# 1,000,000 Books
are available to read at

www.ForgottenBooks.com

Read online
Download PDF
Purchase in print

ISBN 978-1-5278-2301-3
PIBN 10893762

This book is a reproduction of an important historical work. Forgotten Books uses state-of-the-art technology to digitally reconstruct the work, preserving the original format whilst repairing imperfections present in the aged copy. In rare cases, an imperfection in the original, such as a blemish or missing page, may be replicated in our edition. We do, however, repair the vast majority of imperfections successfully; any imperfections that remain are intentionally left to preserve the state of such historical works.

Forgotten Books is a registered trademark of FB &c Ltd.
Copyright © 2018 FB &c Ltd.
FB &c Ltd, Dalton House, 60 Windsor Avenue, London, SW19 2RR.
Company number 08720141. Registered in England and Wales.

For support please visit www.forgottenbooks.com

# 1 MONTH OF FREE READING

at
www.forgottenbooks.com

By purchasing this book you are eligible for one month membership to ForgottenBooks.com, giving you unlimited access to our entire collection of over 1,000,000 titles via our web site and mobile apps.

To claim your free month visit:
www.forgottenbooks.com/free893762

\* Offer is valid for 45 days from date of purchase. Terms and conditions apply.

**English**
Français
Deutsche
Italiano
Español
Português

# www.forgottenbooks.com

**Mythology** Photography **Fiction** Fishing Christianity **Art** Cooking Essays Buddhism Freemasonry Medicine **Biology** Music **Ancient Egypt** Evolution Carpentry Physics Dance Geology **Mathematics** Fitness Shakespeare **Folklore** Yoga Marketing **Confidence** Immortality Biographies Poetry **Psychology** Witchcraft Electronics Chemistry History **Law** Accounting **Philosophy** Anthropology Alchemy Drama Quantum Mechanics Atheism Sexual Health **Ancient History Entrepreneurship** Languages Sport Paleontology Needlework Islam **Metaphysics** Investment Archaeology Parenting Statistics Criminology **Motivational**

# THE PRACTITIONER'S
# MEDICAL DICTIONARY

---

GOULD—SCOTT

## OTHER DICTIONARIES

### By GEORGE M. GOULD, A.M., M.D.

THE POCKET PRONOUNCING MEDICAL DICTIONARY, 35,000 WORDS. **Seventh Edition, Revised.** 1008 Pages. Full Flexible Leather, Gilt Edges, Round Corners, $1.00, Thumb Indexed, $1.25, Postpaid.

THE ILLUSTRATED DICTIONARY OF MEDICINE, BIOLOGY AND ALLIED SCIENCES. **Sixth Edition** with additions and corrections, and a **Supplement** including **38,000** additional words. Numerous Illustrations. Large Square Octavo. 22c4 Pages. Double-Columned. Half Morocco, $14.00, Postpaid.

# THE PRACTITIONER'S MEDICAL DICTIONARY

CONTAINING ALL THE WORDS AND PHRASES GENERALLY USED IN MEDICINE AND THE ALLIED SCIENCES, WITH THEIR PROPER PRONUNCIATION, DERIVATION, AND DEFINITION

BY GEORGE M. GOULD, A.M., M.D.

AUTHOR OF "AN ILLUSTRATED DICTIONARY OF MEDICINE, BIOLOGY, AND ALLIED SCIENCES," "THE STUDENT'S MEDICAL DICTIONARY," "POCKET MEDICAL DICTIONARY," ETC.

THIRD EDITION—REVISED AND ENLARGED

BY R. J. E. SCOTT, M.A., B.C.L., M.D.

FELLOW OF THE NEW YORK ACADEMY OF MEDICINE

EDITOR OF HUGHES' "PRACTICE OF MEDICINE," GOULD AND PYLE'S "CYCLOPEDIA OF MEDICINE AND SURGERY," ETC.

BASED ON RECENT MEDICAL LITERATURE

WITH MANY TABLES

PHILADELPHIA
P. BLAKISTON'S SON & CO.
1012 WALNUT STREET

Copyright 1916 by P. Blakiston's Son & Co.

## PREFACE TO THE THIRD EDITION

The chief feature of this revision is the large number of new words included, probably 20,000. Another point of importance is that the volume is compact and easy to andle, while still sufficiently comprehensive to serve present-day demands. This reat increase in the number of words with a corresponding decrease in the size of the ook has been achieved by the omission of nearly all the illustrations now familiar in he usual text-books and works of reference.

A dictionary is for casual use and brief consultation; therefore, the size of the type need not be larger than legibility demands: that selected is similar to what has been used in Gould's Pocket Dictionary (of which thousands are sold yearly) and is a little larger than in the present edition of the unabridged Webster. The result is a volume of about three-fourths the bulk of the previous edition, or of any of the medical dictionaries of its class with about 71,000 definitions.

The eponymic terms have been placed in their proper alphabetical order, so that one need no longer be in doubt whether to look under "test" or "reaction", "phenomenon" or "syndrome", "sign", or "symptom", for the desired information; this moreover permits the insertion of the nationality and the dates of birth and death of those referred to. The proper name is constant; generic terms vary.

The definitions are based upon the standard literature and authoritative text-books of the day, and are not copied from the older vocabularies.

As to pronunciation: The alphabetical sound of the letter has been the key. This avoids the use of a confusing number of diacritics. Only when there may be any doubt has the proper pronunciation been indicated by a diacritic mark. Over a letter "¯" means that that letter has its usual alphabetical sound.

Proper names and their derivatives only have been capitalized in the title-words; this is in accord with present usage which is a revival of the custom of some of the lexicographers of the last century.

With the exception of a few signs which have no letter-equivalents, all of the matter that preceded the regular pages in the former edition has been placed in alphabetical order in the body of the book, where it will more readily be found.

The critical reader scarcely needs to be reminded that the principal duty of the lexicographer is similar to that of a census-taker; it is his duty to make an inventory of the words and their pronunciation as he finds them among the well-informed; it is not his province to reform or to invent substitutes for the terms held to be undesiable. If words exist and are used, they should be recorded, whether they are well-born, vulgar, hybrid, obsolete, or anomalous. Unfortunately many of our medical terms were coined by men who knew "little Latin and less Greek"; but when once these terms gain currency, usage sanctions them, although philologically incorrect.

We have striven for the fortunate *medio tutissimus ibis*, and if the aim has been successfully carried out, then this Middle-of-the-Road Medical Dictionary contains neither too much, nor too little, and should carry further the popularity of Gould's series of Medical Dictionaries, of which more than one-third of a million volumes are in the hands of the English-speaking practitioners of the world.

Among many, one illustration of the unexpected reach of far-off influence may be permitted:—In the English-Chinese Lexicon of Medical Terms, compiled by Philip B. Cousland, M.B., C.M. (Edin.), the valorous, philanthropic, and learned Editor, in his Preface says: *It is largely based on Gould's Medical Dictionary, and the Nomenclature of the Royal College of Physicians of England, etc.*

In its new form, it is believed that this well-known dictionary will be even more useful than before.

GEORGE M. GOULD,
R. J. E. SCOTT.

# NOTES CONCERNING THE HISTORY OF LEXICOGRAPHY*

*Nomina si nescis perit cognitio rerum*, said Coke with the acumen of the legal mind, and it is generally true that the knowledge of things depends upon the knowledge of their names. Discoveries of new facts, or new standpoints for viewing old facts, demand new tags or "nicking" symbols whereby their status may be fixed and their recognition insured and made more clear for distant or future students. Few philosophic and scientific minds may exhibit an aloofness and a freedom from the tyranny of words to enable them to study things without the aid of words and namings. But nothing, it is admitted, is more blundering in a personal sense, and more harmful to the progress of science, than the exhaustion of interest so soon as a classification and nomenclature have been made. The ridiculous is only needed to end in the absurd, and this is generally supplied by their wrong pigeon-holing and false ticketing. A diagnosis once made, a mere word, long, mysterious, and meaningless, pinned upon the bunched symptoms, and further study of etiology, prophylaxis, or therapeutics is with too many at an end. Over 200 years ago Dr. South tried to check this "fatal imposture and force of words" by showing how "the generality of mankind is governed by words and names," not by things as they are, but as they are called—in a word, by "verbal magic."

And yet in a groping science like medicine, one that inductively, slowly, and tentatively is feeling its way toward the truth, this need of naming every step forward is peculiarly necessary. It is the condition of securing the step in itself, and of guiding the aftercomers. It is the blazing of trails into the wilderness of the unknown. Of course no one can tell what lines of research may finally prove the best and true, and none, therefore, what blazes will be useful or useless. New trails, shorter, easier, and better, may indeed be discovered, and when the wilderness country is settled, all trails will either be abandoned or become well-known roads. But even then good sign-posts and pointing index-fingers will be helpful for strangers, and some of the old names will never be discarded. None can surely foretell what words may die and what ones become a part of the language. Hence the lexicographer may not too recklessly exclude.

The history of lexicography finds its first data about 700 or 800 A. D., in glosses, or the more common explanatory words annexed or superposed over "hard" terms, and made either in Latin or in the glossator's own vernacular. A list of such glosses was called a *glossarium*, or as we say, a glossary. It soon became the custom for children and students to learn by heart the classified lists of the names of things, such as those of the parts of the body, of animals, trades, tools, virtues and vices, diseases, etc. Such a list constituted a *vocabularium* or vocabulary. These glosses and vocabularies were in time thrown together in bundles, at first without any order, and as lists, without losing their individuality. Then came the "first letter order," in which all words and terms beginning with the letter A were bundled together, still without discrimination, so that the entire list of words beginning with A, or B, had to be scanned in order to find a special word. The classification proceeded to an arrangement of the items also according to the second letter, then the third, etc., until after hundreds of years complete alphabetization came into use. At first the aim had been to explain difficult Latin words by easier Latin ones; then by English ones, and in the tenth and eleventh centuries the English equivalents were the rule, and the glossaries were Latin-English. The first book of this kind to be called a *dictionarium*, that is a repertory of *dictiones* or sayings, was that of Sir Thomas Elyot in 1538, and from that time the word *dictionary* has supplanted all others; so much so that it is now the title of any alphabetic gathering not only of words but of any kind of knowledge whatsoever.

* From the preface to Gould's "A Dictionary of New Medical Terms."

Our modern language of medicine is unique in that it is made up of the unchanged and undigested materials and relics used or contributed during its entire history. The persisting substratum is Latin, upon which has been placed a mass of pseudogreek words not physiologicly created nor grown by natural philologic methods, but springing Minervalike from the brains of thousands of modern Jupiters. These largely bear the marks of their parentage in characteristics that do not, or should not, beget a spontaneous pride of lineage. From a highly variegated medievalism that has, indeed, never ended, we have taken over another unassimilable conglomerate, and superadded are thousands of dissimilar terms derived from modern chemistry, biology, bacteriology, and many other sciences. Each single group of contemporaneous nationalities contributes to the others its share of names, and is itself hard at work endeavoring to fuse the whole heritage into homogeneity and unity with the amalgam of the spirit of the general language dominant among its people. The result is a strange hodge-podge of the medical language of two or more thousand years and of many special national tongues, in mechanic, not chemic mixture, with modern sounds and symbols, the whole amazingly heterogeneous and cacophonous. The thirtieth century medical student will probably be compelled to memorize *iter a tertio ad quartum ventriculum*, etc., and to write his orders for drugs in a sad mixture of sorry Latin so far as his knowledge will carry, and then to end it in despair in the vulgar manner of speech of his contemporaries. In general biology the law holds that the ontogeny epitomizes and repeats the phylogeny; but only at the different successive stages of its individual development. In medical language the phylum is always present, and there are no successive stages; there has been no rebirth or inheritance; the ontogeny goes on preserving all the old origins and accretions, and simply adding the new to them. For this sort of evolution there is no name (unless Weissmann's immortality theory is applicable), and its study may be commended to the Darwins and Spencers of the future as a noteworthy exception to hitherto formulated laws. The result is before us: a huge and unassimilated philologic mass, many times greater than it should be, the despair of medical students and of the makers of dictionaries. These word-books, of course, reproduce the phylogenetic history in the same way, and there is no escape from the republication of all the methods and most all the words gathered and found useful in the course of ages. Here with some modifications of detail must be repeated the glosses and vocabularies of a thousand years ago, the foiled attempts together with the partial successes at alphabetic arrangement, and lastly the addition of the modern encyclopedia.

The functions of the dictionary-maker have thus become multiplied and varied. As the gloss-lists and vocable-lists grew into dictionariums and as alphabetization became thoroughgoing, as one after another subject was added to the word-gatherer's work so our technical dictionary has at last become in part encyclopedic and expository, its plan and outworking still somewhat subject to the personality, scholarship, and judgment of the author. It will always remain an open question how far the author should or may go in giving individual color to his dictionary. Johnson's famous definitions of *excise, lexicographer, oats, pension, pensioner, tory, whig*, etc.; Webster's "Americanism" in spelling; the Century's seconding in various ways the obvious trending of philologic progress,—these, and many such illustrate the lexicographer's belief in his own, at least, "limited" free-will.

"Johnson's great work," says Dr. Murray, "raised English lexicography altogether to a higher level. In his hands it became a department of literature." The technical dictionary of to-day may indeed claim a higher office than that, because no monograph or text-book comes near the far-reaching and lasting influence of modern encyclopedic dictionaries. They help more than teacher or text-book to bring order into the student's forming mind, and to systematize and make definite his knowledge. In postgraduate life and practice there is no book that is so frequently consulted, and the teachings of which are so clearly kept in memory. This is because of the validity of the maxim of Coke.

Solely upon condition, however, that the author has put heart, intellect, and labor into his work! If he has been content to repeat, copy, and adopt, it will not be so. And even then only if other repeaters, copiers, and adopters "do not break through and steal." As has often happened since, dictionary-theft is an ancient story. As long as

## PREFACE

250 years ago Phillips plagiarized the *glossographia* of Blount. The robbed author indignantly exposed the shamelessness of the cribber, even of misprints and errors. But he was not ashamed! *More suo* the thief, having no defense, made none, and instead proceeded to correct all the errors pointed out by Blount, and, in many subsequent editions, the quack-lexicographer reaped the reward given by a too careless public.

The ancient injustice would be much manifolded in modern times, with an intensely progressing science which demands that, if to be of the best service, new editions of its word-books shall be made every few years. The system must become systematic and the professing truly professional. No spasmodic, incidental, or amateur methods will nowadays avail. Revisions are required, and continuous labor, not only of one but of many, so that helpers, a large corps of them, must be organized, and paid. Over 300 years ago a great worker in this field, one who "contrived and wrought not onelie for our owne private use, but for the common profet of others," even with the patronage of great men "who encouraged in this wearie worke" was grieved that "the charges were so great and the losse of time" so much that he came near having "never bene able alone to have wrestled against so manie troubles." Finding that "his spiritual substance had vanished," old Simon Browne "took to an employment which did not require a soul, and so became a dictionary-maker," piously adding that we should "thank God for everything and therefore for dictionary makers."

---

## SIGNS AND ABBREVIATIONS

Wt. .................... Weight.
m. .................... Minimum .... Minim.
℥ .................... Drachma .... Dram.
℈ .................... Scrupulum .... Scruple.
℥ .................... Uncia .... Ounce.

= .................... Equal to.
∞ .................... Infinity, 20 ft. distance. } Optics.
○ .................... Combined with.

z. .................... Applied to Zygoma.
∞̄ .................... Heard, but not Understood.

H .................... Intensity of Magnetic Force.
I .................... Intensity of Magnetism.

Z. .................... Contraction (Zuckung).
Z. Z.' Z." .................... Increasing strengths of contraction.
κ. .................... Magnetic Susceptibility.
μ. .................... Magnetic Permeability.
Micron. .................... Unit of Microscopic Measurement.
ω. .................... Ohm.
ρ. .................... Specific Resistance.
Ω. .................... Megohm (one-millionth part of an ohm).
⊣|⊢ .................... Battery.
+ .................... Plus. Anode or Positive Pole.
− .................... Minus. Kathode or Negative Pole.
> .................... Greater than, as K > A.
< .................... Less than.

° .................... Degree.
' .................... Inches.
" .................... Foot. Lines; each one-twelfth of an inch or about two millimeters.
! .................... A mark of affirmation or authentication.
? .................... A mark of doubt.
− .................... Figures or words separated by a short dash indicate the extremes of variation, as 5–10″ long, few–many flowered; *i. e.*, varying from 5 to 10 lines in length, and with few to many flowers.

× .................... Used to express magnification, thus × 1000 indicates a magnification of $\frac{1000}{1}$ diameters. The improper fraction $\frac{1000}{1}$ indicates the same thing, but is rarely used.
① .................... An annual Herb.
② .................... A biennial Herb.
♃ .................... A perennial Herb.
ḃ .................... An Undershrub, deciduous.
ḃ .................... An Undershrub, evergreen.
ƺ .................... A Shrub, deciduous.
ƺ .................... A Shrub, evergreen.
ƻ .................... A Tree, deciduous.
ƻ .................... A Tree, evergreen.
♭ .................... An herbaceous Vine, annual or biennial.
♄ .................... A woody Vine, deciduous.
♄ .................... A woody Vine, evergreen.
ɫ .................... A trailing Herb, annual or biennial.
ɫ .................... A trailing Herb, perennial.
≈ .................... An aquatic plant.
⚥ .................... Flowers perfect.

♂ .................... A male animal, or a plant or flower bearing only stamens or antheridia.
♀ .................... A female animal or a plant or flower bearing only pistils or archegonia.
○ .................... A young animal of undetermined sex, thus ♂ o, young male, or ♀ yg for young female, but ○ *juv* (*juvenis*, young).
⊙ .................... A monocarpic plant.
○= .................... Cotyledons accumbent.
○‖ .................... Cotyledons incumbent.
§ .................... A plant introduced and naturalized.
† .................... A plant cultivated for ornament.
‡ .................... A plant cultivated for use.
8 .................... Monecious.
♂♀ .................... Diecious.
♂ ☿ ♀ .................... Polygamus.
o .................... Wanting or none.
∞ .................... Numerous or indefinite; more than twenty when applied to stamens.
σ .................... The microsecond represents .001 second or the unit of time in experiments or psychophysical reactions.

# THE PRACTITIONER'S MEDICAL DICTIONARY

## A

**A.** Chemical symbol of *argon*.
**a** [ἀ, ἀν, or ἀμ, without]. 1. The Greek letter *alpha*, called alpha privative, equivalent to the prefix *un-* or *in-*. It denotes absence or want of the thing or quality expressed by the root of the word. 2. Abbreviation for *accommodation, ampere, anode, anterior, aqua, arteria, total acidity.*
**aa** [ἀνά, of each]. An abbreviation, written *āā,* used in prescriptions to denote repetition of the same quantity for each item.
**āāā.** Abbreviation for *amalgam.*
**Aaron's sign** (âr′-un) [Charles D. *Aaron*, American physician, 1866– ] In appendicitis, pressure over McBurney's point causes distress in the region of the stomach or heart.
**aasmus** (a-as′-mus) [ἀασμός, a breathing out]. Asthma.
**A.B.** Abbreviation of *Artium Baccalaureus,* Bachelor of Arts.
**ab** [*ab*, from]. A Latin preposition signifying *from*.
**abaca** (ab′-ak-ah; sp. pron., ah-vah-kah′). Manila hemp; also *Musa textilis,* the plant which produces it. See *hemp.*
**abactio** (ab-ak′-she-o) [*abigere*, to drive away]. An abortion, or labor, artificially induced.
**abactus venter** (ab-ak′-tus-ven′-ter) [*abigere,* to drive out; *venter,* the belly]. An abortion procured by artificial means.
**Abadie's sign** (ab-ad-e′) [J. M. *Abadie,* French ophthalmologist, 1842– ]. Spasm of the levator palpebræ superioris in exophthalmic goiter.
**abaissement** (ah-bâs′-mon(g)) [Fr.]. 1. Depression, falling. 2. Couching.
**abalienatio mentis** (ab-āl-yen-a′-she-o) [see *abalienation*]. Insanity.
**abalienation** (ab-āl-yen-a′-shun) [*ab*, away; *alienare*, to transfer]. Decay, especially mental decay, insanity.
**abalienated** (ab-āl′-yen-a-ted) [*abalienatus,* alienated, estranged]. 1. Deranged, or insane. 2. Gangrenous, or so severely injured as to require amputation or extirpation.
**abanet.** See *abnet.*
**abaptiston** (ah-bap-tis′-ton) [ἀ, priv.; βάπτιστος, immersed]. A trephine so shaped that penetration of the brain is impossible.
**abarthrosis** (ab-ar-thro′-sis) [*ab*, from; *arthrosis*, a joint]. Same as *diarthrosis* or *abarticulation.*
**abarticular** (ab-ar-tik′-u-lar) [*ab*, from; *articulus,* joint]. Not connected with or not situated near a joint.
**abarticulation** (ab-ar-tik-u-la′-shun) [*ab*, from; *articulatio*, joint]. 1. Same as *diarthrosis*; sometimes also a synonym of *synarthrosis*. 2. A dislocation.
**abasia** (ah-ba′-ze-ah) [ἀ, priv.; βάσις, a step]. Motor incoordination in walking. See *astasia.* **a. astasia,** inability to walk or stand in a normal manner. **a. atactica,** a form marked by awkwardness and uncertainty of movement. **a., choreic,** that due to choreic cramps in the legs. **a., paralytic,** that form in which the legs give way under the weight of the body and walking is impossible. **a., paroxysmal trepidant,** a form of astasia-abasia (*q. v.*) in which trepidation similar to that of spastic paraplegia stiffens the legs and prevents walking. **a., trembling,** inca acit to walk on account of trembling of the legs.p y
**abasic** (ah-ba′-sik) [see *abasia*]. Pertaining to, or affected with, abasia.
**abatage** (ah-bah-tazj′) [Fr.]. 1. The slaughter of an animal to prevent the infection of others. 2. The art of "casting" an animal preparatory to an operation.
**abatardissement** (ah-bah-tar-dees′-mon(g)) [Fr.]. The gradual degeneration or deterioration of a breed or race.
**abatement** (a-bāt′-ment). Mitigation or decrease in severity of pain, or of any untoward symptom or condition.
**abattoir** (ah-bat-war′) [Fr.]. A slaughter-house or establishment for the killing and dressing of animals.
**abaxial** (ab-ak′-se-al) [*ab*, from; *axis*, an axle]. Not situated in the line of the axis.
**Abbe's catgut rings** (ab′-e) [Robert *Abbe,* New York surgeon, 1851 ]. Rings composed of 8 or 10 turns of heavy catgut in the shape of an oval, with inside diameter of two inches, for use in intestinal anastomosis. **A.'s operation,** lateral anastomosis of intestine with catgut rings. **A.'s string-method,** cutting through an esophageal stricture by the sawing action of a string one end of which passes through the mouth and the other end through an opening in the stomach.
**Abbe's condenser, A.'s illuminator** (ab′-ba) [Ernst *Abbé,* German physicist, 1845–1905). A system of lenses attached to a microscope for condensing the light upon an object. **A.'s lenses, apochromatic,** see *apochromatic lens.* **A.'s test-plate,** an instrument for testing microscopic objectives for spherical and chromatic aberration, composed of a microscopic slide with 6 cover-glasses ranging from 0.09 to 0.024 mm. thick, silvered on one side. Delicate, parallel, ruled lines are cut through the silver film, thus making a kind of micrometer with transparent rulings.
**Abbott's method** (ab′-ot) [Edville G. *Abbott,* American orthopedist, 1872– ]. For treatment of *scoliosis:*—overcorrection by means of plaster jackets and bandages.
**A.B.C. liniment.** Compound liniment of aconite. It contains liniment of aconite 40, liniment of belladonna 40, and chloroform 20.
**A.B.C. process.** A process for the deodorization of sewage by the addition of a mixture of alum, blood, and charcoal.
**Abderhalden's test for pregnancy** (ab′-der-hahlden) [Emil *Abderhalden,* Swiss physiologist and chemist, 1877– ]. During pregnancy microscopic portions of the chorionic villi enter the maternal blood and cause the production of protective ferments which may be detected in the serum by an optical method and a dialyzation method. The ferments disappear within a short time after delivery or abortion.
**abdomen** (ab-do′-men) [*abdere,* to hide]. The large inferior cavity of the trunk, extending from the brim of the pelvis to the diaphragm, and bounded in front and at the sides by the lower ribs and abdominal muscles, and behind by the vertebral column, the psoas and the quadratus lumborum muscles. It is artificially divided into 9 regions by two circular

lines, the upper parallel with the cartilages of the ninth ribs, the lower with the iliac crests, and by two lines drawn vertically upwards from the center of Poupart's ligament. These lines are differently situated by different writers. The regions thus formed are, above, the right hypochondriac, the epigastric, and the left hypochondriac; in the middle, the right lumbar, umbilical, and left lumbar; and below, the right inguinal, the hypogastric, and the left inguinal. a., accordion, Kaplan's term for a swelling of the abdomen attended with flattening of the arch of the diaphragm and increased respiration. It is not due to the presence of gas or to tumor, and disappears under anesthesia; nervous pseudotympany. a., acute, any acute abdominal condition requiring prompt operation. a., boat-shaped, a., carinate, see under *scaphoid*. a. obstipum, congenital shortening of the rectus abdominis muscle. a., pendulous, a relaxed condition of the abdominal walls in which the latter hang down over the pubis. a., scaphoid, see under *scaphoid*. a., uncinate, one in which the terminal segments and those next to them are turned under the others.
 **abdominal** (*ab-dom'-in-al*) [*abdomen*]. Pertaining to or connected with the abdomen. a. aneurysm, see *aneurysm*. a. aorta, the part of the aorta below the diaphragm. a. aponeurosis, see *aponeurosis*. a. bandage, see *a. binder*. a. binder, a broad bandage of muslin or flannel applied to the abdomen for making pressure after delivery or after an operation. Sometimes a many-tailed bandage is used. a. brain, the solar plexus. a. breathing, see *a. respiration*. a. cavity, the cavity within the peritoneum. a. compress, a form of local pack, made by forming folds of a coarse linen towel of sufficient breadth to reach from the ensiform cartilage to the pubis; one of the folds is then wrung out of cold water, applied, and the remainder is rolled around the body so as to retain it in position. a. dropsy, ascites. a. ganglia, the semilunar ganglia. a. gestation, *pregnancy, extrauterine*. a. hysteria, a hysteric condition simulating peritonitis, in which the abdomen becomes extremely painful to the touch, swollen, and distended with gas. a. line, the linea alba. a. lines, muscle tracings on the abdominal walls. a. muscles, the internal and external obliques, the transversalis, rectus, pyramidalis, and quadratus lumborum. a. phthisis, tuberculous disease of the intestines or peritoneum. a. press, see *prelum abdominale*. a. reflex, see *reflexes*. a. regions, see *abdomen*. a. respiration, respiration,

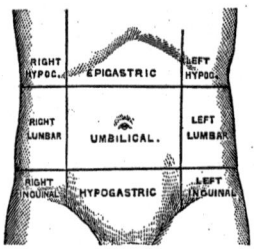

ABDOMINAL REGIONS.

carried on chiefly by the diaphragm and abdominal muscles. a. ring, external, a triangular opening in the fibers of the aponeurosis of the external oblique muscle, transmitting the spermatic cord of the male and the round ligament of the female. a. ring, internal, an oval aperture in the fascia transversalis that transmits the spermatic cord of the male and the round ligament of the female. a. section, see *celiotomy*. a. surgery, the branch of surgery that deals with the lesions of the abdominal viscera and the operations performed upon them through incisions in the abdominal walls. a. typhus, enteric fever. a. viscera, the organs contained in the abdominal cavity.
 **abdominoanterior** (*ab-dom-in-o-an-te'-re-or*). Having the belly forward (used of the fetus in the womb).
 **abdominocystic** (*ab-dom-in-o-sis'-tik*) [*abdomen*;

κύστις, bladder]. Relating to the abdomen and bladder.
 **abdominogenital** (*ab-dom-in-o-jen'-it-al*). Relating to the abdomen and the genitalia. a. nerve, inferior, the ilioinguinal nerve. a. nerve, superior, the iliohypogastric nerve.
 **abdominohysterectomy** (*ab-dom-in-o-his-ter-ek'-to-me*). Removal of the uterus through an abdominal incision.
 **abdominohysterotomy** (*ab-dom-in-o-his-ter-ot'-o-me*). Hysterotomy through an abdominal incision.
 **abdominoposterior** (*ab-dom-in-o-pos-te'-re-or*). Having the belly toward the mother's back (used of the fetus in the womb).
 **abdominoscopy** (*ab-dom-in-os'-ko-pe*) [*abdomen*; σκοπεῖν, to examine]. Examination of the abdomen for diagnostic purposes, by inspection, palpation, measurement, percussion, etc.
 **abdominoscrotal** (*ab-dom-in-o-skro'-tal*). Relating to the abdomen and the scrotum. a. muscle, the cremaster muscle.
 **abdominothoracic** (*ab-dom-in-o-tho-ras'-ik*). Relating to the abdomen and thorax.
 **abdominous** (*ab-dom'-in-us*). Having a large abdomen.
 **abdominouterotomy** (*ab-dom-in-o-u-ter-ot'-o-me*). See *abdominohysterotomy*.
 **abdominovaginal** (*ab-dom-in-o-vaj'-in-al*). Relating to the abdomen and the vagina.
 **abdominovesical** (*ab-dom-in-o-ves'-ik-al*). Relating to the abdomen and the urinary bladder. a. pouch, a fold of the peritoneum in which are comprised the urachal fossæ.
 **abduce** (*ab-dūs'*) [*ab*, away; *ducere*, to lead]. To draw away, as by an abductor muscle.
 **abducens** (*ab-dū'-senz*) [L., "leading away"]. A term applied to certain muscles, or their nerves, that draw the related part from the median line of the body. Also, the sixth pair of nerves supplying the external recti of the eyes. a. labiorum, same as *a. oris*. a. oculi, the external rectus muscle of the eye. a. oris, the levator anguli oris muscle.
 **abduct** (*ab-dukt'*) [*abducere*, to lead away]. To draw away from the median line.
 **abduction** (*ab-duk'-shun*) [*ab*, from; *ducere*, to lead]. 1. The withdrawal of a part from the axis of the body. 2. The recession or separation from each other of the parts of a fractured bone or the sides of a wound.
 **abductor** (*ab-duk'-tor*). See *abducens*. a. auris, the abductor muscle of the ear. a. digiti (*dij'-it-i*) quin'ti, hallu'cis, in'dicis, min'imi digiti, pol'licis, see *muscles, table of*.
 **abenteric** (*ab-en-ter'-ik*) [*ab*, from; ἔντερον, intestine]. Outside the intestine; involving or pertaining to organs or parts other than intestinal. a. typhoid, see under *typhoid*.
 **abepithymia** (*ab-ep-e-thi'-me-ah*) [*ab*, from; ἐπιθυμία longing]. 1. A perverted longing, or desire. 2. Paralysis of the solar plexus (the diaphragm formerly was regarded as the seat of the soul [θυμός], and of the desires).
 **Abernethy's fascia** [John *Abernethy*, English surgeon, 1764-1831]. The subperitoneal areolar tissue that separates the external iliac artery from the iliac fascia overlying the psoas. A.'s operation, for ligation of *the external iliac artery*. In the earlier operation an incision was made in the line of the artery for about three inches, commencing nearly four inches above Poupart's ligament. Later the incision was less nearly vertical and more curved, with the convexity downward and outward, extending from about one inch within and one inch above the anterior superior spine to one and one-half inches above, and external to, the center of Poupart's ligament. A.'s sarcoma, a circumscribed fatty tumor found chiefly on the trunk.
 **aberrant** (*ab-er'-ant*) [*ab*, from; *errare*, to wander]. Deviating from the normal or regular type in appearance, structure, course, etc., as the *aberrant* duct of the testis or liver, *aberrant* arteries, etc. a. arteries, long, slender vessels connected with the brachial or axillary artery.
 **aberratio humorum** (*ab-er-a'-she-o*) [see *aberrant*]. An abnormal tendency or direction of blood or other fluid to a part; as in vicarious menstruation. a. lactis, milk metastasis; see *galactoplania*. a. mensium, a. menstruorum, see *menstruation*, vicarious.
 **aberration** (*ab-er-a'-shun*) [see *aberrant*]. Deviation from the normal; mental derangement; fetal

malformation; vicarious menstruation; escape of the fluids of the body by an unnatural channel. In optics, any imperfection of focalization or refraction of a lens. **a., chromatic,** the dispersion arising from unequal refraction of light of different parts of the spectrum. The violet rays, being more refrangible than the red rays, are brought to a focus nearer the lens, and the image is surrounded by a halo of colors. **a., dioptric,** see *a.*, *spherical*. **a., distantial,** indistinct vision due to distance. **a., lateral,** a deviation of a ray in any direction from the axis measured in the focal plane perpendicularly to the axis. **a., longitudinal,** a deviation of a ray from the focus, measured along the axis above or below the focal plane. **a., mental,** a degree of paranoia that may or may not amount to insanity. **a., Newtonian,** same as *a., chromatic*. **a., spherical,** the excess of refraction of the peripheral part of a convex lens over the central part, producing an imperfect focus and a blurred image.

**abevacuation** (*ab-e-vak-u-a'-shun*) [*ab*, from; *evacuation*]. 1. A morbid evacuation; an excessive or deficient discharge. 2. The passage of matter from one organ or cavity into another; metastasis.

**abeyance** (*ab-a'-ans*) [O. Fr., for "open-mouthed expectation"]. A suspension of activity, or of function; a state of suspended animation, or action.

**Abies** (*a'-be-ēs*) [L.]. A genus of coniferous plants including the fir, hemlock, and spruce. **A. balsamea,** silver fir, balsam-fir, or balm of Gilead, a tree of the nat. ord. *Coniferæ*, from which is derived Canada balsam. A. canadensis, hemlock-spruce; bark of the Canadian fir-tree. It is used as an astringent in various local and internal conditions. It yields Canada pitch. **A. excelsa,** Norway spruce. It yields Burgundy pitch. **A. pectinata,** the European silver fir. Its buds are resinous, balsamic, and sudorific.

**abietene** (*ab-i'-et-ēn*), C₇H₁₆. A hydrocarbon obtained from *Pinus sabiniana*, a California nutpine. It is an aromatic, volatile liquid, agreeing in composition with normal heptane. Syn:, *erasene*.

**abietic, abietinic** (*ab-i-et'-ik, ab-i-et-in'-ik*) [*Abies*]. Pertaining to the genus *Abies*, as *abietic* acid, C₄₄H₆₄O₅ or C₂₆H₄₀O₂, occurring in the resin of *Abies excelsa* and *Larix europæa*.

**abietin** (*ab-i'-et-in*) [*Abies*]. A resinous principle obtained from the turpentine of various species of pine and fir. **a. anhydride,** C₄₄H₆₀O₄, the main constituent of resin.

**abietite** (*ab-i'-et-īt*), C₆H₁₂O₅. A sugar resembling mannite, found in the needles of the European silver fir, *Abies pectinata*.

**abiogenesis** (*ab-i-o-jen'-es-is*) [ά, priv.; βίος, life; *genesis*]. The (theoretic) production of living by nonliving matter. The older term was *spontaneous generation*.

**abiogenetic, abiogenous** (*ab-i-o-jen-et'-ik, ab-i-oj'-en-us*). Pertaining to abiogenesis; characterized by spontaneous generation.

**abiogeny** (*ab-i-oj'-en-e*). See *abiogenesis*.

**abiological** (*ah-bi-o-loj'-ik-al*) [ά, priv.; βίος, life; λόγος, treatise]. Not pertaining to biology.

**abiosis** (*ab-i-o'-sis*) [ά, priv.; βίος, life]. The absence of life.

**abiotic** (*ab-i-ot'-ik*). Opposed to, or incapable of, or incompatible with life.

**abiotrophy** (*ab-i-ot'-ro-fe*) [ά, priv.; βίος, life; τροφή, nourishment]. Degeneration or decay due to defective vital endurance.

**abirritant** (*ab-ir'-it-ant*) [*ab*, from; *irritare*, to irritate]. 1. Tending to diminish irritation; soothing. 2. Relating to diminished sensitiveness. 3. A remedy or agent that allays irritation.

**abirritation** (*ab-ir-it-a'-shun*) [see *abirritant*]. Diminished tissue-irritability; atony or asthenia.

**abjoint** (*ab-joint'*) [*abjungere*, to separate]. To separate by means of a joint or septum.

**abjunction** (*ab-jungk'-shun*) [see *abjoint*]. The separation by means of a joint or septum, as of spores from a growing hypha in some fungi.

**ablactation** (*ab-lak-ta'-shun*) [*ab*, from; *lactare*, to give suck]. The weaning of a child. The end of the suckling period.

**ablastemic** (*ah-blas-tem'-ik*) [ά, priv.; βλάστημα, a shoot]. Non-germinal; in no way related to germination.

**ablastous** (*ah-blas'-tus*) [άβλαστος, not budding, sterile]. In biology, producing no germs or buds. Sterile.

**ablate** (*ab-lāt'*) [*ab*, from; *latum*, from *ferre*, to carry]. To remove; to cut off.

**ablation** (*ab-la'-shun*) [see *ablate*]. Removal of a part, as a tumor, by amputation, excision, etc.

**ablatio retinæ** (*ab-la'-she-o ret'-in-e*). Detachment of the retina.

**ablepharia, ablepharon** (*ah-blef-a'-re-ah, ah-blef'-ar-on*) [ά, priv.; βλέφαρον, the eyelid]. A congenital condition in which there is a total absence either of eyelids or of the interpalpebral fissure. **a., partial,** a congenital defect in one or more of the eyelids. **a., total,** a congenital condition in which there is either a total absence of eyelids or the interpalpebral fissure.

**ablepharous** (*ah-blef'-ar-us*) [see *ablepharia*]. Without eyelids.

**ablepharus** (*ah-blef'-ar-us*). An individual affected with ablepharia.

**ablepsia, ablepsy** (*ah-blep'-se-ah, ah-blep'-se*) [άβλεψία, without sight]. 1. Blindness. 2. Dulness of perception.

**abluent** (*ab'-lu-ent*) [*abluere*, to wash away]. Detergent. That which cleanses or washes away.

**ablution** (*ab-lu'-shun*) [see *abluent*]. Washing or cleansing the body. Separation of chemical impurities by washing.

**abmortal** (*ab-mor'-tal*) [*ab*, from; *mors*, death]. Passing from dead or dying to living muscular fiber (used of electric currents).

**abnerval** (*ab-ner'-val*) [*ab*, from; *nervus*, a sinew]. Passing from a nerve (used of electric currents in muscular fiber).

**abnet** (*ab'-net*) [Hebr., a girdle]. A girdle, or girdleshaped bandage.

**abneural** (*ab-nū'-ral*) [*ab*, from; νεύρον, nerve]. Pertaining to a part remote from the neural or dorsal aspect; ventral.

**abnormal** (*ab-nor'-mal*) [*ab*, away from; *norma*, a rule]. Not normal; not conformable with nature or with the general rule.

**abnormalism** (*ab-nor'-mal-izm*) [*abnormal*]. 1. Abnormality. 2. An abnormal thing or structure.

**abnormality, abnormity** (*ab-nor-mal'-it-e, ab-nor'-mit-e*). The quality of being abnormal; a deformity or malformation.

**aboiement** (*ah-bwah-mon'*(*g*)) [Fr.]. Barking; the involuntary utterance of barking sounds.

**abolition** (*ab-o-lish'-un*) [*abolitio*]. Destruction; cessation; suspension, as of a physiological function.

**abolitionism** (*ab-o-lish'-un-izm*) [*abolitio*, an abolishing]. A movement originating in England to abolish the regulation and control of prostitution by the health-officers. Also applied to the movement to abolish vivisection.

**abomasum, abomasus** (*ab-o-ma'-sum, ab-o-ma'-sus*) [*ab*, away; *omasum*, paunch]. The reed or proper digestive stomach of ruminating mammals; also called "fourth," or "true," stomach.

**aborad** (*ab-o'-rad*) [*ab*, away from; *os*, mouth]. Away from the mouth; in an aboral situation or direction.

**aboral** (*ab-o'-ral*) [*ab*, away from; *os*, the mouth]. Opposite to, or remote from, the mouth.

**aborigines** (*ab-or-ij'-in-ēz*) [*ab*, from; *origo*, origin, beginning]. Primitive, autochthonous, native, indigenous.

**abort** (*ab-ort'*) [*ab*, from; *ortus*, from *oriri*, to grow]. 1. To miscarry; to expel the fetus before it is viable. 2. To prevent full development, as of a disease. 3. To come short of full development.

**aborticide** (*ab-or'-tis-īd*) [*abortus; cædere*, to kill]. 1. The killing of the unborn fetus. 2. The means of killin the fetus. 3. Causing the destruction of a fetus.g

**abortient** (*ab-or'-shent*) [see *abort*]. Abortive; abortifacient.

**abortifacient** (*ab-or-te-fa'-shent*) [*abortus; facere*, to make]. 1. Causing abortion. 2. A drug or agent inducing the expulsion of the fetus.

**abortion** (*ab-or'-shun*) [*abortus*, a miscarriage]. The expulsion of the ovum before the child is viable; that is, any time before the end of the sixth month. By some authors expulsion of the ovum during the first 3 months is termed *abortion;* from this time to viability it is termed *immature delivery,* or *miscarriage,* and from the period of viability to that of maturity, *premature delivery.* **a., accidental,** see *a., spontaneous*. **a., artificial,** that produced intentionally. **a., criminal,** that not demanded for therapeutic reasons. **a., embryonic,** abortion up to

the fourth month. **a., epidemic,** the occurrence of many cases at about the same time, due to widespread distress, excitement, or privation, or to some form of poisoning, such as ergotism. **a., fetal,** abortion after the fourth month. **a., habitual,** repeated abortion in successive pregnancies, usually due to syphilis. **a., incomplete,** when the membranes or the placenta is retained. **a., induced,** see *a., artificial.* **a., inevitable,** when the embryo or fetus is dead, or when there is an extensive detachment or rupture of the ovum. **a., justifiable,** same as *a., therapeutic.* **a., missed,** the death of the fetus and not followed within two weeks by its expulsion. **a., ovular,** abortion within three weeks after conception. **a., partial,** the premature loss of one fetus in a case of multiple gestation. **a., spontaneous,** that not induced by artificial means. **a., therapeu′tic,** induced abortion to save the mother's life. **a., tubal,** the escape of a fertilized ovum through the abdominal opening of the oviduct into the peritoneal cavity.
**abortionist** (*ab-or′-shun-ist*) [see *abortion*]. One who criminally produces abortions; especially one who follows the business of producing abortions.
**abortive** (*ab-or′-tiv*) [see *abortion*]. Prematurely born; coming to an untimely end; incompletely developed; cutting short the course of a disease; abortifacient.
**abortus** (*ab-or′-tus*) [L.]. An aborted fetus; abortion.
**abouchement** (*ab-oosh′-mon(g)*) [Fr.]. The termination of a small vessel in a larger one.
**aboulia** (*ah-boo′-le-ah*). See *abulia.*
**aboulomania** (*ah-boo-lo-ma′-ne′ah*). See *abulomania.*
**ab ovo** (*ab o′-vo*) [L.]. In biology, from the egg; from the beginning.
**abrachia** (*ah-bra′-ke-ah*) [ά, priv.; βραχίων, arm]. The condition of an armless monster.
**abrachiocephalia** (*ah-bra-ke-o-sef-a′-le-ah*) [*abrachia*; κεφαλή head]. Headless and armless.
**abrachiocephalus** (*ab-rak-e-o-sef′-al-us*) [*abrachia*; κεφαλή, head]. A headless and armless monster.
**abrachius** (*ah-bra′-ke-us*). See *abrachia.*
**abrade** (*a-brād′*) [*abradere,* to rub off]. To remove by friction or chafing; to roughen by friction.
**abraham** (*a′-bra-ham*). To sham; to feign sickness or lunacy. **A.-man,** 1. A mendicant lunatic from the Abraham Ward of Bethlehem Hospital, London; they bore a distinctive badge. 2. An impostor who feigned to be a lunatic and begged in the guise of an Abraham man.
**abrasio** (*ab-ra′-ze-o*) [L.]. An abrasion. **a. cor′neæ,** a scraping off of the superficial epithelium of the cornea. **a. dentium,** wearing away of teeth.
**abrasion** (*ab-ra′-shun*) [*ab,* priv.; *radere,* to rub]. Excoriation of the cutaneous or mucous surface by mechanical means. In dentistry, the wearing away of the dentine and enamel, or the cutting edges of the teeth, whether by mechanical or chemical means.
**abrasor** (*ab-ra′-zor*) [L., "abrader"]. A surgeon's rasp or xyster; any file or instrument used in the surgical or dental abrasion of a surface; also, a rasp used in pharmacy.
**abrastol** (*ab-rast′-ol*). See *asaprol.*
**abrin** (*ab′-rin*). A phytotoxin obtained from the *Abrus precatorius;* its action is similar to that of ricin, but is less poisonous.
**abrosia** (*ab-ro′-ze-ah*) [άβρωσία, fasting]. Want of food; fasting.
**abrotanum** (*ab-rot′-an-um*) [άβρότονον, an aromatic plant]. The plant called southern-wood, *Artemisia abrotanum.*
**abruptio** [L.]. Abruption; a tearing away. **a. placentæ,** premature detachment of the placenta.
**abruption** (*ab-rup′-shun*) [*ab,* away, from; and *rumpere,* to break]. 1. A rupture or tearing asunder. 2. A transverse fracture.
**Abrus** (*a′-brus*) [άβρός, pretty]. Jequirity; Indian licorice. The seeds of *A. precatorius,* or wild licorice. Its properties are thought to be due to the presence of certain ferments. See *abrin.* Infusions applied to the conjunctiva or to any mucous surface induce violent purulent inflammation with growth of false membrane. It is used in producing artificial conjunctivitis.
**abscess, abscessus** (*ab′-ses, ab-ses′-us*) [*abscessus,* a departure or separation]. A localized collection of pus surrounded by a wall of lymph. Syn., *ecpyema; gathering.* According to location, abscesses are named *dorsal, mammary, ischiorectal, periyphlitic,*

*retropharyngeal,* etc. **a., acute,** one resulting from an acute inflammation of the part in which it is formed. Syn., *abscessus per fluxum.* **a., alveolar,** abscess in the gum or alveolus. **a. amebic,** a variety of abscess found in the liver and lung and containing amebæ. **a., anorectal,** one of the celluloadipose tissue near the anus. **a., arthrifluent,** a wandering abscess having its origin in a diseased joint. **abscessus arthriticus,** Musgrave's term for intestinal abscesses due to "gouty dysentery." **a., atheromatous,** an area of softening in the wall of a vessel the result of sclerotic endarteritis. **a., bicameral,** one with two pockets. **a., biliary,** one connected with the gallbladder or a bile-duct. **a., bursal,** abscess in a bursa. **a., canalicular,** mammary abscess that communicates with a milk-duct. **abscessus carniformis,** Severinus' name for a hard sarcoma of the joints. **a., chronic,** a., cold, one of slow and apparently non-inflammatory development, generally about a bone, joint, or gland. It is usually tuberculous and contains cheesy material. **a., circumscribed,** one that is limited by an exudation of lymph. **a., cold,** see *a., chronic.* **a., congestive,** one in which the pus appears at a point distant from where it is formed. **a., embolic,** one formed at the seat of a septic embolus. **a., fecal,** one in the rectum or large intestine. **a., fixation,** an abscess produced by the subcutaneous injection of an irritant as a treatment of grave septicemia. **a., glandular,** one formed about a lymph-gland. **a., gravitation,** one in which pus formed in one part of the body tends to migrate, usually to portions deeper or lower down, in the direction gravity would take it. **a., hematic,** one due to an extravasated blood-clot. **a., hemorrhagic,** one containing blood. **a., idiopathic,** one not attributable to any disease. **a., iliac,** a wandering abscess of the iliac region. **a., infecting mitral,** one due to a lymph embolus caused by endocarditis. **a., intramastoid,** one of the mastoid process of the temporal bone. **a., ischiorectal,** one in the ischiorectal fossa. **a., lacunar,** one in the urethral lacunæ. **a., lumbar,** a wandering abscess of the lumbar region. **a., lymphatic,** 1. The suppuration of a lymphatic gland. 2. An enlarged bursa mucosa. **a., mammary,** one in the female breast. **a., marginal,** one located near the anal orifice. **a., mastoid,** suppuration occurring in the cells of the mastoid portion of the temporal bone. **a., metastatic,** an abscess secondary to pyemia and ulcerative endocarditis, but not occurring through septicemia. It is usually of embolic origin and generally located in the lungs and liver. **a., miliary,** a small embolic abscess. **a., milk,** a mammary abscess occurring during lactation. **a., otic cerebral,** a., otitic cerebral, an abscess of the brain following a purulent disease of the inner ear. **a., parametric,** a., parametritic, a form occurring frequently between the folds of the broad ligament of the uterus or in the neighboring cellular tissue. **a., paranephric,** a., paranephritic, one occurring in the tissues about the kidney. **a., perimetric,** a., perimetritic, one within the peritoneum originating from inflammation of the peritoneal covering of the uterus. **a., perinephric,** a., perinephritic, one occurring in the region immediately surrounding the kidney. **a., peripleuritic,** one that occurs beneath the parietal pleura as the result of pleurisy, a diseased rib, or an injury. **a., periproctitic,** one in the loose areolar tissue surrounding the lower part of the rectum. **a., peritoneal,** a collection of softened exudate which has become encysted in cases of peritonitis. **a., peritonsillar,** one that forms in acute tonsillitis around one or both tonsils. **a., phlegmonous,** an acute abscess. **abscessus pneumococcalis,** one due to infection by pneumococci. **a., postcecal,** one located back of the cecum. **a., posttyphoid,** chronic abscess following typhoid. **a., prelacrimal,** an abscess due to caries of the lacrimal or the ethmoid bone, producing a swelling at the inner canthus immediately below the upper margin of the orbit. **a., primary,** one formed at the seat of pyogenic infection. **a., psoas,** one arising from disease of the lumbar or lower dorsal vertebræ, the pus descending in the sheath of the muscle, and usually pointing beneath Poupart's ligament. **a., pyemic,** see *pyemia.* **a., residual,** one formed in or about the residues of former inflammation. **A.-root,** the root of *Polemonium reptans;* alterative, astringent, and expectorant. **a., scrofulous,** one due to tuberculous degeneration of bone or lymph-glands. **a., secondary,** same as *a., embolic.* **a., septicemic,** one resulting from septic infection or accompanying septicæmia. a., shirtstud, two abscesses communicating by means of a sinus. a., spermatic, one involving the seminiferous tubules. a., spinal, one due to necrosis or disease of a vertebra. a., spirillar, Verneuil's name for an abscess containing spirilla from the saliva. a., stitch, one formed about a stitch or suture. a., subaponeurotic, one beneath an aponeurosis or fascia. a., subareolar, one beneath the alveolar epithelium of the nipple. a., subfascial, one beneath a fascia; postfascial abscess. a., submammary, one lying between the mammary gland and the chest-wall. Syn., *postmammary abscess; retromammary abscess.* a., subpectoral, one beneath the chest muscles. a., subperitoneal, one arising between the parietal peritoneum and the abdominal wall. Syn., *preperitoneal abscess.* a., subphrenic, one located beneath the diaphragm. a., sudoriparous, an abscess due to inflammation of obstructed sweat-glands. a., sympathetic, a secondary or metastatic abscess at a distance from the part at which the exciting cause has acted (*e. g.*, a bubo). a., thecal, one in the sheath of a tendon. a., tuberculous, see *a., chronic.* a., tympanitic, one containing gas generated by putrefaction. Syn., *abscessus flatuosus; gas abscess.* a., urethral. 1. Suppuration of a urethral lacuna; a lacunar abscess. 2. One involving the circumurethral tissue. a., urinary, one resulting from extravasation of urine. a., urinous, one containing urine mingled with the pus. a., verminous, a., worm, one containing intestinal worms, from communication with the intestine. a., wandering, one in which the pus has traveled along the connective tissue spaces and points at some locality distant from its origin. Syn., *hypostatic abscess; abscessus per congestum; abscessus per decubitum.*

**abscessed** (*ab'-sesd*). Affected with or caused by an abscess, as "abscessed teeth."

**abscession** (*ab-sesh'-un*) [*abscessio,* departure]. 1. An abscess; a critical discharge. 2. Metastasis.

**abscissæ** (*ab-sis'-se*) [*ab, away; scindere,* to cut]. The transverse lines cutting vertical ones at right angles, to show by a diagram the relations of two series of facts, as, *e. g.*, the number of pulse-beats or the temperature record in given periods of time.

**abscission** (*ab-sish'-un*) [see *abscissa*]. Removal of a part by cutting; or the suppression of a physiological function.

**absconsio** (*ab-skon'-se-o*) [*abscondere,* to hide]. A sinus or cavity whether normal or pathological.

**absence** (of mind) (*ab'-sens*) [*absentia,* absence]. Inattention to surroundings; in marked instances it may be a result of central lesions. It is often seen in epileptics and melancholiacs.

**absentia epileptica** (*ab-sen'-she-ah ep-il-ep'-tik-ah*). Brief losses of consciousness occurring in the mild form of epilepsy.

**abs. feb.** Abbreviation of *absente febre* [L.]. In the absence of fever.

**absinthe** (*ab'-sinth*). See under *absinthium.*

**absinthiate** (*ab-sin'-the-āt*). A salt of absinthic acid.

**absinthiated** (*ab-sin'-the-a-ted*). 1. Mixed with absinthe. 2. Containing wormwood.

**absinthin** (*ab-sinth'-in*) [*absinthium*]. A bitter crystalline principle obtainable from wormwood. See *absinthium.*

**absinthism** (*ab-sinth'-izm*). A disease similar to alcoholism, the result of the excessive use of absinthe. It is characterized by general muscular debility and mental disturbances, that may proceed to convulsions, acute mania, or general paralysis.

**absinthium** (*ab-sinth'-e-um*) [L.]. Wormwood. The leaves and tops of *Artemisia absinthium.* Absinthium contains a volatile oil and an intensely bitter principle, *absinthin,* $C_8H_{16}O_4$, which is a narcotic poison. Absinthium increases cardiac action and produces tremor and epileptiform convulsions. Dose 20–40 gr. (1.3–2.6 Gm.) in infusion. It is used as a stomachic tonic. *Absinthe,* a French liqueur, is an alcoholic solution of the oil exhibited with oils of anise, marjoram, and other aromatic oils.

**absinthol** (*ab-sinth'-ol*), $C_{10}H_{16}O$. The principal constituent of oil of wormwood; it is isomeric with ordinary camphor.

**absolute** (*ab'-so-lūt*) [*absolvere,* to complete]. Perfect, entire, unconditional. a. alcohol, see *alcohol.* a. temperature, see *temperature.* a. zero, see *zero.*

**absorb** (*ab-sorb'*) [*absorbere,* to suck up]. To suck up or imbibe; to take within one's self.

**absorbefacient** (*ab-sorb-e-fa'-shent*) [*absorptio,* absorption; *facere,* to make]. Favoring or tending to produce absorption.

**absorbent** (*ab-sor'-bent*) [see *absorb*]. 1. Absorbing; capable of absorbing. 2. An organ or part that absorbs. 3. A term applied to the lacteals and lymphatics. 4. In materia medica, a drug or medicine that produces absorption of diseased tissue. a. cotton, see *cotton.* a. glands, see *lymphatics.* a. system, the lacteals and lymphatics with their associated glands.

**absorptio** (*ab-sorp'-she-o*) [see *absorb*]. a. morbosa, see *absorption, excrementitial* (2). a. pulmonalis, see *absorption, pulmonary.* a. sana, see *absorption, physiological.*

**absorptiometer** (*ab-sorp-she-om'-et-er*) [*absorption;* μέτρον, a measure]. A device for measuring the thickness of the layer of liquid that is taken up between two glass plates by capillary attraction. Used in conjunction with a spectrophotometer, it serves as a hematoscope.

**absorption** (*ab-sorp'-shun*) [see *absorb*]. The permeation or imbibition of one body by another. a., chylous, the act or process of the entrance of the oil-globules of the chyle into the central canals of the intestinal villi. a., coefficient of, that number which represents the volume of a gas absorbed by a unit volume of water at 0° C. and at a barometric pressure of 760 mm. a., cutaneous, absorption by the skin. a., disjunctive, the removal of living tissue around a necrosed mass, and its consequent separation from its surroundings. a., excrementitial. 1. The absorption of fluid excretions by the mucosa. 2. The absorption of excretions or morbid products (bile, pus) by the blood. Syn., *pathological absorption; absorptio morbosa.* a., external, the taking up by the skin or mucous surfaces of pabulum or medication applied to the exterior of the body or of an organ. a., internal. 1. The absorption of waste-products by the tissues; absorption of decomposition of disassimilation. 2. The taking up of pabulum by the tissues; absorption of nutrition; molecular, nutritive, organic absorption. a., interstitial, the removal by the absorbent system of effete matters. a. lines, Fraunhofer's lines, dark lines of the spectrum, called Fraunhofer's lines, caused by the arrest or absorption of the ethereal waves of certain lengths and rapidities, mainly by vapors of the sun's atmosphere. a., lymphatic, that which occurs in lymphatic vessels. a. method, to determine whether or not hematuria is due to lesion of the bladder. It is based on the fact that the undenuded surface of the bladder will not absorb foreign substances. Fifteen grains of potassium iodide are injected into the bladder, and fifteen minutes later the saliva is examined for iodine. If found, it is an indication of an unhealthy state of the bladder. a., molecular, a., nutritive, a., organic, see *a., internal* (2). a., pathological, see *a., excrementitial* (2). a., physiological, a phenomenon forming an important part of the digestive process, caused in part by the vital activity of the epithelial cells and in part by the physical laws of imbibition, diffusion, and osmosis. Syn., *absorptio sana.* a., progressive, atrophy of a part due to pressure. a., pulmonary, the taking up of oxygen, or of vapors (as of ether), by the lungs. a., purulent, 1. *a., excrementitial* (2). 2. pyemia. a., recrementitial, the absorption of surplus secretions. a., respiratory, see *a., pulmonary.* a. spectrum, a spectrum showing black lines where colors have been absorbed by the transmitting medium. a. tube, see under *tube.* a., ulcerative, that by which an ulcer forms or extends its area. a., venous, absorption by the veins.

**absorptive** (*ab-sorp'-tiv*) [see *absorb*]. Having the power or function of absorbing.

**abstergent** (*ab-ster'-jent*) [*abs,* from; *tergere,* to cleanse]. 1. Cleansing; detergent. 2. A cleansing agent. See *detergent.*

**abstersive** (*ab-ster'-siv*) [*abstersivus*]. Abstergent.

**abstinence** (*ab'-stin-ens*) [*abs,* from; *tenere,* to hold or keep]. Privation or self-denial in regard to food, liquors, etc. See *fasting.*

**abstract** (*ab'-strakt*) [*abstrahere,* to draw away]. In pharmacy, a solid preparation containing the soluble principles of a drug evaporated and mixed with sugar of milk.

**abstraction** (*ab-strak'-shun*) [*abstractio,* a drawing away]. 1. Blood-letting. 2. Attention to one idea to the exclusion of others. 3. In pharmacy, the process of distillation.

**abstractum** (ab-strak'-tum) [pl., abstracta]. An abstract. See *abstract.*

**abterminal** (ab-ter'-min-al) [ab, from; terminus, end]. Passing from tendinous into muscular tissue (used of electric currents).

**abulia** (ah-boo'-le-ah) [ά, priv.; βουλή, will]. Loss or defect of will-power.

**abulic** (ah-boo'-lik) [see *abulia*]. Characterized by or affected with abulia.

**abulomania** (ah-boo'-lo-ma'-ne-ah) [abulia; μανία, madness]. A disease of the mind characterized by imperfect or lost will-power.

**abuse** (ab-ūs') [abusus, a using up]. 1. Misuse or overuse. 2. Rape. **a.,** self-, masturbation.

**abvacuation** (ab-vak-u-a'-shun). Same as *abevacuation, q. v.*

**a.c.** Abbreviation of the Latin *ante cibum,* before meals. Also abbreviation of air-conduction.

**acacanthrax** (ak-ah-kan'-thraks) [ά, priv.; κακός, bad; άνθραξ, a carbuncle: pl., *acacanthraces*]. Non-malignant anthrax.

**Acacia** (ah-ka'-she-àh) [L.]. ° 1. A large genus of leguminous trees, shrubs, and herbs, many of them Australian or African. A number of the species are medicinal, and some are poisonous. The bark is usually very astringent. 2. Gum-arabic, which is produced by various species—*A. lebbek, A. nilotica, A. vera,* and *A. verek. A. senegal* also furnishes gum-arabic, a nearly white, transparent gum, soluble in water. It is used in the manufacture of mucilage, and contains *arabin,* C12H22O11, identical in composition with cane-sugar. **a. anthelmintica,** see *mussanin*. **a. catechu,** see *catechu.* **a., mucilage of** (*mucilago acaciæ*, U. S. P.), acacia, 34; water, to make 100 parts; incompatible with alcoholic tinctures. **a., syrup of** (*syrupus acaciæ*, U. S. P.), mucilage, 25; simple syrup, 75. It is used in various mixtures as a demulcent and to suspend insoluble powders.

**Acalypha** (ah-kal'-if-ah) [άκαλύφής, unveiled]. A genus of euphorbiaceous plants. *A. fruticosa,* of India, is useful in dyspepsia and diarrhea, and is tonic and alterant. *A. hispida* has similar uses. *A. indica* is a plant common in India. The leaves are expectorant, emetic, laxative. *A. virginica,* of North America, is diuretic and expectorant. Dose of the *fluid-extract* 10 min.-1 dr. (0.6-4.0 Cc.); of the juice (*succus acalyphæ*), for an infant, 1 dr. (4 Cc.).

**acampsia** (ah-kamp'-se-ah) [ά, priv.; κάμπτειν, to bend]. Inflexibility of a limb.

**acantha** (ak-an'-thah) [άκανθα, a thorn]. 1. A vertebral process. 2. The spinal column. 3. Spina bifida.

**acanthesthesia, acanthæsthesia** (ak-anth-es-the'-ze-ah) [άκανθα, a prickle; αίσθησις, sensation]. A sensation as of pricking with needles.

**Acanthia lectularia** (ak-an'-the-ah lek-chu-la'-re-ah) [L.]. The common bedbug.

**acanthial** (ak-an'-the-al) [see *acanthion*]. Pertaining to the acanthion.

**acanthion** (ak-an'-the-on) [άκάνθιον, a little thorn]. A point at the base of the nasal spine.

**Acanthocephala** (ak-an-tho-sef'-al-ah) [άκανθα, spine; κεφαλή, head]. An order of parasitic worms, characterized by a thorny armature of the head and proboscis. They are generally grouped in one genus, *Echinorrhynchys.* They infest pigs, birds, and fishes, and in their larval stage live in crustaceans.

**acanthoid** (ak-an'-thoid) [άκανθα, a spine]. Resembling a spine, or spicula; spinous.

**acantholysis** (ak-an-thol'-is-is) [άκανθα, prickle; λύσις, a loosening, a wasting]. Any skin disease in which there is an atrophy of the prickle-layer. **a. bullosa,** see *epidermolysis*.

**acanthoma** (ak-an-tho'-mah) [άκανθα, a spine]. A neoplasm, or localized excessive growth in any part of the prickle-cell layer of the skin.

**acanthopelvis** (ak-anth-o-pel'-vis) [άκανθα, thorn; *pelvis*]. Same as *acanthopelys*.

**acanthopelys** (ak-anth-op'-el-is) [άκανθα, thorn; πέλυς, pelvis]. A pelvis that is encroached upon by exostoses.

**acanthosis** (ak-an-tho'-sis) [άκανθα, a spine]. Any skin disease marked by abnormities in the prickle-cell layer. **a. nigricans,** a general pigmentation of the skin, with papillary, mole-like growths.

**acanthulus** (ak-an'-thu-lus). An instrument for removing thorns from wounds.

**acapnia** (ah-kap'-ne-ah). A condition of diminished carbon dioxide in the blood.

**acapsular** (ah-kap'-su-lar) [ά, priv.; *capsula,* a small box or capsule]. In biology, destitute of a capsule.

**acardia** (ah-kar'-de-ah) [ά, priv.; καρδία, heart]. Congenital absence of the heart.

**acardiac** (ah-kar'-de-ak). 1. Having no heart. 2. A fetus with no heart.

**acardiacus** (ah-kar'-di-ak-us) [see *acardia*]. A synonym employed by German writers for omphalosite. **a. acephalus,** one in which the head is wanting, the thorax rudimentary, the pelvis and contiguous parts perfectly formed. **a. amorphus,** a shapeless lump with only rudiments of organs.

**acardiohemia, or acardiohæmia** (ah-kar-de-o-he'-me-ah) [ά, priv.; καρδία, heart; αίμα, blood]. Lack of blood in the heart.

**acardionervia** (ah-kar-de-o-ner'-ve-ah) [ά, priv. καρδία, heart; *nervus,* a sinew]. Diminished nervous action or nerve-stimulus in the heart.

**acardiotrophia** (ah-kar-de-o-tro'-fe-ah) [ά, priv.; καρδία, heart; τροφή, nutrition]. Atrophy of the heart.

**acardius** (ah-kar'-de-us). Congenital absence of the heart. An acardiac monster.

**acarian** (ah-ka'-re-an). Of or pertaining to the acarids or mites.

**acariasis** (ah-kar-i'-as-is). A disease due to mites. See *mange.*

**acaricide** (ah-ar'-is-īd) [*acarus; cædere,* to kill]. An agent that destroys acarids.

**acarid, acaridan** (ak'-ar-id, ak-ar'-id-an) [άκαρής, small; tiny]. Pertaining to *acarus.*

**Acarina** (ak-ar-i'-na). An order of *Arachnida,* which includes the ticks and mites. They may cause severe symptoms from their bites, apart from the introduction of any parasite such as *Spirochæta.*

**acarinosis** (ak-ar-in-o'-sis) [*acarus,* a mite]. Any disease, as the itch, produced by a mite or acarid.

**acarodermatitis** (ak-ar-o-der-mat-i'-tis) [*acarus,* a mite; *dermatitis*]. Dermatitis caused by acari, or mites.

**acaroid** ⁿ(ak'-ar-oid) [*acarus;* είδος, like]. Mite-like. **a. gum,** Botany Bay gum; resina lutea. An aromatic resin used in Australia as a remedy for gastric troubles, intestinal catarrhs, diarrheas, etc. Dose 8-16 gr. (0.5-1.0 Gm.) in alcoholic solution. Benzoic acid is prepared from it, and it is said to have the properties of storax and balsam of Peru. **a. resin.** See *a. gum.*

**acarophobia** (ak-ar-o-fo'-be-ah) [*acarus;* φόβος, fear]. Morbid fear of the itch.

**acarotoxic** (ak-ar-o-toks'-ik) [*acarus,* a mite; τοξικόν, a poison]. Poisonous, or destructive, to acari.

**acarpæ** (ah-kar'-pe) [ά, priv.; καρπός, fruit]. A name proposed for a group of skin diseases in which there are no papules, tubercles, or elevated points.

**acarpia** (ah-karp'-e-ah) [άκαρπία]. Sterility; barrenness; unfruitfulness.

**acarpous** (ah-karp'-pus) [ά, priv.; καρπός, fruit]. 1. Having no elevations; not nodular. 2. Producing no fruit; sterile, barren.

**Acarus** (ak'-ar-us) [ά, priv.; κείρειν, to cut (because so small)]. The mite, or tick, a parasite of man and animals. **A. scabiei,** *Sarcoptes scabiei,* the itch-mite, a small parasite with numerous sharp tubercles, spines, and hairs on the dorsal surface. See *scabies.*

**acatalepsia, acatalepsy** (ah-kat-al-ep'-se-ah, ah-kat'-al-ep-se) [ά, priv.; καταλαμβάνειν, to understand]. 1. Uncertainty in diagnosis. 2. Mental impairment; dementia.

**acataleptic** (ah-kat-al-ep'-tik) [ά, priv.; καταλαμβάνειν, to understand]. 1. Uncertain; doubtful (used of a prognosis or a diagnosis of a disease). 2. A person affected with acatalepsy.

**acatamathesia** (ah-kat-am-ath-e'-ze-ah) [ά, priv.; καταμάθησις, understanding]. 1. Inability to understand conversation, due to mental disorder. 2. A morbid blunting of the perceptions; as in psychical deafness, or psychical blindness.

**acataphasia** (ah-kat-af-a'-ze-ah) [ά, priv.; κατά, after; φάσις, utterance]. A disorder in the syntactical arrangement of uttered speech, due to some central lesion.

**acataposis** (ah-kat-ap-o'-sis) [ά, priv.; κατά, down; πόσις, a drinking, a swallowing]. A difficulty in swallowing; dysphagia.

**acatastasia** (ah-kat-as-ta'-ze-ah) [άκαταστασία]. Absence of regularity, or of fixed character, in the course of a disease, or in the nature of an excretion.

**acatastatic** (*ah-kat-as-tat'-ik*). Marked or characterized by acatastasia; irregular; not of definite type.

**acatharsia** (*ah-kath-ar'-se-ah*) [ἀκαθαρσία, uncleansed state]. Impurity; foulness; need of purgation, or cleansing.

**acathectic** (*ah-kath-ek'-tik*) [ἀκάθεκτικοs, ungovernable]. Not able to retain. **a. jaundice**, see *jaundice*.

**acaudal, acaudate** (*ah-kaw'-dal, ah-kaw'-dāt*) [ă, priv.; *cauda*, a tail]. Tailless.

**ACC.** Abbreviation for *anodal closure contraction*.

**accelerans nerve** (*ak-sel'-er-ans*) [L.]. A nerve that increases the rate and force of the heart's action.

**acceleration** (*ak-sel-er-a'-shun*) [*accelerare*, to hasten]. Quickening, as of the rate of the pulse or of the respiration.

**accelerator** (*ak-sel'-e-ra-tor*). [see *acceleration*]. 1. That which accelerates. 2. A muscle which hastens a physiological discharge. **a. nerves**, nerves passing from the medulla to the heart and conducting stimuli that cause acceleration of the heart's action. **a. partus**, an abortifacient or ecbolic agent. **a. urinæ**, a muscle of the penis the function of which is to expel the last drops in urination, to expel the semen, and to assist erection. The sphincter vaginæ is its analogue in the female.

**accentuated** (*ak-sent'-u-a-ted*). Abnormally or unusually distinct, as respiratory or heart sounds.

**accentuation** (*ak-sen-tu-a'-shun*) [*accentuare*]. Increased loudness or distinctness.

**access** (*ak'-ses*) [*accessus*, an approach]. 1. An attack of a disease. 2. The return of a fit, or paroxysm. 3d Cohabitation. And see *non-access*.

**accession** (*ak-sesh'-un*) [*ad*, to; *cedere*, to go]. The assault, beginning, or onset of a disease, or of a stage of the same; applied especially to a recurrence of periodical disease.

**accessorius** (*ak-ses-o'-re-us*) [pl., *accessorii*]. 1. Contributory in a secondary degree; accessory. 2. An accessory. **a. ad iliocostalem**, see *muscles, table of.* **a. Willisii**, the spinal accessory nerve.

**accessory** (*ak'-ses-o-re* or *ak-ses'-o-re*) [*accessorius*]. Auxiliary; assisting. A term applied to certain glands, muscles, ducts, nerves, arteries, etc., that are auxiliary in function, course, etc., to the principal. Certain small muscles, as the lumbricales, are regarded as accessory to more important muscles. **a. nu'cleus**, the origin of the spinal accessory nerve. **a. of the parot'id**, the socia parotidis, a small gland.

**accident** (*ak'-se-dent*) [*accedere*, to occur]. 1. In legal medicine, an event occurring to an individual without his expectation, and without the possibility of his preventing it at the moment of its occurrence. 2. An intercurrent or complicating symptom or event, not to be looked for in the regular progression of an attack of disease.

**accidental** (*ak-se-dent'-al*) [*accidentalis*]. 1. Due to, or caused by, an accident. 2. Intercurrent; having no essential connection with other conditions or symptoms. **a. images**, after-images. **a. murmur**, a murmur due to anemia.

**·accipiter** (*ak-sip'-it-er*) [L., "a hawk"]. A facial bandage with tails radiating like the claws of a hawk. **a. quinqueceps**, a five-headed accipiter bandage. **a. triceps**, a three-headed accipiter bandage.

**acclimatation, acclimation, acclimatization** (*ak-li-mat-a'-shun, ak-lim-a'-shun, ak-li-mat-iz-a'-shun*) [*ad*, to; *clima*, climate]. The process of becoming accustomed to the climate, soil, water, etc., of a country to which a plant, animal, person, or a people has removed.

**accommodation** (*ak-om-o-da'-shun*) [*accommodare*, to adjust]. Adaptation or adjustment, particularly the adjustment of the eye for different distances. **a., absolute**, the accommodation of either eye separately. **a., asthenopia of**, subnormal power of the function of accommodation, or the pain or discomfort from accommodative effort. **a., binocular**, the combined accommodation of the two eyes. **a., histological**, the occurrence of changes in the morphology and function of cells following changed conditions. **a., negative**, the opposite of positive accommodation, the refractive power of the eye being lessened. **a. of the eye**, that function of the ciliary muscle and lens whereby objects at different distances are clearly seen. It depends upon the inherent elasticity of the lens, which when the ciliary muscle of an emmetropic eye is at rest, is adapted to the proper focalization of theoretically parallel rays of light. Objects nearer, to be clearly seen, require a greater refracting power on the part of the eye because the rays from such objects are more divergent. This additional refracting power is gained by an increased anteroposterior diameter of the lens, brought about by the contraction of the ciliary muscle, which occasions a loosening of the suspensory ligament and a thickening of the lens by its own elasticity. **a. phosphenes**, the peripheral light-streak seen in the dark after the act of accommodation. **a., positive**, that when the eye being focused for a more distant object is required for fixation upon a nearer point. **a., range of relative**, the range of accommodation at the command of the eye for any particular degree of convergence. **a., reflex**, Argyll Robertson pupil. **a., region of**, the extent controlled by the eye within which it distinguishes objects clearly from the state of rest to that of maximum accommodation. **a., spasm of**, a term used to express excessive or persistent contraction of the ciliary muscle, following the attempt to overcome error of refraction. It stimulates myopia. **a., subnormal**, deficient power of accommodation. **a., supernormal**, excessive power of accommodation. **a., theory of, Helmholtz's**, that the increased convexity of the lens is produced by a relaxation of the suspensory ligament, thus removing the influence which tends to flatten the lens and permitting the latter by its elasticity to become more convex. **a., theory of, Schoen's**, that the contraction of the ciliary muscle produces the same effect on the lens as is produced upon a rubber ball when held in both hands and compressed with the fingers. **a., theory of, Tschernig's**, by the contraction of the anterior part of both the radiating and circular fibers of the ciliary muscle the ciliary processes are drawn backward, and the suspensory ligament pulled backward and outward; pressure of the anterior portion of the muscle causes increased convexity of the lens.

**accommodative** (*ak-om'-o-da-tiv*) [*accommodare*, to adjust]. Pertaining to the function of accommodation, or resulting from it. **a. iridoplegia**, inability of the iris to respond to accommodative effort.

**accouchée** (*ak-koo-shay*) [Fr., à, to; *couche*, a bed]. A woman delivered of a child.

**accouchement** (*a-koosh-mon(g)*) [Fr.]. The French term for childbirth. **a., force**, rapid and forcible delivery with the hand.

**accoucheur** (*a-koo-shur*) [Fr.]. A professional male assistant at childbirth.

**accoucheuse** (*a-koo-shuz*) [Fr.]. A midwife.

**accrementitial** (*ak-re-men-tish'-al*) [*accrescere*, to increase]. In biology, of or pertaining to the process of accrementition.

**accrementition** (*ak-re-men-tish'-un*) [*ad*, to; *crescere*, to grow]. A growth in which increase takes place by interstitial development from blastema, and also, by reproduction of cells by fission. The production or development of a new individual by the separation of a part of the parent; gemmation.

**accrete** (*ak-rēt'*). In biology, grown together.

**accretion** (*ak-re'-shun*) [*ad*, to; *crescere*, to increase]. 1. A term denoting the manner by which crystalline and certain organic forms increase their material substance. 2. The adherence of parts normally separate. 3. An accumulation of foreign matter in any cavity.

**accubation** (*ak-u-ba'-shun*) [*accubare*, to recline]. 1. A reclining posture; the taking to one's bed. 2. The act of lying in bed with another person.

**accumulation** (*ak-u-mu-la'-shun*) [*accumulare*, to heap up]. An amassing or collecting together. **a., fecal**, an excessive aggregation of feces in the large intestine; coprostasis.

**accumulator** (*ak-u'-mu-la-tor*) [*accumulare*, to heap up]. An apparatus to store electricity.

**-aceæ.** A suffix used in botany to designate a family, the name chosen being one of the principal genera. Ex., *Rosa, Rosaceæ, Ranunculus, Ranunculaceæ*.

**acedia** (*ah-se'-de-ah*) [ἀκηδία]. A certain form of melancholia.

**acelia, acœlia** (*ah-se'-le-ah*) [ă, priv.; κοιλία, a cavity]. The absence of a natural cavity. Syn., *acelosis*.

**acelious** (*ah-se'-le-us*) [ă, priv.; κοιλία, the belly]. Without a belly; applied to those extremely emaciated.

**acelosis, acœlosis** (ah-sel-o'-sis). See *acelia*.
**acelous** (ah-se'-lus) [ἀ, priv.; κοῖλος, hollow]. Without intestines; anenterous.
**A. C. E. mixture.** An anesthetic mixture composed of alcohol, 1 part; chloroform, 2 parts; ether, 3 parts. See *anesthetic*.
**acenaphthene** (as-en-af'-thēn) [*aceticus*; *naphthalene*], $C_{12}H_{10}$. A hydrocarbon that occurs in coal tar, and separates on cooling from the fraction boiling at 260–280° C. It crystallizes from hot alcohol in long needles melting at 95° C. and boiling at 277° C.
**acentric** (ah-sen'-trik) [ἀ, priv.; κέντρον, center]. Not eccentric; not originating in, or pertaining to, a nerve-center; peripheric.
**aceognosia** (as-e-og-no'-se-ah) [ἄκος, a remedy; γνῶσις, knowledge]. A knowledge of remedies.
**aceology** (as-e-ol'-o-je) [ἄκος a remedy; λόγος, a discourse]. Therapeutics; medical and surgical treatment of disease; acology.
**acephalemia, acephalæmia** or **acephalhemia, acephalhæmia** (ah-sef-al-e'-me-ah) [ἀ priv.; κεφαλή, head; αἷμα, blood]. Deficiency of blood in the head.
**acephalia** (ah-sef-a'-le-ah) [ἀ, priv.; κεφαλή, head]. Absence of the head.
**acephalism** (ah-sef'-al-izm). See *acephalia*.
**acephalobrachia** (ah-sef-al-o-bra'-ke-ah) [ἀ, priv.; κεφαλή, head; βραχίων, arm]. Absence of the head and arms.
**acephalobrachius** (ah-sef-al-o-bra'-ke-us). A monster with neither head nor arms.
**acephalocardia** (ah-sef-al-o-kar'-de-ah) [ἀ, priv.; κεφαλή, head; καρδία, heart]. Absence of the head and heart.
**acephalocardius** (ah-sef-al-o-kar'-de-us). A monster with neither head nor heart.
**acephalocheiria, acephalochiria** (ah-sef-al-o-ki'-re-ah) [ἀ, priv.; κεφαλή, head; χείρ, hand]. Absence of the head and hands.
**acephalocheirus, acephalochirus** (ah-sef-al-o-ki'-rus) [see *acephalocheiria*]. A monster with neither head nor hands.
**acephalocyst, acephalocystis** (ah-sef'-al-o-sist, ah-sef-al-o-sist'-is) [ἀ, priv.; κεφαλή, head; κύστις, a bladder]. The bladderworm. A headless, sterile hydatid, found in the liver and other organs. **acephalocystis plana**, Laennec's name for certain concretions found in the sheaths of tendons and in muscles. **acephalocystis racemosa**, the hydatid mole of the uterus.
**acephalogaster** (ah-sef-al-o-gas'-ter) [ἀκέφαλος, headless; γαστήρ, belly]. A monster with neither head nor belly.
**acephalogasteria** (ah-sef-al-o-gas-te'-re-ah) [see *acephalogaster*]. Absence of the head and belly.
**acephalophorous** (ah-sef-al-off'-or-us) [ἀ, priv.; κεφαλή, head; φέρειν, to bear]. Destitute of a distinct head.
**acephalopodia** (ah-sef-al-o-po'-de-ah) [ἀ, priv.; κεφαλή, head; πούς, foot]. Absence of the head and feet.
**acephalopodius** (ah-sef-al-o-po'-de-us) [see *acephalopodia*]. A monster with neither head nor feet.
**acephalorrhachia, acephalorachia** (ah-sef-al-or-a'-ke-ah) [ἀ, priv.; κεφαλή, head; ῥάχις, spine]. Absence of the head and vertebral column.
**acephalorrhachus** (ah-sef-al-or-a'-kus) [ἀ, priv.; κεφαλή, head; ῥάχις, spine]. A monster destitute of head and vertebral column.
**acephalostomia** (ah-sef-al-o-sto'-me-ah) [ἀ, priv.; κεφαλή, head; στόμα, mouth]. Absence of the head, with a mouth-like opening on the superior aspect.
**acephalostomus** (ah-sef-al-os'-to-mus) [see *acephalostomia*]. A monster without a head, but with a mouth-like aperture.
**acephalothoracia** (ah-sef-al-o-tho-ra'-se-ah) [ἀ, priv.; κεφαλή, head; θώραξ, chest]. Absence of the head and thorax.
**acephalothorax** (ah-sef-al-o-tho'-raks). A monster destitute of head and thorax. Syn., *acephalothorus*.
**acephalothorus** (ah-sef-al-o-tho'-rus). A monster without head or thorax. See *acephalothoracia*.
**acephalous** (ah-sef'-al-us) [ἀκέφαλος, headless]. Headless.
**acephalus** (ah-sef'-al-us) [see *acephalia*]. A species of omphalositic monsters characterized by complete absence of the head and usually of the upper extremities. It is the commonest condition among the omphalosites. **a. dibrachius**, an acephalus with two upper limbs in a more or less rudimentary state. **a. dipus**, an acephalus with two more or less developed lower extremities. **a. monobrachius**, one with one upper extremity, a cervical vertebra, and one or two more or less developed lower extremities. **a. monopus**, one with only one lower extremity, more or less developed. See *acephalopodius*. **a. sympus**, one in which the trunk ends in a long conic point at the end of which are attached one or two feet.
**acerate** (as'-er-āt) [*acer*, sharp]. 1. A salt of aceric acid. 2. Sharp-pointed, acicular.
**aceratosis** (ah-ser-at-o'-sis) [ἀ, priv.; κέρας, horn]. Deficiency or imperfection of corneous tissue. Akeratosis.
**acerbity** (a-serb'-it-e) [*acerbitas*, sharpness, sourness]. Acidity combined with astringency.
**acercus** (ah-ser'-kus) [ἄκερκος, without a tail]. A monstrosity without a tail or the coccygeal vertebræ.
**acerdol** (as'-er-dol), $MnO_2K_2KOH$. An oxidation-product of potassium and manganese. It is used as an oxidizer and disinfectant.
**aceric** (as-er'-ik) [*acer*, a maple tree]. Pertaining to, or found in the maple; as aceric acid.
**aceride** (as'-er-id) [ἀ, priv.; *cera*, wax]. An ointment or plaster containing no wax.
**acerotous** (ah-ser'-o-tus) [ἀ, priv.; κηρότ, wax]. Containing no wax; said of ointments and plasters.
**acervuline** (as-er'-vu-lin) [*acervulus*, a heap]. Agminated, or aggregated; as certain mucous glands.
**acervuloma** (ah-ser-vu-lo'-mah) [*acervulus*, little heap; pl., *acervulomata*]. See *psammoma*.
**acervulus, a. cerebri** (as-er'-vu-lus ser'-e-bri). Concretionary matter near the base of the pineal gland, consisting of alkaline phosphates and carbonates, with amyloid matter. Syn., *brain-sand*.
**acescence** (as-es'-ens) [*acescere*, to grow sour]. 1. The process of becoming sour; the quality of being somewhat sour. 2. A disease of wines, whereby they become sour, owing to the agency of *Mycoderma aceti*.
**acescent** (as-es'-ent). Somewhat acid or tart; acidulous.
**acesodyne, acesodynous** (ah-ses'-o-dīn, ah-ses-od'-in-us) [ἀκεσώδυνος]. Allaying pain; anodyne.
**acestoma** (as-es'-to-mah) [ἀκεστός, curable]. The mass of young granulation tissue which later forms the cicatrix.
**aceta** (as-e'-tah). Plural of *acetum, q. v*.
**acetabular** (as-et-ab'-u-lar) [*acetabulum*, a vinegar cup]. Pertaining to the acetabulum.
**acetabulum** (as-et-ab'-u-lum) [see *acetabular*]. A cup-shaped depression on the outer aspect of the innominate bone for the reception of the head of the femur. **a. cotyle**, the articular cavity of the innominate bone. **a. humeri**, the glenoid cavity.
**acetal** (as'-et-al) [*acetum*, vinegar]. 1. $C_6H_{14}O_2$. Ethidene diethylate, a colorless liquid with an ethereal odor, produced by the imperfect oxidation of alcohol under the influence of platinum black. It is sparingly soluble in water; boils at 104° C.; sp. gr. at 20° is 0.8304. Its action is that of a soporific. Dose 1 dr. (4 Gm.). 2. A mixture said to consist of acetic ether and oils of cloves, bergamot, lavender, lemon, menthol, orange, rosemary, thyme, and absolute alcohol. **a., dimethyl**, see *methylal*.
**acetaldehyde** (as-et-al'-de-hid). The normal aldehyde; ethaldehyde. See *aldehyde*.
**acetals** (as'-et-als) [*acetum*, vinegar]. Products of the combination of aldehydes with alcohols at 100° C.
**acetamide, acetamid** (as-et'-am-id), $CH_3$. $CO$. $NH_2$. A white, crystalline solid produced by distilling ammonium acetate, or by heating ethyl acetate with strong aqueous ammonia. It combines with both acids and metals to form unstable compounds.
**acetamidoantipyrine** (as-et-am-id-o-an-ti-pi'-rin): A crystalline compound used as antipyrine.
**acetamidophenol** (as-et-am-id-o-fen'-ol). $C_6H_4OH$.$NH$.$C_2H_3O$. An oxidation-product of acetanilide; hydroxyantifebrin.
**acetaminol** (as-et-am'-in-ol), $C_{15}H_{22}NO_4$. A reaction-product of paranitrobenzoyl chloride with eugenol-sodium, followed by reduction and acetylization. It occurs as white scales or crystalline powder, soluble in alcohol and insoluble in water, and melting at 160° C. It is used in pulmonary tuberculosis. Syn., *para-acetamido-benzoyleugenol*; *acetamido-benzoyl*.
**acetanilide** (as-et-an'-il-id), $C_8H_9NO$. Phenylacetamide. A white, crystalline solid, produced by boiling anilin and glacial acetic acid together for several hours, the crystalline mass being then dis-

tilled. It melts at 114° and boils at 259°. It is soluble in hot water, alcohol, and ether. Under the name *antifebrin* it is prescribed as an antipyretic. Dose 2-10 gr. (0.13-0.65 Gm.), not exceeding 30 gr. (2 Gm.) in the 24 hours; of the *compound powder* (*pulvis acetanilidi compositus*, U. S. P.) 7½ gr. (0.5 Gm.). **a.,** **ammoniated,** a mixture of acetanilide, 25 parts; ammonium carbonate, 10 parts; sodium bicarbonate, 5 parts; sugar of milk, 60 parts. It is recommended as causing less depression than acetanilide alone. **a.,** **monobromated,** see *antisepsin*.

**acetas** (*as'-et-as*). An acetate or salt of acetic acid.

**acetate** (*as'-et-āt*) [see *acetic*]. Any salt of acetic acid. **a. of lead,** plumbi acetas; see *plumbum*.

**acetated** (*as'-et-a-ted*). Treated with or containing an acetate, acetic acid, or vinegar.

**acetic** (*as-e'-tik*) [*acetum*, vinegar]. Pertaining to *acetum* or vinegar; sour. See *acid, acetic*. **a. acid amide,** see *acetamide*. **a. acid esters,** see *methyl acetate* and *ether, acetic*. **a. acid salts.** 1. Readily soluble crystalline salts formed from the bases. 2. Basic salts formed from iron, aluminum, lead, and copper; sparingly soluble in water. 3. Alkali salts, which have the property of combining with a molecule of acetic acid to produce acid salts. **a. aldehyde,** see under *aldehyde*. **a. anhydride,** $C_4H_6O_3$, a colorless, mobile liquid, highly refractive, and with an odor of acetic acid. Sp. gr. 1.080 at 15° C.; boils at 136°-138° C. Syn., *acetyl oxide; acetic oxide;* socalled *anhydrous acetic acid*. **a. ether,** see under *ether*. **a. fermentation,** the development of acetic acid by the activity of the *Mycoderma aceti*. **a. fungus,** any one of several minute fungoid organisms capable of inciting and maintaining acetic fermentation, as first proved by Pasteur in 1864.

**acetification** (*as-et-e-fi-ka'-shun*) [*acetum*, vinegar; *facere*, to make]. The production of vinegar by acetic fermentation.

**acetimeter, acetimetric, acetimetry.** See *acetometer; acetometric; acetometric*.

**acetin** (*as'-et-in*) [*acetum*, vinegar], $C_3H_5(C_2H_3O_2)_3$. A chemical compound formed by the union of glycerol and acetic acid.

**acetoacetate** (*as-et-o-as'-et-āt*). A salt of acetoacetic acid.

**acetoacetic acid** (*as-et-o-as-e'-tik*). A monobasic acid formed from acetic acid by replacing one of the hydrogen atoms of the acid radical with the aceticacid radical, acetyl. See *Gerhardt*. **a. esters,** $CH_3 . CO . CH_2 . CO_2R$, liquids possessing an ethereal odor, produced by the action of metallic sodium upon acetic esters; they dissolve with difficulty in water and can be distilled without decomposition.

**acetoarsenite** (*as-et-o-ar'-sen-īt*). A salt composed of an acetate and an arsenite of the same base.

**acetobromide** (*as-et-o-bro'-mid*). An acetic-acid salt in which part of the hydrogen of the acid radicle has been replaced by bromine.

**acetochloride** (*as-et-o-klor'-īd*). A salt composed of an acetate and a chloride of the same base.

**acetoglycocoll** (*as-et-o-gli'-ko-kol*),

$$CH_3 < \frac{NH . C_2H_3O .}{CO_2H .}$$

A substance resembling a monobasic acid, obtained from the action of acetyl chloride on glycocoll silver and of acetamide on monochloracetic acid; it is soluble in alcohol; melts at 206° C. Syn., *acetamidoacetic acid; aceturic acid*.

**acetoiodide** (*as-et-o-i'-o-did*). A double salt containing the acetate and iodide of the same radical.

**acetol** (*as'-et-ol*). 1. See *acetyl carbinol*. 2. A remedy for toothache, said to consist of acetic acid, 8.46%; alum, 3.07%; water, 88.5%; with a small proportion of essential oils of sage, clove, and peppermint.

**acetoluid** (*as-e-tol'-u-id*), $C_7H_7NH . C_2H_3O$. Acetoorthotoluid. An antipyretic resembling acetanilid. The dose is not accurately determined.

**acetomel** (*as-et'-o-mel*). See *oxymel*.

**acetometer** (*as-et-om'-et-er*) [*acetum*, vinegar; μέτρον, a measure]. An instrument used in the quantitative determination of acetic acid.

**acetometric** (*as-et-o-met'-rick*). Pertaining to acetometry.

**acetometry** (*as-et-om'-et-re*) [*acetum*, vinegar; μέτρον, measure]. The quantitative estimation of the amount of acetic acid in vinegar. Usually made by an *acetometer*.

**aceton.** 1. See *acetone*. 2. A proprietary remedy for headache and influenza.

**acetonasthma** (*as-et-on-az'-mah*) [*acetone; asthma*]. Attacks of dyspnea similar to uremic asthma, accompanied, with restlessness, headache, nausea, vomiting, transient amaurosis, and acetonuria.

**acetone, aceton** (*as'-et-ōn*) [*acetum,* vinegar], $CH_3 . CO . CH_3$. Dimethylketone. A colorless, mobile liquid, of peculiar odor and burning taste, present in crude wood-spirit; it occurs in small quantities in the blood and in normal urine, and in considerable quantities at times in the urine of diabetic patients. It is miscible with ether, alcohol, and water. It, is used as an anesthetic and anthelmintic. Dose 15-20 min. (0.9-1.2 Cc.). Syn., *mesitic alcohol; mesityl alcohol; methyl acetyl; acetyl methyl*. See *Chautard, Gunning, Legal, Lieben, Malerba, le Nobel, Penzoldt, Reynolds*. **a. chloroform,** $HO . C(CH_3)_2CCl_3$, a compound formed by the addition of potash to equal weights of acetone and chloroform. It occurs as white crystals, sparingly soluble in water, more freely in alcohol and glycerol. Its 1 % aqueous solution is called *aneson*. It is used as a hypnotic and anesthetic. Dose 15-20 gr. (1.0-1.3 Gm.). Syn., *chloretone; trichlortertiary butyl alcohol; trichlorpseudobutyl alcohol*. **a. diethylsulphons,** see *sulphonal*. **a., monochlorated,** $C_3H_5ClO$, a colorless liquid having a pungent odor, obtained by chlorinating acetone. **a. phenylhydrazone,** $(CH_3)_2C : N_2HC_6H_5$, one of the nitrogen derivatives of ketone. **a. resorcinol,** $C_9H_{10}O_4 + H_2O$, a combination of resorcinol with acetone and fuming hydrochloric acid added hot. It occurs in small anhydrous prisms, soluble in alkaline solutions, insoluble in water, alcohol, ether, and chloroform. It melts at 212°-213° C. It is used in the same manner as resorcinol.

**acetonemia, acetonæmia** (*as-et-on-e'-me-ah*) [*acetone; αἷμα*, blood]. The presence of acetone in the blood.

**acetones** (*as'-et-ōnz*). A class of compounds that may be regarded as consisting of two alcoholic radicals united by the group CO, or as aldehydes in which hydrogen of the group COH has been replaced by an alcoholic radical.

**acetonin** (*as-et'-on-in*). 1. A body produced by the action of ammonia on acetone. 2. Dihydrotriacetonamine.

**acetonitrate** (*as-et-o-ni'-trāt*). A double salt, the acetate and nitrate of the same radical.

**acetonitril** (*as-et-on-i'-tril*), $CH_2CN$ or $C_2H_3N$. Methyl cyanide. It is a colorless liquid, having an agreeable odor, and is prepared by distilling acetamide with $P_2O_5$. It may also be produced from prussic acid and diazomethane. It melts at $-41°$ C., boils at 81.6° C., and has a sp. gr. of 0.789 at 15° C. Syn., *carbamine*.

**acetonoresorcinol.** See *acetone resorcinol*.

**acetonuria** (*as-et-o-nu'-re-ah*) [*acetone; οὖρον,* urine]. The presence of acetone in the urine.

**acetonyl** (*as-et'-on-il*), $CH_2—CO—CH_3$. A univalent radical obtained from acetone by taking away one atom of hydrogen.

**acetophenetidine** (*as-et-o-fen-et'-id-ēn*). See *phenacetine*.

**acetophenone** (*as-et-o-fe'-nōn*), $C_6H_5(CO)(CH_3)$. Phenyl methyl ketone; also called hypnone; a hypnotic and antiseptic. It results from the action of zinc methyl upon benzoyl chloride and crystallizes in large plates, melts at 20.5° and boils at 202°. It is without satisfactory action. Dose 4-15 min. (0.26-1.0 Cc.).

**acetophenoneorthooxyquinolin** (*as-et-o-fē-non-ortho-oks-e-kwin'-ol-in*), $C_8H_6NO . CH_3 . CO . C_6H_5$. A base obtained by interaction between a halogen compound of acetophenone and orthoquinolin in the presence of solvents and an alkali. It forms well-defined salts, is soluble in volatile solvents, and melts at 130° C. It is said to have hypnotic and antineuralgic properties; is odorless, tasteless, and nonirritating.

**acetophenophenetidine** (*as-et-o-fē-nōn-fen-et'-id-in*). A condensation-product of acetophenone and paraphenetidine. **a. citrate,**

$$C_6H_4 < \frac{OC_2H_5}{N=C(CH_3)(C_6H_5) . H_3C,}$$

lemon-yellow needles, soluble in ether and hot alcohol, insoluble in water. It melts at 88° C.; is antipyretic and antineuralgic. Dose 8-15 gr. (0.5-1.0 Gm.). Syn., *malarin*.

**acetopyrine, acetopyrin** (as-et-o-pi'-rēn, -rin). A mixture of antipyrine and acetyl salicylic acid, occurring as a whitish, crystalline powder, soluble with difficulty in cold water, ether, and petroleum ether; readily soluble in warm water, alcohol, chloroform, and warm toluol. It is antipyretic. Dose 7 gr. (0.4 Gm.) 6 times daily. Syn., *antipyrine acetylsalicylate*. **a. acetosalicylate**, antipyretic, analgesic, sedative; employed in influenza, bronchitis, rheumatic headache, sciatica, hemicrania, and acute articular rheumatism.

**acetous** (as-e'-lus) [*acetum*, vinegar]. Resembling vinegar; pertaining to or charged with vinegar or acetic acid.

**acetozone** (as-et'-o-zōn). See *benzoylacetylperoxide*.

**acetparaphenetidine** (as-et-par-a-fe-net'-id-ēn). Same as *phenacetine*.

**acetparatoluid** (as-et-par-ah-tol'-u-id), $C_9H_{11}NO$. Antipyretic, colorless crystals, slightly soluble in water, moderately soluble in alcohol; it melts at 149° C. Dose 15–30 gr. (1–2 Gm.). Syn., *acetparamidotoluol; paratolylacetamide*.

**acetphenetidin** (as-et-fe-net'-id-in) [*acetum*; *phenol*]. A compound derived from phenol, having antipyretic and antineuralgic properties. It is crystalline, tasteless, and almost insoluble in water. Dose 4–30 gr. (0.26–2.0 Gm.). Syn., *phenacetine*.

**acetum** (as-e'-tum) [L.; gen., *aceti*; pl., *aceta*]. Vinegar. An impure, dilute acetic acid produced by acetous fermentation of wine, cider, or other fruit-juice. In pharmacy, a solution of the active principles of certain drugs in dilute acetic acid. See *vinegar*. **a. aromaticum** (N. F.) ["aromatic vinegar"], a mixture of alcohol, water, and acetic acid, aromatized with the oils of rosemary, lavender, juniper, peppermint, cassia, lemon, and cloves. **a. britannicum**, an aromatic vinegar consisting of glacial acetic acid, 600; camphor, 60; oil of cloves, 2; oil of cinnamon, 1; oil of lavender, 0.5.

**acetyl** (as'-et-il) [*acetum*, vinegar], $C_2H_3O$. A univalent radical supposed to exist in acetic acid and its derivatives. Aldehyde may be regarded as the hydride, and acetic acid as the hydrate, of acetyl. Syn., *acetosyl; acetoyl; acetoxyl; othyl*. **a.-anhydride**, see *acetic anhydride*. **a.-atoxyl**, an atoxyl substitution product, better known as *arsacetin*, q. v. **a. benzene**, see *acetophenone*. **a. bioxydamide**, see *acetamide*. **a. bromide**, $C_2H_3BrO$, a reaction-product of acetic acid with phosphorus pentabromide; it is a fuming liquid which turns yellow in the air; it boils at 81° C. It is used as a reagent. **a. carbinol**, $CH_3 . CO . CH_2OH$, a saturated ketol produced by the action of water and barium carbonate upon chloracetone, also by fusing cane-sugar and grape-sugar with caustic potash. It is a colorless oil with a feeble, peculiar odor; boils at 145°–150° C. Syn., *pyroracemic alcohol; acetone alcohol; oxyacetone; methyl ketol; acetol*. **a. chloride**, $C_2H_3ClO$, a reaction-product of acetic acid with phosphorus trichloride; it is a colorless, hi_hl_ refracting, fuming liquid; sp. gr. 1.1305 at 0° C.; boils at 55° C. It is used as a reagent. **a. ethylphenylhydrazin**, $C_{14}H_{18}N_2O_2$, colorless needles obtained by heating a solution of ethylenephenylhydrazin with an excess of acetic anhydrid. It is recommended as an antipyretic. Syn., *phenylhydrazinacetylethyl*. **a. formyl**, see *aldehyde, pyroracemic*. **a. hydrate**, acetic acid. **a. hydride**, same as *acetic aldehyde*. See under *aldehyde*. **a. iodide**, $C_2H_3OI$, a reaction-product of acetic acid with iodine and phosphorus; it is a brown, fuming liquid; sp. gr. 1.98 at 17° C.; boils at 105°–108° C. **a. isocyanide**, $(C_2H_3O)-N \equiv C$, a liquid in its simple form, but capable of polymerization as a crystalline solid. It boils at 93° C. Syn., *acetic isocyanide; cyanacetyl*. **a. isoeugenol**, the direct antecedent of vanillin in the manufacture of the synthetic product, and is used as a substitute for vanillin. **a. leuco-methylene-blue**, a colorless form of methylene-blue for internal use. **a. methyl**, see *acetone*. **a. oxide**, same as *acetic anhydride*. **a.-paraamidophenylsalicylate**, see *salophen*. **a. peroxide**, $(C_2H_3O)_2O_2$, a thick liquid, insoluble in water, but readily dissolved by ether and alcohol. It is a powerful oxidizing agent. It is decomposed in sunlight and explodes violently when heated. **a. phenylhydrazid, a. phenylhydrazin**, same as *hydracetin* and *pyrodin*. **a. tannin**, a grayish-yellow, slightly hygroscopic, odorless, tasteless powder, soluble in alcohol, dilute sodium phosphate, sodium carbonate, or sodium borate; slightly soluble in hot water and ether;

insoluble in cold water; melting at 190° C. It is an astringent and is used internally in chronic diarrhea. Externally, it is used in chronic pharyngitis. Dose 3–7½ gr. (0.2–0.5 Gm.). Application, 3 % solution in 5 % sodium phosphate. Maximum dose 60 gr. (4 Gm.) daily. Syn., *tannigen*. **a. thymol**, $C_{11}H_{16}O_2$, a colorless antiseptic liquid with a pungent taste having a s_eci_c gravity of 1.009 at 0° C. and boiling at 244.4° C. Syn., *thymol acetate*. **a. tribromsalol**, fine, white acicular crystals which melt at 108.5°; insoluble in water; soluble in alcohol. Syn., *cordyl*. **a. urethane**, see *urethane*.

**acetylene** (as-et'-il-ēn). [*acetum*, vinegar], $C_2H_2$. A colorless gas, with a characteristic, unpleasant odor, burning with a luminous, smoky flame. It is formed by the imperfect combustion of illuminating gas and other hydrocarbons. The *acetylene* series of hydrocarbons has the general formula $C_nH_{2n-2}$.

**acetylization** (as-et-il-i-za'-shun). The act of combining with or producing compounds of acetic acid or acetyl.

**Achalmé's bacillus** (ak-al'-ma). An anaerobic bacillus, probably identical with Welch's *Bacillus aerogenes capsulatus*; it has been regarded as the cause of acute articular rheumatism.

**ache** (āk) [AS., *acan*, to ache]. Any continuous or throbbing pain.

**acheilia** (ah-ki'-le-ah) [á, priv.; χεῖλος, a lip]. The congenital absence of lips.

**acheilous** (ah-ki'-lus) [see *acheilia*]. Lipless.

**acheilus** (ah-ki'-lus) [á, priv.; χεῖλος, a lip]. A person affected with acheilia.

**acheir** (ah'-kīr) [á, priv.; χείρ, the hand]. 1. Acheirous. 2. Said of fishes lacking pectoral fins.

**acheiria** (ah-ki'-re-ah) [á, priv; χείρ, a hand]. The congenital absence of hands.

**acheirous** (ah-ki'-rus) [see *acheiria*]. Affected with acheiria.

**acheirus** (ah-ki'-rus). An acheirous person, or fetus; one who was born without hands.

**achene** (a-kēn'). Same as *achenium*.

**achenium** (ah-ke'-ne-um) [á, priv.; χαίνειν, gape; pl., *achenia*]. In biology, a small, dry, one-seeded, indehiscent fruit.

**achilia** (ah-ki'-le-ah). See *acheilia*.

**Achillea** (ak-il-e'-ah) [*Achilles*, its reputed discoverer]. Milfoil; yarrow. The herb A. *millefolium*. Its properties are due to a bitter, aromatic, astringent, tonic extractive, *achillein*, and a volatile oil. It has long been used as a vulnerary, and has been highly recommended for intermittent and low exanthematous fevers. Dose 1 oz.–1 pint infusion *ad lib*.; of the *extractive*, 1–3 dr. (4–12 Gm.); of the *volatile oil*, 5–15 min. (0.3–1.0 Cc.). To the genus *Achillea* belong various other unofficial medicinal plants, as *A. moschata*, of the Alps, used in preparing cordials and a diaphoretic medicine, and *A. ptarmica*, or sneezewort, a strong sialagogue.

**achillein, achilleinin** (ak-il-e'-in, -i'-num), $C_{20}H_{38}N_2O_{15}$. A glucoside obtained from *Achillea millefolium*. Occurs as a brownish-red, amorphous mass, of a strongly bitter taste, soluble in water, less soluble in alcohol, insoluble in ether. It is stated that divided doses up to 30–75 gr. (2–5 Gm.) cause marked irregularity of the pulse.

**Achilles tendon** (ak-il'-ēs ten'-don). The tendon of the gastrocnemius and soleus muscles, inserted into the back of the heel. **A. t. reflex**, contraction of the calf of the leg on tapping the tendo Achillis.

**achillobursitis** (ak-il-o-bur-si'-tis) [*Achilles tendon*; *bursitis*]. Inflammation of the bursae lying approximate to the Achilles tendon.

**achillodynia** (ak-il-o-din'-e-ah) [*Achilles tendon*; ὀδύνη, pain]. Pain referred to the insertion of the Achilles tendon.

**achillorrhaphy** (ak-il-or'-af-e) [*Achilles tendon*; ῥαφή, suture]. Suture of the Achilles tendon; practised by C. Bayer instead of achillotomy for the sake of lengthening the tendon. This is exposed, the length divided in half, the upper end of one side, the lower end of the other, cut across, and both the cut surfaces united by a suture.

**achillotomy** (ak-il-ot'-o-me) [*Achilles tendon*; τομή, a cutting]. The subcutaneous division of the Achilles tendon.

**achillotenotomy** (ak-il-o-ten-ot'-o-me). Same as *achillotomy*.

**achilous** (ah-ki'-lus). See *acheilous*.

**achiria** (ah-ki'-re-ah). See *acheiria*.

**achirous** (ah-ki'-rus). See *acheirous*.

**achirus** (ah-ki'-rus). See *acheirus*.
**achlorhydria** (ah-klor-hi'-dre-ah) [ά, priv.; *chlorhydric* (acid)]. Absence of free hydrochloric acid from the gastric juice.
**achloropsia** (ah-klor-op'-se-ah) [ά, priv.; χλωρός, green; ὄψις, vision]. Green-blindness.
**acholia** (ah-ko'-le-ah) [ά, priv.; χολή, bile]. 1. Absence of biliary secretion. 2. Any condition obstructing the escape of the bile into the small intestine. 3. Asiatic cholera. 4. A mild temperament. a., **pigmentary**, that in which there are deficiency of bile and lack of color in the feces, but no jaundice.
**acholic** (ah-kol'-ik). 1. Affected with acholia. 2. Able to cure jaundice. 3. Due to acholia.
**acholous** (ah'-ko-lus). Pertaining to or affected with acholia.
**acholuria** (ah-kol-u'-re-ah) [ά, priv.; χολή, bile; οὖρον, urine]. The absence of bile-pigment in the urine.
**achondroplasia** (ah-kon-dro-pla'-ze-ah) [ά, priv.; χόνδρος, cartilage; πλάσσειν, to form]. Lack of development in a cartilaginous structure; the absorption of cartilage during its transformation into bone. 2. Parrot's term for a form of fetal rickets in which the limbs are short, the curves of the bones exaggerated, and there is an absence of the proliferating zone of cartilage at the junction of the epiphyses. The children are generally still-born. This condition is very much like a fetal cretinism. Also known as *chondrodystrophia fetalis*.
**achondroplastic** (ah-kon-dro-plas'-tik). Pertaining to achondroplasia.
**achor** (a'-kor) [ἄχωρ, chaff, scurf, or dandruff: pl., **achores** (a-kor'-ēz)]. Crusta lactea, a small pustule, followed by a scab, upon the heads of infants; milkcrust. a. **barbatus**, barber's itch.
**achordal** (ah-kor'-dal) [ά, priv.; χορδή, cord]. Not derived from the notochord.
**achoresis** (ah-kor-e'-sis) [ά, priv.; χωρεῖν, to make room; pl., **achoreses**]. Grossi's term for the diminished capacity of a hollow organ, as of the bladder. Syn., *achoria*. Cf. *stenochoria*.
**Achorion** (a-ko'-re-on) [dim. of ἄχωρ, chaff]. A genus of fungous organisms, including several species (possibly modified forms of *Penicillium glaucum*) found in the skin, especially the hair-follicles. A. **keratophagus**, the form causing *onychomycosis*. A. **lebertii**, the parasite of *Tinea tonsurans*. A. **schoenleinii**, the species occurring in ringworm, or *Tinea favosa*.
**Achras** (ak'-ras) [ἀχράς, the wild pear]. A genus of arboraceous plants of the order *Sapotaceæ*. A. **sapota** [*cochitzapotl*, Mex.], the sapodilla plum; a species indigenous to South America. The fruit is edible, sweet, cloying; said to be beneficial in strangury. The seeds are laxative and diuretic; they are exhibited in emulsion in cases of gravel and renal colic. The bitter astringent bark (*cortex jamaicensis*) has been used as a substitute for cinchona bark. The bark and seeds yield the glucoside *sapotin*. The sap yields chicle-gum.
**achreocythemia** (ah-kre-o-si-the'-me-ah) [ἀχρεῖος, colorless; κύτος, cell; αἷμα, blood]. Lack of coloring matter in the blood.
**achroa, achroia** (ah-kro'-ah, ah-kroi'-ah). Same as *achroma*, q. v.
**achroacyte** (ah-kro'-as-īt) [ἄχροιος, colorless; κύτος, cell]. A colorless cell, or lymphocyte.
**achroacytosis** (ah-kro-ah-si-to'-sis) [ἄχροιος, colorless; κύτος, cell]. Abnormal development of lymphcells.
**achroiocythemia** (ah-kroi-o-si-the'-me-ah), or **achrœocythæmia** (ah-kre-o-si-the'-me-ah) [ἄχροιος, colorless; κύτος, cell; αἷμα blood]. A deficiency of hemoglobin in the red corpuscles; also, the diseased state that is associated with such deficiency.
**achroiocytosis** (ah-kroi-o-si-to'-sis). Same as *achroacytosis*.
**achroma** (ah-kro'-mah) [ά, priv.; χρῶμα, color]. Absence of color; albinism. Syn., *achromasia; achromatia; achromausis; achromoderma; vitiligo.* a., **congenital**, see *albinism*. a. **cutis**, see *leukoderma*.
**achromacyte** (ah-kro'-mas-īt) [ά, priv.; χρῶμα, color; κύτος, cell]. A degenerated, decolorized erythrocyte; a "phantom" or shadow corpuscle. Syn., *Ponfick's shadow corpuscle; Bizzozero's blood-platelet; Hayem's corpuscle or hematoblast*.
**achromasia** (ah-kro-ma'-ze-ah) [ά, priv.; χρῶμα, color]. (1) An absence of color in the body; cachectic pallor. (2) Loss of stain from a cell, a phenomenon occurring in the *in vitro* method of staining living cell. See *achroma*.
**achromatia** (ah-kro-ma'-she-ah). See *achroma*.
**achromatic** (ah-kro-mat'-ik) [ά, priv.; χρῶμα, color]. 1. Without color, 2. Colorblind. 3. Relating to achromatin. a. **lens**, one the dispersing power of which is exactly neutralized by another lens with the same curvature, but having a different refractive index. a. **spindle**, see *nuclear spindle*.
**achromatin** (ah-kro'-mat-in) [ά, priv.; χρῶμα, color]. The groundwork of the nucleus of a cell; it is so called because it is not readily stained by coloring agents.
**achromatism** (ah-kro'-mat-izm) [ά, priv.; χρῶμα, color]. 1. Absence of chromatic aberration. 2. Absence of color.
**achromatophil** (ah-kro-mat'-o-fil) [ά, priv.; χρῶμα, color; φιλεῖν, to love]. 1. Showing no affinity for stains. 2. A microbe or histologic element which does not stain readily.
**achromatophilia** (ah-kro-mat-o-fil'-e-ah) [*achromatophil*]. The condition of being refractory to staining.
**achromatopsia** (ah-kro-mat-op'-se-ah) [ά, priv.; χρῶμα, color; ὄψις, sight]. Color-blindness; daltonism. a., **partial**, a form in which only one pair of colors, which to the normal eye are complementary, appear gray or white. a., **total**, that in which all the colors appear as white or gray.
**achromatosis** (ah-kro-mat-o'-sis) [ά, priv.; χρῶμα, color]. Any disease characterized by deficiency of pigmentation in the integumentary tissues.
**achromatous** (ah-kro'-mat-us) [ά, priv.; χρῶμα, color]. Deficient in color.
**achromaturia** (ah-kro-ma-tu'-re-ah) [ά, priv.; χρῶμα, color; οὖρον, urine]. A colorless state of the urine.
**achromia** (ah-kro'-me-ah). See *achroma*.
**achromodermia** (ah-kro-mo-der'-me-ah) [ά, priv.; χρῶμα, color; δέρμα, skin]. An albinotic or colorless state of the skin.
**achromophilous** (ah-kro-mof'-il-us) [ά, priv.; χρῶμα, color; φιλεῖν, to love]. Not readily stained; not chromophilous.
**achromotrichia** (ah-kro-mo-trik'-e-ah) [ά, priv.; χρῶμα, color; θρίξ, hair]. Absence of pigment from the hair.
**achromous** (ah-kro'-mus) [ά, priv.; χρῶμα, color]. Pale, colorless; having no color.
**achronizoic** (ah-kron-e-zo'-ik) [ά, priv.; χρονίζειν, to hold out]. A term applied to drugs which are incapable of remaining unchanged for any length of time.
**achronychous** (ak-ron'-ik-us). See *acronychous*.
**achroodextrin** (ah-kro-o-deks'-trin) [ἄχροος, colorless; *dexter*, right]. A reducing dextrin formed by the action of the diastatic ferment of saliva upon starch. It is a modification of dextrin and may be precipitated by alcohol; it is not converted into sugar by ptyalin, nor colored by iodine.
**achylia** (ah-ki'-le-ah) [ά, priv.; χυλός, juice]. Absence of chyle. Syn., *achylosis*. a. **gastrica**, Einhorn's term for a condition of the stomach marked by destruction of the glandular structures with resulting absence of chyme ferment, and even mucus; called *anadenia gastrica* by Ewald.
**achylosis** (ah-ki-lo'-sis) [ά, priv.; χυλός, juice]. Deficient chylification. See *achylia*.
**achylous** (ah-ki'-lus) [see *achylia*]. Deficient in chyle or in one of the digestive juices.
**achymia, achymosis** (ah-ki'-me-ah, ah-ki-mo'-sis) [ά, priv.; χυμός, chyme]. Deficient formation of chyme.
**achymous** (ah-ki'-mus). Deficient in chyme.
**acicular** (as-ik'-u-lar) [*acus*, a needle]. Needlelike.
**acid, acidum** (as'-id, -um) [*acere*, to be sour]. 1. A name applied to any substance having a sour taste. 2. A compound of an electronegative element with one or more atoms of hydrogen which can be replaced by electropositive atoms, when a salt is formed. The majority of acids contain oxygen, and are known as *oxyacids*; those not containing oxygen are termed *hydrogen acids* or *hydracids*. Acids vary in their terminations according to the quantity of oxygen or other electronegative constituent. Those having the maximum of oxygen end in -*ic*; those of a lower degree, in -*ous*. When there are more than two combinations, the prefix *per-* is joined to the highest, and *hypo-* to the lowest. Acids that end in -*ic*, as sulphuric acid, form salts

ACID 12 ACID

terminating in *-ate*; those ending in *-ous* form salts terminating in *-ite*. **a., abietic, abietinic,** see *abietic*. **a., abric,** $C_{12}H_{24}N_8O$, a crystallizable acid, said to exist in jequirity. **a., absinthic,** an acid obtained from wormwood; said to be identical with succinic acid. **a., acetic,** an acid solution composed of 36 parts of absolute acetic acid, $C_2H_4O_2$, and 64 parts of water. It has strongly acid properties. **a., acetic, dilute,** contains 6% of absolute acid. Dose 1–2 dr. (4–8 Cc.). An impure form, obtained by the destructive distillation of wood, is known as woodvinegar, or pyroligneous acid. **a., acetic, glacial,** the absolute acid occurring in crystals melting at 22.5° C. It is an escharotic. **a., acetoacetic,** same as *a., diacetic*. **a., achilleic,** same as *a., aconitic*. **a., aconitic,** $C_6H_6O_6$, occurs in different plants, as *Aconitum napellus*, sugar-cane, and beet-root. It crystallizes in small plates that dissolve readily in alcohol, ether, and water, and melt at 186°–187°. **a., acrylic.** 1. $CH_2=CH.CO.OH=C_3H_4O_2$. A monobasic acid which may be considered as the oxide of acrolein, a colorless liquid. 2. A general term for organic acids of the group $C_nH_{2n-2}O_2$, comprising two groups, the normal acrylic and the isocrylic acids. Normal acrylic acids occur in vegetable or animal organisms or are derived from natural products. Isoacrylic acids are formed synthetically by the abstraction of the elements of water from certain acid ethers, which in turn are derived from oxalic acid by substituting 2 molecules of an alcohol radical of the series $C_nH_{2n+1}$ for an atom of hydrogen. **a., adipic,** $C_6H_{10}O_4$, obtained by oxidizing fats with nitric acid. It crystallizes in shining leaflets or prisms; is soluble in 13 parts of cold water; melts at 148°. It is dibasic. **a., agaricic,** $C_{16}H_{30}.O_5+H_2O$, a resin acid obtained from the fungus *Polyporus officinalis*, growing on larch trees. The acid has been recommended for checking night-sweats. It also checks the other excretions and diminishes thirst. It is mildly cathartic. **a.s, alcohol,** $CnH_{2n}<^{OH}_{COOH}$, monobasic acid having the properties of the monohydric alcohols. They are distinguished as primary, secondary, and tertiary, according as they contain, in addition to the carboxyl group, the group $—CH_2OH$, the radical $=CHOH$, or the group $\equiv C.OH$. Syn., *oxyacids; hydroxy-fatty acids*. Cf. *a., glycollic*. **a.s, aldehyde,** bodies which combine the properties of a carboxylic acid and of an aldehyde. **a., aldepalmitic,** $C_{16}H_{30}O_3$ the chief component of the butter of the cow. **a., alginic,** an organic substance from algæ that combines with bases to form soluble and insoluble compounds. **a., aliphatic,** same as *a., fatty*. **a., allanturic,** $C_7H_{10}N_8O_4$, from allantoin, by the action of dilute nitric acid. **a., alloxantic,** $C_8H_2N_2O_4$, a crystalline acid obtained by treating alloxan with alkalies. **a., alloxyproteic,** a neutral sulphur compound found in the urine. **a., amidoacetic,** see *glycin*. **a., amidobenzoic,** $C_7H_7NO_2$, occasionally found in the urine. **a., amidosuccinamic,** same as *asparagin*. **a. s., amino,** a large group of nitrogen-holding substances derived from the decomposition of proteins. **a., aminoacetic,** same as *glycocoll*, *q. v.* **a., anacardic,** $C_{22}H_{32}O_3$, a tetratomic acid obtained by Städler from the cashew-nut. It is used as an anthelmintic in the butter of ammonium anacardate. **a., angelic,** $C_5H_8O_2$, a crystalline monobasic acid. It exists free along with valeric and acetic acids in the roots of *Angelica archangelica*, and as butyl and amyl esters in Roman oil of cumin. It crystallizes in shining prisms, melts at 45°, and boils at 185°. It has a peculiar odor and taste. **a., anisic,** $C_8H_8O_3$, obtained by oxidizing anisol and anethol with $HNO_3$, and from aniseed by the action of oxidizing substances. It is antiseptic and antipyretic, and is used in the treatment of wounds and acute articular rheumatism. Dose of the sodium salt 15 gr. (1 Gm.). Syn., *methylparaoxybenzoic acid*. **a., anisuric,** $C_{10}H_{11}NO_3$, an acid formed by the action of anisyl chloride on the silver compound of glycocoll; it also occurs in the urine after the ingestion of anise. **a., anticyclic,** a white fragrant powder with pleasant, acid taste, readily soluble in water, alcohol, and glycerol; it is used as an antipyretic. Dose 1⁄60 gr. (0.0006 Gm.). **a., antirrhinic,** an acid from the leaves of digitalis. **a., apiolic,** decomposition product of apiol. **a., apocrenic,** Berzelius' term for a brown, amorphous substance obtained from the sediment of chalybeate waters. **a., arabic,** see *arabin*. **a., arachic,** **a.,** **arachidic, a., arachinic,** $C_{20}H_{40}O_2=C_{19}H_{39}.COOH$, a monobasic fatty acid obtained from oil of peanut, *Arachis hypogæa*. **a., argentic,** silver monoxide. **a., aromatic,** a name applied to certain organic acids occurring in the balsams, resins, and other odoriferous principles. Also, in pharmacy, a dilute mineral acid reinforced by aromatic substances in order to modify its flavor. **a., arsenic, a., arsenous,** see *arsenic trioxide*. **a., arsinic,** any one of a class of acids formed by the oxidation of arsins or arsonium compounds. **a., aseptic,** an antiseptic solution consisting of an aqueous solution of 5 Gm. of boric acid in 1000 Gm. of hydrogen dioxide (1.5%); 3 Gm. of salicylic acid may be added. **a., asparagic, a., asparaginic, a., asparamic,** same as *a., aspartic*. **a., aspartic,** $C_4H_7NO_4$, occurs in the vinasse obtained from the beet-root, and is procured from albuminous bodies in various reactions. It is prepared by boiling asparagin with alkalies and acids; crystallizes in rhombic dibasic prisms or leaflets, and dissolves with difficulty in water. **a., aspartic, inactive,** $NH_2C_2H_3(CO_2H)_2$, formed by heating aspartic acid with water or with alcoholic ammonia to 140°–150° C., or with HCl to 170°–180° C. Syn., *asparacemic acid*. **a., atrolactic,** $C_9H_{10}O_3$, a monobasic acid obtained from acetophenone by means of prussic acid and $H_2SO_4$ or dilute HCl. **a., auric,** $Au(OH)_3$, gold trihydroxide. **a., azelaic, a., azelainic,** $C_9H_{16}O_4$, an oxidation-product of oleic acid, Chinese wax, castor oil, or cocoanut oil; soluble in water, alcohol, and ether, melts at 106°–107° C., and boils at 360° C. Syn., *anchoic acid; lepargylic acid; azelic acid; azeloinic acid*. **a., azotic,** nitric acid. **a., benzamic,** see *a., amidobenzoic*. **a., benzoic,** $C_7H_6O_2$, occurs free in some resins, chiefly in gum benzoin and in coaltar. It crystallizes in white, shining needles or leaflets, melts at 120°, and distils at 250°. It volatilizes readily, its vapor possessing a peculiar odor. **a., blattic,** see *antihydropin*. **a., boric, a., boracic,** see *boron*. **a. of borax,** orthoboric acid. **a., borocitric,** a combination of boric and citric acid forming a white powder which is used as a solvent for urates and phosphates in urinary calculi, gout, etc. Dose 5–20 gr. (0.3–1.3 Gm.). **a., borophenylic,** $C_6H_7BO_3$, obtained by the action of phosphorus oxychloride upon a mixture of boric acid and phenol. It is an antiseptic white powder with a mild aromatic taste, not easily soluble in water, melting at 204° C. It is fatal to lower forms of life, but does not affect the higher forms. Syn., *phenylboric acid*. **a., borasalicylic,** $B(OH)(OC_7H_4.CO_2H)_2$, a combination of boric and salicylic acids in molecular proportion. It is used externally instead of salicylic acid. **a., brom-,** one in which bromine has replaced one or more atoms of hydrogen in the acid radical. **a., bromacetic,** see *a., monobromacetic*. **a., bromhydric, hydrobromic acid*. **a., bromic,** $HBrO_3$, a colorless, acid liquid. **a., bursic, a., bursinic,** a yellow, hygroscopic mass obtained from an aqueous extract of *Capsella bursapastoris* by the action of lead acetate and ammonia and evaporating. Its aqueous solution is used in the same manner as ergotin, hypodermatically and also internally. **a., butic, a., butinic,** see *a., arachic*. **a., butyric,** $C_4H_8O_2$, an acid having a viscid appearance and rancid smell. It is obtained commercially by the fermentation of a mixture of sugar and butter or cheese in the presence of an alkaline carbonate, but occurs in various plants, in codliver oil, in the juice of meats, and in the perspiration. Combined with glycerol as glyceryl butyrate, it is essentially butter. **a., cacodylic,** see *a., dimethylarsenic*. **a., caffeic,** $C_9H_8O_4$, obtained when the tannin of coffee is boiled with potassium hydroxide. **a., cahincic or caincic,** see *cahincin*. **a., camphoric,** $C_{10}H_{16}O_4$, a dibasic acid, obtained by boiling camphor with $HNO_3$; it crystallizes from hot water in colorless leaflets; melts at 178°, and decomposes into water and its anhydride, $C_8H_{14}(CO)_2O$. It is used in nightsweats of phthisis. Dose 10–30 gr. (0.65–2.0 Gm.). **a., capric or caprinic,** $C_9H_{19}CO.OH$, occurs in small quantity as a glycerid in cow's butter. It crystallizes in fine needles, melting at 30° C., and is very insoluble in boiling water. **a., caproic,** $C_5H_{11}O_2$, the sixth in the series of fatty acids; a clear, mobile oil, colorless, inflammable, and with a very acid and penetrating taste. **a., caprylic or caprillic,** $C_7H_{15}CO.OH$, an acid combined with glycerol, forming a glycerid existing in various animal fats; it is liquid at ordinary temperatures. **a., carbamic,** $H_2N.CO.OH$, carbonic acid in which $NH_2$ replaces OH; it is not known

# ACID 13 ACID

in the free state; its ammonium salt is contained in commercial ammonium carbonate. The esters of carbamic acid are called urethanes. **a., carbazotic,** see *a., picric*. **a., carbolic,** $C_6H_5OH$, phenol,—the official designation of this substance,—is procured from coal-tar by fractional distillation. It has a very peculiar and characteristic odor, a burning taste, is poisonous, and has antiseptic properties. The sp. gr. at the melting-point is 1.060–1.066; it crystallizes in colorless rhombic needles that melt at about 40° C., boiling at about 180°, and it is not decomposed upon distillation. At ordinary temperatures it dissolves in water with difficulty (1 : 19.6 at 25° C.), but is soluble in alcohol, ether, glacial acetic acid, and glycerol in all proportions. It unites with bases to form salts, known as *carbolates*. Upon exposure to light and air it deliquesces and acquires a pinkish color. It is used in the manufacture of many of the artificial coloring-matters, *e. g.*, picric acid. It is a powerful antiseptic and germicide. Internally it is useful in vomiting, fermentation in the stomach, and as an intestinal antiseptic; locally, as a caustic. Dose, internally, ½–2 gr. (0.03–0.13 Gm.). **a., carbolic, camphorated,** a mixture of phenol 1 part and camphor 3 parts. **a., carbolic, chlorinated,** see *trichlorphenol*. **a., carbolic, iodized,** a solution of 20 parts of iodine in 76 parts of phenol with the addition of 4 parts of glycerol. It is used as an antiseptic and escharotic. **a. carbolic, liquefactum** (B. P.). Dose .1–2 min. (0.06–0.13 Cc.). **a., carbolsulphuric,** a mixture of equal parts of phenol and concentrated sulphuric acid. It is used as a disinfectant in 2 to 3 % solution. **a., carbonaceous,** see *carbon dioxide*. **a., carbonaphtholic,** see *a., oxynaphthoic*. **a., carbonic,** $CO_2$, carbon dioxide; an ultimate product of the combustion of carbon compounds; a colorless, odorless gas, heavier than air, incapable of sustaining respiration. **a., carminic,** $C_{17}H_{18}O_{10}$, a coloring-matter found in the buds of certain plants, and especially in cochineal, an insect inhabiting different varieties of cactus. It is an amorphous, purple-red mass, readily soluble in water and alcohol, and yields red salts with the alkalis. **a., carthamic,** see *carthamin*. **a., caseic,** lactic acid (*q. v.*). **a., catechinic, or catechuic,** same as *catechin*. **a., cathartic, a., cathartinic,** an active principle from several species of *cassia*. **a., cerebric, or cerebrinic,** $C_{26}H_{131}$-$NO_2$, from brain tissue. **a., cerotic, or cerotinic,** $C_{27}H_{54}O_2$, a fatty acid existing in beeswax and in Chinese wax. **a.-characteristic,** the replaceable hydrogen and the elements immediately bound to it in the molecule of an acid, as the CO . OH of organic acids. **a., chloracetic** [*chlorine and acetic*], an acid, called also *monochloracetic acid* produced by the substitution of chlorine for the hydrogen of the radical in acetic acid. It is sometimes used as a caustic. **a., chloric,** $HClO_3$, an acid known only in its compounds (*chlorates*) and its aqueous solution. **a., cholalic,** see *a., cholic*. **a., choleic,** $C_{24}H_{40}O_4$, from ox-bile. **a., cholesteric,** $C_{12}H_{16}O_7$, an acid obtained by Tappeiner from the oxidation of cholic acid with potassium dichromate and sulphuric acid. This must not be confounded with cholesterinic acid. **a., cholesterinic,** $C_8H_{10}O_5$, a dibasic acid obtained from cholesterin and from cholic acid by action of nitric acid; it occurs as a gum-like, yellow, hygroscopic body with an acid taste. **a., cholic,** $C_{24}H_{40}O_5$, from glycocholic and taurocholic acids; it crystallizes from out of a hot solution in small anhydrous prisms, sparingly soluble in water, and melting at 195°. **a., cholodic,** derived from cholalic acid. **a., chondroitic,** $C_{18}H_{27}SNO_{17}$, from cartilage. **a., chromic** (*chromii trioxidum*, U. S. P.), strictly, the compound $H_2CrO_4$; it forms salts called chromates. It is a crystalline solid; escharotic. **a., chrysophanic,** $C_{15}H_{10}O_4$, exists in the lichen, *Parmelia parietina*, in senna leaves, and in the rhubarb root. It crystallizes in golden-yellow needles or prisms, melting at 162°. Syn., *rheinic acid*. See *chrysorobin*. **a., cinchotannic,** see *cinchotannin*. **a., cinnamic, a., cinnamylic,** $C_9H_8O_2$, occurs in peru and tolu balsams, in storax, and in some benzoin resins. It has been used in tuberculosis, both internally and externally. Dose 1–10 min. (0.06–0.65 Cc.) hypodermatically. **a., citric,** $C_6H_8O_7$, occurs free in lemons, black currants, bilberries, beets, and in various other acid fruits. It crystallizes with one molecule of water in large rhombic prisms that melt at 100°, are colorless, inodorous, and extremely sharp in taste. It is refrigerant, antiseptic,

and diuretic. **a., colopholic, a., colophonic,** an acid obtained from turpentine; it is used in plasters. **a., copahuvic, a., copaivic,** $C_{20}H_{30}O_2$, an almost colorless, coarsely crystalline powder, obtained from copaiba; it is soluble in alcohol, ether, and benzene. Sometimes written *copaibic A*. **a., cresolsulphuric,** $C_7H_7O$ . $SO_2$ . OH, exists in the urine in small traces. **a., cresotic, a., cresotinic,** $C_8H_8O_3$, an aromatic acid of which 3 isomeric compounds may be formed by the action of sodium and carbonic anhydride on the 3 modifications of cresol. They all occur in alkaline urine. The para compound, melting at 151° C. is used as an antipyretic in the form of sodium cresolate. Dose 2–20 gr. (0.13–1.3 Gm.); maximum dose 60 gr. (4 Gm.). Syn., *oxytoluic acid; homosalicylic acid*. **a., cresylic,** see *cresol*. **a., cryptophanic,** $C_{10}H_{18}N_2O_{10}$, said to exist in small quantities in human urine. **a., cubebic,** $C_{13}H_{14}O_7$(?), a white, waxy mass, turning brown on exposure, obtained from cubeb berries, the unripe fruit of *Piper cubeba*, soluble in alcohol, ether, and alkaline solutions, and used as a diuretic. Dose 5–10 gr. (0.3–0.6 Gm.) in pills several times daily. **a., cumic,** $C_{10}H_{12}O_2$, produced by the oxidation of cuminic alcohol with dilute $HNO_3$. Very soluble in water and alcohol; crystallizes in colorless needles or leaflets; melts at 116° and boils at about 290°. **a., cyanic,** CONH, obtained by heating polymeric cyanuric acid. **a., cyanuric,** see *a., tricyanic*. **a., cynureinic,** $C_{30}H_{14}N_2O_6$. decomposition product of proteids found in dogs' urine. **a., damaluric,** $C_7H_{12}O_2$, found in urine. **a., dextrotartaric,** tartaric acid. **a., diacetic,** $C_4H_6O_3$, an acid present in the urine in certain stages of diabetes and other diseased conditions. **a., dichloracetic,** $CHCl_2$ . $CO_2H$, produced when hydrated chloral is heated with CNK or potassium ferrocyanide and water. At ordinary temperature it occurs as a caustic, colorless liquid, but crystallizes at a low temperature. Sp. gr. 1.522 at 15° C.; boils at 189°–191° C.; soluble in water and alcohol. It is used as an escharotic in skin diseases. **a., diiodosalicylic,** $C_7H_4I_2O_3$, a white, crystalline powder, soluble in alcohol and ether, slightly soluble in water, and melting at 220°–230° C. It is antipyretic, analgesic, and antiseptic, and is used in rheumatism and gout. Dose 8–20 gr. (0.5–1.3 Gm.) 3 or 4 times daily in wafers; maximum dose 30 gr. (2 Gm.). **a., dimethylarsenic,** $As(CH_3)_2OOH$, a substance formed by the oxidation of cacodyl, occurring in large, permanent prisms, odorless and slightly sour. It is soluble in water and alcohol and melts at 200° C. It is considered not to be toxic, and because of its solubility is easily absorbed. Syn., *cacodylic acid*. **a., dithiochlorsalicylic,** $SC_6H_3$ - Cl . OH . COOH, a reddish-yellow powder obtained by heating a mixture of salicylic acid and sulphur chloride to 140° C. It is recommended as an antiseptic. **a., dithiosalicylic,** $C_{14}H_{10}S_2O_6$, obtained from salicylic acid and sulphur chloride heated to 150° C., and existing in two modifications differing in the solubility of their salts. It is an antiseptic, analgesic, antipyretic, yellowish-gray powder, partly soluble in water. Its lithium and sodium salts only are used in medicine as substitutes for salicylic acid. **a., doeglic,** $C_{19}H_{36}O_2$, a crystalline monobasic acid obtained from the oil of the doegling, or bottle-nosed whale. **a., dracic, a., draconic, a., draconylic,** see *a., anisic*. **a., ethylenelactic,** $CH_2$ (OH) . $CH_2$ - $CO_2H = C_3H_6O_3$, an acid isomeric with ethidene lactic acid or the lactic acid of fermentation; is obtained from acrylic acid by heating with aqueous sodium hydroxide to 100° C. and in various other ways. It is a thick, uncrystallizable syrup; on heating it loses water and is converted into acrylic acid. Syn., *hydracrylic acid; β-oxypropionic acid; β-hydroxypropionic acid*. **a., ethylenephenylhydrazinsuccinic,** $C_{20}H_{22}N_4O_6$, an acid obtained from an alcoholic solution of ethylenephenylhydrazin and succinic anhydride by boiling. It occurs in acicular crystals, soluble in alcohol and ether. It is used as an antipyretic. **a., ethylidenelactic,** lactic acid. **a., excretolic,** fatty acid from feces. **a., fatty,** a monobasic acid formed by the oxidation of a primary alcohol. The fatty acids have a general formula of $C_nH_{2n}O_2$. Syn., *aliphatic acid*. **a., fellic,** $C_{23}H_{40}O_4$, a crystalline cholic acid obtained by Schotten from human bile; it is due to admixture with this acid that cholic acid from human bile differs in appearance from that obtained from other sources. **a., filicic,** $C_{14}H_{18}O_5$, from rhizome of *Dryopteris filix-mas*. **a., fluoric,** hydrofluoric acid in aqueous solution; a strong escharotic. a., formic, $CH_2O_2$, an acid obtained from a fluid emitted by ants when irritated; it is also found in stinging nettles, in shoots of the pine, and in various animal secretions. It is prepared by heating oxalic acid and glycerol. It is a colorless, mobile fluid, with a pungent odor; it is a vesicant. a., gallic, $C_7H_6O_5$, occurs free in nutgalls, in tea, and in the fruit of various other plants. It is obtained from ordinary tannic acid by boiling it with dilute acids. It crystallizes in fine, silky needles containing one molecule of water. It dissolves slowly in water and readily in alcohol and ether; has a faintly acid, astringent taste; melts at near $220°$. It is astringent and disinfectant; useful in night-sweats, diabetes, and chronic diarrhea. a., gallotannic, the tannin of nut-galls. a., gaultheric, see *methyl salicylate*. a., gentianic, *gentisin, q. v.* a., gluconic, $C_6H_{12}O_7$, formed by the oxidation of dextrose, cane-sugar, dextrin, starch, and maltose with chlorine or bromine water. Most readily obtained from glucose. It is dextrorotatory, but does not reduce Fehling's solution. Melts at $200°$. a., glutamic, a., glutaminic, $C_5H_9NO_4$, decomposition product of proteids. a., glutaric, $C_5H_8O_4$, found in decomposing pus. a., glycerinophosphoric, a., glycerinphosphoric, $C_3H_9PO_6$, a dibasic acid in combination with the fatty acids and cholin as lecithin in the yolk of eggs, in bile, in the brain, and in the nervous tissue. It is formed by mixing glycerol with metaphosphoric acid. It is a pale yellow, oily liquid, without odor, having a sour taste; soluble in water and alcohol; is used in the treatment of neurasthenia, tabes, etc. Dose $1\frac{1}{2}$–5 gr. (0.1–0.3 Gm.) 3 times daily. a., glycerinsulphuric, $C_3H_8SO_6$, a monobasic body forming a series of salts called glycerosulphates. Syn., *sulphoglyceric acid*. a., glycerophosphoric, a decomposition product of lecithin. a., glycerosulphuric, see a., glycerinsulphuric. a., glycocholic, $C_{26}H_{43}NO_6$, a monobasic acid found in bile; sparingly soluble in water and crystallizing in minute needles. a., glycollic, $C_2H_4O_3$, oxyacetic acid, produced by the action of nascent hydrogen upon oxalic acid. It is a thick syrup that gradually crystallizes on standing over sulphuric acid; the crystals melt at $80°$ and deliquesce in the air. It dissolves readily in alcohol, water, or ether. a., glycosuric, an acid sometimes occurring in urine. a., glycuronic, $C_6H_{10}O_7$. This acid has been found in urine; it probably does not exist there normally, but appears after taking certain drugs, as benzol, indol, nitrobenzol, and the quinine derivatives. a., guaiacolcarbonic, a., guaiacolcarboxylic, $C_8H_8O_4$, a monobasic crystalline acid, melting at $150°$ C. It is antiseptic and antipyretic. a., gummic, see *arabin*. a., gymnemic, $C_{32}H_{55}O_{12}$, a greenish-white, amorphous powder with a harsh acid taste, soluble in alcohol and chloroform and slightly soluble in water and ether. It is obtained from the leaves of *Gymnema sylvestre*, and obtunds the taste for bitter or sweet things, but not for sour, pungent, or astringent ones. It is used as a mouthwash in 12 % hydroalcoholic solution before taking nauseous medicines. a., helvellaic, an acid which destroys red blood-corpuscles, obtained by Böhm from juice of the mushrooms belonging to the genus *Helvella*. a., helvellic, $C_{12}H_{20}O_7$, an acid obtained from fresh belladonna, occurring as a yellow, transparent, syrupy liquid of strong acid reaction. a., hippuric, $C_9H_9NO_3$, occurs in considerable amount in the urine of herbivorous animals, sometimes in that of man. It crystallizes in rhombic prisms, and dissolves readily in hot water and alcohol. Syn., *benzoyl glycocoll*. a., homogentisic, $C_8H_8O_4$, same as a., *oxymandel*, an acid separated by Baumann from highly-colored urine, believed to be formed by the action of bacteria on the tyrosin normally found in pancreatic digestion. a., hydra-, see *hydrogen acids under acid*. a., hydracrylic, $C_3H_6O_3$, an acid isomeric with lactic acid. See q., *ethylenelactic*. a., hydriodic, HI, a gaseous acid. Its solution (*acidum hydriodicum dilutum*, U. S. P.) and a syrup prepared from it, *syrupus acidi hydriodici* (U. S. P.), are used as alteratives, with the general effects of iodine. Dose of the *syrup* 1–4 dr. (4–16 Cc.). a., hydriodic, dilute, a 10 % solution of hydriodic acid in 90 % of water; an alterative of especial value in scrofulosis of children. a., hydrobromic, HBr; the dilute acid, which is the chief form used, consists of 10 parts acid and 90 parts water. It is a solvent for quinine, is useful in hysteria, congestive headaches, and neuralgia, and is recommended as a substitute for potassium and sodium bromides. Dose 20 min.–2 dr. (1.3–8.0 Cc.). a., hydrochloric, HCl, a liquid consisting of 31.9 % by weight of HCl gas in 68.1 % of water. It is colorless, pungent and intensely acid. Syn., *muriatic acid*. a., hydrochloric, dilute, a 10 % solution of absolute acid in water. Valuable as an aid to digestion. Dose 3–10 min. (0.19–0.65 Cc.). a., hydrocyanic, aqueous, the hydrocyanic acid obtained by distillation, which contains a certain percentage of water before removal by fractional distillation and desiccation. a., hydrocyanic, dilute, HCN, a liquid consisting of 2 % of the acid with 98 % of water and alcohol. It possesses an odor like that of bitter almonds. Prussic acid is found in the bitter almond, the leaves of the peach, and in the cherry-laurel, from the leaves of which it is distilled. It is one of the most active poisons known, death from complete asphyxia being almost instantaneous. It is valuable for its sedative effects in vomiting, whooping-cough, and spasmodic affections. Dose 1–3 min. (0.06–0.2 Cc.). Syn., *prussic acid*. a., hydrocyanic, vapor, 1 part of dilute acid in 4–6 parts of water, warmed, and the vapor inhaled to relieve irritable coughs. a., hydrofluoric, HF, a compound of hydrogen and fluorine; powerfully corrosive, used for etching on glass. a., hydroparacumaric, $C_9H_{10}O_3$, occurs in urine in minute quantities. a., hydrosulphuric, $H_2S$, a gas formed during the putrefaction of albuminous substances; it occurs in sulphur mineral waters, and is produced by the action of mineral acids on metallic sulphides. It has the odor of rotten eggs. Syn., *hydrogen sulphide; sulphureted hydrogen; sulphhydric acid*. a., hypochlorous, HClO, an unstable compound, important as a disinfecting and bleaching agent. a., hypogeic, a., hypogæic, $C_{16}H_{30}O_2$, a monobasic acid found in peanut (*Arachis hypogæa*) oil, occurring as fine, colorless, stellate groups of needles which melt at $33°$ C. and solidify again at $28°$–$30°$ C.; soluble in alcohol and ether; insoluble in water. a., hyponitrous, HNO, forms hyponitrites. a., hypophosphorous, $H_3PO_2$; its salts (hypophosphites), also the dilute acid, and a syrup prepared from it, are used as remedial agents. a., ichthyolsulphonic, $C_{28}H_{38}S_4O_6$, an acid produced from Tyrolean bituminous mineral by the action of sulphuric acid; it is strongly acid and contains about 16.4 % of sulphur. It is antiphlogistic and astringent, and is used in the form of its salts, chiefly "ichthyol," the ammonium salt. a., igasuric, from seeds and surrounding pulp of nux vomica. a., indigosulphuric, $C_{16}H_8S_2N_2O_8$, from indigo by the action of sulphuric acid. a., indoxylsulphonic, $C_8H_7NSO_4$, found in urine. a., indoxylsulphuric, an acid that, combined with potassium, occurs in the urine as indican. a., inorganic, a mineral acid or one in which the carboxyl group CO.OH is absent. a., inosic, a., inosinic, $C_{10}H_{14}N_4O_{11}$, found in muscle tissue. a., iodic, $HIO_3$, a monobasic acid. Its solution (2 %) has been recommended as an alterative by subcutaneous injection. a., iodosobenzoic, $C_6H_4 . OI .- COOH_2$, a compound analogous in action to iodoform. a., isobutylcarbonic, a., isobutylcarboxylic, see a., *valeric, normal*. a., isobutylformic, a., isopropylacetic, see a., *isovaleric*. a., isovaleric, $(CH_3)_2 .- CH . CH_2 . CO_2H$, an isomer of valeric acid, obtained from oil of valerian or from oxidation of amyl-alcohol; occurs as a transparent, colorless, oily liquid with odor of valerian and old cheese; melts at $51°$ C.; boils at $174°$ C. Sp. gr., 0.9470 at $0°$ C. Used in nervous affections. Maximum dose 10 drops; a day, 40 drops. Syn., *monohydrated valeric acid; valeric acid; primary pentoic acid; isobutyl carboxyl; isopropylacetic acid*. a., jecoleic, an acid forming one of the essential constituents of cod-liver oil and isomeric with doeglic acid. a., kombic, a compound obtained by Fraser in the lead precipitate from an aqueous solution of alcoholic extract of strophanthin. It is freely soluble in water and of strongly acid reaction. a., kynureic, see a., *cynureic*. a., lactic, $HC_3H_5O_3$, a liquid containing 75 % of absolute acid in 25 % of water, produced in the fermentation of milk. It is useful in aiding digestion, in diabetes, in tuberculosis of the larynx, and as a solvent of false membrane in diphtheria. Dose $\frac{1}{2}$ dr.–$\frac{1}{2}$ oz. (2–16 Cc.) in the 24 hours. a., lactic, diluted (B. P.), lactic acid, 3 oz., distilled water, sufficient to make one pint. Dose $\frac{1}{2}$–2 dr. (2–8 Cc.). a., lactolactic, a., lactyllactic, $C_6H_{10}O_5$, a monobasic acid obtained from a solution of lactic acid heated to $130°$ to

140° C. Syn., *lactyl lactate; lactic anhydride; lactyl anhydride.* a., **lanoceric,** $C_{30}H_{60}O_4$, an acid resulting from the saponification of lanolin; it melts at 104° C. a., **lanopalmitic,** $C_{16}H_{32}O_3$, resulting from the saponification of lanolin. It melts at 87°. a., **leucamic,** see *leucin.* a., **levulinic,** $C_5H_8O_3$, obtained from levulose, cellulose, cane-sugar, etc.; a very hygroscopic crystalline substance, soluble in water, ether, or alcohol, and melting at 33.5° C. a., **linoleic,** $C_{16}H_{28}O_2$, occurs as a glycerid in drying oils, such as linseed oil, hemp oil, poppy oil, and nut oil. a., **lupamaric,** the bitter acid of hops. a., **lysuric,** $C_8H_{12}(COC_6H_5)_2N_5O_3$, a substance obtained by Drechsel from lysin by action of benzoyl chloride. a., **maleic,** a., **maleinic,** $C_4H_4O_4$, obtained from malic acid by distillation; it occurs in prisms, soluble in water, alcohol, and ether, melting at 130° C., boiling at 160° C. a., **malic,** $C_4H_6O_5$, a bibasic acid, occurring free or in the form of salts in many plant-juices, in unripe apples, in grapes, and in mountain-ash berries. It forms deliquescent crystals that dissolve readily in alcohol, slightly in ether, and melt at 100°; it has a pleasant acid taste. a., **malonic,** $C_3H_4O_4$, occurs in the deposit found in the vacuum pans employed in beet-sugar manufacture; it may be obtained by the oxidation of malic acid with chromium trioxide. a., **mandelic,** $C_6H_5 . CH(OH) . CO_2$H, formed from benzaldehyd by the action of prussic acid and HCl. a., **mannitic,** $C_6H_{12}O_7$, from sugars by oxidation. a., **margaric,** a., **margarinic,** $C_{17}H_{34}O_2$, a monobasic acid existing in nearly all animal fats and occurring as a solid substance melting at about 60° C. It is believed by some to be a mere mixture of palmitic and stearic acids. a., **marine,** hydrochloric acid. a., **meconic,** $C_7H_4O_7$, a tribasic acid, occurring in opium in union with morphine. It crystallizes with 3H₂O in white laminæ. a., **mephitic,** carbon dioxide. a., **mesotartaric,** inactive tartaric acid obtained by heating 30 parts of tartaric acid with 4 parts of water for 2 hours to 165° C. a., **metaphosphoric,** $HPO_3$, a glassy solid, freely soluble in cold water, and converted by boiling into orthophosphoric acid. It is used as a test for albumin in the urine. a., **mineral,** see a., *inorganic.* a., **monobromacetic,** $C_2H_3BrO_2$, produced by heating acetic acid with bromine; it is escharotic and antiseptic. Syn., *bromacetic-acid.* a., **monochloracetic,** $C_2H_3ClO_2$, from chlorine by action of boiling acetic acid containing sulphur and iodine; used in xanthoma. a., **monoiodosalicylic,** $C_7H_5IO_3$, produced by boiling salicylic acid with iodine and alcohol. It is used in acute articular rheumatism. Dose 15-45 gr. (1-3 Gm.) a day. a., **mononitrosalicylic,** $C_6H_3(NO_2)OH .- CO_2H$, an acid obtained by action of nitric acid on indigo or on salicylic acid. Syn., *indigotic acid; nitrospiroylic acid; nitroanilic acid; anilic acid.* a., **morphoxylacetic,** $C_{17}H_{22}NO_5 . C . H_2CO_2H$, a narcotic similar to morphine but weaker. a., **mucic,** $C_6H_{10}O_8$, from gums and sugars. a., **muriatic,** see a., *hydrochloric.* a., **muriatic, dephlogisticated,** a., **muriatic, oxygenated,** chlorine. a., **muriatic, superoxygenated,** chloric acid. a., **myotonic,** an acid obtained from *Palicourea marcgrafii,* occurring as a yellowish, oily, narcotic, and extremely poisonous liquid. a., **myristic,** $C_{14}H_{28}O_2$, from nutmegs. a., **myrionic,** $C_{10}H_{12}N_2O_{10}$, an acid that occurs as a potassium salt in the seeds of black mustard. a., **β-naphthalinsulphonic,** $C_{10}H_7 . SO_3H$, an acid occurring in white, opalescent scales with generally a tinge of red; freely soluble in water and alcohol, slightly in ether. It is a sensitive reagent for albumin. a., **naphthionic,** $C_{10}H_6(NH_2) . SO_3H$, an acid obtained from naphthylamine by action of ammonium sulphite. It is recommended as an antidote for nitrite poisoning; also in the treatment of acute iodism and in troubles of the bladder originating in the alkalescence of the urine. Dose 40-60 gr. (2.5-4.0 Gm.) daily. Syn., *α-naphthylaminsulphonic acid.* a., **naphthoic,** $C_{11}H_8O_2$, a crystalline substance of which 2 isomeric compounds may be formed by saponification of the 2 modifications of naphthonitril. a., **narceic,** see *narcotin.* a., **neurostearic,** $C_{19}H_{36}O_2$, from brain-tissue. a., **nicotinic,** $C_6H_5NO_2$, from tobacco. a., **nitric,** $HNO_3$, a liquid consisting of 68 % absolute acid in 32 % of water. The pure acid is colorless, fuming, and highly caustic. It is used in cauterization of chancres and phagedenic ulcers and as a reagent. a., **nitric, anhydrous,** nitrogen pentoxide. a., **nitric, dilute,** contains 10 % absolute acid. It is used internally to aid digestion, to stimulate the hepatic function, etc. Dose 3-15 min. (0.2-1.0 Cc.), well diluted. a., **nitric, monohydrated,** pure nitric acid. a., **nitro-,** an acid produced from another acid by replacing the hydrogen with nitryl ($NO_2$). a., **nitroanilic,** same as a., *mononitrosalicylic.* a., **nitrohydrochloric,** a., **nitromuriatic,** a golden-yellow, fuming mixture of 4 parts of nitric and 15 of hydrochloric acid. It is a solvent of gold; it is valuable in affections of the liver. Dose 1-7 min. (0.06-0.45 Cc.), very dilute. Syn., *aqua regia.* a., **nitrohydrochloric, dilute,** consists of 4 parts nitric acid, 18 parts hydrochloric acid, and 78 parts water. Dose 5-20 min. (0.3-1.3 Cc.), well diluted. a., **nitrosonitric,** fuming nitric acid. a., **nitrospiroylic,** see a., *mononitrosalicylic.* a., **nitrous,** $HNO_2$, from decomposing nitrites. a., **Nordhausen,** brown, fuming sulphuric acid, first manufactured at Nordhausen. a., **nucleic,** a., **nucleinic,** any one of a group of organic acids containing C, H, O, N, and a large proportion of P. The nucleic bases are present in the nucleic acid radicals as organic compounds. The nucleic acids occur in nature, free or in combination with albumins, when they are called primary acids. On decomposition they yield nucleic bases, and according to their origin are termed *spermonucleic acid, thymono-nucleic acid, yeast-nucleic acid,* etc. According to Kossel, there are in reality only 4 true nucleic acids, viz., adenylic acid, guanylic acid, sarcylic (hypoxanthylic) acid, and xanthylic acid. On decomposition the primary acids give rise to secondary acids which contain more phosphorus than the primary acids, and may or may not give rise to xanthin bases on further decomposition; according to Simon, they may be divided into acids of the type of *plasminic acid* and of *thyminic acid* respectively. a., **oleic,** a., **oleinic,** $C_{18}H_{34}O_2$, an acid present in many fats and oils. It is a colorless oil, crystallizing on cooling, soluble in alcohol, benzol, and the essential oils; insoluble in water. It saponifies when heated with alkaline bases. It is used in making the oleates. a., **organic,** an acid characterized by the presence of the carboxyl group, CO . OH. a., **orthoamidosalicylic,** $C_6H_3(NH_2)(OH)- COOH$, a gray, amorphous, slightly sweet, inodorous powder obtained by reduction of orthonitrosalicylic acid and insoluble in water, alcohol, and ether. It is employed in chronic rheumatism. Dose 3-7 gr. (0.25-0.5 Gm.). a., **orthoboric,** see *boron.* a., **orthophosphoric,** $H_3PO_4$, ordinary phosphoric acid, as distinguished from metaphosphoric and pyrophosphoric acids. a., **osmic,** $OsO_4$, the oxide of *osmium,* one of the rarer elements; it occurs as yellow, acrid, burning crystals, yielding an intensely irritating vapor; has been recommended for hypodermatic use in sciatica, strumous glands, and cancer; is used in histology as a fixing agent and stain for fats. a., **otoic,** same as a., *caprylic.* a., **oxalic,** $C_2H_2O_4$, a colorless, crystalline solid, obtained by treating sawdust with caustic soda and potash. Occurs in many plants, chiefly as potassium oxalate; with 2 parts of water it crystallizes in fine, transparent monoclinic prisms. Is soluble in 9 parts of water at moderate temperature and quite easily in alcohol. Has been recommended in amenorrhea. In large doses it is a violent poison. a., **oxaluric,** $C_3H_4N_2O_4$, oxidation product of uric acid. a., **oxuric,** Vauquelin's name for impure alloxanic acid. a., **oxybutyric,** see *oxybutyric.* a., **oxygen,** an acid containing more oxygen than is requisite for saturation. a., **oxymandelic,** $C_8H_8O_4$, occurs in urine in acute yellow atrophy of the liver. a., **oxymuriatic.** 1. Hydrochloric acid. 2. Chloric acid. 3. Chlorine. a., **β-oxynaphthoic,** $C_{11}H_8O_3$, obtained from sodium betanaphthol by the action of carbon dioxide with heat. It is a surgical antiseptic. Syn., *β-naphtholcarboxylic acid; β-carbonaphthoic acid.* a., **oxypropionic,** lactic acid. a., **oxyproteic,** a neutral sulphur compound found in the urine. a., **palmitic,** $C_{16}H_{32}O_2$, an acid existing as a glycerol ether in palm-oil and in most solid fats. a., **paracresotic,** $C_8H_8O_3$, an intestinal antiseptic. a., **parafumaric,** see a., *maleic.* a., **paralac'tic,** see *sarcolactic.* a., **paraoxyphenylacetic,** $C_8H_8O_3$, found in small quantities in the urine. a. of **pearls,** acid phosphate of sodium. a., **pectic,** $C_{14}H_{22}O_{15}$, from pectin. a., **perchloric,** $HClO_4$, a volatile liquid; it forms perchlorates. a., **periodic,** $HIO_4+3H_2O$, an acid obtained from iodine by the action of concentrated perchloric acid; is soluble in water and alcohol, slightly in ether, and melts at 130°-133° C. Is a powerful oxidizer. Syn., *hepta-*

*iodic acid.* a., **permanganic**, $HMnO_4$, a monobasic acid. a., **perosmic**, see *a., osmic.* a., **phenacetu′ric**, found in the urine of herbivorous animals, sometimes in human urine. a., **phenic**, carbolic acid. a., **phenolsulphonic**, see *a., sulphocarbolic.* a., **phenylic**, phenol. a., **phenylsalicylic**, $C_{13}H_{10}O_4$, a white, antiseptic powder, soluble in alcohol, ether, and glycerol, but very slowly in water; is used as a surgical dressing like iodoform. Syn., *orthooxydiphenylcarbolic acid; phenylorthooxybenzoic acid.* a., **phenylsulphuric**, see *a., sulphocarbolic.* a., **phocenic**, see *a., valeric.* a., **phosphoantimonic**, a yellowish, very acid substance, obtained from antimonium pentachloride by the action of concentrated aqueous solution of sodium phosphate. Used as an alkaloid reagent. a., **phosphocarn′ic**, $C_{10}H_{17}N_3O_8$, a nitrogenous extraction of muscle. a., **phosphoric**, $H_3PO_4$, contains 50 % each of acid and of water; is obtained from bones or by oxidation of phosphorus. Syn., *orthophosphoric acid.* a., **phosphoric, anhydrous**, $P_2O_5$, obtained from phosphorus by complete combustion, occurring as a bulky, light, white, deliquescent powder, soluble in water. Is used as a chemical agent. a., **phosphoric, dilute**, contains 10 % of absolute acid. Employed in digestive disturbances, in strumous diseases, and to dissolve phosphatic deposits. Dose 5–30 min. (0.32–2.0 Cc.). a., **phosphoric, glacial**, a., **phosphoric, monobasic**, see *a., metaphosphoric.* a., **phosphorous**, $H_3PO_3$, a dibasic oxyacid of phosphorus, containing one atom of oxygen less than phosphoric acid. a., **phosphotungstic**, $H_3PO_4 \cdot 12WO_3$, an acid used as an alkaloid and peptone test. a., **picric**, $C_6H_2(NO_2)_3OH$, obtained by the nitration of phenol. Forms pale yellow, shining, prismatic, laminar, or columnar crystals, which possess a very bitter taste. Is readily soluble in hot water; its solution dyes silk and wool a beautiful yellow color. It is recommended as an antiperiodic and anthelmintic. Used as a test for albumin and sugar. Dose 5–15 gr. (0.32–1.0 Gm.) a day. Syn., *carbazotic acid; trinitrophenol.* a., **pimentic**, see *eugenol.* a., **pipitzahoic**, a., **pipitzahoimic**, $C_{14}H_{20}O_5$, a purgative principle discovered by Rio de la Loza in species of *Perezia*, and also obtained from *Trixis radiale*. Used as a mild drastic. Dose 3–5 gr. (0.2–0.3 Gm.). a., **pivalic**, see *a., valeric, tertiary.* a., **plasminic**, a secondary nucleic acid obtainable from yeast. Is soluble in water and precipitates albumins in acid solution. Its phosphoric acid radical is capable of forming a true organic iron compound containing 1 % of iron. On decomposition with mineral acids by boiling it yields nucleic bases and phosphoric acid. a., **plumbic**, $PbO_2$, lead dioxide. a., **polybasic**, acids containing several carboxyl groups. a., **polychromic**, see *a., aloetic.* a., **propionic**, $C_3H_6O_2$, an oxidation-product of propylic alcohol; it is a clear, colorless liquid, with an odor like butyric and acetic acids, and a specific gravity of 1.013 at 0° C.; is miscible with water and boils at 141° C. a., **propionylsalicylic**, a compound obtained from salicylic acid by action of anhydrous propionic acid. Used in gout and rheumatism. a., **prussic**, see *a., hydrocyanic.* a., **pyridintricarboxylic**, a., **pyridintricarbonic**, $C_8H_5NO_6$, an oxidation-product of cinchona alkaloids; it is a white, crystalline powder, soluble in water and alcohol, and melting at 250° C. Is antipyretic, antiseptic, and antiperiodic; used in whooping-cough, typhoid and intermittent fevers, etc., and externally as an injection in urethral inflammation. Dose 10 gr. (0.6 Gm.) 5 times daily. Syn., *carbocinchomeronic acid.* a., **pyro-**, an acid formed from another acid by action of heat. a., **pyroboric**, $H_2B_4O_7$, from boric acid by heat. a., **pyrogallic**, $C_6H_6O_3$, pyrogallol, formed by heating gallic acid with water to 210°. It forms white leaflets or needles, is readily soluble in water, less so in alcohol and ether. Useful in the treatment of certain skin diseases; is poisonous and must be used with caution. a., **pyroligneous**, crude acid obtained in the destructive distillation of wood. It is a clear liquid, of reddish-brown color and strong acid taste, with a peculiar penetrating odor described as empyreumatic, due largely to the furfurol it contains. It contains from 4 to 7 % of real acetic acid. a., **pyrophosphoric**, the dihydric phosphate, $2H_2O \cdot P_2O_5$, one of the forms of phosphoric acid. It is poisonous. Its iron salt is used in medicine. The pure acid is a soft glassy mass. a., **pyrosorbic**, see *a., maleic.* a., **quinic**, $C_7H_{12}O_6$, from cinchona bark. a., **rheinic**, see *a., chrysophanic.* a., **ricinoleic**, $C_{18}H_{34}O_3$, the

active principle of castor oil. a., **rosolic**, $C_{20}H_{16}O_3$, from rosanilin by action of nitric acid used as a dye and test for acids. a., **rutic**, same as *a., capric.* a., **rutinic**, $C_{25}H_{28}O_{15}$, the coloring principle of rue. a., **salicylacetic**, a., **salicyloacetic**, $C_9H_8O_5$, a reaction-product of sodium salicylate in a soda solution with sodium monochloracetate; soluble in boiling water and alcohol, slightly in cold water, ether, chloroform, and benzene. It is antiseptic and used in the same manner as salicylic acid. Syn., *acetosalicylic acid; salicyloxyacetic acid; salicylhydroxyacetic acid.* a., **salicylic**, $C_7H_6O_3$, occurs in the buds of *Spiræa ulmaria*, in the oil of wintergreen, and in other varieties of gaultheria. It forms either a white crystalline powder, or white prismatic and acicular prisms without odor or taste. It is soluble in water and in chloroform, and is antiseptic; it is used in the treatment of acute articular rheumatism and myalgia. Dose 5–20 gr. (0.3–1.3 Gm.), not exceeding 1 dr. (4 Gm.). Syn., *orthooxybenzoic acid.* a., **salicylsulphonic**, a., **salicylsulphuric**, see *a., sulphosalicylic.* a., **salicyluric**, $C_9H_9(OH)NO_3$, a compound found in urine after taking salicylic acid. a. **of salts**, hydrochloric acid. a., **sarcolactic**, $C_3H_6O_3$, occurs in blood and in muscles, to which it gives their acid reaction, especially after the muscles have been in a state of activity. It is also found in urine in phosphorus-poisoning. a., **sclerotic**, a., **sclerotinic**, an acid found in ergot, of which it is one of the active principles. a., **scoparic**, see *scoparin.* a. **of sea-salt**, hydrochloric acid. a., **septic**, nitric acid. a., **sphacelinic**, an acid, regarded as the constituent of ergot, which causes gangrene and develops the cachexia of that disease. a., **stearic**, a., **stearinic**, $C_{18}H_{36}O_2$, associated with palmitic and oleic acids as a mixed ether, in solid animal fats, the tallows. a., **stibious**, $SbCl_3$, a colorless, transparent mass, soluble in alcohol and carbon disulphate, and melting at 73.2° C. It is a caustic. Syn., *antimonious oxide of antimony; antimony trichloride.* a., **stibous**, $C_{14}H_{12}O_5$ (Gmelin), a crystalline substance obtained from oil of bitter almonds by action of fuming sulphuric acid. a., **succinic**, $C_4H_6O_4$, an acid obtained in the distillation of amber, and also prepared artificially. a., **sulphanilic**, $C_6H_4(NH_2) \cdot SO_3H$, obtained by heating anilin (1 part) with fuming $H_2SO_4$ (2 parts) to 180° until $SO_3$ appears. It crystallizes in rhombic plates which effloresce in the air. It is used as a reagent. a., **sulphazotized**, a class of acids formed from potassium nitrite by action of sulphurous acid. a., **sulphocarbolic**, $C_6H_5HSO_4$, phenyl bisulphate, formed by the union of phenol and sulphuric acid. Its salts, the sulphocarbolates, are used in medicine as intestinal antiseptics, etc. a., **sulphoindigotic**, a., **sulphoindylic**, see *a., indigosulphuric.* a.s **sulphonic**, a class of acids of the general formula $Rn \cdot (SO_2 \cdot OH)_n$ when Rn is a radical whose quantivalence is N. Such acids are derived from sulphuric acid by the substitution of a radical for hydroxyl; or they may be regarded as acid sulphites derived from sulphurous acid, $H_2SO_3$, by the replacement of half of its hydrogen by a basic radical. a., **sulphonilic**, see *a., sulphanilic.* a., **sulphophenic**, see *a., sulphocarbolic.* a., **sulphosalicylic**, $C_7H_6SO_5$, an acid obtained from salicylic acid by the action of sulphuric anhydride, occurring as white crystals, soluble in water and alcohol, melting at 120° C., and colored an intense violet-red by ferric chloride. It is used as a test for albumin in urine. Syn., *salicylsulphonic acid.* a., **sulphothiocarbonic**, see *a., xanthogenic.* a., **sulphuric**, $H_2SO_4$, a heavy, oily, corrosive acid, consisting of not less than 92.5 % sulphuric anhydride and 7.5 % of water. It is used as a reagent and as a caustic. Syn., *oil of vitriol.* a., **sulphuric aromatic**, contains 20 % acid, diluted with alcohol and flavored with cinnamon and ginger. It is used as an astringent in diarrhea and in night-sweats; also in hemoptysis. Dose 5–15 min. (0.32–1.0 Cc.). a., **sulphuric, dilute**, contains 10 % strong acid to 90 % of water. It is used as an astringent. Dose 10–15 min. (0.65–1.0 Cc.), well diluted. a., **sulphuric, fuming**, $H_2SO_4 \cdot SO_2$, an oily liquid, fuming in the air, obtained by roasting ferrous sulphate. Syn., *Nordhausen oil of vitriol; Nordhausen acid.* a., **sulphurous**, $H_2SO_3$, a colorless acid containing about 6.4 % of sulphurous anhydride in 93.6 % of water. The gas, $SO_2$, is a valuable disinfectant. The acid is used as a spray or lotion in diphtheria, stomatitis, and as a wash for indolent and syphilitic ulcers. The various hyposulphites are mainly valuable in that they decompose and give off sulphur dioxide. Dose 5 min.-1 dr. (0.32-4.0 Cc.). a., **sulphydric**, see *a., hydrosulphuric*. a., **sumbulic**, a., **sumbulolic**, see *a., angelic*. a., **sylvic**, $C_{20}H_{30}O_2$, from resin. a., **tannic**, $C_{14}H_{10}O_9$, an astringent acid obtained from nutgalls, and occurring in yellowish, scaly crystals. It is soluble in water and alcohol. It is an antidote in poisoning by alkaloids and tartar emetic, and is used as an astringent in catarrh of mucous membranes, and externally in many skin diseases. Dose 1-20 gr. (0.065-1.3 Gm.). Syn., *tannin*. (For preparations of *tannic acid* see respective headings.) a., **tanningenic**, a., **tanningic**, see *catechin*. a., **tartaric**, $H_2C_4H_4O_6$, an astringent acid widely distributed in the vegetable world, occurring principally in the juice of the grape, from which it deposits after fermentation in the form of acid potassium tartrate (argol). It is chiefly employed in refrigerant drinks and in baking-powders; 20 grains neutralize 27 of potassium bicarbonate, 22 of sodium bicarbonate, and 15½ of ammonium carbonate. Dose 10-30 gr. (0.65-2.0 Gm.). a., **tartaric, inactive**, see *a., mesotartaric*. a., **taurocholic**, $C_{26}H_{44}NSO_7$, occurs in bile; it is very soluble in water and alcohol and crystallizes in fine needles. a., **telluric**, $H_4TeO_4+2H_2O$, the dibasic acid of tellurium. a., **tetraboric**, $H_2B_4O_7$, boric acid heated to 160° C., forming a glassy mass. Syn., *pyroboric acid*. a., **tetrathiodichlorsalicylic**, ($S_2$: $C_6HCl[OH]COOH)_2$, obtained from salicylic acid by the action of sulphuryl chloride, and heat; it occurs as a reddish-yellow powder, soluble in aqueous alkalies. It is antiseptic and used as a dusting-powder. a., **thiacetyienic**, see *a., thioacetic*. a., **thio-**, an acid in which sulphur is substituted for oxygen. a., **thioacetic**, $C_2H_4OS$, a clear, pungent, sour liquid with a sulphureted hydrogen odor, obtained from glacial acetic acid and phosphorus pentasulphide. It is used as a substitute for sulphureted hydrogen in analysis. Syn., *ethanethiolic acid; thiacetylenic acid; thiacetic acid; acetosulphuric acid*. a., **thiolinic**, a dark mass, consisting of linseed oil and sulphur dioxide, used in skin diseases. Syn., *sulphurated linseed oil; thiolin*. a., **thioncarbonthiol**, see *a., xanthogenic*. a., **thiosalicylic**, $C_7H_6SO_2$, a brownish-yellow mass obtained from amidobenzoic acid by the successive action of nitrous acid and sulphureted hydrogen; a surgical antiseptic. a., **trichloracetic**, $HC_2Cl_3O_2$, an acid formed from acetic acid, 3 atoms of the hydrogen of which are, in the new acid, replaced by chlorine. It is used as a reagent for the detection of albumin in the urine and as a caustic. a., **trichlorcarbolic**, a., **trichlorphenic**, see *trichlorphenol*. a., **tricyanic**, $H_3C_3N_3O_3$, obtained from tricyanogen chloride by boiling it with water and alkalies. It crystallizes from aqueous solution with two molecules of water in large rhombic prisms; soluble in 40 parts of cold water; easily soluble in hot water and in alcohol. Syn., *cyanuric acid*. a., **trimethacetic**, a., **trimethylacetic**, a., **trimethylcarbincarbonic**, see *a., valeric, tertiary*. a., **tropic**, $C_9H_{10}O_3$, from atropine. a., **tumenolsulphonic**, a substance obtained from tumenol by action of fuming sulphuric acid; used as a dusting-powder. a.s, **uramic**, a series of carbamide—CONH— compounds occurring in the urine after the ingestion of amido-acids. They comprise methylhydantoic acid, taurocarbamic acid, uramidobenzoic acid, and tyrosinhydantoinic acid or hydantoin hydroparacumaric acid. They are found after the ingestion of sarcosin or methylglycocoll, of taurin, amidobenzoic acid, and tyrosin respectively. a., **ureous**, see *xanthin*. a., **uric**, $C_5H_4N_4O_3$, an acid found in the urine of all animals, especially man and the carnivora,—rarely in the herbivora,—abundantly in the excrement of birds, reptiles, and mollusks. It exists usually in combination with the metals of the alkaline group. It is separated from urine by adding hydrochloric acid and allowing the crystals to settle. a. **of urine.** 1. Phosphoric acid. 2. Uric acid. a., **urobenzoic**, see *a., hippuric*. a., **urocanic**, a., **urocalulic**, $C_6H_8N_2O_2 + 11_2O$, from dogs' urine. a., **uroproteic**, $C_{66}H_{116}N_{20}SO_{34}+nH_2O$, from dogs' urine. a., **valeric**, a., **valerianic**, $C_5H_{10}O_2$, is formed by oxidizing normal amyl-alcohol. It is a mobile liquid with caustic acid taste and the pungent smell of old cheese. a., **valeric, active**, see *a., methylethylacetic*. a., **valeric, normal**, $CH_3(CH_2)_3CO_2H$, an isomer of valeric acid, first prepared by Lieben and Rossi from pentonitril ($C_4H_7CN$); it is a liquid with odor of normal butyric acid, boiling at 186° C.,

melting at 59° C. Sp. gr., 0.9568 at 0° C.; Syn., *pentoic acid; normal propylacetic acid; isobutyl carbonic acid*. a., **valeric, tertiary**, $(CH_3)_3C \cdot CO_2H$, a fatty crystalline acid containing a tertiary alcohol radical, discovered by Butlerow, who obtained it synthetically from tertiary butyl alcohol; melts at 35° C.; boils at 163° C. Syn., *pivalic acid; trimethylacetic acid; pseudovaleric acid; trimethacetic acid; pivalic acid; trimethylcarbincarbonic acid*. a., **veratric**, $C_9H_{10}O_4$, occurs with veratrine in sabadilla seeds; soluble in water and alcohol. a., **viburnic**, ordinary valeric acid discovered in *Viburnum opulus*. a.s, **vinic**, acids obtained from alcohol by action of acids. a., **vitriolic**, sulphuric acid. a., **xanthogenic**, HO . CS . SH, an acid not existing in the free state; the xanthates are obtained from it. Syn., *sulphothiocarbonic acid; thioncarbonthiol acid*. a.s, **xanthoproteic**, nitrogenous substances obtained from solutions of proteids by action of nitric acid. a., **xanthylic**, a primary nucleic acid yielding xanthin on decomposition. a., **yeast-nucleic**, $C_{40}H_{59}N_{16}O_{27}$.- $2P_2O_5$, a primary nucleic acid occurring in yeast; it contains a carbohydrate group, as Kossel was able to obtain from it a hexose and a pentose.

**acida** (as'-id-ah) [L]. Plural of *acidum, q. v.*

**acidalbumin** (as-id-al-bu'-min). A proteid acted upon or dissolved in the stronger acids, and yielding an acid reaction.

**acidemia** (as-id-e'-me-ah). A condition of decreased alkalinity of the blood.

**acid-fast** (as'-id-fast). Not easily decolorized by acids when stained.

**acidifiable** (as-id-i-fi'-a bl) [*acidum*, acid; *fieri*, to become]. Capable of becoming sour.

**acidifiant** (as-id-if'-i-ant). See *acidifiable*.

**acidification** (as-id'-if-ik-a-shun) [*acidum*, acid; *facere*, to make]. Conversion into an acid; the process of becoming sour.

**acidify** (as-id'-if-i). 1. To convert into an acid. 2. To render sour, to acidulate.

**acidimeter** (as-id-im'-et-er) [*acidum*, acid; μέτρον, a measure]. An instrument for performing acidimetry.

**acidimetric** (as-id-e-met'-rik). Pertaining to acidimetry.

**acidimetry** (as-id-im'-et-re) [see *acidimeter*]. Determination of the free acid in a solution by an acidimeter or by chemical reactions.

**acidism** (as'-id-izm). Same as *acidosis, q. v.*

**acidity** (as-id'-it-e) [*acidum*, acid]. The quality of being acid; sourness; excess of acid.

**acidity of the stomach**, sourness of the stomach due to oversecretion of acid or to fermentation of the food.

**acidol** (as'-id-ol). Trade name of betaine hydrochloride, $C_5H_{12}NO_2Cl$.

**acidology** (as-id-ol'-o-je) [ἀκίς, a bandage, a point; λόγος, a treatise]. The science of surgical appliances.

**acidometer** (as-id-om'-et-ur). See *acidimeter*.

**acidometric** (as-id-o-met'-rik). See *acidimetric*.

**acidometry** (as-id-om'-et-re). See *acidimetry*.

**acidophil**, **acidophile** (as-id'-o-fil) [*acidum*, acid; φίλος, loving]. 1. Susceptible of imbibing acid stains. 2. A substance having an affinity for acid stains.

**acidosis** (as-id-o'-sis) [*acidum*, acid]. Acid intoxication caused by an abnormal production of acids in the body and their faulty elimination.

**acidosteophyte** (as-id-os'-te-o-fit) [ἀκίς, a point; *osteophyte*]. A sharp, or needle-shaped, osteophyte.

**acidoxyl** (as-id-oks'-il). A compound of an acidyl or acid radical with oxygen.

**acid-proof** (as'-id-pruf). Same as *acid-fast, q. v.*

**acidulate** (as-id'-u-lāt) [*acidulare*, to make sour]. To render acid or sour.

**acidulated** (as-id'-u-la-ted). Somewhat sour or acid.

**acidulous** (as-id'-u-lus) [see *acidulated*]. Moderately sour.

**acidulum** (as-id'-u-lum) [L. dim. of *acidum*]. An acid salt.

**acidum** (as'-id-um) [L.]. See *acid*.

**acidyl** (as'-id-il). The radical of an organic acid, particularly those hydrocarbons of the formula $C_nH_{n-1}$.

**acidylated** (as-id'-il-a-ted). Combined with the residue of a fatty acid (acidyl).

**acinesia** (as-in-e'-ze-ah). See *akinesia*.

**acinesic**, **acinetic**. See *akinetic*.

**acinetatrophia** (as-in-et-at-ro'-fe-ah) [*acinesis*; *atrophia*]. Atrophy due to lack of exercise.

**acini** (*as'-in-i*) [L.]. Plural of *acinus, q. v.*
**aciniform** (*as-in'-if-orm*) [*acinus*, a grape]. Grapelike.
**acinose** (*as'-in-ōs*). See *acinous*.
**acinotubular** (*as-in-o-tu'-bu-lar*) [*acinus*, a grape; *tubulus*, a tube]. Applied to a gland or other structure having tubular acini or secreting sacs.
**acinous** (*as'-in-us*) [*acinus*, a grape]. 1. Relating to an acinus or having acini. 2. Resembling a grape or a cluster of grapes; composed of granular concretions.
**acinus** (*as'-in-us*) [*acinus*, a grape; pl., *acini*]. Any one of the smallest lobules of a compound gland, as an *acinus* of the liver.
**aciurgia** (*as-e-er'-je-ah*), or **aciurgy** (*as'-e-er-je*) [ἀκίς, point; ἔργειν, to work]. Operative surgery.
**aclastic** (*ak-las'-tik*). Not refracting.
**acleidian** (*ah-kli'-de-an*) [ἀ, priv.; κλείς, the collarbone]. Without clavicles.
**acleitocardia** (*ah-kli-to-kar'-de-ah*) [ἀ, priv.; κλείειν, to close; καρδία, the heart]. Imperfect closure of the foramen ovale.
**acmastic** (*ak-mas'-tik*) [*acme*]. Pertaining to disease with regular increase and decrease. (*Epacmastic;* first period. *Paracmastic;* period of decline.)
**acme** (*ak'-me*) [ἀκμή, a point]. 1. The highest point. 2. The crisis or critical stage of disease. 3. Acne; an acne papule; a wart.
**acmon** (*ak'-mon*) [ἄκμων, an anvil]. The incus.
**acne** (*ak'-ne*) [ἀκμή, a point]. A common, usually chronic, inflammatory disease of the sebaceous glands, occurring mostly about the face, chest, and back. The lesions may be papular, pustular, or tubercular. It occurs usually between the ages of puberty and 24 years, is generally worse in winter, and is associated with menstrual and gastrointestinal troubles. The individual lesions consist of minute pink, acuminate papules or pimples, in the center of which is a black-topped comedo (*a. punctata, a. papulosa*). Syn., *acne varus; acne vulgaris; whelk; stone pock; acne boutonneuse; acné éruptive*. **a. adenoid**, see *lupus, disseminated follicular*. **a. adolescentium**, synonym of *a. vulgaris*. **a. albida**, synonym of *milium*. **a., arthritic**, a form common in adults, especially in women at the climacteric, and thought to be connected with the arthritic diathesis. **a. artificialis**, that form that disappears when the cause is removed. **a. atrophica**, synonym of *a. varioliformis*. **a., bromine**, see *a. coagminata*. **a. cachecticorum**, a form occurring in debilitated, cachectic persons after prolonged wasting diseases, as phthisis. The eruption occurs usually on the trunk or legs, and is characterized by flat, dull-red papules and pustules of the size of a pinhead to that of a lentil. **a. cheloidienne**, see *dermatitis papillaris capillitii*. **a., chlorine**, a form occurring among men engaged in manufacturing hydrochloric acid. The skin of the face is pigmented, comedones and pustules of varying size are thickly scattered over the face, brow, scalp, neck, back, upper thorax, genitals, and inner surface of the thighs. Atheromata and curious cornifications resembling those of Darier's disease are present on the scalp. **a. ciliaris**, acne at the edges of the eyelids. **a. coagminata**, a form in which the lesions occur in clusters. The name is generally applied to the acne due to the internal use of bromine or its compounds; the groups of closely aggregated pustules form thick patches covered with scabs of dried pus, presenting beneath a dusky red and often moist surface. **a., concrete**, see *seborrhœa sicca*. **a., congestive**, see *a. rosacea*. **a. contagiosa**, an inoculable pustular disease of horses, said to differ from horse-pox. **a. cornea**, conic, discolored outgrowths, grouped or solitary, consisting of hard plugs of sebaceous matter projecting from the follicles. Syn., *ichthyosis follicularis*. **a. decalvans**, an inflammatory disease of hair-follicles with destruction of the hairs and atrophy or cicatrization of the skin. **a. disseminata**, synonym of *a. vulgaris*. **a., elephantiasic**, see *a., hypertrophica*. **a. erythematosa**, see *a. rosacea*. **a., fluent**, see *seborrhœa oleosa*. **a. frontalis**, see *a. varioliformis*. **a. generalis**, acne that has become general over the surface of the body. **a. granulosa**, see *a. cachecticorum*. **a. hordeolans**, **a. hordeolaris**, a form with the pustules arranged in linear groups. **a. hypertrophica**, a stage of acne rosacea in which there is a permanent, intensely red, non-inflammatory, nodulated thickening of the tips and sides of the nose, expanding it both laterally and longitudinally. **a. indurata**, a variety of acne vulgaris characterized by chronic, livid indurations, the result of extensive perifollicular infiltration. It is especially seen in strumous subjects. **a., iodine**, acne due to the prolonged use of an iodide. **a. keratosa**, a rare form in which a horny plug takes the place of the comedo, and by its presence excites inflammation. **a. luposa**, see *a. telangiectodes*. **a. medicamentosa**, acne due to the internal administration of certain drugs—as iodine, bromine, etc. **a. mentagra**, see *sycosis*. **a. miliaris**. 1. Milium. 2. A pustular variety of acne rosacea. **a., miliary arthritic**, see *a. cachecticorum*. **a., miliary scrofulous**, a variety of the disease usually occurring on the forehead; the pustules are small, discrete, or confluent, and often arranged in geometric figures. **a. molluscoidea, a. molluscum**, see *molluscum. contagiosum* **a. necrotica**, see *a. varioliformis*. **a. pancreat'ica**, small cysts in the pancreas due to obstructions of the smaller ducts. **a. papulosa**, see *acne*. **a., penicilliform**, see *tinea asbestina*. **a. piocalis**, a form of dermatitis common in fiber-dressers who work with paraffin and in persons otherwise brought in contact with tar or its vapor. It involves chiefly the extensor surfaces of the limbs. Syn., *tar acne*. **a., pilous**, a variety in which the pustules involve the hair-bulbs. **a., pilous, umbilicated**, a variety in which each pustule is umbilicated and pierced by a hair. **a. punctata**, a variety of acne vulgaris. **a. punctata albida**, see *milium*. **a. pustulosa**, a variety of acne vulgaris characterized by abscesses. **a. rhinophyma**, same as *a. hypertrophica*. **a. rodens**, synonym of *a. varioliformis*. **a. rosacea**, a chronic hyperemic or inflammatory affection of the skin, situated usually upon the face, especially the nose, cheeks, forehead, and chin. Syn., *rosacea; telangiectasis faciei; nævus araneus; brandy nose; whisky nose; spider nevus; spider cancer*. **a. rosacea congestiva**, see *a. hypertrophica*. **a., scorbutica**, acne associated with scurvy. **a. scrofulosa**, a variety of acne cachecticorum, occurring in strumous children. **a. sebacea**, synonym of *seborrhœa*. **a. sebacea cornea**, see *Darier's disease*. **a. sebacea molluscum**, see *atheroma*. **a., sebacea, crusty**, see *seborrhœa sicca*. **a., sebaceous, dry**, **a. sebacea exsiccata**, see *xeroderma*. **a., sebaceous, fluent**, see *seborrhœa oleosa*. **a., simplex**, a variety of acne vulgaris. **a. solaris**, a form due to exposure to the sun, marked by red papules that seldom suppurate, occurring on the nose, lower eyelids, and cheeks. **a. sycosiformis**, same as *sycosis non-parasitica*. **a., syphilitic, a. syphilitica**, a form with inflammation in the follicles, appearing in the scattered, pointed pustules with copper-colored base. Syn., *acneiform syphiloderm*. **a. tarsi**, an inflammatory affection of the large sebaceous glands of the eyelashes (Meibomian glands). **a. telangiectodes**, **a. telangiectodes**, Kaposi's name for a nonpustular disease having its origin in the hair-follicles and presenting smooth, shining, circumscribed, hemispheric nodules, pale-pink to brownish-red in color, from a pinhead to a cherry-stone in size. Epithelial cyst formation and degeneration of the hair-follicle attend it. Syn., *disseminated follicular lupus simulating acne; acne luposa; lupus miliaris; lupus follicularis acneiformis; acute disseminated nodular tuberculous lupus:* \ **a. tuberata, a. tuberculosa**, see *a. indurata*. **a., tuberculoid, a., tuberculous, umbilicated, a. umbilicata**, see *molluscum contagiosum*. **a., varicose**, a form characterized by dilated superficial capillaries. **a. varioliformis**, a somewhat rare disease, situated chiefly about the forehead, at the junction with the hairy scalp, and extending into the hair. The pustules appear in groups. Its etiology is unknown. **a. vulgaris**, see *acne*.
**acneform, acneiform** (*ak'-ne-form, ak-ne'-e-form*). Resembling acne.
**acnemia** (*ak-ne'-me-ah*) [ἀ, priv.; κνήμη, leg]. 1. Deficiency in the calf of the leg. 2. A condition marked by total absence of legs.
**acnemous** (*ak'-ne-mus*) [ἀ, priv.; κνήμη, leg]. Having recumbent calves; having no legs.
**acnitis** (*ak-ni'-tis*) [ἀκμή, a point; ιτις, inflammation]. See *hydrosadenitis phlegmonosa*.
**Acocanthera** (*ak-o-kan-the'-ra*) [ἀκωκή, a point; ἀνθηρός, blooming]. A genus of plants of the order *Apocynaceæ. A. abyssinica* yields an African arrow-poison, *mshangu*, secured from a decoction of the branches, the toxic property being due to a crystalline glucosid, $C_8H_{14}O_4$. *A. deflersii* and *A. schimperi* are used as arrow-poisons in Africa. The poisonous principles are crystalline glucosides. *A. venenata*

is a species indigenous to southern Africa; a decoction of the bark is used by the natives to poison arrows. The poisonous principle is a glucoside, *acocanthetin*, similar to or identical with *ouabain*.

**acœlius** (*ah-se'-le-us*). See *acœlius*.

**acœsis** (*ak-o-e'-sis*). See *audition*.

**acognosia** (*ah-kog-no'-ze-ah*). See *acognosia*.

**acography** (*ak-og'-raf-e*) [ἄκος, a remedy; γράφειν, to write]. A description of remedies.

**acoin** (*ak'-o-in*). Hydrochloride of diparaany-silmonoparaphenetylguanidin, a white powder, used in infiltration anesthesia by Schleich's method in a 1 : 1000 solution of 0.8 % solution of sodium chloride; also in 1 % aqueous solution in ophthalmology.

**acolasia** (*ak-o-la'-ze-ah*) [ἀκολασία, intemperance]. Unrestrained self-indulgence; lust; intemperance.

**acolastic** (*ak-o-las'-tik*) [ἀκολασία, intemperance]. Due to, or characterized by, acolasia.

**acology** (*ak-ol'-o-je*). [ἄκος, remedy; λόγος, a discourse]. *Aceology, q. v.*

**acolous** (*ah-ko'-lus*) [ἀ, priv.; κῶλον, limb]. Having no limbs.

**acomia** (*ah-ko'-me-ah*) [ἀ, priv.; κόμη, hair]. Baldness. A deficiency of hair arising from any cause.

**acomous** (*ah-ko'-mus*) [ἀ, priv.; κόμη, hair]. Hairless, blad.

**aconine** (*ak'-o-nin*). $C_{25}H_{39}NO_{11}$. A decomposition product of aconitine.

**aconite** (*ak'-on-it*). See *aconitum*.

**aconitia** (*ak-o-nish'-e-ah*). Aconitine or aconitina.

**aconitic acid** (*ak-on-it'-ik*). See *acid, aconitic*.

**aconitin** (*ak-on'-it-in*). See *aconitine*.

**aconitina** (*ak-on-it-i'-nah*. 1. See *aconitine*. 2. An impure aconitine or combination of principles obtained from the root of *Aconitum napellus*, as prepared by Morson. Its salts do not crystallize, but form gum-like masses.

**aconitine** (*ak-on'-it-ēn*), $C_{33}H_{45}NO_{12}$. *aconitina* (U. S. P.). An intensely poisonous alkaloid from *Aconitum napellus* and other species; it occurs as white, flat crystals of slightly bitter taste. Dose ⅟₁₆₀ gr. (0.0003 Gm.). Syn., *aconiticum; aconitinum*, a., amorphous, a mixture of several bases found in the bulbs of *Aconitum napellus*. Its principal constituents are aconitine and picroaconitine. It is 15 or 20 times less poisonous than pure crystallized aconitine. a., **British**, $C_{33}H_{43}NO_{12}$ (Wright), the alkaloid prepared by Morson from *Aconitum ferox*. It is a yellowish-white, crystalline powder. Dose ⅟₃₂₀ gr. (0.0002 Gm.). Also called *English aconitine*; *acraconitine*; *Morson's napelline or pure aconitine*; *Hübschmann's pseudaconitine*; *Flückiger's nepaline*. a., **Duquesnel's**, see *a. nitrate*. a. **hydrobromide**, $C_{33}H_{45}NO_{12}HBr + 2½H_2O$ (Jürgens), from crystalline aconitine, occurring as small white tablets, soluble in water and alcohol. Dose the same as the crystalline alkaloid. a. **hydrochloride**, $C_{33}H_{45}NO_{12}HCl + 3H_2O$ (Jürgens), a white, crystalline powder from crystalline aconitine, soluble in water and alcohol. Dose about the same as the alkaloid. Syn., *aconitine chlorhydrate; aconitine hydrochlorite*. a. **nitrate**, $C_{33}H_{45}NO_{12}HNO_3$, fine white prisms or rhombic crystals; it is highly poisonous and is used in neuralgia and rheumatism. Dose about the same as the alkaloid. Syn., *Duquesnel's aconitine*. a. **phosphate**, a salt of aconitine. It occurs as a white, crystalline powder or as a yellowish-white, amorphous powder. a. **salicylate**, a salt of aconitine occurring as a white, crystalline powder or as a yellowish-white, amorphous powder. a. **sulphate**, $(C_{33}H_{45}NO_{12})_2H_2SO_4$, a salt of aconitine occurring as a crystalline powder, in glass-like lumps, or as a yellowish-white, amorphous powder.

**aconitum** (*ak-on-i'-tum*) [L.]. The root of *Aconitum napellus*. It possesses a bitter, pungent taste, and produces numbness and persistent tingling in the tongue and lips. Is very poisonous. It depresses the heart, respiration, circulation, and paralyzes the sensory nerves. Is antipyretic, diaphoretic, and diuretic. The active principle is *aconitine*. As a diaphoretic and depressant to the circulation it is highly beneficial in fevers, acute throat affections, and inflammation of the respiratory organs. Dose ½-2 gr. (0.03-0.13 Gm.). aconiti, **abstractum**, has double the strength of the powdered drug or its fluid extract. Dose ¼-1 gr. a., **extractum**. Dose ⅙-⅓ gr. (0.011-0.022 Gm.). a., **fluidextractum** (U. S. P.), has a strength of 1 drop to the grain of the powdered drug. Dose ½-2 min.

0.03-0.13 Cc.). a., **linimentum** (B. P.), aconite root, camphor, and rectified spirit. a., **oleatum**, a 2 % solution of aconite in oleic acid. a., **tinctura** (U. S. P.), contains aconite 10, alcohol and water each sufficient to make 100 parts. Dose 10 min. (0.6 Cc.). a., **unguentum** (B. P.), 8 grains to the ounce.

**aconuresis** (*ah-kon-u-re'-sis*) [ἀ, priv.; *conari*, to strive; οὔρησις, urination]. Involuntary discharge of urine.

**acoprosis** (*ah-kop-ro'-sis*) [ἀ, priv.; κόπρος, excrement]. Deficient formation of feces.

**acoprous** (*ah-kop'-rus*). Characterized by the absence of excrement in the bowels.

**acopyrine** (*ak-o-pi'-rin*). A combination of aspirin and antipyrine; it is used in rheumatism. Dose, 0.5 gm. 5 or 6 times daily.

**acor** (*a'-kor*) [L.]. Acrimopy: acidity, as of the stomach.

**acorea** (*ah-ko-re'-ah*) [ἀ, priv.; κόρη, pupil]. Absence of the pupil.

**acoria** (*ah-ko'-re-ah*) [ἀ, priv.; κόρος, satisfaction]. 1. A greedy or insatiable appetite. 2. Temperance in eating. 3. A nervous stomach affection characterized by a sense of fulness.

**acorin** (*ak'-o-rin*). A bitter glucoside obtained from *Acorus calamus*, or sweet flag.

**acormus** (*ah-kor'-mus*) [ἀ, priv.; κορμός, the trunk]. A monster without a trunk or body.

**acorus** (*ak'-o-rus*). See *calamus*.

**acosmia** (*ah-kos'-me-ah*) [ἀ, priv.; κόσμος, order]. 1. Poor health. 2. Irregularity in the course of a disease. 3. Ataxia. 4. Baldness. 5. Any deformity causing irregularity of the features. Syn., *acosmy*.

**acouiation** (*ah-koo-la'-le-on*). An instrument used in teaching speech to deaf-mutes.

**acoumeter, acouometer** (*ah-koo'-me-ter, ah-koo-om'-e-ter*) [ἀκούειν, to hear; μέτρον, a measure]. 1. An instrument for measuring the acuteness of hearing. 2. An instrument arranged to give a typical sound of a vowel, which may be used as a standard to which other sounds may be referred.

**acoumetric, acoumometric** (*ah-koo-met'-rik, ah-koo-mo-met'-rik*). Pertaining to the auditory sense or to the power of estimating the relative distance of sounds. Syn., *acusmetricus; acusmometricus*.

**acoumetry** (*ah-koo'-met-re*) [ἀκούειν, to hear; μέτρον, a measure]. The measurement or testing of the acuteness of the hearing.

**acouophony** (*ah-koo-of'-on-e*) [ἀκούειν, to hear; φωνή, sound]. Same as auscultatory percussion.

**acouoxylon** (*ah-koo-oks'-il-on*) [ἀκούειν, to hear; ξύλον, wood]. A wooden (pine) stethoscope.

**acouphone** (*ah-koo-fōn*) [ἀκούειν, to hear; φωνή, sound]. A mechanism to aid defective hearing.

**acousia** (*ah-koo'-se-ah*) [ἀκουσία, constraint]. 1. Involuntary action. 2. The faculty of hearing; audition.

**acousma** (*ah-koos-* or *kowz'-mah*) [ἄκουσμα, thing heard; pl., *acousmata*]. An auditory hallucination; a condition in which imaginary sounds are noticed by the patient, are believed by him to be real.

**acousmatagnosis** (*ah-koos-mat-ag-no'-sis*). Inability to memorize sounds.

**acousmatamnesia** (*ah-koos-mat-am-ne'-se-ah*). Inability to remember sounds.

**acousmetric** (*ah-koos-met'-rik*). See *acoumetric*.

**acoustic** (*ah-koos'-tik* or *a-kows'-tik*) [ἀκουστικός]. Relating to the ear or science of sound. a. **duct**, the external meatus of the ear. a. **nerve**, the eighth cranial nerve. a. **tetanus**, the rapidity of the induction shocks in a frog's nerve-muscle preparation, as measured by the pitch of a vibrating rod. a. **tubercle**, a rounded elevation on either side of the floor of the fourth ventricle.

**acousticon** (*ah-koos'-tik-on*). An ear-trumpet.

**acoustics** (*ah-koos'-tiks* or *a-kows'-tiks*). The science of sound.

**acquired movements** (*ak-wi'-erd moov'-mentz*). Those brought under the influence of the will only after conscious and attentive effort and practice, in distinction from reacquired movements, those reinstated in their former proficiency after injury to the motor regions of the brain.

**acracholia** (*ak-ra-ko'-le-ah*) [ἀκραχολία]. A fit of passion; passionateness.

**acraconitine**. See *pseudaconitine*.

**acrania, acranial** (*ah-kra'-ne-ah, ah-kra'-ne-al*) [ἀ, priv.; κρανίον, skull]. The condition of a monster with partial or complete absence of the cranium.

**acranius** (*ah-kra'-ne-us*) [ἀ, priv.; κρανίον, cranium]. A monster wholly or partly destitute of cranium.
**acrasia** (*ah-kra'-ze-ah*) [ἀ, priv.; κράσις, moderation]. 1. Intemperance; lack of self-control. 2. Acratia.
**acratia** (*ah-kra'-she-ah*) [ἀκράτεια: ἀ, priv.; κράτος, force]. Impotence, loss of power.
**acraturesis** (*ah-krat-u-re'-sis*) [ἀκράτεια, lack of strength; οὔρησις, micturition]. Inability to micturate from atony of the bladder.
**Acree-Rosenheim formaldehyde reaction** in testing for proteins. Put a few drops of a solution of formaldehyde (1 : 5000) in a solution of protein and mix well. After 2-3 minutes allow a little concentrated sulphuric acid to flow into the test-tube slowly, so that the two solutions do not mix. A violet color appears at the line of contact.
**acribometer** (*ak-re-bom'-et-ur*) [ἀκριβής, accurate; μέτρον, a measure]. A device for measuring minute objects.
**acrid** (*ak'-rid*) [*acer*, sharp]. Pungent; irritating.
**acridine** (*ak'-rid-in*) [*acrid*], C₁₃H₉N. A substance produced by heating anilin and salicylic aldehyde to 260° with ZnCl₂. It dissolves in dilute acids with a beautiful green fluorescence, and has a very pungent odor.
**acrimony** (*ak'-rim-o-ne*) [*acrimonia*]. Irritating quality, pungency, corrosiveness: an acrid quality or state.
**acrinia** (*ah-krin'-e-ah*) [ἀ, priv.; κρίνειν, to separate]. Diminution or suppression of a secretion or excretion.
**acrinyl sulphocyanate** (*ak'-rin-il*). An acrid and vesicating substance found in white mustard.
**acrisia** (*ah-kris'-e-ah*) [ἀ, priv.; *crisis*]. The absence of a crisis from a disease; an unfavorable crisis or turn in the course of an attack of disease.
**acritical** (*ah-krit'-ik-al*) [ἀ, priv.; κρίσις, a crisis]. Without a crisis; not relating to a crisis.
**acritochromacy** (*ah-krit-o-kro'-mas-e*) [ἄκριτος, undistinguished; χρῶμα, color]. Color-blindness, achromatopsia.
**acroæsthesia.** See *acroesthesia.*
**acroanesthesia** (*ak-ro-an-es-the'-ze-ah*) [ἄκρον, extremity; ἀναίσθησια, want of feeling]. Anesthesia of the extremities.
**acroarthritis** (*ak-ro-ar-thri'-tis*). Arthritis of the extremities.
**acroasphyxia** (*ak-ro-as-fiks'-e-ah*) [ἄκρον, extremity; ἀ, priv.; *σφίξις*, pulse]. Asphyxia of the extremities. Phenomenon of Raynaud.
**acroblast** (*ak'-ro-blast*) [ἄκρον, extremity; βλαστός, a germ]. Kollmann's term for that part of the germinal membrane of the embryo which gives rise to blood-vessels filled with blood and probably connective tissue.
**acrobystia** (*ak-ro-bis'-te-ah*) [ἀκροβυστία, the foreskin]. 1. The prepuce. 2. Circumcision.
**acrobystiolith** (*ak-ro-bis'-te-o-lith*) [ἀκροβυστία, the prepuce; λίθος, a stone]. A preputial calculus.
**acrobystitis** (*ak-ro-bis-ti'-tis*). Inflammation of the prepuce.
**acrocarpous** (*ak-ro-kar'-pus*) [ἄκρον, extremity; καρπός, fruit]. In biology, fruiting at the tips, as mosses.
**acrocephalia** (*ak-ro-sef-a'-le-ah*) [ἄκρον, the summit; κεφαλή, the head]. Deformity of the head, the top of which is more or less pointed.
**acrocephalic, acrocephalous** (*ak-ro-sef'-al-ik, ak-ro-sef'-al-us*). Characterized by or affected with acrocephalia.
**acrocephaly** (*ak-ro-sef'-al-e*) [ἄκρον, a point; κεφαλή, the head]. Same as *acrocephalia*.
**acrocheir** (*ak'-ro-kir*) [ἄκρον, point; χείρ, hand]. The ends of the fingers considered together; the forearm and hand.
**acrochordon** (*ak-ro-kor'-don*) [ἀκροχορδών, literally, the end of a catgut cord]. A pedunculated or pensile wart. Synonym of *molluscum fibrosum*.
**acrocinesis, acrocinetic.** See *akrokinesis, akrokinetic*.
**acrocyanosis** (*ak-ro-si-an-o'-sis*) [ἄκρον, extremity; κύανος, blue]. Blueness of the extremities due to vasomotor disturbance.
**acrodermatitis** (*ak-ro-der-mat-i'-tis*) [ἄκρον, extremity; δέρμα, skin; ιτις, inflammation]. Inflammation of the skin of an extremity. **a., perstans,** acrodermatitis which constantly recurs.
**acrodigitalins** (*ak-ro-dij'-it-al-ins*). Digitalis substances which do not possess the general characteristics of glucosides.

**acrodynia, acrodyny** (*ak-ro-din'-e-ah, ak'-ro-din-e*). [ἄκρον, extremity; ὀδύνη, pain]. 1. Epidemic erythema; a disease closely allied to pellagra. Characterized mainly by pricking pains in the palm and soles, hyperesthesia followed by anesthesia of these parts, and an erythematous eruption, preceded by bullæ, chiefly on hands and feet. Followed by exfoliation and dark-brown or black pigmentation. Syn., *pedionalgia epidemica; erythema epidemicum*. 2. Clarus' term for a rheumatic disorder of the nerves.
**acroesthesia** (*ak-ro-es-the'-ze-ah*) [ἄκρος, extreme; ἄκρον, extremity; αἴσθησις, sensation]. 1. Exaggerated sensitiveness or sensibility. 2. Pain in the extremities.
**acrokinesis** (*ak-ro-kin-e'-sis*) [ἄκρον, extreme; κίνησις, movement]. Abnormal freedom of action, as in certain cases of hysteria.
**acrokinetic** (*ak-ro-kin-et'-ik*). Characterized by acrokinesis.
**acrolein** (*ak-ro'-le-in*) [*acer*, sharp; *oleum*, oil], C₃H₄O. Acrylic aldehyde. A colorless, mobile liquid, of pungent odor, derived from the decomposition of glycerol.
**acromania** (*ak-ro-ma'-ne-ah*) [ἄκρος, extreme; μανία, madness]. Incurable insanity.
**acromastitis** (*ak-ro-mas-ti'-tis*) [ἄκρον, extremity; μαστός, breast; ιτις, inflammation]. Inflammation of the nipple.
**acromastium** (*ak-ro-mas'-te-um*) [ἄκρον, extremity; μαστός, breast]. The nipple.
**acromegalia** (*ak-ro-meg-a'-le-ah*). See *acromegaly*.
**acromegaly** (*ak-ro-meg'-al-e*). Abnormal development of the extremities associated with disease of the pituitary body or thyroid gland. Also known as Marie's disease.
**acromelalgia** (*ak-ro-mel-al'-je-ah*). See *erythromelalgia*.
**acromial** (*ak-ro'-me-al*) [ἄκρον, the summit; ὦμος, the shoulder]. Relating to the acromion. **a. process,** the acromion.
**acromicria** (*ak-ro-mik'-re-ah*) [ἄκρον, extremity; μικρός, small]. Abnormal smallness of the extremities. A condition in which there is a reduction in the size of the nose, ears, and face, as well as hands and feet.
**acromioclavicular** (*ak-ro-me-o-kla-vik'-u-lar*) [*acromion; clavicle*]. Relating to the acromion and the clavicle.
**acromiocoracoid** (*ak-ro-me-o-kor'-ak-oid*). Pertaining to the acromion and the coracoid process.
**acromiohumeral** (*ak-ro-me-o-hu'-mer-al*) [*acromion; humerus*]. Relating to the acromion and the humerus. **a. muscle,** the deltoid.
**acromion** (*ak-ro'-me-on*) [ἄκρον, the summit; ὦμος, the shoulder]. The triangular-shaped process at the summit of the scapula.
**acromiothoracic** (*ak-ro-me-o-tho-ras'-ik*) [*acromion*; θώραξ, thorax]. Relating to the shoulder and thorax.
**acromphalus** (*ak-rom'-fal-us*) [ἄκρον, point; ὀμφαλός, the navel]. 1. The center of the umbilicus, to which the cord is attached. 2. The first stage of umbilical hernia, marked by a pouting of the navel. 3. The remains of the umbilical cord attached to the child.
**acromyle** (*ak-rom'-il-e*) [ἄκρον, point; μύλη, patella]. The patella.
**acronarcotic** (*ak-ro-nar-kot'-ik*) [*acer*, sharp; *narcotic*]. 1. Both acrid and narcotic. 2. An agent which combines an irritating and obtunding effect; acting directly upon the peripheral nerves when applied externally, or upon the brain and spinal cord, producing paralysis, convulsions, and narcosis.
**acroneurosis** (*ak-ro-nū-ro'-sis*) [ἄκρον, extremity; νεῦρον, a nerve]. Any neurosis manifesting itself in the extremities.
**acronychus** (*ak-ron'-ik-us*) [ἀκρόνυχος]. Having claws, nails, or hoofs; achronychous.
**acronyx** (*ak'-ro-niks*) [ἄκρον, extremity; ὄνυξ, a nail]. Ingrowing of the nail.
**acroparalysis** (*ak-ro-par-al'-is-is*) [ἄκρον, extremity; παράλυσις, palsy]. Paralysis of the extremities.
**acroparesthesia** (*ak-ro-par-es-the'-ze-ah*) [ἄκρον, extremity; παρά, around; αἴσθησις, sensation]. 1. Abnormal or perverted sensation in the extremities. 2. Extreme or confirmed paresthesia.
**acropathology** (*ak-ro-path-ol'-o-je*) [ἄκρον, extremity; πάθος, disease; λόγος, treatise]. The pathology of the extremities.
**acropathy** (*ak-rop'-a-the*) [ἄκρον, extremity; πάθος, disease]. Any disease of the extremities.

**acrophobia** (ak-ro-fo'-be-ah) [ἄκρον, a height; φόβος, fear]. Morbid dread of being at a great height.
**acroposthia** (ak-ro-pos'-the-ah) [ἄκρος, extreme; πόσθη, foreskin]. The distal part of the prepuce.
**acroposthitis** (ak-ro-pos-thi'-tis) [ἄκρος, extreme; πόσθη, foreskin]. Inflammation of the prepuce. Posthitis.
**acrorrheuma** (ak-ro-ru'-mah) [ἄκρον, an extremity; ῥεῦμα, a flux]. Rheumatism of the extremities.
**acroscleriasis** (ak-ro-skle-ri'-as-is) [ἄκρον, extremity; σκληρός, hard]. Sclerotic changes in the extremities.
**acroscleroderma** (ak-ro-sklet-o-der'-mah). See *sclerodactylis*.
**acrose** (ak'-rōz). A substance isolated from condensation-products of glycerose (an oxidation-product of glycerol) and formaldehyde, forming the starting-point for the synthesis of fruit-sugar, grape-sugar, and mannose.
**acrosome** (ak'-ro-sōm) [ἄκρον, extremity; σῶμα, body]. A small body at the front part of the head of the spermatozoon.
**acrosphacelus** (ak-ro-sfas'-el-us). Gangrene of the digits.
**Acrostichum** (ak-ros'-tik-um) [ἄκρον, a point; στίχος, a line of writing]. A genus of ferns of the order *Polypodiaceæ*. *A. aureum*, a tropical species; the rhizome is used in decoction for dysentery and disease of the spleen. A salt prepared from the leaves is applied to ulcers. *A. dichotomum*, an Arabian species [*medjabesa* or *mejahoese*]; the leaves are applied to burns. *A. flavens*, a South American species, used as a laxative. *A. furcatum*, an Australian species having edible rhizomes. *A. huacsaro*, a Peruvian species. It is said to be sudorific and anthelmintic. *A. sorbifolium*, a West Indian species. The juice is mixed with oil, ginger, and pepper, and used as a cataplasm in sick headache.
**acrotarsium** (ak-ro-tar'-se-um) [ἄκρον, the summit; ταρσός, the tarsus]. The instep.
**acroteria** (ak-ro-te'-re-ah) [ἀκρωτήρια]. The extremities.
**acroteriasis** (ak-ro-te-ri'-a-sis) [ἀκρωτηριάλειν, to cut off the extremities]. Mutilation by the loss of an extremity, especially a hand or foot. In teratology, the absence of such a part.
**acroteriasmus** (ak-ro-te-ri-az'-mus). Same as *acroteriasis*.
**acroteric** (ak-ro-ter'-ik) [ἀκρωτήρια, the extremities]. Relating to the extremities; applied to conditions in which the extremities are most affected.
**acrothymion**, or **acrothymum** (ak-ro-thi'-me-on, ak-ro-thi'-mum) [ἄκρον, summit; *thyme*]. A rugose wart with a broad top.
**acrotic** (ah-krot'-ik) [ἀ, priv.; κρότος, a striking]. 1. Any defective beating of the pulse; failure of the pulse. 2. [ἄκρος, extreme, outmost]. Relating to the glands of the skin; affecting the surface.
**acrotizm** (ah'-krot-izm). See *acrotic* (1).
**acrotrophoneurosis** (ak-ro-trof-o-nu-ro'-sis) [ἄκρον, an extremity; τροφή, nourishment; νεῦρον, nerve]. A trophic disturbance of the extremities of central origin.
**acrylaldehyde** (ak-ril-al'-de-hīd). See *acrolein*.
**act** (akt) [*agere*, to put in motion]. The fulfilment of a purpose or function. **a., imperative**, the act of an insane person in response to an imperative morbid impulse. **a., sexual**, see *coitus*.
**Actæa** (ak-te'-ah) [ἀκτῆ, the elder]. A genus of ranunculaceous plants having active medicinal qualities. *A. alba*, the white cohosh, has much the same qualities as *A. spicata*. *A. cimicifuga* and *A. racemosa* are more important. See *cimicifuga*. *A. rubra*, red cohosh, and *A. spicata* are purgative and emetic.
**actinic** (ak-tin'-ik) [ἀκτίς, a ray]. Those rays of the spectrum capable of producing chemical changes; found in the violet and ultraviolet parts.
**actinism** (ak'-tin-ism) [ἀκτίς, a ray]. 1. The chemical quality of light, or of the sun's rays. 2. The radiation of heat or light, or that branch of science which treats of it.
**actinium** (ak-tin'-e-um) [see *actinic*]. A radio-active substance, thought to be an element, found in pitchblende.
**actinobacillosis** (ak-tin-o-bas-il-o'-sis) [ἀκτίς, ray; *bacillus*]. A disease of cattle and other domestic animals due to a bacillus which produces radiate structures in the affected tissues.
**actinobolia** (ak-tin-o-bo'-le-ah) [ἀκτινοβολεῖν, to radiate]. 1. A term formerly used to express the process by which the impulses of the will are conveyed to the different parts of the body. 2. Von Helmont's term for the phenomena now included under hypnotism.
**actinobolism, actinobolismus** (ac-tin-ob'-o-lizm, ak-tin-ob-o-lis'-mus). See *actinobolia*.
**actinocerate, actinocerous** (ak-tin-os'-er-āt,¹ -us) [ἀκτίς, a ray; κέρας, a horn]. Having horn-like processes radiately arranged.
**actinochemistry** (ak-tin-o-kem'-is-trē) [ἀκτίς, a ray; χημεία, chemistry]. Chemistry dealing with decomposition of substances by light.
**actinocongestin** (ak-tin-o-kon-jes'-tin). A substance derived from the tentacles of *Actinia*; it consists of a toxin and a proteid and when injected into animals causes congestion of the viscera.
**actinodermatitis** (ak-tin-o-der-mat-i'-tis) [ἀκτίς, a ray; *dermatitis*]. Cutaneous lesions produced by application of the röntgen-rays. Syn., *radiodermatitis*.
**actinogram** (ak-tin'-o-gram) [ἀκτίς, a ray; γράφειν, to write]. The record made by the actinograph. Skiagram.
**actinograph** (ak-tin'-o-graf). An apparatus to measure the actinism of sunlight. Skiagraph.
**actinography**. See *actinology*.
**actinology** (ak-tin-ol'-o-je) [ἀκτίς, a ray; λόγος, a discourse]. 1. In biology, that kind of homological relation that exists between the successive segments, regions, or divisions of a part or organ, in that they radiate or spring from it. 2. The science of the chemical action of radiant light: actinography. 3. The part of zoology which treats of the *radiata*.
**actinolyte** (ak-tin'-o-līt) [ἀκτίς, a ray; λύειν, to loose]. An apparatus designed for use in actinotherapy.
**actinometer** (ak-tin-om'-et-er) [ἀκτίς, a ray; μέτρον, measure]. An apparatus for determining the intensity of actinic rays.
**actinomyces** (ak-tin-om'-i-sēz) [ἀκτίς, a ray; μύκης, a fungus; pl., *actinomycetes*]. A vegetable parasite, the cause of the disease actinomycosis. It is also called the *ray-fungus*. It probably belongs to the cladothrix group of schizomycetes. In some tissues it presents itself in the form of a roset of fine filaments clubbed at their outer ends; in the center are numerous coccus-like bodies, the spores of the organism.
**actinomycoma** (ak-tin-o-mi-ko'-mah) [ἀκτίς, a ray; μύκης, a fungus; pl., *actinomycomata*]. A tumor such as is characteristic of actinomycosis.
**actinomycosis** (ak-tin-o-mi-ko'-sis) [ἀκτίς, a ray; μύκης, a fungus]. A parasitic, infectious, inoculable disease, first observed in cattle, and also occurring in man, and characterized by the manifestations of chronic inflammation, with or without suppuration, often resulting in the formation of granulation tumors, especially about the jaws. The disease is due to the presence of a parasite, the *ray-fungus*, or *actinomyces*. Syn., *lumpy-jaw; holdfast; wooden tongue*.
**actinomycotic** (ak-tin-o-mi-kot'-ik). Pertaining to actinomycosis.
**actinotherapy** (ak-tin-o-ther'-ap-e) [ἀκτίς, a ray; θεραπεία, therapy]. The therapeutic use of actinic rays.
**action** (ak'-shun) [*agere*, to do or perform]. A doing; a working; especially the performance of a function. **a., after-**, the brief persistence of negative variation of the electric current in a tetanized muscle. **a.s, animal**, voluntary movements. **a. of arrest**, see *inhibition*. **a., automatic**, see *a., reflex*. **a., capillary**, see *attraction, capillary*. **a., catalytic, a., contact**, see *catalysis*. **a., chemical**, see *reaction*. **a., diastaltic**, see *a., reflex*. **a., electro-capillary**, electric phenomena resulting from chemical reaction between dissimilar fluids connected by a capillary medium. **a., inhibitory**, see *inhibition*. **a., local**, the production of currents between different parts of the same cell of a galvanic battery. **a.s, natural**, the vegetative functions. **a.s, pseudomotor**, Heidenhain's term for phenomena resulting from stimulation of the chorda tympani after section of the hypoglossal nerve; movements due to vascular or lymphatic engorgement. **a., reflex**, an involuntary movement of part of the body resulting from an impression carried by a sensory or afferent nerve to a center, and then sent back by an efferent nerve to the part, usually at or near the source of irritation. **a., safety-valve**, the incomplete closure of the tricuspid valve, especially in cases of resistance in the

pulmonary circulation. a., sexual, functioning of the generative apparatus. a.s, vital, those essential to the continuance of vitality, as of the heart and lungs.
**activate** (*ak'-tiv-āt*). To render active.
**activation** (*ak-tiv-a'-shun*). The process of activating.
**activator** (*ak'-tiv-a-tor*). 1. An agent which renders active some other chemical agent such as an enzyme. Also known as *kinase*, or *coenzyme* in the case of ferments. The term is generally applied to biochemical reactions. 2. The internal secretion of the pancreas.
**active** (*ak'-tiv*) [see *action*]. 1. Energetic; decisive; as *active* treatment. 2. Due to an intrinsic force as distinguished from passive—e. g., *active* hyperemia. a., optically, possessing optic rotatory power.
**activity** (*ak-tiv'-it-e*) [*agere*, to do or perform]. Capacity for acting; sensibility; vitality; potency; energy. a., optic, the property of certain chemical molecules to rotate the plane of polarization, due to the presence of one or several asymmetric carbon atoms in the molecule of every optically active body. Cf. *rotatory power*. a., sense of muscular, see *muscular sense*, under *muscular*.
**actol** (*ak'-tol*). Trade name for silver lactate.
**actual** (*ak'-chu-al*) [*agere*, to do or perform]. Real; effective. a. cautery, see *cautery*.
**actuation** (*ak-chu-a'-shun*). The mental function that is exercised between the impulse of volition and its performance.
**acuclosure** (*ak-u-klo'-zhūr*) [*acus*, a needle; *claudere*, to close]. A method of arresting hemorrhage by the aid of a needle which holds the artery closed for a day. It embraces *acupressure* and *acutorsion*.
**acuductor** (*ak-u-duk'-tor*) [*acus*, a needle; *ducere*, to lead]. A needle carrier.
**acufilopressure** (*ak-u-fi'-lo-pres-ur*) [*acus*, needle; *filum*, a thread; *pressure*]. A combination of acupressure and ligation.
**acuition** (*ak-u-ish'-un*) [*acuere*, to sharpen]. Increased effect of a drug's action by the addition of another drug.
**acuity** (*ak-u'-it-e*) [see *acuition*]. Acuteness or clearness, as *acuity* of vision.
**aculeate** (*ak-u'-le-āt*) [*aculeus*, a sting, prickle]. In botany, armed with prickles, *i. e., aculei*; as the rose and brier. In biology, having a sting.
**acumeter** (*ak-u'-me-ter*). An instrument for testing hearing. See *acoumeter*.
**acuminate** (*ak-u'-min-āt*) [*acuminatus*, pointed; acute]. Sharp-pointed.
**acupression, acupressure** (*ak-u-presh'-un, ak'-u-presh-ūr*) [*acus*, a needle; *pressura*, pressure]. The operation to stop hemorrhage by compressing the artery with a needle inserted into the tissues upon either side.
**acupuncture** (*ak'-u-punk-chur*) [*acus*, a needle; *pungere*, to prick]. Puncture of the skin or tissue by one or more needles for the relief of pain, the exit of fluid, the coagulation of blood in an aneurysm, etc.
**acus** (*a'-kus*) [L.]. A (surgical) needle.
**acusia** (*ah-koo'-ze-ah*). See *acousia* (2).
**acusimeter, acusiometer** (*ah-koo-sim'-et-er, ah-koo-se-om'-et-er*). Same as *acoumeter*.
**acustica** (*ah-koos'-tik-ah*). See *acoustics*.
**acusticus** (*ah-koo'-stik-us*) [L.]. The auditory, or eighth cranial, nerve.
**acute** (*ak-ūt'*) [*acutus*, sharp]. Having a rapid onset, a. short course, and pronounced symptoms and termination. Sharp, severe.
**acutenaculum** (*ak-u-ten-ak'-u-lum*) [*acus*, a needle; *tenaculum pl., acutenacula*]. A needle-holder.
**acuteness** (*ak-ūt'-nes*) [*acutus*, sharp]. The quality of being acute, rapid or sharp. Referring to vision, used as a synonym of keenness or acuity.
**acuticostal** (*ak-ūt-i-kos'-tal*) [*acutus*, sharp; *costa*, a rib]. Having projecting ribs.
**acutorsion** (*ak-u-tor'-shun*) [*acus*, a needle; *torsion*]. The twisting of an artery with a needle as a means of controlling hemorrhage.
**acyanoblepsia** (*ah-si-an-o-blep'-se-ah*) [ἀ, priv.; κύανος, blue; βλέπειν, to look. Same as *acyanopsia*.
**acyanoblepic** (*ah-si-an-o-blep'-tik*). Affected with or pertaining to acyanoblepsia.
**acyanopsia** (*ah-si-an-op'-se-ah*) [ἀ, priv.; κύανος, blue; ὄψις, sight]. Inability to distinguish blue colors.
**acyclia** (*ah-sik'-le-ah*) [ἀ, priv.; κυκλεῖν, to circulate]. Arrested circulation of body-fluids.

**acyclic** (*ah-sik'-lik*) [ἀ, priv.; κυκλικός, circular]. 1. In botany, not whorled. 2. Not characterized by a self-limited course. Cf. *Cyclic*. 3. In chemistry, aliphatic, having the structure of the open chain compounds.
**acyesis** (*ah-si-e'-sis*) [ἀ, priv.; κύησις, pregnancy]. 1. Sterility of the female. 2. Non-pregnancy. 3. Incapacity for natural delivery. Syn., *aciesis*.
**acyeterion** (*ah-si-et-e'-re-on*) [see *acyesis*]. An agent to prevent conception.
**acyetic** (*ah-si-et'-ik*) [ἀ, priv.; [κύησις, pregnancy]. Relating to acyesis.
**acyl** (*as'-il*). An acid organic radical derived from an organic acid by the removal of a hydroxyl group (OH).
**acyoblepsia** (*as-i-o-blep'-se-ah*). Same as *acyanoblepsia*.
**acystia** (*ah-sis'-te-ah*) [ἀ, priv.; κύστις, bladder]. Absence of the bladder.
**acystinervia** (*ah-sis-tin-er'-ve-ah*) [ἀ, priv.; κύστις, bladder; *nervus*, a nerve]. Paralysis or lack of nerve stimulus in the bladder.
**acystonervia, acystoneuria** (*ah-sis-to-nur'-ve-ah, -nu'-re-ah*). See *acystinervia*.
**a.d.** Abbreviation for Latin *auris dextra*, right ear.
**ad** [*ad*, to]. A Latin preposition signifying *to, toward*, at, etc.; as, *ad libitum*, at pleasure or according to discretion.
**ad., or add.** A contraction of *adde*, or *additur*, meaning, add, or let there be added; used in prescription writing.
**adacrya** (*ah-dak'-re-ah*) [ἀ, priv.; δάκρυον, tear]. Absence or deficiency of the secretion of tears.
**adactyl** (*ah-dak'-til*) [ἀ, priv.; δάκτυλος, digit]. 1. Without fingers or without toes. 2. A monstrosity that has an absence of digits.
**adactylia** (*ah-dak-til'-e-ah*) [ἀ, priv.; δάκτυλος, a finger]. Absence of the digits.
**adactylism** (*ah-dak'-til-ism*) [ἀ, priv.; δάκτυλος, a finger]. The absence of the digits.
**adactylous** (*ah-dak'-til-us*), see *adactylism*.
**adalin** (*ad'-al-in*). A proprietary preparation used as a sedative and hypnotic. It is said to be bromodiethylacetyl urea.
**adamantin** (*ad-am-an'-tin*) [ἀδάμας, adamant]. Pertaining to adamant. a. cement, a substance used for filling teeth, consisting of finely powdered silex or pumice stone mixed with an amalgam of mercury and silver. See *amalgam*. a. substance, the enamel of the teeth.
**adamantinoma** (*ad-am-an-tin-o'-mah*) [ἀδάμας, adamant; ὄμα, tumor]. An epithelial tumor resembling in structure the enamel organ of a developing tooth.
**adamantoblast** (*ad-am-an'-to-blast*). An enamel-cell; a columnar epithelial cell from which the enamel of the teeth is developed. Ameloblast.
**Adamkiewicz, dentiline cells of** (*ad-ahm'-ke-a-vits*) [Albert *Adamkiewicz*, Austrian pathologist, 1850- ]. A peculiar form of nerve-corpuscle lying below the neurilemma of medullated nerve-fibers; it is stained yellow by safranin. A.'s reaction for proteins. To a mixture of one volume concentrated sulphuric acid and two volumes glacial acetic acid add the protein. At the ordinary temperature a reddish-violet color is obtained slowly but more quickly on heating. The liquid has also a feeble fluorescence, and gives an absorption band between the lines B and F in the solar spectrum.
**adamon** (*ad'-am-on*). A preparation used as a substitute for valerian; it is a sedative.
**Adams's operation** (Sir William *Adams*, English surgeon, 1760-1829; William *Adams*, English surgeon, 1820- ). *Osteotomy* for ankylosis of the hip-joint, the neck of the femur being divided subcutaneously, within the capsule. 2. *Corectopy*; the iris is drawn into a small, corneal incision, in order to change the position to the natural pupil. 3. For *deviated nasal septum*; the bent cartilaginous septum is forcibly straightened by means of special flat, parallel-bladed forceps. 4. For *Dupuytren's contraction*, when the bands extend far down the sides of the finger. It consists in multiple subcutaneous section of the palmar fascia from without inward. 5. For *ectropion*; a triangular wedge is removed from the whole thickness of the lower lid, and the edges are united by sutures. 6. *Iliac colotomy*; a modification of Cripps' operation, in which a vertical incision is made external to the epigastric artery. 7. For *prolapsus uteri*, see *Alexander's operation*.

# ADAM'S APPLE 23 ADENO-

**Adam's apple.** See *Pomum adami*.

**Adams-Stokes syndrome or disease** [Robert *Adams*, Scotch physician, 1794-1861; William *Stokes*, Irish physician, 1804-1878]. A symptom-complex consisting of bradycardia in association with epileptiform or apoplectiform seizures. Heart-block is often present.

**Adams's disease.** See *Adams-Stokes' disease*.

**Adansonia digitata** (*ad-an-so'-ne-ah dij-it-a'-tah*) [Michel *Adanson*, French naturalist, 1727-1806]. The baobab-tree, a native of Africa. The bark is used in the form of an infusion, 1 oz. to 1 pint, as a remedy for intermittent fever.

**adansonine** (*ad-an'-so-nin*). A febrifugal alkaloid from the leaves and bark of *Adansonia digitata*.

**adanto blaka.** A malady common among the negroes of the Gold Coast and of frequent prevalence in the tropic zone; it is due to an animal parasite.

**adaptation** (*ad-ap-ta'-shon*) [*adaptare*, to adjust]. In biology, favorable organic modifications suiting a plant or animal to its environment. **a.** of the retina, the faculty possessed by the retina of accommodating the power of vision to a diminished amount of light, as in a darkened room.

**adapter** (*ad-ap'-ter*) [*adaptare*, to adjust]. 1. Anything which serves the purpose of fitting one thing to another. An instrument by means of which the direct electric current may be adapted to the various forms of electrotherapeutic treatment. 2. A piece of tubing used to connect the neck of a retort with a receiver. 3. A microscope attachment for centering or decentering the illuminating apparatus. 4. A collar used to fit an objective to a different nosepiece than that for which it was made.

**adarticulation** (*ad-ar-tik-u-la'-shun*) [*ad*, to; *articulatio*, a jointing]. See *arthrodia*.

**adde** (*ad'-e*) [imperative sing. of *addere*, to add]. Add; a direction used in prescription writing.

**ad deliq.** Abbreviation of *ad deliquium* [L.]. To the point of fainting.

**addephagia** (*ad-e-fa'-je-ah*) [L.]. See *bulimia*.

**addiment** (*ad'-im-ent*) [*addere*, to add]. Ehrlich's and Morgenroth's term (1899) for an active thermolabile substance (destroyed by a temperature of 56° C.) contained in normal serum and capable of rendering active the immune body of Ehrlich and setting up bacteriolysis and hemolysis. See *complement*.

**addimentary** (*ad-im-ent'-ar-e*). Pertaining to addiment.

**Addison's anemia** [Thomas *Addison*, English physician, 1793-1860]. Pernicious anemia. **A.'s disease,** a disease of the suprarenal capsules, first described by Addison, and characterized by tuberculous infiltration of the capsules, discoloration of the skin, progressive anemia, and asthenia, ending in death from exhaustion. Bronzed skin may occur without disease of the suprarenal capsules, and the latter have been the seat of morbid processes without an accompanying change in the skin. Syn., *melasma suprarenale; dermatomelasma suprarenale; cutis ærea; bronzed skin.* **A.'s keloid,** morphea. **A.'s pill,** Guy's pill.

**additamentum** (*ad-it-am-en'-tum*) [L.]. Any appendix, as an epiphysis. **a. ad sacrolumbalem,** see *muscles*. **a. coli,** the appendix vermiformis. **a. necatum,** the olecranon. **a. suturæ lambdoidalis,** the occipitomastoid suture. **a. ulnæ,** the radius. **a. uncatum ulnæ,** the olecranon.

**addition** (*ad-ish'-un*) [*addere*, to add]. The formation of a molecule by the direct union of two or more different molecules without decomposition. **a. compound,** see under *compound*. **a. product,** see under *product*. **a. reaction;** see under *reaction*.

**adducens** (*ad-du'-senz*) [*adducere*, to bring toward]. An adductor, a term applied to certain muscles. **a. oculi,** the internal rectus muscle of the eye.

**adducent** (*ad-du'-sent*) [see *adducens*]. Performing adduction.

**adduct** (*ad-ukt'*) [*adducere*, to bring forward]. To draw toward the median line of a body.

**adduction** (*ad-uk'-shun*) [see *adducens*]. Any movement whereby a part is brought toward another or toward the median line of the body.

**adductor** (*ad-duk'-tor*) [*adducere*, to bring forward]. Any muscle effecting adduction. **a. brevis, hallucis, longus, magnus, minimus, obliquus hallucis, obliquus pollicis, transversus hallucis, transversus pollicis;** see *muscles*, *table of*.

**adelodermatous, adelodermous** (*ad-el-o-der-mat-us, ad-el-o-der'-mus*) [ἄδηλος, not seen; δέρμα, skin]. Having concealed integument, as invaginated tracts.

**adelomorphous** (*ad-el-o-mor'-fus*) [ἄδηλος, not seen; μορφή, form]. Not clearly defined; applied to certain cells in the gastric glands.

**adelphia** (*ad-el'-fe-ah*). A form of monstrosity characterized by the union of two organisms above, the lower portions being separated.

**adelphotaxy** (*ad-el-fo-taks'-e*) [ἀδελφός, brotherhood; τάσσειν, to arrange]. The tendency of motile cells to arrange themselves into definite positions.

**ademonia** (*ad-e-mo'-ne-ah*) [ἆ, priv.; δημονία, trouble, distress]. Mental distress.

**ademosyne** (*ad-e-mos'-in-e*) [ἀδημοσύνη, trouble, distress]. Depression of spirits; home-sickness.

**aden** (*a'-den*) [ἀδήν, an acorn, a gland]. A gland; a bubo.

**adenalgia** (*ad-en-al'-je-ah*) [*āden*; ἄλγος, pain]. Glandular pain.

**adenase** (*ad'-en-ās*). An enzyme which converts adenin to hypoxanthin.

**adenasthenia** (*ad-en-as-the'-ne-ah*) [*aden*; ἀσθένεια, weakness]. 1. Functional weakness of a gland. 2. A disorder of the stomach characterized by diminished and enfeebled secretion without anatomic lesion. **a. gastrica,** see *adenasthenia* (2).

**adendric** (*ah-den'-drik*) [ἆ, priv.; δένδρον, tree]. Unprovided with dendrons.

**adendritic** (*ah-den-drit'-ik*) [ἆ, priv.; δένδρον, tree]. Without dendrites.

**adenectomy** (*ad-en-ek'-to-me*) [*aden*; ἐκτομή, excision]. The excision of a gland.

**adenectopia** (*ad-en-ek-to'-pe-ah*) [*aden*; ἔκτοπος, away from a place]. A condition in which the gland does not occupy its proper position.

**adenectopic** (*ad-en-ek-top'-ik*). Pertaining to adenectopia.

**adenemphratic** (*ad-en-em-frat'-ik*). Pertaining to adenemphraxis.

**adenemphraxis** (*ad-en-em-fraks'-is*) [*aden*; ἔμφραξις, a stoppage]. Glandular obstruction.

**Aden fever.** See *dengue*. **A. ulcer.** See *phagedena tropica*.

**adenia** (*ad-e'-ne-ah*) [*aden*]. A hyperplasia of the tissue of lymphatic glands leading to the formation of tumors. See *lymphadenoma*. **a.s, anabiromic,** Piorry's term for diseases of the glandular adnexa of the digestive tract. **a., leukemic,** adenia associated with a leukemic condition of the blood. **a., simple,** that form which is unaccompanied by any increase in the number of the white blood-corpuscles. A synonym of *Hodgkin's disease*.

**adenic** (*ad-en'-ik*) [*aden*]. Relating to or of the nature of a gland.

**adeniform** (*ad-en'-e-form*) [*aden*; *forma*, resemblance]. Shaped like a gland.

**adenin** (*ad'-en-in*). See *adenine*.

**adenine** (*ad'-en-ēn*) [*aden*], C₅H₅N₅. 6 aminopurin. The simplest member of the uric-acid group of leukomaines, apparently formed by polymerization of hydrocyanic acid, first discovered in the pancreas. It occurs, with other bases, as a decomposition-product of nuclein, and may be obtained from all animal and vegetable tissues rich in nucleated cells. It crystallizes in leaflets with pearly luster. It exists abundantly in the liver and urine of leukocythemic patients. Adenine is not poisonous.

**adeninehypoxanthine** (*ad-en-ēn-hi-po-zanth'-ēn*). C₅H₅N₅+C₅H₄N₄O. A compound of adenine and hypoxanthine first observed by Kossel and isolated by Bruhns, occurring in thick, starch-like, semitransparent masses, becoming white and chalky.

**adenitis** (*ad-en-i'-tis*) [*aden*; *ιτις*, inflammation]. Inflammation of a gland. Syn., *phlegmasia adenosa;* *phlegmasia glandulosa.* **a. cervicalis syphilitica,** an engorgement of the cervical lymphatic glands; a sign of syphilitic infection. **a. cubitalis,** Grünfeld's term for inflammation of the epitrochlear lymphatic gland. **a. hyperplastica,** Grünfeld's term for a bubo in which plastic exudation predominates. **a. pubica,** bubo of the public region, often accompanied by suppurative lymphangitis of the dorsum of the penis. **a., syphilitic, primitive,** see *bubo, syphilitic.* **a. universalis,** a widespread induration of the lymphatic glands accompanying primary syphilis.

**adenization** (*ad-en-i-za'-shun*) [*aden*]. 1. The assuming of a glandular appearance. 2. Adenoid degeneration.

**adeno-** [ἀδήν, a gland]. A prefix denoting relation to glands.

**adenoblast** (*ad'-en-o-blast*) [*adeno-*; βλαστός, a germ]. 1. Any functionally active gland-cell; a cell that assists in the glandular action. 2. Haeckel's name for an embryonic cell which forms a gland.

**adenocarcinoma** (*ad-en-o-kar-sin-o'-mah*) [*adeno-*; *carcinoma*]. Adenoma blended with carcinoma.

**adenocele** (*ad'-en-o-sēl* [*adeno-*; κήλη, a tumor]. A cystic tumor containing adenomatous elements. See *adenoma*.

**adenocellulitis** (*ad-en-o-sel-u-li'-tis*) [*adeno-*; *cellulitis*]. Inflammation of a gland and the surrounding cellular tissue.

**adenochirapsology** (*ad-en-o-ki-rap-sol'-o-je*) [*aden*; χείρ, hand; ἅπτειν, to touch; λόγος, treatise]. The obsolete doctrine of the healing of scrofula by the touch of a king's hand.

**adenochondroma** (*ad-en-o-kon-dro'-mah*) [*aden*; χόνδρος, cartilage: *pl.*, *adenochondromata*]. A tumor consisting of both glandular and cartilaginous tissue.

**adenocyst** (*ad'-en-o-sist*) [*adeno-*; κύστις, a cyst]. A cystic lymphatic gland; a glandular cyst. Cf. *adenocystoma*.

**adenocystoma** (*ad-en-o-sis-to'-mah*) [*adeno-*; κύστις, a cyst; ὄμα, a tumor]. A cystic adenoma.

**adenodermia** (*ad-en-o-dur'-me-ah*) [*aden*; δέρμα, skin]. Disease of the glands of the skin.

**adenodiastasis** (*ad-en-o-di-as'-tas-is*) [*aden*; διάστασις, separation]. 1. Displacement of a gland. 2. Abnormal separation of a gland into distinct parts.

**adenodynia** (*ad-en-o-din'-e-ah*) [*aden*; ὀδύνη, pain]. See *adenalgia*.

**adenofibroma** (*ad-en-o-fi-bro'-mah*) [*adeno-*; *fibroma*]. A combination of adenoma and fibroma.

**adenofibrosis** (*ad-en-o-fi-bro'-sis*) [*adeno-*; *fibrosis*]. Fibroid degeneration of a gland, particularly the inflammatory neoplasms involving sudoriparous glands, due to infection with *Botryomyces*. Cf. *botryomycosis*.

**adenogenesis** (*ad-en-o-jen'-es-is*) [*adeno-*; γένεσις, a creation]. The development of a gland.

**adenographer** (*ad-en-og'-ra-fur*). A writer on glands.

**adenography** (*ad-en-og'-ra-fe*) [*adeno-*; γράφειν, to write]. 1. That part of descriptive anatomy which treats of the glandular system. 2. A treatise on glands and the glandular system.

**adenohypersthenia** (*ad-en-o-hi-per-sthe'-ne-ah*) [*adeno-*; ὑπέρ, over; σθένος, strength]. Excessive activity of the glands. a. **gastrica**, a condition characterized by the secretion of gastric juice abnormally rich in hydrochloric acid or excessive in quantity.

**adenoid** (*ad'-en-oid*) [*adeno-*; εἶδος, resemblance]. 1. Resembling a gland. 2. In the plural, the same as *adenoid vegetations*. a. **acne**, see *lupus*, *disseminated follicular*. a. **body**. 1. The prostate gland. 2. A melanotic tumor. a. **disease**, synonym of *Hodgkin's disease*. a. **muscle**, see *thyroadenoideus under muscle*. a. **tissue**, lymphadenoid tissue. a. **tumor**, see *adenoma*. a. **vegetations**, a term applied to a hypertrophy of the adenoid tissue that normally exists in the nasopharynx.

**adenoidectomy** (*ad-en-oi-dek'-to-me*) [*adenoid*; ἐκτομή, excision]. An operation for the removal of adenoids.

**adenoids**. See *adenoid vegetations*.

**adenolipoma** (*ad-en-o-lip-o'-mah*) [*adeno-*; *lipoma*]. A combination of adenoma and lipoma.

**adenolipomatosis** (*ad-en-o-lip-o-mat-o'-sis*) [*adenolipoma*]. A diseased condition of the lymphatic system characterized by fatty deposits in the neighborhood of the neck, axillæ, and groins. It is generally unattended with pain. Syn., *multiple lipomata*.

**adenologaditis** (*ad-en-o-log-ad-i'-tis*) [*adeno-*; λογάδες, whites of the eyes; ἶτις, inflammation]. 1. Ophthalmia neonatorum. 2. Inflammation of the glands and conjunctiva of the eyes.

**adenology** (*ad-en-ol'-o-je*) [*adeno-*; λόγος, a discourse]. The science of or a treatise on the glandular system.

**adenolymphocele** (*ad-en-o-limf'-o-sēl*) [*adeno-*; *lymph*; κήλη, tumor]. Dilatation of the lymph-vessels and enlargement of the lymphatic glands.

**adenolymphoma** (*ad-en-o-lim-fo'-mah*) [*adeno-*; *lymphoma*]. A combined adenoma and lymphoma. See *lymphadenoma*.

**adenom** (*ad'-en-om*). A preparation used as a genitourinary sedative and anaphrodisiac.

**adenoma** (*ad-en-o'-mah*) [*adeno-*; ὄμα, a tumor: pl., *adenomata*]. 1. An epithelial tumor constructed after the type of a secreting gland. 2. Any tumor which has as its characteristic feature tubes or spaces lined with epithelium, whether or not it arises from or is connected with a gland. a. **carcinomatodes renis**, a renal neoplasm probably derived from aberrant adrenal tissue in the kidney. a. **destruens**, a destructive form of adenoma. a. **diffusum**, hyperplasia of the mucous membrane with predominance of glandular elements. a. **fibrosum**, a fibrous growth in the stroma of a gland. a., **heteropodous**, one arising from the metastasis of normal glandular tissue. a., **lupiform**, see *lupus erythematosus*. a., **malignant**, an adenomatous carcinoma. a., **papillary**, a. **papilliferum**, a form arising from either the alveolar or the tubular adenoma through stronger growth of the epithelium and the formation of papillæ of connective tissue. a., **racemose**, an adenoma after the type of a racemose gland. a., **renal**, glandular carcinoma of the kidney. a. **sebaceum**, a fatty tumor of the face composed of sebaceous glands. a. **simplex**, a tumor-like hyperplasia of a gland. a. **sudoriparum**, a cutaneous tumor involving hyperplasia of the sweat-glands. Cf. *hidrosadenitis*. a., **tubular**, an adenoma after the type of a tubular gland. a., **umbilical**, a tumor at the navel originating through the coalescence of Meckel's diverticulum with the umbilical ring, through which the intestinal mucosa appears in the navel. Syn., *intestinal ectropia*.

**adenomalacia** (*ad-en-o-mal-a'-she-ah*) [*adeno-*; μαλακία, softening]. Abnormal softening of a gland.

**adenomatome** (*ad-en-o'-mat-ōm*) [*adenoma*; τομή, a cutting]. Cutting forceps or scissors for use in the removal of adenomatous growths.

**adenomatosis** (*ad-en-o-mat-o'-sis*). A condition characterized by diffuse overgrowth of glandular tissue.

**adenomatous** (*ad-en-o'-mat-us*). Pertaining to an adenoma; characteristic of glandular hyperplasia.

**adenomeningeal** (*ad-en-o-men-in'-je-al*) [*adeno-*; μῆνιγξ, a membrane]. Pertaining to or affecting the glands of a membrane.

**adenomesenteritis** (*ad-en-o-mes-en-ter-i'-tis*) [*adeno-*; *mesentery*; ἶτις, inflammation]. Inflammation of the mesenteric glands.

**adenomyoma** (*ad-en-o-mi-o'-mah*) [*adeno-*; μῦς, a muscle; ὄμα, a tumor: *pl.*, *adenomyomata*]. A tumor composed of glandular and muscular tissues. a., **branchiogenic**, cyst-formation in consequence of inflammation of the mucous bursa in the median line of the neck.

**adenomyxoma** (*ad-en-o-miks-o'-mah*) [*adeno-*; μύξα; mucus; ὄμα, a tumor]. A growth having the characters of adenoma and myxoma.

**adenomyxosarcoma** (*ad-en-o-miks'-o-sar-ko-mah*). A rare combination of malignant tumor forms (observed in the cervix uteri); a primary adenoma with secondary sarcoma and finally myxomatous degeneration of the stromas.

**adenoncosis** (*ad-en-on-ko'-sis*) [*adeno-*; ὄγκωσις, swelling]. The enlargement of a gland.

**adenoncus** (*ad-en-ong'-kus*) [*adeno-*; ὄγκος, a mass]. A glandular tumor.

**adenopathy, adenopathia** (*ad-en-op'-a-the, ad-en-o-pa'-the-ah*) [*adeno-*; πάθος, disease]. Any disease of a gland. a., **angiobromic**, see *adenias, angiobromic*. a., **primary**, the lymphadenitis resulting from primary syphilitic infection. a., **syphilitic**, the enlarged and indurated cervical, inguinal, and cubital glands symptomatic of syphilitic infection. a., **tracheobronchial**, a., **tracheobronchic**, hypertrophy of the peribronchial lymphatic glands observed in the course of various diseases, causing spasmodic cough. a., **tracheolaryngeal**, inflammation and hypertrophy of the tracheolaryngeal lymphatic glands.

**adenopharyngeal** (*ad-en-o-far-in'-je-al*) [*adeno-*; φάρυγξ, pharynx]. Pertaining to the thyroid gland and the pharynx.

**adenopharyngitis** (*ad-en-o-far-in-ji'-tis*) [*adeno-*; φάρυγξ, pharynx; ἶτις, inflammation]. Inflammation of the tonsils and pharynx.

**adenophlegmon** (*ad-en-o-fleg'-mon*) [*adeno-*; φλέγμονη, inflammation]. Suppurative inflammation of a gland. Phlegmonous lymphadenitis.

**adenophthalmia** (*ad-en-of-thal'-me-ah*) [*adeno-*; ὀφθαλμός, the eye]. Inflammation of the Meibomian glands.

**adenophyma** (*ad-e-no-fi'-ma*) [*adeno-*; φῦμα, a tumor or growth]. A soft swelling of a gland.

**adenosarcoma** (*ad-en-o-sar-ko'-mah*) [*adeno-*; sar-

*coma*]. A tumor with the characters of adenoma and sarcoma combined.
**adenosarcorhabdomyoma** (*ad-en-o-sar-ko-rab-do-mi-o'-mah*). A neoplasm composed of the elements of sarcoma; adenoma, and rhabdomyoma.
**adenoscirrhus** (*ad-en-o-skir'-us*) [*adeno-; scirrhus*]. Adenoma with scirrhous or carcinomatous elements.
**adenosclerosis** (*ad-en-o-skle-ro'-sis*) [*adeno-; σκληρός*, hard]. A hardening of a gland, with or without swelling.
**adenose** (*ad'-en-ōs*) [*ἀδήν*, gland]. Glandular; abounding in glands; gland-like.
**adenosis** (*ad-en-o'-sis*) [*ἀδήν*, a gland]. 1. Any glandular disease. 2. Any chronic glandular disorder. **a. scrofulosa**, see *scrofula*.
**adenosynchitonitis** (*ad-en-o-sin-ki-ton-i'-tis*) [*adeno-; σύν*, with; *χιτών*, a covering; *ιτις*, inflammation]. 1. Inflammation of the Meibomian glands. 2. Ophthalmia neonatorum.
**adenotome** (*ad'-en-o-tōm*) [*adeno-; τομή*, a cutting]. An instrument for incising a gland or for removing adenoids.
**adenotomy** (*ad-en-ot'-o-me*) [*adeno-; τομή*, a cutting]. The anatomy of the glands; dissection or incision or removal of a gland.
**adenous** (*ad'-en-us*) [*ἀδήν*, gland]. See *adenose*.
**adenyl** (*ad'-en-il*). The radical, C₅H₄N₄, contained in adenin.
**adephagia** (*ad-e-fa'-je-ah*) [*ἀδηφάγος*, eating one's fill; gluttonous]. Voracious appetite; bulimia.
**adeps** (*ad'-eps*) [L.; gen., *adipis*]. 1. Lard. The fat obtained from the abdomen of the hog, composed of 38 % of stearin and margarin and 62 % olein. It forms 70 % of ceratum and 80 % of unguentum. 2. Fatness. 3. Animal fat. **a. anserinus, a. anseris**, goose-grease. **a. benzoinatus** (U. S. P.), benzoinated lard; contains 2 % of benzoin. A preparation of lard, 48 parts, and 1 part of Peruvian balsam. **a. ex fele**, cat's grease. **a. lanæ** (U. S. P.), lanolin. **a. lanæ hydrosus** (U. S. P.), hydrous wool-fat, the purified fat of the wool of the sheep. **a. ovillus, a. ovis**, mutton suet. A fixed oil (*oleum adipis*) is expressed from lard. **a. præparatus** (U. S. P.), purified fat of the hog. **a. suillus**, hog's lard; adeps.
**adepsin** (*ad-ep'-sin*) [*adeps*, lard]. A petrolatum much like vaselin.
**adermia** (*ah-der'-me-ah*) [*ά*, priv.; *δέρμα*, skin]. Absence or defect of the skin.
**adermogenesis** (*ah-der-mo-jen'-es-is*) [*ά*, priv.; *δέρμα*, skin; *γένεσις*, generation]. Deficient cutaneous development.
**adermotrophia** (*ah-der-mo-tro'-fe-ah*) [*ά*, priv.; *δέρμα*, skin; *τροφή*, nutrition]. Atrophy of the skin.
**adesmosis** (*ah-des-mo'-sis*) [*ά*, priv.; *δεσμός*, a band]. Atrophy of the cutaneous connective tissue.
**adgenic, adgenicus** (*ad-jen'-ik, ad-jen'-ik-us*) [*ad*, to; *gena*, the chin]. Attached to the genial tubercles or apophyses.
**Adhatoda** (*ad-ha-to'-da*) [from the Tamil name]. A genus of plants of the order *Acanthaceæ. A. hysopifolia*, a species native of South Africa; the willow-leaved Malabar nut; bitter, aromatic. *A. vasica*, a species native of tropical Asia; the Malabar nut. The juice of the leaves is used as an expectorant. The leaves, flowers, and root are considered antispasmodic and are given in asthma, intermittent fever, and rheumatism. The fresh flowers are bound over the eyes in cases of ophthalmia. In decoction the leaves with other remedies are used as an anthelmintic. The nut is emmenagogue and used to expel the dead fetus.
**adhesion** (*ad-he'-shun*) [*adhærere*, to stick to]. 1. The attractive force between two dissimilar bodies that are in contact. 2. Abnormal union of two surfaces as a result of inflammation, etc. **a., primary**, called also *healing by first intention* and by *immediate union*, a method of healing of wounds by the production of lymph, followed by the vascularization and cicatrization of the exudate. **a., secondary**, or *healing by second intention*, or *by granulation*, is that mode of healing attended by the production of pus and the formation of granulations.
**adhesive** (*ad-he'-siv*) [see *adhesion*]. 1. Sticky; tenacious. 2. Resulting in or attended with adhesion. **a. inflammation**, inflammation accompanied by plastic exudation, and tending to the union of apposed surfaces. **a. plaster**, resin plaster, see *resin* and *emplastrum*.
**adhesol** (*ad-he'-sol*). A surgical dressing said to contain copal resin, 350 parts; benzoin, 30 parts;

oil of thyme, 20 parts; alphanaphthol, 3 parts; tolu balsam, 30 parts; ether, 1000 parts.
**adhyoid** (*ad-hi'-oid*). Adherent to the hyoid bone.
**adiadochokinesis** (*ah-di-ad-o-ko-kin-e'-sis*) [*ά*, priv.; *διάδοχος*, succeeding]. Inability to perform rapidly alternating movements, such as pronation and supination.
**Adiantum** (*ad-e-an'-tum*) [*ά*, priv.; *διαντός*, capable of being wetted]. A genus of ferns; the maiden-hair. *A. capillus-veneris* and *A. pedatum*, of North America, are serviceable in coughs and as demulcents.
**adiaphoresis** (*ah-di-af-o-re'-sis*) [*ά*, priv.; *διαφορεύω*, to perspire]. Deficient sweat.
**adiaphoretic** (*ah-di-af-o-ret'-ik*) [*ά*, priv.; *διαφορεύω*, to perspire]. Reducing the sweat; anidrotic.
**adiaphorous** (*ad-i-af'-or-us*) [*αδιάφορος*, indifferent]. Neutral; inert; doing neither harm nor good.
**adiapneustia** (*ah-di-ap-nūs'-te-ah*) [*ά*, priv.; *διαπνευστέειν*, to perspire]. A stoppage of perspiration.
**adiarthrotos** (*ah-di-ar-thro'-tos*) [*αδιάρθρωτος*, not jointed]. 1. Without joints; unjointed. 2. Inarticulate (applied to speech).
**adiathermancy** (*ah-di-ath-er'-man-se*) [*ά*, priv.; *διά*, through; *θέρμη*, heat]. Impermeability to radiant heat.
**adiathermic** (*ah-di-a-thur'-mik*) [*ά*, priv.; *διά*, through; *θερμή*, heat]. Impervious to radiant heat.
**adiathesia** (*ah-di-ath-e'-se-ah*) [*ά*, priv.; *διάθεσις*, condition]. A condition or particular disease that is not congenital.
**adiathetic** (*ah-di-ath-e'-sik*) [*ά*, priv.; *διάθεσις*, condition]. Not connected with any diathesis.
**adiathetic** (*ah-di-ath-et'-ik*) [*ά*, priv.; *διάθεσις*, condition]. Adiathesic.
**adiemorrysis, adiæmorrhysis** (*ah-di-e-mor'-e-sis*) [*ά*, priv.; *διά*, through; *αἷμα*, blood; *ῥύσις*, flowing]. Failure of the circulation of the blood through the veins, due to some obstruction.
**adietetic** (*ah-di-et-et'-ik*). 1. Unwholesome for food. 2. Unmindful of dietetic requirements.
**adigan** (*ad'-ig-an*). A digitalis preparation which has been freed from digitonin and other saponin-like constituents; it is said to be effective and nontoxic.
**adipatum** (*ad-ip'-a-tum*). An ointment-base said to consist of lanolin, vaselin, paraffin, and water.
**adipic** (*ad-ip'-ik*) [*adeps*, lard]. Of or belonging to fat. **a. acid**, see *acid, adipic*.
**adipocele** (*ad'-ip-o-sēl*) [*adeps; κήλη*, hernia]. A true hernia with hernia sac, containing only fatty tissue.
**adipocellular** (*ad-ip-o-sel'-u-lar*). Made up of fat and connective tissue.
**adipoceration** (*ad-ip-os-er-a'-shun*) [*adeps*; fat; *cera*, wax]. The formation of adipocere.
**adipocere** (*ad'-ip-o-sēr*) [*adeps; cera*, wax]. A wax-like substance formed by the exposure of fleshy tissue to moisture, with the exclusion of air; *i. e.*, in the earth or under water. It consists of the fatty acids in combination with the alkaline earths and ammonium. Human bodies in moist burial places often undergo this change.
**adipofibroma** (*ad-ip-o-fi-bro'-mah*) [*adeps; fibroma*]. A combined fatty and fibrous tumor.
**adipogenous** (*ad-ip-oj'-en-us*) [*adeps*, fat; *gignere*, to produce]. Producing fat and adipose tissue.
**adipol** (*ad'-ip-ol*). Trade name of a mineral substance used as a base for ointments.
**adipolysis** (*ad-ip-ol'-is-is*) [*adeps; λύσις*, dissolution]. The cleavage or hydrolysis of fats in the process of digestion by the action of a fat-splitting enzyme.
**adipolytic** (*ad-ip-o-lit'-ik*). 1. Efficacious in the digestion or cleavage of fats. 2. An agent efficient in fat-digestion. Cf. *steapsin*.
**adipoma** (*ad-ip-o'-mah*) [*adeps; ὄμα*, a tumor]. A fatty tumor; lipoma.
**adipometer** (*ad-ip-om'-et-ur*) [*adeps*, fat; *μέτρον*, a measure]. An instrument for the estimation of fat.
**adipose** (*ad'-ip-ōs*) [*adeps*]. Fatty. **a. tissue**, fatty tissue distributed extensively through the body. Consists of areolar connective tissue, the cells of which contain fat-globules.
**adiposis** (*ad-ip-o'-sis*) [*adeps*]. Corpulence; fatty infiltration. **a. dolorosa**, Dercum's disease, characterized by the formation of soft nodules throughout the connective tissue of the body, accompanied by neuralgic pains. **a. hepatica**, fatty degeneration or infiltration of the liver.
**adipositas** (*ad-ip-os'-it-as*) [L.]. Fatness; corpulency. **a. cordis**, a fatty condition of the heart. **a. universa'lis**, obesity.

**adiposity** (*ad-ip-os'-it-e*). Fatness; corpulency.
**adiposuria** (*ad-ip-o-su'-re-ah*). The presence of fat in the urine. Lipuria.
**adipsa** (*ad-ip'-sah*) [neut. pl. of *adipsus*, without thirst]. 1. Remedies to allay thirst. 2. Foods which do not produce thirst.
**adipsia** (*ah-dip'-se-ah*) [ά, priv.; δίψα, thirst]. Absence of thirst.
**adipsous** (*ah-dip'-sus*) [ά, priv.; δίψα, thirst]. Quenching thirst.
**aditus** (*ad'-it-us*) [*adire*, to go to]. In anatomy, an entrance. **a. ad antrum**, the outer side of the attic, opening upward, backward, and outward into the mastoid antrum. It gives lodgment to the head of the malleus and the greater part of the incus. **a. ad aquæductum Sylvii**, the entrance to the ventricular aqueduct situated at the lower posterior angle of the third ventricle of the brain. **a. ad infundibulum**, a smaller canal extending from the third ventricle into the infundibulum; it is also called *vulva*. **a. ad laryngem**, **a. laryngis**, the entrance to the larynx. **a. glottidis**, one of the openings (superior or inferior) of the glottis.
**adjuster** (*ad-jus'-ter*) [Fr., *adjuster*, to adjust]. 1. A device formerly used for forcible reduction of dislocations. 2. One for holding together the two ends of a silver wire suture, to secure approximation of the parts without strain on the tissues.
**adjustment**, coarse. Commonly, the rack and pinion for raising or lowering the tube of a microscope a considerable distance without lateral deviation. **a., fine**, the micrometer screw generally at the top of the column of a microscope for raising or lowering the tube slowly through a short distance.
**adjuvant** (*ad'-ju-vant*) [*adjuvare*, to assist]. A medicine that assists the action of another to which it is added.
**Adler's benzidine reaction** for blood. Mix equal parts of a saturated solution of benzidine in alcohol or glacial acetic acid and of hydrogen dioxide (3 %). Add to this 1 Cc. of an aqueous solution of blood: a green or blue color develops. The blood solution should be acid in reaction.
**ad lib.** Abbreviation of *ad libitum* [L.]. At pleasure; as much as you please.
**admaxillary** (*ad-maks'-il-a-re*). Pertaining to maxillary structures. Cf. *gland, admaxillary*.
**adminiculum lineæ alʹbæ**. See *Cooper's ligament*.
**admortal** (*ad-mor'-tal*) [*ad*, to; *mors, mortis*, death]. Moving from living muscular tissue toward that which is dead or dying, as electric currents.
**admove, admoveatur** (*ad'-mo-ve, ad-mo-ve-a'-tur*) [imper. sing. and 3d pers. sing., subj., pass., of *admovere*, to apply]. Apply; let there be applied; directions used in prescription-writing.
**adnasal** (*ad-na'-sal*) [*ad*, near to; *nasus*, the nose]. Pertaining to the nose.
**adnata** (*ad-na'-tah*) [*ad*, to; *nasci*, to be born, to grow]. 1. The *tunica adnata*; the conjunctiva; more correctly, a tendinous expansion of the muscles of the eye; it lies between the sclerotic and the conjunctiva. 2. One of the coats of the testicle.
**adnate** (*ad'-nāt*) [*adnatus*, grown to]. Congenitally attached or united.
**adnephrin** (*ad-nef'-rin*). Trade name of a preparation similar to epinephrin.
**adnerval** (*ad-ner'-val*) [*ad*, to; *nervus*, a nerve]. Moving toward a nerve; said of electric currents in muscular fiber.
**adneural** (*ad-nū'-ral*) [*ad*, to; νεῦρον, a nerve]. 1. A term used to describe a nervous affection in which the disease is at the very point where the symptoms appear. 2. Adnerval.
**adnexa** (*ad-neks'-ah*) [*ad*, to; *nectere*, to join]. Adjunct parts, as the *adnexa* of the uterus. **a. bulbi**, the appendages of the bulb of the eye. **a. oculi**, the appendages of the eye, as the lids and lacrimal apparatus. **a. uteri**, the Fallopian tubes and the ovaries.
**adnexitis** (*ad-nek-si'-tis*). Inflammation of the adnexa uteri.
**adnexopexy** (*ad-neks'-o-pek-se*). The operation of raising and fixing the uterine adnexa to the abdominal wall.
**adolescence** (*ad-o-les'-ens*) [*adolescere*, to grow]. The period between puberty and maturity, in males from about 14 to 25 years; in females, from 12 to 21 years.
**adonidin** (*ad-on'-id-in*) [*Adonis*]. A glucoside derived from *Adonis vernalis*, a plant indigenous in Europe and Asia. It is recommended in cardiac dropsy. Dose $\frac{1}{4}-\frac{1}{2}$ gr. (0.008–0.016 Gm.). **a. tannate**, a yellowish-brown powder, soluble in alcohol, slightly soluble in water; it is used in the same manner as the glucoside.
**Adonis** (*ad-o'-nis*). A genus of European herbs belonging to the order *Ranunculaceæ*. **A. æstivalis**, a plant much used in Italy as a cardiac tonic. Dose of *fluidextract* 1–2 min. (0.06–0.12 Cc.); of the *tincture* 10–30 min. (0.6–2.0 Cc.). **A. vernalis**, is used as a cardiac stimulant, antipyretic, and diuretic. Dose of the *tincture* 3–20 min. (0.2–1.3 Cc.).
**adoral** (*ad-o'-ral*) [*ad*, near to; *os*, the mouth]. Situated near the mouth.
**adorbital** (*ad-orb'-it-al*) [*ad*, near to; *orbita*, orbit]. Pertaining to the orbit. **a. bone**, see *lacrimal bone*.
**adosculation** (*ad-os-ku-la'-shun*) [*ad*, to; *osculari*, to kiss]. 1. Impregnation by external contact without intromission. 2. An articulation in which one part is inserted into the cavity of another.
**adrenal** (*ad-re'-nal*) [*ad*, near to; *ren*, the kidney]. 1. Adjacent to the kidney. 2. The suprarenal capsule.
**adrenalin** (*ad-ren'-al-in*), C₁₀H₁₅NO₃. Trade name for a preparation containing the active principle of the suprarenal gland. **a. chloride**, used in solution of 1 : 10,000 to 1 : 1000 in surgical operations on the eye, ear, nose, urethra, etc.; it is a powerful astringent, hemostatic, and heart tonic.
**adrenalinemia** (*ad-ren-al-in-e'-me-ah*) [*adrenalin*; αἷμα, blood]. Presence of adrenalin in the blood.
**adrenalitis** (*ad-ren-al-i'-tis*). Inflammation of the suprarenal glands.
**adrenals** (*ad-re'-nalz*) [*ad*, near to; *ren*, the kidney]. The suprarenal capsules.
**adrenine** (*ad-ren'-ēn*). A preparation of the medulla of the suprarenal gland.
**adrenitis** (*ad-ren-i'-tis*). Inflammation of the adrenals.
**adrenol** (*ad-re'-nol*). An oily solution of adrenalin.
**adrenoxidase** (*ad-ren-oks'-sid-ās*). Oxygenized adrenal secretion, said to be present in blood plasma.
**adrenoxin** (*ad-ren-oks'-in*) [*adrenal*; *oxygen*]. An organic compound or oxidizing substance formed in the lungs by the internal secretion of the adrenals combined with the atmospheric oxygen. This substance endows the blood-plasm with its oxidizing properties (Sajous).
**Adrian's mixture**. A hemostatic mixture containing chloride of iron 25 parts, chloride of sodium 15 parts, and water 60 parts.
**adrin** (*ad'-rin*). Epinephrin hydrate, an active principle of the suprarenal gland; used as a local hemostatic and vasomotor stimulant.
**adrue** (*ad-rū'-e*). Antiemetic root. The root of *Cyperus articulatus*; it is anthelmintic, aromatic, stomachic. Dose of the *fluidextract* 20–30 min. (1.3–2.0 Cc.).
**adscititious** (*ad-si-tish'-us*). Additional; added from without.
**adsorption** (*ad-sorp'-shun*). 1. The power possessed by certain substances of taking up fluids (apart from capillary attraction). 2. The process whereby a substance becomes a part of another and remains in a state midway between mechanical mixture and chemical combination.
**adsternal** (*ad-stern'-al*) [*ad*, near to; *sternum*]. Pertaining to or situated near the sternum.
**adstrictio** - (*ad-strik'-she-o*) [*adstringere*, to draw together; pl., *adstrictiones*]. 1. The retention of any natural excretion. 2. The action of an astringent. 3. The ligation of a blood-vessel. **a. alvei**, constipation.
**ADTe.** Abbreviation of *anodal duration tetanus*; symbol for tetanic contraction, produced by an application of the positive pole with the circuit closed.
**adterminal** (*ad-ter'-min-al*) [*ad*, near to; *terminus*, the end]. Moving toward the insertion of a muscle; said of electric currents in muscular fiber.
**adult** (*ad'-ult*) [*adultus* from *adolescere*, to grow]. Mature; of full legal age. One of mature age. **a. sporadic cretinism**, see *myxedema*.
**adulterant** (*ad-ul'-tur-ant*). 1. The substance used in the process of sophistication. 2. One who adulterates.
**adulteration** (*ad-ul-ter-a'-shun*) [*adulterare*, to corrupt or falsify]. The admixture of inferior, impure, inert, or less valuable ingredients to an article for gain, deception, or concealment.

**adustion** (*ad-us'-chun*) [*adustus*, burned up]. 1. The quality of being scorched or parched. 2. Cauterization.

**advancement** (*ad-vans'-ment*) [Fr., *avancer*, to advance]. An operation to remedy strabismus, generally in conjunction with tenotomy, whereby the opposite tendon from the overacting one, having been cut, is brought forward, so that, growing fast in a more advanced position, it shall have more power to act upon the globe of the eye. **a.**, capsular, an operation similar to that on the tendon upon Tenon's capsule. It differs from advancement in that the tendon itself is not divided. **a. of the round ligaments**, an operation for replacement of the uterus by taking up "the slack of the round ligaments." See *Alexander's Operation*. **a. of Tenon's capsule**, see *a., capsular*.

**adventitia** (*ad-ven-tish'-e-ah*) [*adventitius*, foreign]. The external coat of a blood-vessel.

**adventitious** (*ad-ven-tish'-us*) [*adventitius*, foreign]. Accidental, foreign, acquired, as opposed to natural or hereditary; occurring out of the ordinary or normal place or abode.

**adynamia, adynamy** (*ah-din-a'-me-ah, ah-din'-a-me*) [ἀ, priv.; δύναμις, power]. Loss of vital or muscular power; prostration.

**adynamic** (*ah-din-am'-ik*). See *adynamia*.

**adynamicoataxic** (*ad-in-am-ik-o-at-aks'-ik*). Pertaining to adynamia and ataxia.

**adynatus** (*ad-in'-at-us*). Weakly, sickly.

**æ-**. See **e** for English words beginning with æ.

**Aeby**, plane of. In craniometry, one passing through the nasion and basion perpendicular to the median plane.

**Aëdes** (*ah-e'-dēz*) [ἀηδής, annoying]. A genus of mosquitoes. **A. calopus**, the mosquito of yellow fever, also called *Stegomyia calopus*.

**ædœagra** (*e-de-a'-grah*). See *edeagra*.

**ædœitis** (*e-de-i'-tis*). See *edeiitis*.

**ædœodynia** (*e-de-o-din'-e-ah*). See *edeodynia*.

**ædœology** (*e-de-ol'-o-je*). See *edeology*.

**ædœomania** (*e-de-o-ma'-ne-ah*). See *edeomania*.

**ædœoscopy** (*e-de-os'-ko-pe*). See *edeoscopy*.

**ædœotomy** (*e-de-ol'-o-me*). See *edeotomy*.

**ægagropilus** (*e-gag-rop'-il-us*) [αἴγαγρος, a wild goat; πίλος, felt]. An intestinal concretion formed of hair, found in animals and occasionally in man. A bezoar.

**ægilops** (*e'-jil-ops*). See *egilops*.

**ægobronchophony** (*e-go-brong-koff'-o-ne*). See *egobronchophony*.

**ægonia** (*e-go'-ne-ah*) [L.]. A minor or slight egophony.

**ægophony** (*e-goff'-o-ne*). See *egophony*.

**æluropsis** (*el-u-rop'-sis*) [αἴλυρος, cat; ὄψις, appearance]. Obliquity of the eye or of the palpebral fissure.

**æquabiliter justo major, or minor pelvis** (*e-kwa-bil'-it-er*). See *pelvis*.

**aer** (*a'-er*). 1. See *atmos*. 2. See *air*. **a. dephlogisticus**, oxygen. **a. fixus**, carbon dioxide.

**aerate** (*a'-er-āt*). To supply with air; to charge with gas; to oxygenate, carbonate, etc.; to arterialize.

**aerated** (*a'-er-a-ted*) [ἀήρ, atmosphere]. Charged with gas or air; arterialized. **a. waters**, waters charged with a greater amount of carbon dioxide than they will absorb under ordinary conditions.

**aëration** (*a-er-a'-shun*) [ἀήρ, air]. Charging with air or gas, such as carbon dioxide; the state of being supplied with air or gas.

**aerator** (*a'-er-a-tor*). A machine for forcing gas or air into liquids.

**aerendocardia** (*a-er-en-do-kar'-de-ah*) [ἀήρ, air; ἔνδον, within; καρδία, heart]. The existence of air within the heart.

**aerenterasic** (*a-er-en-tur-a'-sik*) [ἀήρ, air; ἔντερον, the intestine]. Flatulent, tympanitic.

**aerenterectasia** (*a-er-en-ter-ek-ta'-se-ah*) [ἀήρ, air; ἔντερον, intestine; ἔκτασις, distention]. Flatulent distention of the abdomen by gas within the intestines.

**aerial** (*a-e'-re-al*). Pertaining to the air. **a. conduction**, hearing through air-vibrations.

**aerhemoctonia** (*a-er-hem-ok-to'-ne-ah*) [ἀήρ, air; αἷμα, blood; κτόνος, killing]. Death [by the entrance of air into the veins.

**aericolous** (*a-er-ik'-ol-us*) [*aer*, air; *colere*, to inhabit]. Inhabiting the air. Living in the open air.

**aeriferous** (*a-er-if'-er-us*) [ἀήρ, air; *ferre*, to bear]. Conveying air, as the trachea and its branches.

**aerification** (*a-er-if-ik-a'-shun*) [ἀήρ, air; *facere*, to make]. 1. The process of charging with air; the state of being charged with air. 2. Emphysema.

**aerifluxus** (*a-er-if-luks'-us*) [ἀήρ, air; *fluxus*, flow]. Any abnormal escape of air, as by belching, flatulence, etc.

**aeriform** (*a-e'-re-form*) [ἀήρ, air; *forma*, form]. Resembling air or gas.

**aerify** (*a-er'-e-fi*) [ἀήρ, air; *facere*, to make]. 1. To fill with air; to combine with air. 2. To change to a gaseous state.

**aeroanaerobic** (*a-er-o-an-a-er-o'-bik*). Applied to organisms which are both aerobic and anaerobic.

**aerobe** (*a'-er-ōb*) [ἀήρ, air; βίος, life]. One of the aerobia. See *aerobic*.

**aerobia** (*a-er-o'-be-ah*) [ἀήρ, air; βίος, life]. Plural or *aerobion*. Organisms that require air or free oxygen for the maintenance of life. **a.**, **facultative**, organisms normally or usually anaerobic, but under certain circumstances acquiring aerobic power. **a.**, **obligate**, organisms dependent upon free oxygen at all times; never anaerobic.

**aerobic** (*a-er-o'-bik*) [ἀήρ, air; βίος, life]. Requiring oxygen (air) in order to live. A term applied to bacteria requiring free oxygen. Those which do not grow in oxygen are called *anaerobic*. There are forms that are able to grow without oxygen under favorable conditions, though they make use of it when present; others that may grow in its presence, but flourish best without; these are called respectively *facultative aerobic* or *facultative anaerobic*, while those first mentioned are called *obligatory aerobic* or *obligatory anaerobic*.

**aerobion** (*a-er-o'-be-on*) [ἀήρ, air; βίος, life]. An aerobe. See *aerobia*, and *aerobic*.

**aerobioscope** (*a-er-o-bi'-o-skōp*) [ἀήρ, air; βίος, life; σκοπεῖν, to examine]. An apparatus for collecting and filtering bacteria from the air.

**aerobiosis** (*a-er-o-bi-o'-sis*) [ἀήρ, air; βίος, life]. Life that requires the presence of air, or free oxygen.

**aerobiotic** (*a-er-o-bi-ot'-ik*) [ἀήρ, air; βιωτικός, pertaining to life]. Thriving only in the presence of air.

**aerocele** (*a-er'-o-sēl*) [ἀήρ, air; κήλη, tumor]. A tumor varying with respiration, found in the thyroid region, usually unilateral, with walls resembling mucosa and containing mucous or mucopurulent matter. Sometimes congenital, but oftener the result of violent coughing or straining. When acquired, it may disappear spontaneously. Syn., *aerial bronchocele; aerial goiter; pneumatocele; tracheocele; hernia of the trachea*.

**aerocolpos** (*a-er-ō-kol'-pos*). Distention of the vagina with air or gas.

**aerocystoscope** (*a-er-o-sist'-o-skōp*). Same as *aerourethroscope*.

**aerocystoscopy** (*a-er-o-sist-os'-ko-pe*). Examination of the bladder with the aerourethroscope, the bladder being distended with air.

**aerodermectasia** (*a-er-o-der-mek-ta'-se-ah*) [ἀήρ, air; δέρμα, skin; ἔκτασις, distention]. Surgical emphysema; distention of the subcutaneous connective tissue by air.

**aeroductor** (*a-er-o-duk'-tor*) [ἀήρ, air; *ducere*, to lead]. An apparatus to prevent asphyxia of the fetus if the after-coming head is retained.

**aerodynamics** (*a-er-o-di-nam'-iks*) [ἀήρ, air; δύναμις, power]. The branch of physics that deals with gases in motion.

**aeroenterectasia** (*a-er-o-en-ter-ek-ta'-ze-ah*) [ἀήρ, air; ἔντερον, intestine; ἔκτασις, dilatation]. Distention of the bowels with gas.

**aerogen** (*a'-er-o-jen*) [ἀήρ, air; γεννᾶν, to produce]. Any gas-producing microorganism.

**aerography** (*a-er-og'-ra-fe*) [ἀήρ, air; γραφή, a writing]. Description of air and its qualities.

**aerohydropathy** (*a-er-o-hi-drop'-a-the*) [ἀήρ, air; ὕδωρ, water; πάθος, disease]. Pneumatic treatment of disease, combined with hydropathy.

**aerohydrotherapy**. See *aerohydropathy*.

**aerology** (*a-er-ol'-o-je*) [ἀήρ, air; λόγος, treatise]. The science of the air and its qualities.

**aerometer** (*a-er-om'-et-er*) [ἀήρ, air; μέτρον, a measure]. An instrument for ascertaining the density of gases.

**aeromicrobe, aeromicrobion** (*a-er-o-mi'-krōb, -kro'-be-on*). See *aerobe*.

**aeropathy** (*a-er-op'-ath-e*). Caisson disease, *q. v.*

**aeroperitonia** (*a-er-o-per-it-o'-ne-ah*) [ἀήρ, air; *peritoneum*]. Air or gas in the peritoneal cavity.

**aerophagia**, **aerophagy** (a-er-o-fa'-je-ah, a-er-of'-a-je) [ἀήρ, air; φαγεῖν, to eat]. The imbibing and swallowing of air, especially observed in hysterical patients. **a.**, **rectal**, aspiration of air by the rectum.
**aerophil** (a-er'-o-fil) [ἀήρ, air; φιλεῖν, to love]. 1. An open-air-loving person or creature. 2. Aerophic.
**aerophobia** (a-er-o-fo'-be-ah) [ἀήρ, air; φόβος, fear]. Dread of a current of air.
**aerophone** (a'-er-o-fōn) [ἀήρ, air; φωνή, sound]. An instrument for increasing the amplitude of sound-waves.
**aerophore** (a'-er-o-fōr) [ἀήρ, air; φέρειν, to carry]. 1. A device for inflating the lungs of a still-born child with air. 2. A breathing apparatus, used by firemen and others, to prevent the inhalation of noxious gases.
**aerophysic** (a-er-o-fiz'-ik) [ἀήρ, air; φυσᾶν, to inflate]. Inflated; distended with air; flatulent.
**aerophyte** (a-er-o-fīt) [ἀήρ, air; φύτον, plant]. A plant living exclusively in the air.
**aeroplethysmograph** (a-er-o-pleth-iz'-mo-graf) [ἀήρ, air; πληθυσμός, an enlargement; γράφειν, to write]. An apparatus for registering graphically the expired air; the latter raises a very light and carefully equipoised box placed over water, and this moves a writing-style.
**aeroporotomy** (a-er-o-por-ot'-o-me) [ἀήρ, air; πόρος, a pore; τομή, a cutting]. The operation of admitting air to the lungs, as by intubation or tracheotomy.
**aeroscope** (a'-er-o-skōp) [ἀήρ, air; σκοπεῖν, to observe]. An instrument for estimating the purity of the ai ; also an instrument for the examination of air-dust.
**aeroscopy** (a-er=os'-ko-pe) [see *aeroscope*]. The investigation of atmospheric conditions.
**aerostatics** (a-er-o-stat'-iks) [ἀήρ, air; στατικός, standing]. The branch of physics that treats of the properties of gases at rest.
**aerotaxis** (a-er-o-taks'-is) [ἀήρ, air; τάξις, order]. A form of taxis in which living organisms are attracted or repelled by oxygen.
**aerotherapeutics**, **aerotherapy** (a-er-o-ther-a-pū'-tiks, a-er-o-ther'-ap-e) [ἀήρ, air; θεραπεύειν, to heal]. A mode of treating disease by varying the pressure or the composition of the air breathed.
**aerothermotherapy** (a-er-o-ther-mo-ther'-ap-e) [ἀήρ, air; θέρμη, heat; θεραπεία, therapy]. Treatment with hot air.
**aerothorax** (a-er-o-tho'-raks). See *pneumothorax*.
**aerotonometer** (a-er-o-ton-om'-et-er) [ἀήρ, air; τόνος, tension; μέτρον, a measure]. An instrument for estimating the tension of gases in the blood.
**aerotonometry** (a-er-o-ton-om'-et-re). Measurement of the tension or pressure of gases in the blood.
**aerotropism** (a-er-ot'-ro-pizm) [ἀήρ, air; τρέπειν, to turn]. 1. In biology, the deflection of roots from the normal direction of growth by the action of gases. 2. The tendency of certain protozoa, to mass around a bubble of air.
**aerotympanal** (a-er-o-tim'-pan-al) [ἀήρ, air; τύμπανον, a drum]. Pertaining to the air and the tympanum. Cf. *air*, *innate*.
**aerourethroscope** (a-er-o-u-rēth'-ro-skōp) [ἀήρ, air; οὐρήθρα, urethra; σκοπεῖν, to examine]. An instrument modified from the endoscope used in aerourethroscopy. Syn., *aerocystoscope*.
**aerourethroscopy** (a-er=o-u-rē-thros'-ko-pe) [ἀήρ, air; οὐρήθρα, urethra; σκοπεῖν, to examine]. Urethroscopy conjoined with inflation of the urethra with air.
**aerozol** (a'-er-o-zol) [ἀήρ, air; ὄζειν, to smell]. A mixture of essential oils said to contain 75% of ozone; it is used by inhalation in catarrhal affections.
**aerteriversion** (a-er-ter-iv-er'-shun). See *arterioversion*.
**aerteriverter** (a-er-ter-iv-er'-ter). See *arterioverter*.
**ærugo** (e-ru'-go) [L., gen., *æruginis*]. 1. Rust of a metal. 2. Copper rust; verdigris. **æ. ferri**, the subcarbonate of iron. **æ. plumbi**, lead carbonate or subcarbonate.
**ærumna** (e-rum'-nah) [L.]. Mental distress, or mental and physical distress combined.
**Aerva** (a-er'-vah) [Ar.]. A genus of plants of the order *Amarantaceæ*. **A. lanata**, a species native of tropical Asia and Arabia. It furnishes chaya-root, which contains a mucilaginous principle and has been used as a diuretic, in strangury, and as a depurative.
**æs** (ēs) [L.]. Copper or brass. See *copper*.
**æsculetin**. See *esculetin*.

**æsculin** (es'-ku-lin). See *esculin*.
**Æsculus** (es'-ku-lus) [L.]. A genus of sapindaceous shrubs and trees; buckeye. **Æ. glabra**, Ohio buckeye. The bark is tonic, astringent, and antiperiodic. Dose of fluid-extract 10–20 min. (0.6–1.2 Cc.). **Æ. hippocastanum**, horse-chestnut. The bark is tonic, astringent, antiperiodic. Dose of fluid-extract 20–60 min. (1.2–3.7 Cc.). **Æ. pavia**, red buckeye. The bark has been used as a febrifuge. The fruit is said to be an active convulsant.
**æstates** (es-ta'-tēs) [L., pl.]. Freckles or sunburn.
**æsthema** (es-the'-mah) [αἴσθημα; pl. *æsthemates*]. A perception. sensation. sense.
**æsthematology** (es-the-mat-ol'-o-je). See *esthematology*.
**æsthesia** (es-the'-ze-ah). See *esthesia*.
**æsthesin** (es'-the-sin). See *esthesin*.
**æsthesiogen** (es-the'-se-o-jen). See *esthesiogen*.
**æsthesiography** (es-the-se-og'-ra-fe). See *esthesiography*.
**æsthesiology** (es-the-se-ol'-o-je). See *esthesiology*.
**æsthesiomania** (es-the-se-o-ma'-ne-ah). See *esthesiomania*.
**æsthesiometer** (es-the-se-om'-et-er). See *esthesiometer*.
**æsthesiometry** (es-the-se-om'-et-re). See *esthesiometry*.
**æsthesioneurosis** (es-the-se-o-nu-ro'-sis). See *esthesioneurosis*.
**æsthesis** (es-the'-sis). See *esthesis*.
**æsthetica** (es-thet'-ik-ah) [αἴσθησις, perception by the senses]. Diseases characterized by impairment or abolition of any of the senses.
**æsthophysiology** (es-tho-fiz-e-ol'-o-je). See *esthophysiology*.
**æstival** (es'-tiv-al). See *estival*.
**æstivation** (es-tiv-a'-shun). See *estivation*.
**æstuarium** (es-tu-a'-re-um). See *estuarium*.
**æstuation** (es-tu-a'-shun). See *estuation*.
**æstus** (es'-tus) [L.]. Heat; especially a flushing, or sudden glow of heat. **æ. volaticus**, wildfire rash; strophulus.
**ætas** (e'-tas) [L.]. Age; a period of life. See *age*.
**æther** (e'-ther). See *ether*.
**ætherism** (e'-ther-ism). See *etherism*.
**æthiopification** (e-the-op-if-ik-a'-shun). See *ethiopification*.
**æthiopiosis** (e-the-op-e-o'-sis). See *ethiopiosis*.
**æthiops** (e'-the-ops) [Αἰθίοψ, Ethiopian]. An old term for any black mineral powder used in medicine. **æ. antimonialis**, a black triturate of mercury, antimony, and sulphur, made after several distinct formulæ. **æ. martialis**, black oxide of iron. **æ. mineralis**, black amorphous triturate of mercury with sulphur, in various proportions.
**æthomma** (eth-om'-ah) [αἰθός, of a burnt color; ὄμμα, the eye]. 1. Paré's term for a pigmented condition of the humors and tunics of the eye. 2. Kühn's term for a morbid condition marked by flashes of light and flame appearing before the eye.
**ætiology** (e-te-ol'-o-je). See *etiology*.
**afebrile** (ah-feb'-ril) [ā, priv.; *febrilis*, feverish]. Without fever.
**afetal** (ah-fe'-tal) [ā, priv.; *fetus*, an offspring]. Without a fetus.
**affection** (af-ek'-shun) [*afficere*, to affect]. Disease. **a., parainfectious**, one in which the symptoms or conditions are only indirectly related to the disease named; a by-condition or accessory infection of certain diseases characterized by the appearance of symptoms attributable to an intercurrent or secondary infection, as in the case of noma occurring in cases of measles and due to infection with diphtheria. **a., pneumogastropituitous**, see *pertussis*. **a., polyuric**, see *lithuria*. **a., primary**, one independent of any preceding disease. **a., secondary**, one that is a complication or sequel of a preexisting disease. **a., vaporous**, see *vapors*.
**affective** (af-ek'-tiv) [see *affection*]. Exciting emotion. **a.** faculties, the emotions and propensities, especially those peculiar to man. **a. insanity**, emotional or impulsive insanity.
**affenspalte** (af'-fen-spal-ter) [German for ape's split]. The parietöoccipital fissure; ape-fissure.
**afferent** (af'-er-ent) [*afferens*, carrying to]. Carrying toward the center. Of *nerves;* conveying impulses toward the central nervous system; sensory; centripetal. Of *blood-vessels:* those, as the arteries, conveying blood to the tissues. Of *lymphatics.* those conveying lymph to a lymphatic gland.

**afferentia** (*af-er-en'-she-ah*). See *vasa*.

**affiliation** (*af-il-e-a'-shun*) [*ad*, to; *filius*, son]. In medical jurisprudence, the act of imputing or affixing the paternity of a child in order to provide for its maintenance.

**affinity** (*af-in'-it-e*) [*affinis*, akin to]. 1. Relationship. 2. Attraction. 3. In biology, morphologic, physiologic, and phylogenetic relationship between organisms. a. of aggregation, cohesive attraction; the mechanical affinity of similar molecules tending to the formation of masses. Syn., *quiescent affinity; affinitas quiescens*. a., chemical, the force, exerted at inappreciable distances, that unites atoms. a. of composition, the tendency of substances to unite directly without previous decomposition. Syn., *affinitas compositionis; simple affinity; single affinity; compound affinity; mixing affinity*. a., developed, that exhibited by compounds, but which is not possessed by the constituents separately. Syn., *affinitas producta; resulting affinity; secondary affinity*. a., divellent, the tendency to form new compounds at the expense of decomposition of those previously existing. Syn., *affinitas divellens; separating affinity*. a., elective, the preference of one substance for another over a second or third. a., elementary. 1. That which exists between the elements of two or more compounds. 2. Physicochemical relationship of elementary substances. a., mediating, that by virtue of which a substance lacking the power of combination with a certain substance secures it by preliminary combination with another. Syn., *appropriate affinity; imparted affinity; intermediate affinity; inducing affinity; inductive affinity; affinity of an intermedium; affinitas adjuta; affinitas appropriata; affinitas approximata*. a., morbid, the tendency of certain affections to exist synchronously or as sequels. a., reciprocal, chemical attraction between the elements of a secondary compound, tending, under altered conditions, to the reformation of the primary compound. Syn., *alternating elective affinity; affinitas reciproca*. a., simple elective, that exhibited by a simple body for a single element of a compound. Syn., *single elective affinity*. a. of solution, that existing between a dissolved substance and its solvent. a., vital, the selective action or chemiotaxis exhibited by the several tissues of an organism for their peculiar pabulum.

**affion, affioni** [Turkish]. Crude opium; it contains regularly 10 % of morphine. Syn., *offium*.

**affixion** (*af-ik'-shun*) [*affigere*, to fasten]. Adhesion.

**afflatus** (*af-la'-tus*) [L., "a blowing upon"]. 1. A draft or blast of air. 2. A sudden attack. 3. A supposed inspiration or divine influence.

**affluence** (*af'-lu-ens*) [*affluentia*, from *affluere*, to flow to]. A determination or influx, as of blood to a part.

**affluent** (*af'-lu-ent*) [*affluens*, flowing to]. Producing a congestion; determinant; flowing in or upon.

**afflux, affluxion** (*af'-luks, af-fluk'-shun*) [*affluere*, to flow toward]. The flow of the blood or other liquid to a part.

**affuse** (*af-ūz'*) [*affundere*, to pour upon]. To sprinkle or pour upon from a height; to shower.

**affusio** (*af-u'-se-o*) [L.; pl., *affusiones*]. 1. An affusion. 2. Suffusion. 3. Infusion. 4. Cataract.

**affusion** (*af-u'-zhun*) [*affundere*, to pour upon]. The pouring of water upon an object, as upon the body in fever, to reduce temperature and calm nervous symptoms. a., cold. Currie's method of treating fevers by pouring cold water over the patient. a., *affusio frigida*.

**afibroma** (*ah-fi-bro'-mah*) [á, priv.; *fibroma*]. A mass of fibrous tissue which is not arranged so as to form a tendon or fascia.

**African arrow-poison.** See *strophanthus*. **A. fever**, synonym of *dengue*. **A. gum**, gum-arabic. **A. lethargy**, a "sleeping-sickness" affecting west African coast negroes. Increasing somnolence is the characteristic symptom. It is very fatal—death from exhaustion follows in from 3 to 6 months. Syn., *nélavan*.

**afridol** (*af'-rid-ol*). An antiseptic, said to be an orthotoluate of mercury and sodium.

**afrodyn** (*af'-ro-din*) [ἀφροδίσια, venery]. An aphrodisiac, the principal ingredient of which is said to be the tincture of myorapuama.

**aftannin** (*af'-tan-in*). An infusion of herbs with formaldehyde and glycerin used in veterinary practice.

**after** (*af'-ter*) [AS., *æfter*, back]. 1. The anus; the buttocks. 2. Next in succession.

**after-action**, the negative variation in an electrical current continuing for a short time in a tetanized muscle. a., inner, that involving the whole muscle or muscular fiber. a., terminal, that affecting only the ends of the muscular fibers.

**after-birth**, the popular designation of the placenta, cord, and membranes, sometimes called the *secundines*.

**after-brain.** See *hindbrain* and *metencephalon*.

**after-care**, the care or nursing of convalescents; specifically, the treatment of patients discharged as cured from lunatic asylums.

**after-cataract**, *cataracta secundaria*; an opacity of the media of the eye after operation for cataract due to opacification of the capsule or to non-absorption of the remains of the lens-substance.

**after-current** (*af-ter-kur'-ent*). See under *current*.

**after-damp**, a poisonous mixture of gases, such as carbon monoxide and carbon dioxide, found in coal mines after an explosion of inflammable gases.

**after-gilding** (*af-ter-gild'-ing*). A term introduced by Apathy to designate the process of treating nerve-tissues with salts of gold after fixation and hardening. Cf. *foregilding*.

**after-hearing**, a neurotic condition in which sounds are heard after the wave-motion that produces them has ceased.

**after-images**, continued retinal impressions after the stimulus of the light or image has ceased to act. A *positive after-image* is a simple prolongation of the sensation; a *negative after-image* is the appearance of the image in complementary colors. After sensations may be also experienced with other senses.

**afterings** (*af'-ter-ings*). See *after-milk*.

**after-milk**, the strippings; the last milk taken from the teat at any one milking. It is peculiarly rich in butter, as compared with the fore-milk.

**after-pains.** See *pains*.

**after-perception**, the perception of a sensation after the stimulus has passed away.

**after-production** (*af-ter-pro-duk'-shun*). A new growth; neoplasm.

**after-sensation**, a sensation lasting longer than the stimulus producing it.

**after-taste**, a gustatory sensation produced some time after the stimulus has been removed.

**after-treatment.** See *after-care*.

**after-vision** (*af-ter-vish'-on*). The perception of an after-image.

**Ag.** Abbreviation for *argentum*, Latin for silver.

**agalactia** (*ah-gal-ak'-te-ah*) [á, priv.; γάλα, milk]. Non-secretion or imperfect secretion of milk after child-birth.

**agalactous** (*ah-gal-ak'-tus*) [á, priv.; γάλα, milk]. 1. Without milk. 2. Not suckled; not nourished with milk. 3. Capable of diminishing the secretion of milk.

**agal-agal** (*ah'-gal-ah'-gal*). See *agar-agar*.

**agalasia** (*ah-gal-a'-ze-ah*). See *agalactia*. **a. contagiosa**, an epidemic, contagious disease of sheep and goats, marked by drying-up of the milk.

**agalorrhea, or agalorrhœa** (*ah-gal-o-re'-ah*) [á, priv.; γάλα, milk; ῥέειν, to flow]. A cessation of the flow of milk.

**agamic** (*ah-gam'-ik*) [á, priv.; γάμος, marriage]. In biology, not sexual; not pertaining to the sexual relation; asexual reproduction; parthenogenesis.

**agamogenesis** (*ah-gam-o-jen'-es-is*) [á, priv.; γάμος, marriage; γένεσις, generation]. Reproduction without fecundation, as, e. g., by gemmation. See *parthenogenesis*.

**agamogenetic** (*ah-gam-o-jen-et'-ik*) [á, priv.; γάμος, marriage; γένεσις, generation]. Pertaining to agamogenesis.

**agamospore** (*ah-gam'-o-spor*) [á, priv.; γάμος, marriage; σπορά, offspring]. In biology, an asexually produced spore.

**aganactesis** (*ag-an-ak-te'-sis*) [ἀγανάκτησις, physical pain]. Irritation; physical pain or uneasy sensation.

**aganoblepharon** (*ag-an-o-blef'-ar-on*) [ἀγανοβλέφαρος, mil-eyed]. Adhesion of the eyelids to each other.

**agar-agar** (*ag'-ar-ag'-ar*) [Ceylon]. A kind of glue made from certain sea-weeds, such as *Gracilaria lichenoides* and *Gigartina speciosa*, used in medicine to make suppositories, and in bacteriological studies to make a solution in which microorganisms are bred or kept. See *gelose*.

**agar hanging block** (*ag'-ar*). In bacteriology a small block of nutrient agar cut from a poured plate, and placed on a cover-glass, the surface next the glass having been first touched with a loop from a young fluid culture or with a dilution from the same. It is examined upside down, the same as a hanging drop.

**agaric** (*ag-ar'-ik*). Touchwood; spunk; tinder; the product of different species of *Boletus*, a genus of mushrooms. *Boletus laricis*, *Polyporus officinalis*—is the *white* or *purging agaric*. Agaric or *agaricinic acid*, in doses of $\frac{1}{16}-\frac{1}{4}$ gr. (0.004–0.02 Gm.), is also useful in night-sweats. Dose of the *extract* 3–6 gr. (0.19–0.38 Gm.); of the *tincture* 3–20 min. (0.18–1.2 Cc). *Agaricus chirurgorum*, *Boletus chirurgorum*, surgeon's agaric, a parasitic fungus formerly used for *moxa*. Soaked in solution of potassium nitrate it forms *spunk*. *Agaricus muscarius*, fly agaric, poisonous mushroom, contains an alkaloid, *muscarine*. Dose of the *alkaloid* $\frac{1}{8}-2$ gr. (0.008–0.13 Gm.), *Muscarine nitrate* is used hypodermically. Dose $\frac{1}{10}-\frac{3}{4}$ gr. (0.006–0.048 Gm.).

**agaricin** (*ag-ar'-is-in*) [see *agaricus*]. 1. C₁₆H₂₀O₅ +H₂O. A white, crystalline substance, the active principle of *Agaricus albus*. It has proved useful in the night-sweats of pulmonary tuberculosis. Dose $\frac{1}{20}-\frac{1}{10}$ gr. (0.003–0.006 Gm.). Unof. 2. An impute alcoholic extract of the agaric, *Polyporus officinalis*; used as an anhidrotic.

**Agaricus** (*ag-ar'-ik-us*) [ἀγαρικόν, of Dioscorides, from *Agaria*, a former district of Poland or Sarmatia, whence the Greeks derived the larch agaric]. A large genus of hymenomycetous fungi; mushrooms and toadstools. Cf. *Polyporus amanita*. **A. chirurgorum**, see under *agaric*. **A. rubra**, **A. sanguinea**, these species, indigenous to France, were formerly included under *A. rubra*. They yield the alkaloid *agarythrine* and the rose-red coloring-matter *ruberin*.

**agarythrine** (*ag-ar'-ith-rin*). A yellowish-white alkaloid extracted by ether from *Agaricus rubra* and *A. sanguinea*. It has a bitter taste and leaves a burning sensation in the mouth.

**agaster** (*ah-gas'-ter*) [ἀ, priv.; γαστήρ, the stomach]. One without a stomach.

**agastric** (*ah-gas'-trik*) [see *agaster*]. Without an intestinal canal, as the tape-worms.

**agastronervia** (*ah-gas-tro-ner'-ve-ah*) [ἀ, priv.; γαστήρ, the stomach; *nervus*, a nerve or sinew]. See *agastroneuria*.

**agastroneuria** (*ah-gas-tro-nū'-re-ah*) [ἀ, priv.; γαστήρ, the stomach; νεῦρον, a nerve]. Deficiency in the nerve-stimulus sent to the stomach.

**agathin** (*ag'-ath-in*) [ἀγαθός, good]. C₆H₄(OH).- CH.N.N. (CH₃). C₆H₅. A greenish-white, crystalline substance, obtained by the interaction of salicylic aldehyde and α-methylphenylhydrazine. It is used as an antineuralgic in doses of 8 gr. (0.52 Gm.) 2 or 3 times daily. Its action is cumulative.

**Agave** (*a-ga'-ve*) [ἀγαυή, noble]. A large genus of amaryllidaceous plants, natives of North America. *A. americana*, American aloe, the leaves of a plant growing in North America. It is diuretic and antisyphilitic. Dose of the *fluidextract* ½–1 dr. (2–4 Cc.). The fresh juice is also similarly employed. The fermented juice, called *pulque*, is a moderately stimulant drink, very popular in Mexico.

**age** (*āj*). The length of time a being has existed; also, a certain stage in life. **a. of consent**, in medical jurisprudence the age at which a minor is considered capable of consenting to sexual intercourse; it is usually placed at 16 years. **a. critique**, the climacteric. **a., marriageable, a., nubile**, see *nubility*. **a. of puberty**, see *puberty*.

**agenesia, agenesis** (*ah-jen-e'-se-ah, ah-jen'-es-is*) [ἀ, priv.; γένεσις, generation]. 1. Incomplete and imperfect development. 2. Impotence, barrenness.

**agenosomia** (*ah-jen-o-so'-me-ah*) - [ἀ, priv.; γεννᾶν, to beget; σῶμα, body]. Defective development of the genitals.

**agenosomus** (*ah-jen-o-so'-mus*) [ἀ, priv.; γεννᾶν, to beget; σῶμα, body]. A variety of single autositic monsters, of the species *Celosoma*, in which there is a lateral or median eventration occupying principally the lower portion of the abdomen, while the genital and urinary organs are either absent or very rudimentary.

**agent** (*a'-jent*) [*agere*, to act, to do]. A substance or force that by its action effects changes in the human body.

**agerasia** (*aj-er-a'-ze-ah*) [ἀγηρασία, eternal youth].

Vigorous old age; age without its wonted feebleness and decay.

**ageusia, ageustia** (*ah-gu'-se-ah, ah-goost'-e-ah*) [ἀ, priv.; γεῦσις, taste]. Abolition of the sense of taste. **a., central**, that due to lesion of the cerebral centers of the gustatory nerves. **a., conduction**, that due to lesion in the nerves between their origin and distribution. **a., peripheral**, that due to disorder of the ends of the nerves of taste.

**agger** (*aj'-er*) [L.]. In anatomy, a pile or mound. **a. nasi**, an oblique ridge on the inner surface of the nasal process of the maxilla; also called *crista ethmoidalis*. **a. valvulæ venæ** [*pl.*, *aggeres valvularum venarum*], the eminence of a venous valve; a projection within the lumen of a vein at the junction of a valve.

**agglomerate** (*ag-lom'-er-āt*) [*agglomerare*, to wind into a ball]. Grouped or clustered.

**agglutinant** (*ag-lu'-tin-ant*). See *agglutinative*

**agglutinate** (*ag-lu'-tin-āt*) [see *agglutination*]. To glue together; to unite by adhesion.

**agglutinatio** (*ag-lu-tin-a'-she-o*). Agglutination. **a. maxillæ inferioris**, trismus.

**agglutination** (*ag-lu-tin-a'-shun*) [*agglutinate*, to paste to]. 1. A joining together. 2. A copulative phenomenon accompanying hemolysis or bacteriolysis, thought by Gruber to be due to some deleterious effect on the membrane of the bacteria or blood-corpuscles which makes it sticky. **a. test**, see *Widal's test*.

**agglutinative** (*ag-lu'-tin-a-tiv*) [see *agglutination*]. 1. Favoring agglutination; adhesive. 2. Any substance with adhesive properties, fitted to retain the edges of wounds in apposition. 3. A remedy promoting the repair of wounds by favoring nutrition.

**agglutinin** (*ag-lu'-tin-in*) [see *agglutination*]. A specific principle occurring in the blood-serum of an animal affected with a disease of microbic origin and capable of causing the clumping of the bacteria peculiar to that disease, as exemplified in the Widal reaction.

**agglutinogen** (*ag-lu-tin'-o-jen*). A substance which when introduced into the body is capable of causing the formation of an agglutinin.

**agglutinoid** (*ag-lu'-tin-oid*). An agglutinin with the xymotoxic group deficient or absent.

**agglutinophore** (*ag-glu-tin'-o-fōr*). Same as *zymophore*, *q. v*.

**agglutitio** (*ag-lu-tish'-e-o*) [*ad*, against; *glutire*, to swallow]. Difficult deglutition; an obstruction to swallowing.

**agglutogen** (*ag-lu'-to-jen*). See *agglutinogen*.

**agglutogenic** (*ag-lu-to-jen'-ik*) [*agglutinin*; *generare*, to produce]. Relating to substances from which agglutinins originate.

**agglutometer** (*ag-lu-tom'-et-er*). An apparatus used in performing the agglutination or Widal test.

**aggregate** (*ag'-re-gāt*) [*ad*, to; *gregare*, to collect into a flock]. Grouped into a mass. **a. glands**, Peyer's patches.

**aggregation** (*ag-re-ga'-shun*) [*ad*, to; *gregare*, to collect into a flock]. 1. The massing of materials together. 2. A congeries or collection of bodies, mostly of such as are similar to each other.

**aggressin** (*ag-res'-in*) [*aggressio*, an attack]. A substance produced in the body by bacteria, having the property of weakening the normal protective substances of the body. By some it is held that this substance increases the virulence of the bacteria.

**aggressinogen** (*ag-res-in'-o-jen*). An antigen which gives rise to aggressins.

**aggressivity** (*ag-res-iv'-it-e*). The degree of activity displayed by an invading microorganism against the protective forces of the host.

**agitation** (*aj-it-a'-shun*) [*agitare*, to excite, arouse]. 1. Fatiguing restlessness with violent motion; mental disturbance. 2. A stirring or shaking, as in pharmacy.

**agitator** (*aj-it-a'-tor*) [*agitare*, to excite]. Any apparatus for stirring or shaking substances; a glass rod used for stirring.

**aglandular** (*ah-glan'-du-lar*) [ἀ, priv.; *glandula*, a gland]. Having no glands; without glands.

**aglaukopsia** (*ag-law-kop'-se-ah*) [ἀ, priv.; γλαυκός, green; ὄψις, vision]. Green-blindness.

**aglia** (*ag'-le-ah*) [L.]. A speck or spot upon the cornea or on the white of the eye.

**aglobulia** (*ah-glo-bu'-le-ah*) [ἀ, priv.; *globulus*, a globule]. A decrease in the quantity of red blood-corpuscles.

**aglobulism** (*ah-glob'-u-lizm*) [ἀ, priv.; *globulus*, a globule]. Aglobulia; oligocythemia.

**aglossia** (*ah-glos'-e-ah*) [ἀ, priv.; γλῶσσα, the tongue]. 1. Absence of the tongue. 2. Dumbness; senile impairment of speech.

**aglossostomia** (*ah-glos-o-sto'-me-ah*) [ἀ, priv.; γλῶσσα, the tongue; στόμα, mouth]. The condition of a mouth without a tongue.

**aglossus** (*ah-glos'-us*) [see *aglossia*]. A person without a tongue.

**aglutition** (*ah-glu-tish'-un*) [ἀ, priv.; *glutire*, to swallow]. Difficulty in swallowing; inability to swallow.

**agmatology** (*ag-mat-ol'-o-je*) [ἀγμός, a fracture; λόγος, a discourse]. The science or study of fractures.

**agminate, agminated** (*ag'-min-ăt, ag'-min-a-ted*) [*agmen*, a multitude]. Aggregated; clustered. **a. glands,** see *gland, Peyer's*.

**agnail** (*ag'-nāl*). 1. Hangnail. 2. A whitlow. 3. A corn.

**agnathia** (*ah-gna'-the-ah*) [ἀ, priv.; γνάθος, a jaw]. Absence or defective development of the jaws.

**agnathus** (*ag'-na-thus*) [ἀ, priv.; γνάθος, a jaw]. A monster with no lower jaw.

**agnea,** or **agnœa** (*ag-ne'-ah*) [ἄγνοια, want of perception]. A condition in which the patient does not recognize things or persons.

**Agnew's splint** (*ag'-nū*) [David Hayes *Agnew*, American surgeon, 1818–1892]. *For hip-joint disease;* a long splint with a perineal band (fitted closely against the tuber ischii) and a foot-piece; used after the disappearance of acute symptoms and designed to support the weight of the trunk.

**agnin** (*ag'-nin*) [*agnus*, a lamb]. A fatty substance derived from sheep's wool.

**agnina membrana** (*ag-ni'-nah mem-bra'-nah*) [L.]. "The lamb-like, or woolly, membrane,"—the amnion.

**agnolin** (*ag'-no-lin*). Purified wool fat; adeps lanæ.

**agnosia** (*ah-gno'-se-ah*) [ἀ, priv.; γνῶσις, a recognizing]. Loss of the perceptive faculty which gives recognition of persons and things.

**-agoga, -agogue** [ἀγωγός, one who leads]. A suffix, denoting agents that drive out other substances, as emmenagogues, lithagogues, etc.

**agomphiasis** (*ah-gom-fi'-as-is*) [ἀ, priv.; γομφίος, a tooth]. Same as *agomphosis*.

**agomphious** (*ah-gom'-fe-us*) [ἀ, priv.; γομφίος, a tooth]. Without teeth.

**agomphosis** (*ah-gom-fo'-sis*) [see *agomphious*]. 1. Absence of the teeth. 2. A loosening of the teeth.

**agonal** (*ag'-on-al*) [ἀγωνία, a struggle]. Struggling; relating to the death-struggle.

**agonia** (*ag-o'-ne-ah*) [ἀγωνία, a contest or struggle]. 1. Distress of mind; extreme anguish. 2. The death struggle. [ἄγονος, barren]. Barrenness; sterility; impotence. **a. bark,** see *agoriada*.

**agoniadin** (*ag-on-i'-ad-in*), $C_{10}H_{14}O_6$. A glucoside found in Agonia bark, and used as an antiperiodic.

**agonous** (*ag'-o-nus*) [ἄγονος, unfruitful]. Barren; impotent.

**agony** (*ag'-o-ne*) [see *agonal*]. Violent pain; extreme anguish; the death-struggle.

**agopyrine** (*ag-o-pi'-rin*). An influenza remedy said to contain salicin, 4 gr.; ammonium chloride ⅓ gr.; cinchonine sulphate, ⅓ gr.

**agoraphobia** (*ag-o-ra-fo'-be-ah*) [ἀγορά, a market-place, assembly; φόβος, fear]. A morbid fear of open places or spaces.

**agoriadin** (*ag-o-ri'-ad-in*) [Sp.]. $C_{10}H_{14}O_6$. A glucoside, probably the active principle of *agoriada*.

**Agostini's reaction for glucose.** To 5 drops of the urine add 5 drops of 0.5 % solution of gold chloride and 3 drops of 20 % potash solution, and heat gently. In the presence of glucose a red color will be produced.

**-agra** (ἄγρα, a seizure]. A Greek word added as a suffix to various roots to denote *seizure, severe pain;* as podagra, etc.

**agraffe** (*ag-raf'*) [Fr. *agrafe*, a hook, clasp]. An instrument to keep the edges of a wound together.

**agrammatism** (*ah-gram'-at-izm*) [ἀ, priv.; γράμμα, a word]. A phenomenon of aphasia, consisting in the inability to form words grammatically, or the suppression of certain words of a phrase; a form of aphasia.

**agraphia** (*ah-graf'-e-ah*) [ἀ, priv.; γράφειν, to write]. Inability to express ideas by writing. **a., absolute,** a variety in which no letters can be formed. Syn., *literal agraphia.* **a., acoustic,** loss of capacity to write from dictation. **a. amnemonica,** a form in which letters can be written, but without conveying any meaning. **a. atactica,** that form in which letters cannot be formed from lack of muscular coordination. **a., literal, a. literalis,** see *a., absolute.* **a., motor,** inability to recall the movements of the hand necessary in writing. **a., musical,** pathological loss of the ability to write musical notes. **a., optic,** inability to copy writing, but ability to write from dictation. **a., verbal,** a variety in which a number of words without meaning can be written. Cf. *paragraphia.*

**agraphic** (*ah-graf'-ik*) [see *agraphia*]. Affected with or pertaining to agraphia.

**agremia** (*ag-re'-me-ah*) [ἄγρα, seizure; αἷμα, blood]. The condition of the blood in gout; the gouty diathesis.

**agria** (*ag'-re-ah*) [ἄγριος, wild]. A pustular eruption; malignant pustule; herpes.

**agridinium** (*ag-rid-in'-e-um*). A dye-stuff used with arsenophenylglycin, for its trypanocidal properties.

**agrielcosis** (*ag-re-el-ko'-sis*) [ἄγριος, wild; ἕλκωσις, ulceration]. A malignant or uncontrollable ulceration.

**agrimony** (*ag'-rim-o-ne*) [ἀγρός, a field; μόνος, alone]. The root of *Agrimonia eupatoria*, a mild astringent. Dose of *fluidextract* ½–2 dr. (2–8 Cc.).

**agriopsoria** (*ag-re-op-so'-re-ah*) [ἄγριος, wild; ψώρα, itch]. An incurable or severe attack, or variety, of itch.

**agriothymia** (*ag-re-o-thi'-me-ah*) [ἄγριος, wild; θυμός, mind; will]. Maniacal fury.

**agrippa** (*ag-rip'-ah*) [L.]. One born with the feet foremost.

**agromania** (*ag-ro-ma'-ne-ah*) [ἀγρός, a field; μανία, madness]. A mania for living in the country.

**agron** [East Indian]. A disease which occurs in India, marked by roughening of the tongue, with fissures.

**agrypnetic** (*ah-grip-net'-ik*) [ἀ, priv.; ὕπνος, sleep]. 1. Sleepless; wakeful. 2. Preventing sleep; agrypnotic.

**agrypnia** (*ah-grip'-ne-ah*) [ἀ, priv.; ὕπνος, sleep]. Loss of sleep; insomnia.

**agrypnocoma** (*ah-grip-no-ko'-mah*) [ἀγρυπνος, sleepless; κῶμα, coma]. Coma vigil; wakeful lethargy, with low-muttering delirium.

**agrypnotic** (*ah-grip-not'-ik*) [ἀ, priv.; ὕπνος, sleep]. 1. Preventing sleep; causing wakefulness. 2. A medicine that prevents sleep.

**aguamiel** (*ah-goo-ah-me-el'*) [Sp.]. The sap of the pulque magueys, *Agave atrovirens*, and *A. Mexicana*. From it is made the fermented drink pulque. It is said to have diuretic, laxative, galactagogue, and nutrient properties.

**ague** (*a'-gū*) [*acutus*, sharp; acute; Fr., *aigu*]. 1. Malarial or intermittent fever; characterized by paroxysms consisting of chill, fever, and sweating, at regularly recurring times, and followed by an interval or intermission the length of which determines the epithets quotidian, tertian, etc. In some cases there is a double paroxysm, and hence these are called double quotidian, double tertian, etc. The duration of each paroxysm varies from 2 to 12 hours. Syn., *fever and ague; intermittent fever; periodic fever; malarial fever; marsh fever; paludal fever; miasmatic fever.* 2. A chill. **a., Aden,** see *dengue.* **a., brass-founders',** a disease common among brass-founders, characterized by symptoms somewhat resembling an imperfect attack of intermittent fever, the recurrence of the paroxysms, however, being irregular. The direct cause is generally thought to be the inhalation of the fumes of deflagrating zinc or "spelter." **a., brow-,** intermittent neuralgia of the brow. **a.-cake,** chronic enlargement of the spleen in diseases of malarial origin. **a., catenating,** ague associated with other diseases. **a.-drop,** see *Fowler's solution.* **a., dumb,** ague without well-marked chill, and with at most only partial or slight periodicity. Syn., *dead ague; irregular ague; latent ague; masked ague.* **a., face,** tic douloureux. **a., partial,** ague attended with pain which is limited to some part or organ. **a.-tree,** common sassafras. **a.-weed.** 1. See *Gentiana.* 2. *Eupatorium perfoliatum*, or thoroughwort.

**aguish** (*a'-gu-ish*). Resembling or relating to ague; affected with ague.

**agurin** (*ag'-ū-rin*). A compound of sodium theobromate and sodium acetate; it is recommended as a diuretic in doses of 24 gr. (1.5 Gm.).

**Ah.** Abbreviation of hypermetropic astigmatism.

**Ahlfeld's sign** (*ahl'-felt*) [Johann Friedrich *Ahlfeld*, German obstetrician, 1843– ]. Irregular tetanic contractions affecting localized areas of the uterus, observed after the third month of pregnancy.

**ahypnia** (*ah-hip'-ne-ah*) [ἀ, priv.; ὕπνος, sleep]. Sleeplessness.

**ahypnosis** (*ah-hip-no'-sis*) [ἀυπνία, sleeplessness]. Entire absence of the capacity to sleep, most marked in insanity.

**aichmophobia** (*āk-mo-fo'-be-ah*) [αἰχμή, a spear point; φόβον, to fear]. An extravagant dread of sharp or pointed instruments.

**aidoio-** (*a-doi'-o*). See *edeo-*.

**aidoiomania** (*a-doi-o-ma'-ne-ah*). See *edeomania*.

**ail** (*āl*) [ME., *eyle*]. 1. To be out of health. 2. A slight indisposition. 3. Garlic. **a., Wetherbee,** a popular name for progressive muscular atrophy, from the fact that several successive generations of a Massachusetts family of that name were affected with the disease.

**Ailanthus** (*a-el-an'-thus*). See *Ailantus*.

**ailantus** (*a-el-an'-tus*) [Malacca, *ailanto*, "tree of heaven"]. The bark of *A. glandulosa*, commonly known as "tree of heaven." Its properties are due to an oleoresin and a volatile oil. It is a nauseant and drastic purgative and an excellent anthelmintic against tape-worm. Dose of *fluidextract* 10 min.–1 dr. (0.6–4.0 Cc.); of *tincture* 10 min.–2 dr. (0.6–8.0 Cc.).

**ailing** (*āl'-ing*). Indisposed; out of health; not well.

**ailment** (*āl'-ment*) [ME., *eyle*]. A disease; sickness; complaint.

**ailurophobia** (*a-lu-ro-fo'-be-ah*) [αἴλουρος, a cat; φόβος, fear]. A morbid fear of cats.

**ainhum** (*in'-hoom*) [negro word, meaning *to saw*]. A disease of Guinea and Hindustan, peculiar to negroes, in which the little toes are slowly and spontaneously amputated at about the digitoplantar fold. The process is very slow, is unaccompanied by any constitutional symptoms, and its cause is unknown. It sometimes attacks the great toe.

**aiodine** (*ah-i'-o-din*). A preparation of the thyroid gland and tannin. It is a tasteless powder, of which each gram is said to represent 10 Gm. of the fresh glands and to contain 0.4% of iodine. It is used in myxedema.

**air** [ἀήρ, the lower, dense air as distinguished from αἰθήρ, the upper and purer air]. The atmosphere. *Atmospheric air* consists of a mixture of 77 parts by weight, or 79.19 by volume, of nitrogen, and 23 parts by weight, or 20.81 by volume, of oxygen, with 0.03 to 0.06 parts by volume of $CO_2$. It also contains traces of ammonia, argon, nitrites, and organic matter. By virtue of its oxygen it is able to sustain respiration. One hundred cubic inches weigh 30,935 grains. The pressure of the air at sea-level is about $14\frac{3}{4}$ pounds upon the square inch. **a., alkaline,** free or volatile ammonia. **a., azotic,** nitrogen. **a.-bag,** see *a.-cushion*. **a.-bath,** therapeutic exposure to air, which may be heated, condensed, or variously medicated. **a.-bed,** an airtight rubber mattress, inflated with air, employed in conditions requiring prolonged confinement to bed. **a.-bladder,** see *a.-vesicles*. **a.-cell,** an air-sac; an air-vesicle of the lung. **a., complemental,** the amount of air that can still be inhaled after an ordinary inspiration. **a. conduction,** a method of testing the hearing-power by means of a watch held at varying distances from the ear, or by the employment of a number of tuning-forks of varying pitch. **a.-cure,** the therapeutic employment of air. **a.-cushion,** a cushion filled with air, and usually made of soft india-rubber. **a., dephlogisticated,** an old name for oxygen. **a.-douche,** the inflation of the middle ear through the nose. **a.-embolism,** the entrance of free air into the blood-vessels during life. **a., expired,** that driven from the lungs in expiration. **a., factitious,** carbon dioxide. **a., fixed,** an old name for carbon dioxide. **a., hepatic,** hydrogen sulphide. **a.-hunger,** dyspnea on both inspiration and expiration. **a., inspired,** that taken into the lungs on inspiration. **a., liquid,** air which has been liquefied by intense pressure; an extreme cold is produced by its evaporation. **a., mephitic,** carbon dioxide. **a.-passages,** the nares, mouth, larynx, trachea, and bronchial tubes. **a., phlogisticated,** an old name for nitrogen. **a.-pump,** an apparatus for exhausting or compressing air. **a., reserve, a., supplemental,** the air that can still be exhaled after an ordinary expiration. **a., residual,** that remaining in the lungs after the most complete expiration possible. **a.-sac,** see *a.-vesicles*. **a., solid,** of **Hales,** carbon dioxide; so called because of its property of forming solid carbonates with metallic oxides. **a.-space,** a space in tissues filled with air or other gases. **a., stationary,** that remaining in the lungs during normal respiration. **a., supplemental,** see *a., reserve*. **a.-tester,** an instrument for testing the purity of the air. **a., tidal,** that taken in and given out at each respiration. **a.-trap,** a trap to prevent the escape of sewer gas. **a.-vesicles,** the alveoli of the lung, the ultimate division of the air-passages. **a., vital,** an old name for oxygen.

**air-break wheel, air-breaking wheel.** An arrangement by means of which the sparks may be promptly extinguished when using a 110-volt continuous current to excite a coil; the spark formed at the contact-brushes when the coil is energized is blown out instantaneously by the air-blast.

**airoform** (*ār'-o-form*). Same as *airol*.

**airogen** (*ār'-o-jen*). See *airol*.

**airol** (*ār'-ol*). Trade name for *bismuth iodosubgallate*. *q. v.*

**akamathesia, akamathesis.** See *akatamathesia*.

**akamushi disease** (*ah-kah-mu'-she*) [Jap. *aka*, red; *mushi*, bug, or insect]. Japanese river fever.

**akanthesthesia** (*a-kan-thes-the'-ze-ah*) [ἄκανθα, a thorn; αἴσθησις, sensation]. A form of paresthesia or perverted sensation in which there is a feeling as of a sharp point.

**akanthion** (*a-kan'-the-on*). See *acanthion*.

**akaralgia** (*ak-ar-al'-je-ah*). A proprietary "headache cure." It contains sodium salicylate, sodium sulphate, magnesium sulphate, lithium benzoate, and nux vomica.

**akarkine** (*ak-ar'-kin*). Trade name for arsenic albuminate; it is used as a cancer cure.

**akatamah** (*ak-ah-tah'-mah*). The native West Central African name for an endemic peripheral neuritis of obscure origin marked by numbness and intense prickling and burning in the presence of cold or damp.

**akatamathesia** (*ah-kat-am-ath-e'-ze-ah*) [ἀ, priv.; καταμάθησις, understanding]. Inability to understand.

**akathisia** (*ah-kath-e'-ze-ah*) [ἀ, priv.; καθίζειν, to be seated]. A name given by Lad Haskovec to a form of rhythmic chorea in which the patient is unable to remain seated; the affection resembles astasia-abasia.

**akidopeirastic** (*ak-id-o-pi-ras'-tik*) [ἀκή, ἀκίς, needle; πειραστικός, proving]. Relating to the exploratory puncture of a diseased area by means of a stout needle.

**akidopeirastica** (*ak-id-o-pi-ras'-tik-ah*) [ἀκίς, a point; πειράζειν, to make a trial of]. Exploratory incision or puncture.

**akinesia, akinesis** (*ah-kin-e'-se-ah, ah-kin-e'-sis*) [ἀ, priv.; κίνησις, motion]. Lack of or imperfect motion; motor paralysis. **a. algera,** an affection characterized by abstinence from voluntary movement on account of pain, which any active muscular effort causes. The condition is probably a form of neurasthenia. **a., crossed,** a motor paralysis on the side opposite that in which the lesion exists. **a., reflex,** impairment or loss of reflex action.

**akinesis,** cerebral, that in which the lesion is in the cerebrum. **a. iridis,** rigidity or immobility of the iris. **a., spinal,** motor impairment due to a lesion of the cord.

**akinetic** (*ah-ki-net'-ik*) [*akinesia*]. 1. Relating to or affected with akinesia. 2. An agent lessening muscular action.

**akoulalion** (*ah-koo-la'-le-on*) [ἀκούειν, to hear; λάλος, speech]. A mechanical contrivance to aid defective audition, used in training the deaf and dumb to speak.

**akouphone** (*ah'-koo-fōn*) [ἀκούειν, to hear; φωνή, sound]. A mechanism to aid defective hearing.

**akromegaly, akromegalia** (*ak-ro-meg'-a-le, ak-ro-me-ga'-le-ah*) [ἄκρον, extremity; μεγάλη, large]. A disease characterized by an overgrowth of the extremities and of the face, including the bony as well as the soft parts. The etiology is unknown. In a number of cases the pituitary body has been enlarged; disease of the thyroid gland has also been found in some instances.

**Al.** Chemical symbol of aluminum.

**al.** 1. The Arabic definite article *the*, prefixed to many words to designate preeminence, etc., as alkali, alcohol. 2. A chemical suffix denoting similarity to or derivation from an aldehyde, as chlor*al*, butyr*al*, etc.

**ala** (*a'-lah*) [L., "a wing": *pl., alæ*]. 1. A wing. 2. Any wing-like process. 3. The arm or shoulder; in animals, the shoulder-blade. **a. alba lateralis**, the nucleus of the glossopharyngeal nerve. **a. alba medialis**, the hypoglossal nucleus. **a. auris**, the pinna of the ear. **a. cinerea**, a triangular space of gray matter in the fourth ventricle of the brain, probably giving origin to the pneumogastric nerves. **a. descendens**, the pterygoid process of the sphenoid bone. **a. ethmoidalis**, the alar process of the ethmoid. **alæ laterales**. 1. The great wings of the sphenoid bone. 2. Wing-like processes on each side of the nasal spine of the frontal bone. **a. lob'uli antra'lis**, the lateral part of the median cerebellar lobe. **a. mag'na**, the great wing of the sphenoid. **a. par'va**, the small wing of the sphenoid. **a. pon'tis**, the posterior part of the roof of the fourth ventricle. **alæ majores**. 1. The greater wings of the sphenoid. 2. The external labia pudendi. **alæ minores**. 1. The lesser wings of the sphenoid. 2. The labia minora pudendi. **a. nasi**, the lateral cartilage of the nose. **alæ parvæ**, the lesser wings of the sphenoid. **alæ pulmonum**, the lobes of the lung. **a. of sacrum**, the flat, triangular surface of bone extending outward from the base of the sacrum, supporting the psoas magnus muscle. **a. uvulæ**, a medullary layer running from the posterior part of the uvula of the cerebellum to the amygdalæ. **a. vespertilionis**, the broad ligament of the uterus. **alæ vulvæ**, the labia of the pudendum.

**alabaster** (*al-a-bas'-ter*). 1. Hydrous calcium sulphate. 2. Calcium carbonate.

**alabastrine** (*al-a-bas'-trēn*). 1. Relating to or resembling alabaster. 2. Naphthalene.

**alalia** (*al-a'-le-ah*) [ἀ, priv.; λαλιά, talk]. 1. Impairment of articulation from paralysis of the muscles of speech or from local laryngeal disease. 2. Aphasia due to a psychic disorder. **a., mental**, a form observed in children, which consists in inability to speak through excessive stammering. Cf. *dyslalia, lalophobia, mogilalia, paralalia*. **a., relative**, same as *a., mental*.

**alalic** (*al-a'-lik*) [ἀ, priv.; λαλιά, talk]. Characterized by or pertaining to alalia.

**alangine, alanginum** (*al-an'-jin, -um*). An alkaloid obtained from *Alangium lamarkii*, soluble in alcohol, in ether, and in chloroform; it is used as a febrifuge and emetic.

**alanin** (*al'-an-in*), $C_3H_7NO_2$. Lactamic acid. An organic base obtained by heating aldehyde ammonia with hydrocyanic acid in the presence of an excess of HCl. It occurs in aggregated hard nodules with a sweetish taste. It is soluble in 5 parts of cold water; less soluble in alcohol; insoluble in ether. **a., mercury**, mercury amidopropionate.

**alant camphor.** A camphor from elecampane. See *helenin*.

**alantic** (*al-an'-tik*) [Ger., *alant*, elecampane]. Pertaining to or derived from elecampane. **a. anhydride**, $C_{15}H_{20}O_3$, a crystalline substance derived from the root of elecampane, melting at 66° C.

**alantin** (*al-an'-tin*). Same as *inulin*.

**alantol** (*al-an'-tol*), $C_{20}H_{24}O$. Inulol. An aromatic liquid obtained from elecampane; used in the same manner as creosote in pulmonary tuberculosis.

**alar** (*a'-lar*) [*ala*, a wing]. 1. Wing-like. 2. Relating to the shoulder, or axilla. **a. ligaments**, lateral synovial folds of the ligament of the knee-joint. **a. ligaments**, odontoid, lateral ligaments of the odontoid process.

**alares** [pl. of *alaris*]. 1. The pterygoid muscles. 2. The wings of the sphenoid.

**alaris** (*al-a'-ris*) [*ala*, a wing]. Wing-shaped. See *alar*.

**alate** (*a'-lāt*) [*ala*]. Winged.

**alatus** (*al-a'-lus*). 1. Winged. 2. An individual in whom there is a marked backward projection of the shoulder blades.

**alaxa** (*al-ak'-ser*). Trade name of an aperient preparation, the chief constituent of which is cascara sagrada.

**alba** (*al'-bah*) [L., "white"]. The white fibrous tissue of the brain and nerves. **a.**, reticular, the reticulated layer of alba on the anterior half of the uncinate gyrus. Syn., *substantia reticularis alba*.

**albaras, albarras** [Ar.]. A skin disease characterized by the formation of white, shining patches. Syn., *white leprosy; baras; barras*.

**albargin** (*al-bar'-jin*) [*album*, white; *argentum*, silver]. A compound of silver (15 %) and gelatose (a transformation-product of glue). A yellow powder, freely soluble in water, used in treatment of gonorrhea in injections of 0.2 % solution 4 or 5 times daily.

**albedo** (*al-be'-do*) [L., "whiteness"]. Whiteness. **a. retinæ**, retinal edema. **a. unguis**, or **unguium**, the lunula of the nail.

**albefaction** (*al-be-fak'-shun*) [*albus*, white; *facere*, to make]. The act or process of blanching or rendering white.

**Albert's disease** [Eduard *Albert*, Austrian surgeon, 1841–1900]. Achillodynia; inflammation of the retrocalcanean bursa, generally secondary to osteitis of the os calcis.

**albescent** (*al-bes'-ent*) [*albescere*, to become white]. Whitish.

**albicans** (*al'-be-kanz*) [*albicare*, to grow white]. 1. One of the corpora albicantia of the brain. 2. White; whitish.

**albicantia** (*al-be-kan'-she-ah*) [L.]. (Plural of *albicans* (1).

**albiduria** (*al-bid-u'-re-ah*) [*albidus*, white; οὖρον, urine]. White urine. Chyluria.

**Albini's nodules** (*ahl-be'-ne*) [Giuseppe *Albini*, Italian physiologist, 1830–      ]. Small nodules found on the free edge of the auriculoventricular valves in some infants.

**albinism, albinismus** (*al'-bin-izm, āl-bin-iz'-mus*) [*albus*, white]. That condition of the skin in which there is a congenital absence of pigment involving its entire surface, including the hair and the choroid coats and irises of the eyes. It is usually associated with nystagmus, photophobia, and astigmatism. Syn., *alphosis; congenital achroma; congenital leukoderma; leukæthiopia; achromatosis; leukopathia; albitudo*. **a., acquired, a. acquisita**, see *vitiligo*. **a., partial**, congenital absence of pigmentation in certain parts of the skin, appearing in irregular, white, sharply defined spots. Especially characteristic are the changes of color in the hair, often observed in negroes. The hairs are white and grow upon skin devoid of pigment, or normally colored. Syn., *poliosis circumscripta*.

**albino** (*al-be'-no*) [Sp.]. A person affected with albinism.

**albinotic** (*al-bin-ot'-ik*). Affected with albinism.

**albinuria** (*al-bin-u'-re-ah*) [*albus*, white; οὖρον, urine]. 1. Chyluria; whiteness of the urine. 2. Albuminuria.

**albocinereous** (*al-bo-sin-e'-re-us*) [*albus*, white; *cinereus*, gray]. Having both white and gray matter.

**alboferrin** (*al-bo-fer'-in*). An odorless, light-brown powder, readily soluble in cold water. It is said to consist of albumin, 90.14 %; iron, 0.68 %; phosphorus, 0.324 %; amidonitrogen, 0.13 %; and mineral substances, 0.5 %. It is indicated in chlorosis, anemia, etc. Dose 15–45 gr. (1–3 Gm.) for children; 45–75 gr. (3–5 Gm.) for adults, a day.

**albolene** (*al'-bo-lēn*) [*albus*, white; *oleum*, oil]. A hydrocarbon oil, colorless, tasteless, odorless, used as an application to inflamed surfaces.

**albor** (*al'-bor*) [*albus*]. 1. A whiteness. 2. Egg-albumen. 3. [Ar., *al būl*.] Urine. **a. cutis, a. nativus**, albinism. **a. ovi**, white of egg.

**albuginea** (*al-bū-jin'-e-ah*) [*albus*]. 1. White or whitish. 2. A layer of white fibrous tissue investing an organ or part. Syn., *tunica albuginea*. **a. oculi**, the sclerotic coat of the eye. **a. ovarii**, the tunica albuginea of the ovary. **a. testis**, the tunica albuginea of the testicle.

**albugineotomy** (*al-bū-jin-e-ot'-o-me*) [*albuginea*; τομή, cutting]. Incision of any tunica albuginea (*q. v.*).

**albugineous** (*al-bū-jin'-e-us*). 1. Whitish. 2. Belonging to a tunica albuginea.

**albuginitis** (*al-bū-jin-i'-tis*) [*albuginea*; ιτις, inflammation]. Inflammation of a tunica albuginea.

**albugo** (*al-bū'-go*) [L.]. 1. A white spot, as upon the cornea. 2. A whitish, scaly eruption. 3. The white of an egg.

**albukalin** (*al-bū'-kal-in*), $C_8H_{17}N_2O_6$. A substance found in leukemic blood.

**albulactin** (*al-bū-lak'-tin*). See *lactalbumin*.

**album** (*al'-bum*) [*albus*, white]. A substance

characterized by whiteness. a. candiense, bismuth subnitrate. a. canis, see *a. græcum.* a. ceti, spermaceti. a. græcum, the feces of dogs fed upon bones, and whitened by exposure. It was formerly used in medicine. a. hispaniæ, a. hispanicum, blanc d'Espagne, bismuth subnitrate. a. nigrum, the feces of rats and mice, formerly used as a diuretic and purgative. a. ovi, white of egg.
**albumen** (al-bū'-men) [albus]. The white of an egg. See *albumin.*
**albumimeter** (al-bū-mim'-et-er) [albumin; μέτρον, a measure]. See *albuminimeter.*
**albumin** (al-bū'-min) [albus, white]. A proteid substance, the chief constituent of the animal tissues. Its molecule is highly complex. It is soluble in water and coagulable by heat. It contains the following elements: Carbon, 51.5 to 54.5; hydrogen, 6.9 to 7.3; nitrogen, 15.2 to 17.0; oxygen, 20.9 to 23.5; sulphur, 0.3 to 2.0. Albumen, white of egg, often called albumin, is largely composed of it. Other varieties are called after their sources or characteristic reactions, as acid-albumin, alkali-albumin, muscle-albumin, serum-albumin, ovum-albumin, vegetable-albumin, etc. Syn., *coagulable animal lymph; coagulable lymph of the serum.* See *Axenfeld, Barral, Boedeker, Cohen, Furbringer, Heller, Heynsius, Hindenlang, Johnson, MacWilliam, Méhu, Millon, Oliver, Oxyphenylsulphonic Acid, Parnum, Raabe, Rees, Roberts, Spiegler, Tanret, Zouchlos.* a., acid, that changed by the action of acid. a., blood-, see *serum-albumin.* a., caseiform, that variety not coagulated by heat, but precipitated by acids. a., circulating, that found in the fluids of the body. a., derived, a modification of albumin resulting from the action of certain chemicals upon native albumin. a., egg, albumin of which white of egg is the type. a., floating, same as *a. circulating.* a., imperfect, one which fails to give all the ordinary reactions. a., lacto-, see *lactalbumin.* a., milk, see *eiweiss milch.* a., muscle-, a variety found n muscle-juice. a., native, any albumin occurring normally in the tissues. a., organic, that forming an integral part of the tissue. a., serum-, see *serum-albumin.* a., vegetable, that found in various vegetable juices.
**albuminate** (al-bū'-min-āt). A compound of albumin and certain bases, as *albuminate* of iron.
**albuminaturia** (al-bū-min-at-u'-re-ah) [albuminate; οὖρον, urine]. The abnormal presence of albuminates in the urine.
**albuminid** (al-bū'-min-id). Acidalbumin; syntonin.
**albuminiferous** (al-bu-min-if'-er-us) [albumin; ferre, to bear]. Yielding albumin.
**albuminimeter** (al-bū-min-im'-et-er), [albumin; μέτρον, a measure]. An instrument for the quantitative estimation of albumin in urine.
**albuminimetry** (al-bū-min-im'-et-re). The quantitative estimation of the albumin in a liquid.
**albuminiparous** (al-bū-min-ip'-ar-us) [albumin; parere, to produce]. Yielding albumin.
**albuminofibrin** (al-bū'-min-o-fi'-brin). A compound of albumin and fibrin.
**albuminogenous** (al-bū-min-oj'-en-us) [albumin; γεννᾶν, to produce]. Producing albumin.
**albuminoid** (al-bū'-min-oid) [albumin; εἶδος, likeness]. 1. Resembling albumin. Applied to certain compounds having many of the characteristics of albumin. 2. Any nitrogenous principle of the class of which normal albumin may be regarded as the type. a. degeneration, or disease, see *amyloid degeneration.*
**albuminolysin** (al-bū-min-ol'-is-in). A lysin which causes destruction of albumins.
**albuminometer** (al-bū-min-om'-et-er). See *albuminimeter.*
**albuminometry.** See *albuminimetry.*
**albuminone** (al-bū'-min-ōn) [albumin]. A principle derived from certain albuminoids; it is soluble in alcohol and is not coagulable by heat.
**albuminorrhea** (al-bū-min-or-e'-ah) [albumin; ῥοία, a flow]. Excessive discharge of albumins.
**albuminose** (al-bū'-min-ōs) [albumin]. 1. A product of the digestion of fibrin or of any albuminoid in very dilute hydrochloric acid; acidalbumin. 2. Albumose, or one of the products of the digestion of albumin by the gastric juice.
**albuminosis** (al-bū-min-o'-sis) [albumin]. Abnormal increase of the albuminous elements in the blood, or the condition that results from such increase.
**albuminous** (al-bū'-min-us) [albumin]. Containing, or of the nature of, albumin.

**albuminuria** (al-bū-min-u'-re-ah) [albumin; οὖρον, urine]. The presence in the urine of albumin, usually serum-albumin. Albumin in the urine may result from disease of the kidneys or from the admixture of blood or pus with the urine. Its presence is sometimes not accounted for by either of these causes. See *a., cyclic.* a. acetonica, albuminuria due to asphyxia. Syn., *anoxemic albuminuria.* a. of adolescence, see *a., cyclic.* a., adventitious, see *a., pseudo-.* a., cardiac, that due to chronic valvular disease. a., cicatricial, a form in which epithelial desquamation is assumed to be replaced by tissue incapable of restraining the transudation of albumin from the blood. a., colliquative, that due to great disassimilation of the blood-corpuscles or adipose tissue. a., consumptive, see *a., colliquative.* a., cyclic, a condition, also known as physiological, simple, functional, or transient albuminuria, or the albuminuria of adolescence, in which a small quantity of albumin appears in the urine, especially of the young, at stated times of the day; hence the term, "cyclic." a., dietetic, that due to the ingestion of certain forms of food. a., dystrophic, that dependent upon imperfect formation of the blood-corpuscles. a., emulsion, that in which the urine has a milky turbidity due to minute corpuscular elements. a., exudative, Gubler's name for albuminuria partially due to the filtration of albumin through the membranes of the kidney and also to the presence in the urine of products of inflammation, as in cases of nephritis. a., false, a mixture of albumin with the urine during its transit through the urinary passages, where it may be derived from blood, pus, or special secretions that contain albumin. a., febrile, that due to fever, or associated with acute infectious diseases, slight changes occurring in the glomerules without organic lesion. a., functional, see *a., cyclic.* a., globular, that due to destruction of blood-corpuscles or dependent upon the presence of blood in the urine. a., gouty, albumin in the urine of elderly persons, who secrete a rather dense urine containing an excess of urea. a., intrinsic, see *a., true.* a., mixed, the presence of a true with a pseudo-albuminuria. a., nephrogenous, that due to renal disease. a., orthostatic, a form dependent upon an upright posture. a., paroxysmal, same as *a., cyclic.* a., partial, a form in which it is assumed that only certain tubules are affected. Syn., *albuminuria parcellaire.* a., physiological, the presence of albumin in normal urine, without appreciable coexisting renal lesion or diseased condition of the system. a., pretuberculous, a condition observed in young persons as a premonitory stage of tuberculosis, believed to be due to the congestive action of the tuberculous virus upon the renal structure. a.: pseudo-, albuminuria dependent upon the presence of such fluids as blood, pus, lymph, spermatic fluid, or the contents of an abscess cavity, in the urine. Syn., *adventitious albuminuria.* a., residual, a form in which a small amount of albumin may persist following an attack of nephritis. a., true, that due to the excretion of a portion of the albuminous constituents of the blood with the water and salts of the urine. Syn., *intrinsic albuminuria.*
**albuminuretic** (al-bū-min-ū-ret'-ik). 1. Causing albuminuria. 2. A drug which causes albuminuria.
**albuminuric** (al-bū-min-ū'-rik) [see *-albuminuria*]. Associated with, of the nature of, or affected by, albuminuria.
**albumoid** (al'-bū-moid). A protein found in cartilage and in the crystalline lens; it is but slightly soluble in acid and alkaline solutions, and insoluble in neutral solutions.
**albumone** (al-bū'-mōn). A protein found in the blood; it cannot be coagulated by heat.
**albumoscope** (al-bū'-mo-skōp) [albumin; σκοπεῖν, to examine]. An appliance for determining the presence and amount of albumin in urine.
**albumose** (al'-bū-mōs) [albumin]. Any albuminoid substance ranking among the first products of the splitting-up of proteins by enzymes, and intermediate between the food-albumins and the typical peptones. According to Kühne, there are at least two albumoses, *antialbumose* and *hemialbumose.* Hemialbumose yields the following: *Protalbumose, deuteroalbumose, heteroalbumose,* and *dysalbumose.*
**albumosuria** (al-bū-mos-ū'-re-ah) [albumose; οὖρον, urine]. The presence of albumose in the urine. a., Bence-Jones', see *a., myelopathic.* a., myelopathic, a condition marked by persistent occurrence

of albumose in the urine, accompanied by softening of the bones, owing to sarcomatous disease.
**alburnum** (al-ber'-num) [L., "sap-wood"]. In biology, young wood, sap-wood.
**albus** (al'-bus) [L.]. White.
**alcali** (al'-ka-li). See alkali.
**alcaptonuria** (al-kap-ton-ū'-re-ah). See alkaptonuria.
**alcarnose** (al-kar'-nōs). A nutrient preparation containing maltose combined with albumoses.
**alchemy** (al'-kem-e) [Ar., al-kem'l, of doubtful derivation]. The supposed art of the transmutation of metals (into gold) and of finding a remedy for all diseases.
**Alcock's canal** [Thomas Alcock, English anatomist, 1784–1833]. A canal formed by the separation of the layers of the obturator fascia for the transmission of the pudic nerve and vessels.
**alcogel** (al'-ko-jel). A jelly-like combination of alcohol and silicic acid.
**alcohol** (al'-ko-hol) [Ar., al-koh'l, the fine powder for staining eyelids]. 1. Any compound of an organic hydrocarbon radical with hydroxyl. Alcohols are classed as *monacid* (monatomic), *diacid* (diatomic), and *triacid* (triatomic), according to the number of hydroxyl radicals present in the molecules. 2. Ethyl-alcohol, $C_2H_5OH$: A liquid obtained by the distillation of fermented grain or starchy substance. It is inflammable, colorless, and possesses a pungent odor and burning taste. Internally, it is a cerebral excitant and cardiac stimulant; in large doses a depressant, narcotic poison, producing muscular incoordination, delirium, and coma. It exists in wine, whisky, brandy, beer, etc., and gives to them their stimulant properties. Commercial alcohol contains 92.3 % of absolute alcohol with 7.7 % of water. It is valuable as a cardiac stimulant in acute failure of the heart's action and in adynamic conditions. a., **absolute** (*alcohol absolutum*, U. S. P.), ethyl-alcohol deprived of water. a., **benzyl**, $C_7H_8O$, obtained from benzaldehyde by the action of sodium amalgam. a., **caustic**, sodium ethylate. a., **chlorethyl**, $C_2H_5OCl$, a substitution-product of ethyl-alcohol in which one atom of hydrogen is replaced by one atom of chlorine. a., **cinnamic**, a., **cinnamyl**, a., **cinnamylic**, $C_9H_{10}O$, yellowish needles or crystalline masses obtained from the distillation of styracin. It is soluble in alcohol, ether, water, glycerol, and benzine; melts at 30°–33° C.; boils at 250° C. It is antiseptic and is a deodorizer in a 12.5 % glycerol solution. Syn., *styrilic alcohol; crystallized styrone*. a., **denatured**, alcohol into which some other substance has been introduced, rendering it unfit for drinking but still useful for other purposes. a. **deodoratum**, ethyl-alcohol from which odorous and coloring-matters have been removed by filtration through charcoal. a., **dilute** (*alcohol dilutum*, U. S. P.) contains 41.5 %, by weight, of alcohol. a., **ethyl-**, see *alcohol* (2). a., **fatty**, one obtained from a hydrocarbon of the fatty series. a., **iso-**, an alcohol derived from a hydrocarbon containing carbon atoms which unite directly with more than two other carbon atoms. a., **methyl-**, $CH_4O$, commonly known as "wood spirit." a., **phenic**, same as *phenol*. a., **primary**, a., **secondary**, a., **tertiary**, an alcohol produced by the replacement of 1, 2, or 3 hydrogen atoms in carbinol by alkyls. a., **unsaturated**, that derived from the unsaturated alkylens in the same manner as the normal alcohols are obtained from their hydrocarbons. In addition to the general character of alcohols, they are also capable of directly binding two additional affinities. a., **wood-**, see a., methyl-.
**alcoholase** (al'-ko-hol-ās). A ferment which converts lactic acid into alcohol.
**alcoholate** (al'-ko-hol-āt). 1. A chemical compound, as a salt, into which an alcohol enters as a definite constituent. 2. A preparation made with alcohol.
**alcoholature** (al-ko-hol'-at-chur) [Fr., alcoolature]. An alcoholic tincture.
**alcoholic** (al-ko-hol'-ik) [Arabic: al, the; koh l, finely powdered antimony]. 1. Pertaining to, containing, or producing alcohol. 2. One addicted to the use of spirituous drinks. a. **radicals**, the name applied to the univalent hydrocarbon radicals which unite with OH to form alcohols.
**alcoholica** (al-ko-hol'-ik-ah). In pharmacy, alcoholic preparations.
**alcoholimeter** (al-ko-hol-im'-et-er). See *alcoholometer*.

**alcoholism** (al'-ko-hol-izm). The morbid results of excessive or prolonged use of alcoholic liquors. The term *acute alcoholism* has been used as a synonym for inebriety. The *chronic* form is associated with severe disturbances of the digestive and nervous systems.
**alcoholist** (al-ko-hol'-ist). An individual affected with alcoholism.
**alcoholization** (al-ko-hol-iz-a'-shun). The art or process of alcoholizing; the state of being alcoholized; the product of the process of alcoholizing.
**alcoholize** (al'-ko-hol-īz). 1. To impregnate with alcohol. 2. To convert into an alcohol. 3. To reduce to a very subtle powder.
**alcoholomania** (al-ko-hol-o-ma'-ne-ah). Morbid craving for intoxicating beverages.
**alcoholometer** (al-ko-hol-om'-et-er) [alcohol; μέτρον, a measure]. A hydrometer or other instrument used in determining the percentage of alcohol in any liquid.
**alcoholometry** (al-ko-hol-om'-et-re) [alcohol; μέτρον, a measure]. The determination of the proportion of alcohol present in any liquid.
**alcoholophilia** (al-ko-hol-o-fil'-e-ah) [alcohol; φιλεῖν, to love]. The appetite for strong drink; a craving for intoxicants.
**alcolene** (al'-ko-lēn). A mixture of ethyl and methyl alcohols.
**alcometrical** (al-ko-met'-rik-al), Relating to the estimation of the amount of alcohol in a liquid.
**aldehydase** (al-de-hi'-dās). An oxydase, capable of oxidizing certain aldehydes to the corresponding acids.
**aldehyde** (al'-de-hīd) [al, the first syllable ot alcohol; *dehyde*, from *dehydrogenatum*]. 1. A class of compounds intermediate between alcohols and acids, derived from their corresponding primary alcohols by the oxidation and removal of 2 atoms of hydrogen, and converted into acids by the addition of an atom of oxygen. They contain the group COH. 2. $C_2H_4O$. Alcohol deprived of 2 atoms of hydrogen, or *acetic aldehyde*. It is a colorless, limpid liquid with a characteristic odor. a.-**alcoholate**, $C_4H_{10}O_2$, an addition compound of acetic acid and ethyl-alcohol. a.-**ammonia**, $C_2H_4ONH_3$, obtained from aldehyde by action of dry ammonia; soluble in water, slightly soluble in ether. Syn., *ammoniated ethylic aldehyde*; *acetylammonium*; *ammonium aldehydate*; *ethidene hydramine*. a., **anisic**, $C_8H_8O_2$, results on oxidizing various essential oils (anise, fennel, etc.) with dilute $HNO_3$. a., **aromatic**, an aldehyde obtained as an oxidation-product of a primary aromatic alcohol and in turn giving rise by oxidation to a monobasic aromatic acid. a., **benzoic**, $C_7H_6O$, the oil of bitter almonds. Syn., *benzaldehyde*. a. **characteristic**, the univalent radical, $C(H)=O$, common to the aldehydes. a., **cinnamic**, $C_9H_8O$, the chief ingredient of the essential oil of cinnamon and cassia. a., **collidin**, a., **collinic**, an oxidation-product of albuminoids and gelatin; a colorless, viscid oil with odor like oil of cinnamon. a., **formic**, $CH_2O$ or HCHO is microbicidal and antiseptic. Syn., *formaldehyde*. a., **glycolyl**, $CH_2(OH)$.-CHO. an oxidation-product of tartaric acid when digested with water at 50°–60° C. a., **isobutylic**, a., **isobutyryl**, $C_4H_8O$, a transparent, colorless, highly refractive, pungent liquid; sp. gr., 0.797 at 15° C.; soluble in alcohol; boils at 61° C. a., **isovaleral**, a., **isovaleric**, $C_5H_{10}O$, a pungent, oily liquid, with an odor of apples, obtained from oxidation of amyl-alcohol; sp. gr., 0.804 at 15° C.; miscible in alcohol and ether; boils at 92.5° C. a., **pyroracemic**, $CH_3$.-CO . CHO, a yellow volatile oil obtained by boiling isonitrosoacetone with dilute sulphuric acid. Syn., *acetylformyl*; *methylglyoxal*; *propanalon*. a., **thio-**, an aldehyde in which the oxygen in the aldehyde characteristic is replaced by sulphur. a., **toluic**, a., **toluylic**, $C_8H_8O$, a substance occurring in 3 isomeric forms, all of which are liquids.
**alder** (al'-der). See *alnus*.
**aldin** (al'-din) [see *aldehyds*]. An amorphous basic chemical substance, formed from an ammonia compound of aldehyde. Several aldins are known.
**aldol** (al'-dol) [see *aldehyde*], $C_4H_8O_2$. A colorless, odorless liquid, obtained by the action of dilute HCl on crotonaldehyde and acetaldehyde. It is miscible with water, and at 0° has a sp. gr. of 1.120; upon standing, it changes to a sticky mass that cannot be poured.
v. **Aldor's method** of testing for proteose in urine. Use 10 Cc. of urine; acidify with hydro-

chloric acid, and add phosphotungstic acid until no more precipitate occurs. Centrifugate the solution; wash the precipitate with absolute alcohol until the latter is free from color. Dissolve the precipitate in water to which is added a little potassium hydroxide. If the solution turns blue, heat gently until colorless. When cool apply the biuret test; if positive, proteoses are present.

**aldoses** (al'-do-sez) [see aldehyde]. Carbohydrates which contain the aldehyde group, CHO. The aldehyde alcohols, containing the atomic group CH(OH).CHO.

**aldoxim**, or **aldoxime** (al-doks'-im) [see aldehyde]. Products derived from aldehydes by the substitution of the oxim group N . OH for oxygen.

**ale** (āl) [AS., ealu]. An alcoholic beverage brewed from malt and hops. It contains from 3 to 7 % of alcohol.

**alecithal** (ah-les'-ith-al) [å, priv.; λεκιθος, yolk]. A term applied to certain ova having the food-yolk absent, or present only in very small quantity.

**alegar** (a'-le-gar). Vinegar made of ale.

**aleipsis** (al-īp'-sis) [ἄλειψις, an anointing]. Steatosis; fatty degeneration.

**alembic** (al-em'-bik) [Ar., al, the; ἄμβιξ, a cup]. A vessel used for distillation.

**alembroth** (al-em'-broth) [origin unknown]. An old name for a compound of the chlorides of ammonium and mercury. Its solution has been used as an antiseptic.

**Aleppo boil**, A. button, A. evil, A. pustule, A. ulcer. See *furunculus orientalis*.

**alepton P** (al-ep'-ton). Colloidal ferromanganese peptonate.

**alepton S**. Colloidal ferromanganese saccharate.

**aletocyte** (al-e'-to-sīt) [ἀλήτης, wanderer; κυτίς, a small box, a cell]. A wandering cell.

**aletrin** (al'-et-rin). See aletris.

**aletris** (al'-et-ris). Star-grass; unicorn-root; starwort; colic root. The root of *A. farinosa*. It is tonic, diuretic, and anthelmintic, and was formerly a popular domestic remedy in colic, dropsy, and chronic rheumatism. Dose of *fluidextract* 10–30 min. (0.65–2.0 Cc.); of *tincture* (1 in 8 proof spirit) 1–2 dr. (4–8 Cc.); of *aletrin*, the extractive, ¼–4 gr. (0.016–0.26 Gm.).

**aleudrin** (a-lū'-drin). A white crystalline substance, used as a hypnotic and sedative. It is sparingly soluble in water, but dissolves readily in alcohol, chloroform, ether, and fatty oils.

**aleukæmia** (ah-lū-ke'-me-ah) [å, priv.; λευκός, white; αἷμα, blood]. Deficiency in the proportion of white cells in the blood.

**aleukocytic** (ah-lū-ko-sit'-ik) [å, priv.; leukocyte]. Absence of leucytosis.

**aleukocytosis** (ah-lū-ko-si-to'-sis) [å, priv.; λευκός, white; κύτος, cell]. A diminished or insufficient formation of leukocytes.

**aleurometer** (al-ū-rom'-et-er) [aleuron; μέτρον, a measure]. An instrument used for the examination of crude gluten as to its power of distending under the influence of heat, as a means of judging of the value of a flour for bread-making.

**aleuron** (al-ū'-ron) [ἄλευρον, flour]. 1. Wheat flour. 2. Small, round protein particles found in seeds.

**aleuronat** (al-ū'-ro-nat) [aleuron]. A vegetable albumin used as a substitute for bread in cases of diabetes.

**aleuroscope** (al-ū'-ro-skōp). See aleurometer.

**Alexander's operation** or **Alexander-Adams's operation** [William *Alexander*, English surgeon; James A. *Adams*, Scotch surgeon]. A shortening of the uterine round ligaments through an inguinal incision, to cure retrodisplacement.

**alexanderism** (al-eks-an'-der-izm) [*Alexander the Great*]. The insanity of conquest; agriothymia ambitiosa.

**alexeteric** (al-eks-e-ter'-ik) [ἀλεξητήρ, defender]. Good against poison, venom, or infection.

**alexeterium** (al-eks-e-te'-re-um) [ἀλεξητήρ, a defender]. An external defensive remedy against poison or infection, as distinguished from *alexipharmac*, an internal remedy. The plural *alexeteria* was formerly used to designate remedies in general, but applied later to those used against the poisonous bites of animals.

**alexia** (ah-leks'-e-ah) [å, priv.; λέξις, word]. Word-blindness. A form of aphasia in which the patient is unable to recognize written or printed characters. a., cortical, a variety of Wernicke's sensory aphasia produced by lesions of the left gyrus angularis. a., motor, inability to read aloud what is written or printed, although it is comprehended. a., musical, loss of the ability to read music. a., optic, inability to comprehend written or printed words. a., subcortical, that due to interruption of the direct connection between the optic center and the gyrus angularis.

**alexin** (al-eks'-in) [ἄλεξις, help]. 1. A defensive proteid existing normally in the blood; any phylaxin or sozin. 2. Any antibacterial substance, found in the blood of certain animals and giving immunity to certain toxins. See *immunity*.

**alexipharmac**, **alexipharmic** (al-eks-e-far'-mak, -mik) [ἀλέξειν, to repel; φάρμακον, a poison]. 1. A medicine neutralizing a poison. 2. Acting as an internal antidote. See *alexiterium*.

**alexipharmacon** (al-eks-e-far'-mak-on) [see *alexipharmac*]. Any alexipharmic medicine.

**alexipyretic** (al-eks-e-pi'-ret-ik) [ἀλέξειν, to ward off; πυρετός, a fever]. 1. A febrifuge. 2. Acting as a febrifuge.

**alexiterium** (al-eks-it-e'-re-um). See *alexeterium*.

**alexocyte** (al-eks'-o-sīt) [ἀλέξειν, to ward off; κύτος, a cell]. 1. Hankin's name for an amphophile leukocyte. 2. A protective cell of the body which is said to secrete alexins.

**aleze** (ah-lāz) [Fr., alèze]. A cloth to protect the bed from becoming soiled by excreta, etc.

**alga** (al'-ga) [alga, a seaweed; pl., algæ]. A seaweed; one of a group of acotyledonous plants living mostly in the water.

**algæ** (al'-je) [alga, a seaweed]. Plural of *alga*, *q. v.*

**algaroth** (al'-gar-oth) [Victor *Algarotus*, Veronese physician]. Oxychloride of antimony.

**algefacient** (al-je-fa'-shent) [algor, cold; *facere*, to make]. Cooling, refrigerant.

**algesia** (al-je'-ze-ah) [ἄλγος, pain]. 1. Pain; suffering. 2. Hyperesthesia as regards the sensation of pain; also neuralgia.

**algesichronometer** (al-je-ze-kro-nom'-et-er) [ἄλγος, pain; χρόνος, time; μέτρον, a measure]. An instrument used to note the lapse of time before a nerve center responds to a painful stimulus.

**algesimeter** (al-jes-im'-et-er) [ἄλγος, pain; μέτρον, a measure]. An instrument for determining the acuteness of the sense of pain. a., Björnström's, one to test the sensibility of the skin. a., Boas', an instrument consisting of a pad and spring, used to determine the relative sensitiveness over the epigastrium. The normal tolerance is 9 to 10 kilograms; in cases of gastric ulcer, 1 to 2 kilograms.

**algesthesis** (al-jes-the'-sis) [ἄλγος, pain; αἴσθησις, feeling]. The perception of pain; painful disease.

**algetic** (al-jet'-ik) [ἀλγεῖν, to have pain]. Pertaining to, or producing, pain.

**-algia** (al'-je-ah) [ἄλγος, pain]. A suffix denoting pain, as odontalgia, neuralgia, etc.

**algid** (al'-jid) [algidus, cold]. Cold; chilly. a. cholera, the cold stage of Asiatic cholera. a. fever, a pernicious intermittent fever, with great coldness of the surface of the body. a. state, the cold stage of a disease.

**algidism**, **algidity** (al'-jid-izm, al-jid'-it-e) [see algid]. A marked sense of coldness; chilliness. a., progressive, see *sclerema neonatorum*.

**algiomotor** (al-je-o-mo'-tor) [ἄλγος, pain; *movere*, to move]. Causing movements attended with pain.

**algiomuscular** (al-je-o-mus'-ku-lar) [ἄλγος, pain; *musculus*, a muscle]. Causing pain in the muscles.

**algogenic** (al-go-jen'-ik) [ἄλγος, pain; γεννᾶν, to produce]. 1. Causing neuralgic pain. 2. [*algidus*, cold; γεννᾶν, to produce]. Lowering the body-temperature below the normal.

**algolagnia** (al-go-lag'-ne-ah) [algos; λαγνεία, venery]. Sexual perversion in which pain enjoined or endured plays a part.

**algometer** (al-gom'-et-er) [algos; μέτρον, a measure]. An instrument for testing the sensibility of a part to pain.

**algometry** (al-gom'-et-re) [ἄλγος, pain; μέτρον, a measure]. The testing of pain. a., electric, a comparative estimation of the pain produced by an induced electric current.

**algophobia** (al-go-fo'-be-ah) [ἄλγος, pain; φόβος, dread]. Unreasonable or morbid dread of pain.

**algopsychalia** (al-go-si-ka'-le-ah). See *psychoalgalia*.

**algor** (al'-gor) [L.]. A sense of chilliness or coldness.

**algos** (*al'-gos*) [ἄλγος, pain]. Pain; a painful disease, or attack.
**algoscopy** (*al-gos'-ko-pe*) [*algor*, cold; σκοπεῖν, to see]. Same as *cryoscopy*.
**algospasm** (*al'-go-spazm*) [ἄλγος, pain; σπασμός, spasm]. Painful spasm or cramp.
**algospastic, algospasticus** (*al-go-spast'-ik, -us*) [ἄλγος, pain; σπαστικός, a pulling]. Resembling or of the nature of painful cramps.
**Alibert's disease** (*al-e-bār'*) [Jean Louis Alibert, French physician, 1766–1837]. Mycosis fungoides. A.'s keloid, true keloid.
**alible** (*al'-e-bl*) [*alibilis*, nutritive]. Nutritive; absorbable and assimilable.
**alices** (*al'-is-ēs*) [L.]. Red spots preceding the pustulation in smallpox.
**alicyclic** (*al-i-si'-klik*) [ἄλειφαρ, fat; κύκλος, a circle]. Having the properties of both aliphatic (open-chain) and cyclic (closed-chain) compounds.
**alienatio** (*āl-yen-a'-she-o*). See *alienation*. a. partis, gangrene.
**alienation** (*āl-yen-a'-shun*) [*alienus*, strange]. Mental derangement.
**alienism** (*āl'-yen-izm*) [*alienare*, to deprive of reason]. The study and treatment of mental disorders.
**alienist** (*āl'-yen-ist*) [see *alienation*]. One who treats mental diseases; a specialist in the treatment of insanity.
**aliform** (*al'-if-orm*) [*ala*, wing; *forma*, shape]. Wing-shaped. a. process, the wing of the sphenoid.
**alima** (*al-i'-mah*) [ἄλιμος, without hunger]. Alimentary substances.
**aliment** (*al'-im-ent*) [*alimentum*, from *alimentare*, to nourish]. Nourishment; food. a., accessory, a., adjective, a condiment. a., substantive, a food with nutritive value as distinguished from a condiment.
**alimentary** (*al-im-en'-ia-re*) [see *aliment*]. Nourishing. a. bolus, the food after mastication and just prior to swallowing. a. canal, a. system, a. tract, a. tube, the digestive tube, from the lips to the anus, with its accessory glands. a. duct, the thoracic duct.
**alimentation** (*al-im-en-ta'-shun*) [*alimentare*, to nourish]. The act of supplying with food. The process of nourishment. a., artificial. See *feeding, artificial*. a., artificial-, forced, see *feeding, forced*. a., iodic-, the administration of iodine with the food. a., rectal, the nourishing of a patient by the administration of small quantities of concentrated food through the rectum. a., voluntary-, the nourishment of those who are willing to be fed, but are incapacitated.
**alimentotherapy** (*al-im-ent-o-ther'-ap-e*). The treatment of disease by systematic feeding.
**alinasal** (*al-i-na'-sal*) [*ala*, a wing; *nasus*, the nose]. Pertaining to the *ala nasi*, or wing of the nose.
**alinjection** (*al-in-jek'-shun*) [*alcohol*; *inicere*, to inject]. A process of preserving anatomical specimens by repeated injections of alcohol.
**aliphatic** (*al-e-fat'-ik*) [ἄλειφαρ, fat]. 1. Pertaining to a fat. 2. Belonging to the open-chain series of organic compounds. a. acid, see *acid, fatty*. a.-cyclic. See *alicyclic*.
**aliptic** (*al-ip'-tik*). 1. Relating to inunction. 2. Gymnastic; pertaining to physical culture.
**aliquot** (*al'-e-kwot*) [*alius*, some; *quot*, how many]. A part of a number or quantity which will measure it without a remainder, as 4 is an aliquot of 12.
**alisphenoid** (*al-is-fē'-noid*) [*ala*, a wing; *sphenoid*]. 1. Pertaining to the greater wing of the sphenoid bone. 2. The bone that in adult life forms the main portion of the greater wing of the sphenoid.
**alizaramid** (*al-iz-ar'-am-id*), $C_{14}H_6O_2\{^{OH}_{NH_2}$. A brown, crystalline substance obtained from boiling a dilute solution of alizarin in ammonia. Syn., *amidoanthraquinon*.
**alizarimid** (*al-iz-ar'-im-id*), $C_{14}H_7NO_2$. A violet-red substance obtained from flocculent precipitated alizarin by action of ammonia with heat; it becomes nearly black on drying.
**alizarin** (*al-iz'-ar-in*) [Ar., *al*, the; *'açārah*, to extract], $C_{14}H_8O_4$; dihydroxyanthraquinone. The red coloring principle occurring in *Rubia tinctorum* and in anthracene. It occurs in red, prismatic crystals, readily soluble in ether and alcohol. The alizarins form a group of the anthracene colors. a.-blue, a crystalline blue coloring-matter formed by heating nitroalizarin in combination with $H_2SO_4$ and glycerol.

**alkadermic** (*al-ka-der'-mik*) [*alkali*; δέρμα, skin]. Pertaining to or containing an alkaloid used in subcutaneous injection.
**alkalescence** (*al-ka-les'-ens*) [Ar., *al-qalīy*, soda-ash]. Slight or commencing alkalinity.
**alkalescent** (*al-ka-les'-ent*) [see *alkalescence*]. Somewhat alkaline.
**alkali** (*āl'-ka-li*) [see *alkalescence*]. The term includes the hydroxides of the alkali metals; these are electropositive, are strong bases, uniting with acids to form salts, turn red litmus blue, and saponify fats. a.-albumin, a derived albumin; a proteid that has been acted upon by dilute alkalies and yields an alkaline reaction. a.-albuminate, a soluble powder used as a culture-medium in bacteriology. a., caustic, the solid hydroxide of potassium or sodium. a., fixed, potassium or sodium hydroxide. a. metals, sodium, potassium, lithium, cesium, and rubidium. a., organic, one forming an essential constituent of an organism. a., vegetable, potash or potassium carbonate; also applied to the alkaloids. a., volatile, ammonium hydroxide, which is decomposed by heat with the evolution of ammonia; also ammonium carbonate.
**alkaligenous** (*al-ka-lig'-en-us*) [*alkali*; γενής, producing]. Affording or producing an alkali.
**alkalimeter** (*al-ka-lim'-et-er*) [*alkali*; μέτρον, a measure]. An instrument for estimating the alkali in a substance.
**alkalimetry** (*al-ka-lim'-et-re*) [see *alkalimeter*]. The measurement of the amount of an alkali in a substance.
**alkaline** (*al'-ka-lin*) [*alkali*]. Having the qualities of or pertaining to an alkali. a. air, ammonia. a. earths, the oxides of calcium, barium, strontium, and magnesium. a. metals, those whose hydroxides are alkalies. a. reaction, one in which red litmus paper is turned blue.
**alkalinity** (*al-ka-lin'-i-te*) [*alkali*]. The quality of being alkaline.
**alkalinuria** (*al-ka-lin-u'-re-ah*) [*alkali*; οὖρον, urine]. Alkalinity of the urine.
**alkalithia** (*al-ka-lith'-e-ah*). A proprietary effervescent preparation used in rheumatism, said to contain 1 gr. (0.065 Gm.) caffeine, 5 gr. (0.32 Gm.) lithium bicarbonate, 10 gr. (0.65 Gm.) sodium bicarbonate, in each heaping teaspoonful. Dose 1 heaped teaspoonful 3 times daily in a large glass of water.
**alkalization** (*al-ka-li-za'-shun*) [*alkali*]. The act of rendering a thing alkaline; the state or quality of being rendered alkaline.
**alkaloid** (*al'-ka-loid*) [*alkali*; εἶδος, likeness]. Any one of the nitrogenous compounds occurring in plants, and resembling ammonia in being basic and in their method of forming salts with acids. Alkaloids are believed to be substituted ammonias. Alkaloids are, as a rule, the most active parts of plants; many are used in medicine. a. s, animal, substances chemically like alkaloids, formed in the decomposition of animal tissues. See *leukomaine*. a., artificial, one produced synthetically. a., cadaveric, a., putrefactive, see *ptomaine*. a., fixed, the solid alkaloids; they contain carbon, hydrogen, nitrogen and oxygen. a., glucoside, a substance which exhibits the characteristics of an alkaloid, but is capable of decomposition into sugar and another substance when acted upon by dilute acid. a., volatile, the liquid alkaloids; they contain no oxygen.
**alkaloidal** (*al-ka-loid'-al*) [*alkali*; εἶδος, likeness]. Having the qualities of an alkaloid.
**alkalometry** (*al-kal-om'-e-tre*). Administering alkaloids. See *dosimetry*.
**alkaluretic** (*al-ka-lu-ret'-ik*) [*alkali*; οὖρον, urine]. 1. Causing or tending to cause a flow of alkaline urine. 2. A drug rendering the urine alkaline.
**alkamin** (*al'-kam-in*). See *alkine*.
**alkane** (*al'-kān*). See *paraffin* (2).
**alkanet** (*al'-kan-et*) [Sp., dim. of *alcana*, henna]. The root of the herb, *Alkanna* (*Anchusa*) *tinctoria*, yielding a red dye that is used in staining wood, coloring adulterated wines, and in pharmacy to give a red color to salves, etc.
**alkanin** (*al'-kan-in*). See *alkannin*.
**alkanna-red**. See *alkannin*.
**alkannin** (*al'-kan-in*) [see *alkanet*]. Alkanna-red; a valuable coloring-matter obtained from alkanet.
**alkapton** (*al-kap'-ton*). A yellowish, resinous, nitrogenous body occasionally found in urine.
**alkaptonuria** (*al-kap-ton-'-ū'-re-ah*) [*alkapton*; οὖρον,

urine]. The presence of alkapton in the urine. It has been found in cases of pulmonary tuberculosis and in other instances in which there were no local lesions or general disease. Urine containing alkapton turns dark on standing or on the addition of an alkali.

**alkargen** (al-kar'-jen) [alkarsin; γεννᾶν, to produce]. Dimethylarsenic acid, obtained from alkarsin. by the action of water.

**alkarhein** (al-kar-e'-in). A proprietary alkaline preparation of rhubarb and pancreatin.

**alkarsin** (al-kar'-sin) [alcohol; arsenic]. "Cadet's fuming liquid"; an extremely poisonous liquid containing cacodyl. It is of a brown color, and on exposure to the air ignites spontaneously.

**alkasal** (al'-ka-sal). See aluminum-potassium salicylate.

**alkatrit** (al'-ka-trit) [alkali; triturare, to rub together]. A triturate made from an alkaloid.

**alkeins** (al'-ke-inz). A collective name for the ethers formed from the alkines.

**alkermes** (al'-kur-mēz). See kermes.

**alkine** (al'-kīn). Any member of the acetylene series of hydrocarbons. Syn., alkomin.

**alkyl** (al'-kil) [alkali]. The name applied to any of the univalent alcohol radicals, $C_nH_{2n+1}$; methyl, ethyl, etc., are alkyls. **a.-sulphides**, thioethers; sulphur analogues of the ethers. They are colorless liquids, generally insoluble in water, and possessing a disagreeable odor resembling that of garlic.

**alkylamine** (al-kil'-am-in). A body having the constitution of ammonia in which an alkyl replaces hydrogen; 1, 2, or 3 hydrogen atoms of the ammonia molecule may suffer this replacement, thus yielding primary or monalkylamines, having the general formula $NH_2(C_nH_{2n+1})$; secondary or dialkylamines, having the general formula $NH(C_nH_{2n+1})(C_pH_{2p+1})$; and tertiary or trialkylamines, of the general formula $N(C_nH_{2n+1})(C_pH_{2+p1})(C_qH_{2q+1})$.

**alkylate** (al'-kil-āt). A compound derived from a montaomic alcohol by replacement of the hydroxyl hydrogen by a metal.

**alkylation** (al-kil-a'-shun). The exchange of hydroxylic hydrogen atoms for alkyls.

**alkylene** (al'-ki-lēn). See olefin.

**alkylogen** (al-kil'-o-jen). A haloid salt of an alcohol radical.

**allachesthesia, allachæsthesia** (ăl-ah-kes-the'-ze-ah) [ἀλλαχή, in another place; αἴσθησις, sensation]. Erroneous localization of tactile impressions, differing from allocheiria in the respect that the sensation is felt on the same side of the body, but in a different place from that in which the irritation occurs.

**allantiasis** (al-an-ti'-as-is) [ἀλλᾶς, a sausage]. Sausage-poisoning, due to the ingestion of sausages in which putrefactive changes have taken place.

**allantoic** (al-an-to'-ik) [ἀλλᾶς, a sausage; εἶδος, resemblance]. Pertaining to the allantois. **a. circulation**, the fetal circulation through the cord and the umbilical vessels. **a. vesicle**, the hollow allantois of certain animals.

**allantoid** (al-an'-toid) [see allantoic]. 1. Resembling a sausage. 2. Relating to the allantois. **a. liquid**, see liquor amnii spurius.

**allantoides** (al-an-to'-id-ēz). 1. Allantoid. 2. A sausage. 3. The great toe. 4. The allantois.

**allantoin** (al-an'-to-in) [see allantoic]. $C_4H_6N_4O_3$. A crystalline substance occurring in traces in normal urine, and prepared from uric acid by oxidation. Also the characteristic constituent of the allantoic fluid, and likewise found in fetal urine and amniotic fluid.

**allantois** (al-an'-to-is) [see allantoic]. One of the fetal membranes derived from the mesoblastic and hypoblastic layers. Its function is to convey the blood-vessels to the chorion. The lower part finally becomes the bladder, the upper, the urachus.

**allantotoxicon** (al-an-to-toks'-ih-on) [ἀλλᾶς, a sausage; τοξικόν, a poison]. A poisonous substance, probably a ptomaine, that develops during the putrefactive fermentation of sausage.

**ailaxis** (al-aks'-is) [ἀλάσσειν, to vary]. Metamorphosis, transformation; the act or process of conversion into some other condition or thing. Syn., allagē.

**allelomorph** (al-e'-lo-morf) [ἀλλήλων, of one another; μορφή, form]. In Mendelian inheritance one of a pair of contracted characters which become segregated in the formation of germ cells.

**allelomorphic** (al-e-lo-mor'-fik). Pertaining to, or characteristic of an allelomorph, q. v.

**allelomorphism** (al-e-lo-mor'-fizm). The presence, in Mendelian inheritance, of allelomorphic characters.

**allelotaxis** (al-e-lo-tak'-sis) [ἀλλήλων, of one another; τάξις, arrangement]. The development of a part from different embryonic structures.

**Allen's iodine test** [Charles Warrenne Allen, American physician, 1854-1906]. See under linea versicolor.

**Allen's reaction** for phenol. Add to one or two drops of the liquid to be tested a few drops of hydrochloric acid and then one drop of nitric acid. A cherry-red coloration is produced.

**allene** (al-ēn'). $CH_2=C=CH_2$. An isomere of allylene. Syn., β-allylene; isoallylene.

**allentinesis** (al-en'-the-sis) [ἄλλος, other; ἔνθεσις, insertion]. The presence in or the introduction of foreign bodies into the organism.

**alleosis**, or **allœosis** (al-e-o'-sis) [ἀλλοίωσις, change]. 1. Change; alterative effect; recovery from illness. 2. Mental disorder.

**alleotic**, or **allœotic** (al-e-ot'-ik) [ἀλλοίωσις, change]. 1. Alterative. 2. A remedy or agent having an alterative action.

**allergen** (al'-er-jen) [allergy; γεννᾶν, to produce]. A hypothetical substance of a toxic nature, supposed to produce allergy.

**allergy, allergia** (al'-er-je, al-er'-je-a) [ἄλλος, other; ἐνέργεια, energy (from ἔργον, work)]. A form of acquired immunity, in which a person reinfected reacts differently from the way in which he reacted after the primary infection. It is associated with anaphylaxis.

**allesthesia** (al-es-the'-ze-ah) [ἄλλος, other; αἴσθησις, feeling]. Synonym of allocheiria.

**allevation** (al-e-va'-shun) [ad, to; levare, to lift up]. 1. The relief or palliation of pain. 2. The raising or lifting of a patient from the bed or from the reclining posture.

**alleviator** (al-e'-ve-a-tor) [allevare, to lighten]. A device for raising or lifting a sick person from the bed.

**allex** (al'-eks) [L.]. Same as hallux.

**alliaceous** (al-e-a'-shus) [allium, garlic]. Resembling garlic, or pertaining to the same.

**alligator-forceps** (al'-e-ga-tor-for'-seps). A surgeons' toothed forceps, one of the jaws of which works with a double lever.

**Allingham's operation** [William Allingham, English surgeon, 1830-1908]. 1. For excision of the rectum; the patient in the lithotomy position, an oval incision is made into both ischio-rectal fossæ, around the bowel, and prolonged backward to the coccyx; the bowel is isolated, and separated with the écraseur, scissors, or Paquelin cautery. 2. For hemorrhoids; the pile is dissected off from the muscular tissue with scissors, the pedicle ligated, and the mass cut off. 3. For inguinal colotomy; the incision is from one and one-half to three inches long, and is made parallel with the outer third of Poupart's ligament, and about one-half inch above. **A.'s painful ulcer**, anal fissure. **A.'s rectal plug**, an appliance for controlling hemorrhage from the rectum.

**Allis' sign** [Oscar H. Allis, American surgeon]. Relaxation of the fascia lata between the iliac crest and the trochanter major is indicative of fracture of the neck of the femur.

**alliteration** (al-it-er-a'-shun) [ad, to; litera, letter]. A form of dysphrasia in which the patient arranges his words according to the sound.

**allium** (al'-e-um) [L.]. Garlic. The undried bulb of A. sativum. It contains a pungent, volatile oil that is found also in the leek and the onion. In small amounts garlic acts as a condiment and aids in the digestion and absorption of food. In chronic bronchitis garlic applied as a poultice to the chest and internally in boiled milk is beneficial. Poultices of garlic applied to the spine are recommended in infantile convulsions and may be applied over the abdomen in gastrointestinal catarrh. A. cepa, the common onion, and A. porrum, the leek, have similar qualities. **A.**, syrup of (syrupus allii), contains fresh garlic, 20 Gm.; sugar, 80 Gm.; dilute acetic acid, a sufficient quantity to make 100 Cc. Dose 1-4 dr. (4-16 Cc.).

**allo-**. A prefix used in chemistry to designate a body which has been rendered more stable by heat; also used to represent isomerism when there is "relative asymmetry."

**allocheiria**. See allochiria.

**allochesthesia** (al-ok-es-the'-ze-ah). Same as allachesthesia.

# ALLOCHEZIA 39 ALLYL

**allochezia, allochetia** (*al-o-ke'-ze-ah, al-o-ke'-she-ah*) [ἄλλος, other; χέζειν, to desire to go to stool]. 1. The passage of feces from the body through an abnormal opening. 2. The passing of non-fecal matter from the bowels.

**allochiria** (*al-o-ki'-re-ah*) [ἄλλος, other; χείρ, hand]. An infrequent tabetic symptom, in which, if one extremity be pricked, the patient locates the sensation in the corresponding member of the other side.

**allochroic** (*al-lo-kro'-ik*) [ἄλλος, another; χρῶμα, color]. Of changeable or diversified color.

**allochroism** (*al-ok'-ro-izm*) [ἄλλος, other; χρῶμα, color]. 1. Variation in color. 2. A change of color.

**allochromasia** (*al-o-kro-ma'-ze-ah*) [ἄλλος, other; χρῶμα, color]. 1. Change of color in a part or tissue. 2. Color-blindness.

**allocinetic** (*al-o-sin-et'-ik*). See *allokinetic*.

**allogamy** (*al-og'-am-e*) [ἄλλος, other; γάμος, marriage]. In biology, cross fertilization.

**allogotrophia** (*al-o-go-tro'-fe-ah*) [*allos;* τρέφειν, to nourish]. The nourishment of one part of the body at the expense of some other part.

**alloisomerism** (*al-o-i-som'-er-ism*) [ἄλλος, other; ἰσομερής, having equal parts]. The application of the same structural formula to many different compounds; a variety of isomerism.

**allokinetic** (*al-o-kin-et'-ik*) [ἄλλος, other; κίνησις, motion]. Moved or set in motion by external impressions or forces; not autokinetic.

**allolalia** (*al-o-la'-le-ah*) [*allos;* λαλεῖν, to speak]. Any perversion of the faculty of speech. See *alalia*.

**allolalic** (*al-o-lal' ik*) [ἄλλος, other; λαλεῖν, to speak]. Affected with allolalia.

**allomerism** (*al-om'-er-ism*) [ἄλλος, other; μέρος, shape]. In chemistry, the property of retaining a constant crystalline form while the chemical constituents present, or their proportions, vary.

**allomorphic, allomorphous, allomorphus** (*al-o-mor'-fic, -us*). Affected with allomorphism.

**allomorphism** (*al-o-morf'-ism*) [ἄλλος, other; μορφή, shape]. The property possessed by certain substances of assuming a different form while remaining unchanged in constitution.

**allopath, allopathist** (*al'-o-path, al-op'-ath-ist*) [ἄλλος, other; πάθος, affection]. One who practises allopathy. A common, but incorrect designation for a regular practitioner.

**allopathy** (*al-op'-a-the*) [ἄλλος, other; πάθος, affection]. According to Hahnemann, the inventor of the term, that method of the treatment of disease consisting in the use of medicines the action of which upon the body in health produces morbid phenomena different from those of the disease treated; erroneously used of the regular medical profession; opposed to homeopathy.

**allophasis** (*al-off'-as-is*) [ἄλλος, other; φάσις, speech]. Incoherency of speech; delirium.

**allophemy** (*al-off'-e-me*) [ἄλλος, other; φήμι, to speak]. See *heterophemy*.

**alloplast** (*al'-o-plast*) [ἄλλος, other; πλαστός, form, mold]. In biology, a plastid composed of several tissues; the opposite of homoplast.

**allorrhythmia** (*al-or-rith'-me-ah*) [*allos;* ῥυθμός, rhythm]. Variation in intervals of the pulse.

**allosan** (*al'-o-san*). The allophanic acid ester of santalol. It is a white, crystalline powder, used as santalol.

**allosteatodes** (*al-o-ste-at-o'-dēz*) [ἄλλος, other; στεατώδης, fat-like]. Marked by perversion or morbidity of the sebaceous secretion.

**allotherm** (*al'-o-therm*) [ἄλλος, other; θέρμη, heat]. An organism whose temperature is directly dependent on its culture-medium.

**allotoxin** (*al-o-toks'-in*) [ἄλλος, other; τοξικόν, poison]. Any substance, produced by tissue-metamorphosis within the organism, that tends to shield the body by destroying microorganisms or toxins that are inimical to it.

**allotriodontia** (*al-ot-re-o-don'-she-ah*) [ἀλλότριος, strange; ὀδούς, tooth]. 1. The transplanting of teeth from one person to another. 2. The existence of teeth in abnormal situations, as in tumors.

**allotriogeustia** (*al-ot-re-o-gūs'-te-ah*) [ἀλλότριος, strange; γεῦσις, taste]. Perversion of the sense of taste; abnormity of the appetite.

**allotriolith** (*al-ot'-re-o-lith*) [ἀλλότριος, strange; λίθος, stone]. A calculus composed of unusual material or formed in an abnormal situation.

**allotriolithiasis** (*al-ot-re-o-lith-i'-as-is*) [ἀλλότριος, strange; λίθος, a stone]. The formation or existence of a calculus of unusual material, or composed entirely or in part of a foreign body.

**allotriophagy** (*al-ot-re-off'-a-je*) [ἀλλότριος, strange; φαγεῖν, to eat]. Depraved or unnatural appetite.

**allotriotexis** (*al-ot-re-o-teks'-is*) [ἀλλότριος, strange; τέξις, birth]. 1. Abnormality in delivery. 2. The birth or delivery of a monstrosity.

**allotriuria** (*al-ot-re-u'-re-ah*) [ἀλλότριος, strange; οὖρον, urine]. Abnormality of the urine.

**allotrope** (*al'-o-trōp*) [see *allotropic*]. One of the forms in which an element capable of assuming different forms may appear.

**allotrophic** (*al-o-troff'-ik*) [ἄλλος, other; τροφή, nourishment]. Having perverted or modified characters as a nutrient.

**allotropic** (*al-o-trop'-ik*) [ἄλλος, other; τρόπος, manner]. 1. Characterized by allotropism. 2. Relating to or marked by isomerism.

**allotropism** (*al-ot'-rop-ism*) [see *allotropic*]. 1. The term expresses the fact of certain elements existing in two or more conditions with differences of physical properties; thus, carbon illustrates allotropism by existing in the forms of charcoal, plumbago, and the diamond. 2. Appearance in an unusual or abnormal form.

**allotropy** (*al-ot'-ro-pe*). Allotropism.

**allotrylic** (*al-o-tril'-ik*) [ἀλλότριος, foreign; ὕλη, matter]. Due to the presence of a foreign principle or material; enthetic. a. affections, morbid states caused by the lodgment of foreign substances in the organism. The foreign substance may be animate or inanimate, organic or inorganic.

**alloxamide** (*al-oks'-am-id*) [*alloxan; amide*]. A substance, $C_8H_4N_2O_4$, obtained from alloxan by the action of ammonia.

**alloxan** (*al-oks'-an*) [*allantoin; oxalic*], $C_4H_2N_2O_4$. A crystalline substance produced by the oxidation of uric acid.

**alloxantin** (*al-oks-an'-tin*) [*alloxan*], $C_8H_4N_4O_7$ $+3H_2O$. A substance obtained by reducing alloxan with $SnCl_2$, zinc, and HCl, or $H_2S$ in the cold. It occurs in small, hard, colorless prisms that turn red when treated with ammonia.

**alloxin** (*al-oks'-in*) [*allantoin*]. Any of a series of xanthin bases, the result of the splitting-up of chromatin, and which on oxidation produce uric acid.

**alloxur, alloxuric** (*al-oks'-ur, al-oks-u'-rik*) [ἄλλος, other; ὀξύς, sharp]. A term applied by Kossel and Krüger to the xanthin bases, from the fact that these, like uric acid, contain alloxan and urea groups. a. bases, a. bodies, xanthin, hypoxanthin, guanin, paraxanthin, adenin.

**alloxuremia** (*al-oks-u-re'-me-ah*) [*alloxur; uremia*]. Toxemia due to the resorption of the xanthin or alloxur bases.

**alloxuria** (*al-oks-u'-re-ah*) [*alloxur;* οὖρον, urine]. The pathological secretion of alloxur bodies (uric acid, xanthin, hypoxanthin, paraxanthin, adenin, carnin, etc.) in the urine.

**alloy** (*al-oi'*) [Fr. *aloyer,* from L. *alligare,* to combine]. 1. A compound of two or more metals by fusion. 2. The least valuable of two or more metals that are fused together.

**allspice** (*awl'-spīs*). The fruit of *Eugenia pimenta*. a., Carolina, the leaves of *Calycanthus floridus*, having the properties of an aromatic stimulant. See *pimenta*.

**allus** (*al'-us*) [L.]. The great toe. a. pollex, the thumb.

**allyl** (*al'-il*) [*allium*, garlic], $C_3H_5$. A univalent alcohol radical. Syn., *allylum; acryl.* a. acetate. 1. $C_5H_8$. $C_2H_3O$, an aromatic liquid with sharp taste, boiling at 103°–104° C. 2. A salt of allylacetic acid. a. alcohol, $C_3H_5HO$. A colorless, inflammable liquid, with pungent odor, boiling at 97° C. a. aldehyde, $C_3H_5O$. A synonym of *acrolein.* a. borate, $(C_3H_5)_3BO_3$, a liquid giving off pungent, irritating vapors which cause a flow of tears; it boils at 168°–175° C. a. bromide, $C_3H_5Br$, a liquid with pungent odor; sp. gr., 1.436 at 15° C., soluble in alcohol and ether; boils at 70°–71° C. Syn., *bromopropylene.* a. carbamine, $CN.C_3H_5$, a liquid obtained by heating allyl iodide with silver cyanide; it has an extremely foul and penetrating odor; boils at 96°–106° C. Syn., *allyl cyanide; allyl isocyanide.* a. carbimide, $CO.NC_3H_5$, a foul liquid causing flow of tears, formed by the action of potassium pseudocyanate upon allyl iodide. Syn., *allyl isocyanate; allyl carboxylamine; allyl pseudocyanate.* a. chloride,

$C_3H_5Cl$, a pungent liquid; sp. gr., 0.937 at 20° C.; boils at 45° C. Syn., *chlorotrivylen*. **a. cyanamide.** See *sinamine*. **a. dioxide**, $C_6H_{10}O_3$, a colorless liquid obtained from allyl alcohol by action of glycerol and oxalic acid; sp. gr., 1.16 at 16° C.; boils at 171°–172° C.; soluble in water, alcohol, and chloroform. Syn., *diallyl oxide*. **a. iodide**, $C_3H_5I$, a pungent liquid; sp. gr., 1.848 at 12° C.; soluble in alcohol; boils at 100°–102° C. It is a reaction-product of phosphorus, iodine, and allyl alcohol. **a. mustard oil**, $CS.N.C_3H_5$. The principal constituent of ordinary mustard oil. Syn., *allyl pseudosulphocyanate; allyl pseudothiocyanate; allyl isothiocyanate; allyl isosulphocyanate; allyl thiocarbimide*. **a. nitrate**, $C_3H_5.$-$NO_3$, a mobile liquid of pungent odor, boiling at 106° C., formed from silver nitrate by action of allyl bromide. **a. phenol**, $C_9H_{10}O$, a body obtained from anisic aldehyde by action of potash; it forms laminar crystals. **a. sulphate**, $C_6H_5HSO_4$, a substance acting as a monobasic acid and forming salts called allyl sulphates. Syn., *allyl-sulphuric acid; allyl and hydrogen sulphate*. **a. sulphide**, $(C_3H_5)_2S$, the essential oil of garlic. It is stomachic and sedative. **a. thiocyanate**, $NC.SC_3H_5$, a colorless, strongly refracting, oily liquid, with odor of garlic and hydrocyanic acid, isomeric with allyl mustard oil and producing headache, nervous excitement, and nausea when inhaled. Syn., *artificial oil of mustard; allyl sulphocyanide*. **a. tribromide**, $C_3H_5Br_3$, a colorless liquid used as an antispasmodic. Dose 5 drops.

**allylamine** (*al-il′-am-in*) [*allium; amide*], $NH_2$-$(CH_5)$. Ammonia in which a hydrogen atom is replaced by allyl. It is a caustic liquid.

**almatein** (*al-mat′-e-in*). A compound of hematoxylin and formaldehyde: it has no odor, and has been recommended as a substitute for iodoform.

**Almén's reagent for blood** [August Almén, Swedish physiologist, 1833– ]. A liquid containing blood or blood-coloring matters, if well shaken with a mixture of equal parts of tincture of guaiacum and oil of turpentine, becomes blue. **A.'s test for glucose,** heat the liquid with a solution of bismuth subnitrate dissolved in caustic soda and Rochelle salts; if it contains glucose, the liquid becomes cloudy, dark brown, or nearly black in color, and finally a black deposit appears.

**almond** (*ah′-mond*) [ME., *almonde*]. See *amygdala*. **a.-bran,** a cosmetic powder consisting of perfumed powdered almonds and borax. **a.-bread,** a variety of bread made from almond flour, for use in diabetes as a substitute for ordinary bread. **a.-eyed,** applied to the Mongolian race on account of the peculiar elliptical form and slanting appearance of the eyelids. **a. of the ear, a. of the throat,** the tonsil. **a. mixture,** see under *amygdala*. **a. oil,** oleum amygdalæ. See *amygdala*. **a. oil, bitter,** oleum amygdalæ amaræ. See under *amygdala*. **a.-paste,** a magma of bitter almonds, alcohol, white of egg, and rose-water, used to soften the skin and prevent the hands and lips from chapping.

**alnuin** (*al′-nū-in*) [Celtic, *al*, near; *law*, a riverbank]. A precipitate from the tincture of *Alnus rubra*. Said to be alterative and resolvent. Dose gr. ii–x.

**Alnus** (*al′-nŭs*) [L.]. 1. Alder-bark. 2. A genus of shrubs and trees of the order *Cupuliferæ*. *A. glutinosa,* common European alder, has astringent bark and leaves, which are used in intermittent fever and as an application in wounds and ulcers. *A. serrulata* contains tannic acid. The decoction of bark and leaves is astringent and used as a gargle and as a lotion for wounds and ulcers. Dose of *powdered bark* 10 gr. (0.65 Gm.); of the *fluidextract* 30–60 min. (2–4 Cc.). *A. incana* has qualities similar to *A. serrulata*. It is recommended as a hemostatic.

**alochia** (*ah-lo′-ke-ah*) [ἀ, priv.; λόχια, the lochia]. Absence of the lochia.

**Aloe** (*al′-o*). A genus of liliaceous plants. See *aloes*. **a. americana,** see *agave*. **a.-resin,** an amorphous resinous constituent of aloes obtained as a deposit from a hot aqueous solution of aloes on cooling.

**aloedary** (*al′-o-ed-a-re*). A compound aloetic purgative medicine.

**aloeretin** (*al-o-e-re′-tin*). See *aloe-resin*.

**aloes** (*al′-oz*) [ἀλόη, the aloe]. The inspissated juice of several species of aloe, of which *Aloe socotrina, A. barbadensis,* and *A. capensis* are most commonly used. Its properties are due to a glucoside, *aloin,* $C_{17}H_{18}O_7$. It is a tonic astringent, useful in amenor- rhea, chronic constipation, and atonic dyspepsia. It is also an emmenagogue and anthelmintic. Dose 2–5 gr. (0.13–0.32 Gm.). **a.-bitter,** a bitter principle obtained from aloes by evaporation of the aqueous extract from which the aloe-resin has been extracted. **a.-bitter, artificial,** a body obtained from aloes by action of nitric acid. **a., decoctum, compositum** (B. P.), Socotrine aloes, myrrh, and saffron, of each, 2 parts; potassium carbonate, 4 parts; licorice-juice, 24 parts; water, 768 parts; reduce by boiling to 642 parts and add 192 parts of compound tincture of cardamom. Dose ½–2 gr. (0.032–0.13 Gm.). **a., enema** ($B_t$ P.), aloes, potassium carbonate, and mucilage of starch. **a. et asafœtidæ, pilulæ** (B. P.), aloes and asafetida, of each, 1⅓ gr. (0.1 Gm.). **a. et ferri, pilulæ** (U. S. P., B. P.), contain 1 gr. (0.065 Gm.) each of aloes, ferrous sulphate, and powdered powder, incorporated with confection of roses. **a. et mastiches, pilulæ** (U. S. P., B. P.), "Lady Webster's pills," contain aloes, 2 gr. (0.13 Gm.); mastic and red rose, ½ gr. (0.032 Gm.). **a. et myrrhæ, pilulæ** (U. S. P., B. P.), each contains aloes, 2 gr. (0.13 Gm.); myrrh, 1 gr. (0.065 Gm.); aromatic powder, ½ gr. (0.032 Gm.), mixed with syrup. **a. et myrrhæ, tinctura** (U. S. P., B. P.), aloes, 10; myrrh, 10; alcohol, 100 parts. Dose ½–2 dr. (2–8 Cc.). **a., extractum** (U. S. P.). Dose 2 gr. (0.12 Gm.). **a., extractum, aquosum,** prepared by mixing aloes 1 part with 10 parts boiling water, straining and evaporating. Dose ½–5 gr. (0.032–0.32 Gm.). **a., hepatic,** dark, liver-colored aloes, mostly Barbadian. **a., pilulæ** (U. S. P., B. P.), aloes and soap, of each, 2 gr. (0.13 Gm.). **a. purificata** (U. S. P.), the common drug purified by solution in alcohol and evaporation. Dose 1–5 gr. (0.065–0.32 Gm.). **a. socotrina, pilula** (B. P.), contains Socotrine aloes, hard soap, oil of aniseed, and confection of roses. Dose 5–10 gr. (0.32–0.65 Gm.). **a., tinctura** (U. S. P., B. P.), consists of aloes, 10; licorice, 10; dilute alcohol, 100 parts. Dose ½–2 dr. (2–8 Cc.). **a., vinum** (B. P.), has aloes, 6; cardamom, 1; ginger, 1; white wine, 100 parts. Dose 1–4 dr. (4–16 Cc.).

**aloetic** (*al-o-et′-ik*) [*aloes*]. Containing or pertaining to aloes.

**aloetin** (*al-o-e′-tin*). 1. Aloe-resin. 2. A yellow, crystalline principle obtainable from aloes.

**alogia** (*ah-lo′-je-ah*) [ἀ, priv.; λόγος, word, reason]. 1. Inability to speak, due to some psychical defect. 2. Stupid or senseless behavior.

**alogotrophy** (*al-o-got′-ro-fe*) [ἄλογος, strange, absurd; τροφή, nutrition]. Irregular and perverted nutrition, leading to deformity.

**aloin** (*al′-o-in*) [*aloes*]. A bitter principle found in aloes. It forms fine needles, possesses a very bitter taste, and acts as a strong purgative. Several glucosides of this name are described, as, *barbaloin, nataloin, zanaloin, socaloin*. Dose ½–2 gr. (0.032–0.13 Gm.).

**aloisol** (*al-o-is-ol′*). An oily liquid obtained from the distillation of aloes with quicklime.

**alopecia** (*al-o-pe′-she-ah*) [ἀλωπεκία, a disease of foxes resembling mange]. Deficient hair; baldness. It may be universal or partial, congenital or acquired. It follows a large number of systemic affections. Syn., *lapsus capillorum; defluxio capillorum; vulpis morbus*. **a. adnata,** see *a., congenital*. **a. areata,** that condition in which, suddenly or slowly, one or several, usually asymmetrically distributed, patches of baldness appear upon the hairy regions of the body, more often upon the scalp and parts covered by the beard. Syn., *area Celsi; tinea decalvans; porrigo decalvans; alopecia circumscripta*. **a., cachectic,** that due to general malnutrition. **a. circumscripta,** see *a. areata*. **a., congenital,** a rare form, seldom complete, due to absence of hair-bulbs. **a. furfuracea,** a form of baldness associated with a disorder of the scalp, marked by hyperemia, itching, and exfoliation of dry or fatty scales from its surface. It may be acute or chronic, and produce a dryness, brittleness, and lack of luster in the hair. Syn., *alopecia pityroides capillitii; pityriasis capitis; seborrhœa capillitii; pityriasis simplex*. **a. localis,** that form occurring in one or more patches at the site of an injury or in the course of a nerve. Syn., *alopecia neuritica*. **a. neurotica,** a name given to baldness of trophoneurotic origin. **a. orbicularis,** same as *a. circumscripta*. **a. pityroides capillitii,** see *a. furfuracea*. **a. pityroides universalis,** a rapid and general denudation of hair occurring in debilitated states, preceded by abundant desquamation of fatty

**scales.** **a. senilis,** that occurring in old age. **a. simplex,** the idiopathic premature baldness of young adults. It is most common in males, and is often associated with premature grayness. **a. syphilitica,** that due to syphilis. **a. unguis, a. unguium,** the falling-off of the nails. Syn., *onychoptosis.* **a.-universalis,** that in which there is a general falling-out of the hairs of the body.

**aloxanthin** (*al-oks-an'-thin*), $C_{18}H_{16}O_8$. A yellow substance obtained from barbaloin and socaloin by the action of potassium bichromate.

**alpenstich** (*alp'-en-stik*) [Ger.]. A form of severe pleurisy or pleuropneumonia with typhoid symptoms peculiar to mountainous regions.

**alpha** (*al'-fah*) [ἄλφα, the first letter of the Greek alphabet]. The Greek letter α, used in combination with many chemical terms to indicate the first of a series of isomeric bodies, as alphanaphthol. **a.-eigon,** a compound of iodine and albumin containing 15 % of iodine and soluble in water. **a.-leukocyte,** one disintegrating during blood-coagulation.

**alphanaphthol** (*al-fah-naf'-thol*). A variety of naphthol.

**alphasol** (*al'-fa-sol*). Trade name of a preparation used as an antiseptic in rhinology and laryngology.

**alphenols** (*al'-fe-nolz*). A class of compounds having the characteristics of both alcohols and phenols.

**alphodeopsoriasis** (*al-fo-de-o-so-ri'-a-sis*) [ἀλφώδης, leprous; ψωρίασις, psoriasis]. A form of psoriasis resembling leprosy.

**alphodermia** (*al fo-der'-me-ah*) [ἀλφός, white; δέρμα, the skin]. Achromatosis; any disease marked by lack of pigmentation.

**alphol** (*al'-fol*), $C_{17}H_{15}O_3$. The salicylic ether of alphanaphthol, a white, crystalline powder, soluble in alcohol, in ether, and in fatty oils, and insoluble in water; melts at 83° C. It is an internal antiseptic. Dose 8-15 gr. (0.52-1.0 Gm.) 3 times daily.

**alphos** (*al'-fos*) [ἀλφός, vitiligo]. 1. An old name for leprosy. 2. Psoriasis.

**alphosis** (*al-fo'-sis*) [see *alphos*]. Albinism; leukoderma.

**alphozone** (*al'-fo-zōn*). Succinic dioxide. A white crystalline powder derived from hydrogen dioxide by action of succinic acid. It is used as a germicide in dilute aqueous solutions.

**alphus** (*al'-fus*). 1. See *alphos*. 2. A scrofulous pustular disease of the skin attended with the formation of white crusts. **a. confertus,** a scrofulous form of impetigo with clustered lesions attended with formation of white crusts. **a. leuce,** Plenck's name for a skin disease marked by white spots, which penetrate the skin deeply and involve the hairs, and if pricked, a milky fluid exudes. Syn., *vitiligo leuce; leuce.* **a. simplex,** Plenck's name for a skin disease marked by white patches not involving the hairs and wandering from one part to the other, with roughening of the skin. **a. sparsus,** a scrofulous disseminated ecthyma attended with formation of white crusts.

**Alpinia** (*al-pin'-e-ah*) [Prosper *Alpinus,* Italian botanist, 1553-1617]. A genus of zingiberaceous tropical plants. *A. chinensis, A. officinarum,* and other species furnish galangal.

**Alquié's operation** (*al-ke-a'*) [Alexis Jacques *Alquié,* French surgeon, 1812-1865]. Alexander's operation.

**Alsace gum** (*al-sās'*). See *dextrin.*

**alsol** (*al'-sol*). A preparation of aluminum acetate and tartaric acid; used as an astringent and disinfectant.

**Alstonia** (*al-sto'-ne-ah*) [Charles *Alston,* Scotch physician, 1683-1760]. A genus of apocynaceous trees and shrubs. *A. constricta,* the Australian fever-tree, yields the alkaloid *alstonine.* The bark is tonic, antiperiodic, and antipyretic, and is used in intermittent fevers. Dose of *fluidextract* 30-60 min. (2-4 Cc.). *A. scholaris,* the devil-tree, a native of the East Indies, furnishes dita-bark; it is tonic, astringent, antiperiodic, and anthelmintic.

**alstonidin** (*al-ston'-id-in*). An amorphous substance contained in a variety of dita-bark.

**alstonin** (*al-sto'-nin*). An amorphous substance contained in a variety of dita-bark.

**alstoninine.** A crystalline alkaloid, $C_{21}H_{20}N_2O_4$, obtained from *Alstonia constricta.*

**alt. dieb.** Abbreviation for the Latin *alternis diebus,* every other day.

**alter** (*awl'-ter*). To castrate or spay.

**alterant** (*awl'-ter-ant*). Same as *alterative.*

**alterative** (*awl'-ter-a-tiv*) [*alterativus*]. 1. A medicine that alters the processes of nutrition, restoring, in some unknown way, the normal functions of an organ or of the system. The most important alteratives are arsenic, iodine, the iodides, mercury, and gold. 2. Changing; alterant; re-establishing healthy nutritive processes.

**alternate** (*awl'-ter-nāt*) [*alternare,* to do by turns]. Occurring successively in space or time. **a. hemiplegia,** see *hemiplegia.*

**alternating** (*awl'-ter-na-ting*) [see *alternate*]. Occurring successively. **a. currents,** electric currents the direction of which is constantly changing. **a. insanity,** a form of insanity in which there are regular cycles of exaltation and depression.

**alternation** (*awl-ter-na'-shun*) [see *alternate*]. Repeated transition from one state to another. **a., of generations.** 1. In biology, a generative cycle in which the young do not resemble the parent, but like forms are separated by one or more unlike generations. 2. That form of reproduction in which some of the members of the cycle can produce new beings non-sexually, while in the final stage reproduction is always sexual. Tenia or tapeworm, is an example.

**alternator** (*awlt'-er-na-tor*). An apparatus for converting the direct dynamo current into an alternating current.

**Althaus' oil.** An oil made as follows: Metallic mercury, 1 part; pure lanolin, 4 parts; 2 % phenol, 5 parts. It is used in the treatment of syphilis in injections of 5 min. (0.3 Cc.) at a dose.

**Althea, Althæa** (*al-the'-ah*) [L.]. Marshmallow. The peeled root of *Althæa officinalis,* a plant of the mallow family. It consists of about one-third of vegetable mucus and starch, together with the alkaloids *asparagine* and *altheine* (latterly regarded as identical). Its decoction is employed as a mucilaginous drink. **a., ointment of** (*unguentum althææ*), an ointment composed of marshmallow root, 2 parts; turmeric, flaxseed, and fenugreek, each, 1 part; water, 70 parts; lard, 44 parts; yellow wax, 6 parts. **a., syrup of** (*syrupus althææ*), contains 4 % althea. Dose indefinite. *Asparagine* possesses sedative and diuretic properties, and is useful in ascites and gout. Dose 2-3 gr. (0.13-0.19 Gm.).

**alt. hor.** Abbreviation for the Latin *alternis horis,* every other hour.

**althose** (*al'-thōs*). Trade name of a preparation containing senega, squill, and codeine; used as an expectorant.

**altitude** (*al'-te-tūd*) [*altitudo,* height]. The height, as of an individual. In climatology, the elevation of a place above the sea-level. **a.-staff,** a device employed for measuring the exact height of recruits. It consists of a rigid upright with a vertex-bar moving without play at right angles to the upright.

**Altmann's granules.** Round bodies staining readily with carbolfuchsin, and regarded as cell-derivatives which have grown through the assimilation of fat. Their absence is supposed to indicate cancer. They are probably allied to Russell's bodies. According to Ross, the substance which forms chromosomes.

**altricious** (*al-trish'-us*) [*altrix,* a nurse]. Requiring a long nursing; hence, slow of development (the reverse of precocious).

**alum** (*al'-um*) [*alumen,* alum]. Any one of a class of double sulphates formed by the union of one of the sulphates of certain non-alkaline metals with a sulphate of some alkaline metal. The standard (or common commercial) alum, the official *alumen* (U. S. P.), is the aluminum-and-potassium sulphate, $AlK(SO_4)_2+12H_2O$. It is a powerful astringent and styptic, and is also extensively used in the arts. **a., alumina-,** a mixture of alum and aluminum sulphate. **a., aluminum-,** an alum composed of a double sulphate of aluminum and another radical. **a., ammonia,** the same as the standard, except that the potassium is replaced by ammonium. It is official in Great Britain, and is extensively used on account of its cheapness. What is known as *concentrated* or *patent* alum is the normal aluminum sulphate (*alumini sulphas,* U. S. P.), which is not a true alum. **a., ammonioferric** (*ferri et ammonii sulphas,* U. S. P.), is strongly styptic, and is useful in leukorrhea. Dose 5-10 gr. (0.32-0.65 Gm.). **a., burnt,** alum dried by heat; a spongy, pulverizable substance. It is used as an astringent and on fungous growths. Dose 5-30 gr. (0.333-2.0 Gm.). Syn., *calcined alum; alumen exsiccatum; alumen ustum.* **a.,** feather, **a.,** feathered. 1. Alum occurring in a fibrous form. 2. Asbestos. **a.-hematoxylin, a,** purple stain for tissues, obtained from an alcoholic solution of hematoxylin by addition of an aqueous solution of potash alum. **a.,** potash, **a.,** potassa, **a.,** potassic, **a.,** potassium, an alum containing potassium, particularly ordinary alum, or aluminum-and-potassium sulphate. **a.,** potassioferric, is similar to ammonioferric alum. **a.,** soda, double sulphate of sodium and aluminum; it is too soluble for ordinary uses. **a.-whey,** a preparation obtained by boiling 2 dr. of alum in a pint of milk and straining. It is used as an astringent and internal hemostatic in wineglassful doses.

**alumen** (al-ū'-men) [L., gen., *aluminis*]. See *alum*. **a. exsiccatum** (U. S. P.), burnt or dehydrated alum. See *alum, burnt*.

**alumil** (al'-ū-mil). Alumina in combination with acids.

**alumina** (al-ū'-min-ah) [L.], Al₂O₃. Aluminum oxide: the principal ingredient of clay and of many stones, earths, and minerals.

**aluminated** (al-ū'-min-a-ted). Combined with alum, alumina, or aluminum.

**aluminated copper.** See *lapis divinus*.

**aluminic, aluminicus** (al-ū-min'-ik, -us). Relating to or having the nature of alum.

**aluminiferous** (al-ū-min-if'-er-us) [*alum; ferre,* to bear]. Yielding alum.

**aluminium.** See *aluminum*.

**aluminol, alumnol** (al-ū'-min-ol, al-um'-nol) [*aluminum*]. The aluminum salt of betanaphthol sulphonic acid. It is an astringent and antiseptic; and is used in gonorrhea, endometritis, and diseases of the ear, nose, skin, etc.

**aluminosis** (al-ū-min-o'-sus) [*alum;* νόσος, disease]. A chronic catarrhal inflammation of the lungs found in pottery workers.

**aluminous** (al-ū'-min-us). Relating to or containing alum, alumina, or aluminum. **a. chalybeate,** a term applied to mineral waters containing alum and iron.

**aluminum, aluminium** (a-lū'-min-um, a-lū-min'-e-um) [L.], Al = 27. Quantivalence II, IV. A silver-white metal distinguished by its low sp. gr.—about 2.6. It is largely used in the arts and for certain surgical instruments. **a.** acetate, AlO . 4C₂H₃O₂ +4H₂O. Used as an internal and external disinfectant. Dose 5-10 gr. (0.3-0.6 Gm.) 3 times daily. **a. acetoborate,** antiseptic and disinfectant. **a. acetoglycerinate,** glycerite of aluminum acetate. It has one-fifth the strength of aluminum acetotartrate; used in 50 % solution in diseases of the nose, throat, and ear. **a. acetotartrate,** an energetic nontoxic disinfectant and astringent. It is applied in 0.5 to 2 % solutions in diseases of the air-passages; for chilblains, in 50 % solution. **a. boroformate,** prepared from freshly precipitated aluminium hydroxide dissolved in 2 parts of formic acid, 1 part of boric acid, and 7 parts of water. It is used as an astringent and antiseptic. **a. borotannate,** a reaction-product from tannic acid with borax and aluminium sulphate, containing 76 % tannin, 13.23 % alumina, 10.71 % boric acid; used as a disinfectant and astringent in skin diseases, applied pure or attenuated in ointment or dusting-powder. Syn., *cutal; cutol.* **a. borotannotartrate,** a compound of aluminum borotannate and tartaric acid; is used externally in skin diseases and in gonorrhea in 0.5 to 10 % solution. Syn., *soluble cutal* or *cutol.* **a. borotartrate,** an energetic, astringent, nonirritant antiseptic, used externally in inflammatory diseases of the throat and nose, and applied in substance or in solution with the addition of glycerin. Syn., *boral.* **a.** bromide, Al₂Br₆. In combination with aluminum chloride it is used as a gargle in diphtheria or taken internally. **a. caseinate,** an intestinal astringent. Dose 4-5 gr. (0.25-0.3 Gm.). **a. chloride,** Al₂Cl₆, colorless hexagonal plates which fume in moist air. It is astringent and antiseptic, and is also used in bleaching teeth. **a. gallate,** basic, a brown, antiseptic dusting-powder made by precipitating a solution of aluminum sulphate with a solution of gallic acid to which sodium hydroxide has been added. **a. hydroxide** (*alumini hydroxidum,* U. S. P.), Al₂(HO)₆, a tasteless white powder, feebly astringent. Dose 3-20 gr. (0.2-1.3 Gm.). Syn., *aluminum hydrate.* **a. oleate,** Al(C₁₈H₃₃O₂)₃, a yellowish mass, soluble in alcohol, in ether, in benzene, and in oleic acid. It is used as an antiseptic in skin diseases. **a.-and-potassium sulphate,** AlK(SO₄)₂+12H₂O, a valuable astringent, used in catarrh, leukorrhea, gonorrhea. Dose 10-20 gr. (0.65-1.3 Gm.). In teaspoonful doses it is an emetic. Syn., *alum.* **a.-and-potassium sulphocarbolate,** Al₂K₂(C₆H₄HSO₄)₈, an antiseptic, astringent, and styptic; it is used externally in a 5 to 20 % aqueous solution in cases of cancer and putrid ulcerations, and as a mouth-wash. **a. salicylate,** Al(C₇H₅O₃)₃, a reddish-white antiseptic powder used in nasal catarrh and ozena. Syn., *salumin.* **a. salicylate, ammoniated,** a yellowish-white powder used as an antiseptic and astringent in inflammation of the nose and throat by dry insufflations or painting with a 20 % solution in 50 % of glycerol and 30 % of water. Syn., *soluble salumin.* **a.-and-sodium silicate,** Na₂SiO₄Al₄(SiO₄)₃, obtained by adding aluminium hydroxide to a boiling solution of sodium silicate and sodium hydroxide. It is used in surgical dressings. **a. sozoiodolate,** is used as an antiseptic wash in 2 to 3 % solution. **a. sulphate** (*alumini sulphas,* U. S. P.), Al₂(SO₄)₃, an antiseptic and astringent used as a lotion in 5 % solution. **a. sulphocarbolate,** Al₂(C₆H₄HSO₄)₆, white crystals, soluble in water, in glycerol, and in alcohol. It is recommended as an antiseptic in cystitis and suppurating sores. Syn., *soxal.* **a. tannate,** a compound of aluminum and tannic acid. **a. tannotartrate,** yellowish-white plates or powder, soluble in water; used as an astringent and antiseptic insufflation or gargle in laryngeal or catarrhal troubles. Syn., *soluble tannal.* **a.-and-zinc sulphate,** Al₂(SO₄)₃ZnSO₄, a white, crystalline powder, soluble in water. It is used as a caustic.

**alumroot.** The root of *Heuchera americana.* Its properties are due to gallic and tannic acids. It is very astringent. Dose of the *fluidextract* 10-20 min. (0.65-1.3 Cc.). Also the root of *Geranium maculatum*, a mild astringent.

**alundum** (al-un'-dum). A preparation of alumina used for making appliances which are to be subjected to severe heat in the laboratory.

**alusia** (al-ū'-se-ah) [ἀλύειν, to wander] Hallucination; morbid state of mind.

**alv.** deject. Abbreviation for the Latin *alvi dejectiones*, the intestinal evacuations.

**alv.** adstrict. Abbreviation for the Latin *alvo adstricta*, the bowels being confined.

**alvearium** (al-ve-a'-re-um) [L.]. The external auditory canal or meatus.

**alveated** (al'-ve-a-ted) [*alveatus*, hollowed out like a trough]. Honeycombed; channeled; vaulted like a beehive.

**Alvegniat's pump** (al-vān'-yah). A mercurial air-pump used in estimating the gaseous constituents of the blood.

**alveloz** (al-vel-ōth') [Sp.]. An extractive from *Euphorbia icterodoxa*, having diuretic properties. It is highly recommended as a topical application in cancer.

**alveola** (al-re'-o-la) [*alveolus,* a small hollow]. A little depression.

**alveolar** (al-ve'-o-lar) [see *alveola*]. Pertaining to an alveolus. **a. abscess,** a gum-boil. **a. arch,** the alveolar surface of the jaw. **a. artery,** a branch of the internal maxillary artery. **a. border,** the margin of the jaws. **a. index,** in craniometry, the gnathic index; the ratio of the distance between the basion and alveolar point, to the distance between the basion and the nasal point, multiplied by 100. (Sometimes the basilar index is called the alveolar index.) **a. passages,** the ultimate division of the bronchi, emptying into the infundibula. **a. points,** see *craniometric points.* **a. process,** the border of the superior maxilla, in which the alveoli are placed. **a. sarcoma,** see *sarcoma.* **a. structure,** having small, superficial cavities, as in the mucous membrane of the stomach.

**alveolarium** (al-ve-o-la'-re-um) [*alveus,* a bee-hive]. A name sometimes applied to the external meatus of the ear. It is so called because the wax of the ear gathers in that place.

**alveolate** (al-ve'-o-lāt, or al'-ve-o-lāt) [*alveolatus,* hollowed out like a little tray]. In biology, pitted, honeycombed.

**alveoli** (al-ve'-o-li). Genitive and plural of *alveolus*.

**alveolin** (al-ve'-o-lin). A chemical substance obtained from the alveolar network in the deuto-merites of gregarines.

**alveolitis** (*al-ve-o-li'-tis*) [*alveolus;* ιτις, inflammation]. Inflammation of the alveolus of a tooth.
**alveolocondylean** (*al-ve'-o-lo-kon-dil'-e-an*) [*alveolus*, a hollow; κόνδυλος, a knuckle]. In craniometry, pertaining to the alveolus and condyle. a. plane. See *plane.*
**alveolodental** (*al-ve'-o-lo-den'-tal*) [*alveolus; dens,* a tooth]. Pertaining to the teeth and their sockets.
**alveololabial** (*al-ve'-o-lo-la'-be-al*). Pertaining to the alveolar processes and the lips.
**alveololabialis** (*al-ve-o-lo-lab-e-a'-lis*) [*alveolus,* a hollow; *labium,* the lip]. The buccinator muscle.
**alveolomaxillary** (*al-ve'-o-lo-maks-il'-a-re*). The buccinator muscle.
**alveolosubnasal** (*al-ve'-o-lo-sub-na'-sal*) [*alveolus,* a hollow; *sub,* under; *nasus,* nose]. In biology, pertaining to the alveolar and subnasal points of the skull. a. **prognathism,** see *prognathism.*
**alveolus** (*al-ve'-o-lus*) [L.]. 1. The bony socket of a tooth. Syn., *phatne; phainia; phatnion.* 2. A cell. 3. An air-cell of the lung. 4. A cavity, depression, pit, cell, or recess. a. **of a gland,** the terminal lobule of a racemose gland. a. **laryngeus,** see *pouch, laryngeal.* a. **of the stomach,** one of the honeycomb-like depressions found in the stomach.
**alveus** (*al'-ve-us*) [*alveus,* a trough]. 1. A trough, tube, or canal; applied to ducts and vessels of the body. 2. A cavity or excavation. a. **ampullascens,** a. **ampullescens,** a. **ampullosus,** see *receptaculum chyli.* a. **communis,** the utricle of the ear. a. **cornu ammonis,** see *a. hippocampi.* a. **hippocampi,** a certain structure in the cerebral hemisphere investing the convexity of the hippocampus major. a. **urogenitalis,** see *uterus masculinus.*
**alvine** (*al'-vin* or *al'-vīn*) [*alvus,* belly]. Pertaining to the belly. a. **concretion,** an intestinal calculus. a. **dejections,** a. **discharges,** the feces. a. **obstruction,** constipation.
**alvus** (*al'-vus*) [L., pl. and gen., *alvi*]. 1. The belly or its contained viscera. 2. Diarrhea. a. **adstricta,** a. **astricta,** an extreme degree of constipation. a. **dura,** constipation. a. **renis,** the pelvis of the kidney.
**alymphia** (*ah-limf'-e-ah*) [ά, priv.; *lympha,* lymph]. A deficiency of lymph.
**alypin** (*al'-e-pin*). The hydrochloride of tetramethyl-diamino-dimethyl-ethyl-carbinol-benzoate. It is a synthetic preparation, similar to cocaine and stovaine, and is used as a local anesthetic. It is less toxic than cocaine. For the eye and urethra, a 2 per cent stolution is used; elsewhere, a stronger solution.
**alyssus** (*al-is'-us*) [ά, priv.; λύσσα, madness]. Preventing or curing rabies.
**Alzheimer's disease** (*als'-hi-mer*). A mental disorder generally occurring in middle life; it is characterized by insidious onset, a rapidly progressive course, and final dementia.
**Am.** Abbreviation for *ametropia,* and for *mixed astigmatism.*
**am-.** A prefix indicating the group NH₂.
**A. M.** Abbreviation of *Artium Magister,* Master of Arts.
**ama**-(*ah'-ma*) [ἄμη, a water-pail]. An enlargement at the end opposite the ampulla of a bony canal of the labyrinth of the internal ear.
**A. M. A.** Abbreviation for *American Medical Association.*
**amaas** (*ah'-mahs*) [Kaffir, soured milk]. A mild form of small-pox prevalent in South Africa and elsewhere; milk-pox.
**amacrine** (*am'-ak-rin*) [ά, priv.; μακρός, long; ίς, a fiber]. Applied to nerve-cells entirely devoid of axiscylinder processes.
**amadou** (*am'-a-doo*) [Fr., *amadouer,* to coax]. German tinder or touchwood; *Boletus igniarius,* a fungus found on old tree-trunks, used to stanch local hemorrhage and as a dressing for wounds, etc. a. **de Panamá,** a hemostatic prepared from the leaf-hairs of *Micronia mucronata.*
**amalgam** (*am-al'-gam*) [μάλαγμα, a soft mass]. 1. A combination of mercury with any other metal. 2. Any soft alloy. **a.-carrier** and **-plugger,** an instrument designed for carrying and introducing amalgam into the cavity of a tooth. a., **dental,** compounds of a basal alloy of silver and tin with mercury, used for filling teeth. Gold, platinum, copper, zinc, or bismuth is frequently added as a third metal to the basal alloy. **a.-manipulator,** an instrument used by dentists for preparing amalgam fillings.

**amalgamate** (*am-al'-gam-āt*). To unite a metal in an alloy with mercury. To unite two dissimilar substances. To cover the zinc elements of a galvanic battery with mercury.
**amalgamation** (*am-al-gam-a'-shun*) [see *amalgam*]. In metallurgy, the process of combining mercury with some other metal, as practised in separating silver and gold from ores.
**amandin** (*am-an'-din*) [Fr., *amande,* almond]. A proteid contained in sweet almonds.
**amanitin** (*am-an'-it-in*) [ἀμανῖται, a kind of fungi]. 1. A principle identical with cholin, obtained from the fly-agaric. 2. A poisonous glucoside obtainable from various species of agaric.
**amara** (*am-a'-ra*) [*amarus,* bitter]. 1. Bitters. 2. The bitter alkaloids. 3. [ἀμάρα, a trench.] A sewer, drain, or stream. In the plural, *amarœ,* the hollows of the outer ear.
**amaril** (*am'-ar-il*) [Sp., *amarillo, yellow*]. The poison induced by *Bacillus icteroides.*
**amarillic** (*am-ar-il'-ik*). Pertaining to yellow fever. Cf. *amarylism.*
**amarin** (*am'-ar-in*) [see *amara*], C₂₁H₁₈N₂, triphenyldihydroglyoxalin. It results from boiling hydrobenzamide with caustic potash. It has a poisonous effect on animals.
**amaroids** (*am'-ah-roids*). All distinctly bitter vegetable extractives of definite chemical composition other than alkaloids and glucosides. Their names end in *-in* or *-inum*. Also called "bitter principles."
**amarthritis** (*am-gr-thri'-tis*) [ἄμα, together; ἄρθρον, a joint; ιτις, inflammation]. Arthritis affecting many, or several joints at once.
**amarum** (*am-a'-rum*) [see *amara*]. 1. A bitter. 2. Magnesium sulphate. a., **genuine,** magnesium sulphate. a. **purum,** any simple bitter.
**amarylism** (*am'-ar-il-izm*) [see *amaril*]. Yellow fever.
**amasesis** (*ah-mas-e'-sis*) [ά, priv.; μάσησις, chewing]. Inability to chew.
**amasthenic** (*am-as-then'-ik*) [ἄμα, together; σθένος, strength]. Uniting the chemical rays of light in a focus, as a lens.
**amastia** (*ah-mas'-ti-ah*) [ά, priv.; μαστός, breast]. Congenital absence of the mammæ or nipples.
**amativeness** (*am'-at-iv-nes*) [*amare,* to love]. The sexual passion.
**amatory** (*am'-at-o-re*) [*amator,* a lover]. Pertaining to love. a., **fever,** love-sickness; chlorosis. a. **muscles,** the oblique muscles of the eye, used in ogling.
**amaurosis** (*am-au-ro'-sis*) [ἀμαυρόειν, to darken]. Partial or total blindness, especially that occurring without demonstrable lesion of the eye. Syn., *paropsis amaurosis; gutta serena; cataracta nigra.* a., **albuminuric,** that due to renal disease. a. **arthritica,** that due to gout. a. **atonica,** that due to physical debility. a. **centralis,** that due to disorder of the central nervous system. a., **cerebral,** that due to disease of the brain. a. **compressionis,** cerebral amaurosis caused by pressure upon the optic nerve. a., **congenital,** that existing from birth. a. **congestiva,** that due to cerebral congestion. a., **diabetic,** that associated with diabetes. a. **dimidiate,** that occurring in only one half of the visual field. a., **epileptiform,** a., **epileptoid,** sudden blindness not confined to epileptics, but considered by some to be epileptic in its nature. Dilatation of the retinal veins has been noted, but no changes in the retinal arteries have been observed. Syn., *retinal epilepsy; ophthalmemicrania.* a. **ex hæmorrhagia,** a. **ex hyperopsia,** an incurable, inexplicable blindness occurring suddenly after hemorrhages, especially of the stomach. a., **hysterical,** that accompanying hysteria. a. **intermittens larvata,** a blindness, often unilateral, occurring with mild intermittent fever, which is frequently followed by atrophy of the optic nerve. a., **intermittent,** bilateral amaurosis occurring as a complication of intermittent fever. It usually begins with the chill and continues until the sweating stage. a., **progressive,** the progressive atrophy of the intraocular optic nerve-endings. a., **reflex,** that resulting from a reflex action upon the optic nerve from some remote source of irritation. a., **saburral,** sudden temporary blindness occurring in an attack of acute gastritis. a., **spasmodic,** blindness due to convulsions. a., **spinal,** that caused by atrophy of the optic nerve, due to lateral or multiple sclerosis. Syn., *rachialgic amaurosis.* a.

**sympathica**, a., sympathetic, functional disorder of one eye from reflex transmission of disease of the other eye. a., uremic, that due to uremia.
**amaurotic** (am-au-rot'-ik) [see *amaurosis*]. Relating to or affected with amaurosis. a. cat's-eye, a light-reflex through the pupil in suppurative choroiditis.
**amaxophobia** (am-aks-o-fo'-be-ah) [ἅμαξα, a car; φόβος, fear]. Morbid dread of being in, or riding upon, a car or wagon.
**amazia** (ah-ma'-ze-ah) [ἀ, priv.; μαζός, the breast]. Congenital absence of the mammary gland.
**Ambard's coefficient** (ahm-bar') [Leo *Ambard*, French physician]. For estimating renal activity: it showe the relation between the amount of urea in the blood and that excreted by the kidneys.

$$K = \frac{Ur}{\sqrt{D \times \frac{70}{P} \times \frac{\sqrt{C}}{25}}}$$

Ur = the quantity of urea in a liter of blood; D = the total urea excreted in 24 hours; C = the amount of urea in the urine; P = the weight of the patient in kilograms.
**amber** (am'-ber). See *succinum*.
**ambergris** (am'-ber-gris) [*amber*; Fr., *gris*, gray]. A biliary or intestinal concretion of the sperm-whale, *Physeter macrocephalus*. It exhales a fragrant, musky odor when warmed, and is used in adynamic fevers, chronic catarrh, and nervous diseases. Dose 1-3 gr. (0.065-0.2 Gm.).
**ambidexter** (am-be-deks'-ter) [*ambo*, both; *dexter*, the right hand]. An ambidextrous person.
**ambidexterity** (am-be-deks-ter'-it-e). Ability to use both hands equally well; ambidextrousness.
**ambidextrous** (am-be-deks'-trus) [see *ambidexter*]. Able to use both hands equally well.
**ambilateral** (am-be-lat'-er-al) [*ambo*, both; *latus*, side]. Relating to or affecting both sides.
**ambilevous** (am-be-le'-vus) [*ambo*, both; *lævus*, on the left side]. Unskilful in the use of both hands.
**ambiopia** (am-bi-o'-pe-ah). See *diplopia*.
**ambitus** (am'-bit-us) [*ambire*, to surround]. A circumference. a. cerebelli, Burdach's term for the cerebellum, pons, and oblongata taken together.
**ambloma** (am-blo'-mah) [ἄμβλωμα, an abortion; *pl.*, *amblomata*]. An amblosis or abortion; an aborted fetus.
**amblosis** (am-blo'-sis) [ἄμβλωσις, an abortion]. An abortion.
**ambliotic** (am-blot'-ik) [ἀμβλωτικός]. Abortifacient.
**amblyaphia** (am-ble-a'-fe-ah) [ἀμβλύς, dull; ἀφή, touch]. Dulness of the sense of touch.
**amblygeustia** (am-ble-jūs'-te-ah) [ἀμβλύς, dull; γεῦσις, taste]. A diminution or blunting of the sense of taste.
**amblyope** (am'-ble-ōp). A person affected with amblyopia.
**amblyopia** (am-ble-o'-pe-ah) [ἀμβλύς, dulled; ὤψ, eye]. Dimness of vision, especially that not due to refractive errors or organic disease of the eye. It may be *congenital* or *acquired*, the *acquired* being due to the use of *tobacco* (*amblyopia nicotinica*), *alcohol*, or other *toxic* influences; to *traumatism*; or it may be *hysterical*. *Nyctalopia* and *hemeralopia* are other forms; it may arise from *entoptic phenomena*, such as *musca volitantes*, *micropsia*, *megalopsia*, *metamorphopsia*, etc. It may take the form of contracted fields of vision, of color-blindness, or anesthesia of the retina. Syn., *obfuscatio*; *offuscatio*. a., crossed, a. cruciata, amblyopia occurring through lesion of the brain, in which a dimness of vision with contraction of the field of vision exists in the eye on the side opposite to the lesion. a. ex anopsia, amblyopia from disuse or from nonuse. a., postmarital, that due to sexual excess, called also *Burn's amaurosis*.
**amblyopiatrics** (am-ble-o-pe-at'-riks) [*amblyopia*; ἰατρικός, belonging to medicine]. The therapeutics of amblyopia.
**amblyoscope** (am'-ble-os-kōp) [*amblyopia*; σκοπεῖν, to look]. An instrument by means of which an amblyopic eye is trained to take its share in vision.
**ambo** (am'-bo). See *ambon*.
**amboceptoid** (am-bo-sep'-toid). A degenerated amboceptor which has lost its binding group (haptophore) on the one hand for the cell, or, on the other hand, for the complement.

**amboceptor** (am-bo-sep'-tor) [*ambo*, both; *capere*, to receive]. A hypothetical thermostabile substance found in blood-serum after inoculation. It possesses two haptophore groups, viz., a cytophile and a complementophile. Synonyms: *immune body, reparative, sensitizer, desmon, fixative, fixator, philocytase, receptor of the third order*. a. unit, the smallest quantity of amboceptor in the presence of which a given quantity of red blood corpuscles will be dissolved by an excess of complement.
**amboceptorgen** (am-bo-sep'-tor-jen). An antigen giving rise to amboceptors.
**ambon** (am'-bon) [ἄμβων, the lip of a cup]. The fibrocartilaginous ring that surrounds a socket in which the head of a large bone is received, such as the acetabulum, or the glenoid cavity.
**ambos** (am'-bos) [Ger.]. The incus, or anvil bone.
**Amboyna button** (am-boi'-nah but'-un). See *frambesia*.
**ambra** (am'-bra) [L.]. 1. Amber. 2. Ambergris. 3. Spermaceti. a. alba. 1. Spermaceti. 2. A light-colored amber obtained in Brazil. a. atra, see *a. nigra*. a. cineracea, a. cineria, a. cineritia, see *ambergris*. a. flava, a. fulva, see *succinum*. a. grisea, see *ambergris*. a. nigra, general name for any dark-colored amber or ambergris or dark, resinous substance; also lignite and jet.
**ambrein** (am'-bre-in) [Fr., *ambre*]. A substance much resembling cholesterin; it is obtained from ambergris by digestion in hot alcohol.
**ambrosia** (am-bro'-zhe-ah) [ἀμβροσία, the food of the gods]. A genus of composite-flowered herbs. A. artemisiæfolia, common hog-weed of North America; stimulant, tonic, antiperiodic, and astringent. A. trifida has properties similar to A. artemisiæfolia. The pollen of these plants is by some regarded as a cause of hay-fever.
**ambulance** (am'-bu-lans) [*ambulare*, to walk about]. 1. In Europe the term is applied to the surgical staff and arrangements of an army in service. 2. In the United States the word is restricted to a vehicle for the transference of the sick or wounded from one place to another. 3. In Europe a portable military hospital and its equipments accompanying the army in its movements. a. chaser, a "shyster" lawyer who drums up accident damage cases against firms and corporations.
**ambulant, ambulating, ambulatory** (am'-bu-lant, am'-bu-lā-ting, am'-bu-la-to-re). Relating to walking or changing location; not confined to bed. a. blister, a blister that changes its location. a. clinic, a clinic for patients that can walk; a dispensary. a. erysipelas, erysipelas that shifts from place to place. a. tumor, a pseudotumor. a. typhoid, walking typhoid; enteric fever in which the patient does not, or will not, take to his bed.
**ambulatorium** (am-bu-la-to'-re-um) [L.]. A dispensary.
**ambustial** (am-bust'-she-al) [*amburere*, to scorch]. Caused by a burn.
**ambustion** (am-bus'-chun) [*ambustio*, a burn]. A burn or scald.
**ameba, amœba** (am-e'-bah) [ἀμοιβή, a change]. A colorless, single-celled, jelly-like, protoplasmic organism found in sea and fresh waters, constantly undergoing changes of form and nourishing itself by engloping surrounding objects. a. bucca'lis, found in dental caries. a. coli, the ameba of dysentery. This is a protoplasmic mass, resembling the water ameba, 20 to 30 μ in diameter, and composed of a nucleus and a highly granular protoplasm containing vacuoles. It is found in large numbers in the stools of certain forms of dysentery, in the intestinal mucous membrane, and at times in the socalled dysenteric abscess of the liver. Whether it is the real cause of the disease is not definitely established. a. dysenter'iæ, the organism responsible for amebic dysentery. a.-enteritis, chronic enteritis due to invasion of *amœba coli*. a. gingiva'lis, one species found about the gums. a. histolyt'ica, same as the *a. dysenteriæ*.
**amebaphobia** (am-e-bah-fo'-be-ah) [*ameba*; φόβος, fear]. A morbid fear of becoming infected with amebæ.
**amebiasis** (am-e-bi'-as-is). The state or condition of being infected with amebæ.
**amebic** (am-e'-bik) [see *ameba*]. Pertaining to or characterized by amebæ. a. dysentery, dysentery associated with the presence in the bowel of *amœba coli*.

**amebicide** (am-e'-bis-īd) [ameba; cædere, to kill]. 1. Destructive of amebæ. 2. A remedy that destroys amebæ.
**amebiform** (am-e'-be-form). See ameboid.
**amebism, amœbism, amebaism, amœbaism** (am'-e-bism, am-e'-ba-izm). A pathological condition due to the invasion of the system by amebæ.
**amebocyte** (am-e'-bo-sīt). A leukocyte.
**ameboid** (am-e'-boid) [ameba; εἶδος, resemblance]. 1. Resembling an ameba in form or in movement, as the white blood-cells. 2. In bacteriology, of cultures which assume various shapes.
**amebula, amœbula** (am-e'-bu-lah). A merozoite having the power of ameboid movement.
**ameburia** (am-e-bu'-re-ah). The occurrence of amebæ in the urine.
**ameleia** (am-el-i'-ah) [ἀμέλεια, indifference]. Morbid apathy; indifference.
**amelia** (ah-me'-le-ah) [ά, priv.; μέλος, limb]. Congenital absence of the limbs.
**amelification** (am-el-if-ik-a'-shun). The formation of the enamel of the teeth by means of the enamel cells—ameloblasts.
**amelioration** (am-ēl-yo-rā'-shun) [ad, to; melior, better]. Improvement.
**ameloblast** (am-el'-o-blast) [Anglo-French, amel, enamel; βλαστός, a germ]. An enamel-cell, one of the cylindrical cells covering the papilla of the enamel organ of the teeth, and forming a beautifully regular epithelial layer that produces the enamel.
**amelus** (am'-el-us) [ά, priv.; μέλος, limb]. A monstrosity without limbs.
**amenia** (ah-me'-ne-ah). See amenorrhea.
**amenomania** (am-en-o-ma'-ne-ah) [amanus, agreeable; μανία, madness]. A mild form of mania in which the symptoms are manifested under the form of gaiety, fondness of dress, exaggeration of social condition, etc.; a cheerful, or joyous delirium; a morbid elevation of the spirits.
**amenorrhea, amenorrhœa** (ah-men-or-e'-ah) [ά, priv.; μήν, month; ῥεῖν, to flow]. Abnormal absence of menstruation. Syn., paramenia obstructionis; amenia. a., ovarian. a., paramenia, radical, that due to nonovulation. a., physiologic, absence of menstruation during pregnancy. a., primitive, a term applied to those cases in which the catamenia have not appeared at the proper time. a., secondary, that in which the discharge has been arrested after it has existed during the reproductive period.
**amenorrheal** (ah-men-or-e'-al) [see amenorrhea]. Pertaining to amenorrhea.
**ament** (am'-ent) [ab, from; mens, mentis, the mind]. A person affected with amentia; an idiot.
**amentia** (ah-men'-she-ah) [ά, priv.; mens, mind]. Defective intellect; idiocy.
**amenyl** (am'-en-il). Methylhydrastimide. It is a vasodilator and is used as an emmenagogue. Dose gr. ¾ (0.05 gm.) twice daily.
**amerism** (am'-er-ism) [ά, priv.; μέρος, a part]. The quality or condition of not dividing into segments or fragments.
**ameristic** (ah-mer-is'-tik) [ά, priv.; μέρος, a part]. Not segmented.
**amesiality** (ah-me-ze-al'-it-e). The throwing of a part, as the pelvis, to one side of the mesial line of the figure.
**ametabolic** (ah-met-ab-ol'-ik) [ά, priv.; μεταβόλη, changeable]. Not due to, or causing, or undergoing, metabolism.
**ametamorphosis** (ah-met-ah-mor'-fo-sis) [ά, priv.; μεταμόρφωσις, change]. The absence of metamorphosis.
**ametria** (ah-met'-re-ah) [ά, priv.; μήτρα, womb]. 1. Congenital absence of the uterus. 2. [ά, priv.; μέτρον, a measure.] Immoderation; asymmetry.
**ametrohemia, ametrohæmia** (ah-met-ro-he'-me-ah) [ά, priv.; μήτρα, womb; αἷμα, blood]. A defective uterine blood-supply.
**ametrometer** (ah-met-rom'-et-er) [ά, priv.; μέτρον, a measure]. An instrument for measuring ametropia.
**ametrope** (ah'-met-rōp) [ά, priv.; μέτρον, a measure; ὄψις, sight]. An individual affected with ametropia.
**ametropia** (ah-met-ro'-pe-ah) [ά, priv.; μέτρον, a measure; ὄψις, sight]. The condition when an imperfect image is formed upon the retina, due to defective refractive power of the media or to abnormities of form of the eye. In myopia the anteroposterior diameter is too great or the power of the refractive media is too great; hyperopia (or hypermetropia) is the exact reverse; astigmatism is due to imperfect curvature of the cornea or of the retina, or to inequality of refracting power in different parts of the lens; presbyopia is due to inelasticity of the lens, producing insufficient accommodation; aphakia, or absence of the lens, produces both insufficient refracting power and loss of accommodation.
**ametropic** (ah-met-rop'-ik) [see ametropia]. Affected with or pertaining to ametropia.
**ametrous** (ah-met'-rus) [ά, priv.; μήτρα, womb]. Lacking a uterus.
**amianthinopsy** (am-e-an-thin-op'-se) [ά, priv.; ἰάνθινος, violet-colored; ὄψις, sight]. Violet-blindness; incapacity to distinguish violet rays.
**amic** (am'-ik) [ammonia]. Pertaining to or having the nature of ammonia, or of an amine.
**Amici's disc, A.'s stria.** See Krause's disc.
**amicrobic** (ah-mi-kro'-bik) [ά, priv.; microbion, microbe]. Not due to, or associated with, microbes.
**amicron** (ah-mik'-rōn) [ά, priv.; μίκρον, small]. A particle which is too small to be observed with the ultramicroscope.
**amicroscopic** (ah-mi-kro-skop'-ik). Too small to be observed by the ultramicroscope.
**amide** (am'-id) [ammonia]. A chemical compound produced by the substitution of an acid radical for one or more of the hydrogen atoms of ammonia. The amides are primary, secondary, or tertiary, according as 1, 2, or 3 hydrogen atoms have been so replaced. They are white, crystalline solids, often capable of combining with both acids and bases. a., acid. 1. An amido-acid. 2. An amide as distinguished from amine or alkamide. a., allophanic, see biuret. a. bases, see amine, primary.
**amidine** (am'-id-in) [Fr., amidon, starch]. 1. Starch altered by heat into a horny, transparent mass; soluble starch; the part of starch that is soluble in water. 2. [ammonia.] One of a class of monacid bases produced from the nitrites by heating with ammonium chloride. In the free condition they are quite unstable. They contain the group C . NH . NH₂.
**amido-** (am'-id-o). A prefix denoting a chemical compound containing the univalent radical NH₂.
**amidoacetic acid** (am-id-o-as-e'-tik). See glycocoll and glycin.
**amidoacetophenetidin** (am-id-o-as-et-o-fe-net'-id-in) See phenocoll.
**amidoacid** (am-id-o-as'-id) [ammonia; acetum, vinegar]. An acid containing the amido-group NH₂.
**amidoazotoluol** (am-id-o-az-o-tol'-u-ol). A reddish-brown powder, allied to scarlet-red; it is soluble in alcohol, ether, and fatty oils, but is insoluble in water. It is used as an ointment to promote the growth of epithelium on granulating surfaces.
**amidobenzene, amidobenzol** (am-i-do-ben'-zēn, -sol). See aniline.
**amidocaffeine** (am-id-o-kaf'-e-in), C₈H₉(NH₂)N₄O. Fine acicular crystals obtained by heating bromcaffeine with alcoholic ammonia.
**amidocaproic acid.** Same as leucin.
**amidogen** (am-id'-o-jen) [amide; γενᾶν, to produce]. The hypothetical univalent radical, NH₂, replacing one atom of H in amido-compounds. See amide.
**amidoguaiacol** (am-id-o-gwi'-ak-ol). A product of acetoanisidin by nitration and reduction. It melts at 184° C. The salts are employed in the preparation of colors and medicines.
**amidosuccinamic acid** (am-id-o-suk-sin-am'-ik). Same as asparagin.
**amidosulphonal** (am-id-o-sul'-fon-al). Amidoacetone ethyldisulphone, a sedative.
**amidoxim, or amidoxime** (am-id-oks'-im). A substance derived from an amidine (2) by the substitution of an OH group for an atom of hydrogen.
**amidulin** (am-id'-u-lin) [see amidine]. Soluble starch; prepared by the action of H₂SO₄ on starch, thus removing the starch-cellulose.
**amimia** (ah-mim'-i-ah) [ά, priv.; μῖμος, a mimic]. Loss of the power of imitation or of making gestures.
**amine, amin** (am'-in) [ammonia]. The amines are chemical compounds produced by the substitution of a basic atom or radical for one or more of the hydrogen atoms of ammonia, or basic derivatives of carbon, containing nitrogen and viewed as ammonia derivatives. They are called monamines, diamines, triamines, etc., according to the number of amidogen molecules, NH₂, substituted for H. a., primary, an amine in which one hydrogen atom is replaced by a univalent alkyl. a., secondary, an amine in which

two hydrogen atoms are replaced by univalent alkyls. **a., tertiary,** an amine in which three hydrogen atoms are replaced by univalent alkyls.
**amino-.** A prefix denoting a chemical compound containing the univalent radical $NH_2$. **a.-acid,** an organic acid in which one of the hydrogen atoms is replaced by $NH_2$.
**aminoform** (*am-in'-o-form*). See *urotropin*.
**aminol** (*am'-in-ol*) [*amine*]. A gaseous substance derived from the methylamine of, herring-brine mixed with milk of lime. It is disinfectant, and has been used in the purification of sewage.
**aminopurin** (*am-in-o-pu'-rin*). Any compound derived from purin by substitution of one of the hydrogen atoms by the amino group, $NH_2$; adenin.
**aminosuria** (*am-in-o-sū'-re-ah*) [*amine*; οὖρον, urine]. The presence of amines in the urine when voided.
**amitosis** (*ah-mit-o'-sis*) [ά, priv.; μίτος, a thread]. Cell-multiplication by direct division or simple cleavage.
**amitotic** (*ah-mit-ot'-ik*) [see *amitosis*]. Of the nature of, or characterized by, amitosis. **a. cell-division,** direct cell-division, as distinguished from karyokinesis.
**amma** (*am'-ah*) [ἄμμα, a tie: *pl., ammata*]. A truss or girdle for hernia.
**ammeter** (*am'-et-er*) [*ampère*; μέτρον, a measure]. A form of galvanometer in which the value of the current is measured directly in ampères.
**ammic** (*am'-ik*). See *ammoniacum*.
**ammism** (*am'-izm*) [ἄμμός, sand]. Ammotheraphy; psammism.
**ammonemia** (*am-o-ne'-me-ah*). The supposed presence of ammonium carbonate in the blood.
**ammonia** (*am-o'-ne-ah*) [from the name of Jupiter *Ammon*, from the neighbourhood of whose temple in Libya ammonium chloride was obtained]. A colorless, pungent gas, $NH_3$, very soluble in water. The preparations of ammonia are used as antacids and as gastric and cardiac stimulants, in headache, hysteria, etc. It is a stimulant to the heart, and, in its elimination through the lungs, stimulates and liquefies the bronchial secretion. **ammoniæ, aqua** (U. S. P.), water of ammonia; a solution containing 10 % of the gas in water. Dose 3 min.-½ dr. (0.3-2.0 Cc.), well diluted. **ammoniæ, aqua, fortior** (U. S. P.), contains 28 % of the gas in solution. **ammoniæ, linimentum** (U. S. P.), ammonia-water, 35; cottonseed oil, 60; alcohol, 5 %. **ammoniæ, spiritus** (U. S. P.), a 10 % solution of ammonia-water in alcohol. Dose 10 min.-1 dr. (0.65-4.0 Cc.), diluted. **ammoniæ, spiritus, aromaticus** (U. S. P.), aromatic spirit of ammonia, an alcoholic solution of ammonium carbonate flavored with lemon, lavender, and pimenta. Dose ½-2 dr. (2-8 Cc.).
**ammoniac** (*am-o'-ne-ak*). 1. See *ammoniacum*. 2. Relating to ammonia. 3. Relating to ammoniacum.
**ammoniacal** (*am-o-ni'-ak-al*) [*ammonia*]. Containing or relating to ammonia.
**ammoniacum** (*am-o-ni'-ak-um*) [*ammonia*]. Ammoniac. A gum obtained from a Persian plant, *Dorema ammoniacum*. It is a stimulating expectorant and laxative, resembling asafetida, employed in chronic bronchial affections. Dose 10-30 gr. (0.65-2.0 Gm.). **ammoniaci cum hydrargyro, emplastrum,** ammoniac, 72; mercury, 18 %, with sulphur, acetic acid, and oil, q. s. **ammoniaci, emplastrum,** 100 parts of ammoniac digested with 140 parts of acetic acid, diluted, strained, and evaporated. **ammoniaci, emulsum,** a 4 % emulsion in water. Dose ½-1 oz. (15-30 Cc.).
**ammoniameter** (*am-o-ne-am'-et-er*) [*ammonia*; μέτρον, a measure]. An instrument for testing the strength of ammonia solutions.
**ammoniated** (*am-o'-ne-a-ted*) [*ammonia*]. Combined with ammonia.
**ammoniemia,** or **ammoniæmia** (*am-o-ne-e'-me-ah*) [*ammonia*; αἷμα, blood]. The theoretical decomposition of urea in the blood, yielding ammonium compounds.
**ammonin** (*am'-o-nin*). A soda deposit used in the making of soap.
**ammonionitrometry** (*am-o-ne-o-ni-trom'-et-re*) [*ammonium*; *nitrogen*; μέτρον, a measure]. An analytic method of estimating separately the amount of ammonia, nitrogen, and nitric acid contained in a compound.
**ammonium** (*am-o'-ne-um*) [*ammonia*]. A hypothetic univalent alkaline base, having the composition $NH_4$. It exists only in combination. **a. acetate, solution of** (*liquor ammonii acetatis*, U. S. P.), spirit of Mindererus, dilute acetic acid neutralized with ammonia. Dose 1 dr.-1 oz. (3.75-30.0 Cc.). **a. anacardate,** an ammonium compound of the resinous acids of cashew-nut. It is a doughy mass, soluble in alcohol, and used as a hair-dye. **a. arsenate,** $(NH_4)_2HAsO_4$. It is used as an alterative in skin diseases. Dose ½ gr. (0.03 Gm.), gradually increased, 3 times daily. **a. benzoate** (*ammonii benzoas*, U. S. P.), $NH_4C_7H_5O_2$. Dose 5-15 gr. (0.32-1.0 Gm.). **a. bisulphate,** $NH_4HSO_4$. Dose 10-30 gr. (0.65-2.0 Gm.). **a. bisulphite,** $NH_4HSO_3$. It is antiseptic and used internally in fermentative dyspepsia, externally in skin diseases. Dose 10-30 gr. (0.65-2.0 Gm.). **a. bitartrate,** $NH_4HC_4H_4O_6$, a white, crystalline acid powder. It is used in the manufacture of baking-powder. **a. borate,** $2(NH_4HB_2O_4)+3H_2O$, used in renal colic; in combination with codeine it is used in tuberculosis of the lungs. Dose 10-20 gr. (0.65-1.3 Gm.) every hour in water with licorice. **a. borobenzoate,** an intestinal antiseptic. **a. bromide** (*ammonii bromidum*, U. S. P.), $NH_4Br$, used in epilepsy, cough, and rheumatism. Dose 10 gr.-½ dr. (0.65-2.0 Gm.). **a. carbamate,** $NH_4NH_2CO_2$, a white, crystalline, volatile powder, stimulant, a reaction-product of carbon dioxide and ammonia gas. Syn., *ammonium carbonate anhydrine*. **a. carbazotate,** see *a. picrate*. **a. carbonate,** $C_2H_{11}N_2O_5$.- $NH_4$, antiseptic and antipyretic. Dose 2-6 gr. (0.13- 0.4 Gm.). Syn., *ammonium phenate; ammonium phenylate*. **a. carbonate** (*ammonii carbonas*, U. S. P.), $C_2H_{11}N_3O_5$, a compound of ammonium and carbonic acid. It is a stimulant expectorant and cardiac stimulant. Dose 5-10 gr. (0.32-0.65 Gm.). **a. chloride** (*ammonii chloridum*, U. S. P.), $NH_4Cl$, sal ammoniac, is used in bronchitis, rheumatism, and liver disease. Dose 1-20 gr. (0.065-1.3 Gm.). **a. chloride, troches of** (*trochisci ammonii chloridi,* U. S. P.), each lozenge contains 2 gr. (0.13 Gm.) of the salt. **a. embelate,** the ammonium salt of embellic acid, $NH_4C_{17}H_{19}O_3$. It is a teniacide. Dose for children 3 gr. (0.2 Gm.); for adults 6 gr. (0.4 Gm.). **a. fluoride,** used in enlargement of the spleen. Dose ⅛-1 gr. (0.003-0.032 Gm.). It is recommended in dyspeptic flatulence, 16 gr. (1 Gm.) dissolved in 10 oz. (300 Cc.) of distilled water; 1 tablespoonful after each meal. **a. formate,** $NH_4CHO_2$, used in chronic paralysis. Dose 5-10 gr. (0.32 Gm.). **a. glycerinophosphate,** $(NH_4)_2PO_4C_3H_5(OH)_2$, soluble in water. It is used in neurasthenia, Addison's disease, etc. Dose 3-4 gr. (0.2-0.26 Gm.) several times daily. **a. glycyrrhizate,** an expectorant. **a. hypophosphite,** $NH_4PH_2O_2+H_2O$, white, laminate crystals, soluble in water. Dose 10-30 gr. (0.65- 2:0 Gm.) 3 times daily. **a. iodide** (*ammonii iodidum,* U. S. P.), $NH_4I$. Dose 2-10 gr. (0.13-0.65 Gm.). **a. nitrate,** $NH_4NO_3$, used in preparing nitrous oxide. **a. persulphate,** $(NH_4)_2S_2O_8$, colorless crystals, soluble in water with turbidity. It is a disinfectant and deodorizer. Application, 0.5 to 2 % solution. **a. phosphate,** $(NH_4)_3HPO_4$. Dose 5-20 gr. (0.32- 1.3 Gm.). **a. phosphate, dibasic,** $(NH_4)_2HPO_4$. Used in rheumatism and gout. Dose 5-20 gr. (0.32-1.3 Gm.) 3 or 4 times daily in ½ oz. water. **a. picrate,** $C_6H_2(NH_4)(NO_2)_3O$, a salt in yellow needles, of bitter taste; like other picrates, it is explosive, and must be handled with care. It is antipyretic and antiperiodic, and tends to correct gastric disturbances. Dose 5 gr. (0.32 Gm.) in 24 hours. Syn., *Ammonium carbazotate*. **a. salicylate** (*ammonii salicylas*, U. S. P.), $NH_4C_7H_5O_3$, an antirheumatic, antipyretic germicide and expectorant. Dose 2-10 gr. (0.13-0.65 Gm.). **a. silicofluoride,** $2NH_4F$. $SiF_4$, an energetic antiseptic and reconstituent. It is used by inhalation in diseases of the nose and throat. **a. succinate,** $(NH_4)_2C_4H_4O_4$; recommended, 1 part in 120 parts of water, as a specific in colic. Dose 1 tablespoonful every 15 minutes. **a. sulphate,** $(NH_4)_2SO_4$, used in the preparation of other ammonium salts. **a. sulphite,** $(NH_4)_2SO_3$, an antiseptic used in fermentative dyspepsia. Dose 5-20 gr. (0.3-1.3 Gm.). Applied externally in skin diseases, 1 part in 10 parts of water. **a. sulphocarbolate,** $NH_4C_6H_4HSO_4$, antiseptic. Dose 1-5 gr. (0.06-0.3 Gm.). **a. sulphoricinate,** brown, ointment-like masses, soluble in alcohol and water. It is antiseptic and deodorant, and applied in 20 % solution in skin diseases or on ulcerated mucous membranes. **a. tartrate,** $(NH_4)_2C_4H_4O_6$, clear crystals,

soluble in water. It is an expectorant. Dose 5–30 gr. (0.3–2.0 Gm.). **a. thiosulphate,** $(NH_4)_2S_2O_3$, soluble in water; antiseptic. Dose 5–30 gr. (0.3–2.0 Gm.) in water. **a. tungstate,** fine white crystalline powder or needles, soluble in water. **a. urate,** $(NH_4)C_5H_3N_4O_3$, white crystalline powder, slightly soluble in water. It is antiseptic and used in 4 % ointment in chronic eczema. Ammonium urate occurs in alkaline urine and at times in urinary calculi. **a. valerate,** $NH_4C_5H_9O_2$, is used as a sedative in hysteria. Dose 1–5 gr. (0.065–0.32 Gm.).
**ammoniuria** (am-o-ne-ū'-re-ah) [ammonia; οὖρον, urine]. A condition marked by excess of ammonia in the urine.
**Ammon's fissure** (am'-on) [Friedrich Agust von *Ammon*, German ophthalmologist. 1799–1861]. A pyriform fissure, occurring during the early fetal period in the lower portion of the sclerotic coat of the eye. **A.'s operation.** 1. *Blepharoplasty;* removal of all cicatricial tissue and freeing of the remains of the lid, followed by transplantation of a flap from the cheek. 2. For *destruction* of the *lactimal sac;* incision into, and excision of, a portion of the anterior wall of the sac; closure by adhesive inflammation. 3. For *ectropion* (from caries); an incision is made around the cicatrix, the tissues are dissected free, and after closing the lid the wound is closed over the cicatrix. 4. For *symblepharon;* the lid is divided by two converging incisions, into three portions—two lateral and a central wedge-shaped portion; the former are united by sutures, and after union the central, wedge-shaped part is dissected out. **A.'s posterior scleral protuberance,** a variety of posterior ectasia of the sclera of the eye.
**Ammon's horn** (am'-on) [*Ammon*, an Egyptian deity, represented with a ram's head]. The hippocampus major of the brain.
**ammonol** (am'-on-ol), $C_8H_8NH_2$. A proprietary remedy said to be ammoniated phenylacetamide; pale-yellow crystals, said to be analgesic and antipyretic. Dose 5–20 gr. (0.3–1.3 Gm.). **a. salicylate,** a remedy for headache. Dose 8 gr. (0.5 Gm.).
**ammotherapy** (am-o-ther'-a-pe) [ἄμμος, sand; θεραπεύειν, to heal]. The use of sand-baths in the treatment of disease.
**amnemonic** (am-ne-mon'-ik) [ἀ, priv.; μνημονικός, relating to the memory]. Accompanied by or resulting in impairment of the memory.
**amnesia** (am-ne'-se-ah) [ἀμνησία, forgetfulness]. Loss of memory, especially of the ideas represented by words. **a., auditory,** word-deafness. **a., retroanterograde,** a perversion of memory in which recent events are referred to a far-removed past, while the occurrences of the remote past seem recent. **a., retrograde,** loss of memory for incidents and events which occurred a shorter or longer time before the attack of the disease. Besides that which may result from severe infectious disease or from epilepsy, it may be due to trauma or to hysteria. **a., visual,** word-blindness, or inability to recognize printed or written words.
**amnesic** (am-ne'-sik). Relating to amnesia. **aphasia,** see *amnesia.*
**amnestia** (am-nes'-te-ah) [ἀμνηστεία, forgetfulness]. Amnesia.
**amnestic** (am-nes'-tik) [ἀμνηστεία, forgetfulness]. 1. Amnesic. 2. Causing amnestia.
**amnia** (am'-ne-ah) [ἀμνίον, a young lamb]. Plural of *amnion,* q. v.
**amnial** (am'-ne-al). See *amniotic.*
**amniochorial** (am-ne-o-ko'-re-al) [amnion; χόριον, a membrane]. Pertaining to both amnion and chorion.
**amnioclepsis** (am-ne-o-klep'-sis) [ἀμνίον, amnion; κλέπτειν, to steal away]. The slow and unnoticed escape of the liquor amnii.
**amniocleptic, amniocleptious** (am-ne-o-klep'-tik, -us) [amnion; κλέπτειν, to steal away]. Relating to the unmarked escape of the liquor amnii.
**amnion** (am'-ne-on) [ἀμνίον, a young lamb]. The innermost of the fetal membranes; it is continuous with the fetal epidermis at the umbilicus, forming a complete sheath for the umbilical cord and a sac or bag in which the fetus is inclosed. It contains one or two pints of *liquor amnii.* It is a double, nonvascular membrane, the inner layer or *sac* derived from the epiblast, the outer from the mesoblast. The cavity of the inner folds is called the *true amnion,* that of the outer, the *false.* Syn., *agnina membrana; agnina pellicula; membrana agnina; agnina tunica;*

**abgas; abghas. a., dropsy of,** excessive secretion of liquor amnii.
**amnionic** (am-ne-on'-ik) [amnion]. Relating to the amnion.
**amniorrhea** (am-ne-o-re'-ah) [amnion; ῥοία, a flow]. The discharge of the liquor amnii.
**amnios** (am'-ne-os). 1. The liquor amnii. 2. The amnion.
**amniota** (am-ne-o'-tah) [ἀμνίον, a young lamb]. Animals with an amnion and allantois, comprising mammals, birds, and reptiles. Those without an amnion are called *anamnia.*
**amniotic** (am-ne-ot'-ik) [amnion]. Relating to the amnion. **a. cavity,** the sac of the amnion. **a. fluid,** the liquor amnii. See *amnios.*
**amniotitis** (am-ne-o-ti'-tis) [ἀμνίον, a young lamb; ιτις, inflammation]. Inflammation of the amnion.
**amniotome** (am'-ne-o-tōm) [ἀμνίον, a young lamb; τομή, a cut]. An instrument for puncturing the fetal membranes.
**amnitis** (am-ni'-tis). Same as *amniotitis.*
**amoeba** (am-e'-bah). See *ameba.*
**amoebiasis** (am-e-bi'-as-is). See *amebiasis.*
**amoebiform.** Same as *ameboid.*
**amoebism, amoeboism.** See *amebism.*
**amoeboid** (am-e'-boid). See *ameboid.*
**amoebula.** See *amebula.*
**amok, amuck** (am-ok', a-muk') [A Malay word denoting "an impulse to murder."]. In a state of murderous frenzy; in Oriental regions persons, mostly hashish eaters, often attack and kill those whom they meet while in a state of wild fury. In some cases the infuriated persons take this method of s kin death, for they are shot down at sightee g
**Amomum** (am-o'-mum) [ἄμωμον, an Eastern spice plant]. A genus of scitaminaceous plants to which the cardamom (*A. cardamomum*) and "grains of paradise" (*A. granum paradisi*) belong.
**amor** (am'-or) [L.]. Love. **a. insanus,** see *erotomania.* **a. sui,** love of self; vanity. **a. veneris,** Columbus' term for the clitoris.
**amorpha** (ah-morf'-ah) [ἀ, priv.; μορφή, shape]. 1. A cutaneous eruption having no definite form. 2. A macula. 3. Apparent diseases in which no lesions can be discovered. 4. Intertrigo. **a. infantilis,** a. lactantium, infantile intertrigo. **a. vulgaris,** intertrigo.
**amorphia** (ah-mor'-fe-ah) [see *amorpha*]. Shapeless condition.
**amorphinism** (ah-mor'-fin-izm) [ἀ, priv.; *morphine*]. The condition resulting from the withdrawal of morphine from one habituated to the drug.
**amorphism** (ah-mor'-fizm) [see *amorpha*]. The state of being amorphous or without shape; want of crystalline structure.
**amorphous** (ah-mor'-fus) [see *amorpha*]. Formless; shapeless; not crystalline.
**amorphus** (ah-mor'-fus) [ἀ, priv.; μορφή, a form]. An acardiacus without head or extremities. See also *anideus.* **a. globulus,** see *anideus.*
**amotio** (am-o'-she-o) [L.]. A detachment. **a. retinæ.** See *ablatio retinæ.*
**amp.** Abbreviation for *ampère.*
**ampelopsin** (am-pel-op'-sin). A tonic extract made from *Ampelopsis quinquefolia,* Virginia creeper.
**ampelotherapy** (am-pel-o-ther'-a-pe) [ἄμπελος, a grape-vine; θεραπεύειν, to heal]. The grape-cure (q. v.).
**amperage** (am-pār'-ahj) [André Marie *Ampère,* French physicist, 1775–1836]. The number of ampères passing in a given circuit.
**ampère** (am'-pār) [see *amperage*]. A unit of measurement of an electric current. It is the electromotive force of one volt produced in a circuit having one ohm of resistance. **A.'s law,** same as Avogadro's law, q. v.
**amperemeter** (am-pār'-me-ter) [ampère; μέτρον, a measure]. An instrument for estimating the strength of the current of an electric circuit in ampères.
**amphamphoterodiplopia** (am-fam-fo-ter-o-dip-lo'-pe-ah). See *amphodiplopia.*
**ampharkyochrome** (am-far-ke'-o-krōm). Same as ampharkyochrome.
**amphauxesis, amphauxis** (am-fawks-e'-sis, am-fawks'-is) [ἀμφί, around; αὔξησις, increase]. Growth or increase by concentric circles. Syn., *amphiphya.*
**amphemeros, amphemerus** (am-fem'-er-os, -us). 1. Quotidian. 2. A quotidian fever.

**amphi-** (am'-fe) [ἀμφί, around]. A prefix signifying *about, on both sides, around,* etc., as amphiarthrosis, amphibia, etc.

**amphiarkyochrome** (am-fe-ar'-ke-o-krōm) [ἀμφί, both; ἄρκυς, net; χρῶμα, color]. A term applied by Nissl to a nerve-cell the stainable portion of whose cell-body is in the form of a pale network, the nodal points of which are joined by an intensely staining network.

**amphiarthrodial** (am-fe-ar-thro'-de-al). Relating to *amphiarthrosis.*

**amphiarthrosis** (am-fe-ar-thro'-sis) [amphi-; ἄρθρον a joint]. A form of mixed articulation in which the surfaces of the bones are connected by broad discs of fibrocartilage or else are covered with fibrocartilage and connected by external ligaments. It is distinguished by limited motion, as, *e. g.,* between the vertebræ.

**amphiaster** (am'-fe-as-ter) [amphi-; ἀστήρ, a star]. The figure formed in indirect cell-division by the achromatin threads and chromatin granules united to form the socalled nuclear spindle, together with the threads of cell-protoplasm radiating from a rounded clear space at each end of the spindle, known as the stars or suns.

**amphibia** (am-fib'-e-ah) [amphi-; βίος, life]. A class of the *Vertebrata,* living both in the water and upon the land, as the frog, newt, etc.

**amphibious** (am-fib'-e-us) [see *amphibia*]. Living both on land and in water.

**amphiblastic** (am-fe-blas'-tik) [amphi-; βλαστός, a germ]. Pertaining to that form of complete segmentation that gives rise to an amphiblastula.

**amphiblastula** (am-fe-blas'-tu-lah) [amphi-; *blastula,* dim. of βλαστός, a germ]. The mulberry-mass- or morula-stage in the development of a holoblastic egg. It follows the stage known as amphimorula.

**amphiblestritis** (am-fe-bles-tri'-tis) [ἀμφιβλήστρον, a net; ιτις, inflammation]. Inflammation of the retina.

**amphiblestroid** (am-fe-bles'-troid) [ἀμφιβλήστρον, a net; εἶδος, form]. Net-like. *a.* **apoplexia,** Apoplexy of the retina. *a.* **membrane,** the retina.

**amphibolia** (am-fe-bo'-le-ah) [ἀμφιβολία, uncertainty]. The vacillating period of a fever or disease.

**amphibolic** (am-fe-bol'-ik) [see *amphibolia*]. Uncertain; doubtful. Applied to a period in the febrile process occurring between the fastigium and the defervescence, and marked by exacerbations and remissions.

**amphicelous** (am-fe-se'-lus) [ἀμφί, at both ends; κοῖλος, hollow]. In biology, biconcave, as the center of the vertebræ of fishes.

**amphicentric** (am-fi-sen'-trik) [ἀμφί, both; κέντρον, a point]. Originating and ending in the same vessel.

**amphicrania** (am-fe-krā'-ne-ah) [amphi-; κρανίον, the skull]. Headache affecting both sides of the head.

**amphicreatine** (am-fe-kre'-at-in) [amphi-; κρέας, flesh], C₉H₁₉N₇O₄. One of the muscle-leukomaines. It crystallizes in brilliant oblique prisms of a yellowish-white color, and is faintly basic.

**amphicreatinine** (am-fe-kre-at'-in-in) [see *amphicreatin*], C₉H₁₉N₇O₄. A member of the creatinin group of leukomaines derived from muscle.

**amphicroic** (am-fe-kro'-ik) [amphi-; κρούειν, to test]. Having the power to turn blue litmus-paper red and red litmus-paper blue.

**amphicytula** (am-fe-sit'-u-lah) [ἀμφί, on both sides; κύτος, cell]. The parent cell of an amphiblastic ovum.

**amphidesmic, amphidesmous** (am-fe-des'-mik, -mus) [ἀμφί, on both sides; δεσμός, a bond, a fetter]. Furnished with a double ligament.

**amphidiarthrosis** (am-fe-di-ar-thro'-sis) [amphi-; διάρθρωσις, articulation]. A mixed articulation such as that of the lower jaw, which partakes of the nature both of amphiarthrosis and diarthrosis.

**amphigastrula** (am-fe-gas'-tru-lah) [ἀμφί, on both sides; γαστήρ, belly]. The gastrula of an amphiblastic ovum.

**amphigony** (am-fig'-o-ne) [amphi-; γόνος, offspring]. The sexual process in its broadest sense; gamogenesis.

**amphimicrobian** (am-fe-mi-kro'-be-an) [amphi-; μίκρος, small; βίος, life]. Both aerobian and anaerobian.

**amphimixis** (am-fi-miks'-is) [amphi-; μίξις, mixing]. The mingling of two individuals or their germs; sexual reproduction.

**amphimorula** (am-fe-mor'-u-lah) [amphi-; *morula,* a mulberry]. The morula, or globular mass of cleavage cells resulting from unequal segmentation, the cells of the hemispheres being unlike in size.

**Amphioxus** (am-fe-oks'-us) [ἀμφί, both; ὀξύς, sharp]. A genus of fishes tapering at both ends, the lancelet.

**amphipyrenin** (am-fe-pi'-ren-in) [ἀμφί, around; πυρήν, mass]. The nuclear membrane of a cell.

**amphismela** (am-fis-me'-lah) [ἀμφί, both; μήλη, a probe]. A double-edged surgical knife.

**Amphistoma** (am-fis'-to-mah) [amphi-; στόμα, mouth]. A genus of trematode worms, named from the mouth-like apparatus at either end, also called *amphistomum.* One species, *A. hominis,* has been found in the large intestine of man.

**amphistomiasis** (am-fis-to-mi'-as-is). The condition of being infested with Amphistoma.

**amphitrichous** (am-fit'-rik-us) [amphi-; θρίξ, a hair]. Applied to the type of flagellation in certain bacteria having a flagellum or flagella single at each pole.

/ **amphodiplopia** (am-fo-dip-lo'-pe-ah) [ἀμφώ, both; διπλόος, double; ὤψ, eye]. Double vision affecting each of the eyes.

**amphogenous** (am-foj'-en-us). See *amphoteric.*

**amphopeptone** (am-fo-pep'-tōn). A mixture of hemipeptone and antipeptone.

**amphophil, amphophilous** (am'-fo-fil, am-fof'-il-us) [ἀμφώ, both; φιλεῖν, to love]. Readily stainable alike with acid and with basic dyes.

**amphoric** (am-for'-ik) [amphora, a vase with two handles]. Resembling the sound produced by blowing across the mouth of a bottle. *a.* **breathing,** breath-sounds with musical quality heard in diseased conditions of the lung, especially in pulmonary tuberculosis with cavity-formation. *a.* **resonance,** in auscultation, a metallic sound like that of blowing into a bottle, caused by the reverberation of sound in a cavity of the lung. *a.* **respiration,** see *a. breathing.*

**amphoricity** (am-for-is'-i-te) [amphoric]. The quality of being amphoric; the giving forth of amphoric sounds.

**amphoriloquy** (am-for-il'-o-kwe) [amphoric, *loqui,* to speak]. The production of amphoric sounds in speaking.

**amphorophony** (am-for-of'-o-ne) [amphoric; φωνή, a sound]. An amphoric resonance or sound.

**amphoteric, amphoterous** (am-fo-ter'-ik, am-fot'-er-us) [ἀμφότεροι, both of two]. Double-sided; having the power of altering the color of both red and blue litmus test-paper; a condition sometimes presented by the urine. *a.* **elements,** elements whose oxides unite with water, some to form acids, others to form bases.

**amphoterodiplopia** (am-fot-er-o-dip-lo'-pe-ah) [L.]. Amphodiplopia.

**amphotropin** (am-fo-tro'-pin). Hexamethylenetetramine camphorate ((CH₂)₆N₄)₂ . C₈H₁₄(COOH)₂. It acts as a urinary antiseptic, is said to promote the regeneration of sloughing epithelium, and to increase diuresis and the elimination of uric acid in pathological conditions.

**amplexation** (am-pleks-a'-shun) [*amplexatio,* an embrace]. The treatment of a fractured clavicle by an apparatus that fixes the shoulder and covers a part of the chest and neck.

**amplexus** (am-pleks'-us) [L., an embrace]. 1. An embracing; coitus. 2. Embraced, surrounded.

**ampliation** (am-ple-a'-shun) [*ampliare,* to increase]. Dilatation or distention of a part or cavity.

**amplification** (am-plif-ik-a'-shun) [*amplificare,* to enlarge]. 1. In microscopy, increase of the visual area. 2. Enlargement, as of a diseased organ.

**amplifier** (am'-ple-fi-er) [see *amplification*]. An apparatus used in microscopy for increasing the magnification. It consists of a diverging lens or combination placed between the objective and the ocular, and gives to the image-forming rays from the objective an increased divergence.

**amplitude** (am'-ple-tūd) [*amplus,* broad]. The range or extent, as of vibrations and undulations, the pulse-wave, etc.

**ampoule** (am-pool') [see *ampulla*]. A small, sealed, glass capsule, usually holding one dose of a hypodermic solution, sterile and ready for use.

**ampul** (am-pool'). See *ampoule.*

**ampulla** (am-pul'-ah) [L., "a Roman wine-jug": pl., *ampullæ*]. 1. The trumpet-mouthed or dilated

extremity of a canal, as of the lacrimal canal, the receptaculum chyli, the Fallopian tubes, mammary ducts, semicircular canals, vas deferens, etc. 2. A bulla or blister. **a.** chyli, the receptaculum chyli. **a.** of rectum, the portion above the perineal flexure. **a. vitrea,** a glass bottle.
**ampullaceous** (*am-pul-a'-shus*). 1. Flask-shaped; big-bellied; gibbous. 2. Relating to an ampulla. 3. Attended with the formation of bullæ or blebs.
**ampullar, ampullate** (*am-pul'-ar, am'-pul-āt*). Relating to an ampulla; shaped like an ampulla.
**ampullitis** (*am-pul-i'-tis*). Inflammation of an ampulla, more especially that of the vas deferens.
**ampullula** (*am-pul'-ū-lah*) [dim. of *ampulla*]. A small ampulla, as in the lymphatic or lacteal vessels.
**amputation** (*am-pu-ta'-shun*) [*amputare,* to cut away]. The removal of a limb or any projecting part of the body. Amputation may be by the knife, ligature, or other means, or it may be the result of pathological processes, as gangrene, constriction (*e. g.,* of the cord in the fetus). **a.,** accidental, the separation of a limb by some form of accident. **a.,*aperiosteal,** one in which the periosteum is completely removed from the end of the cut bone or bones. **a.,** bloodless, one in which there is but slight loss of blood, on account of the circulation being controlled by mechanical means. **a.,** central, one in which the scar is situated at or near the center of the stump. **a.,** circular, that performed by making a single flap, by circular sweeps of a long knife, through skin and muscles, in a direction vertical to the long axis of the limb. **a.,** circular skin-flap, a modification of the circular, in which the skin-flap is dissected up, and the muscles divided at a higher level. **a.,** coat-sleeve, a modification of the circular, in which the cutaneous flap is made very long, the end being closed by being gathered together by means of a tape. **a.,** congenital, amputation of fetal portions, due to constriction by amniotic bands. **a.,** consecutive, an amputation during the period of suppuration or later. **a. in contiguity,** amputation at a joint. **a. in continuity,** amputation of a limb elsewhere than at a joint. **a.,** cutaneous, one in which the flaps are composed exclusively of the integuments. **a.,** diclastic, one in which the bone is broken with an osteoclast and the soft tissues divided by means of an écraseur. Its object is to avoid hemorrhage and purulent infection. **a.,** double flap, one in which two flaps are formed from the soft tissues. **a.,** dry, see *a.,* bloodless. **a.,** eccentric, one in which the scar is situated away from the center of the stump. **a.,** elliptic, one that may be performed by a single sweep, as in the circular method; the wound, however, having an elliptic outline, on account of the oblique direction of the incision. **a.** of expediency, one performed for cosmetic effect. **a., flap,** one in which one or more flaps are made from the soft tissues, the division being made obliquely. **a., flap-less,** one in which, on account of destruction of the soft parts, flaps cannot be formed, the wound healing by granulation. **a., galvanocaustic,** one in which the soft parts are divided with the galvanocautery, followed by division of the bone by the saw. **a., immediate,** one done within 12 hours after the injury, during the period of shock. **a., intermediary, a., intermediate, a., intrapyretic,** one performed during the period of reaction and before suppuration. **a., intrauterine,** see *a., congenital.* **a., major,** amputation of an extremity above the wrist- or ankle-joint. **a., mediate,** see *a., intermediary.* **a., mediotarsal.** 1. Chopart's amputation. 2. An amputation through the tarsus, preserving the scaphoid bone. **a., minor,** amputation of a small part, as a finger. **a., mixed,** a combination of the circular and flap methods. **a., multiple,** amputation of two or more members at the same time. **a., musculocutaneous,** one in which the flaps consist of skin and muscle. **a., musculotegumentary,** see *a., musculocutaneous.* **a., natural,** see *a., congenital.* **a., oblique,** see *a., oval.* **a., osteoplastic,** one in which there are section and apposition of portions of bone in addition to the amputation. **a., oval,** a modification of the elliptic, in which the incision consists of two reversed spirals instead of the one oblique. **a., partial.** 1. One in which but a portion of the extremity is removed. 2. An incomplete congenital amputation. **a., pathological,** one done for tumor or other diseased condition. **a., primary,** one done after the period of shock and before the occurrence of inflammation. **a., racket,**

a variety of the oval amputation in which there is a single longitudinal incision continuous below with a spiral incision on either side of the limb. **a., secondary,** one performed during the period of suppuration. **a., spontaneous,** see *a., congenital;* it also occurs in the disease, ainhum. **a., sub-astragalar,** a partial amputation of the foot, leaving only the astragalus. **a., subperiosteal,** one in the continuity, the cut end of the bone being covered by periosteal flaps. **a., supracondylar,** see *Gritti's operation.* **a., synchronous,** see *a., multiple.* **a., tertiary,** that performed after the inflammatory reaction stage has passed. **a. by transfixion,** one done by thrusting a long knife completely through a limb and cutting the flaps from within out.
**amuck.** See *amok.*
**amusia** (*ah-mū'-se-ah*) [ἀ, priv.; μοῦσα, muse]. Loss of the ability to produce or comprehend music or musical sounds; an abnormity as regards music analogous to aphasia as regards the faculty of speech. **a.,** motor, that in which music is understood, but the power of singing or otherwise reproducing music is lost. **a.,** sensory, musical deafness, or the loss of the power of comprehension of musical sounds.
**Amussat's operation** (*am-oo-sah'*) [Jean Zuléma *Amussat,* French surgeon. 1796–1856]. 1. A method of arresting hemorrhage by torsion of the arteries by means of two forceps. 2. For *atresia vaginæ;* dilatation by the use of the finger or a dull instrument, without cutting. 3. For *castration;* by incision upon the posterior surface of the scrotum. 4. For *enterorrhaphy;* in cases of completely divided intestine, each end is invaginated and passed over a cork, with a groove at either end, and the intestine is tied in the grooves. 5. For *imperforated rectum;* the formation of an artificial anus in the perineum, with or without excision of the coccyx. 6. For *lumbar colotomy;* a transverse incision is made, crossing the outer border of the quadratus lumborum muscle. **A.'s valves,** see *Heister's valves.*
**amussis** (*am-us'-is*) [L., "a carpenter's rule or level"; pl., *amusses*]. One of two portions into which a median fissure divides the posterior commissure of the brain.
**amyasthenia** (*am-i-as-the'-ne-ah*). Same as *amyosthenia.*
**amyctic** (*am-ik'-tik*) [ἀμυκτικός, mangling]. 1. Caustic; irritating. 2. A caustic or corrosive drug.
**amydriasis** (*ah-mid-ri'-ah-sis*). See *mydriasis.*
**amyelencephalia** (*ah-mi-el-en-sef-a'-le-ah*) [ἀ, priv.; μυελός, marrow; κεφαλή, the head]. Absence of both brain and spinal cord.
**amyelencephalus** (*ah-mi-el-en-sef'-al-us*) [ἀ, priv.; μυελός, marrow; κεφαλή, the head]. A fetal monster having neither brain nor spinal cord.
**amyelia** (*ah-mi-e'-le-ah*) [ἀ, priv.; μυελός, marrow]. Congenital absence of the spinal cord.
**amyelic** (*ah-mi-e'-lik*) [see *amyelia*]. Relating to amyelia.
**amyelinic** (*ah-mi-el-in'-ik*). Without myelin.
**amyelonervia** (*ah-mi-el-o-ner'-ve-ah*). See *amyeloneuria.*
**amyeloneuria** (*ah-mi-el-o-nū'-re-ah*) [ἀ, priv.; μυελός, marrow; νεῦρον, a nerve]. Paresis of the spinal cord.
**amyelonic** (*ah-mi-el-on'-ik*). 1. Amyelic. 2. Without marrow.
**amyelotrophy** (*ah-mi-el-ot'-ro-fe*) [ἀ, priv.; μυελός, marrow; τροφή, nourishment]. Atrophy of the spinal cord.
**amyelus** (*ah-mi'-el-us*). See *amyelic.*
**ámyelus** (*ah-mi'-el-us*) [ἀ; priv.; μυελός, marrow]. A fetal monstrosity with partial or complete absence of the spinal cord.
**amyencephalus** (*ah-mi-en-sef'-al-us*). See *amyelencephalus.*
**amygdala** (*am-ig'-dal-ah*) [ἀμυγδάλη, almond]. 1. The tonsil. 2. A small lobule on the lower surface of each cerebellar hemisphere, projecting into the fourth ventricle. 3. Almond. The seeds of *A. amara* and *A. dulcis,* containing the principle *emulsin.* The former contains *amygdalin.* The expressed oil of the sweet almond is a demulcent and is useful in skin affections; in doses of 1–2 dr. (4–8 Gm.), a mild laxative; that of *A. amara* is used in cosmetics. **a. amara** (U. S. P.), the bitter almond. **a. dulcis** (U. S. P.), the sweet almond. **amygdalæ amaræ, aqua** (U. S. P.), a 1:1000 solution of the oil of bitter almonds in water. Dose 1 dr. (4 Cc.). **amygdalæ amaræ,** oleum (U. S. P.), contains 3–14 % of

**hydrocyanic** acid and has similar uses. Dose ½–1 min. (0.016–0.065 Cc.). **amygdalæ amaræ, spiritus** (U. S. P.), the spirit of bitter almonds. **amygdalæ, emulsum** (U. S. P.), oil of sweet almonds 6 %; sugar, water, and acacia q. s. **amygdalæ expressum, oleum** (U. S. P.), expressed oil of almonds. Dose 1 cc. (30 Cc.). **amygdalæ, syrupus** (U. S. P.), syrup of almond; demulcent and slightly sedative. Dose 1–2 dr. (4–8 Cc.).
**amygdalæ** (am-ig'-dal-e) [L., pl. of amygdala]. The tonsils.
**amygdalectomy** (am-ig-dal-ek'-to-me) [amygdala; ἐκτομή, a cutting-out]. Excision of a tonsil.
**amygdalin** (am-ig'-dal-in) [see amygdala], $C_{20}H_{27}NO_{11}+3H_2O$. A glucoside formed in bitter almonds, in various plants, and in the leaves of the cherry-laurel. Under the influence of emulsin, contained in the almond, it splits up into glucose and hydrocyanic acid.
**amygdaline** (am-ig'-dal-ēn) [see amygdala]. 1. Almond-like. 2. Pertaining to the tonsil.
**amygdalitis** (am-ig-dal-i'-tis) [amygdala; ιτις, inflammation]. Tonsillitis.
**amygdaloid** (am-ig'-dal-oid) [amygdala; εἶδος, form]. Resembling an almond. a. **fossa**, the depression for the lodgment of the tonsil. a. **tubercle**, a projection of gray matter at the end of the descending cornu of the lateral ventricle of the brain. It is attached to the temporal lobe, and appears to be nearly isolated by white substance.
**amygdalolith** (am-ig-dal'-o-lith) [amygdala; λίθος, a stone]. A concretion or calculus found in the tonsil.
**amygdaloncus** (am-ig-dal-ong'-kus) [amygdala; ὄγκος, a mass]. Any tumor or swelling of the tonsil.
**amygdalopathy** (am-ig-dal-op'-ath-e) [amygdala; πάθος, a disease]. Any disease of the tonsils.
**amygdalotome** (am-ig'-dal-o-tōm) [amygdala; τέμνειν, to cut]. An instrument used in cutting the tonsils.
**amygdalotomy** (am-ig-dal-ot'-o-me) [see amygdalotome]. Tonsillotomy. Partial or complete abscission of a tonsil.
**amygdophenin** (am-ig-dof'-en-in), $C_6H_4(OC_2H_5)NH.OC.CH(OH)C_6H_5$. A grayish-white, crystalline powder, derived from paramidophenol. It is antirheumatic. Dose 15 gr. (1 Gm.) from 1 to 6 times daily in powder. Syn., *phenylglycolphenetidin*.
**amygmus** (am-ig'-mus) [ἀμυγμός]. Scarification.
**amykos** (ah-mi'-kos) [ἀ, priv.; μύκος, a fungus]. An antiseptic fluid composed of boric acid, glycerin and infusion of cloves. Of reputed service in gonorrhea, dental caries, and catarrhs.
**amyl** (am'-il) [ἄμυλον, starch]. The radical, $C_5H_{11}$, of amylic alcohol, the fifth member of the series of alcohol radicals, $C_nH_{2n+1}$. **a.-alcohol**, see *amylic alcohol*. **a. bromide**, $C_5H_{11}Br$, a transparent, colorless liquid, soluble in alcohol. It is antiseptic and germicidal. **a. colloid**, a fluid preparation consisting of amyl hydride, 480 parts; gasoline, 1 part; veratrine, 6 parts; collodion, to 960 parts. It is painted on the skin in neuralgia, sciatica, etc. Syn., *anodyne colloid*. **a. hydrate**, see *amylic alcohol*. **a. hydride**, a fractional product of petroleum ether; it is an antiseptic. Syn., *hydramyl; pentylene; pentylhydride*. **a. iodide**, $C_5H_{11}I$, the reaction-product of isoamylic alcohol, iodine, and phosphorus. It is sedative and antiseptic, and is used as an inhalation in dyspnea. **a. nitrite**, $C_5H_{11}NO_2$, a clear, yellowish, volatile liquid, of a penetrating odor. It produces vascular dilation and stimulates the heart's action, and is useful in angina pectoris, respiratory, neuroses, etc. Dose, *internally*, 1 min. (0.016–0.065 Cc.) dissolved in alcohol; by *inhalation*, 2–5 min. (0.12–0.3 Cc.). **a. nitrite, carbureted**, amyl nitrite saturated with carbon monoxide. It is suggested as a substitute for pure amyl nitrite, to obviate pressure in the head and other secondary objectionable properties. **a. salicylate**, a compound obtained from the action of chlorine on a saturated solution of salicylic acid in amylic alcohol. It is said to have the sedative properties of the amylic derivatives as well as antirheumatic qualities. Dose in acute rheumatism 10 capsules of 3 gr. (0.2 Gm.) each, daily. **a. valerate**, **a. valerianate**, $C_{10}H_{20}O_2$. It is a cholesterin solvent and is used as a sedative in gall-stone colic. Dose 2–3 gr. (0.13–0.2 Gm.). Syn., *apple oil*.
**amylaceous** (am-il-a'-se-us) [see amyl]. Containing starch; starch-like. See *corpora amylacea*.

**amylamine** (am-il'-am-in). See *isoamylamine*. **a. hydrochlorate**, $C_5H_{14}NCl$, a reaction-product of amyl cyanate, potassium hydrate, and hydrochloric acid, occurring as deliquescent scales or crystals. It is an antipyretic. Dose 7–15 gr. (0.45–1.0 Gm.).
**-amylase** (am'-il-ās) [ἄμυλον, starch; -ase]. Any amylolytic enzyme, causing hydrolytic cleavage of the molecules of starch.
**amylate** (am'-il-āt). 1. A combination formed by the replacement of the hydrogen of the hydroxyl molecule in amylic alcohol with a metal or basic radical. 2. A compound of starch with a radical.
**amylene** (am'-il-ēn) [see amyl], $C_5H_{10}$. A liquid hydrocarbon having dangerous anesthetic properties. **a.-chloral**, $CCl_3.CH.OH.O.C.(CH_2)_2C_2H_5$, dimethyl-ethyl-carbinol-chloral. It is hypnotic. Syn., *dormiol*. **a. hydrate**, $C_5H_{11}O$, a tertiary alcohol used as a hypnotic. Dose 30 min.–1 dr. (2–4 Cc.).
**amylenization** (am-il-en-iz-a'-shun). The production of anesthesia by means of amylene.
**amylenol** (am-il'-en-ol). Amyl salicylate; used externally in rheumatism.
**amylic** (am-il'-ik) [see amyl]. Pertaining to amyl. **a. alcohol, fusel oil; potato-starch alcohol; amyl hydrate.** An alcohol having the composition $C_5H_{12}O$, produced in the continued distillation of fermented grain. It was formerly used to adulterate whisky. It is a solvent and reagent.
**amylin** (am'-il-in) [see amyl]. The insoluble wall of the starch-grain.
**amylism** (am'-il-izm). The toxic condition produced by amyl alcohol.
**amylobacter** (am-il-o-bak'-tur) [ἄμυλον, starch; βακτήριον, a little rod]. A genus of schizomycetes characterized by a period of development in which it contains starch in its interior.
**amylodextrin** (am-il-o-deks'-trin). Same as *erythrodextrin*. See *soluble starch*.
**amyloform** (am-il'-o-form). An odorless white powder produced by the chemical combination of starch with formaldehyde. It is non-toxic, quite insoluble, and is not decomposed under 180° C. It is recommended as a surgical antiseptic.
**amylogen** (am-il-o-jen) [ἄμυλον, starch; γεννᾶν, to produce]. Soluble starch.
**amylogenic** (am-il-o-jen'-ik) [ἄμυλον, starch; γεννᾶν, to produce]. Starch-producing.
**amylohydrolysis** (am-il-o-hi-drol'-is-is) [ἄμυλον, starch; ὕδωρ, water; λύσις, solution]. The hydrolysis of starch.
**amylohydrolytic** (am-il-o-hi-dro-lit'-ik). Relating to the hydrolysis of starch.
**amyloid** (am'-il-oid) [ἄμυλον, starch; εἶδος, form]. 1. Starch-like. 2. A starchy substance. 3. Glycogen. 4. Virchow's name for a waxy body found in animal tissue as a result of disease and resembling starch only in the one particular that it is stained by iodine. Cf. *amyloid degeneration*. **a. bodies**, bodies resembling starch-grains, found in the nervous system, the prostate, etc. They are the result of a localized amyloid degeneration. **a. degeneration, waxy or lardaceous degeneration.** A degeneration characterized by the formation of an albuminous substance, resembling starch in its chemical reactions. The process affects primarily the connective tissue of the blood-vessels of various organs, and is connected with or due to chronic suppuration in the body. Amyloid substance gives a brown color with iodine, a red color with gentian-violet, and turns blue on being treated with iodine and sulfuric acid. **a. kidney**, see *Bright's disease*.
**amyloidosis** (am-il-oid-o'-sis). See *amyloid degeneration*.
**amylolysis** (am-il-ol'-is-is) [ἄμυλον, starch; λύσις, solution]. The digestion of starch, or its conversion into sugar.
**amylolytic** (am-il-o-lit'-ik) [see *amylolysis*]. Pertaining to or effecting the digestion of starch, as the ferments in the saliva and pancreatic juice that convert starch into sugar.
**amylon** (am'-il-on) [ἄμυλον, starch]. 1. Starch. 2. Glycogen. 3. A principle found in grape-juice.
**amyloplast** (am'-il-o-plast) [ἄμυλον, starch; πλάσσειν, to form]. A leukoplast; a starch-forming protoplasmic granule.
**amylopsin** (am-il-op'-sin) [ἄμυλον, starch; ὄψις, appearance]. A ferment found in the pancreatic juice which changes starch into sugar.
**amylose** (am'-il-ōs) [ἄμυλον, starch]. Any one of

the group of carbohydrates, comprising starch, glycogen, dextrin, inulin, gum, cellulose, and tunicin.
**amylum** (am'-il-um) [L.], $C_6H_{10}O_5$. Starch.
**amyli, glyceritum** (U. S. P.), contains starch, 10; water, 10; glycerol, 80 %; used for external application. a. **iodatum**, contains starch, 95 %; iodine, 5 %, triturated with distilled water and dried. Dose 1. dr.—½ oz. (4–16 Gm.). **amyli, mucilago** (B. P.), used in making enemas.
**amyluria** (am-il-u'-re-ah) [ἄμυλον, starch; οὖρον, urine]. Presence of starch in the urine.
**amyocardia** (am-i-o-kar'-de-ah) [ἀ, priv.; μῦς, muscle; καρδία, the heart]. Lack of muscular power in the heart's contractions.
**amyostasia** (am-i-os-ta'-ze-ah) [ἀ, priv.; μῦς, muscle; στάσις, standing]. An abnormal trembling of the muscles while in use, often seen in locomotor ataxia.
**amyosthenia** (am-i-os-the'-ne-ah) [ἀ, priv.; μῦς, muscle; σθένος, force]. Deficient muscular power.
**amyosthenic** (am-i-o-sthen'-ik) [ἀ, priv.; μῦς, muscle; σθένος, force]. Pertaining to amyosthenia. Also, a medicine or agent depressing muscular action.
**amyotaxia** (ah-mi-o-taks'-e-ah) [ἀ, priv.; μῦς, muscle; τάξις, arrangement]. Motor disturbance of the muscles, of spinal or cerebral origin. Muscular ataxia.
**amyotonia** (am-i-o-to'-ne-ah) [ἀ, priv.; μῦς, muscle; τόνος, tone]. Lack of muscular tone; myatonia.
**amyotrophia** (am-i-o-tro'-fe-ah) [ἀ, priv.; μῦς, muscle; τροφή, nourishment]. Atrophy of a muscle.
**amyotrophic** (am-i-o-trof'-fik) [see amyotrophia]. Characterized by muscular atrophy a lateral sclerosis, lateral sclerosis combined with muscular atrophy. The lesion is in the pyramidal tracts and in the ganglion-cells of the anterior gray horns of the spinal cord. The disease has a marked tendency to involve the medulla. a. **paralysis**, that due to muscular atrophy.
**amyotrophy** (am-i-ot'-ro-fe). See amyotrophia.
**amyous** (am'-i-us) [ἀ, priv.; μῦς, muscle). Weak, deficient in muscle or muscular strength.
**amyrin** (am'-e-rin) [amyris], $C_{40}H_{48}O$. A resinous principle derived from Mexican Elemi.
**Amyris** (am'-e-ris) [L.]. A genus of tropical trees and shrubs producing fragrant resins and gums, such as Elemi, etc.
**amyxia** (am-iks'-i-ah) [ἀ, priv.; μύξα, mucus]. Absence or deficiency of mucous secretion.
**amyxis** (ah-miks'-is) [ἀμύσσειν, to scarify]. Scarification.
**amyxodes** (ah-miks-o'-dēz). 1. Deficient in mucus; relating to amyxia. 2. Scarified; relating to amyxis.
**amyxorrhea** (am-iks-o-re'-ah) [ἀ, priv.; μύξα, mucus; ῥοία, flow]. Absence of the normal mucous secretion.
**-an,** a suffix applied to a class of bodies related to the starch and sugar group.
**ana** (an'-ah) [ἀνά, so much each]. A Greek preposition signifying through, up, again, etc. In prescriptions contracted to āā, meaning of each.
**-ana.** A termination preferably used as a suffix to the name of a species around which others naturally cluster, in the naming of subsections or groups of species; e. g., the group of species of Helix related to H. pomatia may be indicated by the term pomatiana.
**anabasis** (an-ab'-as-is) [ἀναβαίνειν, to go up]. The increasing stage of acute disease.
**anabatic** (an-ab-at'-ik) [see anabasis]. Increasing; growing more intense; as the anabatic stage of a fever.
**anabiosis** (an-ab-i-o'-sis) [ἀναβιοῦν, to come to life again]. The reappearance of vitality in an apparently lifeless organism. Resuscitation; reanimation.
**anabiotic** (an-ab-i-ot'-ik) [ἀνά, again; βίος, life]. 1. Relating to anabiosis. 2. Restoring the strength or activity.
**anabole** (an-ab'-o-le) [ἀναβάλλειν, to throw up]. A throwing up; what is thrown up; vomit; vomiting; expectoration; regurgitation.
**anabolergy** (an-ab-ol'-er-je) [ἀναβάλλειν, to throw up; ἔργον, work]. The force expended or work performed in anabolism or in anabolic processes.
**anabolic** (an-ab-ol'-ik) [ἀναβάλλειν, to throw up]. Pertaining to or characterized by anabolism.
**anabolin** (an-ab'-o-lin) [see anabolic]. Any substance formed during the anabolic process.
**anabolism** (an-ab'-o-lizm) [see anabolic]. Synthetic or constructive metabolism. Activity and repair of function; opposed to katabolism.

**anabrosis** (an-ab-ro'-sis) [ἀνάβρωσις, an eating up]. Corrosion, or superficial ulceration.
**anabrotic** (an-ab-rot'-ik) [ἀνάβρωσις, an eating up]. Pertaining to anabrosis; corrosive.
**anacampsis** (an-ah-kamp'-sis) [ἀνακάμπτειν, to bend back]. A flexure.
**anacamptic** (an-ah-kamp'-tik) [see anacampsis]. Reflected, as sound or light; pertaining to or causing a reflection.
**anacamptometer** (an-ah-kamp-tom'-et-er) [ἀνακάμπτειν, to bend back; μέτρον, a measure]. An apparatus for measuring reflexes.
**Anacardium** (an-ah-kar'-de-um) [ἀνά, up; καρδία, the heart, from its heart-shaped seeds]. 1. A genus of tropical trees. A. occidentale yields cashew-gum and the cashew-nut. 2. The oil of the pericarp of the cashew-nut, known as cardol, and used as an escharotic. It is said to be of value in leprosy. a., **ointment of,** 1 part of the tar to 8 of lard or vaselin, used as a blistering ointment. a., **tincture of,** 1 to 10 of rectified spirit. Dose 2–10 min. (0.12–0.6 Cc.).
**anacatadidymus** (an-ak-at-ad-id'-im-us) [ἀνά, up; κατά, down; δίδυμος, a twin]. Divided above and below, but jointed centrally into one; said of certain twin monsters.
**anacatadidymus** (an-ak-at-ad-id'-im-us) [ἀνά, up; κατά, down; δίδυμος, a twin]. An anacatadidymous monstrosity.
**anacatharsis** (an-ak-ath-ar'-sis) [ἀνά, up; κάθαρσις, purgation]. Expectoration; vomiting.
**anacathartic** (an-ak-ath-ar'-tik) [ἀνά, up; κάθαρσις, purgation]. 1. Causing anacatharsis. 2. An expectorant, emetic, or sternutatory drug or agent.
**anachlorhydria** (an-ah-klor-hid'-re-ah). The lack of hydrochloric acid in the gastric juice.
**anacid** (an-as'-id) [ἀν, priv.; acidum, acid]. Slightly acid; subacid; not having the normal amount of acidity.
**anacidity** (an-as-id'-it-e). The lack of normal acidity; subacidity; inacidity.
**anaclasimeter** (an-ak-las-im'-et-er) [anaclasis; μέτρον, measure]. An instrument for measuring the refraction of the eye.
**anaclasis** (an-ak'-las-is) [ἀνάκλασις, a breaking-off or back]. 1. Reflection or refraction of light or sound. 2. A fracture. 3. Forcible flexion of a stiff joint.
**anaclastic** (an-ak-las'-tik) [ἀνάκλασις, a breaking-off, or back]. Pertaining to refraction, or to anaclasis.
**anaclisis** (an-ak'-lis-is) [ἀνάκλισις, reclining]. Decubitus; the reclining attitude.
**anacroasia** (an-ak-ro-a'-ze-ah) [ἀν, priv.; ἀκρόασις, hearing]. Inability to understand words that are heard, while the same words if read by the patient are understood.
**anacrotic** (an-ak-rot'-ik) [ἀνά, up; κρότος, a stroke]. Relating to or characterized by anacrotism.
**anacrotism** (an-ak'-ro-tizm) [see anacrotic]. The condition in which there is one or more notches on the ascending limb of the pulse-curve.
**anacusia, anacusis** (an-ak-oo'-se-ah, an-ak-oo'-sis) [ἀν, priv.; ἀκούειν, to hear]. Complete deafness.
**anadenia** (an-ad-e'-ne-ah) [ἀν, priv.; ἀδήν, gland]. Insufficiency of glandular function. Chronic want of gastric secretion. a. **gastrica,** Ewald's name for achylia gastrica. a. **ventriculi,** see achylia gastrica.
**anadesma** (an-ah-dez'-mah) [ἀναδέσμη, a fillet]. A band or fascia.
**anadicrotic** (an-ah-di-krot'-ik) [ἀνά, up; δίς, twice; κρότος, a stroke]. Characterized by anadicrotism.
**anadicrotism** (an-ah-di'-krot-izm) [see anadicrotic]. Dicrotism of the pulse-wave occurring in the upward stroke.
**anadidymous** (an-ad-id'-im-us) [ἀνά, up; δίδυμος, a twin]. Cleft upward into two, while single below—said of certain joined twins.
**anadidymus** (an-ad-id'-im-us) [see anadidymous]. An anadidymous monster.
**anadiplosis** (an-ad-ip-lo'-sis) [ἀνά, up; διπλοῦν, to double]. The reduplication or redoubling of a fever paroxysm.
**anadiplotic** (an-ah-dip-lot'-ik) [ἀνά, up; διπλοῦν, to double]. Characterized by anadiplosis.
**anadipsia** (an-ah-dip'-se-ah) [ἀνά, intensive; δίψα, thirst]. Intense thirst.
**anadrome** (an-ad'-ro-me) [ἀναδρομή, a running up]. 1. An upward determination of the blood. 2. A pain ascending from the lower to the higher portion

**anæmatopoiesis** (an-e-mat-o-poi-e'-sis). See *anematopoiesis*.

**anæmatosis** (an-e-mat-o'-sis). See *anematosis*.

**anæmia** (an-e'-me-ah). See *anemia*.

**anæmic** (an-e'-mik). See *anemic*.

**anaerobe** (an-a'-er-ōb). See *anaerobion*.

**anaerobia** (an-a-er-o'-be-ah) [ἀν, priv.; ἀήρ, air; βίος, life]. Plural of *anaerobion*. Microorganisms having the power of living without air or free oxygen. a., facultative, applied to organisms normally or usually living in the presence of oxygen, but capable of becoming anaerobic.

**anaerobic** (an-a-er-o'-bik) [see *anaerobia*]. Living in the absence of the oxygen or air. See *aerobic*.

**anaerobion** (an-a-er-o'-be-on). See *anaerobia*.

**anaerobiosis** (an-a-er-o-bi-o'-sis) [see *anaerobia*]. Life sustained in the absence of free oxygen; the power of living where there is no free oxygen.

**anaerobiotic, anaerobious** (an-a-er-o-bi-ot'-ik, an-a-er-o'-be-us) [see *anaerobia*]. Capable of existing without free oxygen.

**anaerophyte** (an-a'-e-ro-fīt) [ἀν, priv.; ἀήρ, air; φυτόν, a plant]. In biology, a plant capable of living without a direct supply of oxygen.

**anaeroplastic** (an-a-er-o-plas'-tik) [ἀν, priv.; ἀήρ, air; πλάσσειν, to shape]. Pertaining to anaeroplasty.

**anaeroplasty** (an-a'-er-o-plas-te) [ἀν, priv.; ἀήρ, air; πλάσσειν, to shape]. The treatment of wounds by immersion in warm water, so as to exclude the air.

**anæsthesia** (an-es-thē'-ze-ah). See *anesthesia*.

**anæsthesin** (an-es'-thes-in). Same as *anesthesin*.

**anæsthetic** (an-es-thet'-ik). See *anesthetic*.

**anagenesis** (an-aj-en'-e-sis) [ἀναγέννησις, regeneration]. Reparation or reproduction of tissues.

**anagnosasthenia** (an-ag-nos-as-the'-ne-ah) [ἀνάγνωσις, reading; *asthenia*]. Neurasthenia in which any attempt to read is accompanied by distressing symptoms.

**anagoge, anagogia** (an-a-go'-je, an-a-go'-je-ah) [ἀναγωγή, a bringing up]. Vomiting. a. **hæmatis, a. sanguinis**, a rush of blood to the head.

**anagraph** (an'-a-graf) [ἀναγραφή, a writing out]. A physician's prescription or recipe.

**anagyrine** (an-aj-i'-rin) [ἀνά, backward]; γύρος, a circle], CHN₃O₂. An alkaloid from the seeds of *Anagyris fœtida*, a leguminous shrub of Southern Europe. Its hydrochloride is poisonous, slowing the respiration, and interfering with the heart's action. a. **hydrobromide**, C₁₄H₁₈N₂O₂HBr. Small, white, shining scales, soluble in water and alcohol, melting at 265° C. It is used as a heart stimulant.

**anakhre** (an-ak'-er). Synonym of *goundou* (q. v.).

**anakroasia** (an-ak-ro-a'-ze-ah). See *anacroasia*.

**anakusis** (an-ak-oo'-sis). See *anacusia*.

**anal** (a'-nal) [*anus*, the fundament]. Pertaining to the anus.

**analdia** (an-al'-de-ah). See *marasmus*.

**analepsia** (an-al-ep'-se-ah). See *analepsis*.

**analepsis** (an-al-ep'-sis) [see *analeptic*]. 1. Recovery of strength after disease. 2. Suspension, as in a swing. 3. Epilepsy with gastric aura.

**analeptic** (an-al-ep'-tik) [ἀναληπτικός, restorative]. 1. Restorative. 2. Any agent restoring health after illness.

**analeptol** (an-al-ep'-tol). A tonic preparation said to contain phosphorus, ₁₈₀ gr.; nux vomica extract, ½ gr.; cinchona, 2 gr.; coca leaves, 1 gr., and the addition of aromatics.

**analgen** (an-al'-jen) [ἀν, priv.; ἀλγος, pain], C₂₀H₁₄N₂O₄. A white, tasteless, crystalline powder, almost insoluble in water, soluble with difficulty in cold alcohol, but more readily in hot alcohol and dilute acids. It melts at 406.4° F. It is employed as an analgesic, antineuralgic, and antipyretic. Dose 10–30 gr. (0.65–2.0 Gm.).

**analgesia** (an-al-je'-se-ah) [see *analgen*]. Insensibility to or absence of pain. a. **algera**, a. **dolorosa**, severe pain in a part with loss of general sensibility. a. **panaris**, synonym of *Morvan's disease*.

**analgesic** (an-al-je'-sik) [see *analgen*]. 1. Anodyne; relieving pain. 2. Affected with analgesia. 3. A remedy that relieves pain.

**analgesin** (an-al'-je-sin). See *antipyrine*.

**analgetic** (an-al-je'-tik). See *analgesic*.

**analgia** (an-al'-je-ah). [ἀν, priv.; ἀλγος, pain]. Absence of pain.

**analgic** (an-al'-jik) [see *analgen*]. Analgesic.

**analgin** (an-al'-jin). Synonym of *creolin*.

**analogue analog**, (an'-al-og) [ἀνάλογος, conformable]. A part or organ having the same function as another, but with a difference of structure. The correlative term, *homologue*, denotes identity of structure with difference of function. The wing of the butterfly and that of the bird are *analogous*, but the wing of a bird and the arm of a man are *homologous*.

**analogous** (an-al'-o-gus) [see *analogue*]. Conforming to, proportionate, answering to.

**analogy** (an-al'-o-je) [ἀνάλογος, conformable]. Similarity in function or origin between parts or organs, without identity.

**analosis** (an-al-o'-sis) [ἀνάλωσις, expenditure]. A wasting away; atrophy.

**analysis** (an-al'-is-is) [ἀναλύειν, to unloose]. The resolution of a compound body into its constituent parts. a., **absorptiometric**, the determination of the composition of gaseous bodies by observation of the amount of absorption which occurs on exposure to a liquid in which the coefficient of absorption of different gases is already known. a., **clinical**, a thorough examination of symptoms, lesions, and history to determine the nature of a disease and its cause. a., **densimetric**, analysis of a subject by means of determining the specific gravity of the solution and thus estimating the amount of dissolved matter. a., **dry**, that by means of blowpipe, etc.; also spectral analysis. a., **eudiometric**, see a., **gasometric**. a., **gasometric**, the determination of the constituents of gaseous compounds, especially the determination of the amount of oxygen in specimens of atmospheric air. a., **gravimetric**, the quantitative determination, by weight, of the elements of a body. a., **immediate**, see a., **proximate**. a., **indirect**, a quantitative estimation of the elements of a compound obtained not by isolating them, but by causing them to form new combinations and observing the relation of the molecular weight of these to that of the original body. a., **inorganic**, that of inorganic matter. a., **microchemical**, chemical analysis with the aid of a microscope. a., **organic**, the determination of the elements of matter found under the influence of life. The analysis of animal and vegetable tissues. a., **polariscopic**, analysis conducted with the polariscope. a., **prismatic**, spectral analysis. a., **proximate**, the determination of the simpler compound into which a substance may be resolved. a., **qualitative**, the determination of the nature of the elements that compose a body. a., **quantitative**, the determination of the proportionate parts of the various elements of a compound. a., **radiation**, a method of analysis based upon discoveries of Becquerel and taking advantage of the comparative radioactivity of various metals. a., **spectral**, the determination of the composition of a body by means of the spectroscope. a., **thermometric**, analysis by means of observation of the varying temperature produced by the interaction of substances mixed or combined together. a., **ultimate**, the resolution of a compound into its ultimate elements. a., **volumetric**, the quantitative determination of a constituent by volume. a., **wet**, analysis conducted by means of solutions and precipitations.

**analyst** (an'-al-ist). The person who makes an analysis; analyzer.

**analyzer** (an'-al-i-zer) [see *analysis*]. 1. An analyst. 2. In a polariscope, the Nicol prism, which exhibits the properties of light after polarization. 3. An apparatus for recording the excursions of tremor movements.

**Anam ulcer**. A form of phagedena, common in hot countries.

**Anamirta** (an-am-er'-ta). A genus of *Menispermaceæ*. *A. paniculata*, or *Menispermum cocculus*, is the source of cocculus indicus.

**anamirtin** (an-am-er'-tin) [*Anamirta*, a genus of plants], C₁₉H₃₄O₇. A glycerid derived from cocculus indicus, the berry-like fruit of *Anamirta paniculata*.

**anamnesia** (an-am-ne'-ze-ah). See *anamnesis*.

**anamnesis** (an-am-ne'-sis) [ἀνάμνησις, a recalling to mind]. 1. The faculty of memory; recollection. 2. That which is remembered; information gained from the patient and others regarding the past history of a case.

**anamnestic** (an-am-nes'-tik) [see *anamnesis*]. 1. Pertaining to the anamnesis, or history of a case. 2. Remembering. 3. Restorative of the memory.

**anamnionic** (*an-am-ne-on'-ik*). Same as *anamniotic*.
**anamniotic** (*an-am-ne-ot'-ik*) [ἀν, priv.; ἀμνίον, amnion]. Without an amnion.
**anamorphosis** (*an-am-orf-o'-sis*) [ἀνά, again; μορφοέιν, to form]. 1. Distortion or anomaly of development. In biology, gradual change of form in successive members of a group. 2. In optics, that process by which a distorted image is corrected by means of a curved mirror. **a.**, catoptric, correction of a distorted image by means of a conic or cylindric mirror. **a.**, dioptric, correction of a distorted image by means of a pyramidal glass.
**ananabasia** (*an-an-ab-a'-ze-ah*) [ἀν, priv.; ἀνάβασις, an ascending]. A form of abulia manifested by incapacity to ascend heights.
**ananaphylaxis** (*an-an-ah-fil-ak'-sis*). A condition which neutralizes anaphylaxis; it is wrongly termed antianaphylaxis.
**ananastasia** (*an-an-as-ta'-ze-ah*) [ἀν, priv.; ἀνάστασις, a rising up]. An abulistic inability to rise from a sitting posture.
**anandria** (*an-an'-dre-ah*) [ἀν, priv.; ἀνήρ, man]. Lack of virility; impotence.
**anangioplasia** (*an-an-je-o-pla'-se-ah*) [ἀν, priv.; ἀγγεῖον, a vessel; πλάσσειν, to form]. Congenital narrowing of the caliber of the blood vessels.
**anangioplasm** (*an-an'-je-o-plazm*) [ἀν, priv.; ἀγγεῖον, a vessel; πλάσμα, something formed]. Imperfect vascular development.
**anapeiratic** (*an-ap-i-rat'-ik*) [ἀναπειρᾶσθαι, to do again]. A condition due to excessive exercise, as in writers' cramp.
**anaphalantiasis** (*an-af-al-an-ti'-as-is*) [ἀνά, up; φάλανθος, bald in front]. The falling out of the eyebrows.
**anaphase** (*an'-af-āz*) [ἀνά, up; φάσις, a phase]. The phenomenon of karyokinesis immediately preceding the formation of the daughter-stars, and up to the formation of the resting daughter-nuclei.
**anaphia** (*an-a'-fe-ah*) [ἀν, priv.; ἀφή, touch]. 1. Defective sense of touch. 2. A state of abnormal sensitiveness to touch. 3. A state in which nothing is learned by palpation.
**anaphora** (*an-af'-or-ah*) [ἀναφορά, a bringing up]. 1. A bringing up, as by coughing. 2. Recovery from illness. 3. Rush of blood to the head. 4. A violent inspiration or respiration.
**anaphoresis** (*an-af-or-e'-sis*) [ἀν, priv.; φέρειν, to carry]. A diminution in the activity of the sweat-glands.
**anaphoretic** (*an-ah-for-et'-ik*). 1. Checking perspiration. 2. An agent that checks the secretion of sweat.
**anaphoria** (*an-af-o'-re-ah*). An upward tendency of the eyes and of the visual axes.
**anaphrodisia** (*an-af-ro-diz'-e-ah*) [ἀν, priv.; Ἀφροδίτη, Venus]. Absence or impairment of sexual appetite.
**anaphrodisiac** (*an-af-ro-diz'-e-ak*). 1. Relating to, affected by, or causing anaphrodisia. 2. An agent that allays the sexual desire.
**anaphrodite** (*an-af'-ro-dīt*). An individual affected with anaphrodisia.
**anaphylactic** (*an-ah-fil-ak'-tik*) [ἀν, priv.; φύλαξ, a guardian]. 1. Having the property of diminishing immunity instead of reinforcing it. 2. A serum which diminishes immunity. **a. shock**, the general condition produced by the repeated injections of foreign serum.
**anaphylactin** (*an-ah-fil-ak'-tin*). A substance supposed to produce anaphylaxis; "a toxic, or irritating nonassimilable substance, assumed to be part of the proteid introduced on first injection, which renders the tissue cells abnormally susceptible to reinjections of the same substance."
**anaphylatoxin** (*an-ah-fil-ah-tok'-sin*), The poisonous substance which produces the symptoms in anaphylaxis; it is non-specific, and is supposed to be formed by anaphylactin and the newly injected protein.
**anaphylaxis** (*an-ah-fil-ak'-sis*). Induction of disease; specifically, an intoxication due to the union of a foreign substance with antibodies produced by previous introduction of the same substance; opposed to prophylaxis.
**anaphylaxy** (*an-ah-fil-aks'-e*). See *anaphylaxis*.
**anaplase** (*an'-ap-lāz*) [ἀνά, up; πλάσσειν, to build]. The stage of growth and development; the period before full maturity.

**anaplasia** (*an-ah-pla'-ze-ah*). 1. The tendency of certain tissues toward reversion to an earlier or embryonal type. 2. A similar tendency in cells to revert to a less differentiated condition, prior to cell division.
**anaplasis** (*an-ah-pla'-sis*). See *anaplasty*.
**anaplasm** (*an'-ah-plazm*). See *anaplasty*.
**anaplast** (*an'-ap-last*) [ἀναπλάσσειν, to shape]. See *leukoplast*.
**anaplastic** (*an-ap-las'-tik*) [ἀναπλάσσειν, to build up]. 1. Relating to anaplasty; restoring a lost or defective part. 2. Agent that facilitates repair. **a. surgery**, anaplasty.
**anaplasty** (*an'-ap-las-te*). An operation for the restoration of lost parts; plastic surgery.
**anaplerosis** (*an-ap-le-ro'-sis*) [ἀνά, up; πληροῦν, to fill]. The restoration, or repair of a wound, sore, or lesion in which there has been a loss of substance.
**anaplerotic** (*an-ap-le-rot'-ik*) [ἀνά, up; πληροῦν, to fill]. 1. Promotive of repair, favoring granulation. 2. A remedy or application that promotes repair.
**anapnograph** (*an-ap'-no-graf*) [ἀναπνοή, respiration; γράφειν, to write]. An apparatus registering the movements of (1) inspiration and expiration, (2) the quantity of air inhaled.
**anapnoic** (*an-ap-no'-ik*) [ἀνά, against; ἄπνοια, want of breath]. 1. Relieving dyspnea. 2. Favoring respiration.
**anapnometer, anapneometer** (*an-ap-nom'-et-er, an-ap-ne-om'-et-er*) [ἀναπνοή, respiration; μέτρον, a measure]. An anapnograph.
**anapophysis** (*an-ap-of'-is-is*) [ἀνά, back; ἀπόφυσις, an offshoot]. An accessory process of a lumbar or dorsal vertebra, corresponding to the inferior tubercle of the transverse process of a typical dorsal vertebra.
**anaptic** (*an-ap'-tik*) [ἀν, priv.; ἀφή, touch]. Pertaining to or marked by anaphia: loss of the tactile sense.
**anarcotine** (*ah-nar'-ko-tin*) [ἀ, priv.; narcotic]. Narcotine, which from its lack of narcotic power is mis-named.
**anarithmia** (*an-ar-ith'-me-ah*). An inability to count.
**anarrhea**, or **anarrhœa** (*an-ar-e'-ah*) [ἀνά, up; ῥοία, flow]. Afflux to an upper part, as of blood to the head.
**anarrhexis** (*an-ar-eks'-is*) [ἀνά, up; ῥῆξις, fracture]. Surgical refracture of a bone.
**anarthria** (*an-ar'-thre-ah*) [ἀν, priv.; ἄρθρον, articulation]. 1. Defective articulation. 2. Absence of vigor. 3. Without joints. **a. centralis**, partial aphasia due to central lesion. **a. literalis**, stammering.
**anarthrous** (*an-ar'-thrus*) [ἀν, priv.; ἄρθρον, a joint]. Jointless. So corpulent that no joints are visible. 2. Lacking vigor. 3. Inarticulate.
**anasarca** (*an-ah-sar'-kah*) [ἀνά, through; σάρξ, the flesh]. An accumulation of serum in the subcutaneous areolar tissues of the body. Syn., *catasarca; episarcidium; hydrodermus; intercus; hydrops cellularis*. **a.**, acute, a form in which the flesh preserves its normal color and the depression made by the finger disappears quickly. **a. a fluxu**, that due to loss of body-fluids, as in diarrhea or diabetes. **a. americana**, South American disease marked by sleepiness, headache, debility, and swelling of the abdomen, said to be due to the ingestion of sea-crabs. **a. essential**, that due to malnutrition. **a. exanthematica**, that attributed to the suppression of an exanthem, especially erysipelas. **a. urinosa**, that due to suppression of urine. Syn., *utinary leukophlegmasia*.
**anasarcin** (*an-ah-sar'-sin*). A remedy for dropsy, said to consist of the active principles of *Oxydendron arboreum, Sambucus nigra*, and *Urginea scilla*. Trade name of a remedy claiming to be a heart tonic and diuretic.
**anasarcous** (*an-ah-sar'-kus*) [see *anasarca*]. Affected with anasarca.
**anasomia** (*an-ah-so'-me-ah*) [ἀνά, up; σῶμα, body]. A deformed condition in which the limbs are abnormally adherent to the body.
**anaspadiac** (*an-ah-spa'-di-ak*)]. A person affected with anaspadias.
**anaspadias** (*an-as-pa'-de-as*) [ἀνά, up; σπάειν, to draw]. A urethral opening upon the upper surface of the penis.
**anaspasis** (*an-ah-spa'-sis*) [see *anaspadias*]. 1. A contraction. 2. Revulsion.
**anastalsis** (*an-as-tal'-sis*). A term suggested by Cannon for the upward moving wave of contraction

occurring in the first part of the colon during digestion. There is no preceding wave of inhibition.

**anastaltic** (an-as-tal'-tik) [ἀνασταλτικός, checking; putting back]. 1. Strongly astringent. 2. Centripetal; afferent.

**anastasis** (an-as'-tas-is) [ἀνάστασις, a setting up]. 1. Recovery; convalescence. 2. An upward afflux of the body humors. 3. Resuscitation of one apparently dead.

**anastate** (an'-as-tāt) [ἀνάστατος, caused to rise]. Any substance that appears in or is characteristic of an anabolic process.

**anastatic** (an-as-tat'-ik) [see anastasis]. Tending to recovery; restorative.

**anastigmatic** (an-ah-stig-mat'-ik). Free from astigmatism; said especially of photographic objectives which are corrected for astigmatism as well as for spheric and chromatic aberration.

**anastole** (an-as'-to-le) [ἀναστολή, retracted]. Retraction; shrinking away, as of the lips of a wound.

**anastomose** (an-as'-to-mōz) [see anastomosis]. To produce anastomosis; to communicate by anastomosis.

**anastomosis** (an-as-to-mo'-sis) [ἀναστομόειν, to bring to a mouth]. 1. The intercommunication of blood-vessels. 2. The establishment of a communication between two hollow parts or between two distinct portions of the same organ. See a., intestinal. 3. A whetting of the appetite. a., crucial, an arterial anastomosis in the upper part of the thigh, formed by the anastomotic branch of the sciatic, the first perforating, the internal circumflex, and the transverse branch of the external circumflex arteries. a., entero-. See enteroanastomosis. a., intestinal, an operation consisting in establishing a communication between two parts of the intestine.

**anastomotic** (an-as-to-mot'-ik) [see anastomosis]. 1. Pertaining to anastomosis. 2. Sharpening the appetite. 3. Aperient. 4. Causing dilation of the peripheral blood-vessels. 5. A communicating artery or vein. See under artery and under vein.

**anastomotica** (an-as-to-mot'-ik-ah). 1. A communicating artery or vein. 2. Tonic, aperient, or deobstruent medicines. a. magna, see under artery.

**anastomotris** (an-as-to-mo'-tris) [L.; pl., anastomotridēs]. Any kind of a dilating instrument.

**anastrophe** (an-as'-tro-fe) [ἀναστρέφειν, to turn upside down]. Inversion, particularly of the viscera.

**anatherapeusis** (an-ath-er-ap-u'-sis) [ἀνά, up; θεραπεύσις, medical treatment]. Treatment by increasing doses.

**anathrepsis** (an-ath-rep'-sis) [ἀναθρεψις, a fresh growth]. A renewal of lost flesh after recovery.

**anathreptic** (an-ath-rep'-tik) [ἀναθρεψις, a fresh growth]. Restorative of lost flesh; nutritive.

**anatomical, anatomic** (an-at-om'-ik-al, an-at-om'-ik) [anatomy]. Pertaining to anatomy. a. tubercle, see verruca necrogenica.

**anatomicochirurgical** (an-a-tom-ik-o-ki-rur'-jik-al). Relating to anatomy and surgery.

**anatomicomedical** (an-at-om-ik-o-med'-ik-al). Relating to medicine and anatomy or to medical anatomy.

**anatomicopathological** (an-at-om-ik-o-path-o-loj'-ik-al). Pertaining to anatomy and pathology.

**anatomicophysiological** (an-at-om-ik-o-fiz-e-o-loj'-ik-al). Relating to anatomy and physiology.

**anatomicosurgical** (an-at-om-ik-o-sur'-je-kal). Relating to anatomy and surgery.

**anatomist** (an-at'-om-ist) [see anatomy]. One who is expert in anatomy. a.'s snuff-box, the triangular space between the tendons of the extensor of the metacarpal bone of the thumb and the extensor of the first phalanx on the back of the hand.

**anatomize** (an-at'-om-īz). To dissect.

**anatomy** (an-at'-o-me) [ἀνά, up; τέμνειν, to cut]. The science of the structure of organs or of organic bodies. a., applied, anatomy as concerned in the diagnosis and treatment of pathological conditions. a., artistic, that branch of anatomy treating of the external form of men and animals, their osseous and muscular systems, and the relative size of different parts and members of their bodies. a., comparative, the investigation and comparison of the anatomy of different orders of animals or of plants, one with another. a., descriptive, a study of the separate and individual portions of the body, apart from their relationship to surrounding parts. a., general, that branch of descriptive anatomy treating of the structure and physiological properties of the tissues and their arrangement into systems without regard to the disposition of the organs of which they form a part. a., gross, anatomy dealing with the naked-eye appearance of tissues. a., homological, the study of the correlations of the several parts of the body. a., medical, the application of anatomy to a study of the causation and symptomatology of nonsurgical diseases. a., microscopical, a., minute, that studied under the microscope. a., morbid, a., pathological, a study of diseased structures. a., physiognomonical, the study of expressions depicted upon the exterior of the body, especially upon the face. a., physiological, an anatomical study of tissues in respect to their functions. a., practical, dissection. a., regional, a study of limited parts or regions of the body, the divisions of which are collectively or peculiarly affected by disease, injury, operations, etc. a., surgical, the application of anatomy to surgery. a., topographical, the anatomy of a part in its relation to other parts. a., transcendental, the study of the general design of the body, and of the particular design of the organs. Anatomy as related to theories of type, and evolution. a., vegetable, the branch of botany which treats of the relative position, form, and structure of the organs of plants. a., veterinary, the anatomy of domestic animals.

**anatresis** (an-at-re'-sis) [ἀνατιτρᾶν, to bore through]. Perforation; trephining.

**anatricrotic pulse** (an-ah-tri-krot'-ik). A pulse wave with three breaks on the ascending curve.

**anatripsis** (an-at-rip'-sis) [ἀνάτριψις, a rubbing]. 1. Rubbing; the removal of a part or growth by scraping or rubbing; inunction. 2. An upward or centripetal movement in massage. 3. A crushing, as of calculi. 4. Itching; scratching to allay itching.

**anatriptic** (an-at-rip'-tik) [see anatripsis]. A medicine to be applied by rubbing.

**anaxon, anaxone** (an-aks'-on) [ἀν, priv.; axis]. A neuron devoid of axis-cylinder processes. Syn., amacrine cell.

**anazotic** (an-az-ot'-ik) [ἀν, priv.; azotum, nitrogen]. Without azote or nitrogen.

**anazoturia** (an-az-o-ū'-re-ah) [ἀν, priv.; azotum, nitrogen; οὖρον, urine]. A condition of deficient excretion of nitrogen in the urine, the urea being chiefly diminished.

**anazyme** (an'-a-zīm). The commercial name for a combination of carbolic and boric acids; it is a substitute for iodoform.

**AnCC.** Abbreviation for anodal closure contraction.

**anchilops** (ang'-kil-ops). See anchylops.

**anchone** (ang'-ko-ne) [ἀγχειν, to strangle]. A spasmodic constriction of the throat observed in hysteria.

**anchorage** (ang'-kor-aj). 1. The fixation of a floating or displaced viscus, whether by a natural process or by surgical means. 2. In dentistry, the means adopted for the retention of a dental filling, particularly its initial portion.

**anchoralis** (an-ko-ra'-lis) [ancora, an anchor]. The coronoid process of the ulna.

**anchusin** (ang'-kū-sin) [ἄγχουσα, alkanet]. C₁₅H₁₄O₅. The red coloring-matter found in alkanet-root. See alkanet.

**anchyloblepharon** (ang-kil-o-blef'-ar-on). See ankyloblepharon.

**anchyloglossia** (ang-kil-o-glos'-e-ah). See ankyloglossia.

**anchylops** (ang'-kil-ops) [ἄγχι, near; ὤψ, the eye]. Abscess at inner angle of eye, prior to rupture.

**anchylosis** (ang-kil-o'-sis). See ankylosis.

**anchylostomiasis** (ang-kil-o-sto-mi'-as-is). See ankylostomiasis.

**anchylostomum** (ang-kil-os'-to-mum). See ankylostoma.

**anciptal** (an-sip'-it-al) [anceps, double]. Two-edged.

**ancistrum** (an-sis'-trum) [ἄγκιστρον, a fish-hook]. A surgical hook.

**ancon** (ang'-kon) [ἀγκών, the elbow]. Originally the olecranon process; applied to the elbow generally.

**anconad** (ang'-ko-nad) [ἀγκών, the elbow]. Toward the olecranon, or elbow.

**anconagra** (ang-kon-a'-grah) [ἀγκών, the elbow; ἄγρα, a seizure]. Arthritic pain at the elbow.

**anconal, anconeal** (ang'-kon-al, ang-ko'-ne-al) [ἀγκών, the elbow]. Pertaining to the elbow.

**anconen** (*ang'-kon-en*) [ἀγκών, the elbow]. Belonging to the ancon in itself.
**anconeus** (*ang-ko-ne'-us*). See under *muscle.*
**anconoid** (*ang'-kon-oid*) [ἀγκών, the elbow]. Resembling the elbow.
**ancyloglossum** (*an-sil-o-glos'-um*). See *tongue-tie.*
**ancylomele** (*an-sil-o-me'-le*). See *ankylomele.*
**Ancylostoma.** See *Ankylostoma.*
**ancyra** (*an'-si-rah*) [ἄγκυρα, an anchor]. A hook.
**ancyroid** (*an'-sir-oid*) [ἄγκυρα, anchor; εἶδος, form]. Shaped like an anchor.
**Andernach's ossicles.** See *Wormian bones.*
**Andersch's ganglion** [Carl Samuel *Andersch,* German anatomist]. The petrosal ganglion. A.'s nerve, see *Jacobson's nerve.*
**Anderson's pill.** The compound gamboge pill. A.'s reaction for distinguishing between quinoline and pyridine salts, the chloroplatinates of the latter, when boiled with water, are changed into insoluble double salts with the elimination of hydrogen chloride, whereas the former remain in solution.
**andolin** (*an'-do-lin*). Trade name for a mixture of anesthetics for spinal analgesia. It is said to contain eucaine, stovaine, adrenalin hydrochloride and saline solution.
**Andral's decubitus** [Gabriel *Andral,* French physician, 1797–1876]. The position usually assumed in the early stage of pleurisy by the patient, who seeks to alleviate the pain by lying on the sound side.
**andranatomy** (*an-dran-at'-o-me*) [ἀνήρ, a man; ἀνατομία, anatomy]. Human anatomy; the anatomy or dissection of the male human subject.
**Andreasch's reaction** for cystein. To the hydrochloric acid solution add a few drops of dilute ferric chloride solution and then ammonia. The liquid will become a dark purplish red.
**androgalactozemia** (*an-dro-gal-ak-to-ze'-me-ah*) [ἀνήρ, man; γάλα, milk; ζημία, loss]. The oozing of milk from the male mamma.
**androgenous** (*an-droj'-en-us*) [ἀνήρ, a man; γεννᾶν, to bear]. Giving birth to males.
**androgyna** (*an-droj'-in-ah*) [ἀνήρ, a man; γυνή, woman]. A hermaphrodite; a female in whom the genital organs are similar to those of the male.
**androgyneity** (*an-droj'-in-e-it-e*) [see *androgyna*]. Hermaphroditism.
**androgynism** (*an-droj'-in-izm*) [ἀνήρ, man; γυνή, woman]. Hermaphroditism.
**androgynous** (*an-droj'-in-us*) [ἀνήρ, man; γυνή, woman]. Hermaphrodite. Having the characteristics of both sexes.
**androgynus** (*an-droj'-in-us*) [see *androgyna*]. A hermaphrodite. A male with genital organs similar to those of the female.
**androlepsia** (*an-dro-lep'-si-ah*) [ἀνδρολημψία, a seizure; of men]. The process of fecundation in the female.
**andrology** (*an-drol'-o-je*) [ἀνήρ, man; λόγος, science]. 1. The science of man, especially of the male sex. 2. The science of the diseases of the male genitourinary organs.
**andromania** (*an-dro-ma'-ne-ah*) [ἀνήρ, a man; μανία, madness]. Nymphomania.
**andrometoxin** (*an-drom-et-oks'-in*) [*Andromeda;* τοξικόν, poison]. A poisonous anodyne principle found in *Andromeda japonica,* occurring in *Kalmia latifolia* and some other ericaceous plants and found in poisonous honey from Trebizond.
**andromorphous** (*an-dro-mor'-fus*) [ἀνήρ, man; μορφή, form]. Shaped like a man.
**androphobia** (*an-dro-fo'-be-ah*) [ἀνήρ, a man; φόβος, fear]. Fear or dislike of the male sex.
**androphonomania** (*an-dro-fo-no-ma'-ne-ah*) [ἀνδρόφονος, man-killing; μανία, madness]. Homicidal insanity.
**androsymphysia, androsymphysis** (*an-dro-sim-fiz'-e-ah, an-dro-sim'-fiz-is*) [ἀνήρ, a man; σύν, together; φύειν, to grow]. 1. A monstrosity formed by the fusion of two male fetuses. 2. The growing together of the male genitalia.
**-ane.** A suffix indicating a saturated hydrocarbon.
**anebous** (*an-e'-bus*) [ἄνηβος]. Not come to man's estate; not having reached puberty; immature.
**anecpyetous** (*an-ek-pi-e'-tus*) [ἀνεκπύητος]. 1. Not suppurating. 2. Preventing suppuration; insuppurable.
**anectasia** (*an-ek-ta'-se-ah*). See *anectasis.*
**anectasin** (*an-ek'-ta-sin*) [ἀν, priv.; ἐκ, out of; τείνειν, to stretch]. A product of bacterial action with an influence on the vasomotor nerves contrary to *ectasin* (*q. v.*).

**anectasis** (*an-ek'-tas-is*) [ἀν, priv.; ἔκτασις, extension]. Deficient size of an organ or part.
**anedemin** (*an-e-de'-min*). Trade name of a dropsy remedy. It is said to contain squill, strophanthus, apocynum, and sambucus.
**anedeus** (*an-e'-de-us*) [ἀ, priv.; αἰδοῖα, the genitals]. Lacking genital organs.
**aneilema** (*an-i-le'-mah*) [ἀνά, up; εἰλεεῖν, to roll]. Flatulence; air or wind in the bowels; colic.
**aneilesis** (*an-i-le'-sis*) [ἀνειλεεῖν, to roll together]. 1. See *aneilema.* 2. Twisting of the body in athletics. 3. Evolution.
**Anel's operation** for **aneurysm** [Dominique *Anel,* French surgeon, 1628-1725]. Ligation on the cardiac side close to the aneurysm. A.'s probe, A.'s sound, a fine probe used for exploring or dilating the lacrimal puncta and lacrimal canals. A.'s syringe, a syringes used in injecting fluids into the lacrimal passages.
**anelectric** (*an-el-ek'-trik*) [ἀν, priv.; ἤλεκτρον, amber]. 1. Readily giving up electricity. 2. A good conductor; a substance which readily parts with electricity.
**anelectrode** (*an-el-ek'-trōd*) [ἀνά, upward; *electrode*]. The positive pole of a galvanic battery; anode.
**anelectrotonic** (*an-el-ek-tro-ton'-ik*) [ἀν, priv.; ἤλεκτρον, electricity; τόνος, tension]. Relating to anelectrotonus.
**anelectrotonus** (*an-el-ek-trot'-o-nus*) [see *anelectrotonic*]. The decreased irritability that is present in a nerve in the neighborhood of the anode.
**anematosis, anæmatosis** (*an-e-ma-to'-sis*). 1. General anemia. 2. Idiopathic anemia.
**anemia** (*an-e'-me-ah*) [ἀν, priv.; αἷμα, blood]. Deficiency of blood as a whole, or deficiency of the number of the red corpuscles or of the hemoglobin. It may be *general* or *local.* Local anemia, or *ischemia,* is the result of mechanical interference with the circulation of the affected part. General anemia is either idiopathic or symptomatic. **a., aplastic,** anemia in which the formative processes in the bone marrow do not take place. **a., cytogenic,** synonym of *a., idiopathic.* **a., essential,** synonym of *a., idiopathic.* **a., idiopathic,** a form in which the lesion is in the blood or in the blood-making organs. **a., infantum pseudoleukæmica,** a form of primary anemia described by von Jaksch as peculiar to the young child. Morse holds that chlorosis is a condition wholly foreign to infantile life and that von Jaksch's disease does not represent a distinct clinical entity. **a., lymphatic,** synonym of *Hodgkin's disease;* see *lymphadenoma.* **a., malignant,** see *pernicious anemia.* **a., miners,** see *uncinariasis.* **a., myelogenous,** anemia attended with hyperplasia of myelogenous tissue. **a., paludal,** anemia associated with or caused by malaria. **a., pernicious,** see *pernicious anemia.* **a., primary,** see *a., idiopathic.* **a., secondary,** that due to a distinct cause, as hemorrhage, cancer, wasting discharges, poisons, etc. Syn., *symptomatic anemia.* **a., septic,** one which is septic to secondary conditions, usually about the mouth. **a., splenic,** chronic anemia with enlarged spleen, blood-changes, chloranemia, leukopenia, hemorrhages from the stomach, and pigmentation of the skin. **a., symptomatic,** see *a., secondary.* **a., tumid,** see *uncinariasis.*
**anemic** (*an-em'-ik*) [see *anemia*]. Pertaining to anemia. **a. infarct,** a wedge-shaped area of coagulation-necrosis occurring in organs possessing terminal arteries. It is the result of the sudden stopping of such an artery by a thrombus or an embolus. **a. murmur,** a murmur heard in anemic conditions, soft and blowing in character, and disappearing with the anemia. It is generally heard over the base of the heart. **a. necrosis,** the coagulation-necrosis of tissues resulting from the sudden stoppage of the supplying artery.
**anemometer** (*an-e-mom'-et-er*) [ἄνεμος, wind; μέτρον, a measure]. An instrument for measuring the velocity of the wind.
**Anemone** (*an-em'-o-ne*) [ἀνεμώνη, the wind-flower]. A genus of ranunculaceous herbs, most of which have active medicinal and poisonous qualities. See *pulsatilla.*
**anemonin** (*an-em'-o-nin*) [see *anemone*], $C_{15}H_{12}O_6$. The active principle of the anemone. It is given in bronchitis, asthma, and spasmodic cough. Dose $\frac{1}{4}-\frac{1}{2}$ gr. (0.016-0.048 Gm.) twice daily.
**anemonol** (*an-em'-on-ol*) [ἀνεμώνη, wind-flower; *oleum,* oil]. The volatile oil extracted from anemone; it is a powerful vesicant.

**anemopathy** (an-em-op'-ath-t) [ἄνεμος, wind; πάθος, disease]. Therapeutic treatment by inhalation.
**anemophobia** (an-e-mo-fo'-be-ah) [ἄνεμος, wind; φόβος, fear]. Morbid dread of draughts or of winds.
**anemotrophy,** or **anæmotrophy** (an-em-ot'-ro-fe) [ἀν, priv.; αἷμα, blood; τροφή, nourishment]. A deficiency of blood nourishment; an impoverished state of the blood.
**anemydria, anæmydria** (an-em-id'-re-ah) [ἀ, priv.; αἷμα, blood; ὕδωρ, water]. Insufficiency of the watery element in blood.
**anencephalia** (an-en-sef-a'-le-ah) [ἀν, priv.; ἐγκέφαλος, brain]. Congenital absence of the brain.
**anencephalic** (an-en-sef-al'-ik) [ἀν, priv.; ἐγκέφαλος, brain]. Pertaining to or characterized by anencephalia.
**anencephalohemia** (an-en-sef-al-o-he'-me-ah) [ἀν, priv.; ἐγκέφαλος, brain; αἷμα, blood]. Insufficiency of blood in the brain.
**anencephaloid** (an-en-sef'-al-oid) [ἀν, priv.; ἐγκέφαλος, brain]. Pertaining to anencephalia.
**anencephaloneuria** (an-en-sef-al-on-ū'-re-ah). [ἀν, priv.; ἐγκέφαλος, brain; νεῦρον, a nerve]. Imperfect nerve-action of the brain.
**anencephalotrophia,** or **anencephalotrophy** (an-en-sef-al-o-tro'-fe-ah or -lot'-ro-fe) [ἀν, priv.; ἐγκέφαλος, brain; τροφή, nutrition]. Atrophy, or lack of nutrition of the brain.
**anencephalus** (an-en-sef'-al-us). [see *anencephalia*]. A species of single autositic monsters in which there is no trace of the brain.
**anenergia** (an-en-er'-je-ah) [ἀν, priv.; ἐνέργεια, energy]. Lack of vigor or power.
**anenteremia** (an-en-ter-e'-me-ah) [ἀν, priv.; ἔντερον, an intestine; αἷμα, blood]. Bloodless condition of the bowels.
**anenteroneuria** (an-en-ter-o-nu'-re-ah) [ἀν, -priv.; ἔντερον an intestine; νεῦρον, a nerve]. Intestinal atony.
**anenterotrophia** (an-en-ter-o-tro'-fe-ah) [ἀν, priv.; ἔντερον, an intestine; τροφή, nourishment]. Defective intestinal nutrition.
**anenterous** (an-en'-ter-us) [ἀν, priv.; ἔντερον, intestine]. In biology, having no intestine, as a tapeworm or a fluke.
**anepia** (an-ep'-e-ah) [ἀνεπής, speechless]. Inability to speak.
**anepiploic** (an-ep-ip-lo'-ik) [ἀν, priv.; ἐπίπλοον, the caul]. Having no epiploon or omentum.
**anepithymia** (an-ep-e-thim'-e-ah) [ἀν, priv.; ἐπιθυμία, desire]. Loss of any natural appetite.
**anerethisia** (an-er-eth-iz'-e-ah) [ἀν, priv.; ἐρεθίζειν, to excite]. Imperfect irritability, as of a muscle or nerve.
**anergasis** (an-er'-ga-sis) [see *anergia*]. Absence of functional activity.
**anergia** (an-er'-je-ah) [ἀν, priv.; ἔργον, work]. Sluggishness; inactivity.
**anergic** (an-er'-jik) [see *anergia*]. Characterized by sluggishness; as, *anergic* dementia.
**aneroid** (an'-er-oid) [ἀ, priv.; νηρός, wet; εἶδος, form]. Working without a fluid. a. barometer, see *barometer*.
**anerythroblepsia** (an-er-ith-ro-blep'-se-ah). Same as *anerythropsia*.
**anerythropsia** (an-er-ith-rop'-se-ah) [ἀν, priv.; ἐρυθρός, red; ὄψις, sight]. Impaired color-perception of red.
**anesin, aneson** (an'-es-in, an'-es-on). A proprietary aqueous solution of acetone-chloroform; used as a hypnotic and local anesthetic.
**anesis** (an'-es-is) [ἄνεσις, remission]. An abatement or relaxation in the severity of symptoms.
**anesthecinesis, anæsthecinesis** (an-es-the-sin-e'-sis) [ἀ, priv.; αἴσθησις, feeling; κίνησις, movement]. A condition marked by loss of sensibility and motor capacity.
**anesthesia, anæsthesia** (an-es-the'-se-ah) [ἀναισθησία, want of feeling]. A condition of total or partial insensibility, particularly to touch. **a. angiospas'-tica**, loss of sensibility due to spasm of blood-vessels. **a., bul'bar,** that due to a lesion in the medulla oblongata. **a., central,** due to disease in the nerve-centers. **a., cerebral,** that due to disease of the cerebrum. **a., crossed,** anesthesia on one side of the body, due to a central lesion of the other side. **a., disso'ciated,** loss of pain and temperature sensations, the tactile sense being still present. **a. dolorosa,** severe pain experienced after the occurrence of complete motor and sensory paralysis, a symptom observed in certain diseases of the spinal cord. **a., dolorous** (of Liebreich), the transient but painful anesthesia produced by the injection of water in sufficient quantity to edematize the papillary layer of the derma and subjacent layers. The pain is due to the inhibitory swelling of the cells. **a., efferent,** that due to disorder of the nerve-terminations, disturbing their conductivity. **a., electric,** anesthesia caused by the passage of an electric current through a part. **a., facial,** anesthesia of those parts to which the sensory branches of the fifth cranial nerve are distributed. **a., general,** anesthesia of the entire body, including the abolition of all perceptive power with consequent loss of consciousness. **a., girdle,** a zone of anesthesia encircling the body, due to circumscribed disease of the spinal cord. **a., infiltration-,** local anesthesia effected by subcutaneous injections. **a., intraneural,** local anesthesia effected by injection into a nerve trunk. **a., Javanese,** that produced by pressure upon the carotids. **a., local,** that limited to a part of the body. **a., mixed,** that partially produced and prolonged by the administration of morphine or other cerebral anodyne before the anesthetic is given. **a., muscular,** loss of the muscular sense. **a., olfactory,** anosmia. **a., optic,** amaurosis. **a., partial,** anesthesia in which some degree of sensibility is still present. **a., peripheral,** that depending upon changes in the peripheral nerves. **a., primary,** a temporary insensibility to slight pain occurring in the beginning of anesthesia and during which minor operations can be performed. **a., rectal,** that produced by the injection of an anesthetic agent into the rectum. **a., regional,** that limited to a part of body supplied by an afferent nerve which has been cocainized. **a., sexual,** anaphrodisia. **a., spinal,** (1) that due to a lesion of the spinal cord; (2) that produced by the injection of an anesthetic into the spinal subarachnoid space. **a., surgical,** that induced by the surgeon by means of anesthetics for the purpose of preventing pain, producing relaxation of muscles, or for diagnostic purposes. **a., tactile,** loss of sense of touch. **a., thermic,** loss of temperature sense. **a., unilateral,** hemianesthesia.
**anesthesimeter** (an-es-thes-im'-et-er) [*anesthesia*; μέτρον, a measure]. An instrument to measure the amount of an anesthetic administered in a given time.
**anesthesin** (an-es'-thes-in). Paramidobenzoic acid ester; it is used as a local anesthetic, also, internally, for gastralgia.
**anesthesiology** (an-es-the-ze-ol'-o-je) [*anesthesia*; λόγος, science]. The science of anesthesia and anesthetics.
**anesthetic** (an-es-thet'-ik) [see *anesthesia*]. 1. Without feeling; insensible to touch or pain. 2. A substance that produces insensibility to touch or to pain, diminished muscular action, and other phenomena. Anesthetics may be general, local, partial, and complete. **a., general,** one used for securing general anesthesia. **a. (general) mixtures,** contain combinations of substances for producing anesthesia. **a., local,** an anesthetic that, locally applied, produces absence of sensation in the organ or tissue so treated.
**anestheticism** (an-es-thet'-is-izm) [*anesthetic*]. The quality of being anesthetic.
**anesthetization** (an-es-thet-iz-a'-shun) [ἀναίσθητος, insensible]. The act of placing under the influence of an anesthetic.
**anesthetize** (an-es'-thet-iz) [see *anesthetization*]. To put under the influence of an anesthetic.
**anesthetist, anesthetizer** (an-es'-thet-ist, an-es'-thet-i-zer) [see *anesthetization*]. One who administers an anesthetic.
**anesthol** (an-es'-thol). A trade name for a mixture of ether, chloroform and ethyl chloride. The proportions of ether and chloroform vary; the ethyl chloride is 17 per cent. It is used as a general anesthetic.
**anesthyl** (an-es'-thil). A local anesthetic said to consist of ethyl chloride, 5 parts; methyl chloride, 1 part.
**anethol** (an'-eth-ol) [*anethum;* oleum, oil]. $C_{10}H_{12}O$. The chief constituent of the essential oils of anise and fennel. It is employed in preparing the *elixir anethi* (N. F.), being more fragrant and agreeable than the anise oil. **a., liquid,** an isomeric modification of anethol; it is an antiseptic, oil-like liquid. Syn., *isanethol*.
**anethum** (an-e'-thum) [ἀνά, up; αἴθειν, to burn, from the pungency of the seeds]. Dill; the dried

fruit of *Peucedanum graveolens*, indigenous to southern Europe. It is aromatic, carminative, and stimulant. Dose of the *oil* (*oleum anethi*, B. P.) 1-4 min. (0.06-0.24 Cc.); of the *water* (*aqua anethi*, B. P.) 1-2 oz. (30-60 Cc.).

**anetic** (*an-et'-ik*) [ἀνετικός, relaxing]. Soothing; calmative; anodyne.

**anetiological** (*an-e-te-o-loj'-ik-al*) [ἀν, priv.; αἰτία, cause; λόγος, word]. Having no known cause; dysteleological.

**anetodermia** (*an-et-o-der'-me-ah*) [ἀνετός, relaxed; δέρμα, skin]. Relaxation of the skin.

**anetus** (*an'-et-us*) [ἀνετός, loosened]. Any intermittent fever.

**aneuria** (*ah-nū'-re-ah*) [ἀ, priv.; νεῦρον, a nerve]. Lack of nervous power.

**aneuric** (*ah-nū'-rik*) [see *aneuria*]. Characterized by aneuria.

**aneurism** (*an'-ū-rizm*). See *aneurysm*.

**aneuros** (*ah-nū'-ros*) [ἄνευρος, without sinews]. Feeble, inelastic, relaxed.

**aneurosis** (*ah-nū-ro'-sis*) [ἀ, priv.; νεῦρον, a nerve]. A lack of nerves.

**aneurysm** (*an'-ū-rizm*) [ἀνεύρυσμα, a widening]. A circumscribed dilatation of the walls of an artery. Syn., *Abscessus spirituosus*. **a., abdominal**, an aneurysm of the abdominal aorta. **a., active**, cardiac dilation with hypertrophy. **a., acute**, an ulceration of the heart-wall which, by communicating with one of the chambers of the heart, forms an aneurysmal pouch. **a., ampullary**, a small saccular aneurysm; it is most common in the arteries of the brain. **a.** by **anastomosis**, a dilatation of a large number of vessels,—small arteries, veins, and capillaries,—the whole forming a pulsating tumor under the skin. This form of aneurysm is especially seen upon the scalp. **a., arteriovenous**, the simultaneous rupture of an artery and a vein, the blood from both being poured out into the cellular tissue and forming a false aneurysm. A *varicose aneurysm* is produced by the rupture of an aneurysm into a vein. An *aneurysmal varix* results from the establishment of a communication between an artery and a vein, the latter becoming dilated and pulsating. **a., cardiac**, an aneurysm of the heart. **a., circumscribed**, an aneurysm, either true or false, in which the contents are still within the artery though there may be rupture of one or two of its coats. **a., cirsoid**, a tortuous lengthening and dilatation of a part of an artery. **a., compound**, one in which one or several of the coats of the artery are ruptured and the others merely dilated. **a., consecutive**, **a., diffused**, follows rupture of all the arterial coats, with infiltration of surrounding tissues with blood. **a., dissecting**, one in which the blood forces its way between the coats of an artery. **a., ectatic**, an expansion of a portion of an artery due to yielding of all the coats. **a., endogenous**, one formed by disease of the vessel-walls. **a., exogenous**, one due to traumatism. **a., external**. 1. One remote from the great body-cavities. 2. One in which the cavity of the tumor is entirely or chiefly outside of the inner coat of the artery. **a., false**, **a., spurious**, one due to a rupture of all the coats of an artery, the effused blood being retained by the surrounding tissues. **a., fusiform**, a spindle-shaped dilatation of an artery. **a., hernial**, one in which the internal coat of the artery, with or without the middle coat, forms the aneurysmal sac which has forced its way through an opening in the outer coat. **a., lateral**, an aneurysm projecting on one side of a vessel, the rest of the circumference being intact. **a., miliary**, a sac-like dilatation of an arteriole, often the size of a pin's head. **a., mycotic**, one due to the growth of bacteria in the vessel-wall. **a., osteoid**, a pulsating tumor of a bone. **a., partial**. 1. See **a., *lateral*. 2. An aneurysmal dilatation of a portion of the heart. **a., passive**, **a., passive cardiac**, cardiac dilatation with thinning of the heart-wall. **a., peripheral**, **a., peripheric**, one involving the whole circumference of an artery. **a., racemose**, see **a., *cirsoid*. **a., sacculated**, a sac-like dilatation of an artery communicating with the main arterial trunk by an opening that is relatively small. **a., spurious**, see **a., *false*. **a., subclavicular**, an aneurysm of the axillary artery at a point too high to admit of ligation below the clavicle. **a., surgical**, see **a., *external*. **a., true**, one in which the sac is formed of one, two, or all of the arterial coats. **a., varicose**, see under **a., *arteriovenous*.

**aneurysmal** (*an-ū-riz'-mal*) [see *aneurysm*]. Of the nature of or pertaining to an aneurysm. **a., diathesis**, a body-condition favoring the development of aneurysms. **a., varix**, see under **aneurysm, *arteriovenous*.

**aneurysmatic** (*an-ū-riz-mat'-ik*) [ἀνεύρυσμα, a widening]. Affected with or of the nature of aneurysm.

**aneurysmectomy** (*an-ū-riz-mek'-to-me*) [ἀνεύρυσμα, aneurysm; ἐκτομή, excision]. Excision of the sac of an aneurysm.

**aneurysmoplasty** (*an-ū-ris'-mo-plas-te*). Restoration of the artery in aneurysm; reconstructive endo-aneurysmorrhaphy.

**aneurysmorrhaphy** (*an-ū-riz-mor'-af-e*). The suturing of an aneurysm.

**aneurysmotomy** (*an-ū-riz-mot'-o-me*). Incision into the sac of an aneurysm.

**aneurysmus** (*an-ū-ris'-mus*). 1. Dilatation; for formation of an aneurysm. 2. Aneurysm.

**aneuthanasia** (*an-ū-than-a'-se-ah*) [ἀ, priv.; εὐθανασία, an easy death]. A painful or difficult death.

**an. ex.** (*an'-eks*). An abbreviation of *anode excitation*.

**anfract** (*an'-frakt*) [*anfractus*, a winding]. An anfractuosity or sinuosity; an anfractuous organ or structure.

**anfractuosity** (*an-frak-tū-os'-it-e*) [*anfractus*, a bending round]. 1. Any one of the furrows or sulci between the cerebral convolutions. 2. Any spiral turn or winding; an interruption; a detour. **a., ethmoidal**, an ethmoidal cell.

**anfractuous** (*an frak' tū us*) [*anfractus*, a bending round]. Characterized by windings and turnings; sinuous.

**angeial** (*an-je'-al*) [ἀγγεῖον, a vessel]. Vascular.

**angeio-** (*an-je-o-*). See *angio-*.

**Angelica** (*an-jel'-ik-ah*) [L.]. The seeds and root of *Angelica archangelica*. It is an aromatic stimulant and emmenagogue. Dose of the *seeds* or *roots* 30 gr.-1 dr. (2-4 Gm.).

**angel's wing** (*ān'-jelz wing*). A deformity of the scapula in which it turns forward and then backward, giving the shoulder a peculiar dorsal bulge.

**angi** (*an'-jē*). Inguinal buboes.

**angiectasis** (*an-je-ek'-tas-is*) [ἀγγεῖον, a vessel; ἔκτασις, dilation]. Abnormal dilatation of a vessel; enlargement of capillaries.

**angiectopia** (*an-je-ek-to'-pe-ah*) [ἀγγεῖον, a vessel; ἔκτοπος, displaced]. Displacement or abnormal position of a vessel.

**angielcosis** (*an-je-el-ko'-sis*). See *angielcus*.

**angielcus**, or **angeielcus** (*an-je-el'-kus*) [ἀγγεῖον, a vessel; ἕλκος, an ulcer]. An ulcer in the walls of a vessel.

**angiemphraxis** (*an-je-em-fraks'-is*) [ἀγγεῖον, a vessel; ἔμφραξις, obstruction]. Obstruction of a vessel or of vessels.

**angiitis, angeitis** (*an-je-i'-tis*) [ἀγγεῖον, a vessel; ιτις, inflammation]. Inflammation of a lymph-vessel or of a blood-vessel.

**angileucitis** (*an-je-lū-si'-tis*). Same as *angioleucitis*.

**angina** (*an'-jin-ah* or (incorrectly) *an-ji'-nah*) [*angere*, to strangle]. Any disease attended by a sense of choking or suffocation, particularly an affection of the fauces or pharynx presenting such symptoms. **a. abdominis**, a condition due to aneurysm or arteriosclerosis of the celiac plexus, and accompanied by severe paroxysms of abdominal pain. **a. acuta**, simple sore throat. Syn., *angina simplex*. **a. aphthosa**, **a., aphæthous**, a form attended with the formation of aphthæ in some part of the throat. **a. canina**, croup. **a. cardiac**, angina pectoris. **a. cruris**, intermittent lameness. **a. exsudativa**, croup. **a. externa**, synonym of *mumps*. **a., fibrinous**, a noninfectious disease of the throat simulating diphtheria, marked by the formation of a layer of fibrinous exudation which is chiefly confined to the tonsils. The constitutional symptoms are slight. **a., follicular**, clergyman's sore throat; see *pharyngitis, granular*. **a., herpetic**, angina observed in connection with smallpox and herpes, marked by formation of vesicles in the throat which may be attended with patches of exudation. **a. laryngea**, synonym of *laryngitis*. **a. lingualis**, same as *glossitis*. **a. Ludovici**, **a., Ludwig's**, see *Ludwig's angina*. **a. maligna**, diphtheria. **a. maxillaris**, mumps. **a. membranacea**, synonym of *diphtheria*. **a. parotidea**, the mumps, or parotitis. **a. pectoris**, a paroxysmal neurosis with intense pain and oppression about the heart. It usually occurs in the male after 40 years of age, and is generally associated with diseased conditions of the heart and aorta. There is a sense of impending death, and frequently there is a fatal termination. a. pectoris vasomotoria, a term given by Nothnagel and Landois to an angina associated with vasomotor disturbances, coldness of the surface, etc. a., phlegmónous. 1. An inflammation of the mucous and submucous tissues of the throat, with a tendency to extend more deeply, attended by edematous swelling. 2. Acute inflammation of the deep-seated structures of the throat, with a tendency to pus-formation. a., pseudo-, a neurosis occurring in anemic females, simulating angina pectoris, but characterized by a less grave set of symptoms and never resulting fatally. a., pultaceous, an affection of the throat marked by the presence of whitish or grayish patches which are easily detached, as they are not true exudations. a., rheumatic, a form of catarrhal angina in rheumatic persons, marked by sudden onset of intense pain on swallowing. a. serosa, a., serous. 1. Catarrhal angina. 2. Edema of the glottis. a. simplex, see *a.*, *acuta*. a. suffocativa, diphtheria. a., thymic. 1. Laryngismus stridulus. 2. Bronchial asthma. a. tonsillans, quinsy. a. trachealis, croup. a., ulceromembranous, see *tonsillitis*, *herpetic*. a. varicosa, dyspnea due to enlarged tonsillar tissues. a. vera, a. vera et legitima, quinsy.

**anginal** (an'-jin-al). Relating to angina.
**anginoid** (an'-jin-oid) [see *angina*]. Resembling angina.
**anginophobia** (an-ji-no-fo'-be-ah) [*angina*; φόβος, fear]. Morbid fear of angina pectoris.
**anginose** (an'-jin-ōs) [see *angina*]. Pertaining to angina; characterized by symptoms of suffocation.
**angio-** (an-je-o-). A prefix signifying relating to a vessel.
**angioasthenia** (an-je-o-as-the'-ne-ah) [*angio-*; ἀσθένεια, weakness]. Atony of the blood-vessels.
**angioataxia** (an-je-o-at-aks'-e-ah) [*angio-*; ἀταξία, want of order]. An irregularity in the tension of the blood-vessels.
**angioblast** (an'-je-o-blast) [*angio-*; βλαστός, a germ]. An embryonic cell developing into vascular tissue.
**angiocardiokinetic** (an-je-o-kar-de-o-kin-et'-ik). [*angio-*; καρδία, heart; κινεῖν, to move]. 1. Stimulating or affecting the action of movements of the heart and blood-vessels. 2. A drug which stimulates or affects the movements of the heart and blood-vessels.
**angiocarditis** (an-je-o-kar-di'-tis) [*angio-*; καρδία, the heart; ιτις, inflammation]. An inflammation of the heart and blood-vessels (hypothetical).
**angiocavernous** (an-je-o-kav'-er-nus). Relating to cavernous angioma.
**angioceratodeitis.** See *angiokeratoditis*.
**angiochalasis**, or **angieochalasis** (an-je-o-kal'-as-is) [*angio-*; χάλασις, relaxation]. Dilatation or relaxation of the blood-vessels.
**angiocheiloscope** (an-je-o-ki'-lo-skōp) [*angio-*; χεῖλος, a lip; σκοπεῖν, to look]. An instrument by means of which the blood-circulation in the capillaries of the mucosa of the lips is magnified for observation.
**angiocholitis** (an-je-o-ko-li'-tis) [*angio-*; χολή, bile; ιτις, inflammation]. Inflammation of the biliary ducts.
**angioderma pigmentosum** (an-je-o-der'-mah pig-men-to'-sum). See *atrophoderma*.
**angiodermatitis** (an-je-o-der-mat-i'-tis). Inflammation of the vessels of the skin.
**angiodiastasis** (an-je-o-di-as'-tas-is) [*angio-*; διάστασις, a separation]. 1. Displacement or dilatation of a vessel. 2. Retraction of the severed ends of a blood-vessel.
**angiodystrophia, angiodystrophy** (an-je-o-dis-tro'-fe-ah, -dis'-tro-fe) [*angio-*; δυς, bad; τροφή, nourishment]. Defective nutrition of the vessels.
**angioelephantiasis** (an-je-o-el-e-fan-ti'-as-is). See *elephantiasis telangiectodes*.
**angiofibroma** (an-je-o-fi-bro'-mah). A fibrous degenerating angioma.
**angiogenesis, angiogeny** (an-je-o-jen'-es-is, an-je-og'-en-e) [*angio-*; γεννᾶν, to produce]. The development of the vessels.
**angioglioma** (an-je-o-gli-o'-mah) [*angio-*; *glioma*]. A glioma rich in blood-vessels.
**angiograph** (an'-je-o-graf) [*angio-*; γράφειν, to write]. A variety of sphygmograph.

**angiography** (ah-je-og'-ra-fē) [see *angiograph*]. A description of the vessels; angiology.
**angiokeratoditis** (an-je-o-ker-at-o-di'-tis) [*angio-*; κέρας, cornea; ιτις, inflammation]. Vascular keratitis.
**angiokeratoma** (an-je-o-ker-at-o'-mah) [*angio-*; κέρας, horn; ὄμα, tumor]. Lymphangiectasis; telangiectatic wart; a very rare disease of the extremities, characterized by warty-looking growths that develop on dilated vessels in persons with chilblains, etc. Dark vascular spots the size of pin-points or pin-heads develop as an attack of chilblains is subsiding. The disease is peculiar to childhood.
**angiokinesis** (an-je-o-kin-e'-sis) [*angio-*; κινεῖν, to move]. Excitation or action of the blood-vessels.
**angioleucitis** (an-je-o-lū-si'-tis) [*angio-*; λευκός, white; ιτις, inflammation]. Inflammation of the lymphatic vessels.
**angioleukasia** (an-je-o-lū-ka'-zhe-ah) [*angio-*; λευκός, white; ἔκτασις, dilation]. Dilation of the lymphatics.
**angiolith** (an'-je-o-lith) [*angio-*; λίθος, stone]. A venous calculus, phlebolith.
**angiolithic** (an-je-o-lith'-ik) [*angio-*; λίθος, a stone]. A term applied to neoplasms in which crystalline or mineral deposits take place, with hyaline degeneration of the coats of the vessels.
**angiology** (an-je-ol'-o-je) [*angio-*; λόγος, science]. The science of the blood-vessels and lymphatics.
**angiolymphitis** (an-je-o-limf-i'-tis). Same as *angioleucitis*.
**angiolymphoma** (an-je-o-limf-o'-mah) [*angio-*; lympha, lymph; ὄμα, tumor]. A tumor formed of lymphatic vessels.
**angioma** (an-je-o'-mah) [*angio-*; ὄμα, a tumor]. A tumor formed of blood-vessels. a., cavernous, an angioma with communicating blood-spaces, like the cavernous tissue of the penis, originating chiefly from the distended veins. Syn., *angioma cavernosum*; *angioma circumscriptum*. a., fissural, Virchow's name for a nevus which he judged, from its location, corresponding to that of a fetal fissure, might be due to a disposition to form anomalies on the part of the region adjacent to the fissures. a., plexiform, one consisting of enlarged, tortuous capillaries forming a patch varying in color from claret to steel-blue; if there is great increase of blood-vessels, the growth has the character of a tumor, and large examples of this variety are lobular in structure. a. serpiginosum, infective angioma; nevus, lupus. a., telangiectatic, an angioma composed of dilated blood-vessels. a., tuberose, a., tuberous, one occurring in subcutaneous tissue and presenting the appearance of a lipoma as it gradually replaces the adipose tissue, or it may be accompanied by a true fatty growth.
**angiomalacia** (an-je-o-mal-a'-she-ah) [*angio-*; μαλακία, a softening]. Softening of the blood-vessels.
**angiomatosis** (an-je-o-mat-o'-sis). A condition favoring the production of angiomata.
**angiometer** (an-je-om'-et-er). See *sphygmograph*.
**angiomyces** (an-je-o-mi'-sēz) [*angio-*; μύκης, a fungus; an excrescence]. A fungoid or spongy dilation of the capillaries.
**angiomyocardiac** (an-je-o-mi-o-kar'-de-ak) [*angio-*; μῦς, muscle; καρδία, the heart]. Pertaining to the blood-vessels and the muscle of the heart.
**angiomyoma** (an-je-o-mi-o'-mah) [*angio-*; μῦς, a muscle; ὄμα, a tumor: *pl.*, *angiomyomata*]. A vascular and erectile muscular tumor.
**angiomyopathy** (an-je-o-mi-op'-a-the) [*angio-*; μῦς, muscle; πάθος, disease]. Any affection of the vessels involving the musculature.
**angiomyosarcoma** (an-je-o-mi-o-sar-ko'-mah). A tumor containing elements of angioma, myoma, and sarcoma.
**angioneoplasma** (an-je-o-ne-o-plas'-mah) [*angio-*; νέος, new; πλάσμα, moulded substance; *pl.*, *angioneoplasmata*]. A neoplasm made up of blood-vessels or lymph-vessels.
**angioneurectomy** (an-je-o-nū-rek'-to-mē) [*angio-*; νεῦρον, nerve; ἐκτομή, excision]. Resection of all the cord-elements of the prostate except the vas, with its artery and vein.
**angioneuredema** (an-je-o-nū-red-e'-mah). Same as *angioneuroedema*.
**angioneuroedema** (an-je-o-nū-ro-e-de'-mah). See *angioneurotic edema*.
**angioneurosis** (an-je-o-nū-ro'-sis) [*angio-*; *neurosis*]. A neurosis of the blood-vessels; a disturbance of the

## ANGIONEUROTIC 59 ANGLE

**vasomotor system,** either of the nature of a spasm of the blood-vessels (*angiospasm*) or of paralysis (*angioparalysis*).
**angioneurotic** (*an-je-o-nū-rot'-ik*) [see *angioneurosis*]. Pertaining to angioneurosis. a. **edema,** an acute circumscribed swelling of the subcutaneous or submucous tissues, probably due to vasomotor lesion. The disease often runs in families. It is at times periodic, and is associated with colic and gastric disturbances.
**angioneurotomy** (*an-je-o-nū-rot'-o-me*) [*angio-;* νεῦρον, nerve; τομή, a cutting]. Division of the nerves and vessels of a part.
**angionoma** (*an-je-on-o'-mah*) [*angio-;* νομή, ulcer]. Ulceration of a vessel.
**angionosis** (*an-je-o-no'-sis*) [*angio-;* νόσος, a disease]. See *angiopathy*.
**angiopancreatitis** (*an-je-o-pan-kre-at-i'-tis*). Inflammation of the vascular tissue of the pancreas.
**angioparalysis** (*an-je-o-par-al'-is-is*) [*angio-;* παράλυσις, paralysis]. Vasomotor paralysis.
**angioparalytic** (*an-je-o-par-al-it'-ik*) [see *angioparalysis*]. Relating to or characterized by angioparalysis.
**angioparesis** (*an-je-o-par'-es-is*) [*angio-;* πάρεσις, paresis]. Partial paralysis of the vasomotor apparatus.
**angiopathy** (*an-je-op'-a-the*) [*angio-;* πάθος, disease]. Any disease of the vascular system.
**angiophorous** (*an-je-of'-or-us*) [*angio-;* φερεῖν, to bear]. Applied to tissue which accompanies and supports vessels.
**angioplania** (*an-je-o-pla'-ne-ah*) [*angio-;* πλάνη, a wandering]. Irregularity or abnormality in the course of a vessel.
**angioplasty** (*an'-je-o-plas-te*) [*angio-;* πλάσσειν, to form]. Plastic surgery upon blood-vessels.
**angioplerosis** (*an-je-o-pler-o'-sis*) [*angio-;* πλήρωσις, a filling-up]. Engorgement of the vessels.
**angiopressure** (*an-je-o-presh'-ur*). The production of hemostasis by means of angiotribe and forceps without ligation.
**angiorhigosis** (*an-je-o-ri-go'-sis*) [*angio-;* ῥίγος, cold]. Rigidity of the vessels.
**angiorrhagia,** or **angeiorrhagia** (*an-je-or-a'-je-ah*) [*angio-;* ῥηγνύναι, to break]. Bleeding from a vessel.
**angiorrhaphy** (*an-je-or'-af-e*) [*angio-;* ῥάφη, suture]. Suture of a vessel or vessels. a., **arteriovenous,** the suturing of an artery to a vein, so as to turn the arterial blood into the vein.
**angiorrhea** (*an-je-or-e'-ah*) [*angio-;* ῥεῖν, to flow]. An oozing of blood.
**angiorrhexis** (*an-je-or-eks'-is*) [*angio-;* ῥῆξις, a bursting]. Rupture of a blood-vessel.
**angiosarcoma** (*an-je-o-sar-ko'-mah*) [*angio-;* σάρξ, flesh; ὄμα, a tumor]. A vascular sarcoma.
**angiosclerosis** (*an-je-o-skle-ro'-sis*) [*angio-;* σκληρός, hard]. The induration and thickening of the walls of the blood-vessels.
**angioscope** (*an'-je-o-skōp*) [*angio-;* σκοπεῖν, to inspect]. An instrument for examining the capillary vessels.
**angiosialitis** (*an-je-o-si-al-i'-tis*) [*angio-;* σίαλον, saliva; ιτις, inflammation]. Inflammation of the duct of a salivary gland.
**angiosis** (*an-je-o'-sis*) [ἀγγεῖον, a vessel]. Any disease of blood-vessels or of lymphatics.
**angiospasm** (*an'-je-o-spazm*) [*angio-;* σπασμός, a spasm]. A vasomotor spasm.
**angiospastic** (*an-je-o-spas'-tik*) [see *angiospasm*]. Characterized by or of the nature of angiospasm.
**angiosperm** (*an'-je-o-sperm*) [*angio-;* σπέρμα, seed]. In biology, a plant the seeds of which are produced within a closed vessel.
**angiostegnosis** (*an-je-o-steg-no'-sis*) [*angio-;* στέγνωσις, stoppage]. Stoppage or constriction of a vessel.
**angiostenosis** (*an-je-o-sten-o'-sis*) [*angio-;* στένωσις, a narrowing]. Narrowing of a vessel.
**angiosteogenic, angiosteogenous** (*an-je-o-ste-oj'-en-ik, -us*) [*angio-;* ὀστέον, a bone; γεννᾶν, to produce]. Relating to, producing, or produced by calcification of the vessels.
**angiostrophe, angiostrophy** (*an-je-os'-tro-fe*) [*angio-;* στροφή, a twist]. Torsion of a vessel for the arrest of hemorrhage.
**angiosymphysis** (*an-je-o-sim'-fiz-is*) [*angio-;* σύμφυσις, a growing together]. The growing together of vessels.

**angiosynizesis** (*an-je-o-sin-e-ze'-sis*) [*angio-;* συμίζανειν, to collapse]. The collapse of the walls of a vessel and subsequent growing together.
**angiotasis** (*an-je-ot'-as-is*) [*angio-;* τάσις, tension]. The tension of the vessels.
**angiotatic** (*an-je-ot-at'-ik*) [*angio-;* τάσις, tension]. Relating to angiotasis.
**angiotelectasia, angiotelectasis** (*an-je-o-tel-ek-tā'-ze-ah, an-je-o-tel-ek'-ta-sis*). See *telangiectasis*.
**angiotenic** (*an-je-o-ten'-ik*) [*angio-;* τείνειν, to stretch]. Due to or marked by distention of the blood-vessels.
**angioteria** (*an-je-o-te'-re-ah*) [*angio-;* τέρας, a wonder]. An abnormal development of the vascular system.
**angiothlipsis** (*an-je-o-thlip'-sis*) [*angio-;* θλίβειν, to rub; to gall]. The abrasion of a vessel.
**angiotitis** (*an-je-o-ti'-tis*) [*angio-;* otitis]. Inflammation of the blood-vessels of the ear.
**angiotome** (*an'-je-o-tōm*) [*angio-;* τομή, a cutting] The vascular tissue of an embryonic metamere.
**angiotomy** (*an-je-ot'-o-me*) [see *angiotome*]. 1. Incision into a vessel. 2. That branch of anatomy relating to the vascular system.
**angiotribe** (*an'-je-o-trīb*) [*angio-;* τρίβειν, to grind or bruise]. A clamp furnished with powerful jaws used by Tuffier to occlude arteries in vaginal hysterectomy.
**angiotripsy** (*an-je-o-trip'-sē*) [see *angiotribe*]. Vascular torsion and compression by means of the angiotribe.
**angiitis** (*an-ji'-tis*) See *angiitis*.
**angle, angulus** (*ang'-gl, ang'-gu-lus*) [*angulus,* an angle]. 1. A corner. 2. The degree of divergence of two lines or planes that meet each other; the space between two such lines. a. **of aberration,** see *a. of deviation.* a., **acromial,** that formed between the head of the humerus and the clavicle. a., **alpha,** in optics, that formed by the intersection of the visual line and optic axis. a., **alveolar,** that formed between a line passing through a spot beneath the nasal spine and the most prominent point of the lower edge of the alveolar process of the superior maxilla and the cephalic horizontal line. a. **of aperture,** in optics, that included between two lines joining the opposite points of the periphery of a lens and the focus. a., **biorbital,** in optics, that formed by the intersection of the axes of the orbits. a., **cardio-hepatic,** the angle formed by the junction of the upper limit of hepatic dullness with the right lateral line of cardiac dullness. a., **carrying,** angle between the longitudinal axis of the forearm and that of the arm, when the forearm is extended. a., **costal,** the angle formed by the meeting of ribs at the ensiform cartilage. a., **critical,** that made by a beam of light passing from a rarer to a denser medium, with the perpendicular, without being entirely reflected. a. **of deviation.** 1. In magnetism, the angle traversed by the needle when disturbed by some magnetic force. 2. In optics, that formed by a refracted ray and the prolongation of the incident ray. a.s, **distal,** the angles formed by the union of the other surfaces of the tooth crown with the distal surface. a. **of elevation,** in optics, that made by the visual plane with its primary position when moved upward or downward. a., **epigastric,** same as *a., costal*. a., **great,** of the eye, the inner angle of the eye. a. **of incidence,** in optics, the angle at which a ray of light strikes a denser medium and undergoes reflection or refraction. a.s, **incisal,** in dentistry, the angles of the various lateral surfaces of the tooth crowns at their junction with the incisal surface. a. **of inclination** (of pelvic canal), in obstetrics, that formed by the anterior wall of the pelvis with the conjugate diameter. a. **of inclination** (of pelvis), in obstetrics, that formed by the pelvis with the general line of the trunk, or that formed by the plane of the inferior strait with the horizon. a. **of jaw,** the junction of the lower border of the ramus of the mandible with its posterior border. a.s, **labial.** 1. See *a.s of the lips*. 2. In dentistry, the angles of the labial surface of the tooth crown which join the other surfaces. a., **limiting,** see *a., critical*. a. **of the lips,** that formed by the union of the lips at each extremity of the mouth. a., **Louis',** that between the manubrium and gladiolus of the sternum. a., **Ludwig's,** see *a., Louis'*. a., **mesial,** the angles formed at the junction of the mesial surfaces of a tooth crown with the other surfaces. a., **meter-,** in optics, the degree of convergence of the eyes when centered on an object

one meter distant from each. **a., nasal** (of the eye), the inner angle of the eye. **a., optic,** that included between lines joining the extremities of an object and the nodal point. The smallest is about 30 seconds. **a., pelvivertebral,** same as *a. of inclination* (of pelvis). **a.** of polarization, in optics, the angle of reflection at which light is most completely polarized. **a., principal,** the angle formed by that side of a prism receiving the incident ray with the side from which the refracted ray escapes. **a. of pubes,** that formed by the junction of the pubic bones at the symphysis. **a.** of reflection, in optics, that which a reflected ray of light makes with a line drawn perpendicular to the point of incidence. **a.. of refraction,** in optics, that which exists between a refracted ray of light and a line drawn perpendicular to the point of incidence. **a., Rolandic,** the acute angle formed by the fissure of Rolando with the superior border of the cerebral hemisphere. **a., sacrovertebral,** that which the sacrum forms with the last lumbar vertebra. **a.,** **sigma,** one between the radius fixus and a line from the hormion to the staphylion. **a., sternoclavicular,** that existing between the clavicle and the sternum. **a., subcostal,** see *a., costal.* **a., subpubic,** that formed at the pubic arch. **a.** of supination of the hand, **a.** of supination of the radius, the extent to which the hand is capable of being supinated; about 180°. **a., Sylvian,** the angle formed by the posterior limb of the Sylvian fissure with a line perpendicular to the superior border of the hemisphere. **a., temporal** (of the eye), the outer canthus of the eye. **a., visual,** see *a., optic.* **a., xiphoid,** that formed by the sides of the xiphoid notch.

**Anglesey leg** (*an'-gl-se*) [Marquis of *Anglesey*, 1768-1854]. An artificial limb formed from a solid piece of wood hollowed out to receive the stump and provided with a steel joint at the knee. The ankle-joint was made of wood, to which motion was communicated by strong catgut strings posteriorly and a spiral spring anteriorly.

**anglicus sudor** (*ang'-lik-us su'-dor*) [L.]. English sweating fever. A contagious malignant fever, also known as *ephemera maligna*, characterized by black or dark-colored sweat.

**angophrasia** (*an-go-fra'-ze-ah*) [ἄγχειν, to choke; φράσις, utterance]. A speech-defect consisting of a choking, drawling utterance, occurring in paralytic dementia.

**angor** (*an'-gor*) [*angor*, a strangling]. Syn., *angina.* **a. animi,** a sense of imminent dissolution. **a. pectoris,** angina pectoris.

**angostura** (*an-gos-tu'-rah*) [Sp., *Angostura*, a S. A. town]. Cusparia bark. The bark of *Galipea cusparia.* It is a stimulant tonic and febrifuge, used in malignant bilious fever, intermittent fever, and dysentery. In large doses it is emetic. Dose of *fluidextract* 10-30 min. (0.6-2.0 Cc.); of the *bark* 10-40 gr. (0.6-2.5 Gm.); of the *infusion* (*infusum cuspariæ*, B. P.) 1-2 oz. (30-60 Cc.).

**Angstroem's unit** (*awng'-strēm*) [Anders Jonas *Ångstroem*, Swedish physicist, 1814-1874]. A unit of length equal to one one-hundred-millionth of a centimeter or one ten-thousandth of a micron: used for measuring wave lengths.

**Anguilula** (*an-gwil'-ū-lah*) [dim. of *anguilla*, an eel: *pl., anguillulæ*]. A genus of very small nematode worms. **A. aceti,** the common vinegar eel. **A. intestinalis et stercoralis,** *Strongyloides intestinalis*, a worm found in the intestines and feces of persons in tropical and subtropical countries.

**angular** (*an'-gū-lar*) [*angulus*, an angle]. Pertaining to an angle. **a. artery,** the terminal branch of the facial artery. **a.' gyrus,** a. convolution, a convolution of the brain; see *convolution.* **a. movement,** the movement between two bones that may take place forward and backward or inward and outward. **a. processes,** the external and internal extremities of the orbital arch of the frontal bone.

**angulation** (*an-gū-la'-shun*). The formation of angular loops in the intestine.

**angulus** (*an'-gū-lus*) [L.]. See *angle.*

**angustura.** See *angostura.*

**anhalonine** (*an-hal-o'-nēn*) [*Anhalonium*, a genus of cacti], $C_{12}H_{15}NO_3$. A poisonous alkaloid from *Anhalonium lewinii.* It forms salts with the ordinary acids. **a. hydrochloride,** $C_{12}H_{15}NO_3HCl$, is a cardiac and respiratory stimulant and is used as is strychnine in angina pectoris, asthma, and pneumothorax.

**anhaphia** (*an-ha'-fe-ah*). See *anaphia.*

**anhedonia** (*an-hed-o'-ne-ah*). Complete loss of the sensation of pleasure.

**anhelation** (*an-hel-a'-shun*) [*anhelare*, to pant]. Shortness of breath; dyspnea.

**anhelitus** (*an-hel'-it-us*) [L.]. 1. Respiration. 2. Difficult respiration; asthma.

**anhelose, anhelous** (*an'-hel-ōs, -us*). Panting, out of breath.

**anhematosis** (*an-hem-at-o'-sis*) [ἀν, priv.; αἱματόειν, to make bloody]. Defective formation of the blood.

**anhepatogenic** (*an-hep-at-o-jen'-ik*) [ἀν, priv.; ἧπαρ, liver; γεννᾶν, to produce]. Not originating in or produced by the liver.

**anhidrosis** (*an-hid-ro'-sis*) [ἀν, priv.; ἱδρώς, sweat]. Partial or complete absence of sweat secretion.

**anhidrotic** (*an-hid-rot'-ik*) [see *anhidrosis*]. 1. Tending to check sweating. 2. An agent that checks sweating.

**anhistic, anhistous** (*an-his'-tik, an-his'-tus*) [ἀν,priv.; ἱστός, aweb]. Structureless; not organized;plasmic.

**anhydration** (*an-hi-dra'-shun*) [ἀν, priv.; ὕδωρ, water]. 1. See *dehydration.* 2. The state or condition of not being hydrated.

**anhydremia** (*an-hi-dre'-me-ah*) [ἀν, priv.; ὕδωρ, water; αἷμα, blood]. The opposite of hydremia. A diminution of the watery constituents of the blood.

**anhydric** (*an-hi'-drik*). See *anhydrous.*

**anhydride** (*an-hi'-drid*) [ἀν, priv.; ὕδωρ, water]. A chemical compound, derived from an acid by the withdrawal of a molecule of water; or an oxide, which on combination with water forms an acid. Carbon dioxide and sulphur dioxide are examples.

**anhydrite** (*an-hi'-drīt*). Anhydrous calcium sulphate.

**anhydromyelia** (*an-hi-dro-mi-e'-le-ah*) [ἀν, priv.; ὕδωρ, water; μυελόν, marrow]. A deficiency of the fluid that normally fills the spinal cavity.

**anhydrous** (*an-hi'-drus*) [see *anhydride*]. In chemistry, a term used to denote the absence of water, especially of water of crystallization.

**anhypnia** (*an-hip'-ne-ah*) [ἀν, priv.; ὕπνος, sleep]. Sleeplessness, insomnia.

**anhypnosis** (*an-hip-no'-sis*) [ἀν, priv.; ὕπνος, sleep]. Insomnia.

**anhysteria** (*an-his-te'-re-ah*) [ἀν, priv.; ὑστέρα, the uterus]. Absence of the uterus.

**anianthinopsy** (*an-e-an-thin-op'-se*) [ἀν, priv.; ἰάνθινος, violet-colored; ὄψις, sight]. An inability to recognize violet tints.

**anideus** (*an-id'-e-us*) [ἀν, priv.; εἶδος, form]. The lowest form of omphalosite, in which the parasitic fetus is reduced to a shapeless mass of flesh covered with skin.

**anidous** (*an-i'-dus*) [ἀν, priv.; εἶδος, form]. Formless, from general arrest of development; used of fetal monsters.

**anidros, anidrus** (*an-id'-ros, -rus*). Marked by the absence of perspiration.

**anidrosis** (*an-id-ro'-sis*). See *anhidrosis.*

**anidrotic** (*an-id-rot'-ik*). See *anhidrotic.*

**anile** (*an'-il*) [*anus*, an old woman]. Imbecile; like an old woman.

**anilide anilid,** (*an'-il-id*) [Ar., *al*, the; *nīl*, dark blue]. A compound formed by the action of acid chloride or acid anhydride upon aniline. The anilides are very stable derivatives.

**anilidmetarsenite** (*an-il-id-met-ar'-sen-it*), $C_6H_5$-$NO_2AsC_6H_4NHAsO_2$. A white, odorless powder containing 37.69% of arsenic, about half as much as arsenic trioxide. It dissolves in water up to 20%, and is used by subcutaneous injection in skin diseases. Dose $\frac{1}{4}$-3 gr. (0.05-0.2 Gm.) of 20% solution a day. Syn., *atoxyl.*

**anilin, aniline** (*an'-il-in*) [see *anilide*], $C_6H_7N$. Amidobenzene; phenylamine; formed in the dry distillation of bituminous coal, bones, indigo, isatin, and other nitrogenous substances. It is made by reducing nitrobenzene. It is a colorless liquid with a faint, peculiar odor, boiling at 183°; its sp. gr. at 0° is 1.036. When perfectly pure, it solidifies on cooling, and melts at −8°. It is slightly soluble in water, but dissolves readily in alcohol and ether. Combined with chlorine, the chlorates, and hypochlorites, it yields the various aniline dyes known by the names of *a. purple, a. green, a. black, a. blue,* etc. It is used in chorea and epilepsy in $\frac{1}{2}$ gr. (0.03 Gm.) doses. Syn., *phenylamine; crystallin; cyanol.*

**anilinophile** (*an-il-in'-o-fil*) [*anilin;* φιλεῖν, to love]. 1. Readily stained with anilin. 2. A tissue or element staining readily with anilin.

**anilipyrine** (*an-il-i-pi'-rin*). A feebly toxic white powder, consisting of acetanilide, 1 part; antipyrine, 2 parts, melted together. It is more soluble in water than either of its constituents. Dose, 8–16 gr.

**anilism** (*an'-il-izm*) [*anilin*]. An acute or chronic disease produced in workmen in aniline factories by the poisonous fumes. The symptoms are debility, vertigo, gastrointestinal disturbance, and cyanosis.

**anility** (*an-il'-it-e*) [*anilis*, an old woman]. The state of being imbecile or childish.

**anima** (*an'-im-ah*) [L.; spirit]. 1. The soul; the vital principle. 2. Formerly, the active principle of a drug or medicine. 3. A current of air; the breath; the mind; consciousness. In the plural, *animæ*, the swimming-bladders of herring, used as a diuretic. a. aloes, refined aloes. a. brutalis, the blood. animæ deliquium, syncope. animæ gravitas, an offensive breath. a. hepatis, iron sulphate, from its supposed efficacy in liver disease. animæ pathemata, mental affections. a. stahliana, a., Stahl's, the vital principle of plants or animals.

**animal** (*an'-i-mal*) [*anima*, the spirit, breath, or life]. An organism capable of ingesting and digesting food. No sharp line of distinction exists between the lowest animals and certain vegetables. The higher animals are distinguished by the power of locomotion and the possession of a nervous system. a., charcoal, bone-black, ivory-black, etc., is the product of the calcining of bones in closed vessels. a. chemistry, that concerned with the composition of animal bodies. a. electricity, electricity generated in the body. a.-gum, $C_{11}H_{10}O_{10}$ | $8H_2O$. A substance prepared from mucin by Landwehr, and so named on account of its resemblance to the gum of commerce. It occurs in many tissues of the body, is soluble in water, and in alkaline solution readily dissolves cupric oxide, the solution not being reduced on boiling. It yields no coloration with iodine, and is very feebly dextrorotatory. a. heat, the normal temperature of the body in man—about 98.5° F. (37° C.). a. magnetism, mesmerism; hypnotism. a. starch, see *glycogen*. a. tissue, the textures of the body.

**animalcule** (*an-im-al'-kūl*) [*animalculum*, a minute animal]. An animal organism so small as to require the microscope for its examination.

**animality** (*an-im-al'-it-e*). The state of having an animal nature.

**animalization** (*an-im-al-iz-a'-shun*) [*animalis*, animate]. The process of assimilating food to the tissues of the body.

**animation** (*an-im-a'-shun*) [*animare*, to have life or existence]. To be possessed of life. Formerly used to denote the effect of the vital principle by which the fetus acquires the power of continuing its existence. a., suspended, a condition marked by interrupted respiration and consciousness; caused by strangulation, the inhalation of carbon dioxide or other gases, etc.

**anime** (*an'-im-e*) [Fr., *animé*, origin doubtful]. A name of various resins, especially that of *Hymenæa courbaril*, a tree of tropical America; sometimes used in plaster, etc.

**animism** (*an'-im-izm*) [*anima*, soul]. Stahl's theory of life and disease, namely, that the soul is the source of both normal and pathological activities.

**aniodol** (*an-i'-o-dol*). A glycerol solution of trioxymethylene, useful as an antiseptic in 1 % solution. The univalent radical of anisic alcohol. a. hydrate, anisic alcohol.

**anisamide** (*an-is'-am-id*), $C_8H_9NO_2$. The amide of anisic acid; anisyl amide.

**anisate** (*an'-is-āt*) [*anisum*, anise]. A salt of anisic acid.

**anisated** (*an'-is-a-ted*) [*anisum*, anise]. Containing anise.

**anischuria** (*an-is-kū'-re-ah*) [*ἀν*, priv.; *ἰσχουρία*, retention of urine]. Enuresis or incontinence of urine.

**anise** (*an'-is*). See *anisum*.

**aniseed** (*an'-is-ēd*). Anise-seed. The seed of *Pimpinella anisum*. See *anisum*.

**anisette** (*an'-is-et*) [*anisum*, anise]. A liqueur prepared by the distillation of the seeds of star anise, fennel, and coriander with water and alcohol and the addition of sugar.

**anisic acid.** See *acid, anisic*.

**anisidin** (*an-is'-id-in*), $N(C_7H_7O)H_2$. A base obtained from nitranisol by action of ammonium sulphide in alcoholic solution; with acids it forms crystalline compounds. Syn., *Methylphenidin; Methylamidophenol*. a. citrate, an analgesic similar to phenetidin citrate.

**anisine** (*an'-is-in*) [*anisum*, anise], $C_{12}H_{14}N_2O_3$. A crystalline alkaloid, a derivative of anise.

**aniso-** (*an'-is-o*) [*ἄνισος*, unequal]. In combination, unequal, unsymmetrical.

**anisochromatic** (*an-is-o-kro-mat'-ik*) [*aniso-*; *χρῶμα*, color]. Not having the same color throughout; said of solutions containing two pigments used in testing for color-blindness.

**anisocoria** (*an-is-o-ko'-re-ah*) [*aniso-*; *κόρη*, pupil]. Inequality of the diameter of the pupils.

**anisocytosis** (*an-i-so-si-to'-sis*) [*aniso-*; *κύτος*, cell]. Abnormal inequality in the size of the red blood-corpuscles.

**anisodactylus** (*an-is-o-dak'-til-us*) [*aniso-*; *δάκτυλος*, a finger]. With unequal digits.

**anisodont** (*an-i'-so-dont*) [*aniso-*; *ὀδούς*, tooth]. Having irregular teeth of unequal length.

**anisognathous** (*an-is-og'-na-thus*) [*aniso-*; *γνάθος*, jaw]. Having the two jaws unlike as to the molar teeth.

**anisol** (*an'-is-ol*) [see *anisine*], $C_7H_8O$. Methylphenyl ether, produced by heating phenol with potassium and methyl iodide or potassium methyl sulphate in alcoholic solution. It is an ethereal-smelling liquid, boiling at 152°; its sp. gr. at 15° is 0.991.

**anisomelia** (*an-is-o-me'-le-ah*) [*aniso-*; *μέλος*, limb]. An inequality between corresponding limbs.

**anisomelous** (*an-is-om'-el-us*) [*aniso-*; *μέλος*, a limb]. Having limbs of unequal length.

**anisomeria** (*an-is-o-me'-re-ah*) [*aniso-*; *μέρος*, part]. The condition of having unequal organs or parts in successive series.

**anisometrope** (*an-is'-o-me-trōp*). A person with dissimilar refractive power of the two eyes.

**anisometropia** (*an-is-o-met-ro'-pe-ah*) [*aniso-*; *μέτρον*, a measure; *ὤψ*, [the eye]. A difference in the refraction of the two eyes.

**anisometropic** (*an-is-o-met-rop'-ik*) [see *anisometropia*]. Affected with anisometropia.

**anisonormocytosis** (*an-is-o-nor-mo-si-to'-sis*) [*aniso*; *norma* a rule; *κύτος*, cell]. The presence in the blood of the normal number of leukocytes, but with an abnormal proportion of the various kinds of leukocytes among themselves.

**anisopia** (*an-is-o'-pe-ah*) [*aniso-*; *ὤψ*, eye]. Inequality of visual power in the two eyes.

**anisosthenic** (*an-is-o-sthen'-ik*) [*aniso-*; *σθένος*, strength]. Not of equal power; used of pairs of muscles.

**anisotachys** (*an-is-ot'-a-kis*) [*aniso-*; *ταχύς*, quick]. Applied to an accelerated pulse of varying rapidity.

**anisotropal,** anisotropic, anisotropous (*an-is-o-trop'-al, an-is-o-trop'-ik, an-is-ot'-ro-pus*) [*aniso-*; *τρόπος*, turning]. Not possessing the same light-refracting properties in all directions; a term applied to doubly refracting bodies. In biology, varying in irritability in different parts or organs.

**anisotrophy** (*an-is-ot'-ro-fē*) [see *anisotropal*]. The quality of being doubly refractive or unequally refractive in different directions; or of being unequally responsive to external influences.

**anisum** (*an'-is-um*) [L.]. Anise. The fruit of *Pimpinella anisum*. Its properties are due to a volatile oil. It is slightly stimulant to the heart action. It liquefies bronchial secretions, and is therefore a favorite ingredient in cough-mixtures. Dose 10–20 gr. (0.65–1.3 Gm.). **anisi, aqua** (U. S. P.), oil of anise, 1; water, 500 parts. Dose indefinite. **anisi, essentia** (B. P.). Dose 10–20 min. (0.6–1.2 Cc.). **anisi, oleum** (U. S. P.), an ingredient in tinctura opii camphorata. Dose 1–5 min. (0.06–0.3 Gm.). **anisi, spiritus** (U. S. P.), a 10 % solution of the oil in alcohol. Dose 1–2 dr. (4–8 Cc.).

**anisuria** (*an-is-ū'-re-ah*) [*aniso-*; *urine*]. A condition characterized by alternate polyuria and oliguria.

**anisyl** (*an'-is-il*) [*anisum*], C₈H₇O₂. A hypothetic radical supposed to be found in anise and its derivatives.

**anitin** (*an'-it-in*). A brownish powder obtained from ichthyol. In 33 % aqueous solution it combines with phenols, etc., to form anitols. Syn., *ichthyosulphonic acid.*

**anitol** (*an'-it-ol*). Any one of the soluble compounds formed by anitin with phenols, cresols, etc.; they possess germicidal properties.

**anitrogenous** (*ah-ni-troj'-en-us*) [ἀ, priv.; *nitrogen*]. Nonnitrogenous.

**ankle** (*ang'-kl*) [ME., *ancle*]. The joint between the leg and the foot. It is a ginglymus joint, with four ligaments, the anterior, posterior, internal, and external. a.-bone, the astragalus. a.-clonus, the succession of a number of rhythmic muscular contractions in the calf of the leg when the foot is suddenly flexed by pressure upon the sole. It is a symptom of various diseases of the spinal cord, especially those involving the lateral pyramidal tracts. a.-jerk, see *a.-clonus.* a.-joint, see *ankle.* a. reflex, see *a.-clonus.* a., tailors', a ganglion or synovial sac over the external malleolus in tailors, due to their constrained posture when at work. a. valgus, a debilitated condition of the ankle-joint due to laxity of the internal lateral ligament, permitting the foot to act as in talipes valgus.

**ankola** (*an-kog-lah*) [Hind.]. The bitter, emetic root-bark of *Alangium lamarkii*, a tree of tropical Asia and Africa. It is used in India in skin diseases and leprosy.

**ankyla, ankyle** (*ang'-kil-ah, -e*) [ἀγκύλη, anything bent]. 1. An angular part, particularly the elbow. 2. Ankylosis of a joint with flexion. 3. Abnormal adhesion of parts.

**ankylenteron** (*ang-kil-en'-ter-on*) [ἀγκύλη, a coil; ἔντερον, an intestine]. An adhesion between intestinal coils.

**ankyloblepharon** (*ang-kil-o-blef'-ar-on*) [*ankyle*; βλέφαρον, the eyelid]. The adhesion of the ciliary edges of the eyelids.

**ankylocheilia, ankylochilia** (*ang-kil-o-ki'-le-ah*) [*ankyle*; χεῖλος, lip]. Adhesion of the lips.

**ankylocolpos** (*ang-kil-o-kol'-pos*), [*ankyle*; κόλπος, the vagina]. Atresia of the vagina or vulva.

**ankylodactylia** (*ang-kil-o-dak-til'-e-ah*) [*ankyle*; δάκτυλος, finger]. Adhesion of fingers or toes to one another.

**ankylodeire, ankylodere, ankyloderis** (*ang-kil-o-di'-re, -de'-re, ang-kil-od'-er-is*) [*ankyle*; δειρή, the neck]. Wry-neck; torticollis.

**ankylodontia** (*ang-kil-o-don'-she-ah*) [*ankyle*; ὀδούς, a tooth]. Irregularity in the position of the teeth.

**ankyloglossia, ankyloglossum** (*ang-kil-o-glos'-e-ah, ang-kil-o-glos'-um*) [*ankyle*; γλῶσσα, the tongue]. Tongue-tie.

**ankylomele** (*ang-kil-om'-el-e*) [*ankyle*; μήλος, a limb]. 1. The abnormal growing together of limbs (as of the fingers or toes). 2. [μήλη, a probe] A curved probe.

**ankylomerism** (*ang-kil-om'-er-ism*) [*ankyle*; μέρος, a part]. Abnormal adherence of parts to each other.

**ankylopodia** (*ang-kil-o-po'-de-ah*) [*ankyle*; πούς, a foot]. Ankylosis of the ankle-joint.

**ankyloproctia** (*ang-kil-o-prok'-te-ah*) [*ankyle*; πρωκτός, the anus]. Atresia of the anus.

**ankylorrhinia** (*ang-kil-o-rin'-e-ah*) [*ankyle*; ῥίς, the nose]. Marked adhesion between the walls of a nostril.

**ankylose** (*ang'-kil-ōz*) [*ankyle*]. To be, or to become, consolidated or firmly united.

**ankylosed** (*ang'-kil-ōzd*). Fixed by ankylosis.

**ankylosis** (*ang-kil-o'-sis*) [see *ankylose*]. Union of the bones forming an articulation, resulting in a stiff joint. a., capsular, that due to cicatricial shrinking of the joint-capsule. a., cartilaginous, a form observed as a sequel of subacute coxitis in the young, marked with great muscle tension and absence of suppuration; the cartilages may remain intact for a long time, although the shrunken synovial membrane has ceased to secrete. a., central, that due to causes present within the joint. a., extracapsular, that due to rigidity of the parts external to the joint. a., false, a., spurious, that due to the rigidity of surrounding parts. a., generalized, ankylosis affecting many joints, or a tendency toward it. a., intracapsular, that due to rigidity of the structures within a joint. a., ligamentous, when the medium is fibrous. a., muscular, that due to muscular contraction. a., true,

a., bony, that in which the connecting material is bone.

**Ankylostoma, Ankylostomum** (*ang-kil-os'-to-mah, -mum*) [ἀλκύλος, crooked; στόμα, a mouth]. A genus of nematode worms, one species of which, *A. duodenale* (hook-worm), is sometimes found in the human intestine. It produces a condition analogous to pernicious anemia. See *uncinariasis.*

**ankylostomiasis** (*ang-kil-os-to-mi'-as-is*) [see *Ankylostoma*]. The morbid condition produced by the presence of the parasite *Ankylostoma duodenale* in the human intestine. It is especially prevalent among brickmakers and other workmen in Europe. Syn., *dochmiasis; brickmakers' anemia; tunnel anemia; miners' cachexia; Egyptian chlorosis; uncinariasis; hookworm disease.*

**ankylotia** (*ang-kil-o'-she-ah*) [*ankyle*; οὖς, ear]. Union of the walls of the meatus auditorius.

**ankylotome** (*ang-kil-o-tōm*) [*ankyle*; τόμη, a cut]. A knife for operating on tongue-tie. 2. Any curved knife.

**ankylotomy** (*ang-kil-ot'-o-me*) [*ankyle*; τόμη, cut]. A cutting operation for the relief of tongue-tie.

**ankylourethria** (*ang-kil-ū-re'-thre-ah*). See *ankylurethria.*

**ankylurethra, ankylurethria** (*ang-kil-ū-re'-thrah, -re'-thre-ah*) [*ankyle*; *urethra*]. Urethral stricture or atresia.

**ankyrism** (*ang'-kir-izm*) [ἀγκύρισμα, a hooking]. Articulation or suture by one bone hooking upon another.

**ankyroid** (*ang'-kir-oid*) [ἄγκυρα, a hook]. Hook-shaped. a. cavity, in the brain, the posterior or descending cornu of the lateral ventricle. a. process, the coracoid process.

**anlage** (*ahn-lahg-eh*) [German]. Pl. *anlagen* or *anlages.* 1. The primitive undifferentiated mass of cells or rudiment of a part in a developing embryo. 2. The place in the embryo where differentiation first appears.

**Annandale's operation** (*an'-an-dāl*) [Thomas *Annandale*, Scotch surgeon, 1838-1907]. 1. For *dislocated cartilages;* incision into the knee-joint and stitching of the dislocated cartilages into their proper position. 2. For *genu valgum;* partial excision of both condyles of the femur. 3. For *nasopharyngeal polypus;* division of the alveolar margin and palatal portions of the upper jaw along their center, from before backward, and perforation of the bony septum of the nose, thus permitting separation of the two portions of the bone and exposure of the polypus. 4. For *varicocele;* a modification of Lee's operation, the veins only being excised, the scrotum being left intact. 5. For *webbed-fingers;* the longitudinal incisions are made along the sides of each finger. A.'s **triangle**, the space bounded in front by the patella, above by the articular surface of the femur, and below by the margin of the tibia.

**annatto** (*an-at'-o*). See *annotto.*

**anneal** (*an-ēl'*) [Saxon, *annelan*, to heat]. To heat and cool slowly, as gold or other metals.

**annectant** (*an-ek'-tant*) [ad, to; *nectere*, to bind]. Linking or binding together. a. **convolutions**, see *convolution.*

**annelism** (*an'-el-izm*) [*anellus*, dim. of *annulus*, a ring]. Possessing a ringed structure.

**annexa** (*an-neks'-ah*). See *adnexa.*

**annexopexy** (*an-neks'-o-peks-e*). See *adnexopexy.*

**annidalin** (*an-id'-al-in*). Dithymoltriiodide. A substitute for iodoform and aristol, *q. v.*

**annotto** (*an-ot'-o*) [native American]. A coloring-matter obtained from the pellicles of the seeds of *Bixa orellana*. It is used to color plasters and butter. Syn., *annatto; arnotto.*

**annuens** (*an'-ū-ens*) [*annuere*, to nod]. The rectus capitis anticus minor muscle.

**annular** (*an'-ū-lar*) [*annulus*, a ring]. Ring-like. a. **cartilage**, the cricoid cartilage. a. **finger**, the ring-finger. a. **ligament**, the ligament surrounding the wrist and the ankle. a. **muscle of Mueller**, the circular fibers of the ciliary muscle. a. **process**, a. **protuberance**, the pons Varolii. a. **reflex**, a ring-like reflection sometimes seen with the ophthalmoscope around the macula.

**annulate** (*an'-ū-lāt*). Characterized by, made up of, or surrounded by rings.

**annulorrhaphy** (*an-ū-lor'-af-e*) [*annulus*, ring; ῥαφή, suture]. Closure of a hernial ring or sac by suture.

**annulose** (*an'-ū-lōs*) [*annulus*, a ring]. Possessing rings.

**annulus** (*an'-ŭ-lus*) [see *annular*]. A ring-shaped or circular opening. **a. abdominalis**, the external or internal abdominal ring. **a. abdominis**, the inguinal ring. **a. ciliaris**, the boundary between the iris and the choroid. **a.**, femoralis, femoral ring (O. T. crural ring). **a.**, **fibrocartilagineus**, fibrocartilaginous ring. **a. fibrosus**. 1. The external part of the intervertebral discs. 2. Firm connective tissue containing elastic fibers surrounding the auriculoventricular openings of the heart. Syn., *annulus fibrosus attioventricularis*. 3. The circular fibrous attachment of the tympanic membrane to the tympanic plate. **a.**, **hæmorrhoidalis**, hemorrhoidal ring. **a.**, **inguinalis abdominalis**, abdominal inguinal ring (O. T. internal abdominal ring). **a. inguinalis** cutaneus, the external abdominal ring. **a.**, **inguinalis** subcutaneus, subcutaneous inguinal ring (O. T. external abdominal ring). **a.**, iridis major, greater ring of iris. **a.**, iridis minor, lesser ring of iris. **a. membranæ tympani**, an incomplete bony ring that forms the fetal auditory process of the temporal bone. **a. migrans**, a disease of the tongue marked by crescentic bands of a light-colored rash which spread over its dorsal surface and sometimes over the sides and under surface. Syn., *annulus errans*. **a. osseus**, the tympanic plate. **a. ovalis**, the rounded or oval margin of the foramen ovale. **a.**, tendineus communis (Zinni), common tendinous ring of Zinn. **a. tracheæ**, a tracheal ring. **a.**, tympanicus, tympanic ring. **a. umbilicus**, the umbilical ring. **a. urethralis**, urethral ring. **a. ventriculi**, the pylorus.

**AnOC.** Abbreviation for anodal opening contraction.

**anocathartic** (*an-o-kath-ar'-lik*) [ἄνω, upward; καθαρτικός, purging]. Emetic.

**anocavernosus** (*an-o-kav-ur-no'-sus*). See *bulbocavernosus*.

**anocelia**, **anocœlia** (*an-o-se'-le-ah*) [ἄνω, upward; κοιλία, a cavity]. The thorax.

**anoceliadelphous** (*an-o-se-le-ah-del'-fus*) [ἄνω, upward; κοιλία, a cavity; ἀδελφός, a brother]. United by the thorax or upper part of the abdomen.

**anochilon, anocheilon, anochilos** (*an-o-ki'-lon, -los*) [ἄνω, upward; χεῖλος, a lip]. 1. The upper lip. 2. An individual having a large upper lip.

**anochiloschisis, anocheiloschisis** (*an-o-ki-los'-kis-is*) [ἄνω, upward; χεῖλος, a lip; σχίζειν, to split]. An operation of splitting the upper lip for reducing its size.

**anoci-association** (*ah-no'-se-as-o-se-a'-shun*). The condition in which pain, fear, shock, and neuroses are blocked, and so excluded, in surgical cases. See *noci-association*.

**anococcygeal** (*an-o-kok-sij'-e-al*) [*anus*, the fundament; κόκκυξ, the coccyx]. Pertaining to the anus and the coccyx. **a. ligament**, a ligament that connects the tip of the coccyx with the external sphincter ani muscle.

**anodal** (*an'-o-dal*) [ἀνά, up; ὁδός, a way]. Relating to the anode; electropositive. **a. closure**, the closure of an electric circuit with the anode placed in relation to the muscle or nerve which is to be affected. **a. closure clonus**, **a. closure contraction**, see *contraction, anodal closure*. **a. diffusion**, same as *cataphoresis*. **a. duration**, the duration of an anodal closure contraction. **a. opening contraction**, see *contraction*.

**anode** (*an'-ōd*) [see *anodal*]. The positive pole of a galvanic battery. **a.**, **soluble**, Sprague's term for an anode formed of the metal which is deposited.

**anodermous** (*an-o-der'-mus*) [ἀ, priv.; δέρμα, the skin]. Without the appearance of an epidermis.

**anodic** (*an-od'-ik*) [ἄνω, upward; ὁδός, way]. 1. In biology, applied to the upper edges of leaves arranged in ascending spirals. 2. Ascending. 3. Anodal.

**anodinia** (*an-o-din'-e-ah*) [ἀ, priv.; ὠδίς, the pain of childbirth]. Absence of labor-pains.

**anodinous** (*an-od'-in-us*). Without labor pains.

**anodmia** (*an-od'-me-ah*) [ἀν, priv.; ὀδμή, smell]. Absence of the sense of smell.

**anodont, anodontous, snodous** (*an'-o-dont, an-o-dont'-us, an'-od-us*) [ἀν, priv.; ὀδούς, a tooth]. Toothless.

**anodontia** (*an-o-don'-she-ah*) [ἀν, priv.; ὀδούς, tooth]. Absence of the teeth.

**anodyne** (*an'-o-dīn*) [ἀν, priv.; ὀδύνη, pain]. 1. A medicine that gives relief from pain. 2. Relieving pain. **a.**, **Hoffmann's**, compound spirit of ether.

**anodynia** (*an-o-din'-e-ah*) [see *anodyne*]. 1. Freedom from pain. 2. Loss of sensation. Cf. *anodinia*.

**anœdochium** (*an-e-o-do'-ke-um*). [ἄνος, without understanding; δοχός, a receptacle]. A lunatic asylum.

**anoesia** (*an-o-e'-ze-ah*) [ἀνοησία, a want of sense]. Want of understanding.

**anogon** (*an'-o-gon*). The mercurous salt of diiodoparaphenolsulphonic acid. It is said to contain nearly 50 per cent. of mercury and 30 per cent. of iodine. It is insoluble in the ordinary solvents, and is used in the treatment of syphilis.

**anoia** (*an-oi'-ah*) [ἄνοια, idiocy]. Synonym of *idiocy*.

**anomalism** (*an-om'-al-izm*) [ἀνώμαλος, strange]. Deviation from the normal order or standard.

**anomalology** (*an-om-al-ol'-o-je*) [ἀνώμαλος, strange; λόγος, science]. The science of anomalies.

**anomalonomy** (*an-om-al-on'-o-me*) [ἀνωμαλία, irregularity; νόμος, a law]. The science of the laws governing anomalism.

**anomalotrophy** (*an-om-al-ol'-ro-fe*). An anomaly of nutrition.

**anomalous** (*an-om'-al-us*) [see *anomaly*]. Irregular; characterized by deviation from the common or normal order.

**anomalus** (*an-om'-al-us*) [ἀνώμαλος, not ordinary]. A muscle or muscular slip sometimes occurring beneath the levator labii superioris alæque nasi.

**anomaly** (*an-om'-al-e*) [ἀνωμαλία, irregularity]. A marked deviation from the normal; an abnormal thing or occurrence.

**anome** (*an-o'-me-ah*). 1. See *anomaly*. 2. [ἀ, priv.; ὄνομα, name]. Loss of ability to name objects or to recognize names.

**anomous** (*an-o'-mus*) [ἀ, priv.; ὦμος, the shoulder]. Without shoulders.

**Anona** (*an-o'-nah*) [Malay, *menona*]. A genus of shrubs and trees of the order *Anonaceæ*, native of tropical America. *A. ambotay* is a native of French Guiana. The bark is applied to malignant ulcers. *A. glabra* is a West Indian species. The juice of the unripe fruit is applied to ulcers. *A. muricata*, soursop, rough anona, is an American tree, but cultivated in all tropical countries, where the ripe fruit is a favorite food and used in a cooling drink for fevers. The astringent unripe fruit is used in intestinal atony. The bark is astringent and irritant; the root-bark is used in cases of disease resulting from ingestion of poisonous fish; the leaf is anthelmintic and externally a suppurant. The edible fruit of *A. obtusifolia* is used in South America and in the West Indies by the natives as a narcotic. *A. reticulata*, custard-apple, is a West Indian tree, but cultivated throughout the tropics. The unripe dried fruit and seeds are used as an intestinal astringent; the kernels of the seeds are very poisonous; the leaves are anthelmintic. *A. spinescens*, of Brazil; the seeds are used to poison vermin; the fruit as a poultice. *A. squamosa*, sweet-sop, bullock's-heart, is an American tree cultivated throughout the tropics for its fruit, which is used medicinally as is *A. muricata*. The seeds are used to destroy insects; the bark is employed by the Malays and Chinese as a tonic.

**anonychia** (*an-o-nik'-e-ah*) [ἀν, priv.; ὄνυξ, nail]. Absence of the nails.

**anonyma** (*an-on'-im-ah*) [ἀν, priv.; ὄνομα, name]. The innominate artery.

**anonymos** (*an-on'-im-us*) [see *anonyma*]. The cricoid cartilage.

**anonymous** (*an-on'-im-us*) [see *anonyma*]. Nameless. **a. bone**, see *innominatum*.

**anoopsia** (*an-o-op'-se-ah*) [ἀνώ, upward; ὄψις, vision]. Strabismus in which the eye is turned upward.

**anoperineal** (*a-no-per-in-e'-al*). Relating to the anus and the perineum.

**Anopheles** (*an-of'-el-ēz*) [ἀνωφελής, harmful]. A genus of dipterous insects (mosquitoes), belonging to the family *Culicidæ*. *A. christophersé*, of India, harbors sporozoits, and in districts where present the endemic index of malaria varies from 40 to 72 %. *A. maculipennis*, is the common form of northern and central Europe and America, and the common agent in the transmission of the malaria parasite. Syn., *Anopheles quadrimaculatus*. *A. rossii*, the most widely distributed species in India, breeding in foul water; does not carry the parasite of benign nor of malignant tertian fever, and in Calcutta, where this is the prevalent species, the endemic index of malaria is zero.

**anophelicide** (*an-of-el'-is-īd*) [*anopheles*; *cædere* to kill]. An agent which is destructive to anopheles.

**anophelifuge** (an-of-el'-if-ūj) [anopheles; fugare, to put to flight]. An agent which prevents the bite or attack of anopheles.
**anophelism** (an-of'-el-izm). Infestation of any region, with anopheles.
**anophoria** (an-o-fo'-re-ah) [ἄνω, upward; φέρειν, to bear]. See anotropia.
**anophthalmia** (an-of-thal'-me-ah) [ἄν, priv.; ὀφθαλμός, eye]. Congenital absence of the eyes. a. cyclopica, a congenital malformation in which the eye-socket is very ill-developed and the orbit rudimentary or altogether absent.
**anophthalmos** (an-off-thal'-mus) [ἄν, priv.; ὀφθαλμός, eye]. 1. Congenital absence of the eyes. 2. A person born without eyes.
**anophthalmus** (an-of-thal'-mus). See anophthalmos.
**anopia** (an-o'-pe-ah) [ἄν, priv.; ὤψ, the eye]. Absence of sight, especially that due to defect of the eyes.
**anopsia** (an-op'-se-ah) [ἄν, priv.; ὄψις, vision]. See amblyopia.
**anopubic** (a-no-pu'-bik). Relating to the anus and the pubes.
**anorchia** (an-or'-ke-ah). See anorchism.
**anorchism** (an'-or-kizm) [ἄν, priv.; ὄρχις, the testicle]. Absence of the testicles.
**anorchous** (an-or'-kus) [ἄν, priv.; ὄρχις, the testicle]. Without testicles.
**anorchus** (an-or'-kus) [ἄν, priv.; ὄρχις, the testicle]. A person in whom the testicles are absent or not descended.
**anorectal** (a-no-rek'-tal). Pertaining to the anus and the rectum.
**anoretic, anorectous** (an-o-rek'-tik, an-o-rek'-tus) [ἄν, priv.; ὄρεξις, appetite]. Without an appetite.
**anorexia** (an-or-ek'-se-ah) [ἄν, priv.; ὄρεξις, appetite]. Absence of appetite. a. nervosa, an hysterical affection occurring chiefly in young neurotic females, and characterized by a great aversion to food.
**anoria** (an-or'-e-ah) [ἀνωρία, untimeliness]. Immaturity.
**anormal** (ah-nor'-mal) [ἄν, priv.; norma, a rule]. Abnormal.
**anorrhoea** (an-or-or-e'-ah) [ἄν, priv.; ὀρρός, serum; ῥοία, a flow]. A diminished or defective secretion of serous substance.
**anorthography** (an-or-thog'-ra-fe) [ἄν, priv.; ὀρθός, straight; γράφειν, to write]. Incapacity to write correctly; motor agraphia.
**anorthopia** (an-or-tho'-pe-ah) [ἄν, priv.; ὀρθός, straight; ὄψις, vision]. 1. A defect in vision in which straight lines do not seem straight, and parallelism or symmetry is not properly perceived. 2. Squinting; obliquity of vision.
**anorthoscope** (an-or'-tho-skōp) [ἄν, priv.; ὀρθός, straight; σκοπεῖν, to look]. An apparatus for connecting in one perfect visual image disconnected and incomplete pictures.
**anorthosis** (an-or-tho'-sis) [ἄν, priv.; ὀρθωσις, a making straight]. Absence or defect of erectility.
**anoscope** (a'-no-skōp) [anus; σκοπεῖν, to look]. An instrument for examining the rectum.
**anoscopy** (an-os'-kop-e). Inspection of the anus by means of the anoscope.
**anosia** (an-o'-se-ah) [ἄ, priv.; νόσος, disease]. Without disease; normal health.
**anosmabic** (an-oz-mab'-ik). See anosmatic.
**anosmatic** (an-oz-mat'-ik) [ἄν, priv.; ὀσμή, smell]. 1. With small olfactory lobes. 2. Not having a keen sense of smell.
**anosmia** (an-oz'-me-ah) [ἄν, priv.; ὀσμή, smell]. Absence of the sense of smell. a. afferent, that due to the loss of the conductivity of the olfactory nerves. a. central, that due to cerebral disease. a., organic, that due to disease of the nasal pituitary membrane. a., peripheral, that due to disease of the peripheral ends of the olfactory nerves.
**anosmic, anosmous** (an-oz'-mik, -mus). 1. Without odor. 2. Having no sense of smell.
**anosphrasia** (an-os-fra'-se-ah). Defect or absence of the sense of smell.
**anosphresis** (an-os-fre'-sis) [ἄν, priv.; δσφρησις, odor]. Same as anosphrasia.
**anospinal** (a-no-spi'-nal) [anus; spine]. Relating to the anus and the spinal cord. a. center, a center that controls the anal sphincters. It is situated in the lumbar portion of the spinal cord.
**anostomosis** (an-os-tom-o'-sis). See anastomosis.
**anostosis** [(an-os-to'-sis) [ἄν, priv.; ὀστέον, bone]. Defective development of bone.

**anotia** (an-o'-she-ah) [see anotous]. Congenital absence of the ears.
**anotous** (an-o'-tus) [ἄ, priv.; οὖς, ear]. Devoid of ears; earless.
**anotropia** (an-o-tro'-pe-ah) [ἄνω, upward; τρέπειν, to turn]. A condition in which the visual axes have a tendency to rise above the object looked at.
**anotus** (an-o'-tus) [ἄν, priv.; οὖς, ear]. Destitute of ears.
**anovarthyroid serum** (an-o-var-thi'-royd) [ἄν, priv.; ovum; thyroid]. A serum from sheep whose ovaries and thyroid gland have been removed. It has been used in osteomalacia.
**anovesical** (a-no-ves'-ik-al) [anus; vesica, the bladder]. Pertaining conjointly to the anus and urinary bladder.
**anoxemia, anoxaemia, anoxyemia** (an-oks-e'-me-ah, an-oks-e-e'-me-ah) [ἄν, priv.; oxygen; αἷμα, blood]. 1. A lack of oxygen in the blood. 2. An abnormal condition due to the breathing of an insufficient amount of oxygen; mountain sickness; balloon sickness.
**anoxoluin, anoxolyin** (an-oks-ol'-ū-in, -i-in) [ἄν, priv.; ὀξύς, sharp; λύειν, to dissolve]. The substance opposed to oxolyin, which, according to Le Conte, exists with it in fibrin, albumin, globulin, and casein, and which is not soluble in glacial acetic acid.
**anoxycausis** (an-oks-e-kaw'-sis) [ἄν, priv.; ὀξύς, sharp; καυσις, a burning]. Combustion without the presence of oxygen.
**anozol** (an'-o-zol). A combination of iodoform and thymol; deodorous iodoform.
**ansa** (an'-sah) [L., "a handle"]. A loop. a. atlantis, the uppermost cervical ansa. a. capitis, the zygomatic arch. a., cervical, one of the intercommunicating branches of the anterior cervical nerves. a., coccygeal. See A. sacralis. a., galvanocaustic, the wire loop of a galvanic cautery. Syn., ligatura candens. a., Haller's, the loop formed by the nerve joining the facial and glossopharyngeal nerves. a., Henle's, a part of the uriniferous tubule. a. hypoglossi, a loop formed at the side of the neck by the junction of the descendens noni nerve, with branches of the second and third cervical nerves. a., intergenicularis, fibers connecting the geniculate bodies. a., intestinalis, any loop of the small intestine. a. lenticularis, a bundle of fibers proceeding from the neural laminæ between the divisions of the lenticular nucleus. Syn., ansa lentiformis; lenticular loop. a. lumbalis, a. lumbaris, one of the connecting ramifications between the branches of the lumbar nerves. a., peduncularis, the ventral stalk of the thalamus. a. sacralis, a loop joining the ganglion impar with the sympathetic trunks of the two sides. a., sternal, the interclavicular notch. a. subclavialis, a., subclavian, see a. of Vieussens. a. supramaxillaris, one of communication between the ventral and dorsal superior dental nerves. a. of Vieussens, a loop extending from the third cervical ganglion and surrounding the subclavian artery. a. of Wrisberg, the nerve joining the right great splanchnic and right pneumogastric nerves.
**Anschuetz's chloroform.** See chloroform.
**anserine** (an'-ser-ēn) [anser, a goose]. Resembling a goose. a. disease, a wasting of the muscles of the hands, rendering the tendons unduly prominent, and suggesting the appearance of a goose's foot. a. skin, goose-skin.
**ansiform** (an'-si-form) [ansa, a handle; forma, shape]. Loop-shaped.
**Anstie's rule** (an'-ste) [Francis Edmund Anstie, English physician, 1833–1874]. No healthy man can take without injury more than the equivalent of 1½ ounces of absolute alcohol per diem. A.'s test for alcohol in urine, a mixture of potassium bichromate 1, and strong sulphuric acid 300 is added drop by drop to the urine; an emerald green color denotes the presence of a toxic amount of alcohol.
**ant-, anti-** (ant-, or ant'-te) [ἀντί, against]. Prefixes to compound words signifying opposed to, against, counteracting, etc.
**antacid** (ant-as'-id) [anti, against; acidus, acid]. 1. Neutralizing acidity. 2. A substance counteracting or neutralizing acidity. An alkali.
**antacidin** (ant-as'-id-in). Saccharate of lime.
**antacrid** (ant-ak'-rid) [anti-; acer, sharp]. Correcting acridity in the secretions. a. tincture, see guaiac mixture, Fenner's.
**antagonism** (an-tag'-on-izm) [see antagonist]. Opposition; opposed action, as of two sets of muscles or of two remedial agents.

**antagonist** (an-tag'-o-nist) [ἀνταγωνιστής, counteracting]. A term applied to a drug that neutralizes the therapeutic effects of another. In anatomy, a muscle that acts in opposition to another. a., associated, a name given to that muscle of a healthy eye that turns the globe in the same direction as the affected muscle of the opposite eye would, if normal, turn the eye to which it belongs.
**antagonistic** (an-tag-o-nis'-tik) [see *antagonist*]. Opposing.
**antalgesic** (ant-al-je'-sik) [see *antalgic*]. Antalgic.
**antalgic** (ant-al'-jik) [anti-; ἄλγος, pain]. 1. Relieving pain. 2. A remedy that relieves pain.
**antalkaline** (ant-al'-kal-īn) [anti-; alkali]. 1. Neutralizing alkalies. 2. An agent neutralizing alkalies, as acids.
**antanacathartic** (ani-an-ah-kath-ar'-tik) [anti-; ἀνά, up; κάθαρσις, purgation]. 1. Checking expectoration. 2. An agent which checks expectoration.
**antanemic** (ant-an-e'-mik) [anti-; anemic]. 1. Correcting anemia. 2. A remedy efficient in anemia.
**antaphrodisiac** (ant-af-ro-dis'-e-ak) [anti-; ἀφροδίσια, sexual desire]. 1. Lessening the venereal desires. 2. An agent that lessens the venereal impulse; an anaphrodisiac.
**antaphroditic** (ant-af-ro-dit'-ik). See *antaphrodisiac*.
**antapoplectic** (ant-ap-o-plek'-tik) [anti-; ἀποπληξία, apoplexy]. Efficient in preventing or treating apoplexy.
**antarthritic** (ant-ar-thrit'-ik) [anti-; ἀρθριτικός, gouty]. 1. Relieving gout. 2. A medicine for the relief of gout.
**antasphyctic** (ant-as-fik'-tik) [anti-; ἀσφυκτος, pulseless]. 1. Efficient in preventing asphyxia. 2. An agent efficacious in preventing asphyxia.
**antasthenic** (ant-as-then'-ik) [anti-; ἀσθένεια, weakness]. Tending to correct debility and restore the strength.
**antasthmatic** (ant-az-mat'-ik) [anti-; ἄσθμα, shortdrawn breath]. 1. Relieving asthma. 2. A medicine serving for the relief of asthma.
**antatrophic** (ant-at-rof'-ik) [ant-; ἀτροφία, wasting]. 1. Preventing atrophy. 2. A drug that will prevent wasting or atrophy.
**antebrachial** (an-te-bra'-ke-al). Pertaining to the forearm.
**antebrachium** (an-te-bra'-ke-um). See *antibrachium*.
**antecardium** (an-te-kar'-de-um) [anti-; καρδία, the heart]. The scrobiculus cordis, or pit of the stomach; the infrasternal depression; the precordium.
**ante cibum** (ante si'-bum). Latin for "before a meal."
**antecornu** (an-te-kor'-nu). See *precornu*.
**antecubital** (an-te-ku'-bit-al) [ante, before; cubitum, the elbow]. Situated in front of the elbow.
**antecurvature** (an-te-kur'-va-tūr) [ante, forward; curvatus, bent]. A forward curvature.
**antedisplacement** (an-te-dis-plās'-ment) [ante, forward; O. F., *desplacer*, to put out of place]. Forward displacement of a part or organ.
**antefebrile** (an-te-feb'-ril) [ante, before; febris, fever]. The period before a fever.
**antefixatio uteri** (an-te-fiks-a'-she-o ū'-ter-i). The operative suturing of the uterus in retroflexion.
**anteflexion** (an-te-flek'-shun) [ante, before; flectere, to bend]. A bending forward. a. of uterus, a condition in which the fundus of the uterus is bent forward.
**antehelix** (an-te-he'-liks). See *anthelix*.
**antelocation** (an-te-lo-ka'-shun) [ante, before; locus, a place]. The forward displacement of an organ or viscus.
**antemetic** (ant-em-et'-ik). See *antiemetic*.
**antemortem** (an'-te-mor'-tem) [L.]. Before death.
**antenarial** (an-te-na'-re-al) [ante, before; nares, the nostrils]. Situated in front of the nostrils.
**antenatal** (an-te-na'-tal) [ante, before; natus, born]. Occurring or existing before birth.
**anteneasmum, anteneasmus** (an-ten-e-az'-mum, -mus). P. Zacchias' term for a form of dementia marked by restlessness and a suicidal tendency.
**ante partum** (an'-te par'-tum) [L.]. Before delivery.
**antepileptic** (an-tep-il-ep'-tik) [anti-; ἐπίληψις, epilepsy]. Relieving epilepsy.
**anteprostatic** (an-te-pros-tat'-ik) [ante, before; προστάτης, one who stands before]. Situated before the prostate. a. glands. 1. Cowper's glands. 2. Certain small accessory glands sometimes found between Cowper's gland and the prostate.
**antepyretic** (an-te-pi-ret'-ik) [ante; πυρετός, fever]. Prior to the development of fever.
**antereisis** (ant-er-i'-sis) [ἀντέρεισις, resistance]. The resistance opposed by a dislocation during its reduction.
**anterethic** (an-ter-eth'-ik) [anti-; ἐρεθισμός, irritation]. Soothing; allaying irritation.
**anterior** (an-te'-re-or) · [L., "before"]. Situated before or in front of; pertaining to the part or organ situated toward the ventral aspect of the body. a. poliomyelitis, inflammation of the anterior horns of the spinal cord, giving rise to a characteristic paralysis, common in children. a. rotation, the forward turning of the presenting part in labor.
**antero-** (an'-te-ro-) [anterior, before]. A prefix signifying position in front.
**anterodorsal** (an-te-ro-dor'-sal) [antero-; dorsum, the back]. Pertaining to the ventral aspect of the dorsum.
**anterograde** (an'-te-ro-grād) [antero-; gredi, to go]. Proceeding forwards.
**anteroinferior** (an-te-ro-in-fe'-re-or) [antero-; inferior, lower]. Situated in front and below.
**anterointerior** (an-te-ro-in-te'-re-or). Located ventrally and internally.
**anterointernal** (an-te-ro-in-tur'-nal). Situated in front to the inner side.
**anterolateral** (an-te-ro-lat'-er-al) [antero-; latus, a side]. In front and to or on one side; from the front to one side.
**anteromedian** (an-te-ro-me'-de-an) [antero-; medius, the middle]. In front and toward the middle.
**anteroparietal** (an-te-ro-par-i'-et-al) [antero-; parietal]. Anterior and also parietal. a. area, the anterior part of the parietal area of the cranium.
**anteroposterior** (an-te-ro-pos-te'-re-or) [antero-; posterior, backward]. Extending from before backward.
**anterosuperior** (an-te-ro-su-pe'-re-or) [antero-; superior, upper]. Situated in front and above.
**anterotic** (ant-e-rot'-ik) [anti-; ἐρωτικός, pertaining to love]. Anaphrodisiac.
**anteversion** (an-te-ver'-shun) [ante, forward; vertere, to turn]. A turning forward. a. of uterus, a tilting forward of the uterus.
**anthectic** (an-thek'-tik) [anti-; ἐκτικός, hectic]. 1. Efficacious against tuberculosis. 2. An agent or remedy efficient against tuberculosis.
**antheline** (an-thel'-is-in) [ἀνθέλιξ, the inner curvature of the ear]. Pertaining to the anthelix.
**anthelix** (an'-the-liks) [ἀνθέλιξ, the inner curvature of the ear]. The ridge surrounding the concha of the external ear posteriorly.
**anthelmintic** (an-thel-min'-tik) [anti-; ἕλμινς, a worm]. 1. Efficacious against worms. 2. Avermicide.
**anthema** (an'-the-mah) [ἀνθεῖν, to bloom]. An exanthem; a skin eruption.
**anthemis** (an'-them-is) [ἄνθεμις, a flower]. Camomile. The flower-heads of *A. nobilis*, the properties of which are due to a volatile oil, a camphor, and a bitter principle. It is useful in coughs and spasmodic infantile complaints, and is an excellent stomachic tonic. Infusion of 4 dr. to 1 pint, given in doses of 1–2 oz. (30–60 Cc.). **anthemidis, extractum** (B. P.). Dose 2–10 gr. (0.13–0.65 Gm.). **anthemidis, infusum** (B. P.). Dose 1–4 oz. (30–120 Cc.). **anthemidis, oleum**, the volatile oil of camomile. Dose 2–10 min. (0.12–0.6 Cc.).
**anthemorrhagic** (ant-hem-or-aj'-ik) [anti-; αἷμα, blood; ῥαγία, a bursting]. Checking or preventing hemorrhage.
**anther** (an'-ther) [ἀνθηρός, in full bloom]. In biology, the male sexual organ in plants; the summit and essential part of the stamen. It contains the pollen or fecundating substance of the flower.
**antherpetic** (ant-her-pet'-ik) [anti-; herpes]. 1. Efficient against herpes. 2. An efficacious remedy for herpes.
**anthocephalous, anthocephalus** (an-tho-sef'-al-us) [ἄνθος, a flower; κεφαλή, a head]. Having a flowershaped head; *e. g.*, tænia anthocephala.
**Anthony's fire, St.** See *erysipelas*.
**anthorism, anthorisma** (an'-thor-izm, an-thor-iz'-mah) [anti-; ὅρισμα, a boundary]. A diffuse swelling.
**anthracemia** (an-thras-e'-me-ah) [anthrax; αἷμα, blood]. 1. Woolsorter's disease; splenic fever of animals; a disease due to the presence in the blood of *Bacillus anthracis*. 2. Asphyxia due to carbon monoxide poisoning.

**anthracene** (*an'-thra-sēn*) [*anthrax*], $C_{14}H_{10}$. A hydrocarbon formed from many carbon compounds when they are exposed to a high heat; also from coal-tar. It crystallizes in colorless, monoclinic tables, showing a beautiful blue fluorescence; dissolves with difficulty in alcohol and ether, but easily in hot benzene; melts at 213°. It is the base from which artificial alizarin is prepared.

**anthracia** (*an-thra'-se-ah*) [*anthrax*]. A name for diseases characterized by the formation of carbuncles. a. **pestis**, the plague. a. **rubula**, synonym of *frambesia*.

**anthracic** (*an'-thras-ik*) [*anthrax*]. Pertaining to or of the nature of anthrax.

**anthracin** (*an'-thras-in*) [*anthrax*]. A toxic ptomaine derived from pure cultures of the bacillus of anthrax.

**anthracina** (*an-thras-e'-nah*). Melanotic carcinoma.

**anthracion** (*an-thras'-e-on*) [*anthrax*]. Contagious anthrax.

**anthracite** (*an'-thras-īt*) [ἄνθραξ, a coal]. A variety of mineral coal containing but little hydrogen, and therefore burning almost without flame.

**anthracnosis** (*an-thrak-no'-sis*) [ἄνθραξ, a coal; νόσος, disease]. Black rot, a fungus disease of vines, caused by the *Phoma uvicola*, or *Sphaceloma ampelium*.

**anthracoid** (*an'-thrak-oid*) [*anthrax*; εἶδος, likeness]. Resembling carbon, anthrax, or the gem carbuncle.

**anthracolemus, anthracoloemus** (*an-thra-kol-e'-mus*) [*anthrax*; λοιμός, a plague]. Contagious anthrax.

**anthracoma** (*an-thrak-o'-mah*) [*anthrax*]. A carbuncle.

**anthracometer** (*an-thrak-om'-et-er*) [*anthrax*; μέτρον, a measure]. An instrument for estimating the amount of carbon dioxide in the air.

**anthracometry** (*an-thrak-om'-et-re*) [ἄνθραξ, carbon; μέτρον, a measure]. The determination of the amount of carbon dioxide in air.

**anthraconecrosis** (*an-thrak-o-ne-kro'-sis*) [*anthrax*; νέκρωσις, death]. The necrotic transformation of a tissue into a black mass, as in dry gangrene.

**anthracopestis** (*an-thrak-o-pes'-tis*) [*anthrax*, *pestis*, a plague]. Malignant anthrax.

**anthracophlyctis** (*an-thrak-o-flik'-tis*) [*anthrax*; φλύκτις, a pustule]. The same as *anthracopestis*.

**anthracosis** (*an-thrak-o'-sis*) [*anthrax*; νόσος, disease]. 1. "Miners' lung." A diseased condition of the lung produced by the inhalation of coal-dust. It is a form of pneumokoniosis. 2. A malignant or corroding ulcer; a carbuncle.

**anthraflavon** (*an-thra-flav'-on*) [*anthracene; flavus*, yellow], $C_{14}H_8O_4$. A substance acting as a dibasic acid, forming yellow needles subliming without fusion at temperatures above 300° C.

**anthragallol** (*an-thra-gal'-ol*), $C_{14}H_8O_5$. A reaction-product of benzoic, gallic, and sulphuric acids. It occurs as a dark-brown paste or orange-red acicular crystals, soluble in alcohol; melts at 310° C. Sublimes at 290° C. It is used in dyeing. Syn., *trioxyanthraquinone*.

**anthrapurpurin** (*an-thra-pur'-pu-rin*) [*anthracene; purpurin*], $C_{14}H_8O_5$. A derivative of anthraflavic acid and an isomer of purpurin and of flavopurpurin, almost identical with the latter; it forms orange-colored needles. a. **acetate**, a. **diacetate**, a fine yellow, tasteless powder, freely soluble in glacial acetic acid and xylol, sparingly so in alcohol; insoluble in water. It is used as an aperient and laxative (it colors the urine red). Dose 7½ gr. (0.5 Gm.). Syn., *purgatin; purgatol*.

**anthraquinolin** (*an-thra-kwin'-ol-in*) [*anthracene; quina*, bark], $C_{17}H_{11}N$. A crystalline substance melting at 170° C., boiling at 446° C.; its solutions exhibit an intensely blue fluorescence.

**anthraquinone** (*an-thra-kwin'-ōn*) [*anthracene; quinone*], $C_{14}H_8O_2$. A substance produced by oxidizing anthracene with $HNO_3$. It sublimes in yellow needles, melting at 277° C., and is soluble in hot benzene and $HNO_3$.

**anthrarobin** (*an-thra-ro'-bin*), $C_{14}H_{10}O_3$. A derivative of alizarin, similar to chrysarobin. It is a yellowish-white powder, insoluble in water, but soluble in alcohol and dilute alkaline solutions. It is useful in psoriasis, herpes, pityriasis versicolor.

**anthrasol** (*an'-thra-sol*). A proprietary coal-tar preparation; used in the form of an ointment for pruritus and for skin affections.

**anthrax** (*an'-thraks*) [ἄνθραξ, a coal or a carbuncle]. 1. A carbuncle. 2. An acute infectious disease due to *Bacillus anthracis*. Syn., *milzbrand; charbon; woolsorter's disease; splenic fever; splenic apoplexy; Siberian cattle plague; plaga ignis; acacanthrax; mal de Chabert; abscessus gangrænescens; abscessus gangrænosus*. a., **apoplectic**, a very acute and virulent form of malignant anthrax coming on without premonitory symptoms and chiefly affecting horses and cattle. a., **contagious**, malignant anthrax. a., **hemorrhoidal**, a contagious form affecting the rectum of animals and marked by evacuations of dark-colored blood. a., **malignant**, see *anthrax* (2). a., **pulmonary**, gangrene of the lungs. a., **symptomatic**, see *black-leg*.

**anthropo-** (*an-thro-po-*) [ἄνθρωπος, a man; a human being]. A prefix signifying relating to man or to the human race.

**anthropogenesis** (*an-thro-po-jen'-es-is*) [*anthropo-*; γένεσις, generation]. The development of man, as a race (*phylogenesis*) and as an individual (*ontogenesis*).

**anthropogeny** (*an-thro-poj'-en-e*) [*anthropo-*; γεννᾶν, to produce]. The study or science of the descent of man.

**anthropoglot** (*an'-thro-po-glot*) [*anthropo-*; γλῶσσα, the tongue]. Human-tongued, as a parrot.

**anthropography** (*an-thro-pog'-ra-fe*) [*anthropo-*; γράφειν, to write]. A treatise upon the human structure or organism.

**anthropoid** (*an'-thro-poid*) [*anthropo-*; εἶδος, like]. Man-like.

**anthropology** (*an-thro-pol'-o-je*) [*anthropo-*; λόγος, discourse]. The science of the nature, physical and psychological, of man and of mankind.

**anthropometallism** (*an-thro-po-met'-al-ism*) [*anthropo-*; *metal*]. "Hypnotism or the like condition, induced by looking at a metallic disc.

**anthropometer** (*an-thro-pom'-et-er*) [*anthropo-*; μέτρον, a measure]. An instrument used in anthropometry.

**anthropometry** (*an-thro-pom'-et-re*) [*anthropo-*; μέτρον, a measure]. The determination of the measurement, weight, strength, and proportions of the parts of the human body.

**anthropomorphic** (*an-thro-po-mor'-fik*) [*anthropo-*; μορφή, form]. Man-like.

**anthropomorphism** (*an-thro-po-mor'-fizm*) [*anthropo-*; μορφή, form]. 1. Anthropomorphosis (*q. v.*). 2. The theory which ascribes human attributes to the Deity.

**anthropomorphosis** (*an-thro-po-mor-fo'-sis*) [*anthropo-*; μορφή, form]. The development of the human figure; a change into the shape of a man.

**anthropomorphous** (*an-thro-po-mor'-fus*) [*anthropo-*; μορφή, form]. Resembling a man in shape or character.

**anthroponomy** (*an-thro-pon'-om-e*) [*anthropo-*; *man*; νόμος, a law]. The sum of what is known concerning the laws which control the formation and functions of the human body.

**anthropophagy** (*an-thro-pof'-a-je*) [*anthropo-*; φαγεῖν, to devour]. 1. Cannibalism. 2. Sexual perversion leading to rape, mutilation, and cannibalism.

**anthropophobia** (*an-thro-po-fo'-be-ah*) [*anthropo-*; φόβος, fear]. A symptom of mental disease consisting in fear of society.

**anthroposomatology** (*an-thro-po-so-mat-ol'-o-je*) [*anthropo-*; σῶμα, body; λόγος, science]. The sum of what is known regarding the human body.

**anthropotomy** (*an-thro-pot'-o-me*) [*anthropo-*; τομή, section]. Human anatomy, or dissection of the human body.

**anthropotoxin** (*an-thro-po-toks'-in*) [*anthropo-*; τοξικόν, poison]. The toxic substance supposed to be excreted by the lungs of human beings.

**anthydropic** (*ant-hi-drop'-ik*) [*anti-*; ὕδρωψ, dropsy]. Effective against dropsy.

**anthypnotic** (*ant-hip-not'-ik*) [*anti-*; ὕπνος, sleep]. 1. Preventive of sleep. 2. An agent that tends to induce wakefulness.

**anthypochondriac** (*ant-hip-o-kon'-dre-ak*) [*anti-*; *hypochondriac*]. Efficient in overcoming hypochondriasis.

**anthysteric** (*ant-his-ter'-ik*) [*anti-*; ὑστέρα, the uterus]. 1. Overcoming hysteria. 2. A remedy against hysteria.

**anti-** (*an-ti-*) [ἀντί, against]. A prefix meaning against.

**antiabrin** (an-ti-a'-brin) [anti-; abrin]. Ehrlich's term for a hypothetic alexin in the blood of animals rendered immune against abrin.
**antiacid** (an-te-as'-id). Antacid.
**antiades** (an-ti'-ad-ēz) [pl. of ἀντιάς, tonsil]. The tonsils.
**antiaditis** (an-ti-ad-i'-tis) [ἀντιάς, tonsil; ιτις, inflammation]. Tonsillitis.
**antiadoncus** (an-ti-ad-ong'-kus) [ἀντιάς, a swollen tonsil; ὄγκος, a heap]. Any tumor or swelling of the tonsils.
**antiagglutinin** (an-te-ag-lu'-tin-in). A substance having the power of neutralizing the corresponding agglutinin. q. v.
**antiaggressin** (an-te-ah-gres'-in) [anti-; aggressin]. A substance having the power of neutralizing the corresponding aggressin.
**antialbumate, antialbuminate** (an-te-al'-bū-māt, an-te-al-bū'-min-āt) [anti-; albumen, white of egg]. Parapeptone; a product of the imperfect digestion of albumin. It is changed by the pancreatic ferment into antipeptone.
**antialbumid** (an-te-al'-bū-mid). See antialbumate.
**antialbumin** (an-te-al'-bū-min) [see antialbumate]. One of the products of the action of the digestion of albumin; it is probably one of the preformed substances existing in the proteid molecule.
**antialbumose** (an-te-al'-bū-mōs) [see antialbumate]. One of the albumoses produced by the action of pancreatic juice on albumin. It resembles syntonin or acidalbumin, and is convertible into antipeptone.
**antialexin** (an-te-al-ek'-sin). A substance which has the power of neutralizing the corresponding alexin.
**antiamboceptor** (an-te-am-bo-sep'-tor). A substance which inhibits the action of an amboceptor.
**antianaphylactin** (an-te-an-ah-fi-lak'-tin). A substance which inhibits the action of an anaphylactin.
**antianaphylaxis** (an-te-an-ah-fi-lak'-sis). A condition neutralizing anaphylaxis: a state of absolute insusceptibility; see ananaphylaxis.
**antiantibody** (an-te-an'-te-bod-e). An antibody to an antibody.
**antiantitoxin** (an-te-an-te-toks'-in). An antibody which is formed in immunization with an antitoxin and which inhibits its action.
**antiaphrodisiac** (an-te-af-ro-diz'-e-ak). See anaphrodisiac.
**antiapoplectic** (an-te-ap-op-lek'-tik). An agent which affords relief in, or prevents apoplexy.
**antiar** (an'-te-ar). See antiarin.
**antiarin** (an-te'-ar-in) [Javanese, antiar or antjar], $C_{14}H_{20}O_5+2H_2O$. The active principle of Antiaris toxicaria or Upas antiar, Javanese poison-tree. Intensely poisonous and used as an arrow-poison. Is cardiac depressant. Dose ₁⁄₁₀₀ gr. (0.00065 Gm.).
**antiarsenin** (an-te-ar'-sen-in). An antitoxin produced as the result of the administration of arsenic.
**antiarthrin** (an-te-ar'-thrin). The commercial name for a preparation said to consist chiefly of the extractives of horse chestnut, with salicin, saligenin, dextrose, and hydrochloric acid. It is said to be a specific for gout. Dose, 1 gm.
**antiarthritic** (an-te-ar-thrit'-ik) [anti-; arthritis]. A remedy against gout.
**antiasthmatic** (an-te-as-mat'-ik). See antasthmatic.
**antiautolysin** (an-te-aw-tol'-is-in) [anti-; αὐτός, self; λύσις, solution]. A substance developed in the blood having the power to restrain the solvent action of autolysin.
**antibacterial** (an-te-bak-te'-re-al) [anti-; bacterial]. 1. Opposed to the germ theory of disease. 2. Opposed to or restraining bacterial action.
**antibacterin** (an-te-bak'-ter-in). 1. A pale yellow fluid said to consist of boric acid, 6.25 parts; iron chloride solution, 1.5 parts; chloric ether, to make 100 parts. It is used by inhalation in tuberculosis, beginning with 150 gr. (10 Gm.) daily, and increasing to 10 times that quantity. 2. Crude aluminum sulphate mixed with soot.
**antibechic** (an-te-bek'-ik) [anti-; βήξ, a cough]. 1. Alleviating or curing cough. 2. A remedy for cough or hoarseness.
**antibilious** (an-te-bil'-yus) [anti-; bilious]. Effective against bilious disorders.
**antibiosis** (an-te-bi-o'-sis) [anti-; βίος, life]. An association between two or more organisms which is harmful to one of them. It is the opposite of symbiosis.

**antibiotic** (an-te-bi-ot'-ik) [anti-; βίος, life]. 1. Pertaining to antibiosis. 2. Tending to destroy life.
**antiblennorrhagic** (an-te-blen-or-aj'-ik) [anti-; βλέννα, mucus; ῥήγνυναι, to burst]. Efficient in preventing or curing gonorrhea.
**antibodies** (an-te-bod'-ēz). Characteristic constituents of the blood and fluids of the immune animal; antagonistic to the harmful action of bacteria; e. g., antitoxins, agglutinins, precipitins, etc. Cf. antitoxin.
**antibrachial** (an-te-bra'-ke-al) [anti-; βραχίων, the arm]. Pertaining to the forearm.
**antibrachium** (an-te-bra'-ke-um). The forearm.
**antibromic** (an-te-bro'-mik) [anti-; βρῶμος, a stench]. 1. Deodorant. 2. A drug that destroys offensive smells. A deodorizer.
**antibrule** (an'-ti-brūl). A proprietary analgesic, antiseptic, and keratoplastic.
**anticachectic** (an-te-kak-ek'-tik) [anti-; κακός, bad; ἕξις, habit]. Effective in destroying cachexia. 2. A remedial agent against cachexia.
**anticacochymic** (an-te-kak-o-kim'-ik) [anti-; κακός, bad; χυμός, juice]. Anticachectic.
**anticalculous** (an-te-kal'-ku-lus) [anti-; calculus]. Good against calculus; antilithic.
**anticancrin** (an-te-kang'-krin). See cancroin.
**anticarcinomatous** (an-te-kar-sin-o'-mat-us) [anti-; καρκίνωμα, cancer]. Preventing carcinoma.
**anticardium** (an-te-kar'-de-um) [anti-; καρδία, the heart]. The scrobiculus cordis, or pit of the stomach; the infrasternal depression.
**anticarious** (an te ka' re us) [anti ; caries, decay] Preventing decay, as of the teeth.
**anticatarrhal** (an-te-kat-ar'-al) [anti-; catarrh]. Counteracting catarrh.
**anticathode** (an-te-kath'-ōd). The part of a Crookes' tube opposite the cathode; it is that part on which the cathode rays impinge.
**anticaustic** (an-te-kaws'-tik). Arresting the action of a caustic agent.
**anticausticon** (an-te-kaws'-tik-on) [anti-; καυστικός, burning]. A preparation of soluble water glass.
**antichirotonus, anticheirotonus** (an-te-ki-rot'-o-nus) [anti-; χείρ, hand; τόνος, tension]. Forcible and steady inflection of the thumb, seen at times in or before attacks of epilepsy.
**antichlor** (an'-te-klor). 1. Sodium thiosulphate. 2. Potassium sulphate.
**antichlorin** (an-te-klo'-rin). A preparation used in anemia and said to consist of glucose, basic bismuth formate, and sodium bicarbonate.
**antichlorotic** (an-te-klo-rot'-ik) [anti-; χλωρότης, greenness]. Counteracting chlorosis.
**anticholerin** (an-te-kol'-er-in) [anti-; χολέρα, cholera]. A product isolated from cultures of cholera bacilli, and used in the treatment of cholera.
**anticipating** (an-tis'-ip-a-ting) [anticipare, to take before]. Occurring before the regular or expected time, as an anticipating intermittent fever, one in which the paroxysms occur earlier on successive days.
**anticlinal** (an-te-kli'-nal) [anti-; κλίνειν, to slope]. Sloping in opposite directions. a. vertebra, in man, the tenth thoracic vertebra, where the thoracic vertebræ begin to assume the characters of the lumbar.
**anticloudine** (an-te-klow'-din). Trade name of a paste for preventing moisture from precipitating on eyeglasses, mirrors, or glass or nickel instruments.
**anticnemion** (an-tik-ne'-me-on) [anti-; κνήμη, leg]. The shin or front of the leg.
**anticnesmatic** (an-tik-nes-mat'-ik) [anti-; κνησμός, itching]. 1. Efficient against itching. 2. A remedy for itching.
**anticoagulant** (an-te-ko-ag'-u-lant) [anti-; coagulum]. 1. Opposed to or preventive of coagulation. 2. A substance preventing coagulation.
**anticoagulin** (an-te-ko-ag'-u-lin). A substance formed in the body antagonistic in its action to that of a coagulin (q. v.).
**anticomplement** (an-te-kom'-ple-ment) [anti-; complement]. A substance held by Ehrlich in his lateralchain theory to enter into the composition of an antihemolysin (q. v.). It is capable of neutralizing the action of a complement. Cf. antiimmune body under body.
**anticomplementary** (an-te-kom-ple-men'-tar-e). Capable of lessening of abolishing the action of a complement.
**anticontagious** (an-te-kon-ta'-jus). Counteracting contagion.

**anticonvulsive** (*an-te-kon-vul'-siv*). Effective against convulsions.
**anticope** (*an-tik'-op-e*) [ἀντικοπή, a beating back]. Resonance; reaction; repercussion; counterstroke.
**anticornutin** (*an-te-kor-nū'-tin*). 1. Topasol G. II, an antiseptic combination of zinc and copper ferrosulphates. 2. Topasol G. IV, a combination of iron, zinc, and calcium sulphate.
**anticoroin** (*an-te-ko'-ro-in*). Topasol G. V, an antiseptic combination of zinc, and magnesium sulphates.
**anticreatinine** (*an-te-kre-at'-in-in*). A leukomaine derived from creatinine.
**anticrisis** (*an-te-kri'-sis*) [*anti-*; *crisis*]. An agent or phenomenon preventing a crisis.
**anticritical** (*an-te-krit'-ik-al*) [*anti-*; κρίσις, a crisis]. Preventing the crisis of a disease.
**anticteric** (*ant-ik'-tur-ik*) [*anti-*; *icterus*]. 1. Efficient against jaundice. 2. An efficient agent against jaundice.
**anticus** (*an-ti'-kus*) [*anticus*, that in front]. Anterior; in front of.
**anticyclic acid.** See *acid, anticyclic*.
**anticytolysin** (*an-te-si-tol'-is-in*). A substance opposing a cytolysin.
**anticytotoxin** (*an-te-si-to-toks'-in*). A substance antagonistic in its action to a cytotoxin.
**antidiabetic** (*an-te-di-ab-et'-ik*) [*anti*; *diabetes*]. 1. Efficient against diabetes. 2. A remedy for diabetes.
**antidiabeticum** (*an-te-di-a-bet'-ik-um*). A preparation recommended for diabetes, said to consist of wheat starch, sugar of milk, sulphur, powdered senna leaves, and fennel. Syn., *glycosolveol; glycosolvol*.
**antidiabetin** (*an-te-di-ab-e'-tin*). A mixture of saccharin and mannite, used instead of sugar by diabetics.
**antidiarrheal** (*an-te-di-ar-e'-al*) [*anti-*; *diarrhea*]. Preventing or overcoming diarrhea.
**antidiastase** (*an-te-di'-as-tās*). An antibody to diastase.
**antidiastole** (*an-te-di-as'-to-le*) [ἀντιδιαστολή, distinction]. Differential diagnosis.
**antidigestive** (*an-te-di-jes'-tiv*) [*anti-*; *digestion*]. Preventing the proper digestion of the food.
**antidinic** (*an-te-din'-ik*) [*anti-*; δῖνος, a whirl]. Relieving or preventing vertigo.
**antidiphtherin** (*an-te-dif'-ther-in*). A solution containing cultures of *Bacillus diphtheriæ*, used against diphtheria. a., **Klebs'**, a preparation obtained by precipitation with alcohol from the culture-fluid of *Bacillus diphtheriæ* after removal of the bacilli.
**antidolorin** (*an-te-do'-lor-in*) [*anti-*; *dolor*, pain]. A proprietary preparation of ethyl chloride, used for the relief of superficial pain.
**antidotal** (*an-te-do'-tal*) [*anti-*; δοτός, given]. Having the nature of an antidote.
**antidote** (*an'-te-dōt*) [see *antidotal*]. An agent preventing or counteracting the action of a poison. a., **arsenical**, is prepared by dissolving 100 parts of the hydrated sulphate of iron in 250 parts of water, to which 15 parts of burnt magnesia and 250 parts of water are added. a., **chemical**, one that changes the chemical nature of the poison so that it becomes insoluble or harmless. a., **mechanical**, one that prevents absorption by holding the poison in mechanical suspension or by coating the stomach. a., **physiological**, one that counteracts the physiological effects of a poison. a., **universal**, a mixture of 1 part of dissolved iron sulphate in 2 parts of magnesia water.
**antidotism** (*an'-te-do-tizm*) [see *antidotal*]. Therapeutic or physiologic antagonism; the possession of antidotal properties; the act of giving antidotes.
**antidromic nerve impulses** (*an-te-drom'-ik*) [*anti-*; δρόμος, a running]. Nerve impulses passing in the opposite direction to the normal, such as occurs when vasodilatation follows peripheral stimulation of an afferent nerve.
**antidynamic** (*an-te-di-nam'-ik*) [*anti-*; δύναμις, force]. Weakening, depressing.
**antidyne, antidynous** (*an'-te-dīn*, *an-tid'-in-us*) [*anti-*; ὀδύνη, pain]. Anodyne.
**antidyscratic** (*an-te-dis-krat'-ik*) [*anti-*; δυσκρασία; bad temperament]. Tending to overcome, as a dyscrasia.
**antidysenteric** (*an-te-dis-en-ter'-ik*) [*anti-*; *dysentery*]. 1. Serviceable against dysentery. 2. A remedy for dysentery.
**antidysentericum** (*an-te-dis-en-ter'-ik-um*). A proprietary remedy for dysentery and chronic diarrhea, said to consist of myrobalans, pelletierine, extract of rose and gum arabic.
**antidysuric** (*an-te-dis-ū'-rik*) [*anti-*; δυσουρία, difficult micturition]. Relieving dysuria.
**antiemetic** (*an-te-em-et'-ik*) [*anti-*; *emetic*]. Preventing emesis; relieving nausea.
**antiendotoxic** (*an-te-en-do-toks'-ik*). Preventing or counteracting the effect of endotoxins.
**antiendotoxin** (*an-te-en-do-toks'-in*). An antibody which counteracts a bacterial endotoxin.
**antienzyme** (*an-te-en'-zīm*) [*anti-*; *enzyme*]. An agent which neutralizes the action of an enzyme.
**antiephialtic** (*an-ti-ef-e-al'-tik*). See *antephialtic*.
**antiepilectic** (*an-te-ep-il-ek'-tik*). See *antepilectic*.
**antifebrile** (*an-te-feb'-ril*) [*anti-*; *febris*, a fever]. An agent reducing a fever; a febrifuge.
**antifebrin** (*an-te-feb'-rin*) [*anti-*; *febris*, a fever], $C_6H_5.C_2H_3O.NH$. The proprietary name of acetanilide or phenylacetamide. A white, crystalline powder, insoluble in water, freely soluble in alcohol, ether, and chloroform. It is antipyretic and analgesic. The drug's official name is *acetanilidum*. Dose 5–10 gr. (0.3–0.6 Gm.).
**antiferment** (*an-te-fer'-ment*) [*anti-*; *fermentum*, leaven]. An agent that prevents fermentation.
**antifermentative** (*an-te-fer-men'-ta-tiv*) [*antiferment*]. Preventing fermentation.
**antiflatulent** (*an-te-flat'-u-lent*). 1. Efficient against flatulence. 2. A remedy for flatulence.
**antiformin** (*an-te-for'-min*). Trade name of a disinfectant preparation containing solution of potassium or sodium hypochlorite and of sodium hydrate. It has a powerful solvent action on certain organic substances; and is used in the laboratory for the separation of tubercle bacilli from sputum, urine, and other pathological products which contain these bacilli.
**antifungin** (*an-te-fun'-jin*). Magnesium borate, used as a gargle.
**antigalactagogue** (*an-te-gal-ak'-ta-gog*) [*anti-*; γάλα, milk; ἀγωγός, leading]. Same as antigalactic.
**antigalactic** (*an-te-gal-ak'-tik*) [*anti-*; γάλα, milk]. 1. Lessening the secretion of milk. 2. A drug that lessens the flow of milk.
**antigen** (*an'-te-jen*) [*anti-*; γεννᾶν, to produce]. Any bacterium or substance which, when injected into an organism, is capable of causing the formation of an antibody.
**antigermin** (*an-te-jer'-min*). A compound of copper and an acid, forming a yellowish-green, tenacious mass, soluble in 200 parts of water. It is said to be disinfectant, deodorant, and bactericidal.
**antigerminal** (*an-te-jer'-min-al*) [*anti-*; *germ*]. Relating to the pole of the ovum opposed to the germinal pole.
**antigonorrheic** (*an-te-gon-o-re'-ik*). A substance which is capable of aiding in the cure of gonorrhea.
**antihelix** (*an-te-he'-liks*). See *anthelix*.
**antihemagglutinin** (*an-te-hem-ag-glū'-tin-in*). A substance opposed in action to the hemagglutinins (*q. v.*).
**antihemolysin** (*an-te-he-mol'-is-in*) [*anti-*; αἷμα, blood; λύσις, solution]. A complex substance in the blood-serum developed by inoculations with hemolysins. It is an antibody to hemolysin; and is composed of anticomplements and antiimmune bodies.
**antihemolytic** (*an-te-hem-o-lit'-ik*). Relating to an antihemolysin; not capable of dissolving blood-corpuscles. Preventing hemolysis.
**antihemorrhoidal** (*an-te-hem-or-oid'-al*). 1. Effective against hemorrhoids. 2. A remedy for hemorrhoids.
**antiherpetic** (*an-te-her-pet'-ik*) [*anti-*; *herpes*]. Preventing herpes.
**antihidrotic** (*an-te-hi-drot'-ik*) [*anti-*; ἱδρώς, sweat]. 1. Diminishing the secretion of sweat. 2. An agent that lessens perspiration.
**antihormone** (*an-te-hor'-mōn*) [*anti-*; *hormone*]. A hormone which counteracts another hormone; an antagonistic hormone; a chalone.
**antihydropic** (*an-te-hi-drop'-ik*). See *anthydropic*.
**antihydropin** (*an-te-hi'-dro-pin*) [*anti-*; ὕδωρ, water]. A crystalline principle obtainable from the common cockroach, *Blatta (Periplaneta) orientalis*, and said to be diuretic. Dose 10–20 gr. (0.6–1.3 Gm.).
**antihysterical** (*an-te-his-ter'-ik-al*). Relieving or inhibiting hysteria.
**anticteric** (*an-te-ik-ter'-ik*) [*anti-*; *icteric*]. Serviceable against jaundice.
**antiimmune bodies.** See under *body*.

**antiisolysin** (an-te-i-sol'-is-in). A substance which is capable of counteracting the action of an isolysin.

**antikamnia** (an-te-kam'-ne-ah) [anti-; κάμνειν, to suffer pain]. A proprietary remedy said to be composed of sodium bicarbonate, acetanilide, and caffeine. It is used as an analgesic in doses of 5-10 gr. (0.32-0.65 Gm.).

**antikathode** (an-te-kath'-ōd) [anti-; kathode]. A piece of platinum foil so placed in a Crookes tube as to intercept the kathode rays; being thus rendered fluorescent, it becomes a source of roentgen-rays.

**antiketogen** (an-te-ke'-to-jen) [anti-; ketogen]. A substance which produces antiketogenesis.

**antiketogenesis** (an-te-ke-to-jen'-es-is) [anti-; ketone (acetone); genesis]. The dimination of acidosis by the oxidation of sugar and allied substances in the body.

**antiketogenic** (an-te-ke-to-jen'-ik) [anti-; ketone (acetone); γεννᾶν, to produce]. 1. Pertaining to antiketogenesis. 2. Preventing the formation of acetone.

**antikinase** (an-te-ki'-nās). An antibody to kinase.

**antikol** (an'-tik-ol). A proprietary antipyretic mixture said to contain acetanilide, sodium bicarbonate, and tartaric acid.

**antilabium** (an-te-la'-be-um). See antelabium.

**antilactase** (an-te-lak'-tās). Antibody which counteracts lactase.

**antilactic** (an-te-lak'-tik). See antigalactic.

**antilactoserum** (an-te-lak-to-se'-rum). A substance antagonistic in its action to lactoserum (q. v.).

**antilemic, antilœmic, antiloimic** (an-te-le'-mik, an-te-loi'-mik) [anti-; λοιμός, the plague]. Efficacious against the plague or other pestilence.

**antileprol** (an-te-lep'-rol). The ethyl ester of chaulmoogra acid, recommended in place of chaulmoogra oil in treatment of leprosy.

**antilepsis** (an-til-ep'-sis) [ἀντίληψις, a receiving in return]. 1. The treatment of disease by the application of the remedy to a healthy part; revulsive treatment. 2. A taking root. 3. A taking effect. 4. A seizure; an attack. 5. The support of a bandage.

**antileptic** (an-til-ep'-tik) [ἀντιληψις, a receiving in return]. 1. Revulsive. 2. Supporting, assisting.

**antilethargic** (an-te-leth-ar'-jik). 1. Arresting lethargy; hindering sleep. 2. An agent efficacious against lethargy.

**antileukocidin** (an-te-lu-ko'-si-din). The antibody for the leukocytic poison of the streptococcus.

**antileukotoxin** (an-te-lu-ko-tok'-sin). The antibody to a leukotoxin.

**antilipase** (an-te-lip'-ās). A substance inhibiting or counteracting a lipase.

**antilithemic** (an-te-lith-e'-mik) [anti; lithemia]. Correcting lithemia.

**antilithic** (an-te-lith'-ik) [anti-; λίθος, a stone]. 1. Efficacious against calculus. 2. An agent preventing the deposit of urinary sediment.

**antilobium** (an-te-lo'-be-um) [anti-; λοβός, the lobe of the ear]. The tragus or part of the ear opposite the lobe.

**antilœmic** (an-ti-le'-mik). See antilemic.

**antiluetic** (an-te-lu-et'-ik) [anti-; lues, the plague; syphilis]. Efficacious against syphilis.

**antilypyrin** (an-te-le-pi'-rin). An antipyretic and analgesic substance obtained by heating acetanilide, 1 part, with antipyrine, 2 parts. Dose 7-8 gr. (0.45-0.52 Gm.).

**antilysin** (an-til'-is-in) [anti-; λύσις, a loosing]. A substance opposed to the activity of a lysin.

**antilysis** (an-til'-is-is). The condition due to the activity of antilysins.

**antilyssic** (an-te-lis'-ik) [anti-; λύσσα, rabies]. 1. Tending to cure rabies. 2. A remedy for rabies.

**antilytic**. Relating to the action of an antilysin.

**antimalarial** (an-te-mal-a'-re-al). Preventing or curing malaria.

**antimaniacal** (an-te-ma-ni'-ak-al) [anti-; μανία, madness]. Overcoming insanity.

**antimellin** (an-te-mel'-in). A remedy employed in diabetes purporting to be a glucoside separated from the fruit of Eugenia jambolana.

**antimephitic** (an-te-mef-it'-ik) [anti-; mephitis, a pestilential exhalation]. Efficacious against foul exhalations or their effects.

**antimere** (an'-te-mēr) [anti-; μέρος, a part]. 1. Any one of the segments of the body that are bounded by planes typically at right angles to the long axis of the body. 2. A homotype.

**antimerology** (an-te-mer-ol'-o-je) [anti-; μέρος, a part; λόγος, science]. The science of homotypic parts.

**antimetropia** (an-te-met-ro'-pe-ah) [anti-; metropia]. A condition characterized by opposing states of refraction in the two eyes, as, for example, the existence of myopia in one eye and of hyperopia in the other.

**antimiasmatic** (an-te-mi-as-mat'-ik) [anti-; μίασμα, exhalation]. Preventive of malaria.

**antimicrobic** (an-te-mi-kro'-bik) [anti-; microbe]. Arresting the development of microbes; antibacterial.

**antimicrophyte** (an-te-mik'-ro-fit) [ἀντί, against; μικρός, small; φυτόν, plant]. A germicide.

**antimigraine** (an-te-mig'-rān). A proprietary preparation said to consist of caffeine, antipyrine and sugar. Dose, 1.5 gm. Syn., antihemicranin.

**antimonial** (an-te-mo'-ne-al) [antimonium, antimony]. Containing antimony.

**antimonic** (an-te-mon'-ik) [see antimonial]. A term applied to those compounds of antimony that correspond to its higher oxide.

**antimonide** (an'-te-mo-nid). Any binary combination of antimony.

**antimonious** (an-te-mo'-ne-us) [see antimonial]. A term denoting those compounds of antimony that correspond to its lower oxide.

**antimonium** (an-te-mo'-ne-um). See antimony.

**antimony** (an'-te-mo-ne) [L., antimonium]. Sb = 120.2; quantivalence III and V. A metallic, crystalline element possessing a bluish-white luster. The symbol Sb is derived from the old name, stibium. Antimony is found native, as the sulphide, $Sb_2S_3$, as the oxide, and is a constituent of many minerals. It is used commercially chiefly for making alloys. Type-metal, Britannia metal, and Babbitt antifriction metal are alloys of antimony. In medicine salts of antimony are used less frequently than formerly. The salts are cardiac and arterial depressants, diaphoretic and emetic, and in large doses powerful gastrointestinal irritants, producing symptoms resembling those of Asiatic cholera. Antimony has been used as an antiphlogistic in sthenic inflammation, as a diaphoretic and expectorant, and as an emetic. a. arsenate, a heavy white powder; it is used in syphilitic affections of the skin. Dose $\frac{1}{10}$ gr. (0.001 Gm.) 4 times daily. a. arsenite, a fine white powder; it is used in skin diseases. a. chloride, $SbCl_3$, the "butter" of antimony; a strong caustic. a. iodide, $SbI_3$, red crystals, decomposed by water, soluble in carbon disulphide; melts at 167° C. It is alterative. Dose $\frac{1}{4}$-1 gr. (0.016-0.065 Gm.) in pills. a. oxychloride, the "powder of algaroth"; now little used. a. pentoxide, $Sb_2O_5$, antimonic acid, combines with bases to form antimoniates. a., pills of, compound (pilulæ antimonii compositæ, B. P.), Plummer's pills, contain calomel and sulphureted antimony, of each, $\frac{1}{2}$ gr. (0.032 Gm.). a. and potassium tartrate (antimonii et potassii tartras, U. S. P.; antimonium tartaratum, B. P.), $2KSbOC_4H_4O_6 \cdot H_2O$, "tartar emetic." Dose $\frac{1}{16}$-$\frac{1}{4}$ gr. (0.004-0.016 Gm.). a., powder of (pulvis antimonialis, B. P.), antimonial powder, James' powder, consists of antimonious-oxide 33, and calcium phosphate 67 parts, and is diaphoretic; in large doses, emetic and cathartic. Dose 3-8 gr. (0.2-0.5 Gm.). a. sulphide, $SbS_3$, black sulphide of antimony. Dose $\frac{1}{4}$ gr. (0.016-0.065 Gm.). a. sulphide, golden, $Sb_2S_5$, a fine, odorless, orange-yellow powder, soluble in alkaline solutions. It is alterative, diaphoretic, emetic, and expectorant. Dose $\frac{1}{6}$-$1\frac{1}{2}$ gr. (0.01-0.1 Gm.) several times daily. a., sulphureted (antimonium sulphuratum, B. P.), the sulphide with a small but indefinite amount of the oxide. Dose 1-5 gr. (0.065-0.32 Gm.). a. tartrate, $(SbO)_2C_4H_4O_6 + H_2O$, a white, crystalline powder. Used internally as a substitute for arsenic in affections of the skin. Dose $\frac{1}{10}$ gr. (0.0065 Gm.) 3 to 5 times daily. a. trioxide, antimonious acid, $Sb_2O_3$; soluble in hydrochloric and tartaric acids. Dose 1-2 gr. (0.065-0.13 Gm.). It is an ingredient of James' powder. a., vegetable, honeset, a., wine of (vinum antimonii, U. S. P.), boiling water, 60; tartar emetic, 4; stronger white wine, 1000 parts. It contains about 2 gr. of tartar emetic to the ounce. Dose 5-15 min. (0.3-1.0 Cc.).

**antimonyl** (an'-tim-on-il). SbO. The univalent radical of antimonous compounds.

**antimucorin** (an-te-mū'-kor-in). Topasol G. III, an antiseptic preparation of iron and zinc sulphate.

**antimycetic** (an-te-mi-se'-tik) [anti-; μύκης, fungus]. 1. See actinomycotic. 2. A fungicide.

**antimycotic** (*an-te-mi-kot'-ik*) [*anti-*; μύκης, a fungus]. Destructive to microorganisms.

**antimydriatic** (*an-te-mid-re-at'-ik*) [*anti-*; μυδρίασις, mydriasis]. 1. Opposed to or arresting dilatation of the pupils. 2. A drug efficacious against mydriasis.

**antinarcotic** (*an-te-nar-kot'-ik*) [*anti-*; νάρκωσις, a benumbing]. Preventing narcosis.

**antinausea** (*an-te-naw'-se-ah*). A remedy for seasickness, said to consist of cocaine and antipyrine.

**antinephritic** (*an-te-nef-rit'-ik*) [*anti-*; νεφρός, the kidney; ιτις, inflammation]. Preventing or curative of renal disease.

**antinervin** (*an-te-ner'-vin*) [*anti-*; *nervus*, a tendon or nerve]. Salbromalide, a mixture of bromacetanilide and salicylanilide; used for the relief of neuralgia.

**antineuralgic** (*an-te-nu-ral'-jik*) [*anti-*; νεῦρον, a nerve; ἄλγος, pain]. Overcoming neuralgia.

**antineuritic** (*an-te-nu-rit'-ik*). 1. Efficient in neuritis. 2. A remedy against neuritis.

**antineuropathic** (*an-te-nu-ro-path'-ik*) [*anti-*; νεῦρον, nerve; πάθος, a disease]. 1. Efficient against nervous disorders. 2. A remedy efficient in nervous diseases.

**antineurotic** (*an-te-nu-rot'-ik*) [*anti-*; νεῦρον, a nerve]. A remedy of service in nervous diseases.

**antineurotoxin** (*an-te-nu-ro-tok'-sin*). A substance which inhibits or counteracts a neurotoxin.

**antiniad** (*an-tin'-e-ad*) [*anti-*; ἰνίον, the nape of the neck]. Toward the antinion; glabellad.

**antinial** (*an-tin'-e-al*) [*anti-*; ἰνίον, the nape of the neck]. Pertaining to the antinion.

**antinien** (*an-tin'-e-en*) [*anti-*; ἰνίον, the nape of the neck]. Belonging to the antinion in itself.

**antinion** (*an-tin'-e-on*) [*anti-*; ἰνίον, the nape of the neck]. See *craniometrical points*.

**antinonnin** (*an-te-non'-in*), C₆H₅.(NO₂)₃.CH₂OK, potassium orthodinitrocresylate. See *dinitrocresol*.

**antinosin** (*an-te-no'-sin*) [*anti-*; νόσος, disease]. Tetraiodophenolphthalein, the soluble sodium salt of nosophen; it is a greenish-blue antiseptic powder, used in powder or in solutions of 1 : 1000, for irrigations or gargle.

**antiobesic** (*an-te-o-be'-sik*) [*anti-*; *obesity*]. 1. Efficient against corpulence. 2. A remedy for corpulence.

**antiodontalgic** (*an-te-o-don-tal'-jik*) [*anti-*; ὀδούς, tooth; ἄλγος, pain]. Curative of toothache.

**antiopsonin** (*an-te-op'-son-in*). A substance retarding or destroying the action of an opsonin.

**antiorgastic** (*an-te-or-gas'-tik*) [*anti-*; ὀργασμός, swelling, excitement]. Anaphrodisiac.

**antiotomia, antiotomy** (*an-te-o-to'-me-ah, an-te-ot'-om-e*) [ἀντίας, a tonsil; τέμνειν, to cut]. Excision of the tonsils.

**antipaludean** (*an-te-pal-u'-de-an*) [*anti-*; *palus*, a marsh]. Efficient against malarial diseases.

**antiparalytic** (*an-te-par-al-it'-ik*) [*anti-*; *paralysis*]. 1. Efficient against paralysis. An agent or remedy efficacious in paralysis.

**antiparasitic** (*an-te-par-as-it'-ik*) [*anti-*; παράσιτος, a parasite]. 1. Destroying parasites. 2. An agent destroying parasites.

**antiparastata** (*an-te-par-as'-tat-ah*) [*anti-*; παραστάτης, testicle]. Cowper's glands.

**antiparastatitis** (*an-te-par-as-tat-i'-tis*) [*anti-*; παραστάτης, a testicle]. Inflammation of Cowper's glands.

**antipathic** (*an-te-path'-ik*) [*anti-*; πάθος, disease]. 1. A synonym of *allopathic*, both terms alike being rejected by the advocates of rational medicine. 2. Producing contrary symptoms. 3. Antagonistic. 4. Anodyne.

**antipathy** (*an-tip'-a-the*) [*anti-*; πάθος, affection]. 1. Aversion; an opposing property or quality. 2. Morbid disgust or repugnance for particular objects. 3. Allopathy (*q. v.*). 4. An object exciting morbid dislike or aversion. 5. Chemical incompatibility. **a., insensile**, morbid repugnance excited by the presence of some object which was not perceived by any of the senses. **a., sensile**, morbid aversion aroused by some appreciable quality of the exciting object.

**antipeptone** (*an-te-pep'-tōn*) [*anti-*; πέπτειν, to cook; digest]. A variety of peptone not acted upon by trypsin.

**antiperiodic** (*an-te-pe-ri-od'-ik*) [*anti-*; περίοδος, a going round]. 1. Preventing periodic attacks of a disease. 2. A remedy against periodic disease. **a. tincture**, see *Warburg's tincture*.

**antiperistalsis** (*an-te-per-is-tal'-sis*) [*anti-*; περί, around; στάλσις, compression]. Reversed peristalsis; inverted or upward peristaltic action.

**antiperistaltic** (*an-te-per-is-tal'-tik*) [see *antiperistalsis*]. Relating to antiperistalsis.

**antiperonosporin** (*an-te-per-o-nos'-por-in*). Topasol G. I, an antiseptic preparation of zinc and copper sulphates.

**antiphagin** (*an-tif'-a-jin*) [*anti-*; *phagocyte*]. A substance formed in virulent bacteria which protects them against phagocytosis.

**antiphagocytic** (*an-te-fag-o-sit'-ik*). Protecting against or preventing phagocytosis.

**antiphialtic** (*ant-if-e-al'-tik*) [*anti-*; ἐφιάλτης, nightmare]. Preventive of nightmare.

**antiphlogistic** (*an-te-flo-jis'-tik*) [*anti-*; φλόγωσις, inflammatory heat]. 1. Counteracting fever. 2. An agent subduing or reducing inflammation or fever. 3. Applied to the pneumatic theory of Lavoisier as having supplanted Stahl's phlogistic theory. **a. treatment**, bloodletting, the application of cold, the administration of antipyretics, etc.

**antiphlogistine** (*an-te-flo-jis'-tin*) [see *antiphlogistic*]. Trade name of a paste said to consist of kaolin or purified clay, glycerol, and antiseptics; it is a substitute for poultices.

**antiphlogosis** (*an-te-flo-go'-sis*) [see *antiphlogistic*]. 1. The reduction of inflammation. 2. Inflammation purposely excited to counteract other inflammation.

**antiphone** (*an'-te-fōn*) [*anti-*; φωνή, sound]. An appliance worn in the auditory meatus, and intended to protect the wearer from noises.

**antiphthiriac, antiphtheiriac** (*an-te-thi'-re-ak*) [*anti-*; φθείρ, a louse]. 1. Efficient against lice or the condition caused by them. 2. An agent effective against lice.

**antiphthisic** (*an-te-tiz'-ik*) [*anti-*; φθίσις, a wasting]. Efficient against phthisis. An agent checking in phthisis.

**antiphthisin** (*an-te-ti'-sin*). A modified tuberculin, made from the slight residue after precipitation with sodium bismuth iodide.

**antiphymin** (*an-te-fi'-min*) [*anti-*; φῦμα, a tubercle]. Trade name of a preparation used in tuberculosis. It is said to consist of formaldehyde, ozone, sulphur dioxide. Used by inhalation in tuberculosis.

**antiphytosin** (*an-te-fi-to'-sin*). A preparation resembling tuberculin.

**antipilus** (*an-te-pi'-lus*) [*anti-*; *pilus*, a hair]. Trade name of a preparation for removing hair.

**antiplasis** (*an-te-pla'-sis*). See *antiplasm*.

**antiplasm** (*an'-te-plasm*) [*anti-*; πλάσμα, a thing molded]. 1. Formation according to a pattern. 2. Remolding into the normal form.

**antiplastic** (*an-te-plas'-tik*) [*anti-*; πλάσσειν, to form]. 1. Unfavorable to granulation or to the healing process. 2. An agent impoverishing the blood. 3. Preventing or checking plastic exudation.

**antipleuritic** (*an-te-plu-rit'-ik*) [*anti-*; πλευρῖτις, pleurisy]. Overcoming pleurisy.

**antipneumonic** (*an-te-nu-mon'-ik*) [*anti-*; *pneumonia*]. Of value in treating pneumonia.

**antipneumotoxin** (*an-te-nu-mo-toks'-in*). An antitoxin opposing pneumotoxin.

**antipodagric** (*an-te-po-dag'-rik*) [*anti-*; ποδαγρά, gout]. Efficacious against gout.

**antipodal** (*an-tip'-od-al*) [*anti-*; πούς, a foot]. Situated directly opposite. **a. cells**, a term applied to a group of four cells formed in the lower end of the embryo-sac opposite to the cells constituting the egg-apparatus. **a. cone**, the cone of astral rays opposite to the spindle-fibers.

**antipraxia** (*an-te-praks'-e-ah*) [*anti-*; πράσσειν, to do]. Antagonism of functions or of symptoms.

**antiprecipitin** (*an-te-pre-sip'-it-in*). A substance antagonistic to a precipitin (*q. v.*).

**antiprostate** (*an-te-pros'-tāt*). See *anteprostate*.

**antipruritic** (*an-te-pru-rit'-ik*) [*anti-*; *pruritus*, itching]. 1. Relieving the sensation of itching. 2. A drug that relieves the sensation of itching.

**antipsoric** (*an-tip-so'-rik*) [*anti-*; ψώρα, the itch]. Effective against itching or the itch.

**antiputrefactive** (*an-te-pu-tre-fak'-tiv*). See *antiseptic*.

**antipyic** (*an-te-pi'-ik*) [*anti-*; πύον, pus]. Checking or restraining suppuration.

**antipyogenic** (*an-te-pi-o-jen'-ik*) [*anti-*; πύον, pus; γενᾶν, to form]. Preventing or counteracting suppuration.

**antipyonin** (*an-te-pi'-on-in*). Sodium tetraborate.

**antipyresis** (*an-te-pi-re'-sis*) [*anti-*; πυρετός, fever]. The reduction of fever by means of antipyretics.

**antipyretic** (*an-te-pi-ret'-ik*) [see *antipyresis*].

1. Cooling; lowering the temperature. 2. An agent reducing temperature. The most important antipyretic agents are cold, diaphoretics, and the newer remedies, many of which are coal-tar products, such as antipyrine, acetanilide, phenacetin, etc.

**antipyrine, antipyrin** (*an-te-pi'-rin*) [*anti-*; πῦρ, fever heat], $C_{11}H_{12}N_2O$. Phenazone. The scientific name is dimethyloxychinicin-phenyldimethylpyrazolon, or dihydrodimethylphenylpyrazine. An alkaloidal product of the destructive distillation of coaltar. It may be produced by heating acetoacetic ester with methylphenylhydrazine. It is a grayish or reddish-white, crystalline powder, slightly bitter, soluble in water, alcohol, and chloroform, and crystallizes from an ethereal solution in shining leaflets melting at 113°. It reduces temperature, causes sweating, at times vomiting, peculiar eruptions, pruritus, coryza, etc. Not rarely a cyanotic condition of the face and hands is produced. Antipyrine is incompatible with nitrous compounds. It is a powerful antipyretic and analgesic. Dose 5–15 gr. (0.3–1.0 Gm.). a. **bichloral**, a trituration-product of 94 parts of antipyrine with 165.5 parts of chloral hydrate; it is hypnotic and analgesic. Maximum dose 45 gr. (3 Gm.). Syn., *dichloralantipyrine*. a. **mandelate**, a crystalline compound of antipyrine and amygdalic acid, used as a remedy for whoopingcough. Dose ½–8 gr. (0.05–0.5 Gm.). Syn., *tussol;* *phenylglycollate.* a. **salicylate**, a. **salol**, a brown liquid obtained by fusing together equal parts of phenyl salicylate and antipyrine. It is recommended as an antiseptic, and as a hemostatic in uterine hemorrhage, applied by means of cotton tampons. Syn., *salipyrine.* a., test for, see *Fieux*.

**antipyrinomania** (*an-te-pi-rin-o-ma'-ne-ah*) [*anti-*; πυρετός, fever; μανία, madness]. A condition similar to morphinism, due to excessive use of antipyrine. It is marked by nervous excitement.

**antipyrotic** (*an-te-pi-rot'-ik*) [*anti-*; πύρωσις, a burning]. 1. Efficacious against burns. 2. An agent curative of burns.

**antirabic** (*an-te-ra'-bik*) [*anti-*; rabies, madness]. Preventing or curing rabies.

**antirennene** (*an-te-ren'-ēn*). Morgenroth's name for the principle which appears in the blood of an animal following the introduction of rennet. It has the power of impeding the action of rennet on milk.

**antirheumatic** (*an-te-ru-mat'-ik*) [*anti-*; *rheumatism*]. Preventing or curing rheumatism.

**antirheumaticum** (*an-te-ru-mat'-ik-um*). A compound of sodium salicylate and methylene-blue. It occurs in blue, prismatic crystals, soluble in water and alcohol. Dose 1–1½ gr. (0.06–0.09 Gm.).

**antirheumatin** (*an-te-ru'-mat-in*). An ointment used in treatment of rheumatism, and said to contain fluorphenetol, 1 part; difluordiphenyl, 4 parts; vaselin, 10 parts; woolfat, 85 parts.

**antirheumol** (*an-te-ru'-mol*). A solution of the glycerin ester of salicylic acid in glycerin and alcohol. It is used as a liniment in rheumatism.

**antiricin** (*an-te-ris'-in*). The antibody to ricin. Its action is inhibited by cold and accelerated by heat.

**antirrhachitic** (*an-te-rak-it'-ik*) [*anti-*; ῥάχις, the spine]. Serviceable against rickets.

**antirrheoscope** (*an-te-re'-o-skōp*) [ἀντίρροια, a flowing back; σκοπεῖν, to view]. J. J. Oppel's device for observing the manifestations of visual vertigo.

**Antirrhinum** (*an-te-ri'-num*) [L.]. A genus of scrophulariaceous plants. A. *linaria*, called also *Linaria vulgaris*, toadflax, ramsted, "butter-and-eggs," is a herbaceous plant of Europe and North America; diuretic, cathartic, and irritant; used as a poultice and fomentation.

**antiscabin** (*an-te-ska'-bin*). A preparation said to consist of beta-naphthol, balsam of Peru, soap, glycerin, boric acid, and alcohol. It is used in the treatment of scabies.

**antiscabious** (*an-te-ska'-be-us*) [*anti*, against; *scabies*]. Effective against the itch.

**antiscarlatinal** (*an-te-skar-lat'-in-al*) [*anti*, against; *scarlatina*]. Efficient against scarlet fever.

**antiscirrhous** (*an-te-skir'-us*). Efficient against scirrhus.

**antisclerosin** (*an-te-skle-ro'-sin*). Trade name of a preparation of various inorganic salts, similar to Truncek's serum, used in arteriosclerosis to lessen the intra-arterial pressure.

**antiscolic** (*an-te-skol'-ik*) [ἀντί, against; σκώληξ, a worm]. Vermifuge. See *anthelmintic*.

**antiscorbutic** (*an-te-skor-bū'-tik*) [*anti-*; *scorbutus*, scurvy]. 1. Effective against scurvy. 2. A remedy useful in scurvy.

**antisecosis** (*an-te-sek-o'-sis*) [*anti-*; σηκοειν, to weigh, balance]. 1. A restoration of health, strength, etc. 2. Regulation of the food.

**antisensibilisin** (*an-te-sen-sib-il'-is-in*). One of the substances in an antigen.

**antisensitizer** (*an-te-sen'-sit-i-zer*). In Ehrlich's side-chain theory, a substance antagonistic in its action to that of the intermediary body or sensitizer.

**antisepsin** (*an-te-sep'-sin*) [*anti-*; σῆψις, putrefaction], $C_6H_4BrNHC_2H_3O$. Asepsin; bromated acetanilide; soluble in alcohol and ether, insoluble in water. It is antipyretic, analgesic, and antiseptic. Dose 6–7 gr. (0.39–0.45 Gm.).

**antisepsis** (*an-te-sep'-sis*) [see *antisepsin*]. Exclusion of the germs that cause putrefaction.

**antiseptic** (*an-te-sep'-tik*) [see *antisepsin*]. 1. Having power to prevent the growth of the bacteria upon which putrefaction depends. 2. An agent that prevents development of bacteria. Among the principal antiseptics are mercuric chloride, creolin, phenol, iodoform, thymol, salicylic acid, boric acid, formaldehyde, and potassium permanganate. a. **gauze**, open cotton cloth charged with an antiseptic. a. **ligature**, catgut or other material rendered aseptic by soaking in antiseptic solutions. a. **treatment of wounds**, this looks to thorough antisepsis as regards the wound, the instruments, the operator's hands, the dressings, etc.

**antisepticin** (*an-te-sep'-tis-in*). Trade name of an antiseptic mixture containing benzoic and boric acids, thymol and eucalyptol.

**antisepticism** (*an-te-sep'-tis-izm*) [see *antisepsin*]. The theory or systematic employment of antiseptic methods.

**antisepticize** (*an-te-sep'-tis-īz*) [see *antisepsin*]. To render antiseptic; to treat with antiseptics.

**antisepticol** (*an-te-sep'-tik-ol*). Trade name of a liquid antiseptic said to contain boric acid, sodium borate, benzoic acid, thymol, eucalyptol, menthol and oil wintergreen.

**antiseptin** (*an-te-sep'-tin*) [see *antisepsin*]. 1. Zinc borothymoliodide. It consists of 85 parts zinc sulphate, 2.5 parts each of zinc iodide and thymol, and 10 parts boric acid. It is an antiseptic. 2. A proprietary preparation said to consist of sodium or potassium silicate, 2 parts, and a 0.1 % solution of mercuric chloride, 1 part.

**antiseptol** (*an-te-sep'-tol*) [see *antisepsin*]. Cinchonine iodosulphate, an odorless and fairly effective substitute for iodoform.

**antiserum** (*an-te-se'-rum*). A serum having the power of agglutinating and precipitating another serum. a. **method**, a method of differentiating human from other blood; modified Uhlenhuth's antiserum method. Human blood-serum is injected into the peritoneal cavity of rabbits in doses of 10 Cc. every 8 or 10 days. After 6 injections their blood is collected and preserved on ice; the serum is pipeted off after 24 hours. Some rabbits, as control-animals, are not injected. The blood to be tested is, if dried, first dissolved, and then, as in fluid blood, diluted with ordinary water and salt solution. Several drops of the test-serum are added and the tubes placed at a temperature of 35°. If the blood to be tested is human, a turbidity appears invariably; if not human, it remains clear.

**antisialagogue, antisialagog** (*an-te-si-al'-a-gog*) [*anti-*; σίαλον, saliva; ἀγωγός, leading]. 1. Preventing or checking salivation. 2. A remedy that is effective against salivation.

**antisialic** (*an-te-si-al'-ik*) [*anti-*; σίαλον, saliva]. 1. Checking the flow of saliva. 2. An agent that checks the secretion of saliva.

**antisideric** (*an-te-sid-er'-ik*) [*anti-*; σίδερος, iron]. 1. Incompatible with iron and counteracting its effects; impoverishing the blood. 2. An agent or drug opposed to the action of iron; one which impoverishes the blood.

**antispasmin** (*an-te-spaz'-min*), $C_{29}H_{36}NO_8Na + 3NaC_7H_5O_3$. A compound of 1 molecule of narceine sodium united with 3 molecules of sodium salicylate, occurring as a white, slightly hygroscopic powder containing about 50 % of narceine. It is sedative and hypnotic. Dose ¼–1½ gr. (0.01–9.1 Gm.).

**antispasmodic** (*an-te-spaz-mod'-ik*) [*anti-*; σπασμός, a spasm]. 1. Tending to relieve spasm. 2. An agent relieving convulsions or spasmodic pains, as the narcotics, the nitrites, etc.

**antispastic** (an-te-spas'-tik) [anti-; σπαστικός, drawing]. 1. Revulsive; counterirritant. 2. Antispasmodic. 3. A revulsive agent. 4. An antispasmodic.

**antispermotoxin** (an-te-spur-mo-toks'-in). A substance opposed in its action to spermotoxin.

**antispirochetic** (an-te-spi-ro-ke'-tik) [anti-; spirochæte, a genus of bacteria]. 1. Arresting the action of spirochetes. 2. An agent having this power.

**antisplenetic** (an-te-splen-et'-ik) [anti-; splen, the spleen]. Remedial in diseases of the spleen.

**antistaphylolysin** (an-te-staf-il-ol'-is-in) [anti-; staphylococci, a genus of bacteria; λύσις, a loosing]. A substance antagonistic to the toxic products of staphylococci, contained in healthy blood-serum.

**antistasis** (an-tis'-tas-is) [anti-; στάσις, a standing]. Opposition; opposing effect.

**antistatic** (an-tis-tat'-ik), Antagonistic.

**antisternum** (an-te-stur'-num). The part of the back opposite the breast.

**antistreptococcic** (an-te-strep-to-kok'-sik) [anti-; streptococci, a genus of bacteria]. Antagonistic to or preventing the action of streptococci.

**antistreptococcin** (an-te-strep-to-kok'-sin). 1. The streptococcus-antitoxin. 2. A serum used in erysipelas.

**antistrumous** (an-te-stru'-mus) [anti-; struma, a scrofulous tumor]. Effective against struma or scrofula.

**antisudoral** (an-te-sū'-dor-al) [anti-; sudor, sweat]. Checking the secretion of sweat.

**antisudorific** (an-te-sū-dor-if'-ik) [anti-; against; sudor, sweat; facere, to make]. Checking the excretion of sweat.

**antisudorin** (an-te-sū'-dor-in) [anti-; sudor, sweat]. A proprietary mixture said to consist of boric, citric, and salicylic acids, borax, glycerin, alcohol, distilled water, and several ethers; it is used to diminish sweating of the feet.

**antisyphilitic** (an-te-sif-il-it'-ik) [anti-; syphilis]. 1. Effective against syphilis. 2. A remedy used in the treatment of syphilis.

**antitabetic.** An agent used to mitigate or aid in the cure of tabes dorsalis.

**antitetanic** (an-te-tet-an'-ik). Noting an agent used to mitigate or aid in the cure of tetanus.

**antithenar** (an-te-the'-nar) [anti-; θέναρ, the flat of the hand or the sole of the foot]. 1. Opposite to thenar. 2. A muscle that extends the thumb or opposes it to the hand; an antithenar muscle. a. eminence, the border of the palm of the hand from the base of the little finger to the wrist. a. muscles, of the toe and of the thumb; the abductor pollicis pedis and the flexor brevis pollicis manus; also, the first dorsal interosseous muscle.

**antithermic** (an-te-ther'-mik) [anti-; θέρμη, heat]. Cooling; antipyretic.

**antithermin** (an-te-ther'-min) [see antithermic], C₁₁H₁₄O₂N₂. Phenylhydrazinlevulinic acid, a coaltar derivative used as an antipyretic, analgesic, and antiseptic. Dose 5 gr. (0.3 Gm.).

**antithermolin** (an-te-ther'-mo-lin). Trade name of clay preparation used as an anodyne and antiphlogistic.

**antithrombin** (an-te-throm'-bin). A substance of the nature of a ferment, having the power of retarding or preventing coagulation.

**antithyroidin** (an-te-thi-roid'-in). See serum, thyroid.

**antitonic** (an-te-ton'-ik). 1. Counteracting the effects of a tonic. 2. A drug having opposite effects to those of a tonic. 3. Diminishing tone or tonicity.

**antitoxic** (an-te-toks'-ik) [anti-; τοξικόν, poison]. Antidotal; counteracting poisons.

**antitoxigen** (an-te-tok'-sij-en) [antitoxin; γεννάν, to produce]. Any substance which induces the production or increase of antitoxin in the blood.

**antitoxin** (an-te-toks'-in) [see antitoxic]. A counterpoison or antidote elaborated by the body to counteract the toxins of bacteria. According to some authorities, antitoxins are, like the toxins, bacterial products. Antitoxins are used in the treatment of certain infectious diseases and also to confer immunity against these diseases. 2. The commercial name for a fine white powder said to be a coal-tar product and used as an analgesic and antipyretic. Dose 10–15 gr. (0.65–1.0 Gm.) in from 1 to 4 hours. a., artificial, an antitoxin prepared by passing an electric current through a toxic bouillon.

a., diphtheria, one prepared from the blood-serum of an animal inoculated with Bacillus diphtheriæ. a., tetanus, one prepared from the blood-serum of an animal inoculated with Bacillus tetani. a. unit, 10 times the amount of serum requisite to neutralize completely 10 times the minimum fatal dose of diphtheria toxin in a half-grown guinea-pig; or the amount of antitoxin which, when inoculated into a guinea-pig of 250 Gm. weight, will neutralize 100 times the minimum fatal dose of toxin of standard weight.

**antitragic** (an-te-traj'-ik) [anti-; τράγος, the tragus]. Pertaining to the antitragus. a. muscle, a mere rudiment in man; it arises from the antitragus, and extends to the cauda of the helix.

**antitragus** (an-te-tra'-gus). An eminence of the external ear opposite the tragus.

**antitrismus** (an-te-tris'-mus) [anti-; τρισμός, a creaking]. A condition of tonic spasm in which the open mouth cannot be closed.

**antitrope** (an'-te-tröp) [anti-; τρέπειν, to turn]. Organs arranged to form a symmetrical pair. Thus the right eye is an antitrope to the left. 2. An antibody.

**antitropin** (an-te-tro'-pin). An antibody.

**antitrypsin** (an-te-trip'-sin). An antibody inhibiting the action of trypsin.

**antitryptic** (an-te-trip'-tik). 1. A ferment inimical to bacteria. 2. Antagonistic to proteolysis.

**antitryptic index.** The power of any given serum to inhibit tryptic digestion as compared with that possessed by a normal standard serum. It is said to be raised in cancerous conditions, and it is used to differentiate gastric cancer from gastric ulcer.

**antituberculin** (an-te-tū-ber'-kū-lin). Antibodies found in the sera of individuals who have been treated with tuberculin.

**antituberculotic** (an-te-tū-ber-kū-lot'-ik) [anti-; tuberculum, a tubercle]. Good against tuberculosis.

**antitulase** (an-te-tū'-lās). An immunizing serum for tuberculosis obtained from animals which have been injected with tulase.

**antituman** (an-te-tū'-man). Trade name of a cancer remedy containing sodium chondroitin sulphate. Dose 1 to 2 grains.

**antitussin** (an-te-tus'-in) [anti; tussis, cough]. An ointment consisting of difluordiphenyl 5 parts; vaselin, 10 parts, and lanolin, 85 parts; used as an application in catarrh.

**antitussive** (an-te-tus'-iv) [anti-; tussis, cough]. 1. Relieving or preventing cough. 2. A remedy for cough.

**antityphoid** (an-te-ti'-foid). Opposed to typhoid. a. extract, a preparation obtained by injecting repeatedly cultures of typhoid bacilli of increasing virulence into the peritoneal cavity of rabbits. The animals are killed as soon as they do not react to poisonous doses, and extracts are made of the thymus, spleen, bone-marrow, brain, and spinal cord, by soaking these organs in a solution of salt, glycerol, and alcohol, with the addition of some pepsin. The filtrate is injected in typhoid cases.

**antitypic** (an-te-tip'-ik) [anti-; τύπος, a type]. 1. Efficient against the periodic recurrence of a paroxysm or fever. 2. Irregular; not conformable to a type. 3. An antiperiodic.

**antiuratic** (an-te-ū-rat'-ik). 1. Efficacious against the deposition of urates. 2. An agent that prevents the deposit of urates.

**antiurease** (an-ti-ū'-re-ās). An antibody to urease.

**antivenene, antivenin** (an-te-ven'-ēn, -in) [anti-; venenum, poison]. A serum perfected by Calmette by injecting cobra venom mixed with solutions of calcium hypochlorite into horses. It is used in doses of 2½–5 dr. (10–20 Cc.) in bites of venomous serpents.

**antivenereal** (an-te-ven-e'-re-al) [anti-; venereus, pertaining to Venus, or to sexual intercourse]. Antisyphilitic; anaphrodisiac.

**antivenomous** (an-te-ven'-om-us). Antagonistic to venom; a term applied to immunized animals, to certain serums, and to antitoxins.

**antivermicular** (an-te-vur-mik'-ū-lar) [anti-; vermis, a worm]. Anthelmintic.

**antiverminous** (an-te-vur'-min-us). See antivermicular.

**antivirulent** (an-te-vir'-ū-lent) [anti-; virus, a poison]. Effective against viruses.

**antivivisection** (an-te-viv-is-ek'-shun). Opposition to vivisection or animal experimentation.

**antivivisectionist** (*an-te-viv-is-ek'-shun-ist*) [*anti-;* *vivus*, living; *sectio*, a cutting]. One who opposes the practice of vivisection.

**antizymotic** (*an-te-zi-mot'-ik*) [*anti-;* ζύμωσις, fermentation]. 1. Preventing or checking fermentation. 2. An agent preventing the process of fermentation; an antiferment.

**antlia** (*ant'-le-ah*) [ἀνά, up; τλαείν, to lift]. A syringe or pump. **a. gastrica,** a stomach pump. **a. lactea,** a pump for drawing milk from the breast. **a. mammaria,** same as *a. lactea.*

**antocular** (*ant-ok'-u-lar*) [*ante*, before; *oculus*, the eye]. Situated in front of the eye.

**antodontalgic** (*an-to-don-tal'-jik*). See *antiodontalgic.*

**antodyne** (*an'-to-dīn*): Trade name of an analgesic and sedative, derived from phenol. Dose 7½ grains (0.5 gr.).

**antophthalmic** (*ant-off-thal'-mik*) [*anti-;* ὀφθαλμία, ophthalmia]. Preventive or curative of ophthalmia.

**antorbital** (*ant-orb'-it-al*) [*ante*, before; *orbita*, the orbit]. Located in front of the orbit.

**antorgastic** (*ant-or-gas'-tik*). See *antiorgastic.*

**antozenic** (*ant-o-ze'-nik*) [*anti-;* ὄζειν, to smell]. Curative of ozena.

**antozone** (*ant-o-zōn*) [*anti-;* ὄζειν, to smell]. An imaginary allotropic modification of oxygen, now known to be only hydrogen dioxide.

**antozostomatic** (*ant-o-zos-to-mat'-ik*) [*anti-;* ὀζόστομος, having a foul breath]. Corrective of a foul breath.

**antra** (*an'-trah*). Plural of *antrum, q. v.*

**antracele** (*an'-tra-sēl*) [*antrum;* κήλη, a tumor]. Dropsy of the antrum; an accumulation of fluid in the maxillary sinus.

**antral** (*an'-tral*) [*antrum*]. Relating to an antrum.

**antrectomy** (*an-trek'-to-me*) [*antrum;* ἐκτομή, excision]. Surgical removal of the walls of an antrum, especially the mastoid antrum.

**antritis** (*an-tri'-tis*) [*antrum;* ιτις, inflammation]. Inflammation of an antrum, especially the antrum of Highmore.

**antroatticotomy** (*an-tro-at-ik-ot'-o-me*). The operation of opening the mastoid antrum and the attic of the tympanum.

**antrocele** (*an'-tro-sēl*). See *antracele.*

**antronalgia** (*an-tron-al'-je-ah*) [*antrum;* ἄλγος, pain]. Pain in the antrum.

**antronasal** (*an-tro'-na-sal*). Pertaining to the antrum of Highmore and the nasal fossa.

**antrophore** (*an'-tro-for*). Cacao-butter bougies, containing tannin, 5 %; resorcinol, 5 %; thallin sulphate, 2 to 5 %; zinc sulphate, 0.5 %.

**antrophose** (*an'-tro-fōs*) [ἄντρον, a cavity; φῶς, light]. A phose having its origin in the central ocular mechanism.

**antrorse** (*an-trōrs'*) [*ante*, before; *versus*, turned]. In biology, directed upward or forward.

**antroscope** (*an'-tro-skōp*) [*antrum;* σκοπεῖν, to look]. An instrument for examining the maxillary sinus.

**antroscopy** (*an-tros'-ko-pe*). Inspection of the antrum by means of an antroscope.

**antrotome** (*an'-tro-tōm*) [*antrum;* τέμνειν, to cut]. An instrument for the performance of mastoid antrotomy.

**antrotomy** (*an-trot'-o-me*). Incision of an antrum.

**antrotympanic** (*an-tro-tim-pan'-ik*) [*antrum;* τύμπανον, a drum]. Relating to the cavity of the tympanum and to the tympanic antrum.

**antrotympanitis** (*an-tro-tim-pan-i'-tis*) [ἄντρον, a cave; τύμπανον, a drum]. Chronic purulent otitis media.

**antrum** (*an'-trum*) [L.]. 1. A cavity or hollow space, especially in a bone. 2. The antrum of Highmore. **a., cardiac,** Luschka's name for a dilatation sometimes found in the esophagus immediately above its passage through the diaphragm. **a., dental,** pulp-cavity of a tooth. **a., duodenal,** the normal dilatation presented by the duodenum near its origin. **a. ethmoidal,** the ethmoid sinus. **a. of Highmore,** a cavity in the superior maxillary bone. Syn., *antrum genæ.* **a. Highmori testis,** see *mediastinum testis.* **a., mastoid,** the hollow space beneath the roof of the mastoid process. **a., maxillary,** see *a. of Highmore.* **a. pyloricum Willisii,** the cavity of the pylorus. **a. tubæ,** a sac-like dilatation of the Fallopian tube about an inch from the fimbriated extremity, regarded by some as occurring only in pregnancy. **a. tympanicum,** the mastoid antrum.

**Antyllus' method** for aneurysm [Antyllus, a Greek physician of the third century A. D.]. It consists in ligation above and below the sac, followed by opening of the aneurysm and evacuation of its contents.

**anuresis** (*an-u-re'-sis*) [ἀν, priv.; οὖρον, urine]. Anuria.

**anuretic** (*an-u-ret'-ik*) [see *anuresis*]. Pertaining to or affected with anuria.

**anuria** (*an-u'-re-ah*) [see *anuresis*]. Suppression of the urine.

**anuric** (*an-u'-rik*) [see *anuresis*]. Pertaining to anuria.

**anurous** (*an-u'-rus*) [ἀν, priv.; οὐρά, a tail]. Without a tail.

**anus** (*a'-nus*) [L., "the fundament"]. The extremity of the rectum; the lower opening of the alimentary canal. **a., artificial,** an opening established from the bowel to the exterior at a point above the normal anus, most commonly from the colon, either in the lumbar or in the iliac region. **a., fissure of,** a slight tear in the mucous membrane at the anus, usually due to passage of hardened feces. It is very painful. **a., fistula of,** fistula in ano, a sinus opening from the rectum into the connective tissue about the rectum or discharging externally. **a., imperforate,** absence of the anus, the natural opening being closed by a membranous septum. **a., infundibuliform,** a relaxed condition of the anus with destruction of the natural folds. **a., preternatural,** an abnormal aperture serving as an anus, whether congenital, made by operation, or due to disease or injury. Syn., *fecal fistula; anus preternaturalis.* **a., preternatural ileovaginal, a., preternatural vaginal, a. præternaturalis vestibularis,** the rare abnormity of the rectum opening through the vulva. **a., Rusconi's,** the blastopore. **a., umbilical,** a preternatural anus located in the umbilical region. **a. vulvovaginalis,** an anal opening communicating with the vulva.

**anusol** (*an'-us-ol*). Trade name for suppositories of bismuth iodoresorcinsulphonate; used in rectal diseases.

**anvil** (*an'-vil*). See *incus.*

**anxietas** (*ang-zi'-et-as*). See *anxiety.* **a. tibiæ, a. tibiarum.** 1. An annoying sensation of restlessness in the muscles of the legs noted in neurasthenia. 2. An irregular movement of the legs. Syn., *fidgets.*

**anxiety** (*ang-zi'-et-e*) [*anxius*, anxious]. Restlessness, agitation and general malaise, or distress, often attended with precordial pain, and a noticeable appearance of apprehension or worry visible in the features.

**anydremia, anydræmia** (*an-id-re'-me-ah*). See *anhydremia.*

**anypnia** (*an-ip'-ne-ah*) [ἀν, priv.; ὕπνος, sleep], Sleeplessness.

**anytin** (*an'-it-in*). See *anitin.*

**anytol.** See *anitol.*

**AOC.** Abbreviation of anodic opening contraction.

**aochlesia** (*ah-ok-le'-ze-ah*) [ἀ, priv.; ὄχλησις, disturbance]. Rest; tranquillity; catalepsy.

**aol** (*a'-ol*). Trade name of a derivative of santalum album.

**A. O. M.** Abbreviation for Master of Obstetric Art.

**aorta** (*a-ort'-ah*) [ἀορτή, aorta]. The large vessel arising from the left ventricle and distributing, by its branches, arterial blood to every part of the body. It ends by bifurcating into the common iliacs at the fourth lumbar vertebra. The **arch,** that extending from the heart to the third dorsal vertebra, is divided into an *ascending*, a *transverse*, and a *descending* part. The *thoracic* portion extends to the diaphragm; the *abdominal*, to the bifurcation. **a., cardiac,** that part of the embryonic vascular system giving rise to the aortic arches. **a. chlorotica,** a narrowing of the aorta, sometimes found in chlorotic patients. **a., dorsal.** 1. The embryonic vessel formed by the junction of the two primitive aortas. Syn., *primordial aorta; subvertebral aorta.* 2. The thoracic aorta. **a., inferior,** the abdominal aorta. **a., left,** the embryonic division of the vascular system which finally becomes the aorta. **a., main,** the embryonic vessel formed by the junction of the two primitive aortas. **a., pectoral,** the thoracic aorta. **a., pelvic,** the middle sacral artery. **a., pericardiac,** the part of the aorta within the pericardial cavity. **a., primitive.** 1. That part of the aorta extending from its origin to the point where it first branches. 2. Two embryonic branches of the cardiac aorta extending through the first visceral arch and uniting to form the dorsal aorta. **a.**, **right**, the embryonic division of the aortic bulb which finally forms the pulmonary artery. **a.**, **root of**, the origin of the aorta at the heart. Syn., *radix aortæ*. **a.**, **superior**, the thoracic aorta. **a.**, **systemic**, see *a.*, *left*. **a.**, **thoracic**, see under *aorta*.

**aortal** (*a-ort'-al*) [see *aorta*]. Relating to the aorta.
**aortarctia** (*a-ort-ark'-she-ah*) [ἀορτή, aorta; *arctare*, to constrict]. A constriction or stenosis of the aorta.
**aortectasia** (*a-or-tek-ta'-ze-ah*) [ἀορτή, aorta; ἐκ, out; τάσις, a stretching]. Aortic dilatation.
**aorteurysma** (*a-ort-ū-ris'-mah*) [ἀορτή, aorta; εὔρυσμα, a widening: *pl.*, *aorteurysmata*]. Aortic aneurysm or dilatation.
**aortic** (*a-ort'-ik*) [see *aorta*]. Pertaining to the aorta. **a. arch**, see *aorta* and *arch*. **a. area**, the part of the thorax about the second right costal cartilage, where the aortic murmurs and sounds are best heard. **a. foramen**, see *a. opening of diaphragm*. **a. murmur**, a murmur produced by disease of the aortic valves. **a. opening of diaphragm**, the aperture in, or really behind, the diaphragm, through which the aorta passes. **a. opening of heart**, the opening between the heart and the aorta. **a. plexus**, the plexus of sympathetic nerves, situated on the front and sides of the aorta, between the origins of the superior and inferior mesenteric arteries. **a. sinus**, a deep depression between the leaflets of the aortic valve and the aortic wall. **a. valve**, the three semilunar valves closing the aortic opening during the cardiac diastole.
**aortism** (*a-or'-tizm*). A liability to disease of the aorta.
**aortitis** (*a-ort-i'-tis*) [*aorta*; ιτις, inflammation]. Inflammation of the aorta. **a.**, **nummular**, that characterized by white, circular patches in the inner coat.
**aortoclasia, aortoclasis** (*a-or-to-kla'-ze-ah, -sis*) [*aorta*; κλάσις, a breaking]. Rupture of the aorta.
**aortolith, aortolite** (*a-or'-to-lith, -līt*) [*aorta*; λίθος, a stone]. A calculus formed in the aorta.
**aortolithia** (*a-or-to-lith'-e-ah*). A calcareous deposition in the aorta.
**aortomalacia, aortomalaxia** (*a-ort-o-mal-a'-se-ah, -aks'-e-ah*) [*aorta*; μαλακία, softening]. Softening of the aorta.
**aortopathy** (*a-ort-op'-athe*) [*aorta*; πάθος, disease]. Any disease of the aorta.
**aortoptosis, aortoptosia** (*a-or-top-to'-sis, -to'-se-ah*) [*aorta*; πτώσις, a falling]. A drooping of the abdominal aorta associated with visceroptosis.
**aortosclerosis** (*a-ort-o-skle-ro'-sis*) [*aorta*; σκληρόs, hard]. Induration of the aorta.
**aortostenosis** (*a-ort-o-sten-o'-sis*) [*aorta*; στενός, narrow]. Stenosis or narrowing of the aorta.
**aosmic** (*a-os'-mik*) [ἀ, priv.; ὀσμή, smell]. Having no odor.
**apaconitine** (*ap-ak-on'-it-in*). See *apoaconitine*.
**apagma** (*ap-ag'-mah*) [ἀπό, from; ἀγνύμαι, to break; *pl.*, *apagmata*]. 1. Separation, as of a fractured bone. 2. The part separated.
**apallagin** (*ap-al'-aj-in*) [ἀπαλλαγή, deliverance]. An antiseptic mercury salt of nosophen (*q. v.*).
**apandria** (*ap-an'-dre-ah*) [ἀπό, from; ἀνήρ, a man]. Morbid dislike of the male sex.
**apanthropia** (*ap-an-thro'-pe-ah*). See *apanthropy*.
**apanthropy** (*ap-an'-thro-pe*) [ἀπό, from; ἀνθρωπος, man]. Aversion to society; morbid desire for solitude.
**aparthrosis** (*ap-ar-thro'-sis*) [ἀπό, from; ἄρθρον, a joint]. 1. Dislocation; luxation of a joint. 2. In anatomy, diarthrosis.
**apastia** (*ap-as'-te-ah*) [ἀπαστία, fasting]. Abstinence from food, as a symptom of mental disorder.
**apathetic** (*ap-ath-et'-ik*) [ἀ, priv.; πάθος, feeling]. Affected with apathy; listless; without emotion.
**apathy** (*ap'-ath-e*) [ἀ, priv.; πάθος, feeling]. Insensibility; want of passion or feeling.
**apatropine** (*ap-at'-ro-pēn*) [ἀπό, from; *atropine*], C₁₇H₂₁NO₂. A compound derived from atropine by the action of nitric acid. It is said to produce peculiar convulsions.
**ape** (*āp*) [ME.]. A man-like monkey. **a. fissures**, those fissures of the human brain that are also found in apes. **a.-hand**, a peculiar shape of the hand produced by the wasting of the thumb-muscles; it is seen in some cases of progressive muscular atrophy.
**apella** (*ap-el'-lah*) [ἀ, priv.; πέλλα, skin]. A circumcised person; one with a short prepuce.
**apellous** (*ah-pel'-us*) [ἀ, priv.; πέλλα, skin]. 1. Skinless. 2. Without a prepuce; circumcised.

**apenta** (*ah-pen'-tah*). A Hungarian aperient water.
**apepsia** (*ah-pep'-se-ah*) [ἀ, priv.; πέπτειν, to digest]. Cessation or absence of the digestive function. **a.**, **hysterical**, apepsia due to hysteria. Syn., *hysterical anorexia*. **a. nervosa**, see *anorexia nervosa*.
**apepsinia** (*ah-pep-sin'-e-ah*) [ἀ, priv.; *pepsin*]. Absence of pepsin or pepsinogen from the gastric juice.
**apeptic** (*ah-pep'-tik*) [see *apepsia*]. Affected with apepsia.
**apeptous** (*ah-pep'-tus*) [ἀ, priv.; πέπτειν, to digest]. 1. Crude, indigestible, uncooked. 2. Aseptic.
**aperception** (*ap-ur-sep'-shun*). See *apperception*.
**apergol** (*ap-er'-gol*). A preparation containing apiol, ergotin, oil of savine, and aloin.
**aperient** (*ap-ē'-ri-ent*) [*aperire*, to open]. 1. Laxative; mildly purgative. 2. A mild purgative; a laxative.
**aperiodic** (*ah-pe-re-od'-ik*) [ἀ, priv.; περίοδος, a circuit]. Not periodic.
**aperistalsis** (*ah-per-is-tal'-sis*) [ἀ, priv.; περί, around; στάλσις, constriction]. Cessation of the peristaltic movements of the intestine.
**aperitive** (*ap-er'-it-iv*) [*aperire*, to open]. 1. Aperient. 2. Deobstruent. 3. Stimulating the appetite; an appetizer. **a.**, **hygienic**, hygienic measures for stimulating the appetite.
**aperitol** (*ap-er'-it-ol*). A proprietary purgative consisting of the acetate and valerate of phenolphthalein. Dose 6 gr. (0.4 gm.).
**apertometer** (*ap-ur-tom'-et-er*) [*aperture*; μέτρον, a measure]. An optic device for determining the angle of aperture of microscopic objectives.
**apertor** (*ap-er'-tor*) [L., an opener or beginner]. In anatomy, anything that opens. **a. oculi**, the levator palpebræ muscle.
**apertura** (*ap-er-tu'-rah*). An opening. **a. anterior ventriculi tertii cerebri**, the vulva cerebri. **a. aquæductus cochleæ**, opening of aqueduct of cochlea on the petrous bone. **a. chordæ**, the internal opening of the canal for the chorda tympani nerve. **a. canalis inguinalis**, the inguinal ring. **a. declivis**, the anus. **a. externa aquæductus vestibuli**, external opening of the aqueduct of the vestibule. **a. externa canaliculi cochleæ**, external opening of the canaliculus of the cochlea. **a. inferior canaliculi tympanici**, inferior opening of tympanic canaliculus. **a. lateralis ventriculi quarti**, the foramen of Key and Retzius. **a. medialis ventriculi quarti**, the foramen of Magendie. **a. narium**, same as nares. **a. pelvis (minoris) inferior**, lower opening of lesser pelvis (O. T. pelvic outlet). **a. pelvis (minoris) superior**, upper opening of lesser pelvis (O. T. pelvic inlet). **a. pelvis superior**, the superior strait of the pelvis. **a. piriformis**, piriform opening (O. T. anterior nares). **a. scalæ vestibuli cochleæ**, an opening between the vestibule and the scala vestibuli of the cochlea. **a. sinus sphenoidalis**, opening of sphenoidal sinus. **a. spinalis**, the vertebral foramen. **a. superior canaliculi tympanici**, opening for the smaller petrosal nerve. **a. thoracis inferior**, lower thoracic opening. **a. thoracis superior**, upper thoracic opening. **a. tympanica canaliculi chordæ**, opening of the iter chordæ posterius into the tympanum. **a. uterina**, opening of the Fallopian tube into the uterus.
**aperture** (*ap'-er-chūr*) [*apertura*, an opening]. An opening. **a.**, **angular**, in the microscope, the angle formed between a luminous point placed in focus and the most divergent rays that are capable of pasing through the entire system of an objective. **a.**, **numerical**, the capacity of an objective for admitting rays from the object and transmitting them to the image.
**apex** (*a'-peks*) [L., "the extreme end of a thing"; *pl.*, *apices*]. The summit or top of anything; the point or extremity of a cone. **a. auriculæ** (Darwini), tip of the auricle of the ear. **a.-beat**, the impulse of the heart felt in the fifth intercostal space, about 3½ inches from the middle of the sternum. **a. capituli fibulæ**, apex of the head of the fibula; the styloid process of the fibula. **a. cartilaginis arytænoideæ**, tip of the arytenoid cartilage. **a. columnæ posterioris**, apex of the posterior column. **a. cordis**, the apex of the heart. **a. linguæ**, tip of the tongue. **a. of the lung**, the upper extremity of the lung behind the border of the first rib. **a. murmur**, a murmur heard over the apex of the heart. **a. nasi**, the tip of the nose. **a. radicis dentis**, apex of the root of a tooth.
**aphacia** (*ah-fa'-se-ah*). See *aphakia*.
**aphacic** (*ah-fa'-sik*). See *aphakic*.

# APHAGIA 75 APHTHA

**aphagia** (ah-fa'-je-ah) [ἀ, priv.; φαγεῖν, to eat]. Inability to eat or to swallow.

**aphakia** (ah-fa'-ke-ah) [ἀ, priv.; φακός, a lentil; the crystalline lens]. The condition of an eye without the lens.

**aphakic** (ah-fa'-kik) [see *aphakia*]. Not possessing a crystalline lens.

**aphalangiasis** (ah-fa-lan-ji'-as-is) [ἀ, priv.; φάλαγξ, phalanx]. The loss or absence of fingers and toes, as in leprosy. Cf. *ainhum*.

**aphasia** (ah-fa'-ze-ah) [ἀ, priv.; φάσις, speech]. Partial or complete loss of the power of expressing ideas by means of speech or writing. Aphasia may be either motor or sensory. Motor or *ataxic aphasia* consists in a loss of speech owing to inability to execute the various movements of the mouth necessary to speech, the muscles not being properly coördinated, owing to disease of the cortical center. It is usually associated with *agraphia*, "aphasia of the hand," inability to write, and right-sided hemiplegia. Some aphasiacs can write, but are unable to articulate words or sentences; this variety is variously named *aphemia*, *alalia*, or *anarthria*, according as the impairment of speech is more or less marked. Charcot supposes the center for articulate language divided into 4 subcenters—a visual center for words, an auditory center for words, a motor center of articulate language, and a motor center of written language. Lesions of one or more of these centers produce the characteristic forms of aphasia, all of which have clinical exemplifications. *Sensory aphasia*, or *amnesia*, is the loss of memory for words, and may exist alone or in association with motor aphasia. Amnesia appears clinically in 3 distinct forms: 1. Simple loss of memory of words. 2. *Word-deafness*, or inability to understand spoken words (there is usually some paraphasia connected with this form). 3. *Word-blindness*, or inability to understand written or printed words. a., Broca's, motor aphasia. a., conduction, such as is due to defect in some commissural connection between centers. a., cortical, a., pictorial, a., true, destruction of the junction of the auditory speech-center. a., functional, that in which there is no manifest lesion, but it occurs as a result of excitement in hysteria or in severe constitutional disorders. a., gibberish, a form of transcortical aphasia in which the speech is confused, words or syllables being transposed or jumbled together, due to disruption of the tracts associating cortical speech-centers. Syn., *jargon aphasia*. a., Kussmaul's, see *Kussmaul*. a., mixed, combined motor and sensory aphasia. a., optic, inability to give the names for objects seen, due to interrupted connection between the centers for vision and speech. a., pure, a., isolated, a., subcortical, a., subpictorial; aphasia arising from a lesion interrupting impulses toward the afferent tracts proceeding to the auditory speech-center. a., supracortical, a., suprapictorial, that form of lesion completely severing the connection of the auditory center with the cortical center, but not destroying the auditory speech-center, the afferent tracts proceeding to it or the efferent tracts passing from it to the motor speech-center. a., tactile, inability to recognize objects by the sense of touch, due to lesion in the central parietal lobule. a., total, a., universalis, inability to utter a single word. a., Wernicke's, cortical sensory aphasia.

**aphasiac** (ah-fa'-ze-ak) [see *aphasia*]. One who is aphasic.

**aphasic** (ah-fa'-zik) [see *aphasia*]. Relating to or affected with aphasia.

**aphelexia** (af-el-eks'-e-ah). An incorrect form of the word *aphelxia*, q. v.

**aphelotic** (af-el-ot'-ik) [ἀφέλκειν, to draw away]. Absent-minded; lost in reverie.

**aphelxia** (af-elks'-e-ah) [ἀφέλκειν, to draw away]. Absence of mind; inattention to external impressions.

**aphemesthesia** (ah-fem-es-the'-ze-ah) [ἀ, priv.; φήμη, voice; αἴσθησις, sensation] Word-blindness; word-deafness.

**aphemia** (ah-fe'-me-ah) [ἀ, priv.; φήμη, voice]. Motor aphasia; inability to articulate words or sentences from centric and not from peripheral disease. See *aphasia*.

**aphemic** (ah-fe'-mik) [see *aphemia*]. Relating to or affected with aphemia.

**aphephobia** (af-e-fo'-be-ah) [ἀφή, touch; φόβος, fear]. Hyperesthetic dread of contact with other persons.

**apheter** (af'-et-er) [ἀφέτηρ, one who lets go or sends away]. A supposed impulse-carrying, or trigger-material, probably a catastate, which communicates to the inogen the nerve impulse that causes its destruction, and the consequent muscular contraction. In a larger sense, any trigger-material that takes part in any functional process may be called an apheter.

**aphilanthropy** (ah-fil-an'-thro-pe) [ἀ, priv.; φιλεῖν, to love; ἀνθρωπος, man]. Absence of social feeling; a frequent sign of approaching melancholia.

**aphlogistic** (ah-flo-jist'-ik) [ἀ, priv.; φλόξ, a flame]. 1. Noninflammable. 2. Burning without flame.

**aphonia** (ah-fo'-ne-ah) [ἀ, priv.; φωνή, voice]. 1. Loss of speech, due to some peripheral lesion. 2. Hysterical, or paralytic absence of the power of speech. 3. Voicelessness. a. clericorum, clergyman's sore-throat. a., paralytic, see *paralysis*, *phonetic*. a. paranoica, stubborn silence in the insane. a., spastic, see *dysphonia spastica*.

**aphonic** (ah-fon'-ik) [see *aphonia*]. Speechless; voiceless.

**aphorama, aphorema** (af-o-ra'-mah, -re'-mah) [ἀφορᾶν, to have in full view]. The state of having projecting eyes, enabling one to see at a distance on each side without moving the head.

**aphoresis** (ah-for-e'-sis) [ἀ, priv.; φόρησις, bearing]. 1. Separation or ablation of a part, either by excision or amputation. 2. Lack of the power of endurance, as of pain.

**aphoria** (ah-fo'-re-ah) [ἀ, priv.; φέρειν, to bear]. Sterility of the female; unfruitfulness. a. impercita, that attributed to aversion. a. impotens, that due to impairment of conceptive power. a. incongrua, that attributed to nonresponsive condition of the conceptive power to the seminal fluid. a. paramenica, that due to menstrual disorder. a. polyposa, that attributed to the existence of a uterine polyp.

**aphoric, aphorous** (af'-or-ik, af'-or-us) [ἄφορος, sterile]. 1. Relating to causing, caused by, or affected with sterility. 2. Unbearable, insufferable.

**aphose** (ah'-fōz) [ἀ, priv.; φῶς, light]. A subjective sensation of shadow or darkness. Cf. *phose*.

**aphrasia** (ah-fra'-ze-ah) [ἀ, priv.; φράξειν, to utter]. Absence of the power to utter connected phrases. a., paralytic, that due to paralysis of the ideation faculty. a., superstitious, the voluntary avoidance of certain words from scruples of nicety or religion.

**aphrenic, aphrenous, aphrænous** (ah-fren'-ik, ah'-fren-us, ah-fre'-nus) [ἀ, priv.; φρήν, the mind]. Insane.

**aphrodescin, aphrodæscin** (af-ro-des'-in) [ἀφροδίτη, foamy], C₈H₁₂O₂₃. A glucoside constituent of the cotyledons of horse-chestnut. It is a colorless, amorphous powder, soluble in alcohol and water, its watery solution frothing like soap.

**aphrodisia** (af-ro-diz'-e-ah) [Ἀφροδίτη, Venus]. Sexual desire, especially when morbid or immoderate; sexual congress.

**aphrodisiac** (af-ro-dis'-e-ak) [see *aphrodisia*]. 1. Stimulating the sexual appetite; erotic. 2. An agent stimulating the sexual passion.

**aphronesis** (ah-fro-ne'-sis) [ἀ, priv.; φρόνησις, good sense]. Foolishness, silliness, madness.

**aphronia** (ah-fro'-ne-ah) [ἀ, priv.; φρήν, the mind]. Apoplexy.

**aphtha** (af'-tha) [ἄφθα, an eruption; pl., *aphthæ*]. A form of stomatitis characterized by the presence of small white vesicles in the mouth, occurring chiefly in children under 3 years, and supposed to be due to a special microorganism. Syn., *acacos*; *acacus*; *ophlyctis*; *morbus aphthosus*; *thrush*; *stipue*; *angina aphthosa*; *aphthous stomatitis*. a. **anginosa**, a form of sore throat attended by slight fever, redness, and enlargement of the fauces, with the formation of small whitish specks on the tongue and mucosa of the throat. It usually occurs in cold, damp weather and in women and children. **aphthæ, Bednar's**, two symmetrically placed ulcers seen at times on the hard palate of cachectic infants, one on each side of the mesial line. **aphthæ, cachectic**, those appearing beneath the tongue, and associated with grave constitutional symptoms; Riga's disease. Syn., *Cardarelli's aphtha*. a. **epizootica**, see *foot-and-mouth disease*. a. **febrilis**, ulceration of the mouth, extending to the esophagus and stomach, and accompanied by fever. a. **serpens**, **aphthæ serpentes**, see *cancrum oris*. **aphthæ tropicæ**, a disease of the tropics marked by epigastric fulness, pain, vomiting, diarrhea, and redness of the tongue, with the formation of small, white, painful spots on it.

Syn., *tropical sprue; psilosis; gastroenteritis aphthosa indica; phlegmasia membranæ mucosæ gastropulmonalis*. **aphthæ,** Valleix's, see *aphtha, Bednar's.*
**aphthæ** (*af'-the*). Plural of *aphtha, q. v.*
**aphthenxia** (*af-lhengks'-e-ah*) [ά, priv.; φθέγξις, utterance]. A form of aphasia with impaired expression of articulate sounds.
**aphthoid** (*af'-thoid*) [see *aphtha*]. Resembling aphthæ.
**aphthongia** (*af-thon'-ge-ah*) [ά, priv.; φθόγγος, a sound]. A peculiar form of aphasia due to spasm of the muscles supplied by the hypoglossal nerve.
**aphthous** (*af'-thus*) [see *aphtha*]. 1. Pertaining to or affected with aphthæ. 2. Presenting the appearance of a surface covered with little ulcers.
**apical** (*a'-pik-al*) [*apex*, the top]. Pertaining to the apex.
**apices** (*a'-pis-ez*) [L.]. Plural of *apex*. Summits.
**apicifixed** (*a-pis'-e-fikt*). Attached by the apex.
**apiciform** (*a-pis'-e-form*) [*apex*, the top; *forma*, form]. Sharp-pointed.
**apicilar** (*a-pis'-il-ar*) [*apex*, the top]. Attached to or located upon an apex.
**apiin** (*ap'-e-in*) [*apium*, parsley], $C_{27}H_{32}O_{16}$. A glucoside obtained from the leaves, stems, and seeds of parsley, *Apium petroselinum*. It is a yellowish-white, crystalline powder, soluble in hot water and alcohol, slightly soluble in cold water, insoluble in ether.
**apinoid** (*ah'-pin-oid*) [ά, priv.; πίνος, dirt; εἶδος, form]. Clean; not foul. a. **cancer,** scirrhus; so called from its cleanly section.
**apiol** (*ap'-e-ol*) [*apium*, parsley; *oleum*, oil], $C_{18}H_{14}O_4$. A principle occurring in parsley-seeds; it crystallizes in long white needles, with a slight odor of parsley; melts at 30° C. (86° F.), and boils at 294° C. (572° F.). It is used in dysmenorrhœa and in malaria. In large doses it produces ringing in the ears and frontal headache. Syn., *parsley-camphor*. Dose 10–15 gr. (0.65–1.0 Gm.). a., green, crude ethereal oil from seeds of parsley, *Apium petroselinum*. It is used as an emmenagogue and antiperiodic. Dose, in dysmenorrhœa, 5–10 min. (0.3–0.6 Cc.) 2 or 3 times daily; in malaria, 15–30 min. (1–2 Cc.). a., liquid, an alcoholic extract of parsley-seeds.
**apiolin** (*ap'-e-ol-in*). Rectified essential oil of parsley, a yellow neutral liquid boiling at about 300° C., soluble in alcohol. It is used as an emmenagogue. Dose, 0.2 gm. 2 or 3 times daily.
**apion** (*ap'-e-on*) [*apium*, parsley]. A substance obtained from apiolic acid by heating with dilute sulphuric acid; melts at 69° C.
**apiosoma** (*ap-e-o-so'-mah*). A protozoon said to be found in the blood of patients suffering from typhus fever; it is related to *Piroplasma bigonicum*.
**apiphobia** (*ap-e-fo'-be-ah*) [*apis*, a bee; φόβος, fear]. Morbid terror of bees and of being stung by them.
**Apis** (*a'-pis*) [L., a bee]. A genus of hymenopterous insects. A. **mellifica,** the honey-bee; in homeopathy the poison of the honey-bee's sting, or a preparation thereof.
**apisin** (*ap'-is-in*) [*apis*, a bee]. Bee-poison.
**apisinasion** (*ap-is-in-a'-shun*). Poisoning from the stings of bees.
**apituitarism** (*ah-pit-ū'-it-ar-izm*). The condition of absence of the function of the pituitary body, owing to removal of that body.
**Apium** (*a'-pe-um*) [L.]. A genus of umbelliferous plants. A. **graveolens,** see *celery*. A. **petroselinum,** is the common garden parsley; aperient, diuretic, somewhat antiperiodic; useful in dysmenorrhœa. Dose of the fluidextract (of the root) gtt. xv–ʒj.
**aplacental** (*ah-plas-en'-tal*) [ά, priv.; *placenta*]. Destitute of placenta.
**aplanasia** (*ah-plan-a'-ze-ah*) [ά, priv.; πλανᾶν, to wander]. Entire or nearly entire absence of spherical aberration.
**aplanatic** (*ah-plan-at'-ik*) [see *aplanasia*]. Not wandering; rectilinear. a. **focus,** that focus of a lens the rays from which do not undergo spherical aberration in their passage through the lens. a. **lens,** a lens corrected for aberration of light and color; a rectilinear lens.
**aplanatism** (*ah-plan'-at-izm*). See *aplanasia*.
**aplasia** (*ah-pla'-ze-ah*) [ά, priv.; πλάσσειν, to form]. Incomplete or defective development. Syn., *agenesis*.
**aplastic** (*ah-plas'-tik*) [see *aplasia*]. 1. Structureless; formless. 2. Incapable of forming new tissue. 3. Relating to aplasia. 4. Defective in fibrin. 5. Applied to inflammations unattended with organizable exudation. a. **lymph,** a nonfibrinous material incapable of coagulation or organization.
**aplestia** (*ah-ples'-te-ah*) [ἀπληστία, insatiate desire]. Insatiable hunger; acoria.
**apleuria** (*ah-plū'-re-ah*) [ά, priv.; πλευρά, a rib]. Congenital absence of the ribs.
**apnea, apnœa** (*ap-ne'-ah*) [ά, priv.; πνεῖν, to breathe]. 1. A transient cessation of respiration from an overabundance of oxygen, as, *e. g.*, after forcible respiration. 2. Asphyxia. a., **cardiac,** the period of apnea in Cheyne-Stokes respiration. a., **nervous,** that due to disorders of the centers of respiration. a., **placental,** placental tuberculosis. a., **uterine,** a form of dyspnea observed in hysterical patients, due to no manifest disease. Syn., *uterine asthma*.
**apneumatic** (*ap-nū-mat'-ik*) [ά, priv.; πνεῦμα, breath]. 1. Collapsed; uninflated, not inflatable; said of parts of the lung. 2. Carried on with the exclusion of air, as an apneumatic operation or process.
**apneumatosis** (*ah-nū-mat-o'-sis*) [ά, priv.; πνευμάτωσις, inflation]. Collapse or non-inflation of the air-cells.
**apneumia** (*ap-nū'-me-ah*) [ά, priv.; πνεύμων, lung]. Congenital absence of the lungs.
**apnœa** (*ap-ne'-ah*). See *apnea*.
**apo-** (*ap'-o*) [ἀπό, from]. A prefix denoting *from, away, separation.*
**apoaconitine** (*ap-o-ak-on'-it-ēn*) [ἀπό, from; *aconitum*, aconite], $C_{33}H_{43}NO_{11}$. An alkaloid prepared from aconitine by dehydration.
**apoatropine** (*ap-o-at'-ro-pēn*) [ἀπό, from; *atropine*], $C_{17}H_{21}NO_2$. An alkaloid obtained by the action of $HNO_3$ on atropine.
**apobiosis** (*ap-o-bi-o'-sis*) [ἀπό, from; βίος, life]. Local death of a part.
**apoblema** (*ap-o-ble'-mah*) [ἀποβλημα; ἀπό, away; βάλλειν, to throw]. The product of abortion.
**apobole** (*ap-ob'-o-le*) [ἀποβολή, a throwing away]. Expulsion; abortion.
**apocamnosis** (*ap-o-kam-no'-sis*) [ἀποκάμνειν, to grow utterly weary]. Intense and readily induced fatigue.
**apocatastasis** (*ap-o-kat-as'-tas-is*) [ἀποκατάστασις, restoration]. 1. Return to a previous condition. 2. The subsidence of an abscess or tumor.
**apocatharsis** (*ap-o-kath-ar'-sis*) [ἀπό, away; κάθαρσις, purgation]. Purgation; abevacuation.
**apocathartic** (*ap-o-kath-ar'-tik*). Same as *cathartic*.
**apocenosis** (*ap-o-sen-o'-sis*) [ἀποκενοῦειν, to drain]. 1. An increased flow or evacuation of blood or other humors. 2. A partial evacuation. In the plural, *apocenoses*, Cullen and Swediaur's term for diseases marked by fluxes and unattended by fever.
**apochromatic** (*ap-o-kro-mat'-ik*) [ἀπό, away; χρῶμα, color]. Without color. a. **lens,** a lens of a special variety of glass, corrected for spherical and chromatic aberration.
**apocodeine** (*ap-o-ko'-de-in*) [ἀπό, from; *codeine*], $C_{18}H_{19}NO_2$. An alkaloid prepared from codeine by dehydration. It is emetic and expectorant, with other qualities much like those of codeine, and is recommended in chronic bronchitis. The hydrochloride is generally used. Dose 3–4 gr. (0.2–0.25 Gm.).
**apocope** (*ap-ok'-o-pe*) [ἀπό, from; κοπή, a cutting]. Amputation or abscission; an operation or a wound that results in loss of substance.
**apocopous** (*ap-ok'-o-pus*) [ἀπόκοπος, cut off]. Castrated.
**apocoptic** (*ap-o-kop'-tik*) [ἀποκόπτειν, to cut off]. Affected by or occurring from the removal of a part.
**apocrustic** (*ap-o-krus'-tik*) [ἀποκρούειν, to beat off]. Repellent; defensive; astringent.
**apocynein** (*ap-o-sin'-e-in*). A glucoside from *Apocynum cannabinum*, similar in character to digitalein.
**apocynin** (*ap-os'-in-in*) [see *apocynum*]. The precipitate from a tincture of *Apocynum cannabinum*; tonic, alterative, and cathartic. Dose ⅛–1 gr. (0.016–0.065 Gm.).
**apocynum** (*ap-os'-in-um*) [*apocynon*, dogbane]. Canadian hemp. The root of *A. cannabinum*, the properties of which are due to *apocynin*. It is a good expectorant; in full doses it is emetic and cathartic. Dose 5–20 gr. (0.3–1.2 Gm.); of *tincture* 5–40 min. (0.3–2.5 Cc.). Another American species, *A. androsæmifolium*, has similar properties. a., **fluidextract of** (*fluidextractum apocyni*, U. S. P.). Dose 5–20 min. (0.3–1.2 Cc.).

**apodal** (ap'-od-al). See *apodous*.
**apodemialgia** (ap-o-de-me-al'-je-ah) [ἀποδημία, journey; ἄλγεω, to grieve]. A morbid dislike of home-life with a desire for wandering.
**apodia** (ah-po'-de-ah) [ἀ, priv.; πούς, a foot]. Congenital absence of feet.
**apodous** (ap'-o-dus) [ἀ, priv.; πούς, a foot]. Footless; characterized by apodia.
**apogamy** (ap-og'-am-e) [ἀπό, away from; γάμος, marriage]. In biology, 1. Asexual reproduction where the opposite usually occurs. 2. The total and normal absence of sexual reproductive power.
**apokenosis** (ap-o-ken-o'-sis). See *apocenosis*.
**apolar** (ah-po'-lar) [ἀ, priv.; πόλος, the end of an axis]. Not possessing a pole. a. cells, nerve-cells without processes.
**apolepsis** (ap-o-lep'-sis) [ἀπόληψις, a leaving off]. Suppression or retention of a secretion or excretion; cessation of a function.
**apolexis** (ap-o-leks'-is) [ἀπόληξις, a declining]. The decline of life; the stage of catabolism or decay.
**Apollinaris** water (ap-ol-in-a'-ris). A German alkaline mineral water, highly charged with carbonic acid, and largely used as a diluent in gout, rheumatism, etc.
**apolysin** (ap-ol'-is-in), $C_6H_4(OC_2H_5)NH_2C_6O_6$. A compound of citric acid and phenetidin. It is antipyretic and analgesic. Dose 8–90 gr. (0.5–5.0 Gm.) daily. Syn., *monophenetidin citric acid*.
**apomorphine** (ap-o-mor'-fēn) [ἀπό, from; morphine], $C_{17}H_{17}NO_2$. An artificial alkaloid, derived from morphine by the abstraction of a molecule of water. a. hydrochloride (apomorphinæ hydrochloridum, U. S. P.), is the salt used, and is a grayish, crystalline powder. It acts as a centric emetic. Dose $\frac{1}{30}$–$\frac{1}{10}$ gr. (0.003–0.0065 Gm.), hypodermatically, or $\frac{1}{30}$–$\frac{1}{6}$ gr. (0.0065–0.01 Gm.) by the mouth. It is expectorant in small doses.
**apomorphosis** (ap-o-mor-fo'-sis) [ἀπομορφοῦν, to change the form]. A chemical change by which one substance acting upon another takes something away from it.
**apomyelin** (ap-o-mi'-el-in) [ἀπό, from; μυελός, marrow]. A peculiar phosphatized principle reported to exist in the brain tissue and containing no glycerol.
**apomyttosis** (ap-o-mit-o'-sis) [ἀπομύσσειν, to blow the nose]. Any disease marked by stertor; a sneezing.
**aponal** (ap'-o-nal). The carbamic acid ester of tertiary amyl alcohol; it is used as a hypnotic in doses of 15 to 30 grains (1–2 grammes).
**apone** (ap-ōn') [Fr.: ἀ, priv.; πόνος, pain]. An anodyne; especially the concentrated tincture of capsicum; used externally for the relief of pain, and internally in small doses, diluted, for hemorrhoids, dyspepsia, and mania. Dose gtt. iij–x.
**aponeurography** (ap-o-nū-rog'-ra-fe) [ἀπονεύρωσις, aponeurosis; γράφη, a writing]. A description of the fasciæ, or aponeuroses.
**aponeurology** (ap-o-nū-rol'-o-je) [ἀπονεύρωσις, aponeurosis; λόγος, an account]. The science of the fasciæ or aponeuroses.
**aponeurosis** (ap-o-nū-ro'-sis) [ἀπό, from; νεῦρον, a tendon]. A fibrous, membranous expansion of a tendon giving attachment to muscles or serving to inclose and bind down muscles. a. of occipitofrontalis muscle, the aponeurosis that separates the two slips of the occipitofrontalis muscle. a. of soft palate, a thin, firm, fibrous layer, attached above to the hard palate, and becoming thinner toward the free margin of the velum. a., subscapular, a thin membrane attached to the entire circumference of the subscapular fossa, and affording attachment by its inner surface to some of the fibers of the subscapularis muscle. a., supraspinous, a thick and dense membranous layer that completes the osseofibrous case in which the supraspinatus muscle is contained, affording attachment by its inner surface to some of the fibers of the muscle. a., vertebral, a thin aponeurotic lamina extending along the whole length of the back part of the thoracic region, serving to bind down the erector spinæ, and separating it from those muscles that unite the spine to the upper extremity.
**aponeurositis** (ap-on-ū-ro-si'-tis) [aponeurosis, ιτις, inflammation]. Inflammation of an aponeurosis.
**aponeurotic** (ap-on-ū-rot'-ik) [aponeurosis]. Pertaining to an aponeurosis. a. fascia, a deep fascia.
**aponeurotome** (ap-on-ū'-ro-tōm) [ἀπονεύρωσις, aponeurosis; τομή, cutting]. An instrument for dividing fasciæ.

**aponeurotomy** (ap-on-ū-rot'-o-me) [ἀπονεύρωσις, aponeurosis; τομή, cutting]. The incision, dissection, or anatomy of the fasciæ; fasciotomy.
**apophlegmatic** (ap-o-fleg-mat'-ik) [ἀπό, away; φλέγμα, phlegm]. Promoting the expulsion of mucus from the air passages.
**apophysary** (ap-off'-is-a-re) [ἀποφύειν, to put forth]. Pertaining to or of the nature of an apophysis.
**apophysate** (ap-off'-is-āt) [ἀπό, from; φύσις, growth]. Furnished with an apophysis.
**apophyseal, apophysial** (ap-o-fiz'-e-al). Same as *apophysary*.
**apophysis** (ap-off'-is-is) [ἀπό, from; φύσις, growth; pl. *apophyses*]. A process, outgrowth, or swelling of some part or organ, as of a bone. a., basilar, the basilar process of the occipital bone. a., cerebral, the pineal body. **apophyses, false**, see *epiphyses*. a. lenticularis, the orbicular process of the temporal bone. a. raviana, the processus gracilis of the malleus. **apophyses, true**, those which have never been epiphyses.
**apophysitis** (ap-of-is-i'-tis) [see *apophysis*; ιτις, inflammation]. 1. Inflammation of an apophysis. 2. Appendicitis.
**apoplasmia** (ap-o-plaz'-me-ah) [ἀπό, away; πλάσμα, plasm]. Deficiency of the blood-plasm.
**apoplectic** (ap-o-plek'-tik) [apoplexy]. Pertaining to or affected with apoplexy. a. equivalents, a name given to the premonitory symptoms of apoplexy, indicating that the brain is subject to alterations in blood-pressure.
**apoplectiform** (ap-o-plek'-tif-orm) [apoplexy; forma, form]. Resembling apoplexy.
**apoplectigenous** (ap-o-plek-tij'-en-us) [apoplexy; γεννᾶν, to produce]. Producing apoplexy or cerebral hemorrhage.
**apoplectoid** (ap-o-plek'-toid). Same as *apoplectiform*.
**apoplexy** (ap'-o-pleks-e) [ἀποπλήσσειν, to strike down, to stun]. The symptom-complex resulting from hemorrhage or the plugging of a vessel in the brain or spinal cord. The term is sometimes also applied to the bursting of a vessel in the lungs, liver, etc. a., asthenic, that due to vital depression. a., atonic, that which comes on gradually and does not attain a high degree of development. Syn., *imperfect apoplexy*. a., atrabilious, deep melancholy attributed to resorption of bile. a., bulbar, that due to hemorrhage into the substance of the oblongata, causing paralysis of one or both sides of the body, inability to swallow, difficulty in protruding the tongue, dyspnea, gastric disorders, and tumultuous action of the heart. a., capillary, one resulting from rupture of capillaries. a., consecutive, that due to the arrest of some habitual discharge or eruption. a., cutaneous. 1. See *purpura hæmorrhagica*. 2. A sudden effusion of blood to the skin and subcutaneous tissue. a., dysarthritic, a form accompanying arthritic diseases, in which the pain disappears from the joints, and vertigo, pain in the head, etc., appear. a., epileptic, coma with epileptoid symptoms, sometimes observed in cerebral and acute inflammatory diseases. a., febrile, paroxysmal fever attended with deep sleep and stertor. Syn., *apoplexia febricosa*. a., fulminant, a sudden and fatal apoplexy. a., ingravescent, a term applied to a form of apoplexy in which there is a slowly progressive loss of consciousness, due to a gradual leakage of blood from a ruptured vessel. a., muscular, an escape of blood into the muscular tissue. a., nervous. 1. Acute anemia of the brain. 2. A condition marked by symptoms of cerebral congestion and hemorrhage which are due to functional disturbance of the nervous system. a. of the ovary, a., ovarian, hemorrhage into the stroma of the ovary, through the rupture of a follicle, converting the organ into a cyst or hematoma. The blood is gradually absorbed, though it gives rise to great pain; the cause is unknown. a., phlegmonous, a condition attributed to inflammation of the brain and its membranes; it is marked by delirium, fever, severe headache, conjunctival injection, lacrimation, and a hard pulse. a., pituitous, serous apoplexy. a., placental, a., placentary, escape of blood into the placental substance. a., pontile, apoplexy due to a rupture of a blood-vessel in the pons Varolii. a., progressive, that in which there is a very gradual increase of the paralysis and other symptoms. a., pulmonary, escape of blood into the pulmonary parenchyma. a., pulmonary,

vascular, very acute and extensive congestion of the lungs, leading to apoplectic symptoms and a fatal termination. a., **sanguineous,** hemorrhage into or upon the brain. a., **serous,** that due to an effusion of serious matter into or upon the brain. a., **simple,** the name given to those cases of death from coma in which no cerebral lesion is found. a., **spinal,** rupture of a blood-vessel of the spinal cord. a., **splenic,** (1) flow of blood into the splenic substance; (2) contagious anthrax. a., **suppurative,** that due to purulent processes and fever. a., **symptomatic,** that attributed to another disease or to the arrest of some habitual evacuation. a., **uterine,** escape of blood into the muscular tissue of the uterus. a., **venous,** that due to congestion of the veins.

**apopsychia** (*ap-op-sik'-e-ah*) [ἀπό, away; ψυχή, spirit]. Syncope; fainting; a faint.

**apoptosis** (*ap-op-to'-sis*) [ἀπό, away; πτῶσις, a falling]. A falling off, as of a crust, or of the hair; loosening of a scab or crust.

**apoquinamine** (*ap-o-kwin'-am-ēn*), C₁₉H₂₂N₂O. An artificial alkaloid occurring as a white, amorphous substance derived from quinamine, conquinamine, or quinamidine by action of hydrochloric acid.

**aporetin** (*ap-o-re'-tin*) [ἀπό, from; ῥητίνη, a resin]. A purgative resin derived from rhubarb.

**aporocephalous** (*ap-o-ro-sef'-al-us*) [ἄπορος, difficult to distinguish; κεφαλή, the head]. Having a head scarcely distinguishable.

**aporrhegma** (*ap-o-reg'-mah*) [ἀπο-; ῥηγνύναι, to break in pieces]. A substance split off from another substance by biological action.

**aporrhinosis** (*ap-or-in-o'-sis*) [ἀπό, from; ῥίς, nose]. A discharge from the nostril.

**aporrhipsis** (*ap-or-ip'-sis*) [ἀπό, away from; ῥίπτειν, to throw]. The throwing off of the clothes or the bed clothes; a symptom seen in some cases of insanity and in delirium.

**aposepsis** (*ap-o-sep'-sis*) [ἀπόσηψις, putrefaction; see *sepsis*]. Complete putrefaction.

**aposia** (*ah-po'-ze-ah*) [ἀ, priv.; πόσις, drinking]. Absence of thirst; adipsia.

**apositia** (*ap-o-sit'-e-ah*) [ἀπό, from; σῖτος, food]. Aversion to or loathing of food.

**apositic** (*ap-o-sit'-ik*) [ἀπό, from; σῖτος, food]. Impairing the appetite; affected with apositia.

**apostasis** (*ap-os'-tas-is*) [ἀπόστασις, a standing away from]. 1. An abscess. 2. The end or the crisis of an attack of disease; termination by crisis. 3. An exfoliation.

**apostatic** (*ap-os-tat'-ik*) [ἀπόστασις, a standing away from]. Relating to or of the nature of an apostasis.

**apostaxis** (*ap-o-staks'-is*) [ἀπό, from; στάξις, a dropping]. A discharge of fluid by drops; epistaxis.

**apostem** (*ap'-o-stem*), or **apostema** (*ap-o-ste'-mah*) [ἀπόστημα, an abscess]. An abscess.

**apostematic** (*ap-os-tem-at'-ik*) [ἀπόστημα, an abscess]. Relating to or of the nature of an abscess.

**apostemation** (*ap-os-tem-a'-shun*) [*apostematio,* abscess formation]. The formation of an apostem or abscess.

**aposthia** (*ah-pos'-the-ah*) [ἀ, priv.; πόσθη, prepuce; penis]. Congenital absence of the prepuce or penis.

**Apostoli's method** [Georges *Apostoli,* French physician, 1847-1900]. The use of strong electrolytic or chemical galvanocaustic currents in the treatment of diseases of the female generative organs, especially uterine fibroids.

**apostrophe** (*ap-os'-tro-fe*) [ἀπό, away; στρέφειν, to turn]. The arrangement of chlorophyll bodies along the side walls of the cells as a result of excess or deficiency of light. Cf. *epistrophe* and *dystrophe*.

**apothecaries' weight.** A system of weights and measures used in compounding medicines. The troy pound of 5760 grains is the standard. It is subdivided into 12 ounces. The ounce is subdivided into 8 drams, the dram into 3 scruples, and the scruple into 20 grains. For fluid measure the quart of 32 fluidounces is subdivided into 2 pints, the pint into 16 fluidounces, the ounce into 8 fluidrams, and the fluidram into 60 minims. The following symbols and abbreviations are used:

℥, *minim.*  
℈, *scrupulus,* a scruple (20 grains).  
℥, *drachma,* a dram (60 grains).  
ʒ, *uncia,* an ounce (480 grains).  
℔, *libra,* a pound.  
O., *octarius,* a pint.  
gr., *granum,* a grain.  
ss., *semissis,* one-half.  

See *Weights and Measures.*

**apothecary** (*ap-oth'-e-kā-re*) [ἀποθήκη, a storehouse]. 1. A druggist or pharmaceutical chemist, one who prepares and sells drugs, fills prescriptions, etc. 2. In Great Britain a physician filling his own prescriptions; especially one licensed by the Society of Apothecaries of London, or by the Apothecaries' Hall of Ireland.

**apothem, apothema** (*ap'-o-them, ap-oth'-em-ah*) [ἀπό, from; θέμα, a deposit]. A brown powder deposited from vegetable infusions or decoctions exposed to the air.

**apothesis** (*ap-oth'-es-is*) [ἀπόθεσις, a putting back]. The reduction of a fracture or luxation. a. *funiculi umbilicalis,* the reposition of an abnormally protruded umbilical cord.

**apotheter** (*ap-oth'-et-er*). A navel-string repositor devised by Braun, consisting of a staff with a sling attached in which the prolapsed funis is placed and carried up into the uterine cavity.

**apous** (*ah'-pus*). See *apodous.*

**apozem, apozema** (*ap'-o-zem, ap-oz'-em-ah*) [ἀπό, away; ζεῖν, to boil]. A decoction, especially one to which medicines are added.

**apparatotherapy** (*ap-ar-at-o-ther'-ap-e*). Treatment by mechanical apparatus.

**apparatus** (*ap-ar-a'-tus*) [*apparatus,* preparation]. 1. A collection of instruments or devices used for a special purpose. 2. Anatomically the word is used to designate collectively the organs performing a certain function. 3. A collection of pathological phenomena. 4. Cystotomy. a., **absorbent,** the blood-vessels and lymphatics. a., **acoustic,** a., **auditory,** the external and internal ear, the auditory canal, the tympanum, and the Eustachian tube. a., **chirurgicus,** surgical apparatus. a., **digestorius,** digestive apparatus. a., **lacrimalis,** lacrimal apparatus. a. **ligamentosus colli,** the occipitoaxoid ligament, a broad band at the front surface of the spinal canal that covers the odontoid process. a. **magnus,** a. **major,** median cystotomy. a. **minor,** lateral lithotomy. a. **respiratorius,** respiratory system. a., **segmental,** see *nephridia.* a., **sound-conducting,** a collective term for the auricle, external auditory canal, tympanum, Eustachian tube, and mastoid cells. a., **sound-perceiving,** that part of the organism concerned in the perception of sound consisting of the auditory nerve, and its center of origin and peripheral distribution, or the organs of the labyrinth. a., **urinary,** the kidneys, ureters, bladder, and urethra. a. **urogenitalis,** urogenital system. a., **uropoietic,** the kidneys.

**apparent** (*ap-a'-rent*) [*apparere,* to appear). Seeming; appearing to be like. a. **death,** see *death.*

**apparition** (*ap-ar-ish'-un*) [*apparitio,* an appearance]. 1. A visual delusion or hallucination. 2. The sudden aggregation of scattered principles into an element or corpuscle.

**appendage** (*ap-en'-dāj*) [*appendere,* to weigh; hang]. Anything appended, usually of minor importance. a., **auricular.** 1. The projecting part of the cardiac auricle. 2. Virchow's name for a round or elongated cartilaginous prominence in front of the tragus. a. **cecal,** the appendix vermiformis. a.s, **cutaneous,** a.s, **dermal,** the nails, hair, sebaceous glands, and sweat-glands. a.s, **epiploic,** see *appendices epiploicae* under *appendix.* a.s **of the eye,** the eye-lashes, eyebrows, lacrimal gland, lacrimal sac and ducts, and conjunctiva. a.s, **fetal,** the placenta, amnion, chorion, and umbilical cord. a.s, **moss-like,** short processes seen on some nerve fibres in the granular layers of the cerebellum. a., **ovarian,** the parovarium. a., **pineal,** the epiphysis. a., **pituitary,** th hypophysis. a.s, **uterine,** the ovaries and oviducts.

**appendalgia** (*ap-end-al'-je-ah*) [*appendix;* ἄλγος, pain]. Pain in the appendicular region.

**appendectomy** (*ap-en-dek'-to-me*). See *appendicectomy.*

**appendiceal, appendical** (*ap-en-di-se'-al, ap-en'-di-kal*). See *appendicular.*

**appendicectomy** (*ap-en-dis-ek'-to-me*) [*appendix;* ἐκτομή, excision]. Excision of the vermiform appendix.

**appendices epiploicae** (*ap-en'-dis-ēz ep-ip-lo'-is-e*). See *appendix.*

**appendicitis** (*ap-en-dis-i'-tis*) [*appendix;* ἰτις, inflammation]. Inflammation of the vermiform appendix. Syn., *paratyphlitis; epityphlitis; abscess of iliac form.* a., **gangrenous,** that in which the vermiform appendix is found gangrenous and sloughing, usually with one or more perforations and free leakage, a large section of the right groin full of lemon-colored, septic fluid, a puddle of filth underneath the cecum and ileum, the omentum fixed with a cluster of bowel adhesions beneath. Syn., *green groin*. a. **larvata**, an incipient or latent form of appendicitis. a. **obliterans**, an inflammation characterized by the progressive obliteration of the lumen of the appendix, by the disappearance of the epithelial lining and glandular structure. The symptoms are acute attacks of brief duration, moderate swelling at the seat of disease, and persistent tenderness in the region of the appendix during the intermissions.

**appendicostomy** (*ap-en-dik-os'-to-me*). The operation of opening the vermiform appendix, previously anchored in an incision in the anterior abdominal wall, for the purpose of irrigating the cecum and colon; employed in amebic dysentery and constipation. Syn., *Weir's operation*.

**appendicular** (*ap-en-dik'-u-lar*) [*appendicula*, a small appendix]. 1. Pertaining to the vermiform appendix. 2. Pertaining to the limbs. a. **colic**, a spasmodic colicky pain originating in the appendix.

**appendiculate** (*ap-en-dik'-u-lāt*). Having appendages or protruding accessory parts.

**appendix** (*ap-en'-diks*) [pl., *appendices; appendere*, to hang upon or to]. An appendage. a. **auricularis**, see *appendage, auricular* (1). a. **cerebri**, the pituitary body. a., **ensiform**, see *xiphoid*. a. **epididymidis**, the vas aberrans. **appendices epiploicæ**, fatty projections of the serous coat of the large intestine. a. **lobularis**, the flocculus. a., **supraphonoid**, a. **ventriculi**, the hypophysis. a., **vermiform**, a. **vermiformis**, the small, blind gut projecting from the cecum. a., **xiphoid**, see *xiphoid*.

**apperception** (*ap-er-sep'-shun*) [*appercipere*, to perceive]. The conscious reception of a sensory impression; the power of receiving and appreciating sensory impressions.

**appetence, appetency** (*ap'-e-tens, ap'-e-ten-se*) [*appetentia*, appetite]. An appetite or desire; the attraction of a living tissue for those materials that nourish it.

**appetite** (*ap'-e-tīt*) [*appetere*, to desire]. The desire for food; also any natural desire; lust. a.-**breakfast**, more tasty and desirable than the ordinary test meal, and calculated to excite a more natural flow of gastric juice. a.-**juice**, flow of gastric juice provoked by the mere sight and taste of food (without swallowing it). a., **perverted**, that for unnatural and indigestible things, frequent in disease and in pregnancy.

**appetizer** (*ap-e-ti'-zer*). [*appetere*, to desire]. A medicine, or dose, taken to stimulate the appetite.

**applanate** (*ap'-lan-āt*) [*ad*, to; *planus*, flat]. Horizontally flattened.

**applanatio, applanation** (*ap-lan-a'-she-o, ap-lan-a'-shun*) [L.]. A flattening. a. **corneæ**, flattening of the entire surface of the cornea from disease.

**apple** (*ap'-l*) [AS., *æppel*, an apple]. The fruit of the tree, *Pyrus malus*. a., **Adam's**, see *pomum Adami*. a.-**brandy**, an alcoholic spirit distilled from cider; cider-brandy. a. **extract**, see *extractum ferri pomatum* under *extract*. a. **eye**, synonym of *exophthalmos*. a. **head**, a term for the broad thick skull of dwarfs. a. **oil**, amyl valerate.

**applicator** (*ap'-lik-a-tor*) [L.]. An instrument used in making applications.

**Appolito's operation** (*ap-ol-e'-to*). Enterorrhaphy by means of a form of right-angle continuous suture.

**apposition** (*ap-o-zish'-un*) [*apponere*, to apply to]. 1. The act of fitting together; the state of being fitted together. 2. An addition of parts. 3. Development by accretion.

**approximal** (*ap-roks'-im-al*) [*ad*, to; *proximus*, next]. That which is next to; contiguous. In dentistry, pertaining to contiguous surfaces, as approximal fillings.

**apraxia** (*ah-praks'-e-ah*) [ἀ, priv.; πράσσειν, to do]. Soul-blindness; mind-blindness; object-blindness; an affection in which the memory for the uses of things is lost, as well as the understanding of the signs by which the things are expressed.

**aprication** (*ap-re-ka'-shun*) [*apricatio*, a basking in the sun]. The sun-bath; sun-stroke.

**aproctia** (*ah-prok'-she-ah*) [ἀ, priv.; πρωκτός, anus]. Absence or imperforation of the anus.

**aproctous** (*ah-prok'-tus*) [ἀ, priv.; πρωκτός, the anus]. Having imperforation of the anus.

**apron** (*a'-pron*) [O. F., *naperon*]. 1. A cloth or rubber covering to prevent the clothing from becoming soiled. 2. The omentum. a., **Hottentot**, artificially or abnormally elongated labia minora. Syn., *pudendal apron*. a., **masonic**, a name sometimes given to a support, attached to the waist, for the penis and testicles in gonorrheal cases. a. **of succor**, a canvas stretcher for carrying the wounded.

**aprosexia** (*ah-pro-seks'-e-ah*) [ἀπροσεξία, want of attention]. A mental disturbance consisting in inability to fix the attention upon a subject. An inability to think clearly and to comprehend readily what is read or heard; a condition sometimes observed in the course of chronic catarrh of the nose or of the nose and pharynx.

**aprosopia** (*ah-pro-so'-pe-ah*) [ἀ, priv.: πρόσωπον, the face]. A form of fetal monstrosity with absence of part or all of the face.

**aprosopous** (*ap-ros'-o-pus*) [ἀ, priv.; πρόσωπον, the face]. Exhibiting aprosopia.

**aprosopus** (*ap-ros'-o-pus*) [ἀ, priv.; πρόσωπον, the face]. An aprosopous fetus.

**apselaphesia** (*ap-sel-af-e'-ze-ah*) [ἀ, priv.; ψηλάφησις, touch]. Loss of the tactile sense.

**apsithyria, apsithurea** (*ah-psith-i'-re-ah, -ū'-re-ah*) [ἀ, priv.; ψιθυρίζειν, to whisper]. Hysterical aphonia, in which the patient loses the voice and is also unable to whisper.

**apsychia** (*ah-si'-ke-ah*) [ἀ, priv.; ψυχή, spirit]. Unconsciousness; a faint or swoon.

**aptyalia, aptyalism** (*ap-ti-a'-le-ah, ap-ti'-al-ism*) [ἀ, priv.; πτυαλίζειν, to spit]. 1. Deficiency or absence of saliva. 2. Psychic salivation; debility and general disorder from loss of oxydases due to excessive expectoration.

**apulosis** (*ap-ū-lo'-sis*) [οὐλειν, to cicatrize]. Cicatrization, or a cicatrix.

**apulotic** (*ap-ū-lot'-ik*) [οὐλειν, to cicatrize]. Promoting cicatrization, or apulosis.

**apus** (*a'-pus*) [ἀ, priv.; πούς, foot]. 1. A monstrosity consisting in absence of the lower limbs, or feet. 2. An apodous fetus.

**apyknomorphous** (*ah-pik-no-morf'-us*) [ἀ, priv.; πυκνός, compact; μορφή, form]. Applied by Nissl to feebly staining cells, or those in which the stainable portions are not arranged in close proximity.

**apyonin** (*ah-pi'-on-in*) [ἀ, priv.; πύον, pus]. A remedy introduced as a substitute for pyoktanin in ophthalmic practice. It is said to be identical with yellow pyoktanin.

**apyous** (*ah-pi'-us*). Having no pus.

**apyretic** (*ah-pi-ret'-ik*) [ἀ, priv.; πυρετός, fever]. Without fever.

**apyrexia** (*ah-pi-reks'-e-ah*) [ἀ, priv.; πυρετός, fever]. The non-febrile stage of an intermittent fever; intermission or absence of fever.

**apyrexial** (*ah-pi-rek'-se-al*) [see *apyrexia*]. Pertaining to, or the nature of, or characterized by apyrexia.

**aq.** Abbreviation for *aqua* [L.], water; also for water of crystallization.

**aqua** (*ak'-wah*) [L., gen., and pl., *aquæ*]. Water. An oxide of hydrogen, having the composition $H_2O$. It is a solid below $32°$, a liquid between $32°$ and $212°$, vaporizes at $212°$ at the sea-level (bar. 760 mm.), giving off vapor of tension equal to that of the air. Water is an essential constituent of all animal and vegetable tissues. In the human body it forms 2 % of the enamel of the teeth, 77 % of the tissues, 78 % of the blood, and 93 % of the urine. Water is a valuable antipyretic; internally, it is diuretic. It is the most useful of all the solvents. *Aquæ*, in pharmacy, designates various medicated waters. a. **ammoniæ** (U. S. P.), ammonia-water. Dose 10-30 min. (0.6-2.0 Cc.). a. **ammoniæ fortior** (U. S. P.), stronger ammonia-water, used externally. a. **amygdalæ amaræ** (U. S. P.), bitter almond water. Dose 2 dr. (8 Cc.). a. **anethi** (B. P.), dill-water. Dose ½-2 oz. (15-60 Cc.). a. **anisi** (U. S. P.), anise water. Dose ½-2 oz. (15-60 Cc.). a. **aurantii florum** (U. S. P.), orange-flower water. Dose ½-2 oz. (15-60 Cc.). a. **aurantii florum fortior** (U. S. P.), triple orange-flower water. a. **bulliens**, boiling water. a. **calcis**, lime water. a. **camphoræ** (U. S. P.), camphor-water. Dose ½-1 oz. (15-30 Cc.). a. **carbolisata**, 22 parts of liquefied phenol in 978 parts of distilled water. Dose 1 dr.-½ oz. (4-16 Cc.). a. **chlori** (*liquor chlori compositus*, U. S. P.), chlorine water. Dose 1-4 dr. (3.7-15.0 Cc.). a. **chloroformi** (U. S. P.), chloroform-water. Dose ½-2 oz. (15-60 Cc.). a. **cinnamomi** (U. S. P.), cinnamon-water.

Dose ½-2 oz. (15-60 Cc.). a. **communis,** common water. a. creosoti (U. S. P.), creosote-water. Dose 1-4 dr. (3.7-15.0 Cc.). a. **destillata** (U. S. P.), distilled water. a. ferrata, a chalybeate water. a. **fervens,** hot water. a. **fluvialis,** river-water. a. **fœniculi** (U. S. P.), fennel-water. Dose 1-2 oz. (30-60 Cc.). a. fontana, well- or spring-water. a. fortis, see *acid, nitric.* a. hamamelidis (U. S. P.), hamamelis water. Dose 2 dr. (8 Cc.). a. hydrogenii dioxidi (U. S. P.), solution of hydrogen dioxide used chiefly locally. a. **labyrinthi,** the clear fluid existing in the labyrinth of the ear. a. laurocerasi (B. P.), cherry-laurel water. Dose 5-30 min. (0.3-2.0 Cc.). a. **Levico,** water from springs at Levico in the Tyrol, containing arsenic, iron, and copper. a. **marina,** sea water. a. **menthæ piperitæ** (U. S. P.), peppermint-water. Dose 1-2 oz. (30-60 Cc.). a. **menthæ viridis** (U. S. P.), spearmint water. Dose 1-2 oz. (30-60 Cc.). a. **oculi,** the aqueous humor. a. **omnium florum,** a liquid distillation-product of cow-dung collected during the month of May; formerly used in pulmonary tuberculosis. a. **pimentæ** (B. P.), allspice water. Dose ½-2 oz. (15-60 Cc.). a. **pluvialis,** rain-water. a. **puteana,** well-water. a. regia, see *acid, nitrohydrochloric.* a. rosæ (U. S. P.), rose-water. Dose 1-2 oz. (30-60 Cc.). a. rosæ fortior (U. S. P.), used for making rose-water. a. vitæ, brandy or spirit.

**aquacapsulitis** (ak'-wah-kap-sū-li'-tis) [aqua; capsula; ιτις, inflammation]. Inflammation of the membrane of Descemet; serous iritis.

**aquæ** (ak'-we) [pl. of aqua]. Waters; medicated waters.

**aquæductus** (ak-we-duk'-tus), see *aqueduct.*

**aquapuncture** (ak'-wah-pungk'-chūr) [aqua; punctura, a puncture]. 1. Counterirritation by means of a very fine jet of water impinging upon the skin; it is useful in neuralgic disorders. 2. The hypodermatic injection of water as a placebo.

**aquatic** (a-kwat'-ik) [aqua]: Pertaining to water. a. cancer, synonym of *cancrum oris.*

**aqueduct, aquæductus** (ak'-we-dukt, ak-we-duk'-tus) [aqua; ductus, a leading]. A canal for the passage of fluid; any canal. **aquæductus cerebri,** see *a. Sylvii.* **aquæductus cochleæ,** aqueduct of the cochlea. a., **communicating, aquæductus communicationis,** a small canal sometimes found at the junction of the mastoid part of the temporal bone with the petrosa, which transmits a venous branch to the end of the transverse sinus. a. of Cotunnius, the aqueduct of the vestibule, extending from the utricle to the posterior wall of the pyramid in the brain. **aquæductus Fallopii,** see under *Fallopian.* **aquæductus Sylvii,** the aqueduct of Sylvius, the passageway from the third to the fourth ventricle, the *iter a tertio ad quartum ventriculum.* Syn., *ventricular aqueduct.* a., **temporal,** an inconstant canal at the dorsal part of the superior angle of the petrosa, for passage of the squamosopetrosal sinus. **aquæductus vestibuli,** the aqueduct of the vestibule of the ear.

**aqueous** (a'-kwe-us) [aqua]. Watery. a. **chamber** of the eye, the space between the cornea and the lens; the iris divides it into an anterior and a posterior chamber. a. **extract,** a solid preparation of a drug made by evaporation of its aqueous solution. a. **humor,** the fluid filling the anterior chamber of the eye.

**aquiducous** (a-kwe-dū'-kus) [aquiducus; aqua, water; ducere, to lead]. Hydragogue.

**aquiferous** (ak-wif'-ur-us) [aqua, water; ferre, to bear]. Carrying water or lymph.

**aquocapsulitis** (ak-wo-kap-sū-li'-tis). See *aquacapsulitis.*

**aquosity** (a-kwos'-it-e) [aquositas, watery]. The state or condition of being watery; moisture.

**aquozon** (ak'-wo-zōn). Ozonized, distilled, and sterilized water, containing 3 % by volume of ozone.

**arabate** (ar'-ab-āt). A salt of arabic acid.

**arabic acid** (ar'-ab-ik). See *arabin.* a., **gum-,** see *acacia.*

**arabin** (ar'-ab-in) [arabic], (C₆H₁₀O₅)₂+H₂O. Arabic acid. A transparent, glassy, amorphous mass, an exudate from many plants. It is soluble in water, and is the principal constituent of gum-arabic.

**arabinose** (ar'-ab-in-ōs) [arabic], C₅H₁₀O₅. One of the glucoses made from gum-arabic on boiling with dilute H₂SO₄. It crystallizes in shining prisms that melt at 100°; is slightly soluble in cold water, has a sweet taste, and reduces Fehling's solution, but is not fermented by yeast.

**arabite** (ar'-ab-īt) [arabinose], C₅H₁₂O₅. A substance formed from arabinose by the action of sodium amalgam. It crystallizes from hot alcohol in shining needles, melting at 102°. It has a sweet taste, but does not reduce Fehling's solution.

**Arachis** (ar'-ak-is) [ἄραχος, a leguminous plant]. A genus of leguminous plants. **A. hypogæa,** see *ground nut.* a. **oil,** peanut oil.

**arachnida** (ar-ak'-nid-ah) [ἀράχνη, a spider]. A class of arthropods to which belong ticks (*acari*), mites (*linguatulidæ*), spiders (*araneida*), and scorpions (*scorpionida*).

**arachnitis** (ar-ak-ni'-tis) [arachnoid; ιτις, inflammation]. Inflammation of the arachnoid membrane of the brain. Syn., *leptomeningitis externa; arachnodeitis; arachnoiditis; arachnoideitis; arachnoitis.* a., **rhachidian,** a., **spinal,** spinal meningitis.

**arachnoid** (ar-ak'-noid) [ἀράχνη, a spider's web; εἶδος, form]. 1. Resembling a web. 2. The arachnoid membrane. Syn., *membrana media cerebri; meningion; meningium; meninx arachnoidea; meninx media; meninx serosa.* 3. Pertaining to a membrane. 4. Thready; feeble; said of the pulse. a. **cavity,** the space between the arachnoid and dura mater. a. **membrane,** the delicate membrane of the brain and cord between the dura and pia mater. It is separated from the pia by the subarachnoid space, and passes over the convolutions without dipping down into the fissures between them.

**arachnoidal** (ar-ak-noid'-al) [see *arachnoid*]. Pertaining to the arachnoid membrane.

**arachnoidea** (ar-ak-noid'-e-ah) [see *arachnoid*]. The arachnoid membrane; see *arachnoid.* a. **encephali,** arachnoid of brain. a. **oculi,** outer layer of choroid. a. **spinalis,** arachnoid of spine.

**arachnoiditis** (ar-ak-noi-di'-tis). See *arachnitis.*

**arachnoidism** (ar-ak'-noi-dizm) [ἀράχνη, spider] The condition produced by the bite of poisonous spiders.

**arachnoiditis** (ar-ak-noid-i'-tis). Same as *arachnitis.*

**arachnolysin** (ar-ak-nol'-is-in) [ἀράχνη, a spider; λύσις, dissolution]. A very active hemolytic substance extracted from spiders.

**arachnophia** (ar-ak'-no-pi'-ah) [arachnoid; pia]. The arachnoid and the pia considered together.

**arachnorrhinitis** (ar-ak-no-rin-i'-tis) [ἀράχνη, spider; ῥίς, nose; ιτις, inflammation]. A disease of the nasal passages supposed to be due to the presence of a spider.

**arachnotis** (ar-ak-no-ti'-tis) [ἀράχνη, spider; οὖς, ear; ιτις, inflammation]. Inflammation said to be caused by a spider in the auditory canal.

**arack** (ar'-ak). See *arrack.*

**aræometer** (ar-e-om'-et-er). See *areometer.*

**araiocardia** (ar-i-o-kar'-de-ah) [ἀραιός, thin; καρδία, heart]. Bradycardia.

**Aralia** (ar-a'-le-ah) [L.]. A genus of plants, order *Araliaceæ,* embracing several species, having aromatic, diaphoretic, and resolvent properties. Ginseng, wild sarsaparilla, petty-morrel, and other plants esteemed in popular medicine belong here; few have active qualities of high value in any disease.

**Aran's green cancer** (ar-ahn') [François Amilcar Aran, French physician, 1817-1861]. Chloroma; malignant lymphoma of the orbital cavity associated with grave leukemia, and tending to form metastases through the lymphatic system. Syn., *cancer vert d'Aran.* **A.'s law,** fractures of the base of the skull are the result of injury to the vault, the extension taking place by irradiation along the line of the shortest circle. The fractures of the base which occur by *contre-coup* are exceptions to this law.

**Aran-Duchenne's disease** (ar-ahn'-doo-shen') [see *Aran;* Guillaume Benjamin Amand Duchenne de Boulogne, French physician, 1806-1875]. Progressive muscular atrophy.

**araneous** (ar-a'-ne-us) [aranea, a spider's web]. 1. Full of webs; resembling a cobweb. 2. Applied to a thready, feeble pulse. 3. Consisting of separate filaments. a. **membrane,** the arachnoid membrane.

**Arantius, bodies of** (ar-an'-she-us) [Julius Cæsar Arantius (*Arantio,* or *Aranzio*), Italian anatomist, 1530-1589]. The fibrous tubercles in the center of each segment of the semilunar valves. **A., canal of, A.,** duct of, the ductus venosus. The smaller of the two branches into which the umbilical vein divides after entering the abdomen; it empties into the ascending vena cava and becomes obliterated after birth. **A., ligament of,** the obliterated ductus venosus of Arantius. **A.**, ventricle of, a small culdesac in the medulla oblongata, forming the lower termination of the fourth ventricle.

**araroba** (ar-ar-o'-bah) [Brazil]. Goa powder. An oxidation-product of the resin found deposited in the wood of the trunk of *A. andira*, of Brazil. Its active principle is *chrysarobin* or *chrysophanic acid*. It is largely used in skin affections.

**arbor** (ar'-bor) [L.]. A tree. A name for the arbor vitæ of the cerebellum. **a. vitæ** [tree of life]. 1. A term applied to the arborescent appearance of a section of the cerebellum, and also to a similar appearance of the folds of the interior of the cervix uteri. 2. The *Thuja occidentalis*.

**arborescent** (ar-bor-es'-ent) [arbor]. Branching like a tree.

**arborization** (ar-bor-iz-a'-shun) [arbor]. 1. A form of nerve-termination in which nerve-fiber is brought into contact with muscle-fiber by means of an expansion. 2. A group of crystals showing a tree-like appearance. **a.**, **terminal**. 1. A branched end of a sensory nerve. 2. A motor end-plate. **a.**, **vascular**, a tree-like branching of blood-vessels.

**arbulith** (ar'-bu-lith). Trade name of a mixture of lithium benzoate and arbutin; it is used as a urinary antiseptic and antilithemic.

**arbutin** (ar'-bu-tin) [arbutus], $C_{12}H_{15}O_7$. A bitter glucoside, obtained from *Arctostaphylos uva-ursi*, or bearberry. It is neutral, crystalline, and resolvable into glucose and hydroquinone. It is diuretic. Dose 15-30 gr. (1-2 Gm.). See *Uva ursi*.

**Arbutus**, (ar-bū'-tus) [L.]. A genus of ericaceous shrubs and trees. *A. menziesii*, the madroño of California, has an astringent bark, useful in diarrhea. *A. unedo*, the European arbutus, is astringent and narcotic. **A.**, trailing, see *Epigæa*.

**arc** (ark) [arcus, a bow]. A part of the circumference of a circle; a more or less curved passageway. **a.**, **bigonial** (of lower jaw), a measurement around the anterior margin of the jaw. **a.**, **binauricular**, a measurement from the center of one auditory meatus to the other, directly upward across the top of the head. **a.**, **bregmatolambdoid**, a measurement along the sagittal suture. **a.**, **diastaltic nervous**, Marshall Hall's term for the nerves concerned in a reflex action. **a.**, **frontal**, the measurement from the nasion to the bregma. **a.**, **maximum transverse**, the measurement across the face from a point on each side just anterior to the external auditory meatus. **a.**, **nasobregmatic**, a line measured from the root of the nose to the bregma. **a.**, **nasomalar**, measurement between the outer margins of the orbits over the nasion. **a.**, **naso-occipital**, measurement from the root of the nose to the lowest occipital protuberance. **a.**, **occipital**, measurement from the lambda to the opisthion. **a.**, **parietal**, measurement from the bregma to the lambda. **a.**, **reflex**, the pathway for a reflex act, comprising the center, the afferent and efferent nerve. **a.**, **voltaic**, the band of light formed by the passage of a strong electric current between two adjacent carbon points.

**arcade** (ar-kād') [see arc]. 1. A series of arches; an arch. 2. The bow of a pair of spectacles. **a.**, **crural**, Poupart's ligament. **a.**, **Flint's**, the arteriovenous arch about the base of the renal pyramids. **a.**, **temporal**, **a.**, temporal, inferior, the zygoma. **a.**, temporal, superior, the orbital arch.

**arcanum** (ar-ka'-num) [L., "a secret"]. A secret medicine.

**arcate** (ar'-kāt) [arcatus, bow-shaped]. Bow-shaped; curved; arcuate.

**arcatura** (ar-ka-tū'-rah) [arcus, a bow]. A condition of horses marked by the undue outward curvature of the forelegs.

**arcein** (ar'-se-in). Arecolin hydrobromide; it is an active miotic.

**arch** (arch) [arcus, a bow]. 1. A structure having a curved outline resembling that of an arc or a bow. 2. A part of a circle. **a.**, **abdominothoracic**, the lower boundary of the front of the thorax. **a.**, **alveolar**, that marking the outlines of the alveolar processes of the jaw. **a.**, **anastomotic**, one uniting two veins or arteries. **a.**, **anterior hyoid**, a general term which includes the tympanohyal, epihyal, stylohyal, and ceratohyal arches. **a.**, **aortic**, see *aorta*. **a.s**, **aortic**, five pairs of vascular arches existing in the fetus. **a.s**, **axillary**, twigs of the latissimus dorsi, sometimes passing over the vessels and nerves to the anterior part of the axilla, where they disappear in the tissues. **a.s**, **branchial**, the cartilaginous arches that support the gills of fishes. They are also present in the human fetus. **a.s**, **cervical**, the fourth and fifth postoral arches. **a.**, **cortical**, that portion of the renal substance which stretches from one column to another and surrounds the base of the pyramids. **a.**, **costal**, the arch of the ribs. **a.**, **cotylosacral**, one formed by the sacrum and the osseous structures extending to the coxofemoral joints. Syn., *standing arch*. **a.**, **crural**, Poupart's ligament. **a.**, **dental**. 1. The parabolic curve formed by the cutting-edges and masticating surfaces of the teeth. 2. The alveolar arch. **a.**, **epencephalic**, the bones lying over the epencephalon, uniting in man to form the occipital bone. Syn., *neuroccipital arch*. **a.**, **facial**, the first postoral arch. **a.**, **femoral**, same as *a.*, *crural*. **a.**, **femoral**, deep, a band of fibers originating apparently in the transverse fascia, arching across the crural sheath and attached to the middle of Poupart's ligament and the pectineal line. Syn., *deep crural arch*. **a.s of the foot**, certain arches formed by the bones of the foot; the most distinct is the transverse in the line of the tarsometatarsal articulations. The inner longitudinal is composed of the os calcis, the astragalus, the navicular, the 3 cuneiforms, and the first 3 toes, and the outer longitudinal is made up of the os calcis, the cuboid, and the fourth and fifth toes **a.**, **gluteal**, an opening in the gluteal fascia transmitting the gluteal vessels and nerves. **a.**, **hemal**, Owen's term for the inferior loop of the typical vertebra. It is so called because it surrounds the essential portion of the vascular system. It is formed dorsally by the centrum, laterally by the pleurapophyses and hemapophyses, and inferiorly by the hemal spine. Syn., *infravertebral arch; subcentral arch; vertebral ventral arch*. **a.**, **hyoid**, the second branchial arch of vertebrates. Syn., *lingual arch; arch of tongue; parietohemal arch*. **a.**, **inguinal**, Poupart's ligament. **a.**, **ischiopubic**, that formed by the pubis and the ischiopubic branches. **a.**, **ischiosacral**, one formed by the sacrum, the descending branches of the ischia, and the ilia lying between. Syn., *sitting arch*. **a.**, **laryngeal**, Callender's term for one in the embryo composed of a membranous plate extending from the lower portion of the skull and developing into the inferior constrictor muscle, the cartilages of the larynx, the superior portion of the trachea, and the thyroid body. **a.s**, **lateral inferior** (of the skull), the bones encircling the mouth, nose, and larynx. **a.s**, **lateral superior**, the bones encircling the cerebrum, the cerebellum, and the oblongata. **a.**, **mandibular**, the first branchial arch, developing into the lower jaw. Syn., *maxillary arch*. **a.**, **maxillary**. 1. See *a.*, *mandibular*. 2. See *a.*, *palatomaxillary*. **a.**, **mesencephalic**, one formed by the basisphenoid, alisphenoid, parietal, and mastoid bones. Syn., *neuroparietal arch*. **a.**, **nasal**, one uniting the two frontal veins. **a.**, **neural**, the superior loop of the typical vertebra inclosing the neural canal. Syn., *dorsal vertebral arch; supravertebral arch*. **a.**, **occipitohemal**. See *girdle, shoulder-*. Syn., *pectoral arch; scapular arch; scapuloclavicular arch; scapulocoracoid arch*. **a.**, **osteoblastic**, those formed imperfectly or completely by the osteoblasts, arising from the bony trabeculæ, already developed and finally becoming bone. **a.**, **palatal**, the concavity of the hard palate when seen in transverse section. **a. of the palate**, **posterior**, that formed by the posterior pillars of the fauces. Syn., *palatopharyngeal arch*. **a.**, **palatine**, that formed by the anterior pillars of the fauces. Syn., *anterior arch of the palate*. **a.**, **palatomaxillary**, one formed by the palatine, maxillary, and premaxillary bones or their analogue; it is looked upon as the hemal arch of the nasal vertebra. Syn., *maxillary arch*. **a.**, **palmar**, the arch formed by the radial artery and ulnar arteries in the palm of the hand; there are two—a superficial and a deep. Syn., *radial arch*. **a.**, **palmar**, **superficial**, the continuation of the ulnar artery across the palm. **a.**, **pelvic**, the bones of the pelvis considered as the hemal arches of the sacral vertebræ. **a.**, **pharyngeal**, the fifth pair of branchial arches. **a.**, **plantar**, the arch made by the external plantar artery. **a.s**, **postoral**, arches in the fetus, five in number, that develop into the lower jaw and throat. See *a.*, *branchial*. Syn., *cephalic, poststernal, skeletal, subaxial, vascular, visceral arches*. **a.**, **prosencephalic**, one considered as the neural arch of the frontomandibular vertebra; it is formed by the frontal, presphenoid, and orbito-

**sphenoid bones.** Syn., *neurofrontal arch*. **a.** of **pubes,** that part of the pelvis formed by the convergence of the rami of the ischium and pubis on each side. Syn., *subpubic arch.* **a., radial.** See *a., palmar.* **a., rhinencephalic,** the neural arch of the nasal vertebra, formed by the vomer and the prefrontal and nasal bones. Syn., *neuronasal arch.* **a., Riolan's,** the arch of the mesentery which is attached to the transverse mesocolon. **a., stylohyoid,** the hemal arch of the parietal vertebra formed by the stylohyal, epihyal, ceratohyal, basihyal, glossohyal, and urohyal bones. **a., supraorbital,** the curved and prominent margin of the frontal bone that forms the upper boundary of the orbit. **a.s, tarsal,** the arches of the palpebral arteries. **a., thyrocartilaginous,** a communicating branch between the superior thyroid arteries of the two sides, lying at about the level of the angle of the thyroid cartilage. **a., thyrohyal, a., thyrohyoid,** the third of the postoral arches; it develops into the hyoid body and the greater cornua of the hyoid bone. **a., tonsillar.** See *isthmus of the fauces.* **a., trabecular,** one formed by the junction of the middle trabeculæ of the skull, containing the hypophysis and the infundibulum. **a.** of a vertebra, the part of a vertebra, formed of two pedicles and two laminæ, inclosing the spinal foramen. **a., vertebral.** 1. A neural arch. 2. A hemal arch. **a., zygomatic,** the arch formed by the malar and temporal bones. Syn., *subocular arch; suborbital arch; temporal arch.*

**arch-, archi** [ἀρχή, primitive]. Prefixes denoting first, chief, or principal.

**archæocyte** (ar-ke′-o-sīt) [ἀρχαῖος, ancient; κύτος, a cell]. A wandering or free ameboid cell.

**archæus** (ar-ke′-us) [ἀρχαῖος, ancient]. In spagiric medicine, the invisible counterpart of the visible body; solar heat as a source of life. 2. v. Helmont's name for the vital principle of an organism.

**archameba** (ark-am-e′-bah) [arch-; ἀμοιβή, change]. Haeckel's hypothetic progenitor of all amebas and of all higher forms of life.

**archamphiaster, archiamphiaster** (ark-am-fe-as′-ter, ar-ke-am-fe-as′-ter) [arch-; ἀμφί, afound; ἀστήρ, star]. In biology; those amphiasters concerned in the production of the polar globules.

**Archangelica** (ark-an-jel′-ik-ah) [ἀρχάγγελος, archangel]. A genus of umbelliferous plants. See *Angelica.*

**archebiosis** (ark-e-bi-o′-sis) [arch-; βίος, life]. Spontaneous generation.

**archegenesis** (ark-e-jen′-es-is). The same as *archebiosis.*

**archegonium** (ark-e-go′-ne-um) [ἀρχή, beginning; γόνος, race]. The female reproductive organ of the higher cryptogams.

**archegony** (ar-keg′-o-ne) [ἀρχέγονος, first of a race]. The doctrine of spontaneous generation.

**archelogy** (ar-kel-o′-je) [ἀρχή, a beginning; λόγος, science]. The study of the foundation principles of anthropology.

**archenteric** (ark-en-ter′-ik). Relating to the archenteron.

**archenteron** (ark-en′-ter-on) [arch-; ἔντερον, intestine]. The embryonic alimentary cavity.

**archeocyte** (ar′-ke-o-sīt). Same as archæocyte.

**archepyon** (ar-ke-pi′-on) [ἀρχή, a beginning; πύον, pus]. Pus that has become caseated, or so thick that it does not flow.

**archespore, archesporium** (ar′-ke-spōr, -e-um) [ἀρχή, a beginning; σπορά, a seed]. In biology, the cells that give rise to the lining of the anther-cell and to the mother-cells of the pollen.

**archetype** (ar′-ke-tīp) [arch-; τύπος, a type]. In comparative anatomy, an ideal type or form with which the individuals or classes may be compared. A standard type; original type; prototype.

**archi-.** See *arch-.*

**archiater** (ar-ke-a′-ter) [ἀρχιατρός, a chief physician]. 1. The head physician in a court. 2. The chief physician of an institution.

**archiblast** (ar′-ke-blast) ┤archi-; βλαστός, germ]. In embryology, the granular areola surrounding the germinal vesicle. In pathology, the important tissues of the body as contrasted with the *parablast*, or connective tissues.

**archiblastic** (ar-ke-blas′-tik) [see *archiblast*]. Derived from the archiblast. The parenchymatous tissues are regarded as archiblastic.

**archiblastoma** (ar-ke-blas-to′-mah) [archiblast; ὄμα, a tumor]. A tumor composed of archiblastic tissue, such as myoma, neuroma, papilloma, adenoma, carcinoma, etc.

**archiblastula** (ar-ke-blas′-tu-lah) [see *archiblast*]. In embryology, a ciliated, vesicular morula, resulting from complete and regular yolk-division and by invagination forming the archigastrula.

**archicytula** (ar-ke-sit′-u-lah) [archi-; κύτος, a cell]. A fertilized egg-cell in which the nucleus is discernible.

**archigaster** (ar-ke-gas′-ter) ‚archi-; γαστήρ, belly]. The primitive, perfectly simple intestine; archenteron.

**archigastrula** (ar-ke-gas′-tru-lah) [see *archigaster*]. The gastrula as it is observed in the most primitive types of animal development; called also bellgastrula, from its shape.

**archigenesis** (ar-ke-jen′-es-is). See *archebiosis.*

**archil** (ar′-kil) [ME., *orchell*]. A violet coloringmatter similar to litmus, chiefly obtained from the lichen, *Roccella tinctoria*; used for staining animal tissues.

**archimonerula** (ar-ke-mon-er′-ū-lah) [archi-; μονήρης, single; solitary]. In embryology, a special name given by Haeckel to the monerula stage of an egg undergoing primitive and total cleavage.

**archimorula** (ar-ke-mor′-ū-lah) [archi-; μόρον, a mulberry]. In embryology, the solid mass of cleavage cells, or mulberry mass, arising from the segmentation of an archicytula, and preceding the archiblastula and archigastrula.

**archinephric** (ar-ke-nef′-rik) [ἀρχή, first; νεφρός, the kidney]. Pertaining to the archinephron.

**archinephron** (ar-ke-nef′-ron) [archi-; νεφρός, kidney]. The primitive or embryonic stage of the kidney or renal apparatus. The Wolffian body.

**archineuron** (ar-ke-nū′-ron) [archi-; νεύρον, a nerve]. 1. A primitive neuron. 2. The neuron at which the impulse starts in any physiological act involving the nervous system.

**archipallium** (ar-ke-pal′-e-um) [archi-; *pallium*, a cloak]. The olfactory pallium, the rhinencephalon.

**archistome** (ar′-kis-tōm). See *blastopore.*

**archītis** (ar-ki′-tis) [ἀρχός, anus; ιτις, inflammation]. Proctitis; inflammation of the anus.

**archocele** (ar′-ko-sēl) [ἀρχός, anus; κήλη, hernia]. Rectal hernia.

**archocystocolposyrinx,** or **archocolpocystosyrinx** (ar-ko-sis-to-kol-po-sir′-ingks), or ar-ko-kol-po-sis-to-sir′-ingks) [ἀρχός, anus; κύστις, bladder; κόλπος, vagina; σύριγξ, fistula]. Recto-vesico-vaginal fistula.

**archocystosyrinx** (ar-ko-sist-o-sir′-inks) [ἀρχός, anus; κύστις, bladder; σύριγξ, fistula]. A rectovesical fistula.

**archometrum** (ar-ko-met′-rum) [ἀρχός, anus; μέτρον, measure]. A device for ascertaining the caliber of the anus, or for dilating its sphincters.

**archoplasm, archoplasma** (ar′-ko-plazm, ar-ko-plaz′-mah) [ἀρχών, a ruler; πλάσμα, a thing formed]. Boveri's term for the substance from which the attraction-sphere, the astral rays, and the spindlefibers of mitosis are derived and of which they consist. Syn., *kinoplasm.*

**archoptoma** (ar-kop-to′-mah) [ἀρχός, anus; πτῶμα, a fall]. A prolapse of the rectum.

**archoptosis** (ar-kop-to′-sis) [ἀρχός, anus; πτῶσις, a falling]. Rectal prolapse.

**archoptotic** (ar-kop-tot′-ik) [ἀρχός, anus; πτῶσις, a falling]. Relating to archoptoma or archoptosis.

**archorrhagia** (ar-ko-ra′-je-ah) [ἀρχός, anus; ῥήγνυναι, to break out]. Rectal hemorrhage.

**archorrhea** (ar-ko-re′-ah) [ἀρχός, anus; ῥεῖν, to flow]. A discharge of blood or of any pathological fluid from the anus.

**archos** (ar′-kos) [ἀρχός, the anus]. The anus.

**archostegnoma** (ar-ko-steg-no′-mah) [ἀρχός, anus; στεγνόειν, to consolidate]. Archostenosis; a rectal stricture.

**archostegnosis** (ar-ko-steg-no′-sis) [ἀρχός, anus; στέγνωσις, a stopping]. A rectal stricture.

**archostegnotic** (ar-ko-steg-not′-ik) [ἀρχός, anus; στέγνωσις, a stopping]. Relating to a rectal stricture.

**archostenosis** (ar-ko-sten-o′-sis) [ἀρχός, anus; στενός, narrow]. Stricture of the rectum.

**archostenotic** (ar-ko-sten-ot′-ik) [ἀρχός, anus; στενός, narrow]. Relating to rectal stricture.

**archosyrinx** (ar-ko-sir′-ingks) [ἀρχός, anus; σύριγξ, a pipe]. 1. A syringe for the rectum. 2. Fistula in ano.

**archyle** (ar-ki′-le) [ἀρχή, a beginning; ὕλη, matter]. See *protyle.*

**arciform** (ar′-se-form) [*arcus*, bow; arch; *forma*, form]. Arcuate, bowshaped; especially used to

designate certain sets of fibers in the medulla oblongata.
**arctation** (*ark-ta'-shun*) [*arctatio*, to draw close together]. Contraction of an opening or canal.
**Arctium** (*ark'-te-um*) [L.]. Burdock. See *Lappa*.
**arctostaphylos** (*ark-to-staf''-il-os*). See *Uva ursi* and *Manzanita*.
**arcual** (*ar'-kū-al*) [*arcualis*, arched]. Arched; bent or curved.
**arcuate** (*ar'-kū-āt*) [*arcuatio*, a bowing]. Arched; curved; bow-shaped. **a. fibers of the cerebellum,** associating fibers connecting one lamina with another. **a. fibers of the cerebrum,** associating fibers connecting adjacent convolutions.
**arcuation** (*ar-kū-a'-shun*) [*arcuatio*, a bowing]. Curvature, especially of a bone.
**arcula** (*ark'-ū-lah*) [*arcula*, a casket]. The orbit. **a. cordis,** the pericardium.
**arculus** (*ar'-kū-lus*) [dim. of *arcus*, a bow]. An arching support for bed-clothes.
**arcus** (*ar'-kus*) [L., "a bow"]. A bow or arch. **a. aortæ,** the arch of the aorta, or transverse aorta. **a. arteriarum,** the arteriæ arciformes of the kidney. **a. arteriosus manus,** the palmar arch. **a. arteriosus palpebræ,** an arterial arch along the edge of the eye-lid. **a. arteriosus pedis,** the plantar arch. **a. atlantis,** the arch of the atlas. **a. axillaris,** arch formed by the axillary artery. **a. carpidorsalis,** the posterior carpal arch. **a. cartilaginis cricoideæ,** arch of the cricoid cartilage. **a. coli intestini,** the transverse colon. **a. corneæ,** see *a. senilis*. **a. costarum, arch of the ribs. a. cruralis,** Poupart's ligament **a. cruralis profundus,** the deep crural arch. **a. dentalis,** the dental arch. **a. faucium,** the palatine arch. **a. glossopalatinus,** the anterior pillar of the fauces. **a. jugalis,** the zygomatic arch. **a. juvenalis,** a white ring around the cornea occurring in young individuals and resembling the arcus senilis. **a. lumbocostalis lateralis (Halleri),** ligamentum arcuatum externum. **a. lumbocostalis medialis (Halleri),** ligamentum arcuatum internum. **a. major ventriculi,** the great curvature of the stomach. **a. medullaris,** the fornix. **a. minor ventriculi,** the lesser curvature of the stomach. **a. occipitoparietalis,** an annectant gyrus between the superior parietal lobule and the occipital lobe. **a. palatini,** the pillars of the fauces. **a. palatoglossus.** Same as *a. glossopalatinus*. **a. palatopharyngeus,** the posterior pillar of the fauces. **a. palmaris,** the palmar arch. **a. pharyngoepiglotticus,** folds of mucous membrane passing from the pharynx to the epiglottis. **a. pharyngopalatinus,** posterior pillar of fauces. **a. plantaris,** the plantar arch. **a. popliteus,** the arcuate popliteal ligament. **a. senilis,** a ring of opacity at the edge of the cornea seen in the aged. **a. senilis lentis,** an opaque ring in the equator of the crystalline lens; it sometimes occurs in the aged. **a. spiralis,** the zona arcuata in the organ of Corti. **a. subpubicus,** the pubic arch. **a. superficialis volæ,** the superficial palmar arch. **a. supraorbitalis,** the supraorbital arch. **a. tarseus,** the tarsal arch. **a. tarsi oculi.** Same as *a. arteriosus palpebræ*. **a. thyrocartilagineus,** the arch formed by the superior thyroid arteries and the thyroid cartilage. **a. trachealis anterior,** the arch formed by the inferior thyroid arteries in passing over the trachea. **a. unguium.** See *lunula*. **a. vasculosi,** arches formed by branches of the renal artery in the kidney. **a. vasculosi renales,** arches at the bases of the Malpighian pyramids, formed by anastomoses of tiny ramifications of the renal artery. They give off vessels supplying the cortex of the kidneys, the Malpighian corpuscles, and the capillary plexuses about the uriniferous tubules. Syn., *fornices vasculosi renum*. **a. venosus,** (1) an arch joining the anterior jugular veins; (2) the venous arch in the palm of the hand; (3) a venous arch on the back of the fingers. **a. vertebralis,** a vertebral arch. **a. volaris,** the palmar arch. **a. zygomaticus,** the zygomatic arch.
**ardent** (*ar'-dent*) [*ardens*, *ardens*, to burn]. Burning; fiery; glowing; accompanied by a sensation of burning. **a. fever,** heat fever or thermal fever. **a. spirits,** alcoholic liquors.
**ardor** (*ar'-dor*) [L., "heat"]. Violent heat; burning. **a. urinæ,** burning pain in the inflamed urethra during micturition. **a. venereus,** sexual desire. **a. ventricoli,** pyrosis, heart-burn.
**are** (*ar*). French metric unit of square measure; it is a square whose side is 10 metres.
**area** (*a'-re-ah*) [L., "an open space"]. A limited extent of surface. **a. acustica,** or **a., auditory,** (1) the receptive center for audition in the superior temporal gyrus; (2) an area in the lateral angle of the floor of the fourth ventricle. **a., Broca's.** Same as *a. parolfactoria*. **a. Celsi,** alopecia areata. **a., cord,** that part of the cortex in which lesions would produce degeneration of the spinal cord. **a. cribrosa,** small perforated space in the internal auditory meatus through which pass filaments of the auditory nerve. **a., crural,** a space at the base of the brain included between the pons and chiasm. **a. diffluens,** alopecia areata. **a., diffraction,** a clear area seen in the microscopic image around all bodies of greater or less refractive power. **a. embryonalis.** Same as *a. germinativa*. **a. germinativa,** or embryonic spot, the oval germinating spot of the embryo. **a. hypoglossi.** Same as *Trigonum hypoglossi*. **a. intercruralis,** or interpeduncularis, an area at the base of the brain between the crura cerebri. **a.s, motor,** the convolutions in front of the Rolandic fissure, containing the centers for voluntary motion. **a., non-nucleated,** one of the clear spaces found at times between the endothelial cells of blood-vessels; they have no nuclei, are smaller than endothelial cells, and are considered to be due to the removal of parts of the surrounding endothelium. **a., occipital,** (1) that part of the occipital bone above the superior curved line; (2) the portion of the brain beneath the occipital bone. **a. opaca,** the opaque circle about the *a. pellucida*. **a. paraterminalis,** a space on the mesial aspect of the embryonic cerebral hemisphere. **a. parolfactoria (Brocæ),** a small vertical gyrus beneath the corpus callosum and continuous with the gyrus cinguli. **a. pellucida,** the light central portion of the *a. germinativa*. **a., postpontile,** that of the metencephalon comprising the olivary bodies and the lower lateral portion of the cerebellum. **a. postrema,** on the floor of the fourth ventricle between the ala cinerea and the tænia ventriculi quarti. **a., Rolandic,** the excitomotor area of the cerebral hemispheres, comprising the ascending frontal and ascending parietal convolutions. **a., sensor, sensory,** or **sensorial,** the general area of the cerebral cortex in which sensation is perceived. **a., septal,** the inner surface of each of the laminæ which make up the septum lucidum. **a., somesthetic,** the area for body feelings or tactile sensation in the postcentral convolution; the entire receptive and psychic sensory area. **a. vagi,** the trigonum vagi or ala cinerea. **a. vasculosa,** the space in the area opaca where blood-vessels first develop. **a.s, viscerocutaneous,** areas of skin and viscera corresponding to different spinal segments. **a., visual,** the area of the cortex cerebri, *viz.,* the cuneus and superior occipital gyrus, where vision is perceived. **a. vitellina,** yolk-area outside the area vasculosa in mesoblastic eggs. **a., vocal;** the portion of the glottis lying between the vocal bands.
**Areca** (*ar-e'-kah*). A genus of East Indian palms. *A. catechu* is extensively distributed throughout the tropics of Asia, where it has been cultivated from the earliest times. It furnishes the betel-nut (*q. v.*); the powdered nut is used as a vermifuge.
**arecaidin** (*ar-e-ka'-id-in*), $C_7H_{11}NO_2$. An acid contained in areca-nut, of which arecoline is its methyl ether.
**arecaine** (*ar-e'-ka-ēn*), $C_7H_{11}NO_2 + H_2O$. A poisonous teniacidal alkaloid obtained from areca-nut, forming colorless crystals soluble in water, insoluble in alcohol, in ether, and in benzol.
**arecane, arekane** (*ar'-ek-ān*). An oily and volatile basic substance obtainable from areca-nut; said to be a purgative and sialagogue, and to slow the pulse.
**areca-nut** (*a-re'-kah-nut*). See *betel*.
**arecin** (*ar'-es-in*) [Sp., *areça*], $C_{25}H_{26}N_2O$. 1. An organic base isomeric with brucine, derived from cinchona-bark. 2. A brown-red coloring-matter obtained from areca-nuts. Syn., *areca red*.
**arecoline** (*ar-e'-ko-lēn*) [*areca*], $C_8H_{13}NO_2$. A liquid alkaloid isolated from the seeds of *areca catechu*. It is a powerful poison, affecting the heart similarly to muscarine. It has anthelmintic properties. Dose $\frac{1}{16} - \frac{1}{10}$ gr. (0.004–0.006 Gm.). **a. hydrobromide,** is used as a miotic, applied in 1 % solution, and in the treatment of glaucoma. In veterinary practice it is used as a cathartic and anthelmintic. Injection for horse, $\frac{1}{2}$–1 gr. (0.032–0.065 Gm.).
**arefaction** (*ar-e-fak'-shun*) [*arefactio*; *arefacere*, to make dry]. 1. Exsiccation or desiccation. The removal of the structural or constitutional water from a substance. 2. The drying of drugs before

powdering them. 3. Dryness, as of the skin. 4. Withering, as of a paralyzed limb.

**areflexia** (*ar-e-fleks'-e-ah*) [ἀ, priv.; *reflectere*, to bend back]. The failure of a reflex; areflexion.

**areflexion.** See *areflexia*.

**arena** (*ar-e'-nah*) [*arena*, sand]. 1. Brick-dust deposit from urine; gravel. 2. Sabulous matter; brain-sand.

**arenaceous** (*ar-e-na'-se-us*) [*arenaceus; arena*, sand]. Of the nature of sand or gravel; sabulous.

**arenation** (*ar-e-na'-shun*) [*arena*, sand]. A sand-bath. The application of hot sand to a limb or part of the body. See *ammotherapy*.

**areocardia** (*ar-e-o-kar'-de-ah*) [ἀραιός, thin, rare; καρδία, heart]. Bradycardia.

**areola** (*ar-e'-o-lah*) [dim. of *area*, an open space]. 1. The brownish space surrounding the nipple of the breast. This is sometimes called *areola papillaris*. A second areola, surrounding this, occurs during pregnancy. The pigmentation about the umbilicus is called the *umbilical areola*. 2. Any interstice or minute space in a tissue. a., primary, cell-spaces still containing cartilage-cells in the matrix of ossifying cartilage-bone. Syn., *primary marrow cavities; medullary spaces*.

**areolar** (*ar-e'-o-lar*) [see *areola*]. Relating to or characterized by areolæ. a. tissue, cellular tissue; loose connective tissue.

**areolate, or areolated** (*ar-e'-o-lāt*, or *ar-e'-o-la-ted*) [*areola*, dim. of *area*, an open space]. Marked or characterized by areolæ.

**areometer** (*ar-e-om'-et-er*) [ἀραιός, rare, thin; μέτρον, a measure]. An instrument for measuring the specific gravity of liquids.

**areosis, aræosis** (*ar-e-o'-sis*). The process of becoming less compact; dilution.

**arevareva** (*ar-a-var-a'-vah*) [Tahitian]. A scaly skin-disease said to be caused by the habitual use of the drug *Kava*, q. v. It is accompanied by eye-disease, with dimness of vision.

**argal** (*ar'-gal*). See *argol*.

**argamblyopia** (*ar-gam-ble-o'-pe-ah*) [ἀργός, idle, disused; *amblyopia*]. Amblyopia due to disuse of the eye.

**Argand burner** (*ar'-gand*) [Ami Argand]. A burner that uses gas or oil, and contains an inner tube for supplying the flame with air.

**argas** (*ar'-gas*). The dove-tick. Found in dove-cots and pigeon roosts; it may give rise to edema or urticaria in man.

**Argasidæ** (*ar-gas'-e-de*). A family of ticks, practically all members of which are pathogenic to man.

**argema** (*ar'-jem-ah*) [ἄργεμα, an ulcer; pl., *argemata*]. A white ulcer of the margin of the cornea, following phlyctenula.

**argentamid** (*ar-jen'-tam-id*). An antiseptic liquid preparation of silver.

**argentamine** (*ar-jen'-tam-ēn*). A colorless alkaline liquid consisting of an 8 % solution of silver phosphate in a 15 % aqueous solution of ethylenediamide. It is applied in gonorrhea and conjunctivitis in 1 : 4000 solution. Syn., *ethylenediamide silver phosphate*.

**argentation** (*ar-jen-ta'-shun*) [*argentum*]. 1. Staining with a preparation of silver. 2. The act of silvering. 3. The process of injecting mercury into the vessels of an anatomic specimen. 4. Argyria.

**argentic** (*ar-jen'-tik*). Containing silver.

**argentine** (*ar-jen-tēn*). Containing or resembling silver.

**argentol** (*ar'-jen-tol*), C₆H₅N . OH . SO₄Ag. Silver quinaseptol, a yellow powder, sparingly soluble in water; used as a surgical antiseptic and astringent in ointment 1 : 100 or 2 : 100, in solution 1 : 1000 to 3 : 1000.

**argentous** (*ar jen'-tus*). Containing silver; applied to a compound containing a relatively larger amount of silver than an ordinary silver compound (argentic compound).

**argentum** (*ar-jen'-tum*) [L.]. Silver. Ag=107.88; quantivalence, 1; specific gravity, 10.4 to 10.5. A malleable and ductile metal of brilliant white luster. It tarnishes only in the presence of free sulphur, sulphur gases, and phosphorus. **argenti cyanidum** (U. S. P.), AgCN, silver cyanide, is used in the preparation of diluted hydrocyanic acid. **argenti nitras** (U. S. P.), AgNO₃, silver nitrate, argentic nitrate, "lunar caustic," a powerful astringent and an escharotic of moderate strength. It stains skin and other tissues black. If too long administered, it leaves a slate-colored, insoluble deposit of silver under the skin (*argyria*). It is used in gastric catarrh, in gastric ulcer, in intestinal ulceration, and as an alterative in scleroses of the nervous system. Dose ⅛–⅓ gr. (0.01–0.032 Gm.). **argenti nitras fusus** (U. S. P.), "stick caustic," contains 4 % of silver chloride. It is used locally. **argenti nitras mitigatus** (U. S. P.), the mitigated caustic, or diluted stick, is fused with an equal amount of potassium nitrate. **argenti oxidum** (U. S. P.), Ag₂O, explosive when treated with ammonia. Used internally for the same conditions in which the nitrate is used. Dose ½–2 gr. (0.032–0.13 Gm.). **argentum vivum**, an old name for mercury or quicksilver.

**argiamblyopia** (*ar-je-am-ble-o'-pe-ah*) [ἀργία, disuse; *amblyopia*]. See *argamblyopia*.

**argilla** (*ar-jil'-ah*) [ἄργιλλος, potter's clay]. White or potter's clay; alumina.

**argillaceous** (*ar-jil-a'-shus*) [ἄργιλλος, white clay]. Clay-like; composed of clay.

**arginase** (*ar'-jin-ās*). A ferment which has the power of splitting arginin into urea and ornithin.

**arginin** (*ar'-jin-in*), C₆H₁₄N₄O₂. Guanidin diamino-valeric acid, a substance formed in the cleavage of the protein molecule. It is hydrolyzed to urea and ornithin.

**argol** (*ar'-gol*) [ἀργός, white]. The impure tartar derived from wine; crude potassium bitartrate. See *tartar*.

**argon** (*ar'-gon*) [ἀργός, idle; inactive]. An inert gaseous element present in the atmosphere. Its symbol is A; atomic weight, 39.88. Argon may be obtained by freeing air, which has been deprived of its carbon dioxide and water, from oxygen by means of red-hot copper, and then absorbing the nitrogen by means of metallic magnesium. The residual gas, the passage of the gases being repeated a number of times, is argon. Chemically, it is the most inert element known.

**argonin** (*ar'-go-nin*). Silver casein; a soluble, antiseptic silver salt.

**Argyll Robertson pupil** (*ar'-gĭl*) [Douglas Argyll Robertson, Scotch physician, 1837–1909]. A pupil that acts to accommodation but not to light; it is seen in tabes dorsalis, paretic dementia, in some cases of encephalomalacia, senile brain atrophy, syphilis, hydrocephalus, etc.

**argyria** (*ar-ji'-re-ah*) [*argentum*]. A form of discoloration of the skin and mucous membranes produced by the prolonged administration of silver, the granules of silver being deposited in much the same position as those of the natural pigment of the skin.

**argyriasis** (*ar-jir-i'-as-is*). See *argyria*.

**argyrism** (*ar'-jir-izm*). Argyria.

**argyritis** (*ar-jir-i'-tis*). Yellow or silver litharge; lead monoxide of a yellow color.

**argyrol** (*ar'-jir-ol*). A soluble silver salt obtained by combining a proteid of wheat with 30 % of silver. It is used in gonorrhea. Syn., *silver vitelline*.

**argyrosis** (*ar-ji-ro'-sis*). Same as *argyria* (q. v.).

**arhinencephalia.** See *arrhinencephalia*.

**arhinia** (*ah-rin'-e-ah*) [ἀ, priv.; ῥίς, nose]. Congenital absence of the nose.

**arhovin** (*ar'-o-vin*). Addition-product of diphenylamine and thymylbenzoic acid ester, used in the treatment of gonorrhea.

**arhythmia** (*ar-ith'-me-ah*). See *arrhythmia*.

**Arica bark.** Calisaya bark exported from Arica, Chili. It contains the alkaloid, aricine.

**aricine** (*ar'-is-ēn*) [*Arica*], C₂₃H₂₆N₂O₄. An alkaloid obtained from several varieties of cinchona-bark.

**aridura** (*ar-id-ū'-rah*) [L.]. Dryness; a drying up, withering, or wasting of a part, or of the organism as a whole. 2. Hectic fever.

**aristocardia** (*ar-is-to-kar'-de-ah*) [ἄριστερός, left; καρδία, heart]. Deviation of the heart to the left side.

**aristochin** (*ar-is'-to-kin*). The ester of diquinine carbonic acid; it is a white, tasteless powder containing 96 per cent. of quinine.

**aristol** (*ar'-is-tol*) [ἄριστος, best], (C₆H₃CH₃OI .-C₃H₇)₂. Dithymoliodide. It is also called annidalin, although this is dithymoltriiodide. An iodine compound used as a substitute for iodoform as an antiseptic dressing. It has the advantage of being odorless, and is used either in the powder form or as a 5 to 10 % ointment with vaselin or lanolin.

**Aristolochia** (*ar-is-to-lo'-ke-ah*) [ἄριστος, best; λοχεία, the lochia]. A genus of exogenous herbs, many species of which have active medicinal qualities. *A. clematitis*, of Europe, has been used as a tonic, stimulant, and diaphoretic. *A. cymbifera*, of South America, furnishes a part of the drug called *guaco*, and is a good tonic and stimulant, *A. rotunda*, a species of southern Europe, with offensive odor and bitter taste, is employed as an emmenagogue and in gout. *A. serpentaria*, Virginia snakeroot, is at present more used in medicine than any other species. See *serpentaria*.

**aristolochic** (*ar-is-to-lo'-kik*) [ἄριστος, best; λοχεία, the lochia]. 1. Having the property of expelling the placenta, or of exciting or promoting the lochial discharge. 2. A medicine used for expelling the secundines or for exciting the lochial flow.

**aristolochin** (*ar-is-to-lo'-kin*) [see *Aristolochia*]. A bitter principle found in Virginia snakeroot. See *serpentaria*.

**aristoquin** (*ar-is'-to-kwin*). Same as *aristochin*.

**Aristotle's experiment** [*Aristotle*, Greek philosopher, 384-322 B. C.]. The eyes being closed, when a small spherical object is placed between two crossed fingers of one hand so that it touches the radial side of one and the ulnar side of the other, the sensation produced is that of two objects.

**arithmomania** (*ar-ith-mo-ma'-ne-ah*) [ἀριθμός, a number; μανία, madness]. An insane anxiety with regard to the number of things that fall under the observation. Sometimes it consists in constant or uncalled-for counting of objects, sometimes in the mere repeating of consecutive numbers.

**arkyochrome** (*ar'-ke-o-krōm*) [ἄρκυς, a net; χρῶμα, a color]. A somatochrome nerve-cell in which the stainable portion of the cell-body appears in the form of network.

**arkyostichochrome** (*ar-ke-o-stik'-o-krōm*) [ἄρκυς, a net; στίχος, a row or rank; χρῶμα, a color]. Applied by Nissl to a nerve-cell in which the chromophilic particles of its cell-body present a combination of both the striated (stichochrome) and network (arkyochrome) arrangements, so that it is difficult to decide which dominates; *e. g.*, the Purkinje cells of the cerebral cortex.

**arico-urease** (*arl'-ko-u'-re-ās*). A preparation of the urealytic enzyme obtained from the soy bean, *Soja hispida*. It decomposes urea into ammonia and carbon dioxide; and is used to determine the amount of urea in the urine, blood and other body fluids.

**Arlt's ointment** [Carl Ferdinand Ritter von *Arlt*, Austrian physician, 1812-1887]. An ointment containing 7½ gr. (0.5 Gm.) of belladonna to 1¼ dr. (5 Gm.) of blue ointment. **A.'s recess, A.'s sinus**, a small depression, directed forward and outward, in the lower portion of the lacrimal sac; it is not constant. **A.'s trachoma**, granular conjunctivitis; trachoma.

**Arlt-Jaesche's operation** [see *Arlt's ointment*.] For *distichiasis*; the edge of the lid and the contained ciliary bulbs are dissected from the tarsus, a crescent-shaped piece of skin is removed from the lid above the flap, and the edges of the wound are united, thus transplanting the ciliary bulbs farther away from the edge of the lids.

**arm.** 1. The upper extremity from the shoulder to the elbow. 2. The upper extremity from the shoulder to the wrist. 3. That portion of the stand connecting the body or tube of a microscope with the pillar. **a.**, center, the cortical center for the movement of the arm; it is situated in the middle third of the ascending frontal and ascending parietal convolutions. **a., milk**, phlegmasia alba dolens in the arm.

**armamentarium** (*ar-ma-men-ta'-re-um*) [L, an arsenal]. The outfit of medicines or instruments of the physician or surgeon. **a. chirurgicum**, surgical instruments and appliances. **a. lucinæ**, an outfit of obstetrical instruments.

**Armanni-Ehrlich's degeneration.** Hyaline degeneration of the epithelial cells of Henle's looped tubes in diabetes.

**armarium** (*ar-ma'-re-um*) [L.]. 1. Same as *armamentarium*. 2. The literary outfit of a physician or surgeon, his library.

**armature** (*ar'-mat-chūr*) [*armatura*, equipment]. 1. A mass of soft iron at the extremity of a magnet. Also, the core of iron around which coils of insulated wire are wound. 2. Any protective investment of an organism. 3. A condenser.

**Armenian** (*ar-me'-ne-an*) [Armenia]. Of or belonging to Armenia. **A. blue.** Same as *ultramarine*. **A. bole**, a reddish, unctuous earth or clay formerly much used in medicine, now used in tooth-powders and in veterinary practice. It is absorbent and astringent.

**armilla** (*ar-mil'-ah*) [*armilla*, a bracelet, ring]. 1. The annular ligament of the wrist. 2. The Gasserian ganglion.

**armpit** (*arm'-pit*) [*armus*, shoulder; *puteus*, a well]. The axilla.

**army itch.** A distressing, chronic form of itch prevalent in the United States at the close of the civil war. The itching was intense. The eruption was seen especially on the arms, forearms, chest, abdomen, and lower extremities, particularly on the ulnar side of the forearm and inner aspect of the thigh. It resembled prurigo associated with vesicles, pustules, and eczema.

**Arneth's classification of neutrophiles** (*ar'-nāt*) [Joseph *Arneth*, German physician, 1873– ]. The polynuclear neutrophiles are classified according to the number of nuclear lobes which they contain. The normal is said to be: 1 lobe, 5 per cent.; 2 lobes, 35 per cent.; 3 lobes, 41 per cent.; 4 lobes, 17 per cent.; and 5 lobes, 2 per cent.

**Arnica** (*ar'-nik-ah*) [L.]. A genus of composite-flowered plants. The *arnica* of the U. S. P. is the dried flower-heads of the plant commonly known as "leopardsbane," *A. montana*. The root (*arnicæ radix*) is official in the B. P. Its properties are probably due to an alkaloid, *trimethylamine*, $C_3H_9N$. In small doses it is a cardiac stimulant; in larger doses, a depressant. It is a popular remedy, when locally applied, for sprains, bruises, and surface wounds. Dose 15 gr. (1 Gm.). **a., infusion of** (20 parts flowers, 100 parts water), superior to the tincture for local use. **a. plaster**, contains extract of root, 33; lead-plaster, 67 parts. **a. root, extract of.** Dose 3–5 gr. (0.2–0.3 Gm.). **a. root, fluidextract of.** Dose 5–10 min. (0.3–0.65 Cc.). **a. root, tincture of**, 10 %. Dose 5–30 min. (0.3–2.0 Cc.). **a., tincture of** (*tinctura arnicæ*, U. S. P.), 20 %. Dose 15–30 min. (1–2 Cc.).

**arnicin** (*ar'-nis-in*) [*arnica*], $C_{20}H_{30}O_4$. A brownish, bitter neutral principle extracted from the root of *Arnica montana*.

**Arnold's bundle** [Friedrich *Arnold*, German anatomist, 1803–1890]. The fibers which form the inner third of the crusta of the cerebral peduncles. **A.'s canal**, a small canal in the petrous portion of the temporal bone, transmitting Arnold's nerve. **A.'s fold.** See *Béraud's valve*. **A.'s ganglion**, the otic ganglion. **A.'s ground plexus**, a plexus formed by the axis-cylinders of nonmedullated nerve-fibers in smooth muscular tissue. **A.'s innominate canal**, a nonconstant canal in the base of the skull, internally to the foramen rotundum; it transmits the superficial and deep petrosal branches that have become fused into one nerve. **A.'s ligament**, the ligament connecting the body of the incus with the roof of the tympanic cavity. **A.'s membrane**, the pigmentary layer of the iris. **A.'s nerve**, the auricular branch of the pneumogastric nerve. **A.'s operculum**, the operculum of the island of Reil. **A.'s recurrent nerve**, a sensory branch of the ophthalmic division of the trigeminus that anastomoses with the trochlear nerve and is distributed to the tentorium cerebelli and the posterior part of the falx cerebri. **A.'s stratum reticulatum**, the network formed by the fibers connecting the occipital lobe with the optic thalamus before they enter the latter.

**Arnott's bed** (*ar'-not*) [Neil *Arnott*, Scotch physician, 1788–1874]. A rubber mattress filled with water, designed to prevent bedsores.

**Arnoux's sign** (*ar-noo'*). In case of twins, a stethoscope applied over the mother's abdomen at a point about midway between the two fetal hearts will sometimes enable the physician to hear an apparent unison, and at other times a distinct rhythm of four beats; the double and quadruple rhythm alternate with great regularity.

**aroma** (*ar-o'-mah*) [ἄρωμα, spice]. The fragrance or odor emanating from certain vegetable substances, especially those used for food and drink.

**aromatic** (*ar-o-mat'-ik*) [see *aroma*]. 1. Having a spicy odor. 2. A substance characterized by a fragrant, spicy taste and odor, as cinnamon, ginger, the essential oils, etc. 3. A qualification applied to any carbon compound originating from benzene, $C_6H_6$. Their stability is relatively great as compared with that of the fatty bodies. **a. acids**, those derived from the benzene group of hydrocarbons. **a. compound,** any benzyl derivative. **a. fluidextract,** aromatic powder, 100; alcohol, sufficient to make 100 Cc. **a. powder.** See *cinnamomum*. **a. sulphuric acid.** See *acid, sulphuric*. **a. tincture,** an alcoholic solution of aromatic powder. **a. vinegar,** any mixture of aromatic oils in vinegar. **a. wine,** a wine containing in each 100 parts 1 part each of lavender, origanum, peppermint, rosemary, sage, and wormwood.

**aromatin** (ar-o'-mat-in). A succedaneum for hops.

**aromatize** (ar-o'-mat-īz) [ἄρωμα, spice]. To make aromatic; to spice.

**aromin** (ar-o'-min) [see aroma]. A substance derived from urine. When heated, it emits a fragrant odor.

**Aronson's serum** (ar'-on-sun) [Hans Aronson, German bacteriologist, 1865– ]. An antistreptococcus serum obtained from horses that have been treated with highly virulent cultures.

**arophene** (ar'-o-fēn). A proprietary dental anesthetic.

**arphoaline** (ar-fo'-al-ēn). A preparation containing arsenic, phosphorus and albumin; it is used as a local application for cancers and ulcers.

**arrachement** (ar-ash-mon(g)') [Fr., a tearing out]. Tearing out; extraction.

**arrack** (ar'-ak) [Ind.]. A liquor distilled from malted rice. Any alcoholic liquor is called arrack in the East.

**arrector** (ar-ek'-tor) [L., "an erector"]. An erector. **a. pili,** a fan-like arrangement of a layer of smooth muscular fibers surrounding the hair-follicle, the contraction of which erects the follicle and produces *cutis anserina*, or "goose-skin."

**arrectores pilorum** (ar-ek-to'-rēz pi-lo'-rum) [L.]. Plural of *arrector pili, q. v.*

**arrest** (ar-est') [ad, to; restare, to withstand]. 1. Stoppage, detention. 2. A disease of a mangy character affecting the hind leg of horses between the ham and postern. **a., action of, inhibition** (*q. v.*). **arrested development,** is when an organ or organism fails in its normal evolution, stopping at the initial or intermediate stages of the process. **arrested head,** when in parturition the child's head is hindered but not impacted in the pelvic cavity.

**arrhea** (ah-re'-ah) [ἀ, priv.; ῥοία, a flow]. The cessation or suppression of any discharge.

**arrhenal** (ar'-en-al). A monomethyl sodium arsenate, recommended in treatment of tuberculosis. Dose ¾ gr. (0.05 Gm.) daily.

**Arrhenius' law** (ah-ra'-ne-oos) [Svante August Arrhenius, Swedish chemist, 1859– ]. A solution must have a high osmotic pressure in order to be electrically conductive.

**arrhinencephalia** (ar-in-en-sef-al'-e-ah) [ἀ, priv.; ῥίς, nose; ἐγκέφαλον, the brain]. A form of partial anencephalia in which there is malformation of the nose.

**arrhinia** (ah-rin'-e-ah). Congenital absence of the nose.

**arrhythmia** (ah-rith'-me-ah) [ἀ, priv.; ῥυθμός, rhythm]. Absence of rhythm.

**arrhythmic** (ah-rith'-mik) [see *arrhythmia*]. Without rhythm; irregular.

**arrosion** (ar-o'-shun) [*arrodere,* to gnaw]. The gnawing or destruction of vessel-walls by ulcerous processes.

**arrow-poison** (ar-o-poi'-zun). See *curara*.

**arrowroot** (ar'-o-root) [ME., arow; roote]. A variety of starch derived from *Maranta arundinacea* of the West Indies, southern United States, etc. It is a popular remedy for diarrhea, and is widely used as a food. Many other starchy preparations are sold as arrow-root.

**arsacetin** (ars-as'-et-in). Sodium acetyl arsanilate; acetyl atoxyl. It is an organic arsenic compound used in sleeping-sickness, syphilis, and skin diseases.

**arsamine** (ars'-am-ēn). Same as *atoxyl*.

**arsan** (ar'-san). An organic arsenic preparation consisting of vegetable protein with about 4 per cent. of arsenic.

**arsanilate** (ar-san'-il-āt). A salt of arsanilic acid.

**arsenate, arseniate** (ar'-sen-āt, ar-se'-ne-āt) [*arsenic*]. Any salt of arsenic acid. **a., acid,** a monohydric or dihydric arsenate. **a., basic,** an arsenate combined with the oxide or hydrate of a base. **a., dihydric.** 1. An acid arsenate containing two atoms of hydrogen. 2. See *pyroarsenic acid*. **a., monohydric.** 1. An acid arsenate containing one atom of hydrogen. 2. Metarsenic acid, $HAsO_3$, a crystalline substance obtained from arsenic trioxide by heating above 200° C. **a., neutral.** 1. A normal arsenate. 2. A pyroarsenate.

**arsenauro** (ar-sen-aw'-ro). A double bromide of gold and arsenic; 10 min. contain $\frac{1}{12}$ gr. each of gold and arsenic bromides. It is an alterative and a tonic. D s 5–15 min. (0.3–1.0 cc.) in water three times daily.

**arsenglidin** (ar-sen-gli'-din). Same as *arsan*.

**arsenhemol** (ar-sen-hem'-ol). A compound of hemol and 1 % of arsenic trioxide, forming a brown powder. It is used as a substitute for arsenic as an alterative and hematinic. Dose 2 gr. (0.1 Gm.) 3 times daily.

**a$_r$s$_e$niasis** (ar-sen-i'-as-is). Same as *arsenism*.

**arsenic, arsenicum, arsenum** (ar'-sen-ik, ar-sen'-i-kum, ar-se'-num). 1. As = 74.96; quantivalence III, V. A brittle, crystalline metal, of a steel-gray color, tarnishing on exposure to the air. Sp. gr. 5.73. It sublimes at 180° C., and gives off a garlicky odor. In medicine arsenic is used as an alterative in anemia, chronic malaria, asthma, pulmonary tuberculosis, as a gastric sedative, and in chorea. 2. Arsenic trioxide. 3. Pertaining to arsenic. **a. bromide,** $AsBr_3$, is used in diabetes. Dose $\frac{1}{60}$ gr. (0.001 Gm.). **a. bromide, solution of** (*liquor arseni bromidi*), Clemens' solution, a 1 % solution of arsenic bromide. Dose 1–4 min. (0.06–0.24 Cc.). **a.,** butter of. See **a. chloride. a. caseinate,** a soluble arsenic compound for internal administration. **a. chloride,** $AsCl_3$, a colorless, oily liquid decomposed by water. Dose $\frac{1}{60}$ – $\frac{1}{16}$ gr. (0.001–0.004 Gm.). Syn., *butter of arsenic; chloride of caustic arsenic*. **a. disulphide,** $As_2S_2$, occurs native as realgar. Syn., *sandaraca; red sulphide of arsenic; red arsenic.* An artificial disulphide of arsenic is prepared in the arsenic works and contains about 15 % of arsenic and 27 % of sulphur. Syn., *red arsenic glass; ruby sulphur; red orpiment*. **a.,** flowers of, a fine white powder, formed by the sublimation of arsenic trioxide. **a. glass,** a term applied to the vitreous mass obtained either by heating arsenic pyrites with sulphurous ores, or by the resublimation of the "flowers of arsenic" obtained by subliming arsenic pyrites. Syn., *white arsenic glass*. **a. iodide** (*arseni iodidum,* U. S. P.), arsenous iodide, $AsI_3$. Dose $\frac{1}{20}$ – $\frac{1}{4}$ gr. (0.003–0.008 Gm.). **a. iodide, solution of** mercuric and of (*liquor arseni et hydrargyri iodidi,* U. S. P.), Donovan's solution; contains arsenous iodide, 10 Gm.; red mercuric iodide, 10 Gm.; distilled water, q. s. to make 1000 Gm. **a., test for.** See *Bettendorf, Marsh, Reinsch*. **a. trioxide** (*arseni trioxidum,* U. S. P.), $As_2O_3$, arsenous acid; "ratsbane." Dose $\frac{1}{60}$ – $\frac{1}{10}$ gr. (0.002–0.006 Gm.). Syn., *white arsenic*. **a. trioxide, solution of** (*liquor acidi arsenosi,* U. S. P.; *liquor arsenici hydrochloricus,* B. P.), a 1 % solution of the trioxide in hydrochloric acid and distilled water. Dose 2–5 min. (0.12–0.3 Cc.). **a. trisulphide,** $As_2S_3$, translucent, lemon-colored, rhombic prisms, occurring in nature; sp. gr. 3.46; a corrosive and depilating agent recommended for removal of warts. Syn., *orpiment; yellow sulphide of arsenic; arsenicum* (Pliny); *arsenii sulphidum citrinum;* King's yellow. **a., white.** See **a. trioxide**.

**arsenical** (ar-sen'-ik-al) [*arsenum,* arsenic]. Pertaining to arsenic.

**arsenicalism,** (ar-sen'-ik-al-izm,) [*arsenic*]. Chronic arsenic poisoning.

**arsenicate** (ar-sen'-ik-āt). To impregnate with arsenic.

**arsenicophagy** (ar-sen-ik-off'-a-je) [*arsenum,* arsenic; φαγεῖν, to eat]. The habitual eating of arsenic.

**arsenide** (ar'-sen-īd). A compound of arsenic with another element.

**arsenionization** (ar-sen-i-on-i-za'-shun). The electrical administration of arsenic ions into the tissues.

**arseniophosphate** (ar-sen-e-o-fos'-fāt). A compound of a base with both arsenic and phosphoric acids.

**arsenism** (ar'-sen-izm) [*arsenum,* arsenic]. Chronic arsenical poisoning; arsenicalism.

**arsenite** (ar'-sen-īt) [*arsenic*]. Any salt of arsenous acid.

**arseniureted** (ar-sen'-yu-ret-ed). Combined with arsenic so as to form an arsenide. **a. hydrogen,** arsine, $AsH_3$.

**arsenization** (ar-sen-iz-a'-shun) [*arsenum,* arsenic]. Treatment with arsenical remedies.

**arseno–** (ar'-se-no). Arsenic combined in the form —As = As —. Thus arsenobenzol is $C_6H_5$ — As = As — $C_6H_5$.

**arsenobenzol** (*ar-sen-o-ben'-zol*). 1. See *arseno-*. 2. See *salvarsan*.

**arsenoblast** (*ar-sen'-o-blast*) [ἄρσην, male; βλαστός, germ]. In biology, the male element of the sexual cell, capable of multiplication by division; the opposite of the *thelyblast* or female element.

**arsenocerebrin** (*ar-sen-o-ser'-e-brin*). A proprietary preparation said to contain cerebrin and sodium cacodylate; it is suggested for use in epilepsy.

**arsenophenylglycin** (*ar-sen-o-fen-il-gli'-sin*). A synthetic arsenic compound, a substitution-product of atoxyl, used in syphilis, trypanosomiasis and protozoan diseases in general.

**arsenous** (*ar-se'-nus*) [*arsenic*]. Containing arsenic. **a. acid.** See *arsenic trioxide*.

**arsine** (*ar'-sin*). Arseniureted hydrogen, AsH₃.

**arsins** (*ar'-sins*) [*arsenic*]. Peculiar volatile arsenic bases found by Selmi to be produced by the contact of arsenic trioxide and albuminous substances.

**arsinyl** (*ar'-sin-il*). The proprietary name for disodium-methylarsenate, a nontoxic substance allied to cacodyl and free from its garlicky odor. It is said to be a powerful tonic. Dose ½ gr. (0.03 Gm.) twice daily.

**arsonate** (*ar'-so-nāt*). A salt of arsonic acid.

**arsonic acid** (*ar-son'-ik*). An acid derived from arsenic acid by the substitution of an organic radical for one of the hydroxyl groups.

**arsonium** (*ar-so'-ne-um*) [*arsenic; ammonium*], AsH₄. A univalent radical in which arsenic replaces the nitrogen of ammonium.

**Arsonvalization** (*ar son val is a' shun*) [d'*Arsonval*, French physiologist and physicist, 1851– ]. The therapeutic application of Tesla currents.

**arsycodile** (*ar-sik'-od-il*). 1. A chemically pure cacodylate of sodium, a nontoxic salt used in emaciation. Dose ½ gr. (0.03 Gm.) 4 times daily. 2. The trade name of a number of preparations containing sodium cacodylate.

**artarine** (*ar'-tar-ēn*). An alkaloid, C₂₀H₁₇NO₄, from artar root; it is a cardiac stimulant, with action similar to veratrine.

**artar root** (*ar'-tar root*). A drug from West Africa, probably the root of *Xanthoxylum senegalense*.

**artefact** (*ar'-te-fakt*) [*arte*, by art; *factum*, made]. In microscopy and histology, a structure that has been produced by mechanical, chemical, or other artificial means; a structure or tissue that has been changed from its natural state.

**Artemisia** (*ar-tem-is'-e-ah*) [''Αρτεμις, the goddess Diana]. A genus of plants of the order *Composita*. **A. abrotanum**, southern-wood, is stimulant, tonic, and vermifuge, and is popularly used as a vulnerary. It is similar in properties to wormwood. Dose of *fluidextract* 10–20 m̄in. (0.6–1.2 Cc.). **A. absinthium.** See *absinthium*. **A. abyssinica**, an African species yielding the drug *zerechiti*, applied to relieve cramps in the final stages of malaria. **A. arborescens**, of southern Europe, is stomachic and tonic, and is used as is *A. absinthium*. **A. chinensis**, of Asia, is employed by the Chinese as a tonic and emmenagogue, and the down covering the leaf-surface in the preparation of moxa. **A. frigida**, wild sage, mountain sage, sierra salvia. An herb of western United States, introduced as a substitute for quinine in the treatment of periodic fevers. Also of service in diphtheria, rheumatism, and scarlatina. Dose of the *fluidextract* 1–2 dr. (4–8 Cc.). **A. maritima**, affords pure wormseed. **A. mexicana**, an American species, is said to be a stimulant, emmenagogue, and anthelmintic. **A. pontica**, Roman wormwood; it grows in Europe and Asia, has a pleasant odor and taste, and is used as a tonic and stimulant; it is burned in Egypt during the plague to ward off contagion. **A. santonica**, a species of Persia and Tartary, a variety of wormseed sometimes imported from Russia. **A. spicata**, an Alpine species with strong aromatic properties. **A. tridentata**, sagebrush, a shrub of the elevated portion of western North America, containing a pungent volatile oil. It is diaphoretic and stimulant. The Indians use an infusion of the plant as remedial for colds and headache and as a vermifuge. **A. trifida**, is found in the valleys of Utah and Wyoming, and has properties similar to *A. tridentata*. **A. vulgaris**, mugwort, a popular remedy in various diseases.

**arterenol** (*ar-te-re'-nol*). A proprietary drug, said to have an action similar to suprarenal preparations.

**arteria** (*ar-te'-re-ah*) [ἀρτηρία trachea; artery.] A hollow tube. See *artery*. **a. acetabuli**, artery of the acetabulum. **a. alveolaris inferior**, inferior dental artery. **a. alveolaris superior anterior**, anterior superior dental artery. **a. alveolaris superior posterior**, posterior dental artery. **a. angularis**, angular artery. **a. anonyma**, innominate artery. **a. appendicularis**, appendicular artery. **a. arcuata**, arcuate or metatarsal artery. **a. auditiva interna**, auditory artery. **a. auricularis posterior**, posterior auricular artery. **a. auricularis profunda**, deep auricular artery. **a. axillaris**, axillary artery. **a. basilaris**, basilar artery. **a. brachialis**, brachial artery. **a. bronchialis**, bronchial artery. **a. buccinatoria**, buccal artery. **a. bulbi urethræ**, artery of the bulb of the urethra. **a. bulbi vestibuli (vaginæ)**, artery of the vestibular bulb of the vagina. **a. canalis pterygoidei (Vidii)**, Vidian artery. **a. carotis communis**, common carotid artery. **a. carotis externa**, external carotid artery. **a. carotis interna**, internal carotid artery. **a. centralis retinæ**, central artery of the retina. **a. cerebelli inferior anterior**, anterior inferior cerebellar artery. **a. cerebelli inferior posterior**, posterior inferior cerebellar artery. **a. cerebelli superior**, superior cerebellar artery. **a. cerebri anterior**, anterior cerebral artery. **a. cerebri media**, middle cerebral artery. **a. cerebri posterior**, posterior cerebral artery. **a. cervicalis ascendens**, ascending cervical artery. **a. cervicalis profunda**, deep cervical artery. **a. cervicalis superficialis**, superficial cervical artery. **a. chorioidea**, anterior choroidal artery. **a. ciliaris anterior**, anterior ciliary artery. **a. ciliaris posterior brevis**, short posterior ciliary artery. **a. ciliaris posterior longa**, long posterior ciliary artery. **a. circumflexa femoris lateralis**, external circumflex artery. **a. circumflexa femoris medialis**, internal circumflex artery. **a. circumflexa humeri anterior**, anterior circumflex artery. **a. circumflexa humeri posterior**, posterior circumflex artery. **a. circumflexa ilium profunda**, deep circumflex iliac artery. **a. circumflexa ilium superficialis**, superficial circumflex iliac artery. **a. circumflexa scapulæ**, dorsalis scapulæ artery. **a. clitoridis**, artery of the clitoris. **a. cœliaca**, celiac artery or axis. **a. colica dextra**, right colic artery. **a. colica media**, middle colic artery. **a. colica sinistra**, left colic artery. **a. collateralis media**, middle collateral artery. **a. collateralis radialis**, articular branch of superior profunda artery. **a. collateralis ulnaris inferior**, anastomotica magna artery. **a. collateralis ulnaris superior**, inferior profunda artery. **a. comitans nervi ischiadici**, comes nervi ischiadici, or companion artery of the sciatic artery. **a. communicans anterior**, anterior communicating artery. **a. communicans posterior**, posterior communicating artery. **a. conjunctivalis anterior**, anterior conjunctival artery. **a. conjunctivalis posterior**, posterior conjunctival artery. **a. coronaria (cordis) dextra**, right coronary artery. **a. coronaria (cordis) sinistra**, left coronary artery. **a. cystica**, cystic artery. **a. deferentialis**, deferential artery. **a. digitalis dorsalis**, dorsal digital artery. **a. digitalis plantaris**, collateral digital branch artery. **a. digitalis volaris communis**, palmar digital artery, or collateral digital artery. **a. dorsalis clitoridis**, dorsal artery of the clitoris. **a. dorsalis nasi**, dorsal artery of the nose. **a. dorsalis pedis**, dorsal artery of the foot. **a. dorsalis penis**, dorsal artery of the penis. **a. epigastrica inferior**, deep epigastric artery. **a. epigastrica superficialis**, superficial epigastric artery. **a. epigastrica superior**, superior epigastric artery. **a. episcleralis**, episcleral artery. **a. ethmoidalis anterior**, anterior ethmoidal artery. **a. ethmoidalis posterior**, posterior ethmoidal artery. **a. femoralis**, femoral artery. **a. frontalis**, frontal artery. **a. gastrica dextra**, pyloric artery. **a. gastrica sinistra**, gastric or coronary artery. **a. gastroduodenalis**, gastroduodenal artery. **a. gastroepiploica dextra**, right gastro-epiploic artery. **a. gastroepiploica sinistra**, left gastro-epiploic artery. **a. genu inferior lateralis**, inferior external articular artery. **a. genu inferior medialis**, inferior internal articular artery. **a. genu media**, azygos articular artery. **a. genu superior lateralis**, superior external articular artery. **a. genu superior medialis**, superior internal articular artery. **a. genu suprema**, anastomotica magna (of knee). **a. glutæa inferior**, sciatic artery. **a. glutæa superior**, (superior) gluteal artery. **a. hæmorrhoidalis inferior**, inferior hemorrhoidal artery. **a. hæmorrhoidalis media**, middle hemorrhoidal artery. **a. hæmorrhoidalis superior**, superior hemorrhoidal artery. **a. helicina**, a spiral artery. **a.**

hepatica, hepatic artery. a. hyaloidea, hyaloid artery. a. hypogastrica, internal iliac artery. a. ilea, ileal artery, one of the rami intestini tenuis arteries. a. ileocolica, ileocolic artery. a. iliaca communis, common iliac artery. a. iliaca externa, external iliac artery. a. iliolumbalis, iliolumbar artery. a. infraorbitalis, infraorbital artery. a. intercostalis, intercostal artery. a. intercostalis suprema, superior intercostal artery. a. interlobaris renis, interlobar artery of kidney. a. interossea communis, common interosseous artery. a. interossea dorsalis, posterior interosseous artery. a. interossea recurrens, posterior interosseous recurrent artery. a. interossea volaris, anterior interosseous artery. a. jejunalis, jejunal artery. a. labialis anterior, anterior scrotal or labial artery. a. labialis inferior, inferior labial or coronary artery. a. labialis posterior, posterior labial artery. a. labialis superior, superior labial or coronary artery. a. lacrimalis, lacrimal artery. a. laryngea inferior, inferior laryngeal artery. a. laryngea superior, superior laryngeal artery. a. lienalis, splenic artery. a. ligamenti teretis uteri, artery of the round ligament of the uterus. a. lingualis, lingual artery. a. lumbalis, lumbar artery. a. lumbalia ima, lowest lumbar artery. a. malleolaris anterior lateralis, external malleolar artery. a. malleolaris anterior medialis, internal malleolar artery. a. malleolaris posterior lateralis, posterior peroneal artery. a. malleolaris posterior medialis, internal malleolar artery. a. mammaria interna, internal mammary artery. a. masseterica, masseteric artery. a. maxillaris externa, facial artery. a. maxillaris interna, internal maxillary artery. a. mediana, median artery. a. mediastinalis anterior, anterior mediastinal artery. a. meningea anterior, anterior meningeal artery. a. meningea media, middle meningeal artery. a. meningea posterior, posterior meningeal artery. a. mentalis, mental artery. a. mesenterica inferior, inferior mesenteric artery. a. mesenterica superior, superior mesenteric artery. a. musculophrenica, musculophrenic artery. a. nutritia femoris inferior, inferior nutrient artery of femur. a. nutritia femoris superior, superior nutrient artery of femur. a. nutritia fibulæ, nutrient artery of fibula. a. nutritia humeri, nutrient artery of the humerus. a. nutritia pelvis renalis, nutrient artery of renal pelvis. a. nutritia tibiæ, nutrient artery of tibia. a. obturatoria, obturator artery. a. occipitalis, occipital artery. a. œsophagea, esophageal artery. a. ophthalmica, ophthalmic artery. a. ovarica, ovarian artery. a. palatina ascendens, ascending palatine artery. a. palatina descendens, descending palatine artery. a. palatina major, greater palatine artery. a. palatina minor, lesser palatine artery. a. palpebralis lateralis, lateral palpebral artery. a. palpebralis medialis, middle or internal palpebral artery. a. pancreaticoduodenalis inferior, inferior pancreaticoduodenal artery. a. pancreaticoduodenalis superior, superior pancreaticoduodenal artery. a. penis, artery of penis. a. perforans prima, first perforating artery. a. perforans secunda, second perforating artery. a. perforans tertia, third perforating artery. a. pericardiacophrenica, the comes nervi phrenici. a. perinei, superficial perineal artery. a. peronæa, peroneal artery. a. phrenica inferior, inferior phrenic artery. a. phrenica superior, superior phrenic artery. a. plantaris lateralis, external plantar artery. a. plantaris medialis, internal plantar artery. a. poplitea, popliteal artery. a. princeps pollicis, principal artery of thumb. a. profunda brachii, superior profunda artery. a. profunda clitoridis, deep artery of clitoris. a. profunda femoris, deep femoral artery. a. profunda linguæ, ranine artery. a. profunda penis, artery to the corpus cavernosum. a. pudenda externa, external pudic artery. a. pudenda interna, internal pudic artery. a. pulmonalis, pulmonary artery. a. radialis, radial artery. a. recurrens radialis, radial recurrent artery. a. recurrens tibialis posterior, posterior recurrent tibial artery. a. renalis, renal artery. a. recurrens ulnaris, recurrent ulnar artery. a. sacralis lateralis, lateral sacral artery. a. sacralis media, middle sacral artery. a. scrotalis anterior, anterior scrotal artery. a. sigmoidea, sigmoid artery. a. spermatica externa, cremasteric artery. a. spermatica interna, (internal) spermatic artery. a. sphenopalatina, sphenopalatine or nasopalatine artery. a. spinalis anterior, anterior or ventral spinal artery. a. spinalis posterior, posterior or dorsal spinal artery. a. sternocleidomastoidea, sternomastoid artery. a. stylomastoidea, stylomastoid artery. a. subclavia, subclavian artery. a. sublingualis, sublingual artery. a. submentalis, submental artery. a. subscapularis, subscapular artery. a. supraorbitalis, supraorbital artery. a. suprarenalis inferior, inferior suprarenal artery. a. suprarenalis media, middle capsular artery. a. tarsea lateralis, lateral tarsal artery. a. tarsea medialis, medial tarsal artery. a. temporalis media, middle temporal artery. a. temporalis profunda anterior, anterior deep temporal artery. a. temporalis profunda posterior, posterior deep temporal artery. a. temporalis superficialis, superficial temporal artery. a. testicularis, testicular artery. a. thoracalis lateralis, long thoracic artery. a. thoracalis suprema, superior thoracic artery. a. thoracoacromialis, acromiothoracic artery or thoracic axis. a. thoracodorsalis, thoracodorsal artery. a. thymica, thymic artery. a. thyreoidea ima, lowest thyroid artery. a. thyreoidea inferior, inferior thyroid artery. a. thyreoidea superior, superior thyroid artery. a. tibialis anterior, anterior tibial artery. a. tibialis posterior, posterior tibial artery. a. transversa colli, transversalis colli. a. transversa faciei, transverse artery of face. a. transversa scapulæ, suprascapular artery. a. tympanica anterior, anterior tympanic artery. a. tympanica inferior, inferior tympanic artery. a. tympanica posterior, posterior tympanic artery. a. tympanica superior, superior tympanic artery. a. ulnaris, ulnar artery. a. umbilicalis, umbilical artery. a. urethralis, urethral artery. a. uterina, uterine artery. a. vaginalis, vaginal artery. a. vertebralis, vertebral artery. a. vesicalis inferior, inferior vesical artery. a. vesicalis superior, superior vesical artery. a. volaris indicis radialis, radialis indicis artery. a. zygomaticoorbitalis, zygomatico-orbital artery.

**arteriac** (ar-te'-re-ak) [arteria]. 1. Pertaining to the trachea, or to the arteries. 2. A remedy used in diseases of the trachea or of the arteries.

**arteriæ** (ar-te'-re-e) [L., plural of arteria]. Arteries. a. arciformes, arciform arteries of renal arches. a. gastricæ breves, the vasa brevia. a. interlobulares, interlobular arteries. a. intestinales, intestinal arteries or vasa intestini tenuis. a. metacarpeæ dorsales, dorsal interosseous arteries. a. renis, renal arteries. a. metacarpeæ volares, volar or palmar interosseous arteries. a. metatarseæ dorsales, dorsal interosseous arteries. a. metatarseæ plantares, digital branches of the plantar arch. a. scrotales posteriores, superficial perineal arteries. a. surales, inferior muscular arteries.

**arteriagra** (ar-ter-e-a'-grah) [arteria; ἄγρα, a seizure]. Neuralgia of an artery.

**arterial** [see arteria]. Pertaining to an artery.

**arterialization** (ar-te-re-al-iz-a'-shun) [see arteria]. 1. The process of making or becoming arterial; the change from venous blood into arterial. 2. Vascularization.

**arteriarctia** (ar-te-re-ark'-te-ah) [arteria; arctus, bound]. Constriction or stenosis of an artery.

**arteriasis** (ar-te-ri'-as-is) [see arteria]. Degeneration of an artery; it may be either calcareous or fatty.

**arteriectasis, arteriectasia** (ar-te-re-ek'-tas-is, ar-te-re-ek-ta'-se-ah) [arteria; ἔκτασις, a stretching out]. Arterial dilatation.

**arteriectopia** (ar-te-re-ek-to'-pe-ah) [arteria; ἔκτοπος, out of place]. Displacement or abnormality in the course of an artery.

**arteriitis** (ar-te-re-i'-tis). See arteritis.

**arterin** (ar'-ter-in) [see arteria]. Hoppe-Seyler's term for the arterial blood-pigment contained in the red corpuscles.

**arterioarctia** (ar-te-re-o-ark'-te-ah). See arteriarctia.

**arteriocapillary** (ar-te-re-o-kap'-il-a-re) [arteria; capillary]. Pertaining to arteries and capillaries. a. fibrosis, a chronic inflammatory process characterized by an overgrowth of connective tissue in the walls of the blood-vessels. It is known also as arteriocapillary fibrosis of Gull and Sutton.

**arteriochalasis** (ar-te-re-o-kal'-as-is) [arteria; χάλασις, a slackening]. Arterial atony.

**arteriococcygeal gland** (ar-te-re-o-kok-sij'-e-al). Luschka's gland.

**arteriodialysis** (ar-te-re-o-di-al'-is-is) [arteria; διάλυσις, dissolution]. Attenuation of the arterial walls with or without rupture.

**arteriodiastasis** (ar-te-re-o-di-as'-tas-is) [arteria; διάστασις, separation]. 1. The retraction of the two ends of a divided artery. 2. See *arterioectopia*. 3. The divergence of two arteries that lie near each other normally.

**arteriodiplopiesmus** (ar-te-re-o-dip-lo-pi-ez'-mus) [*arteria*; διπλοῦς, twofold; πιεσμός, pressure]. D'Etiolles' procedure for obtaining rapid coagulation of the blood in that part of an artery lying between two points upon which simultaneous pressure is made.

**arteriofibrosis** (ar-te-re-o-fi-bro'-sis). See *arteriocapillary fibrosis*.

**arteriogram** (ar-te'-re-o-gram). See *sphygmogram*.

**arteriograph** (ar-te'-re-o-graf) [*arteria*; γράφειν, to record]. A form of sphygmograph.

**arteriography.** (ar-te-re-og'-ra-fe) [*arteria*; γράφη, a writing]. 1. A description of the arteries. 2. The graphic representation of the pulse-waves.

**arteriola** (ar-te-re-o'-lah) [L.: *pl.*, *arteriolæ*]. An arteriole. **a. recta,** one of the arterioles going to the pyramids in the cortex of the kidney.

**arteriolæ** (ar-te-re o'-le) [L.]. Arterioles. **a. auricularis cordis,** coronary arteries of the heart. **a. rectæ,** vasa recta of the kidney.

**arteriole** (ar-te'-re-ōl) [*arteriola*]. A very small artery. **a., straight,** one of the small blood-vessels supplying the medullary pyramids of the kidneys.

**arteriolith** (ar-te'-re-o-lith) [*arteria*; λίθος, a stone]. A calculus in an artery from calcification of a thrombus.

**arteriology** (ar-te-re-ol'-o-je) [*arteria*; λόγος, science]. The science of the arteries; the anatomy, physiology, and pathology of the arteries.

**arteriomalacia** (ar-te-re-o-mal-a'-se-ah) [*arteria*; μαλακία, softness]. Softening of the wall of an artery.

**arteriomalacosis** (ar-te-re-o-mal-ak-o'-sis). See *arteriomalacia*.

**arteriometer** (ar-te-re-om'-et-er) [*arteria*; μέτρον, measure]. An instrument for measuring the changes in the caliber of a pulsating artery.

**arterionecrosis** (ar-te-re-o-ne-kro'-sis) [*arteria*; *necrosis*]. Necrosis of an artery or arteries.

**arteriopalmus** (ar-te-re-o-pal'-mus) [*arteria*; παλμός, palpitation]. Throbbing of the arteries.

**arteriopathy** (ar-te-re-op'-a-the) [*arteria*; πάθος, illness]. Any disease of an artery or of arteries.

**arterioperissia, arterioperittia** (ar-te-re-o-per-is'-e-ah, -it'-e-ah) [*arteria*; περισσός, excessive]. Abnormal or excessive arterial development. Syn., *perittarteria*; *perissoarteria*.

**arteriophlebotomy** (ar-te-re-o-fle-bot'-o-me) [*arteria*; φλέψ, a vein; τέμνειν, to cut]. Local bloodletting.

**arteriopituitous** (ar-te-re-o-pit-ū'-it-us) [*arteria*; *pituita*, mucus]. Applied to the blood-vessels of the nasal passages.

**arterioplania** (ar-te-re-o-pla'-ne-ah) [*arteria*; πλανᾶσθαι, to wander]. Deviation or tortuousness in the course of an artery.

**arterioplasty** (ar-te'-re-o-plas-te) [*arteria*; πλάσσειν, to form]. Matas' operation for aneurysm.

**arterioplegmus** (ar-te-re-o-pleg'-mus) [*arteria*; πλέγμα, anything twined or plaited]. Perplication. Syn., *arterioploce*.

**arteriorenal** (ar-te-re-o-re'-nal) [*arteria*; *ren*, the kidney]. Pertaining to the renal blood-vessels.

**arteriorrhagia** (ar-te-re-or-a'-je-ah) [*arteria*; ῥήγνυναι, to break forth]. Arterial hemorrhage.

**arteriorrhaphy** (ar-te-re-or'-af-e) [*arteria*; ῥαφή, suture]. Suture of an artery.

**arteriorrhexis** (ar-te-re-o-reks'-is) [*arteria*; ῥῆξις, a bursting]. Rupture of an artery.

**arteriosclerosis** (ar-te-re-o-skle-ro'-sis) [*arteria*; σκληρός, hard]. A chronic inflammation of the arterial walls, especially of the intima.

**arteriosclerotic** (ar-te-re-o-skle-rot'-ik) [see *arteriosclerosis*]. Pertaining to arteriosclerosis. **a. kidney,** a kidney the seat of chronic interstitial inflammation affecting primarily the blood-vessels.

**arteriosity** (ar-te-re-os'-it-e) [*arteria*]. The quality of being arterial.

**arteriostenosis** (ar-te-re-o-stem-o'-sis) [*arteria*; στενός, narrow]. The narrowing of the caliber of an artery in any part.

**arteriosteogenesis** (ar-te-re-os-te-o-jen'-e-sis) [*arteria*; ὀστέον, a bone; γένεσις, production]. Calcification of an artery.

**arteriosteosis, arteriostosis** (ar-te-re-os-te-o'-sis, ar-te-re-os-to'-sis). See *arteriosteogenesis*.

**arteriostrepsis** (ar-te-re-o-strep'-sis) [*arteria*; στρέψις, a twisting]. The twisting of an artery for the purpose of staying a hemorrhage.

**arteriotome** (ar-te'-re-o-tōm) [*arteria*; τέμνειν, to cut]. A knife for use in arteriotomy.

**arteriotomy** (ar-te-re-ot'-o-me) [*arteria*; τέμνειν, to cut]. 1. The cutting or opening of an artery for the purpose of bloodletting. 2. Dissection or anatomy of the arteries.

**arteriotrepsis** (ar-te-re-o-trep'-sis) [*arteria*; στρέψις, torsion]. See *arteriostrepsis*.

**arterious** (ar-te'-re-us) [*arteria*]. Relating to the arteries; arterial.

**arteriovenous** (ar-te-re-o-ve'-nus) [*arteria*; *vena*, vein]. Both arterial and venous; involving an artery and a vein, as an *arteriovenous* aneurysm.

**arterioversion** (ar-te-re-o-ver'-shun) [*arteria*; *vertere*, to turn]. Weber's method of arresting hemorrhage by turning vessels inside out by means of an instrument called the arterioverter.

**arterioverter** (ar-te-re-o-ver'-ter). An instrument for performing arterioversion.

**arteritis** (ar-te-ri'-tis) [*arteria*; ιτις, inflammation]. 1. Inflammation of an artery. 2. Inflammation of the external coat of an artery. **a. deformans.** See *endarteritis, chronic*. **a. obliterans.** See *endarteritis obliterans*. **a. syphilitica,** endarteritis deformans caused by syphilis. **a. umbilicalis,** septic inflammation of the umbilical arteries in the newborn.

**artery.** (*ar'-ter-e*) [see *arteria*]. One of the tube-like vessels through which the blood is propelled by the heart to all parts of the body. Arteries end in arterioles and capillaries. They are composed of 3 coats: the outer, or *tunica adventitia;* the middle, or *tunica media*, the muscular coat; the internal, or *intima*, composed of endothelial cells, fibrous and elastic tissue. **a., abdominal.** See *a., circumflex iliac, deep*. **a., abdominal,** external or subcutaneous. See *a., epigastric, superficial;* **a., pudic, external superior**. **a., abdominal, posterior**. See *a., epigastric, deep*. **a., acetabular,** a branch of the internal circumflex artery distributed to the hip-joint. **a., acromiothoracic** (thoracic axis), origin, second branch of first part of axillary; distribution, shoulder, arm, upper anterior part of chest, and mammary gland; branches, acromial, humeral, pectoral, clavicular. **a., alar thoracic,** origin, second part of axillary; distribution, lymphatic glands in axilla. **a., anastomotic,** those which connect other arteries more or less remote from each other. **a., anastomotic** (of external plantar), origin, external plantar; distribution, outer border of foot; it anastomoses with the tarsal and metatarsal branches of the dorsalis pedis. **a., anastomotic** (of internal plantar), origin, internal plantar; distribution, inner side of foot; it anastomoses with internal tarsal branch of the dorsalis pedis. **a., anastomotica magna** (of brachial), origin, brachial; distribution, elbow; branches, posterior and anterior. **a., anastomotica magna** (of superficial femoral), origin, superficial femoral (in Hunter's canal); distribution, knee; branches, superficial and deep. **a., angular,** origin, the termination of the facial; distribution, lacrimal sac and lower part of orbicularis palpebrarum; it anastomoses with infraorbital. **aorta, abdominal,** origin, thoracic aorta; termination, two common iliacs; branches, phrenic (right and left), celiac axis, suprarenal or capsular (right and left), superior mesenteric, lumbar (4 pairs), renal (right and left), spermatic (right and left), inferior mesenteric, right and left common iliac, middle sacral. **aorta, arch,** origin left ventricle of heart; distribution, thoracic aorta; branches, two coronary, innominate, left common carotid, left subclavian. **aorta,** primitive, that portion from the origin to the point at which the first branch is given off. **aorta, thoracic,** origin, arch of aorta; termination, abdominal aorta; branches, 2 or 3 pericardiac, 3 bronchial, 4 or 5 esophageal, 20 intercostal, subcostal (or twelfth dorsal), diaphragmatic, aberrans. **a., articular, middle** (of knee), origin, popliteal; distribution, crucial ligaments and joint. **a., articular, superior external** (of knee), origin, popliteal; distribution, crureus and knee. **a., articular, superior internal,** origin, popliteal; distribution, knee. **a., auditory, external,** a division of the first part of the internal maxillary artery; it enters the tympanum by the glaserian fissure and is distributed to the tympanum. **a., auricular, posterior,** origin, fifth branch of external carotid; distribution, back of auricle, scalp, and part of neck; branches, parotid, muscular, stylomastoid, anterior terminal or auricular, and posterior terminal or mastoid. **a., axillary,** origin, subclavian; distribution, brachial and seven branches; branches, superior thoracic, acromiothoracic, long thoracic, alar thoracic, subscapular, anterior and posterior circumflex. **a., azygos** (of the tongue), a small artery formed by the junction of branches of the dorsal arteries of the tongue; it extends along the median line of the dorsum of the tongue. **a., basilar,** origin, by confluence of right and left vertebral; distribution, brain; branches, transverse (or pontile), internal auditory, anterior cerebellar, superior cerebellar, two posterior cerebral. **a., brachial,** origin, axillary; distribution, arm and forearm; branches, superior and inferior profunda, anastomotica magna, nutrient, muscular, radial, and ulnar. **a., cardiac,** origin, gastric; distribution, cardiac end of stomach. **a., carotid, common,** origin, *right side,* innominate; *left side,* arch of aorta; distribution, external and internal carotid; branches, external and internal carotid. **a., carotid, external,** origin, common carotid; distribution, anterior part of neck, face, side of head, integuments, and dura mater; branches, ascending pharyngeal, superior thyroid, lingual, facial, occipital, posterior auricular, temporal, internal maxillary. **a., carotid, internal,** origin, common carotid; distribution, greater part of brain, the orbit, internal ear, forehead, and nose; branches, tympanic, vidian, arteria receptaculi, pituitary, gasserian, meningeal, ophthalmic, posterior communicating, anterior choroid, anterior cerebral, middle cerebral. **a., carotid, primitive.** See *a., carotid, common.* **a., celiac,** origin, abdominal aorta; distribution, stomach, duodenum, spleen, pancreas, liver, and gall-bladder; branches, gastric, hepatic, splenic. **a., central** (of retina), origin, ophthalmic; distribution, retina. **a.s, central system of,** Heubner's and Duret's term for the primary or secondary branches of the circle of Willis; they are distributed to the central ganglia of the brain. **a., cerebellar, anterior,** origin, basilar; distribution, anterior inferior surface of cerebellum. **a., cerebellar, inferior,** origin, vertebral; distribution, vermiform process and cortex of cerebellum; branches, inferior vermiform and the hemispheral. **a., cerebellar, superior,** origin, basilar; distribution, superior vermiform process and circumference of cerebellum; branches, superior vermiform and hemispheral. **a., cerebral, anterior,** origin, internal carotid; distribution, anterior portion of cerebrum; branches, anterior communicating, ganglionic (or central), commissural, hemispheral (or cortical). **a., cerebral, middle,** origin, internal carotid; distribution, middle portion of cerebrum; branches, ganglionic (or central), hemispheral (or cortical). **a., cerebral, posterior,** origin, basilar; distribution, temporosphenoid and occipital lobes; branches, ganglionic (or central) and hemispheral (or cortical). **a., cervical,** origin, uterine; distribution, cervix uteri. **a., cervical, ascending,** origin, inferior thyroid; distribution, deep muscles of neck and spinal canal; branches, muscular, spinal, and phrenic. **a., cervical, deep,** origin, superior intercostal; distribution, deep muscles of neck and spinal canal; branches, muscular, anastomotic, vertebral (or spinal). **a., cervical, superficial,** origin, transverse cervical; distribution, trapezius, levator anguli scapulæ, splenius muscles, and posterior chain of lymphatic glands. **a., cervical, transverse** (transversalis colli), origin, thyroid axis; distribution, posterior cervical and scapular regions; branches, posterior scapular and superficial cervical. **a., circumflex, anterior** (of axillary), origin, axillary; distribution, pectoralis major, biceps, and shoulder-joint; branches, bicipital and pectoral. **a., circumflex iliac, deep,** origin, external iliac; distribution, upper part of thigh and lower part of abdomen; branches, muscular and cutaneous. **a., circumflex, posterior** (of axillary), origin, axillary; distribution, deltoid, teres minor, triceps, and shoulder-joint; branches, nutrient, articular, acromial, muscular. **a., colic, left,** origin, inferior mesenteric; distribution, descending colon. **a., colic, middle,** origin, superior mesenteric; distribution, transverse colon. **a., colic, right,** origin, superior mesenteric; distribution, ascending colon. **a., colic, transverse,** origin, colic, middle; distribution, transverse colon. **a., comes nervi phrenici.** See *a., phrenic, superior.* **a., communicating.** 1. One establishing communication between two arteries. 2. An artery having as origin the dorsalis pedis; it enters into the formation of the plantar arch and has two digital branches. **a., communicating** (or perforating), origin, deep palmar arch; distribution, joins proximal ends of metacarpal and second and third dorsal interosseous arteries. **a., communicating, anterior,** origin; anterior cerebral; it assists in formation of anterior boundary of the circle of Willis; sends branches to caudate nucleus. Syn., *communicans willisii.* **a., communicating, posterior,** origin, posterior cerebral; it enters into formation of circle of Willis; sends branches to uncinate convolution and optic thalamus; branches, uncinate, middle thalamic. **a. compressor, a. constrictor,** an instrument for occluding an artery for the purpose of arresting or preventing hemorrhage. **a., coronary, inferior,** origin, facial; distribution, lower lip. **a., coronary, left,** origin, left anterior sinus of Valsalva; distribution, heart; branches, left auricular, anterior interventricular, left marginal, terminal. **a., coronary, right,** origin, right anterior sinus of Valsalva; distribution, heart; branches, right auricular, preventricular, right marginal, posterior interventricular, transverse. **a., coronary, superior,** origin, facial; distribution, upper lip. **a.s, cortical system of,** Heubner and Duret's term for the arteries distributed to the cerebral cortex and the parts immediately beneath it. **a., diaphragmatic,** origin, thoracic aorta; distribution, diaphragm. **a., digital,** origin, external plantar; distribution, outer side of the second and third, fourth, and fifth toes. **a., digital, palmar,** origin, superficial palmar arch; distribution, both sides of little, ring, and middle finger and ulnar side of index-finger. **a., dorsal** (of penis), origin, pudic; distribution, penis. **a., dorsalis hallucis,** a continuation of dorsalis pedis; distribution, great and second toes. **a., dorsalis pedis,** origin, continuation of anterior tibial; distribution, assists to form plantar arch; branches, tarsal, metatarsal, dorsalis hallucis, communicating. **a., end,** **a., terminal,** an artery that does not anastomose with other arteries by means of large branches; there is usually a capillary anastomosis. **a., epigastric, deep** (or inferior), origin, external iliac; distribution, abdominal wall; branches, cremasteric, pubic, muscular, cutaneous, terminal. **a., epigastric, superficial,** origin, common femoral; distribution, inguinal glands, skin, superficial fascia, and abdominal wall. **a., epigastric, superior,** origin, internal mammary; distribution, abdominal wall and diaphragm, liver, and peritoneum, branches, phrenic, xiphoid, cutaneous, muscular, hepatic, and peritoneal. **a., epiploic,** origin, right and left gastroepiploic; distribution, omentum. **a., esophageal.** 1. Origin, gastric; distribution, esophagus. 2. Origin, inferior thyroid; distribution, esophagus. 3. Origin, left phrenic; distribution, esophagus. 4. (4 or 5). Origin, thoracic aorta; distribution, esophagus. **a., esophageal, inferior,** origin, coronary (of stomach); distribution, esophagus. **a., facial,** origin, third branch of external carotid; distribution, pharynx and face; branches, ascending, or inferior palatine, tonsillar, glandular, muscular, submental, masseteric, buccal, inferior labial, inferior and superior coronary, lateralis nasi, angular. **a., femoral, common,** origin, continuation of external iliac; distribution, lower part of abdominal wall, upper part of thigh and genitalia; branches, superficial epigastric, superficial circumflex iliac, superficial external pudic, deep external pudic, profunda. **a., femoral, deep,** see *a., femoral, profunda.* **a., femoral, profunda,** origin, common femoral; distribution, muscles of thigh; branches, external circumflex, internal circumflex, and three perforating. **a., femoral, superficial,** origin, continuation of common femoral; distribution, muscles of thigh and knee-joint; branches, muscular, saphenous, anastomotica magna. **a. forceps,** a forceps for catching or twisting an artery, as a hemostat. **a., frontal,** a branch of the ophthalmic artery; it ascends the inner part of the orbital arch and supplies the periosteum, muscles, and integument of the middle forehead. **a., gastric** (or coronary), origin, celiac axis; distribution, stomach, liver, and esophagus; an hes, esophageal, cardiac, gastric, and hepatic. **a., gastroduodenal,** a branch of the hepatic artery given off near the pyloric orifice of the stomach; branches, right gastroepiploic and superior pancreaticoduodenal. **a., gluteal,** a branch of the internal iliac which runs backward between the lumbosacral cord and the first sacral nerve, turns around the upper margin of the great sacrosciatic foramen, and divides opposite the interval between the gluteus medius and pyriformis muscles, into the deep and superficial gluteal arteries. **a., gluteal, deep,** origin, gluteal; distribution, deep muscles of posterior gluteal

region. a., gluteal, inferior, origin, sciatic; distribution, gluteus maximus. a., gluteal, superficial, origin, gluteal; distribution, gluteus maximus and integument over sacrum. a., gluteal, superior, origin, deep gluteal; distribution, muscles adjacent. a.s, helicine, the arteries found in cavernous tissue, as in the testicle, uterus, ovary, etc. a., hemorrhoidal, inferior (or external), origin, pudic; distribution, sphincter muscle, levator ani. a., hemorrhoidal, middle, origin, internal iliac, anterior division; distribution, middle part of rectum. a., hemorrhoidal, superior, origin, inferior mesenteric; distribution, upper part of rectum. a., hepatic, origin, celiac axis; distribution, liver, pancreas, part of duodenum, and stomach; branches, pancreatic, subpyloric, gastroduodenal, right and left terminal. a., iliac, common, origin, terminal branch of abdominal aorta; distribution, peritoneum, subperitoneal fat, ureter, and terminates in external and internal iliac; branches, peritoneal, subperitoneal, ureteric, external and internal iliac. a., iliac, external, origin, common iliac; distribution, lower limb; branches, deep epigastric, deep circumflex iliac, muscular, and continues as femoral. a., iliac, internal, origin, common iliac; distribution, pelvic and generative organs and inner side of thigh; branches, anterior and posterior trunk. a., iliac, internal (anterior trunk), origin, internal iliac; distribution, pelvic and generative organs and thigh; branches, hypogastric, superior, middle, and inferior vesical, middle hemorrhoidal, uterine, vaginal, obturator, sciatic, internal pudic. a., iliac, internal (posterior trunk), origin, internal iliac; distribution, muscles of hip and sacrum; branches, iliolumbar, lateral sacral and gluteal. a., innominate, origin, arch of aorta; distribution, right side of head and right arm; branches, right common carotid, right subclavian, occasionally thyroidea ima. a., intercostal, anterior, origin, internal mammary; distribution, intercostal muscles, ribs (upper five or six), and pectoralis major. a., intercostal, anterior, origin, musculophrenic; distribution, lower five or six intercostal spaces. a., intercostal, superior, origin, subclavian; distribution, neck and upper part of thorax; branches, deep cervical, first intercostal, arteria aberrans. a., interosseous, anterior, origin, interosseous (common); distribution, muscles of forearm. a., interosseous, common, origin, ulnar; distribution, interosseous membrane and deep muscles of the forearm; branches, anterior and posterior interosseous. a., interosseous, posterior, origin, ulnar; distribution, muscles of forearm. a., labial, superior. See a., coronary, superior. a., laryngeal, superior, origin, superior thyroid; distribution, intrinsic muscles and mucous membrane of larynx. a., lenticulostriate, origin, middle cerebral; distribution, lenticular and caudate nuclei. a., lingual, origin, external carotid; distribution, tongue; branches, byoid, dorsalis linguæ, sublingual, ranine. a., mammary, external. See a., thoracic, long. a., mammary, internal, origin, subclavian; distribution, structures of thorax; branches, superior phrenic, mediastinal (or thymic), pericardiac, sternal, anterior intercostal, perforating, lateral intercostal, superior epigastric, internal mammary. a., maxillary, external. See a., facial. a., maxillary, internal (maxillary group), origin, external carotid; distribution, structures indicated by names of branches; branches, deep auricular, tympanic, middle meningeal, mandibular, small meningeal. a., maxillary, internal (pterygoid group), origin, external carotid; distribution, structures indicated by names of branches; branches, masseteric, posterior deep temporal, internal and external pterygoid, buccal, anterior deep temporal. a., maxillary, internal (sphenomaxillary group, origin, external carotid; distribution, structures indicated by names of branches; branches, posterior dental (or alveolar), infraorbital, posterior (or descending) palatine, Vidian, pterygopalatine, nasal, or sphenopalatine. a., median (arteria comes nervi mediani), origin, anterior interosseous; distribution, median nerve and superficial palmar arch. a., mediastinal, anterior (or thymic), origin, internal mammary; distribution, connective tissue, fat, and lymphatics in superior and anterior mediastinums, thymus gland. a.s, medullary.
1. Those supplying the medullary substance of the brain. 2. The nutrient arteries. a., meningeal.
1. Origin, ascending pharyngeal; distribution, membranes of brain. 2. Origin, posterior ethmoid; distribution, dura mater. a., meningeal, anterior, origin, internal carotid; distribution, dura mater. a., meningeal, middle or great, origin, internal maxillary; distribution, cranium and dura mater; branches, anterior and posterior. a., meningeal, posterior. 1. Origin, occipital; distribution, dura mater. 2. Origin, vertebral; distribution, dura mater. a., meningeal, small, origin, internal maxillary; distribution, Gasserian ganglion, walls of cavernous sinus, and dura mater. a., mesenteric, inferior, origin, abdominal aorta; distribution, lower half of large intestine; branches, left colic, sigmoid, superior hemorrhoidal. a., mesenteric, superior, origin, abdominal aorta; distribution, whole of small intestine and upper half of large; branches, inferior pancreaticoduodenal, colica media, colica dextra, ileocolic, vasa intestini tenuis. a., musculophrenic, origin, internal mammary; distribution, diaphragm, fifth and sixth lower intercostal spaces, oblique muscles of abdomen; branches, phrenic, anterior intercostals, muscular. a., nasal, origin, ophthalmic; distribution, lacrimal sac and integuments of nose; branches, lacrimal and transverse nasal. a., nasopalatine. See a., sphenopalatine. a., obturator, origin, anterior division, internal iliac; distribution, pelvis and thigh; branches, iliac (or nutrient), vesical, pubic, external and internal pelvic. a., obturator, external, origin, obturator; distribution, muscles about obturator foramen. a., occipital. 1. Origin, fourth branch of external carotid; distribution, muscles of neck and scalp; branches, sternomastoid, posterior meningeal, auricular, mastoid. princeps cervicis, communicating, muscular, terminal. 2. A branch of the posterior cerebral artery distributed to the occipital gyri and surrounding parts. a., omphalomesenteric, origin, primitive aorta; distribution, subsequently becomes the umbilical. a., ophthalmic, origin, internal carotid; distribution, the eye, adjacent structures, portion of face; branches, lacrimal, supraorbital, central artery of retina, muscular, ciliary, posterior and anterior ethmoid, palpebral, frontal, nasal. a., ovarian, origin, abdominal aorta; distribution, ovary, ureter, Fallopian tube, uterus; branches, ureteral, Fallopian, uterine, ligamentous. a., palatine, origin, ascending pharyngeal; distribution, soft palate and its muscles. a., palatine, ascending (or inferior), origin, first branch of facial; distribution, upper part of pharynx, palate, and tonsils; branches, palatine, tonsillar. a., palatine, descending, origin, internal maxillary; distribution, to soft and hard palate; branches, anterior and posterior. a., palmar arch, deep, origin, radial and communicating of ulna; distribution, palm and fingers; branches, princeps pollicis, radialis indicis, palmar interosseous (3), recurrent carpal, posterior perforating. a., palmar arch, superficial, origin, ulnar and superficialis volæ; distribution, palm and fingers; branches, digital (4), muscular, cutaneous. a., pancreatic. 1. Origin, hepatic; distribution, pancreas. 2. Origin, splenic; distribution, pancreas. a., pancreaticoduodenal, inferior, origin, superior mesenteric; distribution, pancreas and duodenum. a., pancreaticoduodenal, superior, origin, gastroduodenal; distribution, duodenum and pancreas. a., perforating (or posterior communicating) (3), origin, deep palmar arch; distribution, interosseous spaces. a.s, pericardiacophrenic, the pericardiac divisions of the internal mammary artery connecting with sternal ramifications of the same artery and with branches of the superior phrenic, bronchial, and intercostal arteries to form the subpleural mediastinal plexus. a., pharyngeal. 1. Origin, pharyngopalatine; distribution, roof of pharynx. 2. Origin, sphenopalatine; distribution, roof and contiguous portions of pharynx. a., pharyngeal, ascending, origin, first branch external carotid; distribution, pharynx, soft palate, tympanum, posterior part of neck, and membranes of brain; branches, prevertebral, pharyngeal, palatine, tympanic, meningeal. a., phrenic, origin, ascending cervical; distribution, phrenic nerve. a., phrenic, superior (comes nervi phrenici), origin, internal mammary; distribution, pleura, pericardium, and diaphragm. a., plantar arch, origin, external plantar artery; distribution, anterior part of foot and toes; branches, articular and plantar digital. a., plantar, deep, origin, metatarsal; distribution, assists in formation of plantar arch. a., plantar, external, origin, posterior tibial; distribution, sole and toes; branches, muscular, calcaneal, cutaneous, anastomotic, posterior per-

forating, plantar arch. **a., plantar, internal,** origin, posterior tibial; distribution, inner side of foot; branches, muscular, cutaneous, articular, anastomotic, superficial digital. **a., popliteal,** origin, continuation of femoral; distribution, knee and leg; branches, cutaneous, muscular (superior and inferior) or sural, articular, superior and inferior external, superior and inferior internal and azygos, terminal (anterior and posterior tibial. **a., profunda (deep femoral),** origin, femoral; distribution, thigh; branches, external and internal circumflex, three perforating. **a., profunda, inferior,** origin, brachial; distribution, triceps, elbow-joint. **a., profunda, superior,** origin, brachial; distribution, humerus, muscles and skin of arm; branches, ascending, cutaneous, articular, nutrient, muscular. **a., pterygopalatine (pterygopharyngeal),** origin, internal maxillary; distribution, pharynx, Eustachian tubes, and sphenoid cells; branches, pharyngeal, Eustachian, sphenoid. **a., pudic, external, deep (inferior),** origin, femoral, common; distribution, skin of scrotum (or labium in female). **a., pudic, external, superficial (superior),** origin, common femoral; distribution, integument above pubes and external genitalia. **a., pudic, internal,** origin, internal iliac, anterior division; distribution, generative organs; branches, external (or inferior) hemorrhoidal, superficial perineal, muscular, arteries of bulb, crus, and dorsal of penis. **a., pulmonary,** origin, right ventricle; distribution, lungs; branches, right and left. **a., pyloric, inferior,** origin, gastroduodenal or right gastroepiploic; distribution, pyloric end of stomach. **a., pyloric, superior,** origin, hepatic; distribution, pyloric end of stomach. **a., radial,** origin, brachial; distribution, forearm, wrist, hand; branches, radial recurrent, muscular, anterior and posterior carpal, superficial volar, metacarpal, dorsalis pollicis, dorsalis indicis. deep palmar arch. **a., ranine,** origin, lingual; distribution, tongue and mucous membrane of mouth. **a., renal,** origin, abdominal aorta; distribution, kidney; branches, inferior suprarenal, capsular, ureteral. **a.s, retinal,** the central artery of the retina and the upper and lower arteries on the nasal side and on the temporal side of the optic nerve. **a., sacral,** See .a., sacral, middle. **a., sacra media.** See .a., sacral, middle. **a., sacral, middle,** origin, continuation of aorta; distribution, sacrum and coccyx. **a., scapular, dorsal,** origin, subscapular; distribution, muscles of infraspinous fossa; branches, infrascapular. **a., scapular, posterior,** origin, continuation of transverse cervical; distribution, muscles of scapular region; branches, supraspinous and infraspinous, subscapular, muscular. **a., sciatic,** origin, internal iliac, anterior division; distribution, pelvic muscles and viscera, and branches; branches, coccygeal, inferior gluteal, muscular, anastomotic, articular cutaneous, comes nervi ischiadici, vesical, rectal, prostatic, etc. **a., spermatic,** origin, abdominal aorta; distribution, scrotum and testis; branches, ureteral, cremasteric, epididymal, testicular. **a., sphenopalatine (nasopalatine),** origin, internal maxillary; distribution, pharynx, nose, and sphenoid cells; branches, pharyngeal, sphenoid, nasal, ascending septal. **a., spinal.** 1. Origin, ascending cervical; distribution, spinal canal. 2. Origin, intercostals; distribution, spinal canal and spine. 3. Origin, lateral sacral; distribution, spinal membranes and muscles and skin over sacrum. **a., spinal, anterior,** origin, vertebral; distribution, spinal cord. **a., spinal, lateral,** origin, vertebral; distribution, vertebræ and spinal canal. **a., spinal, posterior,** origin, vertebral; distribution, spine. **a., splenic.** 1. Origin, celiac axis; distribution, spleen, pancreas, part of stomach, omentum; branches, small and large pancreatic, left gastroepiploic, vasa brevia, terminal. 2. Origin, left phrenic; distribution, spleen. **a., subclavian,** origin, right, innominate; left, arch of aorta; distribution, neck, thorax, arms, brain, meninges, etc.; branches, vertebral, thyroid axis, internal mammary, superior intercostal. **a., subscapular,** origin, axillary; distribution, subscapularis, teres major, latissimus dorsi, serratus magnus, axillary glands; branches, dorsal and infrascapular. **a., suprascapular (transversalis humeri),** origin, thyroid axis; distribution, muscles of shoulder; branches, inferior sternomastoid, subclavian, nutrient, suprasternal, acromial, articular, subscapular, supraspinous, and infraspinous. **a., Sylvian,** the middle cerebral artery. **a., temporal,** origin, external carotid; distribution, forehead, parotid gland, masseter muscle, ear; branches,

parotid, articular, masseteric, anterior auricular, transverse facial, middle, anterior and posterior temporal. **a., temporal, deep, anterior,** origin, internal maxillary; distribution, anterior part of temporal fossa. **a., termatic,** origin, anterior communicating; distribution, lamina cinerea and corpus callosum. **a., thoracic, acromial,** origin, axillary; distribution, muscles of shoulder, arm, and chest; branches, acromial, humeral, pectoral, clavicular. **a., thoracic, alar,** origin, axillary; distribution, axillary glands. **a., thoracic, external.** See .a., thoracic, long. **a., thoracic, internal.** See .a., mammary, internal. **a., thoracic, long (external mammary),** origin, axillary; distribution, pectoral muscles, serratus magnus, mammary and axillary glands. **a., thymic,** origin, internal mammary; distribution, connective tissue, fat, and lymphatics of mediastinum and thymus. **a. of the thyroid axis,** origin, subclavian; distribution, shoulder, neck, thorax, spine, cord; branches, inferior thyroid, suprascapular, and transverse cervical. **a., thyroid, inferior,** origin, thyroid axis; distribution, larynx, esophagus, and muscles of neck; branches, muscular, ascending cervical, esophageal, tracheal, and inferior laryngeal. **a., thyroid, superior,** origin, external carotid; distribution, omohyoid, sternohyoid, sternothyroid, thyroid gland; branches, hyoid, sternomastoid, superior laryngeal, cricothyroid. **a., thyroidea ima,** origin, innominate (usually); distribution, thyroid body. **a., tibial, anterior,** origin, popliteal; distribution, leg; branches, posterior and anterior tibial recurrent, muscular, internal and external malleolar. **a., tibial, posterior,** origin, popliteal; distribution, leg, heel, and foot; branches, peroneal, muscular, medullary, cutaneous, communicating, malleolar, calcanean, internal and external plantar. **a., tonsillar.** 1. Origin, ascending palatine; distribution, tonsil and Eustachian tube. 2. Origin, facial; distribution, tonsil and root of tongue. **a., transversalis colli.** See .a., cervical, transverse. **a., ulnar,** origin, brachial; distribution, forearm, wrist, and hand; branches, anterior and posterior ulnar, recurrent, common interosseous, muscular, nutrient, anterior and posterior ulnar, carpal, palmar arch. **a., uterine.** 1. Origin, internal iliac, anterior branch; distribution, uterus; branches, cervical, vaginal, azygos. 2. Origin, ovarian; distribution, uterus. **a., vasa brevia,** origin, splenic; distribution, stomach. **a., vertebral,** origin, subclavian; distribution, neck and cerebrum; branches, lateral spinal, muscular, anastomotic, posterior meningeal, posterior and anterior spinal, posterior cerebellar. **a., vesical, inferior,** origin, internal iliac, anterior division; distribution, bladder, prostate, seminal vesicles, and vagina (in female). **a., vesical, middle,** origin, superior vesical; distribution. bladder. **a., vesical, superior,** origin, internal iliac, anterior division; distribution, bladder; branches, deferential, ureteric, middle vesical (occasionally). **a., Vidian,** origin, internal maxillary; distribution, roof of pharynx, Eustachian tube, and tympanum; branches, pharyngeal, Eustachian, tympanic. **a., vitelline.** See a., omphalomesenteric.

**arthragra** (ar-thra'-grah) [arthron; ἄγρα, seizure]. Gout.

**arthragrosis** (ar-thrag-ro'-sis) [arthron; ἄγρα, seizure (pl., arthragroses)]. Gout. In the plural, gouty disorders affecting the skin.

**arthral** (ar'-thral) [arthron]. Articular; relating to an arthron.

**arthralgia** (ar-thral'-je-ah) [arthron; ἄλγος, pain]. Neuralgic pain in a joint. Syn., arthroneuralgia; articular neuralgia. **a. saturnina.** pain in the joints and rigidity and cramps in the approximate muscles; it is symptomatic of lead-poisoning.

**arthralgic** (ar-thral'-jik) [see arthralgia]. Relating to arthralgia.

**arthrectasia, arthrectasis** (ar-threk-ta'-ze-ah, ar-threk'-ta-sis) [arthron; ἔκτασις, dilation]. Dilation of a joint-cavity.

**arthrectomy** (ar-threk'-to-me) [arthron; ἐκτομή, a cutting-out]. Excision of a joint.

**arthredema, arthrœdema** (ar-thred-e'-mah) [arthron; οἴδημα, a swelling, tumor]. Edema affecting a joint.

**arthrelcosis** (ar-threl-ko'-sis) [arthron; ἕλκωσις, ulceration]. Ulceration of a joint.

**arthremia** (ar-thre'-me-ah) [arthron; αἷμα, blood]. A congested condition of a joint.

**arthrempyema** (ar-threm-pi'-e-mah) [arthron; ἐμπύημα, suppuration]. Suppuration or abscess of a joint.

**arthrempyesis** (ar-threm-pi-e'-sis). Suppuration in a joint.
**arthrentasis** (ar-thren'-ta-sis) [arthron; ἔντασις, distortion]. Distortion of the limbs due to gout.
**arthric** (ar'-thrik). See *arthritic* (2).
**arthrifluent** (ar-thrif'-lu-ent) [arthron; fluere, to flow]. Applied to abscesses proceeding from a diseased joint.
**arthrifuge** (ar'-thrif-ūj) [arthron; fugare to put to flight]. A remedy for gout.
**arthritic** (ar-thrit'-ik) [arthritis]. Relating—1. To arthritis or to gout. 2. To a joint.
**arthritis** (ar-thri'-tis) [arthron; ιτις, inflammation]. Inflammation of a joint. **a.**, acute, acute joint-inflammation, particularly that due to gout. Syn., *arthritis vera*. **a.**, acute serous, acute synovitis. **a. arthrodynia**, gout. **a. asthmatica**, a form observed in elderly persons subject to asthma, and mitigated by an attack of the latter. **a.**, atrophic. Synonym of *Charcot's joint disease*. **a.**, blennorrhagic, gonorrheal rheumatism. **a.**, chronic, a form in which the joints are not so much affected as are other parts of the body. **a. deformans**, chronic inflammation of a joint with deformity; rheumatoid arthritis. **a.**, diaphragmatic, angina pectoris. **a.**, erratic, retrocedent or metastatic gout. **a. fungosa**, tuberculous disease of the joints; white swelling. **a.**, gonorrheal, gonorrheal synovitis. **a.**, gouty, that due to gout. **a. hiemalis**, winter gout, a form occurring less frequently in summer than in other seasons. **a.**, internal. See *a.*, *visceral*. **a.**, intervertebral. See *spondylarthritis*. **a. ischias**, gout in the hip. **a. larvata**, **a.**, latent, a masked form not manifested by the usual symptoms. **a. maxillaris**, rheumatoid arthritis of the temporomaxillary joint. **a. nodosa**. See *osteoarthritis*. **a. pauperum**. Synonym of *a.*, *rheumatoid*. **a. podagra**, gout in the feet. **a.**, proliferating. See *a. deformans*. **a.**, retrograde, suppressed gout. **a.**, rheumatoid, a chronic joint affection characterized by inflammatory overgrowth of the articular cartilages and synovial membranes, with destruction of those parts of the cartilages subject to intraarticular pressure; there is progressive deformity. Syn., chronic rheumatoid arthritis; osteoarthritis; rheumatic gout; nodular rheumatism; arthritis deformans. **a. sicca**, rheumatoid arthritis. **a.**, strumous. See *a. fungosa*. **a.**, subdiarthroidal, a form of fungous arthritis in which fleshy granulations occur between the bone and the cartilage of the joint. **a. syphilitica**, gonorrheal rheumatism; also the nocturnal pains of syphilis. **a. typica**, acute arthritis. **a.**, urethral, gonorrheal rheumatism. **a. urica**, gout attributed to excessive formation of uric acid. Syn., *arthritis uratica; panarthritis urica; uarthritis*. **a. vertebralis**, a breakdown of the intervertebral discs. **a.**, visceral, gout affecting an internal organ, with alternating attacks in the joints.
**arthritism**. (ar'-thrit-izm) [arthron]. Gout or the gouty diathesis.
**arthritolith** (ar-thrit'-o-lith) [arthron; λίθος, a stone]. Gouty calcareous deposit or concretion in or around a joint.
**arthro-** (ar-thro-) [arthron]. A prefix denoting relating to the joints.
**arthrobacterium** (ar-thro-bak-te'-re-um) [arthro-; bacterium]. A bacterium forming arthrospores.
**arthrocace** (ar-throk'-as-e) [artho-; κακός, ill]. Fungous, strumous, or tuberculous arthritis; caries of a joint. **a. coxarum**, see *coxalgia*. **a.**, senile, changes in the joints occurring in the aged.
**arthrocarcinoma** (ar-thro-kar-sin-o'-mah) [arthro-; carcinoma]. Carcinoma affecting a joint.
**arthrocele** (ar'-thro-sel) [arthro-; κήλη, a tumor]. Swelling of a joint.
**arthrocenchriasis** (ar-thro-sen-kri'-as-is) [arthro-; κεγχρίας, like a grain of millet]. A miliary eruption occurring about a joint.
**arthrochondritis** (ar-thro-kon-dri'-tis) [arthro-; χόνδρος, a cartilage; ιτις, inflammation]. Inflammation of the cartilaginous parts of a joint.
**arthroclasia** (ar-thro-kla'-se-ah) [arthro-; κλάειν, to break]. The breaking-down of ankyloses in order to produce free movement of a joint.
**arthrocleisis** (ar-thro-kli'-sis) [arthro-; κλείειν, to shut]. See *arthrodesis*.
**arthrodesis** (ar-throd'-es-is) [arthro-; δέσις, a binding]. Surgical fixation of paralyzed joints.
**arthrodia** (ar-thro'-de-ah) [ἀρθρωδία, a kind of articulation]. A form of joint admitting of a gliding movement.
**arthrodial** (ar-thro'-de-al) [arthrodia]. Pertaining to or of the nature of arthrodia.
**arthrodynia** (ar-thro-din'-e-ah) [arthro-; ὀδύνη, pain]. Pain in a joint; arthralgia.
**arthrodynic** (ar-thro-din'-ik) [arthro-; ὀδύνη, pain]. Relating to or affected by arthrodynia.
**arthroempyesis** (ar-thro-em-pi-e'-sis) [arthro-; ἐμπύησις, suppuration]. Suppuration in a joint.
**arthrogenous** (ar-throj'-en-us) [arthro-; γεννᾶν, to produce]. Forming an articulation. **a. spore**, an arthrospore.
**arthrography** (ar-throg'-ra-fe) [arthro-; γράφειν, to write]. A description of the joints.
**arthrogryposis** (ar-thro-grip-o'-sis) [arthro-; γρύπωσις, flexure]. 1. Permanent flexure of a joint; ankylosis. 2. Persistent idiopathic contracture of a joint. 3. Tetany or tetanilla.
**arthrokleisis**. See *arthrocleisis*.
**arthrolith** (ar'-thro-lith) [arthro-; λίθος, a stone]. One of the free bodies which occur in joints, arising from the segmentation of warty outgrowths of joint cartilage or of synovial membrane. Syn., *arthrophyte; arthremphyte; joint-bodies; joint-mice; mures articulares; corpora mobilia articulorum; corpora libera articulorum; tophus arthriticus; arthrotophus*.
**arthrolithiasis** (ar-thro-lith-i'-as-is) [see *arthrolith*]. Gout.
**arthrology** (ar-throl'-o-je) [arthro-; λόγος, science]. The science of joints.
**arthrolysis** (ar-throl'-is-is) [arthro-; λύσις, a solution]. The division or removal of adhesions and bone from an ankylosed joint.
**arthromeningitis** (ar-thro-men-in-ji'-tis) [arthro-; μήνιγξ, membrane; ιτις, inflammation]. Synovitis.
**arthron** (ar'-thron) [ἄρθρον, a joint]. A joint or an articulation.
**arthronalgia** (ar-thron-al'-je-ah). See *arthralgia*.
**arthroncus** (ar-throng'-kus) [arthro-; ὄγκος, a swelling]. 1. A cartilaginous body such as occasionally forms within the knee-joint. 2. Swelling of a joint.
**arthroneuralgia** (ar-thro-nū-ral'-je-ah) [arthro-; νεῦρον, nerve; ἄλγος, pain]. Neuralgic pain in a joint.
**arthropathology** (ar-thro-path-ol'-o-je) [arthro-; πάθος, disease; λόγος, science]. The branch of pathology dealing with joint-diseases.
**arthropathy** (ar-throp'-a-thē) [arthro-; πάθος, disease]. 1. Any joint disease. 2. A peculiar trophic disease of the joints, sometimes occurring in locomotor ataxia and syringomyelia; rarely in general paralysis of the insane and in disseminated sclerosis. Syn., *Charcot's joint*. **a.**, Charcot's, see *arthropathy* (2). **a. osteopulmonary**, Marie's disease, an enlargement of the ends of long bones in long standing pulmonary disease. **a.**, vertebral, arthropathy with depressions and rugosities of the vertebræ.
**arthrophlogosis** (ar-thro-flo-go'-sis) [arthro-; φλέγειν, to burn]. Inflammation of a joint.
**arthrophysis** (ar-throf'-lis-is) [arthro-; φλύσις, an eruption]. Gout accompanied with a cutaneous eruption. **a. cardiaca**. See *miliaria arthritica*. **a. vulgaris**. See *eczema arthriticum*.
**arthrophyma** (ar-thro-fi'-mah) [arthro-; φῦμα, a swelling]. Swelling of a joint.
**arthrophyte** (ar'-thro-fīt) [arthro-; φύτον, a growth]. A growth in a joint.
**arthroplasty** (ar'-thro-plas-te) [arthro-; πλάσσειν, to form]. The making of an artificial joint.
**Arthropod** (ar'-thro-pod) [arthro-; πούς, foot]. A member of the phylum arthropoda which embraces crustaceans, insects and spiders.
**arthropodous** (ar-throp'-o-dus) [arthro-; πούς, a foot]. In biology, having jointed legs.
**arthropyosis** (ar-thro-pi-o'-sis) [arthro-; πύωσις, suppuration]. Pus-formation in a joint.
**arthrorheumatism** (ar-thro-ru'-mat-izm) [arthro-; rheumatism]. Articular rheumatism.
**arthrorrhagia** (ar-thro-ra'-je-ah) [arthro-; ῥηγνύναι, to burst forth]. Hemorrhage into a joint.
**arthrosia** (ar-thro'-ze-ah) [arthron]. Painful inflammatory or other affection of a joint.
**arthrosis** (ar-thro'-sis) [ἄρθρωσις, to fasten by a joint]. Articulation or jointing; a suture.
**arthrospore** (ar'-thro-spōr) [arthro-; σπόρος, a seed]. A spore formed by fission, as opposed to an *endospore*.
**arthrosteitis** (ar-thro-ste-i'-tis) [arthro-; ὀστέον, bone; ιτις, inflammation]. Inflammation of the bony parts of a joint.
**arthrostenosis** (ar-thro-sten-o'-sis) [arthro-; στένωσις, a narrowing]. Contraction of a joint.

**arthrosteophyma** (ar-thro-ste-o-fi'-mah) [arthro-; ὀστέον, bone; φῦμα, tumor]. A tumor of the bone in a joint.

**arthrosyrinx** (ar-thro-sir'-ingks) [arthro-; σῦριγξ, a pipe]. A fistulous opening into a joint.

**arthrotome** (ar'-thro-tōm) [arthro-; τομή, a cutting]. A stout knife used in the surgery of the joints.

**arthrotomy** (ar-throt'-o-me) [arthro-; τέμνειν, to cut]. Incision of a joint.

**arthrotrauma** (ar-thro-traw'-mah) [arthro-; τραῦμα, an injury]. An injury to a joint.

**arthrotropia** (ar-thro-tro'-pe-ah) [arthro-; τροπή, a turning]. Torsion of a limb.

**arthrotyphoid** (ar-thro-ti'-foid). Typhoid fever with articular involvement.

**arthrous** (ar'-thrus) [arthron]. Pertaining to a joint or joints; jointed.

**arthroxerosis** (ar-thro-zer-o'-sis) [arthro-; ξέρωσις, a dry state]. Chronic osteoarthritis.

**arthroxesis** (ar-throks'-es-is) [arthro-; ξέσις, a scraping]. The surgical treatment of an articular surface by scraping.

**Arthus phenomenon** (ar'-toos) [Maurice Arthus, French bacteriologist]. A rabbit treated with horse's serum at intervals of six days shows a soft infiltrate after the fourth injection, a hard infiltrate after the fifth injection, and gangrene after the sixth or seventh injection; this last is followed by death.

**artiad** (ar'-te-ad) [ἄρτιος, even]. In chemistry, a term designating an element or radical having an even quantivalence.

**article** (art'-ikl) [articulus, a little joint]. A joint; a segment of a jointed series.

**articular** (ar-tik'-ū-lar) [articularis; of the joints]. Pertaining to an articulation or joint.

**articularis** (ar-tik-ū-la'-ris). Articular. **a. genu.** See subcrureus, in table of muscles.

**articulate** (ar tik'-ū-lāt) [articulare, to divide in joints]. 1. Divided into joints. 2. Distinct, clear. **a. speech,** the communication of ideas by spoken words.

**articulatio** (ar-tik-ū-la'-she-o) [L., a joint]. A joint; see articulation. **a. acromioclavicularis,** acromioclavicular joint. **a. atlantoepistrophica,** joint between atlas and epistropheus or axis. **a. atlantooccipitalis,** joint between atlas and occipital bone. **a. carpometacarpea pollicis,** carpometacarpal joint of the thumb. **a. calcaneocuboidea,** calcaneocuboid joint. **a. cochlearis,** spiral joint. **a. composita,** compound joint. **a. coxæ,** hip-joint. **a. cricoarytænoidea,** arycornicula te synchondrosis. **a. cricothyreoidea,** cricothyreoid articulation. **a. cubiti,** elbow-joint. **a. cuneonavicularis,** cuneonavicular joint. **a. ellipsoidea,** elliptical joint. **a. genu,** knee joint. **a. humeri,** shoulder-joint. **a. humeroradialis,** humeroradial articulation. **a. humeroulnaris,** humero-ulnar articulation. **a. incudomalleolaris,** joint between anvil and hammer. **a. incudostapedia,** joint between anvil and stirrup. **a. intercarpea,** intercarpal articulation, carpal joints. **a. mandibularis,** jaw-joint. **a. manus,** joint of the hand. **a. ossis pisoformis,** joint of the pisiform bone. **a. radioulnaris distalis,** inferior radio-ulnar joint. **a. radioulnaris proximalis,** superior radio-ulnar joint. **a. sacroiliaca,** sacro-iliac joint. **a. sellaris,** saddle joint. **a. simplex,** simple joint. **a. sphæroidea,** spherical joint. **a. sternoclavicularis,** sternoclavicular joint. **a. talocalcanea,** talocalcanean joint. **a. talocalcaneonavicularis,** talocalcaneonavicular joint. **a. talocruralis,** ankle-joint. **a. talonavicularis,** talonavicular joint. **a. tarsi transversa** (Choparti), Chopart's transverse articulation of the tarsus. **a. tibiofibularis,** superior tibiofibular articulation. **a. trochoidea,** trochoid or pivot joint.

**' articulation** (ar-tik-ū-la'-shun )[articulus, a joint]. 1. A joint; a connection between two or more bones, whether or not allowing movement between them. The articulations are divided into: (1) Synarthroses, immovable, subdivided into schindyleses, or grooved joints; gomphoses, in sockets, as the teeth; and suturæ, as in the bones of the skull; (2) diarthroses, or movable joints, subdivided into the arthrodia, or gliding joints; the ginglymus, or hinge-like; the enarthroses, or ball-and-socket joints; (3) amphiarthroses, or those of a mixed type. 2. The enunciation of spoken speech. 3. The articulating contact of the cusps in the positions of mastication. **a., false,** one formed between the end of a dislocated bone and the contiguous parts or between the parts of a broken bone. Syn., pseudarthrosis. **a., supplementary,** a false articulation in which the ends of the fragments become rounded and covered with a fibrous capsule.

**articulationes** (ar-tik-ū-la-she-o'-nēz) [L. pl., of articulatio]. Joints. **a. capitulorum,** capitular joints or articulations between the heads of the ribs and the vertebræ. **a. carpometacarpeæ,** carpometacarpal joints. **a. costotransversariæ,** costotransverse joints. **a. costovertebrales,** joints between ribs and vertebræ. **a. digitorum manus,** joints of the fingers. **a. digitorum pedis,** joints of the toes. **a. interchondrales,** interchondral joints. **a. intermetacarpeæ,** intermetacarpal joints. **a. intermetatarseæ,** intermeta.arsal joints. **a. intertarseæ,** intertarsal joints. **a. metacarpophalangeæ,** metacarpophalangeal joints. **a. metatarsophalangeæ,** metatarsophalangeal joints. **a. ossiculorum auditus,** joints of the auditory ossicles. **a. manus,** joints of the hand. **a. pedis,** joints of the foot. **a. sternocostales,** sternocostal joints.

**articulator** (ar-tik'-ū-la-tor) [axiculus, a joint]. An instrument used in mechanical dentistry for holding the models in position while the artificial teeth are being arranged and antagonized upon the plates.

**articulatory** (ar-tik'-ū-la-tor-e). Relating to articulation.

**articulo mortis, in** (ar-tik'-ū-lo mor'-tis) [L.]. At the moment of death. In the act of dying.

**articulus** (ar-tik'-ū-ius) [dim. of artus, a joint; pl. and gen. articuli]. 1. A joint; a knuckle. 2. A segment; a part; a limb. 3. A moment of time.

**artifact** (ar'-te-fakt). See artefact.

**artificial** (ar-te-fish'-al) [artificialis]. Made or imitated by art. **a. anus,** an opening in the abdomen or loin to give exit to the feces. **a. eye,** a film of glass, celluloid, rubber, etc., made in imitation of the front part of the globe of the eye, and worn in the socket or over a blind eye for cosmetic reasons. **a. feeding,** the feeding of an infant by other means than mother's milk. **a. leech.** See leech, artificial. **a. palate.** See palate, artificial. **a. pupil,** the result of removal of a piece of the iris (iridectomy, iridodialysis, etc.) to allow the light to pass through the opening. **a. respiration,** the aeration of the blood by artificial means—a method of inducing the normal function of respiration, as in asphyxia neonatorum, drowning, etc. The chief methods are:—Bain's, Byrd's, Calliano's, Dew's, Forest's, Hall's, Howard's, Laborde's, Pacini's, Rosenthal's, Satterthwaite's, Schafer's, Schroeder's, Schultze's, and Sylvester's, q. v.

**artistomia** (ar-te-sto'-me-ah) [ἄρτι, exactly; στόμα, a mouth]. 1. Distinctness in utterance. 2. The condition of an aperture, especially in surgical incisions, in which the size is perfectly adapted to the purpose.

**artiyls** (ar'-te-ils) [ἄρτιος, complete]. Loewig's name for hydrocarbons of the general formula $C_nH_{2n}$.

**Artocarpus** (ar-to-kar'-pus) [ἄρτος, bread; καρπός, a fruit]. A genus of trees of the order Urticaceæ, including the breadfruit-tree, A. incisa. A. blumei is an East Indian species which an edible fruit, the oil of which is used in diarrhea; an ointment from the buds and leaves is applied to buboes. A. integrifolia, native in India, is prized for its wood; the root is used in diarrhea and as an external application in leprosy; the root-bark is used as a vermifuge.

**artus** (ar'-tus) [L.: pl., artus]. A joint; a limb; the joints collectively.

**aryepiglottic** (ar-e-ep-e-glot'-ik). Same as arytenoepiglottic.

**aryl** (ar'-il). A chemical prefix denoting an organic radical belonging to the aromatic series.

**arylarsonates** (ar-il-ar'-so-nāts). Aromatic organic salts of arsenic, such as atoxyl, soamin and six hundred and six.

**arytenectomy** (ar-e-ten-ek'-to-me). See arytenoidectomy.

**arytenoepiglottic** (ar-it-en-o-ep-e-glot'-ik) [ἄρυταινα, a pitcher; εἶδος likeness; ἐπί, upon; γλωττίς, glottis]. Relating to an arytenoid cartilage and to the epiglottis; as the arytenoepiglottic fold (or folds), consisting of a fold of mucous membrane that extends from each arytenoid cartilage to the epiglottis.

**arytenoid** (ar-it'-en-oid) [ἄρυταινα, a pitcher; εἶδος, likeness]. 1. Resembling the mouth of a pitcher. 2. Pertaining to the arytenoid cartilages. **a. cartilages,** two cartilages of the larynx regulating, by means of the attached muscles, the tension of the vocal bands. **a. glands,** muciparous glands, found in large numbers along the posterior margin of the arytenoepiglottic fold in front of the arytenoid cartilages. **a. muscle,** a muscle arising from the posterior surface of one arytenoid cartilage and inserted into the corresponding parts of the other. It is composed of three planes of fibers, two oblique and one transverse. It draws the arytenoid cartilages together.

**arytenoidectomy** (*ar-e-ten-oid-ek'-to-me*) [*arytenoid;* ἐκτομή, a cutting-out]. Removal of an arytenoid cartilage.

**arytenoiditis** (*ar-e-ten-oid-i'-tis*). Inflammation of the arytenoid cartilage or muscles.

**arythmia** (*ar-ith'-me-ah*). See *arrhythmia*.
**arythmic** (*ar-ith'-mik*). See *arrhythmic*.
**A. S.** Abbreviation for Latin *auris sinistra*, the left ear.
**As.** 1. Chemical symbol for arsenic. 2. Abbreviation for astigmatism.

**asa** (*a'-sah*) [Pers., *āsā*, mastic]. A gum. **a. dulcis,** benzoin; also the drug called *laser*.

**asab** [Ar.]. An African venereal disease said to differ from syphilis.

**asafetida, asafœtida** (*as-a-fet'-id-ah*) [*asa,* gum; *fœtida,* stinking]. A gum-resin obtained from the root of *Ferula fœtida*. It is slightly soluble in alcohol and forms an emulsion with water. Its properties are due to a light volatile oil. It is antispasmodic, stimulating, expectorant, and is used in hysteria and in bronchial affections. Dose 5-20 gr. (0.32-1.3 Gm.). **a., emulsion of** (*emulsum asafœtidæ,* U. S. P.), a 4 % emulsion of asafetida. Dose ½-2 oz. (15-60 Cc.). Syn., *milk of asafetida.* **a., pills of** (*pilula asafœtidæ,* U. S. P.), composed of asafetida, soap, and water. Dose 1-3. **a., tincture of** (*tinctura asafœtidæ,* U. S. P.), strength, 20 %. Dose 10-30 min. (0.6-2.0 Cc.). Dewees' carminative (*mistura magnesiæ et asafœtidæ*) is an unofficial preparation composed of magnesium carbonate, 5; tincture of asafetida, 7; tincture of opium, 1; sugar, 10; distilled water, sufficient to make 100 parts. Dose ½ dr.-½ oz. (1-15 Cc.).

**asaphia** (*as-a'-fe-ah*) [ά, priv.; σαφής, clear]. Indistinctness of utterance, especially that due to cleft palate.

**asaprol** (*as'-ap-rōl*), CaC₂₀H₁₄S₂O₈+3H₂O. Calcium betanaphthol-α-monosulphonate, a substance readily soluble in water and alcohol, and recommended in asthma, tonsillitis, and acute articular rheumatism, in doses of from 15-60 gr. (1-4 Gm.).

**asarcia** (*ah-sar'-se-ah*) [ά, priv.; σάρξ, flesh]. Emaciation; leanness.

**asarin** (*as'-ar-in*). Same as *asarone*.

**asarol** (*as'-ar-ol*) [*asarum;* oleum, oil], C₁₀H₁₈O. A camphor-like body derived from asarum.

**asarone** (*as'-ar-ōn*) [ἄσαρον, asarabacca], C₂₀H₃₆O₅. Asarin. The solid component of the oil from *Asarum europæum.* It forms monoclinic prisms, has an aromatic taste, and smells like camphor.

**Asarum** (*as'-ar-um*) [ἄσαρον, asarabacca]. A genus of aristolochiaceous plants. *A. canadense,* called wild ginger, Canada snakeroot, with other North American species, is used chiefly in domestic practice. It is a fragrant, aromatic stimulant. Dose of *fluidextract* 15 min.-½ dr. (1-2 Cc.). *A. europæum* has diaphoretic, emetic, purgative, and diuretic qualities, but is now little used except in veterinary practice.

**asbestiform** (*as-best'-e-form*) [*asbestos*]. Fibrous in structure.

**asbestos** (*as-bes'-tos*) [ἄσβεστος, unquenchable]. A soft fibrous mineral made up of flexible or elastic filaments, and the best nonconductor of heat known. Mixed with plaster it is used in mechanical dentistry as a substitute for sand to form the investment preparatory to soldering. It has also a limited use in surgery.

**asbolic, asbolicous, asbolicus** (*as-bol'-ik,* ' -us) [ἄσβολος, soot]. Sooty; due to soot; *e. g., carcinoma scroti asbolicum.*

**asbolin** (*as'-bol-in*) [see *asbolic*]. A bitter, acrid, yellow oil extracted from soot; it is used in tuberculosis.

**ascariasis** (*as-kar'-i-as-is*) [*ascaris*]. The symptoms produced by the presence of ascarides in the gastro-intestinal canal.

**ascaricide** (*as-kar'-is-īd*) [*ascaris; cædere,* to kill]. A medicine that kills ascarides.

**Ascaridæ** (*as-kar'-e-de*) [*ascaris*]. A family of nematode worms, to which belongs the round-worm (*Ascaris lumbricoides*) and the threadworm (*Oxyuris vermicularis*).

**ascarides** (*as-kar'-id-ēz*). Plural of *ascaris.*
**ascaridiasis** (*as-kar-id-i'-as-is*). The presence of ascarides in the intestine.

**Ascaris** (*as'-kar-is*) [ἀσκαρίς, a species of intestinal worm; pl. *ascarides*]. A genus of parasitic worms inhabiting the intestine of most animals. **A. alata,** a variety that has rarely been found in man. **A. lumbricoides,** a variety found in the ox, hog, and man. It inhabits the small intestine, especially of children. **A. mystax,** the roundworm of the cat, rarely found in man. **A. trichiuris,** the whip-worm. **A. vermicularis.** Synonym of *Oxyuris vermicularis*.

**ascending** (*as-end'-ing*) [*ascendere,* to rise]. Taking an upward course; rising (as parts of the aorta and colon, and as one of the venæ cavæ). **a. aorta,** the first part of the aorta. **a. colon,** the first part of the colon. **a. current,** in electricity, one going from the periphery to a nerve-center. **a. degeneration,** a degeneration of the nerve-fibers extending from the periphery to the center, or, in the spinal cord, from below upward toward the brain. **a. metamorphosis,** same as *anabolism*. **a. paralysis.** See *paralysis, ascending.* **a. tracts,** the centripetal tracts of the spinal cord, carrying afferent impulses.

**Asch's operation** [*ash*) [Morris J. Asch, American physician]. For deviation of nasal septum; it consists in a crucial incision over the deflection, taking up the segments, reduction of the deflection, and insertion of a tube to hold the segments in place.

**Ascherson's vesicles** (*ash'-er-sun*) [Ferdinand Moritz Ascherson, German physician, 1798-1879]. The peculiar small globules formed when oil and an albuminous fluid are agitated together; formerly thought to be cells.

**ascheturesis** (*as-ket-ū-re'-sis*) [ἄσχετος, resistless; οὔρησις, urination]. An uncontrollable desire to urinate; irrepressible urination.

**aschistodactylism** (*as-kis-to-dak'-til-izm*) [ἀσχίστος, uncloven; δάκτυλος, finger]. A synonym of *syndactylism.*

**aschistodactylous** (*as-kis-to-dak'-til-us*). Affected with syndactylism.

**Aschoff bodies** (*ah'-shoff*) [Ludwig Aschoff, German pathologist, 1886- ]. Nodular bodies found in the myocardium in patients who have suffered from rheumatism.

**ascia** (*ah'-se-ah* or *as'-she-ah*) [ά, priv.; σκιά, shadow]. A spiral bandage applied without reverses, each turn of which overlaps the preceding for about one-third of its width. *Dolabra repens* is the same as the preceding, but the spirals are formed more obliquely and do not overlap each other, but are separated by a greater or less interval. Syn., *dolabra currens; fascia spiralis.*

**ascites** (*as-i'-tēz*) [ἀσκίτης, a kind of dropsy; from ἀσκός, a bag]. An abnormal collection of serous fluid in the peritoneal cavity; dropsy of the peritoneum. It is either local in origin or part of a general dropsy. The ascitic fluid is usually clear, yellow, and coagulates on standing. It may be turbid, blood-stained, and contain lymph-particles or shreds. There are uniform enlargement of the abdomen, fluctuation, percussion-dulness. Its usual cause is cirrhosis of the liver. Syn., *abdominal dropsy; hydroperitoneum; hydrops peritonæi.* See *Duparque's method for detecting ascites.* **a., active, a., acute,** that in which there is a sudden large effusion due to exposure or cold. **a., adiposus,** ascites characterized by a fluid, milky appearance, due to the presence of it of numerous cells that have undergone fatty degeneration and solution. It is seen in certain cases of carcinoma, tuberculosis, and other chronic inflammations of the peritoneum. Syn., *ascites oleosus.* **a. chylosus,** the presence of chyle in the peritoneal cavity. It follows rupture of a chyle-duct. **a. intercus,** an effusion occurring between the skin and the peritoneum. **a. intermuscularis,** edema of the abdominal muscles. **a., mechanical, a., passive,** that due to diseases which retard the blood-current in the portal vein. **a. saccatus.** 1. A form in which the effusion is prevented by adhesions or inflammatory exudate from entering the general peritoneal cavity. Syn., *encysted dropsy of the peritoneum.* 2. An ovarian cystoma. **a., sanguineous,** a bloody form affecting sheep and lambs. Syn., *diarrhemia.* **a. vaginalis,** a collection of liquid within the sheath of the rectus abdominis muscle. **a. vulgatior,** a form apparently due to diseased kidneys, and preceded by scanty, highly colored urine.

**ascitic** (as-it'-ik) [see *ascites*]. Pertaining to or affected with ascites.
**asclepiadin** (as-kle-pi'-ad-in) [*asclepias*]. A bitter glucoside obtainable from various species of *Asclepias*. It is poisonous, and has emetic, purgative, and sudorific properties.
**Asclepias** (as-kle'-pe-as) [ἀσκληπιάς]. 1. Pleurisy-root. The root of *Asclepias tuberosa*. A popular remedy in the Southern States for pleurisy. It is diaphoretic, emetic, and cathartic. The infusion recommended has a strength of 1 oz. of the powdered root to 32 oz. of water. Dose a teacupful every 3 or 4 hours. 2. A genus of plants of the order *Asclepiadaceæ*. *A. curassavica*, blood-flower, is an herb common to tropical America; astringent, styptic, and anthelmintic against the tape-worm. Dose of fluidextract 20 min.–1 dr. (1.3–4.0 Cc.). *A. longifolia*, of the western United States, is diaphoretic.
**asclepin** (as'-kle-pin) [*asclepias*]. 1. A poisonous principle obtainable from asclepiadin by the separation of glucose from the latter. 2. The precipitate from a tincture of *Asclepias tuberosa*; alterative, evacuant, tonic, sedative. Dose 2–4 gr. (0.13–0.26 Gm.).
**asclepion** (as-kle'-pe-on). A resinous substance, $C_{20}H_{34}O_3$, obtained from *Asclepias syriaca*.
**Ascococcus** (as-ko-kok'-us) [*ascus*; κόκκος, a kernel]. A genus of the family of *Schizomycetes*. The Ascococci are microorganisms made up of round or ovoid cells, united in massive colonies, and surrounded by tough, thick, gelatinous envelopes. a. **Billrothii**, a form found in putrid meat; its natural habitat is the air; it is probably not pathogenic.
**Ascoidium** (as-ko-id'-e-um) [*ascus*; εἶδος, likeness]. A genus of *Infusoria* found in the urine and feces of typhoid fever patients, in sewage, in the excrement of cattle, and in the cecum of swine.
**Ascomycetes** (as-ko-mi-se'-tēz). A group of fungi, including *aspergillus* and *oidium*.
**ascospore** (as'-ko-spōr) [*ascus*; σπόρος, spore]. A spore produced by or in an ascus.
**ascus** (as'-kus) [ἀσκός, a bag or bladder]. The characteristic spore-case of some fungi and lichens, usually consisting of a single terminal cell containing eight spores.
**-ase** (ās). A termination denoting an enzyme; thus *lipase*, a fat-splitting enzyme.
**asecretory** (ah-se'-kret-o-re) [ά, priv.; *secretus*, separate]. Dry; without secretion.
**Aselli's glands or pancrease** [Gaspar *Aselli*, Italian anatomist, 1581–1626]. A group of lymphatic glands situated at the root of the mesenęery.
**aselline** (as-el'-ēn). A poisonous leukomaine found in cod-liver oil.
**asemasia** (ah-sem-a'-ze-ah) [ά, priv.; σημασία, a signaling]. Absence of the power to communicate either by signs or by language.
**asemia** (ah-se'-me-ah) [ά, priv.; σῆμα, sign]. Inability to form, express, or understand any sign, token, or symbol of thought or feeling, whether speech, writing or gesture. a. **mimica**. See *amimia*. a. **spuria**. See *parusemia*.
**asepsin** (ah-sep'-sin). See *antisepsin*.
**asepsis** (ah-sep'-sis) [ά, priv.; σῆψις, putrefaction]. Absence of pathogenic microorganisms.
**aseptic** (ah-sep'-tik) [ά, priv.; σηπτός, septic]. Free from pathogenic bacteria, as *aseptic* wounds. a. **surgery**, the mode of surgical practice in which everything that is used, as well as the wound, is in a germ-free condition.
**asepticism** (ah-sep'-tis-izm) [see *aseptic*]. The doctrine or principles of aseptic surgery.
**asepticize** (ah-sep'-tis-īz) [see *aseptic*]. To render aseptic.
**aseptin** (ah-sep'-tin) [see *aseptic*]. A proprietary preparation containing boric acid, used for preserving articles of food.
**aseptol** (ah-sep'-tol) [see *aseptic*], $C_6H_6SO_4$. A reddish liquid, with an odor of phenol, recommended as a disinfectant and antiseptic. It is used externally (1 to 10 % solution) and internally in about the same dose as phenol. Syn., *sozolic acid*; *phenolsulphonic acid*.
**aseptolin** (ah-sep'-tol-in). A preparation of pilocarpine (0.018 %) in an aqueous solution of phenol (2.74 %); it is used in tuberculosis and in malaria. Dose 50–70 min. (3–4 Cc.) daily, injected subcutaneously.
**asexual** (ah-seks'-ū-al) [ά, priv.; *sexus*, sex]. Without sex; nonsexual.

**asexualization** (ah-seks-ū-al-iz-a'-shun). Removal of the testicles in the male, or of the ovaries in the female.
**As. H.** Abbreviation for *hyperopic astigmatism*.
**ash** [ME., *asch*]. 1. The incombustible mineral residue that remains when a substance is incinerated. 2. See *manna*. a. **manna**. See *manna*. a., **prickly**. See *xanthoxylum*.
**asialia** (as-e-a'-le-ah) [ά, priv.; σίαλον, spittle]. Deficiency or failure of the secretion of saliva.
**Asiatic** (a-zhe-at'-ik) [*Asia*]. Pertaining or belonging to Asia. A. **cholera**. See *cholera, Asiatic*. A. **pill**, a pill composed of arsenic trioxide, black pepper, powdered licorice, and mucilage.
**Asimina** (as-im-e'-nah) [L.]. A genus of trees. *A. triloba* is the papaw tree of North America.
**asitia** (ah-sit'-e-ah) [ά, priv.; σῖτος, food]. The want of food; also a loathing for food.
**askelia** (ah-skel'-e-ah) [ά, priv.; σκέλος, leg]. Nondevelopment of the legs.
**As. M.** Abbreviation for *myopic astigmatism*.
**asoma** (ah-so'-mah) [ά, priv.; σῶμα, body]. A species of omphalositic monster characterized by an absence of the trunk. The head is never well formed, and the vessels run from it to the placenta in the membranes. Beneath the head is a sac in which rudiments of body-organs may be found. This is the rarest form of omphalosites.
**asomus** (ah-so'-mus) [ά, priv.; σῶμα, body]. A monster with only a rudimentary body.
**asonia** (ah-so'-ne-ah) [ά, priv.; *sonus*, a sound]. Tone-deafness.
**aspalasoma** (as-pal-as-o'-mah) [ἀσπάλαξ, mole; σῶμα, body]. A variety of single autositic monsters of the species *Celosoma*, in which there is a lateral or median eventration occupying principally the lower portion of the abdomen, with the urinary apparatus, the genital apparatus, and the rectum opening externally by three distinct orifices.
**asparagine** (as-par'-aj-ēn) [*asparagus*], $C_4H_8N_2O_3$. An alkaloid found in the seeds of many plants, in asparagus, beet-root, peas, and beans. It forms shining, four-sided, rhombic prisms, readily soluble in hot water, but not in alcohol or ether. It is an amid of aspartic acid, and forms compounds with both acids and bases. It is diuretic. Asparagine hydrargyrate has been used as an antisyphilitic, in doses of ⅙ gr. (0.01 Gm.) hypodermatically.
**asparaginic acid**. See *acid, asparaginic*.
**Asparagus** (as-par'-ag-us) [ἀσπάραγος, asparagus]. 1. The green root of *Asparagus officinalis*, a mild diuretic. Dose of *fluidextract* ½–1 dr. (2–4 Cc.). Unof. 2. A genus of plants belonging to the order *Liliaceæ*. *A. acutifolius*, a species of southern Europe, is said to be more efficient medicinally than *A. officinalis*. *A. racemosus* and *A. sarmentosus*, of the old world tropics, are employed in the same manner as salep; an infusion of the root of *A. sarmentosus* is used to prevent the confluence of smallpox pustules.
**asparamide** (as-par'-am-id). See *asparagine*.
**asparolin** (as-par'-ol-in). A brown liquid said to consist of guaiac, asparagus, parsley, black haw, and henbane. It is used as an antispasmodic uterine tonic. Dose, 2–4 drams in hot water.
**aspartic acid** (as-par'-tik). See *acid, aspartic*.
**aspastic** (ah-spas'-tik). Not spastic.
**aspergillin** (as-per-jil'-in) [*aspergillus*]. A pigment obtained from the spores of *Aspergillus niger*. Syn., *vegetable hematin*.
**aspergillosis** (as-per-jil-o'-sis). Pseudotuberculosis; morbid lesions due to some species of *Aspergillus*.
**Aspergillus** (as-per-jil'-us) [*aspergere*, to sprinkle]. A genus of fungi. A. **auricularis**, a fungus found in the wax of the ear. A. **fumigatus**, found in the ear, nose, and lungs. A. **glaucus**, the bluish mold found upon dried fruit. A.-**keratitis**, corneal inflammation due to invasion by a fungus belonging to the genus *Aspergillus*. Syn., *Keratomycosis aspergillina*. A. **mucoroides**, a species found in tuberculous or gangrenous lung tissue. A.-**mycosis**. See *otomycosis*.
**aspermatic** (ah-sper-mat'-ik) [ά, priv.; σπέρμα, seed]. Affected with or relating to aspermatism.
**aspermatism** (ah-sper'-mat-izm) [ά, priv.; σπέρμα, seed]. 1. Non-emission of semen, whether owing to non-secretion or non-ejaculation. 2. Defective secretion of semen or lack of formation of spermatozoa.
**aspermia** (ah-sper'-me-ah). Same as *aspermatism*.

**aspermous** (*ah-sper'-mus*) [see *aspermatic*]. Without seed.

**asperous** (*as'-per-us*) [*asper*, rough]. Uneven; having a surface with distinct minute elevations.

**aspersion** (*as-per'-shun*) [*aspergere*, to sprinkle]. Treatment of disease by sprinkling the body or the affected part with a medicinal agent.

**aspersus** (*as-per'-sus*) [see *aspersion*]. Covered with scattered dots or punctures.

**asphalgesia** (*as-fal-je'-ze-ah*) [ἀσφί, their own; ἄλγησις, pain]. Pitres' term for a condition observed in hypnotism, in which intense pain follows the touching of certain articles, and prolonged contact produces convulsions.

**asphyctic, asphyctous** (*as-fik'-tik, -tus*) [*asphyxia*]. 1. Affected with asphyxia. 2. Pulseless.

**asphyxia** (*as-fiks'-e-ah*) [ἀ, priv.; σφίξις, the pulse]. Suffocation; the suspension of vital phenomena resulting when the lungs are deprived of oxygen. The excess of carbon dioxide in the blood at first stimulates, then paralyzes, the respiratory center of the medulla. Artificial respiration is therefore required in cases of asphyxia. a. **cataphora**, that with brief incomplete remissions. a., **lethargic**, deep sleep accompanying mental and physical torpor. a., **local**, that stage of Raynaud's disease in which the affected parts are dusky red from intense congestion. a. **neonatorum**, the asphyxia of the newborn from any cause. a. **sideratorum**, loss of consciousness from lightning-stroke. a., **solar**, a. **solaris**, sunstroke. a., **syncopal**, a form of asphyxia in which the heart-cavities are found vacant. a. **valsalviana**, syncope due to disturbance of cardiac functions.

**asphyxial** (*as-fik'-se-al*). Relating to or characterized by asphyxia.

**asphyxiant** (*as-fiks'-e-ant*) [see *asphyctic*]. 1. Producing asphyxia. 2. An agent capable of producing asphyxia.

**asphyxiate** (*as-fiks'-e-āt*) [see *asphyctic*]. To produce or cause asphyxia.

**aspic** (*as'-pik*) [*a* and *spic*, lavender spike]. The great lavender, or spike lavender, *Lavandula spica*. Its oil is at present used in veterinary practice and occasionally in liniments.

**aspidin** (*as'-pid-in*) [*Aspidium*, a genus of ferns]. C₂₃H₂₇O₇. An active principle obtained from malefern.

**aspidiopsoriasis** (*as-pid-e-o-so-ri'-as-is*) [ἀσπίδιον, a little shield; *psoriasis*]. A form of psoriasis marked by the formation of scutiform scales.

**Aspidium** (*as-pid'-e-um*) [L.; gen., *aspidii*]. 1. A genus of ferns known as shield-ferns. 2. The rhizome of *Aspidium filix-mas* and of *A. marginale*, or malefern. Its properties are due to a resin containing filicic acid. It is valuable chiefly against tape-worm. Dose ⅓ dr.-½ oz. (2-15 Cc.). a., **liquid extract of** (*extractum filicis liquidum*; B. P.). Dose 15 min.-1 dr. (1-4 Cc.). a., **oleoresin of** (*oleoresina aspidii*, U. S. P.), an ethereal extract. Dose ½-1 dr. (2-4 Cc.).

**aspidol** (*as'-pid-ol*) [*Aspidium*, a genus of ferns]. C₂₀H₂₄O. A substance isolated from malefern.

**aspidosamin** (*as-pid-os'-am-in*), C₂₂H₂₈N₂O₂. A basic principle from quebracho bark. It is emetic.

**Aspidosperma** (*as-pid-o-sper'-mah*) [ἀσπίς, a shield; σπέρμα, a seed]. A genus of apocynaceous trees, of which the quebracho is the most important.

**aspidospermatin** (*as-pid-o-sper'-mat-in*) [ἀσπίς, a shield; σπέρμα, seed]. A basic substance, from quebracho bark, said to be isomeric with aspidosamine and to depress the temperature when administered.

**aspidospermine** (*as-pid-o-sper'-mēn*) [see *aspidosperma*]. C₂₂H₃₀N₂O₂. An alkaloid extracted from quebracho (*Aspidosperma quebracho*). It is a respiratory stimulant and antispasmodic. Dose 1-2 gr. (0.065-0.13 Gm.).

**aspiration** (*as-pir-a'-shun*) [*ad*, to; *spirare*, to breathe]. 1. The act of sucking up or sucking in; inspiration; imbibition. 2. The act of using the aspirator. 3. A method of withdrawing the fluids and gases from a cavity. a. **pneumonia**. See *pneumonia, aspiration*.

**aspirator** (*as'-pir-a-tor*) [see *aspiration*]. An apparatus for withdrawing liquids from cavities by means of suction.

**aspirin** (*as'-pir-in*). The acetic-acid ester of salicylic acid; small needles without color or taste, used as an antipyretic and analgesic, as is sodium salicylate. Dose 15 gr. (1 Gm.). Syn., *acetyl salicylic acid*.

**aspirolithin** (*as-pir-o-lith'-in*). A proprietary combination of aspirin and lithium.

**aspirophen** (*as-pi'-ro-fen*). A mixture containing salicylic acid and monacetyl phenocoll; it is said to be antirheumatic and antipyratic. Dose 10-15 gr. (0.6-1.0 gm.).

**Asplenium** (*as-ple'-ne-um*). A genus of ferns called spleen-worts, or miltwastes.

**asporogenic** (*ah-spor-o-jen'-ik*) [ἀ, priv.; σπόρος, seed; γενής, producing]. Not reproducing by means of spores; not producing spores.

**asporogenous** (*as-por-oj'-en-us*). Same as *asporogenic*.

**asporous** (*ah-spo'-rus*) [ἀ, priv.; σπόρος, seed]. Without spores.

**assafetida** (*as-a-fet'-id-ah*). See *asafetida*.

**assanation** (*as-an-a'-shun*) [*ad*, to; *sanare*, to make sound]. The improvement of sanitary conditions.

**assault** (*as-awlt'*) [*assalire*, to assail]. An attack. a., **criminal**, in medical jurisprudence, the touching or attempting to touch, on the part of a male, any of the sexual organs (the breasts included) of a female against her will, even though they be covered by clothing.

**assay** (*as-a'*) [Fr., *assayer*]. 1. The testing or analysis of a metal or drug to determine the relative proportion of its constituents. 2. The substance thus tested. 3. The process of assaying.

**Assègat, triangle of.** See under *Assézat*.

**asselline** (*as-el'-ēn*). A poisonous leukomaine obtained from cod-liver-oil.

**Assézat, triangle of.** (*ah-sa-zah'*) [Jules *Assézat*, French anthropologist, 1832- ]. A triangle formed by lines uniting the projection of the nasion on the alveolo-condylar plane and the alveolar and nasal points and one uniting the two latter.

**assident** (*as'-id-ent*) [*assidere*, to sit by]. Usually, but not always, accompanying a disease; as, *assident symptoms*. Opposed to *pathognomonic*.

**assideration** (*as-id-er-a'-shun*) [*ad*, intensive; *sideratio*, an evil influence]. In forensic medicine, infanticide by immersing in ice-cold water.

**assimilable** (*as-im'-il-a-bl*) [*assimulare*, to make like]. Capable of being assimilated; nutritious.

**assimilation** (*as-im-il-a'-shun*) [see *assimilable*]. The process of transforming food into so nutrient a condition that it is taken up by the circulatory system, to form an integral part of the économy; synthetic or constructive metabolism; anabolism. a.-**limit**, the amount of starchy or saccharine food which a person can ingest without the appearance of glycosuria. a., **mental**, the mental reception of impressions and their assignment by the consciousness to their proper place. a., **primary**, that concerned in the conversion of food into chyle and blood. a., **secondary**, that relating to the formation of the organized tissues of the body.

**associated** (*as-o'-se-a-ted*) [*associatus*, united]. Joined. a. **movements**, coincident or consensual movements of muscles other than the leading one, and which, by habit or unity of purpose, are involuntarily connected with its action; both eyeballs move alike in reading, though one be a blind eye. Movement of the normal arm will sometimes produce slight motion of the opposite paralyzed arm. Uniformity of innervation is usually the cause of these movements. a. **paralysis**, a. **spasm**, a common paralysis or spasm of associated muscles. **association center** (*as-so-se-a'-shun*). The center controlling associated movements. a. c., **anterior**, that part of the frontal cortex anterior to the motor area. a. c., **middle**, the island of Reil. a. c., **posterior**, that part of the cortex situated between the sensory area at the equator and the area for vision in the occipital lobe.

**association test**. A word is mentioned to the patient, and the physician observes what other words the patient will give as the ones suggested to him by the first word. The time consumed in this process is also noted.

**assonance** (*as'-o-nans*) [*assonare*, to respond to]. A morbid tendency to employ alliteration.

**assuetude** (*as'-su-e-tūd*). Habituation to disturbing influences; the condition of the organism in which it has acquired such tolerance for a drug or poison that the effect it once had is lost.

**assurin** (*as'-u-rin*), C₄₆H₉₁N₂P₂O₉. A name given by Thudichum to a complex substance occurring in brain tissue.

**astasia** (*ah-sta'-se-ah*) [ἀ, priv.; στάσις, standing].

Motor incoordination for standing. **a.-abasia,** a symptom consisting in inability to stand or walk in a normal manner. The person affected seems to collapse when attempting to walk.

**astatic** (*ah-stat'-ik*). Having no directive tendency. **a. needle,** an apparatus consisting of two needles of equal magnetic moments and exactly reversed in direction.

**asteatosis** (*as-te-at-o'-sis*) [ά, priv.; στέαρ, tallow; ώδης, fulness]. 1. A deficiency or absence of the sebaceous secretion. 2. Any skin disease (as xeroderma) characterized by scantiness or lack of the sebaceous secretion.— **a. cutis,** a condition of diminished sebaceous secretion, as the result of which the skin becomes dry, scaly, and often fissured.

**aster** (*as'-ter*) [ἀστήρ, a star]. 1. The stellate structure surrounding the centrosome. 2. The stellar group of chromosomes during karyokinesis.

**astereognosis** (*ah-ste-re-og-no'-sis*) [ά, priv.; στερεός, solid; γνῶσις, knowledge]. Inability to recognize objects by the sense of touch, due to lesion in the central parietal lobule. Syn., *stereoagnosis.* Cf., *aphasia, tactile.*

**asterion** (*as-te'-re-on*) [*aster*]. A point on the skull corresponding to the junction of the occipital, parietal, and temporal bones.

**asternal** (*ah-ster'-nal*) [ά, priv.; στέρνον, the breast-bone]. 1. Without a sternum. 2. Not connected with the sternum. **a. ribs,** the five lower pairs, because not joined directly to the sternum.

**asternia** (*ah-ster'-ne-ah*) [see *asternal*]. Absence of the sternum.

**asteroid** (*as'-ter-oid*) [*aster*; εἶδος, likeness]. 1. Stellate. 2. See *astrocyte.*

**asterol** (*as'-ter-ol*). Trade name of a preparation of paraphenolsulphonate of mercury and ammonium tartrate; it is used as a surgical antiseptic and bactericide.

**asthenia** (*ah-sthe'-ne-ah*) [ά, priv.; σθένος, strength]. Absence of strength; adynamia. Syn., *lipopsychia.*

**asthenic** (*ah-sthen'-ik*) [see *asthenia*]. Characterized by asthenia.

**asthenogenia, asthenogenesis** (*ah-sthen-o-je'-ne-ah, ah-sthen-o-jen'-es-is*) [*asthenia*; γεννᾶν, to produce]. The production of asthenia.

**asthenometer** (*ah-sthen-om'-et-er*) [*asthenia*; μέτρον, a measure]. An instrument for detecting and measuring asthenia; especially, a device for measuring muscular asthenopia.

**asthenope** (*as'-then-ōp*). A person suffering from asthenopia.

**asthenopia** (*ah-sthen-o'-pe-ah*) [*asthenia*; ὤψ, eye]. Weakness of the ocular muscles or of visual power, due to errors of refraction, heterophoria, overuse, anemia, etc. **a.,** accommodative, that due to hyperopia, astigmatism, or a combination of the two, producing strain of the ciliary muscle. **a., muscular,** that due to weakness, incoordination (heterophoria), or strain of the external ocular muscles. **a., nervous, a., retinal,** a rare variety, caused by retinal hyperesthesia, anesthesia, or other abnormity, or by general nervous affections. **a. tarsal,** that due to pressure of the eyelids on the cornea.

**asthenopic** (*ah-sthen-op'-ik*) [see *asthenopia*]. Characterized by asthenopia.

**asthenoxia** (*as-then-ok'-se-ah*) [*asthenia*; *oxygen*]. Insufficient oxidation of the waste products of metabolism.

**asthma** (*az'-mah*) [ἄσθμα, panting]. A paroxysmal affection of the bronchial tubes characterized by dyspnea, cough, and a feeling of constriction and suffocation. The disease is probably a neurosis, and is due to hyperemia and swelling of the bronchial mucous membrane, with a peculiar secretion of a mucin-like substance. The attacks may be caused by direct irritation of the bronchial mucous membrane or by indirect or reflex irritation, as from the nose, the stomach, the uterus. When dependent upon disease of the heart, the kidneys, stomach, thymus, etc., it has been designated *cardiac, renal, peptic, thymic,* etc. **a., arthritic.** 1. That due to gout. 2. Angina pectoris. **a., bronchial.** Same as *asthma.* **a., cardiac,** paroxysmal dyspnea due to heart disease. **a. convulsivum,** Synonym of *asthma.* **a. crystals,** acicular crystals (Charcot-Leyden crystals) contained in the sputum of asthmatic patients. They are generally associated with eosinophil cells. **a. cultrariorum.** See *fibroid phthisis.* **a. dyspepticum,** asthma due to nervous reflexes through the vagus. **a., fuller's, a. fullorum,** a pulmonary affection due to inhaling particles of wool and dust in the manufacture of cloth. **a., grinders'.** See *fibroid phthisis.* **a., hay-.** See *hay-fever.* **a., intrinsic,** that due to direct irritation of the lungs. **a., marine.** See *beriberi.* **a., miller's.** See *laryngismus stridulus.* **a., miner's.** See *anthracosis.* **a. nervosum.** Synonym of *asthma.* **a., organic,** asthma of cardiac origin. **a.-paper, niter-paper. a., paralytic bronchial,** a rare form attributed to a relaxed condition of the bronchioles. **a., pneumobulbar,** Sée's term for a form attributed to pulmonary irritation transmitted to the bronchioles by reflexes through the vagus. **a. purulentum,** that due to an abscess in the respiratory passages. **a., renal,** a paroxysmal dyspnea sometimes occurring in the course of Bright's disease. **a., spasmodic.** See *asthma.* **a., thymic.** Synonym of *laryngismus stridulus.*

**asthmatic** (*az-mat'-ik*) [see *asthma*]. Relating to or affected with asthma.

**asthmaticoscorbutic** (*az-mat-ik-o-skor-bū'-tik*). Relating to asthma and scurvy.

**asthmatophthisis** (*as-mat-o-tis'-is*). Pulmonary tuberculosis attended with asthma. Syn., *asthmatic phthisis.*

**asthmatorthopnea, asthmatophopnea** (*az-mat-or-thop'-ne-ah, az-mor-thop'-ne-ah*) [*asthma*; *orthopnea*]. Orthopnea due to asthma or respiratory obstruction located in the chest.

**asthma-weed.** Lobelia inflata.

**asthmogenic** (*az-mo-jen'-ik*) [*asthma*; γεννᾶν, to produce]. Causing asthma.

**asthmolysin** (*as-mol'-is-in*). A mixture of the extracts of the suprarenal glands and of the hypophysis with some preservative; said to be serviceable in asthma. It is administered by hypodermic injection.

**astigmagraph** (*as-tig'-ma-graf*) [ά, priv.; στίγμα, a point; γράφειν, to write]. An instrument for illustrating the phenomena of astigmatism.

**astigmatic** (*ah-stig-mat'-ik*) [*astigmatism*]. Pertaining to or affected with astigmatism.

**astigmatism** (*ah-stig'-mat-izm*) [ά, priv.; στίγμα, a point, because rays of light from a point are not brought to a point by the refractive media of the eye]. That condition of the eye in which rays of light from a point do not converge to a point on the retina. It is usually due to inequality of curvature of the different meridians of the cornea (*corneal astigmatism*), but may be caused by imperfections of the lens (*lenticular astigmatism*), unequal contraction of the ciliary muscle, or may perhaps be due to retinal imperfection. It may be *acquired* or *congenital,* and may complicate hyperopia or myopia, producing either *simple hyperopic astigmatism,* in which one principal meridian is emmetropic, the other hyperopic, or *compound hyperopic astigmatism,* in which both meridians are hyperopic, but one more so than the other. Complicating myopia we may in the same way have *simple myopic* or *compound myopic astigmatism.* In *mixed astigmatism* one principal meridian is myopic, the other hyperopic. *Regular astigmatism* is when the two principal meridians are at right angles to each other; *irregular astigmatism* when different parts of a meridian have different refracting powers.

**astigmatometer** (*ah-stig-mat-om'-et-er*) [*astigmatism*; μέτρον, a measure]. An instrument for measuring the degree of astigmatism.

**astigmia** (*ah-stig'-me-ah*). See *astigmatism.*

**astigmic** (*ah-stig'-mik*). See *astigmatic.*

**astigmometer** (*ah-stig-mom'-et-er*). See *astigmatometer.*

**astigmometry** (*ah-stig-mom'-et-re*). The measurement of astigmatism.

**astigmoscope** (*ah-stig'-mo-skōp*). An instrument for measuring astigmatism.

**astigmoscopy** (*ah-stig-mos'-kop-e*). The measurement of astigmatism by the astigmoscope; the use of the astigmoscope.

**astomatous** (*ah-sto'-mat-us*) [ά, priv.; στόμα, mouth]. In biology, without a mouth or aperture.

**astomia** (*ah-sto'-me-ah*) [ά, priv.; στόμα, a mouth]. The condition of having no mouth.

**astomous** (*ah-sto'-mus*). See *astomatous.*

**astragalar** (*as-trag'-al-ar*). Relating to the astragalus.

**astragalectomy** (*as-trag-al-ek'-to-me*) [*astragalus*; ἐκτομή, excision]. Excision of the astragalus.

**astragalocalcanean** (*as-trag-al-o-kal-ka'-ne-an*). Relating to the astragalus and calcaneum.

**astragaloscaphoid** (*as-trag-al-o-skaf'-oid*). Relating to the astragalus and the scaphoid bone.
**astragalotibial** (*as-trag-al-o-tib'-e-al*). Relating to the astragalus and the tibia.
**astragalus** (*as-trag'-al-us*) [ἀστράγαλος, a die; the analogous bones of the sheep were used by the ancients as dice]. The ankle-bone, upon which the tibia rests. **Astragalus.** A genus of leguminous plants from some varieties of which gum tragacanth is derived. *A. mollissimus* is the loco-plant. The active principle of this plant has mydriatic properties.
**astral** (*as'-tral*). Pertaining to an aster.
**astraphobia, astrapaphobia** (*as-trah-fo'-be-ah, as-trap-af-o'-be-ah*) [ἀστραπή, lightning; φόβος, fear]. Morbid fear of lightning.
**astriction** (*as-trik'-shun*) [*asctrictio; ad,* to; *stringere,* to bind]. Constipation or any condition resulting from the use of astringents.
**astringency** (*as-trin'-jen-se*) [*ad,* to; *stringere,* to bind]. The quality of being astringent.
**astringent** (*as-trin'-jent*) [*ad,* to; *stringere,* to bind]. 1. Causing contraction; binding. 2. An agent producing contraction of organic tissues, or that arrests hemorrhages, diarrhea, etc.
**astro-** (*as-tro-*) [ἄστρον, a star]. A prefix meaning star or star-shaped.
**astroblast** (*as'-tro-blast*) [*astro-;* βλαστός, a germ]. A variety of glia-cell less differentiated than the endymal cell and astrocytes.
**astrocyte** (*as'-tro-sīt*) [*astro-;* κύτος, cell]. 1. One of the cells derived from the endyma of the embryonic cerebrospinal canal that, in the course of development, wander toward the periphery, undergo modification, and form one of the two chief divisions of glia-cells, the other divisions being the original endymal cells. Syn., *Deiters' cells.* 2. A stellate bone-corpuscle.
**astroid** (*as'-troid*) [ἄστρον, a star; εἶδος, resemblance]. 1. Star-shaped. 2. An astrocyte.
**astrokinetic** (*as-tro-kin-et'-ik*) [*astro-;* κινεῖν, to move]. Applied to the phenomena of motion as exhibited by the centrosomes of cells.
**astrophobia** (*as-tro-fo'-be-ah*) [*astron,* a star; φόβος, fear]. Fear of the stars and celestial space.
**astrophorous** (*as-trof'-or-us*) [*astro-;* φέρειν, to bear]. Having stellate processes.
**astrosphere** (*as'-tro-sfēr*) [*astro-;* σφαῖρα, sphere]. 1. The radially arranged protoplasmic filaments surrounding the centrosome in a dividing cell. 2. The central mass of the aster, exclusive of the filaments or rays, in which the centrosome lies. 3. The entire aster exclusive of the centrosome. See *centrosphere* and *sphere of attraction.*
**astrostatic** (*as-tro-stat'-ik*) [*astro-;* ἱστάναι, to stand]. Applied to the resting condition of the centrosomes of cells.
**Asturian** (*as-tū'-re-an*). Relating to Asturia, an old province of Spain. A. rose. 1. Pellagra. Syn., *rosa asturica; rosa asturiensis.* 2. Leprosy.
**astyclinic** (*as-ti-klin'-ik*) [*astу,* city; *clinic*]. Same as *policlinic.*
**astysia** (*ah-stis'-e-ah*) [ἀ, priv.; στύειν, to make erect]. Incomplete power to erect the penis.
**asurol** (*as'-u-rol*). A preparation containing mercury and sodium amido-oxybutyrate; it is used in syphilis.
**asyllabia** (*ah-sil-a'-be-ah*) [ἀ, priv.; συλλαβή, a syllable]. A condition in which individual letters are recognized, but the formation of syllables and words is impossible.
**asylum** (*as-i'-lum*) [L., "a place of refuge"]. An institution for the support, safe-keeping, cure, or education of those incapable of caring for themselves, such as the insane, the blind, etc. a. ear. See *hamatoma auris* under *hematoma.*
**asymbolia** (*ah-sim-bo'-le-ah*) [ἀ, priv.; σύμβολον, symbol]. The loss of all power of communication, even by signs or symbols.
**asymmetric carbon atom** (*as-im-et'-rik*). In stereochemistry, a carbon atom to which four different univalents are attached.
**asymmetry** (*ah-sim'-et-re*) [ἀ, priv.; συμμετρία, symmetry]. 1. Unlikeness of corresponding organs or parts of opposite sides of the body that are normally of the same size, etc., *e. g., asymmetry* of the two halves of the skull or brain. 2. The linking of carbon atoms to four different groups; the combination of carbon atoms with different atoms or atomic groups. a., meridional. See under *astigmatism.* a., unilateral. See *hemihypertrophy.*

**asymphytous** (*ah-sim'-fit-us*). Distinct; not grown together.
**asynclitism** (*ah-sin'-klit-izm*) [ἀ, priv.; σύν, together; κλίσις, an inclination]. The condition of obliquity of two or more objects to each other; *e. g.,* an oblique presentation of the fetal head at the superior strait of the pelvis.
**asynechia** (*ah-si-ne'-ke-ah*) [ἀ, priv.; συνέχειν, to hold together]. Absence of continuity in structure.
**asynergia** (*ah-sin-er'-je-ah*). Same as *asynergy.*
**asynergic** (*ah-sin-ur'-jik*). Not acting simultaneously or in harmony.
**asynergy** (*ah-sin'-er-jē*) [ἀ, priv.; συνεργία, cooperation]. Faulty coordination of the different organs or muscles normally acting in unison. a., progressive locomotor. a., motorial. See *ataxia, locomotor.* a., verbal, defective coordination of speech, as in aphasia. a., vocal, faulty coordination of the muscles of the larynx due to chorea.
**asynesia** (*ah-sin-e'-ze-ah*) [ἀσυνεσία, stupidity]. Stupidity; loss or disorder of mental power.
**asynetic, asynetous** (*ah-sin-et'-ik, ah-sin'-et-us*). Affected with asynesia.
**asynodia** (*ah-sin-o'-de-ah*) [ἀ, priv.; συνοδία, a traveling together]. Sexual impotence.
**asynovia** (*ah-sin-o'-ve-ah*) [ἀ, priv.; *synovia*]. A deficiency of the synovial fluid.
**asynthesis** (*ah-sin'-the-sis*) [ἀ, priv.; σύνθεσις, a putting together]. A faulty connection of parts.
**asyntrophy** (*ah-sin'-tro-fe*) [ἀ, priv.; συντροφία, a growing up together]. Absence of symmetry in growth and development.
**asystematic** (*ah-sis-tem-at'-ik*) [ἀ, priv.; σύστημα, system]. Diffuse; not restricted to any one or several systems of nerve fibers; applied to nervous diseases that are general.
**asystole, asystolia** (*ah-sis'-to-le, ah-sis-to'-le-ah*) [ἀ, priv.; συστολή, a shortening]. Imperfect contraction of the ventricles of the heart. a., cardiataxic, transitory asystole due to accelerated heart-action. a., cardioplegic. See *amyocardia.*
**asystolic** (*ah-sis-tol'-ik*) [see *asystole*]. Characterized by asystole.
**asystolism** (*ah-sis'-tol-izm*) [ἀ, priv.; συστολή, a shortening]. Inability of the right ventricle of the heart to empty itself of its contents, a condition encountered in the last stages of mitral incompetence. See *asystole.*
**atactic** (*at-ak'-tik*) [ἄτακτος, irregular]. Irregular; incoordinate. Pertaining to muscular incoordination, especially in aphasia.
**atactilia** (*ah-tak-til'-e-ah*). Inability to recognize tactile impressions.
**atavic** (*at'-av-ik*) [*atavus,* a forefather]. Relating to or characterized by atavism.
**atavism** (*at'-av-izm*) [see *atavic*]. The reappearance of a peculiarity in an individual whose more or less remote progenitors possessed the same peculiarity but whose immediate ancestors did not present it.
**atavistic** (*at-av-is'-tik*). Same as *atavic.*
**ataxaphasia** (*at-aks-a-fa'-ze-ah*). Inability to arrange words synthetically into sentences.
**ataxia** (*at-aks'-e-ah*) [ἀταξία, want of order]. Incoordination of muscular action. a., bulbar, tabes due to a lesion in the pons or oblongata. a., cerebellar, a., cerebral, a., spinal, that due to disease of the cerebellum of the brain or of the spinal cord. a. cordis. See *delirium cordis.* a., diphtheritic, a sequel of diphtheria preceding diphtheritic paralysis, and in which the chief phenomena of locomotor ataxia are present. a., family, a., Friedreich's, a., hereditary. See *Friedreich's disease.* a., hereditary cerebellar (of Marie), a form of ataxia that resembles Friedreich's disease in being hereditary, occurring in families; the gait, however, is not the staggering gait of tabes, but the reeling gait of cerebellar disease; the knee-jerk is increased instead of being diminished, and there are no deformities. a., locomotor, a disease of the posterior columns of the spinal cord, characterized by static and motor ataxia, by fulgurant pains, girdle-sensation, Argyll Robertson pupil, disturbances of sensation and of the sphincters, and loss of the patellar reflex. Syn., *posterior spinal sclerosis; tabes dorsalis.* a., moral, the inconstancy of ideas and will, attended with convulsions and pain, observed in hysterical subjects. a., motor, inability to coordinate the muscles in walking. a., paralytic, of the heart, a condition marked by dyspnea, weakness of cardiac sounds, palpitation, edema, and dropsy, without any organic

heart disease. a., **sensory**, a form regarded as due to disturbance of the nerve-tracts lying between the periphery and the centers of coordination; its existence is denied by some authorities. a., **spinal**. See a., *cerebellar*. a., **static**, the failure of muscular coordination in standing still, or in fixed positions of the limbs. a., **thermal**, peculiar large and irregular fluctuations of the body-temperature, due to a condition of incoordination or a disordered or weakened thermotaxic mechanism. This may give rise to the socalled *paradoxic* or *hysterical temperatures*, rising occasionally to 108° or 110° F., without grave or permanent injury. a., **vasomotor**. See *vasomotor ataxia*.

**ataxiagram** (at-aks'-e-a-gram) [ataxia; γράμμα, a marking]. 1. A line drawn by a patient suffering with an ataxial disease. The patient's eyes are open or closed and he attempts to make a straight line. The character of the deviations from a straight line that result are conceived to have a certain diagnostic value. 2. The record made by an ataxiagraph.

**ataxiagraph** (at-aks'-e-a-graf) [ataxia; γράφειν, to write]. An instrument for recording the swaying in ataxia.

**ataxiamnesia** (at-aks-e-am-ne'-ze-ah) [ataxia; amnesia]. Muscular ataxia with loss of or impairment of memory.

**ataxiamnesic** (at-aks-e-am-ne'-zik). Affected with ataxia and amnesia.

**ataxic** (at-aks'-ik) [see *ataxia*]. 1. Pertaining to or affected with ataxia. 2. A person affected with ataxia. a. **aphasia**. See under *aphasia*. a. **fever**. See *typhus*.

**ataxoadynamia** (at-aks-o-ah-di-nam'-e-ah). Adynamia combined with ataxia.

**ataxodynamy** (at-aks-o-din'-am-e) [ataxia; δύναμις, power]. Abnormity in the movements of a part of organ.

**ataxophemia** (at-aks-o-fe'-me-ah) [ataxia; φήμι, to speak]. Lack of coordination in speech.

**ataxophobia** (at-aks-o-fo'-be-ah) [ἀταξία, want of order; φόβος, fear]. 1. Excessive dread of disorder. 2. Morbid dread of suffering from locomotor ataxia.

**ataxospasmodic** (at-aks-o-spas-mod'-ik). Affected with choreic ataxia or relating to it.

**ataxy** (at-aks'-e). See *ataxia*.

**-ate**. A suffix to nouns in chemistry signifying any salt of an oxyacid having the termination *-ic*; as, *sulphate*, *phosphate*.

**atelectasis** (at-el-ek'-tas-is) [ἀτελής, imperfect; ἔκτασις, expansion]. Imperfect expansion or collapse of the air-vesicles of the lung. It may be present at birth, or may be acquired from diseases of the bronchi or lungs. a., **absorption**, acquired atelectasis in which the air has been removed by absorption from within, resulting from the plugging of the bronchial tubes. a., **compression**, acquired atelectasis due to pressure. a., **obstructive**, that due to obstruction of a bronchial tube. See a., *absorption*.

**atelectatic** (at-el-ek-tat'-ik) [see *atelectasis*]. Relating to or characterized by atelectasis.

**ateleiosis** (at-el-i-o'-sis) [ἀτελείωσις, not arriving at perfection]. A disease characterized by abrupt onset, the absence of any perceptible cause, conspicuous infantilism with retention of unimpaired intelligence, and marked tardiness in development of the sexual system. Cf. *progeria*.

**atelencephaly** (at-el-en-sef'-al-e) [ἀτελής, ἐγκέφαλος, brain]. Imperfect development of the brain.

**atelia, ateleiosis** (at-e'-le-ah, at-e-li-o'-sis) [ἀτέλεια, imperfection]. Persistence of the child's characteristics in the adult. Imperfect development. The word is compounded with others to designate the part affected, as *atelocardia*, etc., imperfect development of the heart, etc. a., **asexual**, that type in which the sexual organs are implicated. a., **sexual**, that type in which the sexual organs develop normally.

**atelic** (at'-el-ik) [ἀτελής, incomplete]. Functionless.

**atelo-** (at-el-o-). A prefix signifying imperfect development.

**atelocardia** (at-el-o-kar'-de-ah) [atelo-; καρδία, heart]. An imperfect or undeveloped state of the heart.

**atelocephalous** (at-el-o-sef'-al-us) [atelo-; κεφαλή, head]. Having the skull or head more or less incomplete.

**atelocheilia** (at-el-o-ki'-le-ah) [atelo-; χεῖλος, lip]. Defective development of a lip.

**atelocheiria** (at-el-o-ki'-re-ah) [atelo-; χείρ, hand]. Defective development of the hand.

**ateloencephalia** (at-el-o-en-sef-a'-le-ah) [atelo-; ἐγκέφαλος, brain]. Incomplete development of the brain.

**ateloglossia** (at-el-o-glos'-e-ah) [atelo-; γλῶσσα, tongue]. Congenital defect in the tongue.

**atelognathia** (at-el-og-na'-the-ah) [atelo-; γνάθος, jaw]. Imperfect development of a jaw, especially of the lower jaw.

**atelomyelia** (at-el-o-mi-e'-le-ah) [atelo-; μυελός, marrow]. Congenital defect of the spinal cord.

**atelopodia** (at-el-o-po'-de-ah) [atelo-; ποῦς, foot]. Defective development of the foot.

**ateloprosopia** (at-el-o-pro-so'-pe-ah) [atelo-; πρόσωπον, face]. Incomplete facial development.

**atelorrhachidia** (at-el-o-rak-id'-e-ah) [atelo-; ῥάχις, spine]. Imperfect development of the spinal column, as in spina bifida.

**atelostomia** (at-el-o-sto'-me-ah) [atelo-; στόμα, mouth]. Incomplete development of the mouth.

**athelasmus** (ah-thel-az'-mus) [ἀ, priv.; θηλασμός, a suckling]. Inability to suckle, from defect or want of the nipples.

**athelia** (ah-the'-le-ah) [ἀ, priv.; θηλή, a nipple]. Absence of a nipple.

**athermal** (ah-thur'-mal) [ἀ, priv.; θέρμη, heat]. Cool; applied to spring-water of a temperature below 15° C.

**athermancy** (ah-thur'-man-se) [ἀθέρμαντος, not heated]. The state of being impervious to radiant heat.

**athermanous** (ah-ther'-man-us). Impervious to radiant heat.

**athermic, athermous** (ah-ther'-mik, -mus). 1. Without fever. 2. See *athermanous*.

**athermosystaltic** (ah-ther-mo-sist-al'-tik) [ἀ, priv.; θέρμη, heat; συσταλτικός, drawing together]. Applied to muscles which do not contract under the influence of heat.

**atheroma** (ath-er-o'-mah) [ἀθήρη, gruel; ὄμα, tumor]. 1. A sebaceous cyst containing a cheesy material. Syn., *acne sebacea molluscum; sebaceous cyst; steatoma*. 2. The fatty degeneration of the walls of the arteries in arteriosclerosis; by common usage the word is also applied to the whole process of arteriosclerosis. Arterial atheroma is also termed *atherosis*. a., **capillary**, the formation of fatty granules in the walls of the capillaries.

**atheromasia** (ath-er-o-ma'-ze-ah) [see *atheroma*]. Atheromatous degeneration; the condition of atheroma.

**atheromatosis**. A more or less generalized atheromatous condition of the arteries.

**atheromatous** (ath-er-o'-mat-us) [see *atheroma*]. Characterized by or affected with atheroma. a. **abscess**. See *abscess, atheromatous*. a. **ulcer**, an ulcer formed by the abscess breaking through the intima.

**atherosclerosis** (ath-er-o-skle-ro'-sis) [atheroma; *sclerosis*]. A form of arteriosclerosis in which there is hyperplasia of the outer layers of the involved arteries and degeneration of the elastic layer.

**atherosis** (ath-er-o'-sis) [ἀθήρη, gruel]. A synonym of *atheroma* (2).

**athetoid** (ath'-et-oid) [*athetosis*]. Pertaining to or affected with athetosis. a. **spasm**, a spasm in which the affected member performs athetoid movements.

**athetosis** (ath-et-o'-sis) [ἄθετος, unfixed; changeable]. A condition most frequently occurring in children, and characterized by continual slow change of position of the fingers and toes. It is usually due to a lesion of the brain. It is also called "posthemiplegic chorea," from its occurrence after hemiplegia. a., **general**. See *paraplegia, infantile spasmodic*.

**athlete's heart** [ἀθλεῖν, to contend with]. A slight incompetency of the aortic valves, a condition sometimes found in athletes.

**athrepsia** (ah-threp'-se-ah) [ἀ, priv.; τρέφειν, to nourish]. Malnutrition.

**athreptic** (ah-threp'-tik). Relating to or affected with athrepsia.

**athymia** (ah-thi'-me-ah) [ἀ, priv.; θυμός, spirit]. 1. Despondency. 2. Loss of consciousness. 3. Insanity. 4. Absence of the thymus gland.

**athymic** (ah-thi'-mik) [ἀ, priv.; θυμός, mind]. Affected with athymia.

**athyrea, athyria** (ah-thi'-re-ah) [ἀ, priv.; *thyroid*]. The condition arising from absence of the thyroid gland or suppression of its function. Syn., *myxedema*. Cf. *thyreoprivus*.

**athyreosis** (*ah-thi-re-o'-sis*). Atrophy or absence of the thyroid gland and the pathological condition consequent upon elimination of its function.

**athyria.** Same as *athyrea*.

**athyroidea** (*ah-thi-roid'-e-ah*). Absence of the thyroid gland.

**athyroidation.** Same as *athyrea*, q. v.

**athyroidea.** Same as *athyrea*, q. v.

**athyroidemia** (*ah-thi-roid-e'-me-ah*). Davel's name for myxedema.

**athyroidism** (*ah-thi'-roy-dizm*). Same as *athyreosis* or *athyrea*.

**athyrosis** (*ah-thi-ro'-sis*). See *athyreosis*.

**atlantad** (*at-lan'-tad*) [See *atlas*]. Toward the atlas in situation or direction.

**atlantal** (*at-lan'-tal*) [See *atlas*]. Relating to the atlas.

**atlanten** (*at-lan'-ten*) [See *atlas*]. Belonging to the atlas in itself.

**atlanto-** (*at-lan'-to*) [See *atlas*]. A prefix signifying relation to the atlas; seen in the words atlantoaxial (relating to the atlas and the axis), atlantooccipital, atlanto-odontoid, etc.

**atlantoaxial** (*at-lant-o-aks'-e-al*). See *atloaxoid*.

**atlas** (*at'-las*) [ΆτΛας, a mythological Greek hero who was supposed to carry the earth on his shoulders]. The first of the cervical vertebrae. It articulates with the occipital bone of the skull and with the axis.

**atloaxoid** (*at-lo-aks'-oid*). Relating to the bones termed the atlas and the axis.

**atlodidymus** (*at-lo-did'-im-us*). Same as *atlodymus*.

**atlodymus** (*at-lod'-im-us*) [See *atlas*; δίδυμος, double]. A monstrosity with two heads on one neck and a single body.

**atloido-** (*at-loi'-do*). In composition, the same as *atlanto*; seen in such examples as atloido-axoid, atloido-odontoid, etc.

**atmiatria, atmiatrics.** See *atmiatry*.

**atmiatry** (*at-mi'-at-re*) [ἀτμίς, vapor; ἰατρεία, medical treatment]. Treatment of diseases of the lungs or mucous membranes by inhalation, fumigation, or by directing a current of vapor or gas upon the part.

**atmic** (*at'-mik*) [ἀτμίς, vapor]. Relating to, due to, or consisting of vapor.

**atmidalbumin** (*at-mid-al'-bū-min*). A substance standing between the albuminates and the albumoses, obtained by Neumeister at the same time with atmidalbumose.

**atmidalbumose** (*at-mid-al'-bū-mōs*). Neumeister's name for a body obtained by the action of superheated steam on fibrin.

**atmidiatrics** (*at-mid-re-at'-riks*). Treatment of disease by vapor.

**atmidometer** (*at-mid-om'-et-er*). See *atmometer*.

**atmidometrograph** (*at-mid-o-met'-ro-graf*) [ἀτμός, vapor; μέτρον, a measure; γράφειν, to write]. A self-registering atmidometer.

**atmidoscope** (*at-mid'-o-skōp*) [ἀτμός, vapor; σκοπεῖν, to view]. See *atmometer*.

**atmiometer** (*at-mi-om'-et-er*). A closed cabinet with apparatus for treating diseases by means of atmiatry.

**atmismometer** (*at-mis-mom'-et-ur*). See *atmometer*.

**atmisterion** (*at-mis-te'-re-on*). See *vaporarium*.

**atmo-** (*at-mo-*) [ἀτμός, vapor; breath]. A prefix meaning vapor or breath.

**atmocausia, atmocausis** (*at-mo-kaw'-se-ah, -sis*) [*atmo-*; καῦσις, a burning]. Therapeutic cauterization with steam by means of an atmocautery.

**atmocautery** (*at-mo-kaw'-ter-e*). An apparatus used in practising atmocausis.

**atmograph** (*at'-mo-graf*) [*atmo-*; γράφειν, to record]. A form of self-registering respirometer.

**atmography** (*at-mog'-raf-e*) [*atmo-*; γράφειν, to write]. A description of vapors and evaporation.

**atmokausis** (*at-mo-kaw'-sis*). See *atmocausis*.

**atmology** (*at-mol'-oj-e*) [ἀτμός, vapor; λόγος, science]. The science of vapors and evaporation.

**atmolysis** (*at-mol'-is-is*) [*atmo-*; λύσις, loosing]. A method of separating the ingredients of mixed gases or vapors by means of their different diffusibility through a porous substance.

**atmolyzer** (*at-mol-i'-zur*). An apparatus for separating gases by diffusion.

**atmometer, atmidometer** (*at-mom'-et-er, at-midom'-et-er*) [*atmo-*; μέτρον, a measure]. An instrument for measuring the amount of water exhaled by evaporation from a given surface in a given time, in order to determine the humidity of the atmosphere.

**atmos** (*at'-mos*) [abbreviation of atmosphere]. A proposed unit of air pressure, being the pressure of one dyne on one square centimeter.

**atmosphere** (*at'-mos-fēr*) [*atmo-*; σφαῖρα, a sphere]. 1. The mixture of gases surrounding the earth to the height of about 200 miles. 2. The pressure exerted by the atmosphere at the level of the sea; it is about 15 pounds to the square inch, or 1 kilogram to the square centimeter. 3. In chemistry, any special gaseous medium encircling a body. 4. The climatic state of a locality.

**atmospheric** (*at-mos-fer'-ik*) [see *atmosphere*] Pertaining to the atmosphere. **a. moisture,** the vapor of water mingled with the atmosphere. It varies in quantity according to the temperature. **a. tension,** the pressure of the air per square inch on the surface of a body. Normally, at the sea-level, it is about 15 pounds per square inch, or equal to that of a column of mercury about 30 inches in height. It decreases about $\frac{1}{30}$ inch or $\frac{1}{30}$ pound per square inch for every 90 feet of altitude. Above 10,000 feet the rarity of the atmosphere is usually noticeable in quickened breathing and pulse-rate.

**atmospherization** (*at-mos-fer-iz-a'-shun*). The conversion of venous into arterial blood by the absorption of oxygen. Cf. *dearterialization*.

**atmotherapy** (*at-mo-ther'-ap-e*) [*atmo-*; θεραπεία, therapy]. 1. A name given by Pitres to the treatment of certain tics by methodic reduction of respiration. 2. The treatment of disease by vapor.

**atocia** (*at-o'-se-ah*) [ἄτοκος, barren]. Sterility of the female.

**atom** (*at'-om*) [ἀ, priv.; τέμνειν, to cut]. The ultimate unit of an element; that part of an element incapable of further division, or the smallest part capable of entering into the formation of a chemical compound, or uniting with another to form a molecule—which last is the smallest quantity of a substance that can exist free or uncombined.

**atomic** (*at-om'-ik*) [see *atom*]. Pertaining to atoms. **a. heat,** the specific heat of an atom of an element multiplied by its atomic weight. **a. theory,** the theory of Dalton that all matter is composed of atoms, the weight of each atom differing for the different elements. **a. valence,** the saturating power of the atom of an element as compared with an atom of hydrogen. Syn., *equivalence*. **a. volume,** the atomic weight of an element divided by the density. **a. weight,** the weight of an atom of an element as compared with the weight of an atom of hydrogen.

**atomicity** (*at-om-is'-it-e*) [see *atom*]. 1. Chemical valence; quantivalence. 2. The number of OH groups is an alcohol or a base.

**atomism** (*at'-om-izm*) [ἀ, priv.; τέμνειν, to cut]. 1. The science of atoms. 2. The theory that the universe is composed of atoms.

**atomist** (*at'-om-ist*). One who believes in atomism.

**atomistic** (*at-om-is'-tik*). 1. Relating to or consisting of an atom. 2. Relating to atomism.

**atomization** (*at-om-iz-a'-shun*) [see *atom*]. The mechanical process of breaking up a liquid into fine spray.

**atomizer** (*at'-om-i-zer*) [see *atom*]. An instrument for transforming a liquid into a spray.

**atomology** (*at-om-ol'-o-je*) [*atom*; λόγος, science]. The science of atoms; atomism.

**atonia** (*at-o'-ne-ah*) [ἀτονία, want of tone]. Atony.

**atonic** (*at-on'-ik*) [*atony*]. Relating to or characterized by atony.

**atonicity** (*at-on-is'-it-e*). Lack of tone, atony.

**atony** (*at'-o-ne*) [ἀ, priv.; τόνος, tone]. Want of tone. Debility. Loss of diminution of muscular or vital energy.

**atophan** (*at'-o-fan*). Phenylcinchoninic acid, or phenylchinolin carboxlic acid. It is said to increase the elimination of uric acid in cases of gout and rheumatism.

**atopomenorrhea** (*at-o-po-men-or-e'-ah*) [ἄτοπος, out of place; μήν, month; ῥεῖν, to flow]. Vicarious menstruation.

**at pi** (*ah-top'-ik*) [ἀ, priv.; τόπος, place]. Out of place. c

**atoxic** (*ah-toks'-ik*) [ἀ, priv.; τόξικον, poison]. Not venomous; not poisonous.

**atoxogen** (*ah-toks'-o-jen*) [ἀ, priv.; τόξικον, poison; γενᾶν, to produce]. A defensive substance resembling the enzymes and chemically allied to toxins and antitoxins prepared from the adrenals and spleen of the horse.

**atoxyl** (*at-oks'-il*), $C_6H_4.NH_2.AsO(OH)_2$, a compound of arsenic acid and aniline, used in skin-diseases and in sleeping-sickness.

**atrabiliary** (*at-rah-bil'-e-a-re*) [*atra*, black; *bilis*, bile]. Pertaining to black bile, gloomy, melancholic. **a. capsules**, an old name for the suprarenal capsules.

**atrabilin** (*at-rah-bil'-in*). A preparation of suprarenal capsule; it has a hemostatic and vasoconstrictor action.

**atrachelia** (*ah-trak-e'-le-ah*) [ἀ, priv.; τράχηλος, the neck]. Absence or exceeding shortness of the neck.

**atrachelocephalus** (*ah-trak-el-o-sef'-al-us*) [ἀτράχηλος, without a neck; κεφαλή, the head]. 1. Affected with atrachelia. 2. A monster with no neck or an abnormally short one.

**atrachelous** (*ah-trak'-el-us*). Having no neck or only a very short one; also, beheaded.

**atractenchyma** (*ah-trakt-en'-ki-mah*) [ἄτρακτος, a spindle; ἐγχεῖν, to pour in]. A tissue consisting of spindle-cells.

**atractoid** (*ah-trakt'-oid*). Spindle-shaped.

**atramental** (*at-ram-en'-tal*) [*atramentum*, ink]. Of an inky-black color.

**atremia** (*ah-tre'-me-ah*) [ἀ, priv.; τρέμειν, to tremble]. 1. An absence of tremor. 2. Hysterical inability to walk, stand, or sit without general discomfort and paresthesia of the head and back, all movements being readily executed in the recumbent posture. Syn., *Neftel's disease.*

**atrepsy** (*ah'-trep-se*) [ἀ, priv.; τρέφειν, to nourish]. Ehrlich's term for immunity to tumor cells produced by the absence of the particular nourishment needed for the growth of tumors.

**atresia** (*ah-tre'-ze-ah*) [ἀ, priv.; τρῆσις, perforation]. Imperforation or closure of a normal opening or canal, as of the anus, vagina, meatus auditorius, pupil, etc.

**atresic** (*ah-tre'-zik*) [see *atresia*]. Characterized by atresia.

**atretic** (*ah-tret'-ik*). Same as *atresic.*

**atreto-** (*ah-tre-to-*) [ἄτρητος· imperforate]. A prefix meaning imperforate.

**atretoblepharia** (*at-ret-o-blef-a'-re-ah*) [*atreto-*; βλέφαρον, eye-lid]. Symblepharon, q. v.

**atretocephalus** (*ah-tret o-sef'-al-us*) [*atreto-*; κεφαλή, the head]. A monster with imperforate nostrils or mouth.

**atretocormus** (*ah-tret-o-korm'-us*) [*atreto-*; κορμός, the trunk]. A monster having one or more imperforate openings on the trunk.

**atretocystia** (*ah-ret-o-sis'-te-ah*) [*atreto-*; κύστις, bladder]. Atresia of the bladder.

**atretogastria** (*ah-tret-o gas'-tre-ah*) [*atreto-*; γαστήρ, stomach]. Imperforation of the cardiac or pyloric orifice of the stomach.

**atretolemia** (*ah-tret-o-le'-me-ah*) [*atreto-*; λαιμός, the gullet]. Imperforation of the esophagus or pharynx.

**atretometria** (*at-ret-o-me'-tre-ah*) [*atreto-*; μήτρα, womb]. Atresia of the uterus.

**atretopsia** (*at-ret-op'-se-ah*) [ἄτρητος, imperforate; ὤψ, eye]. Imperforation of the pupil.

**atretorrhinia** (*at-tret-or-rin'-e-ah*) [*atreto-*; ῥίς, the nose]. Nasal atresia.

**atretostomia** (*ah-tret-o-sto'-me-ah*) [*atreto-*; στόμα, the mouth]. Imperforation of the mouth.

**atreturethria** (*ah-tret-u-re'-thre-ah*) [*atreto-*; οὐρήθρα, the urethra]. Imperforation of the urethra.

**atria**. Plural of *atrium.* **a. mortis**, the halls of death (*i. e.*, the heart, lungs, and brain).

**atrial** (*a'-tre-al*) [*atrium*, the fore-court, or hall]. Relating to an atrium.

**atricha** (*ah'-trik-ah*) [ἀ, priv.; θρίξ, hair]. A group of bacteria having no flagella.

**atrichia, atrichiasis** (*ah-trik'-e-ah, ah-trik-i'-as-is*) [ἀ, priv.; θρίξ, hair]. Absence of the hair.

**atrichosis** (*ah-trik-o'-sis*) [see *atrichia*]. A condition characterized by absence of hair.

**atrioventricular** (*a-tre-o-ven-trik'-u-lar*) [*atrium*, hall; *ventriculus*, ventricle]. Relating both to the atrium (or auricle) and to the ventricle of the heart.

**atriplicism** (*at-rip'-lis-izm*) [*Atriplex*, a genus of plants]. A form of poisoning from eating uncooked spinach, *Atriplex littoralis*. It is characterized by painful infiltration of the backs of the hands and forearms and a sensitiveness to light.

**atrium** (*a'-tre-um*) [L., "the forecourt or hall"]. 1. The auricle of the heart. 2. The part of the tympanic cavity of the ear below the head of the malleus. **a. anterius**, the right auricle of the heart. **a. cordis**, the auricle of the heart. **a. cordis dextrum**, the right auricle of the heart. **a. cordis posterius**, the left auricle of the heart. **a., infection-**, the point of entrance of the bacteria in an infectious disease. **a. vaginæ**, the vestibule of the vulva.

**atrolactyl** (*at-ro-lak'-til*), $C_9H_9O_3$. The radical of atrolactic acid. **a.-tropein**. See *aconitine, British.*

**Atropa** (*at'-ro-pah*) ["Ἄτροπος, "she who turns not"; undeviating; one of the three Fates who cut the thread of life—in allusion to the poisonous effects of the plant]. A genus of the natural order *Solanaceæ. A. belladonna* is the deadly nightshade, from which *atropine* is obtained. See *belladonna.*

**atrophia** (*at-ro'-fe-ah*). See *atrophy.*

**atrophic** (*at-ro'-fik*) [*atrophy*]. Pertaining to or affected with atrophy.

**atrophied** (*at' ro-fid*) [ἀ, priv.; τροφή, nourishment]. Wasted; affected with atrophy.

**atrophoderma** (*at-ro-fo-der'-mah*). See *atrophy of the skin.* **a. pigmentosum**. See *xeroderma pigmentosum.*

**atrophodermatosis** (*at-ro-fo-der-mat-o'-sis*) [*atrophy*; δέρμα, the skin]. A class of skin diseases, including atrophoderma, ulodermitis, and scleroderma, characterized by atrophy of the cutis.

**atrophodermia**. Atrophoderma.

**atrophoderma**. Atrophoderma.

**atrophodermia**. Atrophoderma.

**atrophoderma**. Atrophoderma.

**atrophoderma**. Atrophoderma.

**atrophoderma**. Atrophoderma.

**atrophoderma**. Atrophoderma.

**atrophodermia**. Atrophoderma.

**atrophodermia**. Atrophoderma.

**atrophodermia**. Atrophoderma.

**atrophodermia**. Atrophoderma.

**atrophoderma**. Atrophoderma.

**atrophoderma**. Atrophoderma.

**atrophoderma**. Atrophoderma.

**atrophoderma**. Atrophoderma.

**atrophoderma**. Atrophoderma.

**atrophoderma**. Atrophoderma.

**atrophoderma**. Atrophoderma.

**atrophodermia**. Atrophoderma.

**atrophodermia**. Atrophoderma.

**atrophodermia**. Atrophoderma.

**atrophodermia**. Atrophoderma.

**atrophodermia**. Atrophoderma.

**atrophodermia**. Atrophoderma.

**atrophodermia**. Atrophoderma.

**atrophodermia**. Atrophoderma.

**atrophodermia**. Atrophoderma.

**atrophodermia**. Atrophoderma.

**atrophodermia**. Atrophoderma.

**atrophodermia**. Atrophoderma.

**atrophodermia**. Atrophoderma.

**atrophodermia**. Atrophoderma.

**atrophodermia**. Atrophoderma.

**atrophodermia**. Atrophoderma.

**atrophodermia**. Atrophoderma.

**atrophodermia**. Atrophoderma.

**atrophodermia**. Atrophoderma.

**atrophodermia**. Atrophoderma.

**atrophodermia**. Atrophoderma.

**atrophodermia**. Atrophoderma.

**atrophodermia**. Atrophoderma.

**atrophodermia**. Atrophoderma.

**atrophodermia**. Atrophoderma.

**atrophodermia**. Atrophoderma.

**atrophodermia**. Atrophoderma.

**atrophodermia**. Atrophoderma.

**atrophodermia**. Atrophoderma.

**atrophodermia**. Atrophoderma.

**atrophodermia**. Atrophoderma.

**atrophodermia**. Atrophoderma.

**atrophodermia**. Atrophoderma.

**atrophodermia**. Atrophoderma.

**atrophodermia**. Atrophoderma.

**atrophodermia**. Atrophoderma.

**atrophodermia**. Atrophoderma.

**atrophodermia**. Atrophoderma.

**atrophodermia**. Atrophoderma.

**atrophodermia**. Atrophoderma.

**atrophodermia**. Atrophoderma.

**atrophodermia**. Atrophoderma.

**atrophodermia**. Atrophoderma.

**atrophodermia**. Atrophoderma.

**atrophodermia**. Atrophoderma.

**atrophodermia**. Atrophoderma.

**atrophodermia**. Atrophoderma.

**atrophodermia**. Atrophoderma.

**atrophodermia**. Atrophoderma.

**atrophodermia**. Atrophoderma.

**atrophodermia**. Atrophoderma.

**atrophodermia**. Atrophoderma.

**atrophodermia**. Atrophoderma.

**atrophodermia**. Atrophoderma.

**atrophodermia**. Atrophoderma.

**atrophodermia**. Atrophoderma.

**atrophodermia**. Atrophoderma.

**atrophodermia**. Atrophoderma.

**atrophodermia**. Atrophoderma.

**atrophodermia**. Atrophoderma.

**atrophodermia**. Atrophoderma.

**atrophodermia**. Atrophoderma.

**atrophodermia**. Atrophoderma.

**atrophodermia**. Atrophoderma.

**atrophodermia**. Atrophoderma.

**atrophodermia**. Atrophoderma.

**atrophy** of a part with destruction of some of its elements. **atrophia nervea,** atrophy of the nerves. **atrophia nervosa,** gradual emaciation, with loss of appetite, due to unwholesome and depressing environment. **a., pigmentary, a., pigmented,** a form of atrophy so called from a deposit of pigment (yellow or yellowish-brown) in the atrophied cells. **atrophia pilorum propria,** atrophy of the hair, either symptomatic or idiopathic in origin. **a., progressive facial,** a condition characterized by progressive wasting of the skin of the face. Syn., *atrophia nova facialis.* **a., progressive muscular,** a chronic disease characterized by progressive wasting of individual muscles or physiological groups of muscles, and by an associated and proportional amount of paralysis. It is due to a degeneration and atrophy of the multipolar cells in the anterior gray horns of the cord, with consecutive degeneration of the anterior nerve-roots and muscles. The right hand is usually the part first attacked, and takes on a peculiar claw-like form (*main-en-griffe*). The disease is most frequent in males of adult life, and follows excessive muscular exertion. Syn., *chronic anterior poliomyelitis; wasting palsy.* **a., progressive nervous,** Jaccoud's name for atrophy of the spinal nerve-roots due to pressure from a deposit of fibrous substance on the spinal arachnoid. **a., progressive unilateral facial,** a disease characterized by progressive wasting of the skin, connective tissue, fat, bone, and more rarely the muscles of one side of the face. It is most common in females; its course is slow and generally progressive. **a., qualitative, degeneration a., quantitative.** See **a., simple. a., red,** a form of atrophy due to chronic congestion, as seen in the liver in mitral and tricuspid valvular lesions. **a., sclerotic,** a name for connective tissue found at times deposited in the heart-substance after myocarditis. **a., senile,** the physiological atrophy of advanced life. It affects the lungs, the sexual and other organs. **a., senile, of the skin,** an atrophy of the skin usually associated with general signs of senile degeneration. Syn., *atrophia cutis senilis; senile atrophoderma.* **a., serous,** atrophy associated with an infiltration of fluid into the atrophic tissues. **a., simple,** that due to a decrease in the size of individual cells. **a., simple brown,** a condition of the heart in which the muscle-fibers retain their striated appearance, but the muscle-cells are small and contain yellow granules of pigment. **a. of the skin,** atrophy characterized by diminution or disappearance of certain of the elements of the skin: especially seen in advanced age. The skin becomes thin, loose, wrinkled, and discolored. Syn., *atrophia cutis; atrophoderma.* **a., sympathetic,** atrophy of the second member of a pair of organs, following that of the first. **a., trophoneurotic,** that dependent upon abnormality of the nervous supply of an organ or tissue, best illustrated in muscular atrophy from disease of the anterior horns of the spinal cord. **atrophia verminosa,** emaciation due to intestinal worms. **a., white,** nerve atrophy, leaving only white connective tissue.

**atropia** (*at-ro'-pe-ah*). See *atropine.*

**atropic** (*at-rop'-ik*). Relating to the genus *Atropa* or to atropine.

**atropine, atropina** (*at'-ro-pēn, at-ro-pi'-nah*) ["Ατροπος, one of the Fates who cut the thread of life], $C_{17}H_{23}NO_3$. The *atropina* of the U. S. P. is a crystalline alkaloid derived from *Atropa belladonna.* It is a mydriatic, antispasmodic, and anodyne; in small doses a cardiac, respiratory, and spinal stimulant; in large doses a paralyzant of the cardiac and respiratory centers, the spinal cord, motor nerves, and involuntary and voluntary muscles. In full doses it produces dryness of the throat, flushing of the face, dilatation of the pupils, a rise of temperature, and, sometimes an erythematous rash. It is extensively used in ophthalmic practice to dilate the pupil, to paralyze accommodation, and also in various corneal, iritic, and other ocular diseases. Its therapeutic use in general medicine is also manifold; *e. g.*, in inflammatory affections and the pain of cerebral and spinal hyperemia, atonic constipation, cardiac failure, hypersecretions, especially of the sweat, to relieve local spasms, as in intestinal and biliary colic, in asthma, whooping-cough, etc., and as a physiological antagonist in opium-poisoning. **a. borate,** $(C_{17}H_{23}NO_3)_3B_4O_7$, is used in ophthalmic practice. **a. hydrobromide,** $C_{17}H_{23}NO_3HBr$, white crystals, soluble in water and in alcohol. It is used as is atropine. **a. hydrochloride,** $C_{17}H_{23}NO_3HCl$, white crystals, soluble in water and alcohol, slightly in ether. Used in the same manner as atropine. Dose $\frac{1}{150} - \frac{1}{24}$ gr. (0.0006–0.001 Gm.). **a., lamellæ of** (*lamellæ atropinæ*, B. P.), each contains $\frac{1}{5000}$ gr. (0.000013 Gm.) atropine. **a. oleate** (*oleatum atropinæ*, U. S. P.), a 2 % solution of atropine in oleic acid; it is a mydriatic, sedative, and anodyne, and is used as an inunction in cases in which remedies cannot be administered by the mouth. **a. salicylate,** $C_{17}H_{23}NO_3C_7H_6O_3$, a colloidal mass, used as is atropine. **a. santonate,** a compound of atropine and santonic acid, recommended as a mydriatic. **a. santoninate,** $C_{17}H_{23}O_3C_{15}H_{18}O_3$, is used in ophthalmic practice. **a. stearate,** $C_{17}H_{23}NO_3C_{17}H_{35}CO.OH$, fine white needles, greasy to the touch, melting at 120° C., beginning to decompose at 170° C., and containing 50.43 % of atropine. It is soluble in ether and in alcohol. Applied in 1 : 500 oily solution as a substitute for oil of belladonna or oil of hyoscyamus. **a. sulphate** (*atropinæ sulphas*, U. S. P.), the most frequently used preparation of atropine, is a white powder, of bitter taste and neutral reaction, and is soluble in water. Dose $\frac{1}{150} - \frac{1}{50}$ gr. (0.00036–0.008 Gm.). **a., sulphate, solution of** (*liquor atropinæ sulphatis*, B. P.). Dose 1–6 min. (0.065–0.4 Cc.). **a. tartrate,** $(C_{17}H_{23}NO_3)_2C_4H_6O_6$, is used as is atropine.

**atropinism** (*at'-ro-pin-ism*). See *atropism.*

**atropinization** (*at-ro-pin-i-za'-shun*). The production of the physiological effect of belladonna.

**atropinize** (*at'-ro-pin-īz*) [*atropine*]. To bring under the influence of, or to treat with, atropine.

**atropism** (*at'-ro-pism*). Poisoning with, or the morbid condition induced by, atropine.

**atroscine** (*at'-ros-ēn*), $C_{17}H_{21}NO_4$. An alkaloid isomeric with hyoscine, obtained from *Scopolia carniolica.* It has a higher rotatory power than hyoscine, and is from 2 to 4 times stronger in mydriatic action. Syn., *atrosia.*

**attaint** (*at-aint'*) [*attingere*, to touch by striking]. An injury to a horse's leg caused by overreaching.

**attar** (*at'-ar*) [Ar., '*itr*, perfume]. A general name for any of the volatile oils. **a. of rose,** oil of rose. The volatile oil distilled from the fresh flowers of the Damascene rose. It comes mainly from eastern Rumelia, and is generally adulterated with other volatile oils. It is used as a perfume.

**attendant** (*at-en'-dant*) [*attendere*, to wait upon]. A nonprofessional attaché of an asylum or hospital.

**attention** (*at-ten'-shun*). The direction of the will or thought upon an object or to a particular sensation. **a., central,** the "imagination" or mental remaking of the image by the mind when the peripheral visual attention is abrogated. **a., compound synchronous,** in this the consciousness recognizes and correlates or combines multiple streams of synchronous and diverse stimuli, visual, auditory, etc. **a., multiple synchronous auditory,** two or more synchronous tones or sounds or lines of such tones or sounds are recognized by consciousness. **a., multiple synchronous central visual,** the imagining or mental reproduction of multiple synchronous visual trains without the objectively formed images. **a., multiple synchronous visual,** that when the attention recognizes two or more discrete sets of retinal images at the same time. **a., single-stream auditory,** that when a monotone, a sound, or series of single notes or sounds, is listened to, exclusive of others. **a., single-stream central,** that when the central visual attention, without objectively forming images, follows the passing of imagined single or unitary images in single file. **a., single-stream central auditory,** that without the objective audition. **a., single-stream visual,** that form of visual attention existing when the eyes follow a linear concatenation of single or unitary macular images to the exclusion of all others. **a., visual,** that existing when the eyes, consciously, observe a fixed or moving object.

**attenuant** (*at-en'-ū-ant*) [*attenuare*, to make thin]. 1. A medicine or agent increasing the fluidity or thinness of the blood or other secretion. 2. A diluent. 3. Lessening the effect of an agent.

**attenuated** (*at-en'-ū-a-ted*). Wasted; thinned. **a. virus,** a weakened virus.

**attenuating** (*at-en'-u-a-ting*) [see *attenuant*]. Making thin.

**attenuation** (*at-en-u-a'-shun*) [see *attenuant*]. The act of making thin; a thinning, narrowing, or reduction of the strength or size of a substance, especially the weakening of the pathogenic virulence of microorganisms by successive cultivation, by exposure to light, air, heat, or other agency, or by passing through certain animals, so that they may be used as a vaccine to confer immunity from future attacks of the disease. **a., Sanderson's** method of, the passing of virus through the system of another animal (*e. g.*, the guinea-pig, in anthrax) so that it becomes modified in virulency.

**attic** (*at'-ik*). Part of the tympanic cavity situated above the atrium. **a. disease,** chronic suppurative inflammation of the attic of the tympanum.

**atticoantrotomy** (*at-ik-o-an-trot'-o-me*) [*attic; an-trum; τέμνειν,* to cut]. The opening of the attic and mastoid process.

**atticomastoid** (*at-ik-o-mas'-toid*). Relating to the attic and the mastoid.

**atticotomy** (*at-ik-ot'-om-e*) [*attic; τέμνειν,* to cut]. Surgical incision of the attic.

**attitude** (*at'-e-tūd*). See *posture.* **a., crucifixion,** in hysteroepilepsy, a rigid state of the body, the arms stretched out at right angles. **a. of fetus,** the relation of its parts to one another. **a., frozen,** a peculiar stiffness of the gait characteristic of disease of the spinal cord, especially of amyotrophic lateral sclerosis. **a., passionate,** the assumption of a dramatic or theatrical expression, a position assumed by some hysterical patients.

**attollens** (*at-ol'-enz*) [*attollere,* to rise up]. Raising. **a. aurem,** a muscle raising the external ear.

**attraction** (*at-rak'-shun*) [*attrahere,* to draw to]. The tendency of one particle of matter to approach another; affinity. As existing between masses, it is termed *gravitation,* while *molecular attraction* or *cohesion* expresses the force aggregating molecules. **a., capillary,** the force that causes liquids to rise in fine tubes or between two closely approximated surfaces, or on the sides of the containing vessel. **a., chemical,** the attraction of affinity, relates to the attraction of atoms of one element to those of others, resulting in chemical compounds. **a., electric,** the tendency of bodies toward each other when charged with opposite electricities. **a., magnetic,** the influence of a magnet upon certain metallic substances, chiefly iron. **a. sphere,** the central mass of the aster in karyokinesis.

**attrahens** (*at'-ra-henz*) [L., "drawing"]. Drawing forward, as *atrahens aurem,* a muscle drawing the ear forward and upward.

**attrahent** (*at'-ra-hent*) [*attrahens,* drawing]. 1. Drawing to; adducent. 2. A drawing application; an epispastic or rubefacient.

**attrition** (*at-rish'-un*) [*attercre,* to rub against]. 1. An abrasion or chafing of the skin. 2. Any rubbing or friction that breaks or wears the surface.

**at. wt.** Abbreviation of *atomic weight.*

**atypical,** (*ah-tip'+ik-al*) [*ά,* priv.; *τύπος,* a type]. Irregular; not conformable to the type. **a. fever,** an intermittent fever with irregularity of the paroxysm.

**A. u.** Abbreviation of *Ångstrom's unit.*

**Au.** Chemical symbol of the element gold. See *aurum.*

**auante** (*aw-an'-te*) [*αὐαίνειν,* to dry]. A wasting or atrophy.

**auantic** (*aw-an'-tik*) [*αὐαντικός,* wasted]. Characterized by wasting; atrophic.

**Aubert's phenomenon** (*o-bair'*). An optical illusion by which, when the head is inclined to one side, a vertical line is made to appear oblique toward the opposite side.

**auchen** (*aw'-ken*) [*αὐχήν,* the neck]. The neck or throat, or the constricted part of any organ.

**aucheniatria** (*aw-ken-e-at'-re-ah*) [*αὐχήν,* the throat; *ιατρεία,* a healing]. The therapy of throat diseases.

**audiclave** (*aw'-dik-lav*), An instrument for aiding hearing.

**audiometer** (*aw-de-om'-et-er*) [*audire,* to hear; *μέτρον,* a measure]. An instrument for measuring the acuteness of hearing.

**audiometry** (*aw-de-om'-et-re*) [*audire,* to hear; *μέτρον,* a measure]. The measurement, or testing, of the sense of hearing.

**audiphone** (*aw'-dif-ōn*) [*audire,* to hear; *φωνή,* a sound]. An instrument for improving the power of hearing by conveying sounds through the bones of the head to the labyrinth.

**audition** (*aw-dish'-un*) [*audire,* to hear]. The act of hearing. Syn., *acoesis; acousia; acusis.* **a. colorée, color-hearing,** a peculiar association between the auditory and optic nerves, by which a certain sound or musical note will give rise to a subjective sensation of color, the same note in the same person being always associated with the same color. Syn., *chromatic audition.* **a. contre,** the perception by one ear of the vibrations of a tuning-fork placed on the mastoid process on the other side.

**auditory** (*aw'-dit-o-re*) [see *audition*]. Pertaining to the act or the organs of hearing. **a. after-sensations,** the sensations of sounds continuing or occurring after the cessation of the stimulus. **a. amnesia.** See *mind-deafness.* **a. area,** the cerebral center for hearing, probably located in the temporosphenoidal lobe. **a. aura,** an auditory sensation preceding an attack of epilepsy. **a. capsule,** the primitive auditory organ, formed by the invagination of the nervous stratum of the epiblast. **a. center.** Same as *a. area.* **a. dysesthesia.** Same as *dysacusis; q. v.* **a. eminence,** the prominent part of the floor of the fourth ventricle, lying between the inferior and superior fovea. **a. field,** the area within which a sound may be heard. **a. hairs,** the processes of the crista acustica. **a. meatus** (external and internal), the external and internal canals or openings of the ear. **a. nerve,** the eighth cranial nerve, supplying the internal ear; formerly the *portio mollis* of the seventh pair of cranial nerves. **a. nuclei,** the nuclei in the oblongata giving rise to the auditory nerves. **a. ossicles,** the chain of small bones of the middle ear. **a. pit,** the depression in the epiblast on both sides of the embryonic after-brain, destined to form the labyrinth of the ear. **a. teeth,** tooth-like tubercles in the cochlea of the ear. **a. vertigo,** dizziness due to pathological conditions of the ear. See *Ménière's disease.* **a. vesicle,** the ectodermal sac from which is developed the membranous labyrinth.

**auditus** (*aw-di'-tus*) [L.]. Hearing; the sense or power of hearing.

**Auenbrugger's sign** (*ow'-en-broog-er*) [Leopold *Auenbrugger,* Austrian physician, 1722–1809]. Bulging of the epigastric region in cases of extensive pericardial effusion.

**Auer's bodies** (*ow'-er*) [John *Auer,* American physician, 1875– ]. Rod-like bodies seen in the lymphocytes in leukemia.

**Auerbach's ganglia** (*ow'-er-bakh*) [Leopold *Auerbach,* German anatomist, 1828–1897]. The ganglionic nodes in Auerbach's plexus. **A.'s plexus,** plexus myentericus, a nerve-plexus found between the circular and longitudinal muscular coats of the stomach and intestine, and consisting of a network of pale nerve-fibers, at the nodal points of which minute ganglia exist.

**Aufrecht's sign** (*ow'-frekht*) [Emanuel *Aufrecht,* German physician, 1844– ]. Short and feeble breathing heard just above the jugular fossa on placing the stethoscope over the trachea; it is noted in tracheal stenosis.

**augment,** (*awg'-ment*) [*augmentum,* increase]. The increasing stage of a fever or other acute disease.

**augmentation** (*awg-men-ta'-shun*) [*augmentatio,* an increasing]. 1. Same as *augment.* 2. Increase in the violence of symptoms.

**augmentor** (*awg-men'-tor*). An agent which increases or accelerates the action of auxetics; by itself it is unable to produce cell division. See *auxetic.*

**augnathus** (*awg-nā'-thus*) [*αὖ,* besides; *γνάθος,* the jaw]. A monster with two lower jaws.

**aula** (*aw'-lah*) [*αὐλή,* a hall or open court]. The common mesal cavity of the cerebrum, it being also the anterior portion of the third ventricle.

**aulatela** (*aw-lat-e'-lah*) [*aula,* a hall; *tela,* a web]. The roof or covering membrane of the aula.

**aulic** (*aw'-lik*) [*aula,* a hall]. Belonging or pertaining to the aula. **a. recess,** a triangular depression between the precommissure and the two forniculoms of the brain.

**auliplexus** (*aw-le-pleks'-us*) [*aula,* hall; *plexus,* a network]. The choroid plexus of the aula.

**aulix** (*aw'-liks*) [*αὖλιξ,* a furrow]. The sulcus of Monro, a groove on the mesal surface of the thalamus just ventrad of the medicommissure.

**aulophyte** (*aw'-lo-fīt*) [*αὐλός,* pipe or tube; *φυτόν,* a plant]. A symbiotic plant; one that lives within another, but not as a parasite.

**aura** (*aw'-rah*). [*αὔρα,* a breath]. A breath of wind; a soft vapor. The phenomenon preceding an attack of epilepsy. It may be motor, sensory, vasomotor, secretory, or psychic. It is also applied to the symptom preceding an attack of any disease or paroxysm, as the *aura hysterica, aura vertiginosa,* etc. **a., electric,** the current of air that attends the discharge of electricity from a point. **a., epigastric,** a localized epileptic aura.

**aurade, auradin** (*aw'-rād, aw'-rad-in*). A fatty body obtained from oil of orange-flowers. It crystallizes in tasteless, pearly, odorless scales, melting at 131° F.; soluble in water, insoluble in alcohol. Syn., *Neroli camphor.*

**aural** (*aw'-ral*) [*auris,* the ear]. 1. Relating to the ear or to hearing. 2. [*aura.*] Relating to the air or to an aura. **a. vertigo.** See *Ménière's disease.*

**auramine** (*aw'-ram-ēn*) [*aurum,* gold; *amine*]. Yellow pyoktanin, a yellow aniline color used to some extent as an antiseptic.

**aurantia** (*aw-ran'-she-ah*) [*aurantium*]. 1. An orange coal-tar dye; an ammonium salt of hexanitrodiphenylamine. 2. An orange or oranges.

**aurantiamarin** (*aw-ran-te-am'-ar-in*). A bitter glucoside obtained from orange peel.

**aurantin** (*aw-ran'-tin*). See *heptane.*

**aurantium** (*aw-ran'-she-um*) [L.; gen., *aurantii*]. Orange. The fruit of *Citrus vulgaris* and *C. aurantium.* Both the flowers and the rind of the fruit are employed in medicine. **aurantii amari cortex** (U. S. P.), bitter orange-peel. **aurantii amari, fluidextractum** (U. S. P.), bitter orange-peel, alcohol, and water. It is used as a flavor. Dose ½–1 dr. (2–4 Cc.). **aurantii amari, tinctura** (U. S. P.), bitter orange-peel, 20; dilute alcohol, q. s. ad 100. Dose 1–2 dr. (4–8 Cc.). **aurantii corticis, oleum** (U. S. P.), the volatile oil expressed from the rind of the orange; it is aromatic and a mild tonic, but is used mainly as a flavor. Dose 1–5 drops. **aurantii dulcis cortex** (U. S. P.), sweet orange-peel. **aurantii dulcis, tinctura** (U. S. P.), sweet orange-peel, 20; dilute alcohol, q. s. ad 100. Dose 1–2 dr. (4–8 Cc.). **aurantii, elixir,** oil of orange-peel, 1; sugar, 100; alcohol and water, q. s. ad 300. **aurantii florum, aqua** (U. S. P.), stronger orange-flower water and distilled water, of each, 1 volume, **aurantii florum fortior, aqua** (U. S. P.), water saturated with the volatile oil of fresh orange-flowers. **aurantii florum, oleum,** oil of neroli, a volatile oil distilled from fresh orange-flowers. Dose 1–5 drops. **aurantii florum, syrupus** (U. S. P.), sugar, 85; orange-flower water, sufficient to make 100 parts. A common flavoring agent. **aurantii, infusum** (B. P.). Dose 1–2 oz. (30–60 Cc.). **aurantii, infusum, compositum** (B. P.). Dose 1–2 oz. (30–60 Cc.). **aurantii, spiritus,** oil of orange-peel, 5; deodorized alcohol, 95. Dose according to quantity of alcohol desired. **aurantii, spiritus, compositus** (U. S. P.), oil of orange-peel, 20; oil of lemon, 5; oil of coriander, 2; oil of anise, 5; deodorized alcohol, sufficient to make 100 parts. **aurantii, syrupus** (U. S. P.), tincture of sweet orange-peel, 5; citric acid, 0.5; magnesium carbonate, 1; sugar, 82; water sufficient to make 100 parts. **aurantii, tinctura** (B. P.). Dose 1–2 dr. (4–8 Cc.). **aurantii, tinctura, recentis** (B. P.), tincture of fresh orange-peel. Dose 1–2 dr. (4–8 Cc.). **aurantii, vinum** (B. P.), contains 12 % of alcohol.

**aureol** (*aw-re'-ol*). The commercial name of a hair-dye said to contain menthol, 1 %; amidophenolchlorhydrate, 0.3 %; monoamido-diphenylamine, 0.6 %; dissolved in 50 % alcohol which contains 0.5 % sodium sulphite.

**aureola** (*aw-re'-o-lah*). See *areola* (1).

**aureolin** (*aw-re'-o-lin*) [*aurum,* gold]. A yellow pigment obtained by heating paratoluidin with sulphur and treating with fuming sulphuric acid. Syn., *carnotine; polychromin; primulin yellow; sulphine; thiochromogen.*

**auric** (*aw'-rik*) [*aurum,* gold]. 1. Pertaining to aurum or gold. 2. Referring to gold in chemical combination as a triad. **a. acid.** See *acid, auric.*

**auricle** (*aw'-rik-l*) [*auricula,* the ear]. 1. The expanded portion or pinna of the ear. 2. One of the upper chambers of the heart receiving the blood from the lungs (*left auricle*) or from the general circulation (*right auricle*). 3. An ear-shaped appendage. 4. A kind of ear-trumpet. **a., cervical,** congenital cartilaginous remains of the neck, arising about the middle of the sternomastoid as symmetrical bodies, occurring in man occasionally and almost constantly present in the goat.

**auricoammonic** (*aw-rik-o-am-on'-ik*). Containing gold and ammonium.

**auricobarytic** (*aw-rik-o-bar-it'-ik*). Containing gold and barium.

**auricula** (*aw-rik'-ū-lah*) [dim. of *auris,* ear].

1. *Auricle, q. v.* 2. The auricular appendix, a pouch-like appendage to the auricles of the heart.

**auricular** (*aw-rik'-ū-lar*) [see *auricle*]. 1. Relating to the auricle of the ear. 2. Pertaining to the auricles of the heart, as *auricular* appendix. 3. Relating to the auricular nerve, arteries, veins, etc. **a. appendix,** the anterior prolongation of the cardiac auricle. **a. finger,** the little finger. **a. point,** the central point of the external auricular meatus.

**auriculare** (*aw-rik-ū-la'-re*) [*auricularis,* pertaining to the ear]. The *auricular point, q. v.*

**auricularis** (*aw-rik-ū-la*″*-ris*) [see *auricle*]. 1. Auricular. 2. The extensor minimi digiti. See under *muscle.* **a. magnus,** a branch of the cervical plexus of nerves.

**auriculate, auriculated** (*aw-rik'-ū-lat, -ed*). Furnished with ears or ear-like appendages; auricled.

**auriculocranial** (*aw-rik-ū-lo-kra'-ne-al*). Pertaining to both the auricle and the cranium.

**auriculooccipital** (*aw-rik-ū-lo-ok-sip'-it-al*) [*auricula,* the ear; *occiput,* the back of the head]. Pertaining both to the ear and the back of the head. **a., triangle.** See *triangle.*

**auriculotemporal** (*aw-rik-ū-lo-tem'-po-ral*) [*auricle; tempus,* the temple]. Relating to the auricle and to the temporal region. **a. nerve,** a branch of the inferior maxillary, supplying superficial parts about the auricle and temple.

**auriculoventricular** (*aw-rik-ū-lo-ven-trik'-ū-lar*) [*auricle; ventriculus,* the ventricle]. Relating to an auricle and a ventricle of the heart. **a. bundle,** the bundle of His. **a. opening, the opening between the auricles and the ventricles of the heart.

**auriform** (*aw'-rif-orm*) [*auris,* the ear; *forma,* shape]. Ear-shaped.

**auriginous** (*aw-rij'-in-us*). 1. Having the color of gold. 2, Relating to jaundice.

**aurilave** (*aw'-ril-av*) [*auris,* the ear; *lavare,* to wash]. An appliance for cleansing the ears. An ear-brush or ear-sponge mounted upon a handle.

**aurinasal** (*aw-re-na'-sal*) [*auris; nasus,* nose]. Pertaining to the ear and the nose.

**auripuncture** (*aw'-re-punk-chur*) [*auris; punctura*]. Puncture of the membrana tympani.

**auris** (*aw'-ris*) [L.]. The ear. **a. externa,** the outer ear, auricle, pinna. **a. interna, a. intima,** the internal ear, labyrinth. **a. media,** the middle ear, tympanum.

**auriscalp** (*aw'-ris-kalp*) [*auris,* the ear; *scalpare,* to scrape]. An instrument for cleansing the ear. An ear-pick, or probe for the ear.

**auriscope** (*aw'-ris-kōp*) [*auris;* σκοπεῖν, to examine]. An instrument for examining the ear, and especially the Eustachian passage: an otoscope.

**aurist** (*aw'-rist*) [*auris*]. A specialist in diseases of the ear.

**aurobromide** (*aw-ro-bro'-mīd*). Gold and potassium bromide.

**aurous** (*aw'-rus*) [*aurum,* gold]. 1. Pertaining to gold and its compounds. 2. Referring to gold in chemical combination as a monad.

**aurum** (*aw'-rum*) [L.; gen., *auri*], Gold. Au =197.2; quantivalence III. A brilliant yellow metal, having a specific gravity of 19.3. It is soluble in a mixture of nitric and hydrochloric acids. **auri bromidum,** AuBr₃, used in epilepsy and migraine. Dose ⅛–⅓ gr. (0.003–0.01 Gm.). **auri chloridum,** gold chloride. Dose $\frac{1}{60}$–$\frac{1}{30}$ gr. (0.001–0.002 Gm.). Also used as a stain for nerve tissue. **auri et sodii chloridum** (U. S. P.), the double chloride of gold and sodium. It is used as an alterative in chronic inflammations, diabetes, in the treatment of the alcohol habit, etc. Dose $\frac{1}{30}$–$\frac{1}{10}$ gr. (0.002–0.006 Gm.). **a. vegetabile,** saffron.

**auscult, auscultate** (*aws-kul', aws'-kul-tāt*) [*auscultare,* to listen to]. To perform or practise auscultation; to examine by auscultation.

**auscultation** (*aws-kul-ta'-shun*) [see *auscult*]. A method of investigation of the functions and conditions of the respiratory, circulatory, digestive, and other organs by the sounds they themselves give out or that are elicited by percussion. It is called *immediate,* when the ear is directly applied to the part, and *mediate,* if practised by the aid of the stethoscope. *Obstetric auscultation* is practised in pregnancy to detect or study the fetal heart-sounds or the placental murmur. **a.-tube,** in otology, an instrument for listening to the forced passage of air through the ear of another.

**auscultatory** (*aws-kul'-ta-to-re*) [see *auscult*]. Relating to auscultation. a. percussion, the practice of listening with the stethoscope to the sounds produced by percussing a part.

**auscultoscope** (*aws-kult'-o-skōp*). Stethoscope, or phonendoscope.

**autacoid** (*aw'-tak-oid*) [*auto-;* ἄκος, remedy]. A general term for all internal secretions, it includes *hormones* and *chalones, q. v.*

**autan** (*aw'-tan*). Trade name of a preparation said to be a mixture of paraformaldehyde and barium dioxide. It is used to disinfect rooms.

**autechoscope** (*aw-tek'-o-skōp*) [αὐτός, self; ἦχος, sound; σκοπεῖν, to inspect]. A device for enabling a person to listen to sounds produced within his own body.

**autecic, autœcic** (*aw-te'-sik*). See *autecious.*

**autecious, autœcious** (*aw-te'-shus*) [αὐτός, self; οἶκος, dwelling]. Applied to parasitic fungi that pass through all the stages of their existence in the same host.

**autemesia** (*aw-tem-e'-zhe-ah*) [αὐτός, self; ἐμεῖν, to vomit]. Vomiting without manifest cause.

**auto-** (*aw-to-*) [αὐτός, self]. A prefix meaning self, of itself.

**autoactivation** (*aw-to-ak-tiv-a'-shun*) [*auto-; activate*]. The activation of a gland by an enzyme or hormone derived from itself.

**autoanticomplement** (*aw-to-an-te-kom'-ple-ment*). An anticomplement, formed within the body, which is capable of neutralizing its own complements.

**autoaudible** (*aw-to-awd'-i-bl*) [*auto-; audire,* to hear]. Applied to cardiac sounds audible to the patient.

**autoblast** (*aw'-to-blast*) [*auto-;* βλαστός, a germ]. An independent bioblast.

**autocatheterism** (*aw-to-kath'-et-er-izm*) [*auto-; catheter*]. The passage of a catheter by a person upon himself.

**autochthon** (*aw-tok'-thon*) [αὐτόχθων, sprung from the land]. An aboriginal inhabitant.

**autochthonous** (*aw-tok'-thon-us*) [see *autochthon*]. Aboriginal; formed (as, *e. g.,* a clot) in the place where it is found.

**autocinesis** (*aw-to-sin-e'-sis*). See *autokinesis.*

**autocinetic** (*aw-to-sin-et'-ik*). See *autokinetic.*

**autoclasis** (*aw-tok'-la-sis*) [*auto-;* κλάσις, breaking]. A breaking up of a part due to causes developed within itself.

**autoclave** (*aw'-to-klāv*) [*auto-; clavis,* a key]. 1. Self-fastening; closing itself. 2. An apparatus for sterilizing objects by steam-heat at high pressure. 3. To sterilize in an autoclave.

**autoconduction** (*aw-to-kon-duk'-shun*) [*auto-; conduction*]. A term used in electrotherapy for a method of using high-frequency currents. The patient or part to be acted upon is placed inside of the solenoid, without any direct connection with any part of the circuit.

**autocystoplasty** (*aw-to-sis'-to-plas-te*) [*auto-;* κύστις, bladder; πλάσσειν, to form]. Plastic surgery of the bladder with grafts from the patient's body.

**autocytolysin** (*aw-to-si-tol'-is-in*). Same as *autolysin.*

**autocytotoxins** (*aw-to-si-to-toks'-ins*) [*auto-; cytotoxin*]. Cytotoxins produced in the body of the individual by abnormal retention and absorption of the products of degenerated and dead cells.

**autodidact** (*aw'-to-di-dakt*) [*auto-;* διδακτός, taught]. One who is self-taught.

**autodigestion** (*aw-to-di-jes'-chun*) [*auto-; digerere,* to digest]. Digestion of the walls of the stomach by the gastric juice, from disease of the stomach.

**autofundoscope** (*aw-to-fun'-do-skōp*) [*auto-; fundus,* the bottom; σκοπεῖν, to look]. An instrument for self-examination of the vessels about the macular region of the eye.

**autogamous** (*aw-tog'-am-us*) [*auto-;* γάμος, marriage]. In botany, a name applied to flowers that are habitually self-fertilizing.

**autogamy** (*aw-tog'-am-e*). [See *autogamous.*] Self-fertilization.

**autogenesis** (*aw-to-jen'-es-is*) [*auto-;* γένεσις, production]. Spontaneous generation; self-production.

**autogenetic** (*aw-to-jen-et'-ik*) [see *autogenesis*]. Produced within the organism.

**autogenous** (*aw-toj'-en-us*) [see *autogenesis*]. 1. Pertaining to diseases or conditions self-produced within the body and not derived from external sources; applied to poisons generated in the body by its inherent processes. 2. Having a distinct center of development, as parts of bones. a. hemorrhage, hemorrhage due to causes residing within the body; not traumatic. a. vaccine, one derived from the microorganism infecting the person to be immunized, as opposed to *stock vaccines* which are made from standard cultures.

**autognosis** (*aw-tog-no'-sis*) [*auto-;* γνῶσις, knowledge]. Knowledge obtained by self-observation.

**autogony** (*aw-tog'-o-ne*) [αὐτόγονος, self-produced]. The rise of the simplest protoplasmic substance in a formative fluid.

**autographic** (*aw-to-graf'-ik*) [*auto-;* γράφειν, to write]. Self-registering. a. skin, a condition of vasomotor paralysis, usually in hysterical patients, in which markings made upon the skin form quite persistent and intensely red traces. a. woman, one with an autographic skin.

**autographism** (*aw-tog'-raf-izm*) [*auto-;* γράφειν, to write]. The condition observed in the so-called autographic skin; dermographism. See *urticaria factitia.*

**autohypnotic** (*aw-to-hip-not'-ik*). 1. Relating to autohypnotism. 2. An individual who can put himself into a hypnotic state.

**autohypnotism** (*aw-to-hip'-not-izm*) [*auto-;* ὕπνος, sleep]. Mental stupor induced by dwelling intensely upon some all-absorbing thought.

**autoimmunization** (*aw-to-im-u-ni-za'-shun*) [*auto-; immunization*]. Immunization obtained by natural processes at work within the body.

**autoinfection** (*aw-to-in-fek'-shun*) [*auto-; infection*]. Infection by virus originating within the body or transferred from one part of the body to another.

**autoinfusion** (*aw-to-in-fu'-shun*) [*auto-; infundere,* to pour in]. Compulsion of the blood to the heart by bandaging the extremities, compression of the abdominal aorta, etc.

**autoinoculable** (*aw-to-in-ok'-u-la-bl*) [*auto-; inoculare,* to implant]. Capable of being inoculated upon the person already infected. Chancroid is autoinoculable.

**autoinoculation** (*aw-to-in-ok-u-la'-shun*) [see *autoinoculable*]. Inoculation in one part of the body by virus present in another part; self-inoculation.

**autointoxication** (*aw-to-in-toks-ik-a'-shun*) [*auto-;* τοξικόν, a poison]. Poisoning by faulty metabolic products elaborated within the body; autoinfection. a., endogenous, that due to the action of excessive unneutralized or modified discharges from the cells of any tissue acting upon the other tissues without previous discharge from the body; or that due to the action of products of decomposition and necrosis of any tissue acting in a similar manner; or that due to microendoparasites or macroendoparasites. a., exogenous, that due to the action of poisons entering the system from without, through the skin, the digestive, the respiratory or genitourinary tract, as by the absorption of retained excreta, or of decomposition- and fermentation-products developed in the external secretions through the action of those secretions. a., indirect, that caused by the absorption of retained excrements.

**autoisolysin** (*aw-to-is-ol'-is-in*) [*auto-;* ἴσος, equal; λύσις, a loosing]. A serum which dissolves the corpuscles of the individual from which it was obtained and also those of another individual of the same species.

**autokinesis** (*aw-to-kin-e'-sis*) [*auto-;* κίνησις, movement]. Voluntary movement.

**autokinetic** (*aw-to-kin-et'-ik*) [see *autokinesis*]. Pertaining to, or of the nature of, autokinesis.

**autolaryngoscopy** (*aw-to-lar-ing-gos'-ko-pe*) [*auto-;* λάρυγξ, the larynx; σκοπεῖν, to examine]. The examination of one's own larynx.

**autolavage** (*aw-to-lav'-ahj*) [*auto-; lavage*]. The washing out of one's own stomach.

**autolysate** (*aw-tol'-is-āt*) [see *autolysin*]. That which results from or is produced by autolysis.

**autolysin** (*aw-tol'-is-in*) [*auto-;* λύσις, a loosing]. A lysin capable of dissolving the red blood-corpuscles of the animal in the serum of which it circulates.

**autolysis** (*aw-tol'-is-is*) [see *autolysin*]. 1. Self-digestion of tissues within the living body. 2. The chemical splitting-up of the tissue of an organ by the action of an enzyme peculiar to it. 3. The hemolytic action of the blood-serum of an animal upon its own corpuscles.

**autolytic** (*aw-to-lit'-ik*). Relating to autolysis.

**automatic** (*aw-to-mat'-ik*) [αὐτομαΐζειν, to act spontaneously]. Performed without the influence of the will.

**automatism** (aw-tom'-at-izm) [see *automatic*]. The performance of acts without apparent volition, as seen in certain somnambulists and in some hysterical and epileptic patients. a., **epileptic**. See *automatism*.

**automatograph** (aw-to-mat'-o-graf) [αὐτοματίζειν, to act spontaneously; γραφεῖν, to record]. An instrument for registering involuntary movements.

**automaton** (aw-tom'-at-on) [αὐτόματος, spontaneous]. One who acts in an involuntary or mechanical manner.

**automixis** (aw-to-miks'-is) [auto-; ˆμίξις, mixture]. Same as *autogamy*.

**automysophobia** (aw-to-mis-o-fo'-be-ah) [auto-; μύσος, filth; φόβος, fear]. Insane dread of personal uncleanliness.

**autonephrectomy** (aw-to-nef-rek'-to-me) [auto; νέφρος, kidney; ἐκτομή, excision]. Complete stricture of the ureter so that no urine flows from the kidney to the bladder.

**autonomic, autonomous** (aw-ton-om'-ik, -ton'-om-us) [auto-; νόμος, law]. Independent in origin, action, or function; self-governing. a., **nervous system,** the sympathetic nervous system supplying involuntary muscle fibers, secreting glands, and arterioles.

**autonomy** (aw-ton'-o-me) [see *autonomous*]. Independence.

**autoophthalmoscope** (aw-to-of-thal'-mo-skōp). See *autophthalmoscope*.

**autopathic** (aw-to-path'-ik) [auto-; πάθος, suffering]. The same as endopathic or idiopathic.

**autopepsia** (aw-to-pep'-se-ah) [auto-; πέπτειν, to digest]. Autodigestion.

**autophagia** (aw-to-fa'-je-ah) [auto-; φαγεῖν, to eat]. 1. Self-consumption; emaciation. 2. The biting of one's own flesh.

**autophagy** (aw-tof'-a-je). See *autophagia*.

**autophilia** (aw-to-fil'-e-ah) [auto-; φιλεῖν, to love]. Morbid self-esteem.

**autophobia** (aw-to-fo'-be-ah) [auto-; φόβος, fear]. A morbid dread of one's self or of solitude.

**autophonia** (aw-to-fo'-ne-ah). 1. See *autophony*. 2. [auto-; φόνος, murder]. Suicide.

**autophonomania** (aw-to-fo-no-ma'-ne-ah) [αὐτοφονία, suicide; μανία, madness]. Suicidal mania.

**autophonous** (aw-tof'-on-us) [auto-; φωνή, voice]. Having the character of autophony.

**autophony** (aw-tof'-o-ne) [see *autophonous*]. 1. The auscultation of the physician's own voice through the patient's chest. 2. The condition in which one's own voice appears changed. It may be due to chronic inflammation of the ear or to other causes.

**autophthalmoscope** (aw-toff-thal'-mo-skōp). An ophthalmoscope for examining one's own eye.

**autophthalmoscopy** (aw-tof-thal-mos'-ko-pe) [auto-; ὀφθαλμός, the eye; σκοπεῖν, to see]. Examination of one's own eye with the ophthalmoscope.

**autoplasty** (aw'-to-plas-te) [auto-; πλάσσειν, to form]. A method of repairing the effects of a wound or lesion involving loss of tissue by grafting or implanting fresh parts taken from other portions of the patient's body.

**autopsy** (aw'-top-se) [auto-; ὄψις, a seeing]. The postmortem examination.

**autopsychorrhythmia** (aw-to-si-kor-rith'-me-ah) [auto-; ψυχή, mind; ῥυθμός, rhythm]. A morbid rhythmic activity of the brain; it is a symptom of grave insanity.

**autoscope** (aw'-to-skōp) [auto-; σκοπεῖν, to see]. An instrument arranged for the examination of one's own organs by one's self.

**autoscopy** (aw-tos'-ko-pe) [see *autoscope*]. The examination of one's own organs by means of an autoscope.

**autoserotherapy** (aw-to-se-ro-ther'-ap-e) [auto-; serum; therapy]. Treatment of a disease (such as pleurisy) by means of a serum obtained from the patient himself.

**autoserum** (aw-to-se'-rum) [auto-; serum]. A therapeutic serum which is obtained from the patient on whom it is used.

**autosite** (aw'-to-sīt) [auto-; σῖτος, food]. 1. A monster capable of an independent existence after birth. 2. That member of a double fetal monstrosity that nourishes itself by its own organs and also the other member, which is called the parasite.

**autositic** (aw-to-sit'-ik) [see *autosite*]. Of the nature of an autosite.

**autospermotoxin** (aw-to-spurm-o-toks'-in) [auto-; σπέρμα, seed; τοξικόν, poison]. A specific substance produced in the blood-serum of an animal by intravenous injection of spermatozoa of another animal, and which renders the serum of the treated animal toxic for the spermatozoa of both.

**autosterilization** (aw-to-ster-il-iz-a'-shun) [auto-; sterilization]. Sterilization effected by the normal fluids of the body.

**autostethoscope** (aw-to-steth'-o-skōp) [auto-; στῆθος, the chest; σκοπεῖν, to examine]. A stethoscope so arranged that by it one may listen to his own chest-sounds.

**autosuggestibility** (aw-to-suj-es-tib-il'-it-e). That mental state with loss of will, in which auto suggestion easily occurs.

**autosuggestion** (aw-to-suj-es'-chun) [auto-; suggestio, an intimation]. A peculiar mental condition, often developing after accidents, especially railway accidents; it is intimately associated with the hypnotic state. In both of these conditions the mental spontaneity, the will, or the judgment is more or less suppressed or obscured, and suggestions become easy. Thus the slightest traumatic action directed to any member may become the occasion of a paralysis, of a contracture, or of an arthralgia. Syn., *traumatic suggestion*.

**autotemnous** (aw-to-tem'-nus) [auto-; self; τέμνειν, to cut]. Capable of spontaneous division.

**autotherapy** (aw-to-ther'-a-pe). [auto-; θεραπεία, treatment]. The spontaneous or self-cure of a disease.

**autotomy** (aw-tot'-o-me) [auto-; τομή, a cutting]. 1. Self-division; fission. 2. The performance of a surgical operation upon one's own body.

**autotoxemia** (aw-to-toks-e'-me-ah) [auto-; τοξικόν, a poison; αἷμα, blood]. Toxemia from poisons derived from the organism itself.

**autotoxicosis** (aw-to-toks-ik-o'-sis) [auto-; τοξικόν, poison]. The symptoms due to autotoxemia.

**autotoxin** (aw-to-toks'-in) [auto-; τοξικόν, a poison]. Any poisonous product of tissue-metamorphosis.

**autotoxis** (aw-to-toks'-is) [auto-; τοξικόν, poison]. Self-poisoning through the absorption of noxious products of katabolism, as in uremia. Cf. *autointoxication*.

**autotransfusion** (aw-to-trans-fū'-shun) [auto-; *transfusio*, a pouring-out or forth]. The transfer of the blood to the brain and other central organs by elevating the hips and legs and by the use of elastic bandages compressing the limbs.

**autotransplantation** (aw-to-trans-plan-ta'-shun) [auto-; transplantation]. The operation of transplanting to a part of the body tissue taken from another part of the same body.

**autotrophic** (aw-to-trof'-ik) [αὐτός, self; τροφή, nutrition]. Self-nourishing. A term applied to those forms of bacteria which do not require organic carbon and nitrogen, but are able to form carbohydrates and protein out of carbon dioxide and inorganic salts.

**autotuberculin** (aw-to-tū-ber'-kū-lin). Tuberculin prepared from a patient's own sputum.

**autotyphization** (aw-to-ti-fiz-a'-shun) [auto-; typhoid]. The production of a condition resembling typhoid fever from faulty elimination of waste-material.

**autovaccination** (aw-to-vaks-in-a'-shun) [auto-; *vaccinare*, to vaccinate]. The reinsertion of fresh vaccine lymph upon the same person from whom it is taken.

**autumn catarrh.** Synonym of *hay-fever*, since it is apt to occur in the autumn or the fall of the year.

**autumnal** (aw-tum'-nal) [autumn]. Pertaining to the fall of the year. a. **fever.** Synonym of *typhoid fever*.

**auxanogram** (awks-an'-o-gram) [αὔξανειν, to grow; γράφειν, to write]. A pure plate culture of microbes which has been prepared by Beyerinck's auxanographic method in which the colonies indicate which one of several nutrient media is best suited to their growth.

**auxanographic** (awks-an-o-graf'-ik). Pertaining to auxanography.

**auxanography** (awks-an-og'-ra-fe). A method devised by Beyerinck for ascertaining the nutrient mediums suitable for a growing microbe. Plate cultures of poor mediums (e. g., 10 % gelatin or 2 % agar in distilled water) are stippled with drops of solutions the nutrient properties of which are to be tested. The species of microbe under examination will then develop strong colonies only on those spots where the requisite pabulum is present.

**auxanology** (*awks-an-ol'-o-je*) [αὔξανειν, to grow; λόγος, science]. The study of growth.

**auxanometer** (*awks-an-om'-et-er*) [αὔξάνειν, to grow; μέτρον, a measure]. An instrument used in biological study for measuring the growth of young organisms.

**auxe** (*awks'-e*) [αὔξη, increase]. Enlargement in bulk or volume.

**auxesis** (*awks-e'-sis*) [αὔξησις, enlargement]. Increase in size or bulk. Hypertrophy is a word often incorrectly used where auxesis is meant.

**auxetic** (*awks-et'-ik*) [See *auxesis*]. 1. Characterized by auxesis. 2. Increase in size or bulk. 3. An exciter of reproduction; an agent which causes proliferation of human cells, especially leukocytes. See *in vitro*.

**auxiliary** (*awks-il'-e-a-re*) [*auxilium*]. 1. Aiding. 2. An adjuvant. **auxiliaries of respiration**, those muscles brought into action in difficult respiration.

**auxilium** (*awks-il'-e-um*) [L., "help"]. A wheeled vehicle or ambulance with couch and mattresses, for use in the service of field military hospitals.

**auxocardia** (*awks-o-kar'-de-ah*) [αὔξη, an increase; καρδία, the heart]. The normal increase of the volume of the heart during diastole, in distinction from *meiocardia*, the diminution during systole.

**auxochrome** (*awks'-o-krōm*) [αὔξειν, increase; χρῶμα, color]: 1. That which increases color. 2. A term applied to a chemical group which, if added to a chromophore group will produce a dye.

**auxocyte** (*awks'-o-sīt*) [αὔξειν, to increase; κύτος, a cell]. A cell which is concerned in growth or reproduction.

**auxometer** (*awks-om'-et-er*) [αὔξειν, to grow; μέτρον, a measure]. 1. A device for estimating the magnifying power of lenses. 2. See *auxanometer*. 3. A dynamometer. Syn., *auxemeter; auxenometer; auxesimeter; auxiometer; auzometer*.

**auxospore** (*awks'-o-spōr*) [αὔξειν, to grow; σπόρος, seed, offspring]. A large spore produced, either asexually, or by conjugation, in the *Diatomaceæ*.

**auxotonic** (*awks-o-ton'-ik*) [αὔξειν, to grow; τόνος, tension]. Determined by growth. a. movements, movements due to growth rather than to stimulation.

**auzometer** (*aw-zom'-et-ur*). See *auxometer*.

**Av.** Abbreviation for avoirdupois weight; see *weights and measures*.

**ava, ava-kava** (*ah'-vah, ah-vah-kah'-vah*). See *kava-kava*.

**avaism** (*ah'-vah-izm*). A malady from abuse of kava, resembling absinthism.

**avalanche theory**. Pflüger's theory that nerve-energy gathers intensity as it passes toward the muscles.

**avalent** (*ah-va'-lent*) [ā, priv.; *valence*]. Without valency.

**avalvular** (*ah-val'-vū-lar*) [ā, priv.; *valvula*, a valve]. Lacking valves.

**avascular** (*ah-vas'-kū-lar*) [ā, priv.; *vas*, a vessel]. Without blood; not possessing blood-vessels.

**avascularization** (*ah-vas-kū-lar-iz-a'-shun*). The act of rendering a part bloodless, as by compression or bandaging.

**avascularize** (*ah-vas'-kū-lar-īz*). To render bloodless.

**Avellis' symptom-complex** (*ah-vel'-lis*) [Georg *Avellis*, German laryngologist, 1864— ]. Paralysis of one-half of the soft palate, associated with a recurrent paralysis on the same side.

**Avena** (*av-e'-nah*) [L.]. A genus of plants. Oats. **Avena farina**, oatmeal. **A. sativa**, the embryo of the seed of the common oat-plant. It contains starch, gluten, a ferment called diastase, and a small amount of alkaline phosphates, and is a nutritious food. Dose of the concentrated *tincture* or *fluidextract* 10 min.–2 dr. (0.65–8.0 Cc.). The pericarp contains an alkaloid possessed of slight narcotic powers.

**avenin** (*av-e'-nin*) [*avena*]. 1. A precipitate made from a tincture of *Avena sativa*, or the oat. It is a nerve-stimulant and tonic. 2. A nitrogenous principle obtained from the oat, and nearly identical with legumin; the gluten-casein of oats.

**avenious, avenous** (*ah-ve'-ne-us, ah-ve'-nus*) [ā, priv.; *vena*, vein]. Lacking veins.

**avenolith** (*av-en'-o-lith*) [*avena*; λίθος, stone]. An intestinal calculus formed around a grain of oats.

**aversion** (*av-ur'-shun*) [*avertere*, to turn aside]. 1. A turning aside, as in the displacement of an organ or in metastasis. 2. Nausea.

**avidity** (*av-id'-it-e*) [*avidus*, greedy]. In chemistry, the tendency of certain weak acids, in suitable conditions, to dispossess even the strongest acids and to unite with their bases.

**avirulent** (*ah-vir'-ū-lent*) [ā, priv.; *virus*, a poison]. Without virulence.

**avitaminosis** (*ah-vi-tam-in-o'-sis*) [ā, priv.; *vitamine*]. A disease resulting from deficiency of vitamines in the diet.

**Avogadro's law** [Amadeo *Avogadro*, Italian physicist, 1776–1856]. Equal volumes of all gases and vapors, at like temperature and pressure, contain an equal number of molecules.

**avoirdupois weight** (*av-or-du-pois'*). See *weights and measures*.

**avulsio, avulsion** (*av-ul'-se-o, -shun*) [*avellere*, to tear away]. A tearing or wrenching away of a part, as a polyp, a limb, etc. **a. bulbi**, avulsion of the bulb, separation of the pupil from its attachments in consequence of complete or almost complete rupture of the tendons of the optic muscles and nerves.

**axanthopsia** (*ah-zan-thop'-se-ah*) [ā, priv.; ξάνθος, yellow; ὄψις, vision]. Yellow-blindness.

**Axenfeld's test for albumin in urine** (*ahks'-en-felt*) [David *Axenfeld*, German physiologist]. Acidulate with formic acid and add, drop by drop, a 0.1 % solution of gold chloride, and warm. If albumin is present, the solution becomes red, then purplish, and on the addition of more gold chloride, blue. The blue color is also produced by glucose, starch, tyrosin, uric acid, urea, leucin, etc., but the red color is characteristic of albumin.

**axial** (*aks'-e-al*) [*axis*]. Pertaining to or situated in an axis. **a. current**, the column of red corpuscles which, by reason of the weight of the cells, occupies the center or axis of the blood-stream. **a. hyperopia**. See *hyperopia, axial*. **a. neuritis**, inflammation of a nerve axis. **a. stream**. See *a. current*.

**axifugal** (*aks-if'-ū-gal*) [*axis*; *fugere*, to flee]. Centrifugal.

**axilemma** (*aks-il-em'-ah*) [*axis*; λέμμα, husk; skin]. An elastic sheath composed of neurokeratin, inclosing the axis-cylinder of medullated nerve-fibers.

**axilla** (*aks-il'-ah*) [L.]. 1. The armpit. 2. The prominence of the shoulder.

**axillary** (*aks'-il-a-re*) [*axilla*]. Pertaining to the axilla. **a. artery**, the continuation of the subclavian artery, extending from the lower border of the first rib to the insertion of the pectoralis major muscle, where it becomes the brachial. See under *artery*. **a. glands**, the lymphatic glands in the axilla. **a. plexus**, the brachial plexus, formed by the last three cervical and the first dorsal nerves. **a. region**, **a. space**, the irregular conical space of the axilla. **a. vein**, a continuation of the brachial vein, corresponding with the artery and terminating in the subclavian vein.

**axin** (*aks'-in*) [*axinus*]. A fatty and varnish-like substance produced in Mexico by an insect, *Coccus axinus*. It is used in the arts and locally in medicine, being regarded as a good vulnerary and resolvent.

**axioplasm** (*aks'-e-o-plazm*). See *axoplasm*.

**axipetal** (*aks-ip'-et-al*) [*axis*; *petere*, to seek]. Centripetal; applied to the transmission of impulses toward an axone.

**axis** (*aks'-is*) [L., "axletree"]. 1. An imaginary line passing through the center of a body. 2. The second cervical vertebra. 3. A short artery which breaks up into several branches, e. g., thyroid *axis*, celiac *axis*. See under *artery*. **a., basicranial**, in craniometry, a line drawn from the basion to the middle of the anterior border of the cerebral surface of the sphenoid bone. **a., basifacial**, in craniometry, a line drawn from the anterior border of the cerebral surface of the sphenoid to the alveolar point. **a., binauricular**, in craniometry, the imaginary line joining the two auricular points. **a., brain**, the isthmus. **a. celiac**, same as *celiac artery*; see table of *arteries*. **a., cerebrospinal**, the central nervous system. **a.-cord**. See *primitive streak*. **a.-corpuscle**. See *corpuscle, axile*. **a., craniofacial**, in comparative anatomy the bones making the floor of the cranial cavity. **a.-cylinder**, the conducting or essential part of a nerve. Syn., *axis-cylinder of Purkinje*. **a.-cylinder process**, that one of the protoplasmic processes of a nerve-cell which becomes an axis-cylinder. **a., electric**, a line connecting the two poles of an electric body. **a., frontal** (of the eye), an imaginary line running through the eyeball from right to left, and corresponding with the movements of elevation and depression of the eyeball.

**a., hemal**, the aorta. **a., magnetic**, a line connecting the two poles of a magnet. **a. neural**, the cerebrospinal axis. **a., optic**. 1. The line from the center of the cornea to the macula lutea. 2. An imaginary line passing from the center of the eye-piece of a microscope through the body, objective, stage, and substage, to the mirror. **a., pelvic**, an imaginary line passing through all the median anteroposterior diameters of the pelvic canal at their centers. **a., sagittal** (of the eye), an imaginary line running through the eyeball from before backward, and coinciding with the line of vision. **a.-traction**, traction on the fetus in the axis of the pelvis. **a.-traction forceps**, a forceps for performing axis-traction. **a. uteri**. 1. The long diameter of the uterus. 2. A line imagined to pass transversely through the uterus near its junction with the cervix, on which it is said to turn in retroversion. **a., visual**, the line from the object through the nodal point to the macula.

**axite** (aks'-it) [axis]. Gowers' name for the terminal filaments of the axis-cylinder.

**axle teeth** (aks'-l tēth). See *azzle teeth*.

**axo-** (aks-o-) [axis]. A prefix meaning **axis**.

**axodendrite** (aks-o-den'-drīt) [axo-; δένδρον, a tree]. Lenhossék's term for a nonmedullated, axopetally conducting side fibril on the axons, as distinguished from a cytodendrite or one of the true medullated, cellulifugal collaterals.

**axofugal** (ak-so-fu'-gal) [axo-; *fugere*, to flee from]. Directed away from an axis cylinder process.

**axoid** (aks'-oid) [axo ; εἶδος, likeness]. 1. Shaped like a pivot. 2. Relating to the second cervical vertebra.

**axolemma**. See *axilemma*.

**axolysis** (aks-ol'-is-is) [axon; λύσις, solution]. Destruction of an axis cylinder.

**axometer** (aks-om'-et-ur) [axo-; μέτρον, measure]. An instrument used to adjust properly the axes of spectacles to the eyes.

**axon, axone** (aks'-on) [axis]. 1. The body-axis. 2. An unbranched nerve-cell process of the second order. See *dendrite*. 3. The cerebrospinal axis. 4. Kölliker's term for neurite. **a. degeneration**, disintegration and loss of function of the axis-cylinder. **a. hillock**, the pyramidal projection of the nerve-cell protoplasm from which the axon issues.

**axoneuron** (aks-o-nū'-ron) [axo-; νεῦρον, nerve]. A neuron the cell-body (nerve-cell) of which lies in the interior of the brain or the spinal cord. The axoneurons are classified as rhizoneurons and the endaxoneurons.

**axonometer** (aks-o-nom'-et-er) [axo-; μέτρον, a measure]. 1. An instrument used for locating the axis of astigmatism. 2. An apparatus for determining the axis of a cylinder.

**axopetal** (aks-op'-et-al). See *axipetal*.

**axoplasm** (aks'-o-plazm) [axis; πλάσμα, a thing molded]. Waldeyer's term for the delicate stroma of reticular substance holding together the fine fibrillæ of the axis-cylinders. Syn., *neuroplasm*.

**axospongium** (aks-o-spun'-je-um) [axo-; σπόγγος, a sponge]. Held's term for the reticular structure of the axis-cylinder.

**axungia** (aks-un'-je-ah) [L.]. Fat; lard; adeps. **axungiæ lunæ**, a variety of calcium carbonate. **axungiæ vitri**, salt of glass; a scum forming on the surface of molten glass. It is applied as a desiccative and detergent.

**ayapana, ayapano**. The South American name for the leaves of the herb *Eupatorium triplinerve*, of tropical America. It is stimulant, diaphoretic, and tonic, and is used in infusion externally for wounds and abscesses, internally for gastric disorders, and is recommended as a substitute for tea, coffee, and cocoa.

**azalein** (az-a'-le-in). See *fuchsin*.

**azedarach** (az-ed'-ar-ak) [Pers., āzād, free; *dirakht*, a tree]. Pride of China, the bark of *Melia azedarach*, an Asiatic tree naturalized in the southern United States. It occurs in curved pieces or quills, having a sweetish taste. A decoction, ½ oz. to 1 pint, is used as an anthelmintic against the roundworm. Dose ½-1 oz. (15-30 Cc.). Dose of the *fluidextract* 1 dr. (4 Gm.); of the *tincture*, 1 to 8, ½-2 dr. (2-8 Cc.).

**azerin** (az'-er-in) [ἀ, priv.; ξηρός, dry]. A ferment analogous to ptyalin and found in the digestive secretions of *Drosera, Nepenthes*, and probably all other insectivorous plants.

**azoamyly** (ah-zo-am'-il-e) [ἀ, priv.; ζῶον, animal; ἄμυλον, starch]. The inability of the cell (hepatic) to store up the normal amount of glycogen.

**azobenzene** (az-o-ben'-sēn) [*azote*, nitrogen; *benzene*], $C_{12}H_{10}N_2$. A compound formed by the action of sodium amalgam upon the alcoholic solution of nitrobenzene. It forms orange-red, rhombic crystals, readily soluble in alcohol and ether, but sparingly soluble in water. It melts at 68° and distils at 293°.

**azobenzoid** (as-o-ben'-zo-id). An amorphous white powder derived from oil of bitter almonds by action of ammonia.

**azo-compound**. In chemistry, a compound containing the group $-N=N-$ united to two hydrocarbon groups; a compound intermediate between the nitro-compounds and the amido-compounds, and made from the former by partial reduction, or from the latter by partial oxidation.

**azoic** (ah-zo'-ik) [ἀ, priv.; ζωή, life]. 1. Destitute of living organisms. 2. Relating to nitrogen; azotic; nitric.

**azolitmin** (az-o-lit'-min) [ἀ, priv.; ζωή, life; *litmus*], $C_7H_7NO_4$. A deep blood-red coloring-matter obtained from litmus.

**azomethane** (ah-zo-meth-ān'), Hydrocyanic acid.

**azoospermia** (ah-zo-o-sper'-me-ah) [ἀ, priv.; ζωή, life; σπέρμα, seed]. Absence of, or deficient vitality of, the spermatozoa.

**azoresorcin** (az-o-rez-or'-sin) [ἀ, priv.; ζωή, life; *resorcinol*], $C_{12}H_9NO_4$. A derivative of resorcinol, occurring as dark-red and greenish crystals.

**azotation** (az-o-ta'-shun). The assimilation of nitrogen from the air by organisms.

**azote** (az'-ōt) [ἀ, priv.; ζωή, life]. A synonym of *nitrogen*.

**azotemia** (az-o-te'-me-ah) [*azote*; αἷμα, blood]. The presence of nitrogenous compounds in the blood; uremia.

**azotenesis** (az-o-ten-e'-sis) [*azote*]. Any one of a class of diseases said to be due to a superabundance of nitrogen in the system, such as scurvy.

**azotic acid**. Nitric acid.

**azotiodic** (az-ot-i-o'-dik). Containing nitrogen and iodine.

**azotized** (az'-ot-īzd) [*azote*]. Nitrogenized; containing nitrogen.

**azotobacter** (az-o'-to-bak-ter). A class of large aerobic bacteria, capable of fixing free nitrogen from the air. They are found in the soil.

**azotometer** (az-o-tom'-et-er) [*azote*; μέτρον, a measure]. A device for the measurement of nitrogen.

**azotorrhea** (az-o-to-re'-ah) [*azote*; ῥοία, flow]. Excess of nitrogenous matter in the urine or feces.

**azoturia** (az-o-tū'-re-ah) [*azote*; οὖρον, urine]. An increase of the urea and urates in the urine.

**azoxybenzene** (az-oks-e-ben'-sēn) [*azote*; ὀξύς, sharp; *benzene*], $C_{12}H_{10}N_2O$. A compound obtained by the reduction of nitrobenzene. It forms long yellow needles, easily soluble in alcohol and ether, but not in water.

**azulene** (az'-ū-lēn). Same as *cerulein*.

**azyges** (az'-ij-ēs) [ἄζυγής, unwedded]. The sphenoid bone.

**azygos** (az'-ig-os) [ἀ, priv.; ζυγόν, a yoke]. Applied to parts that are single, not in pairs. **a. uvulæ**, a small muscle of the uvula. **a. veins**. See *veins*.

**azygous** (az'-ig-us) [see *azygos*]. Not paired.

**azymia** (ah-zi'-me-ah) [ἀ, priv.; ζύμη, a ferment]. Absence of ferment.

**azymic** (ah-zi'-mik) [ἀ, priv.; ζύμη, a ferment]. Not giving rise to fermentation.

**azymous** (az'-į-mus) [ἀ, priv.; ζύμη, a ferment]. Unfermented.

**azzle teeth** (az'-l) [E. dial., *usual tooth*]. A name given to the molar teeth.

# B

**B.** 1. The chemical symbol of *boron.* 2. Abbreviation for Beaumé's hydrometer; also of *Bacillus*, and *Bacterium.*
**Ba.** The chemical symbol of *barium.*
**B.A.** Abbreviation of Bachelor of Arts.
**Babbit metal** (*bab'-it*). An antifriction alloy composed of 8 parts of tin, 2 of antimony, and 1 of copper. Also used occasionally in dentistry.
**Babes-Ernst's bodies** [Victor *Babes*, Roumanian bacteriologist, 1854– ; Paul *Ernst*, German pathologist, 1859– ]. Bodies found in bacteria, especially those derived from animal bodies or secretions; they stain more deeply than the rest of the cytoplasm.
**Babesia** (*ba-be'-se-ah*) [Victor *Babes*, Roumanian bacteriologist, 1854– ]. Same as *Piroplasma.*
**babesiosis** (*ba-be-se-o'-sis*). Infection with babesia. Same as *piroplasmosis.*
**Babinski's phenomenon, B.'s reflex** (*ba-bin'-ske*) [Jules *Babinski*, French neurologist, 1857– ]. Extension, instead of flexion, of the toes on exciting the sole of the foot; it is connected with a lesion of the pyramidal tract, and is found in organic, but not in hysterical, hemiplegia. Syn., *phénomène des orteils*. **B.'s sign,** diminution or absence of the Achilles tendon reflex in true sciatica as distinguished from hysterical sciatica.
**bablabs, bablah** (*bab'-labz, -lah*). The pods of *Acacia arabica* and several other species; they are used in coughs; the seeds contain 20 % of tannin.
**babool, babul bark** (*ba-bool'*). The astringent, tonic bark of the babul tree, *Acacia arabica*, of India.
**baby** (*ba'-be*). An infant, a newborn child. **b.-farm.** An institution for raising orphan and pauper infants. **b.-farming**, the business of receiving and caring for the infants of those who, for any reason, may be unable or unwilling to bring up their own children.
**bacca** (*bak'-ka*) [L.]. A berry.
**Baccelli's method** (*bat-chel'-le*) [Guido *Baccelli*, Italian physician, 1832– ]. 1. In echinococcus cysts of the liver: aspiration is done on several consecutive days, and washings made with a 1 : 1000 solution of mercury bichloride and a 1 : 100 salt solution. 2. In tetanus: hypodermatic injection of a solution of phenol. **B.'s sign,** aphonic pectoriloquy. The whispered voice is transmitted through a serous, but not through a purulent, pleuritic exudate.
**baccharine** (*bak'-ar-in*). A poisonous alkaloid obtained from *Baccharis coridifolia.*
**Baccharis** (*bak'-ar-is*) [βάκκαρις, a fragrant herb]. A genus of composite trees. *B. halimifolia*, the groundsel-tree, is a shrub of North America. A decoction of the leaves and bark is a popular demulcent and pectoral medicine. *B. pilularis*, kidney plant, a native of the Pacific coast of the United States, is used in cystitis.
**bacchia** (*bak'-e-ah*, or *bak-i'-ah*) [*Bacchus*, the god of wine]. A synonym of *acne rosacea*, a condition often found in drunkards. **b. rosacea.** Synonym of *acne rosacea.*
**bacciform** (*bak'-si-form*) [*bacca; forma,* form]. Berry-shaped.
**Bach's reagent for hydrogen dioxide.** This consists of two solutions: (*a*) 0.03 potassium dichromate and 5 drops of aniline in 1 liter of water; (*b*) 5 % oxalic acid solution. Shake 5 Cc. of the solution to be tested with 5 Cc. of solution *a* and 1 drop of solution *b*; in the presence of hydrogen dioxide a violet-red color results.
**bacilac** (*bas'-il-ak*). Trade name of a preparation of milk which has been soured by the lactobacillus.
**bacillar,** or **bacillary** (*bas'-il-ar; bas'-il-a-re*) [*bacillus*]. 1. Relating to bacilli or to a bacillus. 2. Consisting of or containing rods.
**bacilemia, bacillæmia** (*bas-il-e'-me-ah*) [*bacillus;* αἷμα, blood]. The presence of bacilli in the blood.
**bacilli** (*bas-il'-i*) [*bacillus*]. 1. Plural of *bacillus,*

*q. v.* 2. In pharmacy, cylindrical lozenges made by cutting the lozenge mass, and rolling it into a soft cylinder, on a pill-machine.
**bacilli-carrier.** A person who is apparently in good health but who has pathogenic bacteria (such as typhoid) in his tissues or secretions and so is able to spread the disease.
**bacillicidal** (*bas-il-is-id'-al*) [*bacillus,* a. rod; *cædere,* to kill]. Destructive to bacilli.
**bacillicide** (*bas-il'-is-id*) [*bacillus,* a rod; *cædere,* to kill]. 1. Destructive to bacilli. 2. An agent that destroys bacilli.
**bacilliculture** (*bas-il-e-kul'-chur*) [*bacillus,* a rod; *cultura*, cultivation]. 1. The artificial culture of bacilli for the purpose of studying their nature and life. 2. A culture containing bacilli.
**bacilliform** (*bas-il'-if-orm*) [*bacillus; forma,* form]. Having the shape or appearance of a bacillus.
**bacilliparous** (*bas-il-ip'-ar-us*) [*bacillus; parere,* to produce]. Producing bacilli.
**bacillogenous** (*bas-il-oj'-en-us*) [*bacillus; generare,* to beget]. Due to bacilli; producing bacilli.
**bacillol** (*bas'-il-ol*). A coal-tar distillation-product resembling lysol, its active property being due to cresols, of which it contains 52 %. It is an oily fluid, of faint alkaline reaction, dark-brown color, and odor of pitch, readily soluble in water, with sp. gr. of 1.100, and bactericidal in dilute solution. In veterinary practice it is used in 2 % solution.
**bacillophobia** (*bas-il-o-fo'-be-ah*) [*bacillus;* φόβος, fear]. Morbid fear of microbes.
**bacillosis** (*bas-il-o'-sis*) [*bacillus*]. The condition caused by infection with bacilli.
**bacilluria** (*bas-il-u'-re-ah*) [*bacillus;* οὖρον, urine]. The discharge of urine containing bacilli.
**bacillus** (*bas-il'-us*) [dim. of *baculus,* a rod; pl., *bacilli*]. 1. A genus of the *Schizomycetes* comprising the rod-shaped forms of bacteria. 2. An individual of the genus *Bacillus.* 3. A medicated rod or bougie. 4. Any rod-like body, or, specifically, one of the retinal rods. See *Bacilli table of,* page 111.
**bacillus carrier.** See *bacilli-carrier.*
**back** (*bak*) [ME., *bak*]. Dorsum; posterior aspect. **b.-airing,** a term used in hygiene to designate the admission of fresh air to traps by means of a separate ventilating pipe of small diameter. **b.-rest,** a cloth-covered frame adjusted to any height by means of braces and ratchets, designed to relieve bedridden patients.
**backache** (*bak'-ak*). Pain in the back.
**backbone** (*bak'-bōn*). The vertebral column.
**backset.** A relapse of a disease.
**bacony infiltration** (*ba'-kon-e in-fil-tra'-shun*). Same as *amyloid degeneration.*
**bacteria** (*bak-te'-re-ah*). Plural of *bacterium* (*q. v.*).
**Bacteriaceæ** (*bak-te-re-a'-se-e*) [*bacteria*]. The *Schizomycetes.*
**bacterial** (*bak-te'-re-al*). Resembling, of the nature of, or derived from bacteria. **b. vaccine.** See *bacterine.*
**bactericidal** (*bak-te-ris-id'-al*) [*bacteria; cædere,* to kill]. Destructive to bacteria.
**bactericide** (*bak-te'-ris-id*) [*bacteria; cædere,* to kill]. 1. Destructive to bacteria. 2. An agent that destroys bacteria.
**bacteridium** (*bak-ter-id'-e-um*) [βακτήριον, a little stick]. A genus of *Bacteriaceæ* characterized by immobility of the elements at all periods of their existence (Davaine). The distinction does not now obtain.
**bacteriemia** (*bak-te-re-e'-me-ah*). The presence of bacteria in the blood.
**bacteriform** (*bak-te'-re-form*) [*bacterium; forma,* form]. Shaped like a bacterium.
**bacterination** (*bak-ter-in-a'-shun*). Inoculation with bacterial vaccines.
**bacterine** (*bak'-ter-ēn*). Any vaccine prepared

## TABLE OF BACILLI.

| Name. | Where Found. | Characters. |
|---|---|---|
| B. abortus.................... | Uterus of cow................ | Pathogenic for mammals. |
| B. aceti or aceticus............. | Air, vinegar................. | Zymogenic. |
| B. of Achalme................. | Blood in acute rheumatic fever... | Specificity disputed. |
| B. acidi lactici (Hueppe)........ | Milk....................... | Zymogenic. |
| B. acidi lævolactici (Schardinger).... | Well water.................. | Zymogenic. |
| B. acidificans longissimus (Lafar).... | Distillery yeast-mash.......... | Zymogenic. |
| B. acidiformans (Sternberg)........ | Liver, yellow-fever cadaver...... | Pathogenic. |
| B. adhæsioformans ............. | Peritoneum.................. | Causes pericolonic adhesions. |
| B. aerogenes, I, II, III (Miller)..... | Healthy alimentary tract........ | Zymogenic. |
| B. aerogenes capsulatus (Welch and Nuttall). | Blood and viscera in cases of infectious emphysema. | Zymogenic. |
| B. aerogenes meningitidis (Cantini)... | Meningitis................... | Zymogenic, pathogenic. |
| B. aerophilus (Liborius)........... | Air and water................ | Chromogenic (greenish-yellow). |
| B. albicans pateriformis.......... | Skin in seborrhea............. | Saprophytic. |
| B. albuminis (Bienstock).......... | Feces....................... | Zymogenic. |
| B. albus (Eisenberg)............. | Water...................... | Saprophytic. |
| B. albus anaerobiescens (Vaughan).. | Water...................... | Saprophytic. |
| B. albus cadaveris (Strassmann and Stricker). | Blood of cadaver.............. | Pathogenic. |
| B. albus putidus (Maschek)........ | Water...................... | Zymogenic |
| B. alkaligenes................... | Feces....................... | Nonpathogenic. |
| B. of Allantiasis (Müller).......... | Poisonous sausage, "Blunzen"... | Pathogenic, zymogenic. |
| B. allantoides (Klein)............. | Air........................ | Saprophytic. |
| B. allii (Griffiths)................ | Decaying onions.............. | Zymogenic, chromogenic (green) |
| B. of Alopecia Areata (Kasauli and Sabouraud). | Hair and scalp............... | Pathogenic, chromogenic (brick red). |
| B. alvei (Cheshire and Cheyne)..... | Bee larvæ, foul brood.......... | Pathogenic. |
| B. amylobacter (Grueber).......... | Flour....................... | Zymogenic. |
| B. amylobacter (Van Senus)........ | Fermenting cellulose........... | Symbiotic-zymogenic. |
| B. amylobacter (Van Tieghem)..... | Arable soil, manure............ | Zymogenic. |
| B. amylovorus (Burrill)............ | Pear blight.................. | Zymogenic, phyto-pathogenic. |
| B. "amylozyme" (Perdrix)......... | Water (Paris)................ | Zymogenic. |
| B. anaerobicus liquefaciens (Sternberg) | Intestines, yellow-fever cadaver.. | Pathogenesis undetermined. |
| B. antenniformis (Ravenel)........ | Soil........................ | Saprophytic. |
| B. anthracis (Rayer and Davaine).... | Blood in cases of anthrax, water, soil. | Pathogenic, zymogenic. |
| B. anthracis claviformis (Chauveau and Phisalix). | Anthrax, soil, etc............. | Pathogenic, zymogenic. |
| B. aquaticus liquefaciens (Podrowsky) | Water...................... | Saprophytic. |
| B. aquatilis (Lustig).............. | Water (Aosta)................ | Zymogenic. |
| B. aquatilis (P. and G. C. Frankland). | Well water (Kent)............. | Saprophytic. |
| B. aquatilis fluorescens (Lustig)..... | Water...................... | Saprophytic. |
| B. aquatilis graveolens (Tataroff).... | Water (Dorpat).............. | Chromogenic (yellowish). |
| B. aquatilis radiatus (Zimmermann). | Water...................... | Saprophytic. |
| B. aquatilis solidus (Lustig)........ | Water...................... | Zymogenic. |
| B. aquatilis sulcatus (Weichselbaum). | Water (Vienna).............. | Saprophytic. |
| B. arborescens (P. and G. C. Frankland). | Water (Thames)............. | Zymogenic, chromogenic (orange). |
| B. arborescens nonliquefaciens (Ravenel). | Soil........................ | Saprophytic. |
| B. argenteo-phosphorescens (Katz)... | Sea-water, decaying fish........ | Photogenic. |
| B. argenteo-phosphorescens liquefaciens (Katz). | Sea-water................... | Photogenic. |
| B. arómaticus (Pammel).......... | On cabbage leaves............ | Zymogenic. |
| B. aurantiacus (P. and G. C. Frankland). | Well water.................. | Chromoparous (red-orange). |
| B. aurescens (Ravennel)........... | Soil........................ | Saprophytic. |
| B. aureus (Adametz).............. | Air and water................ | Chromogenic (golden-yellow). |
| B. avisepticus................... | Blood of chickens............. | Pathogenic for birds, rabbits. |
| B. "B" (Hoffmann).............. | Diseased larvæ of Liparis monacha. | Pathogenic. |
| B. baccarinii (Machiati)........... | "Mal nero," or gummosis of grape-vines. | Phytopathogenic. |
| B. of Bang..................... | Same as B. abortus. | |
| B. beri-bericus (Lacerda).......... | Blood in cases of beri-beri...... | Saprophytic. |
| B. berolinensis indicus (Claessen).... | Water (Spree)................ | Chromoparous (indigo-blue). |
| B. bienstockii (Bienstock)......... | Human feces................. | Pathogenic. |
| B. bifidus...................... | Feces of nurselings........... | |
| B. boocopricus (Emmerling)........ | Cow dung................... | Zymogenic. |
| B. botulinus.................... | Pork, sausage, and other meat... | Pathogenic through its toxin. |
| B. of Bovet..................... | Intestine in case of enteritis..... | Pathogenic. |
| B. brassicæ (Pommer)............ | Infusions of cabbage.......... | Saprophytic. |
| B. brevis (Møri)................. | Sewage (Berlin).............. | Pathogenic. |
| B. bronchicanis ................. | In cases of canine distemper.... | |
| B. bronchitidis putridæ (Lumnitzer). | Cases of putrid bronchitis...... | Pathogenic. |
| B. brunneus (Adametz and Wichmann). | Water...................... | Saprophytic. |
| B. buccalis (Vignal).............. | Normal human saliva.......... | Chromogenic (golden-yellow). |
| B. buccalis maximus (Miller)....... | Mouth of man, common....... | Saprophytic. |
| B. bulgaricus................... | Milk....................... | |
| B. butylicus (Fitz)............... | On cereals, common........... | Zymogenic. |
| B. butyri fluorescens (Lafar)....... | Milk, butter................. | Chromoparous (green). |

TABLE OF BACILLI.—(Continued.)

| Name. | Where Found. | Characters. |
|---|---|---|
| B. butyricus (Prazmowski) | Ropy milk, water, soil | Zymogenic. |
| B. "C" (Foutin) | Hailstones | Chromoparous (reddish-yellow). |
| B. cadaveris (Sternberg) | Yellow-fever cadaver | Pathogenic. |
| B. canalis capsulatus (Mori) | Sewage (Berlin) | Pathogenic. |
| B. canalis parvus (Mori) | Sewage (Berlin) | Pathogenic. |
| B. candicans (Frankland) | Soil | Saprophytic. |
| B. of Canestrini | Diseased bees | Pathogenic, chromogenic (pink). |
| B. capsulatus (Mori) | Sewage (Berlin) | Pathogenic. |
| B. capsulatus mucosus (Fasching) | Nasal secretions, influenza (man) | Pathogenic. |
| B. capsulatus smithii (Theobald Smith) | Intestines of swine | Saprophytic. |
| B. carabiformis (Kaczynsky) | Stomach of dog | Saprophytic. |
| B. carnicolor (Tils) | Water (Freiburg) | Chromogenic (flesh-color). |
| B. carotarum (A. Koch) | Carrots and beets | Saprophytic. |
| B. caucasicus (Kern) (Syn. B. Kephir, Sorokin). | Kephir granules | Symbiotic-cymogenic with Saccharomyces kephir. |
| B. caulivorus (Galloway) | Potatoes and pelargoniums | Phytopathologic. |
| B. caviæ fortuitus (Sternberg) | Guinea-pigs, exudates after inoculation with liver of yellow-fever cadaver. | Saprophytic. |
| B. cavicidus havaniensis (Sternberg) | Intestine of yellow-fever cadaver. | Pathogenic. |
| B. cavicidus (Brieger) | Human feces | Pathogenic. |
| B. of Cazal and Vaillard | Cheesy nodules of peritoneum and pancreas. | Pathogenic. |
| B. of Chancroid (Ducrey) | Soft chancres | No growth in artificial cultures. |
| B. chauvæi (Bollinger and Feser) | Tissues of animals with "quarter evil." | Pathogenic, symbiotic-zymogenic with Micrococcus acidi paralactici. |
| B. of Cholera in Ducks (Cornil and Toupet). | Blood of ducks | Pathogenic. |
| B. choleroides (Bujwid) | Water | Pathogenic, methyl mercaptan odor. |
| B. chromo-aromaticus (Galtier) | Diseased pig | Pathogenic. |
| B. of Chyluria (Wilson) | Chylous urine | Pathogenesis undetermined. |
| B. cinctus (Ravenel) | Soil | Zymogenic, chromogenic (bright yellow). |
| B. circulans (Jordan) | Water | Zymogenic. |
| B. cloacæ (Jordan) | Water and in corn affected with "Burrill's disease." | Zymogenic. |
| B. cœruleus (A. J. Smith) | Water (Schuylkill) | Chromogenic (blue). |
| B. cœruleus (Voges) | Water | Chromoparous (blue). |
| B. coli communior | A form of B. coli communis, but with different cultural properties. | |
| B. coli communis (Escherich) | Intestines of men and animals (common). | Pathogenic. |
| . coli concentricus (Fitzpatrick) | Alimentary tract in yellow-fever cases. | Pathogenic. |
| B. coli icteroides (Fitzpatrick) | Yellow-fever cadaver | Pathogenic. |
| B. coli similis (Sternberg) | Human liver | Saprophytic. |
| B. of Colomiatti | Conjunctivitis and xerotic masses in eye. | Saprophytic. |
| B. of Conjunctival Catarrh (Koch) | Cases of "pink eye" | Pathogenic. |
| B. constrictus (Zimmermann) | Water (Chemnitz) | Chromogenic (cadmium-yellow). |
| B. coprogenes fœtidus (Schottelius) | Earth and intestines of hogs | Pathogenic, zymogenic. |
| B. coprogenes parvus (Eisenberg) | Human feces | Pathogenic. |
| B. corallinus (Slater) | Atmospheric dust | Chromoparous (coral-red). |
| B. corticallis (Haenlein) | Sour pine-bark liquor | Zymogenic. |
| B. crassus aromaticus (Tataroff) | Water (Dorpat) | Zymogenic, fruit-like odor. |
| B. crassus sputigenus (Kreibohm) | Human sputum | Pathogenic. |
| B. cuneatus (Rivolta) | Carcasses of domestic animals | Pathogenic. |
| B. cuniculicidus (Koch and Gaffky) | Water (Panke) | Pathogenic. |
| B. cuticularis (Tils) | Water (Freiburg) | Chromogenic (yellow). |
| B. cuticularis albus (Tataroff) | Water (Dorpat) | Saprophytic. |
| B. cyaneo-fluorescens (Zangemeister) | Blue milk | Chromoparous (blue). |
| B. cyaneo-fuscus (Beyerinck) | Blue cheese; glue | Chromoparous (blue). |
| B. cyaneo-phosphorescens (Katz) | Sea-water | Photogenic, chromoparous (green). |
| . cyanogenus (Ehrenburg-Hueppe) | Blue milk | Chromoparous (blue). |
| . cyanogenus (Jordan) | Water | Chromoparous (blue). |
| . cystiformis (Clado) | Urine in case of cystitis | Pathogenesis undetermined. |
| B. "D" (Foutin) | Hailstones | Saprophytic. |
| . of Dantec | Salt codfish which has turned red. | Chromogenic (red). |
| . delicatulus (Jordan) | Water (Lawrence) | Zymogenic, thermophilous. |
| . of Demme | Blood in cases of erythema nodosum. | Pathogenic. |
| . dendriticus (Bordoni-Uffreduzzi) | Water (Turin) | Saprophytic. |
| . denitrificans (Giltray and Aberson) | Soil and air | Zymogenic. |
| B. dentalis viridans (Miller) | Carious dentine | Pathogenic. |
| . devorans (Zimmermann) | Well-water | Saprophytic. |
| . dianthi (Arthur and Bolley) | Bacteriosis of carnations | Phytopathogenic. |
| . diatrypeticus casei (Baumann) | Cheese | Zymogenic. |
| B. diffusus (P. and G. C. Frankland) | Soil and water | Chromogenic (greenish-yellow). |

TABLE OF BACILLI.—(*Continued.*)

| Name. | Where Found. | Characters. |
|---|---|---|
| B. diphtheriæ (Klebs and Loeffler) | Diphtheric membranes. | Pathogenic. |
| B. diphtheriæ columbarum (Loeffler). | Diphtheric exudates in pigeons. | Pathogenic. |
| B. diphtheriæ vitulorum (Loeffler). | Diphtheric exudates in calves. | Pathogenic. |
| B. dysenteriæ (Chantemesse and Widal). | Intestines in dysentery cadavers. | Pathogenic. |
| B. dysodes (Zopf). | Bread. | Zymogenic. |
| B. Eberth's. See *B. typhi abdominalis.* | | |
| B. endocarditidis capsulatus (Weichselbaum). | Viscera in cases of endocarditis. | Pathogenic. |
| B. enteritidis (Gaertner). | Intestines in allantiasis. | Zymogenic, pathogenic. |
| B. entomotoxicon (Duggar). | Diseased squash-bugs (Anasatristis). | Pathogenic. |
| B. epidermidis (Bizzozero). | Epidermis between toes. | Saprophytic. |
| B. erodens (Ravenel). | Soil. | Saprophytic. |
| B. erysipelatos leporis (Koch). | Erysipelas in rabbit. | Pathogenic. |
| B. erysipelatos suis (Koch). | Erysipelas in hogs. | Pathogenic. |
| B. erythrosporus (Eidam). | Putrefying egg-albumen, water. | Chromoparous (red). |
| B. ethaceticus (P. Frankland, Fox, and Macgregor). | Sheep-dung. | Zymogenic. |
| B. ethaceto-succinicus (P. Frankland and Frew). | In a solution of ammonio-ferric citrate. | Zymogenic. |
| B. expneumo-enteritide suis (Klein). | Swine with hog cholera. | Pathogenic. |
| B. facultatus (Sadebach and Fraenkel) | In nonmalignant pharyngeal mycosis. | Pathogenesis undetermined. |
| B. figurans (Vaughan) | Water. | Saprophytic. |
| B. filiformis (Tils). | Water (Freiburg). | Saprophytic. |
| B. filiformis havaniensis (Sternberg). | Liver of yellow-fever cadaver. | Saprophytic. |
| B. of Fiocca. | Saliva of cats and dogs. | Pathogenic. |
| B. fissuratus (Ravenel). | Soil. | Chromoparous (yellow). |
| B. fitzianus (Zopf). | Hay-dust, manure, soil. | Zymogenic. |
| B. flavescens (Pohl). | Marsh water. | Chromogenic (yellow). |
| B. flavocoriaceus (Adametz and Wichmann). | Water. | Chromoparous (sulphur-yellow). |
| B. flavus (Macé). | Water. | Chromogenic (golden-yellow). |
| B. fluorescens albus. | Water. | Chromoparous (yellow). |
| B. fluorescens aureus (Zimmermann). | Water. | Chromoparous (pale yellow). |
| B. fluorescens liquefaciens (Fluegge). | Air and water. | Zymogenic, chromoparous (fluorescent green). |
| B. fluorescens longus (Zimmermann). | Water. | Chromoparous (yellowish-green). |
| B. fluorescens minutissimus. | Water, decomposing infusions. | Zymogenic, chromogenic (bluegreen). |
| B. fluorescens nivalis (Schmolck). | Glacier ice and water. | Chromoparous (green). |
| B. fluorescens nonliquefaciens (Eisenberg and Krueger). | Water and in butter. | Zymogenic, chromoparous (fluorescent-green). |
| B. fluorescens ovalis (Ravenel). | Soil. | Saprophytic. |
| B. fluorescens putidus (Fluegge). | Water. | Zymogenic, chromoparous (yellow). |
| B. fluorescens tenuis (Zimmermann). | Water. | Chromoparous (greenish-yellow). |
| B. fluorescens undulatus (Ravenel). | Soil. | Saprophytic. |
| B. fœtidus (Passet). | Soil. | Saprophytic. |
| B. fœtidus lactis (Jensen). | Milk in Jutland dairies. | Zymogenic. |
| B. fœtidus ozænæ (Hajek). | Nasal secretions in ozena. | Pathogenic. |
| B. formosus (Ravenel). | Soil. | Saprophytic. |
| B. of Fulles. | Soil. | Saprophytic. |
| B. fulvus (Zimmermann). | Air and water. | Chromogenic (gamboge-yellow). |
| B. fuscus (Schroetter). | Putrid infusions of maize. | Zymogenic, chromogenic (yellow). |
| B. fuscus (Zimmermann). | Air and water. | Chromogenic (chrome-yellow). |
| B. fuscus limbatus (Scheibenzuber). | Rotten eggs, water. | Chromogenic (brown). |
| B. gangliformis (Ravenel). | Soil. | Zymogenic. |
| B. gasoformans (Eisenberg). | Water. | Zymogenic, chromogenic (darkyellow). |
| B. gaytoni (Cheshire). | Diseased honey-bees. | Pathogenic. |
| B. geminus major (Ravenel). | Soil. | Saprophytic. |
| B. geminus minor (Ravenel). | Soil. | Saprophytic. |
| B. ginglymus (Ravenel). | Soil. | Saprophytic. |
| B. glaucus (Maschek). | Water. | Chromogenic (gray). |
| B. "Golden-yellow Water" (Adametz and Wichmann). | Water. | Chromogenic (shining yellow). |
| B. gossypinus (Stedman). | Bacteriosis of cotton plant. | Phytopathogenic. |
| B. gracilis (Zimmermann). | Water. | Saprophytic. |
| B. gracilis anaerobiescens (Vaughan). | Water. | Saprophytic. |
| B. gracilis cadaveris (Sternberg). | Human liver. | Pathogenic. |
| B. granulatus (Babes). | Air. | Chromoparous. |
| B. granulosus (Russell). | Sea-mud. | Saprophytic. |
| B. graveolens (Bordoni-Uffreduzzi). | Epidermis between toes. | Saprophytic. |
| B. of Grouse Disease (Klein). | Viscera of diseased grouse. | Pathogenic. |
| B. of Guillebeau (Freudenreich). | Ropy milk and inflamed udders of cows. | Zymogenic, pathogenic. |
| B. gummosus (Happ). | Ropy infusions of digitalis. | Zymogenic. |
| B. guttatus (Zimmermann). | Water. | Saprophytic. |
| B. "h" (Rosenberg). | Water (Main). | Chromoparous (violet). |

TABLE OF BACILLI.—*(Continued.)*

| Name. | Where Found. | Characters. |
|---|---|---|
| . halophilus (Russell) | Sea-mud | Saprophytic. |
| . hansenii (Raspmussen) | Air and water | Chromogenic (yellow). |
| . havaniensis (Sternberg) | Water | Chromogenic (blood-red). |
| B. havaniensis liquefaciens (Sternberg) of Havelburg. | Epidermis | Chromogenic (blood-red). |
| B. helvolus (Zimmermann) | Stomach of yellow-fever cadaver | Zymogenic. |
| . heminecrobiophilus (Arloing) | Water | Chromogenic (Naples-yellow) |
|  | Callous lymphatic glands in guinea-pig. | Pathogenic. |
| . hepaticus fortuitus (Sternberg) | Exudate of guinea-pig after inoculation with liver of yellow-fever cadaver. | Saprophytic. |
| B. hominis capsulatus (Bordoni-Uffreduzzi). | Cadaver of a rag-picker | Pathogenic. |
| B. of Horse-pox (Dieckerhoff and Grawitz). | Pustules of horses having acne contagiosa. | Pathogenic. |
| B. hyacinthi septicus (Heinz) | White rust of hyacinth bulbs and onions. | Phytopathogenic. |
| B. hyalinus (Jordan) | Water, sewage | Saprophytic. |
| B. hydrophilus fuscus (Sanarelli) | Well water (Sienna) | Pathogenic. |
| B. of Ice-cream Poisoning (Vaughan and Perkins). | Ice cream and cheese | Pathogenic. |
| B. icteroides (Sanarelli) | Alimentary tract, yellow-fever cadaver. | Pathogenic, zymogenic, produces amaril. |
| B. of Icterus (Karlinsky and Ducamp). | Blood in case of infectious icterus. | Pathogenesis undetermined. |
| B. ilidzensis capsulatus (Karlinsky) | Hot sulphur springs (Ilidze, Bosnia). | Thermophilous. |
| B. implexus (Zimmermann) | Water | Saprophytic. |
| B. incanus (Pohl) | Swamp-water | Saprophytic. |
| B. indicus (Koch) | Stomach, E. Indian ape | Chromoparous (red-yellow). |
| B. indigoferus (Claessen) | Water | Chromogenic (indigo-blue). |
| B. indigoferus (Voges) | Water | Chromoparous (blue). |
| B. indigogenes (Alparez) | Infusion of indigo-plant leaves | Pathogenic, zymogenic, chromoparous (indigo-blue). |
| B. inflatus (A. Koch) | Air | Saprophytic. |
| B. influenzæ (Pfeiffer) | Air; nasal secretions in influenza | Pathogenic. |
| B. of Intestinal Diphtheria of Rabbits (Ribbert). | Rabbits | Pathogenic. |
| B. intestinus motilis (Sternberg) | Intestine, yellow-fever cadaver | Saprophytic. |
| B. inunctus (Pohl) | Swamp-water | Saprophytic. |
| B. invisibilis (Vaughan) | Water | Saprophytic. |
| B. iridescens (Tataroff) | Water | Chromogenic (greenish-yellow). |
| B. janthinus (Zopf) | Water (Panke) | Zymogenic, chromoparous (violet). |
| B. of Jefferies | Alvine discharges in summer diarrhea. | Saprophytic. |
| B. of Jequirity Ophthalmia (de Wecker and Sattler). | Infusions of jequirity seed and in jequirity ophthalmia. | Saprophytic. |
| B. of Kartulis | Conjunctiva in Egyptian catarrhal conjunctivitis. | Pathogenesis undetermined. |
| B. "Kiel." See B. ruber kielensis. |  |  |
| B. of Kitasato. See B. pestis bubonicæ. |  |  |
| B. of Koubasoff | Carcinoma of stomach | Pathogenic. |
| B. lacmus (Schroeter) | Water | Chromogenic (blue). |
| B. lactis acidi (Marpmann) | Milk | Zymogenic. |
| B. lactis aerogenes (Abelous) | Alimentary tract in healthy persons! | Zymogenic. |
| B. lactis albus (Loeffler) | Milk | Saprophytic. |
| B. lactis cyanogenus (Hueppe) | Blue milk | Chromoparous (blue, triphenylrosanilin). |
| B. lactis erythrogenes (Hueppe and Baginsky). | Red milk | Chromoparous (red). |
| B. lactis peptonans (Sterling) | Pasteurized milk | Zymogenic. |
| B. lactis pituitosi (Loeffler) | Slimy milk | Zymogenic. |
| B. lactis saponacei (Weigmann and Zirn). | Soapy milk | Zymogenic. |
| B. lactis viscosus (Adametz) | Water and ropy milk | Zymogenic. |
| B. of Laser | Diseased mice | Pathogenic. |
| B. latericeus (Adametz and Wichmann). | Water | Chromogenic (brick-red). |
| B. Lemon-yellow (Maschek) | Water | Chromogenic (lemon-yellow). |
| B. leporis lethalis (Gibier and Sternberg). | Intestines of yellow-fever cadaver. | Pathogenic. |
| B. lepræ (Armauer and Hansen) | Leprous tubercles | Pathogenic. |
| B. leptosporus (L. Klein) | Air | Saprophytic. |
| B. of Lesage | Green alvine discharges in infants | Pathogenic. |
| B. lethalis (Babes) | Tissues in case of septicemia | Pathogenic. |
| B. of Letzerich | Urine in nephritis | Pathogenic. |
| B. of Lichen ruber (Laser) | Lymph in Lichen ruber | Pathogenesis undetermined. |
| B. limbatus acidi lactici (Marpmann). | Milk | Saprophytic. |
| B. limosus (Russell) | Sea-water and mud | Saprophytic. |
| B. liodermos (Loeffler) | Water and milk | Zymogenic. |
| B. liquefaciens (Eisenberg) | Water, frequent | Saprophytic. |

TABLE OF BACILLI.—(Continued.)

| Name. | Where Found. | Characters. |
|---|---|---|
| B. liquefaciens bovis (Arloing) | Lungs of diseased ox | Pathogenic. |
| B. liquefaciens communis (Sternberg) | Yellow-fever feces | Saprophytic. |
| B. liquefaciens lactis amari (Freudenreich). | Bitter cream | Zymogenic. |
| B. liquefaciens magnus (Luederitz) | Mice inoculated with soil | Zymogenic. |
| B. liquefaciens parvus (Luederitz) | Mice inoculated with soil | Saprophytic. |
| B. liquidus (P. and G. C. Frankland) | Water (Thames, common) | Saprophytic. |
| B. litoralis (Russell) | Sea-mud | Saprophytic. |
| B. lividus (Plagge and Proskauer) | Water (Berlin) | Chromoparous (blue). |
| B. lucens (Van Tieghem) | Water | Saprophytic. |
| B. of Lucet | Dysentery of fowls | Pathogenic. |
| B. of Lungs of Cattle | Cattle | Pathogenesis undetermined. |
| B. lupuliperda (Behrens) | Hops that had become "warm" | Thermogenic, zymogenic, odor of trimethylamin. |
| B. of Lustgarten | Syphilitic lesions | Specific pathogenesis disputed. |
| B. luteus (Dobrzyniecki) | Carious teeth | Chromoparous (yellow). |
| B. luteus (Fluegge) | Air | Chromoparous (yellow). |
| B. luteus suis (Salmon and Smith) | Perivisceral fluid of hogs | Chromogenic (yellowish-red). |
| B. of Lymph in Fishes (Oliver and Richet). | Fishes | Pathogenesis undetermined. |
| B. lyssæ (Pasteur) | Hydrophobic saliva | Specific pathogenesis disputed. |
| B. magenta (Pearmain and Moor) | Water | Chromogenic (carmin or magenta). |
| B. maidis (Cuboni) | Feces of pellagra patients | Saprophytic. |
| B. malariæ (Klebs and Tommasi Crudeli). | Air and soil; Roman campagnia | Saprophytic |
| B. mallei (Loeffler) | Cases of glanders | Pathogenic. |
| B. marsiliensis (Rietsch and Jobert) | Swine and ferrets affected with plague. | Pathogenic. |
| B. martinez (Sternberg) | Liver of yellow-fever cadaver | Saprophytic. |
| B. of Measles (Canon and Pielicke) | Blood in cases of measles | Pathogenesis undetermined. |
| B. of Meconium | Meconium | Saprophytic. |
| B. megaterium (de Bary) | Water and soil | Zymogenic. |
| B. megatherium (Ravenel) | Soil | Chromogenic (brown). |
| B. melanosporus (Eidam) | Air | Chromogenic (black). |
| B. melitensis | Spleen | Pathogenic for Malta fever. |
| B. melochloros (Winkler and Schroeter) | Wormy apples | Pathogenic, chromogenic (emerald-green). |
| B. membranaceus amethystinus (Eisenberg). | Well-water (Spolato) | Chromoparous (dark violet). |
| B. meningitidis purulentæ (Naumann and Schaeffer). | Pus in case of purulent meningitis | Pathogenic. |
| B. merismopoedioides (Zopf) | Sewage, soil | Saprophytic. |
| B. mesentericus fuscus (Fluegge) | Air, water, soil, hay-dust | Zymogenic. |
| B. mesentericus niger (Biel and Lunt) | Potatoes | Chromoparous (black). |
| B. mesentericus ruber (Globig) | Water, and on potatoes | Zymogenic, chromogenic (pink to red). |
| B. mesentericus vulgatus (Fluegge) | Air, water, milk, potatoes; frequent. | Zymogenic. |
| B. of Miller | Intestinal tract of healthy persons | Zymogenic. |
| B. mirabilis (Hauser) | Decaying animal matter | Pathogenic. |
| B. mollusci (Domenico) | Molluscum contagiosum | Pathogenesis disputed. |
| B. mucosis capsulatis (Friedländer) | Lungs in pneumonia infrequent | Pathogenic. |
| B. mucosus ozænæ (Lowenberg) | Mucous membrane of nostrils | Saprophytic. |
| B. multiformis trichorrhexidis (Hodara). | Diseased hairs in trichorrhexis nodosa barba. | |
| B. multipediculosus (Fluegge) | Air and water | Saprophytic. |
| B. murisepticus (Gaffky) | Water (Panke) | Pathogenic. |
| B. murisepticus pleomorphus (Karlinsky). | Uterine discharges | Pathogenic. |
| B. muscoides (Liborius) | Water, soil, cow-dung | Saprophytic. |
| B. mycoides (Fluegge) | Soil, water, hail | Zymogenic. |
| B. mycoides roseus (Scholl) | Soil | Chromogenic (red). |
| B. necrophorus (Loeffler) | Eye of rabbit inoculated with condyloma. | Pathogenic. |
| B. of Necrosis of Liver in Badgers (Eberth). | Badger | Pathogenesis undetermined. |
| B. of Necrosis of Liver in Guinea-pigs (Eberth). | Guinea-pigs | Pathogenesis undetermined. |
| B. of Nocard | Abscesses in cattle having farcy | Pathogenic. |
| B. No. 41 (Conn) | Butter | Zymogenic. |
| B. nodosus parvus (Lustgarten) | Healthy human urethra | Saprophytic. |
| B. nubilus (P. and G. C. Frankland) | Water (Thames) | Saprophytic. |
| B. ochraceus (Zimmermann) | Water | Saprophytic. |
| B. œdematis aerobicus (Klein) | Exudates of guinea-pigs inoculated with garden soil. | Pathogenic. |
| B. œdematis maligni (Pasteur, Joubert, and Chamberlain). (Vibrion septique of the French.) | Soil, dust, intestines of man and mammals, also in musk. | Zymogenic. |
| B. oleæ (Prillieux and Bioletti) | Disease of olive tree ("olive-knot"). | Phytopathogenic. |
| B. oleæ tuberculosis (Savartane) | Disease of olive tree | Phytopathogenic. |

TABLE OF BACILLI.—(Continued.)

| Name. | Where Found. | Characters. |
|---|---|---|
| B. oogenes fluorescens (Zoerkendoerfer). | Rotten eggs. | Zymogenic, chromoparous (pale green). |
| B. oogenes hydrosulphuricus (Zoerkendoerfer). | Rotten eggs. | Zymogenic. |
| B. "Orange-red" (Adametz and Wichmann). | Water. | Chromoparous (orange-red). |
| B. orthobutylicus (Grimbert). | Fermenting leguminous seeds. | Zymogenic. |
| B. of Osteomyelitis (Kraske and Becker). | Cases of osteomyelitis. | Pathogenic. |
| B. ovatus minutissimus (Unna). | Skin in eczema, seborrhœicum. | Saprophytic. |
| B. oxalaticus (Zopf). | Air. | Chromogenic (ocherous). |
| B. oxytocus perniciosus (Wyssokowitsch). | Milk. | Pathogenic. |
| B. panificans (Laurent). | Bread. | Zymogenic. |
| B., Paracolon (Gwyn). | Blood in infections resembling typhoid fever. | Pathogenic. |
| B. paratyphosus (Archard and Bensaud). | In case of paratyphoid. | Pathogenic. |
| B. parvus ovatus (Loeffler). | Carcass of hog. | Pathogenic. |
| B. (saccharo-bacillus) pastorianus (Van Laer). | Beer-wort. | Zymogenic. |
| B. peptofaciens (Bernstein). | Milk. | Zymogenic |
| B. of Perez. | In cases of ozena. | Pathogenic. |
| B. pestifer (Frankland). | Air. | Saprophytic. |
| B. pestis bubonicæ (Kitasato and Yersin). | Blood and lymphatics in bubonic plague. | Pathogenic. |
| B. phaseoli (E. F. Smith). | Parasitic on legumes. | Zymogenic. |
| B. phlegmonis emphysematosi (Fraenkel). | Pus in emphysema. | Pathogenic. |
| B. phosphorescens (Fischer). | Sea-water. | Photogenic. |
| B. phosphorescens gelidus (Foerster). | Luminous sea-fish. | Photogenic. |
| B. phosphorescens indicus (Fischer). | Sea-water. | Photogenic. |
| B. phosphorescens indigenus (Fischer). | Sea-water, and on fishes. | Photogenic. |
| B. phosphoreus (Cohn). | Sea-water, and on fishes. | Photogenic. |
| B. phylloxericidus (Dubois). | Soil, manure. | Pathogenic to phylloxera. |
| B. pinnatus (Ravenel). | Soil. | Zymogenic. |
| B. pituitosi (Loeffler). | See B. lactis pituitos. | |
| B. plicatus (Zimmermann). | Water. | Chromogenic (grayish). |
| B. pneumoniæ friedländeri (Friedländer). | Pulmonary exudates in croupous pneumonia. | Pathogenic, zymogenic. |
| B. pneumonicus agilis (Schou). | Pneumonia of rabbit. | Pathogenic. |
| B. pneumosepticus (Babes). | Blood in case of septic pneumonia. | Pathogenic. |
| B. polymyxus (Prazmowski). | Vegetable infusions. | Saprophytic. |
| B. polypiformis (Liborius). | Cow-dung. | Saprophytic. |
| B. prausnitzii. | Water, soil. | Saprophytic. |
| B. prodigiosus (Ehrenberg). | Food materials, etc. | Zymogenic, chromogenic (red). |
| B. proteus fluorescens (Jaeger). | Viscera of diseased fowls. | Pathogenic. |
| B. of Pseudodiphtheria (Belfanti). | Human mouth and throat. | Saprophytic. |
| B. pseudoedema (Liborius). | Mice inoculated with garden soil. | Pathogenic. |
| B. Pseudopneumonicus (Fluegge). | Pus. | Pathogenic. |
| B. pseudosepticus (Bienstock). | Exudates in mice inoculated with feces. | Pathogenic. |
| B. pseudotuberculosis (Pfeiffer). | Viscera of horse. | Pathogenic. |
| B. pseudotuberculosis (Rabinowitsch). | Butter. | Pathogenic. |
| B. pseudotuberculosis in Rabbits (Eberth). | Tuberculosis nodule in rabbits. | Pathogenesis undetermined. |
| B. pseudotuberculosis rodentium (Pfeiffer). | Found in rats. | Closely allied to B. pestis. |
| B. psittacosis (Widal and Sicard). | Blood of parrots and human beings having psittacosis. | Pathogenic. |
| B. puerperalis (Engel and Spillmann). | Cases of puerperal sepsis. | Pathogenic. |
| B. pulpæ pyogenes (Miller). | Gangrenous tooth pulp. | Pathogenic. |
| B. punctatus (Zimmermann). | Water (Chemnitz). | Saprophytic. |
| B. of Purpura hæmorrhagica (Babes and Kolb). | Viscera of purpura cadaver. | Pathogenic. |
| B. putrificus coli (Bienstock). | Water, feces. | Zymogenic. |
| B. of Pyemia (Beltzow). | Blood in pyemia. | Pathogenic. |
| B. pyocyaneus (Gessard). | Air, dust, water, pus. | Pathogenic, zymogenic, chromoparous (blue to verdigris-green, pyocyanin). |
| B. pyogenes fœtidus (Passet). | Pus. | Pathogenic, zymogenic. |
| B. pyogenes soli (Bolton). | Exudates of rat inoculated with garden soil. | Saprophytic. |
| B. radiatus (Luederitz). | Exudates of mice and guinea-pigs inoculated with garden soil. | Saprophytic. |
| B. radiatus aquatilis (Zimmermann). | Water. | Chromogenic (ochre-yellow). |
| B. radicicola (Byerinck). | Tubercles of leguminous plants, arable soil. | Zymogenic. |
| B. radiciformis (Tataroff). | Water. | Saprophytic. |
| B. radicosus (Zimmermann). | Water. | Saprophytic. |
| B. ramosus (P. and G. C. Frankland). | Soil, water (Thames). | Zymogenic. |

TABLE OF BACILLI.—(*Continued.*)

| Name. | Where Found. | Characters. |
|---|---|---|
| B. ramosus (Eisenberg and Fraenkel). | Water. | Saprophytic. |
| B. ramosus liquefaciens (Fluegge). | Air, water. | Saprophytic. |
| B. ranicida (Ernst). | Water; frogs dead of septicemia. | Pathogenic. |
| B. reticularis (Jordan). | Sewage. | Saprophytic. |
| B. rheumarthritidis (Kuessmaul). | Effusions in joints in articular rheumatism. | Pathogenesis undetermined. |
| B. "Rhine water" (Burri). | Water (Rhine). | Saprophytic. |
| B. rhinitis atrophicus. | Nasal secretions. | Saprophytic. |
| B. rhinoscleromatis (Cornil and Alvarez). | Tubercles in rhinoscleroma. | Pathogenic. |
| B. rodonatus (Ravenel). | Soil. | Chromoparous (brown to yellow). |
| "Der rothe Bacillus" (Lustig). | Water. | Chromoparous (raspberry red). |
| B. rubefaciens (Zimmermann). | Water. | Chromogenic (pale pink). |
| B. rubellus (Okada). | Guinea-pigs after inoculation with street dust. | Chromoparous (red). |
| B. ruber (Frank). | Water. | Chromoparous (blood-red). |
| B. ruber kielensis (Breunig). | Water (Kiel). | Chromoparous (blood-red). |
| B. ruber ovatus (Bruyning). | Blighted sorghum. | Phytopathogenic. |
| B. rubescens (Jordan). | Sewage. | Chromogenic (pale pink). |
| B. rubidus (Eisenberg). | Water. | Chromogenic (brownish-red). |
| B. saccharo-butyricus (von Klecki). | In "Quargelkase". | Zymogenic. |
| B. sanguinis typhi. (Brannan and Cheesman). | Blood in typhus fever. | Pathogenesis undetermined. |
| B. saprogenes (Rosenbach). | Decaying animal matter, fetid feet, etc. | Pathogenic, zymogenic (trimethylamin). |
| B. saprogenes vini (Kramer). | Wine. | Zymogenic. |
| B. satellitis. | In intestinal ulcers in typhoid fever. | |
| B. of Scarlet Fever (Crooke). | Throat in anginose scarlet fever. | Pathogenesis undetermined. |
| B. schafferi (Freudenreich). | "Puffy" and "Nissler" cheese. | Zymogenic. |
| B. of Scheurlen. | Mammary epithelia. | Saprophytic. |
| B. of Schimmelbusch. | Necrotic tissues in noma. | Pathogenic. |
| B. schutzenbergii. | Sewage. | Zymogenic. |
| B. scissus (Frankland). | Soil. | Saprophytic. |
| B. of Seborrhea (Sabouraud). | Hair and scalp. | Chromogenic (brick-red). |
| B. secalis (Burrill). | See B. zea. | |
| B. "Seidenglanzender" (Tataroff). | Well-water (Dorpat). | Saprophytic. |
| B. of Senile Gangrene (Tricomi). | Blood and tissues in cases of senile gangrene. | Pathogenic. |
| B. septicæmiæ hæmorrhagicæ (Sternberg). | Blood in septicemia. | Pathogenic. |
| B. septicus acuminatus (Babes). | Blood in septic infection. | Pathogenic. |
| B. septicus agrigenus (Nicolaier). | Garden soil. | Pathogenic. |
| B. septicus keratomalaciæ (Babes). | Cadaver; septicemia following kerato-malacia. | Pathogenic. |
| B. septicus sputi (Kreibohm). | Human saliva. | Pathogenic. |
| B. septicus sputigenus (Fluegge). | Healthy and pneumonic sputum. | Pathogenic. |
| B. septicus ulceris gangrænosi (Babes). | Cadaver; septicemia following gangrene. | Pathogenic. |
| B. septicus vesicæ (Clado). | Urine in cystitis. | Pathogenic. |
| B. sessilis (Klein). | Blood of cow. | Saprophytic. |
| B. smaragdinus phosphorescens (Katz). | On luminous fishes. | Photogenic, chromogenic (emerald-green). |
| B. smaragdinus fœtidus (Reimann). | Nasal secretions in ozena. | Pathogenic, chromogenic (green). |
| B. of Smegma (Bunge and Trautenroth). | Smegma. | Saprophytic. |
| B. solanacearum (E. F. Smith). | Brown rot of solanaceous plants. | Phytopathogenic. |
| B. solidus (Luederitz). | Mice after inoculation with garden soil. | Saprophytic. |
| B. solitarius (Ravenel). | Soil. | Saprophytic. |
| B. sorghi (Kellermann and Swingle). | Sorghum blight. | Phytopathogenic. |
| B. of Southern Cattle Plague (F. S. Billings). | Blood of cattle with Texas fever. | Pathogenic. |
| B. spiniferus (Unna). | Skin in eczema seborrhœicum. | Chromogenic (grayish-yellow). |
| B. stoloniferus (Adametz and Wichmann). | Water. | Saprophytic. |
| B. stoloniferus (Pohl). | Marsh-water. | Saprophytic. |
| B. striatus albus (von Besser). | Healthy nasal secretions. | Saprophytic. |
| B. striatus flavus (von Besser). | Healthy nasal secretions. | Chromogenic (sulphur-yellow). |
| B. striatus viridis (Ravenel). | Soil. | Saprophytic. |
| B. stuetzeri (Lehmann and Neumann). | Soil. | Zymogenic. |
| B. suaveolens. | Water. | Zymogenic. |
| B. subflavus (Zimmermann). | | Chromogenic (pale-yellow). |
| B. subtilis (Ehrenberg). | Air, water, soil; frequent. | Zymogenic. |
| B. subtilis simulans (Bienstock). | Human feces. | Saprophytic. |
| B. of Sugar-beet Disease (Arthur and Golden). | Sugar beets. | Pathogenesis not established. |
| B. sulph-hydrogenus (Miquel). | Water. | Zymogenic; evolves $H_2S$. |
| B. sulphureum, I (Holschewnikoff). | Urine. | Saprophytic. |
| B. sulphureum, II (Holschewnikoff). | Mud. | Chromogenic (reddish-brown). |
| B. superficialis (Jordan). | Sewage. | Saprophytic. |

TABLE OF BACILLI.—(Continued.)

| Name. | Where Found. | Characters. |
|---|---|---|
| B. of Swine Plague, Marseilles | See *B. marsiliensis*. | |
| B. sycosiferus fœtidus (Jordan) | Hair and scalp in sycosis. | Pathogenic. |
| B. syncyanus (Ehrenberg) | Water | Chromoparous (blue). |
| B. synxanthus (Schroetter) | Milk | Chromogenic (citron-yellow). |
| B. syphilidis (Lust-garten) | Syphilitic new-growths and secretions. | Pathogenic. |
| B. tardigradus (Detmers) | Water | Saprophytic. |
| B. tartaricus (Grimbert and Ficquet) | Fermenting solution of calcium tartrate. | Zymogenic. |
| B. tenuis sputigenus (Pansini) | Sputum | Pathogenic. |
| B. termo (Mace) | Water | Saprophytic. |
| B. terrigenus (Frank) | Soil | Zymogenic. |
| B. tetani (Nicolaier) | Arable soil, horse-dung, and tissues of persons dead of tetanus. | Pathogenic, zymogenic. |
| B. thalassophilus (Russell) | Sea-mud | Saprophytic. |
| B. thermophilus (Miquel) | Air, water, soil, feces, sewage | Thermophilous. |
| B. tholoideum (Gessner) | Water, sewage, intestinal tract | Pathogenic. |
| B. tracheiphilus | The cause of Cucurbit wilt | Phytopathogenic. |
| B. tremelloides (Schottelius) | Water (Freiburg) | Chromogenic (golden-yellow). |
| B. tremulus (Koch) | Vegetable infusions | Saprophytic. |
| B. of Trichorrhexis nodosa (Markusfeld) | Diseased hair | Zymogenic, pathogenic. |
| B. "Trommelschlagel" (Ravenel) | Soil | Saprophytic. |
| B. tuberculosis (Koch) | Sputum and tissues in tuberculosis. | Pathogenic. |
| B. tuberculosis gallinarum (Maffucci) | Tuberculosis in fowls | Pathogenic. |
| B. of Tuberculosis of Vines | Diseased grape-vines | Phytopathogenic. |
| B. tumescens (Zopf) | Beets | Saprophytic. |
| B. tussis convulsivæ (Afanassiew) | Sputum in cases of pertussis | Pathogenic. |
| B. typhi abdominalis (Eberth) | Water, milk, sewage; and blood, urine, feces, and tissues of typhoid-fever patients. | Pathogenic, zymogenic. |
| B. typhi exanthematici (Plotz) | | Pathogenic for typhus fever. |
| B. typhi murium (Loeffler) | Diseased mice | Pathogenic. |
| B. ubiquitus (Jordan) | Air, water, sewage | Pathogenic. |
| B. ulna (Cohn) | Egg-albumen | Saprophytic. |
| B. ulna (Vignal) | Normal saliva | Saprophytic. |
| B. of Uptadel (Gessner) | Intestinal contents (man) | Pathogenic. |
| B. ureæ (Leube) | Soil, water, manure, old urine, etc. | Zymogenic. |
| B. ureæ (Miquel) | Air | Saprophytic. |
| B. vacuolatus (Ravenel) | Soil | Chromoparous (yellow). |
| B. vaginalis (Doderlein) | Normal vaginal secretions | Saprophytic. |
| B. varicosus conjunctiva (Gombert) | Healthy conjunctival sac in man | Pathogenic. |
| B. vascularis (Sternberg) | Viscera of yellow-fever cadaver | Saprophytic. |
| B. vascularum (Cobb) | Gummosis of sugar cane | Phytopathogenic. |
| B. venenosus (Vaughan) | Water | Pathogenic. |
| B. venenosus brevis (Vaughan) | Water | Pathogenic. |
| B. venenosus invisibilis (Vaughan) | Water | Pathogenic. |
| B. venenosus liquefaciens (Vaughan) | Water | Pathogenic. |
| B. ventriculi (Raczynssky) | Stomach of dog | Saprophytic. |
| B. vermicularis (P. and G. C. Frankland). | Water (Lea) | Chromogenic (flesh-colored). |
| B. vermiculosus (Zimmermann) | Water | Saprophytic. |
| B. of Verruga peruana (Izquierdo) | Nodules in cases of Peruvian wart | Pathogenesis undetermined. |
| B. verticillatus (Ravenel) | Soil | Zymogenic. |
| B. violaceus (Becker) | Water | Chromoparous (deep-violet). |
| B. violaceus (Frankland) | Water | Chromoparous (violet). |
| B. violaceus laurentius (Jordan) | Water | Chromoparous (violet). |
| B. virens (Van Tieghem) | Water | Chromoparous (green). |
| B. virescens (Frick) | In green sputum | Chromogenic (green). |
| B. viridans | Water | Chromoparous (green). |
| B. viridescens liquefaciens (Ravenel) | Soil | Saprophytic. |
| B. viridescens nonliquefaciens (Ravenel). | Soil | Zymogenic. |
| B. viridis (Van Tieghem) | Water | Chromogenic (green). |
| B. viridis flavus (Frick) | Water | Chromogenic (yellowish-green). |
| B. viridis pallescens (Frick) | Air, water (Freiburg) | Zymogenic, chromogenic (yellowish-green). |
| B. viscosus (Van Laer) | Ropy beer | Zymogenic. |
| B. viscosus cerevisiæ (Van Laer) | Ropy beer and milk | Zymogenic. |
| B. viscosus sacchari (Kramer) | Viscous saccharine fluids | Zymogenic. |
| B. viscosus vini (Kramer) | Ropy wine | Zymogenic. |
| B. vulgaris (Hauser) | Putrefying matter | Pathogenic. |
| B. of Weigmann | Bitter milk | Zymogenic. |
| B. "Weissen-" (Eisenberg) | Water | Saprophytic. |
| B. "Weisser-" (Tataroff) | Well-water (Dorpat) | Saprophytic. |
| B. "White" (Maschek) | Water | Saprophytic. |
| B. "X" (Sternberg) | Yellow-fever cadavers | Pathogenic. |
| B. xerosis | In conjunctiva | Non-pathogenic. |
| B. "Y" | In cases of dysentery | |
| B. "Yellow" (Lustig) | Water | Saprophytic. |
| B. of Yersin. See *B. pestis bubonicæ*. | | |
| B. zea (Burrill) | Bacteriosis of Indian corn | Phytopathogenic. |
| B. zurnianus (List) | Air and water | Zymogenic. |

## TABLE OF BACTERIA.

| Name. | Where Found. | Primary Characters. |
|---|---|---|
| B. accidentalis tetani (Belfanti and Pescarolo). | Pus in a case of tetanus | Pathogenic. |
| B. aceti (Hansen) | Sour beer and wine | Zymogenic. |
| B. aceti (Peters) | Sour dough | Zymogenic. |
| B. aceticum (Baginsky) | Beer-wort | Zymogenic. |
| B. aceticum (Zoilder) | Beer-wort | Zymogenic. |
| B. acidi lactici (Grotenfeld) | Feces, water, milk | Zymogenic. |
| B. acne contagiosæ (Dieckerhoff and Grawitz). | Acne contagiosa in horses | Pathogenic. |
| B. aeris minutissimus (Bey) | Air | Chromoparous (canary-yellow). |
| B. amabilis (Dyar) | Air | Chromoparous (bright-yellow). |
| B. ambiguus (Wright) | Water | Saprophytic. |
| B. amethystinus (Eisenberg) | Water | Chromogenic (dark-blue). |
| B. amethystinus mobilis (Germano) | Air | Chromogenic (blue-violet). |
| B. anaerobicum (Fluegge) | Milk | Zymogenic. |
| B. annulatus (Wright) | Water | Chromoparous (yellow). |
| B. apii (Brizi) | Bacteriosis of celery plants | Phytopathogenic. |
| B. apthosus (Siegel) | Liver and kidneys in cases of "Maul-" and "Klauenseuche." | Pathogenic. |
| B. aquatilis communis (Zimmermann) | Water | Saprophytic. |
| B. aquatilis sulcatus quartus (Weichselbaum). | Soil | Saprophytic. |
| B. aurantiacum (Trelease) | Water | Chromogenic (orange). |
| B. aureo-flavus (Adametz) | Water | Chromogenic (chrome-yellow) |
| B. betæ (Arthur and Golden) | Bacteriosis of sugar-beets | Phytopathogenic. |
| B. bovisepticus (Kitt) | "Buffelseuche" | Pathogenic. |
| B. brassicæ (Lehn and Conrad) | Sauerkraut | Zymogenic. |
| B. breslaviensis (Van Ermenghem) | Poisonous meat | Zymogenic, pathogenic. |
| B. brunneo-flavus (Dyar) | Air | Chromogenic (brown to orange). |
| B. brunneum (Schröter) | Putrid infusion of maize | Chromogenic (brown). |
| B. buccalis fortuitus (Vignal) | Healthy saliva | Saprophytic. |
| B. buccalis minutus (Vignal) | Healthy saliva | Chromogenic (golden-yellow). |
| B. of Buffalo Plague (Ratz) | Buffaloes having an infectious disease. | Pathogenic. |
| B butyri colloideum (Lafar) | Butter (frequent) | Saprophytic. |
| B. campestris (Pammel) | Decayed turnips | Zymogenic, chromogenic (cadmium-yellow). |
| B. of Canary-bird Septicemia (Rieck) | Canaries | Pathogenic. |
| B. capitatum (Davaine) | Infusion of albuminous substances. | Saprophytic. |
| B. carlsbergense (Hasen) | Air | Saprophytic. |
| B. carneus (Tils) | Water | Chromogenic (flesh-color). |
| B. catenulus (Dujardin) | Putrid urine, blood in typhoid fever. | Zymogenic. |
| B. caudatus (Wright) | Water | Chromogenic (yellowish). |
| B. centrifugans (Wright) | Water | Zymogenic, chromogenic (greenish). |
| B. chlorinum (Engelmann) | Water | Chromophorous (green). |
| B. choleræ columbarum (Leclancler) | Wild pigeons | Pathogenic. |
| B. choleræ gallinarum (Perroncito) | Chicken cholera | Pathogenic. |
| B. chologenes (Stern) | Case of angiocholitis with meningitis. | Pathogenic. |
| B. chrysogloia (Lafar) | Air, water | Chromoparous (yellow, lipoxan thin). |
| B. citreus (Unna and Tomassoli) | Epidermis in eczema | Chromoparous (citron-yellow). |
| B. citreus adaveris (Strassmann and Streckerb | Human cadaver | Chromogenic (citron-yellow). |
| B. coadnutus (Wright) | Water | Saprophytic. |
| B. coherens (Wright) | Water | Saprophytic. |
| B. coli aerogenes (Lembke) | Dog-dung | Pathogenic. |
| B. coli anindolicum (Lembke) | Dog-dung | Zymogenic, pathogenic. |
| B. coli commune (Escherich) | Abundant in human feces and those of domestic animals. | Pathogenic. |
| B. coli immobilis (Germano and Maurea). | Feces | Symbiotic-zymogenic with Bacillus denitrificans, pathogenic. |
| B. coli mobilis (Messea) | Typhoid stools | Zymogenic. |
| B. colorabilis (Naunyn) | Yellow-fever cadaver | Pathogenic. |
| B. conjunctivitis (Morax) | Conjunctival catarrah and chronic inflammation of conjunctiva. | Pathogenic. |
| B. convolutus (Wright) | Water | Saprophytic. |
| B. of Corn-stalk Disease (Billings) | "Corn-stalk disease" and broncho-pneumonia of cattle. | Pathogenic. |
| B. cuniculi pneumonicus (Beck) | Lung plague of rabbits | Pathogenic. |
| B. cuniculicidus immobilis (Smith) | Cause of a spontaneous rabbit plague. | Pathogenic. |
| B. cuniculicidus septicus (Lucet) | Epizootic of rabbits | Pathogenic. |
| B. cuniculicidus thermophilus (Lucet) | Epidemic of rabbits and guinea-pigs. | Pathogenic. |
| B. decidiosus (Wright) | Water | Chromogenic (yellow). |
| B. decolorans major (Dyar) | Air | Saprophytic. |
| B. decolorans minor (Dyar) | Air | Saprophytic. |
| B. delabens (Wright) | Water | Saprophytic. |

TABLE OF BACTERIA.—(*Continued.*)

| Name. | Where Found. | Primary Characters. |
|---|---|---|
| B. delta (Dyar) | Water | Chromogenic (red). |
| B. dendriticus (Lustig) | Water | Saprophytic. |
| B. denitrificans (Gayon and Dupetit). | Air, soil, straw, horse-dung, etc. | Symbiotic-zymogenic with *Bacterium coli commune*. |
| B. denitrificans (Stuetzer and Burri) | Horse manure | Zymogenic. |
| B. denitrificans agilis (Ampola and Garino). | Manure | Zymogenic. |
| B. diphtheriæ avium (Loir and Duclaux). | Epizootic of fowls | Pathogenic. |
| B. diphtheriæ cuniculi (Ribbert) | Intestinal diphtheria of rabbits | Pathogenic. |
| B. discissum (Dinwiddie) | Milk | Zymogenic. |
| B. domesticus (Dyar) | Air | Chromogenic (yellow). |
| B. dormitator (Wright) | Water | Chromogenic (bright-yellow). |
| B. dubius (Bleisch) | Feces | Pathogenic. |
| B. dubius pneumoniæ (Bunzl and Federn). | Rusty sputum of pneumonia | Pathogenic. |
| B. duplicatus (Wright) | Water | Saprophytic. |
| B. dysenteriæ liquefaciens (Ogata) | Cases of dysentery in Japan | Pathogenic. |
| B. dysenteriæ vitulorum (Jensen) | Dysentery of calves | Pathogenic. |
| B. egregium | Atmospheric dust | Chromogenic (yellow, lipoxanthin). |
| B. ellenbachensis (Stuetzer and Hartleb). | Soil and roots of small grains | Zymogenic. |
| B. emphysematosus (Fraenkel) | Gaseous phlegmon | Pathogenic, zymogenic. |
| B. enchelys (Ehrenberg) | Water | Saprophytic. |
| B. endometritidis (Kaufmann) | Liver abscess | Pathogenesis undetermined. |
| B. epsilon (Dyar) | Air | Chromogenic (pink). |
| B. equi-intestinalis (Dyar and Keith). | Horse manure | Saprophytic. |
| B. eta (Dyar) | Air | Zymogenic, chromogenic (yellow). |
| B. exanthematicus (Babes and Oprescu). | Hemorrhagic infection in man | Pathogenic. |
| B. exiguus (Wright) | Water | Chromogenic (salmon-pink). |
| B. fæcalis alcaligenes (Petruschky) | Feces | Zymogenic, pathogenic. |
| B. fairmontensis (Wright) | Water | Saprophytic. |
| B. farinaceum (Wigand) | Sour dough | Zymogenic. |
| B. felis septicus (Fiocca) | Septicemia, cats | Pathogenic. |
| B. ferrugineus (Dyar) | Air | Chromogenic (brick-red). |
| B. fimbriatus (Wright) | Water | Zymogenic. |
| B. finitimus ruber (Dyar) | Air | Chromogenic (pink-red). |
| B. fischeri (Beyerinck) | Air | Chromogenic (yellowish). |
| B. flexuosus (Wright) | Water | Saprophytic. |
| B. fluorescens (Lepierre) | Water | Chromogenic (yellow-green). |
| B. fluorescens convexus (Wright) | Water | Saprophytic. |
| B. fluorescens crassus (Frick) | Air, water | Pathogenic. |
| B. fluorescens foliaceus (Wright) | Water | Saprophytic. |
| B. fluorescens immobilis | Air, water | Saprophytic. |
| B. fluorescens incognitus (Wright) | Water | Saprophytic. |
| B. fluorescens mutabilis (Wright) | Water | Chromogenic (yellowish-green). |
| B. fluorescens schuylkilliensis (Wright) | Water | Saprophytic. |
| B. friedbergensis (Ebert and Mandry) | Poisonous sausages | Pathogenic. |
| B. furfuris (Wood and Wilcox) | In tanner's bran-plump soak | Zymogenic. |
| B. fuscus liquefaciens (Dyar) | Air | Chromogenic (bright-orange). |
| B. fuscus pallidior (Dyar) | Air | Chromogenic (pink-orange). |
| B. fusiforme (Warming) | Sea-water | Saprophytic. |
| B. gallinarum (Klein) | Enteritis in fowls | Pathogenic. |
| B. gamma (Dyar) | Air | Chromogenic (ocherous). |
| B. gelatinosum betæ (Glazer) | Mucigenous beet-juice | Zymogenic, phytopathogenic. |
| B. geniculatus (Wright) | Water | Saprophytic. |
| B. gingivæ pyogenes (Miller) | Diseased teeth | Pathogenic, chromogenic (yellowish). |
| B. gliscrogenum (Malerba and Sanna-Salaris). | Mucinous viscid urine | Zymogenic, produces gliscrin. |
| B. gummis (Comes) | Gummosis of tomatoes, figs, almonds, oranges, etc. | Phytopathogenic. |
| B. hæmatoldes (Wright) | Water | Chromogenic (blood-red). |
| B. hæmorrhagicus (Kolb) | Septicemia cadaver | Pathogenic. |
| B. hæmorrhagicus nephritidis (Vassale). | Hemorrhagic nephritis | Pathogenic. |
| B. hæmorrhagicus septicus (Babes) | Septicemia in man | Pathogenic. |
| B. hæmorrhagicus venenosus (Tizzoni and Giovannini). | Purpura hæmorrhagica | Pathogenic. |
| B. of Hemorrhagic Septicemia of Swans (Fiorentini). | Swans | Pathogenic. |
| B. hessii (Guillebeau) | Ropy milk | Zymogenic. |
| B. hudsonii (Dyar) | Air | Chromogenic (ocherous-orange). |
| B. hyacinthi (Wakker) | "Yellows" of hyacinth bulbs | Phytopathogenic. |
| B. hydrosulphureum ponticum (Zelinsky). | Ooze (Black Sea) | Zymogenic ($H_2S$) chromogenic (coffee-brown). |
| B. icterogenes (Guarnieri) | Liver and blood in acute yellow atrophy of liver; typhoid stools. | Pathogenic. |
| B. indigonaceus (Schneider) | Water | Chromogenic (indigo-blue). |

TABLE OF BACTERIA.—*(Continued.)*

| Name. | Where Found. | Primary Characters. |
|---|---|---|
| B. inutilis (Dyar) | Air | Pathogenic. |
| B. javaniensis (Eijikmann) | Air | Zymogenic. |
| B. kochii (Hansen) | Air | Saprophytic. |
| B. krallii (Dyar) | Air | Saprophytic. |
| B. kutzingianum (Hansen) | Sour beer | Zymogenic. |
| B. lacticus (Gunther and Thierfelder) | Milk | Zymogenic. |
| B. lactis (Lister) | Milk | Zymogenic. |
| B. lactis aerogenes (Escherich) | Milk and intestines of milk-fed animals. | Zymogenic. |
| B. lactis innocuus (Wilde) | Milk | Saprophytic. |
| B. lacunatus (Wright) | Water | Chromogenic (grayish-yellow). |
| B. larvicida (Dyar) | Diseased larvæ of silkworm (Clisiocampa fragilis). | Pathogenic. |
| B. leucæmiæ canis (v. Lucet) | Dog with leukocythemia | Pathogenic. |
| B. levans (Lehmann and Wolffin) | Sour dough | Zymogenic. |
| B. lindolum (Fodor) | Soil | Zymogenic. |
| B. lineola (Mueller) | Water, soil | Saprophytic. |
| B. litoreum (Warming) | Sea-water | Saprophytic. |
| B. of Liver Abscess (Korn) | Case of liver abscess | Chromogenic (yellow). |
| B. lucens (Nuesch) | Phosphorescent meat | Photogenic. |
| B. ludwigii (Karlinsky) | Hot sulphur springs of Ilidze, Bosnia. | Thermophilous. |
| B. luminosum (Giard) | Abdominal cavity of Amphipod crustacea (Talitrus). | Photogenic, pathogenic. |
| B. luminosus (Beyerinck) | Sea-water | Photogenic. |
| B. luteum (List) | Water | Chromogenic (orange-yellow). |
| B. maddoxii (Miquel) | Fermenting urine | Zymogenic. |
| B. martinezii (Sternberg and Dyar) | Liver of yellow-fever cadaver; air | Pathogenic. |
| B. monachæ (Tubeuf) | Diseased larvæ of the "nun" moth (Liparis monache). | Pathogenic. |
| B. monadiformis (Messea) | Typhoid stools | Pathogenic. |
| B. morbificans bovis (Basenau) | Cow with puerperal fever | Pathogenic. |
| B. morbilli (Lanzi) | Urine in case of measles | Saprophytic. |
| B. mori (Boyer and Lambert) | Diseased mulberry trees | Phytopathogenic. |
| B. multistriatus (Wright) | Water | Saprophytic. |
| B. muripestifer (Laser) | Plague of field mice | Pathogenic. |
| B. murisepticus (Fluegge) | Mice | Pathogenic. |
| B. navicula (Reinke and Berthold) | Wet rot of potatoes | Phytopathogenic. |
| B. nebulosus (Wright) | Water | Saprophytic. |
| B. nexibilis (Wright) | Water | Saprophytic. |
| B. nitrificans (Burri and Stuetzer) | Soil | Zymogenic. |
| B. nitroso, formæ novæ (Rullmann) | Soil | Zymogenic. |
| B. oblongum (Boutroux) | Beer | Zymogenic. |
| B. œdematis thermophilus (Novy) | Guinea-pig inoculated with contaminated nuclein solution. | Pathogenic, zymogenic. |
| B. oleæ (Archangeli) | Tuberculosis of olive trees | Phytopathogenic. |
| B. orchiticus (Kuetscher) | Glandered horse | Pathogenic. |
| B. ovalis (Wright) | Water | Chromogenic (bright-yellow). |
| B. oxylacticus (Dyar) | Air | Chromogenic (ocherous). |
| B. pallescens (Henrici) | Cheese | Saprophytic. |
| B. paradoxus (Kruse and Pasquale) | Liver in case of dysentery | Pathogenic. |
| B. pasteurianus (Hansen) | Beer wort | Zymogenic. |
| B. periplanetæ (Tichomirow) | Diseased cock-roaches (Periplaneta orientalis). | Pathogenic. |
| B. pfluegeri (Ludwig) | Luminous meat and fish | Photogenic. |
| B. phasiani septicus | Diseased pheasants | Pathogenic. |
| B. phosphorescens (Cohn) | Fish and sea-water | Photogenic. |
| B. phosphorescens pfluegeri (Foerster) | Luminous fish | Photogenic. |
| B. photometricum (Engelmann) | Water | Chromophorous (green), photophilous. |
| B. pini (Vuillemin) | Galls on alpine pine | Saprophytic. |
| B. of Pneumonia in Turkeys (Mac-Fadyean). | Turkeys | Pathogenic. |
| B. pneumonicus liquefaciens (Arloing) | Exudates in lung-plague of cattle | Pathogenic. |
| B. pneumosepticus (Klein) | Rusty sputum in pneumonia | Pathogenic. |
| B. porri (Tommasi-Crudeli) | Warts | Saprophytic. |
| B. of Potato Scab (Bolley) | Potatoes | Phytopathogenic. |
| B. primus fullesii (Dyar) | Water | Zymogenic. |
| B. pseudo-conjunctivitis (Kartulis) | Conjunctival secretions | Chromogenic (canary-yellow). |
| B. pseudo-influenzæ (Pfeiffer) | Secretions in broncho-pneumonia, otitis media. | Pathogenesis undetermined. |
| B. pseudotyphosus (Loesener) | Water; liver abscess | Pathogenic. |
| B. pullulans | Water | Chromogenic (yellow). |
| B. putidum (Frick and Dyar) | Air, water | Zymogenic, chromoparous (yellow). |
| B. putredinis (Davaine) | Decaying plants | Saprophytic. |
| B. pyocinnabareus (Ferchmin) | Pus | Chromogenic (red-yellow), odor of trimethylamin. |
| B. pyogenes anaerobicus (Fuchs) | Stinking pus of rabbit | Pathogenic. |
| B. pyogenes fœtidus liquefaciens (Lanz). | Brain abscess after otitis media | Pathogenic, chromoparous (citron-yellow). |
| B. pyogenes minutissimus (Kruse) | Pus in man | Pathogenic. |

TABLE OF BACTERIA.—(Continued.)

| Name. | Where Found. | Primary Characters. |
|---|---|---|
| B. pyriforme (Hansen) | Air | Saprophytic. |
| B. recuperatus (Wright) | Water | Saprophytic. |
| B. refractans (Wright) | Water | Saprophytic. |
| B. rhodochrous (Dyar) | Air | Chromogenic (rose-color). |
| B. rhusiopathiæ suis (Kitt) | Erysipelas of swine | Pathogenic. |
| B. ruber sardinæ (Du Bois Saint Sévrin). | Sardine oil | Chromogenic (carmin-red), zymogenic, odor of trimethylamin. |
| B. rubescens (Lankester) | Water | Chromophorous (bacterio-purpurin). |
| B. rugosus (Wright) | Water | Saprophytic. |
| B. salivæ minutissimus (Wilde) | Saliva | Pathogenic. |
| B. salmoneus (Dyar) | Air | Chromogenic (salmon-pink). |
| B. salmonica (Emmerich and Weibel) | Trout disease | Pathogenic. |
| B. sanguinarium (Smith and Moore) | Infectious leukemia in fowls | Pathogenic. |
| B. (photobacterium) sarcophilum (Dubois). | Phosphorescent flesh | Photogenic. |
| B. sarraceni cola (Dyar) | Leaf of pitcher-plant | Zymogenic. |
| B. secundus fullesii (Dyar) | Air | Saprophytic. |
| B. septicus putidus (Roger) | Cholera cadaver | Pathogenic. |
| B. of Sheep-pox | In cases of sheep-pox or "Schafblättern." | Pathogenic. |
| B. sinuosus (Wright) | Water | Saprophytic. |
| B. of Sporadic Pneumonia in Cattle (Smith). | Cattle | Pathogenic. |
| B. sputigenes crassus (Kreibohm) | Sputum | Pathogenic. |
| B. sputigenes tenuis (Pansini) | In phthisis and catarrhal pneumonia. | Pathogenic. |
| B. subochraceus (Dyar) | Air | Chromogenic (orange). |
| B. suipestifer (Salmon and Smith) | Hog cholera | Pathogenic. |
| B. suisepticus (Schutz) | Swine plague | Pathogenic. |
| B. sulcatus liquefaciens (Kruse) | Water | Chromogenic (yellow-brown). |
| B. sulphureum (Rosenheim) | Wine | Zymogenic (evolves H₂S). |
| B. tachyctonum (Fischer) | Stools in cholera nostras | Pathogenic. |
| B. termo (Dallinger and Drysdale) | Putrefactive material | Zymogenic. |
| B. termo (Vignal) | Normal human saliva | Chromogenic (yellowish-gray). |
| B. theta (Dyar) | Air | Ocherous. |
| B. tiogensis (Wright) | Water | Saprophytic. |
| B. tracheiphilus (Smith) | Diseased melons | Saprophytic. |
| B. trambustii (Trambusti and Galcotti). | Water | Saprophytic. |
| B. tularense | In conjunctivitis and lymphadenitis. | Pathogenic for rodents and transmissible to man. |
| B. ureæ (Jaksch) | Ammoniacal urine | Zymogenic. |
| B. uvæ (Cugini and Macchiati) | Diseased grapes | Chromogenic (honey-yellow). |
| B. vacuolatus (Dyar) | Bladders of Utricularia vulgaris | Zymogenic. |
| B. vaginæ (Doederlein) | Vaginal secretions | Pathogenesis undetermined. |
| B. of Variola (Cose and Feltz) | Vesicles in case of small-pox | Pathogenic. |
| B. vermiforme (Ward) | Ginger-beer | Symbiotic-zymogenic with Saccharomyces pyriformis. |
| B. vernicosum (Zopf) | Water | Zymogenic. |
| B. violaceum (Bergonzini) | Putrefying egg-albumen | Chromogenic (violet). |
| B. violaceus sacchari ( ger and Dyar) | Air | Chromogenic (violet). |
| B. viridis (Lesage)...A | In "green diarrhea" of children | Chromophorous (green). |
| B. xylinum (Brown) | Solutions of carbohydrates | Zymogenic. |
| B. zeta (Dyar) | Air | Chromogenic (orange-red). |
| B. zopfii (Kurth) | Intestinal tract of fowls | Saprophytic. |

SARCINÆ.  TETRADS.

 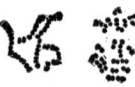 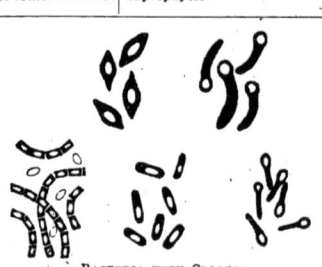

STAPHYLOCOCCI.  STREPTOCOCCI.  DIPLOCOCCI.   BACTERIA WITH SPORES.

from a specific bacterium; a bacterial vaccine. See *vaccine, bacterial.* It is a preparation of killed bacteria suspended in a normal saline solution. **b.,  autogenous,** one prepared from bacteria obtained from the patient. **b., polyvalent,** one containing bacteria of the same species but derived from many different sources. **b., stock,** one made from bacteria not obtained from the patient.

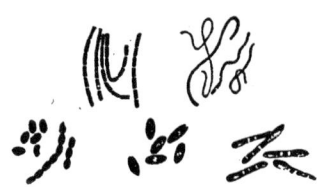

BACILLI OF VARIOUS FORMS.

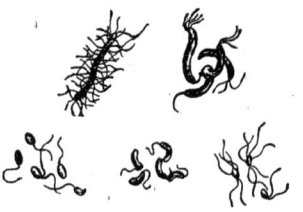

BACTERIA SHOWING FLAGELLA.

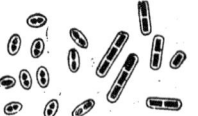

BACTERIA WITH CAPSULES.

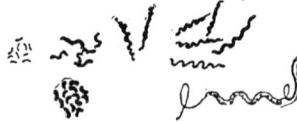

TYPES OF SPIRILLA.

**bacterio-** (bak-te-re-o-) [bacterium]. A prefix meaning relating to bacteria.
**bacteriofluorescin** (bak-te-re-o-flu-o-res'-in). A fluorescent coloring-matter produced by the action of certain bacteria.
**bacteriogenic** (bak-te-re-o-jen'-ik) [bacterio-; generare, to produce]. Caused by bacteria.
**bacteriogenous** (bak-te-re-oj'-en-us). 1. Producing bacteria. 2. Bacteriogenic.
**bacteriohemagglutinin** (bak-te-re-o-hem-ag-lu'-tin-in). A hemagglutinin produced in the body by the action of bacteria; it is very unstable, being destroyed at 58° C.
**bacteriohemolysin** (bak-te-re-o-hem-ol'-is-in). A very unstable hemolysin formed in the body by bacteria; it is destroyed at 58° C.
**bacterioid** (bak-te'-re-oid) [bacterio-; εἶδος, form]. Resembling bacteria; bacteriform.
**bacteriological** (bak-te-re-o-loj'-ik-al). Pertaining to bacteriology.
**bacteriologist** (bak-te-re-ol'-o-jist) [bacteriology]. One versed in bacteriology.
**bacteriology** (bak-te-re-ol'-o-je) [bacterio-; λόγος, science]. The science and study of bacteria.
**bacteriolysin** (bak-te-re-ol'-is-in). A specific antibody developed in the blood by the action of any one bacterium and capable of causing the disintegration of the same bacterium.
**bacteriolysis** (bak-te-re-ol'-is-is) [bacterio-; λύσις, a loosing]. The disintegration of bacteria, usually by means of a specific antibody.
**bacteriolytic** (bak-te-re-o-lit'-ik). Possessing a disintegrating action upon living bacteria. Pertaining to bacteriolysis.
**bacteriopathology** (bak-te-re-o-path-ol'-o-je) [bacterio-; pathology]. The science of diseases due to bacteria.
**bacteriophobia** (bak-te-re-o-fo'-be-ah) [bacterio-; φόβος, fear]. A morbid dread of bacteria or other microorganisms.
**bacteriophytoma** (bak-te-re-o-fi-to'-mah) [bacterio-; φυτόν, a growth]. A newgrowth caused by bacteria.
**bacterioplasmin** (bak-te-re-o-plaz'-min) [bacterio-; πλάσμα, anything formed or molded]. One of several toxic principles or toxalbumins extracted from pathogenic organisms, as of cholera or typhoid fever, by pressure.
**bacterioprotein** (bak-te-re-o-pro'-te-in) [bacterio-; πρῶτος, first]. A protein contained in bacteria. A toxalbumin.
**bacteriopurpurin** (bak-te-re-o-pur'-pu-rin) [bacterio-; purpura, purple]. A peach-colored pigment found in the protoplasm of Beggiatoa roseopersicina.
**bacterioscopic** (bak-te-re-o-skop'-ik) [bacterio-; σκοπεῖν, to view]. Pertaining to bacterioscopy.
**bacterioscopist** (bak-te-re-os'-ko-pist) [bacterio-; σκοπεῖν, to look]. A person devoted to the investigation of bacteria.
**bacterioscopy** (bak-te-re-os'-ko-pe) [bacterio-; σκοπεῖν, to inspect]. The microscopic study of bacteria.
**bacteriosis** (bak-te-re-o'-sis) [bacterium]. The action of bacteria in the system; infection by bacteria.
**bacteriotherapeutic** (bak-te-re-o-ther-ap-ū'-tik) [bacterio-; therapeutic]. Relating to bacteriotherapy.
**bacteriotherapy** (bak-te-re-o-ther'-ap-e) [bacterio-; therapy]. The treatment of disease by the introduction of bacteria or their products into the system.
**bacteriotropic** (bak-te-re-o-trop'-ik) [bacterio-; τροπή, a turning]. A generic term used by Wright to denote all substances in the blood which turn toward or are attracted to bacteria. Said of agglutinins.
**bacteriotropin** (bak-te-re-ot'-ro-pin) [bacterio-; τροπή, a turning]. A principle in the blood which, by its action on bacteria, aids the phagocytic action of the leukocytes.
**bacteriotoxin** (bak-te-re-o-toks'-in). A specific substance poisonous to bacteria.
**bacteritic** (bak-ter-it'-ik). Relating to or due to bacteria.
**Bacterium** (bak-te'-re-um) [βακτήριον, a little stick; pl., bacteria]. 1. A synonym of Schizomyces or Microorganism. 2. The word was formerly restricted to a genus of schizomycetous fungi established by Ehrenberg (1838) and Dujardin (1841), characterized by short, linear, inflexible, rod-like forms, without tendency to unite into chains or filaments. Morphologically, bacteria are spherical (cocci); in the form of straight rods (bacilli); or of twisted rods (spirilla). Bacteria are either aerobic, requiring free oxygen or anaerobic, not requiring free oxygen. Again, certain forms appear to possess the ability to flourish in either condition, and are known as facultative aerobic or facultative anaerobic. Bacteria are either motile or nonmotile; they may exist as saprophytes, facultative parasites, strict, obligate, or true parasites. Those that produce pigment are known as chromogenic; those that produce fermentation, as zymogenic; those that affect adversely the health of plants or animals, as pathogenic. See Bacteria, Table of (page 119), and special terms under appropriate headings.
**bacteriuria** (bak-te-re-ū'-re-ah) [bacterium; οὖρον, urine]. The discharge of urine containing bacteria.
**bacteroid** (bak'-ter-oid) [bacteria; εἶδος, form]. Resembling bacteria.
**baculiform** (bak-ū'-le-form) [baculum, a stick; forma, form]. Rod-shaped.
**Badal's operation** (bad-al') [Antoine Jules Badal, French ophthalmologist, 1840– ]. Rupture of the infratrochlear nerve for the relief of pain in glaucoma.
**bael** (bā'-el). See bel.
**Baelz's disease** (bailts) [Erwin Baelz, German physician, 1845–1913]. Progressive ulceration and destruction of the mucous glands of the lips.
**(von) Baer's law** [Carl Ernst von Baer, Russian embryologist, 1792–1876]. The more special forms of structure arise out of the more general, and that by a gradual change. B.'s vesicle, the ovule.
**Baeyer's reaction for glucose.** Indigo is formed on boiling a glucose solution with orthonitrophenyl-

propionic acid and sodium carbonate. When the glucose is in excess, this blue is converted into indigo white. B.'s reaction for indol, a watery solution of indol to which have been added 2 or 3 drops of fuming nitric acid and then a 2 % solution of potassium nitrite drop by drop, yields a red liquid and then a red precipitate of nitrosoindol nitrate, $C_{16}H_{13}(NO)$-$N_{21}HNO_3$.

**bag** [AS., *bælg*]. 1. A sac. 2. The scrotum. **b., intragastric**, an elastic rubber bag which, when folded over a tube which runs through it, occupies less space than in ordinary stomach-tube, and which has the exact shape of the stomach when it is inflated within that organ. It is employed to obtain the contents of the duodenum. **b.** of waters, the fetal membranes inclosing the liquor amnii and projecting through the os uteri early in labor. The sac usually ruptures when the cervix is dilated.

**bagnio** (*ban'-yo*) [It., *bagno*, from Latin *balneum*, a bath]. 1. A bath-house. 2. A house of prostitution.

**Bagot's local anesthesia mixture.** Cocaine hydrochloride, 0.04 gm. and sparteine sulphate, 0.05 gm.; this is dissolved in 1 or 2 c.c. of boiled water.

**Baillarger's outer band, line, or layer** (*bi'-yar-gha*) [Jules Gabriel François *Baillarger*, French neurologist, 1806–1891]. A white band in the layer of large pyramidal cells of the cortex cerebri. See also *Gennari* and *Vicq d'Azyr*. B.'s **internal band or line**, a white band between the layer of large pyramidal cells and the polymorphous layer of the cortex. **B.'s sign**, difference in the size of the pupils in paralytic dementia.

**bain-marie** (*ban-mah-re'*) [Fr.]. An instrument for immersing solutions, microorganisms, etc., in water or chemical solutions, thus keeping them at a desired temperature.

**Bain's method of artificial respiration.** A modification of Sylvester's method (*q. v.*); the structures surrounding the axilla are seized so that traction is made directly upon the pectoral muscles.

**Baker's cysts** [William Morrant *Baker*, English surgeon, 1839–1896]. Hernial protrusions of the synovial membrane of the joints through the fibrous capsule.

**bakers' itch.** An eczematous affection of the hands, caused by the irritation of the yeast. **b.'s leg**, knock-knee, or genu valgum. **b.'s salt.** A synonym of *smelling-salts* or *ammonium bicarbonate*; it is sometimes used by bakers in leavening cakes. **b.'s stigmata**, corns on the fingers from kneading dough.

**bakkola** (*bak'-o-lah*). A fungus which grows on birch trees in Finland; it is used in the form of a decoction as a cancer-cure.

**balance** (*bal'-ans*) [L., *bilanx*, having two scales]. 1. An instrument for weighing. 2. The harmonious adjustment of related parts. **b., electromagnetic**, an apparatus for estimating the intensity of electric currents. **b., thermic.** See *bolometer*. **b., torsion**, an instrument for estimating magnetic attraction and repulsion.

**balanic** (*bal'-an-ik*) [βάλανος, the glans penis]. Pertaining to the glans of the penis or of the clitoris.

**balanism** (*bal'-an-izm*) [βάλανος, acorn; pessary; the glans penis]. The application of a pessary or suppository.

**balanitis** (*bal-an-i'-tis*) [βάλανος, the glans penis; ιτις, inflammation]. Inflammation of the glans penis.

**balano-** (*bal-an-o-*) [βάλανος, the glans penis]. A prefix meaning relating to the glans penis.

**balanoblennorrhea** (*bal-an-o-blen-or-e'-ah*) [*balano-;* βλέννα, mucus; ῥοία, a flow]. Gonorrheal balanitis.

**balanocele** (*bal-an'-o-sēl*) [*balano-;* κήλη, a hernia]. The protrusion of the glans through an opening in the prepuce, as occurs in gangrenous phimosis.

**balanoplasty** (*bal-an'-o-plas-te*) [*balano-;* πλάσσειν, to form]. Plastic surgery of the glans penis.

**balanoposthitis** (*bal-an-o-pos-thi'-tis*) [*balano-;* πόσθη, prepuce; ιτις, inflammation]. Inflammation of the glans penis and of the prepuce.

**balanopreputial** (*bal-an-o-pre-pu'-she-al*) [*balano-;* *præputium*, prepuce]. Relating to the glans penis and the prepuce.

**balanorrhagia** (*bal-an-or-a'-je-ah*) [*balano-;* ῥήγνυναι, to burst out]. Gonorrheal balanitis, with copious discharge of pus.

**balanorrhea** (*bal-an-or-e'-ah*) [*balano-;* ῥοία, a flow]. Purulent balanitis.

**Balantidium** (*bal-an-tid'-e-um*) [βαλαντίδιον, dim. of βαλάντιον, a bag]. A genus of protozoa. **B. coli**, a protozoan parasite found in the intestine of man and other vertebrates. **B. minutum**, a protozoan found in the feces, smaller than the *B. coli*.

**balanus** (*bal'-an-us*) [βάλανος, an acorn]. 1. The glans of the prepuce or of the clitoris. 2. A pledget, suppository, or pessary.

**balata** (*bal-at'-ah*) [native Guiana]. The dried milky juice of the bully-tree, *Mimusops balata*, and of several other sapotaceous trees of Guiana. It is intermediate between caoutchouc and guttapercha, and is used chiefly in England as a substitute for these.

**balatin** (*bal'-at-in*). The creamy sap from a South American tree, *Mimusops kauki*; it is used as a varnish and vehicle in skin diseases.

**Balbiani, body of** (*bal-be-ah'-ne*) [*Balbiani*, Italian physician,         ]. The yolk-nucleus or idiosome, a small body seen near the nucleus of the oocyte.

**balbuties** (*bal-bū'-she-ēz*) [*balbutire*, to stammer]. Stammering.

**bald** (*bawld*) [ME., *balde*, bald]. Wanting hair. A term applied to one who has lost the hair of the scalp. **b. ringworm.** See *tinea tonsurans*.

**baldness** (*bawld'-nes*). Loss of hair; alopecia.

**Baldy's operation** (*bawl'-de*) [John Montgomery *Baldy*, American gynecologist, 1860–         ]. For *prolapse of the uterus*. The uterus is removed at the internal os, and the cervical stump is fixed to the abdominal wall at the lower end of the incision by means of two silkworm-gut sutures; these latter are made to transfix the cervical stump from side to side and the free ends are brought through the peritoneum, muscles, and deep fascia. They are then securely tied and cut off short and the knots are buried, and the abdominal incision is closed.

**Balfour's disease** [George William *Balfour*, English physician, 1822–1903]. A fatal disease of childhood, with postmortem findings of greenish-yellow or greenish-gray fibrosarcoma in various parts of the body, especially the periosteum. Syn., *chloroma*, *chlorosarcoma*.

**ball** (*bawl*) [ME., *ball*]. 1. An object having a round or spherical shape. 2. In anatomy, any globular part. 3. In veterinary medicine a pill or bolus. **b. of foot**, the rounded part at the base of the great toe. **b. of thumb**, the rounded part at the base of the thumb. **b., martial**, balls made of 2 parts of cream of tartar and 1 part of iron filings; they were used in the preparation of ferruginous baths. Syn., *boli martis*; *globuli martis*. **b.-and-socket joint.** See *diarthrosis* and *enarthrosis*. **b.-thrombus**, a non-attached antemortem clot in the heart.

**Ball's operation.** 1. *Iliac colotomy*, in which the incision is made in the left semilunar line, and the bowel secured above and below the future artificial opening by means of two special clamps, which are removed after suturing and opening of the bowel. 2. *For inguinal hernia;* the sac is separated up to the abdominal portion of the neck, and then twisted around its own axis, after which the fundus is cut away and the stump is secured in the ring.

**ballet-dancer's cramp.** See *cramp*.

**Ballet's disease** [Gilbert *Ballet*, French ophthalmologist, 1853–         ]. Ophthalmoplegia externa. B.'s **sign**, the loss of all voluntary movements of the eyeball, with preservation of the automatic movements and integrity of the movements of the ophthalmic goiter.

**Ballingal's disease.** Mycetoma.

**ballismus** (*bal-iz'-mus*) [βαλλισμός, a leaping]. 1. Chorea. 2. Paralysis agitans.

**balloon** (*bal-oon'*) [Mod. E., *baloon*, a large bag]. 1. In chemistry, a spherical glass receiver with a short neck. 2. To distend a body-cavity by means of air-bags or water-bags. **b. sickness**, an abnormal condition due to the breathing of an insufficient amount of oxygen.

**ballonnement** (*bal-lon-mon*(*g*)) [Fr.]. The ballooning or distending of a part for operative or diagnostic purposes.

**ballooning** (*bal-oon'-ing*). Surgical distention of the vagina or other cavity of the body, by air-bags or water-bags.

**ballottement** (*bal-ot-mon*(*g*)) [Fr., *ballottement*, tossing; shaking about]. A method of diagnosing pregnancy from the fourth to the eighth month. A push is given the uterus by the finger inserted into the vagina, and if the fetus is present, it will rise and

fall again like a heavy body in water.  b., **cephalic**, the rebound of the fetal head against the hand when depressed through the abdominal wall.  b., **ocular**, the falling of opaque particles in a fluid vitreous humor after movements of the eyeball.
**balm** (*bahm*) [*balsamum*, a balsam].  1. A popular synonym of *balsam*.  2. Any soothing application or ointment.  See *melissa*.  b. **of Gilead**, an oleoresin obtained from the *Balsamodendron gileadense*.
**balmony** (*bal'-mo-ne*) [origin uncertain]: The herb *Chelone glabra*; cathartic and anthelmintic.  Dose of the fldext. ℨ ss-j.  *Chelonin*, the concentrated ext.  Dose gr. j-iv.
**balneal** (*bal'-ne-al*).  Relating to baths.
**balneation** (*bal-ne-a'-shun*) [*balneum*, a bath].  1. The act of bathing.  2. Balneotherapy.
**balneography** (*bal-ne-og'-ra-fe*) [*balneum*; γράφη, a writing].  A treatise on bathing, baths, and mineral springs.
**balneology** (*bal-ne-ol'-o-je*) [*balneum*; λόγος, science].  The science of baths and mineral waters, and their effects upon the system.
**balneophysiology** (*bal-ne-o-fiz-e-ol'-o-je*) [*balneum*; *physiology*].  The physiology of bathing; the science of the effects of baths upon the system.
**balneotechnics** (*bal-ne-o-tek'-niks*) [*balneum*, a bath; τέχνη, an art].  The art of properly preparing baths as to constituents and temperature and the administration of them.
**balneotherapeutics** (*bal-ne-o-ther-ap-ū'-tiks*).  See *balneotherapy*.
**balneotherapy** (*bal ne o ther' ap e*) [*balneum;* θεραπεία, treatment].  Systematic bathing for therapeutic purposes, or the treatment of disease by baths.
**balneum** (*bal'-ne-um*) [L.].  A bath.  See *bath*. b. **arenæ**, a sand-bath.  See *ammotherapy*. b. **lacteum**, a milk-bath. b. **luteum**, a mud-bath. b. **pneumaticum**, an air-bath.
**balsam** (*bawl'-sam*) [βάλσαμον, the resin of the balsam-tree]. The resinous, volatile, aromatic substance, liquid or solid, obtained from certain trees by natural exudation or by artificial extraction. Balsams are divided into two classes—those with, and those without, benzoic and cinnamic acids.  In general they are mixtures of various essential oils, resins, and acids.  b.-**apple**, the plant *Momordica balsamina*, and its warty, gourd-like fruit.  It is purgative, but its tincture is used in domestic medicine chiefly as a vulnerary.  b.-**bog**, a singular stonelike, woody, umbelliferous plant, *Bolax*, or *Azorella glebaria*, of the Falkland Islands and Patagonia.  Its aromatic gum is locally prized as a vulnerary, desiccative, and antigonorrheal remedy. b., **Canada**, a turpentine gathered from the natural blisters of the bark of *Abies balsamea*.  It is used as a mounting-medium by microscopists.  See under *turpentine*. b. **of copaiba**.  See *copaiba*. b. **of fir**.  Same as *b.*, *Canada*. b., **friars'**.  See *benzoin*. b., **Houmiri**, b., **Humiri**, the fragrant exudate from the trees *Humiria balsamifera* and *H. floribunda*, natives of South America.  It is used as an expectorant and vermifuge.  Syn., *Umire*. b. **of Peru** (*balsamum peruvianum*, U. S. P.), the balsam obtained from *Toluifera pereiræ*; antiseptic, stimulant to the circulation, and sedative to the nervous system, tonic and expectorant.  Applied locally, it is useful in chronic inflammatory skin diseases.  Dose of the *emulsion* 10–25 min. (0.6–1.5 Cc.).  b.-**root**, a popular name for certain composite-flowered plants of the genus *Balsamorrhiza*. *B. hookerii, B. macrophylla*, and *B. sagittata* are common in the Pacific States and abound in a resinous balsam. b., **Samaritan**, a mixture of equal parts of oil and wine, heated together, and a tenth part of rosemary leaves. b., **stimulant**, a mixture of 8 parts of turpentine and 1 part of mustard flour. b., **sulphur**, a mixture of 8 parts of olive oil and 1 part of sublimed sulphur heated together. b., **tagulaway**, b., **tagulavay**, a yellow oil prepared in the Philippines by boiling the bark and twigs of the çebu, *Parameria vulneraria*, in cocoanut oil; it is used as a vulnerary and in skin diseases.  Syn., *cebur; Jagulaway balsam*. b. **of Tolu**, b. **of Tolutan** (*balsamum tolutanum*, U. S. P.), obtained from *Toluifera balsamum*.  Its properties are due to a volatile oil, *toluene*.  It possesses an agreeable odor, and is a basis for many cough-mixtures.  It is expectorant.  See also *Tolu*.
**balsamation** (*bawl-sam-a'-shun*) [see *balsam*].  1. Embalmment with balsamic or aromatic spices.  2. The act of rendering balsamic.

**balsamic** (*bawl-sam'-ik*).  Pertaining to or having the nature or qualities of a balsam.  b. **tincture**, compound tincture of benzoin.
**balsamodendron** (*bawl-sam-o-den'-dron*).  See *bdellium* and *myrrh*.
**balsamum** (*bawl'-sam-um*) [*balsam*].  A balsam. b. **dipterocarpi**.  See *Gurjun balsam*.
**Balser's fat-necrosis** (*bol'-ser*) [August *Balser*, German surgeon].  An acute disease of the pancreas with areas of fat-necrosis in the interlobular tissue of that organ, in the omentum and mesentery, at times also in the pericardial fat and bone-marrow.
**Bamberger's bulbar pulse** (*bam'-ber-ger*) [Heinrich von *Bamberger*, Austrian physician, 1822–1888].  Pulsation of the jugular vein,—the bulbus venæ jugularis,—synchronous with the systole, in tricuspid insufficiency.  B.'s **disease**, saltatory spasm.  B.'s **fluid**, an albuminous mercuric compound used in the treatment of syphilis.  It is made as follows: To 100 Cc. of a filtered solution of white of egg (containing 40 Cc. of albumin and 60 Cc. of water) there are added 60 Cc. of a solution of mercuric chloride (containing 5 %, or 3 Gm., of mercuric chloride and 60 Cc. of a solution of sodium chloride (containing 20 %); finally, 80 Cc. of distilled water is added, which brings the bulk of the solution up to 300, containing 0.010 mercuric chloride in every cubic centimeter.  B.'s **hematogenic albuminuria**, albuminuria occurring during the later stages of severe anemia.  B.'s **sign**, allocheiria; perception of a stimulus applied to the skin of one extremity at the corresponding place on the other extremity.  B.'s **type of hypertrophic pulmonary osteopathy**, a form in which painful thickenings of the long bones, especially of the forearm and leg, are a prominent symptom.
**bamboo** (*bam-boo'*) [E. Ind., *bambu*].  A popular name for many tree-like, woody-stemmed grasses, especially those of the genus *Bambusa*. *Bambusa arundinacea* is employed as an alterative, anthelmintic, and depurative. b.-**brier**, the root of *Smilax sarsaparilla*, habitat, southern United States.  Its properties are identical with those of sarsaparilla.  Dose of the *fluidextract* ½–2 dr. (2–8 Cc.).
**banana** (*ban-an'-ah*) [Sp.].  The fruit of the common banana, *Musa sapientium*, said to be a valuable alterative, and useful in strumous affections.  Dose of the *fluidextract* 10–30 min. (0.6–2.0 Cc.).
**bananina** (*ban-an-in'-ah*).  Banana flour, plantain flour; the fruit of *Musa sapientium*, dried and pulverized.
**banausea** (*ban-aw'-shur*) [βαναυσία, handicraft].  Mechanical work as opposed to mental achievement; Hippocrates' term for the practice of medicine regarded from a commercial standpoint rather than as an art; quackery.
**band** [ME., *bande*].  That which binds.  A stripe.  A ligament.  b.s, **amniotic**, bands formed by drawn-out adhesions between the fetus and the amnion where the cavity has become distended through the accumulation of fluid.  Syn., *Simonart's bands*. b., **anogenital**, the rudiment of the perineum; a transverse band of integument completing the division of the cloaca in the embryo. b. **auriculoventricular**, the bundle of His. b., **axis**, the primitive streak. b., **belly**-, a flannel band wound around the abdomen. b. **of the colon, inner**, a band-like thickening of the muscular coat running along the inner surface of the ascending and descending colon and the inferior aspect of the transverse colon.  b.s, **fetoamniotic**, amniotic bands producing deformities or intrauterine amputation. b., **furrowed**, a small band of cinerea uniting the uvula cerebelli with the cerebellar tonsils. b., **head-**, a strap for securing a mirror to the forehead. b., **horny** (of Tarinus), the fore part of the tænia semicircularis. b., **mesoblastic**, a band of mesoblastic cells which extends the entire length of the embryo. b., **moderator**, a fibromuscular fillet that frequently extends across the right ventricle of the heart.  Syn., *Moderator band*. b. **of Remak**.  See *fiber*, *axial*; also *Purkinje*, *axis-cylinder of*. b.s, **supraorbital**, the embryonal thickenings above the eyes and to the outer side of them.
**bandage** (*ban'-dāj*) [Fr., *bande*, a strip].  Bandages are usually strips of muslin or other material, of varying widths and lengths, used in surgery for the purpose of protecting, compressing, etc., a part, or for the retention of dressings and applications.  A *simple bandage* or *roller* consists of one piece; a *compound*, of two or more pieces. Starch, plaster-of-paris, silica, dextrin, tripolith, etc., are used for

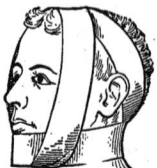
Barton's Bandage.

Recurrent of Head.

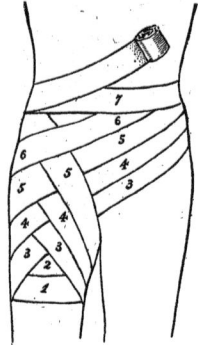
Spica of Groin.

Gibson's Bandage.

Borsch's Eye Bandage.

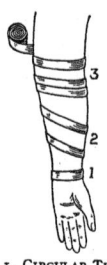
1. Circular Turns.
2. Oblique Turns.
3. Spiral Turns.

Spiral Reversed of the Forearm.

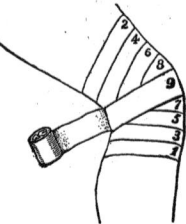
Figure of 8 of Knee.

Spiral Reversed Bandage.

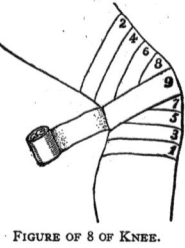
Spica of Foot.

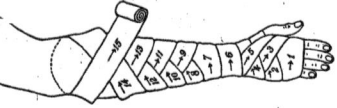
Figure of 8 of Upper Extremity.

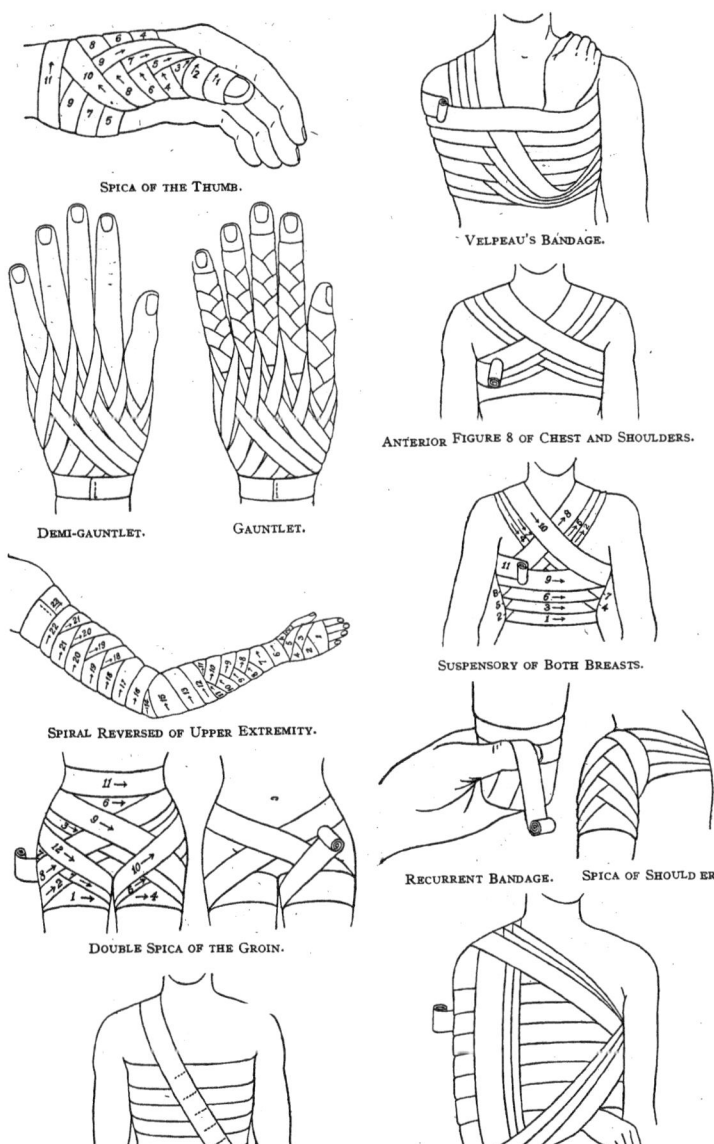

Spica of the Thumb.

Demi-Gauntlet. Gauntlet.

Spiral Reversed of Upper Extremity.

Double Spica of the Groin.

Spiral of the Chest.

Velpeau's Bandage.

Anterior Figure 8 of Chest and Shoulders.

Suspensory of Both Breasts.

Recurrent Bandage. Spica of Shoulder.

Desault's Bandage.

making stiff and *immovable dressings or bandages*. According to their direction, bandages are classed as: 1. *Circular*, circular turns about the part. 2. *Figure-of-*8, the turns crossing one another like the figure 8. 3. *Oblique*, covering the part by oblique turns. 4. *Recurrent*, the turns returning successively to the point of origin. 5. *Spica*, the turns resembling the arrangement of the husks of an ear of corn. 6. *Spiral*, each turn covering one-half of the preceding. 7. *Spiral reverse*, the bandage is reversed in order better to adapt it to the part. Bandages are also classed according to the part to which they are applied. (*See illustrations on pp.* 126, 127.) b., **abdominal**. See *binder*. b., **recurrent**, a bandage used after amputations, to support the flaps of the stump. b.s, **T-**, bandages shaped like the letter T.
**bandager** (*ban'-da-jer*). One skilled in the application of bandages. See *dresser*.
**bandagist** (*ban'-da-jist*). A maker of trusses, bandages, and other surgical appliances to be worn upon the person.
**bandelette** (*ban-dl-let'*). A small bundle.
**Bandeloux's bed** (*band-loo*). An air-bed furnished with a vessel for urine and surmounted with a gauze-covered cradle.
**Bandl's ring** [Ludwig *Bandl*, German obstetrician, 1842–1892]. The line of depression sometimes felt on digital pressure just above the pubes during labor-pains; it corresponds to the site of the internal os uteri.
**bandolin** (*ban'-do-lin*) [origin uncertain]. A mucilage made usually of quince-seeds, or of gum-tragacanth, used mainly as a paste for confining the hair and making it lustrous; it is called also *fixature*. See *cydonium*.
**bandy leg** (*ban'-de leg*). Bow-leg, genu valgum.
**baneberry** (*bān'-ber-e*). See *Actaa*.
**bang, bhang, bangue** (*bang*). See *cannabis indica*.
**Bang's bacillus** [Bernhard L. F. *Bang*, Danish physician, 1848–    ]. The *Bacillus abortus*.
**banian** (*ban'-yan*) [Ar., *banyān*, a trader]. A South Asiatic fig-tree, *Ficus bengalensis*, remarkable for the development of roots and secondary trunks from its branches. The bark and seeds are tonic, diuretic, and antipyretic.
**bant** [see *Banting treatment*]. To try the Banting treatment.
**Banti's disease** (*ban'-te*) [Guido *Banti*, Italian physician]. Enlargement of the spleen with progressive anemia, followed by hepatic cirrhosis.
**Bantingism** (*ban'-ting-izm*) See *Banting treatment*.
**Banting treatment** (*ban'-ting*) [William *Banting*, English undertaker, 1797–1878]. A treatment of obesity; the withdrawal of carbohydrates from the food, and the use of a diet of nitrogenous food.
**banyan** (*ban'-yan*). See *banian*.
**B. A. O.** Abbreviation for *Bachelor of the Art of Obstetrics*.
**baobab tree.** See *Adansonia digitata*.
**baptin** (*bap'-tin*) [*baptisia*]. A purgative glucoside obtainable from *Baptisia tinctoria*.
**baptisia** (*bap-tiz'-e-ah*) [βάπτισις, a dipping]. Wild indigo. The root-bark of *B. tinctoria*, the properties of which are due to an impure glucoside, the socalled *baptisin*. It is laxative and stimulant in moderate doses, emetic and cathartic in large doses, and is used in amenorrhea, typhus and typhoid fevers, and as a local application to indolent ulcers and gangrenous sores. Dose of the *resin* 1–5 gr. (0.065–0.32 Gm.). b., **extract** of. Dose 1–10 gr. (0.065–0.65 Gm.). b., **fluidextract** of. Dose 2–20 min. (0.13–1.3 Cc.). b., **tincture** of. Dose 5–30 min. (0.32–2.0 Cc.).
**baptisin** (*bap'-tiz-in*) [*baptisia*]. 1. A precipitate from the tincture of *Baptisia tinctoria;* antiseptic, purgative, ecbolic, resolvent. D$_{ose}$ 1–3 gr. (0.065–0.2 Gm.). 2. A bitter glucoside obtainable from the plant *Baptisia tinctoria;* it has little medicinal activity.
**baptitoxine** (*bap-tit-oks'-in*) [*baptisia;* τοξικόν, a poison]. A poisonous alkaloid obtained from *Baptisia tinctoria*. It hastens respiration and exaggerates vasomotor activity.
**baptothecorrhea** (*bap-to-the-kor-e'-ah*) [*baptisia*, infected; θήκη, vagina; ῥεειν, to flow]. Gonorrhea in women.
**baptorrhea** (*bap-tor-e'-ah*) [βάπτος, infected; ῥεειν, to flow]. Any infectious discharge from a mucous surface, as gonorrhea.
**bapturethrorrhea** (*bap-tū-reth-ro-re'-ah*) [βάπτος,

infected; *urethra;* ῥεειν, to flow]. Urethral gonorrhea; gonorrhea of men.
**bar** [OF., *barre*]. 1. A band or stripe. 2. The part of the upper jaw of a horse destitute of teeth. 3. An arch. 4. A prominence of the symphysis pubis projecting into the pelvic cavity. 5. A unit of atmospheric pressure representing one megadyne per square centimeter. b.s, **hyoid**, the pair of cartilaginous plates forming the second visceral arch. b., **interureteric**, same as b. *of bladder*. b. **of bladder**, the transverse curved ridge joining the openings of the ureters on the inner surface of the bladder; it forms the posterior boundary of the trigone. Syn., *bar of the bladder; interureteric bar; plica ureterica*.
**Bárány's sign**, B.'s **nystagmus** (*bah'-rah-ne*) [Robert *Bárány*, Austrian otologist, 1876–    ]. If the labyrinth of the ear is diseased injection of water into the external auditory canal causes no nystagmus; if the labyrinth is not diseased, similar injection of water causes nystagmus, such nystagmus being towards the same side if the temperature of the water is above that of the body, and towards the opposite side if the temperature of the water is below that of the body. See *Past pointing*.
**barba** (*bar'-bah*) [L.]. 1. The beard. 2. A hair of the beard.
**Barbados. aloes** (*bar-ba'-dōz al'-ōz*) [*Barbados*, an island in the West Indies]. See *aloes*. B. **distemper**. Synonym of *yellow fever*. B. **leg**. See *elephantiasis arabum*. B. **nut**. See *purging nut*.
**barbaloin** (*bar-bal'-o-in*) [*Barbados;* ἀλόη, aloe], C$_{17}$H$_{20}$O$_7$. The aloin derived from Barbados aloes.
**barbel** (*bar'-bel*) [*barba*, a beard]. The fish *Cyprinus barbus;* the roe is used as a purgative in some countries, and causes vomiting and purging if eaten to excess. b., **cholera**, an epidemic of fish-poisoning from eating diseased barbels. The symptoms are identical with those of cholera nostras, and are due to a ptomaine. Syn., *gastric ichthyotoxism*.
**barberry** (*bar'-ber-e*). See *berberis*. b. **gum**, a commercial name of certain varieties of gum-arabic.
**barber's itch**. Sycosis parasitaria.
**Barberio's test for semen** (*bar-ba'-re-o*). A drop of spermatic fluid or an aqueous extract of a spermatic stain when treated with a saturated aqueous solution of picric acid, shows a precipitate of yellow refractile crystals which increase in size.
**barbiers** (*bar'-bērs*) [E. Ind.]. A paralytic affection common in India, and probably a myelitis. It is often confounded with beriberi.
**barcoo** (*bar'-koo*) [Australian]. A peculiar disease, marked by nausea and vomiting, common in Australia. Its cause is unknown. It is also called "Fly-sickness."
**Bard's sign** [Louis *Bard*, Swiss physician, 1857–    ]. To differentiate between organic and congenital nystagmus. In the former the oscillations of the eyeball increase when the patient follows the physician's finger, moved before his eye alternately from right to left and from left to right. In the latter the oscillations disappear under these conditions.
**bardana** (*bar-da'-nah*) [L.]. The burdock. See *lappa*.
**Bardinet's ligament** (*bar-de-na*) [Barthélemy Alphonse *Bardinet*, French physician, 1809–1874]. The posterior fasciculus of the internal lateral ligament of the elbow-joint; it is attached above to the postero-inferior portion of the internal condyle, and below, by its expanded border, to the inner side of the olecranon process.
**Bareggi's reaction or sign** (*bar-ej'-e*). Twenty or 30 drops of blood collected in a small test-tube and allowed to stand for 24 hours will show a nonretracted clot and a small amount of serum if the blood has been taken from a typhoid-fever patient. In tuberculosis, on the other hand, the clot will retract and an abundance of serum will be formed.
**baregin** (*bar-a'-zhin*). See *glairin*.
**baresthesiometer** (*bar-es-the-se-om'-et-er*) [βάρος, weight; αἰσθησις, perception; μέτρον, a measure]. An instrument for estimating the sense of weight or pressure.
**baresthesiometric** (*bar-es-the-ze-o-met'-rik*). Relating to the baresthesiometer.
**Barfoed's reagent for dextrose** (*bar'-fēd*) [Christen Thomsen *Barfoed*, Swedish physician, 1815–1889]. One part copper acetate dissolved in 15 parts of water; 5 Cc. of acetic acid containing 38 % of glacial

acetic acid, added to 200 Cc. of this solution. Heat this reagent with a dextrose solution, and a reduction of copper suboxide is produced, but not when heated with lactose or maltose.
**baric** (ba'-rik) [barium]. Pertaining to or containing barium.
**barilla** (bar-il'-ah) [Fr., barille, impure soda]. Impure sodium carbonate; being ashes obtained by burning various chenopodiaceous plants of the genera *Salicornia* and *Salsola*. They contain about 30 % of sodium carbonate.
**barium** (ba'-re-um) [βάρος, weight]. Ba = 137.37; quantivalence II: A metal of the group of alkaline earths, of pale yellow color, characterized by a strong affinity for oxygen. The salts are poisonous. **b. arsenate**, Ba₃(AsO₄)₂, a white powder, almost insoluble in water. It is used in tuberculosis and in skin diseases. Dose $\frac{1}{16} - \frac{1}{4}$ gr. (0.004-0.016 Gm.). **b. benzoate**, Ba(C₇H₅O₂)₂+2H₂O, small, colorless plates; it is used instead of digitalis as a heart-stimulant. **b. carbonate**, BaCO₃, used in the preparation of the chloride. **b. chloride**, BaCl₂·2H₂O, soluble, used as a test for sulphates, which it precipitates as barium sulphate, and also as a cardiac and vasomotor stimulant. Dose $\frac{1}{2}$–5 gr. (0.032–0.32 Gm.). **b. chromate**, BaCrO₄, yellow crystals, insoluble in water; it is used as a pigment. Syn., *yellow ultramarine*. **b. hydrate**, **b. hydroxide**, Ba(OH)₂, caustic baryta; a crystalline substance, soluble in water, and used as a test for sulphates, which it precipitates as barium sulphate. **b. iodide**, formerly used as an alterative. Dose $\frac{1}{8}$ gr. (0.008 Gm.). It is employed in ointments. **b. manganate**, Ba-MnO₄, emerald-green powder of microscopic prisms or six-sided plates; it has been used as a pigment. **b. meconate**, a compound of barium and meconic acid; it is used as an anthelmintic. **b. oxide**, BaO, baryta. **b. sulphate**, BaSO₄. **b. sulphide**, BaS, a white, phosphorescent powder, soluble in water; it is used as an alterant. Dose $\frac{1}{2}$–1 gr. (0.032–0.065 Gm.) in keratin-coated pills. **b. sulphocarbolate**, Ba(C₆H₅-SO₄)₂, colorless crystals, soluble in water; it is antiseptic.
**bark** [ME., *barke*]. 1. The cortex or covering surrounding the wood of exogenous trees. 2. Synonym of *Cinchona* or *Calisaya bark*. **b., Jesuits'**, cinchona. **b., Peruvian**. See *cinchona*. (For other barks, see the names of the plants or trees that produce them.)
**Barker's operation** (bar'-ker) [Arthur Edward James Barker, English surgeon, 1850– ]. 1. *For excision of the astragalus*; the incision runs from just above the tip of the external malleolus forward and a little inward, curving toward the dorsum of the foot. 2. *For excision of the hip*; by an anterior incision, commencing on the front of the thigh, one-half inch below the anterior superior iliac spine, and running downward and inward for three inches.
**Barker's post-partum pill** (bar'-ker) [Fordyce Barker, American obstetrician, 1817–1891]. A laxative pill recommended for use during the puerperium. It contains compound extract of colocynth, 1$\frac{3}{8}$ gr., purified aloes $\frac{5}{8}$ gr., extract nux vomica $\frac{3}{16}$ gr., resin of podophyllum $\frac{1}{12}$ gr., ipecac $\frac{1}{12}$ gr., extract of hyoscyamus 1$\frac{1}{4}$ gr.
**Barkow's ligament** (bar'-ko) [Hans Carl Leopold Barkow, German anatomist, 1798–1873]. Ligamentous bundles lying in the fatty tissue of the olecranon fossa; they pass from the upper border of the fossa vertically downward to join the deeper fibers of the posterior ligament of the elbow-joint.
**barley** (bar'-le) [ME., *barly*]. A cereal belonging to the order *Gramineæ*; the most common variety, *Hordeum distichon*, is used as a food, and also in the preparation of malt. **b., decoction of**. See *b.-water*. **b., pearl-**, the decorticated grain, rounded and polished. **b.-water** (*decoctum hordei*, B. P.), a decoction consisting of 2 ounces of pearl-barley boiled in 1$\frac{1}{2}$ pints of water and afterward strained. It is used as a demulcent and food in the diarrheas of children. Dose 1–4 oz. (30–120 Cc.).
**Barlow's disease** (bar'-lo) [Sir Thomas Barlow, English physician, 1845– ]. Infantile scurvy, generally associated with rickets, and characterized by subperiosteal hemorrhages, especially of the long bones, with painful swellings.
**barm** (barm). Yeast.
**Barnes' cervical zone** (barnz) [Robert Barnes, English obstetrician, 1817–1907]. The lowest fourth of the internal surface of the uterus. **B.'s curve**, in obstetrics, the segment of a circle, having for its center the sacral promontory, its concavity looking backward. **B.'s dilators, or bags**, graduated rubber bags used for dilating the cervices uteri in the induction of abortion or premature labor.
**baro-** (bar'-o-) [βάρος, weight]. A prefix implying heaviness.
**baroelectroesthesiometer** (bar-o-e-lek-tro-es-the-ze-om'-et-er) [baro-; electric; αἴσθησις, perception; μέτρον, a measure]. An apparatus to determine the amount of pressure when electric sensibility to pain is felt.
**barograph** (bar'-o-graf) [baro-; γράφειν, to record]. A self-registering barometer.
**barology** (bar-ol'-o-je) [baro-; λόγος, science]. The branch of physics dealing with gravitation.
**baromacrometer** (bar-o-mak-rom'-et-er) [baro-; μακρός, long; μέτρον, measure]. An apparatus for ascertaining the weight and length of new-born infants.
**barometer** (bar-om'-et-er) [baro-; μέτρον, a measure]. An instrument for determining the weight and tension of the atmosphere. It consists essentially of a glass tube, about 36 inches long, closed at one end, filled with mercury, and inverted in a basin of mercury. The mercury will sink in the tube until it rests at a height of about 30 inches (760 mm.) at the sea-level, the height varying as the atmospheric pressure increases or diminishes. **b., air**, a barometer with air or gas imprisoned above the column of liquid; the variations of atmospheric pressure are indicated by the changes in the volume of this air. **b., aneroid**, a metallic box from which the air has been exhausted, the tension being indicated by the collapsing or bulging of the thin, corrugated cover, which is connected with a movable index. **b., boiling-point**, an instrument for determining the atmospheric pressure through observation of the boiling-point of water. Syn., *barothermometer; hypsometer; thermobarometer*. **b.-maker's disease**, a form of chronic mercurial poisoning among the workmen who make barometers. It is due to the inhalation of the fumes of mercury.
**barometric** (bar-o-met'-rik). Relating to atmospheric pressure, to a barometer, or to barometry. **b. light**, the glow produced by the mercury in a barometer tube when shaken.
**barometrograph** (bar-o-met'-ro-graf). See *barograph*.
**barometry** (bar-om'-et-re). The science of atmospheric pressure and the use of the barometer.
**baroscope** (bar'-o-skōp) [baro-; σκοπεῖν, to observe]. An instrument used for determining the loss of weight of a body in air, compared with its weight in a vacuum. A form of baroscope has been invented by Esbach for the quantitative determination of urea.
**Barosma** (bar-os'-mah) [baro-; ὀσμή, smell]. A genus of plants of the order *Rutaceæ*, native to the Cape of Good Hope and vicinity, several species of which yield the buchu of commerce.
**barosmin** (bar-oz'-min) [barosma]. A precipitate from the tincture of *Barosma crenulata*. Dose 2–3 gr. (0.13–0.2 Gm.).
**barotaxis** (bar-o-taks'-is) [baro-; τάξις, orderly arrangement]. Reaction of living matter to changes in pressure.
**barotropism** (bar-ot'-ro-pizm) [baro-; τροπή, turning]. See *barotaxis*.
**Barral's test for albumin and bile-pigments**. Cover the urine in a test-tube with a layer of 20 % solution of aseptol (orthophenolsulphonic acid), and in the presence of albumin a white ring will form at the zone of contact of the two fluids; $\frac{1}{8}$ of a grain of albumin in 1 liter of urine may be thus detected. Mucin causes a similar appearance, but it disappears on boiling. A green ring will indicate the presence of bile-pigments. This is much more sensitive than the color test with fuming nitric acid.
**barrenness** (bar'-en-nes). Sterility in the female.
**Barringtonia** (bar-ing-to'-ne-ah) [D. Barrington, an English naturalist]. A genus of plants of the order *Myrtaceæ*. *B. acutangula* is a tree growing in Australia and India. The juice from the leaves mixed with oil is used in skin diseases; the root is bitter, cooling, and aperient, and is said to be similar to cinchona; the seeds, prepared with sago and butter, are used in diarrhea. *B. butonica* is a tree of Australia and India. The outer portion of the fruit is used to stupefy fish. *B. racemosa* has properties similar to *B. acutangula*.
**Barry, retinacula of** (bar'-e) [Martin Barry, English scientist, 1802–1855]. Tense filaments running

from the thickened portion of the cellular membrane lining the Graafian follicle to other parts of the membrane.

**barsati** (*bar-sat-e'*). A disease affecting horses, considered analogous to cancer. Syn., *atrophic carcinoma*.

**Bartholin, duct of** (*bar'-tho-lin*) [Caspar *Bartholin*, Danish anatomist, 1655-1738]. The larger and longer of the sublingual ducts,' opening into the mouth near to, or in common with, Wharton's duct. **B., foramen of,** the obturator foramen. **B., glands of,** the vulvovaginal glands, a pair of glands situated at the entrance of the vagina, one on each side, and corresponding to Cowper's glands in the male.

**Bartholinian abscess.** An abscess of Bartholin's glands.

**bartholinitis** (*bar-to-lin-i'-tis*). Inflammation of Bartholin's glands.

**Barton's bandage** [John Rhea *Barton*, American surgeon, 1794-1871]. A bandage for the lower jaw. **B.'s fracture,** the separation of the posterior portion of the lower articular surface of the radius.

**Baruch's sign** (*bar'-ook*) [Simon *Baruch*, American physician, 1840-    ]. The resistance of the rectal temperature to a bath of 75° for 15 minutes, with friction; it is pathognomonic of typhoid fever.

**baruria** (*bar-u'-re-ah*) [βαρύς, heavy; οὖρον, urine]. The passage of urine having a high specific gravity; also the condition of the body associated therewith.

**Barwell's operation** (*bar'-wel*). Osteotomy at genu valgum: division of the lower and upper end of the tibia above and below their respective epiphyses.

**bary-** (*bar-e-*) [βαρύς, heavy]. A prefix meaning heavy, dull, or hard.

**baryecoia** (*bar-e-ek-oi'-ah*) [βαρυηκοΐα, hardness of hearing]. Hardness of hearing; partial deafness.

**baryencephalia** (*bar-e-en-sef-a'-le-ah*) [βαρύ-; ἐγκέφαλος, brain]. Dulness of intellect.

**baryencephalus** (*bar-e-en-sef'-al-us*) [see *baryencephalia*]. A person with dull intellect.

**baryglossia** (*bar-e-glos'-e-ah*) [βαρύ-; γλῶσσα, tongue]. Thick, slow utterance.

**baryglossus** (*bar-e-glos'-us*) [βαρύ-; γλῶσσα, the tongue]. An individual suffering from baryglossia.

**baryglottic, baryglotticus** (*bar-e-glot'-ik, -us*). 1. Relating to baryglossia. 2. A person affected with baryglossia.

**barylalia** (*bar-e-la'-le-ah*) [βαρύ-; λαλιά, speech]. Thickness of speech.

**barymastia** (*bar-e-mas'-te-ah*) [βαρύ-; μαστός, breast]. Same as *barymazia*.

**barymazia** (*bar-e-ma'-ze-ah*) [βαρύ-; μαζός, breast]. Heaviness or large size of the mammae.

**baryodmia** (*bar-e-od'-me-ah*) [βαρύ-; ὀδμή, odor]. A heavy, disagreeable odor; also a subjective sensation as of a disagreeable smell.

**baryodynia** (*bar-e-o-din'-e-ah*) [βαρύ-; ὀδύνη, pain]. Severe pain.

**baryphonia** (*bar-e-fo'-ne-ah*) [βαρύ-; φωνή, a voice]. A heaviness or difficulty of speech.

**barysomatia** (*bar-e-so-ma'-she-ah*) [βαρύ-; σῶμα, body]. Excess in the weight of the body.

**baryta, barytes** (*bar-i'-tah, bar-i'-tes*) [*barium*]. Barium oxide. **b., caustic.** See *barium sulphate*. **b., synthetic.** See *barium sulphate*.

**barythymia** (*bar-e-thi'-me-ah*) [βαρύ-; θυμός, mind]. A melancholy, gloomy, or sullen state of mind.

**baryticoargentic** (*bar-it-ik-o-ar-jen'-tik*) [*baryta*; *argentum*]. Containing baryta and silver.

**barytıcosodic** (*bar-it-ik-o-so'-dik*) [*baryta*; *sodium*]. Containing baryta and sodium.

**barytin** (*bar'-it-in*). Native barium sulphate.

**basad** (*ba'-sad*) [βάσις, a foundation]. Toward the basal aspect.

**basal** (*ba'-sal*) [βάσις, a foundation]. Pertaining to or located at the base. **b. ganglia,** the optic thalamus and corpus striatum of the brain.

**basalia** (*ba-sa'-le-ah*) [*basis*]. Huxley's term for the metacarpal bones.

**basculation** (*bas-ku-la'-shun*) [Fr., *basculer*, to swing]. 1. The movement by which retroversion of the uterus is corrected when the fundus is pressed upward and the cervix drawn downward. 2. See *bascule movement*.

**bascule movement** (*bas'-kūl*) [Fr., *bascule,* a swing]. The recoil of the heart in its systolic motion.

**base** (*bās*) [*basis*]. The lowest part of a body or the foundation upon which anything rests. In chemistry, an element or radical that combines with an acid to form a salt. The electropositive molecule or radical of a compound. In pharmacy, the most important part of a prescription. **b., acid-forming, b., acidifiable,** one which forms an acid by uniting with water. **b., aldehyde.** See *aldin*. **b., animal,** a ptomaine.

**base-ball pitcher's arm,** a peculiar condition of the arm arising from over-use, characterized by loss of strength, tenderness, neuralgic pains, and inflammation and hypertrophy of the bone.

**Basedow's disease** (*baz'-e-do*) [Karl Adolphus von *Basedow*, German physician, 1799-1854]. Exophthalmic goiter or Graves' disease. See *goiter, exophthalmic*. **B.'s syndrome,** tachycardia, flashes of heat, and sweating crises.

**Basella** (*bas-el'-ah*) [Malabar. name]. A genus of plants of the order *Chenopodiaceæ*. *B. rubra*, Malabar nightshade, is an esculent herb cultivated throughout India, where the juice of the leaves is given in infantile catarrh and an infusion of the leaves is used as tea.

**basement** (*bās'-ment*) [Fr., *bassement*]. The fundamental portion. **b.-membrane.** See *membrane, basement-*.

**bas-fond** (*bah-fon(g)*) [Fr.]. The floor or lowest oprtion of the urinary bladder.

**Basham's mixture** (*bash'-am*) [William Richard *Basham*, English physician, 1804-1877]. A mixture of iron and ammonium acetate; liquor ferri et ammonii acetatis.

**basi-** (*ba-si-*) [*basis*]. A prefix meaning basis or base. **basial** (*ba'-se-al*) [*basis*]. Relating to a base or to the basion.

**basialveolar** (*ba-se-al-ve'-o-lar*) [*basi-*; *alveolus*]. Relating to the basion and the alveolar point.

**basiarachnitis** (*ba-se-ar-ak-ni'-tis*) [*basi-*; ἀράχνη, a spider; ἶτις, inflammation]. Inflammation of that part of the arachnoid that corresponds to the base of the skull.

**basiator** (*ba-se-a'-tor*) [*basiare*, to kiss]. The orbicularis oris muscle. See *muscles, table of*.

**basic** (*ba'-sik*). 1. Having properties of a base; *i. e.,* capable of neutralizing acids. 2. Basal or basilar. **b. salt,** a salt in which part of the hydroxyl of the base is retained.

**basichromatin** (*bas-e-kro'-mat-in*) [*basi-*; χρῶμα, color]. According to Heidenhain, that portion of the nuclear reticulum stained by basic aniline dyes.

**basicity** (*bas-is'-it-e*). 1. The quality of being basic. 2. The combining power of an acid.

**basicranial** (*ba-se-kra'-ne-al*) [*basi-*; κρανίον, the skull]. Relating to the base of the skull. **b. axis,** a line running from a point midway between the occipital condyles through the median plane to the junction of the ethmoid and presphenoid.

**basidigital** (*ba-se-dij'-it-al*) [*basi-*; *digitus*, a finger]. Pertaining to the bases of the digits. **b. bone,** any metacarpal, or metatarsal bone.

**basidiogenetic** (*bas-id-e-o-jen-et'-ik*) [*basidium*, a spore-producing cell; γένεσις, origin]. In biology, produced on a basidium.

**basidiomycetes** (*bas-id-e-o-mi-se'-tēz*) [*basidium*; μύκης, a fungus]. A division of fungi comprising genera which produce spores upon basidia.

**basidiophore** (*bas-id'-e-o-fōr*) [*basidium,* a spore-producing cell; φερεῖν, to bear]. Furnished with basidia.

**basidiospore** (*bas-id'-e-o-spōr*) [*basidium,* a spore-producing cell; σπορά, seed]. One of the spores produced upon a basidium.

**basidium** (*bas-id'-e-um*) [dim. of βάσις, a base; pl., *basidia*]. In biology, a large cell in the higher fungi, borne on the hymenium and producing spores at its free end.

**basifacial** (*ba-se-fa'-shal*) [*basi-*; *facies,* face]. Pertaining to the lower portion of the face. **b. axis,** in craniometry, a line from the anterior point of the premaxilla to the anterior point of the basicranial axis.

**basifier** (*ba'-sif-i-er*) [*basi-*; *facere,* to make]. A substance capable of converting a body into a base.

**basifugal** (*bas-if'-u-gal*) [*basi-*; *fugere,* to flee]. In biology, derived from, or proceeding away from the base; acropetal; centrifugal.

**basigenic** (*ba-se-jen'-ik*) [*basi-*; γεννᾶν, to produce]. Producing bases.

**basihyal** (*ba-se-hi'-al*) [*basi-*; ὑοειδής, shaped like the letter υ, upsilon]. Either one of the two bones, one on each side, that form the principal part of the body of the hyoid arch.

**basihyoid** (ba-se-hi'-oid). See *basihyal*.
**basilad** (bas'-il-ad) [βάσις, foundation]. Toward or to the base or basilar aspect.
**basilar** (bas'-il-ar) [basis]. Pertaining to the base. **b. artery.** See under *artery*. **b. membrane,** a membranous division-wall separating the scala vestibuli from the scala tympani, extending from the base to the apex of the cochlea, and supporting the organ of Corti. **b. process,** a strong quadrilateral plate of bone forming the anterior portion of the occipital bone, in front of the foramen magnum. **b. suture,** the suture formed by the junction of the basilar process of the occipital bone with the posterior surface of the body of the sphenoid.
**basilateral** (bas-e-lat'-er-al) [basi-; latus, side]. Both basilar and lateral.
**basilemma** (bas-il-em'-ah) [basi-; λέμμα, a husk]. A basement-membrane.
**basilen** (bas'-il-en) [basis, base]. Belonging to the basilar portion in itself.
**basilic** (ba-sil'-ik) [βασιλικός, royal]. Important; prominent; said of a structure or a drug. **b. vein,** a large vein of the arm on the inner side of the biceps.
**basilicon ointment** (bas-il'-ik-on oint'-ment). Rosin cerate. It consists of rosin, 35 Gm.; yellow wax, 15 Gm.; lard, 50 Gm.
**basilobregmatic** (bas-il-o-breg-mat'-ik). Pertaining to the base of the skull and the bregma.
**basilomental** (bas-il-o-men'-tal). Pertaining to the base of the skull and to the chin.
**basilopharyngeal** (bas-il-o-far-in'-je-al). Relating to the basilar process of the occipital bone and to the pharynx.
**basilosubnasal** (bas-il-o-sub-na'-zal). Relating to the basion and the nasion.
**basilysis** (bas-il'-is-is) [basi-; λύσις, a loosening]. The breaking-up of the fetal skull in craniotomy.
**basilyst** (bas'-il-ist) [see *basilysis*]. An instrument for use in perforating the cranial vault and breaking up the base of the skull.
**basin** (ba'-sin) [ME.]. 1. The third ventricle of the brain. 2. The pelvis. **b.-trap,** a trap or seat in the outlet of the basin of a water-closet, placed there to prevent the escape into the apartment of noxious and offensive vapors and gases.
**basinasal** (bas-e-na'-sal) [basion; nasus, the nose]. Relating to the basion and the nasion.
**basioalveolar** (bas-e-o-al-ve'-o-lar). Relating to the basion and to the alveolar point.
**basioccipital** (bas-e-ok-sip'-it-al) [basi-; occiput, the back of the head]. A bone, separate in many of the lower vertebrate animals, forming the central axis of the skull. In adult human life it is the basilar process of the occipital bone.
**basioccipitosphenoidal** (bas-e-ok-sip-it-o-sfe-noid'-al). 1. Relating to the basioccipital bone and the sphenoid.
**basioglossus** (bas-e-o-glos'-us) [basi-; γλῶσσα, the tongue]. That part of the hyoglossus muscle that is attached to the base of the hyoid bone.
**basion** (ba'-se-on) [basis]. A point located at the middle of the anterior margin of the foramen magnum.
**basiotic** (bas-e-ot-ik) [basi-; οὖς, ear]. Relating to the base of the ear.
**basiotribe** (bas'-e-o-trīb) [basion; τρίβειν, to grind or crush]. An instrument used for perforating or crushing the fetal head.
**basiotripsy** (bas-e-o-trip'-se) [see *basiotribe*]. The operation of crushing the fetal head.
**basipetal** (bas-ip'-et-al) [basis, a base; petere, to seek]. In biology, applied to growth in plants from the apex toward the base.
**basipresphenoid** (bas-e-pre-sfe'-noid). 1. Relating to the basisphenoid and presphenoid bones. 2. The basipresphenoid bone.
**basirrhinal** (bas-e-ri'-nal) [basi-; ῥίς, nose]. Relating to the base of the brain and to the nose. Applied to a cerebral fissure located at the base of the olfactory lobe.
**basis** (ba'-sis) [βάσις, foundation]. Base. **b. cranii,** the base of the skull. **b. pedunculi,** the base of the peduncle; the crusta or pes.
**basisphenoid** (bas-e-sfe'-noid) [basi-; σφήν, wedge; εἶδος, form]. The lower part of the sphenoid bone.
**basisylvian** (bas-e-sil'-ve-an) [basi-; *Sylvian*]. Applied to the transverse basilar portion or stem of the Sylvian fissure.
**basitemporal** (bas-e-tem'-po-ral) [basi-; tempora, the temples]. Relating to the base or lower part of the temporal bone.

**basivertebral** (bas-e-ver'-te-bral) [basi-; vertebra, joint]. Relating to the basis or centrum of a vertebra.
**basket cell.** A cell surrounded by a network of fibrils derived from the axis cylinder process of another cell.
**basophil, basophile** (bas'-o-fil) [basis; φιλεῖν, to love]. Cells and tissue elements showing an affinity for basic rather than for acid dyes.
**basophilia** (bas-o-fil'-e-ah) [basophile]. Increase in the number of basophiles in the circulating blood.
**basophilic** (bas-o-fil'-ik) [basophile]. 1. Combining readily with bases; stainable by means of basic dyes. 2. Any histological structure which stains with basic dyes.
**basophilous** (bas-of'-il-us) [basophile]. Stained by basic rather than by acid dyes.
**basophobia** (bas-o-fo'-be-ah) [basis; φόβος, fear]. Complete inability to walk or stand erect, due to emotional causes. The muscles concerned are not appreciably impaired.
**basophobiac** (bas-o-fo'-be-ak). 1. A person affected with basophobia. 2. Relating to or affected with basophobia.
**bass-deafness** (bas'-def-nes). Deafness to certain bass-notes, the higher notes being heard.
**Bassini's operation** (bah-se'-ne) [Edoardo *Bassini*, Italian surgeon, 1847– ]. For the radical cure of inguinal hernia: the sac is exposed, twisted, and the neck ligated and removed; the spermatic cord is lifted, and the border of the rectus is stitched to the edge of the internal oblique, the transversalis muscle, and the transversalis fascia to Poupart's ligament under the cord. The cord is placed upon this layer and the border of the external oblique is stitched to Poupart's ligament over the cord.
**bassora gum** (bas'-o-rah gum). An inferior gum, much used in adulterating tragacanth.
**bassorin** (bas'-o-rin) [*Bassora*, an Asiatic town], $C_8H_{18}O_8$, or $2C_4H_{10}O_8$. 1. A tasteless, odorless, vegetable mucilage, insoluble in cold water, but rendered soluble by alkalies; it is found in gum tragacanth (of Bassora) and in cherry and plum gums. 2. A term for all vegetable mucilages.
**bast.** The inner bark of exogenous plants. The fibrous parts of the bark are used in making cordage, and have a limited use in surgery.
**bastard** (bas'-tard) [OF., *fils de bas*, son of a bast or of a pack-saddle]. 1. An illegitimate child. 2. Illegitimate. 3. A hybrid species. 4. See *bastards*.
**bastards** (bas'-tards) [*bastard*]. The name given to an impure sugar procured by concentrating molasses and allowing it to crystallize slowly in moulds.
**Bastedo's sign** (bas-te'-do) [Walter Arthur *Bastedo*, American physician, 1873– ]. If the colon is distended with air there will be pain and tenderness at McBurney's point in cases of chronic appendicitis.
**Bastian's law, B.-Brun's law** [Henry Charlton *Bastian*, English scientist, 1837– ; Ludwig *Bruns*, German neurologist, 1858– ]. When there exists a complete transverse lesion of the spinal cord above the lumbar enlargement, the tendon reflexes of the lower extremities are abolished.
**basyl** (ba'-sil) [βάσις, foundation]. 1. The electropositive constituent of a compound. 2. A body which unites with oxygen to form a base.
**basylous** (bas'-il-us). See *basigenic*.
**Bateman's disease** [Thomas *Bateman*, English physician, 1778–1831]. Molluscum contagiosum.
**B.'s drops,** the tinctura pectoralis, a weak tincture of opium, camphor, and catechu; a popular remedy in coughs.
**bath** [AS., *bæth*]. 1. A bathing-place or room. 2. The medium in which the body is wholly or partly immersed. As therapeutic agents, baths are classified according as water, vapor, air, etc., is used; according to the temperature, as hot, temperate, cold, etc.; according to the end desired, as nutritional, medicinal, stimulant, etc. Special forms of bath are the moor, peat, mud-, slime, pine-leaf, herb (hay, gentian, camomile, juniper, marjoram, etc.), brine, sand, tan, bran, malt, glue, soup, milk, whey, wine, guano, starch, soap, acid, iron, sulphur, carbonic acid, compressed air, mustard, electric, etc. 3. A medium, such as sand, water, oil, or other substance, interposed between the fire and the vessel to be heated, in chemical manipulations. **b., acid,** add 1½ oz. of nitric acid and 1 to 3 oz. of hydrochloric acid to 30 gallons of warm water in a wooden or

earthenware vessel, and immerse the patient in this for from 10 to 20 minutes. For a foot-bath, add ½ oz. of nitric acid and 1 oz. of hydrochloric acid to 4 gallons of warm water. This is said to be useful in cases of dyspepsia with sluggish liver and constipation. **b.s, acratothermal,** baths prepared from natural mineral waters of high temperature, but in which the gaseous and saline constituents are small in quantity and of feeble therapeutic action. Syn., *indifferent thermal baths; simple thermal baths; unmixed thermal baths.* **b., air-,** a bath in which but little water is employed, the body being exposed freely to the air. It is employed in those cases in which there is a tendency to catch cold on slight exposure. **b., alcohol,** one in dilute alcohol, used to reduce temperature in fever. **b., alkaline,** add 3 oz. of potassium carbonate, or 6 of sodium carbonate, to 25 or 30 gallons of hot water. It is used in chronic squamous skin diseases, chronic rheumatism, and lithemia. It should be taken in a wooden, earthenware, or enameled tub. **b., animal.** 1. One prepared from dung of cattle or the contents of the first stomach of a freshly slaughtered ox. 2. The introduction of the whole body of the part affected with rheumatism into the body-cavity of an animal just slaughtered. **b., antimonial,** one containing 1–2 oz. of tartar emetic; it is used in skin diseases. **b., antirheumatic,** one containing, in sufficient water for the purpose, 100 Gm. of oil of turpentine, 10 Gm. of oil of rosemary, 50 Gm. of sodium carbonate. **b., antisyphilitic,** a solution of 15 Gm. of mercury bichloride in 500 Gm. of water, to be added to the bath at the time of using. **b., astringent,** one prepared with tannin or other astringents, to control sweating or in the treatment of skin diseases. **b., astringent, Most's,** a bath for extensive burns, consisting of a solution of 200 Gm. of alum in 6 to 8 pailfuls of cold water and 1 pailful of curdled milk. **b., balsamic,** one containing tar, turpentine, or the buds and bark of terebinthaceous plants. **b., bog-,** a bath made by mixing bog-earth (produced by the decomposition of plants in the presence of water and found at iron and sulphur springs) with warm water to form a pulpy or mushy consistence. This is used as a mud-bath. **b., borax,** borax, 4 oz.; glycerol, 3 oz.; water, 30 gallons. It is used in the same class of cases as the bran-bath. **b., bran-,** boil 1 pound of bran in 1 gallon of water, strain, and add 30 gallons of water. This is a soothing and emollient bath, and is of service in squamous and irritable conditions of the skin. **b. Brand,** see *Brand.* **b., buff-,** one in which the bather is nude. **b.-chair** [*Bath,* town in England, where the inventor James Heath, lived]. A vehicle for the conveyance of invalids and others; it is mounted on three or four wheels, and may be pushed by hand or drawn by pony or donkey. **b., chemical,** in chemistry, an apparatus for regulating the temperature of chemical processes by surrounding the substance with water, sand, oil, or mercury, through which the heat is communicated. **b., cold,** a bath of cold water, the temperature of the latter varying from 32°–70° F. (0°–21° C.). It is used to reduce fever and as a general stimulant. **b., dipolar,** a hydroelectric bath in which the patient does not come in contact with either of the electrodes, but these are immersed in the water at each end of the tub. **b., effervescent,** a liquid bath containing a free gas, which is given off with effervescence. **b., Egyptian,** a modification of the Turkish bath, with rise of temperature to the maximum point, followed by lowering of temperature to the initial point. **b., electric.** 1. One in which the medium of the bath and the bather's person are included in the circuit of a galvanic current. 2. A bath in which an electric current is generated by the decomposition of the chemical constituents of the medium. **b., electrothermal,** a hot bath combined with exposure to the influence of electricity. **b., foot-,** a bath for the feet, used as a derivative agent in cases of cold, etc. **b., fucus-,** one containing seaweed, or a decoction of it, imparting sodium chloride and a small percentage of iodine. **b., full,** one in which the patient's body is entirely covered by water, so that his chin just clears it. **b., gas-,** one in which a gas is applied to the entire body or an affected part by means of a closed cabinet. **b., graduated,** one in which the temperature of the water is gradually lowered by the addition of cold or iced water. **b., herb-,** made by using the extract of pine-needles or of some aromatic herbs; used as a tonic. **b., hip-,**

See *b., sitz-*. **b., hot,** one in which the temperature of the water ranges from 104° to 110° F. (40°–43.3° C.). It acts upon the skin, producing free perspiration, and accelerates the pulse and respiration. **b., hot-air,** a Turkish bath. A bath in hot air. It is useful as a diaphoretic, and in catarrhal, neuralgic, and rheumatic conditions, but is contraindicated in fatty degeneration of the heart. **b., hydroelectric,** a water-bath charged with electricity. **b., hydrostatic,** a variety of permanent water-bath in which the patient is supported without total immersion. **b., Indian,** massage in combination with a Turkish bath. **b., internal,** lavage of the stomach or rectum. **b., iodine-,** one to which is added a solution of 8 Gm. of iodine and 16 Gm. of potassium iodide in 600 Gm. of water. **b., kinetotherapeutic,** a water-bath in which specified movements are carried out. **b., light-.** 1. See *b., sun-*. 2. Exposure of affected parts to rays of light by means of various apparatus; also of the whole body for inducing perspiration. **b., lime-,** a bath to which slaked lime is added at the time of using; it is used in gout and in treatment of itch. **b., medicated,** a bath in which medicinal substances, as mineral salts, sulphur, etc., are dissolved or held in suspension. **b., mercurial,** a bath for the treatment of syphilis, given in the vapor of mercury, usually prepared by vaporizing calomel over an alcohol lamp. **b., milk-,** a bath in milk, given for nutritive purposes. **b., mineral.** 1. The water of a mineral spring used as the medium. 2. One to which a solution of mineral substances has been added. **b., monopolar,** a hydroelectric bath in which the wall of the metal tub is utilized as a large electrode. The current entering here is conducted to the entire surface of the body in contact with the water, and passes out by means of a large metal electrode the edges of which are covered by a rubber pillow so placed that the patient can lie upon it without coming in contact with the metal. Cf. *b., dipolar.* **b., moor-,** a bath in water mixed with the earth of moors. **b., mud-,** a bath prepared by mixing well-seasoned earths, containing more or less mineral matter, with water containing the same substances. It is useful in chronic rheumatism. **b., mustard-,** made by inclosing from 2 to 4 oz. of ordinary mustard in a piece of muslin or thin linen and hanging it in about four gallons of hot water until the latter becomes yellow, or simply by adding mustard to water. It is used as a general bath for infants in collapse, convulsions, or severe bronchitis, the child being left in until the skin becomes distinctly reddened. It is also used as is the foot-bath or sitz-bath in amenorrhea. **b., Nauheim,** a natural thermal effervescent (gaseous muriated) bath. **b., needle.** See *b., rain-*. **b., nutritive,** one containing wine, milk, or any nutritive ingredient. **b., oxygen-,** an inhalation of oxygen to correct inadequate aeration of the blood; also a local application of oxygen gas to gangrenous ulcers. **b., ozone-.** See *b., fucus-*. **b., pack-,** one in which the body is wrapped in wet cloths. See *b., sheet-*. **b., peat-,** an application of bog-earth containing much vegetable matter and used in gout and rheumatism. **b., pine-,** prepared by adding a decoction of pine-needles, or some pine-extract, to hot water. It is mildly stimulating, and is employed in hysteria, gout, and rheumatism. **b. of Plombières,** a bath to which a solution of 100 Gm. of powdered gelatin in hot water is first added and afterward a mixture of 100 Gm. of sodium carbonate, 20 Gm. each of sodium chloride and sodium bicarbonate, and 60 Gm. of sodium sulphate. **b., plunge-,** a bath, hot or cold, into which the patient plunges. **b., rain-,** consists of from four to six three-fourths circles of pipes secured together at a distance of 2 to 3 inches. Each pipe has three lines of fine perforations, from which the stream issues under pressure, striking the body of the patient at all points with considerable force. **b., Russian,** a vapor-bath, the vapor being generated by throwing water upon heated mineral or metallic substances. Syn., *vapor-bath.* **b., sand-.** 1. One in which the body of the patient is placed in a layer of dry sand that has been heated. 2. In chemistry, the immersion of a crucible, etc., containing a chemical compound, in a vessel containing fine sand, the latter being heated gradually to a high temperature. **b., seaweed.** See *b., fucus-*. **b., sheet-,** the application of cold water to the body through the medium of a wet sheet or towel placed upon the skin. It is used to reduce temperature. **b., shower-,** a douche in which the water is delivered against the body from a nozzle with numerous perforations. **b., sitz-,** one in which only the buttocks and hips are immersed in water. It is useful in pelvic inflammations, amenorrhea, and retention of urine. **b., slime-,** a bath in water mixed with the slimy deposit of organic matter found in rivers and ponds. **b., sponge-,** one in which the patient's body is rubbed with a wet sponge. It is used to reduce temperature. **b., sulphur,** potassium sulphide 4 to 8 oz. in 30 gallons of water; a little sulphuric acid may be added. It is used in certain skin diseases, scabies, lead colic, and lead palsy. **b., sun-,** the exposure of the naked body to the sun's rays. **b., sweat-, b., sweating,** a bath to induce a free flow of perspiration; *e. g.*, a Turkish bath. **b., tepid,** the temperature of the water ranges from 85°–95° F. (29.4°–35° C.). It acts as a sedative, cleansing, and detergent agent; the pulse, respiration, excretion, secretion, and temperature are practically unaffected. **b., Turkish,** one in which the bather is placed successively in rooms of higher temperature, then shampooed or rubbed, and finally stimulated by a douche of cold water. See *b., hot-air.* **b., vacuum,** the treatment of parts by subjecting them to a partial vacucum. **b., vapor-.** See *b., Russian.* **b., warm,** a bath in water having a temperature of from 90°–100° F. (32.2°–38.5° C.). It is used to calm the nervous system, produce sleep, and allay reflex irritability. **b., water-.** 1. A bath of water. 2. In chemistry, a bath of water for immersing vessels containing substances that must not be heated above the boiling-point of water. **b., zinc-chloride,** a chemical bath of molten zinc chloride for immersion of substances not to be heated beyond 700° C.

**bathmic** (*bath'-mik*) [βαθμός, a threshold]. Relating to bathmism.

**bathmism** (*bath'-mism*) [βαθμός, a threshold]. That supposed modification of chemical force which is active in the processes of nutrition.

**bathmos** (*bath'-mos*) [βαθμός, a little fossa]. A shallow depression or fosset.

**bathmotropic** (*bath-mo-trop'-ik*) [βαθμός, threshold; τρέπειν, to turn]. Applied by T. W. Engelmann to a supposed set of fibers in the cardiac nerves which affect the excitability of the muscle. Cf. *dromotropic; inotropic.*

**bathycardia** (*bath-e-kar'-de-ah*) [βαθύς, deep; καρδία, heart]. A condition in which the heart is in a lower position than normal; the condition is an anatomical one, and is not the result of disease.

**bathycentesis** (*bath-e-sen-te'-sis*) [βαθύς, deep; κέντησις, puncture]. A deep surgical puncture; deep acupuncture.

**bathyesthesia** (*bath-e-es-the'-ze-ah*) [βαθύς, deep; αἴσθησις, sensation]. Oppenheim's term for the muscle-sensations.

**bathymetry** (*bath-im'-et-re*) [βαθύς, deep; μέτρον, a measure]. The measurement of the depths of the sea or of any body-cavity, natural or abnormal.

**bathystixis** (*bath-e-stiks'-is*) [βαθύς, deep; στίξις, puncture]. Deep acupuncture.

**batophobia** (*bat-o-fo'-be-ah*) [βατός, a height; φόβος, fear]. 1. Acrophobia; dread of being at a great height. 2. Dread of high objects; fear of passing near a high building or of going through a deep valley.

**batrachocephalus** (*bat-rak-o-sef'-al-us*) [βάτραχος, a frog; κεφαλή, the head]. Having a frog-like head.

**batrachoid** (*bat'-rak-oid*) [βάτραχος, a frog; εἶδος, form. Frog-like.

**batrachoplasty** (*bat'-rak-o-plas-te*) [βάτραχος, a frog; ranula; πλάσσειν, to mold]. Plastic surgical operation for the cure of ranula.

**batracin** (*bat'-ra-sin*) [βάτραχος, a frog]. A poisonous secretion obtained from the cutaneous pustules of toads. According to Calmeil, the poison of toads contains methyl-carbylamine and isocyanacetic acid.

**battalism** (*bat'-al-izm*). See *battarism.*

**battarism** (*bat'-ar-ism*) [βατταρισμός, a stuttering]. Stuttering or stammering.

**battery** (*bat'-er-e*) [*batteria*, a beating; battery]. A series of two or more pieces of apparatus arranged to produce increased effect, as *battery* of boilers, prisms, lenses, galvanic cells. It is usually applied to a series of cells producing electricity (voltaic or galvanic battery); also, frequently, to a single cell. **b., cautery,** a galvanic battery with high electromotive force for heating a wire used as a cautery. **b., faradic, b., faradaic,** one giving an induced or faradic current. **b., galvanic,** one giving a galvanic or chemical current. **b., Hare's,** a battery of cells marked by low resistance. Two large plates of zinc and copper, separated from each other by cloth or some indifferent substance, are rolled on a wooden cylinder and immersed in acidulated water. See *deflagrator.* **b., primary,** the combination of a number of primary cells so as to form a single battery. **b., secondary,** the combination of a number of storage-cells to form a single electric source. **b. storage,** an apparatus consisting of a secondary battery for storing electricity.

**Battey's operation** (*bat'-e*) [Robert Battey, American surgeon, 1828–1895]. Removal of the ovaries in order to eliminate their physiological influence and so produce the menopause.

**Battle's incision** (*bat'-l*). An incision to the right or left of the median line of the abdomen down to the rectus; the inner edge of this muscle is then retracted and the posterior layer of its sheath incised as well as the peritoneum.

**battledore** (*bat'-el-dōr*) [ME., *batyldoure*, a bat for beating clothes]. An instrument shaped like a racket. **b. placenta,** one in which there is a marginal insertion of the cord.

**Battley's sedative drops** (*bat'-le*) [Richard *Battley*, English pharmacist, 1770–1856]. A preparation consisting of opium 3, water 30, alcohol 6; dose ♏. 5–15 (0.3–1.0).

**Baudelocque's diameter** (*bo-d'l-ok*) [Jean Louis *Baudelocque*, French obstetrician, 1746–1810]. In obstetrics, the externalc onjugate diameter of the pelvis.

**Bauer's qualimeter** (*bow'-ers kwol-im'-et-ur*). A static electrometer with pendulum and an index, used to indicate the quality of the x-rays issuing from a Roentgen tube.

**Bauhin, gland of** (*bo'-hin*) [Caspar *Bauhin*, Swiss anatomist, 1560–1624]. See *Blandin's gland.* **B. valve of,** the ileocecal valve.

**Bauhinia** (*bo-hin'-e-ah*). A genus of plants of the order *Leguminosa.* B. *variegata* is a tree of India; the bark is astringent and used as a tonic in fevers; the buds are used in diarrhea and as a vermifuge.

**Baumann's coefficient.** The ratio existing between the total sulphuric acid and the ether sulphuric acids of the urine; it amounts to 10 %. B.'s reaction for **dextrose,** to an aqueous solution of grape-sugar add benzoyl chloride and an excess of sodium hydroxide and shake until the odor of benzoyl chloride disappears. A precipitate of benzoic-acid ester of dextrose will be produced which is insoluble in water and alkalies.

**Baumann and Goldmann's test for cystin.** If a solution of cystin is shaken in caustic soda with benzoyl chloride, a voluminous precipitate of benzoyl cystin will be produced. The sodium salt occurs as silky plates, readily soluble in water, but nearly insoluble in an excess of caustic soda.

**Baumé** (*bo-ma'*) [Antoine *Baumé*, French chemist, 1728–1804]. Abbreviation for Baumé's hydrometer.

**Baumes' law** (*bo-mes'*) [Jean Baptiste Timothée *Baumes*, French physician, 1756–1828]. Same as *Colles's law.* **B.'s sign or symptom,** retrosternal pain in angina pectoris.

**Baunscheidtism** (*bown'-shi-tizm*) [Carl *Baunscheidt*, a German mechanic]. A mode of treating rheumatism and chronic neuralgias by counterirritation, the latter being produced by pricking the exterior of the part affected with fine needles dipped in oil of mustard, formic acid, or other irritant.

**Bavarian dressing, B. splint.** A variety of immovable dressing in which the plaster is applied between two flannel cloths.

**bavarol** (*bav'-ar-ol*). A proprietary brown, aromatic liquid used in 5 % solution as a disinfectant.

**bawchan, bauchee.** Names given in India to the seeds of *Psoralea coryifolia*, used as a tonic and in skin diseases. Syn., *bauchan; bawchwan.*

**bay, lacrimal.** The depression at the inner canthus of the eye, holding the lacrimal canaliculi. **b., oil of.** See *myrcia.* **b.-rum.** See *myrcia.* **b.-tree,** the *Laurus nobilis;* also *Prunus laurocerasus*, commonly called the laurel or the cherry-laurel.

**Bayard's ecchymoses** [Henri Louis *Bayard*, French physician, 1812–1852]. Small capillary hemorrhages found in the pleura and pericardium of infants who, as the result of asphyxia, have made premature efforts at breathing in the uterus.

**bayberry** (*ba'-ber-e*) [ME., *bay; bery*]. 1. The berry of *Laurus nobilis*, bay, or noble laurel. 2. The wax-myrtle, *Myrica cerifera*, and its fruit. 3. The pimenta, or allspice.

**baycuru** (*bi-koo-roo'*) [native S. A.]. The root of a S. American plant, *Statice braziliensis*, one of the most powerful astringents known. It is used for ulcers of the mouth and for glandular enlargements. Dose of the fluidextract ɱ. v-xxx.

**Bayle's disease** [Gaspard Laurent *Bayle*, French physician, 1774–1816]. Progressive general paralysis of the insane. B.'s **granulations**, miliary tubercles.

**bayonet-leg** (*ba'-on-et leg*). A backward displacement of the leg-bones.

**bazin** (*ba'-zin*). Synonym of *molluscum contagiosum*.

**Bazin's disease** (*baz-an'*) [Pierre Antoine Ernest *Bazin*, French dermatologist, 1807–1878]. 1. Psoriasis buccalis. 2. Scrofulous ulcer of the leg. B.'s **erythema**, erythema induratum scrofulosorum, a form most commonly seen in strumous individuals; it attacks the calf, or the leg immediately below, more frequently than the front of the leg, occurring in diffuse, ill-defined patches or in nodules, bright red at first and gradually assuming a violet hue. The nodules may be superficial or deep, a quarter of an inch to an inch or more in diameter, and may be slowly absorbed, and necrose or slough out, leaving a very indolent ulcer.

**B. C.** Abbreviation for Bachelor of Chemistry. Also for *Baccalaureus Chirurgiæ*, Bachelor of Surgery.

**B. Ch.** Abbreviation for *Baccalaureus Chirurgiæ*, Bachelor of Surgery.

**B. C. L.** Abbreviation for Bachelor of Civil Law.

**bdella** (*del'-ah*) [βδέλλα, a leech]. 1. A leech. 2. A varicose vein.

**bdellatomy** (*del-at'-o-me*). See *bdellotomy*.

**bdellepithecium** (*del-ep-e-the'-se-um*) [*bdella*; ἐπιτίθναι, to put on]. A tube for applying leeches.

**bdellium** (*del'-e-um*) [Heb., *d'bŏlakh*]. A resinous gum exuding from various species of *Balsamodendron*. It resembles myrrh. b., **Indian**, has been recommended as an emmenagogue.

**bdellometer** (*del-om'-et-er*) [βδέλλα, a leech; μέτρον, a measure]. A mechanical substitute for the leech, consisting of cupping glass, scarificator, and exhausting syringe.

**bdellotomy** (*del-ot'-o-me*) [βδέλλα, a leech; τομή, a cut]. The opening with the knife of the body of a leech that is drawing blood; by this means the activity of the leech may be considerably prolonged.

**Be.** Chemical symbol of beryllium, now called *glucinum*.

**B. E.** Bacillary emulsion. See *tuberculin*.

**beaded** (*be'-ded*). In bacteriology denoting disjointed semi-confluent colonies along the line of inoculation in a stab culture.

**beads, rhachitic.** "Rhachitic rosary"; the socalled "beading of the ribs" in rickets; a succession of visible and palpable swellings at the points where the ribs join their cartilages.

**beak** (*bēk*) [ME., *beeke*]. 1. The mandibular portion of a forceps. 2. The lower end of the calamus scriptorius. 3. The pad or splenium of the corpus callosum. b., **coracoid**, the beak of the coracoid process of the scapula. b. of the **encephalon**. See *beak* (3). b. of the **sphenoid bone**. See *rostrum sphenoidale*.

**beaker** (*be'-ker*) [Germ. *becher*, a cup, bowl]. A wide-mouthed glass vessel used in chemical laboratories. b.-**cell**, the goblet-cell found in mucous membranes.

**Beale's fiber** (*bēl*) [Lionel Smith *Beale*, English physician, 1828–1906]. The fine spiral fiber surrounding the process of some of the sympathetic ganglion-cells of the frog.

**bean** (*bēn*) [ME., *bene*]. The seeds of several plants, mostly *Leguminosæ*, especially that of the common bean, *Faba vulgaris*. b., **Calabar**. See *physostigma*. b., **castor**. See *ricinus*. b. of St. **Ignatius**. See *ignatia*.

**bearberry.** See *Uva ursi*.

**Beard's disease** [George Miller *Beard*, American physician, 1839–1883]. Nervous exhaustion; neurasthenia.

**Beard-Valleix's points.** See *Valleix's points douloureux*.

**bearing-down.** The feeling of weight or pressure in the pelvis in certain diseases. b.-d. **pains**, uterine pains in labor.

**bear's-foot.** Leafcup. A composite-flowered plant, *Polymnia uvedalia*, of North America. A popular remedy for enlargement of the spleen or the 'ague-cake'' of malarious regions.

**beat** (*bēt*) [ME., *beten*]. The pulsation of the arteries or the impulse of the heart. b., **apex**-, the stroke of the heart-apex against the chest-wall. b., **heart-**, a pulsation of the heart. b., **pulse-**, an arterial pulsation which can be felt.

**Beatty-Bright's friction-sound.** The friction-sound produced by inflammation of the pleura.

**Beau's disease** (*bo*) [Joseph Honoré Simon *Beau*, French physician, 1806–1865]. Asystole; cardiac insufficiency. B.'s **lines**, the transverse rings seen on the finger-nails after convalescence from exhausting diseases.

**Beaumès' sign** (*bo'-mā*). See *Baumes' sign*.

**Beaumês-Colles' law.** See *Colles' law*.

**Beauperthuy's method** (*bo'-per-thwē*) [Louis Daniel *Beauperthuy*]. A method of treating leprosy by good hygiene, bathing with olive oil, the internal administration of mercury bichloride, and abstinence from salted meats.

**Beauvais' disease** (*bo'-vā*). Chronic articular rheumatism.

**bebeerine** (*beb'-e-rēn*). See *nectandra*. b. **hydrochloride**, $C_{19}H_{21}NO_3HCl$, reddish-brown scales, soluble in alcohol and water, and used as an antipyretic and tonic. Dose $\frac{1}{3}-1\frac{1}{2}$ gr. (0.005–0.097 Gm.) 3 or 4 times daily. b. **sulphate**, $(C_{19}H_{21}NO_3)_2H_2SO_4$, reddish-brown scales, soluble in water and alcohol; uses and dose as in b. *hydrochloride*.

**bebeeru bark** (*be-be'-ru*). See *nectandra*.

**Beccaria's sign** (*bak-kah'-re-ah*). Painful pulsating sensations in the occipital region during pregnancy.

**bechesthesis** (*bek-es'-thes-is*) [βήξ, cough; αἴσθησις, sensation]. The sensation in the throat or air-passages that prompts one to cough.

**bechic** (*bek'-ik*) [βήξ, a cough]. Relieving cough; a remedy against cough.

**Bechterew's disease** (*bek'-ter-u*) [Vladimir Mikhailovich von *Bechterew*, Russian neurologist, 1857–   ]. Ankylosis of the vertebral column, associated, as a rule, with muscular atrophy and sensory symptoms. B.'s **layer**, the layer of fibers between and parallel to the tangential fibers and Baillarger's layer in the cerebral cortex. B.'s **nucleus**, the nucleus of the vestibular portion of the auditory nerve; an ill-defined group of ganglion-cells lying dorsad of Deiters' nucleus. B.'s **reaction**, the minimum strength of the electric current necessary to provoke muscular contraction requires a gradual diminution at every interruption of the current or change in density, to prevent tetanic contraction which will occur if the inital strength is maintained. It is observed in tetany. B.'s **sign**, anesthesia of the popliteal space in tabes dorsalis. B.'s **tract**, the central tract of the tegmentum that passes between the mesial side of the superior olivary body and the fillet.

**Beck's method** [Emil G. *Beck*, American surgeon, 1866–   ]. Tuberculous cavities in bone are curetted and then filled with a paste containing 30 per cent. of bismuth subnitrate.

**Beck's operation.** To cure epilepsy due to adhesions resulting from fracture or operations on the skull. The skull defect is plugged with temporal fascia and muscle.

**Becker's reaction for picrotoxin.** Picrotoxin reduces Fehling's solution upon the application of gentle heat. B.'s **sign** [Otto Heinrich Enoch *Becker*, German ophthalmologist, 1828–1890]. Spontaneous pulsation of the retinal arteries in exophthalmic goiter. B.'s **test**, a test for astigmatism made by means of a set of parallel lines in triplets placed in various meridians.

**Béclard's hernia** (*ba'-klar*) [Pierre Augustin *Béclard*, French anatomist, 1785–1825]. Hernia occurring through the saphenous opening. B.'s **nucleus**, a vascular, bony nucleus, of lenticular shape, appearing in the cartilage of the lower epiphysis of the femur during the thirty-seventh week of fetal life.

**Becquerel's disc** (*bek'-er-el*). An apparatus for estimating the difference in temperature between a sound limb and a paralyzed one. B.'s **pills**, pills containing quinine, extract of digitalis, and colchicum seeds; they are used in gout. B.'s **rays**, invisible radiations of electrified particles or ions projected from radioactive bodies such as uranium, radium, polonium, or their salts, without evident cause, and persisting over long periods.

**bed** [AS., *bedd*]. The couch or support on which

the body may rest in sleep or in sickness; usually a mattress of straw, hair, or similar substance. b., air-, a mattress of rubber or leather that can be inflated with air. b.-case, a form of hysteria in which the patient persistently lies in bed. b., fracture-, an especial device for the use of a patient confined with a fracture, composed of sections forming a double or triple inclined plane with an aperture to allow of the ejection of urine and feces. b.-hoist, a device for lifting a patient from bed. b.-pan, a large shallow vessel for receiving the alvine discharges of bedridden patients. b., protection-, a bed arranged for the confinement of maniacs in a recumbent posture. b.-rest, an apparatus for propping up patients in bed. b.-sore, a sore produced on any part of the body by prolonged pressure against the bed or by trophic changes in paralyzed parts. b.-swing, an appliance like a hammock for swinging a patient clear of the bed. b., water-, a rubber mattress containing water; it is used to prevent the formation of bed-sores.

**bedbug** (*bed'-bug*). An apterous insect, *Cimex lectularius*, that infests bedsteads, and at times is parasitic upon the human body.

**bedlam** (*bed'-lam*) [ME., *bedlem*, a corruption of *Bethlehem*]. An insane asylum.

**bedlamism** (*bed'-lam-ism*) [*bedlam*]. Insanity.

**bedlamite** (*bed'-lam-īt*) [*bedlam*]. A madman; an insane person.

**Bednar's aphthæ** [Alois *Bednar̆*, Austrian physician]. Two symmetrically placed ulcers seen at times on the hard palate of cachectic infants, one on each side of the mesial line.

**bedouin itch**. A synonym of the vesicular variety of *lichen tropicus*.

**bedridden** (*bed'-rid-n*). Confined to bed; applied especially to those who seem permanently so affected.

**bed-wetting.** Nocturnal enuresis.

**beef** (*bēf*) [ME., *befe*]. The flesh of domestic cattle. It consists of water, 73; fibrin, 15; gelatin, 4; albumin, 3; fat and other substances, 5 %. b.-extract, the soluble fibrin of lean meat partly desiccated. b.-measles. See *Cysticercus bovis*. b.-tea, the soluble extractive matter of beef, made by steeping finely cut lean beef in its weight of water and straining.

**Beeley's square and plumb-line.** An instrument to measure degrees of deformity.

**beer** (*bēr*). See *malt liquors*.

**Beer's cataract knife** (*bēr*) [Georg Joseph *Beer*]. A knife with a triangular-shaped blade, for making section of the cornea in the removal of the crystalline lens. B.'s operation, extraction of cataract by the flap method.

**beestings** (*bēst'-ings*). First milk after parturition; colostrum.

**beeswax** (*bēs'-waks*). *Cera flava* (U. S. P.), wax secreted by bees, of which their cells are constructed. It is used in making candles, ointments, and pomades.

**beet** (*bēt*). The genus *Beta*, especially *B. vulgaris*, the common beet.

**Begbie's disease** (*beg'-be*) [James *Begbie*, Scotch physician, 1798–1869]. 1. Exophthalmic goiter. 2. Localized rhythmic chorea.

**Beggiatoa** (*bej-e-at-o'-ah*) [J. *Beggiato*, Italian botanist]. A genus of *Schizomycetes*, consisting of swinging or gliding, milk-white, gray, rosy, or violet threads. They decompose sulphur compounds and store up sulphur granules in their protoplasm. They are found in stagnant, fresh, or salt water, particularly in that contaminated with sewage or factory waste.

**begma** (*beg'-mah*) [βῆγμα, a cough; pl., *begmata*]. A cough; the matter expectorated by coughing.

**behen, behmen, behn, ben.** Arabian names for roots of various plants.

**behenic** (*be-hen'-ik*). Derived from behen.

**behen-nut** (*be'-hen-nut*). The seed of *Moringa pterygosperma*, and *M. aptera*, trees of tropical countries. They afford *oil of behen*, or *oil of ben*.

**Béhier-Hardy's symptom** (*ba'-he-a*) [Louis Jules *Béhier*, French physician, 1813–1876]. Aphonia, an early symptom in pulmonary gangrene.

**Behring's law** (*ba'-ring*) [Emil Adolph *Behring*, German physician, 1854– ]. The blood and blood-serum of an individual who has been artificially rendered immune against a certain infectious disease may be transferred to another individual with the effect of rendering the other also immune. B.'s serum, serum containing diphtheria-antitoxin. B.'s tulase, an immunizing remedy against tuberculosis.

**Beigel's disease** (*bi'-gel*) [Hermann *Beigel*, German physician, 1829–1879]. Trichorrhexis nodosa.

**Beissenhirtz's reaction for aniline** (*bi'-sen-hairtz*). On the addition of a grain of potassium dichromate to a solution of aniline in concentrated sulphuric acid the solution becomes first red, then blue, the color gradually disappearing.

**bel, bela** (*be'-lah*) [Hind.]. The dried, half-ripe fruit of *Ægle marmelos*, or Bengal quince. It is recommended as a remedy for chronic diarrhea and dysentery. The ripe fruit is slightly laxative. Dose ½–1 dr. (2–4 Gm.). b., liquid extract of (*extractum belæ liquidum*, B. P.). Dose 1–2 dr. (4–8 Cc.).

**belæ fructus** (*be-le fruk-tus*) [L.]. Bael-fruit; see *bel*.

**belching** (*belch'-ing*) [ME., *belchen*]. The expulsion of gas from the stomach through the mouth; eructation.

**belemnoid** (*bel-em'-noid*) [βέλεμνον, a dart; εἶδος, form]. 1. Dart-shaped; styloid. 2. The styloid process of the ulna or of the temporal bone.

**Belfield's operation** (*bel'-fēld*) [William Thomas *Belfield*, American surgeon, 1856– ]. Making an artificial opening into the vas deferens.

**bell** (*bel*) [ME., *bel*]. A hollow, metallic, sounding instrument. b.-gastrula. See *archigastrula*. b.-metal, an alloy of copper, zinc, tin, and antimony. b. sound, b. tympany, the sound produced in pneumothorax by striking a coin, placed flat upon the chest, with another coin. It can be heard through the stethoscope placed over the affected side.

**bell-crowned** (*bel'-krownd*). Applied to a toothcrown which is largest at the occlusal surface and tapers to the gum.

**Bell's disease** [1. Sir Charles *Bell*, Scotch physiologist, 1774–1842; 2. John *Bell*, Scotch anatomist, 1763–1820]. 1. See *Bell's mania*. 2. See *Bell's paralysis*. B.'s law, [1]. the anterior spinal nerve-roots are motor; the posterior, sensory. B.'s mania, [1]. acute delirium; acute periencephalitis. B.'s muscle, [2]. The short muscular ridge on the inner surface of the bladder, passing forward from the ureteral openings and ending in the uvula vesicæ. B.'s nerves, [2]. The external and internal respiratory nerves—*i. e.*, the posterior thoracic and phrenic nerves. B.'s palsy, B.'s paralysis, [1], peripheral paralysis of the facial nerve. B.'s phenomenon, [1]. upward and outward rolling of the eyeball when an attempt is made to close the eye of the affected side in peripheral facial paralysis. B.'s respiratory nerve, [2]. The long thoracic nerve. B.'s spasm, [1]. convulsive facial tic.

**belladonna** (*bel-ah-don'-ah*) [It., *bella donna*, beautiful lady]. Deadly nightshade. A perennial plant of the order *Solanaceæ*, indigenous to southern Europe and Asia, and cultivated in the United States. Its properties are due to the alkaloids atropine and belladonnine, the latter thought to be identical with hyoscyamine. Both leaves and root are employed. It is used as an antispasmodic, as a cardiac and respiratory stimulant, and to check secretions, as those of the sweat and milk. b. leaves (*belladonnæ folia*, U. S. P.), the dried leaves of *Atropa belladonna*. b. leaves, extract of (*extractum belladonnæ foliorum*, U. S. P.). Dose ⅛ gr. (0.01 Gm.). b. leaves, tincture of (*tinctura belladonnæ foliorum*, U. S. P.). Dose 1–30 min. (0.06–2.0 Cc.). b. liniment (*linimentum belladonnæ*, U. S. P.), made of camphor, 5; fluidextract of belladonna root, 95. b. ointment (*unguentum belladonnæ*, U. S. P.), contains extract of belladonna leaves, 10; diluted alcohol, 6; benzoinated lard, 65; hydrous wool-fat, 20. b. plaster (*emplastrum belladonnæ*, U. S. P.), made of adhesive plaster and extract of belladonna leaves. b. root (*belladonnæ radix*, U. S. P.), the dried root of *Atropa belladonna*. b. root, fluidextract of (*fluidextractum belladonnæ radicis*, U. S. P.). Dose 1–5 min. (0.065–0.3 Cc.).

**belladonnine** (*bel-ah-don'-in*) [see *belladonna*], C₁₇H₂₃NO₃. An alkaloid found in belladonna. It resembles atropine, hyoscyamine, and hyoscine. It occurs with atropine, and is likewise decomposed into tropic acid and oxytropine, C₈H₁₅NO₂. See *belladonna*.

**Bell-Bernhardt's phenomenon.** See *Bell's phenomenon*.

**Bellini's duct** (*bel-e'-ne*). [Lorenzo *Bellini*, Italian anatomist, 1643–1704]. One of the excretory ducts of the kidney. B.'s ligament, a ligamentous band extending from the capsule of the hip-joint to the greater trochanter of the femur. B.'s tubes, the straight uriniferous tubules.

**bellite** (*bel'-īt*) [*bellum*, war]. An explosive employed both in war and in blasting. A principal element in its manufacture is nitrobenzol. The most prominent symptoms induced by its inhalation and absorption are headache, mental confusion, dyspnea, pallor, blueness of the lips, general lividity, coma, insensibility.

**Bell-Magendie's law.** See *Bell's law*.

**Bellocq's cannula** (*bel'-oks*). An instrument used in plugging the nares.

**Bellonia** (*bel-o'-ne-ah*) [Peter *Belon*, a French naturalist, 1499-1564]. A genus of plants of the order *Gesneraceæ*. B. *aspera* is a shrub of the West Indies; the bark is used in intermittent fever and in dysentery.

**bellows** (*bel'-ōz*) [ME., *belowes*]. An instrument for propelling air through a tube or small orifice. **b. sound,** the *bruit de souffle*, q. v.

**belly** (*bel'-e*) [ME., *bely*]. 1. See *abdomen*. 2. Any belly-like enlargement of a part. **b -bound,** constipated. **b.-button,** the navel. **b. of a muscle,** the fleshy part of a muscle.

**bellyache** (*bel-e-āk*). See *colic*.

**beloid** (*bel'-oid*) [βέλος, an arrow]. Arrow-shaped or styloid.

**belonephobia** (*bel-o-ne-fo'-be-ah*) [βέλονη, a pin; φόβος, fear]. A morbid dread of pins and needles, and of sharp-pointed objects in general.

**belonoid** (*bel'-on-oid*) [βέλονη, a needle, bodkin]. Styloid; needle-shaped.

**belonospasis** (*bel-on-os'-pa-sis*) [βέλονη, a point; σπάσις, a drawing]. Irritation by means of needles or metallic tractors.

**belt** (*belt*). A girdle about the waist. **b., abdominal,** a broad, elastic belt worn about the abdomen as a support during pregnancy. **b., magnetic,** a belt consisting of plates of metal fastened upon a strip of felt moistened with dilute acid. It is a cure-all largely used by empirics. **b., test,** a patient suffering from enteroptosis experiences a sensation of relief when firm upward pressure is made on the lower part of the abdomen.

**ben, oil of, benne oil** (*ben, ben'-e*). Oleum ciatinum; it is obtained by expression from the seeds of the several species of *Moringa*. It is a colorless, odorless oil, not readily turning rancid. It is used for extracting odors and for lubricating clocks and light machinery.

**Benario's method** (*ben-ar'-e-o*). For the fixation of blood-films. It consists in the use of a 1 % alcoholic solution of formalin for one minute.

**Bence-Jones' bodies** [Henry *Bence-Jones*, English physician, 1814-1873]. Peculiar bodies, consisting of albumose, found in the urine in certain affections of the bone-marrow, especially neoplasms. **B.-J.'s cylinders,** long cylindrical formations derived from the seminiferous tubules, sometimes seen in the urine. **B.-J.'s reaction,** the urine is acidified with acetic acid and gently heated, in the presence of albumose a precipitate is formed, which is dissolved on boiling and reappears on cooling.

**bends** (*bendz*) [ME., *bend*]. A term used by miners and caisson-laborers for a condition produced by too sudden reduction of the high air-pressure; it is indicated by swellings or small bubbles under the skin; see *caisson disease*.

**Benedict's test for HCl in the stomach.** This consists in auscultation over the stomach after the patient has swallowed a small quantity of saturated solution of sodium bicarbonate. Normal hydrochloric acidity is indicated by a fine crepitation, superacidity by an earlier and louder sound, and in anacidity the crepitation is absent.

**benedictine** (*ben-e-dik'-tin*) [*benedicere*, to bless]. A liqueur or cordial originally prepared by Benedictine monks, and distilled at Fécamp, in France. It much resembles chartreuse and trappistine.

**Benedikt's syndrome** (*ben'-e-dikt*) [Moritz *Benedikt*, Austrian physician, 1835- ]. Paralysis of the motor oculi of one side and tremor of the arm of the opposite side; attributed by Charcot to a lesion in the ventrointernal part of the crura cerebri.

**beng** (*beng*). See *cannabis indica*.

**bengalin** (*ben'-gawl-in*) [Hind., *Bengāl*]. A blue pigment derived from the benzene colors.

**Benger's food.** Partially digested and solidified beef-tea, used as a food for invalids.

**Bengue's anodyne balsam** (*ben'-ga*). A compound consisting of menthol, ʒ2½; methyl salicylate, ʒ2½; wool-fat, ʒ3.

**benign, benignant** (*be-nīn', be-nig'-nant*) [*benignus*, kind]. Not endangering health or life; not malignant, applied to certain tumors.

**Benincasa** (*ben-in-ka'-zah*) [*Benincasa*, an Italian nobleman]. A genus of plants of the order *Cucurbitaceæ*. B. *cerifera* is a perennial species of the East Indies, with large, greenish fruit, the seeds of which are used in dysuria and colic, the rind in tuberculosis, asthma, and chills, and the plant in fevers, vertigo, etc.

**Béniqué's sound** (*ba-ne-ka'*) [Pierre Jules *Béniqué*, French physician, 1806-1851]. A urethral sound with a wide curve.

**benne, oil of** (*ben'-e*). Oil of sesame seeds. See *sesame*.

**Bennett's corpuscles.** Large epithelial cells filled with fatty detritus found in the contents of some ovarian cysts.

**Bennett's fracture** [Edward Hallaran *Bennett*, Irish surgeon, 1837-1907]. A longitudinal fracture of the first metacarpal bone, extending into the carpometacarpal joint and complicated by subluxation.

**benzacetin** (*ben-zas'-et-in*), $C_6H_3(OC_2H_5)(NH.-CH_3CO)COOH$. Colorless crystals, soluble in alcohol, slightly soluble in water; melt at 205° C. It is used in neuralgia. Dose 8-15 gr. (0.52-0.97 Gm.). Syn., *acetamidomethyl-salicylic acid*.

**benzaconine** (*ben-zak'-on-ēn*). An alkaloid from aconite with action similar to aconitine, except that it lacks its antipyretic power and has little effect on the sensory nerves, while it depresses the motor group and also the muscle-fibers.

**benzaldehyde** (*ben-zal'-de-hīd*) [*benzoin; aldehyde*], $C_7H_6O$, *benzaldehydum* (U. S. P.). Bitter-almond oil; a compound that results from the oxidation of benzyl alcohol.

**benzamide** (*ben'-zam-id*) [*benzoin; amide*], $C_7H_7NO$. A compound resulting from the action of benzoyl chloride upon alcoholic ammonia.

**benzamil** (*ben'-zam-il*), $C_9H_{10}NO_3$. A distillation-product of oil of bitter almonds.

**benzanalgen** (*ben-zan-al'-jen*), $C_8H_6OC_2H_5$ . $HN-CO_6CH_3N$. A derivative of chinolin. It is antiseptic, antipyretic, and antineuralgic, and has the power of dissolving uric acid. It is used in rheumatism, tabes dorsalis, and chronic gout. Dose 7½-45 gr. (0.5-3.0 Gm.).

**benzanilide** (*ben-zan'-il-id*) [*benzoin; aniline*], $C_{13}H_{11}NO$. Benzoyl anilide, prepared by the action of benzoyl chloride on aniline. It is used as an antipyretic in children in doses of 3-8 gr. (0.2-0.5 Gm.).

**benzaurin** (*ben-zaw'-rin*), $C_{19}H_{16}O_2$. Red crystals melting at 100° C. Syn., *Phenyldiphenolcarbinol*. **b. anhydride,** a colorless substance dissolving in alkaline solutions with a violet color.

**benzene** (*ben'-zēn*), $C_6H_6$. A hydrocarbon contained in coal-tar. It is formed by the dry distillation of all benzene acids. It is a mobile, ethereal-smelling liquid, of specific gravity 0.899 at 0°. It solidifies at about 0°, melts at +6°, boils at 580°, and burns with a luminous flame. It readily dissolves resins, fats, sulphur, iodine, and phosphorus. Aniline and the aniline colors are derived from it. Syn., *benzol; phene; phenyl hydril*. **b.-sulphonic acid,** $C_6H_5 . SO_3H$, is prepared by boiling together equal parts of benzene and $H_2SO_4$. It occurs in small plates, readily soluble in alcohol and water, and which deliquesce in the air.

**benzenyl** (*ben'-zen-il*). See *phenyl*. **b.-amido-thiophenol,** $C_{13}H_9NS$, acicular crystals with fragrance of roses, obtained from amidophenylmercaptan by benzoic aldehyde and heat; it is soluble in alcohol, ether, carbon disulphide, and dilute hydrochloric acid. Syn., *benzenylamidophenylmercaptan*.

**benzhydrol** (*ben'-hi'-drol*), $C_6H_5-CH(OH)-C_6H_5$. An alcohol occurring as silky, acicular crystals, slightly soluble in water, obtained from an alcoholic solution of benzophenone by action of sodium amalgam. Syn., *diphenylcarbinol*. **b. acetate,** a thick liquid obtained by heating benzhydrol with acetic acid.

**benzidine** (*ben'-zid-ēn*). Diaminobiphenyl, $NH_2.-C_6H_4 . C_6H_4 . NH_2$. A colorless, crystalline substance, formed by reaction of acids upon hydrazo-benzene. Used in test for blood. See *Adler's benzidine reaction for blood*.

**benzil** (*benz'-il*), $C_{14}H_{10}O_2$. A compound produced by the action of nitric acid on benzoin.
**benzilimide** (*ben-zil'-im-id*), $C_{28}H_{22}N_2O_2$. White silky needles obtained from an alcoholic solution of benzil by action of dry ammoniacal gas.
**benzimide** (*ben'-zim-id*) [*benzoyl;* \*amide*], $C_{28}H_{18}N_2O_2$. A compound formed by the action of hydrocyanic acid on benzoyl hydrate. It occurs also in the resinous residue of the rectification of the oil of bitter almonds.
**benzin, benzinum, benzine** (*ben'-zin, -zi'-num, ben'-zēn*). Petroleum ether. The *benzinum* of the U. S. P. is a purified distillate from American petroleum, having a specific gravity of 0.77 to 0.79, boiling at 80° to 90° C., colorless, of ethereal odor, and a slightly peppermint-like taste. It is a valuable solvent for oils, fats, resins, caoutchouc, and some alkaloids. It has been used against tape-worm. It should be distinguished from *benzene*. Dose 5–10 min. (0.33–0.66 Cc.) on sugar or in mucilage. **b., coal-tar**, benzin obtained as a by-product in preparing benzene and toluene from coal-tar oil by action of acids and alkalies. It differs chemically and physically from petroleum benzin, and is used as a cleansing fluid and as a solvent for resin, caoutchouc, etc.
**benzinin** (*ben'-zin-in*). A toxin extracted by Auclair from tubercle bacilli. Syn., *benzinobacillin*.
**benzinobacillin** (*ben-zin-o-bas'-il-in*). See *benzinin*.
**benzinomania** (*ben-zin-o-ma'-ne-ah*). The habit of taking benzin, by inhalation. And see *benzolism*.
**benzite** (*ben' zīt*). A solution of sulphur in two or three parts of hot coal-tar.
**benzoate** (*ben'-zo-āt*) [*benzoin*]. Any salt of benzoic acid. **b., naphthol**. See *benzonaphthol*.
**benzoated** (*ben'-zo-a-ted*) [*benzoin*]. Impregnated with benzoin or with benzoic acid or a benzoate.
**benzodiureid** (*ben-zo-di-u'-re-id*), $C_9H_{12}N_4O_2$. Tiny needles obtained from benzoic aldehyde by action of urea.
**benzoglycollate** (*ben-zo-gli'-kol-āt*). A salt of benzoglycollic acid.
**benzohelicin** (*ben-zo-hel'-is-in*), $C_{20}H_{26}O_8$. A combination of benzoyl and helicin. Syn., *benzoyl helicin*.
**benzoic** (*ben-zo'-ik*) [*benzoin*]. Pertaining to or derived from benzoin. **b. acid**. See *acid, benzoic*.
**benzoin** (*ben'-zoin* or *-zo-in*) [origin obscure].
1. *Benzoinum* (U. S. P.), a resin obtained from *Styrax benzoin*, a tree native to Sumatra and Siam. It is a ketone alcohol, $C_{14}H_{12}O_2$, and may be produced by oxidizing hydrobenzoin with concentrated $HNO_3$. It is antiseptic and disinfectant, and is used mainly as a stimulant expectorant in chronic bronchitis.
2. $C_{14}H_{12}O_2$, a reaction-product of an alcoholic solution of potassium cyanide on benzoic aldehyde, forming yellowish, fragrant prisms, soluble in hot alcohol, melting at 135°–137° C. It is used as an external antiseptic, 1 part in 5 of lard. Syn., *bitter-almond oil camphor; phenylbenzoyl carbinol*. **b., flowers of**, benzoic acid obtained by the sublimation of benzoin. **b., tincture of** (*tinctura benzoini*, U. S. P.), 20 % of the resin in alcohol. Dose ½–1 dr. (2–4 Cc.). **b., tincture of, compound** (*tinctura benzoini composita*, U. S. P.), friars' balsam; Turlington's balsam; it consists of benzoin, 12; aloes, 2; storax, 8; balsam of tolu, 4; alcohol, sufficient to make 100 parts. Dose ½–2 dr. (2–8 Cc.).
**benzoinam** (*ben-zo'-in-am*), $C_{28}H_{24}N_2O$. A crystalline powder derived from benzoin by action of alcoholic solution of ammonia with heat.
**benzoinated** (*ben-zo'-in-a-ted*). Combined or prepared with benzoin.
**benzoinol** (*ben-zo'-in-ol*). An oily liquid, said to consist of albolene with gum benzoin in solution; it is used as an excipient for menthol, camphor, etc., in diseases of the nose and throat.
**benzoiodohydrin** (*ben-zo-i-o-do-hi'-drin*), $(C_3H_5)$-$ClI(C_7H_5O_2)$. A brownish-yellow, oily mass, soluble in alcohol, ether, and petroleum ether, insoluble in glycerol. It decomposes at 100° C., iodine being liberated. It is a succedaneum for potassium iodide, and is given in the same doses. Syn., *chloroiodobenzoic-glycerinester; glyceryl chloriodobenzoate*.
**benzol, benzole, benzoleum** (*ben'-zol, ben'-zōl, ben-sōl-e'-um*). See *benzene*.
**benzolguaiacol** (*ben-zol-gwi'-ak-ol*). See *benzosol*.
**benzoline** (*ben'-zol-ēn*). Impure benzene, used for removing grease.
**benzolism** (*ben'-zol-izm*). Benzol-poisoning, from inhaling the vapor or swallowing it. It is marked in light cases by dizziness, loss of consciousness, and anesthesia; in severer cases by hallucination, epileptic paroxysms, and coma.
**benzonaphthol** (*ben-zo-naf'-thol*) [*benzoin; naphthol*], $C_{10}H_7O(C_7H_5O)$. The benzoate of betanaphthol, used as an intestinal antiseptic in typhoid fever and other intestinal diseases. Dose 2–8 gr. (0.13–0.52 Gm.).
**benzonitrile** (*ben-zo-ni'-tril*) [*benzoin; nitrum, niter*], $C_7H_5N$. An oil obtained from benzene-sulphonic acid by distillation with potassium cyanide. It has an odor resembling that of oil of bitter almonds, and boils at 191° C.; its specific gravity is 1.023 at 0° C.
**benzoparacresol** (*ben-zo-par-ah-kre'-sol*). An intestinal septic, insoluble in water. Dose, 4 gr. (0.25 gm.).
**benzophenid** (*ben-zo-fen'-id*). Phenyl benzoate.
**benzophenoneid** (*ben-zo-fe-no'-ne-id*). An antiseptic and germicidal compound.
**benzopyrine** (*ben-zo-pi'-rin*). Antipyrine benzoate.
**benzosalicin** (*ben-zo-sal'-is-in*). See *populin*.
**benzosalin** (*ben-zo-sa'-lin*). Trade name for the methyl ester of benzosalicylic acid; used in rheumatism, sciatica, neuralgia. Dose, 20–40 gr. (1–2 gm.).
**benzosol** (*ben'-zo-sol*) [*benzoin*], $C_{14}H_{12}O_3$. The benzoate of guaiacol; it contains 54 % of guaiacol. Its chief uses are as an intestinal antiseptic and in pulmonary tuberculosis. Dose 3–12 gr. (0.2–0.8 Gm.).
**benzosulphate** (*ben-zo-sul'-fāt*). A salt of benzosulphuric acid.
**benzosulphinide** (*ben-zo-sul'-fin-id*). Benzosulphinidum (U. S. P.); saccharin.
**benzotrichloride** (*ben-zo-tri-klo'-rid*), $C_7H_5Cl_3$. A colorless, transparent, highly refractive liquid, with penetrating odor, obtained from boiling toluene by action of chlorine; sp. gr. 1.38 at 14° C.; boils at 213°–214° C. Syn., *benzenyl trichloride*.
**benzoyl** (*ben'-zo-il*) [*benzoin*], $C_6H_5CO$. The radical of benzoic acid, of oil of bitter almonds, and of an extensive series of compounds derived from this oil, or connected with it by certain relations. **b.-acetylperoxide**, $C_9H_8O_4$, an oxidized product of the mixed anhydride of acetic and benzoic acids, a crystalline body, slightly soluble in water and very unstable. To prevent explosion by sudden heating or grinding it is diluted with an equal quantity of inert absorbent powder and called *acetosone*. It is used as an intestinal antiseptic. Dose 4–5 gr. (0.26–0.32 Gm.) 3 times daily. Syn., *acetyl-benzoyl peroxide*. **b. chloride**, $C_7H_5OCl$, a transparent, colorless, pungent oil with a specific gravity of 1.21 at 19° C.; it boils at 194°–195° C. When acted on by alcoholic ammonia it gives dibenzylamine. It is used as a reagent in organic analysis and synthesis. Syn., *chlorobenzoyl chloride*. **b.-ecgonin**, $C_{18}H_{23}NO_4H$, a substance intermediate in composition between cocaine and ecgonin. **b.-eugenol**, $C_{17}H_{16}O_3$, a substance forming large, transparent, colorless prisms or small white crystals, soluble in alcohol, ether, chloroform, and acetone; melts at 69°–70.5° C. It is used in tuberculosis. Dose 7½–15 gr. (0.5–1.0 Gm.). **b.-glycoin, b.-glycocoll**, hippuric acid. **b.-guaiacol**. See *benzosol*. **b. hydrate**, benzoic acid; also improperly applied to benzoic aldehyde. **b. peroxide**, a bactericide and disinfectant substance. **b.-phenylhydrazin**, $C_{13}H_{12}N_2O$, an antiseptic. **b. pseudotropein**, a local anesthetic. **b. salicin**, see *populin*. **b.-tropein**, $C_8H_{14}(C_6H_5CO)NO$, silky, acicular needles; it is a local anesthetic.
**benzyl** (*ben'-zil*) [*benzoin*], $C_7H_7$. A univalent hydrocarbon radical that does not exist in the free state, but in combination forms a considerable number of compounds. **b. alcohol**. See *alcohol, benzyl*.
**benzylic** (*ben-zil'-ik*). Relating to or prepared with benzyl.
**benzylidene** (*ben-zil'-id-ēn*) [*benzoin*], $C_7H_6$. A bivalent hydrocarbon radical derived from benzoin compounds.
**Beral's apparatus**. In pharmacy, an apparatus for effecting lixiviation.
**Béraneck's tuberculin** (*ba-ran-ek'*) [Edmond Béraneck, Swiss bacteriologist, 1859– ]. A tuberculin made from two filtrates of tubercle bacilli, one extracted with lime, the other with phosphoric acid.
**Bérard's aneurysm** (*ba-rar*) [Auguste Bérard, French surgeon, 1802–1846]. A varicose aneurysm

having its sac in the tissue immediately surrounding the vein.

**Béraud's ligament** (ba-ro'), [Bruno Jean Jacques *Béraud*, French surgeon, 1825–1865]. The suspensory ligament of the pericardium that is attached to the third and fourth dorsal vertebræ. **B.'s valve,** a fold of mucous membrane found occasionally in the lacrimal sac, which it separates from the nasal duct.

**berberine** (ber'-ber-ēn) [*berberis*], $C_{20}H_{17}NO_4$. An alkaloid found in the bark of *Berberis* and in many other plants. It is recommended as a tonic and antiperiodic, and is an ingredient of various lotions for mucous membranes; it is useful in nasal catarrh, etc. Dose 1–10 gr. (0.065–0.65 Gm.). **b. carbonate,** $C_4H_{14}NO_{11}$, crystalline powder, soluble in hot water; it is antiperiodic, stomachic, and tonic. Dose, antiperiodic, 8–15 gr. (0.52–1.0 Gm.); stomachic and tonic, $\frac{1}{2}$–1 gr. (0.032–0.065 Gm.) 3 times daily. **b. hydrochloride,** is used locally in gonorrhea.

**berberis** (ber'-ber-is) [L.]. Barberry. The berberis of the U. S. P. is the root of *Berberis aquifolium* and other species. Its properties are due to an alkaloid, *berberine*, $C_{20}H_{17}NO_4$. It is an astringent, bitter tonic; in large doses, a cathartic. It has been used locally in conjunctivitis, and internally in malarial and typhoid fevers. **b., fluidextract of** (*fluidextractum berberidis,* U. S. P.). Dose 5–30 min. (0.32–2.0 Cc.). **b., tincture of.** Dose 10 min.–1 dr. (0.65–4.0 Cc.).

**bergamot, oil of** (bur'-gam-ot) [*Bergamo,* a town in Italy], $C_{10}H_{16}$. A volatile oil derived from the rind of the *Citrus bergamia*. It is used mainly as a perfume and as a clearing-agent in histological work.

**bergapten** (bur-gap'-ten), $C_{12}H_8O_4$. A solid, greasy compound obtained from bergamot oil, being the lactone of burgaptenic acid. It melts at 188° C. Syn., *bergamilene; bergamot camphor.*

**bergenin** (bur'-jen-in). [*Bergenia,* a genus of plants], $C_6H_6O_3H_2O$. A bitter, crystalline substance, obtained from various species of saxifrage, melting at 140° C. It is said to be a nerve tonic, with action intermediate between that of salicylic acid and of quinine.

**Berger's paresthesia** (bār'-ja) [Emil *Berger,* Austrian physician]. Paresthesia in youthful subjects, of one or both lower extremities, without objective symptoms, accompanied by weakness. **B.'s sign,** an elliptical or irregular shape of the pupil sometimes seen in the early stage of tabes and paralytic dementia and in paralysis of the third cranial nerve.

**Bergeron's disease** (bār-ja-ron) [Étienne Jules *Bergeron,* French physician, 1817–1900]. An affection characterized by abrupt, lightning-like, muscular contractions, independent of the will and limited ordinarily to the head and arms, involving at times the two extremities of one side. Like Dubini's disease, it is also known as "electric chorea."

**Bergeron-Henoch's chorea.** See *Bergeron's disease.*

**Bergmann's fibers, B.-Deiters' fibers.** The processes of certain superficial neuroglia cells of the cerebellum which radiate toward the surface and are connected with the pia. **B. incision.** An oblique incision to expose the kidney. From the outer edge of the erector spinæ at the level of twelfth rib downward and outward toward the junction of the outer and middle third of Poupart's ligament.

**beriberi** (ber'-e-ber-e) [Singhalese, *beri*, weakness]. An infectious disease, endemic in various countries of Asia (Ceylon, India, China, Japan), Africa, and Australia, and presenting the features of a multiple neuritis. Bad food and defective hygienic conditions are predisposing causes; the true etiological factor is probably a microorganism. Syn., *kakké; myelopathia tropica; panneuritis epidemica.* **b., pseudo-,** Gibbs' name for a disease endemic in the Singapore Lunatic Asylum, which prevails during the wet season and attacks Asiatics only. It is not contagious, is marked by slight anemia, considerable soft anasarca, and tendency to sudden death from shock. The softness of the edema, lack of spastic and paralytic conditions, and rapidity of recovery distinguish it from beriberi. The period of incubation is from one to two days. **b., web,** that marked by anemia and dropsy without paralysis.

**Berkefeld filter** (berk'-felt). A filter of diatomaceous earth used to filter out bacteria and so obtain a sterile filtrate.

**Berlin's disease** [Rudolf *Berlin,* German ophthalmologist, 1833–1907]. Commotio retinæ; traumatic edema of the retina.

**Bernard's canal or duct** (bur-nar') [Claude *Bernard,* French physiologist, 1813–1878]. The supplementary duct of the pancreas. **B.'s center,** the "diabetic center" in the floor of the fourth ventricle. **B.'s granular layer,** the deep layer of cells lining the acini of the pancreas. It is granular in appearance and stains but slightly with carmin. **B.'s puncture,** the puncture of a definite spot in the floor of the fourth cerebral ventricle for the production of artificial diabetes.

**Bernays' aseptic sponge** (bur'-nāz) [Augustus Charles *Bernays,* American surgeon, 1854–1907]. Small circular discs of prepared cotton fiber which has been subjected to great pressure; when placed in water, they increase in size 12 to 15 times. They are recommended as controlling agents in nasal hemorrhage.

**Bernhardt's paresthesia** (burn'-hart) [Martin *Bernhardt,* German neurologist, 1844– ]. Abnormal sensations, especially of numbness, with hyperesthesia and pain on exertion, in the region supplied by the external cutaneous nerve of the thigh.

**Bernhardt-Roth's symptom-complex.** See *Bernhardt's paresthesia.*

**Bernheimer's fibers** (burn'-hi-mer) [Stefan *Bernheimer,* Austrian ophthalmologist, 1861– ]. A tract of nerve-fibers extending from the optic tract to Luys' body.

**berry** (ber'-e). An indehiscent fruit with a pericarp that is succulent throughout, as the grape and gooseberry.

**Berthelot's test for phenol** (bur'-thel-o) [Marcellin Pierre Eugène *Berthelot,* French chemist, 1827–1907]. An ammoniacal solution of phenol treated with sodium hypochlorite produces a beautiful blue coloration.

**Berthollet's law** (bur-tol-a') [Claude Louis *Berthollet,* French chemist, 1748–1822]. When two salts in solution can, by double decomposition, produce a salt less soluble than either, this salt will be produced.

**Bertillonage** (ber-til-lon-a(h)j) [Alphonse *Bertillon,* French anthropologist, 1853–1914]. A system of carefully recorded measurements and descriptions of criminals, for the purpose of future identification, introduced into France by Bertillon and adopted by the police of many large cities of the United States.

**Bertin, bones of** (bur-tan') [Exupère Joseph *Bertin,* French anatomist, 1712–1781]. The sphenoid turbinated bones, partly closing the sphenoid sinuses. **B., column of,** a cortical column of the kidney; the part separating the medullary pyramids. **B., ligament of,** the iliofemoral ligament.

**beryllium** (be-ril'-e-um) [βήρυλλος, beryl]. A bivalent metal obtained from the beryl, whence its name. Syn., *glucinum.* See *elements,* table of *chemical.*

**Berzelius' test for albumin** (bur-ze'-le-us) [Johan Jacob *Berzelius,* Swedish chemist, 1779–1848]. All albuminous substances (except peptone) are precipitated from their aqueous solutions by metaphosphoric acid in freshly prepared concentrated solution.

**besetment** (be-set'-ment). An obsession.

**besicIometer** (bes-ik-lom'-et-er) [Fr., *besicles,* spectacle's; μέτρον, measure]. An instrument used by opticians for measuring the forehead to obtain the proper width for spectacle frames.

**Besnier's rheumatism** (ba-ne-a') [Jules *Besnier,* French physician]. Simple chronic articular rheumatism; chronic arthrosynovitis.

**bestiality** (bes-te-al'-it-e) [*bestia,* a beast]. Sexual intercourse with an animal.

**Bestucheff's mixture, B.'s tincture** (bes'-tu-shef) [Alexei Petrovich *Bestucheff,* Russian general, 1693–1766]. The ethereal tincture of iron chloride, used in erysipelas. It is made as follows: Tincture of iron chloride, 1 part; spirit of nitrous ether, 4 parts. Mix and expose to the rays of the sun in well-closed bottles until the brownish color disappears. The dose is from 1 to 2 teaspoonfuls every 3 hours.

**beta** (be'-tah) [L.]. 1. The beet. 2. The second letter of the Greek alphabet, used in chemical nomenclature to indicate the second of two isomeric compounds.

**betacism** (be'-tas-izm) [*beta,* β, the second letter of the Greek alphabet]. The too-frequent use of the *b*-sound in speech, or the conversion of other sounds into it.

**betaine** (*be'-ta-in*). A ptomaine, $C_5H_{11}NO_2$, obtained from certain animal and vegetable substances, including the beet.

**betanaphthol** (*be-tah-naf'-thol*). See *naphthol*.
**b. bismuth,** a preparation containing 80 % of bismuth oxide and 20 % of betanaphthol. It is a brown powder, insoluble in water, and decomposed into its component parts in the intestine, the betanaphthol being absorbed and discharged with the urine, while the bismuth is evacuated with the stools. It is an intestinal antiseptic. Dose 15-45 gr. (1-3 Gm.). **b. carbonate,** $CO(OC_{10}H_7)_2$, a dinaphthyl ester of carbonic acid obtained by the action of phosgene on betanaphthol sodium. It is used as an intestinal antiseptic. **b. lactate,** lactol.

**betel** (*be'-tel*) [E. Ind.]. A masticatory used in the East. A few grains of the nut of the catechu palm, *Areca catechu*, are rolled up with a small amount of quicklime in a leaf of *Piper betel*, and chewed. It is tonic, astringent, stimulant, and aphrodisiac, and seems to increase the powers of endurance. Dose of *fluidextract* 1-3 dr. (4-12 Cc.).

**bethroot** (*beth-root*). The rhizome of *Trillium erecta*, astringent and tonic. Dose of *fluidextract* ℥ xxx-ʒj. *Triliin*, a concentrated ext. Dose gr. ij-iv.

**betin** (*be'-tin*) [*beta*, a beet]. A precipitate prepared from a tincture of the common beet. It has been proposed as a substitute for ergot.

**betol** (*be' tol*), $C_6H_4O \cdot C_7H_5O_3$. A salicylic ether of naphthol, used in rheumatism and cystitis. It resembles salicylic acid in its properties. Dose 10-15 gr. (0.65-1.0 Gm.). Syn., *naphthalol; salinaphthol*.

**betony** (*bet'-on-e*). The leaves of *Stachys betonica*, formerly used as an emetic, expectorant, cathartic and for various other purposes.

**Bettendorff's test for arsenic** (*bet'-en-dorf*). On heating a solution of stannous chloride in concentrated hydrochloric acid, specific gravity 1.19, with a solution of arsenic or arsenous acids in strong hydrochloric acid, a brownish turbidity or precipitate of metallic arsenic and tin is yielded.

**Bettmann's test.** Same as *Bettendorff's test*.
**betula** (*bet'-ū-lah*) [*betula*, birch]. See *birch*.
**betulase** (*bet'-ū-lās*). See *gaultherase*.
**betulin** (*bet'-ū-lin*) [*betula*, birch], $C_{36}H_{60}O_3$. Birchresin, or birch-camphor, derived from the bark of the white birch.
**betulol** (*bet'-u-lol*). An application for the treatment of rheumatism, said to be more quickly absorbed than oil of wintergreen. Syn., *methyl-oleosalicylate*.

**between-brain.** The interbrain; also the midbrain.

**Betz, giant cells of** (*bets*) [Philipp Friedrich Betz, German physician, 1819- ]. Large ganglion-cells found in the deeper layers of the cortex, especially in the ascending frontal convolution and the paracentral lobule. They are usually arranged in small groups of from three to five that are known as *Betz's nests*.

**Bevan's incision.** To expose the gall-bladder. A vertical incision along the outer border of right rectus muscle.

**bex** (*beks*) [βήξ, a cough]. A cough, or disease characterized by coughing. **b. convulsiva,** whooping-cough. **b. theriodes,** synonym of *whooping-cough*.

**bezoar** (*be'-zo-ar*) [Pers., *pādzahr*, the bezoarstone, supposed antidote for poison]. A concretion found in the stomach or intestine of some animals (especially ruminants), formerly believed to be efficacious in preventing the fatal effects of poison. **b., oriental,** a hard, round concretion obtained from the intestine of the gazelle and other ruminants. It consists of concentric layers of resinous matter which burn with an agreeable odor; it is valued in the East for supposed medicinal properties.

**Bezold's ganglion** (*be'-zōlt*) [Friedrich Bezold, German otologist, 1842-1908]. A ganglion in the interauricular septum of the frog's heart. **B.'s mastoiditis,** destruction of the apex of the mastoid process with a tendency to the formation of an abscess in the neck. **B.'s perforation,** perforation on the inner surface of the mastoid, with formation of an abscess under the sterno-mastoid. **B.'s symptom,** the appearance of an inflammatory swelling at short distance below the apex of the mastoid process is evidence of mastoid suppuration. See *Bezold's mastoiditis*. **B.'s perforation.** **B.'s triad,** diminished or delayed bone conduction, diminished appreciation of low tones, and negative Rinne's test, all together indicating otosclerosis.

**BF.** [Fr. *bouillon filtré*, filtered broth or bouillon]. Same as *Denys' tuberculin*.

**bhang** (*bang*). See *cannabis indica*.
**bhel** (*bel*). See *bel*.
**B. Hy.** Abbreviation for Bachelor of Hygiene.
**Bi.** Chemical symbol of *bismuth*.
**bi-** [*bis*, twice]. A prefix meaning two, twice, double.

**biacuminate** (*bi-ak-ū'-min-āt*) [*bi-; acuminatus*, pointed]. Having two diverging pointed ends.
**bialate** (*bi-a'-lāt*) [*bi-; ala*, a wing]. Having two wings or wing-like appendages.

**Bial's test for pentose** (*be'-al*). Reagent: To 500 c.c. of 30 per cent. HCl add 1 gram of orcin and 25 drops of 10 per cent. ferric chloride solution. Four to 5 c.c. of this reagent are heated to the boiling point and withdrawn from the flame. Add the suspected urine drop by drop, up to 1 c.c. or less; a green color will appear if pentose be present. If dextrose be present in the urine it should first be removed by fermentation with a pure culture of yeast.

**bialuminate** (*bi-al-ū'-min-āt*) [*bi-; aluminum*]. A salt of aluminum containing two equivalents of aluminum combined with one of acid.
**biangulate** (*bi-ang'-ū-lāt*) [*bi-; angulus*, an angle]. With two angles.
**biapiculate** (*bi-ap-ik'-ū-lāt*) [*bi-; apex*, the summit]. With two summits.
**biarsenate** (*bi-ar'-sen-āt*) [*bi-; arsenic*]. An acid arsenate containing two atoms of hydrogen.
**biarticulate** (*bi-ar-tik'-ū-lāt*) [*bi-; articulus*, a joint]. Having a double joint.
**biasteriac, biasterial, biasteric** (*bi-as-ter'-e-ak, bi-as-te'-re-al, bi-as-ter'-ik*) [*bi-; asterion*, a craniometric point]. Relating to the asterion on each side of the skull; extending between the two asterions.
**biauricular** (*bi-aw-rik'-ū-lar*) [*bi-; auricula*, the ear]. Relating to two auricles or to two corresponding auricular points.
**biaurite** (*bi-aw'-e-rīt*) [*bi-; auris*, the ear]. Furnished with two ears or ear-like projections.
**biaxial** (*bi-aks'-e-al*) [*bis*, twice; *axis*]. Furnished with two axes.
**bib.** A portion of a red blood-corpuscle adherent to the crescent bodies of the blood, observed in malaria.
**bibasic** (*bi-ba'-sik*) [*bi-; basis*, a base]. Having two hydrogen atoms replaceable by bases, as certain acids; dibasic.
**biberin** (*bib'-ur-in*). Same as *bebeerin*.
**bibiru bark** (*bib-e'-roo*). See *bebeeru*.
**bibliography** (*bib-le-og'-ra-fe*) [βιβλίον, a book; γράφειν, to write]. A classified list of references, books, or authorities on any subject.
**bibliophobia** (*bib-le-o-fo'-be-ah*) [βιβλίον, a book; φόβος, dread]. Morbid dislike of books.
**biborate** (*bi-bo'-rāt*). See *pyroborate*.
**bibromide** (*bi-bro'-mid*). A compound of bromine with a radical or element, containing twice as much bromine as another similar compound.
**bibulous** (*bib'-ū-lus*) [*bibere*, to drink]. Having the property of absorbing moisture. **b. paper,** blotting paper.
**bicalcarate** (*bi-kal'-kar-āt*) [*bi-; calcar*, a spur]. Furnished with two spurs or spur-like projections.
**bicameral** (*bi-kam'-er-al*) [*bi-; camera*, a vault]. Having two compartments.
**bicapitate** (*bi-kap'-it-āt*) [*bi-; caput*, a head]. Having two heads; bicephalous; dicephalous.
**bicapsular** (*bi-kap'-sū-lar*) [*bi-; capsula*, a capsule]. Having two capsules.
**bicarbonate** (*bi-kar'-bon-āt*) [*bi-; carbonate*]. Any salt of carbonic acid in which only one of the hydrogen atoms has been replaced by a base.
**bicaudal, bicaudate** (*bi-kaw'-dal, bi-kaw'-dāt*) [*bi-; cauda*, tail]. Having two tails or appendages.
**bicavitary** (*bi-kav'-it-a-re*) [*bi-; cavitus*, a cavity]. Having two cavities.
**bicellular** (*bi-sel'-ū-lar*) [*bi-; cella*, a cell]. Composed of two cells.
**bicephalic, bicephalous** (*bi-sef-al'-ik, bi-sef'-al-us*). See *dicephalous*.
**bicephalus.** See *dicephalus*.
**biceps** (*bi'-seps*) [*bi-; caput*, head]. Having two heads, a term applied to several muscles, as *b. brachii, b. flexor cruris*.

**biceptor** (*bi-sep'-tor*) [*bi-*; *receptor*]. A receptor having two complementophile groups.
**Bichat, canal of** (*be'-shah*) [Marie François Xavier Bichat, French anatomist, 1771–1802]. A canal which was supposed by Bichat to exist between the subarachnoid space and the third ventricle. **B., fat-ball of**, the buccal fat-pad, a mass of fat lying in the space between the buccinator and the anterior border of the masseter; it is especially well developed in infants. **B., fissure of**, the transverse curved fissure passing below the splenium; its extremities correspond to the beginning of the Sylvian fissure. It affords passage to the pia, which forms within the hemispheres the tela choroidea and choroid plexus. **B., foramen of**, one connecting the subarachnoid space and third ventricle; it transmits the cerebrospinal fluid. **B., membrane of**, the subendothelial fibroelastic layer of the tunica intima of an artery. **B., tunic of**, the intima of the blood-vessels.
**bichlorid.** See *bichloride*.
**bichloride** (*bi-klo'-rid*) [*bi-*; *chloride*]. A salt containing two equivalents of chlorine.
**bicho** (*be'-cho*). Epidemic gangrenous inflammation of the rectum.
**bichromate** (*bi-kro'-māt*) [*bi-*; *chromium*]. A salt containing two equivalents of chromium.
**bicinctus, bicingulatus** (*bi-sink'-tus, bi-sin-gū-la'-tus*) [*bi-*; *cingere*, to gird]. Having two zones or belts.
**bicipital, bicipitous** (*bi-sip'-it-al, -us*) [*biceps*, double-headed]. 1. With two heads. 2. Relating to one of the biceps muscles.
**biclavate** (*bi-kla'-vāt*) [*bi-*; *clava*, a club]. Clubbed at each end. **b.-bihamate**, with the two club-shaped ends bent toward each other. **b.-cylindrical**, cylindrical and with clubbed ends.
**biconcave** (*bi-kon'-kāv*). Hollow on both surfaces.
**biconvex** (*bi-kon'-veks*). Rounded on both surfaces.
**bicornuate** (*bi-kor'-nū-āt*) [*bicornutus*, with two horns]. Having two horns, as a *bicornuate* uterus.
**bicorporal, bicorporate, bicorporated** (*bi-kor'-por-al, -āt, -ā-ted*) [*bi-*; *corpus*, a body]. Consisting of two bodies.
**bicrescentic** (*bi-kres-en'-tik*) [*bis*, twice; *crescere*, to grow]. Applied to a tooth having two ridges in the form of a double crescent.
**bicrural** (*bi-kru'-ral*) [*bi-*; *crus*, a leg]. Having two legs or leg-like processes.
**bicuspid** (*bi-kus'-pid*) [*bi-*; *cuspis*, the point of a spear]. Having two cusps, as *bicuspid* teeth. **b. valve**, the mitral valve of the heart.
**bicyanate** (*bi-si'-an-āt*) [*bi-*; *cyanogen*]. A salt having two equivalents of cyanic acid and one of a base.
**b. i. d.** Abbreviation of the Latin *bis in die*, twice daily.
**bidacryc** (*bi-dak'-rik*) [*bi-*; *dacryon*]. In craniometry, relating to the two dacryons.
**Bidder, ganglion of** (*bid'-er*) [Friedrich Heinrich Bidder, German anatomist, 1810–1894]. An accumulation of ganglion-cells in the interauricular septum and the auriculoventricular groove of the frog's heart.
**bidental** (*bi-den'-tal*) [*bi-*; *dens*, a tooth]. Having two teeth or tooth-like prominences.
**bidet** (*be-det'* or *be'-da*) [Fr., "pony"]. A tub or basin with fixed attachments for the administering of injections; also for use as a sitz-bath or hip-bath.
**bidigital** (*bi-dij'-it-al*) [*bi-*; *digitus*, a finger]. Referring to the tip of a finger of each hand.
**Bieber's reagent.** Equal volumes concentrated sulphuric acid, red nitric acid, and water.
**Biedert's cream mixture** (*be'-dār*). An infant's food made by mixing 4 oz. of cream with 12 oz. of warm water, and adding ½ oz. of milk-sugar. It contains 1 % of casein, 2.5 % of fat, and 3.8 % of sugar. As the child grows older a larger proportion of milk is added.
**Bieg's entotic test** (*bēg*). When words are audible only on being spoken into an ear-trumpet connected with a catheter placed in the Eustachian tube, but not through the ear-trumpet as ordinarily applied, there is a probable lesion of the malleus or incus which interferes with conduction.
**bielectrolysis** (*bi-e-lek-trol'-is-is*) [*bi*; ἤλεκτρον, amber; λύσις, resolution]. The electrolysis of two substances at the same time.
**biennial** (*bi-en'-e-al*) [*bi*; *annus*, a year]. Every two years. In botany, plants that produce foliage and a root-stalk the first year, flowering and maturing the second.

**Biermer's anemia** (*bēr'-mur*) [Anton Biermer, German physician, 1827–1892]. Pernicious anemia. **B.'s change of pitch**, in hydropneumothorax the tympanitic sound is lower in pitch when the patient is sitting than when he is lying down.
**Biernacki's symptom** (*bēr'-nak-e*) [Edmond Adolfovich Biernacki, Russian pathologist, 1866–1912]. Analgesia of the ulnar nerve at the elbow; it is observed in tabes dorsalis and paretic dementia.
**Bier's cups** (*bēr*) [August Karl Gustav Bier, German surgeon, 1861– ]. Glass appliances used to produce Bier's hyperemia. **B.'s hyperemia.** A method of treatment by artificial production of passive congestion in the part diseased. **B.'s local anesthesia.** 1. Anesthesia in a limb produced by intravenous injections of half per cent. cocaine after the part has been rendered bloodless by elevation and constriction. 2. Anesthesia of the lower part of the body produced by the injection of an anesthetic agent into the spinal membranes.
**Biesiadecki's fossa** (*be-es-e-ah-dek'-e*) [Alfred von Biesiadecki, Russian physician, 1839–1888]. A peritoneal recess which is bounded in front by a more or less well-defined fold, the inner surface of which looks upward over the psoas toward the root of the mesentery, the outer extending toward the crest of the ilium. Syn., *fossa iliacosubfascialis*.
**Biett's collar** (*be-et'-*) [Laurent Theodore Biett, Swiss physician, 1781–1840]. A zone of lenticulo-papular syphilide on the neck.
**bifacial** (*bi-fa'-shal*) [*bi-*; *facies*, a face]. Having the opposite surfaces similar.
**bifarious** (*bi-fa'-re-us*) [*bifarius*]. Twofold; arranged in two more or less regular series or rows.
**Biffi's test for bile.** Acidify 150 to 200 Cc. of urine with sulphuric acid; add drop by drop a 5 % barium chloride solution, using about 30 drops to every 100 Cc. of urine. Pour off the liquid and collect the soft precipitate on absorbent cotton and spread evenly. Place a crystal of potassium dichromate upon the surface of the precipitate, and in the presence of bile a green ring will form around the crystal, changing to blue and then to red. A small amount of albumin will not interfere with the test, but if much is present, the use of a saturated solution of sodium sulphate is recommended instead of the sulphuric acid.
**bifid** (*bi'-fid*) [*bifidere*, to cleave]. Divided into two parts; cleft, as *bifid* uvula. **b. spine**, spina bifida. **b. tongue**, one cleft longitudinally.
**bifissile** (*bi-fis'-l*) [*bi-*; *findere*, to split]. Parting naturally into halves.
**bifistular, bifistulous** (*bi-fis'-tu-lar, -lus*) [*bi-*; *fistula*, a pipe]. With two tubes.
**biflagellate** (*bi-flaj'-el-āt*) [*bi-*; *flagellum*, a whip]. Furnished with two flagella.
**bifocal** (*bi-fo'-kal*) [*bi-*; *focus*, a point]. Having a double focus. Applied to a system of lenses or spectacle-glasses with two foci, chiefly used for the correction of presbyopia, when there is at the same time an error of refraction for distant vision. The distance lens is above that for near-work. These are sometimes called *pantoscopic lenses*, and also *Franklin spectacles*, because the device was first made by Benjamin Franklin. They are also called *cement lenses*, because now made by cementing the lower segment to the distance lens.
**biforate** (*bi-fo'-rāt*) [*bi-*; *foratus*, perforated]. Having two foramina.
**biformity** (*bi-form'-it-e*). The condition of being dimorphous.
**biforous** (*bi'-for-us*). 1. See *biforate*. 2. Having two valves.
**bifurcate** (*bi-fur'-kāt*) [*bi-*; *furca*, a fork]. Divided into two, like a fork.
**bifurcation** (*bi-fur-ka'-shun*) [see *bifurcate*]. Division into two branches, as of the trachea for the aorta.
**bigaster** (*bi-gas'-ter*). See *biventer*.
**Bigelovia** (*big-lo'-ve-ah*) [Jacob Bigelow, American botanist, 1787–1879]. A genus of composite-flowered plants. See *Damiana*.
**Bigelow's ligament** (*big'-el-o*) [Henry Jacob Bigelow, American surgeon, 1816–1890]. The Y-ligament of the hip-joint; iliofemoral ligament. **B.'s operation**, litholapaxy. **B.'s septum**, the calcar femorale, a nearly vertical spur of compact tissue in the neck of the femur, a little in front of the lesser trochanter.
**bigeminal, bigeminous** (*bi-jem'-in-al, -us*) [*bi-*; *geminare*, to double]. Occurring in two pairs.

**bigeminum** (*bi-jem′-in-um*) [*bi*, two; *geminus*, a twin]. One of the corpora bigemina of the brain.

**bihastate, bihastatus** (*bi-has′-tāt, bi-has-ta′-tus*) [*bi-; hasta*, a lance]. With two lance-shaped processes.

**bilabe** (*bi′-lāb*) [*bi-; labium*, lip]. A surgical instrument for removing foreign bodies from the bladder through the urethra.

**bilamellar, bilamellate, bilamellated** (*bi-lam-el′-ar, -āt, -el-a′-ted*) [*bi-; lamella*, a plate]. Consisting of two thin plates.

**bilaminar, bilaminate** (*bi-lam′-in-ar, -āt*) [*bi-; lamina*, a sheet]. Composed of two layers.

**bilateral** (*bi-lat′-er-al*) [*bi-; latus*, a side]. Relating to two sides; pertaining to or affecting both sides of the body. **b. symmetry**, the symmetry of right and left halves.

**bilateralism**- (*bi-lat′-er-al-izm*) [see *bilateral*]. Bilateral symmetry.

**bile** (*bīl*) [*bilis*, the bile]. The substance secreted by the liver. It is mucilaginous, golden-brown, and is composed of biliary salts, cholesterin, mucus, and certain pigments. The principal salts are the sodium salts of taurocholic acid and glycocholic acid. The taste of bile is intensely bitter, its reaction feebly alkaline, and its density from 1026 to 1032. **b. acids, tests for.** See *Drechsel, Mylius, Pettenkofer, Strassburg, v. Udransky*. **b., crystallized,** Plattner's name for sodium taurocholate. **b., cystic,** bile contained in the gall-bladder as distinguished from that which is transmitted directly from the liver to the duodenum. **b., glastine,** that of a bluish color, so called from *glastum*, or woad (*Isatis tinctoria*), used for dyeing blue. **b., hepatic,** that which is transmitted directly from the liver to the duodenum without entering the gall-bladder. **b.-pigments,** the coloring-matters of the bile. For tests see *Barral, Capranica, Cunisset, Dragendorff, Dumontpallier, Fleischl, Glusinski, Gmelin, Huppert, Jolles, Maréchal, Le Nobel, Rosenbach, Smith, Stokvis, Trousseau, Ultzmann, Vitalli*.

**Bilharzia** (*bil-har′-ze-ah*) [Theodor Bilharz, German physician, 1825–1862]. A genus of trematode worms, established by Cobbold, characterized by having the sexes separate. **B. hæmatobia.** See *distoma hæmatobium*.

**bilharziasis, bilharziosis** (*bil-har-zi′-as-is, bil-har-ze-o′-sis*) [*Bilharzia*]. The group of symptoms produced by the presence in the intestine of worms of the genus *Bilharzia*.

**bili-** (*bil′-e*). A prefix denoting relating to the bile.

**biliary** (*bil′-e-a-re*) [*bile*]. Pertaining to the bile; conveying the bile. **b. acids,** glycocholic and taurocholic acids. **b. calculus,** a gall-stone. **b. colic,** colic produced by the passage of gall-stones. **b., diabetes,** Hanot's disease, or hypertrophic cirrhosis of the liver with icterus. **b. ducts,** the hepatic and cystic ducts and the ductus communis choledochus, together with the small ducts in the liver itself.

**biliation** (*bil-e-a′-shun*) [*bile*]. The secretion or excretion of bile.

**bilicyanin** (*bil-e-si′-an-in*) [*bili-; κυάνεος*, blue]. A blue pigment obtained from biliverdin. Syn., *cholcyanin; choleverdin*.

**bilification** (*bil-if-ik-a′-shun*) [*bili-; facere*, to make]. The formation of bile.

**biliflavin** (*bil-e-fla′-vin*) [*bili-; flavus*, yellow]. A yellow coloring-matter derivable from biliverdin.

**bilifulvin** (*bil-e-ful′-vin*) [*bili-; fulvus*, reddish yellow]. An impure form of bilirubin; also a yellow bile color from ox-gall, not normally present in human bile.

**bilifuscin** (*bil-e-fus′-in*) [*bili-; fuscus*, brown], $C_{16}H_{20}N_2O_4$. A pigment occurring in bile and in gall-stones.

**biligulate, biligulatus** (*bi-lig′-ū-lāt, bi-lig-ū-lā′-tus*) [*bi-; ligula*, a little tongue]. Formed like two tongues or having two tongue-like processes.

**bilihumin** (*bil-e-hū′-min*) [*bili-; humus*, earth]. An insoluble residue left after treating gall-stones with various solvents.

**bilin** (*bi′-lin*) [*bile*]. A mixture of sodium taurocholate and glycocholate, forming a constituent of the bile.

**bilineurin** (*bil-e-nū′-rin*). Cholin.

**bilious** (*bil′-yus*) [*biliosus*, full of bile]. 1. Pertaining to bile. 2. A term popularly applied to disorders supposed to arise from a too free secretion of bile. **b. fever,** a remittent fever characterized by the vomiting of bile.

**biliousness** (*bil′-yus-nes*) [*bilious*]. A popular name for a condition characterized by anorexia, constipation, coated tongue, lassitude, and headache, and supposed to be due to disorders in the secretion and flow of bile.

**biliphein** (*bil-e-fe′-in*) [*bilis*, bile; *φαιός*, gray]. A supposed bile-color, now regarded as an impure bilirubin; called also *cholophein*.

**biliprasin** (*bil-e-pra′-sin*) [*bili-; πράσινος*, leek-green], $C_{16}H_{22}N_2O_8$. A pigment occurring in gall-stones, icteric urine, and bile.

**bilipurpin,** or **pilipurpurin** (*bil-e-per′-pin, bil-e-per′-pū-rin*) [*bilis*, bile; *purpura*, purple]. A purple coloring-matter derivable from bilirubin.

**bilipyrrhin** (*bil-e-pir′-in*). See *cholepyrrhin*.

**bilirubin** (*bil-e-roo′-bin*) [*bili-; ruber*, red], $C_{16}H_{20}N_2O_3$. A red coloring-matter, the chief pigment of the bile, and also found in the urine in jaundice. It is insoluble in water, and almost so in ether and alcohol, but it is readily soluble in alkaline solutions. It crystallizes in rhombic plates or prisms.

**bilirubinemia** (*bil-e-roo-bin-e′-me-ah*) [*bilirubin; aίμa*, blood], The presence of bilirubin in the blood.

**bilirubinuria** (*bil-e-roo-bin-ū′-re-ah*) [*bilirubin; urine*]. Presence of bilirubin in the urine.

**bilis** (*bi′-lis*). Bile, gall. **b. bubula,** ox-gall.

**biliverdin** (*bil-e-ver′-din*) [*bili-; viridis,* green], $C_{16}H_{20}N_2O_5$, or $C_8N_9NO_2$. A green pigment, the first product of the oxidation of bilirubin. It gives the characteristic color to the bile of herbivora, and occurs in the urine in jaundice and in gall-stones.

**bilixanthin** (*bil-e-zan′-thin*). See *cholelepin*.

**Billroth's anesthetic** (*bil′-rōt*) [Christian Albert Theodor *Billroth*, Austrian surgeon, 1829–1894]. A mixture containing chloroform 3 parts, and alcohol and ether, each 1 part. **B.'s disease.** 1. Spurious meningocele. 2. Malignant lymphoma. **B.'s mixture,** see *B.'s anesthetic.* **B.'s operation,** pylorectomy: a parietal incision is made in almost a transverse direction; the divided walls of the stomach and bowel are brought together and united by sutures on the side of the greater curvature of the stomach. **B.'s suture,** the button hole stitch.

**bilobate** (*bi-lo′-bāt*) [*bi-; λοβός*, a lobe]. With two lobes; divided into two lobes.

**bilobular** (*bi-lob′-ū-lar*) [*bi*, two; *lobulus*, lobule]. Having two lobules.

**bilocular** (*bi-lok′-ū-lar*) [*bi-; loculus*, a little place]. Having two cells; divided into two compartments; biloculate.

**bimaculate** (*bi-mak′-ū-lāt*) [*bi-; macula*, a spot]. Marked with two spots.

**bimalar** (*bi-ma′-lar*). Extending between the two malar bones.

**bimalate** (*bi-mal′-āt*). In a series of malates, that one which contains twice the amount of malic acid that the first one of the series does.

**bimanous** (*bi-ma′-nus*) [*bi-; manus*, a hand]. Having two hands.

**bimanual** (*bi-man′-ū-al*) [*bi-; manus*, a hand]. With both hands; two-handed. **b. palpation,** palpation by means of both hands.

**bimastoid** (*bi-mas′-toid*). Relating to the two mastoid eminences.

**bimaxillary** (*bi-maks′-il-a-re*). Extending between the two maxillæ.

**bimembral** (*bi-mem′-bral*) [*bi-; membrum*, a member]. With two limbs.

**bimestral** (*bi-mes′-tral*) [*bi-; mensis*, month]. Two months old; continuing two months.

**bimolybdate** (*bi-mol′-ib-dāt*). A molybdate containing twice as much molybdic acid as the corresponding normal molybdate.

**bimucous** (*bi-mū′-kus*) [*bi-; mucus*, mucus]. Relating to two mucous surfaces.

**bimuscular** (*bi-mus′-kū-lar*). Having two muscles. Syn., *dimyarious*.

**binary** (*bi′-nar-e*) [*binus*, a couple]. In chemistry, compounded of two elements. In anatomy, separating into two branches.

**binaural** (*bin-aw′-ral*) [*bi-; auris*, ear]. Pertaining to or having two ears; used for both ears. **b. stethoscope,** a stethoscope with two tubes, one for each ear.

**binauricular** (*bin-aw-rik′-u-lar*). See *binaural.*

**binder** (*bind′-er*) [ME., *byndere*]. A wide bandage about the abdomen, worn by women after labor or after celiotomy, to support the abdominal walls. **b., mammary,** a sling or suspensory for the mamma.

**bindweb** (*bind′-web*). The neuroglia.

**Bing's test** [Albert *Bing*, German otologist, 1844–

]. Let a vibrating tuning-fork be held on the vertex until it has ceased to be audible; then close either ear, and the fork will be heard again for a certain period. If this period of secondary perception is shortened, there exists a lesion of the sound-conducting apparatus; if normal and yet deafness is present, the perceptive apparatus is involved.

**biniodid.** See *biniodide*.

**biniodide** (bin-i'-o-dīd). 1. Having two atoms of iodine in the molecule. 2. A binary salt having twice as many atoms of iodine as it has of the other element.

**binocular** (bin-ok'-ū-lar) [bi-; oculus, an eye]. Pertaining to both eyes. In optics, an instrument with two eyepieces for use with both eyes at once. **b. vision,** the faculty of using both eyes synchronously and without diplopia.

**binotic** (bin-o'-tik). See *binaural*.

**binovular** (bin-ov'-ū-lar) [bi-; ovum, an egg]. Pertaining to or derived from two ova.

**binoxide** (bin-oks'-id). See *dioxide*.

**binuclear** (bi-nū'-kle-ar) [bi-; nucleus, a kernel]. Having two nuclei.

**binucleate** (bi-nū'-kle-āt). Having two nuclei.

**binucleolate** (bi-nū-kle'-o-lāt) [bi-; nucleolus, a lit tle kernel]. Having two nucleoli.

**Binz's test for quinine in the urine** (binis) [Karl Binz, German chemist, 1832– ]. The reagent consists of 2 parts of iodine, 1 part of potassium iodide, and 40 parts of water; on being added to the urine, a precipitate is thrown down if quinine is present.

**bio-** [βίος, life]. A prefix meaning life.

**bio-assay** (bi-o-as-sā') [bio-; assay]. Estimation of the strength of a drug as compared with a standard sample.

**bioblast** (bi'-o-blast) [bio-; βλαστός, a germ]. A plastidule or formative cell; a corpuscle that has not yet become a cell.

**bioblastic** (bi-o-blast'-ik). Relating to bioblasts. **b. theory,** Altmann's, according to which leukocyte granules are considered as definite biological entities, which affect, through oxygen-transmission, both reduction and oxygenation, and in this manner accomplish the disunions and the syntheses of the economy without sacrificing their own individuality. Cf. *color-analysis*.

**biochemics** (bi-o-kem'-iks) [bio-; χημεία, chemistry]. The chemistry of life. See *biochemy*.

**biochemistry** (bi-o-kem'-is-tre) [see *biochemics*]. The chemistry of the living tissues or of life; physiological chemistry.

**biochemy** (bi'-o-kem-e) [see *biochemics*]. Chemical force as exhibited in living organisms.

**biocitin** (bi-o-si'-tin). A preparation containing casein, milk sugar and lecithin.

**bioculate, bioculatus** (bi-ok'-u-lāt, bi-ok-u-la'-tus) [see *binocular*]. Marked by two spots of color different from the chief color.

**biod** (bi'-od) [βίος, life]. 1. Animal magnetism. 2. See *protyl*. 3. Vital force.

**biodesmus** (bi-od-es'-mus) [bio-; δεσμός, a bond]. The vital principle regarded as a bond between organisms.

**biodynamics** (bi-o-di-nam'-iks) [bio-; δύναμις, power]. The dynamics of life; dynamic biology. See *bionomy*.

**biogen** (bi'-o-jen) [bio-; γεννᾶν, to produce]. 1. See *protyl*. 2. See *bioplasm*. 3. See *magnesium dioxide*.

**biogenesis** (bi-o-jen'-es-is) [bio-; γένεσις, origin]. The doctrine that living things are produced only from living things—the reverse of abiogenesis.

**biogenetic** (bi-o-jen-et'-ik) [see *biogenesis*]. Pertaining to biogenesis. **b. law,** the fact that a certain tendency directs the drift or trend of development of a being along a line parallel with that of the series of forms ancestral to it. The being in the course of its development briefly recapitulates that of the ancestral series to which it belongs. Syn., *Mueller's law*.

**biogeny** (bi-oj'-en-e) [see *biogenesis*]. In biology, the evolution of organic forms, either considered individually (*ontogeny*) or tribally (*phylogeny*).

**biognosis** (bi-og-no'-sis) [bio-; γνῶσις, knowledge]. The study of life and its phenomena; biology.

**biograph** (bi'-o-graf) [bio-; γράφειν, to write]. An apparatus for securing photographs of animals in motion. Syn., *kinematograph*.

**biokinetics** (bi-o-kin-et'-iks) [bio-; κίνησις, motion]. The kinetics of life; the science of the movements of living organisms, particularly those that are necessary parts of the process of development. See *karyokinesis*.

**biologic, biological** (bi-o-loj'-ik, -al) [see *biology*]. Pertaining or belonging to biology.

**biologist** (bi-ol'-o-jist) [see *biology*]. One who is a student of biology.

**biologos** (bi-ol'-o-gos) [see *biology*]. A designation proposed for the intelligent living power displayed in cellular and organic action and reaction.

**biology** (bi-ol'-o-je) [bio-; λόγος, science]. The science of life and living things; it embraces the structure, function, and organization of living forms. Syn., *organology*; *organomy*; *somiology*; *zoonomy*. **b., dynamic.** See *bionomy*. **b., static.** See *biostatics*.

**biolysis** (bi-ol'-is-is) [bio-; λύειν, to loosen]. The destruction of life. The devitalization of living tissue by the action of living organisms.

**biolytic** (bi-o-lit'-ik) [bio-; λύειν, to loosen]. Destructive to life; relating to biolysis.

**biomagnetism** (bi-o-mag'-net-izm) [bio-; *magnetism*]. Animal magnetism.

**biometer** (bi-om'-et-er) [bio-; μέτρον, a measure]. 1. A table of life expectancy, etc., upon which the science of life-insurance is based. 2. An instrument, of the nature of a tuning-fork, invented by Dr. Collongues for the reproduction and increase of sounds of the body ordinarily perceived by auscultation.

**biometrics** (bi-o-met'-riks) [see *biometer*]. The science of the body-sounds perceived by auscultation.

**biometry** (bi-om'-et-re) [see *biometer*]. Life-measurement; the estimation of the probable duration of any given life-form—in the past or future.

**bion** (bi'-on) [βίος, life]. A definite physiological individual element or organism. Cf. *morphon*.

**Biondi's fluid.** A staining medium used in histologic laboratories. It is a mixture of orange-G, methyl-green, and acid-fuchsin.

**bionergy** (bi-on'-er-je) [bio-; ἔργον, work]. Life-force; force exercised in the living organism.

**bionomics** (bi-o-nom'-iks) [bio-; νόμος, law]. That branch of natural history which treats of the relations of organisms among themselves and to their environment.

**bionomy** (bi-on'-o-me) [see *bionomics*]. Dynamic biology; biodynamics; the science of the laws and functions of life.

**bionosis** (bi-o-no'-sis) [bio-; νόσος, disease]. A disease caused by a living agent, such as a bacterial disease.

**biontic** (bi-on'-tik) [bios]. Individual as opposed to phyletic.

**bionuclein** (bi-o-nū'-kle-in) [bio-; *nuclein*]. A term suggested by Sacharoff (1902) for the hypothetical substance composed of a combination of iron and nuclein which exists in all enzymes, holding that all vital processes depend upon decomposition of living substance set up by them.

**biophagism, biophagy** (bi-off'-aj-ism, -e) [bio-; φαγεῖν, to eat]. The capacity of absorbing living matter.

**biophagous** (bi-off'-ag-us) [bio-; φαγεῖν, to eat]. Feeding upon living organisms or upon living tissue as insectivorous plants.

**biophilia** (bi-o-fil'-e-ah) [bio-; φιλεῖν, to love]. The instinct for self-preservation.

**biophore** (bi'-o-for) [bio-; φέρειν, to bear]. One of Weismann's hypothetical "bearers of vitality," corresponding to the "plasomes" of Wiesner and Brucke and to the "pangenes" of de Vries, the smallest units that exhibit the primary vital forces, the bearers of the cell-qualities.

**biophthorous** (bi-off'-thor-us) [bio-; φθορά, destruction]. Ruinous to life.

**biophysiography** (bi-o-fiz-e-og'-ra-fe) [bio-; φύσις, nature; γράφειν, to write]. Descriptive or structural biology; organography, as distinguished from biophysiology.

**biophysiology** (bi-o-fiz-e-ol'-o-je) [bio-; φύσις, nature; λόγος, science]. The branch of biology including organogenesis, morphology, and physiology.

**Biophytum** (bi-off'-it-um) [bio-; φυτόν, plant]. A genus of plants of the order *Geraniaceæ*. *B. sensitivum* is a native of the East Indies, where the root is used in inflammations, in gonorrhea, and in pulmonary affections.

**bioplasm** (bi'-o-plazm) [bio-; πλάσμα, form]. Any living matter, but especially germinal or forming matter; matter possessing reproductive vitality.

**bioplasmic** (*bi-o-plaz'-mik*) [*bio-*; πλάσμα, form]. Relating to or of the nature of bioplasm.

**bioplasmin** (*bi-o-plaz'-min*). A hypothetical substance supposed to exist in all living cells and believed to be essential to their life and functional activity.

**bioplasson** (*bi-o-plas'-on*) [*bio-*; πλάσσων, forming]. Elsberg's term for living matter. A synonym of protoplasm or bioplasm.

**bioplast** (*bi'-o-plast*) [*bioplasm*]. A mass or cell of bioplasm that is a unit of living matter.

**bioplastic** (*bi-o-plas'-tik*) [*bio-*; πλαστός, formed]. Relating to or of the nature of a bioplast or of bioplasm.

**biopsia, biopsy** (*bi-op'-se-ah, bi'-op-se*) [*bio-*; ὄψις, vision]. 1. Observation of the living subject; opposed to necropsy. 2. A name coined by Besnier for the excision, during life, of an eruptive lesion or fragment of a newgrowth to establish the diagnosis by means of an examination of the excised piece.

**biopsic** (*bi-op'-sik*) [*bio-*; ὄψις, vision]. Pertaining to biopsy.

**biopsychic, biopsychical** (*bi-o-si'-kik, -al*). Pertaining to psychic phenomena regarded from the biological point of view.

**biorbital** (*bi-or'-bit-al*) [*bi-*; *orbita*, a circle]. Relating to both orbits.

**bios** (*bi'-os*) [βίος, life]. Life.

**bioscope** (*bi'-o-skōp*) [*bio-*; σκοπεῖν, to view]. An instrument used in bioscopy.

**bioscopy** (*bi-os'-ko-pe*) [see *bioscope*]. Examination of the body to ascertain whether life is present. **b., electro-**, examination by the aid of the electric current. In about two hours after death the muscular reaction is lost to faradic stimulation in the tongue; after three or four hours in the extremities; after five or six hours in the trunk. The reaction to galvanism persists somewhat longer.

**biose** (*bi'-ōs*). A disaccharide.

**biosis** (*bi-o'-sis*) [βίος, life]. Life; vitality.

**biostatics** (*bi-o-stat'-iks*) [*bio-*; στατικός, causing to stand]. 1. Static biology; the science of the determinate parts of biology, including anatomy and the physics of the living body. 2. Vital statistics.

**Biot's respiration** (*be'-o*) [Camille *Biot*, French physician]. Meningitic respiration: rapid, short breathing, interrupted by pauses lasting from several seconds to half a minute, sometimes observed in healthy subjects during sleep; most frequently in meningitis, in which it is an unfavorable prognostic sign.

**biotaxis, biotaxy** (*bi-o-taks'-is, bi'-o-taks-e*) [*bio-*; τάξις, arrangement]. 1. The selective and arranging function or activity of life, or of living cells. 2. Systematic biology; the classification of living organisms.

**biothalmy** (*bi'-o-thal-me*) [*bio-*; θάλλειν, to be vigorous]. The art of living long and well.

**biotic** (*bi-ot'-ik*) [see *biotics*]. Pertaining to life or to the laws of animal and vegetable progress and evolution; physiological.

**biotics** (*bi-ot'-iks*) [βιοτικός, vital]. The science of vital functions and manifestations.

**biotomy** (*bi-ot'-o-me*) [*bio-*; τέμνειν, to cut]. Vivisection.

**bipalatinoid** (*bi-pal-at'-in-oid*). A gelatin capsule with two compartments.

**biparasitic** (*bi-par-as-it'-ik*) [*bi-*; παράσιτος, a parasite]. Parasitic upon a parasite.

**biparietal** (*bi-par-i'-et-al*) [*bi-*; *paries*, a wall]. Relating to both parietal bones. **b. diameter**, the distance from one parietal eminence of the cranium to the other.

**biparous** (*bip'-ar-us*) [*bi-*; *parere*, to bring forth]. Producing two at a birth.

**bipartite** (*bi-par'-tīt*) [*bi-*; *pars*, a part]. In biology, composed of two parts or divisions.

**bipartition** (*bi-par-tish'-un*) [see *bipartite*]. Separation into two parts.

**biped** (*bi'-ped*) [*bi-*; *pes*, a foot]. 1. Having two feet. 2. An animal with two feet.

**biperforate** (*bi-per'-fo-rāt*) [*bi*, two; *perforatus*, bored through]. Having two perforations; as a biperforate hymen.

**biplumbic** (*bi-plum'-bik*) [*bi-*; *plumbum*, lead]. Containing two atoms of lead.

**bipocillated** (*bi-pos'-il-a-ted*) [*bi-*; *pocillum*, a little cup]. Having two cup-like appendages.

**bipolar** (*bi-po'-lar*) [*bi-*; *polus*, a pole]. Having two poles or extremities. **b. nerve-cells**, nerve-cells that have two prolongations of the cell-matter.

**bipolarity** (*bi-po-lar'-it-e*) [see *bipolar*]. The condition of having two processes from opposite poles, as a nerve-cell; or of having different electric properties existing at the two poles.

**bipotassic** (*bi-po-tas'-ik*). Having two atoms of potassium.

**bipubiotomy** (*bi-pū-be-ot'-o-me*) [*bi*, double; *pubes*, pubes; τέμνειν, to cut]. Double pubiotomy; an obsolete operation in which the pubic bones were both divided. See *ischiopubiotomy*.

**bipunctate** (*bi-punk'-tāt*) [*bi-*; *punctum*, a point]. Having two dots or points.

**bipupillate** (*bi-pū'-pil-āt*) [*bi-*; *pupilla*, pupil of the eye]. Marked with spots which contain two pupil-like dots.

**biramose, biramous** (*bi-ram'-ōs, -us*) [*bi-*; *ramus*, a branch]. Having two branches.

**birch** (*berch*) [AS., *birce*]. Any tree of the genus *Betula*. Birch-tar, or the tarry oil of *Betula alba*, is useful in certain skin diseases. The bark of *B. lenta*, the American black birch, yields a fragrant volatile oil, identical with that of *Gaultheria procumbens*, for which it is extensively substituted. **b. camphor**. Same as *b.-resin*. **b.-resin**. See *betulin*.

**Bird's formula**. The last two figures of the specific gravity of the urine roughly indicate the number of grains of solids to the ounce of urine. The same two figures multiplied by 2 (*Trapp's factor*) give the parts per 1000. **B.'s sign**, a well-defined zone of dulness with absence of the respiratory sound in hydatid cyst of the lung.

**bird-lime**. A viscous vegetable substance used in Japan as a local dressing for wounds.

**bird's nest bodies, or cells**. The cells of certain forms of carcinoma, distinguished by the concentric arrangement of their cell-walls. See also *cancer nests*. **b., edible**, the nest of certain species of swift, used by the Chinese as food. It consists of marine algæ, *Gelidium*, cemented by the salivary mucus of the bird. **b. sternum**, a deformity of the sternum found in lateral curvature of the spine.

**birefractive, birefringent** (*bi-re-frak'-tiv, bi-re-frin'-jent*) [*bis*, twice; *refringere*, to break back]. Doubly refractive; anisotropic.

**birhinia** (*bi-rin'-e-ah*) [*bi*, two; ῥίς, the nose]. A congenital defect in which there is the formation of two noses.

**birimose** (*bi-ri'-mōs*) [*bis*, twice; *rima*, cleft]. Having two clefts or slits.

**Birkett's hernia** [John *Birkett*, English surgeon]. Intraperitoneal inguinal hernia; hernia into the vaginal process of the peritoneum.

**birth** (*berth*) [ME., *byrth*]. 1. The delivery of a child; parturition. 2. That which is brought forth in parturition. **b., cross**. See *transverse presentation*. **b.-mark**. See *nævus pigmentosus*. **b.-palsy**, any paralytic affection due to an injury received at birth; less correctly, a congenital paralytic affection due to a lesion that existed in the fetal state. **b., partial**, the incomplete expulsion of a child in labor; of legal value in lawsuits for property. **b., plural**, the birth of more than a single child. **b., posthumous**, the birth of a child after the death of its father. **b., precocious**, the occurrence of natural labor after a shorter pregnancy than is usual. **b., premature**. See *labor*. **b.-rate**, the proportion of births per thousand. **b.-root**. See *beth root*. **b.-wort**, the plant *Aristolochia climatitis* (see *Aristolochia*), so called from its former employment as a depurant after childbirth. **b., still**. See *still-born*. **b., virgin-**. See *parthenogenesis*.

**bisacromial** (*bis-ak-ro'-me-al*) [*bis*, two; *acromion*]. Relating to the two acromial processes.

**bisalt** (*bi'-salt*). See *salt, acid*.

**bisaxillary** (*bis-ak'-sil-a-re*) [*bis*; *axilla*]. Pertaining to both axillæ.

**bische** (*bish*) [East Indian name]. Endemic dysentery.

**Bischoff's operation** (*bish'-of*) [Johann Jacob *Bischoff*, German gynecologist, 1841– ]. Abdominal section with extirpation of the gravid uterus.

**Bischoff's test for biliary acids** (*bish'-of*) [Carl Adam *Bischoff*, German chemist, 1855– ]. A red coloration results upon heating biliary acids with dilute sulphuric acid and cane sugar.

**bisection** (*bi-sek'-shun*) [*bi*, two; *sectio*, a cutting]. Division into two parts; in obstetrics, embryotomy.

**bisexual** (*bi-seks'-ū-al*) [*bi-*; *sexus*, sex]. Having the reproductive organs of both sexes; hermaphroditic.

**bisferious** (bis-fer'-e-us) [bis; ferire, to strike]. Having two beats; dicrotic.

**bishoping** (bish'-op-ing). In farriery, filing a space between the teeth of a horse.

**bisiliac** (bis-il'-e-ak) [bis; iliacus]. Relating to the two most distant points of the two iliac crests.

**bisischiadic, bisichiatic** (bis-is-ke-ad'-ik, bis-is-ke-at'-ik) [bis, two; ἰσχιαδικός, relating to the hip]. Relating to both ischia, or to corresponding points on the two ischia.

**Biskra boil, B. button** (bisk'-rah) [Biskra, a town in Algeria]. See *furunculus orientalis.*

**bismal** (bis'-mal). Trade name for the bismuth salt of methylene digallic acid; used as intestinal antiseptic and astringent.

**Bismarck-brown.** A brown, basic aniline dye, extensively used as a stain and counterstain in histology; also called vesuvine.

**bismon** (bis'-mon). Colloidal bismuth oxide, containing 20 per cent. of bismuth, and used like the subnitrate.

**bismutal, bismuthol** (bis'-mu-tal, -thol). Bismuth and sodium phosphosalicylate.

**bismutan** (bis'-mu-tan). A yellow, insoluble powder, said to be a mixture of bismuth, resorcin, and tannin; used in the diarrhea of children.

**bismuth, bismuthum** (bis'-muth, bis-mu'-thum) [L.]. Bi = 208; quantivalence I, III, V. A pinkish-white, crystalline metal. Its commercial salts often contain arsenic. The insoluble salts of bismuth are feebly astringent. The derivatives of bismuth are chiefly employed as astringents and sedatives to mucous membranes and as gastrointestinal antiseptics. The soluble salts are irritant in large doses. **b. albuminate,** a powder containing 9 % of bismuth; it is used in stomachic or intestinal cramp. Dose 5–15 gr. (0.32–0.97 Gm.) 3 or 4 times daily. **b. and ammonium citrate** (bismuthi et ammonii citras, U. S. P.), soluble in water. Dose 1–5 gr. (0.065–0.32 Gm.). **b. and ammonium citrate, solution of** (liquor bismuthi et ammoniæ citratis, B. P.). Dose ¼–1 fl. (2–4 Cc.). **b. benzoate,** Bi(C₇H₅O₂)₃, a white powder containing 27 % of benzoic acid. It is an internal and external antiseptic. Dose 5–15 gr. (0.32–0.97 Gm.). **b. betanaphtholate,** 2Bi (C₁₀H₇O)₃+Bi₂O₃ a light brown, odorless, insoluble powder, containing 80 % bismuth trioxide. It is an intestinal antiseptic. Dose 15–30 gr. (0.97–1.94 Gm.). Syn., *betanaphtholate; naphthol bismuth; orphol.* **b. bilactomonotannate,** an odorless yellow powder, used in the diarrhea of infants. Dose 30–45 gr. (2–3 Gm.). Syn., *lactanin.* **b. borate,** BiBO₃, an intestinal antiseptic. Dose 5–40 gr. (0.32–2.6 Gm.). **b. borophenate,** Bi₂O₂B(C₆H₃)(CO₃)+3H₂O. It is recommended as a surgical dressing used as a dusting-powder, or for burns or scalds applied as a paste (25 to 50 % in glycerol) on lint. Syn., *markasol.* **b. carbolate,** Bi(OH)₂C₆H₅O, a grayish-white powder containing 80 % of bismuth oxide and 18 to 19 % of phenol. It is an intestinal antiseptic and is used externally as a substitute for iodoform. Dose 5–15 gr. (0.32–0.97 Gm.). Syn., *bismuth phenate; bismuth phenylate; phenol bismuth.* **b. carbonate** (bismuthi carbonas, B. P.), (BiO)₂CO₃)₂, H₂O. Dose 5–20 gr. (0.32–1.3 Gm.). **b. and cerium salicylate,** an antirheumatic and intestinal antiseptic. Dose 5–15 gr. (0.32–0.97 Gm.). **b. chrysophanate,** Bi(C₁₅H₉O₄)₂·Bi₂O₃, a yellow, amorphous powder, insoluble in ordinary solvents, but soluble in nitric or sulphuric acid; it is used as a siccative in psoriasis. Application, 5 to 20 % ointment. Syn., *dermol.* **b. citrate** (bismuthi citras, U. S. P.), BiC₆H₅O₇, soluble in water of ammonia. Dose 2–5 gr. (0.13–0.32 Gm.). **b. cresolate,** an odorless, tasteless, grayish-white powder, insoluble in water and alcohol; it is an internal and external antiseptic. **b. dithiosalicylate,** a bulky yellow powder without odor, used as a wound antiseptic and in ophthalmic practice, in diseases of the nose and throat, and in dentistry. Syn., *thioform.* **b. iodosubgallate,** C₆H₂(OH)₄COOBiI, an antiseptic used as a dusting-powder on wounds. Syn., *airol; bismuth oxyiodogallate.* **b. lactate,** BiH(C₃H₄O₃)₂, an internal and external antiseptic. Dose 5–15 gr. (0.32–0.97 Gm.). **b. loretinate,** a combination of bismuth and loretin, used as a surgical and intestinal antiseptic and also in ophthalmology. Dose 7½ gr. (0.5 Gm.). **b., magistery of,** same as *b. subnitrate.* **b. metacresol,** an internal antiseptic consisting of a combination of 75 % of bismuth with 17.5 % of metacresol. **b. methylene-digallate,** 4C₁₅H₁₂O₁₆+3Bi(OH)₃, an internal astringent. Dose 3–5 gr. (0.1–0.3 Gm.) every 3 hours. Syn., *bismal.* **b.-naphthalin benzoate,** an intestinal antiseptic. Dose 8–15 gr. (0.5–1.0 Gm.). Syn., *intestin.* **b. naphthoglycerite,** a remedy for gonorrhea. **b. nitrate,** Bi(NO₃)₃+5H₂O, an astringent and antiseptic. Dose 5–10 gr. (0.32–0.65 Gm.). **b. oxide** (bismuthi oxidi, B. P.), Bi₂O₃. Dose 5–15 gr. (0.32–1.0 Gm.). **b. oxybromide,** BiOBr. It is recommended in the treatment of nervous dyspepsia and hysteria accompanied by gastric pains and vomiting. Dose 5–6 gr. (0.3–0.4 Gm.) several times daily. **b. oxychloride,** pearl white. It is used as a cosmetic. **b. oxyiodide.** See *b. subiodide.* **b. oxyiodomethylgallol,** C₆H₅COOCH(OH)₂O · BiOH . I, a dark-gray powder containing 23.6 % of iodine and 38.4 % of bismuth, used as a surgical antiseptic. Syn., *iodogallicin.* **b. oxyiodopyrogallate. b. oxyiodopyrogallol,** a combination of bismuth subiodide with pyrogallol. It is recommended as a surgical antiseptic. **b. oxyiodotannate,** a wound antiseptic. Syn., *ibit.* **b., pancreatinized,** used in dyspepsia. Dose 15–75 gr. (1–5 Gm.). **b. peptonate, b., peptonized,** used in dyspepsia and gastralgia. Dose 15–75 gr. (1–5 Gm.). Syn., *bismuthated peptone.* **b. permanganate,** Bi(MnO₄)₃, a dry dusting-powder for wounds and ulcers. **b. phenate, b. phenylate.** See *b. carbolate.* **b., phenol-,** a compound of bismuth, 27.5 %, with phenol, 22 %; it is used as an intestinal antiseptic. **b. phosphate,** BiPO₄, an intestinal disinfectant. Dose 3–8 gr. (0.2–0.5 Gm.). **b. powder, compound,** Ferrier's snuff, contains 2 grains of morphine hydrochloride in 1 ounce, with bismuth and acacia. It is used in the treatment of coryza. **b. pyrogallate,** (C₆H₃[OH]₃O)₃BiOH, an internal antiseptic in doses of 5–15 gr. (0.32–0.97 Gm.). Applied in skin diseases in 10 to 20 % ointment or dusting-powder. Syn., *helcosol.* **b. resorcinate,** a yellowish-brown powder containing about 4 % of bismuth trioxide. It is used in catarrh of the stomach. **b. salicylate,** (C₇H₅O₃)₃Bi₂O₃, a salt obtained by Thibault from bismuth oxide, instead of the hydroxide, as is customary. It is used as an external and internal antiseptic. Dose 5–15 gr. (0.32–0.97 Gm.). **b. subbenzoate,** basic benzoate of bismuth; used as a wound antiseptic. **b. subcarbonate** (bismuthi subcarbonas, U. S. P.), (BiO)₂CO₃, insoluble. Dose 10 gr.–1 dr. (0.65–4.0 Cc.); feebly astringent and sedative. **b. subgallate** (bismuthi subgallas, U. S. P.). See *dermatol.* **b. subiodide,** BiOI, used as an antiseptic dusting-powder, like iodoform. **b. subnitrate** (bismuthi subnitras, U. S. P.), BiONO₃ . H₂O, the salt chiefly used in medicine as a sedative astringent to the gastro-intestinal mucous membrane. Dose 10 gr.–1 dr. (0.65–4.0 Cc.). **b. subsalicylate** (bismuthi subsalicylas, U. S. P.), a white, amorphous powder. Dose 4 gr. (0.25 Gm.). **b. sulphite,** a combination of sodium sulphite and bismuth nitrate. It is an intestinal antiseptic. Dose 5–40 gr. (0.32–2.6 Gm.). **b. sulphophenylate,** a general intestinal disinfectant. Dose 3½–8 gr. (0.2–0.5 Gm.) 3 or 4 times daily. **b. tannate,** an intestinal antiseptic. Dose 10–30 gr. (0.65–1.94 Gm.). **b. tribromcarbolate, b. tribromphenate,** Bi₂O₃(C₆H₂Br₃OH), an insoluble powder containing about 60 % of Bi₂O₃. It is used as antiseptic in cholera and intestinal disorders. Dose 8–15 gr. (0.52–0.97 Gm.); maximum dose a day 90 gr. (5.85 Gm.). Syn., *xeroform.* **b. trioxide,** Bi₂O₃. It is incompatible with alkalies and water in excess. It is antiseptic and astringent. Dose 5–40 gr. (0.32–2.6 Gm.). Syn., *bismuthous oxide.* **b. troches** (trochisci bismuthi, B. P.), each contains 2 gr. (0.13 Gm.). **b. valerate,** a white powder with the odor of valeric acid, soluble in dilute hydrochloric or nitric acid, insoluble in water or alcohol; it is used as a sedative and antispasmodic in neuralgia, chorea, epilepsy, etc. Dose 1–3 gr. (0.065–0.194 Gm.).

**bismutal** (bis'-mu-thal). Containing bismuth.

**bismutate** (bis'-mu-thāt). A salt of bismuthic acid.

**bismuthic** (bis'-mu-thik). Relating to bismuth; containing bismuth in its higher valency.

**bismuthol.** See *bismutal.*

**bismuthosis** (bis-mu-tho'-sis). Chronic bismuth-poisoning; it may follow the use of the soluble salts.

**bismuthous** (bis'-mu-thus). Containing bismuth as a trivalent radicle.

**bismutyl** (bis'-mu-thil), BiO. A univalent radical. **b. bromide.** See *bismuth oxybromide.* **b. chloride.**

See *bismuth oxychloride*. **b. iodide.** See *bismuth subiodide*. **b. nitrate.** See *bismuth subnitrate*.
**bismutol** (*bis'-mū-tol*). A compound of sodium salicylate and soluble bismuth phosphate.
**bismutose** (*biz'-mū-tōs*). A bismuth and albumin compound, equivalent in action to bismuth subnitrate. Useful in gastrointestinal affections of infectious character. For children under six months the dose is 15–30 gr. (1–2 Gm.); for those over six months it may be given in 60 gr. (1 dr.) doses.
**bisol** (*bi'-sol*). Soluble bismuth phosphate containing about 20 % of bismuth oxide. It is used in gastralgia. Dose 3–7½ gr. (0.19–0.5 Gm.).
**bispep** (*biz'-pep*). A proprietary preparation containing bismuth, pepsin, ammonium carbonate, and aromatics.
**bissa** (*bis'-ah*) [native African]. An affection of man and sheep, common in Egypt, and characterized by the production of edema. **b. bol,** a kind of myrrh, from *Balsamodendron Kafal*, a tree of E. Africa. It is used largely in adulterating the finer grades of myrrh, and is said to stimulate powerfully the flow of milk in cows.
**bistellate** (*bi-stel'-āt*) [*bi*, two; *stella*, a star]. Shaped like a double star.
**bistephanic** (*bi-ste-fan'-ik*) [*bi*, two; στεφάνιον, dim. of στέφανος, a wreath]. Relating to the two stephanions.
**bistort** (*bis'-tort*) [*bis*, twice; *tortus*, twisted]. Snake-weed, adder's wort. The rhizome of *Polygonum bistorta*, an astringent. Dose of fluidextract ♏ ꜱꜱ–ᴊ.
**bistoury** (*bis'-too-re*) [Fr. *bistouri*]. A long, slender (straight or curved) knife used in surgery. **b.-caché,** one that has the blade concealed for passing to the point to be incised, and by pressure on the handle the blade is exposed and the incision made.
**bistratal** (*bi-stra'-tal*) [*bis*; *stratum*, layer]. Arranged in two layers.
**bistriate** (*bi-stri'-āt*) [*bis*; *stria*, a furrow]. Marked with two lines or streaks.
**bisulcate** (*bi-sul'-kāt*) [*bi*, two; *sulcus*, a furrow]. Having two furrows or grooves.
**bisulphate** (*bi-sul'-fāt*) [*bi-*; *sulphur*]. A sulphate in which the base replaces but one of the two hydrogen atoms of the acid.
**bisulphide** (*bi-sul'-fīd*) [*bi*, two; *sulphur*]. In chemistry, a sulphur compound in which there are two atoms of sulphur to one atom of the other substance of the compound.
**bisulphite** (*bi-sul'-fīt*) [*bi*, two; *sulphur*]. A sulphite in which the base replaces but one of the two hydrogen atoms of the acid.
**bitartrate** (*bi-tar'-trāt*) [*bi-*; τάρταρον, tartar]. Any tartrate in which only one replaceable hydrogen atom has been replaced by a base.
**bite** (*bīt*) [AS., *bītan*]. 1. The corrosion of a substance with an acid. 2. The more or less perfect coaptation of the upper and lower teeth. **b., open,** that in which the upper and lower incisors do not close together. **b., underhung,** that in which the upper incisors overreach the lower.
**bitemporal** (*bi-tem'-por-al*) [*bi-*; *tempora*, the temples]. Pertaining to the two temples.
**bitnoben** (*bit-no'-ben*) [Hind. for "black salt"]. An East Indian polychrest remedy composed of salt, myrobalan and iron.
**bitonal** (*bi-to'-nal*) [*bis*, twice; *tonus*, a tone]. Double-toned.
**Bitot's spots** (*be'-tō*). Xerosis conjunctivæ. Silver-gray, shiny, triangular spots on both sides of the cornea, within the region of the palpebral aperture, consisting of dried epithelium, flaky masses, and microorganisms. Observed in some cases of hemeralopia.
**bitrochanteric** (*bi-tro-kan-ter'-ik*) [*bi-*, two; *trochanter*]. Belonging to the two trochanters.
**bitter** (*bit'-er*) [AS., *bītan*, to bite]. A peculiar, acrid, biting taste, of which that of quinin is an example; unpalatable. **b. almond,** the nut of *Amygdalum amarum*. It contains hydrocyanic acid. **b.-almond oil,** oleum amygdalæ amaræ. See *benzaldehyde*. **b. apple,** the fruit of the colocynth, a purgative remedy. **b.-blain,** a West Indian herb, *Vandellia diffusa*, employed in fevers and in hepatic disorders. **b. bugleweed,** the herb *Lycopus europæus*, alterative and tonic. Dose of fluidextract ½–1 oz. (16–32 Cc.). **b. cucumber,** colocynth. **b. cup.** See *cup, bitter*. **b. purging salt,** magnesium sulphate. **b.-root,** the root of *Gentiana lutea*, a tonic. **b. salts,** magnesium sulphate. **b. tincture,** the *tinctura amara* (N. F.), prescribed also in the German pharmacopœia; it is a tincture of gentian, centaury, bitter orange-peel, orange-berries, and zedoary. Syn., *stomach-drops*. **b. water,** a water containing magnesium sulphate. **b. wine of iron,** a solution of white wine, syrup, iron citrate, and quinine.
**bitters** (*bit'-ers*) [see *bitter*]. 1. Medicines characterized by a bitter taste. 2. An alcoholic drink, an appetizer. **b., aromatic,** medicines that unite the properties of aromatics with those of simple bitters. **b., simple,** medicines that stimulate the gastrointestinal tract without influencing the general system. **b., styptic,** medicines that add styptic and astringent properties to those of bitterness.
**bittersweet** (*bit'-er-swēt*). See *dulcamara*.
**bitumen** (*bi-tū'-men*) [L.]. Mineral pitch or oil, composed of various hydrocarbons. In the solid form it is usually called *asphalt*; in the liquid form, *petroleum*. An intermediate form is known as mineral tar or *maltha*. By distillation, bitumen yields benzol, naphtha, paraffin, and various other hydrocarbons, gaseous and liquid.
**bituminization** (*bi-tū-min-iz-a'-shun*). A conversion into bitumen.
**biurate** (*bi-ū'-rāt*). An acid urate; a urate containing twice as much of the uric-acid constituent as an ordinary urate.
**biuret** (*bi'-ū-ret*) [*bi-*; οὖρον, urine], $C_2H_5N_3O_2 + H_2O$. A compound formed by exposing urea to a high temperature for a long time. It is readily soluble in water and in alcohol; it crystallizes with one molecule of water in the form of warts and needles. When anhydrous, biuret melts at 190° C. with decomposition. **b. reaction for proteids.** See *Piotrowski's reaction*. **b. reaction for urea,** melt urea completely in a dry test-tube and continue the heat for some time. When cold, dissolve in water, add abundant caustic soda and a dilute solution of copper sulphate drop by drop. The solution becomes first pink, then reddish-violet, and, finally, bluish-violet, according to the amount of copper sulphate added.
**bivalence** (*biv'-al-ens*) [*bi-*; *valens*, having power]. In chemistry, a valence or saturating power double that of the hydrogen atom.
**bivalent** (*biv'-al-ent*) [see *bivalence*]. In chemistry applied to an element of which an atom can replace two atoms of hydrogen or other univalent element, or to a radical that has the same valence as such an element.
**bivalve** (*bi'-valv*) [*bi-*; *valva*, a valve]. Having two valves or shells, as a speculum.
**biventer** (*bi-ven'-ter*) [*bi-*; *venter*, a belly]. 1. Having two bellies, as a muscle. 2. A digastric muscle. **b. cervicis,** the inner portion of the complexus muscle. **b. maxillæ,** the digastric muscle.
**biventral** (*bi-ven'-tral*) [see *biventer*]. Having two stomachs; having two bellies, as the digastric muscle.
**bivittate** (*bi-vit'-āt*) [*bi-*; *vitta*, a fillet]. Marked by two longitudinal stripes.
**biv₀** (*be'-vo*). Trade name of a beef and iron wine.
**Bixa** (*biks'-ah*) [*biché*, Brazilian name]. A genus of plants of the order *Bixaceæ*. *B. orellana* is the annotto-tree, a native of South America and now dispersed throughout the tropics, furnishing from the pulp surrounding the seeds the annotto of commerce. The pulp is used as a remedy for dysentery and the seeds are said to be astringent and antipyretic.
**bixin** (*biks'-in*) [*bixa*], $C_{16}H_{26}O_2$. An orange-red coloring-matter found in annotto.
**bizincic** (*bi-zink'-ik*). Containing two atoms of zinc.
**bizygomatic** (*bi-zi-go-mat'-ik*). Relating to both zygomas, or to the most prominent points on the two zygomatic arches.
**Bizzozero's blood-platelets** (*bit-sot'-ser-o*) [Giulio Bizzozero, Italian physician, 1846–1901]. Small, round or elliptical, nonnucleated bodies found in the blood of mammals, including man. **B.'s corpuscles.** See *Neumann's corpuscles*. **B.'s crystals.** See *Charcot's crystals*.
**black** (*blak*) [AS., *blæc*]. Characterized by an absence of color. The appearance of an object from the surface of which none of the spectrum colors is reflected. **b. alder.** See *Prinos*. **b. antimony,** antimony tersulphide, $Sb_2S_5$. **b. ash.** 1. The bark of *Fraxinus sambucifolia*, a mild tonic and

astringent. Dose of *fluidextract* ⅓–1 dr. (2–4 Cc.).
2. Impure sodium carbonate Na₂CO₃ mixed with carbon. **b. blood,** venous blood. **b. cancer.** See *melanosis*. **b. cohosh.** See *cimicifuga*. **b.-damp,** carbon dioxide gas, which is found in greater or less quantity in all collieries, being given off by many coals, either mixed with fire-damp or separately, or produced in various other ways, as by the exhalations of the miners, by fires, and by explosions of fire-damp. Syn., *choke-damp*. **b. death.** See *plague*. **b. disease,** a disease of malarial origin and pernicious course, characterized by extreme darkening of the skin, which may be brown or black in color. It occurs in the Garo Hills in Assam. **b. draught.** See *senna*. **b.-drop.** See *opium*. **b. erysipelas.** Synonym of *anthrax*. **b. eye,** livor (or sugillatio) oculi; ecchymosis of the tissues about the eye, usually from injury. **b. haw.** See *viburnum*. **b. hellebore.** See *hellebore*. **b. jaundice,** an excessive jaundice arising from obstruction of the gall-ducts. The color of the skin is greenish-black. **b.-lead,** a form of carbon properly known as the mineral graphite. **b. measles,** rubeola nigra, a grave or malignant form of measles. **b. pepper.** See *piper nigrum*. **b. phthisis.** A synonym of *miner's phthisis*. **b. sickness,** cerebrospinal fever. **b. snakeroot.** See *cimicifuga*. **b. spit.** See *miner's phthisis*. **b. tongue.** 1. A name given to a condition characterized by the formation, upon the dorsum of the tongue, of a hair-like deposit that passes through various stages of coloration from yellow to brown and finally black, ultimately disappearing by desquamation as gradually as it came. Repeated recurrence is the rule. It is probably an undue proliferation of the epithelium, the result of irritation. 2. A term applied to erysipelatous glossitis. **b. vomit,** the coffee-ground vomit of yellow fever, etc. **b. walnut,** the leaves of *Juglans nigra*, a tonic, alterative, and deobstruent. Dose of *fluidextract* 20–30 min. (1.3–2.0 Cc.). **b.-wash,** *lotio hydrargyri nigra* (B. P.). **b. willow,** the buds of *Salix nigra*, a bitter tonic with aphrodisiac properties. Dose of *fluidextract* ₹5 min.–1 dr. (1–4 Cc.).

**blackbain** [OE.]. Synonym of *anthrax*.
**blackberry** (blak'-ber-e). See *rubus*. **b. cordial,** the *cordiale rubi fructus* (N. F.). Its formula is: blackberry juice, 3 pints; cinnamon, in coarse powder, 2 troy oz.; cloves and nutmeg, in coarse powder, each, ½ troy oz.; dilute alcohol, 2 pints; syrup, 3 pints.
**blackhead** (blak'-hed). See *comedo*.
**blackleg** (blak'-leg). A febrile, generally fatal, disease, chiefly affecting cattle and sheep, which is characterized by the rapid appearance of irregular nodules in the skin and muscular tissues, that are at first tense and very painful, but rapidly become painless and crepitating. It is caused by *Bacillus chauvœi*, or the bacillus of symptomatic anthrax. Syn., *symptomatic anthrax*.
**blackwater fever.** A very fatal infectious disease occurring especially on the low coasts of tropical Africa, but also in Sicily, Greece, Central and South America, Java, New Guinea, and the southern portions of the United States. The disease is almost exclusively confined to the white race, and is characterized by a chill, an irregular intermittent or remittent fever, vomiting, dyspnea, jaundice, and hemoglobinuria. An almost invariable complication is nephritis. Studies of the blood have revealed the presence of a small, annular, nonpigmented, intracellular parasite; also forms having a roset or crescent shape,
**bladder** (blad'-er) [AS., *blæddre*, a blister]. 1. A membranous sac serving for the reception of fluids or gases. 2. The hollow organ which serves as a reservoir for the urine. **b., atony of,** inability to expel the urine, from deficient muscular power. **b., bilobed, b., bilocular,** a sacculated bladder having two pouches. **b., brain-,** a cerebral vesicle. **b., catarrh of.** See *cystitis*. **b., encysted,** a urinary bladder with communicating cysts connected with it. **b., exstrophy of, b., extroversion of.** See *exstrophy of bladder*. **b., gall-.** See *gall-bladder*. **b. germ.** See *blastula*. **b., irritable,** a condition characterized by constant desire to urinate. **b., multilocular,** a sacculated bladder having many pouches. **b., neck of,** the constricted portion continuous with the urethra in front. **b., nervous,** a condition in which there is a frequent desire to urinate, with inability at the same time to perform the act perfectly, and consequent slight dribbling at its close.

**b., sacculated,** a condition due to overextension, in which pouches are formed by the forcing out of its mucous coat between the hypertrophied muscular fibers, and in which urine may be held and become decomposed, and in which calculi may be retained. **b.-stammering,** Sir James Paget's name for that condition observed in young males who are unable to micturate when under observation or surrounded by unusual conditions or objects. It is due to spasm of the compressor urethræ muscle. **b.-stammering, false,** a condition in which there is some mechanical or pathological interference with urination. **b., sterile,** a hydatid cyst without secondary cysts, heads, or broad capsules. **b., supplementary,** a diverticulum caused by sacculation of the urinary bladder. Syn., *parurocystis*. **b.-worm.** See *worm-bladder-*. **b. wrack.** See *Fucus vesiculosus*.
**blade** (blād) [ME., blad, a leaf of grass]. In biology, the broad part of a leaf. **b.-bone.** The scapula, or shoulder-blade.
**blain** (blān) [ME., *blane*, a blister]. 1. A blister; an elevation of the cuticle containing serum; a pustule; a blotch. 2. Synonym of *anthrax*.
**Blainville's ears** (blan-vēl') [Henri Marie Ducrotay de *Blainville*, French anthropologist, 1777–1850]. Congenital asymmetry of the ears.
**Blancard's pills** (blang'-kard) [Stephen *Blancard*, Dutch physician, 1650–1702]. Pills of iron iodide.
**bland** (bland) [*blandus*, mild]. Mild; soothing.
**Blandin's ganglion** [Philippe Frédéric *Blandin* French surgeon, 1798–1849]. The sublingual ganglion, a small gangliform enlargement lying between the lingual nerve and the sublingual gland. **B.'s gland,** a mucipalous gland situated near the tip of the tongue in the median line and opening by several ducts on the lower surface of the tongue.
**Blandin-Nuhn's gland.** See *Blandin's gland*.
**Blasius' (Blaes')** duct. See *Stenson's duct*.
**blast** (blast) [AS., *blæst*]. 1. See *breath*. 2. See *blight* (2). 3. Inflammation. 4. A disease of sheep.
**blastema** (blas-te'-mah) [βλάστημα, from βλαστάνειν, to germinate]. 1. The formative lymph or rudimentary tissue, from which, by differentiation, tissues and organs are developed. A synonym of protoplasm. 2. An undifferentiated protoplasmic layer in certain eggs or embryos. **b., ossific, b., ossifying, b., subperiosteal.** See *osteogenetic layer*. **b. pili,** a hair-papilla.
**blastemic** (blas-tem'-ik) [βλαστάνειν, to germinate]. Relating to blastema; rudimentary; bioplasmic.
**blasticle** (blas'-ik-l). The vitelline nucleus.
**blastid** (blas'-tid) [βλαστός, a germ]. In embryology, a very small clear spot on the fecundated ovum marking the place of the nucleus or cytoblast.
**blastidium** (blas-tid'-e-um) [βλαστός, a germ]. An endospore or cell of endogenous origin.
**blasto-** (blas-to-) [βλαστός, a germ]. A prefix meaning germ or bud.
**blastocardia** (blas-to-kar'-de-ah) [blasto-; καρδία, the heart]. The germinal spot.
**blastocele, blastoceloma** (blas'-to-sēl, -o'-ma) [blasto-; κοίλος, hollow]. The central cavity of the blastula or vesicular morula. The nucleolus of the ovum.
**blastocelis** (blas-to-se'-lis) [blasto-; κηλίς, a spot]. Wagner's name for the germinal spot.
**blastochyle** (blas'-to-kīl) [blasto-; χυλός, juice]. The colorless fluid in the blastodermic vesicles.
**blastocœlum, blastocœloma** (blas-to-se'-lom, blas-to-se-lo'-mah). Same as *blastocele*.
**blastocolysis** (blas-to-kol'-is-is) [blasto-; κόλουσις, a cutting off]. The arrest of a developmental process.
**blastocyst, blastocystinx** (blas'-to-sist, blas-to-sist'-inks) [blasto-; κύστις, a bladder]. The germinal vesicle.
**blastocyte** (blas'-to-sīt) [blasto-; κυτός, a cell]. An undifferentiated embryonic cell; a cell which is capable of giving rise to daughter cells of varying characters.
**blastocytoma** (blas-to-si-to'-ma) [blasto-; κυτός, a cell; ὄμα, a tumor]. A tumor in which the elements are blastocytes, or undifferentiated cells.
**blastoderm** (blas'-to-derm) [blasto-; δέρμα, the skin]. In embryology, the germinal membrane formed by the cells of the morula, lying on the internal surface of the vitelline membrane of the impregnated ovum. The whole hollow sphere, with its surrounding cells, is called the *blastodermic vesicle*, and is formed about the tenth day. The *ectoderm* (or *epiblast*) and the *endoderm* (*entoderm* or *hypoblast*) layers are simply

due to a proliferation of the blastodermic cells about the *germinal area*, whereby the blastoderm is doubled, thus forming these outer and inner layers. The *mesoblast* or middle layer is developed after the others, and probably from the hypoblast. Syn., *blastodermic membrane; germ membrane; germinal membrane*. **b.**, **bilaminar**, the blastoderm when it consists only of the ectoderm and the endoderm. **b.**, **trilaminar**, the blastoderm after the formation of the mesoblast.

**blastodermic** (*blas-to-der'-mik*) [see *blastoderm*]. Relating to the blastoderm. **b. vesicle**, see *blastoderm*.

**blastodisc** (*blas'-to-disk*) [*blasto-*; δίσκος, disc]. A blastodermic disc; a mass or disc capping one pole of the yolk.

**blastogenesis** (*blas-to-jen'-es-is*) [*blasto-*; γένεσις, generation]. 1. In biology, Weismann's theory of origin from germ-plasm, in contradistinction to Darwin's theory of pangenesis. 2. Reproduction by buds.

**blastogeny** (*blas-toj'-en-e*) [see *blastogenesis*]. Haeckel's term for the germ-history of an individual organism; a division of ontogeny.

**blastoma** (*blas-to'-mah*) [*blasto-*; ὄμα, a tumor]. 1. A granular growth due to the presence of a germ or microorganism. 2. One of a peculiar group of true tumors which originate from embryonic cellnests; *e. g.*, chondroma, glioma, etc. Syn., *enblastoma*.

**blastomere** (*blas'-to-mēr*) [*blasto-*; μέρος, a part]. Any one of the nucleated cells or segments into which the fecundated vitellus divides.

**blastomyces** (*blas-to-mi'-sēz*) [*blasto-*; μύκης, a fungus; pl., *blastomycetes*]. A genus of budding fungi usually referred to *Torula* or *Saccharomyces*. Single-celled thallophytes, destitute of chlorophyl, which reproduce by yeast-like buds or by endogenous cell-formation. **b. dermatidis** (Gilchrist, 1894), a yeast-like organism producing a scrofuloderma in man. See *dermatitis, blastomycetic*.

**blastomycetic** (*blast-o-mi-se'-tik*). Pertaining to or caused by budding fungi (*blastomycetes*). **b. dermatitis**. See under *dermatitis*.

**blastomycosis** (*blas-to-mi-ko'-sis*). An affection due to budding fungi (*blastomycetes*). Cf. *blastomyces; saccharomyces; torula*.

**blastoneuropore** (*blas-to-nū'-ro-pōr*) [*blasto-*; νεῦρον, a nerve; πόρος, a pore]. In biology, the temporary aperture in certain embryos formed by the coalescence of the *blastopore* and *neuropore*.

**blastosphere** (*blas'-to-sfēr*) [*blasto-*; σφαῖρα, a sphere]. The blastodermic vesicle or blastula.

**blastophore** (*blas'-to-fōr*) [*blasto-*; φέρειν, to bear]. That part of a sperm-cell that does not become converted into spermatozoa.

**blastophyllum** (*blast-o-fil'-um*) [*blasto-*; φύλλον, a leaf]. The endoderm or ectoderm; a primitive germ-layer.

**blastophyly** (*blas-tof'-il-e*) [*blasto-*; φυλή, a tribe]. The tribal history of individual organisms.

**blastopore** (*blas'-to-pōr*) [*blasto-*; πόρος, passage; pore]. The small opening leading into the notochordal canal, or, after the canal has fused with the yolk-cavity, leading into the archenteron. It is situated at the hind end of the primitive axis and is a small portion of the gastrula mouth.

**blastoprolepsis** (*blast-o-pro-lēp'-sis*) [βλαστός, a germ; πρόληψις, an anticipating]. Hastening of development.

**blastosphere** (*blas'-to-sfēr*) [βλαστός, a germ; σφαῖρα; a sphere]. In biology, the "blastula," "germinal vesicle," or "vesicular germ." A hollow sphere composed of a single, simply layer of germinal cells. A vesicular morula.

**blastostroma** (*blas-to-stro'-mah*). See *embryonic area*.

**blastous** (*blast'-us*). Relating to a blastema.

**blastula** (*blas'-tū-lah*) [dim. of βλαστός, a germ]. The blastodermic vesicle; see *blastoderm*.

**blastulation** (*blas-tū-la'-shun*) [dim. of βλαστός, a germ]. In embryology, the conversion of morula or mulberry-germ into a blastular or vesicular germ.

**blastzellen** (*blast'-tsel-en*) [Ger.]. Primitive cells from which are developed all other kinds of cells. They are seen in the embryo before any beginning of differentiation, and are characterized by their large size, richness in cytoplasm, and large nuclei.

**Blatin's suture**, a modification of Gely's suture in which one needle and two threads of different colors are used.

**Blatta** (*blat'-ah*) [L., "blood-colored"]. 1. A genus of *Blattidæ*. 2. A clot of blood. **B. (Periplaneta) orientalis**, the cockroach; the powdered body is a popular remedy for dropsy among the Russian peasants. *Tinctura blattarum orientalium* is used in whooping-cough. Dose, 1 or 2 drops in water at intervals of 2 hours.

**Blaud's pill** (*blawd*) [P. *Blaud*, French physician, 1774-1858]. A pill containing equal parts iron sulphate and potassium carbonate; for use in anemia, etc.

**bleach** (*blēch*) [ME., *blechen*, to make white]. To make white or pale.

**bleacher's eczema** (*blēch'-erz ek'-zem-ah*). Eczema of the hands of bleachers, due to the use of hot water and strong lye.

**bleaching fluid**. A fluid obtained by passing chlorine gas into an emulsion of calcium hydrate. Syn., *Eau de Javelle; Javelle water*. **b. powder**. chlorinated lime, a mixture of calcium chloride and calcium hypochlorite, containing free chlorine gas. It is used as a disinfectant.

**blear eye** (*blēr'-i*). See *blepharitis ulcerosa*.

**bleb** (*bleb*). See *bulla*.

**bleeders** (*ble'-derz*) [AS., *blēdan*, to bleed]. 1. A popular term for those who are subjects of the hemorrhagic diathesis. Syn., *hemophiliacs*. 2. One who practises venesection. **b.'s disease**. See *hemophilia*.

**bleeding** (*ble'-ding*). See *bloodletting* and *hemorrhage*.

**blemmatrope** (*blem'-at-rōp*) [βλέμμα, a glance; τρέπειν, to turn]. An apparatus for showing the various positions of the eye in its orbit.

**blenal** (*blen'-al*). Sanatol carbonate, a yellow, oily liquid, used in gonorrhea.

**blenna** (*blen'-er*) [βλέννα, mucus]. Mucus. **b. narium**, mucus from the nose.

**blennadenitis** (*blen-ad-en-i'-tis*) [*blenna-*; ἀδήν, gland; ιτις, inflammation]. Inflammation of the mucous glands and follicles.

**blennelytria** (*blen-el-it'-re-ah*) [*blenna-*; ἔλυτρον, sheath, vagina]. Vaginal catarrh, leucorrhea.

**blennemesis** (*blen-em'-es-is*) [*blenna-*; ἐμεῖν, to vomit]. The vomiting of mucus.

**blennenteria** (*blen-en-te'-re-ah*) [*blenna-*; ἔντερον, intestine]. A mucous diarrhea or dysentery.

**blennenteritis** (*blen-en-ter-i'-tis*) [*blenna-*; ἔντερον, intestine; ιτις, inflammation]. 1. Enteritis with a copious discharge of mucus. 2. Inflammation of the mucous membrane of the bowel.

**blennisthmia** (*blen-isth'-me-ah*) [βλέννα, mucus ισθμια, a throat]. Pharyngeal catarrh.

**blenno-** (*blen-o-*) [βλέννα, mucus]. A prefix meaning mucus.

**blennocele** (*blen'-o-sēl*) [*blenno-*; κήλη, a tumor]. Gonorrheal epididymitis.

**blennochesia, blennochezia** (*blen-o-ke'-ze-ah*) [*blenno-*; χέζειν, to want to go to stool]. See *blennenteria*.

**blennocystitis** (*blen-o-sis-ti'-tis*) [*blenno-*; κύστις, bladder; ιτις, inflammation]. Catarrh of the urinary bladder.

**blennogenic**, or **blennogenous** (*blen-o-jen'-ik*, or *blen-oj'-en-us*) [*blenno-*; γένεσις, production]. Producing or secreting mucus; muciparous.

**blennoid** (*blen'-oid*) [*blenno-*; εἶδος, form]. Resembling mucus; myxoid; muciform; mucoid.

**blennoma** (*blen-o'-mah*) [*blenno-*; ὄμα, a tumor; pl., *blennomata*]. 1. A mucous polypus. 2. A myxoma.

**blennometritis** (*blen-o-me-tri'-tis*) [*blenno-*; μήτρα, the uterus; ιτις, inflammation]. Catarrhal metritis.

**blennometrorrhea, blennometrorrhœa** (*blen-o-metror-e'-ah*). See *metroblennorrhea*.

**blennophlogisma, blennophlogosis** (*blen-o-flo-jis'-mah, blen-o-flo-go'-sis*) [*blenno-*; φλόγωσις, inflammation]. Inflammation of a mucosa.

**blennophthalmia** (*blen-off-thal'-me-ah*) [*blenno-*; ὀφθαλμία, disease of the eyes]. Catarrhal conjunctivitis.

**blennoptysis** (*blen-op'-tis-is*) [*blenno-*; πτύσις, a spitting]. Bronchial mucous expectoration.

**blennorrhagia** (*blen-or-a'-je-ah*) [*blenno-*; ῥηγνύναι, to burst forth]. 1. An excessive mucous discharge. 2. Gonorrhea.

**blennorrhagic** (*blen-or-aj'-ik*) [see *blennorrhagia*]. Relating to blennorrhagia.

**blennorrhea** (*blen-or-e'-ah*) [*blenno-*; ῥοή, a flow]. Same as *blennorrhagia*.

**blennorrheal, blennorrhoic** (*blen-or-e'-al, -o'-ik*) [see *blennorrhea*]. Relating to blennorrhea.

**blennorrhinia** (*blen-or-in'-e-ah*) [*blenno-*; ῥίς, the nose]. Coryza; nasal catarrh.

**blennosis** (*blen-o'-sis*) [*blenno-*; νόσοs, disease; *pl.*, **blennoses**]. Any disease of a mucous membrane.

**blennostasin** (*blen-os'-tas-in*). The proprietary name for a yellow solid body said to be cinchonidine hydrobromide, $C_{19}H_{22}N_2O(HBr)_2$, a nontoxic vasomotor constrictor and blennostatic. It is used in influenza, colds, night-sweats, etc. Dose 15-60 gr. (1-4 Gm.) hourly.

**blennostasis** (*blen-os'-tas-is*) [*blenno-*; στάσις, a staying]. The checking or suppression of any mucous discharge.

**blennostatic** (*blen-os-tat'-ik*) [*blenno-*; στάσις, a staying]. 1. Checking or suppressing mucous discharges. 2. An agent capable of suppressing mucous discharges.

**blennostrumous** (*blen-o-stru'-mus*). Relating to gonorrhea and to scrofula.

**blennothorax** (*blen-o-tho'-raks*) [*blenno-*; θώραξ, the thorax]. Pulmonary catarrh.

**blennotorrhea** (*blen-ot-or-e'-ah*) [*blenno-*; οὖs, ear; ῥοία, a flow]. A mucous discharge from the ear.

**blennurethria** (*blen-ū-re'-thre-ah*) [*blenno-*; οὐρήθρα, the urethra]. Urethral gonorrhea.

**blennuria** (*blen-ū'-re-ah*) [*blenno-*; οὖρον, urine]. The discharge of mucus in the urine.

**blennymenerysipelas** (*blen-e-men-er-e-sip'-e-las*) [*blenno-*; ὑμήν, membrane; *erysipelas*], Erysipelas attacking a mucosa.

**blennymenitis** (*blen-im-en-i'-tis*) [*blenno-*; ὑμήν, membrane]. Inflammation of any mucous surface.

**blephara** (*blef'-ar-ah*). Plural of *blepharon*, *q. v.*

**blepharadenitis** (*blef-ar-ad-en-i'-tis*) [*blepharon*; ἀδήν, a gland; ἴτις, inflammation]. Inflammation of the Meibomian glands. **b. tarsalis.** See *hordeolum*.

**blepharal** (*blef'-ar-al*) [βλέφαρον, the eyelid]. Relating to an eyelid or to the eyelids.

**blepharanthracosis** (*blef-ar-an-thrak-o'-sis*) [βλέφαρον, the eyelid; ἀνθράκωσις, a charring; carbuncilization]. Carbuncular inflammation of the eyelid.

**blepharedema** (*blef-ar-e-de'-mah*) [βλέφαρον, the eyelid; οἴδημα, a swelling]. Swelling or edema of the eyelids.

**blepharelosis** (*blef-ar-el-o'-sis*) [βλέφαρον, the eyelid; ἔλᾱν, to roll]. Synonym of *entropion*.

**blepharemphysema** (*blef-ar-em-fis-e'-mah*) [βλέφαρον, eyelid; ἐμφύσημα, an inflation]. Emphysema of an eyelid.

**blepharides** (*blef-ar'-id-ēz*). Plural of *blepharis*.

**blepharis** (*blef'-ar-is*) [βλέφαρίς, an eyelash]. 1. An eyelash. 2. A genus of plants of the order *Acanthaceæ*. *B. capensis* is a plant of South Africa used in blood-poisoning from anthrax and in treatment of snake-bites. Dose 3-4 oz. (90-118 Cc.) of a 1 : 100 decoction.

**blepharism** (*blef'-ar-izm*) [βλεφαρίζειν, to wink]. Rapid involuntary winking; morbidly excessive nictitation.

**blepharitic** (*blef-ar-it'-ik*) [βλέφαρον, the eyelid; ἴτις, inflammation]. Relating to or affected with blepharitis.

**blepharitis** (*blef-ar-i'-tis*) [*blepharon*, ἴτις, inflammation]. Inflammation of the eyelids. **b. acarica**, marginal blepharitis in which the *Demodex folliculorum* is present upon or about the eyelashes. **b. ciliaris, b. marginalis,** inflammation of the ciliary or marginal border of the lids. **b. gangrænosa,** carbuncle of the eyelids. **b. glandularis, b. glandulosa,** inflammation of the Meibomian glands. **b. internus,** palpebral conjunctivitis. **b. phlegmonosa,** inflammation of the cellular tissue of the eyelid. **b. scrofulosa.** See *b. simplex.* **b. simplex,** mild inflammation of the borders of the eyelids with formation of moist yellow crusts on the ciliary margins, gluing together the eyelids. **b. squamosa,** that attended with the formation of scabs. **b. ulcerosa,** an ulcerative inflammation of the eyelids.

**blepharo-** (*blef-ar-o-*) [*blepharon*]. A prefix meaning relating to the eyelid.

**blepharodenitis** (*blef-ar-o-ad-en-i'-tis*). See *blepharadenitis.*

**blepharoadenoma** (*blef-ar-o-ad-en-o'-mah*) [*blepharo-*; ἀδήν, a gland; ὄμα, a tumor]. An adenoma of the eyelid.

**blepharoatheroma** (*blef-ar-o-ath-er-o'-mah*) [*blepharo-*; *atheroma*]. A sebaceous cyst of the eyelid.

**blepharoblennorrhea** (*blef-ar-o-blen-or-e'-ah*). See *ophthalmia, purulent.* **blepharoblennorrhœa gonor-**

**rhoica, b. maligna,** gonorrheal ophthalmia. **blepharoblennorrhœa neonatorum,** ophthalmia neonatorum. **blepharoblennorrhœa urethritica,** gonorrheal ophthalmia.

**blepharocarcinoma** (*blef-ar-o-kar-sin-o'-mah*) [*blepharo-*; *carcinoma*]. Carcinoma affecting the eyelid.

**blepharochalasis** (*blef-ar-o-kal'-as-is*) [*blepharo-*; χάλασις, a slackening]. A method of treating trachoma by excising oval slices from the upper and lower palpebral conjunctiva, with incision of the outer canthus.

**blepharochromidrosis** (*blef-ar-o-kro-mid-ro'-sis*) [*blepharo-*; χρῶμα, color; ἱδρώς, sweat]. Colored sweat of the eyelids, usually of a bluish tint.

**blepharocleisis** (*blef-ar-o-klī'-sis*) [*blepharo-*; κλείσις, closure]. Ankyloblepharon; abnormal closure of the eyelids.

**blepharoclonus** (*blef-ar-ok'-lon-us*) [*blepharo-*; κλόνος, commotion]. Spasm of the orbicularis palpebrarum muscle.

**blepharocoloboma** (*blef-ar-o-kol-o-bo'-mah*). See *coloboma palpebræ.*

**blepharoconjunctivitis** (*blef-ar-o-kon-junk-tiv-i'-tis*). See *conjunctivitis, palpebral.*

**blepharodiastasis** (*blef-ar-o-di-as'-tas-is*) [*blepharo-*; διάστασις, separation]. Excessive separation of the eyelids; or inability to close the eyelids completely.

**blepharodyschrea** (*blef-ar-o-dis-kre'-ah*), or **blepharodyschroia** (*blef-ar-o-dis-kroi'-ah*) [*blepharo-*; δυς, bad; χροία, color]. Discoloration of the eyelid from nevus or from any other cause.

**blepharoedema** (*blef-ar-o-e-de'-mah*). See *blepharedema.*

**blepharoemphysema.** See *blepharemphysema.*

**blepharohematidrosis** (*blef-ar-o-hem-at-id-ro'-sis*) [*blepharo-*; αἷμα, blood; ἱδρώς, sweat]. The rare occurrence of sweating blood from the skin of the eyelid.

**blepharolithiasis** (*blef-ar-o-lith-i'-as-is*) [*blepharo-*; λίθος, a stone]. The formation of marginal concretions within the eyelid.

**blepharomelasma** (*blef-ar-o-mel-az'-mah*) *blepharo-*; μέλας, black]. Seborrhœa nigricans occurring on the eyelid.

**blepharomelena** (*blef-ar-o-mel-e'-nah*). See *blepharochromidrosis.*

**blepharon** (*blef'-ar-on*) [βλέφαρον, the eyelid; pl., *blephara*]. The eyelid; palpebra.

**blepharoncosis** (*blef-ar-ong-ko'-sis*) [*blepharoncus*]. The formation of a blepharoncus, or the condition of suffering due to such a growth.

**blepharoncus** (*blef-ar-ong'-kus*) [*blepharo-*; ὄγκος, an enlargement]. A tumor or swelling of the eyelid.

**blepharonysis** (*blef-ar-on-i'-sis*) [*blepharo-*; νύσσειν, to prick]. Operation for entropion by means of Gaillard's suture.

**blepharopachynsis** (*blef-ar-o-pak-in'-sis*) [*blepharo-*; παχύs, thick]. Morbid thickening of the eyelid.

**blepharophimosis** (*blef-ar-o-fi-mo'-sis*) [*blepharo-*; φίμωσις, a shutting-up]. Abnormal smallness of the palpebral aperture.

**blepharophryastic** (*blef-ar-o-frip-las'-tik*) · [*blepharo-*; ὀφρύς, brow; πλαστικός, plastic]. Pertaining to the plastic surgery of the lid and eyebrow.

**blepharophryplasty** (*blef-ar-off'-re-plas-te*) [*blepharo-*; ὀφρύς, eyebrow; πλάσσειν, to form]. Plastic surgery of the eyebrow and eyelid.

**blepharophthalmia** (*blef-ar-of-thal'-me-ah*) [*blepharo-*; ὀφθαλμός, eye]. Combined palpebral and ocular conjunctivitis.

**blepharophthalmic** (*blef-ar-of-thal'-mik*): Relating to the eyelids and the globe of the eye, or to blepharophthalmia.

**blepharophthalmostat** (*blef-ar-of-thal'-mo-stat*). See *blepharostat.*

**blepharophyma** (*blef-ar-o-fi'-mah*) [*blepharo-*; φῦμα, a growth; *pl.*, *blepharophymata*]. A tumor of, or outgrowth from, the eyelid.

**blepharoplast** (*blef'-ar-o-plast*) [*blepharo-*; πλάσσειν, to form]. An individualized centrosome, found in certain protozoa, such as trypanosoma.

**blepharoplastic** (*blef-ar-o-plas'-tik*) [*blepharo-*; πλαστικός, plastic]. Pertaining to blepharoplasty.

**blepharoplasty** (*blef'-ar-o-plas-te*) [see *blepharoplast*]. An operation for the restoration of any part of the eyelid.

**blepharoplegia** (*blef-ar-o-ple'-je-ah*) [*blepharo-*; πληγή, a stroke]. Paralysis of an eyelid.

**blepharoptosis** (*blef-ar-op-to'-sis*) [*blepharo-*; πτῶσις, a fall]. Drooping of the upper eyelid.

**blepharopyorrhea** (*blef-ar-o-pi-or-e'-ah*) [*blepharo-;* πύον, pus; ῥέειν, to flow]. A flow of pus from the eyelid.
**blepharorrhaphy** (*blef-ar-or'-a-fe*) [*blepharo-;* ῥαφή, a seam]. The stitching together of a portion of the edges of the lids.
**blepharorrhea** (*blef-ar-or-e'-ah*) [*blepharo-;* ῥοία, a flow]. A discharge from the eyelid.
**blepharospasm** (*blef'-ar-o-spazm*) [*blepharo-;* σπασμός, a spasm]. Spasm of the orbicularis palpebrarum muscle; excessive winking.
**blepharospath** (*blef'-ar-o-spath*) [*blepharo-;* σπάθη, a blade]. A forceps for taking up or holding an artery; for use in operations on the eyelid.
**blepharosphincterectomy** (*blef-ar-o-sfink-ter-ek'-to-me*) [*blepharo-;* σφιγκτήρ, sphincter; ἐκτομή, incision]. An operation to lessen the pressure of the upper lid upon the cornea; it consists in making an incision the entire length of the lid, about 2 mm. above the lid-border; by a second incision a small oval flap of skin, 2-4 mm. broad, is removed along with all the underlying muscle-fibers. The wound is closed with two or three sutures.
**blepharostat** (*blef'-ar-o-stat*) [*blepharo-;* στατικός, causing to stand]. An instrument for holding the eyelids apart while performing operations upon the eyes or lids.
**blepharostenosis** (*blef-ar-o-ste-no'-sis*) [*blepharo-;* στενός, narrow]. Diminution of the space between the eyelids.
**blepharosymphysis** (*blef-ar-o-sim'-fiz-is*). See *blepharosynesohia.*
**blepharosyndesmitis** (*blef-ar-o-sin-des-mi'-tis*) [*blepharo-;* σύνδεσμος, a bond]. See *conjunctivitis, palpebral.*
**blepharosynechia** (*blef-ar-o-si-nek'-e-ah*) [*blepharo-;* συνέχεια, a holding together]. The adhesion or growing together of the eyelids.
**blepharotomy** (*blef-ar-ot'-o-me*) [*blepharo-;* τέμνειν, to cut]. Incision into the eyelid.
**blepharydatis** (*blef-ar-id'-at-is*) [*blepharo-;* ὑδατίς, a vesicle]. A hydatid affecting the eyelid.
**blessed thistle.** See *Carduus.*
**Blessig's groove** [Robert *Blessig*, Russian physician, 1830-1878]. The slight groove in the embryonic eye that marks off the fundus of the optic cup from the zone that surrounds the periphery of the lens and corresponds in position with the future ora serrata.
**blight** (*blit*). 1. A partial paralysis of certain facial nerves, arising from sudden or extreme cold. 2. A fungus-disease of plants. . **b. of the eye,** an extravasation of blood within the conjunctiva. **b., sandy,** a form of ophthalmia attended with photophobia and a sensation of grittiness, due to the formation of pus in the openings of the Meibomian glands.
**blighted** (*bli'-ted*). Withered, blasted; affected with blight.
**blind** (*blind*) [AS]. Without sight; deprived of sight. **b. gut,** the cecum. **b. spot,** that part of the fundus of the eye where the optic nerve enters.
**blindness** (*blind'-nes*). Want of vision. **b., Bright's.** See *Bright's blindness.* **b., color-,** imperfect color-perception. This condition is found in about 4 % of persons, is more frequent in men than in women, and is probably due to nonexercise of the color-sense. *Complete color-blindness* is very rare, the different colors probably appearing as different intensities or shades of white light. In *partial color-blindness* defective perception of red is the most frequent, green, blue, and yellow, respectively, being next in order. Tests *for color-blindness* usually consist in matching and classifying colored yarns. **b., cortical,** blindness due to lesion of the cortical center of vision. **b., day-.** See *nyctalopia.* **b., electric-light,** a condition similar to snow-blindness, due to exposure of the eyes to intense and prolonged electric illumination. **b., gold,** a form of retinal asthenopia at times affecting dentists, owing to which there is inability to distinguish the filing from the tooth. **b., intellectual, b., mental.** See *b., psychic.* **b., moon-,** a rare condition of retinal anesthesia said to be due to exposure of the eyes to the moon's rays in sleeping. **b., nervous.** See *amaurosis.* **b., night-.** See *hemeralopia.* **b., object-.** See *apraxia.* **b., psychic,** loss of conscious visual sensation from destruction of the cerebral visual center; there is sight but not recognition. **b., snow-,** photophobia and conjunctivitis due to exposure of the eyes to the glare of sunlight upon snow. **b., soul.** See *b. psychic.* **b., word-.** See *aphasia.*
**blinking** (*blink'-ing*). An involuntary winking.
**blister** (*blis'-ter*) [ME.]. A vesicle resulting from the exudation of serous fluid between the epidermis and true skin; also the agent by which the blister is produced. **b., blood-,** a blister which contains blood. **b., fly-,** a blister of cantharides. See *cantharides.* **b., flying,** a blister that remains long enough to produce redness of the skin and not vesication.
**blistering** (*blis'-ter-ing*). Forming a vesicle upon the skin. **b. collodion,** a solution of cantharidin in collodion. **b. liquid.** Same as *b. collodion.* **b. paper,** paper saturated with cantharides; used for producing vesication.
**bloat** (*blōt*) [ME., *blote*, swelling]. 1. Puffiness; edema; turgidity from any cause, as from anasarca. 2. A form of colic in the horse produced by tympanitic distention of the intestines. Also called *wind colic.*
**block** (*blok*) [ME., *blok*, a piece of wood]. 1. In dentistry a mass of gold-foil for filling teeth, made by folding a tape of foil upon itself several times by means of pliers. 2. To obstruct the path of all sensory impressions in the nerve-trunks and roots in the spinal cord which connect the area of surgical operation with the sensorium. **b. heart.** See *heart.*
**blocking** (*blok'-ing*). F. Franck's term for the transitory sensory paralysis of the entire peripheral distribution of a nerve by the infiltration of its sectional area of a nerve-trunk in any part of its course with cocaine or a similar analgesic.
**Blocq's disease** [Paul Oscar *Blocq*, French physician, 1860-1896]. See *Briquet's ataxia.*
**blondine** (*blon'-dēn*) [Fr., *blonde*, light, fair]. A preparation of hydrogen dioxide, used for bleaching the hair.
**blood** (*blud*) [AS., *blōd*]. The fluid that circulates through the heart, arteries, and veins, supplying nutritive material to all parts of the body. In the human being the blood of the arteries is bright red and dichroic; that of the veins, dark red and monochroic. Blood consists of plasma and corpuscular elements, the latter being the red corpuscles, the white corpuscles, and the blood-plaques. In a cubic millimeter there are about 5,000,000 red and 10,000 white corpuscles. The red color of the blood is due to the hemoglobin contained in the red corpuscles. The total amount of blood was formerly estimated at $\frac{1}{13}$ of the body-weight, but now $\frac{1}{20}$ of the body-weight is believed to be a more correct figure. When exposed to the air, blood coagulates, forming a red clot and a yellowish fluid called serum. Healthy blood consists of 78 % of water and 22 % of solids. See *Almén, Bremer, Hayem, Hoppe-Seyler, Kobert, Ladendorff, Pacini, Rubner, Salkowski, Struve, Van Deen, Wetzel, v. Zaleski.* **b.-casts,** tube casts to which red blood-corpuscles are attached. **b.-cell,** a blood corpuscle. **b.-coagulation.** See *Hammarsten, Lilienfeld, Pekelharing, Schmidt.* **b.-corpuscles, red,** circular, biconcave discs floating in the blood. Red corpuscles are circular in mammals (except the camel) and elliptical in birds and reptiles. In man they are about $\frac{1}{3200}$ inch (7 μ) in diameter and $\frac{1}{12800}$ inch thick. **b.-corpuscles, granular,** bodies described by Erb in blood of mammals and supposed to be transitional blood-corpuscles. **b.-corpuscles white,** *colorless corpuscles*, about one-third larger in diameter than the red—$\frac{1}{2300}$ inch (10 μ). They exhibit movements similar to those of the ameba. **b.-crasis,** the mixture of the constituents of the blood. When the blood-crasis is diseased or disordered, the condition is known as *dyscrasia.* **b.-crystals,** crystals of hematoidin. **b., defibrinated,** blood from which the fibrin has been removed by agitating it with twigs. **b.-disease.** A synonym of *dyscrasia.* **b.-dust.** A synonym of *hemokonia.* **b.-fluke.** See *Bilharzia hæmatobia.* **b.-islands,** a term applied to the groups of corpuscles developed during the first days of embryonic life, within the large branched cells of the mesoblast. **b.-pigments.** See *hemoglobin, hematin, and hematocyanin.* **b.-plaque.** See *b.-platelets.* **b.-plasma,** the *liquor sanguinis*, or fluid part of the blood. **b.-platelets,** circular or oval, light-gray bodies found in the blood. They are from 1 to 1.3 μ in size, and number from 18,000 to 300,000 in the cubic millimeter. Their function is not definitely known; they are an important factor in thrombosis.

**b.-poisoning,** a morbid state due to the circulation of bacteria or their products, or both, in the blood, as the result of a local infection. **b.-pressure,** the tension of the blood in the arteries. **b.-root.** See *sanguinaria.* **b.-shot,** redness due to turgescence of blood-vessels; ecchymosed,—*e. g.,* the eye. **b.-stroke,** apoplexy. **b.-tumor.** See *hematoma.* **b.-vessel,** an artery or a vein.

**bloodless** (*blud'-les*) [*blood*]. Without blood. **b. operations,** surgical operations, such as amputations, in which the member is so bandaged by compresses and elastic bands that the blood is expelled and kept from the part to be operated upon.

**bloodletting** (*blud-let'-ing*). The artificial abstraction of blood from the body. **b., general,** venesection or phlebotomy. **b., local, b., topical,** the removal of a small quantity of blood by cupping leeching, or scarification. **b., revulsive,** that performed for arresting internal hemorrhage. **b., spoliative,** bleeding to reduce the number of bloodcorpuscles.

**bloody** (*blud'-e*). Having the nature of, or filled with blood. **b. flux,** dysentery. **b. sweat,** ephidrosis.

**blotch.** A pimple or blain; a small discolored patch of skin; a group of small pustules.

**blow-pipe** (*blo'-pip*). A cylindrical tube, from twelve to eighteen inches long, about half an inch in diameter at one end, and gradually tapering to a fine point or nozzle, which may be straight or bent at a right angle; it is used in directing the flame of a lamp in a fine conical tongue, for the purpose of producing a high temperature by hastening the combustion.

**blucaloids** (*blu'-kal-oids*). Capsules containing methylene-blue and oil of eucalyptus; said to be useful in malaria.

**blue** [ME., *blew*]. One of the colors of the spectrum. **b. baby,** a child with congenital cyanosis. **b.-blindness,** defective color-perception for blue. **b. cardinal-flower.** Synonym of *Lobelia syphilitica.* **b. cohosh.** See *Caulophyllum.* **b. disease,** cyanosis of the newborn, usually due to congenital disease of the heart. **b. edema,** a puffed and bluish appearance of the limb sometimes seen in hysterical paralysis. **b.-flag.** See *Iris.* **b. gentian,** the root of *Gentiana catesbœi,* tonic and stomachic. Dose of *fluidextract* 10-40 min. (0.65-2.65 Cc.). **b.-gum tree.** See *Eucalyptus.* **b.-mass.** See *mercury mass.* **b., methylene-.** See *methylene-blue.* **b. ointment,** *unguentum hydrargyri dilutum* (U. S. P.). **b. pill,** a pill made from blue-mass. **b., Prussian,** ferric ferrocyanide, $(Fe_3)_2(FeC_6N_6)_3$. **b. stick.** Same as *b.-stone.* **b.-stone,** copper sulphate. **b., Turnbull's,** ferrous ferrocyanide, $Fe_3Fe_2(C_6N_6)_4$. **b., vision,** cyanopsia. **b. vitriol,** copper sulphate.

**bluebottle** (*blu'-bot-l*). See *Carduus.*

**blues** (*bloos*). A popular name for short periods of mental depression; they are usually associated with indigestion.

**Blumberg's sign** (*bloom'-berg*). In active peritonitis, pressure of the hand on the abdomen causes less pain than sudden removal of the pressure; during convalescence, pressure causes greater pain than sudden removal of the pressure.

**Blumenau's nucleus** (*bloo'-men-ow*) [Leonid *Blumenau,* Russian neurologist, 1862– ]. The lateral nucleus of the cuneate nucleus.

**Blumenbach's clivus** (*bloo'-men-bakh*) [Johann Friedrich *Blumenbach,* German physiologist, 1752–1840]. The inclined surface of the sphenoid bone which lies behind the posterior clinoid processes and is continuous with the basilar process of the occipital bone. **B.'s process,** the uncinate process of the ethmoid bone.

**blunt-hook** (*blunt'-hook*). An obstetrical instrument, used mainly in embryotomy.

**blushing** (*blush'-ing*) [ME., *blushen,* to glow]. The sudden and temporary reddening of the face due to vasomotor action caused by shame, modesty, or confusion. See *rubor.*

**Blyth's test for lead in potable water** [Alexander Wynter *Blyth,* English chemist]. On the addition of a 1 per cent. alcoholic tincture of cochineal a precipitate is formed.

**BNA.** Abbreviation for *Basle nomina anatomica,* Basle anatomical nomenclature.

**Boas' reagent** (*bo'-as*) [Ismar *Boaz,* German physician, 1858– ]. A solution of tropeolin or paper saturated with such a solution. **B.'s sign,** the presence of lactic acid in the gastric contents in cases of cancer of the stomach. **B.'s test,** in cases of intestinal atony a splashing sound can be obtained on pressure upon the abdominal wall after the injection of a small quantity (200-300 Cc.) of water into the bowel. **B.'s test for hydrochloric acid in the contents of the stomach,** in 100 Gm. dilute alcohol dissolve 5 Gm. pure resorcinol and 5 Gm. white sugar. Spread a few drops of this reagent in a thin layer upon a porcelain dish and heat gently. If a glass rod dipped in the solution is applied to a drop of the filtrate from the stomach, a deep scarlet streak is produced. **B.'s treatment of hemorrhoids,** after extrusion of the hemorrhoids they are treated by Bier's suction until an edematous ring appears; later, they drop off.

**Boas-Oppler bacillus** (*bo'-as op'-ler*) [Ismar *Boas,* German physician, 1858– ; Bruno *Oppler,* German physician]. A strepto-bacillus sometimes found in the gastric contents; when present in large numbers it is said to indicate carcinoma of the stomach.

**boat-belly** (*bot'-bel-e*). See *scaphoid abdomen.*

**bocaral** (*bo'-kar-al*). An antiseptic preparation containing boric acid, carbolic acid, and alum.

**Bocconia** (*bok-o'-ne-ah*) [Paolo, *Boccone,* Italian botanist, 1633-1704]. A genus of plants of the order *Papaveraceæ.* *B. frutescens* is a native of the West Indies; the juice is purgative and anthelmintic.

**Bochdalek's canal** (*bok-dal-ek'*) [*Bochdalek,* Austrian anatomist, –1883]. A minute canal passing obliquely downward and outward through the membrana tympani from the upper edge of the inner surface of the latter to the umbo, where it opens on the outer surface. **B.'s ganglion,** the supramaxillary ganglion; a small gangliform enlargement at the junction of the anterior and middle dental branches of the superior maxillary nerve. **B.'s gap,** the interval existing in the diaphragm between the costal and lumbar attachments of this muscle. **B.'s glands,** cysts developed in the tongue derived from the primitive thyroglossal duct. **B.'s muscle,** the triticeoglossus muscle, a small muscle extending from the cartilago triticea in the thyrohyoid ligament upward and forward to the tongue, which it enters, together with the posterior part of the hyoglossus. **B.'s tubes,** round or tubular cavities opening into the posterior portion of the thyroglossal duct; like the latter, they are inconstant after birth. **B.'s valve,** a small fold of the lining membrane of the lacrimal duct, near the punctum lacrimale.

**Bock's pharyngeal nerve** [August Carl *Bock,* German anatomist, 1782-1833]. The posterior efferent (pterygopalatine) branch of the sphenopalatine ganglion.

**Bockhart's impetigo.** Epidermic abscesses caused by pyogenic micrococci.

**bodik.** A Malay liquor made from rice.

**Bodo** (*bo'-do*) [L.]. A genus of flagellate protozoa. *B. saltans* has been reported as living in great numbers in unhealthy ulcerations. *B. urinarius* has been found in the urine of cholera-patients.

**body** (*bod'-e*) [AS., *bodig*]. 1. The animal frame with its organs. 2. A cadaver or corpse. 3. The important and largest part of an organ, as the body of the uterus. See also *corpus* and *corpora.* 4. A mass of matter. **b., alloxur.** See *alloxur.* **bodies, amylaceous, b., amyloid,** a term applied by Virchow (1856) to bodies found in the central nervous system of adults and young persons dying of various diseases (not alone of diseases of the nervous system). They are concentrically striated, stain deep brown with Lugol's solution, blue with iodine and sulphuric acid, and give the characteristic amyloid color with the aniline stains. **b., antiimmune,** a substance held by Ehrlich in his lateral-chain theory to enter into the composition of an antihemolysin (*q. v.*). Cf. *anticomplement.* **bodies, bigeminal.** See *corpora quadrigemina.* **b.-cavity,** the space contained within the thoracic and abdominal walls. **b., central.** 1. The nucleus. 2. In the plural, peculiar corpuscles which are permanently present near the nucleus in protoplasm during cell-division. **bodies, chromatin,** bodies of various forms found in the reticulum of a cell undergoing mitosis. **bodies, direction, b., directive,** the minute abortive cells extruded by the egg-cell as the final phenomenon in the process of maturation; polar bodies. **b., immune,** a name given by Pfeiffer to one of the two substances of a hemolytic serum. It is thermostabile and has two affinities, a stronger one for the red blood-cell and a weaker one for the complement. Having two uniting processes, it is an ambo-

ceptor. **b., inclusion,** a nucleoid. **b., intermediary, b., intermediate.** See **b., immune** and **amboceptor. bodies, katalytic,** the ferments. **b.-louse.** See under *pediculus.* **bodies, melon-seed,** bits of fibrin, cartilage, or of tuberculous or syphilitic granulation tissue, found in enlarged bursæ and ganglia. **bodies, metachromatic,** small granules in bacterial cells staining differently from the surrounding cytoplasm. **bodies, parenchymal,** the lobules of the lacrimal gland. **b., perineal,** the tissues between the vulva and anus. **b., pituitary.** See *pituitary.* **b.-sarcode,** the protoplasm of the cell-body. **bodies, semilunar.** See *cells, demilune.*
**Boeck's disease** (*bek*) [Carl Wilhelm *Boeck,* Norwegian physician, 1808–1875]. Sarcoid tumor of the skin. **B.'s lotion,** for dry, itching, inflammatory diseases. Talc and starch, each, 50; glycerol, 20; lead-water, 100. The bottle to be well shaken; the lotion diluted with twice the volume of water and applied with a brush. **B.'s scabies,** scabies crustosa; Norwegian itch.
**Boedeker's reaction for albumin** (*bo'-dek-er*). Treat the liquid with acetic acid and add a solution of potassium ferrocyanide drop by drop. White precipitate of albumin will be formed.
**Boerhaave's glands** (*bor'-hāv*) [Hermann *Boerhaave,* Dutch physician, 1668–1738]. The sudoriparous glands. The convoluted glands in the skin that secrete the sweat.
**Boernstein's test for saccharin.** Extract the substance to be tested with ether; remove the ether and heat with resorcinol and sulphuric acid and next add an excess of soda solution. In the presence of saccharin a strong fluorescence is produced. According to Hooker, other substances, *e. g.,* succinic acid, also produce this reaction.
**Boettcher's cells** (*bet'-kher*) [Arthur *Boettcher,* German anatomist, 1831–1889]. Dark-colored cells, with a basally situated nucleus, found between Claudius' cells. **B.'s crystals.** See *crystals, spermin.*
**Boettcher-Cotugno's space** (*ko-tūn'-yo*). The endolymphatic sac of the internal ear.
**Boettger's test for dextrose**—(*bet'-ger*) [Wilhelm Carl *Boettger,* German chemist, 1871–    ]. Take 5 Gm. of basic bismuth nitrate, 5 Gm. of tartaric acid, and 30 Cc. of distilled water. Add to this slowly a strong solution of sodium hydroxide, stirring continually until a clear solution is obtained. To a small quantity of this add some solution of dextrose and boil. A black precipitate of metallic bismuth is formed. Or the test may be performed in this way: Add some solid bismuth subnitrate to the liquid already rendered alkaline with sodium carbonate or potassium and boil. The existence of dextrose will be evinced by the darkening of the bismuth salt or a black precipitate.
**Bogg's sign.** In hypertrophy of the thymus: an upward shifting of the lower level of percussion dulness when the "seated" subject's head is extended backwards towards the spine.
**Bogros' space** (*bog'-ro*) [Jean Annet *Bogros,* French anatomist, 1786–1825]. A space between the peritoneum above and the fascia transversalis below, in which the lower portion of the external iliac artery can be reached without wounding the peritoneum.
**Bogrow's fibers** (*bog'-ro*). A tract of nerve-fibers passing from the optic tract to the optic thalamus.
**boil** (*boil*) [Mod. E., *boile*]. A furuncle; a localized inflammation of the skin and subcutaneous connective tissue attended by the formation of pus. See *furunculus.* **b., Aleppo, b., Delhi,** a peculiar ulcerative affection endemic in India, due to a specific microorganism. See *furunculus orientalis.* **b., blind,** one of brief continuance and not attended by the formation of a core.
**boiled oil.** Linseed oil that has been heated to a high temperature (130° C. and upward), while a current of air is passed through or over the oil, and the temperature increased until the oil begins to effervesce from evolution of products of decomposition.
**boilermakers' deafness.** See *deafness, boilermakers'.*
**boiling** [ME., *boilen,* to stir]. The vaporization of a liquid when it gives off vapor having the same tension as the surrounding air. **b. point,** the temperature at which a liquid begins to boil.
**bolbomelanosis** (*bol-bo-mel-an-o'-sis*) [βολβός, a bulb; μελάνωμα, blackness]. The process of formation of a melanoma.

**boldine** (*bōl'-din*). See under *boldus.*
**boldo** (*bol'-do*). See *boldus.*
**boldoglucin.** An aromatic glucoside obtained from *Peumus boldus* and other species. It is a hypnotic in doses of 20 gr.–1 dr. (1.3–4.0 Gm.).
**boldus** (*bol'-dus*) [L.]. Boldo. The leaves and stems of an evergreen, *Peumus boldus,* native to Chili, sometimes used in anemia and general debility as a substitute for quinine. It contains a bitter alkaloid, *boldine,* a hypnotic in doses of 3 gr. (0.2 Gm.). Dose of the *fluidextract,* 1–4 min. (0.065–0.26 Cc.); of the *tincture,* which contains 20 % of the drug, 5–8 min. (0.32–0.52 Cc.).
**bole** (*bōl*) [βῶλος, a clod of earth]. A translucent, soft variety of clay formerly much used in medicine—internally as an astringent, externally as an absorbent. Several varieties are used, as the *Armenian bole,* the *Lemnian,* and the *French bole.*
**boletiform** (*bo-let'-e-form*). Shaped like a mushroom.
**Boletus** (*bo-le'-tus*). A genus of fungi, some species of which are edible while others are highly poisonous.
**Bolognini's sign** (*bo-lo-ne'-ne*). On pressing with the tips of the fingers of both hands alternately upon the right and left of the abdomen of a patient who is lying on his back and whose abdominal muscles are relaxed by flexion of the thighs, a sensation of friction within the abdomen is perceived; it is noticed in the early stages of measles.
**bolometer** (*bo-lom'-et-er*) [βολή, a ray, a throw; μέτρον, a measure]. A device for measuring minute differences in radiant heat. Syn., *thermis balanoe.*
**bolus** (*bo'-lus*) [L.]. 1. A large pill. 2. The rounded mass of food prepared by the mouth for swallowing. 3. See *bole.*
**bombus** (*bom'-bus*) [βόμβος, a humming sound]. A ringing or buzzing sound in the ears; tinnitus. Also a sonorous movement or rumbling flatus of the intestines: borborygmus.
**Bond's splint.** *For fracture of the lower end of the radius:*—A thin, wooden splint, with sole-leather edges an inch high, and a curved block to rest in the palm of the hand. A pasteboard splint is used on the back of the forearm, both reaching from below the elbow to almost the distal ends of the metacarpal bones.
**bonducin** (*bon'-dū-sin*), $C_{14}H_{18}O_5$. A bitter principle from bonducella; a white powder, soluble in alcohol, chloroform, fats, and oils, used as a febrifuge. Dose 1½–3 gr. (0.1–0.3 Gm.).
**bonduk, bonducella** (*bon'-dūk, bon-dū-sel'-ah*) [L.]. Bonduc-seeds. The seeds of *Cæsalpinia bonducella,* a tropical plant. It is a bitter tonic and antiperiodic in intermittent fevers. Dose 10–15 gr. (0.65–0.97 Gm.).
**bone** (*bōn*) [AS., *bān*]. A hard tissue that constitutes the framework or skeleton of the body. Bone usually consists of a compact outer mass covered with periosteum, surrounding a reticulated inner structure that incloses a central cavity filled with marrow. A transverse section of a long bone shows bone-tissue to be composed of a number of nearly circular zones, each having a central tube, the Haversian canal, through which the blood circulates. Surrounding the Haversian canal are concentrically arranged laminæ, or layers of bone, between which are found irregular spaces called lacunæ, containing bone-corpuscles and communicating with the Haversian canal and each other by means of canaliculi, through which the nutrition is conveyed to all parts of the bone. **b.-ache,** osteocopic pain. **b., ankle-,** the talus or astragalus. **b., articular,** that element of the mandible or lower jaw which is formed from the condylar portion of the cartilaginous rudiment in Meckel's cartilage. Syn., *os articulare; os condyloideum maxillæ inferioris.* **b.-ash,** the calcic phosphate remaining after bones have been incinerated. **b., basilar.** 1. The sphenoid and occipital bones regarded as one. 2. The sacrum. 3. The last lumbar vertebra. 4. The basioccipital bone. 5. The basisphenoid bone in birds. **b., cancellated, b., cancellous,** bone consisting chiefly of spongy tissue. **b.-cartilage.** See *ossein.* **b.-cell,** an osteoblast. **b., cheek-,** the malar bone. **b.-chips.** See *Senn's bone-plates.* **b., collar-,** the clavicle. **b.-conduction,** the transmission of sound-waves to the auditory nerve by means of sonorous vibrations communicated to the bones of the skull. **b.-cyst,** a tumor distending and thinning bone, and filled with serum or bloody fluid; in rare cases bone-cysts

contain hydatids. b.s, elongated, long bones, like the ribs, devoid of a medullary cavity. b., endochondral, such true bone as originates from osteoblastic centers in fetal cartilage, and not from periosteum. b., epipteric, a small Wormian bone sometimes found between the great wing of the sphenoid and the anterior inferior angle of the parietal bone. b., exercise, an ossification in the left arm of soldiers due to constant pressure of a musket. b., flat, a bone more or less in the form of a plate. b.-gelatin. See *gelatin*, *bone-*. b., intermaxillary. See *mesognathion*. b., long, one consisting of a narrow shaft and two expanded ends. b.s, pneumatic, those containing many air-cells or air-sacs. b., puboischiadic, b., puboischiatic, the os pubis and the ischium taken as one. b., rider's, an ossification of the lower tendon of the adductor longus or magnus from pressure. b., sesamoid. See *sesamoid bone*. b.-setter, a specialist at setting bones; usually an uneducated empiric, and often a pretender to hereditary skill in the business. b., short, one the three dimensions of which are approximately equal. b., stirrup. See *stapes*. b.s, sutural, the Wormian bones of the skull. b.s, Wormian, small bones in the sutures of the skull. See *bones, table of*, page 153.

**bonelet** (bŏn′-lĕt). See *ossicle*.

**boneset.** See *Eupatorium*.

**Bonfils' disease.** See *Hodgkin's disease*.

**Bonnaire method** (bon-air′) [—*Bonnaire*, French obstetrician]. A method of bimanual dilatation of the cervix for rapid artificial delivery.

**Bonnet's capsule** (bon-a′) [Amadée *Bonnet*, French surgeon, 1809–1858]. The posterior portion of the sheath of the eyeball; Tenon's capsule. B.'s position, the position assumed by the thigh in coxitis: flexion, abduction, and outward rotation.

**boohoo** (boo-hoo′) [native S. Pacific]. A kind of gastritis with slight fever and with great nostalgia and depression of spirits. It attacks strangers in some of the Pacific Islands.

**Boophilus** (bo-off′-il-us) [βοῦς, ox; φιλεῖν, to love]. A genus of cattle-ticks. B. annulatus, the cattle-tick which carries the piroplasma responsible for Texas fever (a disease of cattle).

**boot,** Junod's. See *Junod's boot*.

**boracic acid** (bo-ras′-ik). See under *boron*.

**boracil** (bo′-ras-il). A preparation composed of boric acid, benzoic acid, acetanilide, and resorcinol. It is used as an antiseptic dusting powder.

**boracite** (bo′-ras-īt). Native magnesium borate.

**borage** (bo′-raj). The plant *Borago officinalis*, a demulcent, mild refrigerant and diaphoretic. Dose of fluidextract, ℥ j.

**boral** (bo′-ral). See *aluminum borotartrate*.

**boralide** (bo′-ral-id). A proprietary wound-antiseptic said to contain equal parts of boric acid and acetanilide.

**borate** (bo′-rāt) [Ar., *būraq*, borax]. Any salt of boric acid.

**borated** (bo′-ra-ted). Containing or combined with borax or boric acid.

**borax** (bo′-raks). See *boron*.

**borborygmus** (bor-bor-ig′-mus) [βορβορυγμός, a rumbling: *pl.*, borborygmi]. A rumbling of the bowels.

**Bordeaux emulsion** (bor-do′). A preparation containing lime 0.3, copper sulphate 1.6, liquid petrolatum 6.0, and water to 100.

**border** (bord′-er) [ME.]. In anatomy, the boundary of an area or surface. b., basal, b., cuticular, b., striated. See *layer, cuticular*. b., bright, the margin of a ciliated cell. b., vermilion, the line of union of the mucosa of the lip with the skin.

**Bordet's phenomenon** (bor′-dā) [J. *Bordet*, Belgian bacteriologist]. See *Pfeiffer's reaction*, from which it differs only in the use of a minute quantity of normal serum instead of fresh peritoneal fluid. B.'s specific test, for human blood: it is based upon the fact that the blood-serum of an animal subcutaneously injected with the blood of another animal of a different species rapidly develops the property of agglutinating and dissolving the erythrocytes similar to those injected, but has no effect upon blood derived from any other source.

**Bordier-Fraenkel's sign.** See *Bell's phenomenon*.

**boric** (bo′-rik). Relating to boron; containing boron. b. acid. See under *boron*.

**boricin** (bo′-ris-in). A proprietary mixture of borax and boric acid used as an antiseptic.

**boride** (bo′-rīd). A compound of boron with a radical or element.

**borine** (bo′-rēn). 1. A compound of 1 atom of boron and 3 atoms or 3 molecules of a univalent radical. 2. A proprietary antiseptic said to contain boric acid and aromatic stearoptens.

**borism** (bo′-rizm). Poisoning with boric acid or borax.

**borneene** (bor′-ne-ēn), $C_{10}H_{16}$. A peculiar volatile oil, the chief constituent of oil of camphor. It is isomeric with oil of turpentine and holds in solution borneol and rosin.

**Borneo camphor** (bor′-ne-o kam′-for). See *borneol*.

**borneol** (bor′-ne-ol) [*Borneo*], $C_{10}H_{18}O$. Borneo camphor; a substance that occurs in *Dryobalanops camphora*, a tree growing in Borneo and Sumatra. It is artificially prepared by treating the alcoholic solution of common camphor with sodium.

**borneyl** (bor′-ne-il), $C_{10}H_{14}$. The radical of borneol.

**bornyval** (bor′-ne-val). Isovalerate of borneol; an aromatic liquid used as a nerve sedative and antispasmodic. Dose 6 mimims.

**boroborax** (bo-ro′-raks). A crystalline combination of borax and boric acid. It is used as an antiseptic.

**borocalcite** (bo-ro-kal′-sīt). Native calcium borate.

**borocarbide** (bo-ro-kar′-bid). A compound of borax and carbon.

**borochloretone** (bo-ro-klor′-et-ōn). A combination of boric acid and chloretone; used as an antiseptic dusting powder.

**borocitrate** (bo-ro-sit′-rāt). A compound of both citric and boric acid with an element or radical.

**borofax** (bo′-ro-fax). Trade name of an emollient and sedative preparation containing boric acid.

**borofluorin** (bo-ro-flu′-or-in). A proprietary antiseptic and germicide said to contain boric acid, sodium fluoride, benzoic acid, and formaldehyde.

**boroform** (bo′-ro-form). A condensation product of boroglycerin and formaldehyde, used as an antiseptic.

**boroformal, boroformalin** (bo-ro-form′-al, bo-ro-form′-al-in). A proprietary antiseptic said to consist of borosalicylic glycerol, benzoresorcinol, menthol, thymol, eucalyptol, and formalin.

**borogen** (bo′-ro-jen). Boric-acid ethyl ester. It is used by inhalation in diseases of the air-passages.

**boroglycerin** (bo-ro-glis′-er-in). A mixture of boric acid with glycerol. b., glycerite of (*glycerinum boroglycerini*, U. S. P.), boroglycerin dissolved in glycerin by heating. Syn., *solution of boroglycerin*.

**boroglycerol** (bo-ro-glis′-er-ol). Boroglycerin.

**borol** (bo′-rol). Borosulphate of sodium or potassium.

**borolyptol** (bo-ro-lip′-tol). A proprietary internal and external antiseptic said to contain acetoboroglyceride, formaldehyde and the antiseptic constituents of *Pinus pumilio*, eucalyptus, myrrh, storax, and benzoin.

**boron** (bo′-ron) [Ar., *būraq*, borax]. B = 11; quantivalence III, V. A nonmetallic element occurring in two allotropic forms—as a powder and as a crystalline substance. It is the base of boric acid and of the mineral borax. *Boracic* or, more properly, *boric acid*, $H_3BO_3$, is a crystalline substance, found native in the volcanic lagoons of Tuscany. It occurs in white, transparent crystals, soluble in water and alcohol; it is antiseptic. Dose 5–20 gr. (0.32–1.3 Gm.). *Unguentum acidi borici* (U. S. P., B. P.) contains hard paraffin, 2; soft paraffin, 4; boric acid, 1; it is used as an antiseptic and in dermatology. *Borax*, $Na_2B_4O_7+10H_2O$, sodium borate (more correctly, disodic tetraborate), occurs as white, transparent crystals, soluble in water, alcohol, and glycerol; it is used as an antiseptic. Dose 5–40 gr. (0.32–2.6 Gm.). *Glycerinum boracis* (B. P.) contains borax, glycerol, and distilled water; used as a local application. *Mel boracis* (B. P.), borax honey, contains borax, clarified honey, and glycerol (about 1 in 7); used as a local application.

**borophenol** (bo-ro-fe′-nol). A soluble disinfectant compound of borax and phenol.

**borosalicyl, borsalyl** (bo-ro-sal′-is-il, bor′-sal-il). Sodium borosalicylate.

**borosol** (bo′-ro-sol). A proprietary liquid used as a wash for perspiring feet and said to contain aluminum tartrate, boric and salicylic acids, glycerin, and free tartaric acid.

**borotartrate** (bo-ro-tar′-trāt). A combination of boric and tartaric acids with a base.

## TABLE OF BONES.

| Name. | Principal Features. | Articulations and Variety. | Muscular and Ligamentous Attachments. |
|---|---|---|---|
| Anvil. | See *Incus*. | | |
| Astragalus. | See *Talus*. | | |
| Atlas [BNA].* | First cervical vertebra; ring-like; anterior and posterior arches and tubercles; articular surfaces. | Occipital bone—*double arthrodia*. Epistropheus, four joints—*diarthrodia rotatoria and double arthrodia*. | Longus colli (*tubercle*), rectus capitis posticus minor (*spinous process*). |
| Axis. | See *Epistropheus*. | | |
| Calcaneus [BNA] (calcaneum) (os calcis). | The heel bone; irregularly cuboid; lesser process [sustentaculum tali], greater process, peroneal ridge or spine, superior groove articular surfaces. | Astragalus, Cuboid, Scaphoid, } *arthrodia*. | Tibialis posticus, tendo Achillis, plantaris, abductor pollicis, abductor minimi digiti, flexor brevis digitorum, flexor accessorius, extensor brevis digitorum. |
| Capitatum (os) [BNA] (os magnum). | Largest bone of carpus; occupies center of wrist; head, neck, body; short bone. | Scaphoid, Semilunar, } *enarthrodia*. Second, Third, Fourth } Metacarpal—*arthrodia*. Trapezoid, Unciform, } *arthrodia*. | Flexor brevis pollicis (part). |
| Carpus [BNA] (ossa carpi). | Consists of os naviculare manus, os lunatum, os triquetrum, os pisiforme, os multangulum majus, os multangulum minus, os capitatum, os hamatum. See *inidvidual bones*. | | |
| Clavicula [BNA] (clavicle). | Collar-bone, resembles the italic "*f*"; conoid tubercle, deltoid tubercle, oblique line. | Sternum, Scapula, Cartilage of first rib, } *arthrodia*. | Sternomastoid, trapezius, pectoralis major, deltoid, subclavius, sternohyoid, platysma. |
| Coccygis (os) [BNA] (coccyx). | The last bone of the vertebral column; resembles a cuckoo's beak; usually composed of four small segments: base, apex, cornua. | Sacrum—*amphiarthrodia*. | Coccygeus, gluteus maximus, extensor coccygis, sphincter ani, levator ani. |
| Concha nasalis inferior [BNA] (inferior turbinate bone). | Situated on the outer wall of the nasal fossa; irregular bone. | Ethmoid, Maxilla, Lacrimal, Os palatinum, } *synarthrodia*. | None. |
| Costæ [BNA] (ribs). | Twelve in number on each side; shaft, head, neck, tuberosity, angle (anterior and posterior), anterior or sternal extremity; first, second, tenth, eleventh and twelfth are peculiar; flat bones. | Vertebræ—*arthrodia*. Sternum, { First rib—*synarthrodia*. Others—*arthrodia*. | Internal and external intercostals, scalenus anticus, medius and posticus, pectoralis minor, serratus magnus, obliquus externus, transversalis, quadratus lumborum, diaphragm, latissimus dorsi, serratus posticus superior and inferior, sacrolumbalis, musculus accessorius ad sacrolumbalem, longissimus dorsi, cervicalis ascendens, levatores costarum, infracostales. |
| Coxæ (os) [BNA] (innominate bone; os innominatum). | Large, 3 parts; flat bone; with its fellow and sacrum and occcyx forms pelvis. *Ilium*—superior broad expanded portion, crest, superior, middle, and inferior curved lines iliopectineal line, venter, auricular surface, anterior and posterior, superior and inferior spinous processes, ⅖ (about) of acetabulum. *Ischium*—lower and back portion, body, tuberosity and ramus, spine, greater and lesser sacrosciatic notches, external and internal lips of tuberosity, lower boundary of obturator foramen, ⅖ (about) of acetabulum. | { With its fellow of opposite side, Sacrum, Femur—*enarthrodia*. } *synarthrodia*. | *Ilium*—tensor vaginæ femoris, external oblique, latissimus dorsi, iliacus, transversalis, quadratus lumborum, erector spinæ, internal oblique, gluteus maximus, medius, and minimus, rectus, pyriformis, multifidus spinæ, sartorius. *Ischium*—obturator externus and internus, gracilis, levator ani, gemelli superior and inferior, coccygeus, biceps, semitendinosus, semimembranosus, quadratus femoris, adductor magnus, transversus perinei, erector penis. |

* The abbreviation [BNA] used throughout this table indicates that the term is in accordance with the Basle anatomical nomenclature.

TABLE OF BONES.—(Continued.)

| Name. | Principal Features. | Articulations and Variety. | Muscular and Ligamentous Attachments. |
|---|---|---|---|
| Coxæ (continued). | Pubis—body, horizontal ramus, descending ramus, spine, iliopectineal line, angle, symphysis, obturator foramen (upper boundary), ⅓ of acetabulum. | | Pubis—internal and external oblique, transversalis, rectus, pyramidalis, psoas parvus, pectineus, adductor magnus, longus, and brevis, gracilis, external and internal obturator, levator ani, compressor urethræ, accelerator urinæ. |
| Cranium BNA]. | Occipital, parietal (two), frontal, temporal (two), sphenoid, ethmoid. See *individual bones.* | | |
| Cuboideum (os) [BNA] (cuboid). | Somewhat pyramidal; tuberosity. | Os calcis, External cuneiform, Fourth and fifth metatarsal bones, } arthrodia. | Flexor brevis pollicis. |
| Cuneiform (of carpus). | See *Triquetrum (os).* | | |
| Cuneiforme (os) primum [BNA] (internal cuneiform). | Irregularly wedge-shaped; the largest of the three. | Os naviculare pedis, Cuneiforme secundum, First and second metatarsal bones, } arthrodia. | Tibialis anticus and posticus. |
| Cuneiforme (os) secundum [BNA] (middle cuneiform). | Wedge-shaped; smallest of the three. | Os naviculare pedis, Cuneiforme primum, Cuneiforme tertium, Second metatarsal, } arthrodia. | Tibialis posticus. |
| Cuneiforme (os) tertium [BNA] (external cuneiform). | Wedge-shaped. | Os naviculare pedis, Cuneiforme secundum, Os cuboideum, Second, third, and fourth metatarsal bones, } arthrodia. | Tibialis posticus, flexor brevis pollicis. |
| Epistropheus [BNA] (axis). | Second cervical vertebra; body, odontoid process, pedicles, laminæ, spinous process, transverse processes, articular surfaces. | First cervical vertebra—*diarthrodia rotatoria* and *double arthrodia.* Third cervical vertebra—*double arthrodia.* | Longus colli, check ligaments |
| Ethmoidale (os) [BNA] (ethmoid). | Irregularly cubic; situated at anterior part of base of skull; horizontal or cribriform plate, perpendicular plate, two lateral masses, crista galli. | Os sphenoidale Two sphenoid—turbinated, Frontal, Two nasal, Two maxillary, Two lacrimal, Ossa palatinum, Conchæ nasalis inferior, Vomer, } synarthrodia. | None. |
| Femur [BNA]. | Cylindrical; longest, largest, and strongest bone in the body; shaft and two extremities, head, neck, greater and lesser trochanters, linea aspera, condyles; a long bone. | Os coxæ—*enarthrodia.* Tibia—*ginglymus.* Patella—*arthrodia.* | Gluteus medius, gluteus minimus, pyriformis, obturator internus, obturator externus, gemellus superior, gemellus inferior, quadratus femoris, psoas magnus, iliacus, vastus externus, gluteus maximus, short head of the biceps, vastus internus, adductor magnus, pectineus, adductor brevis, adductor longus, crureus, subcrureus, gastrocnemius, plantaris, popliteus. |
| Fibula [BNA]. | Long bone; shaft, upper extremity or head, lower extremity or external malleolus. | Tibia—*arthrodia.* Talus, with the tibia and fibula—*ginglymus.* | Biceps, soleus, peroneus longus, extensor longus digitorum, peroneus tertius, extensor proprius pollicis, tibialis posticus, flexor longus pollicis, peroneus brevis. |

TABLE OF BONES.—(Continued.)

| Name. | Principal Features. | Articulations and Variety. | Muscular and Ligamentous Attachments. |
|---|---|---|---|
| Foot. | Composed of tarsus, metatarsus, and phalanges (q. v.). | | |
| Frontale (os) [BNA] (frontal). | The forehead bone; a flat bone; a frontal portion and an orbitonasal portion; frontal eminences, superciliary ridges, supraorbital arches, supraorbital notches or foramina, internal and external angular processes, temporal ridges, nasal notch, nasal spine, nasal eminence or glabella. | Two parietal, Sphenoid, Ethmoid, Two nasal, Two maxillary, Two lacrimal, Ossa zygomatica, } synarthrodia. | Corrugator supercilii, orbicularis palpebrarum, and temporal on each side. |
| Hamatum (os) [BNA] (unciform). | Wedge-shaped; hook-like process; in lower row of carpus; short bone. | Os lunatum, Fourth and fifth metacarpal, Os triquetrum, Os capitatum, } arthrodia. | Flexor brevis minimi digiti flexor ossis metacarpi minimi digiti, anterior annular ligament. |
| Hand. | Composed of carpus, metacarpus, and phalanges (q. v.). | | |
| Humerus [BNA] | Largest bone of upper extremity; long bone; a shaft and two extremities. Upper extremity presents a head, neck, and greater and lesser tuberosities. Lower extremity, trochlea, olecranon fossa, coronoid fossa, external and internal condyles, supratrochlear foramen, supracondyloid ridges. | Scapula (glenoid cavity)— enarthrodia. Ulna, Radius, } ginglymus. | Supraspinatus, infraspinatus, teres minor, subscapularis, pectoralis major, latissimus dorsi, deltoid, coracobrachialis, brachialis anticus, triceps, subanconeus, pronator radii teres, flexor carpi radialis, palmaris longus, flexor digitorum sublimis, flexor carpi ulnaris, supinator longus, extensor carpi radialis longior, extensor carpi radialis brevior, extensor communis digitorum, extensor minimi digiti, extensor carpi ulnaris, supinator brevis, and anconeus. |
| Hyoideum (os) [BNA] (hyoid). | A bony arch; irregular bone; a body, two greater and two lesser cornua. | None. | Sternohyoid, thyrohyoid, omohyoid, digastricus, stylohyoid, mylohyoid, geniohyoid, genioglossus, hyoglossus, middle constrictor of the pharynx. |
| Ilium (os) [BNA] (ilium). | See Coxæ (os). | | |
| Incus [BNA] (anvil). | Resembles a bicuspid tooth with two roots, body, and two processes. The largest bone in the ear. | Malleus—arthrodia (trigger-joint). Stapes—arthrodia. | None. |
| Inferior maxillary. | See Mandibula. | | |
| Inferior turbinated. | See Concha nasalis inferior. | | |
| Innominate. | See Coxæ (os). | | |
| Ischii (os) [BNA] (ischium). | See Coxæ (os). | | |
| Lacrimale (os) [BNA] (lacrimal). | Small; situated at front part of inner wall of orbit; resembles fingernail; crest, lacrimal groove; flat bone. | Frontal, Ethmoid, Superior maxillary, Inferior turbinated, } synarthrodia. | Tensor tarsi. |
| Lingual. | See Hyoideum (os). | | |
| Lunatum (os) [BNA] (semilunar). | Upper row of carpus; four surfaces, crescentic outline. | Radius—condyloid. Os capitatum, Os hamatum, Os triquetrum, Os naviculare manus, } arthrodia. | |
| Magnum (os). | See Capitatum (os). | | |
| Malar. | See Zygomaticum (os). | | |
| Malleus. | Resembles a hammer; head, neck, handle or manubrium, processus gracilis, processus brevis; irregular. The second largest bone of the ear. | Incus—arthrodia (trigger-joint). | Tensor tympani. |
| Mandibula [BNA] (inferior maxillary; mandible). | Body and two rami; contains the teeth of the lower jaw; symphysis, | The two temporal bones— bilateral condyloid—diarthrodia. | Levator menti, depressor labii inferioris, depressor anguli oris, platysma myoides, buc- |

TABLE OF BONES.—(Continued.)

| Name. | Principal Features. | Articulations and Variety. | Muscular and Ligamentous Attachments. |
|---|---|---|---|
| Mandibula (continued). | mental process, mental foramen, coronoid and condyloid process, head, neck, sigmoid notch; irregular bone. | | cinator, masseter, orbicularis oris, geniohyoglossus, geniohyoideus, mylohyoideus, digastric, superior constrictor, temporal, internal pterygoid, external pterygoid. |
| Maxilla [BNA] (superior maxillary). | Forms with its fellow the whole of the upper jaw; somewhat cuboidal; body and malar, nasal, alveolar and palatine processes; hollow (the antrum of Highmore or maxillary sinus), incisive or myrtiform fossa, canine fossa, canine eminence, infraorbital foramen, posterior dental canals, maxillary tuberosity, posterior palatine canal, infraorbital groove; irregular bone. | Frontal, Ethmoid, Nasal, Os zygomaticum, Lacrimal, Concha nasalis inferior, Palate, Vomer, Fellow of opposite side,   } synarthrodia. | Orbicularis palpebrarum, obliquus inferior oculi, levator labii superioris alæque nasi, levator labii superioris proprius, levator anguli oris, compressor nasi, depressor alæ nasi, dilator naris posterior, masseter, buccinator, internal pterygoid, orbicularis oris. |
| Metacarpalia (ossa) [BNA] (metacarpal). | Five in number; shaft, base, and head; long bones. | Second row of carpus—arthrodia. Phalanges—condyloid. | To the thumb—flexor and extensor ossis metacarpi pollicis, first dorsal interosseous. Second metacarpal bone—flexor carpi radialis, extensor carpi radialis longior, first and second dorsal interosseous, first palmar interosseous, flexor brevis pollicis (frequently). Third metacarpal—extensor carpi radialis brevior, flexor brevis pollicis, adductor pollicis, second and third dorsal interosseous. Fourth metacarpal—third and fourth dorsal and second palmar interosseous. Fifth metacarpal—extensor carpi ulnaris, flexor carpi ulnaris, flexor ossis metacarpi minimi digiti, fourth dorsal and third palmar interosseous. |
| Metatarsalia (ossa) [BNA] (metatarsal). | Five in number; shaft, base, and head; long bones. | Tarsus—arthrodia. Phalanges—condyloid. | First—tibialis anticus (part), peroneus longus, first dorsal interosseous. Second—adductor pollicis, first and second dorsal interosseous, tibialis posticus (part). Third—adductor pollicis, second and third dorsal and first plantar interosseous, tibialis posticus (part). Fourth—adductor pollicis, third and fourth dorsal and second plantar interosseous, tibialis posticus (part). Fifth—peroneus brevis, peroneus tertius, flexor brevis minimi digiti, transversus pedis, fourth dorsal, and third plantar interosseous. |
| Multangulum majus [BNA] (trapezium). | In lower row of carpus; very irregular; six surfaces, groove; short bone. | Os naviculare manus, Multangulum minus, First and second metacarpal.   } arthrodia. | Adductor pollicis, flexor ossis metacarpi pollicis, flexor brevis pollicis (part). |
| Multangulum minus [BNA] (trapezoid). | Smallest bone in second row of carpus; wedge-shaped; six surfaces; short bone. | Os naviculare manus, Second metacarpal, Multangulum majus, Os capitatum,   } arthrodia. | Flexor brevis pollicis (part). |
| Nasale (os) [BNA] (nasal). | Oblong; forms with its fellow the bridge of the nose; nasal foramen, spine, crest; flat bone. | Frontal, Ethmoid, Nasal (opposite), Maxillary,   } synarthrodia. | |

TABLE OF BONES.—(Continued.)

| Name. | Principal Features. | Articulations and Variety. | Muscular and Ligamentous Attachments. |
|---|---|---|---|
| Naviculare (os) manus [BNA] (scaphoid of wrist). | Largest bone of first row; boat-shaped; upper and outer part of carpus; four surfaces; tubercle; short bone. | Radius—*condyloid.* Multangulum majus, Multangulum minus, Os capitatum, Os lunatum, } *arthrodia.* | External lateral ligament of wrist. |
| Naviculare (os) pedis [BNA] (scaphoid of ankle). | Or navicular bone; boat-shaped; inner side of tarsus; four surfaces, tuberosity; short bone. | Talus, Cuneiform (three), Cuboid (occasionally), } *arthrodia.* | Tibialis posticus (part). |
| Occipitale (os) [BNA] (occipital). | Back part and base of cranium; trapezoid in shape; outer and inner tables; external protubcrance, external occipital crest, superior and inferior curved lines, foramen magnum, condyles, jugular process, anterior and posterior condyloid foramina, basilar process, pharyngeal spine. 4 fossæ on internal surface, internal occipital protuberance and crest, grooves for the cerebral sinuses, torcular herophili, jugular foramen; flat bone. | Parietal (two), Temporal (two), Sphenoid, } *synarthrodia.* Atlas—*double arthrodia.* | Twelve pairs—occipitofrontalis, trapezius, sternomastoid, complexus, biventer cervicis, splenius capitis, superior oblique, rectus capitis posticus, major and minor, rectus lateralis, rectus capitis anticus, major and minor, superior constrictor of the pharynx. |
| Palatinum (os) [BNA] (palate). | Back part of nasal fossa; helps to form floor and outer wall of nose, the roof of mouth and floor of orbit, also sphenomaxillary and pterygoid fossæ and the sphenomaxillary fissure; L-shaped; inferior or horizontal plate, superior or vertical plate, posterior palatine canal, posterior spine, inferior and superior turbinated crests, maxillary process, pterygoid process, accessory descending palatine canals, orbital process, sphenoid process, sphenopalatine foramen. | Sphenoid, Ethmoid, Maxillary, Concha nasalis inferior, Vomer, Opposite palate, } *synarthrodia.* | Tensor palati, azygos uvulæ, internal and external pterygoid, superior constrictor of pharynx. |
| Parietale (os) [BNA] (parietal). | Form sides and roof of skull; irregular, quadrilateral; two surfaces, four borders, four angles, parietal eminence, temporal ridge, parietal foramen, furrows for cerebral sinuses, depressions for Pacchionian bodies; flat bone. | Opposite parietal, Occipital, Frontal, Temporal, Sphenoid, } *synarthrodia.* | Temporal. |
| Patella [BNA]. | Flat; triangular sesamoid; anterior part of knee-joint; two surfaces, three borders, apex; flat bone. | Condyles of femur—*partly arthrodial.* | Rectus, crureus, vastus internus, vastus externus. |
| Pelvis. | Composed of two ossa coxæ, os sacrum, and os coccygis (*q. v.*). | | |
| Phalanges digitorum pedis [BNA] (phalanges of foot). | Two of great toe, three of each of the others; shaft, base, head; long bones. | First row with metatarsal and second phalanges—*condyloid.* Second of great toe with first phalanx; of other toes, with first and third phalanges, Third row with second row, } *ginglymus.* | *First—great toe*—inner tendon extensor brevis digitorum, abductor pollicis, adductor pollicis, flexor brevis pollicis, transversus pedis. *Second toe*—first and second dorsal interosseous, first lumbrical. *Third toe*—third dorsal and first plantar interosseous, second lumbrical. *Fourth toe*—fourth dorsal and second plantar interosseous, third lumbrical. |

TABLE OF BONES.—(Continued.)

| Name. | Principal Features. | Articulations and Variety. | Muscular and Ligamentous Attachments. |
|---|---|---|---|
| Phalanges digitorum pedis (continued). | | | *Fifth toe*—flexor brevis minimi digiti, abductor minimi digiti, third plantar interosseous, fourth lumbrical.<br>*Second — great toe*—extensor longus pollicis, flexor longus pollicis.<br>*Other toes*—flexor brevis digitorum, one slip of common tendon of extensor longus and brevis digitorum.<br>*Third*—two slips from the common tendon of the extensor longus and extensor brevis digitorum, and flexor longus digitorum. |
| Phalanges digitorum manus [BNA] (phalanges of hand). | Fourteen in number, three for each finger and two for thumb; shaft, head, base; long bones. | First row with metacarpal bones and second row of phalanges—*condyloid*. Second row with first and third rows, Third row with second row, } *ginglymus*. | *Thumb*—extensor primi internodii pollicis, flexor brevis pollicis, abductor pollicis, adductor pollicis, flexor longus pollicis, extensor secundi internodii.<br>*First—index-finger*—first dorsal and first palmar interosseous.<br>*Middle finger*—second and third dorsal interosseous.<br>*Ring finger*—fourth dorsal and second palmar interosseous.<br>*Little finger*—third palmar interosseous, flexor brevis minimi digiti, abductor minimi digiti.<br>*Second—to all*—flexor sublimis digitorum and extensor communis digitorum; in addition.<br>*To index-finger*—extensor indicis.<br>*To little finger*—extensor minimi digiti.<br>*Third*—flexor profundus digitorum, extensor communis digitorum. |
| Pisiforme (os) [BNA] (pisiform). | Anterior and inner side of carpus; small; spherical; one articular facet; short bone. | Os triquetrum—*arthrodia*. | Flexor carpi ulnaris, abductor minimi digiti, anterior annular ligament. |
| Pubis. | See *Coxæ (os)*. | | |
| Pyramidale (os). | See *Triquetrum (os)*. | | |
| Radius [BNA]. | Outer side of forearm; shaft, head, neck, tuberosity, lower extremity, oblique line, sigmoid cavity, styloid process. | Humerus—*ginglymus*.<br>Ulna, { superior—*diarthrodia rotatoria*.<br>middle—*membranous*.<br>inferior—*diarthrodia rotatoria*.<br>Os lunatum—*condyloid*. | Biceps, supinator brevis, flexor sublimis digitorum, flexor longus pollicis, pronator quadratus, extensor ossis metacarpi pollicis, extensor primi internodii pollicis, pronator radii teres, supinator longus. |
| Ribs. | See *Costæ*. | | |
| Sacrum (os) [BNA]. | Large triangular bone at lower part of vertebral column and upper and back part of pelvic cavity; composed of five vertebræ; base, promontory, four surfaces, apex, central canal, anterior and posterior sacral foramina, lateral masses, laminæ, articular processes, sacral cornua, transverse processes, sacral groove, ala; irregular bone. | Last lumbar vertebra, Coccyx, Ossa coxæ (two), } *amphiarthrodia*. | Pyriformis, coccygeus, iliacus, gluteus maximus, latissimus dorsi, multifidus spinæ, erector spinæ, extensor coccygis. |
| Scaphoid of carpus. | See *Naviculare (os) manus*. | | |
| Scaphoid of tarsus. | See *Naviculare (os) pedis*. | | |
| Scapula [BNA]. | Back part of shoulder; triangular; posterior aspect and side of thorax; two surfaces, three borders, three angles, sub- | Humerus—*enarthrodia*.<br>Clavicle—*arthrodia*. | Subscapularis, supraspinatus, infraspinatus, trapezius, deltoid, omohyoid, serratus magnus, levator anguli scapulæ, rhomboideus major |

TABLE OF BONES.—*(Continued.)*

| Name. | Principal Features. | Articulations and Variety. | Muscular and Ligamentous Attachments. |
|---|---|---|---|
| Scapula *(continued)*. | scapular fossa, subscapular angle, dorsum, spine, supraspinous and infraspinous fossæ, acromion process, glenoid cavity, neck, head, coracoid process; flat bone. | | and minor, triceps, teres major and minor, biceps, coracobrachialis, pectoralis minor, platysma, latissimus dorsi. |
| Semilunar. | See *Lunatum (os)*. | | |
| Sesamoid [BNA]. | Small, rounded masses, cartilaginous in early life, osseous in the adult; developed in tendons; inconstant, except patellæ. | | |
| Sphenoidale (os) [BNA] (sphenoid). | Anterior part of base of skull; bat-shaped, with wings extended; body, two greater and two lesser wings, two pterygoid processes, ethmoid spine, optic groove, optic foramen, olivary process, pituitary fossa, anterior, middle, and posterior clinoid processes, sella turcica, carotid or cavernous groove, lingula, ethmoid crest, sphenoid cells or sinuses, sphenoid turbinated bones, rostrum, vaginal processes, pterygopalatine canal, spinous processes, round foramen, oval foramen, foramen Vesalii, foramen spinosum, pterygoid ridge, external orbital foramens, Vidian canal, pterygoid fossa, internal and external pterygoid plates, hamular process, scaphoid fossa. | All the bones of cranium, Malar (two), Palate (two), Vomer, } *synarthrodia.* | Temporal, external and internal pterygoids, superior constrictor, tensor palati, levator tympani, levator palpebræ, obliquus superior, superior, inferior, internal and external recti. |
| Sphenoid turbinated or sphenoid spongy bones. | Situated at anterior and inferior part of body of sphenoid; exist as separate pieces until puberty and occasionally are not joined in the adult. | Ethmoid, Palate, } *synarthrodia.* | None. |
| Stapes [BNA]. | Resembles a stirrup; one of the ossicles of the tympanum; head, neck, two branches (crura), base; irregular bone. | Incus—*arthrodia.* | Stapedius. |
| Sternum [BNA]. | The breast-bone; manubrium, gladiolus, ensiform cartilage, or first, second, and third pieces, anterior and posterior surface, borders. | Clavicles (two)—*arthrodia.* Costal cartilages (seven on each side) first—*synarthrodia;* others, *arthrodia.* | Pectoralis major, sternomastoid, sternohyoid, sternothyroid, triangularis sterni, aponeuroses of the obliquus externus and internus and transversalis muscles, rectus, diaphragm. Internal and external lateral ligaments. |
| Superior maxillary. | See *Maxilla.* | | |
| Talus [BNA] (astragalus). | Irregularly cubic; forms the keystone of arch of foot; head, neck, six articular surfaces. | Tibia, Fibula, } *ginglymus.* Calcaneus, Os naviculare pedis, } *arthrodia.* | |
| Tarsus [BNA] (ankle). | Consist of calcaneus, or os calcis, talus, cuboid, navicular, internal, middle, and external cuneiform. See *individual bones.* | | |
| Temporale (os) [BNA] (temporal). | Situated at side and base of skull; squamous, mastoid and petrous portions, temporal ridge, zygoma or zygomatic process, eminentia articularis, glaserian fissure, tubercle, glenoid fossa, postglenoid pro- | Occipital, Parietal, Sphenoid, Inferior maxillary, Malar, } *synarthrodia.* | Temporal, masseter, occipitofrontalis, sternomastoid, splenius capitis, trachelomastoid, digastricus, retrahens aurem, stylopharyngeus, stylohyoideus, styloglossus, levator palati, tensor tympani, tensor palati, stapedius. |

TABLE OF BONES.—(Continued.)

| Name. | Principal Features. | Articulations and Variety. | Muscular and Ligamentous Attachments. |
|---|---|---|---|
| Temporale (continued). | cess, tympanic plate, mastoid foramen, digastric fossa, occipital groove, sigmoid fossa, mastoid cells, meatus auditorius externus, hiatus Fallopii, meatus auditorius internus, lamina cribrosa, aquæductus vestibuli, styloid process, stylomastoid foramen, auricular fissure; irregular bone. | | |
| Tibia [BNA]. | At front and inner side of leg; next to femur in length and size; prismoid in form; upper extremity or head, tuberosities, spinous process, tubercle, popliteal notch, shaft, crest (the shin), oblique line, internal malleolus; long bone. | Femur—*ginglymus*. Fibula, { superior—*arthrodia*. middle—*membranous*. inferior—*arthrodia*. } Talus with fibula—*ginglymus*. | Semimembranosus, tibialis anticus, extensor longus digitorum, biceps, sartorius, gracilis, semitendinosus, tibialis anticus, popliteus, soleus, flexor longus digitorum, tibialis posticus, ligamentum patellæ. |
| Trapezium. | See *Multangulum majus*. | | |
| Trapezoid. | See *Multangulum minus*. | | |
| Triquetral. | See *Wormian*. | | |
| Triquetrum (os) [BNA] (cuneiform of carpus). | Pyramidal. | Os lunatum, Pisiform, Os hamatum, Interarticular fibrocartilage, } arthrodia. | |
| Turbinate, inferior. | See *Concha nasalis inferior*. | | |
| Turbinate, middle. | The free convoluted margin of the thin lamella that descends from the under surface of the cirbriform plate of the ethmoid bone. | | |
| Turbinate, superior. | The thin curved plate of the ethmoid that bounds the superior meatus of the nose above. | | |
| Tympanic. | Includes the *incus*, *malleus*, and *stapes* (q. v.). | | |
| Ulna [BNA]. | Inner side of forearm, parallel with radius; prismatic; shaft and two extremities, olecranon process, coronoid process, greater and lesser sigmoid cavities, oblique ridge, perpendicular line, head or lower extremity, styloid process; long bone. | Humerus—*ginglymus*. Radius, { superior—*diarthrodia rotatoria*. middle—*membranous*. inferior—*diarthrodia rotatoria*. } | Triceps, anconeus, flexor carpi ulnaris, brachialis anticus, pronator radii teres, flexor sublimis digitorum, flexor profundus digitorum, flexor longus pollicis (occasionally), pronator quadratus, supinator brevis, extensor ossis metacarpi pollicis, extensor secundi internodii pollicis, extensor indicis, extensor carpi ulnaris. |
| Unciform. | See *Hamatum* (os). | | |
| Vertebra [BNA]. | Twenty-six; cervical seven, dorsal twelve, lumbar five, sacrum (composed of five), coccyx (composed of four); each has a body and an arch; latter has two pedicles, two laminæ, and seven processes, viz.: for articular, two transverse, one spinous). Peculiar vertebræ, first, second, and seventh cervical; first, ninth, tenth, eleventh, and twelfth dorsal, sacrum, coccyx; irregular bones. | Articulate with each other, and with occipital and innominate bones and ribs (q. v.). Intervertebral articulations, between the bodies—*amphiarthrodia*. Between articular processes—*arthrodia*. | Attachment of muscles.—To *the atlas* are attached nine pairs: the longus colli, rectus anticus minor, rectus lateralis, rectus posticus minor, obliquus superior and inferior, splenius colli, levator anguli scapulæ, and first intertransverse. To *the axis* are attached eleven pairs: the longus colli, obliquus inferior, rectus posticus major, semispinalis colli, multifidus spinæ, levator anguli scapulæ, splenius colli, scalenus medius, transversalis colli, intertransversales, interspinales. To *the remaining vertebræ* generally are attached thirty-five pairs and a single muscle: anteriorly, the rectus anticus major, longus colli, scalenus anticus, medius, and |

TABLE OF BONES.—(Continued.)

| Name. | Principal Features. | Articulations and Variety. | Muscular and Ligamentous Attachments. |
|---|---|---|---|
| Vertebra (continued). | | | posticus, psoas magnus, psoas parvus, quadratus lumborum diaphragm, obliquus internus and transversalis, posteriorly, the trapezius, latissimus dorsi, levator anguli scapulæ, rhomboideus major and minor, serratus posticus superior and inferior, splenius, erector spinæ, sacrolumbalis, longissimus dorsi, spinalis dorsi, cervicalis ascendens, transversalis colli, trachelomastoid, complexus, biventer cervicis, semispinalis dorsi and colli, multifidus spinæ, rotatores spinæ interspinales, supraspinales, intertransversales, levatores costarum. |
| Vomer [BNA]. | Situated vertically at back part of nasal fossæ; forms part of septum of nose; somewhat like a plowshare; two surfaces and four borders. | Sphenoid, Ethmoid, Superior maxillary (two), Palate (two), Cartilage of the septum, } synarthrodia. | |
| Wormian. | Supernumerary bones; irregular, inconstant, isolated, interposed between the cranial bones, most frequently in the lambdoid suture. | | |
| Zygomaticum (os) [BNA] (os malæ, the malar or cheek bone). | Small; quadrangular; at upper and outer part of face; forms prominence of cheek, part of outer wall and floor of orbit, part of temporal and zygomatic fossæ; frontal, orbital, maxillary and zygomatic processes, malar foramen, four borders; irregular bone. | Frontal, Sphenoid, Temporal, Maxillary, } synarthrodia. | Levator labii superioris proprius, zygomaticus major and minor, masseter, temporal. |

BOROTARTROL 162 BOUGIE

**borotartrol** (*bo-ro-tar'-trol*). A mixture of neutral sodium tartrate and boric acid.
**Borsch's bandage** (*borsh*). A bandage specially arranged to cover one eye only. See page 126.
**Borsieri's line** (*bor-se-a'-re*). In the early stage of scarlatina, a line drawn on the skin with the fingernail leaves a white mark which quickly turns red and becomes smaller in size.
**borsyl** (*bor'-sil*). A proprietary dusting-powder for perspiring feet, said to consist of borax, boric acid, talcum, and spermaceti.
**boss** (*bos*) [ME., *boce*]. A wide, more or less circular protuberance, as on the skull or on a tumor. b., **parietal.** See *eminence, parietal.* b., Pott's. See *Pott's curvature.* b., **sanguineous.** 1. A swelling due to a contusion and containing extravasated blood. 2. See *caput succedaneum.*
**bossed** (*bosd*). Having a prominent center on a circular flat surface.
**bosselated** (*bos'-el-a-ted*) [*boss*]. Covered with bosses or small nodules.
**Bostock's catarrh** (*bos'-tok*) [John Bostock, English physician, 1773–1846]. Hay-fever.
**Boston's reaction for Bence-Jones' albumose.** It depends upon the presence in the albumose of loosely combined sulphur: (1) 15 to 20 Cc. of filtered urine are placed in a test-tube and to it an equal quantity of saturated solution of sodium chloride is added, and the whole shaken; (2) 2 or 3 Cc. of a 30 % solution of caustic soda are now added and shaken vigorously; (3) the upper one-fourth of the column of liquid is gradually heated over the flame of a Bunsen burner to the boiling-point, whereupon a solution of lead acetate (10 %) is added, drop by drop, boiling the upper previously heated stratum of liquid after each additional drop; (4) when the drop of lead acetate comes in contact with the liquid, a copious pearly or cream-colored cloud appears at the surface, which becomes less dense as the boiling-point is reached; and when boiling is prolonged for one-half to one minute, the upper stratum shows a slight browning, which deepens to a dull black. This lessens in intensity toward the bottom of the tube. After standing the reaction becomes intensified, and a black precipitate falls through the clear liquid and collects at the bottom of the tube.
**botryx** (*bos'-triks*) [βόστρυχος, a curl]. In biology, a helicoid cyme.
**bot** (*bot*) [*botus*, a belly-worm]. The larva of certain species of flies of the genus *Œstrus*, which are conveyed into the stomach of man, where they hatch. Also the threadworm, *Oxyuris vermicularis.*
**Botal's duct, Botallo, duct of** (*bo'-tal, bo-tal'-o*) [Leonardo Botallo, Italian anatomist, 1530–  ]. Ductus arteriosus Botalli. A short vessel in the fetus between the main pulmonary artery and the aorta. B., **foramen of,** the foramen ovale in the interauricular septum of the fetal heart. B., **ligament of,** the remains of Botal's duct.
**botalismus** (*bot-al-iz'-mus*). See *botulism.*
**botanic** (*bo-tan'-ik*) [βοτάνη, an herb]. Pertaining to botany. b. **physician,** a title assumed by certain persons who profess to use only vegetable remedies. See *eclectic.*
**botany** (*bot'-an-e*) [βοτάνη, an herb]. The science of plants—their classification and structure.
**bothrenchyma** (*both-ren'-ke-mah*) [βοθρίον, a pit; ἐγχέω, to pour in]. Pitted tissue.
**bothria** (*both'-re-ah*) [L.]. Plural of *bothrion,* or of *bothrium.*
**Bothriocephalus** (*both-re-o-sef'-al-us*) [*bothrion*; κεφαλή, a head]. A genus of tape-worms. B. **latus,** the fish tape-worm, a common parasite of man in certain European localities. It may reach 25 feet in length, with a breadth of three-fourths of an inch. Syn., *Dibothrium latum; Tænia lata.*
**bothrioid** (*both'-re-oid*) [*bothrion*; εἶδος, likeness]. Pitted; foveolated; covered with pit-like markings.
**bothrion** (*both'-re-on*) [βοθρίον, a pit]. 1. A small cavity; the socket of a tooth. 2. A facet, or fosset, such as is seen upon the head of most of the tapeworms. 3. A deep corneal ulcer.
**bothrium** (*both'-re-um*). See *bothrion.*
**botryoid** (*bot'-re-oid*) [βότρυς, a bunch of grapes; εἶδος, likeness]. Resembling in shape a bunch of grapes.
**Botryomyces** (*bot-re-o-mi'-sēs*) [βότρυς, a bunch of grapes; μύκης, a fungus; pl., *botryomycetes*]. A general term for those fungi which occur in grape-like clusters. B. **equi,** the specific microorganism of botriomycosis.

**botryomycoma** (*bot-re-o-mi-ko'-mah*) [βότρυς, a bunch of grapes; μύκης, a fungus]. A tumor occurring in animals or persons affected with botryomycosis.
**botryomycosis** (*bot-re-o-mi-ko'-sis*) [see *Botryomyces*]. A disease of horses in which fibromatous nodules form in the lungs. It is supposed to be caused by a microorganism called *Botryomyces;* it is communicable to man.
**botryomycotic** (*bot-re-o-mi-kot'-ik*). Relating to or affected with botryomycosis.
**botryophyma** (*bot-re-o-fi'-mah*) [βότρυς, a bunch of grapes; φῦμα, a growth]. A vascular, fungus-like growth from the skin. b. **cæruleum,** a form having a blue coloration. b. **rubrum,** a form having a red color.
**botryotherapy** (*bot-re-o-ther'-a-pe*) [βότρυς, a bunch of grapes; θεραπεύειν, to heal]. The grape-cure; treatment by an almost exclusive diet of grapes.
**Botrytis** (*bot-ri'-tis*) [βότρυς, a bunch of grapes]. A genus of fungi. B. **bassiane,** a mold causing muscardine, a disease of silkworms.
**Böttcher's cells, B.'s crystals.** See under *Boettcher.*
**Böttger's test.** See *Boetiger's test.*
**Bottini's operation** (*bot-te'-ne*) [Enrico Bottini, Italian surgeon, 1837–1903]. For *enlarged prostate gland;* a fresh channel is bored through the substance of the gland by means of the galvano-cautery.
**bottle** (*bot'-l*) [ME., *botel;* Fr., *bouteille*]. A vessel, usually of glass, with a narrow neck. b., **feeding,** a flat flask with a nipple of India-rubber attached, used for feeding infants. b. **nose,** a common name for *Acne rosacea.* b., **specific gravity,** a Florence flask graduated to contain 500-1000 grains of water, with the weight of which an equal volume of any other liquid may be compared. b.-**stoop,** in pharmacy, a block so grooved that it serves to hold a wide-mouthed bottle in an oblique position while a powder is being dispensed from it. b.-**wax,** a hard, stiff variety of wax used in sealing bottles.
**botulin** (*bot'-u-lin*). See *botulismotoxin.*
**botulism, botulismus** (*bot'-u-lizm, bot-u-liz'-mus*) [*botulus*, a sausage]. Sausage-poisoning.
**botulismotoxin** (*bot-u-liz-mo-toks'-in*). A toxic albumose of poisonous meat produced by *Bacillus botulinus,* van Ermengem. Syn., **botulin;** botulinic acid.
**Bouchard's coefficient** (*boo-shar*) [Charles Jacques Bouchard, French physician, 1837–  ]. The ratio existing between the amount of urea and the sum-total of the solids in the urine—approximately 50. B.'s **disease,** dilation of the stomach due to deficient function of the gastric muscular fibers. B.'s **nodosities,** enlargement of the second phalangeal joints of the fingers, associated with dilation of the stomach. B.'s **treatment of obesity,** a daily diet of 1250 Gm. of milk and 5 eggs divided into 5 meals.
**Bouchardat's test** (*boo-shar-dah'*) [Apollinaire Bouchardat, French chemist, 1806–1886]. For *alkaloids.*—When potassium triiodide is added to any alkaloid in solution a brown precipitate results; this latter is soluble in alcohol.
**Bouchut's tubes** (*boo-shoo'*) [Jean Eugene Bouchut, French physician, 1818–1891]. A variety of tubes for intubation of the larynx.
**Boudin's law** (*boo-dan'*) [Jean Christian Marc François Joseph Boudin, French physician, 1806-1867]. The poisons of malaria and tuberculosis are antagonistic. This law is not founded upon fact.
**Bougard's paste** (*boo-gar'*) [Jean Joseph Bougard, French physician, 1815-1884]. A caustic paste containing mercury bichloride, zinc chloride, arsenic, cinnabar, starch, and wheat-flour.
**bougie** (*boo-zhe'*) [Fr., "a candle"]. 1. A slender cylindric instrument made of waxed silk, catgut, etc., for introduction into the urethra or other passage, for the purpose of dilation, exploration, etc. 2. A suppository. b. à **boule.** See *b.,* **bulbous.** b. à **empreinte,** one with a waxy substance adherent to its point, by means of which an impression of the stricture may be taken. b., **armed,** a bougie with a piece of silver nitrate or other caustic attached to its extremity. b., **bulbous,** a bougie with a bulbous tip. b., **caustic,** b., **cauterizant.** See *b.,* **armed.** b., **emplastic,** 1. See *b. à empreinte.* 2. A flexible bougie coated with a mixture of wax, diachylon, and olive-oil. b., **filiform,** a whalebone or other bougie of very small diameter. b., **fusiform,** one with a spindle-shaped shaft. b., **medicated.** 1. A bougie charged with some medicament. 2. A medicated suppository. b., **rosary,** a beaded bougie used in a

strictured urethra. **b., soluble,** a suppository composed of substances dissolving at body-temperature. **b., whip-,** one with filiform end gradually increasing in thickness.

**bouginage** (*boo-zhe-nahzh'*). Dilatation with a bougie.

**bouhou** (*boo-hoo'*). See *boohoo*.

**Bouillaud's disease** (*boo-e-yo'*) [Jean Baptiste *Bouillaud*, French physician, 1796-1881]. Infective endocarditis. **B.'s metallic tinkling,** a peculiar clink sometimes heard to the right of the apex-beat in cardiac hypertrophy.

**bouillon** (*boo-e-yon*(*g*)) [Fr.]. 1. A broth made by boiling meat, usually beef, in water. 2. A liquid nutritive medium for the culture of microorganisms, prepared from finely chopped beef or beef-extract.

**boulimia** (*boo-lim'-e-ah*). See *bulimia*.

**Boulton's solution.** A compound tincture of iodine, phenol, glycerol, and distilled water; it is used for spraying in rhinitis.

**bouquet** (*boo-ka'*) [Fr.]. 1. In anatomy, a cluster of nerves, blood-vessels, or muscles. 2. The delicate perfume and flavor of good wine. 3. The odor characteristic of a disease.

**Bourdin's paste** (*boor-dan'*) [Claude Etienne *Bourdin*, French physician, 19th cent.]. An escharotic mixture of nitric acid with flowers of sulphur.

**bourdonet** (*boor-don-a'*). An ovoid mass of lint.

**bourdonnement** (*boor-dun-mon*(*g*)) [Fr., *bourdonner*, to buzz]. Any buzzing sound. The murmur that is heard when the stethoscope is applied to any part of the body. It is thought to result from contraction of the muscular fibrils.

**Bourget's test** (*boor-zha'*) [Louis *Bourget*, Swiss chemist]. *For iodine or iodides.*—Filter paper is impregnated with starch solution and then moistened with a 5 per cent. solution of ammonium sulphate; on being subsequently made wet with a solution containing iodine, it strikes a deep blue color.

**boutonnière** (*boo-ton-ne-air'*) [Fr., buttonhole]. 1. A buttonhole-like incision. 2. External urethrotomy.

**boutons terminals** (*boo'-ton*(*g*) *ter-me-nal*). Small terminal enlargements or tactile-cells of sensory nerves, as in the nose of the guinea pig and mole.

**Bouveret's disease** (*boo'-ver-a*). Paroxysmal tachycardia. **B.'s sign** in intestinal obstruction, this sign is applicable only to the large gut; great distention of the cecum and a large elevation in the right iliac fossa.

**bovillæ** (*bo-vil'-e*) [L.]. Measles.

**bovin** (*bo'-vin*) [bos, ox]. A modified virus derived from the tubercle bacillus, of greater virulence than bovovaccine.

**bovine** (*bo'-vin*) [*bovinus*, of an ox]. 1. Ox-like. 2. Relating to, or derived from a cow or ox or heifer. **b. heart,** the immensely hypertrophied heart of aortic valvular disease. **b. hunger,** bulimia. **b. lymph,** vaccine virus from cows.

**bovinine** (*bo'-vin-in*) [see *bovine*]. A proprietary preparation of beef used as a food for invalids and convalescents.

**Bovista** (*bo-vis'-tah*) [L.]. 1. A genus of fungi closely allied to *Lycoperdon*; some of the species are edible. 2. The *Lycoperdon bovista,* a fungus or puffball. When dry it is a good styptic, and its tincture has been used in nervous diseases.

**bovovaccine** (*bo-vo-vak'-sin*) [*bovine; vaccine*]. An extract of tubercle bacilli used for protection against bovine tuberculosis.

**bovox** (*bo'-vox*). A proprietary essence or extract of beef.

**bovril** (*bov'-ril*). Trade name of a preparation containing extract of beef, peptone, albumin, and fibrin.

**Bowdichia** (*bo-dich'-e-ah*) [Edward *Bowdich*, an English naturalist]. A genus of plants of the order *Leguminosæ*. *B. virgilioides* is a South American tree, of which the bark (alchornoque or alcornoque bark) is diaphoretic, roborant, and antisyphilitic.

**Bowditch's law** (*bo'-ditch*) [Henry Pickering *Bowditch*, American physiologist, 1840-1911]. Any stimulus which is capable of exciting the heart at all will produce as great a response as the strongest stimulus.

**bowel** (*bow'-el*) [OF., *boel,* from L., *botellus,* a sausage]. The intestine. **b.-complaint,** diarrhea. **b., lower,** the rectum.

**bow-leg** [ME., *bowe*]. An arching outward of the lower limbs. See *genu varum*.

**Bowman's capsule** (*bo'-man*) [Sir William *Bowman*, English anatomist and ophthalmologist, 1816-1892]. The expanded portion forming the beginning of a uriniferous tubule. **B.'s discs,** the products of a breaking up of muscle-fibers in the direction of the transverse striations. **B.'s glands,** glands found in the olfactory mucous membrane. **B.'s membrane,** a thin, homogeneous membrane representing the uppermost layer of the stroma of the cornea, with which it is intimately connected. **B.'s muscle,** *origin;* 1. *Longitudinal portion* (Brucke's muscle): junction of cornea and sclera; 2. *Circular portion* (Mueller's muscle): the fibers form a circle; *insertion;* 1. Outer layers of choroid. 2. Ciliary processes; *innervation,* ciliary; it is the muscle of visual accommodation. **B.'s probe,** a probe used in dilating strictures of the lacrimal duct. **B.'s sarcous elements,** muscle-caskets; the small elongated prisms of contractile substance that produce the appearance of dark stripes in voluntary muscle. **B.'s tubes,** artificial tubes formed between the lamellæ of the cornea by the injection of air or colored fluid.

**Bowman-Mueller's capsule.** See *Bowman's capsule*.

**box, boxwood** (*boks, boks'-wood*). See *buxus*.

**Boyer's bursa** (*bwoi-ya'*) [Alexis *Boyer*, French surgeon, 1757-1833]. The subhyoid bursa. **B.'s cyst,** cystic enlargement of the subhyoid bursa.

**Boyle's law** (*boil*) [Robert *Boyle*, English physicist, 1627-1691]. At any given temperature the volume of a given mass of gas varies inversely as the pressure it bears. Syn., *Mariotte's law*.

**Bozeman's catheter** (*bōz'-man*) [Nathan *Bozeman*, American surgeon, 1825-1905]. A double-current catheter.

**Bozzi's foramen** (*bot'-tze*). See *Soemmering's yellow spot*.

**Bozzolo's sign** (*bot-tso'-lo*) [Camillo *Bozzolo*, Italian physician, 1845– ]. Visible pulsation of the arteries of the nares, said to occur in some cases of aneurysm of the thoracic aorta.

**B. P.** or **B. Ph.** Abbreviation for *British Pharmacopœia.*

**B. P. C.** Abbreviation for British Pharmaceutical Codex.

**Br.** Chemical symbol of bromine.

**bracelets** (*brās'-lets*). Transverse lines across the anterior aspect of the wrist.

**Brachet, mesolateral fold of** (*brash-a'*). The right lamella of the primitive mesentery which passes to the dorsal aspect of the right lobe of the liver and whose free edge bounds the foramen of Winslow.

**brachia** (*bra'-ke-ah*). Plural of *brachium*.

**brachia conjunctiva.** White fibers passing from the cerebellum towards the corpora quadrigemina and the cerebral hemispheres.

**brachial** (*bra'-ke-al*) [*brachium*]. Pertaining to the arm. **b. artery,** the continuation of the axillary artery, extending along the inner side of the arm. See under *artery*. **b. glands,** the lymphatic glands of the arm. **b. plexus,** the plexus of the fifth, sixth, seventh, and eighth cervical and the first dorsal nerves. **b. veins,** the veins of the arm that accompany the brachial artery.

**brachialgia** (*bra-ke-al'-je-ah*) [*brachium;* ἄλγος, pain]. Pain or neuralgia in the arm or in the brachial plexus.

**brachialis anticus** (*bra-ke-a'-lis an'-tik-us*). See *muscles, table of*.

**brachiform** (*bra'-ke-form*) [*brachium; forma,* form]. Arm-shaped.

**brachinin** (*brak'-in-in*). A substance obtained from the bombardier-beetle, *Brachinus crepitans,* of Europe. It is said to be efficacious against rheumatism.

**brachio-** (*bra-ke-o-*) [*brachium*]. A prefix meaning pertaining to the brachium.

**brachiocephalic** (*bra-ke-o-sef-al'-ik*) [*brachio-;* κεφαλή, head]. Pertaining to the arm and the head. **b. artery, b. vein,** the innominate artery and vein.

**brachiocrural** (*bra-ke-o-kru'-ral*) [*brachio-; crus,* the leg]. Pertaining to or affecting the arm and leg.

**brachiocubital** (*bra-ke-o-ku'-bit-al*) [*brachio-; cubitus,* forearm]. Relating to the arm and forearm, as, the *brachiocubital* ligament.

**brachiocyllosis** (*bra-ke-o-sil-o'-sis*) [*brachio-;* κύλλωσις, a bending]. A crookedness of the arm; also the paralysis that may accompany it.

**brachiofacial** (*bra-ke-o-fa'-shal*). Pertaining to both arm and face.

**brachiofascialis** (bra-ke-o-fas-e-a'-lis) [brachio-; fascia, a bundle]. See under *muscles*.
**brachioncus** (bra-ke-ong'-kus) [brachio-; ὄγκος, a swelling]. Any hard and chronic swelling of the arm.
**brachioradial** (bra-ke-o-ra'-de-al) [brachio-; radius]. The supinator longus muscle. See *muscles, table of*.
**brachiorrhachidian, brachiorachidian** (brak-e-o-rak-id'-e-an) [brachio-; ῥάχις, the spinal column]. Relating to the arm and the spinal cord.
**brachiorrheuma** (bra-ke-o-ru'-mah) [brachio-; ῥεῦμα, a flux]. Rheumatism of the arm.
**brachiostrophosis** (bra-ke-o-stro-fo'-sis) [brachio-; στρέφειν, to turn]. A twist or twisted deformity of the arm.
**brachiotomy** (bra-ke-ot'-o-me) [brachio-; τομή, a section]. The surgical or obstetrical removal of an arm.
**brachiplex** (bra'-ke-pleks) [βραχίων, the arm; plexus, a twining]. The brachial plexus.
**brachistocephalic, or brachistocephalous** (bra-kis-to-sef-al'-ik, or bra-kis-to-sef'-al-us) [βράχιστος, shortness; κεφαλή, head]. Having an extremely short and very broad head.
**brachium** (bra'-ke-um) [βραχίων, the arm; pl., brachia]. The arm, especially the upper arm; also, any arm-like object or structure. **brachia cerebelli**, the peduncles of the cerebellum. **brachia cerebri, b. of optic lobes**, the bands connecting the nates and testes with the optic thalamus. **b. conjunctivum cerebelli**, same as *b. copulativum*. **b. copulativum**, the superior peduncle of the cerebellum. **b. pontis**, the brachium of the pons, being also the middle peduncle of the cerebellum. **b. quadrigeminum inferius**, the postbrachium, a white band connecting the postgeniculum to the postgeminum. **b. quadrigeminum superius**, the prebrachium, a white band between the pregeminum and the thalamus.
**Brach-Romberg's sign.** See *Romberg's sign*.
**brachy-** (brak-e-) [βραχύς, short]. A prefix meaning short.
**brachycardia** (brak-e-kar'-de-ah). Same as *bradycardia*.
**brachycephalia** (brak-e-sef-a'-le-ah) [brachy-; κεφαλή, a head]. The quality of being brachycephalic.
**brachycephalic, brachycephalous** (brak-e-sef-al'-ik, brak-e-sef'-al-us) [brachy-; κεφαλή, a head]. 1. Applied to skulls of an egg-like shape, with the larger end behind. 2. Having a skull the transverse diameter of which is more than eight-tenths of the long diameter.
**brachycephalism** (brak-e-sef'-al-izm). See *brachycephalia*.
**brachycephaly** (brak-e-sef'-a-le). See *brachycephalia*.
**brachycheirous, brachychirous** (brak-e-ki'-rus) [brachy-; χείρ, the hand]. Having short hands.
**brachycnemic, brachyknemic** (brak-e-ne'-mik) [brachy-; κνήμη, the leg]. A term applied by Sir W. Turner to a leg proportionately shorter than the thigh.
**brachydactylia** (brak-e-dak-til'-e-ah) [brachy-; δάκτυλος, a digit]. Abnormal shortness of the fingers or toes.
**brachydactylous** (brak-e-dak'-til-us) [brachy-; δάκτυλος, a finger]. Pertaining to an abnormal shortness of the fingers or toes.
**brachyglossal** (brak-e-glos'-al) [brachy-; γλῶσσα, tongue]. Having a short tongue.
**brachygnathous, brachygnathus** (brak-e-na'-thus). [brachy-; γνάθος, a jaw]. Having short jaws.
**brachyhieric** (brak-e-hi'-e-rik) [brachy-; ἱερόν, sacrum]. Having a short sacrum.
**brachykerkic** (brak-e-kerk'-ik) [brachy-; κερκίς, a shuttle]. Having the forearm disproportionately short as compared with the upper arm.
**brachymetropia** (brak-e-me-tro'-pe-ah) [brachy-; μέτρον, a measure; ὤψ, the eye]. nearsightedness or myopia. q. v.
**brachymetropic** (brak-e-me-trop'-ik) [brachy-; μέτρον, a measure; ὤψ, the eye]. Nearsighted, or myopic.
**brachynosis, brachynsis** (brak-e-no'-sis, brak-in'-sis) [brachy-; νόσος, disease]. The contraction or shortening of an organ or part by disease.
**brachyntic** (brak-in'-tik). Related to or affected with brachynosis.
**brachyotus** (brak-e-o'-tus) [brachy-; οὖς, the ear]. Short-eared.
**brachypnea** (brak-ip-ne'-ah) [brachy-; πνοίη, breath]. Abnormal shortness of breath.

**brachypodous** (brak-ip'-o-dus) [brachy-; πούς, foot]. In biology, possessing a short foot or stalk.
**brachyrrhinia** (brak-e-rin'-e-ah) [brachy-; ῥίς, a nose]. Abnormal shortness of the nose.
**brachystaphylic** (brak-e-staf'-il-ik) [brachy-; σταφυλή, the palate]. Having a short alveolar arch.
**bracket** (brak'-et) [OF., braguette]. An apparatus for supporting a joint or rendering it immovable.
**bradesthesia** (brad-es-the'-ze-ah). See *bradyesthesia*.
**brady-** (brad-e-) [βραδύς, slow]. A prefix meaning slow.
**bradyarthria** (brad-e-ar'-thre-ah) [brady-; ἄρθρον, articulation]. Abnormally slow articulation of words: bradylalia.
**bradybolism** (brad-ib'-o-lizm) [brady-; βάλλειν, to throw]. Same as *bradyspermatism*.
**bradycardia** (brad-e-kar'-de-ah) [brady-; καρδία, heart]. Slowness of the heart-beat—the opposite of tachycardia.
**bradycauma** (brad-e-kaw'-mah) [brady-; καῦμα, burn: pl., bradycaumata]. Slow cautery, as with the moxa.
**bradycausis** (brad-e-kaw'-sis) [brady-; καῦσις, a burning]. A slow burning; the application of a slow caustic.
**bradycinesia** (brad-e-sin-e'-se-ah) [brady-; κίνησις, movement]. Extreme slowness of movement.
**bradycrote** (brad'-e-krōt) [brady-; κρότος, beating]. Marked by or relating to slowness of the pulse.
**bradycrotic** (brad-e-krot'-ik) [brady-; κρότος, a beating]. Bradycrote.
**bradydiastole** (brad-e-di-as'-to-lē) [brady-; διαστολή, a drawing apart]. A prolongation of the diastolic pause; it is generally associated with myocardial lesions. Syn., *bradydiastolia*.
**bradyecoia** (brad-e-ek-oi'-ah) [brady-; ἀκούειν, to hear]. Hardness of hearing.
**bradyesthesia bradyesthesia** (brad-e-es-the'-ze-ah) [brady-; αἴσθησις, perception]. Dulness of perception.
**bradyglossia** (brad-e-glos'-e-ah). See *bradylalia*.
**bradylalia** (brad-e-la'-le-ah) [brady-; λαλεῖν, to talk]. A slowness of utterance.
**bradylogia** (brad-e-pep'-se-ah) [brady-; λόγος, discourse]. Bradylalia.
**bradypepsia** (brad-e-pep'-se-ah) [brady-; πέψις, digestion]. Slow digestion.
**bradyphagia** (brad-e-fa'-je-ah) [brady-; φάγειν, to eat]. Slowness in eating.
**bradyphasia** (brad-e-fa'-ze-ah). See *bradylalia*.
**bradyphrasia** (brad-e-fra'-ze-ah) [brady-; φράσις, utterance]. Slowness of speech; it occurs in some types of mental disease.
**bradypnea, bradypnœa** (brad-ip-ne'-ah) [brady-; πνεῖν, to breathe]. Abnormal slowness of breathing.
**bradyspermatism** (brad-e-sper'-mat-izm) [brady-; σπέρμα, seed]. Abnormally slow emission of semen.
**bradysphygmia** (brad-e-sfig'-me-ah) [brady-; σφυγμός, pulse]. Abnormal slowness of the pulse.
**bradytocia** (brad-e-to'-se-ah) [brady-; τόκος, birth]. Abnormally slow or protracted parturition.
**bradytrophic** (brad-e-trof'-ik) [brady-; τροφή, nutrition]. Characterized by slowness of trophic changes.
**bradyuria** (brad-e-u'-re-ah) [brady-; οὖρον, urine]. Slow passage of urine.
**Braidism** (brād'-izm) [James Braid, English physician, 1795-1860]. The hypnotic state produced by fixation of the eyes upon a shining object.
**brain** (brān) [AS., brægen]. That part of the central nervous system contained in the cranial cavity, and consisting of the cerebrum, the cerebellum, the pons, and the medulla oblongata. **b., abdominal**, the solar plexus. **b., after-**. See *metencephalon*. **b.-axis**, that portion of the brain-substance including the island of Reil, the basal ganglia, the crura, pons, medulla, and cerebellum. **b.-bladder**, a cerebral vesicle of the embryo. **b., end-**, see *telencephalon*. **b. fag, brain tire**. **b.-fever**. See *meningitis*. **b., fore-**. See *prosencephalon*. **b., great**, the cerebrum. **b., hind-**. See *epencephalon*. **b., mid-**. See *mesencephalon*. **b. pan**, the cranium. **b., railway-**, a condition analogous to railway spine, and characterized by cerebral disturbance. See *Erichsen's disease*. **b.-sand**. See *acervulus*. **b. stem**, the brain axis. **b. storm**, sudden and severe phenomena due to cerebral causes. **b.-tire**, a condition of brain exhaustion due to excessive functional activity. **b., 'twixt-**. See *diencephalon* and *thalamencephalon*. **b., wet**, the cerebral edema caused by alcoholism.
**bran** (bran). [Breton, brenn]. The epidermis or outer covering of the seeds of most cereals. **b.-**

**bath.** See *bath, bran.* **b. dressing,** a dressing formerly used for compound fracture of the leg. The leg was placed in a fracture-box and surrounded with clean bran.

**branch** (*branch*). 1. A name given to the divisions of offshoots of blood-vessels, lymphatics, or nerves, from the trunk or main stem. 2. A primary division of the animal kingdom.

**branchia** (*brang'-ke-ah*) [βράγχια, gills]. The gills of fishes.

**branchial** (*brang'-ke-al*). Pertaining to the branchia. **b. arches.** See *arches, branchial.* **b. cyst,** a cyst formed of embryonic structures in a branchial arch or cleft. **b. fistula,** a congenital fistula in the neck, in connection with the branchial openings. **b. openings.** See *clefts, visceral.*

**branchiogenic.** (*brang-ke-o-jen'-ik*). Same as *branchiogenous.*

**branchiogenous** (*brang-ke-oj'-en-us*) [*branchia*; γεννᾶν, to produce]. Produced or developed from a branchial cleft.

**branchioma** (*brang-ke-o'-mah*) [*branchia*; ὄμα, a tumor]. A tumor developed from remains of the branchial arches.

**branchiomere** (*brang'-ke-o-mēr*) [*branchia*; μέρος, a part]. The segment of the lateral mesoderm between each two branchial (gill) clefts.

**Brand method** [Ernst *Brand*, German physician, 1827–1897]. A system of baths employed in the treatment of typhoid fever.

**Brandt's method.** Treatment of affections of the Fallopian tubes by massage in an endeavor to force out their contents into the uterus.

**brandy** (*bran'-de*). See *Spiritus vini gallici.*

**branks** (*brangks*) [Gael., *brancas*, a kind of pillory]. The mumps.

**Brasdor's method** (*braz'-dor*) [Pierre *Brasdor*, French surgeon, 1721–1797]. Treatment of aneurysm by ligation of the artery immediately beyond the aneurysm.

**brash** (*brash*) [Dutch, *braaken*, to vomit]. A common name indicating almost any disorder of the digestive system; any rash, or eruption; a short fit of illness. **b., water.** See *pyrosis.*

**brass** (*bras*) [ME., *bras*]. An alloy of copper with 25–40 per cent. of zinc. **b.-founder's ague.** See *ague.*

**Brassica** (*bras'-ik-ah*) [AS.]. A genus of plants of the order *Cruciferæ*, including the common cabbage. **B. alba,** white mustard; the powdered seeds a condiment and rubefacient. **B. nigra,** black mustard.

**brassicon** (*bras'-ik-on*). A proprietary local application for headache; said to consist of 2 gm. oil of peppermint, 6 gm. camphor, 4 gm. ether, 12 gm. alcohol, 6 drops mustard oil.

**brassy-eye.** See *chalcitis.*

**Braun's canal** (*brown*) [Gustav von *Braun*, Austrian obstetrician, 1829–1911]. See *canal, neurenteric.* **B.'s hook.** A hook used for decapitation of the fetus.

**Braun's reaction for glucose** (*brown*) [Christopher Heinrich *Braun*, German physician, 1847– ]. Treat the glucose solution with caustic soda, and warm until it is yellow; then add a dilute solution of picric acid and heat to boiling. A deep red color will be produced. Creatinin gives the same reaction, even in the cold, and acetone also, though slightly.

**Braune's canal.** The continuous passage formed by the uterine cavity and the vagina during labor, after full dilation of the os. **B.'s os internum.** See *Bandl's ring.*

**Braun-Fernwald's sign** (*brown-fairn'-volt*) [Karl von *Braun*, Austrian obstetrician, 1823–1891]. An early sign of pregnancy consisting in an increased thickness of one-half of the body of the uterus and in the presence of a longitudinal median groove, these changes being dependent upon unequal consistence of that organ.

**Bravais-Jackson's epilepsy** (*brav'-a*). See *Jacksonian epilepsy.*

**brawn** (*brawn*) [OF., *braon*, flesh]. The flesh of a muscle; well-developed muscles.

**brawny** (*brawn'-e*) [see *brawn*]. Fleshy; muscular. **b. induration,** pathological hardening and thickening of the tissues.

**Braxton Hicks' sign.** See *Hicks.*

**braxy** (*braks'-e*). Anthrax in sheep.

**brayera** (*bra-ye'-rah*). See *cusso.*

**bread** (*bred*) [AS., *brēad*]. A mixture of flour and water rendered porous by carbon dioxide, and baked. The flour may be of wheat, corn, oat, or rye. The carbon dioxide may be introduced by fermenting the starch with yeast. **b., brown,** a kind of bread made from a mixture of corn, rye, and wheat flour. **b., Graham,** bread made from unbolted wheat flour; it contains more gluten, diastase, and mineral phosphates than ordinary bread. **b.-paste,** a culture-medium for bacteria and molds. Stale, coarse bread is dried, ground to powder, and made into a paste with water. **b.-poultice,** bread-crumbs steeped in hot water. **b., pulled;** fresh bread pulled apart longitudinally and rebaked until brittle. **b., white,** bread made from bolted wheat flour, and therefore deficient in diastase, gluten, and mineral phosphates. Other kinds, such as rye (or black), corn, bran, barley, etc., indicate their composition by their name.

**break** (*brāk*) [AS., *brecan*]. 1. In electricity, to open the circuit of a battery. 2. In surgery, a fracture. 3. To change suddenly and involuntarily from the natural voice to a shrill one or to a whisper, as with boys at puberty, or with adults under strong emotion. **b. shock,** a term sometimes employed in electrotherapeutics for the physiological shock produced on the opening or breaking of an electric circuit.

**breakbone fever.** See *dengue.*

**breast** (*brest*) [ME., *brest*]. 1. The anterior part of the chest. 2. The mamma. **b.-bone,** the sternum. **b., broken,** abscess of the mammary gland. **b., chicken-,** a deformity marked by prominence of the sternal portion of the chest. Syn., *pectus carinatum.* **b., funnel-,** a depression of the chest walls at the sternum resembling the bowl of a funnel; it is like shoemaker's breast, only it may occur at any point. Syn., *funnel-chest.* **b., gathered,** mammary abscess. **b., hysterical,** a form of mastodynia due to hysteria. **b., irritable tumor of,** a name given by Astley Cooper to peculiar, sharply defined fibromatous or neuromatous tumors of small size and extreme tenderness. **b.-pang,** angina pectoris. **b., pigeon-.** Same as *b., chicken-.* **b.-pump,** a sunction-apparatus for removing the milk from the breast. **b., shoemaker's,** a depression of the sternum in shoemakers due to the pressure of tools against it and the xiphoid cartilage.

**breath** (*breth*) [AS., *brath*]. The air exhaled from the lungs. **b.-sounds,** the respiratory sounds heard upon auscultation. See *Table* page 166.

**breathing.** See *respiration.* **b., abdominal,** breathing in which the abdominal walls move decidedly and in which the diaphragm is actively engaged. **b., Cheyne-Stokes'.** See *Cheyne-Stokes' respiration.* **b., interrupted,** **b., cog-wheel,** **b., wavy,** a broken or interrupted inspiratory sound produced by nervousness, irregular contraction of the muscles of respiration, or irregular expansion of the lung from disease. **b., mouth-,** habitual respiration through the mouth. **b., puerile,** the breathing normally heard in children, and heard in adults when the respiratory murmur is exaggerated. **b., suppressed,** entire absence of breath-sounds, as in pleuritic effusion and certain solid conditions of the lung. **b., thoracic,** respiration in which the thoracic walls actively move.

**Brecht's cartilages** (*brekt*). The ossa suprasternalia, two small cartilaginous or bony nodules near each sternoclavicular joint, above the sternum. They are regarded as the rudiments of the episternal bone that is well developed in some animals.

**breech** (*brēch*) [ME., *breech*]. The buttocks. **b., frank,** a breech presentation in which the thighs are flexed and the legs extended on the anterior surface of the fetus. **b. presentation,** presentation of the buttocks of the child at the os uteri during labor.

**breeches splint.** A splint that surrounds the leg; oftenest made of woven wire.

**breeze, electric.** See *static breeze.*

**bregenin** (*breg'-en-in*) [Low Ger., *bregen*, brain], $C_6H_8NO_5$. A name given by Thudichum to a viscous principle, soluble in and crystallizable from alcohol, by means of which it has been extracted from brain-tissue. It is fusible like a fat, but is miscible with water.

**bregma** (*breg'-mah*) [βρέγμα, the sinciput]. 1. The part of the skull corresponding to the anterior fontanel. 2. The junction of the coronal and sagittal sutures.

**bregmatic** (*breg-mat'-ik*). Relating to the bregma.

**bregmatodymia** (*breg-mat-o-dim'-e-ah*) [βρέγμα, the sinciput; δίδυμος, twin]. Teratic union of twins by the bregmata.

## TABLE OF BREATH-SOUNDS IN HEALTH AND DISEASE.
*(Altered and enlarged from J. K. Fowler.)*

| Variety of breathing. | Period. | Pitch. | Quality. | Interval. | Duration. | Intensity. | Where heard. | Condition in which heard. |
|---|---|---|---|---|---|---|---|---|
| Vesicular. | 1. Inspiration. | Low. | Vesicular. | None. | .......... | Variable. | Over the lungs in health. | In health. |
|  | 2. Expiration. | Lower. | Blowing. | ...... | Shorter than inspiration or absent. | Faint or absent. |  |  |
| Bronchial. | 1. Inspiration. | High. | Tracheal. | Distinct. | .......... | Variable. | In *health*, in regions of seventh cervical spine. In *disease*, over areas of consolidation. | In health; pulmonary tuberculosis, lobar pneumonia, large pleuritic effusions, thoracic aneurysm, mediastinal tumors. |
|  | 2. Expiration. | Higher. | Tracheal. | ...... | Equal to or longer than inspiration. | Greater. |  |  |
| Bronchovesicular. | 1. Inspiration. | Higher than in vesicular breathing. | More or less tracheal. | Slight. | .......... | Variable. | In *health*, ant., over sternal portion of infraclavicular region; *post.*, upper part of interscapular region. In *disease*, over slight consolidation. | In health; pneumonia, pulmonary tuberculosis. |
|  | 2. Expiration. | Higher than in vesicular breathing. | More or less tracheal. | ...... | About equal to inspiration. | Greater. |  |  |
| Amphoric. | 1. Inspiration. | Low. | Hollow and metallic. | Distinct. | .......... | Variable. | Over a large cavity communicating with an open bronchus. | Pulmonary tuberculosis. |
|  | 2. Expiration. | Lower. | Both characters more marked. | ...... | Longer than inspiration. | Greater. |  |  |
| Cavernous. | 1. Inspiration. | Low. | Blowing and hollow. | Distinct. | .......... | Variable. | Over a cavity communicating with an open bronchus. | Pulmonary tuberculosis. |
|  | 2. Expiration. | Lower. | Both characters more marked. | ...... | Longer than inspiration. | Greater. |  |  |
| Tubular. | 1. Inspiration. | Higher than in bronchial breathing. | Laryngeal or whiffing. | Distinct. | .......... | Variable. | Over consolidated areas. | Lobar pneumonia, pulmonary tuberculosis. |
|  | 2. Expiration. | Higher. | Laryngeal or whiffing. | ...... | Equal to or longer than inspiration. | Greater. |  |  |

**brein** (brē'-in). A glucoside isolated from *Bryonia alba*. It is a powerful stimulant to the arterioles and useful in the treatment of post-partum hemorrhage and other forms of metrorrhagia.

**Breisky's disease** (brī'-ske) [August *Breisky*, German gynecologist, 1832–1889]. Kraurosis vulvæ. **B.'s method**, of measuring the dimensions of the pelvis at its outlet: Measure externally the distance between the tuberosities of the ischia, and the distance from the junction of the sacrum and coccyx to the lower border of the arcuate ligament.

**Bremer's reaction for diabetic blood** (brem'-ur) [John Lewis *Bremer*, American physician, 1874– ]. The blood is prepared as in ordinary staining methods, and, after drying in a hot-air sterilizer, stained with methylene-blue and eosin. The erythrocytes of diabetic blood are stained greenish-yellow, whereas in normal blood they appear brownish.

**Brenner's formula** (bren'-nur) [Rudolf *Brenner*, German physician, 1821–1884]. 1. With the same current strength the kathodal closing contraction is four times as strong as the kathodal opening contraction. 2. The normal auditory nerve reacts to the kathodal closure by a sound sensation which immediately attains its maximum and then gradually diminishes; the anodal opening causes with the same current strength a somewhat weaker sound that is of short duration.

**brenzcain** (brenz'-ka-in). See *guaiacolbenzyl ester*.

**brenzkatechinuria** (brenz-kat-e-kin-u'-re-ah). See *alkaptonuria*.

**brephopolysarcia** (bref-o-pol-e-sar'-ke-ah) [βρέφος, an infant; πολύς, much; σάρξ, flesh]. Excess of flesh in an infant.

**brephotrophium** (bref-o-tro'-fe-um) [βρέφος, infant; τρέφειν, to nourish]. An infant-asylum; a foundling-hospital.

**brephydrocephalus** (bref-id-ro-sef'-al-us) [βρέφος, an infant; *hydrocephalus*]. Hydrocephalus in infants.

**Breschet's canal, B.'s veins** (bresh'-a) [Gilbert *Breschet*, French anatomist, 1784–1845]. The canals and veins of the diploe. **B.'s helicotrema**, the helicotrema, the foramen of communication between the scala vestibuli and the scala tympani. **B.'s sinus**, the sphenoparietal sinus.

**Bretonneau's diphtheria** (bret'-on-o) Pierre *Bretonneau*, French physician, 1778–1862]. Diphtheria of the pharynx.

**breviductor** (brev-e-duk'-tor) [L.]. The adductor brevis muscle of the thigh. See *muscles, table of*.

**breviflexor** (brev-e-fleks'-or) [L.]. Any short, flexor muscle.

**brevissimus oculi** (brev-is'-im-us ok'-u-li) L.]. The shortest muscle of the eye; the obliquus inferior. See *muscles, table of*.

**Brewer's operation** (*broo'-er*) [George Emerson *Brewer*, American Surgeon, 1861– ]. Closure of wounds of arteries with rubber plates. **B.'s point**, the costo-vertebral angle, which is tender in cases of infection of the kidney.
**Brewster's law** (*broo'-ster*). The polarizing angle has such a value that the reflected and the refracted rays are at right angles to each other.
**brick-dust deposit.** A reddish sediment in the urine, consisting of urates.
**bricklayer's itch.** Eczema due to irritation of lime-mortar.
**brickmakers' anemia.** See *ankylostomiasis, dochmiasis*.
**bridge** (*brij*). 1. The upper ridge of the nose formed by the union of the two nasal bones. 2. In electricity, an apparatus for measuring the resistance of a conductor. **b. coloboma.** See *coloboma*. **b., herpetic,** a term for *fascicular keratitis, q. v.* **b., intercellular,** slender protoplasmic processes connecting proximate cells. Syn., *internuclear bundles*. **b., jugal.** See *arch, zygomatic*. **b. of nose,** the prominent ridge formed by the nasal bones. **b. of Varolius,** the *pons Varolii*. **b.-work,** in dentistry, the adaptation of artificial crowns of teeth to and over spaces made by the loss of natural teeth, by connecting such crowns to natural teeth or roots for anchorage by means of a bridge, and thereby dispensing with plates covering more or less of the roof of the mouth and the alveolar ridge.
**bridle** (*brī'-dl*) [AS., *brīdel*]. 1. A band or filament stretching across the lumen of a passage, or from side to side of an ulcer, scar, abscess, etc. 2. A frenum. **b. stricture,** a stricture due to the presence of a delicate band stretched across the lumen of the urethra.
**Brieger, bacillus of** (*bre'-ger*) [Ludwig *Brieger*, German physician, 1849– ]. *Bacillus cavicidus*. **B.'s method,** a method of separating ptomaines from a putrefying mass.
**Bright's blindness** (*brīt*) [Richard *Bright*, English physician, 1789–1858]. Partial or complete loss of sight, which may be temporary, independent of any change in the optic disc or retina; it occurs in uremia. **B.'s disease,** a generic term for acute and chronic diffuse disease of the kidneys, usually associated with dropsy and albuminuria. **B.'s disease, acute,** an acute inflammation of the kidney; it may be parenchymatous, interstitial, or diffuse. **B.'s disease, chronic,** a chronic inflammation of the kidney, affecting the parenchyma, the connective tissue, or both. Amyloid degeneration is also considered a chronic form of Bright's disease. **B.'s friction-sound.** See *Beatty-Bright's friction-sound*. **B.'s granulations,** the granulations of the large white kidney.
**brightic** (*brīt'-ik*). A person suffering from Bright's disease.
**brightism** (*brīt'-izm*). Chronic nephritis.
**Brill's disease** [Nathan E. *Brill*, American physician, 1860– ]. An acute infectious disease of unknown origin, very similar to a mild form of typhus. It is shorter than typhoid, lasting only 12 to 14 days, and is characterized by a short incubation, intense headache, apathy, prostration, a maculo-papular eruption not disappearing on pressure, and a fall of temperature by crisis or rapid lysis. The Widal test is negative. Prognosis is good; the treatment, symptomatic.
**brim** (*brim*) [ME.]. The upper edge or margin, as the *brim* of the pelvis.
**brimstone** (*brim'-stōn*). See *sulphur*. **b., cane,** sublimed sulphur molded into the form of solid cylinders about an inch in diameter. Syn., *roll-sulphur*. **b., vegetable,** the spores of *Lycopodium clavatum*.
**Brinton's disease** (*brin'-tun*) [William *Brinton*, English physician, 1823–1867]. 1. Linitis plastica; hypertrophy and sclerosis of the submucous connective tissue of the stomach. 2. Infantile scurvy.
**Briquet's ataxia** (*bre-kā'*) [Paul *Briquet*, French physician, 1796–1881]. Hysteric ataxia; astasia-abasia.
**brisement** (*brēz-mon(g)*) [Fr.]. A breaking or rupture. **b. forcé,** the forcible breaking up of structures causing ankylosis of a joint.
**brise-pierre** (*brēz'-pe-ār'*) Fr., "stone-crusher"]. An old form of lithotrite.
**Brissaud-Marie's syndrome** (*bre-so-mar-e'*) [Edouard *Brissaud*, French physician, 1852–1909; Pierre *Marie*, French neurologist, 1853– ]. Hysterical glossolabial hemispasm.
**bristle-cell.** Any one of the ciliated cells at the termination of the auditory nerve-filaments.
**British gum.** See *dextrin*.
**broach** (*brōtch*). A five-sided steel instrument used by dentists for enlarging the canal in the root, and the opening into a decayed cavity in the crown of a tooth.
**broad.** Wide; extensive. **b. ligament.** See *ligament, broad*. **b. tapeworm.** See *Bothriocephalus latus*.
**Broadbent's sign** (*brawd'-bent*) Sir William Henry *Broadbent*, English physician, 1835–1907]. A visible retraction, synchronous with the cardiac systole, of the left side and back in the region of the eleventh and twelfth ribs, in adherent pericardium.
**Broca's aphasia** (*bro'-kah*) [Paul *Broca*, French surgeon and anthropologist, 1824–1880]. Cortical motor aphasia. **B.'s area,** the medial portion of the anterior, olfactory lobe. Syn., *area parolfactoria*; *gyrus olfactorius medialis*. **B.'s cape,** the dividing-point of the fossa Sylvii. **B.'s center,** the posterior part of the left third frontal convolution; it is the center of speech. **B.'s convolution,** the third frontal convolution of the left hemisphere. **B.'s diagonal band,** a band of gray matter forming the posterior part of the anterior perforated space and extending from the gyrus subcallosus to the anterior end of the gyrus hippocampi. **B.'s fissure,** the fissure surrounding Broca's lobe. **B.'s olfactory area,** trigonum olfactorium; the posterior end of the *gyrus rectus*, lying anteriorly to the mesial root of the olfactory tract. **B.'s parietal angle,** in craniometry, that included between two lines joining the auricular point and the bregma and lambda. **B.'s point,** the auricular point, the center of the external auditory meatus. **B.'s pouch,** a pear-shaped sac lying in the tissues of the labia majora; it is analogous in structure to the dartos, but contains no muscular fibers. **B.'s visual plane,** a plane drawn through the axes of the two orbits.
**Brodie's abscess** (*bro'-de*) [Sir Benjamin Collins *Brodie*, English surgeon, 1783–1862]. Chronic abscess of bone, most frequently of the head of the tibia. **B.'s disease,** pulpy disease of a joint, more especially the knee-joint; also spinal neuralgia following trauma, and often hysterical. **B.'s joint,** hysterical arthroneuralgia. **B.'s pain,** the pain caused by lifting a fold of the skin in the neighborhood of a joint in articular neuralgia.
**Broesike's fossa** (*bre'-zik-eh*) [Gustav *Broesike*, German anatomist, 1853– ]. The parajejunal fossa; a recess in the peritoneal cavity which is situated in the first part of the mesojejunum and behind the superior mesenteric artery.
**Brokaw ring** (*brok'-aw*). A ring used in intestinal anastomosis, made of segments of rubber drainage-tubing and threaded with catgut strands.
**brom-, bromo-.** A prefix denoting the presence of bromine in a substance.
**bromal** (*bro'-mal*) [*brom-*; *aldehyde*], $CBr_3.CHO$. Tribromaldehyde, analogous to chloral, and produced by the action of bromine on alcohol. It is a colorless, oily fluid, of a penetrating odor and sharp, burning taste, boiling at 172°–173° C.; it has been used in medicine, having properties similar to those of chloral. **b. hydrate,** $CBr_3.CHO+H_2O$, a fluid of oily consistence, having a structure similar to that of chloral hydrate, but more irritating and narcotic than the latter. It is used as a hypnotic and in epilepsy. Dose 1–5 gr. (0.065–0.32 Gm.).
**bromalbacid** (*bro-mal'-bas-id*). A compound of bromine and albumin used as a sedative. Dose 15–30 gr. (1–2 Gm.).
**bromalbumin** (*bro-mal'-bū-min*). See *bromoalbumin*.
**bromaldehyde** (*bro-mal'-de-hīd*). A compound of bromine and aldehyde.
**bromalin** (*bro'-mal-in*), $C_6H_{12}N_4C_2H_5Br$. A substance occurring as a white, crystalline powder, soluble in water, melting at 200° C. It is a nervesedative and antiepileptic, used as a substitute for potassium bromide. Dose 30–60 gr. (2–4 Gm.) several times a day. Syn., *bromethylformin; hexamethylenetetraminbromethylate*.
**bromaloin** (*bro-mal'-o-in*), $C_{17}H_{15}Br_3O_7$. A derivative of barbaloin by the action of bromine. Syn., *tribromaloin*.
**bromamide** (*bro'-mam-id*) [*brom-; amide*]. A bro-

mine compound of the anilin group, with the formula $C_8H_5Br_2NH \cdot HBr$. It contains 75 % of bromine, and is used as an antipyretic in doses of 10–15 grains (0.65–1.0 Gm.).
**bromargyrite** (*bro-mar'-jir-īt*). Native silver bromide.
**bromate** (*bro'-māt*) [*bromin*]. A salt of bromic acid.
**bromatecerisis** (*bro-mat-ek'-ris-is*) [βρῶμα, food; ἔκκρισις, excrement]. The passage of undigested food.
**bromated** (*bro'-ma-ted*). Impregnated with bromine.
**bromatography** (*bro-mat-og'-ra-fe*) [βρῶμα, food; γραφή, a writing]. A description of or treatise on foods.
**bromatology** (*bro-mat-ol'-o-je*) [βρῶμα, food; λόγος, a science]. The science of foods.
**bromatometer** (*bro-mat-om'-et-ur*) [βρῶμα, food; μέτρον, measure]. An instrument used in bromatometry.
**bromatometry** (*bro-mat-om'-et-re*). The estimation of the daily amount of food requisite for an individual.
**bromatotherapy** (*bro-mat-o-ther'-ap-e*) [βρῶμα, food; θεραπεύειν, to heal]. Treatment of diseased conditions by regulation of the diet.
**bromatotoxicon** (*bro-mat-o-toks'-ik-on*). A general term for the active agent in food-poisoning.
**bromatotoxin** (*bro-mat-o-toks'-in*) [βρῶμα, food; τοξικόν, poison]. A basic poison generated in food by the growth of microorganisms.
**bromatotoxism** (*bro-mat-o-toks'-izm*) [βρῶμα, food; τοξικόν, poison]. Poisoning with infected food.
**brombenzoyl** (*brom-ben'-zo-il*), $C_7H_5O \cdot Br$. A crystalline substance obtained from oil of bitter almonds by action of bromine; it is soluble in alcohol and ether. Syn., *brombenzoylic acid*.
**bromcaffeine** (*brom-kaf-e'-in*), $C_8H_9BrN_4O_2$. A compound obtained by mixing 1 part of caffeine with 5 parts of bromine; melts at 206° C.
**bromeigon** (*bro-mi'-gon*). An insoluble compound of bromine and a proteid.
**bromelin** (*bro'-mel-in*) [*Bromelia*, a genus of plants]. A digestive principle, allied to trypsin, found in the juice of pineapples. It will digest 1500 times its weight of proteids.
**bromethyl** (*brōm'-eth-il*). See *ethyl bromide* under *bromine*. **b. formin.** See *bromalin*.
**brometone** (*bro'-met-ōn*) [*brom-*; *acetone*]. A bromine derivative of acetone, analogous to chloretone. It is used as a nerve sedative instead of the bromides.
**bromglidin** (*brom-gli'-din*). A proprietary organic bromine compound: used as a nerve sedative.
**bromhemol** (*brōm'-he-mol*). A compound of hemol and 2.7 % of bromine. It is used when continued effect of bromine is desired. Dose 15–30 gr. (1–2 Gm.). Syn., *bromohemol*.
**bromhidrosis.** See *bromidrosis*.
**bromhydrate** (*brōm-hi'-drāt*). See *hydrobromate*.
**bromhydric** (*brōm-hi'-drik*). See *hydrobromic*. **b. ether,** hydrobromic ether.
**bromic** (*bro'-mik*). Containing or compounded with bromine.
**bromide, bromid** (*bro'-mīd, bro'-mid*). A salt of hydrobromic acid; the bromides of calcium, iron, ammonium, potassium, and sodium are used in medicine. They allay nervous excitement and are employed as sedatives. **b., basic,** a compound of a bromide with the oxide of the same base. **b. of ethyl.** See *ethyl bromide* under *bromine*.
**bromidia** (*bro-mid'-e-ah*). A proprietary hypnotic and anodyne.
**bromidrosiphobia** (*bro-mid-ros-e-fo'-be-ah*), [βρῶμος, a stench; ἱδρώς, sweat; φόβος, fear]. Insane dread of offensive personal smells, with hallucinations as to the perception of them.
**bromidrosis** (*bro-mid-ro'-sis*) [βρῶμος, a stench; ἱδρώς, sweat]. Osmidrosis; an affection of the sweat-glands in which the sweat has an offensive odor.
**bromidum** (*bro'-mid-um*). Bromide; a salt of bromine.
**brominated, brominized** (*bro'-min-a-ted, -īzd*). Combined with bromine.
**bromine, bromum** (*bro'-men, -mum*) [βρῶμος, a stench]. Br = 79.92 quantivalence I. A reddish-brown liquid which, at ordinary temperatures, gives off a heavy, suffocating vapor. It is a very active escharotic and disinfectant and internally a violent poison. The salts of bromine are cerebrospinal and cardiac depressants, and are employed as sedatives, particularly in epilepsy, eclampsia, spasmodic affections, insomnia, hysteria, migraine, etc. The salts of the alkaline metals are those most commonly used. **bromidum ammonii** (U. S. P.), $NH_4Br$. Dose 5–20 gr. (0.32–1.3 Gm.). **b. blocks,** porous blocks of diatomaceous earth incinerated with calcium saccharate and impregnated with three times their weight of bromine, which is gradually given off by them. They are used as disinfectants. **bromidum calcii** (U. S. P.), $CaBr_2$. Dose 5 gr.–1 dr. (0.32–4.0 Gm.). **b. chloride,** BrCl (below 10° C.), a reddish-yellow, mobile, very volatile liquid. It is used as an internal and external caustic in cancer. **bromide, ethyl,** $C_2H_5Br$, useful in spasmodic coughs. Dose 10 min.–1 dr. (0.65–4.0 Cc.). **bromidi, ferri, syrupus,** contains 10 % of the salt. Dose ½–1 dr. (2–4 Cc.). **b. iodide,** $IBr_3$, a dark-brown liquid, soluble in water; it is used as a gargle in diphtheria, in 0.1 % solution. **bromidum lithii** (U. S. P.), LiBr. Dose 5–20 gr. (0.32–1.3 Gm.). **bromidum, nickel.** See *nickel*. **b. pentachloride,** $BrCl_5$, a caustic liquid. **bromidum potassii** (U. S. P.), KBr. Dose 5 gr.–1 dr. (0.32–4.0 Gm.). **bromidum sodii** (U. S. P.), NaBr. Dose 5 gr.–1 dr. (0.32–4.0 Gm.). **bromidum strontii** (U. S. P.), $SrBr_2$. Dose 15–20 gr. (1.0–1.3 Gm.). **bromidum, zinci** (U. S. P.), $ZnBr_2$. Dose ½–2 gr. (0.032–0.13 Gm.).
**brominism.** See *bromism*.
**brominol** (*bro'-min-ol*). A solution of bromine in sesame oil.
**bromiodide** (*bro-mi'-od-īd*). A compound formed from the bromide and the iodide of the same base.
**bromiodoform** (*bro-mi-o'-do-form*), $CHBr_2I$. A substitution-compound of bromine and iodoform.
**bromipin** (*bro'-mip-in*). A liquid compound of bromine and sesame oil containing 10 % of bromine. It is used as a sedative in epilepsy. Dose 1–3 teaspoonfuls daily.
**bromism, brominism** (*bro'-mizm, bro'-min-izm*) [*bromine*]. Certain peculiar phenomena produced by the prolonged administration of the bromides. The most marked symptoms are headache, coldness of the extremities, feebleness of the heart's action, somnolence, apathy, anesthesia of the soft palate and pharynx, pallor of the skin, and a peculiar eruption of acne which is one of the earliest and most constant symptoms. There is also anorexia, and at times there are loss of sexual power and atrophy of the testes or mammæ.
**bromite** (*bro'-mīt*). 1. Native silver bromide. 2. A salt of bromous acid.
**bromium** (*bro'-me-um*). Bromine.
**bromoalbumin** (*bro-mo-al'-bū-min*). A compound of bromine (10 %) and albumin; it is used in epilepsy. Syn., *bromalbumin; bromosin*.
**bromocaffeine** (*bro-mo-kaf'-e-in*) [*bromine; caffeine*]. A proprietary effervescing preparation containing caffeine. It is used for the relief of headaches.
**bromocamphor** (*bro-mo-kam'-for*). See *camphor, monobromated*.
**bromochloralum** (*bro-mo-klo-ral'-um*) [*bromo-; chloral*]. A proprietary antiseptic and disinfecting compound, containing the bromine and chloride of aluminum.
**bromocoll** (*bro'-mo-kol*) [*bromo-; κόλλα*, glue]. A product of the condensation of bromine, tannin, and gelatin; a light-brown, odorless, almost tasteless powder, containing 20 % of bromine, soluble in alcoholic liquids. It is indicated when other bromides are not well borne. Dose 15–75 gr. (1–5 Gm.) a day; in epilepsy, 123 gr. (8 Gm.). Syn., *dibromotannic glue*.
**bromoform** (*bro'-mo-form*), $CHBr_3$. A bromide having a structure like that of chloroform, $CHCl_3$; it is sedative and anesthetic and is used in whooping-cough and in seasickness in doses of 2–5 min. (0.13–0.32 Cc.). Syn., *formobromide; formylbromide; methenyl tribromide; tribrommethane*.
**bromoformin** (*bro-mo-for'-min*). Bromethylate of hexamethylene tetramine; it is used as a sedative.
**bromoformism** (*bro-mo-form'-izm*). Poisoning with bromoform.
**bromography** (*bro-mog'-ra-fe*). Same as *bromatography*.
**bromohematin** (*bro-mo-hem'-at-in*). Hematin hydrobromide.
**bromohemol.** See *bromhemol*.
**bromohydrate.** See *hydrobromate*.

**bromohydric.** See *hydrobromic*.
**bromohyperidrosis** (*bro-mo-hi-per-id-ro'-sis*) [*bromo-*; ὑπέρ, over; ἱδρωσις, a perspiring]. A condition marked by excessive and offensive perspiration.
**bromoiodism** (*bro-mo-i'-o-dism*) [*bromism; iodism*]. Poisoning by bromine and iodine compounds together.
**bromol** (*bro'-mol*), $C_6H_2Br_3OH$. Tribromophenol; an antiseptic substance used in the form of a powder, solution (1 : 30 olive oil), or ointment (1 : 10). Internally it is used in cholera infantum and typhoid fever. Dose $\frac{1}{10}-\frac{1}{2}$ gr. (0.006–0.02 Gm.).
**bromolein** (*brom-o'-le-in*). A combination of bromine (20 per cent.) and unsaturated oil of almonds.
**bromolithia** (*bro-mo-lith'-e-ah*). A proprietary remedy for gout.
**bromomania** (*bro-mo-ma'-ne-ah*). Insanity from excessive use of bromides.
**bromomenorrhea** (*bro-mo-men-or-e'-ah*) [βρῶμος, stench; μήν, month; ῥεῖν, to flow]. Disordered menstruation marked by offensiveness of the flow.
**bromophenol** (*bro-mo-fe'-nol*). 1. See *bromol*. 2. $C_6H_4BrOH$. A violet-colored liquid obtained from phenol by action of bromine. It is used in a 1 to 2 % ointment in treatment of erysipelas. Syn., *orthobromphenol*.
**bromopnea, bromopnœa** (*brom-op-ne'-ah*) [βρῶμος, stench; πνοία, breath]. Fetid breath.
**bromopropylene** (*bro-mo-pro'-pil-ēn*). See *allyl bromide*.
**bromopyrine** (*bro-mo-pi'-rin*). 1. $C_{11}H_{11}BrN_2O$, a substance used as is antipyrine, occurring in white needles, soluble in alcohol, chloroform, and hot water, melting at 114° C. Dose 5–15 gr. (0.3–1.0 Gm.). Syn., *monobromoantipyrin*. 2. A proprietary mixture said to consist of antipyrine, caffeine, and sodium bromides.
**bromoseltzer** (*bro-mo-selt'-zer*). A proprietary headache remedy.
**bromoserum** (*bro-mo-se'-rum*). A solution of 4 parts of sodium bromide and 1.5 parts of sodium chloride in 1000 parts of water. It is used by injection as a substitute for bromides.
**bromosin** (*bro'-mo-sin*). See *bromoalbumin*.
**bromosoda** (*bro-mo-so'-da*). A proprietary remedy for dyspepsia.
**bromous** (*bro'-mus*). Containing bromine united with oxygen in the same proportion as in the chlorous compounds.
**bromphenols** (*brom-fe'-nols*). A series of bromated phenols occurring at times in the precipitates of tested urine.
**bromum** (*bro'-mum*). See *bromine*.
**bromural** (*bro'-mū-ral*). Monobrom-iso-valerylurea; a white crystalline powder, soluble in hot water; it contains about 35 per cent. of bromine, and is used as a hypnotic. Dose gr. 5–10 (0.3–0.6).
**bromurated** (*bro'-mū-ra-ted*). Containing bromine or a bromine salt.
**bronchadenitis** (*brongk-ad-en-i'-tis*) [*broncho-*; ἀδήν, gland; ιτις, inflammation]. Inflammation of the bronchial lymphatic glands.
**broncheopyra** (*brong-ke-o-pi'-rah*) [*bronchus;* πῦρ, fire]. A suffocative cough.
**bronchi** (*brong-ki*) [L.]. The plural of *bronchus*. The two tubes into which the trachea divides opposite the third dorsal vertebra, called respectively the right and the left bronchus.
**bronchia** (*brong'-ke-ah*) [βρόγχος, the windpipe]. The bronchial tubes, especially those that are smaller than the two bronchi.
**bronchiadenoscirrhus** (*brong-ke-ad-en-o-skir'-us*) [*bronchus;* ἀδήν, a gland; σκιρρός, hard]. Scirrhus of the bronchial glands.
**bronchial** (*brong'-ke-al*) [*bronchus*]. Relating to the bronchi. b. **arteries.** See under *artery*. b. **crises,** dyspneic paroxysms occurring in locomotor ataxia. b. **fluke.** See *Distoma Ringeri*. b. **glands,** the chain of lymphatic glands running beside the bronchi. b. **tube,** a bronchus, or one of its subdivisions.
**bronchiarctia** (*brong-ke-ark'-she-ah*). See *bronchostenosis*.
**bronchiectasis** (*brong-ke-ek'-tas-is*) [*bronchus;* ἐκτασις, dilatation]. Dilatation of the walls of the bronchi. It occurs in chronic bronchitis, in fibroid pneumonia, and in tuberculosis of the lung. It may involve a tube uniformly, producing the *cylindrical* form; or it may occur irregularly in sacs or pockets—the *sacculated* form. The characteristic symptom of bronchiectasis is paroxysmal coughing, with the expectoration of large quantities of mucopurulent, often fetid, matter. Cavernous breathing may be heard over the dilated tubes.
**bronchiectatic** (*brong-ke-ek-tat'-ik*). Pertaining to or affected with bronchiectasis.
**bronchiloquy** (*brong-kil'-o-kwe*) [*bronchus; loqui,* to speak]. Bronchophony.
**bronchiocele** (*brong'-ke-o-sēl*) [*bronchiolus,* a little air passage; κήλη, tumor]. A swelling or dilatation of a bronchiole.
**bronchiocrisis** (*brong-ke-o-kri'-sis*) [*bronchus; crisis*]. Paroxysmal coughing in tabes dorsalis.
**bronchiole** (*brong'-ke-ōl*) [dim. of *bronchus*]. One of the smallest subdivisions of the bronchi.
**bronchiolectasis** (*brong-ke-o-lek'-ta-sis*) [*bronchiole;* ἐκτασις, dilation]. A rare form of bronchiectasis diffused to all parts of the lung, making it appear as if riddled with small cavities.
**bronchioli** (*brong'-ke-o-li*). Bronchioles.
**bronchiolitis** (*brong-ke-o-li'-tis*) [*bronchiole,* ιτις, inflammation]. Inflammation of the bronchioles. b., **asthmatic.** See *b., exudative*. b., **exudative,** b. **exudativa,** an inflammation of the bronchioles, with exudation, a condition by some held to be the cause of bronchial asthma. b. **fibrosa obliterans,** b. **obliterans,** b., **obliterating fibrous,** bronchiolitis resulting in obliteration of the finest bronchioles by connective-tissue plugs.
**bronchiolus** (*brong-ki-o'-lus*) [dim. of *bronchus,* pl. *bronchioli*]. A bronchiole.
**bronchiospasmus** (*brong-ke-o-spaz'-mus*) [*bronchus; spasm*]. Spasm of the bronchi.
**bronchiostenosis** (*brong-ke-o-ste-no'-sis*) [*bronchus;* στένω, narrow]. Contraction of a bronchus or of any one or more of the bronchial tubes.
**bronchismus** (*brong-kis'-mus*). Suffocative bronchial spasm due to spinal paralysis.
**bronchitic** (*brong-kit'-ik*) [*bronchitis*]. Relating to, of the nature of, or affected with, bronchitis.
**bronchitis** (*brong-ki'-tis*) [*bronchus;* ιτις, inflammation]. Inflammation of the mucous membrane of the bronchial tubes. Syn., *pleuritis bronchialis*. b., **acute,** is due to exposure to cold, to the inhalation of irritant vapors, to certain infectious agents, etc. It is characterized by fever, cough, substernal pain, and by dry rales in the early, and moist rales in the later, stages. b., **capillary,** an acute bronchitis of the finer bronchioles; it is generally the result of a downward extension of an acute bronchitis. Children are most frequently affected. Dyspnea, nervous depression, and cyanosis are prominent symptoms. Catarrhal pneumonia is a common complication. b., **catarrhal,** a form attended with profuse mucopurulent discharges. b., **chronic,** a form of bronchitis usually occurring in middle or advanced life, characterized by cough and by dry and moist rales. It may be due to repeated attacks of acute bronchitis, to gout, rheumatism, or tuberculosis, or it may be secondary to cardiac and renal disease. b. **convulsiva,** whooping cough. b., **croupous,** b., **fibrinous,** b., **plastic,** a rare variety attended with the expectoration of casts of the bronchial tubes, containing Charcot-Leyden crystals and eosinophile cells, after a paroxysm of dyspnea and violent coughing. b., **dry,** that unattended by expectoration. b., **mechanical,** a form caused by the inhalation of dust, etc. b., **phthinoid,** a consumptive form with purulent sputum. b. **potter's.** Same as *b., mechanical*. b., **putrid,** b., **fetid,** a variety of chronic bronchitis characterized by the discharge of a copious, half-liquid, extremely offensive sputum. b., **secondary,** one which develops as a complication of some preceding disease. b., **suffocative,** b. **suffocans,** acute capillary bronchitis. b., **summer,** hay-fever.
**bronchium** (*brong'-ke-um*) [L.; *pl*. *bronchia*]. A bronchial tube.
**bronchlemmitis** (*brong-klem-i'-tis*) [*bronchus;* λέμμα, a skin]. Croupous bronchitis.
**broncho-** (*brong-ko-*) [*bronchus*]. A prefix meaning relating to the bronchi.
**bronchoægophony** (*brong-ko-e-gof'-o-ne*). See *bronchoegophony*.
**bronchoalveolar** (*brong-ko-al-ve'-o-lar*). Same as *bronchovesicular*.
**bronchocavernous** (*brong-ko-kav'-er-nus*). Both bronchial and cavernous; it is applied to respiration.
**bronchocele** (*brong'-ko-sēl* [*broncho-*; κήλη, a tumor]. Really a tumor of a bronchus, but generally signifying goiter. b., **aerial.** See *aerocele*.

**bronchocephalitis** (*brong-ko-sef-al-i'-iis*). Synónym of *whooping-cough.*
**bronchoconstriction** (*brong-ko-kon-strik'-shun*). The narrowing of the caliber of the pulmonary air-passages.
**bronchoconstrictor** (*brong-ko-kon-strik'-tor*). Constricting the caliber of the air-passages of the lungs.
**bronchodilator** (*brong-ko-di-la'-tor*). Dilating the caliber of the air-passages of the lungs.
**bronchoegophony** (*brong-ko-e-goff'-o-ne*). [*broncho-;* αἴξ, a goat; φωνή, a voice, sound]. Bronchophony combined with egophony.
**bronchoesophagoscopy** (*brong-ko-e-sof-ag-os'-ko-pe*) [*broncho-; esophagus;* σκοπεῖν, to view]. Inspection of the interior of the bronchi and esophagus.
**bronchohemorrhagia** (*brong-ko-hem-or-a'-je-ah*). Extravasation of blood from the lining membrane of the bronchial tubes.
**broncholemmitis** (*brong-ko-lem-i'-tis*). See *bronchlemmitis.*
**broncholith, broncholite** (*brong'-ko-lith, brong'-ko-lit*) [*broncho-;* λίθος, a stone]. A calculus or concretion formed in a bronchial tube.
**broncholithiasis** (*brong-ko-lith-i'-a-sis*). [*broncho-; lithiasis*]. The formation of calculi in the bronchial apparatus.
**bronchomotor** (*brong-ko-mo'-tor*). Affecting the caliber of the bronchial apparatus.
**bronchomycosis** (*brong-ko-mi-ko'-sis*) [*broncho-; mycosis*]. The growth or presence of fungi in a bronchial tube.
**bronchopathy** (*brong-kop'-a-the*) [*broncho-;* πάθος, disease]. Any disease of a bronchus.
**bronchophonic** (*brong-ko-fon'-ik*) [*broncho-;* φωνή, a voice]. Relating to bronchophony.
**bronchophony** (*brong-kof'-o-ne*) [*broncho-;* φωνή, the voice]. The resonance of the voice within the bronchi as heard on auscultating the chest. It is normally present over the lower cervical spines, in the upper interscapular region, and over the sternal portion of the infraclavicular regions. The most frequent pathological cause is consolidation of the lung. **b., accidental,** that due to disease. **b., pectoriloquous.** See *pectoriloquy.* **b., whispered,** bronchophony elicited by causing the patient to whisper.
**bronchophthisis** (*brong-ko-thi'-sis*) [*broncho-;* φθίσις, a wasting]. Pulmonary tuberculosis characterized by extensive lesions of the bronchial tubes.
**bronchophyma** (*brong-ko-fi'-mah*) [*broncho-;* φῦμα, a growth]. Any growth, as a tubercle, in a bronchial tube.
**bronchoplasty** (*brong'-ko-plas-te*) [*broncho-;* πλάσσειν, to form]. The closure of a tracheal or bronchial fistula by operation.
**bronchoplegia** (*brong-ko-ple'-je-ah*) [*broncho-;* πληγή, a blow]. Paralysis of the bronchial tubes.
**bronchopleurisy** (*brong-ko-plu'-ris-e*). Bronchitis existing with pleurisy.
**bronchopleuropneumonia** (*brong-ko-plu-ro-nū-mo'-ne-ah*). Coexistent bronchitis, pleurisy, and pneumonia.
**bronchopneumonia** (*brong-ko-nū-mo'-ne-ah*) [*broncho-; pneumonia*]. Lobular pneumonia, a term applied to inflammation of the lungs, which, beginning in the bronchi, finally involves the parenchyma of the lungs. This disease is most frequently encountered in children, but may occur in old age, and may be a simple catarrhal or a tuberculous process. Syn., *bronchiopneumonia; bronchoalveolitis; bronchopneumonitis; catarrhal pneumonia; microbronchitis.*
**bronchopneumonitis** (*brong-ko-nū-mon-i'-tis*). Bronchopneumonia.
**bronchopulmonary** (*brong-ko-pul'-mon-a-re*). Relating to the bronchi and lungs.
**bronchorrhagia** (*brong-kor-a'-je-ah*) [*broncho-;* ῥηγνύναι, to burst forth]. Hemorrhage from the bronchial tubes.
**bronchorrhea, bronchorrhœa** (*brong-kor-e'-ah*) [*broncho-;* ῥεῖν, to flow]. A form of bronchitis attended with profuse expectoration. Syn., *blennorrhagia pulmonum; bronchoblennorrhea.* **b., serous,** a form in which the sputum is serous.
**bronchorrhoncus** (*brong-kor-ong'-kus*). A bronchial rale.
**bronchoscope** (*brong'-ko-skōp*) [*broncho-;* σκοπεῖν, to look]. An instrument employed for the direct inspection of the interior of a bronchus.
**bronchoscopy** (*brong-kos'-ko-pe*). Inspection of the interior of the bronchial tubes.

**bronchospasm** (*brong'-ko-spazm*) [*broncho-;* σπασμί]. Bronchial spasm.
**bronchostenosis** (*brong-ko-ste-no'-sis*) [*broncho-;* στενός, narrow]. Contraction of a bronchus.
**bronchotetany** (*brong-ko-tet'-an-e*) [*broncho-; tetany*]. A condition characterized by extreme dyspnea caused by spasm of the muscles in the bronchi preventing the access of air to the lungs.
**bronchotome** (*brong'-ko-tōm*) [*broncho-;* τέμνειν, to cut]. An instrument for cutting the trachea or a bronchus in the operation of bronchotomy.
**bronchotomy** (*brong-kot'-o-me*) [see *bronchotome*]. Incision into the trachea, or bronchus.
**bronchotracheal** (*brong-ko-tra'-ke-al*) [*broncho-;* τραχεία, the windpipe]. Relating to a bronchus (or to both bronchi) and to the trachea.
**bronchovesicular** (*brong-ko-ves-ik'-ū-lar*) [*broncho-; vesicula,* a vesicle]. Both bronchial and vesicular. See *breath-sounds, table of.*
**bronchus** (*brong'-kus*) [βρόγχος, bronchus; pl., *bronchi*]. One of the primary divisions of the trachea. **b., eparterial,** the first branch of the right bronchus situated above the right pulmonary artery. **b., hyparterial,** the left bronchus and the remaining branches of the right bronchus, situated below the pulmonary artery.
**bronzed** (*bronzd*). Tanned; of a bronzed color. **b. skin,** a symptom of Addison's disease. **b.-skin disease.** See *Addison's disease.*
**brood-cells,** in cell-division, the mother-cells inclosing the daughter-cells.
**broom.** See *Scoparius.*
**brossage** (*bro-sazh*) [Fr. "brushing"]. The removing of granulations with a stiff brush, as in trachoma.
**Brossard's type of progressive muscular atrophy** (*bros-ar'*). "Type fémoral avec griffe des orteils" (femoral type with a claw-like appearance of the toes). See *Eichhorst's type.*
**broth,** See *bouillon.*
**broussaisism** (*broo-sa-izm*) [François Joseph Victor *Broussais,* French physician, 1772–1838]. The opinion that gastro-intestinal irritation is the prime cause of disease.
**brow** [AS., *brū*]. The forehead; the superciliary ridge; the eyebrow; the upper anterior portion of the head. **b.-ache, b.-ague,** neuralgia of the first division of the fifth cranial nerve, generally due to malaria. **b.-pang.** Synonym of *hemicrania.* **b. presentation,** presentation of the fetal brow in labor.
**brown** (*broun*) [AS., *brūn*]. Having a dark color inclining toward red or yellow, **b. atrophy,** an atrophy of a tissue associated with a deposit therein of a brown or yellow pigment. **b., Bismarck-.** See *Bismarck-brown.* **b. induration of lung,** a state of the lung due to long-continued congestion, usually arising from valvular heart disease. It is characterized by an increase in connective tissue and an excess of pigment. **b. mixture,** mistura glycyrrhizæ composita. **b. ointment,** the unguentum fuscum (N. F.); called also unguentum matris, or "mother's salve." It is composed of "brown plaster," 2 parts; oil, 1 part; suet, 1 part. **b. plaster,** the emplastrum fuscum camphoratum (N. F.); called also emplastrum matris camphoratum, or "camphorated mother's plaster"; official in German pharmacy. It is made of red oxide of lead, 30 parts; olive-oil, 60 parts; yellow wax, 15 parts; camphor, 1 part.
**Brown's movement.** See *Brownian movement.*
**Browne's (Crichton) sign.** Tremor of the labial commissures and outer angles of the eyes in the early stage of paralytic dementia.
**Brownian movement** [Robert *Brown,* English botanist, 1773–1858]. An oscillatory movement observed under the microscope in very fine granules, drops, etc., when suspended in a liquid. The movement is not locomotion, and is to be distinguished from that of the self-motility of living microorganisms. Its cause is not definitely known, but it may be due to heat, light, electricity, osmosis, etc. Same as *pedesis.*
**Brownism** (*broun-ism*). See *brunonian theory.*
**Brown-Séquard's paralysis** (*sa-kar'*) [Charles Edouard *Brown-Séquard,* French physiologist and neurologist, 1817–1897]. Paralysis and hyperesthesia of one side and anesthesia of the other side of the body.
**Bruce and Muir, septomarginal tract of.** A part of the descending posteromedial tract of the spinal cord.

**Brucea** (brū-se'-ah) [*Bruce*, the Abyssinian explorer (1730-1794)]. A genus of plants of the order *Simarubeæ*. *B. ferruginea* is an Abyssinian species; the bark and root are used in dysentery. *B. sumatrana* is a species of the Asiatic tropics and of Australia; all parts of the plant are bitter, tonic, febrifuge, vermifuge, and antidysenteric.

**Bruch's glands** (brook) [Carl Wilhelm Ludwig *Bruch*, German anatomist, 1819-1884]. Lymph-follicles found in the conjunctiva about the inner canthus of ruminants. **B.'s layer, B.'s membrane,** the lamina basalis which forms the inner boundary of the choroid.

**brucine** (brū'-sēn) [*Brucea*], $C_{23}H_{26}N_2O_4$. A poisonous alkaloid found in *Strychnos nux-vomica* and in *Strychnos ignatia*. It crystallizes in prisms containing $4H_2O$, and melts at $178°$. Its taste is exceedingly bitter and acrid. Its action on the animal economy is similar to, but much less powerful than, that of strychnine. Dose $\frac{1}{12}-\frac{1}{3}$ gr. (0.005-0.03 Gm.); maximum dose $\frac{3}{4}$ gr. (0.05 Gm.), single; 3 gr. (0.2 Gm.) a day. Antidotes: chloral, chloroform, tannic acid. Syn., *brucia; brucinum; brucium; pseudoangustin; vomicine*. **b. hydrobromide,** $C_{23}H_{26}N_2O_4$.-HBr, a substitute for strychnine in ophthalmic surgery; it is 40 times less poisonous. **b. hydrochloride,** $C_{23}H_{26}N_2O_4$.HCl, small white crystals, soluble in water; used as is brucine. **b. nitrate,** $C_{23}H_{26}N_2O_4$.$HNO_3+3H_2O_4$, white crystalline powder, soluble in water; used as brucine. **b. phosphate,** ($C_{23}H_{26}N_2O_4$)$.H_3PO_4$, white, crystalline powder, soluble in water, use and dose the same as those of brucine. **b. sulphate,** ($C_{23}H_{26}N_2O_4$)$_2H_2SO_4+3\frac{1}{2}H_2O$, white, microscopic crystals, soluble in water and alcohol; use and dosage the same as those of brucine.

**Bruck's disease.** A syndrome described by Bruck as consisting of multiple fractures and marked deformity of bones, ankylosis of most of the joints, and muscular atrophy.

**Brudzinski's signs** (brood-zin'-ske). 1. In meningitis, if the neck of the patient is bent forwards, flexion occurs at the hip, knee and ankle. 2. In meningitis, when forcible flexion of the lower limb on one side is made, flexion or extension will be observed in the opposite limb; this is also called *contralateral reflex*.

**Bruecke's lines** (brook'-eh) [Ernst Wilhelm *Bruecke*, Austrian physiologist, 1819-1892]. The broad bands which alternate with Krause's membranes in the fibrils of striated muscles. **B.'s muscle.** 1. See *Bowman's muscle*. 2. The muscularis mucosæ of the small and large intestine. **B.'s reagent for proteids,** saturate a boiling 10 % solution of potassium iodide with freshly precipitated mercuric iodide. Filter when cool; the filtrate is used with hydrochloric acid as a precipitant for the proteids. **B.'s tunica nervea,** the layers of the retina, exclusive of the rods and cones.

**Bruenninghausen's method** (brū'-ning-how-zen) [Hermann Joseph *Bruenninghausen*, German physician, 1761-1834]. The induction of premature labor by dilating the cervix uteri.

**Bruggiser's hernia.** See *Kroenlein's hernia*.

**bruise** (brooz). See *contusion*.

**bruissement** (brū-ēs-mon(g)) [Fr.]. A purring sound heard on auscultation.

**bruit** (brū-e) [Fr., "a noise or report"]. An adventitious sound heard on auscultation. For kinds— *amphoric, rotatory,* etc.—see *murmur*. **b., aneurysmal,** the blowing murmur heard over an aneurysm. **b. d'airain,** the ringing note heard through the stethoscope applied to the chest-wall when a coin is struck against another pressed against the surface of the chest on the opposite side. It is pathognomonic of a collection of gas in the pleural cavity. **b. de clapôtement,** a splashing sound often heard in cases of well-marked dilatation of the stomach when pressure is made upon the abdominal walls. **b. de craquement, b. de cuir neuf,** the creaking sound, like that of new leather, sometimes heard in pericarditis. **b. de diable,** a humming, rushing sound heard in the veins in anemia. **b. de drapeau,** a rustling murmur heard in croup and laryngitis. **b. de froissement,** a clashing sound of the lungs or heart. **b. de galop,** a cantering rhythm of the heart-sounds, in which, owing to a reduplication of the second sound, three sounds are heard. It occurs most frequently in mitral stenosis. **b. de lime,** a cardiac sound as if made by a file or saw. **b. de moulin,** the water-wheel sound. **b. de pot fêlé,** the cracked-pot sound. **b. de rappel,** a sound resembling the double beat upon a drum. **b. de scie.** Same as *b. de lime*. **b. de soufflet,** the bellows-murmur. **b., placental,** the uterine souffle, a blowing sound heard over the uterus in pregnancy.

**Brun's law** (broon). Same as *Bastian's law, q. v.*

**Brun's test for uric acid in minute particles.** Examine the particle in naphthalin monobromide, the index of refraction of which is 1.66. Compare this with indexes of refraction of uric acid, 1.73; calcium oxalate, 1.60; calcium phosphate, 1.63. If on raising the tube of the microscope the crystal becomes brilliant, the substance under the microscope has a higher index than the fluid in which it is immersed; it becomes darker if the substance has a lower index than the fluid.

**brunet,** or **brunette** (brū-net') [Fr.]. 1. Of a dark complexion. 2. One with a dark complexion.

**Brunfelsia** (brun-fel'-se-ah) [O. *Brunfels*, a botanist of Metz, 1464-1534]. A genus of plants of the order *Saponaceæ*. A syrup made from the fruit of *B. americana,* a West Indian species, is used as a tonic in recovery from diarrhea. *B. uniflora,* of Brazil, is purgative, emetic, and emmenagogue. Syn., *mercurio vegetal*.

**Brunn's cell-nests, B.'s epithelial nests** (broon) [Albert von *Brunn*, German anatomist, 1849-1895]. Branched or solid groups of flat epithelial cells occurring in all normal ureters. **B.'s glands.** See *Brunner's glands*. **B.'s layer, B.'s membrane,** the stratum of more or less pyramidal epithelial cells forming the deep layer of the nasal mucous membrane.

**Brunner's glands** (broo'-ner) [Johann Conrad *Brunner,* Swiss anatomist, 1653-1727]. The racemose glands found in the wall of the duodenum.

**Brunonian movement.** See *Brownian movement*.

**B. theory,** Brownism; a doctrine, taught by Dr. John Brown (1735-88), that both physiological and pathological phenomena are due to variations in a natural stimulus, its excess causing sthenic, and its deficiency producing asthenic, diseases.

**brush** [OF., *broce*]. An instrument consisting of a collection of some flexible material fastened to a handle. In medicine various forms of brushes are employed, as the *acid brush,* of glass threads; the *electric brush,* an electrode in the form of a brush; the *laryngeal brush;* the *nasal, pharyngeal,* and *stomach brush*. **b.-burn,** the injury produced by violent friction and the resulting heat; it often resembles a burn or scald. **b., terminal.** See *motorial end-plate*.

**Bryant's ampulla** (brī'-ant) [Sir Thomas *Bryant*, English surgeon]. The apparent distention of an artery immediately above a ligature, due to the contraction of the vessel above the ampulla, where it is not completely filled by the clot. **B.'s iliofemoral triangle,** the rectangle formed by a vertical line dropped from the anterior superior iliac spine to the horizontal plane of the body; by a second line drawn from the anterior superior iliac spine to the tip of the trochanter, and by a third, the "test-line," which joins the two at a right angle to the vertical line. Shortening of the neck of the femur will be indicated by a shortening of the test-line. **B.'s line,** the vertical line forming one of the boundaries of the iliofemoral triangle. **B.'s operation,** for *lumbar colotomy;* an oblique incision is made midway between the last rib and the iliac crest. The bowel is fixed in position and opened.

**Bryce's test** [James *Bryce,* Scotch physician]. For vaccinal infection: the inoculation is repeated at a certain period in the evolution of vaccinia, upon the theory that systemic infection does not take place at once, but only after the lapse of a number of days from the time of inoculation.

**brygmus** (brig'-mus) [βρυγμός, biting]. Same as *odontoprisis*.

**bryology** (bri-ol'-o-je) [βρύον, a moss; λόγος, a science]. The science of mosses.

**bryonia** (bri-o'-ne-ah) [βρυωνία, bryony]. Bryony. The root of *B. alba* and *B. dioica,* indigenous to Europe. Its properties are due to an intensely bitter glucoside, *bryonin,* $C_{28}H_{36}O_{13}$, which is a strong irritant when applied to the skin or mucous membrane, often producing vesication. It is used in pleurisy, pleuropneumonia, rheumatic fever, and colds. Dose of the *powdered root* 10-30 gr. (0.65-2.0 Gm.); of the *infusion* (1 : 16) $\frac{1}{2}$-2 oz. (15-60 Cc.); of the *tincture,* a 10 % solution of the root in alcohol, 1-2 dr. (4-8 Cc.).

**bryonidin** (*bri-on'-id-in*). A glucoside isolated from *Bryonia alba*, more active than bryonin.
**bryonin** (*bri'-o-nin*). See under *bryonia*.
**bryony** (*bri'-o-nē*). See *bryonia*.
**bryoplastic** (*bri-o-plas'-tik*) [βρύον, moss; πλάσσειν, to form]. A descriptive term loosely applied to such abnormal growths of tissue as resemble vegetable forms.
**Bryson's sign** [Alexander *Bryson*, English physician, 1802–1869]. Diminished power of expansion of the thorax during inspiration; occasionally observed in exophthalmic goiter and in neurasthenia.
**B. S.** Abbreviation for (1) Bachelor of Science and (2) Bachelor of Surgery.
**B. Sc.** Abbreviation for Bachelor of Science.
**bubo** (*bū'-bo*) [βουβών, the groin]. Inflammation and swelling of a lymphatic gland, particularly of the groin, and usually following chancroid, gonorrhea, or syphilitic infection. Syn., *adēn; inguinal adenitis; sympathetic abscess*. **b., absorption**. See **b., virulent**. **b., indolent**, one with enlargement and hyperplasia without the formation of pus or any tendency to break down. Syn., *adenitis e blennorrhœa; adenitis e sclerosi*. **b., parotid**. See *parotitis*. **b., pestilential**, that associated with the plague. **b., primary**, a slight adenitis of the groin due to mechanical irritation or other cause; formerly supposed to be due to syphilis without a chancre having preceded. **b., rheumatic**, a hard lump, occurring oftenest on the back of the neck, as a sequel of acute articular rheumatism. **b., serpiginous**, an ulcerated bubo which changes its seat or in which the ulceration extends in one direction while healing in another. **b., simple**. See **b., sympathetic**. **b., strumous**, hypertrophied glands forming a large indolent swelling in a scrofulous subject. **b., sympathetic**, one caused by irritation, friction, injury, etc., and not arising from an infectious disease. **b., syphilitic**, that which appears in syphilis a few days after the primary lesion. It runs a slow course of six months or more. Syn., *primitive syphilitic adenitis*. **b., venereal**, that due to venereal disease. **b., virulent**, an ulcerated, suppurating bubo due to absorption of the virus of a chancre. Syn., *chancrous adenitis*.
**bubon d'emblée** [Fr.]. See *bubo, primary*.
**bubonadenitis** (*bū-bon-ad-en-i'-tis*) [*bubo*; ἀδήν, a gland; ιτις, inflammation]. Inflammation of an inguinal gland.
**bubonalgia** (*bū-bon-al'-je-ah*) [*bubo*; ἄλγος, pain]. Pain in the inguinal region.
**bubonic** (*bū-bon'-ik*) [see *bubo*]. Relating to a bubo. **b. plague**. Synonym of *plague*.
**bubonocele** (*bū-bon'-o-sēl*) [βουβών, the groin; κήλη, hernia]. Inguinal hernia when the gut does not extend beyond the inguinal canal.
**bubononcus** (*bū-bon-ong'-kus*) [*bubo*; ὄγκος, a tumor]. A swelling in the groin.
**bubonopanus** (*bū-bon-o-pa'-nus*) [*bubo*; πᾶνος, torch]. An inguinal bubo.
**bubonulus** (*bū-bon'-ū-lus*) [L., dim. of *bubo*]. Lymphangitis of the dorsum of the penis, often with abscesses; due to chancroidal virus.
**bubophthalmia** (*bū-bof-thal'-me-ah*). See *keratoglobus*.
**bucardia** (*bū-kar'-de-ah*) [βοῦς, ox; καρδία, heart]. Cor bovinum; see *bovine heart*.
**bucca** (*buk'-ah*) [L. gen. and pl., *buccæ*]. 1. The cheek; the hollow of the cheek, or its inner surface. 2. The mouth.
**buccal** (*buk'-al*) [*bucca*, the cheek]. Pertaining to the cheek or mouth. **b. coitus**, sexual perversion where gratification is found by mouth.
**buccellation** (*buk-sel-a'-shun*) [*bucella*, a morsel]. Hemostasis by a lint compress.
**buccilingual** (*buk-se-ling'-wal*) [*bucca*; *lingua*, the tongue]. Relating to the cheek and the tongue.
**buccinatolabialis** (*buk-sin-at-o-la-be-a'-lis*). The buccinator and orbicularis oris regarded as one.
**buccinator** (*buk'-sin-a-tor*) [L., "a trumpeter"]. The thin, flat muscle of the cheek. See *muscles, table of*.
**buccobranchial** (*buk-o-brang'-ke-al*). Relating to the mouth and the branchial cavity.
**buccocervical** (*buk-o-ser'-vik-al*) [*bucca*, cheek; *cervix*, neck]. Pertaining to the cheek and the neck.
**buccogingival** (*buk-o-jin'-jiv-al*). Pertaining to the cheek and the gums.
**buccolabial** (*buk-o-la'-be-al*) [*bucca*, cheek; *labium*, lip]. Pertaining to the cheek and the lip.

**buccolingually** (*buk-o-lin'-gwal-e*). From the cheek toward the tongue.
**buccopharyngeal** (*buk-o-far-in'-je-al*) [*bucca*, the mouth; *pharynx*]. Relating to the mouth and to the pharynx.
**buccula** (*buk'-ū-lah*) [L., dim. of *bucca*]. The fleshy fold seen beneath the chin, and forming what is called a double chin.
**Buchner's albuminoid bodies** (*bookh'-ner*). Defensive proteids. **B.'s humoral theory of immunity**. This supposes that a reactive change has been brought about in the integral cells of the body by the primary affection from which there has been recovery, and this change is protective against similar invasions of the same organism.
**buchu** (*bū'-kū*) [native African]. The leaves of several species of *Barosma*, yielding a volatile oil, to which its properties are probably due, and a bitter extractive, barosmin. Dose 1–5 gr. (0.065–0.32 Gm.). It causes a sensation of glowing warmth over the body, stimulates the appetite, and increases the circulation. It is useful in cystitis and other affections of the genitourinary mucous membrane. Dose of the *leaves* 15–30 gr. (1–2 Gm.). **b., fluidextract of** (*fluidextractum buchu*, U. S. P.). Dose 10 min.–1 dr. (0.65–4.0 Cc.). **b., infusion of** (*infusum buchu*, B. P.). Dose 1–4 oz. (30–120 Cc.). **b. resin**, barosma. **b., tincture of** (*tinctura buchu*, B. P.). Dose 1–2 dr. (4–8 Cc.).
**Buchwald's atrophy** (*buk-volt*). Idiopathic, diffuse, progressive atrophy of the skin.
**Buck's extension** [Gurdon *Buck*, American surgeon, 1807–1877]. An apparatus consisting of a weight and pulley for applying extension to a limb. **B.'s fascia**, the sheath of the corpora cavernosa and the corpus spongiosum, which arises from the symphysis pubis by the suspensory ligament of the penis and is continuous with the deep layer of the superficial perineal fascia.
**buck-bean** (*buk'-bēn*). Bog-bean. The rhizome of *Menyanthes trifoliata*, tonic, antiscorbutic, and emmenagogue. It has been recommended as a vermifuge, and has been used in functional amenorrhea. Dose of *fluidextract* 5–30 min. (0.32–2.0 Cc.).
**bucket fever**. *Dengue, q. v.*
**buckeye bark** (*buk'-i bark*). The bark of *Æsculus glabra*, astringent and tonic, used in rectal irritation, prolapse, and various uterine derangements. Dose of *fluidextract* 3–5 min. (0.2–0.32 Cc.).
**buckthorn** (*buk'-thorn*). See *frangula* and *rhamnus*.
**b., California**. See *cascara sagrada*.
**bucnemia** (*buk-ne'-me-ah*) [βοῦς, an ox; κνήμη, the leg]. Inflammation of the leg, characterized by tenseness and swelling; elephantiasis; also phlegmasia dolens. **b. tropica**, elephantiasis arabum.
**Budd's cirrhosis** [William *Budd*, English physician, 1811–1880]. Hepatic cirrhosis due to autointoxication from the gastrointestinal tract. **B.'s jaundice**. See *Rokitansky's disease*.
**Budde's method** (*bood'-deh*) [G. *Budde*, Danish chemist]. A method of preserving milk. To a quart of fresh milk 12 c.c. of a 3 per cent. solution of peroxide of hydrogen is added; the mixture is heated to 124° F., for a few hours, and is then rapidly cooled, when it is supposed to be sterilized.
**buddeise** (*bood'-de-īz*). To treat by Budde's method.
**budding** (*bud'-ing*) [ME., *budden*]. In biology, a form of reproduction or cell-division, occurring among the polyps and infusorians, in which a bud is given off by the parent and comes to resemble the latter. The process is also called *gemmation*.
**Budge's center**. 1. The ciliospinal center in the cervical spinal cord. 2. The genitospinal center in the lumbar spinal cord.
**Buehlmann's fibers** (*bēl-man*). Certain lines on decayed teeth.
**Buetschli's nuclear spindle** (*bet'-she-lē*). The spindle-shaped figure observed during karyokinesis.
**buffers' consumption** (*buf'-erz con-sump'-shun*). The phthisis occurring among metal-polishers.
**buffy coat** (*buf'-e*). A grayish or buff-colored crust or layer sometimes seen upon a blood-clot, after phlebotomy, and once looked upon as a sign of inflammation. It is caused by the partial subsidence of the red blood-corpuscles.
**bufidine** (*bū'-fid-ēn*) [*bufo*, the toad]. Phrynine, a poisonous alkaloid from the venom of the toad.
**bufotalin** (*bū-fo'-tal-in*), $C_{19}H_{17}O_{38}$. A toxic substance isolated by Phisalix and Bertrand from

the parotid gland and skin of the common toad, *Bufo vulgaris;* it is a transparent resin, soluble in chloroform, alcohol, and acetone. It acts on the heart and does not affect the nervous centers.

**bufotenin** (*bū-fo'-ten-in*). A toxic body found with bufotalin (*q. v.*); it exerts a powerfully paralyzing action on the nervous centers.

**bugantia** (*bū-gan'-she-ah*) [L.]. A chilblain.

**buggery** (*bug'-er-e*) [OF., *bougre*, an heretic]. Sodomy; bestiality.

**bugleweed** (*bū'-gl-wēd*). The herb, *Lycopus virginicus*, narcotic and astringent. Dose of fldext. ʒ ss–ij. *Lycopin*, concentrated extract. Dose gr. j–iv.

**Buhl's desquamative pneumonia** (*bool*) [Ludwig von *Buhl*, German physician, 1816–1880]. Caseous pneumonia, in which the exudate consists chiefly of desquamated alveolar epithelium. B.'s **disease**, acute fatty degeneration of the viscera of the newborn, with hemorrhages in various parts of the body.

**Buhl-Dittrich's law.** In every case of acute general miliary tuberculosis an old focus of caseation is to be found somewhere in the body. This law, being based upon the belief in the etiologic nonidentity of diffuse caseous and miliary tuberculosis, is not in conformity with modern views.

**bukardia** (*bū-kar'-de-ah*) [βουκαρδία, ox-heart]. Hypertrophy of the heart.

**Bulam**, or **Bulama boil** (*bū'-lam* or *bū-lam'-ah*). A boil occurring on the African island of Bulam; probably caused by a burrowing-worm or insect-larva. B. **fever**, a West-Africa coast fever, said to be identical with yellow fever.

**bulb** [L., *bulbus*]. 1. An oval or circular expansion of a cylinder or tube. 2. The medulla oblongata. b., **arterial**, the anterior part of the embryonic heart from the division of which the aortic and pulmonary stems have their origin. b., **brachial**, b., **brachiorhachidian**, the expansion of the spinal cord at the place of distribution of the nerves forming the brachial plexus. b. **of the corpus cavernosum**, the swelling at the junction of the corpora cavernosa. b., **crural**, the dilatation of the spinal cord in the lumbar region. b., **dental**, dental papilla. b. of **the eye**, the eyeball. – b.s, **four**, the corpora quadrigemina. b., **gustatory**. See *taste-bulbs*. b., **hair**-, the swelling at the root of a hair. b., **nerve**-. See *end-bud* and *motorial end-plate*. b., **olfactory**, one of the two bulbs of the olfactory nerve situated on each side of the longitudinal fissure upon the under surface of each anterior lobe of the cerebrum. b., **rhachidian**, the oblongata. b. **of spinal cord**, b. **of spinal marrow**, the oblongata. b.s, **tonsillar**, the lobules of the cerebellum. Syn., *bulbi tonsillares*. b. **of urethra**, the posterior expanded part of the corpus spongiosum. b. **of vagina**, a small body of erectile tissue on each side of the vestibule of the vagina, homologous to the bulb of the urethra of the male. b. **of vena jugularis**, the dilatation at the termination of the external jugular vein.

**bulbar** (*bul'-bar*). Bulbous. Pertaining to the medulla. b. **disease**, b. **paralysis**, a term applied to the progressive and symmetrical paralysis of the muscles of the mouth, tongue, pharynx, and sometimes those of the larynx. This paralysis is due to a disease of the motor nuclei in the medulla oblongata; an acute and a chronic form are met. The acute form is due to hemorrhage or softening; the chronic, to degeneration. There is also a pseudobulbar paralysis, due to symmetrical lesions of the motor cerebral cortex. Syn., *labioglossolaryngeal paralysis*.

**bulbi vestibuli** (*bul'-bī ves-tib'-ū-lī*) [L.]. A name sometimes given to the glands of Bartholin.

**bulbocavernosus** (*bul-bo-kav-ern-o'-sus*) [*bulb; caverna*, a cavern]. The accelerator urinæ muscle, corresponding to the sphincter vaginæ of the female.

**bulbonuclear** (*bul-bo-nū'-kle-ar*). Relating to the oblongata and its nerve-nuclei.

**bulbopetal** (*bul-bop'-et-al*) [*bulb; petere*, to seek]. Moving toward the bulb; said of nerve-impulses.

**bulbourethral** (*bul-bo-ū-re'-thral*) [*bulb; oὐρήθρα*, the urethra]. Relating to the bulb of the urethra.

**bulbous** [*bulb*]. Terminating in a bulb.

**bulbus** (*bul'-bus*) [L.]. A bulb. b. **arteriosus**, at a certain stage in the development of the heart the upper aortic enlargement is so called. b. **cinereus**. See *bulb, olfactory*. b. **crinis**, a hair-bulb. b. **oculi**, the globe of the eye. b. **pili**. Same as b. *crinis*. b. **rhachidicus**, the oblongata. b. **urethræ**. See *urethra*. b. **venæ jugularis internæ inferior**, an enlargement of the jugular vein immediately above its union with the subclavian vein. b. **venæ jugularis internæ superior**, an enlargement of the internal jugular vein at the point of exit from the jugular foramen. b. **vestibuli**, the bulb of the vagina; see *bulb of vagina*.

**bulesis** (*bū-le'-sis*) [βούλησις, the will]. The will, or an act of the will.

**bulimia** (*bū-lim'-e-ah*) [βοῦς, an ox; λιμός, hunger]. Excessive, morbic hunger; it sometimes occurs in idiots and insane persons, and it is also a symptom of diabetes mellitus and of certain cerebral lesions. Syn., *bulimiasis; bulimy*.

**bulimic** (*bū-lim'-ik*). See *bulimia*.

**bulla** (*bul'-ah*) [L., "a bubble"]. A bleb or blister, consisting of a portion of the epidermis detached from the skin by the infiltration beneath it of watery fluid, the result of a liquefaction-necrosis. b. **dolentissima**, a small, very painful cutaneous ulcer, which persists for a long time. b. **ethmoidalis**, a rounded projection into the middle meatus of the nose, due to an enlarged ethmoid cell. b. **a frigore**, a blister from the effect of cold. **bullæ gangrænosæ**, those occurring in moist gangrene of the skin. b. **ossea**, the inflated or dilated part of the bony external meatus of the ear.

**bullate** (*bul'-āt*) [*bulla*, a bubble]. 1. Inflated; fornicated and with thin walls; blistered; marked by bullæ. 2. Of bacterial cultures, a growth rising in convex prominences, like a blistered surface.

**bullation** (*bul-a'-shun*) [*bullare*, to bubble]. 1. Inflation. 2. Division into small compartments.

**bull-dog forceps**. Forceps with strong teeth and a clasp to prevent slipping.

**Buller's shield**. A watch-glass in a frame of adhesive plaster or rubber worn in front of the sound eye to protect it from an infected eye.

**bullet forceps**. For extracting bullets.

**bulletin** (*bul'-let-in*) [*bulla*, a seal]. A brief, official statement of a patient's condition.

**bullet-probe**. An instrument for locating bullets.

**bullous** (*bul'-us*) [*bulla*]. Marked by bullæ; of the nature of a bulla.

**bundle** (*bun'-dl*) [AS., *bindan*, to bind]. In biology, a fascicular grouping of elementary tissues, as nerve-fibers or muscle-fibers. b., **aberrant**, a band of nerve-fibers in the isthmus of the gyrus fornicatus. b., **anterior marginal**, the ventral part of the descending cerebellar tract in the spinal cord. b., **atrioventricular**, the bundle of His. b., **ground**, **anterior**, the anterior root zone in the white matter of the spinal cord. b., **ground, lateral**, a part of the lateral column of the cord extending from the ventral to the dorsal horn. b., **hemispheral**, the posterior one of the two bundles composing the anterior commissure. It originates in the pyramidal cells of the temporal lobe and amygdaloid nucleus, passes through the external capsule and lenticula, unites with the mesial part of the commissure at the point of the decussation of its fibers, and radiates to the opposite temporal lobe. b., **longitudinal**, a bundle of fibers outside of the optic radiation, passing from the occipital to the temporal lobe. b., **marginal**, a small fasciculus at the end of the dorsal cornu of the spinal cord. b., **primitive**, b., **Schwann's primitive**, a muscular fiber. b., **respiratory**, the solitary fasciculus. b., **solitary**, b., **trineural**. See under *solitary*. See also *Helweg, His, Krause, Lenhossék, Meynert, Spitzka, Stilling, Turck, Vicq d'Azyr, Weissmann*.

**Bunge's law.** The epithelial cells of the mammary gland (of the rabbit, cat, and dog) select from the mineral salts of the blood-plasma all the inorganic substances exactly in the proportion in which they are necessary for the development of the offspring and for the building-up of the latter's organism.

**bunioid** (*boo'-ne-oid*) [βουνός, a hill; εἶδος, likeness]. Having a round form; applied to tumors.

**bunion** (*bun'-yun*) [origin uncertain]. A swelling of a bursa of the foot, especially of the metatarso-phalangeal joint of the great toe.

**bunodont** (*bū'-no-dont*) [βουνός, a hill, mound; ὀδούς, tooth]. Pertaining to tuberculate molar teeth.

**bunogaster** (*bū-no-gas'-ter*) [βουνός, a little hill; γαστήρ, stomach]. A protruding abdomen.

**Bunsen burner** [Robert Wilhelm *Bunsen*, German chemist and physicist, 1811–1899]. A form of burner in which, before ignition, the gas is mixed with a sufficient quantity of air to produce complete oxidation. B. **cell**. In this the positive element is zinc; the negative element, carbon; exciting agent, dilute sulphuric acid; depolarizing agent, nitric acid; E.M.F., 1.75-1.96 volts.

**Buphane** (*bu'-fan-e*) [L.]. A genus of plants of the order *Amaryllidea*. *B. disticha* is a native of the Cape of Good Hope; the juice of the bulb is used as an arrow-poison by the Hottentots.

**buphthalmia, buphthalmos** (*būf-thal'-me-ah, -mos*). See *keratoglobus*.

**bur, burr** [ME., *burre*]. 1. In botany, a rough, prickly shell or case. 2. The lobe of the ear. 3. In dentistry, an instrument with a rounded, pointed, cylindrical or ovoid head and a cutting blade, used in the dental engine for excavating carious dentine and for other purposes. **b., surgical,** an instrument similar in form to a dental bur, but larger, designed for surgical operations upon the bones.

**Burckhardt's corpuscles,** peculiar angular or roundish bodies of a yellowish color found in the secretion of trachoma.

**burcquism** (*boork'-izm*). See *metallotherapy*.

**Burdach's column** (*boor'-dakh*) [Karl Friedrich *Burdach*, German physiologist, 1776-1847]. The posteroexternal column of the spinal cord. **B.'s fissure,** a small fissure between the insula and the operculum. **B.'s nucleus,** the cuneate nucleus, a small nucleus of gray matter in the funiculus cuneatus of the oblongata, forming the termination of the long fibers of Burdach's column. **B.'s operculum.** See *Arnold's operculum*.

**burdock** (*ber'-dok*). See *Lappa*.

**burette, buret** (*bu-ret'*) [Fr.]. A graduated tube designed for measuring small quantities of a reagent. It is usually held vertically in a stand and is provided with a stopcock.

**Burgundy pitch.** See under *pix*.

**burking** [*Burke*, a noted criminal]. Suffocation produced by a combination of pressure on the chest with closure of the mouth and nostrils. This was the method employed by Burke.

**Burma boil.** A form of endemic ulcer common in Burma. **B. head,** a disease of the Burmese territory marked by loss of memory, idiocy, homicidal mania, and inability to walk.

**burn.** 1. To become inflamed. 2. To char or scorch. 3. To have the sensation of heat. 4. An injury caused by fire or dry heat. 5. A disease in vegetables. 6. In chemistry, to oxygenize. 7. In surgery, to cauterize. **b., brush-.** See *brush-burn*.

**Burnam's test for formaldehyde** (*bur'-nam*) [Curtis Field *Burnam*, American surgeon]. Same as Rimini's test, q. v.

**burner** [see *burn*]. A common name for a lamp or heating apparatus used in laboratories for chemical and pharmaceutical purposes. See *Argand, Bunsen*.

**Burnett's disinfecting fluid** (*bur-net'*) [Sir William *Burnett*, English surgeon, 1779-1861]. A strong solution of zinc chloride with a little iron chloride.

**burning** (*burn'-ing*) [ME., *bernen*, to burn]. Consuming with heat or fire. **b. bush,** euonymus. **b. glass,** a convex lens or concave mirror causing a sufficient concentration of the sun's rays to ignite an object placed at the focus. **b. of the feet,** a neurotic affection of the soles of the feet, common in India. **b. oil,** kerosene. **b. point,** in testing petroleum oils, the temperature at which a spark or lighted jet will ignite the liquid itself, which then continues to burn. This point is usually 6° to 20° C. higher than the flash-point, but there is no fixed relation between the two.

**Burns' amaurosis** [John *Burns*, Scotch physician, 1774-1850]. Postmarital amblyopia. Impaired vision caused by sexual excess. **B.'s ligament** [Alan *Burns*, Scotch anatomist, 1781-1813]. See *Hey's ligament*.

**Burow's solution** (*boo'-ro*) [Karl August *Burow*, German surgeon, 1809-1874]. A solution consisting of alum, 5 parts; lead acetate, 25 parts; in 500 parts of water; used to wash old ulcers. **B.'s vein.** An inconstant venous trunk formed by branches of the inferior epigastric veins and joining the umbilical vein.

**burquism** (*berk'-izm*). Same as *burcquism*.

**burr.** See *bur*.

**burring** (*bur'-ing*) [ME., *borre*, a harshness in the throat]. Rhotacism; in stammering, the mispronunciation of the letter r. **b. engine,** a dental appliance for the use of burs, etc., in forming cavities, etc. See *dental engine*.

**burrow** (*bur'-o*) [ME., *borow*, a hole]. 1. To make a hole or furrow, as in the skin; said of the itch insect. 2. To force a way through, as pus through the tissues.

**burrowing** (*bur'-o-ing*). The term given to the passage of pus through the tissues after the formation of an abscess.

**bursa** (*bur'-sah*) [L., "a purse"]. A small sac interposed between parts that move upon one another. **b., accidental, b., adventitious,** one resembling a bursa mucosa, but due to friction or pressure. **b., acromial, external,** one beneath the acromion, between the coracoid process, the deltoid muscle, and the capsular ligament. **b., acromial, internal,** one lying above the acromion, between the tendon of the infraspinatus and the teres major. **b. anserina,** one under the insertion of the gracilis and sartorius muscles. **b. cordis,** the pericardium. **b., gluteal.** See under *gluteal*. **b., gluteofascial, b., gluteotrochanteric,** one lying between the trochanter major and the gluteus maximus. **b., iliac.** 1. One lying between the tendon of the iliacus muscle and the trochanter minor. 2. One between the pelvic brim and the iliopsoas muscle. **b. mucosa,** a membranous sac secreting synovial fluid. **b., obturator,** one under the tendon of the obturator internus. **b., omental, b. omentalis,** a large cavity formed by the peritoneum back of the stomach and in the great omentum. The lesser peritoneal sac. **b. patellæ, b. patellaris,** one lying between the patellā and the skin. **b. patellæ amplificata,** housemaid's knee. **b. patellaris lateralis externa,** one lying between the patella and the external lateral dilatation of the tendon of the quadriceps extensor cruris; it is rarely found. **b. patellaris lateralis interna,** one between the patella and the inner lateral dilatation of the quadriceps extensor cruris; it may be either deep or superficial. **b. pharyngea,** a blind pouch projecting upward from the pharynx toward the occipital bone. **b., plantar,** one over the instep either above or below a tendon. **b., popliteal,** a bursa situated in the popliteal space between the tendon of the semimembranosus and the tendon of the inner head of the gastrocnemius, where they rub against each other. **b., prepatellar,** a bursa situated over the patella and the upper part of the patellar ligament. **b., riders',** an enlarged bursa due to excessive horseback riding. **b. sacralis,** one found in the aged over the sacrococcygeal articulation or over the spine of the fourth or fifth sacral vertebra. **b., subhyoid.** See *subhyoid bursa*. **b., synovial,** one found between tendons and bony surfaces. **b. testium,** the scrotum.

**bursal** (*bur'-sal*) [*bursa*, a purse]. Pertaining to a bursa, sac, or follicle.

**bursalis** (*bur-sa'-lis*). The obturator internus muscle.

**bursalogy** (*bur-sal'-o-je*) [*bursa*, a purse; λόγος, science]. The science or study of the bursæ; the anatomy, physiology, and pathology of the bursæ.

**Bursera** (*bur-ser-ah*) [*Burser*, German botanist]. A genus of tropical trees, several species of which afford resinous gums. *B. gummifera* is a native of South America; the resin, chibou or cachibou, is used in plasters and salves and internally in diseases of the lungs and kidney. The leaves are vulnerary, the bark is anthelmintic and antigonorrheic, and the root is used in diarrhea.

**burserin** (*bur'-ser-in*) [see *Bursera*]. A resinous constituent of opobalsamum.

**bursiform** (*bur'-sif-orm*) [*bursa*, a purse; *forma*, form]. Resembling a bursa.

**bursine** (*bur'-sēn*). An alkaloid isolated from *Capsella bursa-pastoris*. It is a yellow, deliquescent powder, used as an astringent, tonic, and styptic instead of ergot, and hypodermically in aqueous solution.

**bursitis** (*bur-si'-tis*) [*bursa*, a purse; *itis*, inflammation]. Inflammation of a bursa. **b., omental,** inflammation of the omental bursa. **b., prepatellaris,** housemaid's knee. **b., retrocalcaneal.** See *achillodynia*. **b., Thornwaldt's,** catarrhal inflammation of the anterior portion of the median recess of the nasopharynx.

**bursopathy** (*bur-sop'-ath-e*) [*bursa*; πάθος, suffering]. Any disease of the bursa, particularly dropsy due to some general disease.

**bursula** (*bur'-su-lah*) [dim. of *bursa*, a purse]. A small bursa; the scrotum.

**Burton's line** [Henry *Burton*, English physician, 19th cent.]. A blue line along the margins of the gums in chronic lead-poisoning.

**butane** (bū'-tān), C₄H₁₀. An anesthetic substance isolated from petroleum. Syn., *butyl hydride*.

**Butcher's saw.** [Richard G. *Butcher*, Irish surgeon]. A saw used in amputations and excisions; it has a narrow blade that can be adjusted at any angle, so that it runs easily and in any direction.

**butter** (*but'-er*) [*butyrum*, butter]. 1. The fatty part of the milk obtained by rupturing the cells of the fat-globules by "churning" or mechanical agitation. 2. Various vegetable fats having the consistency of butter. 3. Certain chemical products having the appearance or consistence of butter. **b. of antimony**, chloride of antimony. **b. of cacao.** See *cacao butter*. **b. of tin**, chloride of tin. **b. of zinc**, zinc chloride.

**butterfly patch.** A patch of lupus erythematosus on the cheeks and nose.

**butterine** (*but'-er-ēn*) [*butter*]. An artificial substitute for butter, made principally of beef-fat. Oleomargarine.

**buttermilk** (*but'-er-milk*). The liquid left after extracting the butter from cream. **b.-belly**, a distended abdomen; pot-belly.

**butternut** (*but'-er-nut*). See *juglans*.

**buttocks** (*but'-uks*) [dim. of *butt*, an end]. The nates. The fleshy part of the body posterior to the hip-joints, formed by the masses of the glutei muscles.

**button** (*but'-un*). See *furunculus orientalis*. **b., Amboyna.** See *frambesia*. **b. anastomosis**, anastomosis by means of a Murphy button. **b., belly-, the navel. b., Biskra.** See *furunculus orientalis*. **b.-bush**, the bark of *Cephalanthus occidentalis*, a tonic, febrifuge, and diuretic. Dose of *fluidextract* ½–1 dr. (2–4 Cc.). **b., Chlumsky's**, an intestinal button made of pure magnesium after the pattern of the Murphy button. It remains undissolved for four weeks, only the outer part becoming softer. **b., Corrigan's**, a steel, button-shaped cautery-iron, introduced by Sir J. C. Corrigan (1802–80). **b.-makers' chorea.** See *chorea*, *buttonmaker's*. **b., Murphy**, a device used in gastroenterostomy or intestinal anastomosis. **b.-snakeroot**, the root of *Liatris spicata* and of *Eryngium yuccæfolium*; a stimulant, tonic, diuretic, and emmenagogue. Dose of *fluidextract* ½–1 dr. (2–4 Cc.). See *Chlumsky*, *Corrigan*, *Murphy*.

**buttonhole fracture.** One in which a missile has perforated the bone. **b., mitral**, an advanced degree of constriction of the mitral orifice of the heart.

**b. operation**, boutonnière operation, *q. v.*

**butyl** (bū'-til) [*butyrum*, butter], C₄H₉. A hydrocarbon alcohol radical. **b.-chloral**, **b.-chloral hydrate.** See *chloral butylicum.* **b. hydride.** See *butane.* **b.-hypnal**, a combination of butyl-chloral and antipyrine. It is hypnotic and antipyretic.

**butylamine** (bū-til'-am-in) [*butyl*; *amine*], C₄H₉NH₂. A substance contained in codliver oil, possessing diuretic and diaphoretic properties.

**butylene** (bū'-til-ēn) [*butyrum*], C₄H₈. A hydrocarbon belonging to the olefin series. It exists in three isomeric forms, all of which are gases at ordinary temperatures.

**butyphus** (bū-ti'-fus) [βοῦς, an ox; τῦφος, stupor]. The cattle-plague. Syn., *rinderpest*.

**butyraceous** (bu-tir-a'-se-us) [*butyrum*]. Resembling or containing butter.

**butyrate** (bū'-tir-āt) [*butyrum*, butter]. A salt of butyric acid.

**butyric** (bū-tir'-ik). Contained in butter; derived from butter. **b. acid.** See *acid, butyric*.

**butyrin** (bū'-tir-in) [*butyrum*], C₃H₅(C₄H₇O₂)₃. A constant constituent of butter, together with olein, stearin, and other glycerids. It is a neutral, yellowish, liquid fat, having a sharp, bitter taste.

**butyroid** (bū'-tir-oid) [*butyrum*]. Buttery; having the consistence of butter.

**butyromel** (bū-tir'-o-mel). The proprietary name for a mixture of 2 parts of fresh butter and 1 part of honey, rubbed together until a clear yellow mixture is obtained. It is used in preparing palatable preparations of codliver oil and other nauseous oleaginous substances.

**butyrometer** (bū-tir-om'-et-ur) [*butyrum*, butter; μέτρον, measure]. An apparatus for determining the proportion of fatty matter in milk.

**butyroscope** (bū-tir'-o-skōp) [*butyrum*, butter; σκοπεῖν, to look]. An instrument for estimating the proportion of fat in milk.

**buxine** (buks'-in) [*buxus*, the box-tree]. Berberine, the alkaloid of *Buxus sempervirens*. It is a white, amorphous powder with a persistent bitter taste; very insoluble in water, but easily soluble in alcohol and chloroform. It is used as a febrifuge; dose gr. 3–6.

**buxus** (buks'-us) [L., the box-tree]. A genus of trees affording boxwood. **b. sempervirens**, the common box or box-tree of Europe and Asia. Its leaves, wood, and oil have been employed in medicine.

**buyo cheek cancer** (bu'-yo) [Tagalog term]. A cancer of the cheek found in natives of the Philippine Islands; it is associated with the chewing of a mass made up of buyo leaves, betel-nut, slaked lime and tobacco.

**Bychowski's test for albumin.** Put a drop or two of the urine into a test-tube filled with hot water and shake it; in the presence of albumin a whitish cloud is formed and is diffused through the liquid. Phosphates give the same result, but the cloud disappears on addition of a drop of acetic acid.

**bynin** (bi'-nin) [βύνη, malt]. 1. A proteid, insoluble in water, found in malt. 2. A proprietary liquid extract of malt made in England. **b. amara**, a combination of bynin (2) with the phosphates of iron, quinine, and strychnine.

**bynol** (bi'-nol). A combination of malt extract and cod-liver oil.

**Byrd's method** (*bird*) [Harvey Leonidas *Byrd*, American physician, 1820–1884]. For *artificial respiration in asphyxia neonatorum:* The physician's hands are placed under the middle portion of the child's back, with their ulnar borders in contact and at right angles to the spine. With the thumbs extended, the two extremities of the trunk are carried forward by gentle but firm pressure, so that they form an angle of 45 degrees with each other in the diaphragmatic region. Then the angle is reversed by carrying backward the shoulders and the nates.

**byrolin** (bi'-ol-in). A proprietary remedy said to be a combination of boric acid, glycerol, and lanolin, and recommended for use in skin diseases.

**bysma** (bis'-mah) [βύσμα, a stopper; plug; *pl.*, *bysmata*]. A plug or tampon.

**byssinosis** (bis-in-o'-sis). A pulmonary affection due to the inhalation of cotton-dust.

**byssocausis** (bis-o-kaw'-sis) [βύσσος, cotton; καῦσις, a burning]. Cauterization by the moxa; moxibustion.

**byssoid** (bis'-oid). Consisting of a filamentous fringe of which the strands are of equal length.

**byssophthisis** (bis-o-tis'-is). See *byssinosis*.

**byssus** (bis'-us) [βύσσος, cotton, flax]. 1. Charpie, lint, or cotton. 2. The hairy growth of the pubic region. 3. In biology, a bunch of silky filaments secreted by the foot, in several molluscs. A name formerly given to the mycelium of large fungi.

**bythium** (bith'-e-um) [βύθος, depth]. A supposed new chemical element said to have been obtained from sulphur. ¶The claim has not been accepted.

# C

**C.** 1. The chemical symbol of *carbon*. 2. The abbreviation of *centigrade*, *congius*, *closure*, *contraction*, *cylinder*, or *cylindrical lens*.

**Ca.** 1. Chemical symbol of *calcium*. 2. Abbreviation for *cathode*.

**caballine aloes** (*kab'-al-in al'-ōs*). An inferior quality of aloes, known also as fetid or horse aloes.

**Cabanis' pallet.** A shovel-shaped instrument consisting of two plates of perforated silver, jointed and movable on each other; it is used to seize the extremity of the nasal probe in Méjean's operation for lacrimal fistula.

**cabbage** (*kab'-āj*). See *brassica*. **c.-rose.** See *rose, pale*. **c., skunk-,** a fetid plant of North America, *Symplocarpus fœtidus*. Its tincture and fluidextract are prescribed as antispasmodics and antasthmatics.

**Cabot's ring bodies** (*cab'-ot*) [Richard Clarke Cabot, American physician, 1868-    ]. Intra- and extra-cellular bodies having the general shape of a ring and found in the blood in severe anemia.

**cacaerometer** (*kak-a-er-om'-e-tur*) [κακόs, bad; ἀήρ, air; μέτρον, measure]. An apparatus for determining the impurity of the air.

**cacaine** (*kak-a'-ēn*) [Nahuatl, *cacauatl*, cacao]. Theobromine, *q. v.*

**cacanthrax** (*kak-an'-thraks*) [κακόs, bad; ἄνθραξ, acoal]. Contagious anthrax. See *anthrax*.

**cacao** (*kak-a'-o*). See *theobroma*. **c.-butter** (*oleum theobromatis*, U. S. P.), is obtained from seeds or nibs of *theobroma cacao*. It is a pure white fat, with a pleasant odor and taste; it fuses at 86° F. (30° C.); its specific gravity is from 0.945 to 0.952. It is used in cosmetics and for pharmaceutical preparations. See also *theobroma*.

**cacaphthæ** (*kak-aff'-the*) [κακόs, bad; ἄφθα, an eruption]. Malignant or cachectic aphthæ. See *aphtha*.

**cacation** (*kak-a'-shun*) [*cacatio*, a going to stool]. Defecation; alvine discharge.

**cacatory** (*kak'-at-o-re*) [*cacatio*, a going to stool]. Attended with diarrhea; as a cacatory fever.

**CaCC.** Abbreviation for cathodal closure contraction; also written CCC.

**caccagogue** (*kak'-a-gog*) [κάκκη, dung; ἀγωγόs, leading]. 1. Aperient; laxative. 2. An aperient, especially an ointment or suppository that induces gentle purgation.

**cacemia** (*kas-e'-me-ah*, or †*kak-e'-me-ah*) [κακόs, bad; αἷμα, blood]. An ill-condition of the blood; depravity of the blood.

**cacesthesis** (*kak-es-the'-sis*) [κακόs, bad; αἴσθησις, sensation]. Morbid sensation.

**cachectic** (*kak-ek'-tik*) [see *cachexia*]. Characterized by cachexia.

**cachelcoma** (*kak-el-ko'-mah*) [κακόs, ill; ἕλκωμα, ulcer: pl., *cachelcomata*]. A malignant or foul ulcer.

**cachet** (*kash-a*) [Fr.]. A pharmaceutical preparation consisting of two concave pieces of wafer, varying in size from ¾ to 1½ inches in diameter, round or oblong in shape, in one of which the powder to be administered is placed, and the other, having previously been moistened, is then laid over the powder and the two margins are pressed together, when they adhere and completely inclose the powder.

**cachexia, cachexy** (*kak-eks'-e-ah, kak-eks'-e*) [κακόs, bad; ἕξις, a habit]. A depraved condition of general nutrition, due to some serious disease, as syphilis, tuberculosis, carcinoma, etc. **c. aquosa.** See *Griesinger's disease*. **c., cancerous, c., carcinomatous,** a condition marked by weakness, emaciation, and a muddy or brownish complexion, due to carcinomatous disease. Syn., *cachexia canceratica; cancerous diathesis*. **c., lymphatic.** Synonym of *Hodgkin's disease*. **c., miner's.** See *uncinariasis*. **c., osteal,** profound cachexia seen in children and accompanied by painful swelling of one of the long bones, with hematinuria or extravasation of blood into a tissue, and often by rhachitic phenomena. **c., pachydermic.** See *myxedema*. **c., paludal.** See *malarial cachexia*. **c., periosteal.** See *c., osteal.* **c. splenetica,** that associated with splenic enlargement. **c. strumipriva, c., thyreopriva,** the condition allied to, if not identical with, myxedema, following the extirpation of the thyroid gland. **c., thyroid,** exophthalmic goitre. **c. venerea,** syphilis. **c. virginum,** chlorosis.

**cachexy** (*kak-eks'-e*). See *cachexia*.

**cachibou** (*kash-i-boo'*). See under *bursera*.

**cachinnation** (*katsh-in-a'-shun*) [*cachinnare*, to laugh loudly]. Immoderate laughter, as in the insane.

**cachou** (*kash-oo'*) [Fr. for "*catechu*"]. An aromatic pill or tablet for concealing the odor of the breath.

**caco-** (*kak-o-*) [κακόs, bad]. A prefix meaning bad or diseased.

**cacocholia** (*kak-o-ko'-le-ah*) [*caco-*; χολή, bile]. A morbid condition of the bile.

**cacochroia, cacochrœa** (*kak-o-kroi'-ah, kak-o-kre'-ah*) [*caco-*; χροία, color]. A bad complexion; unnatural color of the skin.

**cacochylia** (*kak-o-ki'-le-ah*) [*caco-*; χυλόs, juice]. Imperfect or disordered digestion.

**cacochymia** (*kak-o-ki'-me-ah*) [*caco-*; χυμόs, juice]. A morbid state of the fluids, humors, blood, or secretions; faulty stomachic digestion.

**cacocnemia** (*kak-ok-ne'-me-ah*) [*caco-*; κνήμη, leg]. Thinness or ill-condition of the leg or shin.

**cacocolpia** (*kak-o-kol'-pe-ah*) [*caco-*; κόλποs, vagina]. A diseased state of the vagina; gangrene of the vulva.

**cacodes** (*kak-o'-dēz*) [κακώδηs, ill smelling]. Having a foul, offensive odor.

**cacodiacol** (*kak-o-di'-ak-ol*). Guaiacol cacodylate.

**cacodontia** (*kak-o-don'-te-ah*) [*caco-*; ὀδούs, tooth]. A bad condition of the teeth.

**cacodyl** (*kak'-o-dil*) [κακώδηs, ill-smelling; ὕλη, matter], As(CH3)2. Dimethylarsine; a radical containing arsenic, hydrogen, and carbon. It is a colorless, heavy liquid, with an extremely offensive odor; it is inflammable when exposed to air. Its protoxide is called alkarsine (*q. v.*).

**cacodylate** (*kak-o'-il-āt*). A salt of cacodylic acid. The sodium and the iron salts are used in medicine.

**cacodyliacol** (*kak-o-dil-i'-ak-ol*). Guaiacol cacodylate: used in tuberculosis; dose, 2 grains (0.04 gm.).

**cacoethes** (*kak-o-e'-thēz*) [*caco-*; ἦθοs, a habit]. 1. Any bad habit, disposition, or disorder. 2. A malignant ulcer.

**cacoethic** (*kak-o-eth'-ik*) [*caco-*; ἦθοs, habit]. Malignant.

**cacogalactia** (*kak-o-gal-ak'-te-ah*) [*caco-*; γάλα, milk]. A bad or abnormal condition of the milk.

**cacogastric** (*kak-o-gas'-trik*) [*caco-*; γαστήρ, the stomach]. Dyspeptic.

**cacogenesis** (*kak-o-jen'-es-is*) [*caco-*; γένεσιs, formation]. Any morbid, monstrous, or pathological growth or product.

**cacogenics** (*kak-o-jen'-iks*) [κακογενήs, ill-born]. The opposite of eugenics. The sum total of the conditions which tend to the deterioration of the human race.

**cacoglossia** (*kak-o-glos'-e-ah*) [*caco-*; γλῶσσα, tongue]. Gangrene of the tongue.

**cacolet** (*kak'-o-la*) [Fr.]. A mule-chair or horse-pannier for the transportation of the wounded.

**cacomelia** (*kak-o-me'-le-ah*) [*caco-*; μέλοs, limb]. A congenital pathological condition or deformity of a limb.

**cacomorphia** (*kak-o-mor'-fe-ah*) [*caco-*; μορφή, form]. Malformation; deformity.

**caconychia** (*kak-o-nik'-e-ah*) [*caco-*; ὄνυξ, nail]. Disease or defect of a nail or of the nails.

**cacopathy, or cacopathia** (*kak-op'-a-the* or *kak-o-path'-e-ah*) [*caco-*; πάθοs, illness]. Any severe, malignant, or untoward condition or disease.

**cacopharyngia** (*kak-o-far-in'-je-ah*) [*caco-*; φάρυγξ, the pharynx]. Gangrene of the pharynx.

**cacophonia** (kak-o-fo'-ne-ah) [caco-; φωνή, voice]. An altered, depraved, or abnormal voice.
**cacophonic** (kak-o-fon'-ik). Affected with cacophonia.
**cacophthalmia** (kak-off-thal'-me-ah) [caco-; ophthalmia]. A malignant inflammation of the eye.
**cacoplasia** (kak-o-pla'-ze-ah) [caco-; πλάσσειν, to form]. The formation of diseased structures.
**cacoplastic** (kak-o-plas'-tik) [see cacoplasia]. 1. Characterized by a low degree of organization. 2. Relating to cacoplasia.
**cacopragia** (kak-o-pra'-je-ah) [caco-; πράσσειν, to do]. Functional derangement, as of nutritive processes, or of organs.
**cacoproctia** (kak-o-prok'-te-ah) [caco-; πρωκτός, anus]. A gangrenous state of the rectum.
**cacorrhachis** (ka-kor'-rak-is) [caco-; ῥάχις, spine]. A diseased state of the vertebral column.
**cacorrhinia** (kak-or-in'-e-ah) [caco-; ῥίς, nose]. Any diseased condition of the nose.
**cacosmia** (kak-os'-me-ah) [caco-; ὀσμή, smell]. A disgusting smell.
**cacosomium** (kak-o-so'-me-um) [caco-; σῶμα, body]. A hospital for leprosy and other incurable diseases.
**cacospermia** (kak-o-sper'-me-ah) [caco-; σπέρμα, seed]. Any diseased state of the semen.
**cacosphyxia** (kak-o-sfiks'-e-ah) [caco-; σφύξις, pulse]. A disordered state of the pulse.
**cacosplanchnia** (kak-o-splangk'-ne-ah) [caco-; σπλάγχνα, the viscera]. Diseased condition of the digestive tract and resulting emaciation.
**cacostomia** (kah o sto' me ah) [caco; στόμα, mouth]. Any diseased or gangrenous state of the mouth.
**cacothanasia** (kak-o-than-a'-se-ah) [caco-; θάνατος, death]. A painful or miserable death.
**cacotheline** (kak-oth'-el-ēn), C₂₁H₂₂NO₂₀. An alkaloid produced from brucine by the action of HNO₃.
**cacothesis** (kak-oth'-es-is) [caco-; θέσις, a placing]. A faulty position of a part or of the entire organism.
**cacothymia** (kak-o-thim'-e-ah) [caco-; θυμός, mind]. A disordered state of the mind or disposition; mental disorder with moral depravity; insane malignity of temper.
**cacotrichia** (kak-o-trik'-e-ah) [caco-; θρίξ, hair]. A diseased condition of the hair.
**cacotrophia** (kak-o-tro'-fe-ah). Same as cacotrophy.
**cacotrophy** (kak-ot'-ro-fe) [caco-; τροφή, nourishment]. Disordered or defective nutrition.
**cacozyme** (kak'-o-zīm) [caco-; ζυμή, a ferment]. A disorganizing, putrefactive, fermentative, or pathogenic microörganism.
**cactin** (kak'-tin). An acrid resinous glucoside obtained from Cereus grandiflorus.
**cactina** (kak-ti'-nah) [κάκτος, a prickly plant]. A proprietary preparation said to be a proximate principle derived from night-blooming cereus (Cactus grandiflorus and C. mexicana). It is a cardiac stimulant, recommended as a substitute for digitalis.
**cactus** (kak'-tus) [κάκτος, a prickly plant]. A genus of plants. c. grandiflorus, a plant indigenous to the West Indies and cultivated in North America and Europe. It bears large white or straw-colored flowers which bloom only at night. The preparations of cactus are stimulant to the spinal cord, the vasomotor center, and the cardiac ganglia. They have been used as substitutes for digitalis. Dose of tincture 15-20 min. (1.0-1.3 Cc.); of fluidextract 5-10 min. (0.32-0.65 Cc.). Syn., night-blooming cereus; Zerus grandiflorus.
**cacumen** (kak-ū'-men) [L.; pl., cacumina]. 1. The top, as of a plant. 2. The culmen of the vermis superior of the cerebellum.
**cadaver** (kad-av'-er or kad-a'-ver) [cadere, to fall; pl., cadavera]. A dead body, especially that of a human being.
**cadaveric** (kad-av'-er-ik) [cadaver]. Pertaining to the cadaver. c. alkaloids, ptomaines. c. ecchymoses, c. lividity, certain postmortem stains, closely resembling in their general appearance the effects of bruises or contusions. They occur on the lowest and most dependent parts of the body. c. rigidity, rigor mortis. c. spasm, the early, at times instantaneous, appearance of rigor mortis, seen after death from certain causes. It is also called instantaneous rigor and tetanic rigidity.
**cadaverine** (kad-av'-er-ēn) [cadaver], C₅H₁₄N₂. A ptomaine, occurring very frequently in decomposing animal tissues. It is a thick, clear, syrupy liquid, having an exceedingly unpleasant odor. An auxetic in cancer.
**cadaverization** (kad-av-er-iz-a'-shun) [cadere, to fall]. The passage of a living body to the state of a cadaver. Applied to the algid and cyanotic stage of cholera.
**cadaverous** (kad-av'-er-us) [cadere, to fall]. Resembling a cadaver; ghastly; of a deathly pallor.
**cade** (kād) [a Languedoc name]. See juniper. c., oil of (oleum cadinum, U. S. P.), a tarry oil from the wood of Juniperus oxycedrus; it is used in skin diseases.
**caderas** (kad-e'-ras). See mal de caderas.
**Cadet's fuming liquid** [Louis Claude Cadet de Gassicourt, French chemist; 1731-1799]. See alkarsin.
**cadinene** (kad'-in-ēn), C₁₅H₂₄. A sesquiterpene boiling at 274° C.
**cadmium** (kad'-me-um) [καδμία, calamine]. Cd = 112.40 quantivalence II; sp. gr. 8.60-8.69. A bluish-white metal resembling zinc in its general properties. In its physiological action it is escharotic and astringent; internally, in large doses, it produces emesis and violent gastritis. c. iodide, CdI₂, used as an ointment, 1 to 8 of lard. c. salicylate, Cd(C₇H₅O₃)₂, white needles, soluble in water and alcohol. It is used in purulent ophthalmia, etc., and is said to be a more active antiseptic than other cadmium salts. c. sulphate, CdSO₄. 4H₂O, an astringent in gonorrhea and in corneal opacities; used as a lotion in strength of ⅜ gr. or 4 to 1 oz. of water, or as an ointment in 1 : 40 of fresh lard. c. sulphocarbolate, Cd(C₆H₅-SO₄)₂, white crystals, soluble in water; it is antiseptic and astringent. Syn., cadmium sulphophenylate.
**caduca** (kad-dū'-kah) [cadere, to fall]. Thickened mucous membrane of the uterus in pregnancy. See decidua.
**caducity** (kad-dū'-sit-e) [caducitas, senility]. Senility; the feebleness of advanced age.
**caducous** (kad-ū'-kus) [caducus, falling off]. Dropping off very early, as compared with other parts. Deciduous. c. morbus. Falling sickness; an old name for epilepsy.
**cæcal** (se'-kal). See cecal.
**cæcitas** (se'-sit-as) [L.]. Blindness. c. diurna, day-blindness. c. nocturna, night-blindness. c. verbalis, word-blindness.
**cæcitis** (se-si'-tis). See cecitis.
**cæcum** (se'-kum). See cecum.
**cænesthesis** (cen-es-the'-sis). See cenesthesis.
**cæruleus** (se-ru'-le-us) [L.]. Sky-blue. c. morbus, blue disease. See cyanosis.
**cærulosis** (se-ru-lo'-sis). See cyanosis.
**Cæsalpinia** (ses-al-pin'-e-ah) [L.]. A genus of tropical leguminous trees. C. bonducella is a prickly, trailing shrub of most tropical coasts. The seeds, Molucca beans, and the whole plant are anthelmintic and emmenagogue; the oil of the seeds is used in rheumatism. It contains bonducin. C. coriaria, American sumac, divi-divi, is a South American shrub cultivated in India, where the dried powdered pods are used as an antiperiodic. Dose 40-60 gr. (2.6-4.0 Gm.). A decoction of the pods is used as an injection in the treatment of bleeding piles. C. echinata, a tree of Brazil, furnishes Brazil-wood; the bark, rich in tannin, is used as an astringent, roborant, and febrifuge. C. sappanis, a tree of India; the brownish-red wood, sappan-wood, contains sappanin and is used as an astringent; it furnishes a red dye and the root a yellow dye.
**cæsarean operation** or **section** (se-sar'-e-an). See cesarean section.
**cæsium** (se'-se-um). See cesium.
**caffea** (kaf'-e-ah) [L.]. The seeds of Coffea arabica. The dried and roasted seeds are almost universally used in infusion as a beverage, forming a cerebral stimulant and stomachic tonic. They are valuable in promoting digestion and allaying hunger and fatigue. The properties are due to an alkaloid, caffeine, identical with theine. See thea and caffeine. The fluidextract of Caffea viridis is intended as a substitute for the fluidextract of guarana. Dose ½-2 dr. (2-8 Cc.). See Guarana.
**caffeic acid** (kaf-e'-ik). See acid, caffeic.
**caffeine** (kaf'-e-in or kaf-ēn) [caffea], C₈H₁₀N₄O₂ +H₂O. See under caffea. An alkaloid found in the leaves and beans of the coffee-tree, in tea, in Paraguay tea, and in guarana, the roasted pulp of the fruit of Paulinia sorbilis. It occurs in long, silky needles, slightly soluble in cold water and alcohol, with a feebly bitter taste. It is a cerebrospinal, circulatory, and renal stimulant. Dose 1-3

gr. (0.06–0.2 Gm.). Syn., *guaranine; methyltheobromine; psoraline; theine; trimethylxanthine.* **c. borocitrate**, (C₈H₁₀N₄O₂)₃BO₃, a white, crystalline powder, soluble in water, alcohol, and chloroform; it decomposes in water. It has the effect of caffeine combined with the antiseptic action of boric acid. **c. bromide.** See *c. hydrobromide.* **c. carbolate**, C₈H₁₀N₄O₂.HOC₆H₅, a white, crystalline mass, soluble in alcohol and water with decomposition. It is an antiseptic, diuretic, and stimulant, having the combined action of caffeine and phenol, and is used subcutaneously. **c. chloral**, C₈H₁₀N₄O₂—CCl₃COH, a molecular combination of caffeine and chloral, occurring in soluble crystals. It is sedative and analgesic. Dose 3–5 gr. (0.2–0.3 Gm.). **c. citrate**, (*caffeina citrata*, U. S. P.), (C₈H₁₀N₄O₂)₂C₆H₈O₇, a true salt, forming a white, crystalline powder, used in the same manner as caffeine. It is soluble in water and alcohol with decomposition. Dose 1–5 gr. (0.065–0.32 Gm.). **c., citrated,** this is improperly called caffeine citrate and is prepared by dissolving equal weights of caffeine and citric acid in double the quantity of hot distilled water. Dose 3–8 gr. (0.2–0.52 Gm.). **c., citrated, effervescent** (*caffeina citrata effervescens*, U. S. P.). Dose 60 gr. (4 Gm.). **c. citrosalicylate**, (C₈H₁₀N₄O₂)₃C₆H₈O₇+(C₈H₁₀N₄O₂.-C₇H₆O₃)₃, a true salt occurring as a white, crystalline powder, decomposing in water. It is antiseptic and is used as is caffeine. **c. diiodide.** See *c. triiodide.* **c. hydrobromide**, caffeine bromide, a true salt, C₈H₁₀N₄O₂HBr, occurring as large crystals, reddish or greenish on exposure, soluble in water on decomposition. It is used as a diuretic in injections of 4–10 min. of a solution of 10 parts caffeine hydrobromide, 1 part hydrobromic acid, and 3 parts distilled water. **caffeine, injectio, hypodermatica,** 1 grain of caffeine in 3 minims. Dose 1–6 min. (0.065–0.30 Cc.). **c. salicylate**, C₈H₁₀N₄O₂.C₇H₆O₃, a true salt occurring as white, crystalline masses, soluble in water and alcohol with decomposition. It is used instead of caffeine with salicylic acid. Dose as caffeine. **c. and sodium benzoate,** a white powder containing 45.8% of caffeine, soluble in 2 parts of water. It is used instead of caffeine by subcutaneous injection. Dose about double that of caffeine. Syn., *caffeinum natriobenzoicum.* **c. and sodium salicylate,** a white powder, soluble in 2 parts of water, and containing 62.5% of caffeine. It is used in rheumatism, etc., instead of caffeine, by subcutaneous injection. Dose about double that of caffeine. Syn., *caffeinum natriosalicylicum.* **c. and sodium sulphonate,** a diuretic. Dose 15 gr. (1 Gm.). Syn., *symphoral.* **c. tannate,** a yellow powder, more astringent than caffeine; uses and dosage as caffeine. **c. triiodide**, (C₈H₁₀N₄O₂I₂.HI)₂+3H₂O, dark-green prisms, of a metallic luster, soluble in alcohol. It is a diuretic and alterative, used instead of potassium iodide, and said to be nondepressing. Dose 2–4 gr. (0.13–0.26 Gm.). Syn., *diiodocaffeine hydriodate.* **c. valerate**, C₈H₁₀N₄O₂.C₅H₁₀O₂, small lustrous needles, soluble in alcohol with decomposition. It is used in nervous headache, whooping-cough, etc. Dose 2–5 gr. (0.13–0.3 Gm.) several times a day.

**caffeinism** (*kaf'-e-in-izm* or *kaf-ēn'-izm*) [*caffeine*]. Chronic coffee-poisoning; a train of morbid symptoms due to excess in the use of coffee.

**caffeism** (*kaf'-e-izm*). See *caffeinism.*

**caffeol** (*kaf'-e-ol*). See *caffeone.*

**caffeone** (*kaf'-e-ōn*) [*caffea*], C₈H₁₀O₂. A volatile aromatic oily principle (empyreumatic oil) produced by the roasting of coffee. Syn., *caffeol.*

**Cagot** (*kag-o*) [Fr.]. 1. A member of an outcast race or clan in the S. W. of France; formerly regarded as lepers. 2. (By error) a cretin. **C. ear.** This is an ear with no lower lobe; but it is asserted that this is a peculiarity of the Lapps, and not of the Cagots.

**cahinca,** or **cainca** (*kah-hing'-kah*) [native S. American]. The diuretic root of *Chiococca racemosa, C. densifolia,* or *C. anguifuga,* rubiaceous shrubs of tropical America.

**cahincin** (*kah-hin'-sin*) [*cainca* or *cahinca;* the South American name for several species of *Chiococca*], C₄₀H₆₄O₁₅. A glucoside from *Chiococca racemosa* and *C. brachiata.* In small doses it is diuretic and cathartic and in large doses emetic. It is used in dropsy. Dose, diuretic and cathartic, 2–4 gr. (0.13–0.26 Gm.); emetic, 8–15 gr. (0.5–1.0 Gm.); maximum dose 15 gr. (1 Gm.). Syn., *cainic acid.*

**caisson-disease** (*ka'-son-diz-ēz*). Diver's disease, tunnel disease, or the bends, a morbid condition due to increased atmospheric pressure, sometimes occurring in divers, caisson-workers, etc. Paraplegia, hemiplegia, anesthesia, and apoplectic attacks are common, coming on only after a return to the normal atmosphere. The nature of the lesion is obscure.

**Cajal's (Ramón y) cells** (*kah'-hal*) [Santiago Ramón y Cajal, Spanish histologist, 1852– ]. Fusiform or triangular ganglion cells lying near the surface of the cerebral cortex and giving off branched processes parallel to it. **C.'s moss-like appendages.** See *appendage.* **C.'s tassel-cells,** pyramidal cells of the cortex cerebri having a large number of dendritic processes hanging from their base.

**cajuput oil, cajeput oil** (*kaj'-u-put*). [Malay, *cajuputi,* white wood]. A volatile oil (*oleum cajuputi,* U. S. P.), distilled from the leaves of *Melaleuca cajuputi.* It resembles oil of turpentine. It is used in flatulent colic, hysteria, cutaneous disorders, and toothache. Dose 1–5 min. (0.065–0.32 Cc.). **c. spirit** (*spiritus cajuputi,* B. P.), contains 2% of the oil. Dose ½–1 dr. (2–4 Cc.).

**cajuputene, cajeputene** (*caj-u-pū-tēn', caj-e-pū-tēn'*) [Malay], C₁₀H₁₆. The principal constituent of cajuput oil; it is a liquid of an agreeable odor.

**cajuputol, cajeputol** (*kaj-u-pū'-tol*) [see *cajuput oil*]. The more limpid part of cajeput oil; it is found also in some other fragrant volatile oils.

**caked** (*kākd*). Compressed or hardened into a solid mass. **c. bag,** in cows, an inflammation of the mammary gland. **c. breast,** a breast in a puerperal woman in which the milk has become hardened and inspissated.

**cal.** Abbreviation of *small calory.*

**Cal.** Abbreviation of *large or great calory.*

**Calabar bean** (*kal'-ab-ar bēn*) [*Calabar,* a region of West Africa]. See *physostigma.* **C. swellings,** an edematous enlargement occurring in the natives of Calabar and other parts of West Africa. It is thought to be produced by the action of *Filaria loa* in the subcutaneous tissues.

**calabarine** (*kal-ab'-ar-ēn*). An alkaloid from Calabar bean.

**calabarization** (*kal-a-bar-iz-a'-shun*). The act of bringing a person under the effects of calabarine.

**calage** (*kal-ahsh'*) [Fr., wedging]. A method of prophylaxis or treatment of sea-sickness by fixation of the viscera by pillows placed between the abdomen and the wall of the cabin and between the back and edge of the berth, the patient lying upon the side.

**Calaguala** (*kah-lah-gwah'-la*) [Peruvian]. The commercial name for several plants of the order *Polypodiaceæ,* especially *Polypodium calaguala,* of Peru, which is esteemed an excellent resolvent and diaphoretic. It is used in chronic affections of the air-passages and in whooping-cough in doses of 30–60 gr. (2–4 Gm.) daily. A decoction is used externally as a dressing for wounds.

**calamine** (*kal'-am-in*) [*calamina,* a corruption of *cadmia*]. Native zinc carbonate. *Calamina præparata,* the prepared calamine, washed and pulverized, is used mainly as an external exsiccant and astringent.

**calamus** (*kal'-am-us*) [L., "a reed"]. Sweet-flag. The rhizome of *Acorus calamus.* It contains a volatile oil and *acorin,* a bitter nitrogenous principle. The root is an aromatic, stomachic tonic, and a common ingredient of many popular "bitters." **C. draco** is a species of ratan palm that affords a part of the so-called dragon's-blood of commerce. **c., fluid-extract of** (*fluidextractum calami,* U. S. P.). Dose 15 min.–1 dr. (1–4 Cc.). **c. scriptorius** [a writing-pen or reed], the groove on the floor of the lower extremity of the fourth ventricle, at the end of which is the ventricle of Arantius.

**calaya** (*kal-a'-yah*). A fluidextract of the fruit of *Anneslea febrifuga* (?), used in malaria. Dose 30 gr. (2 Gm.) every 2 hours.

**calcalith** (*kal'-kal-ith*). A proprietary preparation said to be a rheumatism remedy and uric-acid solvent. It consists of calcium carbonate, lithia and colchicine.

**calcaneal** (*kal-ka'-ne-al*) [*calcaneum,* the heel]. In biology, relating to the heel-bone, or calcaneum, or to a tuberosity in birds, resembling the calcaneum.

**calcaneo-astragalar** (*kal-ka'-ne-o-as-trag'-al-ar*). See *calcaneoastragaloid.* **c.-astragaloid,** relating to the calcaneum and the astragalus. **c.-cavus,** a club-foot that combines the characters of calcaneus and cavus. **c.-cuboid,** belonging to the calcaneum and the cuboid. **c.-navicular.** Relating to the calcaneum and

the scaphoid bone, or os naviculare. **c.-scaphoid**, belonging to the calcaneum and the scaphoid bone.
**c.-talar** (*kal-ka-ne-o-ta'-lar*). See *calcaneo-astragalar*.
**c.-tibial** (*kal-ka-ne-o-tib'-e-al*) [*calcaneum; tibia*]. Relating to the calcaneum and tibia. **c.-valgo-cavus**, club-foot combining the features of calcaneus, valgus and cavus.

**calcaneum, calcaneus** (*kal-ka'-ne-um, -us*). The os calcis or heel bone. See *bones, table of*.

**calcaneus** (*kal-ka'-ne-us*) [*calcaneum*, the heel]. 1. See *calcaneum*. 2. Club-foot in which the heel alone touches the ground, the instep being drawn up toward the shin.

**calcar** (*kal'-kar*) [L., "a spur"]. 1. Any spur or spur-like point, as the hippocampus minor. 2. Ergot of rye. 3. The calcaneum. 4. The styloid process of the temporal bone. **c. avis**, the hippocampus minor. **c. femorale**, a plate of hard tissue around the neck of the femur. **c. pedis**, the heel bone.

**calcarate** (*kal'-kār-āt*) [*calcar*]. Spurred; furnished with spurs or spur-like processes.

**calcarea, calcaria** (*kal-ka'-re-ah*) [*calx*, limestone]. Lime.

**calcareous** (*kal-ka'-re-us*) [*calcarea*]. 1. Pertaining to or having the nature of limestone. 2. Having a chalky appearance or consistence. 3. Growing in chalk.

**calcariform** (*kal-kar'-e-form*) [*calcar; forma*, form]. Spur-shaped.

**calcarine** (*kal'-kar-ēn*) [*calcar*]. Spur-shaped; relating to the hippocampus minor. **c. fissure.** See *fissure, calcarine*.

**calcic** (*kal'-sik*) [*calx*, lime]. Of or pertaining to lime or calcium. **c. inflammation of gums and peridental membrane**, inflammation caused and maintained by deposits of calculus on the necks of the teeth.

**calcicosis** (*kal-sik-o'-sis*) [*calx*]. Marble-cutter's phthisis; a chronic inflammation of the lung due to the inhalation of marble-dust.

**calcidine** (*kal'-sid-ēn*). Trade name of a preparation containing calcium and iodine; used in catarrhal laryngitis; dose gr. ¼-1 (0.02–0.06).

**calciferous** (*kal-sif'-er-us*). Containing lime, chalk, or calcium.

**calcific** (*kal-sif'-ik*) [*calx*]. Forming lime.

**calcification** (*kal-sif-ik-a'-shun*) [*calx; fieri*, to become]. The deposit of calcareous matter within the tissues of the body. **c., metastatic**, that resulting from an excess of lime-salts in the blood, as occurs in the rapid breaking down of bones from osteomalacia.

**calcigerous** (*kal-sij'-er-us*) [*calx*, lime; *gerere*, to bear]. Containing lime or a lime-salt; as the calcigerous cells of the dentine, or calcigerous tubules in bone.

**calcigrade** (*kal'-sig-rād*) [*calx*, heel; *gredi*, to walk]. Walking on the heels.

**calcination** (*kal-sin-a'-shun*) [*calcinare*, to calcine]. The process of driving off the volatile chemical constituents from inorganic compounds. The expulsion of carbon dioxide from carbonates.

**calcine** (*kal'-sin* or *kal-sin'*) [*calcinare*, to calcine]. To separate the inorganic elements of a substance by subjecting it to an intense heat.

**calcinol** (*kal'-sin-ol*). Trade name of calcium iodate.

**calcis, os**. The heel bone.

**calcium** (*kal'-se-um*) [*calx*]. Ca=40.07; quantivalence II. A brilliant, silver-white metal, the basis of limestone, characterized by strong affinity for oxygen, and isolated with great difficulty. It is best known in the form of *calcium oxide*, quicklime; *calcium hydroxide*, slaked lime; and *calcium carbonate*, limestone or chalk. **c. acetate**, Ca(C$_2$H$_3$O$_2$)$_2$, a white, amorphous powder, soluble in water. It is used in tuberculosis and psoriasis. **c. albuminate**, an alterative and nutrient used in rhachitis. **c. benzoate**, Ca(C$_7$H$_5$O$_2$)$_2$, used in nephritis and albuminuria of pregnancy. Dose 5–10 gr. (0.32–0.65 Gm.) **c. bisulphite, liquid**, a solution of calcium sulphite (CaSO$_3$) in an aqueous solution of sulphurous acid. It is used, when diluted with 4 to 8 times its amount of water, as an antiseptic gargle or wash. **c. borate**, a very light white powder. It is used as an antiseptic and astringent, internally in children's diarrhea and externally in fetid perspiration and weeping eczema. Dose 1–5 gr. (0.06–0.32 Gm.) 3 times daily. Application, 10 to 20 % ointment or dusting-powder. **c. boroglyceride**, an antiseptic substance prepared by heating together calcium borate and glycerol. It forms a transparent, hygroscopic mass, soluble in water and alcohol. **c. bromide** (*calcii bromidum*, U. S. P.), a nerve sedative. Dose 10–30 gr. (0.65–2.0 Gm.) twice daily. **c. bromoiodide**, CaI$_2$+CaBr$_2$, a mixture of calcium iodide and bromide in molecular proportions, forming a yellow powder, soluble in water. It is alterative and sedative. Dose 5–10 gr. (0.32–0.65 Gm.) 3 times daily. **c. carbide**, CaC$_2$, obtained from lime with carbon, by the electric furnace. It occurs in gray to bluish-black irregular lumps, decomposing with water, evolving acetylene, and leaving a residue of slaked lime; sp. gr. 2.22. It is used in the palliative treatment of cancer of the vagina and uterus. **c. carbolate**, Ca(OC$_6$H$_5$)$_2$, a reddish, antiseptic powder, used as a disinfectant and internal and external antiseptic. Dose 2–5 gr. (0.13–0.32 Gm.). **c. carbonate**, Ca(OC$_4$H$_5$)$_2$, a reddish powder used as an internal and external antiseptic. Dose 2–5 gr. (0.13–0.32 Gm.). **c. carbonate, precipitated** (*calcii carbonas præcipitatus*, U. S. P.), CaCO$_3$, a fine white powder, without odor or taste. Dose 15 gr. (1 Gm.). **c. chloride** (*calcii chloridum*, U. S. P.), CaCl$_2$, soluble in water; used internally to increase the coagulability of the blood. Dose 10–20 gr. (0.65–1.3 Gm.). **c. citrate**, Ca(C$_6$H$_5$O$_7$)$_2$+4H$_2$O, a crystalline powder, soluble in 1730 parts of water at 90° C.; more soluble in cold water. A solution is recommended for the treatment of burns. **c. cresylate**, a syrupy fluid obtained by treating calcium hydroxide with cresol. It is used as a disinfectant instead of phenol. **c. creosotate**, Ca(C$_6$H$_5$O$_3$)$_2$, a sulphosalt of aliphatic creasote esters containing 25 % of creasote and occurring as a gray powder soluble in 10 parts of water. Dose 4–10 gr. (0.26–0.65 Gm.) 4 or 5 times daily. **c. ferrophospholactate**, used in the treatment of tuberculosis and rhachitis. Dose 3–7½ gr. (0.2–0.5 Gm.). **c. glycoarsenate**, a crumbling white powder, insoluble in water and alcohol, freely soluble in mineral and organic acids, especially in dilute citric acid. It is used in treatment of tuberculosis. Daily dose ⅙ gr. (0.01 Gm.). **c. glyceroborate**, an antiseptic compound of equal parts of calcium borate and glycerol. **c. glycerophosphate**, CaC$_3$H$_7$PO$_6$, a white, crystalline powder, soluble in cold water, almost insoluble in boiling water; it is a nerve tonic. Dose 2–5 gr. (0.13–0.32 Gm.) 3 times daily. In treatment of enuresis, dose, for adults, 8 gr. (0.5 Gm.) twice daily. **c. hippurate**, Ca(C$_9$H$_8$NO$_3$)$_2$, a white, crystalline powder, slightly soluble in hot water. It is alterative. Dose 5–15 gr. (0.32–1.0 Gm.). **c. hydrate**. See *c. hydroxide*. **c. hydroxide**, slaked lime. **c. hypochlorite**, Ca(ClO)$_2$, white cubes decomposing readily. It is an antiseptic and is used as a disinfectant and strong bleaching agent. **c. hypophosphite** (*calcii hypophosphis*, U. S. P.), Ca(PH$_2$O$_2$)$_2$, a white, crystalline powder, lustrous scales, or transparent crystals, soluble in 7 parts of water, decomposing and giving out inflammable gas above 300° C. It is used in tuberculosis, chlorosis, etc. Dose 10–30 gr. (0.65–2.0 Gm.). Syn., *calcium hypophosphorosum*. **c. hyposulphite**. See *c. thiosulphate*. **c. iodate**, Ca(IO$_3$)$_2$+6H$_2$O, a white, crystalline powder, soluble in 400 parts of water, insoluble in alcohol. It is used internally in doses of 4–5 gr. (0.26–0.32 Gm.) to check fermentation and also as a succedaneum for iodoform. Syn., *calcinol*. **c. iodide**, CaI$_2$, a white powder or yellowish-white hygroscopic mass, soluble in water and alcohol. It is an alterative used instead of potassium iodide. Dose 2–5 gr. (0.13–0.32 Gm.) 3 times daily in syrup. Maximum dose, daily, 15 gr. (1 Gm.). **c. and iron lactophosphate**, a yellowish powder used in treatment of rhachitis and tuberculosis. Dose 3–8 gr. (0.19–0.52 Gm.) several times daily. **c. lactate**, Ca(C$_3$H$_5$O$_3$)$_2$+5H$_2$O, white, opaque, granular masses, soluble in water and hot alcohol. It is used in treatment of rhachitis and tuberculosis of children. Dose 3–10 gr. (0.19–0.65 Gm.) in syrup. **c. lactophosphate**, a crystalline compound of calcium lactate and calcium phosphate, containing 1 % of phosphorus; soluble in water. It is stimulant and nutrient. Dose 3–10 gr. (0.19–0.65 Gm.) 3 times daily. **c. oxide** (calx, U. S. P.), CaO, quicklime. **c. oxide, chlorinated**. See *lime, chlorinated*. **c. oxysulphide**, a compound of calcium, oxygen, and sulphur, forming a yellowish powder; used in washing scrofulous ulcers. **c. permanganate**, Ca(MnO$_4$)$_2$+5H$_2$O, deliquescent, brown crystals with violet luster, soluble in water. It is used internally in diarrhea of children and externally as a mouth lotion. Dose $\frac{3}{4}$–2 gr. (0.05–0.13 Gm.). c. **peroxide,** $CaO_2+4H_2O$, an antiseptic, used in acid dyspepsia and summer diarrhea. Dose (children) $\frac{3}{4}$–2 gr. (0.05–0.13 Gm.). c. **phosphate, antimoniated,** a mixture of precipitated calcium phosphate (67 parts) and antimony oxide (33 parts), occurring as a dull white, gritty powder, without odor or taste, soluble in boiling water. It is alterative, purgative, and emetic, and is used in acute rheumatism and febrile diseases. Dose 3–8 gr. (0.19–0.52 Gm.) 4 to 6 times daily. Syn., *antimonial powder; James' febrile powder.* c. **phosphate, dibasic,** $CaH_2(PO_4)_2$ or $CaHPO_4$, a white powder, soluble in acids, insoluble in water. It is used in diseases of bone, chlorosis, etc. Dose 8–20 gr. (0.52–1.3 Gm.). Syn., *bicalcic phosphate; secondary calcium phosphate.* c. **phosphate, monobasic,** $CaH_4(PO_4)_2+H_2O$, the chief constituent of the so-called "superphosphate of lime," a decomposition-product of tricalcic or dicalcic phosphate and sulphuric acid, occurring as white, deliquescent, strongly acid crystals. c. **phosphate, precipitated** (*calcii phosphas præcipitatus,* U. S. P.), $Ca_3(PO_4)_2$, a bulky white powder, odorless and tasteless. Dose 15 gr. (1 Gm.). c. **phosphate, tribasic,** $Ca_3(PO_4)_2$, a light, white, amorphous powder without odor or taste, soluble in acids, insoluble in water. It is used as is the dibasic. c. **propionate,** $Ca(C_3H_5O_2)_2$, a white powder, soluble in water. c. **quinovate,** a compound of calcium and quinovic acid used in malarial fever and dysentery as a tonic. Dose $\frac{1}{5}$–$\frac{1}{2}$ gr. (0.013–0.032 Gm.). c. **salicylate,** $CaC_7H_4O_3+H_2O$, a white, crystalline powder with alkaline reaction, soluble with difficulty in water. It is used in intestinal diseases. Dose 8–20 gr. (0.52–1.3 Gm.). c. **santonate,** c. **santoninate,** $Ca(C_{15}H_{19}O_4)_2$, a white, odorless, insipid powder, insoluble in water or chloroform. It is anthelmintic. Dose $\frac{1}{2}$–$1\frac{1}{2}$ gr. (0.03–0.1 Gm.). c. **sulphate,** $2CaSO_4+H_2O$, a fine white, odorless and tasteless powder, used in making plaster bandages for fractures. Syn., *gypsum; plaster-of-Paris.* c. **sulphate, dried** (*calcii sulphas exsiccatus,* U. S. P.), a fine white powder without odor or taste. c. **sulphydrate,** $CaS.H_2S$, transparent crystals decomposing in the air; it is used as a depilatory. c. **sulphide,** CaS, a compound of calcium and sulphur, a yellow-white substance with odor of hydrogen sulphide and forming a large percentage of calx sulphurata. It is recommended in treatment of influenza (dose 1 gr. (0.06 Gm.) 4 times hourly) and in treatment of diphtheria (dose $\frac{1}{6}$ gr. (0.01 Gm.) every hour, under 1 year of age, every $\frac{1}{2}$ hour between the ages of 1 and 3, and every 15 minutes between the ages of 3 and 5). Syn., *calcium monosulphide.* c. **sulphide, hydrated,** CaS; it is used as a depilatory. c. **sulphite,** $CaSO_3$, a white powder, soluble in sulphurous acid and in 800 parts of water. It is antiseptic and is used in flatulent diarrhea. Dose $\frac{1}{10}$–5 gr. (0.0065–0.32 Gm.). c. **sulphocarbolate,** $Ca(C_6H_5SO_4)_2+6H_2O$, a white, odorless, astringent powder or scales, soluble in water. It is an internal antiseptic and astringent. Dose 5–15 gr. (0.32–1.0 Gm.) in 1 % solution. c. **sulphophenate.** See *c. sulphocarbolate.* c. **thiosulphate,** $CaS_2O_3$, white antiseptic crystals, soluble in water; it is an internal antiseptic. Dose 3–10 gr. (0.194–0.65 Gm.).

**calcoglobulin** (*kal-ko-glob'-ū-lin*). A combination of soluble calcium salts with an albuminous base. It has a distinct and definite form and is probably the basis of all the calcic tissues of the body.

**calcoid** (*kal'-koid*) [*calx,* lime; εἶδος, resemblance]. A neoplasm of the tooth-pulp.

**calcospherites, calcosphærites** (*kal-ko-sfe'-rītz*) [*calx,* lime; *sphæra,* a sphere]. Hartig's term for the granules or globules formed in embryological dental pulp and in tissues like bone and shell by calcium salts brought by the blood into loose proteid combination and modified by the cytoplasm.

**calcspar** (*kalk'-spar*). Calcium carbonate, $CaCO_3$.

**calculary** (*kal'-kū-la-re*) [*calculus,* a stone]. Relating to or of the nature of a calculus.

**calculifragous** (*kal-kū-lif'-rag-us*) [*calculus,* a stone; *frangere,* to break]. Lithotritic; breaking or reducing a stone in the bladder.

**calculoid** (*kal'-ku-loid*). Resembling a calculus; a concretion.

**calculous** (*kal'-kū-lus*) [*calculus*]. Of the nature of a calculus.

**calculus** (*kal'-kū-lus*) [dim. of *calx,* chalk]. A calcareous or stone-like concretion found in the body, particularly in cavities. c., **alternating,** one composed of alternate layers of the substances of which it is made up. c., **arthritic,** a gouty concretion. c., **articular.** See *c., arthritic.* c., **aural,** hardened cerumen in the external auditory canal. c., **biliary,** a gall-stone. c., **blood,** a fibrinous calculus containing remains of blood-corpuscles. c., **bronchial,** a concretion in an air-passage. c., **chalky,** one made up mainly of calcium carbonate and calcium phosphate with small amounts of magnesium carbonate, water, and organic matter, and frequently having a foreign body as a nucleus. c., **cutaneous.** See *milium.* c., **cystic.** 1. A vesical calculus. 2. A gall-stone. 3. One composed of cystin. c., **cystic-oxide,** c., **cystin,** a urinary calculus, rarely found, and composed largely of cystin. c., **dental,** tartar on the teeth or gums. c., **encysted,** a vesical calculus which has become invested in a pouch springing from the wall of the bladder. c., **essential,** one having its origin within the tissue of an organ and not due to a foreign body. c., **fatty,** a vesical calculus having a nucleus of fat or saponaceous matter. c., **fibrinous,** a vesical calculus made up of dried coagulated albumin. c., **fusible,** a urinary calculus composed of phosphates of ammonium, calcium, and magnesium. c., **hemic,** a concretion of coagulated blood. c., **incarcerated.** See *c., encysted.* c., **intestinal.** Same as *enterolith.* c., **lacteal.** Same as *c. mammary.* c., **laminated,** one made up of layers of different materials. c., **mammary,** a calcareous nodule sometimes obstructing the lactiferous ducts. c., **mulberry,** the oxalate-of-lime variety, resembling a mulberry in shape and color. c., **nasal,** one found in the nasal cavities. c., **organic,** one with a nucleus formed of epithelium, blood, etc. c., **pineal,** brain-sand. See *acervulus.* c., **podagric.** See *c., arthritic.* c., **prostatic,** one in the prostate gland. c., **renal,** a calculus found in the kidney. c., **salivary.** 1. One formed in the ducts of the salivary glands. 2. The tartar deposited on teeth. c., **scrotal.** 1. A vesical or prostatic calculus which has made its way to the scrotum. 2. One formed in the scrotum from calcareous degeneration. c., **secondary,** a vesical calculus formed in consequence of a diseased condition of the mucosa of the urinary tract. c. **serumal,** tartar on the teeth from exudation of diseased gums. c., **urinary,** a concretion composed of concentric layers of crystallized substance cemented together by mucus or other organic material, occurring in the bladder. Urinary calculi (sand, gravel, or stones, according to size) may be classified as follows: (1) Those containing a mixture of uric acid with urates, with either little or no phosphates; (2) mixed calculi, those containing more phosphates than uric acid; (3) calcium-oxalate calculi; (4) phosphatic calculi—composed of calcium phosphate, triple phosphate, or a combination of calcium and magnesium phosphates; (5) calcium-carbonate calculi; (6) cystin calculi; (7) xanthin calculi; (8) fibrinous calculi, consisting of fibrin or inspissated albumin. c., **uterine,** an intrauterine concretion; a womb-stone; formed mainly by calcareous degeneration of a tumor. c., **vesical,** a calculus found in the urinary bladder. c. **xanthic,** a urinary calculus of xanthic oxide.

**calcusol** (*kal'-kū-sol*). A proprietary remedy for gout, said to consist of piperidine parasulphaminebenzoate and potassium bicarbonate.

**Caldani's ligament** (*kal-dah'-ne*) [Leopoldo Marco Antonio *Caldani,* Italian anatomist, 1725–1813]. A fibrous band extending from the inner border of the coracoid process to the lower border of the clavicle and upper border of the first rib, where it unites with the tendon of the subclavius muscle.

**caldarium** (*kal-da'-re-um*). A hot bath, or a room used for the administration of a hot bath.

**calefacient** (*kal-e-fa'-she-ent*) [*calidus,* warm; *facere,* to make]. 1. Warming; producing a sensation of heat. 2. A medicine, externally applied, that causes a sensation of warmth.

**calefactor** (*kal-e-fak'-tor*) [*calidus,* warm; *facere,* to make]. A warmer; a little, portable stove; a pocket stove; a chafing-dish.

**calendula** (*kal-en'-dū-lah*) [*calendæ,* the first day of the month]. Marigold. The flowering plant known as the garden-marigold, *C. officinalis.* c., **tincture of** (*tinctura calendulæ,* U. S. P.), contains 20 % of the leaves and stems. It is used as a local application to wounds, bruises, and ulcers, and has been vaunted as a cure for carcinoma.

**calendulin** (*kal-en'-dū-lin*) [*calendula*]. An amorphous principle obtainable from calendula.
**calentura** (*kal-en-tu'-rah*). 1. See *calenture* (1). 2. Applied to an epidemic disease of horses in the Philippines. It is caused by a species of *Spirillum*. c. **amarilla**, c. vomito negro, yellow fever.
**calenture** (*kal'-en-tūr*) [Sp., *calentura*, heat; L., *calere*, to be hot]. 1. A tropical remittent fever with delirium; formerly, a supposed fever of this kind that attacked mariners, leading them to leap into the sea. 2. Sunstroke.
**calf** (*kahf*) [Icel., *kálfr*]. The thick, fleshy part of the back of the leg, formed by the gastrocnemius and soleus muscles. c.-**bone**, the fibula.
**calibrate** (*kal'-ib-rāt*) [Fr., *calibre*, the bore of a gun]. 1. To estimate the exact size of an opening, as of intestines to be united by anastomosis. 2. To graduate the tubes of a thermometer so that it will indicate the temperature correctly, or to determine the errors of the gradation when made; also, to determine the indication of the reading after the correction of the errors.
**calibration** (*kal-e-bra'-shun*). The act, process, or result of calibrating.
**calibrator** (*kal'-e-bra-tor*). An instrument for determining the exact diameter of the lumen of an opening, as of the urethra. It may consist of a truncated cone supplied with a scale or some form of dilating blades.
**caliche** (*kah-le'-che*) [S. A., "a flake of lime from a wall"]. The South American name for crude sodium nitrate; Chili saltpeter.
**California buckthorn.** See *cascara sagrada*.
**caligation, caliginosity** (*kal-ig-a'-shun, kal-ij-in-os'-it-e*). See *caligo*.
**caliginous** (*kal-ij'-in-us*). Relating to or affected with caligo.
**caligo** (*kal-i'-go*) [L.]. Dimness of vision; an opacity of the cornea, lens, or vitreous humor.
**calipers** (*kal'-ip-ers*) [corruption of *caliber*]. Compasses with curved legs.
**calisaya** (*kal-is-a'-yah*) [S. A.]. Cinchona bark, especially that of *cinchona calisaya*. See *cinchona*.
**calisayin** (*kal-e-sa'-yin*). An amorphous base consisting mostly of quinine obtained from cinchona bark.
**calisthenics, callisthenics** (*kal-is-then'-iks*), [καλὸs, beautiful; σθένος, strength]. The practice of various rhythmical movements of the body, intended to develop the muscles and produce gracefulness of carriage; light gymnastics, especially designed for the use of girls and young women.
**calix** (*ka'-liks*). See *calyx*.
**Callaway's test** (*kal'-la-way*) [Thomas *Callaway*, English surgeon, 19th century]. In dislocation of the humerus the circumference of the affected shoulder, measured over the acromion and through the axilla, is greater than that of the sound side.
**Calleja's olfactory islets.** Nests of large stellate cells interspersed with small nests of minute pyramidal cells, found in the cortex of the hippocampal gyrus.
**Calliano's method of artificial respiration.** A modification of Sylvester's: the arms are drawn up so as to expand the thorax, and then fixed above and behind the head by fastening the wrists together; pressing with the hands upon the thorax some 18 to 20 times a minute to induce respiration.
**callisection** (*kal-e-sek'-shun*) [*callus*, insensitive; *sectio*, a cutting]. Vivisection of anesthetized animals.
**Callisen's operation** (*kal'-is-en*) [Heinrich *Callisen*, Danish surgeon, 1740–1824]. For *lumbar colotomy*; a vertical incision is made following the line of the descending colon.
**callomania** (*kal-lo-ma'-ne-ah*) [καλός, beautiful; *mania*]. Delusion in which the patient believes herself endowed with beauty.
**callosal** (*kal-o'-sal*) [*callosum*]. Pertaining to the corpus callosum.
**callosities** (*kal-os'-it-ēs*). See *callosity*.
**callosity** (*kal-os'-it-e*) [*callus*]. A hard, thickened patch on the skin produced by excessive accumulation of the horny layers. Syn., *callositas; keratoma; tyloma; tylosis.*
**callosomarginal** (*kal-o-so-mar'-jin-al*) [*callosum; margo*, margin]. Relating to the callosal and marginal gyri of the brain.
**callososerrate** (*kal-o-so-ser'-āt*) [*callus; serratus*, saw-shaped]. Serrated callous projections.
**callosum** (*kal-o'-sum*). See *corpus callosum*.

**callous** (*kal'-us*) [*callus*]. Hard. See *callus*.
**callus** (*kal'-us*) [L.]. 1. A callosity; hardened and thickened skin. 2. The new growth of incomplete osseous tissue that surrounds the ends of a fractured bone during the process of repair. c., **interior**, c., **internal**, provisional callus of a fractured bone deposited in its medullary canal. c., **permanent**, the permanent bond of bony union after reabsorption of the *provisional callus*, or cartilage-like, plastic material first thrown out. c. **of skin**, induration and thickening of the skin.
**calmant** (*kahm'-ant*) [ME., *calme*]. A calmative medicine.
**calmative** (*ka(h)l'-mat-iv*). 1. Calming; sedative. 2. An agent that produces a calming or sedative effect.
**Calmette's reaction** or **ophthalmo-tuberculin test** (*kal-met'*) [Albert *Calmette*, French bacteriologist, 1863– ]. A diagnostic measure to detect the presence or absence of tuberculosis. It consists in the instillation into the eye of a drop of a one per cent. aqueous solution of an alcoholic precipitate of Koch's old tuberculin. In a positive reaction the conjunctiva becomes congested and a fibrinous exudate forms in the lower conjunctival sac and at the inner canthus.
**calmin** (*kal'-min*). A compound of antipyrine and heroine; it is used in asthma.
**calolactose** (*cal-o-lak'-tōs*). Trade name of an intestinal disinfectant said to consist of calomel, 1 part; bismuth subnitrate, 1 part; lactose, 8 parts.
**calomel** (*hal'-o-mel*) [καλός, fair; μέλας, black]. Mercurous chloride. c., **colloidal**. Same as *calomelol*.
**calomelol** (*kal'-o-mel-ol*). Colloidal calomel; a proprietary preparation containing calomel and albumin; used as a dusting-powder or in ointment.
**calor** (*ka'-lor*) [L.]. 1. Heat. 2. Moderate fever-heat; less than *fervor* and *ardor*. c. **animalis**, animal heat. c. **febrilis**, fever-heat. c. **fervens**, boiling heat. c. **innatus**, natural or normal heat. c. **internus**, inward fever: fever not appreciable on the surface of the body. c. **mordax**, c. **mordicans**, biting or pungent heat. c. **nativus**, native or animal heat; blood-heat; normal heat.
**calorescence** (*kal-or-es'-ens*). Tyndall's name for the conversion of rays of heat into rays of light.
**Calori's bursa** (*ka-lor'-e*) [Luigi *Calori*, Italian anatomist, 1807–1896]. A bursa between the arch of the aorta and the trachea.
**caloric** (*ka-lor'-ik*) [*calor*]. 1. Pertaining to a calory or to heat. 2. Heat.
**caloricity** (*kal-or-is'-it-e*). [*calor*, heat]. The heat-producing power of the living animal body.
**calorie** (*kal'-or-e*). See *calory*.
**calorifacient** (*kal-or-if-a'-she-ent*) [*calor; facere*, to make]. Heat-producing (applied to certain foods).
**calorific** (*kal-or-if'-ik*) [*calor*, heat; *facere*, to make]. Heat-producing. c. **center**, heat-producing center.
**calorimeter** (*kal-or-im'-et-er*) [*calor; μέτρον*, a measure]. An instrument for measuring the heat that bodies produce or absorb. c., **respiration** (Atwater's), an apparatus used to determine the caloric values of various foods and their effect on metabolism.
**calorimetric equivalent.** The amount of heat necessary to raise the temperature of the calorimeter 1° C.
**calorimetry** (*kal-or-im'-et-re*). The estimation of the heat-units by the calorimeter. c., **direct**, a method of estimating the amount of heat produced and given off by an animal incased in a ventilated cabinet, and inclosed in another cabinet filled with air or water, by gaging the amount imparted to the air or water in the second cabinet. c., **indirect**, that arrived at by an estimation of the calorific value of a known quantity of food ingested by an animal in a given time.
**calorimotor** (*kal-or-im-o'-tor*) [*calor*, heat; *motor*, mover]. A galvanic battery that produces heating-effects, generating electricity in large quantity, but not necessarily with a high electro-motive force.
**calorinesis** (*kal-or-in'-es-is*) [*calor*, heat; *pl., calorineses*]. Any disease characterized by an alteration in the quantity of animal heat.
**calory** (*kal'-or-e*) [Fr., *calorie*]. A heat-unit; the amount of heat required to raise the temperature of one kilogram of water from 0° to 1° C. c., **great**. See *calory*. c., **rational**, the amount of heat required to raise the temperature of one gram of water from

0° to 100° C., and is approximately equal to 100 small calories. c., small, the amount of heat required to raise the temperature of one gram of water 1° C.

**Calot's method** (*kal'-o*) [François Calot, French surgeon, 1861– ]. A method of forcible reduction of angular deformity of the spine.

**calox.** A proprietary preparation, containing calcium peroxide; used as a dentifrice.

**calumba** (*kal-um'-bah*) [native Mozambique, *kalumb*]. Columbo. The root of *Jateorrhiza calumba*, native to South Africa and parts of the East Indies. It is an excellent example of a simple bitter, and contains a bitter principle, *calumbin*, $C_{21}H_{22}O_7$, of which the dose is 1–3 gr. (0.065–0.2 Gm.). It is not astringent, and may be prescribed with salts of iron. It is useful in atonic dyspepsia, and as a mild, appetizing tonic in convalescence. c., **extract of** (*extractum calumbæ*, B. P.). Dose 2–10 gr. (0.13–0.65 Gm.). c., **fluidextract** of (*fluidextractum calumbæ*, U. S. P.). Dose 5–30 min. (0.32–2.0 Cc.). c., **infusion of** (*infusum colombæ*, B. P.). Dose 1–2 oz. (30–60 Cc.). c., **tincture of** (*tinctura calumbæ*, U. S. P.), contains 10 % of calumba. Dose ½–2 dr. (2–8 Cc.).

**calumbin** (*kal-um'-bin*). A bitter principle found in *calumba, q. v.*

**calvaria, calvarium** (*kal-va'-re-ah, -um*) [*calva*, the scalp]. The upper part of the skull; the skullcap.

**calvarian** (*kal-va'-re-an*) [*calva*, the scalp]. Relating to the calvaria.

**calvities** (*kal-vish'-e-ēz*), [*calvus*, bald]. Baldness.

**calvitium** (*kal-vish'-e-um*). See *calvities*.

**calvous** (*kal'-vus*) [*calvus*]. Bald.

**calx** (*kalks*) [L.]. 1. The heel. 2. Calcium oxide. c. **chlorinata** (U. S. P., B. P.), chlorinated lime. c. **sulphurata** (U. S. P., B. P.), consists largely of calcium sulphide; used externally and internally in skin diseases. Dose 1/12 gr. (0.006 Gm.). c. **usta**, c. **viva**, burnt lime, unslaked lime, quick lime.

**calyces of the kidneys** (*kal'-is-ēz*). [plural of *calyx*]. The cup-like tubes of the ureter that encircle the apices of the Malpighian pyramids of the kidneys.

**calyciform** (*kal-is'-e-form*) [*calyx*; *forma*, shape]. Cup-shaped; resembling a calyx.

**calycine** (*kal'-is-in*) [*calyx*, calyx]. In biology, pertaining to or resembling a calyx.

**calycle** (*kal'-ik-l*) [*calyculus*, a little calyx]. In biology, applied to parts that resemble a calyx, as the bracts or leaflets of certain plants, or the cup-cells of zoophytes.

**calyculus** (*kal-ik'-u-lus*) [pl., *calyculi*]. See *calycle*.

**calyculi gustatorii.** See *taste-buds*.

**calyx** (*ka'-liks*) [L.; pl., *calyces*]. A cup; especially one of the cup-like divisions of the pelvis of the kidney into which the pyramids project. c. **of ovum**, the wall of the Graafian follicle from which the ovum has escaped.

**camara** (*kam'-ar-ah*). 1. An arched or vaulted chamber. 2. The fornix of the brain. 3. The hollow of the external ear.

**cambium** (*kam'-be-um*) [L.]. A layer of tissue formed between the wood and the bark of exogenous plants.

**cambodia** (*kam-bo'-de-ah*). See *cambogia*.

**cambogia** (*kam-bo'-je-ah*) [*Camboja*, or *Cambodia*, in Siam]. Gamboge. A resinous gum from *Garcinia hanburii*, a tree native to southern Asia. Its properties are due to gambogic acid. It is a drastic, hydragogue cathartic, decidedly diuretic. c., **compound pill of** (*pilula cambogia composita*, B. P.), contains cambogia, aloes, hard soap, compound powder of cinnamon, and syrup. Dose 5–10 gr. (0.32–0.65 Gm.). It is also officially a constituent of compound cathartic pills.

**camera** (*kam'-er-ah*) [καμάρα, an arched roof or chamber]. 1. See *camara*. 2. In optics, the apparatus used for photography. c. **aquosa**, the anterior aqueous chamber of the eye. c. **cordis**, the enveloping membrane of the heart, the pericardium. c. **lucida**, an optical device for superimposing or combining two fields of view in one eye, invented by the chemist, Wollaston. c. **oculi**, the chamber of the eye. c. **septi lucidi**, the fifth ventricle of the brain.

**camisia fœtus** (*kam-is'-e-ah*). The chorion.

**camisole** (*kam-is-ōl'*) [Fr.]. The strait-jacket, used for the restraint of the violently insane.

**Cammidge's test** [Percy John Cammidge, English physician]. A test for the detection of pancreatic disease by examination of the urine for a substance, probably pentose.

**camomile, chamomile** (*kam'-o-mil*). See *anthemis* and *matricaria*.

**Campani's test** (*kam-pah'-ne*). *For glucose:* If a concentrated solution of lead subacetate mixed with a dilute solution of copper acetate is added to urine containing glucose a yellow to orange-red precipitate results.

**camp cure.** Life in camp and in the open air, adopted as a therapeutic measure. c. **fever**, synonym of *typhus fever*. c. **measles**, an epidemic of measles among soldiers.

**Camper's angle** (*kam'-per*) [Peter Camper, Dutch anatomist, 1722–1789]. See *facial angle*. C.'s **chiasm**, the crossing of the inner fibers of the tendons of the flexor sublimis digitorum (see under *muscle*) after they have separated to give passage to the tendons of the deep flexor. C.'s **fascia**, the superficial layer of the lower part of the superficial fascia of the abdomen. C.'s **ligament**, triangular ligament; the deep perineal fascia. C.'s **line**, a line running from the external auditory meatus to a point just below the nasal spine.

**camphacol** (*kam'-fak-ol*). A proprietary preparation said to contain camphoric acid, formaldehyde and guaiacol; used in catarrhal conditions of the respiratory and urinary tract. Dose gr. 5–20 (0.3–1.2 Gm.).

**camphene** (*kam'-fēn*) [*camphor*]. 1. Any one of the volatile oils or hydrocarbons having the general formula $C_{10}H_{16}$, isomeric with oil of turpentine. Many camphenes, as oil of cloves, etc., exist ready formed in plants. They are liquid at ordinary temperatures. 2. Purified oil of turpentine.

**camphenol** (*kam-fe'-nol*). A compound of camphor cresols, and phenols; it is a disinfectant.

**camphin** (*kam'-fin*), $C_{10}H_8$. A colorless oil obtained by the distillation of camphor with iodine.

**camphoid** (*kam'-foid*) [*camphor*]. A substitute for collodion. It is a solution, one in 40, of pyroxylin, in equal parts by weight of camphor and absolute alcohol.

**camphol** (*kam'-fol*). See *borneol*.

**campholyptus** (*kam-fo-lip'-tus*). A proprietary external anodyne said to consist of eucalyptol, camphor and chloral.

**camphophenique** (*kam-fo-fen-ēk'*) [Fr.]. A proprietary preparation combining camphor and phenol; it is recommended as an antiseptic and local stimulant.

**camphor** (*kam'-for*) [*camphora*, camphor], $C_{10}H_{16}O$. A solid, volatile oil obtained from *Cinnamomum camphora*, a tree indigenous to eastern Asia. It yields camphoric and camphonic acids, also camphor cymol when exposed to a high heat in closed vessels. It is antispasmodic, anodyne, diaphoretic, and stimulant. Applied locally, it is rubefacient. It is used in cholera, vomiting, the typhoid state, headache, diarrhea with pain, etc., cardiac depression, and affections requiring an antispasmodic. c., **artificial**, $C_{10}H_{16}HCl$, a terpene hydrochloride obtained from oil of turpentine by action of hydrochloric acid; it is a solid very similar to camphor. c.-**ball**, an English preparation used as an application to chapped skin. Its composition is spermaceti, 4; white wax, 12; oil of almonds, 5; melt in a water-bath, and add flowers of camphor 4; dissolve, and when nearly cold pour into boxes or mold in gallipots. c., **Borneo**. See *borneol*. c., **carbolated**, a mixture of 2⅓ parts of camphor with 1 each of phenol and alcohol; it is a good antiseptic dressing for wounds. c. **cerate** (*ceratum camphoræ*, U. S. P.), consists of camphor liniment, 3; olive-oil, 12; simple cerate, 85. It is used for itching skin affections. c., **chloral**, a fluid prepared by mixing equal parts of camphor and chloral. It is an excellent solvent for many alkaloids, and is used externally as a sedative application. c., **citrated**, a compound of citric acid and camphor; a white powder, antiseptic, antispasmodic, and stimulant. Dose 3–10 gr. (0.2–0.65 Gm.) several times daily. c., **flowers of**, powdered camphor obtained by condensing sublimed camphor. c.-**ice**, a cosmetic preparation made by melting 16 parts of white wax with 48 parts of benzoated suet, and then adding 8 parts of camphor and 1 part of oil of lavender. c., **Japan**, the commercial variety brought from Japan; it is also called *tub camphor*, from the receptacle in which it comes, or *Dutch camphor*, from its introduction into the market by that people. c. **liniment** (*linimentum camphoræ*, U. S. P.), a preparation consisting of camphor, 20 parts; cotton-seed oil, 80 parts. c. **liniment, compound** (*linimentum camphoræ com-*

*positum*, B. P.), contains camphor and oil of lavender dissolved in rectified spirit, and strong solution of ammonia added. c., **liquid,** oil of camphor. c., **liquid artificial,** $C_{10}H_{18}HCl$, a liquid isomere of solid artificial camphor obtained from oil of turpentine by action of gaseous hydrochloric acid at high temperatures. c., **monobromated** (*camphora monobromata*, U. S. P.), $C_{10}H_{15}BrO$, camphor in which one atom of hydrogen has been replaced by an atom of bromine. It resembles the bromides in therapeutic action. Dose 1–10 gr. (0.065–0.65 Gm.). c., **Neroli.** See *auradæ.*¦ c.-**resin,** $C_{20}H_{30}O_2$, a yellow, resinous [body obtained from camphor by heating it with an alcoholic solution of caustic potash. c. **salicylate,** prepared by heating together 14 parts of camphor and 11 of salicylic acid. It is used as an ointment. c., **spirit of** (*spiritus camphoræ*, U. S. P.), contains camphor, 10; alcohol, 90 parts. Dose 5–20 min. (0.3–1.2 Cc.). c., **tincture of, compound** (*tinctura camphoræ composita*, B. P.), contains opium, benzoic acid, camphor, oil of anise, and proof spirit. Dose 15 min.–1 dr. (1–4 Cc.). c., **tincture of, Rubini's,** a saturated solution of camphor in alcohol. Dose 2–5 min. (0.12–0.3 Cc.). c.-**water** (*aqua camphoræ*, U. S. P.), consists of camphor, 8; alcohol, 5; distilled water, sufficient to make 1000 parts. Dose 1–4 dr. (4–16 Cc.).
**camphoraceous** (*kam-for-a'-shus*). Resembling or containing camphor. Syn., *camphoroid; camphorous.*
**camphorate** (*kam'-for-āt*). A salt of camphoric acid.
**camphorated** (*kam'-for-a-ted*) [*camphor*]. Impregnated with camphor. c. **oil,** camphor liniment (camphor 20, cotton-seed oil 80).
**camphoric** (*kam-for'-ik*). Relating to camphor. c. **acid.** See *acid, camphoric.* c.-**acid phenetidin,** a compound of camphoric acid and paraphenetidin. c. **anhydride,** $C_{11}H_{14}O_3$, the anhydride of camphoric acid, a sticky mass obtained by heating the crude acid.
**camphorism** (*kam'-for-izm*). Camphor-poisoning; a condition marked by gastritis, coma, and convulsions, due to excessive ingestion of camphor and its preparations.
**camphorogenol** (*kam-for-oj'-en-ol*), $C_{10}H_{18}O_2$. A constituent of camphor occurring as a tolerably heavy oil with smell somewhat like camphor; boils at 212–213° C.
**camphoromania** (*kam-for-o-ma'-ne-a*) [*camphor*; μανία, frenzy]. The camphor habit; a morbid craving for camphor.
**camphoroxol** (*kam-for-oks'-ol*). A 3 % solution of hydrogen dioxide containing 32 % of alcohol and 1 % of camphor.
**camphrene** (*kam'-frēn*), $C_8H_{14}O$. A volatile product of camphor and sulphuric acid.
**camphyl** (*kam'-fil*). The hypothetical radical of borneol, $C_{10}H_{17}$. c. **alcohol,** borneol.
**campimeter** (*kam-pim'-et-er*). See *perimeter.*
**campsis** (*kamp'-sis*) [κάμψις, a curving]. Any abnormal curvature or flexion.
**camptodactylia** (*kamp-to-dak-til'-e-ah*) [καμπτός, bent; δάκτυλος, finger]. Permanent flexion of one or more fingers.
**campylochirus** (*kam-pil-o-ki'-rus*) [καμπύλος, crooked; χείρ, hand]. Having distorted hands.
**campylorrhachis** (*kam-pil-or'-a-kis*) [καμπύλος, crooked; ῥάχις, backbone]. A fetus with spinal deformity.
**campylorrhinus** (*kam-pil-o-ri'-nus*) [καμπύλος, crooked; ῥίς, nose]. A monstrosity with a deformity of the nose.
**Canada balsam** (*kan'-a-dah*). See *balsam, Canada.*
C. **fleabane.** See *erigeron.* C. **hemp.** See *apocynum.* C. **pitch.** See *pix canadensis.*
**canadine** (*kan'-ad-ēn*), $C_{21}H_{21}NO_4$. An alkaloid from the rhizome of *Hydrastis canadensis*, occurring in pure white needles, soluble in alcohol and melting at 132°–135° C.
**canadol** (*kan'-ad-ol*). A transparent volatile liquid resembling benzene in smell. It is a local anesthetic used in minor surgical operations.
**canal, canalis** (*kan-al'*, *kan-a'-lis*) [L., *canalis*]. A tubular channel or passage. c., **abdominal.** See *c., inguinal.* c., **adductor.** See *Hunter's canal.* c., **Alcock's.** See *Alcock's canal.* c., **alimentary,** the whole digestive tube from the mouth to the anus. c., **alisphenoid,** in comparative anatomy, a canal in the alisphenoid bone, opening anteriorly into the foramen rotundum, and transmitting the external

carotid artery. c., **alveolar, anterior,** one located in the superior maxilla; it transmits the anterior superior dental nerve. c., **alveolar, inferior,** the inferior dental canal. c., **alveolar, median,** one located in the superior maxilla and transmitting the middle superior dental nerve. c., **alveolar, posterior,** one situated in the superior maxilla; it transmits the posterior superior dental nerve. c., **alveolodental,** any of the dental canals. c., **anal,** the third part of the rectum or space between the rectum proper and the anus. c., **arachnoid,** a space formed beneath the arachnoid membrane of the brain; it transmits the venæ magnæ Galeni. c.-**of Arantius,** the ductus venosus. c., **archinephric,** the duct of the archinephron or primitive kidney. c., **Arnold's innominate.** See *Arnold's canal.* c., **arterial.** See *ductus arteriosus.* c., **atrial,** the cavity of the atrium. c., **auditory, external,** that from the auricle to the tympanic membrane. c., **auditory, internal,** that beginning on the posterior surface of the petrous bone, and extending outward and backward for a distance of about four lines; it transmits the auditory and facial nerves and the auditory artery. c., **auricular.** 1. See *c., auditory, external.* 2. The constriction between the auricular and ventricular portions of the fetal heart. c., **avant-,** the anterior portion of the male urethra. c., **Bartholin's,** the duct of Bartholin's gland. c., **Bernard's.** See *Bernard's canal.* c., **Bichat's.** See *Bichat's canal.* c., **biliary.** See *c., hepatic.* c., **blastoporic.** See *c., neurenteric.* c., **Bochdalek's.** See *Bochdalek's canal.* c. **of bone,** a canaliculus of bone. c., **Braun's.** See *c., neurenteric.* c., **Braune's.** See *Braune's canal.* c., **Breschet's.** See *c.s of diploë.* c., **bullular.** See *c. of Petit.* c., **caroticotympanic,** two or three short canals extending from the carotid canal to the tympanum; they transmit branches of the carotid plexus. c., **carotid,** one in the petrous portion of the temporal bone; it transmits the internal carotid artery. c.s **of cartilage,** the canals in ossifying cartilage during its vascularization, intended to receive prolongations of the osteogenetic layer of the periosteum. They radiate in all directions from the center of ossification. c., **central (of the modiolus),** a canal running from the base to the apex of the cochlea. c., **central (of spinal cord),** the small canal that extends through the center of the spinal cord from the conus medullaris to the lower part of the fourth ventricle. It represents the embryonic ectodermal canal. c., **cerebrospinal.** 1. The neural or craniovertebral canal formed by the skull and the spine, and containing the brain and spinal marrow. 2. The primitive continuous cavity of the brain and spinal cord, not infrequently more or less extensively obliterated in the cord, but in the brain modified in the form of the several ventricles and other cavities. c., **cervical.** See *c. of cervix uteri.* c., **cervicouterine.** See *c., uterine.* c. **of cervix uteri,** that portion of the uterine canal that extends between the internal and the external os. **canalis choledochus,** the common bile-duct. c. **of chorda tympani,** a small canal in the temporal bone, between its squamous and petrous portions, parallel with the Glaserian fissure; it transmits the chorda tympani nerve. c., **ciliary.** See *c. of Fontana.* c., **circumpeduncular,** the lateral ventricles of the brain. c. **of Cloquet.** See *c., hyaloid.* c., **cochlear,** the spiral and snail-like cavity of the cochlea, 28 to 30 mm. long. The base is turned inward toward the internal auditory meatus, and the apex outward toward the tympanum. c., **connecting,** the arched or coiled portion of a uriniferous tubule, joining with a collecting tubule. c. **of Corti.** See *Corti's canal.* c., **Cotugno's,** c. **of Cotunnius.** See *aqueduct of Cotunnius.* c., **craniopharyngeal,** a fetal canal perforating the posterior part of the sphenoid bone and extending from the pharynx to the hypophysis and the epiphysis; it sometimes persists in infancy. c., **craniovertebral.** See *c., cerebrospinal,* and *c., vertebral.* c., **crural.** See *c., femoral.* c.s **of Cuvier.** See *Cuvier's canals.* c., **cystic,** the cystic-duct. c., **deferent,** the vas deferens. c.s, **demicircular.** See *c.s, semicircular.* c., **dental, anterior,** one extending into the facial portion of the superior maxilla; it transmits the anterior dental vessels and nerves. c., **dental, inferior,** the dental canal of the inferior maxilla; it transmits the inferior dental nerve and vessels. c.s, **dental, posterior,** two canals in the superior maxilla. They transmit the superior posterior dental vessels and nerves. c.s, **dentinal,** the minute canals in dentine, extending approximately at right angles to the surface of a tooth from the pulp-cavity, into which they open, to the cementum and enamel. c.s of derivation, anastomotic venous branches extending from deep to superficial veins. c., digestive. See c., alimentary. c.s of diploe, canals in the diploe of the cranium transmitting Breschet's veins. c., ejaculatory. See duct, ejaculatory. c. of epididymys, a convoluted tube, about 20 feet long when straightened, forming the epididymis and continuous with the vas deferens. c., ethmoid, anterior, one between the ethmoid and frontal bones; it transmits the nasal branch of the ophthalmic nerve and the anterior ethmoid vessels. c., ethmoid, posterior. See c., orbital, posterior internal. c., Eustachian. See Eustachian canal. c., facial, the aqueduct of Fallopius; it transmits the facial nerve. c., Fallopian. See c., facial. c. of the Fallopian tube. See aquaeductus Fallopii. c., femoral. 1. The inner compartment of the sheath of the femoral vessels behind Poupart's ligament. 2. See c., Hunter's. c. of Ferrein. See Ferrein's canal. c. of Fontana, a series of small spaces formed by the interlacing of the connective-tissue fibers of the framework of the peripheral processes of the iris, situated in the angle of the anterior chamber, and serving as a medium for the transudation of the aqueous humor from the posterior to the anterior chamber of the eye. Syn., canal of Hovius; ciliary canal; Fontana's spaces. c.s, galactophorous, the lactiferous tubules of the mammary gland. canalis ganglionaris, the spinal canal of the modiolus. c. of Gartner. See Gärtner's canal. c., genital, in comparative anatomy, any canal designed for copulation or for the discharge of ova. c. of Guidi. See c., Vidian. c., Hannover's. See Hannover's canal. c.s of Havers. See Haversian canals. c., hemal, the ventral of the two canals, of which, according to R. Owen, the vertebrate animal is composed. It contains the heart and the other viscera, while the neural canal incloses the central nervous system. c. of Henle. See Henle's canal. c., Hensen's. See Hensen's canal. c., hepatic. 1. The excretory duct of the liver. 2. The radicles of the hepatic duct. c., hernial, one transmitting a hernia. c., His', the thyroglossal duct of the fetus, of which the cecal foramen of the tongue is the vestige and which may persist during postnatal life. c. of Hovius. See c. of Fontana. c. of Huguier. See c. of chorda tympani. c., Hunter's. See Hunter's canal. c. of Huschke. See Huschke's canal. c., hyaloid, a canal running anteroposteriorly through the vitreous body, through which, in the fetus, the hyaloid artery passes, to ramify on the posterior surface of the crystalline lens. c. of the hypoglossus, the anterior condylar foramen. c., incisor, a canal that opens into the mouth by an aperture just behind the incisor teeth of the upper jaw; it is formed by a groove on the adjoining surfaces of the superior maxillae, and has two branches that open into the nasal fossae. c., infraorbital, a small canal running obliquely through the bony floor of the orbit; it transmits the infraorbital artery and nerve. c., inguinal, a canal about 1½ inches long, running obliquely downward and inward from the internal to the external abdominal ring, and constituting the channel through which an inguinal hernia descends; it transmits the spermatic cord in the male and the round ligament of the uterus in the female. c., innominate. See Arnold's innominate canal. c., intestinal, that portion of the alimentary canal that is included between the pylorus and the anus. c., intralobular, biliary, the radicals of the bile-ducts, forming a fine network in and around the hepatic cells, and communicating with vacuoles in the cells. c. of Jacobson. See c., tympanic. c. of Kowalewsky. See c., neurenteric. c., lacrimal. 1. The bony canal that lodges the nasal duct. 2. One of the lacrimal canaliculi. c., Landzert's. See c., craniopharyngeal. c., Lauth's. See c. of Schlemm. c. of Loewenberg. See Loewenberg's canal. c., malar, one in the malar bone transmitting the malar division of the temporomalar branch of the superior maxillary nerve. c.s, mandibular. See c., dental, inferior. c., mastoid, one opening just above the stylomastoid foramen and transmitting the auricular branch of the vagus nerve. c., maxillary. See c., dental. c., median. 1. The central canal of the spinal cord. 2. The aqueduct of Sylvius. c., medullary. 1. The hollow cavity of a long bone, containing the marrow. 2. See c., vertebral. 3. The central canal of the spinal cord. 4. A Haversian canal. 5. In embryology, the medullary tube. c., medullary, cerebrospinal, the central canal of the spinal cord. c., membranous, of the cochlea, a canal in the cochlea, following the turns of the lamina spiralis; it is bounded by the basilar membrane, the membrane of Reissner, and the wall of the cochlea. c.s, membranous, semicircular. See c.s, semicircular. c. of modiolus. See c., spiral, of the modiolus. c., myelonal, the central canal of the spinal cord. c., nasal. 1. See c., lacrimal (1). 2. An occasional canal found in the posterior portion of the nasal bone; it transmits the nasal nerves. c., nasolacrimal. See c., lacrimal (1). c., nasopalatine. See c., incisor. canalis nervi petrosi profundi minoris, one in the petrosa transmitting the deep petrosal nerve. canalis nervi petrosi superficialis majoris, one opening into the Fallopian aqueduct and transmitting the great superficial nerve. c., neural. See c., vertebral. c., neurenteric (of Kowalewsky), also called blastoporic canal, in the embryo, a passage leading from the posterior part of the medullary tube into the archenteron. c., neurocentral. See c., vertebral. c. of Nuck. See Nuck, canal of. c.s, nutritive. See c.s, Haversian. c., obstetric. See c., parturient. c., obturator, a canal in the ilium transmitting the obturator nerve and vessels. c.s, olfactory, in the embryo, the nasal fossae at an early period of development. c., omphalomesenteric, in the embryo, a canal that connects the cavity of the intestine with the umbilical vesicle. c., orbital, anterior internal. See c., ethmoid, anterior. c., orbital, posterior internal, the posterior of two canals formed by the ethmoid bone and the orbital plate of the frontal bone. It transmits the posterior ethmoid vessels. c., palatine, accessory posterior, one or two canals in the horizontal plate of the palate bone, near the groove entering into the formation of the posterior palatine canal. c., palatine, anterior, formed by the union of the incisive canals; it opens on the palate behind the incisor teeth. c., palatine, descending. See c., palatomaxillary. c., palatine, external, small, a small canal in the pyramidal process of the palate bone, close to its connection with the horizontal plate. It transmits the external palatine nerve. c., palatine, posterior, c., palatine, smaller. See c., palatomaxillary. c., palatine, superior, one formed by the palate bone and the superior maxilla, transmitting the large palatine nerve and blood-vessels. c., palatomaxillary, one formed by the outer surface of the palate bone and the adjoining surface of the superior maxilla. It transmits the large palatine nerve and blood-vessel. c., parturient, the channel through which the fetus passes in parturition, comprising the cavity formed by the uterus and vagina considered as a single canal. c., pelvic, the canal of the pelvis from the superior to the inferior strait. c., perivascular, the lymph-spaces about the blood-vessels. c. of Pétit. See Pétit's canal. c., petromastoid, a small canal, not always present, situated at the angle of union between the mastoid and petrous bones. It transmits a small vein from the middle fossa of the skull to the transverse sinus. c.s, petrosal, two canals on the upper surface of the petrous portion of the temporal bone, transmitting the large and small superficial petrosal nerves. c., pharyngotympanic, one in the embryo developing into the Eustachian canal and the tympanum. c., plasmatic, a Haversian canal. c., pore. See c., porous. c., porous, a canal in the ovule, supposed to serve for the entrance of the spermatozoids in fecundation. c., portal, the space in the capsule of Glisson of the liver, in which the portal vein, hepatic artery, and bile-duct lie. c., primitive, the vertebral canal of the embryo. c.s, pseudostomatous, the processes of branched cells that extend from a subepithelial or endothelial layer to the free surface, their free ends forming the pseudostomata. c., pterygoid. See c., Vidian. c., pterygopalatine, one formed by the root of the internal pterygoid plate of the sphenoid bone and the sphenoid process of the palate bone. It transmits the pterygopalatine vessels and nerve. c., pulmoaortic. See ductus arteriosus. c. of the quadrigemina, the Sylvian aqueduct. canalis radicis, one in the root of a tooth. c.s of Recklinghausen. See Recklinghausen's canals. c., recurrent, canalis recurrens. c., Vidian. c., Reichert's. See c., Hensen's. c. of Reissner. See c., membranous, of the cochlea. c. of Rivinus, the duct of the sublingual gland. c. of **Rosenthal.** See *c., spiral, of the modiolus.* c., **sacculocochlear,** one connecting the sacculus and the cochlea. c., **sacculoutricular,** one connecting the sacculus and the utricle. c., **sacral,** the continuation of the vertebral canal in the sacrum. c.s, **Saviotti's.** See *Saviotti's canals.* c. of **Schlemm.** See *Schlemm's canal.* c.s, **semicircular,** bony canals of the labyrinth of the internal ear. They are three in number,—the external, superior, and posterior,—and contain the membranous semicircular canals. c., **semicircular, anterior,** c., **semicircular, anterior vertical.** See *c., semicircular, superior.* c., **semicircular, external,** that one of the semicircular canals of the labyrinth having its plane horizontal and its convexity directed backward. c., **semicircular, frontal.** See *c., semicircular, superior.* c., **semicircular, horizontal.** See *c., semicircular, external.* c., **semicircular, inferior,** c., **semicircular inner,** c., **semicircular, internal.** See *c., semicircular, posterior.* c., **semicircular, lateral.** See *c., semicircular, external.* c., **semicircular, osseous.** See *c.s, semicircular.* c., **semicircular, posterior,** that one of the semicircular canals having its convexity directed backward and its plane almost parallel to the posterior wall of the pyramid. c., **semicircular, posterior vertical,** c., **semicircular, sagittal.** See *c., semicircular, posterior.* c., **semicircular, superior,** that one of the semicircular canals having its convexity directed toward the upper surface of the pyramid. c.s, **seminal,** the seminiferous tubules. c., **serous,** any minute canal connected with the lymph-vessels and supposed to be filled with lymph. c., **sheathing,** the communication between the cavity of the tunica vaginalis of the testicle and the general peritoneal cavity. It soon closes in man, leaving the tunica vaginalis a closed sac. c., **spermatic.** 1. The vas deferens. 2. The inguinal canal in the male. c., **sphenopalatine.** See *c., pterygopalatine.* c., **spinal.** See *c., vertebral.* c., **spiral, of the cochlea,** one that runs spirally around the m$_o$di$_o$lus, taking two turns and a half, diminishing in size from the base to the apex, and terminating in the cupola. c., **spiral, of the modiolus,** a small canal winding around the modiolus at the base of the lamina spiralis. c., **spiroid, of the temporal bone.** See *c., facial.* c., **of Steno,** the duct of the parotid gland. c. of **Stilling.** See *c., hyaloid,* and *c., central, of spinal cord.* c., **suborbital.** See *c., infraorbital.* c., **supraorbital,** one at the upper margin of the orbit. It transmits the supraorbital artery and nerve. c., **tarsal,** one between and below the heads of the abductor hallucis, transmitting the vessels and nerves to the sole of the foot. c., **temporal,** c., **temporomalar.** See *c., zygomaticotemporal.* c., **thoracic,** the thoracic duct. **canales tubæformes,** the semicircular canals. **canalis tuberculorum quadrigeminorum,** the aqueduct of Sylvius. c., **tuboovarian,** the oviduct. c., **tubotympanal,** the inner division of the first gill cleft in the embryo. c., **tympanic,** one that opens on the lower surface of the petrous bone, between the carotid canal and the groove for the internal jugular vein. It transmits Jacobson's nerve. c., **uterine,** the cavity of the uterus, including the body and neck. c., **uterocervical,** the cavity of the cervix uteri. c., **uterovaginal.** 1. The common canal formed by the uterus and vagina. 2. In embryology, the duct of Mueller. c., **vaginal,** the canal of the vagina. c., **vaginoperitoneal,** the inguinal canal. c., **vascular.** See *c.s, Haversian.* c., **vector,** the oviduct. c., **venous,** the ductus venosus. c., **vertebral,** the canal formed by the vertebræ; it contains the spinal cord and its membranes. c., **Vidian.** See *Vidian canal.* c.s, **Volkmann's.** See *Volkmann's canals.* **canalis vomeris,** one lying between the vomer and the lower surface of the sphenoid, and transmitting blood-vessels to the nose. c.s, **vomerobasilar,** c.s, **vomerosphenoid,** lateral, small lateral canals lying between the vomer and the sphenoid, transmitting blood-vessels. c., **vulvar,** the vestibule of the vagina. s., **vulvouterine,** the vagina. c., **vulvovaginal.** 1. The vagina and the vulva considered as a single canal. 2. The orifice of the hymen. c. of **Wirsung,** the pancreatic duct. c. of **Wolff.** See *Wolffian duct.* c., **zygomatic.** See *c., zygomaticotemporal.* c., **zygomaticofacial.** See *c., malar.* c., **zygomaticotemporal,** the temporal canal of the malar bone, running from its orbital to its temporal surface. It transmits a branch of the superior maxillary nerve.

**canalicular** (kan'-al-ik'-u-lar) [*canal*]. Canal-shaped; relating to a canaliculus.
**canaliculization** (kan-al-ik-u-liz-a'-shun) [*canaliculus,* a little canal]. The formation of canaliculi, as in bone, or as in calcified cartilage.
**canaliculus** (kan-al-ik'-u-lus) [L.]. 1. A small canal; especially that leading from the punctum to the lacrimal sac of the eye. 2. Any one of the minute canals opening into the lacunæ of bone. **canaliculi accessorii,** inconstant canals at the outer edge of the anterior condylar foramen, for the transmission of veins. c. of **communication,** c. **communicationis,** a canal at the junction of the petrosa and the mastoid portion of the temporal bone, transmitting a vein from the middle fossa of the skull to the transverse sinus. Syn., *aquæductus communicationis.* c. **laqueiformis.** See *Henle's loop.* c. **mastoideus,** canal for the auricular branch of the vagus nerve. **canaliculi medullares.** See *canals, Haversian.* c. **pharyngeus.** See *canal, pterygopalatine.* **canaliculi, serous.** See *canals of Recklinghausen.* c. **tympanicus,** canal for the tympanic branch of the glossopharyngeal nerve. **canaliculi vasculosi,** Haversian canals.
**canalis** (kan-a'-lis). Same as *canal.*
**canalization** (kan-al-iz-a'-shun) [*canal*]. 1. The formation of canals, as in tissues, etc. 2. A system of wound-drainage without tubes.
**canary-seed** (kan-a'-re-sēd). The hulled seeds of *Phalaris canariensis;* it is used in emollient poultices.
**cancellate, cancellated** (kan'-sel-āt, -a'-ted) [*cancellare,* to provide with a lattice]. Reticulated, or characterized by latticed lines, as the spongy tissue of bones, or certain leaves consisting entirely of veins.
**cancellation** (kan-sel-a'-shun) [*cancelli,* latticework]. The quality of being cancellate; cancellous structure.
**cancelli** (kan-sel'-li) [L. *pl.* of *cancellus,* a lattice]. See *cancellus.*
**cancellous** (kan'-sel-us) [*cancellus,* latticework]. Resembling latticework, as the tissue in the articular ends of long bones.
**cancellus** (kan-sel'-us) [L.; *pl., cancelli,* latticework]. A space, or unit of structure, in cancellous bone; any one of the minute divisions in spongy bone. The spongy, latticework texture of bone.
**cancer** (kan'-ser). 1. See *carcinoma.* 2. Any kind of malignant growth. Syn., *malignant disease.* c., **acinous,** c., **acute,** medullary carcinoma or medullary sarcoma. c., **adenoid,** a malignant form chiefly composed of tubules lined with epithelium; adenocarcinoma. c. **à deux,** cancer attacking both husband and wife, or successively a man's first and his second wife. c. **albus.** See *cancrum oris* and *noma.* c. **anthracinus,** one beginning as a black speck and developing into a mulberry-like growth. c., **apinoid,** a hard cancer, so called because of the cleanness of its section. c., **apioid,** a hard cancer resembling an immature pear in section. c. **aquaticus.** Synonym of *gangrenous stomatitis.* c., **Aran's green.** See *Aran's green cancer.* c., **areolar.** See *carcinoma, colloid.* c.-**bandage,** a crab-shaped bandage; a split cloth of eight tails. c., **black,** melanotic cancer. c. **of the blood,** leukocythemia. c., **cavernous,** a colloid carcinoma in which the alveoli have become absorbed. c.-**cell,** an epithelial cell of peculiar, distorted shape, found in the interior of cancer-nests. It is an ordinary epithelial cell altered in outline by pressure. c., **cellular,** c., **cerebriform.** See *carcinoma, soft,* and *sarcoma, encephaloid.* c., **chimney-sweep's.** See *carcinoma, chimney-sweep's.* c., **chondroid,** a hard cancer, which on section exhibits a shining, bluish-white appearance. c., **chronic,** a scirrhous cancer. c., **cicatrizing,** a form of hard cancer marked by atrophy and shrinking. c., **clay-pipe.** See *c., smoker's.* c., **Cohnheim's theory of the embryonic origin of.** See *Cohnheim's theory.* c., **colloid.** See *carcinoma, colloid.* c., **connective-tissue.** See *c., hard.* c., **dendritic,** a papilloma. c., **duct,** a form of columnar epithelioma. c., **eburneous.** See *scleroderma.* c., **embolic,** one due to embolic infection. c., **encephaloid.** See *carcinoma, encephaloid.* c. **en cuirasse,** disseminated cancer of the skin of the thorax. c., **epithelial,** epithelioma. c., **fasciculated,** a spindle-celled sarcoma. c., **fungoid,** c., **fungous,** c., **hematoid.** See *sarcoma, encephaloid; angioma, cavernous;* and *angiomyces.* c., **gelatiniform,** c., **gelatinous,** a colloid cancer. c., **glaucoid,** c., **green.** See *Aran's green cancer.* c., **hard,** one containing an excess of fibrous tissue.

**cancerate** 186 **cannon-bone**

c., inclusion theory of. See *Cohnheim's theory of the embryonic origin of cancer*. c., **jacket**, a continuously spreading cancerous infiltration of the superficial tissues. c.-**juice**, the milky fluid yielded by the cut surface of a cancer on scraping. c., **lipomatous**, one marked by many fat-cells in the stroma. c., **mammary**. 1. One affecting a mammary gland. 2. See c., mastoid. c., **mastoid**, a form of medullary sarcoma presenting on section the appearance of boiled cow's udder. c., **medullary**. See c.; *soft*. c., **melanotic**, a pigmented form. c. **mollis**, soft cancer. c., **nephroid**, a form of encephaloid sarcoma having the appearance of a kidney in section. c.-**nest**, a mass of cancer-cells. c. **occultus**, latent cancer. c., **osteoid**. 1. One containing a deposition of osseous material. 2. See *osteosarcoma*. 3. See *osteochondroma*. c., **osteolytic**, carcinomatous infiltration of bone without distinct tumor-formation. c., **primary**, c., **primitive**, one not due to infection from some preceding cancerous manifestation. c. **pullulans**, an ulcerating cancer which forms granulations. c., **pulpy**, c., **pultaceous**. See c., *colloid*. c., **ramose**, a form of hard cancer which branches. c., **rodent**. 1. Lupus. 2. Rodent ulcer. c., **scirrhous**. See c., *hard*. c., **smoker's**, epithelioma of the lip or mouth attributed to tobacco smoke or the irritation of a pipe. c., **soft**, one in which the cells predominate, the connective tissue being very small in amount. c., **solanoid**, one having the appearance of a potato in section. c., **soot**. See *carcinoma, chimney-sweep's*. c., **stone**, scirrhous carcinoma. c. **terebrans**, an epithelioma. c., **tubular epithelial**. See *cylindroma*. c. **verrucosus**, epithelioma with wart-like epithelial hypertrophy. c., **villous**. See *papilloma*. c. **xanthosus**. See *Aran's green cancer*.

**cancerate** (kan'-ser-āt). To become cancerous; to be developed into a cancer.

**canceration** (kan-ser-a'-shun). Development into a cancer; the assumption of malignant qualities by a tumor.

**cancerine** (kan'-ser-ēn) [*cancer*]. The name given to a ptomaine obtained from the urine in cases of carcinoma of the uterus. It is a white substance, crystallizing in fine needles and soluble in alkaline solutions. Its formula is C₅H₈NO₃.

**cancerism** (kan'-ser-izm), The tendency to cancerous formation.

**canceroderm** (kan'-ser-o-derm). A. T. Brand's name for angiomata, conspicuous in size and number, which appear in certain people who are not aged, and which he believes bear a relation to malignant growths.

**canceromyces** (kan-ser-o-mi'-sēz). See *cladosporium cancerogenes*.

**cancerous** (kan'-ser-us). Having the qualities of a cancer; malignant.

**canchasmus** (kan-kaz'-mus) [καγχασμός], loud laughter]. Hysterical or immoderate laughter.

**Cancriamœba macroglossia** (kang-kre-ah-me'-bah mak-ro-glos'-e-ah). An organism claimed to have been found in epithelial carcinoma, and of which the spores are identical with Plimmer's bodies.

**cancriform** (kang'-krif-orm) [*cancer*, a crab; *forma*, form]. Resembling a cancer in appearance.

**cancrine** (kang'-krēn). Cancerous.

**cancroid** (kang'-kroid) [*cancer*]. 1. Cancer-like. 2. An epithelioma. 3. A variety of keloid. c. **corpuscles**. See *corpuscles, cancroid*. c., **dermic**, epithelioma attacking all the layers of the skin. c., **follicular**, epithelioma arising in the hair-follicles or in the glands of the skin. c., **papillary**, epithelioma affecting the papillary layer and subsequently the other layers of the skin. c. **ulcer**. See *rodent ulcer*.

**cancroin** (kang'-kro-in) [*cancer*]. A substance (said to be identical with neurin) introduced by Adamkiewicz as a material for hypodermatic injection in cases of malignant disease, it being regarded by him as an alexin destructive of cancer tissue.

**cancrum** (kang'-krum) [*cancer*]. A cancer or rapidly spreading ulcer. c. **nasi**, gangrenous rhinitis of children. c. **oris**, a disease of childhood between the ages of one and five, characterized by the formation of foul, deep ulcers of the buccal surfaces of the cheeks or lips. There is but slight pain, but the prostration is great, and death usually results from exhaustion or blood-poisoning. The disease is bacterial, poor hygienic surroundings and a debilitated system being predisposing causes. Syn., *canker of the mouth; gangrenous stomatitis; noma*;

*gangrenous ulceration of the mouth*. c. **pudendi**, ulceration of the vulva.

**candela** (kan-de'-lah) [*candere*, to glow]. 1. A medicated candle for fumigation. 2. A wax bougie.

**canella** (kan-el'-ah) [dim. of *canna*, a reed]. The bark of *C. alba* deprived of its corky layer and dried, It is a native of the West Indies, and is an aromatic tonic and bitter stomachic. Dose of the powdered bark 15-30 gr. (1-2 Gm.). It is official in the B. P.

**cane-sugar**. See *saccharose* (1).

**canicaceous** (kan-e-ka'-shus), [*canicæ*, a kind of bran]. Furfuraceous.

**canine** (ka'-nin) [*canis*, a dog]. Partaking of the nature of, relating to, or resembling a dog or the sharp tearing-teeth of mammals, located between the incisors and the molars. c. **appetite**, bulimia. c. **eminence**, a prominence on the outer side of the maxilla. c. **fossa**. See *fossa, canine*. c. **laugh**, a sardonic smile or grin. c. **madness**, rabies; hydrophobia. c. **muscle**, the levator anguli oris. c. **teeth**, the cuspid teeth next to the lateral incisors; so called from their resemblance to a dog's teeth. Syn., *conoides; cuspidati; cynodontes; dentes angulares; dentes canini; dentes laniarii; eye-teeth*.

**caniniform** (kan-in'-if-orm) [*caninus*, pertaining to the dog; *forma*, shape]. Applied to teeth resembling canines.

**canities** (kan-ish'-e-ēz) [L.]. Poliosis; hoariness; blanching of the hair.

**canker** (kang'-ker) [*cancer*, a crab]. An ulceration of the mouth, or any ulcerous or gangrenous sore; cancrum oris; in farriery, a fetid abscess of the horse's foot. See *cancrum oris*. c.-**rash**. Synonym of *scarlatina*.

**Canna** (kan'-ah) [κάννα, a cane]. A genus of large-leaved marantaceous plants. *C. indica* has an acrid and stimulant root; it is alterative, diuretic, and diaphoretic. The rhizome of *C. speciosa* affords canna-starch, a substitute for arrowroot.

**cannabene** (kan'-ab-ēn). See under *cannabis*.

**cannabin** (kan'-ab-in). A crystalline resin from Indian hemp; it is hypnotic. Dose 1½-4 gr. (0.097-0.26 Gm.). c. **tannate**, a yellow, astringent powder, soluble in alkaline water or alcohol; it is hypnotic and sedative. Dose 2-10 gr. (0.13-0.6 Gm.).

**cannabindon** (kan-a-bin'-don), C₅H₁₂O. A dark, cherry-red syrup obtained from Indian hemp; soluble in alcohol, ether, and oils. It is hypnotic and narcotic. Dose ½-2 gr. (0.03-0.13 Gm.).

**cannabinine** (kan-ab'-in-ēn) [*cannabis*]. A volatile alkaloid from Indian hemp.

**cannabinol** (kan-ab'-in-ol), C₁₈H₂₄O₂. A red oil obtained by fractional distillation from Indian hemp, and supposed to be its most active ingredient.

**cannabinon, cannabinone** (kan-ab'-in-ōn) [*cannabis*]. An amorphous bitter resinoid from Indian hemp, used as a hypnotic. Dose 1-3 gr. (0.065-0.2 Gm.).

**Cannabis** (kan'-ab-is) [L.]. Hemp. Indian hemp. The flowering tops of *C. sativa*, of which there are two varieties, *C. indica* and *C. americana*, the former, being the more potent; they contain a resin, cannabin, and a volatile oil, from which are obtained *cannabene*, C₁₈H₂₀, a light hydrocarbon, and *cannabene hydride*, a crystalline body. It is antispasmodic, narcotic, and aphrodisiac. In large doses it produces mental exaltation, intoxication, and a sensation of double consciousness. It is used in migraine, in paralysis agitans, in spasm of the bladder, in sexual impotence, in whooping-cough, in asthma, and in other spasmodic affections. *Bang, bhang, cunjah, churrus*, and *hashish* are the various Indian names by which the drug is known. Dose 1 gr. (0.065 Gm.). **C. indica**, extract of (*extractum cannabis indicæ*, U. S. P.). Dose ⅛-1 gr. (0.01-0.065 Gm.). **C. indica, fluidextract of** (*fluidextractum cannabis indicæ*, U. S. P.), an alcoholic preparation. Dose 1-5 min. (0.06-0.3 Cc.). **C. indica, tincture of** (*tinctura cannabis indicæ*, U. S. P.), contains 10 % of the drug. Dose 20 min.-1 dr. (1.3-4.0 Cc.).

**cannabism** (kan'-ab-izm) [*cannabis*, hemp]. The habitual use of Cannabis indica; ill-health caused by the misuse of Cannabis indica.

**cannabist** (kan'-ab-ist) [*cannabis*, hemp]. A devotee to the use of Cannabis indica.

**cannon-bone, canon-bone** (kan'-on bōn). One of the functional and complete metacarpal or metatarsal bones of a hoofed quadruped, supporting the weight of the body upon the feet.

**cannula** (*kan'-ū-lah*) [dim. of *canna*, a tube]. A tube used for withdrawing fluids from the body. It is generally fitted with a pointed rod for puncturing the integument. **c., perfusion,** a double cannula, one tube of which is used for the inflow of a fluid and the other tube for the outflow; it is employed in the irrigation of a cavity.

**çannular, cannulate** (*kan'-ū-lar, -lāt*) [*cannula*]. Tubular; channeled.

**canor** (*ka'-nor*) [L.; pl., *canores*]. A musical sound. **c. stethoscopicus.** See *metallic tinkling*.

**Canquoin's paste** (*kan-kwan'*) [Alexandre *Canquoin*, French physician, 1823– ]. A paste of flour, water, and zinc chloride; it is a powerful escharotic.

**Cantani's diet** (*kan-tah'-ne*) [Arnoldo *Cantani*, Italian physician, 1837–1893]. An exclusive meat-diet in diabetes. **C.'s treatment,** a method of treating cholera by high enemata of large quantities of water containing tincture of opium and tannic acid at a temperature of from 100° to 104° F.

**cantering rhythm** (*kan'-ter-ing rithm*). See *bruit de galop*.

**canthal** (*kan'-thal*) [κανθός, a canthus]. Relating to a canthus.

**cantharene** (*kan'-thar-ēn*), C₁₀N₁₂I₂O₃. A compound obtained from cantharidin by action of hydriodic acid.

**canthariasis** (*kan-thar-i'-as-is*) [*cantharis*]. A term proposed by Hope for the diseases that originate from the presence in the body of coleopterous insects or their larvæ.

**cantharidal** (*kan-thar'-id-al*) [*cantharides*]. Relating to or containing cantharides. **c. collodion** (*collodium cantharidatum*, U. S. P.). See *collodion, cantharidal*.

**cantharidated** (*kan-thar'-id-a-ted*). Containing cantharides.

**cantharides** (*kan-thar'-id-ēz*) [κανθαρίς, a blistering Spanish fly]. The dried body of a species of beetle, *C. vesicatoria* (nat. ord. *Coleoptera*). It contains a powerful poisonous principle, cantharidin (*q. v.*). Locally applied, cantharidis is a rubefacient and vesicant; internally it is an irritant, causing pain and vomiting. In toxic doses it produces severe gastro-enteritis, strangury, and priapism. It is used as an external counterirritant in the form of "blisters." Internally it is employed as a stimulant to the genitourinary mucous membrane, especially in cystitis, atony of the bladder, amenorrhea, etc.; also in skin diseases. **c. cerate** (*ceratum cantharidis,* U. S. P.), cantharides, 35; rosin, 20; yellow wax 20; lard, 35 parts; liquid petrolatum, q. s. **c., liniment,** cantharides, 15 parts; oil of turpentine, q. s. ad 100. **c. ointment** (*unguentum cantharidis,* B. P.), cantharides, yellow wax, olive-oil. **c. paper** (*charta epispastica,* B. P.), blistering paper, contains cantharides, 1; Canada turpentine, 1; 10 olive-oil, 4; spermaceti, 3; white wax, 8; water, 10 parts, spread on paper. **c. plaster** (*emplastrum cantharidis,* B. P.), cantharides, yellow wax, prepared suet, prepared lard, rosin. **c., tincture of** (*tinctura cantharidis,* U. S. P.), contains 10 % of the drug. Dose 3–10 min. (0.09–0.3 Cc.). **c., vinegar of** (*acetum cantharidis,* B. P.), of the strength of 1 to 8.

**cantharidic** (*kan-thar-id'-ik*). Relating to or obtained from cantharides. **c. anhydride.** Synonym of *cantharidin*.

**cantharidin** (*kan-thar'-id-in*) [*cantharides*], C₁₀H₁₂O₄. The bitter principle contained in Spanish flies and other insects; it crystallizes in prisms or leaflets, and melts at 218°. It has an extremely bitter taste, and produces blisters on the skin. See *cantharis*.

**cantharidism** (*kan-thar'-id-ism*) [*cantharides*]. Cantharidal poisoning. **c., external,** poisoning by absorption from a cantharidal blister.

**cantharis** (*kan'-thar-is*). See *cantharides*.

**canthectomy** (*kan-thek'-to-me*) [*canthus*; ἐκτομή, a cutting out]. Excision of a canthus.

**canthitis** (*kan-thi'-tis*) [*canthus*; ιτις, inflammation]. Inflammation of a canthus.

**cantholysis** (*kan-thol'-is-is*) [*canthus*; λύσις, a loosening]. Canthotomy with section of the external canthal ligament.

**canthoplastic** (*kan-tho-plas'-tik*) [*canthus*; πλάσσειν, to form]. Relating to canthoplasty.

**canthoplasty** (*kan'-tho-plas-te*) [*canthus*; πλάσσειν, to form]. An operation for increasing the size of the palpebral fissure by cutting the outer canthus.

**canthorrhaphy** (*kan-thor'-a-fe*) [*canthus*; ῥαφή, a seam]. An operation to reduce the size of the palpebral fissure by suture of the canthus.

**canthotomy** (*kan-thot'-o-me*) [*canthus*; τομή, a cutting]. Surgical division of a canthus.

**canthus** (*kan'-thus*) [κανθός, canthus]. The angle formed by the junction of the eye-lids.

**cantus galli** (*kant'-us gal'-i*) [L., "cock-crowing"]. Same as *child-crowing*.

**canula** (*kan'-ū-lah*). See *cannula*.

**CaOC.** Abbreviation for cathodal opening contraction.

**caoutchouc** (*koo'-tshook*) [S, A.]. Rubber. The chief substance contained in the milky juice that exudes upon incision of a number of tropical trees belonging to the natural orders *Euphorbiaceæ, Artocarpaceæ,* and *Apocynaceæ.* The juice is a vegetable emulsion, the caoutchouc being suspended in it in the form of minute transparent globules. When pure, caoutchouc is nearly white, soft, elastic, and glutinous; it swells up in water without dissolving; the best solvents are carbon disulphide and chloroform. It melts at about 150° C. and decomposes at 200° C.

**cap.,** abbreviation for *capiat*, [L.], let him take.

**cap** (*kap*) [AS., *cappe*]. 1. See *tegmentum*. 2. The tissue covering the conical end of a lymph-follicle. **c., enamel,** the concave enamel-organ covering the top of the growing tooth-papilla. **c., nuclear,** a collection of chromophilic substance on one side of the nucleus of a cell.

**capacity** (*kap-as'-it-e*) [*capacitas*, capacity]. 1. The power of holding or containing; mental or physical ability. 2. Cubic extent. **c., testamentary,** a legal term signifying the degree of mental ability requisite for making a valid will. **c., thermal,** the amount of heat absorbed by a body in being raised 1° C. in temperature. **c., vital,** the total amount of air that can be expelled by the most forcible expiration after the deepest inspiration.

**capiat** (*ka'-pe-at*) [L., "let it take"]. An instrument intended for use in removing remnants of the placenta, polypi, or the like, from the uterine cavity.

**capillaire** (*kap-il-ār'*) [Fr.]. The plant *Adiantum capillus veneris,* a species of maiden-hair fern; also a cough-syrup prepared from the same. See *adiantum*.

**capillaraneurysm** (*kap-il-ar-an'-ū-rizm*) [*capillus*; ἀνεύρυσμα, a widening]. Excessive capillarectasia.

**capillarectasia** (*kap-il-ar-ek-ta'-ze-ah*) [*capillus*; ἔκτασις, a stretching-out]. Dilation of the capillaries.

**capillarimeter** (*cap-il-ar-im'-et-ur*) [*capillus,* a hair; μέτρον, a measure]. A device for estimating the diameter of capillary tubes.

**capillaritis** (*kap-il-ar-i'-tis*). Inflammation of the capillaries.

**capillarity** (*kap-il-ar'-it-e*) [*capillary*]. 1. Capillary attraction; the force that causes fluids to rise in fine tubes or bores. 2. The condition of being capillary.

**capillary** (*kap'-il-a-re*) [*capillus*]. 1. Hair-like; relating to a hair, to a hair-like filament, or to a tube with a hair-like bore. 2. A minute blood-vessel connecting the smallest ramifications of the arteries with those of the veins. **c. attraction.** See *capillarity.* **c. bronchitis.** See *bronchitis, capillary.* **c. drainage.** See *drainage.* **c. fissure, c. fracture,** a linear fracture, without displacement. **c. capillaries, Meigs',** the branching capillaries discovered by A. V. Meigs between the muscular fibers of the human heart. **c. nevus.** See *nevus* (2). **c. pulsation** of the capillaries sometimes seen in aortic regurgitation. **c. vessels,** the capillaries.

**capilliculture** (*kap-il'-e-kul-chur*) [*capillus,* hair; *cultura,* culture]. Systematic treatment for the improvement or restoration of the hair.

**capillitium** (*kap-il-ish'-e-um*) [L.]. The hair of the head, or the portion of the scalp thus covered.

**capillose** (*kap'-il-ōs*) [*capillosus*]. Hairy.

**capillurgy** (*kap'-il-ur-je*) [*capillus,* a hair; ἔργον, work]. The art of destroying superfluous hair.

**capillus** (*kap-il'-us*) [L.; *pl., capilli*]. 1. A hair; specifically a hair of the head. 2. A hair-like filament. 3. A hair's breadth (1-10 to 1-12 of a line).

**capistration** (*kap-is-tra'-shun*) [*capistrum,* a halter].

**capistrum** (*kap-is'-trum*) [L., "a muzzle or halter"; pl., *capistra*]. 1. A bandage for the head or lower jaw. Syn., *capelina.* 2. Trismus.

**capital** (*kap'-it-al*) [*caput*]. 1. Pertaining to the head, or to the summit of a body or object. 2. Of great importance, as a *capital* operation in surgery.

**capitalis reflexa** (*kap-it-a'-lis re-fleks'-ah*). A recurrent bandage for a stump.
**capitate** (*kap'-it-āt*) [*caput*, head]. In biology, having a head or a head-like termination.
**capitatum** (*kap-it-a'-tum*). The large bone of the carpus, the os magnum.
**capitellum** (*kap-it-el'-um*) [dim. of *caput*]. 1. A small head or rounded process of bone. 2. The rounded, external surface of the lower end of the humerus. 3. The bulb of a hair.
**capitiluvium** (*kap-it-el-u'-ve-um*) [*caput*, head; *luere*, to wash]. A washing or bathing of the head; a wash for the head.
**capitium** (*kap-ish'-e-um*) [L.]. A bandage for the head; it may be triangular or four-cornered. c. **magnum**, c. **quadrangulare**, c. **quadratum**, a four-cornered head bandage. c. **minus**, c. **triangulare**, a three-cornered head bandage.
**capitones** (*kap'-it-ōn-ez*) [L.]. Fetuses with heads too large for unassisted delivery.
**capitopedal** (*kap-it-o-ped'-al*) [*caput*, a head; *pes*, foot]. In biology, pertaining to or near the junction of the head and foot.
**capitular** (*kap-it'-u-lar*) [*caput*, head]. Pertaining to a capitulum or head. c. **process of a vertebra**, one with which the head of a rib articulates.
**capitulum** (*kap-it'-u-lum*) [L., "a small head"]. A little head. c. of **Santorini**, a small elevation on the apex of the arytenoid cartilage, corresponding in position to the posterior extremity of the vocal band.
**capnomor** (*kap'-no-mor*) [καπνός, smoke; μοῖρα, a part], C₈H₁₀O₂. A transparent, colorless, oily fluid, a constituent of smoke obtained from the heavy oil of tar. It dissolves caoutchouc.
**cappa** (*kap'-ah*). The ectocinereal lamina of the mesencephal.
**Capparis** (*kap'-ar-is*) [L., "the caper-bush"]. A genus of shrubs including the caper-bush, *C. spinosa*. Its flower-buds (*capers*) are pickled or made into sauce. The bark of the root and the flowers are official remedies in some countries. It is diuretic, cathartic, depurative, stimulant. *C. aphylla*, a shrub of India, is esteemed in the treatment of boils and affections of the joints. *C. coriacea* is a native of Peru; the fruit is antiepileptic and antihysteric. The root-bark of *C. jamaicensis*, of South America, is rubefacient, the root diuretic, the leaves and flowers antispasmodic.
**capped hock** (*kapd' hok*). In farriery, the development of a bruise at the point of the hock of a horse, with the formation of a hygroma, the result of rubbing or striking that part against the partition of the stall. c.-**knee**, a dropsical collection in the bursa in front of the knee-joint of the horse.
**Capranica's reaction for bile-pigments**. Add to the solution chloroform containing some bromine, and shake; it becomes first green, blue-violet, yellowish red, and finally colorless. If the green or blue solution is shaken with HCl, the color is destroyed by the acid. **C.'s reaction for guanin**. 1. A warm solution of guanin hydrochloride with a cold saturated solution of picric acid gives a yellow precipitate occurring as silky needles. 2. Add to a guanin solution a concentrated solution of potassium ferricyanide; a yellowish-brown prismatic precipitate is formed. 3. On the addition of a concentrated solution of potassium chromate to guanin solutions an orange-red crystalline precipitate is formed. It is very insoluble in water.
**caprate** (*kap'-rāt*). A salt of capric acid.
**caprenalin** (*kap-ren'-al-in*). Trade name of a preparation from suprarenal capsules; used as a vasoconstrictor, hemostatic, and cardiac stimulant.
**capreolar** (*kap-re'-o-lar*), **capreolary** (*kap'-re-o-la-re*), **capreolate** (*kap'-re-o-lāt*) [*capreolus*, a tendril]. In biology, climbing, furnished with tendrils; in anatomy resembling tendrils, as the spermatic vessels, *vasa capreolaria*. c. **vessels**, the spermatic vessels, from their twined and twisted appearance.
**capric** (*kap'-rik*) [*caper*, a goat]. Relating or belonging to, or having the odor of, a goat. c. **acid**. See *acid*, *capric*.
**capriloquium** (*kap-ril-o'-kwe-um*) [*caper*, a goat; *loqui*, to speak]. Same as *egophony*.
**caprin** (*kap'-rin*]) [see *capric*]. An oily and flavoring constituent of butter; glycerol caprate.
**caprinate** (*kap'-rin-āt*). See *caprate*.
**caprinic** (*kap-rin'-ik*). See *capric*.
**caprizant** (*kap'-ri-zant*) [see *capric*]. Leaping; of irregular motion, applied to the pulse.

**caproate** (*kap'-ro-āt*). A salt of normal caproic acid.
**caproic** (*kap-ro'-ik*). See *capric*. c. **acid**. See *acid*, *caproic*. c. **anhydride**, C₁₂H₂₂O₃, a neutral oily liquid.
**caproin** (*kap'-ro-in*). A fat, resembling caprin, found in goat's butter.
**caprone** (*kap'-rōn*) [see *capric*], C₁₁H₂₂O. Caproic ketone; a clear, volatile oil found in butter, and forming the larger part of the oil of rue.
**caproyl** (*kap'-ro-il*). 1. C₆H₁₁O. A hypothetical radical. Syn., *hexoyl*. 2. C₆H₁₃, a radical. Syn., *hexyl*.
**caproylamine** (*kap-ro-il'-am-in*) [*caproyl*; *amine*], C₆H₁₅N. Hexylamine. A ptomaine formed in the putrefaction of yeast.
**caprylate** (*kap'-ril-āt*). A salt of caprylic acid.
**caprylic acid** (*kap-ril'-ik*). See *acid*, *caprylic*.
**capsaicin** (*kap-sa'-is-in*). See *capsicin*.
**capsella** (*kap-sel'-ah*) [dim. of *capsa*, a box]. The leaves and stems of *C. bursa pastoris*, common in temperate climates. **C. bursa pastoris**, shepherd's purse; the leaves are hemostatic and antiscorbutic.
**capsicin** (*kap'-sis-in*). 1. C₉H₁₄O₂. The active principle of Cayenne pepper, found in the pericarp and placenta of *Capsicum fastigiatum*, and soluble in alcohol, ether, benzene, and fixed oils. It is a thick, yellowish-red substance, and its vapors are intensely acrid. Dose $\frac{1}{10}-\frac{1}{4}$ gr. (0.006–0.016 Gm.). 2. An oleoresin from capsicum, occurring as an oily liquid devoid of pungency.
**capsicol** (*kap'-sik-ol*) [*capsicum*; *oleum*, oil]. A red oil obtainable from the oleoresin of capsicum.
**capsicum** (*kap'-sik-um*) [*capsa*, a box]. Cayenne pepper. The fruit of *C. fastigiatum*, native to tropical Africa and America. Its odor and hot taste are due to a volatile oil, *capsicin*, C₉H₁₄O₂, which is irritant to the skin and mucous membranes. Internally it is a stomachic, tonic, diuretic and aphrodisiac. It is useful in atonic dyspepsia, flatulent colic, and intermittent fever. *C. annuum* is the common red pepper of the garden. c., **fluidextract of** (*fluidextractum capsici*, U. S. P.). Dose 5 min.–1 dr. (0.3–4.0 Cc.). c., **liniment**, 1 in 10, for chest affections, rheumatism, etc. c., **oleoresin of** (*oleoresina capsici*, U. S. P.). Dose $\frac{1}{8}$ gr. (0.03 Gm.). c. **plaster** (*emplastrum capsici*, U. S. P.), prepared from the oleoresin and adhesive plaster. c., **tincture of** (*tinctura capsici*, U. S. P.), contains 10 % of capsicum. Dose 5–30 min. (0.3–2.0 Cc.).
**capsitis** (*kap-si'-tis*). Same as *capsulitis*.
**capsotomy** (*kap-sot'-o-me*). See *capsulotomy*.
**capsula** (*kap'-su-lah*) [L., "a small box"]. 1. The internal capsule of the brain; it is the thick layer of fibers between the caudatum and thalamus mesad and the lenticula laterad; it is continuous with the crura caudad, and its expansion is called the *corona*. 2. See *capsule*. c. **articularis**, capsular ligament. c. **glomeruli**. See *Bowman's capsule*.
**capsular** (*kap'-su-lar*) [*capsule*]. Pertaining to a capsule. c. **artery**, the middle suprarenal artery, see *arteries*, *table of*. c. **cataract**, an opacity of the capsule of the crystalline lens. c. **hemiplegia**, a hemiplegia due to a lesion in the internal capsule. c. **ligament**, the sac or membranous bag that surrounds every movable joint or articulation. It contains the synovial fluid. c. **vein**, the suprarenal vein, see *vein*.
**capsulation** (*kap-su-la'-shun*) [*capsule*]. The act or process of inclosing in capsules.
**capsule** (*kap'-sūl*) [dim. of *capsa*, a chest]. 1. A membranous sac inclosing a part. 2. An envelope for administering medicines. c., **acoustic**. See *c., auditory*. c., **adipose**. See *c. of the kidney*. c., **aqueous**, c. **of the aqueous humor**, Descemet's membrane. c., **articular**. See *capsular ligament*. c., **auditory**, the primitive auditory organ, formed by the invagination of the nervous stratum of the epiblast. See also *vesicle, auditory*. c., **Bonnet's**. See *Bonnet's capsule*. c., **Bowman's**. See *Bowman's capsule*. c., **Bowman-Mueller's**. See *Bowman's capsule*. c., **brain**. See *capsula* (1). c., **cartilage**, c. **of a cartilage-cell**, the lining of cartilage-cavities containing the cartilage-cells. c., **crystalline**. See *c. of the lens*. c., **external**, a layer of white nerve-fibers forming part of the external boundary of the lenticular nucleus. c., **fibrous**. See *ligament*, *capsular*. c. **of Glisson**. See *Glisson's capsule*. c.s, **glutoid**, gelatin

capsules treated with formaldehyde. **c.,** hemorrhoidal, a metal, capsule-shaped device for applying Vienna paste to a hemorrhoid. **c.,** hyaloid. See *membrana limitans.* **c., internal,** a layer of nerve-fibers on the outer side of the optic thalamus and caudate nucleus, which it separates from the lenticular nucleus, and containing the continuation upward of the crus cerebri. **c. of the kidney,** the fat-containing connective tissue encircling the kidney. **c. of the lens,** a transparent, structureless membrane inclosing the lens of the eye. **c., Malpighian,** the commencement of the uriniferous tubules. See *Bowman's capsule.* **c. Mueller's.** See *Bowman's capsule.* **c., nasal,** the embryonic cartilage which becomes the nose. **c. of a nerve-cell,** that portion of the neurilemma which covers a ganglion-cell. **c., optic,** the embryonic structure forming the sclera. **c., periotic,** the structure surrounding the internal ear. **c., renal.** See *c., suprarenal.* **c.s, seminal,** expansions of the vasa deferentia near the seminal vesicles; applied by some authorities to the seminal vesicles. Syn., *capsulares seminales.* **c.s, sense,** the cartilaginous or bony cavities containing the organs of sense. **c., suprarenal,** the ductless, glandular body at the apex of each kidney. **c., suprarenal accessory,** an additional capsule attaining the size of a pea and sometimes attached to the suprarenal capsule by connective tissue. **c., synovial.** See *membrane, synovial.* **c. of Tenon,** the tunica vaginalis of the eye.
**capsulitis** (*kap-su-li'-tis*) [*capsule; itis,* inflammation]. Inflammation of the capsule of the lens or of the fibrous capsule of the eyeball.
**capsulociliary** (*kap-su-lo-sil'-e-a-re*) [*capsule; cilium,* an eyelid]. Relating to the capsule of the lens and to the ciliary organ.
**capsulolenticular** (*kap-su-lo-len-tik'-u-lar*) [*capsule; lenticula,* a lentil]. Relating to the lens and to its capsule.
**capsulopupillary** (*kap-su-lo-pu'-pil-a-re*) [*capsule; pupilla,* the pupil of the eye]. Relating to the capsule of the lens and to the pupil.
**capsulorrhaphy** (*kap-su-lor'-af-e*) [*capsule; ῥαφή,* a suture]. Suture of a capsule, to repair a rent or to prevent dislocation.
**capsulotome** (*kap'-su-lo-tōm*). 1. See *cystiotome.* 2. An instrument used by Buller in capsulotomy to steady the capsule; it consists of two fine needles fixed parallel to each other in a handle.
**capsulotomy** (*kap-su-lot'-o-me*) [*capsule; τέμνειν,* to cut]. The operation of rupturing the capsule of the crystalline lens in cataract-operations.
**captation** (*kap-ta'-shun*) [*captare,* to desire]. The first or opening stage of the hypnotic trance.
**captol** (*kap'-tol*). A product of the condensation of tannin and chloral; it is used in 1 to 2 % solution as an antiseborrheal agent and lotion for the hair. Syn., *tannochloral.*
**Capuron's cardinal points.** Four fixed points of the pelvic inlet, the two iliopectineal eminences anteriorly, and the two sacroiliac joints posteriorly.
**caput** (*kap'-ut*) [L.; pl., *capita*]. The head; also the chief part or beginning of an organ. **c. breve,** the transversus pedis muscle. **c. cæcum coli,** the cecum. **c. caudati,** the base of the corpus striatum. **c. coli,** the head of the colon; the cecum. **c. cordis,** the base of the heart. **c. cornu posterioris.** Same as *c. gelatinosum.* **c. epididymidis,** head of the epididymis, the globus major. **c. galeatum,** a child's head emerging at birth covered with the caul. **c. gallinaginis.** See *verumontanum.* **c. gelatinosum,** the name given to the translucent gray matter covering the dorsomesal periphery of the dorsal horn of the spinal cord. It is a peculiar, striated substance composed of numerous closely crowded cellular elements, in part connective-tissue cells, in part nerve-cells. **c. humerale,** the humeral head. **c. medullæ, c. medullæ oblongatæ,** Bartholin's name for the cerebrum as distinguished from the oblongata. **capita medullæ oblongatæ,** the thalami. **c. Medusæ,** the peculiar plexus of veins surrounding the umbilicus in periportal cirrhosis of the liver. It represents collateral paths for the return of the venous blood from the abdominal viscera. **c. obstipum.** Synonym of *wry-neck.* **c. penis,** the glans penis. **c. quadratum,** the rectangular head of rickets, flattened upon the top and at the sides, with projecting occiput and prominent frontal bosses. **c. succedaneum,** a tumor composed of a serosanguineous infiltration of the connective tissue situated upon the presenting part of the fetus. **c. tali,** the head of the astragalus.

**c. testis,** the epididymis. **c. transversum.** The same as *c. breve.*
**caputin** (*kap'-u-tin*). A proprietary preparation containing acetanilide.
**caraate** (*kah-rah-aht'-a*). Mal de los pintos.
**carageen, caragheen.** See *carrageen.*
**caramel** (*kar'-am-el*) [Fr., "burnt sugar"]. Cane-sugar deprived of two molecules of water. It is a viscid, brown-colored liquid.
**Carapa** (*kar'-ap-ah*) [*caraipi,* the Guiana name]. A genus of tropical meliaceous trees. *C. guianensis* has an antispasmodic and febrifuge bark, and its seeds afford *carap-oil,* a protective against insects and vermin. *C. moluccensis* is an East Indian tree; the bitter bark is used in diarrhea and the seeds in colic. The fruit and bark of *C. procera,* of the tropics of Asia and Africa, are antiperiodic; the oil from the seeds is anthelmintic and expectorant.
**carapine** (*kar'-ap-ēn*). An alkaloid from the bark of *Carapa guianensis.*
**caraway** (*kar'-ah-wā*). See *carum.*
**carbamate** (*kar'-bam-āt*). A salt of carbamic acid.
**carbamic** (*kar-bam'-ik*). Obtained from carbamide. **c. acid.** See *acid, carbamic.*
**carbamide** (*kar'-bam-id*) [*carbo,* a coal; *amide*], CH$_4$N$_2$O. Urea.
**carbamin** (*kar-bam'-in*). See *acetonitril.*
**carbasus** (*kar'-bas-us*) [*κάρβασα,* fine flax]. Gauze; thin muslin used in surgery. **c. carbolisatus** (N. F.), carbolized gauze. **c. iodoformata** (N. F.), iodoform gauze.
**carbazotate** (*har-bas-o'-tāt*) [*carbo,* a coal; *azotum,* nitrogen]. Same as *picrate.*
**carbazotic acid** (*kar-baz-ot'-ik*). See *acid, picric.*
**carbenzyme** (*kar'-ben-zīm*). Trade name of a preparation containing trypsin and charcoal, used in the treatment of tuberculous fistulæ and other tuberculous lesions; it is said to digest dead and disintegrating tissues.
**carbide** (*kar'-bīd*). A compound formed by the direct union of carbon with some radical or element.
**carbimids** (*kar'-bam-ids*). Bodies isomeric with cyanates, but distinguished from true cyanates in that alkalies decompose them into carbon dioxide and amine. Syn., *carbonylamines.*
**carbinol** (*kar'-bin-ol*) [*carbo*]. Methyl-alcohol, CH$_3$OH. Also a generic term for the alcohols formed by substituting hydrocarbon radicals for the hydrogen in the methyl radical of carbinol.
**carbo** (*kar'-bo*) [L.]. A coal; charcoal. **c. animalis** (U. S. P.), animal charcoal; bone-black; it is used in pharmacy and in manufacturing chemistry largely as a decolorizing agent and as a filter. **c. animalis purificatus** (U. S. P.), purified animal charcoal. Dose 20 gr.–1 dr. (1.3–4.0 Gm.). **c. ligni** (U. S. P.), wood-charcoal; an absorbent, disinfectant, and deodorizer, used in poulticing wounds and dressing ulcers. It is used internally in gastrointestinal irritation.
**carbocyclic compounds** (*kar'-bo-si-klik*). Organic compounds of the closed chain series in which the rings consist of carbon atoms exclusively.
**carboformal** (*kar-bo-form'-al*). A combination of carbon and paraformaldehyde in the form of blocks (Glüh blocks) for purposes of disinfection, the formaldehyde being liberated by the burning of the carbon.
**carbohemia** (*kar-bo-he'-me-ah*) [*carbo;* αἷμα, blood]. Imperfect oxidation of the blood.
**carbohydrate** (*kar-bo-hi'-drāt*) [*carbo;* ὕδωρ, water]. An organic substance belonging to the class of compounds represented by the sugars, starches and celluloses; and containing carbon, hydrogen and oxygen. The carbohydrates form a large class of organic compounds, and may be arranged into three groups: the glucoses (*monoses*); the disaccharids, or sugars; and the polysaccharids. The glucoses are the aldehyde derivatives or ketone derivatives of the hexahydric alcohols, into which they may be converted by the absorption of two hydrogen atoms. They are mostly crystalline substances, very soluble in water, but dissolving with difficulty in alcohol. They possess a sweet taste. The disaccharids and polysaccharids are ethereal anhydrides of the glucoses. They may all be converted into the glucoses by hydrolytic decomposition. The disaccharids are ether-like anhydrides of the hexoses.
**carbohydraturia** (*kar-bo-hi-drat-u'-re-ah*) [*carbohydrate;* οὖρον, urine]. The presence of an ab-

normally large proportion of carbohydrates in the urine.
**carbolate** (*kar′-bol-āt*). 1. A salt of phenol. 2. To impregnate with phenol.
**carbolfuchsin** (*kar-bol-fook′-sin*) [*carbo; fuchsin*]. A staining fluid consisting of 90 parts of a 5% aqueous solution of phenol and 1 part of fuchsin dissolved in 10 parts of alcohol.
**carbolic** (*kar-bol′-ik*) [*carbo; oleum*, oil]. Containing or derived from coal-tar oil. c. acid. See *acid, carbolic*, and *phenol*.
**carbolism** (*kar′-bol-izm*) [see *carbolic*]. Phenol poisoning; a diseased state induced by the misuse or maladministration of phenol. c., cutaneous, dry gangrene due to the continued application of a solution of it upon the skin.
**carbolize** (*kar′-bol-īz*) [see *carbolic*]. To impregnate with carbolic acid. To render aseptic or antiseptic by the use of carbolic acid.
**carbomarasmus** (*kar-bol-mar-az′-mus*) [*carbolic*; μαρασμός, decay]. Chronic carbolism: a condition marked by vomiting, vertigo, headache, salivation, nephritis, and general marasmus.
**carboluria** (*kar-bol-ū′-re-ah*) [*carbo*, a coal; *oleum*, oil; οὖρον, urine]. The presence of carbolic acid in the urine, producing a dark discoloration. It is one of the signs of carbolic-acid poisoning.
**carboxylene** (*kar-bol-zī′-lēn*). A clearing mixture composed of phenol, 1 part, and xylene, 3 parts; used for clearing microscopic sections which are to be mounted in Canada balsam or other resinous medium.
**carbometer.** See *carbonometer*.
**carbon** (*kar′-bon*) [*carbo*]. Charcoal. C = 12; quantivalence II, IV. A nonmetallic element occurring in the various forms of diamond, graphite or "black lead," charcoal, and lamp-black. It is the central or characteristic element of organic compounds. c. dioxide, the acid, gaseous product, having the composition of $CO_2$, commonly known as "carbonic-acid gas" or carbonic acid. It is a colorless gas, having a sp. gr. of 1.52, soluble in cold water, and possessing a pungent smell and an acid taste. Inhaled, it destroys animal life by asphyxiation. c. dioxide snow, frozen $CO_2$ used in the treatment of certain skin affections. c. **disulphide** (*carbonei disulphidum*, U. S. P.), carbon bisulphide, $CS_2$, a colorless, transparent liquid, of offensive odor, highly inflammable, very poisonous. It is used as a solvent for caoutchouc and as a reagent. c. **monoxide**, CO, carbonic oxide, a colorless, tasteless, and inodorous gas, one of the products of imperfect combustion. It is actively poisonous. c. **nitride**, CN, cyanogen. c. **oxysulphide**, a body, COS, formed by conducting sulphur-vapor and carbon monoxide through red-hot tubes; it is a colorless gas, with a faint and peculiar odor. It unites readily with air, forming an explosive mixture, and is soluble in an equal volume of water. It is present in the waters of some mineral springs. c. **tetrachloride**, $CCl_4$, anesthetic, used in asthma by inhalation.
**carbonate** [*carbon*]. A salt of carbonic acid. c., acid, a substitution-compound of carbonic acid in which there is replacement of but one of its hydrogen atoms with a base. c., basic, a compound of a carbonate with the oxide of the same base. c., hydric. See c., acid. c., hydrogen. 1. Carbonic acid. 2. Acid carbonate. c., neutral, c., normal, a substitution-compound of carbonic acid in which a base replaces all its hydrogen.
**carbonated** (*kar′-bo-na-ted*). 1. Containing carbonic acid or carbon dioxide. 2. Changed into a carbonate.
**carbone** (*kar′-bōn*). A carbuncle.
**carbonemia** (*kar-bon-e′-me-ah*) [*carbo*; αἷμα, blood]. An accumulation of carbon dioxide in the blood.
**carboneum** (*kar-bon′-e-um*). Carbon.
**carbonic** (*kar-bon′-ik*). Relating to, obtained from, or containing carbon. c. acid. See *carbon dioxide*. c. anhydride, carbon dioxide. c. snow, carbon dioxide in crystal form.
**carbonide** (*kar′-bon-īd*). 1. See *carbide*. 2. A mineral which contains carbon. 3. An oxalate freed from its hydrogen by heat.
**carbonite** (*kar′-bon-īt*). An oxalate.
**carbonization** (*kar-bon-iz-a′-shun*) [*carbon*]. The process of decomposing organic substances by heat without air, until the volatile products are driven off and the carbon remains.
**carbonometer** (*kar-bon-om′-et-er*) [*carbon*; μέτρον, a measure]. An apparatus for indicating the degree to which the air of a room is vitiated by carbon dioxide.
**carbonometry** (*kar-bon-om′-et-re*). The determination of the amount of carbon dioxide present in air, by the aid of the carbonometer.
**carbonous** (*kar′-bon-us*). Containing carbon.
**carbonyl** (*kar′-bon-il*) [*carbon*]. A hypothetic organic radical having the formula CO.
**carbonylamines** (*kar-bon-il′-am-ēns*). See *carbimides*.
**carborundum** (*kar-bo-run′-dum*). Silicon carbide, SiC, a substance of extreme hardness.
**carbosapol** (*kar-bo-sa′-pol*) [*carbo; sapo*, soap]. A clear disinfectant solution obtained by warming together phenol, 50 parts; yellow soda-soap, 25 parts; and soft potash-soap, 25 parts.
**carbostyril** (*kar-bo-stir′-il*) [*carbo*, charcoal; *styrax*, storax], $C_9H_7NO$. Oxyquinoline; a compound prepared by digesting quinoline with a bleaching-lime solution.
**carbosulphide, carbosulphuret** (*kar-bo-sul′-fīd, -fur-et*). A compound of carbon and sulphur with a radical.
**carbothialdin** (*kar-bo-thi-al′-din*), $C_4H_{10}N_3S_2$. White crystals obtained on evaporating carbon sulphide with an alcoholic solution of aldehyde ammonia. It is soluble in acids.
**carbovinate** (*kar-bo-vin′-āt*). An ethyl carbonate.
**carboxyhemoglobin** (*kar-boks-e-hem-o-glo′-bin*) [*carboxyl; hemoglobin*]. The compound of carbon monoxide and hemoglobin formed when CO is present in the blood. The carbon monoxide displaces the oxygen and checks the respiratory function of the red corpuscles.
**carboxyl** (*kar-boks′-il*) [*carbo*; ὀξύς, sharp]. 1. The group, CO . OH, characteristic of the organic acids. The hydrogen of this can be replaced by metals, forming salts. 2. Same as *carbonyl*.
**carboy** (*kar′-boi*) [Turk., *karaboya*]. A large bottle protected by wickerwork and a wooden box, used in the transportation of corrosive and other liquids.
**carbuncle, carbunculus** (*kar′-bung-kl, kar-bung′-kū-lus*) [*carbo*]. A hard, circumscribed, deep-seated, painful suppurative inflammation of the subcutaneous tissue. It differs from a boil in being of greater size, having a flat top, and several points of suppuration. It is erroneously called anthrax.
**carbunculosis** (*kar-bung-kū-lo′-sis*). A condition characterized by the formation of carbuncles.
**carburet** (*kar′-bū-ret*). Carbide.
**Carcassonne's ligament.** See *Colles' fascia*.
**carcea.** A disease of sheep described by Babes in Rumania; it is probably a form of trypanosomiasis.
**carcinelcosis** (*kar-sin-el-ko′-sis*) [*carcinoma*; ἕλκωσις, ulceration]. A cancerous ulcer, c. fungosa. See *cancer verrucosus*.
**carcinolytic** (*kar-sin-o-lit′-ik*) [*carcinoma*; λύσις, solution]. Said of a substance which is destructive to cancer cells.
**carcinoma** (*kar-sin-o′-mah*) [καρκίνωμα; καρκίνος, a crab; ὄμα, tumor]. Cancer. A malignant epithelial tumor composed of a connective-tissue stroma surrounding groups or nests of epithelial cells. Three varieties are generally described—the squamous, the cylindrical, and the glandular. See *Boas' sign; de Morgan's spots; Semon's symptom; Spiegelberg's sign*. c., acinous. See *cancer, acinous*. c., adenoid, c. adenodes, c. adenoides. See *cancer, adenoid*. c. asbolicum. See *c., chimney-sweep's*. c., chimney-sweep's, epithelioma of the scrotum, occurring among chimney-sweepers, and supposed to be caused by the irritant action of soot. Syn., *soot cancer*. c., colloid, one in which the delicate connective-tissue stroma is filled with colloid matter, the result of a colloid degeneration of the epithelial cells. In some cases the degeneration is mucoid instead of colloid. It affects chiefly the alimentary canal, uterus, etc. c., cylindrical, one in which the cells tend to assume a cylindrical or columnar shape. This shape is best seen in the cells nearest the periphery of the nests. c. **durum**, a hard cancer. c., **encephaloid**, one of rapid growth, with a small amount of stroma, large alveoli, and greater amount of cells and blood-vessels. c., **fibromedullary**, one containing about an equal portion of cells and stroma. c. **fibrosum**, c., **fibrous**. See *c., scirrhous*. c., **glandular**, a carcinoma in which the cells are of the glandular or secreting type. c., **hyaline**. See *c., colloid*. c., **lenticular**, a form of scirrhous cancer. c. **melanodes**, a pigmented cancer. c. **molle**, a soft or a medullary cancer. c. **nigrum**.

See *melanocarcinoma*. c. **psammosum**, one in which stratified calcareous concretions differing from those found in psammomata have replaced the epithelial elements. c., **reticulated**, one which has undergone fatty metamorphosis and exhibits its stroma more distinctly. c. **sarcomatodes, c., sarcomatous**, an adenocarcinoma which has undergone sarcomatous degeneration of the connective tissue. c., **scirrhous**, a form which occurs most commonly in the breast; it has a stout, fibrillated stroma, closely packed with large nucleated cells. Syn., *hard carcinoma*. c. **scroti, c. scroti asbolieum**. See c., *chimney-sweep's*. c., **squamous**, one derived from squamous epithelium; the cells are cuboid in shape. c. **ventriculi**, cancer of the stomach. c., **villous**. See *papilloma*.
**carcinomatoid** (kar-sin-o'-mat-oid) [*carcinoma*; εἶδος, appearance]. Resembling a carcinoma.
**carcinomatosis** (kar-sin-o-mat-o'-sis). The pathological condition giving rise to carcinomata.
**carcinomatous** (kar-sin-o'-mat-us) [*carcinoma*]. Relating to or affected with carcinoma.
**carcinomelcosis** (kar-sin-om-el-ko'-sis). See *carcinelcosis*.
**carcinopolypus** (kar-sin-o-pol'-e-pus). A cancerous polyp.
**carcinosarcoma** (kar-sin-o-sar-ko'-mah) [*carcinoma; sarcoma*]. A mixed tumor having the characters of carcinoma and sarcoma; it usually affects the thyroid gland.
**carcinosis** (kar-sin-o'-sis) [*carcinoma*]. 1. A carcinomatous cachexia; a tendency to the development of malignant disease. 2. A form of carcinoma, usually fatal, beginning generally in the uterus or the stomach and spreading to the peritoneum. c., **acute**, rapidly fatal carcinosis. c., **miliary**, one in which there are many secondary nodules the size of miliary tubercles. c., **miliary, acute**, the rapid formation of minute cancerous nodules, either primary or secondary, within an internal organ or upon its surface.
**carcinous** (kar'-sin-us). Cancerous.
**carcinus** (kar'-sin-us) [καρκίνος, crab]. Same as cancer, or carcinoma.
**cardamom, cardamomum** (kar'-dam-om, kar-dam-o'-mum) [L.]. The fruit of *Elettaria cardamomum*, cultivated in Malabar. Its properties are due to a volatile oil, C₂₆H₃₈. It is an aromatic, carminative stomachic, used as an ingredient of several "bitters." When combined with purgatives it is useful to prevent griping. c., **infusion of**. Dose 2 oz. (64 Cc.). c., **tincture of** (*tinctura cardamomi*, U. S. P.), 20% strength. Dose ½-2 dr. (2-8 Cc.). c., **tincture of, compound** (*tinctura cardamomi composita*, U. S. P.), cardamom, 20; cinnamon, 20; caraway, 10; cochineal, 5; glycerol, 60; dilute alcohol, q. s. ad 1000 parts. Dose ½-2 dr. (2-8 Cc.).
**Cardarelli's symptom** [Antonio *Cardarelli*, Italian physician, 19th century]. See *Oliver's symptom*.
**Carden's amputation** (kar'-den) [Henry Douglas *Carden*, English surgeon, —1872]. Amputation through the condyles of the femur just above the articular surface; a single rounded flap is removed from the front of the joint, and the operation is completed by a circular incision.
**cardia** (kar'-de-ah). 1. The heart. 2. The esophageal orifice of the stomach. 3. The fundus of the stomach.
**cardiac** (kar'-de-ak) [*cardia*]. 1. Pertaining to the heart. 2. Pertaining to the cardia of the stomach. 3. A drug acting especially on the heart. c. **cycle**, the period included between the beginning of one heart-beat and the beginning of another. c. **dropsy**, a dropsical effusion due to heart disease with loss of compensation. c. **ganglia**, ganglia lying in the grooves and substance of the heart—the principal ones are Remak's and Bidder's, the first on the surface of the sinus venosus, and the latter (2) at the auriculoventricular groove. c. **impulse**, the elevation caused by the movement of the heart, usually seen in the fifth left intercostal space r **murmur**. See *murmur, cardiac*. c. **orifice** (of the stomach), the esophageal orifice. c. **passion**. See *cardialgia*. c. **plexus**. See *plexus, cardiac*. c. **rhythm**, the term given to the normal regularity in the force and volume of the individual heart-beats.
**cardiactia** (kar-de-ak'-te-ah) [καρδία, heart; *arctus*, bound]. Cardiac stenosis.
**cardiagra** (kar-de-a'-grah) [καρδία, heart; ἄγρα, seizure]. 1. Gouty attack of the heart. 2. Angina pectoris.
**ardiagraphy** (kar-de-ag'-raf-e). See *cardiography* (2).
**cardialgia** (kar-de-al'-je-ah) [*cardia*; ἄλγος, pain]. Pain in the region of the heart, usually due to gaseous distention of the stomach; heartburn. Syn., *morbus cardiacus; morsus stomachi; morsus ventriculi*. c. **icterica**, heartburn with jaundice. c. **inflammatoria**, gastritis. c. **sputatoria**, pyrosis.
**cardiameter** (kar-de-am'-et-er) [*cardia*, μέτρον, measure]. An apparatus for determining the position of the cardiac orifice of the stomach.
**cardiamorphia** (kar-de-am-or'-fe-ah) [καρδία, heart; ἀ, priv.; μορφή, form]. Deformity or malformation of the heart.
**cardianastrophe** (kar-de-an-as'-tro-fe) [καρδία, heart ἀναστροφή, a turning back]. Congenital displacement of the heart to the right side of the chest.
**cardianesthesia** (kar-de-an-es-the'-ze-ah) [*cardia;* ἀναισθησία, want of feeling]. A condition of the heart marked by lack of sensation.
**cardianeuria** (kar-de-ah-nū'-re-ah) [καρδία, heart; ἀ, priv.; νεῦρον, a nerve]. Lack of nerve-stimulus to the heart.
**cardianeurysma** (kar-de-an-ū-riz'-mah) [*cardia;* ἀνεύρυσμα, a widening]. Aneurysm of the heart.
**cardiant** (kar'-de-ant) [*cardia*]. 1. Affecting the heart. 2. A remedy that affects the heart.
**cardiaortic** (kar-de-ah-or'-tik). Relating to the heart and the aorta.
**cardiaplegia** (kar-de-ah-ple'-je-ah). See *cardioplegia*.
**cardiasthenia** (kar de as the' ne-ah) [*cardia;* ἀσθένεια, weakness]. A peculiar weakness of the heart due to neurasthenic conditions.
**cardiasthma** (kar-de-az'-mah) [καρδία, heart; ἄσθμα, asthma]. Dyspnea or so-called asthma due to heart disease.
**cardiataxia** (kar-de-at-ak'-se-ah) [καρδία, heart; *ataxia*]. Incoördination of the contractions of the heart.
**cardiatelia** (kar-de-ah-te'-le-ah). See *atelocardia*.
**cardiatomy**. See *cardiotomy*.
**cardiatrophia** (kar-de-at-ro'-fe-ah) [καρδία, heart; ἀτροφία, wasting]. Atrophy of the heart.
**cardiauxe** (kar-de-awks'-e) [καρδία, heart; αὔξη, increase]. Enlargement of the heart.
**cardicentesis** (kar-de-sen-te'-sis). See *cardiocentesis*.
**cardiechema** (kar-de-ek-e'-mah) [καρδία, heart; ἤχημα, sound; *pl., cardiechemata*]. A sound produced in or by the heart.
**cardiectasis** (kar-de-ek'-tas-is) [καρδία, heart; ἔκτασις, a stretching out]. Dilatation of the heart.
**cardiectomy** (kar-de-ek'-to-me) [*cardia;* ἐκτομή, cutting out]. Excision of the cardiac end of the stomach.
**cardielcosis** (kar-de-el-ko'-sis) [*cardia;* ἕλκωσις, ulceration]. Ulceration of the heart.
**cardiemphraxia** (kar-de-em-fraks'-e-ah) [καρδία, heart; ἔμφραξις, obstruction]. Obstruction to the blood-current in the heart.
**cardiethmoliposis** (kar-de-eth-mo-lip-o'-sis) [*cardia;* ἠθμός, a sieve; λίπος, fat]. A deposit of fat in the connective tissue of the heart.
**cardieurysma** (kar-de-u-riz'-mah) [καρδία, heart; εὐρύς, wide]. Dilatation of the heart.
**cardinal** (kar'-din-al) [*cardo*, a hinge]. Important; preeminent. c.-**flower**, a common name for several species of *Lobelia*, chiefly *Lobelia cardinalis*. c. **points of Capuron**. See *Capuron's cardinal points*. c. **veins**, the venous trunks which, in the embryonic stage, form the primitive jugular veins.
**cardine** (kar'-dēn). A fluid preparation of sheep-hearts digested in glycerol and boric acid, used subcutaneously as a heart-tonic and diuretic. Dose 50 min.–1½ dr. (3–5 Cc.).
**cardio-** (kar-de-o-) [*cardia*]. A prefix meaning relating to the cardia.
**cardioaccelerator** (kar-de-o-ak-sel'-er-a-tor). Hastening the action of the heart. c. **center**. See *center, cardioaccelerator*.
**cardioaortic** (kar-de-o-a-or'-tik). Relating to the heart and the aorta. c. **interval**, the interval between the apex beat and the arterial pulse.
**cardioarterial** (kar-de-o-ar-te'-re-al). Pertaining to the heart and the arteries.
**cardioaugmenter** (kar-de-o-aug-men'-tor). Increasing the vigor or force of the heart-beat.
**cardiocele** (kar'-de-o-sēl) [*cardio-;* κήλη, hernia]. Hernia of the heart. c. **abdominalis**, hernial protrusion of the heart into the abdomen.

**cardiocentesis** (kar-de-o-sen-te'-sis) [cardio-; κέντησις, puncture]. Puncture of one of the chambers of the heart to relieve engorgement.

**cardioclasia** (kar-de-o-kla'-ze-ah) [cardio-; κλάσις, rupture]. Rupture of the heart.

**cardiodemia** (kar-de-o-de'-me-ah) [cardio-; δημός, fat]. Fatty heart; fatty degeneration of the heart.

**cardiodilator** (kar-de-o-di-la'-tor). An instrument for dilating the cardia.

**cardiodynia** (kar-de-o-din'-e-ah) [cardio-; όδύνη, pain]. Pain in or about the heart.

**cardiodysesthesia**, (kar-de-o-dis-es-the'-ze-ah) [cardia; δυs, bad; αἴσθησις, perception]. Defective innervation of the heart.

**cardiodysneuria** (kar-de-o-dis-nū'-re-ah). See cardiodysesthesia.

**cardiogmus** (kar-de-og'-mus) [cardio-; ὀγμός, a furrow]. 1. Cardialgia. 2. Aneurysm of the heart. 3. Angina pectoris. c. strumosus, synonym of exophthalmic goiter.

**cardiogram** (kar'-de-o-gram) [cardio-; γράμμα, a writing]. The tracing of the cardiac impulse made by the cardiograph.

**cardiograph** (kar'-de-o-graf) [cardio-; γράφειν, to write]. An instrument for registering graphically the modifications of the pulsations of the heart.

**cardiographer** (kar-de-og'-ra-fer) [see cardiograph]. An authority upon diseases of the heart.

**cardiographic** (kar-de-o-graf'-ik) [cardio-; γράφειν, to write]. Pertaining to or recorded by the cardiograph.

**cardiography** (kar-de-og'-ra-fe) [cardio-; γράφειν, to write]. 1. The use of the cardiograph. 2. A description of the anatomy of the heart.

**cardioid** (kar'-de-oid) [cardio-; εἶδος, likeness]. Like a heart.

**cardioinhibitory** (kar-de-o-in-hib'-it-o-re) [cardio-; inhibere, to restrain]. Inhibiting or diminishing the heart's action. The cardioinhibitory fibers pass to the heart through the pneumogastric nerves.

**cardiokinetic** (kar-de-o-kin-et'-ik) [cardio-; κινεῖν, to move]. 1. Exciting the heart-action. 2. An agent which excites the action of the heart.

**cardiolith** (kar'-de-o-lith) [cardio-; λίθος, a stone]. A cardiac concretion.

**cardiology** (kar-de-ol'-o-je) [cardio-; λόγος, discourse]. The anatomy, physiology, and pathology of the heart.

**cardiolysis** (kar-de-ol'-is-is) [cardio-; λύσις, loosening]. Resection of the ribs and sternum over the pericardium to free the latter from its adhesions to the anterior chest-wall in adhesive mediastinopericarditis.

**cardiomalacia** (kar-de-o-mal-a'-she-ah) [cardio-; μαλακία, softness]. Softening of the heart.

**cardiomegalia** (kar-de-o-meg-a'-le-ah) [cardio-; μέγας, large]. Cardiac enlargement.

**cardiomelanosis** (kar-de-o-mel-an-o'-sis) [cardio-; melanosis]. Melanosis of the heart.

**cardiometer** (kar-de-om'-et-er) [cardio-; μέτρον, a measure]. An instrument for estimating the force of the heart's action.

**cardiometry** (kar-de-om'-et-re) [cardio-; μέτρον, a measure]. The estimation of the size and dimentions of the heart (as by means of auscultation and percussion).

**cardiomyolipozis** (kar-de-o-mi-o-lip-o'-sis) [cardio-; μῦs, muscle; λίπος, fat]. Fatty degeneration of the heart-muscle.

**cardiomyomalacia** (kar-de-o-mi-o-mal-a'-she-ah). See cardiomalacia.

**cardioncus** (kar-de-ong'-kus) [cardio-; ὄγκος, a tumor]. An aneurysm in the heart or one in the aorta close to the heart.

**cardionecrosis** (kar-de-o-nek-ro'-sis) [cardio-; νέκρωσις, a killing]. Gangrene of the heart.

**cardionosos**, **cardionosus** (kar-de-on-o'-sos, -sus) [cardio-; νόσος, disease]. Any pathological affection of the heart.

**cardiopalmus** (kar-de-o-pal'-mus) [cardio-; παλμός, palpitation]. Palpitation of the heart.

**cardioparaplasis, cardioparaplasmus** (kar-de-o-par-a-pla'-sis, -plas'-mus) [cardio-; παραπλάζειν, to wander from the right way]. Cardiac malformation.

**cardiopath** (kar'-de-o-path) [cardio-; πάθος, disease]. A sufferer from heart disease.

**cardiopathy** (kar-de-op'-a-the) [cardio-; πάθος, disease]. Any disease of the heart.

**cardiopericarditis** (kar-de-o-per-e-kar-di'-tis) [cardio-; pericardium; ιτις, inflammation]. Associated carditis and pericarditis; inflammation of the heart tissues and of the pericardium.

**cardiophone** (kar'-de-o-fōn) [cardio-; φωνή, voice]. An instrument used to aid in hearing the sounds of the heart.

**cardiophtharsis** (kar-de-of-thar'-sis) [cardio-; φθείρειν, to corrupt]. Any affection of the heart causing destruction of its substance.

**cardioplegia** (kar-de-o-ple'-je-ah) [cardio-; πληγή, a stroke]. Paralysis of the heart.

**cardiopneumatic** (kar-de-o-nū-mat'-ik); [cardio-πνεῦμα, breath]. Pertaining to the heart and respiration. c. movements, those movements of the air in the lungs that are caused by the pulsations of the heart and larger vessels.

**cardiopneumograph** (kar-de-o-nū'-mo-graf) [cardio-; πνεῦμα, breath; γράφειν, to write]. An instrument designed for graphically recording cardiopneumatic movements.

**cardioptosis** (kar-de-op-to'-sis) [cardio-; πτῶσις, falling]. Prolapse of the heart. Syn., Rummo's disease.

**cardiopulmonary** (kar-de-o-pul'-mon-a-re). Relating to the heart and lungs; cardiopulmonic.

**cardiopuncture** (kar-de-o-punk'-chūr) cardio-; punctura, a puncture]. 1. Cardiocentesis. 2. Any surgical or vivisectional puncture of the heart.

**cardiopyloric** (kar-de-o-pi-lor'-ik) [cardio-; pyloric]. Referring to both the cardiac and pyloric portions of the stomach.

**cardiorenal** (kar-de-o-re'-nal) [cardio-; ren, kidney]. Relating to the heart and the kidneys.

**cardiorrhaphy** (kar-de-or'-af-e). Suturing of the heart.

**cardiorrheuma** (kar-de-or-u'-mah) [cardio-; rheumatism]. Rheumatism of the heart.

**cardiorrhexis** (kar-de-or-eks'-is) [cardio-; ῥῆξις, a tearing]. Rupture of the heart.

**cardioschesis** (kar-de-os'-kis-is). [cardio-; σχίσις, a cleaving]. The tearing apart of adhesions which exist between the heart and the chest-wall in adhesive pericarditis.

**cardiosclerosis** (kar-de-o-skle-ro'-sis) [cardio-; σκληροεῖν, to harden]. Induration of the tissues of the heart. See fibroid heart.

**cardioscope** (kar'-de-o-skōp) [cardio-; σκοπεῖν, to view]. An instrument for the observation of the movements or of lesions of the heart.

**cardiospasm** (kar'-de-o-spasm) [cardio-; σπασμός, a drawing]. 1. A spasm of the heart. 2. Spasmodic contraction of the esophageal opening of the stomach.

**Cardiospermum** (kar-de-o-sper'-mum) [cardio-; σπέρμα, seed]. A genus of plants of the order Sapindaceæ. C. halicacabum is a climbing tropical annual; the leaves and mucilaginous root are diuretic and diaphoretic.

**cardiosphygmograph** (kar-de-o-sfig'-mo-graf). An instrument for the simultaneous recording of the heart and pulse movements.

**cardiostenosis** (kar-de-o-ste-no'-sis) [cardio-; στένωσις, narrowing]. Constriction of the heart, especially of the conus arteriosus; also the development of such a constriction.

**cardiotomy** (kar-de-ot'-o-me) [cardio-; τομή, cutting]. 1. The anatomy or dissection of the heart. 2. Incision of the heart. 3. Incision of the cardiac end of the stomach.

**cardiotopography** (kar-de-o-to-pog'-ra-fe) [cardio-; τόπος, place; γράφειν, to write]. The topography or topographic anatomy of the heart and the cardiac area.

**cardiotoxic** (kar-de-o-toks'-ik) [cardio-; τόξικον, poison]. Having a poisonous effect upon or through the heart.

**cardiotrauma** (kar-de-o-traw'-mah) [cardio-; τραῦμα, a wound]. Traumatism or wound of the heart.

**cardiotremus** (kar-de-ot'-ro-mus) [cardio-; tremere, to tremble]. Fluttering of the heart.

**cardiotrophia**, **cardiotrophia** (kar-de-ot-ro'-fe-ah, kar-de-ot-ro'-fe-ah) [cardio-; τροφή, nourishment]. 1. Heart-nutrition. 2. The volume of the heart.

**cardiovascular** (kar-de-o-vas'-kū-lar) [cardio-; vasculum, a small vessel]. Pertaining to the heart and the blood-vessels.

**cardipaludism** (kar-de-pal'-u-dizm) [καρδία, heart; paludism]. Disturbance of the heart's action due to malaria.

**cardipericarditis**. See cardiopericarditis.

**carditic** (kar-dit'-ik) [καρδία, the heart; ιτις, inflammation]. Relating to or affected with carditis.

**carditis** (*kar'-di'-tis*) [*cardia*; ιτις, inflammation]. Inflammation of the heart. **c., internal.** Synonym of *endocarditis*.
**cardivalvulitis** (*kar-de-val-vū-li'-tis*). Endocarditis confined to the valves.
**cardol** (*kar'-dol*). See *anacardium*.
**Carduus** (*kar'-du-us*) [L., "a thistle"]. The seeds of *C. marianus*, St. Mary's-thistle, and *C. benedictus*, blessed thistle. A decoction of the former, 2 oz. to 1 pint, constitutes an old and popular remedy for hemoptysis. The latter is also a popular cure-all, used mainly as a tonic bitter. Dose of *decoction* 1 dr.–½ oz. (4–16 Cc.); of *tincture* 10–20 min. (0.6–1.2 Cc.).
**cargentos** (*kar-jen-tos*). Colloidal silver oxide containing about 50 per cent. of silver, and used similarly to silver nitrate.
**Cargile membrane** (*kar'-gil*) [Charles H. *Cargile*, American surgeon, 1853– ]. An animal membrane resembling gold-beaters' foil, made from the peritoneum of the ox and used in surgery for packing and to prevent adhesions; when applied to the raw surface of the bowel it adheres without supporting stitches and forms an artificial peritoneum. Syn., *animal velum*.
**cariated** (*ka'-re-a-ted*). Carious.
**caribi** (*kah-re'-be*). Epidemic gangrenous proctitis.
**carica** (*kar'-ik-ah*) [*carica*, a dry fig, so called from *Caria* in Asia Minor]. A genus of plants of the order *Papayaceæ*. *C. papaya*, the papaw-tree of tropical America, contains in its leaves and fruit the alkaloid carpaine, besides the ferment papain or papayotin; the leaves also contain the glucoside carposid. The milky juice and the seeds are anthelmintic. **c.-cocoa,** a preparation of cocoa containing papain.
**caricin** (*kar'-is-in*). See *papain*.
**caricous** (*kar'-ik-us*) [*carica*, a fig]. Fig-shaped, as a caricous tumor.
**caries** (*ka'-re-ēz*) [L., "rottenness"]. A molecular death of bone or teeth, corresponding to ulceration in the soft tissues. See *Rust's sign*. **c. articulorum.** 1. Caries of a joint. 2. See *c. fungosa*. **c., atonic,** a form described by Billroth, attended with but little swelling and a thin, fetid discharge. **c. callosa,** syphilitic chancre. **c. carnosa,** fungous caries, marked by large granulation-masses. **c. centralis,** circumscribed chronic osteomyelitis, which, working from within, causes disease of the cortical substances. Syn., *osteitis interna*. **c. dentis, c. dentium.** See *c. of teeth*. **c. fungosa,** tuberculosis of a bone with attached sequestrum, the meshes of the latter being filled with granulations growing into them from the inner surface of the cavity. **c. gallica,** syphilitic chancre. **c. granulosa.** See *c. fungosa*. **c. interna.** See *c. centralis*. **c., lacunar,** a form in which the undermined bone is full of lacunæ. **c. mollis.** See *c. fungosa*. **c., necrotic,** a form in which portions of the bone lie in a suppurating cavity. **c. nongallica,** simple chancre. **c. profunda.** See *c. centralis*. **c. sicca,** a form of tuberculous caries characterized by absence of suppuration, obliteration of the cavity of the joint, and sclerosis and concentric atrophy of the articular extremity of the bone. **c. of spine,** tuberculous osteitis of the bodies of the vertebræ and intervertebral fibrocartilage, producing curvature of the spine. Syn., *Pott's disease*. **c. strumosa,** tuberculous caries. **c. of teeth,** a chemical decomposition of the earthy part or any portion of a tooth, accompanied by partial or complete disorganization of the animal framework of the affected part. Syn., *odontonecrosis*. **c. tuberculosa,** tuberculous caries.
**cariesin** (*ka-ri-es'-in*). A medical preparation of carious bone.
**carina** (*kar-i'-nah*) [L., "the keel"]. 1. Any keel-like structure. 2. A mesial ridge on the lower surface of the fornix cerebri. 3. The spinal column. **c. aquæductus Sylvii,** the carinate inferior margin of the Sylvian aqueduct. **c. vaginæ,** the anterior column of the vagina.
**carinal** (*kar'-in-al*). Carinate.
**carinate** (*kar'-in-āt*) [*carina*]. Keeled.
**cariosity** (*kar-e-os'-it-e*). See *caries*.
**carious** (*ka'-re-us*) [*caries*]. 1. Pertaining to or affected with caries. 2. Marked by irregular pits or perforations so as to present the appearance of carious bone.
**Carissa** (*kar-is'-ah*) [L.]. A genus of shrubs. See *ouabain*.

**carissin** (*kar-is'-in*). According to Bancroft, a glucoside from the bark of *Carissa ovata*, resembling ouabain in action.
**Carlsbad salt** (*kahrls'-bahd*). A salt supposed to be prepared from Carlsbad water. **C. water.** A famous mineral water used largely for chronic affections of the gastrointestinal tract, obesity, gout, and diabetes.
**carmalum, Mayer's.** A stain consisting of carminic acid, 1; alum, 10; water, 200 parts. It is well adapted for sections cut on the freezing microtome.
**carmin** (*kar'-min*), $C_{17}H_{18}O_{19}$. A coloring-matter extracted from cochineal.
**carminant** (*kar'-min-ant*) [*carmen*, a charm]. 1. Carminative. 2. A carminative agent or medicine.
**carminative** (*kar-min'-at-iv*) [*carminare*, to card; hence, to cleanse]. Having the power to cure flatulence and colic. Carminatives are generally aromatics.
**carminophile** (*kar-min'-o-fil*) [*carmin*; φιλεῖν, to love]. Readily stainable with carmin.
**carnal** (*kar'-nal*) [*carnalis*, fleshly]. Pertaining to flesh. **c. knowledge,** sexual intercourse.
**carnallite** (*kar'-nal-īt*) [v. *Carnall*, Prussian mineralogist 1804–1874]. Potassium-magnesium chloride.
**carnation** (*kar-na'-shun*) [*carnatio*]. The natural color of flesh.
**carnauba** (*kar-na-oo'-bah*) [Braz.]. 1. The root of *Copernicia cerifera*, a wax-producing palm-tree of tropical America. It is used in Brazil as an alterative and resembles sarsaparilla in its properties. Dose of the *fluidextract* 30 min.–1 dr. (2–4 Cc.). 2. See *C. wax*. **c. wax,** the wax obtained from *Copernicia cerifera*.
**carneoaponeurotic** (*kar-ne-o-ap-on-ū-rot'-ik*). Fleshy and pertaining to an aponeurosis.
**carneopapillosus** (*kar-ne-o-pap-il-o'-sus*). Composed of fleshy papillæ, as the columns of the vagina.
**carneotendinous** (*kar-ne-o-ten'-din-us*). Both muscular and tendinous.
**carneous** (*kar'-ne-us*) [*carneus*, of flesh]. Fleshy. **c. columns.** See *columnæ carneæ*.
**carniferrin** (*kar-ne-fer'-in*). A tasteless meat preparation containing phosphocarnic acid and 30 % of iron. Dose for adults 8 gr. (0.52 Gm.). Syn., *iron phosphosarcolactate*.
**carniferrol** (*kar-nif'-er-ol*). A preparation of meat-peptone with iron; it is used as a stimulant dietetic. Syn., *liquor carnis ferropeptonatus*.
**carnification** (*kar-nif-ik-a'-shun*) [*caro, carnis*, flesh; *facere*, to make]. A term indicating the alteration of tissue, especially the lung, to a dense, fleshy appearance. **c. of bone.** See *osteosarcosis*. **c. of the lung, congestive,** brown induration of the lung. **c. of the lungs, c., pulmonary.** 1. The change of the parenchyma of the lungs into a red material resembling muscle. 2. A consolidation of the lung from action of inflammation.
**carniformis** (*kar-ne-form'-is*) [see *carnification*]. Having a flesh-like appearance, *e. g., abscessus carniformis*.
**carnigen** (*kar'-ne-jen*). A dietetic albumose.
**carnine** (*kar'-nen*) [*caro, carnis*, flesh], $C_7H_8N_4O_2$. A leukomaine isolated from American meat-extract, but not from muscle tissue itself; also obtained from yeast and wine.
**carnivorous** (*kar-niv'-o-rus*) [*caro, carnis*, flesh; *vorare*, to devour]. Flesh-eating.
**Carnochan's operation** (*kar'-no-kan*) [John Murray *Carnochan*, American surgeon, 1817–1887]. 1. For *elephantiasis*: ligation of the main artery of the limb. 2. *For neuralgia*: removal of the second division of the fifth nerve, together with the sphenopalatine ganglion as far back as the foramen rotundum.
**carnogen** (*kar'-no-jen*) [*caro, carnis*, flesh; *generare*, to produce]. Glycerite of bone-marrow, containing 60 % of red marrow and 25 % of unaltered fibrin of ox-blood, with albumin, suspended in glycerol. It is a hematinic and used chiefly in pernicious anemia. Dose 1–2 tablespoonfuls 3 times daily.
**carnolin** (*kar'-nol-in*). A solution of 1:5 % of formaldehyde; it is a food-preservative and disinfectant.
**carnose** (*kar'-nōs*) [*carnosus*, fleshy]. Resembling or having the consistence of flesh.
**carnosin** (*kar'-no-sin*), $C_9H_{14}N_4O_2$. A base, soluble in water, isolated from Liebig's meat-extract; it melts with decomposition at 239° C.

**carnosity** (*kar-nos'-it-e*) [*carnosus*, fleshy]. A fleshy growth or excrescence.

**Carnot's solution** (*kar'-no*). A solution of gelatin, 5 to 10 per cent., in normal saline. It is used as a local hemostatic.

**caro** (*ka'-ro*) [L. *gen.*, *carnis*]. Flesh. c. **luxurians,** exuberant granulation.

**Caroba** (*kar-o'-bah*) [L.]. The leaflets of *Jacaranda procera* and of *Cybistax antisyphilitica*. It is a popular Brazilian remedy as an emetocathartic, alterative, and tonic in syphilis and in yaws. Dose of the *fluidextract* 15 min.–1 dr. (1–4 Cc.).

**carobin** (*kar'-o-bin*). A crystalline body obtained from *Jacaranda procera*.

**caroid** (*kar'-oid*). A digestive ferment obtained from *Carica papaya*; a pale-yellow powder. Dose 1–3 gr. (0.065–0.2 Gm.).

**carolinium** (*kar-o-lin'-e-um*). The provisional name given by Baskerville to a supposed new element obtained from thorium oxide.

**carone** (*kar'-ōn*). A substance obtained from dihydrocarvone by action of hydrobromic acid.

**Carony bark** (*kar-o'-ne*) [*Caroni*, a river in Venezuela]. Angustura bark.

**carotic** (*kar-ot'-ik*) [κάρος, stupor]. 1. Carotid. 2. Stupefying; or of the nature of stupor. 3. A drug to produce sleep.

**caroticoclinoid** (*kar-ot-ik-o-kli'-noid*). Relating to a carotid artery and a clinoid process of the sphenoid bone.

**caroticotympanic** (*kar-ot-ik-o-tim-pan'-ik*). Relating to the carotid canal and the tympanum.

**carotid** (*kar-ot'-id*) [καρόειν, to produce sleep]. 1. The carotid artery, the principal large artery on each side of the neck. See under *artery*. 2. Of or relating to the carotid artery. c. **gland.** See under *gland*. c. **plexus,** the nerve-plexus around the carotid artery. c. **tubercle,** the anterior tubercle of the transverse process of the sixth cervical vertebra.

**carotidaneurysma** (*kar-o-tid-an-ū-riz'-mah*). Aneurysm of the carotid artery.

**carotis** (*kar-ot'-is*) [L.]. The carotid artery. c. **cephalica,** c. **cerebralis,** the internal carotid artery. c. **communis,** the common carotid artery. c. **externa,** c. **facialis,** the external carotid artery. c. **interna,** the internal carotid artery. c. **primitiva,** the common carotid artery.

**carpagra** (*kar-pag'-rah*) [*carpus*; ἄγρα, a seizure]. A sudden attack of pain at the wrist.

**carpaine** (*kar-pa'-ēn*), $C_{14}H_{25}NO_2$. A nalkaloid extracted from the leaves of *Carica papaya*. It is recommended for the subcutaneous treatment of heart disease. Dose $\frac{1}{10}$–$\frac{1}{6}$ gr. (0.006–0.01 Gm.) subcutaneously, every day or every second day. c. **hydrochloride,** $C_{14}H_{25}NO_2HCl$, bitter white crystals, soluble in water. It is used in mitral insufficiency and aortic stenosis. Dose $\frac{1}{5}$–$\frac{1}{3}$ gr. (0.013–0.022 Gm.) daily. Injection, $\frac{1}{10}$–$\frac{1}{6}$ gr. (0.0065–0.011 Gm.) daily.

**carpal** (*kar'-pal*) [καρπός, the wrist]. Pertaining to the carpus or wrist.

**carpale** (*kar-pa'-le*) [*carpus*]. Any one of the wrist-bones.

**carpectomy** (*kar-pek'-to-me*) [*carpus*; ἐκτομή, excision]. Excision of one or more of the carpal bones.

**carpen** (*karp'-en*) [*carpus*]. Belonging to the carpus in itself.

**carphologia** (*kar-fol-o'-je-ah*). See *carphology*.

**carphology** (*kar-fol'-o-je*) [κάρφος, chaff; λέγειν, to collect]. The aimless picking at the bedclothes, seen in grave fevers, particularly in the socalled typhoid state.

**carpitis** (*kar-pi'-tis*) [*carpus*; ιτις, inflammation]. Inflammation of one or more of the carpal joints.

**carpo-** (*kar-po-*) [*carpus*]. A prefix meaning relating to the carpus.

**carpocace** (*kar-pok'-as-e*) [*carpus*; κακός, bad]. A diseased condition of the wrist.

**carpocarpal** (*kar-po-kar'-pal*). Applied to the articulation between the two rows of carpal bones; also to different parts of the carpus in relation to each other.

**carpocervical** (*kar-po-ser'-vik-al*). Relating to the wrist and the neck.

**carpogenic** (*kar-po-jen'-ik*) [καρπός, fruit; γένης, producing]. Applied to the fruit-producing cell or system of cells in certain algæ.

**carpogenous** (*kar-poj'-en-us*) [καρπός, a fruit; γεννᾶν, to produce]. Fertile; fruit-producing.

**carpogonium** (*kar-po-go'-ne-um*) [καρπός, fruit; γόνος, producing]. In biology, the unfertilized female reproductive organ of certain thallophytes.

**carpolith** (*kar'-po-lith*) [καρπός, fruit; λίθος, a stone]. 1. A hard concretion formed in a fruit. 2. A petrified fruit.

**carpometacarpal** (*kar-po-met-ă-kar'-pal*) [*carpo-*; *metacarpus*]. Relating to the carpus and to the metacarpus.

**carpo-olecranal** (*kar-po-o-le-kra'-nal*) [*carpo-*; *olecranon*]. Relating to the wrist and the lower portion of the upper arm.

**carpopedal** (*kar-po-pe'-dal*) [*carpo-*; *pes*, *pedis*, a foot]. Affecting the wrists and feet, or the fingers and toes. c. **contraction.** See *contraction, carpopedal*. c. **spasm,** a spasm of the hands and feet, or of the thumbs and great toes, associated with laryngismus stridulus of children.

**carpophalangeal** (*kar-po-fă-lan'-je-al*). Pertaining to the wrist and the phalanges.

**carpophalangeus, carpophalanginus** (*kar-po-fal-an'-je-us, kar-po-fal-an-ji'-nus*). 1. Relating to the wrist and to a phalanx. 2. See under *muscle*.

**carpophilous** (*kar-pof'-il-us*) [καρπός, fruit; φιλεῖν, to love]. Parasitic upon fruit.

**carpoptosis** (*kar-pop-to'-sis*) [*carpus*; πτῶσις, a fall]. Wrist-drop.

**carposid** (*kar'-po-sid*). A crystalline glucoside from *Carica papaya*.

**carpozyma** (*kar-po-zi'-mah*) [καρπός, fruit; ζύμη, ferment]. A genus of microorganisms producing fermentation.

**carp's-tongue** (*karps'-tung*). An elevator used in the extraction of roots of teeth.

**Carpue's method of rhinoplasty** (*kar'-poo*) [Joseph Constantine *Carpué*, English surgeon, 1764–1846]. A repair of the nose by taking a heart-shaped flap from the forehead.

**carpus** (*kar'-pus*) [L.]. The eight bones collectively forming the wrist.

**carrageen, carragheen** (*kar'-ag-ēn*) [*Carragheen* in Ireland]. Irish moss. See *chondrus*.

**carreau** (*kar'-o*). Scrofulosis and tuberculosis of the digestive organs.

**carrefour sensitif** (*kar-foor' son-set-eef'*) [Fr., sensory crossway]. The posterior part of the posterior limb of the internal capsule of the cerebrum.

**Carrick bend** (*kar'-ik bend*). A form of knot for fastening together two ligatures. The merit of the knot consists in the free end being held firmly between the two long portions.

**carriers** (*kar'-e-ers*). Individuals who are convalescent from an infectious disease but, while showing no signs or symptoms of the disease, harbor and eliminate the microorganism, and so spread the disease. c., **chronic,** carriers who eliminate the microorganisms for an indefinite period. c., **temporary** or **transitory,** convalescents who eliminate the microorganisms only for a short time after recovery.

**Carrion's disease** [Daniel E. *Carrion*, Peruvian student, 19th century]. Verruga peruviana; Peruvian wart.

**Carron oil** (*kar'-on*). An oil consisting of equal or nearly equal parts of linseed-oil and lime-water. It is used as an application to burns, and is named after the Carron iron-works in Scotland, where it was first employed.

**carrotin** (*kar'-o-tin*) [*carota*, carrot]. $C_{18}H_{24}O$. A lipochrome, the coloring-matter of carrots and tomatoes.

**car-sickness.** The symptoms similar to those of sea-sickness produced by journeying in railway cars.

**Carswell's grapes** [Sir Robert *Carswell*, English physician, 1793–1857]. Pulmonary tubercles when they occur in a racemose distribution at the extremities of several adjacent bronchioles.

**Carter's operation** (*kar'-ter*) [William Wesley *Carter*, American laryngologist, 1869– ]. Transplanting a piece of bone from a rib in order to make a new bridge for the nose.

**Carthagena bark.** Cinchona from Carthagena.

**carthamin** (*karth'-am-in*) [Ar., *qartama*, paint], $C_{14}H_{16}O_7$. The coloring-matter in safflower, the blossoms of *Carthamus tinctorius*.

**carthamus** (*karth'-am-us*) [Ar., *qartama*, paint]. American or bastard saffron or safflower. The flowers of *C. tinctorius*. An infusion, "Saffron tea," is a popular domestic remedy as a diuretic in measles and other exanthematous affections.

**cartilage** (*kar'-til-āj*) [*cartilago*, gristle]. Gristle; a white, semiopaque, nonvascular connective tissue composed of a matrix containing nucleated cells which lie in cavities or lacunas of the matrix. When boiled, cartilage yields a substance called chondrin. c., **annular**. 1. Any ring-shaped cartilage. 2. The cricoid cartilage. c., **anonymous**, the cricoid cartilage. c., **aortic**, the second costal cartilage on the right side. c., **arthrodic**, c., **arthrodial**. See *c.*, *articular*. c., **articular**, that lining the articular surfaces of bones. c., **arytenoid**. See *arytenoid cartilage.* ·c., **asternal**, the costal cartilages which are detached from the sternum. c.-**bone**. 1. See *ossein.* 2. See *c.*, *calcified*. c.s, **Brecht's**. See *Brecht's cartilage.* c., **bronchial**, plates of cartilage, in some instances very minute, found in the bronchial tubes. c., **calcified**, that in which a calcareous deposit is contained in the matrix. Syn., *cartilage-bone; crusted cartilage; primary bone.* c. **cells** or **corpuscles**, connective-tissue cells in matrix or cartilage. c., **cellular**. See *c.*, *parenchymatous.* c., **ciliary**. See *c.*, *palpebral.* c., **corniculate**. See *Santorini's cartilage.* c., **costal**, that occupying the interval between the true ribs and the sternum or adjacent cartilages. c., **cricoid**. See *cricoid cartilage.* c.s, **cuneiform**. See *Wrisberg's cartilages.* c., **dentinal**. See *ossein.* c., **diarthrodial**. See *c.*, *articular.* c., **embryonal**. See *c.*, *parenchymatous.* c., **ensiform**, the third piece of the sternum. Syn., *xiphoid appendix; xiphoid cartilage.* c.s, **epactal**, small cartilaginous nodules on the upper edge of the alar cartilages of the nose. c., **epiphyseal**. See *c.*, *intermediary* (2). c., **fetal**. See *c.*, *temporary.* c., **fibro-**. See *fibrocartilage.* c., **floating**. See *arthrolith.* c., **Huschke's**. See *Jacobson's cartilage.* c., **hyaline**, is distinguished by a finely granular or homogeneous matrix. c., **innominate**, the cricoid cartilage. c.s, **interarticular**, flat fibrocartilages situated between the articulating surfaces of some of the joints. Syn., *interarticular fibrocartilages.* c., **interarytenoid**, an inconstant cartilage found between the arytenoid cartilages. c.s, **interhemal**, nodules of cartilage which aid in the formation of the hemal arch of a vertebra. c., **intermediary**. 1. Cartilage-bone in process of transformation into true bone. 2. That interposed between the epiphysis and diaphysis of a bone. c.s, **interneural**, nodules of cartilage which aid in the formation of the neural arch of a vertebra. c.s, **intervertebral**. See *intervertebral discs.* c., **investing**. See *c.*, *articular.* c., **Jacobson's**. See *Jacobson's cartilage.* c., **Luschka's**. See *Luschka's cartilage.* c., **Luschka's subpharyngeal**. See under *Luschka.* c., **Meckel's**. See *Meckel's cartilage.* c.s, **Morgagni's**. See *Wrisberg's cartilages.* c., **palpebral**, the connective tissue forming the framework of the eyelids. c., **parachordal**. See *parachordal cartilage.* c., **parenchymatous**, that in which cells form the main part of the tissue. c.s, **pyramidal**, the arytenoid cartilages. c.s, **quadrate**, several small cartilages passing out from the alar cartilages in the external part of the nostril. c., **reticular**, a peculiar cartilage found in the auricle of the ear, the epiglottis, and Eustachian tubes. Its peculiarity consists in a network of yellow elastic fibers pervading the matrix in all directions. c., **retiform**. See *c.*, *reticular.* c. of **Santorini**. See *Santorini's cartilage.* c., **Seiler's**. See *Seiler's cartilage.* c.s, **semilunar**, two interarticulating cartilages of the knee. c., **sesamoid**. See *sesamoid bone.* c., **sesamoid (of the larynx)**, Luschka's cartilage. c.s, **sesamoid (of the nose)**. See *c.s, epactal.* c.s, **sigmoid**. See *c.s, semilunar.* c., **synarthrodial**, that of any fixed or slightly movable articulation. c., **tarsal**. See *c.*, *palpebral.* c., **temporary**, that which is ultimately replaced by bone. c., **tubal**, a rolled triangular cartilage running from the osseous part of the Eustachian tube to the pharynx. c. of **Weitbrecht**. See *Weitbrecht's cartilage.* c.s of **Wrisberg**. See *Wrisberg's cartilage.* c., **xiphoid**. See *c., ensiform.* c., **yellow**. See *c., reticular.*

**cartilagin** (*kar-til'-aj-in*) [*cartilago*, cartilage]. A characteristic principle of hyaline cartilage. Boiling changes it into chondrin.

**cartilagines** (*kar-til-aj'-in-ēz*) [plural of *cartilago*]. Cartilages. c. **alares minores**, lesser alar cartilages, sesamoid cartilages. c. **corniculatæ**, corniculate cartilages, cartilages of Santorini. c. **cuneiformes**, cuneiform cartilages, cartilages of Wrisberg.

**cartilaginification** (*kar-til-aj-in-if-ik-a'-shun*) [*cartilago*, cartilage; *facere*, to make]. A change into cartilage.

**cartilaginiform** (*kar-til-aj-in'-if-orm*) [*cartilago*, cartilage; *forma*, form]. Resembling cartilage.

**cartilaginoid** (*kar-til-aj'-in-oid*) [*cartilago*, cartilage; εἶδος, form]. Resembling cartilage.

**cartilaginous** (*kar-til-aj'-in-us*) [*cartilago*]. Made up of or resembling cartilage.

**cartilago** (*kar-til-a'-go*) [L.]. See *cartilage.* c. **alaris major**, the lower lateral cartilage of the nose. c. **alaris minor**, one of the lesser alar cartilages of the nose. c. **auriculæ** or **auris**, the cartilage of the pinna of the ear. c. **basilaris**, the cricoid cartilage; the fibrocartilage in the foramen lacerum medium. c. **corniculata**, the corniculum laryngis. c. **cricoidea**, the cricoid cartilage. c. **cuneiformis**, the cuneiform cartilage of the larynx. c. **ensiformis**, the xiphoid cartilage. c. **epiglottica**, the cartilage of the epiglottis. c. **nasi lateralis**, the upper lateral cartilage of the nose. c. **ossescens**, cartilage destined to become bone. c. **septi nasi**, cartilage of the nasal septum. c. **thyreoidea**, the thyroid cartilage. c. **triticea**, a small oblong cartilaginous nodule often found in the lateral thyrohyoid ligament. c. **tubæ auditivæ**, the cartilage forming front part of the Eustachian tube. c. **vomeronasalis**, the vomerine cartilage or cartilage of Jacobson. c. **Wrisbergii**. Same as *c. cuneiformis.* c. **xiphoidea**, the xiphoid or ensiform process.

**carum** (*ka'-rum*) [κάρον, caraway]. Caraway. It is official in the U. S. P. in the form of the dried fruit of *C. carvi*, indigenous to Europe, and an allied species native to the Pacific coast of America. Its odor and taste are due to a volatile oil. It is used chiefly as a flavor. *C. petroselinum*, parsley, is diuretic and sedative. **carui, aqua** (B. P.), caraway water. Dose (ka'-rus). (30–60 Cc.). **carui, infusum**, 2 dr. to 1 pint. Dose ½–2 oz. (15–60 Cc.). **carui, oleum** (U. S. P.), oil of caraway. Dose 1–5 min. (0.06–0.3 Cc.).

**caruncle** (*kar'-ung-kl*) [*caruncula*]. A small, fleshy growth. c., **lacrimal**, one upon the conjunctiva near the inner canthus. c., **urethral**, a small, bright-red growth situated on the posterior lip of the meatus urinarius: a frequent condition in women. The caruncle varies in size from a hempseed to a filbert; it is very painful, especially during micturition and coitus, and bleeds readily.

**caruncula** (*kar'-ung-kū-lah*) [dim. of *caro*, flesh; pl., *carunculæ*]. A caruncle. **carunculæ cuticulares**, the nymphæ. c. **innominata**, the lacrimal gland. c. **major**, a caruncle marking the common orifice of the common bile-duct and the pancreatic duct. c. **mammillaris**. 1. The olfactory tubercle, between the roots of the olfactory nerves. 2. The enlarged ends of the galactophorous ducts in the nipple. c. **minor**, one in the duodenum in the center of which a supplementary pancreatic duct occasionally opens. c. **Morgagnii**, the middle lobe of the prostate. **carunculæ myrtiformes**, the projections of membrane near the orifice of the vagina, thought to be the remains of the hymen after its rupture. **carunculæ papillares**. See *papilla, renal.* c. **salivalis**. See *c. sublingualis.* c. **sublingualis**, one marking the orifice of Wharton's duct. Syn., *papilla salivalis inferior.* c. **urethræ**. See *caruncle, urethral.*

**caruncular** (*kar-ung'-kū-lar*) [*caruncula*, a caruncle]. Like or pertaining to a caruncle.

**carunculate, carunculated** (*kar-ung'-kū-lāt, -ed*). Furnished with a caruncle.

**carus** (*ka'-rus*) [κάρος, stupor]. Deep, lethargic sleep. c. **catalepsia**, catalepsy. c. **ecstasis**, trance, or catalepsy. c. **lethargus**, lethargy.

**Carus' curve** (*ka'-rus*) [Karl Gustav *Carus*, German obstetrician, 1789–1869]. The longitudinal axis of the pelvic canal, which forms a curved line, having the symphysis pubis as its center.

**carvacrol** (*karv'-ak-rol*) [Ital., *carvi*, caraway; ἄκρος, sharp], $C_{10}H_{13}$ . OH. A liquid body occurring in the oil of certain varieties of satureja. Syn., *cymic phenol; cymophenol; metaisocymophenol; oxycymol.* c. **iodide**, $C_{10}H_{13}OI$, a brown powder, slightly soluble in alcohol, readily soluble in olive-oil, ether, and chloroform, melting at 90° C.; it is used as a substitute for iodoform. Syn., *iodocrol.*

**carvene** (*karv'-ēn*) [It., *carvi*, caraway], $C_{10}H_{16}$. A hydrocarbon contained in caraway. It is a light terpene. See also *citrene.*

**carvol** (*karv'-ol*) [It., *carvi*, caraway; *oleum*, oil], $C_{10}H_{14}O$. An aromatic alcohol isomeric with carvacrol, and obtained from oil of cumin. It is an oil with a pleasant odor, boiling at 225° C.

**carvone** (*kar'-vōn*). Same as *carvol*.
**Carya** (*kar'-e-ah*) [καρύα, the walnut-tree]. Hickory; a genus of trees of the order *Juglandaceæ*, indigenous to North America. *C. tomentosa* yields a crystalline principle, *caryin*, believed to be identical with quercitrin. The leaves of most of the species are aromatic and astringent and the bark bitter and astringent. The inner bark is used in dyspepsia and intermittent fever.
**caryenchyma** (*kar-e-en'-ki-mah*) [κάρυον, nut (nucleus); ἐν, in; χυμός, juice]. The more fluid part of the protoplasm of a nucleus.
**caryin** (*kar'-e-in*). See under *Carya*.
**caryinum** (*kar-e-in'-um*). Nut-oil.
**caryoblast** (*kar'-e-o-blast*) [κάρυον, nucleus; βλαστός, a germ]. Any nucleated plastidule.
**caryochrome** (*kar'-e-o-krōm*). See *karyochrome*.
**caryocinesis** (*kar-e-o-sin-e'-sis*). See *karyokinesis*.
**caryocinetic** (*kar-e-o-sin-et'-ik*). 1. See *karyokinetic*. 2. Ameboid.
**caryolysis** (*kar-e-ol'-is-is*). See *karyolysis*.
**caryomitosis** (*kar-e-o-mi-to'-sis*). See *karyomitosis*.
**caryophyllin** (*kar-e-o-fil'-in*) [*caryophyllus*], $C_{10}H_{16}O$ or $C_{20}H_{32}O_2$. The neutral crystalline principle of cloves.
**caryophyllus** (*kar-e-o-fil'-us*) [κάρυον, a nut; φύλλον, a leaf]. Clove. The unexpanded flowers of *Eugenia aromatica*, distinguished by their pungent, spicy taste. Its properties are due to a volatile oil, which is antiseptic, stimulant, and irritant. It also contains a crystalline body, *eugenin*, $C_{10}H_{12}O_2$, and a camphor, *caryophyllin*, $C_{10}H_{16}O$. It is useful as a stomachic and to prevent "griping" when combined with purgatives. **caryophylli, infusum** (B. P.), a strength of 1 to 40 is recommended. Dose 1-2 oz. (30-60 Cc.). **caryophylli, oleum** (U. S. P.), oil of cloves, contains an acid and a phenol compound. Dose 1-4 min. (0.06-0.24 Cc.). It is used also by microscopists to clarify preparations and tissues for mounting.
**caryoplasm** (*kar'-e-o-plazm*). See *karyoplasm*.
**caryorrhexis**. See *karyorrhexis*.
**casanthrol** (*kas-an'-throl*). A mixture of casein ointment with a coal-tar product; it is used as a varnish in skin diseases.
**casca-bark** (*kas'-kah*). Sassy-bark; ordeal-bark. The bark of *Erythrophlæum guineense*, a tree native to Africa. Its properties are due to an alkaloid. It is valuable in intermittent fevers and as a hearttonic; in over-doses it produces nausea and vomiting. *Erythrophleine*, the active alkaloid, is a local anesthetic. Dose of the *aqueous extract* 1 gr. (0.065 Gm.); of the *fluidextract* 5-15 min. (0.3-0.9 Cc.); of the *tincture* (25 % strength) 10 min. (0.6 Cc.).
**cascador** (*kas'-ka-dor*) [*casca*, bark]. A gatherer of cinchona bark.
**cascanata** (*kas-kan-at'-ah*). A proprietary laxative and alterative said to consist of the active principles of cascara sagrada, gentian, rhubarb, and other herbs, holding in solution phosphate of sodium and magnesium.
**cascara** (*kas'-kar-ah*). Spanish for "bark." c. **amarga**, Honduras bark. The bark of a tree native in Mexico, much used as an alterative tonic in syphilis and skin affections. c. **cordial**, a trade preparation. Dose 15 min.-2 dr. (1-8 Cc.). c. **sagrada** (*rhamnus purshiana*, U. S. P.), the bark of *Rhamnus purshiana*, or California buckthorn. Its properties are due to a volatile oil. It is useful in chronic constipation. Syn., *chittem bark; sacred bark*. c. **sagrada, extract** of (*extractum rhamni purshianæ*, U. S. P.). Dose ½-1 dr. (2-4 Cc.). c. **sagrada, fluidextract of** (*fluidextractum rhamni purshianæ*, U. S. P.). Dose 15 min. (1 Cc.). c. **sagrada, fluidextract of,** aromatic (*fluidextractum rhamni purshianæ aromaticum*, U. S. P.). Dose 15 min. (1 Cc.).
**cascarilla** (*kas-kar-il'-ah*) [Sp., dim. of *casca*, bark]. The bark of *Croton eluteria*, native to the Bahama Islands, an aromatic bitter, increasing the natural secretions of the digestive organs. **cascarillæ, infusum** (B. P.). Dose 1-2 oz. (30-60 Cc.). **cascarillæ, tinctura** (B. P.). Dose ½-2 dr. (2-8 Cc.).
**cascarillin** (*kas-kar-il'-in*) [*cascarilla*], $C_8H_{12}O_3$. The active principle of cascarilla; a white, crystalline, bitter substance, scarcely soluble in water.
**cascarin** (*kas'-kar-in*), $C_{24}H_{30}O_{10}$. A substance isolated by Leprince from the bark of *Rhamnus purshiana* (cascara sagrada), and believed by him to contain the active tonic and laxative principles of that bark; it occurs in granular masses or prisms. Dose 1½-3 gr. (0.099-0.198 Gm.). According to Phipson, this is identical with rhamnotoxin.
**case** (*kās*) [*cadere*, to happen]. 1. A single instance or example of a disease. 2. A covering, or box-like structure. c., **brain-**, the calvaria. c., **muscle-**, see *muscle*. c.-**taking**, the collection of memoranda and notes of an individual case for service in diagnosis or prognosis, or for use in a medico-legal inquiry. c., **trial-**, in ophthalmology, a case containing various lenses for refracting the eye, etc.
**Casearia** (*kas-e-a'-re-ah*). [J. Casearius, Dutch botanist]. A genus of tropical trees of the order *Samydaceæ*. *C. esculenta* is a native of the Asiatic tropics and Australia; its bitter roots are said to be a valuable remedy in hepatic torpor. *C. ovata*, the *anavingah* of the Malays, is a large tree, bitter in all its parts. The fruit is diuretic. *C. tomentosa* is a tree of India; the bitter leaves are used by the natives in medicated baths and the fruit is diuretic.
**casease** (*ka'-se-ās*). An enzyme which digests casein, found by Duclaux and produced by bacteria; notably *Tyrothrix tenuis*.
**caseate** (*ka'-se-āt*). 1. A lactate. 2. To undergo cheesy degeneration.
**caseation** (*ka-ze-a'-shun*) [*casein*]. The precipitation of casein during the coagulation of milk. Also a form of degeneration in which the structure is converted into a soft, cheese-like substance.
**caseiform** (*ka'-ze-if-orm*). Resembling cheese or casein.
**casein** (*ka'-se-in*) [*caseus*, cheese]. A derived albumin, the chief proteid of milk, precipitated by acids and by rennet. It is closely allied to alkali-albumin, but contains more nitrogen and a large amount of phosphorus. It constitutes most of the curd of milk. Syn., *caseum; lacterin*. c. **dyspepton**, an insoluble, semigelatinous substance, separated in the first stages of gastric digestion. c., **gluten**. See c., *vegetable*. c.-**mercury**, a compound of casein and mercury bichloride, soluble in water with a trace of ammonia added; it is antiseptic. c. **ointment**, an ointment-base consisting of casein, 14 parts; potassium hydroxide and sodium hydroxide, each, 0.43 part; glycerol, 7 parts; vaselin, 21 parts; borax, 1 part; water, 56 or 57 parts. c.-**peptone**, a lightbrown, soluble powder used as a nutrient. c. **saccharide**, a compound of dry casein, 1 part; cane-sugar, 9 parts, and sodium bicarbonate enough to render it slightly alkaline. It is useful in preparing emulsions of oils, balsams, terpenes, resins, or gum-resins. c. of the **saliva**, ptyalin. c.-**sodium**, a compound of casein and sodium hydroxide, used as a nutrient. c., **vegetable**, a nitrogenous substance resembling the casein of milk; two varieties have been described—*legumin*, in peas, beans, etc., and *conglutin*, in hops and almonds.
**caseinogen** (*ka-se-in'-o-jen*) [*caseum*, cheese; γεννᾶν, to produce]. A peculiar substance occurring in milk, neither an alkali-albumin nor a globulin, but occupying a distinct position among proteids. When acted upon by a digestive ferment it produces casein, or the curd of milk. Caseinogen is a proteid analogous to fibrinogen, myosinogen, etc.
**caseoiodine** (*ka-se-o-i'-o-din*). A compound of casein and iodine (8 or 9 %) forming a white powder, soluble in dilute hot alcohol and in hot alkalies. It is used in myxedema.
**caseose** (*ka'-se-ōs*) [*caseum*, cheese] A product of the gastric digestion of casein.
**caseous** (*ka'-se-us*) [*caseus*, cheese]. Having the nature or consistence of cheese.
**cashew** (*kash'-o*). The cashew-nut, the product of *Anacardium occidentale*. See *Anacardium*.
**Casimiroa** (*kas-im-iv-o'-ah*) [after *Casimiro Gomez*]. A genus of plants belonging to the order *Rutaceæ*. *C. edulis* is the *zapote blanco* of Mexico; the edible fruit is anthelmintic; the bitter seeds with the leaves and seeds are incinerated and used medicinally.
**casogen** (*ka'-so-jen*). Trade name of a milk food said to contain 95 per cent. of milk protein, 4 per cent. glycerophosphates, and 1 per cent. ovolecithin; used in neurasthenia, anaemia and dyspepsia.
**cassareep, cassaripe** (*kas'-a-rēp*) [South American name]. The concentrated juice of the cassava, the root of *Jatropha manihot*, made innocuous by boiling; it is a condiment, and as an ointment (10 %) is recommended in the treatment of purulent conjunctivitis, corneal ulcers, and other diseases of the eye.

**cassava** (*kas-ah'-vah*) [Sp., *casabe*]. 1. The manioc plant (*Jatropha manihot* and other species of *Jatropha*). 2. Tapioca.
**Casselberry position** (*kas'-el-ber-re*) [William Evans Casselberry, American laryngologist, 1858– ]. The patient after an intubation lies with his face downward while drinking so that the fluid may not enter the tube.
**Casser's (Casserius') fontanel** [Giulio *Casserio*, Italian anatomist, 1545–1616]. The fontanel formed by the temporal, occipital, and parietal bones. **C.'s ganglion**, the Gasserian ganglion. **C.'s muscle**. 1. Ligamentous fibers attached to the malleus and formerly described as the laxator tympani minor muscle. 2. The coracobrachialis. **C.'s perforating nerve**, the external cutaneous nerve of the arm.
**Casserian** (*kas-e'-re-an*). See *Gasserian*.
**Cassia** (*cash'-e-ah*) [κασία, a perfume]. 1. A genus of leguminous plants, several species of which afford senna. 2. An old name, still used commercially, for the coarser varieties of cinnamon. See *Cinnamon*. *C. alata*, the ringworm-shrub, is a widely diffused tropical shrub. The juice of the leaves mixed with lime-juice is used in the treatment of ringworm, and the wood and bark are alterant. *C. beareana* is a species of East Africa. A decoction of the root is highly recommended in black-water fever, and the powdered bark is applied as a dressing to ulcers. *C. marilandica*, of North America, produces the leaves called American senna, which are less active as a cathartic than the true senna. **C.-bark**, cassia-lignea. See *cinnamon*. **C.-buds**, the immature fruit of Chinese cinnamon; used chiefly as a spice. **C., oil of**, a variety of oil of cinnamon, used in pharmacy and in perfumery. **C., purging** (*cassia fistula*, U. S. P.), the dried fruit of a tree growing in tropical regions. The pulp (*cassiæ pulpa*, B. P.) is a mild laxative. Dose 1–2 dr. (4–8 Gm.).
**cast** (*kast*) [ME., *casten*, to throw]. 1. A mass of fibrous or plastic material that has taken the form of some cavity in which it has been molded. From their source, casts may be classified as bronchial, intestinal, nasal, esophageal, renal, tracheal, urethral, vaginal, etc. Of these, the renal casts, by reason of their significance in diseases of the kidney, are the most important. Classed according to their constitution, casts are epithelial, fatty, fibrinous, granular, hyaline, mucous, sanguineous, waxy, etc. See *tube-casts*. 2. Strabismus. **c.s, Kuelz's**. See, *Kuelz's casts*. **c.s, tubular exudation** (of the intestine) a pathognomonic symptom of mucous colitis.
**castanea** (*kas-ta'-ne-ah*) [L.]. Chestnut. The leaves of *C. vesca*. They contain tannic and gallic acids and other principles the value of which is not known. They are used in infusion or decoction as a remedy for whooping-cough. Dose of the *fluidextract* 5–60 min. (0.3–3.8 Cc.).
**Castellani's test** (*kas-tel-ah'-ne*) [Aldo *Castellani*, Italian physician]. If an immune serum is mixed with its corresponding bacteria, the agglutinins for these bacteria are absorbed, as are also the partial agglutinins for the heterologous bacteria.
**Castellino's sign**. See *Oliver's symptom*.
**castor** (*kas'-tor*). See *castoreum*. **c.-bean, c.-oil**. See under *ricinus*. **c.-xylene**, a mixture composed of castor-oil, 1 part, and xylene, 3 parts, used for clearing or clarifying the collodion or celloidin of objects embedded in collodion.
**castoreum** (*kas-to'-re-um*) [κάστωρ, the beaver]. The dried preputial follicles and their secretion, obtained from the beaver, *Castor fiber*. It is a reddish-brown substance with a strong odor. It is antispasmodic and stimulant, its action resembling that of musk. Dose of the *tincture* ½–1 dr. (2–4 Cc.).
**castoria** (*kas-to'-re-ah*) [κάστωρ, the beaver]. A proprietary medicine recommended as a substitute for castor oil.
**castorin** (*kas'-to-rin*). A neutral principle obtainable from castoreum.
**castration** (*kas-tra'-shun*) [*castrare*, to cut]. Orchidectomy; the excision of one or both testicles. **c., female**, removal of the ovaries; oophorectomy; spaying.
**castrensis** (*kas-tren'-sis*) [*castra*, a camp]. 1. Relating to camps. 2. Camp-fever or dysentery due to unsanitary living in camps.
**casual** (*kaz'-u-al*) [*casus*, chance]. 1. Fitted or set apart for the treatment of accidental injuries, as a casual ward in a hospital. 2. An occupant of a casual ward in a hospital.

**casualty** (*kaz'-u-al-te*) [*casus*, chance]. An accidental injury; a wound, or loss of life, accidentally incurred; an injury in a battle.
**Casuarina** (*kas-u-ar-e'-nah*) [*casuarius*, the cassowary, from the resemblance of the stems to the heavy feathers of this bird]. A genus of plants of the order *Casuarineæ*. The tonic and styptic bark of *C. equisetifolia*, of Malaya, is used in the treatment of beriberi. *C. montana* is a native of Malaya; the bark is used in beriberi; the leaves in colic; the seeds in a salve in the treatment of headache.
**casuistics** (*kas-u-is'-tiks*) [*casus*, a case]. The study of individual pathologic cases as a means of arriving at the general history of a disease.
**casumen** (*kas'-u-men*). A proprietary dietetic said to contain 93 % of proteid.
**cata-**. A prefix denoting downward, or according to, or against.
**catabasial** (*kat-ah-ba'-se-al*) [κατά, down; *basion*]. Applied to skulls having the basion lower than the opisthion.
**catabasis** (*kat-ab'-as-is*) [κατάβασις, a descent]. The decline of a disease.
**catabatic** (*kat-ah-bat'-ik*). Pertaining to catabasis.
**catabiotic** (*kat-ah-bi-ot'-ik*) [κατά, intensive; βίος, life]. Applied to the power of growing structures which causes the development of approximate cells to be harmonious with the primary structure.
**catabolergy** (*kat-ab-ol'-er-je*) [κατά, down; βάλλειν, to throw; *έργον*, work]. Energy expended in catabolic processes.
**catabolic** (*kat-ab-ol'-ik*) [κατά, down; βάλλειν, to throw]. Of the nature of, or pertaining to, catabolism.
**catabolin, catabolite** (*kat-ab'-o-lin, -līte*) [κατά, down; βάλλειν, to throw]. Any product of catabolism.
**catabolism** (*kat-ab'-ol-izm*) [κατά, down; βάλλειν, to throw]. Destructive metamorphosis; disassimilation; physiological disintegration; movement toward a cata-state.
**catabolite** (*kat-ab'-o-līt*). Same as *catabolin*.
**catabythismomania** (*kat-ab-ith-iz-mo-ma'-ne-ah*) [καταβυθισμός, submergence; μανία, madness]. Insane impulse to suicide by drowning.
**catabythismus** (*kat-ab-ith-is'-mus*) [καταβυθισμός, submergence]. Drowning; especially suicidal drowning.
**catacausis** (*kat-ak-aw'-sis*) [κατά, down; καίειν, to burn]. Spontaneous combustion.
**cataclasis** (*kat-ak'-las-is*) [κατά, down; κλάσειν, to break]. 1. A fracture. 2. See *cataclisis*.
**catacleisis** (*kat-ak-li'-sis*) [κατάκλεισις, a locking]. Closure of the eyelids by adhesion or by spasm.
**cataclysm** (*kat'-ak-lizm*) [κατακλυσμός; a deluge]. 1. An effusion. 2. A sudden shock.
**catacoustics** (*kat-ah-koos'-tiks*) [κατά, after; *ἀκούειν*, to hear]. The science of reflected sound.
**catacrotic** (*kat-ak-rot'-ik*) [κατά, down; *κρότος*, a striking]. Interrupting the line of descent in a sphygmogram.
**catacrotism** (*kat-ak'-rot-izm*) [κατά, down; *κρότος*, a striking]. An interruption or oscillation of the line of descent in a sphygmogram; the quality of being catacrotic or of being marked by oscillation in the sphygmographic line of descent.
**catadicrotic** (*kat-ah-di-krot'-ik*). Having one or more secondary expansions, as a pulse.
**catadicrotism** (*kat-ad-ik'-rot-izm*) [κατά, down; *δίκροτος*, double beating]. The occurrence of a divided or double pulsation in the downward stroke of the sphygmograph.
**catadidymous** (*kat-ad-id'-im-us*) [κατά, down; *δίδυμος*, twin]. Joined into one, as a twin monstrosity, but with downward cleavage, so that the upper parts are double.
**catadidymus** (*kat-ad-id'-im-us*) [κατά, down; *δίδυμος*, twin]. A catadidymous monstrosity.
**catadioptric** (*kat-ah-di-op'-trik*) [κατά, over against; *dioptric*]. Applied to optical instruments which have the power of reflecting and refracting light at the same time.
**catadrome** (*kat'-ad-rōm*) [καταδρέχειν, to run at or over]. 1. The onset of a disease. 2. The decline of a disease.
**catagenesis** (*kat-aj-en'-es-is*) [κατά, down; *γένεσις*, generation]. The process of creation by retrograde metamorphosis of energy, or by the specialization of energy.
**catagma** (*kat-ag'-mah*) [κάταγμα, a fracture; *pl., catagmata*]. A fracture.

**catagmatic** (*kat-ag-mat'-ik*) [κάταγμα, a fracture]. 1. Relating to or serviceable in cases of fracture. 2. A remedy that promotes the union of broken parts.
**catalase** (*kat'-al-ās*). See *milk-catalase*.
**catalepsy** (*kat'-al-ep-se*) [κατά, down; λαμβάνειν, to seize]. A condition of morbid sleep, associated with a loss of voluntary motion and a peculiar plastic rigidity of the muscles, by reason of which they take any position in which they are placed and preserve it for an indefinite time. The condition is associated with hysteria, with forms of insanity, and is a stage of the hypnotic sleep. **c., local,** that affecting a single organ or group of muscles.
**cataleptic** (*kat-al-ep'-tik*) [see *catalepsy*]. 1. Relating to, affected with, or of the nature of, catalepsy. 2. A person affected with catalepsy.
**cataleptiform** (*kat-al-ep'-tif-orm*) [*catalepsy; forma*, form]. Resembling catalepsy.
**cataleptize** (*kat-al-ep'-tīz*). To reduce to a state of catalepsy.
**cataleptoid** (*kat-al-ep'-toid*) [*catalepsy*; εἶδος, likeness]. Like catalepsy. **c. state,** a condition due to neuromuscular excitability and differing from true catalepsy in that the limbs must be held in fixed attitudes for a few seconds before they maintain themselves and friction causes them to become limp.
**cataleptolethargic** (*kat-al-ep-to-leth-ar'-jik*). Having the nature of catalepsy and lethargy.
**Catalpa** (*kat-al'-pah*) [native Am. Indian]. A genus of American and Asiatic bignoniaceous trees. A. **bignonoides** and A. **speciosa,** of North America, have astringent, anthelmintic, and tonic qualities; the leaves and pods are reputed anodyne, emollient, and antasthmatic.
**catalysis** (*kat-al'-is-is*) [καταλύειν, to dissolve]. In chemistry, a reaction that appears to take place owing to the mere presence of another body that apparently undergoes no change.
**catalyst** (*kat'-al-ist*). A substance having the power to produce catalysis.
**catalytic** (*kat-al-it'-ik*) [καταλύειν, to dissolve]. 1. Of the nature of, or characterized by, catalysis. 2. Any medicine that is supposed to break down, destroy, or counteract morbid agencies existing within the economy. See *alterative*. 3. A retrogressive change.
**catalyzation** (*kat-al-i-za'-shun*). The act or process of catalysis.
**catalyzer** (*kat-a-li'-zur*). Any substance that accelerates chemical or physical processes which would occur without it.
**catamenia** (*kat-am-e'-ne-ah*) [κατά, concerning, according to; μήν, month]. The recurrent monthly discharge of blood during sexual life from the genital canal of the female.
**catamenial** (*kat-am-e'-ne-al*) [*catamenia*]. Pertaining to the catamenia.
**catapasm** (*kat'-ap-azm*) [κατάπασμα, powder]. A dry powder to be sprinkled upon the skin or upon a sore.
**cataphasia** (*kat-af-a'-ze-ah*) [κατάφασις, assent]. A condition of imperfect consciousness, in which the patient repeatedly utters the same word or words spontaneously, or in reply to a question.
**cataphonics** (*kat-ah-fon'-iks*) [κατά, after; φωνή, sound]. That branch of physics treating of reflected sounds.
**cataphora** (*kat-af'-o-rah*) [καταφορά, to fall]. Lethargy; imperfect or restless coma, with intervals of coma-vigil.
**cataphoresis** (*kat-af-or-e'-sis*) [κατά, down; φέρειν, to carry]. The introduction of drugs into the system through the skin by means of ointments or solutions applied by the electrode of a battery. **c., anemic,** the application of cataphoresis upon a part from which the blood-supply has previously been cut off by an Esmarch bandage or a rubber ring. **c., static,** a method of introducing into the body gaseous medicaments that have been inclosed within a bell-jar or tube into which enters a brush electrode connected with the positive pole of an influence machine.
**cataphoretic** (*kat-af-o-ret'-ik*). Pertaining to cataphoresis.
**cataphoria** (*kat-af-o'-re-ah*) [κατά, down; φορός, tending]. A tendency of both eyes to assume too low a plane.
**cataphoric** (*kat-af-or'-ik*) [*cataphoresis*]. 1. Passing, or causing a passage, from the anode to the cathode, through a diaphragm or septum. 2. Relating to lethargy or to apoplexy.
**cataplasis** (*kat-ap'-las-is*) [κατα, down; πλάσις, formation]. 1. The stage of decline in the individual life. 2. The application of a plaster or coating.
**cataplasm, cataplasma** (*kat'-ap-lasm, kat-ap-las'-mah*) [κατάπλασμα, a poultice]. A poultice (*q. v.*). **cataplasma carbonis** (B. P.), a poultice made of wood-charcoal, 1; bread-crumb, 1; linseed-meal, 3; boiling water, 20 parts. **cataplasma fermenti** (B. P.), a mixture of beer, yeast, wheat-flour, water at 100° F. It is a stimulant and antiseptic for indolent ulcers. **cataplasma kaolini** (U. S. P.), a mixture of kaolin, boric acid, thymol, methyl salicylate, oil of peppermint, and glycerol.
**cataplectic** (*kat-ap-lek'-tik*) [κατάπληξις, a striking down]. Fulminant; sudden and overwhelming.
**cataplexis** (*kat-ap-leks'-is*) [κατάπληξις, a striking down]. 1. A sudden and overwhelming shock or attack of disease; prostration by the onset of disease or by shock. 2. Hypnotic sleep.
**cataptosis** (*kat-ap-to'-sis*) [κατά, down; πτῶσις, a falling]. Apoplexy; epilepsy; paralysis; ptosis.
**cataract** (*kat'-ar-akt*) [καταρράκτης, a water-fall]. Partial or complete opacity of the crystalline lens or its capsule. **c., adherent,** opacity of the lens, due to disturbed nutrition, in which it is attached by exudates to the adjacent parts, as in cyclitis. **c., aridosiliquose,** an overripe cataract with a dry, wrinkled capsule. **c., capsular,** cataract due to opacity of the capsule. **c., capsulolenticular,** one involving both the capsule and the lens. **c., chalky.** See *c., aridosiliquose.* **c., cholesterin,** one containing what are apparently crystals of cholesterin. **c., concussion,** a soft cataract due to an explosion or some other concussion. **c.,"cortical,** one due to loss of transparency of the outer layers of the lens. **c., cystic.** See *c., Morgagnian.* **c., diabetic,** a form associated with diabetes. **c.-discission,** an operation preliminary to absorption, or extraction by suction, consisting in rupturing the capsule, so that the aqueous humor gains access to the lens. **c.-extraction,** removal of the cataractous lens by surgical operation. **c., fibrinous,** a false cataract consisting of an effusion of plastic lymph on the capsule and into the field of vision. **c., fibroid,** a false cataract consisting of an opacity in the axis of the visual rays though not in the lens. **c., fluid,** the breaking-up of an opaque lens into a milky fluid. **c., green,** a name given to a grayish-green reflex seen in glaucoma; it is also seen when the pupil is dilated and the media are not completely transparent. **c., grumous,** a spurious cataract from hemorrhage into the cornea or into the vitreous. **c., gypseous,** an overripe cataract presenting a white appearance from having undergone degeneration. **c., hard.** See *c., senile.* **c., hyaloid,** a spurious cataract attributed to opacity of the anterior part of the vitreous. **c., immature,** one in which only a part of the lens-substance is cataractous. **c., incipient,** forked/linear opacities in the equatorial region of the lens seen in middle-aged persons and sometimes remaining unchanged for years. Syn., *arcus senilis lentis; gerontoxon lentis.* **c., interstitial.** See *c., lenticular.* **c., lacteal.** See *c., fluid.* **c., lamellar,** one due to opacity of certain layers between the cortex and nucleus, the remaining layers being transparent. **c., lenticular,** one occurring in the lens proper. **c., lymph, c., lymphatic.** See *c., Morgagnian.* **c., mature,** one in which the whole lens-substance is cataractous. **c., membranous,** a fibrinous deposit from the iris upon the capsule, which becomes opaque. **c., mixed,** one which ultimately affects the whole lens, but begins as a cortical opacity in sharply demarcated streaks or triangular patches. **c., Morgagnian,** one in which an overripe cataract shrinks and leaves a nucleus floating in the dissolved outer layers. **c., myelin,** one containing a semitransparent, yellowish, friable substance. **c., nuclear,** one of moderate extent beginning in the nucleus. **c., pigmented, c., pigmentous,** a spurious cataract due to an injury by which the pigment from the posterior surface of the iris has been detached, forming a tree-like appearance. **c., polar** (anterior or posterior), a form in which the opacity is confined to one pole of the lens. **c., pseudomembranous,** a condition marked by white spots on the lens due to iritis. **c., pupillary,** congenital closure of the pupil. **c., pyramidal,** one in which the opacity is at the anterior

pole and is conoid, the apex extending forward. **c., recurrent capsular, c., secondary,** capsular cataract appearing after the extraction of the lens. **c., ripe.** See **c.,** *mature.* **c., senile,** the cataract of old persons, the most frequent form, and that understood when not specified as *congenital, juvenile, traumatic, soft,* etc. **c., siliculose, c., siliquose.** See **c.,** *aridosiliquose.* **c.,** soft, a form occurring especially in the young; the lens-matter is of soft consistence and milky appearance. **c., spontaneous,** one not dependent upon some other lesion or disease. **c., stony,** one that has undergone degeneration and become of stony hardness. **c., tremulous, c., vacillating,** one associated with laceration of the zonule of Zinn, causing trembling of the iris and of the cataract on movement of the eyeball. **c., true,** lenticular cataract. **c., unripe.** See **c.,** *immature.* **c., zonular.** See **c.,** *lamellar.*
**cataractopiesis** (*kat-a-rak-ĭo-pi-e'-sis*). See *couching.*
**cataractous** (*kat-ar-ak'-tus*) [*cataract*]. Of the nature of or affected with cataract.
**catarrh** (*kat'-ahr'*) [καταρρεῖν, to flow down]. Inflammation of a mucous membrane. The term is also applied to certain inflammations of the tubules of the kidney and the air-vesicles of the lung. **c., alveolar,** a condition occurring in bronchopneumonia in which the alveoli of the lungs contain a granular liquid exudate holding modified epithelial cells and blood-corpuscles. **c., atrophic nasal,** chronic nasal catarrh resulting in dryness and atrophy of the membrane. **c., dry,** bronchitis in which there are frequent, severe attacks of coughing, with pain and but little expectoration. **c., epidemic.** See *influenza.* **c., gastric,** gastritis. **c., hemorrhagic,** bronchial catarrh attended with a superficial extravasation of blood into the mucous tissue. **c., infectious,** that caused by pathogenic microorganisms either by direct invasion or by the effect of toxins generated by them. **c., intestinal,** enteritis. **c., intoxication,** that caused by chemical poison—(*a*) introduced with the ingesta; (*b*) developed from the ingesta through putrefaction; (*c*) that developed from the blood. **c., Laennec's.** 1. See **c.,** *dry.* 2. See **c.,** *pituitous.* **c., membranous nasal,** a form of nasal catarrh marked by the formation of a thick-pseudomembrane. **c., mycotic,** that caused by a fungus. **c., nasal, coryza. c., papillary,** catarrh of the renal papillæ. **c., pituitous** (of Laennec), chronic serous bronchorrhea attended with copious secretion discharged by severe paroxysms of coughing. **c., pulmonary,** bronchitis. **c., rarefying dry** (of the nasopharynx), a state of malnutrition marked by pale, dry mucosa and at times the occurrence of varicose veins in the pharyngeal wall and about the orifices of the Eustachian tubes. **c., rose-,** hay-fever. **c., Russian,** influenza. **c., serous,** that marked by secretion consisting chiefly of a serous fluid. **c., suffocative.** Synonym of *capillary bronchitis.* **c., summer-,** hay-fever. **c., uterine,** endometritis. **c., vasomotor,** hay-fever. **c., vernal.** See *conjunctivitis, vernal.* **c., venereal.** Synonym of *gonorrhea.* **c., vesical,** cystitis.
**catarrhal** (*kat-ahr'-al*) [*catarrh*]. Of the nature of, affected with, or relating to catarrh. **c., fever.** Synonym of *influenza.* **c. inflammation,** an inflammation of an archiblastic surface, characterized by proliferation and desquamation of the epithelium.
**catarrhectic** (*kat-ar-ek'-tik*) [καταρρηκτικός]. Purgative.
**catastalsis** (*kat-as-tal'-sis*). A term suggested by Cannon for the downward moving wave of contraction occurring in the stomach, during digestion. There is no preceding wave of inhibition.
**catastaltic** (*kat-as-tal'-tik*) [κατασrέλλειν, to check, to send downward]. 1. Astringent. 2. Passing from above downward (as a nerve-impulse). 3. An inhibitory or sedative agent.
**catastasis** (*kat-as'-tas-is*) [κατάστασις, a settling]. Condition, state, habit; a decline, or quieting of symptoms; restitution, as of a displaced part.
**catastate** (*kat'-as-tāt*) [κατάστασις, settling down]. Any one of a series of successive catabolic states, substances, or conditions, each one of which is less complex, more stable, and exhibits less functional activity than its predecessor.
**catastatic** (*kat-as-tat'-ik*) [κατάστασις, a settling down]. Relating to a catastasis, or to a catastate.
**catatonia** (*kat-at-o'-ne-ah*) [κατά, down; τόνος, tension]. A form of mental derangement progressing from melancholia successively through mania and stupidity to imbecility and tonic convulsions.
**catatony** (*kat-at'-o-ne*). See *catalonia.*
**catatricrotism** (*kat-ah-tri'-krot-izm*) [κατά, down; τρεῖς, three; κροτός, a striking]. The occurrence of a third pulsation in the downward stroke of the sphygmograph.
**catatropia** (*kat-ah-tro'-pe-ah*) [*kata-*; τρόπος, a turn]. An actual turning of both eyes downward.
**catavertebral** (*kat-ah-ver'-te-bral*) [κατα, down; *vertebra*]. Located on the side of a centrum of a vertebra next to the blood-vessels.
**catching** (*katch'-ing*). Contagious, in the popular sense, i. e., directly from one person to another, as scabies, ringworm, syphilis, gonorrhea, typhus, variola, diphtheria, and scarlatina.
**catechin** (*kat'-e-kin*) [*catechu*], $C_{11}H_{10}O_5 + 5H_2O$. Catechinic acid, the active principle of catechu. It crystallizes in shining needles of a snow-white, silky appearance.
**catechol** (*kat'-e-kol*). See *pyrocatechin.* **c. dimethylate, c. dimethyl-ether.** See *veratrol.* **c. monomethylate, c. monomethyl-ether.** See *guaiacol.*
**catechu** (*kat'-e-kū*) [E. Ind.]. An extract prepared from the wood of *Acacia catechu,* a native of the East Indies. It contains 50 % of tannic acid, and hence is a powerful astringent. It is used in the diarrhea of children and as a gargle and mouth-wash. Dose of the *powdered extract* 10 gr.–½ dr. (0.65–2.0 Gm.). The *catechu* of the Pharmacopeia of 1890 has been replaced by *gambir* (*q. v.*). **c., infusum** (B. P.). Dose 1–1⅓ oz. (30–45 Cc.). **c., pulvis, compositus** (B. P.), contains catechu, kino, and rhatany. Dose 20–40 gr. (1.3–2.6 Gm.).
**catelectrode** (*kat-el-ek'-trōd*). See *cathode.*
**catelectrotonus** (*kat-el-ek-trot'-o-nus*) [κατά, down; ἤλεκτρον, amber; τόνος, tension]. The state of increased irritability of a nerve near the cathode. See *anelectrotonus.*
**catenating** (*kat'-en-a-ting*) [*catenare,* to chain together]. Connecting; linking; *e. g., catenating* ague, ague associated with another disease.
**catenoid** (*kat'-en-oid*) [*catena,* chain; εἶδος, resemblance]. Resembling a chain.
**catgut** (*kat'-gut*). The intestine of various animals, particularly the sheep, treated to make ligatures. **c., carbolized,** catgut rendered aseptic by soaking in a solution of phenol. **c., chromicized,** gut treated with chromium trioxide. **c.-plate,** an appliance for uniting intestinal edges in intestinal anastomosis. It is made of a solid catgut sheet, is thin, large, and flat, and resembles the Senn decalcified bone-plates.
**catharma** (*kath-ar'-mah*) [κάθαρμα, refuse; *pl., catharmata*]. That which is removed by purgation; excrement.
**catharsis** (*kath-ar'-sis*) [καθαίρειν, to purge]. Purgation.
**cathartate** (*kath-ar'-tāt*). A salt of cathartic acid.
**cathartic** (*kath-ar'-tik*) [see *catharsis*]. 1. Purgative. 2. A medicine used to produce evacuations of the bowels; a purgative. **c. acid.** See *senna.* **c. pill, compound.** See under *compound.*
**catharthin** (*kath-ar'-tin*). A bitter principle found in rhubarb, senna and jalap.
**cathartogenin** (*kath-ar-toj'-en-in*). A yellow-brown substance obtained from cathartic acid by decomposition with hydrochloric acid. Syn., *cathartogenic acid.*
**cathartomannite** (*kath-ar-to-man'-īt*). See *sennit.*
**cat-head** (*kat'-hed*). A term applied by Rosch to certain skulls the bones of which are too thin, the form rotund, with the occiput markedly projecting, with all prominences and muscular impressions are inconspicuous. Cf. *apple head.*
**cathelectrotonus** (*kath-el-ek-trot'-o-nus*). See *catelectrotonus.*
**Cathelineau's sign.** See *Tourette's disease.*
**catheresis, catheresis** (*kath-er-e'-sis*) [καθαίρειν, to reduce]. 1. Prostration or weakness induced by medication. 2. Caustic action; it often designates a feebly caustic action.
**catheretic** (*kath-er-et'-ik*) [καθαίρειν, to reduce]. 1. Reducing; weakening; prostrating. 2. Caustic. 3. A reducing or caustic agent.
**catheter** (*kath'-et-er*) [καθετήρ, a thing put down]. 1. A hollow tube for introduction into a cavity through a narrow canal. 2. Specifically, one intended to be passed through the urethra into the bladder. **c., Bozeman's,** a double-current uterine catheter. **c., Eustachian,** an instrument for examining, distending, or making applications to

the Eustachian tube. c. fever, systemic disturbance with fever, following the introduction of a catheter into the urethra., c., Gouley's. See *Gouley's catheter*. c.-life, continuous dependence upon the catheter for evacuation of the bladder. c., lung, a soft-rubber tube that may be passed down the trachea. c., Schroetter's. See *Schroetter's catheter*. c., self-retaining, one that will hold itself within the bladder without other appliances to assist it.
**catheterism, catheterization** (*kath'-et-er-izm, kath-et-er-iz-ā'-shun*) [*catheter*]. The introduction of a catheter.
**catheterize** (*kath'-et-er-īz*). To introduce a catheter.
**catheterostat** (*kath'-et-er-o-stat*) [*catheter*; στατός, standing]. A stand with glass tubes for holding sterilized catheters.
**cathion.** Cation.
**cathodal** (*kath'-o-dal*) [κατά, down; ὁδός, way]. Relating to a cathode.
**cathode** (*kath'-ōd*) [κατά, down; ὁδός, way]. The negative electrode or pole of an electric circuit.
**cathodic** (*kath-od'-ik*) [κατά, down; ὁδός, way]. 1. Relating to a cathode. 2. Proceeding downward; efferent or centrifugal (applied to a nerve-current or nerve-impulse).
**catholicon** (*kath-ol'-ik-on*) [καθολικόν, universal]. A universal remedy; a cure-all.
**cathypnosis** (*kath-ip-no'-sis*) [καθύπνωσις, a falling asleep]. Syn. of *African lethargy*.
**cation** (*kat'-e-on*) [κατιών, going down]. An electropositive element, *i. e.*, one which passes to or is evolved at the cathode in electrolysis.
**cativi, cativia** [Carib for manihot-root]. A skin disease of Central America said to be caused by an animal parasite, and resembling in its lesions grated manihot-root.
**catlin, catling** (*kat'-lin, kat'-ling*) [dim. of *cat*]. A long, pointed, two-edged knife used in amputation.
**catnep, catnip** (*kat'-nep, kat'-nip*) [corruption of *catmint*]. The leaves and tops of the herb *Nepeta cataria*, a stimulant and tonic; a popular remedy for chlorosis, hysteria, etc. Dose of *fluidextract* 1–2 dr. (4–8 Cc.).
**catochus** (*kat'-o-kus*) [κάτοχος, a holding down]. 1. Catalepsy; coma-vigil. 2. Apparent death; trance.
**catodont** (*kat'-o-dont*) [κατά, down; ὁδούς, tooth]. Possessing teeth only in the lower jaw.
**catoptric test.** The diagnosis of cataract by means of the reflection of images from the cornea and lens-capsules.
**catoptrics** (*kat-op'-triks*) [κάτοπτρον, a mirror]. The laws of the reflection of light.
**catoptroscope** (*kat-op'-tro-skōp*) [κάτοπτρον, a mirror; σκοπεῖν, to examine]. An instrument for examining objects by reflected light.
**catoteric** (*kat-o-ter'-ik*) [κατωτερικός, a carrying downward]. A purgative or cathartic.
**catramin** (*kat'-ram-in*). A turpentine obtained from *Tsuga canadensis* and other conifers. It is recommended as a stimulant, diuretic, and expectorant in chronic respiratory troubles, and is used in tuberculosis and lupus subcutaneously and as an embrocation.
**cat's ear.** A deformed ear similar to that of a cat. c.'s eye, a morbid yellowish appearance of the fundus of the eye. c.'s-eye pupil, an elongated pupil.
**cat's-purr.** A peculiar purring bruit heard on auscultation, due to a defect of the mitral valve. Syn., *Frèmissement cataire*.
**cattle-plague.** See *Rinderpest*.
**catulotic** (*kat-ul-ot'-ik*) [κατουλωέιν, to cause to cicatrize]. Promoting cicatrization.
**cauda** (*kaw'-dah*) [L.]. 1. A tail: 2. The part of a muscle forming its insertion. c. cerebelli, the vermiform process. c. epididymidis, the inferior part of the epididymis. c. equina, a term applied collectively to the roots of the sacral and coccygeal nerves, from their resemblance to a horse's tail. c. helicis, an appendage of the cartilage of the ear at the union of the helix and antihelix. c. medullæ (of Bartholin), a collective term for the oblongata and spinal cord. c. muliebris, the clitoris. c. salax, the penis. c. striati, the narrow posterior portion of the caudate nucleus.
**caudad** (*kaw'-dad*) [*cauda*; *ad*, to]. Toward the tail or cauda; opposed to cephalad; in man, downward.
**caudal** (*kaw'-dal*) [*cauda*]. 1. Pertaining to a cauda or tail. 2. Referring to a position near the tail end of the long axis of the body.

**caudate** (*kaw'-dāt*) [*cauda*]. Having or resembling a tail. c. lobe of liver, a small lobe of the liver. c. nucleus, the intraventricular portion of the corpus striatum.
**caudation** (*kaw-da'-shun*) [*cauda*]. 1. The condition of being furnished with a tail. 2. Elongation of the clitoris.
**caudatolenticular, caudolenticular** (*kaw-dat-o-len-tik'-ū-lar, kaw-do-len-tik'-ū-lar*). Pertaining to both the caudate and the lenticular nuclei.
**caudatum** (*kaw-da'-tum*). See *corpus striatum*.
**caudex** (*kaw'-deks*) [L., "a tree-stem"]. 1. In biology, applied to the scaly, unbranching trunk of a palm-tree or tree-fern. 2. The main portion of the brain-stem, the fibers running from the spinal cord to the hemispheres of the brain. c. cerebri. See *caudex* (2). c. dorsalis. 1. The spinal cord. 2. The oblongata. c. encephali, the cerebral peduncle. c. encephali communis, the oblongata and crus cerebri. c. encephali pontilis, the pons. c. medullaris, the cerebral peduncle.
**caudiduct** (*kaw'-de-dukt*) [*cauda*, tail; *ducere*, to draw]. In biology, to draw or carry backward toward the tail.
**caudiferous** (*kaw-dif'-er-us*) [*cauda*; *ferre*, to bear]. Having a tail or tail-like appendage.
**caudle** (*kaw'-dl*) [ME., *caudel*, a warm drink]. A nutritious food for invalids. It is made as follows: Beat up an egg to a froth; add a glass of sherry and half a pint of gruel. Flavor with lemon-peel, nutmeg, and sugar.
**caul** (*kawl*) [ME., *calle*, a hood]. 1. A portion or all of the fetal membranes covering the head and carried out in advance of it in labor. 2. The great omentum.
**cauliflower excrescence.** A tumor with an irregular surface resembling the cauliflower.
**cauline** (*kaw'-lin*) [καυλός, a stalk]. In biology, of or pertaining to the stem.
**caulophyllin** (*kaw-lo-fil'-in*) [καυλός, a stalk; φύλλον, a leaf]. A resinoid precipitate from the tincture of caulophyllum. See *caulophyllum*.
**caulophyllum** (*kaw-lo-fil'-um*) [καυλός, stalk; φύλλον, leaf]. Blue cohosh; "squaw-root"; the rhizome and rootlets of *C. thalictroides*, growing in Canada and the northern United States. It contains a glucoside, *saponin*, and two resins, one of which is *caulophyllin*. It produces intermittent contractions of the gravid uterus, and possesses diuretic, emmenagogue, and antispasmodic powers. There are no official preparations. Dose of the *powdered drug* 5–20 gr. (0.32–1.3 Gm.); of *caulophyllin* 2–5 gr. (0.13–0.32 Gm.).
**cauloplegia** (*kaw-lo-ple'-je-ah*) [καυλός, a stalk; πληγή, a stroke]. Paralysis affecting the penis.
**caulosterin** (*kaw-los'-ter-in*) [καυλός, a stalk; στερεός, solid]. C₂₆H₄₄O. An aromatic compound occurring in the root and stem of seedlings of the yellow lupine which have grown in the dark. It is levorotary, forming lustrous plates which melt at 158°–159° C.
**cauma** (*kaw'-mah*) [καῦμα, a burning; *pl.*, *caumata*]. Fever; heat; pyresis; an inflammatory fever; a burn. c. enteritis, synonym of intestinal catarrh, acute.
**caumatic** (*kaw-mat'-ik*) [καῦμα, a burning]. Pertaining to cauma.
**caumesthesia** (*kaw-mes-the'-ze-ah*) [καῦμα, heat; αἴσθησις, sensation]. A condition in which a person experiences a sense of heat, when the temperature is not high.
**causalgia** (*kaw-sal'-je-ah*) [καυσός, a burning; ἄλγος, pain]. The burning pain that is sometimes present in injuries of the nerves.
**cause** (*kawz*) [*causa*, a cause]. The sources, conditions and origins of a result. The preceding factors that unite to produce a given condition. Causes are spoken of as efficient, instrumental, final, primary, secondary, predisposing, controlling, determining, ultimate, exciting, etc. c., antecedent. See c., predisposing. c., determining, a cause that precipitates the action of another or other causes. c., efficient, c., essential, one that secures the effect independent of the action of other causes. c., endopathic, see c., internal. c., exciting, the immediately preceding and conditioning factor. c., exopathic, c., external, one that acts external to the organism. c., immediate, see c., proximate. c., internal, a cause acting within the organism. c., negative, one consisting in the absence of some prophylactic condition. c., predisponent, c., procatarctic, see c., predisposing. c., predisposing, that which tends to

the development of a condition. c., **primary,** c., **proximate,** that one of several causes which takes effect last and acts with rapidity. c., **remote,** c., **secondary,** an ultimate cause. c., **ultimate,** one that eventually comes into play aided by a proximate cause. c., **vital,** a specific pathogenic microorganism.
**causoma** (*kaw-so'-mah*) [καύσωμα; καίειν, to burn; *pl.*, *causomata*]. A burning; usually an inflammation.
**caustic** (*kaws'-tik*) [*causticum*, caustic]. 1. Very irritant; burning; capable of destroying tissue. 2. A substance that destroys tissue. c. **alkali,** a pure alkaline hydrate or oxide. c. **arrows,** conical sticks charged with caustic material. c., **lunar** (*argenti nitras fusus*, U. S. P.), silver nitrate. c., **metallic,** one containing a metal or a metallic salt. c., **mitigated** (*argenti nitras mitigatus*, U. S. P.), silver nitrate made less active by fusion with potassium nitrate or argentic chloride. c., **perpetual,** fused silver nitrate. c. **potash,** potassium hydroxide. c. **soda,** sodium hydroxide.
**causticity** (*kaws-tis'-it-e*) [καίειν, to burn]. Caustic quality; corrosiveness.
**cauter** (*kaw'-ter*) [καυτήρ, a burner]. A searing-iron or cautery-iron; any caustic application.
**cauterant** (*kaw'-ter-ant*) [καυτήρ, a burner]. 1. Caustic; escharotic. 2. A caustic substance.
**cauterism** (*kaw'-ter-izm*). See *cauterization*.
**cauterization** (*kaw-ter-iz-a'-shun*) [see *cautery*]. The application of a cautery; the effect of such an application. c., **distant,** that performed by holding the cautery at some distance from the surface to be cauterized. c., **galvanochemical** (Apostoli's), the destruction of the mucosa by means of electrolytic action. c., **inherent,** deep cauterization by means of the actual cautery. c., **Neapolitan,** deep cauterization through an incision. c., **objective.** See *c., distant*. c. **by points,** c., **punctate,** deep cauterization with a pointed cautery. c., **slow,** that performed with moxa. c., **subcutaneous,** deep cauterization by injection of caustics or by inclosing the cautery in a tube so as not to affect the superficial parts. c., **tubular,** Tripier's operation of charring the walls of an opening made into a cyst by means of an instrument connected with the negative pole of a battery.
**cauterize** (*kaw'-ter-iz*) [see *cautery*]. To sear or burn with a cautery or a caustic.
**cautery** (*kaw'-ter-e*) [καυτήριον, a branding-iron]. A metal instrument heated by the electric current or in a flame, used to destroy tissue or for producing counterirritation. Syn., *inustorium*. c., **actual,** the white-hot iron. c., **button-,** an iron heated in hot water. c., **galvanic,** a platinum wire heated by electricity. c., **nummular,** a cautery iron fitted with a coin-shaped disc. c., **Paquelin's.** See *Paquelin's cautery*. c., **potential,** c., **virtual,** the application of caustic substances. c., **solar,** a lens for concentrating the rays of the sun upon a part to be cauterized. c., **steam.** See *atmocausis*. c., **thermo-.** See *thermocautery*.
**cava** (*ka'-vah*) [L.]. 1. A vena cava. 2. Any external cavity or hollow of the body.
**caval** (*ka'-val*) [*cava*]. 1. Relating to a vena cava. 2. Hollow.
**cavalry-bone** (*kav'-al-re-bōn*). A bony deposit in the adductor muscles of the thigh.
**cavascope** (*kav'-a-skōp*) [*cava*; σκοπεῖν, to view]. An apparatus for illuminating a cavity.
**cavern** (*kav'-ern*) [*caverna*, a hollow]. A cavity in the lung due to necrosis of the parenchyma; also the cavity of a dilated bronchus. c., **brand,** one due to gangrenous destruction of a circumscribed segment of the lung parenchyma.
**cavernitis** (*kav-er-ni'-tis*). Inflammation of the corpora cavernosa.
**cavernoma** (*kav-ern-o'-mah*) [*cavern*; ὄμα, a tumor]. A cavernous tumor; a cavernous angioma.
**cavernosum** (*kav-ur-no'-sum*). The corpus cavernosum.
**cavernous** (*kav'-er-nus*) [*cavern*]. Having hollow spaces. c. **angioma,** an angioma filled with blood-spaces. c. **bodies,** the corpora cavernosa of the penis. c. **breathing,** the breath-sounds heard over a pulmonary cavity. c. **groove,** the carotid groove. c. **plexus.** See *plexus, cavernous*. c. **sinus,** a venous sinus situated at the side of the body of the sphenoid. c. **tissue,** erectile tissue. c. **tumor,** a cavernous angioma.
**caviar,** or **caviare** (*kav-e-ar'*) [Fr.]. The salted hard roe of the sturgeon and other large fish.

**cavitary** (*kav'-it-a-re*) [*cavitarius*, hollow]. Hollow; having, or forming cavities. Applied to any nematode worm; any intestinal worm that has a body-cavity; a worm that is not anenterous.
**cavitas** (*kav'-it-as*) [L.]. A hollow, a cavity. c. **cochleata.** See *duct, spinal*. c. **glenoidališ,** the glenoid cavity. c. **pulpæ,** the pulp-cavity of a tooth. See *dental cavity*.
**Cavité fever** (*kav-e-ta'*). An acute contagious disease confined almost exclusively to Cavité naval station in the Philippines. It is marked by sudden onset, high temperature, severe muscular pain, and extremely tender and painful eyeballs, the incubation period varying from two days to two weeks.
**cavitis** (*ka-vi'-tis*) [*cava*; ιτις, inflammation]. Inflammation of a vena cava; celophlebitis.
**cavity** (*kav'-it-e*) [*cavum*]. A hollow. See under *abdominal, amniotic, cotyloid, glenoid, pulp*, and *serous*. c., **ankyroid,** the descending cornu of the lateral ventricle. **cavities, cerebral,** the ventricles of the brain. c., **cotyloid,** the acetabulum. c., **cranial,** the hollow of the skull. c., **epiploic,** the omentum. c., **nasal,** the nasal fossa. c., **oral,** that of the mouth. Syn., *cavum oris; spatium oris*. c., **pleural,** the closed space of the pleura included between its parietal and visceral layers. c., **pleuro-peritoneal,** the celom or body-cavity. c., **preperitoneal.** See *Retzius' space*. c., **sigmoid,** one of two depressions on the head of the ulna for articulation with the radius and humerus.
**cavo-valgus** (*ka'-vo-val'-gus*) [*cavus*, hollow; *valgus*, bow-legged]. Cavus combined with valgus. See *clubfoot*.
**cavus** (*ka'-vus*) [L.]. 1. A hollow; a cavity. 2. Talipes arcuatus; hollow-foot.
**cavum** (*ka'-vum*) [L.]. Any hollow or cavity, normal or pathological. c. **abdominis,** the cavity of the abdomen. c. **articulare,** joint cavity. c. **conchæ,** the deepest part of the concave surface of the concha. c. **dentis,** the pulp-cavity of a tooth. c. **epidurale,** epidural cavity. c. **laryngis,** cavity of larynx. c. **Meckelii.** See *Meckel's cavity*. c. **mediastinale anterius,** anterior mediastinal cavity. c. **mediastinale posterius,** posterior mediastinal cavity. c. **medullare,** the medullary canal of bones. c. **nasi,** nasal cavity. c. **oris proprium,** the cavity of the mouth proper. c. **pelvis,** pelvic cavity. c. **pericardii,** the pericardial cavity. c. **peritonæi,** the peritoneal cavity. c. **pharyngis,** cavity of pharynx. c. **pleuræ,** the pleural cavity. c. **Retzii.** See *Retzius's space*. c. **septi,** the embryonal fifth ventricle of the brain. c. **septi pellucidi,** cavity of septum pellucidum, the fifth ventricle. c. **subarachnoideale,** the subarachnoid space. c. **thoracis,** thoracic cavity. c. **subdurale,** the subdural space, the interval between the dura mater and the arachnoid. c. **thoracis,** thoracic cavity. c. **tympani,** the tympanic cavity. c. **uteri,** the cavity of the uterus.
**cayaponine** (*ka-ap'-o-nēn*) [*Cayaponia*, Brazilian name]. An alkaloid extracted from *Cayaponia globosa*, a cucurbitaceous plant of Brazil. It is said to purge without griping. Dose 1 gr. (0.06 Gm.).
**Cayenne pepper** (*ki'-en*). See *capsicum*.
**Cazenave's lupus** (*kas-nāv'*) [P. L. A. *Cazenave*, French dermatologist, 1795–1877]. Lupus erythematosus. C.'s **vitiligo.** See *Celsus' area*.
**Cb.** Chemical symbol of *columbium*.
**Cc., c.c.** Abbreviations of *cubic centimeter*.
**C.C.C.** Abbreviation of cathodal closure contraction; also written CaCC.
**c.cm.** Abbreviation for cubic centimeter; more frequently written c.c.
**CCTe.** Abbreviation for cathodal closure tetanus.
**Cd.** Chemical symbol of *cadmium*.
**Ce.** Chemical symbol of *cerium*.
**ceanothin** (*se-an-o'-thin*). 1. A brown powder obtained from *Ceanothus americanus*. It is purgative and alterative, and is used in syphilis, dysentery, and sore throat. Dose 1–2 gr. (0.065–0.13 Gm.). 2. A resinoid isolated from *Ceanothus americanus*; slightly soluble in alcohol and ether.
**cearin** (*se'-ar-in*). An ointment-base consisting of carnauba wax, 1 part; paraffin, 3 parts, melted together and mixed with 4 times its weight of liquid petrolatum.
**ceasma** (*se-az'-mah*) [κέασμα, a chip; pl., *ceasmata*]. A splinter; a fissured state.
**ceasmic** (*se-az'-mik*) [κέασμα, a chip]. Fissured; remaining in the primitive fissured state of the embryo.

**cebocephalia** (*se-bo-sef-a'-le-ah*) [κήβος, a kind of monkey; κεφαλή, the head]. The condition of being cebocephalic.
**cebocephalic** (*se-bo-sef-al'-ik*) [κήβος, a kind of monkey; κεφαλή, the head]. Of the nature or appearance of a cebocephalus.
**cebocephalus** (*se-bo-sef'-al-us*) [κήβος, a monkey; κεφαλή, head]. A variety of single autositic monsters of the species *Cyclocephalus*, in which there is entire absence of the nose, with, however, two orbital cavities and two eyes, the region between the eyes being narrow and perfectly flat.
**cebur** (*se'-bur*). See *balsam, Tagulaway*.
**cecal** (*se'-kal*) [*cecum*]. 1. Pertaining to the cecum. 2. Ending in a cul-de-sac.
**cecectomy** (*se-sek'-tom-e*) [*cecum*; τομή, a cutting]. Excision of the cecum.
**cecitas** (*se'-sit-as*) [L.]. Blindness.
**cecitis** (*se-si'-tis*) [*cecum*; ιτις, inflammation]. Inflammation of the cecum; typhlitis.
**cecity** (*ses'-it-e*) [*cæcus*, blind]. Blindness.
**cecocele** (*se'-ko-sēl*) [*cecum*; κήλη, hernia]. A hernia into the cecum. Syn., *typhlocele*.
**cecograph** (*se'-ko-graf*) [*cæcus*, blind; γράφειν, to write]. A writing-machine for the use of the blind.
**cecopexy** (*se'-ko-peks-e*) [*cēcum*; πῆξις, fixation]. Fixation of the cecum by a surgical operation.
**cecoplication** (*se-ko-pli-ka'-shun*). An operation for the relief of dilated cecum, consisting in taking folds in its wall.
**cecoptosis** (*se-kop-to'-sis*) [*cecum*; πτῶσις, a falling]. A downward displacement of the cecum.
**cecosigmoidostomy** (*se-ko-sig-moid-os'-to-me*) (*cecum*; *sigmoid*; στόμα, a mouth). The making of an anastomosis between the cecum and sigmoid.
**cecostomy** (*se-kos'-to-me*). [*cecum*; στόμα, a mouth]. The formation of an artificial anus in the cecum.
**cecum, cæcum** (*se'-kum*) [*cæcus*, blind]. The large blind pouch or culdesac in which the large intestine begins. c. **mobile**, abnormal mobility of the cecum, so that it can be pushed up out of its normal situation.
**cecutiency** (*se-ku'-shen-se*) [*cǣcutire*, to become blind]. Tendency to, or the commencement of, blindness.
**cedar** (*se'-dar*) [*cedrus*, cedar]. One of the genus of coniferous trees, *Cedrus*. c.-**oil**, a transparent oil obtained from *Juniperus virginiana*, and used as a clearing agent in histology and for oil-immersion lenses.
**Cedrela** (*sed'-re-lah*) [κεδρελάτη, a cedar-fir tree]. A genus of trees found in tropical regions and allied to mahogany. c. **febrifuga**, of Southern Asia; c. **odorata**, bastard cedar, of tropical America (see *Cailcedra*). c. **rosmarinus**, of Indo-China, and c. **toona**, of India, are among the species that afford active medicines.
**cedrene** (*se'-drēn*) [*cedrus*, cedar], C₁₅H₂₄. A volatile liquid hydrocarbon found in oil of red cedar (see *Juniperus virginiana*), oil of cloves and oil of cubebs. c. **camphor**, C₁₅H₂₄O, a camphor that separates from the oil of red cedar.
**cedrin** (*se'-drin*) [*cedrus*, cedar]. A bitter crystalline substance obtained from cedron.
**cedron** (*se'-dron*) [*cedrus*, cedar]. The seeds of *C. simaba*, a popular external remedy in tropical America for the bites of venomous insects and serpents, and of reputed value in malarial fevers. Dose of the fluidextract ♏ j–viij.
**ceke** (*the'-ke*). A Fiji term for elephantiasis of the scrotum.
**Cel.** Abbreviation for Celsius, scale of thermometer.
**celandine** (*sel'-an-dēn*). See *chelidonium*.
**celarium, cœlarium** (*se-la'-re-um*). The epithelium of the celom.
**celastrine** (*se-las'-trēn*). 1. Mosso's name for a poisonous alkaloid obtained from the leaves of *Catha edulis;* it resembles caffeine physiologically, though more energetic and differing essentially from it. Syn., *kathine*. 2. A bitter principle found by Dragendorff in the leaves of *Celastrus serratus*. 3. A substance forming minute white crystals found by Wayne in *Celastrus scandens*.
**Celastrus** (*sel-as'-trus*) [κήλαστρος, an evergreen tree]. A genus of trees and shrubs, nearly allied to *Euonymus*. *C. paniculatus* is a climbing shrub of India; the oil from the seeds (oleum nigrum) is a powerful stimulant and diaphoretic in gout and fever. *C. scandens*, of North America, is cathartic,

diuretic, and alterative. *C. serratus* is a native of Abyssinia; the leaves are used in malaria; they contain tannin, a bitter principle, *celastrin*, and a volatile oil.
**celation** (*sel-a'-shun*) [*celatio*, a hiding]. The concealment of illness, of a birth, or of pregnancy.
**cele** (*sēl*) [κοιλία, cavity]. An encephalic cavity; used instead of ventricle.
**-cele** (*sēl*) [κήλη, a tumor]. A suffix denoting a tumor, or swelling, or hernia.
**celenteron** (*se-len'-ter-on*). Same as *archenteron*, *q. v.*
**celerina** (*sel-e-ri'-nah*). A proprietary remedy said to contain celery, black haw, coca and kola.
**celery** (*sel'-er-e*) [σέλινον, a kind of parsley]. The stalk of *Apium graveolens*, or common garden celery. It contains *apiol, q. v.* It is reputed to be antispasmodic and nervine. Dose indefinite. c. **seed**, used to cover the taste of other drugs.
**celia** (*se'-le-ah*) [κοιλία, belly]. 1. The belly; the stomach. 2. A ventricle of the brain.
**celiac** (*se'-le-ak*) [*celia*]. Abdominal; pertaining to the belly. c. **artery**. Same as *c. axis*. c. **axis**, a branch of the abdominal aorta; it divides into the gastric, hepatic, and splenic arteries. See under *artery*. c. **disease**, a form of chronic indigestion generally occurring in children under five years of age, and characterized by offensive diarrhea. c. **ganglion**. See *ganglion, semilunar*. c. **passion**, painful diarrhea, or dysentery. c. **plexus**, a sympathetic nerve-plexus situated about the origin of the celiac axis.
**celiaca** (*se-li'-ak-ah*) [κοιλία, the abdomen]. Diseases of the abdominal organs.
**celiacomesenteric** (*se-le-ak-o-mez-en-ter'-ik*). Relating to the celiac and mesenteric regions.
**celiadelphus** (*se-le-ad-el'-fus*) [κοιλία, belly; ἀδελφός, brother]. A monstrosity having two bodies joined at the abdomen.
**celiagra** (*se-le-ag'-rah*) [κοιλία, belly; ἄγρα, seizure]. Abdominal gout.
**celialgia** (*se-le-al'-je-ah*) -[*celia;* ἄλγος, pain]. Pain in the abdomen.
**celian, celine** (*se'-le-an, se'-lin*). Same as *celiac*.
**celianeurysm** (*se-le-an'-u-rizm*) [*celia;* ἀνεύρυσμα, a widening]. An abdominal aneurysm.
**celiectasia** (*se-le-ek-ta'-ze-ah*) [*celia;* ἔκτασις, a stretching-out]. Abnormal distention of the abdominal cavity.
**celiectomy** (*se-le-ek'-to-me*) [*celia;* ἐκτομή, a cutting]. Excision of an abdominal organ.
**celiemia** (*se-le-e'-me-ah*) [*celia;* αἷμα, blood]. Hyperemia of the abdominal viscera.
**celiitis, cœliitis** (*se-le-i'-tis*) [*celia;* ιτις, inflammation]. Inflammation of the abdominal organs.
**celiocele, cœliocele** (*se'-le-o-sēl*) [*celia;* κήλη, a hernia]. Abdominal hernia.
**celiocentesis, cœliocentesis** (*se-le-o-sen-te'-sis*) [κοιλία, belly; κεντήσις, puncture]. Puncture of the abdomen.
**celiocyesis** (*se-le-o-si-e'-sis*) [κοιλία, belly; κύησις, pregnancy]. Abdominal extra-uterine gestation.
**celiodynia, cœliodynia** (*se-le-o-din'-e-ah*) [*celia;* ὀδύνη, pain]. Pain in the abdomen.
**celiogastrotomy** (*se-le-o-gas-trot'-o-me*) [*celia;* γαστήρ, stomach; τομή, cutting]. The opening of the stomach through abdominal incision.
**celiohemia, cœliohæmia**. See *celiemia*.
**celiohysterectomy** (*se-le-o-his-ter-ek'-to-me*) [*celia;* ὑστέρα, womb; ἐκτομή, a cutting out]. 1. Excision of the uterus through an abdominal cut. 2. Porrocesarean section.
**celiohysterotomy** (*se-le-o-his-ter-ot'-o-me*). 1. Same as *cesarean section, q. v.* 2. Incision of the uterus through an abdominal cut.
**celiolymph** (*se'-le-o-limf*) [κοιλία, a cavity; *lympha*, water]. The cerebro-spinal fluid.
**celiomyalgia** (*se-le-o-mi-al'-je-ah*) [κοιλία, belly; μῦς, muscle; ἄλγος, pain]. Pain in the abdominal muscles.
**celiomyitis, cœliomyitis** (*se-le-o-mi-i'-tis*) [*celia;* μῦς, muscle; ιτις, inflammation]. Inflammation of the muscles of the abdomen.
**celiomyodynia, cœliomyodynia** (*se-le-o-mi-o-din'-e-ah*). See *celiomyalgia*.
**celioncus** (*se-le-ong'-kus*) [κοιλία, the belly; ὄγκος, a swelling]. A tumor of the abdomen.
**celioparacentesis** (*se-le-o-par-ah-sen-te'-sis*) [κοιλία, belly; *paracentesis*]. Tapping, or paracentesis of the abdomen.

**celiopyosis** (se-le-o-pi-o'-sis) [κοιλία, belly; πύωσις, a suppuration]. Suppuration in the abdominal cavity.

**celiorrhea** (se-le-or-e'-ah) [κοιλία, the belly; ῥέειν, to flow]. Diarrhea.

**celioschisis** (se-le-os'-kis-is). [celia; σχίσις, fissure] Congenital abdominal fissure.

**celioscope** (se'-le-o-skōp) [κοῖλος, a hollow; σκοπεῖν, to examine]. An apparatus for illuminating and inspecting body-cavities.

**celioscopy, cœlioscopy** (se-le-os'-ko-pe). Kelling's method of examining the peritoneal cavity by filling it with sterile filtered air through a hollow needle, plunging a trocar through the distended abdominal wall, and passing through the trocar a cystoscope by means of which the adjacent peritoneal surface may be inspected.

**celiotomy** (se-le-ot'-o-me) [celia; τομή, a cutting]. Surgical opening of the abdominal cavity.

**celitis** (se-li'-tis) [κοιλία, belly; ιτις, inflammation]. Inflammation of the abdominal organs.

**cell** (sel) [cella, a small, hollow cavity]. 1. A granular mass of protoplasm containing a nucleus. The typical adult cell consists of protoplasm or cell-contents, a nucleus, and, within the latter, one or more nucleoli. The cell may or may not have a cell-wall. The *protoplasm* consists of two parts— the *spongioplasm* and the *hyaloplasm*. The *nucleus* is made up of a nuclear membrane, nuclear fibrils (*chromatin*), and nuclear matrix (*achromatin*). The *nucleolus* is a highly refracting body the function of which is not known. a. A galvanic element or single member of a galvanic battery without the connecting wire between the metals. c., **acid.** Same as c., *delomorphous*. c., **acidophil,** one which attracts acid dyes. Syn., *oxyphil cell.* c.s, **adelomorphous,** epithelial cells composing the chief part of the lining of the glands of the stomach, particularly the pyloric region. They are supposed to secrete pepsinogen. Syn., *central cell; peptic cell.* c., **adventitial.** 1. A branched cell peculiar to the perithelium. 2. A stellate cell of the membrana propria of glands. c., **air-.** See *air-cell.* c.s, **amacrine,** spongioblasts of the inner nuclear layer of the retina; they lack long processes, though sometimes axis-cylinder processes are given off which may extend into the nerve-fiber layer. The bodies of these cells are often partly in the inner molecular layer. c., **ameboid,** a cell capable of changing its form and of moving about like an ameba. c., **apolar,** a nerve-cell without processes. c. **basket,** a neuron from whose axis cylinder there project fibrils which surround another cell. c., **beaker-.** See c., *goblet-.* c., **binary nerve-,** two pyriform nerve-cells contained in a single sheath and each provided at its pointed end with a single nerve-fiber; these radiate in opposite directions. c.-**body,** the mass of a cell. c., **bone-.** See *osteoblast.* c., **brush-.** See *Deiters' cells.* c.s, **calcigerous.** 1. Cells containing earthy salts found in dentine. 2. Mueller and Henle's name for the lacunæ of bone and their canaliculi. c.-**capsule,** a thick or unusually strong cell-wall. c., **central.** Same as c., *adelomorphous.* c.s, **centroacinar,** c.s, **centroacinous,** little cellular masses found by Robert Langerhans in the interstitial connective tissue of the pancreas. Syn., *Langerhans' islets; Renaut's follicular points.* c., **chromatophore,** a cavity directly beneath the epidermis containing pigment and changing its shape and color by means of attached radiating muscular bands. c., **ciliated,** one provided with cilia. c., **cleavage,** a segmentation-cell. c., **columnar,** one of the elongated cells forming columnar epithelium. c.s **commissural.** Same as c., *heteromeral.* c.-**cones,** the cancer-nests of a squamous epithelioma—so called from their conical shape. c., **constant,** the galvanic element of a constant battery. c. **of Corti.** See *Corti's cell.* c.-**cover,** the cuticular layer. c., **cover-.** See c., *tegmental.* c., **cylindrical,** a variety of epithelial cell shaped like a miniature cylinder. c., **cytochrome,** a nerve-cell having a cell-body very small in proportion to its nucleus. c., **daughter-,** a cell originating from the division of the protoplasm of a mother-cell. c.s, **decidual,** a proliferation of young connective-tissue cells above the uterine glands taking place after the ovum is impregnated. c. of **Deiters.** See *Deiters' cells.* c.s, **delomorphous,** Rollet's name for granular cells which stain deeply, occurring next the basement-membrane in the glands of the stomach in the cardiac region. They are supposed to secrete acid. c.s, **demilune,** granular protoplasmic cells found in mucous glands, lying between the mucous cells and the basement-membrane. Syn., *cells of Gianuzzi.* c., **dentine,** c., **dentinal.** 1. An odontoblast. 2. One of the lacunæ in dentine similar to those in bone. 3. One of the bodies forming the matrix- in dentine. c.-**division.** See *karyokinesis.* c.-**doctrine,** the theory that the cell is the unit of organic structure, and that cell-formation is the essential process of life and its phenomena. c., **elementary,** an embryonic cell; also a leukocyte. c., **embryo,** c., **embryonic,** one arising from the division of the ovum. Syn., *elementary cell; formative cell; primary cell; primitive cell; primordial cell.* c.s, **embryoplastic,** cells originating from the mesoblast and becoming stellate or fusiform; they comprise the fixed connective-tissue corpuscles in developing connective tissue. Syn., *fibroplastic bodies* or *cells.* c.s, **endothelial,** flat cells found on the inner surface of vessels and spaces that do not communicate directly with the external air. c., **epidermic,** c., **epithelial.** See *epithelium.* c., **epithelioid,** one of the flattened cells forming an epithelial or endothelial covering in forms of membranous connective tissue. c.s, **ethmoid,** the cellular cavities of the lateral masses of the ethmoid bone. Syn., *ethmoid sinuses.* c., **external ciliated.** See *Corti's cell.* c., **fiber-,** a cell elongated into a fiber. c., **fibrillated.** 1. See *Heidenhain's rods.* 2. One of the fibrillated cells lining the interlobular ducts of the salivary glands. c.-**fission,** cell-division. c.s, **follicular.** 1. Those of which the membrana granulosa is composed. 2. See *Sertoli's columns.* c., **formative.** Same as c., *embryonic.* c.s, **Foule's.** See *Foule's cells.* c., **fusiform,** a spindle-cell. c., **ganglion-,** a large nerve-cell, especially that found in the spinal ganglia. c., **germinal,** an epiblastic cell from which a neurone is derived. c., **giant-,** large multinuclear cells occurring in tuberculosis and other infectious granulomata, in bone, in giant-cell tumors, etc. c. of **Gianuzzi.** See *c.s, demilune.* c.s, **glia-,** neuroglia. c.-**globulins,** Halliburton's name for forms of globulin that occur in lymph-corpuscles and can be extracted from them by solutions of sodium chloride. c., **goblet-,** an epithelial cell that has been bulged out like a goblet by the presence of mucin. c.s, **Golgi's.** See *Golgi's cells.* c., **Grove.** See *Grove's cell.* c., **guard.** See *guard.* c., **gustatory,** a taste cell. c., **hair.** See *hair.* c., **hecateromeric,** c., **hecatomeral,** a nerve-cell of the cinerea of the spinal cord whose processes divide into two, one going to each side of the cord. c., **heckle,** a prickle-cell. c.s **Heidenhain's.** 1. See *c.s, delomorphous.* 2. See *c.s, adelomorphous.* c., **heteromeral,** c., **heteromeric,** a nerve-cell in the cinerea of the spinal cord, the axons of which pass through one of the commissures and enter the white matter of the other side of the cord. Syn., *commissural cell.* c., **histogenetic wandering,** a migratory connective-tissue cell or glandular cell; a wandering cell that is not a leukocyte. c.s, **horn-,** c.s, **horny,** those comprising the stratum corneum of the epidermis; they are homogeneous cells containing keratin, and are modified to form nails, hoofs, hair, etc. c.s, **imbricated,** those overlapping like roof-tiles. c., **indifferent,** a cell found in the walls the neural tube. c.-**islets,** the centers of most active growth in young cellular tissues. They contain the stores of nutriment that are gradually dissolved and digested. c.s, v. **Koelliker's.** See *v. Koelliker's cell.* c.s, v. **Koelliker's tract-.** See *v. Koelliker's tract-cells.* c.s, **liver-,** nucleated polyhedral or spheroidal cells containing granules of glycogen and pigment and more or less fat, forming the glandular substance of the liver. c., **locomotive,** one endowed with power of movement, especially a ciliated cell. c., **lymphoid,** a small, round, connective-tissue cell containing a relatively large nucleus. c., **marrow,** an osteoblast. c.s, **mast-,** leukocytes containing coarse basophile granules. They are occasionally present in the peripheral circulation as the result of certain pathological influences, but are totally foreign to the normal blood of man. They are commonly found in the splenomedullary type of leukemia. The granules of the mast-cell show an intense affinity for basic aniline dyes, toward which they react metachromatically. In view of their distinctive behavior toward selective stains for mucin, Harris suggests for the mast-cell the term *mucinoblast.* c., **mastoid,** one of the hollow air-spaces in the structure of the mastoid process. c.s, **medullary.** 1. Marrow-cells. 2. The ameboid cells of cartilage-bone. c., **mother-**, a cell that divides its protoplasm and gives each part a new cell-wall. c., **motor**, a nerve-cell generating impulses. c., **mucin-**, c., **mucous**, c., **mucus-**, a cell which secretes mucus, particularly a kind of salivary cell secreting mucus, but no albumin. c.-**multiplication**, cytogenesis, a name given to the process of reproduction of cells. It may be *direct*, as when a cell constricts and cuts off a part of itself, or *indirect*, when the division is preceded by the cycle of nuclear changes known as karyokinesis. c.s, **muscle-**, a general term for cells the substance of which is contractile. c., **myeloid**. 1. See *myeloplax*. 2. Applied, from its resemblance to a cell of the red marrow of bone, to one of the oval multinuclear cells of myeloid tumors. c., **naked**, one unprovided with a cell-wall. c.-**nests**, a collection of epithelial cells closely packed together and surrounded by a connective-tissue stroma. Cell-nests are found in carcinoma. c., **neuroglia**, one, of the cells of the neuroglia; flat, round cells, especially numerous about blood-vessels and the pia mater. c., **neutrophil**. See $c_1$, *acidophil*. c., **nuclear**, a nucleated dendritic nerve-cell. c., **nucleated**, a cell containing one or more nuclei. Syn., *karyota*. c.-**nucleus**, the cytoblast; the areola. c.s, **oxyntic**. See *c.s, delomorphous*. c., **oxyphil**. See *c., acidophil*. c.s, **palatine**, the cells formed by the junction of the palatine and ethmoid bones. c., **parietal**. Same as *c., delmorphous*. c., **peptic**. Same as *c., adelomorphous*. c., **pigmented**, one containing granules of pigment. c., **pillar**, a peculiar S-shaped cell with a striated body, found in the organ of Corti. c.s, **plasma-**. 1. (Of Unna.) Cubic or rhombic cells, the protoplasm of which stains deeply with methylene-blue, while the nucleus, which has usually an eccentric situation, is readily decolorized (by creosote or styrone). They are probably derived from lymphocytes, and play an important part in inflammatory reactions, especially in granulomatous processes. 2. (Of Waldeyer.) Nucleated cells of varying size and shape, with voluminous, coarsely granular protoplasm, found in connective tissue, especially about the blood-vessels. c.-**plate**. 1. The equatorial plate in which division of the nucleus occurs during karyokinesis. 2. (Of Strasburger.) The equatorial thickening of the spindle-fibers from which the septum arises during the mitosis of plant-cells. c.-**plate**, **subendothelial**, a small granular cell of unknown function occurring in the intima of blood-vessels. c., **polar**. See under *polar*. c., **porous**. 1. One containing an opening in the side. 2. A porous jar containing one of the liquids of a galvanic battery. c., **prickle-**. See *prickle-cell*. c., **primary**. 1. An embryonic cell. 2. Any undifferentiated cell. c., **protective**. See *c., tegmental*. c.s, **pseudoplasma**, cells found in normal human spleen and differing from plasma-cells in that they are larger, and possess twisted nuclei which do not present the characteristic chromatin arrangement. They appear to be a variety of large mononuclear leukocyte, the protoplasm of which has become basophilic. c.s, **Ranvier's**. See *Ranvier's cells*. c., **roof**. 1. See *c., tegmental*. 2. One found on the convexity formed by the junction of the two rows of arches in the organ of Corti. c., **salivary**, one of those forming the lining of the alveoli of the salivary glands. c.-**sap**, the more fluid part of the cell-contents. c., **sarcogenic**, an embryonic cell which develops into a muscular fiber. c., **segmentation**. See *blastomere*. c.s, **sense-**, c.s, **sensory**, those adapted for the reception and transmission of sensory impressions. c.s, **sensory epithelial**, modified epithelial cells in an organ of sense connected with the fibrils of the nerves of that organ. c., **sensory nerve-**, a nerve-cell the axis-cylinder process of which is supposed to be continued as a sensory nerve. c., **septate**, one with a septum across its lumen. c., **serous fat-**, a fat-cell occurring in emaciated individuals, in which the fat is reduced to a few small globules and in this place there is a pale protoplasm mixed with a mucoid fluid; the cell is no longer spherical. c., **simple**, one which has not undergone differentiation. c., **sister-**, one formed simultaneously with another in the progress of the mother-cell. c., **Sorby tubercular**. See *Sorby's cell*. c.-**spaces**. See *Recklinghausen's canals*. c., **sperm**, a spermatozoon; a spermatoblast. c., **spider-**. See *Deiters' cells*. c., **spindle**, a cell having a fusiform shape. c., **spiral fiber-**, a motor cell of the heart, having a spiral fiber coiled around a larger straight one. These separate, after a short distance, proceeding in different directions. The cell constantly disengages the excitation which the spiral fiber transmits to the heart-muscle. Syn., *Beale's cell*; *spiral fiber ganglion-cell*. c.s, **splanchnic**, those of the splanchnic layer of the mesoderm. c.s, **squamous**, a variety of epithelial cells found on the surface of the skin and certain mucous membranes and characterized by their scale-like flatness. c.-**stations**, cells in the sympathetic ganglion around which the nerve-fibers arborize. c., **sterile**, one occurring in a reproductive organ, but not participating in reproductive processes. c., **Stilling's**. See *Stilling's cells*. c., **stroma**, those forming the mass of an organ. c., **swarm**, a naked ciliated cell. c., **sympathetic**, a nerve-cell of the sympathetic nervous system as distinguished from one of the cerebrospinal system. c., **tapetal**, c., **tapetum**, one which forms or aids others in forming an investment over an organ. c., **taste**, a spindle-shaped cell in taste-buds. c., **tegmental**, one covering and protecting another cell of special function, as, *e. g.*, those forming the outer layer of the taste-buds. Syn., *cover-cell*. c.-**theory**, the doctrine that cell-formation is the essential biogenetic element. c., **twin**, a single cell resulting from the fusion of two cells. c., **two-fluid**, a galvanic element in which two fluids are used. c., **vasofactive**, c., **vasoformative**, a cell that anastomoses with other similar cells so as to form blood-vessels. c.s, **vortex**, Meyer's term for cortex cells which show a peculiar whorl-like and very regular arrangement of the chromophilic material. c.-**wall**, the membrane surrounding a cell. c., **wandering**, a leukocyte. c., **whip-**, a cell furnished with flagella. c., **zinc-carbon**, a galvanic cell in which zinc and carbon are the two elements employed. c., **zinc-copper**, a galvanic cell in which zinc and copper are the elements employed.

**cella** (*sel'-ah*) [L.]. 1. A cell. 2. A portion of the paracele extending caudad from the porta. c. **lateralis**, the lateral ventricle of the brain, or one of its cornua. c. **media**, the central cornu of the lateral ventricle, or that part of the ventricle whence the cornua extend. c., **turcica**. See *sella turcica*. c. **of Wilder**. Same as *c. media*.

**cellasin** (*sel'-as-in*). Trade name of a ferment which is said to split sugar, starch, fat and peptone.
**celliferous** (*sel-if'-er-us*) [*cella*, cell; *ferre*, to bear]. Producing, forming, or bearing cells.
**celloid** (*sel'-oid*) [*cella*, a cell; *eidos*, form]. Resembling a cell.
**celloidin** (*sel-oid'-in*) [*cell*; *eidos*, form]. A concentrated form of collodion for use in embedding objects for histological purposes.
**cellotropin** (*sel-ot'-ro-pin*). Monobenzoyl arbutin, obtained from the action of benzoyl chloride upon arbutin in neutral solution. Used in the treatment of tuberculosis and scrofula.
**cellula** (*sel'-u-lah*). See *cellule*.
**cellular** (*sel'-u-lar*) [*cella*, cell]. Relating to or composed of cells. c. **cartilage**, cartilage composed mainly of large cells, with but little intercellular substance. c. **membrane**, c. **tissue**, areolar tissue; bony connective tissue; cancellous tissue. c. **pathology**. See *pathology, cellular*. c. **therapy**, the name applied by Aulde to the method in therapeutics of exhibiting properly-selected medicaments with a view to restoration of cell-function. It aims to apply scientifically those remedies that experience has shown to possess special curative properties in the restoration of disordered functions.
**cellule** (*sel'-ul*) [*cellula*, a small cell]. A small cell.
**cellulic** (*sel'-u-lik*). Relating to cells; derived from cell-walls by action of acids or alkalies.
**cellulicidal** (*sel-u-lis'-id-al*) [*cellule*; *cædere*, to kill]. Destructive to cells.
**celluliferous** (*sel-u-lif'-er-us*) [*cellula*, a little cell; *ferre*, to bear]. In biology, producing small cells.
**cellulifugal** (*sel-u-lif'-u-gal*) [*cellule*; *fugere*, to flee]. Pertaining to the transmission of impulses from a nerve-cell.
**cellulipetal** (*sel-u-lip'-e-tal*) [*cellule*; *petere*, to seek]. Relating to the transmission of impulses toward a nerve-cell.
**cellulitis** (*sel-u-li'-tis*) [*cellule*; *itis*, inflammation]. a diffuse inflammation of cellular tissue. Syn., *ethmyphytis*. c., **ischiorectal**, inflammation of the cellular tissue lying below the anal levator muscle or anal fascia. c., **pelvic**. See *parametritis*. c.,

**pneumo**coccous, that due to the invasion of pneumococci.
**celluloadipose** (sel-ū-lo-ad'-ip-ōs). Relating to loose connective tissue containing fat-cells.
**cellulocutaneous** (sel-ū-lo-ku-ta'-ne-us) [cellule; cutis, skin]. Relating to cellular tissue and the skin.
**cellulofibrinous** (sel-ū-lo-fi'-brin-us). Both cellular and fibrinous.
**celluloid** (sel'-ū-loid) [cellula, a little cell; εἶδος, form]. A product of the action of camphor upon pyroxylin.
**cellulosa** (sel-ū-lo'-sah) [L.]. A cellular coat. c. **chorioideæ**, the external layer of the choroid coat of the eye.
**cellulose** (sel'-ū-lōs) [cellule], $C_{12}H_{20}O_{10}$. Woodfiber; lignose, the principal ingredient of the cellmembranes of all plants. It is a white, amorphous mass, insoluble in most of the usual solvents. c., **reagent for**. See Schultze, Schweitzer.
**cellulosity** (sel-ū-los'-e-te). The condition of being cellular.
**celology** (sel-ol'-o-je) [κήλη, hernia; λόγος, science]. That branch of surgical science that treats of hernia.
**celom, celoma** (se'-lom, se-lo'-mah) [κοίλωμα, a cavity]. The embryonic body-cavity.
**celophlebitis** (se-lo-fle-bi'-tis) [κοῖλος, hollow; φλέψ, vein; ιτις, inflammation]. Inflammation of a vena cava.
**celophthalmia** (se-lof-thal'-me-ah) [κοῖλος, hollow; ὀφθαλμός, eye]. Hollowness of the eyes.
**celoscope** (se'-lo-skōp) [κοῖλος, hollow; σκοπεῖν, to observe]. An instrument for examining a cavity of the body by means of the electric light, enclosed in a flask and mounted upon a glass shank.
**celosis** (se-lo'-sis) [κοῖλος, hollow]. The formation of any cavity. c., **endocytic**, the formation of a cavity within a cell. c., **paracytic**, the formation of a cavity between cells.
**celosoma** (se-lo-so'-mah) [κοῖλος, hollow; σῶμα, body]. A species of single autositic monsters characterized by more or less extensive body-cleft, with eventration, associated with various anomalies of the extremities, of the genitourinary apparatus, of the intestinal tract, and even of the whole trunk.
**celosomia** (se-lo-sō'-me-ah) [κήλη, hernia; σῶμα, body]. Congenital protrusion of the viscera, with defect of the thoracic or abdominal walls.
**celosomus** (se-lo-so'-mus) [κήλη, a hernia; σῶμα, the body]. A monster with fissure or absence of the sternum and hernia of the thoracic or abdominal organ.
**celostomia** (se-lo-sto'-me-ah) [κοῖλος, hollow; στόμα, mouth]. Hollowness of the voice.
**celotome** (se'-lo-tōm) [κήλη, hernia; τέμνειν, to cut]. A hernia-knife.
**celotomy** (se-lot'-o-me) [κήλη, hernia; τέμνειν, to cut]. The operation for strangulated hernia by incision of the stricture.
**Cels**. Abbreviation for Celsius; the Celsius scale of the thermometer.
**Celsius scale** [Anders Celsius, Swedish astronomer, 1701-1744]. A term sometimes (but erroneously) employed for the centigrade scale on the thermometer. It is the reverse of the centigrade scale, having the freezing-point at 100° and the boiling point at 0°.
**Celsus' area** (sel'-sus) [Aulus Cornelius Celsus, Roman physician and writer, 1st century, A.D.]. Alopecia areata. **C.'s chancre**, the soft chancre or chancroid. **C.'s kerion**, suppurating ringworm, a pustular inflammation of the hair-follicles of the scalp in tinea tonsurans. **C.'s papules**, a form of acute papular eczema (lichen agrius).
**cement** (se-ment') [cæmentum, a rough stone]. 1. Any plastic material capable of becoming hard and of binding together the objects that are contiguous to it. 2. Filling-material for the teeth. 3. The crusta petrosa of the teeth. **c.-substance**, the substance holding together the endothelial cells of the intima of blood-vessels.
**cementation** (sem-en-ta'-shun) [cement]. 1. A process of causing a chemical change in a substance by surrounding it with the powde. of other substances and exposing the whole to red heat in a closed vessel for a length of time. 2. In biology, the concrescence of hyphæ.
**cementinification** (se-men-tin-e-fik-a'-shun). The formation of cementum about the dental root.
**cementoblast** (se-ment'-o-blast) [cæmentum, cement; βλαστός, germ]. A cement-corpuscle in tooth-tissue;

more correctly, an osteoblast that takes part in the development of the dental cement.
**cementodentinary** (se-ment'-o-den'-tin-a-re) [cæmentum, cement; dens, dentis, a tooth]. Relating to the cement and dentine of a tooth.
**cementoma** (se-ment-o'-mah) [cæmentum, cement; ὅμα, tumor; pl., cementomata]. A tumor thrown out by the irritated alveolar periosteum.
**cementoperiostitis** (se-men-to-per-e-os-ti'-tis). Same as pyorrhœa alveolaris.
**cementosis** (sem-en-to'-sis). The development of a cémentoma.
**cementum** (se-ment'-um). A layer of bone developed by ossification of the dental follicle over the root of the tooth. It differs from ordinary bone by the greater number of Sharpey's fibers in it. Its development begins on the milk-teeth during the fifth month.
**cenadelphus, cœnadelphus** (sen-a-del'-fus) [κοινός, common; ἀδελφός, a brother]. A double monster with the halves equally developed or having one or more vital organs in common.
**cenencephalocele** (sen-en-sef-al'-o-sēl) [κενός, empty; ἐγκέφαλος, brain; κήλη, tumor]. A protrusion of pure brain substance through a cranial fissure.
**cenesthesia, cenæsthesia** (sen-es-the'-ze-ah) [κενός, destitute; αἴσθησια, perception]. Hysterical loss of consciousness of identity.
**cenesthesis** (sen-es-the'-sis) [κοινός, common; αἴσθησις, feeling]. A sense of existence, either painful or pleasurable. It is the prevailing conscious state of feeling, either of depression or of exaltation, which is the resultant of the subconscious organic sympathies of the whole organism. It does not exceed physiological limits so long as it does not exclude the normal exercise of mental functions.
**cenesthetic, cœnesthetic** (sen-es-thet'-ik). Relating to cenesthesis.
**cenogenesis, cenogeny**. See kenogenesis.
**cenophobia**. See kenophobia.
**cenosis** (sen-o'-sis) [κένωσις, a draining]. 1. Evacuation; apocenosis. 2. Inanition.
**cenotic** (sen-ot'-ik) [κένωσις, a draining]. 1. Causing cenosis; drastic; purgative. 2. A drastic drug or agent.
**Cent**. Abbreviation for centigrade and centimeter.
**Centaurea** (sen-taw'-re-ah) [κένταυρος, centaur]. A genus of composite-flowered herbs. See carduus.
**centaury** (sen'-taw-re) [centaurea]. A popular name for various plants of the genera Centaurea, Erythrœa, Sabbatia, Chlora, etc., especially Erythrœa centaurium, which is used as a simple, bitter tonic. D se ½-1 dr. (2-4 Cc.) in decoction several times a day.
**center** (sen'-ter) [centrum, the center]. 1. The middle point of any surface or of a body. 2. The ganglion or plexus whence issue the nerves controlling a function. c., **accelerating**, a center in the medulla sending accelerating fibers to the heart. These leave the cord through the branches of communication of the lower cervical and upper six dorsal nerves, passing thence into the sympathetic. c., **anovesical**, one in the spinal cord near the point of origin of the third and fourth sacral nerves. Incontinence of urine and feces is due to paralysis of this center. c., **arm**, the cortical center controlling the movement of the arm, supposed to be in the cortex occupying the middle third of the anterior central gyrus as well as the base of the superior and middle frontal gyri. c., **articulate language**, the speech-coordinating center, which is supposed to include Broca's gyrus, the anterior gyri of the insular, the intervening cortical area, the supramarginal gyrus, the first temporal gyrus, and the angular gyrus. c., **association**. See association. c., **auditory**, a center in the first temporosphenoid convolution upon each side. c., **Broca's**. See c., speech. c., **cardiac**. 1. One in the lower cervical and upper dorsal portions of the spinal cord which controls the movements of the heart. 2. That portion of the oblongata embracing the cardio-accelerator and cardioinhibitory centers. c., **cardio-accelerator**, that of the spinal cord which through the cardiac nerves and plexus sends impulses to the heart, causing it to beat more rapidly. These impulses are not constantly emitted, as are the inhibitory impulses which travel by the pneumogastric. c., **cardioinhibitory**, in the medulla, efferent impulses being carried by the vagus. c., **cerebral inspiratory**, one said to exist in the thalamus, which

by direct stimulation causes deeper and more rapid inspirations. c., **cerebrospinal,** the cerebrospinal axis. c., ciliospinal, connected with the dilatation of the pupil; it is in the lower cervical part of the cord, and extends downward from the first to the third dorsal. c., **color,** a center for perception of colors, said to be situated in the occipital cortex anterior to the apical region. c., **convulsional,** a hypothetical center said to lie in the floor of the fourth ventricle. c., **coordinating,** the cerebellum, the ganglia at the base of the brain, and in some degree the cinerea of the spinal cord, are regarded as controlling co-ordination. c., **cortical,** the parts of the cerebral cortex concerned in motor, sensory, and psychic functions. c., **coughing,** in the medulla, above the inspiratory center. c., **deglutition.** See c., *swallowing*. c., **deputy,** a secondary ganglion-cell in the spinal cord; also a nucleus of one of the cranial nerves. c., **diabetic,** in the posterior part of the anterior half of the floor of the fourth ventricle, in the median line. c., **epiotic,** the ossification center of the mastoid portion of the temporal bone. c., **erection.** See c., *genitourinary*. c., **excitomotor,** the sensitive centers of the brain considered as one; these are the crura, the pons, the oblongata, the deeper parts of the cerebellum, and the corpora quadrigemina. c.s, **facial movement,** one in the ascending frontal gyrus and one in the angular gyrus. c., **genitourinary,** one in the lumbar portion of the spinal cord, but controlled from the medulla, controlling erection of the penis and emission of semen. c., **glycogenic,** the diabetic center. c., **half-vision,** one in the apex of the occipital lobe, receiving impressions from corresponding halves of the two retinæ. c., **head and neck movement,** one in the posterior end of the second frontal gyrus and in the corresponding part of the first frontal gyrus. c., **heat-regulating, c., temperature,** the center for the control of body-temperature. See c., *thermotaxic*. c., **higher visual,** one regarded as lying in the angular gyrus, in which there is effected a combination of the impressions received from the half-vision centers, making a complete image. c., **Hitzig's.** See *Hitzig's center*. c.s **of inhibition,** c.s, **inhibitory.** See *c.s of moderation*. c., **inspiratory.** 1. A reflex center in the oblongata forming part of the respiratory center. 2. See *c., cerebral inspiratory*. 3. A reflex center in the post-geminum. c., **intracardiac,** three small nerve-ganglia connected with the cardiac plexus, to which is due the automatic beating of the heart after separation from the body. c., **kinesthetic,** one in the third left frontal convulution presiding over the motor element in speech. c., **laryngeal cortical,** one in the posterior end of the inferior frontal gyrus. c., **leg,** one in the upper portion of the ascending frontal convolution. c. **for mastication and sucking,** one in the medulla. c., **median** (of Luys). See *Luys, nucleus of*. c., **medullary.** 1. The interior white matter of the cerebral hemispheres. 2. See c., *neural*. c.s **of moderation,** c.s, **moderator,** nervous centers in the spinal cord and the cerebral peduncle which restrain, generally by reflex action, various functions of the body. c., **motor,** a nervous center controlling motion. c., **musculotonic,** that which is continually discharging impulses which keep the muscular system in a condition of slight contraction. It is regarded by some as a special center of the cord, but it is questionable whether this condition is attributable to any special center rather than to the action of all those cells whose function it is to send out motor impulses. c., **nerve-,** c., **nervous,** any group of nerve-cells acting in unison for the performance of some function. c., **neural,** in the embryo, that part of the epiblast ultimately developing into the brain and spinal cord. c., **nutrition.** See c., *trophic*. c., **olfactory,** probably in the hippocampal region of the temporal lobe. c., **opisthotic,** the center of ossification of petrous bone. Huxley's name for the part of the periotic cartilage surrounding the fenestra rotunda and the cochlea. c. **of ossification, the** place in bones at which ossification begins. c., **parenchymatous nerve-,** Körner's name for a nerve-cell existing in the substance of an organ and controlling its action. c., **parturition,** in the spinal cord, at the level of the first and second lumbar vertebræ. c., **peristaltic,** one in the oblongata controlling peristalsis. c., **phonation.** See c., *laryngeal cortical*. c., **psychomotor,** that portion of the cortex from which motor impulses originate. c., **psychosmic,** the olfactory center. c., **reflex,** any nerve-cell or group of cells in the brain, cord, or ganglionic system which receives an impression through centripetal nerve-fibers and transforms it into an impulse which is transmitted through centrifugal nerve-fibers. c., **respiratory,** in the medulla, between the nuclei of the vagus and accessorius. c. **for secretion of saliva,** on the floor of the fourth ventricle. c., **Setschenow's.** See *Setschenow's center*. c., **sneezing,** in the medulla. c., **spasm,** in the medulla, at its junction with the pons. c., **speech,** in the third left frontal convulution in right-handed people; probably the island of Reil has some influence also. c., **sudoral.** See c., *sweat*. c., **supreme,** Spitzka's name for the cortical centers of the brain as a whole. c., **swallowing,** on the floor of the fourth ventricle. c., **sweat,** the dominating center is in the medulla, with subordinate centers in the spinal cord. c., **tactile,** one for the sense of touch, located by Ferrier in the hippocampus and the gyrus hippocampus. c., **thermal cortical,** one discovered in the cerebral cortex of the dog, stimulation of which caused a change in the temperature of the opposite limbs. c., **thermoexcito-,** c., **thermogenic.** 1. A hypothetical center of the cord concerned in the changes in body-temperature. 2. The mesial portion of the striatum and the parts directly beneath it. c., **thermotaxic,** six heat-regulating cerebral centers; of the four principal centers, one is located in the caudatum, one in the subjacent cinerea, one in the cinerea surrounding the most anterior portion of the third ventricle, and one at the anterior inner extremity of the thalamus. c., **trophic,** a nerve-center regulating nutrition. c., **upper, for dilator pupillæ,** in the medulla. c.s, **vascular,** c.s, **vasoconstrictor,** centers in the cord controlling the contractility of the smaller blood-vessels. c., **vasodilator,** in the medulla. c., **vasomotor,** in the medulla. c., **visual,** in the occipital lobe, especially in the cuneus. c., **vomiting,** an area in the oblongata concerned in the reflex act of vomiting; stimulation of the terminal filaments of the vagi excites its action. c., **winking,** the reflex center concerned in winking, situated in the oblongata. c.s, **word-.** 1. One in the left superior temporosphenoid gyrus controlling the perception of words heard. 2. A center in the posterior part of the left parietal lobe and one in the second left frontal gyrus governing the perception of printed or written words.

**centering** (*sen'-ter-ing*) [*center*]. In microscopy, the arrangement of an object or an accessory so that its center coincides with the optical axis of the microscope. In optics, having the pupil and the optic center of the refracting lens in the same axis.

**centesimal** (*sen-tes'-im-al*) [*centum*, a hundred]. In the proportion of 1 to 100.

**centesis** (*sen-te'-sis*) [κέντησις, a pricking]. Puncture; perforation.

**centi-** (*sen-ti-*) [*centum*]. A prefix meaning one hundred.

**centifidous** (*sen-tif'-id-us*) [*centi-*; *findere*, to split]. Cleft into many or 100 parts.

**centigrade** (*sen'-te-grād*) [*centi-*; *gradus*, a step]. Having 100 divisions or degrees. Abbreviation, C. c. **thermometer,** a thermometer with zero as the freezing-point and 100° as the boiling-point of water. See under *thermometer*.

**centigram, centigramme** (*sen'-te-gram*) [*centi-*; γράμμα, a small weight]. The hundredth part of a gram, equal to 0.154328 grain troy.

**centiliter** (*sen'-til-e-ter*) [*centi-*; λίτρα, a pound]. The hundredth part of a liter, equal to 0.6102 of a cubic inch.

**centimeter** (*sen'-tim-e-ter*) [*centi-*; μέτρον, a measure]. The hundredth part of a meter, equal to 0.3937 (or about ⅖) of an inch.

**centinormal** (*sen-te-nor'-mal*) [*centi-*; *norma*, normal]. The 1/100 of the normal; applied to a solution the 1/100 of the strength of a normal solution.

**centrad** (*sen'-trad*) [*centrum*; *ad*, to]. 1. Toward the center, or toward the median line. 2. An angular measure, one hundredth of a radian; about 0.57°.

**central** (*sen'-tral*) [*centrum*]. Relating to the center; passing through the center. c. **artery,** an artery in the optic nerve and retina; it passes to the optic papilla and then divides. See under *artery*. c. **fissure,** the fissure of Rolando. c. **ganglia,** the corpora striata and optic thalami. c. **ligament,** the terminal filum of the spinal cord. c. **lobe,** the island of Reil.

**centrality** (sen-tral'-it-e). Applied to the condition of nervous phenomena originating in the central nervous system and not in the peripheral nerves.

**centraphose, centrophose** (sen'-trah-fōz, sen'-trofōz). See under *phose*.

**centraxonial** (sen-traks-o'-ne-al) [κέντρον, center; ἄξων, axis]. In biology, having a central axial line.

**centre** (sen'-ter). See *center*.

**centren** (sen'-tren) [*centrum*, a center]. Belonging solely to a center.

**centric** (sen'-trik) [*centrum*]. Relating to a center, especially to a nerve-center.

**centricipital** (sen-tris-ip'-it-al) [*centrum*, center; *caput*, a head]. Relating to the centriciput; parietal. c. **vertebra**, the second or more central of the three principal cranial vertebræ.

**centriciput** (sen-tris'-ip-ut) [*centrum*, center; *caput*, head]. The mid-head; the second cranial segment situated between the sinciput and occiput.

**centrifugal** (sen-trif'-u-gal) [*centrum*; *fugere*, to flee]. Receding from the center to the periphery. c. **force**, the force by which a revolving body tends to fly off at the periphery. c. **machine**, one by which tubes of liquid are rapidly revolved for the purpose of driving particles floating in the liquid to the distal ends of the tubes.

**centrifugalization** (sen-trif-u-gal-iz-a'-shun) [see *centrifugal*]. The use of a centrifuge.

**centrifugalized milk**. Milk from which the cream has been separated by whirling it in a centrifugal machine.

**centrifuge** (sen'-trif-uj) [see *centrifugal*]. 1. A centrifugal machine; an apparatus for separating substances by centrifugal force. 2. To submit to the action of a centrifuge.

**centriole** (sen'-tre-ōl) [*centrum*]. Boveri's term for a minute body, central horn, contained within the centrosome; in some cases it is not distinguishable from the latter.

**centripetal** (sen-trip'-et-al) [*centrum*; *petere*, to seek]. Traveling toward the center from the periphery.

**centro-** (sen-tro-) [*centrum*]. A prefix meaning central.

**centroacinal, or centroacinar** (sen-tro-as'-in-al, or cen-tro-as'-in-ar) [*centrum*, center; *acinus*, a grape]. Belonging to the center of an acinus. c. **cells** are found in the acini of the pancreas, etc.

**centrodesmus** (sen-tro-des'-mus) [*centro-*; δεσμός, a band]. Heidenhain's term for the band primarily connecting the centrosomes and giving rise to the central spindle.

**centrodontous** (sen-tro-don'-tus) [κέντρον, a sharp point; ὀδούς, a tooth]. Furnished with sharp-pointed teeth.

**centrodorsal** (sen-tro-dor'-sal). Pertaining to the central dorsal region.

**centrolecithal** (sen-tro-les'-ith-al) [*centro-*; λέκιθος, yolk]. In embryology, having the food-yolk located centrally.

**centronucleus** (sen-tro-nū'-kle-us). Same as *amphinucleus*.

**centrophose** (sen'-tro-fōz) [κέντρον, center; φῶς, light]. See under *phose*.

**centroplasm** (sen'-tro-plazm). The protoplasm of the centrosphere; the archoplasm.

**centrosclerosis, centro-osteosclerosis** (sen-tro-skler-o'-sis, sen-tro-os-te-o-skler-o'-sis) [*centro-*; *sclerosis*]. Osteosclerosis of the central cavities of bones.

**centrosome** (sen'-tro-sōm) [*centro-*; σῶμα, body]. 1. A highly refractive body lying in the protoplasm of the ovum and other cells, and taking an active part in cell-division. Syn., *pole-capsule*. 2. An organ of the cell, usually diminutive, lying within the nucleus or near by in the cytoreticulum. It is regarded as the especial organ of cell-division, and in this sense as the dynamic center of the cell. Syn., *attraction-particle; daughter-periplast; polar corpuscle*. c.s, **quadrille of**, the conjugation of paternal with maternal centrosomes, based upon the view that each germ-cell contributes a centrosome that divides into two daughter-centrosomes. Syn., *quadrille of centers*.

**centrosphere** (sen'-tro-sfēr). See *sphere of attraction*.

**centrostaltic** (sen-tro-stal'-tik) [*centro-*; σταλτικός, constriction]. Relating to the action of nervous force in a spinal center. c. **motion**, the motion of nervous force in the spinal center.

**centrostigma** (sen-tro-stig'-mah) [κέντρον, center; στίγμα, a point]. In morphology, having all the axes converging to a central point.

**centrum** (sen'-trum) [L.]. 1. The center or middle part; the body of a vertebra, exclusive of the bases of the neural arches. 2. A spine; a pointed projection. c. **cinereum**, the gray commissure of the spinal cord. c. **commune**, the solar plexus. c. **geminum**, the capsula. c. **ovale majus**, the large mass of white matter appearing when either of the cerebral hemispheres is cut down to the level of the corpus callosum. c. **ovale minus**, the white matter appearing when the upper part of a hemisphere of the brain is removed. c. **ovale Vieussenii**, the central white matter seen on making a section of the brain at the level of the upper surface of the callosum. Syn., *centrum medullare;* *centrum ovale majus et minus; centrum semiovale Vieussenii; centrum ovale of Vicq d'Azyr; medulla; tegmentum ventriculorum*. c. **rubrum**. See *nucleus tegmenti*. c. **tendineum**, the central tendon of the diaphragm.

**cephaeline** (sef-a'-el-in). See *emetine*.

**cephaelis** (sef-a'-el-is). See *ipecacuanha*.

**cephal-** (sef-al-). See *cephalo-*.

**cephalad** (sef'-al-ad) [*cephal-*; *ad*, to]. Toward the head.

**cephalagra** (sef-al-ag'-rah) [κεφαλή, head; ἄγρα, seizure]. Gouty headache.

**cephalalgia** (sef-al-al'-je-ah) [*cephal-*; ἄλγος, pain]. Headache.

**cephalalgic** (sef-al-al'-jik) [κεφαλή, head; ἄλγος, pain]. Relating to headache.

**cephalanthin** (sef-al-an'-thin). See under *cephalanthus*.

**Cephalanthus** (sef al an'-thus) [*cephal-;* ἄνθος, a flower]. A genus of rubiaceous plants. *C. occidentalis* is the button-bush or crane-willow of North America; its bitter bark is laxative and tonic and is used in periodic fevers and paralysis. The bark contains cephalin, cephaletin, and a toxic principle cephalanthin, which, according to Mohrberg, causes destruction of the red blood-corpuscles, vomiting, convulsions, and paralysis.

**cephalea** (sef-al-e'-ah) [κεφαλαία, headache]. Headache; especially severe or chronic headache, with intolerance of light and sound.

**cephaledema, cephalœdema** (sef-al-e-de'-mah) [*cephal-*; οἴδειν, to swell]. Edema of the head; cerebral edema.

**cephalemia** (sef-al-e'-me-ah) [κεφαλή, head; αἷμα, blood]. An abnormal determination of blood to the head.

**cephalhematocele** (sef-al-hem-at'-o-sēl) [*cephal-*; *hematocele*]. A hematocele situated beneath the scalp, and communicating with a dural sinus. c., **Stromeyer's**. See *Stromeyer's cephalhematocele*.

**cephalhematoma** (sef-al-hem-at-o'-mah) [*cephal-*; *hematoma*]. 1. A collection of blood beneath the pericranium, forming a tumor-like swelling. 2. Caput succedaneum. c., **external**, an effusion between the pericranium and the skull. c., **internal**, an effusion between the dura and the skull.

**cephalhematometer** (sef-al-hem-at-om'-et-ur) [κεφαλή, the head; αἷμα, blood; μέτρον, a measure]. An apparatus for the estimation of the increase or diminishment of the amount of blood within an animal's head.

**cephalhydrocele** (sef-al-hi'-dro-sēl) [κεφαλή, head; ὕδωρ, water; κήλη, tumor]. Effusion of cerebral fluid beneath the occipito-frontal aponeuroses in fractures of the skull.

**cephalic** (sef-al'-ik) [κεφαλή, head]. 1. Pertaining to the head. 2. Any remedy for headache. c. **index**. See *index, cephalic*. c. **vein**, a vein of the upper arm. c. **version**. See *version, cephalic*.

**cephalin** (sef'-al-in) [κεφαλή, the head]. An unstable phosphatic substance obtained from brain-substance; it is allied to lecithin; it is called also *kephalin*.

**cephaline** (sef'-al-ēn). A proprietary headache remedy said to consist of antipyrine and pulverized coffee, each 5 parts, and caffeine and sodium salicylate, each 2 parts. Dose, 4 gr.

**cephalitis** (sef-al-i'-tis). See *encephalitis*. c. **Ægyptiaca**, an epidemic form of encephalitis occurring in Egypt during the hot winds of early summer. c. **littriana**, inflammation of the epiphyses. c. **meningica, meningitis**. c. **nervosa**, pertussis.

**cephality** (sef-al'-e-te) [κεφαλή, the head]. Agassiz's term for the preponderance of the head over the remainder of the organism.

**cephalization** (sef-al-iz-a'-shun) [κεφαλή, the head]. In biology, Dana's term for that specialization the

tendency of which is to concentrate important parts and organs at the head region of the trunk.
**cephalize** (*sef'-al-īz*) [κεφαλή, head]. To develop head-organs.
**cephalo-** (*sef-al-o-*) [κεφαλή, head]. A prefix denoting relating to the head.
**cephalocathartic** (*sef-al-o-kath-ar'-tik*) [*cephalo-;* καθαρτικόs, purging]. 1. Purging or relieving the head. 2. A medicine that relieves the head.
**cephalocele** (*sef'-al-o-sēl*) [*cephalo-;* κήλη, tumor]. Hernia of the brain; protrusion of a mass of the cranial contents.
**cephalocentesis** (*sef-al-o-sen-te'-sis*) [*cephalo-;* κέντησις, puncture]. Surgical puncture of the cranium.
**cephalocercal** (*sef-al-o-ser'-kal*) [*cephalo-;* κέρκος, tail]. In anatomy, from head to tail.
**cephalochord** (*sef'-al-o-kord*) [*cephalo-;* χορδή, cord]. The cephalic portion of the chorda dorsalis in embryonic life.
**cephaloclasia** (*sef-al-o-kla'-ze-ah*). See *cephalotripsy.*
**cephaloclast** (*sef'-al-o-klast*). See *cephalotribe.*
**cephalodymia** (*sef-al-o-dim'-e-ah*) [*cephalo-;* δύμεναι, to enter]. Teratologic union of twins by the merging of their heads together.
**cephalodynia** (*sef-al-o-din'-e-ah*) [*cephalo-;* ὀδύνη, pain]. Rheumatism affecting the occipitofrontalis muscle, the pain being chiefly experienced in the forehead or occiput, and at times involving the eyeballs.
**cephalofacial** (*sef-al-o-fa'-shal*). Relating to the skull and to the face.
**cephalogaster** (*sef-al-o-gas'-ter*) [*cephalo-;* γαστήρ, stomach]. The anterior division of the enteric canal, as in certain parasitic worms, where it is continued into a second division, the typhlosole.
**cephalograph** (*sef'-al-o-graf*) [*cephalo-;* γράφειν, to write]. An instrument for recording the contours of the head.
**cephalography** (*sef-al-og'-ra-fe*) [*cephalo-;* γράφειν, to write]. A description of the head.
**cephalohemometer** (*sef-al-o-hem-om'-el-er*) [*cephalo-;* αἷμα, blood; μέτρον, a measure]. An instrument for noting changes in the intracranial blood-pressure.
**cephaloid** (*sef'-al-oid*) [*cephalo-;* εἶδος, likeness]. Resembling the head.
**cephalology** (*sef-al-ol'-o-je*) [*cephalo-;* λόγος, science]. The science of cranial measurements and indications.
**cephaloma** (*sef-al-o'-mah*) [*cephalo-;* ὅμα, tumor; *pl., cephalomata*]. Encephaloid carcinoma; soft carcinoma.
**cephalomelus** (*sef-al-om'-el-us*) [*cephalo-;* μέλος, a limb]. A form of double monster in which there is a supernumerary limb attached to the head.
**cephalomenia** (*sef-al-o-me'-ne-ah*) [*cephalo-;* μήν, a month]. Vicarious menstruation through the nose.
**cephalomeningitis** (*sef-al-o-men-in-ji'-tis*) [*cephalo-;* μῆνιγξ, a membrane; ιτις, inflammation]. Cephalic meningitis; inflammation of the cephalic meninges.
**cephalometer** (*sef-al-om'-et-er*) [*cephalo-;* μέτρον, a measure]. An instrument for measuring the head.
**cephalometry** (*sef-al-om'-et-re*) [*cephalo-;* μέτρον, a measure]. 1. The use of the cephalometer; craniometry. 2. The art of taking measurements of the head to determine the position of the fissures and convolutions of the brain.
**cephalomyitis** (*sef-al-o-mi-i'-tis*) [*cephalo-;* μῦς, a muscle; ιτις, inflammation]. Inflammation of the muscles of the head.
**cephalonasal** (*sef-al-o-na'-sal*). Relating to the skull and the nose.
**cephalonia** (*sef-al-o'-ne-ah*) [κεφαλή, head]. Macrocephaly with hypertrophy of the brain.
**cephalo-orbital** (*sef-al-o-or'-bit-al*) [*cephalo-;* *orbita*, an orbit]. Relating to the cranium and orbits. c. index. See *index.*
**cephalo-orbitonasal** (*sef-al-o-or-bit-o-na'-zal*) [*cephalo-;* *orbita*, an orbit; *nasalis*, of the nose]. Relating to cranium, orbits, and nose.
**cephalopagus** (*sef-al-op'-ag-us*) [*cephalopagy*]. A double monstrosity having the heads united at the top.
**cephalopagy** (*sef-al-op'-ath-e*) [*cephalo-;* πηγνύναι, to join]. That form of monstrosity marked by the development of two individuals having heads united at the top.

**cephalopathic** (*sef-al-o-path'-ik*) [*cephalo-;* πάθος, disease]. Pertaining or belonging to a disease of the head.
**cephalopathy** (*sef-al-op'-ath-e*) [*cephalo-;* πάθος, disease]. Any disease of the head.
**cephalopharyngeus** (*sef-al-o-far-in'-je-us*). 1. Relating to the head and pharynx. 2. See under *muscle.*
**cephalophyma** (*sef-al-o-fi'-mah*). Synonym of *cephalhematoma.*
**cephaloplegia** (*sef-al-o-ple'-je-ah*) [*cephalo-;* πληγή, a stroke]. Paralysis of the muscles about the head and face.
**cephalorrhachidian** (*sef-al-o-rak-id'-e-an*) [*cephalo-;* ῥάχις, spine]. Same as *cerebro-spinal.*
**cephaloscope** (*sef'-al-o-skōp*) [*cephalo-;* σκοπεῖν, to examine]. A stethoscope for use in auscultation of the head or the ear.
**cephaloscopy** (*sef-al-os'-ko-pe*) [*cephalo-;* σκοπεῖν, to examine]. 1. Auscultation of the head. 2. Examination of the head with a view to ascertaining the condition of the mental faculties.
**cephalostat** (*sef'-al-o-stat*) [*cephalo-;* ἱστάναι, to cause to stand]. A vise or clamp for holding a patient's head; a head-rest.
**cephalothoracopagus** (*sef-al-o-tho-rak-op'-ag-us*) [*cephalo-;* θώραξ, thorax; παγείς, joined]. A double-headed monster with united thoraces and necks. These monsters are divided by Veit into *prosopothoracopagus* and *syncephalus.*
**cephalothorax** (*sef-al-o-tho'-raks*) [*cephalo-;* θώραξ, a breastplate]. In biology, the anterior portion of the body of an arthropod formed by the union of the head and thorax.
**cephalotome** (*sef'-al-o-tōm*) [*cephalo-;* τέμνειν, to cut]. The instrument used in performing cephalotomy.
**cephalotomy** (*sef-al-ot'-o-me*) [*cephalo-;* τομή, section]. The opening or division of the head of the fetus to facilitate labor.
**cephalotractor** (*sef-al-o-trak'-tor*) [*cephalo-;* *tractor*]. Obstetric forceps.
**cephalotrib** (*sef'-al-o-trib*) [*cephalo-;* τρίβειν, to crush]. An instrument for crushing the fetal head.
**cephalotridymus** (*sef-al-o-trid'-im-us*) [*cephalo-;* τρίδυμος, triple]. A three-headed monster.
**cephalotripsy** (*sef-al-o-trip-se*) [*cephalo-;* τρίψις, a crushing]. The operation of crushing the fetal head when delivery is otherwise impossible.
**cephalotriptor** (*sef-al-o-trip'-tor*). See *cephalotrib.*
**cephalotrypesis** (*sef-al-o-tri-pe'-sis*) [*cephalo-;* τρύπησις, a boring]. A trephining of the skull.
**cephaloxia** (*sef-al-oks'-e-ah*). Synonym of *torticollis.*
**ceptor** (*sep'-tor*) [*capere*, to take]. A term suggested by Ehrlich in place of intermediary body. According to the theory and manner of action he distinguishes *uniceptors* and *amboceptors.*
**cera** (*se'-rah*) [L.]. Wax. A mixture of cerotic acid, cerolein, and myricin, gathered by the honey-bee from the pollen of flowers and the leaves of plants. c. alba (U. S. P.), white wax, prepared by bleaching yellow wax. It is valuable as an ingredient of cerates and ointments. c. flava (U. S. P.), yellow wax; it possesses an agreeable balsamic odor, and is soluble in ether, in hot alcohol, and in chloroform.
**ceraceous** (*se-ra'-se-us*) [*cera*, wax]. Waxy. Resembling wax.
**ceral** (*se'-ral*). Pasta cerata, a proprietary vehicle for application of medicaments, said to consist of wax, potash, and water.
**ceramuria** (*ser-am-u'-re-ah*) [κέραμος, potter's earth; οὖρον, urine]. Phosphaturia.
**cerasin** (*ser'-as-in*) [*cerasus*, a cherry-tree]. 1. An ingredient of the gum of cherry-, peach-, and plum-trees, apparently identical with bassorin. 2. A crude precipitate from tincture of choke-cherry.
**cerate** (*se'-rāt*) [*cera*]. In pharmacy, an unctuous preparation consisting of wax mixed with oils, fatty substances, or resins, and of such a consistence that at ordinary temperatures it can be readily spread upon linen or muslin, and yet is so firm that it will not melt or run when applied to the skin. c., camphor. See *camphor cerate.* c., cantharides. See *cantharides cerate.* c., Goulard's. See *Goulard's cerate.* c. of lead subacetate. See *Goulard's cerate.* c., rosin. See *rosin cerate.* c., rosin, compound. See *rosin cerate, compound.* c., touch, a lubricant used in vaginal inspection, consisting of spermaceti, white wax, and caustic soda, each, 1 part; olive-oil, 16 parts. Syn., *ceratum pro tactu.*

# CERATED 209 CEREBROSCOPY

**cerated** (*se'-ra-ted*) [*cera*]. Coated with wax.
**ceratiasis** (*ser-at-i'-as-is*). See *keratiasis*.
**ceratin** (*ser'-at-in*). See *keratin*.
**ceratitis** (*ser-at-i'-tis*). See *keratitis*.
**cerato-** (*ser'-a-to*). See *kerato-*.
**ceratocele** (*ser'-at-o-sēl*). See *keratocele*.
**ceratoglossus** (*ser-at-o-glos'-us*). See *keratoglossus*, in *muscles, table of*.
**ceratohyal** (*ser-at-o-hi'-al*). See *keratohyal*.
**Ceratonia** (*ser-at-o'-ne-ah*) [κερατωνία]. 1. A genus of leguminous trees. 2. The fruit of *C. siliqua*, the carob-tree, a native of the regions about the Mediterranean. The falcate, fleshy pods, called carob-pods, sugar-pods, and St. John's bread, are demulcent and pectoral and contain carobin, carobone, and carobic acid. They are used as food and form the chief constituent of much of the patented food for cattle. The seeds are used as a substitute for coffee.
**ceratonosus.** See *keratonosus*.
**ceratoplasty** (*ser'-at-o-plas-te*). See *keratoplasty*.
**ceratoscope** (*ser'-at-o-skōp*). See *keratoscope*.
**ceratotomy** (*ser-at-ot'-o-me*). See *keratotomy*.
**ceratonyxis** (L.). See *keratonyxis*.
**ceratorrhexis** (*ser-at-o-reks'-is*). See *keratorrhexis*.
**ceratose** (*ser'-at-ōs*). See *keratose*.
**ceratosis** (*ser-at-o'-sis*). See *keratosis*.
**ceratotomy** (*ser-at-ot'-o-me*). See *keratotomy*.
**ceraunics** (*ser-aw'-niks*). See *keraunics*.
**cercaria** (*ser-ka'-re-ah*) [κέρκος, tail]. Any trematode worm (fluke) in its second (or tailed) stage of larval life.
**cercarian** (*ser-ku'-re-un*) [κέρκος, a tail]. Any trematode, or fluke-worm, in the *cercaria* stage.
**cerchnus** (*serk'-nus*) [κέρχνος, rough, hoarse]. Hoarseness; noisy respiration.
**ceratum** (*se'-ra-tum*). See *cerate*.
**cercomonad** (*ser-ko-mo'-nad*). A member of the genus *cercomonas*.
**Cercomonas** (*ser-ko-mo'-nas*) [κέρκος, tail; μονάς, monad]. A genus of flagellate infusorians. **C. intestinalis**, a protozoon, occasionally found in the fecal discharges of patients suffering with typhoid fever, chronic diarrhea, or cholera. Its pathological significance has not yet been ascertained.
**cerea flexibilitas** (*se'-re-ah fleks-ib-il'-it-as*). That condition of muscular tension in the insane in which the limbs may be molded into any position.
**cereal** (*se'-re-al*) [*Ceres*, the goddess of agriculture]. 1. Relating to edible grains. 2. Any edible grain.
**cerealin** (*se-re'-al-in*). An enzyme converting starch into glucose, isolated from bran-extract.
**cerebellar** (*ser-e-bel'-ar*) [*cerebellum*]. Relating to the cerebellum. **c. ataxia**, ataxia due to some cerebellar lesion. **c. tonsil.** See *amygdala* (2).
**cerebellic** (*ser-e-bel'-ik*). See *cerebellar*.
**cerebellifugal** (*ser-e-bel-if'-ū-gal*) [*cerebellum*; *fugere*, to flee]. Tending from the cerebellum.
**cerebellipetal** (*ser-e-bel-ip'-e-tal*) [*cerebellum*; *petere*, to seek]. Tending toward the cerebellum.
**cerebellitis** (*ser-e-bel-i'-tis*) [*cerebellum*; *ιτις*, inflammation]. Inflammation of the cerebellum.
**cerebellocortex** (*ser-e-bel-o-kor'-teks*) [*cerebellum*, *cortex*, bark]. The cortex of the cerebellum.
**cerebellospinal** (*ser-e-bel-o-spi'-nal*) [*cerebellum*, *spina*, the spine]. Relating to the cerebellum and the spinal cord.
**cerebellum** (*ser-e-bel'-um*) [dim. of *cerebrum*]. The inferior part of the brain lying below the cerebrum and above the pons and medulla. It consists of two lateral lobes and a middle lobe.
**cerebral** (*ser'-e-bral*) [*cerebrum*]. Relating to the cerebrum. **c. apoplexy.** See *apoplexy*. **c. arteries.** See *arteries, table of*. **c. fornix.** See *fornix, cerebral*. **c. gyri**, the convolutions of the brain. **c. hemiplegia**, hemiplegia due to cerebral apoplexy. **c. index.** See *index, cerebral*. **c. maculæ**, spots on the skin caused by slight irritation, and abnormally persistent. They may indicate disorder of the vaso-motor mechanism. **c. nerves.** See *nerves, table of*. **c. pneumonia.** See *pneumonia, cerebral*. **c. surprise**, the speedy, but not long-persistent stupor that often follows sudden mental shock or grave lesion or injury of the brain. **c. vesicles**, the embryonic vesicles from which the brain is developed. See *brain-bladder*.
**cerebralgia** (*ser-e-bral'-je-ah*) [*cerebrum*, the brain; ἄλγος, pain]. Pain in the head.
**cerebralism** (*ser'-e-bral-izm*) [*cerebrum*, the brain]. The theory that mental operations are due to the activity of the brain; or that thought is a function of the brain.
**cerebrasthenia** (*ser-e-bras-the'-ne-ah*) [*cerebrum*; *asthenia*]. Cerebral asthenia; cerebral neurasthenia; phrenasthenia.
**cerebrasthenic** (*ser-e-bras-then'-ik*) [*cerebrum*, the brain; ἀσθενής, without strength]. Characterized by, or pertaining to, cerebrasthenia.
**cerebration** (*ser-e-bra'-shun*) [*cerebrum*]. Mental activity. **c., unconscious**, mental activity of which the subject is not conscious.
**cerebriform** (*ser-e'-bre-form*). See *cerebroid*.
**cerebrifugal** (*ser-e-brif'-ū-gal*) [*cerebrum*, the brain; *fugere*, to flee]. Centrifugal; efferent; transmitting or transmitted from the brain to the periphery.
**cerebrin** (*ser'-e-brin*) [*cerebrum*]. 1. $C_{17}H_{33}NO_3$. A nitrogenous glucoside obtained from brain-tissue, nerves, and pus-corpuscles. It is a light, colorless, exceedingly hygroscopic powder. 2. A preparation from the gray matter of the brain of sheep and calves, made with equal parts of glycerol and 0.5 % of phenol solution. It has been used in chorea. Dose 5–10 min. (0.3–0.6 Cc.). Syn., *cerebrin-alpha*; *cerebrinin*.
3. A proprietary antineuralgic elixir, said to contain analgesin, ether, caffeine and cocaine.
**cerebrincide** (*ser-e-brin'-as-id*) [*cerebrum*, the brain]. One of certain substances found in brain-tissue, and capable of combining with metallic oxides.
**cerebrinin** (*se-reb'-rin-in*). See *cerebrin* (2).
**cerebripetal** (*ser-e-brip'-et-al*) [*cerebrum*, the brain; *petere*, to seek]. Centripetal; afferent; transmitting or transmitted from the periphery to the brain.
**cerebritis** (*ser-e-bri'-tis*) [*cerebrum*; *ιτις*, inflammation]. Inflammation of the proper substance of the cerebrum. **c., local**, softening of the brain.
**cerebro-** (*se-re-bro-*) [*cerebrum*]. A prefix denoting relating to the cerebrum.
**cerebrocardiac** (*ser-e-bro-kar'-de-ak*) [*cerebro-*; καρδία, the heart]. Applied to diseases characterized by both cerebral and cardiac symptoms.
**cerebrogalactose** (*ser-e-bro-gal-ak'-tōs*). Same as *cerebrose*.
**cerebrohyphoid** (*ser-e-bro-hi'-foid*) [*cerebro-*; ὑφή, tissue; εἶδος, likeness]. Resembling the substance of the brain.
**cerebroid** (*ser'-e-broid*) [*cerebro-*; εἶδος, likeness]. Resembling brain-substance.
**cerebrol** (*ser'-e-brol*) [*cerebrum*, brain; *oleum*, oil]. An oily, reddish fluid obtainable from brain-tissue.
**cerebrology** (*ser-e-brol'-o-je*) [*cerebro-*; λόγος, science]. The science of the brain; encephalology.
**cerebroma** (*ser-e-bro'-mah*) [*cerebrum*, the brain; ὄμα, a tumor; *pl.*, *cerebromata*]. A growth, outside the cranium, that contains cerebral tissue.
**cerebromalacia** (*ser-e-bro-mal-a'-se-ah*) [*cerebro-*; μαλακία, softness]. Softening of the brain tissue.
**cerebromedullary** (*ser-e-bro-med-ūl'-ar-e*) [*cerebro-*; *medulla*, marrow]. Relating to the brain and spinal cord.
**cerebrometer** (*ser-e-brom'-et-er*) [*cerebro-*; μέτρον, a measure]. An instrument for recording cerebral impulses.
**cerebroolein** (*ser-e-bro-o'-le-in*). A compound of olein and lecithin forming a yellow oil; it is obtained from brain tissue.
**cerebropathy** (*ser-e-brop'-a-the*) [*cerebro-*; πάθος, illness]. 1. A train of symptoms following overwork, and approaching the character of insanity. 2. Cerebral disease in general. **c., psychic**, mental disease resulting from primary lesion of the brain or spinal cord, but presenting distinct symptoms of its own.
**cerebrophysiology** (*ser-e-bro-fiz-e-ol'-o-je*). The physiology of the brain.
**cerebropontile** (*ser-e-bro-pon'-til*). Relating to the cerebrum and pons.
**cerebropsychosis** (*ser-e-bro-sik-o'-sis*) [*cerebro-*; ψύχωσις, animating]. Mental disturbance due to a disease of the psychic centers.
**cerebrorrhachidian** (*ser-e-bro-ra-kid'-i-un*) [*cerebro-*; ῥάχις, spine]. Cerebrospinal.
**cerebrosclerosis** (*cer-e-bro-skle-ro'-sis*) [*cerebro-*; σκληρός, hard]. Sclerosis of cerebral tissue.
**cerebroscope** (*ser-e'-bro-skōp*). An ophthalmoscope used in the diagnosis of brain disease.
**cerebroscopy** (*ser-e-bros'-ko-pe*) [*cerebro-*; σκοπεῖν, to inspect]. 1. Ophthalmoscopy in the diagnosis of brain-disease. 2. Encephaloscopy. 3. The postmortem examination of the brain.

**cerebrose** (*ser'-e-brōs*) [*cerebrum*], C₆H₁₂O₆. A crystallized sugar isomeric with glucose, occurring in brain tissue.
**cerebrosensorial** (*ser-e-bro-sen-so'-re-al*) [*cerebro-; sensorium*, the organ of sensation]. Pertaining to the cerebral sensorium.
**cerebroside** (*ser'-e-bro-sid*) [*cerebrum*]. One of a class of substances occurring in brain tissue, containing cerebrose, just as glucosides contain glucose.
**cerebrosis** (*ser-e-bro'-sis*) [*cerebrum*, the brain]. Any cerebral disorder.
**cerebrospinal** (*ser-e-bro-spi'-nal*) [*cerebro-; spina*, the spine]. Pertaining to the brain and spinal cord. **c. axis.** See *axis, cerebrospinal.* **c. fever.** See under *fever.* **c. fluid**, the fluid between the arachnoid membrane and the pia mater. **c. meningitis.** See *fever, cerebrospinal.* **c. sclerosis**, sclerosis of the brain and spinal cord. **c. system**, the brain, spinal cord, and nerves.
**cerebrospinant** (*ser-e-bro-spi'-nant*) [*cerebro-; spina*, the spine]. A medicine that acts upon the brain and spinal cord.
**cerebrosuria** (*ser-e-bro-sū'-re-ah*). The presence of cerebrose in the urine; cerebral diabetes.
**cerebrotomy** (*ser-e-brot'-o-me*) [*cerebro-; τέμνειν*, to cut]. Surgical or anatomical section of brain-tissue.
**cerebrum** (*ser'-e-brum*) [L.]. The chief portion of the brain, occupying the whole upper part of the cranium, and consisting of the right and left hemispheres. **c. abdominale**, the solar plexus. **c. exsiccatum**, the dried and powdered gray substance of the brain of calves; one part represents five parts of the fresh organ. Dose 30-60 gr. (2-4 Gm.) a day. **c. posterius**, the cerebellum.
**cerecloth** (*sēr'-kloth*) [*cera*]. Cloth impregnated with wax and rendered antiseptic; used as a dressing for wounds.
**cerectomy** (*ser-ek'-to-me*). See *kerectomy*.
**cereiform** (*se-re'-e-form*) [*cereus*, a wax taper; *forma*, form]. Shaped like a wax taper.
**cereolus** (*ser-e'-o-lus*) [*cera*, wax; *pl., cereoli*]. A bougie of waxed linen, often medicated.
**cereometer** (*se-re-om'-et-ur*) [*cera*, wax; *μέτρον*, a measure]. An apparatus for the estimation of the quantity of wax in a given mixture by determining the specific gravity.
**cereous** (*se'-re-us*) [*cereus*]. Made of wax.
**ceresin** (*ser'-es-in*) [*cera*, wax]. Ozokerite that has been bleached without distillation; it is used as a substitute for beeswax.
**cereus** (*se'-re-us*) [L., "a wax candle"]. A genus of cactaceous plants. **c. grandiflorus.** See *cactus grandiflorus.*
**cerevisia** (*ser-e-vis'-e-ah*). See *cervisia*.
**cerevisin** (*ser-e-vis'-in*). Dried yeast used internally in furunculosis (dose 1 teaspoonful before each meal) and for application in leukorrhea and gonorrheal vaginitis (15-30 gr. (1-2 Gm.) in suppository of cacao-butter).
**ceric** (*se'-rik*) [*cera*]. 1. Relating to wax. 2. Containing cerium as a quadrivalent radical.
**ceridin.** (*se'-rid-in*). Cerolin. A fatty substance obtained from yeast; used in the treatment of acne.
**cerin** (*se'-rin*) [*cera*]. 1. An ether of cerotic acid; one of the substances found in wax. 2. (Of Chevreul.) A crystalline precipitate from an aqueous extract of cork by action of hot alcohol.
**cerite** (*sē'-rīt*). A Swedish mineral formerly called the heavy stone of Bastnas, from which cerium is obtained.
**cerium** (*se'-re-um*) [named from the planet *Ceres*]. Ce = 140.25; quantivalence II, IV. One of the rarer metals. It forms two series of salts (*cerous* and *ceric* salts) corresponding to the two oxides. See *elements, table of chemical.* **c. nitrate.** 1. Ce(NO₃)₃ . 12H₂O, white crystals, soluble in water; an antiseptic used in solutions of 1 : 1000. Syn., *cerous nitrate*. 2. Ce(NO₃)₄, a reddish-yellow mass of crystals, soluble in water and alcohol. It is used as a nerve-tonic in irritable dyspepsia and chronic vomiting. Dose 1-3 gr. (0.065-0.2 Gm.), Syn., *ceric nitrate.* **c. oxalate** (*cerii oxalus*, U. S. P.), Ce₂(C₂O₄)₃ . 9H₂O, a white, granular powder, insoluble in water or alcohol, but soluble in hydrochloric acid. It is useful in the vomiting of pregnancy. Dose 1-10 gr. (0.065-0.65 Gm.) in pill. **c. valerate**, has been used in the same class of cases as the oxalate. Dose 1¼ gr. (0.1 Gm.).
**cerolein** (*se-ro'-le-in*) [*cera*, wax]. A substance found in beeswax, soluble in alcohol; probably a mixture of fatty acids.

**cerolin** (*se'-ro-lin*). A preparation said to be the active principle of yeast. It consists of the glycerides of fatty acids with cholesterina, lecithin and ethereal oil. It is said to be useful in furunculosis, acne, sycosis and skin affections.
**ceroma** (*se-ro'-mah*) [*cera; δμα*, a tumor]. A cystic tumor the tissue of which has undergone fatty degeneration.
**ceromel** (*se'-ro-mel*) [*cera*, wax; *mel*, honey]. Honey cerate; wax, one part; honey, two or four parts. It is applied to wounds and ulcers, chiefly in Asiatic countries.
**ceroplas'tik**) [*cera*, wax; *πλάσσειν*, to mould]. Modeled, or as if modeled, in wax. **c. catalepsy.** See *catalepsy*.
**ceroplasty** (*se'-ro-plas-te*) [*cera*, wax; *πλάσσειν*, to mould]. The modeling of anatomical preparations in wax.
**cerostroma, cerostrosis** (*ser-o-stro'-mah, -sis*). See *ichthyosis hystrix.*
**cerotate** (*se'-ro-tāt*). A salt of cerotic acid.
**cerotic** (*se-ro'-tik*). Derived from wax. **c. acid.** See *acid, cerotic.*
**cerous** (*se'-rus*). Containing cerium as a bivalent radical.
**certificate** (*ser-tif'-ik-āt*) [*certificare*, to certify]. A written statement, as for insurance, or in case of birth or death.
**cerumen** (*ser-u'-men*) [*cera*]. The wax of the ear. **ceruminosis** (*ser-u-min-o'-sis*). An excessive secretion of cerumen.
**ceruminous** (*ser-u'-min-us*) [*cera*]. Pertaining to cerumen. **c. glands**, glands secreting cerumen.
**ceruse** (*se'-rūs*) [L., *cerussa*]. 1. White lead: basic carbonate and hydrate of lead. 2. A white face-powder. **c.**, antimony, white oxide of antimony; also antimonic acid.
**cerussa** (*se-rus'-ah*). See *ceruse.*
**cervical** (*ser'-vik-al*) [*cervix*, the neck]. Pertaining to the neck or to the cervix uteri. **c. carcinoma**, carcinoma of the neck of the uterus. **c. endometritis.** See *endocervicitis.* **c. pregnancy**, a rare condition in which, from atrophy of the decidual membranes, the impregnated ovum is not properly held in place, and, dropping, lodges in the cervical canal, where it develops until the uterus expels it.
**cervicalis** (*ser-vik-a'-lis*) [*cervix*]. 1. Cervical. 2. A cervical artery, muscle, nerve, or vein.
**cervicem** (*ser'-vis-em*) [*cervix*, the neck]. Belonging solely to the cervix.
**cervicicardiac** (*ser-vis-ik-ar'-de-ak*) [*cervix*, the neck; *καρδία*, the heart]. Relating to the neck and the heart, as the cervicicardiac nerves, branches of the vagus.
**cerviciplex** (*ser-vis'-ip-leks*) [*cervix*, the neck; *plexus*, a network]. The cervical plexus.
**cervicispinal** (*ser-vis-e-spi'-nal*). Relating to the neck and spinal cord.
**cervicitis** (*ser-vis-i'-tis*) [*cervix; ιτις*, inflammation]. Inflammation of the cervix uteri.
**cervico-** (*ser'-vik-o*). Prefix denoting relation to the neck or to the cervix of an organ.
**cervicoauricular** (*ser-vik-o-aw-rik'-ū-lar*). Relating to the back of the neck and the outer ear.
**cervicobasilar** (*ser-vik-o-bas'-il-ar*). Pertaining to the neck and the basilar region.
**cervicobrachial** (*ser-vik-o-bra'-ke-al*) [*cervico-; brachium*, the arm]. Relating to the neck and the arm.
**cervicobregmatic** (*ser-vik-o-breg-mat'-ik*) [*cervico-; βρέγμα*, the sinciput]. Relating to the cervix or nucha and the bregma.
**cervicodynia** (*ser-vik-o-din'-e-ah*) [*cervico-; οδύνη*, pain]. Cramp or neuralgia of the neck.
**cervicofacial** (*ser-vik-o-fa'-shal*) [*cervico-; facies*, face]. Relating to the neck and the face.
**cervicohumeral** (*ser-vik-o-hū'-mer-al*). Relating to the neck and the humerus.
**cervicomuscular** (*ser-vik-o-mus'-kū-lar*). Relating to the muscles of the neck.
**cerviconasal** (*ser-vik-o-na'-zal*). Running from the back of the neck to the nose.

**cervico-occipital** (*ser-vik-o-ok-sip'-it-al*) [*cervico;- occiput*, the back of the head]. Relating to the neck and the back of the head.
**cervico-orbicular** (*ser-vik-o-or-bik'-ū-lar*) [*cervico-; orbicularis*, circular]. Relating to the neck and the orbicular muscle.
**cervicoscapular** (*ser-vik-o-skap'-ū-lar*). 1. Relating to the back of the neck and the scapula. 2. The transverse artery or vein of the neck.
**cervicovaginal** (*ser-vik-o-vaj'-in-al*) [*cervico-; vagina*]. Relating to the cervix uteri and the vagina.
**cervicovesical** (*ser-vik-o-ves'-ik-al*). Pertaining to the bladder and the cervix uteri.
**cervimeter** (*ser-vim'-et-er*) [*cervix*; μέτρον, a measure]. An instrument for measuring the cervix uteri.
**cervisia** (*ser-vis'-e-ah*) [L.]. Ale or beer. **cervisiæ fermentum**, beer-yeast. The ferment obtained in brewing beer, and produced by *Saccharomyces cerevisiæ*.
**cervix** (*ser'-viks*) [L.]. A constricted portion or neck. c. **conical**, c., **conoid**, c., **conoidal**, malformation of the cervix uteri marked by a conical shape and elongation, with constriction of the os externum. c. **cornu**, the constricted portion of the cornu dorsale. c. **obstipa**, c. **rigida**, wryneck. c. **tapiroid**, a cervix uteri with a very elongated anterior lip. c. **uteri**, the neck of the uterus. c. **vesicæ**, the neck of the bladder.
**ceryl** (*se'-ril*) [*cera*, wax], $C_{27}H_{56}$. An organic radical found in combination in beeswax.
**cesarean section**, or operation [*cædere*; to cut]. Extraction of the fetus through an incision made in the abdomen. c. **section, postmortem**, extraction of the child after the mother's death.
**cesarotomy** (*se-sar-ot'-om-e*). Cesarean section.
**cesium, cæsium** (*se'-ze-um*) [L., "bluish-gray"], Cs = 132.81; quantivalence I. A rare alkaline metal resembling potassium in physical and chemical properties. c. **and ammonium bromide**, CsBr . - 3NH₄Cl, sedative used in epilepsy. Dose 15-45 gr. (1-3 Gm.); maximum dose 90 gr. (6 Gm.). c. **bitartrate**, CsHC₄H₄O₆, used in nervous heart-palpitation. Dose 3-5 gr. (0.18-0.3 Gm.). c. **bromide**, is a good sedative, but its cost is very great. c. **carbonate**, Cs₂CO₃; used in epilepsy. c. **chloride**, lowers the pulse-rate and raises arterial pressure. Dose 2-5 gr. (0.13-0.32 Gm.). c. **hydrate**, c. **hydroxide**, CsOH, is used in epilepsy. c. **and rubidium and ammonium bromide**, CsBr.RbBr . 6(NH₄Br), a nervine. Dose 15-45 gr. (1-3 Gm.) once or twice daily; maximum dose 90 gr. (6 Gm.). c. **sulphate**, Cs₂SO₄, used as an antiepileptic.
**cestode, cestoid** (*ses'-tōd, ses'-toid*) [κεστός, a girdle; εἶδος, likeness]. Shaped like a girdle or ribbon; applied to worms, of which *Tænia* is a type.
**cestus** (*ses'-tus*) [L., a girdle]. The fold of the metatela encircling the dorsal part of the brain-tube.
**cetacea** (*se-ta'-se-ah*) [κῆτος, a whale]. An order of mammals living in the sea, as the whale, dolphin, etc.
**cetaceum** (*se-ta'-se-um*) [see *cetacea*]. Spermaceti. A fatty substance somewhat resembling paraffin in its physical properties. It is obtained from the head of the sperm-whale, *Physeter macrocephalus*. It is soluble in ether, in chloroform, and in boiling alcohol, and is employed as an emollient. **Cetacei, ceratum**, contains spermaceti, 10; white wax, 35; olive-oil, 55 parts. **Cetacei, unguentum** (B. P.), contains spermaceti, white wax, almond-oil, and benzoin.
**cetic, cetinic** (*se'-tik, se-tin'-ik*). Pertaining to cetin or to the whale.
**cetin** (*se'-tin*) [see *cetacea*], C₃₂H₆₄O₂. The chief constituent of commercial purified spermaceti. It is a fatty, crystalline substance, soluble in alcohol and ether, insoluble in water, melting at 49° C., and volatilizing at 360° C. Syn., *cetinum*.
**Cetraria** (*se-tra'-re-ah*) [*cætra*, a short Spanish shield]. 1. A genus of lichens. 2. Iceland moss — a lichen, *C. islandica*, found in Iceland and other northern countries. It contains a form of starch, *lichenin*, that gelatinizes when boiled with water. It is a feebly tonic demulcent, sometimes used in pulmonary affections. **Cetrariæ, decoctum** (B. P.), contains 5 % of the lichen. Dose 2-4 oz. (60-120 Cc.).
**cetrarin** (*se-tra'-rin*) [*cætra*, a short Spanish shield]. The bitter principle of Iceland moss, crystallizing in fine needles, and nearly insoluble in water.

**cetyl** (*se'-til*) [*cetus*, a whale], C₁₆H₃₃. An alcoholic radical existing in beeswax, and spermaceti.
**cevadilla** (*sev-ad-il'-ah*). See *sabadilla*.
**cevadine** (*sev'-ad-ēn*), C₃₂H₄₉NO₉. A crystalline alkaloid of cevadilla. See *veratrine*.
**Ceylon sickness**. Beriberi.
**ceyssatite** (*ses'-a-tīt*) [*Ceyssat*, a village of Puy-de-Dôme, France]. A fossil earth from the village of Ceyssat, France, composed almost entirely of pure silica. It is used as an absorbent dusting-powder.
**C.F.** Abbreviation for Canadian Formulary of Unofficial Preparations.
**Cg.** Abbreviation for centigram.
**C.G.S.** Abbreviation for *centimeter, gram, second*; denoting that system of scientific measurements which takes the *centimeter*, the *gram*, and the *second* as the units respectively of distance, mass (or weight), and time.
**Chabert's disease** (*shab-air'*) [Philibert *Chabert*, French veterinarian, 1737-1814]. Symptomatic anthrax; black-leg. C.'s oil, a mixture of crude animal oil and oil of turpentine.
**Chaddock's external malleolar sign** (*chad'-ock*) [Charles Gilbert *Chaddock*, American physician]. Extension of one or more of the toes when the external inframalleolar skin area is irritated; it is found in organic disease of the spino-cortical reflex paths.
**Chadwick's sign** [James Read *Chadwick*, American gynecologist, 1844-1905]. Same as *Jacquemier's sign*.
**chæraphrosyne** (*he-raf-ros'-in-e*) [χαίρειν, to rejoice; ἀφροσύνη, senselessness]. Amenomania.
**chæromania** (*ke-ro-ma'-ne-a'h*) [χαίρειν, to rejoice; μανία, madness]. Amenomania.
**chaffbone** (*chaf'-bōn*). A name for the inferior maxilla.
**Chagas' disease** (*chah'-gahs*) [Carlos *Chagas*, Spanish physician]. Parasitic thyroiditis.
**Chagres fever** (*shag'-ras*) [*Chagres*, a river on the isthmus of Panama]. A malignant form of malaria, endemic on the isthmus of Panama.
**chain** (*chān*). 1. A series of connected links of metal, etc. 2. In (organic) chemistry (a series of atoms linked together by one or more bonds). c. **écraseur**, an écraseur of which a chain forms the cutting part. c.-**saw**, a surgeon's saw, the teeth of which are linked together like a chain.
**chalastic** (*kal-as'-tik*) [χαλαστικός, making supple]. 1. Emollient, softening. 2. An emollient or laxative medicine.
**chalastodermia** (*kal-as-to-der'-me-ah*). Synonym of *dermatolysis*.
**chalaza** (*kal-a'-zah*) [*chalazion*]. One of the twisted cords binding the yolk-bag of an egg to the lining membrane of the shell; or that part of a seed where its coats unite with each other and the nucleus.
**chalazia** (*kal-a'-ze-ah*) [*chalazion*]. 1. The socalled hailstone sputa. 2. A chalazion.
**chalazion** (*kal-a'-ze-on*) [χαλάζιον, a small hailstone]. A tumor of the eyelid from retained secretion of the Meibomian glands; a Meibomian cyst. Syn., *porosis palpebræ*. c. **terreum**, one in which there is degeneration of the contents and change to calcium carbonate and cholesterin. Syn., *lithiasis palpebralis*.
**chalazonephritis** (*kal-a-zo-nef-ri'-tis*) [*chalazion; nephritis*]. Granular nephritis.
**chalcitis** (*kal-si'-tis*) [χαλκός, anything made of metal]. A severe inflammation of the eyes, marked at first by excessive lacrimation and sensitiveness to light, resulting in blurred vision and continued flow of mucus. It is due to rubbing the eyes after the hands have been used on brass, as is done by trolley-car conductors and workmen. Syn., *brassy eye; chalkitis*.
**chalcosis** (*kal-ko'-sis*) [χαλκός, copper]. A deposit of copper particles in the tissues.
**chalice-cell** (*chal'-is*). Goblet cell.
**chalicosis** (*kal-ik-o'-sis*) [χάλιξ, gravel]. A disease of the lungs caused by the inhalation of dust or sand.
**chalinoplasty** (*kal-in-o-plas'-te*) [χαλινός, a bridle or rein; πλάσσειν, to form]. An operation to form a new frenum of the tongue.
**chalk** (*chawk*) [*calx*, limestone]. Calcium carbonate. See *calcium*. c.-**stone**, gout-stone — a deposit beneath the skin in gouty patients.
**chalkitis** (*kal-ki'-tis*). See *chalcitis*.
**chalodermia** (*kal-o-dur'-me-ah*) [χάλασις, a slackening; δέρμα, the skin]. A term for dermatolysis.

**chalone** (*kal'-ōn*) [χαλᾶν, to loosen]. An inhibitory hormone. See *hormone*.
**chalonic** (*kal-on'-ik*). Pertaining to chalone.
**chalybeate** (*ka-lib'-e-āt*) [χάλυψ, steel]. 1. Containing iron. 2. Having the color or taste of iron. 3. A substance or medicine containing iron.
**Chamælirium** (*kam-e-lir'-e-um*) [χαμαί, on the earth; λείριον, a lily]. A genus of plants of the order Liliaceæ. The rhizome of *C. luteum*, devil's-bit, of the United States and Canada, is a uterine tonic, anthelmintic, diuretic, and febrifuge. Dose of *aqueous infusion* (1 oz. to 1 pint) a wineglassful.
**chamber** (*chām'-ber*) [*camera*, a chamber]. A cavity or space. c., anterior (of the eye), the space between the cornea and the iris. c., aqueous (of the eye), the space between the cornea and lens. c., posterior (of the eye), the space between the iris and the lens. The chambers of the eye contain the aqueous humor. c., resonance, a resonant chamber attached to a tuning-fork for acoustic investigation.
**Chamberland filter** (*tshām'-ber-land*) [Charles Edouard *Chamberland*, French bacteriologist, 1851–1908]. A filter made of unglazed porcelain; only ultramicroscopic microorganisms pass through it.
**chamecephalic** (*kam-e-sef-al'-ik*) [χαμαί, low; κεφαλή, head]. Characterized by chamecephaly.
**chamecephalous** (*kam-e-sef'-al-us*). See *chamecephalic*.
**chamecephaly** (*kam-e-sef'-al-e*) [χαμαί, low; κεφαλή, head]. In craniometry, that condition of the skull in which the cephalic index is 70° or less. A flat and receding skull.
**chameconcha** (*kam-e-kong'-kah*) [χαμαί, low; κόγχη, orbit]. In craniometry, an orbital index below 80.01°.
**chameconchous** (*kam-e-cong'-kus*) [χαμαί, low; κόγχη, concha]. In craniometry, having an orbital index of not more than 80°.
**chamecranious** (*kam-e-kra'-ne-us*) [χαμαί, low; κρανίον, skull]. In craniometry, having the greatest length of the skull proportioned to its height.
**chameleon-phenomenon** (*kam-e'-le-on-fen-om'-e-non*). A peculiar reaction shown by *Bacillus pyocyaneus*; when grown on agar, a light-green color is imparted to the medium, which after 48 hours turns very dark green. On potato a yellowish-brown growth is formed, which turns green when the superficial portion is removed by scraping, but it soon resumes its brown color.
**chameprosopic** (*kam-e-pro-so'-pik*) [χαμαί, low; πρόσωπον, face]. Low-faced; having the zygomatic facial index below 90°.
**chamocephalic** (*kam-o-sef-al'-ik*) [χαμαί, on the ground, low; κεφαλή, head]. See *chamecephalic*.
**chamois-skin** (*sham'-wah-, or sham'-e-skin*). Properly the skin or tanned leather of the chamois; now prepared from split sheep-skin. It is used in surgery and for underclothing.
**chamomile** (*kam'-o-mīl*). See *anthemis* and *matricaria*.
**chamoprosopic** (*kam-o-pro-so'-pik*) [χαμαί, on the ground; πρόσωπον, face]. See *chameprosopic*.
**champacol** (*sham'-pa-kol*) [*champaka*, Bengalêse name]. A camphor, $C_{17}H_{30}O$, from the wood of the champak-tree, *Michelia champaca*. Syn., *Champaca camphor*.
**champagne** (*sham-pān*) [Fr.]. An effervescent wine useful as a remedy for nausea and vomiting.
**champignon** (*shawm-pin-e-on(g)'*). A suppurative inflammation of the spermatic cord of a horse, developing sometimes after castration.
**chancebone** (*chans'-bōn*). A name for the ischium.
**chancre** (*shang-ker*) [Fr.]. A term formerly used indiscriminately for any primary venereal ulcer, but now generally applied to the initial lesion of syphilis (*q. v.*). c., arsenical, ulceration resembling a syphilitic chancre, but due to arsenic. c., hard, c., Hunterian, c., indurated, c., infecting; c., non-suppurating, c., true, the ulcer of syphilitic origin, which is followed by constitutional syphilis. c., nonincubatory, c., noninfecting, c., simple, c., soft, a contagious, suppurating, nonsyphilitic venereal ulcer, properly called *chancroid*. c., Sahara, the Aleppo boil.
**chancroid** (*shang'-kroid*) [*chancre*; εἶδος, form]. A local, infective process, transmitted by sexual intercourse, and characterized by ulceration, local glandular involvement, and often suppuration. It has been variously termed the soft, nonindurated, simple, or nonsyphilitic chancre. See *chancre*. c., **phagedenic**, chancroid with a tendency to slough.

c., **serpiginous**, phagedenic chancroid that spreads superficially in curved lines.
**chancroidal** (*shang-kroi'-dal*). Pertaining to or of the nature of a chancroid.
**chancrous** (*shang'-krus*). Of the nature of a chancre.
**change** (*chānj*). The word is colloquially used for either the establishment or the cessation of the menstrual function. c. of life, the menopause.
**channel** (*chan'-el*) [ME., *chanel*]. See *canal*. c.s, **intercellular**. 1. Irregular canals of communication between the intercellular spaces interposed between prickle-cells, and thought to be connected with the lymph-capillaries. 2. Tiny canals between gland-cells. c.s, **intracellular**, the minute canals which connect vacuoles in the cell-body of liver-cells with the biliary canaliculi or intercellular channels. c., **lymphatic**, c., **plasmatic**. See *Recklinghausen's canal*, and *canal, serous*.
**channel-bone** (*chan'-el-bōn*). The clavicle.
**Chantreuil's method** (*shang-treel'*). In pelvimetry, a method of ascertaining the distance between the tuberosities of the ischia (11 cm.) in estimating the size of the pelvic outlet. The two thumbs are placed upon the tuberosities, while an assistant measures the distance between them.
**chap** [ME., *chappen*, to cleave]. A slight or superficial fissure of the skin, usually upon the lips, hands, or nipples.
**chaparra amargosa.** A plant growing in Mexico and Texas; the infusion and the fluid extract are used in amebic dysentery.
**chappa** (*chap'-ah*). The name among the Popo people in the colony of Lagos for a disease believed to be neither tuberculous nor syphilitic, marked by severe initial pains in muscles and joints, followed by swelling and the formation of round multiple nodules the size of a pigeon's egg; without forming abscesses these are exposed by ulceration of the skin. The disease finally attacks the bones.
**characterizing group.** A group of atoms in the molecule of a compound which distinguishes that class of compound from all other classes.
**charbon** (*shar'-bon*) [Fr., "charcoal"]. Anthrax (*q. v.*).
**charcoal** (*char'-kōl*) [ME., *charren*, to turn; *col*, coal]. Coal made by subjecting wood to a process of smothered combustion. See under *carbo*.
**Charcot's artery** (*shar-ko'*) [Jean Martin *Charcot*, French physician, 1825–1893]. The "artery of cerebral hemorrhage," one of the lenticulostriate arteries that passes through the outer part of the putamen. C.'s **cirrhosis**. See *Hanot's disease*. C.'s **crystals**, octahedral crystals of the phosphate of Schreiner's base (spermin), found in the sputum of asthma, in seminal fluid, leukemic blood, and feces. C.'s **disease**. 1. Amyotrophic lateral sclerosis. 2. Arthropathy of tabes dorsalis. 3. Multiple cerebrospinal sclerosis. C.'s **fever**, a septic fever occurring in cases of jaundice due to impacted gall-stones. C.'s **gait**, the gait of Friedreich's ataxia. C.'s **intermittent claudication**, intermittent paresthesia of the legs attended with pain, tremor, and excessive perspiration due to arteriosclerosis; a condition first noted by French writers in apparently healthy horses and afterward observed in man. Syn., *angina cruris; angiosclerotic paroxysmal myasthenia; intermittent lameness; intermittent limping*. C.'s **joint**. See *C.'s disease* (2). C.'s **method**. See *hypnotism*. C.'s **pain**, hysterical pain in the ovarian region. C.'s **posterior foot-zone**. See *Burdach's column*. C.'s **sensory crossway**, the posterior third of the posterior limb of the internal capsule. Syn., *carrefour sensitif*. C.'s **sign**, in facial paralysis the eyebrow is raised; in facial contracture it is lowered. Syn., *signe du sourcil*. C.'s **syndrome**, intermittent claudication, an affection connected with arteriosclerosis of the lower extremities. C.'s **zones**, the syterogenhic zones.
**Charcot-Guinon's disease** (*shar-ko'-gwe-nyong'*). Dementia complicating some cases of progressive muscular dystrophy.
**Charcot-Leyden's crystals** (*shar-ko-li'-den*). See *Charcot's crystals*.
**Charcot-Marie's symptom.** See *Marie's symptom*.
**C.-M.'s type of progressive muscular atrophy**, the neurotic type of progressive muscular atrophy; progressive neural muscular atrophy, commencing in the muscles of the feet and the peroneal group.
**C.-M.-Tooth's type of progressive muscular atrophy.** See *C.-M.'s type of progressive muscular atrophy*.

**Charcot-Neumann's crystals.** See *Charcot's crystals.*
**Charcot-Robin's crystals.** See *Charcot's crystals.*
**Charcot-Vigouroux's sign.** See *Vigouroux's sign.*
**chariot** (*char'-e-ot*). The movable coil of an induction apparatus.
**charlatan** (*shar'-lat-an*) [Ital., *ciarlatano*, a quack]. A quack; a pretender to medical skill; an advertising doctor.
**charlatanism, charlatanry** (*shar'-lat-an-izm, -re*). 1. The state of being a quack. 2. The practices of a quack.
**Charles' law** (*chaŕlz*) [Jacques Alexander Cæsar *Charles*, French physicist, 1746–1823]. Equal increments of temperature add equal amounts to the product of the volume and pressure of a given mass of gas. The increase is $\frac{1}{273}$ of its volume measured at $-273°$ C., which is the zero of absolute temperature.
**charleyhorse** (*char'-le-hors*). Stiffness of the right arm and leg in baseball players.
**charpie** (*shar'-pe*) [*carpere*, to pluck]. Picked or shredded lint; linen shreds for dressing wounds.
**Charrière's guillotine** (*shar-re-ār'*) [Joseph François Benoit *Charrière*, French instrument maker, 1803–1876]. An instrument for excising tonsils. **C.'s scale,** the French scale for measuring the size of urethral sounds or catheters; the consecutive numbers differ by $\frac{1}{3}$ mm.
**charta** (*kar'-tah*) [χάρτης, paper]. A paper. In pharmacy, a strip of paper the fibers of which are impregnated with a medicinal substance. Also a wrapper for holding powders. c. cantharidis, c. **epispastica,** blistering-paper. c. **emporetica,** porous or bibulous paper. c. **exploratoria,** test-paper. c. **sinapis** (U. S. P.), mustard-paper.
**chartreuse** (*shar-trooz'*) [Fr.]. A tonic cordial, obtained by distillation from various plants growing on the Alps.
**chartula** (*kart'-u-lah*) [dim. of *charta*]. A little paper, especially a paper containing a single dose of a medicinal powder.
**chasma, chasmus** (*kaz'-mah, kaz'-mus*) [χασμός, a gaping]. A yawn.
**Chassaignac's axillary muscle** (*shas-a-nyak'*) [Edouard Pierre Marie *Chassaignac*, French surgeon, 1804–1879]. A nonconstant muscular bundle that extends across the axillary hollow from the lower border of the latissimus dorsi to the lower border of the pectoralis minor or to the brachial fascia. **C.'s tubercle,** the carotid tubercle on the transverse process of the sixth cervical vertebra.
**chaudepisse** (*shōd-pēs*) [Fr. "hot urine"]. The scalding and painful urination of the acute stage of gonorrhea.
**chauffeur's fracture** (*sho-fer*). Same as Colles' fracture; caused by "back fire" of the motor which suddenly jerks the crank handle in the opposite direction from which it is being turned.
**chaulmoogra, or chaulmugra oil** (*chawl-moog'-rah*) [E. Ind.]. A fixed oil expressed from the seeds of *Gynocardia odorata,* a tree native to the East Indies. It is soluble in alcohol, and its properties are due to gynocardic acid. It is used in leprosy, in scaly eczema, psoriasis, and syphilitic skin affections. For external use, 1 part of the acid to 24 of petroleum. Internally, 5–10 min. (0.32–0.65 Cc.) of the oil or $\frac{1}{2}$–3 gr. (0.032–0.2 Gm.) of the acid, in capsules.
**Chaussier's areola** (*sho-se-a'*) [François *Chaussier*, French physician, 1746–1828]. The areola of inflammatory induration of a malignant pustule. **C.'s line,** the raphe of the corpus callosum.
**Chautard's test for acetone** (*sho-tar'*). Allow sulphurous acid to pass through a solution of 0.25 Gm. of fuchsin in 500 Cc. of water until the solution becomes yellow. On the addition of a portion of this to the liquid to be tested for acetone it will assume a violet color if acetone is present.
**Chauveau's retention theory** (*sho-vo'*) [Auguste *Chauveau*, French veterinarian, 1827– ]. See under *immunity*.
**chawstick** (*chaw'-stik*). See *chewstik.*
**chaya, c.-root** (*chi'-ah*). The plant, *Aerva lanata.* Syn., *shaya-root.*
**Ch. B.** Abbreviation for *Chirurgiæ Baccalaureus,* Bachelor of Surgery.
**Cheadle-Barlow's disease** [Walter Butler *Cheadle,* English physician]. See *Barlow's disease.*
**check** (*chek*) [OF., *eschec,* from Pers. *shāh,* a-king]. A sudden stop. c.**-experiment.** See *control experiment.* c.**-ligament.** See *ligament, check-.*

**checkerberry** (*chek'-er-ber-e*). A popular name for *Gaultheria procumbens.*
**cheek** (*chēk*) [AS., *cēáce*]. The side of the face; it is composed of fat, areolar tissue, muscles, etc. c. **bone,** the malar bone. c. **teeth,** the molar teeth.
**cheese** (*chēs*) [AS., *cēse*]. A food prepared from the casein of skimmed or unskimmed milk.
**cheesy** (*chēz'-e*) [*cheese*]. Of the nature of cheese. c. **degeneration,** c. **necrosis,** caseous degeneration or caseation; the conversion of the tissues into a substance resembling cheese. c. **tubercle,** a tubercle that has undergone cheesy necrosis.
**cheil-, cheilo-** (*kīl-, ki-lo-*). For words thus beginning see *chil-* or *chilo-.*
**cheiranthin** (*ki-ran'-thin*). A glucoside from the leaves and seeds of *Cheiranthus cheiri,* with action similar to that of digitalis.
**cheiro-** (*ki-ro-*). For words thus beginning see *chiro-.*
**chekan, cheken** (*chek'-en*) [Chilian]. 1. The leaves of *Eugenia cheken,* a South American shrub. It is diuretic and expectorant and similar in action to eucalyptus. It is used in chronic pharyngitis, laryngitis, etc. Dose of the *fluidextract* $\frac{1}{2}$–1 dr. (2–4 Cc.). 2. The crude resin obtained from *Cannabis indica.*
**chelate** (*ke'-lāt*) [χηλή, a claw]. 1. Claw-shaped. 2. Having claw-shaped appendages or processes.
**chelene** (*ke-lēn'*). Ethyl chloride.
**chelerythrine** (*kel-er'-ith-rēn*), $C_{19}H_{17}NO_4$. A poisonous alkaloid obtained from *Chelidonium.*
**chelidonine** (*hel-id'-o-nēn*) [*chelidonium*], $C_{19}H_{17}$-$N_3O_3+H_2O$, or $C_{19}H_{17}NO_4$. A crystalline alkaloid of celandine (*Chelidonium majus*). c. **phosphate,** a white, crystalline powder, soluble in water; it is used as an analgesic. c. **sulphate,** $(C_{20}H_{19}NO_5)_2H_2SO_4,$ a white crystalline substance, soluble in water; it is a narcotic like morphine, but less toxic. Dose 1$\frac{1}{2}$–3 gr. (0.1–0.2 Gm.).
**chelidonism** (*kel-id'-on-izm*). Poisoning by *Chelidonium majus;* it is marked by inflammation of the mouth and gastrointestinal tract and hematuria. It is due to the action of chelerythrine.
**chelidonium** (*kel-id-o'-ne-um*) [χελιδόνιον, celandine]. Celandine. The leaves and stems of *C. majus,* with properties due to a number of alkaloids and acids. It is a drastic cathartic and externally an irritant, and has been used in jaundice, whooping-cough, and catarrhal pneumonia. Dose of the *plant* 10–30 gr. (0.65–2.0 Gm.); of the *juice* 5–20 min. (0.32–1.3 Cc.). Unof.
**chelidoxanthin** (*kel-id-o-zan'-thin*) [*chelidonium*]. One of the bitter, crystalline constituents of celandine.
**cheloid** (*ke'-loid*). See *keloid.*
**cheloma** (*ke-lo'-mah*). Same as *keloid.*
**chelonin** (*kel'-on-in*). See *balmony.*
**chelotomy** (*ke-lot'-o-me*). See *kelotomy.*
**Chelsea pensioner** (*chel'-se*) [*Chelsea,* town in England]. Compound confection of guaiacum; it contains guaiacum resin, 1; rhubarb, 2; acid potassium tartrate, 7.5; nutmeg, 1; sublimed sulphur, 14.5; and clarified honey, 74. It is a popular remedy (in England) for rheumatism and gout.
**chematropism** (*kem-at'-ro-pizm*). See *chemotropism,* and *chemotaxis.*
**chemical, chemic** (*kem'-ik-al, kem'-ik*) [*chemistry*]. Of or pertaining to chemistry. c. **antidote,** an antidote which decomposes a poison. c. **equation,** the formula representing a chemical reaction. c. **food,** compound syrup of phosphates. c. **reflex.** Same as *humoral reflex, q. v.*
**chemicity** (*kem-is'-it-e*). The state of having chemical properties.
**chemicoanalytic** (*kem-ik-o-an-al-it'-ik*). Relating to chemical analysis.
**chemicocautery** (*kem-ik-o-kaw'-ter-e*). Cauterization by means of chemical agents.
**cheminosis** (*kem-in-o'-sis*) [*chemistry;* νόσος, disease]. Any disease caused by chemical agents.
**chemiotaxis, chimiotaxis** (*kem'-e-o-taks-is, kim'-e-o-taks-is*). See *chemotaxis.*
**chemise** (*she-mēz'*). A form of surgical dressing made of muslin and applied after operations upon the rectum or bladder to control or prevent hemorrhage.
**chemism** (*kem'-izm*) [χημεία, chemistry]. 1. Chemical force. 2. Iatrochemistry or chemistry. See *spagirism.* 3. The theory that assumes the development of the universe to be due to chemical processes.

**chemist** (*kem'-ist*). 1. One skilled in chemistry. 2. A druggist. c. **pharmaceutical**, a druggist.

**chemistry** (*kem'-is-tre*) [χημεία, chemistry]. The science of the molecular and atomic structure of bodies. c., **actinic**, c., **actino-**, that treating of decomposition of light. c., **analytic**, that concerned in the determination of the constituents and decomposition-products of substances; also in the estimation of the relative proportion of their elements and the number and interrelation of the atoms contained in a molecule. c., **animal**, that dealing with animal substances. c., **atomic**, that concerned in the structure of molecules, the relations of their contained atoms, and the laws governing their combination. c., **electro-**. See *electrochemistry*. c., **empirical**, c., **experimental**. 1. The sum of chemical knowledge established by experiments. 2. The carrying on of experiments for determining chemical laws and knowledge. c., **forensic**, that concerned in legal investigations. c., **galvano-**. See *galvanochemistry*. c., **inorganic**, the chemistry of substances which do not contain carbon. c., **organic**, the chemistry of organic substances, or of the carbon compounds. c., **physiological**, the chemistry of the vital processes of animals and plants. c., **pneumatic**, the chemistry of vapors and gases. c., **stœchiometric**. See c., *atomic*. c., **synthetic**, that which deals with the building-up of compounds from their elements.

**chemocephalus** (*kem-o-sef'-al-us*) [χαμαί, low; κεφαλή, head]. An individual possessed of a flat head.

**chemoceptor** (*kem-o-sep'-tor*). One of the side chains or receptors in a living cell, having the power of fixing chemical substances in the same way that bacterial toxins are fixed.

**chemoimmunology** (*kem-o-im-u-nol'-o-je*). The study of the chemical processes concerned with the problem of immunity.

**chemolysis** (*kem-ol'-is-is*). Chemical decomposition.

**chemoreceptor** (*kem-o-re-sep'-tor*). See *chemoceptor*.

**chemosis** (*ke-mo'-sis*) [χήμωσις, a gaping]. Swelling of the conjunctiva.

**chemosmosis** (*kem-os-mo'-sis*) [chemistry; ὠσμός, an impulse]. Chemical action resulting from osmosis.

**chemosmotic** (*kem-os-mot'-ik*). Relating to or due to chemosmosis.

**chemosynthesis** (*kem-o-sin'-thes-is*). The building up of compounds by chemical action.

**chemotactic** (*kem-o-tak'-tik*) [*chemotaxis*]. Pertaining or relating to chemotaxis.

**chemotaxis** (*kem-o-taks'-is*) [*chemistry*; τάσσειν, to order, arrange]. The property of cellular attraction and repulsion. It is displayed by the protein constituents of the protoplasm of various species of bacteria, as well as by proteids from a great variety of sources. The qualifications positive and negative are added according as the phenomenon is one of attraction or repulsion.

**chemotherapy** (*kem-o-ther'-ap-e*). Treatment of disease based on the affinity which is supposed to exist between various chemical agents and the tissues of the body or invading microorganisms.

**chemotic** (*ke-mot'-ik*) [*chemosis*]. Pertaining to or marked by chemosis.

**chemotropism** (*kem-ot'-ro-pizm*) [*chemistry*; τροπή, a turning]. The destruction of bacteria by phagocytes; the victory of the phagocytes over bacteria, or of bacteria over phagocytes. In biology, the attraction of leukocytes by certain chemical substances held in solution in the blood. Cf. *chemotaxis*.

**chenopodium** (*ken-o-po'-de-um*) [χήν, a goose; πόδιον, a little foot]. American wormseed; the fruit of *C. ambrosioides*, or *anthelminticum*, a plant native to the United States, with properties due to a volatile oil, which is the only preparation used. It is an efficient anthelmintic against the roundworm. c., **oil of** (*oleum chenopodii*, U. S. P.). Dose 5-15 min. (0.32-1.0 Cc.).

**cheoplastic** (*ke-o-plas'-tik*) [χέειν, to pour; πλαστικός, plastic]. Made soft and yielding by heat. c. **metal**, an alloy composed of tin, silver, and bismuth, with a small trace of antimony.

**Cherchewsky's disease** (*sher-shef'-ske*) [Michael Cherchewsky, Russian physician]. Nervous ileus. An affection, closely simulating intestinal obstruction, that has been observed in neurasthenia.

**cheromania, chæromania** (*ker-o-ma'-ne-ah*). See *amenomania*.

**cherophobia** (*ker-o-fo'-be-ah*) [χαίρειν, to rejoice; φόβος, fear]. Morbid fear of gaiety, or of being happy.

**cherry** (*cher'-e*) [κερασός, cherry-tree]. The bark of the common cherry, *Prunus serotina*, a mild bitter and tonic containing tannin. Dose of *fluidextract* ½-1 dr. (2-4 Cc.). *Prunin*, a concentrated extract; dose 1-3 gr. (0.065-0.2 Gm.). See also **choke-cherry** and *Prunus virginiana*. c.-**laurel**, the European evergreen cherry, *Prunus laurocerasus*. Water distilled from its leaves is used in the same way as dilute hydrocyanic acid. Dose 30 min.-1 dr. (2-4 Cc.). c., **wild**. See *prunus*.

**chervil** (*shur'-vil*) [AS., *cerfille*]. The European potherb *Anthriscus cerefolium*. It is said to be deobstruent, diuretic, and emmenagog.

**Chervin's method** (*sher'-van*) [Claudius Chervin, French teacher, 1824-1896]. A method of treating stuttering.

**chest**. See *thorax*. c., **alar**, c., **paralytic**, c., **phthisical**, c., **pterygoid**, a narrow thorax having a winged appearance from abnormal projection of the wings of the scapula. c., **barrel-**, a peculiar formation of the chest observed in cases of long-standing emphysema of the lungs; it is round, like a barrel, and in respiration is lifted vertically instead of being expanded laterally. c., **emphysematous**. See c., *barrel-*.

**chestnut**. See *castanea*.

**chewstick** (*chu'-stik*). The bark of *Gonania domingensis*, a popular aromatic bitter in the West Indies. It is also used as a dentifrice and masticatory. Dose of the fluidextract j-ij.

**Cheyne's nystagmus**. See *Cheyne-Stokes' nystagmus*. **C.'s symptom**. See *Cheyne-Stokes' respiration*. **Cheyne's operation** (*chain*) [William Watson Cheyne, English surgeon]. For the radical cure of *femoral hernia*; after reducing the hernia, a flap of the pectineus muscle is raised and made to cover the hernial orifice.

**Cheyne-Stokes' asthma** (*chain-stōks*) [John Cheyne, Scotch physician, 1777-1836; William Stokes, Irish physician, 1804-1878]. Dyspnea due to pulmonary congestion in an advanced stage of chronic myocarditis. **C.-S.'s nystagmus**, a variety of nystagmus in which the oscillations of the eyeball have a rhythmic variation similar to the rhythm of Cheyne-Stokes' respiration. **C.-S.'s phenomenon**, **C.-S.'s respiration**, a form or type of breathing in which there is a rhythmic increase of the respirations up to a certain degree of rapidity, then gradually decreasing again to temporary cessation. It occurs in certain grave affections of the central nervous system, heart, and lungs, and in intoxications.

**Chian** (*ki'-an*). Pertaining to Chios, an island in the Ægean Sea. **C. turpentine**. See *terebinthina*.

**chiasm, chiasma** (*ki'-asm, ki-as'-mah*) [χιάζειν, to make a cross, as an X]. 1. The optic commissure. 2. A crossing. c., **Camper's**. See *Camper's chiasm*.

**chiasmal** (*ki'-as-mal*). Pertaining to the optic chiasm.

**chiastometer** (*ki-as-tom'-et-er*) [χιαστός, crossed; μέτρον, a measure]. An instrument for measuring any deviation of the optic axes from parallelism.

**chibou** (*see-boo'*) [Fr.]. The resin or gum of *Bursera gummifera*, a tree in Florida and tropical America; it is locally valued in diseases of the lung and kidneys, and is used in various plasters and ointments; it is called also *cachibou* and *archipin*.

**chickahominy fever** (*chik-a-hom'-in-e*). A synonym of *Typhomalarial fever*.

**chicken** (*chik'-en*) [AS., *cicen*]. The domestic fowl. c.-**breast**, an abnormally prominent condition of the sternum and of the sternal region; pigeon-breast; it is seen in rhachitic persons, etc. c. **cholera**. See *cholera, chicken*. c.-**fat clot**, a clot of blood, yellowish in color, containing largely of fibrin, and containing but few red cells. c.-**pox**. See *varicella*.

**chicle**. See *balata*.

**chicory** (*chik'-or-e*). *Cichorium intybus*, a composite plant of Europe and Asia, naturalized and growing in the United States. Its ground root is used to adulterate coffee.

**chielin** (*ki'-el-in*). A thick, nontoxic, brown, viscous substance isolated from the bulb of the tulip. It is recommended in eczema and in skin diseases in veterinary practice.

**Chiene's line** [John *Chiene*, Scotch surgeon, 19th cent.]. Imaginary lines designed to aid in localizing the cerebral centers in operations upon the brain.

**chignon fungus** (*shēn-yon(g)*) [Fr.]. A fungoid disease of the hair in which oval or roundish masses

surround the hair-shaft at irregular intervals. It is also miscalled *chignon gregarine.*
**chigoe** (*chig'-o*) [Fr., *chique*]. Sand-flea; *sarcopsylla penetrans*, a small parasite of the skin, affecting usually that portion between the toes and fingers; also the red harvest mite, leptus irritans. It is also written *chigo, chegoe, chigga, chiggre, chigger, jigger.*
**chilalgia** (*kil-al'-je-ah*) [χεῖλος, lip; ἀλγός, pain]. Neuralgia affecting the lips.
**chilblain** (*chil'-blān*) [AS., *cēle*, cold; *blĕgen*, a boil]. A congestion and swelling of the skin, due to cold, and attended with severe itching or burning; vesicles and bullæ may form, and these may lead to ulceration. Syn., *erythema pernio; pernio.*
**childbed**. The popular term for the puerperal state. c. fever, puerperal fever.
**childbirth**. Parturition.
**child-crowing**. The crowing sound of the respiration that characterizes laryngismus stridulus.
**chilectropion** (*ki-lek-tro'-pe-on*) [χεῖλος, lip; *ectropion*]. Eversion of the lip.
**Chili saltpeter**. Sodium nitrate.
**chilitis** (*ki-li'-tis*) [χεῖλος, lip; ιτις, inflammation]. Inflammation of the lip.
**chill** (*chil*) [AS., *cēle*, chilliness]. A sensation of cold accompanied by shivering, usually appearing shortly after exposure to cold or wet. It is frequently the initial symptom of acute disorders, as pneumonia, etc. It is a prominent symptom of various forms of malarial fever.
**chills and fever**. A popular term for intermittent fever.
**chilo-** (*ki-lo-*) [χεῖλος, lip]. A prefix meaning relating to the lips.
**chiloangioscope** (*ki-lo-an'-je-o-skōp*) [χεῖλος, lip; ἀγγεῖον, vessel; σκοπεῖν, to look]. An apparatus for observing the circulation of the blood in the human lip.
**chiloangioscopy** (*ki-lo-an-je-os'-ko-pe*) [χεῖλος, lip; ἀγγεῖον, vessel; σκοπεῖν, to look]. The use of the cheiloangioscope.
**chilocace** (*kil-ok'-as-e*) [χεῖλος, lip; κακός, evil]. A firm, reddish swelling of the lip in scrofulous children.
**chilognathopalatoschisis** (*ki-log-nath-o-pal-at-os'-kisis*) [χεῖλος, lip; γνάθος, jaw; *palatum*, the palate; σχίσις, a splitting]. Marchand's term for a malformation marked by fissure of the lip, alveolar process, and palate.
**chilognathus** (*ki-log'-na-thus*) [χεῖλος, lip; γνάθος, jaw]. Harelip.
**chilogramma** (*ki-lo-gram'-mah*) [χεῖλος, lip; γράμμα, a mark; *pl., cheilogrammata*]. Jadelot's labial line.
**chiloncus** (*ki-long'-kus*) [χεῖλος, lip; ὄγκος, tumor]. Tumor of the lip.
**chiloplasty** (*ki'-lo-plas-te*) [χεῖλος, lip; πλάσσειν, to form]. Any plastic operation upon the lip.
**chilorrhagia** (*ki-lor-a'-je-ah*) [χεῖλος, lip; ῥήγνυναι, to burst forth]. Hemorrhage from the lips.
**chiloschisis** (*ki-los'-kis-is*) [χεῖλος, lip; σχίσις, a split]. Hare-lip. c. complicated, harelip attended with fissure of the palate or of the alveolar arch.
**chilostomatoplasty** (*ki-lo-sto-mat-o-plas'-te*) [χεῖλος, lip; στόμα, mouth; πλάσσειν, to form]. Chiloplasty including restoration of the mouth.
**chimaphila** (*ki-maf'-il-ah*) [χεῖμα, winter; φίλος, loving]. Pipsissewa; prince's-pine; the leaves of *C. umbellata*, an evergreen found in the United States, an astringent tonic and excellent diuretic. The bruised leaves are used as a rubefacient. It is valuable in dropsy, in renal disease, and in affections of the urinary passages. c., **decoction of** (*decoctum chimaphilæ*, B. P.). Dose 1–3 oz. (30–90 Cc.). c., **fluidextract of** (*fluidextractum chimaphilæ*, U. S. P.). Dose ½–2 dr. (2–8 Cc.).
**chimaphilin** (*ki-maf'-il-in*). A resinoid from the leaves of *Chimaphila umbellata;* diuretic in dose of from one to four grains (0.065 to 0.26 Gm.).
**chimney-sweep's cancer.** See *carcinoma, chimney-sweep's.*
**chimogene** (*ki'-mo-jen*) [χειμών, cold winter weather; γενᾶν, to produce]. A highly volatile liquid proposed by Vanderweyde as a substitute for rhigolene, ether, etc., producing cold in local anesthesia.
**chin** [AS., *cin*]. The mentum; the lower part of the face, at or near the symphysis of the lower jaw. **c.-cough**, whooping cough. **c.-jerk, c.-reflex**. See *jaw-jerk* and under *reflex.*
**china** (*kin'-ah* or *ke'-nah*). Same as *cinchona.*

**chinaphthol** (*kin-af'-thol*). A yellow, bitter, insoluble powder, used as an intestinal antiseptic. Dose 7½–7.5 gr. (0.5–5.0 Gm.) daily. Syn., *betanaphthol-a-monosulphate.*
**chinaseptol** (*kin-a-sep'-tol*). See *diaphtol.*
**chinine, chininum** (*kin'-ēn, kin-in'-um*). See *quinine.*
**chinoform** (*kin'-o-form*). A compound of formaldehyde with cinchotannin. Syn., *quinoform.*
**chinoidine, chinoidinum** (*kin-oi'-dēn, kin-oi-di'-num*) [Sp., *china*]. Quinoidine. A mixture of amorphous alkaloids obtained in the manufacture of quinine. It has the therapeutic properties of quinine. Dose 1–20 gr. (0.065–1.3 Gm.). c., **animal**, a substance giving, like quinine, a blue fluorescence in solutions of dilute acids, first obtained by Bence-Jones from the liver, but found in all the organs and tissues of the body, especially in the nerves. c. **borate**, yellowish scales, soluble in water and alcohol, used as is chinoidine. Dose 8–15 gr. (0.5–1.0 Gm.). c. **citrate**, reddish scales, soluble in water and alcohol. Dose 5–25 gr. (0.32–1.6 Gm.). c. **tannate**, a yellow or brown powder, slightly soluble in alcohol; antipyretic, astringent, and tonic. Dose 2–12 gr. (0.13–0.8 Gm.). In veterinary practice it is given in hog-cholera in 24 gr. (1.5 Gm.) doses 3 times daily.
**chinol** (*kin'-ol*). Quinoline monohypochlorite, C₉H₆N·ClO, a white, crystalline, odorless powder, with a pungent taste; soluble in alcohol, almost insoluble in cold or hot water. It is antipyretic and analgesic. Dose 3–5 gr. (0.19–0.32 Gm.).
**chinoline, chinolina** (*kin'-o-len, kin-o-li'-nah*). See *quinoline.*
**chinone** (*kin'-ōn*). See *quinone.*
**chinopyrin** (*kin-o-pi'-rin*). See *quinopyrine.*
**chinoral** (*kin'-or-al*). An oily, bitter liquid containing quinine and chloral; antiseptic and hypnotic. Dose 1–15 gr. (0.06–1.0 Gm.). Syn., *quinochloral.*
**chinosol** (*kin'-o-sol*). See *quinosol.*
**chinotoxin** (*kin-o-toks'-in*). A synthetic compound said to possess properties similar to those of curare. Syn., *diquinolin dimethyl sulphate.*
**chinotropin** (*ki-no-tro'-pin*). Quinate of urotropin; used as a urinary antiseptic and uric acid solvent; dose, gr. 5–15 (0.3–10 Gm.).
**chinovin** (*kin'-o-vin*) [*china*, quinine], C₃₀H₃₆O₉. A glucoside obtained from cinchona.
**chiolin** (*ki'-ol-in*). A proprietary remedy for diseases of the skin.
**chionablepsia, chionablepsy** (*ki-on-ab-lep'-se-ah, ki-on-ab-lep'-se*) [χιών, snow; ἀβλεψία, without sight]. Loss of sensibility of the retina resulting from the exposure of the eyes to reflection of the sunlight upon snow; snow-blindness.
**chionanthin** (*ki-o-nan'-thin*) [χιών, snow; ἄνθος, a flower]. A precipitate from the tincture of the rootbark of *Chionanthus virginiana;* it is an aperient, diuretic, tonic, and narcotic. Dose 1 to 3 grains.
**Chionanthus** (*ki-o-nan'-thus*) [χιών, snow; ἄνθος, a flower]. A genus of oleaceous trees and shrubs. C. **virginiana**, fringe-tree or poison-ash. The root is used as a vulnerary.
**chionyphe** (*ki-on'-if-e*). Madura-foot.
**chionyphe** (*ki-on'-if-e*) [χιών, snow; ὑφή, a texture]. 1. See *mycetoma.* 2. A genus of fungi. c. **Carteri**, a parasitic fungus, apparently the cause of the disease known as *fungus foot*, or *madura foot, q. v.*
**chiragra** (*ki-rag'-rah*) [χείρ, the hand; ἄγρα, a seizure]. Gout in the hand.
**chiralgia** (*ki-ral'-je-ah*). See *chiragra.*
**chirapsia** (*ki-rap'-se-ah*) [χείρ, hand; ἁψις, a touching]. Friction with the hand; massage.
**chirarthritis** (*ki-rar-thri'-tis*) [χείρ, hand; ἄρθρον, a joint; ιτις, inflammation]. Rheumatism or arthritis of the hand.
**chirata, chiretta** (*ke-ra'-tah, ke-re'-tah*) [Hind., *chirāetā*, a species of gentian]. The dried plant of *Swertia chirayita.* It resembles gentian in its therapeutic properties, and is an excellent tonic. It does not contain tannin. Dose of the powdered plant 15–30 gr. (1–2 Gm.). c., **fluidextract of** (*fluidextractum chiratæ*, U. S. P.). Dose 15–30 min. (1–2 Cc.). c., **infusion of** (*infusum chiratæ*, B. P.). Dose 2 oz. (64 Cc.). c., **tincture of** (*tinctura chirata*, B. P.) (10 % strength). Dose ½–2 dr. (2–8 Cc.).
**chiratin, chirettin** (*ki-ra'-tin, ki-ret'-in*) [Hind., *chirāetā*, a species of gentian], C₂₆H₄₈O₁₅. A light-yellow, crystalline, bitter glucoside, obtained from chirata.

**chiretta** (*kir-et'-ah*). See *chirata*.
**chirismus** (*ki-riz'-mus*) [χειρισμός, a handling]. 1. Manipulation; a kind of massage. 2. Spasm of the hand.
**chiro-, cheiro-** (*ki-ro-*) [χείρ, the hand]. A prefix meaning hand.
**chirokinesthetic** (*ki-ro-kin-es-thet'-ik*) [χείρ, hand; *kinesthetic*]. Relating to the subjective perception of the motions of the hand, particularly in writing.
**chirol** (*ki'-rol*). A solution of resins and fatty oils in a mixture of ethers and alcohols, used as a protective varnish for the hands in surgery.
**chirology** (*ki-rol'-o-je*) [χείρ, hand; λόγος, science]. A method of communicating with deaf-mutes by means of the hands; it is distinct from dactylology.
**chiromancy** (*ki-ro-man'-se*). See *palmistry*.
**chiromania** (*ki-ro-ma'-ne-ah*) [χείρ, hand; μανία, insane desire]. Masturbation.
**chiromegaly** (*ki-ro-meg'-al-e*) [χείρ, hand; μεγάλη, large]. Enlargement of one, or both hands, but not of akromegalic nature. Syn., *pseudoakromegaly*.
**chirometer** (*ki-rom'-et-ur*) [χείρ, hand; μέτρον, measure]. Osiander's instrument for measuring a distance on the finger in manual pelvimetry.
**chiropelvimeter** (*ki-ro-pel-vim'-et-er*) [*chiro-*; *pelvis*; μέτρον, a measure]. In manual pelvimetry, an instrument for measuring the hand.
**chiroplasty** (*ki'-ro-plas-te*) [χείρ, hand; πλάσσειν, to form]. A plastic operation on the hand.
**chiropodalgia** (*ki-ro-pod-al'-je-ah*). See *acrodynia*.
**chiropodist** (*ki-rop'-o-dist*) [*chiro-*; πούς, foot]. A surgeon or person who professionally treats diseases of the hands and feet, especially corns, bunions, and affections of the nails.
**chiropody** (*ki-rop'-od-e*) [χείρ, hand; πούς, foot]. The business or profession of a chiropodist.
**chiropompholyx** (*ki-ro-pom-fo'-liks*) [χείρ, hand; πόμφολυξ, blister]. Dysidrosis; pompholyx; an ill-defined, inflammatory skin-disease confined to the hands and feet, and characterized by the development of peculiar vesicles or blebs, arranged in groups.
**chiropractic** (*ki-ro-prak'-tik*) [χείρ, hand; πράσσειν, to do]. A method of restoring health by manipulation of the spinal column.
**chirospasm** (*ki'-ro-spazm*) [χείρ, the hand; σπασμός, a drawing]. Writers' cramp.
**chirotheca, cheirotheca** (*ki-ro-the'-ka*) [*chiro-*; θήκη, case]. A long, narrow roller bandage for wrapping the fingers. **c. completa**, one for all the fingers of a hand. **c. incompleta**, one for a single finger only.
**chirurgeon** (*ki-rur'-jon*) [χειρουργός, a surgeon]. A surgeon.
**chirurgery** (*ki-rur'-jer-e*) [*chirurgia*, from χείρ, hand; ἔργον, work]. Same as *surgery*.
**chirurgia** (*ki-rur'-je-ah*) [see *chirurgeon*]. Surgery.
**chirurgical** (*ki-rur'-jik-al*) [χειρουργία, surgery]. Pertaining to surgery.
**chirurgicogynecological** (*ki-rur-je-ko-jin-e-kol-oj'-ik-al*). Pertaining to surgical operations for gynecological conditions.
**chitin** (*ki'-tin*) [χιτών, a tunic], $C_{15}H_{26}N_2O_{10}$. A colorless skeleton; the animal analogue of the cellulose of plants.
**chitinization** (*kit-in-iz-a'-shun*). Transformation into chitin.
**chitinous** (*ki'-tin-us*). Resembling chitin. **c. degeneration**, amyloid degeneration.
**chitonitis** (*ki-ton-i'-tis*) [χιτών, a tunic; ιτις, inflammation]. Inflammation of any investing membrane.
**chitosan** (*kit'-o-san*), $C_{14}H_{26}N_2O_{10}$. A cleavage-product of chitin heated to 180° C. with alkali and a little water; it is soluble in dilute acids.
**chittim-bark** (*chit'-im-bark*). See *cascara sagrada*.
**chlamydobacteria** (*klam-id-o-bak-te'-re-ah*) [χλαμύς, mantle; *bacteria*]. Bacteria surrounded by a thick capsule or sheath.
**chlamydospore** (*klam'-id-o-spor*) [χλαμύς, mantle; σπόρα, seed]. In biology, applied to a spore having its own protective envelope.
**chlamydozoa** (*klam-id-o-zo'-ah*) [χλαμύς, cloak; ζῷον, animal]. Protozoa consisting of a cell or cells surrounded by a capsule.
**chliasma** (*kli-az'-mah*) [χλιαίνειν, to make warm]. A fomentation, a poultice.
**chloasma** (*klo-az'-mah*). [χλοάζειν, to be pale green]. A deposit of pigment in the skin, occurring in patches of various sizes and shapes, and of a yellow, brown, or black color. Syn., *discolorations*; *melanoderma*; *melasma*. **c. hepaticum**, liver-spots;

a form following dyspepsia and popularly associated with hepatic disturbance. **c. phthisicorum**, the brown patches upon the skin of the forehead or upper portions of the cheeks in tuberculous patients. **c. uterinum**, chiefly located on the forehead, temples, cheeks, nipples, and median line of abdomen. They are marked during pregnancy, and often during menstruation.
**chloracetic acid** (*klor-as-e'-tik*). See *acid, chloracetic*.
**chloracetization** (*klor-as-et-iz-a'-shun*). The production of local anesthesia by chloroform and glacial acetic acid.
**chloracetyl** (*klor-as'-et-il*). 1. $C_2Cl_2$. A radical formed from acetyl by the replacement of hydrogen with chlorine. 2. Acetyl chloride.
**chloracne** (*klor-ak'-ne*) [*chlorine*; *acne*]. An acneiform eruption of the face, chest and back, occurring in workers in chlorine, and the chlorides.
**chloral** (*klo'-ral*) [*chlorine*; *aldehyde*], $C_2Cl_3HO$. A pungent, colorless, mobile liquid. The name is often misapplied to chloral hydrate. Syn., *acetochloral*; *trichlorated* or *trichloracetic aldehyde*. **c. anhydroglyco-**, chloralose. **c., anhydrous**, chloral as distinguished from chloral hydrate. **c. antipyrine**. See *hypnal*. **c., butyl-** (*butyl-chloral hydras*, B. P.), croton-chloral, $C_4H_5Cl_3+OH_2O$, a solid occurring in crystalline scales, resembling chloral hydrate, but made with butyl, $C_4H_9$, as a base, instead of ethyl, $C_2H_5$. Its properties are similar to those of chloral, but are much feebler. Dose 5–20 gr. (0.32–1.3 Gm.) in syrup. **c.-caffeine**, the residue upon evaporation of a concentrated aqueous or alcoholic solution of chloral hydrate 7.8 parts and caffeine 10 parts; hypnotic, sedative, and analgesic. Injection, 3–6 gr. (0.2–0.4 Gm.) 2 or 3 times daily. **c., camphorated**, a liquid made by triturating equal parts of chloral hydrate and camphor; it is an anodyne. **c. hydrate** (*chloralum hydratum*, U. S. P.), a colorless, crystalline solid having the composition $C_2HCl_3(HO)_2$; the hydrate of chloral. It is a powerful hypnotic, antispasmodic, and depressant to the cerebral, medullary, and spinal centers, and to a limited extent is an anesthetic. It is serviceable in fevers accompanied by cerebral excitement, in chorea, convulsions, and in delirium tremens, but should be used with great caution. Dose 5–20 gr. (0.32–1.3 Gm.). **c., syrup of** (B. P.). Dose ½–2 dr. (2–8 Cc.). **c. urethane**. See *uralium*.
**chloralacetaldoxime**[*klo-ral-as-et-al-doks'-im*), $C_4H_5-NO_2Cl_3$. A white, crystalline powder, soluble in alcohol and ether, melting at 74° C. It is hypnotic.
**chloralacetophenonoxime** (*klo-ral-as-et-o-fe-non-oks'-im*), $C_8H_8$. $C_3HCl_3$. A substance forming colorless prisms, soluble in alcohol and ether, melting at 81° C. It is used in tetanus and epilepsy.
**chloralacetoxime** (*klo-ral-as-et-oks'-im*), $C_4H_8N_2O$. $Cl_3$. A white, crystalline powder, soluble in alcohol and ether, and melting at 72° C. It is hypnotic.
**chloralamide** (*klo-ral'-am-id*). See *chloralformamide*.
**chloralbacid** (*klo-ral-bas'-id*). A compound of chlorin and albumin. It is used as a tonic in gastric disorders. Dose .7⅔–15 gr. (0.5–1.0 Gm.). **c. sodium**, a compound of chloralbacid and sodium; it is used in gastric and intestinal affections. Dose 15–30 gr. (1–2 Gm.) before meals.
**chloralbenzaldoxime** (*klo-ral-ben-zal-doks'-im*), $C_9-H_8NO_2Cl_3$. A white, crystalline powder, soluble in alcohol and ether, melting at 62° C. It is hypnotic and antiseptic.
**chloralcamphoroxime** (*klo-ral-kam-for-oks'-im*), $C_{12}-H_{18}NO_2Cl_3$. A white, crystalline powder, soluble in alcohol and ether, melting at 98° C. It is hypnotic, stimulant, and antiseptic.
**chloralformamide** (*klo-ral-form'-am-id*). A crystalline solid (*chloralformamidum*, U. S. P.), $C_3H_4Cl_3NO_2$. Used as a hypnotic. Dose 30–45 gr. (2–3 Gm.).
**chloralic** (*klo'-ral-ik*) [*chloral*]. Relating to chloral.
**chloralid** (*klo'-ral-id*) [*chloral*], $C_5H_2Cl_6O_3$. A substance obtained when trichlorlactic acid is heated to 150°, with an excess of chloral; it crystallizes from alcohol and ether, in large prisms, and is insoluble in water.
**chloralimide** (*klo-ral'-im-id*) [*chloral*; *imide*], $CCl_3$.-$CH$. $NH$. A hypnotic allied to chloral, soluble in alcohol. The dose is the same as that of chloral hydrate.
**chloralin** (*klo'-ral-in*). An antiseptic fluid containing monochlorphenol and bichlorphenol. It is

used in 2 to 3 % solution; as a gargle, in 0.5 to 1 % solution.
**chloralism** (klō'-ral-izm) [*chloral*]. 1. Chloral-poisoning, the morbid state caused by the injudicious use of chloral. 2. The habit of using chloral.
**chloralization** (klo-ral-iz-a'-shun). 1. See *chloralism*. 2. Anesthesia by means of hydrated chloral.
**chloralize** (klō'-ral-īz) [*chloral*]. To put under influence of chloral.
**chloralorthoform** (klor'-al-or'-thō-form). Trade name of a combination of chloral and orthoform recommended as a local anæsthetic.
**chloralose** (klō'-ral-ōs) [*calcis chloridi; glucose*], $C_8H_{11}Cl_3O_6$. Anhydroglyco-chloral; a product of the action of anhydrous chloral upon glucose. It is used as a hypnotic. Dose 3–14 gr. (0.2–0.9 Gm.).
**chloraloximes** (klo-ral-oks'-ims). A series of chemical compounds the physiological activities of which are claimed to be due to their splitting up in the system into chloral hydrate and their respective oximes. See *chloralacetoxime, chloralbenzaldoxime,* etc.
**chloraloxylose** (klo-ral-ọ-zi'-lōs) [*chloral; xylose*]. A combination of hydrated chloral and xylose. A convulsivant.
**chloralum** (klo'-ral-um). 1. Chloral. 2. Crude aluminum chloride mixed with various sodium and calcium salts; a disinfectant. c. **hydratum,** the official name of chloral hydrate in the U. S. P.
**chloralurethane** (klo-ral-u'-re-thān). Same as *ural*.
**chloramide** (klō'-ram-id). Same as *chloralamide*.
**chloranemia, chloranæmia** (klor-un-e'-me-ah). Synonym of *chlorosis*.
**chloranodyne** (klor-an'-o-dīn) [*chlorine; anodyne*]. A proprietary remedy, introduced as an improvement on chlorodyne. It contains morphine hydrochloride, tincture of cannabis indica, chloroform, dilute hydrocyanic acid, and aromatics. Dose for an adult 15 min. (1 Cc.).
**chlorargentate** (klor-ar'-jen-tāt). A combination of silver chloride with the chloride of some other radical or element.
**chlorarsenous** (klor-ar-se'-nus). Arsenous and also containing chlorine.
**chlorate** (klō'-rāt) [*chlorine*]. A salt of chloric acid.
**chlorated** (klō'-ra-ted). Containing, combined with, or charged with chlorine.
**chlorazol** (klo'-raz-ol). A highly toxic, oily liquid, obtained from albumin, glutin, or dried muscle by action of strong nitric and hydrochloric acids.
**chlorbenzoyl** (klōr-ben'-zo-il). See *benzoyl chloride*.
**chlorbromide** (klōr-brō'-mid). A combination of a radical with chlorine and bromine.
**chlorcamphor** (klōr-kam'-for). A name for several compounds of chlorine and camphor.
**chloremia, chloræmia** (klo-re'-me-ah) [χλωρός, green; αἷμα, blood]. A blood-disorder, either idiopathic or associated with other ailments, consisting in a diminution of the percentage of the hemoglobin and a decrease in the number of red blood-corpuscles.
**chlorepatitis** (klōr-ep-at-i'-tis) [*chlorine; hepatitis*]. Chronic hepatitis combined with chlorosis.
**chlorephidrosis** (klōr-ef-id-ro'-sis) [*chlorine; ἐφίδρωσις,* perspiration]. A condition characterized by greenish perspiration.
**chlorethyl** (klo-reth'-il). See *ethyl chloride*.
**chlorethylene** (klōr-eth'-il-ēn). A substance formed from ethylene by replacement of one or more atoms of hydrogen with chlorine. c. **chloride,** c. **dichloride,** $C_2H_3Cl_3$, an oil with odor like that of ethene chloride, boiling at 115° C.; employed as an anesthetic.
**chlorethylidene** (klōr-eth'-il-id-ēn). A chlorine substitution-compound of ethylidene. c. **chloride,** c. **dichloride,** $C_2H_4Cl_2$, a liquid used as an anesthetic. Syn., *chlorinated ethyl chloride; monochlorethylidene dichloride.*
**chloretone** (klō'-ret-ōn). See *acetone chloroform*.
**chlorhematin** (klōr-hem'-a-tin). See *hemin*.
**chlorhydria** (klōr-hi'-dre-ah). An excess of hydrochloric acid in the stomach.
**chlorhydric** (klōr-hi'-drik). Composed of chlorine and hydrogen; hydrochloric.
**chloric** (klō'-rik) [*chlorine*]. Pertaining to or containing chlorine. c. **acid.** See *acid, chloric.* c. **ether.** 1. See *ethyl chloride.* 2. See *chloroform, spirit of.*
**chloride, chlorid** (klo'-rid) [*chlorine*]. A binary compound, one of the elements of which is chlorine. c., **methyl-.** See under *anesthetic, local.*

**chloridrometer** (klor-id-rom'-et-ur) [*chloride; μέτρον,* measure]. An apparatus for the estimation of chlorides in the urine.
**chlorimetry** (klor-im'-et-re) [*chlorine; μέτρον,* a measure]. The estimation of the amount of available chlorine in a compound.
**chlorine, chlorin** (klō'-rin) [χλωρός, green]. Chlorum. Cl = 35.46; quantivalence I. A greenish-yellow gas, prepared by decomposing sodium chloride, NaCl. It is highly irritative to the skin and mucous membranes, producing spasmodic closure of the glottis. It is a valuable disinfectant. The *liquor calcis chloridi* (B. P.) contains 1 pound of the salt to a gallon of water. Chloride of lime (*calx chlorinata,* U. S. P.), a hypochlorite of calcium, contains free chlorine and is a valuable disinfectant. Dose, internally, 3–6 gr. (0.2–0.4 Gm.). c.-**hunger,** the condition of the body when chlorine (usually in the form of common salt) is lacking. Among the immediate results of this deficiency are indigestion and albuminuria. c.-**vapor** (B. P.), used for inhalation. c.-**water** (*liquor chlori compositus,* U. S. P.), contains 4 % of the gas in solution. It is a good antiseptic wash. Dose internally 10–30 min. (0.65–2.0 Cc.).
**chlorinated** (klō'-rin-a-ted). Containing chlorine or combined with it.
**chloriodoform** (klo-ri-o'-do-form), $CHCl_2I$. A yellow oil obtained from chloroform by replacement of one atom of chlorine by an atom of iodine; it boils at 131° C. and does not solidify.
**chloriodolipol** (klo-ri-o-do-lip'-ol). A combination of creosote antiseptic; in surgery, 2 to 3 % solution is employed; for inhalation in diseases of the air-passages, 5 % solution.
**chlorite** (klō'-rīt). Any one of the salts of chlorous acid; they are used as bleaching and oxidizing agents.
**chlormethyl** (klor-meth'-il). Methyl chloride.
**chloroalbumin** (klo-ro-al'-bū-min). A derivative of peptone, protogen, or albumoses by action of chlorine.
**chloroanemia, chloroanæmia** (klo-ro-an-e'-me-ah). Same as *chloranemia.*
**chlorobrom** (klō'-ro-brōm) [*chlorine; bromine*]. A solution each ounce of which contains 30 grains of chloralamide and of potassium bromide. It is hypnotic and useful in sea-sickness.
**chlorobromhydrin** (klo-ro-brom-hi'-drin). A substance formed from glycerol by replacement of two molecules of hydroxyl with one atom of chlorine and one of bromine. Syn., *Allyl chlorobromhydrin.*
**chlorodyne** (klō'-ro-dīn) [*chlorine; ὀδύνη,* pain]. An English proprietary remedy supposed to contain chloroform, ether, morphine, cannabis indica, hydrocyanic acid, and capsicum. It is anodyne and narcotic. Dose 10–30 min. (0.65–2.0 Cc.).
**chloroform** (klō'-ro-form) [χλωρός, green; *formyl*]. Methyl trichloride, $CHCl_3$. A heavy, colorless liquid obtained by the action of chlorinated lime on methyl-alcohol. The commercial article, *chloroformum venale,* contains 2 % of impurities. Administered internally in large doses, chloroform produces narcosis and violent gastroenteritis. In small doses it is antispasmodic and carminative. Chloroform has an agreeable odor and a sweetish taste. It solidifies in the cold, boils at 62° C., and has a specific gravity at 15° C. of 1.502. Externally it is much employed as an ingredient of rubefacient and anodyne liniments. Mixed with a large percentage of air and inhaled, it is one of the most valuable of general anesthetics, but occasionally (1 : 3000) causes death by cardiac paralysis. See under *anesthetic.* Deep injections of chloroform in the vicinity of the sciatic nerve have been recommended in sciatica. Syn., *chloroformum; chloroformyl.* c., **alcoholized,** a mixture of chloroform and alcohol. c., **ammoniated,** equal parts of ammonia in alcohol and chloroform; antipyretic and anodyne. c., **Anschuetz's,**

$$(C_6H_4 <^{CO}_{O})_4 \cdot 2CHCl_3$$

a crystalline substance which liberates pure chloroform on application of gentle heat. Syn., *salicylid chloroform.* c., **emulsion of** (*emulsum chloroformi,* U. S. P.), chloroform, 4; expressed oil of almond, 6; tragacanth, 1.5; water sufficient to make 100 parts. c., **gelatinized,** equal parts of chloroform and white of egg shaken together. c. **liniment** (*linimentum chloroformi,* U. S. P.), chloroform, 300 Cc.; soap liniment, 700 Cc. c. **and morphine, tincture of** (B. P.), a substitute for chlorodyne. Each dose of

10 min. (0.65 Cc.) contains chloroform, $1\frac{1}{4}$ min.; ether, $\frac{1}{3}$ min.; alcohol, $1\frac{1}{4}$ min.; morphine hydrochloride, $\frac{1}{48}$ gr.; dilute hydrocyanic acid, $\frac{5}{8}$ min.; oil of peppermint, $\frac{1}{80}$ min.; fluidextract of licorice, $1\frac{1}{4}$ min.; treacle and syrup q. s. **c., Pictet's,** chloroform obtained in a pure state by crystallizing at a low temperature. **c., spirit** of (*spiritus chloroformi*, U. S. P.), pure chloroform, 10; alcohol, 90 parts. Dose 10 min.–1 dr. (0.65–4.0 Cc.). **c., tincture of, compound** (*tinctura chloroformi composita*, B. P.), chloroform, 2; alcohol, 8; compound tincture of cardamom, 10. Dose 20 min.–1 dr. (1.3–4.0 Cc.). **c.-water** (*aqua chloroformi*, U. S. P., B. P.). Dose $\frac{1}{2}$–2 oz. (15–60 Cc.).

**chloroformin** (*klo-ro-form'-in*). A poison extracted by Auclair from tubercle bacilli by means of chloroform. Syn., *chloroformobacillin*.

**chloroformism** (*klo'-ro-form-izm*) [*chloroform*]. 1. The use of chloroform to excess for its narcotic effect. 2. The symptoms produced by this use of the drug.

**chloroformization** (*klo-ro-form-i-za'-shun*). 1. The act of administering chloroform as an anesthetic. 2. The anesthetic results from the inhalation of chloroform.

**chlorogenine** (*klo-roj'-en-ēn*). See *alstonin* (2).

**chlorol** (*klo'-rol*). A solution of sodium chloride, mercury bichloride, and hydrochloric acid, each, 1 part, and 3 % of copper sulphate in 100 parts of water; it is disinfectant and antiseptic.

**chloroleukemia** (*klo-ro-lū-ke'-me-ah*). Leukemia combined with chlorosis.

**chlorolin** (*klo'-ro-lin*). A solution said to consist mainly of the chlorphenols; recommended as a disinfectant and as an antiseptic wash in 2 to 3 % solution.

**chlorolymphoma** (*klo-ro-limf-o'-mah*) [χλωρός, green; *lympha*, water; ὄμα, a tumor]. Another name for chloroma; it is thought by some to be a variety of lymphoma, from its containing lymphocytes.

**chloroma** (*klo-ro'-mah*) [χλωρός, green; ὄμα, a tumor: *pl.*, *chloromata*]. "Green cancer"; a rare variety of sarcoma, of a greenish tint, usually seated upon the periosteum of the bones of the head.

**chlorometer** (*klo-rom'-et-ur*) [χλωρός, green; μέτρον, a measure]. An apparatus for the estimation of the amount of chlorine in a compound.

**chloromorphine** (*klo-ro-mor'-fēn*). An intermediate product between morphine and apomorphine.

**chloronaphthol** (*klo-ro-naf'-thol*). A disinfectant, non-poisonous substitute for carbolic acid; said to be a combination of creosote with an alkali. It is used as a dip for cattle to destroy ticks.

**chloropercha** (*klo-ro-purch'-ah*). A solution of guttapercha in chloroform. It is used in dentistry as nonconducting cavity linings, pulp-cappings, and for filling the roots of pulpless teeth.

**chlorophane** (*klo'-ro-fān*) [χλωρός, green; φαίνειν, to show]. A yellowish-green chromophane. See *chromophane*.

**chlorophthisis** (*klo-ro-ti'-sis*). Pulmonary tuberculosis associated with chlorosis.

**chlorophyl, chlorophyll** (*klo'-ro-fil*) [χλωρός, green; φύλλον, leaf]. The green coloring-matter of plants. It decomposes carbon dioxide, setting free oxygen and forming new organic compounds. This decomposition takes place only or chiefly in the presence of sunlight. The chlorophyl is contained in certain parts of the protoplasm of the plant. It is the substance by the agency of which carbohydrates are formed in green plants.

**chloropia, chloropsia** (*klo-ro'-pe-ah, klo-rop'-se-ah*) [χλωρός, green; ὄψις, vision]. Disordered or defective vision in which all objects appear green.

**chloroplastid** (*klo-ro-plas'-tid*) [χλωρός, green; πλασσεῖν, to form or mould]. In biology, a chlorophyl-granule.

**chloroplastin** (*klo-ro-plas'-tin*) [χλωρός, green; πλαστός, formed]. Schwartz's name for the protoplasm in chlorophyl grains.

**chloroquinone** (*klo-ro-kwin'-ōn*). Any chlorine substitution-compound of quinone.

**chlorosalol** (*klo-ro-sal'-ol*). See *chlorphenyl salicylate*.

**chlorosarcoma** (*klo-ro-sar-ko'-mah*). See *chloroma*.

**chlorosin** (*klo'-ro-sin*). A compound of albumin and chlorine, used in gastric catarrh.

**chlorosis** (*klo-ro'-sis*) [χλωρός, green]. The "green sickness." A form of anemia, most common in young women, and characterized by a marked reduction of hemoglobin in the blood, with but a slight diminution of red corpuscles. In some cases there is a hyperplasia of the sexual organs and the heart and large blood-vessels. The symptoms are those of anemia—a greenish color of the skin, gastric and menstrual disturbances. Syn., *chloranemia; chloremia; green sickness; morbus virgineus; pallor luteus; pallor virginum; parthenosis*. **c. adultarum,** that occurring between 30 and 40 years of age. Syn., *acmæochlorosis; chlorosis tarda*. **c. ægyptiaca, c. æthiopum, c., Egyptian,** uncinariasis. **c. florida,** a rare form of chlorosis in which the color is high. **c. pituitosa.** Synonym of *mucous colitis*. **c. tarda.** See *c. adultarum*. **c., tropical, c. tropica,** uncinariasis.

**chlorosonin** (*klo-ro'-son-in*). A compound of chloral and hydroxylamine; it is hypnotic.

**Chlorostigma** (*klo-ro-stig'-mah*) [χλωρός, green; στίγμα, stigma]. A genus of plants of the order *Asclepiadaceæ*. *C. stuckertianum* is a plant of South America; the root, stem, and leaves are said to possess powerful galactagogic properties. An alkaloid, *chlorostigmine*, has been extracted.

**chlorotic** (*klo-rot'-ik*) [*chlorosis*]. 1. Relating to chlorosis. 2. A person affected with chlorosis.

**chlorous** (*klor'-us*). Containing or combined with chlorine; generally restricted to compounds containing chlorine combined with oxygen, but containing less oxygen than the chloric compounds.

**chlorozone** (*klo'-ro-zōn*) [*chlorine; ozone*]. A yellow liquid assumed to be a mixture of chlorine and ozone; it is formed by passing nascent chlorine through caustic soda; its composition is not known. It is a strong bleaching-agent and disinfectant.

**chlorphenol** (*klōr-fe'-nol*) [*chlorin; phenol*], $C_6H_4Cl.OH$. Monochlorphenol, a substance possessed of antiseptic properties. It is recommended for inhalation in diseases of the respiratory passages.

**chlorphenyl** (*klōr-fen'-il*). A substance obtained from trichlorphenic acid by action of nitric acid. **c. salicylate,** $C_6H_4(OH)CO.OC_6H_4Cl$, a crystalline substance obtained from a mixture of orthochlorphenol and parachlorphenol by action of phosphorus pentachloride. The ortho-compound, *chlorsalol*, is used as a surgical antiseptic; the para-compound, as a substitute for phenyl salicylate. Dose 60–90 gr. (4–6 Gm.) daily. Syn., *salicylic chlorphenol ester*.

**chlorsalol** (*klor-sa'-lol*). See under *chlorphenyl salicylate*.

**chlorum** (*klo'-rum*) [*chlorin*]. The official pharmaceutical name of chlorine. See *chlorine*.

**chloruret** (*klor'-ū-ret*). The same as *chloride*.

**chloryl** (*klo'-ril*). A mixture of ethyl and methyl chlorides; an anesthetic. Syn., *coryl*.

**Chlumsky's button** (*klum'-ske*). An intestinal button made of pure magnesium after the pattern of the Murphy button. It remains undissolved for four weeks, only the outer part becoming softer.

**Ch.M.** Abbreviation for *Chirurgiæ Magister*, Master of Surgery.

**choana** (*ko'-an-ah*) [χοάνη, a funnel; *pl.*, *choanæ*]. 1. A funnel-like opening. 2. A name applied to the posterior nasal orifices.

**choanoid** (*ko'-an-oid*) [*choana*; εἶδος, likeness]. Funnel-shaped.

**chocolate** (*chok'-ol-āt*) [Mex., *chocolatl*, chocolate]. A dried paste prepared from the powder of cacao-seeds (see *Cacao*) with various mucilaginous and amylaceous ingredients. It is used to prepare a beverage, and also as a vehicle, especially for quinine. **c., acorn-,** a mixture of ground acorns with pure chocolate, prepared according to the formula of Liebreich by Stollwerk, of Cologne. It contains nearly 2 % of tannic acid and is used in the dietetic management of diarrhea in enteritis.

**chœradology** (*ker-ad-ol'-o-je*) [χοιράς, scrofula; λόγος, science]. The science of scrofula.

**choke** (*chōk*) [ME., *choken*, to choke]. To suffocate; to prevent access of air to the lungs by compression of or by obstructing the trachea. **c.-cherry,** the fruit of *Prunus virginiana* (not of the Pharmacopeia), common in the United States. It is antispasmodic, tonic, and slightly astringent. **c. damp,** a name given by miners to carbon dioxide gas; called also *black-damp*.

**choked disc** (*chōkd*). See *papillitis*.

**chokes** (*chōks*). The same as caisson disease, *q. v.*

**choking** (*chōk'-ing*). Partial or complete suffocation, whether by the lodgment of food, or any foreign body in the larynx, trachea, pharynx, or eso-

phagus, or by the inhalation of any irrespirable gas or vapor. c. distemper, a name applied in Eastern Pennsylvania to cerebro-spinal meningitis in the horse.
chol-, cholo- [χολή, bile]. Prefixes meaning bile.
cholæmia. Same as *cholemia*.
cholagogic (*kol-ag-oj'-ik*) [*cholagogue*]. 1. Stimulating the flow of bile. 2. A cholagogue.
cholagogue, cholagog (*kol'-ag-og*) [*chol-*; ἀγωγός, leading]. 1. Stimulating the flow or the secretion of bile. 2. Any agent that promotes the flow of bile.
cholalic acid (*kol-a'-lik*). See *acid, cholic*.
cholangeitis (*kol-an'-je-i'-tis*). Same as *cholangitis*.
cholangiostomy (*kol-an-je-os'-to-me*) [*chol-*; ἀγγεῖον, vessel; στόμα, mouth]. The formation of a fistula into the gall-bladder.
cholangiotomy (*kol-an-je-ot'-o-me*) [*chol-*; ἀγγεῖον, vessel; τομή, a cutting]. The incision of an intrahepatic bile-duct for the removal of a calculus.
cholangitis (*kol-an-ji'-tis*) [*chol-*; ἀγγεῖον, vessel; ιτις, inflammation]. Inflammation of a bile-duct. c., obliterative, closure of the bile-ducts the result of inflammation.
cholate (*kōl'-āt*). Any salt of cholic acid.
cholecyanin (*kol-e-si-an'-in*). Synonym of *bilicyanin*.
cholecyst, cholecystis (*kol'-e-sist*, *kol-e-sis'-tis*) [*chol-*; κύστις a bladder]: The gall-bladder.
cholecystalgia (*kol-e-sis-tal'-je-ah*) [*cholecyst*, ἄλγος]. Biliary colic.
cholecystectasia (*kol-e-sis-tek-ta'-se-ah*) [*cholecyst*; ἔκτασις, a distention]. Distention or dilatation of the gall-bladder.
cholecystectomy (*kol-e-sis-tek'-to-me*) [*cholecyst*; ἐκτομή, a cutting off]. Excision of the gall-bladder.
cholecystendesis. See *cholecystendysis*.
cholecystendysis (*kol-e-sis-ten'-dis-is*) [*cholecyst*; ἔνδυσις, an entering]. Cholecystotomy.
cholecystenteroanastomosis (*kol-e-sis-ten'-ter-o-an-as-to-mo'-sis*). Same as *cholecystenterostomy*.
cholecystenterorrhaphy (*kol-e-sist-en-ter-or'-af-e*) [*cholecyst*; *enterorrhaphy*]. The operation of suturing the gall-bladder to the small intestine.
cholecystenterostomy (*kol-e-sist-en-ter-os'-to-me*) [*cholecyst*; *enterostomy*]. The artificial establishment of a communication between the gall-bladder and the intestine.
cholecystic (*kol-e-sis'-tik*) [*cholecyst*]. Relating to the gall-bladder.
cholecystis (*kol-e-sis'-tis*). See *cholecyst*.
cholecystitis (*kol-e-sis-ti'-tis*) [*cholecyst*; ιτις, inflammation]. Inflammation of the gall-bladder. c., Eberth's, that due to *Bacillus typhi abdominalis*.
cholecystocolostomy (*kol-e-sis-to-ko-los'-to-me*) [*cholecyst*; *colostomy*]. The surgical establishment of a passage between the gall-bladder and the colon.
cholecystocolotomy (*kol-e-sis-to-ko-lot'-o-me*) [*cholecyst*; *colotomy*]. Incision into the gall-bladder and colon.
cholecystoduodenostomy (*kol-e-sis-to-du-od-en-os'-to-me*) [*cholecyst*; *duodenostomy*]. The establishment of an artificial communication between the gall-bladder and the duodenum.
cholecystogastrostomy (*kol-e-sis-to-gas-tros'-to-me*) [*cholecyst*; *gastrostomy*]. The formation of an opening between the gall-bladder and the stomach.
cholecystoileostomy (*kol-e-sis-to-il-e-os'-to-me*) [*cholecyst*; *ileostomy*]. The formation of an opening between the gall-bladder and the ileum.
cholecystojejunostomy (*kol-e-sis-to-je-jun-os'-to-me*) [*cholecyst*; *jejunostomy*]. The establishment of a communication between the gall-bladder and the jejunum.
cholecystolithotripsy (*kol-e-sis-to-lith'-ot-rip-se*) [*cholecyst*; λίθος, a stone; τρίβειν, to crush]. The crushing of gall-stones in the gall-bladder.
cholecystoncus (*kol-e-sis-tong'-kus*) [*cholecyst*; ὄγκος, a tumor]. A swelling or tumor of the gall-bladder.
cholecystopexy (*kol-e-sis'-to-pek-se*) [*cholecyst*; πῆξις, fixation]. Suture of the gall-bladder to the abdominal wall.
cholecystorrhaphy (*kol-e-sist-or'-af-e*) [*cholecyst*; ῥαφή, a seam]. Suture of the gall-bladder, especially suture to the abdominal wall.
cholecystostomy (*kol-e-sist-os'-to-me*) [*cholecyst*; στόμα, mouth]. The establishment of an opening into the gall-bladder.
cholecystotomy (*kol-e-sist-ot'-o-me*) [*cholecyst*; τέμ-νειν, to cut]. Incision of the gall-bladder to remove gall-stones, etc.
choledoch (*kol'-e-dok*) [*choledochus*]. 1. Conducting bile. 2. A bile-duct. 3. The common bile-duct.
choledochectomy (*kol-ed-o-kek'-to-me*) [*choledochus*; ἐκτομή, excision]. Excision of a part of the common bile-duct.
choledochendysis (*kol-ed-ok-en'-dis-is*) [*choledochus*; ἔνδυσις, an entry]. See *choledochotomy*.
choledochitis (*kol-ed-o-ki'-tis*) [*choledochus*; ιτις, inflammation]. Inflammation of the common bile-duct.
choledochoduodenostomy (*kol-ed-o-ko-du-od-en-os'-to-me*) [*choledochus*; *duodenum*; στόμα, mouth]. The surgical establishment of a passage between the common bile-duct and the duodenum.
choledochoenterostomy (*kol-ed-o-ko-en-ter-os'-to-me*) [*choledochus*; ἔντερον, bowel; στόμα, mouth]. The surgical establishment of a passage between the cavity of the common bile-duct and the small intestine.
choledocholithiasis (*kol-ed-ok-o-lith-i'-as-is*) [*choledochus*; *lithiasis*]. The formation of a calculus in the common bile-duct.
choledocholithotomy (*kol-ed-ok-o-lith-ot'-o-me*) [*choledochus*; *lithotomy*]. The incision of the common bile-duct for the removal of gall-stones.
choledocholithotripsy (*kol-ed-o-ko-lith'-o-trip-se*) [*choledochus*; λίθος, a stone; τρίβειν, to rub]. The crushing of a gall-stone in the common bile-duct, without opening the duct.
choledocholithotrity (*kol-ed-o-ko-lith-ot'-rit-e*) [*choledochus*; λίθος, a stone; *terere*, to rub]. The crushing of a gall-stone in the common bile-duct.
choledochostomy (*kol-ed-o-kos'-to-me*) [*chol-*; δέχεσ-θαι, to receive; στόμα, mouth]. The formation of a fistula in the common bile-duct through the abdominal wall.
choledochotomy (*kol-ed-o-kot'-o-me*) [*choledochus*; τομή, section]. An incision into the common bile-duct.
choledochus (*kol-ed'-o-kus*) [χολή, bile; δέχεσθαι, to receive]. Receiving or holding bile. c., ductus communis, the common excretory duct of the liver and gall-bladder.
choleglobin (*kol-e-glo'-bin*) [χολή, bile; *globin*]. Latschenberger's name for the antecedent of bile-pigment, resulting, in his estimation, from the decomposition of the coloring-matter of blood.
cholehemia, cholehæmia (*kol-e-hem'-e-ah*, *-he'-me-ah*). See *cholemia*.
choleic (*kol-e'-ik*) [χολή, bile]. Pertaining to the bile.
cholein (*kol'-e-in*) [χολή, bile]. A mixture of several principles of the bile; a fatty principle found in bile.
cholelith (*kol'-e-lith*) [*chol-*; λίθος, a stone]. A biliary calculus or gall-stone.
cholelithiasis (*kol-e-lith-i'-as-is*) [χολή, bile; *lithiasis*]. The presence of, or a condition associated with, calculi in the gall-bladder or in a gall-duct.
cholelithotomy (*kol-e-lith-ot'-o-me*) [*cholelithiasis*; τέμνειν, to cut]. An incision for the removal of gall-stones.
cholelithotripsy (*kol-e-lith-ot-rip'-se*) [*cholelithiasis*; τρίψις, a rubbing]. The operation of crushing a gall-stone.
cholelithotrity (*kol-e-lith-ot'-re-te*). See *cholelithotripsy*.
cholemesis (*kol-em'-es-is*) [*chol-*; ἔμεσις, vomiting]. The vomiting of bile.
cholemia, cholæmia (*kol-e'-me-ah*) [χολή, bile; αἷμα, blood]. The presence of bile in the blood.
cholemic (*ko-le'-mik*) [*chol-*; αἷμα, blood]. Relating to cholemia.
choleplania (*kol-e-pla'-ne-ah*) [χολή, bile; πλάνη, a wandering]. Jaundice.
cholerasin (*ko-le-pra'-sin*). A bile-pigment.
cholepyrrhin (*kol-e-pir'-in*) [χολή, bile; πυρρός, orange-colored]. 1. The brown coloring-matter of bile. Syn., *bilepyrrhin*. 2. Bilirubin.
cholera (*kol'-er-ah*) [χολή, bile; ῥοία, a flow]. 1. A name given to a number of acute diseases characterized mainly by large discharges of fluid material from the bowels, vomiting, and collapse. 2. A synonym of *Asiatic cholera*. c., Asiatic, c., algid, an acute, specific, highly malignant disease, existing in India and the tropics of Asia during the entire year, and occasionally spreading as an epidemic over large areas. It is characterized by vomiting, alvine

discharges resembling flocculent rice-water, severe cramps, and collapse. The rate of mortality varies from 10 to 66 %, the average being over 50 %. The cause is the comma bacillus of Koch, which is always found in the rice-water discharges. The germs commonly gain entrance into the system by means of the drinking-water. **c. asphyctica,** Asiatic cholera marked by early collapse and speedy death. **c., barbel.** See under *barbel.* **c., bilious,** a form of the disease attended with excessive discharge of bile. **c.-blue.** See under *pigment.* **c.-cells, c.-corpuscles,** fungi found in dejecta of cholera patients. **c., chicken,** a very fatal epidemic disease of fowls, marked by tumefaction of the lymphatic glands, with inflammation and ulceration of the digestive organs. **c., English.** See *c. morbus.* **c., epidemic.** Synonym of *Asiatic cholera.* **c.-fever.** 1. Cholera-typhoid. 2. Intermittent cholera. **c., hog,** an infectious disease attacking swine and characterized by a patchy redness of the skin, with inflammation and ulceration of the bowels, enlargement of the abdominal glands, and congestion of the lungs. **c. infantum,** the "summer complaint" of infants and young children; an acute disease occurring in warm weather, and characterized by pain, vomiting, purgation, fever, and prostration. The disease is supposed to be caused by the bacillus of Shiga, and is favored by the prolonged action of heat, together with errors in diet and hygiene. It is most common among the poor and in hand-fed babes. The disease is of short duration, death frequently ensuing in from 3 to 5 days. **c., intermittent,** a form of simple cholera sometimes accompanying the onset of tertian fevers. **c., malignant,** Asiatic cholera. **c. morbus,** an acute catarrhal inflammation of the mucous membrane of the stomach and intestine, with pain, purging, vomiting, spasmodic contractions of the muscles, etc. It is a disease of the heated term and is very similar to Asiatic cholera in its symptomatology. **c. nostras.** Same as *c. morbus.* **c. orientalis,** Asiatic cholera. **c.-red.** See under *pigment.* **c. sicca,** a term applied to those cases of Asiatic cholera in which rice-water liquid is found in the intestine after death, though none had been voided during life. **c., sporadic.** See *c. morbus.* **c. suppressa.** See *c. sicca.* **c.-typhoid,** a soporific condition resembling typhus, lasting from 2 to 7 days, and attributed to uremia resulting from acute nephritis. It frequently follows Asiatic cholera.

**choleragenic** (*kol-er-a-jen'-ik*) [*cholera;* γεννᾶν, to produce]. Tending to produce, or to spread, cholera.

**choleraic** (*kol-er-a'-ik*) [*cholera*]. Pertaining to or resembling cholera. **c. diarrhea,** diarrhea characterized by a profuse, exhausting discharge of watery material.

**cholerase** (*kol'-er-ās*). The special bacteriolytic enzyme of the cholera vibrio. Cf. *pyocyanase* and *typhase.*

**choleric** (*kol'-er-ik*). 1. Having abundant bile. 2. Applied to a temperament easily excited to anger. 3. Choleraic.

**choleriform** (*kol-er'-if-orm*) [*cholera; forma,* form]. Resembling or appearing like cholera.

**cholerigenous** (*kol-er-ij'-en-us*) [*cholera;* γένεσις, production]. Giving origin to cholera.

**cholerine** (*kol-er-ēn'*) [dim. of *cholera*]. A mild form of Asiatic cholera, or the initial stage of a more severe form.

**choleroid** (*kol'-er-oid*) [*cholera;* εἶδος, like]. Resembling cholera; choleriform.

**choleromania** (*kol-er-o-ma'-ne-ah*). See *cholerophobia.*

**cholerophobia** (*kol-er-o-fo'-be-ah*) [*cholera;* φόβος, fear]. Morbid dread of cholera.

**cholerotyphus** (*kol-er-o-ti'-fus*). 1. See *cholera-typhoid.* 2. The most malignant type of Asiatic cholera.

**cholerrhagia** (*kol-er-aj'-e-ah*) [*chole-;* ῥηγνύναι, to burst forth]. 1. Synonym of cholera morbus. 2. A flow of bile.

**cholerythrin** (*kol-er'-ith-rin*). See *bilirubin.*

**cholestearin** (*kol-es-te'-ar-in*). See *cholesterin.*

**cholesteatoma** (*kol-es-te-at-o'-mah*) [*cholesterin;* στεάρωμα, a sebaceous tumor: pl., *cholesteatomata*]. A teratoid tumor containing plates of cholesterin, epithelial cells, hair, and other dermal structures, and occurring most frequently in the brain.

**cholesteatomatous** (*kol-es-te-at-om'-at-us*) [*chole-;* στεάρωμα, a sebaceous tumor.]. Of the nature of, pertaining to, or affected with, cholesteatoma.

**cholestegnosis** (*kol-e-steg-no'-sis*) [χολή, bile; στέγνωσις, a making close]. Thickening of the bile.

**cholesteremia, cholesteræmia, cholesterinemia, cholesterinæmia** (*kol-es-ter-e'-me-ah, kol-es-ter-in-e'-me-ah*) [*cholesterin;* αἶμα, blood]. The morbid state resulting from the retention of cholesterin in the blood. The condition is probably due to the retention of the bile acids.

**cholesterilins** (*kol-es-ter'-il-ins*). Hydrocarbons formed from cholesterin by action of concentrated sulphuric acid, and supposed to stand in close relationship to the terpene group.

**cholesterin** (*kol-es'-ter-in*) [χολή, bile; στέαρ, fat], **cholesteryl** (*kol-es'-ter-il*), $C_{26}H_{44}$. The radical of cholesterin. $C_{27}H_{46}OH$. A monatomic alcohol, a constituent of bile, gall stones, nervous tissue, egg yolk, and blood, and sometimes found in foci of fatty degeneration. It is a glistening, white, crystalline substance, soapy to the touch, crystallizing in fine needles and rhombic plates. It is insoluble in water, soluble in hot alcohol, ether, or chloroform. It is held in solution in the bile by the bile salts; it is levorotatory. The power of immunizing against and neutralizing snake-venom is attributed to it. **c., tests for.** See *Lieber-mann-Burchard, Obermueller, Salkowski, Schiff, Schultze.*

**cholesterinuria** (*ko-les-ter-in-u'-re-ah*) [*cholesterin;* οὖρον, urine]. The presence of cholesterin in the urine.

**cholesterol** (*kol-es'-ter-ol*). Same as *cholesterin.*

**cholesteryl** (*kol-es'-ter-il*), $C_{26}H_{44}$. The radical of cholesterin.

**choletelin** (*kol-et'-el-in*) $C_{14}H_{18}N_2O_6$. An amorphous, soluble, yellow pigment derived from bilirubin. It is the final product of the oxidation of bile pigments. It is readily soluble in alkalies, alcohol, and chloroform.

**choletherapy** (*kol-e-ther'-ap-e*) [χολή, bile; *therapy*]. The remedial use of bile.

**choleuria** (*kol-e-ū'-re-ah*) [χολή, bile; οὖρον, urine]. The presence of bile in the urine.

**choleverdin** (*kol-e-ver'-din*) [χολή, bile; *viridis,* green]. See *bilicyanin.*

**cholic** (*kol'-ik*) [χολή, bile]. Pertaining to the bile. **c. acid.** See *acid, cholic.*

**cholicele** (*kol'-is-ēl*) [χολή, bile; κήλη, a tumor]. A tumor of the gall bladder, due to accumulation of bile.

**choline,** (*kol'-in*) [χολή, bile], $C_5H_{15}NO_2$. A ptomaine occurring in bile, and elsewhere. It is also found in the extracts of the suprarenals, and is a product of the decomposition of lecithin. An auxetic in cancer.

**cholocele** (*kol'-o-sēl*). See *cholicele.*

**cholochrome** (*kol'-o-krōm*) [χολή, bile; χρῶμα, color]. Any bile pigment.

**cholocyanin** (*kol-o-si'-an-in*). Synonym of *bili-cyanin.*

**cholocyst** (*kol'-o-sist*). See *cholecyst.*

**chologestin** (*kol-o-jes'-tin*). A proprietary cholagogue said to contain sodium glycocholate, sodium salicylate, sodium bicarbonate, and pancreatin.

**cholohematin** (*kol-o-hem'-at-in*) [χολή, bile; αἶμα, blood]. A pigment found in the bile of the ox and sheep. It is probably a derivative of hematin.

**choloidinic acid** (*kol-oi-din'-ik*), $C_{24}H_{39}O_4$. A decomposition product of cholic acid.

**chololith** (*kol'-o-lith*) [χολή, bile; λίθος, stone]. A gall stone.

**chololithiasis** (*kol o-lith-i'-as-is*). See *cholelithiasis..*

**chololithic** (*kol-o-lith'-ik*) [χολή, bile; λίθος, a stone] Pertaining to a cholelith.

**choloplania.** See *choleplania.*

**cholopoiesis** (*kol-o-poi-e'-sis*) [χολή, bile; ποίησις, a making]. The formation of bile.

**cholorrhagia** (*kol-or-a'-je-ah*) [χολή, bile; ῥηγνύναι, to burst forth]. A sudden flow of bile.

**cholorrhea** (*kol-or-e'-ah*) [χολή, bile; ῥοία, a flow]. Any excessive discharge of bile.

**cholosis** (*kol-o'-sis*) [χολή, bile]. 1. Any disease caused by or associated with a perversion of the biliary secretion. 2. Lameness. **c. americana,** yellow fever.

**cholostegnosis.** See *cholestegnosis.*

**cholotic** (*kol-ot'-ik*) [*chol-*]. Due to or associated with a cholosis.

**choluria** (*kol-ū'-re-ah*) [*chol-;* οὖρον, urine]. The presence of bile, bile-salts, or bile-pigments in the urine. Also, the greenish coloration of the urine.

**chondral** (*kon'-dral*) [*chondrus*]. Cartilaginous; relating to or composed of cartilage.

**chondralgia** (kon-dral'-je-ah) [chondro-; ἄλγος, pain]. Pain in or about a cartilage.
**chondrectomy** (kon-drek'-to-me) [chondro-; ἐκτομή, a cutting out]. Surgical excision of a cartilage, or of a part of one.
**chondren** (kon'-dren) [chondro-]. Belonging to a cartilage in itself.
**chondric** (kon'-drik). See chondral.
**chondrification** (kon-drif-ik-a'-shun) [chondrus; facere, to make]. The process of being converted into cartilage.
**chondrify** (kon'-drif-i) [chondro-; fieri, to become]. To convert into cartilage; to become cartilaginous.
**chondrigen** (kon'-drij-en) [chondrin; γεννᾶν, to produce]. That material of the hyaline cartilage which on boiling with water becomes chondrin.
**chondrigenous** (kon-drij'-en-us) [chondrin; γεννᾶν, to produce]. Producing chondrin; relating to cartilage that has not hardened.
**chondriglucose** (kon-dre-glu'-kōs) [chondrin-; γλυκύς, sweet]. A material formed by boiling cartilage with mineral acids. It has a sweet taste and the properties of glucose.
**chondrin** (kon'-drin) [chondrus]. A substance obtained from the matrix of hyaline cartilage by boiling. It resembles gelatin in general properties, but differs from it in not being precipitated by tannic acid. **c. balls**, a substance found in cartilage and composed of chondromucoid and chondroitic acid.
**chondriomite** (kon'-dre-o-mīt) [chondro-; μίτος, thread]. A chain of mitochondria.
**chondritis** (kon-dri'-tis) [chondro-; ιτις, inflammation]. Inflammation of a cartilage.
**chondro-** (kon-dro-) [chondrus]. A prefix meaning relating to cartilage.
**chondroarthritis** (kon-dro-ar-thri'-tis). An inflammation of the cartilaginous parts of a joint.
**chondroblast** (kon'-dro-blast) [chondro-; βλαστός, germ]. A cell of developing cartilage.
**chondroblastoma** (kon-dro-blas-to'-mah). Same as chondroma.
**chondrocarcinoma** (kon-dro-kar-sin-o'-mah). A carcinoma containing cartilaginous tissue.
**chondrocele** (kon'-dro-sēl) [chondro-; κήλη, a tumor]. A sarcocele containing masses resembling cartilage.
**chondroclasis** (kon-drok'-las-is) [chondro-; κλάσις, fracture]. The crushing of a cartilage.
**chondroclast** (kon'-dro-klast) [chondro-; κλάειν, to break]. A cell supposed to be concerned in the absorption of cartilage.
**chondrocoracoid** (kon-dro-kor'-ak-oid). Relating to a costal cartilage and to the coracoid process of the scapula.
**chondrocostal** (kon-dro-kos'-tal) [chondro-; costa, a rib]. Relating to the ribs and their cartilages.
**chondrocranium** (kon-dro-kra'-ne-um) [chondro-; cranium]. The cartilaginous cranium, as of the embryo.
**chondrocyte** (kon'-dro-sīt) [chondro-; κύτος, cell]. A cartilage cell.
**chondrodendron** (kon-dro-den'-dron) [chondro-; δένδρον, a tree]. A genus of South American menispermaceous climbing plants. C. glaberrimum and C. tomentosum are among the plants that furnish pareira.
**chondrodialysis** (kon-dro-di-al'-is-is) [chondro-; dialysis]. The decomposition of cartilage.
**chondrodynia** (kon-dro-din'-e-ah) [chondro-; ὀδύνη, pain]. Pain in a cartilage.
**chondrodystrophia** (kon-dro-dis-tro'-fe-ah) [chondro-; δύς, bad; τρέφειν, to nourish]. Fetal rhachitis. **c. fœtalis**. See achondroplasia and achondroplasty.
**chondroepiphysis** (kon-dro-ep-if'-is-is) [chondro-; epiphysis]. A cartilage which later develops into a bony epiphysis.
**chondrofetal** (kon-dro-fe'-tal). Relating to fetal cartilage.
**chondrofibroma** (kon-dro-fi-bro'-mah) [chondro-; fibra, a fiber; ὄμα, a tumor: pl., chondrofibromata]. Chondroma with fibromatous elements.
**chondrofibromatous** (kon-dro-fi-bro'-mat-us) [chondro-; fibra, a fiber; ὄμα, a tumor]. Of the nature of chondrofibroma.
**chondrogen** (kon'-dro-jen) [chondro-; γεννᾶν, to produce]. 1. See chondrigen. 2. A substance found in fetal and early life, forming a part of the tissue of imperfectly developed cartilage.
**chondrogenesis** (kon-dro-jen'-es-is) [chondro-; γένεσις, formation]. The formation of cartilage.
**chondrogenetic** (kon-dro-jen-et'-ik) [chondro-; γένεσις, formation]. Forming cartilage; relating to chondrogenesis.
**chondrogenous** (kon-droj'-en-us) [chondro-; γένεσις, production]. Of the nature of chondrogen; producing cartilage.
**chondroglossus** (kon-dro-glos'-us). See muscles, table of.
**chondroglucose** (kon-dro-glu'-kōs) [chondro-; γλυκύς, sweet]. A glucose obtained from cartilage.
**chondrography** (kon-drog'-ra-fe) [chondro-; γράφειν, to write]. An anatomical description of the cartilages.
**chondroid** (kon'-droid) [chondro-; εἶδος, form]. Resembling cartilage.
**chondroitic acid** (kon-dro-it'-ik). A complex organic acid found in small quantities in normal urine. It is chondroitin-sulphuric acid, and is found in chondromucoid.
**chondrology** (kon-drol'-o-je) [chondro-; λόγος, science]. The science of cartilages.
**chondroma** (kon-dro'-mah) [chondro-; ὄμα, tumor]. A cartilaginous tumor.
**chondromalacia** (kon-dro-mal-a'-se-ah) [chondro-; μαλακία, softening]. Softening of a cartilage. **c. auris**. Same as hæmatoma auris.
**chondromatous** (kon-drōm'-at-us) [chondro-; ὄμα, a tumor]. Relating to or of the nature of cartilage.
**chondromucoid** (kon-dro-mū'-koid). A mucin found in cartilage. Cf. osseomucoid; tendomucoid.
**chondromyoma** (kon-dro-mi-o'-mah) [chondro-; myoma]. A neoplasm presenting the characteristics of both chondroma and myoma.
**chondromyxoma** (kon-dro-miks-o'-mah) [chondro-; μύξα, mucus; ὄμα, a tumor: pl., chondromyxomata]. A chondroma with myxomatous elements.
**chondrophyma** (kon-dro-fi'-mah) [chondro-; φῦμα, a growth]. 1. A tumor of a cartilage. 2. A neoplasm with cartilaginous elements. 3. See chondrophyte.
**chondrophyte** (kon'-dro-fīt) [chondro-; φυτόν, a plant]. A fungous neoplasm springing from a cartilage.
**chondroplast** (kon'-dro-plast). See chondroblast.
**chondroporosis** (kon-dro-por-o'-sis) [chondro-; πόρος, a passage]. The thinning of cartilage by the formation of spaces, occurring during the process of ossification.
**chondroprotein** (kon-dro-pro'-te-in). A protein occurring normally in cartilage.
**chondrosarcoma** (kon-dro-sar-ko'-mah) [chondro-; sarcoma]. A tumor composed of cartilaginous and sarcomatous tissue.
**chondrosarcomatous** (kon-dro-sar-kōm'-at-us). Relating to chondrosarcoma.
**chondrosidin** (kon-dros'-id-in). The hyalin obtained from chondrosin.
**chondrosin** (kon'-dro-sin) [chondrosia, a genus of sponges]. A hyalogen obtained from the sponge, chondrosia reniformis.
**chondrosis** (kon-dro'-sis) [chondrus]. 1. Formation of cartilage. 2. A cartilaginous tumor.
**chondrosteous** (kon-dros'-te-us) [chondro-; ὀστέον, bone]. In biology, having a cartilaginous skeleton.
**chondrosternal** (kon-dro-ster'-nal). Pertaining to the sternum and costal cartilages.
**chondrosyndesmus** (kon-dro-sin-dez'-mus). See synchondrosis.
**chondrotome** (kon'-dro-tōm) [chondro-; τέμνειν, to cut]. An instrument for cutting cartilage.
**chondrotomy** (kon-drot'-o-me) [see chondrotome]. The division of a cartilage.
**chondroxiphoid** (kon-dro-zif'-oid). Pertaining to the costal cartilages and the ensiform cartilage.
**chondrus** (kon'-drus) [χόνδρος, a grain; cartilage]. 1. Irish moss. The substance of the algæ C. crispus and C. mammillosus. These yield, on boiling with water, a soluble colloid consisting mainly of mucilage. This is demulcent and somewhat nutrient. Dose indefinite. 2. A cartilage; the ensiform cartilage.
**Chopart's amputation** (sho-par) [François Chopart, French surgeon, 1743–1795]. An amputation of the foot consisting of a disarticulation through the tarsal bones, leaving only the os calcis and the astragalus. **C.'s joint**, the mediotarsal articulation; the line of articulation which separates the astragalus and os calcis from the remaining tarsal bones.
**chord**. See cord.
**chorda** (kor'-dah) [L.]. A cord, tendon, or nervefilament. **chordæ arteriarum umbilicalium**, the lateral ligaments of the bladder. **c. dorsalis**. See notochord. **c. magna**, tendo Achillis. **c. obliqua**,

the oblique ligament of the superior radio-ulnar articulation. **c. saliva,** saliva produced by stimulation of the chorda tympani nerve. **c. spermatica,** the spermatic cord. **c. tendinea,** any one of the tendinous strings connecting the papillary muscles of the heart with the auriculoventricular valves. **c. tympani.** See under *nerve*. **c. umbilicalis,** the umbilical cord. **c. venæ umbilicalis,** the round ligament of the liver. **c. vertebralis.** See *notochord*. **c. vocalis,** a vocal band.

**chordal** (*kor'-dal*) [*chorda*, a cord]. Relating to a chorda, especially to the notochord.

**chordapsus** (*kor-dap'-sus*) [χορδή, a chord; ἅπτειν, to tie up]. Synonym of *acute intestinal catarrh*.

**chordee** (*kor-de'*) [*chorda*]. A painful curved erection of the penis with concavity downward. The corpus spongiosum being infiltrated from urethral inflammation, does not fill with blood during erection, and so acts like a bow-string.

**chorditis** (*kor-di'-tis*) [χορδή, a cord; ιτις, inflammation]. Inflammation of the vocal bands. **c. tuberosa,** a localized thickening on the vocal bands, often bilateral and situated at the junction of the posterior two-thirds with the anterior third.

**chordoma** (*kor-do'-mah*) [*chorda*, a cord]. 1. Virchow's name for the upper part of a persistent notochord. 2. A tumor composed of tissue of the same nature as that of the notochord. 3. A small tumor occurring in the median line of the clivus, near the articulation of the sphenoid with the occipital bone.

**chordoskeleton** (*kor-do-skel'-et-on*). The portion of the skeleton surrounding the notochord.

**chordotonal** (*kor-do-to'-nal*) [χορδή, chord; τόνος, tone]. In biology, applied to sense-organs or parts of arthropods that are responsive to sound-vibrations.

**chordurethritis** (*kor-dū-re-thri'-tis*). See *chordee*.

**chorea** (*ko-re'-ah*) [χορεία, dancing]. St. Vitus' dance. A functional nervous disorder, usually occurring in youth, characterized by irregular and involuntary action of the muscles of the extremities, face, etc., with general muscular weakness. Frequently a mitral systolic murmur is heard, often hemic, but in a large proportion of cases due to endocarditis, and there seems to be a close relation between the two diseases. Rheumatism often coexists. Chorea may be caused by a number of conditions, among which are fright and reflex irritation. It affects girls about three times as frequently as boys. Occasionally a form of chorea is seen in the adult, and may become a serious complication of pregnancy, resulting in the death of both fetus and mother. When it occurs late in life, it generally resists treatment. Syn., *chorea anglorum; chorea sancti Viti; epilepsia saltatoris; St. John's dance*. **c., bilateral,** that due to cerebral lesions causing development of choreic symptoms on both sides of the body. **c., buttonmaker's,** a form of chorea occurring in persons employed in making buttons. **c., cardiac,** a form marked by palpitation and other cardiac disorders. **c., chronic progressive,** Hoffmann's name for *Huntington's chorea*. **c., congenital.** Synonym of *birth-palsy*. **c. cordis.** See *c., cardiac*. **c.-corpuscles** (Elischer), peculiar cells found in the brain in cases of chorea and regarded as pathognomonic; they have, however, been found in the brains of those who have never had the disease. **c., dancing,** hysterical chorea marked by rhythmic dancing movements. **c.-demonomania,** epidemic chorea. **c., diaphragmatic,** spasm of the diaphragm. **c. dimidiata,** hemichorea. **c., electric.** See *Dubini's disease*. **c., epidemic.** See *choromania*. **c., essential,** that occurring independently and not as a symptom of some other disease. **c., facial,** convulsive tic. **c., false.** See *c., symptomatic*. **c., general,** a form of chorea in which all or almost all of the voluntary muscles are subject to irregular contractions. **c. gravidarum,** intractable chorea occurring during pregnancy, toward its close sometimes aggravated and attended with fever. **c. gravis,** severe and dangerous cases of chorea. **., habit-.** See *habit-spasm*. **c., hammering,** a form marked by coordinated rhythmic spasm in consequence of which persistent hammering with the fist upon some object will be indulged in. **c., hereditary.** See *Huntingdon's chorea*. **c., Huntingdon's.** See *Huntingdon's chorea*. **c., hysterical.** See *c. major*. **c., imaginative,** choromania. **c., imitative,** choreic movements developed in children from association with choreic subjects. **c., infantile.** See *c. minor*. **c. insaniens,** maniacal chorea; a grave form of chorea usually seen in women, and associated with mania, and generally ending fatally. It may develop during pregnancy. **c. laryngea, c., laryngeal, c. laryngis.** 1. A condition attended with clonic spasm of the laryngeal muscles and marked by inability to sustain coordinate action. 2. A condition marked by spasmodic motions of some of the muscles of expiration, causing a cry. **c., limp,** West's name for a sequel of motor paralysis in children marked by very slight choreic movements. **c. major,** a form of hysteria in which there are continual regular oscillatory movements. **c., maniacal.** See *c. insaniens*. **c., metaparalytic, c., methemiplegic.** See *c., postparalytic*. **c., methodic.** See *c. major*. **c., mimetic.** See *c., imitative*. **c. minor,** simple chorea. **c. mollis.** See *hemiplegia, choreic*. **c., Morvan's.** See *Morvan's chorea*. **c., neuralgica,** convulsive tic. **c. nutans,** that attended with nodding motions. **c. oculi,** choreic movements of the eyes due to cerebral lesions. **c. pandemica.** See *Dubini's disease*. **c., paralytic** (Gowers). See *hemiplegia, choreic*. **c., partial,** imperfect choreic movements associated with contractures, due to cerebral lesion. **c., posthemiplegic, c., postparalytic,** a form of involuntary movement seen in patients after an attack of hemiplegia. **c., prehemiplegic, c., prohemiplegic,** choreic spasms of the hands or feet forerunning hemiplegia. **c. procursiva.** Synonym of *paralysis agitans*. **c., rhythmic.** See *c. major*. **c., school-made,** chorea resulting from overstimulation of children at school. **c., secondary.** See *c., symptomatic*. **c., semilateralis,** hemichorea. **c. senilis.** 1. Paralysis agitans. 2. The trembling incident to age. **c., Sydenham's.** See *Sydenham's chorea*. **c., symptomatic,** that dependent upon some organic disease. **c., unilateral.** See *hemichorea*.

**choreal** (*ko-re'-al*). Pertaining to chorea; choreic.

**choreic** (*ko-re'-ik*) [*chorea*]. Relating to, of the nature of, or affected with chorea.

**choreiform** (*ko-re'-if-orm*) [*chorea; forma,* form]. Resembling chorea.

**choremania, choreomania** (*ko-re-ma'-ne-ah, ko-re-o-ma'-ne-ah*) [*chorea;* μανία, madness]. Synonym of *choromania*.

**choreoid** (*ko'-re-oid*) [χορεία, dancing; εἶδος, like]. Pertaining or similar to chorea.

**chorial** (*ko'-re-al*) [χόριον, skin]. Chorionic.

**chorioblastosis** (*ko-re-o-blas-to'-sis*) [χόριον, skin; βλαστάνειν, to germinate]. Any anomaly of growth of the corium and subcutaneous connective tissue.

**choriocapillaris** (*ko-re-o-kap-il-a'-ris*) [*chorion; capillus*, a hair]. The network of capillaries over the inner portion of the choroid coat of the eye.

**choriocele** (*ko'-re-o-sēl*) [χόριον, a skin; κήλη, hernia]. A hernial protrusion of the choroid coat of the eye.

**chorioepithelioma** (*ko-re-o-ep-e-the-le-o'-mah*). A neoplasm apparently due to excessive proliferation of chorionic epithelium; see *deciduoma*. **c. benignum,** degenerated relics of fetal epithelium or epiblast in the maternal tissues. **c. malignum,** Marchand's (1895) name for an epithelioma due to malignant degeneration of fetal epiblast left in the maternal tissues. Syn., *deciduoma malignum; sarcoma deciduocellulare; syncytioma malignum*.

**chorioid** (*ko'-re-oid*). See *choroid*.

**chorioidal** (*ko-re-oid'-al*). See *choroid* (2).

**chorioideal tubercle** (*ko-re-oid'-e-al tū'-ber-kl*). See *choroid tubercle*.

**chorioideremia** (*ko-re-oid-er-e'-me-ah*). See *choroideremia*.

**chorioiditis** (*ko-re-oid-i'-tis*). See *choroiditis*.

**chorioidoretinitis** (*ko-re-oid-o-ret-in-i'-tis*). See *choroidoretinitis*.

**chorioma** (*ko-re-o'-mah*) [*chorion;* ὄμα, tumor]. A neoplasm developed from the chorion. See *chorioepithelioma*.

**chorion** (*ko'-re-on*) [χόριον, skin; fetal membrane]. The outermost of the fetal membranes, formed from the vitelline membrane, the false amnion, and the allantois. The chorion lies between the amnion and the decidua reflexa and vera. **c., cystic degeneration of,** a myxoma of the chorion, producing the socalled "hydatid mole." It is characterized by rapid increase in the size of the uterus, hemorrhage, often profuse, beginning during the second month of pregnancy, and the discharge of small cysts, whitish in appearance, surrounded by bloody clots. These cysts vary in size from a pin-head to a filbert. **c. frondosum, c., shaggy,** the part covered by villi.

It helps to form the placenta. c. **læve**, the membranous portion of the chorion.
**chorionic** (*ko-re-on'-ik*) [*chorion*]. Relating to the chorion.
**chorionin** (*ko-re-on'-in*). A name given by Bronchacourt (1902) to a preparation made from sheep's placenta by submitting it to pressure without heat. The juice thus expressed is made palatable with syrup after being sterilized with ammonium fluoride; employed as a galactagogue.
**chorionitis** (*ko-re-on-i'-tis*). 1. See *scleroderma*. 2. Placentitis.
**chorioretinal** (*ko-re-o-ret'-in-al*). Pertaining to the choroid and retina.
**chorioretinitis** (*ko-re-o-ret-in-i'-tis*). Inflammation of the choroid and retina.
**chorioretinitis** (*ko-re-o-ret-in-i'-tis*). See *choroidoretinitis*.
**chorisis** (*ko-ri'-sis*) [χώρισις, a separation]. In biology, the development of two or more members when but one is expected; a doubling.
**choristoblastoma** (*ko-ris-to-blas-to'-mah*) [*choristoma*; *blastoma*]. An autonomous tumor originating in a choristoma.
**choristoma** (*ko-ris-to'-mah*) [χωρίστος, separated]. A tumor due to hyperplasia of an aberrant primordium.
**choroid** (*ko'-roid*) [*chorion*; εἶδος, likeness]. 1. The vascular tunic of the eye, continuous with the iris in front, and lying between the sclerotic and the retina. 2. Pertaining to the choroid; choroidal. c. **membrane**, the choroid. c. **plexus**, a vascular plexus in the ventricles of the brain. c. **tubercle**, a diagnostic sign of tuberculous meningitis found by ophthalmoscopic investigation.
**choroidal** (*ko-roid'-al*). Pertaining to the choroid.
**choroideremia** (*ko-roid-er-e'-me-ah*) [*choroid*; ἐρημία, desolation]. Absence of the choroid.
**choroiditis** (*ko-roid-i'-tis*) [*choroid*; ιρις, inflammation]. Inflammation of the choroid coat of the eye. It may be anterior, the foci of exudation being at the periphery of the choroid; or central, the exudate being in the region of the macula lutea; diffuse or disseminated, characterized by numerous round or irregular spots scattered over the fundus; exudative or nonsuppurative, when there are isolated foci of inflammation scattered over the choroid; metastatic, when due to embolism; and suppurative, when proceeding to suppuration. c., **areolar**, c. **areolaris**, that in which the first foci occur near the fovea and extend toward the periphery in constantly increasing distances. c. **guttata senilis**. See *Tay's choroiditis*. c. **serosa**. Synonym of *glaucoma*.
**choroidocyclitis** (*ko-roid-o-si-kli'-tis*) [*choroid*; κύκλος, a circle; ιρις, inflammation]. Inflammation of the choroid and of the ciliary body.
**choroidoiritis** (*ko-roid-o-i-ri'-tis*) [*choroid*; *iritis*]. Inflammation of the choroid and the iris.
**choroidoretinitis** (*ko-roid-o-ret-in-i'-tis*) [*choroid*; *retinitis*]. Choroiditis associated with retinitis. c., **ametropic**, that caused by ametropia.
**chorology** (*ko-rol'-o-je*) [χώρος, a place; ιρις, science]. The science of the geographic distribution of animals and plants.
**choromania** (*kor-o-ma'-ne-ah*) [χορός, a dance; μανία, madness]. A nervous disorder characterized by dancing or other rhythmic movements; epidemic chorea; dancing mania.
**choronosologia, choronosology** (*ko-ro-no-sol-o'-je-ah, ko-ro-no-sol'-o-je*) [χώρος, a region; νόσος, a disease; λόγος, science]. The science of the geographic distribution of diseases or of endemic diseases of some region.
**Christian Science**. An alleged system of therapy; a form of faith-cure; Eddyism.
**Christison's formula** (*kris'-tis-on*) [Sir Robert Christison, Scotch physician, 1797–1882]. A formula for estimating the amount of solids in the urine: multiply the last two figures of the specific gravity expressed in four figures by 2.33 (or by 2, Trapp; or by 2.2, *Loebisch*). This gives the amount of solids in every 1000 parts.
**chroatol** (*kro'-at-ol*), $C_{18}H_{18}.2HI$. A greenish-yellow, crystalline substance obtained by action of turpentine on iodine; used in powder or ointment in treatment of skin diseases. Syn., *terpiniodohydrate*.
**chroma-, chromato-** [χρῶμα, color]. Prefixes meaning colored.
**chromaffin cells** (*kro'-maf-fin*) [*chroma-*; *affinis*, akin to]. Cells that have an affinity for chromium, which makes them yellow; the term is applied to the adrenal system. Syn., *phaochrome, paraganglia*. c. **hormone, epinephrine**.
**chromate** (*kro'-māt*) [χρῶμα, color]. Any salt of chromic acid.
**chromatelopsia, chromatelopsis** (*kro-mat-e-lop'-se-ah, -sis*) [*chroma-*; ἀτελής, imperfect; ὄψις, vision]. Color-blindness.
**chromatic** (*kro-mat'-ik*) [χρῶμα, color]. Relating to or possessing color. c. **aberration**. See *aberration, chromatic*. c. **audition**, luminous sensations aroused by sound.
**chromatidrosis**. See *chromidrosis*.
**chromatin** (*kro'-mat-in*) [see *chromatic*]. The portion of the protoplasm of a cell that takes the stain, forming a delicate reticular network or plexus of fibrils permeating the achromatin of a cell. Syn., *karyomitome*.
**chromatism** (*kro'-mat-ism*) [χρωματισμός, coloring]. 1. Abnormal coloration of any tissue. 2. Chromatic aberration.
**chromatoblast** (*kro-mat'-o-blast*) [*chromato-*; βλαστός, a germ]. Same as *chromatophore*.
**chromatodermatosis**. See *chromodermatosis*.
**chromatodysopia** (*kro-mat-o-dis-o'-pe-ah*) [*chromato-*; δύς, ill; ὄψις, vision]. Color-blindness.
**chromatogenous** (*kro-mat-oj'-en-us*) [*chromato-*; γεννᾶν, to beget]. Producing color.
**chromatology** (*kro-mat-ol'-o-je*) [*chromato-*; λόγος, science]. The science of colors. Also the spectroscopic investigation of colors.
**chromatolysis** (*kro-mat-ol'is-is*) [*shromato ?* λύσις, a loosing]. 1. Flemming's term for the breakingdown of the nucleus at the death of the cell. Syn., *karyolysis*. 2. The disintegration and disappearance of the Nissl granules from nerve-cells.
**chromatometer** (*kro-mat-om'-et-ur*) [*chromato-*; μέτρον, a measure]. An instrument for measuring color-perception or the intensity of colors.
**chromatometry** (*kro-mat-om'-et-re*). 1. See *chromatoptometry*. 2. The estimation of the coloring power of a substance.
**chromatopathia** (*kro-mat-o-path'-e-ah*) [*chromato-*; πάθος, disease]. Any pigmentary skin-disease; a chromatosis.
**chromatopathy** (*kro-mat-op'-a-the*). See *chromatopathia*.
**chromatophile** (*kro-mat'-o-fil*). Same as *chromophilous*.
**chromatophobia** (*kro-mat-o-fo'-be-ah*) [*chromato-*; φόβος, dread]. Abnormal fear of colors.
**chromatophore** (*kro-mat'-o-for*) [*chromato-*; φόρος, bearing]. Any colored cell-plastid.
**chromatophorous** (*kro-mat-off'-or-us*) [*chromato-*; φέρειν, to bear]. Containing pigment or pigment-cells.
**chromatoplasm** (*kro-mat'-o-plazm*) [*chromato-*; πλάσμα, anything formed]. The substance of the chromatoplasts as distinguished from the other cell-substances, karyoplasm, cytoplasm, metaplasm, paraplasm, etc.
**chromatoplast** (*kro-mat'-o-plast*). See *chromatophore*.
**chromatopseudopsis** (*kro-mat-o-sū-dop'-sis*) [*chromato-*; ψευδής, false; ὄψις, sight]. Color-blindness.
**chromatopsia** (*kro-mat-op'-se-ah*) [*chromato-*; ὄψις, vision]. A disorder of vision in which color-impressions arise subjectively. It may be due to disturbance of the optic centers, or to drugs, especially santonin.
**chromatopsy** (*kro'-mat-op-se*). See *chromatopsia*.
**chromatoptometry** (*kro-mat-op-tom'-et-re*) [*chromato-*; ὄψις, vision; μέτρον, a measure]. The testing of the sensibility of the eye with respect to color-perception.
**chromatoscope** (*kro-mat'-o-skōp*) [*chromato-*; σκοπεῖν, to observe]. An instrument for determining the refractive index of colored light.
**chromatoscopy** (*kro-mat-os'-ko-pe*) [*chromato-*; σκοπεῖν, to observe]. The determination of the color of objects.
**chromatosis** (*kro-mat-o'-sis*) [χρῶμα, color]. Pigmentation; a pathological process or pigmentary disease consisting in a deposit of coloring-matter in a locality where it is usually not present, or in excessive quantity in regions where pigment normally exists.
**chromatoskiameter** (*kro-mat-o-ski-am'-et-er*) [*chromato-*; σκιά, a shadow; μέτρον, a measure]. Holmgren's apparatus for testing color-sense, consisting

of a lamp and a white screen, on which is cast the shadow of a pencil placed in front of different colored glasses. A scale indicates when the shadows are of equal brightness.

**chromatosome** (kro-mat'-o-sōm) [chromato-; σῶμα, body]. In biology, the "nuclear rods" of the nucleus.

**chromaturia** (kro-mat-ū'-re-ah) [chromato-; οὖρον, urine]. Abnormal coloration of the urine.

**chrome** (krōm). See *chromium*. **c.-alum**, CrK₂(SO₄)₂+12H₂O. A compound of chromium and potassium sulphate, forming large, dark-violet crystals soluble in five parts of water, the solution turning green when heated; used as a pigment.

**chromesthesia** (krom-es-the'-ze-ah) [χρῶμα, color; αἴσθησις, perception by the senses]. The association of colors with words, letters, and sounds.

**chromhidrosis** (krōm-hid-ro'-sis). Same as *chromidrosis*.

**chromic** (kro'-mik) [chromium]. Pertaining to or made from chromium. **c. acid, c. anhydride**, chromium trioxide.

**chromicize** (kro'-mis-īz). To impregnate with chromic acid or a chromium salt.

**chromidium** (kro-mid'-e-um). Any one of the granules of nuclear substance found in the cytoplasm.

**chromidrosis** (kro-mid-ro'-sis) [χρῶμα, color; ἱδρώς, sweat]. A rare condition of the sweat in which it is variously colored, being bluish, blackish, reddish, greenish, or yellowish. *Black sweat* (*seborrhœa nigricans*) occurs usually in hysterical women, the face being most often affected. It is associated with chronic constipation and is due to the presence of indican in the sweat. *Red sweat* (*hematidrosis*) may be due to an exudation of blood into the sweat-glands, or to the presence of a microorganism in the sweat.

**chromism** (kro'-mizm). Excessive or abnormal coloration.

**chromite** (kro'-mīt). 1. A combination of chromium sesquioxide with the oxide of some other metal. 2. Any organic pigment.

**chromium** (kro'-me-um) [χρῶμα, color]. Cr=52.0; quantivalence II and IV. One of the elements of the iron group. The various salts of chromium, especially the derivatives of chromium trioxide, CrO₃, are much used in the manufacture of pigments and as a caustic. All are poisonous. **c. and potassium sulphate.** See *chrome-alum*. **c. sesquioxide**, Cr₂O₃, a green pigment occurring in nature as chrome-ochre; it can be prepared artificially. **c. trioxide.** See *acid, chromic*.

**chromo-** (kro-mo-) [χρῶμα, color]. A prefix meaning colored.

**chromoaromatic** (kro-mo-ar-o-mat'-ik). Applied to microorganisms that are colored and aromatic.

**chromoblast** (kro'-mo-blast). See *chromatophore*.

**chromocrinia** (kro-mo-krin'-e-ah) [chromo-; κρίνειν, to separate]. The secretion of coloring-matter, as in the sweat, etc.

**chromocyte** (kro'-mo-sīt) [chromo-; κύτος, a cell]. Any colored cell.

**chromocytometer** (kro-mo-si-tom'-et-er) [chromocyte; μέτρον, a measure]. An instrument for estimating the proportion of hemoglobin present in the blood.

**chromocytometry** (kro-mo-si-tom'-et-re) [chromo-; κύτος, cell; μέτρον, a measure]. The estimation of hemoglobin by means of the chromocytometer.

**chromocystoscopy** (kro-mo-sis-tos'-ko-pe) [chromo-; *cystoscopy*]. Cystoscopy and inspection of the orifices of the ureters after the administration of a substance that will stain the urine.

**chromodermatosis** (kro-mo-der-mat-o'-sis) [chromo-; *dermatosis*]. A skin disease characterized by discoloration of the surface.

**chromodiagnosis** (kro-mo-di-ag-no'-sis) [chromo-; *diagnosis*]. Sicard's term (1901) to designate the diagnosis of hemorrhages of the neuraxis by yellow discoloration of the cephalorhachidian fluid (xanthochromia), due, according to Tuffier, to the normal lutein of the serum.

**chromogen** (kro'-mo-jen) [chromo-; γεννᾶν, to produce]. Any principle of the animal or vegetable economy which is susceptible, under suitable circumstances, of being changed into a coloring-matter.

**chromogenesis** (kro-mo-jen'-es-is) [chromo-; γεννᾶν, to produce]. The production of pigments or coloring matter, as by bacterial action.

**chromogenic** (kro-mo-jen'-ik) [see *chromogen*]. 1. Producing color or pigment; applied generally to pigment-producing bacteria. 2. Relating to chromogen.

**chromolume** (kro'-mo-lūm) [chromo-; *lumen*, light]. An apparatus for the production of colored rays for therapeutic purposes. **c., electro-arc**, a special device for solarization by means of the arc light.

**chromolysis** (kro-mol'-is-is). Same as *chromatolysis*.

**chromomere** (kro'-mo-mēr) [chromo-; μέρος, a part]. One of the minute granules composing the chromosomes. Syn., *id*.

**chromometer** (kro-mom'-et-ur). See *chromatometer* (2).

**chromometry** (kro-mom'-et-re). See *chromatometry* (2).

**chromoparic, chromoparous** (kro-mo-par'-ik, kro-mop'-ar-us) [chromo-; *parere*, to furnish]. Excreting a colored transformation-product which is diffused out upon and into the surrounding medium. It is said of certain bacteria.

**chromophane** (kro'-mo-fan) [chromo-; φαίνειν, to appear]. The pigment of the inner segments of the retinal cones of certain animals. There are at least three varieties, chlorophane, rhodophane, xanthophane.

**chromophile** (kro'-mo-fil). Same as *chromophilous*.

**chromophilous** (kro-mof'-il-us) [chromo-; φίλος, loving]. Readily stained; easily absorbing color.

**chromophobic** (kro'-mo-fo-bik) [chromo-; φόβος, fear]. Not stainable; not readily absorbing color. Cf. *chromophilous*.

**chromophor** (kro'-mo-for) [chromo-; φέρειν, to bear]. 1. Those chromogenic bacteria that possess pigment as an integral part of their organism. 2. The chemical group in an anilin dye which gives the color.

**chromophoric, chromophorous** (kro-mo-for'-ik, kro-mof'-or-us). Applied to chromogenic bacteria in which the pigment is stored in the cell-protoplasm of the organism.

**chromophose** (kro'-mo-fōz) [chromo-; φῶς, light]. A subjective sensation of color. See *phose*.

**chromophyl** (kro'-mo-fil) [chromo-; φύλλον, a leaf]. A comprehensive term for the coloring-matter of plant cells, including chlorophyl, xanthophyl, erythrophyl, cyanophyl, phycophyl, the various phycochromes, etc.

**chromophytosis** (kro-mo-fi-to'-sis) [chromo-; φυτόν, a plant]. 1. Any microscopic plant-growth that produces a discoloration of the skin in which it grows; any pigmentary skin disease caused by a vegetable parasite. 2. A synonym of *tinea versicolor* or *Eichstedt's disease*.

**chromoplasm** (kro'-mo-plazm) [chromo-; πλάσμα, anything formed]. The network of a nucleus, so called because it stains readily.

**chromoplastid, or chromoplastidule** (kro-mo-plas'-tid, or kro-mo-plas'-tid-ūl) [chromo-; πλάσσειν, to form]. A pigment-granule imbedded in the protoplasm of a plant or animal. It is also called chromoleucite.

**chromoprotein** (kro-mo-pro'-te-in) [chromo-; *protein*]. Any proteid capable of being broken up into albumin and a coloring-matter.

**chromopsia** (kro-mop'-se-ah). See *chromatopsia*.

**chromoptometer** (kro-mop-tom'-et-er) [chromo-; μέτρον, a measure]. A contrivance for determining the extent of development of color-vision.

**chromoradiometer** (kro-mo-ra-de-om'-et-ur) [chromo-; *radius*, ray; μέτρον measure]. An instrument for measuring the penetrative power of the Roentgen rays.

**chromorhinorrhea** (kro-mo-rin-or-e'-ah) [chromo-; ῥίς, nose; ῥοία, a flow]. The discharge of a colored secretion from the nose.

**chromoscope** (kro'-mo-skōp). See *chromatoscope*.

**chromosochromic** (kro-mo-so-kro'-mik). Containing chromium both as a trivalent and as a bivalent radical.

**chromosome** (kro'-mo-sōm) [chromo-; σῶμα, body]. A chromatin-fiber formed during karyokinesis. Syn., *karyomita*. **c., bivalent**, one representing two chromosomes joined end to end. Cf. *c.s, pseudoreduction of*. **c. granules**, granules of lymphocytes seen in the *in vitro* examination of living cells. **c., plurivalent**, one having the value of two or more chromosomes. Cf. *c.s, pseudoreduction of*. **c.s, pseudoreduction of**, apparent reduction of the number of chromosomes through increase of bivalent or

plurivalent chromosomes. c.s, **reduction of,** the halving of the number of chromosomes in the germ-nuclei during maturation.
**chromosot** (kro′-mo-sŏt). A disinfectant said to consist mainly of sodium sulphate and sodium sulphite.
**chromospermism** (kro-mo-sperm′-izm) [chromo-; σπέρμα, seed]. A condition in which the semen is colored. Cf. *cyanospermia*.
**chromostroboscope** (kro-mo-stro′-bo-skŏp) [chromo-; στρόβος, a twisting; σκοπεῖν, to inspect]. A device for showing the persistence of visual impressions of color.
**chromotherapy** (kro-mo-ther′-ap-e) [chromo-; θεραπεία, treatment]. The treatment of disease by colored light.
**chromoureteroscopy** (kro-mo-ū-ret-er-os′-ko-pe). See *chromocystoscopy*.
**chromule** (kro′-mūl) [chromo-; ὕλη, matter]. Coloring-matter in plants, especially when not green, or when liquid.
**chronic** (kron′-ik) [χρόνος, time]. Long-continued; of long duration; opposed to acute.
**chronicity** (kron-is′-it-e) [chronic]. The state of being chronic or long-continued.
**chronograph** (kron′-o-graf) [χρόνος, time; γράφειν, to write]. An instrument for graphically recording intervals of time in physiological and psychophysical experiments.
**chronoscope** (kron′-o-skŏp) [χρόνος, time; σκοπεῖν, to inspect]. An instrument for measuring extremely short intervals of time. c., **A-form,** an apparatus introduced by Galton for measuring the time of certain psycho-physical reactions. It is so called from its outline, which somewhat resembles that of the letter A.
**chronothermal** (kron-o-ther′-mal) [χρόνος, time; θέρμη, heat]. Pertaining to the theory that all diseases are characterized by periods of intermitting chill and heat; relating to periodicity in changes of bodily temperature.
**chronotropic** (kron-o-trop′-ik) [χρόνος, time; τρέπειν, to turn]. Pertaining to influences which modify the rate of a periodically recurring phenomenon (heart-beat).
**chrotoplast** (kro′-to-plast) [χρώς, skin; πλασσεῖν, to form]. A skin-cell; a dermal or epithelial cell.
**chrotopsia, chrupsia** (kro-top′-se-ah, krup′-se-ah). See *chromatopsia*.
**chrysarobin, chrysarobinum** (kris-ar-o′-bin, kris-ar-o-bi′-num) [χρυσός, gold; araroba (nat. East Ind.), bark of a leguminous tree], C₃₀H₂₆O₇. A reduction-product of chrysophanic acid. It occurs in Goa powder and araroba powder. It is a yellow-colored powder, the product of the decay of *Vouacapoua araroba*, a Brazilian tree. It is a gastrointestinal irritant; locally and internally it is useful in psoriasis, but stains the skin a dark yellowish-brown color. Dose, internally, ⅛–¼ gr. (0.008–0.032 Gm.). c. **ointment** (unguentum chrysarobini, U. S. P.), contains 10 % of the drug with 90 % benzoinated lard. c. **oxide,** a brownish-black powder obtained from chrysarobin in boiling water by the action of sodium peroxide. It is recommended in treatment of eczema and acne rosacea in 5 to 10 % solution.
**chrysoidin** (kris-oi′-din) [χρυσός, gold; εἶδος, like], C₁₂H₁₃N₄Cl. A coal-tar color used in dyeing. It is the hydrochloride of diamidoazobenzene. It consists of dark-violet crystals soluble in water. It dyes bright-yellow on silk and cotton. 2. C₇H₂₂O₄. A yellow coloring-substance found in asparagus berries.
**chrysolein** (kris-o′-le-in). Sodium fluoride.
**chrysophan** (kris′-o-fan) [χρυσός, gold; φαίνειν, to show], C₁₄H₁₈O₄. A glucoside found in rhubarb.
**chrysophanic acid** (kris-o-fan′-ik). See *acid, chrysophanic*.
**chrysophyl** (kris′-o-fĭl) [χρυσός, gold; φύλλον, a leaf]. Xanthophyl; a bright golden-yellow crystalline pigment derived from leaves.
**chrysoretin, chrysorrhetin** (kris-o-ret′-in). A yellow pigment found in senna and identical with chrysophan.
**chrysotoxin** (kris-o-toks′-in). See *sphacelotoxin*.
**chthonophagia,** or **chthonophagy** (thon-o-fa′-je-ah, or thon-off′-a-je) [χθών, earth; φαγεῖν, to eat]. Dirt-eating; geophagy.
**chuchuarine** (chu-chu′-ar-in) [*Chuchuara*, Indian name], C₂₀H₁₆N₁₈O₅. An alkaloid obtained from the seeds and wood of *Semecarpus anacardium*, an anacardiaceous aphrodisiac plant of the East Indies.

It is extremely poisonous, acting somewhat like strychnine.
**churning sound** (churn′-ing). A peculiar splashing-sound like that made by a churn, heard in the chest in some cases of pleural effusion.
**Chvostek's symptom** (vos′-tek) [Franz Chvostek, Austrian surgeon, 1835–1884]. Increase of the mechanical irritability of the motor nerves, especially the facial, in post operative tetany. A sudden local spasm is elicited by a slight tap on one side of the face. See also *Weiss' sign*.
**chylaceous** (ki-la′-se-us). Composed of chyle.
**chylangioma** (ki-lan-je-ŏ′-mah) [chyle; ἀγγεῖον, a vessel]. 1. Retention of chyle in lymphatic vessels with dilatation of the latter. 2. A tumor of lymph-vessels containing chyle.
**chylaqueous** (ki-la′-kwe-us) [chyle; aqua, water]. Like water and chyle. c. **fluid,** the digested food of nutritive fluid in the somatic or perigastric cavity of invertebrates; it is never enclosed in distinct vessels and represents the blood of higher animals.
**chyle** (kīl) [χυλός, juice]. The milk-white fluid absorbed by the lacteals during digestion. On standing, it separates into a thin, jelly-like clot and a substance identical with serum. c.-**corpuscle,** any floating cell of the chyle. These cells resemble, and are probably identical with, the colorless blood-corpuscles. c., **granular,** c., **molecular base of,** the minute particles of fat which give the milky appearance to chyle.
**chylemia** (ki-le′-me-ah) [chyle; αἷμα, blood]. The presence of chyle in the blood.
**chylidrosis** (ki-lid-ro′-sis) [chyle; ἵδρωσις, a sweating]. Milkiness of the sweat.
**chylifaction** (ki-le-fak′-shun) [chyle; facere, to make]. The forming of chyle from food.
**chylifactive** (ki-le-fak′-tiv) [chyle; facere, to make]. Chyle-forming.
**chyliferous** (ki-lif′-er-us) [chyle; ferre, to carry]. Containing or carrying chyle.
**chylific** (ki-lif′-ik) [chyle; facere, to make]. Making chyle; pertaining to chylifaction.
**chylification** (ki-le-fik-a′-shun) [chyle; facere, to make]. The process by which chyle is formed, separated, and absorbed by the villi of the small intestine.
**chylificatory** (ki-lif′-ik-at-o-ré) [chyle; facere, to make]. Chyle-making.
**chylivorous** (ki-liv′-or-us) [chyle; vorare, to devour]. Applied to parasitic organisms subsisting on chyle.
**chylocele** (ki′-lo-sēl) [chyle; κήλη, a tumor]. An effusion of chyle into the tunica vaginalis testis. c. **parasitic.** See *filaria sanguinis hominis*.
**chylocyst** (ki′-lo-sist) [chyle; κύστις, bladder]. The chyle-bladder; the reservoir of Pecquet.
**chylocystic** (ki-lo-sis′-tik) [chyle; κύστις, bladder]. Relating to the chylocyst.
**chyloderma** (ki-lo-der′-mah) [chyle; δέρμα, skin]. Scrotal elephantiasis, with accumulation of lymph in the thickened skin and in the enlarged lymphatic vessels; lymph-scrotum.
**chylodochium** (ki-lo-do′-ke-um) [chyle; δοχεῖον, receptacle]. The receptaculum chyli.
**chylogaster** (ki-lo-gas′-ter) [chyle; γαστήρ, the stomach]. The duodenum, so-called because of its being the chief seat of chylous digestion.
**chylogastric** (ki-lo-gas′-trik). Pertaining to the chylogaster.
**chylopericardium** (ki-lo-per-ik-ar′-de-um) [chyle; περί, around; καρδία, the heart]. A rare condition, in which chyle is present in the pericardium, as a consequence of the formation of a channel of communication between a chyle-duct and the cavity of the heart-sac.
**chyloperitoneum,** (ki-lo-per-it-on-e′-um). A condition marked by an effusion of chyle in the peritoneum.
**chylopoiesis** (ki-lo-poi-e′-sis) [chyle; ποιεῖν, to make]. Chylification.
**chylopoietic** (ki-lo-poi-et′-ik) [see *chylopoiesis*]. Making or forming chyle.
**chyloptyalism** (ki-lop-ti′-al-izm) [chyle; πτυαλεῖν, to spit]. Milkiness of the saliva.
**chylorrhea** (ki-lor-e′-ah) [chyle; ῥεῖν, to flow]. An excessive flow of chyle; also, a diarrhea characterized by a milky color of the feces.
**chylosis** (ki-lo′-sis) [χύλωσις, a converting into juice]. Chylification.
**chylothorax** (ki-lo-tho′-raks) [chyle; θώραξ, the chest]. The presence of chyle in the pleural cavity,

**chylous** (*ki'-lus*) [*chyle*]. Relating to or resembling chyle.

**chyluria** (*ki-lū'-re-ah*) [*chyle*; οὖρον, urine]. The passage of chyle in the urine. It is thought to be caused by a disordered condition of the lacteals, and is also connected with the presence in the blood of *Filaria sanguinis hominis*, which blocks up the lymph-channels.

**chyme** (*kīm*) [χυμός, chyme]. Food that has undergone gastric digestion and has not yet been acted upon by the biliary, pancreatic, and intestinal secretions.

**chymiferous** (*ki-mif'-er-us*) [*chyme; ferre,* to bear]. Capable of producing chyme.

**chymification** (*ki-me-fik-a'-shun*) [*chyme; facere,* to make]. The change of food into chyme by the digestive process.

**chymorrhea** (*kim-or-e'-ah*) [*chyme;* ῥεῖν, to flow]. A discharge of chyme.

**chymosin** (*ki'-mo-sin*) [*chyme*]. Rennin; the rennet ferment.

**chymosinogen** (*ki-mo-sin'-o-jen*). The antecedent body from which chymosin is developed.

**Ciaglinski's sensory tract** (*se-a-glin'-ske*). A tract of ascending fibers in the posterior gray commissure of the thoracic part of the spinal cord.

**Ciamician and Magnanini's test for skatol.** Skatol warmed with sulphuric acid produces a purple-red color.

**cibarian** (*sib-a'-re-an*) [*cibus,* food]. Relating to food and the organs concerned in mastication and deglutition.

**cibarious** (*sib-a'-re-us*) [*cibus,* food]. Serving as food; nutritious; edible.

**cibation** (*si-ba'-shun*) [*cibus,* food]. 1. The act of receiving nourishment. 2. The process of condensing a liquid.

**cibisotome** (*si-bis'-ot-ōm*) [κίβισις, pouch; τομή, cut]. An instrument for opening the capsule of the lens.

**cibophobia** (*si-bo-fo'-be-ah*) [*cibus,* food; φόβος, fear]. Morbid aversion to food.

**cicatricial** (*sik-at-rish'-al*) [*cicatrix*]. Pertaining to or of the nature of a cicatrix. **c. deformities,** abnormal contractions caused by cicatrices. **c. tissue,** a form of dense connective tissue seen in cicatrices.

**cicatricose, cicatrisate** (*sik-at'-re-kōs, sik-at'-riz-āt*). Marked with cicatrices or cicatricial impressions.

**cicatrix** (*sik-a'-triks*) [L.: *pl., cicatrices*]. A scar. The connective tissue which replaces a localized loss of substance. Its color is usually whitish and glistening when old, red or purple when newly developed. **c., exuberant, c., hypertrophic, c., keloid,** one that hypertrophies after the healing of a wound and becomes red and prominent. **c., vicious,** one that impairs the function of a part.

**cicatrizant** (*sik'-at-ri-zant*) [*cicatrix,* a scar]. 1. Tending to cicatrize or heal. 2. A medicine that aids the formation of a cicatrix.

**cicatrization** (*sik-at-riz-a'-shun*) [*cicatrix*]. The process of healing of a wound.

**cicatrize** (*sik'-at-riz*). To heal.

**cicatrose** (*sik'-at-rōs*). See *cicatricose*.

**Cichorium** (*sik-o'-re-um*) [κιχώριον]. A genus of plants of the order Compositæ. *C. intybus,* chicory, succory, bunk, is a hardy perennial of Europe; the root of the wild plant is said to be a powerful alterative. **C. glucoside,** C₂₁H₃₄O₁₉+4½H₂O (?), a bitter glucoside obtained from the flowers of *C. intybus*.

**Cicuta** (*sik-u'-tah*) [L., "hemlock"]. A genus of umbelliferous plants. *C. virosa* is a poisonous species of northern Europe. It is never used internally, but has been applied externally in rheumatism.

**cicutine** (*sik'-ū-tēn*) [*cicuta,* hemlock]. 1. An alkaloid obtained from *Cicuta virosa*. 2. The same as coniine.

**cicutism** (*sik'-ū-tizm*) [*Cicuta*]. Poisoning with water-hemlock, *Cicuta virosa*. It is marked by epileptiform convulsions, dilatation of the pupils, cyanosis of the face, and coma.

**cicutoxin** (*sik-ū-toks'-in*) [*cicuta,* hemlock; τοξικόν, poison]. The poisonous active principle of *Cicuta virosa*. It is a viscid, non-crystallizable liquid of unpleasant taste and acid reaction.

**cilia** (*sil'-e-ah*) [pl. of *cilium,* the eyelid or eyelash]. 1. The eyelashes. 2. The locomotor and prehensile organs of certain microorganisms. 3. The hair-like appendages of certain epithelial cells, the function of which is to propel fluid or particles.

**ciliariscope** (*sil-e-ar'-is-kōp*) [*cilium,* eyelid; σκοπεῖν, to look at]. An instrument (essentially a prism) for examining the ciliary region of the eye.

**ciliarotomy** (*sil-e-ar-ot'-o-me*) [*cilia;* τέμνειν, to cut]. Surgical section of the ciliary zone for glaucoma.

**ciliary** (*sil'-e-a-re*) [*cilia*]. 1. Pertaining to the eyelid or eyelash. 2. Relating to ciliary movement. 3. Pertaining to the ciliary apparatus. **c. apparatus,** the structure related to the mechanism of accommodation. **c. arteries,** anterior, posterior long, and posterior short, branches of the ophthalmic artery, supplying the recti muscles, the ciliary apparatus, and the posterior structures of the eye, with the exception of the retina. **c. body,** the ciliary muscle and processes. **c. canal,** the canal of Fontana. **c., ganglion,** the ganglion at the apex of the orbit, supplying the ciliary muscle and iris. **c. ligament.** See *ligamentum pectinatum*. **c. movement,** movement by means of vibratory cilia. **c. muscle,** the muscle of accommodation, the contraction of which lessens the tension upon the suspensory ligament of the lens. **c. nerves,** branches of the ophthalmic ganglion supplying the anterior structures of the eyeball and the accommodative apparatus. **c. neuralgia,** neuralgic pain of the eye, brow, temple, etc. **c. processes,** circularly arranged choroid foldings continuous with the iris in front. **c. region,** the pericorneal or "danger zone," corresponding to the position of the ciliary body. **c. zone,** the ciliary processes collectively.

**ciliata** (*sil-e-ah'-tah*). A class of protozoa characterized by the presence of cilia.

**ciliated** (*sil'-e-a-ted*) [*cilia*]. 1. Having cilia. 2. Of bacteria, having fine hair-like processes, like cilia.

**ciliation** (*sil-e-a'-shun*) [*cilium,* an eyelash]. The condition of having cilia.

**ciliospinal** (*sil-e-o-spi'-nal*) [*cilia; spina,* the spine]. Relating to the ciliary zone and the spine. **c. center,** See *center, ciliospinal*.

**cilium** (*sil'-e-um*). See *cilia*.

**cillo** (*sil'-o*), or **cillosis** (*sil-o'-sis*) [*cilium,* an eyelash]. A spasmodic trembling of the eyelid.

**cillotic** (*sil-ot'-ik*) [*cilium,* an eyelash]. Pertaining to or affected with cillo.

**cimbia** (*sim'-be-ah*) [L.]. The white band seen upon the ventral aspect of the crus cerebri, the tractus pedunculi transversus of Gudden.

**cimbial** (*sim'-be-al*) [*cimbia,* a cincture]. Relating to the cimbia.

**cimex** (*si'-meks*) [L., "a bug"]. A genus of hemipterous insects. **c. lectularius,** the common bedbug.

**cimicifuga** (*sim-is-if'-ū-gah*) [*cimex; fugare,* to drive away]. Black snakeroot; black cohosh. The root of *C. racemosa,* ord. Ranunculaceæ. A stomachic, antispasmodic, aphrodisiac, expectorant, and diuretic. Its action on the heart is similar to that of digitalis. It has been used in cardiac diseases, functional impotence, chorea, and ovarian neuralgia. **c., extract of** (*extractum cimicifugæ,* U. S. P.). Dose 4 gr. (0.25 Gm.). **c., fluidextract of** (*fluidextractum cimicifugæ,* U. S. P.). Dose 5–30 min. (0.32–2.0 Cc.). **c., liquid extract of** (*extractum cimicifugæ liquidum,* B. P.). Dose 3–30 min. (0.2–2.0 Cc.). **c., tincture of** (*tinctura cimicifugæ,* U. S. P.) (20 % strength). Dose 15 min.–1 dr. (1–4 Cc.).

**cimicifugin** (*sim-is-if'-u-jin*) [*cimex,* a bug; *fugare,* to drive away]. The precipitate from a tincture of the root of *Cimicifuga racemosa*; it is an antispasmodic, diaphoretic, nervine, emmenagogue, parturient, and narcotic. Dose 1 to 2 grains.

**cimmol** (*sim'-ol*). See *aldehyde, cinnamic*.

**cina** (*si'-nah*) [L.]. The plant *Artemisia santonica*. See *santonica*.

**cinchamidine** (*sin-kam'-id-ēn*) [*cinchona;* amidin]. C₁₉H₂₄N₂O. An alkaloid found in the mother-liquor from which cinchonidine has been extracted.

**cincholine** (*sin'-ko-lēn*) [*cinchona*]. A pale yellow liquid alkaloid isolated from the mother-liquors of quinine.

**cinchona** (*sin-ko'-nah*) [Countess *Chinchon,* 17th century]. Peruvian bark. The bark of several varieties of cinchona, a tree native to the eastern slopes of the Andes, the most valuable being *C. calisaya*. Other varieties are *C. condaminea,* pale bark, *C. pitayensis,* Pitayo bark, and *C. micrantha*. Cinchona bark contains 21 alkaloids, of which four —quinine, cinchonine, quinidine, and cinchonidine— are the most important. Cinchona has the same

**physiological** action and therapeutic uses as its chief alkaloid, quinine. See *quinine*. It is also an astringent, bitter, and stomachic tonic, stimulating appetite and promoting digestion, beneficial in atonic dyspepsia and adynamia, but especially useful in malarial affections. **c.**, decoction of (*decoctum cinchonæ*, B. P.). Dose 1–2 oz. (30–60 Cc.). **c., fluidextract of** (*fluidextractum cinchonæ*, U. S. P.). Dose 10 min.–1 dr. (0.65–4.0 Cc.). **c., infusion of** (*infusum cinchonæ*, B. P.). Dose 1 dr.–1 oz. (4–30 Cc.). **c., liquid extract of** (*extractum cinchonæ liquidum*, B. P.). Dose 5–10 min. (0.3–0.6 Cc.). **c., red** (*cinchona rubra*, U. S. P.), the dried bark of *C. succirubra*. Dose 15 gr. (1 Gm.). **c., tincture of** (*tinctura cinchonæ*, U. S. P.) (20 % of bark). Dose ½–2 dr. (2–8 Cc.). **c., tincture of, compound** (*tinctura cinchonæ composita*, U. S. P.), Huxham's tincture. Dose 1 dr.–½ oz. (4–16 Cc.).

**cinchonamine** (sin-kon'-am-ēn) [*cinchona; amine*], $C_{19}H_{24}N_2O$. An alkaloid of cuprea bark. It occurs in glistening, colorless crystals that are nearly insoluble in water and but slightly soluble in ether.

**cinchonate** (sin'-ko-nāt). A salt of cinchonic acid.

**cinchonicine** (sin-kon'-is-ēn) [*cinchonin*], $C_{19}H_{22}N_2O$. An artificial alkaloid derived from cinchonine.

**cinchonidine** (sin-kon'-id-ēn), $C_{19}H_{22}N_2O$. An alkaloid derived from cinchona. It is a crystalline substance resembling quinine in general properties. **c. bisulphate, c. disulphate**, $C_{19}H_{22}N_2O$. $H_2SO_4$ $+5H_2O$, prisms soluble in water and alcohol; antiperiodic. Dose 15–30 gr. (0.97–1.94 Gm.). Syn., *acid cinchonidine sulphate*. **c. salicylate**, has decided antiperiodic properties. **c. sulphate** (*cinchonidinæ sulphas*, U. S. P.), $(C_{19}H_{22}N_2O)_2 . H_2SO_4+3H_2O$, less bitter than quinine, and valuable as an antipyretic. Dose 1–20 gr. (0.065–1.3 Gm.) or more. **c. tannate**, a tasteless, yellow, amorphous powder, soluble in alcohol; it is used in intermittent fevers. Dose 8–16 gr. (0.52–1.04 Gm.).

**cinchonine** (sin'-ko-nēn) [*cinchona*], $C_{19}H_{22}N_2O$. An alkaloid derived from cinchona. It is a colorless, crystalline body, similar to quinine in therapeutic effects, but less active. **c. bisulphate**, $C_{19}H_{22}N_2O$.- $H_2SO_4$, used as is cinchonine. **c. dihydrochloride**, is said to contain 60 % of cinchonine; antipyretic and antiseptic. **c. herapathite, c. iodosulphate**. See *antiseptol*. **c. hydrochloride**, used as is cinchonine. **c. nitrate**, $C_{19}H_{22}N_2O$. $HNO_3+H_2O$, used as is cinchonine. **c. salicylate**, $C_{19}H_{22}N_2$. $C_7H_6O_5$; used in rheumatism in malarial regions. Dose 5–20 gr. (0.32–1.3 Gm.). **c. sulphate** (*cinchoninæ sulphas*, U. S. P.), $(C_{19}H_{22}N_2O)_2 . H_2SO_4+2H_2O$. It is soluble with difficulty in water, but soluble in adiculated water. Dose 5–30 gr. (0.32–2.0 Gm.). **c. tannate**, yellow powder, soluble in alcohol; used in the same manner as is cinchonine.

**cinchonism** (sin'-ko-nism) [*cinchona*]. The systemic effect of cinchona or its alkaloids when given in full doses. The symptoms produced are a ringing in the ears, with deafness, headache, giddiness, dimness of sight, and a weakening of the heart's action.

**cinchonize** (sin'-ko-nīz) [*cinchona*]. To bring under the influence of cinchona or its alkaloids.

**cinchonology** (sin-ko-nol'-o-je) [*cinchona*; λόγος, science]. The science of the derivatives of cinchona.

**cinchotannin** (sin-ko-tan'-in), $C_{14}H_{16}O_3$. A glucoside existing in cinchona bark in the proportion of 3 to 4 %: a brownish-red substance, soluble in water and alcohol, and forming white precipitates with tartar emetic and gelatin. Syn., *cinchotannic acid; quinotannic acid*.

**cinclisis** (sin'-klis-is) [κίγκλισις, any quick, repeated motion]. Quick, spasmodic movement of any part of the body, but particularly applied to rapid winking. Hippocrates' term for quick motion of the chest, as in dyspnea. Syn., *cinclesmus*.

**cincture** (singk'-tūr) [*cinctura*, a girdle]. A belt or girdle. **c.-feeling**, a sensation as if the waist were encircled by a tight girdle. See *girdle-pain*.

**cinematics** (sin-e-mat'-iks). See *kinetics*.

**cinematograph**. See *kinematograph*.

**cineol** (sin'-e-ol) [*cina*, wormseed; *oleum*, oil], $C_{10}H_{18}O$. The principal constituent of wormseed, cajuput, and eucalyptus oils.

**cineraceous** (sin-er-a'-shus) [*cinerea*]. Ash-gray in color.

**Cineraria** (sin-er-a'-re-ah) [*cinerarius*, pertaining to ashes]. A genus of composite plants. **C. maritima**, the juice of this plant has been long used in Venezuela in the belief that, dropped in the eye, it would cause the absorption of cataract.

**cinerea** (sin-e'-re-ah) [*cinereus*, ashen]. The gray substance of the brain, spinal cord, and ganglia. **c., lamina**, a thin layer of gray substance extending backward above the optic commissure from the termination of the corpus callosum to the tuber cinereum.

**cinereal** (sin-e'-re-al). Ashy.

**cineritious** (sin-er-ish'-us) [*cineres*, ashes]. Ashlike or pertaining to ashes. **c. substance**, the cortex of the brain, from the color of the same. **c. tubercle**, the tuber cinereum.

**cinesia** (sin-e'-se-ah). See *kinesis*.

**cinesiology** (sin-es-e-ol'-o-je). See *kinesiology*.

**cinesis** (sin-e'-sis). See *kinesis*.

**cinesitherapy** (sin-es-e-ther'-a-pe). See *kinesitherapy*.

**cinetica** (sin-et'-ik-ah) [κινεῖν, to move]. Medicines or diseases that affect the motor apparatus.

**cingula** (sin'-gū-lah). 1. A band, girdle, or zone. 2. Burdach's name for the upper part of the fornicate gyrus.

**cingule** (sin'-gūl) [*cingulum*]. The groove separating the primitive cusp or tubercle frequently found on the lingual face of the upper incisor teeth. Syn., *cingulum; cingulus*.

**cingulum** (sin'-gū-lum) [*cingere*, to gird]. 1. A girdle or zone; the waist. 2. Herpes zoster or shingles. 3. See *cingule*. 4. A fibrous bundle in the fornicate gyrus of the brain. Syn., *bundle of the gyrus fornicatus; fasciculus arcuatus*. **c. extremitatis inferioris**, the pelvic girdle. **c. extremitatis superioris**, the shoulder girdle. **c. Halleri**, the abdominal muscles. **c. veneris**. See *corona veneris*.

**cinnabar** (sin'-ab-ar) [κιννάβαρι, a pigment]. Mercuric sulphide, HgS.

**cinnabarsana** (sin-ab-ar-san'-ah). A preparation said to consist of arsenic trioxide, cinnabar, charcoal, and water; it has been used as a cancer remedy.

**cinnamein** (sin-am-e'-in), $CHO_3(CH)$. A constituent of balsams of Peru and Tolu, and is obtained from sodium cinnamate by heating with benzyl chloride. Syn., *benzyl cinnamate*.

**cinnamene** (sin-am'-ēn). See *styrol*.

**cinnamic** (sin-am'-ik) [*cinnamom*]. Pertaining to or derived from cinnamon. **c. acid**. See *acid, cinnamic*. **c. aldehyde** (*cinnaldehydum*, U. S. P.). See *aldehyde, cinnamic*.

**cinnamol** (sin'-am-ol). 1. See *styrol*. 2. See *aldehyde, cinnamic*.

**cinnamomum** (sin-am-o'-mum). See *cinnamon*.

**cinnamon** (sin'-am-on). The inner bark of the shoots of several species of *Cinnamomum*, native to Ceylon and China, the latter variety being known in commerce under the name of *cassia*. Two varieties are official: *Cinnamomum saigonicum*, Saigon cinnamon, and *C. zeylanicum*, Ceylon cinnamon. Its properties are due to a volatile oil. It is an agreeable carminative and aromatic stimulant, used in flatulence, colic, enteralgia, etc. **c., aromatic, fluidextract of**, contains aromatic powder, 10; alcohol, 8 parts. Dose 10–30 min. (0.65–2.0 Cc.). **c., aromatic, powder of**, made up of aromatic powder, cinnamon, of each, 35; cardamom, nutmeg, of each, 15. Dose 10–30 gr. (0.65–2.0 Gm.). **c., compound powder of** (*pulvis cinnamomi compositus*, B. P.), cinnamon-bark, cardamom-seeds, and ginger. Dose 3–10 gr. (0.2–0.65 Gm.). **c., oil of** (*oleum cinnamomi*, U. S. P.), the volatile oil of cinnamon. Dose 1–5 min. (0.065–0.32 Cc.). **c., spirit of** (*spiritus cinnamomi*, U. S. P.), 10 % of the oil in spirit. Dose 5–30 min. (0.32–2.0 Cc.). **c., tincture of** (*tinctura cinnamomi*, U. S. P.), 20 % of powdered Saigon cinnamon in glycerol, alcohol, and water. Dose ½–2 dr. (2–8 Cc.). **c. water** (*aqua cinnamomi*, U. S. P.), 2 parts of oil in 1000 of water. Dose 1–2 oz. (30–60 Cc.).

**cinnamyl** (sin'-am-il) [*cinnamon*], $C_9H_7O$. The radical believed to exist in cinnamic acid. **c. cinnamate**. See *styracin*. **c.-eugenol**, $C_9H_{18}O_3$. It is antiseptic and is used hypodermatically instead of eugenol in tuberculosis. Injection, 2–8 min. (0.12–0.5 Cc.) of olive-oil solution. **c. hydrate**, cinnamic acid. **c. hydride**, cinnamic aldehyde. **c.-metacresol**, the metacresol ester of cinnamic acid; a nontoxic, nonirritating antiseptic substance recommended in treatment of tuberculosis. Syn., *hetocresol*.

**cionectomy** (si-on-ek'-to-me) [κίων, the uvula; ἐκτομή, a cutting out]. Ablation of the uvula.

**cionitis** (si-on-i'-tis) [κίων, the uvula; ιτις, inflammation]. Inflammation of the uvula.
**cionoptosis** (si-on-op-to'-sis) [κίων, uvula; πτῶσις, a falling]. Prolapse of the uvula.
**cionorrhaphia** (si-on-or-af'-e-ah) [κίων, the uvula; ῥαφή, a suture]. See *staphylorrhaphy*.
**cionotome** (si-on'-o-tōm) [κίων, the uvula; τομή, a cutting]. An instrument for cutting off the uvula.
**cionotomy** (si-on-ot'-o-me) [κίων, uvula; τομή, a section]. Incision of the uvula.
**ciose** (si'-ōs). Trade name of a dry preparation containing the protein substance of lean beef in a soluble form.
**Cipollina's test for glucose** (sip-ol-e'-nah). Mix 4 c.c. of dextrose solution (or urine), 5 drops of phenylhydrazine (base), and ½ c.c. of glacial acetic acid in a test-tube. Heat over a low flame for one minute. Add 4 or 5 drops of sodium hydroxide (sp. gr. 1.16) taking care that the fluid remains acid. Heat the mixture again for a moment and then cool. Crystals of glucosazone usually form at once. If they do not, allow test-tube to stand at least twenty minutes before final decision is reached.
**circellus** (ser-sel'-us) [L.; pl., *circelli*]. A small circle. **circelli cerebelli**, the laminæ of the cerebellum. **c. venosus hypoglossi**, a venous plexus encircling the hypoglossal nerve in the anterior condylar foramen; it communicates with the occipital sinus and with the jugular vein.
**circinate** (sir'-sin-āt) [*circinatus*, circular]. Having a circular outline or a ring formation. **c. eruption**, see *wandering rash*.
**circinus** (sir'-sin-us) [κίρκινος, circle]. Herpes zoster; zona.
**circle** (ser'-kl) [κίρκος, a circle]. A ring; a line, every point of which is equidistant from a point called the center. **c., ciliary**, the ciliary ligament. **c. of diffusion**. See *diffusion-circle*. **c. of Haller**. See *Haller, circle of*. **c., Huguier's**. See *Huguier, circle*. **c. of Willis**. See *Willis, circle*. **c. of Zinn**. See *Haller, circle of* (1).
**circocele** (sur'-ko-sēl). See *cirsocele*.
**circuit** (ser'-kit) [*circuitus*, a going round]. The course of an electric current. **c.-breaker**, an apparatus for interrupting the circuit of an electric current.
**circular** (ser'-kū-lar). 1. Ring-shaped. 2. Pertaining to a circle. 3. Marked by alternations of despondency and excitation, as in circular insanity. **c. amputation**, amputation with an incision surrounding the limb. **c. insanity**. See *insanity, circular*. **c. sinus**. See *sinus, circular*.
**circulation** (ser-kū-la'-shun) [*circulatio*, a circular course]. Passage in a circle, as the circulation of the blood. **c., collateral**, that taking place through branches and secondary channels after stoppage of the principal route. **c., fetal**, that of the fetus, including the circulation through the placenta and umbilical cord. **c., first, c., primitive**, that of the embryo, a closed system, carrying nutriment and oxygen to the embryo. **c., placental**, the period of circulation. **c., portal**, the passage of the blood from the gastrointestinal tract and spleen through the liver, and its exit by the hepatic vein. **., pulmonary**, the circulation of blood through the lungs by means of the pulmonary artery and veins, for the purpose of oxygenation and purification. **c., second**, the fetal circulation, replacing the omphalomesenteric system. **c., systemic**, the general circulation, as distinct from the pulmonary circulation. **c., third**, that of the adult. **c., vitelline**, first or primitive circulation.
**circulatory** (ser'-kū-la-to-re). Pertaining to the circulation.
**circulus** (ser'-kū-lus) [L.]. 1. A circle. 2. See *Willis, circle of*. **c. arteriosus Halleri**, the circle of Haller. **c. arteriosus iridis major**, an arterial circle around the circumference of the iris. **c. arteriosus iridis minor**, one around the free margin of the iris. **c. articuli vasculosus**, that formed by the blood-vessels in the synovial membrane about the cartilages of a joint. **c. arteriosus Willisii**, the circle of Willis, an arterial circle at the base of the brain. **c. ganglionis ciliaris**, a circular nerve-plexus in the ciliary muscle. Syn., *orbiculus gangliosus*. **c. osseus**, the tympanic ring. **c. venosus Halleri, c. venosus mammæ**, an anastomosis of veins around the nipple.
**circum-** (ser-kum-) [L.]. A prefix meaning around, about.
**circumagentes** (sir-kum-aj-en'-tēz) [L., "causing to revolve"]. 1. The oblique muscles of the eye.

2. The infraspinatus and supraspinatus muscle that revolve the arm.
**circumanal** (sir-kum-a'-nal) [*circum; anus*, the fundament]. Periproctous; surrounding the anus.
**circumaxile** (ser-kum-aks'-il) [*circum-; axis*]. Encircling an axis.
**circumaxillary** (sir-kum-aks'-il-ar-e) [*circum-; axilla*]. Surrounding the axilla.
**circumbuccal** (sir-kum-buk'-al) [*circum-; bucca*, the cheek]. Surrounding the mouth.
**circumbulbar** (sir-kum-bul'-bar). Surrounding a bulb, especially the eyeball.
**circumcision** (sir-kum-sizh'-un) [*circum-; cædere*, to cut]. The removal of the foreskin; excision of a circular piece of the prepuce.
**circumclusion** (sir-kum-klu'-shun) [*circum-; cludere*, to close]. A form of acupressure in which the pin is passed beneath the vessel, a wire loop placed over its point, and its ends brought over the artery and made fast.
**circumcorneal** (sir-kum-kor'-ne-al) [*circum-; corneus*, horny]. Around or about the cornea.
**circumduction** (sir-kum-duk'-shun) [*circum-; ducere*, to lead]. The movement of a limb in such a manner that its distal part describes a circle, the proximal end being fixed.
**circumference** (sir-kum'-fer-ens) [*circumferre*, to carry around]. The distance around a part.
**circumferential** (sur-kum-fer-en'-shal). Pertaining to a circumference; peripheral.
**circumflex** (sir'-kum-fleks) [*circum-; flectere*, to bend]. Winding around. The name given to a number of arteries, veins, and nerves, on account of their course.
**circumgyration** (sir-kum-ji-ra'-shun) [*circum-; gyrare*, to turn]. See *vertigo*.
**circuminsular** (sir-kum-in'-sū-lar) [*circum-; insula*, island]. Surrounding the island of Reil.
**circumlental** (sir-kum-len'-tal) [*circum-; lens*, a lentil; lens]. Surrounding the lens. **c. space**. See *space*.
**circumnuclear** (sir-kum-nū'-kle-ar) [*circum-; nucleus*, kernel]. Surrounding the nucleus.
**circumnutation** (sir-kum-nū-ta'-shun) [*circum-; nutare*, to nod]. A bending successively toward all points of the compass. Applied to the movements of young and growing organs.
**circumocular** (sir-kum-ok'-ū-lar) [*circum-; oculus*, eye]. Surrounding the eye.
**circumoral** (sir-kum-o'-ral) [*circum-; os*, mouth]. Surrounding the mouth.
**circumorbital** (sir-kum-or-bit-al) [*circum-; orbita*, orbit]. Around the orbit.
**circumpolarization** (sir-kum-po-lar-i-za'-shun) [*circum-; polus*, pole]. 1. The rotation of a ray of polarized light. 2. The quantitative estimation of sugar in a suspected liquid by the degree of the rotation of polarized light, sugar rotating the ray to the right, albumin to the left.
**circumrenal** (sir-kum-re'-nal) [*circum-; ren*, the kidney]. Around or about the kidney.
**circumscribed** (sir-kum-skrībd) [*circum-; scribere*, to write]. Strictly limited or marked off; well defined; distinct from surrounding parts, as a *circumscribed* inflammation or tumor.
**circumtonsillar** (sir-kum-tons-il'-lar) [*circum-; tonsil*]. Surrounding the tonsil.
**circumvallate** (sir-kum-val'-āt) [*circum-; vallum*, wall]. Surrounded by a wall or prominence. **c. papillæ**, certain papillæ at the base of the tongue.
**circumvascular** (sir-kum-vas'-kū-lar) [*circum-; vasculum*, vessel]. Surrounding a blood vessel, or other vessel.
**cirrholysin** (sir-ol'-is-in). See *fibrolysin*.
**cirrhonosus** (sir-on'-o-sus) [κιρρός, yellow; νόσος, disease]. 1. A fetal disorder, marked by yellowness of the serous membranes. 2. Abnormal postmortem yellowness of any surface or tissue.
**cirrhose** (sir-ōs') [*cirrus*, a tendril]. In biology, provided with tendrils.
**cirrhosis** (sir-o'-sis) [κιρρός, reddish-yellow from the color of the cirrhotic liver]. Chronic inflammation of an organ, characterized by hardening due to an overgrowth of the connective tissue. **c., alcoholic**. See *c., atrophic*. **c., annular**. See *c., multilobular*. **c., atrophic**, a form of cirrhosis of the liver occurring in hard drinkers, characterized by great overgrowth of the interstitial substance, with atrophy of the parenchyma. **c., biliary**, a form of cirrhosis of the liver due to chronic retention of bile from long-

continued obstruction of the bile-ducts. c., Budd's. See *Budd's cirrhosis.* c., cardiac, c. cordis, hypertrophy of the connective tissue between the muscular fibers of the heart. c., cardiotuberculous, that accompanied by tuberculosis and symptoms of cardiac disease. c., Charcot's. See *Hanot's disease.* c., fatty, that in which the hepatic or other cells become infiltrated with fat. c., Glissonian, perihepatitis. c., Hanot's: See *Hanot's disease.* c. hepatis, interstitial hepatitis. c., hypertrophic, a form of cirrhosis in which the liver is permanently enlarged. The disease is probably infectious, and is characterized by an overgrowth of the connective tissue which has no tendency to contract. c., irritative, interstitial hepatitis due to irritation by some toxic substance which has been carried to the liver by the hepatic or portal veins. c., Laennec's. See *Laennec's disease.* c. of the lung, interstitial pneumonia. c. mammæ, chronic interstitial mastitis. c., mixed, that presenting features of both the atrophic and the hypertrophic form. c., multilobular, a form of interstitial hepatitis in which many lobules are surrounded by a fibrous ring. c., muscular, the induration of connective tissue, fatty degeneration, and atrophy of the muscular fibers which take place in muscular contracture. c., obstructive, cirrhosis of the liver due to the obstruction of the passage of blood or bile from the liver. c., periportal, atrophic cirrhosis, so called because the hyperplasia of the connective tissue follows the portal vessels. c., pigmentary diabetic, cirrhosis of the liver with pigmentation of the skin. c., pulmonary, c. pulmonum, interstitial pneumonia. c., renal, c. renum, interstitial nephritis. c. of the spleen, chronic hypertrophy and induration of the spleen, with thickening of the capsule. c. of the stomach, chronic interstitial gastritis. c., Todd's. See *Todd's cirrhosis.* c., tuberculous, cirrhosis of the liver due to tuberculosis. It is rare; the majority of cases have occurred in children. c., turbinated, defective turbinated bodies due to disappearance or diminishment of the erectile structure in cases of atrophic rhinitis. c. ventriculi. See *c. of the stomach.*

cirrhotic (sir-ot'-ik) [cirrhosis].—Affected with, or relating to, cirrhosis. c. kidney, chronic interstitial nephritis.

cirrus (sir'-us) [L.]. 1. A lock or tuft of hair. 2. The male genital organ of *Cestodes,* usually adherent to the anterior end of the cirrus-pouch. c.-pouch, a structure made up of muscle and connective tissue attached to the male genital aperture of the *Cestodes* and serving to protrude the cirrus.

cirsaneurysma (sirs-an-ū-riz'-mah). See *aneurysm, cirsoid.*

cirsectomy (sur-sek'-to-me) [κιρσός, varix; ἐκτομή, a cutting out]. Excision of a piece of a varicose vein.

cirsocele (sir'-so-sēl) [cirsoid; κήλη, tumor]. A varicose tumor, especially of the spermatic cord.

cirsoid (sir'-soid) [κιρσός, a varix; εἶδος, form]. Resembling a varix or dilated vein.

cirsomphalos (sir-som'-fal-os) [cirsoid; ὀμφαλός, navel]. A varicose condition of the navel.

cirsophthalmia (sir-soff-thal'-me-ah) [κιρσός, varix; ὀφθαλμός, the eye]. 1. Ophthalmia, with an apparent varicose condition of the conjunctival vessels. 2. Corneal staphyloma, with an appearance of varicosity of the surface.

cirsotome (sir'-so-tōm) [κιρσός, a varix; τομή, a cutting]. A cutting instrument for the operation of cirsotomy.

cirsotomy (sir-sot'-o-me) [cirsoid; τέμνειν, to cut]. Excision of a varix.

cis-. A prefix proposed by Baeyer to designate relative asymmetry in unsaturated carbon compounds.

cissa (sis-ah) [L.]. See *pica.*

cissampeline (sis-am'-pel-ēn). An alkaloid from pareira root, identical with beberine.

Cissampelos (sis-am'-pel-os) [κισσός, ivy; ἄμπελος, a vine]. A genus of climbing menispermaceous plants. The root of C. capensis, of South America, is cathartic and emetic. C. pareira, of tropical America, false pareira brava (q. v.), is tonic and diuretic.

cistern (sis'-tern) [cisterna, a vessel; receptacle]. 1. A reservoir. 2. Any dilation of the space between the pia and arachnoid. c. of Pecquet, the receptaculum chyli. c., seminal, the posterior culdesac of the vagina. Syn., *receptaculum seminis.*

cisterna (sis-ter'-nah) [L.]. Same as *cistern.* In the plural, cisterna, the subarachnoid spaces. c. ambiens. 1. One of the pockets situated over the optic lobes. 2. See *canal, arachnoid.* c. basilis, that part of the anterior subarachnoid space holding the circle of Willis; it is divided by the chiasm into two parts, the *cisterna anterior* and the *cisterna inferior.* c. cerebellaris, c. cerebellomedullaris, cerebellomedullary cistern, or *cisterna magna;* see *postcisterna.* c. chiasmatis, the interpeduncular space. . c. chyli, see *receptaculum chyli.* c. corporis callosi, the third ventricle. c. fossæ lateralis cerebri, c. fossæ Sylvii, cistern of the lateral fossa of the cerebrum; see *c. Sylviana.* c. intercruralis, c. interpeduncularis, the anterior subarachnoid space at the base of the brain. c. intercruralis profunda, that part of the subarachnoid space lying directly above the space included between the crura cerebri. c. intercruralis superficialis, that part of the subarachnoid space included between the pons and the chiasm. c. lateralis pontis, a small space extending along the outer edge of the pons. c. lumbaris, see *receptaculum chyli.* c. magna. 1. A large cisterna where the arachnoid spreads across from the caudad border of the cerebellum to the oblongata. 2. The fourth ventricle. c. perilymphatica, in the ear, a large space adjacent to the foot-plate of the stapes. c. pontis, the anterior subarachnoid space. cisternæ subarachnoideales, the subarachnoid spaces. c., superior, that included in the angle between the splenium, the superior surface of the cerebellum, and the posterior aspect of the quadrigeminum. c. Sylviana, the part of the subarachnoid space lying immediately above the Sylvian fissure. Syn., *c. fossæ lateralis cerebri.* c. venæ magnæ cerebri, see *canal, arachnoid.*

Cistus (sis'-tus) [κίστος, the rock-rose]. A genus of plants of the order *Cistaceæ,* growing in the old world. C. areticus, C. cyprius, C. ladaniferus, and C. ledon, afford the resinous substance labdanum, or ladanum.

citarin (sit'-ar-in). Trade name of sodium anhydromethylenecitrate. Used in the treatment of gout.

citral (sit'-ral) [citrus, a lemon]. C10H16O. An aldehyde found in oil of lemon and many of the essential oils; a golden-yellow liquid giving aroma and value to oil of lemon.

citrate (sit'-rāt) [citric acid]. Any salt of citric acid.

citric (sit'-rik) [citrus]. Pertaining to or derived from lemons or citrons. c. acid. See *acid, citric.*

citrine (sit'-rēn) [citrus]. Yellow; of a lemon-color. c. ointment (unguentum hydrargyri nitratis, U. S. P.), a preparation consisting of mercury dissolved in nitric acid and mixed with some fatty substance. It is made by adding 7 parts of nitric acid to 76 parts of warmed lard oil, and then mixing it with 7 parts of mercury dissolved in 10 parts of nitric acid.

citrocoll (sit'-ro-kol). Phenocoll citrate: used as an antipyretic and antineuralgic.

citrol (sit'-rol). Itrol citrate, a silver preparation used in gonorrhea.

citronella (sit-ron-el'-ah) [dim. of κίτρον; the citron-tree]. A fragrant grass. c. oil, the essential oil of various grasses, mostly of the genus *Andropogon;* used chiefly as a perfume, and as a protection against insects; antirheumatic.

citronellol (sit-ron-el'-ol), C10H18O. A body isomeric with borneol, obtained from oil of citronella.

citrophen (sit'-ro-fen), C6H4OH—CONH—OC2H4C2H4. Paraphenetidin citrate. It is antipyretic and antineuralgic. Dose 3-15 gr. (0.2-1.0 Gm.)

citrospirine (sit-ro-spi'-rēn). A compound of acetyl-salicylic acid and citrated caffeine.

citrullin (sit-rul'-in). A resinoid from *Citrullus colocynthis.* It is a cathartic extensively used in veterinary practice. Syn., *amorphous colocynthidin; colocynthidin.*

citrullus (sit-rul'-lus) [L.]. A genus of the *Cucurbitaceæ,* comprising but two species, indigenous to tropical Asia and southern Africa. C. colocynthis, the bitter cucumber or gourd, furnishes colocynth.

citrurea (sit-ru'-re-ah). A combination of citric acid, urea, and lithium.

Citrus (sit'-rus) [L.]. A genus of aurantiaceous trees. See *aurantium, bergamot, lime, limo.*

cittosis (sit-to'-sis) [κίττα, κίσσα, pica]. Pica; a longing for strange or improper food.

civet (siv'-et). A semi-liquid, unctuous secretion

from the anal glands of *Viverra civetta, V. zibetha*, and *V. rasse*, carnivorous old-world animals; themselves called civets. It is now used as a perfume; formerly as an antispasmodic and stimulant, like musk.

**Civinini's spine.** A small spine on the outer border of the external pterygoid plate, giving attachment to the pterygospinous ligament.

**Cl.** Chemical symbol for chlorine.

**cl.** Abbreviation for *centiliter*.

**cladode** (*klad'-ōd*) [κλάδος, a branch; εἶδος, form]. In biology, branch-like.

**Cladonia** (*klad-o'-ne-ah*) [κλάδος, a branch]. A genus of lichens. C. **rangiferina**, the reindeer-moss; a lichen that grows extensively in Asia, Europe and N. America. It is used as a food in famine-seasons, and is locally distilled, affording an alcoholic spirit.

**cladosporium cancerogenes** (*klad-o-spo'-re-um kan-ser-oj'-en-ēz*). A fungus said to be the cause of carcinoma. Syn., *canceromyces*.

**Cladothrix** (*klad'-o-thriks*) [κλάδος, branch; θρίξ, a hair]. A genus of *Schizomycetes* having long, apparently branching filaments.

**clairaudience** (*klār-aw'-de-ens*) [Fr. *clair*, clear; *audience*, hearing]. The alleged telepathic hearing of sounds uttered at a great distance.

**clairvoyance** (*klār-voi'-ans*) [Fr. *clair*, clear; *voir*, to see]. The alleged ability (in certain states), to see things not normally visible; the pretended ability to see the internal organs of a patient, and thus diagnosticate his ailments.

**clamp** (*klamp*) [Ger., *Klampe*]. An instrument for compressing the parts in surgical operations to prevent hemorrhage, etc.

**clang** (*klang*). A sharp metallic sound; a hoarse voice. **c.-deafness**, a defect of hearing in which sounds are heard, but their more delicate qualities are not perceived. **c.-tint**, the timbre, or delicate shading of a tone. See *timbre*.

**clap** (*klap*). Gonorrhea. **c.-threads**, slimy threads consisting of mucus and pus-cells in the urine of gonorrheal patients.

**clapotage, clapotement** (*klap-ŏt-ahzh'*, *klap-ōt-mon(g)'*) [Fr.]. The splashing sound of a liquid in succession.

**Clapton's line.** Greenish discoloration of the gums and teeth, especially the incisors, in chronic copper-poisoning.

**claquement** (*klahk'-mon(g)*) [Fr., clapping, slapping]. 1. In massage, percussion with the flat of the hand. 2. The clack or flapping sound caused by sudden closure of the heart-valves.

**claret** (*klar'-et*) [*clarus*, clear]. A light wine of a red color.

**clarificant** (*klar'-if-ik-ant*) [*clarus*, clear; *facere*, to make]. A substance used for the purpose of clearing solutions turbid from insoluble matter.

**clarification** (*klar-if-ik-a'-shun*) [*clarus*, clear; *facere*, to make]. The operation of removing the turbidity of a liquid or naturally transparent substance. It may be accomplished by allowing the suspended matter to subside, by the addition of a clarificant or substance that precipitates suspended matters, or by moderate heating.

**clarify** (*klar'-if-i*) [*clarus*; *facere*, to make]. To free a liquid or solution from insoluble substances; to make clear.

**clarifying** (*klar-if-i'-ing*) [*clarus*, clear; *facere*, to make]. Clearing; purifying. **c. reagent**, any preparation used for purifying microscopical and anatomical preparations that have been mounted in gummy media. Oil of cloves, turpentine, creosote, xylol, and oil of bergamot are the chief.

**Clark's sign** [Alonzo *Clark*, American physician, 1807–1887]. A tympanitic sound over the hepatic region in tympanites due to perforative peritoneal inflammation.

**Clarke's corroding ulcer.** [Sir Charles Mansfield *Clarke*, English physician, 1782–1857]. Progressive ulcer of the cervix uteri. **C.'s tongue**, the hard, fissured, and nodular tongue of syphilitic glossitis sclerosa.

**Clarke's vesicular column** [Jacob Augustus Lockhart *Clarke*, English physician, 1817–1880]. A column of gray substance occupying the region to the outer and posterior side of the central canal of the spinal cord, at the inner part of the base of the posterior cornu, it contains fusiform cells, and is the trophic center for the direct cerebellar tract.

**clasmacytosis** (*klas-mah-si-to'-sis*). Same as *clasmatocytosis*.

**clasmatocyte** (*klas-mat'-o-sīt*) [κλάσμα, fragment; κύτος, cell]. A form of very large connective-tissue corpuscles that tend to break up into granules or pieces.

**clasmatocytosis** (*klas-mat-o-si-to'-sis*) [κλάσμα, fragment; κύτος, a cell]. The breaking up of clasmatocytes, and the formation of islands of granules from their debris.

**clasmatosis** (*klas-mat-o'-sis*). See *clasmatocytosis*.

**clasp** (*klasp*) [ME., *claspen*, to grasp firmly]. **c.-knife rigidity**, a spastic condition of a limb, as a result of which extension is completed with a "spring," as in a knife-blade. It is met in the cerebral palsies of children.

**classification** (*klas-if-ik-a'-shun*) [*classis*, a class; *facere*, to make]. An orderly arrangement of names, objects, diseases, etc., according to their properties and peculiarities.

**clastic** (*klas'-tik*) [κλαστός, broken]. Breaking up into fragments; causing division.

**clastothrix** (*klas'-to-thriks*). Synonym of *Trichorrhexis nodosa*.

**Clathrocystis** (*klath-ro-sis'-tis*) [κλῆθρα, a trellis; κύστις, pouch]. A genus of microorganisms with round or oval cells, forming zoogleæ in the form of circular layers.

**claudication** (*klaw-dik-a'-shun*) [*claudicare*, to limp]. 1. Lameness. 2. An obstruction. **c., Charcot's intermittent, c., intermittent**, intermittent paresthesia of the legs attended with pain, tremor, and excessive perspiration due to arteriosclerosis; a condition first noted by French writers in apparently healthy horses and afterward observed in man. Syn., *angina cruris; angiosclerotic paroxysmal myasthenia; intermittent lameness; intermittent limping*. **c., spontaneous**, the lameness that occurs as an early symptom of coxarthrocace in children.

**Claudius' cells** [Friedrich Matthias *Claudius*, German anatomist, 1822–1869]. Polyhedral or conoidal cells lining the outer angle of the scala media of the cochlea. **C.'s fossa**, the ovarian fossa, a triangular space containing the ovary; it is bounded anteriorly by the round ligament, above by the external iliac vein, and below by the ureter.

**claustral** (*klaws'-tral*). Pertaining to the claustrum.

**claustrophilia** (*klaws-tro-fil'-e-ah*) [*claustrum*; φιλεῖν, to love]. A morbid dread of open places; it is noted in neurasthenia.

**claustrophobia** (*klaws-tro-fo'-be-ah*) [*claustrum*; φόβος, fear]. Morbid distress at being in a room or confined space.

**claustrum** (*klaws'-trum*) [*claudere*, to shut; pl., *claustra*]. A barrier; applied to several apertures that may be closed against entrance. Also, a layer of cinerea (gray nervous matter) between the insula and the lenticula. **c., gutturis**, the opening of the pharynx. **c. oris**, see *velum palati*. **c. virginale, c. virginitatis**, the hymen.

**clausura** (*klaw-sū'-rah*) [L.]. Closure; atresia; as of a passage. **c. tubalis**, closure of a Fallopian tube. **c. uteri**, an imperforate state of the uterine cervix.

**clava** (*kla'-vah*) [L., "a club"]. An enlargement of the funiculus gracilis.

**clavate** (*klav'-āt*) [*clava*]. Club-shaped or becoming gradually thicker toward one end. **c. nucleus**, a collection of nerve cells within the clava.

**clavation** (*klav-a'-shun*) [*clavatio*; *clavus*, a nail]. Same as *gomphosis*.

**clavelization** (*klav-el-iz-a'-shun*) [Fr., *clavellée*; sheep-pox]. Inoculation with sheep-pox virus; ovination.

**claven**, or **claviculen** (*kla'-ven, kla-vik'-ū-len*) [*clavis*, clavicle]. Belonging to the clavicle in itself.

**Claviceps** (*klav'-is-eps*) [*clava*; *caput*, head]. A genus of fungi. **C. purpurea**, the fungus producing the ergot of rye.

**clavicle** (*klav'-ik-l*) [*clavicula*; *clavus*, a key]. The collar-bone. **c.-crutch**, Cole's device for purchase of a broken clavicle; it is so furnished with pads and adjustments as to render bandaging unnecessary.

**clavicotomy** (*klav-ik-ot'-o-me*) [*clavicle*; τομή, a cutting]. Surgical section of the clavicle.

**clavicula** (*klav-ik'-ū-lah*). The clavicle. **c. capitis**, the projection formed by the pterygoid and entopterygoid bones on the pleurapophysis of the hemal arch of the nasal vertebra.

clavicular (*kla-vik'-ū-lar*) [*clavicle*]. Relating to the clavicle.
claviculate (*klav-ik'-ū-lāt*). 1. Having a clavicle. 2. Wrinkled; corrugated.
claviculus (*klą-vik'-ū-lus*) [dim. of *clavus*, a nail; *pl., claviculi*]. One of Sharpey's fibers, *q. v.*
claviform (*klav'-e-form*). See *clavate*.
clavin (*kla'-vin*), C₁₁H₁₈N₂O₄. One of the active principles of ergot; it is said to be nontoxic.
clavipes (*klav'-e-pēz*) [*clava; pes*, a foot]. Having club-shaped feet.
clavis uteri (*kla'-vis u'-ter-i*). Womb-key; an electrotherapeutic intrauterine device, designed for the application of electricity in certain pathological conditions of the uterus and adnexa.
clavus (*kla'-vus*) [L., "a nail; a wart; a corn"]. Corn; a hyperplasia of the horny layer of the epidermis, in which there is an ingrowth as well as an outgrowth of horny substance, forming circumscribed epidermal thickenings, chiefly about the toes. c. hystericus, a pain in the head, as if a nail were being driven in.
claw-foot. A form of talipes due to depression of the heads of the metatarsal bones, with forced extension of the first phalanges and flexion of the last; it is a result of paralysis of the interossei and lumbricales muscles and of those inserted into the sesamoid bone of the great toe.
claw-hand. A condition of the hand characterized by overextension of the first phalanges and extreme flexion of the others. The condition is a result of atrophy of the interosseous muscles, with contraction of the tendons of the common extensor and long flexor. Syn., *main-en-griffe*.
clean, cleaning (*klēn, klēn-ing*). A word used in practical anatomy to denote the complete removal (during the process of dissection) of the fat or connective tissue, from the surface of any structure.
cleansings (*klen'-zings*). The lochia.
clearing agent. A substance used in microscopy to render tissues transparent and suitable for mounting.
cleavage (*kle'-vāj*) [AS., *cleōfan*, to split asunder]. 1. The linear clefts in the skin indicating the general direction of the fibers. They govern to a certain extent the arrangement of the lesions in skin diseases. The lines of cleavage run, for the most part, obliquely to the axis of the trunk, sloping from the spine downward and forward; in the limbs they are mostly transverse to their longitudinal axis. 2. A mode of cell-division. c., egg-. See *segmentation*. c.-nucleus, the nucleus which in the fertilized egg results from the union of the male and female nuclei.
Cleemann's sign, C.'s test. In fracture of the femur with shortening there is a wrinkle above the ligamentum patellæ, which disappears when the shortening is corrected by extension.
cleft (*kleft*). Divided. A fissure. c., branchial. See *c., visceral.* c., genital, a depression in the genital region of the embryo from which the cloaca is developed. c.-hand, a congenital deformity in which some finger or fingers are widely separated from the others. c. palate, a congenital fissure of the palate. c. sternum, congenital fissure of the sternum. c., visceral, the four slit-like openings on each side in the cervical region in the fetus, sometimes called the branchial openings. The slits close (in the human fetus), except the upper, from which are developed the auditory meatus, tympanic cavity, and Eustachian tube.
cleidagra, cleisagra (*kli-dag'-rah, kli-sag'-rah*) [κλείς, clavicle; ἄγρα, seizure]. Gouty pain in the clavicle.
cleidal (*kli'-dal*) [κλείς, clavicle]. Relating to the clavicle; clavicular.
cleidarthritis (*kli-dar-thri'-tis*) [κλείς, clavicle; *arthritis*]. Inflammation of the sternoclavicular articulation.
cleido- (*kli-do-*) [κλείς, clavicle]. A prefix meaning pertaining to the clavicle.
cleidocostal (*kli-do-kos'-tal*). Pertaining to the ribs and the clavicle.
cleidohyoid (*kli-do-hi'-oid*). Relating to the clavicle and the hyoid.
cleidomastoid (*kli'-do-mas'-toid*). Pertaining to the clavicle and to the mastoid process.
cleido-occipital (*kli-do-ok-sip'-it-al*). Relating to the clavicle and occiput. c. muscle. See under *muscle*.
cleidoscapular (*kli-do-skap'-ū-lar*). Relating to the clavicle and the scapula.

cleidosternal (*kli-do-stur'-nal*). Sternoclavicular.
cleidotomy (*kli-dot'-o-me*) [*cleido-*; τέμνειν, to cut]. The operation of dividing the clavicles in cases of difficult labor due to the broad shoulders of the child.
cleidotripsy (*kli-do-trip'-se*). The operative crushing of the clavicle.
clematine (*klem'-at-in*). An alkaloid from *Clematis vitalba.*
Clematis (*klem'-at-is*). A genus of ranunculaceous plants of many species, most of which are acrid or poisonous. C. corymbosa is powerfully irritant and resistant. C. crispa and C. erecta are diuretic and diaphoretic, and are said to be antisyphilitic. C. viorna, C. virginica, and C. vitalba are similar in properties to C. erecta.
Clemens' solution (*klem'-enz*). Liquor potassii arsenatis et bromidi.
cleptomania (*klep-to-ma'-ne-ah*). See *kleptomania*.
cleptophobia (*klep-to-fo'-be-ah*). See *kleptophobia.*
clergyman's sore throat. A chronic hypertrophic form of pharyngitis, with more or less enlargement of the tonsils and lymph-follicles of the posterior wall, due to excessive or improper use of the voice.
Clerodendron (*kler-o-den'-dron*) [κλῆρος, a lot; δένδρον, tree]. A genus of tropical shrubs and trees of the order *Verbenaceæ*. C. infortunatum is a species indigenous to India and Malaya; it is used as a substitute for chirata; the juice of the leaves is a tonic, febrifuge, and vermifuge. C. nereifolium is a species found in Malaya; the root and leaves are antisyphilitic, tonic, and vulnerary; the root and fruit are used to stupefy fish. C. serratum is indigenous to India; the root is tonic and stomachic; the fruit, purgative and diuretic. The leaves and an insect larva found on the branches of C. trichotomum are used as an ascaricide. C. villosum is a species indigenous to Malaya; the root is stomachic, the sap vermifugal.
Clevenger's fissure (*klev'-en-jer*). The inferior occipital fissure; a small fissure between the second and third occipital convolutions.
clicking sounds (*klik'-ing sounds*). Peculiar sharp sounds heard in auscultating the apex of a tuberculous lung. They indicate the commencement of softening in a tuberculous deposit. See *rale*.
clidagra. See *cleidagra*.
clidarthritis. 1. See *cleidagra*. 2. See *cleidarthritis*.
clidocostal. See *cleidocostal.*
clidotomy. See *cleidotomy.*
clidotripsy. See *cleidotripsy.*
climacter (*kli-mak'-tur*). See *climacteric.*
climacteric (*kli-mak'-ter-ik*) [κλιμακτήρ, the round of a ladder]. A period of life at which the system was believed to undergo marked changes. These periods were thought to occur, every seven years. The word is now generally applied to the menopause. c. age, puberty; also in women the time of cessation of the catamenia. c. epoch. Same as *c. age.* c., grand, the sixty-third year.
climate (*kli'-mat*) [κλίμα, a region, or zone, of the earth]. The sum of those conditions in any region or country that relate to the air, the temperature, moisture, sunshine, winds, etc., especially in so far as they concern the health or comfort of mankind.
climatic (*kli-mat'-ik*) [κλίμα, a region or zone of the earth]. Pertaining to climate.
climatology (*kli-mat-ol'-o-je*) [κλίμα, climate; λόγος, science]. The science of climate.
climatotherapy (*kli-mat-o-ther'-a-pe*) [κλίμα, clime; θεραπεία, a waiting on]. The employment of climatic measures in the treatment of disease.
climax (*kli'-maks*) [κλίμαξ, ladder]. The acme, or height of a disease; the period of greatest intensity.
clinic (*klin'-ik*) [κλίνη, a bed]. 1. Medical instruction given at the bedside, or in the presence of the patient whose symptoms are studied and whose treatment is considered. 2. A place where such instruction is given. 3. A gathering of instructors, students, and patients for the study and treatment of disease.
clinical (*klin'-ik-al*) [*clinic*]. 1. Relating to bedside treatment or to a clinic. 2. Pertaining to the symptoms and course of a disease as observed by the physician, in opposition to the anatomical changes found by the pathologist. c. thermometer. See *thermometer.*
clinician (*klin-ish'-an*) [*clinic*]. A physician whose opinions, teachings, and treatment are based upon experience at the bedside; a clinical instructor; one who practises medicine.

**clinicist** (*klin'-is-ist*) [*clinic*]. A clinician.
**clinicopathology** (*klin-ik-o-path-ol'-o-je*) [*clinic; pathology*]. Pathological conditions as open to clinical observation.
**clino-** (*klin-o-*) [κλίνειν, to incline]. A prefix denoting inclination or declination.
**clinocephalia** (*klin-o-sef-a'-le-ah*) [*clino-;* κεφαλή, head]. Abnormal flatness of the top of the head.
**clinocephalous** (*klin-o-sef'-al-us*) [*clino-;* κεφαλή, head]. Having the top of the head abnormally flat.
**clinocephalus** (*klin-o-sef'-al-us*) [*clino-;* κεφαλή, the head]. A variety of dolichocephalus occurring through synostosis of the sphenoparietal suture and resulting in a saddle-formed depression of the skull. Syn., *saddle-head.*
**clinodactylous** (*klin-o-dak'-til-us*) [*clino-;* δάκτυλος, finger]. Pertaining to an abnormal flexure, deviation or curvature of the fingers or toes.
**clinodiagonal** (*klin-o-di-ag'-on-al*) [*clino-; diagonal*]. Inclined and diagonal; obliquely transverse.
**clinoid** (*klin'-oid*) [κλίνη, a bed; εἶδος, likeness]. Resembling a bed; applied to sundry bony structures of the body, as the *clinoid* processes. c. **processes.** See under *process*.
**clinology** (*klin-ol'-o-je*) [*clino-;* λόγος, science]. 1. The science of the decline of animal life after it has reached the meridian. 2. [κλίνη, a bed]. The study of beds for the sick.
**clinometer** (*klin-om'-et-er*) [*clino-;* μέτρον, a measure]. An apparatus to estimate the rotational capacity of the ocular muscles. c., **Duane's.** See *Duane's test.*
**clinoscope** (*klin'-o-skōp*) [*clino-;* σκοπεῖν, to view]. An instrument for measuring the torsion of the eyes when gazing at a fixed object with the axes of vision presumably parallel.
**clinostat** (*klin'-o-stat*) [*clino-;* στατός, placed]. An apparatus for regulating the exposure of plants to the sunlight.
**clinotechny** (*klin-o-tek-ne*) [κλίνη, a bed; τέχνη, an art]. The art of making and preparing beds for the sick.
**cliseometer** (*klis-e-om'-et-er*) [κλίσις, inclination; μέτρον, a measure]. An instrument for measuring the degree of inclination of the pelvic axis.
**clition** (*klit'-e-on*) [κλιτύς, a slope]. A craniometric point located in the middle of the anterior border of the clivus ossis.
**clitoralgia** (*klit-or-al'-je-ah*) [*clitoris;* ἄλγος, pain]. Pain referred to the clitoris.
**clitoridauxe** (*klit-or-id-awk'-se*) [*clitoris;* αὔξη, increase]. Hypertrophy of the clitoris.
**clitoridectomy** (*klit-or-id-ek'-to-me*) [*clitoris;* ἐκτομή, excision]. Excision of the clitoris.
**clitoris** (*klit'-or-is*) [κλειτορίς, clitoris]. The homologue in the female of the penis, attached to the ischiopubic rami by two crura or branches, which meet in front of the pubic joint to form the body, of corpus. It possesses erectility. c. **crises**, paroxysms of sexual excitement in women suffering from tabes.
**clitorism** (*klit'-or-izm*). 1. Enlargement or hypertrophy of the clitoris. 2. Tribadism.
**clitoritis** (*klit-or-i'-tis*) [*clitoris;* ιτις, inflammation]. Inflammation of the clitoris.
**clitorotomy** (*klit-or-ot'-o-me*). Incision of the clitoris.
**clitorrhagia** (*klit-or-a'-je-ah*) [*clitoris;* ῥηγνύναι, to burst forth]. Hemorrhage from the clitoris.
**clivis** (*kli'-vis*). Same as *declivis cerebelli.*
**clivus** (*kli'-vus*) [L., "a slope"]. A slope. c. **ossis**, c. of Blumenbach, the slanting surface of the body of the sphenoid bone between the sella turcica and the basilar process of the occipital bone. c. **moniculi.** Same as *declivis cerebelli.*
**cloaca** (*klo-a'-kah*) [L., "a sewer"]. 1. In early fetal life, the common orifice of the intestine and the allantois. 2. A fistulous tract in bone discharging pus from a sequestrum. 3. A common outlet to the rectum and the bladder. c., **congenital**, a malformation in which the rectum opens into the genitourinary tract. c., **urogenital**, an abnormal common opening of the urethra and vagina due to a defective urethrovaginal septum. c., **vesicorectovaginal**, a common aperture of the bladder, rectum, and vagina, due to deformity or trauma.
**cloacal** (*klo-a'-kal*) [*cloaca,* a sewer]. Pertaining to or serving as a cloaca.
**clonic** (*klon'-ik*) [*clonus*]. Applied to convulsive and spasmodic conditions of muscles characterized by alternate contractions and relaxations.
**clonicity** (*klon-is'-it-e*). The state of being clonic.

**clonism, clonismus** (*klo'-nizm, klo-niz'-mus*) [*clonus*]. A clonic spasm or a succession of clonic spasms; clonospasm.
**clonograph** (*klon'-o-graf*) [*clonus;* γράφειν, to write]. An apparatus for recording the spasmodic movements of the head, extremities, lower jaw, and trunk, as well as the tendon-reflexes.
**clonospasm** (*klon'-o-spazm*) [κλόνος, commotion; σπασμός, a spasm]. A clonic spasm.
**clonus**, (*klo'-nus*) [κλόνος, commotion]. A series of movements characterized by alternate contractions and relaxations; a clonic spasm. Involuntary, reflex, irregular contractions of muscles when put suddenly upon the stretch. According to the part affected, the phenomenon is spoken of as *ankle-, foot-, rectus-,* or *wrist-clonus*, etc. See under *reflex*.
**Cloquet's canal** (*klo-ka'*) [Hippolyte *Cloquet*, French surgeon, 1787–1840; Jules Germain *Cloquet*, French surgeon, 1790–1883]. The hyaloid canal; an irregular canal running anteroposteriorly through the center of the vitreous body and transmitting the hyaloid artery during fetal life. **C.'s fascia**, the crural septum. **C.'s ganglion**, the nasopalatine ganglion. **C.'s hernia**, subpubic hernia; a femoral hernia passing behind and internally to the femoral vessels and resting on the pectineus muscle. **C.'s ligament.** See *Haller's habenula.*
**clostridial** (*klos-trid'-e-al*) [κλωστήρ, a spindle]. Pertaining to, or caused by *Clostridium*.
**Clostridium** (*klos-trid'-e-um*) [κλωστήρ, a spindle]. A genus of bacteria differing from bacilli in the fact that their spores are formed in an enlarged part of the cell.
**closure** (*klo'-zhŭr*) [*clausura*, a closing]. The act of completing or closing an electric circuit.
**clot** (*klot*) [AS., *clāte*, a bur]. A peculiar solidification of the blood, such as takes place when it is shed.
**clotbur** (*klot'-bur*). The leaves of *Xanthium strumarium*, much used as a domestic remedy for bites of poisonous insects and venomous serpents. Also an active styptic. Dose of the fluidextract ʒ j–ij.
**clottage** (*klot'-āj*). The blocking up of a canal (as a ureter) with a blood-clot.
**cloudy swelling.** Parenchymatous degeneration; a swelling-up of the elements of a tissue, with the formation in them of fine granules due to the change of soluble albuminates into insoluble.
**clove** (*klōv*). See *Caryophyllus*. c.-**hitch knot**, a form of double knot in which two successive loops are made close to each other on the same piece of cord or bandage, a half-twist being given to the junction of each loop at the time of making it.
**clownism** (*klown'-izm*). That stage of hysteroepilepsy in which there is an emotional display with a remarkable series of contortions.
**clubbed fingers.** Knobbed deformity of the finger-tips, with curvature of the nails over the finger-ends; seen in some cases of pulmonary and cardiac disease.
**club-foot.** See *talipes*. c., **heel**, talipes calcaneus. c., **inward**, talipes varus. c., **outward**, talipes valgus.
**club-hand.** A deformity of the hand similar to that of club-foot.
**club-moss** (*klub'-mos*). See *Lycopodium*.
**clumping** (*klump'-ing*). See *agglutination* (2). c. **serum.** See under *serum*.
**clunes** (*klu'-nēz*) [pl. of *clunis*, buttock]. The buttocks, nates.
**clupein** (*klu'-pe-in*) [*clupea*, a kind of small riverfish], C₄₈H₈₀N₁₇O₆+4H₂O. A protamine from the herring. Syn., *salmin*.
**Clusia** (*klu'-se-ah*) [Charles de l'*Escluse* (1526–1609)]. A genus of plants of the order *Guttiferæ*, many species of which yield a gum-resin called West Indian balsam. *C. flava*, of the West Indies, yields the milky sap used as a substitute for copaiba. *C. insignis*, of Brazil, yields a milky sap used as a salve. *C. hiliariana*, of the West Indies and South America, yields a gum used as a drastic and vulnerary; the fruit is edible and the astringent bark is employed in diarrhea.
**clysis** (*kli'-sis*) [κλύζειν, to cleanse]. The administration of an enema; the cleansing by means of an enema.
**clysma** (*kliz'-mah*). See *clyster*.
**clysmic** (*kliz'-mik*). Relating to an enema; suitable for cleansing or washing.

clyster (*klis'-ter*) [κλυστήρ, an injection]. An enema. See *alimentation, rectal*. c., **meat-bouillon-wine-** (Fleiner): 80 Gm. of beef-tea and 40 Gm. of mild white wine. Inject 2 or 3 times a day at body-heat. c., **meat-pancreas-** (Leube): 150 Gm. good beef scraped and chopped fine; 80 Gm. fresh pancreas (cow or hog) free from fat; mix with 150 Gm. luke-warm water; inject from 50 to 100 Gm. at a time, by means of a simple funnel, and at blood-heat. c., **nutritive (Boas)**: warm 250 Gm. of milk, stir in 2 egg-yolks, 1 teaspoonful of common salt, and 1 table-spoonful of wheat-starch, and afterward add 1 table-spoonful of red wine. If the mucous membrane is easily irritated, 4 or 5 drops of tincture of opium may be added. c., **nutritive (Ewald)**: wheaten starch, ½ teaspoonful, is boiled with a cup (100 Gm.) of a 20 % solution of grape-sugar, and 1 wineglass (150 Gm.) of red wine added. Then the solution is cooled to 35° C. and 2 or 3 eggs beaten smooth with 1 teaspoonful of cold water and a little salt are stirred in slowly. Inject at blood-heat. c., **nutritive** (Jaccoud): bouillon, 250 Gm.; wine, 120 Gm.; yolks of 2 eggs; and peptone, 5 to 20 Gm. c., **nutritive** (Rosenheim): peptone, 4 to 8 Gm.; 2 eggs; glucose, 15 Gm., and sometimes, if desired, emulsions of cod-liver oil.
**clysterize** (*klis'-tér-īz*). To administer a cluster.
**C.M.** An abbreviation for *Chirurgiæ Magister*, Master of Surgery.
**Cm.** Abbreviation of centimeter.
**cm.** Abbreviation of centimeter.
**CN.** 1. Abbreviation for *cyanogen*. 2. Trade name of a disinfectant.
**cnemial** (*ne'-me-al*) [κνήμη, the leg]. Relating to the tibia or leg; crural.
**cnemis** (*ne'-mis*) [κνήμη, the leg]. The tibia or shin-bine.
**cnemitis** (*ne-mi'-tis*) [κνήμη, shin; ιτις, inflammation]. Inflammation of the tibia.
**cnemoscoliosis** (*ne-mo-sko-le-o'-sis*) [κνήμη, the leg; σκολιός, curved]. Lateral curvature of the leg.
**cnicin** (*ni'-sin*) [κνῆκος, a plant of the thistle kind], C₂H₁₆O₁₅. A crystalline bitter substance found in *Cnicus benedictus*, Blessed thistle.
**cnidosis** (*ni-do'-sis*) [κνίδη, nettle]. Urtication; nettle-rash.
**Co.** Chemical symbol of cobalt.
**coagulable** (*ko-ag'-ū-la-bl*). Capable of coagulation.
**coagulant** (*ko-ag'-ū-lant*) [*coagulare*, to curdle]. 1. Causing the formation of a clot or coagulum. 2. A coagulating agent.
**coagulated** (*ko-ag'-ū-la-ted*) [*coagulare*, to curdle]. Clotted; curdle. c. **proteids**, a class of proteids produced by heating solutions of egg-albumin or serum-albumin up to 70° C. or higher. At the body-temperature they are readily converted into peptones by the action of the gastric juice in an acid medium, or of pancreatic juice in an alkaline medium.
**coagulation** (*ko-ag-ū-la'-shun*) [*coagulum*]. The formation of a coagulum or clot, as in blood or in milk. c. **necrosis**, a peculiar metamorphosis by which cells lose their vitality and change their chemical composition.
**coagulative** (*ko-ag'-ū-la-tiv*) [*coagulum*]. Causing or favoring or marked by coagulation. c. **necrosis**. See *necrosis, coagulative*.
**coagulin** (*ko-ag'-ū-lin*) [*coagulum*]. 1. A substance endowed with capacity to precipitate certain albuminous bodies contained in the culture-fluid injected into an inoculated animal. 2. A proprietary preparation used to check hemorrhage.
**coagulometer** (*ko-ag-ū-lom'-et-er*) [*coagulum*; μέτρον, a measure]. An apparatus for the determination of the rapidity of coagulation of the blood. c., **Wright's**, a cylinder surrounded by pockets for thermometer and coagulation-tubes.
**coagulose** (*ko-ag'-ū-lōs*). Trade name of a blood coagulant obtained by precipitating horse serum.
**coagulum** (*ko-ag'-ū-lum*) [*coagulare*, to curdle]. A clot. The mass of fibrin, inclosing red and colorless corpuscles and serum, that forms from the blood after the latter has been drawn from the body. Also, the curd of milk and the insoluble form of albumin.
**Coakley's operation** (*kōk'-le*) [Cornelius Godfrey Coakley, American laryngologist, 1862— ]. For disease of frontal sinus: the anterior wall of the sinus is removed and the sinus is curetted; the nasal duct is also curetted with a view to procuring its obliteration.

**coalescence** (*ko-al-es'-ens*) [*coalescere*, to grow together]. The union of two or more parts or things previously separate.
**coalescent** (*ko-al-es'-ent*). In a condition of coalescence.
**coalitus** (*ko-al'-it-us*) [L.]. Coalescent; coalescence. c. **artuum**, adhesion of limbs to each other. See *ankylomele*.
**coal-tar** (*kōl'-tar*). A by-product in the manufacture of illuminating gas; it is a black, viscid fluid, of a characteristic and disagreeable odor. The specific gravity ranges from 1.10 to 1.20. Its composition is extremely complex, and its principal constituents are separated, one from the other, by means of fractional distillation. Among the principal products manufactured from coal-tar are anthracene, benzol, naphtha, creosote, phenol, pitch, etc. From the basic oil of coal-tar are manufactured the anilin or coal-tar colors or dyes.
**coaptation** (*ko-ap-ta'-shun*) [*con*, together; *aptare*, to fit]. The proper union or adjustment of the ends of a fractured bone, the lips of a wound, etc.
**coarctate** (*ko-ark'-tāt*) [*coarctare*, to press together]. Crowded together. c. **retina,** a funnel-shaped retina.
**coarctation** (*ko-ark-ta'-shun*) [*coarctate*]. A compression of the walls of a vessel or canal, narrowing or closing the lumen; reduction of the normal or previous volume, as of the pulse; shriveling and consequent detachment, as of the retina. A stricture.
**coarctotomy** (*ko-ark-tot'-o-me*) [*coarctatus*, constricted; τέμνειν, to cut]. The cutting of a stricture.
**coarse** (*hors*). Not fine; gross. c. **adjustment**. See *adjustment, coarse*. c. **features of disease**, macroscopic organic lesions, such as swelling, hemorrhage, etc.
**coarticulation** (*ko-ar-tik-ū-la'-shun*) [*con*, together; *articulare*, to join, articulate]. A synarthrosis.
**coat** (*kōt*) [*cottus*, a tunic]. A cover or membrane covering a part or substance. c., **buffy**, the upper fibrinous layer of the clot of coagulated blood, characterized by its pale color, due to absence of red corpuscles. c., **internal elastic**. See *Henle's fenestrated membrane*. c., **internal fibrous**. See c., *subepithelial*. c., **middle**, the tunica media. c., **subepithelial**, the middle layer of the intima, composed of fusiform and stellate cells and finely granular substances with longitudinal and transverse fibrils. Syn., *innermost longitudinal fibrous coat; intermediary layer; internal fibrous coat; striated layer of the internal coat*. c., **uveal**, the uvea. c., **vaginal**. 1. The fibrous capsule of the eyeball. 2. See *tunica vaginalis*.
**coating** (*kōt'-ing*) [*cottus*, a tunic]. A covering, as of a wound, the tongue, etc. c. **of the tongue**, a condition of the tongue indicative of abnormality of the digestive tract. c. **of pills**, a covering of various substances to conceal the taste in swallowing.
**cobalt** (*ko'-bawlt*) [*Kobold*, a German mythological goblin]. A tough, heavy metal having some of the general properties of iron. Its oxides have been employed in medicine, but are now very little used. Symbol, Co; atomic weight, 58.97. c. **nitrate**, CO-(NO₃)₂+6H₂O. It is said to be a successful antidote in poisoning by hydrocyanic acid and potassium cyanide. c. **and potassium nitrate**, COK₃(NO₂)₆, cobalt yellow, a powder, slightly soluble in water; antispasmodic and antidyspneic. Dose ¼-½ gr. (0.016-0.032 Gm.). Syn., *potassium cobaltonitrite*. c. **salipyrin**, a salicylate of cobalt and antipyrin.
**Cobelli's glands** (*ko-bel'-e*). A ring of mucous glands in the mucosa of the esophagus, just above the cardia.
**cobra** (*ko'-brah*) [Port.]. A venomous snake of India, *Naja tripudians*. c.**-lysin**, Myers' term for the hemolytic poison of cobra venom. It is destroyed by heat and neutralized by antivenin. Cf. *cobra nervine; echidnase; echidnotoxin*. c. **nervine**, one of the principles isolated by Myers from cobra venom. It is not decomposed by heat or neutralized by antivenin.
**coca** (*ko'-kah*). See *Erythroxylon*.
**cocaethylin** (*ko-kah-eth'-il-in*), C₁₈H₂₃NO₄. A white powder obtained from benzoylecgonin by action of ethyl iodide. It is soluble in alcohol and ether and almost insoluble in water; it is a local anesthetic, milder than cocaine. Syn., *benzoylecgoninethylic ester; ethylbenzoylecgonin; homococaine*.
**cocaine** (*ko'-kah-ēn* or *ko-kān'*) [S.A., *coca*], C₁₇H₂₁-NO₄. Cocaine (*cocaina*, U. S. P.) is the chief alkaloid of *Erythroxylon coca*. It is at first stimulant and afterward narcotic, and resembles caffeine in

its action on the nerve-centers, and atropine in its effects on the respiratory and circulatory organs. Its long-continued use (*cocaine-habit*) is followed by insomnia, decay of moral and intellectual power, emaciation, and death. It is a local anesthetic when applied to the surface of mucous membranes or given hypodermatically. Applied to the conjunctiva of the eye, it causes also dilatation of the pupil and paralysis of the function of accommodation. Dose ½–2 gr. (0.008–0.13 Gm.). Syn., *methylbenzoylecgonin*. c. **aluminum citrate**, a double salt consisting of three molecules of albuminum citrate and one of cocaine; it is used as an astringent and as a local anesthetic. c. **aluminum sulphate**, a compound of aluminum sulphate and cocaine. It is used as is. cocaine aluminum citrate. c. **benzoate**, $C_{17}H_{21}NO_4 \cdot C_7H_6O_2$, anodyne and anesthetic. c. **borate**, a white, crystalline powder containing 68.7 % of cocaine. It is used in eye-douches and subcutaneous injections. c. **cantharidate**, $(C_{17}H_{21}NO_4)_2C_{10}H_{12}O_4$. It is used hypodermatically in tuberculosis; injection, $\frac{1}{160}$–$\frac{1}{48}$ gr. (0.0013–0.0016 Gm.) in 500 parts of chloroform-water. c. **carbolate**, a crystalline mass containing 75 % of cocaine; it is analgesic, anticatarrhal, and a local anesthetic. Dose $\frac{1}{16}$–$\frac{1}{6}$ gr. (0.005–0.01 Gm.), once or twice daily in capsules. Injection, 16 min. (1 Cc.) of 1:1250 solution in dilute alcohol. Application, 1 to 3 % solution with 30 % alcohol, 5 % powder, or pure. c. **cerate**, 1:30, for burns, etc. c. **chloride**. See c. *hydrochloride*. c. **citrate**, used to stop toothache. c. **hydrobromide**, $C_{17}H_{21}NO_4HBr$, small white crystals, soluble in water. It is used instead of cocaine hydrochloride and the dosage is the same. c. **hydrochloride** (*cocainæ hydrochloridum*, U. S. P.), $C_{17}H_{21}NO_4 \cdot HCl$, most commonly used for local anesthesia in 2 to 8 % solution, internally, $\frac{1}{4}$–2 gr. (0.008–0.13 Gm.). c. **hydrochloride, solution of** (*liquor cocainæ hydrochloratis*, B. P.). Dose 2–10 min. (0.13–0.65 Cc.). c. **hydroiodide**, $C_{17}H_{21}NO_4 \cdot HI$, a suggested substitute for cocaine hydrochloride in producing electroanesthesia. c. **lactate**, $C_{17}H_{21}NO_4C_3H_6O_3$, a white liquid of the consistence of honey; it is used particularly in tuberculous cysts of the bladder. Injection (into the bladder), 1½ gr. (0.1 Gm.) dissolved in 5 parts each of lactic acid, and distilled water. c. **lamellæ** (*lamellæ cocainæ*, B. P.), each contains $\frac{1}{640}$ gr. (0.0065 Gm.) of cocaine hydrochloride. c. **muriate**. See c. *hydrochloride*. c. **nitrate**, $C_{17}H_{21}NO_4 \cdot HNO_3$. It is used in combination with silver nitrate in treatment of disorders of the genitourinary tract. Dose, as the hydrochloride, maximum dose $\frac{3}{4}$ gr. (0.049 Gm.), single; 2¼ gr. (0.146 Gm.) a day. c. **oleate** (*oleatum cocainæ*, U. S. P.), a 10 % solution in oleic acid, for external use. c. **phenate**, a topical application in catarrhs and in rheumatism, used as a 5 to 10 % alcoholic solution; also internally. Dose $\frac{1}{12}$ gr. (0.005–0.01 Gm.). c. **phthalate**, contains 64.6 % of the alkaloid. It is used hypodermatically instead of cocaine hydrochloride. c. **saccharate**, moist crystalline plates used in diseases of the throat; to a 5 % solution corresponds to a 4 % solution of cocaine hydrochloride. c. **salicylate**, $C_{17}H_{21}NO_4 \cdot C_7H_6O_3$, is used in spasmodic asthma in the same manner as cocaine hydrochloride. c. **tartrate**, $(C_{17}H_{21}NO_4)_2 \cdot C_4H_6O_6$. Uses and dose same as of cocaine hydrochloride.

**cocainism** (*ko-ka'-in-izm*) [*cocaine*]. The cocaine-habit.

**cocainist** (*ko-ka'-in-ist*). One addicted to habitual use of cocaine.

**cocainization** (*ko-ka-in-iz-a'-shun*) [*cocaine*]. The bringing of the system or an organ under the influence of cocaine. c., **endomeningeal**, c., **intraspinal**, c., **spinal-canal**, c., **spinal subarachnoid**, c., **subarachnoid**. See *Corning-Bier* method under *anesthetic*.

**cocainize** (*ko-ka'-in-iz*). To bring under the influence of cocaine.

**cocainomania** (*ko-ka-in-o-ma'-ne-ah*) [*cocaine*; μανία, madness]. The habit of using cocaine; properly, insanity due to the cocaine-habit.

**cocainomaniac** (*ko-ka-in-o-ma'-ne-ak*) [*cocaine*; μανία, madness]. One who is insane from the effects of cocaine.

**cocapyrine** (*ko-ka-pi'-rin*). A mixture of cocaine, 1 part; antipyrine, 100 parts; used as an analgesic and antipyretic. Dose 3½ gr. (0.22 Gm.).

**Coccaceæ** (*kok-ka'-se-e*) [see *coccus*]. A group of schizomycetous fungi or bacteria, including as genera the *micrococcus, sarcina, ascococcus,* and *leuconostoc*.

**coccal** (*kok'-al*) [*coccus*]. Relating to cocci.

**coccidial** (*kok-sid'-e-al*). Relating to, or caused by, coccidia.

**coccidoidal granuloma** (*kok-sid-oi'-dal gran-u-lo'-ma*). Granuloma due to the presence of the *Oidium coccidoides*.

**Coccidioides immitis pyogenes** (*kok-sid-e-o'-id-ēs im-i'-tis pi-oj'-en-ēs*). A pathogenic microorganism discovered by Ophüls and Moffitt (1900). It produces in human beings chronic suppurative processes or caseation.

**coccidiosis** (*kok-sid-e-o'-sis*) [*coccidium; νόσος,* disease]. The group of symptoms produced by the presence of coccidia in the body.

**Coccidium** (*kok-sid'-e-um*) [*coccus;* pl., *coccidia*]. A genus of protozoa, by some referred to as the so-called psorosperms. See *psorosperm*. C. **oviforme**, has been found in intestinal epithelium and in the liver of man, and often in the liver of the rabbit. True coccidia are nonmotile cell-parasites. C. **sarkolytus**, the name given by Adamkiewicz to the so-called parasite of carcinoma.

**coccigenic** (*kok-sij-en'-ik*). [κόκκος, berry; γεννᾶν, to produce]. Caused by micrococcus.

**coccinella** (*kok-sin-el'-ah*). See *cochineal*.

**coccineous** (*kok-sin'-e-us*) [*coccinus,* scarlet]. In color, pure carmine tinged with yellow.

**coccobacteria** (*kok-o-bak-te'-re-ah*) [*coccus;* βακτήριον, a little rod]. The rod-like or spheroidal bacteria found in putrefying liquids, and called *C. septica*.

**coccogenous** (*kok-oj'-en-us*) [*coccus;* γεννᾶν, to produce]. Caused by the presence of cocci.

**coccomelasma** (*kok-o-mel-as'-mah*) [κόκκος, berry; μέλασμα, blackness]. A granular dermal melanosis.

**cocculin** (*kok'-ū-lin*). See *picrotoxin*.

**cocculus indicus** (*kok'-ū-lus in'-dik-us*). The dried fruit of *Anamirta cocculus*. It is an active narcotic poison. It is employed as a destroyer of vermin. See *picrotoxin*.

**coccus** (*kok'-us*) [κόκκος, a berry; pl., *cocci*]. 1. A genus of insects including *C. cacti,* the cochineal insect. 2. A spherical bacterium—a micrococcus.

**coccyalgia** (*kok-se-al'-je-ah*) [κόκκυξ, coccyx; ἄλγος, pain]. Coccygodynia.

**coccycephalus** (*kok-se-sef'-al-us*) [*coccyx;* κεφαλή, the head]. 1. Having a beaked process for a head. 2. A monstrosity with such a head.

**coccydynia** (*kok-se-din'-e-ah*). See *coccygodynia*.

**coccygeal** (*kok-sij'-e-al*) [κόκκυξ, coccyx]. Pertaining to the occcyx.

**coccygectomy** (*kok-sij-ek'-to-me*) [κόκκυξ, coccyx; ἐκτομή, excision]. Surgical excision of the coccyx.

**coccygeomesenteric** (*kok-sij-e-o-mes-en-jer'-ik*). Relating to the caudal and mesenteric areas; applied to an embryonic vein.

**coccygeus** (*kok-sij'-e-us*) [*coccyx*]. One of the pelvic muscles. See under *muscle*.

**coccygodynia** (*kok-sig-o-din'-e-ah*) [*coccyx; ὀδύνη,* pain]. Pain referred to the region of the coccyx; combined almost exclusively to women who have given birth to children.

**coccygotomy** (*kok-sig-ot'-o-me*) [κόκκυξ, coccyx; τομή, a cutting]. Cutting of the coccyx.

**coccyodynia** (*kok-se-o-din'-e-ah*). See *coccygodynia*.

**coccyx** (*kok'-siks*) [κόκκυξ, cuckoo (resembling the bill)]. The last bone of the spinal column, formed by the union of four rudimentary vertebræ.

**cochia** (*kok'-e-ah* or *kotch'-e-ah*) [L.; of Gr. κόκκιον, a pill]. A name for certain drastic and mainly aloetic pills (*pilulæ cochiæ*); pills of aloes and colocynth. The name cochia is now adjectival.

**cochineal** (*kotch'-in-ēl* or *kotch-in-ēl'*) [ME., *cutchaneal*]. The dried insects of a species of plant-lice, *Coccus cacti,* parasitic upon a cactus of Mexico and Central America. It contains a rich red coloring-matter, carmine, used mainly as a dyeing agent. It is thought to be valuable in whooping-cough. Dose ½ gr. (0.02 Gm.).

**cochinilin** (*kotch-in-il'-in*). The same as *carminic acid*.

**Cochin-leg** (*kot'-chin*) [*Cochin-China*]. Synonym of *elephantiasis arabum*.

**cochlea** (*kok'-le-ah*) [κόχλος, a conch-shell]. A cavity of the internal ear resembling a snail-shell. It describes 2½ turns about a central pillar called the modiolus or columella, forming the spiral canal, about 1½ inches in length. See also *ear*.

**cochlear, cochleare** (*kok'-le-ar, kok-le-a'-re*) [L.].

A spoon; a spoonful. c. **amplum,** c. **magnum,** a tablespoon. c. **medium,** a dessertspoon. c. **minimum,** c. **parvum,** a teaspoon.

**cochlear** (*kok'-le-ar*) [κόχλος, a conch-shell]. Pertaining or belonging to the cochlea. c. **nerve,** the nerve supplying the cochlea. See *nerves, table of.*

**Cochlearia** (*kok-le-a'-re-ah*). A genus of plants, containing *C. armoracia,* horse-radish, and *C. officinalis,* scurvy-grass.

**cochleariform** (*kok-le-ar'-e-form*) [*cochlear; forma,* shape]. 1. Spoon-shaped. 2. [κόχλος, a conch-shell.] Having the shape of a snail-shell.

**cochleate** (*kok'-le-āt*) [*cochleatus,* spiral]. Spirally coiled, like a snail-shell.

**cochleitis** (*kok-le-i'-tis*). See *cochlitis.*

**cochlitis** (*kok-li'-tis*) [*cochlea;* ιτις, inflammation] Inflammation of the cochlea.

**cocinin** (*ko'-sin-in*). A peculiar fatty principle, the chief constituent of cocoanut oil. Syn., *cocin; cocostearin; cocostearyl; cocyl.*

**Cock's operation** [Edward *Cock,* English surgeon, 1805–1892]. A method of external urethrotomy; the urethra is opened behind the stricture without a guide, the knife being carried into the median line of the perineum and the incision extended vertically as far as is necessary.

**Cock's peculiar tumor.** Extensive septic ulceration of the scalp, resembling an epithelioma and developed from a neglected sebaceous cyst.

**cockeye** (*kok'-i*). Strabismus.

**cockroach** (*kok'-rōch*). See *blatta.*

coco (*ko'-ko*). See *coko disease.*

**cocoa,** coco (*ko'-ko*). See *cacao* and *theobroma.* **c.-butter.** See *cacao-butter.*

**coco-olein** (*ko-ko-o'-le-in*) [*cacao; oleum,* oil]. A proprietary substitute for cod-liver oil, said to be derived from cocoa-nut oil.

**coctolabile** (*kok-to-la'-bil*) [*coctus,* boiled; *labilis* unstable]. Not able to withstand the temperature of boiling water without change.

**coctoprecipitin** (*kok-to-pre-sip'-it-in*) [*coctus,* cooked; *precipitin*]. A precipitin produced from a serum which has been boiled.

**coctostabile** (*kok-to-sta'-bil*) [*coctus,* boiled; *stabilis,* stable]. Able to withstand the temperature of boiling water without change.

**cod** (*kod*) [ME.]. The *Gadus morrhua,* a fish furnishing cod-liver oil. **c.-liver oil,** an oil derived from the liver of the *Gadus morrhua,* and ranging in color, according to the method of its preparation, from pale straw to dark brown; its specific gravity is 0.923 to 0.924 or even 0.930 at 15° C. See *morrhua.*

**codamine** (*ko'-dam-ēn*) [κώδεια, poppy-head; *amine*], $C_{20}H_{25}NO_4$. A crystalline alkaloid of opium, isomeric with laudanine. When ferric chloride is added to it, it assumes a deep-green color.

**codeia** (*ko-de'-ah*). See *codeine.*

**codeine** (*ko'-de-in*) [κώδεια, the poppy-head], $C_{18}H_{21}NO_3+H_2O$, *codeina* (U. S. P.). A white, crystalline alkaloid of opium resembling morphine in action, but being weaker. It is used in cough and in diabetes mellitus. Dose ½–2 gr. (0.032–0.13 Gm.). Syn., *codeia; methylmorphine.* c. **acetate,** $C_{18}H_{21}NO_3 . C_2H_4O_2$, use and dose same as codeine. c. **citrate,** used as is codeine. c. **hydrobromide,** $C_{18}H_{21}NO_3 . HBr+2H_2O$, used as is codeine. c. **hydrochloride,** $C_{18}H_{21}NO_3 . HCl+2H_2O$, use and dose same as codeine. c. **hydroiodide,** $C_{18}H_{21}NO_3 . HI+H_2O$, use and dose same as codeine. c. **nitrate,** $C_{18}H_{21}NO_3 . HNO_3$, use and dose same as codeine. c. **phosphate** (*codeinæ phosphas,* U. S. P.), soluble in water. It is similar to morphine in action, but less toxic. Dose hypodermatically, ½ gr. (0.032 Gm.). c. **salicylate,** a white powder, soluble in water, used in rheumatism. c. **sulphate** (*codeinæ sulphas,* U. S. P.), the sulphate of the alkaloid. Dose ⅙–¼ gr. (0.01–0.016 Gm.). c. **valerate,** an antispasmodic and sedative. Dose ¼ gr. (0.016 Gm.).

**codeonal** (*ko'-de-on-al*). Trade name of a preparation containing codeine and veronal; it is used as a hypnotic.

**codex** (*ko'-deks*) [L.; *pl., codices*]. A pharmacopœia or book of formulæ; specifically, the French pharmacopœia. c. **medicamentarius,** the French pharmacopœia.

**codol** (*ko'-dol*). See *retinol.*

**codrenin** (*kod-ren'-in*). Trade name of a preparation containing cocaine hydrochloride and adrenalin hydrochloride; used as a local anesthetic and hemostatic.

**coefficient** (*ko-ef-ish'-ent*) [*con,* together; *efficere,* to produce]. A figure indicating the degree of physical or chemical alteration characteristic of a given substance under stated conditions. **c., Baumann's.** See under *Baumann.* **c., biological,** the energy consumed by the body at rest. **c., Bouchard's.** See under *Bouchard.* **c., Haeser's.** See *Christison's formula.* **c., isotonic,** the lowest degree of concentration of a solution of a salt in which laking of blood does not occur. **c. jelly,** a preparation of agar used in Ross's *in vitro* method. c. **of diffusion,** the index of diffusion (*q. v.*) plus the time and temperature required to stain the nucleus. See in *vitro.* **c. of solubility of a gas,** the amount of a gas which is dissolved at a given temperature in 1 c.c. of a liquid, when the pressure of gas on the liquid is 760 mm. Hg. **c., Trapp's.** See *Trapp's formula.* **c., urotoxic,** the number of urotoxic units per kilogram of body weight excreted in twenty four hours. **c., Yvon's.** See under *Yvon.*

**cœlarium.** See *celarium.*

**coelectron** (*ko-e-lek'-tron*) [*con,* together; *electron*]. The matrix which, associated with the electron, forms the atom of ponderable matter. Syn., *atomic core.*

**cœlenteron** (*se-len'-ter-on*) [κοιλός, hollow; ἐντερόν, intestine]. Same as *archenteron.*

**cœlia** (*se'-le-ah*). See *celia.*

**cœliac** (*se'-le-ak*). See *celiac.*

**cœliadelphus** (*se-le-ad-el'-fus*). See *celiadelphus.*

**cœliagra** (*se-le-a'-grah*). See *celiagra.*

**cœlialgia** (*se-le-al'-je-ah*). See *celialgia.*

**cœliocyesis** (*se-le-o-si-e'-sis*). See *celiocyesis.*

**cœliolymph** (*se'-le-o-limf*). See *celiolymph.*

**cœliomyalgia** (*se-le-o-mi-al'-je-ah*). See *celiomyalgia.*

**cœlioncus** (*se-le-ong'-kus*). See *celioncus.*

**cœlioplegia** (*se-le-o-ple'-je-ah*) [κοιλία, the belly; πληγή, a stroke]. A synonym of Asiatic cholera.

**cœliorrhœa** (*se-le-or-e'-ah*). See *celiorrhea.*

**cœlioschisis** (*se-le-os'-kis-is*). See *celioschisis.*

**cœlioscope** (*se'-le-o-skōp*). See *celioscope.*

**cœliotomy** (*se-le-ot'-o-me*). See *celiotomy.*

**cœlitis** (*se-li'-tis*). See *celitis.*

**cœlom, cœloma** (*se'-lom, se-lo'-mah*). See *celom.*

**coelongate** (*ko-e-lon'-gāt*) [*con,* together; *elongatus,* elongated]. Of equal length.

**cœlophlebitis** (*se-lo-fle-bi'-tis*). See *celophlebitis.*

**cœlophthalmia** (*se-loff-thal'-me-ah*). See *celophthalmia.*

**cœlosis** (*se-lo'-sis*). See *celosis.*

**cœlosoma** (*se-lo-so'-mah*). See *celosoma.*

**cœlostomia** (*se-lo-sto'-me-ah*). See *celostomia.*

**cœnesthesis** (*sen-es-the'-sis*). See *cenesthesis.*

**cœnobium** (*se-no'-be-um*) [κοινός, common; βιος, life]. In biology, a composite zoophyte or any colony of independent cells held together by a common investment.

**cœnoblast** (*se'-no-blast*) [κοινός, common; βλαστός, a germ]. In biology the primitive germinal layer, giving rise to the endoderm and mesoderm.

**cœnotype** (*se'-no-tīp*) [κοινός, common; τύπος, type]. The fundamental type-form of a group.

**cœnurus** (*se-nūr'-us*) [κοινός, common; οὐρά, tail]. The larva of *Tænia cænurus,* producing the disease of sheep called staggers. c. **cerebralis,** a hydatid found mainly in the brain and spinal canal of the ox and sheep (mostly in young animals). Occasionally it has been discovered in the muscles of man. It is known to be the larva of the tape-worm, *Tænia cænurus.*

**co-enzyme** (*ko-en'-zīm*). A substance whose presence is essential for the due activity of a certain enzyme.

**coercible** (*ko-ers'-ib-il*) [*coercere,* to curb]. Applied to gases which are capable of being liquefied.

**coercive** (*ko-ers'-iv*). Capable of being rendered magnetic and continuing so.

**coferment** (*ko-fur'-ment*). Same as *co-enzyme.*

**coffea** (*kof'-e-ah*). The coffee tree. *C. arabica,* the common coffee plant, is the original source of most of the coffee cultivation.

**coffee** (*kof'-e*). See *coffea.* **c.-ground vomit,** the material ejected by emesis in gastric carcinoma and other conditions that give rise to a slow hemorrhage into the stomach. It consists of blood changed by the action of the gastric juice, and mixed with other contents of the stomach.

**coffeinism** (*kof'-e-in-izm*). Excessive habitual use of coffee, or the state of ill-health that results from it.

**coffeol** (*kof'-e-ol*). See *caffeol.*

**coffeon** (*kof'-e-on*). A product obtained by condensing the material volatilized when coffee is roasted. The pleasant flavor of coffee is due to it.
**coffer dam.** See *rubber dam.*
**coffeurin** (*kof-e-ū'-rin*) [*caffea*, coffee; οὖρον, urine]. A principle said to sometimes be present in urine after the free use of coffee as a beverage or medicine. The urine then has the odor of coffee, and its color is red, brownish, or deep-brown.
**coffin** (*kof'-in*) [κόφινος, a basket]. 1. A case intended to hold the dead body. 2. In farriery, the hollow portion of a horse's hoof. **c.-birth**, postmortem expulsion of the fetus. **c.-bone**, the last or distal phalanx of a horse's foot.
**coffinism** (*kof'-in-izm*) [after Dr. *Coffin*, who advocated it]. A variety of quackery or professed system of medical practice. It resembles so-called Thomsonianism.
**cognac** (*kōn-yak*) [a district in France]. French brandy distilled from wines produced in the district of Cognac.
**cognominal** (*kog-nom'-in-al*) [*cognomen*, a surname]. A word formed from an individual's surname; e. g., *mackintosh, ampère.*
**cog-wheel breathing, c. respiration.** A type of breathing characterized by a jerky, wavy inspiration.
**cohabitation** (*ko-hab-it-a'-shun*) [*con*, together; *habitare*, to dwell]. 1. The living together of a man and woman, with or without legal marriage. 2. Sexual connection.
**Cohen's test for albumin.** To the acid solution of albumin add a solution of potassium bismuthic iodide and potassium iodide. The albumin is precipitated.
**cohesion** (*ko-he'-shun*) [*cohærere*, to stick together]. The force whereby molecules of matter adhere to one another; the attraction of aggregation.
**Cohn's stomata.** Minute gaps in the interalveolar walls of the normal lung.
**Cohnheim's areas, C.'s fields** (*kōn'-hīm*) [Julius Friedrich *Cohnheim*, German pathologist, 1839–1884]. Small polygonal fields visible in transverse section of muscle fibers. **C.'s frog.** See *salt-frog*. **C.'s terminal arteries.** 1. Terminal arteries without anastomoses. 2. The short arteries supplying the basal ganglia of the cerebrum. **C.'s theory,** a theory that all true tumors are due to faulty embryonal development. The embryonal cells do not undergo the normal changes, are displaced, or are superfluous. When the favorable conditions are presented later in life, they take on growth, with the formation of tumors of various kinds. See *cancer, Cohnheim's theory of*. **C.'s tumor-germs,** small aberrant or heterotopic masses of embryonic tissue from which new growths may originate.
**cohobation** (*ko-ho-ba'-shun*) [*cohobare*, to redistil]. 1. Redistillation. 2. Recurrence of disease.
**cohosh** (*ko'-hosh*) [Am. Ind.]. A name given to several medicinal plants. **c., black.** See *cimicifuga.* **c., red.** See *Actæa rubra*. **c., white.** See *Actæa alba*.
**coil** (*koil*) [*colligere*, to gather together]. A spiral formed by winding. **c.-gland.** See *sweat-gland*. **c., induction-,** rolls of wire used to produce an electric current by induction. **., Leiter's.** See *Leiter's tubes*. **c., primary,** the inner coil of an induction apparatus. **c., resistance-,** a coil of wire of known electric resistance, used for estimating resistance. **c., secondary,** the outer coil of an induction apparatus.
**coin-catcher** (*koin'-katch-er*). An instrument for seizing and removing a coin or other foreign body lodged in the esophagus.
**coindication** (*ko-in-de-ka'-shun*) [*con*, with; *indicare*, to indicate]. A concurrent indication; a collateral and confirmatory indication. Cf. *contraindication*.
**coinosite** (*ko-in'-o-sīt*) [κοινός, common; σιτεῖν, to feed]. An animal parasite capable of separating itself from its host at will; a free commensal organism. **coin-sign, coin-test** (*koin'-sīn, koin'-test*). See *bell-sound*.
**coition** (*ko-ish'-un*). Same as *coitus*.
**coitophobia** (*ko-it-o-fo'-be-ah*) [*coitus;* φόβος, fear]. Morbid dread of coitus from disgust or dyspareunia.
**coitus** (*ko'-it-us*) [*coire*, to come together]. The act of sexual connection. Copulation. **c. disease,** the venereal disease of the horse. **c. interruptus.** See *c. reservatus*. **c. reservatus,** congressus interruptus; incomplete sexual intercourse; the incomplete performance of the sexual act; onanism.

**coko disease** (*ko'-ko*). A name applied in the Fiji Islands to a disease resembling frambesia.
**colalgia** (*ko-lal'-je-ah*) [*colon;* ἄλγος, pain]. Pain in the great intestine.
**cola-nut** (*ko'-lah-nut*). See *kola-nut*.
**colasaya** (*ko-las-a'-ah*). Trade name of a preparation of calisaya bark, cola, iron, and phosphates.
**colation** (*ko-la'-shun*) [*colare*, to strain]. The operation of straining.
**colatorium** (*kol-at-o'-re-um*) [L.]. A sieve, colander, or strainer; used in pharmacy.
**colature** (*ko'-lat-yūr*) [*colatura*, straining]. 1. In pharmacy, a liquid that has been subjected to colation. 2. See *colation*.
**colaxes** (*kol-awks'-e*) [κόλον, colon; αὔξη, increase]. Distention of the colon.
**colchicein** (*kol-chis-e'-in*) [*colchicum*], C₁₇H₂₁NO₅ +2H₂O. A crystalline decomposition-product of colchicine. It is used subcutaneously in treatment of gout. Dose $\frac{1}{60}$–$\frac{1}{30}$ gr. (0.001–0.002 Gm.).
**colchicine** (*kol'-chis-ēn*) [*colchicum*], C₂₂H₂₅NO₆, *colchicina* (U. S. P.). An alkaloid of colchicum; it is a pale, brownish-yellow, exceedingly bitter powder, freely soluble in water. It is a very active poison. Its dose is $\frac{1}{20}$ gr. (0.0032 Gm.) hypodermatically. **c. salicylate.** See *colchisal*.
**colchicum** (*kol'-chik-um*) [κολχικόν, colchicum]. Meadow-saffron. The corm and seed of *C. autumnale*, the properties of which are due to an alkaloid, *colchicine*. It is an emetic, diuretic, diaphoretic, and drastic cathartic. It is valuable in acute gout and in some forms of rheumatism. Dose of the *powdered corm* (*colchici cormus*, U. S. P.) 2–8 gr. (0.13–0.52 Gm.); of the *powdered seeds* (*colchici semen*, U. S. P.) 1–5 gr. (0.065–0.32 Gm.). **c. corm, extract of** (*extractum colchici cormi*, U. S. P.). Dose 1 gr. (0.065 Gm.). **c., extract of, acetic** (*extractum aceticum colchici*, B. P.). Dose ½–2 gr. (0.032–0.13 Gm.). **c. seed, fluidextract of** (*fluidextractum colchici seminis*, U. S. P.). Dose 3 min. (0.2 Cc.). **c. seed, tincture of** (*tinctura colchici seminis*, U. S. P.), 10 % strength. Dose 10–30 min. (0.6–2.0 Cc.). **c. seed, wine of** (*vinum colchici seminis*, U. S. P.), 10 % in strength. Dose 10–30 min. (0.6–2.0 Cc.).
**colchiflor** (*kol'-chi-flor*). A remedy for gout prepared from a tincture made from the fresh flowers of *Colchicum autumnale* and powdered kola. It is said to be free from the drastic properties contained in preparations from the bulb and seeds of colchicum.
**colchisal** (*kol'-chis-al*). Colchicine salicylate. A yellow, amorphous powder, soluble in alcohol, ether, and water. It is used in gout and arthritis. Dose $\frac{1}{100}$ gr. (0.00065 Gm.).
**colcothar** (*kol'-ko-thar*). A crude sequioxide of iron; red oxide of iron; a tonic and hemostatic.
**cold** (*kōld*) [AS., *ceald*]. 1. The comparative want of heat. 2. A term used popularly for coryza and catarrhal conditions of the respiratory tract. Cold is employed largely in various forms as a therapeutic agent, mainly for the purpose of lowering temperature and allaying irritation and inflammation. It may be used in the form of affusion, that is, the sudden application of a considerable volume of cold water to the body. Cold may be used as an anesthetic in baths (see *bath*); in the form of compresses applied over the affected part; in the form of irrigation, especially in the treatment of bruised and injured members; as a lotion, for the purpose of relieving local heat, pain, and swelling; as an injection, in the form of ice-water, into the vagina or rectum, for various conditions; and as the cold pack, which is a valuable means of reducing the body-temperature in cases of hyperpyrexia. Cold may be applied in the dry form by means of the ice-cap or bladder, an india-rubber bag filled with ice, snow, or a freezing mixture. **c. abscess.** See *abscess, cold*. **c. bath.** See *bath, cold*. **c.-blooded.** See *poikilothermic*. **c. coil.** See *Leiter's coils*. **c.-cream** (*unguentum aqua rosæ*, U. S. P.), spermaceti, 125 Gm.; white wax, 120 Gm.; expressed oil of almond, 560 Gm.; stronger rose-water, 190 Gm. in which finely powdered sodium borate 5 Gm. has been dissolved. When used as a vehicle for metallic salts the sodium borate should be omitted. Used for chapping of face and hands, abrasions, etc. **c. on the chest,** bronchial catarrh, *q. v.* **c. pack.** See *pack, cold.* **c., rose-,** hay-fever. **c.-sore,** herpes labialis. **c. spots.** See *temperature sense*. **c., St. Kilda's, c., strangers',** in the Hebrides, a form of

influenza ascribed by the natives to the arrival of a ship and the presence of outsiders. c. **stroke**, a condition, the analogue of heat-stroke, in which from excessive cold the patient is suddenly overcome and falls into a comatose state, shortly followed, as a rule, by death. c. **test**, a test applied, chiefly to lubricating oils to determine the point at which the oil begins to congeal, and ceases to flow.

**colectomy** (*ko-lek'-to-me*) [*colon;* ἐκτομή, cutting out]. Excision of a portion of the colon.

**coleitis** (*kol-e-i'-tis*) [κολεός, sheath; ιτις, inflammation]. Vaginitis.

**Coleman-Shaffer fever diet** (*kōl'-man shaf'-er*) [Warren *Coleman*, American physician, 1869– ; P. A. *Shaffer*, American physician]. Primarily arranged for typhoid, but probably suitable for other fevers. The principle is to supply the patient with sufficient food to maintain nutrition balance. It consists of large amounts of carbohydrate and fat, and relatively little protien. The proportions and total amount are regulated according to the digestive capacity of the patient.

**coleocele** (*kol'-e-o-sēl*) [κολεός, sheath; κηλή, tumor]. Vaginal tumor or hernia.

**coleocystitis** (*kol-e-o-sis-ti'-tis*) [κολεός, sheath; *cystitis*]. Inflammation of vagina and bladder.

**coleoptosis** (*kol-e-op-to'-sis*) [κολεός, sheath; πτῶσις, a fall]. Prolapse of the vaginal wall.

**coleorrhexis** (*kol-e-or-eks'-is*) [κολεός, a sheath; ῥῆξις, a rupture]. Rupture of the vagina.

**coleostegnosis** (*kol-e-o-steg-no'-sis*) [κολεός, a sheath; στέγνωσις, a constriction]. Contraction or atresia of the vagina.

**coleotomy** (*kol-e-ot'-o-me*) [κολεός, a sheath; τομή, a cutting]. A cutting operation upon the vagina; colpotomy.

**Coley's fluid, C.'s mixture** [William Bradley *Coley*, American surgeon, 1862– ]. A combination of the toxins of *Streptococcus erysipelatis* and *Bacillus prodigiosus;* it has been used as a remedy for cancer in the early stage.

**colibacillary** (*ko-li-bas'-il-a-re*). Pertaining to or produced by the *Bacillus coli communis*.

**colibacillosis** (*ko-li-bas-il-o'-sis*). The morbid condition due to infection with *Bacillus coli*.

**colibacilluria** (*ko-li-bas-il-u'-re-ah*). Presence of the *Bacillus coli communis* in the urine.

**colic** (*kol'-ik*) [*colon*]. 1. Pertaining to the colon. 2. A severe griping pain in the bowels, due to spasm of the intestinal walls; also any severe spasmodic pain in the abdomen. c., **appendicular**. Same as c., *vermicular* (1). c., **biliary**, that due to the passage of a gall-stone through the gall-ducts. c., **crapulent**, c., **crapulous**, that due to excess in eating and drinking. c., **cystic**, colicky pain in the urinary bladder. c., **Devonshire**. Synonym of c., lead. c., **hemorrhoidal**, intense pain near the anus and sacrum preceding a discharge from the hemorrhoidal vessels. c., **hepatic**, biliary colic. c., **hernIary**, the pain attending hernia. c., **inflammatory**, the intense pain attending colitis. c., **lead**, c., **saturnine**, intestinal colic due to lead-poisoning. It is characterized by excruciating abdominal pain, a hard and retracted condition of the abdomen, slow pulse, and increased arterial tension. Syn., *colica pictonum; painter's colic*. c., **menstrual**, the pain of menstruation. c., **metastatic**, that due to metastasis of gout or to suppression of the menses or the hemorrhoidal flow. c., **ovarian**, ovaralgia. c., **renal**, that due to the presence of a calculus in the ureter. c., **saburral**, that resulting from over-eating. c., **uterine**, colicky pains experienced at the menstrual epochs, often coming on in paroxysms. c., **vermicular**. 1. Pain in the vermiform appendix, due to catarrhal inflammation resulting from stoppage of its outlet. 2. That due to intestinal worms. Syn., *verminous colic; worm colic*.

**colica** (*kol'-ik-ah*) [L.]. 1. Colic artery, see *arteries, table of*. 2. Colic. c. **æruginis**, copper colic. c. **damnoniensis**, c. **damnoniorum**, c. **figulorum**, c. **hispaniensis**, lead colic. c. **intertropica**, lead colic, formerly supposed to be an endemic disease of the tropics. c. **japonica**. See *colic, crapulent*. c. **mucosa**. Synonym of *mucous colitis*. c. **pictonum** (literally, colic of the people of Poitou), lead-colic, painters' colic. See *colic, lead*. c. **pituitosa**. Synonym of *enteritis, pseudomembranous;* see *colitis, mucous*. c. **pulsatilis**. See *colic, inflammatory*. c. **rhachialgia**, lead colic. c. **scortorum**, a pain in the pelvis of youthful prostitutes, symptomatic of the ovarian disease that results from their unnatural sexual relations. It is due probably to a pyosalpinx.

**colicodynia** (*kol-ik-o-din'-e-ah*). Pain in the large intestine.

**colicolitis** (*ko-le-kol-i'-tis*). See *dysentery*.

**colicoplegia** (*ko-lik-o-ple'-je-ah*) [κολικός, pertaining to the colon; πληγή, a stroke]. Paralysis of the intestines. Also a synonym of *lead-colic*.

**colicystitis** (*ko-le-sist-i'-tis*) [*colon;* κύστις, bladder; ιτις, inflammation]. Cystitis dependent upon the pathogenic activity of the colon bacillus.

**colicystopyelitis** (*ko-le-sist-o-pi-e-li'-tis*). Combined cystitis and pyelitis due to *Bacillus coli communis*.

**coliform** (*kol'-e-form*) [*colum*, a sieve; *forma*, form]. 1. Sieve-like; ethmoid; cribriform. 2. Denoting or pertaining to those microorganisms which resemble the *Bacillus coli communis*.

**coli-group**. A group of pathogenic bacilli including the *Bacillus coli communis*, the typhoid, paratyphoid and paracolon bacilli, and some others.

**coli-infection**. Infection with the *Bacillus coli communis*.

**colilysin** (*ko-lil'-is-in*). A hemolysin formed by *Bacillus coli communis*.

**colipuncture** (*ko-le-punk'-chur*). See *colocentesis*.

**colitis** (*ko-li'-tis*) [*colon;* ιτις, inflammation]. Inflammation of the colon. c., **croupous**, c., **desquamative**, c., **diphtheritic**, c., **follicular**, c., **membranous**, c., **mucomembranous**, c., **plastic**. See *c., mucous*. c., **idiopathic ulcerative**, a specific affection due to microorganisms, beginning in and throughout its course, invariably limited to the colon. c., **mucous**, a clinical combination of symptoms characterized by periodic abdominal pains associated generally with abnormalities of the secretory and absorptive functions, and with the discharge of peculiarly formed mucous masses, sometimes resembling exact casts of the intestine. Syn., *chronic exudative enteritis; chronic mucocolitis; diarrhœa tubularis; fibrinous diarrhea; follicular-colonic dyspepsia; follicular duodenal dyspepsia; intestinal croup; pellicular enteritis; pseudomembranous enteritis*. c., **ulcerative**. Synonym of *dysentery*.

**collacin, collastin** (*kol'-as-in, -tin*). A substance found by Unna, in colloid degeneration of the skin.

**collætina** (*kol-e'-tin-ah*). A proprietary adhesive plaster said to consist of lanolin and caoutchouc.

**collaform** (*kol'-a-form*). A formaldehyde-gelatin preparation intended as a vulnerary.

**collagen** (*kol'-aj-en*) [κόλλα, glue; γεννᾶν, to produce]. A substance existing in various tissues of the body, especially bone and cartilage; it is converted into gelatin by boiling.

**collagenic** (*kol-a-jen'-ik*). Forming or producing collagen or gelatin.

**collagenous** (*kol-aj'-en-us*) [*collagen*]. Containing or resembling collagen or gelatin.

**collapse** (*kol-aps'*) [*collabi*, to fall together]. 1. Extreme depression and prostration from failure of the circulation, as in cholera, shock, hemorrhage, etc. 2. An abnormal sinking or retraction of the walls of an organ. c. **of lung**, return of a portion or the whole of a lung to its fetal or airless condition from some mechanical hindrance to the entrance of air. It is characterized by dyspnea, with more or less cyanosis, and is mainly encountered in bronchopneumonia.

**collapsing** (*kol-aps'-ing*) [*collapse*]. Suddenly breaking down. c. **pulse**. See *Corrigan's pulse*.

**collar-bone** (*kol'-er-bōn*). The clavicle.

**collar-crown** (*kol'-er-krown*). A collar-like device to hold an artificial tooth to a natural root.

**collargol, collargolum** (*kol-lar'-gol, -um*). Colloidal silver; argentum Credé. A nonirritating antiseptic.

**collateral** (*kol-at'-er-al*) [*con*, together; *lateralis*, of the side]. 1. Accessory or secondary; not direct or immediate. 2. One of the first branches of an axis-cylinder of a nerve-cell passing at a right angle.

**collecting tubes of the kidney**. A name given to the ducts discharging into the calices of the kidneys.

**collemia, collæmia** (*kol-e'-me-ah*) [κόλλα, glue; αἷμα, blood]. Haig's term for a condition of capillary obstruction which he attributes to a clogging of the capillaries by urates or colloid deposits.

**collenchyma** (*kol-en'-kim-ah*). The tissue of the primary cortex in plants, just beneath the epidermis.

**Colles' fascia** (*kol'-ēz*) [Abraham *Colles*, Irish surgeon, 1773–1843]. The deep layer of the perineal fascia. It is attached to the base of the triangular ligament, to the anterior lips of the rami of the pubes and ischia laterally, and anteriorly it is continuous with the dartos of the scrotum. **C.'s fracture**, transverse fracture of the lower extremity of the radius, with displacement of the hand backward and outward. **C.'s law**, the child of a syphilitic father will render its mother immune against syphilis. In Colles' original words: "A new-born child affected with congenital syphilis, even[although it may have symptoms in the mouth, never causes ulceration of the breast which it sucks, if it be the mother who suckles it, though continuing capable of infecting a strange nurse." **C.'s ligament**, the fibers which pass from the outer portion of Poupart's ligament behind the internal pillar of the abdominal ring and are inserted into the linea alba, where they interlace with those of the opposite side. **C.'s space**, the space beneath the perineal fascia containing the ischiocavernosus, transversus perinei, and bulbocavernosus muscles, the bulbous portion of the urethra, the posterior scrotal (labial) vessels and nerves, and loose areolar tissue.

**Colles-Beaumès' law** (*kol'-ez-bo'-ma*) [see *Colles; Beaumès*]. See *Colles' law*.

**colliculectomy** (*kol-ik-ū-lek'-to-me*) [*colliculus; ἐκτομή*, excision]. Removal of the verumontanum.

**colliculitis** (*kol-lik-ū-li'-tis*) [*colliculus*, mound, ιτις, inflammation]. Inflammation of the colliculus seminalis.

**colliculus** (*kol-ik'-ū-lus*) [dim. of *collis*, hill: *pl., colliculi*]. 1. A small eminence. 2. The verumontanum. **c. bulbi**, cerebral parietes. 3. The verumontanum. **c. bulbi, c. bulbi intermedius**, the layer of erectile tissue surrounding the male urethra on its entrance into the bulb. **c. cervicalis**, a fold of mucosa extending dorsad from the apex of the trigonum vesicæ. **c. facialis**, the eminentia facialis; see under *eminence*. **c. glandis**, two eminences on the inferior surface of the corona of the glans penis, between which the frenum is attached. **c. inferior**, any one of the posterior quadrigeminal bodies; see *postgeminum*. **colliculi nervi ethmoidalis**, the striatum. **colliculi nervorum opticorum**, the thalamus. **c. papillaris**, the tuberculum or processus papillaris of the liver. **colliculi posteriores**. See *postgeniculum*. **c. rotundus**. See *terete eminence*. **c. rotundus anterior**, the anterior division of the terete eminence. **c. rotundus posterior**, the posterior division of the terete eminence. **c. seminalis**. See *verumontanum*. **c. superior**, any one of the anterior quadrigeminal bodies; see *pregeminum*. **c. urethralis**, the verumontanum.

**collidine** (*kol'-id-in*) [*κόλλα*, glue], $C_8H_{11}N$. A ptomaine obtained from pancreas and gelatin allowed to putrefy together in water. **c. aldehyde**. See *aldehyde, collidine*.

**collier's lung** (*kol'-yers*). Synonym of *anthracosis*. **Collier's ponto-spinal tract** (*kol'-yers*). The descending part of the medial longitudinal bundle, found in the tegmentum.

**colliform** (*kol'-if-orm*). A proprietary preparation containing formaldehyde and gelatin; used as a dressing for wounds.

**colligamen** (*kol-li-ga'-men*) [*colligare*, to bind]. 1. A ligament. 2. A name given to a variety of bandages prepared with glycerol and a glycerol-zinc paste.

**collilongus** (*kol-e-long'-gus*) [*collum*, neck; *longus*, long]. The muscle called longus colli. See *muscles*.

**collimator** (*kol'-im-a-tor*) [*collineare*, to aim at]. The receiving telescope of a spectroscope.

**collin** (*kol'-in*) [*κόλλα*, glue]. Gelatin in soluble form.

**collinic** (*kol-in'-ik*). Relating to or obtained from gelatin.

**Collinsonia** (*kol-in-so'-ne-ah*) [Peter *Collinson*, English botanist, 1694–1768]. A genus of labiate herbs. *C. canadensis*, stoneroot, healall, is a coarse plant with a disagreeable smell; it has tonic, diuretic, and diaphoretic properties. Dose 15-60 gr. (1-4 Cc.) in decoction; of the *fluidextract* 10 min.-1 dr. (0.65–4.0 Cc.); of the *tincture* (1 : 10) ½-2 dr. (2-8 Cc.).

**collinsonin** (*kol-in-so'-nin*) [*Collinsonia*]. A precipitate from the tincture of the root of *Collinsonia canadensis*. It is tonic, diaphoretic, resolvent, and diuretic. Dose 2 to 4 grains.

**colliquation** (*kol-ik-wa'-shun*) [*con*, together; *liquare*, to melt]. The liquefaction or breaking down of a tissue or organ.

**colliquative** (*kol-ik'-wa-tiv*) [*colliquation*]. Profuse or excessive; marked by excessive fluid discharges. **c. diarrhea**, a profuse watery diarrhea. **c. necrosis**. See *necrosis, liquefactive*. **c. sweat**, a profuse clammy sweat.

**colliquefaction** (*kol-ik-we-fak'-shun*) [*colliquation*]. A melting or fusing together.

**collocated** (*kol'-o-ka-ted*) [*collocare*, to place]. Corresponding with in respect to location; applied especially to parts of the brain that are adjacent, one ectal and the other ental; *e. g.*, the calcarine fissure and the calcar.

**collocystis** (*kol-o-sis'-tis*) [*κόλλα*, glue; *κύστις*, cyst]. A gelatin capsule to facilitate the swallowing of a drug.

**collodion** (*kol-o'-de-on*) [*κολλώδης*, glue-like]. *Collodium* (U. S. P.). A dressing for wounds made by dissolving guncotton in ether and alcohol; it is used as a substitute for adhesive plaster. See *pyroxylin*. **c., acetone**, one prepared from guncotton, 5 parts; ether, 10 parts; alcohol, 10 parts; acetone, 20 parts; castor-oil, 6 parts. It is more elastic than ordinary flexible collodion. **c., blistering**. Same as *c., cantharidal*. **c., cantharidal** (*collodium cantharidatum*, U. S. P.), a blistering solution of collodion and cantharides. **c., flexible** (*collodium flexile*; U. S. P.), collodion with the addition of castor-oil and Canada balsam. **c., iodized, flexible** collodion with the addition of 5 % of iodine. It is used in chilblains. **c., iodoform**, flexible collodion with 5 % of iodoform. **c., styptic** (*collodium stypticum*, U. S. P.), a mixture of collodion with tannic acid, ether, and alcohol.

**collodium** (*kol-o'-de-um*). See *collodion*.

**colloid** (*kol'-oid*) [*κόλλα*, glue; *εἶδος*, likeness]. 1. A nondialyzable organic substance. See *dialysis*. 2. A substance formed by colloid degeneration of the epithelium. See *degeneration, colloid*. 3. Having the nature of glue. 4. In chemistry, amorphous and noncrystalline. **c. cancer**. See *cancer, colloid*. **c. cyst**, a cyst with jelly-like contents. **c. degeneration**. See *degeneration, colloid*.

**colloidal** (*kol-oid'-al*). See *colloid* (3 and 4).

**colloidin** (*kol-oid'-in*) [*colloid*], $C_5H_{11}NO_4$. A jelly-like substance obtained from colloid tissue.

**colloidogen** (*kol-oid'-o-jen*). A hypothetical substance which is presumed to be concerned in holding the mineral matter of the body in a colloid state.

**coloma** (*kol-o'-mah*) [*κόλλα*, glue; *ὄμα*, a tumor]. 1. A cystic tumor containing a gelatiniform substance. 2. A colloid cancer which has undergone degenerative changes.

**collonema** (*kol-o-ne'-mah*) [*κόλλα*, glue; *νῆμα*, tissue: *pl., collonemata*]. Myxoma, or myxosarcoma.

**collopexia** (*kol-o-pek'-se-ah*) [*collum*, neck; *πῆξις*, fixing]. The surgical fixation of the neck of the uterus; trachelopexia.

**collosin** (*kol'-o-sin*) [*κόλλα*, glue]. A skin-varnish made by the addition of camphor to acetone collodion.

**colloxylin** (*kol-oks'-il-in*). See *pyroxylin*.

**collum** (*kol'-um*) [L.]. The neck; especially the anterior part of the neck. **c. anatomicum**, the anatomical neck of the humerus. **c. chirurgicum**, the surgical neck of the humerus. **c. distortum**. Synonym of *torticollis*. **c. femoris**, the neck of the femur. **c. pedis**, the instep.

**collunarium** (*kol-lū-na'-re-um*) [L.]. A nasal douche.

**collutorium**, **collutory** (*kol-u-to'-re-um, kol'-u-to-re*) [*colluere*, to rinse]. A mouth-wash; a gargle.

**collyrium** (*kol-ir'-e-um*) [*κολλύριον*, an eye-salve]. 1. A lotion for the eyes. 2. Formerly, a suppository.

**coloboma** (*kol-o-bo'-mah*) [*κολοβοῦν*, to mutilate]. A congenital fissure of the iris, choroid, or eyelids. **c., bridge**, a variety in which the pupil is separated from the coloboma by a narrow thread of iris tissue that stretches like a bridge from one pillar of the disc. **c. palpebræ**, **c. palpebrarum**, a form of partial ablepharia consisting in a fissure of the eyelid—most frequently the upper lid. Syn., *blepharocoloboma*.

**colocentesis** (*ko-lo-sen-te'-sis*) [*colon; κέντησις*, puncture]. Surgical puncture of the colon.

**colocholecystostomy** (*ko-lo-kol-e-sis-tos'-to-me*). See *cholecystocolotomy*.

**colocleisis** (*ko-lo-kli'-sis*) [*colon; κλεῖσις*, closure]. Occlusion of the colon.

**colocyster** (*ko-lo-klis'-ter*) [*colon; clyster*]. An enema in the colon.

**colocolostomy** (*ko-lo-kol-os'-to-me*). The operation of forming a connection between two portions of the colon.

**colocynth** (*kol'-o-sinth*). Same as *colocynthis*.
**colocynthein** (*kol-o-sin'-the-in*) [κολοκυνθίς, colocynth], C₆H₁₄O₁₈. A resinous decomposition-product of colocynthin.
**colocynthidism** (*kol-o-sinth'-id-izm*) [*colocynthis*]. Poisoning from undue use of colocynth. A condition marked by violent inflammation of the digestive tract, watery and bloody stools, bilious vomiting, cramps in the calves of the legs, and collapse.
**colocynthin** (*kol-o-sin'-thin*) [*colocynthis*]. The bitter principle of colocynth. See *colocynthis*.
**colocynthis** (*kol-o-sin'-this*) [κολοκυνθίς, colocynth; *gen., colocynthidis*]. Colocynth. Bitter apple. The fruit of *Citrullus colocynthis*, from which the seeds and rind have been removed. Its properties are due to a bitter glucoside, *colocynthin*, C₅₆H₈₄O₂₃, the dose of which is 1/10–1/2 gr. (0.003–0.013 Gm.). It is a tonic and astringent purgative, and is used mainly as an ingredient of compound cathartic pills. **colocynthidis, extractum** (U. S. P.), alcoholic. Dose 1/2–2 gr. (0.032–0.13 Gm.). **colocynthidis, extractum, compositum** (U. S. P.), contains colocynth extract, 16; aloes, 50; cardamom, 6; resin of scammony, 14; soap, 14; alcohol, 10 parts. Dose 5–20 gr. (0.32–1.3 Gm.). **colocynthidis, pilula, composita** (B. P.), contains colocynth, aloes, scammony, potassium sulphate, and oil of cloves. Dose 5–10 gr. (0.32–0.65 Gm.). **colocynthidis, pilulæ, et hyoscyami** (B. P.), pills of colocynth and henbane. Dose 5–10 gr. (0.32–0.65 Gm.).
**coloenteritis** (*ko-lo-en-ter-i'-tis*) [*colon; enteritis*]. Inflammation of the small and large intestines. Occ. *enterocolitis*.
**colohepatopexy** (*ko-lo-hep-a-to-pek'-se*) [*colon*; ἧπαρ, liver; πῆξις, fixation]. Fixation of the colon to the liver to form adhesions, after operations on the gall-bladder.
**colomba,** colombo (*kol-om'-bah, kol-om'-bo*). See *calumba*.
**colon** (*ko'-lon*) [κῶλον, the colon]. The part of the large intestine beginning at the cecum and terminating at the end of the sigmoid flexure. In the various parts of its course it is known as the *ascending* colon, the *transverse* colon, the *descending* colon, and the *sigmoid flexure;* this latter is sometimes divided into the *iliac* colon, and *pelvic* colon.
**colonalgia** (*ko-lon-al'-je-ah*) [*colon;* ἄλγος, pain]. Pain in the colon; colic.
l ni (*ko-lon'-ik*) [*colon*]. Pertaining to the colmo c.
**colonitis** (*ko-lon-i'-tis*). See *colitis*.
**colonometer** (*kol-on-om'-et-er*) [*colony*; μέτρον, a measure]. An apparatus for estimating the number of colonies of bacteria on a culture-plate.
**colonopexy** (*ko-lon'-o-pek-se*). Same as *colopexy*.
**colonoscope** (*ko-lon'-o-skōp*) [*colon;* σκοπεῖν, to view]. An instrument for examining the colon.
**colonoscopy** (*ko-lon-os'-ko-pe*). Examination by means of a colonoscope.
**colony** (*kol'-o-ne*) [*colonia*, colony]. A collection or assemblage, as of microorganisms in a culture.
**colopexia,** colopexy (*ko-lo-peks'-e-ah, ko'-lo-pek-se*) [*colon;* πῆξις, a fixing]. Suturing of the sigmoid flexure to the abdominal wall.
**colopexotomy** (*ko-lo-peks-ot'-o-me*) [*colon;* πῆξις, a fixing; τέμνειν, to cut]. Incision into and fixation of the colon. Colopexy with the formation of an artificial anus.
**colophonium** (*kol-o-fo'-ne-um*). Rosin; used in differentiating eosinmethylene-blue and other stains.
**colophony** (*kol'-o-fo-ne*) [Κολοφών, a city of Ionia]. Rosin. The solid residue left on distilling off the volatile oil from crude turpentine. See *rosin*.
**coloproctia** (*kol-o-prok'-te-ah*) [*colon;* πρωκτός, anus]. The formation of an artificial colonic anus.
**coloptosis** (*ko-lop-to'-sis*) [*colon;* πτῶσις, a falling]. Descent or displacement of the colon.
**colopuncture** (*kol-o-punk'-chur*). Same as *colocentesis*.

color (*bul'-or*) [Π.,], 1. A visual sensation due to radiated or reflected light. 2. That quality of an object perceptible to sight alone. 3. A pigment. **c.-analysis,** Ehrlich's method of identifying the various forms of leukocytes. It depends upon the distinctive manner in which the protoplasmic granules react toward the acid, basic, and socalled neutral solutions of the anilin dyes. Five varieties of granules are recognized and designated by the Greek letters, α, β, γ, δ, ε: (1) α-*granules* (eosinophil, oxyphil, or coarse oxyphil granules); (2) β-*granules* (amphophil granules); (3) γ-*granules* (mast-cell or coarse basophil granules); (4) δ-*granules* (fine basophil granules); (5) ε-*granules* (neutrophil or fine oxyphil granules). **c.-blindness.** See *blindness*. *color-.* **c.,** complementary, any color that added to another color, or to a mixture of colors, produces white. **c.-gustation.** See *pseudogeusesthesia*. **c.-hearing,** the excitation of the visual center for color through the auditory nerve. **c.-index,** the proportionate amount of hemoglobin contained in each red blood corpuscle. It is normally from .9 to 1.5. **c.-sensation,** the perception of color; it depends on the number of vibrations of the ether.
**colorectitis** (*ko-lo-rek-ti'-tis*) [*colon; rectum; ἴτις,* inflammation]. Inflammation of the colon and rectum. Also a synonym of *dysentery*.
**colorectostomy** (*ko-lo-rek-tos'-to-me*) [*colon; rectum;* στόμα, a mouth]. The surgical establishment of a passage between the colon and the rectum.
**colorimeter** (*kol-or-im'-et-er*) [*color;* μέτρον, a measure]. An instrument for determining the quantity of coloring-matter in a mixture, as in the blood.
**colorimetric** (*kol-or-im-et'-rik*). Relating to methods of color-measuring.
**colostomy** (*ko-los'-to-me*) [*colon;* στόμα, a mouth]. 1. The formation of an artificial anus by an opening into the colon. 2. Any surgical operation upon the colon that makes a permanent opening into it, whether internal or external.
**colostration** (*kol-os-tra'-shun*) [*colostrum*]. A disease or illness of young infants ascribed to the effects of the colostrum.
**colostric** (*kol-os'-trik*) [*colostrum*]. Relating to colostrum.
**colostrorrhea** (*ko-los-tro-re'-ah*) [*colostrum; ῥοία,* flow]. Profuse discharge of colostrum.
**colostrous** (*kol-os'-trus*) [*colostrum*]. Of the nature of or containing colostrum.
**colostrum** (*kol-os'-trum*) [L.]. The first milk from the mother's breasts after the birth of the child. It is laxative, and assists in the expulsion of the meconium. **c. corpuscles,** small microscopic bodies contained in the colostrum. They are the epithelial cells of the mammary glands, full of oil-globules. After about the third day these cells burst and set free the fat-globules before they leave the gland, and in this way the true milk is formed.
**colotomy** (*ko-lot'-o-me*) [*colon;* τέμνειν, to cut]. Incision of the colon, abdominal, lateral, lumbar, or iliac, according to the region of entrance.
**colotyphoid** (*ko-lo-ti'-foid*). Typhoid accompanied with follicular ulceration of the colon and lesions in the small intestine.
**colpalgia** (*kol-pal'-je-ah*) [*colpo-;* ἄλγος, pain]. Vaginal pain or neuralgia.
**colpatresia** (*kol-pat-re'-se-ah*) [*colpo-;* ἀτρητος, not perforated]. Occlusion or atresia of the vagina.
**colpectasia** (*kol-pek-ta'-she-ah*) [*colpo-;* ἔκτασις, a stretching out]. Vaginal dilatation.
**colpedema** (*kol-ped-e'-mah*) [*colpo-;* οἴδημα, swelling]. Edema of the vagina.
**colpemphraxis** (*kol-pem-fraks'-is*) [*colpo-;* ἔμφραξις, a stoppage]. Obstruction of the vagina.
**colpeurynter** (*kol-pu-rin'-ter*) [*κόλπος,* vagina; εὐρύνειν, to widen]. An inflatable bag or sac used for dilating the vagina.
**colpeurysis** (*kol-pu'-ris-is*) [see *colpeurynter*]. Dilation of the vagina, especially that effected by means of the colpeurynter.
**colpitis** (*kol-pi'-tis*) [κόλπος, vagina; ἴτις, inflammation]. Inflammation of the vagina.
**colpo-** (*kol-po-*) [κόλπος, vagina]. A prefix denoting relation to the vagina.
**colpocace** (*kol-pok'-a-se*) [*colpo-;* κακός, bad]. Gangrene of the vagina.
**colpocele** (*kol'-po-sēl*) [*colpo-;* κηλή, hernia]. Hernia or tumor in the vagina.
**colpoceliotomy** (*kol-po-se-le-ot'-o-me*) [*colpo-;* *celiotomy*]. Incision into the abdomen by the vaginal route. c., antero-lateral, Duhrssen's name for a new vaginal operative route into the abdomen. It consists in a combination of vaginal celiotomy with complete division of one broad ligament.
**colpocleisis** (*kol-po-kli'-sis*) [*colpo-;* κλεῖσις, a closure]. The surgical closure of the vagina.
**colpocystic** (*kol-po-sis'-tik*) [*colpo-;* *κύστις,* bladder]. Relating to the vagina and the bladder; vesico-vaginal.
**colpocystitis** (*kol-po-sis-ti'-tis*) [*colpo-;* κύστις,

bladder; ιτις, inflammation]. Inflammation of the vagina and the bladder.

**colpocystocele** (kol-po-sis'-to-sēl) [colpo-; κύστις, a bladder; κηλή, a tumor]. A hernia of the bladder into the vagina, with prolapse of the anterior vaginal wall.

**colpocystoplasty** (kol-po-sist'-o-plas-te) [colpo-; κύστις, bladder; πλάσσειν, to form]. Plastic surgery of the vagina and bladder.

**colpocystosyrinx** (kol-po-sis-to-sir'-ingks) [colpo-; κύστις, cyst; σύριγξ, pipe]. Vesico-vaginal fistula.

**colpocystotomy** (kol-po-sis-tot'-o-me) [colpo-; κύστις, bladder; τομή, a cut]. Surgical incision of the bladder through the vaginal wall.

**colpocystoureterocystotomy** (kol-po-sist-o-ū-re-ter-o-sist-ot'-o-me) [colpo-; κύστις, bladder; ureter; cystotomy]. Exposure of the orifices of the ureter by incision of the walls of the bladder and vagina.

**colpodesmorrhaphy** (kol-po-des-mor'-af-e) [colpo-; δεσμός, a fastening; ῥαφή, a seam]. Suturing of the vaginal sphincter.

**colpohyperplasia** (kol-po-hi-per-pla'-ze-ah) [colpo-; hyperplasia]. Hyperplasia of the vagina. c. cystica, a form of degeneration of the vaginal mucosa, occurring during pregnancy, and characterized by the formation of gas-cysts, due, according to Lindenthal, to an organism which he calls Bacillus emphysematis vaginæ. Syn., colpitis vesiculosa emphysematosa; emphysema vaginæ.

**colpohysterectomy** (kol-po-his-ter-ek'-to-me) [colpo-; hysterectomy]. Removal of the uterus through the vagina.

**colpohysteropexy** (kol-po-his'-ter-o-pek-se) [colpo-; hysteropexy]. Vaginal hysteropexy; supravaginal amputation of the cervix and anastomosis of the uterus and the vaginal mucosa. Syn., hysteropexy vaginalis.

**colpohysterorrhaphy** (kol-po-his-ter-or'-af-e) [colpo-; hysterorrhaphy]. An operation for prolapse of the uterus; colpohysteropexy.

**colpohysterotomy** (kol-po-his-ter-ot'-o-me) [colpo-; ὑστέρα, womb; τομή, section]. Surgical incision of the uterus through the vagina.

**colpomyomectomy** (kol-po-mi-o-mek'-to-me) [colpo-; myomectomy]. Myomectomy through the vagina.

**colpomyomotomy** (kol-po-mi-o-mot'-o-me) [colpo-; μῦς, a muscle; τομή, a section]. Vaginal myomotomy; colpomyometomy.

**colpopathy** (kol-pop'-a-the) [colpo-; πάθος, illness]. Any disease of the vagina.

**colpoperineoplasty** (kol-po-per-in-e'-o-plas-te) [colpo-; perineum; πλάσσειν, to form]. Plastic operation for abnormality of the vagina and perineum.

**colpoperineorrhaphy** (kol-po-per-in-e-or'-af-e) [colpo-; perineorrhaphy]. Repair of a perineal laceration by denuding and in part suturing the posterior wall of the vagina.

**colpopexy** (kol'-po-pek-se) [colpo-; πῆξις, fixation]. Fixation of the vagina; vaginapexy.

**colpoplastic** (kol-po-plas'-tik) [colpo-; πλάσσειν, to form]. Relating to colpoplasty.

**colpoplasty** (kol'-po-plas-te) [colpo-; πλάσσειν, to form]. Plastic surgical operation upon the vagina.

**colpopolypus** (kol-po-pol'-ip-us) [colpo-; polypus]. Polypus of the vagina.

**colpoptosis** (kol-pop-to'-sis) [colpo-; πτῶσις, a falling]. Prolapse of the vaginal walls.

**colporrhagia** (kol-por-a'-je-ah) [colpo-; ῥηγνύναι, to burst forth]. Vaginal hemorrhage.

**colporrhaphy** (kol-por'-a-fe) [colpo-; ῥαφή, a seam]. Suture of the vagina.

**colporrhea** (kol-por-e'-ah) [colpo-; ῥοία, a flow]. Vaginal leukorrhea; a mucous discharge from the vagina.

**colporrhexis** (kol-por-ek'-sis) [colpo-; ῥῆξις, rupture]. Laceration or rupture of the vagina.

**colposcope** (kol'-po-skōp) [colpo-; σκοπεῖν, to view]. An instrument for the visual examination of the vagina; a vaginal speculum.

**colpospasmus** (kol-po-spaz'-mus) [colpo-; σπασμός, spasm]. Spasm of the vagina.

**colpostegnosis** (kol-po-steg-no'-sis) [colpo-; στέγνωσις, closure]. Same as colpostenosis.

**colpostenosis** (kol-po-sten-o'-sis) [colpo-; στένωσις, a narrowing]. Constriction of the vagina.

**colpostenotomy** (kol-po-sten-ot'-o-me) [colpo-; στενός, narrow; τομή, a cutting]. Surgical division of colpostenosis.

**colposynizesis** (kol-po-sin-iz-e'-sis) [colpo-; συνίζησις, a falling in]. Narrowness of the vagina.

**colpotomy** (kol-pot'-o-me) [colpo-; τομή, a cutting]. Surgical incision of the vagina.

**colpoureterocystostomy** (kol-po-ū-re-ter-o-sis-tot'-o-me) [colpo-; ureter; cystotomy]. The exposure of the ureteral openings by cutting through the walls of the vagina and bladder. Colpocysto-ureterotomy.

**colpoureterotomy** (kol-po-ū-re-ter-ot'-o-me). Incision of the ureter through the vagina.

**colpoxerosis** (kol-po-ze-ro'-sis) [colpo-; ξερός, dry]. Morbid dryness of the vagina.

**colt-ill** (kōlt'-il). See strangles.

**coltsfoot**. The leaves of Tussilago farfara, a demulcent and tonic, sometimes prescribed in chronic cough. Dose of a decoction (1 oz. to 1 pint) a teacupful; of fluidextract 1–2 dr. (4–8 Cc.).

**columbin** (kol-um'-bin). See calumbin under calumba.

**columbium** (ko-lum'-be-um). A rare chemical element, symbol Cb; atomic weight 93.5.

**columbo** (kol-um'-bo). See calumba.

**columella** (kol-ū-mel'-ah) [L., "a little column"]. 1. The modiolus or central axis of the cochlea of the human ear. See cochlea and modiolus. 2. A bone in birds which takes the place of the ossicles of the ear in man. c. fornicis, one of the anterior pillars of the fornix. c. nasi, the septum of the nose.

**column** (kol'-um) [columna]. A name given to several parts of the body that furnish support to surrounding parts, or that have the shape of pillars. c., anterior, the layer of white matter in either half of the spinal cord included between the anterior horn and nerve-root and the anterior median fissure. c., anterolateral ascending. See Gowers' c. c., commissural, one of nerve-substance extending along the bottom of the horizontal fissure of the cerebellum and connecting the anterior and posterior superior lobes. c., crossed pyramidal, c., crossed, of Tuerck, c., cuneiform. See c., posteroexternal. c., direct cerebellar, in the spinal cord, is situated outside of the lateral pyramidal tract. c., direct pyramidal, c., direct, of Tuerck. See c. of Tuerck. c. of the external ring, the free border of the aponeurosis of the external oblique muscle, forming the edges of the external abdominal ring. c., gray, anterior, the ventral cornu of the spinal cord. c., gray, posterior, the dorsal cornu of the spinal cord. c. of the intermediolateral tract, a column of motor cells in the intermediolateral tract of the spinal cord; it is best seen in the dorsal region. c., lateral, the layer of white matter in either half of the spinal cord included between the posterior horn and nerve-roots and the anterior horn and nerve-roots. c., posterior, a collection of white matter situated in the spinal cord on either side between the posterior horns and posterior nerve-roots and the posterior median fissure. c., posteroexternal, the outer wider division of the posterior column of the cord; the column of Burdach. c., posteromedian, the median division of the posterior column of the cord; the column of Goll. c., posterovesicular. See Clarke's c. c., respiratory. See fasciculus, solitary. c., vesicular, one of the nerve-cells in the posterior gray horn of the spinal cord.

**columna** (kol-um'-nah) [L.; pl., columnæ]. A column or pillar. c. adiposa, a fat-column. c. anterior, anterior horn or column. c. Bertini, that part of the cortical structure of the kidneys that separates the sides of any two pyramids. c. carneopapillaris anterior, the anterior column of the vagina. c. carneopapillaris posterior, the posterior column of the vagina. c. fornicis, anterior pillar or column of fornix. c. lateralis, lateral horn or column. c. nasi, the antero-posterior septum between the nostrils. c. pliearum, the columns of the vagina. c. posterior, posterior horn or column. c. rugarum, the columns of the vagina. c. vertebralis, the spinal column.

**columnæ carneæ**, the muscular columns projecting from the inner surface of the ventricles of the heart. c. cinereæ, the dorsal and ventral horns of the spinal cord. c. cordis. See columna carneæ. c. cristarum vaginæ, the columns of the vagina. c. griseæ, columns or horns of gray matter. c. papillares. Same as musculi papillares. c. quintæ, a bundle of nerve-fibers apparently originating in the upper nucleus of the fifth cranial nerve and decussating along the Sylvian aqueduct. c. recti. See Morgagni, columns of. c. vaginæ, ridges on the anterior walls of the vagina.

**columnella** (kol-um-nel'-ah). See columella.

**columning** (kol'-um-ing) [column]. The placing of vaginal tampons to support a prolapsed uterus.

**colytic** (*ko-lit'-ik*) [κολυτικός, preventive]. Preventive; inhibitory; caused by an obstruction; antiseptic.

**coma** (*ko'-mah*) [κῶμα, a deep sleep]. Unconsciousness from which the patient cannot be aroused by external stimulus. **c., alcoholic,** that due to poisoning by alcohol. **c., apoplectic,** that due to apoplexy. **c., uremic,** that due to uremia. **c. vigil,** a comatose condition in which the patient lies with open eyes, but is unconscious and delirious. This occurs occasionally in typhoid and typhus fevers and in delirium tremens.

**comatose** (*ko'-mat-ōs*) [*coma*]. In a condition of coma.

**combiner** (*kom-bī'-ner*) [*combinare*, to join]. That which combines. **c., galvanofaradic,** an instrument by means of which the galvanic and faradic currents can be used alternately or in combination in electrotherapeutics.

**combustion** (*kom-bus'-chun*) [*comburere*, to burn up]. The process of oxidation, attended with the liberation of heat and sometimes of light. **c., slow,** same as decay. **c. spontaneous,** that due to heat from chemical changes, such as the spontaneous ignition of oiled waste or shoddy in woolen mills, factories, etc.

**Comby's sign** [Jules *Comby,* French physician, 19th century]. A form of stomatitis involving the buccal mucosa, diagnostic of incipient measles. There is slight swelling, and the mucous membrane becomes reddish, the superficial epithelial cells become whitish as if brushed over by a paint-brush. It may occur evenly or in patches. Cf. *Koplik's spots.*

**comedo** (*kom'-e-do*) [L., "a glutton"; pl., *comedones*]. A chronic disorder of the sebaceous glands characterized by yellowish or whitish elevations, the size of a pinpoint or of a pinhead, containing in their center exposed blackish points. They are found usually on the face, back of the neck, chest, and back, and are often associated with acne. They occur, as a rule, in the young. Occasionally a parasite, *Demodex folliculorum,* is found in each comedo. Syn., *blackheads; grubs; grub-worms.*

**comes** (*ko'-mēz*) [L.; pl., *comites*]. A companion as a vein to an artery, or an artery to a nerve.

**comestible** (*kom-est'-e-bl*) [L., *comestibilis*]. Edible.

**comfrey** (*kum'-fre*). The root of *Symphytum officinale,* a demulcent, slightly astringent and tonic drug. It is a common ingredient of domestic cough-mixtures. Dose of the decoction indefinite; of the fluidextract ʒ j–ij.

**comma bacillus** (*kom'-ah bas-il'-us*). See *spirillum choleræ asiaticæ* under *spirillum.*

**comma-tract** (*kom'-ah-trakt*). A comma-shaped tract in the posterolateral column of the spinal cord.

**commensal** (*kom-en'-sal*) [*com,* together; *mensa,* table]. In biology, the harmonious living together of two organisms, animals or plants. One of two such organisms.

**commensalism** (*kom-en'-sal-ism*). The intimate association of two different living organisms dependent on each other.

**comminuted** (*kom'-in-ū-ted*) [*comminuere,* to break into pieces]. Broken into a number of pieces.

**comminuter** (*kom'-in-ū-ter*) [*comminuere,* to break into pieces]. An apparatus used to produce a spray; an atomizer.

**comminution** (*kom-in-ū'-shun*) [*comminutio; comminuere,* to break in pieces]. The process by which a solid body is reduced to pieces of varying sizes. It includes the various operations of cutting, rasping, grating, slicing, pulverizing, levigating, triturating, elutriating, granulating, etc. See, also, *fracture, comminuted.*

**Commiphora** (*kom-if'-o-rah*) [κόμμι, gum; φέρειν, to bear]. A genus of shrubs and trees of the order *Burseraceæ,* found in Africa and the East Indies. *C. africanum* yields African bdellium. *C. agallocha* yields Indian bdellium. *C. myrrha* yields myrrh. *C. opobalsamum* yields balsam of Mecca or of Gilead.

**commissura** (*kom-is-ū'-rah*). See *commissure.* **c. ansata,** a tract of fibers in the optic chiasm. **c. anterior alba,** anterior white commissure. **c. anterior grisea,** anterior gray commissure. **c. anterior (cerebri),** anterior commissure of the cerebrum. **c. brevis,** a portion of the inferior vermiform process of the cerebellum. **c. habenularum,** commissure of the habenula. **c. hippocampi,** the fornicommissure or lyra of the fornix, *q. v.* **c. inferior (Guddeni),** inferior commissure of Gudden. **c. labiorum,** junction of the lips, the angle of the mouth. **c. labiorum anterior,** anterior commissure of the labia majora. **c. labiorum posterior,** posterior commissure of the labia majora. **c. magna,** the corpus callosum. **c. magna cerebelli,** the superior peduncles of the cerebellum. **c. maxima,** or **maxima cerebri,** the corpus callosum. **c. palpebrarum lateralis,** lateral palpebral commissure, or external canthus. **c. palpebrarum medialis,** medial palpebral commissure, or internal canthus. **c. posterior,** posterior commissure. **c. posterior (cerebri),** posterior commissure of the cerebrum. **c. simplex,** a small cerebellar lobe. **c. superior (Meynerti),** superior commissure of Meynert. **c. ventralis alba,** the band of white matter between the gray matter and the anterior fissure of the spinal cord.

**commissural** (*kom-is-ū'-ral*) [*commissure*]. Having the properties of a commissure; uniting symmetrical parts, as *commissural fibers* of the brain.

**commissure** (*kom'-is-ūr*) [*con,* together; *mittere,* to send]. 1. That which unites two parts. 2. A joint, a seam. 3. The point of union of the lips, eyelids, or labia. 4. Strands of nerve-fibers uniting hemispheres of the brain, or the two sides of the spinal cord. **c., anterior** (of third ventricle), a rounded cord of white fibers placed in front of the anterior crura of the fornix. **c., arcuate.** 1. The posterior optic commissure. 2. Same as Gudden's inferior commissure, *q. v.* **c., gray** (of spinal cord), the transverse band of gray matter connecting the masses of gray matter of the two halves of the spinal cord. **c., gray, anterior,** nerve-fibers in the gray columns of the cord, which, crossing to the opposite side in front of the central canal, decussate in two directions, part of the fibers entering into the opposite cornu ventrale, part into the cornu dorsale. **c., gray, posterior,** that portion of the gray commissure of the spinal cord lying dorsad to the central canal. **c., habenal,** fibers joining one habena to the other; the dorsal stalk of the epiphyses. **c., horseshoe.** See *Wernekink's c.* **c., inferior.** See *Gudden's inferior c.* **c., middle,** a band of soft gray matter connecting the optic thalami. **c., optic,** the union and crossing of the two optic nerves in front of the tuber cinereum. **c., posterior** (of third ventricle), a flattened white band connecting the optic thalami posteriorly. **c., soft** (of the brain). Same as *c., middle.* **c., white, anterior** (of spinal cord), a layer of fibers separating the posterior gray commissure from the bottom of the anterior median fissure. **c., white, posterior** (of spinal cord), a band of fibers separating the gray commissure from the bottom of the posterior median fissure.

**commotio** (*kom-o'-she-o*) [L.]. A concussion; commotion or shock. **c. cerebri,** concussion of the brain. **c. retinæ,** concussion or paralysis of the retina from a blow on or near the eye. It is characterized by sudden blindness, but there is little or no ophthalmoscopic evidence of any lesion. The sight is usually regained, and its loss is supposedly due to disturbance of the retinal elements. **c. spinalis,** railway spine.

**communicable** (*kom-ū'-nik-abl*) [*communicabilis*]. Capable of being transmitted from one person to another.

**communicans** (*kom-ū'-nik-ans*) [L.]. 1. Communicating. 2. Alternating; connecting. **c. hypoglossi, c. noni.** See under *nerve.* **c. Willisii.** The posterior communicating artery of the brain.

**commutator** (*kom'-ū-ta-tor*) [*commutare,* to exchange]. An instrument for automatically interrupting or reversing the flow of an electric current.

**comose** (*ko'-mōs*) [*coma,* hair]. Having much hair.

**comp.** Abbreviation for Latin *compositus, compositum,* compound.

**compact** (*kom'-pakt*) [*compactus,* joined together]. Solid, dense; closely compressed. **c. tissue,** the external, hard part of bone.

**comparascope** (*kom-par'-as-kōp*). An apparatus attached to a microscope for the simultaneous comparison of two different slides.

**comparative anatomy.** See *anatomy, comparative.*

**compatibility** (*kom-pat-ib-il'-it-e*) [Fr., *compatibilité*]. Of medicines, the relation of one substance to another, so that they may be mixed without chemical change or loss of therapeutic power.

**compensating** (*kom'-pen-sa-ting*) [see *compensation*]. Making good a deficiency. **c. ocular.** See under

**ocular.** c. **operation,** in ophthalmology, tenotomy of the associated antagonist in cases of diplopia from paresis of one of the ocular muscles.

**compensation** (*kom-pen-sa'-shun*) [*compensare,* to equalize]. The act of making good a deficiency; the state of counterbalancing a functional or structural defect.

**compensatory** (*kom-pen'-sa-to-re*) [see *compensation*]. Making good a deficiency. Restoring the balance, after failure of one organ or part of an organ, by means of some other organ or part of an organ.

**complaint** (*kom-plānt'*) [*complangere,* to lament]. A disease or ailment. c., **bowel-,** diarrhea. c., **summer-,** summer diarrhea.

**complement** (*kom'-ple-ment*) [*complere,* to complete]. A thermolabile body resembling a ferment and found in serum and cell protoplasm. It acts in conjunction with the amboceptor in causing lysis. "The cell is the lock, the amboceptor the key, and the complement the hand that turns the key." Syn., *addiment; alexin; cytase.* See *alexin, Wassermann's test.* c. **deviation of,** c., **fixation of,** the entering of the complement into combination with an antigen-immune-body, so that hemolysis or bacteriolysis is hindered.

**complemental, complementary** (*kom-ple-men'-tal, kom-ple-men'-ta-re*) [*complement*]. Supplying a deficiency. c. **air.** See *air,* complemental. c. **colors,** a term applied to any two colors which combined produce white light, as, *e. g.,* blue and yellow.

**complementoid** (*kom-ple-ment'-oid*). A complement that has lost its power of causing lysis. It results from the destruction of a complement (*q. v.*); it can go to form an anticomplement.

**complementophil, complementophile** (*kom-ple-ment'-o-fil*). The haptophore group of the intermediary body by means of which it combines with the complement.

**complementophilic** (*kom-ple-ment-o-fil'-ik*). Showing a special affinity for the complement.

**complexus** (*kom-pleks'-us*) [L., "complex"]. The totality of symptoms, phenomena, or signs of a morbid condition. c. **muscle.** See under *muscle.*

**complication** (*kom-plik-a'-shun*) [*complicare,* to fold together]. A disease occurring in the course of some other disease and more or less dependent upon it.

**composition** (*kom-po-zish'-un*) [*compositio,* a putting together]. 1. Compounding; applied to drugs. The constitution of a mixture. 2. The kind and number of atoms which are contained in the molecule of a compound. c. **powder,** a popular name (originally Thomsonian) for the *pulvis myricæ compositus,* N. F. At present it is a mixture of finely powdered bayberry bark (*Myrica cerifera*), 12 parts; ginger, 6 parts; capsicum and cloves, each 1 part. It is useful for sore throat and severe coryza. c. **tea,** a warm drink composed of different spices, and used as a stomachic and to abort an attack of coryza.

**compos mentis** (*kom'-pos men'tis*) [L.]. Of sound mind.

**compound** (*kom'-pound* or *kom-pound'*) [*com,* together; *ponere,* to put]. 1. To mix, as drugs. 2. A mixture composed of several parts. 3. A substance composed of definite proportions of two or more elements in chemical union. c., **addition,** one formed from two other substances by direct union. c., **binary,** a substance composed of two elements or of an element and a compound behaving as an element. c. **cathartic pills** (*pilulæ catharticæ compositæ,* U. S. P.), pills of colocynth, mild mercurous chloride, resin of jalap, gamboge, and diluted alcohol. Dose 2 pills. c., **endothermic,** one absorbing heat in its formation. c., **exothermic,** one in which there is no elevation of temperature attending its formation. c., **explosive,** an unstable organic product containing much oxygen and readily decomposing. c. **fracture.** See *fracture, compound.* c., **quaternary,** a substance composed of four elements. c., **satisfied,** a chemical compound in which the combining capacities of all the elements are satisfied; one in which there are no free valences. c., **saturated,** one in which the elements have their maximum valences all filled. c., **substitution,** a compound formed from another body by replacement of one or more of its elements by another body or bodies. c., **ternary,** a compound composed of three elements.

**compounding** (*kom-pound'-ing*). The mixing, manipulation, and preparation of the drugs ordered in a prescription.

**compress** (*kom'-pres*) [*compressus,* pressed together]. A folded cloth, wet or dry, applied firmly to the part for relief of inflammation or to prevent hemorrhage. c., **electrothermic,** an appliance consisting of flexible pillows and of thin wires isolated by asbestos and covered with canvas. This, when connected with a strong electric current, serves to supply a modified form of dry heat. c. **cribriform,** c., **fenestrated,** a compress with a hole for drainage. c., **graduated,** a compress composed of folds of a gradually increasing size.

**compressed** (*kom-presd'*) [*compressus,* pressed together]. Firmly pressed together; having the lateral diameter reduced. c. **air,** air the density of which has been increased by compression. c.-**air bath.** See *bath.* c.-**air illness.** See *caisson disease.*

**compression** (*kom-presh'-un*) [*compress*]. The state of being compressed. c.-**atrophy,** atrophy of a part from constant compression. c., **digital,** compression of an artery by the fingers. c.-**myelitis.** See *myelitis, compression-.*

**compressor** (*kom-pres'-or*) [*compress*]. 1. An instrument for compressing an artery, vein, etc. 2. It is also applied to muscles having a compressing function, as the *c. naris, c. venæ dorsalis penis,* etc. See *muscles, table of.* 3. One of the light springs inserted on either side of the stage of a microscope for holding the slide in position; a clip. c. **sacculi laryngis,** the inferior aryteno-epiglottideus muscle. See *muscles, table of.* c. **urethræ,** constrictor urethræ. See *muscles, table of.*

**compressorium** (*kom-pres-o'-re-um*) [*compressor,* a compress]. An instrument devised for making pressure on the cover-glass of a microscope-slide in order to favor separation of the elements of the specimen to be examined.

**conalbumin** (*kon-al'-bū-min*). A proteid body obtained by Osborne and Campbell from white of egg, and so designated "on account of its close relation in properties and composition to ovalbumen." Cf. *ovalbumen; ovomucin; ovomucoid.*

**conarial** (*ko-na'-re-al*) [κωνάριον, the pineal gland]. Relating to the conarium. c. **vein.** See *vein.*

**conariohypophyseal** (*ko-na-re-o-hi-po-fis'-e-al*) [κωνάριον, the pineal gland; ὑπόφυσις, an undergrowth]. Relating to the conarium and to the hypophysis of the cerebrum; pineo-pituitary.

**conarium** (*ko-na'-re-um*) [κωνάριον; dim. of κῶνος, a cone]. The pineal gland.

**conation** (*ko-na'-shun*) [*conari,* to endeavor]. The exertive power of the mind, including will and desire; a special act or exercise of the exertive power.

**concassation** (*kon-kas-a'-shun*) [*concassatio; con,* together; *cassare* or *quassare,* to shake, to beat]. 1. The shaking of medicines, as in a bottle; the pulverizing of drugs by beating. 2. Mental distress or affliction.

**concatenate** (*kon-kat'-en-āt*) [L., *con,* together; *catena,* chain]. Linked together; such as enlarged lymph glands.

**Concato's disease** (*kon-kah'-to*) [Luigi Maria *Concato,* Italian physician, 1825–1882]. Tuberculosis affecting successively various serous membranes, terminating usually in pulmonary tuberculosis.

**concave** (*kon-kāv'*) [*con,* together; *cavus,* hollow]. Hollow; incurved, as the inner surface of a hollow sphere.

**concavity** (*kon-kav'-it-e*). A depression or fossa.

**concavoconvex** (*kon-ka-vo-kon-veks'*). Having one surface concave, the other convex, the convexity exceeding the concavity. See *lens, concavoconvex.*

**concavum pedis** [L.]. The hollow, or the arch of the sole of the foot.

**concavoconcave** (*kon-ka'-vo-kon-kāv*) [*concave*]. Concave on both sides. See *lens, biconcave.*

**conceive** (*kon-sēv'*) [*concipere,* to take in]. To become pregnant.

**concentrated** (*kon'-sen-tra-ted*). Made stronger or purer.

**concentration** (*kon-sen-tra'-shun*) [*con,* together; *centrum,* the center]. 1. The act of making denser, as of a mixture, by evaporating a part of the liquid. 2. Afflux toward a part. 3. The strength of a solution. 4. The state of fixed and restricted attention.

**concentric** (*kon-sen-trik*). Having a common center. c. **hypertrophy of the heart,** increase in the muscular texture of the heart, the capacity of the cavities remaining unchanged.

**concept** (*kon'-sept*) [*conceptum,* something understood]. The subject of a mental conception.

**conception** (kon-sep'-shun) [concipere, to conceive]. 1. The fecundation of the ovum by the spermatozoon. 2. The abstract mental idea of anything; the power or act of mentally conceiving. **c., imperative,** a false idea that a person dwells upon and cannot expel from his mind, even when he knows it to be absurd. It dominates his actions and is a symptom of insanity.
**concha** (kong'-kah) [κόγχη, a shell: pl., *conchæ*]. 1. A shell. Applied to organs having some resemblance to a shell, as the naris, vulva, etc. 2. The external ear. 3. The turbinated bone. **c. auris,** the hollow part of the external ear. **c. inferior,** the inferior turbinated bones. **c. labyrinthi,** the cochlea. **c. media,** the middle turbinated bone. **c., Morgagni's,** the superior turbinated bone of the ethmoid. **c. superior,** the superior turbinated bone. **conchæ turbinatæ,** the turbinated bones.
**conchinine** (kon'-kin-ēn);—See *quinidine*.
**conchitis** (kong-ki'-tis) [κόγχη, a shell; *itis*, inflammation]. Inflammation of the concha.
**conchoidal** (kong-koi'-dal). Shell-like; shaped like a cell.
**conchoscope** (kong'-ko-skōp) [concha; σκοπεῖν, to inspect]. A speculum and mirror for inspecting the nasal cavity.
**conchotome** (kong'-ko-tōm) [concha; τομή, a cutting]. An instrument for the surgical removal of the turbinated bone.
**concoctio** (kon-kok'-she-o) [L.]. Digestion. **c. tarda.** Synonym of *dyspepsia*.
**concoction** (kon-kok'-shun) [concoquere, to boil together]. The act of boiling two substances together.
**concomitant** (kon-kom'-it-ant) [concomitari, to accompany]. Accompanying. **c. strabismus.** See under *strabismus*. **c. symptoms,** symptoms that are not in themselves essential to the course of a disease, but that may occur in association with the essential symptoms.
**concrement** (kon'-kre-ment) [concrescere, to grow together]. A concretion.
**concrescence** (kon-kres'-ens) [con, together; crescere, to grow]. 1. See *concretion* (3). 2. Held's term for the plunging of the terminal of one neuron deep into the cell-body of another. **c. of teeth,** a growing together of the roots of two teeth after complete development.
**concretion** (kon-kre'-shun) [see *concrescence*]. 1. The solidification or condensation of a fluid substance. 2. A calculus. 3. A union of parts normally separate, as the fingers.
**concubitus** (kon-kū'-bit-us) [concumbere, to lie together]. Copulation.
**concussion** (kon-kush'-un) [concussio, a violent shock]. Shock; the state of being shaken; a severe shaking or jarring of a part; also, the morbid state resulting from such a jarring. **c. of the brain,** a condition produced by a fall or blow on the head, and marked by unconsciousness, feeble pulse, cold skin, pallor, at times the involuntary discharge of feces and urine; this is followed by partial stupor, vomiting, and headache, and eventually recovery. In severe cases inflammation of the brain or a condition of feeble-mindedness may follow. **c. of the labyrinth,** deafness and tinnitus from a blow or an explosion. **c. of the spinal cord,** a condition caused by severe shock of the spinal column, with or without appreciable lesion of the cord. It leads to functional disturbances analogous to railway spine.
**concussor** (kon-kus'-or) [concutere, to shake]. In massage, an apparatus for gently beating the part to be treated.
**condensability** (kon-dens-ab-il'-it-e) [condensare, to condense]. Capacity for undergoing condensation.
**condensation** (kon-den-sa'-shun) [condensare, to condense]. 1. Making more compact or dense. 2. The changing of a gaseous substance to a liquid, or a liquid to a solid. 3. In chemistry, the union of two or more molecules by the linking of carbon-atoms and the formation of complicated carbon-chains. 4. The pathological hardening, with or without contraction, of a soft organ or tissue.
**condensed** (kon-densd') [condensare, to make thick]. Made compact; reduced to a denser form. **c. milk,** milk that has had most of its watery elements evaporated. Condensed milk prepared with the addition of cane-sugar is a white or yellowish-white product of about the consistence of honey, and ranging in specific gravity from 1.25 to 1.41. It should be completely soluble in 4 or 5 times its bulk of water, without separation of any flocculent residue, and then possess the taste of fresh, sweetened milk. Condensed milk prepared without the addition of cane-sugar is not boiled down to the same degree, and therefore remains liquid.
**condenser** (kon-den'-ser). 1. A lens or combination of lenses used in microscopy for gathering and concentrating rays of light. See *Abbé's condenser*. 2. An apparatus for condensing gases. 3. An apparatus for the accumulation of electricity.
**condiment** (kon'-dim-ent) [condimentum, spice]. Spice, sauce, or other appetizing ingredients used with food.
**condom,** (kon'-dum) [corruption of *Conton*, English physician 18th century, said to be the inventor]. A sheath worn over the penis during copulation for the purpose of preventing conception or infection.
**conductance** (kon-duk'-tans). The ratio of an electric current through a conductor to the electromotive force.
**conductibility** (kon-dukt-i-bil'-e-te) [see *conductor*]. 1. Capacity for being conducted. 2. Conductivity.
**conducting power. c., centrifugal,** the power of carrying centrifugal impulses from the nervous centers to the periphery. **c., centripetal,** the power of conducting centripetal impulses from the periphery to the nervous centers.
**conduction** (kon-duk'-shun) [see *conductor*]. The passage or transfer of force or material from one part to another. **c.-resistance,** the resistance encountered by an electric current in passing through a circuit.
**conductivity** (kon-duk-tiv'-it-e). The capacity for conducting.
**conductor** (kon-duk'-tor) [conducere, to draw together]. 1. A body that transmits force-vibrations, such as those of heat or electricity. 2. A term applied to the electrodes and cords by which they are joined to the battery. 3. An instrument serving as a guide for the surgeon's knife. 4. In physiology, any part of the nervous system that transmits impulses. **c., sonorous,** a term applied to certain nerve-fibers which interlace with the auditory striæ.
**condurangin** (kon-du-ran'-jin) [condurango]. A mixture of glucosides from condurango bark, occurring as an amorphous yellow powder of an aromatic bitter taste, soluble in water, alcohol, and chloroform. It is used as a stomachic and astringent in gastric cancer and chronic dyspepsia. Dose 1/10-1/4 gr. (0.006–0.016 Gm.) 3 times daily.
**condurango bark** (kon-du-ran'-go) [Peruvian]. Bark of *Gonolobus tetragonus* or of *Conglobus condurango*; a remedy much used in South America as an alterative in syphilis. It was introduced into the United States as a remedy for carcinoma of the stomach, but yielded uncertain results. It is a stomachic tonic. Dose of the *fluidextract* ½–1 dr. (2–4 Cc.); of the *tincture* 1–2 dr. (4–8 Cc.)
**Condy's fluid** (kon'-de) [Henry Bollmann *Condy*, English physician, 19th century]. One part of sodium or potassium permanganate dissolved in 500 parts of water; it is a useful disinfectant.
**condylar** (kon'-dil-ar) [condyle]. Pertaining to a condyle.
**condylarthrosis** (kon-dil-ar-thro'-sis) [condyle; ἄρθρον, a joint]. A form of diarthrosis wherein a condyle is set in a shallow and elliptic cavity and free and varied movement of the joint is possible; condylar articulation.
**condyle** (kon'-dīl) [κόνδυλος, a knuckle]. Any rounded eminence such as occurs in the joints of many of the bones, especially the femur, humerus, and lower jaw.
**condylectomy** (kon-dil-ek'-to-me) [condyle; ἐκτομή, excision]. Excision of a condyle.
**condylion** (kon-dil'-e-on) [condyle]. In craniometry, the point at the lateral tip of the condyle of the jaw.
**condyloid** (kon'-dil-oid) [condyle; εἶδος, likeness]. Resembling or pertaining to a condyle.
**condyloma** (kon-dil-o'-mah) [κονδύλωμα, a swelling: pl., *condylomata*]. A wart-like growth or tumor usually near the anus or pudendum. **c. acuminatum,** the pointed condyloma or wart of the genital organs, often of non-syphilitic origin. Syn., *acrothymion; acrothymiosis*. **c. endocysticum, c. endofolliculare, c. porcelaneum, c. subcutaneum.** See *molluscum contagiosum*. **c. latum,** the flat, broad, moist syphilid or mucous patch. Syn., *papula madidans*. Cf. *c. acuminatum*. **c., syphilitic, c. syphiliticum.** See

## CONDYLOMATOUS 244 CONIDIUM

*c. latum.* c., **thymic,** condyloma that assumes the form of a papilloma; so called from its resemblance to thyme-blossoms.

**condylomatous** (*kon-dil-o'-mat-us*). Of the nature of a condyloma.

**condylosis** (*kon-dil-o'-sis*). The formation of a condyloma.

**condylotomy** (*kon-dil-ot'-o-me*) [*condyle;* τέμνειν, to cut]. Extra-articular osteotomy; a division through the condyles of a bone.

**condylus** (*kon'-di-lus*) [L.; *pl., condyli*]. Condyle. c. **lateralis,** lateral condyle, outer condyle, or external tuberosity. c. **medialis,** medial condyle, inner condyle, or internal tuberosity. c. **occipitalis,** occipital condyle.

**cone, conus** (*kōn, ko'-nus*) [κῶνος, a cone]. 1. A solid body having a circle for its base, and terminating in a point. 2. The mechanical element of the tooth-crown. 3. See *conus.* c., **antipodal,** in mitosis the cone of astral rays opposite the spindle-fibers. c.-**bipolars,** bipolar cells of the inner nuclear layer of the retina, connected with the cones of the retina externally and ramifying internally in the middle of the molecular layer. c.-**element,** a cell of the sensory or nerve epithelium of the retina, consisting of a conic tapering external part, the *cone proper,* prolonged into a nucleated enlargement from the farther side of which the *cone-fiber* passes inward to terminate by an expanded arborization in the outer molecular layer. c.-**fiber,** one of the fibers of the retinal cones. c.-**foot,** one of the bulbous processes of the cone-granules of the retina. c., **graduated,** a cone-shaped body used for measuring the size of orifices of vessels, etc., especially in post-mortem examinations. c.-**granules,** those of the outer nuclear layer of the retina, connected with the cones of the ninth layer externally, and internally by a thick process which becomes bulbous (the cone-foot); they terminate in fine fibers in the outer molecular layer. Cf. *rod-granules.* c. **of light,** the triangular reflection from the normal tympanic membrane; also the bundle of light-rays entering the pupil and forming the retinal image. c., **retinal,** one of the rod-like bodies which, with the associated rods, forms one of the outer layers of the retina, the so-called rod-and-cone layer. c. **spermatic,** one of the series of cones forming the head of the epididymis, and composed of the coiled efferent tubules.

**conessi bark** (*kon-es'-e*). The bark of *Wrightia zeylanica,* and of *Holarrhena africana,* apocynaceous shrubs or trees of India and Africa. It is extensively used in India in dysentery and as an antiperiodic. Syn., *tellicherry bark.*

**conessine** (*kon-es'-in*). See *Wrightine.*

**confectio** (*kon-fek'-she-o*) [L.; *gen., confectionis*]. Official name for any confection, *q. v.* c. **Damocratis.** See *mithridate.*

**confection** (*kon-fek'-shun*) [*confectio,* a preparation]. In pharmacy, a mass of sugar and water, or of honey, used as an excipient with a prescribed medicinal substance.

**confectioner's disease.** A disease of the finger-nails occurring in confectioners. The nail loses its polish and becomes black, and the periungual portion becomes loosened and raised.

**confertus** (*kon-fer'-tus*) [*confercire,* to press close together]. Pressed together, dense, crowded; applied to cutaneous eruptions.

**configuration** (*kon-fig-u-ra'-shun*) [*configurare,* to form after something]. In chemistry, a term designating the "constitution" or "structure" of the molecule.

**confinement** (*kon-fīn'-ment*) [*con,* together; *finis,* boundary; limit]. The condition of women during childbirth.

**confirmatory** (*kon-fir'-mat-or-e*) [*confirmare,* to confirm]. Confirming. c. **incision,** an abdominal section, made to confirm a diagnosis, as in case of malignant disease of the ovary, uterus, peritoneum, etc.

**confluens sinuum** (*kon'-flu-enz sin'-u-um*) [L.]. Confluence of the sinuses; the torcular Herophili.

**confluent** (*kon'-flu-ent*) [*confluere,* to flow together]. Running together. The opposite of discrete. In anatomy, coalesced or blended; applied to two or more bones originally separate, but subsequently formed into one.

**confocal** (*kon-fo'-kal*). Having the same focus.

**conformator** (*kon-for-ma'-tor*) [Fr., *conformateur*]. A form of cephalograph used in determining the outlines of the skull in craniometry.

**confrontation** (*kon-fron-ta'-shun*) [*confrontari,* to be contiguous to]. The examination of a person by whom a diseased person may have been infected as a means of diagnosing the disease in the latter.

**confusion** (*kon-fū'-zhun*) [*com,* together; *fundere,* to pour]. Mixing; confounding. c. **colors,** a set of colors so chosen that they cannot be distinguished by one who is color-blind. c. **letters,** test-type letters, such as C, G, O, or F, P, T, liable to be mistaken for one another.

**cong.** Abbreviation for Latin *congius,* a gallon.

**congelation** (*kon-jel-a'-shun*) [*congelatio,* a freezing]. 1. Freezing; frost-bite; intense cold or its effect on the animal economy or any organ or part. 2. The chilling or benumbing effect of any freezing-mixture or application; mainly employed for its local anesthetic effect. 3. Coagulation.

**congener** (*kon'-jen-er*) [L., of the same race]. 1. Belonging to the same genus; closely allied. 2. A congenerous muscle.

**congenerous** (*kon-jen'-er-us*) [*congener,* of the same race]. Of the same genus. c. **diseases,** allied diseases. c. **muscles,** muscles producing one action.

**congenital** (*kon-jen'-it-al*) [*congenitus,* born with]. Existing at birth.

**congested** (*kon-jes'-ted*). In a state of congestion.

**congestion** (*kon-jes'-chun*) [*congerere,* to heap up]. An abnormal collection of blood in a part or organ. Congestion may be active or passive, atonic or inflammatory, functional or hypostatic. It is also named from the parts affected; the most important varieties of morbid congestion are the cerebral, spinal, pulmonary, hepatic, and renal. c., **pleuro-pulmonary,** c., **pulmonary,** Potain's type of, congestion of the lungs marked by symptoms similar to those of pleurisy.

**congestive** (*kon-jes'-tiv*) [*congestion*]. Marked by, due to, or of the nature of congestion. c. **fever,** malarial fever.

**congius** (*kon-je'-us*) [L.: *pl., congii*]. A fluid measure; a gallon.

**conglobate** (*kon-glo'-bāt*) [*con,* together; *globare,* to make into a globe]. Rounded. c. **glands,** the absorbent or lymphatic glands. See *gland.*

**conglomerate** (*kon-glom'-er-āt*) [*conglomerare,* to heap up]. 1. Massed together; aggregated. 2. A mass of units without order. c. **glands,** acinous glands.

**conglutin** (*kon-glu'-tin*) [*con,* together; *gluten,* glue]. One of the proteids found in almonds, peas, beans, and other kinds of pulse.

**conglutinant** (*kon-glu'-tin-ant*) [*conglutinare,* to glue together]. Adhesive; promoting union, as of the edges of a wound.

**conglutination** (*kon-glu-tin-a'-shun*) [*conglutinare,* to glue together]. The abnormal union of two contiguous surfaces or bodies, as of two fingers, or of the opposed surfaces of the pleural or pericardial sac.

**conglutinin** (*kon-glu'-tin-in*). That substance in serum which causes conglutination; it is easily precipitated and is non-specific.

**Congo red.** A red coloring-matter which becomes blue in the presence of free HCl. It is used in the chemical investigation of the gastric juice. c. **root,** the root of *Psoralea melilotoides,* a leguminous herb of the United States. It is an aromatic bitter tonic, recommended in chronic diarrhea.

**congress** (*kong'-gres*) [*congressus,* a meeting together]. An assemblage for deliberative purpose. c., **sexual,** coitus, or carnal intercourse.

**conhydrine** (*kon-hi'-drin*) [κώνειον, hemlock; ὕδωρ, water], $C_8H_{17}NO$. A solid alkaloid of conium, an oxyconine occurring in pearly, iridescent, white, foliaceous crystals, with a faint, conine-like odor, and melting at $121°$ C. and distilling at $226°$. c. **pseudo-,** $C_8H_{17}NO$, white acicular crystals obtained from crude coniine soluble in water, alcohol, benzene, ether, and chloroform; melt at $98°$ C. and boil at $230°-232°$ C.

**coni vasculosi** (*ko'-ni vas-kū-lo'-si*). A series of conical masses of tubules that together form the globus major of the epididymis.

**conic, conical** (*kon'-ik, kon'-ik-al*) [*cone*]. Cone-shaped. c. **cornea.** See *keratoglobus.*

**conicine** (*kon'-is-in*). See *coniine.*

**conidia** (*ko-nid'-e-ah*) [L.]. Plural of *conidium.*

**conidial** (*ko-nid'-e-al*) [*conidium*]. Pertaining to, or of the nature of a conidium.

**conidium** (*ko-nid'-e-um*) [κόνις, dust; ἴδιον, dim. suffix; *pl., conidia*]. In biology, the deciduous

asexual spores of certain fungi. Also called *basidiospore* and *acrospore*.
**conidiophore** (*ko-nid'-e-o-fōr*) [*conidium;* φέρειν, to bear]. The mycelial thread of a fungus which carries the conidia.
**conidiospore** (*ko-nid'-e-o-spōr*) [*conidium; spore*]. Same as conidium.
**coniine** (*ko'-ne-in*). See *conine* and *conium*.
**coniism** (*ko'-ne-izm*) [*conium*]. Poisoning by conium. It begins with paralysis of the legs, which extends to the arms and respiratory muscles, leading to unconsciousness and death.
**conine** (*ko'-nēn*) [*conium*], $C_8H_{15}N$. A liquid alkaloid which is the active principle of *conium*. Dose $\frac{1}{60}-\frac{1}{10}$ gr. (0.001–0.006 Gm.). Unof. **c., animal.** See *cadaverin*. **c. hydrobromide,** $C_8H_{17}$·($C_2H_7$), recommended in spasmodic affections. Dose $\frac{1}{30}-\frac{1}{15}$ gr. (0.002–0.004 Gm.). All preparations are of uncertain strength. **c. hydrochloride,** $C_8H_{17}$·NHCl, the principal salt of conine, is used as is conine hydrobromide.
**coniosis** (*ko-ne-o'-sis*). See *koniosis*.
**conium** (*ko-ne'-um*) [κώνειον, hemlock]. Hemlock. Both the leaves and the fruit are official in the B. P. The *conium* of the U. S. P. is the full-grown but unripe fruit of the spotted hemlock, *C. maculatum*. It contains three alkaloids and a volatile oil. Its properties are mainly due to the alkaloids *conine*, $C_8H_{15}N$, and *methylconine*, $C_9H_{16}NCH_3$. It produces motor paralysis, without loss of sensation or of consciousness. Toxic doses cause death by paralysis of the organs of respiration. It is valuable in acute mania, delirium tremens, tetanus, blepharospasm, asthma, and whooping-cough. **conii, abstractum,** made from conium, 200; dilute hydrochloric acid, 6; sugar of milk and alcohol, q. s. to make 100 parts of abstract. Dose $\frac{1}{4}$–3 gr. (0.032–0.2 Gm.). **conii, cataplasma** (B. P.), made from the leaves; for external use. **conii, fluidextractum** (U. S. P.). Dose 1–2 min. (0.065–0.13 Cc.). **conii, pilula, composita** (B. P.), contains extract of hemlock and ipecac. Dose 5–10 gr. (0.32–0.65 Gm.). **conii, succus** (B. P.), made from the leaves. Dose 30 min.–2 dr. (2–8 Cc.). **conii, tinctura,** 15 % strength. Dose 10 min.–1 dr. (0.65–4.0 Cc.). **conii, vapor** (B. P.), for inhalations.
**conjugal** (*kon'-ju-gal*) [*con*, together; *jugare*, to yoke]. Pertaining to marriage; affecting both husband and wife. **c. diabetes,** diabetes affecting husband and wife together; this is said to be not infrequently observed.
**conjugate** (*kon'-ju-gāt*) [see *conjugal*]. 1. Yoked or coupled. 2. The anteroposterior diameter of the brim of the pelvis, the plane of the brim being regarded as an ellipse. **c., anatomical.** See *conjugate* (2). **c. deviation.** See *deviation, conjugate*. **c. diameter** (of the pelvis). See *conjugate* (2), and *diameter, pelvic*. **c. foci,** two foci which are interchangeable. **c., obstetrical, c., true,** the minimum diameter of the pelvic inlet.
**conjugation** (*kon-ju-ga'-shun*). 1. A form of reproduction or cell-division. 2. The process in which two protozoa come together and exchange materials. **c. nucleus,** the segmentation-nucleus.
**conjunctiva** (*kon-junk-ti'-vah*) [*conjunctivus*, connecting]. The mucous membrane covering the anterior portion of the globe of the eye, reflected upon the lids and extending to their free edges. Its parts are called palpebral and bulbar or ocular. **c., bulbar, c., ocular,** that covering the anterior third of the eyeball, from the retrotarsal fold to the margin of the cornea. **c., palpebral,** the conjunctiva of the eyelid. **c. reaction.** Same as ophthalmo-reaction.
**conjunctival** (*kon-junk'-tiv-al,* or *kon-junk-ti'-val*) [*conjunctivus,* connecting]. Relating to the conjunctiva.
**conjunctivitis** (*kon-junk-tiv-i'-tis*) [*conjunctiva; ιτις,* inflammation]. Inflammation of the conjunctiva. Syn. *ophthalmia externa; ophthalmia mucosa*. **c., catarrhal, c., catarrhal, acute,** the most common form, usually mild, resulting from cold or irritation. See *ophthalmia*. **c. catarrhalis æstiva.** See *vernal conjunctivitis*. **c., contagious, acute,** that due to the presence of *Bacterium ægypticum*. See *trachoma*. **c., croupous,** a variety associated with the formation of a whitish-gray membrane that is easily removed. **c., diphtheritic,** a specific purulent inflammation of the conjunctiva due to the Klebs-Loeffler bacillus. **c., Egyptian.** See *trachoma*. **c., follicular,** a form characterized by numerous round, pinkish bodies found in the retrotarsal fold. **c., gonorrheal,** a severe form of purulent conjunctivitis caused by infection with gonococci. **c. granulosa.** See *trachoma*. **c., hemorrhagic.** See *pink-eye*. **c., hypertrophic, c, hypertrophica,** chronic catarrhal conjunctivitis attended with enlargement of the conjunctval papillæ. **c., lacrimal,** a form due to the presence of irritating secretion from the conducting part of the lacrimal apparatus. **c., lithiasis,** irritation of the conjunctiva due to deposition of calcareous matter in the tissue of the palpebral conjunctiva. **c., phlyctenular,** a form characterized by the presence on the ocular conjunctiva of small vesicles surrounded by a reddened zone. **c., purulent,** conjunctivitis characterized by a thick, creamy discharge. **c., subacute,** redness and thickening of the conjunctiva, largely confined to the conjunctiva of the lids and fornices, a scanty secretion of mucus, with some pus-corpuscles, due to the presence of a bacillus.
**conjunctivoma** (*kon-junk-tiv-o'-mah*) [*conjunctiva;* ὄμμα, tumor]. A tumor consisting of conjunctival tissue; it occurs on the eyelid.
**connate** (*kon'-āt*). 1. Congenital. 2. United; confluent.
**connection** (*kon-ek'-shun*) [*connectere,* to connect]. Sexual intercourse.
**connective** (*kon-ek'-tiv*). Connecting, binding. **c. tissue,** the binding tissue of the body; see *tissue*.
**connectivum** (*kon-nek-ti'-vum*) [L.]. A connective tissue.
**Connell's suture** (*kon'-el*) [F. Gregory *Connell,* American surgeon, 1864– ]. A suture used in circular enterorrhaphy.
**connivens** (*kon-i'-venz*). 1. See *connivent*. 2. Arranged in circular folds; e. g., *valvulæ conniventes*.
**connivent** (*kon-i'-vent*) [*connivere,* to wink at]. In botany converging toward each other; applied to stamens that converge above, as those of the violet.
**conoid, conoidal** (*ko'-noid, ko-noi'-dal*) [κῶνος, cone; εἶδος, shape]. Of a conical shape. **c. body,** the pineal gland. **c. ligament,** the lower and inner part of the coracoclavicular ligament. **c. tubercle,** the eminence on the inferior surface of the clavicle to which the conoid ligament is attached.
**conomyoidin** (*ko-no-mi-oi'-din*) [κῶνος, cone; μῦς, muscle; εἶδος, form]. A contractile protoplasmic material found in the cones of the retina.
**conquassant** (*kon-kwas'-ant*) [*conquassare,* to shake severely]. Very severe (applied chiefly to labor-pains at the acme of their intensity).
**conquinamine** (*kon-kwin'-a-mēn*). An alkaloid of cuprea-bark.
**Conradi-Drigalski medium** (*kon-rah'-de dre-gal'-ske*). 100 c.c. of lactose litmus agar is liquefied by heat and 1 c.c. of a solution of crystal violet added (crystal violet 0.1, gram, distilled water 100 c.c.). Typhoid colonies growing on this medium are pink, while those of colon bacillus are bluish-gray.
**Conradi's line** (*kon-rah'-de*) [Andrew Christian *Conradi*, Norwegian physician, –1869]. A line drawn from the base of the xiphoid process to the point of the apex-beat, marking, under normal conditions, the upper limit of percussion-dulness of the left lobe of the liver.
**consanguineous** (*kon-san-gwin'-e-us*) [*consanguineus,* of the same blood]. Related by a common parentage.
**consanguinity** (*kon-san-gwin'-it-e*) [*con,* together; *sanguinis,* of blood]. The relationship arising from common parentage; blood-relationship.
**consciousness** (*kon'-shus-nes*) [*conscius,* knowing]. The state of being aware of one's own existence, of one's own mental states, and of the impressions made upon one's senses; ability to take cognizance of sensations. **c., double,** that morbid condition in which there are two separate and alternating states of mental consciousness, in either one of which the events that have occurred in the other state are not remembered by the patient.
**consenescence** (*kon-sen-es'-ens*) [*consenescere,* to grow old]. The state or condition of growing old.
**consensual** (*kon-sen'-sū-al*) [*consensus,* agreement]. Excited reflexly by stimulation of another part, usually a fellow organ, as the *consensual* reaction of one pupil when the iris of the other eye is stimulated.
**consensus** (*kon-sen'-sus*) [L.]. Agreement. General harmonious action of different organs in effecting a purpose.

**consent** (*kon-sent'*) [*consentire*, to agree]. In forensic medicine, this term signifies willing participation in unnatural or illegal intercourse.

**conserva**, or **conserve** (*kon-ser'-vah*, or *con-serv'*) [*conservare*, to keep]. A confection, q. v.

**conservancy** (*kon-ser'-van-se*) [*conservare*, to keep]. Public conservation of health or of things that make for health; the sum of hygienic or preservative legislation.

**conservation** (*kon-ser-va'-shun*) [*conservare*, to keep]. Preservation without loss. c. of energy. See under *energy*.

**conservative** (*kon-ser'-vat-iv*) [*conservare*, to keep]. Aiming at the preservation and restoration of injured parts; as conservative surgery or dentistry.

**consistence** (*kon-sis'-tens*). Degree of density, or hardness.

**consolidant, consolidating** (*kon-sol'-id-ant, kon-sol-id-a'-ting*) [*consolidare*, to make firm]. Tending to heal or promoting the healing of wounds or fractures; favoring cicatrization.

**consolidation** (*kon-sol-id-a'-shun*) [*consolidare*, to make firm]. The process of becoming firm or solid, as a lung in pneumonia.

**consommé** (*kon'-so-ma*) [French]. A clear, strong soup of meat and vegetable, freed from fat.

**consonant, consonating** (*kon'-so-nant, kon-so-na'-ting*) [*con*, together; *sonare*, to sound]. Applied to pulmonary sounds heard on auscultation which sound in unison with some other sound.

**constant** (*kon'-stant*) [*constans*, steady]. Fixed, not changing. c. **battery**, c. **cell**, c. **element**, one yielding a constant current. c. **current**, an uninterrupted current, one that goes continuously in one direction.

**constipation** (*kon-stip-a'-shun*) [*constipare*, to crush tightly together]. A condition in which the bowels are evacuated at long intervals or with difficulty; costiveness.

**constitutio** (*kon-sti-tū'-she-o*). See *constitution*. c. **lymphatica**, Paltauf's term for a pathological condition marked by hyperplasia of the entire lymphatic system, including the thymus gland, and frequently by a hyperplasia of the vascular system, and, in females, of the genital organs.

**constitution** (*kon-stit-ū'-shun*) [*constituere*, to dispose]. In chemistry, the atomic or molecular composition of a body, together with the relation which the atoms bear to each other. In pharmacy, the composition of a substance. In physiology, the general temperament and functional condition of the body.

**constitutional** (*kon-stit-ū'-shun-al*). Pertaining to the state of the constitution. c. **diseases**, such diseases as are inherent, owing to an abnormal structure of the body. Also, a condition in which the disease pervades the whole system. General diseases, in contradistinction to local.

**constrict** (*kon-strikt'*). To draw together in one part.

**constrictor** (*kon-strik'-tor*) [*constringere*, to bind together]. Any muscle that contracts or tightens any part of the body. See under *muscle*. c. **isthmi faucium**. See *palatoglossus*, in *muscles, table of*.

**constringent** (*kon-strin'-jent*) [*constringere*, to constrict]. Same as *astringent*, q. v.

**constructive** (*kon-struk'-tiv*) [*construere*, to build up]. Relating to the process of construction; anabolic.

**consultant** (*kon-sul'-tant*) [*consultare*, to take counsel]. A consulting physician; one summoned by the physician in attendance to give counsel in a case.

**consultation** (*kon-sul-ta'-shun*) [*consultare*, to take counsel]. A deliberation between two or more physicians concerning the diagnosis of the disease of a patient and the proper method of treatment.

**consumption** (*kon-sump'-shun*) [*consumere*, to consume or wear away]. A wasting away, especially a wasting disease like tuberculosis, particularly pulmonary tuberculosis or tuberculosis of the bowels.

**consumptive** (*kon-sump'-tiv*) [*consumere*, to consume or wear away]. 1. Of the nature of tuberculosis. 2. One afflicted with pulmonary tuberculosis.

**contact** (*kon'-takt*) [*contactus*, a touching]. 1. A touching. 2. A person who has been exposed to a contagious disease. c.-**action**, catalysis. c.-**bed**, a large open basin containing a layer of coke or cinders, for the purification of sewage by bringing it into contact with bacteria which set up rapid decomposition and destruction of the organic matter. Cf.

**septic tank**. c.-**breaker**, an instrument by means of which a galvanic circuit is broken. c. **lenses**, in optics, a glass shell the concavity of which is in contact with the globe of the eye, a layer of liquid being interposed between the lens and the cornea. c. **series**, a series of metals ranged in such an order that each becomes positively electrified by contact with the one that follows it. c.-**substance**, a catalyst. **contactile**, or **contactual** (*kon-tak'-til*, or *kon-tak'-tū-al*) [*contactus*, contact]. 1. Tactile. 2. Due to or spreading by actual contact.

**contagion** (*kon-ta'-jun*) [*contingere*, to touch]. 1. The process by which a specific disease is communicated from one person to another, either by direct contact or by means of an intermediate agent. 2. The specific germ or virus from which a communicable disease develops.

**contagious** (*kon-ta'-jus*) [*contagion*]. Communicable or transmissible by contagion or by a specific contagium.

**contagium** (*kon-ta'-je-um*) [L.]. Any virus or morbific matter by means of which a communicable disease is transmitted from the sick to the well. c. **animatum**, c. **vivum** ("living contagium"), any living vegetable or animal organism that causes the spread of an infectious disease.

**contemplative** (*kon-tem'-pla-tiv*). A person who induces a sexual orgasm by an act of the imagination.

**contiguity** (*kon-tig-ū'-it-e*) [*contingere*]. Proximity. c., **amputation in the**, one performed at a joint, without section of a bone. c., **solution of**, separation of parts are normally in contact, such as dislocation.

**contiguous** (*kon-tig'-ū-us*). In contact, or adjacent.

**continence** (*kon'-ti-nens*) [*continere*, to hold together]. Self-restraint, especially in regard to the sexual passion.

**continued** (*kon-tin'-ūd*) [*continuere*, to make continuous]. Persisted in. c. **fever**, a fever that is long continued, without intermissions.

**continuity** (*kon-tin-ū'-it-e*) [*com*, together; *tenere*, to hold]. The state of being continuous or uninterrupted. c., **amputation in the**, amputation in which a bone is divided. c., **solution of**, division of a tissue by traumatism, inflammation, or disease; such as fracture, incision.

**contortion** (*kon-tor'-shun*) [*contorquere*, to twist]. A twisting or writhing, as of the body.

**contour** (*kon'-toor*) [Fr., *contour*, circuit]. 1. The line that bounds, defines, or terminates a figure. 2. In operative dentistry, to effect the restoration of lost parts of teeth by building them up with gold, etc. c.-**fillings**, in dentistry fillings in which the material is so built out as to restore the lost portion of the crown of the tooth; distinguished from plane or flush-fillings.

**contoured** (*kon'-toord*). Of a bacterial culture, one that has an irregular but smoothly undulating surface, like that of a relief map.

**contra-aperture** (*kon-trah-ap'-er-chur*). A counter-opening.

**contraceptive** (*kon-trah-sep'-tiv*) [*contra*, against; *conception*]. An agent which prevents conception. **contract** (*kon-trakt'*) [*contrahere*, to draw together]. 1. To draw the parts together; to shrink. 2. To acquire by contagion.

**contractile** (*kon-trak'-til*) [*contrahere*, to contract]. Having the power or tending to contract.

**contractility** (*kon-trak-til'-it-e*) [see *contract*]. That property of certain tissues, especially muscle, of shortening upon the application of a stimulus. c., **faradic**. See *galvanofaradization*. c., **galvanic**. See *galvanocontractility*. c., **idiomuscular**, that peculiar to degenerated muscles. c., **neuromuscular**, normal contractility as distinguished from idiomuscular contractility.

**contractio prævia** (*kon-trak'-she-o pre'-ve-ah*). Narrowing of the lower segment of the uterus in front of the descending head.

**contraction** (*kon-trak'-shun*). [see *contract*]. Approximation of the elements of a tissue or organ, thus diminishing its volume or contents. c., **anodal closing**, c., **anodal opening**, the contraction taking place at the anode on closing or opening the circuit. c., **carpopedal**, a variety of tetany occurring in infants, generally associated with dentition or seat-worms. There is a flexing of the fingers, toes, elbows, and knees and a general tendency to convulsions. c., **cathodal duration**, one occurring at the cathode and continuing during the whole time of closure of the circuit. Syn., *cathodal-closure tetanus*. **c., clonic,** alternate muscular contraction and relaxation. **c., closing,** muscular contraction produced at the instant that the electric current is closed. **c., Dupuytren's.** See under *Dupuytren*. **c., fibrillary,** inordinate contraction of different muscle fibrillæ in a muscle. **c., front-tap,** a phenomenon often observed in cases with exaggerated knee-jerk. When the foot is placed at a right angle to the leg and the muscles of the front of the leg are tapped, the foot is extended. See under *reflex*. **c., Gowers'.** See *c., front-tap*. **c., hour-glass,** a contraction of an organ, as the stomach or uterus, at the middle. **c., idiomuscular,** muscular contraction from direct stimulation. **c. isometric,** one showing mainly the changes in tension in a muscle, without any marked shortening. **c., isotonic,** contraction of a muscle, its tension remaining the same throughout the act. **c., myoclonic,** the convulsive spasmodic contraction of a muscle. **c., opening,** the muscular contraction produced by opening or breaking the circuit. **c., palmar,** Dupuytren's contraction. **c., paradoxic,** a phenomenon that consists in the contraction of a muscle, caused by the passive approximation of its extremities. **c.-remainder,** the stage of elastic aftervibration or residual contraction persisting in a muscle after withdrawal of the stimulus. **c.-ring,** the boundary-line between the upper and lower segments of the parturient uterus. **c., tonic.** See *spasm, tonic*. **c., vermicular,** peristaltic contraction.

**contractor** (*kon-trakt'-or*). A tensor muscle.

**contracture** (*kon-trak'-chur*) [*contraction*]. Contraction; permanent shortening, as of a muscle; distortion or deformity due to the shortening of a muscle or of various muscles. **c., functional,** one that ceases when the person is unconscious. **c., nurse's.** See *nurse's contracture*. **c., organic,** one that persists even when the person is unconscious. **c., Thomsenean,** that occurring in Thomsen's disease. **c., Volkmann's.** See under *Volkmann*.

**contradolin** (*kon-trad'-ol-in*). A compound of acetamide, salicylic acid, and phenol; analgesic. Dose 4–8 gr. (0.25–0.5 Gm.) hourly.

**contrafissura** (*kon-trah-fis-sū'-rah*) [*contra*, opposite to; *fissura*, a fissure]. Cranial fissure or fracture produced by a blow upon the skull at a point distant from or opposite to the seat of the fracture.

**contraindicant** (*kon-trah-in'-dik-ant*) [*contra*, opposed to; *indicare*, to indicate]. 1. Having the effect of a contraindication. 2. A symptom, indication, or condition that forbids the use of a particular remedial measure or set of measures.

**contraindication** (*kon-trah-in-dik-a'-shun*) [*contra*, against; *indicare*, to point out]. That modifying condition in which a remedy or a method of treatment is forbidden.

**contralateral** (*kon-trah-lat'-er-al*) [*contra*; *latus*, side]. Opposite; applied to a muscle acting in unison with another on the opposite side of the body. **c. reflexes.** See under *reflex*.

**contraluesin** (*kon-tra-lū'-es-in*) [*contra*: *lues*]. A preparation consisting of a mixture of sozoiodolate of mercury, quinine, and salicylic acid, the mercury being in such a finely divided state that it can enter directly into the blood-stream; it is given by intramuscular injection in doses of 0.15 gram of mercury every five days.

**contrast stain** (*kon'-trast*) [*contra*, against; *stare*, to stand]. A double stain, in which the special object to be examined takes one color, and everything else another color.

**contrastimulant** (*kon-trah-stim'-ū-lant*) [*contra*, against; *stimulare*, to stimulate]. 1. Counteracting the effect of a stimulus; depressing; sedative. 2. A sedative remedy.

**contrastimulus,** or **controstimulus** (*kon-trah-stim'-ū-lus*, or *kon-tro-stim'-ū-lus*) [*contra*, against; *stimulus*, a stimulus]. An influence that is opposed, or acts in opposition, to a stimulus.

**contratoxin** (*kon-trat'-oks-in*) [*contra*; *toxin*]. A vaccine serum used in the treatment of tuberculosis. It is a mixture of the blood plasma of various animals, mixed in proportions calculated to produce a lytic action on various microorganisms without producing lysis of the human red corpuscles.

**contre-coup** (*kōn-tr-koo*) [Fr.]. Counter-stroke. The transmission of a shock from the point struck to a point on the opposite side of the body or the part.

**contrectation** (*kon-trek-ta'-shun*) [*contrectatio*; *contrectare*, to touch]. 1. Digital examination; palpation; touch; manipulation, as in massage. 2. The impulse to approach and caress a person of the opposite sex (H. Ellis.).

**control** (*kon-trōl'*) [*contra*; *rotula*, a roll]. A standard by which to check observations and insure the validity of their results. Colloquially, the term is sometimes used as a noun for control animal or control experiment. **c. animal,** one used in a control experiment. **c. experiment,** an experiment carried out under normal or common circumstances or conditions, to serve as a standard whereby to test the variation or value of another experiment carried out under peculiar or abnormal circumstances.

**controller** (*kon-trōl'-er*). An apparatus for regulating the electric current to the operation of small lamps, faradic coils, small motors, etc.

**contunding** (*kon-tund'-ing*) [see *contusion*]. Producing a contusion; bruising.

**contusion** (*kon-tū'-zhun*) [*contundere*, to bruise]. A bruise or injury inflicted without the integument being broken. **c.-pneumonia,** a form of pneumonia following traumatism.

**conus** (*ko'-nus*) [κῶνος· a cone]. 1. A cone. 2. A crescentic patch of atrophic choroid tissue near the optic papilla in myopia. **c. arteriosus,** the cone-shaped eminence of the right ventricle of the heart, whence arises the pulmonary artery. **c. cochleæ,** the modiolus. **c. cordis,** the ventricular part of the heart. **c. corporis striati,** the ventral extremity of the corpus striatum. **c. elasticus,** the cricothyroid membrane. **coni Malpighii, c. tubulosi,** the Malpighian pyramids. **c. medullaris,** the cone-like termination of the spinal cord, continuous as the filum terminale. **coni retinæ.** See *cone, retinal*. **c. terminalis.** See *c. medullaris*. **coni testiculi.** See *coni vasculosi*. **coni vasculosi,** a series of conical masses that together form the globus major of the epididymis.

**convalescence** (*kon-val-es'-ens*) [*convalescere*, to become well]. A term applied to the restoration to health after disease.

**convalescent** (*kon-val-es'-ent*) [*convalescere*, to become well]. 1. One recovering from a sickness. 2. Recovering from sickness.

**convallamaretin** (*kon-val-am-ar'-e-tin*), $C_{20}H_{36}O_3$. A crystalline substance obtained by heating convallamarin with dilute sulphuric acid.

**convallamarin** (*kon-val-am'-ar-in*) [*convallaria*; *amarus*, bitter], $C_{23}H_{44}O_{12}$. A glucoside derived from *Convallaria majalis*. It is soluble in water and is used as a cardiac stimulant. Dose ¼ gr. (0.05 Gm.). Syn., *convallamarinum*.

**convallarin** (*kon-val-ar'-e-tin*), $C_{14}H_{26}O_3$. A substance obtained from convallarin by prolonged boiling in dilute acids.

**Convallaria** (*kon-val-a'-re-ah*) [*convallis*, a valley]. A genus of liliaceous plants. *C. majalis* is the lily-of-the-valley. All parts of the plant are used in medicine. Its properties are due to *convallarin*, $C_{34}H_{58}O_{11}$, and *convallamarin*, $C_{23}H_{44}O_{12}$, glucosides. It is a cathartic, diuretic, and cardiac stimulant. **c., extract of.** Dose 2–10 gr. (0.13–0.65 Gm.). **c., fluidextract of** (*fluidextractum convallariæ*, U. S. P.). Dose 2–11 min. (0.13–0.7 Cc.). **c., infusion of,** prepared with three times its weight of water. Dose ½–2 oz. (15–60 Cc.).

**convallarin** (*kon-val'-ar-in*) [*convallaria*], $C_{34}H_{62}O_{11}$. A crystalline purgative glucoside derived from *Convallaria majalis*.

**convection** (*kon-vek'-shun*) [*convehere*, to carry together]. A transmission or carrying, as of heat or electricity. **c.-current,** a current of a liquid or gas heated to a temperature above that of the surrounding medium; it rises to the surface because of its lesser density, and thus the entire fluid or gas acquires the same temperature.

**convergence** (*kon-ver'-jens*) [*con*, together; *vergere*, to incline]. Inclination or direction toward a common point, center, or focus, as of the axes of vision upon the near-point. **c.-stimulus adduction,** the power of adduction of the eyes, provoked by fixation of the gaze upon an object placed at the near-point.

**convergent** (*kon-ver'-jent*) [see *convergence*]. Tending to a common point or center. **c. strabismus.** See *strabismus, convergent*.

**converter** (*kon-vert'-er*). See *alternator*.

**convex** (*kon-veks'*) [*convexus*, vaulted]. Rounded,

as a swelling of a round or spherical form on the external surface; gibbous; opposed to concave.
**convexity** (*kon-veks′-it-e*) [see *convex*]. A surface rounding outward; the quality of being convex.
**convexoconcave** (*kon-veks-o-kon-kāv′*). See *concavoconvex*.
**convexoconvex** (*kon-veks-o-kon-veks′*). Having two convex surfaces; biconvex. See *lens, biconvex*.
**convolute** (*kon′-vo-lūt*) or **convoluted** (*kon-vo-lū′-ted*) [*convolutus*, rolled together]. Rolled together. c. bones. See turbinated bones.
**convolution** (*kon-vo-lū′-shun*) [*convolvere*, to roll together]. A fold, twist, or coil of any organ, especially any one of the prominent convex parts of the brain, separated from each other by depressions or sulci. c., angular, the posterior part of a convolution situated between the intraparietal fissure in front and above, and the horizontal limb of the Sylvian fissure and the hinder part of the first part of the first temporal fissure below. The anterior part is called the supramarginal convolution. c.s, annectant, small convolutions which connect the occipital with the temporosphenoid and parietal lobes. c., anterior central, c., ascending frontal, the convolution in front of the fissure of Rolando. c., ascending parietal, the convolution just behind the fissure of Rolando. c., Broca's, the inferior or third frontal convolution. c., fornicate, a long convolution on the mesial surface of the brain above the corpus callosum. c.s, frontal, the convolutions of the frontal lobe. c., hippocampal, the part of the fornicate convolution that winds around the splenium of the corpus callosum. c.s, insular, the small convolutions composing the island of Reil. c., marginal, the mesial surface of the first frontal convolution. c.s, occipital, the convolutions making up the (occipital lobe. c., paracentral, a convolution on the mesial surface of the brain, representing the junction of the upper ends of the ascending frontal and ascending parietal convolutions. c.s, parietal, the convolutions of the parietal lobe. c., posterior central. See c., *ascending parietal*. c., supramarginal. See c., *angular*. c.s, temporal, the convolutions of the temporal lobe. c., uncinate, the hook-like termination of the fornicate convolution.
**convolvulin** (*kon-vol′-vū-lin*) [*convolvere*, to roll together], $C_{31}H_{50}O_{15}$. A glucoside derived from the roots of jalap (*Convolvulus purga*). It is a gummy mass, with active purgative properties.
**Convolvulus** (*kon-vol′-vū-lus*) [*convolvere*, to roll together]. A genus of plants. C. panduratus, wild potato. The tuber is a mild cathartic. Dose gr. xl. C. purga. See *jalap*. C. scammonia, *scammony*.
**convulsant** (*kon-vul′-sant*) [see *convulsion*]. A medicine that causes convulsions.
**convulsio cerealis** (*kon-vul′-she-o se-re-a′-lis*) [L.]. 1. Convulsion caused by ingestion of a cereal. 2. Convulsion of arms and legs from spoiled corn; ergotism.
**convulsion** (*kon-vul′-shun*) [*convellere*, to convulse]. An involuntary general paroxysm of muscular contraction. It is either tonic (without relaxation) or clonic (having alternate contractions of opposite groups of muscles). c., epileptiform, one characterized by total loss of consciousness. c., hysterical, one due to hysteria; consciousness is only apparently lost. c., infantile, due to a number of causes such as rickets, exhaustion, etc.; sometimes called "screaming fits." c., local, one affecting one muscle, member, or part of a member. c., mimetic, c., mimic, a facial convulsion. c., oscillating, c., oscillatory, one in which the separate fiber-bundles of a muscle are affected successively and not simultaneously. c., puerperal. See *eclampsia*. c., salaam, eclampsia nutans, *q. v.* c., suffocative, laryngismus stridulus. c., tetanic, general tonic convulsions without loss of consciousness. c., tonic. See *spasm, tonic*. c., toxic, one due to the action of some toxic agent upon the nervous system. c., uremic, one that occurs in kidney disease due to retention in the blood of matters that should be eliminated by the kidney.
**convulsionary** (*kon-vul′-shun-a-re*) [*convellere*, to convulse]. One who is subject to convulsions; especially one of a set of patients who are subject to epidemic or imitative convulsions.
**convulsivant** (*con-vul′-siv-ant*) [*convellere*, to convulse]. An agent that causes convulsions.
**convulsive** (*kon-vul′-siv*). Of the nature of, or marked by, convulsions or spasms. c. cerebral typhus. See *Dubini's disease*. c. cough. See *cynobex hebetica*. c. tic. See *habit-spasm*. c. tremor. See *paramyoclonus*.

**coolie-itch** (*koo′-le*). Ankylostomiasis; water-itch.
**Cooper's arsenious ointment**, an ointment composed of arsenious acid and sulphur, each one part, and spermaceti cerate, eight parts. C.'s disease [Sir Astley Paston Cooper, English surgeon, 1768–1841]. See *Reclus' disease*. C.'s fascia. 1. The fascia transversalis. 2. The cellular layer beneath the dartos. C.'s hernia, encysted hernia of the tunica vaginalis. C.'s irritable breast, mastodynia neuralgica; neuralgia of the breast. C.'s irritable testicle, neuralgia of the testis. C.'s ligament. 1. The lower, thickened portion of the fascia transversalis, which is attached to the spine of the pubis and the iliopectineal eminence. 2. Arciform, ligamentous fibers extending from the base of the olecranon to the coronoid process on the inner aspect of the elbow-joint. C.'s operation. For *ligation of the external iliac artery*; an incision four or five inches long is made parallel with Poupart's ligament, and nearly an inch above it, commencing just outside the center of the ligament and extending outward and upward beyond the anterior superior iliac spine. C.'s suspensory ligaments, the fibrous processes that connect the capsule of the convex surface of the mammary gland with the overlying skin.
**coopers' knee.** An enlarged bursa patellæ found in coopers, as the result of pressure exerted by the knee against the barrel.
**coordination** (*ko-or-din-a′-shun*) [*con, together; ordinare*, to regulate]. The harmonious activity and proper sequence of operation of those parts that cooperate in the performance of any function.
**coordinator** (*ko-or′-din-a-tor*) [see *coordination*]. The part of the nervous system regulating coordination. c., oculonuchal, Spitzka's name for the part of the postero-longitudinal fasciculus below the floor of the fourth ventricle.
**coossify** (*ko-os′-if-i*) [*con, together; os*, bone; *facere*, to make]. To grow together as one bone, said of bones or parts of bone usually separate.
**copaiba** (*ko-pa′-e-bah*) [Sp.]. Balsam of copaiba. The oleoresin of *Copaifera officinalis, C. coriacea, C. guianensis, C. multijuga, C. cordifolia, C. laxa, C. nitida, C. oblongifolia*, and *C. langsdorffii*, leguminous trees, native to South America. It is a stimulant, diuretic, diaphoretic, and an expectorant, and is much used in gonorrhea. Syn., *copaiva*. c., balsam of. See *copaiba*. c., East Indian. See *Gurjun balsam*. c., mass of, copaiba, 94; magnesia, 6 parts. Dose 10 gr.-1 dr. (0.65-4.0) Gm. c., mixture of, compound, Lafayette's mixture: copaiba, 7 dr.; oil of cubebs, 1 dr.; glycerite of yolk of egg, 7 dr.; triturate and add syrup, 2½ oz.; then add, with constant stirring, solution of potassium hydroxide, ½ oz.; compound tincture of cardamom, 2 dr.; sweet spirit of niter, ½ oz.; enough peppermint-water to make 8 oz. Dose 1 dr.-½ oz. (4-16 Cc.). c., oil of (*oleum copaibæ*, U. S. P.), a colorless substance constituting about one-half of copaiba, and used for the same purposes. Dose 10-15 min. (0.65-1.0 Cc.). c., resin of, the residue after distilling off the volatile oil of copaiba, mainly copaibic acid. Dose 1-5 gr. (0.065-0.3 Gm.).
**cophosis** (*ko-fo′-sis*) [κωφός, deaf]. Deafness or dulness of hearing.
**copiopia** (*ko-pe-o′-pe-ah*) [κόπος, a straining; ὤψ, eye]. Eye-strain; weariness of the eyes. c. hysterica, a term applied to those symptoms that indicate hyperesthesia of the fifth and optic nerves.
**copodyskinesia** (*kop-o-dis-kin-e′-ze-ah*) [κόπος, toil, fatigue; δυς, hard; κίνησις, motion]. Difficult or faulty motion due to constant repetition of the same act. Professional spasm, or occupation neurosis.
**copos** (*kop′-os*) [κόπος, fatigue]. 1. Lassitude; exhaustion after illness. 2. Cramp in the calves of the legs.
**copper** (*kop′-er*). Cuprum. Symbol, Cu; atomic weight, 63.57. A reddish-brown metal existing in nature chiefly in the form of copper pyrites, which is a double salt of copper and iron sulphide. Various salts are used in medicine. In toxic doses they are gastrointestinal irritants. In therapeutic doses they are used as astringents in inflammation of mucous membranes. They are also employed as emetics, and, externally, as caustics. See *elements, table of chemical*. c. acetate, $Cu(C_2H_3O_2)_2$, verdigris, used in pulmonary diseases and as a lotion in skin diseases. Dose $\frac{1}{20}$-½ gr. (0.0065-0.016 Gm.). c. acetoarsenite,

Paris-green, used as a pigment and an insecticide. c. **acetophosphate,** employed in chlorosis and amenorrhea. **c.-alum.** See *c., aluminated.* **c., aluminated,** a combination of sulphates of copper and aluminum and potassium nitrate, occurring as a green powder; a mild caustic used in ophthalmia. Syn., *copper-alum; lapis divinus.* c. **amalgam,** a metallic filling-material composed of copper and mercury. c. **ammoniate,** ammonium carbonate, 3; copper sulphate, 4 parts; useful in chorea, hysteria, etc. Dose $\frac{1}{6}$–1 gr. (0.01–0.065 Gm.). c. **ammoniosulphate, c. and ammonium sulphate,** obtained by dissolving copper sulphate in ammonia-water and precipitating with alcohol. It is antispasmodic and astringent. Dose $\frac{1}{2}$–2 gr. (0.03–0.13 Gm.) 3 or 4 times daily with tincture of opium after meals; maximum dose 5 gr. (0.3 Gm.) single; 10 gr. (0.6 Gm ) a day. Application for gleet, etc., 0.2 to 1 % solution or ointment. c. **arsenate,** a blue powder obtained from ammonium arsenate with copper sulphate. It is used as an alterative in syphilis. Dose $\frac{1}{32}$–$\frac{1}{8}$ gr. (0.002–0.008 Gm.). c. **arsenite,** a salt valuable in intestinal diseases. Dose $\frac{1}{100}$ gr. (0.00065 Gm.). c. **carbonate,** a compound of copper and carbonic acid. **c. nitrate** $Cu(NO_3)3H_2O$, is used for the same purposes as the sulphate. **c.-nose.** Synonym of *acne rosacea.* c. **nucleinate,** a compound of nucleol and copper oxide containing 6 % of copper; it occurs as a fine powder and is used in chronic conjunctivitis. Syn., *cuprol.* c. **oleate,** $Cu(C_{18}H_{33}O_2)_2$, a mixture of 10 % copper oxide dissolved in oleic acid, forming a greenish-blue, granular powder, soluble in ether. It is applied to indolent ulcers; ointment, 10 to 20 % in lanolin. c. **oxide,** a compound of copper and oxygen. c. **oxide, black,** CuO, a brownish-black, amorphous powder obtained from copper nitrate or copper carbonate by ignition. It is used as a teniafuge. Dose $\frac{3}{4}$–$1\frac{1}{2}$ gr. (0.05–0.11 Gm.) 3 or 4 times daily in pills for two weeks, abstaining from acid food. Externally it is used as an ointment with lard in treatment of chronic glandular induration. It is also employed in organic analysis. Syn., *copper monoxide.* c. **oxide, red,** $Cu_2O$, a dark-brown, crystalline powder. Syn., *copper hemioxide; copper suboxide.* c. **phosphate,** $CuHPO_4$, a bluish-green powder. It is used in tuberculosis. Dose $\frac{1}{8}$–$\frac{1}{2}$ gr. (0.008–0.032 Gm.) several times daily. **c., reaction for.** See *Schoenbein.* c. **sulphate** (*cupri sulphas,* U. S. P.), $CuSO_4 . 5H_2O$, soluble in water, valuable as an emetic tonic, and astringent. Dose, as an emetic, 2–5 gr. (0.13–0.32 Gm.); as a tonic, $\frac{1}{8}$–$\frac{1}{2}$ gr. (0.01–0.032 Gm.). c. **sulphocarbolate,** $CuC_6H(SO_4)_2+6H_2O$, green crystals soluble in water and alcohol. Syn., *cupriaseptol.*

**copperas** (*kop'-er-as*) [*cupri rosa,* rose of copper(?)]. A common name for ferrous sulphate, $FeSO_4+7H_2O$.

**copra** (*kop'-rah*) [Hind.]. The dried and crushed kernel of the cocoanut, from which cocoa-oil is expressed. c. **itch,** a peculiar eruption noticed in persons working in copra mills in Ceylon. The eruption generally begins on the hands, and spreads to the arms, legs, and trunk, but does not affect the face.

**copragogue** (*kop'-rag-og*) [κόπρος, dung; ἀγωγός, leading]. A remedy to carry off feces; a purgative.

**copraol** (*kop'-ra-ol*) [*copra,* the dried kernel of the cocoanut]. A solid fat, derived from the cocoanut, and used as a substitute for cacao-butter in making suppositories. It melts at 30.3° C.

**copremia** (*kop-re'-me-ah*) [κόπρος, dung; αἷμα, blood]. A form of general blood-poisoning arising from chronic constipation. The symptoms are anemia, sallow complexion, anorexia, frontal headache, vertigo, nausea, flatulence, thirst, fetid breath, lassitude, hypochondriasis, and irritability of temper.

**copremesis** (*kop-rem'-es-is*) [κόπρος, dung; ἔμεσις, vomiting]. The vomiting of fecal matter.

**copro-** (*kop-ro-*) [κόπρος, dung]. A prefix meaning relating to the feces or to dung.

**coproctic** (*kop-rok'-tik*) [κόπρος, feces]. Relating to feces; fecal.

**coprolalia** (*kop-ro-la'-le-ah*) [*copro-;* λαλιά, speech]. The use of filthy and offensive language when a manifestation of disease.

**coprolith** (*kop'-ro-lith*) [*copro-;* λίθος, a stone]. A hard mass of fecal matter in the bowels.

**coprophagy** (*kop-rof'-a-je*) [*copro-;* φαγεῖν, to eat]. The eating of feces, a symptom seen in insane and hysterical patients.

**coprophilous** (*kop-roff'-il-us*) [κόπρος, dung; φίλος, loving]. Fond of feeding or growing upon fecal matter; said of certain bacteria.

**coproplanesis** (*kop-ro-plan-e'-sis*) [*copro-;* πλάνησις, wandering]. Escape of feces through a fistula or other abnormal opening.

**coprorrhea** (*kop-ror-e'-ah*) [κόπρος, dung; ῥέειν, to flow]. Synonym of *diarrhea.*

**coprostasia** (*kop-ro-sta'-se-ah*). See *coprostasis.*

**coprostasis** (*kop-ros'-tas-is*) [*copro-;* στάσις, a standing]. The accumulation of fecal matter in the bowel.

**coptine** (*kop'-tēn*) [κόπτειν, to cut]. A colorless alkaloid of *goldthread.* See *coptis.*

**coptis** (*kop'-tis*) [κόπτειν, to cut]. Goldthread. The root of *C. trifolia,* a simple bitter tonic resembling quassia. It contains *coptine,* an alkaloid closely allied to berberine. Dose 10–30 gr. (0.65–2.0 Gm.).

**copula** (*kop'-ū-lah*) [L., "a band"]. 1. The *copula alba cerebri,* an anterior commissure of the cerebrum. 2. A thin lamina joining the rostrum with the terma. 3. Same as *sporont* or *oocyst.* 4. Same as *amboceptor.* 5. Sexual intercourse.

**copulation** (*kop-ū-la'-shun*) [*copulare,* to couple]. The act of sexual intercourse.

**coqueluche** (*kōk-lūsh'*) [Fr., a hood or cowl]. Synonym of *influenza;* also of *whooping-cough.*

**coquilles** (*ko-kēl'*) [Fr.]. A variety of dark eye-glasses curved like shells.

**cor** (*kor*) [L.; gen., *cordis*]. The heart. See *heart.* c. **adiposum,** a heart with a simple excess of the normal subpericardial fat. c. **bovinum.** See *bovine heart.* c. **hirsutum,** c. **hispidum,** c. **tomentosum.** See *c. villosum.* c. **membranaceum,** the auricular part of the heart. c. **mobile,** a heart which changes its position with the change of posture of the individual. c. **villosum,** hairy heart; the peculiar shaggy appearance presented by the heart in acute plastic pericarditis, with the deposited fibrin existing in long shreds.

**coraco-** (*kor-ak-o-*) [κόραξ, a crow]. Pertaining to the coracoid process.

**coracoacromial** (*kor-ak-o-ak-ro'-me-al*). Relating to the coracoid process and the acromion.

**coracobrachialis** (*kor-ak-o-brā-ke-al'-is*). See under *muscle.*

**coracoclavicular** (*kor-ak-o-klav-ik'-ū-lar*). Relating to the coracoid process and the clavicle. Syn., *omoclavicular.*

**coracohumeral** (*kor-ak-o-hū'-mer-al*). Relating to the coracoid process and the humerus.

**coracohyoid** (*kor-ak-o-hi'-oid*). 1. Relating to the coracoid process and the hyoid bone. 2. The omohyoid muscle.

**coracoid** (*kor'-ak-oid*) [*coraco-;* εἶδος, likeness]. 1. Having the shape of a crow's beak. 2. The coracoid process. c. **ligament,** a triangular ligament joining the coracoid process to the acromion. c. **notch,** the notch in the upper border of the scapula. **c. process,** a beak-shaped process of the scapula.

**coracopectoralis** (*kor-ak-o-pek-tor-a'-lis*). The pectoralis minor muscle. See *muscles, table of.*

**coracoscapular** (*kor-ak-o-skap'-ū-lar*). Relating to the coracoid process of the scapula and to some other portion of the scapula.

**coral** (*kor'-al*). The hard substance secreted by marine polyps. c. **calculus,** a peculiar dendritic form of calculus found in the pelvis of the kidney, and forming a complete mold of the infundibula and calices.

**cord** (*chorda,* a string]. 1. A tendon; any string-like body. 2. Used as a synonym for the *umbilical cord,* the vascular, cord-like structure connecting the placenta and fetus. **c., axis-.** See *primitive streak.* **c., bioplasson,** a reticulum formed by branching cells. **c., colic, transverse,** Glénard's term for that portion of the transverse colon which becomes hard and rigid as the result of a stoppage of fecal matter by the kinking of the colon near its attachment by the pylorocolic ligament. **c., dorsal,** the notochord. **c., false, c., superior** (vocal), a fold of mucous membrane on either side of the middle line of the larynx, inclosing the superior thyroarytenoid ligament. **c., genital,** Thiersch's name for an embryonic structure formed from the two Wolffian ducts and the Muellerian ducts. c. **of Hippocrates,** the Achilles tendon. **c., lumbosacral,** a nerve-trunk formed from the divisions of the fourth and fifth lumbar nerves. **c., muscular,** a cord-like prominence of a muscle due to morbid excitability of its fibers. **c., presentation of,** descent of the umbilical cord between the presenting part and the membranes at the beginning of labor. **c., prolapse of,** descent of the umbilical cord at the

rupture of the bag of waters; incomplete, if remaining in the vagina, complete, if protruding therefrom. c.s, sonorous, the semicircular canals of the internal ear. c. spermatic. See *spermatic cord*. c., spinal. See *spinal cord*. c., true, vocal c., vocal, the vocal band. See under *larynx*.
**cordate** (*kor'-dāt*) [*cor*]. Heart-shaped.
**cordein** (*kor'-de-in*). A white, crystalline substance used as an analgesic and antiseptic, Syn., *methyltribromsalol*.
**Cordia** (*kor'-de-ah*) [E. and V. *Cordus*, German physicians (1486-1535 and 1515-1544)]. A genus of shrubs and trees of the order *Boraginaceæ*. C. *aubletii* is indigenous to Guiana; the leaves are used as an application to tumors and skin diseases. C. *myxa* is indigenous to the East Indies, but cultivated in Arabia and Egypt. The fruit is used in coughs, the powdered bark in ringworm, the root as a purgative.
**cordial** (*kord'-yal*) [*cor*]. 1. Pertaining to the heart; exhilarant; stimulant. 2. An aromatic, spirituous stimulant.
**cordiale** (*kor-de-a'-le*) [L.]. A cordial. c. **rubi fructus**. See *blackberry cordial*.
**cordiform** (*kor'-de-form*) [*cor; forma*, form]. Cordate; shaped like a heart.
**cordite** (*kor'-dīt*). A smokeless gun-powder consisting of gun-cotton dissolved in acetone and nitroglycerin.
**cordol** (*kor'-dol*). See *salol tribromide*.
**cordon** (*kor'-don*) [Fr., a cord or rope]. A line of posts to enforce a quarantine against a place infected with an epidemic disease.
**cordyl** (*kor'-dil*). See *acetyl tribromsalol*.
**core** (*kor*). 1. The central slough of a boil or carbuncle. 2. The axial or central portion of the terminal corpuscle in a nerve. 3. A bundle of soft iron wires used as a magnet in the center of a coil. c., atomic. See *coelectron*. 4. (*kor'-eh*) [κόρη, pupil]. The pupil of the eye.
**coreclisis, corecleisis** (*kor-ek-li'-sis*) [κόρη, the pupil; κλείσις, a closure]. Pathological closure or obliteration of the pupil.
**corectasis** (*kor-ek'-ta-sis*) [*core* (4); ἔκτασις, a stretching out]. Dilatation of the pupil.
**corectome** (*kor-ek'-tōm*) [*core* (4); ἐκτέμνειν, to cut out]. An instrument used in iridectomy.
**corectomedialysis** (*kor-ek-to-me-di-al'-is-is*). See *iridectomy* and *corediaIysis*.
**corectomy** (*kor-ek'-to-me*) [*core* (4); ἐκτέμνειν, to cut out]. See *iridectomy*, and *pupil, artificial*.
**corectopia** (*kor-ek-to'-pe-ah*) [*core* (4); ἔκτοπος, misplaced]. An anomalous position of the pupil; displacement of the pupil.
**coredialysis** (*kor-e-di-al'-is-is*) [*core* (4); διάλυσις, dialysis]. The production of an artificial pupil at the ciliary border of the iris.
**corediastasis** (*kor-e-di-as'-tas-is*) [*core* (4); διάστασις, dilatation]. Dilatation of the pupil.
**corelysis** (*kor-el'-is-is*) [*core* (4); λύσις, a loosening]. The detachment of iritic adhesions to the lens or to the cornea.
**coremorphosis** (*kor-e-mor-fo'-sis*) [*core* (4); μόρφωσις, formation]. The operation for establishing an artificial pupil.
**corenclisis** (*kor-en-kli'-sis*) [*core* (4); ἐγκλείσις, inclusion]. The formation of a new pupil by displacement, the iris being drawn aside and in part excised.
**coreometer** (*kor-e-om'-et-er*) [*core* (4); μέτρον, a measure]. An instrument for measuring the pupil of the eye.
**coreometry** (*kor-e-om'-et-re*) [see *coreometer*]. The measurement of the pupil of the eye.
**coreoncion** (*kor-e-on'-se-on*). A double-hooked irisforceps.
**coreoplasty** (*kor'-e-o-plas-te*) [*core* (4); πλάσσειν, to form]. Any operation for forming an artificial pupil.
**coretomodialysis** (*kor-et-o-mo-di-al'-is-is*). See *iridectomy*.
**coretomy** (*kor-et'-o-me*) [*core* (4); τέμνειν, to cut]. Iridotomy or iridectomy; any surgical cutting operation on the iris.
**coriaceous** (*kor-e-a'-she-us*) [*corium*, leather]. 1. Leathery, tough. 2. Of a bacterial culture, one which will not yield to the platinum needle.
**coriamyrtin** (*ko-re-am-er'-tin*) [*Coriaria myrtifolia*, myrtle], $C_{30}H_{36}O_{10}$. An exceedingly poisonous principle, a glucoside, obtained from the fruit of *Coriaria*

*myrtifolia*. A cardiac stimulant. Maximum dose ¼ gr. (0.001 Gm.).
**coriander, coriandrum** (*ko-re-an'-der, ko-re-an'-drum*) [κορίαννον, coriander]. Coriander-seed. The *coriandrum* of the U. S. P. is the dried ripe fruit of *Coriandrum sativum*, an aromatic, carminative, and stimulant, used mainly to give flavor to other remedies and as a corrective to griping purgatives. Dose 10-20 gr. (0.65-1.3 Gm.). c., oil of (*oleum coriandri*, U. S. P.), the volatile oil. Dose 2-5 min. (0.13-0.32 Cc.).
**coriandrol** (*kor-e-an'-drol*), $C_{10}H_{18}O$. The chief constituent of oil of coriander; a liquid isomeric with borneol.
**Coriaria** (*ko-re-a'-re-ah*) [L.]. A genus of poisonous shrubs of several species, having a wide geographical distribution. C. **myrtifolia**, used in dyeing and tanning, has poisonous berries and shoots. The seeds and shoots of C. **sarmentosa** of New Zealand afford what is called *toot-poison*. The memory is said to be impaired after recovery from poisoning by this plant.
**coridin** (*kor'-id-in*), $C_{10}H_{15}N$. A liquid base obtained from the distillation of bones.
**corisol** (*kor'-is-ol*). A preparation of suprarenal capsule said to be useful in catarrh of the nasopharynx.
**corium** (*ko'-re-um*) [L., "a hide; leather"]. The deep layer of the skin. See *skin*. c. **phlogisticum**. Same as *crusta phlogistica, q. v.*
**corm** (*korm*) [κορμός, the trunk of a tree]. The bulbous underground part of certain plants, as the crocus.
**corn** (*korn*) [*cornu*, horn]. A local induration and thickening of the skin from friction or pressure. See *clavus*. c.-silk. See *Zea mays*. c.-smut. See *ustilago*. c.-starch, the commercial name of a starch derived from maize, and extensively used as an article of food, especially for invalids.
**Cornaro's diet**. A diet for indigestion and the results of riotous living, devised by Luigi Cornaro, a Venetian gentleman of the seventeenth cenntury. It consisted of a daily allowance of bread, meat, and yolk of egg, amounting to 12 ounces in all. With this he took 14 ounces of a light Italian wine each day.
**cornea** (*kor'-ne-ah*) [*corneus*, horny]. The transparent anterior portion of the eyeball, its area occupying about one-sixth the circumference of the globe. It is continuous with the sclerotic, and is nourished by lymph from the looped blood-vessels at its peripheral border. c., conical. See *keratoglobus*. c., transplantation of, the operation of engrafting a section of transparent cornea from some animal into the space of an excised portion of human cornea.
**corneal** (*kor'-ne-al*). Pertaining to the cornea. c. **corpuscles**, stellate bodies in the corneal lacunæ. c. **lacunæ, c. spaces**, stellate spaces in the corneal lamellæ between the corpuscles.
**corneitis** (*kor-ne-i'-tis*). See *keratitis*.
**corneoblepharon** (*kor-ne-o-blef'-ar-on*) [*cornea; blepharon*]. Adhesion of the surface of the eyelid to the cornea.
**corneocalcareous** (*kor-ne-o-kal-ka'-re-us*) [*corneus*, horny; *calcareus*, pertaining to lime]. Formed of a mixture of horny and calcareous substances.
**corneoiritis** (*kor-ne-o-i-ri'-tis*). See *keratoiritis*.
**corneosclera** (*kor-ne-o-skle'-rah*) [*cornea; sclera*]. The cornea and sclera taken together.
**corneous** (*kor'-ne-us*) [*corneus*, horny]. Horny or horn-like. c. **tissue**, the substance of the nails.
**cornet** (*kor-net'*) [*cornu*]. 1. A small ear-trumpet worn within the ear and sometimes concealed by the hair of the wearer. 2. A bony layer. c., **Bertin's**, c., **sphenoid**, the anterior part of the body of the sphenoid bone.
**corneum** (*kor-ne'-um*). The stratum corneum or horny layer of the skin.
**corniculate** (*kor-nik'-ū-lāt*) [*cornu*]. Furnished with horns or horn-shaped appendages.
**corniculum** (*kor-nik'-ū-lum*) [L. dim. of *cornu*, a horn; *pl., cornicula*]. A small cornu or horn-like process. c. **laryngis**, a small horn-shaped mass of cartilage on the arytenoid cartilages; called also the *cartilages of Santorini*. **cornicula of the hyoid bone**, c. **interna ossis hyoidei**, the small cornua of the hyoid. **cornicula santoriniana**. See *c. laryngis*.
**cornification** (*kor-nif-ik-a'-shun*) [*cornu; facere*, to make]. The process of hardening or making horny.
**cornin** (*kor'-nin*) [*corneus*, horny]. A precipitate from the tincture of the bark of Dogwood, *Cornus*

*florida;* it occurs in white silky, bitter crystals, and is a tonic, stimulant, and astringent. Dose two to four grains. 2. A bitter crystalline substance from the bark of *Cornus florida.* Syn., *cornic acid.*

**cornstalk disease of cattle and horses.** A disease caused by feeding on cornstalks left standing in the field after the corn has been gathered in the fall. It has been attributed—(1) to corn-smut; (2) to scarcity of salt and water; (3) to "dry murrain," a hard and dry condition of the third stomach, supposed to be morbid, though really normal; (4) to the presence of potassium nitrate in the fodder; (5) to the presence of a bacterium. Syn., *bronchopneumonia bovis.*

**cornu** (*kor'-nu*) [L.; pl., *cornua*]. A horn. A name applied to any excrescence resembling a horn. **c. Ammonis,** the hippocampus major of the brain. **c. anterius,** the anterior horn. **c. cervi,** hartshorn or ammonium hydroxide. **c. cutaneum,** a horn-like excrescence arising from the skin. Syn., *cornu humanum.* **c., dental,** a horn of the dental pulp. These extensions form the body of the dental pulp, which corresponds with the positions of the cusps of the teeth. **c. descendens, c. inferius, c. laterale, c. magnum, c. medium,** the medicornu, that prolongation of the lateral ventricle which, curving outward around the back of the thalamus, descends beneath it, and, extending forward and inward, ends in the anterior extremity of the hippocampal gyrus. **c. dorsale,** the dorsal projection of the mass of cinerea seen upon each half of the spinal cord in transverse section. Syn., *crus posticum; posterior cornu.* **c. humanum.** See *c. cutaneum.* **c. inferius,** the inferior or descending horn. **c. majus,** greater horn. **c. minus,** lesser horn. **c. occipitale, c., posterius** (of the lateral ventricle), the postcornu, a conical prolongation of the lateral ventricle, curving outward, backward, and inward into the occipital lobe. Syn., *cavitas digitata; cornu ancyroide.* **c. superius,** superior horn. **cornua of the uterus.** 1. The lateral infundibuliform prolongations of the uterine cavity into which the Fallopian tubes open. 2. The oviducts. **c. ventrale,** the ventral projection of the mass of cinerea seen upon each half of the spinal cord in transverse section. Syn., *crus anterius.*

**cornua** (*kor'-nū-ah*) [L.]: Plural of *cornu.*
**cornual** (*kor'-nū-al*) [*cornu*]. Relating to a cornu.
**c. myelitis,** myelitis affecting the anterior cornua of the spinal cord.
**cornucopia** (*kor-nū-ko'-pe-ah*) [L., horn of plenty]. An offset of the choroid plexus of the fourth ventricle into the lateral recess of the ventricle.
**cornus** (*kor'-nus*) [L.]. Dogwood. The bark of the root of *C. florida,* the properties of which are due to a crystalline principle, *cornin.* It is a simple stomachic bitter and slightly antiperiodic. Dose of the *fluidextract* 10 min.–1 dr. (0.65–4.0 Cc.).
**cornutine** (*kor'-nū-tēn*). An alkaloid, the active principle of ergot. Dose $\frac{1}{12}$ gr. (0.005 Gm.). **c. citrate.** Dose $\frac{1}{20}$–$\frac{1}{10}$ gr. (0.003–0.006 Gm.) in spermatorrhea.
**cornutol** (*kor'-nū-tol*). Trade name of a fluidextract of ergot, said to be aseptic and adopted for hypodermic use.
**coroclisis, corocleisis** (*ko-ro-kli'-sis*) [see *coreclisis*].
**corodialysis** (*kor-o-di-al'-is-is*) [κόρη, the pupil; διάλυσις, loosening]. See *iridodialysis.*
**corodiastasis** (*kor-o-di-as'-tas-is*) [κόρη, the pupil; διάστασις, separation]. Dilatation of the pupil.
**corometer** (*ko-rom'-et-er*). Same as *coreometer.*
**corona** (*ko-ro'-nah*) [L., a "crown"]. 1. A crown. 2. The corona radiata. **c. capitis,** the crown of the head, the top of the head. **c. ciliaris,** the ciliary ligament. **c. dentis,** the crown of a tooth. **c. glandis,** the ridge of the glans penis. **c. radiata,** a radiating mass of white nerve-fibers ascending from the internal capsule to the cortex cerebri. **coronæ tubulorum.** See *Lieberkühn's crypts.* **c. veneris,** a circle of syphilitic blotches occurring on the forehead.
**coronad** (*kor'-o-nad*) [*corona,* the crown; *ad,* to]. Toward the coronal aspect of the head, or towards any corona.
**coronal** (*kor-o'-nal*) [*corona*]. Encircling like a crown; pertaining to the crown of the head. **c. suture,** the suture joining the frontal with the two parietal bones.
**coronale** (*kor-o-na'-le*) [L.]. 1. The frontal bone. 2. A point on the coronal suture where the frontal diameter is greatest.

**coronamen** (*kor-o-na'-men*) [L., a crowning; *pl., coronamina*]. Same as *coronet.*
**coronaria** (*kor-o-na'-re-ah*). A coronary artery, of the heart, or lips, or stomach. **c. ventriculi,** the coronary artery of the stomach.
**coronary** (*kor'-o-na-re*) [*corona*]. A term applied to vessels, nerves, or attachments that encircle a part or an organ. **c. artery,** one of the arteries around the heart and lips; also the gastric artery. **c. bone,** the small postern or median phalanx of a horse's foot. **c. ligament,** a ligament of the knee; also one of the liver. **c. sinus,** a passage for the blood into the right auricle. **c. valve,** the valve protecting the orifice of the coronary sinus.
**corone** (*ko-ro'-ne*). The coronoid process of the inferior maxilla.
**coronen** (*ko-ro'-nen*) [*corona,* a crown]. Belonging to the corona in itself.
**coroner** (*kor'-o-ner*) [*coranator,* a crown officer]. An officer who inquires by authority of the law into the causes of sudden or violent deaths. **c.'s inquest,** the legal inquiry before a jury into the cause of a sudden or violent death.
**coronet** (*kor'-o-net*) [Fr., dim. of *couronne,* a crown]. 1. In biology, a crowning circle of hairs. 2. In veterinary surgery, the lowest part of the postern of the hoof, also called coronamen.
**Coronilla** (*kor-o-nil'-ah*) [dim. of *corona*]. A genus of leguminous herbs. *C. scorpioides,* an annual of southern France, furnishes *coronillin* (*q. v.*). *C. varia* is diuretic, purgative, and poisonous. It is used as a euœodanæum for digitalis in cardiac disease. An aqueous extract and a powder of the fresh plant are given in doses of 1½ gr. (0.098 Gm.).
**coronillin** (*ko-ro-nil'-in*) [see *coronilla*]. A glucoside, C$_7$H$_{12}$O$_5$, from *Coronilla scorpioides;* it is a cardiac tonic and diuretic. Dose 1–2 gr. (0.06–0.13 Gm.).
**coronion** (*ko-ro'-ne-on*). The apex of the coronoid process of the inferior maxilla.
**coronitis** (*kor-o-ni'-tis*). Inflammation of the coronary substance of the horse's hoof.
**coronium** (*ko-ro'-ne-um*). A supposed element, said to be lighter than hydrogen, and believed to exist in volcanic gases.
**coronobasilar** (*kor-o-no-bas'-il-ar*) [*corona,* a crown; *basis,* the base]. Extending from the coronal suture to the basilar aspect of the head.
**coronofacial** (*kor-o-no-fa'-shal*) [*corona; facies,* face]. Relating to the crown of the head and to the face.
**coronoid** (*kor'-o-noid*) [*corona* or κορώνη, a crow; εἶδος, likeness]. Crown-shaped or crow-shaped, as the coronoid process of the ulna or of the jaw.
**coroparelcysis** (*kor-o-par-el'-si-sis*) [κόρη, the pupil; παρέλκυσις, a drawing aside]. Operative displacement of the pupil to remedy partial opacity of the cornea by bringing it opposite a transparent part.
**coroplasty** (*ko'-ro-plas-te*). Same as *coreplasty.*
**coroplhthisis** (*ko-roff'-this-is*) [κόρη, pupil; φθίσις, a wasting]. Habitual or permanent contraction of the pupil due to a wasting disease of the eye.
**coroscopy** (*kor-os'-ko-pe*). See *retinoscopy.*
**corpora** (*kor'-por-ah*) [pl. of *corpus,* a body]. A general term applied to certain parts of the body having a rounded or ovoid shape. **c. albicantia,** two white masses in the interpeduncular space at the base of the brain, the projections of the anterior pillars of the fornix. **c. amylacea,** certain bodies found in nervous and other tissues after death; they are probably the result of degeneration. **c. aranacea,** a granular substance occurring at times in the masses of papilloma. Syn., *sand-bodies.* **c. Arantii.** See *Arantius.* **c. cavernosa,** the cylindrical bodies of erectile tissue forming the chief part of the penis. Also, the two masses of erectile tissue composing the clitoris. **c. geniculata.** See *geniculate bodies.* **c. globosa cervicis uteri,** cysts of the neck of the uterus, Nabothian cysts. **c. olivaria,** the two oval masses behind the pyramids of the oblongata. **c. pyramidalia,** the two bundles of white matter of the oblongata. **c. quadrigemina,** the optic lobes of the brain, the four rounded eminences situated under the corpus callosum. The anterior pair are called the nates, and the posterior, the testes. **c. restiformia,** the cord-like bodies extending between the oblongata and the cerebrum. **c. sesamoidea.** Same as *c. Arantii.* **c. striata,** two gray bodies in the lateral ventricles of the brain.
**corporeal** (*kor-por'-e-al*) [*corpus,* a body]. Per-

taining to the body. c. **endometritis.** See *endometritis.*

**corpse** (*korps*) [*corpus*, a body]. A cadaver, a dead body.

**corpulence, corpulency** (*kor'-pū-lens, -se*) [*corpulentus*, corpulent]. Obesity; fatness of the body.

**corpulent** (*kor'-pū-lent*) [*corpulentus*, corpulent]. Excessively fat; obese.

**corpulin** (*kor'-pū-lin*). A remedy for obesity said to constis of bladderwrack (*Fucus vesiculosis*), tamarinds, and cascara sagrada.

**corpus** (*kor'-pus*) [L., "a body"; pl., *corpora* (*q. v.*)]. 1. A body; the human body. 2. The body or shaft of a bone or other structure. c. **annulare,** the pons Varolii. c. **bigeminum,** an optic lobe. c. **callosum,** the broad band of white matter uniting the hemispheres of the cerebrum. c. **caudatum,** a ganglion or free ring of gray matter circling around the lenticular nucleus of the brain. It is massive in the frontal portion, but becomes attenuated caudad; the anterior portion is called the head; the posterior, the tail. c. **cavernosum urethræ,** the corpus spongiosum. c. **cavernosum vaginæ,** the spongy tissue of the vagina. c. **ciliare,** the ciliary body, that part of the middle coat of the eye comprising the ciliary muscles and processes. c. **dentatum.** 1. See *olivary body.* 2. The central folded gray nucleus of the cerebellum. c. **fibrosum,** a tough, semiopaque body occurring in the ovary, due to some fibrous change in the corpus luteum. c. **fimbriatum,** the lateral thin edge of the tænia hippocampi. c. **geniculatum,** a tubercle of the lower portion of the optic thalamus. c. **Highmorianum.** See *Highmore, body of.* c. **luteum,** the yellow body formed in the ovary in the site of a Graafian vesicle after the escape of the ovum. c. **luteum, false,** that resulting when pregnancy does not occur. Syn., *corpus luteum of menstruation.* c. **luteum, true,** that resulting when pregnancy takes place. Syn., *corpus luteum of pregnancy.* c. **mammillare.** See *corpora albicantia.* c. **pampiniforme,** the parovarium. c. **phacoide,** the crystalline lens. c. **pyramidale,** the pyramid of the oblongata. c. **rhomboidale.** The same as c. *dentatum.* c. **spongiosum,** the spongy part of the penis encircling the urethra. c. **striatum,** a mass of gray matter extending into the lateral ventricles of the brain and composed of the caudate and lenticular nuclei.

**corpuscle** (*kor'-pus-l*) [dim. of *corpus*]. 1. A small body or particle. 2. A molecule or atom. 3. A cell. 4. A blood cell. c., **axile,** c., **axis-,** the central portion of a tactile corpuscle. Syn., *axile body.* c.s, **Babes-Ernst's.** See *Babes-Ernst's corpuscles.* c.s, **Bennett's.** See *Bennett's corpuscles.* c.s, **Bizzozero's.** See *Bizzozero's blood-platelets.* c.s, **blood-,** red, biconcave, nonnucleated discs, circular in outline, and containing red coloring-matter, termed hemoglobin, to which the color of the blood is due. Red corpuscles have been divided, according to their size, into normocytes (normal in size), megalocytes (of excessive size), microcytes (abnormally small), and poikilocytes (of irregular shape and size). The red corpuscles in the blood of man are about $\frac{1}{3200}$ inch in diameter and $\frac{1}{12000}$ inch thick, and their number is about 5,000,000 to each cubic millimeter of blood. They consist of a colorless stroma infiltrated with the coloring-matter (hemoglobin). c.s, **blood-, white** (or colorless), flattened cells, about $\frac{1}{2500}$ inch in diameter, existing in the ratio of 1 : 500 compared with red corpuscles. Their protoplasm is granular; they have one or more nuclei and no cell-wall. They possess contractile power and alter their shape readily. The colorless corpuscles are variously designated as eosinophil, basophil, neutrophil, mononuclear, polynuclear, lymphocytes, transitional, large, small, etc. c., **bone-,** an osteoblast. c.s, **Burckhardt's.** See *Burckhardt's corpuscles.* c.s, **cancroid,** the pearly bodies of squamous epithelioma. c., **cartilage.** See *cartilage.* c.s, **chorea.** See under *chorea.* c.s, **chromophile.** See *Nissl's bodies.* c.s, **chyle,** lymph-corpuscles. c.s, **colostrum.** See *colostrum corpuscles.* c.s, **concentric.** See *Hassall's bodies.* c.s, **corneal,** connective-tissue corpuscles containing an oval nucleus and furnished with numerous branching processes occurring within the fibrous groundwork of the cornea. Syn., *Toynbee's corpuscles; Virchow's corpuscles.* c., **cytoid,** a leukocyte. c., **Davaine's.** See *Bacillus anthracis* under *Bacilli.* c.s of **Donne.** See *colostrum corpuscles.* c.s, **Drysdale's** ovarian. See under *Drysdale.* c.s, **genital,** special nerve-endings in the external genitalia. c.s, **ghost-, phantom-corpuscles.** c.s, **Gierke's.** See under *Gierke.* c.s, **Golgi's.** See under *Golgi.* c., **Golgi-Mazzoni's.** See under *Mazzoni.* c.s, **Grandry's.** See under *Grandry.* c.s, **Hassall's.** See *Hassall's bodies.* c., **Hayem's.** See *achromacyte.* c.s, **Herbst's.** See under *Herbst.* c.s, **Jaworski's.** See under *Jaworski.* c.s, **Key and Retzius'.** See under *Key.* c.s, **Krause's.** See *Krause's corpuscles.* c.s, **Langerhans'** stellate. See under *Langerhans.* c.s, **Leber's.** See *Gierke's c.s.* c.s, **Lostorfer's.** See under *Lostorfer.* c.s, **lymph-,** nucleated ameboid cells found in lymph and chyle. Upon entering the blood with the lymph they are called white blood-corpuscles. The smaller ones have little if any ameboid movement, and are sometimes spoken of as free nuclei on account of their small cell-body; some of these corpuscles are coarsely granular and are therefore called granular cells. c., **Malpighian (of the kidney).** See *Malpighian corpuscles.* c.s, **Malpighian (of the spleen).** See *Malpighian corpuscles.* c., **Mazzoni's.** See under *Mazzoni.* c.s, **Meissner's.** See *c.s, tactile-* (1). c.s, **Merkel's.** See *Grandry's c.s.* c.s, **Miescher's.** See *Miescher's tubes.* c.s, **milk-,** of v. Kölliker, cells containing fat-globules observed in the acini of the mammary gland and breaking up into milk-globules on reaching the lactiferous ducts. c.s, **Montgomery's.** See *Montgomery's glands.* c., **Morgagni's.** See *Morgagni's globules.* c., **nerve,** nerve-cells. c.s, **Norris' invisible.** See under *Norris.* c.s, **Nunn's.** See *Bennett's c.s.* c.s, **Pacinian.** See *Pacinian's c.s.* c.s, **Patterson's.** See under *Patterson.* c., **pavement-.** See *cells, endothelial.* c.s, **phantom-.** Decolorized blood-corpuscles; and see *blood-platelets.* c., **polar-,** the centrosome. c.s, **Reissner's.** See under *Reissner.* c., **Rollett's nerve-.** See *Mazzoni's c.* c.s, **shadow-.** See *achromacyte.* c.s, **spleen-,** c.s, **splenic.** See *Malpighian corpuscles.* c.s, **tactile-.** 1. (Of *Wagner.*) The small, oval bodies found in the papillæ of the skin and enveloped by nerve-fibers. 2. See *Grandry's c.s.* c., **taste-.** See *taste-bud.* c.s, **terminal.** See *Krause's c.s.* c.s, **touch-.** See *c.s, tactile-.* c.s, **Toynbee's.** See under *Toynbee.* c.s, **transparent,** of Norris. See *Norris' c.s, invisible.* c.s, **Traube's.** See *achromacyte.* c.s, **typhic,** the epithelial cells of Peyer's patches which in typhoid fever have become granulated through degeneration. c.s, **Vater's.** c.s, **Vater-Pacini's.** See *Pacinian c.s.* c.s, **Virchow's.** See *c.s, corneal.* c., **Zimmermann's.** See under *Zimmermann.*

**corpuscula** (*kor-pus'-kū-lah*) [L.]. Plural of *corpusculum.*

**corpuscular** (*kor-pus'-kū-lar*) [*corpuscle*]. Relating to or of the nature of a corpuscle.

**corpusculation** (*kor-pus-kū-la'-shun*) [*corpusculum,* a corpuscle]. A condition in which the corpuscles of the blood have undergone hyperplasia, being larger and more numerous than normal.

**corpusculous** (*kor-pus'-kū-lus*) [*corpusculum,* a corpuscle]. Corpuscular.

**corpusculum** (*kor-pus'-kū-lum*) [L. dim. of *corpus; pl., corpuscula*]. A little body; a corpuscle. c. **articulare mobile.** See *arthrolith.* **corpuscula ossea,** ca. **radiata,** ca. **chalicophora.** See *bone.* c. **lamellosum.** See *Pacinian corpuscles.* c. **triticeum.** See *cartilago triticea.* **corpuscula Wrisbergii,** the cuneiformi cartilages.

**corradiation** (*kor-ra-de-a'-shun*). The act of radiating together, as focused rays.

**correctant, corrective** (*kor-ek'-tant, kor-ek'-tiv*) [*corrigere,* to correct]. 1. Modifying favorably. 2. A substance used to modify or make more pleasant the action of a purgative or other remedy.

**correction** (*kor-ek'-shun*) [*correctus; corrigere,* to amend]. The rectification of any abnormality (as a refractive or muscular defect), or of any undesirable quality (as in a medicine).

**correlated** (*kor'-el-a-ted*) [*correlatus,* related]. Interdependent; related. c. **atrophy.** See *atrophy.*

**correlation** (*kor-el-a'-shun*) [*correlatus,* related]. Interdependence; relationship.

**Corrigan's button,** or **cautery** (*kor'-ig-an*) [Sir Dominic John Corrigan, Irish physician, 1802–1880]. A button-shaped cautery iron fastened in a wooden handle. **C.'s disease.** 1. Aortic insufficiency. 2. Cirrhosis of the lung. **C.'s line,** the purple or brownish-red line on the margin of the gums in chronic copper-poisoning. **C.'s pulse,** "water-hammer pulse"; the abrupt, quickly receding, jerking pulse of aortic insufficiency. **C.'s respiration,**

"nervous or cerebral respiration." Frequent shallow and blowing breathing in low fevers—e. g., in typhus. **C.'s sign,** an expansive pulsation felt in cases of aneurysm of the abdominal aorta.
 **corrigent** (kor'-ij-ent). See *correctant*.
 **corroborant** (kor-ob'-o-rant) [*corroborans*, strengthening]. A tonic invigorating remedy.
 **corrosion** (kor-o'-zhun) [*con*, together; *rodere*, to gnaw]. The process of corroding or the state of being corroded. **c.-anatomy,** that branch of anatomy which demonstrates an anatomical specimen by means of a corrosive process that eats away those parts which it is not desired to preserve. In some cases a resisting-substance is injected, so as to preserve the vessels and ducts from corrosion. **c.-preparation,** one in which the vessels, ducts, or cavities of organs are filled by a fluid that will harden and preserve the shape of the vessel or cavity after the organ itself is corroded, digested, or otherwise destroyed.
 **corrosive** (kor-o'-siv) [see *corrosion*]. 1. Eating away. 2. A substance that destroys organic tissue either by direct chemical means or by causing inflammation and suppuration. **c. chloride, c. sublimate.** See *mercury bichloride*.
 **corrosol** (kor'-o-sol). A proprietary mercurial preparation for hypodermic use; said to contain cacodylate and succinate of mercury, and to be non-irritating and painless.
 **corroval** (kor'-o-val). A variety of curare, or arrow-poison; a cardiac and muscular paralyzant.
 **corrovaline** (kor-o'-val-en). A poisonous alkaloid obtained from corroval.
 **corrugator** (kor'-u-ga-tor) [*corrugere*, to wrinkle]. That which wrinkles. See under *muscle*.
 **corset** (kor'-set). In surgery, an investment for the abdomen or chest, or both; useful in some spinal disorders and deformities, and in fractures or injuries of the thoracic walls. **c.-liver,** a liver characterized by a furrow resulting from the pressure exerted by a corset and situated chiefly in the right lobe. It is due to habitual tight lacing, and hence is found in women and, rarely, among soldiers.
 **Corsican moss.** A mixture of fragments of various seaweeds brought from Corsica. It is said to be alterative, febrifuge, anthelmintic, and nutritious. Syn., *Helminthochorton*.
 **cortex** (kor'-teks) [L., "bark"]. 1. The bark of an exogenous plant. 2. The surface-layer of an organ. 3. The external gray layer of the brain, the substantia corticalis, or cortical substance. 4. The peripheral portion of an organ, situated just beneath the capsule. **c. aurantii,** orange-peel. **c. cerebri.** See *cortex* (3). **c. corticis,** the outer sheath of the kidney. **c. degeneration.** Synonym of *general paralysis of the insane*. See under *paralysis*. **c. renalis,** the cortical substance of the kidney.
 **Corti's arch** (kor'-te) [Marchese Alfonso Corti, Italian histologist, 1822–1876]. The arch formed in the organ of Corti by the two files of rods. **C.'s canal,** the triangular canal formed by the pillars of Corti, the base of which corresponds to the membrana basilaris. It extends over the entire length of the lamina spiralis. **C.'s cells,** the outer haircells of Corti's organ. **C.'s fibers.** See *C.'s rods*. **C.'s ganglion,** the ganglion spirale, an aggregation of ganglion-cells in the spiral canals of the cochlea. **C.'s membrane,** the membrana tectoria of the cochlea. **C.'s organ,** a complicated organ, the product of differentiation of the epithelial lining of the cochlear canal, resting on the basilar membrane of the cochlea and containing the end-organs of the cochlear nerves. **C.'s rods,** the pillars of the arch of the organ of Corti. **C.'s teeth,** the auditory teeth; the tooth-like projections on the edge of the limbus laminæ spiralis of the ear. **C.'s tunnel.** Same as *C.'s canal*.
 **cortical** (kor'-tik-al) [*cortex*]. Pertaining to the cortex or bark or to the cortex of any organ or structure. **c. cataract, c. opacity in the cortex** of the lens. **c. epilepsy, c. paralysis,** such as is due to a lesion of the cortical substance of the brain.
 **corticate** (kor'-te-kāt). Furnished with a bark or cortex.
 **corticifugal** (kor-te-sif'-u-gal) [*cortex; fugere*, to flee]. Conducting away from the cortex.
 **corticipetal** (kor-te-sip'-et-al) [*cortex; petere*, to seek]. Conducting toward the cortex.
 **corticoafferent** (kor-te-ko-af'-er-ent). See *corticipetal*.

 **corticoefferent** (kor-te-ko-ef'-er-ent). See *corticifugal*.
 **corticospinal** (kor-tik-o-spi'-nal). Pertaining to the cortex of the brain and the spinal cord.
 **corundum** (ko-run'-dum) [Hind., *kurand*]. A native crystalline aluminum oxide, Al₂O₃. Mixed with melted shellac, it is formed into wheels for use in the dental laboratory and for grinding in general.
 **coruscation** (kor-us-ka'-shun) [*coruscare*, to glitter]. A glittering or flashing of light, also the subjective sensation of light-flashes.
 **Corvisart's disease** (kor-ve-sar') [Jean Nicolas Corvisart, French physician, 1755–1821]. Idiopathic cardiac hypertrophy. **C.'s facies,** the facies of cardiac insufficiency.
 **corybantism** (kor-e-bant'-ism) [κορυβαντισμός, corybantic frenzy]. Maniacal frenzy, with sleeplessness, choreic excitement and visual hallucinations.
 **corydalin** (kor-id'-al-ēn) [*corydalis*]. 1. An extract from the root of *Corydalis formosa* (*Dicentra canadensis*); it is used in syphilis and scrofula.
 **corydaline** (kor-id'-al-in). An alkaloid, C₂₂H₂₇NO₄ (Freund) from *Corydalis tuberosa*; it is used as a heart-tonic. Dose 1–5 gr. (0.065–0.032 Gm.).
 **corydalis** (kor-id'-al-is) [κορυδαλλίς, the crested lark]. 1. Turkey-corn. The tuber of *C. formosa* (*Dicentra canadensis*), a tonic, diuretic, and alterative. Dose of *fluidextract* 10–40 min. (0.6–2.5 Cc.). 2. A genus of plants of the order *Papaveraceæ. C. tuberosa*, holewort, hollowwort, is an herb indigenous to Europe; the rhizome is anthelmintic and emmenagogue. It contains **corydaline** and corydine.
 **corydine** (kor'-id-ēn). An amorphous alkaloid from *Corydalis tuberosa*.
 **coryfin** (kor'-if-in). Ethyl glycolic acid ester of menthol.
 **coryl** (kor'-il). The name given to an anesthetic composed of ethyl chloride and methyl chloride in such proportions that the boiling-point of the mixture is about 32° F.
 **coryleur** (kor-il-ur'). An apparatus for spraying with coryl.
 **Corynebacterium** (kor-i'-ne-bak-te'-re-um)) [κορύνη, a club; *bacterium*]. A genus of bacilli, club-shaped, granular and gram-positive. **C. commune,** the pseudo-diphtheria bacillus. **C. diphtheriæ,** the bacillus of diphtheria. **C. granulomatis maligni, C. hodgkini,** a bacillus of diphtheroid nature found in Hodgkin's disease. **C. mallei,** the bacillus of glanders.
 **coryza** (ko-ri'-zah) [κόρυζα, a running at the nose]. Catarrh of the mucous membrane of the nasal passages and adjacent sinuses, popularly called a "cold in the head." See *rhinitis*. **c. caseosa,** a term applied by Cozzolino to a disease in which the nostrils are filled with caseous masses. **c. idiosyncratic.** See *hayfever*. **c. maligna,** synonym of *snuffles of the newborn*. **c. periodic vasomotor.** See *hay-fever*. **c. vasomotor.** Synonym of *hay-fever*.

 **cosaprin** (kos'-ap-rin), $C_6H_4 < ^{SO_2Na}_{NH-CO-CH_3}$. A sulphoderivative of acetanilide; it is a whitish-gray powder with a slightly saline taste, freely soluble in water. It is used as an antipyretic instead of acetanilide. Dose 5–8 gr. (0.3–0.5 Gm.) 3 times daily.
 **cosmesis** (koz-me'-sis) [κοσμεῖν, to adorn]. The art of preserving or increasing beauty.
 **cosmetic** (koz-met'-ik) [κοσμεῖν, to adorn]. 1. Beautifying. 2. A remedy designed to hide defects of the skin or other external parts. **c. operation,** a surgical operation to give a natural appearance to a defective or unsightly part.
 **cosmetology** (koz-met-ol'-o-je) [κόσμετος, orderly; λόγος, science]. The science of the proper care of the body with respect to cleanliness, dress, etc.
 **cosmic** (koz'-mik) [κόσμος, the universe]. Worldwide; of wide distribution, as a *cosmic* disease.
 **cosmoline** (koz'-mo-lēn). See *petrolatum*.
 **costa** (kos'-tah) [L.: *pl., costæ*]. A rib. **costæ fluctuantes,** floating ribs. **costæ illegitimæ,** c. **mendosæ, c. spuriæ,** false ribs. **costæ legitimæ,** c. **veræ,** true ribs.
 **costal** (kos'-tal) [*costa*]. Pertaining to the ribs. **c. arch,** the arch of the ribs. **c. cartilages,** the twelve cartilaginous extensions of the ribs. **c. respiration,** respiration carried on chiefly by the chest muscles.

**costalgia** (kos-tal'-je-ah) [costa, a rib; ἄλγος, pain]. Intercostal neuralgia; pain in the ribs.
**costate** (kos'-tāt). Ribbed; furnished with ribs or connecting structures.
**costectomy** (kos-tek'-to-me) [costa; ἐκτομή, excision]. Excision of a rib; costotomy.
**costen** (kos'-ten) [costa, a rib]. Belonging to a rib in itself.
**Coster's paste.** A remedy formerly used in the treatment of tinea tonsurans. It is made of iodine, 2 drams, in one ounce of oil of pitch.
**costicartilage** (kos-te-kar'-til-āj) [costa, a rib; cartilago, gristle]. A costal cartilage or unossified sternal rib.
**costicervical** (kos-te-ser'-vik-al) [costa, a rib; cervix, a neck]. Relating to the neck and ribs.
**costicervicalis** (kos-te-ser-vik-a'-lis) [costa, a rib; cervix, the neck]. The cervicalis ascendens muscle. See *muscles, table of*.
**costiform** (kos'-te-form). Rib-shaped.
**costispinal** (kos-te-spi'-nal) [costa, a rib; spina, the spine]. Relating to the ribs and vertebral column.
c. **muscles**, levatores costarum. See *muscles, table of*.
**costive** (kos'-tiv) [constipare, to be bound]. Constipated; affected with costiveness.
**costiveness** (kos'-tiv-ness) [costive]. An abnormality of digestion characterized by retention and hardness of the feces; constipation.
**costo-** (kos-to-) [costa]. A prefix denoting connection with the ribs.
**costoabdominal** (kos-to-ab-dom'-in-al). Relating to the ribs and the abdomen.
**costocentral** (kos-to-sen'-tral). Pertaining to a rib and the body (or centrum) of a vertebra with which it articulates.
**costochondral** (kos-to-kon'-dral). Pertaining to the ribs and their cartilages.
**costoclavicular** (kos-to-klav-ik'-ū-lar). Pertaining to the ribs and the clavicle.
**costocolic** (kos-to-kol'-ik). Relating to the ribs and the colon.
**costocoracoid** (kos-to-kor'-ak-oid). Pertaining to the ribs and the coracoid process.
**costohumeral** (kos-to-hū'-mur-al). Connected with the ribs and humerus.
**costoinferior** (kos-to-in-fe'-re-or). Relating to the lower ribs; applied to a form of respiration in which the lower ribs move more than the upper.
**costopubic** (kos-to-pū'-bik). Relating to the ribs and the pubis.
**costopulmonary** (kos-to-pul'-mon-a-re). Relating to the ribs and the lungs.
**costoscapular** (kos-to-skap'-ū-lar). 1. Relating to the ribs and the scapula. 2. The serratus magnus muscle.
**costosternal** (kos-to-stur'-nal). Pertaining to the ribs and the sternum.
**costosuperior** (kos-to-sū-pe'-re-or). Relating to the upper ribs.
**costotome** (kos'-to-tōm) [costo-; τέμνειν, to cut]. A strong knife or heavy shears with the under blade in the shape of a hook, for cutting the costal cartilages in dissection, etc.
**costotomy** (kos-tot'-o-me) [costa, τομή, section]. Resection or division of a rib.
**costotrachelian** (kos-to-tra-ke'-le-an). Relating to the ribs and to the transverse processes of the cervical vertebræ.
**costotransverse** (kos-to-tranz'-vers). 1. Pertaining to the ribs and transverse vertebral processes. 2. The scalenus lateralis. See under *muscle*.
**costotransversectomy** (kos-to-tranz-vers-ek'-to-me) [costotransverse; ἐκτομή, a cutting out]. Excision of part of a rib and a transverse vertebral process.
**costovertebral** (kos-to-ver-te'-bral). Pertaining to the ribs and vertebræ.
**costoxiphoid** (kos-to-zif'-oid). Relating to the ribs and to the ensiform cartilage.
**cot** (kot) [AS., cote]. 1. A small bed. 2. The finger of a glove. See *finger-cot*. c., **fever-**, c., **Kibbee's**, a bed devised especially for applying cold-water treatment to fever patients.
**Cotard's syndrome.** A form of paranoia characterized by delusions of negation, with sensory disturbances and a tendency to suicide. Syn., *Délire chronique des négations*.
**cotargit** (ko-tar'-jit). Trade name of a substance composed of cotarnin hydrochloride and ferric chloride; it is used as a hemostatic.

**cotarnine** (ko-tar'-nin) [an anagram of narcotine], $C_{12}H_{15}NO_4$. An oxidation-product of narcotine. c. **hydrochloride**, $C_{12}H_{15}NO_4 \cdot HCl \cdot H_2O$, small yellow crystals, soluble in water and alcohol. It is an internal hemostatic. Dose ½–2 gr. (0.03–0.13 Gm.). Syn., *stypticin*.
**COTe.** An abbreviation for *cathodal opening tetanus*.
**coto** (ko'-to) [Sp. "a cubit"]. Coto bark. The bark of a tree native to Bolivia. It contains a bitter principle, *cotoin*, $C_{22}H_{18}O_6$, irritant to the skin and mucous membranes. It is recommended for diarrhœa and zymotic fevers, and for the night-sweats of pulmonary tuberculosis. Dose of the *powder* 1–15 gr. (0.065–1.0 Gm.); of the *fluidextract* 5–15 min. (0.32–1.0 Cc.); of the *tincture* (1 : 10) 10–30 min. (0.65–2.0 Cc.).
**cotoin** (ko'-to-in) [coto], $C_{22}H_{18}O_6$. An astringent principle from coto (q. v.). It is employed in dysentery and cholera. Dose ½–5 gr. (0.03–0.3 Gm.). *Paracotoin* is one-half as strong as cotoin.
**cottage-hospital** (kot'-āj-hos'-pit-al). A small establishment for the purpose of providing for the sick in a small and isolated community.
**Cotting's operation** (kot'-ing) [Benjamin Eddy Cotting, American surgeon, 1812–1898]. For in-growing toe-nail; all the overlying tissues, together with the sides of the toe, are sliced off freely. The contraction in healing produces a cure.
**cotton** (kot'-n) [Ar., qūtun, cotton]. Gossypium, a white, fibrous seed-hair that envelops the seeds of the cotton-plant. c., **absorbent**, cotton so prepared that it readily absorbs water. See also *gossypium*. c., **gun-**. See *pyroxylin*. c., **-oil**. See *c.-seed oil*. c.**-root**, c.**-root bark.** See under *gossypium*. c.**-seed oil**, an oil obtained by pressure from the hulled seeds of several species of *gossypium* (q. v.). c., **styptic**, cotton saturated with a styptic substance. c. **wool**, absorbent cotton.
**cottonoid** (kot'n'-oid). Trade name of absorbent cotton prepared for surgical use.
**Cotugno's canal** (ko-toon'-yo) [Domenico Cotugno, Italian anatomist, 1736–1822]. The aquæductus vestibuli. Syn., *canalis cotunnii*. C.'s **disease**, sciatica. Syn., *malum cotunnii*. C.'s **liquor**, the perilymph of the osseous labyrinth of the ear. Syn., *liquor cotunnii*. C.'s **nerve**, the naso-palatine nerve. C.'s **space**, the saccus endolymphaticus of the internal ear.
**Cotunnius** (ko-tun'-ne-us). See *Cotugno*.
**cotyle** (kot'-i-le) [κοτύλη, a socket, cup]. The acetabulum.
**cotyledon** (kot-il-e'-don) [κοτυληδών, a socket]. 1. Any one of the enlarged, vascular villi of the chorion which project into depressions of the decidua vera. 2. Any one of the numerous rounded portions into which the uterine surface of the placenta is divided. 3. A genus of plants of the order *Crassulaceæ*. C. *umbilicus*, navelwort of Europe, has been highly recommended in epilepsy, but its medicinal properties are feeble.
**cotyloid** (kot'-il-oid) [κοτύλη, a cup; εἶδος, form]. Cup-shaped. c. **cavity**, c. **fossa**, the acetabulum. c. **ligament**, a ligament surrounding the acetabulum. c. **notch**, a notch in the anterior and lower border of the acetabulum.
**cotylopubic** (kot-il-o-pū'-bik). Relating to the acetabulum and the os pubis.
**cotylosacral** (kot-il-o-sa'-kral). Relating to the acetabulum and the sacrum.
**couch-grass** (kowtch'-gras). See *triticum*.
**couching** (kowtch'-ing) [Fr., *coucher*, to depress]. The operation now fallen into disuse, of depressing a cataractous lens into the vitreous chamber, where it was left to be absorbed.
**cough** (kof, or kawf). A sudden, violent expulsion of air after deep inspiration and closure of the glottis. c., **chin** whooping-cough. c., **dry**, that unattended by expectoration. c., **ear-**, cough excited reflexly from some morbid condition of the ear. c., **moist**, cough with free expectoration. c., **Morton's**. See under *Morton*. c., **pleuritic**, the dry, short, frequent cough of pleurisy, pneumonia, and phthisis, which accompanies the pain and friction-sounds of pleurisy and disappears with effusion or when bronchitis supervenes. c., **reflex**, cough produced by irritation of a remote organ. c., **stomach-**, from gastric irritation. See *tussis*. c., **Sydenham's**. See under *Sydenham*. c., **uterine**, a reflex cough occurring in sufferers from genital disease due to irritation of

the uterovaginal fibers of the hypogastric plexus supplying the fornix vaginæ and cervix uteri and the nerves and ganglia supplying the fundus uteri and ovaries. c., winter-, a short troublesome cough of old people due to chronic bronchitis, and recurring every winter.

**coulomb** (*koo-lom'*) [Charles Augustin de *Coulomb*, French physicist, 1736–1806]. The unit of measurement of electrical quantity; the quantity of electricity that passes during one second through a conductor having a resistance of one ohm, with one volt of electromotive force. The *microcoulomb* is the millionth part of this amount.

**Coulomb's law** (*koo-lom'*) [see *coulomb*]. The force exerted between electrically charged bodies, placed at a distance, is directly proportional to the products of the amounts of charge and inversely proportional to the square of the distance between them.

**coumarin** (*koo'-mar-in*); $C_9H_6O_2$. A vegetable proximate principle found in *Dipteryx odorata*, Tonka bean, and in *Melilotus officinalis*. It conceals the odor of iodoform.

**counteraction** (*kown-ter-ak'-shun*). The action of a drug or agent opposed to that of some other drug or agent.

**counterdies.** See *dies.*

**counterextension** (*kown-ter-eks-ten'-shun*). See under *extension.*

**counterfissure** (*kown-ter-fish'-ur*). [See *contrafissura.*

**counterindication** (*kown-ter-in-dik-a'-shun*). See *contraindication.*

**counterirritant** (*kown-ter-ir'-it-ant*). An agent which produces counterirritation; a drug which attracts blood to the surface.

**counterirritation** (*kown-ter-ir-it-a'-shun*) [*contra,* against; *irritare,* to irritate]. Superficial inflammation produced artificially, in order to exercise a good effect upon some adjacent or deep-seated morbid process.

**counteropening** (*kown'-ter-o-pen-ing*) [*contra; opening*]. An incision made in an abscess or cavity, opposite to another, generally for purposes of drainage.

**counterpoison** (*kown'-ter-poi-zn*) [*contra; poison*]. A poison given as an antidote to another poison.

**counterpressure** (*kown'-ter-presh-ur*). Pressure opposed to pressure.

**counterpuncture** (*kown'-ter-punk-chur*). See *counteropening.*

**counterstain** (*kown'-ter-stān*). 1. A stain used to bring into contrast parts of tissues colored by another stain. 2. To apply a counterstain.

**counterstroke** (*kown'-ter-strōk*). See *contrecoup.*

**Countess's powder.** Synonym of *cinchona bark.*

**coup de fouet** (*koo-der-foo-a'*) [Fr.]. stroke of a whip]. Lawn-tennis leg; rupture of the plantaris muscle.

**coup de soleil** (*koo-der-so-lay'*) [Fr.]. Sunstroke. See *Heat-stroke.*

**coupler** (*kup'-ler*) (*copulare,* to bind]. 1. Used for fastening wire to a tooth to correct dental irregularities. 2. A device for connecting parts of an electric apparatus.

**courbometer** (*koor-bom'-et-er*) [Fr., *courbe,* a curve; μέτρον, a measure]. A device of Chatelain to show the curve of the alternating current.

**courses** (*kors'-ez*). See *menses.*

**court-plaster** (*kort'-plas-ter*). See *plaster.*

**Courvoisier's law** (*koor-vwah-ze-a'*) [Ludwig G. *Courvoisier,* French surgeon, 1843– ]. 1. Tumors of the head of the pancreas almost invariably cause dilatation of the gall-bladder. 2. In the majority of instances of obstruction of the common bile-duct by gall-stone the gall-bladder is contracted; in obstruction from other causes the gall-bladder is dilated.

**cousso** (*koo'-so*). See *cusso.*

**Coutoubea** (*koo-too'-be-ah*) [South American name]. A genus of plants of the order *Gentianeæ. C. spicata,* of Brazil, is emmenagogue and anthelmintic.

**couveuse** (*koo-vus'*) [Fr.]. See *incubator.*

**cover-glass.** In microscopy, the thin slip of glass covering the object mounted on the slide.

**cowhage** (*kow'-haj*), **cowitch** (*kow'-aj, -itch*) [Hind., *kawānch,* cowage]. The external hairs of the pod of *Mucuna pruriens,* formerly used in medicine as a mechanical vermifuge. See also *Mucuna.*

**Cowie's guaiac test for blood in the feces.** To 1 Gm. of moist feces add 4 to 5 Cc. of glacial acetic acid. Extract the mixture with 30 Cc. of ether. Take 1 to 2 Cc. of the extract and add an equal volume of water; agitate; then add a few granules of powdered guaiac resin and all$_{ow}$ it to dissolve. When dissolved gradually add 30 drops of old turpentine or hydrogen peroxide. The presence of blood is indicated by the appearance of a blue color.

**Cowling's rule** (*kow'-ling*). A rule for dosage. The age of the child at the next birthday is the numerator and 24 the denominator. According to this, the dose of a child approaching four years of age would be $\frac{4}{24} = \frac{1}{6}$ of the dose for an adult.

**Cowper's cyst** (*kow'-per*) [William *Cowper,* English surgeon, 1666–1709]. A retention cyst of Cowper's gland. **C.'s glands,** the bulbourethral glands; two compound tubular glands situated between the two layers of the triangular ligament, anteriorly to the prostate gland; they correspond to Bartholin's glands in the female. **C.'s ligament,** the portion of the fascia lata that is attached to the crest of the pubis.

**Cowperian cyst.** A retention cyst formed in Cowper's gland.

**cowperitis** (*kow-per-i'-tis*). Inflammation of the glands of Cowper, usually gonorrheal in origin.

**cowpox, cowpock** (*kow'-poks*). A contagious eruptive fever occurring in the cow, and thought to correspond with smallpox in man.

**coxa** (*koks'-ah*) [L., "the hip"]. The hip-joint or the hip. **c. valga,** a condition, the reverse of coxa vara, in which the angle between the neck and the shaft of the femur is increased above 140 degrees. **c. vara,** a condition in which the neck of the femur is bent downward sufficiently to cause symptoms; this bending may reach such an extent that the neck forms with the shaft a right angle or less, instead of the normal angle of 120 to 140 degrees.

**coxagra** (*koks-a'-grah*) [*coxa,* the hip; ἄγρα, seizure]. 1. Gout in the hip. 2. Coxalgia, sciatica.

**coxal** (*koks'-al*) [*coxa,* the hip]. Relating to the coxa.

**coxalgia** (*koks-al'-je-ah*) [*coxa;* ἄλγος, pain]. Literally, pain in the hip-joint, but generally used synonymously with hip-disease.

**coxalgic** (*koks-al'-jik*) [*coxalgia*]. Relating to coxalgia.

**coxankylometer** (*koks-ang-kil-om'-et-er*) [*coxa;* ἀγκύλος, bent; μέτρον, a measure]. Volkmann's instrument for measuring the deformity in hip-disease.

**coxarthritis** (*koks-ar-thri'-tis*). The same as *coxitis.*

**coxarthrocace** (*koks-arth-rok'-as-e*) [*coxa;* ἄρθρον, joint; κακός, bad]. A fungoid inflammation of the hip-joint.

**coxarum morbus.** Hip-joint disease.

**Coxe's hive mixture.** A mixture of squill and senna, of each, 120; tartar emetic, 3; sugar, 1200; calcium phosphate, 9; dilute alcohol and water, to 2000.

**coxitis** (*koks-i'-tis*) [*coxa;* ιτις, inflammation]. Inflammation of the hip-joint. **c. cotyloidea,** that confined principally to the acetabulum. **c., senile,** a rheumatoid disease of the hip-joint occurring in old people, marked by pain, stiffness, and wasting, without any tendency to suppuration.

**coxodynia** (*koks-o-din'-e-ah*) [*coxa,* hip; ὀδύνη, pain]. Same as *coxalgia.*

**coxofemoral** (*koks-o-fem'-or-al*) [*coxa; femur,* the thigh-bone]. Relating to the hip and the femur, as the *coxofemoral* joint—the hip-joint.

**coxopathy** (*koks-op'-ath-e*) [*coxa;* πάθος, disease]. Any affection of the hip-joint.

**coxotuberculosis** (*koks-o-tū-ber-kū-lo'-sis*) [*coxa,* hip; *tuberculum,* a tubercle]. Tuberculous disease of the hip-joint.

**c.p.** Abbreviation for *chemically pure.*

**Cr.** Chemical symbol of *chromium.*

**crab-louse** (*krab'-lows*). See *Pediculus pubis.*

**crabs'-eyes** (*krabs'-iz*). 1. Flat, calcareous concretions (*Lapides cancrorum*) derived from the stomach of the crab; they have been used as a means of removing foreign bodies from the eye. 2. A name for the seeds of *Abrus precatorius.*

**crab-yaws** (*krab'-yors*). See *frambesia.*

**crachotement** (*kra-shot-mon(g)'*) [Fr.]. A peculiar reflex following operations upon the utero-ovarian organs, marked by a desire to spit, without the ability to do so. It is usually accompanied by a tendency to syncope.

**cracked-pot sound.** A peculiar sound elicited by percussion over a pulmonary cavity communicating with a bronchus.

**cradle** (kra'-dl) [AS., cradol]. In surgery, a wire or wicker frame so arranged as to keep the weight of the bed-clothing from an injured part of the body. It is employed in the treatment of fractures, wounds, etc. **c. cap,** a name given to the scabs composed of dirt and sebum, that form on the scalps of neglected infants and children. **c., ice-,** the suspension over a febrile patient, by means of iron frames, of a number of zinc buckets, kept half-filled with ice, and inclosed in a light covering. **c.-pessary,** a cradle-shaped pessary for treating retrodisplacements of the uterus.

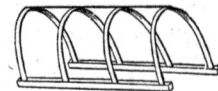

CRADLE FOR SUPPORTING BEDCLOTHES.—*From Fullerton.*

**cram** (kram). To store the memory with information for the mere purpose of passing an examination. **c.-stunt,** arrest in mental development due to overstudy.

**cramp** (kramp) [Teut., krampf]. A spasmodic tonic contraction of a muscle attended with sharp pain. **c., intermittent,** tetany. **c., professional,** spasm of certain groups of muscles, from their continuous use in different occupations, as *writer's, hammerman's, piano-player's, dancer's cramp,* etc. **c.s,** tonic (of fingers and toes in children), tetany.

**Crampton's muscle** [Sir Philip Crampton, Irish surgeon, 1777-1858]. A bundle of striated muscular fibers extending from the annular ligament to the sclera in the eye of birds.

**crane's bill root.** See *geranium.*

**cranial** (kra'-ne-al) [cranium]. 1. Relating to the cranium. 2. Relating to position nearer the head end of the long axis of the body. **c. bosses,** flat, bony elevations of the frontal and parietal bones sometimes seen at the angles of the anterior fontanel. They are said to be a proof of hereditary syphilis. **c.-capacity, modes of measuring.** See under *skull.* **c. nerves.** See *nerve.*

**cranialis** (kra-ne-a'-lis) [L.]. Cranial.

**craniectomy** (kra-ne-ek'-to-me) [*cranium;* ἐκτομή, a cutting out]. The surgical removal of strips or pieces of the cranial bones. It is performed in cases of microcephaly.

**craniencephalometer** (kra-ne-en-sef-al-om'-et-er) [*cranium;* ἐγκέφαλος, the brain; μέτρον, a measure]. An instrument for determining the position of the gyri of the brain from the outer surface of the head.

**cranio-** (kra-ne-o-) [*cranium*]. A prefix meaning relating to the cranium.

**cranioabdominal** (kra-ne-o-ab-dom'-in-al). Relating to the cranium and the abdomen; applied to temperaments showing a predominance of cerebral and abdominal influences.

**cranioacromial** (kra-ne-o-ak-ro'-me-al) Relating to the cranium and the acromion.

**cranioaural** (kra-ne-o-aw'-ral). Relating to the cranium and the ear.

**craniocele** (kra-'ne-o-sel) [*cranio-;* κήλη, a tumor]. Encephalocele, *q. v.*

**craniocerebral** (kra-ne-o-ser'-e-bral). Relating to the cranium and the cerebrum.

**craniocervical** (kra-ne-o-ser'-vik-al). Relating to the cranium and the neck.

**cranioclasis** (kra-ne-ok'-las-is) [*cranio-;* κλάσις, a breaking]. See *cranioclasm.*

**cranioclasm** (kra'-ne-o-klasm) [*cranio-;* κλάειν, to break]. The operation of breaking the fetal head by means of the cranioclast.

**cranioclast** (kra'-ne-o-klast) [see *cranioclasm*]. A heavy forceps for crushing the fetal head.

**cranioclasty** (kra'-ne-o-klas'-te). See *cranioclasm.*

**craniocleidodysostosis** (kra-ne-o-kli-do-dis-os-to'-sis) [*cranio-;* χλεῖς, clavicle; δύς, difficult; *ostosis*]. Congenital defect of the clavicle associated with imperfect ossification of the bones of the cranium.

**craniodiaclast** (kra-ne-o-di'-ak-last) [*cranio-;* διακλάν, to break into pieces]. An instrument for breaking the skull in craniotomy.

**craniodidymus** (kra-ne-o-did'-im-us) [*cranio-;* δίδυμος, double]. Same as *cephalopagus.*

**craniofacial** (kra-ne-o-fa'-shal) [*cranio-; facies,* face]. Relating to the cranium and the face.

**craniognomy** (kra-ne-og'-no-me). See *cephalology.*

**craniograph** (kra'-ne-o-graf) [*cranio-;* γράφειν, to record]. An instrument for recording the outlines of the skull.

**craniography** (kra-ne-og'-ra-fe) [*cranio-; γράφειν,* to write]. That part of descriptive craniology the object of which is to describe the parts or regions of the skull or bones of the face.

**craniohematoncus** (kra-ne-o-hem-at-ong'-kus) [*cranio-;* αἷμα, blood; ὄγκος, a tumor]. Synonym of *cephalhematoma.*

**craniology** (kra-ne-ol'-o-je) [*cranio-;* λόγος, science]. A branch of anatomy comprising the study of skulls.

**craniomalacia** (kra-ne-o-mal-a'-se-ah). See *craniotabes.*

**craniomandibular** (kra-ne-o-man-dib'-u-lar) [*cranio-; mandibula,* the mandible]. Relating to the skull and the lower jaw.

**craniometer** (kra-ne-om'-et-er) [*cranio-;* μέτρον, a measure]. An instrument for measuring the dimensions of the skull.

**craniometric, craniometrical** (kra-ne-o-met'-rik, -al) [see *craniometer*]. Pertaining to craniometry. **c. point,** any one of the points of measurement used in craniometry. The craniometric points are the following: ACANTHION, a point in the median line of the skull at the base of the nasal spine. ALVEOLAR POINT, the point between the two middle incisors of the upper jaw. ANTINION, that point on the glabellum, and in the median line, that is farthest from the inion. ASTERION, the point behind the ear where the parietal, temporal, and occipital bones meet. AURICULAR POINT, the center of the orifice of the external auditory meatus. BASION, the middle point of the anterior margin of the foramen magnum. BREGMA, the point where the coronal and sagittal sutures meet. DACRYON, or DAKRYON, the point beside the root of the nose where the frontal, lacrimal, and superior maxillary bones meet. ENTOMION, the point where the parietal notch of the temporal bone receives the anterior extension of the mastoid angle of the parietal bone. GLABELLA, or GLABELLUM, the point in the median line between the superciliary arches, marked by a swelling, sometimes by a depression. GNATHION. Same as *mental point.* GONION, the point at the angle of the lower jaw. HORMION, the anterior point of the basilar portion of the united sphenooccipital bone, where it is crossed by the median line. INION, the external occipital protuberance. JUGAL POINT, the point situated at the angle that the posterior border of the frontal process of the malar bone makes with the superior border of its zygomatic process. KORONION, the apex of the coronoid process of the inferior maxilla. LAMBDA, the point of meeting of the sagittal and the lambdoid sutures. MALAR POINT, a point situated on the tubercle on the external surface of the malar bone or at the intersection of a line drawn from the external extremity of the frontomalar suture to the tubercle at the inferior angle of the malar bone and a line drawn nearly horizontally from the inferior border of the orbit over the base to the superior border of the zygomatic arch. MAXIMUM OCCIPITAL POINT, or OCCIPITAL POINT, the posterior extremity of the anteroposterior diameter of the skull, measured from the glabella in front to the most distant point behind. MENTAL POINT, the middle point of the lower border of the lower jaw. METOPION, or METOPIC POINT, a point in the middle line between the two frontal eminences. NASION, or NASAL POINT, the middle of the frontal suture at the root of the nose. OBELION, the part of the sagittal suture between the two parietal foramina. OCCIPITAL POINT. See *maximum occipital point* in this table. OPHRYON, the middle of the supraorbital line, which, drawn across the narrowest part of the forehead, separates the face from the cranium. OPISTHION, the middle point of the posterior border of the foramen magnum. PROSTHION, the alveolar point. PTERION, the point where the frontal, parietal, temporal, and sphenoid bones come together. RHINION, the upper median point of the anterior nasal opening. SPINAL POINT. Same as *subnasal point.* STEPHANION, INFERIOR, the point where the ridge for the temporal muscle intersects the coronal suture. STEPHANION, SUPERIOR, the point where the coronal suture crosses the temporal ridge. SUBNASAL POINT, the middle of the inferior border of the anterior nares at the base of the nasal spine. SUPRA-AURICULAR POINT, the point vertically over the auricular point at the root of the zygomatic process. SUPRANASAL POINT, SUPRAORBITAL POINT. Same as ophryon. SYMPHYSION, the median point of the outer border of the alveolus of the lower jaw. VERTEX, the superior point of the skull. In obstetrics, that conical portion of the skull the apex of which is at the posterior fontanel and the base of which is formed by the biparietal and trachelobregmatic diameters.

**craniometry** (kra-ne-om'-et-re) [see *craniometer*]. The ascertainment of the proportions and measurements of skulls.

**craniopagus** (kra-ne-op'-ag-us). See *cephalopagus*.

**craniopathy** (kra-ne-op'-ath-e). See *cephalopathy*.

**craniopharyngeal** (kra-ne-o-far-in'-je-al). Relating to the cranium and the pharynx.

**craniophore** (kra'-ne-o-for) [κρανίο-; φέρειν, to bear]. A device for holding the skull during craniometric study.

**cranioplasty** (kra'-ne-o-plas-te) [cranio-; πλαστός, formed]. The surgical restoration or correction of cranial deficiencies.

**craniorrhachischisis** (kra-ne-or-rak-is'-kis-is) [cranio-; ῥάχις, spine; σχίσις, a cleaving]. Congenital fissure of the skull and spine.

**cranioschisis** (kra-ne-os'-kis-is) [cranio-; σχίσις, a cleaving]. Congenital fissure of the skull.

**craniosclerosis** (kra-ne-o-skle-ro'-sis) [cranio-; σκληρός, hard]. A condition the antithesis of that seen in craniotabes. A thickening of the cranial bones, usually due to rhachitis.

**cranioscopy** (kran-e-os'-ko-pe). See *phrenology*.

**craniospinal** (kra-ne-o-spi'-nal) [cranio-, spine]. Pertaining to the cranium and spinal column.

**craniostegnosis** (kra-ne-o-steg-no'-sis) [cranio-; στέγνωσις, contraction]. Contraction of the skull.

**craniostenosis** (kra-ne-o-sten-o'-sis). See *craniostegnosis*.

**craniostosis** (kran-e-os-to'-sis) [cranio-; ὀστέον, a bone]. Congenital ossification of the cranial sutures.

**craniotabes** (kra-ne-o-ta'-bēz) [cranio-; *tabes*, a wasting]. An atrophy of the cranial bones occurring in infancy, with the formation of small, shallow, conical pits in the bone-substance. Craniotabes results from rhachitis, syphilis, or marasmus.

**craniotabetic** (kra-ne-o-tab-et'-ik) [cranio-; *tabere*, to waste away]. Pertaining or belonging to craniotabes, q. v.

**craniothoracic** (kra-ne-o-thor-as'-ik). Relating to the skull and the thorax; applied to temperaments showing a predominance of cerebral and thoracic influences.

**craniotome** (kra'-ne-o-tōm) [cranio-; τέμνειν, to cut]. An instrument used in craniotomy.

**craniotomy** (kra-ne-ot'-o-me) [cranio-; τομή, a cutting]. 1. The operation of reducing the size of the fetal head by cutting or breaking it up, when delivery is otherwise impossible. 2. The excision of a part of the skull. c., **linear**. See *craniectomy*.

**craniotonoscopy** (kra-ne-o-ton-os'-ko-pe) [cranio-; τόνος, tone; σκοπεῖν, to examine]. An auscultatory method devised by Gabritschewsky for the localization of changes in the bones of the skull (thinning or thickening) by means of the variations in sound transmitted through the bones and a special resonator (pneumatoscope) placed in the mouth.

**craniotractor** (kra-ne-o-trak'-tor) [cranio-; *tractor*, a drawer]. A cranioclast designed to be used also, or mainly, as a tractor.

**craniotripsotome** (kra-ne-o-trip'-so-tōm) [cranio-; τρίψις, a rubbing; τομή, a section. v. Cassagny's instrument for performing cranioclasty.

**craniotrypesis** (kra-ne-o-trip-e'-sis) [cranio-; τρύπησις, a boring]. Trephining.

**craniotympanic** (kra-ne-o-tim-pan'-ik) [cranio-; *tympanum*]. Pertaining to the skull and the tympanum.

**craniovertebral** (kra-ne-o-ver'-te-bral). Same as *cerebrospinal*.

**cranioviscera**l (kra-ne-o-vis'-ur-al). Relating to the cranium and the viscera.

**cranitis** (kra-ni'-tis). Inflammation of one or more of the cranial bones.

**cranium** (kra'-ne-um) [κρανίον, the skull]. The skull. The cavity that contains the brain, its membranes, and vessels. c. **cerebrale**, the cerebral cranium or calvaria. c. **viscerale**, the visceral cranium, or bones of the face.

**crank** (krank) [Ger., sick]. A popular term for an eccentric individual or a hobby-rider. See *paranoiac*.

**cranter** (kran'-ter) [κραντήρ, κραίνειν, to finish,

render perfect: *pl., cranteres*]. A wisdom-tooth. The dentes sapientiæ are sometimes so called because their presence is necessary to a perfect denture.

**crapaudine** (krap'-aw-dēn) [Fr., *crapaud*, a toad]. In veterinary surgery, an ulcer on the coronet of a horse's hoof.

**crapulent, crapulous** (krap'-u-lent, -lus) [*crapula*, drunkenness; surfeit]. Marked by excess in eating and drinking.

**craquement** (krahk'-mon(g)) [Fr.]. Any crackling sound heard in auscultation.

**craseology, crasiology** (kras-e-ol'-o-je) [κρᾶσις, mixture; λόγος, science]. The science of temperaments.

**crasis** (kra'-sis) [κρᾶσις, mixture]. Temperament, constitution. c., **verminous**, an old term used to designate a peculiar dyscrasia of the system due to the presence of worms.

**crassamen** (kras'-am-en). See *crassamentum*. c. **sanguinis**. See *buffy coat*.

**crassamentum** (kras-am-en'-tum) [L., "thickness"]. A clot, as of blood.

**Cratægus** (kra-te'-gus) [κράταιγος, the hawthorn]. A genus of rosaceous shrubs. *C. oxyacantha*, a European shrub, contains a crystallizable principle, *crategin*, in the bark. A strong tincture in doses of 3 drops is used in heart disease.

**crateriform** (kra-ter'-if-orm) [*crater*, a bowl; *forma*, shape]. Goblet-shaped or deep-saucer-shaped. Excavated like a crater.

**cratomania** (krat-o-ma'-ne-ah) [κράτος, power; μανία, madness]. A delirium of exaltation in which the patient conceives himself to possess vast power.

**craurosis**. See *kraurosis*.

**cravat** (kra-vat') [Fr., *cravate*]. A bandage of triangular shape, used as a temporary dressing for a wound or fracture. The middle is applied to the injured part, and the ends are brought around and tied.

**craw-craw** (kraw'-kraw). A variety of filariasis; see *filaria sanguinis hominis*.

**crealbin** (kre-al'-bin). An internal antiseptic said to consist of creolin and albumin. Syn., *creolalbin*.

**cream** (krēm) [*cremor*, thick juice or broth]. The rich fat part of milk. c. **of tartar**, potassium bitartrate, $KHC_4H_4O_6$; it is diuretic and aperient.

**creamometer** (krēm-om'-et-er) [*cremor*, cream; μέτρον, a measure]. An instrument for estimating the amount of cream in milk.

**crease** (krēs) [Celtic]. A line made by folding. c., **gluteofemoral**, c., **iliofemoral**, the crease that bounds the buttock below, corresponding nearly to the lower edge of the gluteus maximus muscle. It is of supposed significance in the diagnosis of hip-disease.

**creasol** (kre'-as-ol) [κρέας, flesh; *oleum*, oil], see *creosol*.

**creasote, creasotum** (kre'-a-sōt, kre-a-so'-tum). See *creosote*.

**creatin** (kre'-at-in) [κρέας, flesh], $C_4H_9N_3O_2$. A neutral organic substance that occurs in the animal organism, especially in the juice of muscles. c., **dehydrated**, creatinin.

**creatinase** (kre-at'-in-ās). An enzyme which converts creatin into creatinin.

**creatinemia** (kre-at-in-e'-me-ah) [*creatin*; αἷμα, blood]. An excess of creatin in the blood.

**creatinin** (kre-at'-in-in) [*creatin*], $C_4H_7N_3O$. An alkaline substance, a normal constituent of urine. It crystallizes in rhombic prisms and is a strong base. It is much more soluble than creatin. c., **reaction for**. See *Jaffé, Kerner, v. Maschke, Weyl*.

**creatoxism** (kre-at-oks'ism). See *kretoxism*.

**crebruria** (kreb-ru'-re-ah) [*creber*, close together; *ouron*, urine]. Frequent micturition.

**creche** (krāsh) [Fr., a crib]. A day nursery or infant shelter.

**credargan** (kre-dar'-gan). A proprietary preparation of colloidal silver.

**Credé's method** (kred-a') [Karl Siegmund Franz Credé, German gynecologist, 1819-1892]. 1. A prophylactic measure against ophthalmia neonatorum by the instillation, into the eyes of new-born children, of a few drops of a 1 or 2 % solution of silver nitrate. 2. A method of expelling the placenta by grasping the uterus firmly through the abdominal walls, kneading it to excite contraction, and then pressing downward toward the sacrum. See *Dublin method*.

**Credé's ointment** (kred-a') [Benno C. Credé, German surgeon, 1847– ]. A soluble silver ointment made from colloidal silver, applied by inunctions in septicemia and pyemia. Dose ¼–1 dr. (2–4 Gm.), repeated every 12 hours until abatement of symptoms.

**creek dots.** Small shining dots, of unknown nature and often hereditary, occurring at times in the retina anterior to the retinal vessels; they were so named by Marcus Gunn, who first described them.

**creeping sickness** (*krēp'-ing sik'-nes*). The gangrenous form of ergotism.

**cremaster** (*kre-mas'-ter*) [κρεμᾶν, to support]. The muscle that draws up the testis. See under *muscle*.

**cremasteric** (*kre-mas-ter'-ik*) [*cremaster*]. Pertaining to the cremaster muscle. c. **reflex.** See under *reflex*.

**cremation** (*kre-ma'-shun*) [*cremare*, to burn]. The destruction of the dead body by burning, as distinguished from interment.

**crematory** (*kre'-mat-or-e*) [*cremare*, to burn]. An establishment for burning the bodies of the dead, or for consuming garbage and other refuse matter.

**cremnophobia** (*krem-no-fo'-be-ah*) [κρημνός, a crag; φόβος, fear]. Morbid fear of precipices.

**cremometer** (*kre-mom'-et-er*) [*cream*; μέτρον, a measure]. A graduated tube for determining the percentage of cream in milk.

**cremor** (*kre'-mor*) [L., "broth"]. Cream. Any thick substance formed on the surface of a liquid. c. **tartari,** cream of tartar.

**cremule** (*krem'-ūl*). A troche of medicated chocolate cream.

**crena** (*kre'-nah*) [L.]. A notch, especially such a notch as is seen on the sutural margins of the cranial bones. c. **ani,** the anal cleft. c. **clunium,** same as *c. ani*.

**crenate** (*kre'-nāt*), or **crenated** (*kre-na'-ted*) [*crena*, a notch]. Notched or scalloped. In botany, leaves that are serrated. See *crenation*.

**crenation** (*kre-na'-shun*) [*crena*, a notch]. A notched or mulberry-like appearance of the red corpuscles of the blood. It is seen when they are exposed to the air or strong saline solutions.

**crenotherapy** (*kre-no-ther'-ap-e*) [κρήνη, a spring; θεραπεία, treatment]. Treatment by water from mineral springs.

**Crenothrix** (*kre'-noth-riks*) [κρήνη, a spring; θρίξ, hair]. A genus of *Schizomycetes* the filaments of which are enveloped in a gelatinous sheath.

**crenulate** (*kren'-ū-lāt*) [*crena*, a notch]. Finely crenate.

**creocamph** (*kre'-o-kamf*). A preparation of creosote and camphoric acid, added to mercurial cream, to lessen the pain after injection of the latter.

**creoform** (*kre'-o-form*). A solid, tasteless antiseptic consisting of guaiacol, creosote, and formaldehyde.

**creolalbin** (*kre-ol-al'-bin*). See *crealbin*.

**creolin** (*kre'-o-lin*) [κρέας, flesh; *oleum*, oil]. A coal-tar product deprived of phenol; it is an antiseptic, used especially as a douche in obstetrical practice. It has also been used in a solution of 5:1000 for irrigation of the bowel in dysentery and enterocolitis.

**creosal** (*kre'-o-sal*). A dark-brown, hygroscopic powder, with odor and taste of creosote, obtained by heating beechwood creosote with tannic acid and phosphorous oxychloride. It is antiseptic and astringent, and is used in bronchial inflammations. Dose 15–135 gr. (1–9 Gm.) daily. Syn., *tannosal*.

**creosin** (*kre'-o-sin*). A compound of creosote, iodine, calcium hypophosphite, and balsam of Peru; it is used like creosote.

**creosoform** (*kre-o'-so-form*). A combination of creosote and formaldehyde, occurring as a greenish powder.

**creosol** (*kre'-o-sol*) [κρέας, flesh; *oleum*, oil], C₈H₁₀O₂. One of the principal phenols contained in creosote. It is formed from guaiacum-resin, and is found in beechwood tar. It is a colorless, oily liquid of an agreeable odor and a burning taste, boiling at 220° C. It is very similar to guaiacol.

**creosolid** (*kre-o-sol'-id*). See *creosote-magnesia*.

**creosomagnesol** (*kre-o-so-mag'-ne-sol*). A dry mixture of potassium hydroxide, creosote, and magnesia; antiseptic. Dose 2 gr. (0.13 Gm.) in pill with honey.

**creosotal** (*kre-o-so'-tal*). See *creosote carbonate*.

**creosote, creosotum** (*kre'-o-sōt, kre-o-so'-tum*) [κρέας, flesh; σώζειν, to preserve]. The product of the distillation of wood-tar, preferably that from the beech, *Fagus sylvatica*, consisting of a mixture of phenol-compounds. It is an inflammable oily liquid, differing in this respect from phenol. It does not coagulate albumin or collodion. Most of the commercial creosote consists of phenol or contains a large percentage of it. It is antiseptic, astringent, styptic, anesthetic, and escharotic. It is used extensively in pulmonary tuberculosis. Dose 1–3 min. (0.06–0.2 Cc.). c., **alpha-,** a preparation containing the constituents of normal creosote mixed in such proportion that it contains 25% of crystalline guaiacol. c., **beechwood,** that obtained from beechwood. c. **benzoate,** an antiseptic used as a spray in diseases of the throat and nose. c.-**calcium chlorhydrophosphate,** a white, syrupy mass used in tuberculosis. Dose 3–8 gr. (0.19–0.52 Gm.) twice daily. c. **carbonate,** guaiacol carbonate with other carbonates containing 90% of beechwood creosote. Maximum daily dose in tuberculosis 80 min. (5 Cc.). It is recommended in treatment of croupous-pneumonia. Dose 15 gr. (1 Gm.) every 2 hours. Syn., *creosotal*. c.-**magnesia,** a mixture of creosote and calcined magnesia, free from odor and taste of creosote. It is a nonirritant antiseptic. Dose 8 gr. (0.52 Gm.). Syn., *creosolid*; *magnesium creosotate*. c. **mixture** (*mistura creosoti*, B. P.), creosote and glacial acetic acid, of each, 16 min. (1 Cc.), dissolved in 15 oz. (55 Cc.) of water to which 1 oz. (30 Cc.) of syrup and ⅓ dr. (a Cc.) of spirit of juniper have been added. Dose 1–2 dr. (4–8 Cc.). c. **ointment** (*unguentum creosoti*, B. P.), creosote, 1; simple ointment, 12; for local application. c. **oleate,** a yellowish, oily liquid used in the same manner as creosote. Dose 40–60 min. (2.5–3.8 Cc.) daily. Syn., *creosoteoleic ether*; *oleocreosote*. c. **phosphate,** PO₄(C₈H₇)₃, a syrupy fluid containing 80% of creosote and 20% of phosphoric acid anhydride; it is used as a substitute for creosote. Syn., *tricreosote phosphate*. c. **tannophosphate,** an amber-colored fluid used in tuberculosis. c. **valerate,** a noncaustic fluid which is used in all forms of tuberculosis. Dose 3 min. (0.2 Cc.), increasing to 18–28 min. (1.1–1.7 Cc.) daily, in milk. Syn., *eosot*. c. **vapor** (*vapor creosoti*, B. P.), for inhalation. c. **water** (*aqua creosoti*, U. S. P.), a 1% solution. Dose 1–4 dr. (4–16 Cc.).

**creotoxin** (*kre-o-tok'-sin*). See *kreotoxin*.

**creotoxism** (*kre-o-tok'-sism*). See *kreotoxism*.

**crepitant** (*krep'-it-ant*) [*crepitare*, to crackle]. Possessing the character of crepitation. c. **rale.** See under *rale*.

**crepitatio, crepitation, crepitus** (*krep-it-a'-she-o, krep-it-a'-shun, krep'-it-us*) [*crepitare*]. 1. The grating of fractured bones. 2. The crackling of the joints. 3. The noise produced by pressure upon tissues containing an abnormal amount of air or gas, as in cellular emphysema. 4. The sound heard at the end of inspiration in the first stage of croupous pneumonia. It closely resembles the sound produced by rubbing the hair between the fingers held close to the ear. **crepitus indux,** a crepitant rale heard in pneumonia at the beginning of hepatization. **crepitus redux,** a crepitant rale heard in pneumonia during the stage of resolution; usually the first manifestation of the recession of the disease. c., **silken,** a sensation such as is produced when two surfaces of silk are rubbed together, felt by the hand when manipulating a joint affected with hydrarthrosis.

**crepitus.** See *crepitation*.

**cresalol** (*kres'-al-ol*) [*cresol*; *salol*]. Cresol salicylate, an intestinal antiseptic.

**cresamine** (*kres'-am-in*). An antiseptic and germicide mixture of ethylenediamine and tricresol.

**cresaprol** (*kres'-ap-rol*). See *cresin*.

**crescent** (*kres'-ent*) [*crescere*, to grow]. 1. Sickle-shaped, or shaped like the new moon. 2. A form of *Plasmodium malariæ*; one of the crescentic, nonflagellate, refractive, pigmented bodies seen in the blood of persons suffering from protracted forms of malarial poisoning; *i. e.*, after the second week in estivoautumnal fever, in malarial remittent fever, and in the cachectic victims of chronic malaria. c.s **of Gianuzzi.** See under *Gianuzzi*. c., **gray,** one lateral half of the gray matter of the spinal cord. c., **myopic.** See *myopic crescent*. c.-**sphere,** Lambertin's term for that phase of development of the malarial parasite when it becomes sausage-shaped or crescent-shaped. It constitutes a sexual phase of the parasite and is destined to be swallowed by *Anopheles* and to carry on the further life-history of the parasite. **c.s of the spinal cord,** the lateral gray bands of the spinal cord as seen in horizontal section.

**crescentic** (*kres-en'-tik*). 1. Shaped like a new moon. 2. Derived from a member of the genus *Crescentia; e. g.,* crescentic acid.

**cresegol** (*kres'-e-gol*). Mercury orthonitro-parasulphonate; a reddish-brown powder used as a surgical disinfectant.

**cresin** (*kre'-sin*). A mixture of cresol, 25 %, and sodium cresoxylacetate; a brown, clear fluid, said to be less poisonous than phenol. It is used in 0.5 to 1 % solution as a wound antiseptic. Syn., *cresaprol.*

**cresochin** (*kres'-o-kin*). A proprietary disinfectant solution containing chinolin, tricresol, and chinolintricresol sulphonate.

**cresoform** (*kres'-o-form*). A mixture of creosote and formaldehyde; used externally as an antiseptic, and internally in tuberculous enteritis. Dose, 10-30 min. (0.65-2.0 c.c.).

**cresol** (*kre'-sol*) [κρέας, flesh; *oleum,* oil]. C₇H₈O. Cresylic acid; a body obtained from the distillation of coal-tar. It is a colorless, caustic liquid, with properties similar to those of phenol, but is superior as an antiseptic. Syn., *paramethyl phenol.* **c.-anitol,** a compound of anitol and cresol, used as a bactericide. **c. iodide.** See *losophan.* **c.-naphthol,** a brown, viscous, tar-like liquid, insoluble in water; it is used as a germicide. **c. salicylate, c.-salol.** See *cresalol.* **c., solution of,** compound (*liquor cresolis compositus,* U. S. P.). See under *solution.*

**cresolin** (*kres'-o-lin*) [κρεας, flesh; *oleum,* oil]. A proprietary preparation used as a disinfectant.

**cresomagnesol** (*kres-o-mag'-nes-ol*). A mixture of caustic potash, creosote, and magnesia.

**crest** (*krest*) [*crista,* a crest]. A ridge or linear prominence, especially of bone. See *crista.* **c., deltoid,** a ridge on the humerus at the attachment of the deltoid muscle. **c., ethmoid.** 1. A transverse ridge on the inner aspect of the nasal process of the superior maxilla. 2. The turbinated crest. **c., frontal,** a ridge along the middle line of the internal surface of the frontal bone. **c., iliac, c. of ilium,** the thickened and expanded upper-border of the ilium. **c., incisor** (of Henle), the forward prolongation of the nasal crest, terminating in the anterior nasal spine; the cartilage of the nasal septum rests upon it. Syn., *crista incisiva.* **c., infratemporal,** one on the outer aspect of the great wing of the sphenoid and separating the part of the bone which partly forms the temporal fossa from that which aids in forming the zygomatic fossa. **c., lacrimal,** a vertical ridge dividing the external surface of the lacrimal bone into two parts. **c., lambdoid.** See *c., occipital.* **c., nasal,** a crest on the internal border of the nasal bone and forming part of the septum of the nose. **c., neural,** a ridge found on either side of the neural tube in the embryo. **c., obturator,** a bony ridge running from the spine of the os pubis to the anterior end of the cotyloid notch. **c., occipital,** a vertical ridge on the external surface of the occipital bone, extending from the occipital protuberance to the foramen magnum. **c., pubic, c. of pubes,** a crest extending from the spine to the inner extremity of the pubes. **c., sacral, c. of sacrum,** a series of eminences forming a longitudinal ridge on the middle line of the posterior surface of the sacrum. **c. sphenoid,** a thin ridge of bone in the median line of the anterior surface of the body of the sphenoid bone. **c., sphenomaxillary,** an arched crest formed in part by the anterior surface of the great wing of the sphenoid and in part by the pterygoid process forming the border of the sphenomaxillary fissure. **c., supramastoid,** a bony ridge above the external auditory meatus. **c. of tibia,** the prominent border or ridge on the front of the tibia; the shin. **c., turbinated,** a prominent horizontal ridge on the internal surface of the palate bone. **c., zygomatic,** the anterior border of the great wing of the sphenoid; it articulates with the malar bone and separates the orbital from the temporal surface.

**cresyl** (*kres'-il*), C₇H₇. The radical of cresol. **c. alcohol,** C₆H₄(OH)CH₃, formed from phenyl alcohol by the substitution of a molecule of methyl for an atom of the hydrogen of the phenyl. **c. hydrate.** See *c. alcohol.*

**cresylate** (*kres'-il-āt*). Any compound of cresol with a metallic radical.

**cresylic acid** (*kres-il'-ik*). See *cresol.*

**creta** (*kre'-tah*) [L.]. Chalk. Native calcium carbonate. **cretæ, mistura** (U. S. P.), consists of compound chalk powder, 20; cinnamon-water, 40; water, 40. It is used in diarrhea. Dose ½ oz. (15 Cc.). **c. præparata** (U. S. P.), prepared chalk; chalk freed from impurities by washing. Dose 5-20 gr. (0.32-1.3 Gm.). **cretæ, pulvis, aromaticus** (B. P.). Dose 10 gr.-1 dr. (0.65-4.0 Gm.). **cretæ, pulvis, compositus** (U. S. P.), compound chalk powder; consists of prepared chalk, 30; acacia in powder, 20; sugar, 50. Dose 5 gr.-1 dr. (0.32-4.0 Gm.). **cretæ, trochisci,** each contains prepared chalk, 4 gr.; acacia, 1 gr.; sugar, 6 gr.; with a little nutmeg.

**cretaceous** (*kre-ta'-shus*) [*creta*]. 1. Chalky. 2. Chalky-white in color.

**Cretan fever** (*kre'-tan*) [*Crete,* an island in the Mediterranean Sea]. Same as Malta fever.

**cretefaction** (*kre-te-fak'-shun*). See *calcification.*

**cretin** (*kre'-tin*) [Fr., *crétin,* a simple-minded person]. A person affected with cretinism.

**cretinism** (*kre'-tin-ism*) [*cretin*]. A congenital disease, characterized by absence of the thyroid gland, diminutiveness of size, thickness of neck, shortness of arms and legs, prominence of abdomen, large size of face, thickness of lips, large and protruding tongue, and imbecility or idiocy. It occurs endemically in the goitrous districts of Switzerland, and sporadically in other parts of Europe and in America. Lack of the secretion of the thyroid gland seems to be the cause. **c., acquired, c., adult.** Synonym of *myxedema.*

**cretinoid** (*kre'-tin-oid*) [*cretin*]. 1. Resembling a cretin; resembling cretinism. 2. A person who resembles a cretin. **c. state, the morbid state** presented by a sufferer from cretinism; cretinism.

**cretinous** (*kre'-tin-us*). Pertaining to cretinism.

**crewels** (*kroo'-els*) [Fr., *écrouelles,* scrofula]. Synonym of *scrofula.*

**crib** (*krib*) [ME., *crib,* a manger]. A small frame with inclosed sides for a child's bed. 2. A stall for cattle. **c. biting.** See *cribbing* and *windsucking.*

**cribbing** (*krib'-ing*) [*crib*]. The peculiar wearing of a horses' teeth, due to a habit of biting his crib or manger, and at the same time sucking air into the stomach.

**cribrate** (*krib'-rāt*) [*cribrum*]. Perforated, sieve-like.

**cribration** (*krib-ra'-shun*) [*cribrum*]. 1. The state of being cribriform or perforate. 2. The act of sifting.

**cribriform** (*krib'-rif-orm*) [*cribrum*]. Perforated like a sieve. **c. fascia,** the portion of the fascia of the thigh covering the saphenous opening. **c. plate,** the upper perforated plate of the ethmoid bone.

**cribrose** (*krib'-rōs*) [*cribrum,* a sieve]. In biology, sieve-like.

**cribrum** (*krib'-rum*) [L.; pl., *cribra*]. A sieve. **c. benedictum,** a perforate septum, supposed by the ancients to separate two hypothetical cavities of the kidneys, by which the blood in the upper one was strained and freed from impurities. **cribra orbitalia,** inconstant porosities behind the edge of the orbit on the inferior surface of the orbital plate of the frontal bone; they may be culdesacs or, when developed more fully, may communicate.

**Crichton Browne's sign.** See *Browne's sign.*

**crick** (*krik*) [ME., *cricke,* a twist in the neck]. Any painful spasmodic affection, as of the back or neck.

**crico-** (*kri-ko-*) [κρίκος, a ring]. A prefix denoting connection with the cricoid cartilage.

**cricoarytenoid** (*kri-ko-ar-it'-en-oid*) [*crico-; arytenoid*]. Pertaining to the cricoid and arytenoid cartilages.

**cricohyoid, cricohyoideus** (*kri-ko-hi'-oid, kri-ko-hi-oid'-e-us*). Relating to the cricoid cartilage and the hyoid bone.

**cricoid** (*kri'-koid*) [*crico-;* εἶδος, form]. Ring-shaped. **c. cartilage,** the ring-shaped cartilage of the larynx.

**cricoidectomy** (*kir-koid-ek'-to-me*) [*cricoid;* ἐκτομή, excision]. The excision of the cricoid cartilage.

**cricopharyngeal** (*kri-ko-far-in'-je-al*). Relating to the cricoid cartilage and the pharynx.

**cricothyreotomy** (*kri-ko-thi-re-ot'-o-me*). Incision through the cricoid and thyroid cartilages.

**cricothyroid** (*kri-ko-thi'-roid*) [*crico-; thyroid*]. Pertaining to the cricoid and thyroid cartilages. **c. artery,** a small branch of the superior thyroid, crossing the cricothyroid membrane. **c. membrane,** a ligamentous membrane that lies between the cricoid and thyroid cartilages. **c. muscle.** See under *muscle.*

**cricothyrotomy** (*kri-ko-thi-rot'-o-me*). Cricotomy with division of the cricothyroid membrane.
**cricotomy** (*kri-kot'-o-me*) [*crico-*; τέμνειν, to cut]. Surgical laryngotomy by cutting through the cricoid cartilage.
**cricotracheal, cricotrachealis** (*kri-ko-tra'-ke-al, -tra-ke-al'-is*). Relating to the cricoid cartilage and to the trachea.
**cricotracheotomy** (*kri-ko-trak-e-ot'-o-me*) [*crico-*; *tracheotomy*]. Tracheotomy through the cricoid cartilage.
**criminal** (*krim'-in-al*) [*crimen*, an accusation, a crime]. Of the nature of crime. **c. abortion**, see *abortion*. **c. assault**, see *assault*.
**criminology** (*krim-in-ol'-o-je*) [*crimen*, crime; λόγος, science]. The science of crime and of criminals; criminal anthropology; the study of crime as a branch of morbid psychology.
**crinate, crinated** (*krin'-āt, krin-a'-ted*) [*crinis*, a hair]. Bearded with long hairs or hair-like processes; crinite.
**criniform** (*krin'-e-form*) [*crinis*, a hair; *forma*, form]. Filiform; resembling horsehairs.
**crino** (*kri'-no*) [*crinis*, hair; *pl.*, *crinones*]. 1. A skin-affection of infants supposed to be due to the presence of a hair-worm. 2. Same as *comedo*.
**crinogenic** (*krin-o-jen'-ik*) [κρίνειν, to separate; γενναν, to produce]. Stimulating the production of secretions generally.
**crinose** (*krin'-ōs*). Hairy.
**crinosin** (*krin'-o-sin*) [*crinis*, hair]. A nitrogenized fat from brain-substance, crystallizing in hair-like threads.
**crinosity** (*krin-os'-it-e*). Hairiness.
**Cripps' operation** (*krips'*) [William Harrison *Cripps*, English surgeon]. *Iliac colotomy;* an imaginary line from the anterior superior iliac spine to the umbilicus is crossed at right angles, 1½ inches from the superior spine, by an incision 2½ inches long. The bowel is fixed in position and opened.
**crisis** (*kri'-sis*) [κρίσις, a decisive point]. 1. A turning-point, as that of a disease or fever; especially, the sudden favorable termination of the acute symptoms of an infectious disease. 2. Paroxysmal disturbance of function accompanied with pain. **c. bronchial**, a paroxysm of dyspnea sometimes occurring in tabes. **c. cardiac**, a paroxysm of cardiac distress or disordered action. **crises, Dietl's.** See *Dietl's crises*. **c., doctrine of,** the theory that the gradual climax of morbid phenomena was announcement of the completion of the union of morbific material—which could then be evacuated by the sweat, urine, or stools—spontaneously or by the administration of diuretics, purgatives, etc. **c., enteralgic,** a paroxysm of pain in the lower part of the abdomen, occurring in tabes. **crises, gastric,** attacks of intense, paroxysmal pain in the abdomen, often attended with vomiting. They occur in loco-motor ataxia. **c., hemati**₀**, c., hemic,** the crisis in a fever marked by increase in the number of blood-plates. **c., nephralgic, c., nephritic,** a ureteral paroxysm of pain observed in tabes. **c., rectal,** paroxysmal rectalgia occurring in tabes dorsalis and in diabetes. **c., tabetic,** paroxysmal pain occurring in the course of tabes dorsalis.
**Crismer's test for glucose** (*kriz'-mer*) [Léon *Crismer*, Belgian chemist, 1858– ]. An alkaline solution of glucose when heated to boiling with a solution of ½ part safranin in 1000 parts water decolorizes the safranin solution or renders it pale yellow. It is not decolorized when heated with uric acid, creatinin, or creatin in an alkaline solution.
**crispation** (*kris-pa'-shun*) [*crispare*, to curl]. 1. See *crispatura*. 2. A slight involuntary quivering of the muscles.
**crispatura** (*kris-pah-tu'-rah*) [L.]. A puckering; a contracture. **c. tendinum,** Dupuytren's contraction.
**crista** (*kris'-tah*) [L.]. Crest. **c. acustica,** a yellow elevation projecting into the equator of the ampulla of the ear. **c. ampullaris.** See *c. acustica.* **c. basilaris.** See *pharyngeal tubercle.* **c. buccinatoria,** a ridge giving origin to the fibers of the buccinator muscle, found in the groove on the anterior surface of the coronoid process of the lower jaw. **c. capituli,** one on the head of a rib dividing its articular surface into two parts. **c, colli costæ,** a crest on the superior border of the neck of a rib. **c. colli inferior,** one on the lower aspect of the neck of a rib. **c. colli superior,** one on the upper aspect of the neck of a rib. **c. conchalis,** the inferior turbinated crest of the maxilla and palate-bone. **c. ethmoidalis,** the superior turbinated crest of the maxilla and palate-bone. **c. falciformis,** a horizontal crest dividing the lamina cribrosa. **c. galli,** cock's-crest, the superior triangular process of the ethmoid bone. **c. helicis,** a projection of the helix above the external auditory meatus. **c. iliaca,** the crest of the ilium. **c. ilii,** the crest of the ilium. **c. infratemporalis,** the pterygoid ridge of the sphenoid bone. **c. interossea,** the interosseous border. **c. intertrochanterica,** the posterior intertrochanteric line. **c. lacrimalis posterior,** the vertical ridge on the orbital surface of the lacrimal bone. **c. obturatoria,** the obturator crest of the os pubis. **c. occipitalis externa,** the external occipital crest. **c. sacralis articularis,** one of the small tubercles of the sacrum representing the articular processes of vertebræ. **c. sacralis lateralis,** one of the rudimentary transverse processes of the sacral vertebræ. **c. sacralis media,** the tubercular ridge of the sacrum. **c. sphenoidalis,** the sphenoidal crest. **c. spiralis,** a ridge on the upper border of the spiral lamina of the cochlea. **c. transversa,** the crista falciformis. **c. tuberculi majoris,** the external or posterior bicipital ridge of the humerus. **c. tuberculi minoris,** the internal or anterior bicipital ridge of the humerus. **c. urethralis,** the crest of the urethra. **c. vestibuli,** an almost vertical bony ridge on the inferior and median walls of the vestibule of the ear, separating the fovea hemielliptica from the fovea hemisphærica. Syn., *pyramis vestibuli.*
**cristallin** (*kris'-tal-in*). A kind of collodion, in which the ether and alcohol employed as solvents for pyroxylin are replaced by methyl-alcohol. It does not dry so readily as ordinary collodion. Syn., *crystallin.*
**cristate** (*kris'-tāt*). Crested.
**crith** (*krith*) [κριθή, barley-corn]. The assumed unit of mass for gases. It is the weight, *in vacuo*, of one liter of hydrogen, at 0° C., which is 0.0896 of a gram, or 1.37 grains.
**critical** (*krit'-ik-al*) [*crisis*]. 1. Pertaining to a crisis in disease, period of life, etc. 2. A qualification applied to temperature and to pressure in relation to gases.
**croated** (*kro'-ka-ted*) [see *crocus*]. Containing saffron.
**croceous** (*kro'-se-us*). Saffron-colored; containing saffron.
**crocidism, crocidismus, crocidixis** (*kro'-sid-ism, kro-sid-is'-mus, -iks'-is*). See *carphology.*
**croconic** (*kro-kon'-ik*). Saffron-colored.
**crocose** (*kro'-kōs*). A dextrorotary sugar obtained from crocin by decomposition.
**crocoxanthin** (*kro-ko-zan'-thin*). A yellow pigment occurring in the petals of *Crocus aureus.*
**Crocq's serum** (*krok*) [Jean *Crocq*, Belgian physician]. A two per cent. solution of sodium phosphate.
**crocus** (*kro'-kus*) [κρόκος, crocus; saffron]. Saffron. The stigma of the flowers of *C. sativus.* It is an aromatic stimulant, emmenagogue, and antispasmodic. Dose of the *tincture* (10 % in strength) 1–2 dr. (4–8 Cc.); of the *drug* 5–20 gr. (0.32–1.3 Gm.) in infusion.
**crocydisma** (*kro-sid-iz'-mus*) [κροκυδίζειν, to pick at]. Same as *carphology.*
**Crombie's ulcer** (*krom'-be*). A small ulcer on the gums, near the last two molar teeth, occurring in sprue.
**Crookes' tube** [Sir William *Crookes*, English physicist, 1832– ]. A highly exhausted vacuum-tube used in producing x-rays.
**Cropper's bodies** (*krop'-er*). Spindle-shaped masses sometimes found in the red blood corpuscles.
**cross** (*kros*). 1. In biology, a cross-breed in plants, the result of cross-fertilization. 2. A structure in which parts cross each other. **c.-birth,** shoulder-presentation, or other presentation requiring version. **c.-education,** E. W. Scripture's term for the curious results that appear in certain cases where exercise of an organ or limb develops not only that particular organ or limb, but the corresponding one on the opposite side. **c.-eye.** See *strabismus.* **c.-fertilization,** in biology, the fertilization of the ovules of one species by the seedgerms of another. **c.-foot,** *pes. varus.* **c.-knee.** See *genu valgum.* **c.-leg, c.-legged,** a deformity that sometimes follows double hip-joint disease; the legs are crossed in walking. **c.-legged progression,** a method of walking in which one foot gets over in front of the other. It is a symptom

of certain cord-lesions. c., occipital. See *occipital protuberance, internal*.

**crossed** (*krōsd*). Having the shape of a cross. Affecting alternate sides of the body. c. amblyopia. See *amblyopia*. c. anesthesia. See *anesthesia, crossed*. c. hemiplegia, c. paralysis. See *paralysis, crossed*. c. reflexes, reflex movements on one side of the body, excited by stimulation of a part on the opposite side.

**crotalin** (*krō'-tal-in*) [κρόταλον, a rattle]. An albuminous body contained in the poison of the cobra. It is not coagulable by heat at 212° C.

**Crotalus** (*krō'-tal-us*) [κρόταλον, a rattle]. A genus of serpents, including the typical rattlesnakes. C. poison, the virus of the rattlesnake, used as a remedy by homeopathists.

**crotaphion** (*kro-taf'-e-on*) [κρόταφος, the temple of the head]. A craniometrical point at the dorsal end of the pterion.

**crotchet** (*kroch'-et*) [ME., *crochett*, a little hook]. A hook used in extracting the fetus after craniotomy.

**crotin** (*krō'-tin*). A mixture of toxic albuminoids contained in croton seeds. It is a yellowish powder containing about 21 % of ashes, soluble in water and in a 10 % solution of sodium chloride; it is a protoplasmic poison.

**Croton** (*krō'-ton*) [κρότων, a tick]. A great genus of euphorbiaceous plants. C. eleuteria yields cascarilla; C. tiglium yields croton oil. c. aldehyde, $C_4H_8O$, a compound obtained by the condensation of acetaldehyde when heated with HCl, with water and zinc chloride. c.-chloral. See *chloral, butyl*. c. oil (*oleum tiglii*, U. S. P.), a fixed oil expressed from the seeds of *C. tiglium*. See under *tiglium*.

**crotonallin** (*kro-ton-al'-in*). A poisonous albuminoid from the seeds of *Croton tiglium*.

**crotonglobulin** (*kro-ton-glob'-ū-lin*). A poisonous albuminoid from the seeds of *Croton tiglium*.

**crotonic** (*kro-ton'-ik*). Belonging to or derived from a plant belonging to the genus *Croton*; e. g., crotonic acid.

**crotonism** (*krō'-ton-ism*). Poisoning by croton oil; a condition marked by hemorrhagic gastroenteritis.

**crotonol** (*krō'-ton-ol*), $C_5H_{14}O_3$. An acid, aromatic principle in croton oil, believed to be the vesicant constituent of the oil.

**crounotherapy** (*kru-no-ther'-ap-e*) [κρουνός, a spring; *therapy*]. Riesman's term for the employment of mineral waters for drinking-cures.

**croup** (*kroop*) [AS., *krōpan*, to cry aloud]. A disease of the larynx and trachea of children, prominent symptoms of which are a harsh, "croupy" cough and difficulty in breathing; it is often accompanied by the development of a membranous deposit or exudate upon the parts. It is usually caused by the diphtheria bacillus, sometimes by other microorganisms. Syn., *membranous croup; pseudomembranous croup; true croup*. c., artificial, traumatic membranous laryngitis. c., bronchial. See *bronchitis, croupous*. c., catarrhal, a simple noncontagious inflammation of the larynx accompanied by the formation of membrane. c., diphtheritic, laryngeal diphtheria. c., false, a spasm of the muscles of the larynx with a slight inflammation. c., intestinal. See *colitis, mucous*. c.-kettle, a small boiler heated by a lamp and contained within a metallic cylinder. The boiler is furnished with an inhaling tube, and water or any medicament may be placed within it and the escaping steam inhaled. c., spasmodic. See *c., false*.

**croupine** (*krōop'-ēn*). Laryngismus stridulus.

**croupous** (*kroop'-us*). Pertaining to croup. c. membrane, the yellowish-white membrane forming in the larynx in croup. c. pneumonia. Same as lobar pneumonia; see under *pneumonia*.

**crowd-poison** (*krowd'-poi-son*). Volatile organic matter recognizable in the air of ill-ventilated places where many persons are congregated.

**crown** (*krown*). See *corona*. c.-bark. See *loxa bark*. c., fibrous, c., radiating. See *corona radiata*. c., French. See *corona veneris*. c., gall, a disease of the peach, apricot, almont, prune, plum, apple, pear, English walnut, grape, raspberry, cherry, poplar, and chestnut, due to a parasite plasmodium, *Dendrophagus globosus*. c. glass, a kind of glass used in optics. c., post and plate, a porelain facing with a backing of gold, and a post fitting the enlarged pulp canal, and a disc covering the exposed surface of the tooth root. c.-setting, the operation of joining an artificial crown to the root of a natural tooth; improperly called "pivoting." c. of a tooth, the exposed part of the tooth above the gums, covered with enamel. c.-work, the adaptation of an artificial crown of porcelain or gold on the cervical portion of the natural root of a tooth.

**crucial** (*kru'-shal*) [*crux*, a cross]. Resembling or pertaining to a cross, as a *crucial* incision.

**crucible** (*kru'-sib-l*) [*crucibulum*, a melting pot]. A vessel of conical shape in which substances are exposed to the heat of a fire or furnace.

**cruciform** (*kru'-se-form*) [*crux; forma*, form]. Crucial; shaped like a cross.

**crude** (*krūd*) [L. *crudus*, raw, unripe]. In the natural form, raw, or unrefined.

**crudivorous** (*kru-div'-or-us*) [*crudus*, raw; *vorare*, to devour]. Applied to savages subsisting entirely upon uncooked food.

**cruels** (*kroo'-els*). Synonym of *scrofula*.

**cruentous** (*kru-en'-tus*) [*cruor*]. Bloody.

**cruenturesis** (*kru-en-tū-re'-sis*) [*cruentus*, bloody; οὖρον, urine]. Hematuria.

**cruor** (*kru'-or*) [L., "blood"]. Blood, especially coagulated blood.

**cruorin** (*kru'-or-in*) [*cruor*, blood]. Synonym of *hemoglobin*.

**crupper** (*krup'-er*) [Fr., *croupe*, the rump]. 1. The buttocks of a horse. 2. The sacrococcygeal region. 3. The base of the tail in mammals. c.-bone, the coccyx.

**crura** (*kru'-rah*) [Plural of *crus*, a leg]. A name applied to certain parts of the body, from their resemblance to legs or roots; see *crus*. c. ampullaria, ampullary limbs. c. anthelicis, c. bifurcata, two ridges on the inner aspect of the external ear, converging at the antihelix. Syn., *radices, anthelicis*. c. cerebelli, the peduncles of the cerebellum, superior, middle, and inferior. c. cerebri, the peduncles of the cerebrum. c. of diaphragm, the muscular bundles arising from the vertebræ, etc., and inserted into the central tendon. c. of fornix. See *pillars, anterior and posterior of the fornix*. c. of penis, the corpora cavernosa.

**crurænus** (*kru-re'-us*) [L.]. See *muscles, table of*.

**crural** (*kru'-ral*) [*crura*]. 1. Pertaining to the thigh. 2. Pertaining to the crus cerebri. c. arch. See *ligament, crural*. c. hernia, femoral hernia. c. ring, the femoral ring; the upper opening of the femoral canal, bounded in front by Poupart's ligament and the deep crural arch, behind by the pubis, internally by Gimbernat's ligament, externally by a fibrous band separating it from the femoral vein. c. septum. See *septum crurale*. c. sheath, the femoral sheath.

**crureus** (*kru-re'-us*) [L.]. One of the muscles of the thigh. See *muscles, table of*.

**crurin** (*kru'-rin*). Edinger's name for quinolinbismuth-sulphocyanide; used as a dressing for ulcers.

**cruritis** (*kru-ri'-tis*). See *phlegmasia alba dolens*.

**crurogenital** (*kru-ro-jen'-it-al*) [*crura; genitalis*, relating to generation]. Relating to the thighs and the genitalia.

**cruroinguinal** (*kru-ro-in'-gwin-al*). Relating to the thigh and the groin.

**crus** (*krus*) [L.; *pl., crura* (*q. v.*)]. A leg, limb, or support. c. anterius, anterior limb. c. breve, short limb. c. cerebelli, any one of the cerebellar peduncles. c. cerebri, either of the two peduncles connecting the cerebrum with the pons. c. clitoridis, crus of the clitoris. c. commune, common limb. c. fornicis, posterior pillar of fornix. c. helicis, limb of the helix. c. inferius, inferior pillar. c. intermedium, intermediate crus. c. laterale, lateral crus. c. longum, long limb. c. mediale, medial crus. c. of the diaphragm, either of the two fibromuscular bands arising in front of the vertebræ and inserted into the central tendon of the diaphragm. c. posterius, posterior limb. c. simplex, simple limb. c. superius, superior pillar.

**crusocreatinin** (*kru-so-kre-at'-in-in*), $C_5H_8N_4O$. A leukomaine isolated from muscle tissue.

**crust** (*krust*) [*crusta*]. A covering, especially a dried exudate on the skin. c., milk. See *crusta lactea*.

**crusta** (*krus'-tah*) [L., "a crust"]. 1. See *crust*. 2. The inferior portion of the crus cerebri. c. adamantina dentium, crust, or enamel, of the teeth. c. lactea, seborrhea of the scalp in infants. c. lamellosa, psoriasis. c. osteoides (radicis), c. petrosa, a thin layer of bone covering the fang of a tooth. c. phlogistica, the yellowish layer of the upper stratum of a blood-clot, coagulating slowly.

**crutch** (*krutch*) [ME., *crutche*]. A staff with a concave cross-piece fitting under the arm-pit, and often with a grip for the hands midway on the staff, used as a support in walking. c. **paralysis**, paralysis of an upper extremity due to the pressure of the crutch-head upon the nerves of the axilla, especially the musculospiral nerve. c., **perineal**, a support or brace of various forms by means of which the leg of a patient in the lithotomy position may be adjusted or held at any height or angle.
**Cruveilhier's atrophy** (*kroo-vāl-yeh'*) [Jean Cruveilhier, French pathologist, 1791–1874]. Progressive muscular atrophy. **C.'s disease**. 1. Ulcer of the stomach. 2. Progressive muscular atrophy. **C.'s fascia**, the superficial layer of the perineal fascia. **C.'s plexus**. 1. A plexus of the posterior cervical region which is derived from the great occipital nerve and the first and second cervical nerves. 2. The plexus of varicose veins in a variety of angioma. **C.'s ulcer**, simple ulcer of the stomach.
**cry** (*krī*) [ME., *crien*, to cry]. The utterance of an inarticulate vocal sound; or the sound so uttered; the sound of the voice in lamentation. c., **epileptic**, see under *epileptic*. c., **hydrencephalic**, see *hydrocephalic cry*.
**cryalgesia** (*kri-al-jē'-ze-ah*) [κρύος, cold; ἄλγησις, pain]. Pain from the application of cold.
**cryesthesia** (*kri-es-the'-ze-ah*) [κρύος, cold; αἴσθησις, sensation]. Undue sensitiveness to cold.
**crymodynia** (*kri-mo-din'-e-ah*) [κρυμός, icy cold; ὀδύνη, pain]. Cryalgesia; pain coming on in cold or damp weather.
**crymotherapy** (*kri-mo-ther'-ap-e*) [κρυμός, icy cold; θεραπεία, therapy]. Ribard's term for the therapeutic use of great cold applied locally. A bag filled with carbonic snow at a temperature of −176° F. is applied daily for half an hour to the pit of the stomach. It is previously surrounded by cotton to prevent injury to the skin.
**cryogenin** (*kri-oj'-en-in*). Metabenzamino-semicarbazide. It is given in treatment of tuberculosis for diminishing the fever, being innocuous and effective.
**cryometer** (*kri-om'-et-er*) [κρύος, cold; μέτρον, measure]. A thermometer for measuring very low temperatures.
**cryoscopic** (*kri-os-kop'-ik*). Relating to cryoscopy.
**cryoscopy** (*kri-os'-ko-pe*) [κρύος, cold; σκοπεῖν, to examine]. The process whereby the freezing-point of certain liquids, blood, urine, etc., may be compared with that of distilled water.
**cryostase** (*kri'-os-tās*). A compound of equal parts of phenol, camphor, saponin, and traces of oil of turpentine. It solidifies when heated, and becomes liquid when cooled to below 0° C. Recommended as an antiseptic.
**crypt** (*kript*) [κρυπτός, hidden]. 1. A small sac or follicle. 2. A glandular cavity. **c.s of Lieberkuehn**. See under *Lieberkuehn*. **c.s, multilocular**. 1. Sharpey's name for simple glands with pouched or sacculated walls. 2. The lobules of a compound gland. **c.s, sebaceous**, the sebaceous glands. **c., synovial**. See *Bursa mucosa*. **c.s, synoviparous**, extensions of the synovial membranes sometimes perforating the capsule of the joints and occasionally becoming shut off from the main sac. **c.s of the tongue**, small pits in the mucosa of the tongue with walls studded with globular projections, each of which contains a vascular loop and is furnished with lymph-follicles.
**crypta** (*krip'-tah*). Same as *crypt*.
**cryptitis** (*krip-ti'-tis*) · [*crypt*; ιτις, inflammation]. Inflammation of a crypt, or of crypts. c., **urethral**, phlegmasia of the mucous follicles of the urethra.
**crypto-** (*krip-to-*) [*crypt*]. A prefix meaning relating to a crypt, or a small sac or follicle.
**cryptobiotic** (*krip-to-bi-ot'-ik*) [*crypto-*; βίος, life]. Having dormant life; applied formerly to calculi, crystals, or any inanimate objects which increase in size. Syn., *lithobiotic*.
**cryptocephalus** (*krip-to-sef'-al-us*) [*crypto-*; κεφαλή, head]. A fetal monster with an imperfectly formed and concealed head.
**Cryptococcus** (*krip-to-kok'-us*). A genus of *Saccharomyces*. Same as *Blastomyces*.
**cryptocrystalline** (*krip-to-kris'-tal-īn*). See *microcrystalline*.
**cryptodidymus** (*krip-to-did'-im-us*) [*crypto-*; δίδυμος; twin]. A teratism in which one fetus is concealed within another.
**cryptogam** (*krip'-to-gam*) [*crypto-*; γάμος, marriage].

In biology, one of the *Cryptogamia*, a division of the vegetable kingdom comprising all plants with concealed sexual organs, without pistils or stamens.
**cryptogamic** (*krip-to-gam'-ik*). See *cryptogamous*.
**cryptogamous** (*krip-tog'-am-us*). Belonging to the cryptogamia; having the processes of the reproductive function obscured or concealed.
**cryptogenetic**, **cryptogenic** (*krip-to-jen-et'-ik, krip-to-jen'-ik*) [*crypto-*; γεννᾶν, to produce]. 1. Obscure as to origin. 2. Parasitic from the outset within another living organism.
**cryptolith** (*krip'-to-lith*) [*crypto-*; λίθος, stone]. A concretion or calculus formed within a crypt.
**cryptolithiasis** (*krip-to-lith-i'-a-sis*) [*crypto-*; lithiasis]. The calcification and ossification of tumors of the skin and subcutaneous tissue.
**cryptomenorrhea**, **cryptomenorrhœa** (*krip-to-men-o-re'-ah*) [*crypto-*; μήν, month; ῥόα, flow]. The occurrence of the subjective symptoms of menstruation without any flow of blood.
**cryptomerorrhachischisis** (*krip-to-mer-o-rak-is'-kis-is*) [*crypto-*; μέρος, a part; ῥάχις, the spine; σχίζειν, to cleave]. Spina bifida occulta, a variety with bony deficiency but without a tumor.
**cryptophthalmos**, **cryptophthalmia** (*krip-tof-thal'-mos, krip-tof-thal'-me-ah*) [*crypto-*; ὀφθαλμός, the eye]. 1. Congenital union of the eyelids, usually over imperfect eyes. 2. A person who has congenital union of the eyelids.
**cryptopine** (*krip'-to-pēn*) [*crypto-*; ὄπιον, opium], $C_{21}H_{23}NO_5$. One of the alkaloids of opium, colorless and odorless. It is said to be anodyne and hypnotic, but it is less safe than morphine. Dose ⅛ gr. (0.008 Gm.).
**cryptoporous** (*krip-top'-or-us*) [*crypto-*; πόρος, a pore]. Having hidden or obscure pores.
**cryptopyic** (*krip-to-pi'-ik*) [*crypto-*; πύον, ᾳ pus]. Characterized by concealed suppuration.
**cryptorchid**, **cryptorchis** (*krip-tor'-kid, -kis*) [*crypto-*; ὄρχις, testicle]. A person with retained testicles, *i. e.*, not descended into the scrotum.
**cryptorchidism** (*krip-tor'-kid-izm*) [see *cryptorchid*]. Retention of the testes in the abdomen or inguinal canal.
**cryptorhetic organs**, **tissues** (*krip-to-ret'-ik*)[*orypto-*; ῥεῖν, to flow]. Organs or tissues which have an internal secretion.
**cryptoscope** (*krip'-to-skōp*) [*crypto-*; σκοπεῖν, to inspect]. See *fluoroscope*.
**cryptozygous** (*krip-toz'-ig-us*) [*crypto-*; ζυγόν, yoke]. Having the dental arches or zygomata concealed from view when the skull is viewed from above.
**crystal** (*kris'-tal*) [κρύσταλλος, clear ice]. In chemistry, a substance that assumes a definite geometric form. **c.s, Bizzozero's**. See *Charcot's c.s.* **c.s, Boettcher's**. See *c.s, spermin*. **c.s, Charcot's**, **c.s, Charcot-Leyden's**. See *Charcot's crystals*. **c.s, Charcot-Neumann's**, **c.s, Charcot-Robin's**. See *Charcot's c.s.* **c.s, coffin-lid**, crystals of triple phosphate found in the urine in dyspepsia and cystitis. **c., dumb-bell**, crystals of calcium oxalate, seen in urine. **c.s, ear-**. See *otolith*. **c., Florence's**. See *Florence's crystals*. **c.-gazing**, gazing into the depths of a crystal globe or the surface of a clear vessel of water in order to produce selfhypnotism or autosuggestion with a view to obtaining socalled telepathic impressions. **c.s, hedgehog**. See *hedgehog crystals*. **c.s, hemin**. See *hemin crystals*. **c.s, knife-rest**, peculiar indented crystals of triple phosphate occasionally found in urine. **c.s, lead-chamber**, those found in the lead-chambers in which sulphuric acid is manufactured, and having the composition $HSO_3(NO_2)$. **c.s, Leyden's**. : See *Charcot's c.s*. **c.s, Lubarsch's**. See *Lubarsch's crystals*. **c.s, Neumann's**. Same as *Charcot's c.s*. **c.s, Schweiner's**. See *Charcot's c.s*. **c.s, spermin**, a combination of phosphoric acid with a base, spermin ($C_5H_{14}N_2$), forming long, monoclinic, prism-like crystals with curved edges, found in spermatic fluid after drying it or allowing it to stand and in desiccated white of egg. A strong solution of iodine and potassium iodide stains them a deep brown or violet. Syn., *Boettcher's crystals*. **c.s, Teichmann's**. See *hemin crystals*. **c.s of Venus**, copper acetate crystals. **c.s, Virchow's**. See *Virchow's c.s.* **c.s, Zenker's**. See *Charcot's c.s.*
**crystalban** (*kris-tal'-ban*). Of Payer; a resinous constituent of gutta-percha.
**crystalbumin** (*kris-tal'-bū-min*). An albuminous body found by Béchamp in the watery extract of crystalline lens.
**crystalfibrin** (*kris-tal-fi'-brin*). An albuminous

body obtained by means of hydrochloric acid from crystalline lens.
**crystallin** (*kris'-tal-in*) [*crystal*]. 1. The globulin of the crystalline lens. 2. See *cristallin*.
**crystalline** (*kris'-tal-īn*) [*crystal*]. Like a crystal. c. **lens.** See *lens, crystalline*.
**crystallitis** (*kris-tal-i'-tis*). See *phakitis*.
**crystallization** (*kris-tal-iz-a'-shun*) [*crystal*]. The process by which the molecules of a substance arrange themselves in geometric forms when passing from a gaseous or a liquid to a solid state. c., **alcohol of,** an alcohol uniting molecule by molecule with a crystalline substance and aiding in the preservation of the crystalline form of the latter. c., **water of,** the water of salts that cannot be extracted without destruction of their crystalline nature.
**crystallographic test for blood.** The use of sodium iodide as a reagent forming characteristic blood-crystals. Its use depends on the formation of iodine-hematin.
**crystallography** (*kris-tal-og'-ra-fe*) [*crystal*; γράφειν, to write]. The science of crystals, their formation, etc.
**crystalloid** (*kris'-tal-oid*): Having a crystalline structure, as distinguished from colloid. In biology, one of the crystal-like proteid bodies found in seeds, tubers, etc.
**crystallomagnetism,** (*kris-tal-o-mag'-net-izm*). That property by which certain crystals point to the north when suspended horizontally.
**crystallometry** (*kris-tal-om'-et-re*) [*crystal*; μέτρον, a measure]. The science of the measurement of the angles of crystals.
**crystallose** (*kris'-tal-ōs*). Sodium saccharinate.
**crystalluridrosis** (*kris-tal-ū-rid-ro'-sis*) [*crystal*; οὖρον, urine; ἱδρώς, sweat]. A condition marked by excretion of urinary elements in the sweat which crystallize on the skin.
**Cs.** Chemical symbol of *cesium*.
**Cu.** Chemical symbol of *copper* (*cuprum*).
**cubeb, cubeba** (*kū'-beb, kū-be'-bah*) [Pers., *kabāba,* cubeb]. The unripe fruit of *Piper cubeba,* cultivated in Java. Its properties are due to a volatile oil, C₁₅H₂₄, and an organic acid. It is an aromatic stimulant, diuretic in small doses, and is useful in affections of the bladder and urethra. It is also employed in catarrh of the air-passages, etc. Dose 10 gr.-2 dr. (0.65-8.0 Gm.). c., **fluidextract of** (*fluidextractum cubebæ,* U. S. P.), alcoholic. Dose 5-30 min. (0.32-2.0 Cc.). c., **oil of** (*oleum cubebæ,* U. S. P.), the volatile oil. Dose 5-20 min. (0.32-1.3 Cc.). c., **oleoresin of** (*oleoresina cubebæ,* U. S. P.), ethereal. Dose 5-30 min. (0.32-2.0 Cc.). c., **tincture of,** 10 % in strength. Dose 10 min.-3 dr. (0.65-12.0 Cc.). c., **troches of** (*trochisci cubebæ,* U. S. P.), oleoresin, ½ gr.; oil of sassafras, 1/20 gr.; extract of licorice, 4 gr.; acacia, 2 gr.; syrup of tolu q. s., in each troche. Dose 1-3.
**cubebene** (*kū'-beb-ēn*) [*cubeb*]. C₁₅H₁₈. The light portion of the essential oil of cubebs.
**cubebic acid** (*kū-beb'-ik*). See *acid, cubebic*.
**cubebin** (*kū-beb'-in*) [*cubeb*], C₁₀H₁₀O₃. An odorless, crystalline substance obtained from cubeb.
**cubebism** (*kū'-beb-izm*). Poisoning by cubeb; it is marked by acute gastroenteritis.
**cubic** (*kū'-bik*). Pertaining to a cube. c. **niter,** sodium nitrate. c. **space** (of air), the amount of space required by persons in health and in disease.
**cubiform** (*kū'-be-form*). Cuboid.
**cubit** (*kū'-bit*) [*cubitus*]. 1. The forearm; cubitus. 2. The ulna. 3. The elbow.
**cubital** (*kū'-bit-al*). Relating to the forearm, to the elbow, or to the ulna. c. **bone,** the cuneiform bone.
**cubitale** (*kū-bit-a'-le*). The cuneiform bone.
**cubitocarpal** (*kū-bit-o-kar'-pal*). Relating to the forearm and to the carpus.
**cubitodigital** (*kū-bit-o-dij'-it-al*). Relating to the forearm or the ulna and to the fingers.
**cubitometacarpal** (*kū-bit-o-met-ah-kar'-pal*). Relating to the forearm or the ulna and the metacarpus.
**cubitoradial** (*kū-bit-o-ra'-de-al*). Relating to both the ulna and the radius.
**cubitosupraphalangeal** (*kū-bit-o-sū-prah-fal-an'-je-al*). Relating to the forearm or the ulna and the bases of the phalanges.
**cubitus** (*kū'-bi-tus*) [L., "the elbow"]. The forearm; elbow; ulna. c. **valgus,** a deformity consisting of an abnormal curvature of the humeral diaphysis. c. **varus.** See *gunstock deformity*.

**cubocuneiform** (*kū-bo-kū-ne'-if-orm*). Relating to the cuboid bone and to one or more of the cuneiform bones.
**cuboid** (*kū'-boid*) [κύβος, cube; εἶδος, like]. Resembling a cube. c. **bone,** a bone of the foot situated at the outer anterior part of the tarsus.
**cuca** (*koo'-kah*). Same as *coca;* see *erythroxylon*.
**cucullaris** (*kū-kū-la'-ris*). The trapezius muscle.
**Cucumis** (*kū'-kū-mis*) [L., "a cucumber"]. A genus of plants. See *colocynthis*. C. *melo,* muskmelon, is a species indigenous to the old-world tropics and widely cultivated. The root is emetic and diuretic and contains melonenemetin. The juice of the fruit of C. *sativus,* cucumber, is purgative, diuretic, and resolvent. It is used in skin diseases and as a cosmetic.
**cucurbit** (*kū-ker'-bit*) [*cucurbita,* a gourd]. 1. A cupping-glass. 2. The body of an alembic. 3. Any plant of the order *Cucurbitaceæ*.
**Cucurbita** (*kū-kurb'-it-ah*) [L., "a gourd"]. A genus of plants of the order *Cucurbitaceæ.* C. *fœtidissima* is a species of North America; the leaf, root, and seeds are edible and the bitter fruit is recommended in the treatment of hemorrhoids. The ripe seed of C. *pepo,* the pumpkin, is the *pepo* of the U. S. P., an anthelmintic. See *pepo*.
**cucurbitation** (*kū-ker-bit-a'-shun*) [*cucurbitatio,* a cupping]. The operation of cupping.
**Cuguillère's serum** (*koo-ge-yār*) [E. *Cuguillère,* French physician]. A preparation containing allyl sulphide and tincture of myrrh, used hypodermatically in cases of tuberculosis.
**Cuignet's method** (*koo-ēn-ya*) [Ferdinand Louis Joseph *Cuignet,* French ophthalmologist, 1823— ]. Retinoscopy.
**cuirass** (*kwe'-ras*) [Fr.]. A close-fitting or immovable bandage for the front of the chest. c. **cancer.** See *cancer en cuirasse*. c., **tabetic,** an anesthetic area encircling the chest in tabetic patients.
**culdesac** (*kul'-de-sak*) [Fr.]. A closed or "blind" pouch or sac. c., **Douglas'.** See under *Douglas*.
**culex** (*kū'-leks*) [L., "a gnat"]. A mosquito. c. **fasciatus.** See *Stegomyia fasciata.* c. **fatigans,** transmits filaria.
**Culicidæ** (*kū-lis'-id-e*). A family of insects, order Diptera, which includes the mosquitoes.
**culicide** (*kū'-lis-īd*) [*culex,* a gnat; *cadere,* to kill]. Any agent which destroys mosquitoes.
**culicifuge** (*kū-lis'-if-ūj*) [*culex; fugare,* to drive away]. An agent to drive away mosquitoes.
**culmen** (*kul'-men*) [L., "summit"]. A part of the cerebellum on the cephalic side of the vermis. c. **monticuli,** the highest lobule of the cerebellum.
**culminal** (*kul'-min-al*) [*culmen,* a summit]. Relating to the culmen.
**cultivation** (*kul-tiv-a'-shun*). Same as *culture*. c., **fractional,** cultivation in which a small piece of a culture containing several species of bacteria is transferred to a new medium and used to form a new culture; and a small piece of this culture is similarly transferred to form another, and so on until a culture of a single microorganism is obtained.
**culture** (*kul'-chūr*) [*colere,* to till]. The growth of microorganisms on artificial media. The act of cultivating microorganisms on artificial media. c., **fractional.** See c., *pure*. c., **hanging-drop,** a culture in which the microorganism is inoculated into a drop of fluid on a cover-glass and the latter is inverted over a glass slide having a central concavity. c.-**medium,** a substance used for cultivating bacteria. Culture-media are either liquid or solid, bouillon and milk being the important liquid media, and gelatin, agar, blood-serum, and potato, the principal solid media. c., **needle-.** See c., *stab-*. c., **plate-,** a culture of bacteria on a medium spread upon a flat plate or in a double dish. c., **pure,** a culture of a single microorganism. c., **slant-,** one made on the slanting surface of a medium, so as to get a greater surface. c., **stab-,** one in which the medium is inoculated by means of a needle bearing the microorganisms, and which is inserted deep down into the medium. c., **thrust-,** same as c., *stab-*.
**Culver's physic,** or root (*kul'-verz*). See *leptandra*.
**cumarin** (*kū'-mar-in*). See *coumarin*.
**cumene** (*kū'-mēn*) [*cumin*], C₉H₁₂. A compound made by distilling cumic acid with lime.
**cumic** (*kū'-mik*) [*cumin*]. Derived from or pertaining to cumin. c. **acid.** See *acid, cumic*.
**cumin** (*kū'-min*) [*cuminum,* cumin]. An umbelliferous plant, *Cuminum cyminum,* native in Egypt

and Syria. The fruit possesses well-marked stimulating and carminative properties. Its active principle is an oil.
**cuminic** (*kū-min'-ik*). See *cumic*.
**cumol** (*kū'-mol*). See *cumene*.
**cumulative** (*kū'-mū-la-tiv*) [*cumulare*, to heap up]. Increasing; adding to. c. **action**, c. **effect**, the production of a marked and sudden result, after the administration of a considerable number of comparatively ineffective doses.
**cumulus** (*ku'-mū-lus*) [L., a heap]. A heap or mound. c. **oophorus**, the ovarian mound, or discus proligerus. c. **ovigerus**. See *ovule*. c. **proligerus**. See *discus proligerus*.
**cundurango** (*kun-dū-ran'-go*). See *condurango*.
**cuneal** (*kū'-ne-al*) [*cuneus*]. See *cuneiform*.
**cuneate** (*kū'-ne-āt*) [*cuneus*]. Wedge-shaped. c. **fasciculus**, c. **funiculus**, the continuation of the posteroexternal column of the cord into the medulla. c. **nucleus**. See *nucleus*.
**cuneiform** (*kū-ne'-if-orm*) [*cuneus; forma*, shape]. Wedge-shaped; cuneate. c. **bones**, a wedge-shaped bone of the carpus, and three wedge-shaped bones at the anterior part of the tarsus. c. **cartilage**, a cartilage beside the arytenoepiglottidean fold. c. **column**. See *Burdach's column*.
**cuneocuboid** (*kū-ne-o-skaf'-oid*). Pertaining to both the cuneiform and cuboid bones.
**cuneohysterectomy** (*kū-ne-o-his-ter-ek'-to-me*) [*cuneus*, wedge; ὑστέρα, the womb; ἐκτομή, a cutting out]. The excision of a wedge-shaped piece of uterine tissue, a procedure that has been advocated in the treatment of anteflexion of the uterus.
**cuneonavicular** (*kū-ne-o-nav-ik'-ū-lar*). Relating to the cuneiform and the navicular bones.
**cuneoscaphoid** (*kū-ne-o-skaf'-oid*). Relating to the cuneiform bones and to the scaphoid bone.
**cuneus** (*kū'-ne-us*) [L., "a wedge"]. A wedge-shaped convolution on the median aspect of the occipital lobe. Syn., *cuneate lobule; internal occipital lobule; lobulus cerebri cuneatus; lobulus cerebri occipitalis (internus); lobulus cuneatus; lobulus cuneiformis; lobus cuneus; lobus pyriformis; triangular lobule*. c. **cinereus**. See *ala cinerea*. c. **thalami optici intergenicularis**, that part of the thalamus lying between the external and internal geniculate bodies.
**cunicular** (*kū-nik'-ū-lar*). Furrowed.
**cuniculus** (*kū-nik'-ū-lus*) [L., "a subterranean passage"]. The burrow of the itch-mite. In the plural, *cuniculi*, the semicircular canals of the ear.
**Cunisset's test for bile-pigments** (*koo-ne-sa'*). Addition to the urine of half its volume of chloroform imparts a yellow color to the latter.
**cunnilinguist** (*kun-e-ling'-gwist*) [*cunnus*, the vulva; *lingere*, to lick]. A sexual pervert who practises licking the vulva.
**cunnus** (*kun'-us*) [L.]. The vulva.
**cuorin** (*kū'-or-in*). A substance which has been isolated from heart muscles; it is a diphosphatide, and is allied to lecithin.
**cup** (*kup*) [AS., *cuppe*]. 1. To bleed. 2. A cupping-glass. 3. A drinking-vessel or the contents of such a vessel. c.**s, antimonial,** cups made of antimony which impart emetic properties to the contained liquid. c., **bitter**, a drinking-cup of quassia wood. c., **Chinese**, a drinking-vessel of red arsenic which imparts cathartic properties to wine left standing in it during the night. c., **dry**, a cup for merely drawing the blood to the surface. c., **favus**, a depression in a favus-scale surrounding a hair. c., **glaucomatous**, a deep depression in the optic papilla seen in cases of glaucoma. c.**s, Montgomery's**. See *Montgomery's cups*. c. **physiological**, the normal concavity of the optic papilla. c., **retinal**, excavation of the optic disc. c., **wet**, a cup for abstracting blood through incisions in the skin.
**Cuphea** (*kū'-fe-ah*) [κῦφος, a hump, from the shape of the calyx]. A genus of plants of the order *Lythraceæ*. *C. antisyphilitica*, *C. balsamona*, *C. ingrata*, and *C. microphylla* are employed in decoction in the American tropics in the treatment of syphilis. *C. viscosissima*, a viscid annual of the United States, is a homeopathic remedy used in the treatment of cholera infantum.
**cuphosis** (*kū-fo'-sis*). See *kyphosis*.
**cupola** (*kū'-po-lah*) [L., "a dome"]. The dome-shaped extremity of the canal of the cochlea; also the summit of a solitary gland of the small intestine.
**cupped** (*kupt*). Having the upper surface depressed; applied to the coagulum of blood after phlebotomy. c. **disc**, excavation of the optic papilla, normally present in slight degree, but pathological if excessive.
**cupper** (*kup'-er*). One who practices cupping.
**cupping** (*kup'-ing*) [*cup*]. 1. A method of blood-derivation by means of the application of cupping-glasses to the surface of the body. 2. The formation of a cup-like depression. c., **dry**, a form of counter-irritation in which the blood is drawn to the surface by means of a cup. This is used mainly in inflammatory affections of the lung. c.-**glass**, a small bell-shaped glass capable of holding three to four ounces, in which the air is rarefied by heat or by exhaustion, and the glass applied to the skin, either with or without scarification of the latter. c., **wet**, the abstraction of blood after scarification.
**cupragol** (*kū'-pra-gol*). A compound of copper and albumin used in solutions of 1 to 5 % as an antiphlogistic and secretory stimulant.
**cupram** (*kū'-pram*) [*cuprum*]. A solution of copper carbonate in ammonia-water used as a fungicide.
**cuprammonic** (*kū-pram-on'-ik*). Containing copper and ammonia.
**cuprargol** (*kū-prar'-gol*). A cupronucleic acid compound, occurring as a gray powder, slowly soluble in water. It is used in the treatment of conjunctivitis by instillation of a 1 to 5 % solution once or several times daily; in trachoma used as an astringent in 20 % aqueous solution.
**cuprate** (*kū'-prāt*). A salt of cupric acid.
**cupratin** (*kū'-pra-tin*). A preparation of copper albuminoid, similar to ferratin.
**cuprea bark** (*kū'-pre-ah*). The bark of certain species of *Remijia*. It affords quinine and the associated alkaloids.
**cupreine** (*kū'-pre-in*), $C_{19}H_{22}N_2O_2$. An alkaloid derived from cuprea bark.
**cuprene** (*kū'-prēn*), $C_7H_8$. A nonvolatile insoluble hydrocarbon; a yellow, bulky solid consisting of matted filaments resembling amadou in appearance, obtained by passing a current of pure acetylene gas over bright copper filings.
**cuprescent** (*kū-pres'-ent*). Having the appearance of copper.
**cupressin** (*kū-pres'-in*). Cypress oil.
**cupriaseptol** (*kū-pre-ah-sep'-tol*). Copper sulphocarbolate.
**cupric** (*kū'-prik*). Containing copper as a bivalent element. c. **oxide**. See *copper oxide, black*.
**cuprocitrol** (*kū-pro-sit'-rol*). A copper and citrol derivative used in trachoma in 5 to 10 % salve.
**cuprohemol** (*kū-pro-he'-mol*). A compound of copper and hemol used in tuberculosis. Dose 1½-2 gr. (0.1-0.13 Gm.).
**cuprol** (*kū'-prol*). See *copper nucleinate*.
**cuprosopotassic** (*kū-pro-so-po-tas'-ik*). Relating to a combination of cuprous salt and potassium.
**cuprosulphate** (*kū-pro-sul'-fāt*). A double sulphate containing copper.
**cuprotartrate** (*kū-pro-tar'-trāt*). A combination of copper and tartaric acid.
**cuprous** (*kū'-prus*). Containing copper as a univalent element. c. **oxide**. See *copper oxide, red*.
**cuprum** (*kū'-prum*) [L.]. Copper. See *copper*.
**cupula** (*kū'-pū-lah*) [L., "a little cup"]. An invisible substance on the cristæ acusticæ that on the application of fixation fluids coagulates and becomes visible. c. **terminalis**. See *membrana tectoria*.
**cupular, cupulate** (*kū'-pū-lar, -lāt*). Cup-shaped.
**cupule** (*kū'-pūl*) [*cupula*, a little cup]. In biology, a cup-shaped organ, as an acorn-cup; or the sucking apparatus of an insect's foot.
**curaçao, curaçoa** (*kū-ra-so'*) [Dutch island of *Curaçao*, north of Venezuela]. A cordial or elixir prepared from brandy, and flavored, principally with orange-peel. It is used as a vehicle for certain medicines.
**curage** (*kū-rahzh*) [Fr.]. 1. Curettage; cleansing of the eye or of an ulcerated or carious surface. 2. A term used by some authorities for clearing the uterine cavity by means of the finger, as distinguished from the use of the curet.
**curangin** (*kū-ran'-jin*). A glucoside obtained from *Curanga amara*. It is used as a febrifuge in India and is similar in action to digitalis.
**curara, curare, curari** (*kū-rah'-rah, -re*) [S. A.]. Woorara. A vegetable extract obtained from *Paullinia curare* and certain members of the *Strychnos*

**family.** It is a powerful paralyzant of the motor nerves and of the voluntary muscles. Its alkaloid is *curarine*, $C_{18}H_{18}N$. Dose $\frac{1}{120}$ gr. (0.0006 Gm.) hypodermatically. It is used in South America and elsewhere as an arrow-poison. Toxic doses cause death by paralysis of the organs of respiration. It has been reported effectual in cases of hydrophobia and tetanus. The initial dose is $\frac{1}{10}$ gr. (0.0065 Gm.) given hypodermatically in a watery solution; this may be gradually increased to $\frac{1}{3}$ gr. (0.02 Gm.), but the latter dose should be given with caution.

**curarine** (*kū'-rah'-rēn*). See *curara*.

**curarization** (*kū-rah-riz-a'-shun*) [*curara*]. The state of one subjected to the full influence of curara by hypodermatic injection. Voice and power of motion are generally abolished, but not sensibility to pain. Syn., *curarism*. **c., spontaneous**, conditions of autointoxication occasioned by the paralyzing influence upon the circulation and upon the end-plates of the neuromuscular system of toxic substances produced in the body by the tetanization of the muscles.

**curarize** (*kū'-rah-rīz*). To bring a subject under the influence of curara.

**curatio** (*kū-ra'-she-o*) [*cura*, care]. The treatment and nursing of a patient.

**curative** (*kū'-rat-iv*) [*cura*, care]. Having a healing tendency.

**curcas** (*ker'-kas*). See *jatropha*.

**curcin** (*ker'-sin*). A toxalbumina analogous to ricin found in the seeds of *Jatropha curcas*.

**curcuma** (*ker'-kū-mah*) [L.]. Turmeric. The rhizome of *Curcuma longa*, of India, a plant of the *Zingiberaceæ*, with action similar to that of ginger. It contains a yellow coloring-matter, *curcumin*, $C_{14}H_{14}O_4$, and is employed as a yellow dye in pharmacy, occasionally, to color ointments and other preparations, and in chemistry its solution in alcohol is used as a test for alkalies, which turn it brown. Paper tinged with an alcoholic solution (*turmeric paper*) may be used instead.

**curcumin** (*ker'-kū-min*) [*curcuma*, saffron], $C_{14}H_{14}O_4$. The coloring-matter of turmeric; it crystallizes in orange-yellow prisms, and melts at 177° C. It dissolves in the alkalies to brownish-red salts.

**curd** (*kerd*) [ME.]. The coagulum of milk that separates on the addition of rennet or an acid to milk.

**cure** (*kūr*) [*cura*, care]. The successful treatment of a disease; also, a system of treatment, as *faith-cure, mind-cure, grape-cure, water-cure* (see *hydropathy*), *hunger-cure, rest-cure*, etc. **c., potato**, a method of treating foreign bodies in the alimentary tract by the ingestion of mashed potatoes. The bulk becomes embedded in the potato-mass that is formed.

**curettage** (*kū-ret-ahj'*). See *curettement*.

**curette, curet** (*kū-ret'*) [Fr.]. An instrument, shaped like a spoon or scoop, for scraping away exuberant or dead tissue.

**curettement** (*kū-ret'-ment*) [Fr., *curettement*]. The removal of vegetations, retained placenta, etc., by means of a curet. Syn., *curetage; cureting; curettage; curettement*.

**curetting** (*kū'-ret-ing*). See *curettement*.

**curie** (*kū'-re*) [Marie Sklodowska Curie, Polish-French physicist, 1867– ]. Unit of measurement of radium emanations, or of radio-active substances.

**curine** (*kū'-rēn*). An alkaloid obtained from curare; a microcrystalline powder slightly soluble in cold water, easily soluble in alcohol, chloroform, and dilute acids. It has no effect upon the motor nerves, but acts upon the heart.

**curled** (*kurld*). Occurring in parallel chains of wavy strands, as in colonies of anthrax bacillus.

**Curling's ulcer.** A duodenal ulcer produced by extensive burns of the skin.

**currant-jelly clot.** A soft, red clot seen post-mortem in the heart and blood-vessels.

**current** (*kur'-ent*) [*currere*, to run]. A term applied to the transference of electric force, which is likened to the flow of a liquid in a tube. **c., abterminal**, the secondary current observed at any point in the nerve or muscle on the passage of a single induction shock. **c., adterminal**, the negative current observed at any point in a nerve or muscle on the passage of a single induction shock. **c., after-**, a current produced in nervous or muscular tissue when a constant current which has been flowing through the same has been stopped. **c., alternating**, a term applied to a current which, by means of an interrupter, is alternately direct and reverse. **c., anelectrotonic**, that observed at the anode on passage of a constant current through a nerve. **c.s, angular**, those which are inclined to each other at some angle. **c., ascending**, the current formed by placing the positive electrode upon the periphery of a nerve and the negative higher up on the trunk of the nerve, or on the surface over the nerve-center in the spinal cord. **c., battery**, a galvanic current. **c., branch.** See *c., derived*. **c.-breaker**, a rheotome. **c., centrifugal**, a descending current. **c., centripetal**, an ascending current. **c.-changer**, a rheotrope. **c., combined**, that obtained by the combined action of the galvanic and faradic currents. **c.-condenser**, an apparatus for collecting the extra currents generated by an induction machine in operation which it combines to form a current of opposite direction to that of the battery current, and which upon being transmitted to the core demagnetizes it and thus increases the rapidity of the interruption and the strength of the induced current. **c., constant.** See *c., continuous.* **c., continuous**, a constant, uninterrupted current in one direction. **c., d'Arsonval**, the high potential discharge of a current-condenser through a large solenoid of wire. **c., derived**, a current drawn off by a derivation-wire from the main current. **c., descending**, one passing through a nerve centrifugally, the anode being placed proximally, the kathode, distally. **c., direct**, a current constant in direction, in contradistinction to an alternating current. **c., faradic**, the current produced by an induction-coil or by a magneto-electric machine. **c., galvanic**, a current generated by the decomposition of acidulated water by means of metallic plates. **c., induced.** See *c.s, secondary.* **c., inducing**, a primary current; one which, by its opening or closure, produces a faradic current in an adjoining circuit. Cf. *c.s, induced.* **c.s, induction, c.s, inductive.** See *c.s, secondary.* **c., interrupted**, a current that is alternately opened and closed. **c.-interrupter**, a rheotome. **c., katelectrotonic**, that observed at the kathode on passage of a constant current through a nerve. **c., labile**, a current applied while moving one or both electrodes over the surface treated. **c., magnetoelectric**, a faradic current generated by a magnet. **c., primary.** See *c., inducing.* **c., primitive**, an electric current from which a derived current has not been drawn off. **c.-regulator**, a rheostat. **c., reversed**, that produced by changing the poles. **c.-reverser**, a rheotrope. **c.s, secondary**, momentary currents produced in a coil of insulated wire, introduced within the field of another coil, when the circuit is made or broken in the second coil. **c., sinuous**, one sent in a curved line. **c., sinusoidal**, an alternating induced electric current in which the electromotive force is so varied that its rise and fall in a positive direction are immediately succeeded without a break by an exactly corresponding fall and rise in the negative direction, and the rise and fall in both directions would, if graphically illustrated, describe a sine curve. **c., spinal-cord**, an electric current applied by placing both the electrodes on the spine. **c., spinal-cord muscle**, an electric current in the application of which one electrode is placed over the spine, the other over a muscle. **c., spinal-cord nerve**, an electric current in the application of which one electrode is placed on the spine, the other on a nerve. **c., stabile**, a current applied with both electrodes in a fixed position. **c., static**, the current from a static machine. **c., uniform**, an electric current which retains the same strength throughout its application. **c., voltaic**, the continuous current.

**curriculum** (*kur-ik'-ū-lum*) [L.]. The regular course of study in a college.

**Curschmann's spirals** (*koorsh'-man*) [Heinrich Curschmann, German physician, 1846–1910]. Spiral threads of mucin contained in the small, thick pellets that are expectorated during an asthmatic paroxysm. They are supposed to be casts of the bronchioles, and contain Charcot-Leyden crystals and eosinophil cells.

**curvature** (*kur'-va-tūr*) [*curvare*, to curve]. A bending or curving. **c., angular.** See *Pott's c.* **c., compensatory**, in spinal curvature, a secondary curve, occurring as the result of the efforts of the trunk to maintain its upright position. **c., lateral**, scoliosis. **c., Pott's.** See *Pott's curvature.* **c. of spine**, a bending of the vertebral column.

**curve** (*kerv*) [*curvus*, bent]. A bending or flexure.

**c., Barnes'.** See *Barnes' curve.* **c.-basis,** a common level from which spring the ascending lines of a sphygmographic tracing. **c.** of **Carus.** See *Carus' curve.* **c., Ellis-Damoiseau's, c., Garland's S-.** See *Ellis' line.* **c., staircase-,** the myographic curve produced by repeated stimulation. **c., temperature-,** a graphic curve showing the variations of the temperature for a given period. **c.s, Traube's, c., Traube-Hering's.** See under *Traube.*
**cuscamidine** (*kus-kam'-id-ēn*) [*cusco; amide*]. An alkaloid found in cinchona bark.
**cuscamine** (*kus'-kam-ēn*). An alkaloid found in cinchona bark.
**cusco bark** (*kus'-ko bark*). A kind of cinchona, the bark of a variety of *Cinchona pubescens.*
**Cusco's speculum** (*kus'-ko*) [Edouard Gabriel Cusco, French surgeon, 1819–1894]. A bivalve vaginal speculum.
**cusconidine** (*kus-kon'-id-ēn*) [*cusco*]. An alkaloid found in cinchona bark.
**cusconine** (*kus'-ko-nēn*) [*cusco*], $C_{23}H_{26}N_2O_4$. A crystalline alkaloid found in Cinchona bark.
**Cushing's suture** (*koosh'-ing*) [Harvey Williams Cushing, American surgeon, 1869– ]. A continuous intestinal suture.
**cushion** (*koosh'-un*). In anatomy, an aggregate of adipose or elastic tissue relieving pressure upon tissues lying beneath. **c., coronary,** the cushion of the upper edge of the hoofs in solipeds. **c. of the epiglottis,** the tubercle of the epiglottis, a median elevation of the mucosa within the larynx below the epiglottis. **c., Eustachian,** a part of the posterior wall of the Eustachian tube. **c., Passavant's,** the bulging of the posterior pharyngeal wall, produced during the act of swallowing, by the upper portion of the superior constrictor pharyngis. **c., plantar,** in solipeds, a cuneiform fibrous body lying between the plantar part of the hoof and the perforans tendon. **c., sucking,** fatty pads found on the buccinator muscles in young infants.
**cusp** (*kusp*) [*cuspis*, a point]. The eminence on the crown of a tooth.
**cusparia bark** (*kus-pa'-re-ah bark*). See *angustura.*
**cuspated, cusped, cuspid, cuspidal** (*kusp-a'-ted, kuspā', kusp'-id, kusp'-id-al*). See *cuspidate.*
**cuspidate** (*kus'-pid-āt*) [*cuspis*, a point]. Tipped with a sharp, rigid point.
**cuspid teeth** (*kus'-pid*). The four teeth that have conical crowns. They are situated, one on each side, in each jaw, between the lateral incisor and first bicuspid.
**cusso** (*koos'-o*). Brayera. The *cusso* of the U. S. P. is the dried panicles of the pistillate flowers of *Hagenia abyssinica.* It contains tannic acid, a volatile oil, and a crystalline principle, *koussin*, $C_{31}H_{38}O_{10}$. It is a valuable anthelmintic against tape-worms. In large doses it produces nausea and emesis. Dose 240 gr. (16 Gm.). Syn., *cusso; cousso; kousso.* **c. infusion** of (*infusum cusso,* B. P.). Dose 4–8 oz. (118–235 Cc.).
**cutal** (*kū'-tal*). See *aluminum borotannate.*
**cutaneous** (*kū-ta'-ne-us*) [*cutis*]. Pertaining to the skin. **c. emphysema.** See *emphysema, cutaneous.* **c. reflex,** a reflex action from irritation of the skin. **c. respiration,** the transpiration of gases through the skin.
**Cuterebra noxialis** (*kū-ter-e'-brah noks-e-a'-lis*). A bot-fly belonging to the *Ostridæ,* whose larva may cause myiasis in man.
**cuticle** (*kū'-tik-l*) [*cutis*]. The epidermis or scarfskin. **c., enamel.** See *Nasmyth's cuticle.*
**cuticolor** (*kū'-ti-kul-or*) [*cutis; color*]. A term descriptive of various ointments and powders simulating the color of the skin and used in the treatment of skin diseases.
**cuticolous** (*kū-tik'-ol-us*) [*cutis; colere,* to inhabit]. Living under the skin; said of parasitic larvæ.
**cuticula** (*kū-tik'-ū-lah*) [dim. of *cutis*, a skin]. The outer, finely lamellated layer of the wall of hydatid cysts. **c. dentis,** the cuticle of a tooth; Nasmyth's membrane; the delicate horny envelope that covers the enamel of young and unworn teeth. **c. pili,** the cuticle of the root-sheath of a hair.
**cuticular** (*kū-tik'-ū-lar*) [*cuticula*]. Pertaining, resembling, or of the nature of cuticle.
**cuticularization** (*kū-tik-ū-lar-iz-a'-shun*) [*cuticula,* dim. of *cutis*, the skin]. The formation of a cuticula.
**cutification** (*kū-tif-ik-a'-shun*) [*cutis; facere,* to make]. The formation of skin.
**cutin** (*kū'-tin*) [*cutis*]. 1. In biology, cork-substance; a modification of cellulose, also called suberin. 2. A preparation of the muscular layer of the intestine of cattle. It is used for dressing wounds and as a substitute for catgut.
**cutipunctor** (*kū-te-punk'-tor*) [*cutis*, skin; *punctor,* puncturer]. An instrument for puncturing the skin.
**cutireaction** (*kū-te-re-ak'-shun*) [*cutis*, skin; *reaction*]. See *Pirquet's reaction, Moro's test.*
**cutis** (*kū'-tis*) [L.]. The derma or true skin. **c. ærea,** the bronze color of the skin in Addison's disease. **c. anserina.** See *goose-flesh.* **c. laxa.** See *dermatolysis.* **c. pendula,** a flabby skin. **c. testacea,** a variety of seborrhea in which the trunk and extensor surfaces of the extremities are covered with large, thick plates of greasy, inspissated sebum, usually greenish or blackish, from accumulation of dirt upon them. **c. unctuosa.** Synonym of *seborrhea.* **c. vera,** the corium.
**cutisector** (*kū-te-sek'-tor*) [*cutis; sector,* a cutter]. An instrument for taking small sections of skin from the living subject.
**cutitis** (*kū-ti'-tis*). Inflammation of the skin; dermatitis.
**cutization** (*kū-tiz-a'-shun*) [*cutis,* skin]. The acquirement of the characters of true skin by exposed mucous membrane.
**cut-off** (*kut'-off*). A device for cutting off the flow of a gas or liquid or electric current. **cut-off muscle,** a popular designation for the compressor urethræ muscle. See *muscles, table of.*
**cutol** (*kū'-tol*). See *aluminum borotannate.*
**Cuvier's canals** (*koo'-ve-a*) [Georges Leopold Chrétien Frédéric Dagobert Cuvier, French scientist, 1769–1832]. In the embryo, two short vessels opening into the common trunk of the omphalomesenteric veins, each being formed by the union of two veins, the anterior cardinal, or jugular, and the posterior cardinal veins. The right one becomes the superior vena cava; the left one disappears.
**Cy.** Abbreviation for cyanogen; sometimes used instead of the chemical symbol CN.
**cyanacetyl** (*si-an-as'-et-il*). See *acetyl isocyanide.*
**cyanalcohol** (*si-an-al'-ko-hol*) [*κύανος,* dark-blue; *alcohol*]. Cyanhydrin; a substance obtained by the union of an aldehyde with hydrocyanic acid.
**cyanaldehyde** (*si-an-al'-de-hīd*) [*κύανος,* dark-blue; *aldehyde*]. A substitution-compound of acetic aldehyde.
**cyanate** (*si'-an-āt*). A salt of cyanic acid.
**cyanemia, cyanæmia** (*si-an-e'-me-ah*). Bluish blood, due to imperfect oxygenation.
**cyanephidrosis** (*si-an-ef-id-ro'-sis*) [*κύανος,* blue; *ἐφίδρωσις,* sweat]. Blue sweat.
**cyanformic** (*si-an-form'-ik*). Containing formic acid and cyanogen.
**cyanhematin** (*si-an-hem'-at-in*) [*κύανος,* blue; *αἷμα,* blood]. A substance produced by adding a solution of cyanide of potassium to a solution of blood, and heating gently for some time.
**cyanhemoglobin** (*si-an-hem-o-glo'-bin*). A compound of hydrocyanic acid with hemoglobin formed in cases of poisoning with this acid. It gives the blood a bright red color.
**cyanhidrosis** (*si-an-hid-ro-sis*). Same as *cyanephidrosis.*
**cyanhydric acid.** Same as *hydrocyanic acid.*
**cyanhydrin** (*si-an-hi'-drin*). See *cyanalcohol.*
**cyanhydrosis** (*si-an-hid-ro'-sis*). See *cyanephidrosis.*
**cyanic** (*si-an'-ik*) [*κύανος,* blue]. 1. Blue or bluish. 2. Relating to or containing cyanogen. **c. acid.** See *acid, cyanic.*
**cyanide** (*si'-an-īd*) [*κύανος,* blue]. Any compound of cyanogen with a metal or a radical. Most of the cyanides are actively poisonous.
**cyanidrosis** (*si-an-id-ro'-sis*). Same as *cyanephidrosis.*
**cyano-** (*si-an-o-*) [*κύανος,* blue]. A prefix meaning blue.
**cyanochroia** (*si-an-o-kroi'-ah*). [*cyano-;* χροία, color]. Cyanosis.
**cyanoderma** (*si-an-o-der'-mah*) [*cyano-;* δέρμα, skin]. Cyanosis.
**cyanoform** (*si-an'-o-form*), $CH(CN)_3$. A compound occurring in small needles obtained by heating chloroform and potassium cyanide and alcohol.
**cyanogen** (*si-an'-o-jen*) [*cyano-;* γεννᾶν, to produce]. A radical having the structure CN, an acid compound of carbon and nitrogen, existing as a colorless, combustible gas; it is exceedingly poisonous. Syn.,

**prussin. c. iodide,** a poisonous crystalline substance produced by combining cyanogen and iodine.

**cyanomycosis** (*si-an-o-mi-ko'-sis*) [*cyano-;* μύκης, fungus]. The production of blue pus, or of pus charged with *micrococcus pyocyaneus.*

**cyanopathy** (*si-an-op'-a-the*). See *cyanosis.*

**cyanophil, cyanophile** (*si-an'-o-fil*) [*cyano-;* φιλεῖν, to love]. Auerbach's term for the blue-staining nuclear substance of cells of plants and animals.

**cyanophilic, cyanophilous** (*si-an-o-fil'-ik, -of'-il-us*). Having an especial affinity for blue or green dyes.

**cyanophyl** (*si-an'-o-fil*) [*cyano-;* φύλλον, leaf]. Fremy's name for a supposed blue constituent of chlorophyl.

**cyanopia, cyanopsia** (*si-an-o'-pe-ah, -op'-se-ah*) [*cyano-;* ὤψ, eye]. A perverted state of the vision rendering all objects blue.

**cyanosed** (*si'-an-ōsd*). Affected with cyanosis.

**cyanosis** (*si-an-o'-sis*) [κύανος, blue]. A bluish discoloration of the skin from deficient oxidation of the blood, caused by local or general circulatory disturbances. **c., congenital,** blue disease; cyanosis due to a congenital lesion of the heart or of the great vessels. **c., local,** the preferred term for local anemia, digiti mortui, regional ischemia, or local syncope.

**cyanospermia** (*si-an-o-sper'-me-ah*) [*cyano-;* σπέρμα, sperm]. Semen of a bluish tint.

**cyanotic** (*si-an-ot'-ik*) [κύανος, blue]. Relating to or affected with cyanosis.

**cyanurea** (*si-an-ū'-re-ah*), $C_2H_3N_3O$. An amorphous body obtained from urea by action of cyanogen iodide with heat.

**cyanuric** (*si-an-ū'-rik*). Relating to or containing cyanurea.

**cyanurin** (*si-an-ū'-rin*) [κύανος, blue; οὖρον, urine]. Uroglaucin or urine-indigo; indigo found in the urine in cystitis and in chronic kidney-diseases; it is also occasionally found in apparent health.

**cyasma** (*si-as'-mah*) [κυεῖν, to be pregnant; pl., *cyasmata*]. The peculiar freckle sometimes seen upon pregnant women.

**cyathus** (*si'-a-thus*) [κύαθος, cup]. 1. A cup or glass. 2. The canal of the infundibulum cerebri.

**cyclarthrosis** (*sik-lar-thro'-sis*) [κύκλος, a circle; ἄρθρωσις, a joint]. A circular or rotatory articulation.

**cyclasterion scarlatinale** (*si-klas-te'-re-on skar-lat-in-a'-le*). A supposed protozoal parasite found in the cutaneous cells in scarlet fever.

**cycle** (*si'-k'l*) [κύκλος, a circle]. A period in which a round of operations or events is repeated; a succession of events or symptoms. **c., aberrant,** the establishment of communication between the pulmonary and bronchial vessels from congestion due to mitral stenosis. **c., cardiac,** the complete cardiac movements embracing the systolic and diastolic movements, with the interval between them. **c., cardiacovascular,** the circuit of the blood through the organism. **c. of generation, c., generations,** Haeckel's term for the successive changes through which an individual passes from its birth to the period when it is capable of reproducing its kind. **c. of Golgi.** See under *Golgi.* **c. of Ross.** See under *Ross.*

**cyclencephalus** (*si-klen-sef'-al-us*). See *cyclocephalus.*

**cyclic** (*si'-klik*) [*cycle*]. 1. Having cycles or periods of exacerbation or change; intermittent. 2. Having a self-limited course, as certain diseases. **c. albuminuria.** See *Albuminuria, cyclic.* **c. compound,** in chemistry, an organic compound belonging to the closed-chain series. **c. insanity.** See *insanity, circular.*

**cyclitic shell** (*si-klit'-ik*) [κυκλικός, circular]. A coherent, solid, fibrous mass of exudate, completely enveloping the crystalline lens.

**cyclitis** (*si-kli'-tis*) [κύκλος, a circular body; ιτις, inflammation]. Inflammation of the ciliary body, manifested by a zone of congestion in the sclerotic coat surrounding the cornea. It may be serous, plastic, or suppurative. *Iridocyclitis* is the involvement of both iris and ciliary body in the inflammatory process.

**cyclo-** (*si-klo-*) [κύκλος, a circle]. A prefix meaning relating to a circle or to the ciliary body.

**cyclocephalus** (*si-klo-sef'-al-us*) [*cyclo-;* κεφαλή, head]. A species of single autositic monsters characterized by a more or less complete absence of the olfactory organs, together with an intimate union of imperfectly developed or rudimentary visual organs, situated in the median line.

**cyclochoroiditis** (*si-klo-ko-roid-i'-tis*) [*cyclo-; choroiditis*]. Combined inflammation of the ciliary body and the choroid.

**cyclodialysis** (*si-klo-di-al'-is-is*). Detachment of the ciliary body from the sclera. It is performed purposely to effect reduction of intraocular tension in certain cases of glaucoma, especially when iridectomy is contraindicated or has failed.

**cyclophoria** (*si-klo-fo'-re-ah*) [*cyclo-;* φερεῖν, to bear]. 1. A term applied to the circulation of the fluids of the body. 2. An insufficiency of the oblique muscles of the eye, giving the eyes a tendency to roll outward or inward, so that the naturally vertical meridians would diverge either at the upper or lower extremities.

**cyclopia** (*si-klo'-pe-ah*). See *synophthalmia.*

**cycloplegia** (*si-klo-ple'-je-ah*) [*cyclo-;* πληγή, a stroke]. Paralysis of the ciliary muscle of the eye.

**cycloplegic** (*si-klo-ple'*-jik). 1. Relating to cycloplegia. 2. A drug which paralyzes the ciliary muscle.

**cyclops** (*si'-klops*) [*Cyclops*, a mythological monster with one eye located in the middle of the forehead]. A congenital malformation consisting in a fusion of the two eyes into one. **c. quadricornis** (*kwad-re-kor'-nis*). A fresh-water crutacean (copepod) which serves as intermediary host in the development of *Filaria medinensis.*

**cyclotherapy** (*si-klo-ther'-a-pe*) [κύκλος, wheel; θεραπεία, therapy]. The use of the bicycle as a therapeutic measure.

**cyclothermia** (*si-klo-thi'-me-ah*) [κύκλος, circle; θυμός, mind]. Cyclic insanity. See *insanity.*

**cyclothymiac** (*si-klo-thi'-me-ak*). Affected with cyclothymia; a person so affected.

**cyclotome** (*sik'-lo-tōm*) [κύκλος, circle; τομός, cutting]. A kind of knife for performing cyclotomy.

**cyclotomy** (*si-klot'-o-me*) [*cyclo-;* τομή, section]. An operation for the relief of glaucoma, consisting of an incision through the ciliary body.

**cydonium** (*si-do'-ne-um*) [κυδώνιον, a quince]. Quince-seed. The seeds of *Cydonia vulgaris,* employed mainly for the mucilage contined in the covering, which consists of a compound of gum and glucose, and is a bland demulcent; it is also used as a hair-dressing.

**cyematocardia** (*si-em-at-o-kar'-de-ah*) [κύημα, a fetus; κάρδια, a heart]. Fetal rhythm of the heart-sounds.

**cyesiognosis** (*si-e-se-og-no'-sis*) [κύησις, pregnancy; γνῶσις, knowledge]. The diagnosis of pregnancy.

**cyesiology** (*si-e-se-ol'-o-je*) [κύησις, pregnancy; λόγος, treatise]. The science of gestation in its medical aspects.

**cyesis** (*si-e'-sis*) [κύησις, pregnancy]. Pregnancy.

**cyetic** (*si-et'-ik*) [*cyesis*]. Relating to pregnancy. **Cyl.** Abbreviation for *cylinder,* or *cylindrical lens.*

**cyltomy** (*sil-ik-ot'-o-me*) [κύλιξ, cup; τέμνειν, to cut]. Surgical incision of the ciliary muscle.

**cylinder** (*sil'-in-der*) [κύλινδρος, a cylinder]. 1. An elongated body of the same transverse diameter throughout and circular on transverse section. 2. See *cast.* 3. A cylindrical lens. **c.s, Bence Jones'.** See *Bence-Jones' cylinders.*

**cylindrenchyma** (*sil-in-dreng'-kim-ah*) [κύλινδρος, a cylinder; ἔγχυμα, an infusion]. A tissue composed of cylindrical cells.

**cylindric, cylindrical** (*sil-in'-drik, sil-in'-drik-al*) [κύλινδρος, a cylinder]. Pertaining to or like a cylinder. **c. lens.** See *lens.*

**cylindroadenoma** (*sil-in-dro-ad-en-o'-mah*). An adenoma containing cylindrical masses of hyaline matter.

**cylindrocephalic** (*sil-in-dro-sef-al'-ik*). Affected with cylindrocephaly.

**cylindrocephaly** (*sil-in-dro-sef'-a-le*) [κύλινδρος, cylinder; κεφαλή, head]. A cylindrical formation of the skull.

**cylindrodendrite** (*sil-in-dro-den'-drīt*). See *paraxon.*

**cylindroid** (*sil'-in-droid*) [*cylinder;* εἶδος, likeness]. A name given to a mucous cast frequently found in the urine in cases of mild irritation of the kidney. Cylindroids are ribbon-like forms, usually of great length, and of about the same diameter as renal casts. They may assume various shapes. One extremity is usually pointed and may be drawn out into a long tail.

**cylindroma** (*sil-in-dro'-mah*) [*cylinder;* ὄμα, a tumor]. A myxosarcoma in which the degeneration is confined to areas surrounding the blood-vessels.
**cylindrosarcoma** (*sil-in-dro-sar-ko'-mah*). A tumor containing both cylindromatous and sarcomatous elements.
**cylindrosis** (*sil-in-dro'-sis*) [κύλενδρουν, to roll]. A variety of bony articulation, the bone being rolled upon itself to form a canal and than a suture, as in the cranium.
**cylindruria** (*sil-in-dru'-re-ah*) [*cylinder;* οὖρον, urine]. The discharge of urine containing hyaline casts or cylindroids.
**cyllin** (*sil'-in*). Trade name of a preparation of creolin, used as a disinfectant.
**cyllopodia** (*sil-o-po'-de-ah*) [κυλλόπους, crook-footed]. The state of having a distorted foot, especially talipes varus.
**cyllosis** (*sil-o'-sis*) [κύλλωσις, crippled]. Club-foot; lameness from deformity.
**cyllosoma** (*sil-o-so'-mah*) [κυλλός, lame; σῶμα, a body; *pl., cyllosomata*]. A variety of single autositic monsters of the species celosoma, in which there is a lateral eventration occupying principally the lower portion of the abdomen, with absence or imperfect development of the lower extremity on that side occupied by the eventration.
**cyllosomus** (*sil-o-so'-mus*) [κυλλός, crooked; σῶμα, body]. A monster characterized by an eventration in the side of the lower abdominal region and imperfect development of the corresponding leg.
**cyllum** (*sil'-um*) [κυλλός, crooked]. Knock-knee.
**cymarin** (*si'-mar-in*). Trade name of a preparation of *Apocynum canabinum;* it is used like digitalis.
**cymba** (*sim'-bah*) [κύμβη, a boat]. In biology, a boat-shaped sponge-spicule. c. conchæ, the upper part of the concha of the ear, above the root of the helix.
**cymbiform** (*sim'-bif-orm*) [*cymba,* boat; *forma,* shape]. In biology, boat-shaped. c. bone, the scaphoid bone.
**cymbocephalic, cymbocephalous** (*sim-bo-sef-al'-ik, sim-bo-sef'-al-us*) [κύμβη, boat; κεφαλή, head]. Exhibiting cymbocephaly.
**cymbocephaly** (*sim-bo-sef'-a-le*) [κύμβη, boat; κεφαλή, head]. The condition of having a boat-shaped head.
**cyme** (*sim*) [κῦμα, a young sprout]. In biology, a loose flower-cluster on the determinate or centrifugal plan.
**cymene** (*si'-mēn*) [*cyminum,* cumin], $C_{10}H_{14}$. A hydrocarbon that occurs, together with cumic aldehyde in Roman caraway oil (from *Cuminum cyminum*), and in other ethereal oils.
**cymenyl** (*si'-men-il*), $C_{10}H_{13}$. The univalent radical found in cymene.
**cymic** (*si'-mik*). Relating to cymene.
**cymol** (*si'-mol*). See *cymene*.
**cymose** (*si'-mōs*) [κῦμα, a young sprout]. In biology, pertaining to a cyme; bearing cymes.
**cynanche** (*sin-ang'-ke*) [κύων, dog; ἄγχειν, to strangle]. An old name for any acute affection of the throat, as diphtheria, croup, tonsillitis, etc., in which the patient struggles for breath (as a panting dog). c. dysarthritica, sore throat sometimes resulting in abscess occurring during the course of an attack of arthritis. Syn., *arthritic angina.* c. maligna, a fatal form of sore throat. c. sublingualis, inflammation of the connective tissue of the floor of the mouth. c. suffocativa, Synonym of *croup*. c. tonsillaris. See *quinsy*.
**cynanthropia** (*sin-an-thro'-pe-ah*) [κύων, dog; ἄνθρωπος, a man]. A mania in which the patient believes himself a dog, and imitates the actions of one.
**cynapine** (*si'-nap-ēn*). A poisonous alkaloid obtained from the seeds of *Æthusa cynapium*.
**cyniatria** (*si-ne-a'-tri-ah*) [κύων, dog; ἰατρεία, medication]. The science or treatise of the diseases of the dog and their treatment.
**cynic** (*sin'-ik*) [κυνικός, dog-like]. Pertaining to a dog. c. spasm, a contraction of the facial muscles upon one side so as to expose the teeth, in the manner of an angry dog.
**cynobex hebetica** (*si'-no-beks he-bet'-ik-ah*) [κύων, dog; βήξ, cough; ἡβητικός, youthful]. The convulsive cough of puberty; a loud, dry, barking cough that often attacks boys and girls at puberty.
**cynocephalous** (*si-no-sef'-al-us*) [κύων, dog; κεφαλή, the head]. With the head dog-shaped.
**cynodontes** (*si-no-don'-tēz*) [κύων, dog; ὀδούς,

tooth]. The canine teeth, so called from their resemblance to the teeth of a dog.
**Cynoglossum** (*si-no-glos'-um*) [κύων, dog; γλῶσσα, the tongue]. A genus of boraginaceous plants. The powdered root, leaves, and flowers of *C. officinale,* hound's-tongue, are recommended as a cure for ulcerated epithelioma; application twice daily.
**Hydrophobia;** rabies.
**cynolyssa** (*si-no-lis'-ah*) [κύων, dog; λύσσα, rabies]. Hydrophobia; rabies.
**cynomania** (*si-no-ma'-ne-ah*) [κύων, dog; μάνια, madness]. Term proposed by Ellis (1899) as a substitute for lyssa or rabies.
**cynophobia** (*si-no-fo'-be-ah*) [κύων, a dog; φόβος, fear]. 1. Morbid fear of dogs. 2. Imaginary hydrophobia.
**cynorexia** (*si-no-reks'-e-ah*) [κύων, dog; ὄρεξις, appetite]. Canine voracity; bulimia.
**cynospasmus** (*si-no-spaz'-mus*) [κύων, dog; σπασμός, spasm]. Same as *cynic spasm*.
**cynurenic acid** (*sin-ū-ren'-ik*) [see *cynurin*], $C_{20}H_{14}N_2O_5 + 2H_2O$. A crystalline acid found in dog's urine. It is a decomposition-product of proteids. On heating it cynurin is evolved.
**cynurin** (*sin'-ū-rin*) [κύων, dog; οὖρον, urine], $C_{13}H_{14}N_2O_3$. A base from cynurenic acid.
**Cyon's nerve** (*se-on(g)*) [Elie de *Cyon,* Russian physiologist, 1843- ]. The depressor nerve of the heart, derived from the pneumogastric.
**cyophoria** (*si-o-fo'-re-ah*) [κύος, fetus; φέρειν, to carry]. Pregnancy; gestation.
**cyophorin** (*si-of'-or-in*). See *gravidin*.
**cyopin** (*si'-o-pin*) [κυάνεος, blue; πύον, pus]. The coloring-matter found in blue-pus.
**cyotrophy** (*si-ot'-ro-fe*) [κύος, fetus; τροφή, nourishment]. Nutrition of the fetus.
**Cyperus** (*si-pe'-rus*) [κύπειρος, a sweet-smelling marsh-plant]. A genus of sedges. *C. articulatus,* adrue, a species of South America, is antiemetic and tonic. Dose of *fluidextract* 10-30 min. (0.6-1.8 Cc.). *C. rotundus* is a tropical species, the tubers of which are tonic and stimulant and are used in treatment of cholera.
**cyphosis** (*si-fo'-sis*). See *kyphosis*.
**cyphotic** (*si-fot'-ik*). See *kyphotic*.
**cypress oil** (*si'-pres*). Oleum cupressi. An oil distilled from the leaves and young branches of *cupressus sempervirens*.
**cyprian** (*sip'-re-an*) [Island of *Cyprus,* the original source of copper]. Containing copper.
**cypridol** (*sip'-rid-ol*). A 1 % solution of nascent mercuric iodide in aseptic oil; it is used in syphilis (dose 3 gr.-0.2 Gm.) and as an application in skin diseases.
**cypridopathy** (*sip-rid-op'-ath-e*) [Κύπρις, Venus; πάθος, disease]. An adenopathy of venereal origin.
**cypripedin** (*sip-rip-e'-din*) [Κύπρις, Venus; πόδιον, a slipper]. A precipitate from the tincture of *Cypripedium pubescens;* antispasmodic, nervine, narcotic. Dose ½ to 3 grains.
**Cypripedium** (*sip-re-pe'-de-um*) [Κύπρις, Venus; πόδιον, a slipper]. Lady's-slipper. The roots of *C. pubescens* and *C. parviflorum,* American valerian, the properties of which are due to a volatile oil and an acid. It is an antispasmodic and stimulant tonic, used instead of valerian, which it resembles. Dose of the *fluidextract* 10-30 min. (0.6-1.8 Cc.).
**cypriphobia** (*sip-rif-o'-be-ah*) [Κύπρις, Venus; φόβος, fear]. 1. Fear of sexual intercourse. 2. Fear of contracting venereal disease.
**Cyprus fever** (*si'-prus fev-er*). Same as *Malta fever.* C. vitriol. Same as *blue vitriol.*
**cyrtocephalus** (*sir-to-sef'-al-us*) [κυρτός, convex; κεφαλή, head]. Having a short head.
**cyrtocoryphus** (*sir-to-kor'-if-us*) [κυρτός, convex; κορυφή, the crown of the head]. Lissauer's term for a skull with a parietal angle between 122° and 132°.
**cyrtograph** (*sir'-to-graf*) [κυρτός, curved; γράφειν, to record]. A recording cyrtometer.
**cyrtoid** (*sir'-toid*) [κυρτός, curved; εἶδος, likeness]. Hump-like.
**cyrtoma** (*sir-to'-mah*) [κύρτωμα, a curving; *pl., cyrtomata*]. A convexity, especially one that is abnormal.
**cyrtometer** (*sir-tom'-et-er*) [κυρτός, curved; μέτρον, a measure]. An instrument adapted for measuring curves. One form is used to locate the fissures of the brain.
**cyrtometopus** (*sir-to-met-o'-pus*) [κυρτός, convex; μέτωπον, the forehead]. Lissauer's term for a skull in which the angle formed by lines connecting the

**nasion** with the bregma and the metopion is between 120° and 130.5°.

**cyrtometry** (*sur-tom'-et-re*) [κυρτός, curved; μέτρον, a measure]. The measurement of the curves of the body.

**cyrtonosus** (*sir-ton'-o-sus*) [κυρτός, curved; νόσος, a disease]. Synonym for rhachitis.

**cyrtopisthocranius** (*sir-to-pis-tho-kra'-ne-us*) [κυρτός, curved; ὀπισθοκράνιον, the occiput]. Lissauer's term for a skull in which the angle of the summit of the occiput is between 117° and 140°.

**cyrtosis** (*sir-to'-sis*) [κυρτός, curved]. 1. Spinal curvature. 2. Any deformity of the bones.

**cyrturanus** (*sir-tū-ra'-nus*) [κυρτός, curved; οὐρανός, the roof of the mouth]. Lissauer's term for a skull in which the angle of the roof of the mouth is between 132° and 147.5°.

**cyst** (*sist*) [κύστις, a pouch]. 1. A bladder. 2. A cavity containing fluid and surrounded by a capsule. **c.s, adgenic,** congenital dermoid cysts adhering to the genial tubercles. **c.s, adhyoid,** dermoid cysts adherent to the hyoid bone. **c., adventitious,** one inclosing a foreign body. **c., air-,** one containing gas. **c., allantoic,** cystic dilation of the urachus. **c., apoplectic,** one inclosing a cerebral effusion of blood. **c., arachnoid,** a meningeal hematoma. **c., atheromatous.** See *atheroma* (1). **c.s, Baker's.** See *Baker's cysts*. **c., blood-.** See *hematoma*. **c., Boyer's.** See *Boyer's cyst*. **c., branchial,** one formed from incomplete closure of a branchial cleft in an embryo. **c.** of the broad ligament, one originating in the ovary and developing between the layers of the broad ligament. **c., butter-,** a cystic tumor of the mammary gland containing semisolid contents of yellowish-brown color and of buttery consistence, that may harden with exposure to the air. **c., chyle,** one in the mesentery containing chyle. **c., colloid,** a cyst with jelly-like contents. **c., compound.** See *c., multilocular.* **c., conjunctival,** a rare congenital cyst, transparent and of the size of a pea, occurring near the corneal margin. **c., Cowperian,** **c., Cowper's.** See under *Cowper*. **c.s, crab's-eye,** small vesicles which appear over Heberden's nodosities. **c., cutaneous.** See *c., dermoid*. **c., daughter-,** any one of the small cysts developed by secondary growth from the walls of a large cyst. **c., dentigerous,** one containing teeth. **c., dermoid,** a congenital cyst containing bone, hair, teeth, etc. **c., distention,** a normal serous cavity distended with a collection of watery fluid. **c., echinococcus-,** a cyst formed in various tissues and organs of man by the larva of the *Tænia echinococcus* of the dog, taken into the stomach. **c., extravasation,** a cyst formed by the encapsulation of a hemorrhage or other fluid into the tissues. **c., exudation.** See *c., extravasation*. **c., false.** See *c., adventitious*, and *c., exudation*. **c., follicular,** one due to the occlusion of the duct of a small follicle or gland. **c., Gartnerian,** a cystic tumor originating in Gartner's duct. **c., hydatid.** See *c., echinococcus-*. **c., intraligamentous.** See *c. of the broad ligament*. **c., involution-,** multiple cystic dilatation of the milk-ducts after the menopause. **c., Kobelt's.** See under *Kobelt*. **c., Meibomian.** See *chalazion*. **c., Morgagnian.** See *Morgagni, hydatid of*. **c., mucoid,** a retention cyst in a mucous follicle or in the duct of a muciparous gland. **c., mucous,** a retention cyst containing mucus. **c., multilocular,** one composed of many separate compartments. **c.s, Nabothian,** small retention cysts formed by the Nabothian follicles. **c., neural,** a cyst-like distention of a lymph-space of the brain or spinal cord. **c., nevoid,** one with vascular walls. **c., pilous,** **c., pilocystic.** See *pilonidal fistula*. **c., proligerous,** **c., recurring proliferous,** a cystic adenosarcoma; one that has undergone cystic degeneration. **c., renal,** a cyst-like dilatation of the kidney. **c., retention,** one that is due to the retention of the secretion of a gland, in consequence of closure of the duct, as in mucous or sebaceous cysts. **c., ricegrain, cystic distention** of a synovial sheath containing rice-seed bodies. **c., sebaceous,** a retention cyst of a sebaceous gland. **c., secondary,** a cyst within a cyst. **c., serous,** one containing transparent watery fluid. **c., softening,** one due to encapsulation of the fluid after liquefactive necrosis. **c., sterile,** a true hydatid cyst which fails to reproduce. **c., sublingual.** See *ranula*. **c., true,** an abnormal cyst not formed by the dilatation of some canal or cavity. **c., umbilical,** a congenital cyst in the umbilical region said to be due to shutting-off of a portion of the fetal stomach. **c., unilocular,** one having but a single cavity. **c., urinary,** a thin-walled cyst of the cortical substance of the kidney, projecting beneath the capsule and containing a clear yellow fluid. **c.-worm.** See *cysticercus*.

**cystadenoma** (*sist-ad-en-o'-mah*) [*cyst; adenoma*]. 1. An adenoma containing cysts. 2. Adenoma of the bladder. **c. papilliferum,** an adenoma containing cysts with papillæ on the inner aspect of the cyst-walls.

**cystadenosarcoma** (*sist-ad-en-o-sar-ko'-mah*). Combined cystadenoma and sarcoma. See *cyst, proligerous*.

**cystalgia** (*sist-al'-je-ah*) [*cyst;* ἄλγος, pain]. Pain in the bladder.

**cystamine** (*sist'-am-ēn*). A compound of formaldehyde and ammonia, used as a bactericide and antiseptic in cystitis and gout. Dose 5–10 gr. (0.33–0.66 Gm.).

**cystanastrophe** (*sist-an-as'-tro-fe*). See *inversion of bladder*.

**cystatrophia** (*sist-at-ro'-fe-ah*) [*cyst; atrophy*]. Atrophy of the urinary bladder.

**cystauchenitis** (*sist-aw-ken-i'-tis*) [*cyst;* αὐχήν, neck; ιτις, inflammation]. Inflammation of the neck of the bladder; trachelocystitis.

**cystauchenotomy** (*sist-aw-ken-ot'-o-me*) [*cyst;* αὐχήν, neck; τέμνειν, to cut]. A surgical incision into the neck of the bladder.

**cystauxe** (*sist-awks'-e*) [*cyst;* αὔξη, increase]. Thickening of the bladder.

**cystectasia, cystectasy** (*sist-ek-ta'-se-ah, sist-ek'-tas-e*) [*cyst;* ἔκτασις, a stretching out]. 1. Dilatation of the bladder. 2. Surgical dilatation of the urethra or of its prostatic portion in certain operations for stone.

**cystectomy** (*sist-ek'-to-me*) [*cyst;* ἐκτομή, excision]. 1. Excision of the cystic duct. 2. Excision of gall bladder, or part of the urinary bladder. 3. Removal of a cyst.

**cystein** (*sist'-e-in*) [*cyst*], $C_3H_7NO_2S$. A compound obtained by reducing cystin; it is a crystalline powder, soluble in water, yielding an indigo-blue color with ferric chloride; in the air it rapidly oxidizes to cystin. **c., reaction for.** See *Andreasch's reaction for cystein*.

**cystelcosis** (*sist-el-ko'-sis*) [*cyst;* ἕλκωσις, ulceration]. Ulceration of the bladder.

**cystencephalia** (*sist-en-sef-al'-e-ah*) [*cyst;* ἐγκέφαλος, the brain]. The state or condition of a cystencephalus.

**cystencephalus** (*sist-en-sef'-al-us*) [*cyst;* ἐγκέφαλος, the brain]. A form of monstrosity in which the brain is replaced by a cyst-like structure.

**cystendesis** (*sis-ten-de'-sis*) [*cyst;* ἔνδησις, suturing]. Suture of a wound in the gall-bladder or in the urinary bladder.

**cysterethism** (*sist-er'-eth-ism*) [*cyst;* ἐρεθισμός; irritation]. Irritability of the bladder.

**cysthitis** (*sis-thi'-tis*) [κύσθος, vulva; ιτις, inflammation]. Inflammation of the vulva.

**cysthus** (*sis'-thus*) [L.]. 1. The vulva. 2. The anus.

**cysthypersarcosis** (*sist-hi-per-sar-ko'-sis*) [*cyst;* ὑπέρ, over; σάρκωσις, the growth of flesh]. Hypertrophy of the muscular walls of the bladder.

**cystic** (*sist'-ik*) [*cyst*]. 1. Pertaining to or resembling a cyst. 2. Pertaining to the urinary bladder or to the gall-bladder. **c. degeneration,** degeneration with cyst-formation. **c. duct,** the duct of the gall-bladder.

**cysticercoid** (*sist-e-ser'-koid*). Resembling *cysticercus;* applied to any encysted tape-worm larva.

**Cysticercus** (*sist-e-ser'-kus*) [*cyst;* κέρκος, a tail]. The embryo of a tape-worm when it has reached the encysted stage. A hydatid. **C. bovis,** the larva of *Tænia saginata,* the beef tape-worm. **C. cellulosæ,** the larval parasite inhabiting the intermuscular connective tissue of the pig, producing the condition known as "measles." It is rarely found in the tissues of man. Its progenitor is the *Tænia solium*. **C. tenuicolis,** that of *Tænia marginata* of the dog.

**cysticolithectomy** (*sist-ik-o-lith-ek'-to-me*) [*cystic duct;* λίθος, a stone; ἐκτέμνειν, to cut out]. Von Greiffenhagen's operation for removal of calculi from the gall-bladder, consisting in opening the cystic duct alone and leaving the gall-bladder intact.

**cysticotomy** (*sist-ik-ot'-o-me*) [*cystic duct;* τομή, a cutting]. Incision into the cystic duct.

**cysticus** (*sis'-tik-us*). Any one member of a

**family** of tape-worms which in the course of development form the cysticercus or bladder-worm.

**cystidolaparotomy** (*sist-id-o-lap-ar-ot'-o-me*) [*cystic; laparotomy*]. An abdominovesical incision.

**cystidomyeloma** (*sist-id-o-mi-el-o'-mah*) [κύστις, bladder; μυελός, marrow; ὄμα, tumor]. A medullary carcinoma of the bladder.

**cystidotrachelotomy** (*sist-id-o-trak-el-ot'-o-me*). See *cystauchenotomy*.

**cystifellotomy** (*sist-if-el-ot'-o-me*) [*cyst; fel*, bile; τομή, a cutting]. See *cholecystotomy*.

**cystiform** (*sist'-if-orm*) [*cystis*, a bladder; *forma*, shape]. Encysted, cystomorphous.

**cystin** (*sist'-in*) [*cyst*], C₆H₇NO₂S. A substance found in the urine. It occurs in regular, colorless, six-sided tables, of very characteristic appearance. c., test for. See Baumann and Goldmann, Liebig, Mueller.

**cystinemia** (*sist-in-e'-me-ah*) [*cystin;* αἷμα, blood]. The occurrence of cystin in the blood.

**cystinuria** (*sist-in-ū'-re-ah*) [*cystin;* οὖρον, urine]. The presence of cystin in the urine.

**cystipathy** (*sist-ip'-a-the*) [*cyst;* πάθος, disease]. Any disease of the bladder.

**cystirrhagia** (*sist-ir-aj'-e-ah*). See *cystohemorrhagia*.

**cystirrhea** (*sist-ir-e'-ah*) [*cyst;* ῥοία, a flow]. Vesical catarrh.

**cystis** (*sist'-is*). 1. A cyst. 2. A bladder. **c. fellea,** the gall-bladder. **c. urinaria,** the urinary bladder.

**cystic** (*sist-it'-ik*) [*cystitis*]. Relating to cystitis.

**cystitis** (*sist-i'-tis*) [*cyst;* ιτις, inflammation]. Inflammation of the bladder.

**cystitome** (*sist'-it-ōm*). See *cystotome*.

**cystitomy** (*sist-it'-o-me*). See *cystotomy* (2).

**cysto-** (*sis-to-*) [κύστις, bladder]. A prefix denoting relation to the bladder.

**cystoadenoma** (*sis-to-ad-en-o'-mah*). Same as *cystadenoma*.

**cystoblast** (*sist'-o-blast*). See *cytoblast*.

**cystobubonocele** (*sist-o-bū-bon'-o-sēl*) [*cysto-;* βουβών, groin; κήλη, hernia]. Hernia of the bladder through the inguinal ring.

**cystocarcinoma** (*sist-o-kar-sin-o'-mah*). Carcinoma with cystic degeneration.

**cystocarp** (*sist'-o-karp*) [*cysto-;* καρπός, fruit]. In biology, a name sometimes applied to the sporocarp of certain algæ.

**cystocele** (*sist'-o-sēl*) [*cysto-;* κήλη, a hernia]. A hernia of the bladder.

**cystochondroma** (*sist-o-kon-dro'-mah*) [*cysto-;* χόνδρος, cartilage; ὄμα, a tumor: pl., *cystochondromata*]. A neoplasm presenting the characters of both chondroma and cystoma.

**cystocolostomy** (*sist-o-kol-os'-tom-e*) [*cysto-; colostomy*]. The surgical establishment of a permanent passage from the gall-bladder to the colon.

**cystodynia** (*sist-o-din'-e-ah*) [*cysto-;* ὀδύνη, pain]. Same as *cystalgia*.

**cystoelytroplasty** (*sis-to-el-it'-ro-plas-te*) [*cysto-;* ἔλυτρον, sheath; πλάσσειν, to form]. Surgical repair of vesicovaginal fistula.

**cystoenterocele** (*sist-o-en'-ter-o-sēl*) [*cysto-;* ἔντερον, an intestine; κήλη, a hernia]. A hernia containing a part of the bladder and intestine.

**cystoepiplocele** (*sist-o-ep-ip'-lo-sēl*) [*cysto-;* ἐπίπλοον, the omentum; κήλη, a hernia]. Hernia of the bladder and of the omentum.

**cystoepithelioma** (*sist-o-ep-ith-e-le-o'-mah*) [*cysto-; epithelioma*]. An epithelioma containing cysts filled with fluid.

**cystofibroma** (*sist-o-fi-bro'-mah*) [pl. *cystofibromata*]. A neoplasm presenting the character of both a fibroma and a cystoma.

**cystogastrostomy** (*sis-to-gas-tros'-to-me*) [*cysto-;* γαστήρ, stomach; στόμα, mouth]. An operation which consists in anastomosing the gall-bladder to the stomach.

**cystogen** (*sist'-o-jen*). See *formin*.

**cystogenia, cystogenesis** (*sist-o-je'-ne-ah, sist-o-jen'-e-sis*) [*cysto-;* γεννᾶν, to produce]. The formation or genesis of cysts.

**cystohemia** (*sist-o-he'-me-ah*) [*cysto-;* αἷμα, blood]. A congested condition of the bladder.

**cystohemorrhagia** (*sist-o-hem'-or-a-je-ah*) [*cysto-; hemorrhage*]. Vesical hemorrhage.

**cystoid** (*sist'-oid*) [*cyst*; εἶδος, likeness]. 1. Having the form or appearance of a bladder or cyst. 2. Composed of a collection of cysts. 3. A pseudocyst.

**cystolith** (*sist'-o-lith*) [*cysto-;* λίθος, a stone]. Vesical calculus.

**cystolithectomy** (*sis-to-lith-ek'-to-me*) [*cysto-;* λίθος, stone; ἐκτομή, excision]. 1. The excision of a calculus from the bladder. 2. The removal of a gallstone from the gall-bladder.

**cystolithiasis** (*sist-o-lith-i'-as-is*) [*cysto-;* λίθος, a stone]. Stone in the bladder; also that condition of the system that is associated with the presence of vesical calculus.

**cystolithic** (*sist-o-lith'-ik*) [*cysto-;* λίθος, a stone]. Pertaining to a vesical calculus.

**cystology** (*sist-ol'-o-je*) [*cysto-;* λόγος, science]. The science of cyst-formations.

**cystolutein** (*sist-o-lū'-te-in*) [*cysto-; luteus*, yellow]. A yellow coloring-matter found in cysts.

**cystoma** (*sist-o'-mah*) [*cyst;* ὄμα, a tumor: pl., *cystomata*]. A newgrowth made up of cysts; applied especially to ovarian cysts. **c. glandulare proliferum, c. proliferum papillare,** proliferating cystoma, a cystic formation derived from gland-ducts and acini. It is the most common form of ovarian and pancreatic cystoma; the lining of the inner wall consists of epithelium showing papillomatous growths or crypts resembling the acini of a gland. Syn., *cylindrocellular adenoma*.

**cystomerocele** (*sist-o-me'-ro-sēl*) [*cysto-;* μηρός, thigh; κήλη, hernia]. A hernia of the bladder through the femoral ring.

**cystomorphous** (*sist-o-mor'-fus*). Having the form of a cyst.

**cystomyoma** (*sist-o-mi-o'-mah*). A myoma containing cysts.

**cystomyxoadenoma** (*sist-o-miks-o-ad-en-o'-mah*). Cystomyxoma with adenoma.

**cystomyxoma** (*sist-o-miks-o'-mah*). A myxoma containing cysts.

**cystoncus** (*sist-ong'-kus*) [*cysto-;* ὄγκος, a swelling]. Any swelling of the bladder.

**cystonephrosis** (*sist-o-ne-fro'-sis*) [*cysto-;* νεφρός, kidney]. A cystic or cystomorphous dilatation of the kidney.

**cystoneuralgia** (*sist-o-nū-ral'-je-ah*) [*cysto-; neuralgia*]. Neuralgia of the bladder; cystalgia.

**cystoparalysis** (*sist-o-par-al'-is-is*) [*cysto-; paralysis*]. Paralysis of the bladder.

**cystopexy** (*sist'-o-peks-e*) [*cysto-;* πῆξις, fixation]. Fixation of the bladder, an operation for the cure of cystocele.

**cystophlegmatic** (*sist-o-fleg-mat'-ik*) [*cysto-;* φλέγμα, phlegm]. Pertaining to vesical mucus.

**cystophotography** (*sist-o-fo-tog'-ra-fe*). Photographing the interior of the bladder for diagnostic purposes.

**cystoplastic** (*sist-o-plas'-tik*). Relating to cystoplasty.

**cystoplasty** (*sist-o-plas'-te*) [*cysto-;* πλάσσειν, to form]. Plastic operation upon the bladder.

**cystoplegia** (*sist-o-ple'-je-ah*) [*cysto-;* πληγή, a blow]. Paralysis of the bladder.

**cystoptosis** (*sist-op-to'-sis*) [*cysto-;* πτῶσις, a fall]. The projection of some portion of the mucous membrane of the bladder into the urethra.

**cystopurin** (*sis-to-pū'-rin*). A proprietary urinary antiseptic; a mixture of hexamethylenamine and sodium acetate.

**cystopyelitis** (*sist-o-pi-el-i'-tis*) [*cysto-; pyelitis*]. Inflammation of the urinary bladder and the pelvis of the kidney.

**cystopyelonephritis** (*sist-o-pi-e-lo-nef-ri'-tis*). See *cystopyelitis*.

**cystopyic** (*sist-o-pi'-ik*). Relating to suppuration of the bladder.

**cystorectostomy** (*sist-o-rek-tos'-to-me*) [*cysto-; rectum;* στόμα, a mouth]. The formation of a fistula between the rectum and bladder.

**cystorrhagia** (*sist-or-a'-je-ah*). See *cystohemorrhagia*.

**cystorrhaphy** (*sist-or'-af-e*) [*cysto-;* ῥαφή, a seam]. Suture of the bladder.

**cystorrhea** (*sist-or-e'-ah*) [*cysto-;* ῥοία, a flow]. 1. Vesical catarrh. 2. Vesical hemorrhage. 3. Polyuria.

**cystorrhexis** (*sist-or-eks'-is*) [*cysto-;* ῥῆξις, rupture]. Rupture of the bladder.

**cystosarcoma** (*sist-o-sar-ko'-mah*) [*cysto-; sarcoma*]. Mueller's name for a sarcoma containing cysts.

**cystoschisis** (*sist-osk'-is-is*) [*cysto-;* σχίσις, a cleaving]. A congenital fissure of the urinary bladder from imperfect development.

**cystoscirrhus** (*sist-o-skir'-us*) [*cysto-;* σκιρρόs, an induration]. Scirrhus of the urinary bladder.

**cystoscope** (*sist'-o-skōp*) [*cysto-;* σκοπεῖν, to examine]. An instrument for inspecting the interior of the bladder.

**cystoscopy** (*sist-os'-ko-pe*) [*cysto-;* σκοπεῖν, to examine]. Examination of the interior of the bladder by means of the cystoscope.

**cystose** (*sist'-ōs*) [*cyst*]. Cystic; full of cysts.

**cystospasm** (*sist'-o-spazm*) [*cysto-; spasm*]. Spasm of the bladder.

**cystospastic** (*sist-o-spas'-tik*) [[*cysto-;* σπαστικόs, pulling]. Relating to spasm of the bladder.

**cystospermitis** (*sist-o-sperm-i'-tis*) [*cysto-;* σπέρμα, a seed; ιτιs, inflammation]. Inflammation of the seminal vesicles.

**cystosteatoma** (*sist-o-ste-at-o'-mah*). See *cyst, sebaceous*.

**cystostomy** (*sist-os'-to-me*) [*cysto-;* στόμα, a mouth or opening]. The formation of a fistulous opening in the bladder-wall.

**cystotome** (*sist'-o-tōm*) [see *cystotomy*]. A knife used in cystotomy; also a knife used in rupturing the capsule of the lens in cataract operations.

**cystotomy** (*sist-ot'-o-me*) [*cysto-;* τομή, a cutting]. 1. Incision of the bladder. 2. Surgical division of the anterior capsule.

**cystotrachelotomy** (*sist-o-trak-el-ot'-o-me*). See *cystauchenotomy*.

**cystoureteritis** (*sis-to-ū-re-ter-i'-tis*). Inflammation of the bladder and ureter.

**cystourethritis** (*sis-to-ū-reth-ri'-tis*). Inflammation of the bladder and urethra.

**cystourethroscope** (*sis-to-ū-re'-thro-skōp*). An instrument for inspecting the bladder and posterior urethra.

**cytameba, cytamœba** (*sit-am-e'-bah*). See *plasmodium malariæ*.

**cytase** (*si'-tās*). Metchnikoff's name for *complement*.

**cytaster** (*si-tas'-ter*). The same as *aster*.

**cyte** (*sit*) [κύτοs, a hollow]. Any cell; especially a nuclear cell (used mostly in composition).

**cythemolysis** (*si-them-ol'-is-is*) [κύτοs, cell; αἷμα, blood; λύσιs, dissolution]. Dissolution of the corpuscles of the blood.

**cytherean** (*si-the'-re-an*) [*Cythera*, an island sacred to Venus]. Venereal. c. **shield,** a condom.

**cytheromania** (*sith-ur-o-ma'-ne-ah*). See *nymphomania*.

**cytisine** (*sit'-is-in*) [κύτισοs, a kind of clover], $C_{20}H_{27}N_3O$. A poisonous alkaloid from *Cytisus laburnum*, the common laburnum, indigenous to the higher mountains of Europe and widely cultivated. c. **hydrochloride,** $C_{11}H_{14}N_2O \cdot HCl$. It is a nervine. Dose $\frac{1}{30} - \frac{1}{12}$ gr. (0.003-0.005 Gm.) subcutaneously. c. **nitrate,** $C_{11}H_{14}N_2O \cdot (HNO_3)_2 + 2H_2O$, used as is cytisine hydrochloride.

**cytisism** (*sit'-is-izm*) [κύτισοs, a kind of clover]. Poisoning by means of *Laburnum anagyroides*, sometimes occurring in children and characterized by pains in the stomach, vomiting, weakness in the legs, meteorism, and collapse.

**cytitis** (*si-ti'-tis*) [*cutis; ιτιs,* inflammation]. Dermatitis.

**cyto-** (*si-to-*) [κύτοs, a cell]. A prefix denoting relating to a cell.

**cytoblast** (*si'-to-blast*) [*cyto-;* βλαστόs, a germ]. 1. In biology, applied to the nucleus of a cell; also one of the amœboid cytodes going to make up the cytoblastema of sponges. 2. One of the hypothetical ultimate vital units of the cell. See *bioblast*. 3. Any naked cell or protoblast.

**cytoblastema** (*si-to-blas-te'-mah*). See *blastema*.

**cytochemism** (*si-to-kem'-izm*) [*cyto-;* χημεία, chemistry]. The reaction of the living cell to chemical reagents, antitoxins, etc.

**cytochemistry** (*si-to-kem'-is-tre*). The chemistry of living cells.

**cytochrome** (*si' to krōm*) [*cyto-;* χρώμα, color]. A term applied by Nissl to nerve-cells deficient in cell-protoplasm, the nucleus not being completely surrounded. The nucleus stains well and is about the size of the leukocyte nucleus.

**cytochylema** (*si-to-ki-le'-mah*) [*cyto-;* χυλόs, juice]. Strasburger's term for the inter-reticular portion of protoplasm; cell-juice.

**cytochyma, cytochyme** (*si-to-ki'-mah, si'-to-kim*) [*cyto-;* χυμόs, juice]. The water-sap in the vacuoles of the cytochylema of protoplasm.

**cytoclasis** (*si-to-kla'-sis*) [*cyto-;* κλᾶν, to break; to weaken]. Cell-necrosis.

**cytode** (*si'-tōd*) [*cyto-;* εἶδοs, form]. The simplest, most primitive form of cell, without nucleus or nucleolus.

**cytodendrite** (*si-to-den'-drīt*) [*cyto-;* δένδρον, a tree]. Lenhossék's term for a true medullated, cellulifugally conducting collateral fibril of a nerve-cell. Cf. *axodendrite*.

**cytoderm** (*si'-to-derm*) [*cyto-;* δέρμα, skin]. In biology, a cell-wall.

**cytodiagnosis** (*si-to-di-ag-no'-sis*). The determination of the nature of a pathogenic liquid by the study of the cells it contains.

**cytodieresis** (*si-to-di-er'-es-is*) [*cyto-;* διαίρεσιs, division]. The process of cell-division.

**cytodistal** (*si-to-dis'-tal*) [*cyto-; distare,* to stand apart]. Applied to that portion of an axon furthest removed from its cell of origin.

**cytogenesis** (*si-to-jen'-es-is*) [*cyto-;* γένεσιs, production]. The formation or genesis of the cell.

**cytogenetic** (*si-to-jen-et'-ik*) [*cyto-;* γένεσιs, production]. Relating to cell-formation.

**cytogenic** (*si-to-jen'-ik*). See *cytogenous*.

**cytogenous** (*si-toj'-en-us*) [*cyto ;* a cell; γένηs, producing]. In biology, producing cells.

**cytogeny** (*si-toj'-en-e*). See *cytogenesis*.

**cytoglobin** (*si-to-glo'-bin*) [*cyto-;* globus, a ball]. An albuminoid, obtainable in the form of a white, soluble powder. It forms about 3 % of the pulp of the lymphatic glands.

**cytography** (*si-tog'-ra-fe*) [*cyto-;* γράφειν, to write]. Descriptive of cells.
• **cytography** (*si-tog'-ra-fe*) [*cyto-;* γράφειν, to write]. A description of cells.

**cytohyaloplasm** (*si-to-hi'-al-o-plazm*) [*cyto-;* ὕαλοs, transparent; πγάσμα, formed matter]. Strasburger's name for the reticulum of protoplasm.

**cytohydrolist** (*si-to-hi'-dro-list*) [*cyto-; hydrolysis*]. An agent producing hydrolysis of cellular substance.

**cytoid** (*si'-toid*) [κύτοs, cell; εἶδοs, likeness]. Resembling a cell.

**cytolergy** (*si-tol'-er-je*) [*cyto-;* ἔργον, work]. Cell-activity.

**cytologist** (*si-tol'-o-jist*) [*cyto-;* λόγοs, science]. One who is versed in cytology.

**cytology** (*si-tol'-o-je*) [*cyto-;* λόγοs, science]. The science of cell-formation and cell-life.

**cytolymph** (*si'-to-limf*) [*cyto-; lympha,* clear water]. The gr und-substance or matrix of the cytoplasm of cells.o

**cytolysin** (*si-tol'-is-in*) [*cyto-; lysin*]. A substance produced in the body through the injection of foreign cells of any kind; it can destroy the same kind of cells as were used in the injection. Cf. *epitheliolysin, leukocytolysin, nephrolysin, spermolysin*.

**cytolysis** (*si-tol'-is-is*). Cell-dissolution.

**cytolytic** (*si-to-lit'-ik*). Relating to or concerned in cell-destruction.

**cytoma** (*si-to'-mah*) [*cyto-;* ὄμα, tumor]. A cell tumor; a tumor consisting of cells artificially arranged.

**cytometer** (*si-tom'-et-er*) [*cyto-;* μέτρον, a measure]. A device for counting cells, especially blood-corpuscles. See also *hemocytometer*.

**cytomicrosome** (*si-to-mik'-ro-sōm*) [*cyto-;* μικρόs, small; σῶμα, body]. A microsome of cytoplasm.

**cytomitoma** (*si-to-mi-to'-mah*) [*cyto-;* μίτοs, a fiber]. The fibrillar part of a cell-body. See *mitome*.

**cytomorphosis** (*si-to-mor-fo'-sis*) [*cyto-;* μόρφωσιs, a shaping]. A term proposed by Minot to designate comprehensively all the structural alterations which cells or successive generations of cells may undergo from the earliest undifferentiated stage to their final destruction.

**cyton** (*si'-ton*) [κύτοs, a cell]. 1. A cell. 2. A nerve cell.

**cytopathy** (*si-top'-ath-e*) [*cyto-;* πάθοs, disease]. Disease of the living cell.

**cytophagous** (*si-toff'-ag-us*) [*cyto-;* φαγεῖν, to devour]. Phagocytic; of the nature of a phagocyte; cell-devouring.

**cytophagy** (*si-tof'-aj-e*) [*cyto-;* φαγεῖν, to devour]. The englobing of cells by other cells.

**cytophil, cytophile** (*si'-to-fil*) [*cyto-;* φιλεῖν, to love]. The haptophorous group of the intermediary body with which it combines with the receptor of the cell. See *amboceptor*.

**cytophysiology** (*si-to-fiz-e-ol'-o-je*) [*cyto-; physiology*]. The physiology of a cell unit.

**cytoplasm, cytoplasma** (*si′-to-plazm, -plas′-mah*) [*cyto-;* πλάσμα, anything formed]. 1. Protoplasm. 2. Cell-plasm other than that of the nucleus; the paraplasm and endoplasm of a cell.
**cytoplastin** (*si-to-plas′-tin*) [*cyto-;* πλάσσειν, to mold]. Schwartz's name for cell-protoplasm.
**cytoproximal** (*si-to-proks′-im-al*) [*cyto-; proximare,* to draw near]. A term applied to that portion of an axon nearest its cell of origin.
**cytoreticulum** (*si-to-ret-ik′-u-lum*) [*cyto-; reticulum,* a little net]. Same as *cytomitoma.*
**Cytorrhyctes, Cytoryctes** (*si-tor-ik′-tēz*) [*cyto-;* ὀρύκτης, a digger]. A genus of protozoa. C. **aphtharum**, a species believed to cause foot-and-mouth disease. **C. luis**, one thought to be the cause of syphilis. **C. scarlatinæ**, one believed to be the exciting agent of scarlatina. **C. variolæ**, one found in variola and, in a modified form, in vaccinia.
**cytoscopy** (*si-tos′-ko-pe*). See *cytodiagnosis.*
**cytose** (*si′-tōs*). A cellulose-dissolving enzyme found in the snail and carp.
**cytosin** (*si′-tos-in*), $C_{11}H_{20}N_{16}O_4 + 5H_2O$. A basic substance obtained as a cleavage product from thymin.
**cytosome** (*si′-to-sōm*) [*cyto-;* σῶμα, a body]. A cell-body as distinguished from the nucleus.
**cytospongium** (*si-to-spun′-je-um*) [*cyto-;* σπόγγος, sponge]. The cell network or spongioplasm containing in its meshes the hyaloplasm. Cf. *mitome* and *paramitome.*
**cytostasis** (*si-tos′-tas-is*) [*cyto-; stasis*]. Stoppage or plugging of the capillaries by the blood-cells.
**cytostome** (*si′-to-stōm*) [*cyto-;* στόμα, mouth]. In biology, the oral aperture of a unicellular organism.
**cytotactic** (*si-to-tak′-tik*). Pertaining to cytotaxis.
**cytotaxis** (*si-to-taks′-is*) [*cyto-;* τάξις, order]. The directive influence which determines the arrangement of cells. The selective, ordering, and arranging function of a living cell.
**cytothesis** (*si-to-the′-sis*) [*cyto-;* θέσις, a placing or arranging]. Cell-repair.
**cytotoxic** (*si-to-toks′-ik*). Same as *cytolytic.*
**cytotoxin** (*si-to-toks′-in*). See *cytolysin.* Cf. *hemotoxin, hepatotoxin, leukotoxin, nephrotoxin, spermatoxin.*
**cytotrophy** (*si-tot′-ro-fe*) [*cyto-;* τροφή, nourishment]. The growth of the cell and sustentation of cell-life.
**cytozoon** (*si-to-zo′-on*) [*cyto-;* ζῷον, animal]. A protoplasmic cell-mass, probably parasitic in nature, with independent movement; found by Gaule in defibrinated blood and other structures.
**cytozyme** (*si′-to-zīm*). A substance, found in various tissues, capable of activating thrombin, the febrin-ferment. Also known as *coagulin, thrombokinase.*
**cytula** (*sit′-u-lah*) [κύτος, a cell]. In biology, an impregnated ovum.
**cytuloplasm** (*sit′-u-lo-plazm*) [*cyto-;* πλάσσειν, to form, mold]. In biology, the mingled ovoplasm and spermoplasm in a cytula.
**Czermak's interglobular spaces** (*cher′-ma(h)k*) [Johann Nepomuk *Czermak,* German physician, 1828-1873]. Irregular branched spaces in the crusta petrosa and enamel of the teeth.
**Czerny Lembert suture.** The application of Lembert sutures after the Czerny sutures are in place.
**Czerny's operation** (*cher′-ne*) [Vincenz *Czerny,* German surgeon, 1842-    ]. For the *radical cure of inguinal hernia;* the sac is exposed and isolated; the neck is tied with a strong catgut ligature, and cut off below this point; the stump is pushed into the abdominal cavity; the borders of the opening are freshened and united by continuous catgut sutures. **C.'s suture.** For *intestinal wounds;* one in which the needle is passed from the serous surface through the wound, down to, but not including, the mucous membrane, and through the wound on the opposite side, and out on the serous surface.
**Czerny-Trunecek's method.** A method of treating cutaneous epithelioma: by cauterization, or by the application, with a brush, after scarification, of the following solution: arsenic trioxid, 1 Gm.; ethyl-alcohol and distilled water, each, 40-50 Cc.

# D

**D.** An abbreviation of *dexter*, right; *diopter*, unit for measuring refractive power of a lens; *dosis*, dose; *detur*, let it be given; *dentur*, let them be given; *da*, give; *duration; density*.

**daboia** (*dab'-oi-ah*). The venom of Russell's viper.

**Da Costa's disease** [Jacob M. *Da Costa*, American physician, 1833–1900]. Retrocedent gout.

**dacry-, dacryo-** (*dak-re-, dak-re-o-*) [δάκρυον, a tear]. Prefixes signifying pertaining to the tears or tear-ducts.

**dacryadenalgia** (*dak-re-ad-en-al'-je-ah*) [*dacry-*; ἀδήν, gland; ἄλγος, pain]. Pain in a lacrimal gland.

**dacryadenitis** (*dak-re-ad-en-i'-tis*). Same as *dacryoadenitis*.

**dacryadenoscirrhus** (*dak-re-ad-en-o-skir'-us*) [*dacry-*; ἀδήν, a gland; σκιρρός, hard]. An indurated tumor of the lacrimal gland.

**dacryagogatresia** (*dak-re-ag-o-gat-re'-ze-ah*) [*dacry-*; ἀγωγός, leading; *atresia*]. Atresia or obstruction of a tear-duct.

**dacryagogue, dacryagog** (*dak'-re-a-gog*) [*dacry-*; ἀγωγός, leading]. 1. Inducing tears. 2. An agent causing a flow of tears.

**dacrycystalgia** (*dak-re₁sis-tal'-je-ah*). See *dacryocystalgia*.

**dacryelcosis** (*dak-re-el-ko'-sis*) [*dacryo-*; ἕλκωσις, ulceration]. Ulceration of the lacrimal apparatus.

**dacryogelosis** (*dak-re-jel-o'-sis*) [*dacry-*; γέλως, laughter]. Alternate weeping and laughing.

**dacryoadenitis** (*dak-re-o-ad-en-i'-tis*) [*dacry-*; ἀδήν, a gland; ιτις, inflammation]. Inflammation of a lacrimal gland.

**dacryoblenorrhea** (*dak-re-o-blen-or-e'-ah*) [*dacryo-*; βλέννος, mucus; ῥοία, a flow]. Chronic inflammation and discharge of mucus from the lacrimal sac.

**dacryocele** (*dak'-re-o-sēl*) [*dacryo-*; κήλη, hernia]. See *dacryocystocele*.

**dacryocyst** (*dak'-re-o-sist*) [*dacryo-*; *cyst*]. The lacrimal sac.

**dacryocystalgia** (*dak-re-o-sist-al'-je-ah*) [*dacryocyst*; ἄλγος, pain]. Pain in the lacrimal sac.

**dacryocystitis** (*dak-re-o-sis-ti'-tis*) [*dacryocyst*; ιτις, inflammation]. Inflammation of the lacrimal sac. **d. blennorrhoica,** purulent inflammation of the lacrimal sac. **d. phlegmonosa,** inflammation of the tissues composing the lacrimal sac and of the surrounding soft parts.

**dacryocystitome** (*dak-re-o-sis'-tit-ōm*). Same as *dacryocystotome*.

**dacryocystoblennorrhea** (*dak-re-o₁sis-to-blen-or-e'-ah*) [*dacryocyst; blennorrhea*]. Chronic inflammation of the lacrimal sac with a muco-purulent discharge.

**dacryocystocele** (*dak-re-o-sist'-o-sēl*) [*dacryo-*; κῆλη, hernia]. Protrusion of a lacrimal sac.

**dacryocystoptosis** (*dak-re-o-sis-top-to'-sis*) [*dacryo-*; κύστις, a cyst; πτῶσις, a falling]. Prolapse or downward displacement of a lacrimal sac.

**dacryocystotome** (*dak-re-o-sis'-to-tōm*) [*dacryocyst*]. An instrument for dividing strictures of the lacrimal passages.

**dacryocystotomy** (*dak-re-o-sist-ot'-o-me*). Incision of the lacrimal sac.

**dacryohemorrhea, dacryohæmorrhœa** (*dak-re-o-hem-or-e'-ah*) [*dacryo-*; αἷμα, blood; ῥοία, a flow]. The weeping of bloody tears.

**dacryohemorrhysis** (*dak-re-o-hem-or'-is-is*) [*dacryo-*; αἷμα, blood; ῥύσις, a flowing]. The weeping of bloody tears; a flow of blood from a lacrimal duct.

**dacryoid** (*dak'-re-oid*) [*dacryo-*; εἶδος, form]. Resembling a tear.

**dacryolin** (*dak'-re-o-lin*). The albuminous material in tears.

**dacryolite** (*dak'-re-o-līt*). See *dacryolith*.

**dacryolith** (*dak'-re-o-lith*) [*dacryo-*; λίθος, a stone]. A calcareous concretion in the lacrimal passages.

**dacryolithiasis** (*dak-re-o-lith-i'-as-is*) [*dacryo-*; λίθος, a stone]. The morbid condition that is attended by the formation of dacryoliths; also, the presence of dacryoliths.

**dacryoma** (*dak-re-o'-mah*) [*dacryo-*; ὄμα, tumor]. 1. A lacrimal tumor. 2. Obstruction of the lacrimal puncta, causing epiphora.

**dacryon** (*dak'-re-on*). 1. A tear. 2. See under *craniometric point*.

**dacryops** (*dak'-re-ops*) [*dacryo-*; ὄψ, eye]. 1 Watery eye. 2. A cyst of the duct of a lacrimal gland.

**dacryoptosis** (*dak-re-op-to'-sis*) [*dacryo-*; πτῶσις, a falling]. 1. The falling or shedding of tears. 2. Dacryocystoptosis.

**dacryopyorrhea** (*dak-re-o-pi-or-e'-ah*) [*dacryo-*; *pyorrhea*]. Purulent lacrimation.

**dacryopyosis** (*dak-re-o-pi-o'-sis*) [*dacryo-*; πύον, pus] Suppuration in the lacrimal apparatus.

**dacryorrhea** (*dak-re-or-e'-ah*) [*dacryo-*; ῥοία, a flow]. An excessive flow of tears.

**dacryosolen** (*dak-re-o-so'-len*) [*dacryo-*; σωλήν, pipe]. A lacrimal duct or canal.

**dacryosolenitis** (*dak-re-o-so-len-i'-tis*) [*dacryo-*; σωλήν, a pipe; ιτις, inflammation]. Inflammation of a lacrimal duct.

**dacryostenosis** (*dak-re-o-ste-no'-sis*) [*dacryo-*; στενός, narrow]. Stenosis or stricture of a lacrimal duct.

**dacryosyrinx** (*dak-re-o-sir'-inks*) [*dacryo-*; σύριγξ, pipe]. 1. A lacrimal fistula. 2. A syringe for use in the lacrimal ducts.

**dacryuria** (*dak-re-u'-re-ah*) [*dacryo-*; οὖρον, urine]. The enuresis which accompanies weeping in cases of hysteria, fright, or nervousness.

**dactyl** (*dak'-til*) [δάκτυλος, a finger]. A digit; a finger or a toe.

**dactylagra** (*dak-til-a'-grah*) [*dactyl*; ἄγρα, seizure]. An attack of gout or rheumatism in the fingers.

**dactylate** (*dak'-til-āt*) [*dactyl*]. Resembling a finger. Possessing five rays or appendages.

**dactylic** (*dak'-til-ik*). Pertaining to a finger or a toe.

**dactyliferous** (*dak-til-if'-er-us*) [*dactyl*; *ferre*, to bear]. 1. Having fingers or finger-like parts, organs, or appendages. 2. [δάκτυλος, a date, so called because shaped like a dactyl.] Date-bearing, as *Phœnix dactylifera*, the date-palm.

**dactylion** (*dak-til'-e-on*). See *syndactylism*.

**dactylitis** (*dak-til-i'-tis*) [*dactyl*; ιτις, inflammation]. Inflammation of a finger or a toe. **d. syphilitica,** a rare tertiary-syphilitic affection of the fingers and toes, consisting of a gummatous infiltration of the subcutaneous connective tissue and of the fibrous portions of the joints and bones. It is accompanied by great deformity, and should be distinguished from a similar affection of the muscular tissues, called by Lewin phalangitis syphilitica. Syn., *paronychia syphilitica*.

**dactylocampsodynia** (*dak-til-o-kamp-so-din'-e-ah*) [*dactyl*; κάμψις, a bending; ὀδύνη, pain]. Painful flexion of the fingers or toes.

**dactylogram** (*dak-til'-o-gram*) [*dactyl*; γράμμα, a mark]. A finger print, generally used for purposes of identification.

**dactylograph** (*dak-til'-o-graf*) [*dactyl*; γράφειν, to write]. 1. A "typewriter"; a writing machine operated by the fingers. 2. A keyboard instrument for the purpose of communication between blind deaf-mutes.

**dactylogryposis** (*dak-til-o-grip-o'-sis*) [*dactyl*; γρυπός, curved]. Abnormal curvature of the fingers or toes.

**dactyloid** (*dak'-til-oid*) [*dactyl*; εἶδος, form]. Resembling a finger.

**dactylology** (*dak-til-ol'-o-je*) [*dactyl*; λόγος, speech]. Conversation or talking by the fingers.

**dactylolysis** (*dak-til-ol'-is-is*) [*dactyl*; λύσις, loosening]. The falling off of a finger or toe. **d. spontanea.** See *ainhum*.

**dactylomegaly** (*dak-til-o-meg'-al-e*) [*dactyl*; μέγας, large]. A condition in which one or more of the fingers or toes is abnormally large.

**dactyloscopy** (*dak-til-os'-ko-pe*) [*dactyl*; σκοπεῖν, to examine]. Examination of finger prints, employed as a means of identification.

**dactylose, dactylous** (*dak'-til-ōs, -us*). See *dactylate*.
**dactylospasm** (*dak-til-o-spazm'*) [*dactyl; σπασμός*, a spasm]. Spasm of a digit.
**dactylosymphysis** (*dak-til-o-sim'-fis-is*) [*dactyl; σύν*, together; *φύειν*, to grow]. Syndactylism.
**dactylotheca** (*dak-til-o-the'-kah*) [*dactyl; θήκη*, a case]. See *finger-cot*.
**dadyl** (*dad'-il*). Balnchet and Sell's name for a camphene produced by the action of lime on artificial camphor.
**dæmonomania** (*de-mo-no-ma'-ne-ah*). See *demonomania*.
**Dæmonorops** (*de-mon'-o-rops*) [*δαίμων*, a devil; *ῥώψ*, a shrub]. A genus of plants of the order *Palmæ*. The inspissated juice of *D. draco*, a palm of Malaya, constitutes the finest dragon's-blood. *D. grandis*, same habitat as *D. draco*, affords a variety of dragon's-blood.
**Daffy's elixir** [Thomas *Daffy*, Englishman, 1680– ]. A compound aromatic tincture of senna.
**dahlia** (*dahl'-ya*) [*Dahl,* a Swedish botanist]. A genus of composite plants. The roots of several species are edible, diuretic, diaphoretic, and carminative, and furnish a purple coloring-matter. The bulbs of *D. variabilis,* a Mexican species, yield white inulin. **d.-paper,** a purple test-paper made from several species of *dahlia;* acids change its color to red and alkalies to green. **d.-violet.** See *pyoktanin, blue.*
**dahlin** (*dah'-lin*). 1. An anilin dye obtained by the action of ethyl iodide on mauvein. It gives a reddish-purple color. 2. A form of inulin obtained from the roots of *Inula helenium.* See *inulin.* Syn., *allantin; menyanthin; sinistrin; syantherin.*
**dakryon** (*dak'-re-on*). See *dacryon.*
**dakryops** (*dak'-re-ops*). Same as *dacryops.*
**Dalbergia** (*dal-bur'-je-ah*). A genus of tropical leguminous, papilionaceous plants. *D. sissoo* is a species of India and Afghanistan. The raspings of the wood are employed as an alterative. *D. sympathetica* is a tree of the East Indies. An infusion of the bark is administered in dyspepsia; the leaves are applied externally in leprosy and other cutaneous diseases, and internally as an alterative. The seeds yield an oil used in rheumatism, and the milky juice of the root is applied to ulcers.
**Dalby's carminative.** An old empirical carminative and mildly opiate mixture, answering nearly to the *mistura carminativa* of the National Formulary. It contains about two and a half minims of tincture of opium to the fluidounce.
**Dallas' operation** (*dal'-as*). For *radical cure of inguinal and femoral hernia;* after a transverse incision through the integument a special instrument is introduced to produce abrasion of the hernial canal; the instrument is then withdrawn, and the external wound sealed with iodoform collodion and a compress applied. The canal is obliterated by the resulting inflammatory action.
**Dalrymple's sign** [John *Dalrymple*, Scotch ophthalmologist, 1804–1852]. See *Stellwag's sign.*
**Dalton's law, Dalton-Henry's law** [John *Dalton*, English physicist and chemist, 1766–1844; Joseph *Henry*, American physicist, 1797–1878]. Although the volume of a gas absorbed by a liquid remains constant, the weight (volume multiplied by the density) of the absorbed gas rises and falls in proportion to its pressure.
**Daltonian** (*dal-ton'-e-an*). 1. Pertaining to John Dalton. 2. A color-blind person.
**daltonism** (*dal'ton-izm*). Color-blindness.
**dam.** See *rubber-dam.*
**damar, damaria** (*dam'-ar, dam-a'-re-ah*). See *dammar.*
**damiana** (*dam-e-an'-ah*). The leaves of *Turnera aphrodisiaca,* found in Mexico and lower California; a stimulant tonic and aphrodisiac. It is the basis of a great number of quack remedies. Dose of the *extract* 2–10 gr. (0.13–0.65 Gm.); of the *fluidextract* 10 min.–1 dr. (0.65–4.0 Cc.); of the *leaves* 1 oz. (3 Cc.) daily.
**dammar** (*dam'-ar*) [Hind., *dāmar,* resin]. A gum or resin produced by various species of *dammara* and other trees. Syn., *damar; dammaria; resina damara.* **d., true,** is obtained from the *Dammara orientalis,* a coniferous tree indigenous in the East Indies, and also from *Dammara australis,* in New Zealand. **d.-varnish,** a mounting medium used in microscopy; it is made by adding 10 parts of white dammar to 20 parts of benzene, decanting after 24 to 48 hours, and adding 4 parts of pure oil of turpentine.
**dammaran** (*dam'-ar-an*). A neutral resin obtained from dammar.
**dammarin** (*dam'-ar-in*). A resin extracted from dammar.
**Damoiseau's curve** (*dam-wah-zo'*) [Louis Hyacinthe Céleste *Damoiseau,* French physician, 19th century]. See *Ellis' sign.*
**damper** (*damp'-er*) [ME., *dampen,* to choke]. 1. A shutter placed in a flue to control draft. 2. A device attached to a galvanometer to control the secondary currents.
**Dana's operation** (*da'-ner*) [Charles Loomis *Dana,* American neurologist, 1852– ]. Resection of the posterior spinal nerve roots for spastic paralysis and other conditions.
**Dance's sign** [Jean Baptiste Hippolyte *Dance,* French physician, 1797–1832]. A depression about the right flank or iliac fossa, regarded by Dance as indicating invagination of the cecum.
**dance, St. Vitus'.** See *chorea.*
**dancing disease.** See *tarantism.*
**dancing mania.** See *choromania.*
**dandelion** (*dan'-de-li-on*). See *taraxacum.*
**dandruff** (*dan'-druf*) [origin unknown]. The scurf or scales formed upon the scalp in seborrhea.
**dandy fever** (*dan'-de*). See *dengue.*
**dangerous area of scalp** (*dān'-jer-us*). The space between the aponeurosis of the occipitofrontalis and the pericranium.
**Daniell** (*dan'-yel*) [John Frederic *Daniell,* English physicist, 1790–1845]. A unit of electrical measurement equal to 1.124 volts. **D. cell,** positive element, zinc; negative element, copper; exciting agent, zinc sulphate; depolarizing agent, cupric sulphate; E. M. F., 1.0–1.14 volts.
**Danielssen's disease** (*dan'-e-el-sen*) [Daniel Cornelius *Danielssen,* Norwegian physician, 1815–1894]. Anesthetic leprosy.
**dansomania** (*dan-so-ma'-ne-ah*). See *choromania.*
**danta** (*dan'-tah*) [Sp.]. The American tapir, *Tapirus americanus;* the powdered hoofs are employed as a sudorific and as a remedy for epilepsy.
**Danysz bacillus** (*dan'-is*). A bacillus which is probably the *Bacillus typhi murium.* **D.'s phenomenon,** when toxin is added to antitoxin in two fractions, a considerable time being allowed to elapse between the additions, the final mixture is more toxic than when the total amount is added all at once.
**Daphne** (*daf'-ne*). See *mezereon.*
**daphnetin** (*daf'-net-in*) [*daphne*]. $C_9H_6O_4 + H_2O$. A substance obtained by the decomposition of the glucoside *daphnin.* It crystallizes in yellow needles or prisms, melting at 255° C.
**daphnin** (*daf'-nin*) [*δάφνη,* laurel], $C_{15}H_{16}O_9 + 2H_2O$. A glucoside from the bark of several species of *daphne.*
**daphnism** (*daf'-nizm*). Poisoning by *Daphne mezereum,* or allied plants. It produces a hemorrhagic gastroenteritis with delirium and collapse.
**D'Arcet's metal** (*dar'-sa*). An alloy employed for filling teeth and in the making of dental plates. It consists of bismuth, 8 parts; lead, 5 parts; and tin, 3 parts. It fuses at 212° F.
**Darier's disease** (*dar-e-a'*) [F. J. *Darier,* French physician, 1856– ]. Psorospermose folliculaire; keratosis follicularis.
**Darkschewitsch's fibers** (*dark'-she-vitsh*) [Livorius *Darkschewitsch,* Russian neurologist, 1858– ]. A tract of nerve-fibers extending from the optic tract to the ganglion of the habenula. **D.'s nucleus,** a nucleus situated on each side of the median line in the gray matter near the junction of the Sylvian aqueduct with the third ventricle.
**d'Arsonvalization** (*dar-son-val-iz-a'-shun*). See *Arsonvalization.*
**dartoic, dartoid** (*dar-to'-ik, dar'-toid*). [*dartos;* *εἶδος,* likeness]. Resembling or consisting of the dartos; having slow, involuntary contractions, like the dartos.
**dartos** (*dar'-tos*) [*δαρτός,* flayed]. The contractile musculofibrous layer beneath the skin of the scrotum.
**d. muliebris,** a similar structure under the skin of the labia majora.
**dartre** (*dar'-tr*) [Fr.]. Any herpetic or other chronic skin-disease; a term vaguely used in French and the older English medical literature.
**dartrous** (*dar'-trus*) [Fr., *dartre*]. Of the nature of

tetter or herpes; herpetic. **d. diathesis,** the predisposition to chronic skin diseases.

**Darwin's ear** [Charles Robert *Darwin*, English naturalist, 1809-1882]. A congenital deformity of the ear in which the helix is absent at the upper outer angle of the ear so that the free border forms a sharp point upward and outward. In another form a blunt point (*Darwin's tubercle*) projects from the upper portion of the helix toward the center of the ear.

**darwinism** (*dar'-win-izm*) [*Darwin*]. The doctrine that higher organisms have been developed from lower forms by the influence of natural selection.

**date-disease.** See *Aleppo boil.* **d.-fever.** Synonym of *dengue.*

**datum-plane** (*da'-tum-plān*). An assumed horizontal plane from which the measurements in craniometry are taken.

**Datura** (*da-tū'-rah*) [Hind., *dhatūrā*, a certain plant]. A genus of *Solanaceæ*, or nightshade family. *D. arborea* is a South American species; the leaves are used as an emollient. *D. ceratocaula* is indigenous to tropical America; its properties are similar to those of *D. stramonium. D. fastuosa* is found throughout the tropics of the old world. The root is administered by Mohammedan physicians in epilepsy. The fruit, seeds, and leaves are used in poultices for boils, carbuncles, and in the treatment of herpetic diseases. A tincture and decoction are given as a remedy for asthma. The plant is poisonous and soporific, and is used in India as an intoxicant, and by professional poisoners for killing newborn female infants. Cf. *Dhatureas. D. metel* possesses qualities similar to *D. fastuosa. D. sanguinea* is a South American species, the *floripondio* of Peru, from the seeds of which an intoxicating beverage called *tonga* is prepared. Taken alone and in large doses it produces furious delirium, but diluted is a soporific. The seeds are used in the preparation of an ointment. *D. stramonium* is the thorn-apple. See *stramonium.*

**daturine** (*dat'-u-rēn* or *da-tu'-rēn*) [see *Datura*], $C_{17}H_{23}NO_3$. A poisonous alkaloid from *Datura stramonium*, identical with hyoscyamine and isomeric with atropine. It is employed in the treatment of mania, epilepsy, neuralgia, rheumatism, syphilis, cancer, pains, spasms, asthma, and as a hypnotic in insanity. Dose $\frac{1}{120}$-$\frac{1}{60}$ gr. (0.0003-0.001 Gm.). Treatment in case of poisoning: emetics, stomach-pump, castor-oil. **d. hydrochloride,** $C_{17}H_{23}NO_3HCl$. Uses and dose same as of daturine. **d., light,** hyoscyamine, obtained by Ladenburg from *Datura stramonium*. **d. sulphate,** $(C_{17}H_{23}NO_3)_2H_2SO_4$. Uses and dose same as of daturine.

**daturism** (*dat'-u-rizm*) [see *Datura*]. Stramonium-poisoning.

**Daubenton's angle** (*do-ban-ton*(*g*)) [Louis Jean Marie *Daubenton*, French physician, 1716-1799]. Occipital angle, in craniometry, that formed by the intersection of the basicranial axis and the plane of the occipital foramen. **D.'s line,** a line joining the opisthion and the projection of the lower border of the orbit. **D.'s plane,** in craniometry, that passing through the opisthion and the inferior borders of the orbits.

**dauciform** (*daw'-si-form*). See *daucoid.*
**daucoid** (*daw'-koid*) [δαῦκον, the carrot; εἶδος, likeness]. Resembling a carrot; dauciform.
**Daucus** (*daw'-kus*) [δαῦκον, the carrot]. A genus of plants of the order *Umbelliferæ. D. carota,* the carrot, is a cultivated biennial indigenous to Europe and the Orient. The root contains sugar, starch, pectin, malic acid, albumin, a volatile oil, and a crystalline coloring-matter (*carotin*). It is a stimulant when applied to indolent ulcers, and is fed to horses to render the coat glossy. The aromatic seeds (fruit) are diuretic and are used in dropsy and nephritic complaints. *D. gingidium,* a species indigenous to Europe and northern Africa, yields a gum-resin, bdellium siculum.

**daughter** (*daw'-ter*). A female child or descendant. **d.-cell.** See *cell*, *daughter-*. **d.-cyst.** See *cyst*, *daughter-*. **d.-nuclei.** See *karyokinesis.* **d.-star,** an amphiaster. See *karyokinesis.*

**Davainea madagascariensis** (*da-va'-ne-ah*) [Casimir Joseph *Davaine*, French physician, 1812-1882]. A tapeworm occurring in man, found in Madagascar and elsewhere.

**Davidsohn's sign** (*da'-vid-son*) [Hermann *Davidsohn,* German physician, 1842- ]. The illumination of the pupil obtained on placing an electric light in the mouth will be less marked on the side on which there is a tumor or empyema of the antrum of Highmore.

**Davy's lever** (*da'-ve*) [Richard *Davy*, English surgeon, 1838- ]. A wooden sound which is inserted into the rectum for the purpose of making pressure on one of the iliac arteries; it is used to arrest hemorrhage.

**Davy's test for phenol.** To 1 or 2 drops of the phenol solution add 3 or 4 drops of a solution of 1 part molybdic acid in 10 or more parts of concentrated sulphuric acid. A pale yellowish-brown coloration is produced, which passes to reddish-brown and then to a beautiful purple.

**day-blindness.** See *nyctalopia* and *hemeralopia.*
**daymare** (*da'-mār*). A state of temporary distress and terror, resembling nightmare, but coming on when the patient is awake. It is thought to be due to a diseased state of the blood-vessels of the brain.

**day-nursery.** See *crèche.*
**daysight** (*da'-sit*). See *hemeralopia.*
**D.D.S.** Abbreviation for *Doctor of Dental Surgery.*
**de-** [*de,* from, away]. A prefix denoting down, away from, occasionally it has an intensive meaning.
**deacidification** (*de-as-id-if-ik-a'-shun*). The act or process of neutralizing an acid.
**deactivation** (*de-ak-tiv-a'-shun*). 1. The process of becoming inactive or of making inactive. 2. Loss of radioactivity.
**dead** (*ded*). 1. Without life; destitute of life. 2. Numb. **d.-born,** still-born. **d. finger.** See *sphacceloderma* and *night-palsy.* **d.-house,** a morgue; an apartment in a public institution for keeping dead bodies. **d. nettle.** See *lamium.* **d. space,** a cavity left after the closure of a wound. **d. voice,** a voice without nasal resonance; the so-called nasal voice.
**deadly** (*ded'-le*). Capable of causing death; fatal; mortal. **d. nightshade.** See *atropa,* and *belladonna.*
**deaf** (*def*) [AS., *deaf*]. Lacking the sense of hearing; in a condition of impaired hearing. **d. fields,** two small triangular planes, converging toward the external auditory meatus, and in which the vibrating tuning-fork is not heard. **d.-mutism,** the state of being both deaf and dumb; the deafness may be congenital or acquired, and prevent the individual from learning to speak. **d.-mutism, hysterical,** a condition of deaf-mutism of sudden development, due to hysteria. **d.-points,** some points near the ear in which a vibrating tuning-fork cannot be heard.
**deafness** (*def'-nes*) [*deaf*]. The state of being deaf. Deafness may be due to disease of the external auditory canal, the middle ear, the internal ear, the auditory nerve, or the brain. **d., bass,** difficulty in hearing low tones. **d., boilermakers',** deafness resulting from working among machinery, and characterized by inability to hear ordinary conversation, while hearing power is increased amid loud noise. **d., cerebral,** that due to a brain-lesion. **d., cortical,** that due to disease of the cortical centers for hearing. **d., mind, d., psychic,** inability to recognize or understand the sounds heard, due to destruction of the central area of the auditory center. **d., paradoxic.** See *Willis's paracusis.* **d., speech-,** a variety of psychic deafness resembling word-deafness, except that the faculty of repeating and writing after dictation is not lost. **d., tone.** See *tone.* **d., word-.** See *d., psychic.*
**dealbate** (*de-al'-bāt*) [*dealbatus,* whitwashed]. In biology, coated with a fine white down or powder.
**dealbation** (*de-al-ba'-shun*) [see *dealbate*]. The process or act of becoming or being made white, as by bleaching.
**dealcoholization** (*de-al-ko-hol-i-za'-shun*). The removal of alcohol from an object or compound used in microscopic technic. **d.-agent,** a liquid employed for the purpose of getting rid of the alcohol in preserved specimens, and to facilitate the penetration of paraffin in microtomy.
**deambulation** (*de-am-bu-la'-shun*) [*deambulare,* to take a walk]. Gentle exercise as by walking.
**deamidation** (*de-am-id-a'-shun*) [*de-; amide*]. The conversion of amido-acids into oxyacids.
**deammoniated** (*de-am-o'-ne-a-ted*) [*de,* from; *ammonium*]. Deprived of ammonia.
**deanesthesiant** (*de-an-es-the'-se-ant*) [*de,* from; *anesthesia*]. A means for arousing the system from a state of anesthesia.
**deaquation** (*de-ak-wa'-shun*) [*de,* from; *aqua,*

# DEARGENTATION 276 DECIPARA

water]. The act or process of removing water from a substance.

**deargentation** (*de-ar-jen-ta'-shun*) [*deargentare*, to plate with silver]. The act or process of silvering.

**dearterialization** (*de-ar-te-re-al-i-za'-shun*) [*de*, from; *arterialization*]. The transformation of the blood from the arterial to the venous state. Cf. *atmospherisation*.

**dearticulation** (*de-ar-tik-u-la'-shun*). See *diarthrosis, disarticulation, dislocation*.

**death** (*deth*) [AS., *dedth*]. The cessation of life. d., **binsical**, death preceded by mania. d., **black**, an exceedingly fatal epidemic called the "plague," which occurred in Europe during the fourteenth century, and during which it is estimated, 20,000,000 persons died. d., **local**, death of a part. d.; **molar**, **necrosis, gangrene**. d., **molecular**, death of individual cells; ulceration. d., **muscular**, a state of the muscles in which they no longer react to stimuli. d.-**rate**; the annual mortality per 1000. d.-**rattle**, the gurgling sound heard in the throat of a dying person. d., **somatic**, death of the organism as a whole.

**deauration** (*de-aw-ra'-shun*) [*deaurare*, to gild]. The act or process of gilding.

**debilitant** (*de-bil'-it-ant*) [*debilitare*, to weaken]. 1. An agent allaying excitement. 2. Weakening.

**debility** (*de-bil'-it-e*). See *asthenia*. d., **nervous**. See *neurasthenia*.

**Debove's disease** (*de-boov'*) [Maurice Georges Debove, French physician, 1845– ]. Splenomegaly. D.'s **membrane**, the basement membrane of the the mucosa of the trachea, bronchi, and intestinal tract.

**débridement** (*da-brēd-mon*(g)) [Fr.]. The enlargement of a wound or hernia in operating; the slitting of any constricting tissue or band.

**deca-** (*dek-a-*) [δέκα, ten]. Ten; prefixed to the units of weight, capacity, and length in the metric system, it signifies a measure ten times as large as the unit. See *metric system*.

**decagram** (*dek'-a-gram*) [δέκα, ten; *gram*]. Ten grams or 154.32349 grains, 0.353 ounce avoirdupois, or 0.3215 ounce troy. See *metric system*.

**decalcification** (*de-kal-sif-ik'-a-shun*) [*de*, priv.; *calx*, lime; *facere*, to make]. The withdrawal of the lime-salts of bone.

**decalcify** (*de-kal'-sif-i*) [see *decalcification*]. To remove lime-salts from tissues.

**decalcifying fluid** (*de-kal'-sif-i-ing flu-id*). A solution used for the purpose of depriving tissue of its calcium salts. Chromic acid 1 gram, water 200 c.c., then add 2 c.c. nitric acid,—is recommended.

**decaliter, decalitre** (*dek-a-le'-ter*) [δέκα, ten; *liter*]. Ten liters, or 2½ imperial gallons, or 2.64 U. S. gallons. See *metric system*.

**decalvant** (*de-kal'-vant*) [*decalvare*, to make bald]. Destroying hair.

**decameter, decametre** (*dek'-a-me-ter*) [δέκα, ten; *meter*]. Ten meters or 393.7 English inches, or 32.8 feet. See *metric system*.

**decane** (*dek'-ān*) [δέκα, ten], $C_{10}H_{22}$. A hydrocarbon of the paraffin series.

**decantation** (*de-kan-ta'-shun*) [*de*, down; *cantus*, a side]. The operation of removing the supernatant fluid from a sediment.

**decapitation** (*de-kap-it-a'-shun*) [*de*, from; *caput*, head]. The act of beheading, especially as performed on the fetus when other means of delivery have failed.

**decapitator** (*de-kap'-it-a-tor*) [*de*, from; *caput*, head]. An instrument used in performing decapitation.

**decapsulation** (*de-kap-su-la'-shun*). Removal of a capsule; especially removal of the capsule of the kidney.

**decarbonated** (*de-kar'-bon-a-ted*). Deprived of carbon dioxide.

**decarbonization, decarburation, decarbonization** (*de-kar-bon-i-za'-shun, de-kar-bū-ra'-shun, de-kar-bū-ri-za'-shun*). The act or process of freeing a substance from carbon.

**decay** (*de-ka'*) [*de*, down; *cadere*, to fall]. 1. Putrefactive change. 2. The ultimate katabolic state; decline of life, of health, or of one or more functions.

**decemcostate** (*de-sem-kos'-tāt*) [*decem*, ten; *costa*, a rib]. Having ten ribs.

**decemfid** (*de'-sem-fid*) [*decem*; *findere*, to divide]. Cut into ten parts.

**decemipara** (*de-sem-ip'-ar-ah*) [*decem*; *parere*, to bring forth]. A woman pregnant for the tenth time.

**decentered** (*de-sent'-erd*) [*de*, from; *center*]. Out of common center; said of lenses as to focus, or of masses as to equilibrium, etc.

**decentration** (*de-sen-tra'-shun*) [see *decentered*]. Removal from a center.

**decerebrated** (*de-ser'-e-bra-ted*). Decerebrized.

**decerebrize** (*de-ser'-e-brīz*) [*de*, from; *cerebrum*]. To remove the brain, as of a frog, in physiological experiments; decerebrate.

**dechloridation** (*de-klo-rid-a'-shun*). The removal of salt from the diet with the object of reducing the quantity of chlorides in the body tissues and fluids.

**dechloruration** (*de-klor-ū-ra'-shun*). The producing of decreased excretion of chlorides in the urine by means of dechloridation.

**deci-** (*des-e-*) [*decem*, ten]. A prefix which, joined to the metric units of length, capacity, and weight, signifies a measure one-tenth as large as the unit. See *metric system*.

**decidua** (*de-sid'-ū-ah*) [*deciduus*, falling off]. The mucous membrane which lines the uterus and surrounds the ovum during pregnancy. Syn., *decidua membrana; decidua tunica*. d. **basalis**. Same as *d. serotina*. d., **catamenial**. See *d., menstrual*. d., **epichorial**. See *d. reflexa*. d. **graviditatis**, the menstrual decidua during pregnancy. d., **inter-uteroplacental**. See *d. serotina*. d. **membrana**, See *decidua*. d., **menstrual**, the outer layer of the uterine mucosa which is shed during menstruation. In membranous dysmenorrhea it is discharged in pieces before disintegration. Syn., *catamenial decidua*. d. **placentalis subchorialis**, the layer of the maternal placenta lying next the chorion. Syn. *decidua subchorialis*. d. **reflexa**, that part of the decidua growing about the ovum and inclosing it as a sac. d. **serotina**, that part of the decidua vera upon which the ovum lies, and from which the placenta is subsequently formed. d. **serotina, glandular**. See *d. serotina, uterine*. d. **placentalis subchorialis**, that portion of decidua serotina which is in contact with the parts of cotyledons of the placenta, as distinguished from the uterine decidua serotina. d. **serotina, uterine**, the outer layer of the decidua serotina; the glandular decidua serotina. d. **subchorialis**. See *d. placentalis subchorialis*. d. **tuberosa et polyposa**, a form of decidual endometritis characterized by a roughened condition and polypoid growths of the uterine mucosa. d. **vera**, the thickened, vascular, spongy mucous membrane of the gravid uterus.

**decidual** (*de-sid'-ū-al*). Belonging to the decidua. d., **cells**, a proliferation of young connective-tissue cells above the uterine glands, taking place after the ovum is impregnated, and producing an hypertrophy of the mucous membrane of the uterus. d. **endo-metritis**, see *endometritis*.

**deciduation** (*de-sid-ū-a'-shun*). The act or process of dropping off or shedding.

**deciduitis** (*de-sid-ū-i'-tis*). Inflammation of the decidual membranes of the gravid uterus.

**deciduoma** (*de-sid-ū-o'-mah*) [*decidua*; δμα, a tumor: *pl., deciduomata*.] An intrauterine tumor containing decidual relics, and believed to arise from some hyperplasia of a retained portion of the decidua. By some it is considered a sarcoma. d. **malignum**, a variety of uterine sarcoma first described by Saenger, which in its microscopic characters strongly resembles decidual tissues. Syn., *choroioepithelioma malignum; sarcoma deciduocellulare; syncytioma malignum*.

**deciduosarcoma** (*de-sid-ū-o-sar-ko'-mah*). See *deciduoma malignum*.

**deciduous** (*de-sid'-ū-us*) [*de*, away from; *cadere*, to fall]. Falling off. d. **teeth**, the temporary teeth or milk-teeth.

**decigram** (*des'-e-gram*) [*decimus*, tenth; *gram*]. One-tenth of a gram or 1.54 grains troy. See *metric system*.

**declin** (*des'-il-an*). A solution of formaldehyde and potassium oleate; used as an antiseptic and disinfectant.

**deciliter** (*des'-e-le-ter*) [*decimus*, tenth; *liter*]. One-tenth of a liter, or 3.52 English fluidounces or 3.38 U. S. fluidounces. See *metric system*.

**decimeter** (*des'-e-me-ter*) [*decimus*, tenth; *meter*]. One-tenth of a meter, or 3.937 inches. See *metric system*.

**decinormal** (*des-e-nor'-mal*) [*deci-*; *norma*, normal]. Having one-tenth the strength of the normal.

**decipara** (*de-sip'-ar-ah*) [*decem*, ten; *parere*, to bring forth]. A woman pregnant for the tenth time.

**declination** (dek-lin-a'-shun) [decline]. The dip of the magnetic needle.
**declinator** (dek'-lin-a-tor) [decline]. An instrument for holding the dura apart during trephining.
**decline** (de-klīn') [declinare, to bend]. 1. A gradual decrease, as of a fever. 2. A wasting away of the bodily strength. 3. A popular term for *pulmonary tuberculosis*.
**declive** (de-klīv') [declivis, sloping]. 1. A lower or descending part. 2. See *declivis cerebelli*.
**declivis cerebelli** (de-kli'-vis ser-e-bel'-i) [L.]. The sloping posterior aspect of the monticulus.
**decoction** (de-kok'-shun) [decoquere, to boil down]. A liquid preparation obtained by boiling vegetable substances in water.
**decoctum** (de-kok'-tum) [L.: *pl.*, *decocta*]. A decoction.
**decollation** (de-kol-a'-shun). See *decapitation*.
**decollator** (de-kol'-a-tor) [decollare, to behead]. An instrument for fetal decapitation.
**decolorant** (de-kul'-or-ant) [de, priv.; color]. An agent for the altering or removal of color.
**decoloration** (de-kul-or-a'-shun) [decolorare, to deprive of color]. Removal of color.
**decolorize** (de-kul'-or-iz) [decolorare, to deprive of color]. To remove the excess of coloring-matter from stained histological preparations, for purposes of differentiation.
**decombustion** (de-kom-bust'-yun). See *deoxidation*.
**decompensation** (de-kom-pen-sa'-shun) [de, priv.; *compensare*, to compensate]. Failure of compensation (as of the circulation or of the heart).
**decomposability** (de-kom-po-za-bil'-it-e) [de, from; *componere*, to compose]. Capability of being decomposed.
**decompose** (de-kom-pōz'). 1. To cause a compound to break up into its simpler constituents. 2. To undergo putrefaction.
**decomposition** (de-kom-po-zish'-un) [decomponere, to decompose]. 1. The separation of the component principles of a body. 2. Putrefactive fermentation.
**decompression** (de-kom-presh'-un). The removal of compression or pressure. **d. injury**, injury from the effects of a sudden vacuum. See *caisson disease*.
**decongestive** (de-kon-jes'-tiv) [de, from; *congerere*, to bring together]. Relieving congestion.
**decortication** (de-kor-tik-a'-shun) [de, from; *cortex*, the bark]. 1. The stripping of the bark or husk of a plant. 2. The stripping off of portions of the cortical substance of the brain from the summits of the gyri. 3. Decapsulation, as of the kidney. 4. Removal of the cortex of any viscus. **d. pulmonary**, pleurectomy.
**decostate** (de-kos'-tāt) [de, from; *costa*, a rib]. Without ribs.
**decrement** (dek'-re-ment). See *decline*.
**decrepit** (de-krep'-it) [decrepitus, old]. Broken down with age.
**decrepitation** (de-krep-it-a'-shun) [decrepitare, to crackle]. A crackling or crepitation.
**decrepitude** (de-krep'-it-ūd) [decrepit]. The state of being decrepit; senility; the feebleness of old age.
**decrustation** (de-krust-a'-shun) [de, from; *crusta*, a crust]. The detachment of a crust.
**decubital** (de-kū'-bit-al). Relating to a decubitus or to a bed-sore.
**decubitus** (de-kū'-bit-us) [decumbere, to lie down]. 1. The recumbent or horizontal posture. 2. A bed-sore. **d., acute**, a form of bed-sore due to cerebral lesions. **d., Andral's**. See under *Andral*.
**decursus fibrarum cerebralium** [L.]. The running down of the cerebral fibers.
**decurtation** (de-kur-ta'-shun) [decurtare, to curtail]. The ablation or shortening of a structure or usual duration of a condition.
**decurvature** (de-kurv'-a-chur) [decurvatus, bent back]. A descending curvature.
**decussate** (de-kus'-āt) [see *decussation*]. To intersect; to cross.
**decussatio** (de-kus-a'-she-o) [L.]. A decussation, or crossing. **d. brachii conjunctivi**, decussation of the brachium conjunctivum. **d. fontinalis**, fountain decussation. **d. lemniscorum**, decussation of the fillet or lemniscus. **d. pontinalis** (more correctly *pontilis*), in the pons, a decussation of tegmental fibers from the thalamus. **d. pyramidum**, the decussation of the pyramids. **d. nervorum trochlearium**, decussation of the trochlear nerves on the upper surface of the valvula.

**decussation** (de-kus-a'-shun) [decussatio, a crossing]. A chiasma or X-shaped crossing, especially of symmetrical parts, as of nerve-fibers, nerve-tracts, or nerve-filaments. The principal decussations are that of the optic nerve and that of the lateral pyramidal tracts in the medulla. **d. of the brachium conjunctivum**, crossing of fibers in the postgeminum to the opposite red nucleus. **d. of the fillet**, the crossing of afferent fibers in the medulla. **d., Forel's**. See under *Forel*. **d., fountain**, Spitzka's term for such a decussation of nerve-fibers as is seen in the cortex of the anterior quadrigemina. **d., motor**. See *d. of the pyramids*. **d. of the optic nerve**, the chiasm. **d., pineal**, Spitzka's term for the crossing of certain fibers of the inner division of the reticular formation. **d., piniform**. See *d. of the pyramids, sensory*. **d., pyramidal, superior**, See *d., pyramidal, upper*. See *d. of the pyramids, sensory*. **d. of the pyramids, d., pyramidal**, the oblique crossing of the bundles of the anterior pyramids of the oblongata from opposite sides of the median fissure. Syn., *inferior decussation; motor decussation; piniform decussation; ventral decussation of the pyramids*. **d. of the pyramids, sensory**, a crossing of certain fibers having their origin in the funiculi cuneati of the spinal cord, which occurs in the upper part of the oblongata, between the anterior pyramids and the gray floor of the fourth ventricle. Syn., *decussation of the fillet; interolivary decussation; pyramidal posterior decussation*. **d. of the pyramids, ventral**. See *d. of the pyramids*. **d., sensory**. 1. See *d. of the pyramids, sensory*. 2. The crossing of the outer bundles of the anterior pyramids of the spinal cord. Syn., *superior decussation*. **d., sensory, middle**, a crossing in the median line of certain fibers between the upper and lower pyramids. **d., tegmental, of Meynert**, the crossing of the fibers in the cortex of the anterior quadrigeminum. **d., ventral**. See *d. of the pyramids*.
**decussorium** (de-kus-o'-re-um) [L.]. An instrument for depressing the dura after trephining.
**dedalous, dædalous** (ded'-al-us) [δαιδάλεος, curiously wrought]. Labyrinthiform; intricately wrought.
**dedasol** (ded'-as-ol). A proprietary tablet containing digitalis.
**dedentition** (de-den-tish'-un) [de, down; *dens*, a tooth]. The shedding of the teeth.
**dedolation** (ded-o-la'-shun) [dedolatio, a hewing off]. A cutting off obliquely.
**deemetinize** (de-em'-et-in-īz). To deprive ipecacuanha of its emetic principle, emetin.
**deep** (dēp). Not superficial. **d. reflexes**. See under *reflex*. **d. water**, water obtained from a porous layer beneath the first impervious stratum.
**defatigatio** (de-fat-ig-a'-she-o) [L.]. Over-fatigue; overstrain, as of the heart-muscle. **d.mentis**, brain-fag.
**defecation** (def-ek-a'-shun) [defæcare, to separate from the dregs]. 1. The evacuation of the bowels. 2. Clarification, as of wine. Cf. *decantation*.
**defect** (de-fect') [defectus, a failure]. A lack or failure; absence of any part or organ; absence or failure of a normal function.
**defemination** (de-fem-in-a'-shun). The loss or diminution of female characteristics, with the assumption of male characteristics by a woman.
**defensive protein** (de-fen'-siv). A globulin variety present in the animal body, possessing germicidal functions.
**deferent** (def'-er-ent) [deferens, carrying away]. Carrying away or down; efferent. See *vas*.
**deferentectomy** (def-er-ent-ek'-to-me). Excision of the vas deferens.
**deferential** (def-er-en'-shal). Pertaining to the vas deferens.
**deferentiovesical** (def-er-en-she-o-ves'-ik-al). Pertaining to both the vas deferens and the bladder.
**deferentitis** (def-er-en-ti'-tis) [deferens; *ιτις*, inflammation]. Inflammation of the vas deferens.
**deferred shock**. The late onset of the symptoms of shock.
**defervescence** (de-fer-ves'-ens) [defervescere, to cease boiling]. Disappearance of fever.
**defibrillation** (de-fi-bril-a'-shun) [de, from; *fibrilla*, a small fiber]. The tearing of the brain-substance in the direction of the least resistance, in order to make cleavage-preparations.
**defibrination** (de-fi-brin-a'-shun) [de, from; *fibra*, a fiber]. The removal of fibrin from blood or lymph.
**defining power** (de-fi'-ning). See *definition*.

**definition** (def-in-ish'-un) [definire, to bound by limits]. In optics, the power of a magnifying lens to show clear outlines of the object examined, free from aberration or distortion.

**definitive** (de-fin'-it-iv). Limiting the extent; final.

**deflagration** (def-lag-ra'-shun) [deflagrare, to be consumed]. A sudden, violent combustion, such as accompanies the oxidation of certain inorganic substances by mixing them with an easily decomposing salt, such as the alkaline chlorates and nitrates.

**deflagrator** (def-la-gra'-tor) [see deflagration]. An apparatus for producing very rapid combustion.

**deflect** (de-flekt'). [deflectere, to bend away]. To turn or bend from a straight course.

**defloration** (def-lo-ra'-shun) [de, from; flos, a flower]. On the part of the female, the first sexual connection. The loss of those marks or features that indicate virginity.

**defluvium capillorum** (de-flū'-ve-um kap-il-or'-um). Alopecia.

**defluxion** (de-fluk'-shun) [de, down; fluere, to flow]. 1. A discharge. 2. A catarrh; a descent of the humors or secretions. 3. A rapid falling, as of the hair or eyebrows.

**deformation** (de-for-ma'-shun) [deformare, to deform]. The process of disfigurement. **d., Sprengel's.** See under Sprengel. **d., Volkmann's.** See under Volkmann.

**deforming** (de-form'-ing). Disfiguring. **d. arthritis.** See arthritis deformans. **d. ostitis.** See ostitis.

**deformity** (de-for'-mit-e). Abnormal shape or structure of a body or part. **d., anterior.** See lordosis.

**defunctionalization** (de-funk-shun-al-iz-a'-shun). The act of destroying a function.

**defurfuration** (de-fur-fur-a'-shun) [de, from; furfur, bran]. Desquamation.

**defuselation** (de-fū-sel-a'-shun). The removal of fusel oil from spirits.

**defusion** (de-fū'-zhun). See decantation.

**deganglionate** (de-gan'-gle-on-āt). To remove a ganglion or ganglia.

**degenerate** (de-jen'-er-āt) [see degeneration]. 1. To revert to a lower type. 2. An individual who has reverted to a lower type.

**degeneration** (de-jen-er-a'-shun) [degenerare, to become base]. 1. A morbid process consisting in the conversion of the elements of a tissue into some inert substance. 2. A term indicating imperfect or abnormal development of the psychic faculties. Syn., degenerescence. **d., Abercrombie's**, amyloid degeneration. **d., albuminoid**, a cloudy and granular swelling of the cell protoplasm. **d., albuminous**, albuminous infiltration. **d., amyloid**, characterized by the formation of an albuminous substance, resembling starch in its chemical reaction. **d., ascending**, a trophic degeneration of nerve-fibers or tracts progressing from the site of the original lesion toward the cerebrum. **d., bacony.** Same as d., amyloid. **d., calcareous.** See infiltration, calcareous. **d., cellulose.** See amyloid degeneration. **d., cheesy.** See caseation. **d., chitinous.** See amyloid degeneration. **d., colloid**, the change of the protoplasm of epithelial cells into a substance that resembles mucus, but is not precipitated by alcohol or acetic acid. **d., cystic**, degeneration with cyst-formation. **d., cystoid, of the retina**, round or oval cystoid spaces surrounded by hypertrophied radial fibers found in the retina at all ages, as described by Iwanoff. **d., descending**, a degeneration of nerve-fibers or tracts extending peripherally from the original lesion. **d., earthy.** See calcification and infiltration, calcareous. **d., fascicular**, that form of atrophy of paralyzed muscles following pathological change in the motor ganglion-cells of the central tube of the gray matter of the spinal cord or their efferent fibers. **d., fatty**, a change of the proteids of the tissues into fat. **d., fibrofatty, of the placenta**, an association of fatty degeneration of the placenta with fibromatous degeneration of the chorionic villi and of the decidua serotina. **d., fibroid**, a change into fibrous tissue. **d., fibrous, of the heart**, hyperplasia of the cardiac connective tissue accompanying chronic interstitial inflammation. **d., gelatiniform.** See d., colloid. **d., granular**, parenchymatous degeneration distinguished by a deposit of albuminoid particles. **d., gray**, in nervous tissue, a gray degeneration due to chronic inflammation. **d., hyaline**, a degeneration affecting particularly the connective tissue of the walls of blood-vessels, and giving rise to a substance resembling amyloid material, but lacking its reactions. See amyloid degeneration. **d., hyaloid.** See amyloid degeneration. **d., hydrocarbonaceous**, Paschutin's term for a special degeneration peculiar to diabetes. Syn., Paschutin's degeneration. **d., lardaceous.** Same as d., amyloid. **d., liquefactive**, a process resembling fatty degeneration, accompanying fibrinous exudations. **d., mineral.** See calcification. **d., mucoid**, the degeneration of tissue into a jelly-like, transparent substance containing mucin. **d., myelin**, a process sometimes occurring in chronic pneumonia in which there is a formation of myelin coincident with fatty degeneration in the pulmonary alveoli. **d., myxomatous.** See d., mucoid. **d., parenchymatous.** See cloudy swelling. **d., parenchymatous, of the kidney**, a degeneration of the parenchyma of the kidney following the acute nephritis of pregnancy, diphtheria, or an acute attack of fever. It is accompanied by 10 to 25 % of albumin in the urine, which remains of normal quantity. **d., Paschutin's.** See d., hydrocarbonaceous. **d., pigmentary, d., pigment**, a pigmentation of the muscles accompanying the atrophy due to cachexia, insufficient food, or the marasmus of old age. **d., putrid.** See hospital gangrene. **d., reaction of.** See reaction of degeneration. **d., secondary.** See Wallerian degeneration. **d., signs of**, physical imperfections, such as asymmetry of corresponding parts, adherent lobules of the ear, stammering, supernumerary or deficient digits, etc., observed in persons presenting psychic degeneration. **d., theroid**, in psychiatry, the lowering or approximation of the human mental faculties and instincts to those of the lower animal. **d., trabecular**, a degeneration of the bronchial wall in which there is a hypertrophy of the elastic and inelastic tissues of the fibrous sheath of the bronchus and its cartilages. **d., uratic**, the deposition of uric acid and the urates in the tissues. **d., Virchow's.** See amyloid degeneration. **d., vitreous.** See albuminoid disease and amyloid degeneration. **d., Wallerian.** See under Wallerian. **d., waxy.** Same as d., amyloid.

**degenerative** (de-jen'-er-a-iv). Of or pertaining to degeneration.

**deglabration** (deg-la-bra'-shun) [deglabrare, to make smooth]. The process of becoming bald.

**deglutible** (de-gloot'-ibl) [deglutitio, a swallowing]. Capable of swallowing, or of being swallowed.

**deglutitio impedita** (de-gloo-tish'-e-o im-ped-i'-tah). Synonym of dysphagia.

**deglutition** (deg-loo-tish'-un) [deglutitio, a swallowing]. The act of swallowing.

**deglutitive** (deg-loo'-tit-iv). Relating to deglutition.

**degote** (de-gōt') [Russ.]. Oil of white birch.

**degradation** (deg-rad-a'-shun) [degradatio, a descent by steps]. Gradual physiological and histological change for the worse; degeneration; retrograde metamorphosis.

**degrease** (de-grēs') [Fr. dégraisser]. To remove fat, as from bones in the preparation of skeletons.

**degreasing** (de-grēs'-ing) [Fr., dégraisser]. Removing the fat, as from bones.

**degree** (de-gre') [de, from; gradus, a step]. 1. Position in a graded series; quality. 2. The units or intervals of thermometric or other scales. Also, a title or testimonial of qualification granted by a university or college.

**degustation** (de-gus-ta'-shun) [degustare, to taste]. The act of tasting.

**degut** (de-gūt') [Russ.]. Birch oil or tar. See birch.

**dehematize** (de-hem'-at-iz) [de, from; αἷμα, blood]. To deprive of blood.

**dehiscence** (de-his'-ens) [de, off; hiscere, to gape or yawn]. The act of splitting open. **d.s, Zuckerkandl's.** See under Zuckerkandl.

**dehumanization** (de-hū-man-iz-a'-shun) [de, from; humanus, human]. 1. The loss of the proper characteristics of humanity, either by insane persons, or by debased criminals. 2. The supposed loss of some quality pertaining to the human species; as in the alleged dehumanization of vaccine virus.

**dehydrate** (de-hi'-drāt) [de, from; ὕδωρ, water]. To remove water from.

**dehydration** (de-hi-dra'-shun) [de, away from; ὕδωρ, water]. The removal of water

**dehydrogenize** (de-hi'-dro-jen-iz). To deprive of hydrogen.

**dehypnotization** (de-hip-no-ti-za'-shun). Waking from hypnotism.

**deintoxication** (de-in-toks-ik-a'-shun) [de, from; intoxication]. The process of overcoming the effects of toxic substances.

**deintoxification** (de-in-toks-if-ik-a'-shun). See detoxification.

**Deiters' cells** (di'-ters) [Otto Friedrich Carl Deiters, German anatomist, 1834-1863]. 1. The branched, flattened cells of the neuroglia. 2. The cylindricoconical cells resting upon the basilar membrane of Corti's organ and supporting the hair-cells. D.'s **nucleus**, a large nucleus situated in the oblongata between the inner portion of the cerebral peduncles and the restiform body. D.'s **phalanges**, the phalangeal processes of Deiters' cells in the organ of Corti. D.'s **process**, the axis-cylinder process of a nerve-cell; the neuraxon.

**dejecta** (de-jekt'-ah) [dejicere, to throw down]. Intestinal evacuations; alvine discharges; fecal matter. Excrementitious matter in general.

**dejection** (de-jek'-shun) [dejecta]. The discharge of fecal matter; the matter so discharged. Also a state of despondency.

**dejecture** (de-jek'-chur) [dejecta]. Matter evacuated from the intestine; feces.

**Déjérine's disease** (da-zher-ēn') [Joseph Jules *Déjérine*, French neurologist, 1849- ]. Hypertrophic interstitial neuritis of infancy.

**Déjérine-Sottas' disease**, D.-S.'s **type of muscular atrophy**. See *Déjérine's disease*.

**delacerate** (de-las'-ur-āt) [delacerare]. To tear to pieces, or lacerate severely.

**delaceration** (de-las-er-a'-shun) [delaceratio]. To tear to pieces or lacerate severely.

**delactation** (de-lak-ta'-shun). See ablactation.

**Delafield's hematoxylin** (del'-a-fēld) [Francis *Delafield*, New York physician, 1841- ]. Dissolve 4 Gm. of hematoxylin in 25 Cc. of absolute alcohol, and add 400 Cc. of a saturated aqueous solution of ammonium alum. Expose to light and air for 3 or 4 days; filter; add to the filtrate 100 Cc. each of glycerol and methyl-alcohol. An excellent nuclear stain.

**delamination** (de-lam-in-a'-shun) [de, away; lamina, a plate]. The splitting into layers.

**deleterious** (del-et-e'-re-us) [δηλητήριος, hurtful]. Hurtful, injurious.

**Delhi boil** (del'-he) [city in India]. See *furunculus orientalis.*

**delicate** (del'-ik-at) [delicatus, delicate]. Of a refined constitution. Feeble. In a condition of poor health.

**deligation** (del-ig-a'-shun) [deligatio, a binding]. Ligation, as of an artery.

**delimitation** (de-lim-it-a'-shun) [delimitare, to mark out]. The determination of the limits of areas, regions, or organs in physical diagnosis.

**deliquation, deliquiation** (del-ik-wa'-shun, del-ik-we-a'-shun). Deliquescence.

**deliquescence** (del-ik-wes'-ens) [deliquescere, to melt away]. A liquefaction by absorption of water from the atmosphere.

**deliquescent** (del-ik-wes'-ent) [see deliquescence]. Dissolving; applied especially to salts that absorb moisture from the air and liquefy.

**deliquium** (del-ik'-we-um) [L.]. An absence. d. **animi**. 1. Failure of the mind; mental decay; melancholy; lowness of the spirits. 2. Syncope or fainting.

**delirament** (de-lir'-am-ent) [deliramentum, delirium]. Delirium.

**delire à Java**. See *lata*.

**deliriant, delirifacient** (de-lir'-e-ant, de-lir'-e-fa'-she-ent) [delirium]. Producing delirium.

**delirious** (de-lir'-e-us) [delirium]. Affected with delirium.

**delirium** (de-lir'-e-um) [L., "madness"]. A condition of mental excitement with confusion and usually hallucinations and illusions. d., **alcoholic**. See *d. tremens*. d. **constantum**, the constant repetition and expression of a single fixed idea, characteristic of the delirium of insane persons having fever. d. **cordis**, a violent, tumultuous beating of the heart. d., **depressive**, a form of general delirium in which there is a marked torpidity as to ideas, feelings, and determinations. d., **Dupuytren's**. See d. nervosum. d., febrile, the delirium of fever. d. **of grandeur**, a condition in which an individual has insanely exaggerated ideas of his own importance or of his possessions. d., **inanition**. That occurring in a person weakened by a febrile affection. d.

**nervosum**, the delirium following severe surgical operations or injuries. d. **of persecution**, that in which the patient imagines himself the object of persecution. d., **primordial**, a form marked by ideas which persistently dominate the mind. d., toxic, delirium caused by poisons. d., **traumatic**. See d. nervosum. d. **tremens**, that arising from alcoholic poisoning. Characterized by constant tremor, insomnia, great exhaustion, distressing illusions, and hallucinations.

**delitescence** (del-it-es'-ens) [delitescere, to lie hid]. The sudden disappearance of inflammation by resolution.

**deliver** (de-liv'-er) [de, from; liberare, to free]. To free from something, especially to deliver a woman of a child or of the after-birth. The word is also applied to the part removed, as to deliver the placenta or a tumor.

**delivery** (de-liv'-er-e) [see deliver]. The act of delivering or freeing from something, especially the relieving of a woman of the contents of the uterus; parturition; childbirth. d., **postmortem**, the birth of a fetus after the death of the mother.

**delomorphous** (de-lo-mor'-fus) [δῆλος, conspicuous; μορφή, form]. Having a conspicuous form. d. **cells of Rollet**. See under *Rollet*.

**delphini oleum** (del-fi'-ni o'-le-um) [L.]. The oil of the common porpoise, *phocæna communis*. It is said to have all the medicinal virtues of cod-liver oil, without the disagreeable qualities of the latter.

**delphine, delphinium, delphinoidine, delphisine** (del'-fin-ēn, del-fin'-e-um, del-fin-oid'-ēn, del'-fis-ēn). See *staphisagria*.

**delta** (del'-tah) [δέλτα, Δ, the fourth letter of the Greek alphabet]. 1. Any triangular space. 2. The vulva, from its triangular shape. d. **fornicis**, a triangular area of the ventral surface of the fornix dorsad of the portæ; lyra fornicis. d. **mesoscapulæ**, the triangular area at the root of the spine of the scapula.

**deltoid** (del'-toid) [δέλτα; εἶδος, likeness]. 1. Delta-shaped. 2. A muscle of the shoulder. See *muscles, table of*. d. **ligament**, the internal lateral ligament of the ankle-joint. d. **ridge**, the ridge on the humerus for the insertion of the deltoid muscle. **deltoideus** (del-toid-e'-us). See *muscles, table of*.

**delusion** (de-lū'-shun) [de, from; lusus, play]. A false belief, the falsity of which is apparent, but out of which the person cannot be reasoned by indubitable evidence. d.s, **expansive**, d.s, **large**, a symptom of the second stage of general paralysis of the insane, in which the patient conceives ideas involving colossal size, magnificent wealth, or extravagant numbers. **delusional** (de-lū'-zhun-al) [delusion]. Of the nature of a delusion; characterized by delusions. d. **stupor**. See *insanity, confusional*.

**demagnetization** (de-mag-net-i-za'-shun). The act of depriving an object of magnetic properties.

**demagnetize** (de-mag'-net-īz). To deprive an object of magnetic properties.

**demarcation** (de-mark-a'-shun) [demarcare, to set the bounds of]. Separation. d., **line of**, a red line forming at the edge of a gangrenous area and marking the limit of the process.

**Demarquay's symptom** (de-mar-ka') [Jean Nicolas *Demarquay*, French surgeon, 1811-1875]. Immobility or lowering of the larynx during deglutition and phonation; it is characteristic of tracheal syphilis.

**demedication** (de-med-ik-a'-shun). The removal of deleterious drugs from the system, as lead, arsenic, or phosphorus, by the reversal of the electric current used in cataphoresis, in a suitably arranged bath.

**demembration** (de-mem-bra'-shun) [demembrare, to deprive of a limb or limbs]. The cutting off of a member; amputation; castration.

**dement** (de'-ment) [dementia]. A person suffering with dementia.

**dementation** (de-men-ta'-shun) [dementia]. Loss of mind; insanity.

**demented** (de-ment'-ed). Deprived of reason.

**dementia** (de-men'-she-ah) [de, from; mens, the mind]. A form of insanity characterized by a deterioration or loss of the intellectual faculties, the reasoning power, the memory, and the will. d., **paralytic**, general paralysis of the insane. See *paresis*. d. **paranoides**, a form of d. præcox, characterized by paranoiac delusions. d. **præcox**, a form which appears at the age of puberty in children previously intellectually bright; there are various

delirious symptoms at the beginning; constant sudden impulses and rapid termination in a dementia which is more or less complete. **d.**, primary, that occurring independently of other forms of insanity. **d.**, secondary, that following another form of insanity. **d.**, senile, that due to the degenerations of old age. **d.**, terminal, that coming on toward the end of other forms of insanity or certain nervous diseases.
**demi-** (*dimidius*, half]. A prefix meaning half.
**demifacet** (*dem-e-fas'-et*) [*demi-*; *facet*]. One-half of an articulation surface adapted to articulate with two bones.
**demilune cells** (*dem'-e-lūn*). See *Adamkiewicz*, *Gianuzzi* and *Heidenhain*.
**demimonstrosity** (*dem-e-mon-stros'-it-e*) [*demi*, half; *monstruosus*, monstrous]. A variety of congenital deformity that does not give rise to appreciable disorder of function.
**demineralization** (*de-min-er-al-iz-a'-shun*). Increase in the elimination of mineral salts. **d., coefficient of**, the quantity of mineral matter as compared with the total solids, in the urine.
**demipenniform** (*dem-e-pen'-e-form*) [*demi-*; *penna*, a wing]. Applied to structures or organs which have one of two margins winged.
**Demodex** (*de'-mo-deks*) [δημός, fat; δήξ, an insect]. A genus of parasitic insects. **D. folliculorum**, the pimple-mite, a minute parasite found in the sebaceous follicles, particularly of the face. It probably does not produce any symptoms.
**demography** (*de-mog'-ra-fe*) [δήμος, the people; γράφειν, to write]. The science of peoples collectively considered; social science, including that of vital statistics and the consideration of questions of state medicine. **d., dynamic**, a study of the activities of human communities, their rise, progress, and fall. **d., static**, a study of the anatomy of a human community, its numbers, the sex, age, wealth, calling, etc., of the people.
**demonomania** (*de-mon-o-ma'-ne-ah*) [δαίμων, a devil; μανία, madness]. A form of madness in which a person imagines himself possessed of a devil.
**demonomaniac** (*de-mo-no-ma'-ne-ak*) [δαίμων, a devil; μανία, madness]. One who suffers with demonomania.
**demonomy** (*de-mon'-om-e*) [δαίμων, the people; νόμος, a law; a custom]. The science of humanity.
**demonopathy** (*de-mon-op'-a-the*) [δαίμων, a demon; πάθος, disease]. Same as *demonomania*.
**demonophobia** (*de-mon-o-fo'-be-ah*) [δαίμων, a devil; φόβος, fear]. Morbid dread of devils and demons.
**demonstrator** (*dem'-on-stra-tor*) [*demonstrare*, to show]. 1. One who instructs in the practical application of the arts and sciences. 2. An assistant or subordinate teacher. 3. The index finger
**De Morgan's spots** (*de-mor'-gan*) [Campbell *De Morgan*, English physician, 1811-1876]. Bright red nevoid spots frequently seen on the skin in cases of cancer.
**demorphinization** (*de-morf-in-i-za'-shun*) [*de*, from; *morphine*]. Treatment of morphinism by gradual withdrawal of the drug.
**Démours' membrane** (*dem-oor'*) [Pierre *Demours*, French ophthalmologist, 1702-1795]. See *Descemet's membrane*.
**demulcent** (*de-mul'-sent*) [*demulcere*, to soothe]. 1. Soothing; allaying irritation of surfaces, especially mucous membranes. 2. A soothing substance, particularly a slippery, mucilaginous liquid.
**(de) Mussey's point**, de **M.'s symptom**. See *Mussey's (de) point*.
**denarcotized** (*de-nar'-ko-tīzd*) [*de*, priv.; *narcotine*]. 1. Deprived of narcotizing qualities. 2. Of opium, deprived of narcotine.
**denatured** (*de-na'-churd*). Changed, made different from normal. **d. alcohol**, ethyl alcohol which has been rendered unfit for drinking by the addition of methyl alcohol and benzine.
**denaturization** (*de-nat-ū-ri-za'-shun*) [*de*, priv.; *natura*, nature]. Alteration in the characteristics of an organic substance by chemical action, boiling, or addition.
**dendraxon** (*den-draks'-on*) [*dendron*; *axon*]. Von Lenhossék's term for a neuron with a short axon, its axonal processes being for the most part devoid of sheaths.
**dendric** (*den'-drik*) [*dendron*]. Provided with dendrons.
**dendrite** (*den'-drīt*). See *dendron*.

**dendritic, dendroid** (*den-drit'-ik*, *den'-droid*) [*dendron*]. Branching like a tree.
**dendron** (*den'-dron*) [δένδρον, a tree]. One of the short, free projections or socalled protoplasmic processes of a nerve-cell.
**dengue** (*dong'-ga*) [West Ind.]. An acute, epidemic, infectious disease, characterized by a febrile paroxysm, severe pains in the bones, joints, and muscles, and, at times, a cutaneous eruption. The period of incubation is from 3 to 5 days; the invasion is sudden, with high fever (106° F.), severe pains in the muscles, bones, and joints, the last being swollen and reddened. After the fever has lasted 3 or 4 days it subsides, but at the end of from 2 to 4 days a second paroxysm accompanied with pain occurs. Convalescence is slow; complications are rare. Syn., *breakbone fever*; *dandy fever*.
**denidation** (*de-ni-da'-shun*) [*de*, priv.; *nidus*, a nest]. The disintegration and ejection of the superficial part of the uterine mucosa.
**Denigès' test for formaldehyde in milk** (*den-e-zha'*) [Georges *Denigès*, French chemist, 1859- ]. Make a solution of 40 Cc. of 0.5 % solution of fuchsin; 250 Cc. distilled water; 10 Cc. of sodium bisulphite, sp. gr. 1.375; 10 Cc. pure sulphuric acid. To 1 Cc. of this solution add 10 Cc. of the suspected milk and let it stand 5 minutes. Then add 2 Cc. of pure hydrochloric acid and shake. In the presence of formaldehyde a violet color will appear; a yellowish-white color in its absence. **D.' test for uric acid**, convert uric acid into alloxan by the action of nitric acid; expel the excess of nitric acid by gentle heat, and treat with a few drops of sulphuric acid and a few drops of commercial benzol (containing thiophen); a blue coloration will result.
**denigration** (*de-ni-gra'-shun*) [*denigrare*, to blacken]. The act or process of rendering black; the state of having become black.
**Denisensko's method** (*den-is-en'-sko*). The subcutaneous injection of a watery extract of *Chelidonium majus* in the treatment of cancer.
**denitration** (*de-ni-tra'-shun*). The process of taking away nitric acid from a compound.
**denitrify** (*de-ni'-tre-fi*) [*de*, priv.; *nitrogen*]. To remove nitrogen.
**denitrifying** (*de-ni'-tre-fi-ing*). Applied to bacteria which reduce nitric acid to nitrous acid and ammonia.
**Denonvillier's fascia** (*de-non-g-vēl-ya*) [Charles Pierre *Denonvillier*, French surgeon, 1808-1872]. The rectovesical fascia between the prostate gland and rectum.
**de novo** (*de no'-vo*) [L.]. Anew.
**dens** (*denz*) [L.: *pl.*, *dentes*]. 1. A tooth. See *teeth*. 2. The tooth-like process on the body of the axis, going through the front part of the ring of the atlas. **d. serotinus**, a wisdom tooth.
**densimeter** (*den-sim'-et-er*) [*densus*, dense; μέτρον, a measure]. An appliance for ascertaining the specific gravity of a liquid.
**densimetric** (*den-sim-et'-rik*). Having reference to the use of the densimeter.
**density** (*den'-sit-e*) [*densitas*; thickness]. Closeness; compactness, especially the degree of closeness of one body compared with an equal volume of another taken as a standard: specific gravity. In electricity, the amount of electricity accumulated on a unit of surface during a given time.
**dentagra** (*den-ta'-grah*) [*dens*, a tooth; ἄγρα, a seizure]. 1. Toothache. 2. A tooth-forceps.
**dental** (*den'-tal*) [*dens*]. Pertaining to the teeth. **d. arch**, the arch in the alveolar process. **d. bulb**, the dental papilla. **d. engine**, a machine worked by a treadle and possessing a flexible cable and adjustable arm and hand-piece, which afford great facility of movement and adaptation. By means of attachments the hand-piece drills can be operated at various angles. **d. germ**, the rudiment of a tooth. **d. sac**, the sac that encloses the developing tooth in the embryo. **d. tubuli**, the minute wavy tubes occurring in the dentine of teeth.
**dentalgia** (*den-tal'-je-ah*) [*dens*; ἄλγος, pain]. Toothache.
**dentalis lapis** (*den-ta'-lis lap'-is*) [L.]. Salivary calculus; tartar of the teeth.
**dentarpaga** (*den-tar'-pa-gah*) [*dens*; ἁρπάγη, hook]. An instrument for the extraction of teeth.
**dentata** (*den-ta'-tah*). See *axis* (2).
**dentate** (*den'-tāt*) [*dens*]. Toothed; having a toothed or serrated edge. **d. body**. See *corpus dentatum*. **d. convolution**, a convolution found in

the hippocampal fissure. **d. fascia,** the serrated free edge of the dentate convolution. **d. fissure,** the hippocampal fissure.
**dentation** (*den-ta'-shun*). The formation of tooth-like structures, as on the margin of a leaf.
**dentatum** (*den-ta'-tum*) [L.]. The dentate nucleus of the cerebellum.
**dentelation** (*den-tel-a'-shun*). The condition of being furnished with tooth-like processes.
**dentes** (*den'-tēz*) [L., plural of *dens*]. Teeth. See *teeth*. **d. acuti,** the incisor teeth. **d. adulti,** the teeth of second dentition. **d. adversi,** the incisor teeth. **d. angulares,** the canine or cuspid teeth, so called probably because they are situated at the angles of the alveolar arch, at the corners of the mouth, or from the angular shape of their crowns. **d. bicuspidati,** bicuspid teeth. **d. canini,** the cuspid or canine teeth; so called from their resemblance to the teeth of a dog. **d. cariosi,** carious teeth. **d. columellares,** the molar teeth. **d. cuspidati,** cuspid teeth. **d. exserti** (*exsertere*), to thrust out, teeth that project or are in front of the dental arch, but applied more particularly to the cuspids. **d. incisores,** incisor teeth. **d. lactei,** the milk, temporary, or deciduous teeth. See *deciduous teeth*. **d. molares,** molar teeth. **d. primores,** the incisor teeth; so called because they occupy the front or anterior part of the dental arch. **d. sapientiæ,** the wisdom-teeth. A name given to the third molar tooth of each half of the jaws. **d. tomici** (*tomicus*, cutting), the incisor teeth.
**dentiaskiascope** (*den-te-ah-ski'-a-skŏp*) [*dens; skiascope*]. An instrument for examining the teeth and alveoli. It consists of a small fluorescent screen within an aluminum case, so situated that the screen image is reflected upon a mirror which the operator sees through a tube.
**denticle** (*den'-tik-l*) [*denticulus*, a small tooth]. A small tooth or projecting point.
**denticulate** (*den-tik'-u-lāt*) [*denticle*]. Having minute dentations; furnished with small teeth or notches.
**denticulus** (*den-tik'-u-lus*) [L.]. A little tooth.
**dentier** (*don(g)-te-a'*) [Fr.]. A French word signifying a base of metal, ivory or any other substance, employed as a support or attachment for artificial teeth. The term is also sometimes applied to a set of artificial teeth.
**dentification** (*den-tif-ik-a'-shun*) [*dens; facere*, to make]. The formation of teeth; dentition.
**dentiform** (*den'-tif-orm*) [*dens; forma*, shape]. Odontoid, tooth-like.
**dentifrice** (*den'-tif-ris*) [*dens; fricare*, to rub]. A substance for cleansing the teeth.
**dentigerous** (*den-tij'-er-us*) [*dens; gerere*, to carry]. Bearing or containing teeth, as a *dentigerous cyst*.
**dentilabial** (*den-te-la'-be-al*) [*dens; labium*, lip]. Relating to the teeth and lips.
**dentilave** (*den'-te-lāv*) [*dens; lavare*, to wash]. A mouth-wash or tooth-wash.
**dentilingual** (*den-ti-lin'-gwal*) [*dens; lingua*, tongue]. Relating to the teeth and tongue.
**dentinal** (*den'-tin-al*). Pertaining to or composed of dentine. **d. fibers,** the protoplasmic substance in the dentinal tubules. **d. papillæ,** the forerunners of the dentinal pulp. **d. tubules,** canals in the matrix of dentine.
**dentinalgia** (*den-tin-al'-je-ah*) [*dentine;* ἄλγος, pain]. Pain in dentine.
**dentine, dentin** (*den'-tēn, den'-tin*) [*dens*]. A modified osseous tissue forming the principal part of a tooth, and consisting, histologically, of dental tubuli and intertubular tissue, chemically, of the phosphates of calcium and magnesium, the carbonate and fluoride of lime, and organic matter, chiefly gelatin. The bony structure of the tooth lying under the enamel of the crown and the cement substance of the root. **d., secondary,** adventitious deposits of dentine which occur in or upon the dental pulp after tooth-formation is complete.
**dentinification** (*den-tin-if-ik-a'-shun*) [*dentine; facere*, to make]. The formation of dentine through the agency of specialized cells, the odontoblasts.
**dentinitis** (*den-tin-i'-tis*) [*dentine;* ιτις, inflammation]. Inflammation of the dentinal fibrils.
**dentinoid** (*den'-tin-oid*). 1. Similar to dentine. 2. Pertaining to an odontoma.
**dentinosteoid** (*den-tin-os'-te-oid*) [*dentine;* ὀστέον, bone]. A tumor of dentine and bone.
**dentiphone** (*den'-tif-ōn*) [*dens;* φωνή, a voice].

A form of audiphone in which the vibrating disc is attached to the teeth.
**dentist** (*den'-tist*) [*dens*]. One who practises dentistry.
**dentistry** (*den'-tis-tre*). Dental surgery, embracing everything pertaining to the prevention and treatment of diseases of the teeth.
**dentition** (*den-tish'-un*) [*dens*]. Teething; the cutting of the teeth. **d., primary,** the cutting of the temporary or milk-teeth. **d., secondary,** the eruption of the 32 permanent teeth.
**dentoid** (*den'-toid*) [*dens;* εἶδος, resemblance]. Tooth-like.
**dentoiletta** (*dent-wah-let'-ah*). A device consisting of two mirrors so arranged that persons may examine their own teeth.
**dentola** (*den'-to-lah*). A solution used on swollen gums, said to consist of cocaine hydrochloride, 1 part; potassium bromide, 10 parts; glycerol and water, each, 200 parts.
**dentolingual** (*den-to-ling'-wal*). Pertaining to the teeth and the tongue or lingual nerve.
**dentoliva** (*den-tol'-iv-ah*) [*dens; oliva*, an olive]. The olivary nucleus.
**dentomental** (*den-to-ment'-al*). Pertaining to the teeth and chin.
**dentonasal** (*den-to-na'-zal*). Pertaining to the teeth and nose.
**denture** (*den'-chur*) [*dens*]. 1. The entire set or group of teeth; the whole assemblage of teeth in both jaws. 2. A set, or plate, of artificial teeth. #
**Denucé's ligament** (*den-oo sa'*) [Maurice *Denucé*, French surgeon, 1859– ]. A short and broad fibrous band in the wrist-joint, connecting the radius with the ulna.
**denucleated** (*de-nu'-kle-a-ted*). Deprived of the nucleus.
**denudation** (*den-u-da'-shun*) [*denudare*, to denude]. A stripping or making bare.
**denuding** (*de-nu'-ding*) [*denudare*, to denude]. A stripping or making bare. **d. of the teeth,** an affection that consists in the gradual destruction of the enamel of the anterior or labial surfaces of the incisors, cuspids, and sometimes of the bicuspids; the molars are rarely affected by it.
**denutrition** (*de-nu-trish'-un*) [*de,* from; *nutrire*, to nourish]. 1. Faulty or absent nutrition. 2. An atrophy and degeneration of tissue arising from lack of nutrition.
**deobstruent** (*de-ob'-stroo-ent*) [*de; obstruere*, to obstruct]. 1. Removing obstruction. 2. A medicine that removes obstruction; an aperient.
**deodorant** (*de-o'-dor-ant*) [*de; odorare*, to smell]. 1. Removing or concealing offensive odors. 2. A substance that removes or conceals offensive odors.
**deodoriferant** (*de-o-do-rif'-er-ant*) [see *deodorant*]. 1. Possessing the power of overcoming bad odors. 2. See *deodorant*.
**deodorized** (*de-o'-dor-īzd*) [see *deodorant*]. Deprived of odor.
**deodorizer** (*de-o'-dor-i-zer*) [*de,* priv.; *odorare,* to smell]. A deodorizing agent; a substance that destroys offensive odors.
**deontology** (*de-on-tol'-o-je*) [δέον, right, binding; λόγος, science]. The science of duty. **d., medical,** medical ethics.
**deoppilant, deoppilative** (*de-op'-il-ant, -at-iv*) [*de; oppilare,* to stop]. The same as *deobstruent*.
**deorsum** (*de-or'-sum*) [L.]. Downward. **d. vergens.** See *vergens*.
**deorsumduction** (*de-or-sum-duk'-shun*). A downward movement, as of the eye.
**deossification** (*de-os-if-ik-a'-shun*) [*de,* away; *os,* bone; *facere,* to make]. The absorption of bony material; the deprivation of any part of its bony character.
**deoxidation** (*de-oks-id-a'-shun*) [*de,* from; *oxygen*]. The removal of the oxygen from a chemical compound.
**deoxygenation** (*de-oks-e-jen-a'-shun*). The process of removing oxygen from a compound.
**deozonize** (*de-o'-zon-īz*) [*de,* from; *ozone*]. To deprive of ozone.
**depancreatize** (*de-pan'-kre-at-īz*). To remove the pancreas.
**dephlegmation** (*de-fleg-ma'-shun*) [*de,* from; φλέγειν, to burn]. The removal of water by distillation.
**dephlogisticate** (*de-flo-jis'-tik-āt*) [*de-,* priv.; φλόγωσις, inflammation]. To lessen inflammation in a part.

**depigmentation** (*de-pig-ment-a'-shun*). The removal of natural pigments from the skin or from microscopic preparations by the action of weak solutions of bleaching or oxidizing solutions.
**depilate** (*dep'-il-āt*) [*depilare*, to remove the hair]. To remove the hair.
**depilation** (*dep-il-a'-shun*) [*depilate*]. The removal or loss of the hair.
**depilatory** (*de-pil'-at-o-re*) [*depilate*]. 1. Having the power to remove the hair. 2. A substance, usually a caustic alkali, used to destroy the hair.
**depilous** (*dep'-il-us*) [*depilate*]. Hairless.
**deplanate** (*dep'-lan-āt*) [*deplanare*, to level]. Leveled; flattened.
**deplethoric** (*de-pleth'-or-ik*) [*de*, priv.; *plethora*]. Marked by absence of plethora.
**depletion** (*de-ple'-shun*) [*deplere*, to empty]. 1. The act of diminishing the quantity of fluid in the body or in a part, especially by bleeding. 2. The condition of the system produced by the excessive loss of blood or other fluids.
**depletive, depletory** (*de-ple'-tiv, de-ple'-tor-e*) [*deplere*, to empty]. 1. Causing or tending toward depletion. 2. A medicine that depletes.
**deplumation** (*de-plu-ma'-shun*) [*de*, down; off; *pluma*, feather]. The loss of the eyelashes.
**depolarization** (*de-po-lar-iz-a'-shun*) [*de*; *polus*, pole]. The neutralization of polarity.
**depolarizer** (*de-po'-lar-i-zer*). A refracting plate used with a polarizer which resolves the polarized ray into ordinary and extraordinary rays.
**deportation** (*de-por-ta'-shun*) [*de*; *portare*, to bear]. Veit's term for the process in which the chorionic fringes are detached and lose all connection with the fetal placenta.
**deposit** (*de-pos'-it*) [*de*; *ponere*, to place]. A sediment; a collection of morbid particles in a body.
**depositive** (*de-pos'-it-iv*) [*deposit*]. A term applied to that state of the skin in which lymph is poured out and papules arise.
**depravation** (*dep-rav-a'-shun*) [*depravare*, to vitiate]. A deterioration or morbid change in the secretions, tissues, or functions of the body.
**depraved** (*de-prāvd'*). Corrupt, perverted or vitiated.
**depressant** (*de-pres'-ant*) [see *depression*]. Lowering. 2. A medicine that diminishes functional activity.
**depressed** (*de-prest'*) [see *depression*]. 1. Referring to a state of lowered vitality; affected with depression. 2. Having the dorsolateral diameter reduced. 3. Flattened from above downward. **d. fracture**, a cranial fracture with sinking of the bone.
**depression** (*de-presh'-un*) [*deprimere*, to depress]. 1. A hollow or fossa. 2. Inward displacement of a part, as of the skull. 3. Lowering of vital functions under the action of some depressing agent.
**depressive** (*de-pres'-iv*). Causing depression.
**depressomotor** (*de-pres-o-mo'-tor*) [*depression; movere*, to move]. 1. Diminishing motion. 2. An agent that diminishes the action of the motor apparatus.
**depressor** (*de-pres'-or*) [*depression*]. 1. A muscle, instrument, or apparatus that depresses. 2. A nerve, stimulation of which lowers the functional activity of a part, as the depressor nerve of the heart. 3. One of two substances found in the infundibulum part of the hypophysis, having distinct physiological properties. It produces a fall of blood-pressure. Cf. *pressor*.
**deprimens** (*dep'-rim-enz*) [L.]. Depressing. A depressing muscle. **d. oculi**, the rectus inferior muscle of the eye.
**depucelation** (*de-pū-sel-a'-shun*) [*depucelatio*]. The act of deflowering.
**depurant** (*dep'-ū-rant*) [*depurare*, to purify]. 1. Purifying; cleansing. 2. A medicine that purifies the animal economy.
**depurated** (*dep'-ū-ra-ted*). Purified; cleansed.
**depurative** (*dep'-ū-ra-tiv*) [see *depurator*]. Purifying or cleansing.
**depurator** (*dep'-ū-ra-tor*) [*depurare*, to purify]. A drug or device for cleansing.
**DeR.** A contraction and symbol of the term *reaction of degeneration*.
**deradelphus** (*der-ad-el'-fus*) [δέρη, neck; ἀδελφός, brother]. A monocephalic dual monstrosity, with fusion of the bodies above the umbilicus, and with four lower extremities and three or four upper.

**deradenitis** (*der-ad-en-i'-tis*) [δέρη, neck; ἀδήν, a gland; ιτις, inflammation]. Inflammation of the cervical glands.
**deradenoncus** (*der-ad-en-ong'-kus*) [δέρη, neck; ἀδήν, a gland; ὄγκος, mass]. Swelling of a neck-gland.
**deranencephalia** (*der-an-en-sef-a'-le-ah*) · [δέρη, neck; ἀν, priv.; ἐγκέφαλος, brain]. Teratism marked by absence of the head and brain, the neck being present.
**derangement** (*de-rānj'-ment*). Disorder of intellect; insanity.
**Derbyshire neck** (*der'-be-sher*). See *goiter*.
**Dercum's disease** (*der'-kum*) [Francis Xavier Dercum, American physician, 1856– ]. **Adiposis dolorosa**, a painful dystrophy of the subcutaneous connective tissue, somewhat resembling myxedema.
**derencephalocele** (*der-en-sef'-al-o-sēl*) [δέρη, neck; *encephalocele*]. Hernia of the brain through a fissure in the cervical vertebræ.
**derencephalus** (*der-en-sef'-al-us*) [δέρη, neck; ἐγκέφαλος, brain]. Affected with derencephalus; of the nature of derencephalus.
**derencephalus** (*der-en-sef'-al-us*)· [δέρη, neck; ἐγκέφαλος, brain]. A variety of single autositic monsters of the species anencephalus; in which the bones of the cranial vault are rudimentary, the posterior portion of the occiput absent, and the upper cervical vertebræ bifid, the brain resting in them.
**deric** (*der'-ik*) [δέρος, the skin]. External; pertaining to the ectoderm.
**derivant** (*der'-iv-ant*) [see *derivation*]. Derivative; a derivative drug.
**derivation** (*der-iv-a'-shun*) [*derivare*, to turn a stream from its banks]. The drawing away of blood or liquid exudates from a diseased part by creating an extra demand for them in some other part.
**derivative** (*de-riv'-at-iv*). 1. Producing derivation. 2. An agent that produces derivation.
**derm, derma** (*derm, der'-mah*) [δέρμα, skin]. The true skin.
**Dermacentor** (*der-ma-sen'-tor*) [δέρμα, skin; κέντωρ, a goader]. A genus of ticks, some species of which are responsible for the spread of the infecting principle of Rocky Mountain fever. D. reticularis, a tick through which is disseminated the *Piroplasma canis* and *P. hominis*.
**dermad** (*der'-mad*) [*derm; ad*, to]. Externally; toward the skin; ectad.
**dermagra** (*der-ma'-grah*). See *pellagra*.
**dermal** (*der'-mal*). Pertaining to the skin. **d. muscle**, a skin-muscle. **d. skeleton**. See *dermoskeleton*.
**dermalaxia** (*der-mal-aks'-e-ah*) [*derm*; μαλακία, softness]. Morbid softening of the skin.
**dermalgia** (*der-mal'-je-ah*). See *dermatalgia*.
**dermanoplasty** (*der-man'-o-plas-te*) [*derm;* ἀναπλάσσειν, to form anew]. Skin-grafting.
**Dermanyssus** (*der-man-is'-us*) [δέρμα, skin; νύσσειν, to prick]. A genus of itch-mites. D. avium is a species found on birds and sometimes on the human subject.
**dermapostasis** (*der-ma-pos'-ta-sis*) [*derm*; ἀπόστασις, a falling away]. A skin disease with focal induration.
**dermatagra** (*der-mat-a'-grah*). 1. Dermatalgia. 2. See *pellagra*.
**dermatalgia** (*der-mat-al'-je-ah*) [*derm*; ἄλγος, pain]. Pain in the skin unaccompanied by any structural change, and caused by some nervous disease or reflex influence.
**dermataneuria** (*der-mat-ah-nū'-re-ah*) [*derm*; **ἀ**, priv.; νεῦρον, a nerve]. Derangement of the nerve-supply of the skin, giving rise to anesthesia or paralysis.
**dermatatrophia** (*der-mat-ah-tro'-fe-ah*) [*derm; atrophy*]. Atrophy of the skin.
**dermatauxe** (*der-mat-awks'-e*) [δέρμα, skin; αὔξη, augmentation]. Thickening or hypertrophy of the skin.
**dermathemia** (*der-mat-he'-me-ah*) [*derm;* αἷμα, blood]. A congestion of the skin. Syn., *dermæmia; dermohemia*.
**dermatic** (*der-mat'-ik*) [δέρμα, the skin]. 1. Relating to the skin. 2. A remedy for diseases of the skin.
**dermatin** (*der'-mat-in*). A preparation used in dermatology. It consists of salicylic acid, 5–7 parts; starch, 7–15 parts; talc, 25–50 parts; silicic acid, 30–60 parts; kaolin, 3–9 parts. It is used as a protective.

**dermatitis** (*der-mat-i'-tis*) [*derm;* ιτις, inflammation]. An inflammation of the skin. **d. ambustionis, d. calorica,** the form due to burns and scalds. **d., blastomycetic,** a skin disease caused by a yeast-like fungus, *Blastomyces dermatitidis.* **d. congelationis.** Same as *frost-bite.* **d. contusiformis,** erythema nodosum. **d., Duhring's.** See *d. herpetiformis.* **d., electroplating,** a form of inflammation due to limedust employed in finishing electroplating, which, coming in contact with an abrasion, forms a caustic paste which in time produces an ulcer. **d. exfoliativa,** an acute or chronic inflammation of the skin, in which the epidermis is shed more or less freely in large or small scales. See *pityriasis rubra.* **d. gangrænosa,** sphaceloderma; gangrenous inflammation of the skin. **d. herpetiformis,** an inflammatory skin disease of a herpetic character, the various lesions showing a tendency to group. It is a protean disease, appearing as erythema, vesicles, blebs, and pustules, and is associated with fever, itching, and burning. **d. hiemalis,** a recurrent inflammation of the skin associated with cold weather and allied to the erythemas; the color is dark blue. It attacks distal extremities first and appears to be due to circulatory disturbance. **d. medicamentosa,** drug-eruptions; inflammatory eruptions upon the skin due to the action of certain drugs taken internally. **d. nodularis necrotica,** a necrosis of the skin and superficial layers of the corium, due primarily to changes and consequent obstruction in the bloodvessels between the cutis and subcutis. **d., oidial.** See *d., blastomycetis.* **d. papillaris, capillitii,** a chronic skin disease affecting the nape of the neck and adjacent parts, and characterized by minute red papules, which occasionally suppurate, and are usually traversed by a hair. They unite to form hard, white or reddish, keloid-looking elevations, from which a bundle of atrophied hairs protrudes. **d., primal,** that caused by contact with *primula obconica;* it is due to the fine hairs on the under surface of the leaves. **d., pustular,** impetigo. **d., Roentgen-ray, d., X-ray,** that due to prolonged exposure to Roentgen-rays. **d. traumatica,** that resulting from traumatism. **d., vegetative,** elevated, vegetating lesions covered with crusts and very prone to bleeding, occurring in remissions of eczema on nurslings, and believed to be due to some infection. A similar affection in adults has been recorded under the names *eczema végétante* and *pyodermite végétante.* **d. venenata,** that produced by the local action of irritant substances. **d., X-ray.** Same as *d., Roentgen-ray.*

**dermato-** [δέρμα, skin]. A prefix signifying pertaining to the skin.

**dermato-autoplasty** (*der-mat-o-aw'-to-plas-te*) [*dermato-; autoplasty*]. Dermatoplasty by means of grafts taken from the patient's body.

**Dermatobia** (*der-mat-o'-be-ah*) [*dermato-;* βίος, life]. A bot-fly of Central America, the eggs of which are not infrequently deposited in the skin and produce a swelling very like an ordinary boil.

**dermatobiasis** (*der-mat-o-bi'-as-is*). Infection with Dermatobia.

**dermatocellulitis** (*der-mat-o-sel-ū-li'-tis*). Inflammation of the subcutaneous connective tissue.

**dermatoconiosis, dermatokoniosis** (*dur-mat-o-kon-i-o'-sis*) [*dermato-;* κονία, dust]. Any skin-disease due to dust. Cf. *enteroconiosis; pneumoconiosis.*

**dermatocyst** (*der-mat'-o-sist*) [*dermato-; cyst*]. A cyst of the skin.

**dermatodynia** (*der-mat-o-din'-e-ah*). See *dermatalgia.*

**dermatodyschroia** (*der-mat-o-dis-kroi'-ah*) [*dermato-;* δύσχροια, a bad color]. Abnormal pigmentation of the skin.

**dermatography** (*der-mat-og'-ra-fe*) [*dermato-;* γράφειν, to write]. 1. A description of the skin. 2. See *dermographia.*

**dermatoheteroplasty** (*der-mat-o-het'-er-o-plas-te*) [*dermato-; heteroplasty*]. Dermatoplasty by means of grafts taken from the body of another than the patient.

**dermatoid** (*der'-mat-oid*). See *dermoid.*

**dermatokelidosis** (*der-mat-o-kel-id-o'-sis*) [*dermato-;* κηλιδοῦν, to stain]. Pigmentation of the skin.

**dermatokeras** (*der-mat-o-ker'-as*) [*dermato-;* κέρας, a horn]. See *cornu cutaneum.*

**dermatol** (*der'-mat-ol*) [*derm*], $C_6H_2$ (OH)$_3$ $CO_2Bi(OH)_2$. Bismuth subgallate, an astringent, antiseptic powder, of yellow color, used in affections of the skin and mucous membranes that are associated with excessive secretion; it is especially recommended for diarrhea in tuberculosis and typhoid fever. Dose internally $\frac{1}{2}$–1$\frac{1}{2}$ dr. (2–6 Gm.) daily.

**dermatologist** (*der-mat-ol'-o-jist*) [see *dermatology*]. A skin specialist.

**dermatology** (*der-mat-ol'-o-je*) [*dermato-;* λόγος, science]. The science of the skin, its nature, structure, functions, diseases and treatment.

**dermatolysis** (*der-mat-ol'-is-is*) [*dermato-;* λύσις, a loosing]. A hypertrophy of the skin and subcutaneous tissue, with a tendency to the formation of folds.

**dermatoma** (*der-mat-o'-mah*) [*dermato-;* ὄμα, a tumor]. A tumor of the skin.

**dermatomalacia** (*der-mat-o-mal-a'-se-ah*) [*dermato-;* μάλακια, softness]. Morbid softening of the skin.

**dermatome** (*der'-ma-tōm*) [*dermato-;* τέμνειν, to cut]. An instrument for incising the skin.

**dermatomelasma** (*der-mat-o-mel-az'-mah*) [*dermato-;* μέλασμα, a black color]. Addison's disease.

**dermatomere** (*dur-mat'-o-mēr*) [*dermato-;* μέρος, a part]. The integumentary portion of the embryonic metamere.

**dermatomucosomyositis** (*der-mat-o-mū-ko-so-mi-o-si'-tis*). Inflammation involving the skin, mucosa, and muscles.

**dermatomycosis** (*der-mat-o-mi-ko'-sis*) [*dermato-; mycosis*]. Any skin disease caused by a vegetable parasite. **d. achorina.** See *favus.* **d. barbæ nodosa.** See *sycosis parasituria.* **d. diffusa.** See *tinea imbricata.* **d. favosa,** favus of the skin, exclusive of that of the hair and nails. **d. furfuracea.** See *tinea versicolor.* **d. maculovesiculosa.** See *tinea trichophytina.* **d. marginata.** See *eczema marginatum.* **d. palmellina,** a parasitic disease described by Pick as affecting the axillæ, the chest, the backs of the hands, the inner surfaces of the thighs, and the pubes. It is characterized by the presence of rounded spores adherent to the hairs. **d. pustulosa.** See *impetigo contagiosa.*

**dermatomyoma** (*der-mat-o-mi-o'-mah*) [*dermato-;* μῦς, muscle; ὄμα, a tumor; *pl., dermatomyomata*]. Myoma seated upon or involving the skin.

**dermatomyositis** (*der-mat-o-mi-o-si'-tis*) [*dermato-; myositis*]. An infectious inflammation of both skin and muscles, accompanied by edema, fever, and general depression. Cf. *myositis; polymyositis.*

**dermatoneuria** (*der-mat-o-nū'-re-ah*). See *dermatoneurosis.*

**dermatoneurology** (*der-mat-o-nū-rol'-o-je*) [*dermato-; neurology*]. Neurology limited to the skin.

**dermatoneurosis** (*der-mat-o-nū-ro'-sis*) [*dermato-; neurosis*]. A neurosis of the skin. **d. indicatrix,** an eruption of the skin due to nervous disease and indicative of more serious symptoms. **d., stereographic,** a form characterized by an elevation, welt, or wheal corresponding in size and shape to the object the application of which produced the elevation. It is due to an extravasation of serum, and is illustrated in the welts that follow light blows of the whip on a nervous horse.

**dermatonosis** (*der-mat-on'-o-sis*) [*dermato-;* νόσος, disease]. Any disease of the skin.

**dermatonosus** (*der-mat-on'-o-sus*) [*dermato-;* νόσος, disease]. Any skin-disease. **d., neuropathic,** any cutaneous disease of nervous origin, as angioneurosis, trophoneurosis, and idioneurosis.

**dermatopathology** (*der-mat-o-path-ol'-o-je*). The pathology of the skin.

**dermatopathy** (*der-mat-op'-ath-e*) [*dermato-;* πάθος, disease]. Any skin-disease.

**Dermatophilus penetrans.** The jigger, chigger, or chigoe; a sand-flea of the West Indies and India; parasitic in man.

**dermatophone** (*der'-mat-o-fōn*) [*dermato-;* φωνή, sound]. A stethoscopic appliance devised by Voltolini-Hueter, by means of which one may perceive the sound of the blood-current in the skin. It also makes perceptible the muscle-tones, and in the tendons and bones it demonstrates the transmitted vibrations. Syn., *myophone; osteophone; tendophone.*

**dermatophyte** (*der'-mat-o-fīt*) [*dermato-;* φυτόν, a plant]. 1. Any species of fungous vegetation that grows upon the skin. 2. A cutaneous appendage, as a hair, feather, scale, nail, or horn.

**dermatoplasia** (*der-mat-o-pla'-ze-ah*) [*dermato-;* πλάσσειν, to form]. The reparative power of the skin to injury.

**dermatoplasty** (*der'-mat-o-plas-te*) [see *dermato-*

*plasia*]. An operative replacement of destroyed skin by means of flaps or skin-grafts.
**dermatorrhagia** (*der-mat-or-a'-je-ah*) [*dermato-*; ῥηγνύναι, to burst forth]. Hemorrhage from the skin.
**dermatorrhea** (*der-mat-or-e'-ah*) [*dermato-*; ῥοία, a flowing]. A morbidly increased secretion from the skin.
**dermatosclerosis** (*der-mat-o-skle-ro'-sis*). See *scleroderma*.
**dermatosis** (*der-mat-o'-sis*) [*dermato-*; νόσος, disease]. Any disease of the skin. d., **angioneurotic**, an infectious, toxic, or essential skin disease, characterized by a general disturbance of the vascular tension, together with inflammatory excitation at the surface of the skin. d., **engorgement**, a skin disease characterized by passive derangement of the circulation, with imperfect venous and lymphatic absorption. d., **hemorrhagic**, a traumatic or essential disease of the skin, characterized by hemorrhage from the cutaneous blood-vessels, without inflammation or stasis; dermatorrhagia. d., **neurotic**, a cutaneous affection due to disease of the sensory or trophic nerves. It may be self-limited in its course (cyclic), as herpes zoster and herpes febrilis; or it may be acyclic, as neurotic edema, neurotic atrophy, and neurotic necrosis of the skin. d., **parasitic**, a papulovesicular disease described by Nielly, in which a filaria-like parasite was found in the vesicles, and believed by him to be identical with craw-craw. d., **postvaccinal**, a dermatosis following vaccination, marked by lesions similar to those of urticaria pigmentosa except that desquamation is present and dermographism is absent.
**dermatosome** (*der'-mat-o-sōm*) [*dermat,-*; σῶμα, body]. 1. A thickening or knot in the equatorial region of each spindle-fiber in the process of cell-division. 2. One of the hypothetical ultimate units that form the membrane of vegetable cells.
**dermatospasm** (*der'-mat-o-spasm'*) [*dermato-*; σπασμός, spasm]. Cutis anserina, or goose-skin.
**dermatosyphilis** (*der-mat-o-sif'-il-is*) [δέρμα, skin; *syphilis*]. The cutaneous manifestations of syphilis. See *syphiloderma*.
**dermatotherapy** (*der-mat-o-ther'-ap-e*). The therapeutics of cutaneous affections.
**dermatotomy** (*der-mat-ol'-o-me*) [*dermato-*; τέμνειν, to cut]. The anatomy or dissection of the skin.
**dermatoxerasia** (*der-mat-o-zer-a'-se-ah*) [*dermato-*; ξηρασία, dryness]. Dryness of the skin. Cf. *xeroderma*.
**dermatozoon** (*der-mat-o-zo'-on*) [*dermato-*; ζῷον, an animal; pl., *dermatozoa*]. Any animal parasitic upon the skin.
**dermatozoonosus** (*der-mat-o-zo-on-o'-sus*) [*dermatozoon*; νόσος, a disease]. A cutaneous disease due to animal parasites, such as *acarus, filaria, ixodes, leptus, pediculus, pulex, sarcoptes*, etc.
**dermatrophia** (*der-ma-tro'-fe-ah*). See *dermatatrophia*.
**dermectasia** (*der-mek-ta'-ze-ah*). See *dermatolysis*.
**dermelminthiasis** (*der-mel-min-thi'-as-is*) [*derm*; ἕλμινς, a worm]. A cutaneous affection due to a parasitic worm. Cf. *dermatozoonosus*.
**dermen** (*der'-men*) [δέρμα, the skin]. Belonging to the derma itself.
**dermenchysis** (*der-men'-kis-is*) [*derm*; ἔγχυσις, a pouring in]. Hypodermatic injection.
**dermepenthesis** (*der-mep-en'-thes-is*) [δέρμα, skin; ἐπένθεσις, insertion]. Synonym of skin-grafting.
**dermexanthesis** (*der-meks-an-the'-sis*) [δέρμα, skin; ἐξάνθησις, eruption; pl., *dermexantheses*]. Any skin-disease marked by a rash-like eruption.
**dermic** (*der'-mik*) [*derm*]. Relating to the skin or formed of skin. d. **graft**, a skin-graft. d. **layer**, the middle layer of the membrana tympani.
**dermis** (*der'-mis*). The corium or true skin.
**dermitis** (*der-mi'-tis*). See *dermatitis*.
**dermo-** (*der-mo-*). The same as *dermato-*.
**dermoabdominalis** (*der-mo-ab-dom-in-a'-lis*). Pertaining to the skin of the abdomen.
**dermoactinomycosis** (*der-mo-ak-tin-o-mi-ko'-sis*). Infection of the skin by actinomyces.
**dermoblast** (*der'-mo-blast*) [*dermo-*; βλαστός, sprout]. The part of the mesoblast which develops into the corium.
**dermocyma, dermocymus** (*der-mo-si'-mah, der-mo-si'-mus*) [*dermo-*; κῦμα, the fetus]. A monster fetus containing another within it.
**dermoepidermal** (*der-mo-ep-e-derm'-al*). Partaking of both the superficial and deep layers of the skin; said of skin-grafts.

**dermogen** (*der'-mo-jen*). Trade name of a dusting powder, the chief constituent of which is zinc oxide.
**dermographia** (*der-mo-graf'-e-ah*) [*dermo-*; γράφειν, to write]. A condition of the skin in which tracings made with the finger-nail or a blunt instrument are followed by elevations at the points irritated. It is common in the condition termed vasomotor ataxia.
**dermographic** (*der-mo-graf'-ik*) [*dermo-*; γράφειν, to write]. Affected with dermographia. d. **pseudourticaria**. Same as *dermographia*.
**dermographism** (*der-mo-graf'-izm*) [*dermo-*; γράφειν, to write]. Autographism. See *autographic skin* and *urticaria factitia*.
**dermography** (*der-mog'-ra-fe*) [*dermo-*; γράφειν, to write]. Dermographia; dermal autographism.
**dermohemal** (*der-mo-he'-mal*) [*dermo-*; αἷμα, blood]. In biology, applied to parts of the ventral or hemal fins of certain fishes.
**dermohemia** (*der-mo-he'-me-ah*). See *dermathemia*.
**dermoid** (*der'-moid*) [*dermo-*; εἶδος, like]. 1. Resembling skin. 2. A dermoid cyst. d. **cyst**, a cyst containing elements of the skin, as hair, teeth, etc.
**dermoidectomy** (*der-moid-ek'-to-me*) [*dermoid*; ἐκτομή, excision]. Excision of a dermoid cyst.
**dermol** (*der'-mōl*). See *bismuth chrysophanate*.
**dermolabial** (*der-mo-la'-be-al*). Having relation to the skin and the lips.
**dermology** (*der-mol'-o-je*). See *dermatology*.
**dermomuscular** (*der-mo-mus'-kū-lar*). Having relation to both skin and muscles, as certain embryonic tissues.
**dermomycosis** (*der-mo-mi-ko'-sis*). See *dermatomycosis*.
**dermoneurosis** (*der-mo-nū-ro'-sis*). See *dermatoneurosis*.
**dermonosology** (*der-mo-no-sol'-o-je*). Same as *dermatonosology*.
**dermopapillary** (*der-mo-pap'-il-a-re*). Having relation to the papillary layer of the true skin.
**dermopathy** (*der-mop'-a-the*) [*dermo-*; πάθος, disease]. Any skin disease.
**dermophlebitis** (*der-mo-fleb-i'-tis*) [*dermo-*; *phlebitis*]. Inflammation of the cutaneous veins.
**dermophyma venereum** (*der-mo-fi'-mah ve-ne'-re-um*). A soft tumor or excrescence, due to syphilis, and generally found on the surface of the genital organs or rectum.
**dermophyte** (*der'-mo-fīt*). See *dermatophyte*.
**dermoplasty** (*der'-mo-plas-te*). See *dermatoplasty*.
**dermorrhagia** (*der-mor-a'-je-ah*). See *dermatorrhagia*.
**dermosapol** (*der-mo-sa'-pol*). A soap said to contain perfumed cod-liver oil, Peruvian balsam, wool-fat, fat, glycerol, and alkali. To these, specifics may be added; it is used as an inunction in skin diseases, tuberculosis, etc.
**dermoskeleton** (*der-mo-skel'-et-on*) [*dermo-*; σκελετόν, a skeleton]. The exoskeleton.
**dermostenosis** (*der-mo-sten-o'-sis*) [*dermo-*; στένωσις, stenosis]. A tightening of the skin, due to swelling of or to disease. Cf. *scleroderma*.
**dermostosis** (*der-mos-to'-sis*) [*dermo-*; ὀστέον, a bone]. Ossification occurring in the true skin.
**dermosynovitis** (*der-mo-si-no-vi'-tis*) [*dermo-*; *synovitis*]. Inflammation of a subcutaneous bursa together with the adjacent skin. d. **plantaris ulcerosa**, a severe suppuration in the sole of the foot which proceeds from inflammation of the bursa beneath a callosity and gives rise to a perforating ulcer.
**dermosyphilopathy** (*der-mo-sif-il-op'-a-the*) [*dermo-*; *syphilis*; πάθος, affection]. A syphilitic skin disease.
**dermotherapy** (*der-mo-ther'-ap-e*). See *dermatotherapy*.
**dermotomy** (*der-mot'-o-me*). See *dermatotomy*.
**derodidymus** (*der-o-did'-im-us*) [δέρη, neck; δίδυμος, double]. A monstrosity with a single body, two necks and heads, two upper and lower extremities, with other rudimentary limbs occasionally present.
**derospasmus** (*der-o-spas'-mus*) [δέρη, neck; σπασμός, spasm]. Spasm or cramp in the neck.
**derrid** (*der'-id*). A highly toxic substance from *Derris elliptica*, a leguminous plant of Malaya, used in Borneo as an arrow-poison; it kills fish in a dilution of 1 : 5,000,000.
**desalgin** (*des-al'-jin*). Colloidal chloroform; an analgesic powder containing 25 per cent. chloroform combined with albumin.
**desalination** (*de-sal-in-a'-shun*) [*de*, from; *sal*, salt]. The process of decreasing the salinity of a substance by the removal of salts.

# DE SALLE'S LINE 285 DESULPHURATION

**De Salle's line.** See *Salle's (de) line.*
**desanimania** (des-an-im-a'-ne-ah) [de-; priv.; animus, mind; μανία, mania]. Mindless insanity; amentia.
**Desault's apparatus, D.'s bandage** (des-o') [Pierre Joseph *Desault*, French surgeon, 1744-1795]. One for the arm, consisting of an axillary pad held by tapes about the neck, a sling for the hand, and two single-headed rollers. **D.'s splint**, one used in treating fracture of the thigh.
**Descartes' laws** (da-kart') [René *Descartes*, French philosopher, 1596-1650]. See *Snell's laws*.
**Descemet's membrane** (des-ma') [Jean *Descemet*, French anatomist, 1732-1810]. The elastic membrane lining the posterior surface of the cornea.
**descemetitis** (des-em-et-i'-tis). Inflammation of Descemet's membrane; serous iritis.
**descemetocele** (des-em-et'-o-sēl) [*Descemet's membrane*; κήλη, hernia]. Hernia of Descemet's membrane.
**descendens** (de-sen'-denz) [*descendere*, to go down]. Downward. **d. noni**, a branch of the hypoglossal nerve. See under *nerve*.
**descending** (de-sen'-ding) [see *descendens*]. Passing downward. **d. current**. See *current, descending*. **d. degeneration**. See *degeneration, descending*. **d. tract**, a collection of nerve-fibers conducting impulses from the centers to the periphery.
**descensus** (de-sen'-sus) [L.]. A descent, fall, prolapse. **d. ventriculi**. See *gastroptosis*.
**descent** (de-sent') [see *descendens*]. The act of going down; downward motion. **d., theory of**, the theory that all higher organisms have descended by evolution from lower forms; as opposed to the theory of spontaneous generation or special creation. Cf. *biogenesis; Darwinism; evolution*.
**Deschamps' needle** (da-shahm') [Joseph François Louis *Deschamps*, French surgeon, 1740-1824]. A needle on a long handle, used for passing sutures in deep tissues.
**Deshler's salve** (desh'-ler). The compound rosin cerate of the pharmacopeia.
**desiccant** (des'-ik-ant) [*desiccare*, to dry up]. 1, Causing desiccation; drying. 2. A drying medicine or application.
**desiccation** (des-ik-a'-shun). Process of drying.
**desiccative** (des-ik'-a-tiv) [*desiccare*, to dry up]. 1. Drying; desiccant. 2. A medicine or application having the property of drying moist tissues, ulcers and running sores.
**desiccator** (des'-ik-a-tor) [see *desiccant*]. A vessel containing some strongly hygroscopic substance, such as calcium chloride or anhydrous sulphuric acid, and used to absorb the moisture from the air of a chamber.
**desichthol** (des-ik'-thol). Deodorized ichthyol; a preparation produced from ichthyol by the removal of about 5 % of volatile oil, to which the disagreeable odor is due.
**desmameba** (des-mah-me'-bah). [δεσμός, a band; *ameba*]. A connective-tissue corpuscle considered as an ameboid element.
**desmatitis** (des-mat-i'-tis). See *desmitis*.
**desmectasia, desmectasis** (des-mek-ta'-se-ah, -mek'-ta-sis) [δεσμός, a band; ἔκτασις, a stretching]. The stretching of a ligament.
**desmepithelium** (des-mep-ith-e'-le-um) [δεσμός, a band; *epithelium*]. The endothelial or epithelial lining of the blood-vessels, lymphatics, and synovial cavities; the epithelial portions of the mesoderm.
**desmognathus** (des-mo-na'-thus) [δεσμός, a band; γνάθος, the jaw]. A monster, the lower jaw of which has a supplementary head joined to it by ligamentous or muscular attachment.
**desmitis** (des-mi'-tis) [δεσμός, a band; ιτις, inflammation]. Inflammation of a ligament.
**desmo-** (des-mo-) [δεσμός, a band]. A prefix, meaning a band, bond, or ligament.
**desmobacteria** (des-mo-bak-te'-re-ah) [*desmo-*; *bacteria*]. A group of bacteria corresponding to the genus *Bacilli*.
**desmoblast** (des'-mo-blast) [*desmo-*; βλαστός, a germ]. Rouber's term for that portion of the area opaca of the blastoderm, especially in mesoblastic ova, which gives rise to the mesenchyma. Syn., *desmohemoblast*.
**desmocyte** (des'-mo-sīt). A general term denoting any kind of supporting tissue cell.
**desmocytoma** (des-mo-si-to'-ma) [*desmocyte*; ὄγκωμα, a tumor]. A tumor composed of desmocytes; sarcoma.

**Desmodium** (des-mo'-de-um) [*desmo-*; εἶδος, form]. A genus of plants of the order Leguminosæ. An infusion of the roots of *D. erythrynæfolium*, of South America, is used in diarrhea and dysentery. The root of *D. incanum*, of the West Indies, is prized as a remedy for dysentery. The root of *D. tortuosum*, of North America and the West Indies, is purgative. *D. triflorum* is found in all tropical countries. The fresh plant is applied to abscesses and wounds.
**desmodynia** (des-mo-din'-e-ah) [*desmo-*; ὀδύνη, pain]. Pain in a ligament.
**desmogenous** (des-moj'-en-us) [*desmo-*; γεννᾶν, to produce]. Of ligamentous origin or causation.
**desmography** (des-mog'-ra-fe) [*desmo-*; γράφειν, to write]. The description of the ligaments.
**desmohemoblast** (des-mo-hem'-o-blast). See *desmoblast*.
**desmoid** (des'-moid) [*desmo-*; εἶδος, likeness]. Like a ligament; fibrous. **d. tumor**, a fibroid tumor.
**desmology** (des-mol'-o-je) [*desmo-*; λόγος, science]. The anatomy of the ligaments. Cf. *syndesmography*.
**desmoma** (des-mo'-mah) [*desmo-*; ὄγμα, tumor]. A connective-tissue tumor; a fibroma.
**desmon** (des'-mon) [δεσμός, a band]. London's name for the intermediary body of Ehrlich; an amboceptor.
**desmoneoplasm** (des-mo-ne'-o-plazm) [*desmo-*; *neoplasm*]. Any neoplasm made up of connective tissue.
**desmonosology** (des-mon-os-ol'-o-je). See *desmopathology*.
**desmopathology** (des-mo-path-ol'-o-je) [*desmo-*; *pathology*]. The pathology of ligaments.
**desmopathy** (des-mop'-a-the) [*desmo-*; πάθος, disease]. Any disease of a ligament.
**desmopexia** (des-mo-peks'-e-ah) [*desmo-*; πῆξις, a putting together]. Fixation of the round ligaments to the abdominal wall or to the wall of the vagina for correction of uterine displacement.
**desmopycnosis, desmopyknosis** (des-mo-pik-no'-sis) [*desmo-*; πύκνωσις, condensation]. Dudley's operation of shortening the round ligaments of the uterus.
**desmorrhexis** (des-mor-eks'-is) [*desmo-*; ῥῆξις, a bursting]. The rupture of a ligament.
**desmosis** (des-mo'-sis) [*desmo-*; *pl., desmoses*]. Any disease of connective tissue, especially of the connective tissue of the skin.
**desmotomy** (des-mot'-o-me) [*desmo-*; τομή, section]. The dissection and anatomy of the ligaments; surgical cutting of a ligament.
**desmotropy** (des-mot'-ro-fe). See *tautomerism*.
**desmurgia, desmurgy** (des-mur'-je-ah, des-mur'-je) [*desmo-*; ἔργειν, to do; to work]. The art of bandaging or applying ligatures.
**Desnos' pneumonia** (da-no') [Louis Joseph *Desnos*, French physician, 1828-1893]. See *Grancher's disease*.
**desolution** (de-so-lu'-shun) [*de*, away from; *solutio*, solution]. The separation from one body of another dissolved in it under certain conditions which remove or diminish the solubility of the latter.
**desquamation** (des-pū-ma'-shun) [*desquamare*, to skim froth]. The purification of a liquid by removal of the scum or froth.
**desquamation** (des-kwam-a'-shun) [*desquamare*, to scale off]. The shedding of the superficial epithelium, as of the skin, mucous membranes, and renal tubules. **d., furfuraceous**, branny desquamation. **desquamatio neonatorum**, the epidermal exfoliation of newborn infants which takes place during the first week of life. **desquamatio siliquosa**, the shedding of the skin of a part in a continuous, husk-like structure.
**desquamative** (des-kwam'-at-iv) . [*desquamation*]. Characterized by desquamation.
**dessertspoon**. A domestic measure equal to about 2 dr. (8 Cc.).
**desternalization** (de-ster-nal-i-za'-shun) [*de*, from; *sternum*]. Separation of the sternum from the costal cartilages.
**destructive** (de-struk'-tiv). Hurtful; tending to destroy. **d. distillation**. See *distillation*. **d. metabolism**. See *katabolism*.
**desudation** (des-ū-da'-shun) [*de*, away; *sudare*, to sweat]. 1. Profuse or morbid sweating. 2. Sudamina.
**desudatory** (de-sū'-dat-or-e) [*desudatio*, a sweating]. A sweating-bath.
**desulphuration, desulphurization** (de-sul-phur-a'-shun, de-sul-phur-i-za'-shun). The act or process of

abstracting sulphur from a compound. Cf. *sulphuration*.

**desumvergence** (*de-sum-ver'-jens*) [*desursum*, from above; *vergere*, to turn]. A downward inclination of the eyes.

**det.** Abbreviation for *detur* [L.]. Let it be given.

**detention** (*de-ten'-shun*) [*detinere*, to detain]. The enforced isolation of one or more individuals to prevent the spread of infectious disease; confinement.

**detergal** (*de-ter'-gal*). A proprietary liquid antiseptic soap containing cresol and thymol.

**detergent** (*de-ter'-jent*) [*detergere*, to cleanse]. 1. Purifying; cleansing; abluent. 2. A drug, compound, or solution used for cleansing wounds, ulcers, etc.

**determinant** (*de-ter'-min-ant*) [*determinare*, to limit]. Weismann's name for one of the particles of germ-plasm corresponding to a group of biophores; a primary constituent of a cell or group of cells.

**determination** (*de-ter-min-a'-shun*) [*determinatio*, a directing]. Of the blood, a tendency to collect in a part, as *determination* of the blood to the head.

**determinism** (*de-ter'-min-izm*) [*determinare*, to limit, prescribe]. In biology, a term introduced by Claude Bernard to indicate the fatality of the reproduction of phenomena under similar conditions, as seen in experimental science.

**detersion** (*de-ter'-shun*) [*detergere*, to cleanse]. The action of a detergent; a cleansing.

**detersive** (*de-ter'-siv*). Same as *detergent*.

**dethyroidism** (*de-thi'-roid-izm*). See *athyrea* and *athyreosis*.

**detonation** (*det-o-na'-shun*) [*detonare*, to thunder]. The loud noise made by sudden chemical decomposition, as of the fulminates. Cf. *fulminate*.

**detorsion** (*de-tor'-shun*) [*detorquere*, to turn]. The correction of an abnormal curvature; the restoration of a deformed part to its normal position.

**detoxication** (*de-toks-e-ka'-shun*) [*de*, priv.; τοξικόν, poison]. 1. See *detoxification*. 2. Recovery from the poisonous effects of any substance.

**detoxification** (*de-toks-if-ik-a'-shun*) [*de*, priv.; τοξικόν, poison]. The power of reducing the poisonous properties of a substance.

**detoxify** (*de-toks'-e-fi*). To deprive a substance of its poisonous attributes.

**detrital** (*de-tri'-tal*). Consisting of or pertaining to detritus.

**detrition** (*de-trish'-un*) [*deterere*, to wear off]. The wearing or wasting of a part.

**detritus** (*de-tri'-tus*) [see *detrition*]. Waste-matter from disorganization.

**detruncation** (*de-trun-ka'-shun*). See *decapitation*.

**detrusion** (*de-tru'-zhun*) [*detrudere*, to drive]. An ejection or expulsion; a thrusting or driving down or out.

**detrusor** (*de-tru'-sor*) [*detrudere*, to push down]. 1. A means or instrument for performing expulsion. 2. A muscle having as its function the forcing down or out of parts or materials. **d. urinæ.** See *table of*.

**detumescence** (*det-u-mes'-ens*) [*detumescentia*, a subsidence of a tumor]. 1. The subsidence of any swelling. 2. The impulse to evacuate the accumulated secretion of the sexual fluid (Ellis).

**deutencephalon** (*du-ten-sef'-al-on*). See *diencephalon*.

**deuter-, deutero-** (*du-ter-, du-ter-o-*) [δεύτερος, second]. Greek prefixes indicating the second of two similar substances or conditions, especially that one which contains more of the substance.

**deuteranopia** (*du-ter-an-o'-pe-ah*) [*deuter-*; *anopia*]. A defect in a second constituent essential for color-vision, as in green-blindness.

**deuteria** (*du-te'-re-ah*). The secundines.

**deuteripara** (*du-ter-ip'-ar-ah*) [*deuter-*; *parere*, to bring forth]. A woman pregnant for the second time.

**deuteroalbumose** (*du-ter-o-al'-bu-mōs*). An albumose soluble in water and not precipitated by saturation with sodium chloride or magnesium sulphate, but by ammonium sulphate.

**deuteroelastose** (*du-ter-o-e-las'-tōs*) [δεύτερος, second; ἐλαστικός, elastic]. Elastin-peptone; one of the products of the digestion of elastin. It is not precipitable by saturation with sodium chloride.

**deuterofibrinose** (*du-ter-o-fi'-brin-ōs*). A product formed from blood-fibrin by digestion.

**deuteroglobulose** (*du-ter-o-glob'-u-lōs*). One of the products formed in the digestion of paraglobulin.

**deuterology** (*du-ter-ol'-o-je*) [*deutero-*; λόγος, science]. The biology of the placenta.

**deuteromyosinose** (*du-ter-o-mi-o'-sin-ōs*). A product of myosin digestion.

**deuteropathic** (*du-ter-o-path'-ik*). Pertaining to a disease dependent on, or secondary to another.

**deuteropathy** (*du-ter-op'-a-the*) [*deutero-*; πάθος, a disease]. A disease that is secondary to another.

**deuteroplasm** (*du'-ter-o-plazm*). See *deutoplasm*.

**deuteroproteose** (*du-ter-o-pro'-te-ōs*). A secondary proteose; a soluble product of proteolysis.

**deuteroscopy** (*du-ter-os'-ko-pe*) [δεύτερος, second; σκοπεῖν, to view]. Synonym of *clairvoyance*.

**deuterostoma** (*du-ter-os'-to-mah*) [*deutero-*; στόμα, mouth]. A secondary blastopore.

**deuterotoxins** (*du-ter-o-toks'-ins*). Dissociation products of toxins.

**deuthyalosome** (*du-thi-al'-o-sōm*) [δεύτερος, second; ὑαλός, glass, crystal; σῶμα, body]. The remains of the germinal vesicle after the polar bodies have been extruded, formed by the union of portions of the chromatic stars or discs with portions of the prothyalosome.

**deutiodide** (*du-ti'-o-did*). A biniodide.

**deutipara** (*du-tip'-ar-ah*) [δεύτερος, second; *parere*, to bear]. A woman pregnant for the second time.

**deutochloride** (*du-to-klo'-rid*). The *bichloride*.

**deutoiodide** (*du-to-i'-o-did*). The biniodide.

**deutoplasm** (*du'-to-plazm*) [*deutero-*; πλάσμα, formed material]. A store of nutrient material in the ovum, from which the protoplasm draws to support its growth.

**deutoscolex** (*du-to-skle'-rus*) [*deutero-*; σκληρός, hard]. Relating to an induration secondary to some pathologic condition.

**deutoscolex** (*du-to-sko'-leks*) [δεύτερος, second; σκώληξ, worm]. In biology, applied to secondary or daughter-cysts or bladder-worms that are derived from a scolex or primary bladder-worm.

**deutospermatoblast** (*du-to-sper'-mq-to-blast*) [*deutero-*; σπέρμα, sperm; βλαστός, germ]. Any one of the cells produced by the division of a protospermoblast.

**deutyl** (*du'-til*). See *ethyl*.

**devalgate** (*de-val'-gāt*) [*de*, intensive; *valgus*, bow-leg]. Bowlegged or bandylegged.

**devaporation** (*de-va-por-a'-shun*). To bring vapor back to the liquid state.

**developer** (*de-vel'-op-er*). A chemical compound employed in photography to reduce the metallic salts and to render visible the image upon an exposed plate.

**development** (*de-vel'-op-ment*) [Fr., *developper*, to unfold]. The sequence of organic changes, by which the fertilized ovum becomes the mature animal or plant.

**Deventer's diameter** [Hendrik à Deventer, Dutch obstetrician, 1651–1724]. The oblique pelvic diameter. **D.'s pelvis**, a simple, nonrhachitic pelvis, flattened from before backward.

**Devergie's attitude** (*de-ver-zhe'*) [Marie Guillaume Alphonse Devergie, French physician, 1798–1879]. A posture of a dead body marked by flexions of the elbows and knees, with closure of the fingers and extension of the ankles. **D.'s disease**, lichen ruber.

**deviation** (*de-ve-a'-shun*) [*deviare*, to deviate]. Turning from a regular course, standard, or position. **c., conjugate**, the forced and persistent turning of eyes and head toward one side, observed with some lesions of the cerebrum. **d. of complement**, see under *complement*. **d., primary**, the deviation of the weaker eye from that position that would make its visual line pass through the object-point of the healthy eye. **d., secondary**, the deviation of the healthy eye from that position that would make its visual line pass through the object-point of the weaker eye. **d. of teeth**, a faulty direction or position of one or more teeth.

**deviometer** (*de-ve-om'-et-er*). A variety of strabismometer.

**devisceration** (*de-vis-er-a'-shun*). See *evisceration*.

**devitalize** (*de-vi'-tal-īz*) [*de*, from; *vita*, 'life]. To destroy vitality.

**devitrifaction, devitrification** (*de-vit-re-fak'-shun, de-vit-re-fi-ka'-shun*) [*de*, priv.; *vitrum*, glass; *facere*, to make]. To change from the glass-like state.

**devolution** (*dev-o-lu'-shun*) [*devolvere*, to roll down]. 1. The reverse of evolution; involution. 2. Catabolism. 3. Degeneration.

**Devonshire colic** (*dev'-on-shir*). Lead-colic.

**devorative** (*de-vor'-at-iv*) [*devorare*, to swallow down]. Intended to be swallowed without chewing.

**Dew's method of artificial respiration** (*dū*). The infant is grasped in the left hand, allowing the neck to rest between the thumb and forefinger, the head falling far over backward. The upper portion of the back and the scapulæ rest in the palm of the hand, the other three fingers being inserted in the babe's left axilla, raising the arm upward and outward. The right hand grasps the knees, and the lower portion of the body is depressed to favor inspiration. The movement is reversed to favor expiration, the head, shoulders, and chest being brought forward and the thighs pressed upon the abdomen.

**dew-cure** (*dū'-kūr*). See *Kneippism*.

**Dewees' carminative** (*dū-ēz'*) [William Potts *Dewees*, American obstetrician, 1768–1841]. See under *asafetida*.

**dewlap** (*dū'-lap*). A longitudinal fold of skin under the neck of bovine animals.

**dew-point**. The temperature at which the air is saturated with vapor; the temperature at which dew forms.

**dexiocardia** (*deks-e-o-kar'-de-ah*) [δεξιός, on the right; καρδία, the heart]. Transposition of the heart to the right side of the thorax.

**dexter** (*deks'-ter*) [L.]. Right; upon the right side.

**dextrad** (*deks'-trad*) [*dexter*; *ad*, to]. Toward the right side.

**dextral** (*deks'-tral*). 1. Pertaining to the right side. 2. Showing preference for the right eye, hand, foot, etc., in certain acts or functions.

**dextrality** (*deks-tral'-it-e*) [*dexter*]. The condition of turning toward, being on, or pertaining to the right side.

**dextran** (*deks'-tran*) [*dexter*], $C_6H_{10}O_5$. A stringy, gummy substance formed in milk by the action of bacteria, and also occurring in unripe beetroot.

**dextraural** (*deks-traw'-ral*) [*dexter*; *auris*, the ear]. Right-eared.

**dextren** (*deks'-tren*) [*dexter*, right]. Belonging to the dextral side in itself.

**dextrin** (*deks'-trin*) [*dexter*], $C_6H_{10}O_5$. A soluble carbohydrate into which starch is converted by diastase or dilute acids. It is a whitish substance, turning the plane of polarization to the right. **d., animal**, glycogen.

**dextrinase** (*deks'-trin-ās*). A ferment which converts starch into isomaltose.

**dextrinate** (*deks'-triu-āt*). To change into dextrin.

**dextrinuria** (*deks-trin-ū'-re-ah*) [*dextrin*; οὖρον, urine]. The presence of dextrin in the urine.

**dextro-** (*deks-tro-*) [*dexter*]. A prefix meaning right.

**dextrocardia** (*deks-tro-kar'-de-ah*) [*dextro-*; καρδία, heart]. Transposition of the heart to the right side of the thorax.

**dextrocardial** (*deks-tro-kar'-de-al*) [see *dextrocardia*]. Having the heart to the right of the median line.

**dextrocerebral** (*deks-tro-ser'-e-bral*) [*dextro-*; *cerebrum*, the brain]. 1. Located in the right cerebral hemisphere. 2. Functionating preferentially with the right side of the brain.

**dextrococaine** (*deks-tro-ko'-kah-ēn*). An artificial alkaloid obtained by heating ecgonine or its derivative with strong alkali. It is a local anesthetic and stimulant, in action similar to cocaine, but more rapid, irritating, and fugitive. Syn., *isococaine*.

**dextrocompound** (*deks-tro-com'-pound*). In chemistry, a compound body that causes a ray of polarized light to rotate to the right, a dextrorotatory compound.

**dextrocular** (*deks-trok'-ū-lar*) [*dextro-*; *oculus*, the eye]. Right-eyed.

**dextrocularity** (*deks-trok-ū-lar'-it-e*). The condition of being right-eyed.

**dextroduction** (*deks-tro-duk'-shun*) [*dexter*, right; *ducere*, to draw]. Movement of the visual axis toward the right.

**dextroform** (*deks'-tro-form*). A combination of formaldehyde and dextrin, soluble in water and glycerol. It is used internally in suppurating cystitis and in the treatment of gonorrhea in applications of 10 to 20 % solutions.

**dextroglucose** (*deks-tro-glū'-kōs*). See *dextrose*.

**dextrogyrate** (*deks-tro-ji'-rāt*). Same as *dextrorotatory*.

**dextrogyre** (*deks-tro-jīr'*) [*dextro-*; *gyrare*, to turn around]. A substance producing rotation to the right.

**dextromanual** (*deks-tro-man'-ū-al*) [*dextro-*; *manus*, hand]. Right-handed.

**dextromanuality** (*deks-tro-man-ū-al'-it-e*). The condition of being right-handed.

**dextromenthol** (*deks-tro-men'-thol*). Menthol oxidized by chromic acid.

**dextropedal** (*deks-trop'-ed-al*) [*dextro-*; *pes*, foot]. Right-footed.

**dextropedality** (*deks-trop-ed-al'-it-e*). The condition of being right-footed.

**dextrophoria** (*deks-tro-fo'-re-ah*) [*dextro-*; φόρος, tending]. A tending of the visual lines to the right.

**dextrorotatory** (*deks-tro-ro'-tat-o-re*) [*dextro-*; *rotare*, to whirl]. Turning the rays of light to the right.

**dextrosaccharin** (*deks-tro-sak'-ar-in*). A mixture of saccharin and dextrose 1 : 2000.

**dextrose** (*deks'-trōs*) [*dexter*], $C_6H_{12}O_6$. Grape-sugar; a sugar-belonging to the glucose group, that rotates polarized light to the right. See *glucose*.

**dextrosinistral** (*deks-tro-sin-is'-tral*) [*dextro-*; *sinister*, left]. Extending from right to left.

**dextrosuria** (*deks-trōs-ū'-re-ah*) [*dextrose*; οὖρον, urine]. The presence of dextrose in the urine. Cf. *levulosuria; pentosuria*.

**dextrotorsion** (*deks-tro-tor'-shun*) [*dextro-*; *torquere*, to twist]. A twisting to the right.

**dextroversion** (*deks-tro-ver'-shun*) [*dextro-*; *vertere*, to turn]. Version to the right side.

**dezymotize** (*de-zī'-mo-tīz*) [*de*, priv.; ζύμη, leaven]. To free from ferments or germs.

**dhatureas** (*dah-tu'-re-as*). Professional poisoners of India who employ the *Datura fastulosa*.

**dhobie itch** (*do'-be*). A Hindu name for ringworm of the body. 2. A popular term in the tropics, to indicate any skin disease transmitted by the clothing.

**dhooley** (*doo'-le*). A covered stretcher used in India.

**dhurrin** (*dur'-in*). A glucoside derived from dhurra and occurring in young plants f *Sorghum vulgare* (dhurra or Guinea corn]).

**D. Hy.** Abbreviation for *Doctor of Hygiene*.

**di—** [δίς, twice]. A prefix signifying *two* or *twice*.

**diabète bronzé** (*de-a-bāt'-bron-za'*) [Fr.]. Same as *bronzed diabetes*.

**diabetes** (*di-ab-e'-tēz*) [διαβήτης; διά, through; βαίνειν, to go]. A disease characterized by the habitual discharge of an excessive quantity of urine; used without qualification, the word indicates diabetes mellitus. Syn., *diarrhæa urinosa*. See *Unschuld's sign*. **d., alimentary**, that due to defective assimilative power over the carbohydrates of food. **d., alternating**, a form of diabetes mellitus alternating with gout. **d., artificial**, that form produced in the physiological laboratory by puncturing the floor of the fourth ventricle of the brain. **d., azoturic**, diabetes without glycosuria accompanied by increase of urea in the urine. **d., bili ry**. See *biliary diabetes*. **d., bronzed**, diabetes in association with hemochromatosis, *q. v.* **d., composite**, that in which sugar and oxybutyric acid and its derivatives are discharged in the urine. **d., conjugal**. See *conjugal diabetes*. **d., decipiens**, diabetes mellitus in which there is no polyuria or polydipsia. **d., gouty**, a form of glycosuria occurring in gouty in i-viduals. **d., hydruric**, polyuria in which the water in the urine is in excess without increase in the solid constituents of the urine. **d., inositus**, diabetes mellitus in which inosit takes the place of grape-sugar. **d., insipidus**, a chronic disease characterized by the passage of a large quantity of normal urine of low specific gravity, associated with intense thirst. **d. mellitus**, a nutritional disease characterized by the passage of a large quantity of urine containing sugar; there is intense thirst, with voracious appetite, progressive loss of flesh and strength, and a tendency to a fatal termination. **d., neurogenic**, that due to disorder of the nervous system. **d., pancreatic**, a variety of glycosuria associated with and probably dependent upon disease of the pancreas. **d., phloridzin**, that form produced in animals by the administration of phloridzin. **d., phosphatic**, a condition characterized by polyuria, polydipsia, emaciation, and excessive excretion of phosphates in the urine. **d., puncture**. See *d., artificial*. **d., renal**, a form due to abnormal permeability of the kidneys to sugar.

**diabetic** (*di-ab-et'-ik*) [*diabetes*]. 1. Pertaining to diabetes. 2. A person suffering from diabetes. **d. cataract**, an opacity of the crystalline lens sometimes found in association with diabetes. d. center. See *center*. d. coma, the coma caused by diabetes mellitus. d. gangrene, a moist gangrene sometimes occurring in persons suffering from diabetes. d. puncture, puncture of the diabetic center, which is followed by glycosuria. d. sugar, $C_6H_{12}O_6$, the glucose present in the urine in diabetes mellitus. It is identical with grape-sugar. d. tabes, a peripheral neuritis occurring in diabetic patients, and causing symptoms resembling tabes dorsalis. d. urine, urine containing sugar.

**diabetico** (*di-ab-et'-ik-o*). A beverage recommended in diabetes, said to consist of alcohol, 8.25 %; extractive, 3.27 %; glycerol, 0.82 %; saccharine, 0.023 %; sulphuric acid, 0.036 %; tartaric acid, 0.56 %; phosphoric acid, 0.025 %.

**diabetide** (*di-ab-e'-tīd*) [*diabetes*]. A cutaneous manifestation of diabetes.

**diabetin** (*di-ab-e'-tin*) [*diabetes*]. A trade name for levulose.

**diabetogenic, diabetogenous** (*di-ab-et-o-jen'-ik, di-ab-et-oj'-en-us*) [*diabetes*; γενᾶν, to produce]. Causing diabetes.

**diabetograph** (*di-ab-et'-o-graf*) [*diabetes*; γράφειν, to write]. An instrument which registers the amount of glucose present in the urine which is dropped into it in boiling Fehling's solution.

**diabetometer** (*di-ab-e-tom'-et-er*) [*diabetes*; μέτρον, a measure]. A polariscope for ascertaining the proportion of sugar in diabetic urine.

**diaboleptic** (*di-ab-o-lep'-tik*) [διάβολος, devil; λαμβάνειν, to seize]. An insane or deluded person who professes to have supernatural communications.

**diabrosis** (*di-ab-ro'-sis*) [see *diabrotic*]. Corrosion; erosion, or ulceration.

**diabrotic** (*di-ab-rot'-ik*) [διαβρωτικός; διά, through; βιβρώσκειν, to eat]. 1. Corrosive. 2. A corrosive substance.

**diacaustic** (*di-ak-aws'-tik*) [διά, through; καυστικός, caustic]. 1. A double convex cauterizing lens. 2. Exceedingly caustic.

**diacele, diacoele** (*di'-as-ēl*) [διά, between; κοίλη, a hollow]. The third ventricle of the brain.

**diacetanilide** (*di-as-et-an'-il-īd*), $C_6H_5N(C_2H_3O_2)_2$. A compound of acetanilide and glacial acetic acid closely resembling, but stronger in physiological action than, acetanilide.

**diacetate** (*di-as'-et-āt*). A salt of diacetic acid.

**diacetemia** (*di-as-e-te'-me-ah*) [*diacetic acid*; αἷμα, blood]. Acidosis due to the presence of diacetic acid in the blood.

**diacetic acid** (*di-as-e'-tik*). See *acid, diacetic*.

**diacetin** (*di-as'-et-in*) $C_3H_5(OH)(C_2H_3O_2)_2$. A liquid derivative of glycerin, with a bitter taste. It is also called *acetidin*.

**diacetonuria** (*di-as-et-on-u'-re-ah*). See *diaceturia*.

**diaceturia** (*di-as-et-u'-re-ah*) [*di*, two; *acetum*, vinegar; οὖρον, urine]. The presence of diacetic acid in the urine.

**diachorema** (*di-ak-o-re'-mah*) [διαχώρημα, excrement]. Fecal matter; excrement.

**diachoresis** (*di-ak-o-re'-sis*) [διαχώρησις, a passing through]. Excretion or passage of feces.

**diachoretic** (*di-ak-o-ret'-ik*) [*diachoresis*]. Laxative; aperient.

**diachylon** (*di-ak'-il-on*) [διά, through; χυλός, juice]. Lead-plaster. See *plumbi oxidum* under *plumbum*.

**diacid** (*di-as'-id*) [*di*, two; *acidus*, acid]. Having two atoms of hydrogen replaceable by a base.

**diaclasia, diaclasis** (*di-ak-la'-ze-ah, di-ak'-la-sis*) [διάκλασις, a breaking in two]. 1. Refraction. 2. Breaking a bone, intentionally.

**diaclast** (*di'-ak-last*) [διακλᾶν, to break apart]. An instrument for breaking the fetal head.

**diaclastic** (*di-ak-las'-tik*). Pertaining to diaclasis.

**diacoele, diacoelia** (*di'-as-ēl, di-as-e'-le-ah*). See *diacele*.

**diacope** (*di-ak'-o-pe*) [διά, through; κοπή, a cut]. A deep, incised wound, especially of the head or skull; a lengthwise fracture or cut, as of a bone.

**diacoustics** (*di-ak-oos'-tiks*) [διά, through; ἀκούειν, to hear]. The department of physics which treats of the refraction of sound. Syn., *diaphon*.

**diacrisiography** (*di-ak-riz-e-og'-ra-fe*). An anatomic description of the secretory organs.

**diacrisis** (*di-ak'-ris-is*) [διά, a part; κρίνειν, to separate or secrete; pl., *diacrises*]. 1. A critical discharge. 2. A change or disorder in a secretion. 3. Any disease marked by altered secretions. d.,

**follicular**, an alteration of the secretion of follicular glands, due to disease.

**diacritic, diacritical** (*di-ak-rit'-ik, di-ak-rit'-ik-al*). Diagnostic, distinctive.

**diactinic** (*di-ak-tin'-ik*). Capable of transmitting actinic rays.

**diad** (*di'-ad*) [*di*, two]. 1. Having a quantivalence of two. 2. An element or radical having a quantivalence of two. 3. A unit made up of primary units which are differentiated into parts, but yet constitute an individual; *e. g.*, a morula.

**diaderm** (*di'-a-derm*) [δίς, two; δέρμα, skin]. The two plates or lamina of the two primitive germ-layers, the ectoderm and entoderm taken as one.

**diadexis** (*di-ad-ek'-sis*) [διάδεξις, metastasis]. Metastasis of a disease, with a change of its character; change in the seat and nature of a disease.

**diadokokinesia** (*di-ah-dok-o-kin-e'-se-ah*) [διαδοκός, succeeding; κίνησις, motion]. The normal power of performing alternating movements in rapid succession, *e. g.*, pronation and supination.

**diagnose** (*di'-ag-nōs*) [*diagnosis*]. To make a diagnosis of; to recognize.

**diagnosis** (*di-ag-no'-sis*) [διά, apart; γνῶσις, knowledge]. The determination of the nature of a disease. d., anatomical. 1. A diagnosis based upon the recognition of definite anatomical alterations lying back of the phenomena. 2. A postmortem diagnosis. d., differential, the distinguishing between two diseases of similar character by comparing their symptoms. d. by exclusion, the recognition of a disease by excluding all other known conditions. d., pathological, the diagnosis of the structural lesions present in a disease. d., physical, the determination of disease by inspection, palpation, percussion, or auscultation. d., topographical, that based upon the seat of a lesion.

**diagnostic** (*di-ag-nos'-tik*) [*diagnosis*]. Serving as evidence in diagnosis.

**diagnosticate** (*di-ag-nos'-tik-āt*) [*diagnosis*]. To make a diagnosis.

**diagnostician** (*di-ag-nos-tish'-an*) [*diagnosis*]. One skilled in making diagnoses.

**diagnostics** (*di-ag-nos'-tiks*) [*diagnosis*]. The science and art of diagnosis.

**diagnostitial** (*di-ag-nos-tish'-al*). Procedure having a diagnostic purpose.

**diagometer** (*di-ag-om'-et-er*). An electroscope for determining the relative conductivity of bodies.

**diagraph** (*di'-ag-raf*) [διά, through; γράφειν, to record]. An apparatus for recording the outlines of a part.

**diahydric** (*di-ah-hi'-drik*) [διά, through; ὕδωρ, water]. Relating to transmission through water, as a percussion-note through a stratum of interposed fluid.

**dialysate** (*di-al'-is-āt*). A product taken from a solution by dialysis.

**dialysis** (*di-al'-is-is*) [διά, through; λύειν, to loose]. 1. The separation of parts in general. 2. A loss of strength; dissolution. 3. The separation of several substances from one another in solution by taking advantage of their differing diffusibility through porous membranes. Those that pass through readily are termed crystalloids, those that do not, colloids.

**dialytic** (*di-al-it'-ik*) [*dialysis*]. 1. Pertaining to or similar to the process of dialysis. 2. Producing relaxation (said of a remedy). 3. A condition of divergent change or evolution.

**dialyzable** (*di-al-i'-za-bl*). Capable of being separated by diffusion.

**dialyzed** (*di'-al-izd*) [*dialysis*]. Separated by dialysis. d. raw meat, a reddish fluid with a slightly acid or bitter taste, prepared from fresh beef or mutton to which are added 200 Gm. of water, 5 Gm. of hydrochloric acid, and 2 Gm. of pepsin; the whole is boiled at 38° C. for 5 or 6 hours.

**dialyzer** (*di'-al-i-ser*) [*dialysis*]. An apparatus for effecting dialysis; also the porous septum or diaphragm of such an apparatus.

**diamagnetic** (*di-ah-mag-net'-ik*) [διά, across; *magnet*]. Taking a position at right angles to the lines of magnetic force. d. bodies, bodies not susceptible of being magnetized.

**diameter** (*di-am'-et-er*) [διά, through; μέτρον, a measure]. A straight line joining opposite points of a body or figure and passing through its center. d., craniometric, one of several lines connecting points on opposite surfaces of the cranium. *biparietal*, that joining the parietal eminences; *bitemporal*, that joining the extremities of the coronal suture; *occipitofrontal*, that joining the root of the nose and the most prominent point of the occiput; *occipitomental*, that joining the external occipital protuberance and the chin; *trachelobregmatic*, that joining the center of the anterior fontanel and the junction of the neck and floor of the mouth. **d., pelvic**, any one of the diameters of the pelvis. The most important are the following: *anteroposterior* (of pelvic inlet), that which joins the sacrovertebral angle and the pubic symphysis; *anteroposterior* (of pelvic outlet), that which joins the tip of the coccyx with the subpubic ligament; *conjugate*, the *anteroposterior diameter* of the pelvic inlet; *conjugate, diagonal*, that connecting the sacrovertebral angle and subpubic ligament; *conjugate, external*, that connecting the depression above the spine of the first sacral vertebra and the middle of the upper border of the symphysis pubis; *conjugate, true*, that connecting the sacrovertebral angle and the most prominent portion of the posterior aspect of the symphysis pubis; *transverse* (of pelvic inlet), that connecting the two most widely separated points of the pelvic inlet; *transverse* (of pelvic outlet), that connecting the ischial tuberosities.

**diametric** (*di-am-et'-rik*). 1. Of, pertaining to, or coinciding with a diameter—extremely opposed. **d. pupil**, one which constitutes a vertical slit, as is the case after two iridectomies, one upward and the other downward.

**diamide** (*di'-am-id*) [*di*, two; *amide*]. A double amide formed by replacing hydrogen in two ammonia molecules by an acid radical. See *hydrazin*.

**diamine** (*di'-am-in*) [*di*, two; *amine*]. An amine formed by replacing hydrogen in two molecules of ammonia by a hydrocarbon radical. See *amine*.

**diaminodihydroxyarsenodibenzene** (*di-am-i'-no-di-hi-drok'-se-di-ar-se'-no-di-ben-zēn*) NH₂ . OH . C₆H₃ . As = As . C₆H₃ . OH . NH₂. The correct name for *arsenobenzol* or "606."

**diaminuria** (*di-am-in-ū'-re-ah*) [*diamine*; οὖρον, urine]. The presence of diamine compounds in the urine.

**diamotosis** (*di-am-o-to'-sis*) [διά, through; μοτός, lint]. The packing of a wound or sore with lint.

**diapason** (*di-ap-ās'-on*) [διαπασῶν, concord]. A tuning fork used to determine deafness, and in the diagnosis of ear diseases.

**diapedesis** (*di-ap-ed-e'-sis*) [διά, through; πηδᾶν, to leap]. The passage of the blood or of its formed elements, particularly the red corpuscles, through the unruptured vessel-walls.

**diapedetic** (*di-ap-ed-et'-ik*). Relating to diapedesis.

**diaphane** (*di'-af-ān*) [διά, through; φαίνειν, to show]. 1. A transparent investing membrane of an organ or cell. 2. A small electric lamp used in transillumination.

**diaphaneity** (*di-af-an-e'-it-e*). Transparency.

**diaphanometer** (*di-af-an-om'-et-er*). An instrument for observing the transparency of fluids. See also *lactoscope*.

**diaphanoscope** (*di-af-an'-o-skōp*) [διαφανής, translucent; σκοπεῖν, to inspect]. An instrument for illuminating the interior of a body-cavity so as to render the boundaries of the cavity visible from the exterior.

**diaphanoscopy** (*di-af-an-os'-ko-pe*) [see *diaphanoscope*]. Examination of body-cavities by means of an introduced incandescent electric light.

**diaphanous** (*di-af'-an-us*) [διά, through; φαίνειν, to shine]. Transmitting light; translucent. **d. test of death**, the normal red color of the finger-tips when held toward the light is not present in death.

**diaphemetric** (*di-af-em;et'-rik*) [διά, through; ἀφή, touch; μέτρον, measure]. Pertaining to measurements of tactile sensibility.

**diaphoresis** (*di-ah-for-e'-sis*) [διά, through; φερεῖν, to carry]. Perspiration, especially perceptible perspiration.

**diaphoretic** (*di-ah-for-et'-ik*) [*diaphoresis*]. 1. Causing an increase of perspiration. 2. A medicine that induces diaphoresis.

**diaphotoscope** (*di-ah-fo'-to-skōp*) [διά, through; φῶς, light; σκοπεῖν, to inspect]. A variety of endoscope.

**diaphragm** (*di'-af-ram*) [διά, across; φράγμα, wall]. 1. The wall, muscular at the circumference and tendinous at the center, that separates the thorax and abdomen. The chief muscle of respiration and expulsion. See *muscles, table of*. 2. A thin septum such as is used in dialysis. 3. In microscopy, an apparatus placed between the mirror and object to regulate the amount of light that is to pass through the object. **d., central stop**, in microscopy, a diaphragm having a circular slit just within its margin the center remaining opaque. **d., condensing**, a diaphragm containing lenses for converging the light-rays. Cf. *Abbe's condenser*. **d., cylindrical**, in microscopy, a piece of substage apparatus fitted with perforated stops, each allowing a different amount of light to pass. **d., graduating**, one which allows a concentric increase or diminution of light. See *d., iris*. **d., inferior**, the vertical part of the diaphragm. Syn., *musculus diaphragmaticus minor*. **d. iris**, a device for changing or regulating the amount of light directed upon an object under the microscope. **d. opening**. The opening in the disc or apparatus of a microscope through which the rays of light pass. **d., pelvic**. 1. See *levator ani*, in table of muscles. 2. The levatores ani and the coccygei muscles combined; also called perineal, and rectal diaphragm. **d., perineal**. See *d., pelvic*. **d.-phenomenon**. See *diaphragmatic phenomenon*. **d., plate**, the ordinary perforated plate or simple shutter diaphragm. **d., rectal**. See *d., pelvic*. **d., superior**, the horizontal part of the diaphragm. Syn., *musculus diaphragmaticus major*. **d., thoracoabdominal**, that separating the thoracic and abdominal cavities.

**diaphragma** (*di-ah-frag'-mah*) [L.]. 1. See *diaphragm*. 2. The velum of the *hydromedusæ*. **d. auris**. See *membrane, tympanic*. **d. cerebri**. See *septum lucidum*. **d. hypophyseos**. See *d. sellæ*. **d. narium**, the septum of the nose. **d. oris**, the mylohyoid muscle. **d. pelvicum, d. pelvis**. See *diaphragm, pelvic*. **d. pharyngis**. See *velum pendulum palati*. **d. sellæ**, a shelf-like process at the base of the skull given off by the dura and forming a roof for the pituitary fossa; it is perforated for the passage of the infundibulum. Syn., *d. hypophyseos*. **d. urogenitale**, the inferior layer of the deep perineal fascia. **d. ventriculorum lateralium**. See *septum lucidum*.

**diaphragmalgia** (*di-af-rag-mal'-je-ah*) [*diaphragm*; ἄλγος, pain]. Pain in the diaphragm.

**diaphragmatalgia** (*di-af-rag-mat-al'-je-ah*) [*diaphragm*; ἄλγος, pain]. Pain in or neuralgia of the diaphragm. See *diaphragmodynia*.

**diaphragmatic** (*di-ah-frag-mat'-ik*) [*diaphragm*]. Relating to the diaphragm. **d. phenomenon**, Litten's sign: in a state of health there can be seen a shadow rising and falling from the vertebral column to the attachment of the diaphragm from the seventh rib to the convexity; this movement can be seen through the thoracic walls, and shows the intensity of respiration and the limit of the diaphragm between its position on inspiration and that on expiration. The upper position corresponds to the liver-margin in the state of rest. Deviations of the extent of movement mark certain pathological states.

**diaphragmatitis** (*di-af-rag-mat-i'-tis*) [*diaphragm*; ιτις, inflammation]. Inflammation of the diaphragm.

**diaphragmatocele** (*di-ah-frag-mat'-o-sēl*) [*diaphragm*; κήλη, hernia]. Hernia of a viscus through the diaphragm.

**diaphragmitis** (*di-af-rag-mi'-tis*). See *diaphragmatitis*.

**diaphragmodynia** (*di-af-rag-mo-din'-e-ah*) [*diaphragm*; ὀδύνη, pain]. Pain in the diaphragm.

**diaphtherin** (*di-af'-ther-in*) [διαφθείρειν, to destroy]. Oxyquinaseptol; a coal-tar derivative composed of two molecules of oxyquinolin and one of aseptol. It is a yellow powder, with a phenol-like odor, and is used as an antiseptic in solutions varying in strength from 1 to 50 %.

**diaphtol** (*di-af'-tol*), C₉H₇O₄SN. Orthooxyquinolin-metasulphonic acid. It is used in internal disinfection of the urinary tract in place of salol. Syn., *chinaseptol; quinaseptol*.

**diaphylactic** (*di-af-il-ak'-tik*) [διαφυλακτικός, preserving]. Same as *prophylactic*.

**diaphysis** (*di-af'-is-is*) [διά, through; φύεσθαι, to grow]. 1. The shaft of a long bone. 2. An interspace. 3. A prominent part of a bony process. 4. A ligament of the knee-joint.

**diaphysitis** (*di-af-iz-i'-tis*) [*diaphysis*; ιτις, inflammation]. Inflammation of a diaphysis.

**diaplasis** (*di-ap'-las-is*) [διάπλασις, the setting of a broken limb]. Reduction, as of a dislocation or fracture.

**diaplastic** (*di-ah-plas'-tik*) [*diaplasis*]. 1. Relating to the setting of a fracture or reduction of a dislocation. 2. Any application for a fracture or dislocation.

**diaplex, diaplexus** (*di'-ap-leks, di-ap-leks'-us*) [διά, between; *plexus*, a network]. The choroid plexus of the third ventricle or diacele.

**diapophysial** (*di-ap-off-is'-e-al*) [διά, apart; ἀπόφυσις, an outgrowth]. Relating to a diapophysis.

**diapophysis** (*di-ap-off'-is-is*) [διά, apart; ἀπόφυσις, an outgrowth]. The superior or articular part of a transverse process of a vertebra.

**diapyema** (*di-ap-i-e'-mah*) [διά, through; πύον, pus: *pl., diapyemata*]. An abscess.

**diapyesis** (*di-ap-i-e'-sis*) [διά, through; πύησις, suppuration]. Suppuration.

**diapyetic** (*di-ap-i-et'-ik*) [*diapyesis*]. 1. Promoting diapyesis or suppuration. 2. A suppurative.

**diarius** (*di-a'-re-us*) [*dies*, a day]. Enduring but a single day.

**diarrhea, diarrhœa** (*di-ar-e'-ah*) [διά, through; ῥεῖν, to flow]. A condition characterized by increased frequency and lessened consistence of the fecal evacuations. **d., atonic.** See *d., camp,* and *d., chronic.* **d., atrophic.** See *d., colliquative.* **d., camp,** a form of diarrhea common among soldiers. The discharges are apt to be purulent, and there is liability to thickening and ulceration of the colon. Syn., *atonic diarrhea.* **diarrhœa carnosa,** dysentery in which flesh-like masses are passed. **d., catarrhal.** 1. A form which is often epidemic at times when catarrhs are prevalent. 2. That of catarrhal gastritis or enteritis. Syn., *diarrhœa acuta serosa mucosa; diarrhœa pituitosa; mucous colitis; phlegmatic diarrhœa; rheumatic diarrhœa.* **d., choleraic,** severe, acute diarrhea with serous stools, and accompanied by vomiting and collapse. **d., chronic,** that characterized by continuous and intractable discharges, often offensive, bloody, or containing undigested food. It occurs as a manifestation of an intestinal lesion or of a constitutional disease. Syn., *atonic diarrhea; diarrhœa habitualis; mucous colitis.* **d., colliquative,** that characterized by excessively frequent and copious discharges and extreme prostration, and occurring, as a rule, toward the close of a chronic disease. Syn., *atrophic diarrhea.* **d., crapulous,** that due to excessive eating or drinking. **d., critical,** that occurring at the crisis of a disease. **d., feculent.** See *d., simple.* **d., green,** a form occurring in infants and marked by green alvine discharges. It is infectious and due to the bacillus of Lesage, or *Bacillus fluorescens nonliquefaciens.* **d., inflammatory,** that caused by congestion of the intestinal mucosa following sudden chilling of the body-surface, suppression of perspiration or menstruation. It frequently constitutes the socalled cholera infantum and sthenic diarrhea. **d., lienteric,** a form of diarrhea characterized by the passage of fluid stools containing masses of undigested food. **d., membranous,** a form characterized by the presence of mucous shreds in the stools. It may be chronic or subacute, and alternate with constipation. Hysterical symptoms, griping, and abdominal tenderness are often present. Syn., *tubular diarrhea.* **d., mucous.** See *colitis, mucous.* **d., pancreatic,** a persistent form in which the discharges are thin and viscid. It is supposed to depend upon disease of the pancreas. **d., parasitic,** a diarrhea incited by the presence of intestinal parasites. Cf. *diarrhœa verminosa.* **diarrhœa pituitosa.** See *d., catarrhal.* **d., simple,** that form in which the evacuations consist of fecal matter only. Syn., *diarrhœa fusa; diarrhœa stercoralis; diarrhœa vulgaris; feculent diarrhea; saburral diarrhea.* **diarrhœa stercoralis, d. stercorea, d. stercorosa.** See *d., simple.* **d., sthenic.** See *d., inflammatory.* **d., strumous,** a chronic form frequently met in underfed, strumous children. **d., summer,** an acute form occurring during the intense heat of summer. **d., summer,** of children, that due to *bacillus dysenteriæ,* Shiga, and etiologically identical with acute bacillary dysentery of adults. **d., tubular.** See *d., membranous.* **diarrhœa urinosa.** See *diabetes.* **diarrhœa verminosa,** that due to intestinal worms. Cf. *d., parasitic.* **d., zymotic.** Synonym of *dysentery.*

**diarrheal** (*di-ar-e'-al*). Relating to, or of the nature of diarrhea.

**diarrhemia, diarrhæmia** (*di-ar-e'-me-ah*) [διά, through; ῥεῖν, to flow; αἷμα, blood]. See *ascites, sanguineous.*

**diarthrodial** (*di-ar-thro'-de-al*) [see *diarthrosis*]. Relating to or of the nature of a diarthrosis.

**diarthrosis** (*di-ar-thro'-sis*) [διά, through; ἄρθρωσις, articulation]. A freely movable articulation. The various forms are: *arthrodia,* in which the bones glide upon plane surfaces; *enarthrosis,* ball-and-socket joint, with motion in all directions; *ginglymus,* or hinge-joint, with backward and forward motion; and *d. rotatoria,* with pivotal movement. Syn., *perarticulation; prosarthrosis.* **d. ambigua.** See *amphiarthrosis.* **d. obliqua,** an inconstant articulation between the spinous processes of adjacent lumbar vertebræ. **d. obliqua accessoria,** a double articulation sometimes formed by the spinous processes of adjacent dorsal or lumbar vertebræ near the basal ends of the processes. **d., planiform,** arthrodia. **d., rotatory, d., synarthrodial, d., trochoid, d. trochoides.** See *cyclarthrosis.*

**diarthrotic** (*di-ar-throt'-ik*). See *diarthrodial.*

**diaschisis** (*di-as'-kis-is*) [διασχίζειν, to split]. An inhibition of functional continuity between different parts of the nervous system.

**diasostic** (*di-as-os'-tik*) [διασώζειν, to preserve]. Pertaining to hygiene or the preservation of health; hygienic; diateretic.

**diaspirin** (*di-as'-pir-in*). The succinic acid ester of salicylic acid. It is used as aspirin, but is more diaphoretic. Dose, 15–30 gr. (1–2 gm.).

**diastalsis** (*di-as-tal'-sis*). A term suggested by Cannon for the downward moving wave of contraction, preceded by a wave of inhibition, occurring in the small intestine during digestion.

**diastaltic** (*di-as-tal'-tik*) [διά, apart; στέλλειν, to send]. Reflex; performed (as are many reflex actions) through the medium of the spinal cord.

**diastase** (*di'-as-tās*) [see *diastasis*]. A nitrogenous vegetable ferment found in malt; it converts starch into glucose. **d., animal,** a general term for the amylolytic enzymes of animals, ptyalin, amylopsin, and the special enzyme of the liver capable of converting glycogen into sugar. **d., pancreatic,** amylopsin. **d., salivary,** ptyalin. **d., vegetable,** the enzyme of germinating seeds; diastase proper.

**diastasemia** (*di-as-tas-e'-me-ah*) [διά, apart; στάσις, settling; αἷμα, blood]. Acute anasarca.

**diastasic, diastatic** (*di-as-tas'-ik, di-as-tat'-ik*) [*diastase*]. 1. Pertaining to diastase. 2. Pertaining to diastasis. **d. action,** the conversion of starch into water-soluble substances by diastase.

**diastasimetry** (*di-as-ta-sim'-et-re*). The estimation of the amount of diastase.

**diastasis** (*di-as'-ta-sis*) [διάστασις, separation]. 1. The separation of an epiphysis from the body of a bone without true fracture. 2. A dislocation of an amphiarthrotic joint.

**diastema** (*di-as-te'-ma*) [διάστημα, a fissure]. A cleft or fissure.

**diastematelytria** (*di-as-tem-at-el-it'-re-ah*) [*diastema*; ἔλυτρον, vagina]. Longitudinal and congenital fissure of the vagina.

**diastematenteria** (*di-as-tem-at-en-te'-re-ah*) [*diastema*; ἔντερον, intestine]. A longitudinal fissure of the intestine. Syn., *diastementeria.*

**diastematia** (*di-as-tem-a'-she-ah*) [*diastema*]. An abnormality in which the body is split or fissured longitudinally. Syn., *diastematocaulia.*

**diastematochilia** (*di-as-tem-at-o-kil'-e-ah*) [*diastema*; χεῖλος, lip]. Congenital longitudinal fissure of the lip; hare-lip.

**diastematocrania** (*di-as-tem-at-o-kra'-ne-ah*) [*diastema*; κρανίον, the skull]. A skull congenitally cleft along the median line.

**diastematocystia** (*di-as-tem-at-o-sis'-te-ah*) [*diastema*; κύστις, bladder]. Congenital longitudinal fissure of the bladder.

**diastematogastria** (*di-as-tem-at-o-gas'-tre-ah*) [*diastema*; γαστήρ, stomach]. A mesial fissure of the ventral wall of the body.

**diastematoglossia** (*di-as-tem-at-o-glos'-e-ah*) [*diastema*; γλῶσσα, the tongue]. A congenital longitudinal fissure of the tongue.

**diastematognathia** (*di-as-tem-at-og-na'-the-ah*) [*diastema*; γνάθος, jaw]. Congenital longitudinal fissure of the jaw.

**diastematometria** (*di-as-tem-at-o-me'-tre-ah*) [*diastema*; μήτρα, womb]. Congenital longitudinal median fissure of the uterus.

**diastematomyelia** (*di-as-tem-at-o-mi-e'-le-ah*) [*diastema*; μυελός, marrow]. A congenital splitting or doubling of the spinal cord.

**diastematopyelia** (*di-as-tem-at-o-pi-e'-le-ah*) [*diastema*; πύελος, a trough]. A mesial fissure of the pelvis.

**diastematorrhachia** (*di-as-tem-at-o-ra'-ke-ah*) [*dias-*

# DIASTEMATORRHINIA 291 DICHLORACETIC

*tema; ῥάχις,* the spine]. A congenital longitudinal fissure of the vertebral column.
**diastematorrhinia** (*di-as-tem-at-or-in'-e-ah*) [*diastema; ῥίς,* the nose]. A congenital mesial fissure of the nose.
**diastematostaphylia** (*di-as-tem-at-o-staf-il'-e-ah*) [*diastema; σταφυλή,* the uvula]. A congenital mesial fissure of the uvula.
**diastematosternia** (*di-as-tem-at-o-ster'-ne-ah*) [*diastema; sternum*]. Median congenital fissure of the sternum.
**diastementeria** (*di-as-tem-en-te'-re-ah*). See *diastematenteria*.
**diaster** (*di-as'-ter*). The karyokinetic figure assumed by the aster of a dividing nucleus before the formation of the stars at the ends of the nuclear spindle. It is the sixth stage of karyokinesis. See *karyokinesis*.
**diastin** (*di-as'-tin*). Trade name for a form of diastase.
**diastoid** (*di'-as-toid*). Trade name for a dry malt preparation.
**diastol** (*di'-as-tol*). A proprietary malt extract.
**diastole** (*di-as'-to-le*) [διαστολή, a drawing apart]. The period of dilatation of a chamber of the heart; used alone it signifies diastole of the ventricles. **d., arterial,** the expansion of an artery following the ventricular systole. **d., auricular,** the dilatation of the cardiac auricle. **d., cardiac,** the period of expansion which follows a cardiac contraction. **d., ventricular,** the dilatation of the cardiac ventricles.
**diastolic** (*di-as-tol'-ik*) [*diastole*]. Pertaining to the diastole of the heart. **d. impulse,** the backstroke. **d. murmur,** a murmur occurring during the diastole. **d. thrill,** the vibration felt in the region of the heart during the diastole of the ventricles.
**diastrephia** (*di-as-tref'-e-ah*) [διά, apart; στρέφειν, to turn]. Insanity marked by acts of cruelty and by gross perversion of the moral sense.
**diastrophometry** (*di-as-tro-fom'-et-re*) [διαστροφή, distortion; μέτρον, a measure]. The measurement of deformities.
**diatactic** (*di-at-ak'-tik*) [διατάσσειν, to make ready]. Preparatory. **d. action,** the supposed molecular establishment of unions between different brain-cells and nerve-centers preparatory to coordinated motor activity.
**diatela, diatele** (*di-at-e'-lah, di'-at-ēl*) [διά, between; *tela,* a web]. The membranous roof of the diacele (third ventricle).
**diateretic** (*di-at-er-et'-ik*) [διατηρεῖν, to watch closely]. Of or pertaining to the practice of hygiene; diasostic.
**diaterma** (*di-at-er'-mah*) [διά, between; τέρμα, end]. A portion of the floor of the diacele. It has a nearly dorso-ventral direction.
**diathermal, diathermanous** (*di-ath-er'-mal, di-ath-er'-man-us*) [διά, through; θέρμη, heat]. Permeable by waves of radiant heat.
**diathermometer** (*di-ah-ther-mom'-et-er*) [*dia; thermometer*]. An appliance for measuring the heat-conducting capacity of substances.
**diathesin** (*di-ath'-es-in*), C₇H₈O₃. A substitute for salicylic acid, the salicylates, and salicin; its use is indicated in gouty diatheses. Dose 7½–15 gr. (0.5–1.0 Gm.).
**diathesis** (*di-ath'-es-is*) [dia; τιθέναι, to arrange]. A state or condition of the body whereby it is especially liable to certain diseases, such as gout, calculus, diabetes, etc. It may be acquired or hereditary. **d., aneurysmal,** inherent predisposition to aneurysms. **d., bilious,** the morbid condition that follows chronic disturbance of the portal circulation and imperfect elimination of bile. **d., calculous,** a constitutional tendency to the formation of calculi. **d., cancerous.** See *cachexia, cancerous.* **d., catarrhal,** a tendency to excessive secretion of mucus. **d., climatic,** a morbid state of body dependent upon local physical conditions, as elevation, soil, water, humidity, etc. **d., congestive,** a constitutional tendency to vasomotor disturbances and local congestions. **d., furuncular.** See *furunculosis.* **d., gouty. d., lithic.** See *d., uric-acid.* **d., hemorrhagic.** See *hemophilia.* **d., lithic,** predisposition to lithemia. **d., osseous, d., osteophytic,** a constitutional tendency to the formation of abnormal ankyloses. **d., psychopathic,** a hereditary predisposition to mental derangement. **d., rheumatic.** See *d., uric-acid.* **d., scorbutic.** See *scurvy.* **d., scrofulous,** a hereditary predisposition to scrofulous affections. Syn., *strumous diathesis.* **d., strumous.** See *d., scrofulous.* **d., syphilitic,** hereditary syphilis. **d., tuberculous,** a constitutional inability to resist tuberculous infection. Syn., *phthisical diathesis.* **d., uratic,** tendency to gout. **d., uric-acid,** a constitutional tendency to the accumulation of uric acid and urates in the fluids of the body and the development of rheumatism, gout, etc.; the arthritic, gouty, rheumatic, or lithic diathesis.
**diathetic** (*di-ath-et'-ik*) [*diathesis*]. Relating to a diathesis.
**diatom** (*di'-at-ōm*). One of the *Diatomaceæ,* a group of microscopic Algæ.
**diatomic** (*di-at-om'-ik*) [δίς, two; ἄτομος, atom]. 1. Consisting of two atoms. 2. Divalent.
**diaxon, diaxone** (*di-aks'-ōn*) [δίς; ἄξων, axis] 1. In biology, having two axes. 2. A neuron having two axons.
**diazo-** (*di-az-o-*) [*dis; azotum,* nitrogen]. A prefix, signifying that a compound contains phenyl, C₆H₅, united with a radical consisting of two nitrogen atoms. **d.-reaction, d.-test,** a urinary test, valuable in the diagnosis of enteric fever. The solutions required are: (*a*) A saturated solution of sulphanilic acid in a 5 % solution of hydrochloric acid; (*b*) a 0.5 % solution of sodium nitrite. Mix *a* and *b* in the proportion of 40 Cc. of *a* to 1 Cc. of *b,* and to a few cubic centimeters add an equal volume of urine, and, after shaking well, allow a few drops of ammonia to flow down the side of the tube. A garnet-red color at the point of contact denotes the reaction, or a rose-pink foam after shaking.
**diazonal** (*di-as'-o-nal*) [διά, through; ζώνη, a zone]. Applied by Fürbringer to nerve-trunks which lie across a sclerozone.
**diazyme** (*di'-az-īm*). Trade name for a diastatic extract of the pancreatic glands.
**dib.** The knee-pan or ankle-bone of a sheep's leg.
**dibasic** (*di-ba'-sik*) [δίς, two; βάσις, base]. Of a salt, containing two atoms of a monobasic element or radical; of an acid, having two replaceable hydrogen atoms.
**dibenzyl** (*di-ben'-zil*), C₁₄H₁₄. A compound prepared by the action of sodium upon benzyl chloride.
**dibenzylamine** (*di-ben-zil'-am-in*), C₁₄H₁₅N. An oily liquid, having the constitution of ammonia in which two atoms of hydrogen are replaced by two molecules of benzyl.
**diblastic** (*di-blas'-tik*) [δίς, double; βλαστός, germ]. Referring to any theory of disease that ascribes it to a double agency.
**diblastula** (*di-blas'-tū-lah*) [δίς, two; *blastula*]. A blastula containing both ectoderm and entoderm.
**diborated** (*di-bo'-ra-ted*). Combined with two molecules of boric acid.
**Dibothrium** (*di-both'-re-um*) [L.]. A genus of cestode worms. D. latum. See *Bothriocephalus latus.*
**dibromated** (*di-bro'-ma-ted*). Containing two atoms of bromine in the molecule.
**dibromethane** (*di-bro-meth-ān'*). See *ethylene bromide.*
**dibromide** (*di-bro'-mid*). A compound consisting of an element or radical and two atoms of bromine.
**dicalcic** (*di-kal'-sik*). Containing two atoms of calcium in each molecule. **d. orthophosphate,** Ca₂H₂(PO₄)₂, a salt occurring in urinary deposits.
**dicamphendion** (*di-kam-fen'-de-on*), (C₆H₁₄O)₂. A reaction-product obtained from bromocamphor by action of metallic sodium; it occurs in flat yellow needles which melt at 193° C.
**dicamphor** (*di-kam'-for*), (C₁₀H₁₈O)₂. A colorless crystalline substance, melting at 166° C., obtained with *dicamphendion* (*q. v.*).
**dicentrine** (*di-sen'-trēn*). An alkaloid derived from *Dicentra pusilla,* a Japanese plant.
**dicephalism** (*di-sef'-al-izm*) [δίς, two; κεφαλή, head]. The condition of having two heads.
**dicephalous** (*di-sef'-al-us*) [*dicephalus*]. Two-headed.
**dicephalus** (*di-sef'-al-us*) [δίς, two; κεφαλή, a head]. A monster with two heads.
**dichastasis** (*di-kas'-tas-is*) [δίχασις, division]. In biology, spontaneous fission.
**dichlamydeous** (*di-klam-id'-e-us*) [δίς, two; χλαμύς, a mantle]. In biology, applied to flowers having both floral envelopes.
**dichloracetic acid** (*di-klor-as-e'-tik*). See *acid, dichloracetic.*

**dichloralantipyrine** (*di-klo-ral-an-te-pi'-rin*). See *antipyrine bichloral*.

**dichlorethane** (*di-klor-eth'-ān*). See *ethene chloride*.

**dichloride** (*di-klo'-rid*). A compound in each molecule of which two atoms of chlorine are combined with an element or radical.

**dichlormethane** (*di-klor-meth'-ān*). See *methylene dichloride*.

**dichotomy** (*di-kot'-o-me*) [δίχα, in two; τέμνειν, to cut]. 1. The state of being bifid; the phenomenon of bifurcation. 2. Division into two parts. 3. Division of a professional fee; the paying of a commission by a consultant or surgeon to the practitioner who refers a case. d., **anterior**, said of a double monster united below the upper limbs. d., **posterior**, said of a double monster in which the two individuals are fused above the posterior extremities.

**dichroic** (*di-kro'-ik*). Pertaining to dichroism.

**dichroism** (*di'-kro-izm*) [δίς, double; χροία, color]. The phenomenon of difference of color in bodies when viewed by reflected or by transmitted light.

**dichromasy** (*di-kro'-mas-e*) [δίς, two; χρῶμα, color]. The condition of a dichromat; inability to distinguish more than two colors.

**dichromat** (*di'-kro-mat*). A person with dichromatopsia. Cf. $m_o n_o chromat$; *trichromat*.

**dichromatopsia** (*di-kro-mat-op'-se-ah*) [dis; χρῶμα, color; ὄψις, sight]. A form of color-blindness in which there are two sharply limited regions at the ends of the spectrum, within which there are no changes of hue, but merely of intensity. All other parts of the spectrum, the "middle region," can be produced by mixtures of the two end regions.

**dichromic** (*di-kro'-mik*). 1. Marked by two colors. 2. Containing two atoms of chromium.

**dichromism** (*di-kro'-mizm*) [*di-*; χρῶμα, color]. The state of presenting one color when seen by reflected light, and another when seen by transmitted light. See also *dichromatopsia*.

**dichromophilism** (*di-kro-mof'-il-izm*) [*di-*; χρῶμα, color; φιλεῖν, to love]. Capability for double staining.

**dichromous, dichroous, dichrous** (*di'-kro-mus, di'-kro-us, di'-krus*). Having two colors; relating to dichroism.

**dicinchonine** (*di-sin'-kon-ēn*) [*di-*; *cinchona*], C₃₈H₄₄N₄O₂. An alkaloid of cinchona-bark.

**diclidostosis** (*di-klid-os-to'-sis*) [δικλίδες, folding doors; ὀστέον, a bone]. Ossification of the venous valves. Syn., *osteoclidis*.

**dicranous** (*di-kra'-nus*) [δίκρανος, two-headed]. Dicephalous.

**dicranus** (*di-kra'-nus*) [δίς, double; κράνον, head]. A dicephalous monster.

**dicrotic** (*di-krot'-ik*) [δίκροτος, double beating]. Having a double beat. d. **pulse**. See *dicrotism*. d. **wave**, the recoil-wave of the sphygmographic tracing, generated by closure of the aortic valves.

**dicrotism** (*di'-kro-tizm*) [see *dicrotic*]. A condition of the pulse in which with every wave there is given to the finger of the examiner the sensation of two beats. It is present when the arterial tension is low.

**dictyitis** (*dik-te-i'-tis*) [δίκτυον, net; retinal]. Retinitis.

**dictyopsia** (*dik-te-op'-se-ah*) [δίκτυον, net; ὄψις, view]. The sensation as if a net were stretched before the eyes.

**didactic** (*di-dak'-tik*) [δίδακτικός, apt at teaching]. Teaching by description and theory; opposed to clinical.

**didactylism** (*di-dak'-til-izm*) [δίς, double; δάκτυλος, a finger]. The condition of having congenitally but two digits on a hand or foot.

**didelphic, didelphous** (*di-del'-fik, -fus*) [δίς, double; δελφίς, the uterus]. Having a double uterus.

**diduction** (*di-duk'-shun*) [*diducere*, to draw apart]. Abduction of two parts; the withdrawal of a part.

**diductor** (*di-duk'-tor*) [*diduction*]. A muscle which in action produces diduction.

**didymalgia** (*did-e-mal'-je-ah*) [δίδυμος, testicle; ἄλγος, pain]. Pain in a testicle.

**didymin** (*did'-im-in*) [δίδυμοι, the testes]. A dry preparation made from the testes of the ox. Aphrodisiac dose 5 gr. (0.3 Gm.). In larger doses it is hypnotic.

**didymitis** (*did-e-mi'-tis*) [δίδυμος, a testicle; *ιτις*, inflammation]. Orchitis affecting mainly the body of the testicle.

**didymium** (*di-dim'-e-um*) [δίδυμος, twin]. A substance formerly thought to be an element, but now recognized as a mixture of the two elements neodymium and praseodymium.

**didymodynia** (*did-im-o-din'-e-ah*) [δίδυμος, testicle; ὀδύνη, pain]. Pain in the testicle.

**didymous** (*did'-im-us*) [δίδυμος, twin]. Twin, arranged in a pair, or in pairs.

**didymus** (*did'-im-us*) [δίδυμος, twin]. A twin; a twin-monstrosity; a testicle.

**die** (*di*). To cease to live; to become dead.

**diechoscope** (*di-ek'-o-skōp*) [δίς, twice; ἠχώ, a sound; σκοπεῖν, to examine]. A kind of stethoscope for the simultaneous perception of two different sounds in two different parts of the body.

**diecious** (*di-e'-shus*) [δίς, two; οἶκος, house]. In biology, having the two sexes in different individuals, or in two households as staminate and pistillate flowers separate and on separate plants.

**Dieffenbach's operation** (*de-fen-bakh'*). For amputation at the hip-joint; an elastic ligature is applied around the limb, a circular incision is made down to the bone, the vessels are secured, and the ligature removed; a knife is inserted two inches above the greater trochanter and the incision is carried down the outer aspect of the bone to meet the circular incision; the joint is then disarticulated.

**dielectric** (*di-el-ek'-trik*) [διά, through; ἤλεκτρον, amber]. Transmitting electricity by induction and not by conduction. d., **pseudo-**, any compound which acts as a dielectric when pure, but as an electrolyte when mixed with other members of its own class.

**dielectrolysis** (*di-e-lek-trol'-is-is*) [διά, through; *electrolysis*]. Galvanic electrolysis of a compound, the current passing at the same time through a diseased portion of the body and carrying one of the elements of the compound with it.

**dien** (*di'-en*). A contraction of diencephalon.

**diencephal** (*di-en'-sef-al*). Same as *diencephalon*.

**diencephalon** (*di-en-sef'-al-on*) [διά, between; ἐγκέφαλος, brain]. That part of the brain between the prosencephalon and the mesencephalon. It includes the thalami and the third ventricle. Syn., *between-brain; thalamencephalon*.

**dieresis** (*di-er'-es-is*) [διαίρεσις, a division]. A solution of continuity, as a wound, ulceration, etc.

**dieretic** (*di-er-et'-ik*) [*dieresis*]. Destructive; escharotic; corrosive.

**dies and counter-dies**. Metallic casts obtained by molding in sand or dipping in molten zinc and lead, or other alloys, such as Babbitt metal, and which are used for stamping up dental plates; they are generally described as male and female castings, between which the plate is swaged.

**diestrous, diœstrous** (*di-es'-trus*). Pertaining to a type of sexual season in female animals in which there is a short period of sexual rest.

**diestrum, diœstrum** (*di-es'-trum*) [διά, between; οἴστρος, gad-fly]. Heape's term for the short period of sexual rest characteristic of some female animals.

**diet** (*di'-et*) [δίαιτα, a system or mode of living]. The food taken regularly by an individual; the food adapted to a certain state of the body, as *fever-diet, convalescent-diet*. d., **Banting's**. See *Banting cure*. d., **bland**, one consisting of food that is free from the ingredients which excite heat, but containing all the nutrients—protein, carbohydrates, and fat—necessary for the maintenance of man. d., **diabetic**, a diet mostly of meats and green-vegetables, starches and sugars being excluded. d., **fever-**, a nutritious, easily digestible liquid or semiliquid diet, usually with milk and meat-broths, as a basis. d., **gouty**, a diet of simple nutritious food, avoiding wines, fats, pastries, and much meat. d., **Tuffnell's**, a highly nutritious diet, including but a small amount of liquids, employed in the treatment of aneurysm.

**dietarian** (*di-et-a'-re-an*). A physician who pays special attention to matters of diet.

**dietary** (*di'-et-a-re*) [*diet*]. A system of food-regulation intended to meet the requirements of the animal economy.

**dietetic** (*di-et-et'-ik*) [*diet*]. Pertaining to diet.

**dietetics** (*di-et-et'-iks*). The science of the systematic regulation of the diet for hygienic or therapeutic purposes.

**diethyl** (*di-eth'-il*), C₄H₁₀. A double molecule of ethyl; in a free state it constitutes normal butane. d. **acetal**. See *acetal* (1). d. **acetone, d. ketone**; C₂H₅C₂H₅ . CO. A hypnotic liquid used in mania. Syn., *propione*. d. **glycocoll-guaiacol hydrochlorate**,

an antiseptic used in pulmonary tuberculosis, ozena, etc. Dose 15-60 gr. (1-4 Gm.). Syn. *gujasanol*.
**diethylamine** (*di-eth-il'-am-in*), NC₄H₁₁. A non-poisonous, liquid ptomaine obtained from putrefying fish.
**diethylenediamine**. See *piperazine*.
**dietist** (*di'-et-ist*). One who is expert in questions of diet; a dietarian or dietitian.
**dietitian** (*di-et-ish'-an*) [*diet*]. See *dietist*.
**Dietl's crises** (*dē'-tl*) [Joseph Dietl, Austrian physician, 1804-1878]. Paroxysms of gastric distress and severe abdominal pain occurring in nephroptosis; probably dependent on acute hydronephrosis from twisting of the ureter.
**dietotherapy** (*di-et-o-ther'-ap-e*) [*diet*; θεραπεία, therapy]. The regulation of diet for therapeutic purposes.
**Dietrich's reaction for uric acid.** A red coloration results from the addition of a solution of sodium hypochlorite or hypobromite to the uric-acid solution. The color vanishes on adding caustic alkali.
**Dieudonné's medium** (*de-ū-don-a'*) [Adolph Dieudonné, German physician, 1864- ]. An alkaline medium for bacterial cultures; it is composed of defibrinated ox blood, solution of potassium hydroxide, and cholera agar.
**Dieulafoy's triad** (*de-ū'-laf-oy*) [Georges Dieulafoy, French physician, 1839-1911]. Muscular contraction, tenderness, and hyperesthesia of the skin at McBurney's point in appendicitis.
**differential** (*dif-er-en'-shal*) [*differentia*, difference]. Pertaining to or creating a difference. **d. blood-count**, an estimation of the number of different kinds of leukocytes in a cubic millimeter of blood. **d. diagnosis.** See *diagnosis, differential*. **d. staining,** a method of staining tubercle and other bacilli, founded upon the fact that they retain the color in the presence of certain reagents that decolorize the surrounding tissues.
**differentiation** (*dif-er-en-she-a'-shun*) [see *differential*]. 1. The act or process of distinguishing or making different. 2. Changing from general to special characters; specialization.
**diffluence** (*dif'-lu-ens*) [*diffluere*, to flow apart]. The condition of being almost liquefied.
**diffraction** (*dif-rak'-shun*) [*dis*, apart; *fractus*, broken]. The deflection or the separation into its component parts that takes place in a ray of light when it passes through a narrow slit or aperture. **d. grating,** a strip of glass closely ruled with fine lines; it is often used in the spectroscope in the place of the battery of prisms.
**diffusate** (*dif'-ū-sāt*) [*diffuse*]. The portion of the liquid which passes through the animal membrane in dialysis, and holds crystalloid matter in solution. Dialysate.
**diffuse** (*dif-ūs'*) [*diffundere*, to spread by pouring]. Scattered; not limited to one tissue or spot; opposed to localized.
**diffusibility** (*dif-ū-si-bil'-it-e*). Capacity for being diffused. **d. of gases,** Dalton's term for that property by which two or more gases confined in an inclosed space expand as if the space were occupied by one gas alone, the elastic force of the mixture being equal to the sum of the elastic forces of all the combined gases.
**diffusible** (*dif-ū'-zib-l*) [*diffuse*]. Spreading rapidly; capable of passing through a porous membrane; applied to certain quickly acting stimulants, usually of transient effect.
**diffusiometer** (*dif-ū-ze-om'-et-ur*) [*diffusion*; μέτρον, a measure]. A device for estimating the diffusibility of gases.
**diffusion** (*dif-ū'-zhun*) [*diffuse*]. 1. A spreading-out. 2. Dialysis. **d.-circle,** the imperfect image formed by incomplete focalization, the position of the true focus not having been reached by some of the rays of light or else having been passed. **d. vacuoles.** See under *vacuoles*.
**difluordiphenyl, difluorodiphenyl** (*di-flu-or-di-fen'-il, -o-di-fen'-il*), C₆H₄Fl—C₆H₄Fl; used as a 10 % dusting-powder or as a 10 % ointment in treating luetic ulcers, etc.
**digalen** (*dij'-al-en*). A soluble preparation of digitalis, suitable for intravenous administration.
**digallic acid** (*di-gal'-ik*). Synonym of *tannic acid*.
**digastric** (*di-gas'-trik*) [*bis*, two; γαστήρ, belly]. Having two bellies, as the *digastric* muscle; see *muscles, table of*. 2. Referring to the digastric muscle. **d. groove,** a groove on the mastoid process which serves as the line of origin of the digastric muscle. **d. muscle.** See *muscles, table of*. **d. nerve.** See *nerves, table of*.
**digastricus** (*di-gas'-trik-us*). The digastric muscle. See *muscles, table of*.
**digenesis** (*di-jen'-es-is*) [*bis*, two; γένεσις, generation]. In biology, the alternation of sexual and asexual generation.
**digenetic** (*di-jen-et'-ik*) [*digenesis*]. Relating to alternate generation.
**digenism** (*di'-jĕn-ism*). 1. See *digenesis*. 2. The combined or concurrent action of two causes.
**digerent** (*di'-er-ent*) [*digerere*, to digest]. A digestant; also a medicine that excites the secretion of pus in wounds.
**digest** (*di-jest'*) [*digerere*, to digest]. 1. To make food capable of absorption and assimilation. 2. In pharmacy, to macerate in a liquid medium.
**digestant** (*di-jest'-ant*) [*digest*]. A substance that assists digestion of the food.
**digester** (*di-jest'-er*) [*digest*]. An autoclave or apparatus for destructive distillation.
**digestibility** (*di-jes-tib-il'-it-e*) [*digestibilis*, that can be digested]. Susceptibility of being digested.
**digestible** (*di-jest'-ib-l*) [*digestibilis*, that can be digested]. Capable of being digested.
**digestion** (*di-jes'-chun*) [*digest*]. Those processes whereby the food taken into an organism is made capable of being absorbed and assimilated by the body-tissues. **d., artificial,** digestion carried on outside of the body. **d., gastric,** digestion by the action of the gastric juice. **d., intestinal,** digestion by the action of the intestinal juices, including the action of the bile and the pancreatic fluid. **d., pancreatic,** digestion by the action of the pancreatic juice. **d., peptic.** See *d., gastric*. **d., primary,** gastrointestinal digestion. **d., salivary,** digestion by the saliva. **d., secondary,** the assimilation by the body-cells of their appropriate pabulum.
**digestive** (*di-jes'-tiv*) [*digestion*]. 1. Relating to or favoring digestion. 2. An agent that promotes digestion. **d. tract,** the whole alimentary canal from the mouth to the anus.
**digestol** (*di-jes'-tol*). Trade name of a combination of bismuth subnitrate, pepsin, and salol.
**digipoten** (*dij-ip-o'-ten*). A preparation containing the digitalis glucosides in soluble form. It is said to be of the same strength as digitalis leaf; its dosage is also the same.
**digipuratum** (*dij-ip-ū-ra'-tum*). A proprietary digitalis preparation.
**digistrophan** (*dij-is-tro'-fan*). Trade name of a preparation of digitalis and strophanthus.
**digit** (*dij'-it*) [*digitus*, finger]. A finger or toe.
**digital** (*dij'-it-al*) [*digit*]. 1. Pertaining to the fingers or toes. 2. Performed with the fingers. 3. Resembling a depression made with a finger-tip; *e. g., digital* fossa. **d. arteries,** the arteries of the hands and feet supplying the digits. See under *artery*. **d. compression,** the stoppage of a flow of blood by pressure with the finger. **d. examination,** examination or exploration with the finger.
**digitalacrin** (*dij-it-al-ak'-rin*), C₂₈H₄₄O₈. A substance obtained from digitalis.
**digitalein** (*dij-it-al'-e-in*). 1. One of the constituents of digitalis. 2. A cardiac, tonic and diuretic. Dose $\frac{1}{64} - \frac{1}{32}$ gr. (0.001-0.002 Gm.) 2 to 4 times daily.
**digitaletin** (*dij-it-al-et'-in*), C₂₁H₃₄O₉. A substance obtained from digitalin by heating with dilute acid.
**digitaliform** (*dij-it-al'-e-form*) [*digit; forma*, form]. Finger-shaped.
**digitalin, digitalinum** (*dij-it-al'-in, dij-it-al-i'-num*) [*digitalis*]. 1. C₅H₈O₂(?). The active principle of *Digitalis purpurea*. Dose $\frac{1}{40} - \frac{1}{20}$ gr. (0.001-0.002 Gm.). 2. A precipitate from a tincture of *Digitalis purpurea*. **d., crystallized.** See *digitin*. **d., French,** a yellowish, odorless, bitter powder, said to consist of digitalin with some digitoxin. It is used as a heart-tonic. Dose $\frac{1}{250}$ gr. (0.00026 Gm.) rapidly increased to $\frac{1}{40}$ gr. (0.0016 Gm.) daily; maximum dose $\frac{1}{50}$ gr. (0.0013 Gm.) daily. Syn., *chloroformic digitalin; Homolle's amorphous digitalin; insoluble digitalin*. **d., German,** a white or yellowish powder, said to consist of digitalein with some digitonic and digitalin. It is a noncumulative heart-tonin and diuretic. Dose $\frac{1}{64} - \frac{1}{32}$ gr. (0.001-0.002 Gm.) 3 or 4 times daily in pills or subcutaneously; maximum dose $\frac{1}{16}$ gr. (0.004 Gm.) single; ⅓ gr. (0.022 Gm.) daily. **d., Homolle's, d., insoluble.** See *d., French*. **d., Kilian's,** a white, amorphous powder, exerting

the characteristic effect of digitalis leaves. Dose $\frac{1}{40}$ gr. (0.00025 Gm.). Syn., *digitalinum verum kiliani*. **d., Nativelle's, d.-nativelle,** $C_{36}H_{56}O_{15}$; said to consist chiefly of digitoxin. It is recommended as a heart-tonic and in pulmonary inflammation. Dose $\frac{1}{120}-\frac{1}{24}$ gr. (0.00065-0.001 Gm.). **d., soluble.** See *d., German*.

**digitaliretin, digitalirrhetin** (*dij-it-al-i-ret'-in*), $C_{16}H_{20}O_3$. A substance obtained from digitalin by action of dilute acid with heat.

**digitalis** (*dij-it-a'-lis*) [*digitalis*, pertaining to the fingers]. Foxglove. The *digitalis* of the U. S. P. is the leaves of *D. purpurea*. It contains an amorphous complex substance, *digitalin*, that does not, however, represent the full properties of the leaves. It is a powerful cardiac stimulant, strengthening the systole and lengthening the diastole of the heart. It also acts as a diuretic; in large doses it causes gastric disturbance. It is employed mainly in diseases of the heart when compensation is lost. **d.,** extract. of (*extractum digitalis*, U. S. P.). Dose $\frac{1}{6}-\frac{1}{3}$ gr. (0.01-0.032 Gm.). **d., fluidextract** (*fluidextractum digitalis*, U. S. P.). Dose 1-3 min. (0.06-1.8 Cc.). **d., infusion of** (*infusum digitalis*, U. S. P.), 1½ parts of the leaves in 100 parts. Dose 2-4 dr. (4-8 Cc.). **d., tincture of** (*tinctura digitalis*, U. S. P.), 15 parts of the leaves in 100 of diluted alcohol. Dose 10-20 min. (0.6-1.2 Cc.).

**digitalism, digitalismus** (*dij'-it-al-izm, dij-it-al-is'-mus*). The condition caused by the injudicious use of digitalis, consisting in paralysis of cardiac action.

**digitalization** (*dij-it-al-i-za'-shun*). Subjection to the effects of digitalin or digitalis.

**digitalone** (*dij'-it-al-ōn*). A non-irritating solution of the digitalis glucosides.

**digitalose** (*dij-it-al-ōs*). A white crystalline constituent of digitalis.

**digitation** (*dij-it-a'-shun*) [*digitatus*, having digits]. A finger-like process, or a succession of such processes, especially of a muscle.

**digiten** (*dij'-it-en*) [*digitus*, a finger]. Belonging to a digit in itself.

**digitiform** (*dij'-it-e-form*) [*digit*]; *forma*, form. Finger-shaped.

**digiti mortui** (*dij'-it-i mor'-tu-i*) [L.]. Dead fingers; a cold and white state of the fingers.

**digitin** (*dij'-it-in*), $(C_4H_2O_3)n$. A therapeutically inert substance occurring as a granular, crystalline powder, isolated from the leaves of *Digitalis purpurea*. Syn., *crystallized digitalin*.

**digitofibular** (*dij-it-o-fib'-u-lar*). Pertaining to the fibular aspect of the toes.

**digitol** (*dij'-it-ol*). Trade name of a fat-free tincture of digitalis.

**digitometatarsal** (*dij-it-o-met-a-tar'-sal*). Pertaining to the metatarsus and the toes.

**digitonin** (*dij-it-o'-nin*) [*digitalis*], $C_{31}H_{52}O_{17}$. A white, amorphous mass obtained from digitalis.

**digitoradial** (*dij-it-o-ra'-de-al*). Relating to or situated upon the radial aspect of the fingers.

**digitotibial** (*dij-it-o-tib'-e-al*). Relating to the tibial aspect of the toes.

**digitoulnar** (*dij-it-o-ul'-nar*). Relating to the ulnar aspect of the fingers.

**digitoxin** (*dij-it-oks'-in*) [*digitalis*; τοξικόν, poison], $C_{21}H_{32}O_7$, or $C_{34}H_{56}O_{14}$. A highly poisonous glucoside from *Digitalis purpurea*. A powerful heart-tonic, used in valvular lesions and myocarditis, etc. Dose $\frac{1}{250}-\frac{1}{120}$ gr. (0.00026-0.0005 Gm.) 3 times daily with 3 min. (0.2 Cc.) chloroform, 60 min. (4 Cc.) alcohol, 1½ oz. (45 Cc.) water. Enema, $\frac{1}{80}$ gr. (0.0008 Gm.) with 10 min. (0.6 Cc.) alcohol, 4 oz. (120 Cc.) water, 1 to 3 times daily. Maximum dose $\frac{1}{32}$ gr. (0.002 Gm.) daily.

**digitus** (*dij'-it-us*) [L.: pl., *digiti*]. A finger or toe. **d. annularis,** ring finger. **d. auricularis,** little finger. **d. clavatus,** club-finger. **d. demonstrativus,** index finger. **d. hippocraticus,** club-finger. **d. manus,** a finger. **d. medicus,** the ring finger. **d. medius,** the middle finger. **d. minimus,** the little finger. **d. pedis,** a toe. **d. recellens,** trigger finger.

**diglossia** (*di-glos'-e-ah*) [δίς, double; γλῶσσα, tongue]. The condition of having a double tongue.

**dignathus** (*dig-na'-thus*) [δίς, twice; γνάθος, jaw]. A monster with two lower jaws.

**dihydrate** (*di-hi'-drāt*) [δίς, twice; ὕδωρ, water]. 1. Any compound containing two molecules of hydroxyl. Syn., *bihydrate*. 2. A compound containing two molecules of water.

**dihydrated** (*di-hi'-dra-ted*). Having absorbed two hydroxyl molecules.

**dihydric** (*di-hi'-drik*). Containing two atoms of hydrogen in the molecule.

**dihydride** (*di-hi'-drid*). A compound of two atoms of hydrogen with an element or radical.

**dihydrocollidine** (*di-hi-dro-kol'-id-ēn*), $C_8H_{13}N$. A liquid substance isomeric with a ptomaine obtained from putrid flesh and fish.

**dihydrocoridine** (*di-hi-dro-kor'-id-ēn*), $C_{10}H_{17}N$. A substance isomeric with a ptomaine found in cultures of the *Bacillus allii*.

**dihydrolutidine** (*di-hi-dro-lu'-tid-ēn*) [δίς, double; ὕδωρ, water; *luteus*, yellow], $C_7H_{11}N$. One of the alkaloidal bodies found in cod-liver oil. It is slightly poisonous, in small doses diminishing general sensibility, in large doses causing tremor, paralysis of the legs, or, in animals, the hind limbs, and death.

**dihydroresorcinol** (*di-hi-dro-re-zor'-sin-ol*). Shining white prisms, soluble in water, alcohol, or chloroform, melting at 104°-106° C., obtained from resorcinol by action of sodium amalgam with carbon dioxide. It is recommended as an antiseptic.

**dihydroxyphthalophenone** (*di-hi-droks-e-thal-o-fe'-nōn*). Phenolphthalein.

**dihydroxytoluene** (*di-hi-droks-e-tol'-u-ēn*). See *orcin*.

**dihysteria** (*di-his-te'-re-ah*) [δίς, double; ὑστέρα, the womb]. The presence of a double uterus.

**diiodide** (*di-i'-o-did*) [*di*, two; *iodum*, iodine]. A compound consisting of a basic element and two atoms of iodine.

**diiodoaniline** (*di-i-o-do-an'-il-in*), $C_6H_5$ . $NH_2$ . $I_2$ [1 : 2 : 4]. A reaction-product of aniline with iodine chloride. It is antiseptic and used as an application in skin diseases. Syn., *metadiiodaniline*.

**diiodobetanaphthol** (*di-i-o-do-be-tah-naf'-thol*), $C_{10}H_6I_2O_2$. A yellowish-green powder obtained from mixed solutions of iodine with potassium iodide and betanaphthol with sodium carbonate and sodium hypochlorite. It is used as an antiseptic in place of aristol. Syn., *naphtholaristol*; *naphtholdiiodide*.

**diiodocarbazol** (*di-i-o-do-kar'-ba-zol*), $C_{12}H_6I_2$ : NH. A substance obtained from carbazol by action of iodine with heat; insoluble in water, soluble in alcohol and chloroform. It is recommended as an antiseptic.

**diiododithymol** (*di-i-o-do-di-thi'-mol*). See *aristol*.

**diiodoform** (*di-i-o'-do-form*), $C_2H_2I_4$. A substance obtained from acetylene iodide by excess of iodine and containing 95.28 % of iodine. It decomposes on exposure to light. Used as a substitute for iodoform. Syn., *tetraethylene iodide*.

**diiodomethane** (*di-i-o-do-meth'-ān*). Methylene iodide.

**diiodonaphthol** (*di-i-o-do-naf'-thol*). See *diiodobetanaphthol*.

**diiodoresorcinol** (*di-i-o-do-re-zor'-sin-ol*). A brown, inodorous powder, used as an antiseptic in place of aristol.

**diiodosalicylic acid** (*di-i-o-do-sal-is-il'-ik*). See *acid, diiodosalicylic*. **d.-methylester.** See *sanoform*. **d.-phenylester.** See *diiodosalol*.

**diiodosalol** (*di-i-o-do-sal'-ol*), $C_6H_3I_2(OH)CO_2C_6H_5$. A condensation-product of diiodosalicylic acid with phenol. It is used in treatment of skin diseases.

**diiodothioresorcinol** (*di-i-o-do-thi-o-re-zor'-sin-ol*), $C_6H_2O_2I_2S_2$. It is used as a dusting-powder and in 10 to 20 % ointment.

**dikamali** (*dik-am-ah'-le*) [E. Ind.]. A fetid gum-resin obtained from *Gardenia gummifera* and *G. lucida*. In decoction it is used as an antiperiodic and in the treatment of chronic skin diseases.

**dilaceration** (*di-las-er-a'-shun*) [*dilaceratio*, a tearing apart]. A tearing apart; division of a membranous cataract by a tearing operation.

**dilatatio cordis** (*dil-at-a'-she-o kor'-dis*) [L.]. Dilatation of the heart.

**dilatation** (*dil-at-a'-shun*) [*dilatare*, to expand]. A spreading apart; the state, especially of a hollow part or organ, of being dilated or stretched. **d., digital,** dilatation of a body-cavity or orifice by means of one or more fingers. **d. of heart,** an increase in the size of one or more of the cavities of the heart, arising from a relaxation or weakening of the heart muscle. It is associated with evidences of failure of circulation, resulting in congestion of the lungs and other viscera. **d., hydrostatic,** dilatation of a cavity or part by means of an introduced elastic bag which is subsequently distended with water. **d. of stomach,** increase in size of the

stomach from relaxation of the walls and expansion with gas.
**dilate** (di-lāt') [*dilatation*]. To spread. To increase in size; to spread apart; to stretch.
**dilator** (di-la'-tor) [*dilate*]. 1. An instrument for stretching or enlarging a cavity or opening; also, 2. A dilating muscle. See under *muscle*. **d. iridis**, the set of muscular fibers dilating the pupil. **d. naris.** See *muscles, table of*. **d. pyloric**, the muscle which dilates the pyloric orifice of the stomach. **d. tubæ.** See *tensor palati*, in *muscles, table of.*
**dill** (dil). See *anethum*.
**diluent** (dil'-ū-ent) [*dilute*]. 1. Diluting. 2. An agent that dilutes the secretions of an organ.
**diluin** (di-lū'-in). A normal physiological saline solution to which ½ per cent. of carbolic acid has been added; used as a diluent for tuberculin.
**dilute** (di-lūt') [*diluere*, to wash away]. To make weaker through increasing the bulk by the addition of liquid.
**diluting fluids.** Solutions for use with the hemocytometer. See *Hayem's solution* and *Sherrington's solution, Toisson's solution*.
**dilution** (di-lū'-shun) [*dilute*]. 1. The process of adding a neutral fluid to some other fluid or substance, in order to diminish the qualities of the latter. 2. A diluted substance; the result of a diluting process.
**dilutionist** (di-lū'-shun-ist) [*dilute*]. One who advocates the dilution of medicines. **d., high,** a homeopathist who advocates the extreme attenuation of medicines.
**dimerous** (dim'-er-us) [δίς, two, μέρος, a part]. In biology, bipartite.
**dimethyl** (di-meth'-il), (CH₃)₂. A double molecule of methyl; in the free state it constitutes ethane. **d.-acetal,** C₆H₁₄O₃, a colorless ethereal liquid obtained from aldehyde, methylalcohol, and glacial acetic acid with heat; it is used as an anesthetic, alone or combined with one-half its volume of chloroform. Syn., *ethylidenedimethyl ether*. **d.-amidoantipyrine.** See *pyramidon*. **d.-amidoazobenzene,** C₁₄H₁₅N₃. It is used as an indicator in alkalimetry and as a fat color. Syn., *butter yellow*. **d.-amidophenyl-dimethyl-pyrazolon.** See *pyramidon*. **d.-arsin.** See *cacodyl*. **d. sulphate,** $\frac{CH_3}{CH_3}$>SO₄, a colorless, oily fluid much used in chemical manipulation and giving rise to poisoning with marked local and pulmonary symptoms, convulsions, coma, and paralysis. **d.-xanthin.** 1. See *paraxanthin*. 2. See *theobromine*.
**dimethylamine** (di-meth-il-am'-in), NC₂H₇. A nontoxic ptomaine found in putrefying gelatin, old decomposing yeast, etc.
**dimethylated** (di-meth'-il-a-ted). Combined with two molecules of methyl.
**dimetria** (di-me'-tre-ah) [δίς, double; μήτρα, the womb]. The condition of having a double uterus.
**dimorphism** (di-morf'-izm) [δίς, double; μορφή, form]. The property of assuming or of existing under two distinct forms.
**dimorphobiotic** (di-mor-fo-bi-ot'-ik) [δίς; μορφή, shape; βίωσις, life]. Relating to an organism which runs through two or more morphologically distinct phases in its life-history—a free stage and a parasitic stage.
**dimorphous** (di-mor'-fus) [δίς; μορφή, form]. Existing in two forms.
**dimple** (dim'-pl). A slight depression.
**dineuric** (di-nū'-rik) [δίς; νεῦρον, nerve]. Provided with two neuraxons; said of a nerve-cell.
**dinic, dinical** (din'-ik, din'-ik-al) [δῖνος, whirl]. Pertaining to or useful in the relief of vertigo.
**dinitrate** (di-ni'-trāt). A compound resulting from the replacement of the hydrogen of two molecules of nitric acid by a base.
**dinitrocellulose** (di-ni-tro-sel'-ū-lōs). See *pyroxylin*.
**dinitrocresol** (di-ni-tro-kre'-sol), C₇H₆N₂O₅. Explosive crystals melting at 85° C. The potassium salt of o-dinitrocresol is an excellent insecticide and an efficient remedy in scabies, and mixed with equal parts of soap is, under the name of *antinonnin*, used as an insecticide and to destroy rats and mice. A mixture of the potassium salts of o-dinitrocresol and p-dinitrocresol, a commercial substitute for saffron, has been employed with fatal results.
**dinitroresorcin** (di-ni-tro-re-zor'-sin), C₆H₄N₂O₅ + 2H₂O. It is employed in histological preparations. Syn., *ordinary dinitroresorcin*.
**dinner pills.** A name applied to various mild cathartic pills taken after meals. See *Lady Webster pill*.
**dinomania** (din-o-ma'-ne-ah) [δῖνος, a whirling dance; μανία, mania]. Dancing-mania, choromania, q. v.
**dinus** (di'-nus) [δῖνος, whirl]. Vertigo or dizziness.
**diœcious** (di-e'-shus). See *diecious*.
**dioform** (di'-o-form). Acetylene dichloride, a volatile narcotic.
**diogmus** (di-og'-mus) [διωγμός, a chase]. Palpitation of the heart.
**diomorphine** (di-o-mor'-fēn). A mixture of dionine and morphine.
**dionin** (di'-o-nin), C₂H₅O . (OH) . C₁₇H₁₇NO . HCl + H₂O. Ethylmorphine hydrochloride. It is analgesic, antispasmodic, and sedative, and is employed in diseases of the respiratory passages and in morphinism. Dose ⅙–⅓ gr. (o.01–0.03 Gm.).
**dioning** (di-o'-ning) [Διώνη, the mother of Aphrodite, afterward applied to Aphrodite herself]. Normal love between the opposed sexes; as distinguished from *urning*, abnormal love between the same sexes.
**dionym** (di'-o-nim) [δίς, two; ὄνομα, name]. A name consisting of two words, as medulla oblongata; corpus callosum.
**diophthalmus** (di-off-thal'-mus). See *diprosopus*.
**diopsimeter** (di-op-sim'-et-er) [δίοψις, clear vision; μέτρον, a measure]. An instrument for exploration of the visual field.
**diopter** (di-op'-ter) [διά, through; ὄψεσθαι, to see]. The unit of measurement of the refractive power of an optic lens. It is the refractive power of a lens that has a focal distance of one meter.
**dioptometer** (di-op-tom'-et-er). An instrument for determining ocular refraction. Same as *optometer*.
**dioptometry** (di-op-tom'-et-re) [*diopter*; μέτρον, a measure]. The measurement of the accommodative and refractive states of the eye.
**dioptoscopy** (di-op-tos'-ko-pe) [*diopter*; σκοπεῖν, to examine]. A method of estimating ocular refraction by means of the ophthalmoscope.
**dioptral** (di-op'-tral). Pertaining to a dioptry; expressed in dioptrics.
**dioptric** (di-op'-trik) [*diopter*]. 1. Pertaining to transmitted and refracted light. 2. A diopter.
**dioptrics** (di-op'-triks). A branch of optics treating of the refraction of light by transparent media, especially by the media of the eye.
**dioptroscopy** (di-op-tros'-ko-pe). Same as *dioptoscopy*. See also *retinoscopy*.
**dioptry** (di-op'-tre). See *diopter*.
**dioradin** (di-o-ra'-din). Trade name of the "radioactive menthol-iodine" remedy proposed for tuberculosis
**diorthosis** (di-or-tho'-sis) [διά, throughout; ὀρθόειν, to straighten]. The surgical correction of a deformity, or repair of an injury done to a limb.
**diorthotic** (di-or-thot'-ik). Relating to or effecting a diorthosis.
**dioscorea** (di-os-ko'-re-ah) [*Dioscorides*], Greek physician and botanist]. Wild yam, colic root. The rhizome of *D. villosa*, a creeping-plant, indigenous to the Eastern U. S. It is claimed to be expectorant, diaphoretic, and stimulant to the intestinal canal, in large doses causing neuralgic pains and erotic excitement. It is used successfully for bilious colic. *fluidextract,* of standard strength, dose ♍ xv–xxx.
**dioscorein** (di-os-ko'-re-in) [*Dioscorides*], a Greek botanist]. A precipitate from a tincture of the root of *Dioscorea villosa*. It is antispasmodic, expectorant, and diaphoretic. Dose ½ to 4 grains.
**diose** (di'-ōs). A monosaccharide containing only two carbon atoms; it is the simplest form of sugar.
**diosmic** (di-oz'-mik). Containing two atoms of osmium as a quadrivalent radical.
**diosmin** (di-os'-min) [δίος, divine; ὀσμή, odor]. An active principle obtained from various species of Buchu. It is an amorphous, bitter substance.
**diosmosis** (di-os-mo'-sis). Same as *osmosis*.
**diosmotic** (di-os-mot'-ik) [διά, through; ὠσμός, impulse]. Pertaining to osmosis.
**Diospyros** (di-os'-pi-ros) [Διός, of Jove; πυρός, grain; fruit]. A genus of trees of the order *Ebenaceæ*. The bark of *D. virginiana*, the persimmon-tree of the United States, is astringent, tonic, antiperiodic, and fever, and uterine hemorrhage. Dose of *fluidextract* 30–60 min. (3–4 Cc.).
**diostosis** (di-os-to'-sis) [διά, away from; ὀστέον, a bone]. Displacement of a bone.

# DIOTIC 296 DIPLOSCOPE

**diotic** (di-ot'-ik) [δίς, two; οὖς, ὠτός, ear]. Binaural; pertaining to both ears.

**dioviburnia** (di-o-vi-bur'-ne-ah). A proprietary combination of equal parts of the fluidextracts of *viburnum opulus*, *v. prunifolium*, *chamælirium carolinianum*, *caulophyllum thalictroides*, *aletris farinosa*, *mitchella repens*, *scutellaria lateriflora*, and *dioscorea villosa*. It is antispasmodic and anodyne, and is used in dysmenorrhea, amenorrhea, etc.

**dioxide** (di-oks'-id) [δίς, two; ὀξύς, sharp]. A molecule containing two atoms of oxygen and one of a base.

**dioxogen** (di-oks'-so-jen). Trade name for a solution of hydrogen dioxide.

**dioxyanthranol, dioxyanthrol** (di-oks-e-an'-thran-ol, -throl). See *anthrarobin*.

**dioxybenzene** (di-oks-e-ben'-zēn). See *hydroquinone*.

**dioxydiamidoarsenobenzol** (di-oks-se-di-am'-id-o-ar-sen-o-ben'-zol). Salvarsan; a synthetic compound considered a specific for syphilis.

**dioxygen** (di-oks'-e-jen); H₂O₂. A term introduced for a preparation of pure hydrogen dioxide.

**dioxynaphthalene** (di-oks-e-naf'-thal-ēn), C₁₀H₈O₂. A toxic compound used as a roborant. The daily dose is 3 gr. (0.19 Gm.).

**dioxytoluene** (di-oks-e-tol'-u-ēn). Same as *orcin*.

**dip.** The deviation from the horizontal position shown by a freely suspended magnetic needle.

**dipentene** (di-pen'-tēn), C₁₀H₁₆. Cinene; a compound produced by heating pinene, camphor, and limonene to 250°–300° C. It is present in the Russian and Swedish turpentine oil. It is a liquid with an agreeable lemon-like odor; sp. gr. 0.853; boils at 175°–176° C.

**dipeptide** (di-pep'-tid). A protein substance consisting of two aminoacids.

**diphenyl** (di-fen'-il), C₁₂H₁₀. A hydrocarbon resulting from the action of sodium upon brombenzene in ether or benzene. It is also present in coal-tar.

**diphtheria** (dif-the'-re-ah) [διφθέρα, a skin or membrane]. An acute infectious disease caused by the Klebs-Loeffler bacillus. It is characterized by the formation, on a mucous membrane, most frequently that of the pharynx, of a false membrane, grayish or buff in color, and quite firmly adherent. Any mucous membrane, as the laryngeal, nasal, conjunctival, and, more rarely, the gastrointestinal, vaginal, and that of the middle ear, may be the seat of the disease. The membrane may also be formed on wounds—*surgical* or *wound-diphtheria*. **d., Brétonneau's**, true diphtheria of the pharynx, first described by P. Brétonneau (1826). **d. toxin**, a toxalbumin produced by *Bacillus diphtheriæ*; it is destroyed by a temperature over 60° C., and is capable of causing in susceptible animals the same phenomena induced by inoculation with the living bacillus.

**diphtherial** (dif-the'-re-al). Pertaining to diphtheria.

**diphtheric** (dif-ther'-ik). See *diphtheritic*.

**diphthericide** (dif-ther'-is-īd). A proprietary prophylactic against diphtheria, said to consist of thymol, sodium benzoate, and saccharin.

**diphtherin** (dif'-ther-in). See *diphtheria toxin*.

**diphtheriolysin** (dif-the-re-ol'-is-in). A lysin having a specific action on diphtheria toxin.

**diphtheritic** (dif-ther-it'-ik) [*diphtheria*]. Of or pertaining to diphtheria.

**diphtheritis** (dif-ther-i'-tis). See *diphtheria*.

**diphtheroid** (dif'-ther-oid) [*diphtheria*]. 1. Resembling diphtheria. 2. A general term for all pseudomembranous formations not due to *bacillus diphtheriæ*.

**diphtherotoxin** (dif-ther-o-toks'-in). See *diphtheria toxin*.

**diphthongia** (dif-thon'-je-ah) [δίς, double; φθόγγος, a voice]. The production of a double tone of the voice, due to incomplete unilateral paralysis of the recurrent laryngeal nerve, or to some lesion of the vocal bands that causes each to produce its own sound.

**diphyodont** (dif'-e-o-dont) [διφυής, twofold; ὀδούς, tooth]. In biology, having two sets of teeth, as the milk-teeth and the permanent teeth.

**diplacusis** (dip-lak-ū'-sis) [διπλόος, double; ἄκουσις, hearing]. 1. The hearing of a tone as higher by one ear than by the other. Syn., *diplacusis binauralis*. 2. The hearing of two tones by one ear when only one tone is produced. Syn., *diplacusis uniauralis*.

**diplasmatic** (di-plaz-mat'-ik) [δίς, two; πλάσμα, something formed]. Containing matter other than protoplasm; said of cells.

**diplastic** (di-plas'-tik) [δίς, two; πλάσσειν, to form]. A term applied to cells having two substances in their constitution.

**diplegia** (di-ple'-je-ah) [δίς, double; πληγή, stroke]. Paralysis of similar parts on the two sides of the body. **d., spastic cerebral, of infancy.** See *Little's disease*.

**diplegic** (di-ple'-jik). [δίς, double; πληγή, stroke]. Relating to or of the nature of diplegia.

**diplo-** [διπλόος, double]. A prefix signifying double.

**diploalbuminuria** (dip-lo-al-bū-min-ū'-re-ah) [*diplo-*; *albuminuria*]. The coexistence or alternation of physiological and pathological albuminuria in the same subject.

**diplobacillus** (dip-lo-bas-il'-us) [*diplo-*; *bacillus*]. A double bacillus.

**diplobacterium** (dip-lo-bak-te'-re-um) [*diplo-*; *bacterium*]. A bacterial form made up of two adherent bacteria.

**diploblastic** (dip-lo-blas'-tik) [*diplo-*; βλαστός, a germ]. Having two germinal layers.

**diplocardiac** (dip-lo-kar'-de-ah) [*diplo-*; καρδία, heart]. Having a double heart.

**diplocephalia** (dip-lo-sef-a'-le-ah) [*diplo-*; κεφαλή, the head]. A two-headed monstrosity.

**diplocephalus** (dip-lo-sef'-al-us). A monster with a single body and two heads.

**diplococcus** (dip-lo-kok'-us) [*diplo-*; κόκκος, a berry]. A micrococcus that occurs in groups of two. See *Micrococci*, table of.

**diplocoria** (dip-lo-ko'-re-ah) [*diplo-*; κόρη, pupil]. Double pupil.

**diploe** (dip'-lo-e) [διπλόη, a fold]. The cancellous bony tissue between the outer and inner tables of the skull.

**diploetic** (dip-lo-et'-ik) [*diploe*]. Relating to the diploe; diploic.

**diplogenesis** (dip-lo-jen'-es-is) [*diplo-*; γένεσις, production]. 1. The development of a double or twin monstrosity. 2. The process described by Pigné in 1846 whereby congenital tumors are formed by the inclusion of embryonic remains.

**diploic** (dip-lo'-ik) [*diploe*]. See *diploetic*.

**diploma** (dip-lo'-mah). A document granted by an authorized body of men, showing that the recipient has performed certain work under the prescribed conditions, and is entitled to a definite professional rank and title.

**diplomellituria** (dip-lo-mel-it-ū'-re-ah) [*diplo-*; *mellituria*]. The coexistence or alternation of diabetic and non-diabetic glycosuria in the same subject.

**diplomeric** (dip-lom'-er-ik) [*diplo-*; μέρος, a part]. Applied to muscles arising from two myotomes; *e. g.*, the supraspinatus and infraspinatus muscles.

**diplomyelia** (dip-lo-mi-e'-le-ah) [*diplo-*; μυελός, marrow]. An apparent doubleness of the spinal cord, produced by a longitudinal fissure.

**diploneural** (dip-lo-nū'-ral) [*diplo-*; νεῦρον, nerve]. Pertaining to a muscle, or other structure, supplied by the nerves.

**diplophonia** (dip-lo-fo'-ne-ah) [*diplo-*; φωνή, voice]. A rare symptom of laryngeal disease in which a double note is produced in the larynx. Cf. *diphthongia*.

**diplopia** (dip-lo'-pe-ah) [*diplo-*; ὄψις, sight]. Double vision, one object being seen by the eye or eyes as two. **d., binocular,** the most frequent, is due to a derangement of the muscular balance, the images of the object being thereby thrown upon nonidentical points of the retinæ. **d., crossed, d., heteronymous,** the result of divergent strabismus, the image of the right eye appearing upon the left side and that of the left eye upon the right side. **d., direct, d., homonymous,** the reverse of crossed diplopia, found in convergent strabismus. **d., monocular,** diplopia with a single eye, usually due to hysteria, to double pupil, or beginning cataract.

**diplopic** (dip-lo'-pik) [*diplo-*; ὄψις, sight]. Relating to or affected with diplopia.

**diplopiometer** (dip-lo-pe-om'-et-er) [*diplo-*; ὄψις, sight; μέτρον, measure]. An instrument for measuring the degree of double vision.

**diplosal** (dip'-lo-sal). A proprietary name for the salicylate of salicylic acid.

**diploscope** (dip'-lo-skōp) [*diplo-*; σκοπεῖν, to examine]. An instrument for the investigation of binocular vision.

**diploteratography** (*dip-lo-ter-at-og'-ra-fe*) [*diplo-*; τέρας, a monster; γράφειν, to write]. The description and diagnosis of special forms of double monsters.

**diploteratology** (*dip-lo-ter-at-ol'-o-je*) [*diplo-*; τέρας, a monster; λόγος, science]. The science of twin monstrosities.

**dipolar** (*di-po'-lar*). See *bipolar*.

**dipotassic** (*di-po-tas'-ik*) [*di*, two; *potassium*]. Containing two atoms of potassium in a molecule.

**Dippel's animal oil** (*dipl*) [Johann Conrad *Dippel*, German alchemist, 1673–1734]. Oleum cornu cervi. An oil obtained in distilling bone and deer's horn. It contains pyridin and lutidin. It is antispasmodic and stimulant.

**dipping** (*dip'-ing*). Palpating the liver by sudden pressure.

**dipping needle.** A magnetic needle so hung that it can move freely in a vertical plane.

**diprosopia** (*di-pro-so'-pe-ah*) [*δίς*, double; πρόσωπον, face]. In teratology, the duplication of the face.

**diprosopus** (*di-pros-o'-pus*) [*δίς*, double; πρόσωπον, face]. A monster characterized by a duplicity of the face and head, frequently associated with hydrocephalus, acrania, defective development of the brain, and spina bifida.

**dipsesis** (*dip-se'-sis*). Extreme thirst.

**dipsetic** (*dip-set'-ik*) [διψητικός, causing thirst]. Causing or attended with thirst.

**dipsomania** (*dip-so-ma'-ne-ah*) [δίψα, thirst; μανία, madness]. The uncontrollable desire for spirituous liquors.

**dipsomaniac** (*dip-so-ma'-ne-ak*). A person affected with dipsomania.

**dipsopathy** (*dip-sop'-ath-e*) [δίψα, thirst; πάθος, disease]. The thirst-cure; the treatment of disease by limiting the liquids ingested.

**dipsorrhexia** (*dip-sor-eks'-e-ah*) [δίψα, thirst; ὄρεξις, appetite]. Thebault's term for that early stage of alcoholism in which no organic lesions have as yet appeared in consequence of the alcoholic poisoning, but when the appetite has been developed.

**dipsosis** (*dip-so'-sis*). See *dipsesis*.

**Diptera** (*dip'-ter-ah*) [δίς, two; πτερόν, wing]. An order of insects including the fleas, flies, and mosquitoes.

**Dipterocarpus** (*dip-ter-o-kar'-pus*) [δίπτερος, two-winged; καρπός, fruit]. A genus of trees, chiefly found in southern Asia, some of which furnish gurjun balsam.

**dipterous** (*dip'-ter-us*) [δίς, two; πτερόν, wing]. In biology, having two wings or wing-like processes.

**Dipteryx** (*dip'-ter-iks*) [δίς, two; πτερυξ, a wing]. A genus of leguminous trees. **d. odorata**, the tree that produces the Tonka bean, *q. v.* The seeds are stimulant, antispasmodic, and antiseptic; used in whooping-cough and for flavoring. Dose of fluid-extract, 5–30 ♏ (0.3–1.8 c.c.).

**dipygus** (*dip'-ig-us*) [δίς, double; πυγή, buttocks]. A monstrosity with more or less duplication of the pelvis and lower parts of the back.

**Dipylidium** (*di-pi-lid'-e-um*) [δίπυλος, with two entrances]. A genus of parasitic platode worms.

**diradiation** (*di-ra-de-a'-shun*). See *actinobolia* (1).

**direct** (*di-rekt'*) [*directus*, straight]. In a right or straight line; without the interposition of some medium. **d. current**, a galvanic current. **d. image.** See *image, direct*. **d. murmur.** See under *murmur*. **d. ophthalmoscopy.** See *ophthalmoscopy*. **d. vision**, the perception of an object the image of which falls upon the macula.

**direction** (*di-rek'-shun*) [*dirigere*, to direct]. Relative position considered without regard to linear distance. **d.-spindle**, a fusiform body of the ovule, stretching from the germinal vesicle toward the surface.

**director** (*di-rek'-tor*) [*direct*]. Anything that guides or directs. **d., grooved**, an instrument grooved to guide the knife in surgical operations.

**dirigomotor** (*dir-ig-o-mo'-tor*) [*dirigere*, to direct; *motor*, a mover]. Controlling motor action.

**dirt** (*dert*). Excrement; feces. **d.-eating.** See *chthonophagia* and *geophagism*.

**dis-** [δίς, twice]. 1. A prefix denoting two or double. 2. A prefix denoting apart from.

**disaccharide** (*di-sak'-ar-id*). A carbohydrate formed by the condensation of two monosaccharide molecules.

**disarthral** (*dis-ar'-thral*) [δίς, twice; ἄρθρον, a joint]. Relating to muscles that pass over two joints, *e. g.*, the biceps.

**disarticulation** (*dis-ar-tik-u-la'-shun*) [*dis-*, *articulum*, a joint]. Separation at a joint; amputation at a joint.

**disassimilation** (*dis-as-sim-il-a'-shun*). The process of transformation of assimilated substances into waste-products. Failure or loss of assimilative power.

**disassociation** (*dis-as-o-se-a'-shun*) [*dis*, apart; *associare*, to unite with]. In chemistry, the decomposition of a compound by heat, the molecules reuniting on the removal of the heat; dissociation.

**disc, disk** [*discus*; δίσκος, a quoit or round plate]. A circular, plate-like organ or structure. **d., anisotropous.** See *d., sarcous*. **d., Becquerel's.** See *Becquerel's disc*. **d., blood-.** See *blood-corpuscle*. **d., choked.** See *papillitis*. **d., contractile, d., dark.** See *d., sarcous*. **d., cupped.** See *cupped disc*. **d.-diameter**, the diameter of the optic disc. **d., epiphyseal**, the broad articular surface with slightly elevated rim on each end of the centrum of a vertebra. Syn., *epiphyseal plate*. **d., equatorial.** See *plate, equatorial*. **d., germinal**, the small disc of the blastodermic membrane, in which the first traces of the embryo are seen. **d., intermediate.** See *Krause's membrane*. **d., interstitial**, the more translucent cementing substance conjoining the sarcous elements, and to which the lighter narrower striæ of the muscle-fiber and the intervals of the fibrils are due. Syn., *Englemann's lateral disc; isotropous disc; lateral disc; light disc*. **d., intervertebral.** See *intervertebral discs*. **d., invisible.** See *Norris' invisible corpuscles*. **d., isotropous, d., lateral, d., light.** See *d., interstitial*. **d., median, d., middle.** See *Krause's membrane*. **d., nuclear.** See *plate, equatorial*. **d., optic**, the circular area in the retina that represents the termination of the optic nerve. **d., overgrown.** See *discus proligerus*. **d., Placido's.** See *Placido's disc*. **d., sarcous**, the dark, broad, transverse stripe of striated muscle-fiber; it is anisotropic and supposed to represent the proper contractile substance of the fiber. Syn., *anisotropic disc; contractile disc; dark disc; principal disc; transverse disc*. **d., Schiefferdecker's intermediate.** See under *Schiefferdecker*. **d., stenopeic**, a lens allowing the passage of light-rays only through a straight narrow slit; it is used for testing astigmatism. **d., tactile, d., terminal.** See *meniscus, tactile*, and *Ranvier's tactile discs*. **d., transverse.** See *d., sarcous*. **d., vitelline.** See *discus proligerus*.

**discharge** (*dis-charj'*) [OF., *descharger*, to unload]. 1. A morbid secretion. 2. Any evacuation; also that which is evacuated. 3. A setting free or escape of pent-up energy. In electricity the restoration to a neutral electric condition by which a highly electrified body gives off its surplus of electricity to surrounding objects less highly electrified. **d., conductive**, an electric discharge taking place through conduction. **d., convective**, an electric discharge in which the charged particles of a fluid convey the electricity. **d., disruptive**, an electric discharge with emission of heat and sound.

**discharger** (*dis-char'-jer*). An instrument for setting free electricity stored in a Leyden jar or other condenser.

**discharging** (*dis-char'-jing*). Unloading; flowing out, as pus, etc. **d. lesion**, a brain-lesion that causes sudden discharges of nervous motor impulses.

**disciform** (*dis'-e-form*). Disc-shaped.

**discission** (*dis-ish'-un*) [*discissio; discindere*, to tear or cut apart]. 1. An operation for soft cataract in which the capsule is lacerated a number of times to allow the lens-substance to be absorbed. 2. See *Ransohoff's operation*.

**discoblastic** (*dis-ko-blas'-tik*) [*disc;* βλαστός, a germ]. Undergoing discoid segmentation of the vitellus.

**discoid** (*dis'-koid*) [*disc*]. 1. Shaped like a disc. 2. An excavator having a blade in the form of a disc.

**discoloration** (*dis-kul-ur-a'-shun*) [*discolor*, of different colors]. A change in or loss of the natural color of a part.

**discophorous** (*dis-kof'-or-us*) [*disc;* φέρειν, to bear]. Furnished with a disciform organ or part.

**discoplacenta** (*dis-ko-pla-sen'-tah*). See *placenta, discoid*.

**discoplasm, discoplasma** (*dis'-ko-plazm, -plaz'-mah*). The plasma of red blood-corpuscles.

**discoria** (*dis-ko'-re-ah*). See *dyscoria*.

**discous** (*dis'-kus*) [*disc*]. Discoid.

**discrete** (dis-krēt') [discretus, separated]. Not running together; separate; the opposite of confluent.

**discus** (dis'-kus) [L., "a disc"]. A disc. See disc. d. **articularis**, interarticular fibrocartilage. d. **proligerus**, the mass of cells of the membrana granulosa of the Graafian vesicle that surround the ovum.

**discussion** (dis-kush'-un) [discussio]. The scattering or driving away of a swelling, effusion, or tumor.

**discutient** (dis-kū'-shent) [discutere, to shake apart]. 1. Capable of effecting resolution. 2. A medicine having the power of causing an exudation to disappear.

**disdiaclast** (dis-di'-ak-last) [δίς, double; διά, through; κλᾶν, to break]. One of the small, doubly refractive elements in the contractile fiber of a muscle-fiber.

**disease** (dis-ēs') [dis, negative; ease, a state of rest]. A disturbance of function or structure of any organ or part of the body. d., **acute**, a disease marked by rapid onset and short course. d., **acute specific**. 1. An infectious febrile disease. 2. Acute syphilis. d., **amyloid**. Same as amyloid degeneration. d., **anserine**, muscular wasting of the hand, the prominent tendons suggesting a goose's foot. d., **autogenous**, one due to failure on the part of some group of body-cells to perform its function. d., **bad**, syphilis. d., **barometer-maker's**. See under barometer. d., **bleeders'**. Synonym of hemophilia. d., **blue**, cyanosis. d., **boiler-maker's**, deafness to high-pitched tones, occurring in boiler-makers. d., **brass-founders'**, chronic poisoning from working in brass. d., **caisson-**. See caisson-disease. d., **choleraic**, a tropical affection resembling cholera, due to the ingestion of poisonous fish. Cf. siguatera. d., **cholesterin**, amyloid degeneration. d., **chronic**, one that is slow in its course. d., **constitutional**, one in which a system of organs or the whole body is involved. d., **cyclic**, a disease following cycles or periods of exacerbation or change. d., **cystic, of the breast**. See Reclus' disease. d., **dancing**, tarantism. d., **divers'**, an affection similar to caisson-disease. d., **elevator**, an affection of the heart occurring in elevator-men. Syn., liftman's heart. d., **enthetic**, one introduced extraneously; an infectious disease. d., **entozootic**, one due to the presence of animal parasites within the body. d.s, **eponymic**, those named after individuals. See under name of the individual for definition. d., **fifth**, erythema infectiosum. d., **fish-skin**, ichthyosis. d., **fish-slime**, a peculiar form of septicemia due to punctured wounds by fish-spines. d., **flax-dresser's**, pneumonia from inhalation of particles of flax. d., **fleshworm**, trichinosis. d., **flint**, Synonym of chalicosis. d., **focal**, a localized disease. d., **fourth**, Clement Dukes' term for a contagious disease resembling measles, scarlatina, and rubella, or roserash, but distinct. Syn., quatrième fièvre éruptive. d., **functional**, a disease without discoverable organic lesion. d., **fungous, of India**, a prevalent endemic disease of India affecting the extremities and disorganizing the tissues, due to the implantation of spores in the tissues. d., **guinea-worm**. See guinea-worm disease. d., **habit**, one that results from long continuation and frequent repetition of an act. d.s, **heterotoxic**, those due to toxic substances introduced from without the body. d., **hook-worm**. See hookworm disease. d., **hydrocephaloid**, a disease of children resembling hydrocephalus, following premature weaning. d., **idiopathic**, one that exists by itself without any connection with another disorder; one of which the cause is unknown. d., **infectious**, one arising from the invasion, growth, and multiplication in the body of specific, pathogenic microorganisms which produce a chemical poison that induces its characteristic effects. d., **intercurrent**, a disease occurring during the progress of another of which it is independent. d., **internal**, one affecting the internal organs. d., **jumping**, a form of choromania. See jumpers. d., **lardaceous**, amyloid degeneration. d., **local**, one confined to some particular region of the body or to one tissue or organ. d., **malignant mold-fungus**, an affection of the skin and mucosa, especially of the mouth, lips, and nose; described by de Hahn as due to a mold-fungus. d., **mitral**, one affecting the mitral valves. d., **mucous** (Starr), a form of chronic gastrointestinal catarrh in children. It consists of a mucous flux, from the whole internal surface of the alimentary canal, which interferes mechanically with the digestion and absorption of food, and so impedes nutrition as to suggest the presence of tubercles. It usually arises between the fourth and the twelfth year, frequently as a sequel of pertussis. d., **mucous** (Whitehead), d., **mucous** (of the colon). See colitis, mucous. d., **occupation**, any one of the nervous affections due to the habitual performance of the duties of some occupation. d., **organic**, one due to structural changes. d., **pandemic**, a disease epidemic over a wide area. d., **parasitic**, one due to an animal or vegetable parasite. d., **parenchymatous**, that affecting the parenchyma of an organ. d., **pearl**, bovine tuberculosis. d.s, **protozoal**, pathological conditions due to the invasion of the body by protozoa. d., **protozoic** (of Posadas, Wernicke, etc.). See dermatitis, blastomycetic. d., **pulpy**, tuberculous arthritis. d., **ragsorter's**. See ragpicker's disease. d., **Scythian**. See Scythian. d., **septic**, one arising from the development of pyogenic or putrefactive organisms within the body. d., **septinous**, a form of septic disease in which there is absorption of the toxic substance through an abrasion of the mucosa of the alimentary canal. d., **seven days'**, trismus. d., **specific**, one caused by the introduction of a specific virus or poison within the body; also used as a synonym of syphilis. d., **straddling**. See quebrabunda. d., **structural**, one involving a change of structure in the part first affected. d., **summer**, cholera infantum. d., **suprarenal-capsule**, Addison's disease. d., **system**, one affecting a number of tissues having a common function. d., **teataster's**, a disorder characterised by extreme neurasthenia, rapid heart action, fibrillary muscular twitching, and paresthesia due to excessive use of tea. d., **tricuspid**, that of the tricuspid valves. d., **tsetse-fly**, an African disease of horses, cattle, and other stock due to the Trypanosoma brucei, which is transmitted by the tsetse-fly, Glossina morsitans. d., **vagabond's**. See vagabond. d., **venereal**, one due to sexual intercourse. d., **woolsorter's**, anthrax. d., **zymotic**, a disease arising from the introduction and multiplication of some living germ within the body.

**disengagement** (dis-en-gāj'-ment) [Fr., desengager, to disengage]. Emergence from a confined state; especially the escape of the head of the fetus from the vaginal canal.

**disfigurement** (dis-fig'-ur-ment) [dis, priv.; figurare, to fashion]. Blemish; deformity.

**disgorgement** (dis-gorj'-ment) [OF., desgorger, to vomit]. 1. Ejection by vomiting. 2. The subsidence of an engorgement.

**disgregation** (dis-greg-a'-shun) [disgregare, to separate]. Dispersion; separation, as of molecules or cells.

**disinfect** (dis-in-fekt'). To destroy or remove pathogenic substances or organisms, or to render them inert.

**disinfectant** (dis-in-fek'-tant) [dis, negative; inficere, to corrupt]. An agent that destroys the germs of disease, fermentation, and putrefaction.

**disinfectin** (dis-in-fek'-tin). A brown liquid obtained from treating 5 parts of the residue of naphtha-distillation with 1 part of concentrated sulphuric acid and the resulting product with 5 parts of .10 % soda solution. Diluted it is used as a disinfectant.

**disinfection** (dis-in-fek'-shun) [see disinfectant]. The destroying or removal of pathogenic germs, especially by means of chemical substances.

**disinfectol** (dis-in-fek'-tol). An antiseptic substance analogous to creolin and lysol. It is used in the form of a two to five per cent. emulsion.

**disinfector** (dis-in-fek'-tor). An apparatus for the purpose of disinfection.

**disintegrate** (dis-in'-te-grāt) [dis; integer, the whole]. To break up or decompose.

**disintoxication** (dis-in-toks-ik-a'-shun). See detoxification.

**disinvagination** (dis-in-vaj-in-a'-shun) [dis, neg.; in, in; vagina, a sheath]. The reduction or relief of an invagination.

**disjoint** (dis-joint'). To disarticulate; to separate, as bones, from their natural relations.

**disk**. See disc.

**dislocation** (dis-lo-ka'-shun) [dis; locare, to place]. The displacement of one or more bones of a joint or of any organ from its natural position. See Callaway's, Dugas', Hamilton's tests. d., **complete**,

one in which the joint-surfaces are entirely separated **d.**, **compound**, one in which the joint communicates with the external air through a wound. **d.**, **consecutive**, one in which the displaced bone is not in the same position as when originally displaced. **d.**, **divergent**, separate dislocation of the ulna and radius. **d.**, **double**, displacement at the same time of corresponding bones on both sides of the body. **d.**, **habitual**, one that recurs repeatedly from a relaxed condition of the ligaments or from incomplete repair of the articular capsule. **d.**, **Monteggia's**. See under *Monteggia*. **d.**, **Nélaton's**. See under *Nélaton*. **d.**, **old**, one in which inflammatory changes have occurred. **d.**, **partial**, **d.**, **incomplete**, one in which the articulating surfaces remain in partial contact. Syn., *subluxation*. **d.**, **pathological**, one the result of disease in the joint or of paralysis of the controlling muscles. **d.**, **primitive**, one in which the bones remain as originally displaced. **d.**, **recent**, one in which no inflammatory changes have ensued. **d.**, **relapsing**. See *d.*, *habitual*. **d.**, **simple**, one in which there is no communication with the air through a wound. **d.**, **Smith's**. See under *Smith*. **d.**, **subclavicular**, one of the head of the humerus beneath the pectoralis major below the clavicle. **d.**, **subglenoid**, one of the humerus directly below the glenoid fossa. **d.**, **subpubic**, dislocation of the hip-joint below the pubes. **d.**, **subspinous**, one in which the head of the humerus is held in the infraspinous fossa. **d.**, **thyroid**, displacement of the head of the femur into the thyroid foramen. **d.**, **traumatic**, that due to injury.
**disodic** (*di-so'-dik*) [*di*, two; *sodium*]. 1. Containing two atoms of sodium in the molecule. 2. (*dis-od'-ik*) [*δίς*, twice; *ὁδός*, a way.] Furnished with or relating to two openings.
**disoma**, **disomus** (*di-so'-mak*, -*mus*) [*di*-; σῶμα, body; pl., *disomata*, *disomi*.] A monster having two trunks.
**disorder** (*dis-or'-der*). See *disease*.
**disorganization** (*dis-or-gan-iz-a'-shun*). Destruction or loss of organic structure; complete pathologic or traumatic change in the minute structure of any tissue.
**disorientation** (*dis-o-re-en-ta'-shun*). The loss of the ability to locate one's position in the environment, or the mental confusion seen in psychic disorders.
**dispar** (*dis'-par*) [L.]. Unequal.
**disparate** (*dis'-par-āt*) [*dispar*]. Not alike; unequal or unmated. **d. points**, nonidentical points of the two retinæ. Diplopia is produced when the images of a single object fall upon such points.
**dispareunia** (*dis-par-oo'-ne-ah*). See *dyspareunia*.
**disparity** (*dis-par'-it-e*) [*dispar*]. Difference; inequality. **d.**, **crossed**, a condition of binocular relief whereby in superimposed similar figures having their points of sharpest vision coinciding, a certain other point in the left field appears to be to the right, and the same point in the right field to the left, of the point of sharpest vision. **d.**, **uncrossed**, a similar condition of binocular relief, but in which a point in the left is seen to the left of a similar point in the right field.
**dispensary** (*dis-pens'-ar-e*) [*dispensare*, to distribute]. A charitable institution where medical treatment is given to the poor and others.
**dispensatory** (*dis-pens'-at-or-e*). A treatise on materia medica and the composition, effects, and preparation of medicines.
**dispensing** (*dis-pens'-ing*) [*dispensare*, to weigh out]. The measuring, weighing, and issuing of the drugs ordered in a prescription.
**dispermine** (*di-sperm'-in*). See *piperazine*.
**dispermy** (*di-sperm'-e*) [*di*-; σπέρμα, a seed]. The entrance of two spermatozoa into the ovum.
**dispersion** (*dis-per'-shun*) [*dispersus*, scattered]. The act of scattering. In physics, the separation of a ray of light into its component parts by reflection or refraction; also, any scattering of light, as that which has passed through ground glass.
**dispirem** (*di-spi'-rem*) [*di*-; *spira*, a spiral]. The two skeins of a dividing nucleus formed from the nuclear loops and in development giving rise to the daughter-nuclei.
**displacement** (*dis-plas'-ment*) [Fr., *desplacer*, to displace]. 1. A putting-out of the normal position. 2. Percolation. **d.**, **backward**, **d.**, **dorsal** (of the arm), a backward displacement, across the neck or occipital region, of one of the arms of the fetus,

causing obstruction to delivery. **d.**, **fish-hook**, a displacement of the stomach in which the pyloric orifice faces directly upward and the duodenum extends upward and to the right, connecting with the pylorus at an angle, which produces a constricting hook.
**disposition** (*dis-po-zish'-un*) [*dispositio*, an arranging]. Tendency, either physical or mental, to certain diseases.
**disruptive** (*dis-rup'-tiv*). Bursting; rending. **d. discharge.** See *discharge*.
**dissect** (*dis-ekt'*) [*dissecare*, to cut up]. To cut tissues apart carefully and slowly, in order to allow study of the relations of a part.
**dissecting** (*dis-ek'-ting*) [*dissect*]. Performing dissection. **d. aneurysm**, an aneurysm in which there occurs a separation of the coats of an artery, with hemorrhage between.
**dissection** (*dis-ek'-shun*). The cutting apart of the tissues of the body for purposes of study. **d. tubercle**, the same as *verruca necrogenica*. **d.- wound**, a septic wound acquired during dissection.
**dissector** (*dis-ek'-tor*) [*dissect*]. 1. One who makes a dissection. 2. Handbook or manual of anatomy and instructions for use in dissection. 3. An instrument used for separating structures in dissection or in a surgical operation.
**disseminated** (*dis-em'-in-a-ted*) [*disseminare*, to scatter seed]. Scattered; spread over a large area. **d. sclerosis**, a disease of the central nervous system in which the areas of sclerosis are irregularly scattered throughout the cord and brain. Syn., *multiple or insular sclerosis*.
**dissemination** (*dis-em-in-a'-shun*) [*dis*, apart; *seminare*, to sow]. The scattering or dispersion of disease or disease-germs.
**dissepiment** (*dis-ep'-e-ment*) [*dis*, apart; *sæpire*, to hedge in]. A partition, septum, or diaphragm.
**dissimilation** (*dis-im-il-a'-shun*). See *catabolism*.
**dissipation** (*dis-ip-a'-shun*) [*dissipare*, to scatter]. A dispersion of matter or of the morbid condition that causes disease.
**dissociation** (*dis-o-se-a'-shun*) [*dis*-; *sociare*, to associate]. Separation, especially the separation of a complex compound into simpler molecules by the action of heat. **d.-symptom**, anesthesia to pain and to heat and cold, with preservation of tactile sensibility and of the muscular sense; it is observed in syringomyelia.
**dissolution** (*dis-o-lū'-shun*) [*dissolutio*; *dissolvere*, to set free]. 1. The separation of a body or compound into its parts. 2. Death; decomposition.
**dissolve** (*diz-olv'*). To make a solution of.
**dissolvent** (*diz-ol'-vent*) [*dissolvere*, to loosen, dissolve]. A solvent or resolvent.
**dissonance** (*dis'-o-nans*) [*dissonare*, to disagree in sound]. The combination of such tones as are so different from each other as to produce discord.
**distad** (*dis'-tad*) [*distare*, to be at a distance; *ad*, to]. In the direction of the free extremity of an appendage or part.
**distal** (*dis'-tal*) [*distare*, to be at a distance]. Extreme; at the greatest distance from a central point; peripheral.
**distally** (*dis'-tal-e*) [*distare*, to be at a distance]. Distad.
**distance** (*dis'-tans*) [*distantia*, distance]. The measure of space between two objects. **d. focal**, the distance between the center of a lens and its focus. **d.**, **working**, in the microscope, the distance from the front lens of an objective to the object, when the objective is correctly focused.
**distemper** (*dis-tem'-per*) [*distemperare*, to dissolve]. 1. Disease; malady; indisposition; most commonly applied to the diseases of animals. 2. A disease of young dogs, commonly considered as a catarrhal disorder.
**disten** (*dis'-ten*) [*distare*, to be at a distance]. Belonging to the distal aspect in itself.
**distention** (*dis-ten'-shun*) [*distendere*, to stretch]. The state of being dilated.
**distichia** (*dis-tik'-e-ah*). See *distichiasis*.
**distichiasis** (*dis-tik-i'-as-is*) [*di*-; στίχος, a row]. The condition in which there is a double row of eyelashes, the inner rubbing against the globe. See also *entropion* and *trichiasis*.
**distillate** (*dis'-til-āt*). The product obtained by distillation.
**distillation** (*dis-til-a'-shun*) [*distillare*, to drop little by little]. The process of vaporizing and collecting

the vapor by condensation. It is used mainly in purifying liquids by separating them from non-volatile substances. **d., destructive,** the decomposition of an organic substance in a closed vessel in such a manner as to obtain liquid products. **d., dry,** distillation of solids without the addition of liquids, conducted within a closed vessel in order to hinder combustion. **d., fractional,** a method of separating substances from each other by distilling the mixture containing them at a gradually increased temperature, the different substances being vaporized and collected in the order of their volatility.

**distobuccal** (*dis-to-buk'-al*) [*distare,* to separate; *bucca,* the cheek]. Relating to the distal and buccal walls of the bicuspid and molar teeth.

**distolabial** (*dis-to-la'-be-al*) [*distare,* to separate; *labium,* lip]. Relating to the portions of the anterior teeth between their distal and labial walls.

**distolingual** (*dis-to-lin'-gwal*) [*distare,* to separate; *lingua,* tongue]. Relating to the portions of teeth between their distal and lingual walls.

**Distoma, Distomum** (*dis'-to-mah, -mum*) [*di-;* στόμα, a mouth]. 1. A genus of trematode worms which have an oral as well as a ventral sucker. 2. A general term applied to various genera of trematode worms. See *Schistosomum.* **D. hæmatobium,** a species which, becoming lodged in the portal vessels and the veins of the mesentery and of the urinary tract causes a disease characterized by hematuria, anemia, and diarrhea. It is endemic in parts of the tropics. **D. Ringeri,** a variety infesting the lungs of man in China and Japan and causing periodic hemoptysis.

**distomatosis** (*di-sto-mat-o'-sis*). See *distomia.*

**distomia** (*di-sto'-me-ah*) [see *distoma*]. Congenital duplication of the mouth.

**distomiasis** (*dis-to-mi'-as-is*) [*distoma*]. The presence in the body of distoma.

**distomus** (*dis-to'-mus*). See *diprosopus.*

**distortion** (*dis-tor'-shun*) [*distorquere,* to distort]. 1. A twisted or bent shape; deformity or malformation, acquired or congenital. 2. A writhing or twisting motion, as of the face; a grimace.

**distortor oris** (*dis-tor'-tor o'-ris*) [L., "the distortor of the mouth"]. The zygomaticus minor muscle. See *muscles, table of.*

**distraction** (*dis-trak'-shun*) [*distrahere,* to draw apart]. A method of treating certain joint diseases and bone-fractures by extension and counterextension.

**distribution** (*dis-tri-bū'-shun*) [*distribuere,* to distribute]. The branching of a nerve or artery, and the arrangement of its branches within those parts that it supplies.

**distrix** (*dis-'triks*) [δίς, two; θρίξ, hair]. The splitting of the distal ends of the hair.

**disulphate** (*di-sul'-fāt*) [*di-; sulphur*]. A sulphate containing one atom of hydrogen that can be replaced by a base; an acid sulphate.

**disulphide** (*di-sul'-fīd*). A compound of an element or radical with two atoms of sulphur.

**disvolution** (*dis-vo-lū'-shun*) [*dis; volvere,* to roll down]. Degeneracy; devolution; extreme catabolism.

**disvulnerability** (*dis-vul-ner-ab-il'-it-e*) [*dis,* neg.; *vulnerare,* to wound]. The power of abnormally rapid recovery from wounds, said to be a peculiarity of many criminals.

**dita-bark** (*dī'-tah-bark*) [L.]. The bark of *Alstonia scholaris,* native to the Philippine Islands. It is employed as a tonic and antiperiodic in intermittent fever. Dose of the *fluidextract* 1–2 dr. (4–8 Cc.); of the *powder* 5 gr. (0.32 Gm.).

**ditaine** (*dit'-ah-in*), $C_{22}H_{28}N_2O_4$. An alkaloid from dita-bark, used hypodermatically in tetanus. Dose 1/12 gr. (0.005 Gm.) once or twice daily or until effectual. Syn., *echitamine.*

**ditamine** (*dit'-am-ēn*). Same as *ditaine.*

**dithan** (*dith'-an*). See *trional.*

**dithion** (*dith'-e-on*). A mixture of the two sodium dithiosalicylates occurring as a gray powder. It is used as an antiseptic wash (5 to 10 %) and dusting-powder in gonorrhea and in foot-and-mouth disease.

**dithymoldiiodide, dithmoliodide** (*di-thi-mol-di-i'-o-did, di-thi-mol-i'-o-did*). Aristol.

**ditokus** (*dit'-o-kus*) [δίς, two; τίκειν, to bring forth]. In biology, giving birth to twins, or laying two eggs.

**Dittel's operation** (*dit'-el*). For enlarged prostate; enucleation of the lateral lobes of the prostate by an external incision.

**Dittrich's plugs** (*dit'-rik*) [Franz *Dittrich,* German physician, 1815–1859]. Dirty white or yellowish masses, consisting chiefly of fatty detritus, microorganisms, and crystals of margarin; they are found in the sputum of putrid bronchitis and pulmonary gangrene. **D.'s stenosis,** stenosis of the conus arteriosus.

**diurazin** (*di-ū'-ra-zin*). A substance containing theobromin, formaldehyde and salicylic acid, used as a urinary antiseptic. Dose 6 gr. (0.4 gm.).

**diureid** (*di-ū'-re-id*). A substance derived from a double molecule of urea, by substituting a radical for hydrogen.

**diuresis** (*di-ū-rē'-sis*) [διά, through; οὐρεῖν, to urinate]. Abnormal increase in the secretion of urine.

**diuretic** (*di-ū-ret'-ik*) [*diuresis*]. 1. Increasing the flow of urine. 2. An agent that increases the secretion of urine. **d.s, alternative,** drugs eliminated by the kidney and used for their local action on the surfaces over which they pass. **d.s, hydragogue,** those that increase the flow of water from the kidneys. **d.s, refrigerant,** those that render the urine less irritating while not greatly increasing its flow.

**diuretin** (*di-ū-re'-tin*), $C_7H_7NaN_4O_2$, $C_6H_4OH$·COONa. Theobromin sodiosalicylate. It has been found useful as a diuretic in pleuritic effusion and cardiac dropsy. Dose 90 gr. (6 Gm.) daily in four doses.

**diurnule** (*di-urn'-ūl*) [Fr.]. A form of medicinal tablet or capsule that contains the maximum quantity of a toxic drug that may be administered in 24 hours.

**divagation** (*di-vag-a'-shun*) [*divagatio,* a wandering]. Incoherence of speech or thought.

**divalent** (*di'-va-lent*). See *bivalent.*

**divaricatio palpebrarum** (*di-var-ik-a'-she-o, pal-pe-bra'-rum*). Synonym of *ectropion.*

**divergence** (*di-ver'-jens*). A separation, as of axes.

**divergent** (*di-ver'-jent*) [*divergere,* to diverge]. Moving in different directions from a common point. **d. strabismus.** See *strabismus, divergent.*

**divers' paralysis.** See *caisson-disease.*

**diverticular** (*di-ver-tik'-ū-lar*) [*diverticulum*]. Relating to or arising from a diverticulum.

**diverticulitis** (*di-ver-tik-ū-li'-tis*). Inflammation of a diverticulum.

**diverticulum** (*di-ver-tik'-ū-lum*) [*divertere,* to turn]. A small pouch or sac springing from a main structure. **d., false,** a sacciform dilatation due to disease or injury. **d., Heister's.** See under *Heister.* **d., Meckel's.** See under *Meckel.* **d., Pertik's.** See under *Pertik.* **d., pulsion,** **d., traction,** a false diverticulum produced by traction on the outside of a hollow organ. **d., Vater's.** See *Vater, ampulla of.*

**divi-divi** (*div-e-div'-e*) [S. A.]. The seed-pods of *Cæsalpinia coriaria,* a tree of South America.

**divulsion** (*di-vul'-shun*) [*divulsio,* a tearing apart]. A tearing asunder.

**divulsor** (*di-vul'-sor*) [L.]. An instrument for dilating a part.

**dizziness** (*dis'-e-nes*). The state in which objects seem to be whirling around; vertigo.

**dizzy.** Giddy; light-headedness.

**D. M. D.** Abbreviation for *Doctor of Dental Medicine.* See *D. D. S.*

**D. O.** Abbreviation for *Doctor of Osteopathy.*

**Doane's sign** (*dōn*). Deafness in one ear in typhoid fever presages death; deafness in both ears is a good prognosis.

**Dobell's solution, D.'s spray** (*do'-bel*) [Horace *Dobell,* English physician, 1828– ]. A solution of borax, sodium bicarbonate, and phenol in glycerol and water; it is used as a spray for nasal and throat troubles.

**Dobie's globule** (*do'-be*) [William Murray *Dobie,* English anatomist, 19th century]. A small, round body rendered visible in the center of the transparent disc of a muscular fibril by staining. **D.'s layer, D.'s line.** See *Krause's membrane.*

**dochmiasis** (*dok-mi'-as-is*) [*Dochmius*]. The diseased condition caused by the presence in the body of parasites belonging to the genus *Dochmius.* See *uncinariasis.* Cf. *ankylostomiasis.*

**Dochmius** (*dok'-me-us*) [δόχμιος, crumpled]. A genus of threadworms of the family *Strongylidæ.* **D. duodenalis.** See *ankylostoma.*

**docimasia** (*dos-im-a'-se-ah*) [δοκιμάζειν, to examine]. Examination; testing or assaying.

**docimasiology** (*dos-im-a-se-ol'-o-je*) [*docimasia*; λόγος, science]. The art or science of investigation, embracing medicine, surgery, chemistry, etc.

# DOCIMASTER 301 DORSICUMBENT

**docimaster** (*dos-im-as'-ter*). An examiner or tester.
**docimastic** (*dos-im-as'-tik*) [δοκιμάζειν, to examine]. Making use of tests; testing; proving.
**doctor** (*dok'-tor*) [*doctor*, a teacher]. A teacher. A title conferred by a university or college. A physician licenced to practise medicine.
**dodecadactylitis** (*do-dek-a-dak-til-i'-tis*) [*dodecadactylon*]. Inflammation of the duodenum.
**dodecadactylon** (*do-dek-a-dak'-til-on*) [δώδεκα, twelve; δάκτυλος, finger]. The duodenum.
**dodging time** (*dodj'-ing*). A popular term among women for the period of irregular menstruation of varying duration proceding the full establishment of the menopause.
**doegling oil** (*do'-eg-ling*). The oil of the doegling, or bottle-nosed whale.
**Doehle's inclusion bodies** (*de'-leh*) [Karl Gottfried Paul *Doehle*, German pathologist, 1855– ]. Bodies found in the leukocytes in scarlet fever and streptococcic infections.
**dog's-bane.** See *apocynum*.
**dog-button.** See *nux vomica*.
**dog-nose.** See *goundou*.
**dogwood** See *cornus*.
**dolabra** (*do-la'-brah*) [L.]. A name applied to various bandages. **d. currens.** See *ascia*. **d. repens.** See *ascia*.
**doliariin** (*do-le-a'-re-in*) [*dolium*, a cask]. A vegetable pepsin, like papain and cradin, obtained from *Ficus doliaria*, a wild fig-tree of Brazil. It is vermifuge, purgative, and digestive.
**doli capax** (*do'-li ka'-paks*) [L.; *pl.*, *doli capaces*]. In legal medicine, one capable of guilt; one able or old enough to distinguish right and wrong; one morally or legally responsible. **d. incapax,** one incapable, or not old enough to distinguish wrong from right, or to be legally responsible for wrongdoing.
**dolicho-** (*dol-ik-o-*). A prefix meaning long.
**dolichocephalia** (*dol-ik-o-sef-a'-le-ah*) [*dolicho-*; κεφαλή, head]. The condition of being dolichocephalic.
**dolichocephalic, dolichocephalous** (*dol-ik-o-sef-al'-ik, -sef'-al-us*) [*dolicho-*; κεφαλή, head]. Long-headed; having a relatively long anteroposterior cephalic diameter.
**dolichocephalus** (*dol-ik-o-sef'-al-us*) [see *dolichocephalic*]. A skull having a relatively long anteroposterior diameter; a skull-formation resulting from the too rapid ossification of the longitudinal suture. The varieties are *leptocephalus, sphenocephalus,* and *clinocephalus*. **d. simplex,** that occurring through synostosis of the sagittal suture.
**dolichochamæcephalus** (*dol-ik-o-kam-e-sef'-al-us*) [*dolicho-*; χαμαί, on the ground; κεφαλή, head]. Applied to a skull which is characterized by both dolichocephalia and chamecephaly.
**dolichocnemic, dolichoknemic** (*dol-ik-o-ne'-mik*) [*dolicho-*; κνήμη, a leg]. Having the lower leg of almost the same length as that of the thigh.
**dolichoderus** (*dol-ik-o-de'-rus*). [*dolicho-*; δέρη, the neck]. Having the neck long.
**dolichouromesocephalus** (*dol-ik-o-ū-ro-mez-o-sef'-al-us*) [*dolicho-*; εὐρύς, broad; μέσος, middle; κεφαλή, head]. Having a dolichocephalic skull which is broad in the temporal region.
**dolichoeuroopisthocephalus** (*dol-ik-o-ū-ro-o-pis-tho-sef'-al-us*) [*dolicho-*; εὐρύς, broad; ὄπισθε, behind; κεφαλή, the head]. Having a dolichocephalic skull, broad in the occipital region.
**dolichoeuroprocephalus** (*dol-ik-o-ū-ro-pro-sef'-al-us*) [*dolicho-*; εὐρύς, broad; πρό, before; κεφαλή, the head]. Having a dolichocephalic skull very broad in the frontal region.
**dolichofacial** (*dol-ik-o-fa'-shal*). With a long face.
**dolichohieric** (*dol-ik-o-hi-er'-ik*) [*dolicho-*; ἱερόν, sacrum]. Having a relatively slender sacrum. See *platyhieric*.
**dolichokerkic** (*dol-ik-o-kerk'-ik*) [*dolicho-*; κερκίς, a shuttle]. Having the angle which is formed by the crest of the scapula with its spinal border over 80 degrees.
**dolicholeptocephalus** (*dol-ik-o-lep-to-sef'-al-us*) [*dolicho-*; *leptocephalus*]. Having a skull both dolichocephalic and leptocephalic.
**dolichopellic** (*dol-ik-o-pel'-ik*) [*dolicho-*; πέλλα, a bowl]. Having a relatively long or narrow pelvis. See *platypellic*.
**dolichopelvic** (*dol-ik-o-pel'-vik*). Same as *dolichopellic*.

**dolichoplatycephalus** (*dol-ik-o-plat-e-sef'-al-us*) [*dolicho-*; *platycephalus*]. Having the skull both long and flat.
**dolichorrhine** (*dol'-ik-or-ēn*) [*dolicho-*; ῥίς, the nose]. Long-nosed.
**dolichos.** Same as *mucuna*.
**dolichouranic** (*dol-ik-o-ū-ran'-ik*) [*dolicho-*; οὐρανος, the palate]. Having a long alveolar arch. Cf. *brachyuranic; measuranic*.
**dolioform** (*dol'-e-o-form*) [*dolium*, a cask; *forma*, form]. Cask-shaped.
**doll's-head anesthesia.** Anesthesia of the head, neck, and chest.
**dolomol** (*dol'-o-mol*). Stearate of calcium and magnesium; it is used as a base for dusting-powders, etc.
**dolor** (*do'-lor*) [*dolere*, to feel pain]. Bodily or mental pain or suffering.
**dolorosus** (*do-lor-o'-sus*) [L.]. Full of pain.
**domatophobia** (*do-mat-o-fo'-be-ah*) [δῶμα, house; φόβος, fear]. Insane dread of being in a house; a variety of claustrophobia.
**dominus morborum** (*dom'-in-us mor-bor'-um*) [L.]. The lord of diseases; gout.
**donda ndugu** [African]. Brother ulcer; a disease common on the east coast of Africa, due to some organism that infests stagnant water. It affects especially the leg, which becomes inflamed and sloughs below the healthy tissue.
**Donders'' glaucoma** (*don'-derz*) [Franz Cornelius *Donders*, Dutch ophthalmologist, 1818–1889]. Simple atrophic glaucoma. **D.'s law**, the rotation of the eyeball about the line of sight is involuntary, and when the eyes are fixed on a distant object, the amount of rotation is determined solely by the angular distance of that object from the horizon and from the median plane. **D.,** reduced eye of, a representation of the eye in such a way that all the distances needed in the calculations are represented by whole numbers. **D.'s rings,** rainbow-colored rings seen in cases of glaucoma and by normal and cataractous eyes when the pupil is dilated. They are attributed to the diffraction of light by the cortex of the crystalline lens. **D.'s schema,** an arrangement of the head thorax with manometers to gage the pressure.
**Donné's corpuscles** (*don-na'*) [Alfred *Donné*, French physician, 1801–1878]. 1. The colostrum corpuscles. 2. See *Bizzozero's blood-platelets*. **D.'s test for pus,** into the mass to be tested stir a small piece of caustic potash. The mass will be converted into a tough, slimy material if pus is present.
**Donovan's solution** (*don'-ov-an*) [Edward *Donovan*, English druggist, 1798–1837]. A solution of mercuric iodide and arsenic iodide, one per cent. of each.
**dope** (*dōp*). A slang term for any drug, particularly a narcotic drug.
**dormiol** (*dor'-me-ol*). The commercial name for amylene chloral; recommended as a soporific. Dose 7½–45 gr. (0.5–3.0 Gm.).
**dormitio** (*dor-mish'-e-o*) [L.]. 1. Sleep. 2. A sedative said to consist of dilute alcohol, oil of anise, extract of lettuce, and sugar.
**dorsad** (*dor'-sad*) [*dorsum; ad,* toward]. Toward the dorsal aspect.
**dorsal** (*dor'-sal*) [*dorsum*]. Pertaining to the back or to the posterior part of an organ. **d. artery.** See under *artery*. **d. decubitus,** recumbency in the supine position. **d. nerves,** the spinal nerves coming through the intervertebral foramina of the dorsal vertebræ. **d. reflex.** See under *reflexes*.
**dorsalgia** (*dor-sal'-je-ah*) [*dorsum*; ἄλγος, pain]. Pain in the back.
**dorsalis** (*dor-sa'-lis*) [*dorsum*, the back]. See *arteries, muscles, nerves, veins, tables of*.
**dorsalis pedis** (*dor-sa'-lis pe'-dis*). See under *artery*.
**dorsen** (*dor'-sen*) [*dorsum,* the back]. Belonging to the dorsum in itself.
**dorsi-** (*dor-si-*). The same as *dorso-*.
**dorsicolumn** (*dor-se-kol'-um*) [*dorsi-*; *columna,* a column]. The dorsal column of the spinal cord.
**dorsicommissura** (*dor-se-kom-is-ū'-rah*) [*dorsi-; commissure*]. Wilder's name for the gray commissure of the spinal cord.
**dorsicornu** (*dor-sik-or'-nū*). The posterior cornu of the spinal cord.
**dorsicumbent** (*dor-se-kum'-bent*) [*dorsi-; cubare,* to lie down]. Supine, or lying on the back.

# DORSIDUCTION 302 DOWEL

**dorsiduction** (dor-se-duk'-shun) [dorsi-; ducere, to lead]. The act of moving toward the back.

**dorsiflexion** (dor-se-flek'-shun) [dorsi-; flectere, to bend]. A flexion, as of toes, toward the back.

**dorsimesad** (dor-si-me'-sad) [dorsi-; μέσον, the middle]. Toward the dorsimeson.

**dorsimesal** (dor-si-me'-sal) [dorsi-; μέσον, the middle]. Pertaining to the dorsimeson.

**dorsimeson** (dor-si-me'-sis) [dorsi-; μέσον, the middle]. The dorsal edge of the meson or median plane of the body. The dorsal, mesal, or median line. Cf. *ventrimeson*.

**dorsiscapular** (dor-si-skap'-u-lar) [dorsi-; scapula, the shoulder]. Relating to the dorsum of the scapula.

**dorsispinal** (dor-se-spi'-nal) [dorsi-; spina, spine]. Relating to the back and the spinal column.

**dorsiventral** (dor-se-ven'-tral). See *dorsoabdominal*.

**dorso-** (dor-so-) [dorsum]. A prefix used to signify pertaining to the back.

**dorsoabdominal** (dor-so-ab-dom'-in-al) [dorso-; abdomen]. Relating to both the dorsal and the abdominal region; extending from the back to the abdomen. Syn., *dorsiventral*.

**dorsoacromial** (dor-so-ak-ro'-me-al). Relating to the back and the acromion.

**dorsoanterior** (dor-so-an-te'-re-or). Applied to a fetus having its back toward the ventral aspect of the mother.

**dorsocephalad** (dor-so-sef'-al-ad) [dorso-; κεφαλή, head]. Toward the dorsal aspect of the head.

**dorsocervical** (dor-so-ser'-vik-al). Relating to the back and the neck.

**dorsocostal** (dor-so-kos'-tal). Relating to the back and the ribs.

**dorsodynia** (dor-so-din'-e-ah) [dorso-; ὀδύνη, pain]. Omodynia; scapulodynia; pain in the dorsal region; rheumatism of the muscles of the shoulders and upper part of the back.

**dorsohumeral** (dor-so-hū'-mur-al). Relating to the back and the humerus.

**dorsointercostal** (dor-so-in-tur-kos'-tal). Relating to the back and the intercostal spaces.

**dorsointerosseal, dorsointerosseous** (dor-so-in-ter-os'-e-al, -us). Located between the metacarpal or metatarsal bones and on the back of the hand or foot.

**dorsointestinal** (dor-so-in-test'-in-al). Situated upon the dorsal aspect of the intestine.

**dorsolateral** (dor-so-lat'-er-al). Relating to the back and the sides.

**dorsolumbar** (dor-so-lum'-bar) [dorso-; lumbus, loin]. Relating to the back and the loins.

**dorsomedian** (dor-so-me'-de-an). Situated in or relating to the middle region of the back.

**dorsonasal** (dor-so-na'-sal). Relating to the back of the nose.

**dorsoposterior** (dor-so-pos-te'-re-or). Applied to the position of a fetus having its back toward the dorsal aspect of the mother.

**dorsoradial** (dor-so-ra'-de-al). Relating to or situated upon the dorsal aspect and radial border of the hand, finger, or arm.

**dorsosacral** (dor-so-sak'-sal). Relating to the back and the sacrum.

**dorsothoracic** (dor-so-tho-ras'-ik) [dorso-; thorax]. Relating to the back and the thorax.

**dorsoulnar** (dor-so-ul'-nar). Relating to or situated upon the dorsal aspect and ulnar border of the arm, hand, or finger.

**dorsoventrad** (dor-so-ven'-trad) [dorso-; venter, belly]. In a direction from back to front.

**dorsoventral** (dor-so-vent'-ral). See *dorsoabdominal*.

**dorsum** (dor'-sum) [L.]. 1. The back. 2. Any part corresponding to the back, as the *dorsum* of the foot, hand, tongue, penis, etc.

**dosage** (do'-sāj) [dose]. The determination of the proper amount of a medicine or other agent for a given case or condition. **d., electrical**, the regulation of the strength of an electric current for therapeutic purposes.

**dose** (dōs) [δόσις, a portion given]. The measured portion of medicine to be taken at one time. **d., divided**, a relatively small quantity of a drug taken at short intervals. **d., lethal**, a dose sufficient to kill. **d., maximum**, the largest dose consistent with safety. **d., minimum**, the smallest quantity of a medicine that will produce physiological effects.

**dosimeter** (do-sim'-et-er) [δόσις, a dose; μέτρον, a measure]. A drop-meter; an instrument for measuring minute quantities of a liquid.

**dosimetric** (do-sim-et'-rik) [see *dosimetry*]. Relating to or characterized by dosimetry.

**dosimetry** (do-sim'-et-re) [dose; μέτρον, a measure]. 1. The dosimetric system; the accurate and systematic measurement of medicinal doses. 2. The system of treatment which consists in the use of granules containing a definite quantity of the active principles of drugs.

**dosis** (do'-sis). See *dose*.

**dossil** (dos'-il) [ME., dosil, a spigot]. A cylindric pledget of lint for cleansing wounds.

**dotage** (dōt'-āg). Feebleness of mind; senility.

**dothienenteritis, dothinenteritis** (doth-e-en-en-ter-i'-tis, doth-in-en-ter-i'-tis) [δοθιήν, a boil; ἔντερον, bowel; ιτις, inflammation]. Enteric or typhoid fever; inflammation of Peyer's patches.

**double** (dub'-l). Twofold; in pairs. **d. consciousness**. See *consciousness*. **d. hearing**. See *diplacusis*. **d., touch**, investigation with a thumb in one cavity and the index-finger in another. **d. uterus**, dihysteria. **d. vision**, the seeing of a single object double; diplopia.

**doublet** (dub'-let). In optics, a system consisting of two lenses.

**doubly** (dub'-le). In a two-fold manner. **d. contoured**, in microscopy, an object is doubly contoured when it is bounded by two, usually parallel, dark lines with a lighter band between them.

**douche** (doosh) [Fr.]. 1. A stream of water directed against a part, or one used to flush a cavity of the body. 2. An apparatus for directing a jet of water or other substance against a part. **d., air**, a current of air directed against some organ for therapeutic purposes. **d., alternating**, a hot and cold current applied in succession. **d., capillary**. See *aquapuncture*. **d., Char**cot, a cold daily spinal douche. **d., galvanic**, a precision douche by means of which the faradic or galvanic current can be communicated to the douching current; especially effective in gastralgia and enteralgia. **d., hot-air** the use of a blast of heated air as a method of treatment, especially in gouty rheumatism and neuralgic conditions. **d., mobile**, one applied successively to different regions of the body. **d., precision**, one by which fluid of any desired temperature can be applied under any desired pressure. **d., rain-**, a shower-bath. **d., Scotch**, di, Scottish, one of alternating temperature. **d., sheet-**, one in which a sheet of water is directed through a slit. **d., transitional**. See *d., alternating*, and *d., Scotch*.

**Douglas' crescentic fold** (dug'-las) [James *Douglas*, Scotch anatomist, 1675-1742]. The lower border of the posterior sheath of the rectus abdominis. **D.'s culdesac, D.'s pouch**, a pouch between the anterior wall of the rectum and the posterior wall of the uterus, formed by the reflection of the peritoneum. **D.'s ligaments**, the rectouterine folds of the peritoneum. **D., line of**, the curved lower edge of the internal layer of the aponeurosis of the internal oblique muscle of the abdomen, where it ceases to cover the posterior surface of the rectus muscle. **D.'s semilunar fold**, a thin curved margin that forms the lower part of the posterior wall of the sheath of the abdominal rectus muscle. **D.'s septum**, in the rectus the septum formed by the union of Rathke's folds and transforming the rectum into a complete canal.

**douglasitis** (dug-las-i'-tis). Inflammation of Douglas' pouch.

**doundaké** (doon-dak-a) [Fr.]. The Guinea peach, *Sarcocephalus esculentus*, yielding a tonic, febrifuge, and astringent bark. Dose of the *wine* (3 %) 1-2 oz. (30–60 Cc.); of the *extract* 2½–3 gr. (0.16–0.2 Gm.); of the *bark* 50–60 gr. (3.2–3.8 Gm.); of the *aqueous extract* 3–4 min. (0.2–0.25 Cc.).

**doundakine** (doon-dak-ēn). An alkaloid from doundaké; a substitute for quinine. Dose 3–4 gr. (0.2–0.25 Gm.).

**dourine** (doo'-rēn). A contagious venereal disease of horses, the prominent signs and symptoms consisting in inflammation of the genital organs and lymph glands, and paralysis of the hind legs. The exciting cause is believed to be the *Trypanosoma equiperdum*. Syn.: *Mal de coit*.

**Dover's powder** (do'-ver) [Thomas *Dover*, English physician, 1660–1742]. A powder containing 10 % each of opium and ipecac.

**dowel** (dow'-el) [Fr., *douille*, a socket]. In dentistry, the piece of wood or metal uniting any artificial crown to the root of a natural tooth.

**Dowieism** (*dow'-e-izm*). A form of faith-cure propagated in America by a charlatan named Dowie (1899).
**Doyère's eminence, D.'s hillock, D.'s papilla, D.'s tuft** (*doy-yār'*) [Louis Doyère, French physiologist, 1811–1863]. The slight elevation in a muscular fiber corresponding to the entrance of a nerve-fiber.
**D. P.** Abbreviation for *Doctor of Pharmacy*.
**D. P. H.** 1. Abbreviation for *Diploma in Public Health*. 2. Abbreviation for *Doctor of Public Health*.
**D. R.** Abbreviation for *reaction of degeneration*.
**Dracæna** (*dra-se'-nah*). A genus of liliaceous trees. *D. cinnabari* and *D. schizantha* of eastern Africa, and *D. draco* of western Africa, afford part of the dragon's-blood of commerce.
**drachm** (*dram*). See *dram*.
**dracontiasis** (*dra-kon-ti'-as-is*) [*dracunculus*]. The skin disease caused by *dracunculus medinensis*.
**dracontium** (*dra-kon'-she-um*) [δράκων, a dragon]. Skunk-cabbage.
**Dracunculus** (*dra-kun'-ku-lus*) [dim. of δράκων, a dragon; a serpent]. A genus of threadworms belonging to the family *Filariidæ*. **D. medinensis**. See *Filaria medinensis*.
**draft** (*draft*). See *draught*.
**dragee** (*drah-zhā'*) [Fr.]. A sugar-coated pill, bolus, or comfit; a sugared confection.
**Dragendorff's test for bile-pigments** (*drag'-en-dorf*) [Johann Georg Noel *Dragendorff*, German chemist, 1836–1898]. Spread a few drops of the urine on an unglazed porcelain surface, and after absorption has taken place add a drop or two of nitric acid. If bile is present, several rings of color will be produced, the green ring, which is characteristic of bile-pigments, being chief among them.
**dragonneau** (*drag-on-o'*). The *Filaria medinensis*.
**dragon's-blood** (*drag'-ons-blud*). 1. The astringent resin of *Calamus ratang* and *C. draco*, East Indian ratan-palms. 2. The resin of various species of *Dracæna*. 3. The resin of *Pterocarpus draco*, a West Indian tree. The various kinds of dragon's-blood are astringent, but are no longer used internally.
**drain** (*drān*) [ME., *drainen*, to drain]. A material that affords a channel of exit for the discharge from a wound or cavity. **d., cigarette**, a drain made of a strip of gauze surrounded by rubber dam or gutta-percha.
**drainage** (*drān'-āj*) [*drain*]. The method of effecting the exit of the discharges from a wound or cavities by means of tubes or strands of fibers or by a free incision. **d., capillary**, that by means of capillary attraction, using loosely woven cloth, thread wisps of hair, etc. **d., funnel**, that effected by means of glass funnels. **d., through-**, drainage accomplished by means of counteropenings and the passage of a tube completely through the part to be drained, so that a cleansing fluid may be injected through one opening and allowed to escape by the counteropening. **d.-tube**, a rubber or glass tube with perforations for draining wounds or cavities.
**dram, drachm** (*dram*) [δραχμή, a Greek weight]. The eighth part of the apothecaries' ounce, equal to 60 grains or 3.9 grams. Also the sixteenth part of the avoirdupois ounce, equal to 27.34 grains. **d., fluid-**, the eighth part of a fluid ounce.
**dramatism** (*dram'-at-izm*) [*drama*]. Insanely stilted and lofty speech or behavior.
**drapetomania** (*drap-et-o-ma'-ne-ah*) [δραπέτης, a runaway; μανία, madness]. A morbid desire to wander from home.
**drastic** (*dras'-tik*) [δρᾶν, to act]. 1. Severe; harsh; powerful. 2. A powerful and irritating purgative.
**draught** (*draft*) [AS., *dragan*, to draw]. A quantity of liquid drunk at one gulp. **d., black**, compound infusion of senna. **d., effervescing**, one containing sodium or potassium bicarbonate and a vegetable acid.
**draw**. To digest and cause to discharge; said of a poultice. In dentistry, to remove a tooth from its socket. In andrology, to remove the urine from the bladder by means of a catheter.
**dream** (*drēm*). 1. An involuntary series of images, emotions, and thoughts presented to the mind during sleep. 2. To be conscious of such manifestations. **d., waking**, an illusion or hallucination. **d., wet**, a term given to the emission of semen during sleep.
**dreamy** (*dre'-me*). Full of dreams. **d. state**, a common psychic aura of epilepsy in which the patient experiences a sensation of strangeness or sometimes of terror. It may be associated with flashes of light or auditory auræ.
**Drechsel's test for bile acids** (*drek'-sel*) [Edmund *Drechsel*, Swiss chemist, 1843–1897]. A beautiful red color is produced if bile acids are present in a substance treated with a little cane-sugar and a few drops of a mixture composed of 5 parts of syrupy phosphoric acid and 1 part water, and warmed on a water-bath.
**drench**. In veterinary practice, a draught of medicine.
**drepanidium** (*drep-an-id'-e-um*) [δρεπάνη, a sickle]. The sickle-shaped young of certain protozoa. **d. ranarum**, a (probably) parasitic cytozoon of frogs, blood.
**dresser** (*dres'-er*). An attendant (in English hospitals, usually a medical student) whose special duty is to dress and bandage wounds, and attend to other ward work.
**dressing** (*dres'-ing*) [ME., *dressen*, to make straight]. 1. The application of various materials for protecting a wound and favoring its healing. 2. The material so applied.
**Dressler's disease**. Paroxysmal hemoglobinuria.
**drill**. A surgical instrument used in perforating bones, calculi, or teeth. **d.-bone**, a flat osteoma. **d.-bow**, a bow and string for rotating a drill-stock, effected by passing the string around it, and moving it backward and forward. **d.-stock**, an instrument for holding and turning a drill, either by the thumb and finger or by a handle.
**drink**. To swallow a liquid. The liquid that is swallowed. A draught.
**drivelling** (*driv'-el-ing*). 1. An involuntary flow of the saliva, as in old age, infancy idiocy, and mental stupor. 2. Senile weakness of mind.
**dromograph** (*drom'-o-graf*) [δρόμος, a course; γράφειν, to write]. An instrument for registering the velocity of the blood-current.
**dromomania** (*dro-mo-ma'-ne-ah*) [δρόμος, a course; μανία, madness]. An insane desire to wander; vagabondage.
**dromotropic** (*dro-mo-tro'-pik*) [δρόμος, a course; τρέπειν, to turn]. Applied by T. W. Englemann to a supposed set of fibers in the cardiac nerves which he holds influences the power of conducting the contraction. Cf. *bathmotropic; inotropic*.
**drop** [AS., *dropa*]. 1. A minute mass of liquid which in falling or in hanging from a surface assumes the spherical form. 2. The falling of a part, as from, paralysis. **d., ague**, Fowler's solution. **d., black** acetum opii. See *opium, vinegar of*. **d.-culture**, in bacteriology, a culture prepared by placing a little of the infected material in a drop of the culture-medium. **d., wrist-**. See under *wrist*.
**dropped** (*dropt'*). In a condition of ptosis. **d. foot**, extreme extension of the foot, especially observed in alcoholic neuritis, and dependent upon weakness of the flexors of the foot. **d. hand**, a form of paralysis from lead-poisoning, consisting in the inability to contract the extensors of the wrist. **d. lid**, ptosis of the upper lid. **d. wrist**. Same as *d. hand*.
**dropper** (*drop'-er*). A bottle, tube, or pipet, fitted for the emission of a liquid drop by drop. See *stopper-dropper*.
**dropsical** (*drop'-sik-al*) [*dropsy*]. Affected with or pertaining to dropsy.
**dropsy** (*drop'-se*) [ὕδρωψ, dropsy]. An infiltration of the tissues with diluted lymph, or the collection of such lymph in the body-cavities. **d., abdominal**, ascites. **d., acute**, dropsy due to congestion of the kidneys from sudden exposure to cold. Syn., *active dropsy; febrile dropsy; plethoric dropsy*. **d. of belly**, ascites. **d. of brain**, hydrocephalus. **d., cachectic**, that occurring in cachexia; it is due to decrease of the albuminous constituents of the blood and an increase of the watery constituents. **d., cardiac**, that due to failure of compensation in cardiac disease. **d. of chest**, hydrothorax. **d., encysted**. See *ascites saccatus*. **d., false**, a retention cyst. **d. of the gall-bladder**, dilatation of the gall-bladder with the secretion of the mucous glands and with epithelium: a very unusual condition due to obliteration of the cystic duct. **d., general**. 1. Dropsy of one or more of the large serous sacs of the body combined with anasarca. 2. Superficial dropsy when it affects the trunk and arms as well as the legs. **d., glandular**, that due to disease of the lymphatics. **d., lymphatic**, infiltration of the tissues with lymph caused by obstruction of the

lymphatics. d., mechanical, that due to mechanical obstruction of the veins or lymphatics. d., passive. 1. That due to obstruction of the veins or lymphatics or to defective absorption. 2. See d., cachectic. d. of pericardium, hydropericardium. d. of peritoneum, ascites. d., renal. 1. Anasarca due to disease of the kidneys. 2. Hydronephrosis. d., symptomatic, that said to be induced when there is 6 % of albumin in the blood with a simultaneous increase of arterial pressure.

**drosera** (*dro'-ser-ah*) [δροσερός, covered with dew]. Sundew. *D. rotundifolia* and *D. longifolia* have been used in pulmonary tuberculosis. It is an antispasmodic, and is used in whooping-cough and other spasmodic coughs.

**droserin** (*dros'-er-in*) [δροσερός, dewy]. A ferment resembling pepsin and found in the digestive secretions of most of the insectivorous plants.

**drown.** To deprive of life by immersion in a fluid.

**drowsy** (*drow'-ze*). Inclined to sleep; sleepy.

**drug** [Fr., *drogue*]. A substance used as a medicine. d., antagonistic, one that neutralizes the action of another.

**drum** [ME., *drumme*]. The tympanum. See under *ear*.

**drumhead** (*drum'-hed*). The tympanic membrane.

**drummers' palsy** (*drum'-erz pawl'-ze*). A form of occupation-neurosis, dependent upon the constrained attitude of the hand in beating a drum.

**Drummond's sign** (*drum'-und*) [David Drummond, English physician, 19th century]. The "oral whiff" heard when the mouth is closed; it disappears on compression of the nostrils; it is observed in cases of aneurysm of the thoracic aorta.

**drumstick** (*drum'-stik*). A stick terminating in a knob. The word is used to describe certain microbes, clubbed organs, etc. d. bacillus, *Bacillus putrificus coli*.

**drunkenness** (*drunk'-en-nes*). Acute or habitual alcoholic intoxication. See *alcoholism, dipsomania, inebriety*.

**druse** (*drūs*) [Ger.]. A rupture of tissues with no surface-lesion.

**dry** (dri) [AS., *dryge*]. Free from moisture. d., amputation, amputation without hemorrhage. d.-cupping, cupping without incising the skin. d. gangrene. See *gangrene, dry*. d. labor, one in which there is but a slight discharge of liquor amnii. d. pleurisy, pleurisy without effusion. d. wine, a wine containing little or no sugar.

**Drysdale's corpuscles** (*driz'-dāl*) [Thomas Murray Drysdale, American gynecologist, 1831-1904]. Granular cells, non-nucleated and of varying sizes, which were regarded by Drysdale as peculiar to ovarian fluid.

**D. S.** Abbreviation for *Doctor of Science*.

**D. Sc.** Abbreviation for *Doctor of Science*.

**D. S. M.** Abbreviation for *Diploma in State Medicine*.

**D. S. S.** Abbreviation for *Diploma in Sanitary Science*.

**D. S. Sc.** Abbreviation for *Diploma in Sanitary Science*.

**D.t.** Abbreviation of *duration tetany*.

**D. T. M.** Abbreviation for *Diploma in Tropical Medicine*.

**D. T. M. and H.** Abbreviation for *Diploma in Tropical Medicine and Hygiene*.

**Duane's clinometer** (*doo-ān'*) [Alexander Duane, American ophthalmologist, 1858– ]. A device for estimating torsional deviations of the eye, and also used in the study of metamorphopsia. D.'s test, a candle having been placed in front of the person to be examined, a screen is held before one eye for a time and then suddenly transferred to the other side. The existence of deviation in the first eye is recognized by a sudden apparent displacement of the candle-flame in the direction opposite to that in which the eye has deviated.

**Dubini's disease** (*doo-be'-ne*) [Angelo Dubini, Italian physician, 19th century]. Rapid rhythmic contractions of one or more groups of muscles, beginning in a finger, an extremity, or a half of the face, and extending over the greater part or the whole of the body. They are generally followed by palsies and often by coma and death. The affection has been observed thus far only in Italy. Syn., *electric chorea; spasmus Dubini*.

**Dublin method of expressing the placenta.** A better name for the *Credé method* (q. v.), as it was practised in the Rotunda Hospital in Dublin long before the time of Credé.

**Dubois' abscess, D.'s disease.** The presence of multiple necrotic foci in the thymus glands of infants affected with hereditary syphilis.

**DuBois-Reymond's inductorium** [Emil *DuBois-Reymond*, Berlin physiologist, 1818–1896]. An induction apparatus with a primary and secondary coil in which the primary current is never opened, it being short-circuited. It is used in physiological laboratories. **D.-R.'s key**, an electric switch by means of which the circuit may be closed or the current short-circuited. Syn., *tetanizing key*. **D.-R.'s law**, it is not the absolute value of current density at a given moment that acts as a stimulus to a muscle or motor nerve, but the variation of density.

**duboisine** (*dū-bois'-ēn*), $C_{17}H_{23}NO_3$. An alkaloid from *Duboisia myoporoides*, a tall shrub of Australia identical with hyoscyamine. It is used as a mydriatic; also as a hypnotic and sedative in epilepsy. Dose $\frac{1}{80}$–$\frac{1}{30}$ gr. (0.0008–0.002 Gm.). Antidotes: emetics, pilocarpine, muscarine. d. hydrobromide, dose and uses the same as duboisine. d. hydrochloride, usage and dose as duboisine. d. salicylate, used as a mydriatic in 0.2 to 0.8 % solution. d. sulphate, used as a mydriatic in aqueous solution—$\frac{1}{12}$ gr. to 1 oz. of water (0.005 Gm. to 30 Cc.).

**Du henne's attitude** (*doo-shen'*) [Guillaume Benjamin Amant Duchenne, French neurologist, 1806–1875]. In paralysis of the trapezius the shoulder droops; the shoulder-blade seesaws so that its internal edge, instead of being parallel to the vertebral column, becomes oblique from top to bottom and from without in. **D.'s disease**, tabes dorsalis. **D.'s paralysis**, progressive muscular dystrophy with pseudohypertrophy. **D.'s sign**, sinking-in of the epigastrium during inspiration in cases of marked hydropericardium or impaired movement of the diaphragm from pressure or paralysis. **D.'s syndrome**, labioglossolaryngeal paralysis.

**Duchenne-Aran's disease.** See *Aran-Duchenne's disease*.

**Duchenne-Erb's paralysis.** See *Erb's paralysis*.

**Duchenne-Landouzy's type of progressive muscular atrophy.** See *Landouzy-Déjérine's type of progressive muscular atrophy*.

**Duckworth's syndrome** [Sir Dyce Duckworth, English physician, 1840– ]. Complete stoppage of respiration several hours before that of the heart in certain cerebral diseases attended by intracranial pressure.

**Ducrey's bacillus** (*doo-kra'*) [Augusto Ducrey, Italian physician, 19th century]. A small oval bacillus occurring in chains, the pathogenic agent in chancroid.

**duct** (*dukt*) [*ducere*, to lead]. A tube or channel, especially one for conveying the secretions of a gland. d., alimentary. See *d., thoracic*. d. of Bartholin. See under *Bartholin*. d., Blasius'. See *d. of Stenson*. d., common bile-, a duct formed by the union of the cystic and hepatic ducts. d., Cowperian, the efferent duct of Cowper's gland. d.s of Cuvier. See under *Cuvier*. d., cystic, the excretory duct of the gall-bladder. d., ejaculatory, a duct formed by the union of the vas deferens and the duct of the seminal vesicle and carrying the semen into the urethra. d., endolymphatic, a tubular process of the membranous labyrinth of the ear, passing through the aqueduct of the vestibule into the cranial cavity, where it terminates below the dura mater in a blind enlargement, the sacculus endolymphaticus. d., galactophorous, one of the milk-ducts of the mammary gland. d., Gartner's. See *Gartner's canal*. d., hepatic, a duct formed at the margin of the transverse fissure of the liver by the junction of the right and left hepatic ducts. d., lacrimal. See *d., nasal*. d., lactiferous. See *d., galactophorous*. d., lymphatic, right, the vessel that receives the lymph from the lymphatics of the right arm, the right side of the head and neck, the chest, lung, and right side of the heart, and also from the upper surface of the liver. It terminates at the junction of the right subclavian and internal jugular veins. d.s of Müller. See under *Müller*. d., nasal, the duct that conveys the tears from the lacrimal sac into the inferior meatus of the nose. d., omphalomesenteric. See *d., vitelline*. d., pancreatic, one that extends from the pancreas to the duodenum at the point where the common bile-duct enters the bowel. d., pancreatic, accessory, the excretory duct of the lesser pancreas, opening into the pancreatic duct or into the duodenum, close to the orifice of the common bile-duct. **d., parotid,** that conveying the secretion of the parotid gland into the mouth. **d., prostatic,** any one of the ducts conveying the secretion of the prostate into the urethra. **d., Rathke's.** See under *Rathke*. **d., Reichel's cloacal.** See under *Reichel*. **d. of Rivini.** See under *Rivini*. **d., salivary,** a duct of any salivary gland. See *Wharton's duct, Stenson's duct,* and *d., sublingual*. **d., salivary, inferior,** Wharton's duct. **d., salivary, superior,** Stenson's duct. **d.s, Schueller's.** See under *Schueller*. **d., segmental,** a tube, on each side of the body of the embryo, situated between the visceral and parietal layers of the mesoblast, opening anteriorly into the body-cavity and posteriorly into the cloaca. **d., spermatic,** the vas deferens. **d. of Steno.** See *Stenson's duct*. **d. of Stenson,** the duct of the parotid gland. **d., sublingual.** See *Bartholin's duct* and *Rivini's duct*. **d., thoracic,** a duct 18 to 20 inches long, beginning in the receptaculum chyli, passing upward, and emptying into the left subclavian vein at its junction with the left internal jugular vein. It receives all the lymph and chyle not received by the right lymphatic duct. **d., umbilical.** See *d., vitelline*. **d., urogenital,** one that receives the urine and genital products. **d., vitelline,** the duct from the umbilical vesicle of the embryo to intestine. **d.s, Walther's.** See under *Walther*. **d. of Wharton.** See under *Wharton*. **d. of Wirsung.** See under *Wirsung*. **d., Wolffian.** See under *Wolffian*.

**ductile** (*duk'-til*) [*duct*]. Capable of being drawn out thin, as a wire or thread.

**duction** (*duk'-shun*) [*duct*]. A colloquialism used to represent one or more of the terms abduction, adduction, or sursumduction.

**ductless glands.** The spleen, thyroid, parathyroid, and thymus glands, suprarenal capsules, carotid and coccygeal glands, pineal and pituitary bodies, which have no excretory duct. **Endocrine g.** *q. v.*

**ductor** (*duk'-tor*) [*ducere*, to lead]. A surgical instrument used as a guide or in making traction.

**ductule** (*duk'-tūl*) [*ductulus*,—a small duct]. A small duct.

**ductulus** (*duk'-tu-lus*) [L.]. A small duct. **d. aberrans,** vas aberrans. **d. alveolaris,** a terminal bronchiole. **d. efferens testis,** one of the seminal ducts. **d. rectus,** one of the straight tubules of the testis.

**ductus** (*duk'-tus*) [L.; *pl., ducti*]. A canal or duct. **d. arteriosus,** a short vessel in the fetus connecting the pulmonary artery with the aorta. **d. auditorius, d. cochlearis,** the scala media of the cochlea. **d. choledochus,** the bile duct. **d. Cuvieri.** See *Cuvier, ducts of*. **d. deferens,** the vas deferens. **d. lingualis,** the persistent remains, in the adult, of the upper portion of the embryonal ductus thyreoglossus. **d. nasofrontalis,** one between the frontal sinus and middle meatus of nose. **d. pancreaticus azygos,** the accessory pancreatic duct. **d. pancreaticus minor, d. pancreaticus recurrens, d. pancreaticus Santorini, d. pancreaticus secondarius, d. pancreaticus superior.** See *duct, pancreatic, accessory*. **ducti papillares,** the uriniferous tubules. **d. perilymphaticus,** the aqueduct of the cochlea. **d. Rosenthalianus,** the accessory pancreatic duct. **ducti seminales, ducti seminiferi,** the seminiferous tubules. **d. thyreoglossus,** a small duct connecting the primitive thyroid gland with the upper surface of the tongue. **d. thyroideus,** the persistent remains, in the adult, of the lower portion of the embryonal ductus thyreoglossus. **d. venosus,** a branch of the umbilical vein in the fetus which empties directly into the ascending vena cava.

**Duddell's membrane.** See *Descemet's membrane*.

**Dudley's operation** (*dud'-le*) [Emilius Clark Dudley, American gynecologist, 1850– ]. For *retroversion of the uterus:* the abdomen is opened, and a strip is doubled from the anterior surface of the uterus, and likewise a strip along the inner side of each round ligament, followed by suturing of the three together.

**Dugas' test** (*doo'-gas*) [Louis Alexander Dugas, American surgeon, 1806–1884]. In dislocation of the shoulder-joint the elbow cannot be made to touch the side of the chest when the hand of the affected side is placed on the opposite shoulder.

**Duhring's disease** (*doo'-ring*) [Louis Adolphus Duhring, American dermatologist, 1845– ]. Dermatitis herpetiformis. **D.'s pruritus,** pruritus hiemalis.

**duipara** (*du-ip'-ar-ah*) [*duo,* two; *parere,* to bring forth]. A woman pregnant for the second time.

**Dukes' disease** (*dūks*) [Clement Dukes, English physician, 19th century]. See *disease, fourth*.

**dulcamara** (*dul-kam-a'-ra*) [*dulcis,* sweet; *amarus,* bitter]. Bittersweet. The young branches of *Solanum dulcamara,* containing an alkaloid, *solanine*. In overdoses it causes nausea, emesis, and convulsive muscular movements, and in toxic doses it is a narcotic poison. It is employed in psoriasis and similar skin diseases. Dose of the *fluidextract* 30 min.–1 dr. (2–4 Cc.).

**dulcamarin** (*dul-kam-a'-rin*) [*dulcamara*], $C_{22}H_{34}O_{10}$. A yellow, amorphous glucoside found in dulcamara, with a sweetish-bitter taste, sparingly soluble in water, freely so in alcohol and acetic acid, insoluble in ether and in chloroform.

**dulcify** (*dul'-sif-i*). To render sweet.

**dulcin** (*dul'-sin*), $C_9H_{12}N_2O_2$. A toxic substance 200 times sweeter than cane-sugar, obtained from paraphenetidin by action of potassium cyanate. It is used as a sweetening medium. Syn., *paraethoxyphenylurea; paraphenetol-carbamide; sucrol; valzin*.

**dulcite, dulcitol** (*dul'-sīt, dul'-sit-ol*) [*dulcis,* sweet], $C_6H_{14}O_6$. Sugar from *Melampyrum nemorosum* and other plants.

**dulcose** (*dul';kōs*). Dulcitol.

**dull** (*dul*). 1. Slow of perception. 2. Not resonant on percussion. 3. Not bright in appearance. 4. Not sharp; blunt.

**dullness, dulness** (*dul'-nes*). The quality of being dull; in any sense, lack of resonance on percussion. **d. wooden,** a percussion note sounding as if given out from wood.

**dumb** (*dum*). Unable to utter articulate speech. **d. ague,** a popular term for ague or malaria marked by obscure symptoms. **d.-bell crystals,** crystals of calcium oxalate, sometimes seen in the urine.

**dumbness** (*dum'-nes*). Inability to utter articulate speech.

**dumdum fever** [*Dum Dum,* a town in India]. Same as *kala-azar, q. v.*

**Dumontpallier's test for bile-pigments.** See *Smith's reaction*.

**Dunbar's serum** (*dun-bar'*) [William Philipps Dunbar, American physician 1863– ]. Pollantin.

**Duncan Bird's sign.** See *Bird's sign*.

**Duncan's folds** [James Matthews Duncan, English gynecologist, 1826–1890]. The folds of the loose peritoneal covering of the uterus seen immediately after delivery. **D.'s position of the placenta,** the on presenting itself at the os uteri for expulsion. **D.'s ventricle,** the fifth ventricle. Syn., *sinus Duncanii,* marginal position generally assumed by the placenta

**duodenal** (*du-o-de'-nal* or *du-od'-en-al*) [*duodenum*]. Relating to the duodenum.

**duodenectomy** (*du-o-de-nek'-to-me*) [*duodeno; εκτομή*, excision]. Excision of part or all of the duodenum.

**duodenitis** (*du-o-den-i'-tis*) [*duodenum; ιτις,* inflammation]. Inflammation of the duodenum.

**duodeno-** (*du-od-en-o-*) [*duodenum*]. A prefix meaning relating to the duodenum.

**duodenocholangitis** (*du-o-de-no-ko-lan-ji'-tis*). Inflammation of the duodenum and the common bile-duct.

**duodenocholecystostomy** (*du-od-en-o-kol-e-sis-tos'-to-me*) [*duodeno-; χολή,* bile; *cystostomy*]. The formation of a fistula between the duodenum and gall-bladder.

**duodenocholedochotomy** (*du-od-en-o-ko-led-o-kot'-o-me*). A modification of choledochotomy consisting in incising the duodenum in order to reach the gall-duct.

**duodenocystostomy** (*du-od-en-o-sist-os'-to-me*) [*duodeno-; cystostomy*]. The establishment of a communication between the bladder and the duodenum.

**duodenoenterostomy** (*du-od-en-o-en-ter-os'-to-me*) [*duodeno-; enterostomy*]. The formation of a fistula between the duodenum and small intestine.

**duodenogastric** (*du-od-en-o-gas'-trik*). See *gastroduodenal*.

**duodenojejunal** (*du-o-de-no-jej-u'-nal*) [*duodeno-; jejunum*]. Pertaining to the duodenum and the jejunum.

**duodenorenal** (*du-o-de-no-re'-nal*) [*duodeno-; ren,* the kidney]. Relating to the duodenum and to the kidney.

**duodenostomy** (*du-od-en-os'-to-me*) [*duodeno-; στόμα,* a mouth]. The operation of forming an opening into the duodenum through the abdominal walls.

**duodenotomy** (*dū-od-en-ot'-o-me*) [*duodeno-*; τέμνειν, to cut]. Surgical incision of the duodenum.

**duodenum** (*dū-o-de'-num*) [*duodeni*, twelve each; so called because it is about 12 fingerbreadths long]. The first part of the small intestine beginning at the pylorus. It is from 8 to 10 inches long, is the most fixed part of the small intestine, consists of an ascending, descending, and transverse portion, and contains the openings of the pancreatic duct and the common bile-ducts.

**duotal** (*dū'-o-tal*). The commercial name for guaiacol carbonate.

**duotonal** (*dū-o-to'-nal*). Trade name for a combination of calcium and sodium glycerophosphates.

**Duparque's method for detecting ascites** (*doo-park'*). When fluctuation is indistinct, the patient is to be placed on one side for a few moments, so that the whole quantity of fluid may gravitate to the depending flank; then quickly turned upon the back, when dulness and temporary fluctuation will be found at the site of accumulation.

**Duplay's bursitis** (*doo-play*) [Emmanuel Simon Duplay, French surgeon, 1836– ]. Subacromial or subdeltoid bursitis. **D.'s operation.** 1. *For epispadias:* the urethra is formed at the expense of the corpus spongiosum and corpora cavernosa instead of by flaps. 2. *For hypospadias:* it is performed in three stages; (1) straightening of the penis and the formation of a meatus; (2) the formation of a canal from the meatus to the hypospadial opening; (3) junction of the old and new canals.

**duplicature** (*dū'-plik-a-chur*) [*duplicare*, to double]. The reflection, or folding of a membrane upon itself.

**duplicity** (*dū-plis'-it-e*) [*duplicitas*, doubleness]. The condition of being duplex or double.

**Dupré's syndrome.** Meningism; pseudomeningitis.

**Dupuytren's contraction** (*doo-pwe-tran*) [Guillaume *Dupuytren*, French surgeon, 1778–1835]. A contraction of the palmar fascia causing the fingers to fold into the palm. **D.'s delirium.** See *delirium nervosum*. **D.'s egg-shell symptom**, the sensation of a delicate crepitant shell (egg-shell crackling) imparted on slight pressure in certain cases of sarcoma of long bones. **D.'s false contraction,** a contraction of the palm and fingers due to injury to the palmar fascia. **D.'s finger.** See *Dupuytren's contraction*. **D.'s fracture,** fracture of the lower end of the fibula, with displacement of the foot outward and backward. **D.'s hydrocele,** bilocular hydrocele of the tunica vaginalis testis. Syn., *hydrocèle en bissac*. **D.'s operation.** For amputation at the shoulder-joint: two rounded flaps are taken from the outer and inner aspects of the arm; the outer flap is made first by transfixion from behind at a point two inches below the acromion; the bone is then disarticulated and a short inner flap cut from within outward. **D.'s phlegmon,** unilateral phlegmonous suppuration occupying the anterolateral portion of the neck. **D.'s powder,** a caustic powder consisting of arsenic trioxide, 1 part; calomel, 200 parts. **D.'s splint,** a splint used in the treatment of Pott's fracture of the leg.

**dura, dura mater** (*dū'-rah ma'-ter*) [*durus*, hard; *mater*, mother]. The fibrous membrane forming the outermost covering of the brain and spinal cord.

**duræmatoma** (*dū-rem-at-o'-mah*). See *durematoma*.

**dural** (*dū'-ral*) [*durus*, hard]. Pertaining to the dura.

**duramatral** (*dū-rah-ma'-tral*). See *dural*.

**durematoma** (*dū-rem-at-o'-mah*) [*dura*; αἷμα, blood; δύα, tumor; *pl.*, *durematomata*]. Hematoma of the dura; an accumulation of blood between the dura and arachnoid.

**Durham's tube** (*dur'-um*) [Arthur Edward *Durham*, English surgeon, 1834–1895]. The lobster-tail tube formerly used in tracheotomy.

**duritis** (*dū-ri'-tis*). Inflammation of the dura; pachymeningitis.

**duroarachnitis** (*dū-ro-ar-ak-ni'-tis*) [*durus*, hard; ἀράχνη, a spider; ιτις, inflammation]. Inflammation of the dura and arachnoid membrane.

**duroleum** (*dū-ro'-le-um*). An ointment-base obtained from petroleum.

**Duroziez's disease** (*doo-ro-zje-a*) [Paul Louis *Duroziez*, French physician, 1826–1897]. Congenital mitral stenosis. **D.'s murmur,** a double murmur heard over the femoral artery on pressure with the stethoscope in cases of aortic insufficiency, mitral stenosis, lead-poisoning, contracted kidney, and some fevers. **D.'s sign.** Same as *D.'s murmur*.

**Dusart's syrup** (*doo-sar*) [Lucien O. *Dusart*, French physician, 19th century]. A preparation having for its chief ingredient ferric phosphate; dose, ½ to 2 fl.dr. (2 to 8 c.c.).

**dusting-powder.** Any fine powder used to dust on the skin to absorb or diminish its secretions or allay irritation.

**dust-occupations.** Those that from the nature of the particles of the dust produced may give rise to pneumoconiosis, tuberculosis, or bronchitis.

**Dutch liquid.** See *ethene chloride*.

**Duval's nucleus.** An aggregation of large multipolar ganglion-cells lying ventrolaterally to the hypoglossal nucleus.

**Duverney's foramen** (*doo-ver-na'*) [Joseph *Guichard Duverney*, French anatomist, 1648–1730]. See *Winslow's foramen*. **D.'s gland.** Same as *Bartholin's gland*.

**D. V. M.** Abbreviation for Doctor of Veterinary Medicine.

**D. V. M. S.** Abbreviation for Doctor of Veterinary Medicine and Surgery.

**D. V. S.** 1. Abbreviation for Doctor of Veterinary Science. 2. Abbreviation for Doctor of Veterinary Surgery.

**dwarfism** (*dwarf'-izm*) [ME., *dwarf*]. An abnormal stature in man, often pathological (microcephalia, rickets, etc.), in which the height falls below 1 m. 25 cm.

**dyad** (*di'-ad*). An atom uniting with two monad atoms.

**dyaster** (*di'-as-ter*). A double group of chromosomes during the anaphases of cell-division.

**dye** (*di*) [ME., *dyen*, to dye]. 1. To color a substance by immersing it in some coloring-matter. 2. The material used as a coloring substance. **d., acid,** one produced by combining a substance having coloring properties and which plays the part of an acid (an anilin) with some ordinary base, as sodium, potassium, etc. **d., basic,** one produced by combining with some acid a coloring principle (an anilin) which plays the part of a base.

**dyestuff** (*di'-stuf*). See *dye* (2). **d.s, indifferent,** a group of histological dyestuffs, neither basic nor acid, obtained from acid which possesses a peculiar affinity for fats; an example is sudan III (benzolazobetanaphthol). **d., Ehrlich's term** for a salt the acid portion of which is an acid dye and of which the basic portion is a basic dye.

**dymal** (*di'-mal*). See *didymium salicylate*.

**dymyarious** (*di-mi-a'-re-us*) [δυάς, two; μῦς, a muscle]. Furnished with two muscles.

**dynactinometer** (*di-nak-tin-om'-et-er*) [δύναμις, power; ἀκτίς, a ray; μέτρον, a measure]. An apparatus for determining the intensity of the photogenic rays and estimating the power of object glasses.

**dynam** (*di'-nam*). See *dyne*.

**dynameter** (*di-nam'-et-er*). See *dynamometer* (2).

**dynamia** (*di-nam'-e-ah*) [δύναμις, power]. 1. Ability to resist disease or to withstand the effects of any strain, physical or mental. 2. The sthenic character of any attack of disease.

**dynamic** (*di-nam'-ik*) [*dynamia*]. Pertaining to energy; sthenic; characterized by energy or great force.

**dynamicity** (*di-nam-is'-it-e*) [*dynamia*]. The greatest capacity for inherent power possessed by a substance or organism.

**dynamics** (*di-nam'-iks*). See *mechanics*. **d., vital,** the science of the inherent power of an organism.

**dynamimeter** (*di-nam-im'-et-er*). See *dynamometer*.

**dynamite** (*di'-nam-īt*) [δύναμις, power]. An explosive consisting of nitro-glycerin incorporated with infusorial earth, to give it consistency. It is eight times as powerful as gunpowder.

**dynamization** (*di-nam-iz-a'-shun*) [δύναμις, power]. The so-called potentizing of medicines by comminution and agitation.

**dynamo** (*di'-nam-o*) [δύναμις, power]. A machine in which an electric current is generated by revolving coils of insulated wire through the field of a magnet intensified by the same current.

**dynamoelectric** (*di-nam-o-e-lek'-trik*) [*dynamo-electric*]. Relating to the motor power of electricity.

**dynamogen** (*di-nam'-o-jen*). A proprietary remedy resembling hematogen; used in anemia.

**dynamogenic** (*di-nam-o-jen'-ik*) [*dynamo-*; γεννᾶν, to produce]. Generating force or power.

**dynamogeny** (*di-nam-oj'-en-e*) [*dynamo-*; γεννᾶν, to

beget]. The production of energy; the physiological generation of force.
**dynamograph** (*di-nam'-o-graf*) [*dynamo*; γράφειν, to write]. An instrument designed to measure and record graphically muscular strength.
**dynamography** (*di-nam-og'-ra-fe*) [see *dynamograph*]. 1. Mechanics. 2. The measurement and graphic record of muscular strength.
**dynamometer** (*di-nam-om'-et-er*) [*dynamo*; μέτρον, a measure]. 1. An instrument for the measurement of muscular strength, particularly of the hand. 2. An instrument for estimating the magnifying power of lenses. Syn., *dynameter*; *optical dynamometer*.
**dynamometry** (*di-nam-om'-et-re*) [see *dynamometer*]. The measurement of force by means of the dynamometer. d., vital, the estimation of the inherent force of an individual.
**dynamoneure** (*di-nam'-o-nūr*) [*dynamo*; νεῦρον, nerve]. A spinal motor neuron.
**dynamoscope** (*di-nam'-o-skōp*) [*dynamo*; σκοπεῖν, to examine]. An apparatus for auscultating the muscles.
**dynamoscopy** (*di-nam-os'-ko-pe*). Auscultation of the muscles by means of the dynamoscope.
**dyne** (*dīn*) [δύναμις, power]. A measure of force; it is the force that, when applied to a mass of one gram for one second, will give it a velocity of one centimeter a second.
**dys-** [δυs-]. A prefix meaning bad, hard, difficult, painful.
**dysacousma** (*dis-ak-oos'-mah*) [*dys-*; ἀκουσμα, hearing]. A sensation of pain or discomfort caused by loud or even moderately loud noises.
**dysacusia, dysacousis** (*dis-ak-oo'-ze-ah, -sis*) [see *dysacousma*]. Difficulty of hearing.
**dysæsthesia** (*dis-es-the'-ze-ah*). See *dysesthesia*.
**dysalbumose** (*dis-al'-bū-mōs*). A variety of albumose, insoluble in hot or cold water or hydrochloric acid.
**dysanagnosia** (*dis-an-ag-no'-se-ah*) [*dys-*; ἀνάγνωσις, recognition, reading]. Word-blindness; dyslexia; difficulty in comprehending written language.
**dysantigraphia** (*dis-an-te-gra'-fe-ah*). Inability to copy writing or print.
**dysaphe** (*dis'-af-e*) [*dys-*; ἀφή, touch]. Morbid state of the sense of touch.
**dysapocatastasis** (*dis-ap-o-kat-as'-tas-is*) [*dys-*; ἀποκατάστασις, re-establishment]. Morbid restlessness and dissatisfaction.
**dysaponotocy** (*dis-ap-o-not'-o-se*) [*dys-*; ἄτονος, painless; τόκος, birth]. Painless, but difficult labor.
**dysarthria** (*dis-ar'-thre-ah*) [*dys-*; ἄρθρον, articulation]. Impairment of articulation.
**dysarthritis** (*dis-ar-thri'-tis*) [*dys-*; *arthritis*]. Anomalous gout.
**dysarthrosis** (*dis-ar-thro'-sis*) [*dys-*; ἄρθρον, a joint]. 1. A deformed joint. 2. A false-joint. 3. A dislocation of a joint.
**dysbasia** (*dis-ba'-ze-ah*). Difficulty in walking. d. intermittens angiosclerotica, intermittent claudication due to arteriosclerosis.
**dysblennia** (*dis-blen'-e-ah*) [*dys-*; βλέννα, mucus]. A disordered state or formation of the mucus.
**dysbulia** (*dis-bū'-le-ah*) [*dys-*; βούλεσθαι, to will]. Impairment of will-power.
**dyscatabrosis** (*dis-kat-ah-bro'-sis*) [*dys-*; καταβρῶσις, a devouring]. Difficulty in swallowing food; dysphagia.
**dyschezia** (*dis-ke'-ze-ah*) [*dys-*; χέζειν, to go to stool]. Painful or difficult defecation, as in cases of prolapse of the ovary.
**dyscholia** (*dis-ko'-le-ah*) [*dys-*; χολή, bile]. A disordered or morbid state of the bile.
**dyschondroplasia** (*dis-kon-dro-pla'-ze-ah*) [*dys-*; χόνδροs, cartilage; πλάσιs, molding]. A disease of unknown etiology, attacking the long bones and the metacarpal and phalangeal skeleton of the hand. It is characterized by cartilaginous tissue developing regularly but ossifying very slowly.
**dyschrea, dyschroia, dyschroma** (*dis-kre'-ah, -kroi'-ah, -kro'-mah*). See *dyschroa* and *parachrea*.
**dyschroa, dyschrœa** (*dis-kro'-ah, -kre'-ah*) [*dys-*; χροιά, color]. Discoloration, especially of the skin.
**dyschromasia** (*dis-kro-ma'-ze-ah*). 1. See *dyschroa*. 2. See *dyschromatopsia*.
**dyschromatodermia, dyschromodermia** (*dis-kro-mat-o-dur'-me-ah, dis-kro-mo-dur'-me-ah*). See *dyschroa*.
**dyschromatope** (*dis-kro'-mat-ōp*) [*dys-*; χρῶμα,

color; ὄψις, vision]. An individual affected with dyschromatopsia.
**dyschromatopsia** (*dis-kro-mat-op'-se-ah*) [see *dyschromatope*]. Partial color-blindness; difficulty in distinguishing colors.
**dyschromia** (*dis-kro'-me-ah*) [*dys-*; χρῶμα, color]. Discoloration, especially of the skin.
**dyschylia** (*dis-ki'-le-ah*) [*dys-*; *chyle*]. Disorder of the chyle.
**dyscinesia** (*dis-sin-e'-ze-ah*) [*dys-*; κίνησις, motion]. Difficult or painful motion.
**dyscoria** (*dis-ko'-re-ah*) [*dys-*; κόρη, pupil]. Abnormality of the form of the pupil.
**dyscrasia** (*dis-kra'-ze-ah*) [*dys-*; κρᾶσις, combination]. A depraved condition of the blood or system due to general disease; a condition of increased susceptibility to disease.
**dyscrasic, dyscratic** (*dis-kra'-sik, dis-krat'-ik*) [*dys-*; κρᾶσις, combination). Of the nature of, or affected with a dyscrasia.
**dysecœa, dysecoia** (*dis-es-e'-ah, dis-ek-oi'-ah*) [δυσηκοία, deafness]. Hardness of hearing.
**dysemesia, dysemesis** (*dis-em-e'-ze-ah, -em'-es-is*) [*dys-*; *emesis*]. Painful vomiting; retching.
**dysemia, dysæmia** (*dis-e'-me-ah*) [*dys-*; αἷμα, blood]. A morbid state of the blood.
**dysenteria** (*dis-en-te'-re-ah*). See *dysentery*. d. splenica. Synonym of *melena*.
**dysenteric** (*dis-en-ter'-ik*) [*dysentery*]. Of the nature of or affected with dysentery.
**dysenteriform** (*dis-en-ter-e'-form*) [*dysentery*; *forma*, form]. Resembling dysentery.
**dysenterioid** (*dis-en-ter'-e-oid*). See *dysenteriform*.
**dysentery** (*dis'-en-ter-e*) [*dys-*; ἔντερον, the bowel]. An inflammation of the large intestine, probably infectious in origin, and characterized by pain, rectal tenesmus, and the frequent passage of small amounts of mucus and blood. Anatomically, three varieties may be distinguished: the catarrhal, the diphtheritic, and the gangrenous. The true cause of dysentery is not known in all cases; but the *Amœba coli* is the cause of amebic dysentery.
**dysepulotic** (*dis-ep-ū-lot'-ik*) [*dys-*; ἐπουλωτικόs, healing]. Cicatrizing slowly and imperfectly.
**dyseresthesia** (*dis-er-e-thiz'-e-ah*) [*dys-*; ἐρεθίζειν, to excite]. Diminished sensibility or irritability.
**dysergasia, dysergasy** (*dis-er-ga'-ze-ah, dis-er-ga'-se*) [δυσεργήs, difficult to effect]. Disturbances of function, especially as manifested in neurasthenia.
**dysesthesia** (*dis-es-the'-ze-ah*) [*dys-*; αἴσθησις, sensation]. 1. Dulness of sensation. 2. Painfulness of any sensation not normally painful.
**dysfunction** (*dis-fungk'-shun*). Any abnormality or impairment of function.
**dysgalactia** (*dis-gal-ak'-te-ah*) [*dys-*; γάλα, milk]. Loss or impairment of milk secretion.
**dysgenesia** (*dis-jen-e'-se-ah*) [*dys-*; γεννᾶν, to produce]. 1. Loss or impairment of procreative power.
**dysgenesis** (*dis-jen'-es-is*) [*dys-*; γένεσις, generation]. Sterility; difficulty in breeding.
**dysgenic** (*dis-jen'-ik*). 1. Pertaining to dysgenesis. 2. A term applied to anything which interferes with eugenics. 3. The opposite of eugenics.
**dysgenitalism** (*dis-jen'-it-al-ism*) [*dys-*; *genital*]. The condition resulting from some abnormality in the development of the genital organs.
**dysgeusia** (*dis-ju'-se-ah*) [*dys-*; γεῦσις, taste]. Morbidity or perversion of the sense of taste.
**dysgrammatism** (*dis-gram'-at-ism*) [*dys-*; γράμμα, a letter]. Inability to make the proper use of words. It is a symptom of certain cerebral diseases.
**dysgraphia** (*dis-graf'-e-ah*) [*dys-*; γράφειν, to write]. 1. Impairment of the power of writing as a result of a brain-lesion. 2. Writer's cramp.
**dyshæmia** (*dis-he'-me-ah*). See *dysemia*.
**dyshaphia** (*dis-haf'-e-ah*). See *dysaphe*.
**dyshidria, dysidria** (*dis-hid'-re-ah, dis-id'-re-ah*) [*dys-*; ἱδρώs, sweat]. A morbid condition of the function of perspiration.
**dyshidrosis** (*dis-hid-ro'-sis*). See *dysidrosis*.
**dyshypophysism** (*dis-hi-pof'-is-ism*) [*dys*; *hypophysis*]. A condition produced by an abnormal condition of the pituitary gland. It is characterized by hypotension, tachycardia, sensation of heat, profuse sweats, oliguria, anorexia, asthma, insomnia, and disturbances of nutrition and growth.
**dysidrosis** (*dis-id-ro'-sis*) [*dys-*; ἱδρωσις, sweating]. Synonym of *pompholyx*.
**dyskatabrosis** (*dis-kat-ah-bro'-sis*). See *dyscatabrosis*.

**dyskinesia** (*dis-kin-e'-ze-ah*) [*dys-;* κίνησις, movement]. Impairment of the power of voluntary motion.

**dyslalia** (*dis-lal'-e-ah*) [*dys-;* λαλεῖν, to talk]. Impairment of the power of speaking, due to a defect of the organs of speech.

**dyslexia** (*dis-leks'-e-ah*) [*dys-;* λέξις, reading]. Impairment of the ability to read.

**dyslochia** (*dis-lo'-ke-ah*) [*dys-;* lochia]. An abnormal condition of the lochial discharge.

**dyslogia** (*dis-lo'-je-ah*) [*dys-;* λόγος, speech]. Difficulty in the expression of ideas by speech.

**dyslysin** (*dis'-lis-in*) [*dys-;* λύειν, to dissolve], C₂₄H₃₆O. A product of cholic acid.

**dysmasesis, dysmassesis, dysmastesis** (*dis-mas-e'-sis, dis-mas-te'-sis*). Difficulty of mastication.

**dysmenorrhea, dysmenorrhœa** (*dis-men-or-e'-ah*) [*dys-;* μήν, month; ῥεῖν, to flow]. Difficult or painful menstruation. **d., congestive,** a form of painful menstruation due to an intense congestion of the pelvic viscera. **d., inflammatory,** that due to inflammation. **d., mechanical.** See **d., obstructive. d., membranous,** a very painful form characterized by the discharge of shreds of menstrual decidua. **d., obstructive,** that due to mechanical obstruction to the free escape of the menstrual fluid. **d., ovarian,** that form due to disease of the ovaries. **d., spasmodic,** that form due to spasmodic uterine contraction. **d., vascular,** pain, congestive symptoms in the genital apparatus, and tenesmus of the bladder and anus appearing some days prior to menstruation.

**dysmimia** (*dis-mim'-e-ah*) [*dys-;* μιμεῖσθαι, to mimic]. Impairment of the power to use signs and gestures; inability to imitate.

**dysmnesia** (*dis-mne'-ze-ah*) [*dys-;* μνῆσις, memory]. Impairment or defect of the memory.

**dysmorphia** (*dis-mor'-fe-ah*) [*dys-;* μορφή, form]. Deformity.

**dysmorphophobia** (*dis-morf-o-fo'-be-ah*) [*dys-;* μορφή, form; φόβος, fear]. Morbid dread of deformity; it is a rudimentary form of paranoia.

**dysmorphosteopalinklast** (*dis-morf-os-te-o-pal'-in-klast*). An instrument for refracturing a bone which has united with deformity.

**dysmorphosteopalinklasy** (*dis-morf-os-te-o-pal-in'-kla-se*) [δυσμορφος, deformed; ὀστέον, a bone; πάλιν, again; κλᾶν, to break]. The operation of refracturing a bone which has healed with deformity after a fracture.

**dysmyotonia** (*dis-mi-o-to'-ne-ah*) [*dys-;* μῦς, a muscle; τόνος, tone]. 1. Atony of the muscles. 2. Excessive tonicity in muscles; myotonia.

**dysneuria** (*dis-nū'-re-ah*) [*dys-;* νεῦρον, nerve]. An impairment of nerve-function.

**dysnoia** (*dis-noi'-ah*) [*dys-;* νόος, mood, disposition]. Heavy, gloomy thought.

**dysnusia** (*dis-nū'-ze-ah*) [*dys-;* νοῦς, mind]. Weakness or impairment of the mind.

**dysodia** (*dis-o'-de-ah*) [*dys-;* ὄζειν, to smell]. Fetor; stench; ill smell. Also a synonym of *rhinitis atrophica, q. v.*

**dysodontiasis** (*dis-o-don-ti'-as-is*) [*dys-;* ὀδοντίασις, dentition]. Difficult dentition.

**dysodynia** (*dis-o-din'-e-ah*) [*dys-;* ὀδύνη, pain]. Ineffective labor-pains.

**dysopia** (*dis-o'-pe-ah*) [*dys-;* ὤψ, eye]. Painful or defective vision.

**dysorexia** (*dis-or-eks'-e-ah*) [*dys-;* ὄρεξις, appetite]. A depraved or unnatural appetite.

**dysosmia** (*dis-oz'-me-ah*) [*dys-;* ὀσμή, odor]. Impairment of the sense of smell.

**dysosphresia** (*dis-os-fre'-ze-ah*) [*dys-;* ὄσφρησις, smell]. Impairment of the sense of smell.

**dysostosis** (*dis-os-to'-sis*) [*dys-;* ὀστέον, bone]. Defective formation of bone. **d., cleidocranial,** a singular congenital malformation compatible with life, intelligence, and purity of the blood, consisting in incomplete ossification of the skull, malformation of the palatine arch, and more or less atrophy of the clavicles.

**dyspareunia** (*dis-par-oo'-ne-ah*) [δυσπάρευνος, ill-mated]. Painful or difficult copulation. **d., climacteric,** pain or difficulty in coitus following the menopause; it is regarded as a symptom of kraurosis vulvæ.

**dyspepsia** (*dis-pep'-se-ah*) [*dys-;* πέπτειν, to digest]. Disturbed digestion. **d., acid,** that attended with hyperacidity of the gastric juice. **d., alkaline,** that accompanied by lack of the normal acidity of the gastric juice. **d., atonic,** a form due to insufficient quantity or impaired quality of the gastric juice or to deficient action of the gastric muscles. **d., bilious,** intestinal dyspepsia due to impaired secretion of bile. **d., catarrhal,** that caused by gastric catarrh. **d., chemical,** that due to some change in the constitution of the digestive secretions. **d., feculent,** that due to excess of starchy food in the diet. **d., flatulent, d., gaseous,** that marked by almost constant generation of gas within the stomach. **d., gastric,** that confined to the stomach. **d., gastrointestinal,** that in which both the stomach and the intestine are concerned. **d., gastrorrheal,** that due to gastric catarrh and hypersecretion. **d., inflammatory,** that due to some form of gastritis. **d., intestinal,** that due to imperfect digestive action of the intestinal juices or to lack of tone in the muscular coat of the bowel. **d., lienteric,** that due to defective nutrition or superalimentation, and indicated by lienteric stools. **d., motor.** See **d., atonic. d., muscular,** that due to atony of the muscular coat of the stomach or intestine. **d., nervosecretory,** neurotic dyspepsia with perverted or excessive gastric secretion. **d., nervous,** that characterized by gastric pains, coming on often when the stomach is empty and relieved by eating, and by various reflex nervous phenomena, especially by palpitation. **d., neuralgic,** that marked by intermittent gastrodynia. **d., pyretic,** Gendrin's name for the forms of inflammatory dyspepsia formerly known as bilious or gastric fever. **d., salivary,** that due to excess, deficiency, or defective quality of the saliva or to insufficient mastication.

**dyspepsodynia** (*dis-pep-so-din'-e-ah*). Synonym of *gastralgia.*

**dyspeptic** (*dis-pep'-tik*) [see *dyspepsia*]. 1. Relating to or affected with dyspepsia. 2. A person suffering from dyspepsia.

**dyspeptone** (*dis-pep'-tōn*) [see *dyspepsia*]. An insoluble and unassimilable peptone.

**dysperistalsis** (*dis-per-e-stal'-sis*). Painful or violent, or perverted peristalsis.

**dyspermatism** (*dis-per'-mat-ism*) [*dys-;* σπέρμα, seed]. Difficulty of depositing the sperm within the vagina.

**dyspermia** (*dis-perm'-e-ah*). See *dysspermia.*

**dysphagia** (*dis-fa'-je-ah*) [*dys-;* φαγεῖν, to eat]. Difficulty in swallowing, or inability to swallow. **d. amyotactica,** a disturbance of the act of deglutition, not due to organic changes in the pharynx or esophagus nor to spasm or paralysis of the muscles concerned in swallowing; it consists in a disturbance of the rhythm of the function of the higher nerve-centers. **d. callosa,** that resulting from the destruction of the muscular layers of the esophagus and the formation of cicatricial tissue causing constriction. **d. constricta,** that due to stenosis of the pharynx or esophagus. **d. globosa,** globus hystericus. **d. lusoria,** a doubtful form ascribed to compression of the esophagus by the right subclavian artery when by a freak of nature this artery springs from the aorta behind the left subclavian artery and turns to the right either before or behind the esophagus. **d. spastica,** that due to hysterical spasm of the esophagus or pharynx.

**dysphasia** (*dis-fa'-ze-ah*) [*dys-;* φάσις, speech]. Difficulty of speech depending on a central lesion.

**dysphemia** (*dis-fe'-me-ah*) [*dys-;* φήμη, a speech]. Stammering.

**dysphonia** (*dis-fo'-ne-ah*) [*dys-;* φωνή, voice]. An impairment of the voice.

**dysphoria** (*dis-fo'-re-ah*) [*dys-;* φέρειν, to bear]. Impatience and restlessness; mental anxiety; fidgets.

**dysphotia** (*dis-fo'-she-ah*) [*dys-;* φῶς, light]. An error of refraction in which only near objects are seen; nearsightedness.

**dysphrasia** (*dis-fra'-ze-ah*) [*dys-;* φράσις, speech]. Imperfect speech due to impairment of mental power.

**dysphrenia** (*dis-fre'-ne-ah*) [*dys-;* φρήν, mind]. Any mental disorder.

**dyspituitarism** (*dis-pit-ū'-it-ar-izm*). A condition due to disease or destruction of the pituitary body.

**dysplasmatic, dysplastic** (*dis-plaz-mat'-ik, dis-plast'-ik*). See *cacoplastic.*

**dyspnea, dyspnœa** (*disp-ne'-ah*) [*dys-;* πνεῖν, to breathe]. Difficult or labored breathing. **d., cardiac,** that due to heart disease. **d., renal,** that due to renal disease. **d., Traube's.** See *Traube's dyspnea.*

**dyspneic** (*disp-ne'-ik*) [*dyspnea*]. Affected with or caused by dyspnea.

**dyspragia** (*dis-pra'-je-ah*) [*dys-;* πράγειν, to do; to perform]. Difficult or painful performance of any function.

**dysproteose** (*dis-pro′-te-ōs*). A modified heteroproteose obtained by treating heteroproteose with water.

**dysspermasia, dysspermatism** (*dis-sperm-a′-ze-ah, dis-sperm′-at-izm*). See *bradyspermatism*.

**dysspermia** (*dis-sper′-me-ah*) [*dys-*; σπέρμα, seed]. An abnormal condition of the semen.

**dysstechiasis, dysstœchiasis, dysstichiasis** (*dis-stek-i′-as-is, dis-stik-i′-as-is*). See *distichiasis*.

**dystasia** (*dis-ta′-ze-ah*) [*dys-*; στάσις, standing]. Difficulty in standing.

**dystaxia** (*dis-taks′-e-ah*) [*dys-*; τάξις, regulation; order]. Ataxia or partial ataxia. **d. agitans**, tremor due to irritation of the spinal cord. Syn., *pseudoparalysis agitans*.

**dysteleologic** (*dis-te-le-o-loj′-ik*) [*dys-*; τέλεος, perfect; λόγος, science]. Pertaining to *dysteleology*.

**dysteleology** (*dis-te-le-ol′-o-je*) [*dys-*; τέλεος, perfect; λόγος, science]. The study of rudimentary and useless organs, such as the vermiform appendix.

**dysthanasia** (*dis-than-a′-ze-ah*) [*dys-*; θάνατος, death]. A slow and painful death.

**dysthelasia** (*dis-thel-a′-ze-ah*) [*dys-*; θηλάζειν, to suck]. Difficulty in sucking or in giving suck.

**dysthermasia** (*dis-ther-ma′-ze-ah*) [*dys-*; θέρμη, heat]. Insufficient production of bodily heat.

**dysthesia** (*dis-the′-ze-ah*) [*dys-*; θέσις, an arranging]. 1. Ill condition, especially ill health due to a non-febrile disorder of the blood-vessels. 2. Impatience; fretfulness; ill-temper in the sick.

**dysthetic** (*dis-thet′-ik*) [*dysthesia*]. Of the nature of a dysthesia; cachectic.

**dysthymia** (*dis-thim′-e-ah*) [*dys-*; θυμός, mind]. Melancholy or mental perversion. **d. algetica**, mental perversion due to peripheral nerve-irritation. **d. neuralgica**, mental perversion due to facial or other neuralgias.

**dysthyreosis** (*dis-thi-re-o′-sis*) [*dys-*; *thyroid*]. Impaired functional activity of the thyroid gland.

**dysthyroid, dysthyroidism** (*dis-thi′-roid, -izm*). Incomplete development and function of the thyroid gland.

**dystithia** (*dis-tith′-e-ah*) [*dys-*; τιθή, a nipple]. Difficulty of nursing or inability to nurse at the breast.

**dystocia** (*dis-to′-se-ah*) [*dys-*; τόκος, birth]. Difficult labor. **d., fetal**, difficult labor due to abnormalities of position or size and shape of the fetus. **d., maternal**, that dystocia the cause of which resides in the mother.

**dystonia** (*dis-to′-ne-ah*) [*dys-*; ′τόνος, tone]. Disorder or lack of tonicity.

**dystopia** (*dis-to′-pe-ah*) [*dys-*; τόπος, place]. Displacement of any organ.

**dystrophia** (*dis-tro′-fe-ah*). See *dystrophy*. **d. adiposogenitalis**, a form of dystrophy characterized by adiposity, aplasia of the genitals, and hypotrichiasis; it is believed to be due to diminished function of the pituitary body.

**dystrophic** (*dis-tro′-fik*) [see *dystrophy*]. Pertaining to dystrophy.

**dystrophoneurosis** (*dis-tro-fo-nū-ro′-sis*) [*dys-*; τροφή, nourishment; νεῦρον, a nerve; νόσος, disease]. A disturbance of nutrition caused by abolition or perversion of nervous influence; or a nervous disease caused by ill-nutrition.

**dystrophy** (*dis′-tro-fe*) [*dys-*; τροφή, nourishment]. Faulty nutrition.

**dystropodextrin** (*dis-tro-po-deks′-trin*) [*dys-*; τρέπειν, to turn; *dexter*, right]. A starchy material existing in normal blood, and but slightly soluble.

**dystrypsia** (*dis-trip′-se-ah*) [*dys-*; *trypsin*]. Dyspepsia from lack of trypsin. **d., intestinal**, a term proposed as a substitute for intestinal dyspepsia, since trypsin is the most important enzyme in intestinal digestion.

**dysulotous** (*dis ū′-lo-tus*) [*dys-*; οὐλή, scar. Healing with difficulty.

**dysuresia, dysuresis** (*dis-ū-re′-se-ah, -sis*) [*dys-*; οὔρησις, micturition]. Any disease of the urinary apparatus.

**dysuria** (*dis-ū′-re-ah*) [*dys-*; οὖρον, urine]. Difficult or painful urination.

**dysuriac** (*dis-ū′-re-ak*). A person affected with dysuria.

**dysuric** (*dis-ū′-rik*) [*dysuria*]. Affected with or relating to dysuria.

**dyszoöamylia** (*dis-zo-o-am-il′-e-ah*) [*dys-*; ζωόαμυλον]. Imperfect transformation of dextrose into glycogen (zoöamylon).

# E

**E.** Abbreviation of *eye*, of *emmetropia*, and of *electromotive force*. Chemical symbol of *erbium*.

**e.** A prefix denoting without, from, etc.

**ear** (ēr) [ME., *ere*]. The organ of hearing, consisting of the external ear, the middle ear or tympanum, and the internal ear or labyrinth. The *outer ear* is made up of an expanded portion, the *pinna*, and the external auditory canal. The *middle ear* consists of the tympanum, with the ear-ossicles, the Eustachian tube, and the mastoid cells. The *tympanum* is lined by mucous membrane, and communicates with the pharynx by means of the *Eustachian tube*. It is divided into three parts—the *atrium*, the *attic*, and the *antrum*. Its outer end is closed by the *tympanic membrane*, from which sound is conducted along the *ear-ossicles* (the *malleus, incus,* and *stapes*) to the *fenestra ovalis*, which communicates with the vestibule of the internal ear. By means of the *fenestra rotunda*, which is closed by the entotympanic membrane, it communicates with the cochlea of the internal ear. The *mastoid cells*, which are also part of the tympanum, are airspaces in the mastoid process of the temporal bone. The *internal ear* consists of the bony and membranous labyrinths, which are separated from each other by a space containing the *perilymph*. Each labyrinth consists of three parts: the vestibule, the semicircular canals, and the cochlea. The *bony vestibule* communicates with the tympanum by the fenestra ovalis, closed by the base of the stapes, and also with the other parts of the internal ear. The *semicircular canals* are three in number—the superior, the posterior, and the inferior. The *cochlea*, so named from its resemblance to a snail-shell, is a cylindrical tube that winds around a central axis, the modiolus, which transmits the cochlear nerves and blood-vessels. The *cochlear canal* is divided by the spiral lamina into the *scala vestibuli*, communicating with the vestibule, and the *scala tympani*, communicating with the fenestra rotunda of the tympanum. The *membranous labyrinth* is made up of parts corresponding to the bony labyrinth. The *vestibule* consists of two small sacs, the *utricle* and the *saccule*, which communicate through the vestibular aqueduct, which in places is largely specialized and receives the terminations of the vestibular nerve. The *membranous cochlea*, or *cochlear duct*, contains the acoustic organ of the cochlea, or *organ of Corti*, which consists of a series of epithelial arches formed by the interlocking of the sides of the *pillars* or *rods of Corti*. Upon the inner rods of Corti are the inner acoustic hair-cells; in relation with the outer rods are the outer hair-cells. The organ of Corti is covered by the *membrana tectoria*, or Corti's membrane. **e.-ache.** See *otalgia*. **e., Blainville's.** *Blainville's ear.* **e.-bones**, the auditory ossicles. **e.-cough**, a reflex cough due to irritation of the ear. **e. Darwin's.** See *Darwin's ear*. **e.-drum**, the tympanum. **e.-mold**, otomycosis. **e., Morel's.** See *Morel's ear.* **e.-trumpet**, an instrument to aid the hearing. **e.-wax.** See *cerumen*.

**earth** (erth) [ME., *erthe*]. A name given to various metallic oxides or silicates not soluble in water and not affected by great heat. **e.s, alkaline**, the oxides and hydrates of calcium, magnesium, strontium, barium, and other metals of the same group. **e.-bath**, application of hot earth or sand to the body of a patient. **e., fuller's**, a clay used as an absorbent application to irritated surfaces.

**earthy phosphate.** See *phosphate, earthy*.

**Eastes' test for sugar.** Place 60 Cc. of filtered urine in a beaker of 100 Cc. capacity, add 1 Gm. of sodium acetate and a little less of phenylhydrazin hydrochloride; stir with a glass rod, which is to remain in the beaker. Evaporate on a water-bath to 10 or 15 Cc., scraping the sediment from the sides of the beaker, if it collects there; cool, and examine under the microscope. If there is 1 part to 1000 of sugar in the urine, osazone crystals will be found.

**Easton's syrup** (ēs-ton) [John Alexander *Easton*, English physician, 1807–1865]. A syrup of the phosphates of quinine, iron, and strychnine.

**eat** (ēt). 1. To masticate and swallow food. 2. To corrode.

**eau** (o) [Fr.]. Water. **e. de Cologne**, Cologne water. **e. de Javelle**, solution of potassium hypochlorite. **e. de vie**, brandy; alcoholic spirit.

**Eberth's bacillus** (a'-bärt) [Carl Joseph *Eberth*, German physician, 1835– ]. *Bacillus typhi abdominalis.* See *bacilli, table of.* **E.'s lines**, dark broken lines seen to separate the cardiac muscular cells on staining with silver nitrate. Syn., *lineæ scalariformes*.

**Ebner's germ reticulum** (eb'-ner) [Victor von *Ebner*, Austrian histologist, 1842– ]. A fine, nucleated reticulum existing between the inner cells of the seminiferous tubules. **E.'s glands**, the acinous glands situated in the region of the circumvallate papillæ of the tongue.

**ebonation** (e-bo-na'-shun) [e, away from; *bone*]. The removal of splinters of bone after injury.

**ebonite** (eb'-on-īt) [*ebon*, ebony]. Black hard rubber. See *vulcanite*.

**ebracteate** (e-brak'-te-āt) [e, priv.; *bractea*, a thin plate]. In biology, destitute of bracts.

**ebriecation** (e-bri-e-ka'-shun) [*ebrietas*, drunkenness]. Mental disorder due to the use of alcoholic stimulants.

**ebrietas** (e-bri'-et-as) [L.]. Synonym of *alcoholism*.

**ebriety, ebriosity** (e-bri'-et-e, e-bre-os'-et-e). Synonym of *alcoholism*.

**Ebstein's lesion** (eb'-stīn) [Wilhelm *Ebstein*, German physician, 1836– ]. Hyaline degeneration and insular necrosis of the epithelial cells of the renal tubules in diabetes.

**ebullition** (eb-ul-lish'-un) [*ebullire*, to boil]. Boiling.

**ebur** (e'-bur) [L.]. Ivory. **e. dentis**, dentine, *q. v.*

**eburnated** (e-bur'-na-ted) [*ebur*, ivory]. A term applied to dentine the tubules of which have been obliterated by a calcareous deposit.

**eburnation** (e-bur-na'-shun) [*ebur*, ivory]. 1. An increase in the density of bone following inflammation. 2. Ossification of a cartilage; calcareous infiltration of a tumor.

**eburneous** (e-bur'-ne-us) [*ebur*, ivory]. In biology, ivory-white.

**ecaudate** (e-kaw'-dāt) [e, priv.; *cauda*, a tail]. Tailless; without a tail-like appendage. See *acaudal*.

**Ecballium** (ek-bal'-e-um). See *elaterium*.

**ecblepharos** (ek-blef'-ar-os) [ἐκ, out; βλέφαρον, eyelid]. An ancient form of artificial eye.

**ecbloma** (ek-blo'-mah) [ἐκ, out; βάλλειν, to cast]. An abortion; an aborted fetus.

**ecbolic** (ek-bol'-ik) [ἐκβολή, a throwing out]. 1. Producing abortion or accelerating labor. 2. Any agent producing this effect.

**ecbolin** (ek'-bol-ēn) [ἐκ, out; βάλλειν, to throw]. An alkaloid said to be one of the active principles of ergot. Little is known about it.

**ecbolium** (ek-bo'-le-um) [ἐκ, out; βάλλειν, to throw: *pl., ecbolia*]. Any abortifacient drug.

**eccentric** (ek-sen'-trik) [ἐκ, out; κέντρον, center]. 1. Situated away from the center. 2. Odd or peculiar in behavior, but free from insanity. **e. amputation.** See *amputation*. **e. convulsion**, one due to peripheral irritation. **e. hypertrophy**, hypertrophy of a hollow organ, as the heart, with dilation.

**eccentricity** (ek-sen-tris'-it-e) [ἐκ, out; κέντρον, center]. 1. Oddness or peculiarity of behavior without true insanity. 2. A peculiarity.

**eccentropiesis** (ek-sen-tro-pi-e'-sis) [*eccentric*; πίεσις, a pressing]. Pressure from within outward; a method proposed for the treatment of anal fistula.

**eccephalosis** (ek-sef-al-o'-sis) [ἐκ, out of; κεφαλή, the head]. Synonym of *cephalotomy* or *excerebration*.

**ecchondroma** (ek-on-dro'-mah) [ἐκ, out; χόνδρος, cartilage; ὄμα, tumor: pl., ecchondromata]. A cartilaginous tumor; a chondroma.
**ecchondrosis** (ek-on-dro'-sis) [ecchondroma; pl., ecchondroses]. A cartilaginous outgrowth.
**ecchondrotome** (ek-kon'-dro-tōm) [ἐκ, out; χόνδρος, cartilage; τομή, cutting]. An instrument for the surgical removal of cartilaginous growths.
**ecchymoma** (ek-e-mo'-mah) [ἐκ, out; χυμός, juice; ὄμα, tumor: pl., ecchymomata]. A tumor-like swelling composed of extravasated blood.
**ecchymosis** (ek-e-mo'-sis) [ἐκ, out; χυμός, juice]. An extravasation of blood into the subcutaneous tissues. It is marked by a purple discoloration of the skin, the color gradually changing to brown, green, and yellow.
**ecchymotic** (ek-e-mot'-ik) [ecchymosis]. Relating to or resembling an ecchymosis.
**ecchysis** (ek'-kis-is) [ἔκχυσις, a pouring out]. Any skin-disease characterized by effusion into the dermal tissue.
**ecclasis** (ek-la'-sis) [ἔκκλᾶν, to break in pieces]. A breaking away, as of a small piece of bone from a larger piece.
**ecclisis** (ek-li'-sis) [ἔκκλίνειν, to turn aside]. 1. Dislocation. 2. The displacement of fractured bones.
**ecclysis** (ek-li'-sis) [ἔκκλύζειν, to wash out]. A washing out by injections.
**eccope** (ek'-op-e) [ἐκκοπή, a cutting out]. Excision of a part, or the vertical division of the cranium by a saw or other means.
**eccorthatic** (ek-or-that'-ik) [ἐκ, out; κόρθυς, a heap]. Producing copious fecal discharge.
**eccrinology** (ek-rin-ol'-o-je) [ἐκρίνειν, to secrete; λόγος, science]. The science of secretion, including its physics, physiology, and pathology.
**eccrisiology, eccrisionomy** (ek-kris-e-ol'-o-je, -on'-om-e). See eccrinology.
**eccrisis** (ek'-ris-is) [ἐκκρίνειν, to expel]. The expulsion of waste or morbid products; excretion.
**eccritic** (ek-rit'-ik) [ἐκκριτικός, secretive]. 1. A medicine promoting excretion. 2. Promoting excretion.
**eccyesis** (ek-si-e'-sis) [ἐκ, out; κίησις, pregnancy]. Extra-uterine gestation.
**eccyliosis** (ek-si-le-o'-sis) [ἐκ, out; κυλίνδειν, to roll]. Any disorder of development.
**ecdemic** (ek-dem'-ik) [ἐκδήμος, away from home]. Applied to diseases originating in a distant locality; not endemic.
**ecdemiomania, ecdemomania** (ek-de-me-o-ma'-ne-ah, ek-de-mo-ma'-ne-ah) [ἐκδήμος, away from home; μανία, madness]. Insanity marked by a desire for wandering.
**ecdemionosus** (ek-de-me-on'-o-sus) [ἐκδημέων, to wander; νόσος, disease]. Ecdemiomania.
**ecderon** (ek'-der-on) [ἐκ, out; δέρος, skin]. 1. That layer of skin or mucous membrane that lies outside of the enderon. 2. The outermost or epithelial layer of skin or mucous membrane.
**ecderonic** (ek-der-on'-ik) [ἐκ, out; δέρος, skin]. Belonging to or the outer layer of the ecderon.
**ecdysis** (ek'-dis-is) [ἐκδύειν, to cast off]. Sloughing or casting off of the skin; desquamation.
**ecgonine** (ek'-go-nēn). An alkaloid produced in the decomposition of cocaine by HCl.
**echafolta** (ek-af-ol'-tah). A proprietary antiseptic and alterative; said to be a purified echinacea.
**echidnase** (ek-id'-nās) [ἔχιδνα, viper]. A phlogogenic principle found in snake-venom.
**echidnin** (e-kid'-nin) [ἔχιδνα, viper]. 1. Serpent-poison; the poison or venom of the viper and other similar serpents. 2. A nitrogenous and venomous principle found in the poison-secretion of various serpents.
**echidnotoxin** (ek-id-no-toks'-in). A principle of snake-venom having a general action and a powerful effect on the nervous system.
**Echinacea** (ek-in-a'-se-ah) [ἔχινος, a hedgehog]. A genus of coarse composite plants of N. America. E. angustifolia. Black Sampson. The root of a perennial herb growing in the U. S. It is claimed that it possesses marked alterative value in strumous and syphilitic conditions. Dose of the fluidextract ♏ xv–xxx.
**echinate** (ek'-in-āt) [echinatus, prickly, from echinus, a hedgehog]. Beset with prickles.
**echinococcosis** (ek-in-o-kok-o'-sis). Infection with echinococci.

**echinococcotomy** (ek-in-o-kok-ot'-o-me) [echinococcus; τέμνειν, to cut]. The Posadas-Bobrow operation, consisting in the evacuation of echinococcus-cysts and closure of the cavity by suture.
**echinococcus** (ek-in-o-kok'-us) [ἐχῖνος, a hedgehog; κόκκος, a berry]. 1. The scolex or larval stage of the Tænia echinococcus. 2. Hydatid. e.-cyst. See cyst, echinococcus.
**echinodermatous, echinodermous** (ek-in-o-der'-mat-us, -mus). Having a spiny surface.
**echinol** (ek'-in-ol). A proprietary alterative containing echinacea.
**Echinops** (ek'-in-ops) [ἐχῖνος, a hedgehog; ὤψ, appearance]. A genus of composite plants. E. sphærocephalus, a European species, is laxative and diuretic, and contains an alkaloid, echinopsine, similar in action to brucine and strychnine.
**Echinorhyncus** (e-ki-no-rin'-kus) [ἐχῖνος, hedgehog; ῥύγχος, beak]. A worm parasitic within certain animals and occasionally found in man. E. gigas is the best-known species. E. hominis is smaller, and is perhaps an immature form of the other.
**echinulate** (e-kin'-ū-lāt) [echinulus, dim. of echinus, a hedgehog]. 1. Beset with prickles of small size. 2. Bacterial cultures showing spinous projections.
**echitamine** (ek-it'-am-ēn). See ditaine.
**echitenine** (e-kit'-en-ēn), C₂₀H₂₇NO₄. An amorphous, brown alkaloid of Dita, soluble in alcohol and water.
**echma** (ek'-mah) [ἔχμα, a stoppage: pl., echmata]. A stoppage or obstruction.
**echmasis** (ek'-mas-is) [ἔχμάζειν, to hinder: pl., echmases]. An obstruction or an obstructive disease.
**echmatic** (ek-mat'-ik) [ἔχμάζειν, to hinder]. Due to or marked by an echmasis or an echma.
**echo** (ek'-o) [ἠχώ, a sound]. A reverberated sound. e., amphoric, a vocal resonance in which the transmitted voice sounds as if it were speaking into a narrow-necked bottle. **e.-sign**, a symptom of epilepsy and other brain-conditions in which there is a repetition of the closing word or words of a sentence. It is regarded as the result of perverted will, or impaired or defective inhibition. **e.-speech**, a peculiar method of utterance in one type of hypnotism.
**echoacousia** (ek-o-ah-koo'-se-ah) [echo; ἀκουσία, hearing]. The subjective sensation of hearing echoes after sounds heard normally.
**echoacousia** (ek-o-ah-koo'-se-ah) [echo; ἀκουσία, hearing]. The subjective sensation of hearing echoes after sounds heard normally.
**echographia** (ek-o-graf'-e-ah) [echo; γράφειν, to write]. A form of aphasia in which printed or written questions submitted to the patient are copied without ability to comprehend the inquiry; also, in writing, the last word or letter is repeated.
**echokinesia, echokinesis** (ek-o-kin-e'-se-ah, -e'-sis) [ἠχώ, ἠχησις, motion]. Imitative unwilled action, like that observed in palmus or latah. See palmus and habit-spasm.
**echolalia** (ek-o-la'-le-ah) [echo; λαλιά, babble]. A meaningless repetition, by a person, of words spoken to him by others.
**echolalus** (ek-o-la'-lus) [see echolalia]. A hypnotized person who repeats words heard without comprehension of their meaning.
**echomatism** (ek-om'-at-izm) [echo; ματίζειν, to strive to do]. The opposite of automatism. The mimicking condition produced in hypnotics when the hand is pressed on the vertex of the head.
**echometer** (ek-om'-et-er) [ἠχώ, sound; μέτρον, a measure]. A stethoscope.
**echopathy** (ek-op'-ath-e) [ἠχώ, echo; πάθος, disease]. Any automatic and purposeless repetition of a word or sound heard or of an act seen.
**echophony** (ek-of'-o-ne) [echo; φωνή, voice]. An echo of a vocal sound in auscultation of the chest.
**echophotony** (ek-o-fot'-o-ne) [echo; φῶς, light; τόνος, one]. The production of the sensation of color by the stimulus of aerial waves, or sound. See phonism, photism.
**ecnophrasia** (ek-o-fra'-ze-ah) [echo, φράσις, speech].
Same as echolalia.
**echopraxis** (ek-o-praks'-is) [echo; πρᾶξις, a doing]. The needless continuance, by an insane patient, of some maneuver initiated by the physician in the course of examining the patient.
**echos** (e'-kos) [echo]. Any subjective sensation, as of a sound that has no objective cause.
**echoscope** (ek'-o-skōp) [echo; σκοπεῖν, to examine]. A stethoscope.

ECHOSCOPIA 312 ECTENTAL

**echoscopia** (ek-o-sko'-pe-ah) [echo; σκοπείν, to view]. Auscultation.

**echo-speech** (ek'-o-spēch). Same as *echolalia*.

**echuja** (ek'-ū-jah). An apocynaceous plant, *Adenium bahmianum*, of Africa. It is extremely poisonous.

**echujin, echugin** (ek'-ū-jin). A poisonous glucoside from the plant called *echuja*; it has much the same effects as strophanthin.

**eciomania** (ek-e-o-ma'-ne-ah). See *oikiomania*.

**Eck fistula** [Gottlieb Wilhelm Eck, German physician, 1795–1848]. An artificially made communication between the portal vein and the vena cava inferior.

**Ecker's gyrus**. [Alexander Ecker, German anatomist, 1816–1887]. The gyrus descendens, the most posterior of the occipital convolutions. **E.'s sulcus**, the anterior or transverse occipital sulcus, usually joined to the horizontal part of the interparietal sulcus.

**eclabium** (ek-la'-be-um) [ἐκ, out; labium, a lip]. An eversion of the lip.

**eclampsia** (ek-lamp'-se-ah) [ἐκλάμπειν, to shine or burst forth]. 1. A convulsive or epileptiform seizure occurring in women during pregnancy, labor, or the puerperium. 2. Any convulsive or epileptiform seizure, especially one in which consciousness is not lost. **e., cerebral**, a form in which the irritation is presumed to originate in the brain, as distinguished from uterine eclampsia. **e., infantilis**, a reflex convulsion of childhood. **e. nutans**, an affection characterized by paroxysms, in which the head and upper part of the body are bowed forward several times in succession; the attacks are accompanied by disordered consciousness. Syn., *nodding spasm*; *salaam convulsion*. **e., puerperal**, a convulsion occurring toward the close of pregnancy or during or after labor, believed to be caused by the irritation of the vasomotor centers by retained excrementitious substances. **e. tardissima**, that occurring several days or as long as eight weeks after parturition.

**eclampsism** (ek-lamp'-sism) [eclampsia]. Bar's name for eclampsia without convulsions.

**eclamptic** (ek-lamp'-tik) [eclampsia]. Relating to, or affected with, or of the nature of, eclampsia.

**eclamptism** (ek-lamp'-tism). The morbid condition produced by the retention of various toxic principles and autointoxication, all dependent upon the state of pregnancy; it is prone to result in convulsions, but may show only prodromes, such as headache, impairment of vision, etc.

**eclectic** (ek-lek'-tik). [ἐκλέγειν, to select]. Pertaining to a choosing or selection. Applied by a certain school of physicians to themselves, to denote their principle or plan of selecting or choosing that which they consider good from all other schools.

**eclecticism** (ek-lek'-tis-izm) [ἐκλέγειν, to select]. The doctrine and practice of the eclectics.

**eclegm, eclegma, ecleigma**, (ek'-lem, ek-leg'-mah, ek-lig'-mah) [ἐκ, out; λείχειν, to lick; pl., eclegmata]. An electuary.

**eclimia** (ek-lim'-e-ah). Same as *bulimia*.

**eclipsis** (ek-lip'-sis) [ἔκλειψις, a dying out]. A sudden failure; trance; catalepsy; a sudden and transient loss or impairment of consciousness.

**eclysis** (ek'-lis-is) [ἔκλυσις, a release; a loosening]. Any loosening, as of the bowels. Also a slight amount of, or merely a tendency to, syncope. It is present in anemia of the brain.

**ecmetropia** (ek-me-tro'-pe-ah). See *ametropia*.

**ecmnesia** (ek-ne'-ze-ah) [ἐκ, out; μνῆσις, remembrance]. A gap in memory; amnesia in which there is normal memory to a certain date and loss of memory for a period after it.

**ecnea** (ek-ne'-ah) [ἐκ, out; νοῦς, mind]. Insanity.

**ecoid** (e'-koid) [οἶκος, house]. A blood-shadow; the colorless stroma or framework of red corpuscles of the blood that have been deprived of their hemoglobin; a shadow-corpuscle.

**ecology** (e-kol'-o-je) [οἶκος, a house, family; λόγος, science]. In biology, the science of vegetable and animal economy and activity as shown by their modes of life, e. g., socialism, parasitism. "The terms biology and ecology are not interchangeable, because the latter only forms part of physiology." (Haeckel.)

**economy** (e-kon'-o-me) [οἶκος, house; νόμος, a law]. A general name for the human being considered as a whole. **e., animal**, that of an animal organism. **e., medical**, the rules regulating the practice of medicine and surgery.

**ecophony** (ek-of'-on-e) [echo; φωνή, sound]. An echo immediately following vocal sounds, heard in acute congestion of the lungs.

**ecostate** (e-kos'-tāt) [e, priv.; costa, a rib]. Without ribs.

**écouvillon** (a-koo-ve'-yon(g)) [Fr.]. See *écouvillonage*.

**écouvillonage** (a-koo-ve-yon-ahzh') [Fr.]. The operation of cleansing, and carrying medicinal agents to, the inside of the uterus by means of a swab or brush.

**ecphlysis** (ek'-flis-is) [ἐκφλύζειν, to burst out: pl., ecphlyses]. Any vesicular eruption.

**ecphractic** (ek-frak'-tik) [ἐκφρακτικός, clearing obstruction]. 1. Removing obstructions. 2. An ecphractic medicine.

**ecphronia** (ek-fro'-ne-ah) [ἐκ, out of; φρήν, mind]. Insanity.

**ecphyadectomy** (ek-fi-ad-ek'-to-me) [ἐκφύας, appendage; ἐκτομή, excision]. Excision of the vermiform appendix.

**ecphyaditis** (ek-fi-ad-i'-tis) [ἐκφύας, appendage; ἴτις, inflammation]. Inflammation of the vermiform appendix; appendicitis. This term has also been used to include typhlitis, perityphlitis, etc.

**ecphyas** (ek'-fi-as) [ἐκφύας, an offshoot]. The vermiform appendix.

**ecphyma** (ek-fi'-mah) [ἐκ, out; φύεσθαι, to grow]. An excrescence on the skin. **e. globulus**, a contagious disease of Ireland marked by the formation, on the skin, of tubercles which soften and form raspberry-like tumors.

**ecphysesis** (ek-fiz-e'-sis) [ἐκ, out; φυσάειν, to blow]. Rapid breathing.

**ecpyesis** (ek-pi-e'-sis) [ἐκ, out; πυεῖν, to suppurate: pl., ekpyeses]. 1. Any suppuration or abscess. 2. Any pustular skin-disease.

**ecpyetic** (ek-pi-et'-ik) [ἐκ, out; πυεῖν, to suppurate]. Promoting suppuration.

**ecptoma** (ek-to'-mah) [ἐκ, out; πίπτειν, to fall: pl., ecptomata]. Any falling of a part or organ.

**écrasement** (ā-krahs-mon(g)) [F., "a crushing"]. The removal of a part by means of an écraseur.

**écraseur** (a-krah-zer) [see *écrasement*]. An instrument consisting of a chain or wire loop which is placed about a projecting part, and, by being tightened, gradually cuts through the tissues. **e., galvanic**, one constructed so that the wire loop can be heated to redness while in use, by the passage through it of an electric current.

**ecrodactylia** (ek-ro-dak-til'-e-ah) [ἐκρος, escape; δάκτυλος, digit]. Same as *ainhum*.

**ecsomatic** (ek-so-mat'-ik) [ἐκ, out; σῶμα, body]. Relating to ecsomatics or to material removed from the body, as pus, urine, etc.

**écsomatics** (ek-so-mat'-iks). That department of medicine included in clinical laboratory methods; so called because all the material dealt with is removed from the body and examined elsewhere.

**ecsomatist** (ek-so'-mat-ist). An individual who is versed in clinical laboratory methods.

**ecstaltic** (ek-stal'-tik) [ἐκ, out; στέλλειν, to send]. Sent out from a nerve-center; applied specially to nerve-impulses originating from the spinal cord.

**ecstasis** (ek-sta'-sis). See *ecstasy*.

**ecstasy** (eks'-ta-se) [ἔκστασις, a trance]. A derangement of the nervous system characterized by an exalted visionary state, absence of volition, insensibility to surroundings, a radiant expression, and immobility in statuesque positions.

**ecstrophy** (ek'-stro-fe). Same as *exstrophy*.

**ectacolia, ectacoly** (ek-ta-ko'-le-ah, ek'-ta-ko-le) [ἐκτακός, capable of stretching; κόλον, the colon]. Congenital dilation of a more or less extensive section of the colon.

**ectad** (ek'-tad) [ἐκτός, external; ad, to]. On or toward the ectal part.

**ectal** (ek'-tal) [see *ectad*]. At some surface or aspect farther from a supposed center than that with which a given object is compared; external; superficial.

**ectasia, ectasis** (ek-ta'-ze-ah, ek'-ta-sis) [ἔκτασις, extension]. Distention; dilatation.

**ectasin** (ek'-tas-in) [see *ectasia*]. A substance isolated from tuberculin, which causes dilatation of the vessels.

**ectatic** (ek-tat'-ik) [see *ectasia*]. Distended or dilated.

**ectental** (ek-ten'-tal) [ἐκτός, outward; ἐντός, inward]. Pertaining to the line of union between the

ectoderm and the entoderm. **e. line**, the line of junction of the ectoderm and entoderm.
**ectethmoid** (*ek-leth'-moid*) [*ecto-*; *ethmoid*]. Either one of the lateral cellular masses of the ethmoid bone.
**ecthol** (*ek'-thol*). A proprietary remedy said to contain the active principles of *Echinacea angustifolia* and *Thuja occidentalis*; it is antipurulent and antimorbific. Dose 1 dr. (4 Gm.) 3 times daily.
**ecthyma** (*ek-thi'-mah*) [ἔκθυμα, a pustule]. An inflammatory skin disease attended with an eruption of large, flat, superficial pustules. They vary in size from a ten-cent to a twenty-five-cent piece, and are surrounded by a distinct inflammatory areola. The eruption appears, as a rule, on the legs and thighs where the hairs are thick; it occurs in crops, and may persist for an indefinite period. **e. gangrænosum, e., gangrenous,** a form marked by the appearance of brown discolorations of the skin, usually surrounded by a halo; the center of these efflorescences rapidly becomes necrotic. It is due to *bacillus pyocyaneus*. **e., syphilitic.** See *rupia*.
**ecthymiform** (*ek-thi'-mif-orm*) [ἔκθυμα, a pustule; *orma*, form]. Resembling ecthyma.
**ecthyreosis, ekthyrosis** (*ek-thi-re-o'-sis*, *ek-thi-ro'-sis*). See *athyreosis*.
**ectillotic** (*ek-til-ot'-ik*) [ἐκ, out; τίλλειν, to pluck]. Depilatory; causing the hairs to fall; removing corns from the feet.
**ectiris** (*ek-ti'-ris*) [*ecto-*; *iris*]. That part of Descemet's membrane that lies in front of the iris.
**ecto-** (*ek-to-*) [ἐκτός, without]. A prefix signifying without, upon the outer side.
**ectobatic** (*ek-to-bat'-ik*) [*ecto-*; βαίνειν, to go]. Efferent; centrifugal; moving ectad or distad.
**ectoblast** (*ek'-to-blast*) [*ecto-*; βλαστός, a bud]. The outside membrane of a cell.
**ectocardia** (*ek-to-kar'-de-ah*) [*ecto-*; καρδία, the heart]. An abnormal position of the heart. **e. abdominalis,** a malformation in which the heart is wholly within the abdomen or within a sac in the precordia. **e. cephalica, e. cervicalis,** a form in which the heart is at the base of the neck. **e. extrathoracica,** that in which the heart is external to the thoracic cavity. **e. intrathoracica,** that in which the heart is inside the thorax. **e. pectoralis,** that in which the heart lies in front of the chest.
**ectocentral** (*ek-to-sen'-tral*) [*ecto-*; *central*]. Near to the center and to the external surface.
**ectochoroidea** (*ek-to-ko-roid'-e-ah*). The outer layer of the choroid.
**ectocinerea** (*ek-to-sin-e'-re-ah*) [*ecto-*; *cinereus*, ashy]. The gray substance of the cortex of the brain. Cf. *entocinerea*.
**ectocnemial** (*ek-to-ne'-me-al*) [*ecto-*; κνήμη, the leg]. Located on the external aspect of the fibula.
**ectocolostomy** (*ek-to-ko-los'-to-me*) [*ecto-*; *colostomy*]. A surgical operation upon the colon to establish an external opening.
**ectocondylar, ectocondyloid** (*ek-to-kon'-dil-ar, -oid*). Relating to an ectocondyle.
**ectocondyle** (*ek-to-kon'-dil*) [*ecto-*; *condyle*]. An external condyle.
**ectocornea** (*ek-to-kor'-ne-ah*) [*ecto-*; *cornea*]. The corneal conjunctiva.
**ectocuneiform** (*ek-to-ku-ne'-e-form*) [*ecto-*; *cuneiform*]. 1. Relating to the outer cuneiform bone of the foot. 2. The outer cuneiform bone.
**ectoderm** (*ek'-to-derm*) [*ecto-*; δέρμα, skin]. The outer of the two primitive layers of the embryo; the epiblast.
**ectodermal, ectodermic** (*ek-to-der'-mal, -mik*) [see *ectoderm*]. Relating to the ectoderm; applied to structures derived from the upper epithelial layers of the derma, as hair, chitin, enamel, etc.
**ectoentad** (*ek-to-en-tad*) [*ecto-*; *ἐντός*, within; *ad*, to]. From without inward.
**ectogastrocnemius** (*ek-to-gas-trok-ne'-me-us*). The gastrocnemius externus muscle. See *muscles, table of*.
**ectogenous** (*ek-toj'-en-us*) [*ecto-*; γενᾶν, to produce]. Capable of growth outside of the body; applied especially to bacteria and other parasites.
**ectoglobular** (*ek-to-glob'-u-lar*). Formed outside the blood-globules.
**ectogluteus** (*ek-to-glu-te'-us*). The external gluteus muscle, or gluteus maximus. See *muscles, table of*.
**ectokelostomy** (*ek-to-kel-os'-to-me*) [*ecto-*; κήλη, hernia; στόμα, a mouth]. Vitrac's operation, by which the sac of an infected inguinal hernia is kept open with drainage, the whole being displaced through a counteropening in the abdominal wall, the hernia being then cured radically.
**ectolecithal** (*ek-to-les'-ith-al*) [*ecto-*; λέκιθος, yolk]. In embryology, applied to such eggs as have the formation-yolk enclosed in a superficial layer of food-yolk.
**ectoloph** (*ek'-to-lof*) [*ecto-*; λοφος, ridge]. The external ridge of the upper molar teeth of the horse.
**ectomarginal** (*ek-to-mar'-jin-al*). Situated on the external aspect and near the margin.
**ectomere** (*ek'-to-mēr*) [*ecto-*; μέρος, a share]. Any one of the cells of the ovum that are destined to take part in forming the ectoderm.
**ectomia** (*ek-to'-me-ah*) [ἐκ, out; τέμνειν, to cut]. Excision, amputation.
**-ectomy** [ἐκ, out; τέμνειν, to cut]. A suffix meaning a cutting out.
**ectopagia** (*ek-to-pa'-je-ah*) [*ecto-*; πάγος, a fixture]. The condition of being ectopagous; an ectopagous monstrosity.
**ectopagous** (*ek-top'-ag-us*) [*ecto-*; πάγος, a fixture]. Of the nature of or pertaining to an ectopagus.
**ectopagus** (*ek-top'-ag-us*) [*ecto-*; παγείς, united]. A twin monstrosity united laterally the full extent of the thorax.
**ectoparasitic** (*ek-to-par-as-it'-ik*) [*ectoparasite*]. Of the nature of or pertaining to an ectoparasite.
**ectoparasite** (*ek-to-par'-as-īt*) [*ecto-*; *parasite*]. A parasite that lives on the exterior of its host.
**ectopectoral** (*ek-to-pek'-tor-al*). The outer of the two pectoral muscles; the pectoralis major.
**ectoperitonitis** (*ek-to-per-it-on-i'-tis*) [*ecto-*; *peritonitis*]. Inflammation of the attached side of the peritoneum.
**ectophyte** (*ek'-to-fīt*) [*ecto-*; φυτόν, a plant]. An external parasitic plant-growth; a vegetable parasite on the skin.
**ectophytic** (*ek-to-fit'-ik*) [*ecto-*; φυτόν, a plant]. Of the nature of or pertaining to an ectophyte.
**ectopia** (*ek-to'-pe-ah*) [ἔκτοπος, displaced]. An abnormality of position, usually congenital. **e. ani,** prolapse of the anus. **e. bulbi.** See *e. oculi.* **e. cordis.** See *ectocardia.* **e. lentis,** dislocation or congenital malposition of the crystalline lens. **e. oculi,** abnormal position of the eyeball in the orbit. **e., pupillæ.** See *corectopia.* **e. renis,** floating kidney. **e. testis,** abnormal position of the testicle. **e. vesicæ,** protrusion of the bladder through the wall of the abdomen.
**ectopic** (*ek-top'-ik*) [*ectopia*]. In an abnormal position. **e. gestation,** extrauterine gestation.
**ectoplasm** (*ek'-to-plasm*) [*ecto-*; πλάσσειν, to form]. The outer, hyaline, more compact layer of protoplasm of a cell or unicellular organism.
**ectoplasmatic** (*ek-to-plaz-mat'-ik*). See *ectoplastic*.
**ectoplastic** (*ek-to-plas'-tik*). Relating to ectoplasm; applied to cells in which the ectoplasm is undergoing changes.
**ectopocystic** (*ek-to-po-sist'-ik*). Relating to ectopocystis.
**ectopocystis** (*ek-to-po-sist'-is*) [*ectopia*; κύστις, the bladder]. Displacement of the bladder.
**ectopotomy** (*ek-to-pot'-o-me*) [*ectopia*; τέμνειν, to cut]. Laparotomy for the removal of the contents of an extrauterine gestation-sac.
**ectopy** (*ek'-to-pe*). Same as *ectopia*.
**ectorbital** (*ekt-orb'-it-al*). Relating to the temporal part of the orbits.
**ectoretina** (*ek-to-ret'-in-ah*) [*ecto-*; *retina*]. The external and pigmentary layer of the retina.
**ectorganism** (*ekt-or'-gan-izm*). An organism, external to another. Cf. *ectoparasite*.
**ectosac** (*ek'-to-sak*) [*ecto-*; σάκκος, a sac]. The limiting membrane of an ovum.
**ectosarc** (*ek'-to-sark*) [*ecto-*; σάρξ, flesh]. The outer layer of protozoa; same as *ectoplasm, q. v.*
**ectoskeletal** (*ek-to-skel'-et-al*). Relating to the exoskeleton; exoskeletal.
**ectoskeleton** (*ek-to-skel'-et-un*). Same as *exoskeleton*.
**ectospore** (*ek'-to-spōr*). See *exospore*.
**ectosteal** (*ek-tos'-te-al*) [*ecto-*; ὀστέον, a bone]. Relating to, situated or occurring outside of, a bone.
**ectosteomyces** (*ekt-os-te-o-mi'-sēz*) [*ecto-*; ὀστέον, a bone; μύκης, a fungus]. A fungous newgrowth from a bone.
**ectostosis** (*ek-tos-to'-sis*) [*ecto-*; ὀστέον, a bone]. The growth of bone from without; ossification that begins at the perichondrium, or future periosteum.
**ectothalamus** (*ek-to-thal'-am-us*) [*ecto-*; *thalamus*]. The external medullary layer of the thalamus.

# ECTOTHRIX 314 EDEMATOSCHEOCELE

**ectothrix** (ek'-to-thriks) [ecto-; θρίξ, hair]. An organism parasitic upon the hair. Cf. trichophyton.
**ectotoxemia** (ek-to-toks-e'-me-ah). Toxemia due to an external cause.
**ectotoxin** (ek-to-toks'-in). Same as exotoxin.
**ectotrochanter** (ek-to-tro-kan'-ter) [ecto-; trochanter]. The greater trochanter.
**ectozoon** (ek-to-zo'-on) [ecto-; ζῶον, an animal: pl., ectozoa]. An external animal parasite; an ectoparasite.
**ectrodactylia, ectrodactylism** (ek-tro-dak-til'-e-ah, ek-tro-dak'-til-izm) [ἔκτρωμα, abortion; δάκτυλος, finger]. Congenital absence of any of the fingers or toes.
**ectrogenic** (ek-tro-jen'-ik) [ἔκτρωμα, abortion; γεννᾶν, to produce]. Due to some loss of tissue, chiefly congenital.
**ectrogeny** (ek-troj'-en-e) [ἔκτρωμα, abortion; γεννᾶν, to produce]. Loss or congenital absence of any part or organ.
**ectroma** (ek-tro'-mah) [ἔκτρωμα, abortion: pl., ectromata]. An aborted ovum or fetus.
**ectromelus** (ek-trom'-el-us) [ἔκτρωσις, abortion; μέλος, a limb]. A single autositic monster characterized by the presence of imperfectly formed limbs.
**ectropia** (ek-tro'-pe-ah). See exstrophy. e., intestinal. See adenoma, umbilical.
**ectropic** (ek-trop'-ik). Turned out or everted.
**ectropion** (ek-tro'-pe-on) [ἐκ, out; τρέπειν, to turn]. Eversion of a part, especially of an eyelid.
**ectropionization** (ek-tro-pe-on-iz-a'-shun). Inversion of the upper eyelid and exposure of the conjunctiva to facilitate therapeutic manipulation.
**ectropionize** (ek-tro'-pe-on-iz) [ectropion]. To produce, by operation, the condition of ectropion.
**ectropium** (ek-tro'-pe-um). Same as ectropion.
**ectropodism** (ek-trop'-od-izm) [ἔκτρωμα, abortion; πούς, foot]. Congenital absence of one or more toes.
**ectrosis** (ek-tro'-sis). An abortion, or the production of an abortion.
**ectrotic** (ek-trot'-ik). Tending to cut short; preventing the development of disease; abortive; abortifacient.
**ectylotic** (ek-til-ot'-ik) [ἐκ, away; τύλος, callus]. Tending to remove warts or indurations.
**eczema** (ek'-ze-mah) [ἐκζεῖν, to boil over]. Tetter; an acute or chronic, noncontagious, inflammatory disease of the skin, characterized by multiformity of lesions, and the presence, in varying degrees, of itching, infiltration, and discharge. The skin is reddened, the redness shading off insensibly into the surrounding unaffected parts. e. arthriticum, a vesicular form occurring about gouty joints. e. erythematosum, the mildest form of eczema, in which the skin is reddened and slightly swollen. e. fissum, a form affecting the hands and skin over the articulations, and characterized by the formation of deep, painful cracks or fissures. e. hypertrophicum, a form characterized by permanent hypertrophy of the papillae of the skin, giving rise to general or limited warty outgrowths. e., lichenoid, that marked by thickening of the epidermis. e. madidans, a form characterized by large, raw, weeping surfaces studded with red points. It follows e. vesiculosum. Syn., eczema rubrum. e. marginatum, the most severe form of ringworm of the body. Its seats are the groin, axilla, crotch, and occasionally the popliteal space. Its lesion is marked by a well-defined, festooned, raised margin. e. papulosum, a variety associated with the formation of minute papules of a deep-red color and firm consistence, and accompanied by intense itching. e. pustulosum, the stage of eczema characterized by the formation of pustules. e. rubrum. See e. madidans. e. seborrhoicum. Synonym of seborrhea. e. solare, that form due to irritation from the rays of the sun. e. squamosum, a variety characterized by the formation of adherent scales of shed epithelium. e. sudamen, e. sudorale, that due to excess of perspiration. e. sycomatosum, e. sycosiforme, a pustular form occurring on the hairy parts and affecting the hair-follicles. e. tyloticum, a form occurring on the palmar aspect of the hands and fingers and attended with callosity. e. vesiculosum, an eczema characterized by the presence of vesicles.
**eczematization** (ek-ze-mat-i-za'-shun). A condition of the skin marked by persistent eczema-like lesions, due to continued injury from scratching.
**eczematoid** (ek-zem'-at-oid). Resembling an eczema.
**eczematosis** (ek-zem-at-o'-sis) [pl., eczematoses]. Any eczematous skin-disease.
**eczematous** (ek-zem'-at-us) [eczema]. Of the nature of or affected with eczema.
**eczemine** (ek'-zem-ēn). A white, crystalline substance, soluble in water, feebly alkaline in reaction, extracted from the urine in eczema; toxic.
**eczemogenous** (ek-zem-oj'-en-us) [eczema; γεννᾶν, to produce]. Giving rise to eczema.
**Eddyism** (ed'-e-izm). A form of faith-cure propagated, under the name of Christian Science, by an American woman, Mary Patterson Baker Glover Eddy, known to her followers as "Mother Eddy."
**edea, ædœa** (e-de'-ah) [αἰδοῖα, the genitals]. The genital organs, particularly the external genitals.
**edeagra, ædœagra** (e-de-a'-grah) [edea; ἄγρα, a seizure]. Pain or gout in the genitalia.
**edeatrophia, ædœatrophia** (e-de-at-ro'-fe-ah) [edea; ἀτροφία, a wasting]. Atrophy or wasting of the genital organs.
**edeauxe, ædœauxe** (e-de-awks'-e) [edea; αὔξη, increase]. Swelling or hypertrophy of the genitals.
**Edebohls' operation** (ed'-e-bōls) [George Michael Edebohls, American surgeon, 1853–1908]. Decapsulation or decortication, of the kidney. E.'s position, or posture. See Simon's posture.
**edeitis, ædœitis** (e-de-i'-tis) [edea; ιτις, inflammation]. Inflammation of the external genitals.
**edema, œdema** (e-de'-mah) [οἴδημα; οἰδεῖν, to swell]. An infiltration of serum in a part. e., acute. See œ. calidum. œ., angioneurotic. See angioneurotic edema. e., blue, edema with cyanosis, seen in hysterical paralysis accompanied with pain. œ. calidum, that due to a serous exudation; it is sudden in its onset and resembles acute inflammation. œ. capitis, a serous effusion into the subcutaneous areolar tissue of the scalp. œ. cardiaca (of the kidney), the change in the kidneys due to passive congestion in consequence of heart disease. e., cerebral. See hydrocephalus. e., collateral, the serous infiltration of the tissue encircling an inflamed part. e., compact (of infants), a variety of scleroderma neonatorum in which the skin is edematous. Syn., scleroderma œdematosa. e., cretinoid. See myxedema. œ. ex vacuo, edema of a part to counteract the tendency to a vacuum caused by atrophy of some neighboring part. e., febrile purpuric, localized edema accompanying an eruption of purpura urticans about the joints, and rheumatic fever. œ. frigidum, a chronic swelling, cold to the touch and painless. œ. fugax, edema due to atmospheric changes occurring in the face, eyelids, and neck of chlorotic patients. e., glottidial. See laryngeal edema. e., infectious. See e., malignant. e., inflammatory, a serous infiltration into inflamed tissue. e., Iwanoff's. See under Iwanoff. e., laryngeal. See laryngeal edema. e., malignant, an edematous inflammation that occurs at times after serious injuries, and is characterized by its rapid spread, the speedy destruction of the tissue involved, and the formation of gas. It is due to the bacillus of malignant edema. e., neuroparalytic, that due to paralysis of the vasomotor nerves or to neuroparalytic congestion. œ. oculi. See hydrophthalmia. œ. œdematodes. See œ. frigidum. e., paroxysmal pulmonary, a rare form of edema of the lungs marked by rapid onset, imminent asphyxia, and copious albuminous expectoration. The attack, lasting from a few minutes to some days, may terminate fatally or the symptoms may disappear. œ. puerperarum, phlegmasia alba dolens. e., purulent, a purulent infiltration in which there is a great deal of fluid. e., retinal, the development of irregular spaces filled with transparent fluid at the periphery of the retina occurring after middle age. œ. scleroticum, edema attended with distention.
**edemamycosis, œdemamycosis** (e-de-mah-mi-ko'-sis) [edema; μύκης, fungus]. The name applied by Edington to an ectogenous infective disease, commonly referred to as African horse-sickness; it is characterized by intense congestion of the blood-vessels with consequent edema of the lungs and at times of the subcutaneous tissues of the head and neck.
**edemania, ædœmania** (e-de-ma'-ne-ah). See nymphomania.
**edematization** (e-dem-at-iz-a'-shun) [edema]. Edema of the tissues produced by the injection of a 2 % salt solution at a temperature lower than that of the body.
**edematoscheocele** (e-dem-at-os-ke'-o-sēl) [edema;

# EDEMATOUS 315 EHRLICH'S ANEMIA

δσχη, the scrotum; κήλη, a tumor]. Edematous oscheocele.

**edematous** (*e-dem'-at-us*). Pertaining to or characterized by edema.

**edemerysipelas** (*e-dem-er-e-sip'-e-las*). Edematous erysipelas.

**edentate** (*e-den'-tate*) [*e*, priv.; *dens*, tooth]. Without teeth.

**edentation** (*e-den-ta'-shun*) [*e*, without; *dens*, a tooth]. A deprivation of teeth.

**edentulous** (*e-den'-tu-lus*) [*e*, without; *dens*, a tooth]. Without teeth (applied to one who has lost his teeth).

**edeodynia, ædœodynia** (*e-de-o-din'-e-ah*) [*edea*; ὀδύνη, pain]. Any pain in the genital organs.

**edeogargalismus, ædœogargalismus** (*e-de-o-gar-gal-is'-mus*) [*edea*; γαργαλισμός, a tickling]. Masturbation.

**edeography, ædœography** (*e-de-og'-ra-fe*) [*edea*; γράφειν, to write]. A description of the genitalia.

**edeology, ædœology** (*e-de-ol'-o-je*) [*edea*; λόγος, science]. A treatise or monograph on the organs of generation.

**edeomycodermatitis, ædœomycodermatitis** (*e-de-o-mi-ko-der-mat-i'-tis*) [*edea*; μύκος, mucus; δέρμα, skin; ιτις, inflammation]. Inflammation of the mucous membrane of any of the genital organs.

**edeomania, ædœomania** (*e-de-o-ma'-ne-ah*) [*edea*; μανία, madness]. Nymphomania; satyriasis.

**edeopsophy, ædœopsophy** (*e-de-op'-so-fe*) [*edea*; ψοφεῖν, to utter a noise]. The emission of sounds from the genital organs (as from the bladder or vagina).

**edeoptosis, ædœoptosis** (*e-de-op-to'-sis*) [*edea*; πτῶσις, a fall]. Prolapse of some portion of the genital apparatus.

**edeoscopy, ædœoscopy** (*e-de-os'-ko-pe*) [*edea*; σκοπεῖν, to inspect]. An inspection or professional examination of the genital organs.

**edeotomy, ædœotomy** (*e-de-ot'-o-me*) [*edea*-; τόμη, a cutting]. The anatomy or dissection of the genital organs.

**edestin** (*ed-est'-in*). The chief and characteristic protein of the seeds of sunflower, hemp, squash, and castor-oil bean.

**edible** (*ed'-ib-l*) [*edibilis*, eatable]. A qualification applied to food, the condition of which is good and wholesome.

**Edinger's law** (*ed'-ing-er*) [Ludwig *Edinger*, German anatomist, 1855– ]. A regular and gradual increase of function of a neuron leads at first to increased growth; if carried to excess, especially if irregular and spasmodic, it results in atrophy and degeneration, and ultimately in proliferation of the surrounding tissue. **E.'s nucleus**, the nucleus of the posterior longitudinal bundle, an aggregation of ganglion-cells in the gray matter of the third ventricle at the beginning of the Sylvian aqueduct.

**Edinger-Westphal's nucleus** (*ed'-ing-er-vest'-fahl*) [*Edinger*; Karl Friedrich Otto *Westphal*, German neurologist, 1833–1890]. One of the nuclei of the third cranial nerve in the region of the anterior corpora quadrigemina below the Sylvian aqueduct.

**edipism** (*ed'-ip-izm*) [*Œdipus*, King of Thebes, who put out his own eyes because he had killed his father unwittingly]. Selfinflicted injury to the eyes.

**Edsall's disease** (*ed'-sal*) [David Linn *Edsall*, American physician, 1869– ]. Heat cramp.

**educt** (*e'-dukt*) [*e*, out; *ducere*, to draw]. A compound that exists in any substance and is extracted from it by a chemical or pharmaceutical process; opposed to *product*.

**edulcorant** (*e-dul'-kor-ant*) [*edulcare*, to sweeten]. Sweetening; corrective of acidity or of acrimony.

**edulcoration** (*e-dul-kor-a'-shun*) [*edulcare*, to sweeten]. In chemistry, the act or process of sweetening; the removal of soluble or saline matters by washing.

**effector** (*ef-ek'-tor*). A name given by Sherrington to nerve endings in organs, glands or muscles, which are consequently called effector organs. The term is opposed to *receptor*.

**effeminacy** (*ef-em'-in-as-e*). See *feminism*.

**effemination** (*ef-em-in-a'-shun*) [*effeminare*, to make womanish]. The state of being effeminate.

**efferent** (*ef'-er-ent*) [*efferens*, carrying from.] Carrying away, as *efferent* nerves, nerves conveying impulses away from the central nervous system; of *blood-vessels*, conveying blood away from the tissues; of *lymphatics*, conveying lymph from the lymphatic glands.

**effervescent** (*ef-er-ves'-ent*) [*effervescere*, to boil up]. Susceptible of being made to effervesce, or to dissolve with foaming and with the escape of a gas, such as carbon dioxide.

**effervescing** (*ef-er-ves'-ing*) [*effervescere*, to boil up]. Giving off gas-bubbles; foaming. **e. powder**. See *Seidlitz powder*.

**effete** (*ef-et'*) [L., *effetus*]. Exhausted, worn out.

**effleurage** (*ef-lur-azh*) [Fr.]. In massage, the stroking movement.

**efflorescence** (*ef-lor-es'²ens*) [*efflorescere*, to bloom]. 1. The spontaneous conversion of a crystalline substance into powder by a loss of its water of crystallization. 2. The eruption of an exanthematous disease.

**effluent** (*ef'-lu-ent*) [*effluere*, to flow out]. An outflow. The fluid discharged from works for the treatment of sewage.

**effluvium** (*ef-lu'-ve-um*) [*effluere*, to flow out; pl., *effluvia*]. Any subtle emanation from a substance or person, especially one that is offensively odoriferous.

**efflux** (*ef'-fluks*), or **effluxion** (*ef-fluk'-shun*) [*effluxio*; *effluxus*, an outflow]. 1. An outflow; that which flows out. 2. Abortion, or the escape of the embryo from the uterus during a very early stage of pregnancy.

**effracture** (*ef-frak'-chūr*). Fracture of the cranium, with depression of one or more fragments.

**effumability** (*ef-u-ma-bil'-i-te*). Capacity for volatilization.

**effuse** (*ef-ūs'*) [*effusion*]. Spread out, said of a bacterial culture that is thin, and widely spreading.

**effusion** (*ef-u'-zhun*) [*effundere*, to pour out]. 1. A pouring-out, especially the pouring-out of blood or serum into the cellular tissues or the serous cavities. 2. The effused fluid. **e., pericardial**, an effusion into the pericardium. For signs of, see *Auenbrugger, Ewart, Rotch, Roth, Sansom*, and *Sibson*. **e., pleural**, an effusion into the pleura. For signs of, see *Baccelli, Kellock, Litten, de Mussey, Pitres, Sieur, Skoda*, and *Williams*.

**egagropilus** (*e-gag-rop'-il-us*) [αἰγάγροι, a goat; πῖλος, felt]. An intestinal concretion formed of hair.

**egertic** (*e-jer'-tik*) [ἐγείρειν, to awaken]. Causing wakefulness.

**egest** (*e-jest'*) [*egerere*, to discharge]. To void, as excrement; to defecate.

**egesta** (*e-jes'-tah*) [*egerere*, to cast out]. The discharges of the bowels or other excretory organs.

**egestion** (*e-jes'-chun*) [*egestio*, that which is voided]. Defecation; expulsion of excrements or excretion.

**egg** (*eg*). See *ovum*. **e.-albumin**, albumin, in white of egg, constituting about 60 % of the egg of the domestic fowl.

**egilops, ægilops** (*e'-jil-ops*) [αἰξ, a goat; ὤψ, eye]. Abscess, with perforation, at the inner canthus of the eye, supposed to be a result of lacrimal fistula. See *anchylops*.

**eglandular** (*e-glan'-dū-lar*) [*e*, priv.; *glandula*, a gland]. Destitute of glands.

**eglandulose** (*e-glan'-dū-lōs*) [*e*, priv.; *glandula*, a gland]. Same as *eglandular*.

**egmol** (*eg'-mol*). Trade name of an emulsion of egg and olive oil.

**egobronchophony** (*e-go-brong-koff'-o-ne*). A combination of egophony and bronchophony.

**egois** (*e-gois'*). Compounds of mercury with parasulphonic acid and a phenol. They are redbrown powders, soluble and emetic.

**egomania** (*eg-o-ma'-ne-ah*) [*ego*, I; μανία, madness]. Abnormal self-esteem.

**egophony** (*e-gof'-o-ne*) [αἰξ, a wild goat; φωνή, the voice]. A modification of bronchophony, in which the voice has a bleating character, like that of a kid. It is heard in pleurisy with slight effusion.

**egregorsis** (*eg-re-gor'-sis*) [ἐγείρειν, to wake]. Morbid wakefulness; insomnia.

**Egyptian chlorosis** (*e-jip'-shun*). See *ankylostomiasis*. **E. ophthalmia**. See *trachoma*.

**Ehrenritter's ganglion** (*air'-en-rit-er*). The jugular ganglion.

**Ehret's paralysis** (*air'-et*) [Heinrich *Ehret*, German physician, 1870– ]. A traumatic neurosis following injury to the inner side of the foot or ankle, consisting in spasmodic contracture of the muscles which raise the inner border of the foot and functional paralysis of the peroneal muscles.

**Ehrlich's anemia** (*air'-likh*) [Paul *Ehrlich*, German bacteriologist, 1854– ]. A plastic anemia. A rapidly progressing anemia with hyperplasia of the bone-marrow and hemorrhages into the mucous membranes. E.'s biochemical theory, the theory that a specific chemical affinity exists between specific living cells and specific chemical substances. E.'s method, the use of a saturated solution of anilin in water, as a mordant for better fixing the anilin dyes used in staining bacteria. E.'s method for the fixation of blood-films consists in boiling the specimen for one minute in a test-tube containing absolute alcohol. E.'s reaction, the treatment of the urine with diazobenzosulphuric acid produces a deep-red color that is due to a combination of the reagent with an aromatic amido-compound found in the urine in typhoid fever and pneumonia; frequently also in pleurisy, measles, tuberculosis, erysipelas, and peritonitis. To produce this reaction, equal parts of the reagent and urine are mixed and about one-eighth of their total volume of ammonia is added. The reagent consists of two solutions: (1) Sulphanilic acid, 1 Gm.; hydrochloric acid, 10 Cc.; distilled water, 200 Cc. (2) Sodium nitrite, 0.5 Gm.; distilled water, 100 Cc. E.'s side-chain theory, a theory based upon the phenomena of immunity, q. v., and of cytolysis, and serving to explain these. In this connection see *receptor, haptophore, haptin, amboceptor, uniceptor, complement, toxophore.* E.'s solution, a solution of a basic anilin dye in anilin oil and water.

**Ehrlich-Hata's "606"** (*air'-likh-hah'-tah*) [*Ehrlich;* S. *Hata,* Japanese physician]. Dioxydiamidoarsenobenzol; salvarsan. A synthetic compound considered a specific for syphilis. It is administered by intravenous or intramuscular injection. Dose 0.3 to 0.6 gm.

**Eichhorst's corpuscles** (*ik'-horst*) [Hermann Ludwig *Eichhorst,* Swiss physician, 1849– ]. Small, spherical blood-corpuscles found in pernicious anemia and formerly regarded as characteristic of this disease. E.'s neuritis, a form of neuritis in which the morbid process involves both the nervesheath and the interstitial tissue of the muscles. Syn., *neuritis fascians.* E.'s type, of progressive muscular atrophy, the femorotibial type.

**Eichstedt's disease** (*ik'-sted*) [Karl Ferdinand *Eichstedt,* German physician, 1816–1892]. Pityriasis versicolor.

**eidoptometry** (*i-dop-tom'-et-re*) [εἶδος, appearance; ὀπτεῖν, to see; μέτρον, measure]. The estimation of the a̱cuity of vision.

**eighth nerve.** The auditory nerve.

**eigon** (*i'-gon*). A compound of iodine and albumin used as a substitute for iodine. α-**eigon,** albuminiodatum, a brown powder, odorless and tasteless; contains 20 % of iodine; soluble in alkalies and acids; insoluble in water. α-**eigon-sodium,** sodium iodoalbuminatum, a white, odorless, nearly tasteless powder containing 15 % of iodine. β-**eigon,** peptone iodate, a yellow powder, odorless and tasteless; contains 15 % of iodine. It is recommended as a substitute for iodine when there is digestive weakness. Dose 45–150 gr. (3–10 Gm.) daily. α- and β-eigons are also used as dusting-powders.

**Eijkman's test for phenol** (*ik'-man*). Add to the phenol solution a few drops of an alcoholic solution of nitrous acid, ethyl ether, and an equal amount of concentrated sulphuric acid. A red coloration is produced.

**eilema** (*i-le'-mah*) [εἴλειν, to twist; *pl., eilemata*]. A pain or colic of the bowels; volvulus; tormina.

**eiloid** (*i'-loid*) [εἴλειν, to coil; εἶδος, form]. Having a coiled structure, as an *eiloid* tumor.

**Eimeria** (*i-me'-re-a*). A genus of protozoa; the same as coccidia in the asexual stage.

**Einhorn's method** (*in'-horn*) [Max *Einhorn,* American physician, 1862– ]. 1. A method of ascertaining the condition of the gastric secretion. An apparatus termed a stomach-bucket, consisting of a small oval silver vessel, 1¾ cm. long and ¾ cm. wide, is attached to a silk thread, in which at a distance of 40 cm. from the bucket a knot is made. The patient swallows the bucket, and when the knot enters the mouth the operator knows that the bucket is in the stomach. It is withdrawn after remaining there five minutes, and its contents are tested. Resistance to its removal may be overcome by. having the patient expire deeply or swallow once. 2. See *gastrodiaphany.*

**Einthoven's string galvanometer** (*int'-ho-fen*) [W. *Einthoven,* Dutch physiologist]. See *electrocardiograph.*

**eisanthema** (*i-san-the'-mah*) [εἰς, into; ἄνθημα, inflorescence; *pl., eisanthemata*]. An exanthem on a mucous membrane. See *enanthema.*

**Eiselt's reaction for melanin in urine** (*i'-selt*). Concentrated nitric acid, sulphuric acid, potassium dichromate, or other oxidizing agents render urine containing melanin dark colored.

**eisenzucker** (*i'-zen-zuk'-er*) [Germ. *eisen,* iron; *zucker,* sugar]. Saccharated ferric oxide; it consists of ferric hydroxide, sugar, and sodium hydroxide.

**eisodic** (*is-od'-ik*). Same as *esodic.*

**esophobia** (*i-so-fo'-be-ah*). Synonym of *agoraphobia.*

**esophoria** (*i-so-fo'-re-ah*). Same as *esophoria.*

**eispnea** (*is-pne'-ah*) [εἴσπνοη, a breathing into]. Inspiration; the inhaling of the breath.

**Eitelberg's test** (*i'-tel-berg*) [Abraham *Eitelberg,* Austrian physician, 1847– ]. If a large tuningfork is held at intervals before the ear during 15 or 20 minutes, the duration of the perception of the vibration, during these periods, increases in case the ear is normal, but decreases when a lesion of the sound-conducting apparatus exists.

**eitnerin** (*it'-ner-in*). A German substitute for yolk of egg.

**eiweiss milch** (*i'-vis milk*) [Ger. *eiweiss,* albumen *milch,* milk]. A preparation used for feeding infants it consists of broken up curd from which the whey has been removed, boiled buttermilk and malt sugar are then mixed with it.

**ejaculation** (*e-jak-u-la'-shun*) [*ejaculatio,* a throwing out]. The ejection of the semen. **ejaculatory** (*e-jak'-u-la-tor-e*) [*ejaculation*]. Throwing or casting out. e. duct. See *duct, ejaculatory.*

**ejecta** (*e-jek'-tah*) [L., pl. of *ejectum*]. Things or materials cast out; excretions or excrementitious matters.

**ejection** (*e-jek'-shun*) [*ejectio,* a casting out]. The casting out of excretions or of excrementitious matters; that which is cast out.

**ekaiodoform** (*ek-ah-i-o'-do-form*). A combination of iodoform and 0.5 % of paraformaldehyde. It is used as a dressing for wounds.

**ekiri** (*ek-e'-re*). A severe type of infantile diarrhea occurring in Japan.

**ektogan** (*ek'-to-gan*). The commercial name for zinc peroxide; it is used externally.

**ekzemin** (*ek'-ze-min*). An ointment consisting of precipitated sulphur with coloring-matter and perfume.

**elaboration** (*e-lab-or-a'-shun*) [*elaborare,* to work out]. In physiology, any anabolic process, such as that of making crude food into higher tissue-products.

**elacin** (*el'-a-sin*). Basophile elastin.

**elæometer, elaiometer** (*el-e-om'-et-ur, el-a-i-om'-et-ur*). See *eleometer.*

**elæomyenchysis.** See *eleomyenchysis.*

**elæoptene.** See *eleoptene.*

**elaidin** (*e-la'-id-in*) [ἔλαις, the olive-tree], C₅₇H₁₀₄O₆. A white, crystalline, fatty substance, isomeric with olein, produced by the action of HNO₃ upon certain oils, especially castor-oil.

**elain** (*e-la'-in*). See *eleoptene.*

**elarson** (*el-ar'-son*). Trade name for the strontium salt of chlorarsenobeholic acid; it contains about 13 per cent. of elementary arsenic and about 6¼per cent. of chlorine. It has been used in anemias, neuralgias, chorea, and various skin diseases.

**elastic** (*e-las'-tik*) [ἐλαύνειν, to urge forward]. Returning to the original form after being stretched or compressed. e. bandage, a rubber bandage exerting continuous compression of a part. e. lamina, Descemet's membrane. e. stocking, a rubber stocking exerting continuous pressure. e. tissue, a variety of connective tissue composed of yellow elastic fibers.

**elastica** (*e-las'-tik-ah*). The official name for rubber. See *caoutchouc.*

**elasticin** (*el-as'-tis-in*). See *elastin.*

**elasticity** (*e-las-tis'-it-e*). The property exhibited by some substances of returning to their original shape after the removal of a deforming force.

**elastin** (*e-las'-tin*) [see *elastic*]. An albuminoid substance forming the basis of elastic tissue.

**elastinase** (*e-las'-tin-ās*). A ferment that dissolves elastin.

**elastoid degeneration** (*e-las'-toid*). Hyaline degeneration of the elastic fibers in the wall of an artery; it occurs during involution of the uterus.

**elastometer** (*e-las-tom'-et-er*) [*elastic;* μέτρον, a

**measure].** An apparatus for determining the elasticity of tissues.
**elastose** (*e-las'-tōs*). One of the forms of peptone resulting from the gastric digestion of elastin.
**elater** (*el'-at-er*) [ἐλατήρ, a driver, hurler]. In biology, (*a*) one of the thread-like, usually spirally-coiled, bodies found in the sporangia of mosses, liverworts, and equisetum. They serve for the dispersion of spores. (*b*) One of the free filaments of the capillitium of the slime-moulds. (*c*) One of the anal bristles of the insect called spring-tail.
**elaterin, elaterinum** (*el-at'-er-in, el-at-er-i'-num*) [*elaterium*], $C_{20}H_{28}O_5$. A neutral principle obtained from *Ecballium elaterium*. It is a powerful hydragogue cathartic. Dose $\frac{1}{20}$ gr. (0.0032 Gm.). e., **powder of,** compound, contains elaterin, 1; sugar of milk, 39 parts. Dose $\frac{1}{3}$–5 gr. (0.032–0.32 Gm.). e., **trituration of** (*trituratio elaterini*, U. S. P.), elaterin, 10; sugar of milk, 90 parts; thoroughly mixed. Dose $\frac{1}{2}$–$\frac{3}{4}$ gr. (0.032–0.04 Gm.).
**elaterium** (*el-at-e'-re-um*) [ἐλατήριος, driving away]. The dried sediment from the juice of the squirting cucumber, *Ecballium elaterium*. It is a powerful hydragogue cathartic. Dose $\frac{1}{8}$ gr. (0.008 Gm.).
**elaterometer** (*e-lat-er-om'-et-er*) [ἐλατήρ, a driver; μέτρον, a measure]. An apparatus for determining the elasticity of gases.
**elayl** (*el'-āl*). See *ethylene*.
**elbow** (*el'-bo*). The region corresponding to the junction of the arm and forearm; the bend of the arm. **e.-bone,** the ulna. **e.-jerk,** a reflex flexion of the elbow on striking the biceps tendon.
**elcoplasty** (*el'-ko-plas-te*). See *helcoplasty*.
**elcosis** (*el-ko'-sis*). See *helcosis*.
**elder** (*el'-der*). See *Sambucus*.
**elecampane** (*el-e-kam'-pān*). See *inula*.
**electrargol** (*e-lek-trar'-gol*). Trade name of a sterile solution of electric colloidal silver.
**electric, electrical** (*e-lek'-trik, e-lek'-trik-al*) [*electricity*]. Having the nature of or produced by electricity. **e. chorea.** See *Dubini's disease*. **e. discharger,** an instrument for liberating stored electricity. **e.-light treatment,** the therapeutic application of electric light by means of cabinets in which the patient sits with the light directed upon the affected part. It is used in rheumatism, neuralgia, etc.
**electrician** (*e-lek-trish'-an*). 1. One skilled in electric science or a manipulator of electric apparatus. 2. One who employs electricity in the treatment of disease.
**electricity** (*e-lek-tris'-it-e*) [ἤλεκτρον, amber]. One of the forces of nature developed or generated by chemism, magnetism, or friction. **e., animal,** free electricity in the body. **e., chemical.** See *e., galvanic.* **e., faradic,** that produced by induction. **e., franklinic,** frictional or static electricity. **e., frictional,** that produced by friction. **e., galvanic,** that which is generated by chemical action in a galvanic cell. **e. induced, or inductive,** that produced in a body by proximity to an electrified body. **e., magnetic,** that developed by bringing a conductor near the poles of a magnet. **.e., static,** frictional electricity. **e., voltaic,** galvanic or chemical electricity.
**electrification** (*e-lek-trif-ik-a'-shun*). See *electrization*.
**electrify** (*e-lek'-trif-i*) [*electric; facere,* to make]. To make electric.
**electrization** (*e-lek-triz-a'-shun*) [*electricity*]. The application of electricity to the body. **e., intragastric,** electrotherapy practised by the introduction of an electrode into the stomach. Cf. *electrode, deglutable.*
**electro-** (*e-lek-tro-*) [*electricity*]. A prefix denoting connection with or relation to electricity.
**electroanesthesia** (*e-lek-tro-an-es-the'-ze-ah*) [*electro-;* ἀναισθησία, want of feeling]. 1. Inability to perceive the sensation made by electricity upon the skin. 2. Local anesthesia induced by the introduction of anesthetizing substances into the tissues by means of the electric current without injury to the skin. It is called the cataphoretic method.
**electrobiology** (*e-lek-tro-bi-ol'-o-je*) [*electro-; biology*]. 1. The science of the electrical relations and laws of organic beings. 2. A modern term for mesmerism or hypnotism.
**electrobioscopy** (*e-lek-tro-bi-os'-ko-pe*) [*electro-;* βίος, life; σκοπεῖν, to view]. The test of the existence of life by means of electricity.

**electrocapillarity** (*e-lek-tro-kap-il-ar'-it-e*). See *action, electrocapillary.*
**electrocardiogram** (*e-lek-tro-kar'-de-o-gram*) [*electro-;* καρδία, heart; γράμμα, a writing]. A registration of electromotive variations in heart-action.
**electrocardiograph** (*e-lek-tro-kar'-de-o-graf*) [*electro-;* καρδία, heart; γράφειν, to write]. An instrument for recording the electromotive variations in the action of the heart muscle.
**electrocatalysis** (*e-lek-tro-kat-al'-is-is*) [*electro-; catalysis*]. Catalysis or chemical decomposition produced by the action of electricity.
**electrocautery** (*e-lek-tro-kaw'-ter-e*). See *galvanocautery.*
**electrochemical** (*e-lek-tro-kem'-i-kal*). Pertaining to electrochemistry.
**electrochemism** (*e-lek-tro-kem'-izm*). The theory that all chemical action is caused by electricity.
**electrochemistry** (*e-lek-tro-kem'-is-tre*) [*electro-;* χημεία, chemistry]. The science treating of the chemical changes produced by electricity.
**electrocoagulation** (*e-lek-tro-ko-ag-u-la'-shun*). The destruction or hardening of tumors or tissues by coagulation induced by the passage of high-frequency currents.
**electroconductivity** (*e-lek-tro-kon-duk-tiv'-it-e*). Capability for transmitting electricity.
**electrocution** (*e-lek-tro-ku'-shun*) [*electro-; execution*]. Judicial execution by electricity.
**electrocystoscope** (*e-lek-tro-sis'-to-skōp*). A cystoscope with electric illumination.
**electrosyntoscopy** (*e-lek-tro-sis-tos'-ko-pe*). Cystoscopy with electric illumination.
**electrode** (*e-lek'-trōd*) [*electro-;* ὁδός, a way]. The pieces of metal or other substance fastened to the conducting cords of a battery through which electricity is applied to the body. **e., Alleman's,** a device for the application of electricity to cause absorption of corneal opacities. **e., colon** (Pennington's), an appliance for hydroelectric applications to the colon. It is a perforated hollow carbon electrode connected with the conducting cord by means of a spiral wire passing through and surrounded by a soft colon tube, through which the colon may be flushed, with warm water or saline solution. **e., deglutable,** an electrode suitable to be passed into the stomach for intragastric electrization. **e., dispersing.** See *e., indifferent.* **e., exciting,** in electrotherapy, the small electrode used in nerve- and muscle-stimulation, immediately over or near the nerve to be examined. Syn., *localizing electrode.* Cf. *e., indifferent.* **e., exciting, Erb's,** a bundle of 400 metal threads separated from one another by insulation and tightly incased in a hard-rubber tube; employed in electrotherapy. **e., indifferent,** the large electrode used in nerve and muscle stimulation at a distance from the nerve to be examined. Syn., *dispersing electrode.* Cf. *e., exciting.* **e., localizing.** See *e., exciting.*
**electrodiagnosis** (*e-lek-tro-di-ag-no'-sis*) [*electro-; diagnosis*]. Diagnosis by examining the reaction of the excitable tissues of the body by means of electric currents.
**electrodiaphane** (*e-lek-tro-di'-af-ān*) [*electro-;* διαφαίνειν, to show through]. An apparatus for illumination of the stomach. Cf. *diaphanoscope.*
**electrodiaphany** (*e-lek-tro-di-af'-an-e*). See *diaphanoscopy* and *transillumination*.
**electrodynamics, electrodynamism** (*e-lek-tro-di-nam'-iks, -din'-am-izm*) [*electro-; dynamic*]. The science of the reciprocal action of electric currents.
**electrodynamometer** (*e-lek-tro-di-nam-om'-et-er*) [*electro-; dynamometer*]. An instrument for measuring the strength of electric currents.
**electroendoscopy** (*e-lek-tro-end-os'-ko-pe*). See *diaphanoscopy* and *transillumination*.
**electrogenesis** (*e-lek-tro-jen'-es-is*) [*electro-;* γένεσις, production]. Production by electricity. Results following the application of electricity.
**electrogram** (*e-lek'-tro-gram*) [*electro-;* γράμμα, a writing]. A skiagram.
**electrograph** (*e-lek'-tro-graf*). See *skiagraph.*
**electrography** (*e-lek-trog'-raf-e*). 1. Skiagraphy. 2. Electroscopy.
**electrohemostasis** (*e-lek-tro-hem-os'-ta-sis*) [*electro-; hemostasis*]. Arrest of hemorrhage in a tissue or vessel by grasping it with a forceps, in the jaws of which heat is generated by an electric current, causing desiccation of the tissue and union of the arteries.
**electrokinetic** (*e-lek-tro-kin-et'-ik*). Electromotive

**electrokinetics** (*e-lek-tro-kin-et'-iks*) [*electro-;* κινεῖν, to move]. 1. The science of galvanism. 2. The science of electricity as applied to mechanic motion.

**electrolepsy** (*e-lek'-tro-lep-se*) [*electro-;* ἐπίληψις]. Electric chorea.

**electrolithotrity** (*e-lek-tro-lith-ot'-rit-e*). Disintegration of a vesical calculus by means of electricity.

**electrolizer** (*e-lek'-tro-li-zer*). An instrument for removing strictures by electricity.

**electrology**(*e-lek-trol'-o-je*) [*electro-;* λόγος, science]. That branch of physics treating of the laws and phenomena of electricity.

**electrolysis** (*e-lek-trol'-is-is*) [*electro-;* λύσις, solution]. The dissolution of a chemical compound by an electric current., e., cupric, electrolysis in which a bulb of chemically pure copper is applied directly to the diseased area; the copper oxychloride generated acts as a germicide.

**electrolyte** (*e-lek'-tro-līt*) [see *electrolysis*]. A substance capable of conducting an electric current and being decomposed by it.

**electrolytic** (*e-lek-tro-lit'-ik*) [see *electrolysis*]. Relating to electrolysis.

**electrolyzer** (*e-lek'-tro-li-zer*). See *electrolizer*.

**electromagnet** (*e-lek-tro-mag'-net*) [*electro-; magnet*]. A mass of soft iron surrounded by a coil of wire. A current passing through the wire will make the iron core magnetic.

**electromagnetics** (*e-lek-tro-mag-net'-iks*). 1. The production of magnetic action by means of electricity. 2. The science of the relation of electricity to magnetism.

**electromassage** (*e-lek-tro-mas-ash'*) [*electro-; massage*]. The transmission of a current of electricity through a kneading instrument. Electric treatment combined with massage.

**electromedication** (*e-lek-tro-med-ik-a'-shun*). The introduction of medicaments into the system by electric means.

**electrometer** (*e-lek-trom'-et-er*) [*electro-;* μέτρον, a measure]. An instrument for measuring electric force.

**electrometry** (*e-lek-trom'-et-re*) [see *electrometer*]. The measurement of electricity.

**electromotive** (*e-lek-tro-mo'-tiv*). 1. Pertaining to or producing electric action. 2. Producing electricity. e. force, the force that produces an electric current. Abbreviated E. M. F.

**electromuscular sensibility** (*e-lek-tro-mus'-kū-lar*). Sensibility of muscles to stimulation by electricity.

**electron** (*e-lek'-tron*) [ἤλεκτρον, amber]. 1. Amber. 2. A term used to represent a separate unit of electricity. According to J. J. Thompson, the mass of an electron is about one seventh-hundredth part of that of the hydrogen atom. 3. The ultimate particle of negative electricity. Cf. *ion* and *co-electron*.

**electronecrosis** (*e-lek-tro-ne-kro'-sis*). See *electrocution*.

**electronegative** (*e-lek-tro-neg'-a-tiv*) [*electro-; negare*, to deny]. Pertaining to or charged with negative electricity.

**electroneurotone** (*o-lek-tro-nū'-ro-tōn*) [*electro-;* νεῦρον, a nerve; τόνος, tone]. An apparatus for applying massage by electricity.

**electrooptics** (*e-lek-tro-op'-tiks*). The department of physics which deals with the optical phenomena of electric light.

**electropathology** (*e-lek-tro-path-ol'-o-je*) [*electro-; pathology*]. The study of morbid conditions by the aid of electric irritation.

**electrophobia** (*e-lek-tro-fo'-be-ah*) [*electro-;* φόβος, fear]. A morbid fear of electricity.

**electrophobist** (*e-lek-tro-fo'-bist*). A person having a morbid fear of electricity.

**electrophone** (*e-lek'-tro-fōn*) [*electro-;* φωνή, sound]. An apparatus used in treating deafness, by means of sonorous vibrations.

**electrophorus** (*e-lek-trof'-or-us*) [*electro-;* φέρειν, to carry]. An instrument used to generate small quantities of static electricity.

**electrophotography** (*e-lek-tro-fo-tog'-raf-e*). Same as *skiagraphy*.

**electrophototherapy** (*e-lek-tro-fo-to-ther'-ap-e*) [*electro-; phototherapy*]. Therapeutic treatment by means of electric light.

**electrophysiology** (*e-lek-tro-fis-e-ol'-o-je*) [*electro-; physiology*]. The study of electric reactions, properties, and relations of organs and organic tissues.

**electropositive** (*e-lek-tro-pos'-it-iv*) [*electro-; ponere*, to place]. Pertaining to or charged with positive electricity.

**electroprognosis** (*e-lek-tro-prog-no'-sis*). The use of electricity in prognosis.

**electropuncture** (*e-lek-tro-pung'-chur*) [*electro-; pungere*, to prick]. The use of needles as electrodes, which are thrust into an organ or a tumor or an aneurysm.

**electroscission** (*e-lek-tro-sish'-un*) [*electro-; scindere*, to cleave]. Division of tissues by an electro-cautery knife.

**electroscope** (*e-lek'-tro-skōp*) [*electro-;* σκοπεῖν, to view]. An instrument for detecting the presence of static electricity and determining whether it is positive or negative.

**electrosensibility** (*e-lek-tro-sen-si-bil'-it-e*). The irritability of a sensory nerve to electricity.

**electroskiagraphy** (*e-lek-tro-ski-ag'-raf-e*). Synonym of *skiagraphy*.

**electrostatics** (*e-lek-tro-stat'-iks*). The science of static electricity, or that developed by friction.

**electrostixis** (*e-lek-tro-stiks'-is*) [*electro-;* στίξις, puncture]. Electropuncture, *q. v.*

**electrosurgery** (*e-lek-tro-sur'-jer-e*). The use of electricity in surgery.

**electrosynthesis** (*e-lek-tro-sin'-the-sis*) [*electro-; synthesis*]. Chemical combination by means of electricity.

**electrotaxis** (*e-lek-tro-tak'-sis*) [*electro-;* τάξις, arrangement]. The reaction (attraction or repulsion) of organisms or cells to electric currents.

**electrothanasia** (*e-lek-tro-than-a'-ze-ah*) [*electro-;* θάνατος, death]. Death due to electricity.

**electrothanasize** (*e-lek-tro-than'-as-īz*) [*electro-;* θανατόω, to kill]. To produce death by electricity, but not as capital punishment.

**electrothanatosis** (*e-lek-tro-than-at-o'-sis*) [*electro-;* θανάτωσις, a putting to death]. Death by electricity.

**electrotherapeutics** (*e-lek-tro-ther-ap-u'-tiks*) [*electro-;* θεραπεία, treatment]. The science and art of the application of electricity for therapeutic purposes.

**electrotherapy** (*e-lek-tro-ther'-ap-e*). See *electrotherapeutics*.

**electrotherm** (*e-lek'-tro-therm*) [*electro-;* θέρμη, heat]. An apparatus for producing heat to relieve pain by the application of electricity to the skin.

**electrothermal** (*e-lek-tro-therm'-al*). Pertaining to heat and electricity or to heat generated by electricity.

**electrotome** (*e-lek'-tro-tōm*) [*electro-;* τέμνειν, to cut]. The circuit-breaker of an electric battery; especially one that acts automatically.

**electrotonic** (*e-lek-tro-ton'-ik*) [*electro-;* τόνος, tension]. Relating to or of the nature of electrotonus. e. effect, an altered condition of excitability of a nerve produced when in the electrotonic state.

**electrotonus** (*e-lek-trot'-on-us*) [*electro-;* τόνος, tension]. The change of condition in a nerve during the passage of a constant current. See *anelectrotonus* and *catelectrotonus*.

**electrotrephine** (*e-lek-tro-tre'-fīn*). A trephine operated by electricity.

**electrotropism** (*e-lek-trot'-ro-pism*) [*electro-;* τροπή, a turning]. Same as *electrotaxis*.

**electrovagogram** (*e-lek-tro-va'-go-gram*) [*electro-; vagus;* γράμμα, a writing]. A record of the electrical changes occurring in the vagus nerve, taken with a string galvanometer.

**electrovection** (*e-lek-tro-vek'-shun*) [*electro-; vehere*, to carry]. Electric endosmosis; the introduction of medicaments into the system by means of the electric current; cataphoresis.

**electrozone** (*e-lek'-tro-zōn*). The proprietary name for a disinfectant fluid produced by the electrolysis of sea-water.

**electuary** (*e-lek'-tū-ar-e*) [ἐλεκτόν, a medicine that melts in the mouth]. A soft or pasty mass, consisting of a medicinal substance, with sugar, honey, water, etc.

**eleidin** (*e-le'-id-in*) [ἐλαία, olive-oil]. A material occurring in the form of granules in the stratum granulosum of the epidermis.

**element** (*el'-e-ment*) [*elementum*, a first principle]. Any one of the ultimate parts of which anything is composed, as the *cellular elements* of a tissue. In chemistry, a body that cannot be decomposed into simpler substances. The recognized elements now number about 80. See *elements, table of chemical*, on. p. 319.

## TABLE OF CHEMICAL ELEMENTS.

Based on one in Funk and Wagnall's Standard Dictionary [copyright].—(Published by permission.) The first two columns have been revised to 1916 from Jour. of Amer. Chem. Soc.

| Name. | Symbol. | Atomic Weight. | Specific Gravity.* | Fusing-point or Melting-point. Degrees C. and F. | Valence. | Where and How Found. |
|---|---|---|---|---|---|---|
| Aluminium | Al | 27.1 | 2.58 | 627° C. (1160° F.). | III | In many rocks. (The most abundant metal.) |
| Antimony (stibium) | Sb | 120.2 | 6.7 | 432° C. (808° F.). | V | Chiefly as sulphide, and in various metallic ores. |
| Argentum. See Silver. | | | | | | |
| Argon | A | 39.88 | 1.5† | −128.6° C.(231.4° F.). | | Free in the atmosphere. |
| Arsenic | As | 74.96 | 5.71 | Ab't 500° C. (932° F.). | V | Native, as sulphide, and in various metallic ores. |
| Aurum. See Gold. | | | | | | |
| Barium | Ba | 137.37 | 3.75 | Above redness. | II | In barite and witherite. |
| Beryllium. See Glucinum. | Be | | | | | |
| Bismuth | Bi | 208.0 | 9.8 | 268° C. (517° F.). | V | Native, as sulphide, and in rare minerals. |
| Boron | B | 11.0 | 2.6 | Very high. | III | In borax and various minerals. |
| Bromine | Br | 79.92 | 3.19 | −7.2° C. (−20° F.). | I or VII | Mainly in sea-water and other natural brines. |
| Cadmium | Cd | 112.40 | 8.65 | 231° C. (600° F.). | II | In small amount in zinc ores. |
| Calcium | Ca | 40.07 | 1.6–1.8 | Bright redness. | II | In limestone, and abundantly in other rocks. |
| Carbon | C | 12.005 | 3.52‡ | Infusible. | IV | In coal, limestone, and all organic matter. |
| Cerium | Ce | 140.25 | 6.7 | Below silver. | III or IV | In cerite and other rare minerals. |
| Cesium | Cs | 132.81 | 1.88 | 26.5° C. (80° F.). | I | In lepidolite, pollucite, and mineral springs. |
| Chlorine | Cl | 35.46 | 1.33† | −75.6° C. (−103° F.). | I or VII | In common salt (NaCl) and other chlorides. |
| Chromium | Cr | 52.0 | 7.3 | Above platinum. | II or VI | Mainly in chrome-iron ore. |
| Cobalt | Co | 58.97 | 8.96 | 1500° C. (2732° F.). | II or VIII | In many metallic ores. |
| Columbium (niobium) | Cb | 93.5 | Above 7 | | V | In columbite and other rare minerals. |
| Copper (cuprum) | Cu | 63.57 | 8.9 | 1054° C. (1931° F.). | I or II | Native, and in many ores. |
| Coronium (hypothetical). | | | | | | |
| Didymium. See Praseodymium. | Di | | | | | |
| Dysprosium | Dy | 162.5 | | | | |
| Erbium | Er | 167.7 | | | III | In rare minerals, as gadolinite, etc. |
| Europium | Eu | 152.0 | | | | |
| Ferrum. See Iron. | | | | | | |
| Fluorine | F | 19.0 | | | I or VII | In fluorite (CaF₂) and other minerals. |
| Gadolinium | Gd | 157.3 | | | III | In rare minerals, as gadolinite, etc. |
| Gallium | Ga | 69.9 | 5.95 | 30.1° C. (86° F.). | III | In certain zinc-blendes. |
| Germanium | Ge | 72.5 | 5.47 | 900° C. (1652° F.). | IV | In argyrodite, a rare mineral. |
| Glucinum (beryllium) | Gl | 9.1 | 1.85 | Above redness. | II | In beryl and several rare minerals. |
| Gold (aurum) | Au | 197.2 | 19.3 | 1045° C. (1913° F.). | I or III | Generally free, rarely combined, in various ores. |
| Helium | He | 3.99 | | | | In cleveite and several other rare minerals. |
| Hydrargyrum. See Mercury. | | | | | | |
| Hydrogen | H | 1.008 | 0.025† | −200° C.†(−328° F.). | I | Mainly in water (H₂O). |
| Indium | In | 114.8 | 7.4 | 176° C. (348° F.). | III | In certain zinc ores. |
| Iodine | I | 126.92 | 4.95 | 114° C. (238° F.). | I or VII | Mainly in ashes of seaweeds. |
| Iridium | Ir | 193.1 | 22.4 | 1950° C. (3542° F.). | II or IV | In iridosmin. |
| Iron (ferrum) | Fe | 55.84 | 8 | 1600° C. (2912° F.). | II or IV | As oxide and sulphide, and in nearly all rocks. |
| Kalium. See Potassium. | | | | | | |
| Krypton | Kr | 82.92 | | | | |
| Lanthanum | La | 139.0 | 6.1 | | III | In cerite and other rare minerals. |
| Lead (plumbum) | Pb | 207.20 | 11.36 | 326° C. (850° F.). | II or IV | In galena (PbS) and other ores. |
| Lithium | Li | 6.94 | 0.585 | 180° C. (356° F.). | I | In lepidolite, spodumene, and some rare minerals. |

TABLE OF CHEMICAL ELEMENTS.—(Continued.)

| Name. | Symbol. | Atomic Weight. | Specific Gravity.* | Fusing-point or Melting-point. Degrees C. and F. | Valence. | Where and How Found. |
|---|---|---|---|---|---|---|
| Lutecium | Lu | 175.0 | | | | |
| Magnesium | Mg | 24.32 | 1.75 | Ab't 430° C. (806° F.). | II | In sea-water, magnesite, and many rocks. |
| Manganese | Mn | 54.93 | 7.2 | Above iron. | II or VII | In pyrolusite and many other minerals. |
| Mercury (hydrargyrum) | Hg | 200.6 | 13.596 | −38.8° C. (−38° F.). | I or II | Native and in cinnabar (HgS). |
| Molybdenum | Mo | 96.0 | 8.6 | Very high. | II or VI | Mainly as molybdenite (MoS$_2$). |
| Natrium. See Sodium. | | | | | | |
| Neodymium | Nd | 144.3 | About 6.5 | | III or IV | In cerite and other rare minerals. |
| Neon | Ne | 20.2 | | | | |
| Nickel | Ni | 58.68 | 8.9 | 1450° C. (2642° F.). | II or VIII | In many metallic ores. |
| Niobium. See Columbium. | Nb | | | | | |
| Niton | Nt | 222.4 | | | | Radium emanation. |
| Nitrogen | N | 14.01 | 0.38 ‖ | | V | In the atmosphere and organic matter. |
| Osmium | Os | 190.9 | 22.48 | Nearly infusible. | II or VII | In iridosmin and native platinum. |
| Oxygen | O | 16.00 | 1.11 § | | II or VI | Free in air. (Forms one-half the earth's crust, combined.) |
| Palladium | Pd | 106.7 | 12.1 | 1500° C. (2732° F.). | II or IV | Native and with platinum and gold. |
| Phosphorus | P | 31.04 | 1.84 | 44.2° C. (112° F.) | V | In bones and in apatite and many minerals. |
| Platinum | Pt | 195.2 | 21.5 | 1775° C. (3225° F.) | II or IV | Mainly as native platinum in river-gravels. |
| Plumbum. See Lead. | | | | | | |
| Potassium (kalium) | K | 39.10 | 0.86 | 62.5° C. (144.5° F.). | I | In wood-ashes and many rocks. |
| Praseodymium (didymium) | Pr | 140.6 | About 6.5 | | III or IV | In cerite and other rare minerals. |
| Radium | Ra | 226.0 | | | II | In pitch-blende. |
| Rhodium | Rh | 102.9 | 12.1 | 2000° C. (3632° F.). | II or VIII | With platinum and iridosmin. |
| Rubidium | Rb | 85.45 | 1.52 | 38.5$^d$ C. (101.5° F.). | I | In lepidolite and some mineral springs. |
| Ruthenium | Ru | 101.7 | 12.26 | Nearly infusible. | II or VII | With platinum and iridosmin. |
| Samarium | Sa | 150.4 | | | III | In samarskite, cerite, and other rare minerals. |
| Scandium | Sc | 44.1 | | | III | In gadolinite and other rare minerals. |
| Selenium | Se | 79.2 | 4.5 | 217° C. (425° F.). | II or VI | Mainly in sulphur as an impurity. |
| Silicon | Si | 28.3 | 2.48 | Above 800° C. (1500° F.). | IV | In quartz (SiO$_2$). (Most abundant element after oxygen.) |
| Silver (argentum) | Ag | 107.88 | 10.5 | 954° C. (1750° F.). | I | Native and in many ores. |
| Sodium (natrium) | Na | 23.00 | 0.97 | 95.6° C. (204° F.). | I | In common salt (NaCl) and many rocks. |
| Stannum. See Tin. | | | | | | |
| Stibium. See Antimony. | | | | | | |
| Strontium | Sr | 87.63 | 2.5 | Red heat. | II | In celestite and strontianite. |
| Sulphur | S | 32.06 | 2.07 | 114.5° C. (235° F.). | II or VI | Native and in many natural sulphides and sulphates. |
| Tantalum | Ta | 181.5 | Above 10 | | V | In tantalite and other rare minerals. |
| Tellurium | Te | 127.5 | 6.23 | 455° C. (851° F.). | II or VI | In several rare minerals. |
| Terbium | Tb | 159.2 | | | III | In rare minerals, as gadolinite, etc. |
| Thallium | Tl | 204.0 | 11.19 | 239.9° C. (561° F.). | I or III | In pyrites and in fluedust of sulphuric-acid works. |
| Thorium | Th | 232.4 | 11.23 | Almost infusible. | IV | In thorite and other rare minerals. |
| Thulium | Tm | 168.5 | | | III | In rare minerals, as gadolinite, etc. |
| Tin (stannum) | Sn | 118.7 | 7.25 | 233° C. (551° F.). | II or IV | Mainly in cassiterite (SnO$_2$). |
| Titanium | Ti | 48.1 | | Not fusible. | V | Widely diffused in rocks and clays, in small amounts. |

## TABLE OF CHEMICAL ELEMENTS.—(Concluded.)

| Name. | Symbol. | Atomic Weight. | Specific Gravity.* | Fusing-point or Melting-point. Degrees C. and F. | Valence. | Where and How Found. |
|---|---|---|---|---|---|---|
| Tungsten (wolframium) ... | W | 184.0 | 19.26 | Very high. | IV or VI | Mainly in wolframite (Mn-FeWO$_4$). |
| Uranium......... | U | 238.2 | 18.69 | Very high. | II or VI | In pitch-blende and other rare minerals. |
| Vanadium........ | V | 51.0 | 5.87 | In oxyhydric flame. | V | In vanadinite and other rare minerals. |
| Wolframium. See Tungsten. | | | | | | |
| Xenon............ | Xe | 130.2 | | | | |
| Ytterbium (neoytterbium)... | Yb | 173.5 | | | III | In rare minerals, as gadolinite, etc. |
| Yttrium.......... | Yt | 88.7 | | | III | In gadolinite and other rare minerals. |
| Zinc (zincum).... | Zn | 65.37 | 7.12 | 433° C. (811.5° F.). | II | In ores, as oxide, silicate, sulphide, and carbonate. |
| Zirconium........ | Zr | 90.6 | 4.15 | Above sulphur. | IV | In zircon and other rare minerals. |

\* The factors in the columns of specific gravities and melting-points naturally vary with the form which the element takes (e. g., in carbon the specific gravity varies as diamond, charcoal, or lampblack is taken), but so far as possible the factor of the most typical form is given.
† Of the liquid element.  ‡ Diamond.  ǁ Of the liquid at 0° C.  § Of the liquid at —181° C.

**elementary** (el-e-men'-ta-re) [element]. Pertaining to or having the characters of an element.

**elemi** (el'-em-e) [Ar.]. A resinous exudation probably derived from the *Canarium commune*, although its botanic source is still undetermined. It contains a crystalline resin, *elemin* or *amyrin*. Its action is similar to that of the turpentines. **e., unguentum** (B. P.), elemi and simple ointment; it is used as an application to indolent sores and boils.

**eleometer** (el-e-om'-et-ur) [ἔλαιον, oil; μέτρον, a measure]. An apparatus for ascertaining the specific gravity of oil.

**eleomyenchysis** (el-e-o-mi-en'-kis-is) [ἔλαιον, oil; μῦς, muscle; ἐγχεῖν, to pour in]. 1. The intramuscular injection and congelation of oils in treatment of chronic local spasm. 2. Surgical prosthesis by injection of paraffin.

**eleoptene** (el-e-op'-tēn) [ἔλαιον, oil; πτηνός, volatile]. The permanent liquid principle of volatile oils. See *stearoptene*.

**eleosaccharum** (e-le-o-sak'-ar-um) [L.: *pl.*, *eleosacchara*]. An oil-sugar; a preparation made by saturating thirty grains of sugar with one drop of volatile oil.

**elephantiac, elephantiasic** (el-e-fant'-i-ak, el-e-fant-i-a'-sik). Relating to or affected with elephantiasis; elephantic.

**elephantiasis** (el-ef-an-ti'-as-is) [ἔλέφας, an elephant]. A chronic affection of the cutaneous and subcutaneous tissues, due to obstruction of lymph-vessels, and characterized by enormous thickening of the affected parts. The disease occurs in successive attacks accompanied by fever and by swelling of the affected parts, usually the lower extremities and genital organs; it is endemic in certain tropical countries, and seems to be connected, in many cases, with the presence in the blood of *Filaria sanguinis-hominis*. **e. anæsthetica**, anesthetic leprosy. **e. Arabum.** See *elephantiasis*. **e. asturiensis**, pellagra. **e. congenita cystica**, a state of malformation marked by skeletal defects, general anasarca, and formation of cysts in the subcutaneous tissue. **e. dura**, **e. scirrhosa**, a variety of elephantiasis marked by density and sclerosis of the subcutaneous connective tissues. **e. Græcorum.** See *leprosy*. **e., nevoid.** See *s. telangiectodes*. **e. telangiectodes**, elephantiasis characterized by a great increase in the blood-vessels; dermatolysis.

**elevator** (el'-ev-a-tor) [*elevare*, to lift]. 1. The same as *levator*. See under *muscle*. 2. An instrument for elevating or lifting a part, or for extracting the roots of teeth.

**eleventh nerve.** The spinal accessory nerve.

**eliminant** (e-lim'-in-ant) [*eliminare*, to expel]. 1. Promoting elimination. 2. A drug causing elimination.

**elimination** (e-lim-in-a'-shun) [see *eliminant*]. The process of expelling or casting out, especially waste-products.

**elinguation** (e-lin-gwa'-shun) [*e*, out; *lingua*, the tongue]. Surgical removal of the tongue.

**elinguid** (e-lin'-gwid). Tongue-tied; without the power of speech.

**eliquation** (el-ik-wa'-shun) [*eliquare*, to melt out]. The separation of one substance from another by fusion or melting.

**elixation** (e-liks-a'-shun). 1. A decoction. 2. Digestion.

**elixir** (e-liks'-er) [Ar., *el iksīr*, the philosopher's stone]. A sweetened, aromatic, spirituous preparation, containing only a small amount of an active ingredient. **e., adjuvant** (*elixir adjuvans*, U. S. P.), one made of fluidextract of glycyrrhiza and aromatic elixir. **e., aromatic** (*elixir aromaticum*, U. S. P.), compound spirit of orange, 1.2 Cc.; syrup, 37.5 Cc.; purified talc, 1.5 Gm.; deodorized alcohol, distilled water, each, a sufficient quantity to make 100 Cc. It is used as a vehicle. **e. of iron, quinine, and strychnine phosphates** (*elixir ferri, quininæ, et strychninæ phosphatum*, U. S. P.). Dose 1 dr. (4 Cc.). **e. of phosphorus**, spirit of phosphorus, 21; oil of anise, 0.2; glycerol, 55; aromatic elixir, a sufficient quantity to make 100 Cc.

**elixiviation** (e-liks-iv-e-a'-shun). See *lixiviation*.

**elixoid** (el-iks'-oid). Trade name of fluid preparations of drugs, which are said to keep perfectly in any climate.

**elkodermatosis** (el-ko-der-mat-o'-sis) [ἔλκος, an ulcer; δέρμα, skin; νόσος, disease]. An ulcerative skin affection.

**elkoplasty** (el'-ko-plas-te). See *helcoplasty*.

**Elliot's position** (el'-e-ot) [John Wheelock Elliot, American surgeon, 1852– ]. The patient is placed in the position of a double inclined plane by means of a pillow or cushion under the small of the back.

**Elliot's sign** (el'-e-ot) [George T. Elliot, American dermatologist, 1855– ]. A skin lesion with an indurated or infiltrated border is syphilitic.

**Elliot-Smith, area paraterminalis** of. A space on the mesial aspect of the embryonic cerebral hemisphere. **E.-S., fasciculus præcommissuralis** of, the peduncle of the corpus callosum in the embryo.

**Ellis' ligament.** That part of the rectovesical fascia that extends to the side of the rectum. **E.'s line**, or **curve** the curved line followed by the upper border of a pleuritic effusion or a hydrothorax. **E.'s sign**, during resorption of a pleuritic exudate, the upper border of dulness forms a curve convex toward the head, the highest point of which lies laterally.

**Ellis-Damoiseau's curve.** See *Ellis' sign* and *line*.

**elm.** See *ulmus*.

**elodes** (e-lo'-dēz) [ἑλώδης, swampy; ἕλος, a swamp].

Marsh or paludal fever; malarial fever. **e. icterodes,** yellow fever.
**elongatio, elongation** (*e-lon-ga'-she-o, -shun*). 1. The process of lengthening. 2. A lengthened condition. **e. colli,** pathological lengthening of the cervix uteri through hypertrophy.
**elosin** (*el'-o-sin*). A remedy said to be a resinoid from the root of *Chamalirium carolinianum*. It is tonic, diuretic, emmenagogue, and a vermifuge.
**Elsner's asthma** (*els'-ner*) [Christopher Friedrich *Elsner*, German physician, 1749–1820]. Angina pectoris.
**Elsner's method of diagnosing typhoid** [Ottomar *Elsner*, German pathologist, 1869– ]. Cultures are made from the stools, upon a special culture-medium composed of Holz's acid potato-gelatin with 1% of potassium iodide. Only a few forms of bacteria will grow upon this medium, and among these are *Bacterium coli* and the typhoid bacillus; these latter are of slow growth and in 24 hours are scarcely visible with low power, whereas the coli colonies have attained considerable growth. After 48 hours the typhoid cultures appear in shining aggregations as drops of water with finely molded structure, and the coli colonies are larger, more granular, and brown in color. The presence of the typhoid bacillus can also be detected in food and water by this procedure.
**elutriation** (*e-lu-tre-a'-shun*) [*elutriare*, to wash out]. A process whereby the coarser particles of an insoluble substance are separated from the finer by decanting the fluid after the coarser particles have settled.
**elytratresia** (*el-it-rat-re'-ze-ah*) [*elytro-*; ἀτρητος, imperforate]. Atresia of the vagina; colpatresia.
**elytreurynter** (*el-it-ru-rin'-ter*) [*elytro-*; εὐρύνειν, to make broad]. Same as *colpeurynter*.
**elytritis** (*el-it-ri'-tis*) [ἔλυτρον, vagina; ιτις, inflammation]. Inflammation of the vagina.
**elytro-** (*el'-it-ro*) [ἔλυτρον, vagina]. A prefix signifying relating to the vagina.
**elytrocele** (*el'-it-ro-sēl*) [*elytro-*; κήλη, hernia]. Colpocele; vaginal hernia.
**elytroclasia** (*el-it-ro-kla'-se-ah*) [*elytro-*; κλᾶειν, to break]. Rupture of the vagina.
**elytrocleisis, elytroclisis** (*el-it-ro-kli'-sis*). See *colpocleisis*.
**elytroid** (*el'-it-roid*) [*elytro-*; εἶδος, likeness]. Like a sheath.
**elytroncus** (*el-it-rong'-kus*). Same as *elytrophyma*.
**elytrophyma** (*el-it-ro-fi'-mah*) [*elytro-*; φῦμα, a tumor]. Swelling or tumor of the vagina.
**elytroplastic** (*el-it-ro-plas'-tik*) [*elytro-*; πλάσσειν, to form]. Relating to elytroplasty.
**elytroplasty** (*el'-it-ro-plas-te*) [*elytro-*; πλάσσειν, to form]. A plastic operation upon the vagina.
**elytropneumatosis** (*el-it-ro-nu-mat-o'-sis*) [*elytro-*; πνεῦμα, air]. A collection of air in the vagina.
**elytropolypus** (*el-it-ro-pol'-ip-us*) [*elytro-*; *polypus*]. Vaginal polypus.
**elytroptosis** (*el-it-rop-to'-sis*) [*elytro-*; πτῶσις, a falling]. Prolapse of the vagina.
**elytrorrhagia** (*el-it-ror-a'-je-ah*) [*elytro-*; ῥηγνύναι, to burst forth]. Hemorrhage from the vagina.
**elytrorrhaphy** (*el-it-ror'-a-fe*) [*elytro-*; ῥαφή, a seam]. Suture of the vaginal wall.
**elytrorrhea** (*el-it-ror-e'-ah*) [*elytro-*; ῥοία, a flow]. A vaginal leukorrhea.
**elytrostenosis** (*el-it-ro-ste-no'-sis*) [*elytro-*; στένωσις, a contraction]. Colpostenosis; vaginal stricture.
**elytrotome** (*el-it'-ro-tōm*) [*elytro-*; τομή, a cutting]. An instrument for performing elytrotomy.
**elytrotomy** (*el-it-rot'-o-me*) [*elytro-*; τομή, a cutting]. Surgical incision of the vaginal wall.
**emaciation** (*e-ma-se-a'-shun*) [*emaciare*, to make lean]. Loss of the fat and fulness of the flesh of the body. Leanness.
**emaculation** (*e-mak-u-la'-shun*) [*emaculare*, to remove spots]. The removal of freckles or other spots from the face.
**emailloid** (*em'-il-oid*) [Fr. *émail*, enamel; εἶδος, resemblance]. A tumor developing from and composed of the enamel of a tooth.
**emanation** (*em-an-a'-shun*) [*emanare*, to issue]. 1. Emission; radiation. 2. That which flows or issues from a substance; effluvium.
**emansio** (*e-man'-she-o*) [L.]. A failing. **e. mensium,** delay in the first appearance of the menses.
**emasculation** (*e-mas-ku-la'-shun*) [*emasculare*, to make impotent]. 1. Removal of the testicles; castration. 2. The removal of both testicles, and total extirpation of the penis.
**emballometer** (*em-bal-om'-et-er*) [ἐμβάλλειν, to throw; μέτρον, a measure]. A percussion instrument employed in connection with a stethoscope.
**embalming** (*em-bahm'-ing*). The treatment of a cadaver with antiseptic and preservative substances to keep it from putrefying.
**embed** (*em-bed'*). In histology, to treat a tissue with some substance, as paraffin or celloidin, which shall give it support during the process of sectioncutting.
**embedding** (*em-bed'-ing*). The fixation of a tissue-specimen in a firm medium, in order to keep it intact during the cutting of thin sections.
**Embelia** (*em-bē'-le-ah*) [*Embel*, a German traveler]. A genus of shrubs. **E. ribes,** a myrtaceous shrub that grows in Asia; it is reputed to have anthelmintic properties. Dose of the powdered fruit, 3 j–iv; of the fluidextract 3 j–iv.
**embolalia** (*em-bo-la'-le-ah*). See *embolalia*.
**embole** (*em'-bo-le*) [ἐμβολή, a throwing in]. 1. The reducing of a dislocation. 2. Emboly. 3. Enarthrosis. 4. Embolism.
**embolemia** (*em-bol-e'-me-ah*) [*embolus*; αἷμα, blood]. 1. A state of the blood in which it is said that emboli are readily formed. 2. The presence of emboli in the blood.
**embolic** (*em-bol'-ik*) [*embolus*]. 1. Relating to or caused by an embolus. 2. Pertaining to emboly.
**emboliform** (*em-bol'-if-orm*) [*embolus*]. Resembling an embolus. **e. nucleus.** See under *nucleus*.
**embolism** (*em'-bo-lizm*) [*embolus*]. The obstruction of a blood-vessel, especially an artery, by a fragment of matter brought from another point. **e., air-,** obstruction of a vessel by a bubble of air. **e., fat-,** obstruction of blood-vessels by globules of fat. **e., infective,** embolism in which the emboli contain microorganisms and cause metastatic abscesses. **e., miliary,** a condition in which many small blood-vessels are the seats of emboli. **e., pigment, e., pigmental, e., pigmentary,** embolism due to melanemia and usually occurring in the spleen, liver, brain, or kidney.
**emboloid** (*em'-bo-loid*) [*embolus*; εἶδος, likeness]. Resembling an embolus.
**embololalia** (*em-bo-lo-la'-le-ah*) [*embolus*; λαλιά, babble]. The intercalation of meaningless words into the speech.
**embolophrasia** (*em-bo-lo-fra'-ze-ah*) [*embolus*; φράσις, speech]. Embololalia.
**embolus** (*em'-bo-lus*) [ἐν, in; βάλλειν, to throw; pl., *emboli*]. A particle of fibrin or other material brought by the blood-current and forming an obstruction at its place of lodgment.
**emboly** (*em'-bo-le*) [ἐμβολή, insertion]. The process of invagination that gives rise to a gastrula from a blastosphere or vesicular morula.
**embrocation** (*em-bro-ka'-shun*) [ἐμβρέχειν, to soak in]. 1. The application, especially by rubbing, of a liquid to a part of the body. 2. The liquid so applied.
**embryectomy** (*em-bre-ek'-to-me*) [*embryo*; ἐκτομή, a cutting out]. The surgical removal of the embryo, especially in extra-uterine pregnancy.
**embryo** (*em'-bre-o*) [ἐν, in; βρύειν, to swell with]. 1. The product of conception up to the fourth month of pregnancy. 2. The fertilized germ of an animal.
**embryocardia** (*em-bre-o-kar'-de-ah*) [*embryo*; καρδία, the heart]. A condition in which the heart-sounds resemble those of the fetus, the first and second sounds being almost identical.
**embryochemical** (*em-bre-o-kem'-ikal*) [*embryo*; χημεία, chemistry]. Relating to the changes in the chemical distribution of nitrogen and phosphorus in the fertilized egg during development.
**embryoctonic, embryoctonous** (*em-bre-ok-ton'-ik, -ok'-ton-us*) [*embryo*; κτείνειν, to kill]. Abortifacient; relating to embryoctony.
**embryoctony** (*em-bre-ok'-to-ne*) [see *embryoctonic*]. The destruction of the living fetus; the procurement of abortion.
**embryogenetic, embryogenic** (*em-bre-o-jen-et'-ik, em-bre-o-jen'-ik*) [*embryo*; γένης, producing]. Giving rise to an embryo.
**embryography** (*em-bre-og'-ra-fe*) [*embryo*; γράφειν, to write]. A description of the embryo.
**embryogenesis, embryogeny** (*em-bre-o-jen'-es-is; em-bre-oj'-en-e*) [*embryo*; γένης, producing]. That

department of biology which deals with the development of the fecundated germ.

**embryogeny** (*em-bre-oj'-en-e*). See *embryogenesis*.

**embryograph** (*em'-bre-o-graf*) [*embryo*; γράφειν, to write]. A form of microscope and camera lucida for use in drawing outlines and figures in embryological study.

**embryoid** (*em'-bre-oid*) [*embryo*; εἶδος, form]. Resembling the embryo.

**embryoism, embryonism** (*em'-bre-o-izm*, *em'-bre-on-izm*). The state of being an embryo.

**embryologic, embryological** (*em-bre-o-loj'-ik*, *em-bre-o-loj'-ik-al*) [*embryo*; λόγος, science]. Relating to embryology.

**embryologist** (*em-bre-ol'-o-jist*) [*embryo*; λόγος, science]. One skilled in the science of embryology.

**embryolemma** (*em-bre-o-lem'-ah*) [*embryo*; λέμμα, a husk; *pl.*, *embryolemmata*]. The special fetal membranes, the amnion, serolemma, etc.

**embryology** (*em-bre-ol'-o-je*) [*embryo*; λόγος, science]. The science dealing with the development of the embryo.

**embryoma** (*em-bre-o'-mah*). A dermoid cyst found in the mammalian ovary and testis; regarded by Wilnis as a rudimentary embryo.

**embryometrotrophia** (*em-bre-o-met-ro-tro'-fe-ah*) [*embryo*; μήτρα the womb; τρέφειν, to nourish]. The nourishment of the embryo.

**embryomorphous** (*em-bre-o-mor'-fus*) [*embryo*; μορφή, shape]. Like an embryo or of embryonic origin.

**embryon** (*em'-bre-on*). Same as *embryo*.

**embryonal** (*em-bre-o'-nal*). Same as *embryonic*.

**embryonate** (*em'-bre-o-nāt*) [*embryo*]. 1. Relating to an embryo. 2. Fecundated; containing an embryo.

**embryonic** (*em-bre-on'-ik*) [*embryo*]. Pertaining to the embryo. **e. abortion,** an early abortion. **e. area,** an opaque circular spot that forms on the blastoderm. **e. spot.** See *e. area*. **e. tissue,** tissue in the undifferentiated state, consisting of small, round cells.

**embryoplastic** (*em-bre-o-plas'-tik*) [*embryo*; πλάσσειν, to form]. Participating in the formation of the embryo; it is said of cells.

**embryoscope** (*em'-bre-o-skōp*) [*embryo*; σκοπεῖν, to examine]. An appliance by means of which the course of development of the embryo in eggs with shells may be observed.

**embryospastic** (*em-bre-o-spas'-tik*) [*embryo*; σπᾶν, to draw]. Relating to fetal extraction with an instrument.

**embryotocia** (*em-bre-o-to'-she-ah*) [*embryo*; τόκος, birth]. An abortion.

**embryotome** (*em'-bre-o-tōm*) [*embryo*; τομή, section]. An instrument for performing embryotomy.

**embryotomy** (*em-bre-ot'-o-me*) [see *embryotome*]. The cutting up of the fetus in the uterus for the purpose of reducing its size.

**embryotoxon** (*em-bre-o-toks'-on*) [*embryo*; τόξον, a bow]. A condition resembling *arcus senilis*, sometimes seen at birth.

**embryotrophy** (*em-bre-ot'-ro-fe*) [*embryo*; τροφή, nourishment]. The nutrition of the fetus.

**embryulcia** (*em-bre-ul'-se-ah*) [*embryo*; ἕλκειν, to draw]. 1. Forcible extraction of the fetus. 2. The operation of embryotomy.

**embryulcus** (*em-bre-ul'-kus*). A blunt hook, or obstetric forceps, used in performing embryulcia or in extracting the fetus.

**emedullate** (*e-med'-ul-āt*) [*e*, out of; *medulla*, marrow]. To remove the marrow or pith from.

**emergency** (*e-mer'-jen-se*) [*emergere*, to rise up]. A sudden, pressing, and unforeseen occasion for action; an accident or condition unlooked for, and calling for prompt decision. **e. ration.** See under *ration*.

**emergent** (*e-mer'-jent*) [*emergere*, to rise up]. Sudden, unforeseen, and urgent; calling for prompt decision and action; as an emergent case.

**emerod** (*em'-er-od*). See *hemorrhoid*.

**emesis** (*em'-es-is*) [ἐμέειν, to vomit]. Vomiting.

**emetatrophia** (*em-et-at-ro'-fe-ah*) [ἔμετος, vomiting; ἀτροφία, wasting]. Atrophy or wasting, due to persistent vomiting.

**emetic** (*em-et'-ik*) [see *emesis*]. 1. Having the power to induce vomiting. 2. An agent causing emesis. **e., direct, e., mechanical,** one acting directly on the nerves of the stomach. **e., indirect, e., systemic,** one acting through the blood upon the vomiting center.

**emeticology.** See *emetology*.

**emetine** (*em'-et-ēn*). An alkaloid from ipecac-root; it is emetic, diaphoretic, and expectorant. Emetic dose 1/8–1/4 gr. (0.008–0.016 Gm.); expectorant, 1/60–1/30 gr. (0.001–0.002 Gm.). Also a specific for amebiasis. Dose, subcutaneous injection 1/3 to 2/3 gr. emetine hydrochloride dissolved in sterile saline solution. See *ipecacuanha*.

**emetism** (*em'-et-izm*) [see *emesis*]. Poisoning from undue use of ipecac, manifested by acute inflammation of the pylorus, attended with hyperemesis and diarrhea and in some instances with paroxysms of coughing and asthmatic suffocation.

**emetized** (*em'-et-īzd*). 1. Prepared with tartar emetic. 2. Nauseated.

**emetocatharsis** (*em-et-o-kath-ar'-sis*) [ἔμετος, vomiting; καθαίρειν, to purge]. Vomiting and purgation at the same time, or produced by a common agent.

**emetocathartic** (*em-et-o-kath-ar'-tik*) [*emesis*; *cathartic*]. Having power to induce vomiting and purgation.

**emetology** (*em-et-ol'-o-je*) [ἔμετος, vomiting; λόγος, science]. The study or science of the physiology, pathology, and therapeutics of vomiting, and of the nature of emetics.

**emetomania** (*em-et-o-ma'-ne-ah*) [ἔμετος, vomiting; μανία, madness]. Insane desire for frequent emetics.

**emetomorphine** (*em-et-o-mor'-fēn*) [ἔμετος, vomiting; *Morpheus*, the god of sleep]. Apmorphine.

**emetophobia** (*em-et-o-fo'-be-ah*) [ἔμετος, vomiting; φόβος, fear]. Morbid dread or fear of vomiting.

**E. M. F.** Abbreviation of *electro-motive force*.

**emiction** (*e-mik'-shun*) [*e*, out; *mingere*, to void urine]. Urination.

**emictory** (*e-mik'-tor-e*) [*e*, out; *mingere*, to void urine]. 1. Promoting the secretion of urine. 2. A diuretic medicine.

**emigration** (*em-ig-ra'-shun*) [*e*, out; *migrare*, to wander]. The outward passage of a wandering-cell or leukocyte through the wall of a blood vessel.

**eminence** (*em'-in-ens*) [*eminentia*, an eminence]. A projecting, prominent part of an organ, especially of a bone. **e., arcuate,** a round protuberance on the upper aspect of the petrosa, marking the location of the superior semicircular canal. Syn., *jugum petrosum*. **e., articular,** the projection upon the zygomatic process which marks the anterior boundary of the glenoid fossa. **e., auditory.** See *auditory eminence*. **e., canine.** See *canine eminence*. **e., collateral,** a projection in the lateral ventricle of the brain between the middle and posterior horns. **e. of Doyère.** See *Doyère's eminence*. **e., frontal,** the two eminences of the frontal bone above the superciliary ridges. **e., iliopectineal,** a ridge on the upper surface of the pubic bone. **e., median,** the anterior pyramids. **e., nasal,** the prominence above the root of the nose, between the superciliary ridges. **e., occipital.** 1. The ridge in the paracele corresponding to the occipital fissure, distinct in the fetus. 2. See *occipital protuberance, external*. **e., parietal,** the eminence of the parietal bone. **e., posterior portal,** the caudate lobe of the liver. **e. of the scapha,** one on the dorsal aspect of the external ear corresponding to the scapha. **e., supracondylar,** that formed by the internal or external epicondyles. **e., thenar,** the eminence on the palm at the base of the thumb.

**eminentia** (*em-in-en'-she-ah*) [L.]. An eminence. **e. abducentis,** a medial ridge on the floor of the fourth ventricle. **e. acustica,** an elevation on the floor of the fourth ventricle at the lateral triangles. **e. annularis,** the pons Varolii. **e. arcuata.** See *eminence, arcuate*. **e. articularis,** a rounded ridge on the temporal bone in front of the glenoid fossa. **e. capitata,** the head of a bone. **e. caudata,** an isthmus connecting the Spigelian lobe with the under surface of the right lobe of the liver. **e. cinerea,** the elevated base of the trigonum vagi. **e. collateralis.** See *eminence, collateral*. **e. conchæ,** the posterior projection on the pinna corresponding to the concha. **e. cruciata,** ridges intersecting in the form of a cross on the superior surface of the occipital bone. **e. cuneatus,** a slight swelling of the internal funicle near the eminence of the clava; it contains the internal cuneate nucleus. **e. facialis,** the colliculus facialis. **e. Fallopii,** a ridge on the internal wall of the tympanum. **e. fossæ triangularis,** the posterior projection on the pinna corresponding to the fossa triangularis. **e. gracilis.** See *pyramid, posterior*. **e. hepatis caudata, e. hepatis longitudinalis, e. hepatis radiata,** the lobus caudata, the Spigelian lobe of the liver. **e. intercondyloidea**, the spinous process of the tibia. **e. jugularis**, the spine-like extremity of the jugular process of the occipital bone. **eminentiæ longitudinales**. See *e. caudata*. **e. mandibularis**, a bony protuberance of the inner surface of the skull, beneath the fossa mandibularis. **e. medialis**, an elevation on either side of the median line on the floor of the fourth ventricle. **e. pyramidalis**, a conical projection in the middle ear. **e., scaphæ**, the posterior projection on the pinna corresponding to the scaphoid fossa. **e. styloidea**, an elevation on the posterior wall of the tympanum. **e. teres**. Same as *e. abducentis*.

**eminential** (em-in-en'-shal) [*eminentia*, an eminence]. Relating to an eminence.

**emissarium** (em-is-a'-re-um) [L.]. A term for any canal or channel conveying a fluid outward. It is applied especially to the veins of the skull.

**emissary** (em'-is-a-re) [*emittere*, to send forth]. 1. An outlet. 2. Furnishing an oulet. **e. veins**, small veins piercing the skull and conveying blood outward.

**emission** (e-mish'-un) [*emittere*, to send forth]. 1. An ejaculation, or sending forth. 2. An involuntary seminal discharge.

**Emmanuel movement** (em-an'-u-el) [Emmanuel Church, Boston, where the cult originated]. A religious scheme to treat certain nervous and mental troubles by means of psychotherapy and religious exercises administered by the church authorities or others.

**emmenagogue, emmenagog** (em-en'-ag-og) [ἔμμηνα, the menses; ἀγωγός, leading]. 1. Stimulating the menstrual flow. 2. An agent that stimulates the menstrual flow. **e., direct**, one acting directly on the generative organs. **e., indirect**, one acting by relieving an underlying condition, as anemia, constipation, etc.

**emmenia** (em-e'-ne-ah) [L.]. The menses.

**emmeniopathy** (em-en-e-op'-a-the) [ἔμμηνα, menses; πάθος, illness]. Any disorder of menstruation.

**emmenology** (em-en-ol'-o-je) [ἔμμηνα, menses; λόγος, science]. That branch of science that treats of menstruation.

**emmenorrhea** (em-en-or-e'-ah) [ἔμμηνα, menses; ῥοία, a flowing]. The menses.

**Emmet's operation** (em'-et) [Thomas Addis Emmet, American gynecologist, 1828- ]. 1. Trachelorrhaphy; suturing of the neck of the womb. 2. A method of repairing lacerated perineum.

**emmetrope** (em'-et-rōp) [*emmetropia*]. A person whose eyes are emmetropic.

**emmetropia** (em-et-ro'-pe-ah) [ἐν, in; μέτρον, a measure; ὤψ, the eye]. Normal or perfect vision. The state of an eye in which, when accommodation is suspended, parallel rays of light are brought to a focus upon the retina.

**emmetropic** (em-et-rop'-ik) [see *emmetropia*]. Characterized by emmetropia.

**emodin** (em'-o-din) [Hind., *emodi*, rhubarb], C₁₅H₁₀O₅. A glucoside that occurs with chrysophanic acid in the bark of wild cherry, in cascara sagrada, and in the root of rhubarb.

**emol** (e'-mol) [*emollire*, to soften]. A fine powder composed of talc, silica, aluminum, and a trace of lime, miscible with water, and used as paste in the treatment of various forms of hyperkeratosis.

**emollient** (e-mol'-yent) [see *emol*]. 1. Softening; relaxing; soothing. 2. A substance used by external application to soften the skin; or, internally, to soothe an irritated or inflamed surface.

**emotiometabolic** (e-mo-she-o-met-ah-bol'-ik). Producing metabolism in consequence of some emotion.

**emotiomotor** (e-mo-she-o-mo'-tor). Inducing some activity in consequence of emotion.

**emotiomuscular** (e-mo-she-o-mus'-kū-lar). Relating to muscular activity which is due to emotion.

**emotion** (e-mo'-shun) [*emotio*, agitation]. Mental feeling, or sentiment, with the associated agitation, and often with more or less bodily commotion. With the emotions, desire, impulse, and will are intimately associated.

**emotional** (e-mo'-shun-al). Pertaining to the emotions. **e. insanity**, insanity characterized by exaggeration of the emotions or feelings.

**emotiovascular** (e-mo-she-o-vas'-kū-lar). Relating to some vascular change brought about by emotion.

**emotivity** (e-mo-tiv'-it-e) [*emotio*, agitation]. The degree of an individual's susceptibility to emotion.

**empasm** (em'-pazm) [ἐν, on; πάσσειν, to strew]. A perfumed powder for dusting the person.

**empathema** (em-path-e'-mah) [ἐν, in; πάθημα, suffering; *pl.*, *empathemata*]. A dominant or ungovernable passion or source of suffering. **e. atonicum**, hypochondriasis. **e. entonicum**, active mania. **e. inane**, harebrained and purposeless passion and excitement.

**emphlysis** (em'-flis-is) [ἐν, in, on; φλύσις, eruption: *pl.*, *emphlyses*]. Any vesicular or exanthematous eruption.

**emphractic** (em-frak'-tik) [ἐμφράττειν, to obstruct]. 1. Obstructive; closing the pores of the skin. 2. Any agent that obstructs the function of an organ, especially the excretory function of the skin.

**emphraxis** (em-fraks'-is) [ἔμφραξις, obstruction]. Obstruction; infarction; congestion.

**emphyma** (em-fi'-mah) [ἐν, in; φῦμα, growth; *pl.*, *emphymata*]. A tumor.

**emphysatherapy** (em-fiz-ah-ther'-aṗ-e) [ἐμφυσᾶν, to inflate; *therapy*]. The therapeutic injection of gas into a body-cavity.

**emphysema** (em-fiz-e'-mah) [ἐμφυσᾶν, to inflate]. A condition in which there is air or gas in normally airless tissues or an excess of air in tissues normally containing a certain quantity of it. **e., atrophic**, senile emphysema of the lung, characterized by a diminution in the size of the lung. **e., compensatory**, **e., complementary**, **e., essential**, pulmonary emphysema due to defective expansion of some other area of the lung in consequence of which the affected alveoli have assumed the function of a number of others and give way under the pressure. **e., cutaneous**, the presence of air or gas in the connective tissues beneath the skin, usually of the face. **e., gangrenous**. See *edema, malignant*. **e., hypertrophic**. See *e. pulmonary*. **e., interstitial**, the presence of gas in the connective tissue of a part, particularly in the connective tissue of the lung. **e., pulmonary**, a condition of the lungs characterized by a permanent dilatation of the alveoli with atrophy of the alveolar walls and the blood-vessels, resulting in a loss of the normal elasticity of the lung tissue. It is associated with dyspnea, with hacking cough and defective aeration of the blood. The chest becomes round or barrel-shaped, the right side of the heart hypertrophies, the abdominal viscera are displaced downward. The causes are a lessened resistance on the part of the lung, which may be inherited or acquired, and a distending force, which is usually expiratory in character, and consists in chronic cough, the blowing of wind-instruments, or other labor throwing a strain on the respiratory function. **e., subcutaneous**. See *e., cutaneous*. **e., substantial**. Synonym of *e., pulmonary*. **e., surgical**, distension of the subcutaneous tissue by air. **e., vesicular**, that due to dilatation of the air-vesicles.

**emphysematous** (em-fiz-em'-at-us). Affected with or of the nature of emphysema. **e. girdle**. See *emphysema*.

**emphysemodyspnea** (em-fiz-e-mo-disp'-ne-ah) [*emphysema*; *dyspnea*]. The dyspnea attending pulmonary emphysema.

**emphytriatreusis** (em-fi-tre-al-ru'-sis) [ἔμφυτος, innate; ἰάτρευσις, treatment]. The treatment of disease by an untrained person, whether clairvoyant, medium, trance-doctor, telepathist, mesmerist, or seventh son of a seventh son.

**empiric** (em-pir'-ik) [ἐμπειρικός, experienced]. 1. Based on practical observation and not on scientific reasoning. 2. One who in pertaining medicine relies solely on experience and not on scientific reasoning; a quack, or charlatan.

**empiricism** (em-pir'-is-izm) [*empiric*]. 1. Dependence upon experience or observation. 2. Quackery.

**Empis' "granulie."** Acute miliary tuberculosis of the lungs.

**emplastic** (em-plas'-tik) [ἐμπλαστικός, clogging]. 1. Suitable for a plaster. 2. A constipating medicine.

**emplastration** (em-plas-tra'-shun) [*emplastrum*]. The act of applying a plaster.

**emplastrum** (em-plas'-trum) [L.; *pl.*, *emplastra*]. A plaster (*q. v.*).

**empodic** (em-pod-is'-tik) [ἐμποδίζειν, to hinder]. 1. Checking; preventing. 2. A preventive remedy.

**emprosthocyrtoma** (em-pros-tho-sir-to'-mah, -sis). [ἔμπροσθεν, forward; κύρτωμα, a bending]. Lordosis.

**emprosthokyphosis** (*em-pros-tho-ki-fo'-sis*) [ἔμπροσθεν, forward; *kyphosis*]. Lordosis.
**emprosthotonia** (*em-pros-thot-o'-ne-ah*). Same as *emprosthotonos*.
**emprosthotonos** (*em-pros-thot'-o-nos*) [ἔμπροσθεν, forward; τόνος, tension]. Tonic muscular spasm in which the body is bent forward.
**emprosthozygosis** (*em-pros-tho-zi-go'-sis*) [ἔμπροσθεν, forward; ζυγοῦν, to join]. The condition of conjoined twins in which the fusion is anterior.
**emptysis** (*emp'-tis-is*) [ἐμπτύειν, to spit upon]. Hemorrhage from the lungs; hemoptysis.
**Empusa** (*em-pū'-zah*) [Ἔμπουσα, a hobgoblin]. A genus of fungi parasitic on living insects and causing their death.
**empyema** (*em-pi-e'-mah*) [ἐν, in; πύον, pus]. Pus in a cavity, especially in the pleural cavity. **e. necessitatis**, an empyema in which the pus burrows between the intercostal spaces and appears as a subcutaneous tumor. **e., pulsating**, one that transmits the pulsations of the heart to the chest-wall.
**empyematous** (*em-pi-em'-at-us*). Of the nature of or affected with empyema.
**empyemic** (*em-pi-em'-ik*). See *empyematous*.
**empyesis** (*em-pi-e'-sis*) [ἐμπύειν, to suppurate]. A pustular eruption, as smallpox; any disease characterized by phlegmonous pimples gradually filling with purulent fluid.
**empyocele** (*em-pi'-o-sēl*) [ἐν, in; πύον, pus; κήλη, tumor]. A purulent scrotal tumor.
**empyomphalus** (*em-pi-om'-fal-us*) [ἐν, in; πύον, pus; ὀμφαλός, navel]. A collection of pus at or about the navel.
**empyreuma** (*em-pi-ru'-mah*) [ἐμπύρευμα, a heating; a burnt flavor]. The odor developed in organic matter by destructive distillation.
**empyreumatic** (*em-pi-ru-mat'-ik*) [ἐμπύρευμα, a live coal]. Obtained from some organic substance by the aid of strong heat; as an empyreumatic oil.
**emulgent** (*e-mul'-jent*) [*emulgere*, to milk out]. 1. Draining; applied to the renal arteries as draining out the urine. 2. An emulgent vessel. 3. Any remedy that stimulates the emunctory organs. **e. veins**. See *vein*.
**emulsic** (*e-mul'-sik*). Relating to emulsin.
**emulsification** (*e-mul-sif-ik-a'-shun*). The process of making or becoming an emulsion.
**emulsify** (*e-mul'-se-fi*) [see *emulsion*]. To make into an emulsion.
**emulsin** (*e-mul'-sin*) [see *emulsion*]. A proteid ferment contained in bitter almonds. It aids in emulsifying almond oil, and, by its action on amygdalin, liberates hydrocyanic acid.
**emulsio** (*e-mul'-se-o*) [L.; *pl., emulsiones*]. See *emulsion*.
**emulsion** (*e-mul'-shun*) [*emulsum*, an emulsion]. A preparation consisting of a liquid, usually water, containing an insoluble substance in suspension.
**emulsionize** (*e-mul'-shun-iz*) [*emulgere*, to milk out]. To transform into an emulsion. The pancreatic juice *emulsionizes* fats.
**emulsive** (*e-mul'-siv*) [see *emulsion*]. 1. Forming or readily entering into an emulsion. 2. Affording oil on pressure, as certain seeds.
**emulsoid** (*e-mul'-soyd*) [*emulsion*; εἶδος, resemblance]. An emulsion colloid.
**emulsum** (*e-mul'-sum*) [L.]. An emulsion. The following emulsions are official: *e. amygdalæ, e. asafœtidæ, e. chloroformi, e. olei morrhuæ, e. olei morrhuæ cum hypophosphitibus, e. olei terebinthinæ*.
**emunctory** (*e-munk'-tor-e*), [*emungere*, to wipe the nose]. 1. Excretory; removing waste-products. 2. An organ that excretes waste-materials.
**emundans, emundant** (*e-mun'-danz, -dant*) [*mundare*, to cleanse]. Cleansing and disinfectant; applied to certain washes.
**emundantia, emundants** (*e-mun-dan'-she-ah, e-mun'-dants*) [see *emundans*]. Detergents.
**emundation** (*e-mun-da'-shun*) [see *emundans*]. 1. The act of cleansing. 2. The rectification of drugs.
**emusculate** (*e-mus'-kū-lāt*) [*e*, out; *musculus*, a muscle]. Without muscles.
**emydin** (*em'-id-in*) [μύς, the fresh-water tortoise]. A white proteid substance procured from the yolk of the eggs of turtles.
**enadelphia** (*en-ah-del'-fe-ah*). See *inclusion, fetal*.
**enamel** (*en-am'-el*) [ME., *enamaile*]. The vitreous substance of the crown of the teeth. **e.-column, e.-fiber, e.-prism, e.-rod**, any one of the minute, six-sided prisms of which the enamel of a tooth is composed. **e. cuticle**. See *Nasmyth's cuticle*. **e.-organ**, the ectodermic epithelial cap or process from which the enamel of a tooth is developed.
**enantesis** (*en-an-te'-sis*) [ἐναντίος, opposite]. The approximation of ascending and descending bloodvessels.
**enanthem** (*en-an'-them*). See *enanthema*.
**enanthema** (*en-an-the'-mah*) [ἐν, in; ἄνθημα, bloom, eruption: *pl., enanthemata*]. An eruption on a mucous membrane, or within the body, in distinction from *exanthema*.
**enanthematous** (*en-an-them'-at-us*) [ἐν, in; ἄνθημα, bloom, eruption]. Of the nature of or accompanied by an enanthema.
**enanthesis** (*en-an-the'-sis*) [ἐν, in; ἀνθεῖν, to bloom]. The process which causes an enanthema.
**enanthin, enanthin** (*e-nan'-thin*). A resinous substance contained in *Œnanthe crocata*, and *Œ. fistulosa*.
**enanthotoxin, œnanthotoxin** (*e-nan-tho-toks'-in*), C₁₇H₂₂O₅. A poisonous resinoid contained in *Œnanthe crocata*. It acts as does picrotoxin in producing violent convulsions.
**enanthrope** (*en'-an-thrōp*) [ἐν, in; ἄνθρωπος, man]. A source of disease originating internally.
**enantiobiosis** (*en-an-ti-o-bi-o'-sis*) [ἐναντίος, opposite; βίος, life]. Commensalism in which the associated organisms are antagonistic to each other's development. Cf. *symbiosis*.
**enantiomorphous, enantiomorphic** (*en-an-te-o-mor'-fus, en-an-te-o-mor'-fik*) [ἐναντίος, opposite; μορφή, form]. Similar but contrasted or reversed in form. Thus the two hands are enantiomorphous.
**enantiopathic** (*en-an-te-o-path'-ik*). 1. Palliative. 2. Pertaining to enantiopathy.
**enantiopathy** (*en-an-te-op'-ath-e*) [ἐναντίος, opposite; πάθος, disease]. A disease antagonistic to another disease.
**enarkyochrome** (*en-ar'-ke-o-krom*) [ἐν, in; ἄρκυν, a net; χρῶμα, color]. Nissl's term for a nerve-cell taking the stain best in the cell-body, the formed part of which is arranged in the shape of a network.
**enarthrodia** (*en-ar-thro'-de-ah*). See *enarthrosis*.
**enarthrodial** (*en-ar-thro'-de-al*) [*enarthrosis*]. Having the character of an enarthrosis.
**enarthrosis** (*en-ar-thro'-sis*) [ἐν, in; ἄρθρον, a joint]. A ball-and-socket joint, like that of the hip.
**enarthrum** (*en-ar'-thrum*) [see *enarthrosis*]. A foreign body lodged in a joint.
**encanthis** (*en-kan'-this*) [ἐν, in; κάνθος, canthus]. A newgrowth in the inner canthus of the eye.
**encapsulation** (*en-kap-sū-la'-shun*) [ἐν, in; *capsula*, a capsule]. The process of surrounding a part with a capsule.
**encarditis** (*en-kard-i'-tis*). Same as *endocarditis*.
**enceinte** (*on(g)-sant'*) [Fr.]. With child; pregnant.
**encelitis, encelitis** (*en-se-li'-tis*) [ἐν, in; κοιλία, belly; ιτις, inflammation]. Inflammation of the abdominal viscera.
**encephal** (*en'-sef-al*). Same as *encephalon, q. v.*
**encephalalgia** (*en-sef-al-al'-je-ah*) [*encephalon*; ἄλγος, pain]. Pain in the head. **e. hydropica**, hydrocephalus.
**encephalanalosis** (*en-sef-al-an-al-o'-sis*) [*encephalon*; ἀνάλωσις, a wasting away]. Cerebral atrophy.
**encephalasthenia** (*en-sef-al-as-the'-ne-ah*) [*encephalon*; *asthenia*]. Althaus' term for the cerebral form of neurasthenia; failure of brain power.
**encephalatrophic** (*en-sef-al-at-rof'-ik*) [*encephalo-*; ἀτροφία, wasting]. Relating to brain-atrophy.
**encephalauxe** (*en-sef-al-awks'-e*) [*encephalon*; αὔξη, increase]. Hypertrophy of the brain.
**encephaledema** (*en-sef-al-e-de'-mah*) [*encephalon*; *edema*]. Edema of the brain.
**encephalelcosis** (*en-sef-al-el-ko'-sis*) [*encephalon*; *helcosis*]. Ulceration of the brain.
**encephalemia, encephalæmia** (*en-sef-al-e'-me-ah*). See *encephalohemia*.
**encephalic** (*en-sef-al'-ik*) [*encephalon*]. Pertaining to the brain.
**encephalin** (*en-sef'-al-in*) [ἐγκέφαλος, the brain]. A nitrogenous glucoside extracted from brain-tissue.
**encephalion** (*en-sef-al'-e-on*) [dim. of ἐγκέφαλος, the brain]. The cerebellum.
**encephalitic** (*en-sef-al-it'-ik*) [*encephalo-*; ιτις, inflammation]. Relating to or affected with encephalitis.
**encephalitis** (*en-sef-al-i'-tis*) [*encephalon*; ιτις, inflammation]. Inflammation of the brain. **e.**

**neonatorum** (Virchow), localized softening consisting of numerous yellow spots surrounded by hemorrhage; these occur most commonly in the brains of syphilitic infants.
**encephalo-** (*en-sef-al-o-*) [*encephalon*]. A prefix meaning relating to the encephalon or brain.
**encephalocele** (*en-sef'-al-o-sēl*) [*encephalo-*; κήλη, tumor]. A hernia of the brain through a cranial fissure.
**encephalocœle** (*en-sef-al-o-se'-le*) [*encephalo-*; κοίλος, hollow]. 1. The cranial cavity. 2. The ventricles of the brain.
**encephalodialysis** (*en-sef-al-o-di-al'-is-is*) [*encephalo-*; διά, through; λύειν, to loose]. Softening of the brain.
**encephalodynia** (*en-sef-al-o-din'-e-ah*) [*encephalo-*; ὀδύνη, pain]. Same as *encephalalgia*.
**encephalohemia** (*en-sef-al-o-he'-me-ah*) [*encephalo-*; αἷμα, blood]. Congestion of the brain.
**encephaloid** (*en-sef'-al-oid*) [*encephalo-*; εἶδος, like]. 1. Resembling brain tissue. 2. Soft carcinoma. See *carcinoma, encephaloid*.
**encephalolith** (*en-sef'-al-o-lith*) [ἐγκέφαλος, brain; λίθος, stone]. A calculus of the brain; a brain-stone.
**encephalolithiasis** (*en-sef-al-o-lith-i'-as-is*) [ἐγκέφαλος, brain; λιθίασις, the formation of calculi]. The formation of brain-stones.
**encephalology** (*en-sef-al-ol'-o-je*) [ἐγκέφαλος, brain; λόγος, science]. The anatomy, physiology, and pathology of the brain.
**encephaloma** (*en-sef-al-o'-mah*) [*encephalo-*; ὄγκωμα, tumor]. 1. A tumor of the brain. 2. Encephaloid carcinoma.
**encephalomalacia** (*en-sef-al-o-mal-a'-she-ah*) [*encephalo-*; μαλακία, softening]. Softening of the brain-substance.
**encephalomeningitis** (*en-sef-al-o-men-in-ji'-tis*) [*encephalo-*; meninges; ιτις, inflammation]. Combined inflammation of the brain and membranes.
**encephalomeningocele** (*en-sef-al-o-men-in'-go-sēl*) [*encephalo-*; *meningocele*]. Hernia of the membranes and brain-substance.
**encephalomere** (*en-sef'-al-o-mēr*) [ἐγκέφαλος, brain; μέρος, share]. Any one of the succession of natural segments of axial parts into which the brain is divisible.
**encephalometer** (*en-sef-al-om'-et-er*) [ἐγκέφαλος, brain; μέτρον, measure]. An instrument for measuring the cranium and locating certain brain-regions.
**encephalomyelitis** (*en-sef-al-o-mi-el-i'-tis*). Encephalitis combined with myelitis.
**encephalomyelopathy** (*en-sef-al-o-mi-el-op'-ath-e*) [*encephalo-*; μυελός, marrow; πάθος, disease]. Any disease affecting both the brain and spinal cord.
**encephalon** (*en-sef'-al-on*) [ἐγκέφαλος, brain]. The brain.
**encephalonarcosis** (*en-sef-al-o-nar-ko'-sis*) [ἐγκέφαλος, brain; ναρκοῦν, to benumb]. Stupor from some brain-lesion.
**encephaloncus** (*en-sef-al-ong'-kus*) [ἐγκέφαλος, brain; ὄγκος, a tumor]. See *encephalophyma*.
**encephalopathy** (*en-sef-al-op'-ath-e*) [*encephalo-*; πάθος, disease]. Any disease of the brain.
**encephalophyma** (*en-sef-al-o-fi'-mah*) [ἐγκέφαλος, brain; φῦμα, a growth]. A tumor of the brain.
**encephalopyosis** (*en-sef-al-o-pi-o'-sis*) [*encephalo-*; *pyosis*]. Abscess of the brain.
**encephalorrhachidian, encephalorachidian** (*en-sef-al-o-ra-kid'-e-an*). Same as *cerebrospinal*.
**encephalorrhagia** (*en-sef-al-o-ra'-je-ah*) [ἐγκέφαλος, brain; ῥηγνύναι, to break forth]. Cerebral hemorrhage.
**encephaloscopy** (*en-sef-al-os'-ko-pe*) [*encephalo-*; σκοπεῖν, to examine]. Examination of the brain.
**encephalosepsis** (*en-sef-al-o-sep'-sis*) [*encephalo-*; σῆψις, decay]. Gangrene of the tissue of the brain.
**encephalosis** (*en-sef-al-o'-sis*). The formation of an encephaloma.
**encephalospinal** (*en-sef-al-o-spi'-nal*) [*encephalo-*; *spina*, the spine]. Pertaining to the brain and spinal cord. **e. axis**, the cerebrospinal axis.
**encephalothlipsis** (*en-sef-al-o-thlip'-sis*) [*encephalo-*; θλίψις, pressure]. Pressure on the brain.
**encephalotome** (*en-sef-al'-o-tōm*) [ἐγκέφαλος, brain; τομή, section]. An instrument for slicing the encephalon for examination or preservation.
**encephalotomy** (*en-sef-al-ot'-o-me*) [ἐγκέφαλος, brain; τομή, section]. 1. The anatomy or dissection of the brain; surgical incision of the brain. 2. Obstetric craniotomy.

**enchondral** (*en-kon'-dral*). See *endochondral*.
**enchondroma** (*en-kon-dro'-mah*) [ἐν, in; χόνδρος, cartilage; ὄμα, tumor; *pl.*, *enchondromata*]. 1. A tumor arising from or resembling cartilage in texture, etc.; chondroma. 2. A cartilaginous growth within an organ or tissue.
**enchondrosarcoma** (*en-kon-dro-sar-ko'-mah*). Sarcoma containing cartilaginous tissue.
**enchondrosis** (*en-kon-dro'-sis*) [ἐν, within; χόνδρος, cartilage; *pl.*, *enchondroses*]. An outgrowth of cartilage from an osseous or cartilaginous structure; the process by which an enchondroma is developed.
**enchylema** (*en-ki-le'-mah*) [ἐν, in; χυλός, juice]. A fluid, granular substance filling the interstices of the cell-body and the nucleus.
**enchyma** (*en'-ke-mah*). [ἐγχεῖν, to pour in]. An organic juice elaborated from chyme, the formative juice of tissues.
**enclave** (*en-klav'*, *on(g)-klahv'*) [Fr.]. Any substance enclosed within a foreign tissue, as an oil-globule in a cell; any exclave (*q. v.*) considered in relation to the part that surrounds it.
**enclavement** (*on(g)-klahv'-mon(g)*) [Fr.]. Retention due to a constriction; impaction, as of the head in the pelvic strait.
**enclitic** (*en-klit'-ik*) [ἐγκλιτικός, leaning on]. Presenting obliquely; not synclitic.
**encoleosis** (*en-ko-le-o'-sis*) [ἐν, in; κόλεος, sheath]. Invagination.
**encolpism, encolpismus** (*en-kol'-pism, en-kol-pis'-mus*) [ἐν, in; κόλπος, the vagina]. 1. A vaginal suppository. 2. Medication by vaginal suppositories.
**encolpitis** (*en-kol-pi'-tis*) [ἐν, in; κόλπος, vagina]. Mucous vaginitis.
**encranial** (*en-kra'-ne-al*). See *intracranial*.
**encraty** (*en'-krat-e*) [ἐγκράτεια, mastery]. Self-control; continence or strict temperance.
**encyesis** (*en-si-e'-sis*) [ἐγκύησις]. Pregnancy.
**encysted** (*en-sist'-ed*) [ἐν, in; κύστις, a cyst]. Inclosed in a cyst or capsule.
**encystment** (*en-sist'-ment*) [ἐν, in; κύστις, a bag]. The process of becoming encysted.
**end** [ME., *ende*]. The terminal point of a thing. **e.-artery**, one that does not communicate with other arteries. **e.-body**, Wassermann's term for that substance which kills the bacteria in the production of immunity to typhoid. Cf. *body, immune*. **e. brain**. See *telencephalon*. **e. brush**, the finely branched terminal expansion of an axone. **e.-bud**, **e.-bulb**, the terminal bulb of a nerve in the skin. **e.-organ**, the terminal part of a sensory nerve-fiber. **e.-plate**. 1. The expanded terminal of a motor nerve upon a bundle of muscular fibers. 2. The achromatic masses at the poles of the spindle in karyokinesis of *Protozoa*. **e.-plate, motorial nerve**. See *motorial end-plate*.
**Endamœba, Endameba**. See *Entamœba, Entameba*.
**endangeitis, endangiitis** (*end-an-je-i'-tis, end-an-ji'-tis*) [*endo-*; ἀγγεῖον, vessel; ιτις, inflammation]. Inflammation of the endangium.
**endangic, endangiic** (*end-an'-jik, end-an-jid'-ik*) [see *endangium*]. Endovascular.
**endangium** (*end-an'-je-um*) [*endo-*; ἀγγεῖον, vessel]. The intima or inmost coat of a blood-vessel.
**endanthem** (*end-anth'-them*) [*endo-*; ἀνθεῖν, to bloom]. A term applied to a mucous exanthem.
**endaortitis** (*end-a-or-ti'-tis*) [*endo-*; aorta; ιτις, inflammation]. Inflammation of the intima of the aorta.
**endarteritis** (*end-ar-ter-i'-tis*) [*endo-*; ἀρτηρία, artery; ιτις, inflammation]. Inflammation of the inner coat of an artery. **e. obliterating, arteritis obliterans**, a form in which the production of new connective tissue obliterates the vessel-lumen.
**endaxoneuron** (*en-daks-o-nū'-ron*) [*endo-*; *axoneuron*]. A neuron whose nerve-process does not leave the spinal cord; the endaxoneurons include the column cells and the internal cells.
**endectoplastic** (*end-ek-to-plas'-tik*) [*endo-*; ἐκτός, outward; πλάσσειν, to form]. Applied to cells which form tissue by a metamorphosis of the protoplasm at both the periphery and the center.
**endeictic** (*en-dīk'-tik*) [ἐν; δεικνύναι, to show]. Symptomatic; serving as an indication.
**endeixis** (*en-dīks'-is*) [ἔνδειξις, a pointing out]. A symptom, sign, or indication.
**endemic** (*en-dem'-ik*) [ἐν, in; δῆμος, a people]. Of a disease, found in a certain place more or less constantly.

**endemicity, endemism** (*en-dem-is'-it-e, en'-dem-izm*) [ἐν, in; δῆμος, a people]. The quality of being endemic.
**endemiology** (*en-de'-me-ol-o-je*) [ἐν, in; δῆμος, a people; λόγος, science]. The science of endemic diseases.
**endemoepidemic** (*en-dem-o-ep-e-dem'-ik*). Endemic, but periodically becoming epidemic.
**endepidermis** (*end-ep-e-der'-mis*) [*endo-*; *epidermis*]. The inner layer of the epidermis.
**endermic, endermatic** (*en-der'-mik, en-der-mat'-ik*) [ἐν, in; δέρμα, the skin]. Situated on or applied to the true skin; within the skin. *e*. **medication,** a method of administering medicines through the skin after removal of the cuticle by means of a blister.
**endermism** (*en'-derm-izm*). The endermatic administration of remedies. See *endermic*.
**endermosis** (*en-der-mo'-sis*) [see *endermic*]. 1. A method of administering medicines through the skin by rubbing. 2. Any herpetic affection of a mucosa.
**enderon** (*en'-der-on*) [ἐν, in; δέρος, skin]. The true skin or derm, together with the non-epithelial portion of the mucous membrane.
**enderonic** (*en-der-on'-ik*) [ἐν, in; δέρος, skin]. Pertaining to or of the nature of the enderon.
**endexoteric** (*en-deks-o-ter'-ik*) [*endo-*; ἐξωτερικός, outer]. Due both to internal and external causes.
**end-lobe** (*end'-lōb*). The occipital lobe of the brain.
**endo-** (*en-do-*) [ἔνδον, within]. A prefix meaning within.
**Endo's medium** (*en'-do*). A culture medium of lactose agar with sodium hydroxide, phenolphthalein, fuchsin and sodium sulphite; it was recommended as an aid in differentiating between *B. coli*, and *B. typhosus*.
**endoabdominal** (*en-do-ab-dom'-in-al*). Within the abdomen.
**endoaneurysmorrhaphy** (*en-do-an-ū-ris'-mor-af-e*). The operation of opening an aneurysmal sac and of suturing all openings inside of it.
**endoaortitis** (*en-do-a-or-ti'-tis*). See *endaortitis*.
**endoappendicitis** (*en-do-ap-en-dis-i'-tis*) [*endo-*; *appendicitis*]. Inflammation of the mucosa of the vermiform appendix.
**endoarteritis** (*en-do-ar-ter-i'-tis*). See *endarteritis*.
**endoauscultation** (*en-do-aws-kul-ta'-shun*) [*endo-*; *auscultare*, to listen to]. A method of auscultation by means of an esophageal tube passed into the stomach.
**endoblast** (*en'-do-blast*) [*endo-*; βλαστός, a germ]. The cell-nucleus; the internal blastema.
**endoblastic** (*en-do-blas'-tik*) [see *endoblast*]. 1. Having an endoblast or nucleus. 2. Pertaining to the nucleus.
**endobronchitis** (*en-do-brong-ki'-tis*) [*endo-*; *bronchitis*]. Inflammation of the bronchial mucosa.
**endocardiac, endocardial** (*en-do-kar'-de-ak, -de-al*) [*endocardium*]. Situated or arising within the heart.
**endocarditic** (*en-do-kar-dit'-ik*) [*endo-*; καρδία, the heart]. Pertaining to or affected with endocarditis.
**endocarditis** (*en-do-kar-di'-tis*) [*endocardium*; ιτις, inflammation]. Inflammation of the endocardium or lining membrane of the heart. The condition may be acute or chronic. *Acute endocarditis* is either warty or ulcerative, both of these being microorganismal in origin. The most frequent causes of the acute form are rheumatism and the infectious fevers. The disease usually affects the valves of the left side of the heart, and gives rise to a murmur, to fever, dyspnea, and rapid pulse. In the ulcerative form the symptoms resemble those of pyemia (hectic fever, chills, sweats, embolic processes). *Chronic* or *sclerotic endocarditis* is either a terminal process following the acute forms, or is a primary affection beginning insidiously. The latter is usually associated with general arteriosclerosis, and is due to gout, rheumatism, alcoholism, syphilis, and to other obscure causes. Both the acute and the chronic form give rise to insufficiency or obstruction of the valvular orifice, or to both combined.
**endocardium** (*en-do-kar'-de-um*) [*endo-*; καρδία, the heart]. The serous membrane lining the interior of the heart.
**endocarp** (*en'-do-karp*) [*endo-*; καρπός, fruit]. The inner hard and stony membranous or fleshy layer of a pericarp; *e. g*., the stone of a peach.
**endocervical** (*en-do-ser'-vik-al*) [*endo-*; *cervix*, neck]. Relating to the inside of the uterine cervix.

**endocervicitis** (*en-do-ser-vis-i'-tis*) [*endo-*; *cervix*, neck; ιτις, inflammation]. Inflammation of the lining membrane of the cervix uteri.
**endochondral** (*en-do-kon'-dral*) [*endo-*; χόνδρος, cartilage]. Situated within a cartilage.
**endochorion** (*en-do-ko'-re-on*) [*endo-*; *chorion*]. The inner chorion; the vascular layer of the allantois.
**endochorionic** (*en-do-ko-re-on'-ik*) [*endo-*; *chorion*]. Relating to the endochorion.
**endochrome** (*en'-do-krōm*) [*endo-*; χρῶμα, color]. The coloring-matter, other than green, of the endoplasm of a cell, or that of diatoms or of flowers.
**endochylema** (*en-do-ki-le'-mah*) [*endo-*; χύλος, juice]. The semi-fluid substance filling the protoplasmic reticulum of a cell.
**endochyme** (*en'-do-kīm*) [*endo-*; χυμός, juice]. The formative cell-sap elaborated from the chyme.
**endocolitis** (*en-do-ko-li'-tis*). See *colitis*.
**endocolpitis** (*en-do-kol-pi'-tis*) [*endo-*; κόλπος, vagina]. Same as *encolpitis*.
**endocomplements** (*en-do-kom'-ple-ments*). A class of intracellular complements.
**endocranial** (*en-do-kra'-ne-al*). 1. Relating to the endocranium. 2. See *intracranial*.
**endocranitis** (*en-do-kra-ni'-tis*) [*endo-*; κρανίον, the skull; ιτις, inflammation]. Inflammation of the endocranium; pachymeningitis externa.
**endocranium** (*en-do-kra'-ne-um*) [*endo-*; κρανίον, the skull]. 1. The cerebral dura. 2. The inner surface of the skull.
**endocrin** (*en'-do-krin*) [*endo-*; κρίνειν, to separate]. Any internal secretion.
**endocrinology** (*en-do-kri-nol'-o-je*) [*endo-*; κρίνειν, to separate; λόγος science]. The study of the endocrinous glands and their secretions; the study of the internal secretions.
**endocrinous, endocrinic** (*en-dok'-rin-us, en-do-krin'-ik*) [see *endocrin*]. Pertaining to an internal secretion or to a gland producing such a secretion. *e*. **glands.** 1. Glands producing an internal secretion. 2. The ductless glands.
**endocritic** (*en-do-krit'-ik*). Same as *endocrinic*.
**endocular** (*end-ok'-ū-lar*). Intraocular.
**endocyma** (*en-do-si'-mah*) [*endo-*; κῦμα, a fetus]. A form of double monstrosity in which the parasite is contained within the body of the autosite.
**endocystitis** (*en-do-sist-i'-tis*). See *cystitis*.
**endocytic** (*en-do-sit'-ik*) [*endo-*; κύτος, a cell]. Relating to the contents of a cell.
**endoderm** (*en'-do-derm*) [*endo-*; δέρμα, skin]. The inner of the two primitive cell-layers of the embryo. It lines the cavity of the primitive intestine and its derivatives. Syn., *hypoblast*. See *blastoderm*.
**endodermal, endodermic** (*en-do-derm'-al, -ik*). Relating to the endoderm; applied to structures originating in the lower layers of the derma, as dentin.
**endodermis** (*en-do-der'-mis*) [*endo-*; δέρμα, skin]. The layer of cells surrounding a fibrovascular cylinder.
**endodiascopy** (*en-do-di-as'-ko-pe*) [*endo-*; δία, through; σκοπεῖν, to examine]. Exploration by means of a Crookes tube introduced into a natural body-cavity.
**endodontitis** (*en-do-don-ti'-tis*) [*endo-*; ὀδούς, a tooth; ιτις, inflammation]. Inflammation of the pulp of a tooth.
**endoenteritis** (*en-do-en-ter-i'-tis*). See *enteritis*.
**endoesophagitis** (*en-do-e-sof-aj-i'-tis*) [*endo-*; *esophagitis*]. Inflammation of the membrane lining the esophagus.
**endoexoteric** (*en-do-eks-o-ter'-ik*) [*endo-*; ἐξωτερικός, external]. Applied to a disease the origin of which is both endopathic and exopathic.
**endogastritis** (*en-do-gas-tri'-tis*) [*endo-*; *gastritis*]. Inflammation of the mucous membrane of the stomach.
**endogenesis, endogeny** (*en-do-jen'-e-sis, en-doj'-en-e*) [*endo-*; γένεσις, production]. Growth within; endogenous formation.
**endogenous** (*en-doj'-en-us*) [see *endogenesis*]. Produced within. Applied to spore-formation or cell-formation inside of a parent-cell.
**endoglobular** (*en-do-glob'-ū-lar*) [*endo-*; *globus*, a ball]. Within the blood-corpuscles.
**endognathion** (*end-og-na'-the-on*) [*endo-*; γνάθος, jaw]. The middle portion of the superior maxilla.
**endogonium** (*en-do-go'-ne-um*) [*endo-*; γόνος, seed]. A gonidium formed inside of a receptacle of parent-cell, as in the *Saprolegnieæ*. *Mucorini*, *Vaucheria*.
**endolaryngeal** (*en-do-lar-in'-je-al*) [*endo-*; *larynx*]. Within the larynx.

**endolemma** (*en-do-lem'-ah*). Synonym of *neurilemma*.

**endolymph** (*en'-do-limf*) [*endo-*; *lympha*, water]. The fluid of the membranous labyrinth of the ear.

**endolymphangeal** (*en-do-lim-fan'-je-al*) [*endo-*; *lympha*, lymph; ἀγγεῖον, vessel]. Situated or belonging within a lymph-vessel; as an endolymphangeal nodule.

**endolymphic** (*en-do-lim'-fik*). Relating to or of the nature of endolymph.

**endolysin** (*en-dol'-is-in*) [*endo-*; *lysin*]. An intracellular leukocytic bactericidal substance.

**endomastoiditis** (*en-do-mas-toid-i'-tis*) [*endo-*; *mastoiditis*]. Inflammation within the mastoid cavity.

**endometrectomy** (*en-do-met-rek'-to-me*) [*endometrium*; ἐκτομή, a cutting out]. The extirpation of the entire mucosa of the uterus through the abdomen and incised uterus.

**endometrial** (*en-do-me'-tre-al*) [*endo-*; μήτρα, the womb]. Pertaining to the endometrium; situated within the uterus.

**endometritis** (*en-do-me-tri'-tis*) [*endometrium*; ἴτις, inflammation]. Inflammation of the endometrium. e., **cervical**. See *endocervicitis*. e. **dissecans**, e., **dissecting**, e. **exfoliativa**. See *dysmenorrhea, membranous*. e. **dolorosa**, painful spasms or continuous pain in the region of the uterus, believed to be due to an inflammatory lesion of the uterine mucosa localized in the fundus at the internal orifice and opening of the tube. e., **fungous**, that in which the lining membrane is hypertrophied, with the formation of vascular granulations. Syn., *hemorrhagic endometritis*. e., **hemorrhagic**. See *e*, *fungous*. e., **simple**, a catarrhal inflammation of the endometrium.

**endometrium** (*en-do-me'-tre-um*) [*endo-*; μήτρα, uterus]. The mucous membrane lining the uterus.

**endometry** (*en-dom'-et-re*) [*endo-*; μέτρον, measure]. The measurement of the interior of an organ or cavity, as of the cranium.

**endometrorrhagia** (*en-do-met-ro-raj'-e-ah*). See *metrorrhagia*.

**endomyocarditis** (*en-do-mi-o-kar-di'-tis*) [*endo-*; μῦς, muscle; καρδία, heart; ἴτις, inflammation]. Inflammation of both endocardium and myocardium.

**endomyisal** (*en-do-mis'-e-al*) [*endo-*; μῦς, muscle]. Pertaining to or of the nature of endomysium.

**endomysium** (*en-do-mis'-e-um*) [*endo-*; μῦς, muscle]. The connective tissue between the fibrils of a muscular bundle.

**endonephritis** (*en-do-nef-ri'-tis*) [*endo-*; νεφρός, kidney; ἴτις, inflammation]. Synonym of *pyelitis*.

**endoneural** (*en-do-nū'-ral*) [*endo-*; νεῦρον, nerve]. Relating to or situated within the interior of a nerve.

**endoneurial** (*en-do-nū'-re-al*) [*endo-*; νεῦρον, a nerve]. Relating to the endoneurium.

**endoneuritis** (*en-do-nū-ri'-tis*). Inflammation of the endoneurium.

**endoneurium** (*en-do-nū'-re-um*) [*endo-*; νεῦρον, a nerve]. The delicate connective tissue holding together the fibrils of a bundle of nerves.

**endoparasite** (*en-do-par'-as-īt*) [*endo-*; παράσιτος, parasite]. A parasite living within its host.

**endoparasitic** (*en-do-par-as-it'-ik*). Of the nature of an endoparasite.

**endopathic** (*en-do-path'-ik*) [*endo-*; πάθος, disease]. Pertaining to the origin of disease from conditions or causes not derived from without. See *exopathic*.

**endopathy** (*en-dop'-ath-e*) [*endo-*; πάθος, disease]. Any disease arising within the body.

**endoperiarteritis** (*en-do-per-e-ar-ter-i'-tis*). Endarteritis combined with periarteritis.

**endopericarditis** (*en-do-per-ik-ar-di'-tis*) [*endo-*; περί, around; καρδία, the heart; ἴτις, inflammation]. Inflammation of both endocardium and pericardium.

**endoperimyocarditis** (*en-do-per-e-mi-o-kar-di'-tis*) [*endo-*; περί, around; μῦς, muscle; καρδία, heart; ἴτις, inflammation]. Inflammation of endocardium, pericardium, and myocardium.

**endoperineuritis** (*en-do-per-e-nū-ri'-tis*) [*endo-*; περί, around; νεῦρον, a nerve; ἴτις, inflammation]. Inflammation of both endoneurium and perineurium.

**endoperitonitis** (*en-do-per-it-on-i'-tis*) [*endo-*; *peritonitis*]. Synonym of *peritonitis*.

**endophlebitis** (*en-do-fle-bi'-tis*) [*endo-*; *phlebitis*]. Inflammation of the inner coat of a vein.

**endophyte** (*en'-do-fīt*) [*endo-*; φυτόν, a plant]. Same as *entophyte*.

**endoplasm** (*en'-do-plazm*) [*endo-*; πλάσμα, a thing formed]. The inner granular protoplasm of a protozoan or of a histologic cell.

**endoplast** (*en'-do-plast*) [*endo-*; πλαστός, formed]. In biology, the nuclear body of a protozoan; the h m l gue of the nucleus of a histologic cell. (Huxley.) o o

**endoplastule** (*en-do-plas'-tūl*) [*endo-*; πλαστός, formed]. In biology, a small oval or rounded body, often found attached to the endoplast of a protozoan; the homologue of the nucleolus of a histologic cell. Cf. *entoblast*.

**endorhinitis** (*en-do-ri-ni'-tis*) [*endo-*; *rhinitis*]. Inflammation of the membrane lining the nasal passages.

**endosalpingitis** (*en-do-sal-pin-ji'-tis*) [*endo-*; σάλπιγξ, trumpet, tube; ἴτις, inflammation]. Inflammation of the lining membrane of a Fallopian tube.

**endosarc** (*en'-do-sark*) [*endo-*; σάρξ, flesh]. The inner protoplasm of a protozoan.

**endoscope** (*en'-do-skōp*) [*endo-*; σκοπεῖν, to observe]. An instrument for the examination of a body-cavity through its natural outlet.

**endoscopy** (*en-dos'-ko-pe*) [*endo-*; σκοπεῖν, to observe]. The practice or process of using the endoscope. The examination of cavities or organs within the body by means of an endoscope.

**endosepsis** (*en-do-sep'-sis*) [*endo-*; σῆψις, decay]. Septicemia arising within the body.

**endoskeleton** (*en-do-skel'-et-on*) [*endo-*; σκελετόν, a dry body]. The internal supporting structure of an animal.

**endosmic** (*en-dos'-mik*). Relating to endosmosis.

**endosmometer** (*en-dos-mom'-et-er*) [*endosmosis*; μέτρον, a measure]. An instrument for measuring endosmosis.

**endosmose** (*en'-dos-mōs*). Same as *endosmosis*.

**endosmosis** (*en-dos-mo'-sis*) [*endo-*; ὠσμός, impulsion]. The passage of a liquid through a porous septum from without inward.

**endosmotic** (*en-dos-mot'-ik*) [see *endosmosis*]. Pertaining to endosmosis. e. **equivalent**, the weight of distilled water that passes into the flask of the endosmometer in exchange for a known weight of the soluble substance.

**endosperm** (*en'-do-sperm*) [*endo-*; σπέρμα, seed]. In biology, the albumin of a seed.

**endospore** (*en'-do-spōr*) [*endo-*; σπόρος, seed]. 1. A spore formed within the parent-cell. 2. The inner coat of a spore.

**endosteal** (*en-dos'-te-al*) [*endo-*; ὀστέον, a bone]. Relating to endosteum.

**endosteitis** (*en-dos-te-i'-tis*) [*endo-*; ὀστέον, a bone; ἴτις, inflammation]. Inflammation of the endosteum.

**endosteoma, endostoma** (*en-dos-te-o'-mah, en-dos-to'-mah*) [*endo-*; ὀστέον, bone; pl., *endostomata*]. A bony tumor within a bone, or in a cavity surrounded by bone.

**endostethoscope** (*en-do-steth'-o-skōp*) [*endo-*; *stethoscope*]. A form of stethoscope for auscultation through the esophagus.

**endostoma** (*en-dos'-te-um*) [*endo-*; ὀστέον, bone]. The vascular membranous layer of connective tissue lining the medullary cavity of bones.

**endostosis** (*end-os'-to-sis*) [*endo-*; ὀστέον, bone: pl., *endostoses*]. Ossification of a cartilage.

**endothelial** (*en-do-the'-le-al*) [*endothelium*]. Pertaining to endothelium.

**endothelioid** (*en-do-the'-le-oid*) [*endothelium*; εἶδος, form]. Resembling endothelium.

**endothelioinoma** (*en-do-the-le-o-in-o'-mah*) [*endothelium*; ἴς, a fiber; ὄμα, a tumor: pl., *endothelioinomata*]. A malignant inoma or fibroma of endothelial origin.

**endothelioleiomyoma** (*en-do-the-le-o-li-o-mi-o'-mah*) [*endothelium*; λεῖος, smooth; μῦς, muscle; ὄμα, tumor: pl., *endothelioleiomyomata*]. Endothelial and malignant leiomyoma; myosarcoma.

**endotheliolysin** (*en-do-the-le-ol'-is-in*) [*endothelium*; λύσις, a loosing]. A cytotoxin endowed with the capacity of dissolving endothelial cells. Syn., *hemorrhagin*.

**endothelioma** (*en-do-the-le-o'-mah*) [*endothelium*; ὄμα, a tumor]. A variety of sarcoma formed by the multiplication of the endothelial cells of lymphatic spaces.

**endotheliomyoma** (*en-do-the-le-o-mi-o'-mah*). A myoma springing from endothelium.

**endotheliomyxoma** (*en-do-the-le-o-miks-o'-mah*) [*endothelium*; μύξα, mucus; ὄμα, a tumor: pl., *endotheliomyxomata*]. Endothelial and malignant myxoma.

**endotheliorhabdomyoma** (en-do-the'-le-o-rab-do-mi-o'-mah) [endothelium; ῥάβδος, a rod; μῦς, muscle; ὄμμα, a tumor: pl., endotheliorhabdomyomata]. Endothelial and malignant rhabdomyoma.
**endotheliotoxin** (en-do-the-le-o-toks'-in). See endotheliolysin.
**endothelium** (en-do-the'-le-um) [endo-; θηλή, a nipple]. The lining membrane of serous, synovial, and other internal surfaces. e., subepithelial. See Débove's membrane. e., vascular, that lining the heart-cavities, the blood-vessels, and lymph-vessels.
**endothermic** (en-do-ther'-mik) [endo-; θέρμη, heat]. Relating to the absorption of heat. e. substances, or compounds, those of which the formation is attended with an absorption of heat.
**endothoracic** (en-do-tho-ras'-ik) [endo-; thorax]. Situated or occurring within the thorax.
**endothyropexy, endothyreopexy** (en-do-thi'-ro-pek-se, en-do-thi'-re-o-pek-se). The operation of separating the thyroid from the trachea, and fixing it to one side.
**endotin** (en'-do-tin). A preparation of tuberculin, said to be pure.
**endotome** (en'-do-tōm) [endo-; τέμνειν, to cut]. Strong shears used in decapitation of the fetus.
**endotoscope** (en-do'-to-skōp) [endo-; οὖς, the ear; σκοπεῖν, to view]. An apparatus designed for examination of the ear and for rendering visible the movements of the tympanum.
**endotoxin** (en-do-toks'-in) [endo-; τοξικόν, a poison]. A toxin found within the organism.
**endotracheitis** (en-du-trak-e-i'-tis) [endo-, tracheitis]. Inflammation of the mucous membrane of the trachea.
**endotrachelitis** (en-do-trak-el-i'-tis). See endocervicitis.
**endotrypsin** (en-do-trip'-sin) [endo-; trypsin]. A digestive ferment resembling trypsin in its action; it is derived from yeasts.
**endovascular** (en-do-vas'-kū-lar). See intravascular.
**endovasculitis** (en-do-vas-kū-li'-tis). See endangeitis.
**endovenous** (en-do-ve'-nus). See intravenous. e. medication, the introduction of medicaments in solution into the veins.
**end-plate** (end'-plāt). A flattened disc at the ending of a motor nerve fiber in muscular tissue.
**endyma** (en'-dim-ah) [ἔνδυμα, garment]. The ependyma, or lining epithelial membrane of the ventricles of the brain, and of the cavity of the spinal cord.
**endymal** (en'-dim-al). Relating to the endyma.
**enecation** (en-ek-a'-shun) [e, out; necare, to kill]. Destruction of life; complete exhaustion.
**enechema** (en-ek-e'-mah) [ἐν, in; ἦχημα, sound, ringing]. Tinnitus aurium.
**enecia** (e-ne'-she-ah) [ἠνεκής, continuous]. A continuous fever.
**enema** (en'-em-ah) [ἐν, in; ἱέναι, to send: pl., enemata]. A rectal injection for therapeutic or nutritive purposes. See alimentation, rectal, and clyster.
**enemose** (en'-em-ōs) [enema]. Trade name of a preparation especially designed for colonic alimentation; it is a concentrated fluid made ready for use by simple dilution.
**enepidermatic, enepidermic** (en-ep-e-der-mat'-ik, -der'-mik) [ἐν, in; ἐπιδέρμις, the epidermis]. Pertaining to the treatment of disease by application to the skin.
**energid** (en-er'-jid) [ἐνεργεῖν, to execute]. Sachs' term for the cell-nucleus and the cytoplasm lying within its sphere of influence.
**energin** (en-ur'-jin). An artificial food prepared from protein.
**energometer** (en-er-gom'-et-er) [energy; μέτρον, measure]. An apparatus for measuring blood pressure.
**energy** (en'-er-je) [ἐν, in; ἔργον, work]. The capacity for doing work. All forms of energy are mutually convertible one into the other, without loss, a principle expressed in the term "conservation of energy." e., kinetic, the power of a body in motion. e., latent, e., potential, the power possessed by a body at rest, by virtue of its position, as the potential energy of a suspended weight.
**enervate** (en'-er-vate) [enervare, weakness]. To weaken.
**enervation** (en-er-va'-shun) [enervatio, weakness].
1. Weakness; lassitude; languor from lack of nerve-stimulus; neurasthenia; the reduction of the strength. 2. The removal of a section of a nerve.
**enesol** (en'-e-sol). Mercury salicylarsenate; an antisyphilitic remedy.
**engastrius** (en-gas'-tre-us) [ἐν, in; γαστήρ, belly]. A monstrosity in which one fetus is included within the peritoneal cavity of another.
**Engelmann's intermediate disc** (eng'-el-mahn) [Theodor Wilhelm Engelmann, German physiologist, 1843– ]. See Krause's membrane. E.'s lateral disc, the narrow zone of transparent homogeneous substance on each side of Krause's membrane.
**English disease.** Synonym of rhachitis.
**sweating fever.** A contagious fever of the sixteenth century.
**englobing** (en-glo'-bing) [ἐν, in; globus, a globe]. The taking in of an object by a monad, ameba, or phagocyte.
**engomphosis** (en-gom-fo'-sis) [ἐν, in; γόμφος, nail]. Gomphosis, q. v.
**engonus** (en'-gon-us) [ἐν, in; γενναν, to produce]. 1. Native. 2. Offspring.
**engorged** (en-gorjd') [see engorgement]. Congested. e. papilla. Choked disc; see papillitis.
**engorgement** (en-gorj'-ment) [Fr., engorgement, a choking up]. Overdistention of the vessels of a part with blood.
**enhematospores** (en-hem-at-o-spors) [ἐν, in; αἷμα, blood; σπόρος, a spore]. Ray Lankester's name for the first spores of the malarial parasite produced within the human body.
**onhemospore** (en hem'o sp'or). Same as enhematospore.
**enkatarrhaphy** (en-kat-ar'-af-e) [ἐγκαταρράπτειν, to sew in]. The method of sewing the two sides of a furrow together to bury an epithelial structure.
**enneurosis** (en-ū-ro'-sis) [ἐν, in; νεῦρον, a nerve]. Innervation.
**enomania** (e-no-ma'-ne-ah) [οἶνος, wine; μανία, madness]. Excessive use of or desire for intoxicating liquors; insanity due to intoxication. Also, delirium tremens.
**enophthalmia** (en-off-thal'-me-ah) [ἐν, in; ὀφθαλμός, eye]. Retraction of the eyeball in the orbit.
**enophthalmin** (en-of-thal'-min). Oxytoluylmethylvinyldiacetonalkamine hydrochloride. A substance clsoely allied to eucaine; it is used as a mydriatic in 2 to 5 % solution.
**enophthalmos** (en-of-thal'-mos) [ἐν, in; ὀφθαλμός, the eye]. Recession of the eyeball within the orbit.
**enorchismus** (en-or-kis'-mus). See cryptorchidism.
**enorganic** (en-or-gan'-ik). Referring to that which is inherent in an organism.
**enosimania** (en-os-e-ma'-ne-ah) [ἔνοσις, trembling; μανία, madness]. Insanity marked by terror, the patient expecting for himself the most dreadful visitations.
**enostosis** (en-os-to'-sis) [ἐν, in; ὀστέον, bone]. A tumor or bony outgrowth within the medullary canal of a bone.
**ens** (ens') [esse, to be]. An entity; an inherent quality or power. e. morbi, the pathology of a disease considered apart from its etiology.
**ensellure** (ahn-sel-yur') [Fr.]. The strongly marked curve of the dorsolumbosacral region; saddleback. It is especially marked among Spanish women.
**ensiform** (en'-sif-orm) [ensis, a sword; forma, form]. Shaped like a sword. e. appendix, e. cartilage, the cartilaginous process at the lower extremity of the sternum.
**ensisternal** (en-sis-ter'-nal). Pertaining to the ensisternum.
**ensisternum** (en-sis-ter'-num) [ensis, sword; sternum]. The xiphisternum or ensiform cartilage.
**ensomphalic** (en-som-fal'-ik). Pertaining to an ensomphalus.
**ensomphalus** (en-som'-fal-us) [ἐν, in; σῶμα, body; ὀμφαλός, navel]. A double monstrosity with practically complete and functionating organs, but united with a more or less superficial bond.
**enstrophe** (en'-stro-fe). [ἐν, in; στρέφειν, to turn]. Inversion, as of the margin of an eyelid.
**entacoustic** (ent-ah-oos'-tik) [ἐντός; ἀκούειν, to hear]. Applied to subjective auditory sensations having their origin within the ear or in its vicinity.
**entad** (en'-tad) [see ental]. From without inward; toward a center; the opposite of ectad.
**ental** (en'-tal) [ἐντός, within]. A surface, aspect, or structure farther from the periphery or nearer the

**entallantoic** (*ent-al-an-to'-ik*) [*ento-*; *allantois*]. Located within the allantoic sac.
**entamniotic** (*ent-am-ne-ot'-ik*) [*ento-*; *amnion*]. Located within the folds of the manion.
**Entamœba, Entameba** (*ent-am-e'-bah*). A genus of *amœbæ*, including the species that are internal parasites. See *ameba*.
**entasia, entasis** (*en-ta'-se-ah, en'-tas-is*) [ἔντασις, a straining]. A generic term for spasmodic muscular action; tonic spasm.
**entatic** (*en-tat'-ik*) [*entasis*]. Causing spasm, or strain; aphrodisiac.
**entcephalic** (*ent-en-sef-al'-ik*) [*entos*; *encephalon*]. Applied to sensations having origin within the brain and not in the external world.
**entepicondylar** (*ent-ep-e-kon'-dil-ar*) [*entos*; *epicondyle*]. Located at the inner aspect of the epicondyle.
**entepicondyle** (*ent-ep-e-kon'-dīl*) [*entos*; *epicondyle*]. Owen's name for the internal condyle of the humerus.
**entèqué** (*ahn-ta'-ka*) [Fr.]. A curious disease of South American animals marked by the occurrence of hard spines of bone in the lungs.
**entera** (*en'-ter-ah*). Plural of *enteron*, q. v.
**enteraden** (*ent-er'-ad-ēn*) [*entero-*; ἀδήν, a gland; pl., *enteradenes*]. Any gland of the intestinal tract.
**enteradenitis** (*en-ter-ad-en-i'-tis*) [*enteraden*; ιτις, inflammation]. Inflammation of the intestinal glands.
**enteradenography** (*en-ter-ad-en-og'-ra-fe*) [*enteraden*; γράφειν, to write]. A treatise on the intestinal glands.
**enteradenology** (*en-ter-ad-en-ol'-o-je*) [*enteraden*; λόγος, science]. The anatomy, physiology, and pathology of the intestinal glands.
**enteragra** (*en-ter-a'-grah*) [*entero-*; ἄγρα, seizure]. Gout in the intestine.
**enteral** (*en'-ter-al*) [ἔντερον, intestine]. Intestinal. See *parenteral*.
**enteralgia** (*en-ter-al'-je-ah*) [*entero-*; ἄλγος, pain]. Pain in the bowels.
**enterangiemphraxis** (*en-ter-an-je-em-fraks'-is*) [*entero-*; ἀγγεῖον, a vessel; ἔμφραξις, a stoppage]. Obstruction of the blood-vessels of the intestine.
**enteratrophia** (*en-ter-at-ro'-fe-ah*) [*entero-*; *atrophy*]. Intestinal atrophy.
**enterauxe** (*en-ter-awks'-e*) [*entero-*; αὔξη, growth]. Hypertrophy of the muscles of the intestinal wall.
**enterectasis** (*en-ter-ek'-tas-is*) [*entero-*; ἔκτασις, dilatation]. Dilatation of some part of the small intestine.
**enterectomy** (*en-ter-ek'-to-me*) [*entero-*; ἐκτομή, excision]. Excision of a part of the intestine.
**enterelcosis** (*en-ter-el-ko'-sis*) [*entero-*; ἕλκωσις, ulceration]. Ulceration of the bowel.
**enterembole** (*en-ter-em'-bo-le*) [*entero-*; ἐμβολή, insertion]. Intussusception of the intestine.
**enteremia** (*en-ter-e'-me-ah*) [*entero-*; αἷμα, blood]. Intestinal congestion.
**enteremphraxis** (*en-ter-em-fraks'-is*) [*entero-*; ἔμφραξις, a stoppage]. Intestinal obstruction.
**enterepiplocele** (*en-ter-ep-ip'-lo-sēl*) [*entero-*; ἐπίπλοον, caul; κήλη, hernia]. Hernia in which both bowel and omentum are involved.
**enterepiplomphalocele** (*en-ter-ep-ip-lom-fal'-o-sēl*) [*entero-*; ἐπίπλοον, caul; ὀμφαλός, navel; κήλη, hernia]. Umbilical hernia, with protrusion of the omentum.
**enteric** (*en-ter'-ik*) [ἔντερον, intestine]. Pertaining to the intestine. **e. fever**, typhoid fever.
**entericoid** (*en-ter'-ik-oyd*) [*enteric*; εἶδος, resemblance]. Resembling typhoid fever. **e. fever**, a fever which resembles typhoid fever but is neither typhoid nor paratyphoid.
**enteritic** (*en-ter-it'-ik*) [*entero-*; ιτις, inflammation]. Relating to enteritis.
**enteritis** (*en-ter-i'-tis*) [*entero-*; ιτις, inflammation]. Inflammation of the intestine. **e., chronic cystic**, that characterized by the formation of cystic dilatations of the intestinal glands due to stenosis of the mouths of the gland. **e., diphtheritic, e. diphtheritica**, a form in which the mucosa is covered by a flaky, whitish-gray deposit. **e. nodularis**, that characterized by hyperplastic enlargement of the lymph-nodules. **e., phlegmonous**, a secondary phenomenon due to other intestinal diseases, particularly carcinoma, ulcers of tuberculous, syphilitic, and embolic origin, and occasionally to strangulated hernia and intussusception. **e. polyposa**, that characterized by polypoid growths in the intestine resulting from proliferation of the connective tissue. **e., pseudomembranous**, a nonfebrile affection of the intestinal mucous membrane marked by periodic formation of viscous, shreddy, or tubular exudates, composed chiefly of mucin.

**entero-** (*en-ter'-o-*) [ἔντερον, intestine]. A prefix denoting relation to the intestine.
**enteroanastomosis** (*en-te-ro-an-as-to-mo'-sis*). Intestinal anastomosis.
**enteroapokleisis** (*en-ter-o-ap-o-kli'-sis*) [*entero-*; ἀπόκλεισις, a shutting off]. The surgical exclusion of a portion of the intestine.
**enterobrosis** (*en-ter-o-bro'-sis*) [*entero-*; βρῶσις, an eating]. Perforation or ulceration of the intestine.
**enterocele** (*en'-ter-o-sēl*) [*entero-*; κήλη, hernia]. A hernia containing a loop of intestine.
**enteroceliac, enteroceliac** (*en-ter-o-se'-le-ak*). Relating to the abdominal cavity.
**enterocentesis** (*en-ter-o-sen-te'-sis*) [*entero-*; κέντησις, puncture]. Surgical puncture of the intestine.
**enteroceptive impulses** (*en'-ter-o-sep-tiv*) [*entero-*; *capere*, to take]. Afferent nerve impulses which derive their stimulation from internal organs.
**enterochirurgia** (*en-ter-o-ki-rur'-je-ah*) [*entero-*; χειρουργία, surgery]. Intestinal surgery.
**enterocholecystostomy** (*en-ter-o-ko-le-sis-tos'-to-me*). Same as *cholecystenterostomy*.
**enterocholecystotomy** (*en-ter-o-ko-le-sis-tot'-o-me*). Same as *cholecystenterotomy*.
**enterocleisis** (*en-ter-o-kli'-sis*) [*entero-*; κλεῖσις, a closing]. Occlusion of the bowel.
**enteroclysis** (*en-ter-ok'-lis-is*) [*entero-*; κλύσις, a drenching]. Injection of a large quantity of fluid into the rectum to reach the small intestine; a high enema.
**enteroclysm** (*en'-ter-o-klizm*) [*entero-*; κλύσμα, a clyster]. 1. A rectal injection. 2. A syringe.
**enteroclyster** (*en-ter-o-klis'-ter*) [*entero-*; *clyster*]. A rectal clyster or enema.
**enterocœle** (*en'-ter-o-sēl*) [*entero-*; κοιλία, a cavity]. The abdominal cavity.
**enterocolitis** (*en-ter-o-ko-li'-tis*) [*entero-*; *colitis*]. Inflammation of the small intestine and of the colon.
**enterocolostomy** (*en-ter-o-ko-los'-to-me*) [*entero-*; *colostomy*]. Operation for the formation of a communication between the small intestine and colon.
**enteroconiosis, enterokoniosis** (*en-ter-o-ko-ne-o'-sis*) [*entero-*; κονία, dust]. Any gastrointestinal affection due to dust.
**enterocyst** (*en-ter'-o-sist*). An intestinal cyst.
**enterocystocele** (*en-ter-o-sis'-to-sēl*) [*entero-*; κύστις, bladder; κήλη, tumor]. Hernia involving the urinary bladder and the intestine.
**enterocystoma** (*en-ter-o-sist-o'-mah*) [*entero-*; *cystoma*]. A cystic tumor formed by the persistence of a part of the vitelline duct, opening neither externally nor into the intestinal canal.
**enterocystoschecele** (*en-ter-o-sist-os'-ke-o-sēl*) [*entero-*; κύστις, a bladder; ὄσχεον, the scrotum; κήλη, hernia]. A hernia of the scrotum containing both intestine and bladder.
**enterodialysis** (*en-ter-o-di-al'-is-is*) [*entero-*; *dialysis*]. Complete division of an intestine by injury.
**enterodynia** (*en-ter-o-din'-e-ah*) [*entero-*; ὀδύνη, pain]. Pain referred to the intestines.
**enteroenterostomy** (*en-ter-o-en-ter-os'-to-me*) [*entero-*; *enterostomy*]. The surgical formation of a fistula between two intestinal loops.
**enteroepiplocele** (*en-ter-o-e-pip'-lo-sēl*). See *enterepiplocele*.
**enterogastritis** (*en-ter-o-gas-tri'-tis*) [*entero-*; *gastritis*]. Inflammation of the stomach and intestine.
**enterogastrocele** (*en-ter-o-gas'-tro-sēl*) [*entero-*; γαστήρ, belly; κήλη, tumor]. A hernia containing the gastric and intestinal walls; abdominal hernia.
**enterogenetic, enterogenous** (*en-ter-o-jen-et'-ik, en-ter-oj'-en-us*) [*entero-*; γεννᾶν, to produce]. Originating in the intestine.
**enterograph** (*en'-ter-o-graf*) [*entero-*; γράφειν, to write]. A myograph arranged for measuring the movements of the intestine.
**enterography** (*en-ter-og'-ra-fe*) [*entero-*; γράφειν, to write]. A treatise on or a description of the intestines.
**enterohemorrhage** (*en-ter-o-hem'-or-aj*) [*entero-*; *hemorrhage*]. Intestinal hemorrhage.
**enterohepatitis** (*en-ter-o-hep-at-i'-tis*) [*entero-*; *hepatitis*]. Combined inflammation of the intestines and liver.

**enterohydrocele** (*en-ter-o-hi'-dro-sēl*) [*entero-*; *hydrocele*]. Hydrocele complicated with intestinal hernia.

**enterokinase** (*en-ter-o-kin'-ās*) [*entero-*; κινεῖν, to move]. Pawlow's name for an enzyme of the succus entericus which awakens proteolytic action by converting trypsinogen into trypsin.

**enterokinesia** (*en-ter-o-kin-e'-se-ah*) [*entero-*; κινεῖν, to move]. The motor function of the bowels, peristalsis.

**enterokinetic** (*en-ter-o-kin-et'-ik*). An agent having an action upon the intestinal movements.

**enterol** (*en'-ter-ol*). A mixture of cresols used as an intestinal antiseptic. Dose 15–75 gr. (1–5 Gm.) of a solution of 0.02 Gm. in 100 Cc. of water, daily.

**enterolith** (*en'-ter-o-lith*) [*entero-*; λίθος, a stone]. A concretion formed in the intestine.

**enterolithiasis** (*en-ter-o-lith-i'-as-is*) [*entero-*; *lithiasis*]. The formation of intestinal calculi.

**enterologist** (*en-ter-ol'-o-jist*) [*entero-*; λόγος, science]. One who concerns himself with the study of the intestines and their disorders.

**enterology** (*en-ter-ol'-o-je*) [*entero-*; λόγος, science]. The science of the intestinal viscera; the anatomy, physiology, pathology, and hygiene of the intestines.

**enteromalacia** (*en-ter-o-mal-a'-se-ah*) [*entero-*; μαλακία, softness]. Pathological softening of the bowel-walls.

**enteromenia** (*en-ter-o-me'-ne-ah*) [*entero-*; μήν, month]. Vicarious menstruation by the bowel.

**enteromere** (*en'-ter-o-mēr*) [*entero-*; μέρος, a part]. One of the primitive transverse divisions of the embryonic alimentary tract.

**enteromerocele** (*en-ter-o-me'-ro-sēl*) [*entero-*; μηρός, thigh; κήλη, hernia]. Femoral hernia involving the intestine.

**enteromesenteric** (*en-ter-o-mes-en-ter'-ik*) [*entero-*; *mesentery*]. Pertaining to the intestine and the mesentery. **e. fever**, typhoid fever.

**enterometer** (*en-ter-om'-et-er*) [*entero-*; μέτρον, a measure]. An instrument to measure the lumen of the small intestine.

**enteromphalus** (*en-ter-om'-fal-us*) [*entero-*; ὀμφαλός, the navel]. An umbilical hernia of intestine.

**enteromycosis** (*en-ter-o-mi-ko'-sis*) [*entero-*; *mycosis*]. Intestinal mycosis.

**enteromyiasis** (*en-ter-o-mi-i'-as-is*) [*entero-*; μυῖα, a fly]. Intestinal disease due to the presence of the larvæ of flies.

**enteron** (*en'-ter-on*) [ἔντερον, intestine: *pl.*, *entera*]. 1. The intestinal or alimentary canal, exclusive of these parts that are of ectodermal origin. 2. The intestine.

**enteroncus** (*en-ter-ong'-kus*) [*entero-*; ὄγκος, mass]. A tumor of the bowel.

**enteroparalysis** (*en-ter-o-par-al'-is-is*) [*entero-*; *paralysis*]. Paralysis of the intestine.

**enteropathy** (*en-ter-op'-ath-e*) [*entero*,; πάθος, disease]. Any disease of the intestine.

**enteropexia** (*en-ter-o-peks'-e-ah*) [*entero-*; πῆξις, fixation]. Fixation of a portion of the intestine to the abdominal wall, for the relief of enteroptosis, splanchnoptosis, etc.

**enteropexy** (*en-ter-o-peks'-e*). See *enteropexia*.

**enterophthisis** (*en-ter-off'-this-is*) [*entero-*; φθίσις, wasting]. Intestinal tuberculosis.

**enteroplastic** (*en-ter-o-plas'-tik*) [*entero-*; πλαστικός, formed]. Pertaining to enteroplasty.

**enteroplasty** (*en'-ter-o-plas-te*) [*entero-*; πλάσσειν, to form]. A plastic operation upon the intestine.

**enteroplegia** (*en-ter-o-ple'-je-ah*) [*entero-*; πληγή, a stroke]. Paralysis of the bowels.

**enteroplex** (*en-ter-o-pleks'*) [*entero-*; πλέξειν, to interlace]. See *enteroplexia*.

**enteroplexia**, **enteroplexy** (*en-ter-o-pleks'-e-ah*, *en'-ter-o-pleks-e*) [*entero-*; πλέξειν, to interlace]. A method of treatment of wounds of the intestine in which union of the bowel is obtained by different means, but particularly by an apparatus, the *enteroplex*, without the use of needle and thread. The *enteroplex* is formed of two aluminum rings so constructed that they will fit the one into the other. These are inserted into the cut ends of the bowel, and then the two are pressed together, the intestinal walls being brought into apposition.

**enteroproctia** (*en-ter-o-prok'-she-ah*) [*entero-*; πρωκτός, anus]. The existence of an artificial anus, or of an opening into the bowel for fecal discharge.

**enteroptosis** (*en-ter-op-to'-sis*) [*entero-*; πτῶσις, a fall]. Prolapse of the intestine. See *Stiller's sign*.

**e., Landau's form of**, that due to relaxation of the abdominal walls and pelvic floor.

**enteropyra** (*en-ter-o-pi'-rah*) [*entero-*; πῦρ, a fire]. 1. Enteritis. 2. Typhoid fever. **e. asiatica**, cholera. **e. biliosa**, bilious fever.

**enterorose** (*en'-ter-or-ōs*). A dietetic recommended in gastrointestinal catarrh; a yellow powder miscible with water. Dose 2 dr. (8 Gm.) severa ltimes daily.

**enterorrhagia** (*en-ter-or-aj'-e-ah*) [*entero-*; ῥηγνύναι, to burst forth]. Intestinal hemorrhage.

**enterorrhaphy** (*en-ter-or'-a-fe*) [*entero-*; ῥαφή, suture]. Suture of the intestine. **e.**, circular, the suturing of a completely divided intestine.

**enterorrhea** (*en-ter-or-e'-ah*) [*entero-*; ῥόια, a flow]. Diarrhea.

**enterorrheuma** (*en-ter-or-ru'-mah*) [*entero-*; ῥεῦμα, a flowing]. Intestinal rheumatism.

**enterorrhexis** (*en-ter-or-eks'-is*) [*entero-*; ῥῆξις, rupture]. Rupture of the bowel.

**enterosarcocele** (*en-ter-o-sar'-ko-sēl*) [*entero-*; σάρξ, flesh; κήλη, a hernia]. Intestinal hernia with sarcocele.

**enterosarcoma** (*en-ter-o-sar-ko-mah*) [*entero-*; *sarcoma*]. Sarcoma of the intestine.

**enteroscheocele** (*en-ter-os'-ke-o-sēl*) [*entero-*; ὄσχεον, scrotum; κήλη, hernia]. A scrotal hernia containing intestine.

**enteroscope** (*en'-ter-o-skōp*) [*entero-*; σκοπεῖν, to examine]. An instrument for examining the inside of the intestine by means of electric light.

**enterosepsis** (*en-ter-o-sep'-sis*) [*entero-*; *sepsis*]. Intestinal toxæmia or sepsis.

**enterosis** (*en-ter-o'-sis*) [*entero-*; νόσος, disease: *pl.*, *enteroses*]. Any intestinal disease.

**enterospasm** (*en'-ter-o-spasm*) [*entero-*; σπασμός, spasm]. Spasmodic colic.

**enterostenosis** (*en-ter-o-ste-no'-sis*) [*entero-*; στένωσις, contraction]. Stricture or narrowing of the intestinal canal.

**enterostomy** (*en-ter-os'-to-me*) [*entero-*; στόμα, mouth]. The formation of an artificial opening into the intestine through the abdominal wall.

**enterotome** (*en'-ter-o-tōm*) [*entero-*; τέμνειν, to cut]. An instrument for cutting open the intestine.

**enterotomy** (*en-ter-ot'-o-me*) See *enterotome*]. Incision of the intestine.

**enterotoxism** (*en-ter-o-toks'-izm*) [*entero-*; τοξικόν, poison]. A pathological condition due to the action of microorganisms on food-materials contained in the intestine.

**enterotyphus** (*en-ter-o-ti'-fus*) [*entero-*; *typhus*]. Typhoid fever.

**enterovaginal** (*en-ter-o-vaj'-in-al*). Intestinovaginal; relating to the intestines and the vagina.

**enterozoon** (*en-ter-o-zo'-on*) [*entero-*; ζῷον, an animal; *pl.*, *enterozoa*]. An animal parasite of the intestine.

**enteruria** (*en-ter-ū'-re-ah*) [*entero-*; οὖρον, urine]. The vicarious occurrence of urinary constituents in the intestine.

**enthelioma** (*en-the-le-o'-mah*) [*entero*, within; ἕλκος an ulcer; ὄμα, a tumor]. A comprehensive term including papilloma and adenoma.

**enthelminth** (*en-thel'-minth*) [ἐντός, within; ἕλμινς, a worm]. A parasitic intestinal worm.

**entheomania** (*en-the-o-ma'-ne-ah*) [ἔνθεος, inspired; μανία, madness]. Mania in which the patient believes himself to be inspired; religious, insanity.

**enthesis** (*en'-thes-is*) [ἐν, in; τιθέναι, to place]. The employment of non-living material to take the place of lost tissue.

**enthetic** (*en-thet'-ik*) [ἐντιθέναι, to put in]. Introduced; coming from without; applied especially to syphilitic and other specific contagious diseases. Exogenous.

**enthlasis** (*en'-thla-sis*) [ἐνθλᾶν, to indent]. A depressed, comminuted fracture of the skull.

**entiris** (*en-ti'-ris*) [ἐντός, within; *iris*]. The uvea of the iris, consisting of its inner and pigmentary layer.

**ento-** (*en-to-*) [ἐντός, within]. A prefix denoting within, inside, inner, internal.

**entoblast** (*en'-to-blast*) [*ento-*; βλαστός, a bud, germ]. The nucleolus or germinal spot. Endoblast.

**entoccipital** (*en-tok-sip'-it-al*) [*ento-*; *occipital*]. Situated entad of the occipital gyrus or fissure.

**entocele** (*en'-to-sēl*) [*ento-*; κήλη, hernia]. Internal hernia (as through the diaphragm); ectopia; morbid displacement of an internal organ.

**entoceliac, entocelian** (*en-to-se'-le-ak, en-to-se'-le-an*) [*ento-*; κοιλία, hollow]. Situated within a brain-cavity or ventricle.
**entocelic, entocoelic** (*en-to-se'-lik*) [*ento-*; κοιλία, a cavity]. Within the intestine.
**entocentral** (*en-to-sen'-tral*) [*ento-*; *center*]. Near the center and on the inner aspect.
**entochoroidea, entochorioidea** (*en-to-ko-roi'-de-ah, -ko-re-oi-de'-ah*) [*ento-*; *choroid*]. The inner lining of the choroid membrane of the eye, made up mainly of capillaries.
**entocinerea** (*en-to-sin-e'-re-ah*) [*ento-*; *cinereus*, ashy]. The gray brain-substance surrounding the cavities of the brain and spinal cord.
**entocondylar** (*en-to-kon'-dil-ar*) [*entocondyle*]. Pertaining to an inner condyle.
**entocondyle** (*en-to-kon'-dīl*) [*ento-*; *condyle*]. An inner condyle, as of the humerus of the femur.
**entoconid** (*en-to-ko'-nid*) [*ento-*; κῶνος, cone]. The inner and posterior cusp of a lower molar tooth.
**entocornea** (*en-to-kor'-ne-ah*) [*ento-*; *cornea*]. That part of Descemet's membrane that lines and adheres to the inner surface of the cornea.
**entocranial** (*en-to-kra'-ne-al*). See *intracranial*.
**entocuneiform** (*en-to-kū-ne'-if-orm*) [*ento-*; *cuneiform*]. The inner cuneiform bone of the foot.
**entocyte** (*en'-to-sīt*) [*ento-*; κύτος, cell]. The contents of a cell, inlcuding nucleus, nucleolus, granulations, etc.
**entoderm** (*en'-to-derm*) [*ento-*; δέρμα, skin]. The simple cell-layer lining the cavity of the primitive intestine; the hypoblast. Endoderm.
**entoectad** (*en-to-ek'-tad*) [*ento-*; ἐκτός, external; *ad*, to]. From within outward.
**entogastric** (*en-to-gas'-trik*). [*ento-*; γαστήρ, stomach]. Relating to the interior of the stomach.
**entogenous** (*en-toj'-en-us*). See *endogenous*.
**entoglossal** (*en-to-glos'-al*) [*ento-*; γλῶσσα, tongue]. Situated within the tongue.
**entogluteus** (*en-to-glu-te'-us*) [*ento-*; *gluteus*]. The gluteus minimus muscle. See *muscles, table of*.
**entohyal** (*en-to-hi'-al*). Hyoid and on the inner aspect.
**entohyaloid** (*en-to-hi'-al-oid*) [*ento-*; *hyaloid*]. Located within the vitreous body.
**entomarginal** (*en-to-mar'-jin-al*). Near the margin and internal.
**entome** (*en'-tōm*) [ἐν, in; τομή, cut]. A knife for dividing a urethral stricture.
**entomere** (*en'-to-mēr*) [*ento-*; μέρος, a part]. In embryology, one of the cells forming the center of the mass of blastomeres in the developing mammalian ovum, the outer blastomeres being called *ectomeres*.
**entomiasis** (*en-to-mi'-a-sis*) [ἔντομον, an insect]. Any pathological condition due to infestation with insects.
**entomion** (*en-to'-me-on*) [ἐντομή, notch]. The point where the parietal notch of the temporal bone receives the anterior extension of the mastoid angle of the parietal.
**entomography** (*en-to-mog'-ra-fe*) [ἔντομον, an insect; γράφειν, to write]. A treatise on insects.
**entomology** (*en-to-mol'-o-je*) [ἔντομον, insect; λόγος, science]. That department of zoology devoted to the description of insects.
**entonia** (*en-to'-ne-ah*) [ἐντονία; tension]. Rigidity or tension of a voluntary muscle; tonic spasm.
**entonic** (*en-ton'-ik*) [*entonia*]. Characterized by entonia, or by violent tonic spasm.
**entoparasite** (*en-to-par'-as-īt*) [*ento-*; *parasite*]. An internal parasite; an entozoan or entophyte.
**entopectoralis** (*en-to-pek-tor-a'-lis*) [*ento-*; *pectoralis*]. The pectoralis minor muscle. See *muscles, table of*.
**entoperipheral** (*en-to-per-if'-er-al*) [*ento-*; *peripheral*]. Originating or situated within, and not upon the periphery.
**entophthalmia** (*en-toff-thal'-me-ah*) [*ento-*; *ophthalmia*]. Inflammation of the internal parts of the eyeball.
**entophyte** (*en'-to-fīt*) [*ento-*; φυτόν, a plant]. A vegetable parasite living within the body of its host, as, *e. g.*, a bacterium.
**entoplasm** (*en'-to-plasm*). See *endoplasm*.
**entoplastic** (*en-to-plas'-tik*) [*ento-*; within; πλάσσειν, to form]. Same as *endoplastic*.
**entopterygoid** (*en-to-ter'-e-goid*) [*ento-*; πτέρυξ, a wing; εἶδος, resemblance]. 1. Like a wing and situated entad. 2. Owen's name for the pterygoid process of the sphenoid. 3. E. Coues' name for the internal pterygoid muscle.

**entoptic** (*ent-op'-tik*) [*ento-*; ὀπτικός, pertaining to vision]. Pertaining to the internal parts of the eye. *e.* **phenomena**, visual sensations generated within the eye.
**entoptoscopic** (*en-top-to-skop'-ik*). Relating to entoptoscopy.
**entoptoscopy** (*en-top-tos'-ko-pe*) [*ento-*; ὤψ, eye; σκοπεῖν, to inspect]. The investigation or observational study of the anterior of the eye, or of the shadows within the eye.
**entorbital** (*ent-orb'-it-al*). Located on the inner part of the orbital lobe or entad of its orbital fissure.
**entoretina** (*en-to-ret'-in-ah*) [*ento-*; *retina*]. The innermost layer of the retina, itself composed of five layers, and an inner limiting membrane.
**entorrhagia** (*en-tor-a'-je-ah*) [*ento-*; ῥηγνύναι, to burst forth]. Internal hemorrhage.
**entosarc** (*en'-to-sark*). See *endosarc*.
**entosphenoid** (*en-to-sfen'-oid*) [*ento-*; *sphenoid*]. Sphenoid and internal.
**entosthoblast** (*en-tos'-tho-blast*) [ἐντοσθε, from within; βλαστός, germ]. The supposed nucleus of a nucleolus.
**entostosis, entosteosis** (*en-tos-to'-sis, -tos-te-o'-sis*) [*ento-*; ὀστέον, bone]. An osseous growth within a medullary cavity.
**entosylvian** (*en-to-sil'-ve-an*). Within the Sylvian fissure.
**entothalamus** (*en-to-thal'-am-us*) [*ento-*; *thalamus*]. Spitzka's name for the inner gray thalamic zone.
**entotic** (*ent-o'-tik*) [*ento-*; οὖς, ear]. Pertaining to the internal parts of the ear.
**entotorrhea** (*ent-o-tor-e'-ah*) [*ento-*; *otorrhea*]. Internal otorrhea.
**entotrochanter** (*en-to-tro-kan'-ter*) [*ento-*; *trochanter*]. The lesser trochanter.
**entotympanic** (*en-to-tim-pan'-ik*) [*ento-*; *tympanum*]. Located within the tympanum.
**entozoal** (*en-to-zo'-al*) [*ento-*; ζῷον, an animal]. Caused by or dependent upon entozoa.
**entozoon** (*en-to-zo'-on*) [*ento-*; ζῷον, an animal; pl., *entozoa*]. An animal parasite living within another animal.
**entrails** (*en'-trels*). The bowels and abdominal viscera.
**entropia** (*en-tro'-pe-ah*) [see *entropion*]. A turning inward.
**entropion** (*en-tro'-pe-on*) [ἐν, in; τρέπειν, to turn]. Inversion of the eyelid, so that the lashes rub against the globe of the eye. *e.* **musculare**, that due to contraction of the ciliary part of the orbicular muscle resulting from senile atony. *e.*, **organic**, that due to contraction of the lid resulting from cicatricial contraction of the conjunctiva or to diphtheritic conjunctivitis. *e.* **spasmodicum**, *e.*, **spastic**, *e.* **spasticum**. See *e. musculare*.
**entropionize** (*en-tro'-pe-on-īz*). To turn inward.
**entropium** (*en-tro'-pe-um*). See *entropion*.
**entropy** (*en'-tro-pe*) [ἐντρέπειν, to turn about]. That part of the activity or energy of a body which cannot be converted into mechanical work.
**entyposis** (*en-ti-po'-sis*). The glenoid fossa of the scapula.
**enucleation** (*e-nū-kle-a'-shun*) [*e*, out of; *nucleus*, a kernel]. The shelling-out of a tumor or organ from its capsule. The excision of the eyeball.
**enucleator** (*e-nū'-kle-a-tor*) [*e*, out of; *nucleus*, a kernel]. An instrument used in performing enucleation.
**enula** (*en'-ū-lah*) [L.]. The inner aspect of the gums.
**enule** (*en'-ūl*). Trade name applied to suppositories, medicated bougies, etc.
**enuresis** (*en-ū-re'-sis*) [ἐνουρεῖν, to be incontinent of urine]. The involuntary emptying of the bladder. *e.*, **nocturnal**, that occurring at night during sleep.
**envenomation** (*en-ven-o-ma'-shun*) [ἐν, in; *venom*]. The introduction and action of snake-venom.
**environment** (*en-vi'-ron-ment*) [Fr. *environner*, to surround]. The totality of influences acting upon the organism from without.
**enzootic** (*en-zo-ot'-ik*) [ἐν, in, or among; ζῷον, animal]. Affecting beasts in a certain district; as an enzootic disease.
**enzooty** (*en-zo'-o-te*) [ἐν, in, among; ζῷον, an animal]. An enzootic disease.
**enzyme, enzym** (*en'-zīm*) [ἐν, in; ζύμη, leaven]. 1. Any ferment formed within the living organism. 2. A chemical ferment, as distinguished from organized ferments, such as the yeasts.

**enzymic** (en-zi'-mik) [ἐν, in; ζύμη, leaven]. The nature of the action of an enzyme.

**enzymol** (en'-zi-mol). A proprietary artificial gastric juice prepared from the glands of the stomach; it is used as a solvent and antiseptic, especially in the external treatment of diseases of the ear and nose.

**enzymosis** (en-zi-mo'-sis) [ἐν, in; ζύμη, leaven]. The action of an enzyme.

**enzymotic** (en-zi-mot'-ik) [ἐν, in; ζύμη, leaven]. Pertaining to enzymes.

**enzymuria** (en-zi-mu'-re-ah) [enzyme; οὖρον, urine]. The presence of enzymes in the urine when voided.

**eolipyle, æolipyle** (e-ol'-e-pīl) [αἴολος, windy; πύλη, a narrow passage]. A form of spirit-lamp used to heat cautery-irons.

**eosin** (e'-o-sin) [ἠώς, the dawn], $C_{20}H_8Br_4O_5$. Tetrabromfluorescein; an acid dye produced by the action of bromine on fluorescein suspended in glacial acetic acid. It occurs in red or yellowish crystals, and is used as a stain in histology.

**eosinophil, eosinophilous** (e-o-sin'-o-fil, e-o-sin-off'-il-us) [eosin; φιλεῖν, to love]. Applied to microbes or histological elements showing a peculiar affinity for eosin stain or for acid stains in general.

**eosinophilia** (e-o-sin-o-fil'-e-ah) [see eosinophil]. 1. An increase above the normal standard in the number of eosinophils in the circulating blood. 2. The condition of microbes or histological elements which readily absorb and become stained by eosin.

**eosote** (e'-o-sōt). The commercial name of creosote vaierate.

**epactal** (e-pak'-tal) [ἐπακτός, brought in]. Intercalated; supernumerary. e. bones, Wormian bones. e. cartilages. See cartilage.

**eparsalgia** (ep-ars-al'-je-ah) [ἐπαίρειν, to lift; ἄλγος, pain]. Any disorder due to overstrain of a part.

**eparterial** (ep-ar-te'-re-al) [ἐπί, upon; ἀρτηρία, artery]. Situated above an artery.

**epauxesiectomy** (ep-awk-se-zi-ek'-to-me) [ἐπαύξησις, increase; ἐκτομή, a cutting-out]. Excision of a growth.

**epaxial** (ep-aks'-e-al) [epi-; axis]. Situated or extending over an axis.

**epencephal** (ep-en'-sef-al). See epencephalon.

**epencephalic** (ep-en-sef-al'-ik). Pertaining to the epencephalon.

**epencephalon** (ep-en-sef'-al-on) [ἐπί, upon; encephalon]. The after-brain or hind-brain; the cerebellum and pons taken together; or the cerebellum, pons and medulla.

**ependyma** (ep-en'-dim-ah) [ἐπένδυμα, an upper garment]. The lining membrane of the cerebral ventricles and of the central canal of the spinal cord.

**ependymal** (e-pen'-dim-al). Pertaining to the ependyma.

**ependymitis** (ep-en-dim-i'-tis) [ependyma; ιτις, inflammation]. Inflammation of the ependyma.

**ependymoma** (ep-en-de-mo'-mah) [ependyma; ὄμα, tumor]. Tumor of the lining membrane of the ventricles of the brain or of the central canal of the spinal cord.

**ephebic** (ef-e'-bik) [ἐφηβικός, belonging to puberty]. Pertaining to youth, adolescence, or puberty.

**ephebology** (ef-e-bol'-o-je) [ἔφηβος, youth; λόγος, science]. The science of youth, adolescence, and puberty.

**Ephedra** (ef'-e-drah) [ἐπί, upon; ἕδρα, a seat]. A genus of plants of the Gnetaceæ. E. antisyphilitica has been used in gonorrhea. Dose of the fluidextract 1-2 dr. (4-8 Cc.). E. nevadensis is used as an alterative and antigonorrheic. Dose of fluidextract 1-2 dr. (4-8 Cc.). E. vulgaris contains the alkaloid ephedrine, which is mydriatic.

**ephedrine** (ef'-e-drēn) [ephedra]. An alkaloid, $C_{10}H_{15}NO$, from Ephedra vulgaris; it is a cardiac depressant and harmless mydriatic. e., pseudo-. See pseudoephedrine.

**ephelis** (ef-e'-lis) [ἐπί, upon; ἥλιος, the sun; pl., ephelides]. A freckle.

**ephemeral** (ef-em'-er-al) [ἐφήμερος, living a day]. Temporary. Applied to fevers that are of short duration.

**ephialtes** (ef-e-al'-tēz) [epi-; ἰαλλεῖν, to leap]. See nightmare.

**ephidrosis** (ef-id-ro'-sis) [ἐπί, upon; ἱδρωσις, sweating]. Excessive perspiration. See hyperidrosis. e. cruenta, bloody sweat.

**ephippium** (ef-ip'-e-um) [epi-; ἵππος, horse: pl., ephippia]. Same as sella turcica.

**epi-** [ἐπί, upon]. A prefix signifying upon.

**epiblast** (ep'-e-blast) [epi-; βλαστός, a germ]. The external or upper layer of the blastoderm; called also the ectoderm, from which are developed the central nervous system and the epithelium of the sense-organs, the mucous membranes of the mouth and anus, the enamel of the teeth, the epidermis and its derivatives (hair, nails, glands, etc.).

**epiblastic** (ep-e-blast'-ik). Pertaining to or derived from the epiblast.

**epiblepharon** (ep-e-blef'-ar-on). See epicanthus.

**epibole, epiboly** (ep-ib'-ol-e) [epi-; βάλλειν, to throw]. The enclosure of the large yolk-mass of an invertebrate ovum by the overgrowth of cleavage-cells.

**epibolic** (ep-ib-ol'-ic) [epibole]. Pertaining to epiboly.

**epibulbar** (ep-e-bul'-bar) [epi-; bulb]. Situated upon the globe of the eye; as an epibulbar tumor.

**epicanthus** (ep-e-kan'-thus) [epi-; κανθός, angle of the eye]. A fold of skin over the inner canthus of the eye.

**epicardium** (ep-e-kar'-de-um) [epi-; καρδία, heart]. The visceral layer of the pericardium.

**epicarin** (ep-e-kar'-in). A condensation-product of cresolinic acid and betanaphthol, occurring as an odorless, tasteless, clear, yellowish-gray powder, easily soluble in alcohol, ether, acetone; insoluble in oil. It is used in scabies in 10 % solution.

**epicarp** (ep'-e-karp) [epi-; καρπός, fruit]. In biology, the outer skin of a fruit.

**epicele, epicœle** (ep'-e-sēl) [epi-; κοιλία, belly]. The fourth ventricle.

**epicerebral** (ep-e-ser-e'-bral) [epi-; cerebrum, the brain]. Situated over or on the cerebrum; as the epicerebral space between the brain and the pia.

**epichordal** (ep-e-kord'-al) [epi-; χορδή, a cord]. Located above or dorsad of the notochord; applied especially to cerebral structures.

**epichorial** (ep-e-ko'-re-al). Relating to the epichorion; located on the chorion or on the derma.

**epichorion** (ep-e-ko'-re-on) [epi-; chorion]. 1. The decidua reflexa. 2. The epidermis.

**epichlorhydrin** (ep-e-klor-hi'-drin), $C_3H_5ClO$. A sweet liquid with odor of chloroform, obtained from dichloropropyl alcohol by action of gaseous hydrochloric acid. It is miscible in alcohol and ether, boils at 118°-119° C. Sp. gr. 1.203 at 0° C.

**epichrosis** (ep-e-kro'-sis) [ἐπίχρωσις, a spot]. A discoloration of the skin. e. alphosis, albinism. e. aurigo, a yellow discoloration of the skin, as in icterus. e. ephelis, e. lenticula, pigmentation of the skin from exposure to the sun. e. leucasmus, e. pœcilia, vitiligo. e. spilus. See nævus pigmentosus under nævus.

**epicœlia** (ep-e-se'-le-ah). See epicele.

**epicolic** (ep-e-kol'-ik) [epi-; colon]. Lying over the colon.

**epicoma, epicomus, epicome** (e-pik-o'-mah, e-pik-o'-mus, e-pik'-o-me) [epi-; κόμη, hair]. A parasitic monstrosity having an accessory head united to the principal fetus by the summit.

**epicondylalgia** (ep-e-kon-dil-al'-je-ah) [epicondyle; ἄλγος, pain]. Pain in the muscular mass about the elbow-joint, following fatiguing work.

**epicondylar** (ep-e-kon-dil-ar). Relating to an epicondyle.

**epicondyle** (ep-e-kon'-dīl) [epi-; κόνδυλος, a knuckle]. An eminence upon a bone above its condyle.

**epicondylus** (ep-e-kon'-dil-us). See epicondyle. e. extensorius, the external condyle of the humerus. e. lateralis, the external condyle of the humerus, or the outer tuberosity of the femur. e. medialis, the internal condyle of the humerus, or the inner tuberosity of the femur. e. flexorius, the internal condyle of the humerus.

**epicophosis** (ep-e-ko-fo'-sis) [epi-; κώφωσις, deafness]. Deafness dependent upon some disease.

**epicoracoid** (ep-e-kor'-ak-oid) [epi-, coracoid]. Located upon or over the coracoid process.

**epicostal** (ep-e-kos'-tal) [epi-; costa, a rib]. Situated upon the ribs.

**epicranial** (ep-e-kra'-ne-al). Relating to the epicranium.

**epicranium** (ep-e-kra'-ne-um) [epi-; cranium]. The structures covering the cranium.

**epicranius** (ep-e-kra'-ne-us). The occipitofrontalis muscle.

**epicrisis** (*ep-e-kri'-sis*) [ἐπίκρισις, determination]. The disease-phenomena succeeding crisis.
**epicritic** (*ep-e-krit'-ik*) [ἐπικρινεῖν, to give judgment upon]. Pertaining to sensory nerve fibers which enable one to appreciate very fine distinctions of temperature and touch. These fibers are found in the skin only.
**epicrusis** (*ep-e-kru'-sis*) [*epi-*; κρούσις, stroke]. 1. Massage by strokes or blows; percussion; therapeutic scourging.
**epicyesis** (*ep-e-si-e'-sis*) [*epi-*; κύειν, to be pregnant]. Superfetation.
**epicystic** (*ep-e-sis'-tik*) [*epi-*; κύστις, bladder]. Suprapubic; situated above the urinary bladder.
**epicystitis** (*ep-e-sis-ti'-tis*) [*epi-*; κύστις, bladder; *ιτις*, inflammation]. Inflammation of the tissues above the bladder.
**epicystotomy** (*ep-e-sis-tot'-o-me*) [*epi-*; κύστις, a bladder; τέμνειν, to cut]. Suprapubic incision of the bladder.
**epicyte** (*ep'-e-sit*) [*epi-*; κύτος, cell]. 1. The cell-wall. 2. A cell of epithelial tissue.
**epicytoma** (*ep-e-si-to'-mah*) [*epicyte*; ὄμα, tumor]. Malignant epithelioma.
**epidemic** (*ep-e-dem'-ik*) [*epi-*; δῆμος, people]. Of a disease, affecting large numbers, or spreading over a wide area.
**epidemicity** (*ep-e-dem-is'-it-e*). The quality of being epidemic.
**epidemiography** (*ep-e-dem-e-og'-ra-fe*). [*epidemic;* γράφειν, to write]. A description of epidemic diseases.
**epidemiologic** (*ep-e-dem-e-o-loj'-ik*). Relating to epidemiology.
**epidemiologist** (*ep-e-dem-e-ol'-o-jist*). One who has made a special study of epidemics.
**epidemiology** (*ep-e-dem-e-ol'-o-je*) [*epidemic;* λόγος, science]. The science of epidemic diseases.
**epiderm** (*ep'-e-derm*). See *epidermis*.
**epiderma** (*ep-e-der'-mah*) [*epidermis;* *pl., epidermata*]. Any abnormal outgrowth from the epidermis.
**epidermal** (*ep-e-der'-mal*) [*epidermis*]. 1. Relating to or composed of epiderm. 2. Trade name of scarlet red. **e.-method**, the application of medicinal substance to the skin.
**epidermatic** (*ep-e-der-mat'-ik*). See *epidermic*.
**epidermatoid** (*ep-e-der'-mat-oid*) [*epidermis;* εἶδος, likeness]. Resembling the epidermis.
**epidermic** (*ep-e-der'-mik*) [*epidermis*]. Relating to the epidermis. **e. method**, a method of administering medicinal substances by applying them to the skin.
**epidermidalization** (*ep-e-derm-id-al-i-za'-shun*) [*epidermis*]. The conversion of columnar into stratified epithelium.
**epidermidolysis** (*ep-e-derm-id-ol'-is-is*). See *epidermolysis*.
**epidermidomycosis** (*ep-e-der-mid-o-mi-ko'-sis*) [*epidermis;* μύκης, fungus]. A disease due to the growth of parasitic fungi upon the skin.
**epidermidophyton** (*ep-e-derm-id-o-fi'-ton*) [*epidermis;* φυτόν, a plant]. A fungus found in psoriasis.
**epidermidosis** (*ep-e-derm-id-o'-sis*) [*epidermis*]. A collective name for anomalous growths of the skin of epithelial origin and type.
**epidermin** (*ep-e-der'-min*). A proprietary base for ointments.
**epidermis** (*ep-e-der'-mis*) [ἐπί, upon; δέρμα, skin]. The outer layer of the skin. The scarf-skin, consisting of a layer of horny cells that protects the true skin.
**epidermization** (*ep-e-der-miz-a'-shun*). 1. The formation of epiderm. 2. Skin-grafting.
**epidermoid** (*ep-id-er'-moid*) [*epidermis;* εἶδος, like]. 1. Resembling epidermis. 2. A tumor formed of epidermal cells.
**epidermolysis** (*ep-e-der-mol'-is-is*) [*epidermis*]. A rare skin disease in which bullæ form on the slightest pressure. It shows itself in infancy and is most pronounced in summer.
**epidermophyton** (*ep-e-durm-o-fi'-ton*). See *epidermidophyton*.
**epidiascope** (*ep-e-di'-ah-skōp*) [*epi-*; διά, through; σκοπεῖν, to look]. A magic lantern arranged for ordinary lantern slides, and also for opaque objects; a combined magic lantern and episcope.
**epididymal, epididymic** (*ep-e-did'-em-al, -im'-ik*). Relating to the epididymis.
**epididymectomy** (*ep-e-did-im-ek'-tom-e*) [*epididymis;* ἐκτομή, a cutting out]. Excision of the epididymis.
**epididymis** (*ep-e-did'-im-is*) [ἐπί, upon; δίδυμος, the testes: pl., *epididymides*]. The small body lying above the testis; the superior end is the globus major; the inferior, the globus minor.
**epididymitis** (*ep-e-did-im-i'-tis*) [*epididymis;* *ιτις*, inflammation]. Inflammation of the epididymis.
**epididymo-orchitis** (*ep-e-did-im-o-or-ki'-tis*). Epididymitis combined with orchitis.
**epididymotomy** (*ep-e-did-im-ot'-o-me*). [*epididymis;* τομή, a cutting]. Incision of the epididymis.
**epididymovasostomy** (*ep-e-did-im-o-vas-os'-to-me*) [*epididymis;* vas; στόμα, mouth]. The formation of a lateral anastomosis between the vas and the epididymis; or the vas may be divided, its end split, and the split end sewn into the epididymis. Performed in cases of sterility.
**epidosis** (*e-pid'-o-sis*) [ἐπίδοσις]. Enlargement; increase; exacerbation.
**epidrome** (*e-pe-drōm*) [ἐπιδρομή, a running upon]. Active, or (more often) passive, congestion.
**epidural** (*ep-e-dū'-ral*) [*epi-*; *durus*, hard]. Situated upon or over the dura. **e. space**, the space outside the dura mater of the spinal cord and brain.
**epifolliculitis** (*ep-e-fol-ik-ū-li'-tis*). Inflammation seated about the hair-follicles of the scalp.
**Epigæa** (*ep-e-je'-ah*) [ἐπί, upon; γαῖα, earth]. A genus of trailing ericaceous plants. *E. repens,* trailing arbutus of North America, has diuretic properties.
**epigaster** (*ep-e-gas'-ter*) [*epigastrium*]. The large intestine; hindgut.
**epigastralgia** (*ep-e-gas-tral'-je-ah*) [*epigastrium;* ἄλγος, pain]. Pain in the epigastrium.
**epigastric** (*ep-e-gas'-trik*) [*epigastrium*]. Relating to the epigastrium. **e. reflex.** See under *reflex*.
**epigastriocele, epigastrocele** (*ep-e-gas'-tre-o-sēl, ep-e-gas'-tro-sēl*) [*epigastrium;* κήλη, a hernia]. A hernia in the epigastrium.
**epigastrium** (*ep-e-gas'-tre-um*) [*epi-*; γαστήρ, stomach]. The upper and middle part of the abdominal surface corresponding to the position of the stomach; the epigastric region. See *abdomen*.
**epigastrius** (*ep-e-gas'-tre-us*) [*epi-*; γαστήρ, the stomach]. A form of double-monstrosity, in which one fetus in an undeveloped condition is contained within the epigastric region of the other.
**epigastrocele** (*ep-e-gas'-tro-sēl*) [*epigastrium;* κήλη, hernia]. Hernia in the epigastric region.
**epigenesis** (*ep-e-jen'-es-is*) [*epi-*; γένεσις, generation]. In biology, the theory that holds the embryo to be the result of the union of the male and female elements, and the fully formed organism the result of a gradual process of differentiation, in distinction to the theory of encasement, preformation, or evolution, which held the embryo to preexist enfolded in a minute form within the germ.
**epigenetic** (*ep-e-jen-et'-ik*). Pertaining to epigenesis.
**epiglottic** (*ep-e-glot'-ik*) [*epiglottis*]. Relating to the epiglottis.
**epiglottidean** (*ep-e-glot-id'-e-an*). See *epiglottic*.
**epiglottidectomy** (*ep-e-glot-id-ek'-to-me*) [*epiglottis;* ἐκτομή, excision]. Excision of the epiglottis.
**epiglottiditis** (*ep-e-glot-id-i'-tis*) [*epiglottis;* *ιτις*, inflammation]. Inflammation of the epiglottis.
**epiglottis** (*ep-e-glot'-is*) [*epi-*; γλωττίς, glottis]. A fibrocartilaginous structure that aids in preventing food and drink from passing into the larynx.
**epiglottitis** (*ep-e-glot-i'-tis*). See *epiglottiditis*.
**epignathus** (*ep-ig'-na-thus*) [*epi-*; γνάθος, jaw]. A monstrosity in which the rudimentary organs of a twin are united to the superior maxillary bone.
**epiguanin** (*ep-e-gwan'-in*), $C_5H_6N_5O_2$. A xanthin base sometimes found in the urine; it is similar to guanin in solubilities.
**epihyal bone** (*ep-e-hi'-al*) [*epi-*; *hyoid*]. The stylohyoid ligament when it is ossified.
**epilating forceps**. Forceps for plucking out hairs.
**epilation** (*ep-il-a'-shun*) [*e,* out of; *pilus,* a hair]. The extraction of hair.
**epilatorium** (*e-pil-at-o'-re-um*) [L.]. An application for permanently removing hair.
**epilatory** (*ep'-il-at-o-re*). Removing hair; a remedy for removing hair.
**epilemma** (*ep-e-lem'-ah*) [*epi-;* λέμμα, husk: *pl., epilemmata*]. The neurilemma of very small branches or funiculi of nerve-filaments.

# EPILEPIDOMA 335 EPIPLOSARCOMPHALOCELE

**epilepidoma** (*ep-e-lep-id-o'-mah*). See under *lepidoma*.
**epilepsia** (*ep-e-lep'-se-ah*). See *epilepsy*. **e. gravis.** See *grand mal*. **e. larvata.** See *epilepsy, masked*. **e. mitis.** See *petit mal*. **e. saltatoria.** Synonym of *chorea*. **e. vertiginosa,** *petit mal, q. v.*
**epilepsy** (*ep'-il-ep-se*) [ἐπίληψις, a laying hold of]. A chronic nervous affection characterized by sudden loss of consciousness, with general tonic and clonic convulsions, the paroxysms lasting but a short time. An epileptic seizure is often preceded by a peculiar sensation, or *aura*, and as the patient falls he sometimes makes an outcry—the *epileptic cry*. Syn., *grand mal*. See *petit mal*. **e., cardiac,** paroxysmal tachycardia. **e., cortical, e., focal, e., Jacksonian,** spasmodic contractions in certain groups of muscles, with retention of consciousness, due to local disease of the cortex. **e., cursive,** a form in which the attack is characterized by running. **e., idiopathic,** typical epilepsy. **e., latent,** a form due to some local irritation, generally in the stomach, which ceases on removal of the irritation, but is liable to recur upon any indulgence. **e., masked,** in this, involuntary actions, often violent, replace the convulsion. Syn., *epilepsia larvata*. **e., motorial,** Jacksonian epilepsy. **e., myoclonus,** the occurrence of myoclonus and epilepsy in the same patient, the so-called association-disease. **e., nocturnal,** epilepsy in which the attack occurs during sleep. **e., procursive,** a form in which the patient runs rapidly forward before falling. **e., reflex,** due to some reflex neurosis. **e., spinal,** paroxysms of clonic spasm in the lower extremities sometimes observed in the course of spastic paraplegia. **e., toxemic,** due to poisonous substances in the blood. **e., vasomotor,** that in which extreme contraction of the arteries precedes the attacks.
**epileptic** (*ep-il-ep'-tik*). 1. Pertaining to or like epilepsy. 2. One affected with epilepsy. **e. aura, e. cry.** See under *epilepsy*. **e. dementia,** the dementia which is frequently the terminal stage of epilepsy. **e. equivalents,** transient psychic disturbances replacing the typical convulsions. **e. mania,** mania following or taking the place of the fit.
**epileptiform** (*ep-il-ep'-tif-orm*) [*epilepsy; forma,* form]. Resembling an epileptic attack.
**epileptisant** (*ep-il-ep'-tis-ant*). 1. Producing epileptoid convulsions. 2. A drug which produces epileptoid convulsions: *e. g.,* absinthe.
**epileptogenic** (*ep-il-ep-to-jen'-ik*). See *epileptogenous*.
**epileptogenous** (*ep-il-ep-toj'-en-us*) [*epilepsy*; γενναν, to produce]. Producing epilepsy.
**epileptoid** (*ep-il-ep'-toid*) [*epilepsy*; εἶδος, likeness]. 1. Resembling epilepsy. 2. A person subject to various nervous attacks of the general nature of epilepsy.
**epilose** (*ep'-il-ōs*) [*e,* priv.; *pilosus,* hairy]. Without hair; bald.
**epilymph** (*ep'-e-limf*) [*epi-; lymph*]. The fluid between the bony and the membranous labyrinths.
**epimandibular** (*ep-e-man-dib'-u-lar*) [*epi-; mandibulum,* jaw]. Upon or above the lower jaw.
**epimysium** (*ep-e-mis'-e-um*) [*epi-; μῦς,* a muscle]. The sheath of areolar tissue surrounding a muscle.
**epinasty** (*ep'-e-nas-te*) [*epi-; ναστός,* pressed close]. In biology, curvature produced by excessive growth on the upper side of an extended organ.
**epinephelos, epinephelus** (*ep-e-nef'-el-os, -us*) [*epi-; νεφέλη,* a cloud]. Cloudy, turbid.
**epinephral** (*ep-e-nef'-ral*) [*epi-; νεφρός,* kidney]. Suprarenal.
**epinephrin** (*ep-e-nef'-rin*) [*epi-; νεφρός,* kidney], C₁₇H₁₅NO₄. The active principle of the suprarenal capsule. **e. hydrate,** adrin.
**epinephritis** (*ep-e-nef-ri'-tis*) [*epi-; νεφρός,* kidney; *itis*]. Inflammation of a supra-renal capsule.
**epinephroma** (*ep-e-nef-ro'-mah*). Same as hypernephroma. See *Grawitz's tumor*.
**epinephros** (*ep-e-nef'-ros*) [*epi-; νεφρός,* kidney]. The suprarenal gland.
**epineural** (*ep-e-nū'-ral*) [*epi-; νεῦρον,* a nerve]. In biology, applied to structures attached to a neural arch.
**epineurial** (*ep-e-nū'-re-al*). Relating to the epineurium.
**epineurium** (*ep-e-nū'-re-um*) [*epi-; νεῦρον,* a nerve]. The connective-tissue sheath of a nerve-trunk.
**epinine** (*ep'-in-ēn*). Trade name applied to 3 : 4-dihydroxyphenylethylmethylamine, a synthetic drug used as a vasoconstrictor.

**epinosic** (*ep-e-no'-sik*) [*epi-; νόσος,* disease]. Unhealthy, sickly.
**epionychium** (*ep-e-o-nik'-e-um*). See *eponychium*.
**epiotic** (*ep-e-ot'-ik*) [*epi-; οὖς,* ear]. Situated above or on the cartilage of the ear.
**epiparasite** (*ep-e-par'-a-sīt*). See *epizoon*.
**epipastic** (*ep-e-pas'-tik*) [ἐπιπάσσειν, to sprinkle]. Having the qualities of a dusting-powder.
**epipedometer** (*ep-e-pe-dom'-et-er*) [ἐπίπεδον, surface; *μέτρον,* measure]. An instrument for use in measuring various complex deformities of the body.
**epiperipheral** (*ep-e-per-if'-er-al*) [*epi-; periphery*]. Exterior; at the periphery.
**epipharyngeal** (*ep-e-far-in'-je-al*) [*epi-; pharynx*]. 1. Located upon or above the pharynx. 2. Pertaining to the nasopharynx.
**epipharynx** (*ep-e-far'-inks*) [*epi-; pharynx*]. 1. In biology, the median projection on the internal surface of the upper lip of an insect. 2. The nasopharynx.
**epiphenomenon** (*ep-e-fe-nom'-en-on*) [*epi-; φαινόμενον,* phenomenon]. An exceptional sequence or unusual complication arising in the course of a disease.
**epiphora** (*ep-if'-or-ah*) [*epi-; φέρειν,* to bear]. A persistent overflow of tears, due to excessive secretion or to impeded outflow.
**epiphyseal, epiphysial** (*ep-e-fiz'-e-al*) [*epiphysis*]. Relating to or of the nature of an epiphysis. **e. plate.** See *disc, epiphyseal*.
**epiphyseitis** (*ep-e-fiz-e-i'-tis*) [*epiphysis*; *itis,* inflammation]. Inflammation of an epiphysis.
**epiphyseolysis** (*ep-e-fis-e-ol'-is-is*) [*epiphysis*; λύσις, a loosing]. The separation of an epiphysis.
**epiphysis** (*ep-if'-is-is*) [*epi-; φύεσθαι,* to grow]. A process of bone attached for a time to another bone by cartilage, but in most cases soon becoming consolidated with the principal bone. **e. cerebri,** the pineal gland.
**epiphysitis** (*ep-if-is-i'-tis*). See *epiphyseitis*.
**epiphyte** (*ep'-e-fīt*) [*epi-; φυτόν,* a plant]. A vegetable parasite growing on the exterior of the body.
**epipial** (*ep-e-pi'-al*) [*epi-; pia*]. Upon or above the pia mater.
**epiplasm** (*ep'-e-plasm*) [*epi-; plasma*]. In the sporangium of many fungi, a part of the protoplasm remaining after formation of the spores.
**epiplerosis** (*ep-e-ple-ro'-sis*) [*epi-; πλήρωσις,* filling]. Engorgement; repletion; distention.
**epipleural** (*ep-e-plu'-ral*) [*epi-; pleura*]. 1. Relating to a pleurapophysis. 2. Located on the side of the thorax.
**epiplexus** (*ep-e-pleks'-us*) [*epi-; plexus,* a network]. The choroid plexus of the epicele.
**epiplocele** (*ep-ip'-lo-sēl*) [*epiploon; κήλη,* hernia]. A hernia containing omentum.
**epiploenterocele** (*ep-ip-lo-en'-ter-o-sēl*) [*epiploon; ἔντερον,* intestine; *κήλη,* hernia]. A hernia containing both omentum and intestine.
**epiploenterooscheocele** (*ep-ip-lo-en-ter-o-os'-ke-o-sēl*) [*epiploon; ἔντερον,* intestine; *ὄσχεον,* scrotum; *κήλη,* hernia]. Intestinal and omental hernia into the scrotum.
**epiploic** (*ep-ip-lo'-ik*) [*epiploon*]. Relating or belonging to the omentum. **e. appendages,** small pouches of peritoneum filled with fat, found on the colon.
**epiploischiocele** (*ep-ip-lo-is'-ke-o-sēl*) [*epiploon; ἰσχίον,* hip; *κήλη,* hernia]. Omental hernia through the sciatic notch or foramen.
**epiploitis** (*ep-pip-lo-i'-tis*) [*epiploon; ιτις,* inflammation]. Inflammation of the omentum; omental peritonitis.
**epiplomerocele** (*ep-ip-lo-mer'-o-sēl*) [*epiploon; μηρός,* thigh; *κήλη,* hernia]. Femoral hernia containing omentum.
**epiplomphalocele** (*ep-ip-lom-fal'-o-sēl*) [*epiploon; ὀμφαλός,* navel; *κήλη,* hernia]. Umbilical hernia with protruding omentum.
**epiploön** (*ep-ip'-lo-on*) [ἐπίπλοον, from ἐπί, upon; πλέω, to float]. The omentum.
**epiplopexy** (*ep-ip'-lo-peks-e*) [*epiploon; πῆξις,* a fixing in]. Talma's operation of suturing the great omentum to the anterior abdominal wall for the purpose of establishing a collateral venous circulation in cirrhosis of the liver.
**epiplorrhaphy** (*e-pip-lor'-af-e*) [*epiploon; ῥαφή,* suture]. Same as *epiplopexy*.
**epiplosarcomphalocele** (*ep-ip-lo-sar-kom-fal'-o-sēl*)

**epiploon**; σάρξ, flesh; ὀμφαλοs, navel; κήλη, hernia]. An epiplomphalocele in which the omentum has become indurated.
**epiploscheocele** (*ep-ip-los'-ke-o-sēl*) [*epiploon;* ὀσχέον, scrotum; κήλη, hernia]. Omental hernia descending into the scrotum.
**epipolic** (*ep-e-pol'-ik*) [ἐπιπολή, at the top]. Relating to fluorescence.
**epipteric** (*ep-ip-ter'-ik*) [*epi-;* pterion]. Upon or above the pterion. e. bone. See under *bone.*
**epipygus** (*ep-ip'-e-gus*) [*epi-;* πυγή, rump]. See *pygomelus.*
**episarkin** (*ep-e-sark'-in*) [*epi-;* σάρξ, flesh], C₄H₉N₅O. A xanthin base which occurs in normal urine of man and dogs and in the urine in leukemia.
**episclera** (*ep-e-skle'-rah*) [*epi-;* σκληρόs, hard]. The loose connective tissue lying between the conjunctiva and the sclera.
**episcleral** (*ep-e-skle'-ral*) [*episclera*]. Situated on the outside of the sclerotic coat.
**episcleritis** (*ep-e-skle-ri'-tis*) [*episclera;* ιτιs, inflammation]. An inflammation of the subconjunctival tissues or of the sclera itself.
**episio-** (*ep-iz-e-o-*). A prefix signifying relation to the pubes.
**episiocele** (*ep-iz'-e-o-sēl*) [*episio-;* κήλη, hernia]. Pudendal hernia; vulvar protrusion.
**episioclisia** (*ep-iz-e-o-klis'-e-ah*) [*episio-;* κλείσιs, locking, closure]. Surgical closure of the vulva.
**episioelytrorrhaphy** (*ep-iz-e-o-el-it-ror'-af-e*) [*episio-;* *elytrorrhaphy*]. The operation of suturing a ruptured perineum and narrowing the vagina for the support of a prolapsed uterus.
**episiohematoma** (*ep-iz-e-o-hem-at-o'-mah*) [*episio-;* αἷμα, blood; ὅμα, tumor: *pl.,* *episiohematomata*]. Hematoma of the vulva or pudenda.
**episioitis** (*ep-iz-e-o-i'-tis*) [*episio-;* ιτιs, inflammation of the pudenda.
**episioperineorrhaphy** (*ep-iz-e-o-per-in-e-or'-af-e*). See *episioelytrorrhaphy.*
**episioplasty** (*ep-iz-e-o-plas'-te*). [*episio-;* πλάσσειν, to form]. A plastic operation upon the pubic region, or on the vulva.
**episiorrhagia** (*ep-iz-e-or-a'-je-ah*) [*episio-;* ῥηγνύναι, to break forth]. Hemorrhage from the vulva.
**episiorrhaphy** (*ep-iz-e-or'-a-fe*) [*episio-;* ῥαφή, seam]. An operation for the repair of tears about the vulva.
**episiostenosis** (*ep-iz-e-o-sten-o'-sis*) [*episio-;* στενόs, narrow]. Contraction or narrowing of the vulva.
**episiotomy** (*ep-iz-e-ot'-o-me*) [*episio-;* τομή, section]. Incision through the vulva in child-birth, to prevent rupture of the perineum and to facilitate labor.
**epispadiac** (*ep-e-spa'-de-ak*). 1. Relating to or affected with epispadias. 2. A person affected with epispadias.
**epispadial** (*ep-e-spa'-de-al*). Relating to an epispadias.
**epispadias** (*ep-e-spa'-de-as*) [*epi-;* σπάειν, to pierce]. A condition in which the urethra opens on the upper part of the penis, either on the dorsum or on the glans.
**epispasis** (*e-pis'-pas-is*) [*epi-;* σπάσιs, a drawing]. An eruption or skin-affection due to medical treatment; a drug-exanthem.
**epispastic** (*ep-e-spas'-tik*) [*epi-;* σπάσιs, a drawing]. 1. Blistering. 2. A substance producing a blister.
**episplenitis** (*ep-e-splen-i'-tis*). Inflammation of the fibrous coat of the spleen.
**epistasis** (*e-pis'-tas-is*) [ἐπίστασιs, scum]. 1. A scum or film of substance floating on the surface of urine. 2. A checking or stoppage of a hemorrhage or other discharge.
**epistaxis** (*ep-is-taks'-is*) [ἐπιστάζειν, to cause to drop]. Hemorrhage from the nose.
**episternal** (*ep-e-ster'-nal*) [*epi-;* στέρνον, the sternum]. Above the sternum.
**episthotonos, episthotonus** (*ep-is-thot'-o-nos,* -*us*). See *emprosthotonos.*
**epistropheus** (*ep-is-tro'-fe-us*) [ἐπιστροφεύs, pivot]. The BNA term for the axis, or second cervical vertebra.
**episyivian** (*ep-e-sil'-ve-an*). Situated above the Sylvian fissure.
**epitela** (*ep-e-te'-lah*) [*epi-;* *tela,* a web]. The delicate tissue of Vieussen's valve.
**epiteric** (*ep-e-ter'-ik*) [*epi-;* *pterion*]. Upon or above the pterion.

**epithalamic** (*ep-e-thal'-am-ik*) [*epi-;* *thalamus*]. Situated upon the thalamus.
**epithalamus** (*ep-e-thal'-am-us*). A term including the habenae, epiphysis cerebri, and postcommissure of the brain.
**epithelia** (*ep-e-the'-le-ah*) [*pl.* of *epithelium*]. The epithelial cells, or epithelial layer.
**epithelial** (*ep-e-the'-le-al*) [*epithelium*]. Pertaining to or made up of epithelium.
**epitheliogenetic** (*ep-e-the-le-o-jen-et'-ik*) [*epithelium;* γένεσιs, generation]. Originating from undue epithelial proliferation.
**epithelioid** (*ep-e-the'-le-oid*) [*epithelium;* εἶδοs, likeness]. Resembling epithelium.
**epitheliolysin** (*ep-e-the-le-ol'-is-in*) [*epithelium;* *lysin*]. A cytolysin produced by inoculation with epithelial cells.
**epitheliolytic** (*ep-e-the-le-o-lit'-ik*). Capable of bringing about the destruction of epithelial cells. Metchnikoff found that the introduction of comminuted epithelium into the blood gave this power to the serum.
**epithelioma** (*ep-e-the-le-o'-mah*) [*epithelium;* ὄμα, a tumor]. Properly, any tumor in which epithelium forms the prominent element; by usage the word is restricted to carcinoma of the skin and mucous membranes. e., **columnar**. See e., *cylindrical*. e., **corneous**, one in which the cells resemble the outer layer of epidermal cells. e., **cylindrical**, one in which the epithelial cells resemble ordinary columnar epithelium and the structure resembles ordinary mucosa. Syn., *cylinder-cell cancer; cylindriform epithelial cancer; columnar-celled carcinoma.* e., **cylindrocellular**. See e., *cylindrical*. e., **cystic**, a form containing pits filled with fluid. e., **diffuse**, a form marked by rapid infiltration of the adjacent connective tissue with epithelial cells. e., **glandular**, a not very malignant form composed of gland-cells occurring in mucous membranes, especially of the nose and palate, and of slow growth. e., **multiple cystic**, a variety in which scattered cysts are formed in consequence of mucoid degeneration and the fusion of adjacent drops of fluid. e., **myxomatodes psammosum**, a tumor of the third ventricle of the brain, of the character of a very soft myxoma, and containing very hard, granular, milk-white contents.
**epitheliomatous** (*ep-e-the-le-om'-at-us*). Having the nature of an epithelioma.
**epitheliomuscular** (*ep-e-the-le-o-mus'-kū-lar*). Resembling epithelium and muscle.
**epithelium** (*ep-e-the'-le-um*) [ἐπί, upon; θηλή, nipple]. A term applied to the cells that form the epidermis, that line all canals having communication with the external air, and that are specialized for secretion in certain glands, as the liver, kidneys, etc. Epithelium is divided, according to the shape and arrangement of the cells, into *columnar, cuboidal, flat, pavement, squamous, stratified,* and *tessellated* epithelium; according to function, into protective and glandular or secreting. e., **ciliated**, a form in which the cells bear vibratile filaments or cilia on their free extremities. e., **columnar**, distinguished by prismatic-shaped or columnar cells. e., **fibrillated**. See e., *rod*. e., **germ**, e., **germinal**, e., **germinative**. 1. See *ridge, genital*. 2. The single layer of columnar epithelial cells covering the free surface of the ovary. e., **glandular**, that composed generally of spheroidal cells and constituting the proper secreting substance of a gland. e., **intestinal**, columnar epithelium. e., **Malpighian**. See e., *mucous*. e., **mucous**. 1. The rete mucosum. 2. The entire embryonic epidermis with the exception of the epitrichium. e., **nerve**, epithelium in which sensory cells combined with ordinary epithelial cells form the peripheral terminations of the nerves in the organs of sense. e., **pavement**, a kind composed of cubic cells. e., **pigmentary**, e., **pigmented**, epithelial cells holding pigment-granules. e., **protective**, that serving for protection, as the epidermis, as distinguished from that serving for secretion or sensation. e., **pyramidal**, columnar epithelium. e., **rod**, columnar cells lining certain glands. e., **sensory**. See e., *nerve*. e., **squamous**, the cells have been reduced to scaly plates. e., **stratified**, the cells are arranged in distinct layers. e., **striated**, that consisting of striated cells. e., **subcapsular**, the epithelial-like lining of the internal surface of the capsule of the nerve-cells of spinal ganglia. e., **tabular**, e. **tabularre**, pavement epithelium. e., **tegumentary**, the epidermis. e., **transitional**, epithelium inter-

mediate between simple and stratified. e., **vascular,** vascular endothelium. e. **vibrans,** e., **vibratile,** e., **vibrating,** e. **vibratorium,** ciliated epithelium.
**epithelization** (*ep-e-the-li-za'-shun*). The growth of epithelium over a raw surface.
**epithem** (*ep'-ith-em*) [ἐπίθημα, a poultice; *pl.,* *epithemata*]. Any local application; as a compress, fomentation, lotion, or poultice; from this definition some writers exclude salves, plasters, and ointments.
**epithema** (*ep-ith-e'-mah*). See *epithem.*
**epithesis** (*e-pith'-es-is*) [ἐπιτιθέναι, to lay on]. 1. The surgical correction of deformed or crooked limbs. 2. A splint, or similar appliance.
**epithymia** (*ep-e-thi'-me-ah*) [ἐπιθύμια, longing]. Any natural longing or desire; a yearning.
**epitonic** (*ep-e-ton'-ik*) —[ἐπιτείνειν, to stretch]. Tightly drawn; on the stretch.
**epitonus, epitonus** (*ep-e-to'-nos, -nus*). 1. See *epitonic.* 2. Anything exhibiting abnormal tension or stretched from one point to another.
**epitoxoid** (*ep-e-toks'-oid*). A toxoid (*q. v.*) having a lesser affinity for the antitoxin than is possessed by the corresponding toxin.
**epitrichial** (*ep-e-trik'-e-al*). Relating to the *epitrichium.*
**epitrichium** (*ep-e-trik'-e-um*) [*epi-;* τρίχιον, a small hair]. Superficial layer of fetal epidermis.
**epitrochanterian** (*ep-e-tro-kan-te'-re-an*) [*epi-;* *trochanter*]. Situated upon the trochanters.
**epitrochlea** (*ep-e-trok'-le-ah*) [*epi-;* τροχαλία, a pulley]. The internal condyle of the humerus.
**epitrochlear** (*ep-e-trok'-le-ar*). Applied to muscles of the forearm which are attached to the epitrochlea.
**epitympanic** (*ep-e-tim-pan'-ik*) [*epi-;* τύμπανον, the tympanum]. Upon or above the tympanum. e. **recess,** the attic.
**epitympanum** (*ep-e-tim'-pan-um*). The attic.
**epityphlitis** (*ep-e-tif-li'-tis*) [*epi-;* τυφλόν, the cecum]. Synonym of *appendicitis.*
**epityphlon** (*ep-e-tif'-lon*) [*epi-;* τυφλόν, the cecum]. Küster's name for the vermiform appendix.
**epivertebral** (*ep-e-ver'-te-bral*) [*epi-;* *vertebra*]. 1. Situated upon a vertebra. 2. A spinous process of a vertebra.
**epizoicide** (*ep-e-zo'-is-id*) [*epi-;* ζῷον, an animal; *cædere,* to kill]. A drug or preparation that destroys external parasites.
**epizoon** (*ep-e-zo'-on*) [*epi-;* ζῷον, an animal: *pl.,* *epizoa*]. An animal parasite living upon the exterior of the body.
**epizootic** (*ep-e-zo-ot'-ik*) [see *epizoon*]. An epidemic disease of the lower animals.
**epoikic** (*ep-oi'-kik*) [*epi-;* οἶκος, a house]. Applied to diseases limited to the household or other circumscribed locality.
**eponychium** (*ep-o-nik'-e-um*) [*epi-;* ὄνυξ, fingernail]. A horny condition of the epidermis from the second to the eighth month of fetal life, indicating the position of the future nail. The thickened epitrichium covering the nail area.
**eponym** (*ep'-o-nim*) [ἐπώνυμος, named after a person]. A term derived from the name of a person.
**eponymic** (*ep-o-nim'-ik*) [ἐπώνυμος, named after a person]. Named after some person.
**epoophorectomy** (*ep-o-o-for-ek'-to-me*) [*epoophoron;* ἐκτομή, excision]. Surgical removal of the epoophoron.
**epoophoron** (*ep-o-of'-or-on*) [*epi-;* ᾠόν, egg; φέρειν, to bear]. The parovarium.
**epoptic** (*ep-op'-tik*) [L., *epopticus*]. Fluorescent.
**epostoma** (*ep-os-to'-mah*) [*epi-;* ὀστέον, bone]. An exostosis.
**Epsom salt** (*ep'-sum*) [*Epsom,* a town in Surrey, England]. See *magnesium sulphate.*
**Epstein's pearls** (*ep'-stīn*) [Alois *Epstein,* German physician, 1849— ]. Small, slightly elevated, yellowish-white masses on each side of the median line of the hard palate at birth.
**epulis** (*ep-ū'-lis*) [*epi-;* οὖλα, the gums]. A fibrous tumor of the alveolar processes of the jaws. e., **malignant,** a giant-cell sarcoma of the jaw.
**epuloid** (*ep'-ū-loid*). Like an epulis.
**epulosis** (*ep-ū-lo'-sis*) [*epi-;* οὐλή, scar]. Cicatrization; a cicatrix.
**epulotic** (*ep-ū-lot'-ik*) [*epi-;* οὐλή, scar]. 1. Promoting epulosis or cicatrization. 2. A remedy or application that promotes the healing of wounds or sores.
**equation** (*e-kwa'-shun*) [*æquare,* to make equal]. In chemistry, a collection of symbols and formulæ so arranged as to indicate the reaction that will take place if the bodies represented by these symbols and formulæ are brought together. e., **personal,** an allowance for individual peculiarity or error in an observer's work.
**equator** (*e-kwa'-tor*) [see *equation*]. An imaginary circle surrounding a sphere so as to divide it into equal halves. e. **of a cell,** the boundary of the plane through which division takes place. e. **of the eye,** the equator oculi; a line joining the four extremities of the transverse and vertical axes of the eye.
**equatorial** (*e-kwa-tor'-e-al*) [*æquare,* to make equal]. Pertaining or belonging to an equator. e. **plate.** See *karyokinesis* and *plate.*
**equilibrating operation** (*e-kwil'-ib-ra-ting*). An operation on the ocular muscles to equalize their action in cases of squint.
**equilibration** (*e-kwil-ib-ra'-shun*) [*equilibrium*]. The maintenance of equilibrium.
**equilibrium** (*e-kwe-lib'-re-um*) [*æquus,* equal; *libra,* balance]. An even balancing of a body or condition. e., **indifferent,** that which is independent of the positions assumed by the body. e., **mobile,** the constant temperature kept by neighboring bodies after a mutual exchange of heat proportionate to their capacities; this constancy is due to the fact that after the attainment of heat equilibrium the subsequent emission is equal to the quantity of heat received. e., **neutral.** See e., *indifferent.* e., **nitrogenous,** the condition of the system in which the amount of nitrogen in the matter discharged from the body exactly equals the amount taken in. e., **physiological,** the state of the system in which the amount of material discharged from the body exactly equals the amount taken in. e., **stable,** when, after slight disturbance, the body will return to its original condition or position. e., **unstable,** when it will not so return.
**equinated** (*e-kwin-a'-ted*). Inoculated with the virus of equinia.
**equination** (*e-kwin-a'-shun*) [*equinus,* of a horse]. 1. Inoculation with the virus of equine smallpox. 2. Inoculation with the virus of equinia.
**equinia** (*e-kwin'-e-ah*) [*equus,* a horse]. Glanders; farcy.
**equinocavus** (*e-kwi-no-ka'-vus*) [*equinus; cavus,* hollow]. Dorsal talipes equinus in which the plantar surface is excessively hollowed and creased.
**equinovarus** (*e-kwi-no-va'-rus*) [*equinus; varus,* bent outward]. A variety of talipes presenting the characteristics of talipes equinus and talipes varus.
**equinus** (*e-kwi'-nus*) [*equus,* a horse]. 1. Talipes equinus. 2. Relating to the horse; equine. e. **dorsalis,** a form of talipes equinus in which the patient walks on the dorsal surface of the flexed toes. e. **plantaris,** the form of talipes equinus in which the toes are extended throughout or only at the metatarsophalangeal joint.
**equipotential** (*e-kwi-po-ten'-shal*) [*æquus,* equal; *potentia,* power]. Of equal power; applied in electricity to bodies with equal dynamic units.
**Equisetum** (*ek-wis-e'-tum*) [*equus,* a horse; *sæta,* a bristle]. A genus of cryptogamous plants. *E. hiemale* is used in dropsy and diseases of genitourinary origin. Dose of *fluidextract* 30–60 min. (1.8–3.7 Cc.).
**equivalence, equivalency** (*e-kwiv'-al-ens, -en-se*) [*æquus,* equal; *valere,* to be worth]. The saturating power of an atom of an element as compared with that of an atom of hydrogen. Valence.
**equivalent** (*e-kwiv'-al-ent*). Of equal valency; having the same value. e., **chemical,** the amount of an element capable of combining with a unit weight of hydrogen; it is the atomic weight of the element divided by its valence. e., **endosmotic,** the ratio obtained by dividing the amount of the replacing liquid in osmotic action by the amount replaced. e., **Joule's.** See under *Joule.* e., **psychic epileptic,** mental disturbance or excitement which may take the place of epileptic attacks. e., **toxic,** the quantity of poison capable of killing, by intravenous injection, one kilogram of animal. e., **weight,** same as *equivalent, chemical.*
**equivocal** (*e-kwiv'-o-kal*) [*æquus,* equal; *vox,* sound]. Of doubtful significance, as equivocal symptoms.
**E. R.** Abbreviation for *external resistance.*
**Er.** The chemical symbol of erbium, also written simply E.
**eradication** (*e-rad-ik-a'-shun*) [*e,* out; *radicare,* to root]. Complete or thorough removal.

**erasion** (*e-ra'-shun*) [*e*, out; *radere*, to scrape]. 1. The act of scraping. 2. Scraping or curetting of a joint. 3. The same as arthrectomy.

**Erb's disease** [Wilhelm Heinrich *Erb*, German physician, 1840– ]. Severe pseudoparalytic myasthenia; asthenic bulbar paralysis. **E.'s juvenile form of progressive muscular atrophy,** the scapulohumeral type. **E.'s myotonic reaction.** See *E.'s waves.* **E.'s palsy, E.'s paralysis,** a paralysis involving the deltoid, biceps, brachialis anticus, and supinator longus; often also the supinator brevis, and at times the infraspinatus; rarely the subscapularis. It is traumatic in origin; it may occur during birth. **E.'s point,** a point about two fingerbreadths above the clavicle and one fingerbreadth external to the sternomastoid. Electrical stimulation at this point produces contraction of the deltoid, biceps, brachialis anticus, and supinator longus. **E.'s symptom.** 1. Increase of the electric irritability of the motor nerves in tetany. 2. Dulness on percussion over the manubrium sterni in acromegaly. **E.'s waves,** undulatory movements produced in a muscle by passing a moderately strong constant current through it and leaving the electrodes in place, the circuit remaining closed. They are sometimes seen in Thomsen's disease.

**Erb-Charcot's disease.** Spastic spinal paralysis; spasmodic tabes dorsalis.

**Erben's phenomenon** (*er'-ben*) [Sigmund *Erben*, Austrian physician, 1863– ]. A temporary slowing of the pulse on bending forward or attempting to sit down; it has been observed in neurasthenia.

**Erb-Goldflam's symptom-complex.** See *Erb's disease.*

**erbium** (*ur'-be-um*). A rare element; symbol Er. See *elements, table of.*

**Erb-Westphal's symptom.** See *Westphal's sign.*

**erect** (*e-rekt'*) [*erigere*, to set up]. 1. To raise through engorgement of the tissues. 2. Upright; in the state of erection.

**erectile** (*e-rek'-til*) [*erect*]. Having the quality of becoming erect. **e. tissue,** a tissue consisting of a network of expansile capillaries that, under stimulus, become engorged with blood and cause erection of the part. **e. tumor,** a tumor of erectile tissue.

**erection** (*e-rek'-shun*) [*erect*]. The state of being erect, as *erection* of the penis or clitoris. Fulness and firmness of the genital organs from congestion.

**erector** (*e-rek'-tor*) [*erect*]. 1. A muscle that produces erection of a part. See under *muscle.* 2. A prism frequently attached to the eye-piece of the microscope, for correcting the inversion of the image. **e. nerves.** See *nervi erigentes.* **e. pili,** the unstriped muscular fibers causing the erection of the hair and the phenomenon called goose-flesh or goose-skin.

**eremacausis** (*er-e-mak-aw'-sis*) [ἠρέμα, slowly; καῦσις, burning]. Slow oxidation or gradual decay; slow combustion.

**eremophobia** (*er-em-o-fo'-be-ah*) [ἐρῆμος, desolate; φόβος, fear]. The same as *agoraphobia, q. v.*

**erepsin** (*er-ep'-sin*) [ἐρείπειν, to destroy]. A ferment produced by the intestinal mucosa, having no effect on unaltered albumin but causing cleavage of peptones.

**erethetic, erethetical** (*er-e-thet'-ik, -al*). See *erethismic.*

**erethin** (*er'-e-thin*) [ἐρεθίζειν, to irritate]. The name given by Klebs to that constituent of tuberculin which occasions fever.

**erethism, erethismus** (*er'-e-thizm, er-e-thiz'-mus*) [ἐρεθισμός, irritation]. An abnormal increase of nervous irritability.

**erethisma** (*er-e-thiz'-mah*). An irritant.

**erethismal** (*er-e-thiz'-mal*). Of the nature of an erethism.

**erethismic, erethistic** (*er-e-thiz'-mik, -this'-tik*). Relating to, or affected with, erethism.

**ereuthophobia** (*e-rūth-o-fo'-be-ah*) [ἔρευθος, redness; φόβος, fear]. Morbid fear of blushing.

**ereuthosis** (*e-rūth-o'-sis*) [ἔρευθος, a redness]. Extreme facility for blushing.

**erg** [ἔργον, work]. A unit of work, representing the work done in moving a body against the force of one dyne through a space of one centimeter.

**ergamine** (*er'-gam-ēn*). Trade name of an organic base, beta-iminazolylethylamine, occurring in ergot and its extracts, and also produced by chemical synthesis. It is a stimulant of unstriped muscle, particularly of the uterus; and is used in cases of postpartum hemorrhage.

**ergasiomania** (*ur-gas-e-o-ma'-ne-ah*) [ἐργασία, work; μανία, madness]. 1. An eager desire for work of any kind. 2. Mania for performing operations.

**ergasiophobia** (*ur-gas-e-o-fo'-be-ah*) [ἐργασία, work; φόβος, fear]. 1. Timidity in operating; a dread of operations. 2. Dread of work of any kind.

**ergastoplasm** (*er-gas'-to-plazm*) [ἔργον, work; *plasm*]. Same as *kinoplasm.*

**ergoapiol** (*er-go-a'-pe-ol*). A proprietary combination of apiol, 5 parts; ergotin, 1 part; savin oil, ⅛ part; aloin, ⅛ part; it is used as an emmenagogue. Dose 7–14 gr. (0.45–0.9 Gm.).

**ergogenesis** (*ur-go-jen'-es-is*) [ἔργον, work; γένεσις, production]. Same as *ergogeny.*

**ergogenetic** (*ur-go-jen-et'-ik*) [ἔργον, work; γένεσις, production]. Of the nature of, or pertaining to, ergogeny.

**ergogeny** (*ur-goj'-en-e*) [ἔργον, work; γένεσις, production]. In biology, the energy, both potential and kinetic, involved in the adaptive processes of living organisms; it includes both kinetogeny and statogeny. (Ryder.)

**ergograph** (*ur'-go-graf*) [ἔργον, work; γράφειν, to write]. A recording ergometer. An instrument for recording the extent of movement produced by a contracting muscle, or the amount of work it is capable of doing.

**ergometer** (*ur-gom'-et-er*) [ἔργον, work; μέτρον, measure]. A variety of dynamometer.

**ergone** (*er'-gōn*). Trade name of a sterile preparation of ergot.

**ergophobia** (*er-go-fo'-be-ah*) [ἔργον, work; φόβος, fear]. Morbid dread of work; ergasiophobia.

**ergophore group** (*ur'-go-fōr*) [ἔργον, work; φέρειν, to bear]. A group of atoms belonging to the molecule of an antibody, and by virtue of which its specific (agglutinative or other) action depends.

**ergostat** (*ur'-go-stat*) [ἔργον, work; ἱστάναι, to stand]. An apparatus for testing muscular strength.

**ergot, ergota** (*er'-got, er-go'-tah*) [Fr., *ergot*, a spur]. The sclerotium of the *Claviceps purpurea*, a fungus growing on rye. It is a vasomotor stimulant and causes contraction of the involuntary muscles. It is used to control hemorrhage and to cause uterine contraction; it is also employed in cerebral and spinal congestion, in diabetes insipidus, and in night-sweats. Dose 10 gr.–1 dr. (0.65–4.0 Gm.). **e., extract of** (*extractum ergotæ*, U. S. P.), ergot. Dose 5–20 gr. (0.3–1.3 Gm.); hypodermatically, ¼–5 gr. (0.016–0.32 Gm.). **e., fluidextract of** (*fluidextractum ergotæ*, U. S. P.). Dose ½ dr.–½ oz. (2–16 Cc.). **e., infusion of** (*infusum ergotæ*, B. P.). Dose 1–2 oz. (30–60 Cc.). **e., injection of** (*injectio ergotinæ hypodermica*, B. P.), ergotin and camphor-water. Dose subcutaneously 3–10 min. (0.2–0.65 Cc.). **e., tincture of** (*tinctura ergotæ*, B. P.). Dose 10 min.–1 dr. (0.6–4.0 Cc.). **e., wine of** (*vinum ergotæ*, U. S. P.). Dose 2 dr. (8 Cc.).

**ergotherapy** (*er-go-ther'-ap-e*) [ἔργον, work; θεραπεία, treatment]. Treatment of disease by physical work.

**ergotin** (*er'-go-tin*). Extract of ergot.

**ergotine** (*er-go-tēn'*). An alkaloid of ergot.

**ergotinine** (*er-got'-in-ēn*), $C_{35}H_{40}N_4O_6$. An alkaloid from ergot of rye.

**ergotinol** (*er-got'-in-ol*). A proprietary ammoniated solution of ergotin.

**ergotinum** (*er-go-ti'-num*). See *ergotin.*

**ergotism** (*er'-got-izm*) [*ergot*]. The constitutional effects following the prolonged use of ergot, or of grain containing the fungus *Claviceps purpurea.* The symptoms are of two types, either a spasmodic form with contractions and cramps of the muscles, or a form characterized by dry gangrene.

**ergotized** (*ur'-go-tīzd*). Systemically affected with ergot.

**ergotol** (*ur'-got-ol*). A proprietary liquid preparation of ergot, recommended for hypodermatic injection.

**ergotoxine** (*er-go-toks'-ēn*). An alkaloid derived from ergot, probably identical with cornutine.

**ergoval** (*er'-go-val*). A proprietary standardized fluid extract of ergot.

**Erichsen's disease** (*er'-ik-sen*) [Sir John Eric *Erichsen*, English surgeon, 1818–1896]. Railway-spine; railway-brain. A train of symptoms following accidents, which may assume the form of traumatic hysteria, neurasthenia, hypochondriasis, or melancholia. **E.'s ligature,** one consisting of a double thread, one-half of which is black, the other white; it is used in the ligation of nevi. **E.'s sign,** to differ-

**entiate coxalgia from sacroiliac disease; compression of the two iliac bones causes pain in the latter, but not in the former, affection.

**ericin** (*er'-is-in*) [ἐρείκη, heather]. A dye obtained from common heath and varieties of poplar wood by treating with a hot solution of alum.

**ericolin** (*er-ik'-o-lin*) [ἐρείκη, heath], C₃₄H₅₆O₂₁. A substance found in *uva ursi*. It is an amorphous, yellowish glucoside with a bitter taste, yielding with diluted acids sugar and an essential oil—*ursone*.

**erigens** (*er'-e-jenz*) [L.]. Producing erection, as the nervi erigentes.

**erigeron** (*er-ij'-er-on*) [ἠριγέρων, groundsel]. Fleabane. The plant *E. canadense*, having physiological actions like those of oil of turpentine, but less irritant. It contains oil of erigeron, and is used as a hemostatic. *E. bellidifolium* and *E. philadelphicum* afford similar oil, and have the same properties. *E. canadense* is used in dropsy and diseases of the genitourinary tract. e., **fluidextract of.** Dose 30–60 min. (1.8–3.7 Cc.). e., **oil of** (*oleum erigerontis*, U. S. P.). Dose 10 min.—½ dr. (0.65–2.0 Cc.).

**eriocome** (*er'-e-o-kōm*) [ἔριον, wool; κόμη, hair]. Haeckel's term for a race having wooly hair that covers the head like a continuous fleece, as in the majority of negroes. Cf. *lophocome*.

**eriocomous** (*er-e-ok'-om-us*). Villous; covered with woolly hair.

**eriodictyon** (*er-e-o-dik'-te-on*) [ἔριον, wool; δίκτυον, a net]. Yerba santa or mountain-balm. The leaves of *E. californicum*, a shrub of California, are expectorant and an excipient for quinin, the taste of which they largely conceal. Dose 15 gr. (1 Gm.). e., **extract of.** Dose 2–10 gr. (0.13–0.65 Gm.). e., **fluidextract of** (*fluidextractum eriodictyi*, U. S. P.). Dose 15 min.–1 dr. (1–4 Cc.).

**Erlenmeyer's mixture** (*er'-len-mi-er*) [Friedrich Albrecht Erlenmeyer, German psychiatrist, 1849– ]. A mixture of equal parts of the bromides of potassium, sodium, and ammonium.

**ernutin** (*er-nū'-tin*). A proprietary preparation of ergot.

**erodent** (*e-ro'-dent*) [e, out; *rodere*, to gnaw]. 1. Caustic; causing erosion. 2. A caustic drug.

**erogenic, erogenous** (*er-o-jen'-ik, er-oj'-en-us*) [ἔρως, love; γεννᾶν, to produce]. Producing or stimulating the sexual appetite.

**eromania** (*er-o-ma'-ne-ah*). See *erotomania*.

**erose** (*er-ōs'*) [*erodere*, to eat out]. Having a margin or border irregularly toothed.

**erosion** (*e-ro'-shun*) [*erodere*, to eat out]. The eating away of tissue. e., **aphthous**, the formation of flat ulcers on a mucosa. e., **dental**, a progressive decalcification, affecting most commonly the labial and buccal faces of the teeth, not due to the causes of dental caries, and usually associated with the gouty diathesis. e., **papillary**, a condition developed from simple erosion; after the destruction of the epithelium the exposed points of the papillæ swell and appear as granular, dark-red, and easily bleeding elevations.

**erosive** (*e-ro'-siv*). 1. Pertaining to or causing or characterized by erosion. 2. An agent which produces erosion.

**erotic** (*er-ot'-ik*) [ἔρως, love]. Pertaining to the sexual passion; lustful.

**eroticism** (*er-ot'-is-izm*) [ἔρως, love]. An erotic disposition; erotic display; tendency to erotomania.

**eroticomania** (*er-o-tik-o-ma'-ne-ah*) [*erotic*; μανία, madness]. Same as *erotomania*.

**erotism** (*er'-o-tizm*). A condition of erotic intoxication.

**erotogenic** (*er-ot-o-jen'-ik*) [ἔρως, love; γεννᾶν, to produce]. Causing erotic feelings.

**erotomania** (*er-ot-o-ma'-ne-ah*) [ἔρως, love; μανία, madness]. Morbid exaggeration of the affections, usually toward the opposite sex.

**erotomaniac** (*er-o-to-ma'-ne-ak*). A patient who is afflicted with erotomania.

**erotopath** (*e-rot'-o-path*). A person who is afflicted with erotopathy.

**erotopathy, erotopathia** (*er-ot-op'-ath-e, er-o-to-path'-e-ah*) [ἔρως, love; πάθος, disease]. Perverted sexual instinct.

**erpiol** (*er'-pe-ol*). A proprietary remedy of ergotin, apiol and gossypiin.

**errabund** (*er'-a-bund*) [*errare*, to wander]. Erratic; wandering.

**erratic** (*er'-at-ik*) [*errare*, to wander]. Moving about from place to place; irregular; strange or unusual; eccentric, peculiar.

**errhine** (*er'-in*) [ἐν, in; ῥίς, the nose]. 1. Causing discharges from the nose. 2. A medicine that increases nasal secretions; a sternutatory.

**errhysis** (*er'-is-is*) [ἔρρειν, to go slowly]. Slow bleeding.

**erseol** (*er'-se-ol*). Trade name for quinoline sulphosalicylate; it is used as a substitute for quinine.

**erubescence** (*er-oo-bes'-ens*) [*erubescentia*, blushing]. Redness of the skin.

**eructation** (*e-ruk-ta'-shun*) [*eructare*, to belch]. Belching.

**erugation** (*er-oo-ga'-shun*) [e, out; *ruga*, wrinkle]. The removal of wrinkles.

**erugatory** (*er-oo'-gat-or-e*) [e, out; *ruga*, a wrinkle]. 1. Tending to remove wrinkles. 2. A remedy for wrinkles.

**eruption** (*e-rup'-shun*) [*erumpere*, to burst out]. A bursting forth, especially applied to the skin-lesions of the exanthematous diseases. e., **drug**, e., **medicinal**. See *dermatitis medicamentosa*. e., **Koch's**. See under *Koch*. e., **miliary**, an eruption of little vesicles occurring in the course of febrile diseases.

**eruptive** (*e-rup'-tiv*) [see *eruption*]. Attended by an eruption, as an *eruptive* fever.

**Eryngium** (*er-in'-je-um*) [ἠρύγγη, a sort of thistle]. A genus of plants of the order *Umbelliferæ*. *E. yuccæfolium* is indigenous to the western prairies and southern barrens of the United States. The root is diaphoretic, expectorant, and refrigerant. Dose of *fluidextract* 30–60 min. (1.8–3.7 Cc.).

**erysipelas** (*er-is-ip'-el-as*) [ἐρυθρός, red; πέλλα, skin]. An acute infectious disease due to *Streptococcus erysipelatis* (which is probably identical with the *Streptococcus pyogenes*), and characterized by an inflammation of the skin and subcutaneous tissues. e., **ambulans**. See e., *wandering*. e., **bullosum**, that attended with formation of bullæ. e., **chronicum**. Synonym of *erysipeloid*. e., **diffusum**, that in which the affected area is not sharply defined, the redness merging gradually with the color of the surrounding skin. e., **facial**, erysipelas of the face, the most common form. After an initial chill the temperature rises very high; there may be vomiting and delirium, and the disease may spread rapidly over a great part of the body. The affected area is swollen, has a deep-red color, an elevated margin, and itches. e. **glabrum**, that in which the skin is tightly stretched and has a smooth, shining appearance. Syn., *erysipelas læve; erysipelas lævigatum*. e., **idiopathic**, erysipelas occurring without any visible wound. e., **internal**, e. **internum**, that affecting the interior of the body, especially the mucosas. e., **læve**, e. **lævigatum**. See e. *glabrum*. e. **medicamentosum**, a medicine rash resembling erysipelas, but marked by rapid development, the absence of well-defined areas, and tenderness on pressure. e. **migrans**. See e., *wandering*. e., **phlegmonous**, a form of erysipelas in which there is pus-formation. e., **pustular**, e. **pustulosum**, a variety of erysipelas bullosum in which the bullæ contain pus. e., **serpiginous**, a form which extends by involving neighboring parts of the skin. e., **spontaneous**, that to which no external cause can be assigned. e., **surgical**, e., **traumatic**, erysipelas occurring in the site of a wound. e., **symptomatic**, that dependent on some constitutional disorder. e., **true** that due to infection with *Streptococcus erysipelatis*, Fehleisen. See under *Micrococci*, *table of*. e., **venous**, that accompanied by venous congestion and marked by a dark-red color which does not entirely disappear on pressure. e., **verrucosum**, that characterized by a warty or lumpy appearance. e., **wandering**, a form in which the erysipelatous process successively disappears from one part of the body to appear subsequently at another part. e., **white**, a variety of erysipelatous edema in which there is no manifest dilatation of the blood-vessels.

**erysipelatous** (*er-is-ip-el'-at-us*) [*erysipelas*]. Of the nature of or affected with erysipelas.

**erysipelococcus** (*er-is-ip-el-o-kok'-us*). A name for *Streptococcus erysipelatis*, to which erysipelas is due.

**erysipeloid** (*er-is-ip'-el-oid*) [*erysipelas*; εἶδος, likeness]. A noncontagious disease resembling erysipelas. It is due to *Cladothrix dichotoma*. Syn., *erysipelas chronicum; erythema migrans*.

**erysipelotoxin** (*er-e-sip-el-o-toks'-in*). The toxin of erysipelas.

**erythema**, (er-ith-e'-mah) [ἐρύθαίνειν, to make red]. A redness of the skin occurring in patches of variable size and shape. **e. æstivum**, an intense itching and burning, attended with swelling and formation of bullæ, attacking the feet and ankles of those who walk barefooted, in hay-fields. **e. a frigore, e. a gelu**, chilblain. **e., amorphous**, that in which the efflorescence is irregular in outline and arrangement. **e. angeiectaticum**, Auspitz's term for rosacea in order to convey the idea of its dependence upon dilatation of the cutaneous blood-vessels. **e. annulare**, a form of erythema multiforme in which the lesions shrink and desquamate at the center, but continue to extend at the periphery by a raised margin. **e., choleraic**, erythema multiforme occurring in cholera patients, chiefly affecting the extremities and marked by papules bluish-red or livid in color. It has been observed as occurring at both the initial and the declining stage of the disease. **e. congestivum**, erythema with congestion of the skin. **e. diffusum**, a form resembling scarlatina, with ill-defined outline, the red color of the affected skin merging gradually into that of the surrounding parts. **e. enematogenes**, an eruption sometimes observed in children on the anterior surface of the knees, backs of the elbows, buttocks, and face, appearing from 12 to 24 hours after the administration of an enema. It lasts from 24 to 48 hours, is rarely followed by desquamation, and gives rise to no constitutional disturbance. **e., infectious**, a name given to erythema multiforme to express the theory of its infectious character. **e. intertrigo**, intertrigo; a hyperemia of the skin occurring where the folds of the integument come in contact. The epidermis may be abraded. **e., Lewin's, of the larynx**, simple syphilitic catarrh of the larynx. **e. migrans**. Synonym of *erysipeloid*. **e. multiforme**, an acute inflammatory skin disease characterized by reddish macules, papules, or tubercles, usually appearing on the legs and forearms. It is often ushered in by gastric distress and rheumatic pains. **e. nodosum**, dermatitis contusiformis, an inflammatory disease characterized by the formation, especially on the tibial surfaces, of rounded, elevated, erythematous nodules. **e. serpens**. Same as *e. migrans*. **e. solare**. See *e., symptomatic*. **e., symptomatic**, a hyperemia of the skin, either diffuse or in nonelevated patches. It is either idiopathic, as when arising from the action of the sun (*erythema solare*), or due to various poisons (*erythema venenatum*), or it is symptomatic of systemic disease or gastrointestinal disorder. **e. variolosa**, a rash occurring sometimes in the first stage of smallpox. **e. venenatum**. See *e., symptomatic*.
**erythematica** (er-ith-e-mat'-ik-ah) [erythema]. A form of idiopathic enteritis, according to Cullen.
**erythematous** (er-ith-em'-at-us) [erythema]. Of the nature of erythema.
**erythemoid, erythematoid** (er-ith'-em-oid, er-ith-em'-at-oid) [erythema; εἶδος, resemblance]. Resembling erythema.
**Erythræa** (er-ith-re'-ah) [ἐρυθραῖος, red]. A genus of gentians. *E. centaurium*, the European centaury, is tonic and antipyretic. Dose of *extract* 5–30 gr. (0.32–1.9 Gm.). *E. venusta*, a California species, is a valuable bitter tonic and stomachic.
**erythrasma** (er-ith-raz'-mah) [ἐρυθρός, red]. A rare skin disease attacking the axillæ or inguinal region or the buttocks. It forms reddish or brownish, sharply defined, slightly raised, desquamating patches that cause no itching or inconvenience. It is due to *Bacillus epidermidis*.
**erythremelalgia** (er-ith-rem-el-al'-je-ah). See *erythromelalgia*.
**erythremia, erythræmia** (er-ith-re'-meah). Same as *erythrocythemia*.
**erythrenteria** (er-ith-ren-te'-re-ah) [ἐρυθρός, red; ἔντερον, intestine]. Hyperemia of the intestine.
**erythrism** (er'-ith-rizm) [ἐρυθρός, red]. 1. In biology, applied to conditions of dichromatism in which the normal colors of the integument are affected by an excess of red pigment, as often shown in the plumage of a bird. 2. Broca's term for the pathological condition exhibited by the individual having red hair in a dark-haired race free from intermixture, as among European Jews.
**erythro-** (er-ith-ro-) [ἐρυθρός, red]. A prefix signifying of a red color.
**erythroblast** (er-ith'-ro-blast) [erythro-; βλαστός, a germ]. A rudimentary red blood-corpuscle.
**erythrocatalysis** (er-ith-ro-kat-al'-is-is) [erythro;

catalysis]. Excessive destruction of the red blood corpuscles by phagocytosis.
**erythrochloropia** (er-ith-ro-klo-ro'-pe-ah) [erythro-; χλωρός, green; ὤψ, eye]. A form of subnormal color-perception in which green and red are the only colors correctly distinguished.
**erythrochloropy** (er-ith-ro-klor'-o-pe) [erythro-; χλωρός, green]. Ability to distinguish red and green colors only.
**erythrocruorin** (er-ith-ro-kru'-or-in) [erythro-; cruor, blood]. Same as *hemoglobin*.
**erythrocyte** (er-ith'-ro-sīt) [erythro-; κύτος, a cell]. A red blood-corpuscle.
**erythrocythemia, erythrocythæmia** (er-ith-ro-si-the'-me-ah) [erythro-; κύτος, a cell; αἷμα, blood]. A condition in which there is an increase of red blood corpuscles in the circulation blood.
**erythrocytoblast** (er-ith-ro-si'-to-blast). Same as *erythroblast*.
**erythrocytolysis** (er-ith-ro-si-tol'-is-is) [erythrocyte; λύσις, a loosing]. The plasmolysis of red blood-corpuscles; the escape of soluble substances and the reduction of the volume of the corpuscle.
**erythrocytometer** (er-ith-ro-si-tom'-et-er) [erythrocyte; μέτρον, a measure]. A heavy, graduated, glass capillary tube, the lumen of which is expanded near the upper end into a bulb containing a small cubic glass bead which serves as a stirrer. It is used in counting erythrocytes. Cf. *leukocytometer*.
**erythrocyto-opsonins** (er-ith-ro-si-to-op'-so-nins). Substances which are opsonic for red blood corpuscles.
**erythrocytorrhexis** (er-ith-ro-si-tor-reks'-is). See *plasmorrhexis*.
**erythrocytoschisis** (er-ith-ro-si-tos'-kis-is) [erythrocyte; σχίσις, cleavage]. The splitting-up of red blood-corpuscles into discs resembling blood-platelets. Cf. *plasmoschisis*.
**erythrocytosis** (er-ith-ro-si-to'-sis) [erythrocyte]. 1. The formation of red blood-corpuscles. 2. The presence in the blood, before birth, of red cells with nuclei and with karyokinetic figures.
**erythrodermia** (er-ith'-ro-der-me-ah) [erythro-; δέρμα, skin]. Abnormal redness of the skin.
**erythrodermitis** (er-ith-ro-der-mi'-tis) [erythro-; δέρμα, skin; ιτις, inflammation]. A chromodermatosis characterized by erythema and superficial dermatitis.
**erythrodextrin** (er-ith-ro-deks'-trin) [erythro-; dexter, right]. A dextrin formed by the action of saliva on starch. It yields a red color with iodine.
**erythrogen** (er-ith'-ro-jen) [erythro-; γεννᾶν, to produce]. A green substance that has been found in unhealthy bile, and which (apparently without good reason) has been regarded as "the base of the coloring-matter of the blood."
**erythroglucin** (er-ith-ro-glu'-sin). See *erythrol*.
**erythrogranulose** (er-ith-ro-gran'-ū-lōs) [erythro-; granulum, a little grain]. A granular substance. found in starch-grains, coloring red with iodine.
**erythroid** (er'-ith-roid) [erythro-; εἶδος, resemblance]. Reddish; of a red color.
**erythrol** (er'-ith-rol) [ἐρυθρός, red]. 1. C₄H₆(OH)₄. A crystalline alkaloid from certain algæ and lichens. 2. A double salt of bismuth and cinchonidine. It is used in rare forms of dyspepsia in which acid reaction of the gastric juice is accompanied by the production of butyric acid. **e. tetranitrite**, (CH₃ONO₂)₂-(CHO . NO₂)₂, large scales, soluble in alcohol, insoluble in water, exploding on percussion; recommended as a substitute for amyl nitrite and nitroglycerin in angina pectoris, asthma, lead colic, and cardiac affections. Dose ½–1 gr. (0.03–0.06 Gm.).
**erythrolysin** (er-ith-rol'-is-in). See *hemolysin*.
**erythrolysis** (er-ith-rol'-is-is). Erythrocytolysis.
**erythromannite** (er-ith-ro-man'-īt). Same as *erythrol*.
**erythromelalgia** (er-ith-ro-mel-al'-je-ah) [erythro-; μέλος, a limb; ἄλγος; pain]. An affection of the distal parts of the extremities, particularly the feet, characterized by redness and neuralgic pain. The disease is very obstinate; its pathology is not well understood. It may be a vasomotor neurosis, a neuritis of the peripheral nerves, or it may be due to changes in the spinal cord.
**erythromelia** (er-ith-ro-me'-le-ah) [erythro-; μέλος, limb]. An affection of the extensor surfaces of the arms and legs, characterized by painless progressive redness of the skin; it is distinct from erythromelalgia.

**erythroneocytosis** (*er-ith-ro-ne-o-si-to'sis*) [*erythro-;* νέος, new; κύτος, cell]. The presence of regenerative forms of red blood corpuscles in the circulating blood.

**Erythronium** (*er-ith-ro'-ne-um*) [ἐρυθρός, red]. A genus of liliaceous plants. E. americanum is a species indigenous to the United States; the bulb and all parts of the plant are emetic. Dose 20–30 gr. (1.3–1.9 Gm.).

**erythropenia** (*er-ith-ro-pe'-ne-ah*) [*erythro-;* πενία, poverty]. Deficiency in the number of red blood-corpuscles.

**erythrophage** (*er-ith'-ro-fāj*) [*erythro-;* φαγεῖν, to eat]. Any one of the phagocytic cells which, lying about a hemorrhagic area, take up the pigment of the blood or even red-corpuscles. They are remarkable for their brilliant color (red to golden).

**erythrophil** (*er-ith'-ro-fil*) [*erythro-;* φιλεῖν, to love]. Auerbach's term for the red-staining nuclear substance of animal and vegetal cells.

**erythrophilous** (*er-ith-rof'-il-us*) [see *erythrophil*]. Having an especial affinity for red dyes.

**erythrophleine** (*er-ith-rof'-le-ēn*) [*erythro-;* φλοιός, bark]. A poisonous alkaloid from casca-bark. **e. hydrochloride**, a local anesthetic and cardiac tonic; used chiefly in ophthalmology in 0.05 to 0.25 % solution. Dose $\frac{1}{33}$–$\frac{1}{16}$ gr. (0.002–0.004 Gm.).

**erythrophleum** (*er-ith-rof'-le-um*). Casca-bark.

**erythrophlogosis** (*er-ith-ro-flo-go'-sis*) [*erythro-;* φλόγωσις, a burning]. Inflammation attended with redness.

**erythrophobe** (*er-ith'-ro-fōb*) [*erythrophobia*]. One fearing or disliking red colors.

**erythrophobia** (*er-ith-ro-fo'-be-ah*) [*erythro-;* red; φόβος, fear]. Morbid intolerance of red colors: sometimes observed after operations for cataract. 2. Fear of blushing; ereuthophobia.

**erythrophose** (*er'-ith-ro-fōz*) [*erythro-;* φῶς, light]. A red phose.

**erythropia** (*er-ith-ro'-pe-ah*). Same as *erythropsia*.

**erythropoiesis** (*er-ith-ro-poi-e'-sis*) [*erythro-;* ποίησις, a making]. The formation of red blood corpuscles.

**erythropsia** (*er-ith-rop'-se-ah*) [*erythro-;* ὄψις, vision]. An abnormality of vision in which all objects appear red; red vision.

**erythropsin** (*er-ith-rop'-sin*) [*erythro-;* ὤψ, vision]. An organic substance of the retina. In the presence of light it is believed to form different combinations, constituting color-perception. It is called *visual purple* and *rhodopsin*, q. v.

**erythropyknosis** (*er-ith-ro-pik-no'-sis*) [*erythro-;* πυκνός, thick]. Degenerative changes in the invaded erythrocyte, characteristic of the estivoautumnal infections. It consists in the development of a brassy appearance of the blood-cell, together with distinct crenation.

**erythrose** (*er'-ith-rōs*) [*erythro-*], C₄H₈O₄. Tetrose. A substance derived from erythrol. It is probably a mixture of an aldose and a ketose. It is next to the lowest glucose.

**erythrosin** (*er-ith-ro'-sin*) [*erythro-;* *tyrosin*], C₂₀H₁₈N₂O₅. A compound product by the action of HNO₃ on tyrosin. It is used as a coloring-matter.

**erythrosinophil** (*er-ith-ro-sin'-o-fil*) [*erythrosin;* φιλεῖν, to love]. Easily stainable with erythrosin.

**erythrosis** (*er-ith-ro'-sis*) [ἐρυθρός, red]. 1. Arterial plethora, or the redness of the skin due to it. 2. An exaggerated tendency to blush.

**erythroxyline** (*er-ith-roks'-il-ēn*) [*erythro-;* ξύλον, wood]. Synonym of *cocaine*.

**erythroxylon** (*er-ith-roks'-il-on*) [see *erythroxyline*]. Coca. The leaves of *E. coca*, a shrub indigenous to the Andes. It contains an alkaloid, *cocaine*, C₁₇H₂₁NO₄, to which its properties are mainly due. It is an aromatic tonic and cerebral stimulant. Dose of *coca* (*erythroxylon coca*, B. P.) 2–15 gr. (0.13–1.0 Gm.); of the *fluidextract* (*fluidextractum cocæ*, U. S. P.) 20 min.–1 dr. (1.3–4.0 Cc.); of the *liquid extract* (*extractum cocæ liquidum*, B. P.) 20 min.–1 dr. (1.3–4.0 Cc.).

**erythruria** (*er-ith-ru'-re-ah*) [*erythro-;* οὖρον, urine]. The passage of reddish urine. Hematuria.

**Esbach's reagent** (*es'-bakh*) [Georges Hubert Esbach, French physician, 1843–1890]. Picric acid 1, citric acid 2, water to 100. It is used as a test for albumin in urine.

**escalin** (*es'-kal-in*). Proprietary preparation of powdered aluminum and glycerin; it is said to be indicated in gastric ulcer.

**eschar** (*es'-kar*) [ἐσχάρα, a scab]. A slough, especially that produced by the thermocautery. **e., neuropathic**, a bed-sore.

**escharodermitis** (*es-kar-o-der-mi'-tis*) [*eschar;* δέρμα, skin; ιτις, inflammation]. A skin-inflammation marked by the formation of eschars.

**escharosis** (*es-kar-o'-sis*). The formation of an eschar; escharotic action.

**escharotic** (*es-kar-ot'-ik*) [*eschar*]. 1. Caustic; producing a slough. 2. A substance that produces an eschar; a caustic.

**Escherich's bacillus** (*esh'-er-ik*) [Theodor Escherich, German physician, 1857–1911]. The *Bacillus coli communis;* see *bacilli*, *table of*.

**eschomelia** (*es-ko-me'-le-ah*) [ἔσχατος, worst; μέλος, a limb]. A monstrosity in which there is a defective limb.

**escholalia** (*es-kro-la'-le-ah*) [αἰσχρός, shameful; λαλία, speech]. Same as *coprolalia*.

**eschromythesis** (*es-kro-mi-the'-sis*) [αἰσχρός, base; μυθίζειν, to utter]. The utterance of obscene language by delirious or insane patients.

**esciorcin, æsciorcin** (*es-e-or'-sin*) [*Æsculus*, a genus of trees; *orcin*], C₉H₈O₄. A product of esculetin by action of sodium amalgam. It dissolves in alkalies, green changing to red, and is used in discovering corneal defects and lesions of conjunctival epithelium, the red color being more distinct on the iris than the green color of fluorescein. Application, 1 drop of 10 to 20 % aqueous solution.

**esciorcinol** (*es-e-or'-sin-ol*). Same as *esciorcin*.

**escorcin, æscorcin** (*es-kor'-sin*). See *esciorcin*.

**esculetin** (*es-kū-lo'-tin*) [*Æsculus*, a genus of trees], C₉H₆O₄. A substance present in the bark of the horse-chestnut, partly free, and partly as the glucoside *esculin*, from which it is prepared.

**esculin** (*es'-kū-lin*) [see *esculetin*], C₁₅H₁₆O₉. A glucoside from horse-chestnut bark.

**eseridine** (*es-er'-id-ēn*). An alkaloid, C₁₅H₂₃N₃O₃. It is a laxative and motor excitant and is recommended as a cathartic in veterinary practice. Its uses are the same as eserine, but it is only one-sixth as powerful. Subcutaneous dose $\frac{1}{8}$–$\frac{1}{3}$ gr. (0.01–0.02 Gm.).

**eserine, eserinum** (*es'-er-ēn, es-er-e'-num*) [*esere*, native name of the plant or bean]. An alkaloid obtained from the Calabar bean, and said to be identical with physostigmine (q. v.). **e. benzoate**, C₁₅H₂₁N₃O₂·C₇H₆O₂, used in the same way as eserine. **e. borate**, is mydriatic; the solutions are permanent and nonirritating, used in same way as eserine. **e.-pilocarpine**, a combination of eserine and pilocarpine forming a white, crystalline, soluble powder. It is anodyne and laxative and used in veterinary practice in colic of horses. Injection, 6 gr. (0.4 Gm.) in 5 Cc. of water. **e. salicylate**, C₁₅H₂₁N₃O₂·C₇H₆O₃, is used in 5 % solutions to contract the pupil; red solutions have lost their power. It is also used in intestinal atony. Dose $\frac{1}{80}$–$\frac{1}{20}$ gr. (0.0016–0.003 Gm.), divided into 2, 3, or 4 doses; other uses and dosage the same as of eserine. **e. sulphate**, used in same way as eserine; also, hypodermatically in veterinary practice for colic. Dose 1½ gr. (0.1 Gm.). **e. tartrate**, (C₁₅H₂₁N₃O₂)₂C₄H₆O₆, uses and dosage the same as of eserine.

**Esmarch's bandage, E.'s apparatus** (*es'-mark*) [Johann Friedrich August von Esmarch, German surgeon, 1823–1908]. An elastic rubber bandage used upon a limb to be amputated, in order to drive the blood out of it by the pressure of progressive turns about the limb toward the trunk. **E.'s operation**. 1. For amputation at the hip-joint: the soft parts of the thigh are divided to the bone by a single sweep of the knife five inches below the tip of the trochanter; the bone is then sawed across and a second incision is made to join the first from a point two inches above the trochanter, when the bone is shelled out. 2. For ankylosis of the lower jaw: an incision about two inches long is made along the lower border of the jaw, and a wedge-shaped piece of bone is removed from the horizontal portion. **E.'s tubes**, tubes on the sides of which agar or gelatin has been solidified in a thin layer, by rapid turning of the tube on ice or under ice-water.

**esocolitis** (*es-o-ko-li'-tis*) [ἔσω, within; *colitis*]. Inflammation of the mucous membrane of the colon; dysentery.

**esodic** (*e-sod'-ik*) [ἐς, into; ὁδός, way]. Afferent.

**esoenteritis** (*es-o-en-ter-i'-tis*) [ἔσω, inward; ἔντερον, bowel; ιτις, inflammation]. Inflammation of the mucous membrane of the intestines.

**esoethmoiditis** (es-o-eth-moid-i'-tis) [ἔσω, within; *ethmoiditis*]. Inflammation of the ethmoid sinuses.

**esogastritis** (es-o-gas-tri'-tis) [ἔσω, inward; γαστήρ, belly; ιτις, inflammation]. Inflammation of the mucous membrane of the stomach.

**esogenetic** (es-o-jen-et'-ik) [ἔσω, within; γεννᾶν, to produce]. Produced or arising within the organism.

**esohyperphoria** (es-o-hi-per-fo'-re-ah). See *hyperesophoria* under *heterophoria*.

**esophagalgia, œsophagalgia** (e-sof-ag-al'-je-ah) [*esophagus*; ἄλγος, pain]. Pain in the esophagus.

**esophageal, œsophageal** (e-sof-aj'-e-al) [*esophagus*]. Pertaining or belonging to the esophagus.

**esophagectomy, œsophagectomy** (e-sof-aj-ek'-to-me) [*esophagus*; ἐκτομή, a cutting out]. Extirpation of cancer of the gullet with resection of the walls of the organ.

**esophagectopy, œsophagectopy** (e-sof-aj-ek'-to-pe) [*esophagus*; ἐκτοπος, away from a place]. Displacement of the esophagus.

**esophageurysma, œsophageurysma** (e-so-faj-ūr-iz'-mah) [*esophagus*; εὐρύνειν, to widen]. Abnormal dilation of the esophagus.

**esophagism, œsophagism, esophagismus, œsophagismus** (e-sof'-aj-izm, e-sof-aj-iz'-mus). Spasmodic contraction of the esophagus.

**esophagitis, œsophagitis** (e-sof-aj-i'-tis) [*esophagus*; ιτις, inflammation]. Inflammation of the esophagus.

**esophago-, œsophago-** (e-sof-a-go-) [*esophagus*]. A prefix meaning relating to the esophagus.

**esophagocele, œsophagocele** (e-sof'-ag-o-sēl) [*esophago-*; κήλη, hernia]. An abnormal distention of a portion of the esophagus.

**esophagodynia, œsophagodynia** (e-sof-ag-o-din'-e-ah) [*esophagus*; ὀδύνη, pain]. Same as *esophagalgia*.

**esophagoectasis, œsophagoectasis** (e-sof-ag-o-ek'-tas-is) [*esophago-*; ἐκτείνειν, to stretch]. Diffuse spindleform dilation of the esophagus, almost always due to stenosis of the cardia.

**esophagoenterostomy, œsophagoenterostomy** (e-sof-ag-o-en-ter-os'-to-me) [*esophago-*; *enterostomy*]. Schlatter's operation for the total extirpation of the stomach; the esophagus is first sutured to the duodenum.

**esophagogastroscopy, œsophagogastroscopy** (e-sof-ag-o-gas-tros'-ko-pe) [*esophago-*; γαστήρ stomach; σκοπεῖν, to inspect]. Examination of the interior of the esophagus and stomach by means of the esophagogastroscope.

**esophagomalacia, œsophagomalacia** (e-sof-ag-o-mal-a'-se-ah) [*esophagus*; μαλακία, softness]. Morbid softening of the esophagus.

**esophagometer, œsophagometer** (e-sof-ag-om'-et-er) [*esophago-*; μέτρον, a measure]. An instrument for measuring the esophagus.

**esophagomycosis, œsophagomycosis** (e-sof-ag-o-mi-ko'-sis) [*esophago-*; *mycosis*]. Disease of the esophagus caused by fungi.

**esophagopathy, œsophagopathy** (e-sof-ag-op'-ath-e) [*esophagus*; πάθος, disease]. Any disease of the esophagus.

**esophagoplasty, œsophagoplasty** (e-sof'-ag-o-plast-e) [*esophago-*; πλάσσειν, to shape]. Plastic surgery of the esophagus.

**esophagoplegia, œsophagoplegia** (e-sof-ag-o-ple'-je-ah) [*esophagus*; πληγή, a stroke]. Paralysis of the esophagus.

**esophagoptosis, œsophagoptosis** (e-sof-ag-op-to'-sis) [*esophago-*; πτῶσις, a falling]. Prolapse of the esophagus.

**esophagorrhagia, œsophagorrhagia** (e-sof-ag-or-a'-je-ah) [*esophagus*; ῥηγνύναι, to break forth]. Hemorrhage from the esophagus.

**esophagorrhea, œsophagorrhea** (e-sof-ag-or-e'-ah) [*esophagus*; ῥεῖν, to flow]. A discharge from the esophagus.

**esophagoscope, œsophagoscope** (e-sof-ag'-o-skōp) [*esophago-*; σκοπεῖν, to view]. An instrument for examining the interior of the esophagus by artificial light.

**esophagoscopy, œsophagoscopy** (e-sof-ag-os'-ko-pe) [see *esophagoscope*]. Examination of the interior of the esophagus by means of the esophagoscope.

**esophagospasm, œsophagospasm** (e-sof'-ag-o-spazm). See *esophagismus*.

**esophagostenosis, œsophagostenosis** (e-sof-ag-o-sten-o'-sis) [*esophago-*; στένωσις, constriction]. Constriction of the esophagus.

**esophagostoma, œsophagostoma** (e-sof-ag-os'-to-mah) [*esophago-*; στόμα, a mouth]. An abnormal aperture or passage into the esophagus.

**esophagostomy, œsophagostomy** (e-sof-ag-os'-to-me) [see *esophagostoma*]. The formation of an artificial opening in the esophagus. e. **externa**, the surgical opening of the esophagus from the surface of the neck for the removal of foreign bodies. e. **interna**, incision of the esophagus from the inside by means of the esophagotome for relief of stricture.

**esophagotome, œsophagotome** (e-sof-ag'-o-tōm). An instrument devised for cutting into the esophagus.

**esophagotomy, œsophagotomy** (e-sof-ag-ot'-o-me) [*esophago-*; τομή, a cutting]. Opening of the esophagus by an incision.

**esophagus, œsophagus** (e-sof'-ag-us) [οἴσω, future of φέρειν, to carry; φαγεῖν, to eat]. The gullet. The musculo-membranous canal, about nine inches in length, extending from the pharynx to the stomach.

**esophoria** (es-o-fo'-re-ah). See *heterophoria*.

**esophenoiditis** (es-o-sfe-noid-i'-tis) [ἔσω, within; *sphenoid*; ιτις, inflammation]. Osteomyelitis of the sphenoid bone.

**esoteric** (es-o-ter'-ik) [ἐσωτερός, inner]. Arising within the organism.

**esothyropexy** (es-o-thi'-ro-peks-e). See *exothyropexy*.

**esotropia** (e-so-tro'-pe-ah) [ἔσω, inward; τρέπειν, to turn]. Convergent strabismus.

**espnoic** (esp-no'-ik) [ἐς, into; πνοή, vapor]. 1. Inspiratory. 2. Relating to the injection of gases or vapors.

**essence** (es'-ens) [*essentia*, essence]. 1. That which gives to anything its character or peculiar quality. 2. The peculiar qualities of a drug extracted and reduced to a small compass. 3. A solution of an essential oil in alcohol.

**essential** (es-en'-shal) [*essence*]. 1. Pertaining to the essence of a substance. 2. Of diseases, occurring without a known cause. e. **oils**, the volatile oils obtained from aromatic plants, by distillation or fermentation. e. **paralysis**, paralysis without characteristic anatomical lesions. e. **vertigo**, vertigo without appreciable cause.

**ester** (es'-ter). A compound ether containing both an acid and an alcohol radical.

**esthematology** (es-them-at-ol'-o-je) [αἴσθημα, a perception; λόγος, science]. The science of the sensations and of the sense-apparatus.

**esthesia** (es-the'-ze-ah) [αἴσθησις, sensation]. 1. Capacity of perception; feeling, or sensation. 2. Any nervous disease that affects the senses or perceptions.

**esthesioblast** (es-the-ze-o-blast) [αἴσθησις, sensation; βλαστός, a germ]. Same as *ganglioblast*.

**esthesiodermia** (es-the-ze-o-der'-me-ah) [αἴσθησις, sensation; δέρμα, skin]. An affection of the skin with disturbance of the sensory function, which may be decreased, increased, or abolished.

**esthesiogen** (es-the'-ze-o-jen) [αἴσθησις, sensation; γεννᾶν, to produce]. Any material, as a metal, which in certain states of the body appears to have a specific effect upon the sensibility of the patient.

**esthesiogenic** (es-the-ze-o-jen'-ik) [αἴσθησις, sensation; γεννᾶν, to produce]. Relating to the production of sensations.

**esthesiogeny** (es-the-ze-oj'-en-e) [αἴσθησις, sensation; γεννᾶν, to produce]. The production of altered or perverted sensations.

**esthesiography** (es-the-ze-og'-ra-fe) [αἴσθησις, sensation; γράφειν, to write]. A description of the organs of sensation and perception.

**esthesiology** (es-the-ze-ol'-o-je) [αἴσθησις, a feeling; λόγος, science]. A treatise on, or the science of, the senses.

**esthesiomania** (es-the-ze-o-ma'-ne-ah) [αἴσθησις, feeling; μανία, madness]. Insanity marked by perverted moral feeling and by purposeless eccentricities.

**esthesiometer** (es-the-ze-om'-e-ter) [αἴσθησις, sensation; μέτρον, a measure]. An instrument for measuring tactile sensibility.

**esthesiometry** (es-the-ze-om'-et-re) [αἴσθησις, sensation; μέτρον, measure]. The measurement or estimation of tactile sensibility.

**esthesioneure** (es-the'-ze-o-nūr) [αἴσθησις, sensation; νεῦρον, a nerve]. A sensory neuron.

**esthesioneurosis** (es-the-ze-o-nū-ro'-sis) [αἴσθησις, perception; νεῦρον, nerve]. Any nervous disease in which there are disorders of sensation.

**esthesiosis** (es-the-ze-o'-sis) [αἴσθησις, sensation; νόσος, disease]. See *esthesiodermia*.

**esthesiophysiology** (es-the-ze-o-fiz-e-ol'-o-je). See *esthesophysiology*.

**esthesodic** (*es-the-sod'-ik*) [αἴσθησις, sensation; ὁδός, a way]. Serving to convey sense-impressions, as to the brain.

**esthesis** (*es-the'-sis*) [αἴσθησις, sensation]. Sensibility; sense-perception; a feeling or sense-impression.

**esthiomene** (*es-the-om'-en-e*). [ἐσθιομένη, eating]. Lupus vulgaris.

**esthiomenous** (*es-the-om'-en-us*) [ἐσθιομένη, eating]. Corroding; phagedenic.

**esthophysiology** (*es-tho-fiz-e-ol'-o-je*) [*esthesis; physiology*]. The physiology of sensation and of the sense-apparatus.

**estival, æstival** (*es'-tiv-al*) [*æstas,* summer]. In biology, produced in summer.

**estivation** (*es-tiv-a'-shun*) [*æstivare,* to pass the summer]. In biology, (*a*) the dormant condition of certain plants and animals during the summer; (*b*) the arrangement of the floral organs in the bud.

**estivoautumnal, æstivoautumnal fever.** See *fever, remittent.* **e. parasite,** the parasite of *e. fever;* it is a protozoan, *Plasmodium præcox.*

**Estlander's operation** (*est'-lan-der*) [Jakob August *Estlander,* Finnish surgeon, 1831–1881]. An excision of portions of one or more ribs for the relief of empyema.

**eston** (*es'-ton*). Aluminum acetate; used as a dusting powder.

**estoral** (*es'-to-ral*). A colorless crystalline powder composed of boric acid and menthol: used by insufflation in chronic nasal catarrh.

**estriasis, œstriasis** (*es-tri'-as-is*) [*Œstrus,* a genus of dipterous insects]. Myiasis due to the larva of the *Œstrus.*

**estrual** (*es'-tru-al*) [οἶστρος, gad-fly]. Pertaining to estruation.

**estruation** (*es-tru-a'-shun*) [*estrum*]. Sexual excitement; the socalled *heat* of animals.

**estrum, œstrum, estrus, œstrus** (*es'-trum, es'-trus*) [οἶστρος, gadfly]. Sexual desire; the orgasm.

**estuarium** (*es-tu-a'-re-um*) [*æstus,* heat]. 1. A vapor-bath; also a stove designed to apply warm, dry air to all parts of the body at the same time. 2. A tube through which a hot cautery-iron can be passed to the part to be operated upon.

**estuation** (*est-u-a'-shun*) [*æstus,* heat]. Heat; boiling; fever; a heated state.

**esuritis** (*es-u-ri'-tis*) [*esuries,* hunger]. Gastric ulceration from inanition.

**état mamelonné** (*a'-tah mah-mel-on-a'*) [Fr.]. A condition of the stomach in chronic gastritis in which there is a projection of small elevations consisting of hyperplastic mucous membrane.

**état vermoulu** (*a-tah vär-moo-loō*) [Fr., worm-eaten state]. Irregular ulcerations found on the surface of the brain in connection with advanced arteriosclerosis.

**Eternod, sinus ensiformis of.** A vascular loop connecting the vessels of the chorion with the vessels on the under aspect of the yolk-sac.

**ethacol** (*eth'-ak-ol*). The ethyl morphine salt of guaiacol-sulphonic acid.

**ethanol** (*eth'-an-ol*). See *alcohol* (2).

**ethene** (*eth'-ēn*). Same as *ethylene.* **e. chloride,** C₂H₄Cl₂. Dutch liquid. An anesthetic resembling chloroform, but less dangerous.

**etheogenesis** (*e-the-o-jen'-es-is*) [ἠθεος, bachelor; γένεσις, production]. Non-sexual reproduction by male gametes of protozoa.

**ether, æther** (*e'-ther*) [αἰθήρ, the upper air]. 1. The subtle fluid filling space and penetrating all bodies, the medium of transmission of light, heat, electricity, and magnetism. 2. A compound formed hypothetically from H₂O by the substitution of two alcohol radicals for the H. 3. Diethylic oxide (C₂H₅)₂O, a thin, colorless, volatile, and highly inflammable liquid. The *ether* of the U. S. P. contains 96 % by weight of absolute ether and about 4 % of alcohol containing a little water; its specific gravity at 15° C. is 0.725–0.728. Its chief use is as an anesthetic, it being less dangerous than chloroform. It is also employed as a cardiac stimulant in sudden heart-failure and as a carminative. Dose by the mouth 30 min.–2 dr. (2–8 Cc.) in ice-water. Syn., *ethyl oxide; ethylic ether; sulphuric ether.* **e., acetic** (*æther aceticus,* U. S. P.); has properties like those of ethylic ether. Dose 10 min.–1 dr. (0.65–4.0 Cc.). **e., anesthetic.** 1. A mixture of ether, 20 parts; rhigolene, 80 parts; and petroleum ether, 80 parts; it is used as a local anesthetic. 2. A mixture of absolute alcohol and ether, each, 1 part, and petroleum ether, 4 parts. **e., chloric,** a mixture of chloroform and alcohol. **e., chlormethylmenthyl-,** C₁₀.-H₁₉–O . CH₂Cl, obtained from the action of formaldehyde upon menthol in the presence of hydrochloric acid. It is used in the treatment of catarrhal affections of the air-passages. Syn., *forman.* **e., compound anesthetic,** a combination of equal parts of rhigolene and anhydrous ethyl-ether employed as a spray to produce local anesthesia. **e. cone,** an apparatus used in the administration of ether. **e. drunkenness,** intoxication produced by drinking ether. **e., ethylic.** See *ether* (3). **e., ethylmethyl,** CH₃O . C₂H₅, obtained from sodium methylate by the action of ethyl iodide; it is said to be an effectual anesthetic, free from baleful effects. **æ. fortior,** the ether of the U. S. P. **e., hydriodic.** See *ethyl iodide.* **e., hydrobromic,** ethyl bromide. Dose 10 min.–1 dr. (0.65–4.0 Cc.). **e.-mentholchloroform,** a combination of ether, 15 parts; chloroform, 10 parts; and menthol, 1 part; it is used as an anesthetic spray. **e., methylethyl,** C₃H₈O, a mixed ether composed of one molecule of ethyl and one of methyl, combined with one atom of oxygen. It is used as an anesthetic. Syn., *three-carbon ether.* **e., ozone, e., ozonic, e., ozonized,** a mixture of ether, hydrogen peroxide, and alcohol. It is used in diabetes and whooping-cough. Dose 30–60 gr. (2–4 Gm.) 3 times daily. It is used also as a local antiseptic in scarlatina. **e., spirit of** (*spiritus ætheris,* U. S. P.), a solution of ether in twice its volume of alcohol. **e., spirit of, compound** (*spiritus ætheris compositus,* U. S. P.). See *Hoffmann's anodyne.* **e., sulphurated,** a mixture of sulphur, 1 part; ether, 10 parts. It is used in cholera in teaspoonful doses mixed with carbonated water. **e., sulphuric.** See *ether* (3). **e., terebinthinated,** a combination of ether, 4 parts, and oil of turpentine, 1 or 2 parts. It is used in the treatment of gall-stone. Dose 10–20 min. (0.6–1.2 Cc.). **e., Wiggers'. anesthetic.** See *ethyl chloride, polychlorated.*

**ethereal** (*e-the'-re-al*) [*ether*]. 1. Pertaining to the ether. 2. Made of ether, as *ethereal* tinctures. 3. Volatile.

**etheride** (*e'-ther-īd*). A comprehensive term for any combination of formyl with a haloid.

**etherification** (*e-ther-if-ik-a'-shun*) [*ether; facere,* to make]. The formation of an ether from an alcohol.

**etherify** (*e-ther'-if-i*) [*ether; facere,* to make]. To convert into ether.

**etherin, etherine** (*e'-ther-in, -ēn*). 1. A solid, crystalline body, obtained from ethylene by distillation. 2. A toxin extracted in ether, by Auclair, from tubercle bacilli. Syn., *etherobacillin.*

**etherion** (*e-the're-on*). A gas believed to exist in the air, with a heat conductivity one hundred times that of oxygen.

**etherioscope** (*e-the're-o-skōp*) [*ether;* σκοπεῖν, to examine]. An apparatus for estimating the proportions of ether or of acetic acid to water in a given solution.

**etherism** (*e'-ther-izm*). The phenomena produced upon the animal economy by the administration of ether.

**etherization** (*e-ther-iz-a'-shun*) [*ether*]. The administration of ether to produce anesthesia. This is effected by inhalation of the vapor.

**etherize** (*e'-ther-īz*) [*ether*]. To administer ether.

**etherobacillin** (*e-ther-o-bas-il'-in*). See *etherin* (2).

**etherochloroform** (*e-ther-o-klo'-ro-form*). A mixture of ether and chloroform employed in long-continued anesthesia.

**etheromania** (*e-ther-o-ma'-ne-ah*) [*ether;* μανία, madness]. The mania for drinking ether; ether intoxication.

**etheryl** (*e'-ther-il*). See *ethylene.*

**ethics** (*eth'-iks*) [ἠθικός, moral]. The science of human feelings, thoughts, and actions relating to duty or morality. **e., medical,** the duties a physician owes to himself, his profession and his fellowmen

**ethidene** (*eth'-id-ēn*) [*ether*], C₂H₄. Ethylidene, a bivalent radital. **e. chloride, e. dichloride,** a colorless fluid, tasting and smelling like chloroform. It has been used as a general anesthetic. See under *anesthetic.*

**ethin, ethine** (*eth'-in, -ēn*). See *acetylene.*

**ethiomopemphigus** (*eth-e-o-mo-pem'-fe-gus*) [ἴθμος, accustomed; πέμφιξ, a pustule]. Continued or habitual pemphigus.

**ethionic** (*eth-e-on'-ik*) [*ethylene;* θεῖον, sulphur]. Made up of ethylene and a sulphur compound.

**ethiopification** (*e-the-op-if-ik-a'-shun*) [αἰθίοψ, an Æthiopian; *facere,* to make]. A darkening of the skin such as sometimes results from the misuse of mercurial, silver or arsenical remedies.

**ethmocarditis** (*eth-mo-kar-di'-tis*) [ἠθμός, a sieve; καρδία, heart; ιτις, inflammation]. Inflammation of the connective tissue of the heart.

**ethmocephalus** (*eth-mo-sef'-al-us*) [ἠθμός, a sieve; κεφαλή, head]. A variety of single autositic monsters in which there is a rudimentary nose in the shape of a proboscis terminating anteriorly in two imperfect nostrils or in a single opening.

**ethmocranial** (*eth-mo-kra'-ne-al*) [*ethmoid;* κρανίον, skull]. Relating to the ethmoid and to the rest of the cranium.

**ethmodermitis** (*eth-mo-derm-i'-tis*) [ἠθμός, a sieve; δέρμα, the skin; ιτις, inflammation]. Inflammation of the connective tissue of the skin.

**ethmofrontal** (*eth-mo-frun'-tal*) [*ethmoid; frontal*]. Relating to the ethmoid and frontal bones.

**ethmoid** (*eth'-moid*) [ἠθμός, a sieve; εἶδος, likeness]. 1. The sieve-like bone of the nose, perforated for the transmission of the olfactory nerve; it forms a part of the base of the skull. 2. Relating to the ethmoid bone.

**ethmoidectomy** (*eth-moi-dek'-to-me*) [*ethmoid;* ἐκτομή, excision]. 1. Excision of the ethmoid cells. 2. Excision of part of the ethmoid bone.

**ethmoiden** (*eth-moi'-den*). Belonging to the ethmoid bone in itself.

**ethmoiditis** (*eth-moi-di'-tis*) [*ethmoid;* ιτις, inflammation]. Inflammation of the ethmoid bone or of the ethmoid sinuses.

**ethmoidofrontal** (*eth-moid-o-frunt'-al*). Relating to the ethmoid and frontal bones.

**ethmolacrimal** (*eth-mo-lak'-rim-al*). Relating to the junction of the ethmoid and lacrimal bones.

**ethmopalatine** (*eth-mo-pal'-a-tin*). Relating to the ethmoid and palatal bones, area, or cartilage.

**ethmophlogosis** (*eth-mo-flo-go'-sis*). See *cellulitis.*

**ethmoplecosis** (*eth-mo-ple-ko'-sis*) [ἠθμός, sieve; πλέκειν, to twine]. Any disease attacking the cellular tissue.

**ethmosphenoid** (*eth-mo-sfe'-noid*). Relating to the ethmoid and sphenoid bones.

**ethmoturbinal** (*eth-mo-tur'-bin-al*). Relating to the turbinal portions of the ethmoid bone, forming what are known as the superior and middle turbinated bones.

**ethmovomerine** (*eth-mo-vo'-mer-in*). Relating to the ethmoid bone and the vomer.

**ethmyphitis** (*eth-mif-i'-tis*). See *cellulitis.*

**ethnic** (*eth-nik*) [ἔθνος, a race]. Pertaining to race. **e. idiocy.** See *idiocy, ethnic.*

**ethnography** (*eth-nog'-ra-fe*) [ἔθνος, nation; γραφειν, to write]. A description of the races of men.

**ethnology** (*eth-nol'-o-je*) [ἔθνος, a nation; λόγος, science]. The comparative study of the races of mankind.

**ethoxide** (*eth-oks'-īd*), R . O : $C_2H_5$. A compound of ethyl, oxygen, and a radical or element; an ethylate.

**ethoxycaffeine** (*eth-oks-e-kaf'-e-in*), $C_{10}H_{14}N_4O_3$. A remedy recommended in herpes zoster and migraine. Dose 4 gr. (0.26 Gm.).

**ethyl** (*eth'-il*) [*ether; ὕλη,* matter]. The alcohol radical, $C_2H_5$. Syn., *deutyl.* **e.-acetanilide,** $C_{10}H_{13}$-NO, obtained from ethyl, anilin, and acetyl chloride by heating. It is analgesic and antipyretic. Syn., *acetethylanilide.* **e.-alcohol,** ordinary alcohol of the pharmacopeia. See under *alcohol.* **e. bisulphide,** $C_4H_{10}S_2$, a highly inflammable, colorless, oily liquid with odor of garlic; soluble in alcohol, ether, and chloroform; slightly soluble in water. **e. bromide,** $C_2H_5Br$, a rapid and transient anesthetic; internally it has been recommended as a soporific in doses of 5–20 min. (0.3–1.3 Cc.) greatly diluted with ice-water. Syn., *bromethyl; monobromethane.* See under *anesthetic.* **e. carbamate** (*æthylis carbamis,* U. S. P.), urethane. **e. carbonate,** $C_5H_{10}O_3$, an inflammable, colorless, fragrant liquid, soluble in alcohol and ether, boils at 126° C.; sp. gr., 0.999 at 0° C. **e.-chloralurethane,** *ethyl chloralum chloridum,* U. S. P.), $C_5H_{10}NCl$, an anesthetic resembling chloroform in action. **e. chloride, polychlorated,** a combination of chlorinated ethyl chloride; a clear, colorless liquid, with aromatic odor, miscible in alcohol and ether. It is a local anesthetic and irritant. **e. formate,** $C_3H_6O_2$, a colorless liquid with fragrance of peach-kernels, soluble in ether, water, and alcohol. It is hypnotic and analgesic. Dose 1–2 dr. (4–8 Cc.). Syn., *formic ether.* **e. hydrate,** ordinary alcohol. **e. hydrocupreine,** a derivative of quinine with the formula $C_{19}H_{22}N_2OH.O.C_2H_5$. It is supposed to have a specific influence on the pneumococcus. **e. iodide,** $C_2H_5I$, hydriodic ether, used to relieve the dyspnea of bronchitic asthma and edematous laryngitis. Dose to be inhaled 5 min. (0.32 Cc.) 3 or 4 times daily. **e. lactate,** $C_5H_{10}O_3$, a yellowish or colorless limpid liquid, soluble in water; it is hypnotic and sedative. Dose 8–16 min. (0.5–1 Cc.). **e. nitrite,** $C_2H_5NO_2$, a very volatile, inflammable, ethereal liquid; it is used in alcoholic solution and called sweet spirit of niter. **e. oxide.** See *ether* (3). **e.-pyoktanin,** is recommended in surgery and ophthalmology as more active than ordinary pyoktanin. **e. sulphide,** $C_4H_{10}S$, an oily liquid with an odor of garlic, soluble in alcohol; melts at 93° C.; sp. gr., 0.837 at 20° C. **e. thiocarbimid,** $C_3H_5NS$; it is used as a local irritant in rheumatism, etc. Syn., *ethyl mustard oil.* **e. valerate,** $C_7H_{14}O_2$, a reaction-product of sodium isovalerate, alcohol, and sulphuric acid; it is antispasmic and sedative. Dose 1–2 min. (0.06–0.12 Cc.) several times daily. Syn., *isovaleric ether.*

**ethylamine** (*eth-il'-am-ēn*) [*ethyl; amin*], $C_2H_5N$. A ptomaine found in putrefying yeast. **e. urate,** a remedy for gout and vesical calculi.

**ethylate** (*eth'-il-āt*). A compound of ethylic alcohol in which the H of the hydroxyl is replaced by a base.

**ethylation** (*eth-il-a'-shun*). The act or process of combining with ethyl.

**ethylchloralurethane** (*eth-il-klo-ral-ū'-reth-ān*). Same as *somnal.*

**ethylene** (*eth'-il-ēn*) [*ethyl*]. Olefiant gas, $C_2H_4$. A colorless, poisonous gas which burns with a bright, luminous flame, and when mixed with air explodes violently. It is one of the constituents of illuminating gas. **e. bichloride, e. chloride.** See *ethene chloride.* **e. bromide,** a light, brownish-colored liquid with the formula $C_2H_4Br_2$. It has been used in epilepsy. Dose ½–2 min. (0.05–0.13 Cc.). **e. chloride, monochlorinated,** $C_2H_3Cl_3$, a colorless liquid with pleasant odor, obtained from vinyl chloride by action of antimony pentachloride; it is used as an anesthetic. Syn., *monochlorethylene chloride; vinyl trichloride.* **e.-guaiacol.** See *guaiacol ethylenate.*

**ethylenediamine** (*eth-il-ēn-di'-a-min*). A non-poisonous base isomeric with ethylidenediamine; a solvent of albumin and fibrin, used in diphtheria. **e.-cresol,** a colorless liquid used as a wound antiseptic. **e.-tricresol,** a mixture of ethylenediamine, 10 parts; tricresol, 10 parts; distilled water, 500 parts; it is used as an antiseptic in 0.1 to 1 % solution. Syn., *kresamin.*

**ethylenethenyldiamine** (*eth-il-ēn-eth-en-il-dī'-am-in*). See *lysidin.*

**ethylenimid, ethylenimin** (*eth-il-en-im'-id, -in*), 1. See *piperazin.* 2. $C_2H_5N$. A non-poisonous base found in cholera cultures and believed to be identical with spermin.

**ethylic** (*eth-il'-ik*). Relating to or obtained from ethyl. **e. alcohol,** ethyl-alcohol. **e. aldehyde,** acetic aldehyde. **e. ether.** See *ether* (3).

**ethylidene** (*eth-il'-id-ēn*). See *ethidene.*

**ethylidenediamine** (*eth-il-id-ēn-di'-a-min*), $C_2H_4$(NH$_2$)$_2$. A poisonous ptomaine obtained from decomposing haddock. Injections into mice and guinea-pigs produce hypersecretion from mouth, nose, and eyes, mydriasis, exophthalmos, great dyspnea, and death.

**ethylism** (*eth'-il-izm*). Poisoning by ethyl alcohol.

**ethylization** (*eth-il-i-za'-shun*). The induction of the physiological effects of ethyl bromide.

**ethylize** (*eth'-il-īz*). To anesthetize with ethyl bromide.

**ethylol** (*eth'-il-ol*). Ethyl chloride.

**ethylphenylcarbamate, ethylphenylurethane** (*eth-il-fen-il-kar'-ba-māt, -il'-re-thān*). See *euphorin.*

**ethylthallin** (*eth-il-thal'-in*). An antipyretic compound derived from phenol.

**ethylurethane** (*eth-il-ū'-reth-ān*). See *urethane.*

**etiolate** (*e'-te-o-lāt*) [F., *étioler,* to blanch]. In biology, to blanch or be whitened by the exclusion of light.

**etiolation** (*e-te-o-a'-shun*) [Fr., *étioler,* to blanch]. 1. The paleness or blanching, in plants or animals, from confinement in darkness. 2. Pallor in patients following a long illness.

**etiological, etiologic** (*e-te-o-loj'-ik-al, e-te-ol-oj'-ik*). Pertaining to etiology.
**etiology** (*e-te-ol'-o-je*) [αἰτία, a cause; λόγος, science]. 1. The causation of disease. 2. The science of the causes of the phenomena of life and their relation to physical laws in general.
**etionymous, ætionymous** (*e-te-on'-im-us*) [αἰτία, a cause; ὄνυμα, name]. A term derived from the name of a cause; it is applied to diseases; *e. g.*, alcoholism, lead-colic.
**etrotomy** (*e-trot'-o-me*) [ἦτρον, belly; τομή, section]. A name proposed for pelvic section.
**eubiol** (*ū'-be-ol*). A preparation of hemoglobin.
**eubiose** (*ū'-be-ōs*). A highly concentrated proprietary hematogenous substance.
**eucaine** (*ū'-ka-in, or ū-kān'*). The commercial name for a local anesthetic used as a substitute for cocaine. α-e., e. a, alpha-e., C₁₅H₂₁NO₄ . HCl+H₂O, a benzoyl-meta-methyltetramethyl-para-oxypiperidincarboxylicmethylester, occurring in glossy prisms melting at 104° C. The hydrochloride is used. Application to nose or throat, 5 to 10 % solution; dental surgery, 10 % solution. β-e., e. b, beta-e., C₁₅H₂₁NO₂ . HCl, benzoylvinyldiacetonalkamine hydrochloride; white crystals soluble in 3½ parts of water, melting at 263° C. It is used in 2 % solution in dental surgery as more active and less toxic than cocaine, for which it is used as a substitute. **e. acetate**, recommended for use in ophthalmology.
**eucalyptene** (*ū-kal-ip'-tēn*), C₁₀H₁₆. A hydrocarbon from eucalyptol; the hydrochloride is used as an intestinal antiseptic. Dose 10 30 gr. (1.33—2.0 Gm.). **e. hydrochloride**. See *eucalypteol*.
**eucalypteol** (*ū-kal-ip'-te-ol*), C₁₀H₁₈₂HCl. It is used as an intestinal antiseptic. Dose 24 gr. (1.6 Gm.) daily. Children, 4—12 gr. (0.26—0.78 Gm.) daily. Syn., *terpilene dihydrochloride*.
**eucalyptol** (*ū-kal-ip'-tol*) [*eucalyptus*], C₁₀H₁₈O. A neutral principle obtained from the volatile oil of *Eucalyptus globulus* and of some other species of *Eucalyptus*. It is used in bronchitis and malaria, and also in ear diseases and in urethritis, and externally in various liniments and washes. Dose 5—10 min. (0.32—0.65 Cc.), in capsules, 3 times daily.
**eucalyptus** (*ū-kal-ip'-tus*) [εὖ, well; καλύπτειν, to cover]. The leaves of *E. globulus*, native to Australia, but now cultivated in California. It contains a volatile oil from which eucalyptol is obtained. The properties largely depend on the volatile oil. Eucalyptus has been used as an antiseptic, as a stimulant to mucous membranes, as an antispasmodic in asthma, in migraine, and, with doubtful success in malaria. *E. rostrata* is recommended in sea-sickness. Dose 1 gr. (0.06 Gm.) 3 or 4 times daily. **e., fluidextract of** (*fluidextractum eucalypti*, U. S. P.). Dose 30 min. (2 Cc.). **e., oil of** (*oleum eucalypti*, U. S. P.), the volatile oil. Dose 5 min. (0.32 Cc.) in capsules or emulsion. **e., ointment of** (*unguentum eucalypti*, B. P.), contains 20 % of the oil.
**eucanthus** (*ū-kan'-thus*) [εὖ, expressive of greatness; *canthus*]. Any enlargement of the fleshy papilla at the inner canthus of the eye.
**eucasin** (*ū'-ka-sin*). A casein food-preparation soluble in warm water, obtained by pouring ammonia over casein.
**eucasol** (*ū'-kas-ol*). Soluble eucalyptolanytol, a preparation containing 25 per cent. of eucalyptol; it is used in dental surgery.
**euchinin** (*ū'-kin-in*), C₂₃H₃₀O . CO . OC₂₀H₂₉O. An ethylcarbonic ester of quinine. It is used in whooping-cough, pneumonia, malaria, etc. Dose .15—20 gr. (1—2 Gm.).
**euchlorhydria** (*ū-klor-hi'-dre-ah*) [εὖ, well; χλωρόs, green; ὕδωρ, water]. The presence of a normal, amount of hydrochloric acid in the gastric juice.
**euchlorine** (*ū-klor'-in*) [εὖ, well; χλωρόs, green]. 1. Chlorine protoxide, an antiseptic. 2. A mixture of potassium chlorate and hydrochloric acid; it is used as a spray and gargle in diphtheria.
**eucholia** (*ū-ko'-le-ah*) [εὖ, well; χολή, bile]. Normal condition of the bile.
**euchromatopsia** (*ū-kro-mat-op'-se-ah*) [εὖ, well; χρῶμα, color; ὄψις, sight]. Capacity for correct recognition of colors.
**euchylia** (*ū-ki'-le-ah*) [εὖ, well; χυλόs, the chyle]. A normal condition of the chyle.
**euchymia** (*ū-ki'-me-ah*) [εὖ, well; χυμόs, juice]. A healthy condition of the fluids of the body.
**eucinesia, eukinesia** : (*ū-kin-e'-se-ah*) [εὖ, well; κίνησις, motion]. Normal power of movement.

**eucol** (*ū'-kol*). A combination of eucalyptol, santal oil, cubeb, oleoresin, creosote, and cod-liver oil; it is used in bronchitis and pulmonary consumption.
**eucrasia** (*ū-kra'-she-ah*) [εὖ, well; κρᾶσις, combination]. A healthy condition of the blood or general system; a condition of diminished susceptibility to disease.
**eucrasic** (*ū-kra'-sik*) [see *eucrasia*]. 1. In a condition of good health. 2. Opposed to dyscrasia or capable of bettering it.
**eucyesia, eucyesis** (*ū-si-e'-ze-ah, ū-si-e'-sis*) [εὖ, well; κύησις, pregnancy]. Normal pregnancy.
**eudermol** (*ū-der'-mol*). The proprietary name of nicotine salicylate; used as an ointment in the treatment of skin diseases.
**eudesmin** (*ū-des'-min*), C₂₀H₂₀O₆. A substance found in the kino of *Eucalyptus hemiphloia*.
**eudiemorrhysis, eudiæmorrhysis** (*ū-di-em-or'-is-is*) [εὖ, well; διά, through; αἷμα, blood; ῥύσις, a flowing]. The normal flowing of the blood through the capillaries.
**eudiaphoresis** (*ū-di-af-o-re'-sis*) [εὖ, well; *diaphoresis*]. A healthy condition of perspiration.
**eudiometer** (*ū-de-om'-et-er*) [εὐδία, calm weather; μέτρον, measure]. An instrument for ascertaining the purity of the air, and for the analysis of gases.
**endiometry** (*ū-di-om'-et-re*). See *analysis*, *gasometric*.
**eudosmol** (*ū-dos'-mol*), C₁₀H₁₈O. A crystalline camphor obtained from various species of *eucalyptus*.
**sudoxin** (*ū-doks'-in*). The proprietary name of the bismuth salt of tetraiodophenolphthalein (nosophen); it is used as an intestinal antiseptic. Dose 3–8 gr. (0.2—0.5 Gm.) 3 times daily.
**eudrenin** (*ū-dren'-in*). Trade name of a local anesthetic composed of eucaine and adrenalin.
**euesthesia** (*ū-es-the'-ze-ah*) [εὖ, well; αἴσθησις, sensation]. The sense of well-being; vigor and normal condition of the senses.
**euformol** (*ū-form'-ol*). A proprietary antiseptic containing oils of eucalyptus and wintergreen, thymol, menthol, boric acid, extract of wild indigo, and formaldehyde.
**eugallol** (*ū-gal'-ol*). Pyrogallol monoacetate; it is used in skin diseases, applied with a brush to the affected part, being a powerful inflammatory irritant upon healthy skin.
**eugatol** (*ū'-gat-ol*). A solution of sodium para-amino-diphenylamine monosulphate and of ortho-amino-phenol-sulphate. Used as a hair-dye.
**eugenesis** (*ū-jen'-es-is*) [εὖ, well; γένεσις, generation]. In biology, fertility.
**Eugenia** (*ū-je'-ne-ah*) [after Prince *Eugene*, of Savoy]. A genus of trees and shrubs, mostly tropical; among which are *E. caryophyllata*, which yields caryophyllus, and *E. pimenta*, which produces pimenta.
**eugenic** (*ū-jen'-ik*). See *eugenol*.
**eugenics** (*ū-jen'-iks*) [εὐγενής, well-born]. The science of generative or procreative development. The doctrine of progress of humanity through improved conditions in the relations of the sexes.
**eugenin** (*ū'-jen-in*). See *caryophyllus*.
**eugenoform** (*ū-jen'-o-form*). The sodium salt of eugenolcarbinol; it is an antiseptic and bactericide. Dose 8–15 gr. (0.5–1.0 Gm.).
**eugenol** (*ū'-jen-ol*), C₁₀H₁₂O₂. Eugenic acid; a phenol-like compound that occurs in clove-oil and in allspice, and is convertible into vanillin. It is used as an antiseptic and as a local anesthetic in dentistry. Dose 15 gr. (1 Gm.) well diluted. **e. acetamide**, C₁₃H₁₅O₃N, used in the form of a fine powder as a local anesthetic and wound antiseptic.
**euglobulin** (*ū-glob'-ū-lin*). A protein which with pseudoglobulin forms serum-globulin.
**eugoform** (*ū'-go-form*). A fine, insoluble powder, recommended as a dusting-powder or in ointments (2.5 to 10 %) in skin diseases. Syn., *acetylized guaiacol-methylene*.
**euknesia** (*ū-kin-e'-se-ah*). See *eucinesia*.
**eulachon, eulachon oil, or eulachoni oleum** (*ū'-lak-on*, or *ū-lak-o'-ni o'-le-um*) [native name in North Pacific Islands]. Candle-fish oil. The oil from the fish *Thaleichthys pacificus*, or candle-fish. It is less disagreeable than cod-liver oil, for which it is often substituted. Dose 3 j–iv.
**eulactol** (*ū-lak'-tol*). A dietetic preparation of milk and eggs.
**eulatin** (*ū'-lat-in*). A proprietary compound of

antipyrine with amidobenzoic and bromobenzoic acid.

**eulexin** (*ū-leks'-in*). A proprietary remedy for diabetes mellitus; it is said to consist of jambul, Paraguay tea, cascara sagrada, aromatics, and glycerol. Dose 16 min.–2 dr. (1–8 Cc.) every 4 hours.

**eulyptol** (*ū-lip'-tol*). An antiseptic preparation composed of salicylic acid, 6 parts; phenol and essence of eucalyptus, of each, 1 part.

**eulysin** (*u'-lis-in*). A greenish-yellow resin found with bilin in bile.

**eumenol** (*ū'-men-ol*). A nontoxic fluid extract of *Aralia cordata*, of China and Japan. It is said to be an efficient emmenagogue. Dose 1 teaspoonful (5 Cc.) 3 times daily.

**eumetria** (*ū-met'-re-ah*) [εὖ, well; μέτρον, measure]. The exact quantity of muscular effort which is required to accomplish a definite result.

**eumictin** (*ū-mik'-tin*). Preparation of santol, salol, and urotropin; used as a remedy for gonorrhea.

**eumycetes** (*ū-mi-se'-tēz*) [εὖ, well; μυκῆς, fungus]. Same as *hyphomycetes*.

**eumydrin** (*ū-mid'-rin*). Trade name of atropine methylnitrate, a white powder, used as a mydriatic and also to control the night-sweats of phthisis.

**eunatrol** (*ū-nat'-rol*). Oleate of sodium, recommended as a cholagogue. Dose 15 gr (1 Gm.) twice daily.

**eunoia** (*ū-noi'-ah*) [εὖ, well; νοῦς, mind]. Normal condition of mind and will.

**eunol** (*ū'-nol*). A preparation of naphthols and eucalyptols used in the treatment of skin diseases.

**eunuch** (*ū'-nuk*) [εὐνοῦχος, guardian of the couch]. A male whose genital organs have been removed or mutilated so as to render him impotent.

**eunuchism** (*ū'-nuk-izm*). The condition of being a eunuch.

**eunuchoid** (*ū'-nuk-oid*) [*eunuch*; εἶδος, resembling]. Having the characteristics of a eunuch.

**eunuchoidism** (*ū'-nuk-oid-izm*) [*eunuchoid*]. Eunuchism in which the testicles are present, but their internal secretion is absent.

**euonymin** (*ū-on'-im-in*). A precipitate from the tincture of euonymus; it is tonic, laxative, and expectorant. Dose ½–3 gr. (0.032–0.2 Gm.).

**euonymit** (*ū-on'-im-it*). See *dulcitol*.

**euonymus** (*ū-on'-im-us*) [εὐώνυμος, having a good name]. Wahoo; the bark of *E. atropurpurea* a mild purgative and cholagogue. **e., extract of** (*extractum euonymi*, U. S. P.). Dose 1–5 gr. (0.065–0.32 Gm.). **e., fluidextract of** (*fluidextractum euonymi*, U. S. P.). Dose 8 min. (0.5 Cc.).

**eupareunia** (*ū-par-ū'-ne-ah*) [εὖ, well; πάρευνος, spouse]. Sexual compatibility.

**eupathia** (*ū-path'-e-ah*) [εὖ, well; πάθος, feeling]. 1. Euphoria. 2. Normal sensation. 3. Sensitiveness to impressions.

**eupatorin** (*ū-pat-o'-rin*). 1. A precipitate from the tincture of thoroughwort, *Eupatorium perfoliatum*; it is aperient, emetic, febrifuge and tonic. Dose 1 to 3 grains. 2. A crystalline glucoside from *Eupatorium cannabinum* of Europe. See also *Eupurpurin*.

**Eupatorium** (*ū-pat-o'-re-um*) [εὐπατόριον, agrimony]. A genus of composite-flowered plants. The leaves and flowering tops of *E. perfoliatum*, thoroughwort or boneset. It is a bitter tonic, diaphoretic, and feeble emetic. Dose of the *powdered leaves* 20–30 gr. (1.3–2.0 Gm.). **e., fluidextract of** (*fluidextractum eupatorii*, U. S. P.). Dose 10 min.–1 dr. (0.65–4.0 Cc.).

**eupepsia** (*ū-pep'-se-ah*) [εὖ, well; πέπτειν, to digest]. Sound or normal digestion.

**eupeptic** (*ū-pep'-tik*) [εὖ, well; πέπτειν, to digest]. Possessing a good digestion; promoting digestion.

**euperistalsis** (*ū-per-is-tal'-sis*) [εὖ, well; *peristalsis*]. The quiet peristaltic movements of the intestines in health. See *peristalsis*.

**euphonia** (*ū-fo'-ne-ah*) [εὖ, well; φωνή, voice]. A normal, good, and clear condition of the voice.

**Euphorbia** (*ū-for'-be-ah*). A genus of trees, shrubs and herbs, yielding a milky juice. *E. corollata*, *E. ipecacuanha*, American species, have been employed in medicine on account of their emetic, diaphoretic, and expectorant properties. *E. pilulifera*, of South America and Australia, is used in asthma and bronchitis. *E. resinifera*, of Africa, affords euphorbium. Dose of the *extract* 1 gr. (0.065 Gm.); of the *fluidextract* ½–1 dr. (2–4 Cc.); of the *tincture* ½–1 dr. (2–4 Cc.).

**euphorbin** (*ū-for'-bin*). A precipitate from a tincture of the root of *Euphorbia corollata*; it is an emetic, expectorant, vermifuge and arterial sedative. Dose ¼ to 3 grains.

**euphorbism** (*ū-forb'-izm*). Poisoning by means of species of *Euphorbia*; it is marked by acute inflammation of the digestive tract and asphyxia.

**euphorbium** (*ū-for'-be-um*). An acrid gum-resin obtained from *Euphorbia resinifera*. It is strongly purgative and vesicant, and is now mainly employed in veterinary medicine.

**euphorbon** (*ū-forb'-on*), $C_{18}H_{22}O$. A neutral substance found as a constituent of euphorbium.

**euphoria** (*ū-for'-e-ah*) [εὔφορος, easily carried]. The sense of well-being; health.

**euphoric** (*ū-for'-ik*). Marked by or pertaining to euphoria.

**euphorin** (*ū'-for-in*) [see *euphoria*], $C_9O_2H_{11}$. Phenylurethane, a white, crystalline powder derived from anilin. It is recommended as an analgesic and antipyretic in neuralgia and rheumatism. Dose 5–30 gr. (0.32–2.0 Gm.) daily.

**euphthalmin** (*ūf-thal'-min*) [εὖ, well; ὀφθαλμός, the eye], $C_{16}H_{25}NO_3HCl$. The hydrochloride of the mandelic acid derivative of β-eucaine; it is used as a mydriatic in 2 to 10 % solutions.

**euplastic** (*ū-plas'-tik*) [εὖ, well; πλάσσειν, to form]. Capable of being transformed into healthy tissue.

**eupnea, eupnœa** (*ūp-ne'-ah*) [εὖ, well; πνεῖν, to breathe]. Normal or easy respiration.

**euporphine** (*ū-por'-fēn*). Apomorphine bromomethylate; it is used similarly to apomorphine hydrochloride.

**eupurpurin** (*ū-pur'-pū-rin*) [εὖ, well; *purpura*, purple]. A precipitate from the tincture of *Eupatorium purpureum*; it is diuretic, stimulant, astringent, and tonic. Dose 1 to 4 grains.

**eupyrexia** (*ū-pi-reks'-e-ah*) [εὖ, well; πῦρ, fire]. A slight rise of temperature in the beginning of an infection.

**eupyrine** (*ū-pi'-rēn*). A compound of vanillinethyl carbonate and paraphenetidin; used as an antipyretic. Dose, adults, 15–24 gr. (1.0–1.5 Gm.); children, 5–8 gr. (0.3–0.5 Gm.).

**euquinine** (*ū-kwin'-ēn*). Quininethylcarbonate, a crystalline, tasteless compound. Dose 5–30 gr. (0.32–2.0 Gm.).

**euresol** (*ū'-re-sol*). The commercial name of resorcinol monacetate; dissolved in acetone it is recommended in skin diseases.

**eurobin** (*ū'-ro-bin*). The commercial name of chrysarobin triacetate; it is insoluble in water, but dissolves freely in acetone, chloroform, and ether. It is used in a 2 to 3 % ointment in skin diseases.

**eurodontia** (*ū-ro-don'-she-ah*) [εὐρώς, decay; ὀδούς, a tooth]. Dental caries.

**europhen** (*ū'-ro-fen*). Diisobutylorthocresol iodide, an amorphous yellow powder, recommended as a substitute for iodoform. It is used hypodermatically in doses of ¼–1½ gr. (0.016–0.09 Gm.) in syphilis.

**europisocephalus** (*ū-ro-pis-o-sef'-al-us*) [εὐρύς, broad; ὀπίσω, behind; κεφαλή, the head]. Having the skull broad in the occipital region.

**europium** (*ū-ro'-pe-um*). The provisional name given to a supposed new element. The atomic weight is 152 and it lies midway between gadolinum and samarium.

**europrocephalus** (*ū-ro-pro-sef'-al-us*) [εὐρύς, broad; πρό, in front; κεφαλή, the head]. Having a skull broad in front.

**eurybin** (*ū'-re-bin*). A yellowish, bitter, amorphous powder, soluble in water and alcohol, obtained from *Olearia moschata*.

**eurycephalic, eurycephalous** (*ū-ris-ef-al'-ik, ū-ris-ef'-al-us*) [εὐρύς, wide; κεφαλή, head]. Broadheaded; having a very wide skull.

**eurychasmus** (*ū-rik-az'-mus*) [εὐρύς, broad; χάσμα, a chasm]. Lissauer's term for a skull in which the angle formed between the lines joining the joint of the wing of the vomer and posterior nasal spine and anterior margin of the foramen magnum is between 153° and 154°.

**eurygenesis** (*ū-re-jen'-es-is*) [εὐρύς, broad; γένεσις, origin]. The theory of the origin of a species by gradual amelioration of a race of precursors having a wide-spread or cosmopolitan distribution.

**eurygnathism** (*ū-re-nath'-izm*). The condition of having large jaws.

**eurygnathus, eurygnathous** (*ū-re-nath'-us*) [εὐρύς, broad; γνάθος, jaw]. Large-jawed.

**eurynter** (ū-rint'-er) [εὐρύνειν, to dilate]. An instrument used in dilating. Cf. *colpeurynter*.

**euryon** (ū'-re-on) [εὐρύς, broad]. The craniometric point at the end of the greatest transverse diameter of the skull.

**eurysma** (ū-ris'-mah) [εὐρύνειν, to dilate]. 1. Dilatation. 2. A structure which has undergone dilatation.

**eurythermal** (ū-rith-er'-mal) [εὐρύς, wide; θέρμη, heat]. Capable of sustaining a great range of temperature.

**eurythermic** (ū-re-ther'-mik) [εὐρύς, broad; θέρμη, heat]. Referring to bacteria capable of growing through a wide range of temperature.

**eurythrol** (ū-rith'-rol). An extract from the spleen of oxen, having a honey-like consistence and of aromatic taste and odor. Dose 1 to 2 teaspoonfuls in soup daily.

**eusapyl** (ū'-sa-pil). A solution of chlormetacresol in potassium ricinoleate; it is used as a disinfectant, and as an antiseptic wash.

**euscopol** (ū'-sko-pol). Proprietary name for scopolamine hydrobromide.

**eusemia** (ū-se'-me-ah) [εὖ, well; σῆμα, a sign]. A favorable sign or prognostic.

**eusemin** (ū'-se-min). A local anesthetic composed of cocaine and adrenalin, used in ophthalmic practice.

**Eustachian artery** (ū-sta'-ke-an) [Bartolomeo *Eustachio*, Italian anatomist, 1500–1574]. 1. A branch of the Vidian artery. 2. A branch of the pterygopalatine artery. **E. catheter**, an instrument for examining, distending, or making applications to the Eustachian tube. **E. muscle**, the laxator tympani. **E. tube**, a canal, partly bony and partly cartilaginous, connecting the pharynx with the tympanic cavity. **E. valve**, the fold of the lining membrane of the right auricle of the heart, situated between the opening of the inferior vena cava and the auriculoventricular orifice.

**eustachitis** (ū-sta-ki'-tis). Inflammation of the Eustachian tube.

**eustachium** (ū-sta'-ke-um). The Eustachian tube.

**eustein** (ūs'-ten-in). The double salt of theobromine sodium and sodium iodide; it is used in arteriosclerosis and angina pectoris.

**Eustrongylus** (ū-stron'-jil-us) [εὖ, well; στρογγύλος, round]. A genus of parasitic nematode worms. **E. gigas**, the largest of the parasitic nematodes. Its habitat is the kidney. It is met in many of the lower animals and rarely in man. The symptoms arising from its presence resemble those of renal abscess or calculus. The finding of the eggs in the urine is the most important diagnostic consideration. Oil of turpentine may cause migration of the worm, but extirpation is indicated.

**eusystole** (ū-sis'-to-le) [εὖ, well; συστολή, contraction]. A normal contraction of the cardiac cavities.

**eutaxia** (ū-taks'-e-ah) [εὖ, well; τάξις, order]. A normal condition of the body.

**eutectic** (ū-tek'-tik) [εὖ, well; τίκτειν, to produce]. 1. Well combined; stable; applied to a chemical combination which, in passing from a liquid to a solid state, acts as a simple body, maintaining a constant temperature and its constituent substances remaining associated during solidification. 2. [εὖ, well; τήκειν, to melt]. Melting easily; said of a compound substance which has a lower fusing-point than its constituents have separately. 3. A eutectic substance.

**euteхia** (ū-teks'-e-ah). 1. The condition of being stable and well becomined. 2. The quality of fusing at a low temperature.

**euthanasia** (ū-than-a'-ze-ah) [εὖ, well; θάνατος, death]. 1. An easy or calm death. 2. The killing of people who are suffering from an incurable or painful disease.

**euthenics** (ū-then'-iks) [εὐθηνία, good state of the body]. "The betterment of living conditions for the purpose of securing efficient human beings; race improvement through environment in contrast with *eugenics* which deals with race improvement through heredity" (Ellen H. Richards).

**euthermic** (ū-therm'-ik) [εὖ, well; θέρμη, heat]. Promoting warmth.

**euthesia** (ū-the'-ze-ah) [εὖ, well; θέσις condition]. Good constitution or state of health.

**euthymia** (ū-thim'-e-ah) [εὖ, well; θυμός, mind]. Tranquility; cheerfulness.

**euthymol** (ū-thi'-mol). A nontoxic liquid antiseptic said to contain oils of eucalyptus and winter-green, extract of wild indigo, boric acid, menthol, and thymol. It is used as a spray or internally in doses of 1 dr. (4 Cc.) in water 3 or more times daily.

**eutocia** (ū-to'-se-ah) [εὖ, well; τόκος, child-birth]. Natural or easy childbirth; normal labor.

**eutocous** (ū-to'-kus) [see *eutocia*]. 1. Having an easy delivery. 2. Prolific.

**eutrichosis** (ū-trik-o'-sis) [εὖ, well; θρίξ, hair]. A healthy, normal development of the hair.

**eutrophic** (ū-tro'-fik) [εὖ, well; τρέφειν, to nourish]. Pertaining to eutrophy; promoting the nutritive process; well-nourished; a drug to improve nutrition.

**eutrophy, eutrophia** (ū'-tro-fe, ū-tro'-fe-ah) [εὖ, well; τρέφειν, to nourish]. A state of normal or healthy nutrition; the condition of being well nourished.

**evacuant** (e-vak'-ū-ant) [*evacuare*, to empty]. 1. Emptying. 2. A medicine that causes the emptying of an organ, especially the bowels; a purgative.

**evacuation** (e-vak-ū-a'-shun) [*evacuate*]. 1. The act of emptying, especially of the bowels. 2. That which is evacuated.

**evacuator** (e-vak'-ū-a-tor) [*evacuate*]. An agent to produce emptying, especially an instrument for removing from the bladder fragments of stone after litholapaxy.

**evagination** (e-vaj-in-a'-shun) [e, out; *vagina*, a sheath]. Protrusion from a sheath or invaginating structure. Cf. *invagination*.

**evalvate** (e-val'-vāt) [*evalvis*, without valves]. Destitute of valves.

**evaporation** (e-vap-or-a'-shun) [e, out; *vaporare*, emit vapor]. The conversion of a liquid into vapor.

**evaporometer** (e-vap-or-om'-et-ur) [*evaporare*, to evaporate; μέτρον, measure]. An apparatus for the study of the evaporation from cultivated plants and soils under the influence of different conditions of meteorology, soil, and culture.

**evectics** (e-vek'-tiks). An old name for hygiene; the science of good health.

**evenimation, evenomation** (e-ven-e-ma'-shun, -o-ma'-shun) [e, from; *venom*]. The process of counteracting the effects of a venom.

**eventration** (e-ven-tra'-shun) [e, out; *venter*, the belly]. Protrusion of the abdominal viscera through the abdominal walls.

**eversion** (e-ver'-shun) [*eversio*, a turning out]. A turning outward. **e. of the eyelid**, a folding of the lid upon itself for the purpose of exposing the conjunctival surface or sulcus. See also *ectropium*.

**évidement** (a-vēd-mon(g)) [Fr.]. Splitting open foci of disease and scraping them clean with the sharp curette.

**evidence** (ev'-id-ens) [*evidens*, clear]. In legal medicine, the means by which the existence or nonexistence of the truth or falsehood of an alleged fact is ascertained or made evident; proof, as of insanity. **e., circumstantial**, evidence the conclusions based upon which are beyond actual demonstration. **e., conclusive** or **positive**, evidence that admits of no doubt. **e., expert**, that given before a jury by an expert in any science, art, profession, or trade.

**eviration** (ev-ir-a'-shun) [*evirare*, to castrate]. 1. Castration. 2. Emasculation. 3. A form of sexual perversion in which thère is a deep and permanent assumption of feminine qualities, with corresponding loss of manly qualities. The opposite of this is termed *defemination*.

**evisceration** (e-vis-er-a'-shun) [e, out; *viscera*, the bowels]. The removal of the viscera. **e. of the eye**, removal of the entire contents of the globe of the eye, leaving the scleortic intact. **e., obstetrical**, removal of the abdominal or thoracic viscera of the fetus to facilitate delivery.

**evisceroneurotomy** (e-vis-er-o-nū-rot'-o-me). Evisceration of the eye with division of the optic nerve.

**evittate** (e-vit'-āt) [e, out; *vitta*, a band]. Destitute of bands or stripes.

**evolution** (ev-o-lū'-shun) [*evolvere*, to unroll]. The process of unfolding or developing from a simple to a complex specialized, perfect form. **e., spontaneous**, a series of changes whereby a shoulder presentation is transformed within the pelvis into a combined breech-and-shoulder presentation and delivery effected without artificial aid. **e., threefold law of** (Huxley): (1) Excess of development of some parts in relation to others. (2) Complete or partial suppression of parts. (3) Coalescence of parts usually distinct.

**evulsion** (*e-vul'-shun*) [*evellere*, to pluck out]. The forcible tearing or plucking away of a part.

**Ewald's test-breakfast** (*a'-valt*) [Carl Anton Ewald, German physician, 1845– ]. A method used in ascertaining the condition of the gastric juice. It consists of 35 to 70 grams of white bread and 300 Cc. of water, or a cup of weak tea without milk or sugar. E.'s **test for hydrochloric acid in contents of stomach**, Dilute 2 Cc. of a 10 % solution of potassium sulphocyanide and 0.5 Cc. of a neutral solution of iron acetate to 10 Cc. with water. This makes a ruby-red solution; if a few drops of it are placed in a porcelain dish, and 1 or 2 drops of the liquid to be tested are allowed to come in contact with it, a faint violet cloud is observed in the presence of HCl. On mixing, the color becomes brown.

**Ewart's sign** (*yoo'-art*) [William Ewart, English physician, 1848– ]. In marked pericardial effusion the left clavicle is so raised that the upper border of the first rib can be felt with the finger as far as the sternum.

**ex** (*ex*) [L. for out of]. A prefix denoting out of, away from.

**exacerbation** (*eks-as-er-ba'-shun*) [*exacerbare*, to be violent]. An increase in the symptoms of a disease.

**exalgin** (*eks-al'-jin*) [ἐξ, out; ἄλγος, pain], $C_9H_{11}NO$. Methylacetanilide; a benzene derivative allied to phenacetin. It is an analgesic and antipyretic in doses of from ½–4 gr. (0.032–0.26 Gm.).

**exaltation** (*eks-awl-ta'-shun*) [*exaltatio*, an uplifting]. 1. Increase of functional activity. 2. Increase of mental activity. 3. A morbid mental state characterized by self-satisfaction, ecstatic joy, abnormal cheerfulness, or optimism, or by delusions of grandeur.

**examination** (*eks-am-in-a'-shun*). Investigation, as for the purpose of diagnosis; it is variously qualified as bimanual, digital, oral, physical, etc.

**exangia**, or **exangeia** (*eks-an'-je-ah*) [ἐξ, out; ἀγγεῖον, vessel]. Any dilatation or distention of a blood-vessel, such as aneurysm, varix, or capillary enlargement.

**exanimation** (*eks-an-im-a'-shun*) [*ex*, out; *anima*, spirit]. Real or apparent death; fainting.

**exanthema**, **exanthem** (*eks-an-the'mah*, *eks-an'-them*) [ἐξάνθημα, eruption; *pl.*, *exanthemata*, *exanthems*]. 1. An eruption upon the skin. 2. Any exanthematous or eruptive fever.

**TABLE OF EXANTHEMATA.**

**cerebrospinal meningitis:** *Period of incubation*, unknown. *Stage of invasion*, sudden. *Eruption*, second to fourth day. *Character of eruption*, herpes labialis, purpuric spots, dusky erythema. *Location*, herpes on lips; purpuric spots over entire body. *Duration of disease*, variable; many sequels. *Convalescence* by lysis. **erysipelas:** *Period of incubation*, few hours to 3 or 4 days. *Stage of invasion*, 1 to 3 days. *Eruption*, within 14 hours. *Character of eruption*, bright-red, shining patches with well-defined raised margin. *Duration of eruption*, 4 to 8 days. *Location*: Begins usually on face, may spread to trunk and arms; in the traumatic form begins at the wound. *Desquamation*, branny or in large flakes. Disease lasts 1 to 3 weeks. *Convalescence* by crisis. **measles:** *Period of incubation*, 10 to 12 days. *Stage of invasion*, 4 days. *Eruption*, fourth day. *Character of eruption*, small dark-red papules arranged in crescentic form; complete in 24 hours. *Duration of eruption*, 4 to 5 days. *Location*, face; then downward over body. *Desquamation*, branny, 8 to 11 days. Disease lasts 2 weeks. *Convalescence* by crisis. **rötheln:** *Period of incubation*, 8 to 17 days. *Stage of invasion*, 24 to 48 hours. *Eruption* appears within 48 hours. *Character of eruption*, rose-colored, rounded, discrete macules. *Duration of eruption*, 3 days. *Location*, face and scalp; then downward over body. *Desquamation*, slightly branny. Disease lasts 4 to 7 days. *Convalescence* by crisis. **scarlatina:** *Period of incubation*, 1 to 21 days. *Stage of invasion*, 1 to 2 days. *Eruption* appears within 24 hours. *Character of eruption*, diffuse, scarlet, punctate. *Duration of eruption*, 7 to 10 days. *Location*, neck, chest, face; then over body. *Desquamation*, scales or large flakes about one week. Disease lasts 2 to 3 weeks. *Convalescence* by lysis. **typhoid fever:** *Period of incubation*, 5 to 35 days. *Stage of invasion*, 6 to 8 days. *Eruption* appears on seventh day. *Character of eruption*, rose-colored, lenticular spots, coming on in successive crops. *Duration of eruption*, each crop 3 to 5 days. Continues 10 to 20 days or throughout the whole course of the fever. *Location*, abdomen, chest, and back. *Desquamation*, slightly branny or none. Disease lasts 3 to 4 weeks. *Convalescence* by lysis. **typhus fever:** *Period of incubation*, 4 to 12 days. *Stage of invasion*, 5 days. *Time of appearance*, from the third to the seventh day, usually the fifth day. *Character of eruption*, dusky spots or papules or petechiæ. *Duration of eruption*, few days or throughout the course of the disease. *Location*, sides of chest and abdomen, arms, back. *Desquamation*, slightly branny. Disease lasts 2 to 4 weeks. *Convalescence* by crisis. **varicella:** *Period of incubation*, 4 to 14 days. *Stage of invasion*, 1 to 2 days. *Time of appearance*, from 12 to 24 hours. *Character of eruption*, vesicles in crops. *Duration of eruption*, 5 to 8 days. *Location*, back, chest, arms. *Desquamation*, crusts, 5 to 8 days. Disease lasts 2 weeks. *Convalescence* by lysis. **variola:** *Period of incubation*, 8 to 14 days. *Stage of invasion*, 3 days. *Time of appearance*, fourth day. *Character of eruption*: first, shot-like papules, then vesicles, then umbilicated pustules. *Duration of eruption*, 21 to 25 days. *Location*, face and over body. *Desquamation*, crusts, 12 to 22 days. Disease lasts 4 to 5 weeks. *Convalescence* by lysis.

**exanthematology** (*eks-an-them-at-ol'-o-je*) [*exanthem*; λόγος, science]. The science of the exanthematous diseases.

**exanthematous** (*eks-an-them'-at-us*) [*exanthem*]. Of the nature of or characterized by exanthem or eruption; of the nature of an eruptive fever.

**exanthesis** (*eks-an-the'-sis*) [ἐξάνθησις, a blossoming; *pl.*, *exantheses*]. 1. The breaking out or the appearance of an exanthem. 2. Any exanthematous disease. **e. rosalia arthrodynia.** Synonym of *dengue*.

**exanthropes** (*eks-an'-thrōps*) [ἐξ, out of; ἄνθρωπος, man]. Sources of disease originating externally.

**exanthropia** (*eks-an-thro'-pe-ah*) [ἐξ, out; ἄνθρωπος, man]. Morbid dislike of human society.

**exanthropic** (*eks-an-throp'-ik*). Situated external to the human body; relating to exanthropes. Syn., *extraanthropic*.

**exarteritis** (*eks-ar-ter-i'-tis*) [ἐξ, out; ἀρτηρία, artery; *itis*, inflammation]. Inflammation of the outer coat of an artery.

**exarthrima** (*eks-ar-thrim'-ah*) [ἐξ, out; ἄρθρον; joint; *pl.*, *exarthrimata*]. Luxation or dislocation of a joint.

**exarticulation** (*eks-ar-tik-u-la'-shun*) [*ex*; *articulus*, joint]. 1. Dislocation of a joint. 2. Amputation at a joint.

**exasperate** (*eks-as'-per-āt*) [*exasperare*, to make rough]. Rough; covered with sharp points.

**excarnation** (*eks-kar-na'-shun*) [*ex*; *caro*, flesh]. Separation of injected vessels from a contiguous part.

**excavatio** (*eks-ka-va'-she-o*) [L.]. Excavation. **e. papillæ nervi optici,** excavation of the papilla of the optic nerve; see under *excavation*. **e. rectouterina,** rectouterine excavation, pouch of Douglas. **e. rectovesicalis,** the fold of peritoneum hanging down between the rectum and the bladder in the male. **e. vesicouterina,** uterovesical pouch, the fold of peritoneum hanging down between the bladder and the uterus.

**excavation** (*eks-kav-a'-shun*) [*excavare*, to hollow out]. A hollow or cavity. **e. of the optic nerve,** a hollowing or "cupping" of the optic disc that may be physiological, congenital, or pathological, the result of glaucoma, optic atrophy, etc.

**excavator** (*eks-kav-a'-tor*) [*excavare*, to hollow out]. 1. An instrument like a gouge or scoop used to scrape away tissue. 2. A dental instrument for opening and forming cavities and removing decayed matter from them.

**excentric** (*eks-sen'-trik*) [*ex*, out; *centrum*, center]. 1. See *eccentric*. 2. Out of the center or median line. **e. pains,** radiating pains, symptomatic of spinal disease, due to irritation of the posterior nerve-roots. The pains are felt to be in the peripheral organs, hence the name.

**excerebration** (*eks-ser-e-bra'-shun*) [*ex*, out; *cerebrum*, brain]. The removal of the fetal brain in the process of embryotomy.

**excern** (*ek-sern'*) [*excernere*, to sift out]. To excrete.

**excipient** (*ek-sip'-e-ent*) [*excipere*, to take up]. Any substance combined with an active drug to give the latter an agreeable or convenient form.

**excision** (*eks-sizh'-un*) [*excisio*, a cutting out]. The cutting out of a part.

**excitability** (ek-si-ta-bil'-it-e) [excitare, to rouse]. The property of reacting to a stimulus.
**excitant** (ek-si'-tant) [see excitability]. 1. Stimulating. 2. A remedy that stimulates the activity of an organ.
**excitation** (ek-si-ta'-shun) [see excitability]. The act of stimulating or irritating. e., direct, the stimulation of a muscle by placing an electrode on the muscle itself. e., indirect, the stimulation of a muscle through its nerve.
**exciting** (ek-si'-ting). Calling forth directly, as an *exciting* cause.
**excitoglandular** (ek-si-to-gland'-ū-lar). Arousing or exciting glandular function.
**excitometabolic** (ek-si-to-met-ah-bol'-ik). Exciting metabolic processes.
**excitomotor** (ek-si-to-mo'-tor). Exciting or arousing motor function; also, a drug or agent that increases the activity of the motor nerve-centers.
**excitomuscular** (ek-si-to-mus'-kū-lar). Exciting muscular activity.
**excitor** (ek-si'-tor) [excitare, to rouse]. 1. One who or that which stimulates or excites. 2. A discharger. 3. An electrode which, placed in contact with a nerve, causes excitation.
**excitosecretory** (ex-si-to-se'-kre-to-re). Tending to produce secretion.
**excitovascular** (ex-si-to-vas'-kū-lar). Exciting vascular changes; increasing the activity of the circulation.
**exclave** (eks'-klāv) [Fr.]. A detached portion of any organ, as of a pancreas or ovary.
**exclusion** (eks-klu'-zhun) [excludere, to shut out]. A shutting-out. e., diagnosis by, the reaching of a diagnosis by excluding one hypothesis after another until only one remains.
**excochleation** (eks-kok-le-a'-shun) [ex, out; cochlea, shell, spoon]. Curetting, or scraping material out of a cavity.
**excoriation** (eks-ko-re-a'-shun) [ex, from; corium, the skin]. Abrasion of a portion of the skin.
**excortication** (eks-kor-tik-a'-shun). See *decortication*.
**excrement** (eks'-kre-ment) [excernere, to separate; to excrete]. An excreted substance; the feces.
**excrementitious** (eks-kre-men-tish'-us). Pertaining to excrement.
**excrescence** (eks-kres'-ens) [excrescere, to grow out]. An abnormal outgrowth upon the body.
**excreta** (eks-kre'-tah) [excrete]. The natural discharges of the body, particularly those of the bowel.
**excrete** (eks-krēt'). To remove useless substances from the body.
**excretin** (eks'-kre-tin), C₂₀H₃₆O. A crystalline substance found in feces.
**excretion** (eks-kre'-shun) [excrete]. 1. The discharge of waste-products. 2. The matter so discharged.
**excretory** (eks'-kre-to-re) [excrete]. Pertaining to excretion.
**excursion** (eks-kur'-shun) [ex, out of; currere, to run]. 1. A wandering from the usual course. 2. The extent of movement, as of the eyes from a central position.
**excurvation** (eks-kur-va'-shun) [ex; curvare, to curve]. 1. Outward curvature. 2. A deformity of the upper eyelid in which the tarsal cartilage becomes turned outward.
**exdermoptosis** (eks-derm-op-to'-sis) [ex; δέρμα, the skin; μύωσις, a falling]. Hypertrophy of the sebaceous glands with retention of the secretion.
**exedent** (eks-e'-dent) [ex, out; edere, to eat]. Rodent; eating away the tissues.
**exelcysmos** (eks-el-sis'-mos) [ἐξ, from; ἑλκύειν, to draw]. Extraction. as of feces.
**exencephalocele** (eks-en-sef'-al-o-sēl) [ἐξ, out; ἐγκέφαλος, brain; κήλη, hernia]. Cerebral hernia; an exencephalous tumor; a mass of brain-substance outside of the cranial cavity.
**exencephalon** (eks-en-sef'-al-on). Same as *exencephalus*.
**exencephalus** (eks-en-sef'-al-us) [ex; ἐγκέφαλος, brain]. A species of monsters characterized by a malformed brain, situated without the cranial cavity.
**exenteration** (eks-en-ter-a'-shun) [ex; ἔντερον, intestine]. 1. Removal of the intestine of the fetus, to allow delivery. 2. Same as *evisceration*.
**exenteritis** (eks-en-ter-i'-tis) [ἐξ, out; ἔντερον, bowel; ιτις, inflammation]. Inflammation of the outer or peritoneal coat of the intestine.

**exercise** (eks'-er-siz) [exercere, to keep busy]. Functional activity of the muscles; often applied to such activity when its purpose is the preservation or restoration of the health. e., active, that exerted by the will of the patient. e. bones, rider's bones; osseous growths occasionally found in the muscles. e., passive, when the part is moved by another, or acted upon, as in massage.
**exeresis** (eks-er'-es-is) [ἐξ, out; αἱρέειν, to take]. Surgical removal, as by excision or extraction; evacuation.
**exesion** (eks-e'-shun) [exedere, to corrode]. The gradual superficial destruction of organic parts, particularly bone, in consequence of abscesses and other destroying agencies.
**exfetation** (eks-fe-ta'-shun) [exfetatio]. Ectopic or extrauterine fetation.
**exflagellation** (ex-flaj-el-a'-shun). The act of extruding actively motile chromatin threads from the body of a male malarial parasite.
**exfoliation** (eks-fo-le-a'-shun) [exfoliare, to shed leaves]. The separation of bone or other tissue in thin layers.
**exhalant** (eks-ha'-lant) [exhalare, to breathe out]. 1. Serving for exhalation; exhaling. 2. A pore or organ of exhalation.
**exhalation** (eks-hal-a'-shun) [exhalare, to breathe out]. 1. The giving off of matters in the form of vapor. 2. The vapor, etc., given off by the body through the skin or lungs. 3. Expiration, or breathing out.
**exhauster** (eg-sawst'-er) [exhaurire, to pour out]. An instrument for the removal of soft cataracts.
**exhaustibility** (eg-sawst-e-bil'-it-e). Capacity for being exhausted. e., Faradic, the cessation of excitability in a muscle under repeated stimulation. Cf. *reaction, myasthenic*.
**exhaustion** (eg-sawst'-yun) [see exhauster]. 1. Loss of vital and nervous power from fatigue or protracted disease. 2. The pharmaceutical process of dissolving out one or more of the constituents of a crude drug by percolation or maceration. e., heat-, e., solar. See *insolation*. e. hypothesis, Pasteur's theory that immunity often afforded to the tissues by an attack of infection or following vaccination against infection is due to an abstraction from the tissues by the organism concerned in the primary attack of something necessary to the growth of the infecting organism. It is opposed to the retention theory of Chauveau.
**exhibit** (ek-zib'-it) [exhibere, to give]. To administer, as a medicine.
**exhibition** (ek-zib-ish'-un) [exhibere, to give]. In legal medicine, the exposing of the genitalia in public places. 2. The administration of a remedy.
**exhibitionism** (ek-zib-ish'-un-izm) [exhibere, to give]. A perversion of the sexual feeling that leads the patient to expose the genital organs.
**exhibitionist** (ek-zib-ish'-un-ist) [exhibere, to give]. An insane person who wilfully and indecently exposes the genitals.
**exhilarant** (ek-zil'-ar-ant) [exhilarare, to cheer]. An agent to enliven and cheer the mind. e. gas, nitrous oxide gas.
**exhumation** (eks-hū-ma'-shun) [ex; humus, the ground]. The removal of a corpse from the ground; disinterment.
**exinanition** (eks-in-an-ish'-un) [exinanitio, an enfeebling]. Excessive exhaustion.
**Exner's plexus** (eks'-ner) [Sigmund *Exner*, Austrian physiologist, 1846—  ]. A layer of nerve-plexuses, probably formed by the junction of sensory and motor fibers, in the cerebral cortex, near the surface.
**exo-** (eks-o-) [ἔξω, without]. A prefix meaning without; outside.
**exocardia** (eks-o-kar'-de-ah) [exo-; καρδία, the heart]. Displacement of the heart.
**exocardiac, exocardial** (eks-o-kar'-de-ak, -al) [exocardia]. Originating or situated outside of the heart.
**exocarditis** (eks-o-kar-di'-tis) [ἐξ, out; καρδία, heart; ιτις, inflammation]. Inflammation of the outer surface of the heart. Also a synonym of *pericarditis*.
**exocataphoria** (ex-o-kat-af-o'-re-ah) [exo-; cataphoria]. The condition in which the visual axis turns outward and downward.
**exoccipital** (eks-ok-sip'-it-al) [ex, out; occiput]. Lying to the side of the foramen magnum. e. bone, the neurapophysial or condyloid part of the occipital bone, with which in adult life it is consolidated.

**exochorion** (eks-o-ko'-re-on) [exo-; chorion]. The external layer of the chorion.
**exocolitis** (eks-o-ko-li'-tis) [exo-; colon; ιτις, inflammation]. Inflammation of the outer or peritoneal coat of the colon.
**exocranium** (eks-o-kra'-ne-um) [exo-; cranium]. The outer surface of the skull; the pericranium.
**exocystis** (eks-o-sist'-is) [exo-; κύστις, the bladder]. Prolapse of the urinary bladder.
**exodic** (eks-od'-ik) [exo-; ὁδός, a way]. Transmitting impulses outward from the central nervous system; efferent; centrifugal.
**exodyne** (eks'-o-dīn) [exo-; ὀδύνη, pain]. A mixture of acetanilide 90, sodium salicylate 5, and sodium bicarbonate 5; it is used as an anodyne.
**exogamy** (ex-og'-am-e) [exo-; γάμος, marriage]. Protozoan fertilization by the union of elements derived from two unrelated cells.
**exogastritis** (eks-o-gas-tri'-tis) [exo-; γαστήρ, belly; ιτις, inflammation]. See *perigastritis*.
**exogenetic** (eks-o-jen-et'-ik) [exo-; γεννᾶν, to produce]. Due to an external cause; not arising within the organism.
**exogenous** (eks-oj'-en-us) [see *exogenetic*]. Growing by accretions to the outer surface. e. **disease.** See *exopathy*.
**exognathion** (eks-og-na'-the-on) [exo-; γνάθος, the jaw]. The alveolar process of the superior maxilla.
**exognosis** (ex-og-no'-sis) [ἐξ, out of; γιγνώσκειν, to know]. Diagnosis by exclusion.
**exohysteropexy** (ex-o-his'-ter-o-pek-se) [exo-; ὑστέρα, the uterus; πῆξις, fixation]. Fixation of a prolapsed uterus to the abdominal wall.
**exol** (ex'-ol). A local dental anesthetic.
**exometra** (eks-o-me'-trah) [exo-; μήτρα, the womb]. Prolapse or inversion of the uterus.
**exometritis** (eks-o-me-tri'-tis) [exo-; μήτρα, womb; ιτις, inflammation]. Perimetritis; less correctly, parametritis. See *metritis*.
**exomphalia** (eks-om-fa'-le-ah). Protrusion of the navel.
**exomphalocele** (eks-om-fal'-o-sēl). An umbilical hernia.
**exomphalos** (eks-om'-fal-os) [exo-; ὀμφαλός, navel]. Undue prominence of the navel; also, umbilical hernia.
**exoncoma** (eks-ong'-ko-mah) [ἐξόγκωμα, anything swollen]. A protruding tumor.
**exoncosis** (eks-on-ko'-sis). The formation of a prominent tumor.
**exopathic** (eks-o-path'-ik) [exo-; πάθος, disease]. Pertaining to those causes of disease coming from without or beyond the organism.
**exopathy** (eks-op'-ath-e) [exo-; πάθος, disease]. A disease having its origin in some cause external to the organism.
**exopexy** (eks'-o-peks-e). The surgical anchoring of an organ normally in a body cavity outside of the latter.
**exophoria** (eks-o-fo'-re-ah). See *heterophoria*.
**exophthalmia** (eks-off-thal'-me-ah). See *exophthalmos*.
**exophthalmic** (eks-of-thal'-mik) [*exophthalmos*]. Pertaining to exophthalmos. e. **goiter.** See *goiter*, *exophthalmic*.
**exophthalmometer** (eks-off-thal-mom'-et-er) [ἐξ, out; ὀφθαλμός, eye; μέτρον, measure]. An instrument for measuring the degree of exophthalmos.
**exophthalmos, exophthalmus** (eks-off-thal'-mos, -mus) [exo-; ὀφθαλμός, eye]. Abnormal prominence or protrusion of the eyeballs. e., **pulsating,** that characterized by a bruit and pulsation, due to an aneurysm that pushes the eye forward.
**exoplasm** (eks'-o-plazm) [exo-; πλάσσειν, to form]. In biology, the outer protoplasm of a unicellular organism or histological cell.
**exorbitism** (eks-orb'-it-izm). See *exophthalmos*.
**exormia** (eks-or'-me-ah) [ἐξορμάειν, to go forth]. Any papular skin-disease.
**exosepsis** (eks-o-sep'-sis) [exo-; sepsis]. Sepsis originating outside the body.
**exoskeleton** (eks-o-skel'-et-on) [exo-; σκελετόν, a dried body]. The rigid outer envelop of many of the lower forms of life for the protection and attachment of organs.
**exosmometer** (eks-oz-mom'-et-ur). See *endosmometer*.
**exosmosis** (eks-oz-mo'-sis) [exo-; ὠσμός, thrust]. Outward osmosis. See *osmosis*.

**exosmotic** (eks-oz-mot'-ik) [ἐξ, out; ὠσμός, thrust]. Pertaining to or characterized by exosmosis.
**exosplenopexia, exosplenopexy** (eks-o-sple-no-peks'-e-ah, /eks-o-sple'-no-peks-e) [exo-; σπλήν, spleen; πῆξις, fixation]. An operation substituted for splenectomy, which consists in attaching the spleen in the abdominal wound and fixing it there by its capsule.
**exospore** (eks'-o-spōr) [exo-; outside; σπόρος, seed]. In biology, the outer coat of a spore.
**exostome** (eks'-os-tōm) [exo-; στόμα, mouth]. In biology, the orifice in the outer coat of the ovule, or the outer peristome of a moss.
**exostosis** (eks-os-to'-sis) [exo-; ὀστέον, bone; pl., exostoses]. A bony outgrowth from the surface of a bone.
**exoteric** (eks-o-ter'-ik) [ἐξωτερικός, external]. Synonymous with *exopathic*.
**exotery** (eks-ot'-er-e). See *exopathy*.
**exothermic** (eks-o-thur'-mik) [exo-; θέρμη, heat]. 1. Relating to the giving out of heat. 2. A substance which gives out heat in its production. Cf. *endothermic*.
**exothyropexy** (eks-o-thi'-ro-peks-e) [exo-; thyroid; πῆξις, fixation]. Exposing, the enlarged thyroid gland by a median incision and drawing it outside.
**exotic** (eks-ot'-ik) [ἐξωτικός, foreign]. Pertaining to plants and products from another country.
**exoticosymphysis** (eks-ot-ik-o-sim'-fis-is) [ἐξωτικός, foreign; σύμφυσις, a growing together]. The union of a substance or body with the organism.
**exotospore** (eks-o'-to-spōr) [ἐξωτικός, outward; σπόρος, seed]. The malarial germ brought by the stab of the mosquito (*Anopheles*) into the human blood-vessels; so named from being formed outside the human body.
**exotoxin** (eks-o-toks'-in) [exo-; toxin]. A toxin which is excreted by a microorganism and can afterwards be obtained in bacteria-free filtrates without death or disintegration of the microorganisms.
**exotropia** (eks-o-tro'-pe-ah). See *strabismus*.
**expansive** (eks-pan'-siv) [*expandere*, to spread out]. Comprehensive; wide-extending. e. **delirium,** insane overestimation of one's mental or bodily powers. See *exaltation*.
**expectant** (eks-pek'-tant) [*expectare*, to look out for]. Awaiting or expecting. e. **treatment,** watching the progress of a disease, and not interfering unless warranted by special symptoms.
**expectation** (eks-pek-ta'-shun) [*expectare*, to expect]. Same as *expectant*. e. **of life,** the average number of years that persons of a given age, taken one with another, live, assuming that they die according to a given table of the probabilities of life. It thus has no relation to the most probable life of a single given individual. e. **of l., complete,** the addition of one-half year to the curtate expectation to allow for that portion of a year lived by each person in the year of his death. e. **of l., curtate,** the average number of whole or completed years lived by each person.
**expectorant** (eks-pek'-to-rant) [see *expectoration*]. 1. Promoting expectoration. 2. A remedy that promotes or modifies expectoration.
**expectoration** (eks-pek-to-ra'-shun) [*ex*, out; *pectus*, breast]. 1. The ejection from the mouth of material brought into it from the air-passages. 2. The fluid or semifluid matters from the lungs and air-passages expelled by coughing and spitting. e., **prune-juice,** a sputum containing altered blood expectorated in gangrene and cancer of the lung and in grave pneumonias in the aged. e., **rusty.** See *sputum, rusty*.
**expellent** (eks-pel'-ent) [*expellere*, to drive out]. A medicine that has power to expel a *materies morbi*.
**experimental ten minutes.** The standard time in Ross's *in vitro* method within which mitosis must be induced in lymphocytes.
**expert** (eks'-pert) [*expertus*, proved]. A person especially qualified in a science or art. e., **medical,** a physician peculiarly fitted by experience or special learning to render an authoritative opinion in medicolegal or diagnostic questions.
**expertness.** Special skill or dexterity. e., **dextro-.** Conjoint and superior expertness of the dextral sensory and muscular organs of the body. e., **mixed dextrosinistral.** Some of the centers of the more expert organs in conjoint action are located in one and some in the opposite cerebral hemisphere. e., **sinistro-.** Conjoint and superior expertness of the sinistral sensory and muscular organs of the body.

**expiration** (*eks-pi-ra'-shun*) [*expirare*, to breathe out]. The act of breathing forth or expelling air from the lungs.
**expiratory** (*eks-pi'-ra-to-re*) [see *expiration*]. Relating to expiration.
**expire** (*eks-pīr'*) [*expirare*, to breathe out]. To breathe out; to die.
**exploration** (*eks-plo-ra'-shun*) [*explorare*, to search out]. The act of exploring; investigation of a part hidden from sight by means of touch, by artificial light, etc.
**explorator, explorer** (*eks-plor-a'-tor, ex-plor'-er*) [*explorare*, to search out]. An instrument for use in exploration. e., electric, an instrument for detecting a bullet by means of the electric current.
**exploratory** (*eks-plor'-at-o-re*). Pertaining to exploration. e. puncture, the puncture of a cavity or tumor and extraction therefrom of some of the contents to learn their nature.
**exploring needle.** A needle with a grooved side to allow the passage of fluid along it after it is plunged into a part where the presence of fluid is suspected.
**explosion** (*eks-plo'-zhun*) [*explodere*, to drive away]. 1. The sudden expansion of a body of small volume into great volume, with the resulting effects. 2. The sudden and violent occurrence of any symptom or function.
**explosive** (*eks-plo'-siv*) [*explodere*, to drive away]. See *consonants*. e.-speech, speech characterized by suddenness and explosiveness of enunciation.
**exposure** (*eks-po'-zhur*) [*exponere*, to expose]. 1. The act of laying bare, as the genitals. 2. The state of being open to some action or influence, as of cold or wet.
**expression** (*eks-presh'-un*) [*expressus; exprimere*, to press out]. 1. A pressing out. 2. The facies. e. of fetus or e. of placenta, assisting the expulsion of fetus or placenta by pressure upon the uterus through the abdominal walls. e., rectal, assisting the expulsion of the fetal head by means of two fingers inserted into the rectum and hooked into the mouth or under the chin.
**expulsion** (*eks-pul'-shun*) [*expellere*, to drive out]. The act of driving out. e., spontaneous, the extrusion of the fetus or the placenta without external aid.
**expulsive** (*eks-pul'-siv*) [*expellere*, to drive out]. Pertaining to the extrusion or driving out of the fetus in childbirth, the voiding of the feces, urine, etc.
**exsanguinate** (*ek-san'-gwin-āt*) [*ex-; sanguis*, blood]. 1. To render bloodless. 2. Bloodless.
**exsanguination** (*ek-san-gwin-a'-shun*) [*exsanguinate*]. The act of making bloodless.
**exsanguine** (*ek-sang'-gwin*) [*ex*, out; *sanguis*, blood]. Bloodless.
**exsanguinity** (*ek-sang-gwin'-it-e*). Bloodlessness; extreme pallor.
**exsection** (*ek-sek'-shun*) [*ex; secare*, to cut]. The act of cutting a part out from its surroundings.
**exsertor** (*ek-sert'-or*) [*exserere*, to protrude]. A muscle which protrudes a part.
**exsiccation** (*ek-sik-a'-shun*) [*ex; siccus*, dry]. The act of drying; especially the depriving of a crystalline substance of its water of crystallization.
**exsiccative** (*ek-sik'-a-tiv*). Drying.
**exsiccator** (*ek'-sik-a-tor*) [*exsiccare*, to dry up]. A closed glass vessel containing a tray of sulphuric acid, used to dry and cool substances preparatory to weighing.
**exspuition** (*eks-pū-ish'-un*) [*ex*, out; *spuere*, to spit]. Expectoration; spitting.
**exstrophy** (*ek'-strof-e*) [*ex*; στρέφειν, to turn]. Eversion; the turning inside out of a part. e. of bladder, a congenital condition in which the lower part of the abdominal wall, the anterior wall of the bladder, and usually the symphysis pubis are wanting, and the posterior wall of the bladder is pressed through the opening.
**exsufflation** (*ek-suf-la'-shun*) [*ex*, out; *sufflare*, to blow]. Forced discharge of the breath.
**ext.** Abbreviation of *extractum* or *extract*.
**exta** (*eks'-tah*) [L., pl.]. The viscera, especially those of the chest.
**extasis** (*ek'-sta-sis*). See *ecstasy*.
**extension** (*eks-ten'-shun*) [*extendere*, to stretch out]. 1. A straightening out, especially the muscular movement by which a flexed limb is made straight. 2. Traction upon a fractured or dislocated limb. *Counterextension* is traction made on a part in a direction opposite to that in which traction is made by another force.
**extensometer** (*eks-tens-om'-et-er*) [*extension; μέτρον*, a measure]. A micrometer to measure the expansion of a body.
**extensor** (*eks-tens'-or*) [*extension*]. That which stretches out or extends, as *extensor* muscles. See under *muscle*.
**extenuation** (*eks-ten-ū-a'-shun*) [*ex*, out; *tenuis*, thin]. Thinness; leanness of body; delicacy.
**exterioration** (*eks-te-re-ōr-a'-shun*) [*exterior*, outer]. The faculty of mind by which the image of an object seen is referred to the real situation of the object.
**extern** (*eks'-tern*) [*externus*; outward]. 1. Outside; outside the gates of a hospital. 2. An out-door patient. 3. A medical student, or graduate, who attends to out-door charity cases. e. **maternity**, lying-in in a private house.
**external** (*eks-ter'-nal*) [*externus*, outward]. On the exterior, or on the side removed from the center or middle line of the body.
**externalize** (*eks-ter'-nal-iz*) [*external*]. 1. In psychology, to transform an idea or impression which is on the percipient's mind into a phantasm apparently outside him. 2. To refer to some outside source, as the voices heard by the subject of psychomotor hallucinations.
**externe** (*eks-tern'*) [Fr.]. Same as *extern*.
**exteroceptive impulses** (*eks'-ter-o-sep-tiv*)[*extero-; capere*, to take]. Afferent nerve impulses which derive their stimulatson from external sources.
**exteroceptor** (*eks-ter-o-sep'-tor*) [*extero; receptor*, receiver]. An end organ, in or near, the skin or a mucous membrane, which receives stimuli from the external world.
**extesticulate** (*eks-tes-tik'-ū-lāt*) [*ex*, out; *testiculus*, a testicle]. To castrate.
**extinction** (*eks-tink'-shun*) [*extinguere*, to extinguish]. Complete abeyance or final loss, as of the voice; also, incomplete loss. e. **of mercury**, the rubbing of mercury with lard or some other substance until the particles of mercury are no longer visible.
**extirpation** (*eks-ter-pa'-shun*) [*extirpare*, to root out]. Complete removal of a part.
**extirpator** (*eks-ter-pa'-tor*) [*extirpare*, to root out]. An instrument for extracting the roots of cuspid teeth.
**extra-** (*eks-trah-*) [L.]. A prefix meaning outside; without.
**extra-amniotic** (*eks-trah-am-ne-ot'-ik*). Outside of the amnion; between the amnion and the chorion.
**extra-articular** (*eks-trah-ar-tik'-ū-lar*). Outside of the proper structures of a joint.
**extracapsular** (*eks-trah-kap'-sū-lar*). Outside of the capsular ligament of a joint.
**extracardiac** (*eks-trah-kar'-de-ak*) [*extra*; καρδία, heart]. Situated or occurring outside of the heart.
**extracellular** (*eks-trah-sel'-ū-lar*). External to the cells of an organism.
**extracostal** (*eks-trah-kos'-tal*) [*extra; costa*, a rib]. Outside of the ribs. e. **muscle**, any external intercostal muscle. See *muscles, table of*.
**extracranial** (*eks-trah-kra'-ne-al*). Outside of the cranial cavity.
**extract, extractum** (*eks'-trakt, -trakt'-um*) [*extrahere*, to extract]. In pharmacy, a solid or semisolid preparation, made by extracting the soluble principles of a drug with water or alcohol and evaporating the solution. e., **alcoholic**, that in which alcohol is the solvent. **e.s, animal**, fluidextracts obtained by prolonged digestion of finely chopped organs of animals in glycerol, boric acid, and alcohol, and believed to contain the active principle of the organ. See *musculin, ovarin, testin, thyroidin*, etc. e., **aqueous**, that prepared by using water as the solvent. e., **aromatic fluid**, fluid extract from aromatic powder. e., **compound**, one prepared from more than one drug. e., **ethereal**, one in which ether is the solvent. e., **fluid**. See *fluidextract*. e., **powdered**, an extract dried and pulverized. e., **soft**, an extract evaporated to the consistence of honey. e., **solid**, one made solid by evaporation. **extractum ferri pomatum** (N. F.), an extract made from iron, in the form of fine, bright wire, 1 part; ripe sour apples, 50 parts; water, a sufficient quantity.
**extraction** (*eks-trak-shun*) [*extract*]. 1. The act of drawing out. 2. The process of making an extract. e. **of cataract**, removal of a cataractous lens by surgical operation.
**extractive** (*eks-trak'-tiv*) [*extract*]. Any organic

**extractor** (*eks-trak'-tor*) [*extract*]. An instrument for extracting bullets, sequestra, etc.
**extracurrent** (*eks-trah-kur'-ent*). The induced electric current.
**extradural** (*eks-trah-du'-ral*) [*extra-*; *durus*, hard]. Situated outside of the dura.
**extra-embryonic** (*eks-trah-em-bre-on'-ik*). Situated without or not forming a part of the embryo; it is said of certain structures of the ovum.
**extra-epithelial** (*eks-trah-ep-e-the'-le-al*). Outside of an epithelium.
**extragenital** (*eks-trah-jen'-it-al*). Not situated upon the genitals; applied to chancres.
**extraligamentous** (*eks-trah-lig-a-ment'-us*). External to a ligament.
**extralobular** (*eks-trah-lob'-ū-lar*). Outside of a lobe.
**extramalleolus** (*eks-trah-mal-e'-o-lus*) [*extra*; *malleolus*]. The outer malleolus of the ankle.
**extramedullary** (*eks-trah-med'-ul-ar-e*) [*extra-*; *medulla*]. Situated or occurring outside of the medulla.
**extraneous** (*eks-tra'-ne-us*) [*extraneus*, external]. Existing or belonging outside the organism.
**extraneural** (*eks-trah-nū'-ral*). Situated or occurring outside of a nerve. A term applied to certain nervous affections of which the true seat is more or less remote from the point which manifests the symptoms of nerve-embarrassment.
**extranuclear** (*eks-trah-nū'-kle-ar*). Outside the nucleus of a cell.
**extra-ocular** (*eks-trah-ok'-ū-lar*). Outside the eye, or eyeball; in biology, applied to those antennæ of insects that are located at a distance from the eyes.
**extra-organismal** (*eks-trah-or-gan-iz'-mal*). External to the organism.
**extrapelvic** (*eks-trah-pel'-vik*). Situated or occurring outside the pelvis.
**extraperitoneal** (*eks-trah-per-it-on-e'-al*). External to the peritoneal cavity.
**extrapial** (*eks-trah-pi'-al*). Pertaining to objects external to the pia.
**extrapolar** (*eks-trah-po'-lar*) [*extra-*; *polus*, a pole]. Not lying in the space between the electrodes of a battery. **e. region,** that lying outside the electrodes, as opposed to the intrapolar region, or area, that lying within or directly beneath the electrodes.
**extraradical** (*eks-trah-rad'-ik-al*). Applied to hydrogen atoms not replaceable by a negative or alcoholic radical, but replaceable by a base.
**extrarenal** (*eks-trah-re'-nal*). External to the kidney.
**extrasystole** (*eks-trah-sis'-to-le*). A heart-contraction occurring earlier than the normal systole if the heart-muscle is irritated during the diastolic period.
**extrathoracic** (*eks-trah-tho-ras'-ik*). External to the thoracic cavity.
**extratriceps** (*eks-trah-tri'-seps*). The outer head of the triceps muscle.
**extra-uterine** (*eks-trah-ū'-ter-in*) [*extra-*; *uterus*]. Outside of the uterus. **e. pregnancy.** See *pregnancy, extrauterine.*
**extravasation** (*eks-trav-as-a'-shun*) [*extra-*; *vas*, a vessel]. 1. The passing of fluid outside of the cavity or space normally containing it. 2. The fluid that has passed out.
**extravascular** (*eks-trah-vas'-kū-lar*) [see *extravasation*]. Outside of the vessels.
**extraventricular** (*eks-trah-ven-trik'-ū-lar*). External to a ventricle.
**extremital** (*eks-trem'-it-al*) [*extremus*, outermost]. Situated towards, or pertaining to, an extremity; distal.
**extremity** (*eks-trem'-it-e*) [*extremus*, outermost]. An arm or leg; the distal or terminal end or part of any organ; a hand or foot.
**extrinsic** (*eks-trin'-sik*) [*extrinsicus*, from without]. External; not directly belonging to a part. **e. muscles,** those situated on the exterior of an organ.
**extroversion** (*eks-tro-ver'-shun*). See *exstrophy.*
**extubation** (*eks-tū-ba'-shun*) [*ex*, out; *tubus*, a pipe]. The removal of a laryngeal tube; opposed to intubation.
**exuberance** (*ex-tū'-ber-ans*) [*ex*, out; *tuber*, mass]. A swelling or protuberance.
**extrumescence** (*eks-tū-mes'-ens*) [*ex*, out; *tumescere*, to swell]. A projection or swelling.
**exudate** (*eks'-ū-dāt*) [*exudare*, to sweat]. The material that has passed through the walls of vessels into the adjacent tissues. **e., fibrinous,** coagulation of fluid soon after its escape from the vessels within the spaces into which it has exuded. **e., sero-fibrinous,** serous fluid in which flocculi of coagulated matter float.
**exudation** (*eks-ū-da'-shun*) [*exudate*]. The passing out of serum or pus; the material that has passed out.
**exudative** (*eks-ū-da'-tiv*) [*exudate*]. Of the nature of or characterized by exudation.
**exulceration** (*eks-ul-ser-a'-shun*) [*ex;* out; *ulcerare*, to ulcerate]. A superficial ulceration.
**exumbilication** (*eks-um-bil-ik-a'-shun*) [*ex, out; umbilicus,* navel]. Marked protrusion of the navel.
**exutoria** (*eks-ū-to'-re-ah*) [*exurere*, to burn]. Substances which cause a superficial ulceration of the skin when applied.
**exuviæ** (*eks-ū'-ve-e*) [L.]. Cast-off matters; shreds of epidermis; also, sloughed materials.
**exuviation** (*eks-ū-ve-a'-shun*) [*exuviæ*]. The shedding of the deciduous teeth, or other epidermal part.
**eye** (*i*) [AS., *eāge*]. The organ of vision. It occupies the anterior part of the orbit, is nearly spherical in outline, and is composed of three concentric coats: the sclerotic and cornea, the choroid and iris, and the retina. The *sclerotic* is an opaque, dense, white, fibrous membrane, into the anterior part of which the transparent *cornea* is fitted. The *choroid* is the vascular tissue, and is continuous with the *iris* in front. The latter is a circular membrane with a central perforation, the *pupil*. Within the choroid is the *retina,* a delicate, transparent membrane containing the terminations of the optic nerve. The greater part of the eyeball is filled with a mucoid substance, the *vitreous humor,* against the anterior surface of which rests the *crystalline lens*. The space between the lens and the cornea is divided by the iris into two compartments, communicating through the pupillary opening, the *anterior and posterior chambers*, which contain the *aqueous humor*. Anteriorly the eye is covered by *conjunctiva,* posteriorly by a fibrous capsule (*capsule of Tenon*). The eyeball is moved by a series of muscles attached on the outer surface. Changes in the curvature of the lens are brought about by the ciliary muscle, while the size of the pupil is modified by the action of dilator and constrictor fibers in the iris. **e.s,** alternating dominance of the, dominance of one eye at one time or for one function, alternating with that of the fellow for another time or function. **e., aphakic,** the eye deprived of its crystalline lens. **e., apappendages of,** the eyelids, brows, and lacrimal apparatus. **e., apple of,** formerly the eyeball; the pupil. **e., artificial,** a thin shell of glass, celluloid, or other substance, colored like the natural eye, placed in the socket after enucleation. **e., compound,** the organ of vision formed of several crystal spheres, as in the lower crabs. **e., dominant,** the eye which is unconsciously and preferentially chosen to guide decision and action. **e. diagrammatic of Listing.** See under *Listing*. **e.s, equidominant,** or **divided dominance** of the, having equal or divided dominance. **e., pineal** or **epiphyseal,** the rudimentary median eye in some lizards. **e., reduced, of Donders.** See under *Donders*. **e.s, reversed dominance of the,** the left, because of ametropia, disease, operation, etc. of the right becoming the dominant eye in the right-handed; or *vice versa* in the left-handed. **e., schematic,** an ideal or normal eye.
**eyeball** (*i'-bawl*). The globe of the eye.
**eyebrow** (*i'-brow*). The supercilium, the connective tissue, skin, and hairs above the eye. The hairs serve chiefly to prevent the sweat from falling into the eye.
**eyecells,** cup-shaped cells of porcelain, enameled black, to place over the eye after operations.
**eyecurrent,** the normal electric current that passes from the cornea (positive) to the optic nerve (negative) under the stimulus of light.
**eyedrops.** See *collyrium.* Also an old name for tears.
**eyeglass,** a lens worn in one eye. **Eye-glasses** *pince-nez,* worn instead of spectacles, and held in position by a spring acting upon the bridge of the nose.
**eyeground,** a synonym of the fundus oculi or internal aspect of the vitreous chamber of the eye.
**eyelashes** (*i'-lash-es*). The hairs of the eyelid.
**eyelens,** the lens of a microscope to which the eye is applied; an eye-piece.

**eyelid** (*ī'-lid*). The protective covering of the eyeball, composed of skin, glands, connective and muscular tissue, the tarsus and conjunctiva, with the cilia at the free edge.

**eyepiece** (*ī'-pēs*). Synonym of *ocular*. **e., Huygenian.** See *Huygenian ocular*.

**eyepoint,** the point above an ocular or simple microscope where the greatest number of emerging rays cross.

**eye-speculum,** an instrument for retracting the eyelids.

**eyestrain** (*ī'-strān*). The excess and abnormalism of effort, with the resultant irritation, caused by ametropia or heterophoria. It is applied also to the effects of excessive use of normal eyes.

**eyeteeth** (*ī'-tēth*). The canine teeth of the upper jaw; dog-teeth.

**eyewash,** a medicated water for the eye; a collyrium.

**eyewater,** a collyrium; also the aqueous humor.

**F. 1.** Abbreviation of *Fahrenheit, field of vision.* **2.** Chemical symbol of *fluorine.*
**fabella** (*fa-bel'-ah*) [dim. of *faba*, a bean]. A sesamoid fibrocartilage or small bone occasionally developed in the gastrocnemius muscle.
**fabism** (*fa'-bizm*) [*faba*, bean]. Lathyrism.
**face** (*fās*) [*facies*, the face]. A name applied to the lower and anterior part of the head, including the eyes, nose, mouth, cheeks, lips, etc. **f., adenoid,** a half-idiotic expression, combined with a long, high nose, flattened at the bridge, narrow nostrils, open mouth displaying irregular upper teeth, a drooping jaw, and broadening between the eyes. **f.-ague.** See *ague, brow-.* **f. grippée,** the pinched face observed in peritonitis. **f., mask-like,** a face frequently seen in alcoholic multiple neuritis, in which an expressionless band stretches across the nose and cheeks between the eyes and lips, the skin remaining motionless while the eyebrows, forehead, and lips may be moving freely.
**facet** (*fas'-et*) [Fr., *facette*, a little face]. A small plane surface, especially on a bone or a hard body, like a calculus.
**facial** (*fa'-shal*) [*face*]. Pertaining to the face. **f. angle,** an angle measured in different ways by different authorities. That of Virchow and Holder is formed by the union of a line joining the frontonasal suture and the most prominent point of the lower edge of the superior alveolar process, and a line joining the superior border of the external auditory meatus and the lower portion of the orbit. That of Camper is formed by the union of Camper's line (a line touching the most prominent points of the upper and lower face) and a line joining the acanthion and the auricular point. **f. center,** one in the frontal gyrus for face movements. **f. hemiplegia.** See *hemiplegia.* **f. nerve.** See *nerves, table of.*
**facies** (*fa'-she-ēz*) [L., "face"]. 1. The appearance of the face. 2. A surface. **f. anterior,** the anterior surface. **f. articularis,** an articular surface. **f. auricularis,** auricular surface. **f. cerebralis,** cerebral surface. **f. contactus,** contact surface. **f. convexa,** convex surface. **f. costalis,** costal surface. **f. diaphragmatica,** diaphragmatic surface. **f. dorsalis,** dorsal surface. **f. frontalis,** frontal surface. **f. gastrica,** gastric surface. **f. hippocratica,** an appearance of the face indicating of the rapid approach of dissolution: the nose is pinched, the temples hollow, the eyes sunken, the ears leaden and cold, the lips relaxed, the skin livid. **f. inferior,** inferior surface. **f. infratemporalis,** zygomatic surface. **f. intestinalis,** intestinal surface. **f. labialis,** labial or buccal surface. **f. lateralis,** lateral surface. **f. leontina.** See *leontiasis.* **f. lingualis,** lingual surface. **f. malaris,** malar surface. **f. malleolaris,** malleolar surface. **f. maxatoria,** chewing surface. **f. maxillaris,** maxillary surface. **f. medialis,** medial surface. **f. mediastinalis,** mediastinal surface. **f. nasalis,** nasal surface. **f. orbitalis,** orbital surface. **f. ossea,** bony portion of the face. **f. ovarina,** the emaciated countenance seen in patients with large ovarian cysts. **f. palatina,** palatine surface. **f. parietalis,** parietal surface. **f. patellaris,** patellar surface. **f. pelvina,** pelvic surface. **f. posterior,** posterior surface. **f. renalis,** renal surface. **f. sphenomaxillaris,** temporal surface. **f. superior,** superior surface. **f. volaris,** volar surface.
**faciobrachial** (*fa-she-o-bra'-ke-al*). Relating to the face and arm; generally referring to a form of juvenile muscular dystrophy.
**faciocervical** (*fa-she-o-ser'-vik-al*). Relating to the face and neck.
**faciolingual** (*fa-she-o-lin'-gwal*). Relating to the face and tongue.
**facioscapulohumeral** (*fa-she-o-skap'-u-lo-hu-me-ral*). Relating to the face, scapula, and arm.
**F. A. C. S.** Abbreviation for Fellow of the American College of Surgeons.

**factitious** (*fak-tish'-us*) [*facere*, to make]. Artificial.
**facultative** (*fak'-ul-ta-tiv*) [*facultas*, capability]. Voluntary; optional. **f. aerobic.** See under *aerobic.* **f. anaerobic.** See *aerobic, facultative.* **f. manifest hyperopia,** that part of the manifest hyperopia that can be concealed by the accommodation. **f. parasite,** an organism that, while usually parasitic, can also live outside of its host.
**faculty** (*fak'-ul-te*) [*facultas*, capability]. 1. A special action of the mind through the instrumentality of an organ or organs; any function, particularly any acquired, modified, or facultative function. 2. The corps of professors and instructors of a university and its colleges. **f., medical,** the corps of professors and instructors of a medical college.
**fæcal, fæces** (*fe'-kal, fe'-sēz*). See *fecal, feces.*
**fæcula** (*fek'-u-lah*). See *fecula.*
**fænum-græcum** (*fen-ūm-gre'-kum*). See *fenugreek.*
**fæx** (*feks*) [L., "lees"; pl., *fæces*]. The dregs or sediment of any liquid; fecula. **f. medicinalis liquida,** liquid yeast. It is used in the treatment of acne in the young. Dose ½ teaspoonful to 1 tablespoonful mixed with water once or twice daily with meals. **f. medicinalis sicca,** dry yeast. It is used internally in smallpox to diminish suppuration; also in bronchopneumonia occurring in measles. Dose 1½ teaspoonfuls mixed with 80 Cc. of boiled water in high rectal injection.
**fag** [origin uncertain]. Exhaustion; tire. See *brainfag.*
**Faget's sign** (*fas-zha'*) [Jean Charles Faget, French physician, 1818– ]. A fall in the pulse rate with a rising or horizontal temperature curve; said to be pathognomonic of yellow fever.
**fagopyrum** (*fag-o-pi'-rum*). Poisoning by buckwheat.
**Fahr.,** abbreviation for *Fahrenheit's scale;* see *thermometer.*
**Fahrenheit's thermometer** (*fah'-ren-hīt*) [Gabriel Daniel Fahrenheit, German physicist, 1686–1736]. See *thermometer . Fahrenheit.*
**faint** (*fānt*). 1. A condition of languor. 2. A state of syncope or swooning.
**fainting** (*fānt'-ing*). A swoon; the act of swooning. **f. sickness.** Synonym of *epilepsy.*
**faith-cure.** The system or practice of attempting to cure disease by religious faith and prayer alone.
**falcate** (*fal'-kāt*) [*falx*, a sickle]. Sickle-shaped.
**falcial** (*fal'-se-al*) [*falx*, a sickle]. Relating to the falx cerebri or falx cerebelli.
**falcicula** (*fal-sik'-u-lah*). See *falcula.*
**falciform** (*fal'-si-form*) [*falx; forma*, form]. Having the shape of a sickle. **f. ligament.** See *ligament, falciform.* **f. process,** a process of the dura mater that separates the hemispheres of the brain; the falx.
**falcula** (*fal'-ku-lah*) [dim. of *falx*, a sickle]. The falx cerebelli.
**falcular** (*fal'-ku-lar*) [*falx*, a sickle]. 1. Sickle-shaped. 2. Pertaining to the falx cerebelli.
**fallacia** (*fal-a'-se-ah*) [L.]. An insane illusion; an hallucination. **f. auditoria,** an illusion as to hearing. **f. optica,** any visual illusion.
**fallectomy** (*fal-ek'-to-me*). Same as *salpingectomy.*
**falling** (*fawl'-ing*) [AS., *feallan*, to fall]. Dropping down. **f.-sickness,** epilepsy. **f. of the womb,** a descent of the uterus into the vagina.
**Fallopian** (*fal-o'-pe-an*). Described by Gabriel Fallopio or Falloppius, Italian anatomist, 1523–1562. **F. aqueduct,** F. canal, a canal in the petrosa, extending from the internal auditory meatus to the stylomastoid foramen and transmitting the facial nerve. **F. gestation,** tubal gestation. **F. hiatus,** an opening on the anterior surface of the petrosa, which serves for the transmission of the petrosal branch of the vidian nerve. **F. ligament.** 1. See *Poupart's ligament.* 2. See *Hunter's ligament.* **F. muscle,**

the pyramidalis. **F. tube**, the oviduct. **F. valvè**. See *Bauhin's valve*.
**Fallopius, aqueduct of.** See *Fallopian aqueduct*.
**F., hiatus of.** See *Fallopian hiatus*.
**fallostomy** (*fal-os'-to-me*). Same as *salpingostomy*.
**fallotomy** (*fal-ot'-o-me*). Same as *salpingotomy*.
**Falret's type of mania of persecution.** A form of paranoia occurring in degenerates. Syn., *"Idées de persécution et de persécuteur."*
**false** (*fawls*) [*falsus*, deceptive]. Not genuine; not real; imitating. **f. aneurysm.** See *aneurysm, false*. **f. ankylosis**, ankylosis due to rigidity of the soft tissues. **f. image**, in diplopia, the image of the deviating eye. **f. joint**, a result of non-union of a fractured bone. **f. membrane**, a fibrinous exudate upon a surface. **f. pains**, pains that precede true labor-pains. **f. passage**, a passage, formed by the laceration of any canal. **f. pelvis**, the portion of the pelvic cavity situated above the iliopectineal line. **f. ribs**, the five lower ribs.
**falsetto** (*fawl-set'-o*) [Ital.]. A voice both high-pitched and peculiarly modified as to quality; the highest register of the human voice. It is more obvious in the male than in most female voices.
**falsification** (*fawl-sif-ik-a'-shun*) [*falsus*, deceptive; *facere*, to make]. The fraudulent adulteration of foods or medicines; counterfeiting.
**falx** (*falks*) [L.]. A sickle; a sickle-shaped structure. **f. aponeurotica**, the conjoined tendon. **f. cerebelli**, a sickle-like process of dura mater between the lobes of the cerebellum. **f. cerebri**, the process of the dura separating the hemispheres of the cerebrum. **f. inguinalis**. See *f. aponeurotica*.
**famelic** (*fam-el'-ik*) [*famelicus*, hungry]. Marked by extreme hunger; effective in overcoming the sensation of hunger.
**famelica** (*fam-el'-ik-ah*) [*famelicus*, hungry]. Febris famelica; fever accompanied by hunger.
**fames** (*fa'-mēs*) [L.]. Hunger. **f. bovina**, bulimia. **f. canina** ("dog-hunger"), bulimia. **f. lupina** ("wolf-hunger"), extreme bulimia.
**familial** (*fam-il'-yal*). Characteristic of a family.
**famine fever.** Same as *relapsing fever*.
**fanaticism** (*fan-at'-is-ism*) [*fanaticus*, pertaining to a temple]. Perversion and excess of the religious sentiment. It often trenches upon the domain of insanity; and is sometimes an outcome, at other times a cause, of mental disease.
**fang** [AS., *fangan*, to seize]. The root of a tooth.
**fango** (*fan'-go*). Clay from the hot springs of Battaglio, Italy; it is used as a local application in gout. **f.-therapy**, the therapeutic application of heat and pressure by means of heated fango or other mud.
**fantascopy** (*fan-tas'-ko-pe*). See *retinoscopy*.
**farad** (*far'-ad*) [Michael *Faraday*, English physicist, 1791–1867]. The unit of electric capacity; a capacity sufficient to hold one coulomb of current having a potential of one volt.
**Faraday's law of electrolysis** [see *farad*]. The amount of an ion liberated at an electrode in a given time is proportional to the strength of the current.
**faradic, faradaic** (*far-ad'-ik, far-ad-a'-ik*). Pertaining to induced electric currents. **f. current**, the induced electric current.
**faradimeter** (*far-ad-im'-et-er*) [*farad*; μέτρον, a measure]. An instrument for measuring the strength of an induced electric current.
**faradipuncture** (*far-ad-i-punk'-chur*). The application of faradic electricity by means of needle electrodes thrust into the tissues.
**faradism** (*far'-ad-ism*) [see *farad*]. 1. The electricity produced in an induced or faradic current. 2. Faradization.
**faradization** (*far-ad-iz-a'-shun*) [see *farad*]. Faradism; the application of the induced current to a diseased part. **f. general**, the therapeutic application of the electric current to the organism as a whole.
**faradocontractility** (*far-ad-o-kon-trak'-til-it-e*). Contractility in response to the stimulus of faradic electricity.
**faradopuncture** (*far-ad-o-punk'-chur*) [*Faraday*; *punctura*, a puncture]. The passage of a faradic current into the tissues by acupuncture.
**farcinia** (*far-sin'-e-ah*). Synonym of *equinia*.
**farcinoma** (*far-sin-o'-mah*) [*farciminum*, farcy; ὄμα, a tumor: *pl.*, *farcinomata*]. A farcy-bud, or glanderous tumor; less correctly, farcy, or glanders.
**farctus** (*fark'-tus*) [L., a "stuffing"]. Emphraxis; congestion; infarction.

**farcy** (*far'-se*) [*farcire*, to stuff]. The form of glanders that attacks the skin and lymphatic glands. **f.-bud**, or **f.-button**, a glanderous tumor. **f.-pipes**, the swollen lymph-vessels in glanders.
**fardel-bound** (*far'-del-bownd*) [*fardel*, a load; ME., *bounden*]. 1. A term applied to neat cattle or sheep affected with inflammation of the abomasum, or of the omasum, with impaction of food in the latter. Fardel-bound cattle are said to have "lost the cud."
**farding-bag** (*far'-ding-bag*). The first stomach of a ruminant animal; the paunch or rumen.
**fareol** (*far'-e-ol*). A proprietary anodyne and antipyretic.
**farfara** (*far'-far-ah*). See *coltsfoot*.
**farina** (*far-e'-nah*) [L., "meal"]. The ground or powdered starchy part of seeds, especially that of corn, barley, rye, and wheat.
**farinaceous** (*far-in-a'-shus*) [*farina*]. Having the nature of or yielding flour.
**far-point.** The most distant point at which an eye can see distinctly when accommodation is completely relaxed.
**Farrant's solution.** Make a saturated solution of arsenic trioxide in water by boiling; let it stand for 24 hours and filter. Then to equal quantities of water, glycerol, and arsenic trioxide solution add picked gum-arabic until a thick, syrupy fluid is obtained. In about a week filter slowly through frequently changed filter-paper.
**Farre's tubercles** (*far*) [John Richard *Farre*, English physician, 1775–1862]. Cancerous masses on the surface of the liver.
**Farre's white line** (*far*) [Frederick John *Farre*, English gynecologist, 1804–1886]. The boundary-line at the hilum of the ovary between the germ epithelium and the squamous epithelium of the broad ligament; it marks the insertion of the mesovarium.
**Farre-Waldeyer's line.** See *Farre's white line*.
**farriery** (*far'-yer-e*) [*farraria*, pertaining to iron]. The art of treating the diseases of horses; veterinary surgery.
**far-sightedness.** Hyperopia.
**fascia** (*fash'-e-ah*) [L., "a band"]. 1. The areolar tissue forming layers beneath the skin (*superficial fascia*) or between muscles (*deep fascia*). 2. A bandage. **f., anal.** See *f., ischiorectal*. **f., cervical, deep, superficial**, that which invests the muscles of the neck and incloses the vessels and nerves. **f., cervical, superficial**, that just beneath the skin. **f., Colles'.** See under *Colles*. **f. colli.** 1. The deep and superficial fasciæ of the neck regarded as one. 2. The deep cervical fascia. **f., cremasteric**, a thin covering of the spermatic cord, formed by the stretched fibers of the cremaster muscle. **f., cribriform**, the sieve-like covering of the saphenous opening. **f. dentata**, a serrated band of gray matter of the hippocampal gyrus of the cerebrum. **f. infundibuliform**, the process of the transversalis fascia extended over the spermatic cord. **f., intercolumnar**, a fascia attached to the margins of the external abdominal ring and forming a sheath for the cord and testis. **f., ischiorectal**, that covering the perineal aspect of the levator ani muscle. **f. lata**, the dense fascia surrounding the muscles of the thigh. **f.s of origin**, those serving for the origin of muscles. **f.s, partial**, aponeuroses covering the muscles of a limb and retaining them in position; they are chiefly attached by their extremities to bones. **f., prevertebral**, a band of connective tissue covering the front of the cervical vertebræ and the prevertebral muscles adherent to the basilar process above and running to the third thoracic vertebra below. It is attached to the esophagus and pharynx by loose connective tissue. **f. propria**, **f. propria** of the scrotum, the infundibuliform fascia together with the underlying areolar tissue. **f., semilunar**, **f. semilunaris**, a fibrous band extending downward and inward from the inner aspect of the biceps humeri and its tendon and connecting with the fascia investing the antibrachial muscles arising from the inner condyle of the humerus. **f., transversalis**, that lying between the transversalis muscle and the peritoneum.
**fascial** (*fash'-e-al*) [*fascia*]. Pertaining to or of the nature of a fascia.
**fasciaplasty** (*fash-e-ah-plas'-te*) [*fascia*; πλάσσειν, to form]. Plastic surgery of fascia.
**fasciation** (*fash-e-a'-shun*) [*fascia*, a bandage]. The art or act of bandaging.
**fascicle** (*fas'-ik-l*) [dim. of *fascis*, a bundle]. A

small bundle of fibers. See *fasciculus*. f., **cuneate**, the continuation of the posteromedian column of the spinal cord. f., **fornicate**, the white matter of the fornicate gyrus, the fibers of which extend longitudinally and ramify upward and backward into its secondary gyri. Syn., *fillet of the corpus callosum*. f., **fundamental**, a part of the anterior column extending into the oblongata. f.s, **gyral**. See *fibers, association*. f., **olivary**. See *fillet, olivary*. f., **posterolongitudinal**, fibers connecting the corpora quadrigemina and the nuclei of the fourth and sixth nerves with the parts below. f., **pyramidal**, a portion of the anterior column of the cord extending to the pyramid. f., **solitary**, fibers connecting the internal capsule and lenticular nucleus with parts below.

**fasciculated** (*fas-ik'-ū-la-ted*). United into bundles or fascicles.

**fasciculus** (*fas-ik'-ū-lus*) [dim. of *fascis*, a bundle; pl., *fasciculi*]. A little bundle, particularly of muscle-fibers; a fascicle (*q. v.*). f. **albicantiothalami**. Same as *bundle of Vicq d'Azyr*. f. **arciformis pedis**, the cimbia, *q. v.* f. **cerebello-spinalis**. See *tract, direct cerebellar*. f., **cuneate**, the continuation of Burdach's column, or the posteromedian column of the spinal cord. f., **gracilis**, the posterior pyramid of the medulla. f. **longitudinalis inferior**, fibers connecting the temporal to the occipital lobe. f. **longitudinalis medialis**, a band of fibers found in the mid-brain ventrad to the central gray matter. f. **longitudinalis pyramidalis**. See *tract, pyramidal*. f. **longitudinalis superior**, a bundle of fibers joining the frontal cortex with the parietal, occipital, and external temporal cortex. f. **longitudinalis ventralis**, the anterior longitudinal bundle of fibers arising in the superior colliculus and descending into the spinal cord. f. **marginalis**. See *Lissauer's tract*. f. **occipitofrontalis**, a bundle of fibers extending from the cortex of the frontal lobe to the cortex of the occipital lobe. **fasciculi occipitothalamici**, the bundles of nerve-fibers uniting the thalamus with the occipital lobe. f., **olivary**. See *fillet, olivary*. f. **perpendicularis**, a vertical bundle of fibers from the inferior parietal and superior occipital gyri to the inferior temporal and occipital and the fusiform gyri. f. **pedunculo-mammillaris**, a bundle arising in the corpus mammillare and passing into the mid-brain. f., **posterolongitudinal**, fibers connecting the corpora quadrigemina and the nuclei of the fourth and sixth nerves with the parts below. f. **rectus**. Same as *f. perpendicularis*. f. **retroflexus**, a bundle of nerve-fibers connecting the ganglion of the habenula with the interpeduncular ganglion of the opposite side. See also *Meynert's bundle*. f., **solitary**, fibers connecting the internal capsule and the lenticular nucleus with parts below. See *solitary bundle*. f., **sphenoid**, the part of the corona radiata which enters the temporosphenoid lobe. f. **subcallosus**, a band of long association-fibers lying under the corpus callosum, and connecting the frontal, parietal, and occipital lobes. f. **teres**, the funiculus teres. f. **thalamo-mammillaris**. Same as *bundle of Vicq d'Azyr*. f. **trineuralis**, f., **trineural**. See *solitary bundle*. f. **unciformis**, f. **uncinatus**, a bundle of medullated axons extending between the uncus and the basal portions of the frontal lobe, and connecting the temporal sense area with the olfactory sense area.

**fascination** (*fas-in-a'-shun*) [*fascinatio*, a bewitching]. A form of incomplete hypnotism, intermediate between somnambulism and catalepsy; the alleged controlling influence of one person over another.

**fasciodesis** (*fas-e-od'-e-sis*) [*fascia*; δέσις, binding]. The operation of suturing a tendon to a fascia.

**fasciola** (*fas-i'-o-lah*) [L., a small bandage]. 1. The dorsal continuation of the fascia dentata of the cerebrum. 2. A genus of trematodes. f. **cinerea**. See *fasciola* (1). F. **hepatica**, the liver fluke.

**fascioliasis** (*fas-e-o-li'-as-is*) [*fasciola*, a small bandage]. A term employed for *distomiasis*, or *distomatosis*.

**fasciolopsis** (*fas-ce-o-lop'-sis*). See *fluke*.

**fasciotomy** (*fash-e-ot'-o-me*). See *aponeurotomy*.

**fascitis** (*fas-i'-tis*). Inflammation of a fascia.

**fastidium** (*fas-tid'-e-um*) [L., "a loathing"]. A loathing for food or drink.

**fastigatum** (*fas-tig-a'-tum*). See *nucleus fastigii*.

**fastigium** (*fas-tij'-e-um*) [L., "summit"]. 1. The acme of a disease. 2. The angle between the superior lamina and the inferior medullary velum in the roof of the fourth ventricle.

**fat** [ME.]. A greasy substance, a compound of oleic, palmitic, or stearic acid with glycerol. f.-**cell**, a connective-tissue cell containing oil-globules. f. **columns**, columnar shaped adipose tissue found in the thicker parts of the cutis vera. f. **emboli**. See *embolus*. f.-**necrosis**, a peculiar form of necrosis of a fatty tissue occurring in pinpoint-sized areas of a dead-white color.

**fatigue** (*fa-tēg'*) [*fatigo*, weariness]. Weariness. f. **diseases**, those caused by constant repetition of certain muscular movements. f. **fever**, fever following excessive exertions, and supposed to be caused by the absorption of waste-products. f.-**stuff**, toxic material from tissue-disintegration due to undue fatigue.

**fatty** (*fat'-e*) [*fat*]. Containing fat or derived from fat. f. **acids**, a series of acids with the general formula $C_nH_{2n}O_2$, some of the members of which combine with glycerol to form fats. f. **casts**. See *cast*. f. **degeneration**. See *degeneration, fatty*. f. **heart**. See *heart*. f. **series**, methane and its derivatives.

**fatuity** (*fat-ū'-it-e*) [*fatuitas*, foolishness]. Amentia or dementia.

**fauces** (*faw'-sēz*) [L., "the upper part of the throat"]. The space surrounded by the palate, tonsils, and uvula. f., **isthmus of the**, the space at the back of the mouth inclosed by the margin of the palate, the back of the tongue, and the pillars of the fauces. f., **pillars of the**, the folds formed by the palatoglossus muscle in front of the tonsils and by the palatopharyngeus behind them.

**Fauchard's disease** (*fo'-shar*) [Pierre *Fauchard*, French dentist, 1680–1761]. Alveolodental periostitis; pyorrhœa alveolaris; progressive necrosis of the dental alveoli.

**faucial** (*faw'-se-al*) [*fauces*]. Pertaining to the fauces.

**faucitis** (*faw-si'-tis*) [*fauces*; ιτις, inflammation]. Inflammation of the fauces.

**fault** (*fawlt*) [ME., *faut*, a lack]. In electricity, any failure in the proper working of a circuit due to ground-contacts, cross-contacts, or disconnections. These may be of three kinds: 1. Disconnections; 2. Earths; 3. Contacts.

**fauna** (*faw'-nah*) [*Faunus*, the god of agriculture; pl., *faunæ*]. The entire animal life of any geographical area or geological period.

**faunorum ludibria** (*fawn-or'-um lu-dib'-re-ah*) [L., "sport of the fauns"]. An old designation for nightmare and for epilepsy.

**Fauvel's granules** (*fo-vel'*) [Sulpice Antoine *Fauvel*, French physician, 1813–1884]. Abscesses in the immediate neighborhood of the bronchi or bronchioles.

**favaginous** (*fav-aj'-in-us*) [*favus*]. Resembling favus; having a honeycombed surface.

**faveolate** (*fav-e'-o-lāt*) [*favus*]. Favose, honeycombed; same as alveolate.

**faveolus** (*fav-e'-o-lus*) [*favus*; pl., *faveoli*]. A pit or cell like that of the honeycomb.

**faviform** (*fav'-if-orm*) [*favus*, honeycomb; *forma*, form]. Resembling a honeycomb; a designation of certain ulcerated surfaces.

**favosoareolate** (*fa-vo-so-ar-e'-o-lāt*) [*favus; areola*]. Pitted with reticular markings.

**favus** (*fa'-vus*) [L., "a honeycomb"]. A parasitic skin disease due to the presence of a vegetable parasite, *Achorion schonleinii*. It is characterized by the presence of round, sulphur-yellow, cup-shaped crusts, having a peculiar musty odor, and which are found on microscopic examination to be composed almost entirely of the elements of the fungus. The disease affects most frequently the scalp, but may occur anywhere. Syn., *tinea favosa*. f.-**cup**, any one of the cup-shaped crusts that characterize favus.

**faxwax** (*faks'-waks*) [*fax*, hair; Ger., *wachsen*, to grow]. The ligamentum nuchæ, or the material of which it is composed. It is also called paxwax.

**F. C. S.** Abbreviation for Fellow of the Chemical Society.

**Fe.** Chemical symbol of *ferrum*, iron.

**fear** (*fēr*). An emotion of dread; apprehension; the feeling which in its intenser manifestations is called terror or fright.

**feature** (*fe'-chur*) [*factura*, a making]. Any single part of lineament of the face.

**febralgene** (*feb-ral'-jēn*). A proprietary antipyretic and sedative. Dose 2½–5 gr. (0.16–0.32 Gm.).

**febricide** (*feb'-ris-īd*) [*febris; cædere*, to kill]. Destructive fever.

**febricity** (*fe-bris'-it-e*) [*febris*, a fever]. Feverishness.
**febricula** (*feb-rik'-ū-lah*) [*febris*]. A slight fever of short duration, most frequently encountered among children.
**febriculose** (*feb-rik'-ū-lōs*) [*febris*, fever]. Slightly feverish.
**febriculosity** (*feb-rik-ū-los'-it-e*). Feverishness.
**febrifacient** (*feb-re-fa'-se-ent*) [*febris*, a fever; *facere*, to make]. Causing fever.
**febriferous** (*fe-brif'-er-us*) [*febris*, a fever; *ferre*, to bear]. Causing or conveying fever.
**febrific** (*fe-brif'-ik*) [*febris*, a fever]. Febrifacient.
**febrifugal** (*fe-brif'-ū-gal*) [*febris*, fever; *fugare*, to dispel]. Removing or dispelling fever.
**febrifuge** (*feb'-rif-ūj*) [*febris*; *fugare*, to dispel]. 1. Dispelling fever. 2. An agent that lessens fever.
**febrile** (*feb'-ril*) [*febris*]. Pertaining to or characterized by fever.
**febrility** (*fe-bril'-it-e*) [*febris*, a fever]. The quality of being febrile; feverishness.
**febrinol** (*feb'-rin-ol*). A proprietary antipyretic and analgesic remedy.
**febris** (*feb'-ris*) [L.]. See *fever*. **f. acmastica**, a continued fever. **f. acuta**, ague. **f. amatoria**, chlorosis. **f. bullosa**. Synonym of *pemphigus*. **f. castrensis**. 1. Typhus fever. 2. Remittent fever. **f. castrensis epidemica**, malarial and typhoid fever. **f. catarrhalis**. Synonym of *influenza*. **f. complicata**, Mediterranean fever. **f. dysenterica**. Synonym of *dysentery*. **f. enterica**, typhoid fever. *q. v.* **f. exanthematica articulosa**. Synonym of *dengue*. **f. famelica**. See *famelica*. **f. flava**. See *yellow fever*. **f. Hungarica**. Synonym of *typhus*. **f. innominata**, a fever in which the clinical signs are lacking. **f. lactea**, milk fever. **f. nervosa**, those febrile conditions which appear to be primarily nervous in pathology. **f. nigra**. Synonym of *fever, cerebro-spinal*. **f. petechialis**, typhus. **f. recidiva**. Synonym of *relapsing fever*. **f. recurrens**. Synonym of *fever, relapsing*. **f. remittens**. See *fever, remittent*. **f. undulans**, Malta fever. **f. variolosa**, a form of smallpox described by Sydenham with alarming initial symptoms but mild in its subsequent course. Called by de Haen *variola sine variolis*.
**fecal** (*fe'-kal*) [*feces*]. Pertaining to, consisting of, or discharging feces.
**fecaloid** (*fe'-kal-oid*) [*fæx*; *εἶδος*, likeness]. Resembling feces.
**feces** (*fe'-sēz*) [*fæx*, sediment; *pl.*, *fæces*]. The excretions of the bowels. The feces consist of excretions and secretions from the intestine and of undigested food, the latter being made up of digestible substances that escaped digestion and of indigestible matters, such as nuclein, cellulose, chlorophyl, and mineral salts.
**Fechner's law** (*fek'-ner*) [Fustav Theodor *Fechner*, German physicist, 1801–1887]. The intensity of a sensation is proportional to the logarithm of the stimulus.
**fecula** (*fek'-ū-lah*) [dim. of *fax*, sediment]. 1. The starchy part of a seed. 2. The sediment subsiding from an infusion.
**feculent** (*fek'-ū-lent*) [*fæculentus*, dreggy]. Having sediment.
**fecundate** (*fe'-kun-dāt*). To impregnate; to render pregnant.
**fecundation** (*fe-kun-da'-shun*) [*fecundity*]. The act of fertilizing. **f., artificial**, fecundation brought about by the injection of semen into the vagina or uterus through a syringe or other instrument.
**fecundity** (*fe-kun'-dit-e*) [*fecunditas*, fruitfulness]. The ability to produce offspring.
**Federici's sign** (*fa-dar-e'-che*) [Cesare *Federici*, Italian physician, 1838–1892]. Perception of the heart-sounds over the whole abdomen in cases of perforative peritonitis with escape of gas into the peritoneal cavity.
**feeble** (*fe'-bl*). Lacking strength; weak. **f.-minded, idiotic**.
**feed** (*fēd*) [ME., *fedan*, to nourish]. 1. To supply with food; to graze; to eat. 2. Food, especially that for lower animals; fodder.
**feeder** (*fē'-der*) [ME., *fedan*, to nourish]. An instrument used in the forcible feeding of insane patients who obstinately refuse to eat.
**feeding** (*fe'-ding*) [ME., *fedan*, to nourish]. The taking of food or aliment. **f., artificial**, the introduction of food into the body by means of artificial devices, such as the stomach tube or in the form of an enema. Also, the nourishing of a child by food other than the mother's milk. **f.-bottle**, a glass flask armed with a rubber nipple, used in feeding liquid food to infants. **f.-cup**, a cup used in the forcible feeding of the insane. **f., extrabuccal**, the introduction of food into the system by other channels than the mouth; by subcutaneous nutritive enema, or intravascular injection of food materials; feeding after gastrotomy, through gastric fistulæ. **f., forcible**, the administration of aliment by compulsion to such patients as refuse to take food in the natural manner. **f. by the rectum**, the introduction of food into the rectum in the form of an enema or suppository.
**feel** (*fēl*). To have a sensation of; to try by touch; to have perception by means of the sense of touch.
**feeling** (*fē'-ling*). The sense of touch; any emotion or sensibility; any conscious state of nervous activity; any sensation. **f.s, entoperipheral**, sensations due to stimulation of the peripheral nerves distributed to the interior of the organism. **f.s, epiperipheral**, sensations due to stimulation of the peripheral nerves distributed to the surface of the body, including all the nerves of special sense. **f.s, presentative**, primary sensations caused by direct stimulation. **f.s, representative**, sensations produced by indirect stimulation; revived feelings; ideas.
**feet** (*fēt*). The plural of *foot*, *q. v.* **f., frosted**. See *chilblain*.
**Fehleisen's streptococcus** (*fa'-li-zen*) [Friedrich *Fehleisen*, German physician, 1854– ]. The *Streptococcus erysipelatis*.
**Fehling's solution** (*fa'-ling*) [Hermann von *Fehling*, German chemist, 1812–1885]. See under *F.'s test for glucose*. **F.'s test for glucose**, two solutions are required to be kept in two distinct parts in well-stoppered bottles. (A) Dissolve 36.64 Gm. of copper sulphate in 500 Cc. of water. (B) Dissolve 173 Gm. of Rochelle salt in 100 Cc. of a solution of caustic soda having a specific gravity of 1.34, and dilute with water to 500 Cc. Mix equal volumes of A and B for use; the result is a dark-blue fluid known as *Fehling's solution*. The solution should always be fresh, as tartaric acid has a tendency to become converted into racemic acid, which reduces cupric salts like sugar. Its absence should always be ascertained by boiling the Fehling solution, which should remain unaltered by this process. On addition of a solution of glucose and then boiling, a red precipitate of the cuprous oxide or hydrate occurs.
**feigned disease**. See *malingering*.
**fel** [L.]. Bile. **f. bovinum**. Same as *f. bovis*. **f. bovis** (U. S. P.), oxgall. **f. bovis purificatum** (U. S. P.), purified oxgall; it is said to be tonic and laxative. Dose 5–10 gr. (0.3–0.6 Gm.).
**fellatio** (*fel-a'-she-o*) [L.]. The act of sexual perversion referred to under *fellatrice, q. v.*
**fellator** (*fel-at-or*). Masculine form of *fellatrice, q. v.*
**fellatrice** (*fel-at-rēs'*) [Fr.]. The female agent in irrumation, who receives the male organ in her mouth and by friction with the lips or tongue produces the orgasm.
**fellifluous** (*fel-if'-loo-us*) [*fel*; *fluere*, to flow]. Flowing with gall.
**fellitin** (*fel'-it-in*). A proprietary preparation of oxgall for use in frost-bite.
**fellmongers' disease**. Anthrax; so called as attacking dealers in fells, or pelts, of skins.
**Fell's method** [George E. *Fell*, American physician, 1850– ]. A method of forced respiration in cases of narcotic poisoning or drowning, by means of an apparatus consisting of a tracheotomy-tube attached to a bellows.
**felo-de-se** (*fe'-lo-de-sē*) [Sp.]. A suicide. Also, any one who commits an unlawful malicious act, the consequence of which is his own death.
**felon** (*fel'-on*). See *paronychia*.
**Felt treatment of sciatica**. Subcutaneous injection of $\frac{1}{75}$ gr. (2 mg.) of atropine, followed in 48 hours by a second dose of $\frac{1}{30}$ gr. (2.6 mg.), 48 hours after, $\frac{1}{20}$ gr. (3 mg.).
**female** (*fe'-māl*) [*femina*, woman]. 1. Pertaining to the sex that conceives and bears young; pertaining to woman. 2. Denoting that part of a double-limbed instrument that receives the complementary part. **f., catheter**, a catheter for emptying the female bladder.
**femininity** (*fem-in-in'-it-e*) [*femina*, a woman]. The sum of those qualities that distinguish the female sex.

**feminism** (*fem'-in-ism*) [*femina*, a woman]. Arrested development of the male organs of generation, accompanied by various mental and physical approximations to the characters of the female sex.

**feminonucleus** (*fem-in-o-nū'-kle-us*) [*femina*, woman; *nucleus*]. The embryonic female nucleus, as distinguished from the corresponding male nucleus.

**femoral** (*fem'-or-al*) [*femur*]. Pertaining to the femur. **f. arch,** Poupart's ligament. **f. artery.** See under *artery*. **f. canal,** a canal in the sheath of the femoral vessels, on the inner side of the femoral vein, through which, at times, a hernia descends. **f. hernia.** See *hernia, femoral*. **f. ligament of Hey.** See *Hey's ligament*. **f. ring,** the abdominal end of the femoral canal, normally closed by the crural septum and the peritoneum. **f. sheath,** a continuation downward of the fasciæ that line the abdomen. It contains the femoral vessels.

**femoralis** (*fem-or-a'-lis*) [L.]. 1. See *femoral*. 2. See *quadriceps extensor femoris*, in *muscles, table of*.

**femoren** (*fem'-or-en*) [*femur*, the thigh-bone]. Belonging to the femur in itself.

**femoro-articular** (*fem-or-o-ar-tik'-u-lar*). Articulating with the femur.

**femorocele** (*fem'-or-o-sēl*) [*femur;* κήλη, hernia]. Femoral hernia.

**femoropopliteal** (*fem-or-o-pop-lit-e'-al*). 1. Relating to or contained in the thigh or popliteal space. 2. Relating to the dorsal aspect of the thigh.

**femoropopliteotibial** (*fem-or-o-pop-lit-e-o-tib'-e-al*). Relating to the femur, popliteal space, and tibia.

**femoropretibial** (*fem-or-o-pre-tib'-e-al*). Relating to the thigh and the anterior part of the leg.

**femorotibial** (*fem-or-o-tib'-e-al*). Relating to the femur and the tibia.

**femorovascular** (*fem-or-o-vas'-kū-lar*) [*femur; vasculum*, a little vase]. Relating to the femoral canal.

**femur** (*fe'-mur*) [L.]. 1. The thigh-bone. 2. The thigh.

**fence** (*fens*) [abbreviation of *defense*]. A line of cross-scarification made on the skin surrounding an erysipelatous area, to which a germicide is applied, to prevent progress of the disease.

**fenestra** (*fen-es'-trah*) [L., "a window"]. 1. In anatomy, a name given to two apertures of the ear, the *f. ovalis* and *f. rotunda*. 2. The open space in the blade of a forceps. 3. An opening in a bandage or dressing for drainage, etc. **f. cochleæ, f.** cochlearis, **f. triquetra,** the fenestra rotunda. See under *ear*. **f. semiovalis, f. vestibularis, f. vestibuli,** the fenestra ovalis. See under *ear*.

**fenestral, fenestrate** (*fen-es'-tral, fen'-es-trāt*) [*fenestra*, a window]. Having apertures or openings.

**fenestrated** (*fen-es'-tra-ted*) [*fenestra*]. Perforated. **f. membrane of Henle,** the layer of elastic tissue in the intima of large arteries.

**fennel** (*fen'-l*). See *fœniculum*.

**fenthozon** (*fen'-tho-zon*). A proprietary disinfectant and deodorant said to consist of acetic acid, 26 Gm.; phenol, 2 Gm.; menthol, camphor, and oil of eucalyptus, of each, 1 Gm., and oils of verbena and lavender, of each, 0.5 Gm.

**fenugreek** (*fen'-ū-grēk*). The *Trigonella fænumgræcum*, a leguminous plant cultivated in France and Germany, the seeds of which contain two alkaloids, *choline* and *trigonelline*. The seeds are employed for the preparation of emollient poultices, enemas, ointments, and plasters. They are not used internally.

**Fenwick's disease** [Samuel Fenwick, English physician, 1821– ]. Primary atrophy of the stomach.

**feral** (*fe'-ral*) [*feralis*, deadly]. Deadly or fatal; as a feral disease.

**feralboid** (*fer-al'-boyd*). A peptonized albuminate of iron. It is used in anemia, neurasthenia, etc. Dose ½–⅔ gr. (0.021–0.042 Gm.) 3 times a day. Syn., *feraldoid*.

**fercremol** (*fer'-kre-mol*). A brown, tasteless compound of iron and hemoglobin, containing 3 % of iron. Dose 3–8 gr. (0.2–0.52 Gm.).

**Féréol's nodosities** (*fa-ra-ol'*) [Louis Henri Félix Féréol, French physician, 1825–1891]. Inconstant subcutaneous nodosities observed in cases of acute articular rheumatism.

**Féréol-Graux's type of ocular palsy** (*fa-ra-ol'-gro'*). Associated paralysis of the internal rectus muscle of one side and of the external rectus of the other; it is of nuclear origin.

**Fergusson's speculum** (*fer'-gus-un*). [Sir William Fergusson, Scotch surgeon, 1808–1877]. A vaginal speculum in the form of a silvered glass tube with a coating of caoutchouc.

**ferine** (*fe'-rin*) [*ferinus*, pertaining to a wild beast]. Noxious, malignant, or violent; as a *ferine* disease.

**ferisol** (*fer'-is-ol*). A derivative of cinnamic acid and guaiacol; a very soluble powder. Dose 15 gr. (1 Gm.); intramuscularly, 15 min. (0.92 Cc.) of a 10 % solution.

**fermang** (*fer'-mang*). A proprietary peptonate of iron and manganese.

**fermanglobin** (*fer-man-glo'-bin*). Hemoglobin combined with iron and manganese; used in anemia. Dose a teaspoonful to a dessert-spoonful (5–10 Cc.).

**ferment** (*fer'-ment*) [*fermentum*, leaven; yeast]. Any substance which, in contact with another substance, is capable of setting up changes (*fermentation*) in the latter without itself undergoing much change. Ferments are classified into *unorganized*, or *soluble*, and *organized*, or *living*, ferments. According to the character of the fermentation, the unorganized ferments are divided into amylolytic, proteolytic, fat-decomposing, milk-curdling, and coagulating ferments. *Amylolytic ferments* (sugar-producing or diastatic ferments) convert starch into sugar. The most important is ptyalin of the saliva; but similar ferments are found in the pancreatic and intestinal juice, bile, blood, milk, urine, etc. *Proteolytic ferments* convert proteins into peptones and albumoses. They are found in the stomach (*pepsin*), in the pancreatic juice (*trypsin*), and elsewhere. A *fat-decomposing ferment* is found in the stomach and in the pancreatic juice. *Milk-curdling ferment* is found in the stomach and the intestinal juice. The best example of a *coagulating ferment* is the fibrin-ferment. The organized or living ferments are the yeasts and bacteria. See also *fermentation*. **f., animal,** one secreted by the animal organism. **f., chemical.** See *enzyme*. **f.s, coagulating,** the milk-curdling ferment and fibrin-forming ferment (*thrombin*). **f., digestive,** an enzyme, either of animal or vegetable production, which acts upon a certain kind of food. The digestive ferments embrace the amylolytic, proteolytic, invertive, emulsive, and the milk-curdling ferments. **f., fibrin-.** See *fibrin-ferment*. **f., glycolytic.** 1. One existing in the liver, which changes starch into sugar; also any ferment capable of decomposing sugar. 2. Lépine's name for the internal secretion of the pancreas which reaches the general circulation without entering the intestinal tract. **f.s, hydrolytic.** See *hydrolytic ferments*. **f., inverting, f., inverting.** See *invertin*. **f.s, oxidation, f.s, oxidizing,** ferments existing in the cells and tissues of the body which act as oxygen-carriers and act on hydrogen dioxide and neutral oxygen as well; their activity is destroyed by heat. Syn., *oxidases; oxydases; tissue ferments*. **f., proteolytic,** one which decomposes fat. **f., soluble.** See *enzyme*. **f., steatolytic,** one that splits fat into fatty acids and glycerol, as steapsin of the pancreatic juice and similar ferments found in seeds of poppy, castor-oil beans, Indian hemp, corn, etc.

**fermentable** (*fur-ment'-a-bl*). Capable of being fermented.

**fermental** (*fur-ment'-al*). Endowed with capacity to produce fermentation.

**fermentation** (*fer-men-ta'-shun*) [*ferment*]. The decomposition of complex molecules under the influence of ferments. **f., acetic,** the fermentation whereby weak alcoholic solutions are converted into vinegar; caused by *Bacillus aceti*, etc. **f., alcoholic, f., spirituous,** the conversion of saccharine substances into alcohol; it is due to yeast-germs. **f., ammoniacal,** that giving rise to ammoniacal gas and carbon dioxide, which combine to form ammonium carbonate. The agent of the ammoniacal fermentation of urine is *Micrococcus urea*. **f., butyric,** the conversion of sugars, starches, milk, etc., into butyric acid, due to various microorganisms, especially *Bacillus butyricus*. **f., caseous,** that by which the conversion of milk into cheese is effected. **f., diastatic,** the conversion of starch into glucose by the action of ptyalin, etc. **f., lactic,** the "souring" of milk, caused by *Bacillus lacticus*. **f., propionic,** the production of propionic acid by *Bacillus cavicida*, which decomposes saccharine solutions. **f. test for glucose,** half fill a test-tube with a solution of dextrose, and add a little dried German yeast. Invert the tube over mercury and allow it to stand in a warm place for 24 hours. The sugar will ferment, carbonic-

acid gas accumulates in the tube, and the liquid gives the tests for alcohol. A control-experiment should be made with yeast and water in another test-tube, as a small yield of carbonic acid is often obtained from impurities in the yeast. See also *Roberts*. **f.-tube**, a glass tube used in the fermentation test for glucose, *q. v.* **f., viscous**, a fermentation characterized by the production of a gummy substance.
**fermentemia** (*fer-men-te'-me-ah*) [*fermentum*, leaven; *αἷμα*, blood]. The presence of a ferment in the blood.
**fermentogen** (*fer-ment'-o-jen*) [*fermentum*, leaven; *γεννᾶν*, to produce]. Any substance (like pepsinogen or trypsinogen) that on the reception of the appropriate stimulus is changed into a ferment.
**fern**. Any cryptogamous plant of the order *Filices*. See *aspidium*. **f. female**. *Asplenium filix fœmina*. **f., male**, *Dryopteris filix mas*.
**ferralbumose** (*fer-al'-bu-mōs*). A meat precipitate treated with artificial gastric juice and ferric chloride.
**ferralia** (*fer-a'-le-ah*) [*ferrum*, iron]. Medicinal preparations of iron.
**ferralum** (*fer'-al-um*) [*ferrum*, iron; *alumen*, alum]. 1. Any chalybeate. 2. A proprietary preparation consisting largely of the sulphate of iron and aluminum.
**ferrated** (*fer'-a-ted*). Combined with iron; containing iron.
**ferratin** (*fer'-at-in*). A chemical compound of iron and albumin, introduced as identical with the organic iron component of food. It is used in anemia and malnutrition. Dose 7½ gr. (0.3 Gm.) 3 times daily. **f., Schmiedeberg's**, a nuclein in combination with iron contained in liver. Syn., *Zaleski's hepatin*.
**ferratogen** (*fer-at'-o-jen*). An iron nuclein obtained by cultivating yeast on a medium impregnated with iron. It is used in treatment of chlorosis, the preparation containing 1 % of metallic iron. Syn., *ferric nuclein*.
**Ferrein's canal** (*fer-rīn'*) [Antoine *Ferrein*, French anatomist, 1692–1769]. A triangular channel, supposed to exist between the free edges of the eyelids when they are closed, and to serve for conducting the tears toward the puncta lacrimalia during sleep. **F.'s cords**, the true vocal cords. **F.'s foramen**. See *Fallopian hiatus*. **F.'s pyramids**, the medullary rays, pyramidal in shape, having their apices at the periphery of the cortex of the kidney and their bases in the boundary layer. **F.'s tubes**, the convoluted uriniferous tubules.
**ferri-** (*fer'-i*) [*ferrum*, iron]. A prefix that indicates the ferric, as distinguished from ferrous compounds. Containing iron as a quadrivalent element.
**ferri** (*fer'-i*). Genitive of *ferrum*, iron.
**ferric** (*fer'-ik*). 1. Pertaining to or of the nature of iron. 2. Containing iron as a trivalent or quadrivalent element. **f. ammonium sulphate** (*ferri et ammonii sulphas*, U. S. P.). Dose 7½ gr. (0.5 Gm.). **f. ammonium tartrate** (*ferri et ammonii tartras*, U. S. P.), iron and ammonium tartrate. Dose 4 gr. (0.25 Gm.). **f. chloride** (*ferri chloridum*, U. S. P.), iron perchloride. Dose 1 gr. (0.065 Gm.). **f. chloride, solution of** (*liquor ferri chloridi*, U. S. P.). Dose 2–10 min. (0.13–0.65 Cc.). **f. chloride, tincture of** (*tinctura ferri chloridi*, U. S. P.). Dose 10–30 min. (0.65–2.0 Cc.). **f. citrate** (*ferri citras*, U. S. P.). Dose 3–10 gr. (0.2–0.65 Gm.). **f. citrate, soluble** (*ferri et ammonii citras*, U. S. P.), ammonioferric citrate; iron and ammonium citrate. Dose 2–5 gr. (0.12–0.32 Gm.). **f. citrate, solution of**. Dose 10 min. (0.65 Cc.). **f. citrate, wine of** (*vinum ferri*, U. S. P.). Dose 1–4 dr. (4–16 Cc.). **f. hydroxide** (*ferri hydroxidum*, U. S. P.), hydrated oxide of iron, an antidote to arsenic. **f. hydroxide with magnesium oxide** (*ferri hydroxidum cum magnesii oxido*, U. S. P.), an antidote to arsenic. Dose 3–5 gr. (0.2–0.3 Gm.). **f. hypophosphite** (*ferri hypophosphis*, U. S. P.). Dose 3 gr. (0.2 Gm.). **f. nuclein**. See *ferratogen*. **f. phosphate, soluble** (*ferri phosphas solubilis*, U. S. P.). Dose 5–10 gr. (0.32–0.65 Gm.). **f. pyrophosphate, soluble** (*ferri pyrophosphas solubilis*, U. S. P.). Dose 2–5 gr. (0.23–0.32 Gm.). **f. subsulphate solution of** (*liquor ferri subsulphatis*, U. S. P.), Monsel's solution. Dose 3 min. (0.2 Cc.).
**ferrichthol** (*fer-ik'-thol*). A form of ichthyol iron sulphonate which contains 3.5 % of organically combined iron together with 96.5 % of ichthyol sulphonic acid. It is odorless and tasteless, and is used in the treatment of anemia and chlorosis. Dose 2 gr. (0.13 Gm.).

**ferricyanide** (*fer-i-si'-an-īd*). A compound of ferricyanogen, with an element or radical.
**ferricyanogen** (*fer-i-si-an'-o-jen*). A hexad radical, (FeCN₂)₂.
**Ferrier's snuff** (*fer'-e-er*). A snuff used in acute rhinitis. Its formula is morphine hydrochloride gr. ij, powdered acacia ʒ ij, and subnitrate of bismuth, ℨ vj.
**ferrinol** (*fer'-in-ol*). Iron nucleid, a compound of nucleol and iron oxide containing 6 % of iron.
**ferripton** (*fer-ip'-ton*). A proprietary preparation said to contain 4 % of iron, 7 % of proteids, and 89 % of water. It is used in anemia and chlorosis.
**ferripyrin** (*fer-e-pi'-rin*). See *ferropyrin*.
**ferrisalipyrine** (*fer-e-sal-e-pi'-rēn*). Antipyrine ferrous salicylate, a yellow-brown powder showing a green fluorescence.
**ferro-** (*fer-o-*) [*ferrum*, iron]. A prefix used with the names of ferrous compounds.
**ferrocyanic** (*fer-o-si-an'-ik*). Composed of iron and cyanogen.
**ferrocyanide** (*fer-o-si'-an-īd*). A compound of ferrocyanogen, with an element or radical.
**ferrocyanogen** (*fer-o-si-an'-o-jen*)₁ A tetravalent radical, Fe(CN)₄.
**ferrocyanuret** (*fer-o-si-an'-u-ret*). See *ferrocyanide*.
**ferroferric** (*fer-o-fer'-ik*). Containing iron in both ferric and ferrous combinations.
**ferrohemol** (*fer-o-hem'-ol*). Hemol containing 3 % of added iron. Dose 8 gr. (0.5 Gm.).
**ferrol, Petroleum** (*fer'-ol, fer-o'-le-um*). A proprietary 30 % emulsion of cod-liver oil containing iron phosphate.
**ferromagnesium sulphate** (*fer-o-mag-ne'-se-um sul'-fāt*), FeSO₄+6H₂O. A greenish powder, used in anemia and chlorosis. Dose 8 gr. (0.5 Gm.).
**ferromagnetic** (*fer-o-mag-net'-ik*). Having iron as a constituent and possessing magnetic properties.
**ferrometer** (*fer-om'-et-er*) [*ferrum*; *μέτρον*, a measure]. An apparatus for estimating quantitatively the iron in the blood.
**ferropyrin** (*fer-o-pi'-rin*), (C₁₁H₁₂N₂O)₃Fe₂Cl₆. A hemostatic containing antipyrine, 64 %; iron, 12 %; chlorine, 24 %. It is styptic, antiseptic, and astringent, and is applied externally in gonorrhea and nosebleed. It is used internally in anemia, chlorosis, neuralgia, in doses of from 8-15 gr. (0.52-1.0 Gm.). Application, 1 to 1.5 % solution for gonorrhea; 20 % solution for nosebleed.
**ferrosin** (*fer'-o-sin*). A granular or fine red powder used as a pigment and said to contain iron oxide, 70 to 75 %; lime and albumin, 10 to 20 %; water, 10 to 15 %.
**ferrosodium-citroalbuminate** (*fer-o-so-de-um-sit-ro-al-bū'-min-āt*). A hematinic containing 30 % of ferric oxide. Dose 23 gr. (1.5 Gm.); children, 4–8 gr. (0.26–0.52 Gm.) in soup or syrup.
**ferrosoferric** (*fer-o-so-fer'-ik*). Containing iron as a bivalent and a trivalent radical.
**ferrosoferrous** (*fer-o-so-fer'-us*). Applied to a salt which is compounded of two ferrous salts.
**ferrosol** (*fer'-o-sol*). A double combination of ferric saccharate and saccharate of sodium chloride occurring as a clear, black-brown liquid; used in chlorosis, anemia, and neurasthenia. Dose 1 teaspoonful (5 Cc.) 3 times daily.
**ferrosomatose** (*fer-o-so'-mat-ōs*). A combination of 2 % of iron with somatose; an odorless, tasteless powder, soluble in water. It is used as a tonic in chlorosis, anemia, and debility. Dose 75–150 gr. (5–10 Gm.) daily; as a laxative, 150 gr. (10 Gm.).
**ferrostip** (*fer-o-stip'-tin*). A preparation of iron and formaldehyde occurring in cubic crystals or crystalline powder, soluble in water, melting at 120° C. It is used as a noncaustic, antiseptic hemostatic in dentistry. Dose 5–8 gr. (0.3–0.5 Gm.).
**ferrous** (*fer'-us*) [*ferrum*]. Containing iron as a bivalent element. **f. carbonate, mass of** (*massa ferri carbonatis*, U. S. P.), Vallet's mass. Dose 3–5 gr. (0.2–0.32 Gm.). **f. carbonate, pills of** (*pilulæ ferri carbonatis*, U. S. P.), Blaud's pills, consist of ferrous sulphate, potassium carbonate, sugar, tragacanth, althea, glycerol, and water. **f. carbonate, saccharated** (*ferri carbonas saccharatus*, U. S. P.). Dose 5–10 gr. (0.32–1.3 Gm.). **f. iodide, pills of** (*pilulæ ferri iodidi*, U. S. P.), Blanchard's pills; each pill contains 1 gr. of ferrous iodide. **f., iodide, syrup of** (*syrupus ferri iodidi*, U. S. P.). Dose 15–30 min. (1–2 Cc.). **f. sulphate** (*ferri sulphas*, U. S. P.). Dose 5 gr. (0.32 Gm.). **f. sulphate, dried** (*ferri*

FERROVIN 360 FEVER

*sulphas exsiccatus*, U. S. P.). Dose 3 gr. (0.2 Gm.). **f. sulphate, granulated** (*ferri sulphas granulatus*, U. S. P.). Dose 3 gr. (0.2 Gm.).
**ferrovin** (*fer'-o-vin*). A readily absorbable iron preparation, used in anemia.
**ferrozone** (*fer'-o-zōn*) [*ferrum*, iron; ὄζειν, to smell]. A material consisting in part of iron protosulphate; it is used as a precipitant for sewage.
**ferruginated** (*fer-u'-jin-a-ted*). Having the properties of iron.
**ferruginous** (*fer-u'-jin-us*) [*ferrum*]. 1. Chalybeate. 2. Having the color of iron-rust.
**ferrule** (*fer'-ūl*) [*ferrum*, iron]. A metallic hoop placed around a broken tooth; an instrument used with an attached lever in aligning irregularly placed teeth.
**ferrum** (*fer'-um*) [L.]. Iron, Fe = 55.84. Quantivalence II, IV. The most familiar and most useful of all metals; it is found in many minerals, in nearly all soils, in many mineral waters, and also occurs pure, especially in the form of meteoric iron. Pure iron is rare, nearly all commercial irons containing carbon in various proportions. In pharmacy, iron is used in the form of fine, bright, non-elastic wire, as reduced iron, a metallic iron with a variable amount of iron oxide, and in the form of salts. The therapeutic properties of iron depend on its power to build up the blood, it being a normal constituent of the red corpuscles; hence it is useful in all forms of anemia and in the diseases depending upon the latter. Externally many of the soluble salts of iron are used as styptic and astringent lotions. **f. dialysatum,** dialyzed iron. Dose of the solution 10–20 min. (0.65–1.3 Cc.). **f. reductum** (U. S. P.), reduced iron; iron by hydrogen; Quevenne's iron. Dose 3–6 gr. (0.2–0.4 Gm.). See also under *ferric, ferrous*, and *iron*.
**fersan** (*fer'-san*). A proprietary food-product made from the red corpuscles of beef-blood.
**fertile** (*fer'-til*) [*fertilis*, fruitful]. Prolific; fruitful.
**fertilization** (*fer-til-i-za'-shun*) [*fertile*]. The art of making fertile; impregnation.
**Ferula** (*fer'-ū-lah*) [L.]. A genus of the order *Umbelliferæ*. See *asafetida* and *galbanum*.
**fervor** (*fer'-vor*) [L., "heat"]. Fever-heat; it is defined as being more than calor and less than ardor.
**fessitude** (*fes'-it-ūd*) [*fessus*, weary]. A sensation of weariness.
**fester** (*fes'-tēr*). 1. Any small or superficial ulceration. 2. To suppurate.
**festination** (*fes-tin-a'-shun*) [*festinare*, to hasten]. A gait that increases in rapidity; it is seen in paralysis agitans.
**fetal** (*fe'-tal*) [*fetus*]. Pertaining to the fetus. **f. markings,** furrows and embryonic markings found in the adult kidney.
**fetalism** (*fe'-tal-ism*). The presence or persistence of certain fetal conditions in the body after birth.
**fetation** (*fe-ta'-shun*) [*fetus*]. 1. The formation of a fetus. 2. Pregnancy.
**feticide** (*fe'-tis-īd*) [*fetus; cædere*, to kill]. The killing of the fetus in the womb.
**fetid** (*fe'-tid*, or *fet'-id*) [*fetere*, to become putrid]. Having a foul odor. **f. stomatitis.** Synonym of *ulcerative stomatitis*.
**fetish, fetich** (*fe'-tish*). Any material object regarded with veneration or awe.
**fetishism, fetichism** (*fe'-tish-ism*) [Fr., *fetich*]. The association of lust with the idea of certain portions of the female person, or with certain articles of female attire.
**fetishist, fetichist** (*fe'-tish-ist*) [Fr., *fetich*]. An individual whose sexual interest is confined exclusively to parts of the female body, or to certain portions of the female attire.
**fetlock** (*fet'-lok*). A tuft of hair growing behind the pastern-joint of horses. **f.-joint,** the joint of a horse's leg next to the foot.
**fetlow** (*fet'-lo*). A kind of whitlow or felon, seen upon cattle.
**fetography** (*fe-tog'-raf-e*) [*fetus*; γράφειν, to write]. Skiagraphy of the fetus in utero; embryography.
**fetometry** (*fe-tom'-et-re*) [*fetus*; μέτρον, measure]. The measurement of the fetus, especially of its cranial diameters.
**fetor** (*fe'-tor*, or *fet'-or*) [L.]. Stench; offensive odor. **f. narium.** Synonym of *ozena*.
**fettmilch** of Gaertner. A preparation obtained by putting equal parts of milk and sterile water into the drum of a centrifuge, which is then revolved 4000 times a minute. The fat in the milk collects at the center, and may be drawn off with a tube inserted. The milk obtained should contain the same amount of fat as mother's milk, and by the addition of 35 Gm. of lactose to the liter a milk is produced which in composition resembles human milk very closely. This should be sterilized.
**fetus** (*fe'-tus*) [*fœtus*, offspring]. The unborn offspring of viviparous animals in the later stages of development. **f. amorphous.** See *anideus*. **f. in fetu** (fetus within the fetus), the name applied to those interesting inclusions in which the stronger fetus in its growth had included within its organism the parts of the weaker fetus. **f. papyraceus,** the name given to the malformation resulting in twin-pregnancy, when, owing to an inequality in the circulation of the embryos, the weaker fetus dies, and by continually increasing pressure of the growing fetus is flattened more and more against the uterine walls, until the mass has a thickness little greater than stout parchment.
**fever** (*fe'-ver*) [*febris*, a fever]. 1. An elevation of the body-temperature above the normal. 2. A disease the distinctive characteristic of which is elevation of temperature, accompanied also by quickened pulse and respirations, increased tissue-waste, and disordered secretions. **f., absorption-,** a fever often occurring during the first 12 hours after parturition. **f., African hemoglobinuric.** See **f.,** *blackwater*. **f., asthenic,** one in which there are a weak circulation, a clammy skin, and a low state of the nervous system. **f., bilious remittent.** 1. A term sometimes used as a synonym of *blackwater fever*. 2. Relapsing fever. **f., blackwater,** a disease of the tropics characterized by sudden onset, fever, chills, vomiting, and dyspnea. Syn., *bilious hematuric fever; bilious remittent fever; hematuric fever; hemoglobinuric fever*. **f.-blister.** See *herpes facialis*. **f., brain-,** fever associated with inflammation of the cerebral meninges; meningitis. **f., breakbone.** Synonym of *dengue*. **f., bubonic typhus,** typhus fever with inflammation, swelling, and suppuration of the inguinal, parotid, axillary, submaxillary, or mammary region. **f., catarrhal,** influenza. **f., catheter.** See **f.,** *urethral*. **f., Cavité,** an acute contagious disease confined almost exclusively to Cavité naval station in the Philippines. It is marked by sudden onset, high temperature, severe muscular pain, and extremely tender and painful eyeballs, the incubation period varying from two days to two weeks. **f., cerebrospinal,** an acute infectious disease characterized by inflammation of the meninges of the brain and cord with involvement of the superficial layers of nerve-substance. See *cerebrospinal meningitis* under *exanthemata, table of*. **f., Chagres,** a malignant form of malaria endemic on the isthmus of Panama. **f., childbed,** puerperal fever. **f., continued,** one the course of which is free from remissions or intermissions. **f., dandy.** Synonym of *dengue*. **f., dum-dum.** Same as *kala-azar; q. v.* **f., enteric.** See *typhoid fever*. **f., eruptive.** See *exanthematous*, one that is accompanied by an eruption on the skin. **f., estivoautumnal.** See **f.,** *remittent*. **f., famine.** See *relapsing fever*. **f., fatigue,** that following excessive muscular exercise. **f., fracture,** fever due to fracture of a bone. **f., Gaspard's putrid,** fever due to putrefaction of the intestinal contents. **f., gastric,** a term used indefinitely to indicate any febrile ailment associated with abdominal symptoms. **f. glandular,** an epidemic fever attacking children, marked by swelling of the cervical lymph-glands. **f., hay-.** See *hay-fever*. **f., hectic,** a diurnally intermittent fever with the highest temperature in the evening and accompanied by sweats and chills. It is found in tuberculosis and other diseases associated with the absorption of septic products. **f., hematuric bilious. f., hemoglobinuric.** See **f.,** *blackwater*. **f., hill-,** the pernicious malarial fever of the hill regions of India. **f., intermittent,** one in which the symptoms intermit, with intermediate periods of freedom from the febrile attacks. **f., littoral,** malarial fever in coast regions. **f., low,** fever of an asthenic type. **f., lung,** croupous pneumonia. **f., malarial.** See *malarial fever*. **f., malignant,** a severe and fatal form of any fever. **f., Malta.** See *f., Mediterranean*. **f., Manila,** a special type of pernicious fever occurring in the hot months (April and May) in Manila and sometimes becoming epidemic. **f., Mediterranean,** a specific febrile disease of the Mediterranean coast, characterized by long, irregular pyrexia, frequent relapses, rheumatic complications, constipation, with

no ulceration of Peyer's patches. The incubation period is from 6 to 9 days. Temperature may rise to 106° F.; in fatal cases to 110° F. It is due to *Micrococcus melitensis*, Bruce. **f., melanuric (remittent)**, blackwater fever. **f., metabolic,** a form of fever common in children during the summer, due to increased metabolism and increased tissue-waste clogging the system, owing to inability of the excretory organs to dispose of the waste rapidly enough. **f., milk-,** a slight form of puerperal septicemia, formerly thought to be due to the formation of milk in the mother's breast. **f., paratyphoid,** a condition clinically identical with typhoid fever, but due to a bacillus differing from *Bacillus typhosus* and *B. coli communis*. **f. post-typhoid,** a fever likely to occur directly after an attack of typhoid. **f., puerperal.** See *puerperal fever*. **f., purulent,** the pyrexia attending suppuration. **f., relapsing.** See *relapsing fever*. **f., remittent,** a paroxysmal fever with exacerbations and remissions, but not intermissions; usually applied specifically to remittent malarial fever, the type caused by the estivoautumnal malarial parasite. **f., rheumatic,** febrile symptoms developed in the course of acute rheumatism. **f., Roman,** a malignant malarial fever occurring in the Roman Campagna. **f., scarlet.** See *scarlatina*. **f., septic,** one due to the entrance of septic matter into the system. **f., simple continued,** a continued, non-contagious fever, varying in duration from 1 to 12 days, and usually ending in recovery. **f., spirillum.** Synonym of *relapsing fever*. **f., splenic.** Synonym of *anthrax*. **f., spotted.** 1. Synonym of *cerebrospinal meningitis*. 2. Synonym of *typhus fever*. 3. The local name, among the eastern foot-hills of the Bitter Root Mountains (western U. S.), for an endemic disease characterized by initial chill, constipation, fever, rapid pulse, enlarged spleen, muscular soreness, severe pain in head and back, and an eruption of macular spots, varying from bright red to purple or brownish-red in color. It is due apparently to a hematozoon to which the name *Piroplasma hominis* has been given. **f., sthenic,** a fever characterized by rapid, full pulse, heat and dryness of the skin, high temperature, scanty urine, and delirium. **f., surgical,** the pyrexia consequent upon a surgical operation. **f., Texas.** See *Texas fever*. **f., thermic.** Synonym of *heatstroke*. **f., tick.** 1. Texas fever. 2. Spotted fever (2). **f., traumatic,** that following traumatism. **f., tropical,** yellow fever. **f., typhoid.** See *typhoid fever*. **f., typhotyphus,** Pepper's name for mild typhus marked by laxity of the bowels, tympanites, epistaxis, and bronchial disturbance, the eruption occurring on the fourth day and the crisis from the tenth to the fourteenth day. **f., typhus.** See *typhus fever*. **f., urban,** a fever enduring about three weeks, and similar to mild typhoid except that specific symptoms are absent. **f., uremic,** one due to poisoning from urinary ptomaines; it has been observed after operations on the urinary tract and in urinary diseases. **f., urethral,** the febrile disturbance that follows the use of the catheter or bougie. **f., walking typhoid.** See under *typhoid fever*. **f., yellow.** See *yellow fever*.

**feverish** (*fe'-ver-ish*) [*febris*, fever]. Somewhat affected with fever; febrile.

**fexism** (*feks'-ism*) [Austrian]. A form of cretinism seen in Styria (Austria); its victims are locally called *fexi*.

**F. F. P. S.** Abbreviation of Fellow of the Faculty of Physicians and Surgeons (of Glasgow).

**fiat, fiant** (*fi'-at, fi'-ant*) [pres. subj., third person, sing. and pl., of *fieri*, to be made]. Let there be made.

**fiber, fibre** (*fi'-ber*) [*fibra*, a thread]. A filamentary or thread-like structure. **f.s, accelerating,** nerve-fibers which convey impulses that hasten the rapidity and increase the force of the heart-beat. **f.s, arciform, f.s, arcuate,** the shaped fibers on the anterior aspect of the oblongata. **f.s, association-,** white nerve-fibers situated just beneath the cortical substance and connecting the adjacent cerebral gyri. **f.s, augmented, f.s, augmentor.** See *f.s, accelerating*. **f.s, auxiliary.** See *f.s, secondary*. **f., axial.** 1. The axial band of a nerve-fiber. 2. The central spiral filament, probably contractile, of the flagellum of the spermatozoon. **f.-cell,** a cell elongated into a fiber. **f.s, collateral,** the delicate lateral branches of the nerve-process of a neuron; the paraxons. **f.s, commissural,** fibers joining an area of the cortex of one hemisphere to a similar area of the other. **f.,** elastic. See *tissue, yellow elastic*. **f. of Gerdy.** See under *Gerdy*. **f.s, involuntary muscular,** straight or slightly bent, elongated, spindle-shaped, nucleated cells, bearing more or less distinct longitudinal striations, which make up involuntary or unstriped muscles. Syn., *nonstriated fibers; unstriated fibers; unstriped fibers*. See *muscular tissue*. **f., muscle-.** See *muscle-fiber*. **f., nerve-.** See *nerve-fiber*. **f.s, osteogenic.** See *Sharpey's intercrossing fibers*. **f.s, projection,** fibers joining the cerebral cortex to lower centers and *vice versa*. **f.s of Remak,** the nonmedullated nerve-fibers. **f.s, rivet-,** protoplasmic processes on the basal surface of the columnar cells of stratified squamous epithelium. **f.s, secondary,** in a fibrous structure, those of secondary importance. Syn., *auxiliary fibers*. **f.s of Sharpey.** See under *Sharpey*. **f., smooth muscular,** a muscular fiber-cell. **f.s, spindle-,** achromatic fibrils. **f., spiral,** the coiled fiber peculiar to spiral fiber-cells. See under *cell, spiral fiber-*. **f., straight,** the coiled fiber in a bipolar ganglion-cell. See under *cell, spiral fiber-*. **f.s, sustentacular,** a supporting connective tissue that unites the various layers of the retina. **f., sweat-,** a nervous fibril which on stimulation produces sweating. **f.s, sympathetic,** those of the sympathetic nerve. **f., t-,** a branch given off at right angles to the axis-cylinder of a nerve-cell. **f.s of Tomes.** See *Tomes, fibers of*.

**fibra** (*fi'-brah*) [L.: pl., *fibrae*]. Same as fiber.
**fibralbumin** (*fi-bral-bū'-min*). Globulin.
**fibration** (*fi-bra'-shun*) [*fibra*, fiber]. Fibrous construction, arrangement of fibers.
**fibremia, fibraemia** (*fi-bre'-me-ah*) [*fibra*, fiber; αἷμα, blood]. The presence of fibrin in the blood.
**fibriform** (*fib'-ri-form*) [*fiber; forma*, shape]. Shaped like a fiber.
**fibril** (*fi'-bril*) [*fiber*]. 1. A small fiber or component filament of a fiber. 2. A name applied to minute nerve filaments. 3. The subdivision of a muscular fiber. **f.s, achromatic,** fibrils of achromatic, nuclear, or cell-substance forming lines which extend from pole to pole in a dividing nucleus so as to form a barrel- or barrel-shaped figure. **f.s, chromatic, f.s, nuclear,** the thread-like fibrils consisting of the chromatin in a cell-nucleus.
**fibrillar** (*fi'-bril-ar*) [*fibril*]. Pertaining to fibrils. **f. contractions,** spontaneous contractions successively taking place in different bundles of muscular fibers; they are seen in progressive muscular atrophy and other diseases.
**fibrillary** (*fi'-bril-a-re*). Same as *fibrillar*.
**fibrillation** (*fi-bril-a'-shun*) [*fiber*]. 1. The formation of fibrils. 2. A localized quivering of muscular fibers.
**fibrin** (*fi'-brin*) [*fiber*]. A protein formed in shed blood, lymph, in other body-fluids, and in tissues when these coagulate. It exists in the shape of fibrils, granules, plates, or as a homogeneous material. Fibrin forms about 0.2 % of the blood. **f.-factors,** the substances necessary for and concerned in the formation of fibrin. They are fibrinogen, fibrin-ferment, and certain salts. **f.-ferment,** a ferment obtained from blood-serum after clotting has occurred. It is one of the fibrin-factors, and is probably derived from the leukocytes. **f.-globulin,** Hammarsten's name for a globulin-like substance which coagulates at about +64° C., in blood-serum, and in the serum from coagulated fibrinogen solutions. **f. of Henle.** See under *Henle*. **f., vegetable.** See *casein, vegetable*.
**fibrination** (*fi-brin-a'-shun*) [*fibra*, a fiber]. The acquirement of an abnormal amount of fibrin.
**fibrinemia, fibrinaemia** (*fi-brin-e'-me-ah*) [*fibrin*; αἷμα, blood]. Same as *fibremia*.
**fibrino-** (*fi-brin-o-*) [*fibrin*]. A prefix meaning relating to fibrin.
**fibrinogen** (*fi-brin'-o-jen*) [*fibrino-*; γεννᾶν, to produce]. A protein of the globulin class, obtained from blood-plasma and serous transudations. It is one of the chief elements in the formation of fibrin.
**fibrinogenic** (*fi-brin-o-jen'-ik*) [*fibra*, a fiber; γεννᾶν, to produce]. Of the nature of fibrinogen.
**fibrinogenous** (*fi-brin-oj'-en-us*) [see *fibrinogen*]. Forming or producing fibrin.
**fibrinoglobulin** (*fi-brin-o-glob'-ū-lin*). See *fibrin-globulin*.
**fibrinolysis** (*fi-brin-ol'-is-is*) [*fibrino-*; λύειν, to loose]. The partial dissolution which takes place in fibrin if allowed to stand in contact with the blood from which it was formed.

**fibrinoplastic** (fi-brin-o-plas'-tik) [fibrino-; πλάσσειν, to form]. Of the nature of fibrinoplastin.
**fibrinoplastin** (fi-brin-o-plas'-tin). See *paraglobulin*.
**fibrinoscopy** (fi-brin-os'-ko-pe) [fibrin; σκοπεῖν, to view]. Examination of fibrin of blood-clot, etc.; See *inoscopy*.
**fibrinosis** (fi-brin-o'-sis). A condition marked by excess of fibrin in the blood.
**fibrinous** (fi'-brin-us) [fibrin]. Of the nature of or containing fibrin.
**fibrinuria** (fi-brin-ū'-re-ah) [fibrin; οὖρον, urine]. The passage of urine containing fibrin.
**fibro-** (fi-bro-) [fiber]. A prefix signifying relation to fibers or to fibrous tissue.
**fibroadenoma** (fi-bro-ad-en-o'-mah). Adenoma having fibrous tissue.
**fibroareolar** (fi-bro-ar-e'-o-lar) [fibro-; areola]. Containing fibrous tissue with an areolar arrangement.
**fibroblast** (fi'-bro-blast) [fibro-; βλαστός, a germ]. A cell that forms new fibrous tissue.
**fibroblastic** (fi-bro-blas'-tik) [fibro-]. Pertaining to fibroblasts. 2. Fibroplastic.
**fibrobronchitis** (fi-bro-brong-ki'-tis) [fibra, fiber; bronchitis]. Bronchitis with the expectoration of fibrinous casts.
**fibrocalcareous** (fi-bro-kal-ka'-re-us). Applied to fibrous tumors which have undergone calcareous degeneration.
**fibrocarcinoma** (fi-bro-kar-sin-o'-mah) [fibro-; carcinoma; pl., fibrocarcinomata]. A carcinoma with fibrous elements.
**fibrocartilage** (fi-bro-kar'-til-āj) [fibro-; cartilage]. Cartilage with an intermixture of fibrous elements.
**fibrocartilaginous** (fi-bro-kar-til-aj'-in-us) [fibro-; cartilago, gristle]. Composed of or containing fibrocartilage.
**fibrocellular** (fi-bro-sel'-ū-lar) [fibro-; cellular]. Both fibrous and cellular; fibroareolar.
**fibrochondritis** (fi-bro-kon-dri'-tis) [fibro-; χόνδρος, cartilage; ιτις, inflammation]. Inflammation of fibrocartilage.
**fibroconnective** (fi-bro-kon-ek'-tiv). Having a fibrous structure and the function of connecting.
**fibrocyst** (fi'-bro-sist) [fibro-; κύστις, a cyst]. A fibroma that has undergone cystic degeneration.
**fibrocystic** (fi-bro-sist'-ik). Fibrous and having undergone cystic degeneration.
**fibrocystoid** (fi-bro-sist'-oid). Having the structure of a fibrocyst.
**fibrocystoma** (fi-bro-sist-o'-mah). Fibroma combined with cystoma.
**fibrocyte** (fi'-bro-sit) [fibro-; κύτος, cell]. A fibrous tissue cell.
**fibroelastic** (fi-bro-e-las'-tik) [fibro-; elastic]. Consisting partly of fibrous elastic tissue.
**fibroenchondroma** (fi-bro-en-kon-dro'-mah) [fibro-; enchondroma; pl., fibroenchondromata]. An enchondroma containing fibrous elements.
**fibrofatty** (fi-bro-fat'-e). Consisting of fibrous tissue and fat-corpuscles.
**fibrogen** (fi'-bro-jen). See *fibrinogen*.
**fibroglia** (fi-bro'-gle-ah) [fibro-; γλία, glue]. The supporting structure of connective tissue; it is analogous to the neuroglia of the nervous system.
**fibroglioma** (fi-bro-gli-o'-mah) [fibro-; glioma]. A tumor having the elements of a fibroma and a glioma.
**fibroid** (fi'-broid) [fiber; εἶδος, likeness]. Resembling fibers or composed of fibers; also, a fibroid tumor. **f. degeneration,** transformation of membranous tissue into fiber-like material. **f. heart,** a chronic form of myocarditis in which there is a development of fibrous connective tissue in the cardiac muscle. **f. induration.** See *induration, fibroid*. **f. phthisis,** chronic phthisis in which there is a formation of fibrous tissue. **f. tumor,** a fibroma.
**fibroidectomy** (fi-broid-ek'-to-me) [fibroid; ἐκτομή, excision]. Excision of a fibroid tumor.
**fibroin** (fi'-bro-in) [fibra, fiber], C₁₅H₂₅N₅O₆. An albuminoid; a white, shining substance, the chief constituent of the cocoons of insects and spider-web.
**fibrolaminar** (fi-bro-lam'-in-ar). Relating to a fibrous layer.
**fibrolipoma** (fi-bro-lip-o'-mah) [fibro-; lipoma]. A tumor of fibrous and fatty tissue.
**fibrolysin** (fi-brol'-is-in) [fibro-; λύσις, solution]. Trade name of a solution of thiosinamine sodium salicylate; used in the treatment of keloids, or excessive new connective tissue formation.
**fibroma** (fi-bro'-mah) [fibro-; ὄμα, a tumor]. A benign tumor composed of fibrous tissue. **f., hard,** one containing few cells, being chiefly composed of fibers. **f. lipomatodes.** Same as *xanthoma*. **f. molluscum.** Synonym of *molluscum fibrosum*. **f., soft,** one rich in cells.
**fibromatoid** (fi-bro'-mat-oid) [fibroma; εἶδος, form]. Resembling a fibroma.
**fibromatosis** (fi-bro-mat-o'-sis). See *fibrosis*.
**fibromatous** (fi-bro'-mat-us). Relating to a fibroma.
**fibromucous** (fi-bro-mū'-kus). Consisting partly of mucosa and partly of fibrous tissue.
**fibromuscular** (fi-bro-mus'-kū-lar). Made up of connective tissue and muscle.
**fibromyitis** (fi-bro-mi-i'-tis) [fibra, fiber; μῦς, muscle; ιτις, inflammation]. Inflammation of a muscle, leading to its fibrous degeneration.
**fibromyoma** (fi-bro-mi-o'-mah) [fibro-; myoma.] A tumor composed of fibrous and muscular tissue.
**fibromyomotomy** (fi-bro-mi-o-mot'-o-me) [fibromyoma; τέμνειν, to cut]. The surgical removal of a fibromyoma.
**fibromyxoma** (fi-bro-miks-o'-mah) [fibro-; myxoma]. A tumor composed of fibrous and myxomatous tissue.
**fibromyxosarcoma** (fi-bro-miks-o-sar-ko'-mah). 1. A tumor containing sarcomatous and myxoid tissue. 2. A fasciculated sarcoma which has undergone myxoid degeneration.
**fibroneuroma** (fi-bro-nū-ro'-mah) [fibro-; neuroma]. A tumor composed of fibrous tissue and nerve-fibers.
**fibronuclear, fibronucleated** (fi-bro-nū'-kle-ar, -nū'-kle-a-ted). Relating to tissue which shows many nuclei and fibers.
**fibropericarditis** (fi-bro-per-e-kar-di'-tis). Fibrinous pericarditis.
**fibroplastic** (fi-bro-plas'-tik) [fibro-; πλάσσειν, to form]. Tending to form fibers. **f. tumor,** small spindle-celled sarcoma.
**fibroplastin** (fi-bro-plas'-tin). Same as *paraglobulin*.
**fibropolypus** (fi-bro-pol'-ip-us). A fibroid polypus.
**fibropsammoma** (fi-bro-sam-o'-mah). A tumor consisting of fibromatous and psammomatous elements.
**fibropurulent** (fi-bro-pū'-roo-lent). Consisting of pus containing flakes of fibrin.
**fibroreticulate** (fi-bro-re-tik'-ū-lāt). Consisting of a fibrous network or marked with interlacing fibers.
**fibrosarcoma** (fi-bro-sar-ko'-mah) [fibro-; sarcoma]. A sarcoma containing fibrous tissue. **f., mucocellular** (of the ovary), a form marked by a layer of large, round, bladdery cells lying between the fibrils of the connective tissue.
**fibroserous** (fi-bro-se'-rus) [fibro-; serous]. Having the qualities of a fibrous and serous structure. **f.-s. membranes,** thin, transparent, glistening structures forming closed sacs, that contain certain organs. They are the peritoneum, the two pleuræ, the pericardium, the tunica vaginalis testis, the arachnoid, and synovial membranes.
**fibrosis** (fi-bro'-sis) [fiber]. The development of fibrous tissue. **f., arteriocapillary,** arteriosclerosis; a primary and general fibroid degeneration of the arterioles and capillaries developing about middle life; the caliber of the vessels becomes diminished and they lose their elasticity; there is atrophy of the adjacent tissue, especially in the kidneys, together with cardiac hypertrophy.
**fibrositis** (fi-bro-si'-tis) [fibro-; ιτις, inflammation]. Inflammatory hyperplasia of the white fibrous tissue such as occurs in chronic rheumatism.
**fibrotic** (fi-brot'-ik). Pertaining to fibrosis.
**fibrous** (fi'-brus) [fiber]. Containing fibers; of the character of fibrous tissue. **f. tissue,** the connective tissue of the body.
**fibula** (fib'-ū-lah) [L., "a buckle"]. 1, The slender bone at the outer part of the leg, articulating above, with the tibia and below with the astragalus and tibia. Syn., *perone*. 2. A clasp serving to unite the edges of a wound or the opening of a canal.
**fibular** (fib'-ū-lar) [fibula]. 1. Relating to the fibula. 2. Relating to the outer border of the leg.
**fibulation** (fib-ū-la'-shun). See *infibulation*.
**fibulen** (fib'-ū-len) [fibula]. Belonging to the fibula in fetalt.
**fibulocalcaneal** (fib-ū-lo-kal-ka'-ne-al) [fibula; calcaneum]. Pertaining to or connecting the fibula and the calcaneum.
**F. I. C.** Abbreviation for Fellow of the Institute of Chemistry.
**ficarin** (fik'-ar-in) [ficus, a fig]. A neutral principle

# FICARY 363 FILOVARICOSIS

obtained from the common ficary; it is used internally and externally for piles.

**ficary** (*ĭk'-ar-e*) [*ficaria; ficus*, a fig; a hemorrhoid]. The *Ranunculus ficaria*, pilewort; a common European herb; long a popular remedy for piles.

**ficiform** (*fĭs'-e-form*) [*ficus; forma*, form]. Fig-shaped.

**Ficker's diagnosticum** (*fĭck'-erz-di-ag-nos'-tik-um*) [Philip Martin *Ficker*, German bacteriologist, 1868– ]. An emulsion of dead typhoid bacillus culture, used in the Widal-Gruber test.

**ficosis** (*fĭ-ko'-sis*). See *sycosis*.

**ficus** (*fĭ'-kus*) [L., "a fig-tree"]. 1. The fig. The *ficus* of the U. S. P. is the partially dried fruit of *F. carica*, native of Asia Minor, and cultivated throughout Europe and tropical America. It is laxative and nutritious, and is a constituent of confectio senfiæ. 2. Old name for a hemorrhoidal or condylomatous tumor.

**fidgets** (*fĭj'-ets*). Uneasiness; restlessness; dysphoria, *q. v.*

**fidicinales** (*fĭ-dis-in-a'-lēs*) [*fidicen*, a player on a stringed instrument]. The lumbrical muscles of the hand.

**field** (*fēld*) [ME., *feeld*]. An open space or area. f. of audition, f., auditory, the area surrounding the ear, in every portion of which a given sound is audible to the ear. f. of fixation. See *fixation, field of*. f.s of innervation, special expansions in which the motor nerves to the voluntary muscles terminate. f., magnetic, the portion of space about a magnet in which its action is felt. f. of a microscope, the area that can be seen through a microscope at one time. f. of vision, the space in which the patient can see when the eye is fixed steadily on the object held in the direct line of vision.

**Fieux's test for antipyrine** (*fe'-u*). Add 2.5 Gm. of sodium metaphosphate and 12 drops of sulphuric acid to the suspected fluid, filter, and to the clear filtrate add a few drops of sodium nitrate. If antipyrine is present, a clear green color will develop.

**fifth disease**, erythema infectiosum. **f. nerve.** See *trifacial nerve.* **f. ventricle.** See *ventricle, fifth.* **fig.** See *ficus.* **f.-wart,** a moist condyloma.

**figurate** (*fĭg'-u-rāt*). Having a fixed and definite shape; arranged in a definite shape: said of skin eruptions.

**figure** (*fĭg'-ūr*) [*figura*, a form]. The visible form of anything; the outline of an organ or part. f., achromatic (spindle), f., achromatin, a fuchsin figure assumed by the achromatic fibrils in a dividing cell. f.s, adhesion, Rindfleisch's term for the pattern produced in living protoplasm by the adhesion of the two interpenetrating substances, the reticular framework and the intervening matrix. f., bistellate. See *amphiaster.* f., chromatic (nuclear), f., chromatin, one of the figures formed by the chromatic fibrils of the nucleus during karyokinesis. f., nuclear (spindle or division). 1. Flemming's name for any one of the forms assumed by the nucleus during karyokinesis. 2. Strassburger's name for the spindle stage of karyokinesis.

**figwort** (*fĭg'-wert*). The herb *Scrophularia nodosa*, an alterative, diuretic, and anodyne. It is sometimes used in the form of an ointment for piles. Dose of the *fluidextract* ½–1 dr. (2–4 Cc.).

**fila** (*fĭ'-lah*) [L.]. Plural of *filum*, *q. v.* **f. lateralia pontis**, a strand of fibers at the upper border of the pons. Also called *tenia pontis*.

**filaceous** (*fĭ-la'-she-us*) [*filum*, a thread]. Consisting of threads or thread-like fibers or parts.

**filament** (*fĭl'-a-ment*) [*filum*]. A small, thread-like structure. f., spermatic, the caudal filament of a spermatozoon.

**filamentation** (*fĭl-a-men-ta'-shun*) [L., *filum*, a thread]. Thread formation. A peculiar reaction produced in certain bacteria (*Bacillus coli communis, Proteus*, etc.) when they are brought in contact with blood-serum, and consisting in the formation of long interlacing threads. The reaction is best obtained when the bacteria are suspended in serum derived from the same individual from whose body the bacteria were obtained (so-called "homologous" serum).

**filamentous** (*fĭl-a-ment'-us*) [*filament*]. 1. Like a thread, or made up of threads or filaments. 2. Capable of being drawn out into filaments, like mucus. 3. Containing a stringy substance, as *filamentous urine*.

**filar**(*fĭ'-lar*) [*filum*, a thread]. Filamentous.

**Filaria** (*fĭl-a'-re-ah*) [*filum*, a thread]. A genus of nematode or threadworms, of the family *Filariidæ*. **F. medinensis**, an animal parasite, the female of which works its way from the intestinal tract to the subcutaneous tissue, where, after developing its embryos, it is sooner or later set free by abscess-formation and discharge. Syn., *Guinea-worm.* **F. sanguinis hominis**, the female adult worm was discovered by Bancroft of Brisbane; the male by Aranjo, and the embryo by Demarquay and Lewis. The embryos are about 0.35 mm. long, and inhabit the lymph-channels of the lower extremities and the scrotum. They lead to dilatation of the lymphatics, to hyperplasia of the tissues, to chyluria, hematuria, abscesses, etc. They are found in the blood at night. Elephantiasis arabum and lymph-scrotum are due to the filaria.

**filarial** (*fĭl-a'-re-al*). Relating to the genus *filaria*.

**filariasis** (*fĭl-ar-i'-as-is*) [*filaria*]. A diseased state due to the presence in the body of *Filaria sanguinis hominis* or allied species.

**Filatow's disease** (*fe'-lat-off*) [Nil *Filatow*, Russian physician, 1847– ]. 1. Acute febrile cervical adenitis of children, probably identical with Pfeiffer's glandular fever. 2. Fourth disease, *q. v.* **F.'s spots**, *Koplik's spots, q. v.*

**file** (*fīl*). See *raspatory, xyster*. **f.-cutter's disease**, a form of pneumonokoniosis. **f., dental,** a tooth-file; an instrument for the removal of a portion of one or more teeth, and for the separation of teeth.

**filices** (*fĭl'-is-ēz*) [*pl.* of *filix*, a fern]. Ferns. See *filix*.

**filicic acid** (*fĭl-is'-ik*). An acid, $C_{14}H_{16}O_5$, extracted from *Aspidium filix-mas*.

**filicin** (*fĭl'-is-in*) [*filix*, a fern]. 1. A yellowish white, sticky, odorless powder extracted from the root of *Dryopteris filix mas*. 2. Filicic anhydride.

**filicism** (*fĭl'-is-izm*). Poisoning from overdosage of extract of male-fern.

**filiform** (*fĭl'-if-orm*) [*filum*, a thread; *forma*, form]. Thread-like. **f. bougie.** See *bougie, filiform.* **f. papillæ**, the smallest and most numerous of the papillæ of the tongue, occurring over its whole surface. **f., pulse,** a small, thready, almost imperceptible pulse.

**Filipowicz's sign** (*fĭl-ip'-o-vitch*) [Casimir *Filipowicz*, Polish physician]. A yellowish discoloration of the prominent portions of the palmar and plantar surfaces, seen in typhoid fever. Syn., *Palmoplantar phenomenon.*

**filipuncture** (*fĭl-e-punk'-chūr*) [*filum*, a thread; *punctura*, a puncture]. A method of treating aneurysm by inserting wire threads, hair, or the like to promote coagulation.

**filix** (*fĭ'-liks*) [L.: *pl., filices*]. A general name for any fern. **f. femina**, or **feminea**, the fern now called *asplenium filix femina*, female fern or spleenwort. See *asplenium.* **f. mas,** male fern. See *aspidium.*

**fillet** (*fĭl'-et*) [Fr., *filet*, a thread]. 1. A loop for the purpose of making traction on the fetus. 2. The lemniscus, a band of nerve-fibers connected below with the nucleus gracilis and nucleus caudatus of the medulla and running upward through the pons and crus cerebri to the cerebrum, a portion of the fibers (*lateral fillet*) entering the posterior corpora quadrigemina, another (*mesal fillet*) passing to the anterior corpora quadrigemina and the optic thalamus. A part of the mesal fillet is continued into the cortex. **f. of the corpus callosum.** See *fascicle, fornicate.* **f., olivary,** a fasciculus of nerve-fibers inclosing the olivary body of the medulla.

**filling** (*fĭl'-ing*). The material used in closing cavities in carious teeth.

**film.** A pellicle or thin skin; an opacity of the cornea.

**filmaron** (*fĭl'-ma-ron*). A proprietary anthelmintic from aspidium. **f. oil,** ten per cent. solution of filmaron in castor oil.

**filmozen** (*fĭl'-mo-ien*). A protective vehicle for applying medicaments in skin diseases, consisting of pyroxylin dissolved in acetone with a small quantity of castor-oil.

**filopodium** (*fĭ-lo-po'-de-um*) [*filum*, thread; *πούς*, foot; *pl., filopodia*]. A slender, thread-like pseudopodium.

**filopressure** (*fĭ-lo-presh'-ur*) [*filum*, a thread *pressure*]. Compression of a vessel by means of a wire or a thread.

**filovaricosis** (*fĭ-lo-var-ik-o'-sis*) [*filum*, a thread;

FILTER 364 FISTULA

*varix*, a dilated vein]. A varicosity of the axiscylinder of a nerve-fiber, or the formation of one.
**filter** (*fil'-ter*) [*filtrum*]. An apparatus for straining water or other liquids to remove any undissolved matters. **f.-paper**, an unglazed paper used for filtration. **f., Pasteur-Chamberland.** See under *Pasteur-Chamberland.*
**filth.** Foul, offensive matter. **f.-disease**, any disease due to filth. **f.-dread.** See *mysophobia* and *rupophobia.*
**filtrate** (*fil'-trāt*) [*filter*]. The liquid that has passed through a filter.
**filtration** (*fil-tra'-shun*) [*filter*]. The operation of straining through a filter.
**filtrum** (*fil'-trum*) [L., "felt"; pl., *filtra*]. 1. Felt. 2. A filter or strainer. **filtra ventriculi**, small vertical channels on the back of the larynx between Morgagni's cartilage and the inner edge of the arytenoid cartilage. They end between the vocal bands at the dorsal end of Morgagni's ventricle.
**filum** (*fi'-lum*) [L.]. Any thread-like or filamentous structure; in surgery, a thread or wire. **f. coronarium**, a thread-like ridge at the side of the auriculo-ventricular opening. **f. terminale**, a long slender thread of pia mater, the termination of the spinal cord.
**fimbria** (*fim'-bre-ah*) [L.]. A fringe. **f. cornu Ammonis**, the fimbria hippocampi. **f. of Fallopian tube**, the fringelike process of the outer extremity of the oviduct. **f. hippocampi**, a white band at the bottom of the hippocampal fissure. **f. ovarica**, the longest of the fimbriæ of the Fallopian tube.
**fimbrial** (*fim'-bre-al*) [*fimbria*, a thread]. Relating to the fimbria or to fimbriæ.
**fimbriate** (*fim'-bre-āt*) [*fimbria*]. Fringed with slender processes which are larger than filaments; said of bacterial cultures.
**fimbriated** (*fim'-bre-a-ted*) [*fimbria*]. Fringed.
**f. body**, the corpus fimbriatum.
**fimbriatum** (*fim-bre-a'-tum*) [*fimbria*]. The corpus fimbriatum.
**fimbriocele** (*fim'-bre-o-sēl*) [*fimbria*, a thread; κήλη, hernia]. Hernia enclosing some or all of the fimbriæ of an oviduct.
**fine** (*fīn*). Opposed to coarse. **f. adjustment.** See *adjustment.*
**finger** (*fing'-ger*) [ME.]. A digit of the hand. **f., clubbed**, a finger with the terminal phalanx of which is short and broad, with overhanging nail. It is seen in cases of pulmonary tuberculosis, congenital heart disease, etc. **f.-cot**, a covering of rubber or other material to protect the finger or to prevent infection. **f., mallet**, a deformity marked by undue flexion of the last phalanx. **f., Morse**, an affection resulting from operating the Morse telegraph key. **f.-stall**, a rubber cap for a finger.
**Finney's operation** (*fin'-e*) [John Miller Turpin *Finney*, American surgeon, 1863– ]. A method of performing of gastroduodenostomy.
**Finsen light** (*fin'-sen*) [Niels Ryberg *Finsen*, Danish physician, 1860–1904]. Light from which the heat rays are excluded and only the blue and violet rays remain; it is used in phototherapy. **F.-l. treatment**, a method of treatment by exposure of the diseased part to the violet and ultraviolet rays of the sun or of the electric arc light.
**fir** (*fur*). See *abies.* **f., balsam-.** See *abies balsamea.*
**fire** (*fīr*). 1. The visible heat of burning bodies. 2. A popular name for inflammation affecting the skin. **f.-damp**, the gas contained in coal (marsh gas), often given off in large quantities, and exploding, on ignition, when mixed with atmospheric air. **f. measles.** Synonym of rotheln. **f., St. Anthony's.**, wild, erysipelas.
**first cranial nerve**, the olfactory nerve.
**first intention.** The healing of the lips of a wound by immediate union without suppuration.
**Fischer's test-meal.** This consists of the bread and water of the Ewald breakfast, and in addition a quarter of a pound of finely chopped lean beef broiled and slightly seasoned. It is to be removed from the stomach in three hours.
**fishberry.** See *cocculus indicus.*
**Fisher's brain-murmur.** A systolic murmur heard over the anterior fontanel or in the temporal region in rhachitic infants. **F.'s sign**, a presystolic murmur heard in cases of adherent pericardium without valvular disease.
**fish-skin disease.** See *ichthyosis.*

**Fiske-Bryson's symptom.** See *Bryson's sign.*
**fission** (*fish'-un*) [*fissus; findere*, to cleave]. Reproduction by splitting into two or more equal parts.
**fissipara** (*fis-ip'-ar-ah*) [*fissus; findere*, to cleave; *parere*, to produce]. In biology, applied in a general way to all organisms that multiply by spontaneous self-division.
**fissiparism, fissiparity** (*fis-ip'-ar-ism, fis-ip-ar'-it-e*) [*findere*, to cleave; *parere*, to produce]. Propagation by fission; fissiparous generation.
**fissiparous** (*fis-ip'-ar-us*) [*fission; parere*, to produce]. Propagating by fission.
**fissura** (*fish-u'-rah*) [L.]. A fissure.
**fissural** (*fish'-u-ral*). Pertaining to a fissure.
**fissuration** (*fish-ur-a'-shun*) [*findere*, to split]. Same as *fission.* Also applied to the arrangement of the fissures of various organs, such as the brain.
**fissure** (*fish'-ūr*) [*fissura*]. A groove or cleft. A term applied to the clefts or grooves in various organs, as the skull, the brain, the liver, the spinal cord; also to cracks in the skin or linear ulcers in mucous membranes. **f. anal**, a linear ulcer at the mucocutaneous junction of the anus, giving rise to intense suffering on defecation. **f., auricular**, one in the petrous bone. **f. of Bichat.** See under *Bichat.*
**f. of Broca.** See under *Broca.* **f., calcarine**, one on the mesal aspect of the cerebrum, between the lingual lobule and the cuneate lobe. **f., callosomarginal**, one on the surface of the cerebral hemisphere, dividing the area between the corpus callosum and the margin into nearly equal parts. **f., central.** See *Rolando, f. of.* **f., collateral**, one on the mesal aspect of the cerebrum, between the subcalcarine and subcollateral gyri. It is collocated with the collateral eminence. **f., dentate**, the hippocampal fissure. **f. of the gall-bladder.** See *fossa cystica.*
**f., hippocampal.** See *hippocampal.* **f., interlobular**, **f., longitudinal**, the deep fissure that divides the cerebrum into two hemispheres. **f., occipital**, a deep fissure situated between the parietal and occipital lobes of the brain. **f., palpebral**, the space between the eyelids extending from the outer to the inner canthus. **f., portal.** See *f., transverse.* **f., posterior median** (of spinal cord), a deep, narrow groove extending the whole length of the spinal cord, in the middle line posteriorly. **f., precentral**, in front of the fissure of Rolando and parallel to it. **f., presylvian**, the anterior branch of the fissure of Sylvius. **f., primary** (of His), a fold extending along the mesal line of the hemisphere, producing an external groove and an internal ridge. It begins at the olfactory lobe, which it divides into a ventral and a dorsal part, and, continuing backward in a curved direction, joins the hippocampal sulcus. **f. of Rolando.** See under *Rolando.* **f., semilunar.** See *f., calcarine.* **f., sphenoidal**, a cleft between the great and small wings of the sphenoid bone. **f., sphenomaxillary**, one between the lateral margin of the superior maxilla and the orbital plate of the sphenoid bone. **f. of Sylvius.** See under *Sylvius.* **f., transverse** (of liver), a fissure crossing transversely the lower surface of the right lobe of the liver. It transmits the portal vein, hepatic artery and nerves, and hepatic duct. **f., umbilical**, the anterior portion of the longitudinal fissure of the liver.
**fistula** (*fis'-tu-lah*) [L., "a pipe"]. A narrow canal or tube left by the incomplete healing of abscesses or wounds, and usually transmitting some fluid—either pus or the secretions or contents of some organ or body cavity. **f., abdominal**, one in the abdominal wall communicating with some of the abdominal viscera. **f., aerial**, a small opening in the neck communicating with the larynx, following imperfect closure of incised wounds of the throat. The voice is defective in consequence. **f., alveolar**, one due to necrosis of an alveolus. **f., anal**, a fistula in the neighborhood of the anus, which may or may not communicate with the bowel. **f. ani congenita**, an anomaly of the anus the derivation of which is attributed to the remains of the posterior part of the blastopore. **f., anoperineal**, an anal fistula opening on the perineum. **f., biliary**, an abnormal channel of communication with a biliary duct of the gallbladder. **f., bimucous**, one making a communication between two mucous surfaces. **f., blind**, a fistula open at one end only. **f., blind, external**, one the only opening of which is on the exterior of the body. **f., blind, internal**, one which opens only upon an internal surface. **f., branchial**, an opening that extends from the surface of the neck to the pharynx;

it is an unclosed branchial cleft. **f.**, **cicatricial**, one lined with a cicatricial membrane. **f.**, **coccygeal**. See *pilonidal fistula*. **f.**, **complete**, one having two openings—an internal and an external. **f.**, **cysticocolic**, one leading from the gall-bladder to the colon. **f. Eck's**. See under *Eck*. **f.**, **fecal**, a fistula communicating with the intestine. **f. gastric**, an opening into the stomach, generally artificial, through the abdominal wall. It is sometimes used for feeding a patient who cannot swallow. **f.**, **horseshoe**, a variety of fistula in ano, the external opening being on one side of the anus and the internal opening on the other. **f.**, **labiform**, one characterized by liplike protrusions at the outer margin. **f.**, **lacteal**. See *f.*, *mammary*. **f.**, **mammary**, or **milk**, a fistula of the mamma or of its ducts. **f.**, **ostial**. See *f.*, *labiform*. **f.**, **rectovesicovaginal**, a double fistula giving rise to communication between the rectum, the vagina and the urinary bladder. **f. sacra**, the Sylvian aqueduct. **f.**, **sacral**, a congenital fistula occurring in the lumbosacral region. **f.**, **vesical**, a fistula of the urinary bladder. **f: vesicovaginal**, an opening from the bladder into the vagina.
**fissured** (*fish'-urd*) [*fissus; findere*, to cleave]. Cleft; split.
**fist**. 1. The firmly-closed hand. 2. Same as *bovista*.
**fistular, fistulate** (*fis'-tū-lar, -lāt*) [*fistula*]. Fistulous; of the form or nature of a fistula.
**fistulatome** (*fis'-tū-lat-ōm*) [*fistula;* τέμνειν, to cut]. A cutting-instrument used in the operative treatment of fistula.
**fistulization** (*fis-tū-li-za'-shun*). The act or process of becoming fistulous.
**fistuloenterostomy** (*fis'-tū-lo-en-ter-os'-to-me*). The operation of making a biliary fistula open permanently into the small intestine.
**fistulous** (*fis'-tū-lus*) [*fistula*]. Of the nature of or affected with a fistula.
**fit** [AS., *fitt*, a struggle]. A term applied to any sudden paroxysm of a disease, but especially to an epileptic convulsion.
**Fitz's syndrome** [Reginald Heber *Fitz*, American physician, 1843–1913]. Intense pain in the epigastric region, with vomiting and collapse, all of sudden onset, [and followed by tympanites; diagnostic of acute pancreatitis.
**fixateur** (*fēks-āt-er*). An ambocepter.
**fixation** (*fiks-a'-shun*) [*fixus*, fixed]. 1. The act of fixing or making firm. 2. The operation of rendering fixed, by means of sutures, a displaced or floating organ. **f. of the complement**. See under *complement*. **f., field of**, in optics, the region bounded by the utmost limits of distinct or central vision, and which the eye has under its direct control through its excursions, without movements of the head. **f.-forceps**, for fixing or holding a part in position during a surgical operation.
**fixative** (*fiks'-a-tiv*). 1. Applied to any substance used to fix tissues in the structural condition and shape found in life or for fastening a microscopic section to a slide. 2. See *body, immune*.
**fixator** (*fiks-a'-tor*). See *ambocepter*.
**fixed** (*fikst*) [*fixus*, firm]. Firm; immovable. **f. idea**, a morbid belief, opinion, or conception, entertained constantly by certain insane patients, and more or less permanently dominating the entire mind.
**fixi dentes** (*fiks'-ī den'-tēz*) [L.]. The teeth of the second dentition.
**fixing** (*fiks'-ing*). The preparation of tissue for microscopic study by means of some agent that hardens it and preserves the form and arrangement of the cells.
**F. K. Q. C. P.** Abbreviation for Fellow of the King and Queen's College of Physicians (of Ireland).
**fl.**, or **fld**. Abbreviation of *fluid*.
**flabby** (*flab'-e*). Lax or flaccid; deficient in firmness.
**flabellate** (*flah-el'-āt*) [*flabellum*, a fan]. In biology, fan-shaped; applied to leaves, antennæ, etc.
**flabellum** (*fla-bel'-um*) [L., "fan"]. A group of divergent fibers in the corpus striatum.
**flaccid** (*flak'-sid*) [*flaccus*, flaccid]. Soft; flabby; relaxed.
**flag, sweet-**. See *calamus*.
**Flagellata** (*flaj-el-la'-lah*) [*flagellum*]. A subclass of mastigosphora possessing one or more flagella. In this class are the trypanosomata and the spirochætæ.

**flagellate** (*flaj'-el-āt*) [*flagellum*]. Furnished with slender, whip-like processes.
**flagellation** (*flaj-el-a'-shun*) [*flagellare*, to whip]. 1. Flogging. 2. A term used by Ross for the extrusion of chromatin granules from leukocytes in response to artificial stimulation. 3. Sexual perversion where gratification is produced by flogging. 4. Massage by strokes or blows. 5. Application of electricity by tapping the surface of the body.
**flagelliform** (*flaj-el'-if-orm*) [*flagellum, forma*, form]. Having the form of a flagellum or whip-lash.
**flagellospore** (*flaj-el'-o-spōr*). See *flagellula*.
**flagellula** (*flaj-el'-ū-lah*) [dim. of *flagellum*, a whip]. A flagellate spore; a zoospore.
**flagellum** (*flaj-el'-um*) [L., "a whip": pl., *flagella*]. A whip-like, mobile process; the organ of locomotion of certain bacteria and infusoria.
**flail** (*flāl*). An arm or leg not under muscular control. **f.-joint**, a condition of preternatural mobility frequently following resection of a joint.
**Flajani's disease** (*fla-yan'-e*) [Giuseppe *Flajani*, Italian surgeon, 1741–1808]. See *goiter, exophthalmic*.
**flank** (*flank*) [ME., *flank*, from L., *flaccus*, soft]. The part of the body between the ribs and the upper border of the ilium.
**flap** [ME.]. A loose and partly detached portion of the skin or other soft tissue. **f.-amputation**, one in which flaps of soft tissues are left to cover over the end of the bone. **f., anaplastic**, a skin-flap aiding in the restoration of a neighboring part. **f., autoplastic**, one to replace a part that is destroyed. **f.-extraction**, a method of extracting the crystalline lens so as to make a flap of the cornea.
**flash-point**. The temperature at which a petroleum oil gives off vapors which, mixing with air, cause an explosion or flash of flame, dying out, however, at once.
**flat**. 1. Lying on one plane. 2. A percussion note that is low pitched and without resonance. **f.-ear**. See *Morel's ear*. **f.-foot**, depression of the plantar arch; it differs from splay-foot or talipes valgus in that the sole is not everted. **f.-worm**. See *tape-worm*.
**Flatau's law** (*flat'-ow*) [Edward *Flatau*, Russian physician, 1863– ]. "Law of the eccentric situation of long tracts." The greater the length of the fibers of the spinal cord, the nearer to the periphery are they situated.
**flatness** (*flat'-nes*). The sound obtained by percussing over an airless organ or large effusion.
**flatulence** (*flat'-ū-lens*) [*flatus*]. A condition marked by the presence of gas in the stomach and intestinal canal.
**flatulent** (*flat'-ū-lent*) [*flatus*]. Characterized by flatulence.
**flatus** (*fla'-tus*) [L.]. 1. Gas, especially gas in the gastrointestinal canal. 2. Expired air. 3. Eructation. **f. vaginalis**, expulsion of gas from the vagina.
**flavedo** (*fla-ve'-do*) [L.]. Yellowness or jaundice.
**flavescent** (*flav-es'-ent*) [*flavēscere*, to become yellow]. Yellowish.
**flavopurpurin** (*flav-o-pur'-pū-rin*) [*flavus*, yellow; *purpura*, purple]. C₁₄H₈O₅. A pigment occurring in golden-yellow, acicular crystals.
**flavus** (*flav'-us*) [L.]. Yellow.
**flax** (*flaks*). See *linum*. **f.-dresser's phthisis**, a fibroid pneumonia resulting from the inhalation of particles of flax.
**flaxseed** (*flak'-sēd*). See *linum*.
**flay** (*fla*). To skin.
**flea** (*flē*). See *pulex*.
**fleabane** (*flē'-bān*). See *erigeron*.
**fleam** (*flēm*) [φλεβοτόμον, a lancet; from φλέψ, a vein; τέμνειν, to cut]. A phlebotome.
**Flechsig's column** (*flek'-sig*) [Paul Emil *Flechsig*, German neurologist, 1847– ]. The direct cerebellar tract of the spinal cord. **F.'s tract, F.'s oval field**, the septomarginal tract of the spinal cord.
**fleece of Stilling**. See under *Stilling*.
**Fleischl's reaction for bile-pigments** (*fli'-shl*). Add, by means of a pipet, concentrated sulphuric acid to urine already treated with a concentrated solution of sodium nitrate. The sulphuric acid sinks to the bottom of the test-tube and produces color-layers, as in Gmelin's test.
**Fleischmann's bursa** (*flīsh'-man*) [Gottfried *Fleischmann*, German anatomist, 1777–1850]. A bursa lying in the sublingual space beneath the lingual frenum. Its existence is disputed. **F's hygroma**, distention or inflammation of F.'s bursa.

**Fleming's modification of Wassermann's test.** A much simplified serum-test for syphilis based upon the same principles as the Wassermann reaction.
**F.'s tincture,** an alcoholic preparation of aconite stronger than the official tincture. Dose 2 min. (0.13 Cc.).
**flemingin** (*flem-in'-jin*). A pigment obtained from warras, occurring in small needles.
**Flemming's fibrillary mass** [Walter *Flemming*, German anatomist, 1843– ]. Spongioplasm.
**F.'s germ-centers.** The areas in the adenoid tissue of the spleen and lymphatic glands in which leukocytes are formed. **F.'s solution,** a mixture used in histological study as a fixing agent for tissues. It consists of 15 parts of 1 % solution of chromium trioxide, 4 parts of a 2 % solution of osmic acid, 1 part of glacial acetic acid.
**flesh** [AS., *flǣsc*]. The soft tissues of the body, especially the muscles. **f., proud,** the soft and exuberant granulations of a wound or ulcer. **f.-quotient,** Argutinsky's term for the relationship of the carbon to nitrogen in flesh; it is, on an average, 3.24 : 1.
**fleshy** (*flesh'-e*). Mainly composed of muscular tissue.
**Fletcherism** (*fletsh'-er-izm*) [Horace *Fletcher*, American dietitian, 1849– ]. The thorough mastication of solid food, until all taste of the food is lost.
**flex** (*fleks*) [*flectere,* to bend]. To bend.
**flexibilitas** (*fleks-ib-il'-it-as*) [L.]. Flexibility. **f. cerea,** a condition of the limbs in catalepsy in which they seem as if made of wax.
**flexible** (*fleks'-e-bl*) [*flex*]. That which may be bent, as a *flexible* catheter, *flexible* collodion.
**flexile** (*flex'-il*) [*flexilis,* pliable]. Easily bent.
**Flexner's bacillus** (*fleks'-ner*) [Simon *Flexner,* American bacteriologist, 1863– ]. A bacillus which is said to cause dysentery. **F.'s serum,** an antimeningococcus serum, used in epidemic cerebrospinal meningitis.
**flexion** (*flek'-shun*) [*flex*]. The act of bending; the condition of being bent.
**flexor** (*fleks'-or*) [*flex*]. A muscle that bends or flexes a limb or a part. See under *muscle.*
**flexuous** (*fleks'-ū-us*) [*flectere,* to bend]. In biology, alternately curved in opposite directions.
**flexura** (*fleks-ū'-rah*) [L., a bending]. A bending or curve in an organ.
**flexure** (*fleks'-ūr*) [*flex*]. A bending. **f., caudal,** the bend at the lower portion of the embryo. **f., cephalic,** the arching over of the cephalic end of the embryo. **f., hepatic** (of the côlon), an abrupt bend in the ascending colon to the right of the gall-bladder at the under surface of the liver. **f., sigmoid.** See *sigmoid flexure.* **f., splenic** (of the colon), an abrupt turn beneath the lower end of the spleen, connecting the descending with the transverse colon.
**flighty** (*flī'-te*) [Dan., *vlugtig,* volatile]. Slightly delirious.
**Flindt-Koplik's sign.** See *Koplik's spots.*
**flint-disease.** Synonym of *chalicosis.*
**Flint's arcade** [Austin *Flint,* American physician, 1812–1886]. The arteriovenous arch around the base of the renal pyramids. **F.'s murmur,** a second systolic murmur heard over the apex in cases of marked dilatation of the ventricle from aortic insufficiency.
**floating** (*flō'-ting*) [AS., *fleótan,* to float]. Swimming; free to move around. **f. albumin.** See *albumin, circulating.* **f. kidney,** one that is detached from its normal position and abnormally movable. **f. liver,** one with abnormal mobility; movable liver. **f. rib.** See *rib, floating.* **f. spleen,** one that is separate from its attachments, and displaced.
**flocci** (*flok'-i*) [L. Plural of *floccus,* a tuft]. **f. volitantes.** Same as *muscæ volitantes.*
**floccilegium** (*flok-sil-e'-je-um*) [*floccus,* a flock of wool; *legere,* to pick out]. Carphology.
**floccillation** (*flok-sil-ā'-shun*) [*floccillatio*]. Same as *carphology.*
**floccose** (*flok'-ōs*) [*floccus,* a flock of wool]. 1. Composed of or bearing tufts of woolly or long and soft hairs. 2. A bacterial growth composed of short curving filaments.
**floccular** (*flok'-ū-lar*) [*flocculus,* a little flock of wool]. Pertaining to the *flocculus.*
**flocculence** (*flok'-ū-lens*) [see *flocculus*]. Flakiness; the state of being flocculent.
**flocculent** (*flok'-ū-lent*) [see *flocculus*]. Flaky, downy, or woolly; coalescing in flocky masses.

**flocculus** (*flok'-ū-lus*)* [dim. of *floccus,* a flock of wool; pl., *flocculi*]. 1. A prominent lobe of the cerebellum situated behind and below the middle cerebellar peduncle on each side of the median fissure. 2. A small flock of wool or something resembling it; a tuft, shred, or flake. **f., accessory,** the paraflocculus.
**Floegel's layer** (*fle'-gel*). The layer of granules in the transparent lateral disc of a muscle-fibril.
**Flood's ligament** (*flud*) [Valentine *Flood,* Irish surgeon, 1800–1847]. The glenohumeral ligament.
**flooding** (*flud'-ing*) [AS., *flōd,* a flood]. A copious bleeding from the uterus.
**floor** (*flor*) [ME.]. The basal limit of any hollow organ or open space. **f.-cells,** those found in the floor of Corti's arch. **f. of the pelvis,** the united mass of tissue forming the inferior boundary of the pelvis.
**flora** (*flō'-rah*) [*Flora,* the goddess of flowers]. The entire plant-life of any geographical area or geological period.
**Florence's crystals** (*flor'-ens*) [Albert *Florence,* French physician, 1851– ]. Brown crystals, in the shape of needles or plates obtained by treating semen with a strong solution of iodine and potassium iodide (Florence's reaction); they are also formed in the secretions of the prostate, uterus, vagina, etc. **F.'s reaction.** See above. This is not wholly reliable as a test for human spermatic fluid, since the crystals can also be found in the spermatic fluid of animals.
**flores** (*flō'-rēz*) [pl. of *flos,* a flower]. 1. The flowers or blossoms of a plant. 2. A flocculent or pulverulent form assumed by certain substances after sublimation, as *flores sulphuris,* flowers of sulphur.
**florid** (*flor'-id*) [*floridus,* abounding with flowers] Bright-red in color; rosy as a florid cheek, or countenance. **f. phthisis.** See *galloping consumption.*
**Florida allspice.** See *allspice, Carolina.*
**flos** [L.]. A flower. Singular of *flores, q. v.*
**floss** (*flos'*). Silk which has not been twisted. **f.-silk.** See *silk.*
**flour** [*flos,* a flower]. The finer part of the ground grain, especially of wheat.
**Flouren's doctrine** (*flu-renz'*) [Marie Jean-Pierre *Flourens,* French physiologist, 1794–1867]. A theory that the whole of the cerebrum is concerned in every psychic process.
**flow** (*flō*) [AS., *flōwan,* to flow]. The free discharge of a liquid, as the blood; the menses.
**flower** (*flow'-er*). See *flores.*
**Flower, angle of** [Sir William Henry *Flower,* English anatomist, 1831–1899]. In craniometry, the naso-malar angle.
**F. L. S.** Abbreviation for Fellow of the Linnæan Society.
**fluavil** (*flū'-av-il*), $C_{20}H_{32}O_2$. A transparent yellowish resin found in gutta percha.
**flucticuli** (*fluk-tik'-ū-li*) [pl. of *flucticulus,* a wavelet]. Bergmann's name for the fine, wave-like markings on the surface of the lateral wall of the third ventricle, ventrad of the anterior commissure.
**fluctuation** (*fluk-tū-ā'-shun*) [*fluctuare,* to float or roll]. The wave-like motion produced when a body containing fluid is tapped between the fingers or hands.
**Fluhrer's probe** (*flu'-rer*) [William Francis *Fluhrer,* American physician, 19th century]. An aluminum probe used in investigating gunshot wounds of the brain.
**fluid** (*flu'-id*) [*fluere,* to flow]. 1. A substance whose molecules move freely upon one another; any liquid secretion of the body. 2. Liquid or gaseous. **f., allantoic,** the fluid contents of the allantois. **f., amniotic,** a serous liquor filling the cavity of the amnion. **f., cerebrospinal,** the fluid between the arachnoid membrane and the pia mater. **f., Coley's.** See under *Coley.* **f., colostric.** See *colostrum.* **f., Darby's prophylactic.** See under *Darby.* **f.-dram.** A liquid measure equalling 56.96 grains of distilled water. **f. extract.** See *extractum fluidum.* **f., Haffkine's prophylactic.** See under *Haffkine.* **f. labyrinthine,** the perilymph. **f., Lang's fixative.** See under *Lang.* **f.-ounce.** A liquid measure, eight fluidrams. **f., Scarpa's,** the endolymph. **f., subarachnoid.** See *cerebrospinal fluid.* **f., van Gehuchten's fixative.** See under *van Gehuchten.* **f. vein,** the name given to the eddies produced in a cavity of the heart by regurgitating blood coming in contact with the

# FLUIDACETEXTRACT 367 FOLIE

current entering the cavity in the normal direction. The oscillation of the particles of blood are attended with a blowing sound or murmur.

**fluidacetextract** (*flu-id-as-et-eks'-trakt*). A fluidextract made with acetic acid instead of alcohol.

**fluidextract** (*flu-id-ek'-strakt*). A solution of the solid principles of a vegetable drug, of such strength that 1 Gm. of the drug is fully represented by 1 Cc. of the fluidextract.

**fluidounce** (*flu-id-owns'*). A liquid measure; eight fluidrams.

**fluidram** (*flu-id-ram'*). A liquid measure equal to 56.96 grains of distilled water.

**fluinol** (*flu'-in-ol*). A proprietary preparation of pine and fir needles with volatile oils; it is used as a sedative addition to baths, or for inhalations or gargles.

**fluke** (*fluk*) [ME., *floke*]. Any trematode worm.

**flumen** (*flu'-men*) [L.; pl., *flumina*]. 1. A flow. 2. A name given by Duret to the principal cerebral fissures.

**fluor** (*flu-or*) [L., a flow]. 1. A liquid state. 2. The menstrual flow. f. **albus**, white flow; an old name for leukorrhea. f. **muliebris**. Synonym of *leukorrhea*.

**fluoram** (*flu'-or-am*). Ammonium bifluoride; used as an application to the gums in pyorrhœa alveolaris.

**fluorescein, fluorescin** (*flu-or-es'-e-in, flu-or-es'-in*), $C_{20}H_{12}O_5 + H_2O$. An anhydride of resorcinol, prepared by heating phthalic anhydride with resorcinol to 200° C. It has the property of coloring abrasions of the cornea greenish, and on this account has been used for diagnostic purposes. f.-**sodium**, a 2% alkaline solution employed in diagnosing corneal lesions and in the detection of minute foreign bodies in that tissue; it is suggested as a means of determining apparent death by injection of 16 gr. (1.03 Gm.); if circulation remains, the mucosæ will be stained yellow within a few minutes. Syn., *uranni*.

**fluorescence** (*flu-or-es'-ens*) [*fluor* · (*fluor-spar*), because first observed in this mineral]. A property possessed by certain substances of converting obscure actinic rays, such as the ultraviolet, into luminous rays.

**fluorescent** (*flu-or-es'-ent*). Having the property of fluorescence. f. **screen**, a screen covered with substances which become fluorescent on exposure to the roentgen-rays.

**fluoride** (*flu'-or-ide*) [see *fluorine*]. A compound of fluorine and a base.

**fluorine** (*flu'-or-ēn*) [*fluor-spar*], F = 19; quantivalence I. An element belonging to the chlorine group. The salts formed with the alkaline metals, *fluoride*, have been used in goiter and in rheumatism. See *elements, table of chemical*.

**fluoroform** (*flu-or'-o-form*) [*fluorine; forma*, form], $CHF_3$. A gas, the fluorine analogue of chloroform. f.-**water** (*aqua fluoroformii*), a watery solution (2.8%) of fluoroform, used in tuberculosis and lupus. Dose 1 tablespoonful 4 times daily. Syn., *fluoroformol*.

**fluoroformol** (*flu-or-o-form'-ol*). See *fluoroformwater*.

**fluorol** (*flu'-or-ol*), NaF. Sodium fluoride, an antiseptic.

**fluorometer** (*flu-or-om'-et-er*) [*fluorescence; μέτρον*, a measure]. A device for adjusting the shadow in skiagraphy; a localizer in roentgen-ray examination.

**fluoroscope** (*flu-or'-os-kōp*) [*fluorescence; σκοπεῖν*, to examine]. The instrument for holding the fluorescent screen in roentgen-ray examination.

**fluoroscopy** (*flu-or-os'-ko-pe*). The process of examining the tissues by means of a fluorescent screen.

**fluorphenytol** (*flu-or-fen'-et-ol*), ($C_6H_4F)_2$. A calmative and hypnotic; it is used in whooping-cough.

**fluorrheumin** (*flu-or-ru'-min*). The commercial name of fluorphenetol-difluorodiphenyl, prepared as an ointment and used in the treatment of rheumatism. Dose 77 gr. (5 Gm.) externally.

**flush**. A temporary redness, as the hectic flush, sometimes due to vasomotor paresis.

**flushing** (*flush'-ing*). 1. A frequent symptom in the subjects of cardiac palpitation, and especially in Graves' disease. It implies a condition of vasomotor irritability with a paresis of the arterioles in certain areas. It is often accompanied by local perspiration. It is seldom a marked symptom of organic disease. 2. The process of cleansing by a rapid flow of liquid.

**flux** (*fluks*) [*fluxus*, flowing]. 1. An excessive flow of any of the excretions of the body, especially the feces. 2. Dysentery. f., **alvine**, diarrhea. f., **bloody**, dysentery.

**fluxion** (*fluk'-shun*) [*fluxus*, a flowing]. A gathering of blood or other fluid in one part of the body; congestion, or hyperemia.

**fly** (*fli*). A dipterous insect. f.-**agaric**. See *agaricus muscarius*. f.-**blister, flying blister**. See *blister*.

**focal** (*fo'-kal*) [*focus*].—Pertaining to or occupying a focus. f. **depth**, the power of a lens to give clear images of objects at different distances from it. f. **disease**, f. **lesion**, one that is limited to a small area. f. **distance**, the distance from the focus to a reflecting or refracting surface, or, in the case of a lens, to the principal point of the lens. f. **epilepsy**, epilepsy, due to a focal lesion of the brain. Syn., *Jacksonian epilepsy*.

**focii** (*fo'-sil*) [*focile*, a spindle]. Any bone of the forearm or leg. f. **majus**, the ulna. f. **majus cruris**, the tibia. f. **minus**, the radius. f. **minus cruris**, the fibula.

**focus** (*fo'-kus*) [L., "a fireplace"; pl., *foci*]. 1. The principal seat of a disease. 2. The point (called principal focus or real focus) at which rays of light converge that pass through a convex lens or are reflected from a concave mirror. f., **negative**, f., **virtual**, the point at which divergent rays would meet if prolonged in a backward direction.

**focusing** (*fo'-kus-ing*) [*focus*]. The mutual arrangement of an object and the optical parts of a microscope so that a clear image may be seen. f. **down**, in microscopy, focusing by moving the objective down or toward the object, but at the risk of damaging it. f. **up**, focusing by moving the objective up or away from the object.

**fœniculum** (*fen-ik'-u-lum*) [L¹]. Fennel. The fruit of *F. vulgare*, the properties of which are due to a volatile oil. It is a mild stimulant and aromatic carminative. **fœniculi, aqua** (U. S. P.), 2 parts of the oil of fennel in 1000 of water. Dose ½ dr.–1 oz. (2–32 Cc.). **fœniculi, oleum** (U. S. P.), oil of fennel. Dose 2–5 min. (0.13–0.32 Cc.).

**Foerster's shifting type** (*fers'-ter*) [Richard *Foerster*, German ophthalmologist, 1825–1902]. Variations in the visual field, the limits of which differ according as they are determined by moving the disc from the center outward or from without toward the center; they are seen in traumatic neuroses.

**fœtal** (*fe'-tal*). See *fetal*.

**fœtor** (*fe'-tor*). See *fetor*.

**fœtus** (*fe'-tus*). See *fetus*.

**fogging maneuver**. In repression treatment of esophoria, the reduction of vision to about ⅖ by combining prisms (varying with the muscular imbalance), bases in, with a convex sphere, with which combination glasses the patient reads a half-hour at night before retiring.

**foil** [*folium*, a leaf]. A thin sheet of metal used for filling teeth. f. **carrier**, f. **plugger**, a kind of tweezers used to convey the foil to the cavity in the tooth. f. **crimpers**, an instrument for folding foil.

**fold** (*fōld*) [ME.]. A plication or doubling of various parts of the body. f. f., **arytenoepiglottidean**. See under *arytenoepiglottic*. f., **costocolic**, that which extends from the diaphragm opposite the tenth and eleventh ribs to the splenic flexure of the colon, and forms a shelf-like structure above which lies the spleen. f., **ileocolic**, a semilunar fold of the peritoneum which is attached to the anterior layer of the mesentery, the anterior aspect of the ascending colon, and the cecum as far as the vermiform appendix. Syn., *Luschka's fold*. f., **palpebral**, that formed by the reflection of the conjunctiva from the eyelids on to the eye. There are two folds—superior and inferior. f., **pituitary**, the two layers of dura inclosing the hypophysis.

**folia** (*fo'-le-ah*) [*pl. of folium*, leaf]. Leaves.

**foliaceous** (*fo-le-a'-se-us*) [*folium*, a leaf]. Leaflike.

**Folian process** [Cæcilius *Folius*, Italian anatomist, 1615–1660]. The processus gracilis of the malleus.

**folders** (*fōl'-derz*). Magnifying glasses that shut like "eye-glasses."

**folie** (*fo-le'*) [Fr.]. Insanity. f. **à deux** (*fo-le'-ah duk'*) [Fr.]. See *insanity, communicated*. f. **alternate**, cyclic insanity, q. v. f. **circulaire** (*fo-le-ser-ku-lār'*), cyclic insanity. f. **de doute**. See *doubt, insanity of*.

**Folin's test** (*fo'-lin*) [Otto K. O. *Folin*, American chemist, 1867– ]. 1. (*Quantitative for urea*): The urine is boiled with magnesium chloride, and the urea is decomposed into carbon dioxide and ammonia; the latter is then estimated. 2. (*Quantitative for uric acid*): The uric acid is precipitated with ammonia, ammonium urate being formed, this latter is oxidized with potassium permanganate.

**folium** (*fo'-le-um*) [L., a leaf: *pl.*, *folia*]. 1. In biology, a leaf. 2. Any lamina or leaflet of gray matter, forming a part of the arbor vitæ of the cerebellum. **f. cacuminis**, a lobule on the upper surface of the vermis. **f. vermis**, the terminal lobule in the superior worm of the cerebellum.

**follicle** (*fol'-ik-l*) [*folliculus*, dim. of *follis*, a bellows]. 1. A small lymphatic gland, the tissue of which is arranged in the form of a little sac; also a small secretory cavity or sac. 2. A simple tubular gland. **f., dental**, the dental sac and its contents, the developing tooth. **f., Graafian**, one of the small vesicular bodies in the ovary, each of which contains an ovum. **f., hair**, the depression containing the root of the hair. **f.s of Lieberkühn**. See *Lieberkühn*, *crypts of*. **f., lymph**, collection of adenoid tissue in mucous membranes. **f.s, sebaceous**, the sebaceous glands of the skin. **f.s, solitary**, small discrete lymph-follicles found in the mucous membrane of the intestine.

**folliclis** (*fol'-ik-lis*). A skin disease of tuberculous subjects characterized by a macular eruption which later becomes nodular and then pustular.

**follicular** (*fol-ik'-u-lar*) [*follicle*]. Pertaining to a follicle. **f. tumor**, a sebaceous cyst.

**folliculitis** (*fol-ik-u-li'-tis*) [*follicle*; ιτις, inflammation]. Inflammation of a group of follicles.

**folliculitis abscedens infantum**, follicular furunculosis of children. **f., agminate**, inflammation of a set of follicles. **f., barbæ**, inflammation of the hair-follicles of the beard; sycosis. **f., decalvans**, inflammatory disease of the hair follicles resulting in patches of baldness.

**folliculoma** (*fol-ik-u-lo'-mah*). A tumor originating in a follicle. **f. ovarii malignum**, a malignant tumor of a Graafian vesicle.

**folliculose** (*fol-ik'-u-los*) [*follicle*]. Full of follicles. **folliculosis** (*fol-ik-u-lo'-sis*). A disease in which there is excessive development of the follicles.

**folliculus** (*fol-ik'-u-lus*) [dim. of *follis*; bag: *pl. folliculi*]. Follicle. **f. oophorus primarius**, a follicle surrounding the undeveloped ovum in the ovary. **f. oophorus vesiculosus**, a Graafian follicle. **f. pili**, a hair-follicle. **f. solitarius**, a solitary follicle.

**fomentation** (*fo-men-ta'-shun*) [*fomentare*, to foment]. 1. The application of heat and moisture to a part to relieve pain or reduce inflammation. It may be made by means of cloths soaked in hot water or medicated solution or by a poultice. 2. The substance applied to a part to convey heat or moisture.

**fomes** (*fo'-mēz*) [L., "tinder"; *pl.*, *fomites*]. Any substance capable of acting as the medium for transmitting contagion.

**fomites** (*fo'-mi-tēs*). Plural of *fomes*.

**fons pulsatilis** (*fons pul-sat'-il-is*) [L.]. The anterior fontanel.

**Fontana's bands** (*fon-tah'-nah*) [Felice *Fontana*, Italian anatomist, 1730–1805]. The wavy arrangement presented by nerve-fibers, which lie alongside one another in loose spirals, in places where considerable mobility is possible. **F.'s canals**, **F.'s spaces**, the minute spaces occupying the angle of the iris and communicating with the aqueous chamber and Schlemm's canal.

**fontanel, fontanelle** (*fon-tan-el'*) [Fr., *fontanelle*, a little fountain]. A membranous space between the cranial bones in fetal life and infancy. **f., anterior**, that at the point of union of the frontal, sagittal, and coronal sutures. **f., Casser's**, **f. of Casserius**. See *f.s, lateral*. **f., Gerdy's**. See under *Gerdy*. **f., great**. See *f., anterior*. **f.s, lateral**, two membranous spaces, one in front between the parietal, frontal, and temporal bones (*the anterior lateral or sphenoid fontanel*), and one behind between the parietal, occipital, and temporal bones (*the posterior lateral, mastoid, or Casser's fontanel*). They usually disappear the year after birth. **f., mastofrontal**, an abnormal one at the union of the nasal and frontal bones. **f., posterior**, that at the point of junction of the lambdoid and the sagittal sutures. **f., small**. See *f., posterior*. **f., supraorbital**, in comparative embryology, a cordate membranous space between the occipital cartilage and the skull.

**fonticulus** (*fon-tik'-u-lus*) [dim. of *fons*, fountain]. 1. The depression at the root of the neck, just cephalad of the sternum; more fully, **fonticulus gutturis**. 2. A small artificial ulcer or issue. 3. Same as *fontanel*. **f. major**, **f. quadrangularis**, the anterior fontanel. **f. minor**, **f. triangularis**, the posterior fontanel.

**food** [AS., *fōda*]. Anything which, when taken into the body, is capable of building up tissue, or, by oxidation, of supplying heat.

**foot** [ME.]. 1. The terminal extremity of the leg. It consists of the tarsus, metatarsus, and phalanges, or toes. 2. The base of a microscope. 3. A measure of length equal to 12 inches, or 30.479 cm. **f.-and-mouth disease**, a febrile affection of sheep, cows, pigs, and horses, rarely of man, manifesting itself by the appearance of vesicles and bullæ in the mouth and on the feet. It is probably due to a special microorganism. **f.-cells**, Sertoli's cells. **f.-clonus**. See *ankle-clonus*. **f.-drop**, a falling of the foot due to a paralysis of the flexors of the ankle. **f., fungus-**. See *Madura-foot*. **f.-pound**, the work equal to that of raising a pound to the height of one foot. **f.-reflex**. See *ankle clonus*. **f., tabetic**. 1. An extension of the foot in preataxic tabes observed when the patient is lying down. 2. An affection of the foot in the beginning of tabes, marked by numbness and formication, followed by hypertrophy of the head of the astragalus, scaphoid, cuneiform, and metatarsal bones.

**footing** (*foot'-ling*). With the foot or feet foremost; as a footling presentation in obstetrics.

**forage** (*for'-aj*) [OF., *fourage*]. **f., poisoning**, the preferred term for the so-called epizootic cerebrospinal meningitis of horses. It is attributed to a fungus upon the ensilage.

**foramen** (*for-a'-men*) [*forare*, to pierce: *pl.*, *foramina*]. A perforation or opening, especially in a bone. **f., aortic**, an opening in the diaphragm transmitting the aorta. **f., apical**, the passage at the end of the root of a tooth for the neural supply to the dental pulp. **f., arachnoid**, an opening in the roof of the fourth ventricle. **f., auditory, external**, one located in the external meatus of the auditory canal; it transmits sound-waves to the tympanic membrane. **f., auditory, internal**, one located in the petrous portion of the temporal bone; it transmits the auditory and facial nerves. **f., Bichat's**. See *Bichat*. **f., Botallo's**. See under *Botallo*. **f., cecal** (of frontal bone), a small foramen formed by the frontal bone and the crista galli of the ethmoid; it transmits a vein occasionally. **f., cecal** (of medulla oblongata), one located in a depression at the termination of the anterior median fissure. **f., cecal** (of pharynx), one located in a depression in the mucous membrane, in the median line of the posterior wall of the pharynx. **f., cecal** (of tongue), one located in the posterior termination of the median raphe of tongue; a number of small glands open into it. **f., condyloid, anterior** (sometimes double), that anterior to and to the outer side of each occipital condyle, passing downward, outward, and forward through the basilar process; it transmits the hypoglossal nerve; occasionally a meningeal branch of the ascending pharyngeal artery. **f., condyloid, posterior**, the fossa behind the occipital condyles; it transmits a vein to the lateral sinus. **f., cotyloid**, a notch in the acetabulum converted into a canal by a ligament; transmits vessels and nerves. **f., dental, inferior**, the external aperture of the inferior dental canal, in the ramus of the inferior maxilla; it transmits inferior dental vessels and nerves. **f., esophageal**, passage through the diaphragm for the esophagus. **f., ethmoid, anterior**, a canal between the ethmoid and frontal bones, transmitting the nasal branch of the ophthalmic nerve and anterior ethmoid vessels. **f., frontal**, the supraorbital notch of the frontal bone when it is converted into a canal by a bony process; it transmits the supraorbital vessels and nerves. **f., Galen's**. See under *Galen*. **f., incisor**, the aperture of the incisor canal in the alveolar margin; it transmits nerves and vessels to the incisor teeth. **f., infraorbital**, in the superior maxilla, the external aperture of the infraorbital canal; it transmits the infraorbital nerve and artery. **f., interclinoid, common**, a canal formed by an anomalous process connecting the anterior, middle, and posterior clinoid processes of the sphenoid bone.

FORAMEN 369 FORDYCE'S DISEASE

f., interventriculare, the foramen of Monro, q. v.
f., intervertebral, anterior, the aperture formed by the notches opposite to each other in the laminæ of adjacent vertebræ; it is a passage for the spinal nervæ and vessels. f., intervertebral, posterior, the space between the articular processes of adjacent vertebræ, except the first cervical. f., jugular. See f., lacerated, posterior. f. jugulare spurium, a foramen in the temporal bone of the embryo transmitting a vein from the lateral sinus to the external jugular. f., lacerated, anterior. See f., sphenoidal. f., lacerated, middle, an irregular aperture between the apex of the petrous portion of the temporal bone and the body and great wing of the sphenoid, and the basilar process of the occipital bone; it is an opening for the carotid artery and the large superficjal petrosal nerve. f., lacerated, posterior, the space formed by the jugular notches of the occipital and temporal bones, divided into two portions: the *posterior portion* transmits the internal jugular vein; the *anterior portion*, the ninth, tenth, and eleventh cranial nerves and the inferior petrosal sinus. f. lacerum. Same as f., lacerated. f., Magendie's. See under *Magendie*. f. magnum, a large oval aperture, centrally placed in the lower and anterior part of the occipital bone; it transmits the spinal cord and its membranes; the spinal accessory nerves; the vertebral arteries. f., mastoid, a small foramen behind the mastoid process. It transmits a small artery from the dura; a vein opening into the lateral sinus. f., medullary. See f., nutrient. f., mental, a foramen in the inferior maxilla, external to the incisive fossa, forming a passage for the mental nerve and vessels. f. of Monro. See under *Monro*. f., nutrient, the canal conveying the nutrient vessels to the medullary cavity of a bone. f., obturator, the large ovoid opening between the ischium and the pubis, internal and inferior to the acetabulum; it is partly closed in by a fibrous membrane; it transmits the obturator vessels and nerves. f., occipital. See f. magnum. f., olfactory, numerous foramina in the cribriform plate of the ethmoid, transmitting the olfactory nerves. f., omental, lesser or small. See *Winslow*, f. of. f., optic, the canal at the apex of the orbit, the anterior termination of the optic groove, just beneath the lesser wing of the sphenoid bone; it transmits the optic nerve and ophthalmic artery. f. ovale (of the heart), a fetal opening between the two auricles of the heart, situated at the lower posterior portion of the septum. f. ovale (of the sphenoid), an ovoid aperture near the posterior margin of the great wing of the sphenoid, transmitting the inferior maxillary division of the trigeminal nerve; the small meningeal artery; occasionally, the small petrosal nerve. f., palatine, anterior, the orifice of the incisor canal in the anterior part of the roof of the mouth, constituting the opening for the nasopalatine nerve and a branch of the posterior palatine artery. f., palatine, posterior, the orifice of the posterior palatine canal upon the posterior part of the hard palate; it transmits the descending palatine artery. f., parietal, is near the posterior superior angle of the parietal bone; inconstant. It conveys an emissary vein of the superior longitudinal sinus; occasionally a small branch of the occipital artery. f., pterygopalatine, the external aperture of the pterygopalatine canal, transmitting the pterygopalatine vessels and pharyngeal nerve. f., quadrate, a foramen in the diaphragm for the inferior vena cava. f. rotundum, a round opening in the great wing of the sphenoid bone for the superior division of the fifth nerve. f., sacral, anterior (*four on each side*), on the anterior surface of the sacrum, connecting with the sacral canal, and transmitting the anterior branches of the sacral nerves. f., sacral, posterior (*four on each side*), on the posterior surface of the sacrum, external to the articular processes, and transmitting the posterior branches of the sacral nerves. f., sacrosciatic, great, the oval space between the lesser sacrosciatic ligament and the innominate bone, conveying the pyriformis muscle, the gluteal, sciatic, and pudic vessels and nerves. f., sacrosciatic, small, the space included between the greater and lesser sacrosciatic ligaments and the portion of the innominate bone between the spine and tuberosity of the ischium; it transmits the internal obturator muscle, the internal pudic vessels and nerves. f.s of Scarpa. See under *Scarpa*. f. of Soemmering. See under *Soemmering*. f., sphenopalatine, the space between the sphenoid and orbital processes of the palate bone; it opens into the nasal cavity and transmits branches from Meckel's ganglion and the nasal branch of the internal maxillary artery. f. spinosum, a passage in the great wing of the sphenoid bone, near its posterior angle, for the middle meningeal artery. f.s of Stenson. See under *Stenson*. f., stylomastoid, one between the styloid and mastoid processes of the temporal bone; it is the external aperture of the Fallopian aqueduct. f., supraorbital, a notch in the superior orbital margin at the junction of the middle with the inner third, sometimes converted into a foramen by a bony process or a ligamentous band it transmits the supraorbital artery, veins, and nerve. f.s of Thebesius. See under *Thebesius*. f., thyroid. 1. One in the ala of the thyroid cartilage. 2. See f., obturator. f.s, transverse accessory, anomalous foramina in the transverse processes of the cervical vertebræ transmitting an inconstant accessory vertebral artery. f., vertebral, the space included between the body and arch of a vertebra, transmitting the spinal cord and its appendages. f.s, vertebrarterial, foramina in the transverse processes of the cervical vertebræ for the vertebral artery and vein. f., Weitbrecht's. See under *Weitbrecht*. f. of Winslow. See under *Winslow*.
foraminated (fo-ram'-in-a-ted) [foramen]. Containing foramina.
foraminiferous (fo-ram-in-if'-er-us). Same as foraminated.
foraminulate, foraminulous, foraminulose (for-am-in'-u-lāt, -lus, -lōs). Furnished with very minute openings.
force (fōrs) [fortis, strong]. That which produces or arrests motion. f., absolute muscular, the maximum capacity of shortening shown by a muscle subjected to maximum stimulus. f., chemical, that form of energy which holds atoms together in a molecule. f., electromotive, the force producing an electric current. f., plastic, the generative force of the body.
forced (fōrst) [fortis, strong]. Accomplished by an exertion of force. f. feeding. 1. Systematic overfeeding as a therapeutic measure. 2. Feeding performed against the will of the patient.
forceps (for'-seps) [L.; "a pair of tongs"]. An instrument with two blades and handles used for purposes of drawing on or compressing an object. 2. The curved bundles of fibers passing from the corpus callosum to the cerebral hemispheres. f., alveolar, forceps used in removing portions of the alveolar process. f., anterior, the forceps minor, q. v. f., axis-traction, an obstetrical forceps specially constructed to enable pulling in the direction of the pelvic axis. f., bone, a forceps used for cutting bone. f., dental, forceps used for the extraction of teeth. f., dressing, forceps used for handling surgical dressings. f., duckbill, forceps furnished with duck-bill-shaped beaks used for extraction of roots of teeth. f., epilating, forceps for pulling out hairs. f., fixation, forceps for holding structures in a fixed position during an operation. f., hemostatic, a forceps for controlling hemorrhage. f., obstetrical, forceps used for extracting the fetus. f. major, a curved band of fibers passing from the splenium to the occipital lobe. f. minor, a curved band of fibers passing from the genu of the callosum to the frontal lobe. f., posterior, the forceps major, q. v. f., rongeur. See *Rongeur*. f., sequestrum, strong forceps with serrated jaws of medium length; used for holding or removing the detached portion of bone in case of a sequestrum.
Forchheimer's exanthem (for'-shi-mer) [Frederick *Forchheimer*, American physician, 1853– ]. A maculopapular rose-red eruption on the soft palate and uvula, regarded by Forchheimer as characteristic of rubeola in the absence of any cutaneous eruption.
forcipal (for'-sip-al). Relating to forceps.
forcipate, forcipated (for'-sip-āt, -a-ted) [forceps]. Shaped like a forceps.
forcipressure (for'-se-presh-ūr) [forceps; pressura, a pressing]. The catching the end of the divided vessel with a pair of spring-forceps, which are left on for some time for the purpose of preventing hemorrhage.
Fordyce's disease (for'-dis) [John Addison *Fordyce*, American dermatologist, 1858– ]. A disease affecting the mucous membrane of the lips, consisting

of patches of milium-like bodies, and characterized by itching and burning.
**fore** (*fōr*) [AS.]. In front; coming first.
**forearm** (*fōr'-ärm*). That part of the arm between the wrist and the elbow.
**forebrain** (*fōr'-brān*). The anterior of the encephalic vesicles into which the primary nervous axis of the embryo divides at an early stage: the prosencephalon.
**forefinger** (*fōr'-fing-er*). The index-finger.
**foregilding** (*fōr'-gild-ing*). A term designating the histological process of treating perfectly fresh nerve tissues with salts. Cf. *aftergilding*.
**foregut** (*fōr'-gut*). The embryonic tube corresponding to the pharynx, esophagus, stomach, and duodenum.
**f rehead** (*fōr'-ed*). That part of the face above the orbits.
**foreign** (*for'-en*). Alien; exotic; not native; irrelevant. **f. body**, a substance in a wound, organ, or cavity not normally present.
**forekidney** (*fōr'-kid-ne*). The pronephron.
**Forel's ventral tegmental decussation** (*for'-el*) [A. Forel, Swiss physician]. Crossing of the tract from the red nucleus and the rubro-spinal tract to the opposite side.
**foremilk** (*fōr'-milk*). Colostrum.
**forensic** (*for-en'-sik*) [*forensis*, belonging to the forum]. Pertaining to a court of law. In medicine, that part of the science connected with judicial inquiry.
**foreskin** (*fōr'-skin*). The prepuce.
**Forest's method of resuscitation.** The child is placed on its face, and quick, firm pressure is made on the back; then it is placed in a pail of hot water, and the hands carried upward until the child is suspended by its arms, and mouth-to-mouth insufflation is practised; the arms are then lowered and the body doubled forward; these movements are repeated at the rate of 40 a minute.
**forewaters** (*fōr'-waw-terz*). Hydrorrhea gravidarum.
**fork.** A name frequently given to the space between the thighs.
**form** [*forma*, shape]. The external shape or configuration of the body, or of a part of a body.
**formacoll** (*for'-mak-ol*). See *formaldehydegelatin*.
**Formad's kidneys** (*for'-mad*) [Henry F. Formad, American physician, 1847-1892]. The elongated and enlarged kidneys of chronic alcoholism.
**formagen** (*for'-maj-en*). A proprietary dental cement.
**formal** (*for'-mal*). See *methylal*.
**formalbumin** (*for-mal'-bū-min*). See *formaldehydecasein*.
**formaldehyde** (*for-mal'-dē-hīd*). Formic aldehyde (*q. v.*). **f. acetate**, $C_8H_8O_4$, an antiseptic. **f. bisulphite**, H.COH.$Na_2S_2O_5$, an antiseptic. **f.-casein**, a condensation-product of casein and formaldehyde; it is an inodorous surgical antiseptic. Syn., *formalbumin*. **f.-cotoin**, used in tuberculous diarrhea and in chronic catarrh of the bowels. Dose 4-8 gr. (0.25-0.5 Gm.). Syn., *fortoin*; *methylenedicotoin*. **f.-gelatin**, a combination of 2 % of formaldehyde added to a warm aqueous solution of gelatin; the resultant mass is powdered and used as a surgical dressing. Syn., *formacoll*; *glutol*. **f., para-**, $(CH_2O)_3$, obtained from formaldehyde by heat; antiseptic and astringent; used internally in cholera nostras and as a surgical dressing. Dose 8-15 gr. (0.52-0.97 Gm.) several times daily. Syn., *polymerized formaldehyde*; *triformal*. **f. phenolsulphonic acid**, $2(CH_7OH).C_6H_3.OH$; a wound antiseptic. **f., solution of** (*liquor formaldehydi*, U. S. P.), an aqueous solution containing not less than 37 % by weight of absolute formaldehyde. It is a powerful antiseptic. By means of heat it may be converted into a gas, which is widely used for the disinfection of rooms and dwellings previously exposed to contagion. **f., test for**, phenylhydrazin has been proposed by Vitali as a test for formaldehyde. A mixture of the two gives rise to a milky discoloration; eventually a yellowish deposit is precipitated upon the sides of the test-tube. In concentrated solutions the turbidity appears immediately. In solutions of a strength of 1 : 100 the reaction occurs after a few seconds; in those of 1 : 1000, in 1 minute; in those of 1 : 10,000, in 5 minutes; in those of 1 : 100,000, in 2 to 3 hours. See also *Kenimann*.
**formalin** (*for'-mal-in*) [*formica*, an ant]. A proprietary substance composed of a 40 % solution of formaldehyde. It is used as an antiseptic and as a fixing-agent in histological studies.
**formalith** (*for'-mal-ith*). The proprietary name for diatomaceous earth saturated with a solution of formaldehyde.
**formalose** (*for'-mal-ōs*). A 40 % solution of formaldehyde.
**formamide** (*for'-mam-id*), $CH_3NO$. One of the amides.
**formamint** (*form'-am-int*). Trade name of a preparation containing formaldehyde and lactose.
**forman** (*for'-man*). See *ether, chlormethylmenthyl-*.
**formanganate** (*for-man'-gan-āt*). A mixture of potassium permanganate and formalin; it is used to disinfect rooms.
**formanilide** (*form-an'-il-id*) [*formica*, an ant; *aniline*], $C_7H_7NO$. A substance obtained on digesting aniline with formic acid, or by rapidly heating aniline with oxalic acid. It consists of colorless prisms, readily soluble in water, alcohol, and ether, melting at 46° C.
**formate** (*for'-māt*). A salt of formic acid.
**formatio** (*for-ma'-she-o*) [L.]. A formation. **f. reticularis**, the intercrossing of the fibers of the anterior columns in the medulla.
**formation** (*for-ma'-shun*) [*formatio*]. A thing formed or the process by which it is formed.
**formative** (*for'-ma-tiv*) [*formatio*]. Concerned in the formation of tissue. **f. cells**, large, spherical cells beneath the hypoblast. **f. yolk**, the part of the ovum forming the embryo.
**formatol** (*for'-mat-ol*). A dusting-powder containing formaldehyde.
**formes frustes** (*form froost*) [Fr.]. Incomplete or atypical forms of a disease.
**formeston** (*for-mes'-ton*). Aluminum acetoformate; used as a dusting powder.
**formic, formicic** (*for'-mik, for-mis'-ik*) [*formica*, an ant]. Relating to or derived from ants, or pertaining to formic acid. **f. acid.** See *acid, formic*. **f. aldehyde**, formaldehyde.
**formicant** (*for'-mik-ant*) [*formicare*, to crawl like an ant]. Creeping, or moving with a small and feeble action; applied to the pulse when it is unequal and scarcely perceptible.
**formication** (*for-mik-a'-shun*) [see *formic*]. A sensation like that produced by ants or other insects crawling upon the skin.
**formicin** (*for'-mis-in*). Formaldehyde acetamide; used as an antiseptic and disinfectant.
**formidin** (*for'-mid-in*). Iodide of methylenedisalicylic acid; a proprietary antiseptic used as a substitute for iodoform in dressing wounds.
**formin** (*for'-min*), $C_6H_{12}N_4$. A condensation-product of formaldehyde and ammonia; it is a uric-acid solvent, diuretic, and vesical antiseptic. Dose 15-24 gr. (1.0-1.5 Gm.) in the morning in aqueous solution. Syn., *hexamethylenetetramine*; *urotropin*. **f. salicylate**. See *saliformin*.
**formochlor** (*form'-o-klor*). A solution of formaldehyde and calcium chloride. It is used as a disinfectant by spraying or vaporizing.
**formoform, formoform** (*form-o-for-in, form'-o-'orm*). A dusting-powder for perspiring feet; said to consist of formaldehyde, 0.13 %; thymol, 0.1 % zinc oxide, 34.44 %; and starch, 65.27 %. If the formaldehyde is omitted, it may be used on purulent sores.
**formol** (*form'-ol*). See *formalin*.
**formomethyial** (*form-o-meth'-il-al*), $C_3H_8O$. An ethereal oil obtained from the distillation of a mixture of methyl-alcohol, sulphuric acid, and manganese peroxide. It is anesthetic.
**formonitril** (*form-o-ni'-tril*). Hydrocyanic acid.
**formopyrine** (*form-o-pi'-rin*). A combination of antipyrine with formaldehyde.
**formose** (*form'-ōs*). A sweetish syrup obtained by Loew on the condensation of formic aldehyde in the presence of bases. It consists of a mixture of a nonfermentable sugar (formose) and a fermentable sugar, a hexose which is the starting-point of further syntheses.
**formosyl** (*for'-mo-sil*). A preparation containing formalin, boric acid, phenol, and essential oils. It is used as a mouth wash and as a nasal douche.
**formula** (*form'-ū-lah*) [dim. of *forma*, a form]. 1. A prescribed method. 2. The representation of a chemical compound by symbols. 3. A recipe or prescription. **f., constitutional**, one that indicates

# FORMULARY 371 FOSSA

by means of symbols the relation to each other of the various elements in a compound. Syn., *rational formula*. **f., dental**, one showing the number and arrangement of teeth. **f., empirical**, one that indicates only the constituents and their proportions in a molecule, as HNO₃, nitric acid. **f., glyptic**, a chemical formula designed to illustrate the arrangement and connection of the atoms of a molecule. **f., graphic**. See *f., structural*. **f., official**, one given in an official publication. **f., officinal**, a pharmaceutical formula which, though not official, is commonly followed by pharmacists. **f., rational**. See *f., constitutional*. **f., structural**, one which shows the arrangement and relation of the elements among themselves as well as the number and kind of elements composing the molecule. One in which the symbols are united by the bonds of affinity according to their quantivalence, as H–O–H. **f., vertebral**, one used to indicate the number and arrangement of the vertebræ.

**formulary** (*form'-ū-la-re*) [*formula*]. A collection of formulæ or recipes. **F., National**, a collection of widely used and well-known preparations, omitted from the United States Pharmacopeia, but collected and published by the American Pharmaceutical Association.

**formyl** (*for'-mil*) [*formic acid*; ὕλη, matter]. I. CHO. The radical of formic acid. Syn., *formoxyl*. 2. The trivalent radical, CH‴. Syn., *formylene*. **f. chloride**, f. perchloride, chloroform. **f. iodide**, Iodoform. **f.-phenetidin**, C₆H₁₁NO₂, a substance obtained from phenetidin hydrochloride by action of formic acid with anhydrous sodium formate; it is antiseptic. **f. sulphide**, sulphoform. **f. tribromide**, bromoform. **f. trichloride**, chloroform. **f. triiodide**, iodoform.

**Fornet's ring test** (*for-na'*) [Julius *Fornet*, German physician]. A precipitation test, of value in typhoid, scarlet fever, measles and syphilis.

**fornical** (*for'-nik-al*). Relating to the fornix.

**fornicate** (*for'-nik-āt*) [*fornix*]. Arched. **f. gyrus**. See *convolution, fornicate*.

**fornication** (*for-nik-a'-shun*) [*fornix*, an arch]. The illicit sexual intercourse of an unmarried person of either sex with another, whether married or not.

**fornicolumn** (*for'-ne-kol-um*) [*fornix*, an arch; *columna*, a column]. The anterior pillar of the fornix, one in each hemicerebrum. It is a bundle of fibers ascending from the albicans and thalamus, passing just caudad of the precommissure, forming the cephalic boundary of the porta, and ending in the temporal lobe.

**fornicommissure** (*for-ne-kom'-is-shūr*) [*fornix*, an arch; *commissura*, a commissure]. A lamina of greater or less thickness uniting the two hemifornices of the brain. It is not a true commissure, nor even fibrous in structure.

**fornicrista** (*for-ne-kris'-tah*). See *crista*.

**fornix** (*for'-niks*) [L., "an arch"]. I. A triangular body of white matter beneath the corpus callosum. From the apex, situated anteriorly, the anterior pillars arise and descend to form the corpora mamillaria. From the extremities of the base the posterior pillars descend into the lateral ventricles. The fornix serves as an anteroposterior commissure between the optic thalamus and the hippocampus major and the uncinate gyrus. 2. An arched body or surface; a concavity or culdesac. **f., cerebral**. See *fornix* (1). **f. conjunctivæ**, the culdesac at the point where the bulbar conjunctiva is reflected upon the lid. **f. vaginæ**, the vault of the vagina, the upper part of the vagina, forming when the passage is distended a V-like structure surrounding the cervix uteri.

**fortification-spectra** (*for-tif-ik-a-shun-spek'-trah*). Scotoma scintillans. See *teichopsia*.

**fortin** (*for'-to-in*). See *formaldehyde-cotoin*.

**fossa** (*fos'-ah*) [L., "a ditch": *pl., fossæ*]. A depression or pit. **f., acetabular**, a depression in the center of the acetabulum. **f., amygdaloid**. See *amygdaloid fossa*. **f., anconeal**. See *f., olecranoid*. **f., antecubital**, the depression in front of the elbow. **f., Broesike's**. See *Broesike's fossa*. **f. cæcalis**, a fold of peritoneum forming a pouch upon the surface of the right iliopsoas muscle, and extending to the apex of the cecum. **f., canine**, a depression on the external surface of the superior maxilla, above and to the outer side of the socket of the canine tooth. **f. capitelli**, one for the head of the malleus. **f.s,

**cerebellar**, two shallow, concave recesses on the lower part of the inner surface of the occipital plate for the reception of the hemispheres of the cerebellum. Syn., *inferior occipital fossæ*. **f.s, cerebral**, two shallow, concave recesses on the upper part of the internal surface of the occipital plate for the reception of the hemispheres of the cerebrum. Syn., *superior occipital fossæ*. **f., Claudius'**. See *Claudius' fossa*. **f., coronoid**, a depression in the humerus into which the apex of the coronoid process of the ulna fits in extreme flexion of the forearm. **f.s, costal**, the facets on the bodies of the vertebræ where articulation occurs with the heads of the ribs. **f.s, costotransverse**, depressions (usually three) on each side, upon the dorsal aspect of the three upper segments of the sacrum. **f., cranial**, any of the three depressions in the base of the skull for the reception of the lobes of the brain. See *f.s, mesocranial*; *f.s, postcranial*; *f.s, precranial*. **f. cystica**, **f. cystidis felleæ**, a depression on the lower surface of the right lobe of the liver, which holds the gall-bladder. **f., digastric**, a deep groove on the inner aspect of the mastoid process. **f., digital**, a depression at the base of the inner surface of the great trochanter of the femur. **f., epigastric**. See *infrasternal depression*. **f., flocculær**, the *f. subarcuata* in the child in whom it is larger. **f., glenoid**, the fossa in the temporal bone that receives the condyle of the lower jaw. **f., Hartmann's**. See *Hartmann's fossa*. **f. helicis**, a furrow between the helix and antihelix. **f. hemieliptica**. See *fovea hemielliptica*. **f. hemisphærica**. See *fovea hemispherica*. **f., hyaloid**, a depression in the anterior surface of the vitreous body for the crystalline lens. **f. hypophyseos**. See *f., pituitary*. **f. ileocæcalis anterior**, an inconstant pouch of the peritoneum upon the upper border of the ileocecal valve; open above and on the left side. Syn., *fossa ileocæcalis superior*. **f. ileocæcalis infima**. See *Hartmann's fossa*. **f., ileocolic**. See *Luschka's fossa*. **f., iliac, external**, the outer surface of the ilium. **f., iliac, internal**, the smooth internal surface of the ilium. **f. iliopectinea**, Scarpa's triangle. **f., infraclavicular**. See *Mohrenheim's fossa*. **f., infraspinous**, the recess on the posterior surface of the scapula occupied by the infraspinous muscle. **f., infrasternal**. See *infrasternal depression*. **f., innominata**, a shallow depression between the false vocal band and the arytenoepiglottic fold. **f. interpeduncularis**, a deep groove in the anterior surface of the mid-brain. **f., intersigmoid**, a depression on the lower surface of the mesosigmoid. **f. ischiorectal**, the depression on either side of the anus, bounded on the outer side by the tuberosity of the ischium. **f., lacrimal**, the depression in the orbital plate of the frontal bone for the reception of the lacrimal gland. **f., Landzert's**. See *Landzert's fossa*. **f., Luschka's**. See *Luschka's fossa*. **f., mandibularis**, the glenoid fossa. **f., mastoid**, the groove extending along the inner surface of the mastoid portion of the temporal bone, and forming part of the lateral sinus. **f.s, mental**, **fossæ mentalis**, shallow depressions, on each side of the mental protuberance of the mandible. **f.s, mesocranial**, **f.s, middle, of the skull**, one of the three pairs (right and left) of fossæ into which the interior base of the cranium is divided; they are deeply concave on a much lower level than the precranial fossæ, and lodge the sphenotemporal lobes of the cerebrum. Cf. *f.s, postcranial*; *f.s, precranial*. **f., Mohrenheim's**. See *Mohrenheim's fossa*. **f., navicularis**. I. The dilated portion of the urethra in the glans penis. 2. In the vulva, the depression between the posterior commissure and the fourchet. See *f., scaphoid* (1 and 2). **f. navicularis auriculæ**, fossa of the helix. **f. navicularis laryngei**, one in the mucous membrane of the larynx. **f., olecranoid**, one at the dorsal side of the distal end of the humerus, for the reception of the olecranon. **f. ovalis**, an oval depression in the right auricle of the heart. **f., ovarian**. See *Claudius' fossa*. **f., paracecal**, **f. cæcalis**, an infrequent peritoneal pouch behind and to one side of the cecum. **f., paraduodenal**. See *Landzert's fossa*. **f., parajejunal**. See *Broesike's fossa*. **f. pararectalis**, a depression in the peritoneum on the side of the rectum. **f. paravesicalis**, one on either side of the bladder. **f., patellar**. See *f., hyaloid*. **f. phrenicohepatica**, a pouch of the peritoneum between the left lateral ligament of the liver and the extremity of the left lobe. **f., pituitary**, a depression in the sphenoid bone lodging the pituitary body. **f.s, postcranial**, the lowest in position of the three pairs (right and left) of the cranial fossæ; they lodge the cerebellum, pons, and oblongata. Each fossa is formed by the posterior surface of the pyramid and inner surface of the mastoid portion of the temporal bone and the inner surface of the occipital bone below the horizontal limb of the occipital cross. f.s, precranial, the most elevated in position of the three pairs (right and left) of fossæ into which the internal base of the skull is divided. They lodge the frontal lobes of the brain and are formed by the orbital plates of the frontal bones, the cribriform plate of the ethmoid bone, and the small wings of the sphenoid bone. Cf. *f.s. mesocranial*; *f.s. postcranial*. f., radial, f. radialis, the depression on the humerus above the capitellum which accommodates the head of the radius in extreme flexion of the forearm. f., rectouterine, Douglas' culdesac. f. rhomboidea, the floor of the fourth ventricle of the brain. f., Rosenmüller's. See *Rosenmüller's fossa*. f., scaphoid, f. scaphoidea. 1. A depression in the base of the internal pterygoid plate of the sphenoid bone. 2. A depression between the helix and antihelix of the auricle. Syn., *fossa navicularis; fossa tensoris palati*. f., subarcuata, f., subarcuate, an orifice situated in the newborn on the superior margin of the petrosa, through which the vessels pass to the temporal bone. This opening disappears after birth and is represented in the adult by a depression beneath the arcuate eminence. f. subauricularis, the depression just below the external ear. f., subclavicular. See *Mohrenheim's fossa*. f., subinguinal. See *Scarpa's triangle*. f., sublingual, a depression on the internal surface of the inferior maxillary bone for containing the sublingual gland. f., submaxillary, the oblong depression on the internal surface of the inferior maxillary bone, containing the submaxillary gland. f., suborbital. See *f., canine*. f., subsigmoid, also called intersigmoid fossa, a pouch of peritoneum between the descending mesocolon and the mesosigmoid. f., sulciform, a shallow furrow in the inner fore part of the cavity of the vestibule of the ear, behind the fovea hemielliptica and the fovea hemisphærica, and into which the vestibular aqueduct opens. f., supraclavicular, f. supraclaviculais major, a depression above the clavicle within which lie the axillary blood-vessels and nerves as they emerge from the chest into the armpit. f. supraclavicularis minor, the area between the sternal and clavicular origins of the sternomastoid muscle. f., suprascapular, f., supraspinous. See *supraspinous fossa*. f., suprasternal, f. suprasternalis, the area between the ventral borders of the sternomastoid muscle, the interscapular ligament, the lower border of the mandible, and lines extended between the angles of the jaw and the mastoid process of the temporal bones. f., supratonsillar, the embryonic space above the tonsil covered by a triangular extension of membrane from the anterior pillar; it sometimes persists to adult life. Syn., *palatal recess*. f., temporal, the depression which holds the temporal muscle. f. tonsillaris. See *amygdaloid fossa*. f., triangularis, the fossa of the antihelix. f., trochanteric, a hollow at the base of the inner surface of the great trochanter of the femur. f., trochlear, a hollow in the frontal bone, below the internal angular process, furnishing attachment to the pulley of the superior oblique muscle. f., urachal, the prevesical space. f., zygomatic, a cavity below and on the inner side of the zygoma.

fosset, fossette (fos-et') [Fr.]. 1. A dimple; a small depression. 2. A small deep ulcer of the cornea.

*fossilin (fos'-il-in) [fossilis, dug up]. A trade-name for a product resembling vaselin.

fossula (fos'-u-lah) [dim. of *fossa*, a ditch]. A small fossa; any one of the numerous slight depressions on the surface of the cerebrum.

Fothergill's disease (foth'-er-gil) [John Fothergill, English physician, 1712–1780]. Neuralgia of the trigeminus. F.'s sore throat, the ulcerative angina of severe scarlatina (scarlatina anginosa).

foudroyant (foo-droi-on(t)) [Fr.]. Sudden and overwhelming; fulminant; fulgurant.

foulage (foo-lahzh') [Fr., "fulling, or pressing"]. In massage, a form of manipulation of the tissues. See *fulling*.

Foule's cells (fool). Large cells containing one or more nuclei as large as or larger than a red blood-corpuscle; they have been supposed to be diagnostic of malignant ovarian cysts.

founder (fown'-der) [*fundere*, to pour]. Laminitis (q. v.) of the horse's fore-feet, with the accompanying disorders of related parts. f., chest, founder in a horse, marked or accompanied by atrophy of the chest-muscles.

foundling (found'-ling). An abandoned infant.

fourchette, fourchet (foor-shet') [dim. of *fourche*, a fork]. 1. A fold of mucous membrane just inside the posterior commissure of the vulva. 2. A forked instrument used in division of the frenum linguæ.

Fournier's method of treating syphilis (foor-ne-a') [Jean Alfred *Fournier*, French syphilographer, 1832– ]. The alternate administration of mercury for two months and rest from mercurial medication for a month or more, the treatment being kept up for several years.

fourth cranial nerve, the trochlear nerve. f. disease. An affection resembling measles and scarlet fever. f. venereal disease, gangrenous balanitis. f. ventricle, a space between the cerebellum and pons and medulla.

fovea (fo'-ve-ah) [L., "a small pit": *pl., foveæ*]. A small depression or pit. Applied to many depressions in the body, but more particularly to the *fovea centralis retinæ*, a small pit in the macula lutea, opposite the visual axis; the spot of most distinct vision. f. centralis retinæ. See *fovea*. f. costalis, costal depression on vertebra, a demifacet for head of rib. f. hemielliptica, a small depression on the inner wall of the labyrinth. It is perforated for the passage of filaments of the auditory nerve. f. hemisphærica, a depression in the roof of the labyrinth. f. inferior, a depression at the apex of the trigonum vagi. f. inguinalis lateralis, the external inguinal fossa. f. inguinalis mesialis, the middle inguinal fossa. f. oblonga, a shallow depression on the external surface of the arytenoid cartilages. f. pharyngis, an abnormal depression in the median line of the pharynx. f. superior, a depression at the end of the sulcus limitans on the floor of the fourth ventricle. f. supravesicalis, the internal inguinal fossa. f. triangularis, a deep depression on the external surface of the arytenoid cartilages. f. trochlearis, a hollow in the orbital plate of the frontal bone for the trochlea of the superior oblique muscle.

foveate (fo'-ve-āt) [*foveæ*]. Pitted.

Foveau-Trouvé apparatus (fo-vo-tru-va'). A parabolic mirror with an incandescent or arc lamp in the focus; the former is joined to a concentrating cone which terminates in two quartz plates with a chamber between them; cold water circulates through this chamber and through the whole apparatus, absorbing the heat-rays. The quartz plate is pressed directly upon the part to be treated.

foveola (fo-ve'-o-lah) [dim. of *fovea*: *pl., foveolæ*]. A small fovea or depression. f. coccygea, a small depression back of the anus. f. granularis, any one of the small pits in the cranial bones produced by the Pacchionian bodies. f. radialis, a depression between the tendons of the extensors of the thumb when those muscles are contracted. f. retroanalis, the foveola coccygea. f. triangularis, a triangular depression between the anterior pillars of the fornix.

foveolate (fo-ve'-o-lāt). Marked with slight depressions, dimples, or pits.

Foville's syndrome (fo-vēl') [Achille Louis *Foville*, French neurologist, 1799–1878]. Alternate hemiplegia. F.'s tract, the direct cerebellar tract of the spinal cord.

Fowler's position (fow'-ler) [George Ryerson *Fowler*, American surgeon, 1848–1906]. Semi-erect position obtained by raising the head of the bed 24 to 30 inches, and by pillows.

Fowler's solution (fow'-ler) [Thomas *Fowler*, English physician, 1736–1801]. A solution containing arsenious acid, potassium bicarbonate, and tincture of lavender; liquor potassii arsenitis.

foxglove (foks'-gluv). See *digitalis*.

fractional (frak'-shun-al) [*fractio*, a breaking]. Divided. f. cultivation, the isolation of microorganisms from one another by diluting the mixture containing them to such a degree that a given quantity contains but few organisms. f. distillation. See *distillation*. f. sterilization, intermittent sterilization.

fractionation (frak-shun-a'-shun) [*fractio*, a breaking]. Chemical separation by successive operations.

fractura (frak-tūr'-ah) [L.]. A fracture. f. dentis, fracture of a tooth. f. surcularia, green-stick fracture.

# FRACTURE 373 FREMITUS

**fracture** (*frak'-tūr*) [*frangere*, to break]. A breaking, especially of a bone. For *signs* and *tests* of, see *Allis*, *Cleemann*, *Hueter*, *Keen*, *Morris*. See, also *Aran's law*, *Teevan's law*. **f.-bed**, a bed designed for patients having fractures. It usually has a hole in the center to transmit the discharges. **f.-box**, a long box, without ends or cover, used in the immobilization of fractured legs. **f., buttonhole-**, one in which a missile has perforated the bone. **f., capillary**, one consisting of only a fine crack or fissure. **f., chauffeur's**, fracture of the lower end of the radius or of the carpus produced by reversal of the starting crank while cranking. **f., Colles'**. See *Colles' fracture*. **f., comminuted**, one in which the bone is splintered. **f., complete**, one in which the bone is entirely broken through. **f., complicated**, one associated with injury of adjacent parts. **f., compound**, one in which the point of fracture is in communication with the external air through a wound of the overlying parts. **f. by contrecoup**, a fracture of the skull caused by transmitted violence, and occurring at a distance from the point struck, usually opposite. **f., dentate**, one in which the ends of the fragments are so toothed and interlocked as to prevent displacement. **f., depressed**, one in which the fractured part is depressed below the normal level, as in fracture of the skull. **f., double**, the existence of two fractures in the same bone. **f. fever**, fever due to fracture of a bone. **f., formed**, a fracture suggesting the instrument which caused it. **f., greenstick**, one side of the bone is broken, the other bent. **f., helicoid**, a spiral fracture from twisting of the long bones. Syn., *spiral fracture*. **f., impacted**, one in which one fragment is driven into the other so as to be held fast. **f., interperiosteal**. Same as *f., greenstick*. **f., Pott's**. See *Pott's fracture*. **f., simple**, one in which the overlying integument is intact. **f., Smith's**. See *Smith's fracture*. **f., spiral**. See *f., helicoid*. **f., spontaneous**, one due to a slight force, as then there is disease of the bone. **f., starred**, **f., stellate**, one in which there are fissures radiating from one point. **f., trophic**, one caused by trophic disturbance. **f., ununited**, one in which bony union has failed to occur. **f., willow-**, a greenstick fracture

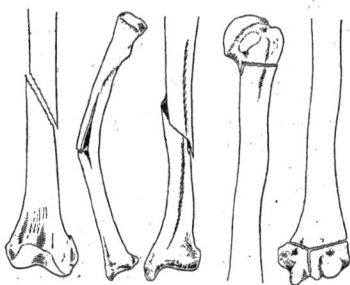

A B C D E
A. Oblique fracture of femur; B. Greenstick fracture of radius; C. Spiral fracture of tibia; D. Transverse impacted fracture of surgical neck of humerus; E. T fracture of lower end of humerus.

**Fraenkel's diplococcus** (*freng'-kel*). The pneumococcus. **F.'s glands**, minute glands opening immediately below the edge of the vocal cords. **F.'s leukemia**, acute leukemia with large mononuclear lymphocytes. **F.'s sign**, diminished tone (hypotonic) of the muscles of the lower extremities in tabes dorsalis.
**frænulum** (*fre'-nū-lum*). See *frenulum*.
**frænum** (*fre'-num*). See *frenum*.
**fragiform** (*fraj'-e-form*) [*fraga*, strawberries; *forma*, form]. Strawberry-shaped.
**fragilitas** (*fraj-il'-it-as*) [L.]. Brittleness. **f. crinium**, an atrophic condition of the hair in which the individual hairs split into numerous fibrils. **f. ossium**, abnormal brittleness of the bones.
**frambesia, frambœsia** (*fram-be'-ze-ah*) [*framboise*, raspberry]. A tropical contagious disease supposed

to be caused by the *Spirochæta pallidula*, and characterized by dirty or bright-red, raspberry-like tubercles, appearing usually on the face, toes, and genital organs. It is most frequent in young negroes. Syn., *pian; yaws*.
**frangible** (*fran'-jib-l*) [*frangere*, to break]. Liable to fracture; breakable.
**frangula** (*fran'-gū-lah*) [L.]. The bark of *Rhamnus frangula*, or alder-buckthorn. The young bark is very irritant; bark at least a year old is laxative, resembling rhubarb in action. **f., fluidextract of** (*fluidextractum frangulæ*, U. S. P.). Dose 10-20 min. (0.65-1.3 Cc.).
**frangulin** (*fran'-gū-lin*), $C_{20}H_{20}O_{10}$. A purgative glucoside from frangula. Dose 1½-3 gr. (0.097-0.19 Gm.).
**Frankenhæuser's ganglion** (*frang'-ken-hoy-zer*) [Ferdinand *Frankenhäuser*, German gynecologist, –1894]. The ganglion cervicale, a cluster of small ganglia at the side of the neck of the uterus.
**frankincense** (*frangk'-in-sens*) [*francum incensum*, pure incense]. An aromatic gum-resin. See *olibanum*. **f., common**, a concrete turpentine obtained from *Pinus palustris* and other species of *Pinus*.
**Franklin plate** (*frank'-lin*) [Benjamin *Franklin*, American physicist, 1706–1790]. A glass plate partly covered on both sides with tin-foil, used as a condenser in frictional electricity. **F. spectacles**, spectacles with each eyeglass divided horizontally into an upper lens, suited for far vision, and a lower, for close work. **Franklinic electricity**. Static or frictional electricity. **F. reaction of degeneration**, a rare form of reaction of degeneration produced by static electricity and similar to that obtained by the faradic current. **F. taste**, an acid taste perceived on applying the positive pole of the electrode of the static machine at a minimum distance of 1 or 2 mm.
**Franklinism** (*frangk'-lin-izm*). Same as *Franklinization*.
**Franklinization** (*frangk-lin-i-za'-shun*). Treatment by static or frictional electricity.
**Frank's operation**. *Of gastrotomy*: consists in forming a valve out of a small cone of the stomach-wall. The oblique incision is made and a cone of the stomach pulled out and its base sutured to the parietal peritoneum. A second incision is then made immediately above the rib-margin, the bridge of the skin is elevated, and the cone pulled up beneath it and the apex sutured into the second wound. The tube is inserted and the first skin-wound closed.
**Frasera** (*fra'-ze-rah*) [John *Fraser*, English botanist, 1750–1817]. The root of American calumba; it is a mild tonic and simple bitter. Dose in powder 1 dr. (4 Gm.).
**fraserin** (*fra'-zer-in*). A precipitate from a tincture of the root of *Frasera carolinensis* it is atonic, stimulant, and mildly astringent. Dose 1 to 3 grains.
**Frauenhofer's lines** (*frown'-hof-er*) [Joseph von *Fraunhofer*, German optician, 1787–1826]. Black lines in the solar spectrum. See *absorption lines*.
**fraxin** (*fraks'-in*) [*fraxinus*, an ash-tree], $C_{16}H_{20}O$. A glucoside from the bark of certain species of ash (*Fraxinus excelsior*, etc.). It forms fine, white, four-sided prisms, of a bitter taste. It has been proposed as a remedy for gout, rheumatism, etc.
**Fraxinus**. A genus of trees, the ash. **F. americana**, white ash; the bark is used in dysmenorrhea. **F. excelsior**, European ash; the bark, leaves, and wood are medicinal.
**F. R. C. P.** Abbreviation of *Fellow of the Royal College of Physicians*.
**F. R. C. S.** Abbreviation of *Fellow of the Royal College of Surgeons*.
**freak** (*frēk*). A popular name for a deformed person.
**freckles** (*frek'-lz*). See *lentigo*.
**Frederici's sign** (*fred-er-e'-che*). Perception of the heart sounds over the whole abdomen in cases of perforative peritonitis, with escape of gas into the peritoneal cavity.
**freezing** (*fre'-zing*). The process of hardening or congealing with cold. **f.-microtome**, a microtome provided with a contrivance for freezing artificially the tissue to be cut. **f.-mixture**, a mixture of salt and snow or ice, which absorbs a great deal of heat in undergoing solution. **f.-point**, the temperature at which a liquid freezes.
**fremitus** (*frem'-it-us*) [L., "a murmur"]. A palpable vibration, as of the chest-walls. **f., friction-**, the vibrations produced by the rubbing to-

gether of two dry surfaces, and felt by the hand. f., hepatic, f., hydatic. See hydatid fremitus. f., rhonchal, vibrations produced by the passage of air through a large bronchial tube containing mucus. f., tactile, the vibratory sensation conveyed to the hand applied to the chest of a person speaking. f., tussive, thrill felt by the hand applied to the chest of a person coughing. f., vocal, the sounds of the voice transmitted to the chest applied to the chest of a person speaking.

**frenal** (*fre'-nal*). Relating to a frenum.

**frenator** (*fre-na'-tor*) [*frenare*, to curb]. 1. Anything that inhibits, curbs, or checks. 2. Dupré's name for any one of the muscles which move the head on the atlas and axis.

**frenetic** (*fren-et'-ik*) [Fr., *frénétique*]. Relating to mental disorder.

**Frenkel's sign** (*freng'-kel*) [H. S. *Frenkel*, Swiss neurologist]. Diminished tone (hypotonia) of the muscles of the lower extremities in tabes dorsalis.

**frenosecretory** (*fre-no-se-kre'-tor-e*) [*frenum*, bridle; *secretio*, a separation]. Exercising a restraining or inhibitory power over the secretions.

**frenotomy, frænotomy** (*fre-not'-o-me*) [*frenum*; τομή, a cutting]. The cutting of any frenum, particularly of the frenum linguæ for tongue-tie.

**frenulum** (*fren'-u-lum*) [dim. of *frenum*, a bridle]. A small frenum; a slight ridge on the upper part of the valve of Vieussens. f. of Giacomini, a narrow band between the uncus and the dentate fascia. f. pudendi, the fourchet. f. valvulæ coli, a narrow membranous ridge on either side of the ileocecal valve....f. veli, the upper thickened part of the valve of Vieussens.

**frenum** (*fre'-num*) [L., "a bridle"]. A fold of integument or mucous membrane that checks or limits the movements of any organ. f. clitoridis, two folds of mucous membrane coming from the nymphæ after having united under the glans of the clitoris. f. labii inferioris and superioris, folds of mucous membrane in the median line uniting each lip to the corresponding gum. f. labiorum pudendi, the fourchet. f. linguæ, the vertical fold of mucous membrane under the tongue. f. præputii, f. of penis, the fold on the lower surface of the glans penis connecting it with the prepuce.

**frenzy** (*fren'-ze*) [φρήν, mind]. Violent mania.

**fret** [ME., *freten*, to eat up]. An abrasion; a chafing; herpes.

**fretum** (*fre'-tum*) [L., "a strait"]. 1. A constriction. 2. A strait; a channel. f. oris, the isthmus of the fauces.

**Freud's theory** (*froyd*). [Sigmund *Freud*, Austrian neurologist, 1856– ]. Hysteria and other neuroses are due to infantile sexual trauma. f. psychoanalysis.

**Freund's law** (*froynt*) [William Alexander *Freund*, German gynecologist, 1833– ]. In the progress of their growth ovarian tumors undergo changes of position: (1) While pelvic, they show a tendency to grow downward behind the uterus; (2) when they have risen out of the pelvis, they tend to fall forward toward the abdominal wall. F's operation, abdominal hysterectomy for carcinoma of the uterus.

**Freund's reaction** (*froynt*) [Hermann Wolfgang *Freund*, German gynecologist, 1859– ]. The serum from a non-cancerous patient causes lysis of cancer cells.

**F. R. F. P. S.** Abbreviation for Fellow of the Royal Faculty of Physicians and Surgeons.

**friable** (*fri'-ab-l*) [*friare*, to break into pieces]. Easily broken or crumbled.

**friars' balsam** (*fri'-arz bawl'-sam*). The compound tincture of benzoin.

**friction** (*frik'-shun*) [*fricare*, to rub]. 1. The act of rubbing. The process, in medicine, called shampooing. Also the inunction of a medicinal substance by rubbing. 2. In massage, firm circular manipulations, always followed by centripetal strokings. Friction may be practised with the fingers, with the tips of the fingers, or with one hand. f.-fremitus. See *fremitus*. f.-sound, the sound observed in auscultation, as a result of the rubbing together of adjacent parts, as of the pleural folds, the pericardium, or the peritoneum, when the layers are dry or roughened.

**frictional electricity** (*frik'-shun-al*). *Static electricity*.

**Friedlaender's bacillus** (*frēd'-len-der*) [Carl *Friedlaender*, German pathologist, 1847–1887]. *Bacillus pneumoniæ*.

**Friedlaender's decidual cells.** The large connective-tissue cells of the uterine mucosa that form the compact layer of the uterine decidua. F.'s disease, obliterative arteritis.

**Friedmann's vasomotor symptom-complex.** A train of symptoms following injury to the head, consisting of headache, vertigo, nausea, and intolerance of mental and physical exertions and of galvanic excitation; it is occasionally associated with ophthalmoplegia and mydriasis. These phenomena may subside and recur with greater intensity, with fever, unconsciousness, and paralysis of the cranial nerves, ending in fatal coma. They are probably due to an encephalitis of slow development with acute exacerbations.

**Friedreich's ataxia** (*frēd'-rik*) [Nicholas *Friedreich*, German physician, 1825–1882]. See *F.'s disease*. F.'s change of pitch, at the height of deep inspiration the tympanitic sound over pulmonary cavities becomes higher in pitch. F.'s disease. 1. Hereditary ataxia. 2. Paramyoclonus multiplex. F.'s foot, talipes cavus, with hyperextension of the toes, observed in hereditary ataxia. F.'s sign, diastolic collapse of the jugular veins in adherent pericardium.

**fright** (*frīt*). Sudden and extreme fear. f.-neuroses, certain neuromimetic disorders following injury; generally considered under the name of "traumatic hysteria." f., precordial, the precordial sensations of anxiety felt immediately before an attack of melancholic frenzy.

**frigidity** (*frij-id'-it-e*) [*frigus*, cold]. Coldness; absence of sexual désire.

**frigorific** (*frig-or-if'-ik*) [*frigus*, cold; *facere*, to make]. Producing extreme cold.

**frigotherapy** (*frig-o-ther'-ap-e*) [*frigus*; θεραπεία, therapy]. The treatment of disease by cold.

**Froehde's reaction for proteids.** A dark-blue coloration is produced by heating a solid proteid with sulphuric acid containing molybdic acid.

**Froehlich's syndrome** (*fre'-lik*) [Alfred *Froehlich*, Austrian neurologist]. Dystrophia adiposogenitalis.

**frog-belly.** The flaccid abdomen seen in children suffering from rickets or from atony of the abdominal cells the result of dyspepsia with flatulent distention.

**f.-face,** a facial deformity due to the growth of polypi or other tumors in the nasal cavities. A temporary condition of this kind may be due to orbital cellulitis or facial erysipeles. f., salt-. See *salt-frog*. f.-unit, the smallest quantity of digitalis which will kill a frog of 30 grammes weight in exactly 30 minutes.

**Frohmann's solution.** A local anesthetic used in dentistry and said to consist of cocaine hydrochloride, 0.2; morphine hydrochloride, 0.25; sodium chloride, 0.2; antipyrine, 1–2; guaiacol, 2 drops; distilled water, 100.

**Frohn's reagent** (*frōn*) [Damianus *Frohn*, German physician, 1843– ]. Add 7 Gm. of potassium iodide and 10 Cc. of hydrochloric acid to 1.5 Gm. of freshly precipitated bismuth subnitrate which has been treated with 20 Cc. of water and heated to the boiling-point.

**frôlement** (*frōl-mon(g)*) [Fr.]. 1. A brushing; in massage, a succession of slow, backward-and-forward movements, as from a center to a periphery, and of reverse. It is done with the palmar surface of the fingers, or with the roulet. 2. A rustling sound heard in auscultation.

**Frommann's lines** (*from'-an*) [Carl *Frommann*, German anatomist]. Transverse lines or striæ appearing on the axis-cylinder of medullated nerve-fibers, near the nodes of Ranvier, after the fibers have been stained with silver nitrate.

**Frommer's test for acetone** (*from'-er*). Render 10 c.c. of urine strongly alkaline by means of potassium hydroxide, add 10 to 12 drops of a 10 per cent. solution of salicylaldehyde in absolute alcohol, and warm the mixture to about 70° C. If acetone be present the fluid becomes yellow, then red, purplish-red, and, on long standing, dark red.

**frons** (*fronz*) [L.: *gen., frontis*]. The forehead.

**frontad** (*front'-ad*) [*front*; *ad*, to]. Toward the frontal aspect.

**frontal** (*front'-al*) [*front*]. 1. Pertaining to the anterior part or aspect of an organ or body. 2. Belonging to the forehead. f. bone, the anterior bone of the skull and superior bone of the face. f. eminence. See *eminence, frontal*. f. lobe. See *lobe, frontal*. f. section. See *section, frontal*. f. sinuses,

# FRONTALIS 375 FUNDUS

the hollow air-spaces in the frontal bone. **f. suture.** See *suture, frontal.*
**frontalis muscle** *(fron-ta'-lis)* [*frontalis*, of the forehead]. The frontal portion of the occipito-frontal muscle. See *muscles, table of.*
**fronten** *(fron'-ten)* [*frons*, the forehead]. Belonging to the frontal bone in itself.
**fronto-** *(fron-to-)* [*frons*, forehead]. A prefix denoting anterior position or expressing a relation with the forehead.
**frontomalar** *(fron-to-mā'-kal),* Relating to the frontal and to the malar bones. **f. suture.** See *suture, frontomalar.*
**frontomaxillary** *(fron-to-mak'-sil-a-re).* Relating to the frontal bone and the upper jaw bones.
**frontomental** *(fron-to-men'-tal).* Running from the top of the forehead to the point of the chin or relating to the forehead and chin.
**frontonuchal** *(fron-to-nū'-kal).* Relating to the forehead and the nape of the neck.
**frontooccipital** *(fron-to-ok-sip'-it-al);* Pertaining to the forehead and the occiput, or on the frontal and occipital bones.
**frontoparietal** *(fron-to-pa-ri'-e-tal).* Relating to the frontal and parietal bones.
**frontotemporal** *(fron-to-tem'-po-ral).* Relating to the frontal and temporal bones.
**front-tap contraction** *(frunt'-tap).* Contraction of the gastrocnemius muscle when the muscles of the front of the leg are tapped.
**Froriep's ganglion** *(fro'-rēp)* [August *Froriep,* German anatomist, 1849- ]. A rudimentary ganglion occasionally found in connection with one or more of the roots of the hypoglossal nerve. **F.'s induration,** myositis fibrosa.
**frost-bite.** The morbid condition of a part the result of extreme cold.
**frosted feet** *(fros'-ted).* See *chilblain.*
**frost-itch.** A name for pruritus hiemalis.
**froth.** Foam, as from the mouth. **f., bronchial,** that coming from the bronchial tubes, as in asthma.
**frottage** *(frot-ahsh')* [Fr.]. 1. Massage, rubbing. 2. A form of sexual perversion in which the subject is induced by simply rubbing against or toying with the clothing of women. An individual so afflicted is called a *frotteur.*
**frotteur** *(frot-ur')* [Fr.]. See *frottage* (2).
**frown.** To scowl. A wrinkling of the brow.
**F. R. S.** Abbreviation for *Fellow of the Royal Society.*
**fructose** *(fruk'-tōs).* See *levulose.*
**fructosuria** *(fruk-to-su'-re-ah)* [*fructus*, fruit; *οὖρον,* urine]. The presence of levulose (fruit-sugar) in the urine.
**frugivorous** *(froo-jiv'-or-us)* [*frux*, fruit; *vorare,* to devour]. Fruit-eating.
**fruit** *(froot)* [*fructus*, fruit]. 1. The developed ovary of a plant, especially the succulent, fleshy parts gathered about the same. 2. The offspring of animals. **f.-sugar.** See *levulose.*
**frumentaceous** *(fru-men-ta'-she-us)* [*frumentum*, grain]. Belonging to or resembling grain.
**frumentum** *(fru-men'-tum)* [L.]. Wheat or other grain. **frumenti, spiritus** (U. S. P.), whisky.
**ft.** Abbreviation of *fiat* or *fiant,* L. "let there be made."
**Fuchs' coloboma** *(fooks)* [Ernst *Fuchs,* German ophthalmologist, 1851- ]. A small crescentic defect of the choroid at the lower border of the optic disc. **F.'s optic atrophy,** peripheral atrophy of the bundles composing the optic nerve.
**fuchsin, fuchsine** *(fook'-sin)* [Leonhard *Fuchs,* German botanist, 1501-1566]. $C_{20}H_{19}N_3 \cdot C_2H_4O_2$. The hydrochloride or acetate of rosanilin, a lustrous, green, crystalline salt, imparting an intense red color to solutions. It is employed as a staining agent in microscopy, and has been used internally in albuminuria. Dose $\frac{1}{10}-\frac{1}{4}$ gr. (0.006-0.01 Gm.). **f. bodies.** See *Russell's bodies.*
**fuchsinophil, fuchsinophile** *(fook-sin'-o-fil)* [*fuchsin;* φιλεῖν, to love]. Stainable with fuchsin.
**Fucus** *(fu'-kus)* [φῦκος, seaweed]. A genus of marine algæ, the rockweeds. *F. vesiculosus,* bladderwrack, sea-wrack, is employed in goiter and glandular enlargements and in obesity, under the name of *antifat.* Dose of the *solid extract* 10 gr. (0.65 Gm.); of the *fluidextract* 1-2 dr. (4-8 Cc.).
**fucusaldehyde,** $C_6H_6O_2.$ Any oily compound from fucus.
**fucusol** *(fu'-kus-sol).* Same as *fucusaldehyde.*

**Fuerbringer's reaction for albumin** *(fūr'-bring-er)* [Paul *Fuerbringer,* German physician, 1849- ]. Gelatin capsules holding the double salt of mercuric chloride and sodium chloride with citric acid, opened at both ends and introduced into the urine, causes a cloudiness or flocculent precipitate in the presence of albumin. **F.'s sign,** a subphrenic abscess may be distinguished from a collection of pus above the diaphragm by the transmission, in case of the former, of the respiratory movements to a needle inserted into the abscess.
**fugacious** *(fu-ga'-shus)* [*fugere,* to flee]. In biology, falling off, or fading early; fleeting; fugitive.
**fugitive** *(fū'-jit-iv).* 1. Wandering, as *e. g.,* a pain. 2. Transient.
**Fuh's test for albumin.** Equal volumes of non-albuminous urine and a mixture of equal parts of phenol and glycerol form an emulsion which clears up on agitation, leaving a transparent and highly refractive liquid. Equal volumes of albuminous urine and this solution when mixed produce a white turbidity which remains in spite of agitation and does not precipitate. The test will show 0.1 % of albumin.
**Fukala's operation** *(fū-kal'-ah)* [Vincenz *Fukala,* German ophthalmologist]. Extraction of the crystalline lens in high degree of myopia.
**fulgurant** *(ful-gū-rant)* [*fulgur,* lightning]. Lightning-like. **f. pains,** pains that are excruciating and come on with lightning-like suddenness, and disappear as quickly.
**fulguration** *(ful-gū-ra'-shun)* [see *fulgurant*]. 1. Lightning-stroke. 2. Sensation of darting pain. 3. Treatment of malignant tumors by means of a high-frequency current of high tension and relatively low amperage applied with a cooled long spark to the area from which the malignant growth has been removed.
**fuliginous** *(fu-lij'-in-us)* [*fuliginosus,* full of soot]. Smoke-like; very dark; soot-colored; applied to lips that are covered with dry, black crusts, as a "fuliginous coating."
**fullers' earth** *(ful'-erz urth).* A siliceous non-fictile clay, used in the fulling of cloth, on account of its soapy quality; it was formerly used as an absorbent, like bole.
**Fuller's operation** *(ful'-er)* [Eugene *Fuller,* American surgeon]. Incision of the seminal vesicles.
**fulling** *(ful'-ing).* In massage, a valuable method of kneading, named from the motion used by fullers in rubbing linen between their hands. It consists in holding the limb between the palms of both hands, with the fingers fully extended, and making a rapid to-and-fro movement with each, the result being that the limb is rolled back and forth between the hands.
**fulminant, fulminating** *(ful'-min-ant, ful'-min-a-ting)* [see *fulgurant*]. Sudden, severe, and rapid in course, as *fulminant* glaucoma.
**fumigation** *(fū-mig-a'-shun)* [*fumigare,* to smoke]. Disinfection by exposure to the fumes of a vaporized disinfectant.
**fuming** *(fū'-ming)* [*fumus,* smoke]. Emitting smoke or vapor.
**function** *(fungk'-shun)* [*fungi,* to perform]. 1. The normal or special action of a part. 2. The chemical character, relationships and general properties of a substance.
**functional** *(fungk'-shun-al).* Pertaining to the special action of an organ. **f. disease,** a derangement of the normal action of an organ without structural alteration.
**funda** *(fun'-dah)* [L., "a sling"; pl., *fundæ*]. A four-tailed bandage.
**fundal** *(fun'-dal)* [*fundus*]. Pertaining to the fundus. **f. placenta,** a placenta normally attached near the fundus of the uterus.
**fundament** *(fun'-dam-ent)* [*fundus*]. 1. The foundation or base of a thing. 2. The anus.
**fundamental** *(fun-dam-ent'-al)* [*fundament*]. Pertaining to the foundation; elementary, essential. **f. tissue,** in biology, unspecialized parenchyma; those tissues of a plant through which the fibro-vascular bundles are distributed.
**fundus** *(fun'-dus)* [L., "the bottom"]. The base of an organ; the part farthest removed from the opening of the organ. **f. glands,** microscopic tubular glands in the cardiac portion of the gastric mucous membrane. **f. oculi,** the posterior portion of the interior of the eye seen by the ophtha moscope. **f. of gall-bladder,** the wide anterior end. **f.-reflex**

**test.** See *retinoscopy*. **f. uteri,** the part of the uterus remotest from the cervix. **f. ventriculi,** the large rounded cul-de-sac cephalad to the cardia of the stomach, when the organ is dilated. **f. vesicæ,** the floor or *bas fond* of the urinary bladder.

**fungal** (*fun'-gal*). Belonging to or like fungi.

**fungate** (*fun'-gāt*) [*fungus*]. 1. To grow up with a fungal appearance; also, to grow rapidly, like a fungus, as certain pathologic growths. 2. A salt of fungic acid.

**fungating** (*fun'-gāt-ing*). Applied to ulcers assuming a fungous appearance.

**fungicide** (*fun'-ji-sīd*) [*fungus; cædere,* to kill]. 1. Destructive to fungi; bactericide. 2. An agent that destroys fungi or bacteria.

**fungiform** (*fun'-jif-orm*) [*fungus; forma,* form]. Having the form of a mushroom, as the *fungiform papillæ* of the tongue.

**fungoid** (*fun'-goid*) [*fungus;* είδος, likeness]. Resembling a fungus.

**fungosity** (*fun-gos'-it-e*) [*fungus*]. A soft excrescence.

**fungous** (*fung'-gus*) [*fungus,* a toadstool]. Having the appearance or qualities of a fungus; excrescent, soft and swiftly-growing.

**fungus** (*fun'-gus*) [L.]. 1. One of the lowest orders of plants, without stems, leaves, or roots, and destitute of chlorophyl. The chief classes of fungi are the molds, or *Hyphomycetes*, the yeasts, or *Saccharomycetes,* and the molds or *Schizomycetes*. 2. A spongy, morbid excrescence, as proud flesh. **f. articuli.** See *arthritis fungosa*. **f. of brain, hernia cerebri. f. of dura mater,** a tumor of malignant nature springing from the dura and perforating the skull and its integuments. **f.-foot.** See *Madura-foot*. **f. hæmatodes,** a bleeding tumor, generally a soft carcinoma. **f., ray-,** the actinomyces.

**funic** (*fū'-nik*) [*funis*]. Pertaining to the funis.
**funicle** (*fū'-nik-l*) [*funis*]. A slender cord.
**funicular** (*fū-nik'-ū-lar*) [*funis*]. Relating to the umbilical or spermatic cord. **f. process,** the portion of the tunica vaginalis that surrounds the spermatic cord.

**funiculate** (*fū-nik'-ū-lāt*). Furnished with a funiculus.

**funiculitis** (*fū-nik-ū-li'-tis*) [*funiculus;* ιτις, inflammation]. Inflammation of the spermatic cord.

**funiculus** (*fū-nik'-ū-lus*) [dim. of *funis,* a cord; *pl., funiculi*]. 1. A cord-like structure, as the spermatic cord or the umbilical cord. 2. A bundle of nerve-fibers in a sheath of perineurium. 3. A name for the different columns of the spinal cord and medulla oblongata, as the *funiculus cuneatus, funiculus gracilis, funiculus of Rolando, funiculus teres.* **f., cuneate lateral, f. of Rolando,** a longitudinal prominence on the surface of the oblongata, between the cuneate funiculus and the line of roots of the spinal accessory nerve. **f. cuneatus,** the continuation into the oblongata of the posterolateral column of the cord; the column of Burdach. **f. gracilis,** the continuation into the oblongata of the posteromedian column of the cord; the column of Goll. **f. solitarius,** a bundle of nerve fibers in the medulla made up of the descending fibers of the glossopharyngeal, facial, and vagus nerves. **f. teres,** a column on each side of the median furrow on the floor of the fourth ventricle. Syn., *corpus teres; fasciculus teres*.

**funis** (*fū'-nis*) [L.]. A cord; the umbilical cord.

**funnel** (*fun'-el*). 1. A wide-mouthed, conical vessel ending in an open tube, used to transfer liquids from one vessel to another, and as a support for paper filters. **f.-breast, f.-chest.** See *breast, funnel-*. **f.-drainage,** drainage of diseased parts by means of funnels. **f.s, Golgi's, f.s, Golgi and Rezzonico's,** funnel-shaped structures composed of spiral threads described by Golgi and others as surrounding the axis-cylinder of a nerve-fiber and supporting the myelin. They appear to be artificially produced in the process of staining. **f., nephridial,** the funnel-shaped beginning of a renal tubule.

**funny bone** (*fun'-e-bōn*). The internal condyle of the humerus. Being crossed superficially by the ulnar nerve, blows upon it give an unpleasant sensation to the hand.

**fur.** A coating of morbid matters, including an increased amount of epithelium, seen upon the tongue in various conditions of disease.

**furca** (*fur'-kah*) [L., a fork]. A fork. **f. orbitalis,** the orbital fork; one of the earliest signs of the orbit seen in the embryo; it is a mere trace of bifurcated bony tissue.

**furcal, furcate** (*fur'-kal, fur'-kāt*) [*furca,* a fork]. Forked; divided into two equal branches.

**furcula, furculum** (*fur'-kū-lah, -lum*). A forked elevation in the floor of the embryonic pharynx; the joined clavicles of a bird; wishbone.

**furfur** (*fur'-fur*) [L.]. Dandruff or porrigo; scurf; the bran of flour. **f., microsporon.** See *tinea versicolor*.

**furfuraceous** (*fur-fū-rā'-she-us*) [*furfur,* bran]. Resembling the scales of bran, as *furfuraceous desquamation*.

**furfural, furfuraldehyde** (*fur'-fū-ral, fur-fū-ral'-de-hīd*). See *furfurol*.

**furfurol** (*fur'-fū-rol*) [*furfur,* bran; *oleum,* oil], $C_5H_4O_2$. A distillation-product from bran, sugar, etc. **f. reaction for proteids,** furfurol is produced on heating proteids with sulphuric acid.

**furfuron** (*fur'-fur-on*). A proprietary liniment for gout and rheumatism said to consist of soap, camphor, salicylic acid, acetic ether, ammonia, and extract of peppermint.

**furfurous** (*fur'-fur-us*). See *furfuraceous*.
**furibund** (*fū'-re-bund*) [*furibundus,* raging]. Raging; maniacal: applied to certain insane patients.

**furor** (*fū'-ror*) [L.]. Madness; fury; a maniacal attack. **f. amatorius,** excessive sexual desire. **f. epilepticus,** epileptic insanity. **f. femininus.** See *nymphomania*. **f. genitalis.** Same as *erotomania*. **f. secandi,** same as tomomania. **f. transitorius.** See *mania, transitory*. **f. uterinus.** See *nymphomania*.

**furred** (*furd*). Having an abnormal coating of granular or epithelial scales and other matter; as a furred tongue.

**furrow** (*fur'-o*) [AS., *furh,* a groove]. A groove. **f., digital,** one of the transverse lines or furrows on the palmar surface of the fingers. **f., genital,** a groove appearing on the genital tubercle of the fetus at the end of the second month. **f.s, interventricular,** two longitudinal grooves separating the two ventricles of the heart. **f.s, Liebermeister's,** depressions on the superior surface of the liver due to pressure of the ribs.

**furuncle** (*fū'-rung-kl*). A boil. See *furunculus*.
**furuncular** (*fū-rung'-kū-lar*) [*furuncle*]. Pertaining to a furuncle. **f. diathesis.** See *furunculosis*.

**furunculin** (*fū-rung'-kū-lin*). A preparation of yeast intended for use in furuncles, gastric and intestinal catarrhs, etc.

**furunculoid** (*fū-rung'-kū-loid*). Resembling a furuncle or boil.

**furunculosis** (*fū-rung-kū-lo'-sis*) [*furunculus*]. A condition associated with the formation of furuncles.

**furunculous** (*fū-rung'-kū-lus*) [*furunculus,* a boil]. Characterized by a continuous production of furuncles.

**furunculus** (*fū-rung'-kū-lus*) [L.]. A furuncle. A local inflammatory affection, commonly involving a skin-gland or hair-follicle, and ending in suppuration. It begins with a painful induration, followed by swelling, suppuration of the corium and subcutaneous connective tissue, and the discharge of a central slough or core. **f. anthracoides,** a small carbuncle. Syn., *anthracoid furuncle*. **f. gangrænescens, f. gangrænosus,** anthrax. **f. malignus,** anthrax. **f. orientalis,** a disease marked by the successive formation of papule, tubercle, scab, and sharply circumscribed ulcer on the face, especially the cheeks and angles of the mouth. It is common along the shores of the Mediterranean Sea. Syn., *Aleppo boil; Biskra-button; Delhi boil; Natal sore; Oriental boil; Pendjeh sore*. **f. vespajus,** a large, indolent furuncle bearing a fancied resemblance to a wasp's nest from its having a number of openings through which it suppurates.

**fuscin** (*fus'-in*) [*fuscus,* dark]. The black pigment of the retina.

**fusel oil** (*fū'-zel*), $C_5H_{11}HO$. Amyl-alcohol. An oily liquid of strong odor. It is an ingredient of crude alcohol obtained by distilling grain and potatoes.

**fusible** (*fūz'-ib-l*) [*fusus,* to melt]. Capable of being melted. **f. calculus,** a calculus that can be liquefied by heat.

**fusiform** (*fūs'-if-orm*) [*fusus,* a spindle; *forma,* shape]. Spindle-shaped. **f. lobule,** the convolution on the median aspect of the brain below the collateral fissure.

**fusion** (*fū'-zhun*) [*fundere*, to pour out]. The process of melting; the act of uniting or cohering.
**fusocellular** (*fū-so-sel'-ū-lar*) [*fusus*, a spindle; *cella*, a cell]. Spindle-celled.
**fustigation** (*fus-tig-a'-shun*) [*fustigare*, to beat]. Flogging. **f., electric,** an application of electricity in which the surface of the body is rapidly tapped with the electrodes of an induced current.

**fustin** (*fus'-tin*) [*fustus*, a knotted stick], $C_{15}H_{10}O_6$. A yellow crystalline coloring-matter obtained from *Rhus cotinus*.
**fututio** (*fū-tū'-she-o*) [L]. Sexual intercourse.
**fututrix** (*fū-tū'-triks*) [L.]. A female who practises tribadism.
**fuzzball** (*fuz'-bawl*). See *Bovista*.

# G

**g.** Abbreviation of *gram*.
**Ga.** Chemical symbol for the element gallium.
**Gabbet's method** (*gab'-et*) [Henry. Singer *Gabbet*, English physician]. *For staining tubercle bacilli:* the dried and fixed preparation is placed for 10 minutes in a solution consisting of fuchsin, 1 part; alcohol, 10 parts; phenol, 5 parts; distilled water, 100 parts; then dried with filter-paper and placed for 5 minutes in a second solution of methylene-blue, 2 parts; sulphuric acid, 25 parts; distilled water, 100 parts; it is then washed and dried. If the stain has been successful, the preparation will have a faint blue color. In the case of negative results, dehydrate with alcohol; clear and mount. A good and quick method.
**Gadberry's mixture.** A combination of iron sulphate, 100 gr.; quinine sulphate, 100 gr.; nitric acid, 100 min.; potassium nitrate, 300 gr.; water enough to make 16 oz. In the N. F. it is called mistura splenetica.
**gadinin** (*gad'-in-in*), $C_7H_{17}NO_2$. A ptomaine obtained from decomposing haddock and also from cultures of the bacteria of human feces.
**gadolinium** (*gad-o-lin'-e-um*). See *elements, table of chemical*.
**gaduin** (*gad'-ū-in*) [*gadus*], $C_{26}H_{46}O_2$. A fatty principle occurring in cod-liver oil.
**gaduol** (*gad'-ū-ol*). See *morrhuol*.
**Gadus** (*ga'-dus*) [γάδος, the whiting]. A genus of fish. *G. morrhua*, the cod; a fish from the livers of which cod-liver oil is obtained.
**Gaertner's bacillus** (*gairt'-ner*) [August *Gaertner*, German bacteriologist, 1848– ]. *Bacillus enteritidis*.
**Gaffky's table** (*gaf'-ke*) [Georg *Gaffky*, German bacteriologist]. A table by means of which the prognosis in cases of tuberculosis may be made by estimating the number of tubercle bacilli in the sputum. The cases are classified by Lawrason Brown as follows: I. Only one to four bacilli in whole preparation; II. Only one on an average in many fields; III. Only one on an average in each field; IV. Two to three on an average in each field; V. Four to six on an average in each field; VI. Seven to twelve on an average in each field; VII. Thirteen to twenty-five on an average in each field; IX. About 100 on an average in each field.
**gafsa button** (*gaf'-sah but'-n*). See *furunculus orientalis*.
**gag** [ME., *gaggen*, to gag]. 1. An instrument placed between the teeth to prevent closure of the jaws. 2. To retch, or attempt to vomit.
**Gage's test for bacterium coli in water.** Dissolve 100 Gm. of dextrose and 50 Gm. of peptone in a liter of boiling water; when cool, filter through paper until clear and add 0.25 % of phenol. To 100 Cc. of the water to be tested add 10 Cc. of the dextrose broth and place in an incubator at 38° C. for 20 hours, when, if *Bacterium coli* is present, there will be a bead on the surface. Give the bottle one vigorous shake and place it before a window. The gas will separate from the liquid and can be seen rising to the surface.
**gaiacyl** (*gi'-as-il*). See *guaiacyl*.
**gaiethol** (*gi'-eth-ol*). See *guaethol*.
**gait** (*gāt*) [Icel., *gata*, a way]. The manner of walking. **g., ataxic,** a gait in which the foot is raised high, thrown forward, and brought down suddenly, the whole sole striking the ground at once. **g., cerebellar,** a gait associated with a staggering movement. **g., cow,** a swaying movement due to knock-knee. **g., equine,** that of peroneal paralysis in which the foot is raised by flexing the thigh on the abdomen. **g., frog,** the hopping gait of infantile paralysis. **g., paraparetic,** that observed in chronic myelitis in which the steps are short and the feet dragged, from inability to lift them. **g., paretic,** a gait in which the steps are short, the feet dragged, the legs held more or less widely apart, and, as the disease progresses, there are uncertainty, shuffling, and staggering. **g., spastic,** a gait in which the legs are held close together and move in a stiff manner, and the toes tend to drag and catch. **g., steppage,** a gait observed in certain neurotic conditions, in which the foot is thrown forcibly forward, the toe lifted high in the air, the heel being first brought down and then the entire foot. **g., tabetic.** See *g., ataxic*. **g., waddling,** that of pseudohypertrophic paralysis, resembling the waddling gait of a duck.
**gala-** (*gal-ah-*) [γάλα, milk]. A prefix denoting relating to milk.
**galactacrasia** (*gal-ak-tak-ra'-ze-ah*) [*gala-*; ά, priv.; κρᾶσις, mixture]. Defect or abnormality in the composition of the milk.
**galactagogue, galactagog** (*gal-ak'-tag-og*) [*gala-*; ἀγωγός, leading]. 1. Inducing the secretion of milk. 2. An agent that increases the secretion of milk.
**galactangioleucitis** (*gal-ak-tan-je-o-lū-si'-tis*) [*gala-*; ἀγγεῖον, vessel; λευκός, white]. Lymphangitis associated with lactation.
**galactapostema** (*gal-ak-tap-os-te'-mah*) [*gala-*; ἀπόστημα, abscess: pl., *galactopostemata*]. Milk-abscess; mammary abscess associated with lactation.
**galactase** (*gal-ak'-tās*) [γάλη, milk]. An enzyme of milk. It is a normal constituent, is soluble, and is active in ripening cheese.
**galactedema** (*gal-ak-te-de'-mah*) [*gala-*; οἴδημα, a swelling]. Swelling of the breast due to accumulation of milk within it.
**galacthemia** (*gal-ak-the'-me-ah*) [*gala-*; αἷμα, blood]. 1. A milky state or appearance of the blood. 2. Bloody milk, or the giving of bloody milk.
**galactic** (*gal-ak'-tik*) [*gala-*]. Relating to or promoting the flow of milk.
**galactidrosis** (*gal-ak-tid-ro'-sis*) [*gala-*; ἱδρώς, sweat]. The sweating of a milk-like fluid.
**galactin** (*gal-ak'-tin*) [*gala-*]. 1. Same as *lactose*. 2. An amorphous substance obtainable from milk.
**galactischia** (*gal-ak-tisk'-e-ah*) [*gala-*; ἴσχειν, to suppress]. Suppression of the secretion of milk; galactoschesis.
**galactoblast** (*gal-ak'-to-blast*) [*gala-*; βλαστός, a germ]. A peculiar fat-containing globule found in the acini of the mammary gland; a colostrum corpuscle.
**galactocele** (*gal-ak'-to-sēl*) [*gala-*; κήλη, tumor]. 1. A cystic tumor of the female breast due to a collection of milk resulting from closure of a milk-duct. 2. Hydrocele with milky contents.
**galactochloral** (*gal-ak-to-klo'-ral*), $C_8H_{14}Cl_3O_6$. A mixture of galactose and chloral heated in presence of hydrochloric acid; similar to chloralose.
**galactoglycosuria** (*gal-ak-to-gli-ko-sū'-re-ah*) [*gala-*; *glycosuria*]. Glycosuria dependent upon lactation.
**galactoid** (*gal-ak'-toid*) [*gala-*; εἶδος, form]. Resembling milk.
**galactoma** (*gal-ak-to'-mah*) [*gala-*; ὄμα, a tumor: pl., *galactomata*]. Same as *galactocele*.
**galactometastasis** (*gal-ak-to-met-as'-tas-is*). Meter.
**galactoplania.**
**galactometer** (*gal-ak-tom'-et-er*). See *lactometer*.
**galactoncus** (*gal-ak-tong'-kus*) [*gala-*; ὄγκος, a swelling]. Same as *galactocele*.
**galactopathy** (*gal-ak-top'-ath-e*) [*gala-*; πάθος, disease]. 1. The application of a wet pack of warm milk (not boiled), used in the Transvaal; South Africa, in zymotic diseases. 2. Galactotherapy.
**galactophagous** (*gal-ak-tof'-ag-us*) [*gala-*; φαγεῖν, to eat]. Subsisting on milk.
**galactophysis** (*gal-ak-tof'-lis-is*) [*gala-*; φύσις, eruption]. 1. A vesicular eruption containing a milk-like fluid. 2. Crusta lactea.
**galactophora** (*gal-ak-tof'-or-ah*) [*gala-*; φέρειν, to bear]. Remedies that increase the secretion of milk.
**galactophoritis** (*gal-ak-tof-or-i'-tis*) [*gala-*; φέρειν, to bear; ιτις, inflammation]. Inflammation of a milk-duct.

**galactophorous** (gal-ak-tof'-or-us) [gala-; φέρειν, to bear]. Milk-bearing. **g.** ducts, the excretory ducts of the mammæ.

**galactophorus** (gal-ak-toff'-or-us) [gala-; φέρειν, to bear]. An artificial nipple placed over the natural organ in order to facilitate sucking and also to protect the natural nipple when abraded.

**galactophthisis** (gal-ak-toff'-this-is) [gala-; φθίσις, consumption]. Emaciation and debility due to excessive secretion of milk.

**galactophyga** (gal-ak-toff'-ig-ah) [gala-; φεύγειν, to shun]. Remedies employed to arrest the secretion of milk.

**galactophygous** (gal-ak-tof'-ig-us) [gala-; φυγή, flight]. Arresting the secretion of milk.

**galactoplania** (gal-ak-to-pla'-ne-ah) [gala-; πλάνη, a wandering]. The metastasis of milk; a disease due to the suppression of lactation and the metastasis of the milk.

**galactoplerosis** (gal-ak-to-ple-ro'-sis) [gala-; πλήρωσις, a filling]. Distention of the breast from hypersecretion of milk, or from closure of the milk-ducts.

**galactopoietic** (gal-ak-to-poi-et'-ik) [gala-; ποιεῖν, to make]. Inducing the secretion of milk; concerned in the secretion of milk.

**galactoposia** (gal-ak-to-po'-ze-ah) [gala-; πόσις, drinking]. The treatment of diseases by the use of a milk-diet; the milk-cure.

**galactopyretic** (gal-ak-to-pi-ret'-ik) [gala-; πυρετός, fever]. Relating to milk-fever.

**galactopyretus** (gal-ak-to-pi-re'-tus) [gala-; πυρετός, fever]. Milk-fever.

**galactorrhea, galactorrhœa** (gal-ak-tor-e'-ah) [gala-; ῥοία, a flow]. An excessive flow of milk.

**galactoschesis** (gal-ak-tos'-kes-is) [gala-; σχέσις, retention]. The retention or suppression of the milk.

**galactoscope** (gal-ak'-to-skōp). See *lactoscope*.

**galactose** (gal-ak'-tōs) [γάλα, milk], $C_6H_{12}O_6$. A sugar formed by boiling milk-sugar with dilute acids. It readily reduces alkaline copper solutions and is fermentable with yeast.

**galactosis** (gal-ak-to'-sis) [γαλάκτωσις]. The secretion of milk.

**galactostasis** (gal-ak-tos'-tas-is). 1. A suppression of the milk-secretion. 2. A stasis of milk in a breast.

**galactosuria** (gal-ak-to-su'-re-ah) [galactose; οὖρον, urine]. The passage of urine containing galactose.

**galactosyrinx** (gal-ak-to-sir'-ingks) [gala-; σύριγξ, a pipe]. Lacteal fistula.

**galactotherapy** (gal-ak-to-ther'-ap-e) [gala-; θεραπεία, treatment]. 1. The treatment of disease in suckling infants by the administration of remedies to the mother or wet-nurse. 2. Milk-cure.

**galactotoxicon** (ga-lak-to-toks'-ik-on). The active agent in poisonous milk.

**galactotoxin** (gal-ak-to-toks'-in). A basic poison generated in milk by the growth of microorganisms. See *tyrotoxicon*.

**galactotoxism** (gal-ak-to-toks'-izm) [gala-; τόξικον, poison]. Milk-poisoning.

**galactotrophy** (gal-ak-tot'-ro-fe) [gala-; τροφή, nourishment]. Nourishing with milk only.

**galactozemia** (gal-ak-to-ze'-me-ah) [gala-; ζημία, loss]. 1. Loss, diminution, or suppression of the milk by failure of secretion. 2. Loss of milk by wastage; oozing of milk from the nipple.

**galactozymase** (gal-ak-to-zi'-māz) [gala-; ζύμη, leaven]. A ferment found in milk capable of liquefying starch.

**galactozyme** (gal-ak'-to-sīm) [gala-; ζύμη, leaven]. A drink made by the fermentation of milk, as kefir, or kumiss; also milk fermented with common yeast.

**galacturia** (gal-ak-tū'-re-ah) [gala-; οὖρον, urine]. Milkiness of the urine; chyluria.

**galanga, galangal** (gal-an'-gah, gal'-an-gal) [Chin.]. The rhizome of *Alpinia officinarum* and of *Kæmpferia galanga* (greater galangal). The active principles are a volatile oil and a resin; the actions are those of a stimulant aromatic. Dose 15–30 gr. (1–2 Gm.).

**galangol** (gal-ang'-gol). The active principle of galangal.

**Galassi pupillary phenomenon.** When the orbicularis palpebrarum is brought into energetic use and the eye closed with vigor, there is a narrowing of the pupil, which dilates when the eye is opened.

**galazyme** (gal'-az-īm) [gala-; ζύμη, leaven]. A fermented drink, made on the continent, from milk by adding to it sugar and a special ferment.

**galbanum** (gal'-ban-um) [Heb., khelb'nāh, white milk]. 1. A gum-resin of *Ferula galbaniflua* and *F. rubricaulis*. It is expectorant, stimulant, and antispasmodic, and is useful in chronic bronchitis, amenorrhea, and chronic rheumatism. Locally it is employed in the form of a plaster for indolent swellings. Dose 10–20 gr. (0.65–1.3 Gm.). **g.,** **compound pills of,** pills of asafetida, galbanum, myrrh, and syrup of glucose. **g. plaster,** galbanum, 16; turpentine, 2; Burgundy pitch, 6; lead-plaster, 76 parts.

**Galbiati knife** (gal-be-ah'-te) [Gennaro *Galbiati*, Italian obstetrician, 1776–1844]. A special knife used in the operation of symphyseotomy.

**galbismin** (gal-biz'-min). A proprietary substitute for iodoform.

**galea** (ga'-le-ah) [galea, helmet: pl., galeæ]. 1. A form of head-bandage. 2. Headache extending all over the head. 3. The aponeurotic portion of the occipito-frontal muscle. 4. The amnion or caul. **g. aponeurotica,** galea (3). **g. capitis,** galea (1) and (3).

**galeamaurosis** (ga-le-am-aw-ro'-sis) [γαλῆ, cat, weasel; ἀμαυροῦν, to darken]. See *amaurotic cat's eye*.

**galeanthropy** (ga-le-an'-thro-pe) [γαλῆ, cat, weasel; ἄνθρωπος, man]. A form of zoanthropy in which the patient believes himself to be transformed into or inhabited by a cat.

**Galeati's glands.** See *Lieberkühn's crypts*.

**galega** (gal-e'-gah) [γάλα, milk; ἄγειν, to lead]. Goat's-rue. A genus of plants belonging in the order Leguminosæ. *G. officinalis* is a European species said to be an efficient galactagogue. Dose of fluid-extract 8–15 gr. (0.52–1.0 Gm.); of *tincture*, containing 6.5 % of extract, 50–100 min. (3.08–6.16 Cc.) 5 times daily.

**galegol** (gal-e'-gol). A proprietary preparation made from galega; used as a galactagogue.

**galena** (ga-le'-nah) [γαλήνη, lead ore]. 1. A remedy for poison. 2. Native lead sulphide.

**Galenic, Galenical** (ga-len'-ik, -al) [*Galen*]. Relating to or consistent with the teachings of Galen. Medicines were said to be galenic (*a*) when they were of vegetable origin, and not chemical or spagyric; (*b*) when they were designed for the use of human patients, and were not veterinary; (*c*) the term was often nearly equivalent to officinal, or official, in the modern sense of the latter word.

**Galen's ampulla** (ga'-len) [Claudius *Galenus*, Greek physician (in Rome), 130–200]. A dilatation of the vena magna Galeni, occurring in the middle of Bichat's fissure, between the splenium and the quadrigeminal bodies. It receives the two basilar and several small cerebral and cerebellar veins. **G.'s anastomosis.** See *G.'s nerve*. **G.'s bandage,** a six-tailed bandage for the head. **G.'s cardiac vein,** the anterior cardiac or right marginal vein. **G.'s cerate,** a cerate composed of white wax 1; oil of almonds 4; rose-water 3. **G.'s chancre.** See *Celsus' chancre*. **G.'s duct.** See *Botal, duct of*. **G.'s foramen,** the opening of the anterior cardiac vein in the right auricle. **G.'s nerve,** a small branch of the superior laryngeal nerve that passes along the posterior surface of the cricoarytenoideus posticus and anastomoses with the ascending branch of the inferior laryngeal nerve. **G.'s veins,** two large venous trunks formed by the deep cerebral veins; they unite to form the vena magna Galeni, which opens into the straight sinus. **G.'s ventricle.** See *Morgagni's ventricle*.

**galeropia,** or **galeropsia** (gal-er-o'-pe-ah, or gal-er-op'-se-ah) [γαλερός, cheerful; ὄψις, vision]. A preternaturally clear and light appearance of objects due to some perversion of the visual apparatus.

**galianconism** (ga-le-an'-ko-nizm) [γαλιάγκων, a short-armed person]. Atrophy of the arm, with shortening.

**Galium.** 1. A genus of herbs. 2. An extract from the leaves of *G. aparine*; it is antiscorbutic and diuretic; used in epilepsy, jaundice, and dropsy. *G. verum,* ladies' bed-straw; it is a refrigerant.

**gall** (gawl) [AS., gealla, bile]. 1. The bile. 2. An excoriation. 3. Nut-gall or galla. **g.-bladder,** the pear-shaped sac on the under surface of the right lobe of the liver, constituting the reservoir for the bile. **g.-cyst,** the gall-bladder. **g.-ducts,** the ducts conveying the bile. **g.-stones,** the concretions occasionally formed in the gall-bladder and bile-ducts. **g.-stones, Courvoisier's law** concerning,

# GALLA 380 GAMBOGE

when the common duct is obstructed by a stone, dilatation of the gall-bladder is rare; when the common duct is obstructed by other causes, dilatation of the gall-bladder is common.

**galla** (*gal'-ah*) [L.]. Nutgall. The *galla* of the U. S. P. is an excrescence on the leaves of *Quercus infectoria*, caused by the deposited ova of an insect. It contains tannic acid, from 10 to 75 %, gallic acid, 5 %. Dose 5–15 gr. (0.32–1.0 Gm.). **gallæ, tinctura** (U. S. P.), 20 %. Dose ½–3 dr. (2–12 Cc.). **gallæ, unguentum** (U. S. P.), nutgall ointment, **gallæ, unguentum, cum opio** (B. P.), an astringent and sedative ointment.

**gallabromol** (*gal-ah-bro'-mol*). See *gallobromol*.
**gallacetophenol** (*gal-as-et-o-fe'-nol*). See *gallacetophenone*.
**gallacetophenone** (*gal-as-et-ò-fe'-nōn*) [*galla*; *acidum, acid*; *phenone*], $CH_3CO \cdot C_6H_2(OH)_3$. A yellow powder prepared from pyrogallic acid; used as a 10 % ointment in dermatology.

**gallai** (*gal'-ai*). Aluminum gallate; it is used as a dusting-powder.
**gallanilide** (*gal-an'-il-id*). See *gallanol*.
**gallanol** (*gal'-an-ol*), $C_{13}H_{11}O_4N + 2H_2O$. The anilide of gallic acid obtained by boiling tannin with aniline; used in skin diseases in 3 to 20 % ointment, or as a dusting-powder when mixed with French chalk.
**gallate** (*gal'-āt*). A salt of gallic acid.
**gallein** (*gal'-e-in*), $C_{13}H_{14}O_7$. Pyrogallol-phthalein; one of the coal-tar colors, obtained on heating pyrogallic acid with phthalic anhydride to 200° C. It dissolves with a dark-red color in alcohol, and with a beautiful blue color in alkalies. It is an important indicator.

**gallianin** (*gal-e'-an-in*). A fluid consisting of 4 parts by volume of ozone dissolved in 1 part of an indifferent vehicle. It is used in veterinary surgery, in heat-stroke, acute pneumonia, etc.
**gallic acid** (*gal'-ik*). See *acid, gallic*.
**gallicin** (*gal'-is-in*), $C_6H_2(OH)_3COOCH_3$. A methyl ether of gallic acid; recommended in conjunctivitis and keratitis.
**gallinol** (*gal'-in-ol*). See *gallanol*.
**gallipot** (*gal'-e-pot*) [OD., *gleypōt*]. An apothecary's pot for holding ointments or confections.
**gallisin** (*gal'-is-in*) [*galla*, an oak-gall], $C_{12}H_{24}O_{10}$. An unfermentable carbohydrate found in starch-sugar.
**gallium** (*gal'-e-um*) [*Gallia*, Gaul]. A rare metal (symbol Ga, sp. gr. 5.935), extremely fusible, and related in chemical properties to aluminum, like which it is capable of forming a series of alums. Some of its compounds are poisonous. See *elements, table of*.
**gallobromol** (*gal-o-bro'-mol*), $C_7Br_2O_5H_4$. A compound obtained from bromine by action of gallic acid; it is sedative, antiseptic, and astringent. Dose 30–45 gr. (2–3 Gm.) a day. Application in 1 to 4 % solution or paste.
**galloformin** (*gal-o-form'-in*). A compound of hexamethylenamine and gallic acid; used as an internal antiseptic.
**gallogen** (*gal'-o-jen*) ellagic or benzoaric acid, $C_{14}H_8O_8$; employed as an intestinal antiseptic. Dose 5–10 gr. (0.3–1.0 gm.).
**Gallois'** test for inosit. Allow the inosit solution to evaporate to incipient dryness; moisten the residuum with a little mercuric nitrate solution and it assumes a yellow color on drying, which becomes a fine red on heating strongly, but disappears on cooling. Tyrosin, sugar, and proteïds must be absent.
**gallon** (*gal'-on*) [ME., *galon*]. A standard unit of volumetric measurement, having in the United States a capacity of 231 cubic inches; four quarts.
**galloping consumption**. The popular term for the very rapidly fatal form of pulmonary tuberculosis, in which there is rapid destruction of the lung-tissue on both sides. It is also called *florid phthisis*. **g. paresis,** rapidly progressive general paralysis.
**gallop rhythm** (*gal'-up rithm*). A peculiar form of cardiac arrhythmia, in which the sounds resemble the footfall of a horse in cantering. It is expressed by the words "rat-ta-tat." It is most frequently heard in interstitial nephritis and arteriosclerosis. It is said to be met with also in healthy persons.
**gallotannic acid** (*gal-o-tan'-ik*). See *acid, tannic*.
**galtah, galtia** (*gal'-tah, gal'-she-ah*) [*gala*, throat, as *galtah* is a form of surra in camels, in which the throat affection is one of the prominent symptoms]. Vernacular term in India for trypanosomiasis.

**Galton's whistle** (*gawl'-ton*) [Francis *Galton*, English anthropologist, 1822–1911]. An instrument used for detecting the perception of high tones by the ear.
**galvanic** (*gal-van'-ik*) [Luigi *Galvani*, Italian scientist, 1737–1798]. Pertaining to galvanism. **c. battery,** a series of cells producing electricity by chemical reaction, and so arranged as to secure the combined effect of the units. **g. electricity,** galvanism.
**galvanism** (*gal'-van-izm*) [see *galvanic*]. Primary electricity produced by chemical action.
**galvanization** (*gal-van-iz-a'-shun*) [see *galvanic*]. The transmission of a current of low electromotive force through any part of the body for the purpose of diagnosing or curing disease.
**galvano-** (*gal-van-o-*) [see *galvanic*]. A prefix denoting a galvanic or primary current of electricity.
**galvanocaustics** (*gal-van-o-kaws'-tiks*). The science of the caustic action of galvanism.
**galvanocautery** (*gal-van-o-kaw'-ter-e*). A form of thermal cautery in which the heat is produced by a galvanic current.
**galvanochemistry** (*gal-van-o-kem'-is-tre*). See *electrochemistry*.
**galvanocontractility** (*gal-van-o-kon-trak-til' i-te*). The property of being contractile under stimulation by the galvanic current.
**galvanofaradaic, galvanofaradic** (*gal-van-o-far-ad-a'-ik, -ad'-ik*). Relating to faradism and to galvanism.
**galvanofaradization** (*gal-van-o-far-ad-i-za'-shun*). The simultaneous excitation of a nerve or muscle by both a galvanic and a faradic current.
**galvanolysis** (*gal-van-ol'-is-is*). See *electrolysis*.
**galvanometer** (*gal-van-om'-et-er*) [*galvano-*; μέτρον, a measure]. 1. An instrument for the qualitative determination of the presence of an electric current. 2. An electrocardiograph.
**galvanoplasty** (*gal'-van-o-plas'-te*) [*galvano-*; πλάσσειν, to form]. Electroplating.
**galvanoprostatotomy** (*gal-van-o-pros-tat-ot'-o-me*) [*galvano-*; *prostate*; τομή, a cutting]. Bottini's operation, *q. v.*
**galvanopuncture** (*gal-van-o-punk'-tūr*). The introduction of fine needles that complete an electric circuit.
**galvanoscope** (*gal-van'-o-skōp*) [*galvano-*; σκοπεῖν, to view]. An instrument for detecting the presence and direction of a galvanic current.
**galvanoscopy** (*gal-van-os'-ko-pe*) [see *galvanoscope*]. The use of the galvanoscope.
**galvanosurgery** (*gal-van-o-sur'-jer-e*). The surgical use of galvanism.
**galvanotaxis** (*gal-van-o-taks'-is*). See *galvanotropism*.
**galvanotherapeutics** (*gal-van-o-ther-a-pū'-tiks*). Treatment by means of the galvanic current.
**galvanothermy** (*gal-van-o-ther'-me*). The galvanic production of heat.
**galvanotonic** (*gal-van-o-ton'-ik*). Both galvanic and tonic; relating to galvanotonus.
**galvanotonus** (*gal-van-ot'-on-us*) [*galvano-*; τόνος, tension]. 1. Electrotonus. 2. The continued tetanus of a muscle between the make and break contraction.
**galvanotropism** (*gal-van-ot'-ro-pizm*) [*galvano-*; πρέπειν, to turn]. The turning movements of living structure or beings under the influence of a current of electricity.
**galyl** (*gal'-il*). Tetraoxydiphosphaminodiarsenobenzene; it contains about 35 per cent. of arsenic, and is used, intravenously, for syphilis.
**galziekte** (*gal-ze-ek'-te*). A South African disease of cattle due to *Trypanosoma theileri*.
**Gambault and Philippe, median triangular tract of.** A part of the descending posteromedial tract of the spinal cord.
**Gambian fever** (*gam'-be-an*) [*Gambia*, on West coast of Africa]. A relapsing fever due to the *Trypanosoma gambiense*.
**gambir** (*gam'-bir*). An extract from the twigs and leaves of *Ourouparia gambir*. Gambir yields the same substances as *catechu*, and its action and uses are the same as those of catechu. **g., compound tincture of** (*tinctura gambir composita*, U. S. P.), used in place of the compound tincture of catechu. Dose 1 dr. (4 Cc.). **g., troches of** (*trochisci gambir*, U. S. P.), made of gambir, sugar, tragacanth, and stronger orange-flower water.
**gamboge** (*gam-booj'*). See *cambogia*.

**gamenomania** (*gam-en-o-ma'-ne-ah*). See *gamomania*.

**gametangium** (*gam-et-an'-je-um*) [γαμέτης, a spouse; ἀγγεῖον, a vessel]. A cell or organ producing or containing sexual elements, or gametes.

**gamete** (*gam'-ēt*). [γαμέτης, a spouse]. In biology, any sexual reproductive body.

**gametocyte** (*gam-et-o-sīt*). A cell different in appearance from the ordinary individuals of the species and from which the gamete is derived.

**gametophyte** (*gam'-et-o-fīt*) [*gamete*; φυτόν, a plant]. In biology, the sexual form of such plants as exhibit alternate generation.

**gametoschizont** (*gam-et-o-skiz'-ont*). A parasite of the sexual type in organisms exhibiting alternation of generation.

**Gamgee tissue** (*gam'-je*). Absorbent cottonwool.

**gamic** (*gam'-ik*) [γάμος, marriage]. In biology, sexual.

**gammacism** (*gam'-as-ism*) [*gammacismus;* γάμμα, the Greek letter Γ, γ, the equivalent of the letter G or g.]. Difficulty in pronouncing the letters "g" and "k."

**gamo-** (*gam'-o*) [γάμος, marriage]. A prefix denoting union, junction.

**gamogenesis** (*gam-o-jen'-es-is*). [γάμος, marriage; γένεσις, generation]. Sexual reproduction.

**gamogenetic** (*gam-o-jen-et'-ik*) [γάμος, marriage; γένεσις, origin]. Relating to gamogenesis.

**gamomania** (*gam-o-ma'-ne-ah*) [γάμος, marriage; μανία, mania]. Insane desire of marriage.

**gamomorphism** (*gam-o-mor'-fizm*) [γάμος, marriage; μορφή, form]. Puberty; sexual maturity.

**gamopetalous** (*gam-o-pet'-al-us*) [γάμος, marriage; πέταλον, a leaf]. In biology, sympetalous; having the petals more or less united.

**gamophyllous** (*gam-o-fil'-us*) [γάμος, marriage; φύλλον, a leaf]. In biology, symphyllous; having the floral envelops united into a single perianth-whorl.

**ganglia** (*gang'-gle-ah*). Plural of *ganglion*.

**gangliac, ganglial, gangliar** (*gang'-gle-ak, gang'-gle-al, gang'-gle-ar*). Same as *ganglionic*.

**gangliasthenia** (*gang-gle-as-the'-ne-ah*) [*ganglion;* ἀσθένεια, weakness]. Ganglionic asthenia; neurasthenia due to defect of ganglionic function.

**gangliate, gangliated** (*gan'-gle-āt, -ed*). 1. Furnished with ganglia. 2. Intertwined or intermixed.

**gangliectomy** (*gang-le-ek'-to-me*) [*ganglion;* ἐκτομή, excision]. Excision of a ganglion.

**gangliform** (*gang'-gle-form*) [*ganglion; forma*, form]. Having the shape of a ganglion.

**gangliitis** (*gang-gle-i'-tis*). See *ganglionitis*.

**ganglioblast** (*gang'-gle-o-blast*) [*ganglion;* βλαστός, a germ]. An embryonic ganglion-cell. Syn., *esthesioblast*.

**gangliocyte** (*gang'-gle-o-sīt*) [*ganglion;* κύτος, cell]. A ganglion-cell.

**ganglioma** (*gang-gle-o'-mah*) [*ganglion;* ὄμα, tumor]. A tumor or swelling of a lymphatic ganglion.

**ganglion** (*gang'-gle-on*) [γάγγλιον, a knot: *pl., ganglia*]. 1. A well-defined collection of nerve-cells and fibers forming a subsidiary nerve-center. 2. A lymph node. 3. An enlarged bursa in connection with a tendon. **g., Andersch's.** See *g., petrous*. **g., Arnold's.** See *g., otic.* **g., auditory,** a prominence on the lateral wall of the fourth ventricle traversed by the auditory striæ. Syn., *tuberculum acusticum*. **g., auricular.** See *g., otic.* **g.s, basal,** those at the base of the cerebrum; they include the corpora striata (caudate and lenticular nuclei) and optic thalami. **g., Bidder's.** See *Bidder's ganglion*. **g., Bochdalek's.** See *Bochdalek's ganglion*. **g., cardiac, superior.** See *Wrisberg's ganglion* (1). **g., carotid,** one in the lower part of the cavernous sinus, beneath the carotid artery; *roots,* filaments from the carotid plexus; *distribution,* carotid plexus. **g., carotid, inferior,** one of the lower portion of the carotid canal; *root,* carotid plexus; *distribution,* filaments to the carotid artery. **g., carotid, superior,** one in the upper portion of the carotid canal; *root,* carotid plexus; *distribution,* filaments to the carotid artery. **g., Casserian.** See *Gasserian ganglion*. **g.-cell,** the large nerve-cell characteristic of the ganglia; similar cells are found in other parts of the nervous system. **g.s, cephalic,** sympathetic ganglia of the head. They include the ophthalmic, sphenopalatine, otic, and submaxillary. **g., cervical, inferior,** that between the neck of the first rib and the transverse process of the last cervical vertebra; *roots,* three lower cervical,

first dorsal; *distribution,* cardiac nerves and plexus, etc. **g., cervical, middle,** or thyroid, that opposite the fifth cervical vertebra, near the inferior thyroid artery; *roots,* fifth and sixth cervical nerves, spinal nerves and ganglia; *distribution,* thyroid gland, cardiac nerve, cavernous plexus. **g., cervical, superior,** that opposite the second and third cervical vertebræ; *roots,* four upper cervical, petrosal, vagus, glossopharyngeal, and hypoglossal nerves; *distribution,* superior, inferior, external, internal branches; carotid and cavernous plexuses. **g., cervical (of uterus),** that near the cervix uteri; *roots,* filaments from the hypogastric plexus, sacral ganglia, and sacral nerves; *distribution,* uterine nerves. **g., ciliary.** See *g., ophthalmic.* **g. of Cloquet.** See *Cloquet's ganglion.* **g., coccygeal.** See *g. impar.* **g., Corti's.** See *Corti's ganglion.* **g., Gasser's.** See *g., Gasserian.* **g., Gasserian, semilunar,** *location,* fossa on the anterior part of the petrous portion of the temporal bone, near the apex; *roots,* fifth cranial nerve, carotid plexus; *distribution,* cervical nerves and various plexuses; maxillary nerves. **g., geniculate,** one in the aqueduct of Fallopius; *roots,* large and small superficial petrosal; *distribution,* facial. **g., hepatic,** one around the hepatic artery; *roots,* hepatic branches of the semilunar ganglion; *distribution,* liver. **g. impar,** **g., coccygeal,** that on the anterior surface of the tip of the coccyx, uniting the two sympathetic nerves; *root,* sympathetic; *distribution,* sympathetic. **g., inferior (of vagus),** one near jugular foramen; *roots,* hypoglossal and cervical nerves and various plexuses; *distribution,* vagus. **g., inframaxillary, anterior,** one near the incisor teeth; *root,* inferior maxillary nerve; *distribution,* filaments to the teeth. **g., inframaxillary, posterior,** one near the last molar tooth; *root,* inferior maxillary nerve; *distribution,* filaments to the teeth. **g., intercarotic,** one connected with the carotid plexus at the bifurcation of the common carotid artery. **g., interpeduncular.** See *Gudden's ganglion.* **g., jugular (Ehrenritter's),** one in the upper part of the jugular foramen; *root,* glossopharyngeal; *distribution,* continuation of the glossopharyngeal. **g., jugular (of vagus),** one in the jugular foramen; *root,* vagus; *distribution,* vagus. **g., lenticular.** See *g., ciliary.* **g., lingual.** See *g., submaxillary.* **g., Lowit's,** the bulbus arteriosus. **g., Ludwig's.** See *Ludwig's ganglion.* **g.s, lumbar** (4 or 5), on each side and behind the abdominal aorta; *root,* sympathetic; *distribution,* sympathetic. **g., lymphatic,** any lymphatic gland. **g., Meckel's.** See *g., sphenopalatine.* **g., mesenteric, inferior,** one in the inferior mesenteric artery; *root,* inferior mesenteric plexus; *distribution,* mesentery and intestine. **g., mesenteric, lateral,** one in connection with superior mesenteric plexus, on left side; *root,* superior mesenteric plexus; *distribution,* mesentery and bowel. **g., mesenteric, superior,** one near the origin of the superior mesenteric artery; *root,* superior mesenteric plexus; *distribution,* sympathetic. **g.. nasal.** See *g., sphenopalatine.* **g., nasopalatine.** See *Cloquet's ganglion.* **g., nodosum,** the ganglion on the trunk of the vagus just before the jugular foramen. **g., ophthalmic,** that in the posterior part of the orbit; *roots,* nasal branch of the ophthalmic, third nerve, cavernous plexus, and Meckel's ganglion; *distribution,* short ciliary. **g. orbital.** See *g., ophthalmic.* **g., otic (Arnold's),** one below the foramen ovale; *roots* inferior maxillary, auriculotemporal, glossopharyngeal, facial, sympathetic, and internal pterygoid; *distribution,* tensor tympani, tensor palati, chorda tympani. **g., petrous (Andersch's),** one on the lower border of the petrous portion of the temporal bone; *root,* glossopharyngeal; *distribution,* tympanic, sympathetic, and vagus. **g., pharyngeal,** one near the ascending pharyngeal artery; *root,* carotid plexus; *distribution,* carotid plexus. **g., phrenic,** one under the diaphragm at the junction of the right phrenic nerve and phrenic plexus; *root,* sympathetic; *distribution,* to the diaphragm, inferior vena cava, suprarenal capsule, hepatic plexus. **g.s, pneumogastric.** See *g., vagus.* **g., prostatic (of Müller),** one on the prostate; *root,* prostatic plexus; *distribution,* filaments to seminal vesicles and cavernous tissue of penis. **g., pterygopalatine.** See *g., sphenopalatine.* **g., Remak's.** See *Remak's ganglion.* **g., renal,** one around the renal artery; *root,* renal plexus; *distribution,* renal artery. **g. of Ribes.** See *Ribes, ganglion of.* **g., sacral,** four or five pairs on the ventral surface of the sacrum; *root,* sympathetic; *distribution,* sympathetic. **g., Scarpa's.**

See *Scarpa's ganglion*. **g., Schacher's.** See *g., ophthalmic*. **g., semilunar.** See *Gasserian's ganglion*.
**g.s, semilunar,** two ganglia, right and left, near the suprarenal bodies, in front of the crura of the diaphragm; *roots*, solar plexus and great splanchnic nerves; *distribution*, solar plexus. **g.s, solar.** See *g.s, semilunar*. **g., sphenopalatine (Meckel's),** one in the sphenomaxillary fossa, near the sphenopalatine foramen; *roots*, superior maxillary, facial, sympathetic; *distribution*, anterior, posterior, and external palatine, nasopalatine, superior nasal, Vidian, •pharyngeal. **g.s, spinal,** those on the spinal nerve near the intervertebral foramina. **g. spirale.** See *Corti's ganglion*. **g.s, splanchnic.** See *g.s, semilunar*. **g., submaxillary,** that above the submaxillary gland; *roots*, gustatory, chorda tympani, submaxillary, sympathetic; *distribution*, mouth, submaxillary gland, and Wharton's duct. **g., superior (of vagus)**, located at the jugular foramen; *roots*, superior cervical and petrous ganglia and spinal accessory; *distribution*, vagus. **g., suprarenal,** that at the junction of the great splanchnic nerves; *root*, solar plexus; *distribution*, suprarenal plexus. **g., thoracic,** twelve pairs between the transverse processes of the vertebræ and the heads of the ribs; *root*, sympathetic; *distribution*, splanchnic nerves and branches to spinal nerves and plexuses. **g., thyroid, inferior.** See *g., cervical, middle*. **g., thyroid, superior.** See *g., cervical, superior*. **g., tympanic,** that in the canal between the lower surface of the petrous portion of the temporal bone and the tympanum; *root*, tympanic branch of the glossopharyngeal; *distribution*, tympanum. **g., vagus.** (1) Of the root. See *g., jugular*. (2) Of the trunk: *location*, below the jugular foramen; *root*, vagus; *distribution*, vagus. **g., Valentin's.** See *Valentin's ganglion*. **g., ventricular.** See *Bidder's ganglion*. **g., vestibular,** that in the aqueduct of Fallopius; *root*, geniculate ganglion; *distribution*, vestibular nerve. **g., Walther's.** See *g. impar*. **g. of Wrisberg.** See *Wrisberg's ganglion*.
**ganglionar** (*gang-gle-on'-er*). Pertaining to, or having the characteristics of, a ganglion.
**ganglionated** (*gang-gle-on-a'-ted*). Same as *ganglionate*.
**ganglionervous system** (*gang-gle-o-ner'-vus*). The sympathetic nervous system.
**ganglioneure, ganglioneuron** (*gang-gle-o-nūr', -nū'-ron*) [*ganglion*; νεῦρον, a nerve]. A neuron the cell-body (nerve-cell) of which lies within the spinal or the cerebral ganglia.
**ganglioneuroma** (*gang-gle-o-nū-ro'-mah*). A neuroma containing ganglion-cells.
**ganglionic** (*gang-gle-on'-ik*) [*ganglion*]. Pertaining to or for the nature of a ganglion. **g. canal,** the canal around the cochlear modiolus, for the spiral ganglion. **g. centers,** masses of gray matter of the brain lying between the floor of the lateral ventricles and the decussation of the anterior pyramids of the cord. They include the optic thalami, corpora striata, and others.
**ganglionica** (*gang-gle-on'-ik-ah*). Drugs affecting the sensibility of the regions supplied by the sympathetic nerve.
**ganglionitis** (*gang-gle-on-i'-tis*) [*ganglion*; *ιτις*, inflammation]. Inflammation of a ganglion.
**gangliopathy** (*gang-gle-op'-ath-e*) [*ganglion*; πάθος, disease]. Any disorder dependent upon a diseased condition of a ganglion; any diseased state of a ganglion.
**Gangolphe's sign** (*gahn-golf'*) [Louis *Gangolphe*, French surgeon]. In intestinal obstruction a serosanguineous effusion in the abdomen soon after strangulation has taken place.
**gangosa** (*gan-go'-sah*) [Sp. *gangoso*, snuffling], A destructive form of nasopharyngitis.
**gangræna oris** (*gan-gre'-nah o'-ris*). Cancrum oris (*q. v.*).
**gangrene** (*gang'-grēn*) [γάγγραινα, a sore, from γραίνειν, to gnaw]. 1. Mortification or death of a part of the body from failure in nutrition. 2. The putrefactive fermentation of dead tissue. **g., atrophic,** that due to embolism or thrombosis. **g., carbolic-acid,** dry gangrene from carbolized dressings. **g., diabetic,** a moist gangrene sometimes occurring in diabetic persons. **g., dry,** shriveling and desiccation from insufficiency of arterial blood. **g., embolic,** that caused by an embolus that cuts off the supply of blood. **g., hospital,** a contagious, rapidly fatal form, arising under crowded conditions, particularly in military hospitals. **g., mixed,** dry gangrene with moist patches. **g., moist,** a form with abundance of serous exudation and rapid decomposition. **g. nosocomial,** hospital gangrene. **g., primary,** that without preceding inflammation of the part. **g., pulpy.** See *g., hospital*. **g., secondary,** a form with preceding inflammation. **g., senile,** that attacking the extremities of the aged. **g., symmetrical,** that attacking corresponding parts of opposite sides. Syn., *Raynaud's disease*. **g., tachetic,** a form marked by the appearance of ecchymotic spots, of greater or less extent, on various parts of the body. It is believed to be due to blood-poisoning. **g., white,** a moist gangrene due to anemia and lymphatic obstruction.
**gangrenopsis** (*gang-gren-op'-sis*) [*gangrene*; ὄψις, the face]. Synonym of *cancrum oris*.
**gangrenosis** (*gang-gren-o'-sis*) [*gangrene*; νόσος, disease]. The condition of being or of becoming mortified or gangrenous.
**gangrenous** (*gang'-gren-us*) [*gangrene*]. Pertaining to or of the nature of gangrene. **g. emphysema.** See *edema, malignant*.
**ganister, gannister** (*gan'-is-ter*) [MHG., *ganster*, a spark]. A very hard silicious fire-clay forming the floor of coal-seams in Yorkshire and Lancashire, England. **g. disease,** the formation of fibroid tissue in the lungs, occurring in ganister miners and grinders, from the irritation produced by breathing the fine dust.
**ganjah** (*gan'-jah*). See *Gunjah*.
**Gant's line** [Frederick James *Gant*, English surgeon]. An imaginary line below the greater trochanter, serving as a guide in section of the femur. G.'s **operation,** *for ankylosis of the hip-joint:* division of the shaft of the femur just below the lesser trochanter.
**gap.** A cleft, fissure, or opening. **g., Bochdalek's.** See under *Bochdalek*.
**gape** (*gāp*). To yawn; the act of yawning.
**gapes** (*gāps*). A disease of young fowls, caused by the presence of a nematode worm, *Syngamus trachealis*, in the trachea.
**gaps, cranial** (*gaps*). Certain occasional congenital fissures of the skull.
**garantose** (*gar'-an-tōs*). Saccharin.
**garbage** (*gar'-bāj*). The refuse materials of kitchen-cookery, etc. **g.-furnace,** a furnace in use in cities and towns to consume the waste material of the place.
**garbled** (*gar'-bld*) [OF., *garbeler*, to inspect closely]. Applied to crude drugs which have been separated from worthless material and made ready for market.
**Garcinia** (*gar-sin'-e-ah*) [Laurent *Garcin*, French botanist, 1752– ]. A genus of guttiferous trees of old-world tropical regions. G. **hanburii,** G. **morella,** G. **pictoria,** and G. **travancorica** afford gamboge. G. **mangostana** yields the palatable fruit called mangosteen.
**gardenin** (*gar-de'-nin*). A compound, $C_{28}H_{30}O_{10}$, obtained from *Gardenia lucida*.
**Gardiner-Brown's test.** In labyrinthine disease the patient ceases to hear the sound of a tuning-fork placed upon the vertex from half a second to several seconds before the examiner ceases to feel its vibrations.
**Garel's sign, Garel-Burger's sign** (*gar-el'*) [Jean *Garel*, French physician, 1852– ]. Luminous perception by the eye of the sound side only when an electric light is placed in the buccal cavity; it is observed in empyema of the antrum of Highmore.
**gargarism, gargarisma** (*gar'-gar-izm, gar-gar-iz'-mah*) [γαργαρισμός, a gargling: *pl.*, *gargarismata*]. A gargle or throat wash.
**garget** (*gar'-get*) [ME., *gargat*, the throat]. 1. A swelling of the throat in swine or cattle. 2. A knotty condition of the udder in cows, attended with inflammation. Syn., *mammitis; weed*.
**gargle** (*gar'-gl*) [OF., *gargouiller*, to gargle]. 1. A solution for rinsing the pharynx and nasopharynx. 2. To rinse the pharynx and nasopharynx.
**Garland's S-curve** (*gar'-land*) [George Minot *Garland*, American physician, 1848– ]. See *Ellis' sign*.
**garlic** (*gar'-lik*). See *allium*.
**garofen** (*gar'-o-fen*). A vegetable analgesic and antipyretic intended as a substitute for morphine and acetanilide.
**Garrod's test for hematoporphyrin in the urine** (*gar'-od*) [Sir Alfred Baring *Garrod*, English physician, 1819–1907.] Add to 100 Cc. of urine 20 Cc. of a

GARROT 383 GASTRITIS

10 % solution of caustic soda and filter. To the filtrate, thoroughly washed in water, add absolute alcohol and enough hydrochloric acid to dissolve perfectly the precipitate. Spectroscopic examination may now be made of the solution for the two absorption bands characteristic of hematoporphyrin. G.'s test for uric acid in the blood, to 30 Cc. of serum add 0.5 Cc. of acetic acid, and immerse a fine thread. The thread becomes incrusted with uric acid crystals. This is obtained especially in gout, but also in leukemia and chlorosis.

**garrot** (*gar'-ot*) [Fr., *garroter*, to bind]. An instrument for compression of an artery by twisting a circular bandage about the part.

**garroting** (*gar-ot'-ing*) [Sp., *garrote*, strangulation by means of an iron collar]. In forensic medicine, a term used in England to signify the forcible compression of a victim's neck by robbers or other criminals.

**garrulitas vulvæ** (*gar-oo'-li-tas vul'-vee*) [L.]. The noisy expulsion of gas from the vagina.

**garrulity** (*gar-oo'-lit-e*) [*garrire*, to prattle]. Talkativeness. **g. of the vulva.** See *garrulitas vulvæ*.

**Gartner's canal, G.'s duct** (*gart'-ner*) [Herman Treschow Gartner, Danish anatomist, 1785–1827]. A tube extending from the broad ligament to the walls of the uterus and vagina during intrauterine life; it is a vestige of the main portion of the Wolffian duct. G.'s cyst. A cystic tumor developed from Gartner's duct.

**gas** [a word coined by the Belgian chemist, van Helmont; it was suggested by χαος, chaos]. An air-like fluid. The word is especially applied to those fluids that, under normal conditions, are aeriform; while those that can be readily condensed to liquids are termed *vapors*. **g., Clayton's,** sulphurous acid gas generated by means of the Clayton furnace, for disinfection and for destroying rats and other vermin. **g.-eye,** a peculiar disease prevalent among the employes of the gas pumping stations in the natural gas regions of the United States. The eyes are inflamed, tender, and sensitive to light. **g., laughing,** nitrous oxide. **g., marsh-, methane. g., olefiant,** ethylene. **g., permanent,** a term formerly applied to those gases, as oxygen, nitrogen, hydrogen, that were thought to be nonliquefiable. **g., phlogisticated,** nitrogen. **g., sewer,** the mixture of gases and vapors which emanate from a sewer.

**Gascoigne's ball** (*gas'-koyn*). Pulverized oriental bezoar formed into balls.

**gaseous** (*gas'-e-us*). Of the nature of a gas.

**gaskaral-H** (*gas'-kar-al*). A proprietary astringent and diuretic remedy. Dose 1½–2 oz. (50–60 Cc.) of the infusion (1 : 20). Syn., *aghara*.

**Gaskell's bridge** (*gas'-kel*) [Walter Holbrook *Gaskell*, English physiologist, 1847–1914]. The atrioventricular bundle.

**gasolene** (*gas'-o-lēn*) [*gas*]. Canadol. A product obtained from petroleum, boiling at 70° to 90° C.; specific gravity, .660 to .690, or 80° to 75° B. It is used in the extraction of oils from oil-seeds and in carburetting coal-gas.

**gasometer** (*gas-om'-et-ur*) [*gas*; μέτρον, a measure]. A device for estimating the amount of gas present.

**gasometric** (*gas-o-met'-rik*). Relating to the measurement of gases.

**gasometry** (*gas-om'-et-re*). See *analysis, gasometric*.

**gasp.** To catch for breath. To breathe spasmodically with open mouth.

**gassed** (*gasd*). Overcome by noxious gas, as by chlorine.

**gasserectomy** (*gas-er-ek'-to-me*) [*Gasserian ganglion*; ἐκτομή, excision]. Excision of the Gasserian ganglion.

**Gasserian** (*gas-e'-re-an*) [referring to Achilles Pirminius *Gasserius*, German surgeon, 1505–1577]. G. artery. 1. A branch given off by the internal carotid to the Gasserian ganglion. 2. A branch of the middle meningeal artery to the Gasserian ganglion. G. fontanel. See *fontanel, Casson's*. G. ganglion, the ganglion of the sensory root of the fifth cranial nerve. See *ganglion, Gasserian*.

**gaster** (*gas'-ter*) [γαστήρ, stomach]. The stomach; the abdomen.

**gasteralgia** (*gas-ter-al'-je-ah*). See *gastralgia*.

**gasterangiemphraxis** (*gas-ter-an-je-em-fraks'-is*) [*gaster;* ἀγγεῖον, vessel; ἔμφραξις, obstruction]. 1. Congestion of the stomachic blood-vessels. 2. Pyloric obstruction.

**gasterasthenia** (*gas-ter-as-the'-ne-ah*). See *gastrasthenia*.

**gasterataxia** (*gas-ter-at-aks'-e-ah*). See *gastroataxia*.

**gasterechema** (*gas-ter-ek-e'-mah*) [*gaster;* ἤχημα, sound]. A sound heard in the auscultation of the stomach.

**gasteremphraxis** (*gas-ter-em-fraks'-is*). 1. See *gasterangiemphraxis* (2). 2. Overdistention of the stomach.

**gasterhysterotomy** (*gas-ter-his-ter-ot'-o-me*). See *gastrohysterotomy*.

**gasteric** (*gas-ter'-ik*). Same as *gastric*.

**gasterin** (*gas'-ter-in*). A preparation of the gastric juice of dogs; it is used as is pepsin.

**gastero-** (*gas'-ter-o-*) [*gaster*]. See *gastro-*.

**Gastou's syndrome** (*gas-too*). Anesthetic prurigo, sometimes observed in cases of alcoholism.

**gastradenitis** (*gas-trad-en-i'-tis*). See *gastroadenitis*.

**gastræmia** (*gas-tre'-me-ah*). See *gastremia*.

**gastral** (*gas'-tral*) [*gaster*]. Pertaining to the gaster or stomach; gastric.

**gastralgia** (*gas-tral'-je-ah*) [*gastro-;* ἄλγος, pain]. Paroxysmal pain in the stomach.

**gastralgokenosis** (*gas-tral-go-ken-o'-sis*) [*gastro-;* ἄλγος; κενός, empty]. A sensory neurosis due to emptiness of the stomach.

**gastraneuria** (*gas-trah-nū'-re-ah*) [*gastro-;* νεῦρον, a nerve]. Impaired or defective action of the nerves of the stomach.

**gastraneurysma** (*gas-tra-nū-riz'-mah*). See *gastrectasis*.

**gastrasthenia** (*gas-tras-the'-ne-ah*) [*gastro-;* ἀσθένεια]. Debility of the stomach.

**gastratrophia** (*gas-tra-tro'-fe-ah*) [*gastro-;* ἀτροφία, atrophy]. Atrophy of the stomach.

**gastrectasis, gastrectasia** (*gas-trek'-tas-is, gas-trek-ta'-ze-ah*) [*gastro-;* ἔκτασις, a stretching out]. Dilatation of the stomach.

**gastrectomy** (*gas-trek'-to-me*) [*gastro-;* ἐκτομή, a cutting out]. Excision of the whole or a part of the stomach.

**gastrelcobrosis** (*gas-trel-ko-bro'-sis*) [*gastro-;* ἕλκος, ulcer; βρῶσις, an eating]. Ulceration of the stomach.

**gastrelcoma** (*gas-trel-ko'-mah*) [*gastro-;* ἕλκος, ulcer]. A gastric ulcer.

**gastrelcosis** (*gas-trel-ko'-sis*) [see *gastrelcoma*]. Ulceration of the stomach.

**gastremia** (*gas-tre'-me-ah*) [*gastro-;* αἷμα, blood]. Congestion of the coats of the stomach.

**gastrenteralgia** (*gas-tren-ter-al'-je-ah*) [*gastro-;* ἔντερον, intestine; ἄλγος, pain]. Neuralgia of the stomach and bowels.

**gastrenteromalacia** (*gas-tren-ter-o-mal-a'-se-ah*) [*gastro-;* ἔντερον, intestine; μαλακία, softening]. Softening of the stomach and intestines (asserted by some to be merely a post-mortem condition).

**gastric** (*gas'-trik*) [γαστήρ, stomach]. Pertaining to the stomach. **g. artery.** See under *artery*. **g. crisis,** a severe paroxysmal attack of pain in the stomach, accompanied by obstinate vomiting, occurring in the course of locomotor ataxia. **g. fever,** acute gastritis. **g. juice,** the secretion of the glands of the stomach. It is a clear, colorless liquid, having an acid reaction and a specific gravity of 1002.5, and containing 5 % of solid matter. A small amount—0.2 to 0.4 %—of hydrochloric acid and a ferment called pepsin are the essential elements. It digests proteids and precipitates the casein of milk. **g. juice, psychic,** Pawlow's term for the gastric juice caused to be secreted by simply showing food to hungry animals. **g. secretion,** same as *gastrin, q. v.*

**gastricism** (*gas'-tris-izm*) [*gaster*]. 1. The theory that ascribes most diseases to some gastric derangement. 2. A gastric disorder.

**gastrin** (*gas'-trin*). A hormone, made in the pyloric glands of the stomach, and supposed to excite secretion of the fundus cells.

**gastritic** (*gas-trit'-ik*) [*gaster*]. Pertaining to or affected with gastritis.

**gastritis** (*gas-tri'-tis*) [*gastro-;* ιτις, inflammation]. Inflammation of the stomach. It may be acute or chronic, catarrhal, suppurative or phlegmonous, or diphtheritic. **g., atrophic,** a chronic form with atrophy of the mucous membrane. **g., croupous, g., diphtheritic, g. membranacea,** a rare form, characterized by formation of a false membrane and recognized by hyaline products upon the mucosa. It may occur as an extension of pharyngeal diphtheria or secondary to other infectious diseases. **g., hypertrophic,** the early stage of chronic gastritis, in which there is a hyperplasia of the mucous membrane. **g., phleg-**

**monous, g., purulent, g. submucosa, g., suppurative,** acute interstitial suppurative inflammation of the stomach-walls; it occurs as a circumscribed abscess or as a diffuse purulent infiltration. **g., polypous, g. polyposa,** a form of chronic gastritis characterized by a great overgrowth of the connective tissue of the organ, giving rise to polypoid projections of fibrous tissue covered by epithelium. **g., pseudomembranous,** a form in which patches of false membrane occur in the stomach. **g., toxic, g. venenata,** acute gastric inflammation due to the ingestion of poisonous or corrosive substances.

**gastro-** (gas-tro-) [γαστήρ, stomach]: A prefix denoting relation to the stomach.

**gastroadenitis** (gas-tro-ad-en-i'-tis) [gastro-; ἀδήν, gland; ιτις, inflammation]. Inflammation of the glands of the stomach.

**gastroadynamic** (gas-tro-ah-din-am'-ik) [gastro-; ἀδύναμος; without strength]. Marked by gastric symptoms and prostration.

**gastroanastomosis** (gas-tro-an-as-to-mo'-sis) [gastro-; anastomosis]. In hour-glass contraction, the formation of a communication between the two pouches of the stomach. Syn., gastrogastrostomy.

**gastroataxia** (gas-tro-ah-taks'-e-ah) [gastro-; ἀταξία, disorder]. Disordered state of the functions of the stomach; that state of the stomach-walls in which some parts are hardened or thickened, while others are softened or inflamed.

**gastroataxic** (gas-tro-ah-taks'-ik). Characterized by gastric symptoms and ataxia.

**gastroatonia** (gas-tro-at-o'-ne-ah). Atonic dyspepsia.

**gastroblennorrhea** (gas-tro-blen-or-e'-ah). An excessive formation of mucus in the stomach.

**gastrobrosis** (gas-tro-bro'-sis) [gastro-; βρῶσις, a gnawing]. Perforating ulcer of the stomach.

**gastrocele** (gas'-tro-sēl) [gastro-; κήλη, hernia]. A hernia of the stomach.

**gastrocnemius** (gas-trok-ne'-me-us). See under muscle.

**gastrocolic** (gas-tro-kol'-ik) [gastro-; κόλον, the colon]. Pertaining to the stomach and the colon. **g. omentum,** the great omentum.

**gastrocolitis** (gas-tro-ko-li'-tis) [gastro-; κόλον, the colon; ιτις, inflammation]. Inflammation of the stomach and colon.

**gastrocoloptosis** (gas-tro-kol-op-to'sis)[gastro-; κόλον, the colon; ptosis]. A prolapse or downward displacement of the stomach and colon.

**gastrocolostomy** (gas-tro-ko-los'-to-me) [gastro-; κόλον, the colon; στόμα, mouth]. The formation of a fistula between the stomach and colon.

**gastrocolotomy** (gas-tro-kol-ot'-o-me) [gastro-; κόλον, colon; τέμνειν, to cut]. 1. Gastrocolostomy. 2. Incision into stomach and colon.

**gastrocolpotomy** (gas-tro-kol-pot'-o-me) [gastro-; κόλπος, vagina; τέμνειν, to cut]. The operation of cesarean section in which the opening is made through the linea alba into the upper part of the vagina.

**gastrocystis** (gas-tro-sis'-tis) [gastro-; κύστις, bladder]. In biology, the single-layered blastodermic vesicle of mammals.

**gastrodialysis** (gas-tro-di-al'-is-is) [gastro-; διαλύσις, a loosening]. A solution of continuity in the gastric wall.

**gastrodiaphane** (gas-tro-di'-af-ān) [gastro-; διά, through; φαίνειν, to show]. An electric apparatus for illuminating the interior of the stomach so that its outlines can be seen through the abdominal wall.

**gastrodiaphanoscopy** (gas-tro-di-af-an-os'-ko-pe) [gastro-; διά, through; φαίνειν, to show; σκοπεῖν, to inspect]. The examination of the stomach by means of the diaphanoscope; gastrodiaphany.

**gastrodiaphany** (gas-tro-di-af'-an-e) [see gastrodiaphane]. A method of exploration of the stomach by means of an electric lamp.

**gastrodidymus** (gas-tro-did'-im-us) [gastro-; δίδυμος, double]. A double monster with one abdominal cavity.

**Gastrodiscus hominis** (gas-tro-dis'-kus hom'-in-is) [gastro-; δίσκος, disc; hominis, of man]. A rare trematode worm; same as Amphistoma hominis.

**gastroduodenal** (gas-tro-dū-od'-en-al) [gastro-; duodenum]. Pertaining to the stomach and the duodenum.

**gastroduodenitis** (gas-tro-dū-od-en-i'-tis) [gastro-; duodenum; ιτις, inflammation]. Inflammation of the stomach and duodenum.

**gastroduodenostomy** (gas-tro-dū-od-en-os'-to-me) [gastro-; duodenum; στόμα, mouth]. The surgical formation of a fistula between the stomach and duodenum.

**gastrodynia** - (gas-tro-din'-e-ah) [gastro-; ὀδύνη, pain]. Pain in the stomach.

**gastroectasis** (gas-tro-ek'-tas-is). Same as gastrectasis.

**gastroelytrotomy** (gas-tro-el-it-rot'-o-me). See gastrocolpotomy.

**gastroenteralgia** (gas-tro-en-ter-al'-je-ah) [gastro-; ἔντερον, bowel; ἄλγος, pain]. Pain in the stomach and bowel.

**gastroenteric** (gas-tro-en-ter'-ik) [gastro-; ἔντερον, bowel]. Pertaining to both stomach and bowel.

**gastroenteritis** (gas-tro-en-ter-i'-tis) [gastro-; ἔντερον, bowel; ιτις, inflammation]. Inflammation of stomach and bowel.

**gastroenteroanastomosis** (gas-tro-en-ter-o-an-as-to-mo'-sis). Anastomosis between the intestine and the stomach.

**gastroenterocolitis** (gas-tro-en-ter-o-kol-i'-tis) [gastro-; ἔντερον, bowel; κόλον, colon; ιτις, inflammation]. Combined inflammation of the stomach, small intestine, and colon.

**gastroenterocolostomy** (gas-tro-en-ter-o-ko-los'-to-me). The formation of a passage between the stomach, small intestine, and colon.

**gastroenterology** (gas-tro-en-ter-ol'-o-je) [gastro-; ἔντερον, intestine; λόγος, treatise]. The study of the stomach and intestine and their diseases.

**gastroenteropathy** (gas-tro-en-ter-op'-ath-e) [gastro-; ἔντερον, bowel; πάθος, disease]. Any disease affecting the stomach and intestine.

**gastroenteroplasty** (gas-tro-en'-ter-o-plas-te). Combined gastroplasty and enteroplasty.

**gastroenteroptosis** (gas-tro-en-ter-op-to'-sis) [gastro-; ἔντερον, bowel; πτῶσις, falling]. Prolapse of the stomach and intestine.

**gastroenterostomy** (gas-tro-en-ter-os'-to-me) [gastro-; ἔντερον, bowel; στόμα, mouth]. The formation of a communication between the stomach and the small intestine.

**gastroenterotomy** (gas-tro-en-ter-ot'-o-me) [gastro-; ἔντερον, bowel; τέμνειν, to cut]. Incision of the intestine through the abdominal wall.

**gastroepiploic** (gas-tro-ep-ip-lo'-ik) [gastro-; ἐπίπλοον, caul]. Pertaining to the stomach and omentum.

**gastroesophageal** (gas-tro-e-so-faj'-e-al) [gastro-; esophagus]. Pertaining to both the stomach and the esophagus.

**gastroesophagitis, gastroœsophagitis** (gas-tro-e-sof-aj-i'-tis). Combined inflammation of the stomach and the esophagus.

**gastrofaradization** (gas-tro-far-ad-iz-a'-shun). The application of faradism to the stomach.

**gastrogalvanization** (gas-tro-gal-van-iz-a'-shun). The application of galvanism to the stomach.

**gastrogastrostomy** (gas-tro-gas-tros'-to-me). The same as gastroanastomosis.

**gastrograph** (gas'-tro-graf) [gastro-; γράφειν, to write]. An apparatus for registering the peristaltic movements of the stomach from the outside. Syn., gastrokinesograph.

**gastrohelcoma** (gas-tro-hel-ko'-mah). See gastrelcoma.

**gastrohelcosis** (gas-tro-hel-ko'-sis). See gastrelcosis.

**gastrohepatic** (gas-tro-he-pat'-ik) [gastro-; ἧπαρ, the liver]. Relating to the stomach and liver.

**gastrohepatitis** (gas-tro-hep-at-i'-tis) [gastro-; hepatitis]. Gastritis and hepatitis occurring simultaneously.

**gastrohyperneuria, gastrohypernervia** (gas-tro-hi-per-nū'-re-ah, -ner'-ve-ah) [gastro-; ὑπέρ, over; νεῦρον, a nerve]. Morbid activity of the nerves of the stomach. Syn., gastryperneuria.

**gastrohypertonic** (gas-tro-hi-per-ton'-ik) [gastro-; ὑπέρ, over; τόνος, tone]. Relating to morbid or excessive tonicity or irritability of the stomach.

**gastrohyponeuria, gastrohyponervia** (gas-tro-hi-po-nū'-re-ah, -ner'-ve-ah) [gastro-; ὑπό, under; νεῦρον, a nerve]. Defective activity of the nerves of the stomach. Syn., gastryponeuria.

**gastrohysterectomy** (gas-tro-his-ter-ek'-to-me) [gastro-; hysterectomy]. Removal of the uterus through the abdominal wall.

**gastrohysteropexy** (gas-tro-his'-ter-o-peks-e) [gastro-; ὑστέρα, the uterus; πῆξις, a fastening]. Abdominal fixatio of the uterus by a surgical operation.

**gastrohysterorrhaphy** (gas-tro-his-ter-or'-af₁e). See *hysteropexy*.
**gastrohysterotomy** (gas-tro-his-ter-ot'-o-me) [*gastro-*; ὑστέρα, the uterus; τέμνειν, to cut]. Incision of the uterus through the abdominal wall, usually for the purpose of removing a fetus; cesarean section.
**gastroid** (gas'-troid) [*gastro-*; εἶδος, likeness]. Like a stomach.
**gastrointestinal** (gas-tro-in-tes'-tin-al). See *gastroenteric*.
**gastrojejunostomy** (gas-tro-jej-u-nos'-to-me) [*gastro-*; *jejunum*; στόμα, mouth]. The formation of a passage from the stomach to the jejunum.
**gastrokinesograph, gastrokynesograph** (gas-tro-kin-es'-o-graf). See *gastrograph*.
**gastrolavage** (gas-tro-lav-ahsh') [*gastro-*; *lavage*]. Washing out of the stomach.
**gastrolienal** (gas-tro-li'-en-al). See *gastrosplenic*.
**gastrolith** (gas'-tro-lith) [*gastro-*; λίθος, a stone]. A calcareous formation in the stomach.
**gastrolithiasis** (gas-tro-lith-i'-as-is) [*gastrolith*]. A morbid condition associated with the formation of gastroliths.
**gastrologist** (gas-trol'-o-jist) [see *gastrology*]. A specialist in gastric disorders.
**gastrology** (gas-trol'-o-je) [*gastro-*; λόγος, science]. 1. A treatise on the stomach. 2. The sum of knowledge regarding the stomach and its diseases.
**gastrolysis** (gas-trol'-is-is) [*gastro-*; λύσις, a loosening]. The breaking-up of adhesions between the stomach and adjacent organs.
**gastromalacia** (gas-tro-mal-a'-she-ah) [*gastro-*, μαλακία, softening]. An abnormal softening of the walls of the stomach.
**gastromalacosis, gastromalaxia** (gas-tro-mal-ak-o'-sis, gas-tro-mal-aks'-e-ah). See *gastromalacia*.
**gastromegaly** (gas-tro-meg'-al-e) [*gastro-*; μεγάλη, large]. Abnormal enlargement of the stomach.
**gastromelus** (gas-trom'-el-us) [*gastro-*; μέλος, a limb]. A monster with accessory limbs attached to the abdomen.
**gastromenia** (gas-tro-me'-ne-ah) [*gastro-*; μήν, month]. Vicarious menstruation by the stomach.
**gastrometritis** (gas-tro-me-tri'-tis) [*gastro-*; *metritis*]. Gastritis concurrent with metritis.
**gastrometrotomy** (gas-tro-met-rot'-o-me). See *laparohysterotomy*.
**gastromucous** (gas-tro-mū'-kus). Characterized by, gastric disturbance and abnormal secretion of mucus.
**gastromycosis** (gas-tro-mi-ko'-sis) [*gastro-*; μύκης fungus]. Gastric disease due to invasion of fungi.
**gastromyeloma** (gas-tro-mi-e-lo'-mah) [*gastro-*; *myeloma*]. A medullary sarcoma of the stomach.
**gastromyxin** (gas-tro-miks'-in). A proprietary preparation of pepsin.
**gastromyxorrhea** (gas-tro-miks-or-e'-ah) [*gastro-*; μίξα, mucus; ῥοία, flow]. Excessive secretion of mucus by the stomach.
**gastronephritis** (gas-tro-nef-ri'-tis). Simultaneous inflammation of the stomach and kidneys.
**gastronesteostomy** (gas-tro-nes-te-os'-to-me). See *gastrojejunostomy*.
**gastroneuria, gastronervia** (gas-tro-nū'-re-ah, -ner'-ve-ah) [*gastro-*; νεῦρον, nerve]. The action of the nerves of the stomach.
**gastro-omental** (gas-tro-o-men'-tal). See *gastroepiploic*.
**gastropancreatic** (gas-tro-pan-kre-at'-ik). Both gastric and pancreatic.
**gastropancreatitis** (gas-tro-pan-kre-at-i'-tis). Simultaneous inflammation of the stomach and pancreas.
**gastroparalysis** (gas-tro-par-al'-is-is) [*gastro-*; *paralysis*]. Paralysis of the stomach.
**gastroparietal** (gas-tro-pa-ri'-et-al) [*gastro-*; *parietal*]. 1. Relating to the stomach-wall. 2. Relating to the stomach and the abdominal wall.
**gastropathy** (gas-trop'-ath-e) [*gastro-*; πάθος, disease]. Any disease or disorder of the stomach.
**gastroperiodynia** (gas-tro-per-e-o-din'-e-ah) [*gastro-*; περίοδος, period; ὀδύνη, pain]. Periodic gastralgia.
**gastroperitonitis** (gas-tro-per-it-on-i'-tis). Simultaneous inflammation of the stomach and peritoneum.
**gastropexis, gastropexy** (gas-tro-peks'-is, gas'-tro-peks-e) [*gastro-*; πῆξις, a fixing]. The fixation of a displaced stomach in its normal position by suturing it to the abdominal wall.
**gastrophore** (gas'-tro-fōr) [*gastro-*; φορός, bearing]. An appliance for fixing the stomach during operations upon that organ.
**gastrophrenic** (gas-tro-fren'-ik) [*gastro-*; φρήν,

14

diaphragm]. Relating to the stomach and diaphragm].
**gastroplasty** (gas-tro-plas'-te). [*gastro-*; πλάσσειν, to form]. Plastic operation on the stomach.
**gastroplegia** (gas-tro-ple'-je-ah) [*gastro-*; πληγή, stroke]. Paralysis of the stomach.
**gastroplication** (gas-tro-pli-ka'-shun) [*gastro-*; *plicare*, to fold]. An operation for relief of chronic dilation of the stomach, consisting in suturing a large horizontal fold in the stomach-wall.
**gastropneumonic** (gas-tro-nū-mon'-ik). 1. Pertaining to the stomach and the lungs. 2. See *pneumogastric*.
**gastroptosia, gastroptosis** (gas-trop-to'-ze-ah, gas-trop-to'-sis) [*gastro-*; *ptosis*]. A prolapse or downward displacement of the stomach.
**gastropyxis, gastroptyxy** (gas-trop-tiks'-is, gas'-trop-tiks-e) [*gastro-*; πτύξις, a folding]. In gastric dilatation, an operation to reduce the size of the stomach.
**gastropulmonary** (gas-tro-pul'-mon-a-re). See *pneumogastric*.
**gastropylorectomy** (gas-tro-pi-lor-ek'-to-me) [*gastro-*; *pylorus*; ἐκτομή, excision]. Excision of the pyloric portion of the stomach.
**gastropyloric** (gas-tro-pi-lor'-ik). Relating to the stomach and the pylorus.
**gastrorrhagia** (gas-tro-ra'-je-ah) [*gastro-*; ῥηγνύναι, to break forth]. Hemorrhage from the stomach.
**gastrorrhaphy** (gas-tror'-a-fe) [*gastro-*; ῥαφή, suture]. 1. Suture of a wound of the stomach or abdominal wall. 2. See *gastroplication*.
**gastrorrhea** (gas-tror-e'-ah) [*gastro-*; ῥοία, a flow]. Excessive secretion of gastric mucus or of gastric juice.
**gastrorrhexis** (gas-tror-eks'-is) [*gastro-*; ῥῆξις, a breaking]. Rupture of the stomach.
**gastrosan** (gas'-tro-san). Trade name of a preparation containing bismuth disalicylate.
**gastroschisis** (gas-tros'-kis-is) [*gastro-*; σχίσις, cleft]. A congenital malformation in which the abdomen remains open.
**gastroscope** (gas'-tro-skōp) [*gastro-*; σκοπεῖν, to see]. An instrument for examining the interior of the stomach.
**gastroscopy** (gas-tros'-ko-pe) [see *gastroscope*]. The inspection of the interior of the stomach by means of the gastroscope.
**gastrosia** (gas-tro'-ze-ah). See *gastroxia*.
**gastrosis** (gas-tro'-sis) [*gastro-*; νόσος, disease: *pl., gastroses*]. A general term for any disease of the abdomen or of the stomach only.
**gastrospasm** (gas'-tro-spazm). A spasm of the stomach.
**gastrosplenic** (gas-tro-splen'-ik) [*gastro-*; *spleen*]. Relating to the stomach and the spleen.
**gastrostaxis** (gas-tro-staks'-is) [*gastro-*; στάξις, an oozing]. The oozing of blood from the mucous membrane of the stomach.
**gastrostegous** (gas-tros'-te-gus) [*gastro-*; στέγος, a roof]. Covering the stomach.
**gastrostenosis** (gas-tro-ste-no'-sis) [*gastro-*; *stenosis*]. A narrowing or stricture of the stomach.
**gastrostomize** (gas-tros'-to-mīz). To perform gastrostomy upon: to make a gastric fistula.
**gastrostomosis** (gas-tros-to-mo'-sis). Same as *gastrostomy*.
**gastrostomy** (gas-tros'-to-me) [*gastro-*; στόμα, mouth]. The establishing of a fistulous opening into the stomach.
**gastrosuccorrhea, gastrosuccorrhoea** (gas-tro-suk-or-e'-ah) [*gastro-*; *succus*, juice; ῥοία, a flow]. Hypersecretion of the gastric juice, Reichmann's disease.
**gastrosuccorrhea mucosa**, an excessive secretion of mucus by the gastric mucosa.
**gastrothoracic** (gas-tro-tho-ras'-ik). Pertaining both to the abdomen and the thorax.
**gastrothoracodidymus** (gas-tro-tho-rak-o-did'-im-us) [*gastro-*; *thorax*; δίδυμος, a twin]. A double monster united at the thorax and abdomen.
**gastrotome** (gas'-tro-tōm) [*gastro-*; τομή, a cutting]. A cutting instrument used in gastrotomy.
**gastrotomy** (gas-trot'-o-me) [see *gastrotome*]. Incision of the abdomen or the stomach.
**gastrotoxin** (gas-tro-toks'-in) [*gastro-*; *toxin*]. A cytotoxin which has a specific action on the cells lining the stomach.
**gastrotrachelotomy** (gas-tro-tra-kel-ot'-o-me) [*gastro-*; τράχηλος, neck; τέμνειν, to cut]. An operation differing from cesarean section only in that the

uterus is opened by a transverse incision of the cervix.
**gastrotubotomy** (*gas-tro-tū-bot'-o-me*). See *laparosalpingotomy*.
**gastrotympanites** (*gas-tro-tim-pan-i'-tēz*). Gaseous distention of the stomach.
**gastroxia, gastroxynsis** (*gas-troks'-e-ah, gas-troksin'-sis*) [*gastro-*; ὀξύς, acid]. Excessive secretion of hydrochloric acid by the stomach, a condition that characterizes a form of dyspepsia. **g. fungosa**, that in which the organic acids are due to mold-vegetation in the stomach.
**gastrozymase** (*gas-tro-zi-mās*) [*gastro-*; *zymase*]. The gastric juice of the pig, obtained from the living animal and carefully dried.
**gastrula** (*gas'-troo-lah*) [γαστήρ, stomach]. The embryo at that stage of its development when it consists of two cellular layers formed by the invagination of the blastula.
**gastrulation** (*gas-troo-lā'-shun*) [*gastrula*]. The process of formation of the gastrula by the invagination of the blastula.
**gastrypalgia** (*gas-trip-al'-je-ah*) [γαστήρ, stomach; ὑπό, under; ἄλγος, pain]. Slight gastralgia.
**gastrypectasia, gastrypectasis** (*gas-trip-ek'-tas-e-ah, -ek'-ta-sis*) [γαστήρ, stomach; ὑπό, under; ἔκτασις, a stretching]. Slight distention of the stomach.
**gastryperneuria** (*gas-trip-er-nū'-re-ah*). See *gastrohyperneuria*.
**gastryperpathia** (*gas-trip-er-path'-e-ah*) [*gastro-*; ὑπέρ, over; πάθος, a disease]. Any severe disease of the stomach.
**gastryponeuria** (*gas-trip-o-nū'-re-ah*). See *gastrohyponeuria*.
**gathering.** A collection of pus beneath the surface.
**gatism** (*gah'-tizm*) [Fr., *gâtisme*, incontinence of feces]. Rectal or vesical incontinence.
**Gaucher's disease** (*go-sha'*). An affection described by Gaucher as "primary epithelioma of the spleen," but probably identical with splenic anemia.
**gaultherase** (*gawl'-ther-ās*). An enzyme found in the bark of *Betula lenta*, in the leaves and berries of *Gaultheria procumbens*, in the root of *Spiræa ulmaria*, and in several species of *Polygala*. It effects the hydrolysis of gaultherin, forming methyl-salicylic acid and glucose.
**gaultheria** (*gawl-the'-re-ah*). The plant, *G. procumbens*, the leaves of which yield a volatile oil. **g.**, oil of (*oleum gaultheria*, U. S. P.), contains 90 % of methyl salicylate, and is used in acute rheumatism and as a local antiseptic. Dose 3–10 min. (0.2–0.65 Cc.). **g.**, oil of, synthetic. See *methyl salicylate*. **g.**, spirit of (*spiritus gaultheria*, U. S. P.), consists of oil of gaultheria, 5; alcohol, 95 parts. It is used chiefly as a flavoring agent. Dose 10–20 min. (0.65–1.3 Cc.). Syn., *teaberry*; *wintergreen*.
**gaultherin** (*gawl'-ther-in*). 1. A glucoside obtained from the bark of the black birch (*Betula lenta*), which, by the action of alkali, is converted into an oil almost identical with the volatile oil of wintergreen. 2. A sodium salt of methyl salicylate (artificial oil of wintergreen) in a nearly pure condition.
**gaultherolin** (*gawl-ther'-ol-in*). Methyl salicylate.
**gaultheromenthol** (*gawl-ther-o-men'-thol*). Trade name for a preparation containing chloroform liniment and wintergreen oil.
**gauntlet** (*gawnt'-let* or *gahnt'-let*) [OF., *gantelet*, dim. of *gant*, a glove]. A bandage that covers the hand and fingers like a glove.
**Gautier's test for carbon monoxide** (*go'-te-a*). Carbon monoxide has the power of decomposing iodic anhydride and forming $CO_2$, the iodine being liberated and absorbed by copper. It is employed by Niclaux to show normal presence of carbon monoxide in blood.
**gauze** (*gawz*) [so called because first imported from Gaza in Syria]. A thin, open-meshed cloth used for surgical dressings. When impregnated with antiseptic substances, it is called *antiseptic gauze*, or, according to the substance used, it is spoken of as *iodoform gauze*, *sublimate gauze*, etc.
**gavage** (*gav-ahzh*) [Fr.]. The administration of liquid nourishment through the stomach-tube.
**Gavard's muscle** (*gav-ar'*) [Hyacinthe Gavard, French anatomist, 1753–1802]. The oblique muscular fibers of the walls of the stomach.
**Gayet's disease** (*ga-ya*) [Prudent Gayet, French surgeon]. A rare and fatal form of narcolepsy somewhat resembling the African sleeping-sickness.
**Gay Lussac's law** (*ga lu-sak'*) [Louis Joseph Gay Lussac, French chemist, 1778–1850]. Same as *Charles' law q. v.*

**Gd.** Chemical symbol of *gadolinium*.
**Ge.** Chemical symbol of *germanium*.
**Geber's glomerules** (*ga'-ber glom'-er-ūl*). The convolutions of the terminal branches of the nerves supplying the epithelial lining of the mouth.
**(von) Gebhart's test for glucose.** To 10 or 15 drops of urine add 10 Cc. of water and a tablet containing sodium carbonate and orthonitrophenylpropionic acid; warm carefully for 2 or 4 minutes, and if sugar is present, the mixture becomes greenish and then dark indigo-blue.
**Gegenbauer's cells** (*ga'-gen-bow-er*) [Carl *Gegenbauer*, German anatomist, 1826–1903]. Osteoblasts.
**Geigel's reflex** (*gi'-gel*) [Richard Geigel, German physician, 1859– ]. The inguinal reflex in the female, corresponding to the cremasteric reflex in the male.
**geisoma, geison** (*gi-so'-mah, gi'-son*) [γεῖσον, anything projecting]. The superciliary ridge of the frontal bone.
**Geissler's tube** (*gis'-ler*) [Heinrich *Geissler*, German physicist, 1814–1879]. A glass tube having a piece of platinum wire sealed into it at each end.
**geissospermine** or **geissine**. (*gi-so-sper'-min*, or *gi'-sin*), $C_{19}H_{24}N_2O_2+H_2O$. An alkaloid from the bark of *Geissospermum læve*; it is a depressant of the respiration and of the cardiac action.
**gel** (*jel*). Graham's name for a colloid which is firm in consistence.
**gelante, gelanthum** (*jel-an'-te, jel-an'-thum*). A mixture of gelatin, tragacanth, rose-water, and thymol recommended as an ointment-vehicle.
**gelasin** (*jel'-as-in*). A preparation of agar-agar.
**gelasma, gelasmus** (*jel-as'-mah, jel-az'-mus*) [γήλασμα, laughter, or *gelasmus*]. Insane or excessive or hysterical laughter.
**gelatification** (*jel-at-if-ik-a'-shun*). 1. The production of gelatin. 2. See *gelification*.
**gelatin** (*jel'-at-in*) [gelare, to congeal]. An albuminoid substance of jelly-like consistence, obtained by boiling connective tissue in water. **g.**, bone, gelatin extracted from osseous tissue. **g. capsules**, capsules of gelatin designed for containing medicines of nauseating taste. **g. culture**, a culture-medium for bacteria containing from 8 to 15 % of gelatin, in order to give it a solid consistence. **g. disc**, a disc of medicated gelatin used in ophthalmology. **g. glycerinated** (*gelatinum glycerinatum*, U. S. P.), a preparation of gelatin, glycerol, and water. Used as a vehicle for suppositories and bougies. Syn., *glycerin-jelly*. **g. medicated**, gelatin discs or lamellæ mixed with medicated substances. The gelatin lamellæ of the British Pharmacopeia contain traces of alkaloids, for introduction into the conjunctival sac. They are dissolved by the tears, and the effects of the alkaloids are thus obtained. **g. peptone**, a substance produced by digesting gelatin.
**g. sugar.** See *glycocoll*.
**gelatination** (*jel-at-in-a'-shun*). See *gelification*.
**gelatiniferous** (*jel-at-in-if'-er-us*) [*gelatin*; *ferre*, to bear]. Producing gelatin.
**gelatiniform** (*jel-at-in'-if-orm*) [*gelatin*; *forma*, form]. Resembling gelatin. **g. degeneration**, waxy or lardaceous degeneration.
**gelatinize** (*jel-at'-in-iz*) [*gelatin*]. To convert into a jelly-like mass.
**gelatinoid** (*jel-at'-in-oid*) [*gelatin*; εἶδος, likeness]. 1. Resembling gelatin. 2. Any member of a class of nitrogenous substances, including chondrin, collagen, elastin, gelatin, etc.
**gelatinosa** (*jel-at-in-o'-sah*). Wilder's term for the substantia gelatinosa.
**gelatinous** (*jel-at'-in-us*) [*gelatin*]. Resembling or having the nature of gelatin.
**gelatio** (*jel-a'-she-o*) [L.]. Synonym of *frostbite*.
**gelation** (*jel-a'-shun*) [*gelatio*, a freezing]. 1. Freezing; also frost-bite, or chilblain. 2. Catalepsy.
**gelatol** (*jel'-at-ol*). An ointment-base consisting of a mixture of oil, glycerol, gelatin, and water.
**gelatose** (*jel'-a-tōs*). A product of the action of gastric juice on gelatin. It is capable of osmosis.
**g. silver.** See *albargin*.
**geld.** To castrate.
**gelding** (*gel'-ding*). 1. Castration. 2. A castrated person or animal.
**gelid** (*jel'-id*) [*gelidus*, cold]. Ice-cold.
**gelification** (*jel-if-i-ka'-shun*). Gelatinization; the conversion of a substance into a jelly-like mass.
Syn., *gelatination*.
**Gellé's test** (*zjel-a'*) [Marie Ernest *Gellé*, French

aurist, 1834– ]. 1. The vibrations of a tuning-fork placed in contact with a rubber tube, the nozzle of which is inserted into the meatus, are distinctly perceived when the air is compressed by pressure upon the bulb attached to the tube. This does not occur when the chain of ossicles is diseased. 2. The air in the external auditory canal is compressed and a vibrating tuning-fork placed upon the vertex of the skull. In the normal ear the vibrations are diminished.

**gelose** (*jel'-ōs*) [*gelare*, to freeze]. 1. The gelatinizing principle of agar. 2. A culture medium for bacteria.

**gelosin** (*jel'-o-sin*). A mucilage from Japanese alga.

**gelotherapy** (*jel-o-ther'-ap-e*) [γελᾶν, to laugh; *therapy*]. Treatment of disease by the induction of laughter.

**gelsemin** (*jel'-sem-in*) [*gelsemium*]. A resinoid from the root of *Gelsemium sempervirens*; it is antipyretic, antispasmodic, emmenagogue, and narcotic. Dose ⅙–1 gr. (0.008–0.065 Gm.).

**gelsemine** (*jel'-sem-ēn*). A poisonous alkaloid, C₁₂H₁₄NO₂, from gelsemium; it is sometimes employed locally in the eye for the production of mydriasis.

**gelseminine** (*jel-sem'-in-ēn*). An alkaloid of gelsemium. It is an amorphous yellowish-white, bitter, very poisonous powder, soluble in ether and chloroform, less so in alcohol and water. Dose gr. $\frac{1}{60}$–$\frac{1}{6}$.

**gelsemism** (*jel'-sem-izm*). Poisoning from the use of *Gelsemium sempervirens*. In light cases it is marked by dizziness, ptosis, and weakness of the legs; in severe cases, by tremor, anesthesia, and dyspnea.

**gelsemium** (*jel-se'-me-um, jel-sem'-e-um*) [*gelseminum*, jasmine]. Yellow jasmine. The root of *G. sempervirens*, the properties of which are mainly due to a bitter alkaloid, *gelsemine*, C₁₂H₁₄NO₂, a powerful motor depressant, antispasmodic, and diaphoretic. In toxic doses it produces diplopia, extreme muscular weakness, and anesthesia, death occurring from asphyxia. Gelsemium is used in neuralgia, especially in migraine, in dysmenorrhea, hysteria, chorea, delirium tremens, and in malarial and typhoid fevers. **g., fluidextract** of (*fluidextractum gelsemii*, U. S. P.). Dose 2–5 min. (0.13–0.3 Cc.). **g., tincture of** (*tinctura gelsemii*, U. S. P.). Dose 5–15 min. (0.3–0.9 Cc.).

**gelsemiumism** (*jel-sem'-e-um-izm*). Habitual poisoning with gelsemium.

**gelsemperin** (*jel-sem'-per-in*). A preparation from *Gelsemium sempervirens*. Dose ⅙–1 gr. (0.008–0.065 Gm.).

**Gely's suture** (*zja-le*) [Jules Aristide Gely, French surgeon, 1806–1861]. An intestinal suture applied by a thread with a needle at each end.

**gemellary** (*jem-el'-ar-e*) [*gemellus*]. Relating to or like twins.

**gemelliparous** (*jem-el-ip'-ar-us*) [*gemellus; parere*, to bring forth]. Bearing twins.

**gemellus** (*jem-el'-us*) [dim. of *geminus*, a twin; pl., *gemelli*]. Applied to one of two muscles, *gemellus superior* and *gemellus inferior*; also to the gastrocnemius muscle, on account of its two heads of origin. See *muscles, table of*.

**gemina** (*jem'-in-ah*) [L., twins]. A name for the corpora quadrigemina, or optic lobes. They constitute the larger part of the mesocœlian roof.

**geminate** (*jem'-in-āt*) [*geminus*, a twin]. In pairs.

**gemination** (*jem-in-a'-shun*) [*geminus*, twin]. The production of twins. The development of two teeth in a single sac. See *geminous teeth*.

**geminous** (*jem'-in-us*) [*geminus*, a twin]. Same as *geminate*. **g. or connate teeth,** twin-formation of two teeth from the occurrence of a double dental germ in a single sac, from which are developed two teeth of the same class, when normally there should be but one.

**gemma** (*jem'ah*) [L.] A bud; a bulb or bulb-like structure.

**gemmation** (*jem-a'-shun*) [*gemmare*, to put forth buds]. Budding; a mode of reproduction seen in low forms of animal and vegetable life, and characterized by the formation of a small projection from the parent-cell, which becomes constricted off and forms an independent individual.

**gemmule** (*jem'-ūl*) [*gemmula*, dim. of *gemma*, a bud]. A small bud.

**genal** (*je'-nal*) [*gēna*, the cheek]. Relating to the cheek. **g. line,** a furrow on the cheek produced by abdominal disease.

**genera** (*jen'-er-ah*) [L.]. Plural of *genus*.

**general** (*jen'-er-al*) [*genus*, race]. Common to a class; distributed through many parts; diffuse. **g. anatomy,** anatomy of the tissues in general, as distinguished from *special anatomy*, that dealing with special organs. **g. paralysis, g. paresis.** See *paralysis, general*. **g. pathology.** See *pathology, general*.

**generalize** (*jen'-er-al-īz*). To make general, as a disease.

**generate** (*jen'-er-āt*) [*generare*, to beget]. To beget; to produce of the same kind.

**generatio æquivoca** (*jen-er-a'-she-o e-kwiv'-o-kah*) [L.]. See *generation, spontaneous*.

**generation** (*jen-er-a'-shun*) [*generare*, to beget]. 1. The act of begetting offspring. 2. A period extending from the birth of an individual to the birth of his offspring, usually estimated at a third of a century. **g., alternate,** the alternation of asexual with sexual generation in the same species of animals or plants, the offspring of one process differing from that of the other. **g., asexual,** reproduction without previous union of two sexual elements; reproduction by fission or gemmation. **g., organs of,** those that are functional in reproduction. **g., sexual,** reproduction by the union of a male and a female element. **g., spontaneous,** the generation of living from non-living matter.

**generative** (*jen'-er-a-tiv*). Pertaining to generation.

**generic** (*jen-er'-ik*) [*genus*, a kind]. 1. Pertaining to the same genus. 2. General.

**genesial, genesiac** (*jen-e'-se-al, jen-e'-se-ak*) [*genesis*]. Pertaining to generation.

**genesic, genetic** (*jen-e'-sik, jen-et'-ik*) [*genesis*]. 1. Pertaining to generation; producing. 2. A drug acting on the genital apparatus. 3. A disease affecting the genital organs. **g. affinity,** relationship by direct descent.

**genesiology** (*jen-e-se-ol'-o-je*) [γένεσις, reproduction; λόγος, science]. The science of reproduction.

**genesis** (*jen'-es-is*) [γένεσις, production]. Begetting; development; origin; formation; generation. **g. affecting sex** (*jen-es'-ik-ah*). Agents affecting the sexual organs.

**geneticist** (*jen-et'-is-ist*) [*genesis*]. One specially interested in genetics.

**genetics** (*jen-et'-iks*). The laws pertaining to generation.

**genetous** (*jen'-et-us*) [*genesis*]. Congenital.

**Genga's bandage** (*gen'-gah*). A form of roller bandage applied from below upward over a graduated compress, to control hemorrhage from a limb.

**Gengou's phenomenon or reaction** (*zjon-goo*) [O. Gengou, French bacteriologist]. Complement fixation.

**genial** (*je'-ne-al*) [γένειον, chin]. Pertaining to the chin. **g. tubercles,** four prominent tubercles on the internal surface of the lower jaw.

**genian** (*je'-ne-an*) [*gena*; γένειον, chin]. Pertaining to the chin.

**geniculate, geniculated** (*jen-ik'-ū-lāt, -ed*) [*geniculatus*, with bended knee]. Abruptly bent. **g. bodies,** the corpora geniculata, two oblong, flattened bodies, the *external* (*pregeniculum*) and the *internal* (*postgeniculum*) *geniculate bodies*, on the posterior inferior part of the optic thalamus. **g. ganglion.** See *ganglion, geniculate*.

**geniculum** (*jen-ik'-ū-lum*) [dim. of *genu*, knee]. 1. A small angular structure. 2. A knot-like structure. 3. One of the two eminences on the latero-caudal aspect of the diencephal; the postgeniculum is mesad and more distinct, and the pregeniculum, laterad and less distinct.

**genio-** (*je-ne-o-*) [γένειον, chin]. A prefix denoting connection with the chin.

**geniohyoglossus muscle.** See under *muscle*.

**geniohyoid muscle.** See under *muscle*.

**genion** (*je'-ne-on*) [γένειον, chin]. 1. The chin. 2. In craniometry, the point at the apex of the lower genial tubercle.

**genioplasty** (*je'-ne-o-plas-te*) [*genio-*; πλάσσειν, to form]. The operation of restoring the chin, or cheek.

**genital** (*jen'-it-al*) [*genitalis*, pertaining to generation; from *gignere*, to beget]. Pertaining to the organs of generation or to reproduction. **g. corpuscle.** See *corpuscle*. **g. eminence, g. tubercle,** an elevation appearing about the sixth week of embryonic life, in front of the cloaca, and from which the penis or

clitoris is developed. **g. folds**, the cutaneous folds around the genital tubercle. **g. furrow**, a groove on the under surface of the genital tubercle. **g. spots**, nasal parts which show increased sensitiveness during menstruation. **g. tubercle**, a fetal eminence giving origin to the genitalia.
**genitalia** (*jen-it-a'-le-ah*) [*genital*]. The organs of generation. In the male these consist of two testicles or seminal glands, with their excretory ducts, the prostate, the penis, and the urethra. The female genitalia include the vulva, the vagina, the ovaries, the Fallopian tubes, and the uterus.
**genitalist** (*jen'-it-al-ist*) [*genitalis*, pertaining to generation]. One who is a specialist in the treatment of diseases and affections of the genital organs.
**genitality** (*jen-it-al'-it-e*) [*genital*]. Capacity for taking part in generation.
**genitals** (*jen-it'-als*). The organs of generation. See *genitalia*.
**genito-** (*jen-it-o-*) [*genitalis*, genital]. A prefix denoting connection with or relation to the genital organs.
**genitocrural** (*jen-it-o-kru'-ral*). See under *nerve*.
**genitourinary** (*jen-it-o-u'-rin-a-re*). Relating to the genitalia and the urinary organs.
**genius** (*je'-ne-us*) [*gignere*, to beget]. Some dominant, distinctive quality. **g. epidemicus**. 1. The predominant characteristic of an endemic or epidemic disease (inflammatory, catarrhal, etc.). 2. The totality of conditions (atmospheric, supernatural, etc.) which favor the prevalence of an endemic or epidemic disease. **g. morbi**, the special or predominant feature of a disease.
**Gennari's layer** (*jen-ah'-re*) [Francisco Gennari, Italian anatomist]. See *Baillarger's layer*.
**genoblast** (*jen'-o-blast*) [γένος, sex; \ βλαστός, germ]. 1. The nucleus of the impregnated ovum, regarded as bisexual. 2. An ovum or spermatozoon.
**genocatachresia** (*jen-o-kat-ak-re'-ze-ah*) [γένος, sex; καταχρησις, misapplication]. Perversion of the sexual instinct.
**genometabole** (*jen-o-me-tab'-o-le*) [γένος, sex; μεταβολή, change]. The modification of sexual character that may follow the menopause.
**genoplasty** (*jen'-o-plas-te*). See *genyplasty*.
**Gentele's test for glucose**. On the addition of a glucose solution to a solution of potassium ferricyanide rendered alkaline with caustic soda or potash it is decolorized, with the formation of potassium ferrocyanide on applying gentle heat. Uric acid gives this same reaction.
**gentian** (*jen'-she-an*) [*gentiana*, gentian]. The common name for species of *Gentiana*. The *gentian* of the U. S. P. is the dried rhizome and roots of *Gentiana lutea*, containing a neutral bitter principle, *gentiopicrin*, $C_{20}H_{30}O_{12}$, and *gentianin*, or *gentisin*, $C_{14}H_{10}O_5$. There are several other species of gentian (*G. purpurea, G. catesbæi*), very similar in action to *G. lutea*. Gentian is an excellent tonic, simple bitter. **g., extract of** (*extractum gentianæ*, U. S. P.). Dose 1-5 gr. (0.065-0.32 Gm.). **g., fluidextract of** (*fluidextractum gentianæ*, U. S. P.). Dose ½-1 dr. (2-4 Cc.). **g., infusion of, compound**. Dose 2 dr.-1 oz. (4-32 Cc.). **g., tincture of, compound** (*tinctura gentianæ composita*, U. S. P.). Dose ½-2 dr. (2-8 Cc.). **g.-violet**, a basic aniline dye, staining tissues violet.
**gentianin** (*jen'-she-an-in*). A crude bitter substance from gentian; it is used as a tonic in dyspepsia, hysteria, etc. Syn., *crude gentianic acid; crude gentisin*.
**gentianophil, gentianophilous** (*jen'-shan-o-fil, jen-shan-of'-il-us*) [*gentian*; φιλεῖν, to love]. Staining readily with dyes of gentian-violet.
**gentianose** (*jen'-she-an-ōs*), $C_{18}H_{66}O_{31}$. A crystallizable polysaccharid obtained from gentian root.
**gentiopicrin** (*jen-she-o-pik'-rin*) [*gentian*; πικρός, bitter]. A bitter, crystalline glucoside obtained from gentian.
**gentisin** (*jen'-tis-in*) [*gentiana*, gentian], $C_{14}H_{10}O_5$. A coloring-matter contained in gentian-root; it crystallizes in yellow needles.
**genu** (*jen'-ū*) [L., "the knee": pl., *genua*]. 1. The knee. 2. Any structure bent like a knee, as the *genu* of the corpus callosum or of the optic tract. **g. extrorsum** out-knee; outward bowing of the knee; bowleg. **g. introrsum**. Same as *g. valgum*. **g. recurvatum**, the backward curvature of the knee-joint. **g. valgum**, inward curving of the knee; knock-knee; in-knee. **g. varum**. Same as *g. extrorsum*.
**genual** (*jen'-ū-al*) [*genu*, the knee]. Pertaining to a knee or to a genu.

**genuclast** (*jen'-ū-klast*) [*genu*; κλᾶειν, to break]. An instrument for breaking adhesions of the knee-joint.
**genucubital** (*jen-ū-kū'-bit-al*) [*genu; cubitus*, elbow]. Relating to or supported by the knees and elbows.
**genufacial** (*jen-ū-fa'-shal*) [*genu*, knee; *facialis*, of the face]. Relating to the knees and face.
**genuflex** (*jen-ū-fleks'*) [*genu*, knee; *flexus*, bent]. Bent at, or like, the knee. Also, bent at any joint.
**genupectoral** (*jen-ū-pek'-to-ral*) [*genu; pectus*, breast]. 1. Relating to the knee and the chest. 2. Pertaining to the knee-chest posture—the patient resting upon the knees and chest.
**genus** (*jēn'-us*) [L.]. A species or collection of species having in common characteristics differing greatly from those of other species.
**geny-** (*jen-e-*) [γένυς, jaw or cheek]. A prefix denoting relation to the jaw or the cheek.
**genyantralgia** (*jen-e-an-tral'-je-ah*) [γένυς, jaw; ἄντρον, cave; ἄλγος, pain]. Pain or neuralgia in the antrum of Highmore.
**genyantritis** (*jen-e-an-tri'-tis*) [*genyantrum*; ιτις, inflammation]. Inflammation of the genyantrum.
**genyantrum** (*jen-e-an'-trum*) [*geny-*; ἄντρον, cave]. The maxillary antrum or antrum of Highmore.
**genycheiloplasty** (*jen-e-ki'-lo-plas-te*) [*geny-*; χεῖλος, lip; πλάσσειν, to form]. Plastic surgery of both cheek and lip.
**genyplasty** (*jen'-e-plas-te*) [*geny-*; πλάσσειν, to form]. An operation for restoring the cheek or the jaw.
**geode** (*je'-ōd*). A lymph-space.
**geoform** (*je'-o-form*). A tasteless, odorless, nontoxic compound of guaiacol and formaldehyde; it is used as an antiseptic.
**geographic tongue** (*je-o-graf'-ik*). An eruption on the dorsum of the tongue, which becomes covered with sinuous, maplike lines.
**geolin** (*je'-o-lin*) [γῆ, earth; *oleum*, oil]. A trade name for a petrolatum resembling vaselin.
**geophagia** (*je-o-fa'-je-ah*). See *geophagism*.
**geophagism** (*je-of'-aj-izm*) [γῆ, earth; φαγεῖν, to eat]. The practice of eating earth or clay.
**geophagist** (*je-of'-aj-ist*) [γῆ, earth; φαγεῖν, to eat]. A dirt-eater.
**geophagous** (*je-of'-ag-us*) [γῆ, earth; φαγεῖν, to eat]. Addicted to dirt-eating.
**geophagy** (*je-of'-aj-e*). See *geophagism*.
**Georget's stupidity** (*jor-jā*). Simple mental confusion without hallucination or delusion.
**geosote** (*je'-o-sōt*). See *guaiacol valerate*.
**geotalose** (*je-o-tal'-ōs*). A compound of creosote with a mucocolloidal base.
**geotropism** (*je-ot'-ro-pizm*) [γῆ, the earth; τρέπειν, to turn]. In biology, the tendency of roots and other parts to grow downward toward the earth.
**geranin** (*je-ra'-ne-in*) [*geranium*]. A precipitate from the tincture of Cranesbill, *Geranium maculatum*. It stimulates and contracts the caliber of the capillary vessels of the mucous membranes, etc. Dose 1 to 3 grains.
**geraniol** (*je-ra'-ne-ol*), $C_{10}H_{17}OH$. A colorless, highly refractive liquid with strong odor of roses, separated from oil of Indian geranium, *Andropogon nardus*. It is an isomerid of borneol.
**geranium** (*jer-a'-ne-um*) [γεράνιον, geranium]. The *geranium* of the U. S. P. is the root of *G. maculatum*, crane's-bill root, the properties of which are due to tannic and gallic acids. It is an astringent, useful in diarrhea, etc. **g., fluidextract of** (*fluidextractum geranii*, U. S. P.). Dose ½-1 dr. (2-4 Cc.).
**geratic** (*jer-at'-ik*) [γῆρας, old age]. Relating to old age.
**geratology** (*jer-at-ol'-o-je*) [γῆρας, old age; λόγος, a discourse]. 1. A department of biology treating of the decadence and gradual extinction of a group of organisms. 2. See *gereology*.
**geriatrics** (*je-re-at'-riks*) [γῆρας, old age; ιατρική, medical treatment]. The branch of medicine dealing with the diseases of old age.
**Gerdy, fibers of** (*ajer-de'*) [Pierre Nicolas *Gerdy*, French surgeon, 1797-1856]. The superficial transverse ligament of the fingers, a fibrous band bounding the distal margin of the palm. **G., fontanel of**, an abnormal or supernumerary fontanel existing between the two parietal bones at the point at which the sagittal suture ceases to be serrated and becomes nearly rectilinear. **G., ligament of**, the suspensory ligament of the axilla that extends along the lower border of the pectoralis major and latissimus dorsi

# GEREOLOGY 389 GIBBER

beneath the skin surrounding the hollow of the axilla. **G.; tubercle of,** a more or less pronounced elevation situated anteroexternally to the tubercle of the tibia, to which it is joined by a short ridge; it serves for the attachment of the tibialis anticus.

**gereology, geræology** (*jer-e-ol'-o-je*). [γῆρας, old age; λόγος, science]. The science of old age; the structural changes and diseases incident to it, its hygiene, etc.

**Gerhardt's change of pitch** (*gār'-hart*) [Carl Adolf Christian Jacob *Gerhardt*, German physician, 1883–1903]. The tympanitic sound heard over partly filled cavities is of a lower pitch when the patient is sitting than when he is lying down. **G.'s disease.** See *erythromelalgia*. **G.'s sign.** 1. A systolic bruit heard between the mastoid process and spinal column in cases of aneurysm of the vertebral artery. 2. Incomplete filling of the external jugular vein on the affected side, occasionally seen in thrombosis of the transverse sinus. 3.-A band of dulness on percussion, superimposed upon the normal precordial dulness, about 3 cm. in width and extending toward the left clavicle; it is observed in cases of the persistence of the ductus arteriosus. 4. The absence of the movement of the larynx in dyspnea due to aneurysm of the aorta. In dyspnea from other causes the excursions of the larynx are extensive.

**Gerhardt's reaction for acetoacetic (diacetic) acid in 'the urine** (*gār'-hart*). A premonitory sign of diabetic coma. Treatment with a solution of ferric chloride causes a gray precipitate, which is removed by filtration; on the further addition of the reagent a deep, Bordeaux-red color is produced, which disappears on adding a few drops of sulphuric acid. **G.'s test for urobilin,** shake the urine with chloroform to extract the urobilin; add to this chloroform extract an iodine solution, and then a solution of caustic potash; a green fluorescence will result.

**Gerhardt-Semon's law** (*gār'-hart-se'mon*) [C. A. C. J. *Gerhardt* (see above); Sir Felix *Semon*, English laryngologist, 1849– ]. Certain central or peripheral lesions of the recurrent laryngeal nerve cause the vocal cord to assume a position midway between adduction and abduction, the lesion of the nerve being insufficient to destroy it and to provoke a complete paralysis of the vocal cord (cadaveric position).

**Gerlach's network** (*ger'-lak*) [Joseph von *Gerlach*, German anatomist, 1820–1896]. The interlacing of the dendritic processes of the ganglion-cells in the gray matter of the spinal cord. The net-work is only apparent, since the processes do not anastomose, but are merely in contact or contiguity. **G.'s tubal tonsil,** a mass of adenoid tissue in the lower part of the Eustachian tube, particularly along its median wall and about the pharyngeal orifice. **G.'s valve,** a circular valve sometimes existing at the orifice of the vermiform appendix in the cecum.

**Gerlier's disease** (*zjer-le-a'*) [Felix *Gerlier*, Swiss physician, 1840– ]. An endemic disease characterized by vertigo, ptosis, paresis of the extremities, and great depression. Syn., *vertige paralysant*.

**germ** (*jerm*) [*germen*, sprig; offshoot]. 1. An ovum, a spore, seed, an undeveloped embryo. 2. A microbe or bacterium. **g.-area,** the spot on an ovum where the development of the embryo begins. **g.-cell,** a cell resulting from a fecundated germinal vesicle. **g.-disease,** any disease of microbic origin. **g.-epithelium,** a thickening on the ventromesial aspect of the Wolffian body, giving rise to the male and female sexual elements." **g.-force,** hypothetic and constructive force. **g.-layer,** any one of the layers of a developing embryo. **g.-plasm,** the reproductive or hereditary substance of living organisms, which is passed on from the germ-cell in which an organism originates in direct continuity to the germ-cells of succeeding generations. **g.-ridge.** Same as *g.-epithelium*. **g. theory,** the doctrine of the origin of every organism from a germ or germ-plasm; also the theory that certain diseases are due to the development of microorganisms in the body.

**German measles.** See *rubella*.

**germander** (*jer-man'-der*) [ME., *germawnder*]. A popular name for plants of the labiate genus *Teucrium*.

**germanium** (*jer-man'-e-um*). See *elements, table of*.

**germicidal** (*jer-mis-i'-dal*) [see *germicide*]. Destroying germs.

**germicide** (*jer'-mis-id*) [*germ*; *cædere*, to kill]. An agent that destroys germs.

**germiculture** (*jer'-me-kul-chur*). [*germ*; *cultura*, culture]. The artificial culture of bacteria.

**germifuge** (*jer'-me-fūj*) [*germ*; *fugare*, to banish]. 1. Having the power of expelling germs. 2. An agent that expels germs.

**germiletum** (*jer-mil-e'-tum*). An antiseptic said to consist of a solution of borohydrofluoric and borosalybenzoic acids, boroglycerol, and formaldehyde with potassium permanganate, menthol, thymol, and aromatics.

**germinal** (*jer'-min-al*) [*germ*]. Pertaining to a germ or to the development of a tissue or organ. **g. area.** See *germ-area*. **g. disc.** See *disc, germinal*. **g. membrane,** the blastoderm. **g. spot,** the nucleolus of the ovule. **g. vesicle,** the blastodermic vesicle.

**germination** (*jer-min-a'-shun*) [*germinatio*, sprouting; budding]. Development or sprouting of a seed or spore.

**germinative** (*jer'-min-at-iv*). Pertaining to germination.

**germol** (*jer'-mol*). A bactericidal preparation analogous to cresol.

**germule** (*jer'-mūl*). An incipient germ.

**gerocomia, or gerocomy** (*jer-o-ko'-me-ah*, or *jer-ok'-o-me*) [γέρων, old man; κομεῖν, to care for]. The hygienic and medical care of old people.

**gerocomium** (*jer-o-ko'-me-um*) [γέρων, old man; κομεῖν, to care for]. A home or institution for aged people.

**geroderma** (*je-ro-der'-mah*) [γέρων, old man; δέρμα, skin]. Dystrophy of the skin producing the wrinkled appearance of old age.

**geromorphia** (*jer-o-der'-me-ah*). See *geromorphism*.

**geromarasmus** (*jer-o-mar-as'-mus*) [γέρων, old man; μαρασμός, a wasting]. Emaciation and weakness characteristic of extreme old age.

**geromorphism** (*jer-o-mor'-fizm*) [γῆρας, old age; μορφή, form]. The appearance of age in a young person.

**gerontic** (*jer-on'-tik*) [γεροντικός, belonging to an old man]. Pertaining to old age.

**gerontin** (*jer-on'-tin*), $C_5H_4N_2$. A poisonous leukomaine from the hepatic cells of dogs.

**gerontopia** (*jer-on-to'-pe-ah*). See *presbyopia*.

**gerontotoxon** (*jer-on-to-toks'-on*) [γέρων, an old man; τόξον, a bow]. The arcus senilis.

**Gerota's capsule** (*jer-o'-tah*). The fascia around the kidney.

**Gerrard's test for glucose.** To a 5 % solution of potassium cyanide add Fehling's solution until the blue color just commences to disappear. On heating this solution to boiling with one containing glucose, no precipitation of cuprous oxide results, but the solution will be more or less decolorized.

**Gersuny's paraffin prosthesis.** The replacement of the cartilaginous portion of the nasal septum by paraffin. **G.'s symptom,** a peculiar sensation of adhesion of the mucosa of the bowel to the fecal mass while pressure is made with the tips of the fingers in cases of coprostasis.

**gestation** (*jes-ta'-shun*) [*gestare*, to bear]. Pregnancy. **g., abdominal,** the form of extrauterine gestation in which the product of conception is developed in the abdominal cavity. **g., double.** 1. Twin pregnancy. 2. The coexistence of uterine and extrauterine pregnancy. **g.; ectopic.** Same as **g., extrauterine. g., extrauterine,** pregnancy in which the product of conception is not contained in the uterine cavity. See *pregnancy*.

**ghee, ghi** [Hind.]. Butter clarified and liquefied by slow boiling and used as a base for ointments and as a dressing for wounds.

**ghost-corpuscles** (*gōst*). See *corpuscle, phantom*.

**Giacomini's band** (*yah-ko-me'-ne*) [Carlo *Giacomini*, Italian anatomist, 1840–1898]. A grayish band, continuous with the gyrus dentatus, which passes from the cleft between the hippocampal and uncinate gyruses transversely over the latter and disappears on its ventricular surface.

**giant** (*ji'-ant*) [γίγας, giant]. A being or organism abnormally large. **g.-cell.** See *cell, giant*. **g.-finger.** Synonym of *macrodactyly*.

**gigantism** (*ji'-ant-izm*). See *gigantism*.

**Gianuzzi, cells or crescents of** (*yan-oot'-se*) [—— *Gianuzzi*, Italian anatomist]. Granular protoplasmic cells found in mucous glands between the mucous cells and the basement-membrane; they play an important part in the functional activity of the gland. They are also called *demilune cells*.

**gibber** (*gib'-er*) [*gibbus*, a hump]. A sac-like en-

largement. **g. inferior thalami.** See *pulvinar*. **g. ulnæ**, the olecranon.
**Gibbon's hydrocele** (*gib'-un*) [Quinton V. *Gibbon*, American surgeon, 1813–1894]. Hydrocele with a voluminous hernia.
**gibbositas** (*gib-os'-it-as*) [L.]. Gibbosity; the condition of being a hunchback.
**gibbosity** (*gib-os'-it-e*) [*gibbus*]. The condition of being humpbacked.
**gibbous** (*gib'-us*) [*gibbus*]. Humpbacked. Swollen, convex, protuberant, especially upon one side.
**gibbus** (*gib'-us*) [L.]. A hump.
**Gibert's pityriasis** (*zhe-bair'*) [Camille Melchior *Gibert*, French physician, 1797–1866]. Pityriasis rosea.
**Gibney's perispondylitis** (*gib'-ne per-e-spon-dil-i'-tis*) [Virgil Pendleton *Gibney*, American orthopedist, 1847– ]. A painful condition of the muscles of the spine.
**Gibraltar fever.** Synonym of *Malta fever*.
**Gibson's bandage** (*gib'-sun*) [Kasson C. *Gibson*, American dentist]. A bandage for fracture of the lower jaw.
**gid** (*gid*) [ME., *gidie*, dizzy]. Staggers in sheep, a disease caused by a cystic worm in the brain, formerly called *Cænurus cerebralis*.
**giddiness** (*gid'-e-nes*) [ME., *gidie*, dizzy]. A sensation of whirling or unsteadiness of the body, usually accompanied by more or less nausea.
**Giemsa's stain** (*gēm'-sah*) [G. *Giemsa*, German bacteriologist]. Azur II, eosin, 3 Gm.; azur II, 0.8 Gm.; glycerin, 250 Gm.; methylalcohol, 250 Gm. This stain is used in the detection of *Spirochæta pallida*.
**Gierke's corpuscles** (*gēr'-keh*) [Hans Paul Bernhard *Gierke*, German anatomist, 1847–1886]. Roundish, colloid bodies, of a significance not yet determined, sometimes found in the central nervous system; they appear to be identical with Hassall's corpuscles. **G.'s respiratory bundle.** See *Krause's respiratory tract*.
**Gifford's reflex** (*gif'-urd*) [Harold *Gifford*, American ophthalmologist, 1858– ]. Contraction of the pupil, occurring when a strong effort is made to close the lids, which are kept apart. **G.'s sign**, inability to evert the upper eyelid in the early stages of exophthalmic goiter.
**gigantism** (*ji-gan'-tizm*) [γίγας, giant]. Abnormal overgrowth or excessive size of the whole or of part of the body.
**gigantoblast** (*ji-gan'-to-blast*) [*gigantism;* βλαστὸς, a germ]. A large nucleated red blood-corpuscle.
**gigantocyte** (*ji-gant'-o-sīt*) [*gigantism;* κύτος, cell]. A large nonnucleated red blood-corpuscle.
**gigantosoma** (*ji-gan-to-so'-mah*) [γίγας, giant; σῶμα, body]. Gigantism.
**gill** (*gil*). One of the respiratory organs of such animals as breathe the air that is mixed with water. Cf. *branchiæ*.
**gill** (*jil*) [*gillo*, a flask]. One-fourth of a pint.
**Gillenia** (*jil-e'-ne-ah*) [Arnold *Gill*, German botanist]. A genus of rosaceous herbs. *G. stipulacea* (bowman's root; Indian physic) and *G. trifoliata*, of North America, are safe and effective substitutes for ipecac. Dose of *fluidextract* of *G. trifoliata*, as expectorant, 3–8 min. (0.2–0.5 Cc.); mild emetic, 20–30 min. (1.2–1.8 Cc.).
**gillenin** (*jil'-en-in*) [see *gillenia*]. The active principle of American ipecac. Dose 4–6 gr. (0.26–0.4 Gm.).
**Gilles de la Tourette's disease.** See under *Tourette*.
**G. de la T.'s sign.** See under *Tourette*.
**gilvor** (*jil'-vor*) [*gilvus*, pale yellow]. The earthy complexion accompanying certain forms of cachexia and dyscrasia.
**Gimbernat's ligament** (*gim'-ber-nat*) [Antonio de *Gimbernat*, Spanish surgeon, 1742–1790]. The triangular portion of the aponeurosis of the external oblique that is attached to the mesal end of Poupart's ligament in front and to the iliopectineal line behind, and externally.
**gin** (*jin*) [OF., *genevre*, juniper]. Common grain-spirit distilled and flavored with juniper-berries. It is a stimulant and diuretic. *Spiritus juniperi compositus* is its official substitute in the U. S. P. **g.-drinker's liver**, the liver of atrophic cirrhosis due to alcoholism.
**gingament** (*jin'-ja-ment*). Trade name of a stomachic; said to contain sodium bicarbonate, ammonium bicarbonate, oil of peppermint, saccharin and ginger.

**ginger** (*jin'-jer*). See *zingiber*.
**gingerol** (*jin'-jer-ol*). A pungent oil from zingiber.
**gingiva** (*jin-ji'-vah*) [L.; *pl.*, *gingivæ*]. The gum; the vascular tissue surrounding the necks of the teeth and covering the alveoli.
**gingival** (*jin'-jiv-al*) [*gingiva*]. Pertaining to the gums. **g. line**, a line along the gums, seen in chronic metallic poisoning, as the blue line of lead.
**gingivalgia** (*jin-jiv-al'-je-ah*) [*gingiva*, gum; ἄλγος, pain]. Neuralgia of the gums.
**gingivitis** (*jin-jiv-i'-tis*) [*gingiva;*- *ιτις*, inflammation]. Inflammation of the gums; ulitis. **g., expulsive**, osteo-periostitis of a tooth, which is gradually expelled from its socket.
**ginglymoarthrodial** (*gin-gle-mo-ar-thro'-de-al*). Both ginglymoid and arthrodial.
**ginglymoid** (*ging'-glim-oid*) [γίγγλυμος, a hinge; εἶδος, likeness]. Resembling a hinge-joint.
**ginglymus** (*gin'-glim-us*). See *diarthrosis*.
**ginseng** (*jin'-seng*) [Chinese, *jin-tsan*, ginseng]. The root of several species of *Panax* or *Aralia*. It has no other medicinal virtues than those of a demulcent, but it has a wonderful reputation in China, to which country most of it is exported.
**Giovannini's disease** (*yo-vah-ne'-ne*) [Sebastiano *Giovannini*, Italian dermatologist]. A rare form of nodular disease of the hair caused by a fungus.
**giraffe** (*jir-af'*). See *dengue*.
**Giraldés' "bonnet à poil"** (*zje-ral-dāz*) [Joachim Albin Cardozo Cazado *Giraldes*, Portuguese surgeon, 1808–1875]. Widening of the cranium in the frontal region in chronic hydrocephalus. **G.'s organ**, the paradidymis, a small, tubular organ found at the junction of the spermatic cord and epididymis. It is a remnant of some of the lower Wolffian tubules, and corresponds to the parovarium in the female.
**Giraldesian organ.** See *Giraldés' organ*.
**Giraud-Teulon's law.** Our binocular retinal images are localized at the point of intersection of the primary and secondary axes of projection.
**girdle** (*gir'-dl*) [AS., *gyrdel*, a waistband]. A band designed to go around the body; a structure resembling a circular belt or band. **g. anesthesia**, an anesthetic ring around the body. **g.-pain**, a sensation as if a girdle were drawn tightly around the body. **g., pelvic**, the bones (the two ossa innominata) forming the support for the lower limbs. **g.-sensation.** Same as **g.-pain.** **g., shoulder-**, the system of bones supporting the upper limbs or arms; it consists of the clavicles, scapulæ, and the manubrium sterni.
**girmir** (*gir'-mir*) [Arab.]. Tartar of the teeth.
**githagism** (*gith'-a-jism*) [*gith*, a black-seeded plant; *agere*, to carry]. The condition of chronic poisoning produced in man and animals attributed to the seeds of the corn-cockle (*Lynchis githago*), which often find their way into cereal foods.
**Giuffrida-Rugieri's stigma of degeneration** (*yoo-fre'-dah-roo-je-er'-e*) [Vincenzo *Giuffrida-Rugieri*, Italian anthropologist]. The absence or incompleteness of the glenoid fossa.
**gizzard** (*giz'-ard*). The strong muscular stomach of birds used for triturating the food. A proprietary substance, *ingluvin*, prepared from it, has been used in dyspepsia.
**Gl.** Chemical symbol of *glucinum*.
**glabella, glabellum** (*gla-bel'-ah*, *-um*) [dim. of *glaber*, smooth]. The smooth triangular space between the eyebrows, just above the root of the nose.
**glabellad** (*gla-bel'-ad*) [dim. of *glaber*, smooth; *ad*, to]. Toward the glabella.
**glabellar** (*gla-bel'-ar*) [dim. of *glaber*, smooth]. Pertaining to the glabella.
**glabellen** (*gla-bel'-en*). Belonging to the glabella in itself.
**glabrate** (*gla'-brāt*) [*glabrare*, to make smooth]. Becoming or tending to smoothness or baldness.
**glabrification** (*gla-bri-fi-ka'-shun*) [*glaber*, smooth; *facere*, to make]. The process of becoming smooth, glistening, and hairless.
**glabrificin** (*gla-brif'-is-in*). An antibody which renders bacteria glabrous.
**glabrous** (*gla'-brus*) [*glaber*, smooth]. Smooth.
**glacial** (*gla'-she-al*) [*glacies*, ice]. Icy; resembling ice in appearance, as *glacial* acetic or phosphoric acid.
**glacialin** (*gla-she-a'-lin*) [*glacies*, ice]. An antiseptic substance used for the preservation of foods. It consists of borax, boric acid, sugar, and glycerin.
**gladiate** (*glad'-e-āt*) [*gladius*, a sword]. Ensiform, or sword-shaped.

**gladioline** (*glad-i'-o-lēn*) [*gladiolus*]. An alkaloid in brain tissue.
**gladiolus** (*glad-i'-o-lus*) [dim. of *gladius*, a sword]. The middle or second piece of the sternum.
**glair** (*glār*) [ME., *glayre*, the white of egg]. The white of egg; any thin, viscous substance, as a mucous discharge.
**glairin** (*glār'-in*) [OF., *glaire*, the white of egg; from *clarus*, clear]. A peculiar organic, gelatinous substance found on the surface of some thermal waters. Syn., *baregin*.
**glairy** (*glār'-e*) [see *glairin*]. Slimy; viscous; mucoid.
**glama** (*gla'-mah*) [L.]., Lippitudo. An accumulation of gummy or hard material at the inner canthus of the eye; the material so accumulated.
**gland** [*glans*, an acorn]. 1. An organ which secretes something essential to the system or excretes waste materials the retention of which would be deleterious to the body. The word is also applied to structures which have no visible, or external secretion, as the suprarenal capsules, the pineal gland, etc. In structure glands may be tubular or racemose, simple or compound. 2. The bulbous end of the penis and clitoris. **g.s, absorbent.** See *g.s, lymphatic.* **g., accessory thyroid,** a small mass of gland-tissue connected with the thyroid gland. **g., acinous.** See *g., racemose.* **g., admaxillary,** an accessory parotid or salivary gland. **g.s, aggregate.** See *Cowper's glands.* **g.s, agminated.** See *Peyer's glands.* **g.s, axillary,** the lymph-glands situated in the axilla. **g. of Bartholin.** See under *Bartholin.* **g.s, blood-.** See *g.s, hematopoietic.* **g.s, bronchial,** the lymph-glands of the root of the bronchi. **g.s, Brunner's.** See under *Brunner.* **g.s, bulbourethral.** See *Cowper's glands.* **g.s, Cabelli's.** See under *Cabelli.* **g., carotid,** a ductless gland at the bifurcation of the common carotid artery. **g.s, ceruminous,** the glands secreting the cerumen of the ear. **g.s, cervical,** the lymph-glands of the neck. **g.s, Clapton-Havers'.** See *Havers' glands.* **g., coccygeal,** a small vascular body at the tip of the coccyx. Syn., *Luschka's gland.* **g., compound,** one composed of a number of small pouches; a gland the duct of which is branched. **g., conglobate.** See *g.s, lymphatic.* **g.s, Cowper's.** See *Cowper's glands.* **g., ductless,** a gland without a duct, as the subrarenal capsule. **g., duodenal.** See *Brunner's g.s.* **g., Duverney's.** See *g. of Bartholin.* **g., endocrinous,** any gland which produces an internal secretion. **g.s, Fraenkel's.** See under *Fraenkel.* **g.s, Gley's.** *g.s, parathyroid.* **g.s, Havers', g.s, Haversian.** See *Havers' glands.* **g.s, hematopoietic,** the socalled glands that are supposed to take part in the formation of the blood, as the spleen, thymus, suprarenal capsules, etc. **g.s, hemolymph,** certain glands occurring chiefly in the retroperitoneal region. They are intermediate between the spleen and ordinary lymph-glands. Their function is mainly hemolytic. There are two types to which the names splenolymph and manolymph or marrow-lymph are applied. **g., Huguier's.** See *g. of Bartholin.* **g.s, integumentary,** the sebaceous and sudoriparous glands. **g.s, intercapsular,** a long, narrow, paired organ, found in the human embryo, corresponding in position and general appearance to the hibernating glands of the lower animals, but from its inner lymphoid structure it is supposed to be a hemolymph gland; no trace of it persists to adult life. **g.s, intestinal, solitary,** the isolated lymph-glands distributed through the intestinal mucous membrane. **g., lacrimal,** a compound racemose gland in the upper and outer portion of the orbit, the function of which is to secrete the tears. **g.s of Lieberkühn.** See under *Lieberkühn.* **g.s of Littré, g.s of Morgagni.** See under *Littré.* **g., Luschka's.** See *Luschka's gland.* **g.s, lymphatic,** small oval masses of lymphatic tissue in the course of lymphatic vessels. Their functions are to act as filters to the blood, retaining foreign particles, and also to form white corpuscles. **g.s, mammary,** the glands that secrete milk. **g., manolymph, g., marrow-lymph,** a variety of hemolymph gland. **g.s, Manz'.** See *Manz's glands.* **g.s, Meibomian,** the minute sebaceous follicles between the cartilage and conjunctiva of the eyelids. **g.s, Montgomery's.** See under *Montgomery.* **g.s, muciparous, g.s, mucous,** the glands in mucous membranes secreting mucus. **g.s, parathyroid,** small lymphatic glands lying near the thyroid, but differing from it in histological structure and not accessory to it. Syn., *Gley's glands; Sandstroem's glands.* **g., parotid,** a large salivary gland situated in front of the ear. **g.s, peptic.** See *peptic glands.* **g.s, Peyer's.** See *Peyer's glands.* **g., pineal.** See *pineal gland.* **g., pituitary,** a term for the hypophysis of the brain. **g., prostate.** See *prostate gland.* **g.s, pyloric,** the glands of the stomach situated near the pylorus and secreting pepsin. **g., racemose,** a gland composed of a number of acini communicating with several excretory ducts, which usually join to form a common duct. **g., Rivini's.** See *g.s, sublingual.* **g., salivary,** a gland that secretes saliva. **g.s, Sandstroem's.** See *g.s, parathyroid.* **g.s, Schüller's.** See *Skene's glands.* **g.s, sebaceous,** the glands in the corium of the skin, secreting sebum. **g., seminal,** the testicle. **g., serous,** a secreting gland, the cells of which are granular and spherical in form, with central nuclei, and which secrete a thin, watery fluid. **g., simple,** a gland having but one secreting sac and a single tube. **g., Skene's.** See *Skene's gland.* **g.s, splenolymph,** certain hemolymph glands intermediate between the spleen and ordinary lymph-glands. **g.s, sublingual,** the smallest of the salivary glands, situated one on each side beneath the tongue. **g., submaxillary,** a salivary gland situated below the angle of the jaw. **g.s, sudoriparous, g.s, sweat,** the convoluted glands in the skin that secrete sweat. **g., Suzanne's.** See *Suzanne's gland.* **g., thymus.** See *thymus.* **g., thyroid.** See *thyroid gland.* **g., Tiedemann's.** See *Bartholin's gland.* **g., tubular,** a gland having a tube-like structure. **g., tubular, compound,** one composed of a number of small tubules with a single duct. **g., urethral.** See *Littré's gland.* **g., vaginal,** one of the glands in the vaginal mucous membrane. **g., Virchow's.** See *Virchow's gland.* **g., vulvovaginal.** See *Bartholin's gland.* **g.s, Waldeyer's.** See under *Waldeyer.* **g.s, Wasmann's.** See under *Wasmann.* **g.s, Willis'.** See under *Willis.*

**glanderous** (*glan'-der-us*) [*glanders*]. Affected with glanders.

**glanders** (*glan'-derz*). [*gland*]. A contagious disease of horses and asses, but communicable to man, and due to the bacillus of glanders or *Bacillus mallei*. It appears in two forms—as glanders proper, when affecting the mucous membranes, and as *farcy*, when limited to the skin and lymphatic glands. On mucous membranes, especially the nasal, it manifests itself as isolated nodules which coalesce and break down into deep ulcers that involve the cartilages and bones. It is apt to extend down to the lungs and give rise to suppuration and pneumonic processes. In man the disease usually runs an acute febrile course typhoid in type, and terminates fatally. Farcy is characterized by nodules (*farcy-buds*) in the skin and lymphatic glands, which break down into irregular chronic ulcers.

**glandiform** (*glan'-de-form*) [*gland; forma*, form 1. Acorn-shaped. 2. Adenoid.
**glandilemma** (*glan-dil-em'-ah*) [*gland;* λέμμα, husk]. The capsule of a gland.
**glandula, glandule** (*glan'-dū-lah, glan'-dūl*) [L. . 1. A little gland. 2. Same as *gland.*
**glandular** (*glan'-dū-lar*) [*glandula*]. Relating to or of the nature of, a gland.
**glandule** (*glan'-ūl*) [*glandula*, a gland]. A small gland.
**glandulen** (*glan'-dū-len*). A preparation of the bronchial glands of sheep, used in the treatment of tuberculosis. Dose 12-20 gr. (0.77-1.3 Gm.) 3 times daily.
**glanduliform** (*gland-dū'-le-form*). Shaped like a gland.
**glandulin** (*gland'-ū-lin*) [*glandula*, a gland]. Extract of gland-tissue.
**glans** (*glanz*) [L., "an acorn"]. 1. An acorn-shaped body. 2. A gland. **g. clitoridis,** the rounded end of the clitoris, analogous to the glans penis of the male. **g. penis,** the conical body forming the head of the penis.
**glaseptic** (*glas-ep'-tik*) [*glass; aseptic*]. Trade name of sterilized solutions in glass ampoules for hypodermic use.
**Glaserian artery** (*gla-se'-re-an*) [Johann Heinrich *Glaserius* (or *Glaser*), Swiss anatomist, 1629–1675]. The tympanic artery. **G. fissure,** the glenoid or petrotympanic fissure, which divides transversely the glenoid fossa of the temporal bone.
**Glasgow's sign** (*glas'-go*). A systolic sound in the brachial artery, heard in latent aneurysm of the aorta.
**glass** (*glas*) [AS., *glas*]. 1. A brittle, hard, transparent substance, consisting usually of the fused amorphous silicates of potassium and calcium; or

# GLASSES 392 GLOBULICIDE

sodium and calcium, with an excess of silica. 2. Any article made of glass. g., crown-, a very hard glass, is a silicate of sodium and calcium. g., flint-, that composed of lead and potassium silicates. g., soluble, potassium or sodium silicate, used as a substitute for plaster-of-Paris. g.-wool, white, silky threads obtained by the action of a powerful blast on a falling stream of molten glass; it is used in draining wounds and in filtering strong acids and alkalies. Syn., *slag-wool*.

**glasses** (*glas'-es*) [see *glass*]. The popular term for spectacles or eye-glasses. g., bifocal, those that have a different refracting power in the upper part from that in the lower; the effect is usually produced by the superposition of segment lenses. g., prismatic, those formed of prisms; used in insufficiency and paralysis of the ocular muscles.

**glassy** (*glas'-e*). 1. Having the appearance of glass; vitreous; hyaline. 2. Expressionless.

**Glauber's salt** (*glow'-ber*) [Johann Rudolf *Glauber*, German chemist, 1603–1668]. Sodium sulphate.

**glaucedo** (*glaw-se'-do*) [L.]. Glaucoma.

**glaucoma** (*glaw-ko'-mah*) [γλαυκός, sea-green]. A disease of the eye characterized by heightened intraocular tension, resulting in hardness of the globe, excavation of the papilla or optic disc, a restriction of the field of vision, corneal anesthesia, colored halo about lights, and lessening of visual power that may proceed to blindness. The etiology is obscure. g. absolutum, g. consummatum, the completed glaucomatous process when the eyeball is exceedingly hard and totally blind. g. acutum, the first of the renewed attack, with the characteristic and inflammatory symptoms, generally intermitting after a few days. g., auricular, that associated with a great increase in the intralabryinthine pressure. g. evolutum, the second stage of glaucoma. g. fulminans, an acute attack coming on with great suddenness and violence. g. hæmorrhagicum, that associated with retinal hemorrhage. Syn., g. *apoplecticum*. g. malignum, a grave form, attended with violent pain and rapidly leading to blindness. g., secondary, that consequent upon other ocular diseases. g. simplex, that form without inflammatory symptoms.

**glaucomatous** (*glaw-ko'-mat-us*) [see *glaucoma*]. Affected with or pertaining to glaucoma.

**glaucosis** (*glaw-ko'-sis*) [*glaucoma*]. The blindness resulting from glaucoma.

**glaucosuria** (*glaw-ko-su'-re-ah*) [γλαυκός, sea-green; οὖρον, urine]. The presence of indican in the urine, which is thereby discolored.

**glaxo** (*glak'-so*). A proprietary food for infants; it is said to consist of pure desiccated milk, with cream and lactose.

**gleet** (*glēt*). The chronic stage of urethritis, characterized by a slight mucopurulent discharge.

**gleety** (*gle'-te*) [*gleet*]. Resembling the discharge of gleet.

**Glénard's disease** (*gla-nar'*) [Franz *Glénard*, French physician, 1819–1894]. Enteroptosis; abdominal ptosis.

**glenohumeral** (*gle-no-hu'-mer-al*) [*glenoid*; *humerus*]. Pertaining to the glenoid cavity and the humerus. g. ligaments, three ligaments of the capsule of the shoulder-joint.

**glenoid** (*gle'-noid*) [γλήνη, a cavity; εἶδος, likeness]. Having a shallow cavity; resembling a shallow cavity or socket. g. cavity, the depression in the scapula for the reception of the head of the humerus. g. fissure. See *Glaserian fissure*. g. fossa, a depression in the temporal bone for articulation with the condyle of the lower jaw.

**Gley's glands** (*gla*) [M. E. *Gley*, French physiologist]. The parathyroid glands.

**glia** (*gli'-ah*) [γλία, glue]. The neuroglia. g.-cells. See *Deiters' cells*.

**gliacoccus** (*gli-ah-kok'-us*) [*glia*; *coccus*]. A micrococcus invested with a gelatinous envelope.

**gliacyte** (*gli'-ah-sīt*) [*glia*; κύτος, cell]. A neuroglia cell.

**gliadin, gliadine** (*gli'-ad-in*) [*glia*]. A proteid found in wheat-gluten.

**glial** (*gli'-al*). Pertaining to glia or neuroglia.

**gliding movement** (*gli'-ding moov'-ment*). The most simple kind of movement that can take place in a joint, one surface gliding or moving over another, without any angular or rotary movement.

**gliobacteria** (*gli-o-bak-te'-re-ah*) [*glia*; *bacteria*]. Bacteria in the zooglea stage, embedded in a gelatinous matrix.

**gliococcus** (*gli-o-kok'-us*) [*glia*; κόκκος, a berry] A micrococcus invested with a gelatinous envelop.

**glioma** (*gli-o'-mah*) [*glia*; ὄμα, a tumor]. A tumor composed of neuroglia cells, and occurring in the brain, spinal cord, retina, nerves, and suprarenal capsules. In the brain it closely resembles the brain-substance, but is usually more gelatinous and darker. In the retina it is often combined with sarcoma (*gliosarcoma*). It may also be combined with fibroma, myxoma, and neuroma. The last combination is known as *neuroglioma ganglionare*.

**gliomatosis** (*gli-o-mat-o'-sis*) [*glioma*]. The development of exuberant masses of glioma-like tissue in the nerve-centers. It is seen in the spinal cord in some cases of syringomyelia.

**gliomatous** (*gli-o'-mat-us*). Of the nature of, or affected with, glioma.

**gliomyoma** (*gli-o-mi-o'-mah*). Glioma combined with myoma.

**gliomyxoma** (*gli-o-miks-o'-mah*). A glioma with a mucoid degeneration.

**glioneuroma** (*gli-o-nū-ro'-mah*). See *neuroglioma ganglionare*.

**gliosarcoma** (*gli-o-sar-ko'-mah*). A tumor having the neuroglia cells of glioma and the fusiform cells of sarcoma.

**gliosis** (*gli-o'-sis*) [*glia*; νόσος, disease]. A brain disease marked by foci of sclerosed gray substance, with the formation of lacunar spaces within the foci. It differs from ordinary diffused sclerosis. g. cervicalis, syringomyelia.

**glischrin** (*glis'-brin*) [γλίσχρος, viscid]. Malerba's name for a nitrogenous mucus formed in urine by *Bacterium gliscrogenum*.

**glischrobacterium** (*glis-kro-bak-te'-re-um*) [γλίσχρος, viscid; *bacterium*]. The microorganism *Bacterium gliscrogenum*, causing mucous degeneration of the urine.

**glischrogenous** (*glis-kroj'-en-us*) [γλίσχρος, viscid; γενναν, to produce]. Giving rise to viscidity.

**glischruria** (*glis-kroó'-re-ah*) [γλίσχρος, viscid; οὖρον, urine]. The presence of glischrin in the urine.

**Glisson's capsule** (*glis'-un*) [Francis *Glisson*, English physician, 1597–1677]. The interlobular connective tissue of the liver, enveloping the portal vein, hepatic artery, and hepatic duct.

**Glissonian cirrhosis.** Perihepatitis.

**glissonitis** (*glis-on-i'-tis*). Inflammation of Glisson's capsule.

**globate** (*glo'-bāt*). Spheroidal; shaped like a globe.

**globe of the eye** (*glōb*). The eyeball.

**globin** (*glo'-bin*) [*globus*]. A proteid derived from hemoglobin.

**globinometer** (*glo-bin-om'-et-er*) [*globin*; μέτρον, a measure]. An instrument devised with special reference to the calculation of the percentage-amount of oxyhemoglobin in a given amount of blood.

**globomyelôma** (*glo-bo-mi-el-o'-mah*) [*globus*; *myeloma*]. A round-celled sarcoma.

**globose** (*glo-bōs'*) [*globosus*, round as a ball]. Spherical in form, or nearly so.

**globular** (*glob'-ū-lar*) [dim. of *globus*, a globe], Having the shape of a globe or sphere.

**globular value.** The relative amount of hemoglobin contained in a red corpuscle. It is a fraction of which the humerator is the percentage of hemoglobin and the denominator the percentage of corpuscle.

**globularetin, globularrhetin** (*glob-ū-lar'-e-tin*), $C_{11}H_{14}O_3$. A decomposition-product of globularin by the action of dilute acids. It is a powerful diuretic, stimulates the secretion of bile, and in large doses causes acute irritation of the intestine. It is used with globularin in gout. Dose ⅔ gr. (0.038 Gm.).

**globularin** (*glob-ū-lar'-in*), $C_{20}H_{16}O_{14}$. A glucoside from the leaves of *Globularia alypum*. Its action upon the heart and nervous system is similar to that of caffeine, while it diminishes the quantity and specific gravity of the urine and its contained urates and uric acid. It is used in connection with globularetin in gout, rheumatism, etc.

**globule** (*glob'-ūl*) [dim. of *globus*]. A small spherical particle, as a blood-corpuscle or lymph-corpuscle; also a small pill or pellet. g.s, directing, g.s, directive, g.s, extrusion, g.s, polar. See *bodies, direction*.

**globulicidal** (*glob-ū-lis-i'-dal*) [*globule*; *cædere*, to kill]. Destructive to the blood-corpuscles.

**globulicide** (*glob-ū'-lis-id*) [*globulus*, a globule; *cædere*, to kill]. 1. Destructive of blood-cells. 2. An agent that destroys blood-cells.

**globuliferous** (*glob-ū-lif'-er-us*) [*globule; ferre*, to bear]. Containing corpuscles, specifically red blood-corpuscles.
**globulimeter** (*glob-ū-lim'-et-er*) [*globulus*, a little ball; μέτρον, a measure]. An instrument for estimating the corpuscular richness of blood.
**globulin** (*glob'-ū-lin*) [*globule*]. 1. A general name for various proteids comprising globulin, vitellin, paraglobulin or serum-globulin, fibrinogen, myosin, and globin, which differ from the albumins in not being soluble in water, but soluble in dilute neutral saline solutions. These solutions are coagulated by heat and precipitated by a large amount of water. 2. Specifically, a proteid found in the crystalline lens. See *Hammarsten, Pohl*.
**globulinuria** (*glob-ū-lin-ū'-re-ah*) [*globulin;* οὖρον, urine]. The presence of globulin in the urine.
**globulism** (*glob'-ū-lizm*) [*globulus*, a little ball]. The administration of medicine in globules; homeopathy.
**globulolysis** (*glob-ū-lol'-is-is*). See *cytolysis*.
**globulose** (*glob'-ū-lōs*). Any product of the peptic digestion of a globulin.
**globulus** (*glob'-ū-lus*) [L.]. The *nucleus globosus* of the cerebellum. It is a mass of gray matter between the fastigatum and the embolus. 2. A globule.
**globus** (*glo'-bus*) [L.]. A ball or globe. **g. hystericus,** the "lump" or choking sensation occurring in hysteria, caused probably by spasmodic contraction of the esophageal and pharyngeal muscles. **g. major,** the larger end or head of the epididymis. **g. minor,** the lower end of the epididymis. **g. pallidus,** the inner and lighter part of the lenticular nucleus.
**glome** (*glōm*) [*glomus*, a ball]. 1. Same as *glomerule*. 2. One of the two rounded prominences which form the backward prolongations of the frog of a horse's foot. Cf. *periople*.
**glomer** (*glo'-mer*) [*glomus*]. A conglomerate gland.
**glomerate** (*glom'-er-āt*) [*glomerare*, to wind around]. Rolled together like a ball of thread.
**glomerular** (*glom-er'-oo-lar*). Relating to a glomerule or the kidney.
**glomerule, glomerulus** (*glom'-er-ūl, glom-er'-oo-lus*) [dim. of *glomus*]. 1. A small rounded mass. 2. A coil of blood-vessels projecting into the expanded end (Bowman's capsule) of each uriniferous tubule, and with it composing the Malpighian body. **g.,** olfactory, a group of nerve-cells, a number of which are embedded in the olfactory nerve-fibers. **g. of the pronephron.** See *glomus* (1). **g.s of the spleen,** round masses of lymphoid tissue developed in the adventitia of the arteries of the spleen.
**glomerulitis** (*glom-er-oo-li'-tis*) [*glomerule; ιτις*, inflammation]. Inflammation of the kidney.
**glomerulonephritis** (*glom-er-ōō-lo-nef-ri'-tis*) [*glomerule;* νεφρός, the kidney; *ιτις*, inflammation]. Inflammation of the Malpighian bodies of the kidney.
**glomus** (*glo'-mus*) [L.], "a ball"; 1. A fold of the mesothelium arising near the base of the mesentery in the pronephron, and containing a ball of blood-vessels. Syn., *glomerule of the pronephron*. 2. The part of the choroid plexus of the lateral ventricle which covers the thalamus. Syn. *glomus chorioideum*. **g. caroticum, g. carotideum,** the carotid gland. **g. coccygeum,** the coccygeal gland.
**glonoin** (*glon'-o-in*) [from Gl = glyceryl; O = oxygen; N = nitrogen, in the formula GlO$_3$(NO$_2$)$_3$, which Gl stands for glyceryl]. Nitroglycerin.
**glonoinism, glonoism** (*glon'-o-in-ism, glon'-o-ism*) [*glonoin*]. Intoxication by nitroglycerin.
**glossa** (*glos'-ah*) [γλῶσσα, the tongue]. The tongue; also the faculty of articulate speech.
**glossagra** (*glos-a'-grah*) [*glossa;* ἄγρα, seizure]. Gouty pain in the tongue.
**glossal** (*glos'-al*) [γλῶσσα, tongue]. Pertaining to the tongue.
**glossalgia** (*glos-al'-je-ah*) [*glossa;* ἄλγος, pain]. Pain in the tongue.
**glossanthrax** (*glos-an'-thraks*). Anthrax, or carbuncle of the tongue.
**glossauxesis** (*glos-awks-e'-sis*) [*glossa;* αὔξησις, increase]. Enlargement of the tongue.
**glossectomy** (*glos-ek'-to-me*) [*glossa;* ἐκτομή, excision]. Amputation or excision of the tongue.
**Glossina** (*glos-si'-nah*) [*glossa*]. A genus of biting flies. **G. morsitans,** a blood-sucking fly which transmits *Trypanosoma brucei*, the parasite of nagana,

a cattle disease of South America. **G. palpalis,** a fly which transmits *Trypanosoma gambiense,* the parasite which causes sleeping sickness.
**glossitic** (*glos-it'-ik*) [*glossa; ιτις*, inflammation]. Pertaining to or affected with glossitis.
**glossitis** (*glos-i'-tis*) [*glossa; ιτις*, inflammation]. Inflammation of the tongue. **g., dissecting,** a form of chronic superficial glossitis characterized by deep furrows upon the tongue that appear to penetrate into the mucous membrane. **g., idiopathic.** Same as **g., parenchymatous. g., interstitial.** Same as **g., parenchymatous. g., parasitic,** an inflammation of the tongue said to be due to parasitic vegetations. It is also called *glossophytia*. **g., parenchymatous,** an inflammation of the tongue involving its substance as well as the mucous membrane. **g. sclerosa,** fibroplastic cellular infiltration of the tongue, producing a sclerosis.
**glosso-** [γλῶσσα, tongue]. A prefix signifying the tongue.
**glossocele** (*glos'-o-sēl*) [*glossa;* κήλη, tumor]. Swelling or edema of the tongue, with consequent extrusion of the organ.
**glossodesmus** (*glos-o-dez'-mus*) [*glossa; δεσμός,* bond]. The frenum linguæ.
**glossodynamometer** (*glos-o-di-nam-om'-et-er*). An apparatus for estimating the capacity of the tongue to resist pressure.
**glossodynia** (*glos-o-din'-e-ah*) [*glossa;* ὀδύνη, pain]. Pain in the tongue, sometimes accompanied by exfoliation of its epithelium.
**glossoepiglottic, glossoepiglottidean** (*glos-o-ep-e-glot'-ik, glos-o-ep-e-glot-id'-e-an*) [*glosso-; epiglottis*]. Pertaining to both tongue and epiglottis.
**glossograph** (*glos'-o-graf*) [*glosso-;* γράφειν, to write]. An instrument for registering the movements of the tongue in speech.
**glossography** (*glos-og'-ra-fe*) [*glosso-;* γράφειν, to write]. A descriptive treatise upon the tongue.
**glossohyal, glossohyoid** (*glos-o-hi'-al, -oid*) [*glosso-; hyoid*]. Pertaining to the tongue and the hyoid bone.
**glossoid** (*glos'-oid*) [*glossa; εἶδος,* appearance]. Resembling a tongue.'
**glossokinesthetic** (*glos-o-kin-es-thet'-ik*) [*glossa; kinesthetic*]. Relating to the subjective perception of the motions of the tongue in speech.
**glossolabial** (*glos-o-la'-be-al*) [*glosso-; labium,* lip]. Relating to the tongue and lips.
**glossolabiolaryngeal paralysis** (*glos-o-la-be-o-lar-in'-je-al*). See *bulbar paralysis*.
**glossolalia** (*glos-o-la'-le-ah*) [*glosso-;* λαλία, speech]. The alleged speaking in foreign or unknown tongues by somnambulists.
**glossology** (*glos-ol'-o-je*) [*glosso-;* λόγος, a treatise]. 1. A treatise concerning the tongue. 2. Nomenclature.
**glossolysis** (*glos-ol'-is-is*) [*glosso-;* λύσις, a loosening]. Paralysis of the tongue.
**glossomantia, glossomantia** (*glos-o-man'-te-ah*) [*glosso-;* μαντεία, divination]. Prognosis of a disease based on the condition of the tongue.
**glossoncus** (*glos-ong'-kus*) [*glossa;* ὄγκος, tumor]. A swelling of the tongue.
**glossopalatine** (*glos-o-pal'-at-in*) [*glosso-; palatum,* palate]. Relating to the tongue and the palate.
**glossopalatinus** (*glos-o-pal-at-i'-nus*). See under *muscle*.
**glossopathy** (*glos-op'-ath-e*) [*glosso-;* πάθος, disease]. Any disease of the tongue.
**glossopeda** (*glos-o-pe'-dah*) [*glosso-; pes, foot*]. Synonym of *foot-and-mouth disease*.
**glossopharyngeal** (*glos-o-far-in'-je-al*) [*glosso-; pharynx*]. 1. Pertaining to the tongue and the pharynx. 2. Pertaining to the glossopharyngeal nerve.
**glossopharyngeus** (*glos'-o-far-in'-je-us*) [*glosso-; pharynx*]. A portion of the superior constrictor muscle of the pharynx. See *muscles, table of*.
**glossophyte, glossophyton** (*glos'-o-fīt, glos-of'-it-on*) [*glosso-;* φυτόν, plant]. A parasitic vegetation growing on the tongue, found in cases of nigrities linguæ.
**glossophytia** (*glos-o-fi'-te-ah*) [*glosso-;* φυτόν, a growth]. A dark discoloration of the tongue, due to the accumulation of spores and dead epithelium. Syn., *black tongue*.
**glossoplasty** (*glos'-o-plas-te*) [*glosso-;* πλάσσειν, to form]. Plastic surgery of the tongue.
**glossoplegia** (*glos-o-ple'-je-ah*) [*glosso-;* πληγή, stroke]. Paralysis of the tongue.

**glossoptosis** (*glos-op-to'-sis*) [*glosso-*; πτῶσις, a falling]. Synonym of macroglossia.
**glossorrhagia** (*glos-or-a'-je-ah*) [*glosso-*; ῥηγνύναι, to burst forth]. Hemorrhage from the tongue.
**glossorrhaphy** (*glos-or'-af-e*) [*glosso-*; ῥαφή, suture]. Surgical suturing of the tongue.
**glossoscopy** (*glos-os'-ko-pe*) [*glosso-*; σκοπεῖν, to inspect]. Diagnostic inspection of the tongue.
**glossosemeiotics** (*glos-o-sem-e-ot'-iks*) [*glosso-*; σημεῖοειν, to mark]. The study or science of the diagnostic and prognostic signs exhibited by the tongue.
**glossospasm** (*glos'-o-spazm*) [*glosso-*; σπασμός, spasm]. Spasm of the tongue.
**glossosteresis** (*glos-o-ster-e'-sis*) [*glosso-*; στέρησις, privation]. Surgical excision of the tongue; absence of the tongue.
**glossotilt** (*glos'-o-tilt*) [*glosso-*; τίλλειν, to pull]. An instrument by which the tongue is drawn forward during the process of artificial respiration.
**glossotomy** (*glos-ot'-o-me*) [*glosso-*; τέμνειν, to cut]. The dissection of the tongue. Also, the excision of the tongue.
**glossotrichia** (*glos-o-trik'-e-ah*) [*glosso-*; θρίξ, hair]. Hairy tongue.
**glossy skin.** A peculiar shining condition of the skin, due to trophic changes following injury or disease of the cutaneous nerves.
**glottagra** (*glot-a'-grah*) [γλῶττα, tongue; ἄγρα, seizure]. Glossagra.
**glottal** (*glot'-al*) [*glottis*]. Pertaining to the glottis.
**glottalgia** (*glot-al'-je-ah*) [γλῶττα, tongue; ἄλγος, pain]. Glossalgia.
**glottic** (*glot'-ik*) [*glotta*; and *glottis*]. 1. Pertaining to the tongue. 2. Pertaining to the glottis.
**glottidean** (*glot-id'-e-an*) [*glottis*]. Pertaining to the glottis.
**glottis** (*glot'-is*) [γλωττίς, glottis]. The rima glottidis. The opening between the arytenoid cartilages, or the interval between the vocal bands; also, the structures collectively that surround that opening.
**glottiscope** (*glot'-is-kōp*) [*glottis*; σκοπεῖν, to inspect]. A form of laryngoscope.
**glottitis** (*glot-i'-tis*). See glossitis.
**glottology** (*glot-ol'-o-je*). See glossology.
**glou-glou** (*gloo'-gloo*). A gurgling sound supposedly produced in the stomach by the respiratory pressure of the diaphragm; heard only when the stomach has become more or less vertical and while a tight corset is on. It is explained as being due to a temporary biloculation of the stomach by the corset.
**glove-area.** (*gluv*). The area of anesthesia of the fingers, hand, and forearm in multiple neuritis. It corresponds to the region of skin covered by gloves of various lengths.
**glovers' stitch** (*gluv'-er*). The continuous suture used especially in repairing wounds of the intestine.
**glucase** (*gloo'-kās*). The enzyme that converts starch into glucose.
**glucin-** (*gloo'-sin*). The sodium salt of amido-triazin-sulphonic acid; a substance resembling saccharin, but less sweet.
**glucinum** (*gloo-si'-num*). See beryllium.
**glucogen** (*gloo'-ko-jen*). See glycogen.
**glucohemia, glucohæmia** (*gloo-ko-he'-me-ah*). See glycohemia.
**glucolysis** (*gloo-col'-is-is*). See glycolysis.
**gluconic** (*gloo-kon'-ik*) [*glucose*]. Of or pertaining to glucose.
**glucoprotein** (*gloo-ko-pro'-te-in*). See glycoprotein.
**glucosamine.** Same as glycosamine.
**glucosazone** (*gloo-kō'-sa-zōn*). See phenylglucosazone.
**glucose** (*gloo'-kōs*) [γλυκύς, sweet]. 1. C₆H₁₂O₆. Grape-sugar; dextrose. A form of sugar found in many fruits, in blood and in lymph, and in the urine in diabetes. It is crystalline; its solution turns the plane of polarized light to the right; it is less soluble and less sweet than cane-sugar, and ferments readily. It can be obtained from starch by the action of diastatic ferments, or by boiling with dilute mineral acids, and crystallizes in nodular masses melting at 86° F. 2. A generic name for a class of carbohydrates having the composition C₆H₁₂O₆, of which ordinary glucose is the type. **g.,** tests for. See *Almén, Baeyer, Barfoed, Baumann; Boettger, Braun, Crismer, Fehling, v. Gebhart, Gentele, Gerrard, Haines, Hassall, v. Jaksch, Knapp, Loewenthal, Maumené, Molisch, Moore, Mulder, Nylander, Pavy,* *Penzoldt, Roberts, Rubner, Saccharimeter, Schiff, Silver, Soldani, Tollen, Trommer, Wender, Worm-Müller.*
**glucoside, glucosid** (*gloo'-ko-sid*) [*glucose*]. Any member of a series of compounds that may be resolved by an acid into glucose and another principle. The more important ones are amygdalin, arbutin, digitalin, and salicin.
**glucosin** (*gloo'-ko-sin*) [*glucose*]. 1. Any one of a series of bases obtained by the action of ammonia on glucose. 2. Trade name of a substitute for sugar.
**glucosuria** (*gloo'-ko-sū'-re-ah*). See glycosuria.
**glucovanillin** (*gloo-ko-van-il'-in*). See glycovanillin.
**glue** (*gloo*) [Low L., *glutem*, accus. of *glus*, glue.] An impure gelatin prepared by boiling the skin, hoofs, and horns of animals. It is a very adhesive substance, and when cold, holds the surfaces between which it is placed firmly together. **g.-like tumor,** a glioma.
**Gluge's corpuscles** (*gloo'-geh*) [Gottlieb *Gluge,* Belgian histologist, 1812–1898]. Compound granular corpuscles; compound granule-cells occurring in tissues that are the seat of fatty degeneration.
**gluside** (*gloo'-sid*). Synonym of saccharin. See glusidum.
**glusidum** (*gloo'-sid-um*) [γλυκύς, sweet],

$$C_6H_4 < \genfrac{}{}{0pt}{}{CO}{SO_2} > NH.$$

Saccharin. Benzoyl-sulphonicimide, a coal-tar derivative. It is an intensely sweet, white powder, 200 times as sweet as cane-sugar. It is antiseptic, and is used to disguise the taste of nauseous medicine. It may be used as a sweetening-agent in diabetes and in the treatment of corpulency.
**glutæus** (*gloo-te'-us*). See gluteus.
**glutamic acid** (*gloo-tam'-ik*), C₅H₉NO₄. A crystalline acid found in gluten and other proteids.
**glutamine** (*gloo-tam'-in*). An amine found in the juice of the beet, gourd, mustard, and other plants.
**glutannin** (*gloo-tan'-in*). Trade name of a combination of tannin and albumin.
**glutannol** (*gloo-tan'-ol*). A proprietary intestinal astringent said to consist of vegetable fibrin and tannic acid. Dose, 4–15 gr. (0.25–1 gm.).
**glutaric acid** (*gloo-tar'-ik*), C₅H₈O₄. A crystalline acid found in decomposed pus.
**gluteal** (*gloo-te'-al*) [*gluteus*]. Pertaining to the buttocks. **g. artery.** See under *artery.* **g. muscles.** See under *muscle.* **g. nerve.** See under *nerve.* **g. reflex.** See under *reflex.*
**glutei** (*gloo'-te-i*). The muscles of the buttocks.
**glutelin** (*gloo'-te-lin*). A class of simple proteins occurring in seeds of cereals; soluble in dilute acids and alkalies insoluble in neutral solutions.
**gluten** (*gloo'-ten*) [L., "glue"]. A nitrogenous substance found in the seed of cereals. It consists mainly of gluten-fibrin, gluten-casein, gliadin, and mucedin. **g.-bread,** bread made from wheat-flour from which all the starch has been removed; it is used as a substitute for ordinary bread in diabetes. **g.-casein,** a nitrogenous substance resembling the casein of milk, and forming about 15 % of the gluten of flour. Syn., *vegetable casein.* **g.-fibrin,** C₃H₅₀N₁₀O₁₁, a brownish mass extracted from gluten.
**glutenin** (*gloo'-ten-in*). A proteid of wheat.
**gluteofascial** (*gloo-te-o-fash'-e-al*) [*gluteus; fascia,* bundle]. Relating to the fascia of the gluteal region.
**gluteofemoral** (*gloo-te-o-fem'-or-al*) [*gluteus; femur*]. Relating to the buttock and the thigh.
**gluteoinguinal** (*gloo-te-o-in'-gwin-al*). Relating to the buttock and groin.
**gluteotrochanteric** (*gloo-te-o-tro-kan-ter'-ik*). Relating to the gluteal region and the trochanter.
**gluteus** (*gloo-te'-us*) [γλουτός, buttock]. One of the large muscles of the buttock. See under *muscle.*
**glutin** (*gloo'-tin*). 1. See *gelatin.* 2. Synonym of *gluten-casein.*
**glutinous** (*gloo'-tin-us*). Viscid; glue-like.
**glutinpeptone sublimate** (*gloo-tin-pep'-tōn*). An antiseptic preparation of mercury containing 25 % of mercuric bichloride and obtained by the action of hydrochloric acid on gelatin. It occurs as a hygroscopic white powder or a noncorrosive liquid used hypodermatically in syphilis. Dose 15 gr. (1 Gm.).
**glutitis** (*gloo-ti'-tis*) [*gluteus;* ιτις, inflammation]. Inflammation of the gluteal muscles.

# GLUTOFORM 395 GLYCOSIDE

**glutoform, glutol** (*gloo'-to-form,* *glu'-tol*). See *formaldehyde-gelatin.*

**glutoid** (*gloo'-toid*). A preparation of gelatin and formaldehyde insoluble in the stomach, but soluble in the intestine. It is used for coating pills or making capsules when intestinal medication is desired.

**glutolin** (*gloo'-to-lin*). An albuminoid body from paraglobulin, supposed to be a constituent of bloodplasm.

**gluton** (*gloo'-ton*). A dietetic substance obtained from gelatin by the action of acids at a high temperature for several hours.

**Gluzinske's test for bile-pigments** (*gloo-zin'-ske*) [Anton *Gluzinski*, Austrian physician]. Boil the solution for a few minutes with formalin, and an emerald-green coloration will result, changing to an amethyst-violet on the addition of a few drops of hydrochloric acid.

**glybolid** (*gli'-bo-lid*). The proprietary name for an antiseptic paste made of equal parts of boralid and glycerin. Syn., *glybrid.*

**glybrid** (*gli'-brid*). Same as *glybolid.*

**glycase** (*gli'-kās*). An enzyme which converts maltose into dextrose.

**glyceleum** (*gli-se'-lē-um*) [γλυκύς, sweet; ἔλαιον, oil]. A mixture of glycerin 2 parts, olive oil 6 parts, almond-meal 1 part; it is used as a base for ointments.

**glycemia, glycæmia** (*gli-se'-me-ah*) [*glucose;* αἷμα, blood]. The presence of glucose in the blood.

**glyceric** (*glis-er'-ik*). Derived from glycerol or glycerol.

**glycerid** (*glis'-er-id*) [*glycerol*]. A compound of glycerol and an acid; the neutral fats are glycerids.

**glycerin, glycerinum** (*glis'-er-in, glis-er-i'-num*) [γλυκύς, sweet]. 1. See *glycerol.* 2. In the British Pharmacopeia, a solution of a medicinal substance in glycerol; a glycerite. **g.-jelly,** a mixture of glycerin and jelly. **g. suppositories** (*suppositoria glycerini,* U. S. P.), each contains 6 Gm. of glycerol; they are used in constipation. **glycerinum pepticum,** trade name of a glycerin extract of concentrated peptic enzyme.

**glycerite, glyceritum** (*glis'-er-īt, glis-er-i'-tum*) [see *glyceri*]. A mixture of medicinal substances with glycerol. The following glycerites are official: *glyceritum acidi tannici; g. amyli; g. boroglycerini; g. ferri quininæ et strychninæ phosphatum; g. hydrastis; g. phenolis.*

**glycerize** (*glis'-er-īz*). To treat or mix with glycerin.

**glyceroborate** (*glis-er-o-bo'-rāt*). A compound made by heating together equal parts of glycerin and a borate.

**glyceroformol** (*glis-er-o-for'-mol*). An antiseptic substance formed by the prolonged action of formaldehyde upon glycerin.

**glycerol** (*glis'-er-ol*) [see *glycerin*], $C_3H_5(HO)_3$. Glycerin. A colorless substance, of syrupy consistence, sweetish to the taste, obtained from fats and fixed oils. Chemically it is a triatomic alcohol, and may be looked upon as propenyl alcohol. It is soluble in water and in alcohol, and has a specific gravity of 1.25 at 15° C. It is used as a vehicle in pharmaceutical preparations, as an emollient application to the skin, as a laxative administered by the mouth or in suppository, for tampons in pelvic congestion, as a substitute for sugar in diabetes, as a mounting-medium in microscopy, and as an addition to bacteriologic culture-mediums.

**glycerolate, glycerolatum** (*glis'-er-o-lāt, glis-er-o-la'-tum*). Same as *glycerite.* **g., aromatic,** a sticky, transparent substance consisting of tragacanth, 4 parts; acetone, 30 parts; glycerol, 46 parts; water, 18 parts; aromatic perfume, 4 parts; it is recommended in the treatment of skin diseases.

**glycerophosphate** (*glis-er-o-fos'-fāt*). A combination of glycerol and phosphoric acid with a base.

**glycerose** (*glis'-er-ōs*), $C_3H_6O_3$. Triose; a substance derived from glycerol. It is the lowest glucose, and is a mixture of glyoeryl aldehyde and dioxyacetone.

**glyceryl** (*glis'-er-il*) [*glycerol*]. The trivalent radical, $C_3H_5$, of glycerol, combining with the fatty acids to form the neutral fats. **g. borate.** See *boroglycerin.* **g. trinitrate, spirit of** (*spiritus glycerylis nitratis,* U. S. P.), spirit of nitroglycerin.

**glycid** (*gli'-sid*) [γλυκύς, sweet], $C_3H_6O_2$. The oxide of hydroxypropene. It is isomeric with acetol and lactic aldehyde.

**glycin** (*gli'-sin*). Synonym of *glycocoll.*

**glycina** (*glis-i'-nah*). Soya-bean. See *soja.*

**glyco-** (*gli-ko-*) [γλυκύς, sweet]. A prefix meaning sweet.

**glycoblastol** (*gli-ko-blas'-tol*). A proprietary hair restorer said to consist of alcohol, glycerin, and capsicum.

**glycocholate** (*gli-ko-ko'-lāt*). A salt of glycocholic acid.

**glycocholic acid** (*gli-ko-kol'-ik*) [*glyco-;* χολή, bile]. An acid found in the bile. See *acid, glycocholic.*

**glycocide** (*gli'-ko-sid*). See *glucoside.*

**glycocin** (*gli'-ko-sin*). See *glycin.*

**glycocoll** (*gli'-ko-kol*) [*glyco-;* κόλλα, glue], $C_2H_5NO_2$. It is obtained when glycocholic acid is boiled with caustic potash, baryta-water, or with dilute mineral acids; also by boiling gelatin with dilute acids. It is capable of acting as a base and as an acid. Syn., *aminoacetic acid; gelatin-sugar; glycin.*

**glycoformal** (*gli-ko-form'-al*). A disinfectant composed of an aqueous solution of formic aldehyde and glycerin.

**glycogelatin** (*gli-ko-jel'-at-in*). An ointment base consisting of glycerol and gelatin.

**glycogen** (*gli'-ko-jen*) [*glyco-;* γεννᾶν, to produce]. A carbohydrate found in the form of amorphous granules in the liver-cells, in all tissues of the embryo, in the testicle, muscles, leukocytes, fresh pus-cells, cartilage, and other tissues. It is formed from carbohydrates and probably also from proteins, and is stored in the liver, where it is converted, as the system requires, into sugar (glucose). It is also known as animal starch. Glycogen is soluble in water, is dextrorotatory, and is colored red by iodine.

**glycogenal, glycogenol** (*gli-ko'-jen-al, -ol*). A substance allied to glycogen. It is used in tuberculosis by inhalation and internally. Dose 15–23 gr. (1.0–1.5 Gm.).

**glycogenesis** (*gli-ko-jen'-es-is*) [γλυκύς, sweet; γένεσις, production]. The formation of sugar in the animal economy, whether normal or pathologic.

**glycogenetic** (*gli-ko-jen-et'-ik*). Pertaining to the formation of sugar or of glycerin.

**glycogenic** (*gli-ko-jen'-ik*). Pertaining to glycogen or to glycogenesis.

**glycogeny** (*gli-koj'-en-e*) [*glycogen*]. Glycogenesis.

**glycohæmia, glycohæmia** (*gli-ko-he'-me-ah*) [*glyco-;* αἷμα, blood]. A saccharine condition of the blood. Syn., *glycemia.*

**glycoheroin** (*gli-ko-her'-o-in*). A proprietary liquid expectorant said to contain heroine and hyoscyamine. Dose 1 dr. (4 Cc.).

**glycol** (*gli'-kol*) [γλυκύς, sweet]. A diatomic alcohol.

**glycolamine** (*gli-kol-am'-in*). Synonym of *glycocoll.*

**glycolic, glycollic** (*gli-kol'-ik*). Derived from glycol.

**glycoline** (*gli'-kol-ēn*). A purified petroleum for use in atomizers.

**glycolysis** (*gli-kol'-is-is*) [*glyco-;* λύσις, dissolution]. The hydrolysis of sugar in the body.

**glycolytic** (*gli-ko-lit'-ik*) [*glucose;* λύσις, dissolution]. Splitting up sugars; pertaining to glycolysis.

**glyconeogenesis** (*gli-ko-ne-o-jen'-es-is*) [*glyco-;* νέος, new; γεννᾶν, to produce]. The formation of carbohydrates from substances which are not carbohydrates, as protein or fat.

**glyconin** (*gli'-ko-nin*). A mixture of yolk of egg, 45, and glycerol, 55 parts. Syn., *glyceritum vitelli.*

**glycopolyuria** (*gli-ko-pol-e-u'-re-ah*) [*glyco-; poly-uria*]. Bouchardat's term for diabetes, attended with a moderate quantity of sugar in the urine and with an increase of lithic acid in the blood.

**glycoproteins** (*gli-ko-pro'-te-ins*) [*glyco-; protein*]. Compound proteins which on decomposition yield a protein on one side and a carbohydrate or derivatives of the same on the other. Some glycoproteins are free from phosphorus (mucins, mucinoids, and hyalogens) and some contain it (phosphoglycoproteins).

**glycorrhea** (*gli-kor-e'-ah*) [*glyco-;* ῥοία, flow]. A discharge of saccharine fluid from the body.

**glycosal** (*gli'-ko-sal*). Monosalicylic glyceryl ester, a white powder readily soluble in hot water and alcohol, and less freely in ether and chloroform. It is antirheumatic. Dose 8–150 gr. (0.52–9.75 Gm.) a day. Applied in 20 % alcoholic solution.

**glycosamine** (*gli-kos'-am-in*). See *chitin.*

**glycosemia, glycosæmia** (*gli-ko-se'-me-ah*). See *glycemia* and *glycohæmia.*

**glycoses** (*gli-ko'-sēs*). Same as *glucoses, q. v.*

**glycoside** (*gli'-ko-sid*). Glucoside.

**glycosolveol** (*gli-ko-sol'-ve-ol*). A proprietary remedy for diabetes said to be obtained from peptone by action of oxypropionic acid and from trypsin by action of a compound of theobromine.

**glycosometer** (*gli-ko-som'-et-er*) [*glyco-*; μέτρον, a measure]. An instrument for use in the estimation of the percentage of sugar in diabetic urine.

**glycosuria** (*gli-ko-sū'-re-ah*) [*glucose;* οὖρον, urine]. The presence of grape-sugar in the urine. For tests see under *glucose*. **g., alimentary,** that due to excessive ingestion of carbohydrates. **g., anxiety** (of v. Noorden), a transitory form due to worry. **g., artificial,** a condition resulting from puncture of the diabetic center in the bulb. Syn., *artificial diabetes; traumatic glycosuria.* **g., diabetic,** that in which sugar and oxybutyric acid and its derivatives are passed in the urine. **g., lipogenic,** the glycosuria of obese subjects which does not amount to true diabetes. **g., persistent.** Synonym of *diabetes mellitus.* **g., toxic,** that observed after poisoning by chloral, morphine, or curara, after inhalation of chloroform or carbonic monoxide, and after the ingestion of phloridzin. See **g.,** *artificial.* **g. of vagrants,** a transient form occurring in underfed vagrants.

**glycothymolin** (*gli-ko-thi'-mol-in*). An antiseptic cleansing solution for the treatment of diseased mucous membrane. Said to consist of glycerol, sodium, boric acid, thymol, menthol, salicylic acid, eucalyptol, and other antiseptics.

**glycovanillin** (*gli-ko-van-il'-in*), $C_6H_3(OCH_3)(OC_6H_{11}O_5)CHO + 2H_2O$. The glucoside of vanillin, formed by the oxidation of coniferin with dilute chromium trioxide. Syn., *glucovanillin.*

**glycozone** (*gli'-ko-zōn*). A combination of pure glycerol with 15 times its own volume of ozone at 0° C. It is a healing agent, used in gastric ulcer, etc., in teaspoonful doses diluted with water.

**glycuronic acid** (*gli-kū-ron'-ik*). See *acid, glycuronic.*

**glycuronuria** (*gli-kū-ron-ū'-re-ah*) [*glyco-;* οὖρον, urine]. The presence of glycuronic acid in the urine.

**glycyltryptophan** (*gli-sil-trip'-to-fan*). A compound of glycin and tryptophan radicals, used as a test for cancer of the stomach.

**glycyrrhea** (*gli-sir-e'-ah*) [γλυκύς, sweet; ῥεῖν, to flow]. Any discharge of glucose from the body.

**glycyrrhoea urinosa.** Synonym of *diabetes mellitus.*

**glycyrrhetin** (*gli-sir'-e-tin*) [γλυκύς, sweet; ῥίζα, root], $C_{18}H_{26}O_4$. An amorphous bitter substance in licorice-root.

**glycyrrhiza** (*glis-ir-i'-zah*) [γλυκύς, sweet; ῥίζα, root], Licorice. The root of *G. glabra,* a demulcent and mild laxative, of sweet taste. It is used in catarrhal affections and as an ingredient of pills. **g., extract of** (*extractum glycyrrhizae,* U. S. P.). Dose 15 gr. (1 Gm.). **g., extract of, pure** (*extractum glycyrrhizae purum,* U. S. P.) used for making pills. **g., fluidextract of** (*fluidextractum glycyrrhizae,* U. S. P.), used as a vehicle for administering quinine. **g., mixture of, compound** (*mistura glycyrrhizae composita,* U. S. P.), brown mixture. Dose 1–4 dr. (4–16 Cc.). **g., powder of, compound** (*pulvis glycyrrhizae compositus,* U. S. P.). Dose 1 dr. (4 Gm.). **g., troches of, and opium** (*trochisci glycyrrhizae et opii,* U. S. P.). Dose 1 or 2.

**glycyrrhizin, glycyrrhizinum** (*glis-ir-i'-zin, -iz-i'-num*) [*glycyrrhiza*]. The active principle of licorice-root. It is in reality an acid, *glycyrrhizic acid,* $C_{44}H_{63}NO_{18}$. **g., ammoniated** (*glycyrrhizinum ammoniatum,* U. S. P.), a sweet preparation used as a substitute for licorice. Dose 5–15 gr. (0.32–1.0 Gm.).

**glykaolin** (*gli-ka'-ol-in*). A compound of aluminium silicate, phenyl salicylate, and glycerol made into a smooth paste; it is indicated in the treatment of wounds, ulcers, sprains, burns, etc.

**glymol** (*gli'-mol*). A proprietary preparation said to be obtained from petroleum; it is used in diseases of the nose and throat.

**glyoxal** (*gli-oks'-al*) [γλυκύς, sweet; ὀξαλίς, sorrel], $C_2H_2O_2$. A substance formed by oxidizing acetaldehyde with $HNO_3$. It is an amorphous non-volatile mass that deliquesces in the air, and is soluble in alcohol and in ether.

**glyoxalin** (*gli-oks'-al-in*) [γλυκύς, sweet; [ὀξαλίς, sorrel], $C_3H_4N_2$. A substance produced by the action of ammonia upon glyoxal. It is easily soluble in water, alcohol and ether, and crystallizes in brilliant prisms, melting at 89° C. and boiling at 255° C.

**gm.** An abbreviation of *gram,* or *gramme.*

**Gmelin's test for bile-pigments in the urine** (*ma'-lin*) [Leopold Gmelin, German physiologist, 1788–1853]. Fuming nitric acid is carefully added, so that it forms a sublayer. At the junction of the two liquids a series of colored layers is formed, in the following order, from above downward: green, blue, violet, red, and reddish-yellow. The green ring must always be present, and the reddish-violet at the same time; otherwise the reaction might be confounded with that for lutein.

**gnat** (*nat*). A dipterous insect, the *Culex pipiens,* differing but slightly from the common mosquito, with which it is popularly confounded. The "bite" consists in a piercing of the skin and the withdrawal of a minute quantity of blood. The gnat has no sting or poison-glands.

**gnathalgia** (*nath-al'-je-ah*) [*gnathion;* ἄλγος, pain]. Pain or neuralgia of the jaw.

**gnathankylosis** (*nath-ang-kil-o'-sis*) [*gnathion;* ankylosis]. Ankylosis of the jaw.

**gnathic** (*na'-thik*) [*gnathion*]. Pertaining to the jaw. **g., index.** See *index, gnathic.*

**gnathion** (*na'-the-on*) [γνάθος, jaw]. The lowest point in the median line of the inferior maxilla.

**gnathitis** (*nath-i'-tis*) [*gnathion; itis,* inflammation]. Inflammation of the jaw or cheek.

**gnatho-** (*nath-o-*) [γνάθος, the jaw]. A prefix signifying the jaw or cheek.

**gnathocephalus** (*nath-o-sef'-al-us*) [*gnathion;* κεφαλή, the head]. A monster lacking all parts of the head except large jaws.

**gnathodynia** (*nath-o-din'-e-ah*) [*gnatho-;* ὀδύνη, pain]. Pain in the jaw, gnathalgia.

**gnathoneuralgia** (*nath-o-nū-ral'-je-ah*) [*gnatho-;* neuralgia]. See *gnathalgia.*

**gnathoparalysis** (*nath-o-par-al'-is-is*) [*gnatho-;* paralysis]. Paralysis of the jaw.

**gnathoplasty** (*nath'-o-plas-te*) [*gnatho-;* πλάσσειν, to shape]. Plastic surgery of the cheek.

**gnathoplegia** (*nath-o-ple'-je-ah*) [*gnatho-;* πληγή, stroke]. Paralysis of the cheek.

**gnathorrhagia** (*nath-or-a'-je-ah*) [*gnatho-;* ῥηγνύναι, to burst forth]. Hemorrhage from the mucous membrane of the cheek or from the jaws.

**gnathoschisis** (*nath-os'-kis-is*) [*gnatho-;* σχίζειν, to split]. Cleft-jaw.

**gnathospasmus** (*nath-o-spaz'-mus*) [*gnatho-;* spasm]. Locked jaw; trismus.

**gnoscopine** (*nos'-ko-pin*), $C_{22}H_{23}N_2O_{11}$. A crystalline alkaloid of opium.

**Goa-powder.** See *araroba.*

**goblet-cells.** Beaker-shaped cells found in mucous membranes.

**Godfrey's cordial.** A non-official preparation of opium containing from half a grain to a grain and a half of opium to the ounce.

**Goggia's sign** (*god'-yah*). If the biceps muscle of the arm is pinched there is a general fibrillation, in a state of health; and a locally limited fibrillation in cases of asthenic disease.

**goggle-eyed** (*gog'l'-īd*). A vulgar synonym of exophthalmos and of strabismus.

**goggles** (*gog'-lz*) [O. E. *goggle,* to roll the eyes]. Spectacles with colored lenses and wire or cloth sides, to protect the eyes from excessive light or dust.

**goiter** (*goi'-tr*) [*guttur,* throat]. Enlargement of the thyroid gland. Syn., *bronchocele;* Derbyshire *neck; tracheocele.* **g., aberrant, g., accessory,** that of an accessory thyroid gland. **g., aerial.** See *aerocele.* **g., amyloid,** a form associated with amyloid degeneration of the small arteries, capillaries, and follicles. **g., anemic.** Synonym of **g.,** *exophthalmic.* **g., carcinomatous,** carcinomatous, carcinoma of the thyroid gland. **g., exophthalmic,** a disease characterized by cardiac palpitation, goiter, exophthalmos, tremor, palpitation being usually the initial symptom. Syn., *Basedow's disease; Graves' disease.* See *Abadie, Dalrymple, Graefe, Stellwag, Vigoroux.* Also see under *Ballet, Becker, Bryson, Joffroy, Marie, Moebius.* **g., perivascular,** one surrounding an important blood vessel. **g., pituitary,** cystic and follicular degeneration in the hypophysis similar to that occurring in the thyroid gland. **g., pneumoguttural.** See *aerocele.* **g., retrovascular,** one traversed anteriorly by a large blood-vessel.

**goitriferous** (*goi-trif'-er-us*) [*goiter; ferre,* to bear]. Giving rise to goiters.

**goitrous** (*goi'-trus*) [*goiter*]. Relating to or affected with goiter.

**gold.** See *aurum*. **g.-beaters' skin,** a thin membrane prepared from the cecum of the ox. **g. cure,** the Keeley cure. **g., fulminating,** Au₂O₃(NH₃)₄, a compound obtained from auric oxide, or auric hydrate by action of ammonia; a greenish-brown powder exploding with great violence on heating or percussion.
**goldenseal.** See *hydrastis*.
**Goldflam's disease** (gŏlt'-flahm) [Sigismund Goldflam, Polish physician]. See *Erb's disease*.
**Goldscheider-Marinesco's law.** The fewer connections a neuron has, and, consequently, the fewer stimuli it receives, the less is its tendency to degeneration.
**Golgi's cells** (gol'-je) [Camillo Golgi, Italian anatomist, 1844– ]. Nerve-cells with very short processes found in the gray matter of the brain and spinal cord. **G.'s corpuscles,** tendon-spindles; small fusiform bodies resembling the Pacinian corpuscles, existing in tendons at the junction of the tendinous fibers with the muscular fibers. They have not been found in the ocular muscles. See *Ross, cycle of*. **G.'s cycle,** that phase of development of the *Plasmodium malariæ* which occurs in human blood. See *Ross, cycle of*. **G.'s funnels,** funnel-shaped structures composed of spiral threads, described by Golgi and others as surrounding the axis-cylinder of a myelinic nerve-fiber and supporting the myelin. They appear to be artificially produced in the process of staining.
**Golgi-Mazzoni's corpuscles.** See *Mazzoni's corpuscles*.
**Golgi-Rezzonico's funnels.** See *Golgi's funnels*.
**Goll's column** [Friedrich Goll, Swiss anatomist, 1829–1903]. The posterointernal column of the spinal cord. **G.'s nucleus,** a small nucleus in the fasciculus gracilis of the oblongata in which the long fibers of Goll's column terminate.
**Golonboff's sign of chlorosis.** An acute pain located directly over the spleen, and pain on percussion over the ends of the long bones, especially the tibia.
**Goltz's experiment** [Friedrich Leopold Goltz, German physician, 1834– ]. Arrest of the heart's action produced in the frog by repeated tapping of the abdomen.
**Gombault-Phillippe's triangle** [Albert François Gombault, French physician, 1844– ]. The triangular area formed in the conus medullaris by the fibers which, higher up, compose the oval field of Flechsig.
**gomenol** (go'-men-ol). A syrup, used in pertussis said to be prepared from the leaves of *Melaleuca leucadendron*. Dose 5–50 min. (0.33–3.33 Cc.) in capsules.
**gomphiasis** (gom-fi'-as-is) [γομφίος, a molar tooth]. Looseness of the teeth.
**gomphosis** (gom-fo'-sis). See *synarthrosis*.
**gonacratia** (gon-ak-ra'-she-ah) [γονή, semen; ἀκράτεια, incontinence]. Spermatorrhea.
**gonad** (gon'-ad) [γονή, semen; pl., *gonades*]. A sexual gland; a testicle or ovary.
**gonades** (gon'-ad-ēs) [gonad]. The reproductive organs; genitalia.
**gonaduct** (gon'-ad-ukt) [γονή, semen; *ductus*, a duct]. The excretory duct of a reproductive gland; an oviduct or a seminal duct.
**gonæ** (go'-ne) [L.]. The genitals.
**gonagra** (gon-a'-grah) [γόνυ, knee; ἄγρα, seizure]. Gout of the knee-joint.
**gonal** (go'-nal). A proprietary remedy for gonorrhea, cystitis, etc., said to be the active principle of sandalwood oil. Dose, 10–20 m. (0.66–1.33 c.c.)
**gonalgia** (gon-al'-je-ah) [γόνυ, knee; ἄλγος, pain]. Pain in the knee-joint.
**gonangiectomy** (go-nan-je-ek'-to-me) [γόνος, generation; ἀγγεῖον, vessel; ἐκτομή, excision]. Excision of a portion of the vas deferens.
**gonarthritis** (gon-ar-thri'-tis) [γόνυ, knee; *arthritis*]. 1. Inflammation of the knee-joint. 2. Synonym of *Conorrhoal synovitis*.
**gonarthrocace** (gon-ar-throk'-as-e) [γόνυ, knee; ἄρθρον, a joint; κάκη, evil]. White swelling of the knee-joint.
**gonarthromeningitis** (gon-ar-thro-men-in-ji'-tis) [γόνυ, knee; ἄρθρον, a joint; μῆνιγξ, membrane; ιτις, inflammation]. Inflammation of the synovial membrane of the knee.
**gonarthrotomy** (gon-ar-throt'-o-me) [γόνυ, knee; ἄρθρον, joint; τομή, incision]. Incision into the knee-joint.

**gonatalgia** (gon-at-al'-je-ah). See *gonalgia*.
**gonatocele** (gon-at'-o-sēl) [γόνυ, knee; κηλή, tumor]. A swelling or tumor of the knee; white swelling.
**gonecyst, gonecystis** (gon'-e-sist, gon-e-sis'-tis) [γονή, semen; κύστις, cyst]. A seminal vesicle.
**gonecystic** (gon-e-sis'-tik). Pertaining to a gonecyst.
**gonecystitis** (gon-e-sis-ti'-tis) [gonecyst; ιτις, inflammation]. Inflammation of the seminal vesicles.
**gonecystolith** (gon-e-sis'-to-lith) [γονή, semen; κύστις, cyst; λίθος, stone]. A concretion or calculus in a seminal vesicle.
**gonecystoncus** (gon-e-sis-tong'-kus) [gonecyst; ὄγκος, tumor]. Any tumor of a seminal vesicle.
**gonecystopyosis** (gon-e-sist-o-pi-o'-sis) [gonecyst; πύωσις, suppuration]. Suppuration of a gonecyst.
**goneitis** (gon-e-i'-tis) [γόνυ, knee; ιτις, inflammation]. Inflammation of the knee.
**gonepoiesis** (gon-e-poi-e'-sis) [γονή, semen; ποιεῖν, to make]. The secretion or elaboration of semen.
**gonepoietic** (gon-e-poi-et'-ik) [γονή, semen; ποιεῖν, to make]. Pertains to secretion of semen.
**gongrona** (gon-gro'-nah) [γογγρώνη, a ganglion]. Synonym of *goiter*.
**gongyloid** (gon'-jil-oid) [γογγύλος, round; εἶδος, likeness]. Having an irregular round shape.
**gonic** (gon'-ik) [γονή, semen]. Pertaining to semen or to generation.
**gonid** (go'-nid), **gonidium** (go-nid'-e-um) [γονή, seed; pl., *gonidia*]. In biology, (*a*) one of the grass-green algal elements of the lichen thallus; (*b*) also applied to various asexually produced reproductive bodies.
**gonidia** (go-nid'-e-ah). Plural of *gonidium*.
**goniocraniometry** (go-ne-o-kra-ne-om'-et-re) [γωνία; angle; *craniometry*]. Measurement of the cranial angles.
**goniometer** (go-ne-om'-et-er) [*gonion*; μέτρον, a measure]. An apparatus for measuring lateral curvatures, adduction and abduction in hip-joint disease, the angle of ankylosed joints, etc., and the angles of crystals. **g., vesical,** an apparatus to measure the angle formed by the long axis of the urethra with a line drawn from the internal urethral orifice to the mouth of the ureter.
**goniometry** (go-ne-om'-et-re). The measurement of angles.
**gonion** (go'-ne-on) [γωνία, an angle]. The outer side of the angle of the inferior maxilla.
**gonioscope** (go'-ne-o-skōp) [γωνία, angle; σκοπεῖν, to see]. An apparatus used in noting the varying angles made by the optical axis with the principal line of action.
**goniozygomatic** (go-ne-o-zi-go-mat'-ik) [γωνία, an angle; *zygoma*]. Relating to the gonion and the zygoma.
**gonitis** (go-ni'-tis) [γόνυ, knee; ιτις, inflammation]. Inflammation of the knee-joint.
**gonoblast** (gon'-o-blast) [γόνος, generation; βλαστός, germ]. A sperm-cell or germ-cell; any cell concerned directly in reproduction; a spermatozoon; an ovum.
**gonobolia** (gon-o-bo'-le-ah) [γονή, semen; βάλλειν, to throw]. 1. The ejaculation of semen. 2. Spermatorrhea.
**gonocele** (gon'-o-sēl) [1] [γόνυ, knee; κήλη, tumor]. A swelling of the knee. [2] [γόνη, semen; κήλη, tumor]. 1. A tumor of a testis, or of a spermatic cord, due to retention of semen. 2. Synonym of *gonorrheal rheumatism*.
**gonochorismus** (gon-o-kor-iz'-mus) [γόνος, generation, sex; χωρισμός, separation]. In biology, the specialization of sex; the differentiation of male or female genitalia in a developing embryo; the separation of the sexes in phylogeny.
**gonocide** (gon'-o-sīd) [*gonococcus; cædere*, to kill]. 1. Destructive to the gonococcus. 2. An agent which kills the gonococcus.
**gonococcal** (gon-o-kok'-al). Relating to the gonococcus.
**gonococcæmia, gonococcsæmia** (gon-o-kok-se'-me-ah) [*gonococcus*; αἷμα, blood]. Gonococci in the blood.
**gonococcic** (gon-o-kok'-sik). Pertaining to the gonococcus; gonococcal.
**gonococcicide** (gon-o-kok'-se-sīd) [*gonococcus; cædere*, to kill]. 1. Destructive to the gonococcus. 2. An agent which kills the gonococcus.
**gonococcus** (gon-o-kok'-us) [γονή, semen; κόκκος, berry]. The specific organism causing gonorrhea. See *Micrococcus gonorrheæ*.

# GONOCYTE 398 GOUT

**gonocyte** (*gon'-o-sīt*) [γονή, semen; κύτος, cell]. Van Beneden's name for the ovum which contains only the female pronucleus, the male part having been expelled as directive bodies. **g.**, **male**, a spermatozoon.

**gonohemia, gonohæmia** (*gon-o-he'-me-ah*) [*gonorrhea*; αἷμα, blood]. Generalized gonorrheal infection.

**gonoid** (*go'-noid*) [γονή, semen; εἶδος, appearance]. Resembling semen.

**gonopepsin** (*gon-o-pep'-sin*). A preparation said to consist of pepsin, boric acid, infusion of cranberries, and water; it is intended as a gonorrheal injection.

**gonophore** (*gon'-o-fōr*) [γονή, seed; φόρος, bearer]. Any structure which conducts or stores up the sexual cells; the vas deferens, seminal vesicles, Fallopian tubes, or uterus.

**gonopoiesis** (*gon-o-poi-e'-sis*). See *gonepoiesis*.

**gonopoietic** (*gon-o-poi-et'-ik*) [γόνος, generation, seed; ποιητικός, productive]. In biology, productive of reproductive elements, as ova or spermatozoa.

**gonorol** (*gon'-or-ol*). A proprietary remedy for gonorrhea said to contain the active principles of sandalwood oil.

**gonorrhea, gonorrhœa** (*gon-or-e'-ah*) [γονή, semen; ῥοία, a flow]. A specific infectious inflammation of the mucous membrane of the urethra and adjacent cavities, due to the gonococcus of Neisser. The disease is characterized by pain, burning urination, a profuse mucopurulent discharge, and a protracted course. It is likely to become chronic, and is frequently accompanied by complications—prostatitis, periurethral abscess, epididymitis, cystitis, purulent conjunctivitis. It may also cause arthritis (gonorrheal rheumatism), endocarditis, and, in women, salpingitis. **g.**, **dry**, a form unassociated with discharge.

**gonorrheal, gonorrhœal** (*gon-or-e'-al*) [*gonorrhea*]. Relating to gonorrhea, as *gonorrheal* ophthalmia. **g. arthritis**, or rheumatism, inflammation of one or more joints as a sequel of gonorrhea.

**gonosan** (*gon'-o-san*). Trade name of a remedy for gonorrhea, said to contain kava-kava and sandalwood oil.

**gonoscheocele** (*gon-os'-ke-o-sēl*) [γονή, semen; ὄσχεον, scrotum; κήλη, tumor]. A distention of the testicle with semen.

**gonotoxemia, gonotoxæmia** (*gon-o-toks-e'-me-ah*). Toxemia attributable to infection with the gonococcus.

**gonotoxin** (*gon-o-toks'-in*). A nondialyzable toxin produced both in the cocci and in the culture-mediums by gonococci.

**gonyagra** (*gon-e-a'-grah*). See *gonagra*.

**gonyalgia** (*gon-e-al'-je-ah*). See *gonalgia*.

**gonybatia** (*gon-e-ba'-she-ah*) [γόνυ, knee; βαίνειν, to go]. Walking upon the knees; a symptom encountered in some paralytic and paretic cases.

**gonycampsis** (*gon-e-kamp'-sis*) [γόνυ, knee; κάμψις, curve]. Deformity of the knee by curvation.

**gonyectyposis** (*gon-e-ek-ti-po'-sis*) [γόνυ, knee; ἐκτύπωσις, a squeezing out]. A bending of the knee outwards; genu varum.

**gonyocele** (*gon'-e-o-sēl*) [γόνυ, knee; κήλη, tumor]. Same as *gonyoncus*.

**gonyoncus** (*gon-e-ong'-kus*) [γόνυ, knee; ὄγκος, tumor]. A tumor or swelling of the knee.

**gonytyle** (*gon-e-ti'-le*) [γόνυ, knee; τύλη, pad]. A prominent thickening of the knee.

**Goodell's law**, **G.'s sign** [William Goodell, American gynecologist, 1829–1894]. When the cervix uteri is as hard as one's nose, pregnancy does not exist; when it is as soft as one's lips, pregnancy is probable.

**goose-flesh, g.-skin**. Skin marked by prominence about the hair-follicles. Syn., *cutis anserina*.

**Gordinier's writing-center**. An area in the cortex cerebri for the educated movements necessary in writing. It is at the posterior end of the medifrontal gyrus of the left cerebral hemisphere in right-handed people.

**Gordon reflex** or **paradoxical reflex** (*gor'-dun*) [Alfred Gordon, American neurologist, 1869– ]. Extension of great toe or all the toes when the deep flexor muscles of leg are being compressed. Found in disease of pyramidal tract anywhere between its origin in the cortex cerebri and its termination at the various levels of the spinal cord.

**gorget** (*gor'-jet*) [*gurges*, a chasm]. A channeled instrument, similar to a grooved director, used in lithotomy.

**gorit** (*gor'-it*). Trade name of a preparation of calcium peroxide.

**gorondou** (*go-ron'-doo*). Same as *goundou*.

**Gosselin's fracture** (*gos-lan*) [Léon Athanase Gosselin, French surgeon, 1815–1887]. A V-shaped fracture of the lower end of the tibia.

**gossypiin** (*gos-ip'-e-in*) [*gossypium*]. A precipitate from a tincture of the root-bark of *Gossypium herbaceum*, recommended as an emmenagogue and diuretic. Dose, 1 to 5 grains.

**Gossypium** (*gos-ip'-e-um*) [L.]. The cotton-tree, *Gossypium herbaceum*, and other species of *Gossypium*, of the order *Malvaceæ*. **gossypii cortex** (U. S. P.), cottonroot bark, is used as an emmenagogue, especially in the form of the fluidextract. Dose ½–1 dr. (2–4 Cc.). **G. purificatum** (U. S. P.), purified cotton; absorbent cotton; cotton-wool; the hairs of the seed of *Gossypium herbaceum* and of other species of *Gossypium*, used as a dressing and as a substitute for sponges in surgery; in pharmacy, as a filtering medium. **gossypii seminis, oleum** (U. S. P.), cotton-seed oil, is used in ointments. See also *cotton*.

**gossypol** (*gos'-ip-ol*). A crystalline compound isolated from cotton-seed, allied to tannin.

**gothic palate**. An enormously high palatal arch.

**Gottstein's basal process** (*got'-stīn*) [Jakob Gottstein, German otologist, 1832–1895]. The attenuated process of an outer hair-cell connecting the latter with the basilar membrane of Corti's organ.

**gouge** (*gowj*) [Fr.]. An instrument for cutting or removing bone or other hard structures.

**Goulard's cerate** (*goo-lar'*) [Thomas Goulard, French physician, 1784– ]. A mixture of lead subacetate, 20, and cerate of camphor, 80. **G.'s extract**, an aqueous liquid containing lead acetate, 180; lead oxide, 110; distilled water, 710. Syn., *liquor plumbi subacetatis*.

**Gould's bowed-head sign** (*goold*) [George Milbry Gould, American ophthalmologist, 1848– ]. In retinitis pigmentosa or other disease destroying the peripheral portion of the retina the patient often bows the head low to see the pavement, in order to bring the image upon the functional portion of the retina.

**Gouley's catheter** (*goo'-le*) [John William Severin Gouley, American surgeon, 1832– ]. A solid curved, steel instrument, grooved on its inferior aspect, for passing over a guide, through a stricture, into the bladder.

**goundou** (*goon'-doo*). An affection occurring among the negroes of the western coast of Africa. It consists of the growth of two bony, ovoid, symmetrical tumors which arise at the root of the nose on each side and which, by their growth, narrow the nasal fossæ and interfere with vision.

**Gouraud's disease** (*goo-ro*) [Vincent Olivier Gouraud, French Surgeon, 1772–1848]. Inguinal hernia.

**Gousset's symptoms of phrenic neuralgia**. A painful point, always present and well defined, to the right of the fourth or fifth chondrosternal articulation; it must not be confounded with the retrosternal pain of chronic aortitis.

**gout** (*gowt*) [*gutta*, a drop]. A disease characterized by a paroxysmal painful inflammation of the small joints, particularly the great toe, accompanied by the deposit of sodium urate. The attack usually comes on at night, is attended by a dusky, glazed swelling of the joint and agonizing pain, and disappears with a sweat in the morning, to recur again at night. In some cases gout presents an atypical form, appearing as dyspepsia, bronchitis, or intestinal catarrh; at times it produces pneumonia and inflammation of the serous membranes. The cause of gout is not definitely known, but is connected with an excess of uric acid or urates in the blood. The disease is most common in high livers. **g.**, **anomalous**, **g.**, **atypical**, that marked by unusual symptoms or at first affecting unusual parts of the body. **g.**, **asthenic**, **g.**, **atonic**, a chronic form marked by enlargement of the joints and thickening and distention of the ligaments and tissues. **g.**, **edematous**, that in which the swelling is not attended with heat, redness, or severe pain. **g.**, **latent**, **g.**, **masked**, lithemia, a condition ascribed to a gouty diathesis, but not presenting the typical symptoms of gout. **g.**, **poor-man's**, gout due to exposure, poor food, and excess in the use of malt liquors. **g.**, **retrocedent**. 1. That form that presents severe internal manifestations, without the customary arthritic symptoms. Syn., *anomalous*, *misplaced*, or *wandering*

**gout;** *arthritis aberrans; arthritis erratica.* 2. Gout which leaves the joints suddenly, to appear in the brain, stomach, or other internal organ. Syn., *abarticular, displaced, extraarticular, flying, metastatic, migrating, recedent, retrograde,* or *transferred gout.* **g., rheumatic.** See *arthritis, rheumatoid.* **g., tophaceous,** a form marked by a deposit of sodium urate on the joint-cartilages and the formation of bony or cartilaginous growths around the ends of the bones.
**gouty** (*gow'-te*) [*g ut*]. Pertaining to or of the nature of gout. **g. diathesis** or **habit,** the peculiar state of the body predisposing to gout. **g. kidney,** a chronically contracted kidney due to gout.
**Gowers' column** (*gow'-erz*) [Sir William Richard Gowers, English neurologist, 1845–1915]. The ascending anterolateral tract of the spinal cord. **G.'s intermediate process,** the lateral horn, a projection of the intermediate gray substance in the dorsal region of the spinal cord. **G.'s paraplegia,** a paraplegia due to vertebral caries. **G.'s symptom,** intermittent and abrupt oscillations of the iris under the influence of light, anterior probably to the total loss of the reflex; it is occasionally seen in tabes dorsalis. **G.'s tract.** Same as *G.'s column.*
**Goyrand's hernia** (*gwar-ahn'*) [Jean Gaspard Blaise *Goyrand,* French surgeon, 1803–1866]. Inguinointerstitial hernia; incomplete inguinal hernia.
**G. Ph.** Abbreviation for German Pharmacopeia.
**gr.** Abbreviation of *granum,* grain; or *grana,* grains.
**Graafian follicles, G. vesicles** (*grah'-fe-an*) [Regnerus de *Graaf,* Dutch anatomist, 1641–1673]. Vesicular bodies found in the cortical layer of the ovary, and each containing an ovum. G. oviduct. See *oviduct.*
**gracile** (*gras'-il*) [L. *gracilis*]. Slender, slight or delicate.
**gracilis** (*gras'-il-is*). See under *muscle.*
**gradatim** (*gra-da'-tim*) [L.]. Gradually.
**gradatory** (*grad'-at-o-re*) [*graduate*]. Adapted for walking.
**graduate** (*grad'-ū-āt*) [*gradus,* a step]. To take a degree from a college or university. Also, a person on whom a degree has been conferred. Also, in pharmacy, a glass vessel upon which the divisions of liquid measure have been marked.
**graduated** (*grad'-ū-a-ted*). Arranged in degrees or steps. **g. compress,** a compress made of pieces decreasing progressively in size, the apex or smallest piece being applied to the focus of pressure.
(**von**) **Graefe's disease** (*gra'-feh*) [Friedrich Wilhelm Ernst Albrecht von *Graefe,* German ophthalmologist, 1828–1870]. Progressive ophthalmoplegia. **v. G.'s knife,** a narrow knife for the performance of the operation for cataract. **v. G.'s operation,** *for cataract:* extraction of the cataract through a scleral incision, with iridectomy and laceration of the capsule. **v. G.'s sign, v. G.'s symptom,** failure of the upper lid to follow the eyeball in glancing downward, elicited in exophthalmic goiter by having the patient alternately rotate the eyes up and down. **v. G.'s spots,** certain spots near the supraorbital foramen, or over the vertebræ, which, when pressed upon, cause a sudden relaxation of the spasm of the eyelids in cases of blepharofacial spasm.
**graft** [ME., *graffe,* A small portion of skin, bone, periosteum, nerve, etc., used to replace a defect in a corresponding structure. **g., autoplastic,** a graft taken from the patient's own body. **g., heteroplastic,** a graft taken from a person other than the subject. **g., homoplastic,** an autoplastic graft. **g., omental;** a small strip of omentum used to strengthen the line of suture in enterorrhaphy. **g., periosteal,** one of periosteum to cover denuded bone or to be placed where bone has been removed to favor new formation. **g., skin-,** a small portion of skin inserted upon a raw surface, such as is produced by a burn, to assist in reproducing the integument. **g., sponge-,** the insertion of a piece of sponge into the tissues to act as a framework for granulations.
**graham bread** (*gra'-ham*) [Silvester *Graham,* English dietitian, 1794–1851]. Brown bread; wheaten bread made from unbolted flour.
**Graham's law** (*gra'-ham*) [Thomas *Graham,* English chemist, 1805–1869]. The rate of diffusion of gases through porous membranes is in inverse ratio to the square root of their weights.
**grain** (*grān*) [*granum,* grain]. 1. Seed, as that of the cereals. 2. A body resembling a seed, as a starch-grain. 3. A small pill. 4. The unit of weight of the troy and the avoirdupois system of weights. See *weights and measures.* **g.s of paradise,** the unripe fruit of *Amomum melegueta* and of *A. granum-paradisi,* brought from West Africa. It is an aromatic stimulant and diuretic, useful in some cases of neuralgia.
**grainage** (*grān-ej*). Weight expressed in grains or fractions of grains.
**gram, gramme** (*gram*) [γράμμα, inscription]. The gravimetric unit of the metric system of weights and measures, equivalent to the weight of a cubic centimeter of distilled water at its maximum density. See *weights and measures.*
**Gram's method.** [Hans Christian Joachim *Gram,* Danish physician, 1853– ]. A method for staining bacteria. The bacteria on the cover-glass or in the section are stained first with Ehrlich's solution, and then are treated with *Gram's solution* (iodine, 1; potassium iodide, 2; water, 300), and then with alcohol. Some bacteria give up the color when washed with alcohol. **G.'s solution.** See under *G.'s method.*
**grammolecular** (*gram-mo-lek'-ū-lar*). Relating to a grammolecule. **g. solution,** a solution in which a grammolecule of the active chemical is contained in each liter.
**grammolecule** (*gram-mol'-e-kūl*). In a solution or mixture, the molecular weight of the active chemical expressed in grams. Syn., *grammole; mol; mole.*
**gramnegative** (*gram-neg'-at-iv*). Incapable of staining by Gram's method.
**grampositive** (*gram-pos'-it-iv*). Capable of staining by Gram's method.
**granatonine** (*gran-at'-on-in*) [*granatum,* the pomegranate]. Pseudopelletierine.
**granatum** (*gran-a'-tum*). Pomegranate. The bark of the stem and root of *Punica granatum.* The bark contains punicotannic acid and mannite, but the active principle is the alkaloid *pelletierine,* C₈H₁₅NO. The chief use of pomegranate and its preparations is as a tæniacide. The *decoction* consists of 2 oz. of bark in 2 pints of water, boiled down to a pint. See also *pelletierine,.* **granati, fluidextractum** (U. S. P.), fluidextract of pomegranate. Dose 30 min. (2 Cc.).
**Grancher's disease** (*grahn-sha'*) [Jacques Joseph *Grancher,* French physician, 1843–1907]. A form of pneumonia with splenization of the lung, the coagulable exudate filling not only the alveoli, but also the larger bronchi. Syn., *Desnos' pneumonia, pneumonie pleuretique; splénopneumonie.* **G.'s sign,** on auscultation in pulmonary condensation the expiratory murmur equals in pitch that of the inspiratory, evidencing obstruction to expired air. **G.'s triad,** the three symptoms characteristic of incipient pulmonary tuberculosis: weakened vesicular murmur, increased vocal fremitus, and Skodaic resonance.
**grandeur: delirium of,** or **delusions of.** Insane exaltation of mind, with false opinions as to one's own greatness and dignity.
**grand mal** (*grong ma(h)l*) [Fr. "great evil"]. A term for fully-developed epilepsy; major epilepsy. See *petit mal.*
**Grandry's corpuscles.** Minute ovoid or spherical taste corpuscles found in the papillæ of the beak and tongue of birds.
**granula** (*gran'-ū-lah*) [*granum,* a grain]. The granules, cytoblasts, or microsomes of protoplasm.
**granular** (*gran'-ū-lar*) [*granule*]. Made up of, or containing, granules. **g. layer.** See under *retina.* **g. lids,** trachoma. **g. pharyngitis,** pharyngitis characterized by the presence of prominent follicles.
**granulase** (*gran'-ū-lās*) [*granum*]. An enzyme found in cereals, converting starch into achroodextrin and maltose.
**granulated** (*gran'-ū-la-ted*) [*granula,* little grains]. Characterized by the presence of granulations or granules.
**granulation** (*gran-ū-la'-shun*) [*granule*]. 1. A capillary loop of blood-vessels surrounded by a group of connective-tissue cells; also the process by which these are formed. 2. The formation of new or cicatricial tissue in the repair of wounds or ulcers, the surface of which has a granular appearance; also, any one of the elevated points of such a surface or formation. **g.s, erethistic, g.s, erethitic,** an acestoma in which severe pain and hemorrhage are caused by slight irritation. **g., exuberant, g., fungous,** an acestoma secreting thin, mucopurulent matter due to local edema or to excessive formative power. **g. tissue,** the material consisting of granu-

**lations** by which the repair of loss of substance or the healing together of surfaces is brought about.
**granulationes arachnoideales** (*gran-ū-la-she-o'-nēz ar-ak-noyd-e-a'-lēs*) [L., arachnoideal granulations]. The Pacchionian bodies.
**granule** (*gran'-ūl*) [*granulum*, a little grain]. A small grain, body, or particle, as the *granules* of à cell; also a small pill. **g.s, alpha, g.s, beta, g.s, delta, g.s, epsilon, g.s, gamma.** See under *color-analysis*. **g.s, Altmann's.** See *Altmann's granules*. **g.s, Bettelheim's.** See *hemokonia.* **g.-cell,** any one of a variety of round cells found in pathologically softened brain-tissue, and densely filled with fat-globules. **g.s, chromophil.** See *Nissl's bodies.* **g.s, Claude Bernard's,** the granules in the secreting cells of the pancreas. **g.s, edematin,** the microsomes forming the mass of the nuclear sap. They have been identified with the "cyanophilous granules" of the nucleus. **g., elementary,** irregular protoplasmic bodies in the blood smaller than ordinary blood-corpuscles. **g.s, interstitial,** those occurring in the sarcoplasm of striated muscle-fibers; they consist of fat and probably also of lecithin. **g.s, leukocyte,** the amphophil, basophil, eosinophil, neutrophil, or oxyphil bodies observed in leukocytes. Cf. *color-analysis*. **g.s, lymph,** lymph-corpuscles. **g.s, Malpighi's,** Malpighian corpuscles. **g.-mass,** a giant-cell. **g.s, Neusser's.** See *Neusser's granules*. **g.s, Nissl's.** See *Nissl's bodies*. **g.s, osseous,** very small granules of inorganic matter which are found in the matrix of bone. **g.s, Schultze's.** See under *Schultze*. **g.,** seminal, any one of the solid particles of the semen, consisting of round, granular corpuscles. **g.s, vitelline, g.s,** yolk. See *spheres, vitelline.* **g.s, Zimmermann's.** Same as *Bizzozero's blood-platelets.* **g.s, zymogen, g.s, zymogenous,** certain granules in the pancreatic cells supposed to give origin to the pancreatic ferments.
**granuliform** (*gran-ū'-le-form*) [*granule; forma,* form]. Resembling small grains.
**granulitis** (*gran-ū-li'-tis*) [*granula,* a little grain; *ιτις,* inflammation]. Acute miliary tuberculosis.
**granulocyte** (*gran'-ū-lo-sīt*). A granular leukocyte.
**granulofatty** (*gran-ū-lo-fat'-e*). Applied to cells in tissue undergoing fatty degeneration, which contain granules of fat. Syn., *granuloadipose.*
**granuloma** (*gran-ū-lo'-mah*) [*granule;* ὅμα, a tumor]. A tumor or tumor-like nodule made up of granulation tissue. **g. annulare,** a peculiar disease midway between inflammation and a neoplasm, characterized by the formation of pale-red or violaceous red nodules on the wrists, backs of the hands, and neck, which develop slowly and form circles by confluence. The lesions are firm; the mucous layer is enormously thickened. **g., infectious,** that due to a specific microorganism; as tubercle, gumma, etc. **g. trichophyticum,** granuloma due to *trichophyton.* **g. tropicum,** frambesia.
**granuloplasm** (*gran'-ū-lo-plazm*) [*granule; plasma,* something formed]. The granular protoplasmic mass in the inner part of a cell.
**granulose** (*gran'-ū-lōs*) [*granule*]. The material that forms the inner and soluble portion of starch-granules.
**granum** (*gra'-num*). See *grain*.
**Granville's lotion** (*gran'-vil*) [Augustus Bozzi *Granville,* English physician, 1783–1871]. A compound liniment of ammonia.
**grape-cure** (*grāp'-kūr*). A treatment of pulmonary tuberculosis consisting in the ingestion of large quantities of grapes.
**grape-sugar.** See *glucose* and *dextrose*.
**graphic** (*graf'-ik*) [γράφειν, to write]. Relating to writing or recording, or to the process of making automatic tracings of phenomena, showing degree, rhythm, etc.
**graphite** (*graf'-īt*) [see *graphic*]. Plumbago or black-lead, an impure allotropic form of carbon. It has been applied externally in skin diseases.
**grapho-** (*graf-o-*) [γράφειν, to write]. A prefix meaning to write.
**graphology** (*graf-ol'-o-je*) [*grapho-;* λόγος, science]. The study of the handwriting for the purpose of diagnosing nerve disease.
**graphomania** (*graf-o-ma'-ne-ah*) [γραφή, writing; μανία, madness]. An insane desire to write.
**graphomaniac** (*graf-o-ma'-ne-ak*) [γραφή, writing; μανία, madness]. One affected with graphomania.
**graphomotor** (*graf-o-mo'-tor*) [*grapho-; movere,* to move]. Relating to graphic movements or to the movements concerned in writing.
**graphorrhea** (*graf-or-e'-ah*) [*grapho-;* ῥοία, flow]. An intermittent condition in certain forms of insanity, marked by an uncontrollable desire to cover pages with usually unconnected and meaningless words.
**graphoscope** (*graf'-o-skōp*) [*grapho-;* σκοπεῖν, to view]. A convex lens devised for the treatment of asthenopia and progressive myopia.
**graphospasm** (*graf'-o-spazm*) [*grapho-;* σπασμός, spasm]. Writers' cramp.
**Grashey's aphasia.** Aphasia due to diminished duration of sensory impressions, with consequent disturbance of perception and association; it is seen in concussion of the brain and in certain acute diseases.
**Grasset-Rauzier's type of syringomyelia.** A form with marked sudoral and vasomotor symptoms.
**grating** (*gra'-ting*) [French, *gratter,* to scratch]. 1. A frame or screen composed of bars. 2. A sound produced by the friction of very rough surfaces against each other. 3. A glass ruled with exceedingly fine parallel lines to produce chromatic dispersion in the rays of light reflected from it.
**Gratiolet's optic radiation** (*gras-e-o-la'*) [Louis Pierre *Gratiolet,* French anatomist, 1815–1865]. Fibers that pass from the optic center in the occipital lobe to the pulvinar and external geniculate body. Syn., *Gratiolet's fibers.*
**gratiolin** (*gra-ti'-o-lin*). $C_{20}H_{34}O_7$. A crystalline, bitter glucoside from *Gratiola officinalis.* It was formerly used as a hydragogue.
**grattage** (*grat-ahzh*) [Fr.]. A method of removing morbid growths, as polypi or trachomatous granulations, by rubbing with a harsh sponge or brush.
**gravative** (*grav'-ah-tiv*) [*gravis*, heavy]. Attended by a sense of weight; said of the pressure-pains of tumors.
**grave** (*grāv*). 1. Serious, severe, dangerous. 2. An excavation in the earth for burying the dead. **g.-wax.** See *adipocere.*
**gravedo** (*grav-e'-do*) [L.]. 1. Muscular rheumatism of the head. 2. Coryza.
**gravel** (*grav'-l*) A granular, sand-like material forming the substance of urinary calculi, and often passed with the urine in the form of detritus.
**graveolent** (*grav'-e-o-lent*) [*gravis,* heavy; *olere,* to smell]. Having a strong, unpleasant odor; fetid.
**Graves' disease** (*grāvz*) [Robert James *Graves,* Irish physician, 1797–1853]. See *goiter, exophthalmic.* **G.'s sign,** an increase of the systolic impulse often noted in the beginning of pericarditis.
**gravid** (*grav'-id*) [*gravidus,* pregnant]. Pregnant. **g. uterus,** the uterus during pregnancy.
**gravida** (*grav'-id-ah*) [*gravid*]. A pregnant woman.
**gravidin** (*grav'-id-in*). See *kyestein*.
**gravidism,** or **gravidity** (*grav'-id-izm,* or *grav-id'-it-e*) [*gravidus,* pregnant]. Pregnancy, or the totality of symptoms presented by a pregnant woman.
**gravidocardiac** (*grav-id-o-kar'-de-ak*) [*gravid; cardiac*]. Relating to cardiac disorders due to pregnancy.
**gravimeter** (*grav-im'-et-er*) [*gravis,* heavy; μέτρον, measure]. An instrument used in determining specific gravities; especially a hydrometer, aerometer, or urinometer.
**gravimetric** (*grav-e-met'-rik*) [*gravis,* heavy; μέτρον, a measure]. Pertaining to measurement by weight. **g. analysis.** See *analysis, gravimetric.*
**gravistatic** (*grav-is-tat'-ik*) [*gravis;* στατικός, causing to stand]. Due to gravitation; applied to a form of congestion.
**gravitation** (*grav-it-a'-shun*). The force by which bodies are drawn to the earth's center.
**gravity** (*grav'-it-e*) [*gravis*]. Weight. **g., specific,** the measured weight of a substance compared with that of an equal volume of another taken as a standard. For gaseous fluids, hydrogen is taken as the standard; for liquids and solids, distilled water at its maximum density.
**Grawitz's granules** (*grah'-vits*) [Paul *Grawitz,* German pathologist, 1850– ]. Minute granules, staining readily with basic dyes, seen in red blood-cells in certain pathological conditions. **G.'s tumor,** hypernephroma; a lipmatoid tumor of the kidney having its origin in aberrant masses of suprarenal tissue, and situated immediately beneath the renal capsule.
**gray** (*grā*) [AS., *græg*]. The color obtained by mixing white and black. **g. atrophy, g. degeneration.**

See *degeneration*. **g. hepatization.** See *hepatization, gray*. **g. matter,** that forming the outer part of the brain and the inner part of the cord, containing the specialized cells of these parts. **g. oil,** mercurial liquid used in syphilis. **g. powder.** See *mercury with chalk*. **g. soap.** See *soap*. **g. softening,** an inflammatory softening of the brain or cord with a gray discoloration. **g. substance.** See *g. matter*.

**grease** (*grēs*). Soft or oily animal fat. In farriery, a swelling and inflammation in a horse's leg, with excretion of oily matter and the formation of cracks in the skin. **g.-trap,** a contrivance employed to prevent clogging of waste-pipes, as well as to save the grease, which has considerable commercial value.

**green** (*grēn*) [ME., *grene*]. Of the color of grass, obtained by mixing yellow and blue. **g.-blindness,** a variety of color-blindness in which green is not distinguished. **g., Paris-,** copper acetoarsenite. **g., Scheele's,** copper arsenite. **g., Schweinfurt.** Synonym of *g., Paris-*.

**Greene's sign** (*grēn*). In percussion of the free cardiac border during full inspiration and again during forced inspiration, the patient either standing or sitting, it will be noticed that the border is displaced outward by the expiratory movement in cases of pleuritic effusion.

**Greenhow's cholera-mixture** (*grēn'-how*) [Thomas Michael *Greenhow*, English physician, 1791–1881]. A mixture containing guaiacum, cloves, cinnamon, each 1 ounce; brandy, 2 pints.

**Greenhow's disease** (*grēn-how*) [Edward Headlam *Greenhow*, English physician, 1814–1888]. Vagabond's disease.

**greensickness** (*grēn'-sik-nes*). Chlorosis.
**green softening.** Purulent softening of nervous matter.
**greenstick fracture.** See *fracture, greenstick*.
**green vitriol.** Ferrous sulphate.
**greffotome** (*gref'-o-tōm*) [Fr., *greffe*, graft; τομή, cutting]. A knife used in cutting slips for surgical grafting.

**Gregarina** (*greg-ar-i'-nah*) [*grex*, a herd]. A genus of *Protozoa*.
**gregarinosis** (*greg-ar-in-o'-sis*) [*grex*, a herd; νόσος, disease]. A morbid condition due to infestation by *Gregarinæ*.

**Gregory's powder** (*greg'-or-e*) [James *Gregory*, Scotch physician, 1753–1821]. Compound powder of rhubarb, consisting of rhubarb, 2 oz.; light magnesia, 6 oz.; ginger, 1 oz.

**Gréhant's method for determining urea in blood and tissues.** It makes use of a solution of mercury, 1 Gm., in 10 Cc. of pure nitric acid, for decomposing the urea; the CO₂ and N are liberated, enabling one to estimate the urea.

**gressorial** (*gres-o'-re-al*) [*gressus*, participle of *gradi*, to walk]. In biology, adapted for walking; gradient.

**Griesinger's disease** (*gre'-zing-er*) [Wilhelm *Griesinger*, German neurologist, 1817–1868]. A form of pernicious anemia connected with the presence of *Dochmius duodenalis* in the intestinal tract. Uncinariasis. G.'s sign. 1. An edematous swelling behind the mastoid process in thrombosis of the transverse sinus. 2. In thrombosis of the basilar artery, compression of the carotids produces symptoms of cerebral anemia (pallor, syncope, convulsions). This sign is of doubtful value, as it may also be caused by disturbances of the cerebral circulation resulting from cardiac and vascular lesions (especially arteriosclerosis).

**Griess' red paper** (*grēs*). Paper charged with sulphanilic acid and naphthylamine sulphate, used in testing for nitrous acid, nitrites, bilirubin, and aldehyde. G.'s test for nitrous acid, an intense yellow color is produced by the addition of a solution of metadiamidobenzol to a dilute solution containing nitrous acid previously acidified with a few drops of sulphuric acid. G.'s yellow paper, paper charged with sulphanilic acid and metadiamidobenzene, used as a sensitive test for nitrites.

**Griffith's mixture** (*grif'-fith*) [Robert Eglesfield *Griffith*, American physician, 1798–1850]. A mixture of iron sulphate, 6; myrrh, 18; sugar, 18; potassium carbonate, 8; lavender, 50; rose-water, 900.

**Grigg's test for proteins.** A precipitate is formed with all proteins except peptones on adding a solution of metaphosphoric acid.

**Grinbert's test for urobilin.** Boil together equal parts of urine and hydrochloric acid and shake with ether. In the presence of urobilin the ether assumes a brownish-red color with a greenish fluorescence.

**grindelia** (*grin-de'-le-ah*) [H. *Grindel*, German botanist, 1776–1836]. The leaves and flowering tops of *G. robusta*, wild sunflower or gum-plant, and *G. squarrosa*. **g., fluidextract** of (*fluidextractum grindeliæ*, U. S. P.). Dose ½–1 dr. (2–4 Cc.). It is used in asthma, bronchitis, and whooping-cough, and locally in rhus-poisoning.

**grindeline** (*grin'-del-ēn*). An alkaloid reported to exist in *Grindelia robusta*.
**grinder** (*grīn'-der*) [AS., *grindan*, to grind]. A molar tooth.
**grinders' asthma.** A fibroid pneumonia, a chronic affection of the lungs resulting from the inspiration of metallic or silicious dust.

**grip, la grippe** (*grip, lah grēp*). See *influenza*. **gripe** (*grip*) [ME., *gripen*, to seize]. 1. To suffer griping pain. 2. A spasmodic pain in the bowel. **g., cutting on the,** an old method of operating for vesical calculus by cutting down directly on the stone in the perineum after having forced it down with the fingers inserted in the rectum. **g.-stick,** a tourniquet.

**gripes** (*grīps*) [see *gripe*]. Colic; tormina.
**grippal** (*grip'-al*). Pertaining to influenza.
**grippotoxin** (*grip-o-toks'-in*). A name for the toxin elaborated by *Bacillus influenzæ*.
**griserin** (*gris'-er-in*). Trade name of a mixture of loretin with alkalies.

**Grisolle's sign** (*gre-zol'*) [Augustin *Grisolle*, French physician, 1811–1869]. The early eruption of smallpox is distinguished from that of measles by the fact that the papules remain distinct to the touch even when the skin is tightly stretched.

**gristle** (*gris'-l*) [AS., *gristel*]. Cartilage.
**Gritti's operation** (*gre'-te*) [Rocco *Gritti*, Italian surgeon]. (*For amputation above the knee-joint*). The patella is preserved in a long anterior flap, and, having had a thin slice removed from its deep surface, is secured in apposition with the femur, the latter having been deprived of its articular surface by being sawn through the condyles.

**groan** (*grōn*). To utter a low, moaning sound, as when in pain. The sound so uttered.
**Grocco's sign** (*grok'-o*) [Pietro *Grocco*, Italian physician]. A paravertebral triangle of dulness in pleural effusion on the side opposite to that of the effusion.
**grocers' itch.** A peculiar psoriasis or eczema of the hands due to irritation from flour, sugar, etc.

**groin** [Icel., *grein*, a branch or arm]. The depression between the abdomen and thigh.
**groove** (*groov*) [D., *groef*]. A furrow or channel. See also *furrow*. **g., alveolingual,** one between the tongue and the lower jaw. **g., bicipital,** the deep groove on the anterior surface of the humerus, separating the greater and lesser tuberosities and containing the long tendon of the biceps. **g., cavernous,** a broad groove on the superior surface of the sphenoid bone lodging the internal carotid artery and the cavernous sinus. **g., dorsal,** the medullary groove. **g., mastoid,** the digastric fossa. **g., medullary,** a long shallow furrow that appears along the dorsal line of the neural tube of the embryo. **g., musculospiral,** one on the external aspect of the humerus which lodges the musculospiral nerve and the superior profunda vessels. **g., neural,** the medullary groove. **g., obturator,** the furrow at the superior and external border of the obturator foramen lodging the subpubic vessels and nerves when they emerge from the pelvic cavity. **g., peroneal,** one on the external aspect of the os calcis lodging the tendon of the peroneus longus. **g., pterygopalatal, g., pterygopalatine.** 1. One in the ventral aspect of the pterygoid process of the sphenoid. 2. A furrow on the vertical part of the palate bone. **g., radial.** See *g., musculospiral*. **g., scapular,** the scapular notch. **g., Schmorl's,** that resulting from emphysematous inflation of those portions of the lungs which lie between the ribs. **g., sigmoid.** See *g., cavernous*. **g., sternal,** one lying between the sternum and the pectoral muscles. **g., subcostal,** a deep furrow lying along the lower border and inner surface of a rib for lodgment of the intercostal vessels and nerves. **g. of Sylvius,** the fissure of Sylvius. **g., ventricular,** two furrows, one on the anterior, one on the posterior, surface of the heart; they indicate the interventricular septum. **g., Verga's lacrimal,** a groove extending downward from the lower orifice of the nasal duct.

**gross** (*grōs*) [Fr., *gros*, great]. Coarse; large. **g. anatomy.** See *anatomy*, *gross*. **g. appearance**, appearance of tissue as seen without a microscope. **g. lesion**, a lesion-perceptible to the eye. **Gross' disease.** See *Physick's encysted rectum*.
**ground** (*grownd*) The bottom; soil; earth. **g.-bundle**, the principal bundle fo nerve-fibers in a group, as the *ground-bundle* of the ventral and lateral columns of the spinal cord. **g.-bundle, anterior, of Flechsig**, that portion of the anterior column outside of the direct pyramidal tracts and running throughout the entire length of the cord; it is made up of fibers having a short course. **g.-bundle, posterior, of Flechsig**, Burdach's column. **g.-itch.** See *itch-coolie*. **g. nut**, a peanut, the fruit of *Arachis hypogæa*.
**group-reaction** (*grūp-re-ak'-shun*). A reaction with an antibody which is characteristic of a whole group of bacteria.
**Grove cell** [Sir William Robert Grove, English physicist, 1811–1896]. A two-fluid battery cell, the fluids being dilute sulphuric and nitric acids, and the metal immersed in them respectively zinc and platinum.
**growing-pains** (*grō'-ing*). A term applied to pains in the limbs occurring during youth, and probably of rheumatic origin.
**growth** (*grōth*). 1. The augmentation of the body that takes place between infancy and adult age. Also, the increase of any part of the body by addition to the number of its cellular elements, without the production of structural abnormality or differentiation into unlike tissues. 2. Any tumor or adventitious structure.
**grub.** See *comedo*.
**Gruber's bursa** (*gru'-ber*) [1. Wenceslaus Leopold Gruber, Russian anatomist, 1814–1890; 2. Josef Gruber, Austrian otologist, 1827–1900; 3. Max Gruber, German bacteriologist, 1853— ], [1.]. The synovial cavity of the tarsal sinus. **G.'s reaction**, [3.]. The addition of some of the culture of *Spirillum choleræ asiaticæ* to the serum of an animal rendered immune to cholera causes these organisms to become nonmotile and to agglutinate. The reaction does not occur with other species. **G.'s test for hearing**, [2.]. If the end of the finger is inserted into the ear after the sound of a vibrating tuning-fork held before the ear has completely ceased, and the tuning-fork is then firmly placed upon the finger, a weakened sound becomes again audible, and remains so for some time.
**Gruber-Widal's reaction.** See *Widal's reaction*.
**Gruby's disease** (*groo'-be*) [David Gruby, Hungarian physician, 1810–1898]. Alopecia areata.
**gruel** (*gru'-el*) [dim. of *grutum*; meal]. A decoction of corn-meal or oat-meal boiled in water to a thick paste.
**gruff** (*gruf*). Any crude drug; also the coarse part of a drug that will not pass through a sieve; the term is used also adjectively, as gruff sulphur or saltpeter.
**grume** (*groom*) [*grumus*, a little heap]. A clot, as of blood; a thick and viscid fluid.
**grumous** (*gru'-mus*) [*grumus*, a little heap]. Clotted; consisting of lumps.
**grutum** (*gru'-tum*). See *milium*.
**Grynfelt's triangle.** A triangular space bounded above by the twelfth rib and the lower border of the serratus posticus inferior, behind by the anterior border of the quadratus lumborum, and anteriorly by the posterior border of the internal oblique. Lumbar hernia may occur in this space.
**gryochrome** (*gri'-o-krom*) [γρῦ, a morsel; χρῶμα, color]. A somatochrome nerve-cell the stainable portion of which consists of minute granules which tend to form threads or heaps.
**gryposis** (*gri-po'-sis*) [γρύπωσις, curvature]. Curvature; abnormal curvature of the nails. See *arthrogryposis* and *onychogryposis*.
**gt.** Abbreviation of *gutta*, drop.
**gtt.** Abbreviation of *guttæ*, drops.
**guacamphol** (*gwah-kam-fol'*). See *guaiacamphol*.
**guacetin** (*gwas'-et-in*). See *guaiacetin*.
**guachamaca** (*gwa-shaw-maw'-kah*). The bark of an apocynaceous tree. *G. toxifera*, or *Malouetia nitida*, furnishes a virulent arrow-poison, somewhat resembling curara; it has been employed in tetanus and hydrophobia.
**guacin** (*gwaw'-sin*) [*guaco*]. A bitter resin from guaco; it is diaphoretic, stimulant, and emetic.
**guaco** (*gwah'-ko*). The *Mikania guaco* and other species of *Mikania* and *Aristolochia*, used in South America, for snake-bites; it has been employed in rheumatism, gout, and in various skin diseases. Dose of a watery extract 3 min. (0.19 Cc.).
**guaconization** (*gwah-kon-i-sa'-shun*). Poisoning and paralysis of the sensory nerve-centers from ingestion of guaco, *Aristolochia cymbifera*.
**guaethol** (*gwa-eth'-ol*). Guaiacol ethyl, $C_8H_4OC_2H_5OH$. It resembles guaiacol in therapeutic action. Dose 2–4 gr. (0.1–0.25 Gm.). Application, 15% ointment. Syn., *Ajacol*; *Pyrocatechin-monoethyl ether*; *Thanatol*.
**guaiac, guaiacum** (*gwi'-ak, -um*). The resin (*guaiacum*, U. S. P.) of the wood (*guaiaci lignum*) of *G. officinale* and *G. sanctum*. It contains guaiacic acid, $C_{13}H_{16}O_5$, guaiac-yellow, guaiacene, $C_8H_8O$, guaiacol, and pyroguaiacin. It is alterative, expectorant, and diaphoretic. It is used in syphilis, chronic rheumatism, and gout. **g., tincture of** (*tinctura guaiaci*, U. S. P.), a solution of the resin in alcohol. Dose ½–2 dr. (2–8 Cc.). **g., tincture of, ammoniated** (*tinctura guaiaci ammoniata*, U. S. P.), a solution of the resin in aromatic spirit of ammonia. This tincture is the preferred one for the administration of guaiac. Dose ½–2 dr. (2–8 Cc.).
**guaiacamphol** (*gwi-ah-kam'-fol*). The camphoric acid ester of guaiacol; employed in treatment of night-sweats of tuberculosis. Dose 3–8 gr. (0.2–0.5 Gm.).
**guaiacene** (*gwi'-as-ēn*), $C_8H_8O$. An oily crystallizable liquid boiling at 118° C., with odor of bitter almonds, obtained from guaiac resin by dry distillation.
**guaiacetin** (*gwi-as'-et-in*). Pyrocatechin-mono-acetate, $C_6H_4.OH.OCH_2COOH$. It is used like guaiacol in tuberculosis. Dose 7½ gr. (0.5 Gm.); 3 times daily and reduced in 3 weeks to 7½ gr. (0.5 Gm.) daily.
**guaiacocaine** (*gwi-ah-ko-ka'-in*). An anesthetic mixture of cocaine and guaiacol used in dentistry.
**guaiacol** (*gwi'-ak-ol*) [S.A.], $C_7H_8O_2$. Methylpyrocatechin, a substance obtained from beechwood creosote and also, synthetically, from pyrocatechin and methylsulphuric acid. It is used as a substitute for creosote in tuberculosis in doses of 3–5 min. (0.19–0.32 Cc.); externally it has been employed as an antipyretic, in doses of 20–40 min. (1.25–2.5 Cc.) painted on the skin. Syn., *methylcatechol*. **g.-benzylester**, $C_6H_5(OCH_3) \cdot CH_2 \cdot C_6H_5$, a local anesthetic. Syn., *brenscain*. **g. biniodide**, $C_7H_6I_2O_2$, is alterative and antituberculous. Dose 2 min. (0.12 Cc.) 3 times daily. **g. cacodylate**, a stable preparation recommended in tuberculosis. Dose ½–2 gr. (0.032–0.13 Gm.). Syn., *cacodiacol*; *cacodyliacol*. **g. carbonate** (*guaiacolis carbonas*, U. S. P.), $(C_7H_7O)_2CO_3$, used in tuberculosis. Dose 3–8 gr. (0.2–0.52 Gm.) 3 times daily and gradually increased to 90 gr. (6 Gm.) daily. Syn., *duotal*. **g. cinnamate.** See *styracol*. **g. ethyl.** See *guaethol*. **g. ethylenate**, $CH_2O.C_6H_4O.C_2H_4O.C_6H_4OCH_3$, a guaiacol ethylene ester, used in tuberculosis. Dose 8–15 gr. (0.5–1.0 Gm.) twice daily. Syn., *ethylene guaiacol*. **g.-glycerylester.** See *guaiamar*. **g. oleate**, a reaction-product of oleic acid, guaiacol, and phosphorus trichloride; it is antiseptic and antituberculous. Dose 5–10 min. (0.3–0.6 Cc.) 3 times daily in capsules. Syn., *oleoguaiacol*. **g.-phosphal**, **g. phosphite**, $P(OC_6H_4OCH_3)_3$, used in tuberculosis. Dose 15–30 gr. (1–2 Gm.) daily. **g. phosphate**, $(C_6H_4OCH_3)_3PO_4$, obtained from guaiacol dissolved in soda solution with addition of phosphorus oxychloride; used in fever of tuberculosis. Dose 4 gr. (0.25 Gm.) every 3 or 4 hours. **g. salicylate**, $C_{14}H_{12}O_4$, is an intestinal antiseptic, and is employed in phthisis, dysentery, rheumatism, etc. Dose 15 gr. (1.0 Gm.) several times daily; maximum dose 150 gr. (10 Gm.). **g. succinate**, $(C_7H_7O_2)_2C_4H_4O_4$, obtained from a mixture of guaiacol and succinic acid with phosphorus oxychloride. Dose 2 min. (0.12 Cc.) 3 times daily in tuberculosis. **g. valerate**, **g. valerianate**, an oily liquid used in pulmonary affections and for hypodermatic injection in tuberculous joints. Dose 3–9 gr. (0.2–0.6 Gm.) 3 times daily. Syn., *geosote*.
**guaiacolate** (*gwi-ak'-ol-āt*). A combination of guaiacol with a base.
**guaiacyl** (*gwi'-as-il*), $C_7H_7O_8SO_3$. The calcium salt of a sulphocompound of guaiacol; used as a therapeutic agent; injected in quantities of 8–25 gr. (0.5–1.5 Gm.) of a 5% solution or 15 gr. (1 Gm.) of a 10% solution.

# GUAIAMAR 403 GUMMATOUS

**guaiamar** (gwi'-am-ar), $C_6H_4.OC_3H_7O_2.OCH_3$, guaiacolglycerylester; employed in tuberculosis and as an intestinal antiseptic. Dose 5–20 gr. (0.33–1.33 Gm.) before meals. It is also applied in arthritis.

**guaiaperol** (gwi-ap'-er-ol). See *piperidin guaiacoate*.

**guaiaquin** (gwi'-ah-kwin), $(C_6H_4O_2CH_3HSO_3)_2C_{22}H_{24}N_2O_2$, the guaiacol bisulphonate of quinine; it is used in malaria, typhoid fever, anemia, etc. Dose 5–10 gr. (0.33–0.65 Gm.) 3 times daily. Syn., *quinine guaiacol bisulphonate*.

**guaiaquinol** (gwi-ah'-kwin-ol). Quinine dibromoguaiacolate.

**guaiasanol** (gwi-as'-an-ol). See *diethyl-glycocollguaiacol hydrochloride*.

**guanase** (gwan'-ās). An enzyme found in the pancreas, thymus and adrenals; it converts guanin into xanthin.

**guanidin** (gwan'-id-in), $CN_3H_5$. Carbondiamidimide; a monacid base forming colorless crystals. See *uramin*.

**guanine** (gwan'-nin) [see *guano*], $C_5H_5N_5O$. A leukomaine found in the pancreas, liver, and in muscle-extract as a decomposition-product of nuclein. It also occurs in guano, and is nonpoisonous. See *Capranica*.

**guano** (gwah'-no) [Peruvian *huanu*, dung]. The excrement of sea-fowl found on certain islands in the Pacific Ocean. It contains guanin and alkaline urates and phosphates, and is used externally in certain skin diseases.

**guarana** (gwah-rah'-nah) [Braz.]. A dried paste prepared from the seeds of *Paullinia cupana*, found in Brazil. It contains an alkaloid, *guaranine*, $C_8H_{10}N_4O_2.H_2O$, nearly identical with caffeine. It is employed in nervous sick-headaches. Dose 1–2 dr. (4–8 Gm.); of *guaranine* 1–3 gr. (0.065–0.19 Gm.), g., fluidextract of (*fluidextractum guaranæ*, U. S. P.). Dose 5–30 min. (0.32–2.0 Cc.).

**guaranine** (gwah-rah'-nēn). An alkaloid, derived from guarana, nearly identical with caffeine. Dose, 1–2 gr. (0.06–0.12 gm.).

**guard** (gard) [*garder*, to keep]. An appliance on a knife to prevent too deep incision. **g.-cell**, in biology, one of the two semilunar epidermal cells, inclosing the opening of a stoma in plants.

**Guarnieri's vaccine-bodies** (gwar-ne-er'-e) [Giuseppi Guarnieri, Italian physician]. See *cytoryctes*.

**guavacine** (gwah'-vas-ēn), $C_6H_9NO_2$. An alkaloid forming colorless crystals soluble in water, obtained from areca-nut, *Areca catechu*. It is used as an anthelmintic.

**gubernaculum** (gū-ber-nak'-ū-lum) [L., "a rudder"]. A guiding structure. **g. dentis**, a bundle of fibrous tissue connecting the tooth-sac of a permanent tooth with the gum. **g., Hunter's**; **g. hunteri**. See *g. testis*. **g. testis**, the cord attached above to the lower end of the epididymis, below to the bottom of the scrotum, and governing the descent of the testes.

**Gubler's hemiplegia** (goob'-ler) [Adolphe Marie Gubler, French physician, 1821–1879]. Hemiplegia of the extremities with crossed paralysis of the cranial motor nerves, especially the facial. **G.'s line**, an imaginary line connecting the superficial points of origin of the trifacial nerves on the lower surface of the pons. A lesion of the pons below this line causes crossed paralysis. **G.'s tumor**, a distention of the synovial sheaths on the dorsum of the hand in palsies of the antibrachial type, and particularly in lead palsy.

**(von) Gudden's inferior commissure** (good'-en) [Bernhard Aloys von *Gudden*, German neurologist, 1824–1886]. Fibers of the optic tract which come from the internal geniculate body and cross in the posterior portion of the chiasma to the opposite tract. **v. G.'s ganglion**, the interpeduncular ganglion, a collection of nerve-cells just above the pons and in the median line. **v. G.'s hemispherical bundle, one in the optic tract passing over to the most lateral portion of the base of the pedun le and thence to the cerebral hemisphere. **v. G.'s daw**, the proximal end of a divided nerve undergoes cellulipetal degeneration. **v. G.'s tractus peduncularis transversus**, the cimbia, a fasciculus passing from the pregeminum and postgeniculum over the crus cerebri to the oculomotor sulcus.

**Guéneau de Mussey's point**. See under *Mussey*.

**Guenz's ligament** (guenīs) [Justus Gottfried *Guenz*, German anatomist, 1714–1784]. The ligamentous fibers of the obturator membrane which form the upper and inner wall of the canal transmitting the obturator vessels and nerves.

**Guenzburg's test for free hydrochloric acid in gastric juice** (*guents'-boorg*) [Alfred *Guenzburg*, German physician]. Two drops of a solution consisting of phloroglucin, 2 Gm.; vanillin, 1 Gm.; alcohol, 30 Cc., mixed with 2 drops of filtered gastric juice, are carefully heated in a porcelain capsule. The presence of free HCl is indicated by the appearance of a bright-red color; if absent, the color will be brown or brownish-red.

**Guérin's fold** (*ga'-ran*) [Alphonse François Marie *Guérin*, French surgeon, 1816–1895]. Same as *G.'s valve*. **G.'s glands**. See *Skene's glands*. **G.'s sinus**, the lacuna magna, situated in the mesial line of the upper wall of the urethra, near the external meatus. **G.'s valve**, a fold of mucous membrane bounding Guérin's sinus.

**guethol** (*gwe'-thol*). An oily liquid allied to guaiacol; analgesic.

**guha** (*goo'-ah*). A form of bronchial asthma, found in the island of Guam, where it is said to be epidemic.

**Guidi's canal**. See *Vidian canal*.

**guillotine** (gil-o-tēn) [Fr.]. A surgical instrument for excision of the tonsils or growths in the larynx, etc.

**Guinea-worm**. *Filaria medinensis*, a nematode worm of the tropics, occasionally parasitic in human tissues. **G. disease**, a disease caused by the presence of *Filaria medinensis* in the subcutaneous cellular tissue of various parts of the body, particularly, the feet and legs.

**Guinon's disease** (*ge'-non(g)*) [Georges *Guinon*, French physician, 1859–    ]. Tic de Guinon. See *Tourette's* (*Gilles de la*) *disease*.

**guipsine** (*gip'-sēn*). A proprietary preparation of mistletoe (*viscus album*); used in arteriosclerosis.

**guja** (*goo'-hah*). A form of epidemic spasmodic bronchial asthma; it occurs in Guam and neighboring islands.

**gujasanol** (*gū-jas'-an-ol*). See *diethyl glycocollguaiacol hydrochloride*.

**gula** (*gū'-lah*) [L.: pl., *gulæ*]. The gullet; the neck and throat, or the pharynx and esophagus.

**gulancha** (*goo-lan'-kah*) [E. Ind.]. The plant, *Tinospora cordifolia*, of India. Its stems and roots are diuretic, tonic, and antiperiodic.

**gular** (*gū'-lar*) [*gula*, the gullet]. Pertaining to the throat, pharynx, or gula.

**Gull's disease** [Sir William Withey *Gull*, English physician, 1816–1890]. Myxedema. **Gull's renal epistaxis**. Essential renal hematuria, or renal hemophilia.

**gullet** (*gul'-et*). See *esophagus*.

**Gullstrand's law** (*gul'-strand*). When the corneal reflex from either of the eyes of the patient, who is made to turn the head while fixing the distant object, moves in the direction in which the head is turning, it moves toward the weaker muscle.

**Gull-Sutton's disease** [*Gull*; Sir John Bland *Sutton*, English surgeon]. Arteriocapillary fibrosis; diffuse arteriosclerosis.

**Gull-Toynbee's law**. In otitis media the cerebellum and lateral sinus are likely to become involved by mastoid disease, while the cerebrum is threatened by caries of the roof of the tympanum.

**gum** [*gummi*, gum]. 1. A concrete vegetable juice exuded from many plants. When treated with nitric acid it yields mucic acid. Gums are either entirely soluble in water or swell up in it into a viscid mass. Various names are given to gums, usually indicating the place whence exported. 2. The gingiva. **g., acacia**, gum from *Acacia senegal*. **g.-Arabic**. See *acacia*. **g., bassora**, a Persian gum said to be from plum and almond trees. **g., Benjamin.-** See *benzoin*. **g., blue**. Synonym of *Eucalyptus globulus*. **g.-boil**, an abscess of the jaw; parulis. **g., British**, dextrin. **g.-resin**, a concrete vegetable juice. **g., spongy**, interstitial infiltration and thickening of the gums with dilation of the capillaries, due to scurvy or analogous conditions. **g. tragacanth**. See *tragacanth*.

**gumma** (*gum'-ah*) [*gum*: pl., *gummata*]. The gummy tumor characterizing the tertiary stage of syphilis. It consists of granulation tissue, with giant-cells, and is the seat of a peculiar degeneration which causes the gummy appearance.

**gummatous** (*gum'-at-us*) [*gumma*]. Of the nature of or affected with gumma.

**gummi** (*gum'-mi*). Latin for *gum* (1).
**gummide** (*gum'-id*). Any compound which yields glucose on decomposition with acids or alkalies.
**gummose** (*gum'-ōs*). A sugar, C₆H₁₂O₆. Obtained from animal gum.
**gummy** (*gum'-e*). 1. Gummatous. 2. Resembling gum.
**gums** (*gumz*). See *gingiva*.
**guncotton** (*gun'-kot-n*). See *pyroxylin*.
**gunjah** (*gun'-jah*). The official part of Indian hemp, consisting of the dried flowering-tops of the female plant, from which the rosin has not been removed. Syn., *ganjah*.
**Gunning's test for acetone.** Add to the liquid to be tested tincture of iodine or Lugol's solution, and then ammonia until a black precipitate is formed. This gradually disappears, leaving a sediment of iodoform.
**Gunn's dots.** Brilliant white dots seen, on oblique illumination, about the macula lutea; they do not seem to be pathologic.
**gunstock deformity.** One caused by fracture of either condyle of the humerus, in which the long axis of the fully extended forearm deviates outwardly from that of the arm.
**gurgling** [Dan., *gurgle*, gargle]. The peculiar sound caused by the passage of gas through a liquid. It is observed upon palpation of the abdomen in enteric fever and other conditions in which the bowel is distended with gas and contains liquid. **g. rale**, a sound heard over the chest when the bronchi or pulmonary cavities contain fluid.
**gurjun balsam** (*ger'-jun*). An oleoresin obtained from several species of *Dipterocarpus*, trees native to southern Asia. It is similar to copaiba, but more decided in therapeutic effects, and is less unpleasant. It is used as an expectorant, and in leprosy and gonorrhea. Dose 15–40 min. (1.0–2.6 Cc.). Syn., *balsamum dipterocarpi; wood-oil*.
**guru** (*goo'-roo*). Same as *kola*.
**gustation** (*gus-ta'-shun*) [*gustare*, to taste]. The sense of taste; the act of tasting.
**gustatory** (*gus'-ta-to-re*) [*gustare*, to taste]. Pertaining to the special sense of taste and its organs. **g. bud**, a taste-bud. **g. cell**, a spindle-cell from the interior of a taste-bud. **g. nerve**. See *nerve*.
**gut** [ME.]. The intestine. See also *catgut*.
**g., blind**, the cecum.
**Guthrie's muscle** (*guth'-re*) [George James *Guthrie*, English surgeon, 1785–1856]. The deep transversus perinæi muscle; *origin*, ramus of pubes; *insertion*, fellow muscle; *innervation*, perineal; it compresses the membranous urethra.
**gutta** (*gut'-ah*) [L.]. A drop. **g. rosacea**, acne rosacea. **g. serena**, amaurosis.
**guttapercha** (*gut-ah-per'-cha*) [Malayan *gutta*, gum; *pertja*, the tree furnishing the gum]. The concrete juice of *Dichopsis gutta* and other species of the natural order *Sapotaceæ*. It is used to make splints, as a dressing for wounds, and as a vehicle for caustic substances.
**guttate** (*gut'-āt*) [*gutta*, a drop]. In biology, spotted as if by drops of something colored.
**guttatim** (*gut-a'-tim*) [L.]. Drop by drop.
**guttiform** (*gut'-e-form*) [*gutta; forma*, form]. Drop-shaped.
**Guttmann's sign** (*goot'-mahn*) [Paul *Guttmann*, German physician, 1834–1893]. A thrill heard over the thyroid in exophthalmic goiter.
**guttur** (*gut'-er*) [L.]. The throat.
**guttural** (*gut'-er-al*) [*guttur*]. Pertaining to the throat.
**gutturotetany** (*gut-er-o-tet'-an-e*) [*guttur; tetanus*]. A form of stuttering in which the pronunciation of such sounds as g, k, q, is difficult.
**Gutzeit's test for arsenic** (*goot'-sīt*). Place a piece of zinc in a test tube with 5 c.c. of diluted sulphuric acid; to this is added 1 c.c. of the suspected liquid; a piece of filter paper moistened with an acid solution of silver nitrate becomes bright yellow in the vapor from the above if arsenic is present.
**Guy's pill** (*gi*) [William Augustus *Guy*, English physician, 1819–1900]. A pill composed of 1 grain each of powdered digitalis leaves, powdered squill, and mercury pill.
**Guye's sign.** Aprosexia occurring in childhood with adenoid vegetations of the nasopharynx.
**Guyon's isthmus** (*ge-yon*(*g*)') [Jean Casimir Félix *Guyon*, French surgeon, 1831– ]. Narrowing and prolongation of the internal os uteri, which thus forms a small canal; it is not pathologic. **G.'s sign**, ballottement in cases of renal tumor.
**gymnasium** (*jim-na'-ze-um*) [γυμνόs, naked]. A place designed and fitted with appliances for the systematic exercise of the muscles and other organs of the body. Also, in Germany, a high-school.
**gymnastic** (*jim-nas'-tik*) [γυμνόs, naked]. Pertaining to bodily exercise, or to the science of preserving health by bodily exercise.
**gymnastics** (*jim-nas'-tiks*) [γυμνόs, naked]. Physical exercise, especially systematic exercise, for the purpose of restoring or maintaining the bodily health. **g., antagonistic, g., resistance-**, physical exercise engaged in by two persons, the one resisting the other, as that adopted in the Schott treatment for cardiac affections. **g., ocular**, regular muscular exercise of the eye by the use of prisms or other means to overcome muscular insufficiency. **g., Swedish**, a system of exercises to restore strength to paretic muscles, consisting in movements made by the patient against the resistance of an attendant.
**gymnobacteria** (*jim-no-bak-te'-re-ah*) [γυμνόs, naked; *bacteria*]. Nonflagellate bacteria.
**gymnocarpous** (*jim-no-kar'-pus*) [γυμνόs, naked; καρπόs, fruit]. In biology, having the fruit naked; applied to lichens and fungi in which the apothecia and hymenia are naked or exposed.
**gymnocyte** (*jim'-no-sīt*) [γυμνόs, naked; κύτοs, cell]. In biology, a naked-celled, unicellular organism.
**gymnoplast** (*jim'-no-plast*) [γυμνόs, naked; πλάσσειν, /to form]. A protoplasmic body without a limiting membrane.
**gymnospore** (*jim'-no-spōr*) [γυμνόs, naked; σπόροs, a seed]. In biology, a naked spore.
**gynæ-** (*jin'-e*). For words beginning thus see *gyne-*.
**gynander** (*jin-an'-der*) [γυνή, woman; ἀνήρ, man]. A man of effeminate or woman-like qualities.
**gynandria** (*jin-an'-dre-ah*) [γυνή, woman; ἀνήρ, man]. The same as *hermaphroditism*.
**gynandrism,** (*jin-an'-drizm*) [γυνή, woman; ἀνήρ, man; μορφή, form]. Hermaphroditism.
**gynandromorphism** (*jin-an-dro-morf'-izm*) [γυνή, female; ἀνήρ, male; μορφή, form]. A combination of both male and female characters.
**gynandrous** (*jin-an'-drus*). In biology, having the stamens and pistils more or less intimately united.
**gynanthropus** (*jin-an'-thro-pus*) [γυνή, a woman; ἄνθρωποs, a man]. A hermaphrodite with predominant male characteristics.
**gynatresia** (*jin-at-re'-ze-ah*) [γυνή, woman; ἀτρησία, atresia]. Imperforation of the vagina.
**gynecatoptron** (*jin-ek-at-op'-tron*) [γυνή, woman; κάτοπτρον, mirror]. A vaginal speculum.
**gynecic** (*jin-e'-sik*) [γυνή, woman]. Relating to women.
**gynecologic** (*jin-e-ko-loj'-ik*) [γυνή, a woman; λόγοs, science]. Relating to gynecology.
**gynecologist** (*jin-e-kol'-o-jist*) [γυνή, woman; λόγοs, science]. One who practises gynecology.
**gynecology** (*jin-e-kol'-o-je*) [γυνή, woman; λόγοs, science]. The science of the diseases of women, especially of those affecting the sexual organs.
**Gynecophorus hematobius.** Same as *Bilharzia hematobia*.
**gynecomania** (*jin-e-ko-ma'-ne-ah*) [γυνή, woman; μανία, madness]. Satyriasis.
**gynecomastia** (*jin-e-ko-mas'-te-ah*) [γυνή, woman; μαστόs, breast]. 1. The excessive development of the breast of a man, either with or without atrophy of the testicles. 2. The secretion of milk by the male.
**gynecomasty** (*jin'-e-ko-mas-te*). See *gynecomastia*.
**gynecomazia** (*jin-e-ko-ma'-ze-ah*) [γυνή, woman; μαζόs, breast]. Same as *gynecomastia*.
**gynecopathy** (*jin-e-kop'-ath-e*) [γυνή, woman; πάθοs, disease]. Any disease of, or peculiar to, women; the study of diseases of women.
**gynecophonus,** (*jin-e-kof'-on-us*) [γυνή, a woman; φόνοs, murder]. 1. Destructive to women. 2. [γυνή, woman; *phone*, the voice]. Having a voice like a woman. 3. A man with an effeminate voice.
**gynephobia** (*jin-e-fo'-be-ah*) [γυνή, woman; φόβοs, fear]. Morbid aversion to the society of women.
**gyniatrics** (*jin-e-a'-triks*) [γυνή, woman; ἰατρεία, therapy]. Gynecology, or gynecologic therapeutics.
**gynocardia** (*jin-o-kar'-de-ah*). See *chaulmoogra*.
**gynocyanauridzarin** (*jin-o-si-an-aw-rid'-za-rin*),

($C_9H_{11}O_7$)$_2$KCNOAu$_2$, used in leprosy, tertiary syphilis, tuberculosis, psoriasis, etc. Dose $\frac{1}{150}-\frac{1}{75}$ gr. (0.03–0.2 mg.) 3 times daily.

**gynoplastic** (*jin-o-plas'-tik*) [γυνή, woman; πλασσεῖν, to mould]. Pertaining to a plastic operation on the female genitals.

**gynophore** (*jin'-o-fōr*) [γυνή, female; φερεῖν, to bear]. In biology: (*a*) The stalk of a pistil raising it above the receptacle; (*b*) The branch of a hydroid gonoblastidium that bears only generative buds containing ova.

**gypsum** (*jip'-sum*) [γύψος, chalk], $CuSO_4 + 2H_2O$. Native calcium sulphate. Deprived of its water of crystallization it constitutes plaster-of-Paris.

**gyral** (*ji'-ral*) [*gyrare*, to turn or whirl]. Pertaining to a gyrus or to gyri.

**gyration** (*ji-ra'-shun*) [*gyrare*, to turn or whirl]. A turning in a circle; also, giddiness.

**gyre** (*jīr*) [*gyrus*, a circle]. A cerebral convolution. And see *gyrus*.

**gyrencephalic** (*ji-ren-sef-al'-ik*). Pertaining to a brain having numerous convolutions.

**gyrencephalus** (*ji-ren-sef'-al-us*). Having a brain with numerous convolutions.

**gyri** (*ji'-ri*) [pl. of *gyrus*]. 1. The convolutions of the brain. 2. The spiral cavities of the internal ear.

**gyroma** (*ji-ro'-mah*) [γῦρος, a circle]. Myoma of the ovary in which the fibrous tissue presents a wavy appearance; it is ascribed to degenerative changes in the fibrous tissue surrounding old contracting corpora lutea.

**gyromele** (*ji'-rom-ēl*) [γῦρος, a circle; μέλε, a kind of cup]. Of Türck, a stomach-tube or probe with a rotating center, which can be fitted with various attachments and used in estimating the size of the stomach, cleansing, massage of the walls, securing cultures, etc.

**gyrosa** (*ji-ro'-sah*) [γῦρος, a circle]. A variety of gastric vertigo in which, when the patient is standing, everything turns around him, and he must close his eyes to avoid falling.

**gyrospasm** (*ji'-ro-spazm*) [γῦρος, circle; σπασμός, spasm]. A peculiar rotary spasm of the head, and sometimes a nodding spasm.

**gyrus** (*ji'-rus*) [γῦρος, a circle; pl., *gyri*]. A convolution of the brain. See *convolution*. **g. ambiens**, or **circumambiens**, a small convolution in the uncus at the end of the lateral olfactory stria. **g. Andreæ Retzii**, any one of a number of ill-defined gyres between the dentate and the hippocampal gyres, in front of the splenium. **g., angular**, the posterior part of that one between the intraparietal fissure in front and above and the horizontal limb of the Sylvian fissure. **g., annectant**, four small convolutions connecting the occipital with the temporo-sphenoid and parietal lobes. **g., ascending frontal**, that in front of Rolando's fissure. **g., ascending parietal**, that just behind Rolando's fissure. **g., callosal**, the convolution immediately above the callosum. **g., dentate**, in man, a rudimentary one in the hippocampal fissure. **g. epicallosus** or **supracallosus**, the indusium, *q. v.* **g. fasciolaris**, the fasciola, *q. v.* **g. fornicatus**, a long convolution on the median surface of the brain above the corpus callosum. **g., frontal**, the convolutions of the frontal lobe. **g., hippocampal**, that part of the fornicate convolution that winds around the splenium of the corpus callosum. **g., insular**, the small gyri composing the island of Reil. **g. intralimbicus**, that part of the uncus caudad of the dentate gyrus. **g. longus insulæ**, the postinsula, a long gyre in the island of Reil. **g., marginal**, the median surface of the first frontal convolution. **g., medifrontal**, the convolution between the superfrontal and subfrontal fissures. **g., mediotemporal**, the convolution between the supertemporal and the meditemporal fissures. **g., mesorbital**, the convolution between the intercerebral and olfactory fissures. **g., occipital**, the convolutions making up the occipital lobe. **g., olfactory**. See *striæ, olfactory*. **g., paracentral**, one on the mesial surface of the brain representing the junction of the upper ends of the ascending frontal and ascending parietal convolutions. **g., parietal**, those of the parietal lobe. **g., postcentral**. Same as *g., ascending parietal*. **g., postparietal**, the convolution between the posterior limb of the meditemporal fissure and the paroccipital fissure. **g., precentral**. Same as *g., ascending frontal*. **g., preinsular**, any one of four or five small gyres in the insula or island of Reil. **g. semilunaris**, a small convolution in the uncus at the end of the lateral olfactory stria. **g., subcalcarine**, a convolution between the calcarine and the collateral fissures. **g., subcollateral**, a convolution between the collateral and the subtemporal fissures. **g., subfrontal**, the convolution between the subfrontal and the Sylvian fissures. **g., subtemporal**, the convolution between the meditemporal and the subtemporal fissures. **g., superfrontal**, the convolution between the callosomarginal and the superfrontal fissures. **g., supertemporal**, the convolution between the Sylvian and the supertemporal fissures. **g., supramarginal**, the anterior part of one between the intraparietal fissure in front and above and the horizontal limb of the Sylvian fissure. **g., temporal**, those of the temporal lobe. **g., transtemporal**, any one of a number of small gyres on the opercular surface of the temporal lobe. **g., uncinate**, the hook-like termination of the fornicate convolution.

# H

**H.** 1. Chemical symbol of *hydrogen*. 2. Abbreviation of *hyperopia*; of *hora*, hour; and of *haustus*, a draught.

**Haab's pupil-reflex** (hahp) [O. *Haab*, Swiss ophthalmologist, 1850– ]. If a bright object already present in the visual field is looked at, the pupils contract, while there is no appreciable change during convergence or accommodation. This points to a cortical lesion. **H.'s sign.** See *H.'s pupil-reflex*.

**habena** (hab-e'-nah) [L., "a rein"]. 1. A frenum. 2. Habenula (2). 3. A bandage or strip of plaster for a wound.

**habenal,** or **habenar** (hab-e'-nal, or hab-e'-nar) [*habena*, a rein]. Relating to the habena.

**habenula** (hab-en'-ū-lah) [*habena*, a rein]. 1. A frenum. 2. A ribbon-like structure; a name applied to different portions of the basilar membrane of the internal ear. 3. A peduncle of the pineal gland. **h. arcuata**, the inner zone of the basilar membrane of the cochlea. **h. conarii**, the peduncle of the pineal gland. **h., tecta.** Same as *h. arcuata*. **habenulæ, ganglion,** a small club-shaped body on the mesial surface of the optic thalamus, in which the corresponding peduncle of the pineal gland terminates.

**habenular** (hab-en'-ū-lar) [*habena*, a rein]. Pertaining to an habenula.

**habit** (hab'-it) [*habere*, to have]. 1. The general condition or appearance of an individual, as a *full habit*, a condition of plethora indicated by congestion of the superficial vessels and obesity. 2. The tendency to repeat an action or condition. **h.-spasm, h.-chorea,** a spasmodic, constantly recurring movement of certain voluntary muscles, usually seen in children.

**habitat** (hab'-it-at) [*habitare*, to dwell]. The natural home of an animal or vegetable species.

**habitus** (hab'-it-us) [*habere*, to have]. Habit; general appearance or expression. **h. apoplecticus,** tendency to apoplexy. **h. enteroptoticus,** the condition characteristic of enteroptosis; the abdomen is long and narrow, and the costal angle is less than 90°.

**habromania** (hab-ro-ma'-ne-ah) [ἁβρος, graceful; μανια, madness]. Insanity with pleasant or agreeable delusions.

**hachement** (hahsh-mon(g)) [Fr., "hacking" or "chopping"]. A form of massage consisting of a succession of strokes performed with the edge of the extended fingers or with the whole hand.

**hacking** (hak'-ing). See *hachement*.

**hadernkrankheit** (hah'-dern-kronk-hīt) [Ger.]. A disease of rag-pickers, by some supposed to be anthrax; others look upon it as malignant edema.

**hæ-.** For words beginning thus, not found below, see *he-*.

**Haeckel's law** [Ernst Heinrich *Hæckel*, German naturalist, 1834– ]. The principle that every organism, in its ontogeny, goes through a series of stages, each of which represents a stage in the evolution of that class of organisms to which it belongs; heredity influencing or securing its palingeny, and the environment causing its kenogeny; in other words, "that the ontogeny is a short repetition of the phylogeny."

**hæma-** (hem-ah-). See *hema-*.
**hæmal** (hem'-al). See *hemal*.
**hæmamœba** (hem-am-e'-bah). See *hemameba*.
**hæmato-** (hem'-at-o-). See *hemato-*.
**hæmatoxylin** (hem-at-oks'-il-in). See *hematoxylon*.
**H. campechianum,** American tropical tree, logwood; the wood is astringent.
**hæmin** (hem'-in). See *hemin*.
**hæmo-** (hem'-o-). For words thus beginning see *hemo-*.
**Hæmogregarinæ** (hem-o-greg-ar-i'-ne). See *Hæmosporidia*.
**hæmorrhage** (hem'-or-āj). See *hemorrhage*.
**Hæmosporidi** m-o-spor-id'-e-ah) alμa, blood; *sporidia*]. An order of sporozoa which live for a part of their life cycle, within the red blood cells of their hosts.

**Haeser's coefficient** (ha'-zer) [Heinrich *Haeser*, German physician, 1811-1884]. The number 2.33, with which the last two figures of the specific gravity of the urine are multiplied in order to obtain the amount of solids in 1000 c.c. of urine.

**Haffkine's method of immunization against cholera** (haf'-kin) [Waldemar Mordecai Wolff *Haff-kine*, Russian bacteriologist, 1860– ]. This consists in the injection of a definite quantity of sterilized culture; 5 days later a small dose, and in 5 days more a larger dose, of the living virulent culture. **H.'s prophylactic fluid,** a preparation of plague bacilli used by inoculation as a preventive of the plague. It is not a serum or a lymph, but a fluid culture of pest-bacilli, grown for several weeks under conditions most favorable to the development of the toxic properties; it is finally deprived of its infective quality by being heated to a temperature that is fatal to the living bacteria, but which does not alter the specific toxin.

**haffkinin** (haf'-kin-in). Haffkine's plague serum. See *Haffkine's prophylactic fluid*.

**Hagedorn needle** (hag'-ed-orn) [Werner *Hagedorn*, German surgeon, 1831-1894]. A curved surgical needle with flat sides.

**hagiotherapy** (ha-je-o-ther'-ap-e) [ἅγιος, sacred; θεραπεία, treatment]. Treatment of disease by means of shrines, relics, the intervention of saints, and other similar observances.

**Hahnemannism** (hahn'-e-man-izm) [Samuel *Christian Frederic Hahnemann*, German physician, the founder of the doctrine of homeopathy, 1755-1843]. See *homeopathy*.

**Haidinger's brushes** (hi'-ding-er) [Wilhelm von *Haidinger*, Austrian mineralogist, 1795-1871]. A brush-like image seen on directing the eye toward a source of polarized light, due to the doubly refractive character of the elements of the macula.

**Haines's coefficient** (hānz) [Walter Stanley *Haines*, American chemist and toxicologist, 1850– ]. The number 1.1 which when multiplied by the last two figures of the specific gravity of urine will give the amount of solids in grains for each fluidounce. **H.'s solution for glucose in urine.** Copper sulphate 3, potassium hydroxide 9, glycerin 100, water 600. Use as Fehling's solution; a red precipitate will be formed.

**hair** [AS., *hær*]. A delicate filament growing from the skin of mammals; collectively, all the filaments forming the covering of the skin. Hair is a modified epidermal structure, and consists of a *shaft* and a *root*, the latter expanded at its end into the *hair-bulb*, which is concave and caps the *hair-papilla*. **h.-bulb,** the expanded portion at the lower end of a hair-root. **h.-cell,** an epithelial cell with delicate, hair-like processes, as, *e. g.,* the *hair-cells* of the organ of Corti. **h.-follicle,** the depression in the corium and subcutaneous connective tissue containing the root of the hair. **h.-papilla,** a portion of the corium projecting upward into the center of a hair-bulb. **h.-salt,** native magnesium sulphate.

**hairy heart.** A heart covered with a rough mass of exudate. **h. tongue,** a tongue covered with hair-like papillæ.

**halakones** (hal'-ak-ōns). Small cones of stiffened gauze, loosely filled with absorbent material which may be medicated: they are designed to fit the nostril so that all air inhaled must pass through the medicated material.

**hale** (hāl). Sound; healthy; robust.

**halide, halid** (hal'-id). Same as *haloid*.

**halimeter** (hal-im'-et-er) [ἅλς, salt; μέτρον, a measure]. An instrument for estimating the proportions of water in milk by means of its power to dissolve common salt.

**halimetry** (*hal-im'-et-re*) [ἅλς, salt; μέτρον, a measure]. The process of determining the quantity of salts in a mixture.
**halisteresis** (*hal-is-ter-e'-sis*) [ἅλς, salt; στέρησις, privation]. The loss of lime-salts of bone.
**halisteretic** (*hal-is-ter-et'-ik*). Pertaining to, or affected with halisteresis.
**halitosis** (*hal-it-o'-sis*) [*halitus*, exhalation]. Foul breath.
**halituous** (*hal-it'-ū-us*) [*halitus*, breath]. Moist, as if from having been breathed upon; applied to the skin.
**halitus** (*hal'-it-us*) [L.]. A vapor, as that expired from the lung.
**Hall's (Marshall) disease** (*hawl*) [Marshall *Hall*, English physician, 1790–1857]. Hydrocephaloid occurring in infants suffering from severe chronic intestinal catarrh. **H.'s facies,** the prominent forehead and small features peculiar to hydrocephalus. **H.'s method of artificial respiration.** The body is turned alternately upon the side or face to compress the chest, and then upon the back to allow the lungs to expand.
**Haller's ansa** (*hal'-ler*) [Albrecht von *Haller*, Swiss anatomist, 1708–1777]. A loop formed in front of the internal jugular vein by a small nerve branching off from the facial just below the stylomastoid foramen, and joining the glossopharyngeal a little below Andersch's ganglion. It is not constant. **H.'s circle.** (1) The plexus of vessels formed by the short ciliary arteries upon the sclerotic, at the entrance of the optic nerve. (2) The circulus venosus mammæ, situated beneath the areola of the nipple. **H.'s colic omentum,** a process of the upper right border of the greater omentum which may become adherent to the testis during fetal life and be included in the sac of an inguinal hernia. **H.'s cones,** the coni vasculosi of the epididymis, small conic masses made up of the convolutions of the efferent tubules of the testicle. They form part of the globus major, and their tubules opening into a common duct form the origin of the vas deferens. **H.'s congenital hernia.** See *Malgaigne's hernia.* **H.'s fretum.** See *H.'s isthmus.* **H.'s habenula,** the slender cord formed by the obliteration of the canal which during early life connects the cavity of the peritoneum with that of the tunica vaginalis. **H.'s isthmus,** the constriction which separates the ventricle from the aortic bulb during early fetal life. Syn., *fretum Halleri.* **H.'s network,** the rete vasculosum of the testis. **H.'s plexus,** the network formed by branches of the external laryngeal and sympathetic nerves on the outer surface of the inferior constrictor pharyngis. Syn., *Haller's laryngeal plexus.* **H.'s splendid line,** the longitudinal fibrous band of the pia corresponding to the site of the anterior median fissure of the spinal cord. Syn., *linea splendens.* **H.'s tripod,** the celiac axis. Syn., *tripus Halleri.* **H.'s tunica vasculosa,** the lamina vasculosa of the choroid. **H.'s vas aberrans,** a small, convoluted duct connected with the tail of the epididymis or the beginning of the vas deferens. **H.'s venous circle,** an incomplete circle of superficial veins frequently seen through the integument of the mammæ, especially during lactation.
**hallex** (*hal'-eks*) [L.: pl., *hallices*]. See *hallux.*
**Hallion's law** (*hal-yon*) [L. *Hallion,* French physician]. "Organic extracts exert on the same organ an exciting influence which lasts for longer or shorter time; when this organ is insufficient, it is conceivable that this influence augments its action and, when it is injured, that it favors its restoration."
**Hallopeau's disease** (*hal-op-o'*) [Henri *Hallopeau*, French dermatologist, 1842– ]. Chronic pustular dermatitis, a form of Neumann's disease.
**hallucal** (*hal'-ū-kal*) [*hallux*]. Pertaining to the hallux, or great toe.
**hallucination** (*hal-lū-sin-a'-shun*) [*alucinari*, to wander in mind]. A false sense-perception; it is the perception of an object or phenomenon which has no external existence, as *hallucination* of sight, sound, smell, taste, or touch.
**hallucinosis** (*hal-lū-sin-o'-sis*). The condition of being possessed by more or less persistent hallucinations.
**hallux** (*hal'-uks*) [L.: pl., *hallices*]. The great toe. **h. dolorosus.** See *h. flexus.* **h. flexus,** a condition allied to and perhaps identical with hammer-toe, in which there is flexion of the first phalanx of the great toe. The second phalanx is usually extended upon the first, and there is more or less rigidity of the metatarsophalangeal joint. **h. rigidus.** See *h. flexus.* **h. valgus,** displacement of the great toe toward the other toes. **h. varus,** displacement of the great toe away from the other toes.
**halo** (*ha'-lo*) [ἅλως, a round threshing-floor]. 1. The areola of the nipple. 2. The luminous circles seen about a light. **h. glaucomatosus,** in glaucoma, a white ring surrounding the optic disc. **h.-symptom,** the colored circles seen around lights in glaucoma.
**halobios** (*hal-o-bi'-os*) [ἅλς, the sea; βίος, life]. The totality of the marine flora and fauna in opposition to *limnobios,* the organic world of fresh water, and *geobios,* the totality of the terrestrial plant and animal world.
**halogen** (*hal'-o-jen*) [ἅλς, salt; γεννᾶν, to produce]. A univalent element that forms a compound of a saline nature by its direct union with a metal. The halogens are chlorine, iodine, bromine, and fluorine. **h. acid,** an acid formed by the combination of a halogen with hydrogen.
**halogenic** (*hal-o-jen'-ik*) [ἅλς, salt; γεννᾶν, to produce]. Salt-producing; producing haloids.
**haloid** (*hal'-oid*) [ἅλς, salt; εἶδος, likeness]. Resembling sea-salt. **h. salts,** any one of those compounds that consist of a metal directly united to chlorine, bromine, iodine, or fluorine.
**halology** (*hal-ol'-o-je*) [ἅλς, salt; λόγος, science]. The chemistry of salts.
**haloscope** (*hal'-o-skōp*) [ἅλς, salt; σκοπεῖν, to examine]. An apparatus for determining the amount of salt in a solution.
**Halsted's operation** (*hol'-sted*) [William Stewart *Halsted,* American surgeon, 1852– ]. For the *radical cure of inguinal hernia:* similar to Bassini's operation, but Halsted makes a new internal ring in addition to transplanting the cord in a new canal.
**Halstern's disease.** Endemic syphilis.
**Halteridium** (*hal-tur-id'-e-um*) [ἀλτήρες, weights held in the hand when leaping]. A genus of parasitic coccidia which infest the blood-corpuscles of birds.
**ham** [AS., *hamm*]. 1. The back part of the knee; the popliteal space. 2. The buttock, hip, and thigh.
**hamamelin** (*ham-am-e'-lin*). A precipitate from a tincture of the bark of witch-hazel, *Hamamelis virginiana;* it is astringent, tonic, and sedative. Dose gr. j-iij.
**hamamelis** (*ham-a-me'-lis*) [ἅμα, together with; μῆλον, apple]. Witch-hazel. **h. bark** (*hamamelidis cortex,* U. S. P.), the bark and twigs of *Hamamelis virginiana,* used in the preparation of hamamelis water. **h. leaves** (*hamamelidis folia,* U. S. P.), the leaves of *Hamamelis virginiana,* used in preparing the fluidextract. **h. leaves,** fluidextract of (*fluidextractum hamamelidis foliorum,* U. S. P.), frequently employed as an astringent gargle in subacute sore throat; internally as a hemostatic, and in suppository or ointment in the treatment of bleeding piles. Dose 5–40 min. (0.3–2.6 Cc.). **h. water** (*aqua hamamelidis,* U. S. P.), an aqueous extract of hamamelis bark; under the name of extract of witch-hazel it is a popular household remedy. Dose ½–1 dr. (2–4 Cc.).
**hamarthritis** (*ham-ar-thri'-tis*) [ἅμα, together; ἄρθρον, a joint; *itis*, inflammation]. Gout involving all of the joints.
**hamartia** (*ham-ar'-she-ah*) [ἁμαρτίον, bodily defect]. An error of development due to defects in tissue-combination.
**hamartoma** (*ham-ar-to'-mah*) [ἁμαρτίον, bodily defect; *oma,* tumor]. 1. A tumor due to a failure of development. 2. A tumor due to a new growth of blood-vessels.
**hamatum** (*ham-a'-tum*) [L. "hooked"]. The unciform bone.
**Hamburger's depot reaction** (*ham'-boor-ger*). ₁⁄₁₀ c.cm. of a 1 : 10,000 dilution of tuberculin is injected just beneath the skin of the forearm or back. If the reaction is positive a subcutaneous infiltration appears within twenty-four hours, and there is a reddening at the site where the point of the needle rested.
**Hamilton Irving apparatus.** An appliance to prevent a patient (after a suprapubic operation) from being wetted by the urine soaking into the dressings.
**Hamilton's test** (*ham'-il-tun*) [Frank Hastings *Hamilton,* American surgeon, 1813–1875]. In dislocation of the shoulder-joint a ruler applied to the

dislocated humerus may be made to touch the acromion and external condyle at the same time.

**Hammarsten's test for globulin** (*ham'-ars-ten*) [Olof *Hammarsten*, Swedish physiologist, 1841– ]. To the neutral solution add powdered magnesium sulphate until no more of the salt dissolves. Separate the globulin thus precipitated by filtration, and wash with a saturated solution of magnesium sulphate.

**hammer** (*ham'-er*) [ME, *hamer*]. 1. In anatomy, the malleus. 2. An instrument for striking. **h.-bone**, the malleus. **h., Mayor's**, one with rounded faces to produce counterirritation on the skin by application when heated. **h., Neef, h., Wagner**, an interrupter or circuit-breaker employed with many induction-coils. **h., percussion-**, a plexor. **h., thermal**, a hammer-shaped cautery-iron. **h.-toe**, a term applied to a condition of the second toe in which the proximal phalanx is extremely extended while the two distal phalanges are flexed.

**hammerman's cramp**. A spasmodic, often painless affection of the muscles of the upper extremity; it is seen in those who use a hammer, and is due to overuse.

**hammock** (*ham'-ok*) [Span., *hamaca*, a hanging mat]. A couch or bed made of netting or canvas, suspended at the ends. It is much used aboard vessels and in tropical regions; it has been used latterly in the transportation of the sick and wounded. Slings for fractured legs, etc., are sometimes called hammocks.

**Hammond's disease** (*ham'-ond*) [William Alexander *Hammond*, American neurologist, 1828–1900]. Athetosis.

**hamose** (*ham'-ōs*) [*hamus*, a hook]. Hooked at the apex.

**hamstring**. The tendons bounding the ham above on the outer and inner side. **h., inner**, the tendons of the semimembranosus, sartorius, gracilis, and semitendinosus muscles. **h., outer**, the tendons of the biceps flexor cruris.

**hamular** (*ham'-ū-lar*) [*hamus*]. Pertaining to or shaped like a hook.

**hamulate, hamulose** (*ham'-ū-lāt, -lōs*) [*hamus*, a hook]. Hooked or hook-shaped.

**hamulus** (*ham'-ū-lus*) [dim. of *hamus*, a hook]. A hook-shaped process, as of a bone. **h. of the cochlea**, the hook-like process of the osseous lamina at the cupola.

**Hancock's operation**. For *amputation through the foot*: a modification of Pirogoff's operation, in which the sawn surface of the os calcis is brought in contact with the transverse section of the astragalus.

**hand** [ME.]. The organ of prehension in bimana and quadrumana, composed of the carpus, the metacarpus, and the phalanges. **h., ape-, h., claw-, h., monkey-**. See *claw-hand*. **h., battledore**, the large hand seen in cases of acromegaly. **h.-electrode**, an electrode for use in the hand. **h., forceps**, a hand which has lost the three middle fingers. **h., trailing**, in synchronous writing of both hands, that upon which the attention, visual or central, is not fixed.

**handkerchief** (*hang'-ker-chif*) [ME., *hand; kerchef*, a kerchief]. A square piece of cloth for wiping the face or nose. **h.-dressing**, a form of temporary dressing for wounds and fractures, made of handkerchiefs.

**hangnail** (*hang'-nāl*). A partly detached piece of epidermis at the root of the nail, the friction against which, as caused inflammation of the abraded surface.

**Hankin's defensive proteids**. Germicidal globulins found by Hankin in the blood of certain animals and giving immunity to certain toxins.

**Hannover's canal** (*han'-o-ver*). The artificial passage produced between the anterior and posterior fibers of the zonules of Zinn by the injection of a viscous fluid. **H.'s intermediate membrane**, the enamel membrane; the inner, cellular layer of the enamel-organ of the dental germ of the fetus.

**Hanot's disease** (*han'-o*) [Victor Charles *Hanot*, French physician, 1844–1896]. Hypertrophic cirrhosis of the liver with icterus.

**Hansen's bacillus** (*han'-sen*) [Gerhard Armauer *Hansen*, Norwegian physician, 1841– ]. The *Bacillus lepræ*.

**haouwa** [E. Ind.]. Synonym, in Bagdad, of *Asiatic cholera*.

**hapalonychia** (*hap-al-o-nik'-e-ah*) [ἀπαλός, soft to the touch; ὄνυξ, nail]. A soft condition of the nails.

**hapantismus** (*hap-an-tis'-mus*) [ἅπας, entire]. Complete adhesion between parts or surfaces.

**haphalgesia** (*haf-al-je'-ze-ah*) [ἁφή, touch; ἄλγος, pain]. A feeling of pain produced by merely touching an object.

**haphemetric** (*haf-e-met'-rik*) [ἁφή, touch; μέτρον, measure]. Relating to esthesiometry. See *esthesiometer*.

**haphephobia** (*haf-e-fo'-be-ah*) [ἁφή, contact; φόβος, fear]. The morbid dread of being touched; mysophobia.

**haphonosus** (*haf-on-o'-sus*) [ἁφή, touch; νόσος, disease]. Any disorder of the sense of touch.

**haplobacteria** (*hap'-lo-bak-te-re-ah*) [ἁπλόος, simple; *bacteria*]. Non-filamentous bacteria.

**haplodermatitis** (*hap-lo-der-mat-i'-tis*) [ἁπλόος, simple; δέρμα, skin; ιτις, inflammation]. A simple or uncomplicated skin-inflammation.

**haplodermitis** (*hap-lo-der-mi'-tis*). Haplodermatitis.

**haplodont** (*hap'-lo-dont*) [ἁπλόος, single; ὀδούς, tooth]. In biology, applied to animals whose molar teeth have simple or single crowns.

**haplolichen** (*hap-lo-li'-ken*) [ἁπλόος, single; λειχήν, lichen]. Same as *lichen simplex, q. v.*

**haplomelasma** (*hap-lo-mel-as'-mah*) [ἁπλόος, simple; μέλασμα, a livid spot]. Simple melasma.

**haplopathy** (*hap-lop'-ath-e*) [ἁπλόος, simple; πάθος, illness]. Any uncomplicated disease.

**haplophyma** (*hap-lo-fi'-mah*) [ἁπλόος, simple; φῦμα, a tumor]. A simple tumor.

**haplopia** (*hap-lo'-pe-ah*) [ἁπλόος, single; ὤψ, vision]. Single vision; used in opposition to diplopia.

**haploscope** (*hap'-lo-skōp*) [ἁπλόος, single; σκοπεῖν, to see]. An instrument for measuring the visual axes. **h., mirror**, an instrument for observing the effects of varying degrees of convergence of the visual axes.

**Haplosporidia** (*hap-lo-spor-id'-e-ah*) [ἁπλόος, single; *sporidia*]. An order of sporozoa, with a simple spore and one nucleus.

**haplotomia, haplotomy** (*hap-lo-to'-me-ah, hapt'-lo-o-me*) [ἁπλόος, simple; τομή, a cut]. A simple incision.

**haptic** (*hap'-tik*) [ἁπτός, subject to the sense of touch]. Pertaining to touch; tactile.

**haptics** (*hap'-tik s*) [ἅπτειν, to touch]. The science of the tactile sense.

**haptine** (*hap'-tēn*) [ἅπτειν, to bind]. 1. In Ehrlich's lateral-chain theory, any thrown-off receptor. These are of three orders: (1) antitoxins; (2) agglutinins, and precipitins; (3) cytotoxins or lysins. 2. An antigen.

**haptodysphoria** (*hap-to-dis-fo'-re-ah*) [ἁπτός, tactile; touched; δύς, difficult; φορός, bearing]. The disagreeable sensation aroused by touching certain objects, as velvet, a peach, or a russet apple.

**haptogen** (*hap'-to-jen*) [ἅπτειν, to bind; γεννᾶν, to produce]. A pellicle forming around fatty matter when brought into contact with albumin. **Syn., haptogenic membrane**.

**haptophil, haptophile** (*hap'-to-fil*) [ἅπτειν, to bind; φιλεῖν, to love]. In Ehrlich's side-chain theory applied to a receptor having an affinity for the haptophore of a toxin.

**haptophore** (*hap'-to-fōr*) [ἅπτειν, to bind; φέρειν, to bear]. That complex of atoms of a toxic unit which unites it to the cell-receptor.

**haptophoric, haptophorous** (*hap-tof-or'-ic, -us*). Combining; pertaining to haptophores. **h. group**. See *haptophore*.

**haramaitism** (*har-am-a'-it-ism*). Anglo-Indian; from *Haram maiti*, the name of an Hindu offender in this way]. Child-marriage in India; also, the collective physical evils that result from that system.

**hardening** (*har'-den-ing*) [AS., *hearde*, hard]. A stage in the preparation of tissues for microscopical examination in which they are rendered firm, so that they may, after embedding, be readily cut.

**Harderian gland** (*har-dē'-re-an*) [Johann Jacob *Harder*, Swiss anatomist, 1656–1711]. A racemose gland located at the inner canthus of the eye of most vertebrates, and especially of those having a well-developed nictitating membrane.

**Hardy-Béhier's symptom**. See *Béhier-Hardy's symptom*.

**hare-eye** (*hār'-i*). See *lagophthalmos*.

**harelip**. Congenital fissure of the lip, due to arrested facial development. **h., complicated**, that with cleft or malformation of the superior maxillary bone also. **h., double**, two clefts of the lip or one of each lip. **h. suture,** a figure-of-8 suture about a pin thrust through the lips of the freshened edges of the cleft.

**harlequin** (*har'-le-kwin*). Variegated; partycolored. **h. fetus,** a fetus with congenital ichthyosis, general seborrhea, or diffuse keratoma. Such subjects are always born prematurely, and have no external ears, eyelids, or lips.

**Harley's disease** (*har'-le*) [George *Harley*, English physician, 1829–1896]. See *Dressler's disease*.

**harmaline** (*har'-ma-lēn*). An alkaloid, $C_{13}H_{14}N_2O$, from harmel; it is used as a stimulant and anthelmintic.

**harmel** (*har'-mel*). Wild rue, *Peganum harmala*, of Turkey; it is a vermifuge.

**harmonia, harmony** (*har-mo'-ne-ah, har'-mo-ne*) [ἁρμονία, harmony]. A form of articulation between two bones that are closely and immovably apposed.

**harpoon** (*har-poon'*) [Fr., *harpon*, a grappling-iron]. An instrument for the removal of bits of living tissue for microscopic examination.

**Harris separator or segregator** (*har'is*) [Malcolm L. *Harris*, American surgeon]. A double catheter is passed into the bladder, and a lever into the rectum or vagina; this lever lifts up the floor of the bladder between the separated ends of the divided catheter; the urine from each kidney is thus collected serarately and flows out through the catheter on the same side.

**Harrison's groove** (*har'-is-on*) [Edward *Harrison*, English physician, 1766–1838]. A curve extending from the level of the ensiform cartilage toward the axilla, and corresponding to the insertion of the diaphragm; it is pronounced in rickets.

**harrowing** (*har'-o-ing*). The action of teasing the fibers of a nerve or tearing them apart with any blunt instrument.

**Hartley-Krause operation** (*hart'-le-krow'-zer*) [Frank *Hartley*, American surgeon; Fedor *Krause*, German surgeon, 1857– ]. The removal of the entire Gasserian ganglion and its roots for relief of facial neuralgia.

**Hartmann's fossa** (*hart'-man*) [Robert *Hartmann*, German anatomist, 1831– ]. A small, infundibular fossa of the peritoneum lying between Tuffier's inferior ligament and the mesoappendix. Syn., *fossa ileocæcalis inflma*.

**hartshorn** (*harts'-horn*). 1. Cornu cervi, the horn of the stag, formerly a source of ammonia, or spirit of hartshorn. 2. A name popularly given to ammonia-water.

**hashish, hasheesh** (*hash'-ēsh*). See *cannabis*.

**Hasner's valve** (*has'-ner*) [Joseph Ritter von *Hasner*, Austrian ophthalmologist, 1819–1892]. An inconstant valvular fold of mucosa at the inferior meatus of the nasal duct.

**Hassall's bodies** (*has'-al*) [Arthur Hill *Hassall*, English physician, 1817–1894]. Concentrically striated corpuscles, apparently of a degenerative character, found in the thymus gland. **H.'s test,** the growth of *Saccharomyces cerevisiæ*, observed under the microscope, is indicative of the presence of sugar in the urine.

**Hastings' stain** (*hās'-tings*) [Thomas Wood *Hastings*, American physician, 1873– ]. A methyleneblue and eosin staining reagent modified from Romanovsky's stain.

**Hata** (*hah'-tah*) [S. , *Hata*, Japanese physician]. See *Ehrlich-Hata*.

**hatters' disease** (*hat'-er*). A form of constitutional mercurial poisoning occurring in the makers of hats; also a skin-disease, arising from the use of mercury and arsenic; also an acute irritation of the respiratory tract caused by the fumes of nitrogen tetroxide; all of these chemicals being used in hatmaking. **h.'s consumption,** a form of pneumonitis occurring in hatters from inhalation of the fur and dust arising during the process of "finishing and pouncing."

**haunch** (*hawnsh*) [Fr., *hanche*]. The part of the body including the hips and the buttocks. **h.-bone,** the ilium.

**haustra coli** (*haws'-trah co'-li*) [L.; pl. of *haustrum*]. Sacculations of the colon.

**haustrum** (*haw'-strum*) [L.]. The pouch or depression of the sacculations of the colon.

**haustus** (*haws'-tus*) [L., a drink or draught]. A draught. **h.-niger,** black draught; the compound infusion of senna.

**haut mal** (*o-mahl'*). See *epilepsy*.

**Havers' canals** (*ha'-verz*) [Clopton *Havers*, English anatomist, 1650–1702]. The canals pervading the compact substance of bone in a longitudinal direction and anastomosing with one another by transverse or oblique branches. They contain bloodvessels and lymphatics. **H.'s glands,** fatty bodies connected with the synovial fringes of most of the joints. They were believed by Havers to secrete the synovia. Syn., *Glandulæ mucilaginosæ*. **H.'s lamellæ,** the concentric lamellæ of bone which form the Haversian canals. **H.'s spaces,** large, irregularly shaped spaces found chiefly in growing bones. **H.'s system,** the concentric arrangement of the bony lamellæ, usually 8 or 10 in number, around a Haversian canal.

**haw** [ME., *haw*, an excrescence in the eye]. 1. The third eyelid, nictitating membrane, or winker of a horse. 2. A diseased or disordered condition of the third eyelid of the horse.

**hawking** (*haw'-king*) [ME., *hauk*, to hawk]. Clearing the throat by a forcible expiration.

**Hay's method.** A method of removing dropsical effusions by producing frequent serous evacuations by means of saline cathartics combined with a dry diet.

**hay-asthma.** See *hay-fever*.

**Hayem's corpuscles, H.'s hematoblasts** (*a-yem'*) [Georges *Hayem*, French physician, 1841– ]. See *Bizzozero's blood-platelets*. **H.'s disease,** apoplectiform myelitis. **H.'s solution,** used in the microscopic examination of blood; it consists of sodium chloride, 1 Gm.; sodium sulphate, 5 Gm.; mercuric chloride, 0.5 Gm.; dissolved in 200 Cc. of distilled water.

**Haygarth's nodes or nodosities** (*ha'-garth*) [John *Haygarth*, English physician, 1740–1827]. Exostoses of the joints of the fingers in arthritis deformans.

**hay-fever.** An acute affection of the conjunctiva and upper air-passages, coming on periodically at certain seasons of the year, especially in summer and autumn, in persons predisposed to the disease. The exciting factor in some cases is the pollen of grasses; in others the disease seems to be caused reflexly by polypi and other diseased conditions of the nose. The chief symptoms are coryza, sneezing, headache, cough, and asthmatic attacks. Syn., *hay-asthma*, *hay-cold*.

**hazeline** (*ha'-zel-ēn*). Trade name applied to preparations of *Hamamelis virginiana*.

**hb.** Abbreviation of *hemoglobin*.

**H. D.** Abbreviation of *hearing distance*.

**He.** Chemical symbol of *helium*.

**head** (*hed*) [ME., *hed*]. 1. The uppermost part of the body; that part of the body containing the brain, the organs of sight, smell, taste, and hearing, and part of the organs of speech. 2. The top, beginning, or most prominent part of anything, as the *head* of the femur, the *head* of the muscle, etc. **h.-drop,** a peculiar disease seen in Japan during the spring and early summer, supposed to be miasmatic in origin. It is attended with inability to hold the head erect, paralytic symptoms in the limbs, and optic disorders. One attack predisposes to others. **h. gut.** See *foregut*. **h.-kidney,** pronephron. **h.-locking,** the entanglements of the heads of twins, at the time of birth. **h.-louse,** pediculus capitis. **h., scald, h., scalled,** any scabby disease of the scalp. **h., swelled,** actinomycosis.

**headache** (*hed'-āk*). Pain in the head. The following varieties of headache are described: local, general; organic, functional; toxemic—alcoholic, caffeinic, diabetic, lithemic, malarial, rheumatic, uremic; reflex—gastric, ocular, nasal, uterine; cardiac, pulmonic, anemic, congestive; hysterical, neurasthenic. Syn., *cephalalgia*. **h., academy.** See *h. panorama*. **h., bilious,** migraine. **h., ocular,** pain in and about the head that results from ametropia, organic disease in, or from impaired function of, any part of the visual apparatus. **h., panorama, h., sightseer's,** headache resulting from the strain of the eyes exposed to brilliant lights or moving objects, etc. **h., sick,** migraine. **h., theater.** See *h., panorama*. **h.-all.** See *Collinsonia*.

**healer** (*hēl'-er*). 1. One who effects cures. 2. One who without medical education claims to cure by some form of suggestion. **h., natural,** one supposed to possess personal magnetism capable of overcoming disease.

**healing** (*he'-ling*) [AS., *helan*, to heal]. The process or act of getting well or of making whole; especially

the getting well of an ulcer or wound. h. by first intention, h., primary, the union of two accurately apposed surfaces without any visible granulating process. h. by second intention, healing through the medium of granulations, which fill up the gap of the wound. h. by third intention, that in which the two granulating surfaces are approximated so as to unite and heal readily.
**health** (helth) [see healing]. That condition of the body in which all the functions are performed normally. h., bill of, the official document issued by quarantine or other public health officers, which grants freedom from sanitary restraint. French, pratique. h., board of, a public body having charge of the sanitation of a stated district.
**hear** (hēr). To perceive by the ear.
**hearing** (hēr'-ing) [AS., hyran, to hear]. The special sense by which the sonorous vibrations of the air are communicated to the mind. The organ of hearing is the ear, whence the vibrations are carried by the auditory nerve to the center of hearing, situated in the temporosphenoid lobe of the cerebrum. h.-distance, the distance at which a certain sound can be heard. h., double, diplacusis. h., Eitelberg's test for. See Eitelberg.
**heart** (hart) [AS., heorte]. A hollow, muscular organ, the function of which is to pump the blood through the vessels. It is enveloped by a serous sac called the *pericardium*, and consists of two symmetrical halves, a *right auricle* and *ventricle* and a *left auricle* and *ventricle*. The right auriculoventricular orifice is guarded by the *tricuspid valve*; the left by a valve with two leaflets—the *mitral*. These valves are broad and thin, consisting of two layers of the lining membrane of the heart, the *endocardium*, separated by a slight amount of connective tissue. To support them, thin *chordæ tendineæ* join their free margins to the muscles in the wall of the ventricle —the *columnæ carneæ*. The outlet of the right ventricle into the *pulmonic artery*, and the left into the *aorta*, are guarded by stout, short, tricuspid valves—the *semilunar valves*. h.-berg, the thoracic portion of the thymus gland of animals. See also *neck-berg*. h., bicycle, cardiac disease due to excessive use of the bicycle. h.-block, dissociation of auricular and ventricular rhythms due to interference with the conduction of the contraction process. h.-b., complete, when the ventricular contractions are independent of the auricular. h.-b., partial or incomplete, when one of the auricular contractions regularly excites the ventricles to contraction. h.-burn, a burning sensation at the epigastrium and lower part of the chest. h.-clot, coagulation of blood in the cardiac cavity. h., fatty, a name given to two distinct pathologic conditions of the heart tissue. In the first there is a true fatty degeneration of the muscular fibers of the heart; in the second there is an increase in the quantity of subpericardial fat—a fatty infiltration. h., hairy. See *cor villosum*. h., icing, Eichhorst's name for a heart the whole surface of which is covered with a dense, thick, marble-white tissue. h., irritable, a peculiar cardiac excitability, marked by pain, palpitation, dyspnea, and rapid pulse; it has been noted especially among soldiers in the field, in whom it has been ascribed to muscular exhaustion. h., low, a low position of the heart due to anatomical conditions, not to disease. Syn., *bathycardia*. h., luxus, a condition in which a primary dilatation of the heart is followed by hypertrophy of the left ventricle; often found in gourmands. h., peripheral, a term applied to the muscular coat of the bloodvessels other than the heart. h., typhoid, overdistention and laceration of the blood-vessels of the heart, with atrophy of the muscle-fibers, due to typhoid fever. h., villous. See *cor villosum*.
**heartburn** (hart'-bern). A burning sensation at the epigastrium and lower part of the chest, caused by gastric fermentation.
**heat** (hēt) [AS., hætu]. 1. A form of kinetic energy communicable from one body to another; it is that form of molecular motion which is appreciated by a special thermal sense. 2. The periodic sexual excitement in animals. h., atomic. See *atomic heat*. h., capacity for, the number of heat-units required to raise the temperature of a body 1° C. h.-centers, centers in the brain for stimulating heat-production or heat-elimination, and for regulating the relation of these. h., latent, the quantity of heat necessary to convert a body into another state without changing its temperature. h., molecular, the product of the molecular weight of a compound multiplied by its specific heat. h., prickly, h.-rash, miliaria. h., specific, the amount of heat required to raise the temperature of a substance a given number of degrees. The unit of specific heat is the Calorie, which is the amount of heat required to raise the temperature of one gram of water from 4° to 5° C. h.-stroke, the symptoms produced by exposure to great heat— either that of the sun or that of heated rooms. h.-unit, the amount of heat required to raise the temperature of one kilogram of water from 0° to 1° C.; it is technically called a calorie.
**Heath's operation** (hēth) [Christopher Heath, English surgeon, 1835– ]. For fixity of the lower jaw: division of the ascending ramus beneath the masseter with a saw introduced through the mouth by means of a small incision above the last molar tooth.
**heaves** (hēvs) [ME., heven, to raise]. A disease of horses, characterized by difficult and laborious respiration. It is also called "broken wind."
**hebeosteotomy** (he-be-os-te-ot'-o-me). See *hebosteotomy*.
**hebephrenia** (he-be-fre'-ne-ah) [ἥβη, puberty; φρήν, mind]. A form of mental derangement occurring in young persons at or soon after the age of puberty, and characterized by mental deterioration and a gradually increasing egotism. It may end in a permanent dementia.
**hebephrenic** (he-be-fre'-ne-ak). One who is affected with hebephrenia.
**hebephrenic** (he-be-fren'-ik). 1. Affected with hebephrenia. 2. One who is affected with hebephrenia.
**Heberden's asthma** (heb'-er-den) [William Heberden, English physician, 1710–1801]. Angina pectoris. H.'s disease, (1) arthritis deformans; (2) angina pectoris. H.'s nodes, H.'s nodosities, deformity of the fingers in arthritis deformans.
**Heberden-Rosenbach's nodes.** See *Heberden's nodes*.
**hebetic** (he-bet'-ik) [ἡβητικός, relating to puberty]. Relating to puberty or to adolescence.
**hebetude** (heb'-e-tūd) [hebetudo, bluntness]. Dulness of the special senses and intellect: a condition present in grave fevers.
**hebetudinous** (heb-e-tū'-din-us) [hebetudo, bluntness]. Affected with hebetude.
**hebosteotomy** (he-bos-te-ot'-o-me) [ἥβη, pubes; ὀστέον, bone; τομή, cutting]. Section through the body of the pubis to facilitate labor; pubiotomy.
**hebotomy** (he-bot'-o-me) [ἥβη, pubes; τέμνειν, to cut]. Van de Velde's operation of sawing the pelvis in cases of obstructed delivery; pubiotomy.
**Hebra's disease** (ha'-brah) [Ferdinand von Hebra, Austrian dermatologist, 1816–1880]. Same as H.'s erythema. H.'s erythema. Polymorphous erythema. H.'s pityriasis, pityriasis rubra. H.'s prurigo, true prurigo.
**hecatomeral, hecatomeric** (hek-at-om'-er-al, hek-at-o-mer'-ik) [ἑκατερόν, each, singly; μέρος, a part]. Applied to a neuron the processes of which divide into two parts, one going to each side of the spinal cord.
**Hecht's test** (hekt) [Hugo Hecht, Austrian physician]. A modification of Wassermann's reaction for syphilis; it is founded on the fact that human blood serum can dissolve ten times its volume of a 2 per cent. solution of sheep's blood.
**hectargyre** (hek'-tar-jīr). A compound of hectine and mercury; used hypodermically, as an antisyphilitic.
**hectic** (hek'-tik) [ἑκτικός, habitual]. 1. Habitual. 2. Pertaining to phthisis. h. fever, a fever caused by absorption of toxic substances formed in the process of suppuration, and characterized by daily intermissions and frequent drenching sweats. It occurs in pulmonary tuberculosis, in pyemia, etc. h. flush, the flushed cheek seen in hectic fever.
**hectine** (hek'-tēn). Sodium benzosulphoparaminophenylarsinate, an arsenical compound said to be less toxic than atoxyl; used in syphilis.
**hecto-** (hek-to-) [ἑκατόν, a hundred]. A prefix signifying one hundred.
**hectogram** (hek'-to-gram) [hecto-; γράμμα, an inscription]. One hundred grams, or 1543.2349 grains.
**hectoliter** (hek'-to-le-ter) [hecto-; λίτρα, a pound]. One hundred liters, equal to 22.009 imperial, or 26.4 United States gallons.

**hectometer** (*hek'-to-me-ter*) [*hecto-*; μέτρον, a measure]. One hundred meters, or 328 feet 1 inch.
**hedeoma** (*he-de-o'-mah*) [ἡδύs, sweet; ὀσμή, smell]. American pennyroyal. The leaves and tops of *H. pulegioides*, the properties of which are due to a volatile oil. It is stimulant, carminative, and emmenagogue. It is used in suppression of the menses, the flatulent colic of children, and, on account of its pungent odor, to drive off fleas and mosquitos. **h.**, **oil of** (*oleum hedeomæ*, U. S. P.). Dose 2–10 min. (0.13–0.65 Cc.).
**hedgehog crystals.** Crystals of ammonium urate, found as a urinary deposit in the form of globular crystals with spiny projections.
**hediosit** (*he'-de-o-sit*). A white crystalline, odorless powder with a sweet taste, easily soluble in water; used in the dietetic treatment of diabetes.
**hedonal** (*he'-don-al*). Methylpropylcarbinol urethane; recommended as a safe hypnotic in the milder forms of insomnia. Dose 20–45 gr. (1.33–2.9 Gm.).
**hedonia** (*he-do'-ne-ah*) [ἡδονή, pleasure]. Abnormal cheerfulness; amenomania.
**hedonism** (*he'-do-nizm*) [ἡδονή, pleasure]. The pursuit of pleasure; in psychic medicine, the unreasoning pursuit of some hobby or whim.
**hedrocele** (*hed'-ro-sēl*) [ἕδρα, breech; κήλη, hernia]. 1. A hernia through the notch of the ischium. 2. Prolapse of the anus.
**hedrosyrinx** (*hed-ro-si'-rinks*) [ἕδρα, anus, fundament; σῦριγξ, pipe]. Fistula in ano.
**heel** (*hēl*) [AS., *hēla*]. The hinder part of the foot. **h.-bone,** the os calcis.
**Hegar's method of diagnosing fibroma** (*ha'-gar*)- [Alfred *Hegar*, German gynecologist, 1830– ]. This consists in drawing downward the uterus with a volsellum while the finger is passed into the rectum and pressed against the tumor; if it is ovarian, it will be immovable; if uterine, there will be great resistance to drawing down the cervix. **H.'s sign. For the relief of cancer of the rectum:** the knife is carried along the sides of the sacrum, making the letter V. He hinges the flap with a chain-saw. **H.'s sign,** compressibility of the lower segment of the uterus and the upper half of the cervix, noticed on bimanual examination during the first two or three months of pregnancy.
**hegemony** (*he-gem'-on-e*) [ἡγεμών, a leader]. The supremacy of one function over a number of others.
**hegonon** (*heg'-on-on*). A substance obtained by treating silver ammonium nitrate with albumose, said to contain about 7 per cent. of organically combined silver. It is a light brown powder soluble in water, and is used as a substitute for silver nitrate.
**hegovia** (*he-go'-ve-ah*). A proprietary remedy for enuresis said to consist of salol, powdered snails, and lithium salicylate.
**Hehner's test for formaldehyde in milk** (*hā'-ner*). Place 5 Cc. of the milk in a test-tube and dilute with an equal volume of water. Carefully pour down the sides of the test-tube strong sulphuric acid containing a trace of ferric chloride, so as to form a layer of acid below the milk. In the presence of formaldehyde a violet ring is formed at the junction of the two liquids; 1 part in 100,000 may be detected. The acid should be of 1.81 to 1.83 sp. gr., and must contain a trace of ferric salt. The charring due to the action of the acid on the milk must not be mist taken for color-reaction.
**Heidenhain's demilunes** (*hi'-den-hīnz dem'-e-lūn*) [Rudolf *Heidenhain*, German physiologist, 1834–1897]. Crescentic bodies lying between the cells and the membrana propria of an acinus of a salivary gland. **H.'s rods, H.'s striæ,** the slender columnar cells of the uriniferous tubules.
**Heim-Kreyssig's sign** (*hīm-kri'-zig*). See *Kreyssig's sign*.
**(von) Heine's infantile paralysis.** Spastic spinal paralysis of infancy.
**Heinecke-Mikulicz operation**(*hi'-neh-eh-mik'-oo-litz*) [Walter Hermann *Heinecke*, German surgeon, 1834– ; Johann von *Mikulicz*, Austrian surgeon, 1850–1905]. Pyloroplasty.
**Heisrath's operation** (*hīs'-rath*). Excision of the tarsus and conjunctiva in cases of trachoma of long standing.
**Heister's diverticulum** (*hi'-ster*) [Lorenz *Heister*, German anatomist, 1683–1758]. The sinus of the jugular vein. **H.'s valves,** the transverse valvular folds of the cystic duct.

**helcodermatosis** (*hel-ko-der-mat-o'-sis*) [*helcoma*; δέρμα, skin]. Skin disease with the formation of ulcers.
**helcoid** (*hel'-koid*) [*helcoma*; εἶδος, likeness]. Resembling an ulcer.
**helcology** (*hel-kol'-o-je*) [ἕλκος, ulcer; λόγος, science]. The pathology and treatment of ulcers.
**helcoma** (*hel-ko'-mah*) [ἕλκος, an ulcer]. An ulcer.
**helcomenia** (*hel-ko-me'-ne-ah*) [ἕλκος, ulcer; μήν, month]. Vicarious menstrual discharge from an ulcer.
**helcoplasty** (*hel'-ko-plas-te*) [*helcoma*; πλάσσειν, to form]. The treatment of ulcers by skin-grafting.
**helcopoiesis** (*hel-ko-poi-e'-sis*) [ἕλκος, ulcer; ποίησις, making]. The surgical formation of an issue, for counter-irritation.
**helcosis** (*hel-ko'-sis*) [ἕλκωσις, ulceration]. The formation and development of an ulcer.
**helcosol** (*hel'-ko-sol*). See *bismuth pyrogallate*.
**Helcosoma tropicum** (*hel-ko-so'-mah trop'-ik-um*) [ἕλκος, ulcer; σῶμα, a body]. A name proposed by Wright for the protozoan parasite of Delhi boil.
**helcotic** (*hel-kot'-ik*) [ἕλκος, an ulcer]. Ulcerative; of the nature of or accompanied by ulceration.
**helenium** (*hel'-en-in*), C₅H₈O. A stearoptene from *Inula helenium*; it is used as an internal and external antiseptic. Dose ⅓–⅔ gr. (0.011–0.022 Gm.).
**helexin** (*hel-eks'-in*). A glucoside, C₂₆H₄₄O₁₁, from *Hedera helix*.
**heliciform** (*hel-is'-e-form*) [*helix*; forma, form]. Spiral; shaped like a snail-shell.
**helicina** (*hel-is'-in-ah*). A mixture of snail mucus and sugar; a white powder, soluble in water, and used as a pectoral remedy. Syn., *saccharated snail-juice*.
**helicine** (*hel'-is-in*) [*helix*]. 1. Spiral in structure. 2. Pertaining to the helix. **h. arteries,** spirally winding arteries supplying the erectile tissue of the penis.
**helicis** (*hel'-is-is*) [*helix*]. Muscle of the helix of the ear. See *muscles, table of*.
**helicoid** (*hel'-ik-oid*) [*helix*; εἶδος, form]. Spiral; coiled like a snail-shell.
**helicopepsin** (*hel-ik-o-pep'-sin*) [ἕλιξ, a spirally coiled snail; πέψις, digestion]. A peptic ferment found by Krukenberg in snails.
**helicopod** (*hel'-ik-o-pod*) [ἕλιξ, a spirally coiled snail; πούς, foot]. A dragging gait in which the foot describes a partial curve.
**helicoprotein** (*hel-ik-o-pro'-te-in*). A phosphoglycoprotein obtained from the glands of the snail, *Helix pomatia*. It is converted by action of alkalies into a gummy, levorotatory carbohydrate called *animal sinistrin*.
**helicotrema** (*hel-ik-o-tre'-mah*) [*helix*; τρῆμα, h₀le]. The opening connecting the scalæ tympani and vestibuli of the spiral canal of the cochlea.
**heliencephalitis** (*he-le-en-sef-al-i-tis*) [ἥλιος, sun; ἐγκέφαλος, brain; ιτις, inflammation]. Encephalitis caused by exposure to the sun's rays.
**Heliodorus' bandage** (*hel-e-o-dōr'-us*) [*Heliodorus*, a Roman surgeon, 1st century]. The T-bandage.
**heliomyelitis** (*he-le-o-mi-el-i'-tis*) [*helios*; μυελόs, marrow; ιτις, inflammation]. Myelitis caused by exposure to the sun's rays.
**helionosus** (*he-le-on'-o-sus*) [*helios*; νόσος, disease]. Sunstroke.
**heliophag** (*he'-le-o-fag*) [*helios*; φαγεῖν, to devour]. A name given to the animal pigment-cell, as being a supposed absorber of the radiant energy of the sun's light and heat.
**heliophilia** (*he-le-of-il'-e-ah*) [ἥλιος, sun; φιλεῖν, to love]. Morbid affinity for the sunlight, resulting in ectsacy and muscular contraction.
**heliophobe** (*he'-le-o-fōb*) [ἥλιος, sun; φόβος, fear]. One who is morbidly sensitive to the effects of the sun's rays.
**heliophobia** (*he-le-o-fo'-be-ah*) [ἥλιος, sun; φόβος, fear]. Morbid fear of exposure to the sun's rays.
**heliosin** (*hel'-e-o'-sin*). An antisyphilitic mixture of various inorganic salts with keratin.
**heliosis** (*he-le-o'-sis*) [ἥλιος, sun]. A sun-bath; also, sunstroke.
**heliostat** (*he'-le-o-stat*) [ἥλιος, sun; στατός, fixed]. A mirror moved by clockwork in such a manner as to reflect continuously the sun's rays in a fixed direction.
**heliotherapy** (*he-le-o-ther'-ap-e*) [ἥλιος, sun; θεραπεία, treatment]. The treatment of disease by exposure of the body to sunlight. Sun-bathing.
**heliotropic** (*he-le-ot'-rop-ik*) [ἥλιος, sun; τρέπειν, to turn]. Relating to the movements of protoplasm under the influence of light.

**heliotropin** (he-le-ot'-ro-pin) [ἥλιος, the sun; τρέπειν, to turn]. 1. See *piperonal.* 2. A bitter, volatile, crystalline, poisonous principle from *Heliotropium europæum,* a European species of heliotrope. Its action is little known.

**heliotropism** (he-le-ot'-ro-pizm) [ἥλιος, the sun; τρέπειν, to turn]. In biology, that property of a plant or plant-organ by virtue of which it bends toward or away from the sunlight.

**heliotropy** (he-le-ot'-ro-pe) [ἥλιος, the sun; τροπή, a turning]. Same as *heliotropism.*

**helium** (he'-le-um) [ἥλιος, sun]. A gaseous body, a supposed atmospheric element, boiling below 264° C., which has resisted all attempts to liquefy it. It forms compounds with hydrogen, carbureted hydrogen, and nitrogen. Symbol He; atomic weight 3.99.

**helix** (he'-liks) [ἕλιξ, a spiral]. 1. The rounded, convex margin of the pinna of the ear. 2. A coil of wire as that of an electromagnet.

**hellebore, helleborus** (hel'-e-bor, hel-leb'-or-us) [ἐλλέβορος, hellebore]. A plant of the genus *Helleborus,* particularly *H. niger,* black hellebore, the root of which contains two glucosides, *helleborin,* $C_{36}H_{42}O_6$, and *helleborein,* $C_{37}H_{40}O_{15}$, to which its properties are due. It is a drastic hydragogue, cathartic, and an emmenagogue, and has been used as a drastic purge in insanity, dropsy, and amenorrhea. Dose of the *powdered root,* as a purge, 10–20 gr. (0.65–1.3 Gm.). **h.,** white. See *veratrum.*

**helleborein** (hel-eb-or'-e-in). A poisonous glucoside, $C_{37}H_{40}O_{15}$, from *Helleborus niger* and *viridis.*

**helleborin,** (hel-eb'-or-in), $C_{36}H_{42}O_6$. A poisonous glucoside from black hellebore.

**helleborism** (hel'-eb-or-izm). 1. The treatment of disease with hellebore. 2. The morbid condition induced by the free exhibition of hellebore.

**Heller's plexus** (hel'-er) [Johann Florian *Heller,* Austrian physician, 1813–1871]. The network of arteries in the deeper layer of the intestinal submucosa. **H.'s test].** 1. A test for albumin in the urine. A little nitric acid is placed in a test-tube and the urine allowed carefully to flow down the side of the tube, so as to form a layer on the acid without mixing. The development of an opaque white ring indicates albumin. 2. A test for the presence of blood-coloring-matter in the urine. The urine is boiled with half its volume of caustic potash, whereby the phosphates are precipitated. The precipitate is colored red if blood is present. 3. For sugar in the urine, see *Moore's test.*

**Hellmund's ointment** (hel'-münt). A narcotic ointment composed of acetate of lead, 10 parts; extract of conium, 30 parts; balsam of Peru, 30 parts; Sydenham's laudanum, 5 parts and cerate, 240 parts.

**Helmerich's ointment** (hel'-mer-ik). An ointment used in the treatment of scabies. It consists of sublimated sulphur, 1 dram; potassium carbonate, ½ dram; and lard, 6½ drams. The ointment should remain in contact with the diseased surface for four or five hours.

**Helmholtz's ligament** (helm'-hōlz) [Hermann Ludwig Ferdinand von *Helmholtz,* German physiologist, 1821–1894]. The anterior ligament of the malleus that encircles the long process of the latter and is inserted into the anterior part of its neck and head. **H.'s line,** the line perpendicular to the plane of the axis of rotation of the eyeballs. **H.'s theory of color-vision.** See *Young-Helmholtz.*

**helminth** (hel'-minth) [ἕλμινς, a worm]. 1. A worm. 2. An intestinal worm.

**helminthagogue, helminthagog** (hel-minth'-ag-og). See *anthelmintic.*

**helminthiasis** (hel-min-thi'-as-is) [helminth]. The diseased condition produced by the presence of worms in the body. **h. elastica,** elastic tumors of the axillæ and groins due to filaria.

**helminthicide** (hel-minth'-is-id) [helminth; cædere, to kill]. See *vermicide.*

**helminthic** (hel-min'-thik). See *anthelmintic.*

**helminthism** (hel-minth'-izm). [helminth]. The existence of intestinal worms in the body.

**helminthochorton** (hel-minth-o-kor'-ton) [helminth; χόρτος, grass]. Corsican moss.

**helminthogenesis** (hel-min-tho-jen'-e-sis). The same as *helminthiasis.*

**helminthoid** (hel-min'-thoid) [helminth; εἶδος, likeness]. Pertaining to or shaped like a worm.

**helminthology** (hel-min-thol'-o-je) [helminth; λόγος, science]. The science of worms, especially those parasitic within the body.

**helminthoma** (hel-min-tho'-mah) [helminth; ὄμα, a tumor: *pl., helminthomata*]. A tumor caused by the presence of a parasitic worm. See *Bulam boil* and *Guinea-worm.*

**helminthoncus** (hel-minth-ong'-kus) [helminth; ὄγκος, a tumor]. An old term for a parasitic skin-disease.

**helminthophobia** (hel-min-tho-fo'-be-ah) [helminth; φόβος, fear]. A nervous state produced by the presence or thought of parasitic worms.

**helminthous** (hel-min'-thus) [helminth]. Wormy.

**helmitol** (hel'-mit-ol). Hexamethylenetetramine anhydromethylene citrate, an analgesic and urinary antiseptic. Dose 10–15 gr. (0.64–0.97 Gm.) 3 times daily in a wineglassful of water.

**Heloderma** (he-lo-der'-mah) [ἥλος, nail; δέρμα, skin]. A genus of lizards. **H. horridum,** of Mexico, and **H. suspectum,** of Arizona (called *Gila Monster*), are said to be the only known species of venomous lizards.

**helodermatous** (he-lo-der'-mat-us) [ἥλος, a nail, wart; δέρμα, skin]. In biology, having a warty or tuberculous skin.

**helodes** (he-lo'-dēz) [ἕλος, a swamp]. 1. Swampy, or marshy. 2. A fever attended with profuse sweating. 3. Marsh-fever.

**helonin** (hel'-o-nin) [ἕλος, a marsh]. A crude precipitate from the tincture of *Chamælirion carolinianum* or *Helonias dioica.* It is tonic anthelmintic, and diuretic. Caution should be observed in its use. Dose, gr. ij-iv.

**helophilous** (hel-of'-il-us) [ἕλος, a marsh; φίλος, loving]. Inhabiting marshes.

**helopyra** (hel-o-pi'-rah) [ἕλος, marsh; πῦρ, fever]. Malarial fever.

**helotic** (hel-ot'-ik) [ἥλος, a nail]. 1. Relating to corns. 2. A vesicant.

**helthin** (hel'-thin). An acidulated solution of sodium parasulphanilate and of sodium or potassium amidonaphthol disulphonate. It is used as a test for nitrites in potable waters.

**Helweg's triangular bundle** (hel'-veg) [Hans Kristian Saxtorph *Helweg,* Danish physician, 1847– ]. The triangular or olivary tract situated in the ventral part of the anterolateral column of the spinal cord.

**hem-, heme-, hemato-** (hem-, hem-ah-, hem-at-o-) [αἷμα, blood]. Prefixes signifying of or pertaining to the blood.

**hemabarometer** (hem-ab-ar-om'-et-er) [αἷμα, blood; *barometer*]. An instrument for the determination of the specific gravity of the blood.

**hemaboloids** (hem-ab'-ol-oids). A proprietary article said to contain a vegetable iron with peptone, bone-marrow and nuclein.

**hemacelinosis** (hem-as-el-in-o'-sis) [hema-; κηλίς, spot; νόσος, disease]. A synonym of the disease, *purpura.*

**hemachroin** (hem-ak-ro'-in) [hema-; χρόα, color]. Same as *hemochroin.*

**hemachromatosis** (hem-ak-rōm-at-o'-sis) [see *hemachrome*]. General hematogenous pigmentation.

**hemachrome** (hem'-ak-rōm) [hema-; χρῶμα, color]. The coloring-matter of the blood; hematin.

**hemachrosis** (hem-ak-ro'-sis) [hema-; χρῶσις, coloring]. 1. Redness of the blood. 2. Any disease in which the blood is abnormally colored.

**hemacyanin** (hem-as-i'-an-in) [hema-; κύανος, blue]. A blue coloring-matter found in the blood and the bile.

**hemacyte** (hem'-as-it). See *hemocyte.*

**hemacytometer** (hem-as-i-tom'-et-er): See *hemocytometer.*

**hemacytozoon** (hem-a-si-to-zo'-on) [hema-; κύτος, cell; ζῷον, animal: *pl., hemacytozoa*]. A protozoon found in the red blood corpuscles.

**hemad** (hem'-ad) [hema-; ad, toward]. 1. Toward the hemal aspect; opposed to *neurad.* 2. A blood-cell or blood corpuscle.

**hemadenology** (hem-ad-en-ol'-o-je) [hem-; ἀδήν, a gland; λόγος, a discourse]. The study of the ductless glands, and their diseases.

**hemadonosos** (hem-ad-on'-o-sos) [hema-; νόσος, disease]. A disease of the blood or of the blood-vessels.

**hemadostenosis** (hem-ad-o-ste-no'-sis) [αἱμάς, bloodstream; στενός, narrow]. Stricture or narrowing of a blood-vessel.

# HEMADOSTEOSIS 413 HEMATINIC

**hemadosteosis** (hem-ad-os-te-o'-sis) [αἱμάς, bloodstream; ὀστέον, bone]. Ossification or calcification of blood-vessels.

**hemadromograph** (hem-ad-rom'-o-graf) [hema-; δρόμος, course; γράφειν, to write]. An instrument for registering changes in the velocity of the bloodstream.

**hemadromometer** (hem-a-dro-mom'-et-er) [hema-; δρόμος, course; μέτρον, a measure]. An instrument for measuring the velocity of the blood-current.

**hemadromometry** (hem-ad-ro-mom'-et-re) [hema-; δρόμος, course; μέτρον, measure]. Measurement of the speed of the blood-current.

**hemadynamics** (hem-ad-i-nam'-iks) [hema-; δύναμις, power]. The science pertaining to the movements involved in the circulation of the blood.

**hemadynamometer** (hem-a-di-na-mom'-et-er) [hema-; δύναμις, strength; μέτρον, a measure]. An instrument for measuring the tension or pressure of blood within the arteries.

**hemafacient** (hem-a-fa'-she-ent) [hema-; facere, to make]. An agent that increases the quantity and quality of the blood.

**hemafecal** (hem-af-e'-kal) [hema-; fæx, dregs]. Characterized by bloody stools. h. jaundice. See jaundice.

**hemagglutination, hemoagglutination** (hem-ag-loo-tin-a'-shun, hem-o-ag-loo-tin-a'-shun) [see hemagglutinins]. The clumping of red blood-corpuscles.

**hemagglutinins, hemoagglutinins** (hem-ag-loo'-tin-ins, hem-o-ag-loo'-tin-ins) [hema-; agglutinin]. Agglutinins which have the power to clump red blood-corpuscles. Syn., erythroagglutinins.

**hemagogue, hemagog** (hem'-ag-og) [hema-; ἀγωγός, leading]. 1. Promoting the menstrual or hemorrhoidal discharge of blood. 2. An agent that promotes the catamenial or hemorrhoidal flow of blood.

**hemal** (he'-mal) [αἷμα, blood]. 1. Pertaining to the blood or vascular system. 2. Pertaining to the ventral aspect of the body, that part containing the heart and blood-vessels. h. arch, the arch formed by the ribs, sternum, and vertebral bodies. h. spine, the sternum or linea alba.

**hemalbumin, hæmalbumin** (hem-al'-bū-min) [hema-; albumin]. 1. A predigested iron albuminate used in anemic conditions. Dose 15 gr. (1 Gm.) several times daily. 2. A preparation of the salts and albuminoid constituents of the blood. 3. A preparation of iron containing hematin, hemoglobin, serum-albumin, paraglobulin, and inorganic constituents of the blood.

**hemaleucin** (hem-al-oo'-sin) [hema-; λευκός, white]. Fibrin; the white portion of a washed blood-clot.

**hemaléukosis** (hem-al-oo-ko'-sis) [hema-; λευκός, white]. The formation of the buffy coat of a clot.

**hemalopia** (hem-al-o'-pe-ah) [hema-; ὤψ, eye]. Effusion of blood in the eye; erythropsia.

**hemalum** (hem-al'-um). A stain for bone-tissue consisting of hematoxylin and alum.

**hemameba, hæmamœba** (hem-am-e'-bah) [hema-; ameba: pl., hemamebæ]. 1. A white blood-cell. 2. A parasitic amœboid microorganism of the blood, as the malarial parasite, **hæmamœba leukemiæ magna**, h. **leukemiæ parva**, bodies once supposed to be of protozoan nature and specific causes of leukemia, shown by Türck to be artefacts resulting from the action of a basic dye upon the mast-cell granules.

**hemamebiasis** (hem-am-e-bi'-as-is). The disease or condition due to infection with hemamebæ.

**hemanalysis** (hem-an-al'-is-is) [hema-; analysis]. Analysis of the blood.

**hemangioendothelioma** (hem-an-je-o-en-do-the-le-o'-mah) [hema-; ἀγγεῖον, vessel; endothelioma]. Epithelial hyperplasia of the capillaries.

**hemangioma** (hem-an-je-o'-mah) [hema-; ἀγγεῖον, vessel; ὄμα, a tumor: pl., hemangiomata]. An angioma made up of blood-vessels.

**hemangiomatosis** (hem-an-je-o-ma-to'-sis). The condition characterized by the presence of multiple hemangiomata.

**hemangiosarcoma** (hem-an-je-o-sar-ko'-mah) [hema-; ἀγγεῖον, vessel; sarcoma]. A vascular sarcoma.

**hemanthine** (hem-an'-thin) [hema-; ἄνθος, a flower]. An alkaloid from Hemanthus coccineus; it is poisonous, with the general properties of atropine.

**Hemanthus** (hem-an'-thus) [hema-; ἄνθος, flower]. A genus of amaryllidaceous plants; blood-flower. **H. coccineus**, of S. Africa, affords an arrow-poison, with the general properties of atropine. The plant is a diuretic and cardiant. It affords hemanthine.

**hemanutrid** (hem-an-ū'-trid). A liquid preparation of hemoglobin, 70 %; glycerol, 20 %; brandy, 10 %.

**hemaphein** (hem-af-e'-in) [hema-; φαιός, dusky]. A brown coloring-matter from blood, regarded as a decomposition-product of hematin.

**hemapheism, hæmaphæism** (hem-af'-e-izm) [hema-; φαιός, dusky]. The passage of reddish-amber colored urine, combined with hepatic disorder.

**hemaphobia** (hem-af-o'-be-ah). See hematophobia.

**hemapoiesis** (hem-ap-oi-e'-sis). See hematopoiesis.

**hemapoietic** (hem-ap-oi-et'-ik). See hematopoietic.

**hemapophysis** (hem-ap-off'-is-is) [hema-; ἀποφύειν, to put forth]. That part of an ideal or perfect vertebra which forms the antero-lateral part of the hemal arch. In man, all the hemapophyses are either cartilaginous or detached.

**hemarthrosis** (hem-ar-thro'-sis) [hema-; ἄρθρον, a joint]. Effusion of blood into a joint.

**hemastatic** (hem-as-tat'-ik). See hemostatic.

**hemastatics** (hem-as-tat'-iks) [hema-; στατικός, standing]. That branch of physiology treating of the laws of the equilibrium of the blood.

**hemasthenosis** (hem-as-then-o'-sis) [hema-; ἀσθένεια, weakness]. A weakening or deterioration of the blood.

**hematachometer** (hem-at-ak-om'-et-er). See hemotachometer.

**hematalloscopy** (hem-at-al-os'-ko-pe) [hema-; ἄλλος, other; σκοπεῖν, to examine]. In medical jurisprudence the examination of the blood to distinguish one kind from another.

**hematangiosus** (hem-at-an-je-ōn'-o-sus) [hema-; ἀγγεῖον, vessel; νόσος, disease]. Disease of bloodvessels.

**hematapostasis** (hem-at-ap-os'-tas-is) [hema-; ἀπόστασις, a standing away from]. Unequal distribution or pressure of blood, with congestion or effusion in some part of the body.

**hematapostema** (hem-at-ap-os-te'-mah) [hema-; ἀπόστημα, abscess: pl., hemapostemata]. An abscess containing extravasated blood.

**hematedema** (hem-at-e-de'-mah) [hema-; οἴδημα, a swelling]. Swelling due to the effusion of blood.

**hematein** (hem-at'-e-in) [αἷμα, blood], C₁₆H₁₂O₆. A crystalline principle derived from, and nearly convertible into, hematoxylin. h.-ammonium, C₁₆H₁₁-O₆. NH₄+4H₂O, a violet-black granular powder, purple in aqueous solution, brown-red in alcoholic solution, used as a stain.

**hemateleum** (hem-at-el'-e-um) [hema-; ἔλαιον, oil]. A yellow or brownish oily fluid obtained by the dry distillation of blood.

**hematemesis** (hem-at-em'-es-is) [hema-; emesis]. The vomiting of blood.

**hematencephalon** (hem-at-en-sef'-al-on) [hema-; ἐγκέφαλος, brain]. A hemorrhage or bleeding within the brain; cerebral apoplexy.

**hematherapy** (hem-ah-ther'-ap-e) [hema-; θεραπεία, therapy]. 1. The therapeutic use of prepared arterial blood of bullocks. 2. Treatment applied to diseases of the blood.

**hemathermous** (hem-ath-er'-mus) [hema-; θερμός, hot]. Having warm blood.

**hemathorax** (hem-ah-tho'-raks). See hemothorax.

**hematic** (hem-at'-ik) [hema-]. Pertaining to, full of, or having the color of, blood. Also, a blood-tonic.

**hematidrosis** (hem-at-id-ro'-sis) [hema-; ἱδρώς, sweat]. A sweating of blood.

**hematimeter** (hem-at-im'-et-er) [hema-; μέτρον, a measure]. An instrument for counting the corpuscles in a given volume of blood.

**hematimetry** (hem-at-im'-et-re) [hema-; μέτρον, measure]. The estimation of the number or proportion of the blood-corpuscles, as by the hemocytometer or hematimeter.

**hematin** (hem'-at-in) [αἷμα, blood], C₃₄H₃₄N₄FeO₅. A decomposition-product of hemoglobin. It is the bluish-black, amorphous, contains iron, and is soluble in dilute alkalies and acids, insoluble in water, in alcohol, and in ether. h.-albumin, a fine brown-red, tasteless, odorless powder obtained by drying blood fibrin. Dose, in anemia, 1-2 teaspoonfuls 3 times daily. h., reduced, hemochromogen. h., vegetable, aspergillin.

**hematinemia, hematinæmia** (hem-at-in-e'-me-ah) [hema-; anemia]. The presence of hematin in the blood.

**hematinic** (hem-at-in'-ik) [hematin]. 1. Same as hematic. 2. Relating to hematin. 3. An agent which tends to increase the proportion of hematin or coloring-matter in the blood.

**hematinometer** (hem-at-in-om'-et-er). Same as hemoglobinometer.
**hematinuria** (hem-at-in-u'-re-ah). Same as hemoglobinuria.
**hemato-** (hem-at-o-) [αἷμα, blood]. A prefix signifying of or pertaining to the blood.
**hematoaerometer** (hem-at-o-a-e-rom'-et-ur) [hemato-; ἀήρ, air; μέτρον, measure]. An instrument for recording the pressure of the gases in the blood.
**hematobious** (hem-at-o'-be-us) [hemato-; βίος, life]. Living in the blood.
**hematobium** (hem-at-o'-be-um) [hemato-; βίος, life]. 1. A blood-corpuscle. 2. A blood-parasite, hematozoon.
**hematoblast** (hem'-at-o-blast) [hemato-; βλαστός, a germ]. A blood-plate; an immature red blood-corpuscle.
**hematocatharsis** (hem-at-o-kath-ar'-sis) [hemato-; κάθαρσις, a cleansing]. The process of expelling toxic substances from the blood.
**hematocathartic** (hem-at-o-kath-ar'-tik) [hemato-; καθαρτικός, cleansing]. 1. Purifying the blood. 2. Any remedy that purifies the blood.
**hematocele** (hem'-at-o-sēl) [hemato-; κήλη, a tumor]. A tumor formed by the extravasation and collection of blood in a part, especially in the tunica vaginalis testis or in the pelvic cavity (pelvic hematocele).
**hematocelia** (hem-at-o-se'-le-ah) [hemato-; κοιλία, a cavity]. An effusion of blood into the peritoneal cavity.
**hematocephalus** (hem-at-o-sef'-al-us) [hemato-; κεφαλή, head]. 1. An effusion of blood into the brain. 2. A monstrosity characterized by an effusion of blood into the cerebral hemispheres.
**hematochezia** (hem-at-o-ke'-ze-ah) [hemato-; χέζειν, to defecate]. The passage of bloody stools.
**hematochlorin** (hem-at-o-klo'-rin) [hemato-; χλωρός, green]. An amorphous green pigment contained in the marginal zone of the placenta.
**hematochrosis** (hem-at-o-kro'-sis) [hemato-; χρῶσις, coloring; pl., hematochroses]. 1. Any disease characterized by discoloration of the skin. 2. A discoloration of the skin.
**hematochyluria** (hem-at-o-ki-lū'-re-ah) [hemato-; χυλός, chyle; οὖρον, urine]. The presence of blood and chylous material in the urine.
**hematocolpos** (hem-at-o-kol'-pos) [hemato-; κόλπος, vagina]. A collection of blood within the vagina.
**hematocrit** (hem'-at-o-krit) [hemato-; κρίνειν, to separate; to judge]. An instrument for making volumetric estimation of the blood-corpuscles by separating, by centrifugal action, the corpuscles from the plasma.
**hematocryal** (hem-at-ok'-re-al) [hemato-; κρύος, cold]. In biology, of or pertaining to the cold-blooded invertebrates.
**hematocrystallin** (hem-at-o-kris'-tal-in). Same as hemoglobin.
**hematocyanosis** (hem-at-o-si-an-o'-sis). Synonym of cyanosis.
**hematocyst** (hem'-at-o-sist) [hemato-; κύστις, bladder]. 1. A cyst containing blood. 2. An effusion of blood into the bladder.
**hematocyte** (hem'-at-o-sīt) [hemato-; κύτος, cell]. A blood-corpuscle.
**hematocytolysis**. See hemocytolysis.
**hematocytometer** (hem-at-o-si-tom'-et-er). See hemocytometer.
**hematocytozoon** (hem-a-to-si-to-zo'-on) [hemato-; κύτος, cell; ζῷον, animal]. A protozoan parasite inhabiting the red blood-corpuscles.
**hematocyturia** (hem-at-o-si-tū'-re-ah) [hemato-; κύτος, cell; οὖρον, urine]. The presence of blood-cells in the urine.
**hematōdes** (hem-at-ō'-dēs) [αἱματώδης, bloody]. Bloody; gorged with or appearing like blood.
**hematodiarrhea** (hem-at-o-di-ar-e'-ah). Synonym of dysentery.
**hematodynamics** (hem-at-o-di-nam'-iks). See hemadynamics.
**hematodynamometer** (hem-at-o-di-nam-om'-et-er). See hemadynamometer.
**hematodyscrasia** (hem-at-o-dis-kra'-se-ah) [hemato-; δυσκρασία, bad temperament]. A diseased or dyscrasic state of the blood.
**hematogaster** (hem-at-o-gas'-ter) [hemato-; γαστήρ, stomach]. Extravasation of blood into the stomach.
**hematogen** (hem-at'-o-jen) [hemato-; γεννᾶν, to produce]. 1. A nucleoalbuminoid preparation of iron 0.3 %. 2. Defibrinated blood with minute percentage of creosote, containing 0.3 % of iron and mixed with glycerol and wine; used in anemia. Dose 1–2 tablespoonfuls. 3. A yellowish powder containing 7 % of iron or a liquid formed by adding ferric citrate and acetic acid to an alkaline solution of albumin. Dose of liquid 1–4 teaspoonfuls. 4. A decomposition-product of vitellin.
**hematogenesis** (hem-at-o-jen'-es-is) [hemato-; γεννᾶν, to produce]. The development of blood or blood-corpuscles.
**hematogenic** (hem-at-o-jen'-ik) [see hematogenesis]. Pertaining to the formation of blood.
**hematogenous** (hem-at-oj'-en-us) [see hematogenesis]. Derived from or having origin in, the blood.
**hematoglobin** (hem-at-o-glo'-bin). See hemoglobin.
**hematoglobulin** (hem-at-o-glob'-ū-lin). Same as hemoglobin.
**hematography** (hem-at-og'-ra-fe) [hemato-; γράφειν, to write]. A description of the blood; hematology.
**hematohidrosis** (hem-at-o-hid-ro'-sis). See hematidrosis.
**hematohiston** (hem-at-o-his'-ton). See globin.
**hematoid** (hem'-at-oid) [hemato-; εἶδος, likeness]. Resembling blood.
**hematoidin** (hem-at-oi'-din) [see hematoid], C₁₆H₁₈N O₃. An iron-free derivative of hemoglobin, occurring in old blood-clots as yellowish-brown rhombohedral crystals.
**hematokolpos** (hem-at-o-kol'-pos). See hematocolpos.
**hematokrit** (hem'-at-o-krit). See hematocrit.
**hematol** (hem'-at-ol). A sterilized hemoglobin mixed with glycerol and brandy.
**hematolin** (hem-at'-o-lin) [hemato-], C₄₈H₇₈N₈O₇. An iron-free derivative of hematin.
**hematologist** (hem-at-ol'-o-jist) [see hematology]. One who makes a special study of the blood and is skilled in the technic of blood-examinations.
**hematology** (hem-at-ol'-o-je) [hemato-; λόγος, science]. The science of the blood, its nature, functions, and diseases.
**hematolymphangioma** (hem-at-o-limf-an-je-o'-mah) [hemato-; lymph; ἀγγεῖον, a vessel; ὄμα, a tumor]. A tumor involving blood-vessels and lymph-vessels.
**hematolysis** (hem-at-ol'-is-is) [hemato-; λύσις, a solution]. 1. Destruction or disorganization of the blood or of the corpuscles. 2. Diminished coagulability of the blood.
**hematolytic** (hem-at-o-lit'-ik) [see hematolysis]. Marked by or tending to blood-impoverishment.
**h. serum**. See serum.
**hematoma** (hem-at-o'-mah) [hemato-; ὄμα, tumor]. A tumor or swelling containing blood. **hæmatoma auris**, insane ear; an effusion of blood or serum between the cartilage of the ear and its covering, occurring in various forms of insanity as the result of injuries or trophic changes. **h. of the dura mater**, an effusion of blood beneath the dura mater, forming membranous layers. **h., pelvic**, an effusion of blood into the cellular tissue of the pelvis. **h.s, valve-**, of the newborn, those due to imperfect development leading to the formation of clefts; they are not produced by hemorrhage and have no relation to the minute hemorrhages found beneath the pericardium and endocardium in cases of death from suffocation.
**hematomancy**, **hematomantia** (hem-at-o-man-se, hem-at-o-man'-she-ah) [hemato-; μαντεία, divination]. The arriving at a diagnosis from examination of the blood.
**hematomatous** (hem-at-o'-mat-us) [hemato-; ὄμα, a tumor]. Relating to or of the nature of a hematoma.
**hematomediastinum** (hem-at-o-me-de-as-ti'-num) [hemato-; mediastinum, the mediastinum]. An effusion of blood into the mediastinal spaces.
**hematometer** (hem-at-om'-et-er). An instrument to estimate the properties or constituents of blood. See hemodynamometer; hemoglobinometer.
**hematometra** (hem-at-o-me'-trah) [hemato-; μήτρα, uterus]. An accumulation of blood within the uterine cavity.
**hematometry** (hem-at-om'-et-re) [hemato-; μέτρον, measure]. The estimation of the number and kind of corpuscles and the quantity of hemoglobin in the blood.
**Hematomonas** (hem-at-om'-on-as) [hemato-; μονάς, monad]. A genus of protozoan parasites inhabiting the blood.
**hematomphalocele** (hem-at-om-fal'-o-sēl) [hemato-; ὀμφαλός, navel; κήλη, hernia]. An umbilical hernia distended with blood.

**hematomyces** (*hem-at-om'-is-ēz*) [*hemato-*; μύκης, a fungus]. A bleeding variety of encephaloid cancer; medullary sarcoma. Syn., *fungus hæmalodes*.

**hematomyelia** (*hem-at-o-mi-e'-le-ah*) [*hemato-*; μυελός, marrow]. Hemorrhage into the spinal cord; an accumulation of blood in the central canal of the spinal cord.

**hematomyelitis** (*hem-at-o-mi-el-i'-tis*) [*hemato-*; *myelitis*]. An acute myelitis attended with an effusion of blood into the spinal cord.

**hematomyelopore** (*hem-at-o-mi'-el-o-por*) [*hemato-*; μυελός, marrow; πόρος, pore]. A cavity in the substance of the myelon resulting from hemorrhage.

**hematoncus** (*hem-at-ong'-kus*) [*hemato-*; ὄγκος, tumor]. Blood-tumor; hemangioma; hematoma.

**hematopathology** (*hem-at-o-path-ol'-o-je*) [*hemato-*; *pathology*]. The science dealing with morbid states of the blood.

**hematopathy** (*hem-at-op'-ath-e*) [*hemato-*; πάθος, disease]. Any disease of the blood.

**hematopedesis** (*hem-at-o-ped-e'-sis*) [*hemato-*; πήδησις, a leaping]. Cutaneous hemorrhage; hematidrosis.

**hematopericardium** (*hem-at-o-per-ik-ar'-de-um*) [*hemato-*; *pericardium*]. An effusion of blood into the pericardium.

**hematopexin** (*hem-at-o-peks'-in*). See *hemopexin*.

**hematopexis** (*hem-at-o-pek'-sis*). Coagulation of the blood.

**hematophagous** (*hem-at-off'-ag-us*) [*hemato-*; φαγεῖν, to eat]. Feeding on blood; blood-sucking.

**hematophilia** (*hem-at-o-fil'-e-ah*). See *hemophilia*.

**hematophobia** (*hem-at-o-fo'-be-ah*) [*hemato-*; φόβος, fear]. Morbid dread of the sight of blood.

**hematophore** (*hem'-at-o-for*) [*hemato-*; φέρειν, to bear]. An instrument used in the transfusion of blood.

**hematophthalmia** (*hem-at-off-thal'-me-ah*). See *hemophthalmia*.

**hematophyte** (*hem'-at-o-fīt*) [*hemato-*; φυτόν, a plant]. A vegetable organism, such as a bacterium, living in the blood.

**hematopinax** (*hem-at-op'-in-aks*) [*hemato-*; πίναξ, tablet: *pl.*, *hematopinaces*]. A blood-plaque.

**hematopisis** (*hem-at-op'-is-is*) [*hemato-*; πνεῖν, to drink]. 1. The drinking of blood. 2. A morbid collection of blood in any cavity of the body.

**hematoplanesis** (*hem-at-o-plan-e'-sis*) [*hemato-*; πλάνησις, wandering]. See *hematoplania*.

**hematoplania** (*hem-at-o-pla'-ne-ah*) [*hemato-*; πλάνη, a wandering]. Vicarious or aberrant course or flow of the blood.

**hematoplasma** (*hem-at-o-plaz'-mah*) [*hemato-*; *plasma*]. The plasma of the blood.

**hematoplast** (*hem'-at-o-plast*). Same as *hematoblast*.

**hematoplastic** (*hem-at-o-plas'-tik*) [*hemato-*; πλαστικός, plastic]. Blood-forming.

**hematoplethora** (*hem-at-o-pleth'-or-ah*). Synonym of *plethora*.

**hematopneumothorax** (*hem-at-o-nū-mo-thor'-aks*). See *pneumothorax*.

**hematopoiesis** (*hem-at-o-poi-e'-sis*) [*hemato-*; ποίησις, a making]. The formation of blood.

**hematopoietic** (*hem-at-o-poi-et'-ik*) [see *hematopoiesis*]. Relating to the processes of blood-making. h. organs, blood-making organs.

**hematoporia** (*hem-at-o-po'-re-ah*) [*hemato-*; ἀπορία, defect]. Deficiency of blood; anemia.

**hematoporphyrin** (*hem-at-o-por'-fir-in*) [*hemato-*; πορφύρα, purple], $C_{68}H_{74}N_8O_{12}$. Iron-free hematin, a decomposition-product of hemoglobin occurring in the urine in conditions associated with destruction of red corpuscles. It is produced by dissolving hematin in concentrated sulphuric acid. h., test for. See Garrod.

**hematoporphyrinuria** (*hem-at-o-por-fir-in-ū'-re-ah*) [*hematoporphyrin*; οὖρον, urine]. The presence of hematoporphyrin in the urine.

**hematoporphyroidin** (*hem-at-o-por-fir-oid'-in*) [*hemato-*; πόρφυρος, purple]. A substance similar in origin and character to hematoporphyrin, but less soluble. Like hematoporphyrin, it is sometimes found in the urine.

**hematoposia** (*hem-at-o-po'-ze-ah*) [*hemato-*; πόσις, a drinking]. The drinking of blood.

**hematopostema** (*hem-at-o-pos'-tem-ah*) [*hemato-*; ἀπόστημα, abscess]. An abscess containing an effusion of blood.

**hematopsia** (*hem-at-op'-se-ah*) [*hemato-*; ὤψ, eye]. An extravasation of blood in the subconjunctival tissues of the eye.

**hematorrhachis, hematorachis** (*hem-at-or'-ak-is*) [*hemato-*; ῥάχις, spine]. Hemorrhage within the vertebral canal.

**hematorrhea** (*hem-at-or-e'-ah*) [*hemato-*; ῥοία, a flow]. A copious flow or discharge of blood.

**hematorrhosis** (*hem-at-or-o'-sis*) [*hemato-*; ὀρρός, serum]. Separation of the serum of the blood.

**hematosac** (*hem'-at-o-sak*) [*hemato-*; *saccus*, a bag]. A blood-cyst.

**hematosalpinx** (*hem-at-o-sal'-pinks*) [*hemato-*; σάλπιγξ, a trumpet]. A collection of blood in a Fallopian tube.

**hematoscheocele** (*hem-at-os'-ke-o-sēl*) [*hemato-*; ὄσχεον, scrotum; κήλη, tumor]. A hemorrhagic tumor or distention of the scrotum.

**hematoscope** (*hem'-at-o-skōp*) [*hemato-*; σκοπεῖν, to view]. An instrument used in the spectroscopic examination of the blood, by means of which the thickness of the layer of blood can be regulated.

**hematoscopy** (*hem-at-os'-ko-pe*) [see *hematoscope*]. Visual examination of the blood; examination of the blood by means of the hematoscope.

**hematose** (*hem'-at-ōs*) [*hemato-*]. Full of blood.

**hematosepsis** (*hem-at-o-sep'-sis*) [*hemato-*; σῆψις, putrefaction]. Septicemia.

**hematosin** (*hem-at'-o-sin*). See *hematin*.

**hematosis** (*hem-at-o'-sis*) [αἷμα, blood]. 1. The process of the formation of blood and the development of blood-corpuscles. 2. The arterialization of the blood.

**hematospectroscope** (*hem-at-o-spek'-tro-skōp*) [*hemato-*; *spectroscope*]. A spectroscope adapted to the study of the blood.

**hematospectroscopy** (*hem-at-o-spek-tros'-ko-pe*) [*hemato-*; *spectrum*, an image; σκοπεῖν, to view]. The use of the hematospectroscope.

**hematospermatocele** (*hem-at-o-sper-mat'-o-sēl*) [*hemato-*; *spermatocele*]. A spermatocele containing blood.

**hematospermia** (*hem-at-o-sper'-me-ah*) [*hemato-*; σπέρμα, seed]. The discharge of bloody semen.

**hematospongus** (*hem-at-o-spun'-gus*) [*hemato-*; σπόγγος, sponge]. Medullary sarcoma.

**hematostatic** (*hem-at-o-stat'-ik*). See *hemostatic*.

**hematotherapy** (*hem-at-o-ther'-ap-e*) [*hemato-*; *therapy*]. The treatment of disease by means of blood or some preparation of blood.

**hematothermal** (*hem-at-o-thur'-mal*) [*hemato-*; θέρμη, heat]. Warm-blooded.

**hematothoracic** (*hem-at-o-thor-as'-ik*). Relating to hematothorax.

**hematothorax** (*hem-at-o-tho'-raks*). See *hemothorax*.

**hematotic** (*hem-at-ot'-ik*). Relating to hematosis.

**hematotoxic** (*hem-at-o-toks'-ik*) [*hemato-*; τοξικόν, a poison]. Pertaining to a poisoned or impure state of the blood.

**hematotympanum** (*hem-at-o-tim'-pan-um*) [*hemato-*; τύμπανον, drum]. Bloody exudation in the drum-cavity.

**hematoxic** (*hem-at-oks'-ik*) [*hemato-*; τοξικόν, a poison]. The same as *hematotoxic*.

**hematoxin** (*hem-at-oks'-in*) [*hemato-*; τοξικόν, poison]. Any blood-poison or poisonous principle developed in the blood.

**hematoxylin** (*hem-at-oks'-il-in*) [*hemato-*; ξύλον, wood], $C_{16}H_{14}O_6$. The coloring-matter of logwood. It is a crystalline substance and is used as a stain in microscopy.

**hematoxylon** (*hem-at-oks'-il-on*) [see *hematoxylin*]. Logwood. The *hæmatoxylon* of the U. S. P. is the heart-wood of *Hæmatoxylon campechianum*; it contains tannic acid and a coloring principle, *hematoxylin*, and is a mild astringent. h., decoction of (*decoctum hæmatoxyli*, B. P.). Dose 1–2 oz. (32–64 Cc.). h., extract of (*extractum hæmatoxyli*, U. S. P.). Dose 5–20 gr. (0.32–1.3 Gm.).

**hematozemia** (*hem-at-o-ze'-me-ah*) [*hemato-*; ζημία, a loss]. A gradual or periodic discharge of blood.

**hematozoic** (*hem-at-o-zo'-ik*) [*hemato-*; ζῷον, an animal]. Pertaining to a hematozoon.

**hematozoon** (*hem-at-o-zo'-on*) [*hemato-*; ζῷον, animal; *pl.*, *hematozoa*]. Any animal parasite in the blood. **Hematozoon malariæ**, a hyaline ameboid body found in the blood of malarial patients.

**hematozymosis** (*hem-at-o-zi-mo'-sis*) [*hemato-*; ξύμωσις, fermentation]. Fermentation of the blood.

**hematozymotic** (*hem-at-o-zi-mot'-ik*) [*hemato-*; ζύμη, leaven]. Relating to a blood-ferment, or to fermentation in the blood of a living organism.

**hematropin** (*hem-at'-ro-pin*). Phenylglycolyltropein; a fluid preparation of hemoglobin.

**hematuresis** (*hem-at-ū-re'-sis*) [see *hematuria*]. The passage of bloody urine.

**hematuria** (*hem-at-ū'-re-ah*) [*hemato-*; οὖρον, urine]. The discharge of urine containing blood. When only the coloring-matter of the blood is found in the urine, it is termed hemoglobinuria or hematinuria. **hæmaturia ægyptica.** See *h., endemic.* **h., chylous,** hematochyluria. **h., endemic,** a form occurring in tropical countries due to parasites peculiar to the particular locality. **h., false,** the discharge of red urine, due to the ingestion of food or drugs containing red pigments.

**hemautogram** (*hem-aw'-to-gram*) [*hemato-*; αὐτός, self; γράμμα, a tracing]. The tracing made in hemautography.

**hemautograph** (*hem-aw'-to-graf*). Same as *hemautogram*.

**hemautography** (*hem-aw-tog'-ra-fe*) [αἷμα, blood; αὐτός, self; γράφειν, to write]. The tracing produced by a jet of blood from a divided artery caught upon paper drawn in front of it.

**hemelytrometra** (*hem-el-it-ro-me'-trah*) [αἷμα, blood; ἔλυτρον, a sheath; μήτρα, uterus]. An accumulation of blood in the uterus and vagina. **h. lateralis,** a collection of menstrual blood in the rudimentary half of a double vagina.

**hemendothelioma** (*hem-en-do-the-le-o'-mah*) [αἷμα, blood; *endothelioma*]. An endothelioma caused by proliferation of the endothelium of the blood-vessels.

**hemeralopia** (*hem-er-al-o'-pe-ah*) [ἡμέρα, day; ὤψ, eye]. Day-vision or night-blindness; a symptom of pigmentary degeneration of the retina, failure of general nutrition, etc. Vision is good by day or in a strong light, but fails at night. See *nyctalopia*.

**hemeraphonia** (*hem-er-af-o'-ne-ah*) [ἡμέρα, day; ά, priv.; φωνή, voice]. Loss of voice during the day, with return of the power of phonation at night.

**hemeropathia** (*hem-er-o-path'-e-ah*) [ἡμέρα, day; πάθος, disease]. 1. Any disease lasting but one day. 2. A disease that is more severe during the day.

**hemi-** (*hem'-e*) [ἡμι-, half]. A prefix signifying half.

**hemiablepsia** (*hem-e-ab-lep'-se-ah*) [*hemi-*; ἀβλεψία, blindness]. See *hemianopsia*.

**hemiacephala** (*hem-e-ah-sef-a'-le-ah*) [*hemi-*; ά, priv.; κεφαλή, head]. A monstrosity having a shapeless tumor representing the head, in which portions of the encephalon are contained. Syn., *acephalia spuria; hypacephalia.*

**hemiacephalus** (*hem-e-ah-sef'-al-us*) [*hemi-*; ά, priv.; κεφαλή, head]. A variety of omphalositic monsters of the species *paracephalus*. It is the lowest grade of development in paracephalus, closely approaching true acephalus.

**hemiachromatopsia** (*hem-e-ah-kro-mat-op'-se-ah*) [*hemi-*; ά, priv.; χρῶμα, color; ὄψις, vision]. Color-blindness in one-half of the field of vision.

**hemiageusia** (*hem-e-ah-gū'-se-ah*) [*hemi-*; ά, priv.; γεῦσις, taste]. One-sided loss or diminution of the sense of taste.

**hemialbumin** (*hem-e-al'-bū-min*). See *antialbumin*.

**hemialbuminose** (*hem-e-al-bū'-min-ōs*). The same as *hemialbumose*.)

**hemialbumose** (*hem-e-al'-bū-mōs*) [*hemi-*; *albumen*, the white of egg]. The most characteristic and most frequently obtained by-product of proteid digestion. It is the forerunner of hemipeptone.

**hemialbumosuria** (*hem-e-al-bū-mos-ū'-re-ah*). The presence of hemialbumose in the urine; propeptonuria.

**hemialgia** (*hem-e-al'-je-ah*) [*hemi-*; ἄλγος, pain]. Unilateral neuralgia.

**hemiamaurosis** (*hem-e-am-aw-ro'-sis*) [*hemi-*; *amaurosis*]. A form of transitory blindness in which hemianopia is combined with amblyopia in the other half of the visual field.

**hemiamblyopia** (*hem-e-am-ble-o'-pe-ah*). See *hemianopsia*.

**hemianalgesia** (*hem-e-an-al-je'-se-ah*) [*hemi-*; ἀν, priv.; ἄλγος, pain]. Insensibility to pain throughout one lateral half of the body and limbs.

**hemianasarca** (*hem-e-an-as-ar'-kah*) [*hemi-*; ἀνά, through; σάρξ, flesh]. Edema of one-half of the body.

**hemianesthesia** (*hem-e-an-es-the'-se-ah*) [*hemi-*; ἀναισθησία, want of feeling]. Anesthesia of one lateral half of the body. **h., alternate,** that affecting one side of the head and the opposite side of the body. **h., bulbar,** that due to disease of the oblongata. **h., cerebral,** that due to lesion in one of the cerebral hemispheres in the part of the capsula between the thalamus and dorsal part of the lenticula. **h., crossed.** 1. That associated with motor paralysis of the opposite half of the body. 2. See *h., alternate*. **h., functional,** that due to functional causes rather than to cerebral or spinal lesion. **h., hysterical,** tactile and thermal hemianesthesia coming on gradually or suddenly and of variable duration in hysterical subjects. **h., organic,** that due to lesion in the central nervous system. **h., saturnine,** that due to lead-poisoning. **h., Türck's,** anesthesia affecting the functions of the posterior spinal roots of one side, at times also those of the nerves of special sense. It is caused by lesions of the posterior portion of the capsula and the contiguous region of the corona radiata.

**hemianopsia, hemianopia** (*hem-e-an-op'-se-ah, hem-e-an-o'-pe-ah*) [*hemi-*; ἀν, priv.; ὄψις, sight). Blindness in one-half of the visual field. It may be bilateral (*binocular*) or unilateral (*unilocular*). **h., binasal,** blindness on the nasal side of the visual field, usually due to disease of the outer sides of the optic commissure. **h., bitemporal,** blindness on the temporal side of the visual field, due to disease of the central parts of the commissure. **h., crossed, h., heteronymous,** a general term for either binasal or bitemporal hemianopia. **h., homonymous,** the form affecting the inner half of one field and the outer half of the other. **h., lateral, h. lateralis,** a form in which the temporal half of one visual field and the nasal half of the other visual field are wanting; a vertical line through the center of vision sharply defining the defect. **h., vertical.** See *h., lateral*.

**hemianoptic** (*hem-e-an-op'-tik*) [*hemi-*; ἀν, priv.; ὄψις, sight]. Affected with hemianopsia.

**hemianosmia** (*hem-e-an-ōs'-me-ah*) [*hemi-*; *dnosmia*]. Loss of smell in one nostril.

**hemiarthrosis** (*hem-e-ar-thro'-sis*) [*hemi-*; ἄρθρωσις, a joining]. A false synchondrosis.

**hemiasynergia** (*hem-e-as-in-ur'-je-ah*) [*hemi-*; ά, priv.; σύν, with; ἔργον, work]. Asynergia affecting only one side of the body.

**hemiataxia** (*hem-e-at-aks'-e-ah*) [*hemi-*; ά, priv.; τάξις, order]. Ataxia limited to one side of the body.

**hemiathetosis** (*hem-e-ath-et-o'-sis*) [*hemi-*; ἄθετος, without fixed position]. Athetosis of one side of the body.

**hemiatonia** (*hem-e-at-o'-ne-ah*) [*hemi-*; ἀτονία, want of tone]. Diminution or loss of muscular or vital energy in one-half of the body. **h. apoplectica,** hemihypertonia postapoplectica.

**hemiatrophy** (*hem-e-at'-ro-fe*) [*hemi-*; ἀτροφία, lack of nourishment]. Atrophy confined to one side of the body.

**hemiazygous** (*hem-e-az'-ig-us*) [*hemi-*; ἄζυγος, unpaired]. Partially paired; imperfectly azygous. **h. veins.** See *vein*.

**hemibranchiate** (*hem-e-brang'-ke-āt*) [*hemi-*; βράγχια, gills]. In biology, having an incomplete branchial apparatus.

**hemic** (*he'-mik*) [αἷμα, blood]. Pertaining to or developed by the blood. **h. calculus,** a concretion of coagulated blood. **h. murmur,** a murmur due to anemia.

**hemicanities** (*hem-e-kan-ish'-e-ēz*) [*hemi-*; *canities*]. Canities on one side only.

**hemicardia** (*hem-e-kar'-de-ah*) [*hemi-*; καρδία, heart]. Half of a four-chambered heart.

**hemicatalepsy** (*hem-e-kat'-al-ep-se*) [*hemi-*; *catalepsy*]. Catalepsy affecting only one lateral half of the subject.

**hemicellulose** (*hem-e-sel'-ū-lōs*) [*hemi-*; *cellula*, a little cell]. A term for all the carbohydrates in the cell-wall which are not colored blue by chlorzinc iodide.

**hemicentrum** (*hem-e-sen'-trum*) [*hemi-*; *centrum*]. Either one of the two lateral elements of the centrum of a vertebra.

**hemicephalia** (*hem-e-sef-a'-le-ah*). Synonym of *hemicrania*.

**hemicephalus** (*hem-e-sef'-a-lus*) [*hemi-*; κεφαλή, head]. A monster in which the cerebral hemispheres and skull are lacking.

**hemicerebrum** (*hem-e-ser-e'-brum*) [*hemi-*; *cerebrum*]. A cerebral hemisphere.

**hemichorea** (hem-e-ko-re'-ah) [hemi-; χορεία, chorea]. A form of chorea in which the convulsive movements are confined to one side of the body.

**hemichromanopsia** (hem-e-kro-man-op'-se-ah). See *hemiachromatopsia*.

**hemichromosome** (hem-e-kro'-mo-sōm) [hemi-; chromosome]. The body formed by the longitudinal splitting of the chromosome.

**hemicollin** (hem-e-kol'-in) [hemi-; κόλλα, glue], $C_8H_{79}N_{14}O_{19}$. A peptone-like substance derived from collagen.

**hemicrania** (hem-e-kra'-ne-ah) [hemi-; κρανίον, head]. 1. Neuralgia of one-half of the head; migraine. 2. Imperfect development or absence of the anterior or posterior part of the skull.

**hemicranic** (hem-e-kra'-nik) [hemi-; κρανίον, skull]. Pertaining to half the skull. **h. equivalents,** a name given to isolated symptoms—flitting scotomata, vomiting, vaso-motor disturbances, etc.—in cases of aborted migraine.

**hemicraniectomy** (hem-e-kra-ne-ek'-to-me) [hemi-; craniectomy]. Doyen's operation of sectioning the cranial vault near the mesial line, from before backward, and pressing the entire side outward to expose one-half of the brain.

**hemicranin** (hem-e-kra'-nin). A proprietary remedy for neuralgia said to consist of phenacetin, 5 parts; caffeine, 1 part; citric acid, 1 part. Dose, 5–15 gr.

**hemicraniosis** (hem-e-kra-ne-o'-sis) [hemi-; κρανίον, head]. Enlargement of one half of the cranium or face.

**hemicyclic** (hem-e-si'-klik) [hemi-; κύκλος, a circle]. In biology, having certain of the floral organs arranged in whorls, and others in a spiral.

**hemidesmus** (hem-e-des'-mus) [hemi-; δεσμός, a band]. Indian sarsaparilla. The dried root of *H. indicus*, imported from India. It is a tonic, alterative, diaphoretic, and diuretic. h., syr. (B. P.). Dose ʒj.

**hemidiaphoresis** (hem-e-di-af-or-e'-sis) [hemi-; διαφόρησις, sweating]. Sweating of one lateral half of the body.

**hemidrosis** (hem-id-ro'-sis). See *hematidrosis*.

**hemidysesthesia** (hem-e-dis-es-the'-ze-ah) [hemi-; δυσ-, difficult; αἴσθησις, sensation]. Dysesthesia of a lateral half of the body.

**hemidystrophia** (hem-e-dis-tro'-fe-ah) [hemi-; δυσ, ill; τροφή, nourishment]. The state of being imperfectly nourished.

**hemiencephalon** (hem-e-en-sef'-al-on) [hemi-; ἐγκέφαλος, brain]. Either lateral half of the brain.

**hemiencephalus** (hem-e-en-sef'-al-us) [hemi-; ἐγκέφαλος, brain]. A monster without organs of sense, but having otherwise a nearly normal brain.

**hemiepilepsy** (hem-e-ep'-il-ep-se) [hemi-; ἐπιληψία, epilepsy]. A form of epilepsy in which the convulsions are confined to one lateral half of the body.

**hemifacial** (hem-e-fa'-shal). Pertaining to one lateral half of the face.

**hemifornix** (hem-e-for'-niks) [hemi-; fornix, an arch, vault]. A name used to designate the irregular, elongated portion of either paracelian floor, composed of hippocamp, fimbria, and fornicolumn.

**hemiglossitis** (hem-e-glos-i'-tis) [hemi-; γλῶσσα, tongue; ιτις, inflammation]. Inflammation of one-half of the tongue.

**hemihidrosis** (hem-e-hid-ro'-sis) [hemi-; ἱδρώς, sweat]. The same as *hemidiaphoresis*.

**hemihypalgesia** (hem-e-hi-pal-je'-ze-ah) [hemi-; hypalgesia]. Hypalgesia limited to one side of the body.

**hemihyperesthesia** (hem-e-hi-per-es-the'-ze-ah) [hemi-; ὑπέρ, over; αἴσθησις, sensation]. Hyperesthesia confined to one lateral half of the body.

**hemihypertonia** (hem-e-hi-per-to'-ne-ah) [hemi-; hypertonia]. Increased muscular tonicity confined to one-half of the body. **h. postapoplectica,** an intermittent tonic spasm distributed over one-half of the body, affecting at times different groups of muscles without loss of power in the muscles affected. It follows an apoplectic attack.

**hemihypertrophy** (hem-e-hi-per'-tro-fe) [hemi-; hypertrophy]. Hypertrophy of half of the body.

**hemihypesthesia** (hem-e-hi-pes-the'-ze-ah) [hemi-; ὑπό, under; αἴσθησις, perception]. Impairment of sensibility in one lateral half of the body.

**hemihypogeusia** (hem-e-hi-po-gu'-se-ah). See *hemiageusia*.

**hemihypothermia** (hem-e-hi-po-ther'-me-ah) [hemi-; ὑπό, under; θέρμη, heat]. Diminution of the temperature limited to one side of the body.

**hemihypotonia** (hem-e-hi-po-to'-ne-ah) [hemi-; ὑπό, under; τόνος, tone]. Partial loss of tonicity of one side of the body.

**hemilateral** (hem-e-lat'-er-al). Pertaining to one lateral half.

**hemilethargy** (hem-e-leth'-ar-je) [hemi-; lethargy]. A state of partial lethargy.

**hemilingual** (hem-e-ling'-wal) [hemi-; lingua, tongue]. Pertaining to one lateral half of the tongue.

**hemilytic** (hem-e-lit'-ik) [hemi-; λύσις, a loosening]. Relating to a condition of retarded change.

**hemimelus** (hem-im'-el-us) [hemi-; μέλος, a limb]. A monster with incomplete or stunted extremities.

**hemimetaboly** (hem-e-met-ab'-o-le) [hemi-; μεταβολή, transformation]. In biology, incomplete or imperfect metamorphosis; hemimetamorphosis.

**hemimetamorphosis** (hem-e-met-am-or'-fo-sis) [hemi-; μεταμόρφωσις, transformation]. In biology, incomplete metamorphosis; hemimetaboly.

**hemimyasthenia** (hem-e-mi-as-the'-ne-ah) [hemi-; μῦς, muscle; ἀσθένεια, weakness]. Myasthenia of one lateral half of the body.

**hemimyoclonus** (hem-e-mi-ok'-lo-nus) [hemi-; μῦς, muscle; κλόνος, commotion]. Clonic spasm of the muscles of one lateral half of the body.

**hemin** (hem'-in) [αἷμα, blood], $C_{34}H_{32}N_4FeO_4 \cdot HCl$. Crystalline hematin chloride, of which Teichmann's crystals, doubly refractive crystals derived from blood by heating a drop of blood on a glass slide with a little glacial acetic acid, are composed. h., test for. See *Teichmann*.

**hemineurasthenia** (hem-e-nū-ras-the'-ne-ah) [hemi-; νεῦρον, a nerve; ἀσθένεια, weakness]. Neurasthenia affecting one lateral half of the body.

**hemiopalgia** (hem-e-op-al'-je-ah) [hemi-; ὤψ, eye; ἄλγος, pain]. Hemicrania with pain in one eye.

**hemiopia** (hem-e-o'-pe-ah). See *hemianopsia*.

**hemiopic** (hem-e-o'-pik). Pertaining to hemianopsia. **h. pupillary reaction,** Wernicke's reaction.

**hemiparanesthesia** (hem-e-par-ah-an-es-the'-ze-ah) [hemi-; paraanesthesia]. Paraanesthesia limited to one side of the body and due to destructive lesion of the lateral half of the spinal cord.

**hemiparaplegia** (hem-e-par-ap-le'-je-ah) [hemi-; παραπληγία, paralysis of the limbs]. Paralysis of a lower limb on one side only.

**hemiparesis** (hem-e-par'-es-is) [hemi-; πάρεσις, impairment of strength]. Paresis of one side of the body.

**hemiparesthesia** (hem-e-par-es-the'-ze-ah) [hemi-; παρά, beside; αἴσθησις, sensation]. Numbness or paresthesia of one lateral half of the body.

**hemiparetic** (hem-e-par-et'-ik) [hemi-; πάρεσις, impairment of strength]. Affected with hemiparesis.

**hemipeptone** (hem-e-pep'-tōn). See *peptone*.

**hemiphalacrosis** (hem-e-fal-ak-ro'-sis) [hemi-; φαλάκρωσις, baldness]. Baldness affecting one lateral half of the head.

**hemiphonia** (hem-e-fo'-ne-ah) [hemi-; φωνή, voice]. Speech having the characteristics of half-voice, half whisper; used by patients in great weakness and exhaustion.

**hemiplegic** (hem-e-plek'-tik). Same as *hemiplegic*.

**hemiplegia** (hem-e-ple'-je-ah) [hemi-; πληγή, stroke]. Paralysis of one side of the body, due usually to a lesion in the internal capsule or corpus striatum, but at times caused by an extensive lesion of the cortex, or a lesion of the crus, pons, medulla, or upper part of the spinal cord. If in the brain, the lesion is on the side opposite to the paralysis. **h., alternate,** paralysis of the facial muscles upon one side, with paralysis of the trunk and extremities upon the opposite side of the body. **h., bilateral spastic.** See *paraplegia, infantile spasmodic*. **h., cerebral,** the ordinary form first described. **h., choreic,** that followed by chorea, especially in young adults. **h., crossed.** Same as *h., alternate*. **h., facial,** motor paralysis of one side of the face. **h., homolateral,** uncrossed hemiplegia. **h., spastic,** a form occurring in infants, in which the affected extremities are spastically contracted. **h., spinal.** See *Brown-Séquard's paralysis*.

15

**hemiplegiac** (hem-e-ple'-je-ak). An individual affected with hemiplegia.
**hemiplegic** (hem-e-ple'-ik). Relating to or affected with hemiplegia; hemiplectic.
**hemiprosoplegia** (hem-e-pro-so-ple'-je-ah). [hemi-; πρόσωπον, face; πληγή, stroke]. Paralysis of one side of the face.
**hemiprotein** (hem-e-pro'-te-in). Same as antialbumin.
**hemirheumatism** (hem-e-ru'-mat-izm). [hemi-; rheumatism]. Rheumatism confined to one lateral half of the body.
**hemisection** (hem-e-sek'-shun) [hemi-; sectio, a cutting]. Bisection; chiefly applied to division into two lateral halves. See mediation.
**hemiseptum** (hem-e-sep'-tum) [hemi-; septum, a partition]. The lateral half of a septum, as of the heart; the lateral half of the septum lucidum.
**hemisine** (hem'-e-sin). Trade name of an active principle of suprarenal gland.
**hemisomus** (hem-e-so'-mus) [hemi-; σῶμα, body]. A monster with one side of the body imperfectly developed.
**hemispasm** (hem'-e-spazm) [hemi-; σπασμός, a spasm]. A spasm affecting only one side of the body.
**hemisphere** (hem'-is-fēr) [hemi-; σφαῖρα, a sphere]. Half a sphere. h., **cerebellar**, either lateral half of the cerebellum. h., **cerebral**, either lateral half of the cerebrum.
**hemisystole** (hem-e-sis'-to-le) [hemi-; συστολή, a contraction]. A peculiar kind of irregular action of the heart-muscle, in which, with every two beats of the heart, only one beat of the pulse is felt.
**hemiterata** (hem-e-ter'-at-ah) [hemi-; τέρας, a monster]. A class of malformations not grave enough to be called monstrous.
**hemiteratic** (hem-e-ter-at'-ik). Pertaining to hemiterata.
**hemithermoanesthesia** (hem-e-ther-mo-an-es-the'-ze-ah) [hemi-; θέρμη, heat; anesthesia]. Insensibility to heat and cold limited to one side of the body.
**hemitis** (hem-i'-tis) [hemi-; ιτις, inflammation]. A condition of the blood associated with inflammation.
**hemitomias** (hem-e-to'-me-as) [hemi-; τομίας, one who has been castrated]. A man who has been deprived of one testis.
**hemitonia** (hem-e-to'-ne-ah) [hemi-; τόνος, tension]. One-sided tonic muscle-contraction in brain disease.
**hemitoxin** (hem-e-toks'-in) [hemi-; toxin]. A toxin deprived of half of its original toxicity.
**hemlock** (hem'-lok). 1. See conium. 2. A tree of the genus Tsuga.
**hemo-, hæmo-** (hem-o-) [αἷμα, blood]. A prefix signifying of or pertaining to the blood.
**hemoagglutination** (hem-o-ag-glu-tin-a'-shun). The clumping of red blood-corpuscles. Cf. hemoagglutinin.
**hemoalkalimeter** (hem-o-al-kal-im'-et-er) [hemo-; alkaline; μέτρον, measure]. An apparatus for estimating the degree of alkalinity of the blood.
**hemobilinuria** (hem-o-bil-in-u'-re-ah) [hemo-; bilis, bile; οὖρον, urine]. The presence of urobilin in the blood.
**hemoblast** (hem'-o-blast). See hematoblast.
**hemocatatonistic** (hem-o-kat-at-on-is'-tik) [hemo-; κατά, down; τόνος, tension]. Tending to diminish the cohesion between the hemoglobin and the red blood-corpuscles.
**hemocelom** (hem-o-se'-lom) [hemo-; κοιλία, a cavity]. A blood-cyst.
**hemochromatosis** (hem-o-kro-mat-o'-sis) [hemo-; χρῶμα, color]. Discoloration of the tissues, particularly the skin, by deposition of a pigment from the blood.
**hemochromogen** (hem-o-kro'-mo-jen) [hemo-; χρῶμα, color; γενᾶν, to produce]. 1. Hemoglobin. 2. A hypothetic substance formed by the decomposition of hemoglobin with acids or alkalies in the absence of oxygen.
**hemochromometer** (hem-o-kro-mom'-et-er) [hemo-; χρῶμα, color; μέτρον, measure]. Colorimeter; an instrument for estimating the amount of oxyhemoglobin in the blood, by comparing a solution of the blood with a standard solution of picrocarminate of ammonium.
**hemoclasis** (hem-ok'-las-is) [hemo-; κλάσις, destruction]. Hemolysis; destruction of the erythrocytes.
**hemoclastic** (hem-o-klas'-tik). Hemolytic.
**hemococcidium** (hem-o-koks-id'-e-um). Same as plasmodium.
**hemoconia.** See hemokonia.
**hemocryoscopy** (hem-o-kri-os'-ko-pe). Cryoscopy applied to blood. See cryoscopy.
**hemocrystallin** (hem-o-kris'-tal-in). See hemoglobin.
**hemocyanin** (hem-o-si'-an-in) [hemo-; κύανος, blue]. A coloring-matter found in the blood of certain invertebrates. It contains copper, and gives to the blood a blue color.
**hemocyte** (hem'-o-sīt). A blood-corpuscle.
**hemocytolysis** (hem-o-si-tol'-is-is) [hemo-; κύτος, a cell; λύειν, to unloose]. The dissolution of blood-corpuscles.
**hemocytometer** (hem-o-si-tom'-et-er) [hemo-; κύτος, a cell; μέτρον, a measure]. An instrument for estimating the number of corpuscles in the blood.
**hemocytotripsis** (hem-o-si-to-trip'-sis) [hemo-; κύτος, a cell; τρίβειν, to rub]. The breaking up of blood-corpuscles under strong pressure.
**hemocytozoon** (hem-o-si-to-zo'-on) [hemo-; κύτος, a cell; ζῷον, an animal]. The plasmodium of malaria.
**hemodia** (hem-o'-de-ah) [αἱμωδεῖν, to set the teeth on edge]. Excessive sensibility of the teeth; the "setting the teeth on edge."
**hemodiagnosis** (hem-o-di-ag-no'-sis) [hemo-; diagnosis]. Diagnosis by examination of the blood.
**hemodiapedesis** (hem-o-di-ap-ed-e'-sis) [hemo-; διαπήδησις, an oozing through]. The transudation of blood through the skin.
**hemodiarrhea** (hem-o-di-ar-e'-ah). Dysentery.
**hemodiastase** (hem-o-di'-as-tās) [hemo-; διάστασις, separation]. The amylolytic enzyme of the blood.
**hemodromograph** (hem-o-dro'-mo-graf). See hemadromograph.
**hemodromometer** (hem-o-dro-mom'-et-er). See hemadromometer.
**hemodynamics** (hem-o-di-nam'-iks). See hemadynamics.
**hemodynamometer** (hem-o-di-nam-om'-et-er). See hemadynamometer.
**hemoferrogen** (hem-o-fer'-o-jen). A dry preparation of blood proposed as a remedy in anemic conditions.
**hemoferrum** (hem-o-fer'-um). 1. See oxyhemoglobin. 2. The iron in the hemoglobin.
**hemofuscin** (hem-o-fus'-in) [hemo-; fuscus, dark]. The yellowish-brown, iron-free pigment found in hemochromatosis.
**hemogallol** (hem-o-gal'-ol) [hemo-; galla, gallnut]. A proprietary substance occurring as a brownish-red powder, and formed by oxidizing the hemoglobin of the blood by pyrogallol. It is used in anemia in doses of gr. jss-vijss.
**hemogastric** (hem-o-gas'-trik) [hemo-; γαστήρ, stomach]. Pertaining to blood in the stomach.
**hemogenesis** (hem-o-jen'-es-is). See hematogenesis.
**hemogenic** (hem-o-jen'-ik). See hematogenic.
**hemoglobic** (hem-o-glo'-bik). Applied to cells containing or generating hemoglobin.
**hemoglobin** (hem-o-glo'-bin) [hemo-; globus, a ball]. The coloring-matter of the red corpuscles. It is an exceedingly complex body, containing iron; it crystallizes in rhombic plates or prisms, and is composed of hematin and a proteid substance, called globulin. It has a strong affinity for oxygen, and the greater part of the oxygen in the blood is in combination with it as oxyhemoglobin. When it gives up the oxygen to the tissues it becomes reduced hemoglobin. It is used in treatment of anemia in daily doses of 75-150 gr. (5-10 Gm.). Syn., purple cruorin. h., **tests for**. See Kobert, Tallqvist.
**hemoglobinemia** (hem-o-glo-bin-e'-me-ah) [hemoglobin; αἷμα, blood]. A condition in which the hemoglobin is dissolved out of the red corpuscles, probably as the result of the destruction of the latter, and is held in solution in the serum. The blood is "lake" colored. It occurs in some infectious diseases and after injecting certain substances into the blood.
**hemoglobiniferous** (hem-o-glo-bin-if'-er-us) [hemoglobin; ferre, to bear]. Yielding or carrying hemoglobin.
**hemoglobinocholia** (hem-o-glo-bin-o-ko'-le-ah) [hemoglobin; χολή, bile]. The presence of hemoglobin in the bile.
**hemoglobinometer** (hem-o-glo-bin-om'-et-er) [hemoglobin; μέτρον, a measure]. An instrument for the quantitative estimation of hemoglobin in the blood. h., **Dare's**, an instrument consisting of a pipet and

two glass plates by means of which the undiluted blood is arranged into a stratum of exact thickness. It is illuminated by direct candle-light, and compared with a glass color-scale which is shifted by means of an adjusting wheel. The percentage of hemoglobin is shown by an indicator. h., **Fleischl's**, in this a certain dilution of the blood is made, and its color then compared with that of different thicknesses of a sliding wedge of red glass. h., **Gowers'**, in this the calculation is made by measuring the amount of dilution necessary to make the blood of the same shade as a standard solution of carmine in gelatin.
**hemoglobinorrhea** (hēm-o-glo-bin-or-e'-ah) [hemoglobin; ῥοία, a flow]. The escape of hemoglobin from the blood-vessels. **hæmoglobinorrhœa cutis**, an effusion of hemoglobin into the skin, due to venous engorgement.
**hemoglobinuria** (hem-o-glo-bin-ū'-re-ah) [hemoglobin; οὖρον, urine]. The presence of hemoglobin in the urine, due either to its solution out of the red corpuscles or to disintegration of the red corpuscles. h., **epidemic**, hemoglobinuria of the new-born associated with jaundice, cyanosis, and nervous symptoms. Syn., *Winckel's disease*. h., **intermittent**, h., **paroxysmal**, a form characterized by recurring periodic attacks. It is related to cold, and is also closely associated with Raynaud's disease. h., **toxic**, that form occurring in consequence of poisoning by various substances.
**hemoglobulin** (hem-o-glob'-ū-lin) [hemo-; globus, a ball]. Same as *hemoglobin*.
**Hemogregarina** (hem-o-greg-ar-i'-nah) [hemo-; Gregarina, a genus of Protozoa]. Gregarine-like bodies found in the blood of persons affected with malaria.
**hemohydronephrosis** (hem - o - hi - dro - nef - ro' - sis) [hemo-; ὕδωρ, water; νεφρός, kidney]. A cystic tumor of the kidney with blood and urine in the contents.
**hemoid** (hem-oid') [hem-; εἶδος, likeness]. Have the appearance of, or resembling, blood.
**hemokelidosis** (hem-o-kel-id-o'-sis) [hemo-; κηλιδοῦν, to stain]. Contamination of the blood; purpura.
**hemokonia** (hem-o-ko'-ne-ah) [hemo-; κονία, dust]. Minute, colorless, highly refractive, spheroidal or dumb-bell-shaped bodies constantly present in normal and pathological blood. They are not more than 1 μ in diameter, and possess active, limited molecular motility, but not true ameboid motion. Syn., *blood-dust; blood-motes*.
**hemol** (hem'-ol) [αἷμα, blood]. A dark-brown powder obtained by the action of zinc upon defibrinated blood, and said to contain 1 % of soluble iron. Dose in anemia 1½-6 gr. (0.1-0.5 Gm.).
**hemoleukocyte** (hem-o-loo'-ko-sīt). A white blood-corpuscle.
**hemolipase** (hem-o-li'-pās). A fat-splitting ferment found in the blood.
**hemology** (hem-ol'-o-je) [hemo-; λόγος, science]. The science treating of the blood.
**hemolutein** (hem-o-loo'-te-in) [hemo-; luteus, yellow]. A yellow coloring-matter obtained from corpora lutea.
**hemolymph** (hem'-o-limf) [hemo-; lympha, water]. 1. Blood and lymph. 2. The circulating nutritive fluid of certain invertebrates. h. **glands**, a variety of glands which are a kind of cross between the hemogenic glands (e. g. the spleen) and the lymphatic glands. The small prevertebral glands are examples.
**hemolysin** (hem-ol'-is-in) [hemo-; λύσις, solution]. A substance produced in the body of one species of animal by the introduction of red blood-corpuscles derived from the body of another species. It is capable of dissolving the red blood-corpuscles of the animal species from which the blood was obtained. Syn., *erythrolysin*. h., **bacterial**, that formed by the action of bacteria.
**hemolysis** (hem-ol'-is-is). See *hematolysis*.
**hemolytic** (hem-ol-it'-ik). An agent causing destruction of the red blood-corpuscles. h. **serum**, a serum which causes hemolysis.
**hemolyze** (hem'-ol-īz). To produce hemolysis.
**hemomanometer** (hem-o-man-om'-et-er) [hemo-; μάνος, thin; μέτρον, a measure]. A manometer used in estimating blood-pressure.
**hemomediastinum** (hem-o-me-de-as-ti'-num). See *hematomediastinum*.
**hemomere** (hem'-o-mēr) [hemo-; μέρος, a part]. A portion of a metamere derived from the vascular system or taking part in its formation.

**hemometer** (hem-om'-et-er). See *hemoglobinometer*.
**hemometra** (hem-o-me'-trah). See *hematometra*.
**hemometrectasia** (hem-o-met-rek-ta'-ze-ah) [hemo-; μήτρα, uterus; ἔκτασις, a stretching]. Dilatation of the uterus from effusion of blood into its cavity.
**hemometry** (hem-om'-et-re) [hemo-; μέτρον, a measure]. Estimation of the amount of hemoglobin or of the number of corpuscles in the blood. *Normal Count*: Erythrocytes (red corpuscles), 5,000,000; leukocytes (white corpuscles), 6000 to 8000; hemoglobin, 100 %. *Differential Count of Leuckoytes*: Polymorphonuclears, 62 to 70 %; small lymphocytes, 20 to 30 %; large lymphocytes, 4 to 8 %; transitionals, 1 to 2 %; eosinophiles, 0.5 to 4 %; mast-cells, 0.25 to 0.5 %.
**hemonervine** (hem-o-nur'-vēn). A proprietary tonic said to consist of calcium phosphoglycerate, hemoglobin, iron, and strychnine.
**hemoophoritis** (hem-o-off-or-i'-tis) [hem-; oophoritis]. Oophoritis with hemorrhage.
**hemopathology** (hem-o-path-ol'-o-je) [hemo-; pathology]. The pathology of the blood.
**hemopericardium** (hem-o-per-ik-ar'-de-um) [hemo-; pericardium]. An effusion of blood into the pericardial cavity.
**hemoperitoneum** (hem-o-per-it-on-e'-um). A bloody effusion into the peritoneal cavity.
**hemopexia** (hem-o-peks'-e-ah) [hemo-; πῆξις, a fixing]. A general name for diseases characterized by a tendency of the blood to coagulate.
**hemopexin** (hem-o-peks'-in). A ferment capable of coagulating blood.
**hemophagic** (hem-o-faj'-ik) [hemo-; φαγεῖν, to eat]. Feeding upon blood; applied to certain animal parasites.
**hemophagocyte** (hem-o-fag'-o-sīt) [hemo-; φαγεῖν, to eat; κύτος, a cell]. Any phagocyte of the blood; a white blood corpuscle.
**hemophilia** (hem-o-fil'-e-ah) [hemo-; φιλεῖν, to love]. Bleeder's disease, an abnormal tendency to hemorrhage. It is usually hereditary, and though it is most common in males, the hereditary influence is transmitted through the mother.
**hemophiliac** (hem-o-fil'-e-ak) [see *hemophilia*]. One who is affected with hemophilia.
**hemophobia**. See *hematophobia*.
**hemophotograph** (hem-o-fo'-to-graf) [hemo-; photograph]. A photograph of blood-corpuscles; it is used in determining the hemoglobin content. Syn., *hemaphotograph*.
**hemophthalmia** (hem-of-thal'-me-ah) [hemo-; ὀφθαλμός, eye]. A hemorrhage into the interior of the eye.
**hemophthalmos** (hem-off-thal'-mos) [hemo-; ὀφθαλμός, the eye]. Blood in the vitreous chamber characterized by a reddish reflex from the pupil.
**hemophthisis** (hem-off-thi'-sis) [hemo-; φθίσις, wasting]. Anemia dependent upon undue depeneration of the red blood-corpuscles.
**hemophysallis** (hem-o-fis-al'-is) [hemo-; φυσαλλίς, a bladder]. A pustule or vesicle filled with blood.
**hemoplanesis, hemoplania** (hem-o-plan-e'-sis, hem-o-pla'-ne-ah). See *hematoplania*.
**hemoplasmodium** (hem-o-plaz-mo'-de-um) [hemo-; *plasmodium*]. The plasmodium of malaria.
**hemoplastic** (hem-o-plas'-tik) [hemo-; πλάσσειν, to form]. Same as *hematoplastic*.
**hemopneumothorax** (hem-o-nū-mo-tho'-raks) [hemo-; πνεῦμα, air; θώραξ, the chest]. A collection of air and blood within the pleural cavity.
**hemopoiesis** (hem-o-poi-e'-sis). See *hematopoiesis*.
**hemoprecipitin** (hem-o-pre-sip'-it-in). See *precipitin*.
**hemoproctia** (hem-o-prok'-te-ah) [hemo-; ἀρχός the anus]. Rectal hemorrhage. Bloody discharge from hemorrhoids.
**Hemoproteus** (hem-o-pro'-te-us). A protozoan parasite of malaria found in the blood-corpuscles of birds.
**hemopsonin** (hem-op'-son-in). A substance which is opsonic for red blood corpuscles.
**hemoptoic, hemoptoe** (hem-op'-tib, hem-op'-to-e). Relating to,or attended by hemoptysis.
**hemoptyic, hemoptysic** (hem-op'-te-ik, hem-op'-tis-ik). See *hemoptic*.
**hemoptysis** (hem-op'-tis-is) [hemo-; πτύειν, to spit]. The spitting of blood from the larynx, trachea, bronchi, or lungs. h., **parasitic**, a disease due to the fluke, *Paragonimus Westermanii*, which lodges in the lungs, and exceptionally in other organs. The diagnosis is made by finding the characteristic ova in the sputum.

## HEMOQUINONE 420 HENLE'S AMPULLA

**hemoquinine** (*hem-o-kwin-ēn'*). Trade name of a preparation said to contain iron, quinine, manganese and arsenic.

**hemorrhage** (*hem'-or-āj*) [*hemo-*; ῥηγνύναι, to burst forth]. An escape of blood from the vessels, either by diapedesis through intact walls or by rhexis through ruptured walls. **h., accidental,** hemorrhage during pregnancy from premature detachment of the placenta when normally situated. **h., capillary,** oozing of blood from the capillaries. **h., concealed,** a variety of accidental hemorrhage in which the bleeding takes place between the ovum and the uterine walls, without escape from the genital tract. **h., consecutive,** one ensuing some time after injury. **h., critical,** occurring at the turning-point of a disease. **h., petechial,** hemorrhage under the surface in the form of minute points. **h., postpartum,** hemorrhage occurring shortly after labor. **h., primary,** that immediately following any traumatism. **h., secondary,** that occurring some time after the traumatism. **h., unavoidable,** hemorrhage from detachment of a placenta prævia. **h., vicarious,** a discharge of blood from a part owing to the suppression of a flow in another part, as *vicarious* menstruation.

**hemorrhagic** (*hem-or-aj'-ik*) [*hemorrhage*]. Relating to or accompanied by hemorrhage. **h. diathesis.** See *hemophilia.* **h. infarct.** See *infarct, hemorrhagic.*

**hemorrhagiferous** (*hem-or-aj-if'-er-us*) [*hemorrhage*; ferre, to bear]. Attended by hemorrhage; giving rise to hemorrhage.

**hemorrhagin, hemorrhagin** (*hem-or-aj'-in*). Flexner's name for endotheliolysin, since it causes extravasations of blood through its direct solvent action upon capillary endothelium.

**hemorrhagiparous** (*hem-or-aj-ip'-ar-us*). See *hemorrhagiferous.*

**hemorrhaphilia** (*hem-or-af-il'-e-ah*) [*hemorrhage*; φίλος, loving]. Synonym of *hemophilia.*

**hemorrhea** (*hem-or-e'-ah*) [*hemo-*; ῥοία, a flow]. A hemorrhage.

**hemorrhelcosis** (*hem-or-el-ko'-sis*) [αἱμοῤῥοίς, a hemorrhoid; ἕλκωσις, ulceration]. The formation of an ulcer upon a hemorrhoid.

**hemorrhinia** (*hem-or-in'-e-ah*) [*hemo-*; ῥίς, nose]. Epistaxis; nose-bleed.

**hemorrhoid** (*hem'-or-oid*) [αἱμόρροος, flowing with blood]. A pile. An enlarged and varicose condition of the veins of the lower portion of the rectum and the tissues about the anus. **h., blind,** one that does not cause bleeding. **h.s, external,** those situated without the sphincter ani. **h.s, internal,** those within the anal orifice.

**hemorrhoidal** (*hem-or-oi'-dal*) [*hemorrhoid*]. 1. Pertaining to or affected with hemorrhoids. 2. Applied to blood-vessels, nerves, etc., belonging to the anus. **h. veins,** the three veins which form a plexus about the lower end of the rectum; through them the general venous system and the portal system communicate.

**hemorrhoidectomy** (*hem-o-roi-dek'-to-me*). Excision of hemorrhoids.

**hemosalpinx** (*hem-o-sal'-pinks*). Same as *hematosalpinx.*

**hemoscope** (*hem'-o-skōp*). Same as *hematoscope.*

**hemosiderin** (*hem-o-sid'-er-in*) [*hemo-*; σίδηρος, iron]. A granular pigment, a product of the decomposition of hemoglobin. It is found where blood is extravasated in contact with active cells, and contains iron.

**hemosiderosis** (*hem-o-sid-er-o'-sis*) [see *hemosiderin*]. A form of hemochromatosis characterized by the deposit of pigments containing iron in the tissues, especially those of the liver and spleen.

**hemosozic** (*hem-o-so'-zik*) [*hemo-*; σώζειν, to save]. Preventing hemolysis; relating to a hemosozin.

**hemosozin** (*hem-o-so'-zin*) [*hemo-*; σώζειν, to save]. Antihemolysin; an antiserum which prevents hemolysis.

**hemospasia** (*hem-o-spa'-ze-ah*) [*hemo-*; σπάειν, to draw]. The drawing of blood to a part, as by drycupping.

**hemospast** (*hem'-o-spast*) [see *hemospasia*]. A device for drawing blood to a part.

**hemospastic** (*hem-o-spas'-tik*) [*hemo-*; σπάειν, to draw]. Effecting or pertaining to hemospasia.

**hemospermatism.** See *hematospermia.*

**Hemosporidium** (*hem-o-spo-rid'-e-um*). Any species of sporozoa living in the blood, e. g., the *Plasmodium malariæ.*

**hemostasin** (*hem-os'-tas-in*). Trade name of a preparation of suprarenal extract.

**hemostasis, hemostasia** (*hem-os'-ta-sis, hem-o-sta'-se-ah*) [*hemo-*; στάσις, a standing]. 1. Stagnation of the blood-current. 2. Arrest of a flow of blood.

**hemostat** (*hem'-o-stat*) [see *hemostasia*]. 1. Hemostatic forceps. 2. A proprietary external remedy for nosebleed said to consist of tannin, quinine sulphate, and benzoated fat.

**hemostatic** (*hem-o-stat'-ik*). 1. Arresting hemorrhage. 2. An agent or remedy that arrests hemorrhage. **h. forceps.** See *forceps, hemostatic.* **h., Martin's.** See *Martin's hemostatic.* **h., Pavesi's,** a mixture of collodion, 100 parts; phenol, 10 parts; pure tannin, 5 parts; benzoic acid, 3 parts. Syn., *Pavesi's styptic collodion.*

**hemostatics** (*hem-o-stat'-iks*). See *hemastatics.*

**hemostatin** (*hem-o-stat'-in*). An extract from the thymus of calves, containing sodium hydroxide and calcium chloride.

**hemosterol** (*hem-os'-ter-ol*). A therapeutic compound from blood of animals.

**hemotachometer** (*hem-o-tak-om'-et-er*) [*hemo-*; τάχος, swiftness; μέτρον, a measure]. An instrument for measuring the rate of flow of blood.

**hemotachometry** (*hem-o-tak-om'-et-re*) [*hemo-*; τάχος, swiftness; μέτρον, measure]. The estimation of the rapidity of blood-circulation.

**hemotelangiosis, hæmotelangeiosis** (*hem-o-tel-an-je-o'-sis*) [*hemo-*; τέλα, a web; ἀγγεῖον, a vessel]. 1. Disease of the finest capillaries. 2. Telangiectasis.

**hemothorax** (*hem-o-tho'-raks*) [*hemo-*; θώραξ; the chest]. An accumulation of blood in a pleural cavity.

**hemotoxic** (*hem-o-toks'-ik*). See *hematoxic.* **h. sensitizer,** Metchnikoff's name for the intermediary body.

**hemotoxin** (*hem-o-toks'-in*) [*hemo-*; τοξικόν, a poison]. 1. A cytotoxin from defibrinated blood. 2. A soluble substance secreted by bacteria and capable of destroying red blood corpuscles. And see *leukocidin.*

**hemotoxis** (*hem-o-toks'-is*). Blood-poisoning.

**hemotropic** (*hem-o-trop'-ik*) [*hemo-*; τρέπειν, to turn]. Applied to the haptophore by which the intermediary body combines with the corpuscle.

**hemp** [ME.]. *Cannabis sativa,* the bast-fiber of which is used for textile purposes. **h., Indian.** See *Cannabis indica.*

**hemuresis** (*hem-u-re'-sis*). Synonym of *hematuria.*

**henbane.** See *hyoscyamus.*

**Henke's retrovisceral space** (*hen'-ker*) [Philipp Jakob Wilhelm *Henke,* German anatomist, 1834–1896]. The prevertebral space of the thorax which is continuous with the cervical space and is filled with areolar and fatty tissue. **H.'s triangle,** or *trigone,* the inguinal triangle, formed by the lateral border of the rectus muscle and the descending portion of the inguinal fold.

**Henle's ampulla** (*hen'-le*) [Friedrich Gustav Jakob *Henle,* German anatomist, 1809–1885]. 1. The fusiform dilatation of the vas deferens near its junction with the seminal vesicle. 2. The expanded outer half of the Fallopian tube. **H. canal of,** a portion of the uriniferous tubules. **H.'s cells,** large cells with granular protoplasm and one or more relatively small nuclei in the seminiferous tubules. **H.'s fenestrated membrane,** the subendothelial fibroelastic layer of the tunica intima of an artery. **H. fibrin of,** a light flocculent precipitate which is separated when semen is diluted with water. **H.'s fissures,** in the muscular fibers of the myocardium. **H.'s glands,** tubular glands found in the palpebral conjunctiva. **H.'s internal cremaster,** the smooth muscular fibers, remains of the gubernaculum, surrounding the vas deferens and the vessels of the spermatic cord. **H.'s ligament,** the inner portion of the conjoined tendon which is chiefly attached to the sheath of the rectus muscle. **H.'s loop,** the U-shaped section of a uriniferous tubule which is formed by a descending and an ascending loop-tube. **H.'s membrane.** See *Bruch's layer.* **H.'s outer fibrous layer,** the zone of cone-fibers at the margin of the fovea centralis. **H.'s sheath.** 1. The perineural sheath. 2. The cellular layer forming the outer portion of the inner root-sheath of the hair. **H.'s sphincter,** the striated muscular fibers which encircle the prostatic and membranous portions of the urethra. **H.'s spine,**

**suprameatal spine;** an inconstant small spine at the junction of the posterior and superior walls of the external auditory meatus. It serves as a landmark in trephining the mastoid process. **H.'s stratum nerveum.** See *Bruecke's tunica nervea.* **H.'s tube,** looped portion of the uriniferous tube of the kidneys.

**henna** (*hen'-ah*) [Arab., *Khanna*, henna]. A cosmetic much used in the Orient; it is prepared from the leaves of *Lawsonia alba*, and is sometimes used externally and internally in leprosy and in skin diseases.

**henocardia** (*hen-o-kar'-de-ah*) [εἶς, ἔν, one; καρδία, heart]. The condition of having but one auricle and one ventricle in the heart; it is normal in some of the lower animals.

**Henoch's purpura** (*hen'-ōks*) [Eduard *Henoch*, German pediatrist, 1820–1910]. A variety of purpura with gastrointestinal symptoms occurring chiefly in young subjects; also a rapidly fatal form of purpura (*purpura fulminans*).

**Henoch-Bergeron's disease.** See *Bergeron's disease.*

**henogenesis** (*hen-o-jen'-es-is*) [εἶς, ἔν, one; γένεσις, origin]. In biology, the developmental history of an individual organism; ontogenesis.

**henosis** (*hen-o'-sis*) [ἔνωσις, uniting]. 1. Healing or uniting. 2. Symblepharon.

**henotic** (*hen-ot'-ik*) [*henosis*]. Tending to heal or to promote union.

**henpuye** (*hen-poo'-ye*) [West African]. See *goundou.*

**henry** (*hen'-re*) [Joseph *Henry*, American physicist, 1797–1878]. The unit of electrical induction. An electromotive force of one volt is induced by a circuit with a variation of current at the rate of one ampère a second.

**Henry's law** (*hen'-re*) [William *Henry*, English chemist, 1775–1836]. See *Dalton's law.*

**Hensen's canal** (*hen'-sen*) [Victor *Hensen*, German physiologist, 1835– ]. The short vertical tube (1 mm. long and 0.5 mm. wide) connecting the blind extremity of the cochlear canal with the saccule. Syn., *canalis reuniens.* **H.'s cells,** columnar epithelial cells found in the organ of Corti. **H.'s disc, H.'s stria,** the colorless transverse band ḥi̇̄ ḥ divides a dark (anisotropic), sarcous element in the middle. **H.'s node,** in the embryo, an accumulation of cells at the anterior end of the primitive streak, through which the neurenteric canal passes from the outside into the blastodermic vesicle.

**Hensing's fold or ligament** (*hen'-sing*) [Friedrich Wilhelm *Hensing*, German anatomist, 1719–1745]. The superior ligament of the cecum. A more or less triangular fold of the peritoneum which is attached to the abdominal wall, from the lower extremity of the kidney to the iliac fossa, by its lower border, and to the posteroexternal aspect of the colon, at times also to the cecum, by its anterior or internal border. The apex is fixed in the lumbar fossa, the lower free border extending from the iliac fossa to the intestine. Syn., *parietocolic fold.*

**hepaptosis** (*hep-ap-to'-sis*). See *hepaloptosis.*

**hepar** (*he'-par*) [ἧπαρ, the liver]. 1. The liver. 2. A substance having the color of liver, as *hepar sulphuris.* **h. induratum,** an affection differing from cirrhosis, occurring after long-continued fevers, frequently characterized by melanemic pigmentary deposits in or near the capillaries and hyperplasia of the interacinous connective tissue. **h. lobatum,** a liver having numerous lobes produced by deep fissures, as in syphilitic hepatitis. **h. siccatum,** the dried and powdered liver of swine freed from fat. Dose, in atrophic cirrhosis of liver, 300 gr. (20 Gm.) daily. **h. sulphuris,** potassium sulphide; formerly much used in medicine, now used mainly by homeopathists.

**heparaden** (*hep-ar'-ad-en*) [*hepar*; ἀδήν, a gland]. A therapeutic preparation of liver-substance, 2 parts; **lactose,** 1 part. It is used in icterus. Dose 0.2–1.54 gr. (6–10 Gm.) daily.

**hepatalgia** (*hep-at-al'-je-ah*) [*hepar*; ἄλγος, pain]. Neuralgic pain in the liver.

**hepatalgic** (*hep-at-al'-jik*) [*hepar*; ἄλγος, pain]. Relating to or affected with hepatalgia.

**hepataposteme** (*hep-at-ap-os-te'-mah*) [*hepar*; ἀπόστημα, abscess: *pl., hepatapostemata*]. An abscess of the liver.

**hepatatrophia** (*hep-at-at-ro'-fe-ah*) [*hepar*; ἀτροφία, atrophy]. Atrophy of the liver.

**hepatauxe** (*hep-at-awk'-se*) [*hepar*; αὔξη, increase]. Enlargement of the liver.

**hepatectomize** (*hep-at-ek'-to-mīz*) [*hepar*; ἐκτομή, an excision]. To excise a part of the liver.

**hepatectomy** (*hep-at-ek'-to-me*) [*hepar*; ἐκτομή, a cutting out]. Excision of the liver, wholly or in part.

**hepatemphractic** (*hep-at-em-frak'-tik*) [*hepar*; ἔμφραξις, obstruction]. Relating to hepatemphraxis.

**hepatemphraxis** (*hep-at-em-fraks'-is*) [*hepar*; ἔμφραξις, obstruction]. Hepatic obstruction.

**hepathelcosis** (*hep-ath-el-ko'-sis*) [*hepar*; ἕλκωσις, ulceration]. Ulceration of the liver.

**hepathemia, hepathæmia** (*hep-ath-e'-me-ah*) [*hepar*; αἷμα, blood]. Sanguineous hepatic congestion. Syn., *hepatohemia.*

**hepatic** (*hep-at'-ik*). Pertaining or belonging to the liver. **h. aloes.** See *aloes.* **h. artery.** See *artery.* **h. duct.** See *duct.* **h. lobes,** the natural anatomical divisions of the liver, usually designated as right, left, quadrate, Spigelian, and caudate lobes. **h. plexus.** See *plexus.* **h. starch,** a synonym of glycogen. **h. zones,** certain areas in an hepatic lobule.

**hepatica** (*hep-at'-ik-ah*). 1. Agents affecting the liver. 2. Liverwort; a genus of ranunculaceous plants. *H. triloba* and *H. acutiloba* were formerly esteemed in the treatment of hepatic, renal, and pulmonary complaints.

**hepatico-** or **hepato-** (*hep-at'-ik-o-* or *hep'-at-o-*) [*hepar*]. Prefixes signifying belonging to or relating to the liver.

**hepaticocholecystostcholecystenterostomy** (*hep-at-ik-o-ko-le-sist-ost-ko-le-sist-en-ter-os'-to-me*). An anastomosis between the gall-bladder and hepatic duct, on one hand, and between the intestine and gall-bladder, on the other.

**hepaticocolic, hepaticogastric.** See *hepatocolic, hepatogastric.*

**hepaticoduodenostomy** (*hep-at-ik-o-du-o-den-os'-to-me*). The formation of an artificial communication between the hepatic duct and the duodenum.

**hepaticoenterostomy** (*hep-at-ik-o-en-ter-os'-to-me*). The formation of an artificial communication between the hepatic duct and the intestine.

**hepaticogastrostomy** (*hep-at-ik-o-gas-tros'-to-me*). The formation of an artificial communication between the hepatic duct and the stomach.

**hepaticolithotripsy** (*hep-at-ik-o-lith'-o-trip-se*). Crushing a stone in the hepatic duct.

**hepaticopancreatic** (*hep-at-ik-o-pan-kre-at'-ik*). Relating to the liver and the pancreas.

**hepaticopulmonary** (*hep-at-ik-o-pul'-mon-a-re*) [*hepatico; pulmo*, lung]. Relating to the liver and the lungs.

**hepaticorenal.** See *hepatorenal.*

**hepaticostomy** (*hep-at-ik-os'-to-me*) [*hepar*; στόμα, m uth]. The formation of a fistula in the hepatic duct.

**hepaticotomy** (*hep-at-ik-ot'-o-me*) [*hepatico-*; τέμνειν, to cut]. Incision of the hepatic duct.

**hepatin** (*hep'-at-in*) [*hepar*]. Glycogen.

**hepatirrhagia** (*hep-at-ir-a'-je-ah*). Same as *hepatorrhagia.*

**hepatirrhea** (*hep-at-ir-e'-ah*). See *hepatorrhea.*

**hepatism** (*hep'-at-ism*) [*hepar*]. Derangement of various functions of the body, due or ascribed to some functional or other disorder of the liver.

**hepatitic** (*hep-at-it'-ik*) [*hepar*; ἶτις, inflammation]. Affected with or relating to, hepatitis.

**hepatitis** (*hep-at-i'-tis*) [*hepar*; ἶτις, inflammation]. Inflammation of the liver. **h., indurative,** a form marked by formation of fibrous tissue causing the liver-cells to atrophy from compression. **h., interstitial,** inflammation of the connective tissue of the liver leading in some cases to the formation of abscesses or to softening and atrophy of the glandular structure, and in others to induration and cirrhosis. **h., interstitial, chronic,** cirrhosis of the liver. **h., parenchymatous, acute, acute yellow atrophy** of the liver. **h., suppurative,** abscess of the liver.

**hepatization** (*hep-at-iz-a'-shun*) [*hepar*]. A change of a tissue into a condition in which it resembles the liver, as *hepatization* of the lung. **h., gray,** that in which the hepatized lung tissue is gray. **h., red,** that in which it is red from an excess of blood.

**hepatizon** (*hep-at-i'-zon*) [ἠπατίζειν, to be like the liver]. Chloasma.

**hepato-** (*hep-at-o-*) [*hepar*]. A prefix denoting relation to the liver.

**hepatocace** (hep-at-ok'-as-e) [hepar; κακόs, ill]. Gangrene of the liver.

**hepatocarcinia** (hep-at-o-kar-sin'-e-ah) [hepar; καρκίνος, cancer]. Malignant disease of the liver.

**hepatocele** (hep'-at-o-sēl) [hepato-; κήλη, a hernia]. Hernia of the liver.

**hepatocholangio-enterostomy** (hep-at-o-ko-lan-je-o-en-ter-os'-to-me) [hepato-; χολή, bile; ἀγγεῖον, vessel; ἔντερον, intestine; στόμα, mouth]. Formation of an artificial communication between the liver and the intestine.

**hepatocirrhosis** (hep-at-o-sir-o'-sis) [hepato-; cirrhosis]. Cirrhosis of the liver.

**hepatocolic** (hep-at-o-kol'-ik) [hepato-; colic]. Relating to the liver and the colon.

**hepatocystic** (hep-at-o-sis'-tik) [hepato-; κύστις, bladder]. Pertaining to the liver and the gall-bladder.

**hepatodidymus** (hep-at-o-did'-im-us) [hepato-; δίδυμος, double]. A monster with a double body from the liver up.

**hepatoduodenal** (hep-at-o-dū-od'-en-al) [hepato-; duodenum]. Relating to the liver and the duodenum.

**hepatoduodenostomy** (hep-at-o-dū-od-en-os'-to-me) [hepato-; duodenum; στόμα, mouth]. The formation of an opening from the liver into the duodenum.

**hepatodynia** (hep-at-o-din'-e-ah) [hepato-; ὀδύνη, pain]. Pain in the liver.

**hepatodysentery** (hep-at-o-dis'-en-ter-e). Inflammation of the liver attended with dysentery; hepatic dysentery.

**hepatoenteric** (hep-at-o-en-ter'-ik) [hepato-; ἔντερον, intestine]. Relating to the liver and the intestine.

**hepatogastric** (hep-at-o-gas'-trik) [hepato-; γαστήρ, stomach]. Pertaining to the liver and the stomach.

**hepatogastritis** (hep-at-o-gas-tri'-tis) [hepato-; γαστήρ, stomach; ιτις, inflammation]. Inflammation of both liver and stomach.

**hepatogen** (hep-at'-o-jen). A proprietary preparation containing desiccated liver substance.

**hepatogenic, hepatogenous** (hep-at-o-jen'-ik, hep-at-oj'-en-us) [hepato-; γεννᾶν, to produce]. Produced by or in the liver. h. **icterus**, jaundice caused by the absorption of bile from the liver.

**hepatography** (hep-at-og'-ra-fe) [hepato-; γράφειν, to write]. A description of the liver.

**hepatohemia** (hep-at-o-hem'-e-ah). See hepathemia.

**hepatoid** (hep'-at-oid) [hepato-; εἶδος, likeness]. Resembling a liver or liver-substance.

**hepatolith** (hep'-at-o-lith) [hepato-; λίθος, stone]. Biliary calculus; gall-stone.

**hepatolithectomy** (hep-at-o-lith-ek'-to-me) [hepato-; λίθος, stone; ἐκτομή, excision]. Surgical removal of one or more gall-stones.

**hepatolithiasis** (hep-at-o-lith-i'-as-is) [hepato-; lithiasis]. A diseased condition characterized by the formation of gall-stones in the liver.

**hepatolithic** (hep-at-o-lith'-ik) [hepato-; λίθος, a stone]. Affected with biliary calculi.

**hepatology** (hep-at-ol'-o-je) [hepato-; λόγος, science]. The science of the nature, structure, functions, and diseases of the liver.

**hepatolysin** (hep-at-ol'-is-in) [hepato-; lysin]. A cytolysin acting on liver cells.

**hepatomalacia** (hep-at-o-mal-a'-she-ah) [hepato-; μαλακία, softness]. Softening of the liver.

**hepatomegalia** (hep-at-o-meg-a'-le-ah) [hepato-; μέγας, large]. Enlargement of the liver.

**hepatomelanosis** (hep-at-o-mel-an-o'-sis). Melanosis affecting the liver.

**hepatomphalocele** (hep-at-om-fal'-o-sēl) [hepato-; ὀμφαλός, the navel; κήλη, a hernia]. An umbilical hernia with part of the liver contained in the sac.

**hepatomyeloma** (hep-at-o-mi-el-o'-mah) [hepato-; μυελός, marrow; ὄμα, tumor; pl., hepatomyelomata]. Medullary carcinoma of the liver.

**hepatoncus** (hep-at-ong'-kus) [hepato-; ὄγκος, a tumor]. A tumor or swelling of the liver.

**hepatonecrosis** (hep-at-o-ne-kro'-sis) [hepato-; νέκρωσις, death]. Gangrene of the liver.

**hepatopathy** (hep-at-op'-ath-e) [hepato-; πάθος, disease]. Any disease of the liver.

**hepatoperitonitis** (hep-at-o-per-it-on-i'-tis) [hepato-; peritonitis]. Inflammation of the peritoneal or serous coat of the liver.

**hepatopexy** (hep'-at-o-peks-e) [hepato-; πῆξις, fixation]. Surgical fixation of a floating liver.

**hepatophage** (hep'-at-o-fāj) [hepato-; φαγεῖν, to eat]. A giant-cell peculiar to the liver, which is said to destroy liver-cells.

**hepatophlebitis** (hep-at-o-fleb-i'-tis) [hepato-; phlebitis]. Inflammation of the veins of the liver.

**hepatophlebotomy** (hep-at-o-fleb-ot'-o-me) [hepato-; φλέψ, a vein; τέμνειν, to cut]. The aspiration of blood from the liver.

**hepatophyma** (hep-at-o-fi'-mah) [hepato-; φῦμα, growth]. Any tumor of the liver.

**hepatoportal** (hep-at-o-por'-tal) [hepato-; portal]. Relating to the portal circulation in the liver; portal as distinguished from reniportal.

**hepatopostema** (hep-at-o-pos-te'-mah) [hepato-; ἀπόστημα, abscess]. Abscess of the liver.

**hepatoptosis** (hep-at-op-to'-sis) [hepato-; πτῶσις, a falling]. Synonym of floating liver.

**hepatopulmonary** (hep-at-o-pul'-mon-a-re). See hepaticopulmonary.

**hepatorenal** (hep-at-o-re'-nal) [hepato-; ren, the kidney]. Relating to the liver and the kidney.

**hepatorrhagia** (hep-at-or-a'-je-ah). Hemorrhage from the liver.

**hepatorrhaphy** (hep-at-or'-a-fe) [hepato-; ῥαφή, suture]. Suture of the liver.

**hepatorrhea** (hep-at-or-e'-ah) [hepato-; ῥοία, flow]. Morbid or excessive secretion of bile by the liver.

**hepatorrhexis** (hep-at-or-eks'-is) [hepato-; ῥῆξις, a rupture]. Rupture of the liver.

**hepatoscirrhus** (hep-at-o-skir'-us) [hepato-; σκίρρος, an induration]. Scirrhous carcinoma of the liver.

**hepatoscopy** (hep-at-os'-ko-pe) [hepato-; σκοπεῖν, to examine]. Examination of the liver.

**hepatosplenitis** (hep-at-o-splen-i'-tis) [hepato-; σπλήν, spleen; ιτις, inflammation]. Inflammation of both liver and spleen.

**hepatostomy** (hep-at-os'-to-me) [hepato-; στόμα, a mouth]. The establishment in the liver-substance of communication between parts of the liver obstructed by concretions.

**hepatotherapy** (hep-at-o-ther'-ap-e) [hepato-; therapy]. The therapeutic use of liver.

**hepatotomy** (hep-at-ot'-o-me) [hepato-; τέμνειν, to cut]. Incision of the liver.

**hepatotoxemia** (hep-at-o-toks-e'-me-ah) [hepato-; toxemia]. Toxemia due to disturbance of the hepatic functions.

**hepatotoxin** (hep-at-o-toks'-in) [hepato-; τοξικόν, a poison]. A cytotoxin found by E. Metchnikoff in the liver.

**hepco flour** (hep'-co). A flour prepared from Soya bean with approximately the following composition: protein 42.9; carbohydrate, 23.4 of which less than one half readily yields sugar; fat 20.8; ash 5.1; fiber 4.2; water 4.6. It is used to be a suitable food material in cases in which carbohydrates are contraindicated, as diabetes, amylaceous dyspepsia and in obesity.

**hephestic** (hef-es'-tik) ['Ήφαιστος, Vulcan]. Prevailing or occurring among hammermen, as hephestic cramp.

**hephestiorrhaphy** (he-fes-te-or'-af-e) ['Ήφαιστος, Vulcan; ῥαφή, a seam]. The application of the actual cautery to the edges of a wound to bring about adhesion.

**hepptine** (hep'-tin) [Maurice Hepp, French physician]. Trade name of pure gastric juice obtained from living pigs.

**heptad** (hep'-tad) [ἑπτά, seven]. An element having a quantivalence of seven.

**heptadicity** (hep-tad-is'-it-e) [see heptad]. Septivalence.

**heptane** (hep'-tān) [see heptad], $C_7H_{16}$. A liquid hydrocarbon of the paraffin group, contained in petroleum and also obtained from the resin of Pinus sabiniana by dry distillation. Syn., abietene.

**heptatomic** (hep-tat-om'-ik) [ἑπτά, seven; ἄτομος, an atom]. Same as heptavalens, q. v.

**heptavalent** (hep-tav-a'-lent) [ἑπτά, seven; valens, having power]. In chemistry, equal to seven atoms of hydrogen in combining or saturating-power; applied to an atom that can be substituted for, or replaced by, seven atoms of hydrogen.

**heptoses** (hep-tōs'-ez) [ἑπτά, seven]. A division of the glucoses, of the composition, $C_7H_{14}O_7$. They are prepared by reducing the corresponding heptonic acids, $C_7H_{14}O_8$ with sodium amalgam.

**herapathite** (her'-ap-ath-īt) [W. B. Herapath, English chemist, 1820–1868]. An iodide of quinine-sulphate, occurring in rhomboidal laminæ. It has been used as a remedy in scrofula and in febrile cases; in microscopy it is employed in polarizing light.

**herb** [*herba*, grass]. A plant the stem of which contains but little wood and dies down to the ground at the end of the season.
**herbaceous** (*her-ba'-se-us*) [*herba*, grass]. In biology. 1. Applied to stems or other organs that have a tender, juicy consistence and perish at the close of the growing-season. 2. Feeding upon herbs.
**herbal** (*her'-bal*) [*herba*, grass]. An old name for a book on herbs; chiefly designating a book on the medicinal virtues of herbs.
**herbalist** (*her'-bal-ist*) [*herba*, grass]. An herb-doctor or simpler; a so-called botanic physician.
**herbarium** (*her-ba'-re-um*) [*herba*, grass]. A collection of dried plants arranged for study; a *hortus siccus*.
**Herbert's operation** (*her'-bert*) [Frederick *Herbert*, American ophthalmologist, 1860– ]. An operation for acute glaucoma, in which a wedge-shaped flap is cut in such a way as to prevent the subsequent cicatrization of the two scleral surfaces.
**herbicarnivorous** (*her-be-kar-niv'-or-us*) [*herba*, grass; *caro*, flesh; *vorare*, to eat]. Omnivorous; living upon both animal and vegetable food.
**herbivora** (*her-biv'-or-ah*) [*herba*, grass; *vorare*, to devour]. A name given to a division of mammalia. Animals that feed on vegetation.
**herbivorous** (*her-biv'-or-us*) [*herb*; *vorare*, to devour]. Living on vegetable food.
**Herbst's corpuscles** (*herpst*) [Ernst Friedrich Gustav *Herbst*, German anatomist, 1803–1893]. A variety of sensory end-organs found in the mucous membrane of the tongue of the duck; they resemble small Vater's corpuscles, but their lamellæ are thinner and closer to each other, while the axis-cylinder within the central core is bordered on each side by a row of nuclei.
**hereditary** (*he-red'-it-a-re*) [*heres*, an heir]. Transmitted from parent to offspring, as *hereditary* disease.
**h. ataxia**, a family disease of the young, depending on combined posterior and lateral-sclerosis of the cord. It differs from tabes and ataxic paraplegia in the early age of its appearance, its hereditary nature, and some other features. **h. syphilis**. See *syphilis, hereditary*.
**hereditation** (*her-ed-it-a'-shun*) [*heredity*]. The effect or influence of heredity.
**heredity** (*he-red'-it-e*) [*hereditas*, heredity]. The transmission of physical or mental qualities or tendencies from ancestor to offspring; the principle or force by reason of which the offspring resembles the parent.
**heredo-** (*her-e-do-*). In composition, hereditary.
**Hering's law** (*hā'-ring*) [Ewald *Hering*, German physiologist, 1834– ]. The distinctness or purity of any sensation or conception depends upon the proportion existing between their intensity and the sum-total of the intensities of all simultaneous sensations and conceptions. **H.'s test**, on looking with both eyes through a tube blackened inside and having a thread across one end, if a small round object is dropped immediately in front of or behind the thread, a subject with binocular vision can at once tell whether it has fallen nearer to his eyes or further away from them than the thread. In the absence of binocular vision a few trials will show that the relative distances of the falling object and the thread cannot be appreciated. **H.'s theory of color-sensation**, this predicates disassimilation and assimilation (decomposition and restitution) of the visual substance in vision—white, red, and yellow representing the sensation of disassimilation; black, green, and blue, that of restitution.
**hermaphrodism, hermaphroditism** (*her-maf'-ro-dizm*, *her-maf'-ro-dit-izm*) ['Ερμῆς, Mercury; 'Αφροδίτη, Venus]. The coexistence, in a single individual, of ovaries and testicles. **h., complex**, a condition in which there are present the internal and the external organs of both sexes. **h., dimidiate or lateral**, a form in which male organs (especially a testicle) are more or less developed on one side, and female organs (especially an ovary) on the opposite side. **h., spurious**, a condition in which the individual is of one sex, but presents the outward signs of the other. **h., transverse**, an instance in which the external organs indicate the one sex, and the internal, the opposite. **h., unilateral**, that in which there are on one side an ovary and a testicle, and, on the other, an ovary or a testicle.
**hermaphrodite** (*her-maf'-ro-dīt*) [see *hermaphrodism*]. An individual affected with hermaphrodism; usually the condition is due to some congenital malformation of the genital organs, such as epispadias, hypospadia, cleft of the scrotum, etc., that makes the determination of sex somewhat doubtful.
**hermaphroditic** (*her-maf-ro-dit'-ik*) ['Ερμῆς, Mercury; 'Αφροδίτη, Venus]. Pertaining to hermaphroditism, *q. v.* See also *teratism*.
**hermetic** (*her-met'-ik*). ['Ερμῆς, Mercury]. Protected from exposure to air; air-tight, as the *hermetic* sealing of a wound.
**hermitine** (*her'-mit-ēn*). The proprietary name for a surgical antiseptic and disinfectant, said to be electrolyzed sea-water. Cf. *electrozone*.
**hermophenol, hermophenyl** (*her-mo-fe'-nol*, *-nil*). A mercuriosodic phenol disulphonate containing 40 % of metallic mercury, used as an antiseptic and antisyphilitic. On wounds, in a solution of 1 : 100; injection in syphilis, 64 min. (4 Cc.) of a solution of 0.5 cg. to the cubic centimeter every 2 or 3 days. Syn., *sodium mercurophenyl disulphonate*.
**hermophilia** (*her-mo-fil'-e-ah*) ['Ερμῆς, Mercury; φιλεῖν, to love]. A predilection for the therapeutic use of mercury.
**hernia** (*her'-ne-ah*) [L.]. A protrusion of a viscus through an abnormal opening in the wall of the containing cavity; used without qualification, the word refers to hernia of the intestine. **h., abdominal**, a protrusion of a portion of the abdominal viscera through some portion of the parietes. **h. adiposa**, a liparocele. **h. adnata**, a congenital hernia. **h. annularis**, umbilical hernia. **h. of the bladder**, protrusion of a part of the bladder through one of the openings of the abdominal cavity. **h. of the brain**, crural. **h. capitis**, **h. cerebri**, a protrusion of the brain through the skull. Syn., *fungus cerebri*. **h., complete**, a hernia in which the sac and its contents have passed the hernial orifice. **h., concealed**, one not perceptible on palpation. **h., congenital**, a form of indirect inguinal hernia in which, the vaginal process of the peritoneum having remained patulous, the bowel descends at once into the scrotum, in direct contact with the testicle. **h., Cooper's**. See *Cooper's hernia*. **h. cordis**, displacement of the heart with encroachment on the diaphragm or mediastinal wall. **h., crural**. **h., cystic**. Same as *cystocele*. **h., diaphragmatic**, a protrusion of a portion of some of the abdominal viscera into the thorax, through a congenital defect in the diaphragm, or through a dilatation or laceration of one of the natural openings. **h., displaced**, one that has been forced from the scrotum into the subperitoneal connective tissue of the abdomen. **h., diverticular**, hernia of a congenital diverticulum of the intestine; hernia of Meckel's diverticulum. Syn., *Littré's hernia*. **h. dolorosa**, a painful incarcerated hernia. **h., duodenojejunal**. See *h., retroperitoneal*. **h., encysted**, a form in which the pouch forming the tunica vaginalis is closed at its upper end, but open below. The hernia in descending along the inguinal canal enters the scrotum behind the tunica vaginalis, and is more or less completely surrounded by its posterior layer. **h., epiploic**, **h. epiploica**. See *epiplocele*. **h., fatty**. See *liparocele* and *steatocele*. **h., femoral**, a hernia through the femoral canal, the tumor appearing on the upper inner aspect of the thigh, below Poupart's ligament. **h., femoroinguinal**, a femoral hernia coexistent with an inguinal hernia. **h., free**, a reducible hernia. **h., funicular**, one with the umbilical cord. **h., Hesselbach's**. See *Hesselbach's hernia*. **h., Holthouse's**. See *Holthouse's hernia*. **h., incarcerated**, a hernia which has become occluded by the accumulation of gas, feces, or undigested food, thus causing obstruction of the bowels. **h., incomplete**, one that has not entirely passed through the hernial orifice. **h., infantile**. See *h., encysted*. **h., inguinal**, a hernia occupying the inguinal canal. **h., inguinocrural**. Same as *Holthouse's hernia, q. v.* **h., intercostal**, a protrusion through the last costal interspaces, due to an intercostal wound. **h., interstitial**, a displaced hernia in which the sac has found a way between two layers of aponeurosis. **h. of iris**, a protrusion of a portion of the iris after iridectomy, trauma, etc. **h., irreducible**, one in which the protruded viscus cannot be returned by manipulation. **h., ischiatic**, a protrusion of the bowel through the great sacrosciatic foramen. **h.-knife**, a probe-pointed knife for incising the constriction of a hernial sac. **h., labial**, a protrusion of the bowel between the vagina and the ramus of the ischium into a labium

majus. h., lacrimal, h. of the lacrimal sac, h. lacrimalis, protrusion of the mucosa of the lacrimal sac through an opening in its anterior wall. h., lateral, h. lateralis, diverticular hernia. h., Lavater's. See *Richter's hernia*. h. lienalis, hernia of the spleen. h. of Littré. See *Littré's hernia*. h., Malgaigne's. See *Malgaigne's hernia*. h., mesocolic, a protrusion of the bowel between two layers of the mesocolon. h., obstructed. See *h., incarcerated*. h., obturator, a protrusion of bowel through the obturator foramen. h., omental, a hernia containing omentum; epiplocele. h., pectineal, one that, having made its way internal to and behind the femoral vessels, rests upon the pectineus muscle. h., perineal, a protrusion of the abdominal contents between the fibers of the levator ani muscle in front of or to one side of the anus. h., properitoneal, one within the abdominal walls in front of the peritoneum. h., pudendal. Same as *h., labial*. h., rectal. 1. See *proctocele*. 2. A protrusion of part of the pelvic or abdominal contents through the anus, held in a sac formed by eversion of the rectum. h., reducible, one that may be returned by manipulation. h., retrocecal, a protrusion of the bowel occupying an inconstant pouch dorsad of the cecum. Syn., *Rieux's hernia*. h., retroperitoneal, a hernia in which the intestine lodges in the fossa duodenojejunalis. h., Richter's. See *Richter's hernia*. h., Rieux's. See *h., retrocecal*. h., sacrorectal, posterior proctocele from defective ossification of the sacrum. h., scrotal, that form of inguinal hernia in which the protrusion has entered the scrotum. h., strangulated, a hernia which is so tightly constricted at its neck as to interfere with its return, with the circulation of blood, and the passage of feces. h., umbilical, a protrusion of the abdominal contents through the umbilicus. h., umbilicovesical, hernia of the bladder through the umbilicus. h., vaginal, one protruding into the vagina. h., ventral, the name applied to protrusions of the abdominal contents through the abdominal walls in situations not usually subject to hernia.
 hernial (*her'-ne-al*). Pertaining to hernia. h. sac, the diverticulum of the peritoneum which the hernia pushes before it or into which it descends.
 herniate (*her'-ne-āt*). To form a hernia.
 herniation (*her-ne-a'-shun*). The formation of a hernia.
, herniocoeliotomy (*her-ne-ō-se-le-ot'-o-me*) [*hernia*, a rupture; κοιλία, belly]. Abdominal section for the relief of hernia.
 hernioenterotomy (*her-ne-o-en-ter-ot'-o-me*) [*hernia*, a rupture; ἔντερον, bowel; τομή, a cutting]. Herniotomy combined with enterotomy.
 herniolaparotomy (*her-ne-o-lap-ar-ot'-o-me*) [*hernia*, a rupture; λαπάρα, the flank; τομή, a cutting]. Same as *herniocoeliotomy*.
 herniology (*her-ne-ol'-o-je*) [*hernia*, hernia; λόγος, science]. That department of surgery which treats of the causes, diagnosis, and treatment of hernia.
 hernioplasty (*her'-ne-o-plas-te*). The operation for the radical cure of hernia.
 herniopuncture (*her-ne-o-punk'-chūr*) [*hernia*; *punctura*, a pricking]. The puncture of a hernia.
 herniotome (*her'-ne-o-tōm*) [*hernia*; τομή, a cutting]. A hernia knife.
 herniotomy (*her-ne-ot'-o-me*) [see *herniotome*]. Operation for the relief of hernia by section of the constriction.
 heroic (*he-ro'-ik*) [ἥρως, a hero]. Bold or daring; rash or unusually severe; applied usually to medical treatment by large doses or by measures involving risk.
 heroin, heroine (*her'-o-in*), C₁₇H₁₇NO(C₂H₃O₂)₂. The diacetic-acid ester of morphine. It is anodyne and sedative, and is used in coughs, dyspnea, and pectoral pains. Dose ¹⁄₁₂–⅓ gr. (0.005–0.032 Gm.).
 heromal (*her'-om-al*). A proprietary remedy for respiratory disorders said to contain malt extract, hypophosphites, and heroine.
 Herophilus, torcular of (*her-of'-il-us*) [*Herophilus*, Greek physician, 335–280 B. C.]. The dilatation at the junction of the superior longitudinal, straight, two lateral, and two occipital sinuses.
 hēroterpine (*her-o-tur'-pēn*). A combination of heroine and terpine hydrate, indicated in bronchitis, asthma, etc.
. herpes (*her'-pēz*) [ἕρπειν, to creep]. An acute inflammatory affection of the skin or mucous membrane, characterized by the development of groups of vesicles on an inflammatory base. h. circinatus.

See *tinea circinata*. h. exedens, a general term for the varieties of herpes and lupus characterized by hard vesicles in thronged clusters and containing dense reddish or yellow fluid. h. facialis, an acute, noncontagious, inflammatory disorder of the skin that appears in the form of one or more groups of vesicles. It is commonly called fever-blisters, a form of herpes appearing especially on the lips. It is frequent in "cold," malaria, croupous pneumonia, and cerebrospinal meningitis. Syn., *herpes febrilis*. h. febrilis. See *h. facialis*. h. gestationis, herpes of the limbs in pregnancy. h. iris, a form of erythema with vesicles growing in a ring. It is usually seen on the backs of the hands and feet. h. labialis. See *h. facialis*. h. præputialis, h. progenitalis, a form of herpes in which vesicles, the size of a pin's head to that of a small pea, occur upon the glans penis and prepuce. h. pyæmicus, impetigo herpetiformis. h. tonsurans. See *tinea tonsurans*. h. zoster, herpes in which the lesions are distributed in relation to the course of a cutaneous nerve, and, as a rule, unilateral. They are usually seen in the line of the intercostal nerves, but may follow the course of any nerve. The outbreak of the eruption is generally preceded by severe neuralgic pain. Syn., *ignis sacer*; *shingles*; *zona*; *zoster*.
. herpetic (*her-pet'-ik*) [*herpes*]. Pertaining to herpes. h. sore throat. See *tonsillitis, herpetic*.
 herpetiform (*her-pet'-if-orm*) [*herpes*; *forma*, form]. Resembling herpes.
. herpetism (*her'-pet-izm*). A constitutional tendency to eruptions of herpes.
 herpetography (*her-pet-og'-ra-fe*). Same as *herpetology*.
 herpetology (*her-pet-ol'-o-je*) [1] [ἑρπετόν, a reptile; λόγος, discourse]. The classified knowledge of reptiles. [2] [ἕρπης, herpes; λόγος, treatise]. The science of skin-diseases, especially those of an herpetic nature.
 Herpetomonas. (*her-pet-om'-on-as*) [ἑρπετόν, a reptile; μονάς, unit]. A genus of flagellated infusorians found in the intestines of insects and in the blood of various animals. H. donovani, the parasite of kala-azar; it is transmitted by the bite of the bedbug.
 Herxheimer's spiral fibers (*hĕrks'-hī-mer*) [Karl *Herxheimer*, German dermatologist, 1861– ]. Spiral fibers found in the rete mucosum of the epidermis. H.'s reaction, the appearance of a maculo-papular eruption, deafness, or blindness, from the sudden onset of neuritis, following the treatment of syphilis by salvarsan, or the cacodylates.
. Heryng's benign ulcer (*her'-ing*) [Théodor *Heryng*, Polish laryngologist, 1847– ]. A solitary ulcer situated on the anterior fauces and resembling a large herpetic vesicle. H.'s sign, an infraorbital shadow observed on introducing an electric light into the mouth in empyema of the antrum of Highmore.
 Herzberg's reagent for free hydrochloric acid. Moisten paper with a solution of Congo red; when dried, it turns blue or bluish-black upon being moistened with hydrochloric acid.
 Hesselbach's hernia (*hes'-el-bakh*) [Franz Kasper *Hesselbach*, German surgeon, 1759–1816]. A lobulated hernia passing through the cribriform fascia. H.'s ligament, the ligamentum interfoveolare; a thin, fibrous, band extending from the posterior surface of the fascia transversalis, near the plica semilunaris, to the pubic bone and Gimbernat's ligament; it forms part of the conjoined tendon. H.'s triangle, a space bounded by Poupart's ligament below, the external border of the rectus abdominis internally, and the deep epigastric artery externally. Direct inguinal hernia occurs in this space.
 heterocephalous (*het-er-o-sef'-al-us*) [ἕτερος, other; κεφαλή, head]. See *heterocephalous*.
 heteradelphia (*het-er-ad-el'-fe-ah*) [*hetero-*; ἀδελφός, brother]. Heteradelphous teratism.
 heteradelphous (*het-er-ad-el'-fus*) [*hetero-*; ἀδελφός, brother]. Relating to an heteradelphus, or to an autosite and its parasite.
 heteradelphus (*het-er-ad-el'-fus*) [*hetero-*; ἀδελφός, brother]. A joined twin monster, consisting of an autositic monster with an attached parasite, the head of the latter being absent.
 heteradenia (*het-er-ad'-ne-ah*) [*hetero-*; ἀδήν, a gland]. 1. Normal glandular structure occurring in a part normally not provided with glands. 2. Glandular structure departing from the normal type.

**heteradenic** (*het-er-ad-e'-nik*) [see *heteradenia*]. Pertaining to or consisting of tissue that is unlike normal glandular tissue, or to glandular tissue occurring in an abnormal place.
**heteradenoma** (*het-er-ad-en-o'-mah*) [*hetero-*; ἀδήν, gland; ὅμα, tumor: *pl.*, *heteradenomata*]. A tumor formed of heteradenic tissue.
**heteralius** (*het-er-a'-le-us*) [*hetero-*; ἅλως, a disc], A double monster in which the parasite is very incomplete, and with no direct connection with the umbilical cord of its host.
**heterauxesis** (*het-er-awks-e'-sis*) [*hetero-*; αὔξησις, increase]. In biology, any unsymmetrical growth, normal or abnormal.
**heterecious** (*het-er-e'-se-us*) [*hetero-*; οἶκος, a house]. Parasitic upon different hosts at different stages of growth.
**heterecism, heterœcism** (*het-er-ē'-sism*) [*hetero-*; οἶκος, a house]. Parasitism upon one host during one stage of growth or generation, and upon another host for the development of another stage or generation.
**hetero-** (*het-er-o-*) [ἕτερος, other]. A prefix denoting unlikeness.
**heteroagglutinin** (*het-er-o-ag-lu'-tin-in*). An agglutinin formed in the blood of an animal as the result of the injection of an antigen from an animal of a different species.
**heteroalbumose** (*het-er-o-al'-bū-mōs*) [*hetero-*; *albumose*]. A variety of albumose soluble in salt solutions, insoluble in water, and precipitated by saturation with sodium chloride or magnesium sulphate.
**heteroalbumosuria** (*het-er-o-al-bū-mo-sū'-re-ah*). The presence of heteroalbumose in the urine.
**heteroblastic** (*het-er-o-blas'-tik*) [*hetero-*; βλαστός, germ; bud]. Arising from tissue of a different kind.
**heterocele** (*het'-er-o-sēl*) [*hetero-*; κήλη, hernia]. A hernia existing in some prolapsed organ, as in a rectocele.
**heterocelous, heterocœlous** (*het-er-o-se'-lus*) [*hetero-*; κοῖλος, hollow]. Convexoconcave.
**heterocentric** (*het-er-o-sen'-trik*) [*hetero-*; κέντρον, center]. Applied to rays that do not meet in a common center.
**heterocephalus** (*het-er-o-sef'-al-us*) [*hetero-*; κεφαλή, the head]. A fetal monstrosity with two heads of unequal size.
**heterochromatosis** (*het-er-o-kro-mat-o'-sis*) [*hetero-*; χρῶμα, color]. 1. Pigmentation of the skin due to substances foreign to the body. 2. See *heterochromia*.
**heterochromia** (*het-er-o-kro'-me-ah*) [see *heterochromatosis*]. A difference in color, as of the irides of the two eyes, or different parts of the same iris.
**heterochromous** (*het-er-o-kro'-mus*) [*hetero-*; χρῶμα, color]. In biology, having different colors.
**heterochronia** (*het-er-o-kro'-ne-ah*) [*hetero-*; χρόνος, time]. The production of a structure or the occurrence of a phenomenon at an abnormal period of time.
**heterochronic, heterochronous** (*het-er-o-kron'-ik*, *het-er-ok'-ron-us*) [see *heterochroma*]. Irregular in occurrence. Occurring at different times, or at other than the proper time.
**heterochylia** (*het-er-o-ki'-le-ah*) [*hetero-*; χυλός, chyle]. A variable condition of the gastric contents, changing suddenly from normal acidity to hyperacidity or anacidity.
**heterocrania** (*het-er-o-kra'-ne-ah*) [*hetero-*; κρανίον, skull]. 1. Asymmetry of the cranium. 2. Headache involving but one side of the head.
**heterocrisis** (*het-er-o-kri'-sis*) [*hetero-*; κρίσις, a crisis]. An abnormal crisis in disease.
**heterocyclic** compound (*het-er-o-si'-klik*) [*hetero-*; κύκλος, a circle]. A closed chain organic compound in which atoms of elements other than carbon enter into the composition of the ring.
**heterodermotrophy** (*het-er-o-der-mot'-ro-fe*) [*hetero-*; δέρμα, skin; τροφή, nutrition]. Disordered or perverted nutrition of the skin.
**heterodesmotic** (*het-er-o-dez-mot'-ik*) [*hetero-*; δεσμός, a bond]. Connecting other parts; applied to nerve-fibers connecting centers of unequal value or associating nervous centers with each other.
**heterodont** (*het'-er-o-dont*) [*hetero-*; ὀδούς, tooth]. In biology, having more than one sort of teeth, as incisors, canines, molars; the opposite of *homodont*.
**heterodymus** (*het-er-od'-im-us*) [*hetero-*; δίδυμος, twin]. A double monster, the accessory part being but an imperfect head, with a neck and thorax by

which it is implanted in the anterior abdominal wall of its host.
**heteroepidermic** (*het-er-o-ep-e-dur'-mik*) [*hetero-*; *epidermis*]. Pertaining to or taken from the skin of some other person; a form of skin-grafting (*q. v.*).
**heterogametous** (*het-er-o-gam'-e-tus*) [*hetero-*; μάμος, marriage]. Pertaining to an individual having both dominant and recessive germ-cells.
**heteroganglionic** (*het-er-o-gan-gle-on'-ik*). Relating to different ganglia; applied to the connecting nerve-fibers between ganglia.
**heterogeneity** (*het-er-o-jen-e'-it-e*) [*hetero-*; γένος, kind]. The condition or quality of being heterogeneous.
**heterogeneous** (*het-er-o-je'-ne-us*) [*hetero-*; γένος, kind]. Differing in kind or nature; composed of different substances; not homogeneous. h. **vaccine**, a vaccine derived from organisms outside of the patient in whose treatment they are to be used. See *autogenous vaccine*.
**heterogenesis** (*het-er-o-jen'-es-is*) [*hetero-*; γένεσις, generation]. A mode of reproduction in which the living parent gives rise to offspring that pass through totally different series of states from those exhibited by the parents, and do not return into the cycle of the parents.
**heterogenetic** (*het-er-o-jen-et'-ik*) [see *heterogenesis*]. Pertaining to heterogenesis.
**heteroglaucous** (*het-er-o-glaw'-kus*) [*hetero-*; γλαυκός, bluish-green]. 1. Having one eye blue and the other black or gray. 2. Relating to the anomalous production of greenish or glaucous spots.
**heterognathous** (*het-er-og'-nath-us*) [*hetero-*; γνάθος, jaw]. Having dissimilar jaws.
**heterogony** (*het-er-og'-o-ne*) [*hetero-*; γόνος, generation]. A form of reproduction that consists in the occurrence, in the cycle of development, of individuals differing in structure from the parent forms and existing under special conditions of nutrition.
**heteroid, heteroideous** (*het'-er-oid*, *-oid'-e-us*) [ἕτερος, other]. Formed diversely; applied to inclosed structures which differ from their investment.
**heteroinfection** (*het-er-o-in-fek'-shun*) [*hetero-*; *infection*]. 1. Infection transmitted by a person who is himself not affected. 2. Infection of any organism by a poison not produced within itself; opposed to *autoinfection*.
**heteroinoculation** (*het-er-o-in-ok-ū-la'-shun*) [*hetero-*; *inoculation*]. Inoculation of one person by another.
**heterolalia** (*het-er-o-la'-le-ah*) [*hetero-*; λαλία, talk]. The utterance of words other than those intended by the speaker; heterophemy.
**heterologous** (*het-er-ol'-o-gus*) [*hetero-*; λόγος, relation]. Differing in structure or form from the normal. h. **tumors**, tumors constituted of a tissue different from that of the part in or on which they are situated.
**heterology** (*het-er-ol'-o-je*) [*hetero-*; λόγος, relation]. Abnormality in nature, form, or structure; development of an abnormal structure.
**heterolopia, heterolopy** (*het-er-o-lo'-pe-ah*, *het-er-ol'-o-pe*) [*hetero-*; λοπός, scale]. The presence of abnormal scales, crusts, or scabs.
**heterolysin** (*het-er-ol'-is-in*) [*hetero-*; λύσις, solution]. A cytolysin produced in the body of one species of animal by the introduction of blood from a different species. Cf. *isolysin*.
**heterolysis** (*het-er-ol'-is-is*) [see *heterolysin*]. The hemolytic action of the blood-serum of one animal upon the corpuscles of another species. Cf. *isolysis*.
**heterolytic** (*het-er-o-lit'-ik*). Pertaining to or produced by heterolysis or a heterolysin.
**heteromeral, heteromeric** (*het-er-om'-er-al*, *-ik*) [*hetero-*; μέρος, part]. Applied to neurons originating in one lateral side of the spinal cord and sending processes to the other side. Cf. *hecatomeral*.
**heteromerous** (*het-er-om'-er-us*) [*hetero-*; μέρος, a part]. 1. Having homologous parts diversely composed. 2. Unlike in chemical composition
**heterometry** (*het-er-om'-et-re*) [*hetero-*; μέτρον, a measure]. Deviation from the normal state in a part, in regard to the amount of its contents.
**heteromorphism** (*het-er-o-mor'-fism*) [*hetero-*; μορφή, form]. A condition marked by difference in form, as compared with the normal form. In chemistry, the property of crystallizing in different forms. In biology: 1. A state of deviation from a type or norm.

2. Exhibiting different forms at different stages in the life-history.
**heteromorphosis** (het-er-o-mor-fo'-sis) [hetero-; μόρφωσις, formation]. Malformation or deformity; any disease characterized by deformity.
**heteromorphous** (het-er-o-mor'-fus) [hetero-; μορφή, form]. Differing from the normal in form.
**heteronephrotrophy** (het-er-e-nef-rot'-ro-fe) [hetero-; νεφρός, kidney; τροφή, nutrition]. Malnutrition or degeneration of any part of the kidney.
**heteronomous** (het-er-on'-o-mus) [hetero-; νόμος, law]. In biology, diversification in any series or set of morphologically related structures through specialization. Abnormal.
**heteronomy** (het-er-on'-o-me) [hetero-; νόμος, law]. 1. Subordination to a law of adaptive modification. 2. The presence of segmentation. Cf. *autonomy* and *homonomy*.
**heteronymous** (het-er-on'-im-us) [hetero-; ὄνομα, name]. On opposite sides; not homonymous; applied to crossed double visual images, such as are seen when there is a relative divergence of the eyes.
**hetero-osteoplasty** (het-er-o-os'-te-o-plas-te) [hetero-; ὀστέον, bone; πλάσσειν, to form]. The surgical grafting of bone, especially with a graft taken from a bone of one of the lower animals.
**heteropagus** (het-er-op'-ag-us) [hetero-; πάγος, fixture]. A double monster in which the parasite, having a head and extremities, is attached to the anterior abdominal wall of its host.
**heteropathic** (het-er-o-path'-ik) [hetero-; πάθος, disease]. Pertaining to or making use of heteropathy.
**heteropathy** (het-er-op'-ath-e) [hetero-; πάθος, suffering]. 1. The treatment of a disease by inducing a different morbid condition to neutralize it. Allopathy. 2. Abnormal reaction to stimulus or irritation.
**heterophasia** (het-er-o-fa'-ze-ah). See *heterophemy*.
**heterophemia,** **heterophemy** (het-er-o-fe'-me-ah, het-er-of'-em-e) [hetero-; φήμη, utterance]. The saying of one thing while another is intended.
**heterophonia** (het-er-o-fo'-ne-ah) [hetero-; φωνή, voice]. Abnormal quality or perversion of the voice.
**heterophoralgia** (het-er-o-for-al'-je-ah) [heterophoria; ἄλγος, pain]. Eye-strain or ocular pain caused by heterophoria.
**heterophoria** (het-er-o-fo'-re-ah) [hetero-; φορός, tending]. A relation of the visual lines of the two eyes other than that of parallelism. *Esophoria* is a tending of the lines inward; *exophoria*, outward; *hyperphoria*, a tending of the right or left visual line in a direction above its fellow; *hyperesophoria*, a tending of the visual lines upward and inward; *hyperexophoria*, upward and outward.
**heterophthalmos** (het-er-of-thal'-mos). See *heterochromia*.
**heterophthongia** (het-er-off-thong'-e-ah) [hetero-; φθόγγος, sound]. Synonym of *ventriloquism*.
**heteroplasia** (het-er-o-pla'-ze-ah) [hetero-; πλάσσειν, to form]. The presence, in a part, of a tissue that does not belong there normally.
**heteroplasm** (het'-er-o-plazm) [see *heteroplasia*]. Abnormal or false tissue.
**heteroplastic** (het-er-o-plas'-tik) [hetero-; πλάσσειν, to form]. 1. Relating to heteroplasia. 2. Differing in structure.
**heteroplastid** (het-er-o-plas'-tid) [hetero-; πλάσσειν, to form]. A surgical graft.
**heteroplasty** (het'-er-o-plas-te) [see *heteroplasia*]. 1. Heteroplasia. 2. The operation of grafting parts taken from another species.
**heteroprosopus** (het-er-o-pro-so'-pus) [hetero-; πρόσωπον, face]. A fetus with two faces; janus or janiceps.
**heteroproteose** (het-er-o-pro'-te-ōs). A product of the digestion of syntonin in the stomach.
**heteropsychology** (het-er-o-si-kol'-o-je) [hetero-; ψυχή, soul; λόγος, treatise]. The study or science of psychology, as based upon facts other than those of one's own subjective experiences.
**heteroptics** (het-er-op'-tiks) [hetero-; ὀπτικός, belonging to sight]. 1. Clairvoyance. 2. Perverted vision.
**heterorexia** (het-er-o-reks'-e-ah) [hetero-; ὄρεξις, desire]. Perversion of the appetite.
**heteroscope** (het'-er-o-skōp) [hetero-; σκοπεῖν, to examine]. An apparatus for the accurate measurement of the various angles at which a deviating eye in strabismus can see.

**heterosexuality** (het-er-o-seks-ū-al'-it-e). Perverted sexual feeling toward one of the opposite sex.
**heterostomy** (het-er-os'-to-me) [hetero-; στόμα, mouth]. Lack of symmetry in the two sides of the mouth.
**heterotaxia** (het-er-o-taks'-e-ah). See *heterotaxis*.
**heterotaxis** (het-er-o-taks'-is) [hetero-; τάξις, order]. The anomalous disposition or transposition of organs.
**heterotonia** (het-er-o-to'-ne-ah) [hetero-; τόνος, tension]. Variable tension.
**heterotopic** (het-er-o-to'-pe-ah) [hetero-; τόπος, place]. A misplacement of normal tissue, especially a congenital malformation of the brain, in which masses of gray matter are found transplanted into the white.
**heterotopic** (het-er-o-top'-ik). See *héterotopous*.
**heterotopous** (het-er-ot'-o-pus) [hetero-; τόπος, place]. Characterized by heterotopia; misplaced.
**heterotoxin** (het-er-o-toks'-in) [hetero-; toxin]. Any poison or toxin introduced into the body from without.
**heterotrichous** (het-er-ot'-rik-us) [hetero-; θρίξ, hair]. Furnished with two kinds of cilia.
**heterotrophia, heterotrophy** (het-er-o-tro'-fe-ah, het-er-ot'-ro-fe) [hetero-; τροφή, sustenance]. Any perversion or disorder of nutrition.
**heterotropia** (het-er-o-tro'-pe-ah). See *strabismus*.
**heterotropical** (het-er-o-tip'-ik, het-er-o-tip'-ik-al) [hetero-; τύπος, pattern]. 1. Differing from type. 2. Applied to a monstrosity consisting of a well-developed fetus from which grows an immature secondary fetus.
**heterotypus** (het-er-o-ti'-pus) [hetero-; τύπος, a type]. A double monster having the parasitic fetus hanging from the ventral wall of the principal subject.
**heterovalvate** (het-er-o-val'-vāt) [hetero-; *valva*, valve]. Having two kinds of valves.
**heteroxanthin** (het-er-o-san'-thin) [hetero-; ξανθός, yellow], $C_5H_6N_4O_2$. A leukomaine that can be isolated in crystalline form from urine.
**heteroxeny** (het-er-oks'-en-e) [hetero-; ξένος, a host]. The quality of living upon different hosts. Cf. *heterecism*.
**hetocresol, hetokresol** (het-o-kre'-sol). See *cinnamyl-metacresol*.
**hetoform** (het'-o-form). Bismuth cinnamate.
**hetol** (he'-tol). See *sodium cinnamate*.
**hetraline** (het'-ral-ēn). A compound of hexamethylene tetramine with dioxybenzene.
**hettocytosis** (het-to-sīr-to'-sis) [ἥττων, less; κύρτωσις, a curvature]. A slight curvature of the spine.
**Heubner's disease** (hoyb'-ner) [Johann Otto Leonhard *Heubner*, German pediatrist, 1843– ]. Syphilitic endarteritis of the brain.
**heurteloup** (hur'-tel-oop) [Charles Louis Stanislas *Heurteloup*, French surgeon, 1793–1864]. An artificial leech for cupping-apparatus.
**hexa-** (heks-ah-) [ἕξ, six]. A prefix signifying six.
**hexabasic** (heks-ah-ba'-sik) [ἕξ, six; βάσις, a base]. Denoting an acid having six replaceable hydrogen atoms.
**hexad** (heks'-ad) [ἕξ, six]. An element the atom of which has a quantivalence of six.
**hexadactylism** (heks-ah-dak'-til-ism) [hexa-; δάκτυλος, a finger]. Having six fingers or toes.
**hexahydrohematoporphyrin** (heks-ah-hi-dro-hem-at-o-por'-fi-rin). A reduction-product of hematin.
**hexamethylenamine** (heks-ah-meth-il-ēn'-am-in), $C_6H_{12}N_4$, the hexamethylenamina of the U. S. P., a condensation-product obtained by the action of ammonia on formaldehyde. Syn., *hexamethyleneiramine; urotropin*.
**hexamethylenetetramine**. Same as *hexamethylenamine*.
**hexane** (heks'-ān) [ἕξ, six], $C_6H_{14}$. The sixth member of the paraffin series of hydrocarbons. It is a liquid, boiling at about 71° C., found in various natural oils.
**hexatomic** (heks-at-om'-ik) [hexa-; ἄτομος, an atom]. Consisting of six atoms; also applied to atoms that are hexavalent, and to alcohols or other compounds having six replaceable hydrogen atoms.
**hexavalent** (heks-av'-al-ent) [hexa-; *valens*, having power]. Having the same combining power as six hydrogen atoms.
**hexhydric** (heks-hi'-drik) [hexa-; ὕδωρ, water]. 1. Containing six atoms of replaceable hydrogen. 2. Containing six molecules of water. h. alcohols, alcohols containing six hydroxyl groups attached

to six' different carbon atoms; they approach the sugars closely in their properties. Moderate oxidation converts them into glucoses.

**hexiology** (*heks-e-ol'-o-je*) [ἕξις, habit; λόγος, science]. The science of the relations of the organism to its environment.

**hexone bases** (*heks'-ōn*). Protein substances containing six atoms of carbon and having basic properties; these are lysin, arginin, histidin.

**hexose** (*heks'-ōs*) [ἕξ, six]. Any monosaccharid which contains six carbon atoms in the molecule.

**hexyl** (*heks'-il*) -[ἕξ, six; ὕλη, substance]. A hypothetical univalent radical C₆H₁₃ occurring in some organic compounds.

**hexylamine** (*heks-il'-am-ēn*), C₆H₁₅N. A ptomaine, found in putrid yeast; it has toxic properties.

**Hey's infantile hernia** (*ha*) [William Hey, English surgeon, 1736–1819]. See *Cooper's hernia*. **H.'s internal derangement**, dislocation of the semilunar cartilages of the knee-joint, especially the internal. **H.'s ligament**, the femoral ligament, a falciform expansion of the fascia lata. **H.'s operation**. 1. For *amputation through the foot:* the same as *Lisfranc's operation*, except that the internal cuneiform bone is sawn through in a line with the articulation of the second metatarsal bone, instead of being disarticulated. 2. For *amputation of the leg:* the amputation is made in the middle of the leg by a long posterior flap, cut by transfixion, and a slightly shorter anterior one.

**Heynsius' test for albumin** (*hīn'-se-uus*) [Adrian Heynsius, Dutch physician, 1831–1885]. Add to the solution acetic acid sufficient to acidify, and a few cubic centimeters of a saturated solution of sodium chloride and boil. A flocculent precipitate is produced by the presence of albumin.

**Hg.** Chemical symbol of *hydrargyrum*, mercury.
**hg.** Abbreviation for hectogram.
**hiant** (*hi'-ant*) [*hiare*, to gape]. Yawning; gaping; opening by a fissure.
**hiation** (*hi-a'-shun*) [*hiare*, to gape]. The act of gaping or yawning. Cf. *pandiculation*.
**hiatus** (*hi-a'-tus*) [L., "a gap"]. 1. A space or opening. 2. The vulva. **h. aorticus**, the aortic opening in the diaphragm. **h. canalis facialis** or **Fallopii**, an oblique opening in the petrous portion of the temporal bone; see *Fallopius, hiatus of*. **h. maxillaris**, one on the inner aspect of the nasal part of the superior maxilla, establishing communication between the nose and the antrum of Highmore. Syn., *hiatus supramaxillaris*. **h. œsophageus**, the esophageal opening in the diaphragm. **h. sacralis**, an opening in the sacral canal posteriorly due to failure of the laminæ of the fifth sacral vertebra to meet in the median line. **h., Scarpa's**. See *Scarpa's hiatus*. **h. semilunaris**, an opening in the deep fascia of the arm for the passage of the basilic vein. **h. subarcuatus**, a depression in the petrosa below the flocculus. **h. tendineus**, the anterior opening of Hunter's canal.

**Hibbs's operation** [Russell Aubra *Hibbs*, American surgeon]. 1. For *Pott's disease:* An osteoplastic operation for the elimination of motion by producing a fusion of the spinous processes, laminæ and lateral articulation of the spine. 2. For *congenital hip disease:* A method of reducing congenital hip dislocation by the aid of a machine, without traumatism to muscle or bone.

**hibernation** (*hi-ber-na'-shun*) [*hibernus*, winter]. The dormant condition or winter-sleep of certain animals, notably bears, hedgehogs, etc., in which animation is almost suspended.

**hiccup, hiccough** (*hik'-up*) [*hic*, a mimic word; *cough*]. A spasmodic contraction of the diaphragm causing inspiration, followed by a sudden closure of the glottis. Syn., *singultus*.

**Hicks' (Braxton) sign** [John Braxton *Hicks*, English gynecologist, 1825–1897]. Intermittent uterine contractions beginning at the end of the third month of pregnancy; they may also be produced by tumors distending the uterus.

**hidden seizure.** A popular name for various forms of slight or sudden epileptiform attacks.

**hidebound disease.** See *scleroderma*.

**hidradenitis, hidroadenitis** (*hi-drad-en-i'-tis, hi-dro-ad-en-i'-tis*). See *hidrosadenitis*. **h. suppurativa**, a condition marked by the formation of tumors the size of a pea which tend to develop into abscesses.

**hidradenoma** (*hi-drad-en-o'-mah*) [ἱδρώς, sweat; *adenoma*]. Hyperplasia of an existing inflammatory tumor of a sweat-gland.

**hidroa** (*hid-ro'-ah*) [ἱδρώς, sweat]. Sudamina; any dermal lesion associated with or caused by profuse sweating.

**hidrocystoma** (*hid-ro-sis-to'-mah*) [ἱδρώς, sweat; κύστις, a cyst; ὄμα, a tumor: *pl.*, *hidrocystomata*]. A variety of sudamina appearing on the face, especially in women in middle and advanced life.

**hidrodermia** (*hi-dro-dur'-me-ah*) [ἱδρώς, sweat; δέρμα, skin]. Anomalies of sweat-secretion.

**hidromancy** (*hi'-dro-man-se*) [ἱδρώς, sweat; μαντεία, divination]. The forming of a prognosis from examination of the perspiration.

**hidronosus** (*hid-ron'-o-sus*) [ἱδρώς, sweat; νόσος, disease]. Any disease of the sweat-glands.

**hidropedesis** (*hid-ro-ped-e'-sis*) [ἱδρώς, sweat; πήδησις, a leaping]. Excessive sweating.

**hidroplania** (*hid-ro-pla'-ne-ah*) [ἱδρώς, sweat; πλάνη, a wandering]. Sweating in an unusual portion of the body.

**hidropoiesis** (*hid-ro-poi-e'-sis*) [ἱδρώς, sweat; ποίησις, formation]. The formation of sweat.

**hidropoietic** (*hid-ro-poi-et'-ik*) [see *hidropoiesis*]. Relating to hidropoiesis.

**hidrorrhea** (*hid-ror-e'-ah*) [ἱδρώς, sweat; ῥοία, a flow]. Excessive flow of sweat.

**hidrosadenitis** (*hid-ros-ad-en-i'-tis*) [ἱδρώς, sweat; ἀδήν, gland; ιτις, inflammation]. Inflammation of the sweat-glands. **h., phlegmonous**, a furuncule beginning in the coil of a sweat-gland. **h., ulcerative**, a variety occurring as superficial ulceration in circular or horse-shoe-shaped areas attaching the palmar or plantar surfaces; it is prone to relapse.

**hidroschesis** (*hid-ros'-kes-is*) [ἱδρώς, sweat; σχέσις, retention]. Retention or suppression of the sweat.

**hidrose** (*hi'-drōs*) [ἱδρώς, sweat]. Relating to sweat.

**hidrosis** (*hid-ro'-sis*) [see *hidrose*]. 1. The formation and excretion of sweat. 2. Abnormally profuse sweating. 3. Any skin disease marked by disorder of the sweat-glands.

**hidrotic** (*hid-rot'-ik*) [ἱδρωτικός, producing sweat]. 1. Diaphoretic or sudorific. 2. A medicine that causes sweating.

**hidrotopathic** (*hid-ro-to-path'-ik*) [ἱδρώς, sweat; πάθος, disease]. Relating to a morbid state of the perspiratory function.

**hieralgia** (*hi-er-al'-je-ah*) [ἱερόν, sacred, sacrum; ἄλγος, pain]. Pain in the sacrum.

**hiera-picra** (*hi-er-ah-pik'-rah*) [L., "sacred bitters"]. Powder of aloes and canella. (Commonly, but incorrectly, called *hicra-picra*.)

**hieromania** (*hi-er-o-ma'-ne-ah*) [ἱερός, sacred; μανία, madness]. Religious frenzy.

**hierotherapy** (*hi-er-o-ther'-a-pe*) [ἱερόν, sacred; θεραπεία, treatment]. The treatment of disease by religious practices.

**high operation.** 1. Supra-pubic lithotomy. See *lithotomy*. 2. Delivery by forceps of a fetus, the instrument being applied at the superior strait.

**Highmore, antrum of** (*hi'-mōr*) [Nathaniel *Highmore*, English anatomist, 1613–1685]. A cavity in the superior maxillary bone communicating with the middle meatus of the nose. **H., body of, H.'s corpus**, a thickening of the tunica albuginea at the posterior part of the testis, from which connective-tissue septa diverge.

**highmoritis** (*hi-mor-i'-tis*), Inflammation of the antrum of Highmore.

**hilar** (*hi'-lar*) [*hilum*, a little thing]. In biology, pertaining to the hilum.

**Hildenbrand's disease** (*hil'-den-brand*) [Johann Valentin von *Hildenbrand*, Austrian physician, 1763–1818]. Typhus fever.

**Hilton's law** (*hil'-tun*) [John *Hilton*, English surgeon, 1804–1878]. The nerve-trunk supplying a joint supplies also the muscles moving the joint, and the skin over the insertion of these muscles. **H.'s line**, a white line marking the junction of the skin of the perineum with the mucosa of the anus. **H.'s muscle**, the arytenoepiglottideus muscle. **H.'s sac**. See *Morgagni's ventricle*.

**hilum, hilus** (*hi'-lum, hi'-lus*) [L., "a little thing"]. A pit, recess, or opening in an organ, usually for the entrance and exit of vessels or ducts.

**hind** (*hīnd*) [AS., *hindan*, after; back]. Pertaining to the rear or posterior extremity. **h.-brain**, division of the brain in the embryo that becomes the

cerebellum and the medulla oblongata. h.-gut, that part of the embryonic intestine from which the cecum, vermiform appendix, colon, and rectum are developed. h.-kidney. See *metanephros*.
**Hindenlang's test for albumin** (*hin'-den-lang*) [C. *Hindenlang*, German physician]. On the addition of solid metaphosphoric acid to the liquid to be tested a precipitate is formed in the presence of albumin.
**hinge-joint** (*hinj'-joint*). See *diarthrosis*.
**hip** [AS., *hype*]. 1. The upper part of the thigh at its junction with the buttocks. 2. The hip-joint. **h.-bath.** See *bath, hip-*. **h.-bone,** the ischium. **h.-girdle,** the pelvic arch. **h.-joint,** the articulation of the femur with the haunch-bone or innominate bone. **h.-joint, disease,** an inflammation of the hip-joint, usually tuberculous, and occurring most commonly in the young, and, according as it begins in the head of the femur, the acetabulum, or in the synovial membrane and proper structures of the joint, divided into femoral, acetabular, and arthritic. The symptoms are shuffling gait, pain often referred to the inner side of the knee, pain in the hip elicited by jarring the heel, deformity, abduction and eversion of the thigh, slight flexion of the knee, and arching of the lumbar spine; later, adduction and inversion of the thigh, with flexion of the knee and shortening of the limb. Suppuration with formation of fistulæ occurs in the advanced stages. Syn., *coxitis*.
**hippanthropy** (*hip-an'-thro-pe*) [ἵππος, horse; ἄνθρωπος, man]. A form of zoanthropy in which the patient believes that he is a horse.
**hippasia** (*hip-a'-se-ah*) [ἱππασία, riding]. Horseback exercise.
**hippiater** (*hip-e-a'-ter*) [ἵππος, a horse; ἱατρός, a physician]. A horse-doctor; a farrier.
**hippiatric** (*hip-e-at'-ric*) [ἵππος, a horse; ἱατρός, a physician]. Pertaining to veterinary surgery.
**hippiatry** (*hip-i'-at-re*) [ἵππος, horse; ἱατρεία, medical art]. Veterinary medicine, in so far as it relates to the horse; farriery.
**hippocamp** (*hip'-o-kamp*) [see *hippocampus*]. The hippocampus major.
**hippocampal** (*hip-o-kam'-pal*) [see *hippocampus*]. Relating to the hippocampus. **h. convolution,** a convolution on the cerebral mesial surface anterior to the lingual lobe. **h. fissure,** a fissure on the cerebral mesial surface above the temporal lobe.
**hippocampus** (*hip-o-kam'-pus*) [ἵππος, horse; κάμπος, a sea-monster: *pl., hippocampi*]. A name applied to two elevations, *hippocampus major* and *hippocampus minor*, the former situated in the middle, and the latter in the posterior, horn of the ventricles of the brain. When the term *hippocampus* is used alone, the *h. major* is meant.
**hippocoryza** (*hip-o-ko-ri'-zah*). Synonym of *equinia*.
**hippocras** (*hip'-o-kras*) [ἵππος, horse; κράσις, strength]. An old-fashioned cordial or liquor, made of red wine, sweetened and spiced.
**Hippocratic** (*hip-o-krat'-ik*) [*Hippocrates*, a Greek physician of the fifth century B. C.]. Described by Hippocrates. **H. expression,** H. facies, an anxious, pinched expression of the countenance, described as characteristic of peritonitis, cholera and other fatal diseases. **H. finger,** hypertrophy of the ungual phalanx and nail in phthisis and other wasting diseases. **H. sound,** the succussion sound. See *Hippocratis succussio*.
**Hippocratis chorda, H. funis.** The Achilles tendon. **H. morbus sacer,** epilepsy. **H. succussio,** succussion employed to obtain a splashing sound in seropneumothorax and pyopneumothorax.
**Hippocratism** (*hip-ok'-rat-izm*) [*Hippocrates*, a Greek physician]. Hippocrates' doctrine of imitating nature in the treatment of disease.
**hippol** (*hip'-ol*). Methylene hippuric acid; it is a colorless, crystalline body, recommended as a urinary antiseptic.
**hippolith** (*hip'-o-lith*) [ἵππος, horse; λίθος, stone]. A calculus or bezoar found in the stomach of the horse.
**hippology** (*hip-ol'-o-je*) [ἵππος, horse; λόγος, knowledge]. The anatomy, pathology, etc., of the horse.
**Hippomane** (*hip-om'-an-e*) [ἵππος, horse; μανία, madness; the ancients believed that horses were madly fond of a plant called by this name]. A genus of euphorbiaceous trees. **H. mancinella** and **H.** **spinosa,** the manchineel trees of tropical America, are extremely acrid and poisonous, even to the touch. They are used locally in medicine, especially in skin-diseases.
**hippomelanin** (*hip-o-mel'-an-in*) [ἵππος, horse; μέλας, black]. A pigment found in melanotic tumors in horses.
**hippomyxoma** (*hip-o-miks-o'-mah*) [ἵππος, a horse; μύξα, mucus]. The swelling attending farcy and glanders. Syn., *hippocoryzoma*.
**hippopathology** (*hip-o-path-ol'-o-je*) [ἵππος, horse; πάθος, disease; λόγος, science]. The science of the diseases of the horse.
**hippophagy** (*hip-off'-a-je*) [ἵππος, horse; φαγεῖν, to eat]. The eating of horse-flesh.
**hippoosteology** (*hip-os-te-ol'-o-je*) [ἵππος, horse; ὀστέον, bone; λόγος, science]. The science of osteology as applied to the horse.
**hippotomy** (*hip-ot'-o-me*) [ἵππος, horse; τομή, a cutting]. The anatomy or dissection of the horse.
**hippurate** (*hip'-u-rāt*) [ἵππος, horse; οὖρον, urine]. Any salt of hippuric acid, *q. v.*
**hippuria** (*hip-u'-re-ah*) [ἵππος, horse; οὖρον, urine]. Excess of hippuric acid in the urine.
**hippuric acid** (*hip-u'-rik*). See *acid, hippuric*. **h. acid, reaction for.** See *Luecke's reaction for hippuric acid*.
**hippuris** (*hip-u'-ris*) [ἵππουρις, horse-tail]. The cauda equina, *q. v.*
**hippus** (*hip'-us*) [ἵππος, horse, from analogy to the movements of this animal]. Spasmodic pupillary movement, independent of the action of light.
**hircismus** (*her-sis'-mus*) [*hircus*, a goat]. The goat-like odor sometimes emitted by the human axilla.
**hircus** (*her'-kus*) [L., "goat"]. 1. The tragus. 2. *Hircismus, q. v.* 3. A hair growing in the axilla.
**Hirschberg's test** (*hērsh'-berg*) [Julius *Hirschberg*, German ophthalmologist, 1843–    ]. A rough estimate of the position of the corneal reflection of a candle-flame held one foot in front of the eye to be tested, the examiner placing his own eye near the candle and looking just over it.
**Hirschfeld's disease** (*hērsh'-felt*). A form of diabetes of rapid march, which usually ends in death in three months, by progressive cachexia or by complication. **H.'s ganglion,** the gyrus hippocampi.
**Hirschsprung's disease** (*hērsh'-sprung*) [Harold *Hirschsprung*, Danish physician, 1830–    ]. Congenital hypertrophic dilatation of the colon.
**hirsute** (*her-sūt'*) [*hirsutus*, shaggy]. Shaggy; hairy.
**hirsuties** (*her-sū'-te-ēz*) [see *hirsute*]. Excessive growth of hair.
**Hirtz's rale.** A moist, subcrepitant rale, of a somewhat metallic character, pathognomonic of tuberculous softening.
**hirudin** (*hir'-u-din*) [*hirudo*, leech]. The active principle of a secretion derived from the buccal glands of the pond-leech, *Sanguisuga medicinalis*; it is said to be a secondary albumose, and has the property of preventing the coagulation of blood.
**hirudiniculture** (*hi-roo'-din-e-kul-chūr*) [*hirudo*, leech; *cultura*, culture]. The artificial breeding and rearing of leeches.
**hirudo** (*hi-roo'-do*) [L.: *pl., hirudines*]. The leech, *q. v.*
**His' canal** [1. Wilhelm *His*, German anatomist, 1831–1904; 2. Wilhelm *His*, Jr., German physician, 1863–    ]. [1] The thyroglossal duct of the fetus, of which the cecal foramen of the tongue is the vestige and which may persist during postnatal life. [2] **H.'s germinal cell,** any epiblastic cell in the neural tube from which a neurone is developed. [2] **H.'s muscle bundle,** a neuromuscular band joining the right auricle to the ventricles in the mammalian heart. [1] **H.'s peripheral veil,** the spongy felt-work formed by the ectal ends of the spongioblasts of the neural tube. [1] **H.'s perivascular spaces,** lymph-spaces surrounding the blood-vessels of the brain and spinal cord. [1] **H.'s stroma,** the trabecular framework of the mammary gland. [2] **H.'s sulcus terminalis,** a furrow on the surface of the right auricle; it corresponds in position to the crista terminalis of His, a vertical ridge in the interior of the right auricle.
**histaffine** (*his'-taf-in*) [*histo-*; *affinis*, related]. 1. Having affinity for tissues. 2. A substance supposed to be present in the blood-serum in certain

diseases such as syphilis and trypanosomiasis; and which is said to produce complement fixation.

**histic** (*his'-tik*) [ἱστός, a web]. Relating to tissue.

**histidin** (*his'-tid-in*), $C_6H_9N_3O_2$. A base present among the hydrolytic products of casein, albumin, blood-serum, and horn, and a constant cleavage-product of the more complex plant and animal proteids.

**histin** (*his'-tin*) [ἱστίον, a web]. Fibrin.

**histioid** (*his'-te-oid*). See *histoid.*

**histioma** (*his-ti-o'-mah*) [ἱστίον, a web; ὄμα, a tumor]. A tissue tumor in which distinct tissues may be recognized, but which do not arrange themselves to form organs.

**histo-** (*his-to-*) [ἱστός, tissue]. A prefix denoting relation to tissue.

**histoblast** (*his'-to-blast*) [*histo-*; βλαστός, a germ]. A cell engaged in the formation of tissue.

**histochemistry** (*his-to-kem'-is-tre*) [*histo-*; *chemistry*]. The chemistry of the histological elements of the body.

**histodialysis** (*his-to-di-al'-is-is*) [*histo-*; διάλυσις, dissolution]. The dissolution of organic tissue.

**histofluorescence** (*his-to-floo-or-es'-enz*). The administration of fluorescing drugs during Roentgen-ray treatment.

**histogenesis** (*his-to-jen'-es-is*) [*histo-*; γένεσις, generation]. The formation of tissues.

**histogenetic** (*his-to-jen-et'-ik*) [see *histogenesis*]. Relating to histogenesis.

**histogenol** (*his-toj'-en-ol*). A compound of phosphorus and arsenic, each dessertspoonful containing ½ gr. (0.032 Gm.) of disodic methyl arsenate and 1½ gr. (0.1 Gm.) of nucleic acid.

**histogeny** (*his-toj'-en-e*). See *histogenesis.*

**histography** (*his-tog'-ra-fe*) [*histo-*; γράφειν, to write]. A description or written account of the tissues.

**histohematin** (*his-to-hem'-at-in*) [*histo-*; αἷμα, blood]. A pigment found in muscles, suprarenal capsules, and other organs, and believed to have a respiratory function.

**histoid** (*his'-toid*) [*histo-*; εἶδος, likeness]. 1. Resembling tissue. 2. Composed of only one kind of tissue.

**histokinesis** (*his-to-kin-e'-sis*) [*histo-*; κίνησις, movement]. Movement that takes place in the minute structural elements of the body.

**histologic, histological** (*his-to-loj'-ik, -al*) [see *histology*]. Relating to histology.

**histologist** (*his-tol'-o-jist*) [*histo-*; λόγος, science]. One who is expert in histology.

**histology** (*his-tol'-o-je*) [*histo-*; λόγος, science]. The minute anatomy of tissues. **h., normal,** the study of sound tissues. **h., pathological,** the study of diseased tissues. **h., topographical,** the study of the minute structure of the organs and especially of their formation from the tissues.

**histolysis** (*his-tol'-is-is*) [*histo-*; λύσις, dissolution]. Disintegration and dissolution of organic tissue.

**histolytic** (*his-tol-it'-ik*) [*histo-*; λύσις, dissolution]. Pertaining to histolysis.

**histomorphology** (*his-to-morf-ol'-o-je*) [*histo-*; μορφή, form; λόγος, science]. The morphology of the histological elements of the body.

**histon** (*his'-ton*) [ἱστός, tissue]. A protein prepared from the nuclei of cells. It belongs to the group of proteins known as albumoses or propeptones.

**histonomy** (*his-ton'-o-me*) [*histo-*; νόμος, a law]. The laws of the development and arrangement of organic tissue.

**histonuria** (*his-ton-ū'-re-ah*). The presence of histon in the urine.

**histopathology** (*his-to-path-ol'-o-je*) [*histo-*; *pathology*]. The study of minute pathological changes in tissues.

**histophysiology** (*his-to-fiz-e-ol'-o-je*) [*histo-*; *physiology*]. The science of the functions of the various tissues.

**histopin** (*his'-to-pin*). A staphylococcus extract used in the treatment of furunculosis.

**Histoplasma capsulatum** (*his-to-plaz'-mah cap-su-la' tum*). A protozoon parasitic in man in Central and South America

**histoplasmosis hominis** (*h s-to-plaz-mo'-sis hom'-in-is*). The diseased state caused by the invasion of the human body by the *Histoplasma capsulatum.*

**histopsyche** (*his-to-si'-ke*) [*histo-*; ψυχή, soul]. The tissue soul; according to Haeckel, the higher psychological function which gives psychological individuality to the compound multicellular organism as a true cell commonwealth.

**history** (*his'-tor-e*) [ἱστορία, a learning by inquiry]. A narrative; story. **h., biological,** the life-story of any animal. **h., medical,** the account obtained from a patient as to his health, past and present, and the symptoms of his disease.

**histosan** (*his'-to-san*). Trade name of guaiacol-albuminate; said to be useful in tuberculosis and other diseases of the respiratory system. Dose ℥ i–iv (4–16 gm.).

**histotherapeutics, histotherapy** (*his-to-ther-ap-ū'-tiks, his-to-ther'-ap-e*) [*histo-*; θεραπεία, therapy]. The remedial use of animal tissues.

**histotome** (*his'-to-tōm*) [*histo-*; τέμνειν, to cut]. An apparatus for cutting tissue for the study of its minute structure; a microtome.

**histotomy** (*his-tot'-o-me*) [see *histotome*]. The dissection of tissues.

**histotripsy** (*his-to-trip'-se*) [*histo-*; τρίψις, a crushing]. The crushing of tissue by an ecraseur.

**histotromy** (*his-tot'-ro-me*) [*histo-*; τρόμος, tremor]. Fibrillary contraction.

**histotrophic** (*his-to-trof'-ik*) [*histo-*; τροφή, nourishment]. Concerning the nutrition of the tissues.

**histotropic** (*his-to-trop'-ik*) [*histo-*; τρόπος, a turn]. The property of entering into chemical combination with the tissues.

**histozoic** (*his-to-zo'-ik*) [*histo-*; ζωή, life]. Living on or within the tissues; denoting certain protozoan parasites.

**histozyme** (*his'-to-zim*) [*histo-*; ζύμη, leaven]. A ferment found in the kidneys of pigs, and concerned in splitting up hippuric acid.

**histrionic** (*his-tre-on'-ik*) [*histrio*, an actor]. Dramatic. **h. mania,** insanity with affectation and lofty manner. **h. muscles,** the muscles of expression of the face. **h. spasm,** spasm of the muscles of expression.

**histrionism** (*his'-tre-on-izm*) [*histrio*, a player]. Dramatic action in insanity or in hysteria.

**Hitzig's center** (*hit'-zig*) [Julius Edward *Hitzig*, German physician, 1838–    ]. A center in the supramarginal gyrus which is supposed to govern the voluntary movements of the eyeballs. **H.'s zone,** a hypesthetic zone extending around the trunk in tabes dorsalis.

**hives** (*hīvz*) [origin uncertain]. 1. Urticaria. 2. In Great Britain the term is also applied to croup, laryngitis, and chicken-pox.

**Hl.** Abbreviation for latent hypermetropia; and for hectoliter.

**Hm.** Abbreviation for manifest hypermetropia; and for hectometer.

**H. M. C.** Abbreviation for hyoscine-morphine-cactine anesthesia.

**hoang-nan, or hwang-nao** (*ho-ang-nan', hwang-now'*]) [Chinese]. A Chinese preparation obtained from the bark of *Strychnos gaultheriana.* Its properties are due to a small percentage of strychnine. It is recommended as an alterative in syphilis, leprosy, and similar diseases, and is an alleged preventive of hydrophobia if given in large doses (gr. xv) during the period of incubation. Dose of the powdered drug gr. iij–v; of the aceto-alcoholic extract gr. ⅛–½; of the tinct. ♏ j–v.

**hoarhound, horehound** (*hōr'-hownd*). See *marrubium.*

**hoarse** (*hors*) [ME., *hoors*, harsh]. Harsh; grating; discordant; applied to the voice.

**hoarseness** (*hors'-nes*) [ME., *hoorsnesse*, hoarseness]. Harshness of the voice depending on some abnormal condition of the larynx or throat.

**hobnail liver.** The liver of advanced atrophic cirrhosis, so called on account of the small projections on the surface. Syn., *gin-drinker's liver.*

**Hoboken's valves.** The secondary windings of the vessels of the umbilical cord that form grooves externally and valve-like projections internally.

**Hoche, bandelette of** (*hoh*). A small bundle of nerve-fibers, a part of the fasciculus posterior proprius.

**Hochsinger's sign** (*hōkh'-zing-er*) [Carl *Hochsinger*, Austrian pediatrist]. The existence of indicanuria in tuberculosis of childhood.

**hock, hough** (*hok*) [ME., *houz*, heel]. The joint on the hind-leg of a quadruped between the knee and the fetlock, corresponding to the ankle-joint in man. In man, the back part of the knee-joint; the ham.

**Hodara's disease** (*ho-dah'-rah*) [Menahem *Hodara*, Turkish physician]. A form of trichorrhexis nodosa that has been observed by Hodara in women in Constantinople.

**Hodge's plane** (*hodj*) [Hugh Lenox *Hodge*, American gynecologist, 1796–1873]. A plane parallel to that of the pelvic inlet, passing through the upper border of the os pubis and the middle of the second sacral vertebra.

**Hodgen's apparatus** (*hod'-gen*) [John Thompson *Hodgen*, American surgeon, 1826–1882]. A modification of Smith's anterior splint.

**Hodgkin's disease** (*hodj'-kin*) [Thomas *Hodgkin*, English physician, 1798–1866]. Pseudoleukemia; progressive hyperplasia of the lymphatic glands associated with anemia.

**Hodgson's disease** (*hodj'-sun*) [Joseph *Hodgson*, English physician, 1788–1869]. Senile atheroma of the aorta with consequent lesion of aortic valves.

**hodograph** (*hod'-o-graf*) [ὁδός, a path; γράφειν, to write]. 1. An instrument for recording locomotor movements. 2. Of Sir Wm. Hamilton, a curve demonstrating the velocity of a moving particle; it is employed in the study of central forces.

**hoe** (*ho*) [ME., *howe*, a hoe]. A scraping-instrument used in operations for cleft-palate, or in dentistry.

**Hoen's degeneration** (*ho'-en*). Degenerative change in striated muscles with nuclear proliferation.

**Hoffa's operation** (*hof'-fer*) [Albert *Hoffa*, German orthopedist, 1859–1907]. A "bloody" method of reducing congenital dislocation of the hip.

**Hoffmann's anodyne** (*hof'-man*). A compound of ether, 30; alcohol, 67; ethereal oil, 3. It is anodyne, stimulant, and antispasmodic, and is used in nervous irritation, angina pectoris, and asthma. Dose 30 min.–2 dr. (2–8 Cc.). H.'s symptom, increase of the mechanical irritability of the sensory nerves in tetany. H.'s test for tyrosin, add to the solution to be tested mercuric nitrate and boil; then add nitric acid containing some nitrous acid. If tyrosin is present, a beautiful red coloration is produced and a red precipitate is formed. H.'s type of progressive muscular atrophy. See *Charcot-Marie's type of progressive muscular atrophy*.

**Hofmeister's test for leucin** (*hof'-mi-ster*). A deposit of metallic mercury is formed on warming a solution of leucin with mercurous nitrate. H.'s test for peptones, prepare phosphotungstic acid by dissolving commercial sodium tungstate in boiling water and adding phosphoric acid until acid in reaction; acidify strongly with hydrochloric acid after cooling, and filter when it has stood 24 hours. On adding this to a peptone solution entirely free from albumin it yields a precipitate.

**hog cholera**. A contagious, febrile, disease of hogs, due to *Bacillus choleræ suis*.

**Hohl's method** (*hōl*) [Anton Friedrich *Hohl*, German physician, 1789–1862]. A method of preserving the perineum in labor. It consists in applying resistance to the presenting part, the thumb being applied anteriorly to the occiput and the index and middle fingers posteriorly upon that portion of the head lying nearest the commissure.

**holadin** (*hol'-ad-in*) [ὅλος, entire; ἀδήν, gland]. Trade name of a preparation of the entire pancreas. It is sold in 3 gr. capsules.

**holagogue** (*hol'-ag-og*) [ὅλος, whole; ἀγωγός, leading]. A medicine or remedy that expels or drives out the whole of a morbid substance. A radical remedy.

**holarthritis** (*hol-ar-thri'-tis*). See *polyarthritis*.

**Holden's line** (*hōl'-den*) [Luther *Holden*, English surgeon, 1815–1905]. A sulcus below the fold of the groin, starting from the femoroscrotal furrow, and fading away between the great trochanter and the anterior superior iliac spine; it crosses the middle of the capsule of the hip.

**holder** (*hōld'-er*). A device for holding instruments, sponges, etc., in surgical operations.

**hold-fast** (*hōld'-fast*). A "lumpy-jaw" tumor. See *actinomycosis*.

**holgin** (*hol'-jin*). Trade name of an antiseptic compound of menthol, formaldehyde and methyl alcohol.

**hollow** (*hol'-o*). 1. Empty within; not solid. 2. A depression; a vacuity. h.-back, lordosis. h.-foot, same as *talipes, cavus, q. v.* h.-horn, h.-tail. Synonym of *Texas fever*.

**hollyhock** (*hol'-e-hok*). See *althea*.

**Holmes' operation** (*hōlmz*) [Timothy *Holmes*, English surgeon, 1825–1907]. For excision of the os calcis: an incision is made from the inner edge of the Achilles tendon along the upper border of the os calcis and the outer border of the foot to the calcaneocuboid joint, and this is joined by another incision running across the sole, the peroneal tendons being divided.

**Holmgren's test** (*holm'-gren*) [Alarik Frithiof *Holmgren*, Swedish physiologist, 1831–1897]. A test for color-blindness. The patient is requested to match skeins of different colored worsted, and if color-blind, he always selects characteristic shades.

**holo-** (*hol'-o-*) [ὅλος, entire]. A prefix signifying entirety.

**holoblast** (*hol'-o-blast*) [*holo-*; βλαστός, germ]. In biology, an ovum that undergoes complete segmentation while germinating.

**holoblastic** (*hol-o-blas'-tik*) [ὅλος, whole; βλαστός, germ]. Applied to ova in which the entire yolk is included in the process of segmentation; one in which there is no separate food-yolk.

**holocaine** (*hol-o-ka'-in*). A crystalline combination of paraphenetidin and acetphenetidin. The hydrochloride is employed as an anesthetic in ophthalmic practice in 1 % solution.

**holocrine** (*hol'-o-krēn*) [*holo-*; κρίνειν, to separate]. Applied to a gland the cell of which, after having elaborated the material of secretion, falls into disuse and disappears. Cf. *merocrine*.

**holodiastolic** (*hol-o-di-as-tol'-ik*) [*holo-*; *diastole*]. Relating to the entire diastole.

**holometabolic** (*hol-o-met-ab-ol'-ik*) [*holo-*; μεταβολή, change]. In biology, applied to animals that undergo complete metamorphosis or transformation, as insects.

**holonarcosis** (*hol-o-nar-ko'-sis*) [*holo-*; νάρκωσις, stupor]. Complete narcosis.

**holopathy** (*hol-op'-ath-e*) [*holo-*; πάθος, disease]. 1. A general or constitutional disease of which a local disorder is but a manifestation. 2. The theory that local diseases are manifestations of a general disorder.

**holoplexia** (*hol-o-pleks'-e-ah*) [*holo-*; πλῆξις, a stroke]. Complete or general paralysis.

**holorhachischisis** (*hol-o-rak-is'-kis-is*) [*holo-*; ῥάχις, spinal column; σχίζειν, to cleave]. A congenital absence of the vertebral canal.

**holoschisis** (*hol-os'-kis-is*) [*holo-*; σχίσις, cleavage]. Amitotic or indirect cell-division; amitosis.

**holosteosclerosis** (*hol-o-ste-o-skler-o'-sis*) [*holo-*; ὀστέον, bone; σκληρός, hard]. General osteosclerosis.

**holosteous** (*hol-os'-te-us*) [*holo-*; ὀστέον, a bone]. In biology, having a completely bony skeleton.

**holosteric** (*hol-os'-ter-ik*) [*holo-*; στερεός, solid]. Not liquid; composed entirely of solids.

**holostomatous** (*hol-o-sto'-mat-us*) [*holo-*; στόμα, mouth]. In biology, having the mouth entire, neither notched nor parts missing.

**holosymphysis** (*hol-o-sim'-fiz-is*) [*holo-*; σύμφυσις, a growing together]. Complete union.

**holosystolic** (*hol-o-sis-tol'-ik*) [*holo-*; *systole*]. Relating to the entire systole.

**holotetanus** (*hol-o-tet'-an-us*) [*holo-*; τέτανος, tetanus]. General tetanus; called also *holotonia*.

**holotomy** (*hol-ot'-o-me*) [*holo-*; τέμνειν, to cut]. Complete surgical excision of a part or organ.

**holotonia**, or **holotony** (*hol-o-to'-ne-ah, hol-ot'-o-ne*) [*holo-*; τείνειν, to stretch]. Same as *holotetanus*.

**holotonic** (*hol-o-ton'-ik*) [*holo-*; τείνειν, to stretch]. Relating to, or characterized by, holotetany.

**holotopic** (*hol-o-top'-ik*) [*holo-*; τόπος, place]. Pertaining to the relation of a part to the entire organism.

**holotopy** (*hol-ot'-o-pe*). Waldeyer's term for the relation of a part or organ to the whole organism. Cf. *idiotopy; skeletotopy; syntopy*.

**holozoic** (*hol-o-zo'-ik*) [*holo-*; ζῷον, an animal]. In biology, entirely resembling animals in mode of nutrition.

**Holthouse's hernia** (*hōlt'-hows*) [Carsten *Holthouse*, English surgeon, 1810–1890]. An oblique inguinal hernia in which, owing to the nondescent of the testis or from other causes, the hernia protrudes outward along the fold of the groin.

**holting** (*hōlt'-ing*) [Barnard *Holt*, English surgeon, 19th century]. The divulsion of a urethral stricture by Holt's dilator.

**Holtz machine** (*hōltz*) [Wilhelm *Holtz*, German physicist, 1836– ]. A particular form of electrostatic induction-machine.

**holzin** (*holt'-zin*). Formaldehyde in a 60 % solution in methyl-alcohol. An antiseptic and disinfectant.
**holzinol** (*holt'-zin-ol*). A solution of formaldehyde in methyl-alcohol containing a small proportion of menthol. Antiseptic and disinfectant.
**homagra** (*hom-a'-grah*). See *omagra*.
**homalocephalus** (*hom-al-o-sef'-al-us*) [ὁμαλός, flat; κεφαλή, the head]. Lissauer's term for "flat headed."
**homalocoryphus** (*hom-al-o-kor'-if-us*) [ὁμαλός, flat; κορυφή, the head]. Lissauer's term for a skull in which the angle formed by two lines drawn from the bregma and the occipital point to the highest point above is between 132° and 142°.
**homalodermatous, homalodermous** (*ho-mal-o-dur'-mat-us, -dur'-mus*) [ὁμαλός, smooth; δέρμα, skin]. Having a smooth skin.
**homalographic** (*hom-al-o-graf'-ik*) [ὁμαλός, level; γράφειν, to write]. Pertaining to homolography.
**h. method**, a method of showing the structure of the body by means of plane sections of a frozen body.
**homalography** (*hom-al-og'-ra-fe*) [ὁμαλός, level; γράφειν, to record]. Anatomy by sections; the representation of structure by means of sketches of various sections.
**homalometopus** (*hom-al-o-met-o'-pus*) [ὁμαλός, flat; μέτωπον, the space between the eyes]. Lissauer's term for a skull having a frontal angle between 130.5° and 141°.
**homalopisthocranius** (*hom-al-o-pis-tho-kra'-ne-us*) [ὁμαλός, flat; ὄπισθεν, behind; κρανίον, the skull]. Lissauer's term for a skull in which the angle formed by lines joining the external occipital protuberance and the occipital point with the highest point of the skull is between 140° and 154°.
**homalosternal** (*hom-al-o-ster'-nal*) [ὁμαλός, even, level; στέρνον, sternum]. In biology, having a raftlike or keelless sternum; as certain birds.
**homaluranus** (*hom-al-ū-ra'-nus*) [ὁμαλός, flat; οὐρά, a tail]. Lissauer's term for a skull in which the angle formed by lines joining the occipital point and the bregma with the highest point of the skull is between 147.5° and 163.5°.
**homatropine** (*ho-mat'-ro-pēn*), $C_{16}H_{21}NO_3$. An alkaloid derived from atropine. It causes dilation of the pupil and paralysis of accommodation as does atropine, but its effects pass off more quickly—usually in two or three days. The hydrobromide is the salt generally employed. **h. hydrobromide** (*homatropinæ hydrobromidum*, U. S. P.), $C_{16}H_{21}NO_3$.·HBr, white crystals used as a mydriati, and in the night-sweats of tuberculosis. Dose $\frac{1}{187}-\frac{1}{65}$ gr. (0.0005–0.0011 Gm.); maximum dose, single, $\frac{1}{65}$ gr. (0.001 Gm.). Application, 1 % solution.
**homaxonial, homaxonic** (*hom-aks-on'-e-al, hom-aks-on'-ik*) [*homo-*; ἄξων, axis]. Having equal axes.
**Home's lobe** [Sir Everard *Home*, English surgeon, 1763–1832]. A small, glandular structure sometimes seen between the caput gallinaginis and the sphincter vesicæ. It represents the third lobe of the prostate and may become considerably enlarged in old people.
**homedric** (*hom-ed'-rik*) [*homo-*; ἔδρα, a base]. Having equal facets.
**homeo-** (*ho-me-o*) [ὅμοιος, like]. A prefix signifying likeness.
**homeochronous** (*ho-me-ok'-ro-nus*) [*homeo-*; χρόνος, time]. 1. Similar in time or periodicity. 2. In true onteogenetic sequence; appearing in proper order or time.
**homeocyte** (*ho'-me-o-sīt*) [*homeo-*; κύτος, cell]. Same as *lymphocyte*.
**homeomerous** (*ho-me-om'-er-us*) [*homeo-*; μέρος, part]. In biology, having given organs or parts distributed uniformly throughout.
**homeomorphous** (*ho-me-o-mor'-fus*) [*homeo-*; μορφή, form]. Like or similar in form and structure.
**homeo-osteoplasty** (*ho-me-o-os'-te-o-plas-te*) [*homeo-*; ὀστέον, bone; πλάσσειν, to mold]. The grafting of a piece of bone similar to that upon which it is grafted.
**homeopath** (*ho'-me-o-path*) [see *homeopathy*]. Homeopathist.
**homeopathic** (*ho-me-o-path'-ik*) [see *homeopathy*]. Relating to homeopathy.
**homeopathist** (*ho-me-op'-ath-ist*) [see *homeopathy*]. A practitioner of homeopathy.
**homeopathy** (*ho-me-op'-ath-e*) [*homeo-*; πάθος, ailment or disease]. A system of treatment of disease by the use of agents that, administered in health, would produce symptoms similar to those for the relief of which they are given.

**homeoplasia** (*ho-me-o-pla'-ze-ah*) [*homeo-*; πλάσσειν, to shape]. The growth of tissue resembling the normal tissue, or matrix, in its form and properties; also the tissue so formed.
**homeoplastic** (*ho-me-o-plas'-tik*) [*homeo-*; πλάσσειν, to form].· Pertaining to a neoplasm resembling its matrix-tissue in texture. One differing widely in this respect is heteroplastic. If separated in position, it is said to be heterotopic; in date, heterochronic.
**homeoplasty** (*ho'-me-o-plas-te*). See *homeoplasia*.
**homeosemous** (*hom-e-o-se'-mus*) [*homeo-*; σημεῖον, a sign]. Similar in import: applied to symptoms.
**homeosis**, or **homoiosis** (*ho-me-o'-sis*, or *ho-moi-o'-sis*) [*homeo-*]. The assimilation of nutrient material.
**homeotherapeutics** (*ho-me-o-ther‹ap-ū'-tiks*) [*homeo-*; θεραπεύειν, to treat]. The homeopathic doctrine of therapeutics.
**homeothermal**, or **homoiothermal** (*ho-me-o-ther'-mal, ho-moi-o-ther'-mal*) [*homeo-*; θέρμη, heat]. Pertaining to animals that are "warm-blooded," or that maintain a uniform temperature despite variation in the surrounding temperature.
**homeothermy** (*ho-me-o-ther'-me*) [*homeo-*; θέρμη, heat]. The condition of having a temperature which is not affected by environment.
**homesickness** (*hōm'-sik-nes*). Nostalgia. An urgent desire to return to one's home. It may be accompanied by a morbid sluggishness of the functions of the various organs of the body, and may develop into profound melancholy.
**homicidal** (*hom-is-i'-dal*) [*homo*, a man; *cædere*, to kill]. Pertaining to homicide. **h. mania**, insanity characterized by murderous impulses.
**homicide** (*hom'-is-īd*) [*homo*, a man; *cædere*, to kill]. The killing of a human being without malice or intent, as distinguished from murder or manslaughter. Also, the taking of human life in general by another. Also, one who takes the life of another.
**homiculture** (*ho'-mik-ul-chur*) [*homo*, man; *cultura*, culture]. The improvement of the human species by attention to the laws of breeding; stirpiculture.
**homo-** (*ho-mo-*). See *homeo-*.
**homoarecoline** (*ho-mo-ar-ek'-o-lēn*); $C_7H_{19}(C_2H_5)$-$NO_2$. The ethyl ether of arecaidine (*q. v.*). A yellowish liquid soluble in water or alcohol. The hydrobromide forming colorless soluble crystals melting at 119° C. is recommended as a substitute for arecoline.
**homoblastic** (*ho-mo-blas'-tik*) [*homo-*; βλαστός, a bud, germ]. In biology, derived from like germs or cells.
**homocentric** (*ho-mo-sen'-trik*) [*homo-*; κέντρον, center]. Concentric; having the same center. **h. rays**, light rays that have a common focus or are parallel.
**homocerebrin** (*ho-mo-ser'-e-brin*) [*homo-*; *cerebrum*]. A substance derived from brain tissue, closely resembling cerebrin, but more soluble in alcohol.
**homochronous** (*ho-mok'-ro-nus*) [*homo-*; χρόνος, time]. Occurring at the same age or period in successive generations.
**homocladic** (*ho-mo-klad'-ik*) [*homo-*; κλάδος, branch]. Referring to an anastomosis between twigs of the same artery.
**homodont** (*ho'-mo-dont*) [*homo-*; ὀδούς, tooth]. In biology, having the teeth alike throughout.
**homœo-** (*ho-me-o-*). For words thus beginning see *homeo-*.
**homogeneity** (*ho-mo-jen-e'-it-e*) [*homo-*; γένος, a kind]. The condition of being homogeneous.
**homogeneous** (*ho-mo-je'-ne-us*) [*homo-*; γένος, kind]. Having the same nature or qualities; of uniform character in all parts.
**homogenesis** (*ho-mo-jen'-es-is*) [*homo-*; γένεσις, birth]. Reproduction in which the offspring passes through the same cycle of changes as the parent itself.
**homogenization** (*ho-mo-jen-iz-a'-shun*) [*homo-*; γεννᾶν, to produce]. The act or process of rendering homogeneous; reduction to a common standard; the process of rendering the objects of microscopic study transparent and fixed.
**homogenous** (*ho-moj'-en-us*). Pertaining to homogeny.
**homogentisic acid** (*hom-o-jen-tis'-ik*). See *acid*.
**homogeny** (*ho-moj'-en-e*) [ὁμογενής, of the same race or family]. In biology, an agreement among organisms depending on the inheritance of a common part or having a common ancestor. See *homogenesis*.

**homoio-** (ho-moi-o-). For words thus beginning, see *homeo-*.
**homoiosis** (ho-moi-o'-sis). See *homeosis*.
**homoiothermal** (ho-moi-o-ther'-mal). 1. Warm-blooded. 2. Maintaining a uniform temperature.
**homolateral** (ho-mo-lat'-er-al) [homo-; latus, side]. On or pertaining to the same side.
**homologue, homolog** (ho'-mo-log) [homo-; λόγος, proportion]. An organ which has the same relative structure, position, or development as another. The same organ in different organisms under every variety of form and function.
**homologous** (ho-mol'-o-gus) [see *homologue*]. Corresponding in structure, either directly or as referred to a fundamental type. In chemistry, being of the same type or series; differing by a multiple or an arithmetical ratio in certain constituents. **h. tissues**, those identical in type of structure. **h. tumor**, a tumor consisting of tissue identical with that of the organ whence it springs. **h. vaccine**, one derived from the microorganism infesting the person to be immunized; autogenous vaccine.
**homology** (ho-mol'-o-je) [see *homologue*]. The quality of being homologous; also, the morphological identity of parts or organs in different animals.
**homomerous** (ho-mom'-er-us) [homo-; μέρος, a part]. Having the parts alike.
**homomorphism** (ho-mo-mor'-fizm) [homo-; μορφή, form]. In biology, superficial resemblance, without true homology; mimicry or adaptive resemblance.
**homomorphous** (ho-mo-mor'-fus) [homo-; μορφή, form]. In biology, exhibiting superficial resemblance, but not truly homologous.
**homonomous** (ho-mon'-o-mus) [homo-; νόμος, law]. Governed by or under the same law.
**homonym** (hom'-o-nim) [homo-; όνομα, name]. That which is homonymous.
**homonymous** (ho-mon'-im-us) [homo-; όνομα, a name]. 1. Having the same sound or name; having the same relative position. **h. diplopia**, a form of diplopia in which the image seen by the right eye is on the right side and that seen by the left eye is on the left side. **h. hemianopia**. See *hemianopia, homonymous*.
**homophonous** (ho-mof'-on-us) [homo-; φωνή, a sound]. Relating to words spelled differently but indistinguishable in sound; it is applied to different conceptions.
**homoplasmic** (ho-mo-plaz'-mik). Same as *homoplastic*.
**homoplasmy** (ho'-mo-plaz-me) [homo-; πλάσμα, a thing moulded]. In biology, homoplastic or homomorphic, i. e., showing mimetic resemblances.
**homoplast** (ho'-mo-plast) [homo-; πλαστός, formed, moulded]. In biology. 1. One of any aggregate or fusion of plastids. 2. An organ or part showing mere superficial or mimetic resemblance to another.
**homoplastic** (ho-mo-plas'-tik) [homo-; πλαστός, formed]. Applied to new growths in which there has been no cytomorphosis, the cells resembling those of the parent tissue, as in angioma and glioma.
**homoplastid** (ho-mo-plas'-tid) [homo-; πλάσμα, form]. An organism each cell of which is endowed with the power of reproducing the species.
**homoquinine** (ho'-mo-kwin-ēn), $C_{19}H_{22}N_2O_2$. A crystalline alkaloid soluble in alcohol or chloroform, found in the bark of *Cinchona pedunculata*, and *Remijia purdieana*.
**homosexual** (ho-mo-seks'-ū-al) [homo-; sexus, sex]. Pertaining to the same sex.
**homosexuality** (ho-mo-seks-ū-al'-it-e) [homo-; sexualis, of a sex]. That form of sexual perversion, acquired or congenital, in which the individual conceives a violent sexual passion for one of the same sex, and gratifies it either by sodomy, by titillation, or platonically.
**homostimulant** (ho-mo-stim'-ū-lant). A term used to indicate the particular action which organic extracts and lipoids exert upon the organs to which they correspond. See *Hallion's law*.
**homothermal** (ho-mo-ther'-mal) [homo-; θερμή, heat]. Warm-blooded.
**homothermic** (ho-mo-ther'-mik) [homo-; θερμή, heat]. Having a uniform temperature.
**homotonic** (ho-mo-ton'-ik) [homo-; τόνος, tone; tension]. Having a uniform or even course.
**homotype** (ho'-mo-tīp) [homo-; τύπος, a pattern]. A part corresponding and similar to another part, as the humerus to the femur.

**homotypical** (ho-mo-tip'-ik-al) [homo-; τύπος, type]. In biology, showing serial correspondence or bilateral symmetry.
**homunculus** (ho-mun'-kū-lus) [L., dim. of *homo*, man]. The fetus; a dwarf.
**Honduras bark** (hon-dū'-ras). Cascara amarga.
**honey** (hun'-e). See *mel*.
**honorarium** (on-or-a'-re-um) [L.]. A professional fee; especially one that is in theory a gift, no formal professional charge having been made.
**honthin, hontin** (hon'-thin, -tin) [named from the town of the discoverer]. A proprietary, odorless, tasteless preparation, said to consist of tannin, albumin, and keratin; an intestinal astringent. Dose 8–20 gr. (0.5–1.3 Gm.) 2 or 3 times daily; infants, 4–5 gr. (0.25–0.32 Gm.) 4 times daily.
**hoof**. The casing of hard, horny substance that sheathes the ends of the digits or in cases the foot in many animals. **h.-bound**, in farriery, having a dryness and contraction of the hoof, resulting in pain and lameness. This condition is also called *contracted heels*.
**hoof-and-mouth disease**. See *foot-and-mouth disease*.
**hook** [AS., hōc]. A curved instrument. **h., blunt**, an instrument for exercising traction upon the fetus in an arrested breech presentation. **h.s, Malgaigne's**. See *Malgaigne's h.s*. **h., Tyrrel's**. See *Tyrrel's h*.
**hook-worm**. Same as *Ankylostoma duodenale* and *Uncinaria americana, q. v*. **h. disease**, uncinariasis, ankylostomiasis.
**Hooper's pill** (hoop'-er) [John Hooper, English apothecary, 18th century]. A pill containing aloes, crystallized sulphate of iron, extract of hellebore, myrrh, soap, canella, and ginger.
**hooping-cough** (hoop'-ing-kof). See *whooping-cough, pertussis*.
**hoose** (hoos) [ME., *hose*, hoarse]. Sheep-cough. A disease of sheep, lambs, etc., due to the presence of *Strongylus filaria*, a nematode worm, in the lungs and air-passages, and characterized by a husky cough, anorexia, dry muzzle, constipation, and dyspnea.
**hoove, hooven** (hoov, hoov'-en) [dial., *hooven*]. Distention of the stomach of a ruminant animal with gas, caused by the fermentation of food. It is generally due to eating too much green food.
**hop**. See *humulus*.
**Hope's camphor mixture** [John Hope, English physician, 1725–1791]. A mixture containing nitric acid, camphor-water, and tincture of opium. It is used in the treatment of serous or choleraic diarrhea. **H.'s sign**, double cardiac beat noted in aneurysm of the aorta.
**Hopmann's polyp**. Papillary hypertrophy of the nasal mucous membrane, presenting the appearance of a papilloma.
**hopogan** (hop'-o-gan). The commercial name for a peroxide of magnesium.
**Hoppe-Goldflam's symptom-complex** (hop'-er-gold'-flam) [Johann Ignaz Hoppe, Swiss physiologist, 1811–1891; S. Goldflam]. Myasthenia gravis; see *Erb's disease*.
**Hoppe-Seyler's test for carbon monoxide in blood** (hop'-er-zi'-ler) [Ernst Felix Immanuel Hoppe-Seyler, German physiologist, 1825–1895]. Add to the blood twice its volume of caustic soda solution of 1.3 specific gravity. Ordinary blood thus treated is a dingy brown mass which, when spread out on porcelain, has a shade of green. Blood containing carbon monoxide, under the same conditions, appears as a red mass which, if spread on porcelain, shows a beautiful red color. **H.S.'s test for xanthin**, add the xanthin to a mixture of a solution of sodium hydroxide and chloride of lime in a porcelain dish; at first a dark-green ring, which quickly turns brown and disappears, forms about each xanthin grain.
**hordein** (hor'-de-in) [hordeum, barley]. A mixture of a protein with starch-cellulose; it exists in barley-starch, but is not soluble.
**hordeolum** (hor-de'-o-lum) [hordeum]. A sty; a furuncular inflammation of the connective tissue of the lids, near a hair-follicle. **h. externum**. See *Zeissian sty*. **h. internum**. See *Meibomian sty*.
**hordeum** (hor'-de-um) [L.]. Barley.
**horehound** (hōr'-hownd). See *marrubium*.
**horismascope** (hor-is'-mah-skōp) [όρισμα, a determination; σκοπεῖν, to examine]. An instrument designed for the detection of albumin, peptones, biliary constituents, etc., in urine.

**horizontal** (*hor-iz-on'-tal*) [ὀριζων, the horizon]. Parallel to the horizon. Referring to planes at right angles to vertical planes.

**Horlick's food** (*hor'-lik*). A food for infants. Its composition is: Water, 3.39; fat, 0.08; grape-sugar, 34.09; cane-sugar, 12.45; no starch; soluble carbohydrates, 87.20; albuminoids, 6.71; ash, 1.28.

**hormion** (*hor'-me-on*) [ὁρμή, the first]. See *craniometric points*.

**hormonadin** (*hor-mon'-ad-in*). Trade name for a pancreatic solution without the enzymes.

**hormonal** (*hor'-mo-nal*). Trade name of a preparation made from the spleen, and said to stimulate intestinal peristalsis. It has been used in constipation.

**hormone** (*hor'-mōn*) [ὁρμάω, I set in motion, arouse]. A chemical substance produced in a more or less distant organ which, passing into the blood-stream and reaching a functionally associated organ, is capable of exciting the latter to activity.

**hormonopoiesis** (*hor-mo-no-poi-e'-sis*) [*hormone-*; ποίησις, a making]. Hormone producing.

**hormonopoietic** (*hor-mo-no-poi-et'-ic*) [*hormone-*; ποιητικός, productive]. Pertaining to hormonopoiesis.

**hormotone** (*hor'-mo-tōn*). A preparation containing hormones of the thyroid, pituitary, ovary, testis, pancreas and spleen.

**horn.** 1. A substance composed chiefly of keratin. 2. Cornu.

**Horner's disease** (*hor'-ner*) [William Edmunds *Horner*, American anatomist, 1793–1853]. A slight ptosis accompanied by miosis, retraction of the eyeball, and flushing of the face of the same side, in destructive lesions of the cervical sympathetic. **H.'s muscle**, the tensor tarsi. **H.'s ptosis.** See *H.'s disease*. **H.'s teeth**, incisor teeth presenting horizontal grooves that are due to a deficiency of enamel.

**horny** (*hor'-ne*) [*horn*]. Composed of or resembling horn. **h. epithelium**, horny granulations in trachoma. **h. layer**, the stratum corneum of the skin.

**horopter** (*hor-op'-ter*) [ὅρος, boundary; ὀπτήρ, an observer]. The sum of all the points seen singly by the two retinæ while the fixation-point remains stationary.

**horopteric** (*hor-op-ter'-ik*) [ὅρος, boundary; ὀπτήρ observer]. Pertaining to an horopter.

**horrida cutis** (*hor'-id-ah kū'-tis*) [L.]. Goose-skin. Cutis anserina.

**horripilation** (*hor-ip-il-a'-shun*) [*horrere*, to stand on end; *pilus*, the hair]. Erection of the hairs of the skin produced by the contraction of the arrectores pili muscles.

**horror autotoxicus** (*hor'-or aw-to-toks'-ik-us*). Ehrlich's term for the non-production of antibodies by an animal against its own tissue cells.

**horrors** (*hor'-ors*) [*horror*, a shaking, terror]. A popular name for *delirium tremens*, *q. v.*

**horse-chestnut.** See *Æsculus hippocastanum*.

**horse-distemper.** Influenza.

**horse-doctor.** A farrier; a veterinary surgeon.

**horse-dose.** A dose of physic for a horse.

**horse-foot.** See *talipes equinus*.

**horse-leech.** A large leech, the *Hæmopis* of S. Europe and N. Africa; also a horse-doctor.

**horse-power.** See *unit*.

**horse-pox.** 1. A pustular disease of horses, which, communicated to cows, produces cow-pox. It is also called *pustular grease*. 2. See *coitus disease*.

**horse-radish** (*hors'-rad-ish*). The plant, *Cochlearia armoracia*, of the order *Cruciferæ*. The root (*armoraciæ radix*, B. P.) contains a volatile oil, and is a gastric stimulant and diuretic. It is chiefly used as a condiment, but has been employed in medicine in dropsy, chronic rheumatism, and scurvy. Dose of the root ½ dr. (2 Gm.) or more.

**horse-shoe fistula.** A name applied to a fistulous tract surrounding the rectum in a semicircle, either in front or behind.

**horse-shoe hymen.** See *hymen*.

**horseshoe-kidney.** A kidney having somewhat the shape of a horseshoe, due to a fusion of the two kidneys at one of their ends, usually the lower.

**horseshoe-magnet.** A magnet bent in the shape of a horseshoe.

**horse-sickness.** See *anthrax*.

**horse-tail.** See *Equisetum*.

**horseweed.** 1. *Callinsonia canadensis*, an indigenous plant, the root of which (in decoction) is used in cystitis, leukorrhea, dropsy, gravel, etc. 2. *Erigeron canadense*.

**horsikin** (*hor'-sik-in*). A model used in teaching the anatomy and surgery of the horse.

**horsine** (*hors'-in*). A French preparation said to be made from the juice of the muscle fibres of the horse. It has been given in tuberculosis.

**Horsley's method** (*hors'-le*) [Sir Victor Alexander Haden *Horsley*, English surgeon, 1857– ]. A method of determining the position of the fissure of Rolando by means of an instrument called a cyrtometer, encircling the head, and having an arm fixed at an angle of 67°, which indicates the position of the fissure. **H.'s test**, a test for glucose. The urine rendered alkaline is boiled with potassium bichromate; if sugar is present, a green color is developed. **H.'s wax**, a compound of phenol 1, oil 2, and wax 7, used to plug the diploe in case of hemorrhage from the skull.

**hospital** (*hos'-pit-al*) [*hospes*, a guest]. A building for the care and treatment of sick or infirm persons. **h. fever**, fever in hospitals due to unsanitary conditions. **h. gangrene**, a contagious, phagedenic gangrene occasionally attacking wounds or open sores. It is confined mainly to military hospitals, and is of microbic origin.

**hospitalism** (*hos'-pit-al-izm*) [*hospital*]. The morbid conditions arising from the gathering of diseased persons in a hospital.

**host** (*hōst*) [*hostis*, a stranger; a landlord]. The organic body upon which parasites live.

**hot** [ME.]. Having or yielding the sensation of heat; stimulating; biting. **h.-air bath.** See *bath, hot-air*. **h.-air treatment**, the local application of superheated dry air, the affected part being introduced into a cylinder or chamber. **h. bath.** See *bath, hot*.

**h. drops**, a term for the tincture of capsicum and myrrh. **h. eye**, congestion of the eye attending gout.

**Hottentot apron.** See *apron, Hottentot*. **H. deformity.** See *steatopygia*.

**hottentotism** (*hot'-en-tot-izm*). An extreme form of congenital stammering.

**hough** (*hok*). Hock; the lower part of the thigh.

**hour-glass contraction.** A contraction of a hollow organ, as the uterus or stomach, near the middle, producing a condition resembling an hour-glass.

**house disease.** Consumption.

**housemaid's knee.** A chronic inflammation of the bursa in front of the patella with an accumulation of serous fluid.

**house-physician.** The resident physician in a hospital.

**house-surgeon.** The resident surgeon in a hospital.

**Houston's folds** or **valves** (*hows'-tun*) [John *Houston*, Irish surgeon, 1802–1845]. Oblique folds, three in number, of the mucous membrane of the rectum. **H.'s muscle**, the compressor venæ dorsalis penis, a fasciculus of the ischiocavernosus, which passes over the dorsum of the penis to join its fellow of the opposite side.

**hove, hoven** (*hōv, ho'-ven*). See *hoove*.

**Hovius' canal.** See *Fontana's canal*. **H.'s membrane.** See *membrana ruyschiana*. **H.'s plexus, H.'s vascular circle.** See *Leber's plexus*.

**Howard's method of artificial respiration** (*how'-ard*) [Benjamin Douglas *Howard*, American physician, 1840–1900]. The patient is placed on his back, with his head lower than his abdomen, and pressure is exerted upon the lower ribs every few seconds.

**Howship's lacunæ** (*how'-ship*) [John *Howship*, English surgeon, died 1841]. **H.'s pits**, minute depressions or pits in bone undergoing absorption, produced by the action of osteoclasts.

**Howship-Romberg's sign.** See *Romberg's sign*.

**h. s.** Abbreviation for *hora somni* [L.]. Bedtime.

**Ht.** Abbreviation for *total hyperopia*.

**Huchard's disease** (*hoo-shar'*) [Henri *Huchard*, French physician, 1844–1910]. Excessive arterial tension due to a spasm of the vasoconstrictors, and which, according to Huchard, causes general arteriosclerosis. **H.'s sign**, the difference in the pulse between the standing and recumbent posture is less in persons with arterial hypertension, and may even be the reverse of that of the normal condition. **H.'s treatment**, a method of treating dilatation of the stomach by almost excluding liquids from the diet.

**huckle-bone** (*huk'-l-bōn*) [ME., *huccle-bone*, the astragalus]. The astragalus, *q. v.*

**Hudson's apparatus.** An apparatus to support the fingers and hands in the treatment of wrist-drop.

**Huebl's sign.** An early sign of pregnancy, consisting in an abnormal thinness and compressibility of the lower segment of the uterus as compared with that part above the insertion of the sacrouterine ligaments, the bimanual examination being carried out with one finger in the rectum.

**Hueck's ligament** (*hŭk*) [Alexander Friedrich *Hueck*, German anatomist, 1802–1842]. The pectinate ligament of the iris.

**Hueter's bandage** (*he'-ter*) [Carl *Hueter*, German surgeon, 1838–1882]. A spica bandage for the perineum. **H.'s sign**, absence of transmission of osseous vibration in cases of fracture with fibrous interposition between the fragments.

**Huguier's canal** (*hoo-ge-a'*) [Pierre Charles *Huguier*, French surgeon, 1804–1873]. A small canal in the temporal bone running parallel to the Glaserian fissure and transmitting the chorda tympani. **H.'s circle,** the anastomosis formed by the branches of the uterine arteries around the uterus, at the junction of the body with the cervix. It is not constant. **H.'s disease.** 1. Hypertrophic elongation of the supravaginal portion of the cervix uteri. 2. Lupus of the vulva. Syn., *Esthiomene de la vulve*. **H.'s glands.** See *Bartholin, glands of*. **H.'s operation.** *A method of performing colotomy:* the right lumbar operation.

**hum, venous.** A peculiar sound heard in the large veins of the neck in some cases of anemia. Syn., *bruit de diable*.

**humanized** (*hū'-man-īzd*). Applied to viruses which have passed through a human being.

**humectant** (*hū-mek'-tant*) [*humectare*, to make moist]. 1. Moistening; like a poultice; diluent. 2. A diluent; a substance used to moisten.

**humectation** (*hū-mek-ta'-shun*) [see *humectant*]. The act of moistening.

**humeral** (*hū'-mer-al*) [*humerus*]. Pertaining to the humerus.

**humeren** (*hū'-mer-en*) [*humerus*]. Belonging to the humerus in itself.

**humerus** (*hū'-mer-us*) [L.]. The bone of the upper arm.

**humid** (*hū'-mid*) [*humidus*, moist]. Moist; damp. **h. gangrene.** See *gangrene*. **h. tetter.** See *eczema*.

**humidity** (*hū-mid'-it-e*) [*humor*]. The state or quality of being moist; moisture; dampness. **h., absolute,** the actual amount of water present in the air at any moment. **h., relative,** the relative amount of water present in air as compared to what the air would contain at the existing temperature were it in condition that of saturation.

**humming-top murmur or sound.** See *hum, venous*.

**humor** (*hū'-mor*) [L., "moisture"]. 1. Any fluid or semifluid part of the body. 2. Disposition; temperament, as the four humors of Galen—the choleric, melancholic, phlegmatic, and sanguine. **h., aqueous,** the transparent fluid of the anterior chamber of the eye. **h., crystalline.** See *lens, crystalline*. **h., vitreous,** the transparent, gelatinlike substance filling the greater part of the globe of the eye.

**humoral** (*hū'-mor-al*) [*humor*]. Pertaining to the natural fluids of the body. **h. pathology,** that system of pathology according to which all diseases result from a disordered or abnormal condition of the fluids or humors of the body. **h. reflex,** functional activity due to the action of a hormone. **h. theory,** that theory which ascribes the production of immunity to the antitoxic or bactericidal action of the fluids of the body.

**humoralism, humorism** (*hū'-mor-al-izm, hū'-mor-ism*). Same as *humoral pathology*.

**humpback, hunchback** (*hump'-bak, hunch'-bak*). See *kyphosis*.

**humulin** (*hū'-mū-lin*) [*humulus*, hop]. 1. The same as *lupulin*. 2. A concentrated preparation from the tincture and decoction of hops.

**humulus** (*hū'-mū-lus*) [L.]. Hops. The *humulus* of the U. S. P. is the fruit-cones or strobiles of *H. lupulus*, which yield a powder, *lupulin*, a volatile oil, and tannin. Hops are tonic and slightly narcotic, and are used internally in dyspepsia, delirium tremens, and insomnia; locally, as emollient poultices. Dose of the *tincture* 1–2 dr. (4–8 Cc.).

**humus** (*hū'-mus*) [L., the earth, ground, soil]. A dark material from decaying vegetable matter.

**Hungarian disease** (*hung-ga'-re-an*) [Hungary]. Synonym of *typhus fever*.

**hunger** (*hung'-ger*) [AS., *hungor*]. A condition marked by a sensation of emptiness of the stomach, with a longing for food. **h., air-,** severe dyspnea or breathlessness. **h.-cure,** treatment by restricted diet.

**Hunt's syndrome** [James Ramsay *Hunt*, American neurologist, 1872– .]. 1. A combination of facial paralysis, earache and herpes, found when both the motor and the sensory fibers of the seventh cranial nerve are diseased. 2. An intention tremor beginning in one extremity, and gradually spreading to other parts of the body, at the same time increasing in intensity, denotes progressive cerebellar disturbance.

**Hunter's canal** (*hun'-ter*) [1. John *Hunter*, English surgeon, 1728–1793; 2. William *Hunter*, English anatomist, 1718–1783]. [1]. A triangular canal formed in the adductor magnus muscle of the thigh; it transmits the femoral artery and vein and internal saphenous nerve. **H.'s gubernaculum.** [2]. See *gubernaculum testis*. **H.'s ligament,** the round ligament of the uterus. **H.'s line,** [2] the linea alba. **H.'s method,** [1] a method of treating aneurysm by ligating the artery on the proximal side of the sac. **H.'s operation,** [2] *for aneurysm:* ligation of the artery on the cardiac side of the aneurysm at some distance from it.

**Hunteri membrana caduca.** The decidua.

**Hunterian chancre.** See *chancre, Hunterian*.

**Huntington's chorea** (*hun'-ting-tun*) [George *Huntington*, American physician, 1850– ]. A hereditary affection of adult or middle life, characterized by irregular movements, disturbance of speech, and gradual dementia.

**Hunyadi Janos** (*hoon-yah'-de yah'-nos*) [from the name of the Hungarian national hero, otherwise called *John Corvinus*]. An aperient mineral water from Buda-Pesth, in Hungary, containing sulphates of magnesium, potassium and sodium, sodium chloride and carbonate, iron oxide, and alumina. It is an effective laxative or cathartic.

**Huppert's reaction for bile-pigments.** After the solution has been treated with milk of lime or with a solution of calcium chloride, precipitate with ammonia; filter and wash the precipitate, treat with alcohol acidified with sulphuric acid, and boil; the liquid will assume a green color.

**Huschke's canal** (*hoosh'-ker*) [Emil *Huschke*, German anatomist, 1797–1913]. A canal formed by the junction of the tubercles of the annulus tympanicus. This is generally obliterated after the fifth year, but may persist through life. **H.'s cartilage.** See *Jacobson's cartilage*. **H.'s foramen,** a perforation often found near the inner extremity of the tympanic plate; it results from an arrest of development. **H.'s teeth,** the serrated projections on the inner wall of the lamina spiralis of the cochlea, roofing over the internal spiral sulcus. Syn., *crista spiralis*. **H.'s valve.** See *Hutchinson's valve*.

**Hutchinson's disease** (*hutsh'-in-sun*) [Sir Jonathan *Hutchinson*, English surgeon, 1828–1913]. See *Tay's choroiditis*. **H.'s facies,** the peculiar facial expression caused by immobility of the eyeballs in ophthalmoplegia externa. **H.'s patch,** a reddish (salmon-colored) patch of the cornea in syphilitic keratitis. **H.'s prurigo,** the prurigo of dentition. **H.'s pupil,** a dilated pupil on the injured side in traumatic meningeal hemorrhage. **H.'s teeth,** peg-shaped incisor teeth, notched at the cutting-edge, frequently seen in congenital syphilis. **H.'s theory,** attributes the origin of leprosy to the eating of fish too continuously or in too great quantities. **H.'s triad,** pathognomonic of hereditary syphilis—(1) diffuse interstitial keratitis; (2) disease of the labyrinth; (3) Hutchinson's teeth.

**huttoning** (*hut'-on-ing*) [after Hutton, the inventor]. A method of manipulating a luxated joint, introduced by one Hutton, a bone-setter.

**Huxham's tincture** (*huks'-ham*) [John *Huxham*, English physician, 1692–1768]. See *cinchona, tincture of, compound*.

**Huxley's layer, H.'s membrane, H.'s sheath** [Thomas Henry *Huxley*, English biologist, 1825–1895]. A layer of nucleated, elongated, polygonal cells lying within Henle's layer of the inner root-sheath of hairs.

**Huygenian ocular** (*hi-ge'-ne-an*) [Christian *Huygens*, Dutch physicist, 1629–1695]. A lens consisting of two planoconvex lenses, the convexities

being directed toward the objective; the lower lens is the *field-lens*, the upper, the *eye-lens*.
**hyal** (*hi'-al*). See *hyoid*.
**hyalin** (*hi'-al-in*) [ὕαλος, glass]. 1. The generic term for the soluble substances obtained from hyalogens by the action of alkalies or superheated water. 2. A translucent substance forming the walls of hydatid cysts.
**hyaline** (*hi'-al-in*) [see *hyalin*]. Resembling glass, crystalline, translucent. **h. cartilage.** See *cartilage, hyaline*. **h. cast, h. cylinder,** a clear, nearly transparent urinary tube=cast. **h. degeneration.** See *degeneration, hyaline*.
**hyalinosis** (*hi-al-in-o'-sis*) [ὕαλος, glass]. Hyaline or waxy degeneration.
**hyalinuria** (*hi-al-in-ū'-re-ah*) [*hyalin*; οὖρον, urine]. The presence of hyalin or hyaline casts in the urine.
**hyalitis** (*hi-al-i'-tis*) [*hyaloid*; ιτις, inflammation]. Inflammation of the hyaloid membrane of the vitreous humor.
**hyalo-** (*hi-al-o-*) [ὕαλος, glass]. A prefix meaning: 1. Transparent. 2. Relating to hyalin.
**hyalogen** (*hi-al'-o-jen*) [*hyalo-*; γεννᾶν, to produce]. 1. Generic term for insoluble substances resembling mucin, found in the walls of hydatid cysts, the vitreous humor, tubercles, etc.; the mother-substance of hyalin. 2. An albuminoid found in cartilage. It is readily changed into hyalin.
**hyaloid** (*hi'-al-oid*) [*hyalo-*; εἶδος, like]. Transparent; glass-like. **h. artery,** a branch of the arteria centralis retinæ in the embryo, traversing the vitreous humor to the posterior capsule of the lens. **h. canal,** the canal in the vitreous humor transmitting the hyaloid artery. **h. fossa,** the depression for the crystalline lens. **h. membrane,** a delicate, transparent membrane surrounding the vitreous humor, except in front, where it becomes fibrous and strong and forms a leaflet of the zonule of Zinn.
**hyaloiditis** (*hi-al-oid-i'-tis*). See *hyalitis*.
**hyaloma** (*hi-al-o'-mah*) [*hyalo-*; ὅμα, a tumor]. The conversion of the eye into a hyaline mass.
**hyalomitome, hyalotome** (*hi-al-om'-it-ōm*), *hi-al'-o-tōm*). See *paramitome*.
**hyalomucoid** (*hi-al-o-mū'-koid*) [*hyalo-*; *mucus*; εἶδος, likeness]. A mucoid found in the fluid of the vitreous humor.
**hyalonyxis** (*hi-al-o-niks'-is*) [*hyalo-*; νύξις, a pricking]. Puncture of the vitreous body of the eye.
**hyalophagia** (*hi-al-o-fa'-je-ah*) [*hyalo-*; φαγεῖν, to eat]. The practice of eating glass, sometimes seen among insane persons.
**hyaloplasm** (*hi'-al-o-plazm*) [*hyalo-*; πλάσμα, plasm]. The fluid portion of the protoplasm of a living cell.
**hyaloserositis** (*hi-al-o-se-ro-si'-tis*) [*hyalo-*; *serosa*; ιτις, inflammation]. Chronic inflammation of the serous membranes with formation of a dense, fibrous hyaline investment in certain regions.
**hyalosome** (*hi'-al-o-sōm*) [*hyalo-*; σῶμα, body]. A body resembling a nucleolus, but staining slightly by either nuclear or plasmatic dyes.
**hyboma** (*hi-bo'-mah*) [ὕβος, humpbacked]. Humpback, gibbosity.
**hybometer** (*hi-bom'-et-er*) [ὕβος, humpbacked; μέτρον, a measure]. An apparatus for measuring pathological gibbosities.
**hybrid** (*hi'-brid*) [ὕβρις, insult]. The offspring of two individuals of distinct but closely related species.
**hybridism,** or **hybridity** (*hi'-brid-ism, hi-brid'-it-e*) [ὕβρις, insult]. Cross-breeding; mixture of races; the combination of various diseases.
**hybridization** (*hi-brid-is-a'-shun*). Cross-breeding.
**hydaleous** (*hi-da'-le-us*) [ὑδαλέος, watery]. Dropsical.
**hydantoin** (*hi-dan-to'-in*), C₃H₄N₂O₂. Glycolyl urea. A crystalline substance derived from allantoin and related to urea.
**hydatenterocele** (*hi-dat-en-ter'-o-sēl*) [ὕδωρ, water; ἔντερον, bowel; κήλη, tumor]. Hydrocele with intestinal hernia.
**hydatic** (*hi-dat'-ik*). Containing hydatids.
**hydatid** (*hi-dat'-id*) [ὑδατίς, vesicle]. 1. A cyst-like body with clear contents, especially that formed by the larva of the *Tænia echinococcus*. 2. Bulbous remnants of embryonic structures. See *Morgagni's hydatid*. **h. disease,** a disease characterized by the presence in various portions of the body of cysts containing the embryo of the *Tænia echinococcus*. **h. fremitus, h. thrill,** a fremitus occasionally obtained on palpating a hydatid cyst. **h. mole.** See *chorion,*

*cystic degeneration of,* and *mole, hydatid.* **h. of Morgagni.** See *Morgagni, hydatid of.*
**hydatidiform** (*hi-dat-id'-if-orm*) [*hydatid*; *forma,* form]. Having the form of a hydatid. **h. mole.** See *mole, hydatidiform.*
**hydatidocele** (*hi-dat-id'-o-sēl*) [*hydatid*; κήλη, tumor]. Oscheocele with hydatid cysts.
**hydatidoma** (*hi-dat-id-o'-mah*) [*hydatid*; ὅμα, a tumor]. Any hydatid cyst or tumor.
**hydatidosis** (*hi-dat-id-o'-sis*) [*hydatid*]. The condition of being affected with hydatids.
**hydatidostomy** (*hi-dat-id-os'-to-me*) [*hydatid*; στόμα, mouth]. The opening and evacuation of a hydatid cyst.
**hydatism** (*hi'-dat-izm*) [ὕδωρ, water]. The sound caused by the moving of pathological fluid in a body cavity.
**hydatogenesis** (*hi-dat-o-jen'-es-is*) [ὕδωρ, water; γένεσις, production]. The formation of water within the tissues or cavities of the body.
**hydatoid** (*hi-dat'-oid*) [*hydatid*; εἶδος, likeness]. 1. Hydatidiform. 2. Watery. 3. The aqueous humor.
**hydatoncus** (*hi-dat-ong'-kus*) [ὕδωρ, water; ὄγκος, a tumor]. Any cyst, or watery tumor.
**hydatorrhea** (*hi-dat-or-e'-ah*) [ὕδωρ, water; ῥοία, flow]. A copious flow or discharge of water.
**hydra-, hydro-** (*hi-drah-, hi-dro-*) [ὕδωρ, water]. Prefixes signifying the presence of water or of hydrogen.
**hydracid** (*hi-dras'-id*) [*hydra-*; *acidum*, acid]. An acid containing hydrogen but not oxygen.
**hydradenitis** (*hi-drad-en-i'-tis*) [ὕδωρ, water; ἀδήν, gland; ιτις, inflammation]. Same as *hidrosadenitis*. 2. Lymphadenitis. **h. destruens suppurativa.** See *hydrosadenitis phlegmonosa*.
**hydradenoma** (*hi-drad-en-o'-mah*). Same as *hidradenoma*.
**hydradenomes** (*hi-drad'-en-ōmz*) [*hydra-*; ἀδήν, gland]. A skin disease marked by the formation of papules, varying in size from that of a pin-head to that of a pea, without subjective symptoms. It is attributed to adenoid epithelioma of the sweat-glands.
**hydræmia** (*hi-dre'-me-ah*). See *hydremia*.
**hydræroperitoneum, hydraeroperitonia** (*hi-drah-er-o-per-it-o-ne'-um, -o'-ne-ah*) [*hydra-*; ἀήρ, air; *peritoneum*]. A collection of gas and fluid in the peritoneal cavity.
**hydragogin** (*hi-drag'-oj-in*). A diuretic and cardiac tonic containing tincture of digitalis, tincture of strophanthus, scillipicrin, scillitoxin, and oxysaponin. Dose 10–15 min. (0.66–1.0 Cc.).
**hydragogue, hydragog** (*hi'-drag-og*) [*hydra-*; ἀγωγός leading]. 1. Expelling water. 2. A purgative that causes copious liquid discharges.
**hydramnios, hydramnion** (*hi-dram'-ne-os, -on*) [*hydra-*; ἀμνίον, fetal membrane]. An abnormal amount of amniotic fluid.
**hydrangea** (*hi-dran'-je-ah*) [*hydra-*; ἀγγεῖον, vessel]. A genus of saxifragaceous shrubs. The root of *H. arborescens* contains *hydrangin* and is employed in lithiasis.
**hydrangeion** (*hi-dran'-je-on*) [ὕδωρ, water, lymph; ἀγγεῖον, vessel; *pl.*, *hydrangeia*]. A lymphatic vessel.
**hydrangeitis** (*hi-dran-je-i'-tis*) [ὕδωρ, water; ἀγγεῖον, a vessel; ιτις, inflammation]. Inflammation of a lymphatic vessel; lymphangitis.
**hydrangin** (*hi-dran'-jin*). A crystalline glucoside from the root of *Hydrangea arborescens*. It melts at 235° C. and by action of dilute acids decomposes into glucose and a resin-like mass.
**hydrangiography, hydrangeiography** (*hi-dran-ji-og'-raf-e*) [ὕδωρ, water; ἀγγεῖον, vessel; γράφειν, to write]. A description of the lymphatic vessels, their anatomy, nature, functions, diseases, etc. Syn., *angiohydrography*.
**hydrangiology** (*hi-dran-ji-ol'-o-je*) [ὕδωρ, water; ἀγγεῖον, vessel; λόγος, science]. The science of the nature, functions, and diseases of the lymphatics.
**hydrangiotomy, hydrangeiotomy** (*hi-dran-je-ot'-o-me*) [ὕδωρ, water; ἀγγεῖον, vessel; τέμνειν, to cut]. The dissection of the lymphatics.
**hydrargyrate** (*hi-drar'-je-rāt*). Relating to mercury; containing mercury.
**hydrargyria, hydrargyriasis, hydrargyrism** (*hi-drar-ji'-re-ah, hi-drar-je-ri'-as-is, hi-drar'-je-rizm*). Chronic mercurial poisoning. See *mercurialism*.

**hydrargyric** (*hi-drar'-je-rik*). Relating to mercury; mercuric.

**hydrargyrol** (*hi-drar'-je-rol*). Mercury paraphenylthionate. **h.-septol.** See *mercury quinoseptolate.*

**hydrargyromania** (*hi-drar-jir-o-ma'-ne-ah*). Insanity due to the unwise use of mercury.

**hydrargyrophobia** (*hi-drar-jir-o-fo'-be-ah*). Morbid dread of mercurial medicines.

**hydrargyrophthalmia** (*hi-drar-jir-off-thal'-me-ah*). Ophthalmia due to mercurial poisoning.

**hydrargyrosis** (*hi-drar-ji-ro'-sis*). 1. See *hydrargyriasis.* 2. Mercurial friction or fumigation.

**hydrargyrum** (*hi-drar'-je-rum*). See *mercury.*

**hydrarsan** (*hi-drar'-san*). A compound of phenacetin, mercuric chloride, arsenic chloride, and potassium iodide; used in syphilis.

**hydrarthrosis** (*hi-drar-thro'-sis*) [*hydra-;* ἄρθρον, a joint]. An accumulation of fluid in a joint.

**hydrarthrus** (*hi-drar'-thrus*). See *hydrarthrosis.*

**hydrastine** (*hi-dras'-tēn*) [*hydrastis*]. 1. An alkaloid (*hydrastina,* U. S. P.) from the root of *Hydrastis canadensis.* 2. A resinous extract from the root of *H. canadensis;* it is a cholagogue, laxative, alterative, antiseptic, etc. **h. hydrochloride,** an astringent, alterative, and hemostatic. Dose $\frac{1}{2}$–1 gr. (0.032–0.065 Gm.) every two hours if needed. Application as astringent 0.1 to 0.5 % solution; in skin diseases 1 % ointment; in chronic bronchitis 10–20 drops of a solution of 15 gr. (1 Gm.) in 5 dr. (18 Cc.) of water 4 times daily.

**hydrastinine** (*hi-dras'-tin-ēn*) [*hydrastis*]. An artificial alkaloid, $C_{11}H_{13}NO_3$, from hydrastine. It is used as a hemostatic, vasoconstrictor, cardiac stimulant, etc. **h. hydrochloride** (*hydrastininæ hydrochloridum,* U. S. P.), $C_{11}H_{11}NO_2 \cdot HCl + H_2O$, used as a uterine hemostatic, emmenagogue, and vasoconstrictor. Dose $\frac{1}{4}$–$\frac{1}{2}$ gr. (0.015–0.032 Gm.) 3 or 4 times daily; maximum daily dose 2 gr. (0.13 Gm.). Injection, 8–16 min. (0.5–1.0 Cc.) of 10 % solution.

**hydrastis** (*hi-dras'-tis*). Goldenseal, a plant of the order *Ranunculaceæ.* The *hydrastis* of the U. S. P. is the rhizome and roots of *H. canadensis,* which contain the alkaloids *hydrastine* and *berberine.* It is tonic, antiperiodic, cholagogue, and diuretic, and has been employed in leukorrhea, cystitis, constipation, menorrhagia, gonorrhea, dyspepsia, etc. Dose of the *hydrastine of commerce,* an impure body, 5–10 gr. (0.32–0.65 Gm.); of the *alkaloid* $\frac{1}{8}$–$\frac{1}{3}$ gr. (0.008–0.022 Gm.). **h., fluidextract of** (*fluidextractum hydrastis,* U. S. P.). Dose $\frac{1}{2}$–1 dr. (2–4 Cc.). **h., glycerite of** (*glyceritum hydrastis,* U. S. P.). Dose $\frac{1}{2}$–1 dr. (2–4 Cc.). **h., tincture of** (*tinctura hydrastis,* U. S. P.). Dose 1 dr. (4 Cc.).

**hydrastol** (*hi-dras'-tol*). A proprietary remedy said to consist of liquid albolene and the active ingredients of hydrastis and cinnamon oil.

**hydrate** (*hi'-drāt*) [ὕδωρ, water]. A compound containing water in chemical combination.

**hydrated** (*hi'-dra-ted*) [see *hydrate*]. Chemically combined with water.

**hydration** (*hi-dra'-shun*) [see *hydrate*]. The process of combining chemically with water, or of converting into a hydrate.

**hydraulics** (*hi-draw'-liks*) [ὕδωρ, water; αὐλός, a pipe]. The science of liquids in motion.

**hydrazine** (*hi'-draz-in*) [*hydra-; azotum,* nitrogen]. 1. $H_4N_2$. Diamine; a colorless, stable gas, soluble in water, having a peculiar odor and a strong alkaline reaction. 2. One of a class of bodies derived from hydrazine by replacing one or more of its hydrogen atoms by a radical.

**hydrazones** (*hi'-draz-ōns*) [ὕδωρ, water; *azotum,* nitrogen]. Phenylhydrazones; a group of chemical bodies produced by the action of phenylhydrazine upon carbonyl compounds. They are usually crystalline compounds, insoluble in water. They are yellow or brown in color, and almost invariably decompose upon fusion.

**hydrectasis** (*hi-drek'-tas-is*) [*hydra-;* ἔκτασις, a stretching out]. Distention by water or by a watery fluid.

**hydremesis** (*hi-drem'-es-is*) [*hydro-;* ἔμεσις, vomiting]. The vomiting of a watery material.

**hydremia** (*hi-dre'-me-ah*) [*hydra-;* αἷμα, blood]. A watery condition of the blood.

**hydrencephal** (*hi-dren'-sef-al*) [*hydra-;* ἐγκέφαλος, brain]. A hydrocephalic brain.

**hydrencephalitis** (*hi-dren-sef-al-i'-tis*) [*hydra-;* ἐγκέφαλος, brain; ιτις, inflammation]. Inflammatory hydrocephalus.

**hydrencephalocele** (*hi-dren-sef'-al-o-sēl*) [*hydra-;* ἐγκέφαλος, brain; κήλη, hernia]. Hernia of the brain, in which the tumor is in part composed of a watery fluid.

**hydrencephalus** (*hi-dren-sef'-al-us*). See *hydrocephalus.*

**hydrenterocele** (*hi-dren-ter'-o-sēl*) [*hydra-;* ἔντερον, intestine; κήλη, hernia]. Intestinal hernia, the sac of which contains some extravasated fluid.

**hydrenterorrhea** (*hi-dren-ter-or-e'-ah*) [*hydra-;* ἔντερον, intestine; ῥοία, flow]. A watery diarrhea.

**hydrepigastrium** (*hi-drep-e-gas'-tre-um*) [*hydra-; epigastrium*]. A collection of fluid between the abdominal muscles and the peritoneum.

**hydriatics** (*hi-dre-at'-iks*). See *hydrotherapeutics.*

**hydriatric** (*hi-dre-at'-rik*) [*hydra-;* ἰατρός, physician]. Relating to the treatment of disease with water.

**hydriatry** (*hi'-dre-at-re*). See *hydrotherapeutics.*

**hydric** (*hi'-drik*). Containing water.

**hydride** (*hi'-drid*). A chemical compound containing hydrogen united to an element or radical.

**hydriodate** (*hi-dri'-o-dāt*). A compound of hydriodic acid with an element or radical.

**hydriodic** (*hi-dre-od'-ik*) [*hydra-; iodine*]. Containing hydrogen and iodine. **h. acid,** HI, a heavy, colorless gas, with a suffocating odor and an acid reaction. **h. acid, syrup of** (*syrupus acidi hydriodici,* U. S. P.), contains 1 % by weight of absolute HI. Dose $\frac{1}{2}$–2 dr. (2–8 Cc.). It is used as an alterative in scrofula, rickets, etc.

**hydro-** (*hi'-dro-*) [ὕδωρ, water]. A prefix signifying water, or that water forms a structural part; also denoting hydrogen.

**hydro** (*hi'-dro*) [ὕδωρ, water]. A hydropathic establishment.

**hydroa** (*hi-dro'-ah*) [*hydro-;* ᾠόν, egg]. A chronic inflammatory disease of the skin characterized by erythema, papules, pustules, vesicles, bullæ, or combinations of these, and by intense itching. Syn., *dermatitis herpetiformis; pemphigus pruriginosus.*

**hydroabdomen** (*hi-dro-ab-do'-men*). See *ascites.*

**hydroadenitis** (*hi-dro-ad-en-i'-tis*). See *hydradenitis.*

**hydroadipsia** (*hi-dro-ad-ip'-se-ah*) [*hydro-;* ἀ, priv.; δίψα, thirst]. Absence of thirst, or of desire for water.

**hydroaeric** (*hi-dro-a-er'-ik*) [*hydro-;* ἀήρ, air]. Applied in auscultation to the sound given by cavities filled with air and water.

**hydroappendix** (*hi-dro-ap-en'-diks*). The dilatation of the vermiform appendix with a watery fluid.

**hydroargentic** (*hi-dro-ar-jen'-tik*) [*hydro-; argentum,* silver]. Containing hydrogen and silver.

**hydrobilirubin** (*hi-dro-bil-e-ru'-bin*) [*hydro-; bilirubin*], $C_{32}H_{40}N_4O_7$. A brown-red pigment formed by treating a solution of bilirubin with sodium amalgam. It is probably identical with stercobilin, the coloring-matter of the feces, and urobilin, the pigment of the urine.

**hydrobiosis** (*hi-dro-bi-o'-sis*) [*hydro-;* βίος, life]. In biology, the origin and maintenance of life in fluid media.

**hydrobromate** (*hi-dro-bro'-māt*) [*hydro-;* βρῶμος, a stench]. A salt of hydrobromic acid.

**hydrobromic** (*hi-dro-bro'-mik*) [see *hydrobromate*]. Composed of hydrogen and bromin. **h. acid.** See *acid, hydrobromic.* **h. ether.** See *ethyl bromide.*

**hydrobromide, hydrobromid** (*hi-dro-bro'-mid*). Same as *hydrobromate.*

**hydrocarbon** (*hi-dro-kar'-bon*) [*hydro-; carbo,* charcoal]. Any compound composed only of hydrogen and carbon. **h., satisfied,** one that has no free valences. **h., saturated,** one that has the maximum number of hydrogen atoms.

**hydrocarbonism** (*hi-dro-kar'-bon-izm*). Poisoning with hydrocarbons, principally observed among miners and workers in petroleum refineries and in those who have used petroleum internally. It is marked by dizziness, cyanosis, loss of consciousness, anesthesia, convulsions, loss of reflexes, weakness of pulse and of breathing.

**hydrocardia** (*hi-dro-kar'-de-ah*). See *hydropericardium.*

**hydrocele** (*hi'-dro-sēl*) [*hydro-;* κήλη, tumor]. A collection of serous fluid about the testicle or spermatic cord. The term is also applied to serous tumors in other locations. **h. colli.** See *Maunoir's hydrocele.* **h. hernialis,** a collection of fluid in a hernial sac due to obstruction of its neck.

**hydrocelia** (*hi-dro-se'-le-ah*) [*hydro-;* κοιλία, belly]. Dropsy of the belly or abdominal region.
**hydrocelodes** (*hi-dro-sel-o'-dēz*) [*hydro-;* κήλη, tumor; εἶδος, form]. A tumor resembling hydrocele, but due to an extravasation of urine.
**hydrocenosis** (*hi-dro-sen-o'-sis*) · [*hydro-;* κένωσις, evacuation] An evacuation of water either by the use of hydragogue cathartics or by the operation of "tapping" the cavity containing the accumulation of fluid. See *paracentesis*.
**hydrocenotic** (*hi-dro-sen-ot'-ik*) [*hydro-;* κένωσις, evacuation]. Relating to or causing hydrocenosis.
**hydrocephalic** (*hi-dro-sef-al'-ik*) [see *hydrocephalus*]. Pertaining to or affected with hydrocephalus. h. cry, the loud cry of a child, indicating pain in the head.
**hydrocephalocele** (*hi-dro-sef'-al-o-sēl*). See *hydrencephalocele*.
**hydrocephaloid** (*hi-dro-sef'-al-oid*) [*hydrocephalus;* εἶδος, like]. 1. Pertaining to or resembling hydrocephalus. 2. Marshall Hall's disease; a disease of infants resembling hydrocephalus. It is a condition of nervous exhaustion generally consequent on prolonged illness or premature weaning. It is sometimes observed in poorly nourished infants just after weaning. The pulse is irregular, the fontanelles depressed, and there is little tendency to vomiting.
**hydrocephalus** (*hi-dro-sef'-al-us*) [*hydro-;* κεφαλή, head]. A collection of fluid in the cerebral ventricles (*internal hydrocephalus*) or outside the brain-substance (*external hydrocephalus*). The symptoms are progressive enlargement of the head, bulging of the fontanels, prominent forehead, thinness of hair and scalp, distention of the superficial veins, mental impairment, muscular weakness, convulsions. *Acute external hydrocephalus* is due to inflammation of the meninges, usually tuberculous; *acute internal hydrocephalus* is caused by ependymitis; chronic hydrocephalus may be congenital or acquired.
**hydrochezia** (*hi-dro-ke'-se-ah*) [*hydro-;* χέζειν, to defecate]. Watery or serous diarrhea.
**hydrochinone** (*hi-dro-kin'-ōn*). See *hydroquinone*.
**hydrochinonuria** (*hi-dro-kin-on-ū'-re-ah*) [*hydroquinone;* οὖρον, urine]. The presence in the urine of hydroquinone due to ingestion of salol, resorcin, etc.
**hydrochlorate** (*hi-dro-klo'-rāt*). Any salt of hydrochloric acid.
**hydrochloric acid** (*hi-dro-klo'-rik*). See *acid, hydrochloric*. h. acid in contents of stomach. See *Boas, Ewald, Guenzburg, Herzberg,* v. *Jaksch, Luttke, Maly, Mohr, Rabuteau, Reoch, Sjoeqvist, Szabo, Uffelmann, v. d. Velden, Winkler, Witz*. h. acid test for formaldehyde in milk, heat in a test-tube 1 Cc. of milk with 4 Cc. of strong hydrochloric acid containing a trace of ferric chloride. In the presence of formaldehyde a purple color appears, varying from a delicate tint to a deep violet. If a yellow color appears, repeat the test, using milk that has been diluted, 1 to 10, with water.
**hydrochloric ether**. See *ethyl chloride*.
**hydrochloride, hydrochlorid** (*hi-dro-klo'-rid*). Same as *hydrochlorate*.
**hydrocholecystis** (*hi-dro-ko-le-sis'-tis*) [*hydro-;* χολή, bile; κύστις, bladder]. Dropsy of the gall-bladder.
**hydrocinnamic acid** (*hi-dro-sin-am'-ik*). See *acid*.
**hydrocirsocele** (*hi-dro-sir'-so-sēl*) [*hydro-;* κιρσός, venous enlargement; κήλη, tumor]. Hydrocele accompanied with varicose veins of the spermatic cord.
**hydrocollidine** (*hi-dro-kol'-id-ēn*) [*hydro-;* κόλλα, glue], C₈H₁₁N. A highly poisonous ptomaine obtained from putrefying mackerel, horse-flesh, and ox-flesh, and said to be identical with one obtained from nicotine.
**hydrocolpos** (*hi-dro-kol'-pos*) [*hydro-;* κόλπος, vagina; κήλη, tumor]. A serous tumor of the vagina.
**hydrocolpos** (*hi-dro-kol'-pos*) [*hydro-;* κόλπος, vagina]. A vaginal retention-cyst containing a watery fluid.
**hydroconion, hydrokonion** (*hi-dro-ko'-ne-on*) [*hydro-;* κονίεν, to fill with dust]. An atomizer; a spraying apparatus.
**hydrocotarnine** (*hi-dro-ko-tar'-nēn*) [*hydro-;* *cotarnine*], C₁₂H₁₅NO₃+½H₂O. A crystalline alkaloid occurring in small amount in opium. It melts at 50° C., and is readily soluble in alcohol, ether, and chloroform.

**hydrocotoine** (*hid-ro-ko'-to-ēn*) [*hydro-;* Sp., *coto*, a cubit], C₁₅H₁₄O₄. An alkaloid of coto-bark, occurring in yellowish crystals without taste.
**Hydrocotyle** (*hi-dro-kot'-il-e*) [*hydro-;* κοτύλη, a cup]. Pennywort; a genus of umbelliferous herbs. The leaves of *H. asiatica* serve as a bitter tonic and alterative, and are very serviceable in skin-diseases, syphilitic sores, and leprosy. *H. centella*, of S. Africa, *H. umbellata*, of America, and *H. vulgaris*, of Europe, have been employed in medicine, but their properties are little known. *H. bonariasis*, of S. America, is diuretic.
**hydrocrania** (*hid-ro-kra'-ne-ah*). Same as *hydrocephalus*.
**hydrocyanic** (*hi-dro-se-an'-ik*). See *cyanogen*, and *acid, hydrocyanic*.
**hydrocyanism** (*hi-dro-si'-an-izm*). Poisoning with hydrocyanic acid; in acute cases marked by loss of consciousness and a sudden fall, generally by cramp, cyanosis, and paralysis.
**hydrocyst** (*hi'-dro-sist*) [*hydro-;* κύστις, bladder]. A cyst containing a water-like liquid. Sometimes it is synonymous with *hydatid*.
**hydrocystoma** (*hi-dro-sis-to'-mah*). See *hidrocystoma*.
**hydroderma** (*hi-dro-der'-mah*) [*hydro-;* δέρμα, skin]. Dropsy of the skin.
**hydrodiarrhea** (*hi-dro-di-ar-e'-ah*) [*hydro-;* *diarrhea*]. Serious diarrhea.
**hydrodiascope** (*hi-dro-di'-as-kōp*) [*hydro-;* διά, through; σκοπεῖν, to view]. A device, shaped like a pair of spectacles, consisting of two chambers filled with physiological salt solution, worn to correct keratoconus and astigmatism.
**hydrodictyotomy** (*hi-dro-dik-te-ot'-o-me*) [*hydro-;* δίκτυον, net; τομή, section]. Surgical incision of the retina for the relief of edema.
**hydrodiffusion** (*hi-dro-dif-ū'-shun*). The physical admixture of two fluids of different densities.
**hydrodiuresis** (*hi-dro-di-ū-re'-sis*) [*hydro-;* διά, through; οὐρεῖν, to urinate]. A copious flow of watery urine.
**hydrodynamics** (*hi-dro-di-nam'-iks*) [*hydro-;* δύναμις, power]. The branch of mechanics treating of fluids in motion.
**hydroelectric** (*hi-dro-e-lek'-trik*) [*hydro-;* *electric*]. Pertaining to electricity developed in connection with water. h. bath, a bath in which the metallic lining of the tub is connected with one pole of a battery, the other pole being in contact with the person of the patient.
**hydr₂electrization** (*hi-dro-e-lek-tri-za'-shun*). Electrization in which water is used as an electrode.
**hydroencephalocele** (*hi-dro-en-sef-al'-o-sēl*). See *hydrencephalocele*.
**hydroenterocele** (*hi-dro-en-ter'-o-sēl*). See *hydrenterocele*.
**hydroepigastrium**. See *hydrepigastrium*.
**hydroepiplocele** (*hi-dro-ep-ip'-lo-sēl*). An epiplocele with water in the sac.
**hydroexostosis** (*hi-dro-eks-os-to'-sis*). An exostosis accompanied by an accumulation of water.
**hydroferrocyanate, hydroferrocyanide** (*hi-dro-fer-o-si'-an-āt, -īd*). A compound of hydroferrocyanic acid with a base.
**hydrofluoric acid** (*hi-dro-flu-or'-ik*). See *acid, hydrofluoric*, and *fluorine*.
**hydrofluosilicate** (*hi-dro-flu-o-sil'-ik-āt*). A salt of hydrofluosilicic acid.
**hydrogalvanic** (*hi-dro-gal-van'-ik*). Relating to galvanism developed by action of fluids.
**hydrogaster** (*hi-dro-gas'-tur*) \ [*hydro-;* γαστήρ, stomach]. Ascites.
**hydrogastria** (*hi-dro-gas'-tre-ah*). A gastric disorder from fluid due to constriction of the esophageal and pyloric orifices.
**hydrogel** (*hi'-dro-jel*). An aqueous colloidal solution in the gelatinized state.
**hydrogen** (*hi'-dro-jen*) [*hydro-;* γεννᾶν, to produce]. Symbol H; atomic weight 1.008; quantivalence 1. A gaseous element, feebly basic, and occurring in nature in greatest abundance combined with oxygen in the form of water, H₂O. It is present in nearly all organic compounds, and is a constant constituent of acids. h. acid. See *hydracid*. h. dioxide, h. peroxide, H₂O₂. an unstable liquid which readily yields up an atom of oxygen, and hence is a powerful oxidizer. It is strongly antiseptic. The solution, *aqua hydrogenii dioxidi* (U. S. P.), contains 3 % of pure H₂O₂. It is useful as an antiseptic application

to inflamed mucous membranes in diphtheria, scarlatina, gonorrhea, etc., and as a cleansing agent of suppurating cavities. See also *Wurster's test for hydrogen dioxide*. **h. monoxide,** $H_2O$, water. **h. persulphide,** $H_2S_2$, a heavy, yellow oil, with a foul, pungent odor and acrid taste, giving off irritating vapors. It bleaches organic coloring-matters and is a powerful antiseptic.

**hydrogenation** (*hi-dro-jen-a'-shun*). The process of causing a combination with hydrogen.

**hydrogenesis** (*hi-dro-jen'-es-is*) [*hydro-*; γεννᾶν, to produce: *pl.*, hydrogeneses]. 1. The collection or formation of a watery fluid. 2. Any disease in which there is a predominance or degeneration of mucous secretion, fat, bile, or milk.

**hydrogenoid** (*hi-droj'-en-oid*) [*hydro-*; γεννᾶν, to produce; εἶδος, likeness]. Applied to a constitution or temperament intolerant of moisture.

**hydrogenous** (*hi-droj'-en-us*). Relating to hydrogen.

**hydrogerous** (*hi-droj'-ur-us*) [*hydro-*; *gerere*, to bear]. Containing or bearing water.

**hydroglossa** (*hi-dro-glos'-sah*) [*hydro-*; γλῶσσα, tongue]. Ranula.

**hydrogol** (*hi'-dro-gol*). A proprietary preparation said to be an aqueous solution of colloidal silver. It is used in gonorrhea.

**hydrohematocele** (*hi-dro-hem'-at-o-sēl*) [*hydro-*; αἷμα, blood; κήλη, tumor]. Hematocele associated with hydrocele.

**hydrohemia** (*hi-dro-hem'-e-ah*). Same as *hydremia*.

**hydrohemostat** (*hi-dro-hem'-o-stat*) [*hydro-*; αἷμα, blood; στατός, stopped]. A device for stopping hemorrhage by means of hydrostatic pressure.

**hydrohemothorax** (*hi-dro-hem-o-tho'-raks*) [*hydro-*; αἷμα, blood; θώραξ, thorax]. An effusion of hemorrhagic fluid into the pleural cavity.

**hydrohymenitis** (*hi-dro-hi-men-i'-tis*) [*hydro-*; ὑμήν, membrane; ιτις, inflammation]. Any inflammation of a serous membrane or surface.

**hydrohystera** (*hi-dro-his'-ter-ah*). See *hydrometra*.

**hydroid** (*hi'-droid*) [*hydro-*; εἶδος, form]. 1. Like water. 2. Living in water.

**hydrokinetics** (*hi-dro-kin-et'-iks*) [*hydro-*; κινεῖν, to set in motion]. The science of the motions of fluids and the causative forces.

**hydrolactometer** (*hi-dro-lak-tom'-et-er*) [*hydro-*; *lac*, milk; μέτρον, measure]. An instrument used in estimating the percentage of water in any given sample of milk.

**hydrolatum** (*hi-dro-la'-tum*) [*hydro-*; *gen.*, hydrolati; *pl.*, hydrolata]. A medicated water.

**hydrolein** (*hi-dro'-le-in*) [*hydro-*; *oleum*, oil]. A proprietary emulsion of cod-liver oil with pancreatin and borax.

**hydrology** (*hi-drol'-o-je*) [*hydro-*; λόγος, science]. A treatise on the nature and uses of water.

**hydrolymph** (*hi'-dro-limf*) [*hydro-*; *lympha*, water]. A term applied to the blood of certain animals which is composed largely of the salt or fresh water in which they live, and containing a small amount of corpuscular elements.

**hydrolysis** (*hi-drol'-is-is*) [*hydro-*; λύειν, to loose]. The decomposition of water.

**hydrolyst** (*hi'-dro-list*) [*hydro-*; λύσις, dissolving]. A substance that, like sulphuric acid, diastase, emulsin, etc., induces hydrolysis; an hydrolytic agent.

**hydrolyte** (*hi'-dro-līt*) [*hydro-*; λύσις, dissolving]. The substance hydrolyzed.

**hydrolytic** (*hi-dro-lit'-ik*) [see *hydrolysis*]. Pertaining to the decomposition of water or the liberation of water during a chemical reaction. h. ferments, those causing a combination with the elements of water in the substances which they decompose.

**hydrolyze** (*hi-dro-līs*). To subject to hydrolysis.

**hydroma** (*hi-dro'-mah*) [*hydro-*; ὅμα, a tumor: *pl.*, hydromata]. A tumor containing water. A cyst or sac filled with water or serous fluid. Also, an edematous swelling. Also, the cystic dilatation of a lymphatic vessel of the neck.

**hydromania** (*hi-dro-ma'-ne-ah*) [*hydro-*; *mania*, madness]. 1. Intense or maddening thirst. 2. Mania with desire for suicide by drowning.

**hydromediastinum** (*hi-dro-med-e-as-ti'-num*) [*hydro-*; *mediastinum*]. A serous effusion into the mediastinum.

**hydromel** (*hi'-dro-mel*) [*hydro-*; μέλι, honey]. A mixture of honey and water with or without a medicinal substance.

**hydromeningitis** (*hi-dro-men-in-ji'-tis*) [*hydro-*; μῆνιγξ, a membrane; ιτις, inflammation]. 1. Inflammation of the membranes of the brain or cord, accompanied by effusion of serous fluid. 2. Inflammation of the membrane of Descemet.

**hydromeningocele** (*hi-dro-men-in'-go-sēl*) [*hydro-*; μῆνιγξ, a membrane; κήλη, a tumor]. 1. A cystic tumor of the meninges protruding through the skull. 2. A form of spina bifida in which the sac contains cerebrospinal fluid.

**hydromeninx** (*hi-dro-men'-inks*) [*hydro-*; μῆνιγξ, a membrane]. Dropsy of the membranes of the brain.

**hydrometer** (*hi-drom'-et-er*) [*hydro-*; μέτρον, a measure]. An instrument for determining the specific gravity of liquids.

**hydrometra** (*hi-dro-me'-trah*) [*hydro-*; μήτρα, the womb]. A collection of watery fluid in the uterus.

**hydrometrectasia** (*hi-dro-me-trek-ta'-se-ah*) [*hydro-*; μήτρα, womb; ἔκτασις, a stretching out]. Hydrometra causing distention of the uterus.

**hydromicrenocephalia, hydromicrencephaly** (*hi-dro-mi-kren-sef-a'-le-ah, hi-dro-mi-kren-sef'-al-e*) [*hydro-*; μικρός, small; κεφαλή, the head]. Micrencephaly leading to, or complicated by, a serous effusion within the cranial cavity.

**hydromphalocele** (*hi-drom-fal'-o-sēl*) [*hydro-*; ὀμφαλός, navel; κήλη, hernia]. Cystic tumor in the sac of an umbilical hernia.

**hydromphalus** (*hi-drom'-fal-us*) [*hydro-*; ὀμφαλός, navel]. A tumor at the navel, distended with water.

**hydromyelia, hydromyelus** (*hi-dro-mi-e'-le-ah, hi-dro-mi'-el-us*) [*hydro-*; μυελός, marrow]. Dilatation of the central canal of the spinal cord with an accumulation of fluid.

**hydromyelitis** (*hi-dro-mi-el-i'-tis*) [*hydro-*; μυελός, marrow; ιτις, inflammation]. Same as *hydrorrhachis interna*.

**hydromyelocele** (*hi-dro-mi'-el-o-sēl*) [*hydro-*; μυελός, marrow; κοιλία, cavity]. 1. Excessive accumulation of fluid in the central canal of the spinal cord. 2. A variety of spina bifida in which remains of the spinal cord cover the tumor.

**hydromyoma** (*hi-dro-mi-o'-mah*) [*hydro-*; *myoma*]. A cystic myoma containing serous fluid.

**hydromyringa, hydromyrinx** (*hi-dro-mir-in'-gah, hi-dro-mi'-rinks*) [*hydro-*; *myrinx*, the tympanic membrane]. 1. The distention of the membrana tympani with water effused within its substance. 2. Less correctly, the same as *hydrotympanum*.

**hydronal** (*hi'-dro-nal*). A preparation obtained by the action of chloral on pyridin; it is used as a hypnotic. Also called *viferral*.

**hydronaphthol** (*hi-dro-naf'-thol*) [*hydro-*; *naphthol*]. A substance, derived from naphthol, and used as an intestinal antiseptic. Dose 3–4 gr. (0.2–0.26 Gm.).

**hydroncus** (*hi-drong'-kus*) [*hydro-*; ὄγκος, mass]. A distention or swelling caused by an accumulation of water. See, also, *edema* and *anasarca*.

**hydrone** (*hi'-drōn*) [*hydro-*]. Armstrong's term for $H_2O$, the molecule of water.

**hydronephrectasia** (*hi-dro-nef-rek-ta'-se-ah*) [*hydro-*; νεφρός, a kidney; ἔκτασις, distention]. Dropsical enlargement of the kidney.

**hydronephros** (*hi-dro-nef'-ros*) [*hydro-*; νεφρός, kidney]. A dropsical kidney. Syn., *hydronephrectasia*.

**hydronephrosis** (*hi-dro-nef-ro'-sis*) [see *hydronephros*]. A collection of urine in the pelvis of the kidney from obstructed outflow. The pressure of the fluid causes in time atrophy of the kidney-structure, and the whole organ is converted into a large cyst.

**hydronephrotic** (*hi-dro-nef-rot'-ik*) [*hydronephrosis*]. Relating to, affected with, or of the nature of, hydronephrosis.

**hydronosos** (*hi-dron-o'-sos*) [*hydro-*; νόσος, disease]. A disease attended with dropsy.

**hydro-oligocythemia** (*hi-dro-ol-ig-o-si-the'-me-ah*) [*hydro-*; ὀλίγος, few; κύτος, cell; αἷμα, blood]. A form of secondary anemia in which there is an increase in the proportion of the serum to the corpuscles of the blood.

**hydroparasalpinx** (*hi-dro-par-ah-sal'-pinks*) [*hydro-*; παρά, beside; σάλπιγξ, tube]. An accumulation of water in the accessory tubes of the oviduct.

**hydroparesis** (*hi-dro-par'-es-is*) [*hydro-*; πάρεσις, paralysis]. A paretic affection characterized by watery effusions, such as are seen in beriberi.

**hydropathic** (*hi-dro-path'-ik*) [*hydro-*]. Pertaining to hydropathy.

# HYDROPATHY 439 HYDROSCOPY

**hydropathy** (*hi-drop'-ath-e*) [*hydro-*; πάθος, disease]. The treatment of diseases by the use of water, externally and internally.

**hydropedesis** (*hi-dro-ped-e'-sis*). See *hidropedesis*.

**hydropericarditis** (*hi-dro-per-ik-ar-di'-tis*) [*hydro-*; *pericarditis*]. Pericarditis accompanied by serous effusion into the pericardium.

**hydropericardium** (*hi-dro-per-ik-ar'-de-um*) [*hydro-*; *pericardium*]. A collection of serum within the pericardial cavity.

**hydroperididymia** (*hi-dro-per-e-did-im'-e-ah*) [*hydro-*; περί, around; δίδυμος, testicle]. Hydrocele.

**hydroperion** (*hi-dro-per'-e-ŏn*) [*hydro-*; περί, around; ᾠόν, egg]. A seroalbuminous liquid existing between the decidua vera and the decidua reflexa and believed to nourish the embryo at an early period.

**hydroperipneumonia** (*hi-dro-per-e-nū-mo'-ne-ah*) [*hydro-*; περί, around; πνεύμων, a lung]. Pneumonia with pleural effusion.

**hydroperitoneum** (*hi-dro-per-it-on-e'-um*) [*hydro-*; *peritoneum*]. Ascites.

**hydroperitonitis** (*hi-dro-per-it-on-i'-tis*). Peritonitis attended with watery effusion.

**hydrophallus** (*hi-dro-fal'-us*) [*hydro-*; φαλλός, penis]. A dropsical swelling of the penis.

**hydrophilism** (*hi-droff'-il-izm*) [*hydro-*; φίλειν, to love]. The property of colloids, cells, tissues, etc., to attract and hold water.

**hydrophilous** (*hi-droff'-il-us*) [*hydro-*; φίλος, loving]. 1. In biology, applied to plants that are fertilized through the agency of water. 2. Absorbing water.

**hydrophlegmasia** (*hi-dro-fleg-ma'-ze-ah*) [*hydro-*; φλεγμασία, inflammation]. Any phlegmasia or inflammation characterized by serous effusion.

**hydrophlogosis** (*hi-dro-flo-go'-sis*) [*hydro-*; φλόγωσις, a burning]. Inflammation attended with serous effusion.

**hydrophobe** (*hi'-dro-fōb*) [*hydro-*; φόβος, fear]. A person who is affected with rabies.

**hydrophobia** (*hi-dro-fo'-be-ah*) [*hydro-*; φόβος, fear]. 1. Fear of water; a symptom of rabies, *q. v.* 2. Used as a synonym of *rabies*.

**hydrophobic** (*hi-dro-fo'-bik*). Pertaining to, or of the nature of, hydrophobia. **h. tetanus**: See *tetanus, hydrophobic*.

**hydrophobin** (*hi-dro-fo'-bin*). The virus of hydrophobia; same as *lyssin*.

**hydrophobophobia** (*hi-dro-fo-bo-fo'-be-ah*) [*hydrophobia*; φόβος, fear]. An intense dread of hydrophobia; a condition producing a state simulating true hydrophobia.

**hydrophone** (*hi'-dro-fōn*) [*hydro-*; φωνή, the voice]. An instrument used in auscultatory percussion, the sound being conveyed to the ear through a column of water.

**hydrophore** (*di'-dro-fōr*) [*hydro-*; φέρειν, to bear]. An apparatus consisting of a short grooved catheter used as an irrigating dilator of the urethra.

**hydrophthalmia** (*hi-droff-thal'-me-ah*) [*hydro-*; ὀφθαλμός, eye]. An increase in the fluid contents of the eye, causing the organ to become distended, resulting in glaucoma, keratoglobus, staphyloma, etc.

**hydrophthalmos** (*hi-droff-thal'-mos*). See *keratoglobus*.

**hydrophyr** (*hi'-dro-fēr*) [*hydro-*]. A variety of peptone insoluble in alcohol.

**hydrophysocele** (*hi-dro-fi'-zo-sēl*) [*hydro-*; φῦσα, air; κήλη, tumor]. Hernia containing both serous fluid and a gas.

**hydrophysometra** (*hi-dro-fi-zo-me'-trah*) [*hydro-*; φῦσα, wind; μήτρα, womb]. An abnormal collection of water, or other fluid, and gas, in the womb.

**hydropic** (*hi-drop'-ik*) [*hydrops*]. Pertaining to dropsy; dropsical.

**hydropica** (*hi-drop'-ik-ah*) [ὑδρωπικόs, dropsical]. Medicines useful in dropsy; especially diuretic remedies.

**hydroplasm** (*hi'-dro-plazm*) [ὕδωρ; plasm] A fluid constituent of protoplasm, or of any plasma.

**hydropleuritis** (*hi-dro-plu-ri'-tis*). Pleurisy attended with effusion.

**hydropneumatic** (*hi-dro-nū-mat'-ik*) [*hydro-*; πνεῦμα, air]. Relating to water and air.

**hydropneumatosis** (*hi-dro-nū-mat-o'-sis*) [*hydro-*; πνευμάτωσις, inflation]. A collection of fluid and air or other gas within the tissues.

**hydropneumonia** (*hi-dro-nū-mo'-ne-ah*) [*hydro-*; πνεύμων, the lung]. A disease thought to consist of a serous infiltration into the lung; pulmonary edema; also, an effusion within the pleura sometimes accompanying pneumonia.

**hydropneumopericardium** (*hi-dro-nū-mo-per-ik-ar'-de-um*) [*hydro-*; πνεῦμα, air; *pericardium*]. A collection of serum and air or other gas within the pericardium.

**hydropneumothorax** (*hi-dro-nū-mo-tho'-raks*) [*hydro-*; πνεῦμα, air; θώραξ, thorax]. The presence of serous fluid and air or gas in the pleural cavity.

**hydropoid** (*hi'-dro-poid*) [ὕδρωψ, dropsy; εἶδος, likeness]. Dropsical.

**hydroposia** (*hi-dro-po'-ze-ah*) [ὕδωρ, water; πόσις, a drinking]. Water drinking; the use of water alone as a beverage.

**hydropotherapy** (*hi-dro-po-ther'-ap-e*) [ὕδρωψ, dropsy; therapy]. The therapeutic use of ascitic fluid.

**hydrops** (*hi'-drops*) [ὕδρωψ, dropsy]. Dropsy. **h. articuli**, a watery effusion into the synovial cavity of a joint. **h. capitis**, hydrocephalus, *q. v.* **h. cystidis felleæ**, dropsy of the gall-bladder. **h. ex vacuo**, a condition following inflammation, in which the lung is unable to expand and the space between the contracted lung and the chest-wall is filled with fluid. **h. tubæ profluens**, hydrosalpinx in which the watery discharges are said to occur from the uterus and vagina, in gushes, the uterine end of the tube from time to time allowing passage of the fluid. **h. vesicæ felleæ**, dropsy of the gall-bladder.

**hydropyonephrosis** (*hi dro pi o nef ro' sis*) [*hydro-*; πύον, pus; νεφρός, kidney]. Distention of the pelvis of the kidney with urine and pus.

**hydropyopneumothorax** (*hi-dro-pi-o-nū-mo-tho'-raks*) [*hydro-*; πύον, pus; πνεῦμα, air; θώραξ, chest]. Hydropneumothorax, associated with the presence of pus.

**hydropyosalpinx** (*hi-dro-pi-o-sal'-pingks*) [*hydro-*; πύον, pus; σάλπιγξ, tube]. An accumulation of serous fluid and pus in an oviduct.

**hydropyretic** (*hi-dro-pi-ret'-ik*) [*hydro-*; πυρετός, heat]. Pertaining to, or affected with, hydropyretos.

**hydropyretos** (*hi-dro-pir'-et-os*) [*hydro-*; πυρετός, heat]. Sweating fever.

**hydroquinine** (*hi-dro-kwin'-ēn*) [*hydro-*; *quinine*], $C_{20}H_{26}N_2O_3$. An alkaloid obtained from cinchona, and frequently contaminating quinine.

**hydroquinone** (*hi-dro-kwin'-ōn*), $C_6H_6O_2$. An isomer of resorcinol and pyrocatechin; found in arbutin, and also obtained from quinine and quinone, etc. It is antipyretic and antiseptic. Dose 15-20 gr. (1.0-1.3 Gm.). See *Uva ursi*.

**hydrorenal** (*hi-dro-re'-nal*) [*hydro-*; *ren*, kidney]. Relating to dropsy of the kidney.

**hydrorheostat** (*hi-dro-re'-o-stat*). A rheostat in which the resistance is furnished by water.

**hydrorrhachiocentesis** (*hi-dror-rak-e-o-sen-te'-sis*) [*hydro-*; ῥάχις, spine; κέντησις, puncture]. Puncture of the spinal meninges in the treatment of hydrorrhachis.

**hydrorrhachis** (*hi-dror'-ak-is*) [*hydro-*; ῥάχις, spine]. A serous effusion within the spinal canal. **h. interna**, syringomyelia, *q. v.*

**hydrorrhachitis** (*hi-dror-rak-i'-tis*) [*hydro-*; ῥάχις, spine; ιτις, inflammation]. Hydrorrhachis with inflammation. See *spina bifida*.

**hydrorrhea** (*hi-dro-re'-ah*) [*hydro-*; ῥοία, a flow]. A flow of watery liquid. **hydrorrhœa gravidarum**, a discharge, from the pregnant uterus, of the mucus that accumulates as a result of excessive secretion of the uterine glands.

**hydrosadenitis** (*hi-dros-ad-en-i'-tis*) [*hydro-*; ἀδήν, gland; ιτις, inflammation]. See *hidrosadenitis*. **h. phlegmonosa**, a furunculus beginning in a sweat-coil. It is also called *hydradenitis destruens suppurativa, acnitis*, and *folliculitis exulcerans*.

**hydrosalpinx** (*hi-dro-sal'-pinks*) [*hydro-*; σάλπιγξ, trumpet]. A distention of the Fallopian tube with fluid.

**hydrosarca** (*hi-dro-sar'-kah*). See *anasarca*.

**hydrosarcocele** (*hi-dro-sar'-ko-sēl*) [*hydro-*; σάρξ, flesh; κήλη, hernia]. Sarcocele with hydrocele.

**hydroscheocele** (*hi-dros'-ke-o-sēl*) [*hydro-*; ὀσχέον, scrotum; κήλη, hernia]. Dropsical hernia of the scrotum.

**hydroscopy** (*hi-dros'-ko-pe*) [*hydro-*; σκοπεῖν, to examine]. The investigation of water or watery fluids.

**hydrosol** (*hi'-dro-sol*). An aqueous colloidal solution.

**hydrosoma** (*hi-dro-so'-mah*) [*hydro-*; σῶμα, body; *pl.*, *hydrosomata*]. In biology, the entire double-walled body of a hydrozoon.

**hydrospermatocyst** (*hi-dro-sper'-mat-o-sist*) [*hydro-*; σπέρμα, seed; κύστις, cyst]. A hydrocele whose fluid contains spermatozoa.

**hydrosphygmograph** (*hi-dro-sfig'-mo-graf*) [*hydro-*; σφυγμός, pulse; γράφειν, to write]. A sphygmographic apparatus, in which the registering device is actuated by the fluctuations of a body of water.

**hydrospirometer** (*hi-dro-spi-rom'-et-er*). A spirometer in which a column of water acts as an index.

**hydrostat** (*hi'-dro-stat*) [*hydro-*; στατός, standing]. An apparatus for preventing the spilling of the fluid of electric batteries during transportation.

**hydrostatic** (*hi-dro-stat'-ik*) [see *hydrostat*]. Relating to hydrostatics. **h. exploration,** M. Sée's method of diagnosing pelvic disease by palpation of the abdomen while the patient is extended in a bath covering its surface. **h. test,** a test for live birth in which the fetal lungs are floated upon water.

**hydrostatics** (*hi-dro-stat'-iks*) [see *hydrostat*]. The science treating of the properties of liquids in a state of equilibrium.

**hydrostomia** (*hi-dro-sto'-me-ah*) [*hydro-*; στόμα, mouth]. Excessive excretion of water from the mouth.

**hydrosudopathy** (*hi-dro-sū-dop'-ath-e*) [*hydro-*; *sudor*, sweat; πάθος, disease]. The treatment of disease by sweating and the use of water internally or externally, or both.

**hydrosulphuric acid** (*hi-dro-sul-fū'-rik*). See *acid, hydrosulphuric*.

**hydrosyntasis** (*hi-dro-sin'-tas-is*) [*hydro-*]. The swelling of tissues, membranes, or protoplasm, etc., by the penetration of water.

**hydrosyringomyelia** (*hi-dro-sir-in-go-mi-e'-le-ah*) [*hydro-*; σῦριγξ, tube; μυελός, marrow]. Dilatation of the central canal of the spinal cord by watery effusion, attended with degeneration and the formation of cavities.

**hydrotaxis** (*hi-dro-taks'-is*) [*hydro-*; τάξις, arrangement]. The determination of the direction of movement by moisture. Cf. *hydrotropism*.

**hydrotherapeutics** (*hi-dro-ther-ap-ū'-tiks*) [*hydro-*; θεραπεύειν, to heal]. The treatment of disease by means of water, or the use of water in the treatment of disease.

**hydrotherapy** (*hi-dro-ther'-ap-e*). See *hydrotherapeutics*.

**hydrothermal** (*hi-dro-ther'-mal*) [*hydro-*; θέρμη, heat]. Pertaining to warm water; said of springs.

**hydrothermostat** (*hi-dro-ther'-mo-stat*) [*hydro-*; *thermostat*]. An apparatus for providing a continuous degree of heat for therapeutic purposes.

**hydrothion** (*hi-dro-thi'-on*) [*hydro-*; θεῖον, sulphur]. Hydrogen sulphide.

**hydrothionammonemia** (*hi-dro-thi-on-am-o-ne'-me-ah*) [*hydro-*; θεῖον, sulphur; *ammonia*; αἷμα, blood]. The condition produced by the presence of ammonium sulphide in the blood.

**hydrothionemia** (*hi-dro-thi-on-e'-me-ah*) [*hydro-*; θεῖον, sulphur; αἷμα, blood]. The condition produced by the presence of hydrogen sulphide in the blood.

**hydrothionuria** (*hi-dro-thi-on-ū'-re-ah*) [*hydro-*; θεῖον, sulphur; οὖρον, urine]. The presence of hydrogen sulphide in the urine.

**hydrothoracic** (*hi-dro-tho-ras'-ik*) [*hydro-*; *thorax*]. Pertaining to hydrothorax.

**hydrothorax** (*hi-dro-tho'-raks*) [*hydro-*; θώραξ, chest]. The presence of serous fluid in the pleural cavity, due to a passive effusion, as in cardiac, renal, and other diseases.

**hydrotimeter** (*hi-dro-tim'-et-er*) [*hydro-*; μέτρον, measure]. An apparatus to determine the amount of calcareous salts in water by means of soap.

**hydrotis** (*hi-dro'-tis*) [*hydro-*; οὖς, ear]. Dropsy of, or effusion into, the ear.

**hydrotomy** (*hi-drot'-o-me*) [*hydro-*; τέμνειν, to cut]. A method of dissecting tissues by the forcible injection of water into the arteries and capillaries, whereby the structures are separated.

**hydrotropism** (*hi-drot'-ro-pizm*) [*hydro-*; τρόπος, a turn]. In biology, that state of a growing plant or organ which causes it to turn either away from or toward, moisture.

**hydrotympanum** (*hi-dro-tim'-pan-um*) [*hydro-*; *tympanum*, the ear-drum]. Dropsical effusion into the cavity of the middle ear.

**hydroureter** (*hi-dro-ū'-re-ter*). Dropsy of the ureter.

**hydrovarium** (*hi-dro-va'-re-um*) [*hydro-*; *ovarium*, ovary]. Ovarian dropsy, or cystoma.

**hydroxide** (*hi-droks'-id*) [*hydro-*; ὀξύς, sharp]. A metallic or basic radical combined with one or more hydroxyl groups. Hydroxides may be regarded as formed from water (HOH) by the substitution for one of its hydrogen atoms of a metal or basic radical.

**hydroxyl** (*hi-droks'-il*) [see *hydroxide*]. The univalent radical, OH, the combination of which with basic elements or radicals forms the hydroxide.

**hydroxylamine** (*hi-droks-il'-am-in*) [*hydroxide*; *amine*], NH₂OH. A basic substance, known only in solution in water or in combination with acids. Its hydrochloride has been used as a substitute for chrysarobin in skin diseases.

**hydrozone** (*hi'-dro-zōn*) [*hydro-*; *ozone*]. An aqueous solution of chemically pure hydrogen dioxide; it is used as a bactericide and healing agent.

**hydruresis** (*hi-dru-re'-sis*) [*hydro-*; οὖρον, urine]. The passage of a relatively large proportion of water in the urine.

**hydruret** (*hi'-dru-ret*). See *hydride*.

**hydruria** (*hi-dru'-re-ah*) [*hydro-*; οὖρον, urine]. The discharge of a large quantity of urine of low specific gravity.

**hydrymenitis** (*hi-dri-men-i'-tis*). See *hydrohymenitis*.

**hygeia, hygieia** (*hi-je'-ah, hi-je-i'-ah*) [ὑγίεια, health]. The state or condition of health.

**hygiama** (*hi-je-am'-ah*). A dietetic said to consist of milk, cereals, and cacao.

**hygieinism** (*hi'-je-in-izm*) [ὑγίεια, health]. Sanitation.

**hygieinization** (*hi-je-in-i-za'-shun*). The establishment of sanitary conditions.

**hygiene** (*hi'-je-ēn*) [ὑγιεινός, good for the health]. The science that treats of the laws of health and the methods of their observance.

**hygienic** (*hi-je-en'-ik*) [see *hygiene*]. Pertaining to hygiene, as *hygienic treatment*, that which simply guards against infraction of the laws of health.

**hygienist** (*hi-je-en'-ist*). One who is a student of, or an expert in, hygiene.

**hygieology** (*hi-je-ol'-o-je*) [ὑγίεια, health; λόγος, science]. The science of health; hygiene.

**hygiology** (*hi-je-ol'-o-je*). See *hygieology*.

**hygrechema** (*hi-grek-e'-mah*) [ὑγρός, moist; ἤχημα, sound; *pl.*, *hygrechemata*]. The peculiar sound produced by a liquid, as heard upon mediate or immediate auscultation.

**hygric** (*hi'-grik*) [ὑγρός, moist]. Pertaining or relating to moisture.

**hygrine** (*hi'-grin*) [ὑγρός, moist]. C₁₄H₁₉N. A liquid alkaloid derived from coca.

**hygro-** (*hi-gro-*) [ὑγρός, moist]. A prefix denoting moist or relating to moisture.

**hygroblepharic** (*hi-gro-blef'-ar-ik*) [ὑγρός, moist; βλέφαρον, eyelid]. Serving to moisten the eyelid, as a hygroblepharic duct.

**hygrocele** (*hi'-gro-sēl*) [ὑγρός, moist; κήλη, a tumor]. Same as *hydrocele*.

**hygrodermia** (*hi-gro-der'-me-ah*) [ὑγρός, moist; δέρμα, skin]. An edematous non-inflammatory skin-affection.

**hygrol** (*hi'-grol*). Colloidal mercury.

**hygrology** (*hi-grol'-o-je*) [ὑγρός, moist; λόγος, science]. The science of the fluids, or so-called humors, of the body.

**hygroma** (*hi-gro'-mah*) [ὑγρο-; ὄμα, tumor]. A bursa, or newly formed sac, distended with fluid.

**hygromatous** (*hi-gro'-mat-us*) [ὑγρος, moist; ὄμα, tumor]. Pertaining to, or characterized by, a hygroma.

**hygrometer** (*hi-grom'-et-er*) [*hygro-*; μέτρον, a measure]. An instrument for determining quantitatively the amount of moisture in the air.

**hygrometric** (*hi-gro-met'-rik*) [see *hygrometer*]. 1. Pertaining to hygrometry. 2. Readily absorbing water; hygroscopic.

**hygrometry** (*hi-grom'-et-re*) [see *hygrometer*]. The measurement of the moisture of the air.

**hygrophanous** (*hi-grof'-an-us*) [ὑγρός, moist; φάνης, show]. In biology, applied to such plant-structures as are transparent when wet, but opaque when dry.

**hygrophobia** (*hi-gro-fo'-be-ah*) [ὑγρός, moist; φόβος, fear]. Insane dislike of water or of moisture.
**hygroscope** (*hi'-gro-skōp*) [ὑγρός, moist; σκοπεῖν, to inspect]. An instrument that indicates variations in the moisture of the air.
**hygroscopic** (*hi-gro-skop'-ik*) [*hygro-*; σκοπεῖν, to see]. Having the property of absorbing moisture from the air.
**hygrostomia** (*hi-gro-sto'-me-ah*) [*hygro-*; στόμα, mouth]. Chronic salivation.
**hyla** (*hi'-lah*). See *paraqueduct*.
**hyle** (*hi'-le*) [ὕλη, matter]. The primitive, undifferentiated matter, mass, or body in nature. Cf. *protyle*.
**hylephobia** (*hi-le-fo'-be-ah*) [*hyle;* φόβος, fear]. Insane dread of materialistic doctrines.
**hylic** (*hi'-lik*). 1. Relating to primitive matter. 2. Adami's name for primal pulp-tissue.
**hylogenesis** (*hi-lo-jen'-es-is*) [ὕλη, matter; γένεσις, formation]. The formation of matter.
**hylogeny** (*hi-loj'-en-e*). Same as *hylogenesis*.
**hylology** (*hi-lol'-o-je*) [ὕλη, matter; λόγος, science]. The science of elementary or crude material.
**hyloma** (*hi-lo'-mah*) [ὕλη, matter; ὄμα, tumor]. A tumor originating in one of the primal pulp-tissues; subdivided into epihyloma, hypohyloma, and mesohyloma.
**hylopathism** (*hi-lop'-ath-izm*) [ὕλη, matter; πάθος, disease]. Any disease arising from defect or disorder of the body-substance.
**hylozoism** (*hi-lo-zo'-izm*) [ὕλη, matter; ζωή, life]. The theory that all matter is endowed with life.
**hymen** (*hi'-men*) [ὑμήν, membrane]. The fold of mucous membrane that partially occludes the vaginal entrance. h., **imperforate**, a congenital abnormality, the hymen not having an opening, and thus closing the vaginal outlet or inlet. h., forms of. 1. *Bifenestrate*, or *hymen biforis*, with two openings. 2. *Bilobate*, with two lobes. 3. *Circular*, with a small foramen. 4. *Cribriform*, with many holes. 5. *Denticular*, with a serrate edge. 6. *Double* (rare). 7. *Fimbriate*, with fringed edges. 8. *Horseshoe*, with its convexity downward. 9. *Imperforate*. 10. *Normal*. 11. *Semilunar*, same as 10, or normal. 12. *Hymen septus*, divided across by a slit. 13. *Hymen subseptus*, covering only the anterior and posterior portions of the passage.
**hymenal** (*hi'-me-nal*). Pertaining to the hymen. h. **tubercles**, the *carunculæ myrtiformes*, q. v.
**hymenitis** (*hi-men-i'-tis*) [*hymen;* ιτις, inflammation]. Inflammation of the hymen or of any membranous structure.
**hymenography** (*hi-men-og'-ra-fe*) [ὑμήν, membrane; γράφειν, to write]. Same as *hymenology*.
**Hymenolepis** (*hi-men-ol'-ep-is*). A genus of Cestoda or tape worms. **H. diminuta**, a tapeworm of rats and mice, occasionally found in man. **H. lanceolata**, a tapeworm of ducks and geese; rarely found in man. **H. nana**, a diminutive tapeworm about 1 inch long sometimes found in the human intestines; also called *Tænia nana*.
**hymenology** (*hi-men-ol'-o-je*) [*hymen;* λόγος, science]. The science of the nature, structure, functions, and diseases of membranes.
**hymenomalacia** (*hi-men-o-mal-a'-se-ah*) [ὑμήν, membrane; μαλακία, softness]. An abnormal softening of membranous tissues.
**Hymenoptera** (*hi-men-op'-ter-ah*) [ὑμήν, membrane; πτερόν, wing]. An order of insects distinguished by two pairs of membranous wings. It includes ants, bees, wasps, ichneumons, flies, etc.
**hymenorrhaphy** (*hi-men-or'-a-fe*) [ὑμήν, hymen; ῥαφή, suture]. 1. Closure, more or less complete, of the vagina by suture at the hymen. 2. Suture of any membrane.
**hymenotome** (*hi-men'-o-tōm*) [ὑμήν, membrane; τέμνειν, to cut]. A cutting instrument used in operations upon membranes.
**hymenotomy** (*hi-men-ot'-o-me*) [ὑμήν, membrane; τέμνειν, to cut]. 1. Surgical incision of the hymen. 2. Dissection of anatomy of membranes.
**hyo-** (*hi'-o-*) [ὑοειδής, hyoid]. A prefix denoting attachment to or connection with the hyoid bone.
**hyobasioglossus**. (*hi-o-ba-se-o-glos'-us*). See *basioglossus*.
**hyocholaic** (*hi-o-ko-la'-lik*) [ὑς, a pig; χολή, bile]. Derived from pig's bile, as hyocholalic acid.
**hyoepiglottic, hyoepiglottidean** (*hi-o-ep-e-glot'-ik, hi-o-ep-e-glot-id'-e-an*) [*hyoid; epiglottis*]. Relating to the hyoid bone and the epiglottis.

**hyoglossal** (*hi-o-glos'-al*) [*hyoid;* γλῶσσα, tongue]. 1. Pertaining to the hyoglossus. 2. Extending from the hyoid bone to the tongue.
**hyoglossus** (*hi-o-glos'-us*). See under *muscle*.
**hyoid** (*hi'-oid*) [T, or υ the Greek letter upsilon; εἶδος, form]. Having the form of the Greek letter upsilon. h. **bone**, a bone situated between the root of the tongue and the larynx, supporting the tongue and giving attachment to its muscles.
**hyolaryngeal** (*hi-o-lar-in'-je-al*). Related to or connected with the hyoid bone and the larynx.
**hyomandibular** (*hi-o-man-dib'-u-lar*). Relating to the hyoid bone and the inferior maxilla.
**hyomental** (*hi-o-ment'-al*). Relating to the hyoid bone and the chin.
**hyopharyngeus** (*hi-o-far-in'-je-us*). The middle pharyngeal constrictor.
**hyoscine** (*hi'-o-sin*) $C_{17}H_{21}NO_4$. A liquid alkaloid found in hyoscyamus. It is a powerful depressant of the cerebrum and the motor centers of the cord, and is employed in insomnia, mania, and excessive sexual excitement. h. **hydrobromide** (*hyoscinæ hydrobromidum*, U. S. P.), the most commonly administered form. Dose $\frac{1}{120}$-$\frac{1}{60}$ gr. (0.0005-0.0008 Gm.).
**hyoscyamine** (*hi-o-si'-am-in*) $C_{17}H_{23}NO_3$. An alkaloid occurring in hyoscyamus. It is isomeric with atropine, is a mydriatic, narcotic, and sedative. Dose $\frac{1}{120}$ gr. (0.0005 Gm.). h. **hydrobromide** (*hyoscyaminæ hydrobromidum*, U. S. P.), yellowish-white, amorphous masses, with nauseous taste and odor of tobacco. It is mydriatic, hypnotic, and sedative. Dose $\frac{1}{120}$-$\frac{1}{60}$ gr. (0.0005-0.001 Gm.), several times daily; as hypnotic for insane $\frac{1}{8}$-$\frac{1}{4}$ gr. (0.008-0.016 Gm.). h. **hydriodide**, $C_{17}H_{23}NO_3$, that obtained from *Duboisia myoporoides*. It is sedative and antispasmodic. Dose for the sane $\frac{1}{120}$-$\frac{1}{60}$ gr. (0.0005-0.001 Gm.); injection for insane $\frac{1}{30}$-$\frac{1}{10}$ gr. (0.002-0.006 Gm.). h. **sulphate** (*hyoscyaminæ sulphas*, U. S. P.), the neutral sulphate of an alkaloid obtained from hyoscyamus. Dose $\frac{1}{120}$ gr. (0.0005 Gm.).
**hyoscyamus** (*hi-o-si'-am-us*). Henbane; a plant of the order *Solanaceæ*. The leaves and flowering tops of *H. niger* yield the alkaloids *hyoscyamine* and *hyoscine*. It is sedative to the nervous system, and has been employed in hysteria, cough, and colic, and to relieve pain in rheumatism, headache, and malignant tumors. h., **extract of** (*extractum hyoscyami*, U. S. P.). Dose $\frac{1}{4}$-2 gr. (0.0165-0.13 Gm.). h., **fluidextract of** (*fluidextractum hyoscyami*, U. S. P.). Dose 5-10 min. (0.32-0.65 Cc.). h. **juice** (*succus hyoscyami*, B. P.). Dose $\frac{1}{2}$-1 dr. (2-4 Cc.). h., **tincture of** (*tinctura hyoscyami*, U. S. P.). Dose $\frac{1}{2}$-1 dr. (2-4 Cc.).
**hyospondylotomy** (*hi-o-spon-dil-ot'-o-me*) [*hyo-;* σπόνδυλος, vertebra; τέμνειν, to cut]. In veterinary practice, puncture of the laryngeal pouch.
**hyosternal** (*hi-o-ster'-nal*) [*hyo-;* στερνόν, sternum]. 1. Relating to the hyoid bone and the sternum. 2. In biology, the second lateral piece of the plastron of a turtle; it is also called *hyoplastron*.
**hyovertebrotomy** (*hi-o-vur-te-brot'-o-me*). See *hyospondylotomy*.
**hyp** (*hip*) [*hypo*, under]. A popular name for hypochondriasis, or persistent depression of spirits; hypo.
**hypacidia** (*hi-pas-id-e'-me-ah*) [*hypo*, under; *acid; αἷμα*, blood]. Deficiency of acid in the blood.
**hypacidity** (*hip-as-id'-it-e*) [*hypo-; acidity*]. Sub-acidity; deficiency in acid constituents.
**hypacusia, hypacusis, hypacusis** (*hip-ah-koo'-sis, hip-ah-kū'-ze-ah*) [*hypo-;* ἄκουσις, hearing]. Impairment of hearing.
**hypactic** (*hi-pak'-tik*) [ὑπάγειν, to carry down]. Slightly purgative. Syn., *hypagogue*. Cf. *lapactic*.
**hypæsthesia** (*hip-es-the'-se-ah*). See *hypesthesia*.
**hypagogue** (*hi'-pah-gog*). See *hypactic*.
**hypalbuminosis** (*hip-al-bū-min-o'-sis*) [*hypo-; albumin*]. Diminution in the proportion of albumin in the blood.
**hypalgesia** (*hip-al-je'-ze-ah*) ὑπό, under; ἄλγησις, painfulness]. Diminished sensitiveness to pain.
**hypalgia** (*hip-al'-je-ah*) [ὑπό, under; ἄλγος, pain]. Slight or moderate pain; diminished sensibility to pain.
**hypalgic** (*hip-al'-jik*) [ὑπό, under; ἄλγος, pain]. Slightly painful; experiencing slight pain.
**hypamnios, hypamnion** (*hip-pam'-ne-os, hi-pam'-ne-on*) [*hypo-; amnion*]. Diminution in the amount of amniotic fluid.

**hypanakinesis** (hi-pan-ak-in-e'-sis) [hypo-; ἀνακίνησις, a swaying to and fro]. Diminution in the movements of the stomach or intestines.

**hypanisognathism** (hip-an-is-og'-nath-izm) [hypo-; ἄνισος, unequal, uneven; γνάθος, the jaw]. In biology, a lack of correspondence between the teeth of the opposite jaws. Cf. *anisognathism* and *epanisognathism*.

**hypanisognathous** (hip-an-is-og'-na-thus) [hypo-; ἄνισος, unequal, uneven; γνάθος, the jaw]. In biology, having the upper teeth broader than the lower.

**hypaphorine** (hi-paf'-or-ēn). A crystalline alkaloid derived from the seeds of *Erythrina lithosperma*.

**hypapophysis** (hip-ap-off'-is-is) [hypo-; apophysis: pl., hypapophyses]. An anterior or ventral apophysis from the centrum of a vertebra in the human skeleton; the atlas is held by some to have an hypophysis, which is blended with the ring of that vertebra.

**hypapoplexia** (hip-ap-o-pleks'-e-ah) [hypo-; ἀποπλήξια, apoplexy]. A slight apoplexy.

**hyparterial** (hi-par-te'-re-al) [hypo-; artery]. Situated beneath an artery.

**hypasthenia** (hip-as-the'-ne-ah) [hypo-; ἀσθένεια, weakness]. Loss of strength in a slight degree.

**hypatmism** (hi'-pat-mizm) [hypo-; ἀτμός, vapor]. Fumigation.

**hypatonia** (hip-at-o'-ne-ah) [hypo-; atony]. A slight amount of atony.

**hypaxial** (hi-paks'-e-al) [hypo-; axis]. Situated beneath or ventrad of the body-axis. Cf. *epaxial*.

**hypectasia, hypectasis** (hi-pek-ta'-ze-ah, hi-pek'-ta-sis) [hypo-; ἔκτασις, a stretching]. Slight or moderate distention.

**hypemia** (hip-e'-me-ah). See *hyphemia*.

**hypencephalon** (hi-pen-sef'-al-on) [hypo-; ἐγκέφαλος, brain]. The corpora quadrigemina, pons and medulla.

**hypendocrisia** (hi-pen-do-kris'-e-ah). Same as *hypoendocrinism*.

**hypeosinophil** (hi-pe-o-sin'-o-fil) [hypo-; eosinophil]. 1. A histological element which does not stain completely with eosin. 2. Staining imperfectly with eosin.

**hyper-** (hi-per-) [ὑπέρ, over]. A Greek prefix signifying above, beyond, or excessive.

**hyperabduction** (hi-pur-ab-duk'-shun). See *superabduction*.

**hyperacanthosis** (hi-per-ak-an-tho'-sis) [hyper-; ἄκανθα, thorn]. Abnormal growth of the prickle-cell layer of the epidermis, as in warts and condylomata.

**hyperacid** (hi-per-as'-id). Excessively acid.

**hyperacidity** (hi-per-as-id'-it-e) [hyper-; acidity]. Excessive acidity.

**hyperacousis, hyperacusia, hyperacusis** (hi-per-ah-koo'-sis, hi-per-ah-koo'-ze-ah) [hyper-; ἄκουσις, hearing]. Morbid acuteness of the sense of hearing; auditory hyperesthesia.

**hyperactivity** (hi-per-ak-tiv'-it-e) [hyper-; activity]. Excessive or abnormal activity.

**hyperacuity** (hi-per-a-kū'-e-te) [hyper-; acuitas, sharpness of vision]. Abnormal or morbid acuity.

**hyperadenoma** (hi-pur-ad-en-o'-mah) [hyper-; over; ἀδήν, gland]. An enlarged lymph-gland.

**hyperadenosis** (hi-per-ad-en-o'-sis) [hyper-; ἀδήν, gland; νόσος, disease]. Enlargement of the lymph-glands; Hodgkin's disease.

**hyperæmia** (hi-per-e'-me-ah). See *hyperemia*.

**hyperaeration** (hi-per-a-er-a'-shun) [hyper-; ἀήρ, air]. The condition of being furnished with excess of ozone.

**hyperæsthesia** (hi-per-es-the'-ze-ah). See *hyperesthesia*.

**hyperalbuminemia** (hi-per-al-bū-min-e'-me-ah). Pernicious anemia characterized by more than the normal percentage of albumin in the blood.

**hyperalbuminosis** (hi-per-al-bū-min-o'-sis) [hyper-; albumin]. An increase in the amount of albumin in the blood.

**hyperalgesia** (hi-per-al-je'-ze-ah) [hyper-; ἄλγησις, pain]. Excessive sensibility to pain. **h., acoustic, h., auditory,** a painful sensation in the ear caused by noises. Syn., *hyperæsthesia acustica*. **h., muscular,** muscular fatigue and exhaustion attending certain diseases. **h., olfactory,** painful sensitiveness of the olfactory apparatus to certain odors.

**hyperalgesic** (hi-per-al-je'-zik) [hyper-; ἄλγησις, pain]. Exhibiting or appertaining to hyperalgesia.

**hyperalgia** (hi-per-al'-je-ah) [hyper-; ἄλγησις, pain]. Excessive pain. **h., acoustic,** excessive hyperacousis.

**hyperalimentation** (hi-per-al-e-men-ta'-shun). See *superalimentation*.

**hyperalimentosis** (hi-per-al-e-men-to'-sis). A morbid condition due to superalimentation.

**hyperalkalescence** (hi-per-al-kal-es'-ens). Excessive alkalinity.

**hyperalonemia** (hi-per-al-on-e'-me-ah) [hyper-; ἅλς, salt; αἷμα, blood]. Excess of blood-salts.

**hyperamnesia** (hi-per-am-ne'-se-ah). See *hypermnesia*.

**hyperanabolism** (hi-per-an-ab'-o-lizm) [hyper-; ἀναβάλλειν, to build up]. Hypertrophy or excess of construction of a tissue or part.

**hyperanakinesis** (hi-per-an-ak-in-e'-sis) [hyper-; ἀνακίνησις, a moving upward; excitement]. Excessive activity of a part. **h. ventriculi,** exaggerated activity of the gastric functions.

**hyperanarthric** (hi-per-an-ar'-thrik) [hyper-; " ἁ, priv.; ἄρθρον, a joint]. Excessively defective in the joints.

**hyperaphia** (hi-per-a'-fe-ah) [hyper-; ἀφή, touch]. Excessive sensitiveness to touch.

**hyperaphic** (hi-per-af'-ik) [hyper-; ἀφή, touch]. Having morbid sensitiveness to touch.

**hyperaphrodisia** (hi-per-af-ro-diz'-e-ah) [hyper-; aphrodisia]. An over-strong venereal appetite.

**hyperapophyseal** (hi-per-ap-o-fis'-e-al) [hyper-; ἀπόφυσις, apophysis]. Pertaining to a hyperapophysis.

**hyperapophysis** (hi-per-ap-off'-is-is) [hyper-; apophysis]. A process projecting backward from a neural spine.

**hyperarithmous** (hi-pur-ar-ith'-mus) [hyper-; ἀριθμός, a number]. Supernumerary.

**hyperarthric** (hi-pur-ar'-thrik) [hyper-; ἄρθρον, joint]. Having supernumerary joints.

**hyperarthritic** (hi-pur-ar-thrit'-ik). Relating to hyperarthritis.

**hyperarthritis** (hi-per-arth-ri'-tis) [hyper-; ἄρθρον, joint; ἴτις, inflammation]. Severe arthritis.

**hyperasthenia** (hi-per-as-the'-ne-ah) [hyper-; ἀσθένεια, weakness]. Extreme weakness.

**hyperauxesis** (hi-per-awks-e'-sis) [hyper-; αὔξησις, increase]. Extreme increase in the size of a part.

**hyperazoturia** (hi-per-az-o-tu'-re-ah) [hyper-; azote; οὖρον, urine]. Excess of nitrogenous matter in the urine.

**hyperbolic** (hi-per-bol'-ik) [hyper-; βάλλειν, to throw]. Exaggerated.

**hyperbrachycephalic** (hi-per-brak-e-sef-al'-ik) [hyper-; βραχύς, short; κεφαλή, head]. Extremely brachycephalic.

**hyperbrachycephaly** (hi-per-brak-e-sef'-al-e) [hyper-; βραχύς, short; κεφαλή, head]. Extreme brachycephaly.

**hyperbulia** (hi-per-bū'-le-ah) [hyper-; βουλή, will]. Exaggerated wilfulness; abnormal development of will-power.

**hypercardia** (hi-per-kar'-de-ah) [hyper-; καρδία, the heart]. Cardiac hypertrophy.

**hypercardiotrophy** (hi-pur-kar-de-ot'-ro-fe). [hyper-; καρδία, heart; τροφή, nourishment]. Hypertrophy of the heart.

**hypercatabolism** (hi-per-kat-ab'-o-lizm) [hyper-; καταβάλλειν, to throw]. Wasting or excess of destruction of a tissue or part.

**hypercatharsis** (hi-per-kath-ar'-sis) [hyper-; κάθαρσις, cleansing]. Excessive purging.

**hypercedemonia** (hi-per-se-de-mo'-ne-ah) [hyper-; κηδεμονία, anxiety]. 1. Extreme anxiety or grief. 2. Excessive care on the part of the physician.

**hypercele** (hi'-per-sēl) [hyper-; κοιλία, belly]. The dorsal portion of the epicele.

**hypercementosis** (hi-pur-sem-en-to'-sis). See *exostosis, dental*.

**hypercenosis** (hi-per-sen-o'-sis) [hyper-; κένωσις, evacuation]. Excessive evacuation, as by purging or bleeding.

**hyperchlorhydria** (hi-per-klor-hi'-dre-ah) [hyper-; chlorhydria]. Excess of hydrochloric acid in the gastric secretion. Rossbach's disease.

**hypercholesteremia** (hi-per-ko-les-ter-e'-me-ah) [hyper-; χολή, bile; στέαρ, fat; αἷμα, blood]. Excess of cholesterin in the blood.

**hypercholia** (hi-per-ko'-le-ah) [hyper-; χολή, bile]. An excessive secretion of bile.

**hyperchondroma** (hi-per-kon-dro'-mah) [hyper-; chondroma]. A cartilaginous tumor.

**hyperchroma** (hi-per-kro'-mah) [hyper-; χρῶμα, color]. The excessive formation of the pigment of the skin, as in phthiriasis or syphilis.

**hyperchromasia** (hi-per-kro-ma'-ze-ah) [hyper-; χρῶμα, color]. A condition characterized by excess of pigment.
**hyperchromatic** (hi-per-kro-mat'-ik) [hyper-; χρῶμα, color]. Exhibiting hyperchroma.
**hyperchromatism** (hi-per-kro'-mat-izm). Same as *hyperchroma*.
**hyperchromatosis** (hi-per-kro-mat-o'-sis) [hyper-; χρῶμα, color]. Excessive pigmentation, as of the skin.
**hyperchylia** (hi-per-ki'-le-ah) [hyper-; χυλός, juice]. Excess of secretion; excessive formation of chyle.
**hypercinesia** (hi-per-sin-e'-ze-ah). See *hyperkinesia*.
**hypercinesis** (hi-pur-sin-e'-sis). See *hyperkinesis*.
**hypercompensation** (hi-per-kom-pen-sa'-shun). The formation of more plastic material than is necessary to compensate for loss.
**hypercrinia** (hi-per-krin'-e-ah) [hyper-; κρίνειν, to separate]. Abnormal or excessive secretion.
**hypercritical** (hi-pur-crit'-ik-al). Relating to a crisis of excessive severity.
**hypercryalgesia** (hi-pur-kri-al-je'-ze-ah) [hyper-; κρίος, cold; ἄλγησις, pain]. Abnormal sensitiveness to cold.
**hypercusia** (hi-per-kū'-ze-ah). See *hyperacusia*.
**hypercyesis** (hi-per-si-e'-sis) [hyper-; κίησις, conception]. 1. Superfetation. 2. The condition in which conceptions follow each other rapidly.
**hypercyrtosis** (hi-per-sir-to'-sis) [hyper-; κύρτωσις, curvature]. Extreme curvature, as of the back or a limb.
**hypercythemia** (hi-per-si-the'-me-ah) [hyper-; κύτος, cell; αἷμα, blood]. Increase in the blood-corpuscles compared with the serum.
**hyperdacryosis** (hi-pur-dak-re-o'-sis) [hyper-; δάκρυ, a tear]. An excessive secretion of tears.
**hyperdactylia** (hi-per-dak-til'-e-ah) [hyper-; δάκτυλος, finger]. Polydactylism.
**hyperdermatosis** (hi-per-der-mat-o'-sis) [hyper-; δέρμα, skin]. Hypertrophy of the skin.
**hyperdesmosis** (hi-per-dez-mo'-sis) [hyper-; δεσμός, a bond]. Hypertrophy of the connective tissue.
**hyperdiacrisis** (hi-pur-di-ak'-ris-is) [hyper-; διάκρισις, a separating]. An abnormally severe crisis.
**hyperdicrotic** (hi-per-di-krot'-ik) [hyper-; δίκροτος, a double beat]. Affected with marked or delayed dicrotism; a condition in which the aortic notch is below the base-line.
**hyperdicrotism** (hi-per-dik'-rot-izm) [hyper-; δίκροτος, a double beat]. Strongly marked or excessive dicrotism.
**hyperdiemorrhysis** (hi-per-di-e-mor'-e-sis) [hyper-; διά, through; αἷμα, blood; ῥύσις, flowing]. Excessive circulation of the blood through the veins.
**hyperdistention** (hi-per-dis-ten'-shun) [hyper-; distendere, to stretch]. Forcible or extreme distention.
**hyperdiuresis** (hi-per-di-ū-re'-sis) [hyper-; διουρεῖν, to pass urine]. Excessive secretion of urine.
**hyperdontogeny** (hi-per-don-toj'-en-e) [hyper-; ὀδούς, tooth; γεννᾶν, to produce]. The occurrence of a third dentition in mature life.
**hyperdynamia** (hi-per-di-nam'-e-ah) [hyper-; δύναμις, energy]. Excessive strength of exaggeration of nervous or muscular function.
**hyperdynamic** (hi-per-di-nam'-ik) [hyper-; δύναμις, energy]. Pertaining to, or marked by, hyperdynamia.
**hypereccrisia, hypereccrisis** (hi-per-ek-kris'-e-ah, hi-per-ek'-kris-is) [hyper-; ἐκ, out; κρίνειν, to separate]. Excessive excretion.
**hypereccritic, hypercritic** (hi-per-ek-rit'-ik). Relating to hypereccrisis.
**hyperechema** (hi-per-ek-e'-mah) [hyper-; ἤχημα, sound; pl., hyperechemata]. A normal sound abnormally exaggerated.
**hyperechesis** (hi-per-ek-e'-sis) [hyper-; ἤχησις, sound]. Abnormal loudness of voice.
**hyperemesis** (hi-per-em'-es-is) [hyper-; ἔμεσις, vomiting]. Excessive vomiting. **h. gravidarum.** See *morning sickness*. **h. lactantium**, intractable vomiting of nurslings.
**hyperemetic** (hi-per-em-et'-ik) [hyper-; ἔμεσις, vomiting]. Pertaining to, or characterized by, excessive vomiting.
**hyperemia** (hi-per-e'-me-ah) [hyper-; αἷμα, blood]. Excessive blood in a part. **h., active**, that caused by an excessive supply of blood going to a part. **h., arterial**, that due to increase of the blood-current from dilatation of the arterioles. **h., Bier's passive.** See *Bier's h.* **h., collateral, h., compensatory**, congestion, either arterial or venous, in one part, compensatory to anemia in another part through transferred blood-pressure. **h., latent**, the condition following continued light-treatment, marked by distinct redness appearing after very slight stimulation and consisting in dilatation of the cutaneous vessels, exudation, and local leukocytosis. **h., passive**, that caused by an impediment to the removal of the blood. **h. of stasis.** See *h., passive.* **h., venous**, that due to diminution of the velocity of the current, from obstruction of the outflow through the veins. Syn., *hyperæmia venosa*.
**hyperencephalus** (hi-per-en-sef'-a-lus) [hyper-; ἐγκέφαλος, brain]. A variety of single autositic monsters in which the upper portion of the skull is entirely lacking.
**hyperendocrinism** (hi-per-en-dok'-rin-izm) [hyper-; ἔνδον, within; κρίνειν, to separate]. Abnormal increase of an internal secretion; the opposite of hypoendocrinism.
**hyperenergy** (hi-per-en'-er-je). Excessive energy or action.
**hyperenteritis** (hi-per-en-ter-i'-tis) [hyper-; ἔντερον, intestine; ιτις, inflammation]. Acute intestinal inflammation; severe enteritis.
**hyperenterosis** (hi-per-en-ter-o'-sis) [hyper-; ἔντερον, bowel]. Hypertrophy of the intestines.
**hyperephidrosis** (hi-per-ef-hid-ro'-sis) [hyper-; ἐφίδρωσις, perspiration]. Excessive or long-continued sweating.
**hypererethisia, hypererethism** (hi-per-er-eth-is'-e-ah, hi-per-er'-eth-izm). [hyper-; ἐρεθισμός, irritation]. Excessive nervous or mental irritability.
**hypergasia** (hi-per-er-ga'-se-ah) [hyper-; ἐργασία, work]. Increased work or functional activity.
**hyperergia** (hi-per-er'-je-ah) [hyper-; ἔργον, work]. Increased functional activity.
**hypererythrocythemia** (hi-per-er-ith-ro-si-the'-me-ah) [hyper-; ἐρυθρός, red; κύτος, cell; αἷμα, blood]. Excess of red corpuscles in the blood.
**hyperesophoria** (hi-per-es-o-fo'-re-ah). See *heterophoria*.
**hyperesthesia** (hi-per-es-the'-ze-ah) [hyper-; αἴσθησις, sensation]. Excessive sensibility.
**hyperesthetic** (hi-per-es-thet'-ik) [hyper-; αἴσθησις, sensation]. Pertaining to hyperesthesia.
**hyperexophoria** (hi-per-eks-o-fo'-re-ah). A turning of the eyes upward and outward.
**hyperextension** (hi-per-eks-ten'-shun). Excessive extension for the correction of orthopedic deformities.
**hyperfecundation** (hi-per-fe-kun-da'-shun). Same as *superfecundation*.
**hyperflexion** (hi-per-flek'-shun) [hyper-; flexio, a bending]. Overflexion, as of a limb. This is one method of treating aneurysm.
**hypergasia** (hi-per-ga'-se-ah) [hypo-; ἐργασία, work]. Diminished work or functional activity.
**hypergastritis** (hi-per-gas-tri'-tis). Very severe gastritis.
**hypergenesis** (hi-per-jen'-es-is) [hyper-; γένεσις, generation]. Excess or redundancy of the parts or organs of the body.
**hypergenetic** (hi-per-jen-et'-ik) [hyper-; γένεσις, generation]. Marked by enlargement or increase of size. See also *teratism*.
**hypergenitalism** (hi-per-jen'-it-al-izm) [hyper-; genital]. Abnormal activity of the internal secretions of the genital organs, producing undue development of the genitals, and precocious puberty.
**hypergeusesthesia** (hi-per-gū-ses-the'-ze-ah). See *hypergeusia*.
**hypergeusia** (hi-per-gū'-se-ah) [hyper-; γεῦσις, taste]. Abnormal acuteness of the sense of taste.
**hypergigantosoma** (hi-per-ji-gant-o-so'-mah) [hyper-; γίγας, a giant; σῶμα, body]. Extraordinary gigantism.
**hyperglobulia** (hi-per-gloo-bū'-le-ah) [hyper-; globus, a ball]. An increase in the number of red blood-corpuscles.
**hyperglucosic** (hi-per-gloo-ko'-sik). A term applied to any diabetic diet containing an amount of carbohydrates larger than the patient's tolerance.
**hyperglycemia** (hi-per-gli-se'-me-ah) [hyper-; γλυκύς, sweet; αἷμα, blood]. Excess of sugar in the blood.
**hyperglycistia** (hi-per-gli-sis'-te-ah) [hyper-; γλυκύς, sweet; ἰστός, tissue]. Excess of glucose in the tissues.

**hyperglycogenia** (hi-per-gli-ko-je'-ne-ah) [hyper-; γλυκύς, sweet; γενής, producing]. The excessive production of glycogen.

**hyperhedonia** (hi-per-hed-o'-ne-ah) [hyper-; ἡδονή, pleasure]. Extreme delight in the gratification of a desire.

**hyperhematosia, hyperhematosis** (hi-per-hem-at-o'-ze-ah, hi-per-hem-at-o'-sis). 1. Extraordinary activity or pressure of the blood. 2. Excess of blood. Syn., *hyperematosis; hyperemosis.*

**hyperhemia, hyperhæmia** (hi-per-he'-me-ah). See *hyperemia.*

**hyperhidrosis, hyperidrosis** (hi-per-hid-ro'-sis, hi-per-id-ro'-sis) [hyper-; ἱδρώς, sweat]. Excessive sweating.

**hyperhydremia** (hi-per-hi-dre'-me-ah) [hyper-; ὕδωρ, water; αἷμα, blood]. Excess of water in the blood.

**hyperhypnosis** (hi-per-hip-no'-sis) [hyper-; ὕπνος, sleep]. Excessive or frequent drowsiness and sleep.

**Hypericum** (hi-per'-ik-um). St. John's wort; a genus of plants, mostly herbs or shrubs, with a resinous juice. H. **perforatum,** one of the commonest of the 160 species, has styptic, stimulant, and diuretic properties, but is now used mainly in domestic practice. Dose of the ext., gr. x-xx; of the fld. ext., ʒ ʃ-ij.

**hyperideation** (hi-per-i-de-a'-shun) [hyper-; idea]. Excessive or morbid mental activity.

**hyperidrosis** (hi-per-id-ro'-sis). See *hyperidrosis.*

**hyperino-epithelioma** (hi-per-in-o-ep-ith-e-le-o'-mah) [hyper-; ἴς, a fiber; ἐπί, upon; θηλή, nipple; ὄγμα, tumor; pl., *hyperino-epitheliomata*]. Scirrhous carcinoma with an abundance of fibrous elements.

**hyperinosemia** (hi-per-in-o-se'-me-ah) [hyper-; ἴς, fiber; αἷμα, blood]. An exaggerated tendency to the formation of fibrin in the blood.

**hyperinosis** (hi-per-in-o'-sis) [hyper-; ἴς, fiber]. An excessive increase in the fibrin-factors in the blood.

**hyperinotic** (hi-per-in-o'-ik) [hyper-; ἴς, fiber]. Characterized by hyperinosis.

**hyperinvolution** (hi-per-in-vo-lu'-shun) [hyper-; *involvere*, to roll around]. Excessive involution of an organ after enlargement, as of the uterus after pregnancy, resulting in a reduction below the normal size.

**hyperisotonia** (hi-per-is-o-to'-ne-ah) [hyper-; ἴσος, equal; τόνος, tone]. Unusual equality of tone or tension; applied to muscles.

**hyperisotonic** (hi-per-is-o-ton'-ik) [hyper-; ἴσος, equal; τόνος, tone]. Applied to a solution of greater density than the blood or some other fluid taken as a standard.

**hyperisotonicity** (hi-per-is-o-ton-is'-it-e). The condition of infusions having too great a saline percentage.

**hyperkeratinization** (hi-per-ker-at-in-i-za'-shun). A hypertrophy of the epithelium seen in the palms and soles in chronic arsenical poisoning.

**hyperkeratomycosis** (hi-per-ker-at-o-mi-ko'-sis) [hyper-; κέρας, horn; μύκης, fungus]. Hyperkeratosis caused by a parasitic fungus.

**hyperkeratosis** (hi-per-ker-at-o'-sis) [hyper-; κέρας, cornea; horn]. 1. Hypertrophy of the cornea. 2. Hypertrophy of the horny layer of the skin. h. **lacunaris pharyngis,** a condition characterized by numerous hard white masses sometimes developing into long horny spines, projecting from the follicles of the lymphoid ring about the pharynx.

**hyperkinesia** (hi-per-kin-e'-ze-ah) [hyper-; κίνησις, energy]. Excessive movement, as that associated with muscular spasm.

**hyperkinetic** (hi-per-kin-et'-ik) [hyper-; κίνησις, energy]. Pertaining to, or marked by, hyperkinesia.

**hyperkoria** (hi-per-ko'-re-ah) [hyper-; κόρος, satiety]. The condition of being quickly satisfied.

**hyperlactation** (hi-per-lak-ta'-shun) [hyper-; *lactare*, to give milk]. Prolongation of lactation beyond the ninth month.

**hyperleukocythemia** (hi-per-lū-ko-si-the'-me-ah). See *hyperleukocytosis.*

**hyperleukocytosis** (hi-per-lū-ko-si-to'-sis) [hyper-; *leukocyte*]. An increase in the number of leukocytes in the blood.

**hyperlipemia** (hi-per-lip-e'-me-ah) [hyper-; λίπος, fat; αἷμα, blood]. Excess of fat in the blood; lipemia.

**hyperliposis** (hi-per-lip-o'-sis) [hyper-; λίπος, fat]. An excess of fat-splitting ferment (lipase) in the blood.

**hyperlithuria** (hi-per-lith-ū'-re-ah) [hyper-; λίθος, stone; οὖρον, urine]. Excess of lithic acid in the urine. Same as *lithuria.*

**hyperlogia** (hi-per-lo'-je-ah) [hyper-; λόγος, speech]. Excessive or maniacal loquacity.

**hyperlymphia** (hi-per-limf'-e-ah) [hyper-; *lympha*, water]. An increase in the amount of lymph in the body.

**hypermastia** (hi-per-mas'-te-ah) [hyper-; μαστός, breast]. Excessive development of the mammary gland.

**hypermature** (hi-per-ma-tūr'). Overmature, overripe, as a cataract.

**hypermedication** (hi-per-med-e-ka'-shun). Excessive employment of drugs.

**hypermegalia, hypermegaly** (hi-per-meg-a'-le-ah, -meg'-al-e) [hyper-; μέγας, large]. Excessive enlargement.

**hypermegasoma** (hi-per-meg-as-o'-mah). See *hypergigantosoma.*

**hypermegasthenic** (hi-per-meg-as-then'-ik) [hyper-; μέγας, great; σθένος, strength]. Abnormally or excessively strong.

**hypermesosoma** (hi-per-mes-o-so'-mah) [hyper-; μέσος, middle; σῶμα, body]. A stature measurably in excess of the ordinary.

**hypermetamorphic** (hi-per-met-ah-morf'-ik). Undergoing frequent transformations.

**hypermetamorphosis** (hi-per-met-am-or-fo'-sis) [hyper-; *metamorphosis*]. In biology, applied to insects that undergo transformation more completely or having more stages than ordinary.

**hypermetrope** (hi'-per-met-rōp). See *hyperope.*

**hypermetropia** (hi-per-me-tro'-pe-ah). Same as *hyperopia.*

**hypermetropic** (hi-per-me-trop'-ik) [hyper-; μέτρον, a measure; ὤψ, sight]. Affected with, or pertaining to, hyperopia.

**hypermicrosoma** (hi-per-mik-ro-so'-mah) [hyper-; μικρός, small; σῶμα, body]. Extreme dwarfishness.

**hypermnesia, hypermnesis** (hi-perm-ne'-ze-ah, hi-perm-ne'-sis) [hyper-; μνῆσις, memory]. Abnormal exaltation of the power of memory.

**hypermotility** (hi-per-mo-til'-it-e) [hyper-; *motilis*, motile]. Excessive action. h., **gastric,** excessive churning action of the gastric walls, often accompanied with hyperacidity due to increased secretion of HCl.

**hypermyelohemia** (hi-per-mi-el-o-he'-me-ah) [hyper-, μυελός, marrow; αἷμα, blood]. Hyperemia of the spinal marrow. Syn., *myelyperemia.*

**hypermyotonia** (hi-per-mi-o-to'-ne-ah) [hyper-; μῦς, muscle; τόνος, tone]. Excessive tonicity of the muscles.

**hypermyotrophy, hypermyotrophia** (hi-per-mi-ot'-ro-fe, hi-per-mi-ot-ro'-fe-ah) [hyper-; μῦς, muscle; τροφή, nourishment]. Hypertrophy of the muscular tissue.

**hypernanosoma** (hi-per-nan-o-so'-ma) [hyper-; νᾶνος, a dwarf; σῶμα, body]. A person of low stature, but larger than a dwarf.

**hypernea, hypernoia** (hi-per-ne'-ah, hi-per-noi'-ah) [hyper-; νοέων, to think]. Excessive or abnormal mental activity; also, the uncontrolled and incongruous activity of the imagination in some cases of insanity.

**hypernephroid** (hi-per-nef'-roid) [hyper-; νεφρός, kidney; εἶδος, likeness]. Suprarenal. h. **tumors,** such as are derived from aberrant suprarenal tissue. See *hypernephroma.*

**hypernephroma** (hi-per-nef-ro'-mah). See *Grawitz's tumor.*

**hypernephrotrophy** (hi-per-nef-rot'-ro-fe) [hyper-; νεφρός, kidney; τροφή, nourishment]. Hypertrophy of the kidney. Syn., *nephrypertrophia.*

**hyperneuria** (hi-per-nū-re-ah) [hyper-; νεῦρον, nerve]. Excessive nerve-action.

**hyperneuroma** (hi-per-nū-ro'-mah) [hyper-; νεῦρον, nerve; ὄγμα, tumor]. An exuberant growth of nervous tissue.

**hyperneurosis** (hi-per-nū-ro'-sis). Excessive development of nervous tissue.

**hypernidation** (hi-per-nid-a'-shun). See *supernidation.*

**hypernormal** (hi-per-nor'-mal). Exceeding the normal state.

**hypernutrition.** See *supernutrition.*

**hyperodontogeny** (hi-per-o-do-don-toj'-en-e) [hyper-; ὀδούς, tooth; γενᾶν, to beget]. The phenomenon of a third dentition late in life.

**hyperoic** (hi-per-o'-ik) [ὑπερῷα, the palate]. Relating to the palate.

**hyperoitis** (hi-per-o-i'-tis) [ὑπερῷα, palate; ιτις, inflammation]. Inflammation of the palate.

**hyperoncosis** (hi-per-on-ko'-sis) [hyper-; ὄγκος, a tumor]. Excessive swelling.

**hyperonychia** (hi-per-o-nik'-e-ah) [hyper-; ὄνυξ, nail]. Hypertrophy of the nails.

**hyperonychosis** (hi-per-on-ik-o'-sis). See *hyperonychia*.

**hyperope** (hi'-per-ōp) [hyperopia]. One who is affected with hyperopia.

**hyperopia** or. **hypermetropia** (hi-per-o'-pe-ah or hi-per-me-tro'-pe-ah) [hyper-; ὤψ, sight]. The condition of the refractive media of the eye in which, with suspended accommodation, the focus of parallel rays of light is behind the retina. It is due to an abnormally short anteroposterior diameter of the eye, or to a subnormal refractive power of its media. **h., absolute**, that which cannot be corrected completely by accommodation, so that there is indistinct vision even for distance. **h., axial**, that due to abnormal shortness of the anteroposterior diameter of the eye, the refractive power being normal. **h., curvature**, a form often combined with astigmatism, due to changes in curvature of the cornea or lens. **h., facultative**, that which may be corrected by the accommodation, so that there is distinct vision at a distance. **h., index**, that developing in old age from sclerosis of the lens. **h., latent**, that part of the total hyperopia that cannot be overcome by the accommodation, or the difference between the manifest and the total hyperopia. **h., manifest**, the amount of hyperopia represented by the strongest convex lens which a person will accept without paralysis of the accommodation. **h., relative**, a high hyperopia in which distinct vision is possible only when excessive convergence is made. **h., total**, the entire hyperopia, both latent and manifest.

**hyperorexia** (hi-per-or-ek'-se-ah) [hyper-; ὄρεξις, appetite]. Bulimia.

**hyperorthognathy** (hi-per-or-thog'-na-the) [hyper-; ὀρθός, straight; γνάθος, the jaw]. Excessive orthognathy; the condition of having a cranial index greater than 91°.

**hyperosmia** (hi-per-os'-me-ah) [hyper-; ὀσμή, smell]. An abnormally acute sense of smell.

**hyperosphresis** (hi-per-os-fre'-sis) [hyper-; ὄσφρησις, the power of smelling]. Exaggeration of the sense of smell.

**hyperosteogeny** (hi-per-os-te-oj'-en-e) [hyper-; ὀστέον, bone; γεννᾶν, to produce]. Excessive development of bone.

**hyperosteopathy** (hi-per-os-te-op'-ath-e) [hyper-; ὀστέον, bone; πάθος, disease]. An excessively diseased condition of the bones.

**hyperostosis** (hi-per-os-to'-sis) [hyper-; ὀστέον, bone]. Exostosis or general hypertrophy of bone tissue.

**hyperoxemia** (hi-per-oks-e'-me-ah) [hyper-; ὀξύς, sharp; αἷμα, blood]. Extreme acidity of the blood.

**hyperparasite** (hi-per-par'-as-īt) [hyper-; parasite]. In biology, a parasite including in itself another parasite.

**hyperparasitism** (hi-per-par'-as-īt-izm) [hyper-; parasite]. The infestation of parasites by other parasites.

**hyperpathia** (hi-per-path'-e-ah) [hyper-; πάθος, disease, sensibility]. 1. Extreme illness. 2. Extreme sensibility.

**hyperpelvic** (hi-per-pel'-vik). Located above the pelvis.

**hyperpepsia** (hi-per-pep'-se-ah) [hyper-; πέψις, digestion]. Dyspepsia characterized by an excess of chlorides in the gastric juice, without an excess of free hydrochloric acid.

**hyperpepsinia** (hi-per-pep-sin'-e-ah). Excessive secretion of pepsin in the stomach.

**hyperperistalsis** (hi-per-per-is-tal'-sis). Peristaltic unrest; a condition characterized by persistent rapid contractions of the stomach in close succession appearing after meals. Syn., *tormina ventriculi nervosa*.

**hyperperitonitis** (hi-per-per-it-on-i'-tis). Very severe or acute peritonitis.

**hyperphagia** (hi-per-fa'-je-ah) [hyper-; φαγεῖν, to eat]. Excess in eating. See *bulimia*.

**hyperphalangia** (hi-per-fa-lan'-je-ah) [hyper-; phalanx]. Abnormal length of one or several of the phalanges.

**hyperpharyngeal** (hi-per-far-in'-je-al). See *suprapharyngeal*.

**hyperphasia** (hi-per-fa'-se-ah) [hyper-; φάσις, saying]. Hyperlogia; insane volubility; lack of control over the organs of speech.

**hyperphenomenal** (hi-per-fe-nom'-en-al). Real.

**hyperphlebectasy** (hi-per-fleb-o-ek'-ta-se) [hyper-; φλέψ, vein; ἔκτασις, a stretching]. Excessive dilatation of the veins.

**hyperphlebosis** (hi-per-fleb-o'-sis). See *hypervenosity*.

**hyperphlogosis** (hi-per-flo-go'-sis) [hyper-; φλόγωσις, a burning]. Violent inflammation.

**hyperphoria** (hi-per-fo'-re-ah). See *heterophoria*.

**hyperphoric** (hi-per-fo'-rik). 1. Relating to hyperphoria. 2. One who is affected with hyperphoria.

**hyperphrasia** (hi-per-fra'-se-ah) [hyper-; φράσις, utterance]. The incoherent and exaggerated utterance of an insane person.

**hyperphrenia** (hi-per-fre'-ne-ah) [hyper-; φρήν, mind]. Passionate mental exaltation of the insane.

**hyperphysemia** (hi-per-fi-se'-me-ah) [hyper-; φῦσα, air; αἷμα, blood]. Excess of gases in the blood, or the abnormal conditions associated with such excess.

**hyperpicrous** (hi-per-pik'-rus) [hyper-; πικρός, bitter]. Excessively bitter.

**hyperpiesis** (hi-per-pi-e'-sis) [hyper-; πίεσις, a pressing]. An abnormally high pressure, as of the blood.

**hyperpigmentation** (hi-per-pig-men-ta'-shun). Excessive pigmentation.

**hyperpimelic** (hi-per-pim'-el-ik) [hyper-; πιμελής, fat]. Relating to obesity.

**hyperpituitarism** (hi-per-pit-u'-it-ar-izm) [hyper-; pituitary]. A condition due to excessive activity of the pituitary gland, and marked by gigantism and hypertrichosis.

**hyperplasia** (hi-per-pla'-ze-ah) [hyper-; πλάσις, molding]. Excessive formation of tissue; an increase in the size of a tissue or organ owing to an increase in the number of cells. Syn., *numerical hypertrophy*.

**hyperplastic** (hi-per-plas'-tik). Pertaining to hyperplasia.

**hyperplerosis** (hi-per-ple-ro'-sis) [hyper-; πλήρωσις, fulness]. Excessive repletion or fulness.

**hyperplexia** (hi-per-pleks'-e-ah) [hyper-; πλῆξις, stroke]. 1. Ecstasy. 2. Melancholia with stupor.

**hyperpnea** (hi-per-pne'-ah) [hyper-; πνοή, breathing]. Exaggeration of respiration.

**hyperporosis** (hi-per-po-ro'-sis) [hyper-; πώρωσις, cementing or uniting]. An excessive formation of callus in the reunion of fractured bones.

**hyperpraxia** (hi-per-praks'-e-ah) [hyper-; πρᾶξις, exercise]. The restlessness of movement characterizing certain forms of mania.

**hyperpresbyopia** (hi-per-pres-be-o'-pe-ah) [hyper-; πρέσβυς, old; ὤψ, eye]. Excessive presbyopia.

**hyperprochoresis** (hi-per-pro-ko-re'-sis) [hyper-; προχωρεῖν, to advance]. Excessive motor action of the stomach. Cf. *hyperperistalsis*.

**hyperpromethia** (hi-per-prom-e'-the-ah) [hyper-; προμήθεια, foresight]. Supernormal power of foresight.

**hyperprosexia** (hi-per-pro-seks'-e-ah) [hyper-; πρόσεξις, attention]. Entire absorption of the attention by a single process.

**hyperselaphesia** (hi-per-sel-af-e'-se-ah) [hyper-; ψηλάφησις, touch]. Abnormal increase of tactile sensibility.

**hyperpyretic** (hi-per-pi-ret'-ik). Pertaining to hyperpyrexia.

**hyperpyrexia** (hi-per-pi-reks'-e-ah) [hyper-; πυρετός, fever]. Excessively high body temperature. By some the term is used only when the temperature is above 106° F.

**hyperresonance** (hi-per-res'-o-nans) [hyper-; resonance]. Increased resonance on percussion.

**hyperrhinencephalia** (hi-per-rin-en-sef-al'-e-ah) [hyper-; ῥίς, the nose; ἐγκέφαλος, the brain]. A congenital deformity characterized by undue frontal predominance and excessive development of the olfactory bulbs.

**hypersarcosis**, **hypersarcoma** (hi-per-sar-ko'-sis, hi-per-sar-ko'-ma) [hyper-; σάρξ, flesh; ὅμα, a tumor]. 1. Excessive granulation, fungosity, or proudflesh. 2. Obesity; hypertrophy.

**hypersecretion** (hi-per-se-kre'-shun) [hyper-; secretion]. Excessive secretion.

**hypersensitive** (hi-per-sen'-sit-iv). Abnormally sensitive.

**hypersomnia** (hi-per-som'-ne-ah) [hyper-; somnus, sleep]. Excessive sleep.

**hyperspasmia** (*hi-per-spaz'-me-ah*). Synonym of *convulsions*.
**hypersplenia** (*hi-per-sple'-ne-ah*). See *splenomegalia*.
**hypersplenotrophy** (*hi-per-splen-ot'-ro-fe*); See *splenomegalia*.
**hypersteatosis** (*hi-per-ste-at-o'-sis*) [*hyper-*; *steatosis*]. Excessive secretion of fat. Syn., *aleipsis acuta*.
**hypersthenia** (*hi-per-sthen'-e-ah*) [*hyper-*; σθένος, strength]. A condition of exalted strength or tone of the body.
**hypersthenic** (*hi-per-sthen'-ik*) [*hyper-*; σθένος, strength]. Characterized by hypersthenia.
**hypersusceptibility** (*hi-per-sus-sep-tib-il'-it-e*). 1. Extreme liability to infection. 2. Anaphylaxis.
**hypersynergia** (*hi-per-sin-ur'-je-ah*) [*hyper-*; συνεργία, a working together]. Excessive coordination; excessive energy in the organs in spreading disease throughout the system.
**hypersystole** (*hi-per-sis'-to-le*) . [*hyper-*; *systole*]. An excessively strong systole.
**hypertension** (*hi-per-ten'-shun*). Excessive tension; supertension.
**hyperthelia** (*hi-per-the'-le-ah*) [*hyper-*; θηλή, a nipple]. The presence of supernumerary nipples.
**hyperthermalgesia** (*hi-per-therm-al-je'-se-ah*) [*hyper-*; θέρμη, heat; ἄλγος, pain]. Abnormal sensitiveness to heat.
**hyperthermia** (*hi-per-ther'-me-ah*). Elevation of temperature above the normal.
**hyperthermoesthesia** (*hi-per-therm-o-es-the'-ze-ah*). See *hyperthermalgesia*.
**hyperthymia** (*hi-per-thi'-me-ah*) '[*hyper-*; θυμός, mind]. 1. Mental hyperesthesia; morbid oversensitiveness. 2. Vehement cruelty or foolhardiness as a symptom of mental disease. 3. Moral insanity.
**hyperthymization** (*hi-per-thi-miz-a'-shun*). Exaggerated activity of the thymus gland and the pathological condition resulting from it.
**hyperthyrea** (*hi-per-thi'-re-ah*). The condition arising from excessive functional activity of the thyroid gland.
**hyperthyroidation** (*hi-per-thi-roid-a'-shun*). Abnormal action or overaction of the thyroid gland.
**hyperthyroidism** (*hi-per-thi'-roid-izm*). An abnormal condition brought about by an excessive or depraved functional activity of the thyroid gland.
**hyperthyrosis** (*hi-per-thi-ro'-sis*). The condition in which there is excess of thyroid substance in the body.
**hypertonia** (*hi-per-to'-ne-ah*) [*hyper-*; τόνος, tone]. 1. Excess of muscular tonicity. 2. Increased intraocular tension.
**hypertonic** (*hi-per-ton'-ik*). Exceeding in strength or tension. h. **salt solution**, one whose osmotic tension exceeds that of the blood-serum.
**hypertonicity** (*hi-per-ton-is'-it-e*). See *hypertonia*.
**hypertoxicity** (*hi-per-toks-is'-it-e*). The quality of being excessively toxic.
**hypertrichiasis**, **hypertrichosis** (*hi-per-trik-i'-as-is*, *hi-per-trik-o'-sis*). Excessive growth of hair of a part or the whole of the body.
**hypertromos** (*hi-per'-tro-mos*) [*hyper-*; τρόμος, tremor]. Excessive tremor or fear.
**hypertrophia** (*hi-per-tro'-fe-ah*) [*hyper-*; τροφή, nourishment]. Same as *hypertrophy*. h. **cordis**, hypertrophy of the heart.
**hypertrophic** (*hi-per-trof'-ik*) [*hyper-*; τροφή, nourishment]. Marked by hypertrophy or excessive size.
**hypertrophous** (*hi-per'-trof-us*). Marked by or exhibiting hypertrophy.
**hypertrophy** (*hi-per'-tro-fe*) [*hyper-*; τροφή, nourishment]. An increase in the size of a tissue or organ independent of the general growth of the body. h., **compensatory**, that resulting from the increased activity of an organ to make up some deficiency in a paired organ or in itself. h., **concentric** (of the heart), increase in the thickness of the walls, without increase in the size of the organ, but with diminution in the capacity of its chambers. h., **eccentric** (of the heart), hypertrophy with dilatation. h., **false**, an increase in some one constituent tissue of an organ, usually the connective tissue. h., **muriform**, a mulberry-like enlargement, as of the posterior ends of the lower turbinals. h., **numerical**, hypertrophy due to an increase in the number of cells. h., **physiological**. Same as h., *compensatory*. h., **simple**. 1. That in which there is increase in the size of the individual cells. 2. Of the heart, increased thickness of the walls, the size of the cavities remaining unchanged. h., **true**, an increase of all the component tissues of an organ, giving increased power.
**hypertropia** (*hi-per-tro'-pe-ah*). See *strabismus*.
**hypertypic** (*hi-per-tip'-ik*). Exceeding the type; excessively atypic.
**hyperuremia** (*hi-per-u-re'-me-ah*) [*hyper-*; οὖρον, urine; αἷμα, blood]. Excess of urea in the blood.
**hyperuresis** (*hi-per-u-re'-sis*). Same as *polyuria*.
**hyperuricemia** (*hi-per-u-ris-e'-me-ah*) [*hyper-*; οὖρον, urine; αἷμα, blood]. Excess of uric acid in the blood.
**hypervaccination** (*hi-per-vak-sin-a'-shun*). A second or subsequent inoculation of an immunized person or animal; it is done with the idea of obtaining (in the case of an animal) a powerful antitoxin.
**hypervenosity** (*hi-per-ve-nos'-it-e*). 1. Excessive development of the venous system. 2. See *supervenosity*.
**hyperventilation** (*hi-per-ven'-til-a-shun*) [*hyper-*; *ventilare*, to fan]. A method of treating some diseases by exposing the body to drafts of air.
**hyperviscosity** (*hi-per-vis-kos'-it-e*). Exaggeration of adhesive properties; observed in the erythrocytes in inflammatory diseases, in anemias, or when they are subjected to the action of poisons, notably snake-poison.
**hypesthesia** (*hip-es-the'-ze-ah*) [ὑπό, under; αἴσθησις, sensation]. Impairment of sensation; lessened tactile sensibility.
**hypesthetic** (*hip-es-thet'-ik*). Pertaining to or affected with hypesthesia.
**hypha** (*hi'-fah*) [ὑφή, a weaving; web; pl., *hyphæ*]. The filament or thread of a fungus; the matted hyphæ form the mycelium.
**hyphedonia** (*hip-hed-o'-ne-ah*) [*hypo-*; ἡδονή, pleasure]. Morbidly diminished pleasure in the gratification of desires. Cf. *hyperhedonia*.
**hyphemia** (*hi-fe'-me-ah*) [ὑπό, under; αἷμα, blood]. 1. Oligemia; deficiency of blood. 2. Hemorrhage in the anterior chamber of the eye.
**hyphidrosis** (*hip-hid-ro'-sis*) [*hypo-*; ἱδρωσις, sweating]. Deficiency of perspiration.
**hyphogenous** (*hi-foj'-en-us*) [ὑφή, a web; γεννάν, to produce]. Due to the hyphæ of some parasitic fungus.
**hyphology** (*hi-fol'-o-je*) [ὑφή, web; λόγος, science]. Same as *histology*.
**Hyphomycetes** (*hi-fo-mi-se'-tēs*) [ὑφή, web; μύκης, fungus]. A group of fungi having the spores on prominent threads; the molds.
**hyphostroma** (*hi-fo-stro'-mah*) [ὑφή, a web; στρῶμα, a bed]. Same as *mycelium*.
**hyphotomy** (*hi-fot'-o-me*) [ὑφή, a web; τομή, a cutting]. The dissection of tissues.
**hypinosis** (*hip-in-o'-sis*) [*hypo-*; ἴς, fiber]. A deficiency of fibrin-factors in the blood.
**hypinotic** (*hip-in-ot'-ik*) [*hypo-*; ἴς, fiber]. Pertaining to, or affected with, hypinosis.
**hypisotonic** (*hip-is-o-ton'-ik*). See *hypoisotonic*.
**hypisotonicity** (*hip-is-o-ton-is'-it-e*) . The quality of having a diminished saline percentage.
**hypnacetin** (*hip-nas'-et-in*), C . H₃CO-NH-C₆H₄-OCH₂-CO-C₄H₅. Acetophenonacetylparaamidophenol ether. It is hypnotic and antiseptic. Dose 3-4 gr. (0.2-0.25 Gm.). Syn., *hypnoacetin*.
**hypnagogic** (*hip-nag-oj'-ik*) [ὕπνος sleep; ἀγωγός, leading]. 1. Inducing sleep; pertaining to the inception of sleep. 2. Induced by sleep.
**hypnagogue** (*hip'-nag-og*) [ὕπνος, sleep; ἀγωγός, leading]. Hypnotic.
**hypnal** (*hip'-nal*) [ὕπνος, sleep]. A drug composed of antipyrin and chloral hydrate and used as a hypnotic. Dose 15 gr. (1 Gm.).
**hypnalgia** (*hip-nal'-je-ah*) [ὕπνος, sleep; ἄλγος, pain]. Pain recurring during sleep.
**hypnapagogue** (*hip-nep'-ag-og*) [ὕπνος, sleep; ἐπαγωγός, enticing]. A medicine that induces sleep; an hypnotic.
**hypniater** (*hip-ne'-a-ter*) [ὕπνος, sleep; ἰατήρ, practitioner]. A somnambulistic or clairvoyant doctor; a hypnotizer or mesmerist.
**hypnic** (*hip'-nik*) [ὑπνικός, producing sleep]. 1. Pertaining to or inducing sleep. 2. An agent that induces sleep.
**hypno-** (*hip-no-*) [ὕπνος, sleep]. A prefix denoting relation to sleep or to hypnotism.
**hypnobat**, or **hypnobate** (*hip'-no-bāt*) [*hypno-*; βατός, walking]. A sleep-walker; somnambulist.

**hypnobatia** (*hip-no-ba'-she-ah*) [*hypno-*; βατόs, walking]. Somnambulism.
**hypnocyst** (*hip'-no-sist*) [*hypno-*; κύστις, bladder]. In biology, an encysted unicellular organism not undergoing sporulation.
**hypnogenetic, hypnogenic, hypnogenous** (*hip-no-jen-et'-ik, hip-no-jen'-ik, hip-noj'-en-us*) [*hypno-*; γεννᾶν, to produce]. 1. Producing or inducing sleep. 2. Inducing hypnotism. **h. spots**, surface-areas of the body, stimulation of which produces sleep.
**hypnography** (*hip-nog'-ra-fe*). Same as *hypnology*.
**hypnolepsy** (*hip'-no-lep-se*) [*hypno-*; λῆψις, seizure]. Excessive or morbid sleepiness; narcolepsy.
**hypnology** (*hip-nol'-o-je*) [*hypno-*; λόγος, science]. The science dealing with sleep or with hypnotism.
**hypnone** (*hip'-nōn*). See *acetophenone*.
**hypnopathy** (*hip-nop'-ath-e*) [*hypno-*; πάθος, disease]. Sleep due to a diseased or morbid condition of the body, brain, or mind; also, narcolepsy; sleepy disease.
**hypnophobia** (*hip-no-fo'-be-ah*) [*hypno-*; φόβος, fear]. Morbid dread of sleep; also, nightmare or night-terror.
**hypnopompic** (*hip-no-pomp'-ik*) [*hypno-*; πομπή, a procession]. Applied to visions seen at the moment of awakening from sleep or prior to complete awakening, as when a dream figure persists in waking life.
**hypnopyrine** (*hip-no-pi'-rēn*). A proprietary hypnotic and antipyretic preparation said to be a chlorine derivative of quinine. Dose, xv gr. (0.25 gm.) 3 or 4 times daily.
**hypnoscope** (*hip'-no-skōp*) [*hypno-*, σκοπεῖν, to examine]. An apparatus to determine if a patient is hypnotized.
**hypnosia** (*hip-no'-ze-ah*) [ὕπνος, sleep]. A condition of morbid drowsiness.
**hypnosis** (*hip-no'-sis*) [see *hypnosia*]. 1. The condition produced by hypnotizing. 2. The production of sleep; also, the gradual approach of sleep.
**hypnotherapy** (*hip-no-ther'-ap-e*) [*hypno-*; θεραπεία, healing]. Hypnotic treatment of disease.
**hypnotic** (*hip-not'-ik*) [*hypnosis*]. 1. Inducing sleep. 2. Pertaining to hypnotism. 3. A remedy that causes sleep.
**hypnotism** (*hip'-not-izm*) [*hypnosis*]. A state of artificial somnambulism or trance, induced in certain persons by concentrating the gaze on a small object or on a revolving mirror, or by complete subjection of their will to that of another, at whose command the hypnotic state develops. Three stages are described—the cataleptic, the lethargic, and the somnambulistic.
**hypnotization** (*hip-not-iz-a'-shun*) [*hypnotism*]. The induction of hypnotism.
**hypnotize** (*hip'-not-iz*). To bring into a hypnotic condition.
**hypnotoid** (*hip'-not-oid*) [*hypno-*; εἶδος, form]. Resembling hypnotism.
**hypnoval** (*hip'-no-val*). Amido-chloral-bromo-isovalerate; said to be a hypnotic.
**hypo** (*hi'-po*). 1. A common abbreviation of hypochondriasis. 2. Sodium thiosulphite.
**hypo-** (*hi-po-*) [ὑπό, under]. A prefix denoting: 1. Deficiency or lack. 2. Below or beneath, opposed to *epi-*, upon. 3. Of acids and salts, denoting those having a smaller number of atoms of oxygen than other compounds of the same elements.
**hypoacidity** (*hi-po-as-id'-it-e*). See *hypacidity*.
**hypoactivity** (*hi-po-ak-tiv'-it-e*). Diminished activity.
**hypoalonemia** (*hi-po-al-o-ne'-me-ah*) [*hypo-*; ἅλς, salt; αἷμα, blood]. A deficiency of the salts of the blood.
**hypoazoturia** (*hi-po-az-ot-ū'-re-ah*) [*hypo-*; *azoturia*]. A diminished amount of urea in the urine.
**hypoblast** (*hi'-po-blast*) [*hypo-*; βλαστός, sprout]. The internal layer of the blastoderm, also called the endoderm, endoblast, or entoderm. From it is developed the intestinal epithelium (except that of the mouth and anus) and that of the glands opening into the intestine, and the epithelium of the air-passages.
**hypoblastic** (*hi-po-blas'-tik*). Pertaining to the hypoblast.
**hypoblepharon** (*hi-po-blef'-ar-on*) [*hypo-*; βλέφαρον, lid]. 1. An artificial eye. 2. A swelling under the eyelid.
**hypobromite** (*hi-po-bro'-mīt*) [*hypo-*; βρῶμος, stench]. A salt of hypobromous acid. **h. method**, a method of estimating the quantity of urea in urine,
based upon the fact that when urea is acted upon by sodium hypobromite it is decomposed into nitrogen, carbon dioxide, and water. From the volume of nitrogen evolved the quantity of urea can be determined.
**hypobulia** (*hi-po-bū'-le-ah*) [*hypo-*; βουλή, will]. Deficiency of will-power.
**hypocardia** (*hi-po-kar'-de-ah*) [*hypo-*; καρδία, heart]. Downward displacement of the heart.
**hypocatalepsis** (*hi-po-kat-al-ep'-sis*) [*hypo-*; κατά, down; λαμβάνειν, to seize]. Slight catalepsy, or epilepsy.
**hypocatharsis** (*hi-po-kath-ar'-sis*) [*hypo-*; καθαίρειν, to purge]. A gentle purgation.
**hypocelom, hypocoelom** (*hi-po-se'-lōm*) [*hypo-*; κοίλωμα, a cavity]. The ventral part of the celom.
**hypochlorhydria** (*hi-po-klor-hi'-dre-ah*) [*hypo-*; χλωρός, green; ὕδωρ, water]. A condition in which there is a diminished amount of hydrochloric acid in the gastric juice.
**hypochlorite** (*hi-po-klo'-rīt*) [*hypo-*; χλωρός, green]. Any salt of hypochlorous acid, HClO. The most important are those of calcium and sodium.
**hypochlorization** (*hi-po-klo-riz-a'-shun*). A method of treating epilepsy and nephritis by reduction of the sodium chloride consumed by the patient to one-half.
**hypocholesteremia** (*hi-po-ko-les-ter-e'-me-ah*) [*hypo-*; χολή, bile; στέαρ, fat; αἷμα, blood]. Decrease or deficiency of the cholesterin of the blood.
**hypochondria** (*hi-po-kon'-dre-ah*) [*hypo-*; χόνδρος, cartilage]. The regions below the costal arches on either side. 2. Same as *hypochondriasis*.
**hypochondriac** (*hi-po-kon'-dre-ak*) [*hypochondrium*]. 1. Pertaining to the hypochondrium. 2. A person who is affected with hypochondriasis.
**hypochondriasis** (*hi-po-kon-dri'-as-is*) [*hypochondrium*]. A condition in which the patient believes himself suffering from grave bodily diseases.
**hypochondrium** (*hi-po-kon'-dre-um*) [*hypo-*; χόνδρος, cartilage]. The upper lateral region of the abdomen beneath the lower ribs.
**hypochromatemia** (*hi-po-kro-mat-e'-me-ah*) [*hypo-*; χρῶμα, color; αἷμα, blood]. Deficiency of the coloring-matter of the blood.
**hypochromatic** (*hi-po-kro-mat'-ik*) [*hypo-*; χρῶμα, color]. Deficient in coloring-matter.
**hypochromatism** (*hi-po-kro'-mat-izm*) [*hypo-*; *chromatin*]. Deficiency of chromatin in the nucleus of a cell.
**hypochromatosis** (*hi-po-kro-mat-o'-sis*). The pathological diminution of the chromatin in a cell-nucleus.
**hypochromemia** (*hi-po-kro-me'-me-ah*) [*hypo-*; χρῶμα, color; αἷμα, blood]. Anemia with an abnormally low color index.
**hypochromia** (*hi-po-kro'-me-ah*) [*hypo-*; χρῶμα, color]. Abnormal pallor or transparency of the skin, occurring in certain skin-diseases.
**hypochrosis** (*hi-po-kro'-sis*) [*hypo-*; χρῶσις, coloring]. Abnormal paleness; lack of normal coloration.
**hypochylia** (*hi-po-ki'-le-ah*) [*hypo-*; χυλός, juice]. Deficiency of secretion; deficiency of chyle.
**hypochyma** (*hi-po-ki'-mah*) [*hypo-*; χέειν, to pour]. An old name for cataract.
**hypocinesia, hypocinesis** (*hi-po-sin-e'-ze-ah, -e'-sis*). See *hypokinesis*.
**hypoclysis** (*hip-ok'-lis-is*) [*hypo-*; κλυσμός, a clyster]. The administration of an enema.
**hypocratous** (*hip-ok'-rat-us*) [*hypo-*; κράτος, strength]. Lacking in strength.
**hypocrinia** (*hip-o-krin'-e-ah*) [*hypo-*; κρίνειν, to separate]. Deficiency of secretion.
**hypocyrtosis** (*hi-po-sir-to'-sis*) [*hypo-*; κύρτωσις, curvature]. A slight amount of curvature.
**hypocystotomy** (*hi-po-sis-tot'-o-me*) [*hypo-*; κύστις, bladder; τομή, a cut]. Perineal cystotomy.
**hypocytosis** (*hi-po-si-to'-sis*) [*hypo-*; κυτός, cell]. Diminution of the number of blood-corpuscles.
**hypoderm** (*hi'-po-derm*) [*hypo-*; δέρμα, skin]. 1. Subcutaneous tissue. 2. A hypodermic injection. In biology, applied to the epithelial membrane lining the cuticular, crustaceous, or chitinous investment of arthropods; also called *hypodermis*.
**hypoderma** (*hi-po-der'-mah*) [*hypo-*; δέρμα, skin]. In biology: 1. A layer of cells, usually collenchyma, just beneath the epidermis of a leaf or stem. 2. A genus of dipterous insects, the bot-flies.
**hypodermatic, hypodermic** (*hi-po-der-mat'-ik, hi-po-der'-mik*) [*hypo-*; δέρμα, skin]. Placed or introduced beneath the skin, as *hypodermatic* injection.

**hypodermatoclysis, hypodermoclysis** (*hi-po-der-mat-ok'-lis-is, hi-po-der-mok'-lis-is*) [*hypodermatic;* κλύσις, injection]. The introduction into the subcutaneous tissues of large quantities of fluids, especially of normal saline solution.

**hypodermatomy** (*hi-po-der-mat'-o-me*) [*hypo-;* δέρμα, skin; τέμνειν, to cut]. Subcutaneous surgical section of parts.

**hypodermotherapy** (*hi-po-der-mo-ther'-ap-e*) [*hypo-;* δέρμα, skin; θεραπεία, therapy]. Subcutaneous medication.

**hypodicrotous** (*hi-po-dik'-ro-tus*) [*hypo-;* δίκροτος, double beat]. Dicrotic in a small degree.

**hypodipsia** (*hip-o-dip'-se-ah*) [*hypo-;* δίψις, drinking]. The drinking of too little water or fluid.

**hypodynamic** (*hi-po-di-nam'-ik*). See *adynamic*.

**hypodynia** (*hi-po-din'-e-ah*) [*hypo-;* ὀδύνη, pain]. Slight or trifling pain.

**hypoemia** (*hi-po-e'-me-ah*) [*hypo-;* αἷμα, blood]. Insufficiency of blood.

**hypoendocrinism** (*hi-po-en-dok'-rin-izm*) [*hypo-;* ἐνδόν, within; κρίνειν, to separate], Deficiency of internal secretion.

**hypoeosinophilia** (*hi-po-e-o-sin-o-fil'-e-ah*). Decrease in the number of eosinophil leukocytes in the blood.

**hypoepinephry** (*hi-po-ep-e-nef'-re*) [*hypo-;* ἐπι, upon; νεφρός, kidney]. Insufficiency of the adrenal secretion.

**hypoerythrocythemia** (*hi-po-er-ith-ro-si-the'-me-ah*) [*hypo-;* ἐρυθρός, red; κύτος, cell; αἷμα, blood]. Deficiency in the normal number of red corpuscles in the blood.

**hypoesophoria** (*hi-po-es-o-fo'-re-ah*) [*hypo-;* *esophoria*]. A tendency of the visual axis of one eye to deviate downward and inward.

**hypoexophoria** (*hi-po-eks-o-fo'-re-ah*) [*hypo-;* *exophoria*]. A tendency of the visual axis of one eye to deviate downward and outward.

**hypofunction** (*hi-po-funk'-shun*). Insufficiency of function.

**hypogastralgia** (*hi-po-gas-tral'-je-ah*) [*hypo-;* γαστήρ, stomach; ἄλγος, pain]. Pain in the hypogastrium.

**hypogastrectasia, hypogastrectasis** (*hi-po-gas-trek-ta'-ze-ah, -trek'-ta-sis*) [*hypogastrium;* ἔκτασις, stretching]. Dilatation of the hypogastrium.

**hypogastric** (*hi-po-gas'-trik*) [*hypogastrium*]. Pertaining to the hypogastrium. h. **artery**, same as internal iliac artery; see under *artery*. h. **plexus**, a sympathetic nerve-plexus in the pelvis. h. **region**, the hypogastrium.

**hypogastrium** (*hi-po-gas'-tre-um*) [*hypo-;* γαστήρ, the belly]. The lower median anterior region of the abdomen.

**hypogastrocele** (*hi-po-gas'-tro-sēl*) [*hypo-;* γαστήρ, stomach; κήλη, hernia]. A hernia in the hypogastric region.

**hypogastrodidymus** (*hi-po-gas-tro-did'-im-us*) [*hypo-;* under; γαστήρ, stomach; δίδυμος, twin]. A double monstrosity in which the two fetuses are united at the hypogastrium.

**hypogastrohemia** (*hi-po-gas-tro-he'-me-ah*) [*hypo-;* γαστήρ, stomach; αἷμα, blood]. Hemorrhage in the hypogastrium.

**hypogastropagus** (*hi-po-gas-trop'-ag-us*) [*hypogastrium;* πάγος, anything solid]. A genus of twin monsters characterized by having the union in the region of the hypogastrium.

**hypogastrorrhagia** (*hi-po-gas-tror-a'-je-ah*). Same as *hypogastrohemia*.

**hypogastrorrhea** (*hi-po-gas-tror-e'-ah*) [*hypo-;* γαστήρ, stomach; ῥοία, a flow]. A slight amount of gastrorrhea.

**hypogastrorrhexis** (*hi-po-gas-tror-eks'-is*) [*hypo-;* γαστήρ, stomach; ῥῆξις, rupture]. Eventration.

**hypogenesis** (*hi-po-jen'-es-is*) [*hypo-;* γένεσις, production, generation]. In biology, direct development, without alternation of generations.

**hypogenous** (*hi-poj'-en-us*) [*hypo-;* γενής, produced]. In biology, growing below the surface or on the under side.

**hypogeusia** (*hi-po-gū'-se-ah*) [*hypo-;* γεῦσις, taste]. Diminution in the sense of taste.

**hypogigantosoma** (*hi-po-ji-gant-o-so'-ma*) [*hypo-;* γίγας, large; σῶμα, body]. A condition of great physical development not amounting to true gigantism.

**hypoglobulia** (*hi-po-glob-ū'-le-ah*). See *oligocythemia*.

**hypoglossal** (*hi-po-glos'-al*) [see *hypoglossus*]. Situated under the tongue. h. **nerve**. See under *nerve*. h. **nucleus**. See *nucleus*.

**hypoglossiadenitis** (*hi-po-glos-e-ad-en-i'-tis*) [*hypo-;* γλῶσσα, tongue; ἀδήν, gland; ἴτις, inflammation]. Inflammation of the sublingual gland.

**hypoglossis** (*hi-po-glos'-is*). See *hypoglottis*.

**hypoglossitis** (*hi-po-glos-i'-tis*) [*hypo-;* γλῶσσα, tongue; ἴτις, inflammation]. Inflammation of the tissue under the tongue.

**hypoglossus** (*hi-po-glos'-us*) [*hypo-;* γλῶσσα, tongue]. The hypoglossal nerve.

**hypoglottis** (*hi-po-glot'-is*) [see *hypoglossus*]. 1. The under part of the tongue. 2. A swelling at the under part of the tongue, as a ranula.

**hypoglucosic** (*hi-po-gloo-ko'-sik*). A term applied to any diabetic diet containing an amount of carbohydrates lower than the patient's tolerance.

**hypoglycemia** (*hi-po-gli-se'-me-ah*) [*hypo-;* γλυκύς, sweet; αἷμα, blood]. Deficiency of sugar in the blood.

**hypognathadenitis** (*hi-pog-nāth-ad-en-i'-tis*) [*hypo-;* γνάθος, jaw; ἀδήν, gland; ἴτις, inflammation]. Inflammation of the submaxillary gland.

**hypognathous** (*hi-pog'-na-thus*) [see *hypognathus*]. Having the lower mandible longer than the upper.

**hypognathus** (*hi-pog'-na-thus*) [*hypo-;* γνάθος, jaw]. A double monstrosity in which the parasite is attached to the inferior maxillary bone.

**hypohæmia** (*hi-po-he'-me-ah*). See *hypohemia*.

**hypohematosis** (*hi-po-hem-at-o'-sis*) [*hypo-;* αἷμα, blood]. A diseased condition marked by hyphemia, or deficiency in the amount of blood; also, the production, or process of inducing such a condition.

**hypohemia** (*hi-po-he'-me-ah*). An extravasation of blood in the eye.

**hypohemoglobinemia** (*hi-po-hem-o-glob-in-e'-me-ah*). See *oligochromemia*.

**hypohepatic** (*hi-po-hep-at'-ik*). Relating to hepatic insufficiency.

**hypohidrosis** (*hi-po-hid-ro'-sis*) [*hypo-;* ἴδρωσις, sweating]. Scanty perspiration.

**hypohydremia** (*hi-po-hi-dre'-me-ah*) [*hypo-;* ὕδωρ, water; αἷμα, blood]. Deficiency of water in the blood.

**hypohyloma** (*hi-po-hi-lo'-mah*). See under *hyloma*.

**hypohypnosis** (*hi-po-hip-no'-sis*) [*hypo-;* ὕπνος, sleep]. Imperfect or partial sleep.

**hypoinosemia** (*hi-po-in-o-se'-me-ah*) [*hypo-;* ἴς, fiber; αἷμα, blood]. Decrease in the tendency to the formation of fibrin in the blood.

**hypoisotonic** (*hi-po-is-o-ton'-ik*) [*hypo-;* *isotonic*]. Applied to a solution the osmotic pressure of which is lower than blood-plasma or some other solution taken as a standard.

**hypokinesia, hypokinesis** (*hi-po-kin-e'-ze-ah, hi-po-kin-e'-sis*) [*hypo-;* κίνησις, motion]. Deficiency in motor reaction under stimulation.

**hypolepidoma** (*hi-po-lep-id-o'-mah*). See under *lepidoma*.

**hypoleukocythemia** (*hi-po-lū-ko-si-the'-me-ah*) [*hypo-;* λευκός, white; κύτος, a cell; αἷμα, blood]. Deficiency of white corpuscles in the blood.

**hypoleukocytosis** (*hi-po-lū-ko-si-to'-sis*) [*hypo-;* *leukocyte*]. A diminution of the number of leukocytes in the blood.

**hypolipemia** (*hi-po-lip-e'-me-ah*) [*hypo-;* λίπος, fat; αἷμα, blood]. Deficiency of fat in the blood.

**hypolipsis** (*hi-po-lip'-sis*). A deficiency of fat-splitting ferment (lipase) in the blood-serum. Cf. *hyperlipsis*.

**hypologia** (*hi-po-lo'-je-ah*) [*hypo-;* λόγος, work]. Poverty of speech as a symptom of cerebral disease.

**hypolympha** (*hi-po-lim'-fah*) [*hypo-;* *lympha*, water]. An extravasation of chyliferous lymph into the anterior chamber of the eye.

**hypolymphia** (*hi-po-lim'-fe-ah*). Insufficiency of lymph.

**hypomania** (*hi-po-ma'-ne-ah*) [*hypo-;* μανία, madness]. A moderate degree of maniacal exaltation.

**hypomastia, hypomazia** (*hi-po-mas'-te-ah, hi-po-ma'-ze-ah*) [*hypo-;* μαστός, the breast]. Abnormal smallness of the mammary gland.

**hypomegasoma** (*hi-po-meg-as-o'-mah*) [*hypo-;* μέγας, great; σῶμα, body]. A tall stature, but quite below gigantism. Cf. *megasoma*.

**hypomelancholia** (*hi-po-mel-an-ko'-le-ah*) [*hypo-;* *melancholia*]. Moderate melancholia; melancholia without delusions.

**hypomenous** (*hi-pom'-en-us*) [*hypo-;* μένειν, to remain]. Same as *hypogenous*.

**hypomesosoma** (*hi-po-mes-o-so'-ma*) [*hypo-;* μέσος, middle; σῶμα, body]. A stature slightly below the medium.

**hypometropia** (*hi-po-me-tro'-pe-ah*) [*hypo-*; μέτρον, a measure; ὤψ, vision]. Myopia.

**hypomicrone** (*hi-po-mik'-rōn*) [*hypo-*; *microne*]. A particle capable of being recognized by the ultramicroscope, but not by the ordinary microscope.

**hypomicrosoma** (*hi-po-mik-ro-so'-mah*) [*hypo-*; μικρός, small; σῶμα, body]. The lowest stature which is not dwarfism.

**hypomoria** (*hi-po-mo'-re-ah*) [*hypo-*; μωρία, folly]. Slight mental disorder.

**hypomyosthenia** (*hi-po-mi-os-the'-ne-ah*) [*hypo-*; μῦς, muscle; σθένος, strength]. Deficiency in muscular power.

**hypomyotonia** (*hi-po-mi-o-to'-ne-ah*) [[*hypo-*; μῦς, muscle; τόνος, tone]. Deficiency in muscular tonicity.

**hyponanosoma** (*hi-po-nan-o-so'-mah*) [*hypo-*; νᾶνος, dwarf; σῶμα, body]. Extreme dwarfishness.

**hyponeuria** (*hi-po-nū'-re-ah*) [*hypo-*; νεῦρον, nerve]. Slight or diminished nerve-power.

**hyponoetic** (*hi-po-no-et'-ik*) [*hypo-*; νοητικός, understanding]. Under the control of the will.

**hyponomous** (*hi-pon-o'-mus*) [*hypo-*; νέμειν, to feed]. Spreading or eating below the surface; applied to certain ulcers, etc.

**hyponychium** (*hi-pon-ik'-e-um*). See *nail-bed*.

**hypoparathyreosis** (*hi-po-par-ah-thi-re-o'-sis*). A pathological state brought about by partial loss or insufficiency of parathyroid tissue.

**hypopepsia** (*hi-po-pep'-se-ah*) [*hypo-*; πέψις, digestion]. Subnormal digestive power.

**hypopepsinia** (*hi-po-pep-sin'-e-ah*). Diminution in the amount of pepsin secreted by the stomach.

**hypophoria** (*hi-po-fo'-re-ah*) [*hypo-*; φορός, tending]. A tendency of the visual axis of one eye to deviate below that of the other

**hypophosphite** (*hi-po-fos'-fīt*) [*hypo-*; *phosphorus*]. A salt of hypophosphorous acid. Those of calcium, iron, manganese, potassium, and sodium are official. **h.s; emulsion of cod-liver oil with** (*emulsum olei morrhuæ cum hypophosphitibus*, U. S. P.), an emulsion made of cod-liver oil, acacia, calcium, potassium, and sodium hypophosphites, syrup, oil of gaultheria, and water. Dose 2 dr. (8 Cc.). **h.s, syrup of** (*syrupus hypophosphitum*, U. S. P.), contains calcium, potassium, and sodium hypophosphites. Dose 2 dr. (8 Cc.). **h.s, syrup of, compound** (*syrupus hypophosphitum compositus*, U. S. P.), contains five hypophosphites, hypophosphorous acid, quinine, and strychnine. Dose 2 dr. (8 Cc.). Both syrups are used in wasting diseases, in scrofula, rickets, etc.

**hypophosphorous acid** (*hi-po-fos-for'-us*). See *acid, hypophosphorous*. **h. acid, dilute** (*acidum hypophosphorosum dilutum*, U. S. P.). Dose 10–30 min. (0.65–2.0 Cc.).

**hypophrasia** (*hi-po-fra'-ze-ah*) [*hypo-*; φράσις, phrase, utterance]. Meagerness or poverty of speech, as a sign of cerebral disease.

**hypophrenic** (*hi-po-fren'-ik*). See *subdiaphragmatic*.

**hypophysectomy** (*hi-pof-is-ek'-to-me*) [*hypophysis*; ἐκτομή, excision]. Surgical removal of the hypophysis cerebri or pituitary body.

**hypophysemia** (*hi-po-fi-se'-me-ah*) [*hypo-*; φῦσα, air; αἷμα, blood]. Deficiency of gaseous elements in the blood.

**hypophysin** (*hi-pof'-is-in*). An organotherapeutic remedy from the hypophysis of the ox. It is used as an adjuvant to iodothyrin in the treatment of akromegaly. Dose 2–5 gr. (0.1–0.3 Gm.) several times daily.

**hypophysis** (*hi-pof'-is-is*) [*hypo-*; φύειν, to grow]. An outgrowth. **h.** cerebri, the pituitary body.

**hypopituitarism** (*hi-po-pit-ū'-it-ar-izm*). A condition due to decreased activity of the pituitary body, and marked by increase of fat, atrophy of the genitals, and loss of sexual power.

**hypoplasia** (*hi-po-pla'-ze-ah*) [*hypo-*; πλάσσειν, to mold]. Defective development of any organ or tissue.

**hypoplasty** (*hi'-po-plas-te*) [*hypo-*; πλάσσειν, to form]. Diminished plastic power.

**hypopraxia** (*hi-po-praks'-e-ah*) [*hypo-*; πρᾶξις, doing]. Inactivity; listlessness; inefficiency as a sign or result of cerebral disorder.

**hyposelaphesia** (*hi-pop-sel-af-e'-ze-ah*) [*hypo-*; ψηλάφησις, touch]. Diminution of sensitiveness to tactile impressions.

**hypoptyalism** (*hi-pop-ti'-al-izm*) [*hypo-*; *ptyalism*]. A mild or slight ptyalism.

**hypopyon** (*hi-po'-pe-on*) [*hypo-*; πύον, pus]. A collection of pus in the anterior chamber of the eye.

**hypoquebrachine** (*hi-po-kweb'-rak-ēn*) [*hypo-*; *quebracho*], C₂₁H₂₆N₂O₃. An alkaloid of quebracho, occurring in yellow masses, melting at about 80° C., and soluble in alcohol and ether.

**hypoquinidol** (*hi-po-kwin'-id-ol*). A proprietary preparation of quinine and phosphorus.

**hyporrhea** (*hi-por-e'-ah*) [*hypo-*; ῥεῖν, to flow]. A slight hemorrhage.

**hyposarca** (*hi-po-sark'-ah*). See *anasarca*.

**hyposcheotomy** (*hi-pos-ke-ot'-o-me*) [*hypo-*; ὄσχεον, scrotum; τέμνειν, to cut]. The surgical puncturing of a hydrocele at the lower part of the tunica vaginalis.

**hyposialadenitis** (*hi-po-si-al-ad-en-i'-tis*) [*hypo-*; σίαλον, saliva; ἀδήν, a gland]. Inflammation of the submaxillary salivary gland.

**hyposmia** (*hi-pos'-me-ah*) [*hypo-*; ὀσμή, smell]. Diminution of the sense of smell.

**hypospadia, hypospadias** (*hi-po-spa'-de-ah, hi-po-spa'-de-as*) [*hypo-*; σπᾶν, to draw]. A condition in which the urethra opens upon the under surface of the penis.

**hypospkyxia** (*hi-po-sfik'-se-ah*) [*hypo-*; σφύξις, pulse]. Diminished blood pressure with venous stasis and general circulatory sluggishness.

**hypostaphylitis** (*hi-po-staf-il-i'-tis*) [*hypo-*; σταφυλίς, uvula; ῖτις, inflammation]. Slight inflammation of the uvula.

**hypostasis** (*hi-pos'-tas-is*) [*hypo-*; *stasis*]. 1. Feces. 2. A settling, also, the sediment. 3. Deposit. 4. The settling of blood in the dependent parts of the body.

**hypostatic** (*hi-po-stat'-ik*) [see *hypostasis*]. Due to, or of the nature of, hypostasis. **h. congestion.** See *hypostasis* (2). **h. pneumonia.** See *pneumonia, hypostatic*.

**hyposthenia** (*hi-po-sthe'-ne-ah*) [*hypo-*; σθένος, strength]. Weakness; subnormal strength.

**hyposthéniant** (*hi-po-sthe'-ne-ant*) [*hyposthenia*]. Reducing the strength; lowering the vital forces.

**hyposthenic** (*hi-po-sthen'-ik*) [*hyposthenia*]. 1. Tendency to, or characterized by, hyposthenia; applied to diseases that are more than ordinarily enfeebling. 2. A medicine that reduces the action of the heart without affecting its rhythm.

**hyposthenuria** (*hi-po-sthen-ū'-re-ah*) [*hyposthenia*; οὖρον, urine]. 1. Suppression of the urine from inability of the kidney to eliminate. 2. Diminution of solids in the urine.

**hypostyptic** (*hi-po-stip'-tik*) [*hypo-*; στυπτικός, astringent]. 1. Moderately or mildly styptic. 2. A mildly styptic medicine.

**hyposynergia** (*hi-po-sin-er'-je-ah*) [*hypo-*; συνεργία, cooperation]. Defective coordination.

**hyposystole** (*hi-po-sis'-to-le*) [*hypo-*; *systole*]. Deficiency of the cardiac systole.

**hypotaxia** (*hi-po-taks'-e-ah*) [*hypo-*; τάξις, arrangement]. A condition of weakened or imperfect coordination.

**hypotension** (*hi-po-ten'-shun*) [*hypo-*; *tensio*, a stretching]. Diminished or abnormally low tension; hypotonia.

**hypothalamus** (*hi-po-thal'-am-us*) [*hypo-*; θάλαμος, thalamus]. A group of prominences and aggregations of ganglia lying on the ventral side beneath the thalamus. Syn., *subthalamus*.

**hypothenar** (*hi-poth'-en-ar*) [*hypo-*; θέναρ, palm]. The fleshy eminence on the palm of the hand over the metacarpal bone of the little finger. Also, the prominences on the palm at the base of the fingers.

**hypothermal** (*hi-po-ther'-mal*) [see *hypothermia*]. Slightly hot; tepid.

**hypothermia, hypothermy** (*hi-po-ther'-me-ah, hi-po-ther'-me*) [*hypo-*; θέρμη, heat]. Subnormal temperature; deficiency in the heat of the body.

**hypothesis** (*hi-poth'-es-is*) [*hypo-*; θέσις, a position]. A supposition set forth for discussion or demonstration. A theory assumed as true.

**hypothyreosis, hypothyroidation, hypothyroidea, hypothyroidism.** Deficient functional activity of the thyroid gland.

**hypothyroidism** (*hi-po-thi'-roid-izm*). A morbid condition attributed to deficient activity of the thyroid gland.

**hypothyrosis** (*hi-po-thi-ro'-sis*). Reduced functional activity of the thyroid gland.

**hypotonia, hypotonus** (*hi-po-to'-ne-ah, hi-pot'-o-nus*) [*hypo-*; τόνος, tension]. Decrease of normal

# HYPOTONIC 450 HYSTERODYNIA

**tonicity** or tension; especially diminution of intraocular pressure.
**hypotonic** (*hi-po-ton'-ik*). 1. Below the normal strength or tension. 2. Less than isotonic.
**hypotoxicity** (*hi-po-toks-is'-it-e*) [*hypo-*; τοξικόν, poison]. A reduced toxicity.
**hypotrichosis** (*hip-o-trik-o'-sis*) [*hypo-*; θρίξ, hair]. A rare congenital anomaly, of entire absence of hair or growth delayed beyond the normal time.
**hypotrophy** (*hi-pot'-ro-fe*) [*hypo-*; τρέφειν, to nourish]. Defective nutrition.
**hypotympanic** (*hi-po-tim-pan'-ik*). Located beneath the tympanum.
**hypouremia** (*hi-po-u-re'-me-ah*) [*hypo-*; οὖρον, urine; αἷμα, blood]. Diminution of the urea normally present in the blood.
**hypourocrinia** (*hi-po-u-ro-krin'-e-ah*) [*hypo-*; οὖρον, urine; κρίνειν, to separate]. A deficient or too scanty secretion of urine.
**hypovenosity** (*hi-po-ven-os'-it-e*) [*hypo-*; venosus, venous]. A condition in which there is incomplete development of the venous system in a given area, resulting in atrophy and degeneration in the muscles.
**hypoxanthine** (*hi-po-zan'-thin*) [*hypo-*; xanthin], $C_5H_4N_4O$. A nonpoisonous leukomaine. It occurs, accompanying adenin and guanin, in nearly all the animal tissues and organs rich in nucleated cells. In minute quantities it is a normal constituent of urine. It has also been found in plants, seeds, ferments, and wines. It is a crystalline body, soluble in cold and boiling water, insoluble in cold alcohol or ether. Hypoxanthine appears to be one of the products formed by the decomposition and successive oxidation of proteid matter previous to the formation of uric acid and urea. It is produced from adenin by the action of nitrous acid. Syn., *sarcine*. **h., test for.** See *Kossel's test for hypoxanthine*.
**hypoxemia** (*hi-poks-e'-me-ah*) [*hypo-*; oxygen; αἷμα, blood]. Insufficient oxygenation of the blood.
**hypsicephalic** (*hip-sis-ef-al'-ik*) [ὕψι, high; κεφαλή, head]. Having a skull with a cranial index over 75.1°.
**hypsicephaly** (*hip-sis-ef'-al-e*) [ὕψι, on high; κεφαλή, head]. The condition of a skull with a cranial index of over 75.1°.
**hypsiloid** (*hip'-sil-oid*) [ὕψιλον, the Greek letter υ, u; εἶδος, resemblance]. Hyoid.
**hypsistenocephalic** (*hip-sist-en-o-sef-al'-ik*). See *hypsicephalic*.
**hypsocephalic, hypsocephalous** (*hip-so-sef-al'-ik, -us*)). See *hypsicephalic*.
**hypsocephaly** (*hip-so-sef'-al-e*). See *hypsicephaly*.
**hypsonosus** (*hip-son'-o-sus*) [ὕψος, height; νόσος, illness]. Mountain-sickness; balloon-sickness; characterized by nausea, headache, epistaxis, etc.
**hypsophobia** (*hip-so-fo'-be-ah*) [ὕψος, height; φόβος, fear]. Morbid dread of being at a great height; aerophobia.
**hypsopisthius** (*hip-so-pis'-the-us*) [ὕψι, on high; ὀπίσθιον, occiput]. Lissauer's term for a skull in which the angle included between the radius fixus and the line joining the hormion and lambda is between 33° and 41°.
**hypurgia, hypurgesis** (*hi-pur'-je-ah, hi-pur-je'-sis*) [ὑπουργία, ὑπουργήσις]. Medical attendance.
**hypurgic, hypurgous** (*hi-pur'-jik, hi-pur'-jus*). Helping, aiding, administering.
**hyrgol, hyrgolum** (*hur'-gol, hur'-gol-um*). Colloidal mercury, an allotropic form of solid mercury.
**Hyrtl's anastomosis or loop** [Joseph *Hyrtl*, Austrian atanomist, 1811–1894]. The transection of two arteries. Syn., *dehiscentia decussantium*. H.'s **sphincter.** See *Nélaton's sphincter*.
**hyssop** (*his'-op*) [ὕσσωπος, an aromatic plant]. The leaves and tops of *Hyssopus officinalis*, an aromatic stimulant, carminative, and tonic, employed in chronic catarrh of the respiratory tract. Dose of the *fluidextract* 1–2 dr. (4–8 Cc.).
**hyster-** (*his-ter-*) [ὑστέρα, uterus]. Prefix signifying relation to the uterus or to hysteria.
**hystera** (*his'-ter-ah*) [ὑστέρα, womb]. The uterus or womb.
**hysteralgia** (*his-ter-al'-je-ah*) [*hyster-*; ἄλγος, pain]. Neuralgic pain in the uterus.
**hysteranesis** (*his-ter-an'-es-is*) [*hyster-*; ἄνεσις, relaxation]. Relaxation and atony of the uterus.
**hysteratresia** (*his-ter-at-re'-ze-ah*) [*hyster-*; ἄτρητος, imperforate]. An imperforate, or impervious condition of the mouth of the womb.
**hysterauxesis** (*his-ter-awks-e'-sis*) [*hyster-*; αὔξησις, pain]. Pain in the womb.

enlargement]. Enlargement of the uterus, normal (as in pregnancy) or abnormal.
**hysterectomy** (*his-ter-ek'-to-me*) [*hyster-*; ἐκτομή, a cutting out]. Excision of the uterus through the abdomen (*abdominal hysterectomy*) or the vagina (*vaginal hysterectomy*).
**hysteredema, hysterœdema** (*his-ter-e-de'-mah*) [*hyster-*; οἴδημα, edema]. Edema of the womb-substance.
**hysterelcosis** (*his-ter-el-ko'-sis*) [*hyster-*; ἕλκωσις, ulceration]. Ulceration of the uterus.
**hysteremphysema** (*his-ter-em-fiz-e'-mah*). See *physometra*.
**hysterergia** (*his-ter-ur'-je-ah*) [ὕστερος, later; ἔργον, work]. The after-results of a remedy or method of treatment.
**hysteria** (*his-te'-re-ah*) [ὑστέρα, womb, from the ancient belief that the condition depended upon uterine disease]. A diseased state of the mind manifesting itself in countless disturbances of the psychic, sensory, motor, and vasomotor functions. The etiology is not definitely known; heredity and mental shock play an important part. All ages and both sexes are subject to the disease, but it is most common in young women. The psychic disturbances consist in increased irritability, tendency to exaggeration, a heightened imagination, hallucinations, and somnambulistic and hypnotic states. Among sensory symptoms are various neuralgias, as clavus, hemicrania, and coccygodynia; anesthesias; hyperesthesias; diminution of the visual field; diplopia; deafness; loss of the sense of taste, etc. The motor symptoms comprise paralyses and contractures of the limbs, tremor, convulsions, catalepsy, aphonia, etc. The chief vasomotor phenomena are cyanosis, cutaneous hemorrhages, and edema. In addition to these symptoms many others are at times noted, as anorexia, vomiting, salivation, polyuria, anuria, etc. **h. major,** hysteroepilepsy. **h. minor,** a mild form of the disease.
**hysteric, hysterical** (*his-ter'-ik, his-ter'-ik-al*). Pertaining to hysteria. **h. ataxia,** a hysterical state marked by loss of sensation in the skin and in the leg muscles. **h. chorea,** a form of hysteria with choreiform movements.
**hystericism** (*his-ter'-is-izm*) [*hyster-*]. The hysterical diathesis or temperament; proneness to the exhibition of hysterical symptoms.
**hystericoneuralgic** (*his-ter-ik-o-nu-ral'-jik*). Like neuralgia but of hysterical origin.
**hysterics** (*his-ter'-iks*). A popular term for the hysterical attack.
**hysteriencephalitis** (*his-ter-e-en-sef-al-i'-tis*) [*hyster-*; ἐγκέφαλον, brain; ιτις, inflammation]. Encephalitis and meningitis following repeated attacks of hysteria.
**hysteritis** (*his-ter-i'-tis*). See *metritis*.
**hystero-** (*his'-ter-o-*) [ὑστέρα, uterus]. A prefix signifying relation to the uterus, or to hysteria.
**hysterobubonocele** (*his-ter-o-bu-bon'-o-sēl*) [*hystero-*; βουβωνοκήλη, an inguinal hernia]. An inguinal hysterocele.
**hysterocatalepsy** (*his-ter-o-kat'-al-ep-se*) [*hystero-*; catalepsy]. A form of hysteria accompanied by catalepsy.
**hysterocele** (*his'-ter-o-sēl*) [*hystero-*; κήλη, hernia]. A hernia containing all or part of the uterus.
**hysterocleisis** (*his-ter-o-kli'-sis*) [*hyster-*; κλεῖσις, closure]. The closure of the uterus by suturing the edges of the os.
**hysterocyesis** (*his-ter-o-si-e'-sis*) [*hystero-*; κύησις, pregnancy]. Uterine pregnancy.
**hysterocystic** (*his-ter-o-sist'-ik*) [*hystero-*; κύστις, the bladder]. Relating to the uterus and bladder.
**hysterocystocele** (*his-ter-o-sist'-o-sēl*) [*hystero-*; κύστις, bladder; κήλη, hernia]. Hysterocele complicated by cystocele; hernia of the womb and the bladder, or parts of them.
**hysterocystocleisis** (*his-ter-o-sist-o-kli'-sis*) [*hystero-*; κύστις, bladder; κλεῖσις, closure]. Bozeman's operation for relief of vesicouterovaginal fistula, consisting in turning the cervix uteri into the bladder and suturing it.
**hysterocystopexy** (*his-ter-o-sist'-o-peks-e*). See *ventrovesicofixation*.
**hysterodynamometer** (*his-ter-o-din-am-om'-et-er*) [*hystero-*; δύναμις, power; μέτρον, a measure]. An apparatus to record the number, intensity, and variations of uterine contractions.
**hysterodynia** (*his-ter-o-din'-e-ah*) [*hystero-*; ὀδύνη, pain]. Pain in the womb.

**hysteroepilepsy** (*his-ter-o-ep'-e-lep-se*) [*hystero-*; *epilepsy*]. A form of hysteria accompanied by convulsions resembling those of epilepsy.
**hysteroepileptogenous** (*his'-ter-o-ep-il-ep-toj'-en-us*) [*hystero-*; *epilepsy*; γεννᾶν, to produce]. Producing hysterical epilepsy.
**hysterofrenic** (*his-ter-o-fren'-ik*) [*hysteria*; *frænum*, a curb]. Capable of checking an attack of hysteria; opposed to hysterogenic.
**hysterogastrorrhaphy** (*his-ter-o-gas-tror'-af-e*). See *hysteropexy*.
**hysterogenic, hysterogenous** (*his-ter-o-jen'-ik, his-ter-oj'-en-us*) [*hystero-*; γεννᾶν, to beget]. Causing or producing a hysterical attack, as *hysterogenic* zones, certain regions pressure upon which excites a hysterical paroxysm.
**hysterogeny** (*his-ter-oj'-en-e*) [*hystero-*; γεννᾶν, to produce]. The induction of the hysterical state or paroxysm.
**hysteroid** (*his'-ter-oid*) [*hystero-*; εἶδος, like]. 1. Resembling hysteria. 2. Pertaining to hystero-epilepsy.
**hysterokataphraxis** (*his-ter-o-kat-ah-fraks'-is*) [*hystero-*; καταφράκτης, a coat of mail]. An operation for including the uterus within supporting metal structures as a medium of replacement.
**hysterolaparotomy** (*his-ter-o-lap-ar-ot'-o-me*) [*hystero-*; λαπάρα, the abdominal wall; τέμνειν, to cut]. Abdominal hysterectomy.
**hysterolith** (*his'-ter-o-lith*) [*hystero-*; λίθος, stone]. Calculus or stone in the womb.
**hysterolithiasis** (*his-ter-o-lith-i'-as-is*) [*hystero-*; λιθίασις, the formation of calculi]. The formation of hysteroliths.
**hysterology** (*his-ter-ol'-o-je*) [*hystero-*; λόγος, science]. The anatomy, physiology, and pathology of the uterus.
**hysteroloxia** (*his-ter-o-loks'-e-ah*) [*hystero-*; λοξίος, oblique]. Oblique displacement or position of the uterus.
**hysterolysis** (*his-ter-ol'-is-is*) [*hystero-*; λύειν, to loose]. Severing the attachments of the uterus. h., vaginal, the operation of detachment of the uterus first from its posterior adhesions and then from its anterior adhesions by posterior and anterior colpotomy and concluding with vaginofixation.
**hysteroma** (*his-ter-o'-mah*). A fibroid tumor of the uterus.
**hysteromalacia** (*his-ter-o-mal-a'-se-ah*) [*hystero-*; μαλακία, softness]. Softening of the tissues of the womb.
**hysteromalacoma** (*his-ter-o-mal-ak-o'-mah*) [*hystero-*; μαλακός, soft]. Softening of the womb or of any part of it.
**hysteromania** (*his-ter-o-ma'-ne-ah*) [*hystero-*; μανία, madness]. Hysterical insanity; also, nymphomania.
**hysterometer** (*his-ter-om'-et-er*) [*hystero-*; μέτρον, measure]. An instrument for measuring the length of the intrauterine cavity.
**hysterometry** (*his-ter-om'-et-re*). [*hystero-*; μέτρον, measure]. The measurement of the size of the uterus.
**hysteromyoma** (*his-ter-o-mi-o'-mah*) [*hystero-*; *myoma*: *pl.*, *hysteromyomata*]. Myoma or fibro-myoma of the uterus.
**hysteromyomectomy** (*his-ter-o-mi-o-mek'-to-me*) [*hystero-*; *myoma*; ἐκτομή, excision]. Removal of a fibroid uterus.
**hysteromyotomy** (*his-ter-o-mi-ot'-o-me*) [*hystero-*; μῦς, muscle; τέμνειν, to cut]. Incision into the uterus for removal or enucleation of a solid tumor.
**hysteroncus** (*his-ter-ong'-kus*) [*hystero-*; ὄγκος, a tumor]. A tumor or swelling of the uterus.
**hysteroneurasthenia** (*his-ter-o-nū-ras-the'-ne-ah*) [*hystero-*; *neurasthenia*]. 1. Neurasthenia resulting from womb-disease. 2. The stage where neurasthenia ceases and hysteria begins.
**hysteroneurosis** (*his-ter-o-nū-ro'-sis*) [*hystero-*; *neurosis*]. A reflex neurosis resulting from irritation of the uterus.
**hystero-oophorectomy** (*his'-ter-o-o-off-or-ek'-to-me*) [*hystero-*; ὠοφόρος, egg-bearing; ἐκτομή, a cutting out]. The surgical removal of the uterus and ovaries together.
**hystero-ovariotomy** (*his-ter-o-o-va-re-ot'-o-me*). See *hystero-oophorectomy*.
**hysteroparalysis** (*his-ter-o-par-al'-is-is*) [*hystero-*; *paralysis*]. Paralysis or weakness of the walls of the womb.
**hysteropathic** (*his-ter-o-path'-ik*) [*hystero-*; πάθος, disease]. Of the nature of, or pertaining to, hysteropathy.

**hysteropathy** (*his-ter-op'-ath-e*) [*hystero-*; πάθος, disease]. Any disease or disorder of the uterus.
**hysteropexy** (*his'-ter-o-peks-e*) [*hystero-*; πῆξις, a fastening]. Fixation of the uterus by a surgical operation to correct displacement.
**hysterophore** (*his'-ter-o-fōr*) [*hystero-*; φορός, bearing]. A form of uterine pessary.
**hysterophrenic** (*his-ter-o-fren'-ik*) [*hystero-*; φρήν, the mind]. Opposed to hysterogenic.
**hysteroplegia** (*his-ter-o-ple'-je-ah*) [*hystero-*; πληγή, a stroke]. Same as *hysteroparalysis*.
**hysteropnix** (*his-ter-op'-niks*) [*hystero-*; πνίξ, suffocation]. Globus hystericus.
**hysterosophy** (*his-ter-op'-so-fe*) [*hystero-*; ψόφος, sound, utterance]. The escape of air from the uterus with an audible sound.
**hysteropsychopathy** (*his-ter-o-si-kop'-ath-e*) [*hystero-*; ψυχή, the mind; πάθος, disease]. Mental disorder secondary to disease of the uterus.
**hysteropsychosis** (*his-ter-o-si-ko'-sis*) [*hystero-*; ψυχή, the mind]. Mental disorder associated with uterine disease.
**hysteroptosis** (*his-ter-op-to'-sis*) [*hystero-*; πτῶσις, a falling]. Falling or inversion of the uterus.
**hysterorrhagia** (*his-ter-or-aj'-e-ah*). See *metrorrhagia*.
**hysterorrhaphy** (*his-ter-or'-a-fe*) [*hystero-*; ῥαφή, suture]. 1. The closure of a uterine incision or rent by suture. 2. Hysteropexy.
**hysterorrhea** (*his-ter-o-re'-ah*) [*hystero-*; ῥοία, flow]. A discharge from the uterus.
**hysterorrhexis** (*his-ter-or-eks'-is*) [*hystero-*; ῥῆξις, rupture]. Rupture of the womb.
**hysterosalpingo-oophorectomy** (*his-ter-o-sal-pin-go-o-of-or-ek'-to-me*). Excision of the uterus, oviducts, and ovaries.
**hysterosalpinx** (*his-ter-o-sal'-pingks*) [*hystero-*; σάλπιγξ, a pipe]. A Fallopian tube or oviduct.
**hysteroscope** (*his'-ter-o-skōp*) [*hystero-*; σκοπεῖν, to view]. A uterine speculum, with a reflector.
**hysteroscopy** (*his-ter-os'-ko-pe*) [*hystero-*; σκοπεῖν, to examine]. Inspection of the uterus.
**hysterospasm** (*his'-ter-o-spazm*) [*hystero-*; *spasm*]. Uterine spasm.
**hysterostomatome** (*his-ter-o-sto'-mat-ōm*) [*hystero-*; στόμα, mouth; τομή, section]. A knife for use in hysterostomatomy.
**hysterostomatomy, or hysterostomatotomy** (*his-ter-o-sto-mat'-o-me, or his-ter-o-sto-mat-ot'-o-me*) [*hystero-*; στόμα, mouth; τομή, section]. Surgical incision of the os uteri, or its enlargement by a cutting operation.
**hysterosyphilis** (*his-ter-o-sif'-il-is*). Hysterical manifestation due to syphilis.
**hysterotabetism** (*his-ter-o-ta'-bet-izm*). Combined tabes and hysteria.
**hysterotokotomy** (*his-ter-o-to-kot'-o-me*) [*hystero-*; τόκος, birth; τομή, section]. Cesarean operation, or delivery through an incision into the womb.
**hysterotome** (*his'-ter-o-tōm*) [*hystero-*; τομή, a cutting]. A hysterotomy-knife or cutting-instrument for use in hysterotomy.
**hysterotomotocia** (*his-ter-o-mo-to'-se-ah*) [*hystero-*; τομή, cutting; τόκος, birth]. An incision into the womb for the removal of a fetus; hysterotokotomy.
**hysterotomy** (*his-ter-ot'-o-me*) [*hystero-*; τομή, a cutting]. Incision of the uterus.
**hysterotrachelorrhaphy** (*his-ter-o-trak-el-or'-a-fe*) [*hystero-*; τράχηλος, neck; ῥαφή, suture]. A plastic operation for the restoration of a lacerated cervix uteri.
**hysterotrachelotomy** (*his-ter-o-trak-el-ot'-o-me*) [*hystero-*; τράχηλος, neck; τομή, section]. Surgical incision of the neck of the womb.
**hysterotraumatism** (*his-ter-o-traw'-mat-izm*) [*hystero-*; τραῦμα, wound]. Hysterical symptoms due to or following traumatism.
**hysterotrismus** (*his-ter-o-triz'-mus*) [*hystero-*; τρισμός, a creaking]. Spasm of the uterus.
**hystrichiasis** (*his-trik-i'-as-is*) [ὕστριξ, a hedgehog]. A disease of the hair in which it stands out stiffly like the spines of the hedgehog.
**hystriciasis, hystricism** (*his-tris-i'-as-is, his'-tris-izm*) [ὕστριξ]. A disease of the hairs in which they stand erect. 2. Ichthyosis hystrix.
**hystrix** (*his'-triks*) [ὕστριξ, porcupine]. Same as *ichthyosis hystrix*.
**hyther** (*hi'-thur*) [ὕδωρ, water; θερμή, heat]. The combined effect of moisture and temperature of the atmosphere upon human beings.

# I

**I.** The chemical symbol for *iodine*.
**i.** Abbreviation for *optically inactive*.
**iamatology** (*i-am-at-ol'-o-je*) [ἴαμα, remedy; λόγος, science]. The science of remedies, or of therapeutics; aceology.
**iasis** (*i-a'-sis*) [ἴασις, treatment, cure]. Medical or surgical treatment.
**-iasis.** A termination denoting a process or its result (as *lithiasis* from λίθος a stone). And see *-osis*.
**iatraliptic** (*i-at-rah-lip'-tik*) [ἰατρός, physician; ἀλείπτης, an anointer]. Curing by using ointments and frictions.
**iatreusiology** (*i-at-roo-se-ol'-o-je*) [ἰάτρευσις, practice of medicine; λόγος, science]. The science of medical or surgical treatment; therapeutics.
**iatreusis** (*i-at-roo'-sis*) [ἰάτρευσις, treatment]. Medical or surgical treatment.
**iatric** (*i-at'-rik*) [ἰατρικός medical]. Pertaining to the physician or to the science of medicine.
**iatro-** (*i-at'-ro-*) [ἰατρός, physician]. A Greek prefix signifying relation to medicine or to physicians.
**iatrochemical, iatrochemic** (*i-at-ro-kem'-ik-al, i-at-ro-kem'-ik*) [*iatro-*; *chemistry*]. Pertaining to the obsolete chemical school of therapeutics; spagiric.
**iatrochemist** (*i-at-ro-kem'-ist*) [*iatro-*; *chemist*]. A follower of iatrochemical or spagiric doctrines.
**iatrochemistry** (*i-at-ro-kem'-ist-re*) [ἰατρός, physician; χημεία, chemistry]. 1. The application of chemistry to therapeutics; the treatment of disease by chemical means. 2. The theory that disease and its treatment are explicable on a chemical basis.
**iatrol** (*i'-at-rol*). Oxy-iodo-methyl-anilide; an odorless and non-toxic antiseptic agent, said to be three times as effective, weight for weight, as iodoform.
**iatroleptica** (*i-at-ro-lep'-tik-ah*) [*iatro-*; ἀλείφειν, to oil the skin]. The treatment of disease by anointing, friction, and exercise.
**iatroliptic** (*i-at-ro-lip'-tik*). See *iatraliptic*.
**iatrology** (*i-at-rol'-o-je*) [*iatro-*; λόγος, science]. The science of medicine; an account of, or treatise on, physicians.
**iatrophysics** (*i-at-ro-fiz'-iks*) [ἰατρός, physician; φυσικός, pertaining to nature]. 1. The treatment of disease by physical measures. 2. The theory that disease and its treatment are explicable on a materialistic or physical basis. The materialistic explanation of disease; applied especially to an obsolete theory of the seventeenth century that sought to explain physiological and therapeutic facts by means of the principles of physics (dynamics and statics).
**iatrosophist** (*i-at-ro-sof'-ist*) [*iatro-*; σοφός, skilful]. A physician skilled in the theory of medicine.
**iatrotechnics** (*i-at-ro-tek'-niks*) [ἰατρός, physician; τέχνη, art]. The art of healing.
**ibit** (*ib'-it*). See *bismuth oxyiodotannate*.
**-ic.** A suffix denoting the higher of two valencies assumed by an element, and incidentally in many cases a larger amount of oxygen.
**icaja, icaya** (*ik'-aj-ah*). An ordeal poison with action similar to nux vomica obtained from the stem of *Strychnos icaja*.
**icajine** (*ik'-aj-in*) [African, *icaja*]. A poisonous alkaloid derived from an African ordeal-drug called *icaja*. It somewhat resembles brucine and is probably derived from some species of *Strophanthus*.
**ice** (*is*) [AS., *is*]. Water in its solid state, which it assumes at a temperature of 0° C., or 32° F. It is used externally in the form of applications, and internally as a refrigerant and to combat nausea. **i.-bag, i.-cap, i.-compress,** measures for applying ice to reduce temperature, to lessen inflammatory action, to check hemorrhage, and to relieve pain.
**Iceland moss** (*is'-land*). See *cetraria* (2). **I. spar,** a crystalline form of calcium carbonate, having doubly refracting properties, and used in instruments for studying polarized light.

**ichnogram** (*ik'-no-gram*) [ἴχνος, a track, footstep; γράφειν, to write]. In forensic connection, the record of a footprint.
**ichor** (*i'-kor*) [ἰχώρ, serum or pus]. An acrid, thin, discharge from an ulcer or wound.
**ichoremia, ichoræmia.** See *ichorrhemia*.
**ichorization** (*i-kor-iz-a'-shun*) [*ichor*]. The conversion of tissue into ichor.
**ichorous** (*i'-kor-us*) [*ichor*]. Resembling or relating to ichor.
**ichorrhea** (*i-kor-e'-ah*) [ἰχώρ, pus; ῥοία, a flow]. A copious flow of ichor.
**ichorrhemia, ichorrhæmia** (*i-kor'-e-me-ah*) [*ichor*; αἷμα, blood]. The presence of septic matter in the blood.
**ichthalbin** (*ik-thal'-bin*). Ichthyol albuminate; used in gastrointestinal diseases. Dose 15-30 gr. (1-2 Gm.) 3 times daily.
**ichthargan** (*ik-thar'-gan*). A combination of silver and ichthyol-sulphonic acid containing 30 % of the former. It is used in acute gonorrhea in injections containing 1-1½ gr. in 8 oz.; irrigation with solutions of 1 : 4000-1 : 750.
**ichthidin** (*ik'-thid-in*) [ἰχθύς, fish]. A substance resembling lardacein, but obtained from the eggs of cyprinoid fishes.
**ichthin** (*ik'-thin*) [ἰχθύς, fish]. An albuminous substance obtained from the eggs of some fishes, and also from cartilaginous fishes and frogs.
**ichthoform** (*ik'-tho-form*). Ichthyol formaldehyde; it is used as an intestinal disinfectant, antiphlogistic, and as a vulnerary. Dose 15-20 gr. (1.0-1.3 Gm.) daily.
**ichthosin** (*ik'-tho-sin*). A compound of ichthyol and eosin used in skin diseases.
**ichthulin** (*ik'-thu-lin*) [ἰχθύς, a fish; ὕλη, matter]. A lardaceous substance, found in the eggs of fishes and in salmon. It is akin to ichthin and ichthidin.
**ichthyiasis** (*ik-the-i'-as-is*). Same as *ichthyosis*.
**ichthyism** (*ik'-the-ism*). See *ichthysmus*.
**ichthyo-** (*ik-the-o-*) [ἰχθύς, fish]. A prefix meaning fish.
**ichthyocolla** (*ik-the-o-kol'-ah*) [*ichthyo-*; κόλλα, glue]. Isinglass. The air-bladder of the sturgeon, *Acipenser huso*, occurring in horny, translucent, white sheets that form a jelly with hot water. It is a form of gelatin, and is used as a food, for clarifying liquids, and as a test for tannic acid. It forms the basis of English court-plaster (*emplastrum ichthyocollæ*).
**ichthyography** (*ik-the-og'-ra-fe*) [*ichthyo-*; γράφειν, to write]. A description of fishes; ichthyology.
**ichthyoid** (*ik'-the-oid*) [*ichthyo-*; εἶδος, like]. Fishlike.
**ichthyol** (*ik'-the-ol*) [*ichthyo-*; *oleum*, oil]. 1. The ammonium or sodium salt of a tarry substance obtained in the distillation of a bituminous mineral containing fossil fish. The chemical formula is $C_{28}H_{36}S_3O_6(HN_4)_2$, or $C_{28}H_{36}S_3O_6Na_2$. It contains about 15 % of sulphur, and is used as an alterative and antiphlogistic, especially in eczema, acne, lupus, and other dermal diseases. Internally it has been employed in rheumatism, syphilis, leprosy, tuberculosis, etc. Dose 10-30 gr. (0.65-2.0 Gm.) in 24 hours. 2. Ammonium ichthyol sulphonate, $(NH_4)_2C_{28}H_{36}S_3O_6$, soluble in water, glycerol, and a mixture of alcohol and ether; freely miscible in oils. Dose as alterative 3-10 min. (0.2-0.65 Gm.) 3 times daily. Application, 5 to 50 % ointment; 2 % solution in gonorrhea. **i. albuminate.** See *ichthalbin*. **i. formaldehyde.** See *ichthoform*. **i., lithium,** $LiC_{28}H_{36}S_3O_6$, a dark-brown mass used in 50 % ointment. Syn., *Lithium ichthyol sulphonate*. **i., silver,** ichthargan. **i., sodium.** See *ichthyol* (1). **i., zinc,** zinc ichthyol sulphonate, a black, tarry mass, used in eczema.
**Ichthyology** (*ik-the-ol'-o-je*) [*ichthyo-*; λόγος, science]. The science of fishes, their anatomy, distribution, and biology.

**ichthyophagous** (ik-the-of'-ag-us) [ichthyo-; φαγεῖν, to eat]. Fish-eating.
**ichthyosis** (ik-the-o'-sis) [ἰχθύs, fish]. A chronic skin disease characterized by the development of epidermal plates somewhat resembling the scales of a fish. **i. follicularis**, a form in which the sebum and epithelium are heaped around the orifices of the hair follicles. **i. hystrix**, a form characterized by warty growths, consisting of elongated and hypertrophied papillæ, covered by greatly thickened epidermis. **i. sebacea**, seborrhea. **i. simplex**, the common form of ichthyosis, in which the surface has a tessellated appearance, from being covered with large, finely corrugated, papery scales.
**ichthyotic** (ik-the-ot'-ik) [ichthyo-]. Relating to or affected with ichthyosis.
**ichthyotomy** (ik-the-ot'-o-me) [ichthyo-; τομή, a cutting]. The dissection or anatomy of fishes.
**ichthyotoxicon** (ik-the-o-toks'-ik-on) [ichthyo-; τόξικον, poison]. 1. The toxin present in the serum of certain fishes, as in that of the eel. 2. A general term for the active agent in poisoning by eating fish.
**ichthyotoxicum** (ik-the-o-toks'-ik-um) [ichthyo-; τόξικον, poison]. A name loosely given to poisoning from eating the flesh of certain fish, and also to the poisonous principle. It may be a natural poisonous principle or the result of putrefactive fermentation of the substance of the fish. It is probably a ptomaine.
**ichthyotoxin** (ik-the-o-toks'-in). A basic poison generated in fish by growth of bacteria or fungi.
**ichthyotoxism** (ik-the-o-toks'-izm). See *ichthysmus*. Cf. *siguatera*.
**ichthysmus** (ik-this'-mus) [ἰχθύs, fish]. Poisoning due to the absorption of mytilotoxin from fish.
**ichtol** (ik'-tol). A proprietary mixture said to consist of lanolin, iodoform, glycerol, phenol, oil of lavender, and oil of eucalyptus, used in skin diseases.
**icing-liver**. Chronic perihepatitis resulting in the formation of an exudate resembling the icing on a cake.
**icon** (i'-kon) [εἰκών, image]. An image or model.
**iconography** (i-kon-og'-ra-fe) [εἰκών, image; γραφειν, to write]. A description by means of pictorial illustration.
**icterencephalotyphus** (ik-ter-en-sef-al-o-ti'-fus) [icterus; ἐγκέφαλος, the brain; τύφος, typhus]. Typhoid fever with marked cerebral symptoms combined with jaundice.
**icterepatitis** (ik-ter-ep-a-ti'-tis) [icterus; ἧπαρ, liver; ιτις, inflammation]. Inflammation of the liver with jaundice.
**icteric** (ik-ter'-ik) [icterus]. Pertaining to or characterized by jaundice. **i. fever**, a form of remittent or relapsing malaria in which jaundice is a marked symptom.
**icteritious** (ik-ter-ish'-us) [icterus]. 1. Affected with or resembling icterus. 2. Yellow, as the skin in jaundice.
**icterode** (ik'-ter-ōd) [ἰκτερώδης, jaundiced]. Icteroid.
**icterogenic, icterogenous** (ik-ter-o-jen'-ik, ik-ter-oj'-en-us) [icterus; γεννᾶν, to produce]. Causing icterus.
**icterohematuria** (ik-ter-o-hem-at-ū'-re-ah) [icterus; hematuria]. Jaundice with hematuria.
**icterohemoglobinuria** (ik-ter-o-hem-o-glo-bin-ū'-re-ah). Combined icterus and hemoglobinuria.
**icteroid** (ik'-ter-oid) [icterus; εἶδος, form]. Resembling the color of, or having the nature of, jaundice.
**icterophthisis** (ik-ter-off'-this-is) [icterus; φθίσις, wasting]. Pulmonary tuberculosis with yellow discoloration of the skin.
**icterus** (ik'-ter-us) [ἴκτερος, jaundice]. Jaundice. **i., acholuric**, a condition characterized by more or less pigmentation of the skin in certain areas or over the whole surface, absence of bile-pigments in the urine and their presence in the blood-serum. **i. diffusion**, that due to the hepatic cells having lost their power of holding back the bile, which consequently diffuses into the fluids of the body. **i. febrilis**. See *Weil's disease*. **i. gravis**, acute yellow atrophy of the liver, an acute disease characterized by jaundice, marked nervous symptoms, diminution in size of the liver, and a rapidly fatal termination. The urine contains bile and crystals of leucin and tyrosin. **i. neonatorum**, that which is sometimes observed in infants during the first few days after birth. The causes are obscure, particularly in the mild form; it may be due to the absorption of biliary pigment from the meconium and its entrance into the circulation through an open ductus venosus; a severe form is due to absence of the large bile-ducts or to septic infection. **i., pancreatic**. 1. A condition arising from stenosis of the pancreatic duct, when the system becomes deluged with pancreatic secretions. 2. That which results from removal of the pancreas. **i. saturninus**, jaundice from lead poisoning.
**ictometer** (ik-tom'-et-er) [ictus, stroke; μέτρον, measure]. An instrument to measure the cardiac impulse.
**ictus** (ik'-tus) [L., "a stroke"]. A sudden attack. **i. epilepticus**, an epileptic fit. **i., laryngeal**, an apoplectiform attack occurring during a severe paroxysm of coughing and passing off in a few seconds. **i. paralyticus**, a paralytic stroke. **i. sanguinis**, apoplexy. **i. solis**, sun-stroke.
**id** [ἴδιος, one's own]. In biology, according to Weismann's theory of the germ-plasm, a vital unit of the third degree, having a definite structure of determinants, which in their turn are made up of biophores; an "ancestral germ-plasm," or unit containing all the primary constituents of the species. Each *id* represents an individuality, and is probably identical with the "microsome" of the nuclear rod. Cf. *idioplasm: idant*.
**idant** (i'-dant) [see *id*]. In biology, according to Weismann's theory of the germplasm, a vital unit of the fourth degree; a group of *ids*, differing from the latter in not being perfectly invariable quantities, but only *relatively* constant, their constitution being modified from time to time, so that the lds which previously belonged to the idant A may later take part in the composition of the idant B or C. A *chromosome* (chromatosome) or nuclear rod, composed of *vital units* or ancestral *plasms*, termed *ids* or microsomata (*microsomes*).
**-ide, -id**. A suffix used in chemistry to denote a combination of two elementary substances, or a radical and an element.
**idea** (i-de'-ah) [ἰδέα; form or semblance]. A mental representation of something perceived. **i., chase**, a condition in acute mania in which disconnected ideas and fancies flow rapidly through the mind. It is a term much used by German authors. **i., fixed**, that form of mania in which one dominant idea controls all actions. **i., imperative**, a morbid idea or insane suggestion imperiously demanding notice, the patient often being painfully conscious of its domination over his will.
**ideal** (i-de'-al). Pertaining to an idea. **i. paraplegia**, reflex emotional paraplegia.
**ideation** (i-de-a'-shun) [ἰδέα, form or semblance]. 1. The formation of a mental conception; the perceptual action by which, or in accord with which, an idea is formed. 2. An impression which conveys some distinct notion, but not of a sensory nature.
**ideational insanity** (i-de-a'-shun-al). A form of insanity characterized by perversion of ideation.
**identical** (i-den'-tik-al) [idem, the same]. Being the same; corresponding exactly. **i. points**, corresponding points of the two retinæ, upon which the rays from an object must be focused in order that it may be seen as one.
**identification** (i-den-tif-ik-a'-shun) [see *identical*]. A method of so describing and registering a person by certain physical peculiarities that he or his body may be identified. **i., anthropometric**. See *i., Bertillon system of*. **i., Bertillon (Alphonse) system of**, consists in the use of those measurements which depend on skeletal parts remaining practically unchanged after adult life is reached. Syn., *anthropometric identification*. **i., Galton system of**, is based upon imprints of the epidermic patterns found upon the balls of the thumbs and fingers. The records used are the printed impressions of the ten digits placed in definite order upon a card. **i., palm and sole system of**, an extension of the Galton system to the palmar and plantar surfaces.
**ideodynamism** (i-de-o-di'-nam-izm) [ἰδέα, idea; δύναμις, force]. The domination of an idea; the control exercised by a suggested idea over the subsequent acts of a person who is, or has been, hypnotized.
**ideoglandular** (i-de-o-glan'-dū-lar). Relating to glandular activity as evoked by a mental concept.
**ideography** (i-de-og'-raf-e) [ἰδέα, an idea; γράφειν, to write]. 1. A description of ideas. 2. An expression of ideas by writing, printing, or hieroglyphics.
**ideology** (i-de-ol'-o-je) [ἰδέα, idea; λόγος, science]. The science of thought.

**ideometabolic** (*i-de-o-met-ab-ol'-ik*). Relating to metabolic action induced by some idea.

**ideomotion** (*i-de-o-mo'-shun*) [ἰδέα, idea; *motio*, a moving]. Motion or action due to some idea, and neither purely voluntary nor reflex.

**ideomotor** (*i-de-o-mo'-tor*) [*idea*; *movere*, to move]. Pertaining conjointly to ideation and movement. **i. center**, that part of the cortex which, influenced by ideation, excites muscular movement. **i. movements**, unconscious movements due to impulses of the mind when the attention is otherwise absorbed.

**ideomuscular** (*i-de-o-mus'-kū-lar*). Relating to influence exerted upon the muscular system by a mental concept.

**ideopegma** (*i-de-o-peg'-mah*) [ἰδέα, idea; πῆγμα, a thing fixed; *pl.*, *ideopegmata*]. A fixed or dominant idea that colors all the thoughts of the patient and thus creates a monomania.

**ideophrenia** (*i-de-o-fre'-ne-ah*) [*idea*; φρήν, mind]. Insanity with marked perversion of ideas.

**ideophrenic** (*i-de-o-fren'-ik*) [ἰδέα, form; φρήν, mind]. Relating to, or marked by ideophrenia.

**ideoplastic** (*i-de-o-plas'-tik*) [ἰδέα, idea; πλάσσειν, to form]. Giving shape to the ideas; that stage of hypnotism in which the idea impressed on the brain of the subject is translated into action.

**ideosynchysia, ideosynchysis** (*i-de-o-sin-ki'-ze-ah, i-de-o-sin'-kis-is*) [ἰδέα, idea; σύγχυσις, a pouring together]. Confusion of ideas; delirium.

**ideovascular** (*i-de-o-vas'-kū-lar*). Relating to a vascular change resulting from a dominant idea.

**idiempresis** (*id-e-em-pre'-sis*) [ἴδιος, own; ἐμπρησις, burning]. Spontaneous combustion. Also, spontaneous inflammation.

**idio-** (*id-e-o-*) [ἴδιος, one's own]. A prefix signifying pertaining to one's self, peculiar to the individual.

**idioagglutinin** (*id-e-o-ag-gloo'-tin-in*) [*idio-*; *agglutinin*]. An agglutinin having a spontaneous origin.

**idioblast** (*id'-e-o-blast*) [*idio-*; βλαστός, offshoot]. 1. In biology, a histological cell having a character different from that of the surrounding cells, owing to a difference either in its form or its contents, e. g., the stellate hair in the interior of the tissue of *Nymphaeaceæ*. 2. See *biophore*.

**idiocrasia, idiocrasis** (*id-e-o-kra'-ze-ah, -kra'-sis*) [*idio-*; κρᾶσις, temperament]. Idiosyncrasy.

**idiocrasy** (*id-e-ok'-ras-e*). Same as *idiosyncrasy*.

**idiocratic** (*id-e-o-krat'-ik*). Relating to an idiosyncrasy.

**idioctonia** (*id-e-ok-to'-ne-ah*) [*idio-*; κτόνος, killing]. Self-murder; suicide.

**idiocy** (*id'-e-o-se*) [ἰδιώτης, a private person]. A congenital condition of mental deficiency, usually accompanied by physical defects, and characterized by an almost total absence of intelligence.

**idiogenesis** (*id-e-o-jen'-es-is*) [*idio-*; γεννᾶν, to produce]. The origin of idiopathic diseases.

**idioglossia** (*id-e-o-glos'-e-ah*) [*idio-*; γλῶσσα, tongue]. Extremely defective utterance, but one in which the same sound is used to express the same idea, even though the sounds used belong to no known language.

**idioheteroagglutinin** (*id-e-o-het-er-o-ag-gloo'-tin-in*) [*idio-*; ἕτερος, other; *agglutinin*]. An agglutinin in normal blood having the property of agglutinating foreign cells and the blood-corpuscles of other species of animals.

**idioheterolysin** (*id-e-o-het-er-ol'-is-in*) [*idio-*; ἕτερος, other; λύειν, to loose]. A lysin in normal blood capable of dissolving foreign cells and the blood-corpuscle of another species of animal.

**idiohypnotism** (*id-e-o-hip'-no-tizm*). Self-induced hypnotism.

**idioisoagglutinin** (*id-e-o-is-o-ag-gloo'-tin-in*) [*idio-*; ἴσος, equal; *agglutinin*]. An inborn nonhereditary substance present in normal blood, due to interchangeable immunization between mother and fetus.

**idioisolysin** (*id-e-o-is-ol'-is-in*) [*idio-*; *isolysin*]. An inborn, nonhereditary isolysin due to an interchangeable immunization between mother and fetus.

**idiologism** (*id-e-ol'-o-jizm*) [*idio-*; λόγος, utterance]. A characteristic expression or form of phraseology peculiar to any person, especially to an insane person.

**idiolysin** (*id-e-ol'-is-in*). A lysin found normally in the blood and having a spontaneous origin.

**idiometritis** (*id-e-o-me-tri'-tis*) [*idio-*; *metritis*]. Inflammation of the parenchymatous substance of the uterus.

**idiomiasma** (*id-e-o-mi-az'-mah*) [*idio-*; μίασμα, stain, defilement; *pl.*, *idiomiasmata*]. A term for any noxious exhalation from the body.

**idiomology** (*id-e-o-mol'-o-je*) [ἰδίωμα, a peculiar phraseology, idiom; λέγειν, to speak]. The study of the peculiarities of speech of various races.

**idiomuscular** (*id-e-o-mus'-kū-lar*) [*idio-*; *musculus*, muscle]. Peculiar to muscular tissue; not involving any nerve-stimulus or any function of the organism except those of the muscle itself. **i. contraction**, contraction of a tired or weakened muscle under certain conditions of extraneous stimulus.

**idioneurosis** (*id-e-o-nū-ro'-sis*) [*idio-*; *neurosis*]. An affection due to some disturbed or abnormal condition of the nerves supplying the affected part; a simple neurosis.

**idiopathic** (*id-e-o-path'-ik*) [*idio-*; πάθος, disease]. Not dependent upon another disease or upon a known or recognized cause. **i. anemia**, pernicious anemia. **i. disease**, a self-existing disease.

**idiopathy** (*id-e-op'-ath-e*) [*idio-*; πάθος, disease]. 1. An idiopathic disease or condition. 2. The fact or quality of being idiopathic.

**idiophrenic** (*id-e-o-fren'-ik*) [*idio-*; φρήν, mind]. Due to disease of the brain; applied to certain forms of insanity.

**idioplasm** (*id'-e-o-plazm*) [*idio-*; πλάσμα, a thing formed]. A reproductive substance not contained in the body of the cell, but in the chromosomes of the nucleus, controlling and determining the actual characters of the particular cell, and also those of all of its descendants.

**idiopsychology** (*id-e-o-si-kol'-o-je*) [*idio-*; ψυχή, soul]. Psychology based upon introspective study of one's own mental acts. Cf. *heteropsychology*.

**idioretinal** (*id-e-o-ret'-in-al*) [*idio-*; *retina*]. Peculiar or proper to the retina.

**idiosome** (*id'-e-o-sōm*). See *idioblast*.

**idiospasm** (*id'-e-o-spazm*) [*idio-*; σπασμός, spasm]. A spasm confined to one part.

**idiospastic** (*id-e-o-spas'-tik*) [*idio-*; σπαστικός, stretching]. Pertaining to *idiospasm*.

**idiosthenia** (*id-e-o-sthe'-ne-ah*) [*idio-*; σθένος, strength]. Having innate or spontaneous strength or power.

**idiosyncrasia** (*id-e-o-sin-kra'-ze-ah*). Same as *idiosyncrasy*.

**idiosyncrasy** (*id-e-o-sin'-kra-se*) [*idio-*; σύν, together; κρᾶσις, a mingling]. 1. Any special or peculiar characteristic or temperament by which a person differs from other persons. 2. A peculiarity of constitution that makes an individual react differently from most persons to drugs or other influences.

**idiosyncratic** (*id-e-o-sin-krat'-ik*). Pertaining to idiosyncrasy. **i. coryza**, a synonym of *hay-fever*.

**idiot** (*id'-e-ot*) [ἰδιώτης, a private person]. A person congenitally almost destitute of intelligence.

**idiotcy, idiotism, idiotry** (*id'-e-ot-se, id'-e-ot-izm, id'-e-ot-ry*). Same as *idiocy*, q. v.

**idiotia** (*id-e-o'-she-ah*). A state of idiocy; idiotism.

**idiotopy** (*id-e-ot'-op-e*) [*idio-*; τόπος, place]. Topographic description pertaining to the relation of different parts of the same organ.

**idolum** (*id-o'-lum*) [εἴδωλον, an image]. An illusion or halucination.

**idorgan** (*id-or'-gan*) [ἴδιος, own; ὄργανον, an organ]. Haeckel's name for a morphological unit made up of two or more plastids and not possessing the positive characteristics of the stock.

**idromania** (*id-ro-ma'-ne-ah*) [ὕδωρ, water; *mania*]. Hydromania; insane desire for water, or to commit suicide by drowning.

**idrosis** (*id-ro'-sis*). See *hidrosis*.

**igasurine** (*ig-as-u'-rēn*) [Malay, *igasur*; *ignatia*]. An alkaloid from ignatia, said to be more poisonous than strychnine. Its existence as a definite compound has been denied.

**igazol** (*ig'-az-ol*). A proprietary gaseous antiseptic said to contain iodoform and formaldehyde. It is used in tuberculosis.

**ignatia** (*ig-na'-she-ah*). St. Ignatius' bean. The seed of *Strychnos ignatii*, containing the alkaloids strychnine and brucine. Its therapeutic effects are similar to those of nux vomica. Dose of the *abstract* ½–1 gr. (0.032–0.065 Gm.); of the *tincture* 2–10 min. (0.13–0.65 Cc.).

**igniextirpation** (*ig-ne-eks-ter-pa'-shun*) [*ignis*; *extirpare*, to root out]. Hysterectomy by cauterization.

**ignipedites** (*ig-ne-ped-i'-tēs*) [*ignis*; *pes*, foot]. Hot-foot; a disorder marked by an intense burning sensation in the soles of the feet.

# IGNIPUNCTURE 455 ILLINITION

**ignipuncture** (ig'-ne-punk-tūr) [ignis; punctura, puncture]. Puncture with platinum needles heated to whiteness by the electric current.
**ignis** (ig'-nis) [L.]. Fire. i. sacer, erysipelas. i. sancti Antonii, Saint Anthony's fire, an old name for erysipelas and for anthrax.
**ignition** (ig-nish'-un) [ignis, fire]. The process of heating solids, especially inorganic compounds, until all volatile matter has been driven off.
**ikota** (ik-o'-tah). A religious mania occurring among women in Siberia.
I. K. therapy. [German immunkörper, immune bodies]. Spengler's method of treating tuberculosis.
**ileac** (il'-e-ak) [ileum]. Pertaining to ileus. i. passion, a disorder marked by severe griping pain, fecal vomiting, with spasm of the abdominal muscles. Syn., ileus.
**ileadelphus** (il-e-ad-el'-fus). See iliadelphus.
**ileectomy** (il-e-ek'-to-me) [ileum; ἐκτομή, a cutting out]. Excision of the ileum.
**ileitic** (il-e-it'-ik). Pertaining to or affected with ileitis.
**ileitis** (il-e-i'-tis) [ileum; ιτις, inflammation]. Inflammation of the ileum.
**ileo-** (il-e-o'-). A prefix signifying relation to the ileum.
**ileocecal** (il-e-o-se'-kal) [ileo-; cecum]. Pertaining to both ileum and cecum. i. fossa, a depression in the lower part of the small intestine at the base of the vermiform process. i. valve, a structure, consisting of two folds of mucosa, that guards the passage between the ileum and cecum.
**ileocecum** (il-e-o-se'-kum). The ileum and cecum regarded as one.
**ileocleisis** (il-e-o-kli'-sis) [ileo-; κλειειν, to lock]. Obstruction or closure of the ileum.
**ileocolic** (il-e-o-kol'-ik) [ileo-; colon]. Pertaining conjointly to the ileum and the colon.
**ileocolitis** (il-e-o-ko-li'-tis) [ileo-; colon; ιτις, inflammation]. Inflammation of the ileum and the colon.
**ileocolonic** (il-e-o-kol-on'-ik). See ileocolic.
**ileocolostomy** (il-e-o-kol-os'-to-me) [ileo-; colon; στόμα, mouth]. The establishment of an artificial communication between the ileum and the colon.
**ileocolotomy** (il-e-o-ko-lot'-o-me) [ileo-; colon; τέμνειν, to cut]. A surgical operation on the ileum and colon.
**ileodicliditis** (il-e-o-dik-lid-i'-tis) [ileo-; δίκλις, valve; ιτις, inflammation]. Inflammation of the ileocecal valve.
**ileoileostomy** (il-e-o-il-e-os'-to-me) [ileo-; στόμα, mouth]. The operation of establishing an artificial communication between two different parts of the ileum.
**ileology** (il-e-ol'-o-je) [εἴλειν, to roll; λόγος, science]. The anatomy, physiology, and pathology of the ileum.
**ileoparietal** (il-e-o-par-i'-et-al). Relating to the walls of the ileum.
**ileopisolitis** (il-e-o-pis-o-li'-tis) [ileo-; pisum, a pea; ιτις, inflammation]. Inflammation of Peyer's patches or glands.
**ileoproctostomy** (il-e-o-prok-tos'-to-me) [ileo-; πρωκτός, rectum; στόμα, mouth]. The surgical formation of a fistula between the ileum and rectum.
**ileorectostomy** (il-e-o-rek-tos'-to-me). See ileoproctostomy.
**ileosigmoidostomy** (il-e-o-sig-moid-os'-to-me). The surgical formation of a fistula between the ileum and sigmoid flexure.
**ileostomy** (il-e-os'-to-me) [ileo-; στόμα, mouth]. The surgical formation of a passage through the abdominal wall into the ileum, or from the ileum to some other hollow organ.
**ileotomy** (il-e-ot'-o-me) [ileo-; τέμνειν, to cut]. Incision of the ileum through the abdominal wall.
**ileotyphus** (il-e-o-ti'-fus) [ileo-; typhus]. Enteric or typhoid fever.
**ileum** (il'-e-um) [εἴλειν, to roll]. The lower portion of the small intestine, terminating in the cecum.
**ileus** (il'-e-us) [ἴλιός, a severe kind of colic]. Ileac passion; intestinal obstruction.
**ilia** (il'-e-ah) [L.: pl. of ilium]. The iliac bones; the flanks or loins.
**iliac** (il'-e-ak) [ileum]. Pertaining to the ilium or to the flanks. i. artery. See under artery. i. crest, the upper free margin of the ilium to which the abdominal muscles are attached. i. fascia, the fascia lining the posterior part of the abdominal cavity and covering the psoas and iliacus muscles. i. fossa,

See fossa, iliac. i. muscle. See iliacus under muscle. i. passion. See ileac passion. i. region, the region external to the hypogastric region. See abdomen.
**iliaco-** (il-i'-ak-o-) [ilia, the loins]. A prefix signifying relation to the loins. i.-femoral, relating to the ilium and the femur. i.-trochanteric, relating to the ilium and the great trochanter of the femur.
**iliacus** (il-i'-ak-us). See muscles, table of.
**iliadelphus** (il-e-ad-el'-fus) [ilia; ἀδελφός, brother]. A monstrosity double from the pelvis upward.
**ilial** (il'-e-al). See iliac.
**ilicin** (il'-is-in) [ilex, the holm-oak]. A crystalline febrifugal principle from the leaves of Ilex aquifolium.
**ilien** (il'-e-en) [ilia, the loins]. Belonging to the ilium in itself.
**ilio-** (il-e-o-). A prefix denoting relation to the ilium.
**iliocolotomy** (il-e-o-ko-lot'-o-me) [ilio-; τομή, a cutting]. Incision of the colon in the iliac region.
**iliocostal** (il-e-o-kos'-tal) [ilio-; costa, rib]. Pertaining to the ilium and ribs. See muscles, table of.
**iliocostalis** (il-e-o-kos-ta'-lis). Iliocostal. See muscles, table of.
**iliodorsal** (il-e-o-dor'-sal) [ilio-; dorsum, the back]. Relating to the dorsal surface of the ilium.
**iliofemoral** (il-e-o-fem'-or-al) [ilio-; femur]. Pertaining conjointly to the ilium and the femur. i. ligament. See ligament, iliofemoral.
**iliohypogastric** (il-e-o-hi-po-gas'-trik) [ilio-; hypogastric]. Pertaining conjointly to the ilium and the hypogastrium. i. nerve. See nerves, table of.
**ilioinguinal** (il-e-o-in'-gwin'-al) [ilio-; inguen, groin]. 1. Pertaining to the ilium and the groin. 2. Lying partly within the iliac and partly within the inguinal region.
**iliolumbar** (il-e-o-lum'-bar). Pertaining to the iliac and lumbar regions.
**iliolumbocostoabdominal** (il-e-o-lum-bo-kos-to-ab-dom'-in-al). Pertaining to the iliac, lumbar, costal, and abdominal regions.
**iliopectineal** (il-e-o-pek-tin-e'-al) [ilio-; pecten, comb]. Pertaining conjointly to the ilium and the pubes. i. line. See line, iliopectineal.
**iliopelvic** (il-e-o-pel'-vik). Pertaining to the iliac region and the pelvis.
**ilioperoneal** (il-e-o-per-o-ne'-al). Relating to the ilium and the peroneal region.
**iliopsoas** (il-e-o-so'-as) [ilio-; ψόα, loin]. Pertaining conjointly to the ilium and the loins. i. muscle, the psoas and iliacus muscles considered as a single muscle.
**iliosacral** (il-e-o-sa'-kral). Relating to the ilium and the sacrum.
**iliosciatic** (il-e-o-si-at'-ik). Relating to the ilium and the ischium. i. notch, the sacrosciatic notch.
**ilioscrotal** (il-e-o-skro'-tal). Relating to the ilium and the scrotum.
**iliospinal** (il-e-o-spi'-nal). Pertaining to the ilium and the spinal column.
**iliotibial** (il-e-o-tib'-e-al) [ilio-; tibia]. Pertaining to or connecting the ilium and the tibia. i. band, a thickened portion of the fascia lata extending from the outer tuberosity of the tibia to the iliac crest.
**iliotrochanteric** (il-e-o-tro-kan-ter'-ik). Pertaining to the ilium and the great trochanter of the femur.
**ilium** (il'-e-um) [L.]. 1. The flank. 2. The superior broad portion of the os innominatum, properly the os ilii.
**ill** (il). Sick; diseased; unwell.
**illacrimation** (il-lak-rim-a'-shun) [lacrima, a tear]. Same as epiphora.
**illaqueation** (il-ak-we-a'-shun) [illaqueare, to insnare]. A method of changing the direction of misplaced cilia by withdrawing them by means of a ligature, with an opening in the tissue of the lid.
**illegitimacy** (il-e-jit'-im-a-se) [in, not; legitimus, according to law]. The condition of being unlawful, or not legitimate.
**illegitimate** (il-e-jit'-im-at) [in, not; legitimus, according to law]. Not in accordance with statutory law. i. child, one born out of wedlock; a bastard.
**illicium** (il-is'-e-um) [illicere, to entice]. Staranise. The fruit of Illicium verum or Illicium anisatum, of the natural order Magnoliaceæ. It is the source of star-anise. I. religiosum and I. parviflorum are poisonous.
**illinition** (il-in-ish'-un) [illinere, to smear]. Inunction.

## ILLNESS 456 IMMUNITY

**illness** (*il'-nes*). Sickness; disease; an attack of disease.

**illumination** (*il-ū-min-a'-shun*) [*illuminare*, to make light]. 1. The act of illuminating or lighting up. 2. The quantity of light thrown on an object. i., **axial**, illumination by light conveyed in the direction of the axis of the microscope. i., **central**, in microscopy, an illumination producted by the rays of light reflected from the mirror passing perpendicularly through the object on the stage. i., **critical**, in microscopy, an illumination in which the lampflame is focused on the object. i., **direct**, illumination of an object by light thrown upon it from in front. i., **focal**, that in which the light is concentrated on an object by means of a lens or mirror. i., **lateral**. See *i. oblique*. i., **oblique**, illumination of an object by throwing light upon it obliquely, usually by means of a lens.

**illuminator, Abbé's.** See *Abbé's condenser*.

**illusion** (*il-ū'-zhun*) [*illusio*, a mocking]. A false interpretation by the mind of a real sensation.

**illusional** (*il-ū'-zhun-al*). Of the nature of an illusion.

**illutation** (*il-lū-ta'-shun*) [*in*, in; *lutum*, mud]. Treatment of disease by the mud-bath.

**im-**. A prefix used in chemistry to indicate the bivalent group NH.

**ima.** (*i'mah*) [*im-us, -a, -um*, lowest, deepest]. The lowest, as *thyroidea ima*, the lowest thyroid artery.

**image** (*im'-āj*) [*imago*, a likeness]. 1. A more or less accurate representation of an object. 2. The picture of an object formed by rays of light reflected, refracted, or passed through a small aperture. **i.**, **after-**. See *after-images*. i., **direct**, i., **erect**, a picture obtained from rays that have not yet come to a focus. i., **false**. See under *false*. i., **inverted**, one turned upside down. Nearly all real images are inverted. i., **real**, that formed at the place where the rays meet. i., **virtual**, an apparent image formed in the direction in which the rays enter the eye, the rays not actually converging at the point where the image is seen. The images formed by plane or convex mirrors and by concave lenses, when the object is placed within the principal focus, are virtual.

**imagination** (*im-aj-in-a'-shun*) [*imaginatio*, imagination]. The picture-making power of the mind. The faculty by which one creates ideas or mental pictures by means of the data derived from experience, ideally revivified, extended, and combined in new forms.

**imago ▼** (*im-a'-go*) [L.; *gen., imaginis*]. 1. The image. 2. The final, adult, or reproductive stage of an insect.

**imapunga** (*im-ap-ung'-gah*) [South African]. A disease occurring to a limited extent among South African cattle closely related in pathology to South African horse-sickness.

**imbalance** (*im-bal'-ans*) [*in*, not; *bilanx*, a balance]. Lack of balance; lack of the power of keeping the erect position; lack of muscular balance (as between the muscle of the eyes).

**imbecile** (*im'-bes-il*) [*imbecillis*, weak]. Feeble in mind.

**imbecility** (*im-bes-il'-it-e*). Mental weakness or defect, similar to that of idiocy, but of less degree. i., **acquired**. Synonym of *dementia præcox*.

**imbed** (*im-bed'*). In histology, to treat a tissue with some substance, as paraffin or celloidin, which shall give it support during the process of sectioncutting.

**imbedding** (*im-bed'-ing*) [*imbed*]. The fixation of a tissue-specimen in a firm medium, in order to keep it intact during the cutting of thin sections.

**imbibe** (*im-bīb'*) [*imbibere*, to drink in]. To drink or suck in.

**imbibition** (*im-bi-bish'-un*) [*in*, in; *bibere*, to drink]. The act of sucking up moisture; the absorption of fluids.

**imbricated** (*im'-brik-a-ted*) [*imbrex*, a tile]. Overlapping, like shingles or tiles on a roof.

**imide, imid** (*im'-id*). Any compound of the radical NH united to a divalent acid radical.

**imidiode** (*im-id'-e-ōd*). Glossy crystals obtained by interaction of paraethoxyphenyl succinimide, iodine, and potassium iodide in the presence of acetic acid; it is used as a wound antiseptic.

**imido** (*im'-id-o*). A prefix denoting an imide.

**imidoxanthin** (*im-id-o-san'-thin*). See *guanine*.

**imitation** (*im-it-a'-shun*) [*imitari*, to imitate]. A production that is similar, to, or a copy of, another object or process. i., **morbid**, the occurrence of a convulsive or mental affection brought about by observing a similar affection in another; mental contagion.

**Imlach's fat-plug** (*im'-lak*) [Francis Imlach, Scotch physician]. A mass of yellowish fat frequently found at the mesial angle of the external inguinal ring, for which it constitutes a landmark during operations.

**immaculate** (*im-ak'-ū-lāt*) [*in*, not; *macula*, a spot]. Pure; spotless.

**immature** (*im-at-ūr'*) [*in*, not; *maturus*, ripe]. Unripe; not yet of an adult age or growth.

**immediate** (*im-e'-de-āt*) [*in*, not; *mediatus*, mediate]. Direct; without the intervention of anything. i. **agglutination**. See *i. union*. i. **auscultation**, auscultation performed with the ear against the surface. i. **contagion**, that from personal contact. i. **union**, union by first intention.

**immedicable** (*im-med'-ik-ab-l*) [*in*, not; *medicare*, to cure]. That which does not yield to medicine or treatment. Incurable.

**immersion** (*im-er'-shun*) [*in*, in; *mergere*, to dip]. The plunging of a body into a liquid. i.-**bath**, a plunge-bath. i., **homogeneous**, a fluid between the objective of a microscope and the cover-glass, having about the same refractive and dispersive power as the glass. i.-**lens**, a lens, usually of high power, the lower end of which is immersed in a drop of some liquid, such as water or oil, that has nearly the same refractive index as glass, and is placed on the cover-glass of the object under examination.

**imminence** (*im'-in-ens*) [*imminere*, to overhang]. An impending or menacing. i., **morbid**, the period immediately preceding the incubation stage of a disease.

**immiscible** (*im-is'-ib-l*) [*in*, not; *miscere*, to mix]. Not capable of being mixed.

**immissio** (*im-ish'-e-o*) [L.]. Insertion. i. **catheteris**. See *catheterism*. i. **penis**, introduction of the penis into the vagina.

**immobilization** (*im-o-bil-iz-a'-shun*) [*in*, not; *movere*, to move]. The act of making firm or of rendering motionless, as *immobilization* of a joint.

**immune** (*im-ūn'*) [*in*, not; *munis*, serving]. 1. Safe from attack; protected against a disease by a natural or an acquired peculiarity. 2. A person who is protected against any special virus. i. **animal**, an animal in control experiment rendered immune by inoculation with some antispecific agent. i. **body**. See *body, immune*. i. **proteids**, substances resulting from combination in the living body of the enzymes of pathogenic bacteria with certain albuminous bodies, probably those derived from the leukocytes. i. **system**, the combination of antigen, amboceptor and complement.

**immunity** (*im-ū'-nit-e*) [see *immune*]. Exemption from disease; the condition of the body wherein it resists the development of morbid processes; resistance to infection. See *Behring's law*. i., **acquired**. See *i. active*. i., **active**, that possessed by an individual after recovering from certain infectious diseases; or that induced by direct treatment with filtered or unfiltered cultures resulting in the production in the body of anti-bacterial or antitoxic substances. i., **antitoxic**, immunity against toxins. i., **congenital**, i., **natural**, that with which the individual is born. i., **passive**, that conferred by the introduction of antitoxins or vaccines. i., **theory of, Buchner's humoral**, this supposes that a reactive change has been brought about in the integral cells of the body by the primary affection from which there has been recovery, and this change is protective against similar invasions of the same organism. i., **theory of, Chauveau's retention**, proposed that bacteria, instead of removing certain essential food-principles from the body, left within the body certain excretory products, and that the accumulation of these products tended to prevent the subsequent invasion of the same species of bacteria. i., **theory of, Ehrlich's side-chain**, considers the individual cells of the body analogous in a certain sense to complex organic substances, and that they consist essentially of a central nucleus to which secondary atom-groups having distinct physiological functions are attached by side-chains such as chemists represent in their attempts to illustrate the reactions which occur in the building up or pulling down of complex organic substances. A cell-equilibrium is supposed to be disturbed by injury to any of the physiological atom-groups, as

by a toxin, and this disturbance results in an effort at compensatory repair during which plastic material in excess of the amount required is generated and finds its way into the blood. This Ehrlich regards as the antitoxin which is capable of neutralizing the particular toxin to which it owes its origin, if this is subsequently introduced into the blood. In this theory a specific combining relation is assumed to exist between various toxic substances and the secondary atom-groups of certain cellular elements of the body. The atom-groups which, in accordance with this theory, combine with the toxin of any particular disease-germ Ehrlich calls the toxiphoric side-chain. **i., theory of, Emmerich and Löw's,** based upon the conclusion that many bacteria generate enzymes capable of digesting the organism by which they were generated and sometimes other organisms as well. **i., theory of, exhaustion hypothesis,** Pasteur's theory that immunity often afforded to the tissues by an attack of infection or following vaccination against infection is due to an abstraction from the tissues by the organism concerned in the primary attack of something necessary to the growth of the infecting organism. It is opposed to the retention theory of Chauveau. **i., theory of, lateral bond.** See *i., theory of, Ehrlich's side-chain.* **i., theory of, Metchnikoff's phagocytic.** See under *Metchnikoff.*
▶ **immunization** (im-ū-niz-a'-shun) [see *immune*]. The act of rendering immune. **i., Haffkine's method of** (against cholera), consists in the injection of a definite quantity of sterilised culture; 5 days later a small dose, and in 5 days more a larger dose, of the living virulent culture.
**immunizator** (im-ū-niz-a'-tor). That which renders immune.
**immunize** (im'-ū-nīz). To give immunity.
**immunizing unit.** See *unit, serum-.*
**immunochemistry** (im-ū-no-kem'-is-tre). That branch of chemistry which treats of immunity and the reactions connected therewith.
**immunologist** (im-ū-nol'-o-jist). One versed in the science of immunity.
**immunology** (im-ū-nol'-o-je). That branch of science which is concerned with the study of immunity.
**immunoprotein** (im-ū-no-pro'-te-in). A protein with bacteriolytic power, formed when attenuated bacterial cultures are injected into animals.
**immunotoxin** (im-ū-no-toks'-in). Any antitoxin.
**impact** (im'-pakt) [*impingere,* to drive into or against]. A forcible striking against.
**impacted** (im-pak'-ted). Driven against and retained, as a wedge. **i. fracture.** See *fracture, impacted.*
**impaction** (im-pak'-shun) [see *impact*]. 1. Concussion. 2. The state of being impacted or fixed in a part, as *impaction* of the feces or *impaction* of a fragment of bone into another fragment.
**impalpable** (im-pal'-pa-bl) [*in,* not; *palpare,* to feel]. Not capable of being felt; unappreciable by touch. **i. powder,** a powder so fine that its separate particles cannot be felt.
**impaludism** (im-pal'-ū-dizm) [*in,* in; *palus,* a marsh]. Chronic malarial poisoning.
**impar** (im'-par) [*in,* not; *par,* equal]. Odd or unequal, or without a fellow. **i., ganglion,** a small ganglion on the coccyx.
**imperative** (im-per-'at-iv) [*imperare,* to command]. Peremptory; absolute; compulsory; binding. **i. conception,** a conception or thought that dominates the actions of an individual, although the falsity of the conception may be recognized.
**Imperatoria** (im-per-at-o'-re-ah) [*imperatorius,* belonging to a commander]. A genus of umbelliferous plants; masterworts. **I. ostruthium,** false pellitory of Spain, is an aromatic stimulant, once prized as a polychrest remedy, but now little used.
**imperfection** (im-per-fek'-shun) [*imperfectus,* imperfect]. A physical defect.
**imperforate** (im-per'-for-āt) [*in,* not; *perforare,* to pierce]. Without opening; not open or pervious, as **imperforate anus.**
**imperforation** (im-per-for-a'-shun) [*in,* not; *perforare,* to bore through]. Occlusion; applied especially to the anus, hymen, vagina, etc.
**imperial** (im-pe'-re-al) [*imperialis,* pertaining to an empire]. Sovereign; commanding. **i.-blue.** Same as *spirit-blue.* **i. drink** or **draft,** potus imperialis, a solution of a half-ounce of potassium bitartrate in three pints of hot water, to which are added four ounces of sugar and half an ounce of fresh lemon-peel. It is a good diuretic and refrigerant drink. **i. granum,** a farinaceous food for infants. Its composition is: Water 5.49, fat 1.01, a trace each of grape-sugar and cane-sugar, starch 78.93, soluble carbohydrates 3.56, albuminoids 10.51, gum, cellulose, etc., 0.59, ash 1.16. **i.-green.** Same as *Schweinfurth-green.* **i.-violet.** Same as *rosanilin-violet.* **i. weights and measures,** those adopted in Great Britain; the old weights and measures in opposition to the metric system. See *Weights and Measures.*
**impermeable** (im-per'-me-a-bl) [*in,* not; *per,* through; *meare,* to go]. Not permitting passage: not capable of being traversed.
**impervious** (im-per'-ve-us) [*in,* not; *pervius,* capable of passage]. Not permitting passage, especially passage of fluids.
**impetiginoid** (im-pet-ij'-in-oid). Same as *impetiginous.*
**impetiginous** (im-pet-ij'-in-us) [*impetigo*]. Affected with or resembling impetigo.
**impetigo** (im-pet-i'-go) [*impetere,* to attack]. An acute inflammatory disease of the skin characterized by discrete, rounded pustules, unattended, as a rule, by itching or other subjective symptoms. **i. adenosa,** an acute contagious, febrile, cutaneous disease characterized by glistening pustules containing a yellow fluid and surrounded by a bright yellow zone. It is attended by pain in the joints, protracted vomiting, chills, and enlarged lymph-glands. **i. contagiosa,** an acute inflammatory contagious disease, characterized by the appearance of vesicles or blebs that dry into flat, straw-colored crusts. **i. herpetiformis,** a rare disease of the skin, characterized by the formation of superficial miliary pustules that may be discrete, but tend to form circular groups. It is most common in pregnant women. **i. syphilitica,** a syphilitic eruption having the characters of small flat pustules. **i. variolosa,** that occurring among the pustules of smallpox when they are drying up.
**impetus** (im'-pe-tus) [L.]. 1. Force or momentum. 2. The onset or attack of a disease, or of a paroxysm.
**implacental** (im-pla-sen'-tal) [*in,* not; *placenta*]. Without a placenta.
**implantation** (im-plan-ta'-shun) [*in,* in; *plantare,* to set]. The act of setting in, as the transplantation of a tooth from the jaw of one person to that of another; the ingrafting of epidermis from the skin of one person upon the body of another; the repair of a wounded intestine by uniting the divided ends. **i., hypodermic,** the introduction of a medicine under the skin. **i., parenchymatous,** the introduction of remedial agents into a neoplasm. **i., teratologic,** a monstrosity consisting of an imperfect, joined to a perfect, fetus.
**imponderable** (im-pon'-der-a-bl) [*in,* not; *pondus,* weight]. Incapable of being weighed; without weight. **i., fluids,** an ancient term, formerly applied to light, heat, and electricity.
**importation** (im-por-ta'-shun) [*in,* in; *portare,* to carry]. Transference from another locality or foreign country. **i. of disease,** the carrying of the contagion of disease.
**impotence** (im'-po-tens) [*in,* nŏt; *potens,* powerful]. Lack of power, especially lack of sexual power in the man.
**impotency** (im'-po-ten-se). See *impotence.*
**impregnate** (im-preg'-nāt) [*impregnare,* to make pregnant]. 1. To render pregnant. 2. To saturate or charge with.
**impregnation** (im-preg-na'-shun) [see *impregnate*]. 1. The act of rendering pregnant; fecundation. 2. The process of saturating with or charging with.
**impressio** (im-presh'-e-o) [L.]. An impression. **i. cardiaca,** a shallow depression on the upper surface of the liver for the heart. **i. colica,** an impression on the under surface of the right lobe of the liver for the hepatic flexure of the colon. **i. duodenalis,** an impression on the liver made by the duodenum. **i. gastrica,** an impression made on the liver by the stomach. **i. pylorica,** an impression made on the liver by the pyloric end of the stomach. **i. renalis,** an impression on the under surface of the liver for the right kidney and suprarenal capsule. **i. suprarenalis,** a depressed area on the liver made by the suprarenal gland.
**impression** (im-presh'-un) [*imprimere,* to press upon]. 1. A hollow or depression. 2. The effect

# IMPRESSIONABLE 458 INCISURA

produced upon the mind, the body, or a disease by external influence. **i.s, digital,** small roundish pits on the inner surface of the bones of the skull. Syn., *impressiones digitata*. **i.s, maternal,** the effects supposed to be produced upon the fetus in the uterus by mental impressions received by the mother during pregnancy. **i.-preparation,** a cover-glass upon which an entire bacterial colony has been fixed by pressing the glass lightly upon the colony. Syn., *Klatschpräparat*.
**impressionable** (*im-presh'-un-a-bl*). Readily susceptible to impressions. **i. heart,** the condition of the heart in which it is very liable to functional disturbance.
**impressorium** (*im-pres-o'-re-um*) [L.]. The seat of impressions; sensorium.
**improcreance** (*im-pro'-kre-ans*) [*in,* not; *procreare,* to beget]. The natural or acquired condition of being unable to procreate, *e. g.,* after the menopause; removal of the ovaries; lack of spermatozoa.
**improcreant** (*im-pro'-kre-ant*). Incapable of procreating.
**impuberal** (*im-pū'-ber-al*) [*in,* not; *pubes,* pubes]. Destitute of hair on the pubes. Not of adult age.
**impulse** (*im'-puls*) [*impellere,* to drive against]. 1. A push or communicated force. 2. A sudden mental feeling that urges onward to an action. **i., cardiac,** the beat of the heart felt in the fifth intercostal space to the left of the sternum. **i., morbid,** a sudden, almost uncontrollable desire to do an unlawful act.
**impulsion** (*im-pul'-shun*) [*impellere,* to impel]. The act of driving or urging onward, either mentally or physically.
**impunctate** (*im-punk'-tāt*) [*in,* not; *punctate*]. Not pricked with dots; not punctate.
**impurity** (*im-pū'-rit-e*) [*in,* not; *purus,* pure or clean]. 1. Want of purity or cleanliness. 2. Adulteration in chemistry; the condition of containing some substance other than that desired. In medicine, a want of clearness in the sounds of the heart, but not sufficient to cause a murmur. 3. The substance which causes uncleanness or adulteration by its presence. **i., respiratory,** the excess of carbon dioxide in the air of a room over that in the outside air.
**imputability** (*im-pū-tab-il'-it-e*). In legal medicine, that degree of mental soundness that makes one responsible for his own acts.
**In.** Chemical symbol of *indium*.
**in-** [L.]. 1. A prefix signifying in or within. 2. A prefix signifying negation. 3. A prefix signifying intensive action. 4. [Is, fiber], a prefix denoting fibrin or fibrous tissue.
**-in.** A termination of no precise significance, mostly applied to bodies the structure of which is not yet known. In materia medica the names of *glucosides* and *neutral principles* terminate in *-in,* and are thus distinguished from *alkaloids* which have the termination *-ine*.
**inacidity** (*in-as-id'-it-e*) [*in-; acidity*]. Want of acidity; applied to deficiency of hydrochloric acid in the gastric juice.
**inaction** (*in-ak'-shun*). Diminution or lack of response to a stimulus.
**inactivate** (*in-ak'-tiv-āt*). To render inactive; usually applied to a hemolytic or immune serum the complement of which has been destroyed by heat.
**inactivation** (*in-ak-tiv-a'-shun*). The destruction of the activity of a body fluid, such as serum.
**inactose** (*in-ak'-tōs*). An optically inactive vegetable sugar.
**inadequacy** (*in-ad'-e-kwa-se*) [*in-; adæquare,* to make equal]. Insufficiency. **i., renal,** that state of the kidney in which it is unable to remove from the blood a sufficient proportion of the effete matters that are normally excreted by it.
**inalimental** (*in-al-im-en'-tal*) [*in-; alimentum,* food]. Not nourishing; not suitable for food.
**inanagenesis** (*in-an-aj-en'-es-is*) [Is, fiber; ἀνά, again; γένεσις, production]. The renewal or regeneration of muscular fiber.
**inanaphysis** (*in-an-af'-is-si*) [Is, fiber; ἀνά, again; φύσις, growth]. Same as *inanagenesis*.
**inangulate** (*in-ang'-ū-lāt*). Having no angles.
**inanimate** (*in-an'+im-āt*) [*in-; animus,* life]. Not animate; dead; without life.
**inanition** (*in-an-ish'-un*) [*inanire,* to make empty]. Emptiness; want of food; wasting of the body from starvation.

**inappetence** (*in-ap'-et-ens*) [*in-; appetere,* to desire]. Loss of appetite.
**inarticulate** (*in-ār-tik'-ū-lāt*) [*in-; articulus,* a joint]. 1. Not jointed or articulated. 2. Vocal sounds not capable of arrangement into syllables, or of being understood.
**in articulo mortis** (*in ar-tik'-ū-lo mor'-tis*). [L.]. At the point of death.
**inassimilable** (*in-as-im'-il-a-bl*) [*in-; ad,* to; *similare,* to make like]. Incapable of assimilation.
**inaxon, inaxone** (*in-aks'-on*) [Is, fiber; ἄξων, axis]. A neuron with a long axon; its axis-cylinder processes for the most part are inclosed within a sheath.
**inca bone.** The interparietal bone. Syn., *incarial bone, q. v.*
**incallosal** (*in-cal-ō'-sal*) [*in;* priv.; *callosum*]. Without a callosum.
**incanate, incanous** (*in'-kan-āt, in'-kan-us*) [*incanus,* hoary]. Hoary white.
**incandescent** (*in-kan-des'-ent*) [*incandescere,* to become white-hot]. Glowing; emitting luminous heat-rays; heated to the degree of emitting light. **i. light,** one in which light is produced by the passage of an electric current through a strip of carbon or platinum suspended in a vacuum.
**incapsuled** (*in-kap'-sūld*) [*in-,* in; *capsula,* a small box]. Inclosed in a capsule; capsulated.
**incarcerated** (*in-kar'-ser-a-ted*) [*incarcerare,* to imprison]. Imprisoned; held fast, as *incarcerated* hernia.
**incarceration** (*in-kar-ser-a'-shun*). The imprisonment of a part, as of the placenta.
**incarial bone** (*in-ka'-re-al*) [Peruvian, *inca,* a prince]. The interparietal bone; usually in adult man a part of the occipital bone. It is called incarial, because, in the skeletons of ancient Peru, the land of the Incas, it is often persistent as a distinct bone.
**incarnant** (*in-karn'-ant*) [*incarnare,* to make flesh]. 1. Flesh-forming; promoting granulation. 2. A remedy or agent which produces flesh or promotes granulation.
**incarnatio** (*in-kar-na'-she-o*) [L.]. Conversion into flesh. **i. unguis,** the ingrowing of a nail. See *onychogryphosis*.
**incarnation** (*in-kar-na'-shun*). 1. Becoming flesh. 2. Granulation.
**incarnification** (*in-kar-nif-ik-a'-shun*). Same as *incarnation*.
**inceal** (*in'-se-al*). See *incudal*.
**incest** (*in'-sest*) [*incestus,* not chaste]. Sexual intercourse between persons of near relationship.
**inch.** The twelfth part of a foot; it equals 25.39954 millimeters.
**incidence** (*in'-sid-ens*) [*incidere,* to fall upon]. A falling upon. The direction in which one body strikes another. **i., angle of,** in optics, the angle at which a ray of light strikes a reflecting or refracting surface. **i., line of,** the path of a ray or a projectile. **i., point of,** the point upon which a ray or projectile strikes a reflecting or refracting surface.
**incident** (*in'-sid-ent*) [see *incidence*]. 1. Falling upon. 2. Same as *afferent*.
**incineration** (*in-sin-er-a'-shun*) [*in,* in; *cineres,* ashes]. The process of heating organic substances until all organic matter is driven off and only the ash remains; cremation.
**incipient** (*in-sip'-e-ent*) [*incipiens,* beginning]. Beginning to exist.
**incisal** (*in-si'-sal*) [*incisio,* a cutting]. Applied to the cutting-edge of incisors.
**incised** (*in-sīzd'*). Cut or notched. **i. wound,** a cleanly cut wound, one made by a sharp-edged instrument.
**incision** (*in-sizh'-un*) [*incisio,* a cutting]. 1. The act of cutting into anything. 2. A wound made with a cutting instrument. **i., confirmatory, i., diagnostic, i., exploratory,** section for diagnostic purposes. **i., crucial,** a cross-shaped incision, consisting of two incisions crossing each other at right angles.
**incisive** (*in-si'-siv*) [see *incision*]. 1. Cutting. 2. Pertaining to the incisor teeth. **i. bone,** that part of the superior maxilla between the two clefts in double harelip. Syn., *intermaxillary bone*.
**incisor** (*in-si'-zor*) [see *incision*]. 1. Anything that cuts, especially an incisor tooth. See under *tooth*. 2. That which supplies the incisor teeth, as the *incisor nerve*.
**incisura** (*in-si-sū'-rah*) [see *incision*]. A notch; an incision. **i., acetabuli,** the cotyloid notch. **i. cardiaca,** a notch in the anterior border of the left lung.

**i. cerebelli,** the sulcus dividing the cerebellar hemispheres. **i. cerebelli anterior,** the notch separating the hemispheres of the cerebellum in front. **i. cerebelli posterior,** the notch separating the hemispheres of the cerebellum behind. **i. intertragica,** the notch between the tragus and antitragus. **i. Rivini,** notch at the upper border of the inner end of the external auditory meatus. **i. Santorini,** either one of the two notches in the cartilaginous portion of the external auditory meatus. **i. temporalis,** the ectorhinal sulcus, a notch half-way between the temporal pole and the uncus. **i. tentorii,** a deep notch in the tentorium cerebelli for the midbrain.

**incisure** (*in-si'-zhur*) [see *incision*]. A slit or notch. **i.s of Lantermann, i.s of Schmidt,** oblique lines running across the white substance of the internodal segments of medullated nerve-fibers.

**inclination** (*in-klin-a'-shun*) [*inclinare*, to incline]. 1. A propensity; a leading. 2. The deviation of the long axis of a tooth from the vertical. **i. of uterus,** obliquity of the uterus.

**inclinometer** (*in-klin-om'-et-er*) [*inclinare*; μέτρον, a measure]. A device for determining the diameter of the eye from the horizontal and vertical lines.

**inclusio fœtalis** (*in-kloo'-ze-o fe-ta'-lis*). See *inclusion, fetal*.

**inclusion** (*in-kloo'-zhun*) [*inclusio*, a shutting up]. 1. The state of being shut in. 2. The act of shutting in. 3. That which is shut in. **i. body,** the granular substance of a red blood-corpuscle, said to be the remnant of a nucleus. **i. fetal,** a monstrosity in which one fetus is included in and overgrown by the tissues of the other fetus.

**incoagulable** (*in-ko-ag'-u-la-bl*) [*in*, not; *coagulare*, to curdle]. That which will not curdle or coagulate.

**incoercible** (*in-ko-er'-sib-l*). Uncontrollable. **i. vomiting,** pernicious vomiting.

**incoherence** (*in-ko-hēr'-ens*) [*in*, not; *cohærere*, to cling together]. The quality of being incoherent; absence of connection of ideas or of language; incongruity or inconsequence of diction.

**incoherent** (*in-ko-he'-rent*) [*in-*; *cohærere*, to stick together]. Not connected; without proper sequence.

**incombustible** (*in-kom-bus'-tib-l*) [*in*, not; *comburere*, to burn up]. Incapable of burning.

**incombustibility** (*in-kom-bus-tib-il'-it-e*) [*in*, not; *comburere*, to burn up]. The state of being incombustible.

**incompatibility** (*in-kom-pat-ib-il'-it-e*). [see *incompatible*]. The state of being incompatible. It may be chemical or physiological or therapeutic.

**incompatible** (*in-kom-pat'-ib-l*) [*in-*; *compatible*]. Of two substances, not miscible without chemical change that destroys the usefulness of either or both; nor capable of being administered together on account of antagonistic properties.

**incompetence, incompetency** (*in-kom'-pe-tens, in-kom'-pe-ten-se*) [*in-*; *competens*, sufficient]. Incapacity; inadequacy; inability to perform the natural functions. **i. of the cardiac valves,** an imperfect state of the valves of the heart in which they permit the return of blood into the cavity from which it came.

**incongruence** (*in-kon'-groo-ens*) [*incongruens*, inconsistent]. Lack of congruence. **i., retinal,** lack of correspondence in the situation of the percipient elements of the two retinæ.

**incongruity** (*in-kon-groo'-it-e*) [*in*, not; *congruere*, to go together]. Absence of agreement or of needful harmony.

**inconscient** (*in-kon'-she-ent*) [*in*, priv.; *conscius*, aware of]. Done without consciousness; applied to impulsive muscular action.

**inconstant** (*in-kon'-stant*) [*in*, priv.; *constare*, to stand together]. Changeable; not constant.

**incontinence** (*in-kon'-tin-ens*) [*in-*; *continere*, to contain]. 1. Inability to control the escape of anything, as of the feces or the urine; involuntary evacuation. 2. Venereal indulgence; lewdness.

**incoordination** (*in-ko-or-din-a'-shun*). Inability to produce voluntary muscular movements in proper order or sequence.

**incorporation** (*in-kor-por-a'-shun*) [*in-*; *corpus*, a body]. The process of intimately mixing the particles of different bodies into a practically homogeneous mass.

**incrassate** (*in-kras'-āt*) [*in*, in; *crassare*, to make thick]. Thickened or swollen.

**incrassation** (*in-kras-a'-shun*) [*in*, in; *crassus*, thick]. The process of making thick, as by inspissation; enlargement of a part, due to fatness.

**increment** (*in'-kre-ment*) [*in-*; *crescere*, to grow]. Increase or growth.

**incremental** (*in-kre-men'-tal*) [*incrementum*, growth]. Pertaining to increment or growth. **i. lines.** See *Salter's lines*.

**incrustation** (*in-krus-ta'-shun*) [*in-*; *crusta*, crust]. The formation of a crust, especially a crust-like deposit of mineral salts.

**incubation** (*in-kū-ba'-shun*) [*in-*; *cubare*, to lie]. 1. The process of sitting upon eggs to favor hatching. 2. The period of a disease between the implanting of the contagium and the development of the symptoms. 3. The process of development of a fecundated ovum. 4. The keeping of a culture in an incubator to obtain the maximum bacterial growth.

**incubator** (*in'-kū-ba-tor*) [see *incubation*]. A device for the artificial hatching of eggs or for the cultivation of bacteria; a contrivance for rearing prematurely born children.

**incubus** (*in'-kū-bus*) [L., "nightmare"]. 1. Nightmare. 2. Anciently, a male demon supposed to have sexual connection with women in their sleep.

**incudal** (*ing'-kū-dal*) [*incus*]. Relating to the incus.

**incudectomy** (*ing-kū-dek'-to-me*) [*incus*; ἐκτομή, a cutting out]. The surgical removal of the incus.

**incudiform** (*in-kū'-dif-orm*) [*incus*, anvil; *forma*, form]. Shaped like an anvil.

**incudomalleal** (*ing-kū-do-mal'-e-al*) [*incus*; *malleus*]. Relating to the incus and the malleus.

**incudostapedial** (*ing-kū-do-sta-pe'-de-al*) [*incus*; *stapes*]. Relating to the incus and the stapes.

**incuneation** (*in-kū-ne-a'-shun*) [*incuneatio*, a wedging]. 1. The impaction of a fracture or of the fetal head. 2. The same as *gomphosis*.

**incurable** (*in-kū'-ra-bl*) [*in-*; *curabilis*, curable]. Not curable.

**incurvation** (*in-ker-va'-shun*) [*incurvare*, to bend]. The state of being bent or curved in.

**incurvorecurved** (*in-ker-vo-re-kervd'*). Curved inward and then backward.

**incus** (*ing'-kus*) [L., "an anvil"]. The middle one of the chain of bones in the middle ear, so termed from its resemblance to an anvil.

**incustapedic** (*ing-kū-stap-e'-dik*). See *incudostapedial*.

**in d.** Abbreviation for *in dies* [L.] daily.

**indagation** (*in-da-ga'-shun*) [*indagare*, to trace out]. 1. Close investigation. 2. Digital examination.

**indecent** (*in-de'-sent*) [*indecens*, unbecoming]. Not decent; obscene. **i. exposure.** See *exhibitionism*.

**indecision** (*in-de-sizh'-un*). Morbid irresolution; want of firmness or of will; abulia or hypobulia.

**indehiscent** (*in-de-his'-ent*) [*in*, not; *hiscere*, to gap]. In biology, not opening spontaneously.

**indentation** (*in-den-ta'-shun*) [*in-*; *dens*, a tooth]. 1. A notch, dent, or depression. 2. A condition of being notched or serrated. **i. of tongue,** the notching of the borders of the tongue made by the teeth.

**index** (*in'-deks*) [L.]. 1. The first or fore-finger. 2. The relation or ratio of one part to another taken as a standard: **i., alveolar,** the degree of prominence of the jaws, measured by the basalveolar length multiplied by 100 and divided by the basinasal length. When the alveolar index is less than 98, the skull is *orthognathic*; when more than 103, *prognathic*; when intermediate, *mesognathic*. **i., cephalic,** the breadth of a skull multiplied by 100 and divided by its length. When this is below 75, the skull is called *dolichocephalic*; when above 80, it is called *brachycephalic*; between these limits, *mesaticephalic*. **i., cerebral,** the ratio of the greatest transverse to the greatest anteroposterior diameter of the cranial cavity, multiplied by 100. **i., color,** the amount of hemoglobin contained in each red blood corpuscle; the quotient of the hemoglobin percentage divided by the percentage of red cells. **i. of diffusion,** as applied to agar jelly containing salt and other substances—the sum of its diffusion-delaying ingredients subtracted from its diffusion-accelerating constituents added to the quantity of stain in the jelly. **i.-finger,** the first finger. **i., gnathic,** the ratio of the distance between the basion and the alveolar point to the distance between the basion and the nasal point, multiplied by 100. **i., hemorenal salt,** the ratio of the amount of inorganic salts in the urine to that in the blood. **i., length-breadth.** See *i., cephalic*. **i. movement.** See under *movement, forced.* i., **obturator** (of the pelvis), the transverse diameter multiplied by 100 and divided by the vertical diameter. i., **opsonic.** See *opsonic.* i., **palatine,** the ratio of the maximum breadth of the palatine arch to its maximum length. i., **pelvic.** See *pelvic index.* i. **of refraction,** the ratio of the sine of the angle of incidence to the sine of the angle of refraction when a ray of light passing from one medium to another is refracted. i., **refractive,** the coefficient of refraction. i., **thoracic,** the ratio of the anteroposterior diameter to the transverse, expressed in percentage. i., **vertical,** the ratio of the vertical diameter of the skull to the maximum anteroposterior diameter, multiplied by 100. i., **volume,** the relation of the volume of the red corpuscles to their number.

**indexometer** (*in-deks-om'-et-ur*) [*index;* μέτρον, measure]. An instrument to determine the index of refraction of liquids.

**Indian corn** (*in'-de-an*). See *zea mays.* **I. hemp.** See *cannabis indica.* **I. poke.** Veratrum viride, *q. v.* **I. tobacco.** See *lobelia.*

**India ink method.** A method of making the *Spirochæta pallida* visible under the microscope by means of India ink.

**India-rubber.** See *caoutchouc.*

**indican** (*in'-dik-an*) [*indigo*]. 1. $C_{26}H_{31}NO_{17}$. A glucoside occurring in indigo-plants, and by the decomposition of which indigo is produced. 2. Potassium indoxyl-sulphate, $C_8H_6NSO_4K$, a substance occurring in urine and sweat, and formed from indol. For tests, see *Jaffé, MacMunn, Obermeyer, Weber.*

**indicanidrosis** (*in-dik-an-id-ro'-sis*) [*indican;* ἱδρώς, sweat]. The presence of indican in the perspiration.

**indicant** (*in'-dik-ant*) [*indicare,* to indicate]. 1. Serving as an index or as an indication. 2. A fact or symptom that indicates a certain treatment; an indication.

**indicanuria** (*in-dik-an-u'-re-ah*) [*indican;* οὖρον, urine]. Morbid excess of indican in the urine. See *Jaffé, MacMunn, Obermeyer, Weber.*

**indication** (*in-dik-a'-shun*) [*indicare,* to point out]. That which points out; a guide, especially that which points out the course of treatment.

**indicator** (*in'-dik-a-tor*). 1. The index-finger. 2. The extensor indicis muscle. 3. In chemistry, a substance used to show by a color-change when a change of reaction has taken place or a chemical affinity has been satisfied. 4. A mechanism like the hand of a dial to register movements or processes.

**indicium** (*in-dish'-e-um*) [L.: *pl., indicia*]. A symptom or sign; a discrimination or diagnostic mark.

**indicophose** (*in'-dik-o-fōz*). A blue-colored phose.

**indifferent** (*in-dif'-er-ent*) [*in-; differens,* different]. 1. Not differentiated; not tending to build up tissue, as *indifferent* cells. 2. Not readily acted upon by agents. 3. Neutral.

**indifferentism** (*in-dif'-er-ent-ism*) [*in,* not; *differens,* different]. Lack of special differentiation.

**indigenous** (*in-dij'-en-us*) [*indu,* within;. *gignere,* to beget]. Native; originating or belonging to a certain locality or country.

**indigestion** (*in-di-jes'-chun*) [*in-; digerere,* to digest]. Imperfect digestion.

**indigitation** (*in-dij-it-a'-shun*) [*in-; digitus,* a finger]. 1. A displacement of a part of the intestine by intussusception. 2. Invagination.

**indiglucin** (*in-de-gloo'-sin*) [*indigo;* γλυκύς, sweet], $C_8H_{10}O_5$. A yellow syrup, one of the decomposition-products of indican.

**indigo** (*in'-dig-o*) [ἰνδικόν, indigo], $C_{16}H_{10}N_2O_2$. A blue pigment formed by the decomposition of the indican contained in various species of *Indigofera* (*Indigofera tinctoria, I. añil, I. argentea*), or in the urine and sweat. i.-**blue,** $C_{16}H_{10}N_2O_2$, a blue pigment from indigo. Syn., *indigotin.* i.-**carmin,** potassium or sodium sulphindigotate, used as a stain in microscopy and as a test for sugar. i.-**carmin paper,** paper charged with indigo-carmin and sodium carbonate used as a test for sugar in urine. i.-**red.** See *indirubin.* i. **white,** indigogen, a substance obtained by the reduction of indigo-blue.

**indigogen** (*in'-dig-o-jen*). See *indigo-white* and *uroxanthin.*

**indigotin** (*in-dig-o'-tin*). See *indigo-blue.*

**indigouria** (*in-dig-o-u'-re-ah*) [*indigo;* οὖρον, urine]. The presence of indigo in the urine; it is due to a decomposition of indican. See *cyanuria.*

**indirect** (*in'-di-rekt*) [*in-; directus,* straight]. Not direct; not in a direct line; acting through an intervening medium. i. **cell-division.** See *karyokinesis.* i. **vision, vision** by some other part of the retina than the macula.

**indirubin** (*in-di-roo'-bin*) [*indigo; rubrum,* red], $C_{16}H_{10}N_2O_2$. A substance isomeric with indigo-blue and very similar to it. It is produced by condensing indoxyl with isatin by means of a dilute soda solution. Syn., *indigo-red.* See *Rosenbach, rosin.*

**indisposition** (*in-dis-po-zish'-un*) [*in-; dispositio,* disposition]. A slight illness not confining the patient to bed.

**indium** (*in'-de-um*) [*indicum,* indigo, so-called from its indigo-blue spectral line]. A rare metal. Symbol In, atomic weight, 114.8. It is very soft, and resembles lead in its properties. See *elements, table of.*

**indol** (*in'-dol*) [*indigo*], $C_8H_7N$. A substance produced in pancreatic digestion, in intestinal putrefaction, and in certain bacterial cultures. It occurs in the feces, giving to them in part their odor, and is eliminated in the urine in the form of indican, being especially increased in intestinal obstruction. See *Baeyer, Nencki, Salkowski.*

**indolaceturia** (*in-dol-as-e-tū'-re-ah*). Presence of indolacetic acid in the urine.

**indolemia** (*in-do-le'-me-ah*) [*indol,* Indian; λοιμός, pestilence]. Asiatic cholera.

**indolent** (*in'-do-lent*) [*in,* not; *dolere,* to feel pain]. Sluggish, without pain; applied to ulcers, tumors, etc.

**indoxyl** (*in-doks'-il*) [*indigo;* ὀξύς, sharp], $C_8H_6$(OH)N. The product derived from indol by oxidation, as it takes place in the liver. i.-**sulphate,** indican, $C_8H_6NSO_4K$, a combination of indoxyl with a sulphate and found in the urine.

**indoxylsulphuric acid** (*in-doks-il-sul-fū'-rik*). See *acid, indoxylsulphuric,* and *indican.*

**indoxyluria** (*in-doks-il-u'-re-ah*) [*indoxyl;* οὖρον, urine]. Excess of indoxyl in the urine.

**induced** (*in-dūsd'*) [see *induction*]. 1. Produced by induction, as **induced electricity.** 2. Produced artificially, as **induced labor.**

**induction** (*in-duk'-shun*) [*inducere,* to lead in]. 1. The act of bringing on. 2. The process of drawing general conclusions from special facts. 3. The production of electricity or magnetism in a body by proximity to another body, which is electrified or magnetized, but not in direct contact with it. i.-**balance,** an instrument used for detecting the presence of metallic bodies by the electric disturbance which they cause. i.-**coil,** a wire wound around a bobbin, used for conducting a galvanic current, by means of which electricity is induced in a second coil.

**inductogram** (*in-duk'-to-gram*). See *skiagram.*

**inductometer** (*in-duk-tom'-et-er*) [*induction;* μέτρον, a measure]. An apparatus for estimating the degree of electric induction.

**inductorium** (*in-duk-to'-re-um*) [L., "a covering"]. An apparatus for producing induced currents. i., **DuBois-Reymond's,** an induction apparatus with a primary and secondary coil in which the primary current is never opened, it being short-circuited. It is used in physiological laboratories.

**indulin** (*in'-du-lin*). A coal-tar dye, used as a tissue stain in histology.

**induinophil** (*in-du-lin'-o-fil*) [*indulin;* φιλεῖν, to love]. Staining with indulin.

**indurated** (*in'-du-ra-ted*) [see *induration*]. Hardened, as *indurated* chancre.

**induration** (*in-du-ra'-shun*) [*in-; durus,* hard]. Hardening of a tissue or part; the state of being or becoming hard; a hardened mass or lump. i., **black,** the hardened, pigmented condition of the lung in anthracosis. i., **brown,** a form of interstitial pneumonia in which there is, in addition to the new-growth of fibrous tissue, a deposit of altered blood-pigment. i., **fibroid.** See *i., gray,* and *cirrhosis.* i., **gray,** the appearance of the lung in chronic pneumonia, the cut surface being smooth, glistening, gray, and dense. i., **red,** an interstitial pneumonia in which the lung is red from congestion.

**indurative** (*in'-du-ra-tiv*). Pertaining to induration.

**indurescent** (*in-du-res'-ent*). Gradually becoming hardened.

**indusium** (*in-dū'-ze-um*) [*induere,* to put on; *pl., indusia*]. 1. A membranous covering. 2. The amnion. 3. A marginal layer of gray matter on the corpus callosum, also called *i. griseum.*

**-ine.** A termination used in forming (1) the names of the elements bromine, chlorine, fluorine,

and iodine; (2) the names of the alkaloids. And see -*in.*
**inebriant** (*in-e'-bre-ant*) [see *inebriety*]. 1. Intoxicant; causing inebriation. 2. An agent that causes inebriation.
**inebriation** (*in-e-bre-a'-shun*) [see *inebriety*]. The condition of drunkenness.
**inebriety** (*in-e-bri'-et-e*) [*inebriare*, to make drunk]. Habitual drunkenness.
**inedia** (*in-e'-de-ah*) [*in*, not; *edere*, to eat]. Synonym of *fasting*.
**inenucleable** (*in-e-nū'-kle-a-bl*) [*in*, not; *enucleare*, to shell out]. Not removable by enucleation.
**inertia** (*in-er'-she-ah*) [*iners*, inactive]. Sluggishness; inability to move except by means of an external force. In physics, that property of matter by virtue of which it is incapable of changing its condition of rest or motion. i., **uterine**, sluggishness of uterine contractions during labor.
**in extremis** (*in eks-tre'-mis*) [L.]. At the end; at the last; at the point of death.
**infancy** (*in'-fan-se*) [*infans*, not able to speak, a little child]. Early childhood. i., **diseases of**, those to which infants are peculiarly liable.
**infant** (*in'-fant*) [*infans*, not able to speak, a little child]. 1. A babe. 2. According to English law, one not having attained the age of 21.
**infanticide** (*in-fant'-is-īd*) [*infant; cædere*, to kill]. 1. The murder of an infant. 2. The murderer of an infant.
**infantile** (*in'-fan-til*) [*infant*]. Pertaining to infancy. i. **hernia**, oblique inguinal hernia behind the funicular peritoneal process. i., **paralysis**, acute anterior poliomyelitis. i. **uterus**, an undeveloped womb.
**infantilism** (*in-fant'-il-izm*) [*infantile*]. The persistence of childish characteristics into adult life. i., **Lorain's type of**, represented by an individual small in stature but of the adult type, with pubic and axillary hair wanting, and with fair intelligence. i., **myxedematous**, a type characterized by chubby face, prominent lips and abdomen, rudimentary genitals, high-pitched voice, second dentition retarded or absent, and infantile mental state.
**infarct** (*in'-farkt*) [*infarcire*, to stuff in]. A wedge-shaped area, either of hemorrhage into an organ (*hemorrhagic infarct*), or of necrosis in an organ (*anemic infarct*), produced by the obstruction of a terminal vessel. i., **uric-acid**, the deposition of crystals of uric acid in the renal tubules of the newborn.
**infarction** (*in-fark'-shun*) [see *infarct*]. The production of an infarct; also the infarct itself.
**infect** (*in-fekt'*) [*inficere*, to put in, or corrupt]. To communicate or transmit the specific virus or germs of disease.
**infecting** (*in-fek'-ting*) [*infection*]. Causing infection, as an *infecting* embolus.
**infection** (*in-fek'-shun*) [*infectio*, from *in*, into; *facere*, to make]. 1. The communication of disease from one body to another, or from one part to another part of the same individual (*autoinfection*). 2. The material conveying the disease; the disease-producing agent. i.-**atrium**, the point of entrance of an infection. i., **consecutive**, septic infection implanted upon an already established morbid process. i., **mixed**, infection by more than one kind of bacterium at the same time. i., **secondary**, same as *i.*, *consecutive.*
**infectious** (*in-fek'-shus*) [see *infection*]. 1. Communicating disease. 2. Caused by an infection.
**infective** (*in-fek'-tiv*) [*infectio*, infection]. Infectious. i. **angioma**. See *angioma serpiginosum.*
**infectivity** (*in-fek-tiv'-it-e*) [*inficere*, to infect]. Infectiousness; the quality of being infectious.
**infecundity** (*in-fe-kun'-dit-e*) [*in-*; *fecundus*, fruitful]. Sterility; barrenness.
**inferent** (*in'-fer-ent*). Same as *afferent.*
**inferior** (*in-fe'-re-or*) [comp. of *inferus*, low]. Lower.
**inferocostal** (*in-fer-o-kos'-tal*) [*inferior; costa*, a rib]. Relating to the lower border of a rib or the region beneath it.
**inferofrontal** (*in-fer-o-front'-al*) [*inferior; frons*, the forehead]. Relating to the inferior part of the frontal lobe.
**inferolateral** (*in-fer-o-lat'-er-al*) [*inferus*, low; *latus*, side]. Situated below and to one side.
**inferoposterior** (*in-fer-o-pos-te'-re-or*) [*inferus*, low; *posterius*, posterior]. Situated backward and below.

**infertility** (*in-fer-til'-it-e*) [*in*, not; *fertilis*, fertile]. Same as *sterility.*
**infestation** (*in-fes-ta'-shun*). The state or condition of being infested. The term is used with reference to the presence of animal parasites in or on the human body.
**infibulation** (*in-fib-ū-la'-shun*) [*in-; fibula*, a clasp]. The operation of fastening the prepuce over the glans penis, or of fastening together the labia of the vagina.
**infiltrate** (*in'-fil-trāt*) [see *infiltration*]. 1. To ooze into the spaces of a tissue. 2. The substance that has oozed out.
**infiltration** (*in-fil-tra'-shun*) [*in-; filtrare*, to strain]. 1. The entrance into the tissue-spaces or into the tissue-elements of some abnormal substance or of a normal substance in excess. 2. The material thus deposited. i.-**anesthesia**, local anesthesia from cocaine-injections. i., **calcareous**, the deposit of lime and magnesium salts in the tissues. i., **cellular**, an infiltration of the tissues with round-cells. i., **circumferential**, in surgery, cutting off the area of operation from all nerve communication with surrounding parts by a wall of anesthetizing edema. i., **fatty**, the deposit of fat in the tissues; the presence of oil- or fat-globules in the interior of a cell. i., **glycogenic**, the deposit of glycogen-granules in the cells. i., **pigmentary**, the deposit of pigment in the tissues, derived either from without or from within. i., **purulent**, the presence of scattered pus-cells in a tissue. i., **serous**, an infiltration of the tissues with diluted lymph. i., **tuberculous**, a confluence of tuberculous nodules. i., **urinous**, the effusion of urine into a tissue. i., **waxy**, a deposit of waxy substance.
**infinite** (*in'-fin-it*) [*in-; finis*, boundary]. Immeasurable or innumerable; unlimited, when compared with any known or conceivable quantity. i. **distance**, a term in optics practically taken as twenty feet. Rays from an object at that distance and entering the eye are practically parallel, as they would be completely if coming from a point at a really infinite distance.
**infinitesimalism** (*in-fin-it-es'-im-al-izm*) [*infinitus*, boundless]. The doctrine that favors the infinitesimal dilution of drugs with the view of potentizing them, or developing their power over disease.
**infirm** (*in-ferm'*) [*in-; firmus*, firm]. Weak or feeble.
**infirmary** (*in-fer'-ma-re*) [*infirmarium*, an infirmary]. A hospital; an institution where ill and infirm persons are maintained during the period of treatment.
**infirmity** (*in-fer'-mit-e*) [*infirm*]. 1. Weakness; feebleness. 2. A disease producing feebleness.
**inflame** (*in-flām'*) [*inflammare*, to set on fire]. To undergo inflammation. To become unduly heated and turgid with blood, owing to a morbid condition.
**inflammation** (*in-flam-a'-shun*). [*inflammare*, to set on fire; to inflame]. A morbid condition with hyperemia, pain, heat, swelling, and disordered function. It is accompanied by overfilling of the blood-vessels, alteration in the blood-vessel walls, outwandering of leukocytes, exudation of plasma, and multiplication of the cells of the surrounding connective tissue. i., **acute**, that in which the processes are active; usually this form is characterized by the cardinal symptoms of inflammation—heat, redness, swelling, and pain. i., **catarrhal**, one occurring on a mucous surface and causing the shedding of its epithelium. i., **chronic**, that in which there is a building-up of new connective tissue. i., **interstitial**, one affecting chiefly the connective tissue of an organ. i., **parenchymatous**, one affecting chiefly the parenchyma of an organ. i., **plastic**. Same as *i., productive*. i., **productive**, that accompanied by the formation of new tissue. i., **reactive**, an inflammation set up around a focus of degeneration to limit the spread of the degenerative process; also the inflammation around a foreign body. i., **specific**, one due to a special microorganism, and characterized by the formation of a tumor-like nodule that tends to degenerate. i., **suppurative**, that attended by the formation of pus. i., **toxic**, that due to poison.
**inflammatory** (*in-flam'-at-o-re*) [*inflammare*, to inflame]. Pertaining to inflammation.
**inflation** (*in-fla'-shun*) [*inflare*, to blow up]. Distention with air.
**inflected** (*in-flek'-ted*) [*in*, in; *flectere*, to bend]. Bent inward or downward.

**inflection, inflexion** (*in-flek'-shun*). 1. A bending inward. 2. Modification of the pitch of the voice in speaking.
**inflexed** (*in-flekst'*) [*inflectere*, to bend]. Bent. Same as *inflected*.
**influenza** (*in-floo-en'-zah*) [Ital., "an influence"]. An epidemic affection characterized by catarrhal inflammation of the mucous membrane of the respiratory tract, accompanied by a mucopurulent discharge, fever, pain in the muscles, and prostration. At times symptoms referable to the gastrointestinal system predominate; at others the symptoms are mainly referred to the nervous system. The cause of the disease is *Bacillus influenzæ*, discovered by Pfeiffer and Canon. Complications are common, pneumonia being the most frequent; pleurisy, otitis media, and neuritis also occur. An occasional sequel is insanity. Syn., *grip; la grippe*.
**influenzin** (*in-floo-en'-zin*). A proprietary remedy said to be a mixture of phenacetin, caffeine, quinine, salicylate, and sodium chloride; used in influenza.
**influx** (*in'-fluks*) [*in*, in; *fluere*, to flow]. An inflow. The act of flowing in.
**infra-** (*in-frah-*) [*infra*, below]. A prefix meaning below or beneath.
**infraaxillary** (*in-frah-aks'-il-a-re*) [*infra-; axilla*, the armpit]. Below the armpit.
**infrabranchial** (*in-frah-brang'-ke-al*) [*infra-; branchia*, gills]. Beneath or below the gills.
**infrabuccal** (*in-frah-buk'-al*) [*infra-; bucca*, the cheek]. In molluscs beneath the buccal mass.
**infraclavicular** (*in-frah-klav-ik'-u-lar*) [*infra-; clavicula*, the collar-bone]. Below the collar-bone. i. **region**, the space on the chest between the clavicle and the third rib.
**infracommissure** (*in-frah-com'-is-ūr*) [*infra-; committere*, to unite]. The inferior commissure of the brain.
**infraconscious** (*in-frah-kon-shus*) [*infra-; conscius*, aware of]. Subconscious.
**infraconstrictor** (*in-frah-kon-strik'-tor*) [*infra-; constringere*, to bind together]. The inferior constrictor of the pharynx. See *muscles, table of*.
**infracortical** (*in-frah-kor'-tik-al*) [*infra-; cortex*, a bark]. Lying beneath the cortical substance of the brain or kidney.
**infracostal** (*in-frah-kos'-tal*) [*infra-; costa*, a rib]. Below the ribs.
**infracostales** (*in-frah-kos-ta'-lēz*). See *muscles, table of*.
**infracotyloid** (*in-frah-kot'-il-oid*). Below the cotyloid cavity or acetabulum.
**infraction** (*in-frak'-shun*) [*in-; fractio*, breaking]. Incomplete fracture of a bone.
**infradiaphragmatic** (*in-frah-di-af-rag-mat'-ik*) [*infra-; diaphragm*]. Situated below the diaphragm.
**infragenual** (*in-frah-jen'-u-al*) [*infra-; genu*, the knee]. Subpatellar.
**infraglenoid** (*in-frah-gle'-noid*) [*infra-;* γλήνη, cavity; εἶδος, likeness]. Located below the glenoid cavity.
**infraglottic** (*in-frah-glot'-ik*) [*infra-; glottis*]. Below the glottis.
**infrahyoid** (*in-frah-hi'-oid*) [*infra-; hyoid*]. Situated below the hyoid bone.
**infrainguinal** (*in-fra-in'-gwin-al*) [*infra-; inguen*, groin]. Below the inguinal region.
**infralemniscus** (*in-frah-lem'-nisk*) [*infra-; lemniscus*, a pendent ribbon]. Wilder's term for the lower lamina of the lemniscus.
**inframammary** (*in-frah-mam'-ar-e*) [*infra-; mamma*, the breast]. Situated beneath the mamma. i. **region**, the area on the chest below the sixth rib.
**inframarginal** (*in-frah-mar'-jin-al*). See *submarginal*. i. **convolution**, the inferior temporal gyrus.
**inframaxillary** (*in-frah-maks'-il-a-re*) [*infra-; maxilla*, the jaw]. Below or under the jaw.
**infraoccipital** (*in-frah-ok-sip'-et-al*). See *suboccipital*.
**infraocclusion** (*in-frah-ok-loo'-shun*). Failure of apposition of one or more teeth when the jaws are closed.
**infraorbital** (*in-frah-or'-bit-al*) [*infra-; orbita*, orbit]. Beneath or below the floor of the orbit. i. **canal**, the canal in the superior maxillary bone that transmits the infraorbital vessels and nerve. i. **foramen**. See *foramen, infraorbital*.
**infrapatellar** (*in-frah-pa-tel'-ar*). Pertaining to parts below the patella.
**infraprotein** (*in-frah-pro'-te-in*). See *metaprotein*.
**infrapubic** (*in-frah-pu'-bik*). Pertaining to parts below the pubis.

**infrarectus** (*in-frah-rek'-tus*) [*infra-; rectus*, straight]. The inferior rectus muscle of the eye. See *muscles, table of*.
**infrascapular** (*in-frah-skap'-u-lar*) [*infra-; scapula*]. Below the shoulder-blade.
**infraspinatus** (*in-frah-spi-na'-tus*). See *muscles, table of*.
**infraspinous** (*in-frah-spi'-nus*) [*infra-; spina*, a spine]. Beneath a spine, as of the scapula or a vertebra. i. **fascia**, the dense membranous fascia covering the infraspinous muscle. i. **fossa**, the shallow depression on the dorsal surface of the scapula, below the spine, and lodging the infraspinatus muscle. i. **muscle**. See *infraspinatus* under *muscle*.
**infrastapedial** (*in-frah-sta-pe'-de-al*) [*infra-; stapes*]. Below the stapes.
**infrasternal** (*in-frah-ster'-nal*) [*infra-; sternum*]. Below the sternum. i. **depression**, the depression of the ensiform cartilage.
**infratemporal** (*in-frah-tem'-po-ral*) [*infra-; tempora*, the temple]. Situated beneath the temporal bone.
**infrathoracic** (*in-frah-tho-ras'-ik*) [*infra-; thorax*]. Below the thorax.
**infratonsillar** (*in-frah-ton'-sil-ar*) [*infra-; tonsilla*, the tonsil]. Below the tonsil.
**infratrochlea** (*in-frah-trok'-le-ah*) [*infra-; trochlea*, a pulley]. Below the trochlea. i. **nerve**. See *nerves, table of*.
**infratrochlear** (*in-frah-trok'-le-ar*). See *subtrochlear*.
**infraturbinal** (*in-frah-tur'-bin-al*) [*infra-; turbo*, a wheel, top]. Inferior turbinal. See *bones, table of*.
**infraumbilical** (*in-frah-um-bil'-ik-al*). Situate below the umbilicus.
**infravaginal** (*in-frah-vaj'-in-al*) [*infra-; vagina*]. Situated below the vaginal vault.
**infriction** (*in-frik'-shun*) [*infrictio*, a rubbing in]. The rubbing of a surface with an ointment or liniment.
**infundibula** (*in-fun-dib'-u-lah*) [L.]. Plural of *infundibulum, q. v.*
**infundibular** (*in-fun-dib'-u-lar*) [*infundibulum*]. Pertaining to a funnel or resembling one; infundibuliform.
**infundibuliform** (*in-fun-dib-u'-le-form*) [*infundibulum; forma*, a form]. Funnel-shaped. i. **fascia**, the funnel-shaped membranous layer that invests the spermatic cord.
**infundibulin** (*in-fun-dib'-u-lin*). An extract of the posterior lobe of the pituitary body.
**infundibuloovarian** (*in-fun-dib-u-lo-o-va'-re-an*) [*infundibulum; ovarium*, ovary]. Relating to the oviduct and to the ovary.
**infundibulopelvic** (*in-fun-dib-u-lo-pel'-vik*). Relating to the oviduct and the pelvis.
**infundibulum** (*in-fun-dib'-u-lum*) [L., "a funnel"]. A funnel-shaped passage or part. i. **of brain**, a tubular mass of gray matter attached to the pituitary body. i. **of cochlea**, a small cavity at the end of the modiolus. i. **of ethmoid bone**, a canal connecting the anterior ethmoid cells with the middle meatus of the nose. i. **of heart**, the arterial cone from which the pulmonary artery arises. i. **of kidney**, one of the primary divisions of the pelvis of the kidney. i. **of lung**, one of the air-spaces into which a terminal bronchiole divides, and which is composed of an aggregation of air-vesicles. i. **of oviduct**, the cavity formed by the fringes at the ovarian end of an oviduct.
**infundin** (*in-fun'-din*). Trade name of a preparation of the posterior lobe of the pituitary body.
**infused** (*in-fūzd'*) [*infundere*, to pour in]. Extracted; steeped. i. **oils**. See *olea infusa*.
**infusible** (*in-fūz'-sib-l*). Incapable of being fused.
**infusion** (*in-fū'-shun*) [*infusum*]. 1. The process of extracting the active principles of a substance by means of water, but without boiling. 2. The product of such a process, known in pharmacy as *infusum* (*q. v.*). 3. The slow injection of liquid into a vein. i. **dural**, the use of the lumbar puncture for immediate applications in cerebrospinal diseases.
**infusodecoction, infusodecoctum** (*in-fū-zo-de-kok'-shun, -tum*). A combination of a decoction of a substance with an infusion of it.
**infusor** (*in-fū'-zor*) [L.]. An instrument by means of which water or a medicated liquid may be made to flow slowly into a vein, or into the parenchymatous tissues.
**Infusoria** (*in-fū-zo'-re-ah*) [*infusum*]. A class of Protozoa so called because they often develop in great numbers in organic infusions.

**infusum** (in-fū'-zum) [in-; fundere, to pour]. An infusion. i. **digitalis** (U. S. P.), infusion of digitalis. Dose 2 dr. (8 Cc.). i. **pruni virginianæ** (U. S. P.), infusion of wild cherry. Dose 2 oz. (60 Cc.). i. **sennæ compositum** (U. S. P.), compound infusion of senna; black draught. Dose 4 oz. (128 Cc.).
**ingesta** (in-jes'-tah) [in-;-gerere, to carry]. Substances introduced into the body, especially foods.
**ingestion** (in-jes'-chun) [ingesta]. 1. The act of taking substances, especially food, into the body. 2. The process by which a cell takes up foreign matters, such as bacilli or smaller cells.
**ingestol** (in-jes'-tol). A proprietary remedy for use in gastric and intestinal diseases. Syn., amarol.
**ingluvial** (in-gloo'-ve-al) [ingluvies, the crop]. Pertaining to the ingluvies.
**ingluvies** (in-gloo'-ve-ēz) [L.]. 1. The crop or craw of birds. 2. The paunch or rumen of ruminating mammals.
**ingluvin** (in'-gloo-vin) [ingluvies]. A preparation obtained from the gizzard of a fowl, *Pullus gallinaceus*, used as a substitute for pepsin and pancreatin, and also in the vomiting of pregnancy. Dose 10–20 gr. (0.65–1.3 Gm.).
**Ingrassias, processes of, I., wings of** (in-gras'-e-as) [Giovanni Filippo Ingrassias, Italian physician, 1510–1580]. The lesser wings of the sphenoid bone.
**ingravescent** (in-grav-es'-ent) [ingravescere, to become heavier]. Increasing in weight or in severity, as *ingravescent* apoplexy.
**ingravidation** (in-grav-id-a'-shun). See *impregnation*.
**ingredient** (in-gre'-de-ent) [ingredi, to step into]. Any substance that enters into the formation of a compound.
**ingrowing nail.** See *nail, ingrowing*.
**inguen** (in'-gwen) [L.]. The groin.
**inguinal** (in'-gwin-al) [inguen]. Pertaining to the groin. i. **canal**, the canal transmitting the spermatic cord in the male and the round ligament in the female. It is situated parallel to and just above Poupart's ligament. i. **glands**, the superficial and the deep glands of the groin. i. **hernia.** See *hernia, inguinal*. i. **ligament**, *Poupart's ligament, q. v.*
**inguino-** (in'-gwin-o-) [inguen, the groin]. In composition, pertaining to the groin.
**inguinoabdominal.** Pertaining conjointly to the groin and the abdomen.
**inguinocrural.** Relating to the groin and the thigh.
**inguinocutaneous.** Relating to the integument of the groin.
**inguinodynia** (in-gwin-o-din'-e-ah) [inguen; ὀδύνη, pain]. Pain in the groin.
**inguinointerstitial** (in-gwin-o-in-ter-stish'-al) [inguen; interstitial]. Within the tissues of the inguinal region.
**inguinolabial** (in-gwin-o-la'-be-al). Relating to the groin and a labium majus.
**inguinoscrotal** (in-gwin-o-skro'-tal). Relating to the groin and the scrotum.
**ingulation** (in-gū-la'-shun) [in, into; gula, the throat]. The introduction of anything into the throat.
**ingurgitation** (in-gur-jit-a'-shun) [ingurgitatio, a swallowing]. 1. The act of swallowing; deglutition. 2. Excess in eating or drinking.
**inhalant, inhalent** (in-ha'-lant, -lent). 1. See *inhalation* (2). 2. Useful for inhalation.
**inhalation** (in-ha-la'-shun) [inhalare, to draw in]. 1. The breathing in of air or other vapor. 2. A medicinal substance to be used by inhalation. i.-**diseases**, those due to the inspiration of air containing dust or any finely divided matter. i.-**therapy**, treatment of a disease by inspiration of medicated vapors.
**inhale** (in'-hāl) [inhalare, to breathe in]. To inspire or draw air or other vapor into the lungs.
**inhaler** (in-ha'-ler). An instrument for inhaling a gas or vapor.
**inherent** (in-her'-ent) [in, to; hærere, to cleave]. Innate; natural to the organism.
**inheritance** (in-her'-it-ans). 1. The act of inheriting. 2. Transmitted characteristics. i., **amphigonous**, of Haeckel, characteristics transmitted from both parents. i., **homochronous**, Haeckel's name for a characteristic in an offspring, shown at an age identical with that in which it was manifest in the parent. i., **homotópic**, of Haeckel, the inheritance of acquired characteristics. i., **particulate**, the reappearance of single peculiarities in the offspring. i., **use**, the inheritance of acquired characters; the acquisition by the offspring of changes in the body-cells of the parent.
**inherited** (in-her'-it-ed) [in, in, to; *heres*, heir]. Derived from an ancestor. i. **disease**, a disease that has been transmitted to a child by its parent.
**inhibit** (in-hib'-it) [inhibere, to check]. To check, restrain, or suppress.
**inhibition** (in-hib-ish'-un) [inhibere, to check]. The act of checking or restraining; a restraint.
**inhibitory** (in-hib'-it-o-re) [see *inhibition*]. Checking; restraining.
**inhibitrope** (in-hib'-it-rōp) [inhibere, to check]. An individual in whom certain stimuli cause a partial arrest of function.
**inhumation** (in-hū-ma'-shun) [inhumare, to put in the ground]. Burial of the dead in the ground.
**iniac, inial** (in'-e-ak, in'-e-al) [inion]. Pertaining to the inion.
**iniad** (in'-e-ad) [ἰνίον, the occiput]. Toward the inial aspect.
**inien** (in'-e-en) [ἰνίον, occiput]. Belonging to the inion in itself.
**iniencephalus** (in-e-en-sef'-al-us) [inion; ἐγκέφαλος, brain]. A fetal monstrosity in which there is a posterior fissure of the skull, with protrusion of the brain-substance, combined with spinal fissure.
**iniodymus** (in-e-od'-im-us) [ἰνίον, occiput; δίδυμος, double]. A teratism with one body and two heads joined at the occiput.
**iniofacial** (in-e-o-fa'-shal). Relating to the inion and the face.
**inioglabellar** (in-e-o-gla-bel'-ar) [ἰνίον, occiput: *glaber*, smooth]. Relating to or joining the inion and the glabella.
**iniomesial** (in-e-o-me'-ze-al). Relating to the inion and to the meson.
**inion** (in'-e-on) [ἰνίον, occiput]. The external protuberance of the occipital bone. See under *craniometric point*.
**iniops** (in'-e-ops) [ἰνίον, occiput; ὤψ, face]. A catadidymous monstrosity with the parts below the navel double, the thoraces joined into one, and one head with two faces, the one incomplete.
**inirritative** (in-ir'-it-a-tiv) [in, not; *irritare*, to irritate]. Not irritant; soothing.
**initial** (in-ish'-al) [in, into; *ire*, to go]. Beginning; early; primary, as the *initial* lesion of syphilis—the chancre. i. **cells**, germ-cells. i. **sclerosis**, the hard chancre.
**initis** (in-i'-tis) [ἴς, muscle;, fiber; ιτις, inflammation]. 1. Inflammation of fibrous or muscular tissue. 2. Inflammation of a tendon; tenontitis.
**inj.** Abbreviation for *injectio* [L.], injection.
**inject** (in-jekt') [see *injection*]. To throw or force in, as to *inject* fluids into the tissues; also, to fill the vessels of an organ. In pathology, to produce a condition of distention of the capillaries with blood.
**injecta** (in-jek'-tah) [L.]. Things introduced, as into the alimentary canal (correlated with *ejecta*).
**injection** (in-jek'-shun) [in, into; *jacere*, to throw]. 1. The act of injecting or throwing in. 2. The substance injected. According to the organ into which the injection is made, different terms are employed, as *urethral, intramuscular, uterine, vaginal* injection, etc. In the British Pharmacopoeia *injectiones* are solutions of active substances used for hypodermatic injection. i., **anatomical**, filling the vessels of a cadaver or of an organ with preservative or coagulating solutions, for purposes of dissection. i., **coagulation**, injection of coagulating solutions into the cavity of an aneurysm. i., **hypodermic**, into the subcutaneous connective tissue, by means of a syringe. i., **nutrient**, injection of nutritive fluids into the rectum or other cavity of the body. i., **opaque, naked-eye**, for anatomical or microscopical purposes, made of plaster of Paris, tallow, vermilion and gelatin, plumbic acetate and potassium bichromate (yellow injection), or plumbic acetate and sodium carbonate (white). i., **transparent microscopical**, made with carmine for red, potassium bichromate for yellow. i.-**pneumonia**, a condition of the lung somewhat resembling pneumonia, following injections of Koch's tuberculin.
**injector** (in-jek'-tor). An apparatus used in injecting.
**injury** (in'-joo-re) [in, not; *jus*, law]. A harm or hurt to the body.
**in-knee** (in'-ne). Knock-knee, or genu valgum, *q. v.*

inlay (in'-la). In dentistry applied to fillings first made and then inserted into a cavity with cement; also applied to any filling occupying but one surface of a tooth.
inlet (in'-let). The place where the air of ventilation is admitted into a room. i. of the pelvis, the heart-shaped space within the brim of the pelvis; the superior pelvic strait.
Inman's disease (in'-man) [Thomas Inman, English physician, 1820-1876]. Myalgia.
innate (in-nāt' or in'-nāt) [innatus, born in]. Congenital; native to the organism; intrinsic.
innervation (in-er-va'-shun) [in, in; nervus, nerve]. 1. Nerve-supply. 2. A discharge of nervous force.
innocent (in'-o-sent) [in, not; nocere, to harm]. Benign, not harmful.
innocuous (in-ok'-ū-us) [innocuus, harmless]. Not injurious.
innominata (in-nom-in-a'-tah). 1. [Plural of innominatum]. The innominate bones. 2. [Feminine of inomnominatus]. The innominate artery.
innominate (in-om'-in-āt) [see innominatum]. Unnamed; unnamable. i. artery. See under artery. i. bone, the irregular bone forming the sides and anterior wall of the pelvic cavity, and composed of the ilium, ischium, and pubis.
innominatum (in-om-in-a'-tum) [L., "nameless"; os, bone, understood]. The innominate bone.
innoxious (in-ok'-shus) [in, not; noxius, harmful]. Harmless, not injurious; same as innocent, q. v.
innutrition (in-ū-trish'-un) [in, not; nutrition]. Want of nutrition or nourishment.
inoblast (in'-o-blast) [is, fiber; βλαστός, germ]. Any one of the cells from which connective tissue is derived.
inocarcinoma (in-o-kar-sin-o'-mah) [is, fiber; καρκίνος, a crab; ὄμα, tumor: pl., inocarcinomata]. A carcinoma with a preponderance of fibrous tissue.
inoccipitia (in-ok-sip-it'-e-ah) [in, negative; occiput]. Deficiency of the occipital lobe of the brain.
inochondritis (in-o-kon-dri'-tis) [is, fiber; χόνδρος, cartilage; ιτις, inflammation]. Conjoined inflammation of tendons and cartilages.
inoculability (in-ok-ū-la-bil'-it-e) [see inoculation]. The quality of being inoculable.
inoculable (in-ok'-ū-la-bl) [see inoculation]. Capable of being inoculated; communicable by inoculation.
inoculation (in-ok-ū-la'-shun) [in, into; oculus, a bud]. 1. The act of introducing the virus of a disease into the body. 2. Specifically, the intentional introduction of a virus for the purpose of producing a mild form of a disease which is severe when spontaneously introduced, as the inoculation of smallpox virus. This is known as preventive inoculation.
inoculator (in-ok'-ū-la-tor) [inoculatio, an ingrafting]. One who or that which inoculates; an instrument used in inoculation.
inocyst (in'-o-sist) [ις, a fiber; κύστις, bladder]. A fibrous capsule.
inoendothelioma (in-o-en-do-the-le-o'-mah) [is, fiber; endothelioma; pl., inoendotheliomata]. Round-celled fibrosarcoma.
inoepithelioma (in-o-ep-ith-e-le-o'-mah) [is, fiber; epithelioma]. Medullary carcinoma containing fibrous tissue.
inogen (in'-o-jen) [is, fiber; γεννᾶν, to produce]. A hypothetical substance believed to occur in muscular tissue and to be decomposed, during contraction, into carbon dioxide, sarcolactic acid, and myosin.
inogenesis (in-o-jen'-es-is) [is, a fiber; γένεσις, generation]. The formation of fibrous or muscular tissue.
inohymenitis (in-o-hi-men-i'-tis) [is, fiber; ὑμήν, membrane; ιτις, inflammation]. Inflammation of fibrous tissue.
inoleiomyoma (in-o-li-o-mi-o'-mah) [is, fiber; λεῖος, smooth; myoma]. Myoma made up of or containing unstriped muscular fibers.
inolith (in'-o-lith) [is, fiber; λίθος, a stone]. A fibrous concretion.
inoma (in-o'-mah) [is, a fiber; ὄμα, a tumor: pl., inomata]. Same as fibroma.
inomyxoma (in-o-miks-o'-mah) [is, fiber; μύξα, mucus; ὄμα, tumor: pl., inomyxomata]. Same as fibro-myxoma.
inopectic (in-o-pek'-tik). Pertaining to inopexia.
inoperable (in-op'-ur-a-bl). That which should not be operated upon.

inopexia (in-o-peks'-e-ah) [is, fiber; πῆξις, coagulation]. A tendency in the blood toward spontaneous coagulation.
inophlogosis (in-o-flo-go'-sis) [is, fiber; φλόγωσις, inflammation]. Inflammation of any fibrous tissue.
inopolypous (in-o-pol'-ip-us) [is, fiber; polypus]. Of the nature of an inopolypus.
inopolypus (in-o-pol'-ip-us) [is, fiber; polypus]. A fibrous polyp.
inorganic (in-or-gan'-ik). Not organic; not produced by animal or vegetal organisms, as an inorganic compound. i. chemistry, chemistry dealing with inorganic compounds. i. compound, a compound not containing carbon.
inorrhabdomyoma (in-or-ab-do-mi-o'-mah) [is, fiber; ῥάβδος, rod; μῦς, muscle; ὄμα, tumor: pl., inorrhabdomyomata]. A fibrous rhabdomyoma.
inoscleroma (in-o-skle-ro'-mah) [is, fiber; σκλήρωμα, induration]. Hardened fibrous tissue.
inosclerosis (in-o-skle-ro'-sis) [is, fiber; σκληρός, hard]. Sclerosis or hardening of fibrous tissue.
inoscopy (in-os'-ko-pe) [is, fiber; σκοπεῖν, to examine]. A method of bacterial investigation designed to set free microorganisms which have become entangled in the fibrin of organic fluids by dissolving the fibrin with a pepsin digestive mixture, when they may be centrifuged for examination.
inosculate (in-os'-kū-lāt) [see inosculation]. To unite by small openings; to anastomose.
inosculation (in-os-kū-la'-shun) [in, in; osculum, a small mouth]. The joining of blood-vessels by direct communication.
inose (in'-ōs). Inosite.
inosemia (in-o-se'-me-ah). 1. [is, fiber; αἶμα, blood]. An excess of fibrin in the blood. 2. [inose; αἶμα, blood]. The presence of inosite in the blood.
inosis (in'-o-sis). See inogenesis.
inosite, inosit (in'-o-sit) [ἰνός, fiber], C₆H₁₂O₆+2H₂O. Muscle-sugar; a saccharine substance occurring in muscles, rarely in urine. For tests for, see Gallois, Scherer, Seidel.
inosituria (in-o-sit-ū'-re-ah) [inosite; οὖρον, urine]. The presence of inosite in the urine.
inosteatoma (in-o-ste-at-o'-mah) [is, fiber; steatoma]. A steatoma with fibrous elements.
inosuria (in-o-sū'-re-ah). See inosituria.
inotagmata (in-o-tag'-mat-ah) [is, strength, force; τάγμα, a regular arrangement, as of soldiers]. The contractile elements that generate the force of protoplasmic motion; they are held to be uniaxial and doubly refractive; also written isotagmata.
inotropic (in-o-trop'-ik) [in, in; τρέπειν, to turn]. Pertaining to influences which modify the contractility of the heart.
inquest (in'-kwest) [in, into; quærere, to ask]. A judicial inquiry, especially one for the purpose of determining the cause of death of one who has died by violence or in some unknown way.
inquination (in-kwin-a'-shun) [inquinatio, pollution]. Pollution; infection; corruption.
inquisition (in-kwiz-ish'-un) [see inquest]. An inquiry, especially one into the sanity or lunacy of a person.
insaccation (in-sak-a'-shun) [in; saccus, sac]. Encystment; enclosure in a sac.
insalivation (in-sal-iv-a'-shun) [in, in; saliva, the spittle]. The mixture of the food with saliva during mastication.
insalubrious (in-sal-ū'-bre-us) [in, not; saluber, healthful]. Unhealthy.
insalubrity (in-sal-ū'-brit-e) [in, not; salubris, wholesome]. Unwholesomeness of air or climate.
insanability (in-san-ab-il'-it-e) [in, not; sanabilis, curable]. Incurableness; the quality or state of being incurable.
insane (in-sān') [in, not; sanus, sound]. Deranged or diseased in mind. i. ear. See hæmatoma auris under hematoma.
insanitary (in-san'-it-a-re) [in, not; sanitas, health]. Not sanitary; not in a proper condition as respects the preservation of health.
insanitation (in-san-it-a'-shun) [in, not; sanitas, health]. Lack of proper sanitary conditions; defect of sanitation.
insanity (in-san'-it-e) [see insane]. A derangement of the mental faculties, with or without loss of volition and of consciousness. Insanity may be due to defective development, to acquired disease, or to natural decay. It is characterized, according to its form, by a variety of symptoms, the most common

of which are change of character and habits, moroseness, confusion, elation, melancholia, mania, delusions, and hallucinations. Melancholia, mania, delusional insanity, and dementia are the four principal types of the affection. i., acquired, that arising after a long period of life of mental integrity. i., affective, a form affecting only the emotions, as melancholia. i., alcoholic, that induced by alcoholic excess, usually a result of hereditary tendencies. i., alternating. See i., circular. i., anemic, that due to anemia. i., arthritic, that due to rheumatism or gout. i., circular, a form of insanity recurring in cycles varying in length from a few days to many months. The arrangement of the cycle varies in different individuals, but is constant in a given case. Thus melancholia may be followed by mania, and this by a lucid interval, the passage from one mental condition to the other being abrupt or gradual. Syn., alternating insanity; cyclothymia. i., climacteric, insanity occurring at or near the menopause. i., communicated, that transmitted by association with an insane person. i., compound, that in which two or more groups of mental faculties are involved. i., concurrent, that, caused by general diseases. i., confusional, an acute insanity produced by nervous shock or exhausting disease, without distinct emotional depression or exaltation, with marked failure of mental power or complete imbecility, often accompanied by hallucinations and loss of physical power. Recovery is usually complete. i., congenital, that existing from birth. i., consecutive, that following some disease or injury not of the brain. i., constitutional, insanity due to some pathological or physiological condition affecting the general system. i., cyclic. Same as i., circular. i., depressive, melancholia. i., deuteropathic, that caused by disorders of or developmental changes in organs other than the brain. i., diabetic, that due to diabetes. i., diathetic, inherited insanity. i., doubting, a form closely allied to delusional insanity, consisting in an uncontrollable doubt and indecision regarding the occupations, duties, or events of the day, of religion, etc. i., egressing, that growing out of a former disease. i., emotional, insanity characterized by derangement of the emotions, either depressing or exalting in character. i., epidemic, a form occasionally manifested among a number of persons in common association, as in convents or schools. i., erotic. See nymphomania. i., general, a general term for mania and melancholia. i., hereditary, that transmitted from parent to child, and not induced by other apparent cause. i., homicidal, that marked by a desire to destroy human life. i., hysterical, chronic insanity secondary to hysteria and preserving the simulative tendencies of hysteria. i., ideal, a general term embracing all the forms in which ideas dependent upon the senses are perverted. i., imitative, a form of communicated insanity marked by mimicry of the insane characteristics of another. i., imposed, delirious ideas imposed by one maniac upon another individual weaker than himself. i., impulsive, a form in which the patient possesses an uncontrollable desire to commit acts of violence. i., intermittent. The same as i., recurrent. i., ischemic, that due to persistent cerebral anemia. i., melancholic, melancholia. i., menstrual, that occurring at the menstrual period; and see i., periodic. i., moral, a form marked by perversion and depravity of the moral sense, apparently without impairment of the reasoning and intellectual faculties. i., notional, a form in which the patient sees objects as they exist, but conceives grossly erroneous ideas concerning them. i., paroxysmal, that marked by temporary paroxysms of mental aberration. i., perceptional, a form characterized by illusions and hallucinations. i., periodic, a condition dependent upon original or acquired psychopathy, in which attacks of insanity occur at regular or irregular intervals. If occurring in women at the menstrual epoch, it is called menstrual insanity. i. of pregnancy, a form occurring during pregnancy, characterized by melancholia, suicidal intent, and abhorrence of friends and relatives. i., primary, a form, often congenital, that arises with the development of the body. It may also proceed from injury or disease of the brain in early life. i. of puberty. See hebephrenia. i., puerperal, a term sometimes applied to the delirium of childbirth, but more properly to the insanity occurring after delivery. i., recurrent, that marked by recurrent attacks of mental aberration with intervening lucid intervals. i., religious, that associated with religious subjects. i., senile, that due to old age. i., stuporous, a primary acute form of dementia, characterized by a tendency to stupor; a disease chiefly met in youth and early maturity. i., surgical, that coming on after surgical operations. i., toxic, an acute form due to systemic poisoning by certain drugs. i., traumatic, insanity marked by perversity, violence, and brief spells of maniacal self-exaltation, progressing slowly with remissions to dementia; it is attributed to injury.
insatiable (in-sa'-she-ab-l) [insatiabilis, that which cannot be satisfied]. Inordinately greedy; incapable of being satisfied. i. appetite. See bulimia.
insatiability (in-sa-she-ab-il'-it-e) [in, not; satiare, to satisfy]. The property of being insatiable. See acoria.
inscription (in-skrip'-shun). The body or main part of a prescription, which contains the drugs and amounts to be used.
inscriptiones tendineæ (in-skrip-she-o'-něs ten-din'-e-e). The lineæ transversæ of the rectus abdominis muscle.
insect (in'-sekt) [in, into; secare, to cut]. Any member of the class of animals called Insecta. i.-powder, a powder employed to destroy or ward off insects, and consisting usually of the powdered flowers of species of Pyrethrum.
insecticide (in-sek'-tis-īd) [insect; cædere, to kill]. A substance that is destructive to insects.
insectiform (in-sek'-tif-orm) [insectum, an insect; forma, form]. Resembling an insect.
insemination (in-sem-in-a'-shun) [inseminare, to plant seed]. 1. The planting of seed. 2. The introduction of semen. 3. Impregnation.
insenescence (in-sen-es'-ens) [insenescentia; in, not, or upon; senex, old]. 1. Vigorous age; old age without its ordinary infirmities. 2. The approach of old age.
insensible (in-sen'-sib-l) [in, not; sentire, to feel]. 1. Incapable of being perceived or recognized by the senses. 2. Unconscious.
insensibility (in-sens-ib-il'-it-e) [in, not; sentire, to feel]. The condition or state of being insensible; absence of consciousness; anesthesia.
insertion (in-ser'-shun) [inserere, to set in]. 1. The act of setting or placing in. 2. That which is set in. 3. The point at which anything, as a muscle, is attached; the place or the mode of attachment of an organ to its support. i., velamentous, the attachment of the umbilical cord to the margin of the placenta.
insidious (in-sid'-e-us) [insidiæ, ambush]. Coming on stealthily or imperceptibly. i. disease, one the onset of which is gradual or inappreciable.
insipid (in-sip'-id) [insipidus, unsavory]. Tasteless.
insitio dentis (in-sish'-e-o den'-tis). Implantation of a tooth.
in situ (in si'-tū) [in, in; situs, position]. In a given or natural position.
insolation (in-so-la'-shun) [in, in; sol, sun]. 1. Exposure to the rays of the sun. 2. Sunstroke or heatstroke; a condition of prostration and fever due to exposure to the direct rays of the sun or to extreme heat.
insoluble (in-sol'-ū-bl) [in, not; solubilis, that can be loosed]. Incapable of being dissolved.
insolubility (in-sol-ū-bil'-it-e). The quality of being insoluble; lack of solubility.
insomnia (in-som'-ne-ah) [in, not; somnus, sleep]. Want of sleep; inability to sleep.
inspection (in-spek'-shun) [inspicere, to look]. In medicine, the examination of the body or any part of it by the eye.
inspergation (in-sper-ga'-shun) [inspergere, to sprinkle upon]. The act of sprinkling or dusting with fine powder.
inspersion (in-spur'-shun). See inspergation.
inspiration (in-spir-a'-shun) [in, in; spirare, to breathe]. The drawing in of the breath.
inspirator (in'-spir-a-tor) [in, in; spirare, to breathe]. An inhaler.
inspiratory (in-spi'-ra-to-re) [see inspiration]. Pertaining to the act of inspiration.
inspirometer (in-spi-rom'-et-er) [see inspiration; μέτρον, measure]. An instrument for measuring the amount of air inspired.
inspissant (in-spis'-ant) [in, in; spissare, to thicken]. 1. Tending to thicken; thickening the blood

or other fluids. 2. An agent that tends to increase the thickness of the blood or some other fluid.
**inspissate** (*in'-spis-āt*) [*inspissare*, to thicken]. To make thick by evaporation or by absorption of fluid.
**instauration** (*in-staw-ra'-shun*) [*instauratio*, renewal]. The first appearance of a physiological condition; the establishment of a new function.
**instep** (*in'-step*) [*in*, in; *step*]. The arch on the upper surface of the foot.
**instillation** (*in-stil-a'-shun*) [*instillare*, to put in little by little]. The pouring of a liquid into a cavity drop by drop.
**instillator** (*in'-stil-a-tor*) [*instillare*, to pour in by drops]. An instrument for pouring a liquid by drops.
**instinct** (*in'-stingkt*) [*instinguere*, to impel]. A natural impulse, which, though unassociated with reason, prompts a useful act.
**instinctive** (*in-stingk'-tiv*) [see *instinct*]. Prompted or determined by instinct; of the nature of instinct.
**institutes of medicine.** The philosophy of the science of medicine, of physiology, pathology, therapeutics, and hygiene, or the general and elementary principles of the same. The term is used sometimes as a synonym of physiology.
**instrument** (*in'-stroo-ment*) [*in*, in; *struere*, to build]. Any mechanical tool or device used to assist in the performance of a certain act.
**instrumental** (*in-stroo-men'-tal*) [*instrument*]. Pertaining to or performed with instruments, as *instrumental* labor.
**instrumentarium** (*in-stroo-men-ta'-re-um*) [L.]. A supply or collection of surgical, dental, or other instruments.
**instrumentation** (*in-stroo-men-ta'-shun*) [*instrument*]. The care or employment of instruments.
**insuccation** (*in-suk-a'-shun*) [*in*, into; *succus*, juice]. The steeping of a drug for a considerable time in water before using it in any pharmaceutical process.
**insufficiency** (*in-suf-fish'-en-se*) [*insufficiens*, insufficient]. The state of being inadequate; incapacity to perform a normal function. i. **of the cardiac valves**, imperfect closure of the valves, permitting regurgitation. Depending upon the valve affected, the insufficiency may be aortic, mitral, tricuspid, or pulmonary. i. **of a muscle**, inability on a part of a muscle to contract sufficiently to produce the normal effect. The term is applied especially to the eye muscles. *Insufficiency of the externi*, a condition in which the contraction of the externi muscles of the eye is weak and is overbalanced by that of the interni, producing esophoria. *Insufficiency of the interni*, defective power on the part of the interni muscles, producing exophoria.
**insufflation** (*in-suf-fla'-shun*). [*in*, in; *sufflare*, to puff]. The act of blowing into, as the *insufflation* of a powder into a cavity; also, the blowing of air into a cavity, as *insufflation* of the middle ear. i., **mouth-to-mouth**, the blowing of air into the mouth of a person, usually a newborn infant, to distend the lungs and counteract asphyxia.
**insufflator** (*in-suf-la-tor'*) [see *insufflation*]. An instrument for blowing air or powders into a cavity.
**insula** (*in'-su-lah*) [L. "an island"]. I. In anatomy, the island of Reil. 2. Any detached part or exclave of an organ. 3. A blood-islet, or island. See *island*.
**insular** (*in'-su-lar*) [*insula*]. I. Pertaining to the island of Reil. 2. Isolated; occurring in patches. i. **sclerosis.** See *sclerosis, multiple*.
**insulate** (*in'-su-lāt*) [*insula*]. To isolate or separate from surroundings. In electricity, to surround a conductor with a nonconducting substance.
**insulation** (*in-su-la'-shun*) [*insulare*, to make like an island]. The process of insulation; the state or quality of being insulated.
**insulator** (*in'-su-la-tor*) [*insulatus*, made into an island]. A nonconducting substance by means of which insulation is effected.
**insusceptibility** (*in-sus-sep-tib-il'-it-e*) [*insusceptibilis*, not susceptible]. Absence of contagious quality; want of susceptibility; immunity.
**integral** (*in'-te-gral*) [*integer*, whole]. Entire; essential.
**integration** (*in-te-gra'-shun*) [*integratio*, a renewing]. The blending of separate parts into one, as in the embryo; assimilation; anabolism.
**integrity** (*in-teg'-rit-e*) [*integer*, whole]. Wholeness; entirety. Also, virginity.

**integument** (*in-teg'-u-ment*) [*in*, upon; *tegere*, to cover]. A covering, especially the skin. i., **fetal**, the fetal membranes.
**integumentary** (*in-teg-u-men'-ta-re*). Pertaining to the skin.
**intellect** (*in'-tel-ekt*) [*intellectus*, understanding]. The mind or the reasoning power.
**intelligence** (*in-tel'-ij-ence*). The understanding that comes from the perception of qualities and attributes of the objective world, and is manifested in the purposive employment of means to attain an end.
**intemperance** (*in-tem'-per-ans*) [*in*, not; *temperare*, to moderate]. Want of moderation; immoderate indulgence, especially in alcoholic beverages.
**intemperant** (*in-tem'-per-ant*). An intemperate person.
**intensification** (*in-ten-sif-ik-a'-shun*) [see *intensity*]. The act of making intense or of increasing the strength of anything.
**intensity** (*in-ten'-sit-e*) [*intensus*, stretched tight]. I. The state of being intense or high-strung. 2. The degree to which a force is capable of rising. 3. A high degree of energy or power.
**intensive** (*in-ten'-siv*) [see *intensity*]. Gradually increased in force or intensity, as the *intensive* method of inoculation.
**intention** (*in-ten'-shun*) [*intentus*, intend]. The end or purpose. See under *healing*. i.-**tremor**, a tremor coming on when attempts at voluntary motion are made.
**inter-** (*in-ter-*) [*inter*, between]. A prefix signifying between.
**interaccessory** (*in-ter-ak-ses'-or-e*) [*inter-*; *accedere*, to go to]. Situated between accessory processes of the vertebræ. i. **muscles**, short lumbar muscles connecting the accessory processes of the vertebræ.
**interacinous** (*in-ter-as'-in-us*) [*inter-*; *acinus*, a berry]. Situated between acini.
**interangular** (*in-ter-ang'-gū-lar*). Occurring between.
**interannular** (*in-ter-an'-ū-lar*). Located between rings or constrictions.
**interarticular** (*in-ter-ar-tik'-ū-lar*) [*inter-*; *articulus*, a joint]. Situated between joints. i. **fibrocartilage**, the flattened cartilaginous plates between the articular cartilages of certain joints.
**interarytenoid** (*in-ter-ar-it'-en-oid*). Between the two arytenoid cartilages. i. **muscle**. See *muscles, table of*.
**interauricular** (*in-ter-aw-rik'-ū-lar*) [*inter-*; *auricula*, auricle]. Situated between the auricles.
**interbody**. (*in'-ter-bod-e*). A substance found in blood-serum, corresponding to the amboceptor of a specific serum.
**interbrain** (*in'-ter-brān*). See *thalamencephalon*.
**intercadence** (*in-ter-ka'-dens*) [*inter-*; *cadere*, to fall]. An irregular beating of the pulse, in which an additional beat is interposed between two pulsations.
**intercadent** (*in-ter-ka'-dent*). Exhibiting the quality of intercadence.
**intercalary, intercalated** (*in-ter'-kal-a-re, in-ter'-kal-a-ted*) [*inter-*; *calare*, to insert]. Placed or inserted between.
**intercalatum** (*in-ter-kal-a'-tum*) [*inter-*; *calare*, to insert]. Substantia nigra; locus niger; a dark mass of crescentic outline situated between the ventral crusta and the dorsal tegmentum of the crus cerebri.
**intercapillary** (*in-ter-kap'-il-a-re*) [*inter-*; *capillus*, a hair]. Between capillaries.
**intercarotic** (*in-ter-kar-ot'-ik*). Same as *intercarotid*.
**intercarotid** (*in-ter-kar-ot'-id*) [*inter-*; *carotid*]. Situated between the external and internal carotid arteries, as the *intercarotid* ganglion; see under *ganglion*.
**intercartilaginous** (*in-ter-kar-til-aj'-in-us*). See *interchondral*.
**intercavernous** (*in-ter-kav-er'-nus*) [*inter-*; *caverna*, a cave]. Situated between two antra, or between the two cavernous sinuses.
**intercellular** (*in-ter-sel'-ū-lar*) [*inter-*; *cellula*, a small cell]. Between cells, as *intercellular* substance of tissue.
**intercentral** (*in-ter-sen'-tral*) [*inter-*; *centrum*, a center]. Between nerve-centers.
**intercerebral** (*in-ter-ser'-e-bral*) [*inter-*; *cerebrum*, cerebrum]. Between the right and left cerebral hemispheres.

**interchondral** (in-ter-kon'-dral) [inter-; χόνδρος, cartilage]. Between cartilages.
**intercidence** (in-ter'-sid-ens). See *intercadence*.
**intercident** (in-ter'-sid-ent). 1. See *intercalary*. 2. See *intercadent*.
**intercilium** (in-ter-sil'-e-um). See *glabella*.
**interclavicular** (in-ter-klav-ik'-u-lar) [inter-; *clavicula*, the collar-bone]. Between the clavicles.
**interclinoid** (in-ter-kli'-noid) [inter-; κλινή, bed; εἶδος, resemblance]. Between the clinoid processes of the sphenoid bone.
**intercoccygeal, intercoccygean** (in-ter-kok-sij'-e-al, -an). Interposed between the coccygeal vertebræ.
**intercolumnar** (in-ter-kol-um'-nar) [inter-; *columna*, column]. Between pillars, as the *intercolumnar fascia*, between the pillars of the external abdominal ring.
**intercondylar, intercondyloid** (in-ter-kon'-dil-ar, in-ter-kon'-dil-oid) [inter-; *condyle*]. Between condyles. **i. eminence,** the spine or knob separating the two condylar portions of the tibia. **i. fossa,** the notch between the condyles of the femur. **i. line,** a transverse line crossing above the intercondyloid fossa, and joining the condyles. **i. notch.** Same as **i. fossa.**
**intercoronoideal** (in-ter-kor-o-noid'-e-al). Lying between the coronoid processes.
**intercostal** (in-ter-kos'-tal) [inter-; *costa*, a rib]. Between the ribs. **i. arteries,** the arteries of the intercostal spaces. See under *artery*. **i. muscles.** See under *muscle*. **i. nerves,** the anterior divisions of the dorsospinal nerves. **i. spaces,** spaces between adjacent ribs.
**intercostales** (in-ter-kos-ta'-lez). See *muscles*, table of.
**intercostohumeral** (in-ter-kos-to-hu'-mer-al) [inter-; *costa*, rib; *humerus*]. Pertaining to the arm and the space between the ribs, as the *intercostohumeral* nerve.
**intercourse** (in'-ter-kors) [*intereursus*, commerce]. Communication. **i., carnal,** sexual connection; coitus. **i. sexual.** Same as **i., carnal.**
**intercoxal** (in-ter-koks'-al) [inter-; *coxa*, hip]. Situated between the coxæ or hips.
**intercranial** (in-ter-kra'-ne-al). Endocranial, relating to the interior of the skull or to the endocranium.
**intercricothyrotomy** (in-ter-kri-ko-thi-rot'-o-me). A cut into the larynx by transverse section of the cricothyroid membrane.
**intercristal** (in-ter-kris'-tal) [inter-; *crista*, crest]. Between the surmounting ridges of a bone, organ, or process.
**intercrural** (in-ter-kroo'-ral) [inter-; *crus*, the leg]. Situated between the legs or the crura.
**intercuneal, intercuneiform** (in-ter-ku-ne'-al, in-ter-ku'-ne-e-form) [inter-; *cuneus*, a wedge]. Between the cuneiform bones.
**intercurrent** (in-ter-kur'-ent) [inter-; *currere*, to run]. Occurring or taking place between. **i. disease,** a disease arising or progressing during the existence of another disease in the same person.
**intercus** (in-ter'-kus) [L.; pl., *intercules*]. 1. See *anasarca*. 2. Subcutaneous.
**intercutaneomucous** (in-ter-ku-ta-ne-o-mū'-kus). Between the skin and mucosa.
**intercutaneous** (in-ter-ku-ta'-ne-us). Subcutaneous.
**interdeferential** (in-ter-def-er-en'-shal). Between the vasa deferentia.
**interdental** (in-ter-den'-tal) [inter-; *dens*, a tooth]. 1. Between the teeth. 2. An interdentium. **i. splint,** a splint used in fracture of the jaw, consisting of a metallic frame at the neck of the teeth, held by wire sutures passing between the teeth.
**interdentium** (in-ter-den'-she-um) [inter-; *dens*, a tooth]. The space between any two of the teeth.
**interdiction** (in-ter-dik'-shun) [*interdictio*, a prohibiting]. A judicial or legal process that deprives an insane person or one suspected of insanity of the management of his own affairs or of the affairs of others.
**interdigital** (in-ter-dij'-it-al) [inter-; *digitus*, a finger]. Between the fingers.
**interdigitation** (in-ter-dij-it-a'-shun) [see *interdigital*]. The locking or dovetailing of similar parts, as the fingers of one hand with those of the other; or of the ends of the obliquus externus muscle with those of the serratus magnus. In dentistry, denoting that in closure of the buccal teeth the cusps of one denture strike fairly into the occluding sulci of the other denture.

**intereruptive** (in-ter-e-rup'-tiv). Between two outbreaks of eruption.
**interfascicular** (in-ter-fas-ik'-u-lar) [inter-; *fasciculus*, a bundle]. Situated between fasciculi.
**interfemoral** (in-ter-fem'-or-al) [inter-; *femur*, the thigh]. Between the femora or thighs.
**interfere** (in-ter-fēr') [inter-; *ferire*, to strike]. In horses, to strike one hoof or the shoe of one hoof against the opposite leg or fetlock.
**interference** (in-ter-fe'-rens) [inter-; *ferire*, to strike]. The act of interfering or preventing. **i. of light,** the mutual neutralization of waves of light, when the crest of one wave falls upon the trough of another. **i. of sound,** the neutralization of two sound-waves, one by the other.
**interfibrillar** (in-ter-fi'-bril-ar) [inter-; *fibrilla*, a small fiber]. Situated between the fibrils of tissues.
**interfilar** (in-ter-fi'-lar) [inter-; *filum*, a thread]. Existing between the filaments of a reticulum.
**interfollicular** (in-tur-fol-ik'-u-lar) [inter-]. Between two follicles.
**interganglionic** (in-ter-gan-gle-on'-ik) [inter-; γάγγλιον, a ganglion]. Connecting one ganglion with another; lying between ganglia.
**interglandular** (in-ter-gland'-u-lar) [inter-; *glandula*, a gland]. Situated between glands.
**interglobular** (in-ter-glob'-u-lar) [inter-; *globulus*, a ball]. Situated between globules. **i. spaces,** irregular cavities seen in a section of dentine, after the earthy matter has been removed by putting a tooth in dilute acid. They are so called because surrounded by minute globules of dentine.
**intergluteal** (in-ter-gloo'-te-al). Between the buttocks.
**intergonial** (in-ter-go'-ne-al) [inter-; γωνία, an angle]. Between the two gonia (angles of the lower jaws).
**intergranular** (in-tur-gran'-ū-lar) [inter-; *granulum*, a small grain]. Between granules.
**intergyral** (in-ter-ji'-ral) [inter-; *gyrus*, a gyre]. Situated between two or more gyri.
**interhemal** (in-ter-hem'-al) [inter-; αἷμα, blood]. Between the hemal arches or spines.
**interhemicerebral** (in-ter-hem-e-ser'-e-bral) [inter-; ἡμι, half; *cerebrum*, cerebrum]. Situated between the cerebral hemispheres.
**interhemispheric** (in-ter-hem-is-fer'-ik) [inter-; ἡμι, half; σφαίρα, sphere]. Situated between hemispheres.
**interhuman** (in-tur-hū'-man) [inter-; *humanus*, human]. Applied to infection transmitted from one human being to another.
**interinhibitive** (in-ter-in-hib'-it-iv) [inter-; *inhibere*, to inhibit]. Mutually inhibitory.
**interjected** (in-ter-ject'-ed) [*interjicere*, to cast between]. Same as *interposed*.
**interjectional** (in-ter-jek'-shun-al) [inter-; *jacere*, to throw]. Interjected; thrown between. **i. speech,** the expression of emotions by inarticulate sounds.
**interjugal** (in-ter-joo'-gal) [inter-; *jugum*, a yoke]. Between the jugal processes of the skull.
**interlabial** (in-ter-la'-be-al) [inter-; *labium*, lip]. Between the lips, or between the labia.
**interlamellar** (in-ter-lam-el'-ar) [inter-; *lamella*, a layer]. Between lamellæ.
**interlaminar** (in-ter-lam'-in-ar) [inter-; *lamina*, a leaf]. Situated between laminæ.
**interligamentous** (in-ter-lig-a-ment'-us) [inter-; *ligamentum*, a ligament]. Between ligaments.
**interlobar** (in-ter-lo'-bar) [inter-; *lobus*, a lobe]. Situated between lobes, as *interlobar* pleurisy.
**interlobular** (in-ter-lob'-ū-lar) [inter-; *lobulus*, a lobule]. Between lobules.
**intermalar** (in-ter-ma'-lar) [inter-; *mala*, the cheekbone]. Situated between the malar bones.
**intermalleolar** (in-ter-mal-e'-o-lar). Between the malleoli.
**intermammary** (in-ter-mam'-ar-e) [inter-; *mamma*, breast]. Between the breasts.
**intermammillary** (in-ter-mam'-il-ar-e) [inter-; *mammilla*, nipple, breast]. Between the nipples; between the breasts.
**intermarginal** (in-ter-mar'-jin-al) [inter-; *margo*, a margin]. Lying between two margins.
**intermarriage** (in-ter-mar'-āj) [inter-; *marriage*]. 1. Marriage between persons related by consanguinity. 2. Marriage between persons of different races.
**intermastoid** (in-ter-mas'-toid) [inter-; μαστός, breast; εἶδος, likeness]. Situated between or connecting the two mastoid processes.
**intermaxilla** (in-ter-maks-il'-ah) [inter-; *maxilla*, jaw]. The premaxilla; the intermaxillary bone.

**intermaxillary** (in-ter-maks'-il-a-re) [inter-; maxilla, jaw-bone]. Between the maxillary bones. i. bone, the small bone that receives the incisors, situated between the superior maxillary bones of the fetus.
**intermediary amputation** (in-ter-me'-de-a-re). Amputation during the inflammatory fever. i. body. See *intermediate body*. i. hemorrhage, hemorrhage following a primary hemorrhage. i. nerve, the nerve of Wrisberg; see under *nerve*.
**intermediate** (in-ter-me'-de-āt) [inter-; medius, middle]. Situated between. i. body, the complementary substance essential to the proper performance of a physiological function—such as enabling the "end-body" or bactericidal substance to combine with the bacteria in typhoid in the production of immunity. Cf. *body, immune*.
**intermediolateral** (in-ter-me-de-o-lat'-er-al) [inter-; medius, middle; latus, side]. Both lateral and intermediate, as the *intermediolateral* tract of the spinal cord, lying between the anterior and posterior horns.
**intermedium** (in-ter-me'-de-um) [inter-; medius, the middle; pl., intermedia]. 1. Any intermediary substance. 2. In pharmacy, an emulsifying or suspending ingredient.
**intermembral** (in-ter-mem'-bral) [inter-; membrum, a limb]. Existing between the members or limbs of an organism.
**intermembranous** (in-ter-mem'-bran-us). Lying between membranes.
**intermeningeal** (in-ter-men-in'-je-al) [inter-; μῆνιγξ, membrane]. Between the dura and the arachnoid, or between the latter and the pia. i. hemorrhage, a hemorrhage between the meninges of the brain or spinal cord.
**intermenstrual** (in-ter-men'-stroo-al) [inter-; mensis, month]. Between the menstrual periods.
**interment** (in-ter'-ment) [in, in; terra, the earth]. The burial of the body.
**intermesenteric** (in-ter-mes-en-ter'-ik). Between the mesenteries.
**intermesoblastic** (in-ter-mes-o-blast'-ik). Between the layers or between the lateral plates of the mesoblast.
**intermetacarpal** (in-ter-met-a-kar'-pal) [inter-; metacarpus]. Between the metacarpal bones.
**intermetatarsal** (in-ter-met-at-ar'-sal). Between the metatarsal bones.
**intermission** (in-ter-mish'-un) [see intermittent]. An interval, as between the paroxysms of a fever or between the beats of the pulse.
**intermittent** (in-ter-mit'-ent) [inter-; mittere, to send, or occur]. Occurring at intervals; characterized by intermissions or intervals, as *intermittent* fever, *intermittent* insanity, *intermittent* pulse, *intermittent* sterilization.
**intermural** (in-ter-mū'-ral) [inter-; murus, a wall]. Occurring or lying between the walls of an organ.
**intermuscular** (in-ter-mus'-kū-lar) [inter-; musculus, a muscle]. Situated between muscles.
**intern** (in'-tern) [Fr., interne]. An in-door or resident physician in a hospital; a member of an in-door staff of physicians.
**internal** (in-ter'-nal) [internus, inward]. Situated within or on the inside. i. capsule, the band of white nerve-matter between the optic thalamus and caudate nucleus on the inner, and the lenticular nucleus on the outer, side. It is the continuation of the crus cerebri, and consists of an anterior and a posterior limb joined at an angle, termed the knee. It is composed of fibers coming from and going to the cortex cerebri. i. ear, the labyrinth, q. v. i. medicine, that branch of medicine which treats of diseases which can not be treated surgically. i. oblique. See *obliquus internus* under *muscle*. i. rectus. See *rectus internus* under *muscle*. i. resistance. See *resistance, internal*.
**internarial** (in-ter-na'-re-al) [inter-; nares, nostrils]. Situated between the nostrils.
**internasal** (in-ter-na'-zal) [inter-; nasus, the nose]. Between the nasal bones.
**interne** (in'-tern). Same as *intern*, q. v.
**interneural** (in-ter-nū'-ral) [inter-; νεῦρον, nerve]. Situated between neural spines.
**interneuronal** (in-tur-nū'-ron-al). Between neurons.
**internist** (in-ter'-nist). A physician, in contradistinction to a surgeon.
**internodal** (in-ter-no'-dal). Situated between two nodes. See *internode*.

**internode** (in'-ter-nōd) [inter-; nodus, a knot]. The space between two nodes of a nerve-fiber, as the *internode* between the nodes of Ranvier. Syn., *internodal segment*.
**internuclear** (in-ter-nū'-kle-ar) [inter-; nucleus, a kernel]. Situated between nuclei.
**internuncial** (in-ter-nun'-she-al) [inter-; nuncius, a messenger]. Serving as a connecting or announcing medium, as *internuncial* fibers, nerve-fibers connecting nerve-cells.
**internus** (in-ter'-nus). 1. See *internal*. 2. The rectus internus muscle of the eye.
**interol** (in'-ter-ol). Trade name of a mineral oil used to lubricate the intestinal tract.
**interolivary** (in-tur-ol'-iv-a-re). Between the olives.
**interoptic** (in-ter-op'-tik) [inter-; opticus, optic]. Between the optic lobes, tracts or nerves of the brain.
**interorbital** (in-ter-or'-bit-al) [inter-; orbita, the orbit]. Situated between the orbits.
**interosseal** (in-ter-os'-e-al). Same as *interosseous*.
**interosseous** (in-ter-os'-e-us) [inter-; os, a bone]. Between bones, as *interosseous* arteries, membranes, muscles, or nerves.
**interpalpebral** (in-ter-pal'-pe-bral) [inter-; palpebra, the eyelid]. Between the palpebræ. i. spot. See *pinguecula*. i. zone, that part of the cornea and of the scleral conjunctiva that ordinarily is not covered by the lids.
**interpapillary** (in-ter-pap'-il-a-re) [inter-; papilla, a papilla]. Between papillæ.
**interparietal** (in-ter-par-i'-e-tal) [inter-; paries, a wall]. Between walls; between the parietal bones, as *interparietal* suture; between parts of the parietal lobe, as *interparietal* fissure. i. bone, a name sometimes given to the upper, squamous, and noncartilaginous part of the occipital bone. i. fissure. See *fissure*. i. suture, the sagittal suture, or that formed by the approximation of the parietal bones.
**interparoxysmal** (in-tur-par-oks-iz'-mal). Between paroxysms.
**interpeduncular** (in-ter-pe-dung'-kū-lar) [inter-; pedunculus, a little foot]. Situated between the cerebral or cerebellar peduncles. i. space, the pons Tarini, or posterior perforated space that forms the posterior portion of the floor of the third ventricle.
**interpellated, interpellatus** (in-ter-pel-a'-ted, in-ter-pel-a'-tus) [interpellare, to interrupt]. Applied by Paracelsus to diseases marked by irregular paroxysms.
**interphalangeal** (in-ter-fa-lan'-je-al) [inter-; φάλαγξ, a finger]. Between the fingers or the toes.
**interpial** (in-ter-pi'-al). Between the layers of the pia.
**interplacental** (in-ter-plas-en'-tal) [inter-; placenta]. Between the placental lacunæ.
**interpleuricostal** (in-ter-ploo-re-kos'-tal). Between the pleura and ribs. i. muscles, the internal pleuricostal muscles.
**interpolar** (in-ter-po'-lar). 1. See *intrapolar*. 2. Between the poles of an electric battery; applied to the effect of the current acting through the whole region of the body between the two poles as contrasted with the polar effect taking place at the point of application of the electrode.
**interpolated** (in-ter'-po-la-ted). See *intercalary*.
**interpolation** (in-ter-po-la'-shun) [interpolare, to furbish up]. The surgical transfer of tissue to a new part, or from one subject to another.
**interposition** (in-ter-po-zish'-un) [interpositio, a placing between]. The development of anatomical structures between existing ones.
**interpositum** (in-ter-pos'-it-um). Same as *velum nterpositum, q. v*.
**interprotometamere** (in-ter-pro-to-met'-a-mēr) [inter-; πρῶτος, first; μετά, among; μέρος, a part]. The part lying between the primary segments of the embryo.
**interproximal** (in-ter-prok'-sim-al) [inter-; proximus, next]. Between adjoining surfaces.
**interproximate** (in-ter-prok'-sim-āt). In dentistry, the space between two adjacent teeth.
**interpterion** (in-ter-te'-re-on) [inter-; pterion, a. craniometric point]. Between the pteria.
**interpterygoid** (in-ter-ter'-e-goid). Lying between the pterygoid processes.
**interpubic** (in-ter-pū'-bik) [inter-; pubis]. Situated between the pubic bones.
**interpyramidal** (in-ter-pir-am'-id-al). Between the pyramids.
**interradial** (in-ter-ra'-de-al) [inter-; radius, a ray]. Situated between two rays.

**interramal** (in-ter-ra'-mal). Between the rami of the mandible.
**interrenal** (in-ter-re'-nal) [inter-; ren, the kidney]. Situated between the kidneys.
**interrupted** (in-ter-up'-ted) [see interrupter]. Discontinuous; broken; irregular.
**interrupter** (in-ter-up'-ter) [interrumpere, to break apart]. That which interrupts; specifically, a device for breaking an electric current.
**interscapular** (in-ter-skap'-ū-lar) [inter-; scapula, the shoulder-blade]. Between the shoulder-blades.
**interscapulum** (in-ter-skap'-ū-lum) [inter-; scapulæ, shoulder-blades: pl., interscapula]. 1. The region of the back between the shoulder-blades. 2. Bartholin's name for the spine of the scapula. 3. A fossa on the dorsal aspect of the scapula.
**intersigmoid** (in-ter-sig'-moid). Pertaining to the space in the loop of the sigmoid.
**interspace** (in'-ter-spās) [inter-; spatium, a space]. An interval between the ribs, or between the fibers or lobules of a tissue or organ.
**interspinal** (in-ter-spi'-nal) [inter-; spina, a spine]. Between spines; specifically, between the spines of the vertebræ. i. **muscles**. See muscles, table of.
**interspinous** (in-ter-spi'-nus) [inter-; spina, a spine]. Situated between spinous processes, as of the vertebræ.
**interstices** (in-ter'-stis-ēz) [interstitium, a space between]. Spaces or intervals; also, pores.
**interstitial** (in-ter-stish'-al) [see interstices]. 1. Situated between important parts; occupying the interspaces or interstices of a part. 2. Pertaining to the interstitial or connective tissue. i. **inflammation**, inflammation of the interstitial or connective tissue. i. **keratitis**. See keratitis, interstitial. i. **kink**, an abnormal angulation in or constriction of the intestine, resulting in a narrowing of the lumen and delay in the progress of the feces. i. **nephritis**. See nephritis, interstitial. i. **pneumonia**. See pneumonia, interstitial. i. **pregnancy**, pregnancy in the Fallopian tube where it passes through the uterine wall. i. **tissue**, the intercellular connective tissue.
**intersuperciliary** (in-ter-sū-per-sil'-e-a-re). Between the superciliary ridges.
**intersystole** (in-ter-sis'-to-le). The interval between the end of the auricular systole and the beginning of the ventricular one.
**intertarsal** (in-ter-tar'-sal). Located between adjacent tarsal bones.
**intertragicus** (in-ter-traj'-ik-ius). See muscles table of.
**intertransversales** (in-ter-trans-ver-sa'-lēs) [see intertransverse]. Short bundles of muscular fibers extending between the transverse processes of contiguous vertebræ.
**intertransverse** (in-ter-trans-vers') [inter-; transversus, turned across]. Connecting the transverse processes of contiguous vertebræ.
**intertriginous** (in-ter-trij'-in-us) [inter-; terere, to rub]. Of the nature of or affected with intertrigo.
**intertrigo** (in-ter-tri'-go) [inter-; terere, to rub]. An erythematous eruption of the skin produced by friction of adjacent parts. i. **ani**, i. **podicis**, chafing of the anus.
**intertrochanteric** (in-ter-tro-kan-ter'-ik) [inter-; trochanter]. Between the trochanters. i. **line**. See line, intertrochanteric.
**intertubular** (in-ter-tū'-bū-lar) [inter-; tubulus, a tube]. Between tubes. i. **substance**, the translucent, granular substance of the dentine of the tooth.
**interureteric** (in-ter-ū-re-ter'-ik) [inter-; οὐρητήρ, ureter]. Situated between the ureters.
**interuteroplacental** (in-ter-ū-ter-o-plas-ent'-al). Between the uterus and the placenta.
**intervaginal** (in-ter-vaj'-in-al) [inter, between; vagina, sheath]. Between sheaths. i. **space**, that found within the sheaths of the optic nerve.
**interval** (in'-ter-val) [inter-; vallum, a rampart]. A space or lapse, either of time or distance, as the interval between the paroxysms of a fever. i., **cardioaortic**, the interval between the apex-beat and the arterial pulse. i. **focal**, the distance between the anterior and posterior focal points.
**intervallary** (in-ter-val'-a-re) [see interval]. Occurring between paroxysms of a disease.
**intervascular** (in-ter-vas'-kū-lar) [inter-; vasculum, a small vessel]. Located between vessels.
**interventricular** (in-ter-ven-trik'-ū-lar) [inter-; ventriculum, a ventricle]. Situated between ventricles. i. **septum**, the partition between the ventricles of the heart.

**interversion** (in-ter-ver'-shun) [inter-; vertere, to turn]. Evolution.
**intervertebral** (in-ter-ver'-te-bral) [inter-; vertebra, a bone of the spine]. Between the vertebræ. i. **discs**, the discs of fibrocartilage between the adjacent surfaces of the bodies of the vertebræ. i. **foramen**. See foramen, intervertebral. i. **notch**, the notch at the base of the pedicle on the sides of the body of each vertebra. i. **substance**. Same as i. discs.
**intervillous** (in-ter-vil'-us) [inter-; villus, a tuft of hair]. Situated between villi.
**interzonal** (in-ter-zo'-nal) [inter-; zona, zone]. In dentistry, applied to the line between enamel and dentine, at the periphery of the latter.
**intestin** (in-tes'-tin). See bismuthnaphthalin benzoate.
**intestinal** (in-tes'-tin-al) [see intestine]. Pertaining to the intestine. i. **absorption**, the absorption of the products of digestion by the capillaries, veins, and lacteals of the mucous membrane of the intestine. i. **anastomosis**. See anastomosis, intestinal. i. **canal**, the entire intestinal passage from the beginning of the duodenum to the anus. i. **concretion**. See enterolith. i. **juice**, the secretion of the intestinal glands, a pale-yellow fluid, alkaline in reaction, having a specific gravity of 1011, and possessing diastasic and proteolytic properties. It also, to a certain extent, emulsifies and decomposes fats. Syn., succus entericus. i. **obstruction**, arrest of or interference with the progress of the feces. i. **stasis**, constipation.
**intestine** (in-tes'-tin) [intestinum, intestine, from intus, within]. The part of the digestive tube extending from the beginning of the pylorus to the anus. It consists of the small and large intestine. The former is about 6½ meters (20 feet) in length, and extends from the pylorus to its junction with the large intestine at the cecum. Three divisions are described—the duodenum, 22 cm. long, is the most important; the jejunum, 2.2 meters long, and the ileum, 4 meters long. The large intestine is about 1.6 meters (5 feet) long, and consists of the cecum (with the vermiform appendix), the colon, and the rectum. The wall of the intestine is made up of four coats—a serous, muscular, submucous, and mucous. Embedded in the wall are minute glands, and projecting from the surface, in the small intestine, are the villi. The function of the intestine is to continue and complete the changes begun in the mouth and stomach and to remove the waste-matter, or feces.
**intestinum** (in-tes-tin'-ū-lum) [L-, a small intestine; pl., intestinula]. 1. The small intestine. 2. A cerebral convolution or gyrus. 3. The umbilical cord or navel-string.
**intestinum** (in-tes-ti'-num) [see intestine]. i., **cæcum**, the cæcum. i. **crassum**, the large intestine. i. **ileum**, the ileum. i. **jejunum**, the jejunum. i. **rectum**, the rectum. i. **tenue**, the small intestine. i. **tenue mesenteriale**, that portion of the small intestine which has a mesentery, namely the jejunum and ileum.
**intima** (in'-tim-ah) [intimus, inmost]. The innermost of the three coats of a blood-vessel.
**intimal** (in'-tim-al). Pertaining to the intima.
**intimitis** (in-tim-i'-tis). Inflammation of an intima.
**intoe** (in'-tō). Hallux valgus.
**intoeing** (in-to'-ing). A condition in which a person (usually a child) walks, with the toes pointing inward, so that when he attempts to run he trips over his feet or falls.
**intolerance** (in-tol'-er-ans) [in, not; tolerare, to bear]. Want of endurance or ability to stand pain. Impatience. Also, the inability to endure the action of a medicine. i. **of light**. See photophobia.
**intonation** (in-to-na'-shun) [intonare, to thunder]. 1. The tone of the voice. 2. The rumbling or gurgling sound produced by the movement of flatus in the bowels.
**intort** (in'-tort). To turn inward.
**intoxicant** (in-toks'-ik-ant) [intoxicare, to intoxicate]. 1. Intoxicating; capable of producing intoxication or poisoning. 2. A drug or agent capable of producing intoxication.
**intoxication** (in-toks-ik-a'-shun) [in, in; τοξικόν, poison]. 1. Poisoning. 2. The acute state produced by overindulgence in alcohol.
**intoxications** (in-toks-ik-a'-shuns) [see intoxication]. A general name for the group of diseases due to the administration of poisons generated entirely outside of the body.
**intra-** (in-trah-). A prefix signifying within or during.

**intra-abdominal** (*in-trah-ab-dom'-in-al*) [*intra-; abdomen*]. Within the cavity of the abdomen.
**intra-acinar, intra-acinous** (*in-trah-as'-in-ar, -us*) [*intra-; acinus*, a berry]. Situated or occurring within an acinus.
**intra-arachnoid** (*in-trah-ar-ak'-noid*). Within or underneath the arachnoid.
**intra-arterial** (*in-trah-ar-te'-re-al*). See *endarterial*.
**intra-articular** (*in-trah-ar-tik'-ū-lar*) [*intra-; articulus*, a joint]. Within a joint.
**intrabronchial** (*in-trah-brong'-ke-al*). Within a bronchus.
**intracapsular** (*in-trah-kap'-sū-lar*) [*intra-; capsula*, a capsule]. Within the capsular ligament of a joint, as *intracapsular* fracture.
**intracardiac** (*in-trah-kar'-de-ak*) [*intra-; καρδία*, heart]. Situated or produced within the heart cavity.
**intracartilaginous** (*in-trah-kar-til-aj'-in-us*) [*intra-; cartilago*, cartilage]. Within a cartilage, as *intracartilaginous* ossification; endochondral.
**intracellular** (*in-trah-sel'-ū-lar*) [*intra-; cellula*, a little cell]. Within a cell.
**intracerebellar** (*in-trah-ser-e-bel'-ar*). Within the cerebellum.
**intracerebral** (*in-trah-ser'-e-bral*) [*intra-; cerebrum*]. Within the cerebrum.
**intracervical** (*in-trah-ser'-vik-al*) [*intra-; cervix*]. Within the cervical canal of the uterus, or any other cervical canal.
**intraciliary** (*in-trah-sil'-e-a-re*) [*intra-; cilium*, an eye-lash]. Situated within the ciliary region.
**intraincisor** (*in-trah-si'-zor*) [*intra-; cadere*, to cut]. A hemostatic forceps which cuts or crushes the intima of an artery and removes the necessity of a ligature.
**intracolic** (*in-trah-kol'-ik*). Within the colon.
**intracranial** (*in-trah-kra'-ne-al*) [*intra-; κρανίον*, the skull]. Within the skull.
**intracutaneous** (*in-trah-kū-ta'-ne-us*) [*intra-; cutis*, the skin]. Within the skin-substance.
**intracystic** (*in-trah-sis'-tik*) [*intra-; κύστις*, a cyst]. Situated or occurring within a cyst or bladder.
**intrad** (*in'-trad*). See *entad*.
**intradermic** (*in-trah-der'-mik*) [*intra-; δέρμα*, skin]. Within the skin.
**intradural** (*in-trah-dū'-ral*) [*intra-; dura*]. Situated or occurring within the dura.
**intraepidermal, intraepidermic** (*in-trah-ep-i-derm'-al, -ik*). Within the substance of the epidermis.
**intra-epithelial** (*in-tra-ep-e-the'-le-al*). Within the epithelium.
**intrafaradization** (*in-trah-far-ad-i-za'-shun*). Faradization applied to the inner surface of a body-cavity.
**intrafascicular** (*in-tra-fas-ik'-ū-lar*). Within a fascicle.
**intrafetation** (*in-trah-fe-ta'-shun*) [*intra-; fetus*]. The formation of a fetus within another fetus.
**intrafilar** (*in-trah-fī'-lar*) [*intra-; filum*, thread]. Situated within the meshes of a network. i. mass. The paramitome, *q. v.*
**intrafistular** (*in-tra-fis'-tū-lar*). Within a fistula.
**intrafusal** (*in-trah-fū'-sal*) [*intra-; fusus*, a spindle]. Pertaining to the striated muscular fibers contained in a muscle-spindle.
**intragalvanization** (*in-trah-gal-van-i-za'-shun*). The application of galvanism to the inner surface of an organ.
**intragastric** (*in-trah-gas'-trik*). Located or occurring within the stomach.
**intraglandular** (*in-trah-glan'-dū-lar*). Within a gland.
**intraglobular** (*in-trah-glob'-ū-lar*). Within a blood-corpuscle; intracorpuscular.
**intragyral** (*in-trah-jī'-ral*) [*intra-; gyrus*, a gyre]. Within a gyre of the brain.
**intrahepatic** (*in-trah-he-pat'-ik*) [*intra-; ἧπαρ*, liver]. Within the liver-substance.
**intrajugular** (*in-trah-joo'-gū-lar*). Within or internal to the jugular foramen, vein, or process.
**intralamellar** (*in-trah-lam-el'-ar*) [*intra-; lamella*, a thin plate]. Within the lamellæ.
**intralaryngeal** (*in-trah-lar-in'-je-al*). Within the larynx.
**intraligamentous** (*in-trah-lig-am-en'-tus*) [*intra-; ligamentum*, a ligament]. Within or between the folds of a ligament, as an *intraligamentous* cyst.
**intralingual** (*in-trah-ling'-gwal*) [*intra-; lingua*, tongue]. Within the substance of the tongue.

**intralobular** (*in-trah-lob'-ū-lar*) [*intra-; lobulus*, a little lobe]. Within a lobule, as the *intralobular* vein of the liver.
**intralocular** (*in-trah-lok'-ū-lar*). Within the loculi of a structure.
**intramammary** (*in-trah-mam'-ar-e*). Within the breast.
**intramarginal** (*in-trah-mar'-jin-al*). Located within a margin.
**intramastoiditis** (*in-trah-mas-toid-i'-tis*). See *endomastoiditis*.
**intramatrical** (*in-trah-mat'-rik-al*). Inside of a matrix.
**intramedullary** (*in-trah-med'-ū-lar-e*) [*intra-; medulla*, marrow]. Within the medulla.
**intramembranous** (*in-trah-mem'-bran-us*) [*intra-; membrana*, a membrane]. Developed or taking place within a membrane, as *intramembranous* ossification.
**intrameningeal** (*in-trah-men-in'-je-al*) [*intra-; μῆνιγξ*, membrane]. Situated within the substance of the membranes of the brain or spinal cord.
**intramural** (*in-trah-mū'-ral*) [*intra-; murus*, a wall]. Within the substance of the walls of an organ, as *intramural* fibroid of the uterus.
**intramuscular** (*in-trah-mus'-kū-lar*) [*intra-; musculus*, a muscle]. Within the substance of a muscle.
**intramolecular** (*in-trah-mo-lek'-ū-lar*) [*intra-; molecula*, a molecule]. Within the molecules of a substance.
**intramyocardial** (*in-trah-mi-o-kar'-de-al*). Within the myocardium.
**intranasal** (*in-trah-na'-sal*) [*intra-; nasus*, nose]. Within the cavity of the nose.
**intranatal** (*in-trah-na'-tal*). Occurring during birth, or at the time of birth.
**intraneural** (*in-trah-nū'-ral*). Within a nerve.
**intranuclear** (*in-trah-nū'-kle-ar*) [*intra-; nucleus*, a kernel]. Within a nucleus.
**intraocular** (*in-trah-ok'-ū-lar*) [*intra-; oculus*, eye]. Within the globe of the eye, as *intraocular* hemorrhage.
**intraoral** (*in-trah-ōr'-al*) [*intra-; os, oris*, a mouth]. Within the mouth.
**intraorbital** (*in-trah-or'-bit-al*) [*intra-; orbita*, orbit]. Within the orbit.
**intraosseous** (*in-trah-os'-e-us*) [*intra-; os, ossis*, a bone]. Within the substance of a bone.
**intraovarian** (*in-trah-o-va'-re-an*) [*intra-; ovarium*, ovary]. Within the ovarian stroma.
**intraparenchymatous** (*in-trah-par-en-ki'-mat-us*) [*intra-; parenchyma*]. Within the parenchyma; between the elements of a tissue.
**intraparietal** (*in-trah-par-i'-e-tal*) [*intra-; paries*, a wall]. 1. Within the wall of an organ. 2. Within the parietal region of the cerebrum, as the *intraparietal* fissure.
**intra partum** (*in'-trah par'-tum*) [L.]. During childbirth or delivery.
**intrapelvic** (*in-trah-pel'-vik*) [*intra-; pelvis*, basin]. Within the pelvic cavity.
**intraperitoneal** (*in-trah-per-it-on-e'-al*) [*intra-; peritoneum*]. Within the peritoneum.
**intrapial** (*in-trah-pī'-tal*). Within the pia mater.
**intraplacental** (*in-trah-plas-en'-tal*) [*intra-; placenta*]. Within the placental tissue.
**intrapleural** (*in-trah-ploo'-ral*) [*intra-; pleura*]. Within the pleural cavity.
**intrapolar** (*in-trah-po'-lar*) [*intra-; polus*, pole]. Between two poles.
**intrapontine** (*in-trah-pon'-tīn*) [*intra-; pons*]. Situated within the pons Varolii.
**intrapulmonary** (*in-trah-pul'-mon-a-re*) [*intra-; pulmonary*]. Within the substance of the lung.
**intrapyretic** (*in-trah-pi-ret'-ik*) [*intra-; πυρετός*, feverish]. Occurring during the febrile stage.
**intrarachidian** (*in-trah-rak-id'-e-an*). Intraspinal.
**intrarectal** (*in-trah-rek'-tal*). Within the rectum.
**intrarenal** (*in-trah-re'-nal*) [*intra-; ren*, kidney]. Within the kidney.
**intraretinal** (*in-trah-ret'-in-al*) [*intra-; retina*]. Within the substance of the retina.
**intrarrhachidian** (*in-trar-rak-id'-e-an*). Intraspinal.
**intrascleral** (*in-trah-skle'-ral*) [*intra-; sclera*]. Situated, occurring, or performed within the sclera.
**intrascrotal** (*in-trah-skro'-tal*) [*intra-; scrotum*]. Within the scrotal sac.
**intraspinal** (*in-trah-spī'-nal*) [*intra-; spina*, spine]. Within the spinal canal.

**intratesticular** (in-trah-tes-tik'-ū-lar). Within the testicle.

**intrathoracic** (in-trah-tho-ras'-ik) [intra-; θώραξ, chest]. Situated or occurring within the thorax.

**intratonsillar** (in-trah-ton-sil'-ar). Situated within the tonsil.

**intratubal** (in-trah-tū'-bal) [intra-; tuba, a trumpet]. Within a Fallopian tube.

**intratympanic** (in-trah-tim-pan'-ik) [intra-; tympanum]. Within the tympanic cavity.

**intraurethral** (in-trah-ū-rē'-thal) [intra-; urethra]. Within the urethra.

**intrauterine** (in-trah-ū'-ter-in) [intra-; uterus]. Within the uterus.

**intravaginal** (in-trah-vaj'-in-al). Within the vagina.

**intravasation** (in-trav-as-a'-shun) [intra-; vas, vessel]. The entrance of extraneous matter, as pus, into a blood-vessel. Cf. *extravasation*.

**intravascular** (in-trah-vas'-kū-lar) [intra-; vasculum, a small vessel]. Within the blood-vessels.

**intravenous** (in-trah-ve'-nus) [intra-; vena, a vein]. Within or into the veins. **i. injection**, the introduction of a solution directly into a vein.

**intraventricular** (in-trah-ven-trik'-ū-lar). Located or occurring within a ventricle.

**intravertebral** (in-trah-ver-te'-bral) [intra-; vertebra]. Same as *intraspinal*.

**intravesical** (in-trah-ves'-ik-al) [intra-; vesica, bladder]. Within the bladder.

**intravillous** (in-trah-vil'-us). Situated within a villus.

**intravital, intra vitam** (in trah vi' tal, tam) [intra ; vita, life]. Occurring during life. **i. stain**, one that will act upon living material.

**intravitreous** (in-trah-vit'-re-us). Within the vitreous humor.

**intrinsic** (in-trin'-sik) [intrinsecus, on the inside]. Inherent; situated within; peculiar to a part, as the *intrinsic* muscles of the larynx.

**intro-** (in-tro-). A prefix signifying within.

**introcession** (in-tro-ses'-shun) [introcedere, to go into]. A depression or sinking in, as of a surface.

**introducer** (in-tro-dū'-sur) [introducere, to lead into]. An instrument used in inserting anything; an intubator.

**introflexion** (in-tro-flek'-shun) [intro-; flexio, a bending]. A bending in; inward flexion.

**introgastric** (in-tro-gas'-trik) [intro-; γαστήρ, stomach]. Conveyed or passed into the stomach.

**introitus** (in-tro'-it-us) [intro-; ire, to go]. An aperture or entrance. **i. pelvis**, the inlet of the pelvis. **i. vaginæ**, the entrance to the vagina.

**intromission** (in-tro-mish'-un) [intro-; mittere, to send]. The introduction of one body into another, as of the penis into the vagina.

**introspection** (in-tro-spek'-shun) [intro-; spicere, to look]. The act of looking inward. **i., morbid**, the morbid habit of self-examination;· insane, or quasi-insane, dwelling upon one's own thoughts, feelings, impulses, fears, or conduct.

**introsusception** (in-tro-sus-sep'-shun) [intro-; suscipere, to receive]. Intussusception.

**introversion** (in-tro-ver'-shun) [intro-; vertere, to turn]. A turning within, as a sinking within itself of the uterus.

**intubation** (in-tū-ba'-shun) [in, in; tuba, a pipe]. **1.** The introduction of a tube into a part, particularly of a tube into the larynx, to allow the entrance of air into the lungs, as in diphtheria. **2.** Catheterism.

**intubationist** (in-tū-ba'-shun-ist). One who is expert in performing intubation of the larynx.

**intubator** (in'-tū-ba-tor). An instrument used in introducing a tube in intubation.

**intumescence** (in-tū-mes'-ens) [intumescere, to swell]. A swelling, of any character whatever. Also, an increase of the volume of any organ or part of the body.

**intumescentia** (in-tū-mes-en'-she-ah) [L.]. A swelling. **i. cervicalis**, the cervical enlargement of the spinal cord. **i. gangliformis**, the reddish, gangliform swelling of the facial nerve in the aqueduct of Fallopius. **i. lumbalis**, the lumbar enlargement of the spinal cord. **i. semilunaris**. See *ganglion, Gasserian*.

**intussusception** (in-tus-sus-ep'-shun) [intus, within; suscipere, to receive]. Invagination or slipping of one part of the intestine into the part beyond. It is most frequent in the young, occurring, as a rule, on the right side, the ileum slipping into the ascending colon, carrying the ileocecal valve in front of it. The condition is characterized by pain, tenesmus, frequent small bloody stools, the presence of a sausage-shaped tumor in the flank, and often, on rectal examination, of a mass in the rectum. **i., ascending, i., regressive, i., retrograde**, that form in which the lower part of the intestine is invaginated in the upper. **i., descending, i., progressive**, that form in which the upper part of the intestine is invaginated in the lower.

**intussusceptum** (in-tŭs-sus-sep'-tum) [see *intussusception*]. In intussusception, the invaginated portion of intestine.

**intussuscipiens** (in-tus-sus-sip'-e-enz). In intussusception, the invaginating segment of the intestine.

**inula** (in'-ū-lah) [L.]. Elecampane. The root of *I. helenium*, a plant of the natural order *Compositæ*, containing a principle resembling starch and termed *inulin*, a crystalline body, alantic acid ($C_{15}H_{20}O_3$), alantol ($C_{10}H_{16}O$), and helenin ($C_6H_8O$). Elecampane is tonic, stimulant, diaphoretic, diuretic, emmenagogue, and expectorant, and has been used in amenorrhea, dropsy, and in scaly skin diseases. Dose 20 gr.-1 dr. (1.3-4.0 Gm.).

**inulase** (in'-ū-lās). An enzyme found in the roots of *Inula helenium* and in the bulb of squill. It decomposes inulin, but has no action on starch, and is destroyed by excessive alkalinity or acidity.

**inulin** (in'-ū-lin) [inula]. **1.** $C_6H_{10}O_5$. A carbohydrate from elecampane and other plants. **2.** A precipitate from the tincture of the root of *Inula helenium*; an aromatic stimulant, tonic, and expectorant. Dose 1-3 gr. (0.06-0.2 Gm.).

**inulol** (in'-ū-lol). See *alantol*.

**inunction** (in-ungh' shun) [inungere, to anoint]. The act of rubbing an oily or fatty substance into the skin; also, the substance used.

**inustion** (in-us'-chun) [inustio, a burning in]. A burning in; deep or thorough cauterization.

**invaccination** (in-vak-sin-a'-shun). Accidental inoculation with some other disease during vaccination.

**invagination** (in-vaj-in-a'-shun) [in, in; vagina, a sheath]. The act of insheathing or becoming insheathed.

**invalid** (in'-val-id) [in, not; validus, strong]. **1.** Not well. **2.** One who is not well, especially one who is chronically ill or whose convalescence is slow. **3.** Suitable for an invalid person, as *invalid* diet, *invalid* chair.

**invaliding** (in'-val-id-ing) [in, not; valere, to be well]. The placing of a soldier or officer on the list of invalids, and thus excusing him from active duty.

**invalidism** (in'-val-id-izm) [in, not; valere, to be well]. Chronic ill-health; the state or condition of being an invalid.

**invasion** (in-va'-shun) [in, upon; vadere, to go]. The onset, especially that of a disease; also, the manner in which the disease begins its attack.

**invermination** (in-ver-min-a'-shun) [in, in; vermis, worm]. A condition of having intestinal worms; ill-health due to parasitic worms; helminthiasis.

**inverse temperature.** A complete reversal of the usual course of the temperature, so that the morning temperature exhibits the maximum, and the evening temperature the minimum. It is not infrequently seen in acute tuberculous affections, and rarely in typhoid fever.

**inversion** (in-ver'-shun) [in, in; vertere, to turn]. **1.** The act of turning inward. **2.** A turning upside down. **3.** In chemistry, the conversion of a dextrorotatory compound into one that is levorotatory. **i. of bladder**, a condition, occurring only in females, in which the bladder is in part or completely pushed into the dilated urethra. **i., sexual**, sexual instinct and attraction towards one of the same sex; homosexuality. *q. v.*

**inversive** (in-ver'-siv). Applied to ferments which convert canesugar into glucose.

**invert** (in'-vert). A person addicted to homosexuality.

**invertase** (in-ver'-tās). Same as *invertin*.

**invertebral** (in-vurt'-e-bral) [in, not; vertebra, backbone]. Without a spinal column.

**invertebrata** (in-ver-te-bra'-tah) [in, not; vertebra]. Animals that have no spinal column.

**invertebrate** (in-vert'-e-brāt) [in, not; vertebra, backbone]. **1.** Without a spinal column; invertebral. **2.** An animal without a vertebra.

**invertin** (in'-ver-tin) [in, not; vertere, to turn]. A ferment found in the intestinal juice, and produced by several species of yeast-plant; it converts cane-sugar in solution into invert-sugar.

**invertor** (in-ver'-tor). A muscle which rotates a part inward.
**invertose.** (in'-ver-tōs). Invert-sugar.
**invert-sugar.** A sugar that turns rays of polarized light to the left. The term is usually applied to levulose or to a mixture of dextrose and levulose.
**investing** (in-ves'-ting) [investire, to invest]. Ensheathing, surrounding. In dentistry, embedding a denture in what is known as investing-material, for the purpose of soldering the linings or backings of the teeth to the plate.
**investiture, investment** (in-vest'-i-chur, in-vest'-ment) [in, in; vestire, to clothe]. A sheath; a covering. i., **fibrous,** (of the suprarenal capsule), a sheath of connective tissue composed of an outer loose portion and an inner part adhering closely to the capsule; its deeper layers contain unstriped muscle-fiber. Syn., involucrum renis succenturiati. i., **myelin,** the medullary sheath.
**inveterate** (in-vet'-er-at) [in, with an intensive force; vetus, old]. Long established; chronic; resisting treatment; obstinate; as an inveterate skin disease.
**invious** (in'-ve-us) [in, not; via, a way]. Impenetrable; impervious.
**invirility** (in-vir-il'-it-e) [in, not; virilis, of a man]. Lack of manly qualities, especially lack of virile power; male impotency.
**inviscation** (in-vis-ka'-shun) [in, in; viscum, birdlime]. Insalivation.
**in vitro** (in vit'-ro) [L.]. Within glass, especially within test-tubes. Also applied to a method of observing under the microscope reproduction and life processes in living cells on a prepared glass slide.
**in vivo** (in vi'-vo) [L.]. Within the living body, as distinguished from occurrences observed in vitro.
**involucre** (in'-vo-lū-ker). Same as involucrum, q. v.
**involucrum** (in-vo-lū'-krum) [involvere, to inwrap; pl., involucra]. The covering of a part. The sheath of bone-enveloping a sequestrum.
**involuntary** (in-vol'-un-ta-re) [in, not; voluntarius, willing]. Performed or acting independently of the will. i. **muscles,** those that are not governed by the will.
**involute** (in'-vo-lūt) [involvere, to roll up, to wrap up]. In biology, rolled up, as the edges of certain leaves in the bud.
**involution** (in-vo-lu'-shun) [involvere, to roll upon]. 1. A turning or rolling inward. 2. The retrogressive change to their normal condition that certain organs undergo after fulfilling their functional purposes. i., **buccal,** the folding in of the epiblast which forms the cavity of the mouth. i.-**forms,** a term applied to microorganisms that have undergone degenerative changes as a result of unfavorable environment. i., **pituitary,** the ingrowth of the epiblast of the mouth cavity which forms the hypophysis. i., **senile,** senile atrophy. i. **of the uterus,** the return of the uterus after gestation to its normal weight and condition.
**inward** (in'-ward). Toward the center. i. **convulsions.** Synonym of laryngismus stridulus.
**inyloma** (in-il-o'-mah) [ἴς, fiber; ὕλη, matter]. A fibrous tumor; inhyloma.
**inymenitis** (in-im-en-i'-tis). See inohymenitis.
**iodacetanilide** (i-o-das-et-an'-il-īd). See iodoacetanilide.
**iodacetyl** (i-o-das'-et-il). See acetyl iodide.
**iodal** (i'-o-dal), C₂I₃HO. A compound formed of the type of chloral, but containing iodine instead of chlorine. It is reported to resemble chloral in its sedative qualities, but is seldom used as a remedy.
**iodalbacide** (i-o-dal'-bas-īd). Iodine (10 %) combined with albumin. In treatment of syphilis, dose, 15 gr. (1 Gm.) 3 to 6 times daily.
**iodalbin** (i-o-dal'-bin). Trade name of a compound of iodine and blood albumin, used like the iodides. Dose, 5–8 gr. (0.3–0.5 gm.).
**iodamyl** (i-o-dam'-il). 1. See amyl iodide. 2. iodamylum. i.-**formol,** a combination of formaldehyde, starch, thymol, and iodine.
**iodamylum** (i-od-am'-il-um). Iodized starch; employed internally in the proportion of 5 parts of iodine to 95 parts of starch. Dose 3–10 gr. (0.2–0.6 Gm.); externally in sluggish ulcers.
**iodanisol** (i-o-dan'-is-ol), C₆H₄(OCH₃)I. A proposed antiseptic forming a yellow or red crystalline mass soluble in alcohol and ether, melting at 47° C.
**iodanitin, iodanitol** (i-o-dan'-it-in, -ol). A combination of iodine and anitin; a bactericide.
**iodanthrak** (i-o-dan'-thrak). An absorption product of iodine and animal charcoal, containing 20 per cent., of iodine; used as an antiseptic.
**iodantifebrin.** See iodoacetanilide.
**iodantipyrin** (i-o-dan-tip-i'-rin). See iodopyrin.
**iodate** (i'-o-dāt). Any salt of iodic acid.
**iodated** (i'-o-da-ted). Charged with iodine.
**iodatum** (i-o-da'-tum). Charged with iodine.
**iodcaffeine.** See iodocaffeine.
**iodethane, iodethyl** (i-od'-eth-ān, -il). See ethyl iodide.
**iodethylformin** (i-od-eth-il-form'-in), C₆H₁₂N₄(C₂H₅I). A proposed substitute for iodides for internal use.
**iodhydrate** (i-od-hi'-drat). Synonym of hydriodate.
**iodia** (i-o'-de-ah). A proprietary remedy said to contain stillingia, menispermum, etc., with five grains of potassium iodide and two grains of iron phosphate in each fluidram. Dose as an alterative, 5 i-ij.
**iodic** (i-od'-ik). Obtained from or containing iodine; also due to the use of iodine. i. **acid.** See acid, iodic.
**iodid.** See iodide.
**iodide, iodid** (i'-o-did) [see iodine]. A compound of iodine with another element or radical.
**iodidum** (i-o-di'-dum). An iodide.
**iodimetry** (i-o-dim'-et-re) [iodine; μέτρον, measure]. The determination of the quantity of iodine in a compound or mixture.
**iodin** (i'-o-din). See iodine.
**iodine, iodum** (i'-o-din, i-o'-dum) [ἰώδης, violet-colored, from ἴον, a violet; εἶδος, like]. Symbol I; atomic weight 126.92; quantivalence I; specific gravity 4.948 at 17° C. (62.6° F.). A nonmetallic element with metallic luster, volatilizing at a low temperature, and giving off an irritating, crimson-purple vapor. It occurs in most marine plants, in shell-fish, and in cod-liver oil. It is soluble in alcohol and in solutions of potassium iodide and of sodium chloride. It is a powerful irritant, and is used chiefly as an alterative in scrofula and rickets; as an absorbent in goiter and lymphatic enlargements; as a counterirritant, and to produce inflammatory reaction in hydrocele and other cysts. The long-continued use of iodine and its preparations produces a form of poisoning termed iodism. See iodism. i.-**green,** a green pigment derived from coal-tar, used in histological work. i. **liniment** (linimentum iodi, B. P.), is used locally. i. **ointment** (unguentum iodi, U. S. P.), used locally as an absorbent. i. **pentoxide.** See acid, iodic. i.-**phosphor,** a combination of phosphorus, ₁/₂₀ in 20 min. of iodipin of 25 % strength. It is used hypodermatically in neurasthenia, spinal sclerosis, gout, etc. Dose 20 min. (1.2 Cc.). i. **reaction, iodophilia** developed by exposing a dried blood-smear to the action of a solution containing 3 parts of potassium iodide and 1 part of iodine in 100 parts of water, brought to syrupy consistence by adding lumps of gum-arabic. i., **solution of, compound** (liquor iodi compositus, U. S. P.), Lugol's solution. Dose 1–10 min. (0.065–0.65 Cc.). i., **tincture of** (tinctura iodi, U. S. P.). Dose 5–15 min. (0.32–1.0 Cc.). It is chiefly used locally. i. **tribromide,** IBr₃, a dark-brown fluid recommended for spraying in diphtheritic sore throat of children: 1 part in 300 of water. i. **vapor** (vapor iodi, B. P.), is used for inhalation.
**iodinophil** (i-o-din'-o-fil) [iodin; φιλεῖν, to love]. Having an affinity for iodine stain. A histological element staining readily with iodine.
**iodinophilia** (i-o-din-o-fil'-e-ah). See iodophilia.
**iodiodoformin** (i-o-di-o-do-form'-in), C₆H₁₃N₄·CHI₃. A light-brown, insoluble powder obtained from hexamethylentetramine by action of iodine; it is recommended as a substitute for iodine.
**iodipin** (i'-o-di-pin). An addition-product of iodine, 10 to 25 %, and sesame oil; used in syphilis, sciatica, etc. Dose, by mouth, 1–4 dr. (3.7–15.0 Cc.) of 10 % solution; enema, 5–7 oz. (148–207 Cc.); subcutaneous injection, 1–2 dr. (3.7–7.5 Cc.) of 25 % solution.
**iodipsol** (i-o-dip'-sol). A compound of iodine used as a substitute for iodoform.
**iodism** (i'-o-dizm) [iodine]. A condition arising from the prolonged use of iodine or iodine compounds, marked by frontal headache, coryza, ptyalism, and various skin-eruptions, especially acne; rarely by a cachexia with atrophy of the sexual organs and marked nervous symptoms.
**iodite** (i'-o-dīt). A salt of iodous acid.

# IODIZED 473 IODOSERUM

**iodized** (i'-o-dīzd) [iodine]. Impregnated with iodine.
**iodoacetanilide** (i-o-do-as-et-an'-il-id), $C_8H_8INO$. An inert substance forming white, tasteless, flaky crystals, insoluble in water, obtained from acetanilide in acetic acid by action of iodine chloride. Syn., *iodantefebrin; iodacetanilide.*
**iodoalbumin** (i-o-do-al'-bū-min). A compound of iodine and albuminoids used in myxedema.
**iodoamylene** (i-o-do-am'-il-ēn), $C_5H_9I$. A reaction-product of valerylene with fuming hydroiodic acid; a clear liquid soluble in alcohol, boils at 142° C. Syn., *valerylene hydroiodide.*
**iodoamylum.** See *iodamylum.*
**iodocaffeine** (i-o-do-kaf'-e-in). Colorless crystals, soluble in water, decomposing in hot water, obtained from a solution of potassium iodide and caffeine by action of sulphureted hydrogen.
**iodocasein** (i-o-do-ka'-se-in). An antiseptic yellow powder with odor of iodine, prepared from iodine and casein; it is used as a vulnerary.
**iodochloroxyquinolin** (i-o-do-klor-oks-e-kwin'-ol-in). A bactericide used in surgery as a substitute for iodoform. Syn., *vioform.*
**iodocin** (i-o'-do-sin). A proprietary antiseptic, analgesic, and styptic.
**iodocol, iodokol** (i-o'-do-kol). A compound of iodine and guaiacol; used in tuberculosis, croupous pneumonia, etc. Dose 3–6 gr. (0.2–0.4 Gm.) 4 or 5 times daily.
**iodocresol** (i-o-do-krē'-sol), $C_7H_7IO$. A compound of iodine and cresol; an odorless yellow powder proposed as a substitute for iodoform. Syn., *traumatol.*
**iodocrol** (i-o'-do-krol). See *carvacrol iodide.*
**iodocyanide** (i-o-do-si'-an-īd). A double salt made up of a cyanide and an iodide of the same base.
**iododerma** (i-o-do-derm'-ah) [iodine; δέρμα, skin]. Skin diseases due to use of iodine and its preparations.
**iodoeugenol** (i-o-do-ū'-jen-ol), $C_{10}H_{11}IO_2$. A compound of iodine and eugenol-sodium; a yellow, inodorous, insoluble powder, melting at 150° C. It is used as an antiseptic.
**iodoform, iodoformum** (i-o'-do-form, i-o-do-form'-um) [iodine; forma, form], $CHI_3$. A yellow, finely crystalline substance having a peculiar penetrating odor, and containing about 96.7 % of iodine by weight. It is readily soluble in chloroform and ether, less readily in alcohol, and but slightly in water. Iodoform is antiseptic and anesthetic, and is used as a dressing to wounds and syphilitic and chancroidal ulcers, either in powder or in the form of iodoform gauze. In tuberculous affections, when it can be directly introduced, it has yielded good results, being in such cases usually employed in the form of an emulsion in olive-oil or as an ethereal solution. Internally it has been used as an alterative in goiter, rickets, pulmonary tuberculosis, and syphilis. Dose 1–3 gr. (0.065–0.2 Gm.). The use of large quantities locally has led to the production of toxic symptoms, which resemble those of meningitis, and to fatty degeneration of the internal organs. Syn., *formyl triiodide; triiodomethane.* i., **deodorous**, a combination of iodoform and thymol. Syn., *anosol.* i. **gauze**, gauze impregnated with iodoform. i. **ointment** (*unguentum iodoformi*, U. S. P.), used as a local antiseptic and stimulant. i. **oleate**, a mixture of 2 % of iodoform in oleic acid. It is used as an external antiseptic on ulcers and abrasions. i.-**salol**, a mixture of iodoform and phenyl salicylate; used as an antiseptic in old wounds and in cavities. i. **suppositories** (*suppositoria iodoformi*, B. P.), used after rectal operations and in fissure of the anus.
**iodoformagen** (i-o-do-form'-maj-en). See *iodoformogen.*
**iodoformal** (i-o-do-form'-al), $C_8H_{12}N_4 \cdot C_2H_5I$. $CHI_3$. A yellow powder produced by the combination of ethyl-hexamethylentetramine hydriodide and iodoform. It is used as a substitute for iodoform.
**iodoformin** (i-o-do-form'-in), $(CH_2)_6N_4 \cdot CHI_3$. An inodorous compound of iodoform, 75 %, with hexamethylenetetramine. i.-**mercury**, a yellowish, insoluble powder recommended as an antiseptic.
**iodoformism** (i-o'-do-form-izm). Poisoning with iodoform.
**iodoformize** (i-o'-do-form-īz). To impregnate with iodoform.
**iodoformogen** (i-o-do-form'-o-jen). A compound of iodoform, 10 %, and albumin, forming a bright yellow, very light powder, insoluble in water and sterilizable at 100° C.
**iodoformum** (i-o-do-form'-um). Iodoform.
**iodogallicin** (i-o-do-gal'-is-in). See *bismuth-oxyiodomethyl gallol.*
**iodogene, iodogenin** (i-o'-do-jēn, i-o-doj'-en-in). A disinfectant, said to be a mixture of charcoal and potassium iodate, molded into cones. Iodine is liberated on combustion.
**iodogenol** (i-o-doj'-en-ol). A compound of iodine and peptonized albumin proposed as a succedaneum for the iodine preparations ordinarily employed internally.
**iodoglandin** (i-o-do-gland'-in). A preparation of thyroid gland said to contain no thyroidin.
**iodoglobulin** (i-o-do-glob'-ū-lin). A substance derived from the thyroid gland, said to be more soluble than thyroidin.
**iodohemol** (i-o-do-he'-mol). A compound consisting of iodine and hemol. See *hemol.*
**iodohydrargyrate** (i-o-do-hi-drar'-ji-rāt). A combination of mercuric iodide with the iodide of another metallic element.
**iodol, iodolum** (i'-o-dol, i-o'-do-lum), $C_4I_4 \cdot NH$. An odorless, grayish-brown powder, soluble in alcohol and in ether, and used as a substitute for iodoform and also in the treatment of diabetes mellitus. Dose ½–5 gr. (0.032–0.32 Gm.). Syn., *tetraiodopyrrol.* i.-**caffeine**, $C_8H_{10}N_4O_2 \cdot C_4I_4NH$, a light-gray, crystalline powder, insoluble in water, containing 74.6 % of iodol and 25.4 % of caffeine. It is a surgical antiseptic. 3yı., *cuffeinated iodol.* i.-**menthol**, a mixture of 1 part of menthol with 99 parts of iodol.
**iodolen, iodolene, iodoline** (i-o'-do-len, -lēn). An iodol albumin compound said to contain 36 % of iodine. It is a yellowish, granular powder, without odor or taste, soluble in hot alkaline solutions. It is a succedaneum for iodides internally and a non-irritant external antiseptic. In tertiary syphilis, dose, 30 gr. (2 Gm.) 6 to 10 times daily.
**iodomethane** (i-o-do-meth'-ān). Methyl iodide.
**iodometric** (i-o-do-met'-rik). 1. Relating to iodometry. 2. In chemical analysis relating to the process or act of determining the quantity of a substance by its reaction with a standard solution of iodine.
**iodometry** (i-o-dom'-et-re) [iodine; μέτρον, a measure]. The estimation of the iodine-content in a compound.
**iodomuth** (i-o'-do-muth), $Bi_5C_7H_7I_5O_5$. A reddish-brown powder containing bismuth and 25 % of iodine. It is siccative, antiseptic, and alterative. Dose 1–10 gr. (0.06–0.6 Gm.).
**iodonaftan** (i-o-do-naf'-tan). An ointment-base containing 3 % of iodine.
**iodonaphthol** (i-o-do-naf'-thol). See *naphthol-aristol.*
**iodone** (i'-o-dōn). Trade name of an antiseptic; it is a periodide of phthalic anhydride.
**iodophen** (i'-o-do-fen). See *nosophen.*
**iodophenacetin** (i-o-do-fe-nas'-et-in). See *iodo-phenin.*
**iodophenin** (i-o-do-fe'-nin) [iodine; phenyl], $C_{10}H_{21}I_2N_2O_4$. A combination of iodine and acetphenetidin. It is an effective bactericide.
**iodophenochloral** (i-o-do-fe-no-klor'-al). A brown fluid used in skin diseases due to parasites; it is said to consist of equal parts of phenol, tincture of iodine and chloral hydrate.
**iodophenol** (i-o-do-fe'-nol). A solution of 20 parts of iodine in 76 parts of fused phenol with 4 parts of glycerol.
**iodophil** (i-o'-do-fil). See *iodinophil.*
**iodophilia** (i-o-do-fil'-e-ah) [iodine; φιλεῖν, to love]. A pronounced affinity for iodine; the term is applied to the protoplasm of leukocytes in purulent conditions.
**iodophosphide** (i-o-do-fos'-fīd). A combination of an iodide with a phosphide.
**iodophthisis** (i-o-dof'-this-is). The emaciation, or local or general wasting that may result from free use of iodine as a remedy.
**iodopyrin** (i-o-do-pi'-rin). Iodantipyrin. A chemical compound of iodine and antipyrine with the composition, $C_{11}H_{11}IN_2O$. It is an antipyretic in doses of from one to five grains.
**iodoserum** (i-o-do-se'-rum). A solution of sodium chloride, 6 parts, and potassium iodide, 2 parts, in 1000 Cc. of water; it is used as a sedative and in syphilis.

**iodosin, iodosinum** (*i-o'-do-sin, i-o-do-si'-num*). A compound of iodine, 15 %, and albumin; it is proposed as a succedaneum for iodohyrin.

**iodospongin** (*i-o-do-spon'-jin*). A substance containing iodine and possessing thyroid properties, isolated from bath-sponge.

**iodostarin** (*i-o-do-star'-in*). An organic preparation of iodine containing 47.5 % of iodine. It is insoluble in water, and is odorless and tasteless; it is said to be less toxic than iodoform.

**iodotannin** (*i-o-do-tan'-in*). An aqueous mixture of iodine and tannin.

**iodoterpin** (*i-o-do-ter'-pin*), $C_{10}H_{16}I$. A combination of iodine, 50 %, and terpin; a dark-brown liquid with the odor of turpentine. It is used as a substitute for iodoform.

**iodotheine** (*i-o-do-the'-in*). Colorless crystals or white powder obtained from sodium iodide with theine by action of sulphureted hydrogen; it is soluble in water, decomposes in hot water. It is used to increase systolic action and arterial pressure of the heart. Dose 2–8 gr. (0.13–0.52 Gm.) 2 to 6 times daily in cachets.

**iodotheobromine** (*i-o-do-the-o-bro'-min*). A reaction-product of theobromine, a solution of potassium iodide and sulphureted hydrogen. It is diuretic, stimulant, and alterative, and is used in cardiac affections. Dose 5–8 gr. (0.32–0.52 Gm.).

**iodotherapy** (*i-o-do-ther'-ap-e*). The treatment or cure of disease by the use of iodine or its compounds.

**iodothymoform** (*i-o-do-thi'-mo-form*). Iodothymol formaldehyde, a condensation-product of thymol and formaldehyde; it is used as a wound antiseptic.

**iodothymol** (*i-o-do-thi'-mol*). See *aristol*.

**iodothyrin** (*i-o-do-thi'-rin*). A lactose trituration of the active constituents of thyroid glands of sheep. One grain contains 0.3 mg. of iodine. It is alterative and discutient. Dose 15–30 gr. (1–2 Gm.) daily. Syn., *thyrein; thyreoiodine; thyroiodine*.

**iodothyroglobulin** (*i-od-o-thi-ro-glob'-ū-lin*). An iodine-containing globulin found in the thyroid gland.

**iodovasogen** (*i-o-do-vas'-o-jen*). A solution of iodine in vasogen; it is recommended in infiltrated and spreading ulcers of the cornea.

**iodovasol** (*i-o-do-vas'-ol*). A combination of vasol and 7 % of iodine.

**iodozen** (*i-o'-do-zen*), $C_6H_5I_3(COOCH_3 . ONa)$. An iodine derivative of methyl salicylate used as an external antiseptic and discutient.

**iodozone** (*i-o'-do-zōn*). A combination of iodine and ozone; it is used as a mouth-wash and as an inhalation in tuberculosis.

**iodum** (*i-o'-dum*). See *iodine*.

**ioduret** (*i-od'-ū-ret*) [*iodine*]. An iodide.

**iodyloform** (*i-o-dil'-o-form*). Trade name of a combination of iodine and gelatin used as a substitute for iodoform.

**iolin** (*i'-o-lin*). A preparation for external use containing 20 % of iodine; it is said not to stain the skin.

**ion** (*i'-on*) [*ἰόν*, going]. An atom or group of atoms set free by electrolysis, and classified as an anion or kation, according as it is set free at the positive or negative pole.

**ionic medication.** See *medication*.

**ionium** (*i-o'-ne-um*) [*ion*]. A recently discovered element of radio-active properties.

**ionization** (*i-on-i-za'-shun*) [*ion*]. Electrolytic dissociation; the production of ions.

**ionize** (*i'-on-iz*). To dissociate into ions; said of an electrolyte.

**ionone** (*i'-on-ōn*). A hydroaromatic ketone prepared synthetically from citral, the odorous principle of lemon oil. It has the odor of violets.

**ionophose** (*i-on'-o-fōz*). A violet phose.

**iontophoresis** (*i-on-to-fo-re'-sis*) [*ion-; φόρησις*, a carrying]. The introduction of ions into the body by the electric current, for therapeutic purposes. Medical ionization; cataphoresis.

**iophobia** (*i-o-fo'-be-ah*) [*ἰός*, poison; *φόβος*, fear]. A morbid dread of poisons.

**iotacism** (*i-o'-tas-izm*) [*ἰῶτα*, the Greek letter I.] Inability to pronounce distinctly the proper sound of the letter *i*.

**iothion** (*i-o-thi'-on*). Diiodhydroxypropane, a yellow, oily fluid, containing 80 per cent. of iodine; used in place of the iodides.

**ipecac, ipecacuanha** (*ip'-e-kak, ip-e-kak-ū-an'-ah*) [Braz., *ipecaaguen*]. The dried root of *Cephaelis*

*ipecacuanha*, a plant of the order *Rubiaceæ*, containing an alkaloid, *emetine*, $C_{15}H_{22}N_2O_4$, and ipecacuanhic acid. In large doses ipecac is emetic; in small doses, diaphoretic and expectorant; and in minute doses, a gastric stimulant. It is used as an emetic, especially in narcotic poisoning, and, in children, to dislodge membranes and secretions in croup and capillary bronchitis; as a diaphoretic in acute colds, as an expectorant in bronchitis, as a sedative (in minute doses in vomiting), and in dyspepsia as a stimulant. It is said to be a specific in tropical dysentery. Emetine is emetic in doses of ⅙–⅓ gr. (0.008–0.016 Gm.), but irritant in large doses. 1., fluidextract of (*fluidextractum ipecacuanhæ*, U. S. P.). Dose 5–30 min. (0.32–2.0 Cc.). 1., opium, powder of (*pulvis ipecacuanhæ et opii*, U. S. P.; *pulvis ipecacuanhæ compositus*, B. P.), Dover's powder. Dose 2–15 gr. (0.13–1.0 Gm.). 1. and opium, tincture of (*tinctura ipecacuanhæ et opii*, U. S. P.). Dose 5–10 min. (0.32–0.65 Cc.). 1., syrup of (*syrupus ipecacuanhæ*, U. S. P.). Dose, as an emetic, 30 min.–1 dr. (2–4 Cc.) for a child; ½–1 oz. (16–32 Cc.) for an adult; as an expectorant, 5 min.– 1 dr. (0.32–4.0 Cc.). 1., wine of (*vinum ipecacuanhæ*, U. S. P.). Dose 1 min.–1 dr. (0.065–4.0 Cc.).

**ipomein** (*ip-o-me'-in*). A glucoside, $C_{17}H_{22}O_{16}$, from the root of *Ipomœa fastigiata*.

**ipsilene** (*ip'-sil-ēn*). A gas used as a disinfectant, obtained from iodoform by action of ethyl chloride with heat and pressure.

**ipsolateral** (*ip-so-lat'-er-al*) [*ipse*, same; *latus*, side]. Situated on the same side, indicating paralytic or similar symptoms which occur on the same side as the cerebral lesion causing them.

**Ir.** Chemical symbol of *iridium*.

**ir.** Abbreviation for *internal resistance*.

**iralgia** (*i-ral'-je-ah*). See *iridalgia*.

**irascibility** (*i-ras-ib-il'-it-e*) [*irasci*, to be angry]. The quality of being choleric, irritable, or of hasty temper. It is a frequent symptom in some varieties of insanity and in neurasthenia, and in some cases it amounts to a species of insanity.

**iretol** (*i'-ret-ol*) [*Iris*, a genus of plants], $C_7H_8O_4$. A phenol obtained by fusing irigenin with potash; it melts at 186° C. Syn., *methoxyphloroglucin*.

**iridadenosis** (*ir-id-ad-en-o'-sis*) [*iris; ἀδήν*, gland]. A glandular affection of the iris.

**iridæmia** (*ir-id-e'-me-ah*). See *iridemia*.

**iridal** (*ir'-id-al*) [*iris*]. Relating to the iris.

**iridalgia** (*ir-id-al'-je-ah*) [*irido-; ἄλγος*, pain]. Pain referable to the iris.

**iridauxesis** (*ir-id-awks-e'-sis*) [*irido-; αὔξησις*, increase]. Auxesis or tumefaction of the iris.

**iridavulsion** (*ir-id-av-ul'-shun*) [*irido-; avellere*, to tear away]. Surgical avulsion of the iris.

**iridectome** (*ir-id-ek'-tōm*) [*irido-; ἐκτομή*, a cutting out]. A cutting instrument used in iridectomy.

**iridectomize** (*ir-id-ek'-tom-iz*) [*iridectomy*]. To excise a part of the iris; to perform iridectomy.

**iridectomy** (*ir-id-ek'-to-me*) [*iris; ἐκτομή*, excision]. The cutting out of a part of the iris.

**iridectropium** (*ir-id-ek-tro'-pe-um*) [*irido-; ἐκτρόπιον*, eversion]. Eversion of a part of the iris.

**iridemia** (*ir-id-e'-me-ah*) [*irido-; αἷμα*, blood]. Hemorrhage from the iris.

**iridencleisis, iridenkleisis** (*i-rid-en-kli'-sis*). See *iridodesis*.

**iridentropium** (*ir-id-en-tro'-pe-um*) [*irido-; ἐντροπή*, a turning in]. Inversion of a part of the iris.

**irideremia** (*i-rid-er-e'-me-ah*) [*iris; ἐρημία*, lack]. Total or partial absence of the iris.

**iridescence** (*ir-id-es'-ens*) [*iridescere*, to shine with rainbow-colors]. The property of breaking up light into the spectral colors.

**iridesis** (*ir-id-e'-sis*). See *iridodesis*.

**iridian** (*ir-id'-e-an*) [*iris*]. Relating to the iris; iridal.

**iridic** (*ir-id'-ik*) [*iris*]. Pertaining to the iris.

**iridicolor** (*ir-id'-ic-ul-or*) [*iris*, a rainbow; *color*]. In biology, iridescent; exhibiting prismatic colors.

**iridin** (*i'-rid-in*) [*iris*]. 1. A precipitated extract of blue flag. See *iris* (2). 2. A glucoside from the rhizome of *Iris florentina*.

**iridium** (*i-rid'-e-um*) [*iris*]. A platinoid metal; alloyed in small percentage with platinum it confers rigidity upon the latter. The alloy is used as plate in mechanical dentistry. Symbol, Ir.; atomic weight, 193.1.

**iridization** (*ir-id-iz-a'-shun*) [*irido-*]. The subjective appearance, as of an iridescent halo seen by persons affected with glaucoma.

**irido-** (*i-rid-o-*) [*iris*]. A prefix meaning relating to the iris.

**iridocapsulitis** (*ir-id-o-kap-sū-li'-tis*). Inflammation involving the iris and the capsule of the lens.

**iridocele** (*ir-id'-o-sēl*). [*irido-*; κήλη, hernia]. Protrusion of part of the iris through a wound or ulcer.

**iridochoroiditis** (*ir-id-o-ko-roid-i'-tis*) [*irido-*; *choroid*; ιτις, inflammation]. Inflammation of both the iris and the choroid of the eye.

**iridocinesis** (*ir-id-o-sin-e'-sis*). See *iridokinesis*.

**iridocoloboma** (*ir-id-o-kol-o-bo'-mah*) [*irido-*; κολόβωμα, a mutilation]. 1. The portion of iris removed in iridectomy. 2. See *coloboma*.

**iridocyclectomy** (*ir-id-o-si-klek'-to-me*) [*irido-*; κύκλος, circle; ἐκτομή, excision]. Excision of the iris and of the ciliary body.

**iridocyclitis** (*ir-id-o-sik-li'-tis*) [*irido-*; κύκλος, a circle; ιτις, inflammation]. Inflammation of the iris and the ciliary body.

**iridocyclochoroiditis** (*ir-id-o-sik-lo-ko-roid-i'-tis*) [*irido-*; κύκλος, circle; χόριον, chorion; εἶδος, likeness; ιτις, inflammation]. Combined inflammation of the iris, the ciliary body, and the choroid.

**iridocystectomy** (*ir-id-o-sist-ek'-to-me*) [*irido-*; κύστις, bladder; ἐκτομή, excision]. Knapp's operation for making a new pupil when iridocyclitis or iridocapsulitis following cataract operations or trauma has closed the cold.

**iridocyte** (*ir'-id-o-sīt*) [*irido-*; κύτος, cell]. Any cell that produces color, either by means of its structure or its contents.

**iridodesis** (*ir-id-od'-es-is*) [*irido-*; δέσις, a binding together]. An operation for the purpose of altering the position of the pupil by drawing the iris into one or two small openings in the cornea and preventing its return by a loop of silk placed around it.

**iridodialysis** (*ir-id-o-di-al'-is-is*). 1. See *corediàlysis*. 2. The separation of the iris from its attachments.

**iridodonesis** (*ir-id-o-do-ne'-sis*) [*irido-*; δόνησις, a trembling]. Tremulousness of the iris; hippus.

**iridokinesis** (*ir-id-o-kin-e'-sis*) [*irido-*; κίνησις, movement]. Any movement of the iris, normal or otherwise.

**iridol** (*i'-rid-ol*) [*Iris*, a genus of plants], C₇H₆(OCH₃)₂OH. A phenol obtained from distillation of iridic acid (C₁₆H₁₈O₆) from orris-root.

**iridoleptynsis** (*ir-id-o-lep-tin'-sis*) [*irido-*; λέπτυνσις, attenuation]. Attenuation or atrophy of the iris.

**iridomalacia** (*ir-id-o-mal-a'-se-ah*) [*irido-*; μαλακία, softness]. Morbid softening of the iris.

**iridomotor** (*ir-id-o-mo'-tor*) [*irido-*; *movere*, to move]. Promoting the motion of the iris.

**iridoncosis** (*ir-id-ong-ko'-sis*) [*irido-*; -ὄγκωσις, a puffing out]. Thickening of the iris.

**iridoncus** (*ir-id-ong'-kus*) [*irido-*; ὄγκος, mass]. A tumor or swelling of the iris.

**iridoparalysis** (*ir-id-o-par-al'-is-is*) [*irido-*; *paralysis*]. Paralysis of the iris.

**iridoparelkysis** (*ir-id-o-par-el'-kis-is*) [*irido-*; παρέλκειν, to draw aside]. An induced prolapse of the iris to effect displacement of the pupil.

**iridoparesis** (*ir-id-o-par'-es-is*) [*irido-*; πάρεσις, a letting go]. A slight or partial paralysis of the iris.

**iridoplania** (*ir-id-o-pla'-ne-ah*) [*irido-*; πλάνη, a wandering]. Same as *hippus*.

**iridoplasma** (*ir-id-o-plaz'-mah*) [*irido-*; πλάσμα, anything formed]. A form of degeneration of the iris.

**iridoplatinum** (*ir-id-o-plat'-in-um*). An alloy of iridium and platinum; used in making electrodes, etc.

**iridoplegia** (*ir-id-o-ple'-je-ah*) [*irido-*; πληγή, stroke]. Paralysis of the sphincter of the iris.

**iridoptosis** (*ir-id-op-to'-sis*) [*irido-*; πτῶσις, a falling]. Prolapse of the iris.

**iridopupillary** (*ir-id-o-pū'-pil-a-re*) [*irido-*; *pupilla*, the pupil of the eye]. Pertaining to the iris and the pupil.

**iridorhexis** (*ir-id-o-reks'-is*) [*irido-*; ῥῆξις, a breaking]. Rupture of the iris.

**iridoschisis, iridoschisma** (*ir-id-os'-kis-is, ir-id-os-kis'-mah*) [*irido-*; σχίσις, σχίσμα, cleft]. Coloboma of the iris.

**iridoscleretomy** (*ir-id-o-skler-ot'-o-me*) [*irido-*; σκληρός, hard; τέμνειν, to cut]. Puncture of the sclera with division of the iris.

**iridosis** (*ir-id-o'-sis*). See *iridodesis*.

**iridosteresis** (*ir-id-o-ster-e'-sis*). See *aniridia* and *irideremia*.

**iridotome** (*ir'-id-o-tōm*) [*irido-*; τομή, a cutting]. A cutting-instrument employed in iridotomy.

**iridotomy** (*ir-id-ot'-o-me*) [*irido-*; τομή, section]. An incision into the iris.

**iridotromos** (*ir-id-ot'-ro-mos*) [*irido-*; τρόμος, tremor]. Hippus; tremor of the iris.

**iregenin** (*i-rij'-en-in*), C₁₆H₁₆O₈. A resolution-product of the glucoside iridin by action of dilute suphuric acid. It has the properties of a phenol, forms crystals melting at 186° C., and gives an intense violet color with ferric chloride.

**iris** (*i'-ris*) [ἶρις, a halo or rainbow]. 1. A colored circular membrane placed between the cornea and the lens, and having a central perforation; the *pupil*. It is about half an inch in breadth, and consists principally of two sets of unstriped muscular fibers, the sphincter of the iris, or *spincter pupillæ*, a narrow zone of circular fibers surrounding the pupil, and the dilator of the iris, or *dilator pupillæ*, a radiate band of fibers extending from the pupil to the border of the iris. 2. A genus of plants of the natural order *Irideæ*. The rhizome of *Iris versicolor* (blue flag) is cathartic, emetic, and diuretic. Dose 10–20 gr. (0.65–1.3 Gm.). **i. bombé**, a condition in which the iris bulges forward due to an increase in the intraocular fluid in the posterior chamber. **i.-contraction**. See under *reflex*. **i., Florentine**, orris-root, the root of *Iris florentina*, emetocatharic and diuretic. At present it is used chiefly as an ingredient of toothpowders. **i.-pigment**, the chemically pure pigment of the bovine eye; triturated with water it is used for tattooing corneal opacities. **Irish ague**. Synonym of *typhus fever, q. v.* **I. button**, syphilis. **I. moss**. See *chondrus*.

**irisin** (*i'-ris-in*). Same as *iridin*.

**irisol** (*i'-ris-ol*). A proprietary disinfectant, said to contain iodoform, 50 %, and boric acid, 45 %.

**iritic** (*ir-it'-ik*) [*iritis*]. Of the nature of, pertaining to, or affected with iritis.

**iritis** (*ir-i'-tis*) [*iris*; ιτις, inflammation]. Inflammation of the iris.

**iritoectomy** (*ir-it-o-ek'-to-me*) [*iris*; ἐκτομή, a cutting out]. The removal of a portion of the iris and iritic membrane for occlusion of the pupil.

**iritomy** (*ir-it'-o-me*). See *iridotomy*.

**iron** (*i'-ern*). See *ferrum*. **i. albuminate**, contains 5 % ferric oxide or 10 % ferric chloride. Dose 10–30 gr. (1.3–2.0 Gm.). **i. and ammonium acetate**, solution of (*liquor ferri et ammonii acetatis*, U. S. P.). Dose 4 dr. (16 Cc.). **i. and ammonium alum**. See *ferric ammonium sulphate*. **i. and ammonium citrate**. See *ferric citrate, soluble*. **i. and ammonium tartrate**. See *ferric. ammonium tartrate*. **i. arseniate** (*ferri arsenias*, B. P.), chiefly valuable for the arsenic it contains. Dose 1/10–1/2 gr. (0.006–0.008 Gm.). **i. bromide**, used in solution in doses of 20 min. (1.2 Cc.) 3 times daily, as an alterative tonic in chorea and scrofula. **i. cacodylate**, used in chlorosis, etc. Dose 2–4 gr. (0.13–0.26 Gm.) daily; hypodermatically 1/2–1 1/2 gr. (0.03–0.09 Gm.). **i. casein, i. caseinate**, a flesh-colored precipitate without taste or odor, obtained from casein of milk with iron lactate, containing 5.2 % of ferric oxide; it is used as a nutritive. Dose 2–10 gr. (0.13–0.65 Gm.) 3 times daily. Syn., *ferrum caseinatum*; *iron nucleoalbuminate*. **i. ferrocyanide**, dark-blue powder or lumps obtained from ferric salts with potassium ferrocyanide; it is used as a tonic, antiperiodic, and cholagogue. Dose 2–5 gr. (0.13–0.32 Gm.). Syn., *Berlin blue*; *ferric ferrocyanide*; *insoluble iron cyanide*; *Prussian blue*. **i. glycerinophosphate**, FePO₄C₃H₅(OH)₂+2H₂O, yellow scales, soluble in water. It is used in neurasthenia, phosphaturia, Addison's disease, etc. Dose 2 gr. (0.13 Gm.) 3 times daily. **i. lactate**, ferrous lactate, a salt occurring in minute whitish-green crystals. Dose 5 gr. (0.32 Gm.). **i. mixture, compound** (*mistura ferri composita*, U. S. P.), Griffith's mixture. Dose 1–2 oz. (30–60 Cc.). **i. nucleoalbuminate**. See *i.-casein*. **i. oleate**, a brownish-green, sticky substance containing Fe(C₁₈H₃₃O₂)₃. It is soluble in ether and is used as a tonic inunction. **i. oxalate**. Dose 2–3 gr. (0.13–0.2 Gm.). **i. oxide, hydrated**. See *ferric hydroxide*. **i. oxide of, magnetic** (*ferri oxidum magneticum*, B. P.). Dose 5–20 gr. (0.32–1.3 Gm.). **i. paranucleinate**, a nutritive preparation of casein of cows' milk, containing 22 % of iron and 2.5 % of phosphorous. Dose 1 gr. (0.3 Gm.) 3 times daily. Syn., *iriferrin*. **i. peptonate**, contains 5 % of red iron oxide with peptone;

a fine yellow-brown powder. Dose 2-8 gr. (0.13-0.52 Gm.). **i. perchloride.** See *ferri chloride*. **i. persulphate.** See *i. sulphate, ferric*. **i. phosphate** (*ferri phosphas*, B. P.). Dose 5-10 gr. (0.32-0.65 Gm.). **i. phosphate, ferric,** $Fe_2(PO_4)_2$, white powder, soluble in acids; used externally in solution with dilute phosphoric acid for carious teeth. Ointment 10 to 20 % in carcinoma. **i. phosphate, soluble.** See *ferric phosphate, soluble*. **i. phosphosarcolactate.** See *carniferrin*. **i. and potassium tartrate** (*ferri et potassii tartras*, U. S. P.). Dose 4 gr. (0.25 Gm.). **i. and quinine citrate** (*ferri et quininæ citras*, U. S. P.). Dose 5-15 gr. (0.32-1.0 Gm.). **i. and quinine citrate, soluble** (*ferri et quininæ citras solubilis*, U. S. P.). Dose 5-10 gr. (0.32-0.65 Gm.). **i., quinine, and strychnine, glycerite of the phosphates of** (*glyceritum ferri, quininæ et strychninæ phosphatum*, U. S. P.). Dose 15 min. (1 Cc.). **i., quinine, and strychnine phosphates, elixir of** (*elixir ferri, quininæ et strychninæ phosphatum*, U. S. P.). Dose 1 dr. (4 Cc.). **i., quinine, and strychnine phosphates, syrup of** (*syrupus ferri, quininæ et strychninæ phosphatum*, U. S. P.). Dose 1 dr. (4 Cc.). **i., reduced.** See *ferrum reductum*. **i. and strychnine citrate** (*ferri et strychninæ citras*, U. S. P.). Dose 1-3 gr. (0.065-0.2 Gm.). **i. succinate,** $Fe(OH)C_4H_4O_4$, amorphous, reddish-brown powder, tonic and alterative, used as solvent in biliary calculi. Dose 1 teaspoonful of the salt with 10 drops of chloroform 4 to 6 times daily. Syn., *ferric succinate*. **i. sulphide, ferric,** $Fe(SO_4)_3$, a grayish-white powder used as a disinfectant and bactericide. Syn., *iron persulphate; iron sesquisulphate; iron tersulphate; normal ferric sulphate*. **i. tannate,** a salt in crimson scales. Dose 8-30 gr. (0.52-2.0 Gm.) in 24 hours. **i. valerianate,** a dark-red, amorphous powder. Dose 2-5 gr. (0.13-0.32 Gm.). **i. vitellinate,** a preparation of iron and yolk of egg. **i., wine of** (*vinum ferri*, U. S. P.). Dose 2 dr. (8 Cc.). **i., wine of, bitter** (*vinum ferri amarum*, U. S. P.), made from the soluble citrate of iron and quinine. Dose 1-4 dr. (4-16 Cc.). See also under *ferric, ferrous*, and *ferrum*.
**ironal, ironol** (*i'-ron-al, -ol*). A preparation said to contain 80 % of absorbable iron.
**ironcosis** (*i-ron-ko'-sis*). See *iridoncosis*.
**irone** (*i'-rōn*) [*Iris*, a genus of plants; *-one*, suffix signifying ketone], $C_{13}H_{20}O$. A substance isolated from *Iris florentina*, believed to be the mother-substance of the odorous constituents of orris-root.
**irotomy** (*i-rot'-o-me*). Same as *iridotomy*.
**irradiating** (*ir-ra'-de-a-ting*) [*irradiare*, to emit rays in every direction]. Radiating from a center, as a pain arising from a definite focus of irritation.
**irradiation** (*ir-a-de-a'-shun*) [*in, on; radiare*, to radiate]. 1. A phenomenon in which, owing to the difference in the illumination of the field of vision or its background, objects appear much larger than they really are. 2. Diffusion in all directions from a common center; applied to nerve impulses, stellate fractures, pains felt in some position in undemonstrable anatomical connection with an affected organ, etc.
**irreducible** (*ir-e-du'-si-bl*) [*in*, not; *reducere*, to lead back]. Not reducible; not capable of being replaced in a normal position, as an *irreducible hernia*.
**irregular** (*ir-eg'-u-lar*) [*in*, not; *regula*, rule]. Not regular; not normal or according to rule; not rhythmic; not recurring at proper intervals, as an *irregular pulse*.
**irreinoculability** (*ir-e-in-ok-u-la-bil'-it-e*) [*in*, not; *re*, again; *inoculatio*, an engrafting]. Insusceptibility to contagion due to previous inoculation.
**irrespirable** (*ir-es-pi'-ra-bl*) [*in*, not; *respirare*, to breathe]. Not capable of being breathed.
**irrhythmia** (*ir-ith'-me-ah*). See *arrhythmia*.
**irrigant** (*ir'-ig-ant*) [*irrigare*, to lead water to]. A substance or wash used in or by irrigation.
**irrigation** (*ir-ig-a'-shun*) [*irrigare*, to lead water to]. The act of washing out by a stream of water, as *irrigation* of the bladder. **i., continuous,** the continuous passage of a stream of water over a surface in order to reduce or limit inflammation.
**irrigator** (*ir-ig-a'-tor*) [*irrigare*, to lead water to]. An apparatus, or device, for accomplishing the irrigation of a part, surface, or cavity.
**irritability** (*ir-it-ab-il'-it-e*) [*irritare*, to excite]. 1. The state of being irritable or of responding to stimuli. 2. A functional disturbance of a part on account of which it reacts excessively to slight stimu-

lation; as *irritability* of the bladder, a condition in which the urine is voided in small quantities at short intervals. **i., contact,** a phenomenon shown by Loeb in muscular movement by action of various salts, e. g., a frog's muscle previously treated with a Na salt the anion of which precipitates Ca is excited by contact with such substances as oil, water, air, etc., unlike a normal muscle. **i., faradic,** the state in which the faradic current will cause muscular contraction. **i., galvanic,** the state in which the galvanic current will cause muscular contraction. **i., muscular,** the inherent contractile quality of a muscle. **i., nervous,** the property of a nerve to transmit impulses upon stimulation.
**irritable** (*ir'-it-a-bl*) [see *irritability*]. 1. Reacting to stimuli. 2. Easily excited. **i. bladder,** a condition of the bladder marked by constant desire to void urine. **i. breast,** a neuralgic condition of the mammary gland, usually associated with uterine affections, or with intercostal neuralgia. **i. heart,** a peculiar condition of the heart characterized by precordial pain, dyspnea on exertion, palpitation, and irregularity of the heart's action. **i. spine,** a condition of spinal anemia frequently occurring in young hysterical females. There is pain along the spine with tenderness on pressure, and vertigo, nausea, palpitation and neuralgia. **i. testicle.** See *testicle*. **i. tongue,** the clean, very red tongue, with enlarged red papillæ about its tip, seen in the dyspepsia of drunkards.
**irritant** (*ir'-it-ant*) [see *irritability*]. 1. Causing or giving rise to irritation. 2. An agent that induces irritation.
**irritation** (*ir-it-a'-shun*) [see *irritability*]. 1. A condition of undue excitement. 2. The act of irritating or stimulating. 3. The stimulus necessary to the performance of a function.
**irritative** (*ir'-it-a-tiv*) [see *irritability*]. Characterized by or dependent on irritation. **i. fever,** a febrile condition dependent upon the presence in the body of irritating substances.
**irruation** (*ir-oo-ma'-shun*) [*irrumare*, to give suck]. Sexual perversion where gratification is found by mouth; same as *fellatio*.
**isadelphia** (*is-ad-el'-fe-ah*) [ἴσος, equal; ἀδελφός, a brother]. A twin monstrosity in which each body is normal in the development of all essential organs but united by unimportant tissues.
**isadelphous** (*is-ad-el'-fus*) [ἴσος, equal; ἀδελφός, brother]. In biology, having an equal number of stamens in each bundle or brotherhood.
**Isambert's disease** (*e-zahm-bair*) [Émile *Isambert*, French physician, 1827-1876]. Tuberculous ulceration of the mouth, fauces, and pharynx.
**isapiol** (*is-ap'-e-ol*) [ἴσος, equal; *apiol*], $C_{12}H_{14}O_4$. An isomer of apiol obtained from it by action of alcoholic solution of potassium hydroxide with heat. In physiological properties it differs but slightly from apiol.
**isatin** (*i'-sat-in*) [*Isatis*, woad], $C_8H_5NO_2$. A substance obtained by the oxidation of indigo with $HNO_3$.
**isatropylcocaine** (*is-at-ro'-pil-ko-ka'-in*), $C_{19}H_{25}NO_4$. An amorphous alkaloid from coca leaves. It has no anesthetic properties, but is said to be an active cardiac poison.
**ischemia** (*is-ke'-me-ah*) [ἴσχειν, to check; αἷμα, blood]. Local anemia.
**ischemic** (*is-hem'-ik*) [see *ischemia*]. Affected with or relating to ischemia.
**ischéocele** (*is-ke'-o-sēl*). See *ischiocele*.
**ischesis** (*is'-kes-is*) [ἴσχειν, to check]. Retention or suppression of a discharge or secretion.
**ischia** (*is'-ke-ah*). Plural of *ischium*.
**ischiadic** (*is-ke-ad'-ik*). Same as *ischiatic*.
**ischiagra** (*is-ke-ah'-grah*) [ἰσχίον, hip; ἄγρα, seizure]. Gout in the hip.
**ischial** (*is'-ke-al*). Pertaining or belonging to the ischium.
**ischialgia** (*is-ke-al'-je-ah*) [*ischium*; ἄλγος, pain]. Sciatica; neuralgia of the hip.
**ischialgic** (*is-ke-al'-jik*) [ἰσχίον, hip; ἄλγος, pain]. Relating to or affected with ischialgia.
**ischias, ischiasis** (*is'-ke-as, is-ki'-a-sis*). See *ischiálgia*.
**ischiálgia... ischias scoliotica** (*sko-le-ot'-ik-ah*). A transitory scoliosis due to a painful affection of the muscles or nerves about the back.
**ischiatic** (*is-ke-at'-ik*) [*ischium*]. Pertaining to the ischium.

# ISCHIATITIS 477 ISODYNAMIC

**ischiatitis** (*is-ke-at-i'-tis*) [ἰσχίον, hip; ιτις, inflammation]. Inflammation of the sciatic nerve.
**ischiatocele** (*is-ke-at'-o-sēl*). See *ischiocele*.
**ischidrosis** (*is-kid-ro'-sis*) [ἰσχειν, to suppress; ἱδρώς, sweat]. Suppression of sweat.
**ischidrotic** (*is-kid-rot'-ik*) [ἰσχειν, to suppress; ἱδρώς, sweat]. Causing a retention or suppression of the sweat.
**ischien** (*is'-ke-en*) [ἰσχίον, hip]. Belonging to the ischium in itself.
**ischigalactic** (*is-ke-gal-ak'-tik*) [ἰσχειν, to restrain; γάλα, milk]. 1. Tending to check the flow of milk. 2. An antigalactic agent or medicine.
**ischio-** (*is-ke-o-*) [*ischium*]. A prefix indicating relationship to the ischium or the hip.
**ischioanal** - (*is-ke-o-a'-nal*) [*ischio-*; *anus*]. Pertaining to the ischium and anus.
**ischiobulbar** (*is-ke-o-bul'-bar*) [*ischio-*; βόλβος, a bulb]. Pertaining to the ischium and the bulb of the urethra.
**ischiocapsular** (*is-ke-o-kap'-sū-lar*) [*ischio-*; *capsula*, a capsule]. Pertaining to the ischium and the capsular ligament of the hip.
**ischiocavernosus** (*is-ke-o-kav-er-no'-sus*) [*ischio-*; *caverna*, cavern]. The erector penis (or erector clitoridis).
**ischiocele** (*is'-ke-o-sēl*) [*ischio-*; κήλη, hernia]. Hernia through the sciatic notch.
**ischiococcygeus** (*is-ke-o-kok-sij'-e-us*) [*ischio-*; *coccyx*]. The coccygeus muscle.
**ischiodidymus** (*is-ke-o-did'-im-us*) [ἰσχίον, hip; δίδυμοι, double]. A double monstrosity united at the hips.
**ischiofemoral** (*is-ke-o-fem'-o-ral*) [*ischio-*; *femur*]. 1. Pertaining to the ischium and the femur. 2. The adductor magnus muscle.
**ischiomenia** (*is-ke-o-me'-ne-ah*). See *ischomenia*.
**ischiomyelitis** (*is-ke-o-mi-el-i'-tis*) [ἰσχίον, hip, loins; μυελός, marrow; ιτις inflammation]. Lumbar myelitis; osphyomyelitis.
**ischioneuralgia** (*is-ke-o-nū-ral'-je-ah*)[*ischio-*; *neuralgia*]. Sciatica.
**ischiopagus** (*is-ke-op'-ag-us*) [ἰσχίον, hip; πάγος, united]. A monomphalic monstrosity united by the coccyges and the sacra. i. tetrapus, one with four legs. i. tripus, one with three legs.
**ischiopagy** (*is-ke-op'-aj-e*) [*ischio-*; πάγος, anything which has become solid]. A double monstrosity in which the two bodies are united at the coccyx and sacrum.
**ischioperineal** (*is-ke-o-per-in-e'-al*) [*ischio-*; *perineum*]. 1. Pertaining to both ischium and perineum; pertaining to the space between the anus and the scrotum. 2. See *Transversus perinæi* under *muscle*.
**ischiophthisis** (*is-ke-off'this-is*) [ἰσχίον, hip; φθίσις, a wasting]. Emaciation associated with or due to hip-joint disease.
**ischioprostatic** (*is-ke-o-pros-tat'-ik*). 1. Relating to the ischium and the prostate. 2. The transversus perinæi muscle.
**ischiopubic** (*is-ke-o-pū'-bik*). Relating to the ischium and the pubes.
**ischiopubiotomy** (*is-ke-o-pū-be-ot'-o-me*). Division of the ischial and pubic rami in otherwise impossible labor.
**ischiorectal** (*is-ke-o-rek'-tal*) [*ischio-*; *rectum*]. Pertaining to both ischium and rectum. i. abscess, an inflammation of the areolar tissue of the ischiorectal fossa. i. fossa. See *fossa, ischiorectal*.
**ischiosacral** (*is-ke-o-sa'-kral*). Pertaining to the ischium and sacrum.
**ischium** (*is'-ke-um*) [ἰσχίον, hip]. The inferior part of the os innominatum; the bone upon which the body rests in sitting.
**ischnogyria** (*isk-no-ji'-re-ah*) [ἰσχνός, feeble; γύρος, curve]. A condition attended with slight development of the cerebral convolutions.
**ischo-** (*is'-ko-*) [ἰσχειν, to suppress]. A prefix meaning suppressed, stopped, or checked.
**ischobiennia** (*is-ko-bien'-e-ah*) [ἰσχειν, to suppress; βλέννος, mucus]. The suppression of a mucous discharge.
**ischocenosis** (*is-ko-sen-o'-sis*) [ἰσχειν, to suppress; κένωσις, discharge]. The suppression of any established or normal discharge.
**ischochezia** (*is-ko-ke'-ze-ah*) [ἰσχειν, to suppress; χέζω, bile]. A suppression of the flow of bile.
**ischochymia** (*is-ko-ki'-me-ah*) [ἰσχειν, to suppress; χυμός, juice]. Dilatation of the stomach.

**ischogalactia** (*is-ko-gal-ak'-te-ah*) [ἰσχειν, to suppress; γάλα, milk]. Suppression of the natural flow of milk.
**ischogyria** (*is-ko-ji'-re-ah*) [ἰσχειν, to suppress; γύρος, a curve]. The small convolutions produced by senile atrophy. Cf. *ischnogyria*.
**ischolochia** (*is-ko-lo'-ke-ah*) [ἰσχειν, to suppress; λόχεια, lochia]. Suppression of the lochial flow.
**ischomenia** (*is-ko-me'-ne-ah*) [ἰσχειν, to suppress; μήν, month; menses]. Suppression of the menstrual flow.
**ischophonia** (*is-ko-fo'-ne-ah*) [ἰσχειν, to suppress; φωνή, voice]. Aphonia. An impediment in speech.
**ischopyosis** (*is-ko-pi-o'-sis*) [ἰσχειν, to suppress; πύον, pus]. The checking of any discharge of pus.
**ischospermia** (*is-ko-sper'-me-ah*) [ἰσχειν, to check; σπέρμα, seed]. Suppression of semen.
**ischuretic** (*is-kū-ret'-ik*) [see *ischuria*]. 1. Relating to or relieving ischuria. 2. A remedy or agent that relieves retention or suppression of urine.
**ischuria** (*is-kū'-re-ah*) [ἰσχειν, to suppress; οὖρον, urine]. Retention or suppression of urine.
**ischyomyelitis** (*is-ke-o-mi-el-i'-tis*). See *ischiomyelitis*.
**isinglass** (*i'-zing-glas*). See *ichthyocolla*. i., **vegetable**. See *agar*.
**island** (*i'-land*). See *insula* and *blood-islands*. i.s, **Langerhans'**. See under *Langerhans*. i. of **Reil**. See under *Reil*.
**isletin** (*iz'-let-in*). Trade name for a preparation containing internal secretions of pancreas and spleen with spermin and sodium cacodylate.
**iso-** (*i'-so-*) [ἴσος, equal]. A prefix signifying equality, or, in chemical nomenclature, isomeric.
**isoagglutinin** (*i-so-ag-gloo'-tin-in*) [*iso-*; *agglutinin*]. An agglutinin in the blood of an individual capable of agglutinating the blood-corpuscles of another individual of the same species.
**isoamylamine** (*i-so-am-il'-am-in*) [*iso-*; *amylum*, starch; *amin*]. A ptomaine obtained in the distillation of horn with potassium hydroxide. It also occurs in the putrefaction of yeast.
**isoamylene** (*i-so-am'-il-ēn*). See *pental*.
**isoapiol** (*i-so-a'-pe-ol*) [*iso-*; *apium*, parsley; *oleum*, oil]. A substance obtained from apiol, exercising a powerful influence upon the vasomotor system.
**isobar** (*i'-so-bar*) [*iso-*; βάρος, weight]. In meterology, a term denoting a line drawn through points having the same synchronous barometric pressure.
**iso-bodies** (*i-so-bod'-ēs*). See *isoagglutinin*, '*isocytolysin*, *isohemolysin*, etc.
**isocholesterin** (*i-so-ko-les'-ter-in*) [*iso-*; *cholesterin*], C₂₆H₄₄O. A substance isomeric with cholesterin, and found in distilled sheeps' fat; it melts at 138° C.
**isocholin** (*i-so-ko'-lin*) [*iso-*; χολή, bile], C₅H₁₅NO₂. A body isomeric with cholin; it is found in flyagaric, and may be formed by oxidizing cholin.
**isochromatic** (*i-so-kro-mat'-ik*) [*iso-*; χρῶμα, color]. Having the same color throughout.
**isochronism** (*i-sok'-ro-nism*) [*iso-*; χρόνος, time]. The quality of occurring at equal intervals of time, or lasting for equal periods of time.
**isochronous** (*i-sok'-ro-nus*) [*iso-*; χρόν s, time]. Occurring at or occupying equal intervals of time.
**isocoria** (*i-so-ko'-re-ah*) [*iso-*; κόρη, pupil]. Equality in diameter of the two pupils.
**isocreatinin** (*i-so-kre-at'-in-in*), C₄H₇N₃O. An isomer of creatinin isolated from decomposing flesh.
**isocytolysin** (*i-so-si-tol'-is-in*) [*iso-*; κύτος, cell; λύειν, to loose]. A cytolysin from the blood of an animal, capable of acting against the cells of other animals of the same species. Cf. *isohemolysin*.
**isodactylous** (*i-so-dak'-til-us*) [ἴσος, equal; δάκτυλος, digit]. In biology, having the fingers and toes alike.
**isodiametric** (*i-so-di-am-et'-rik*) [*iso-*; διά, through; μέτρον, a measure]. Having equal diameters.
**isodimorphism** (*i-so-di-morf'-ism*) [ἴσος, equal; δίμορφος, two-formed]. A form of dimorphism, characterized by the appearance of a substance in two similar but incompatible forms.
**isodont** (*i'-so-dont*) [ἴσος, equal; ὀδούς, tooth]. Having teeth of the same size and shape.
**isodulcite** (*i-so-dul'-sit*). See *rhamnose*.
**isodynamia** (*i-so-di-nam'-e-ah*) [ἴσος, equal; δύναμις, force]. The property of being isodynamic.
**isodynamic** (*i-so-di-nam'-ik*) [*iso-*; δύναμις, force]. Having or generating equal amounts of force. i. **foods**, those that produce an equal amount of heat in undergoing the chemical changes of digestion.

**isoelectrical** (*i-so-e-lek'-trik-al*) [*iso-*; ἤλεκτρον, amber]. Having the same electrical properties throughout.

**isoerythroagglutinin** (*i-so-er-ith-ro-ag-gloo'-tin-in*). See *isoagglutinin*.

**isoform** (*i'-so-form*). An antiseptic dusting powder composed of equal parts of para-iodoxyanisol and calcium phosphate.

**isogamous** (*i-sog'-am-us*) [ἴσος, equal; γάμος, marriage]. A term used to designate that mode of reproduction in which the uniting gametes are of equal size. The same as conjugating.

**isogamy** (*is-og'-am-e*) [ἴσος, equal; γάμος, marriage]. The production of gametes of uniform size and incapable of being distinguished as macrogametes or microgametes. In biology, conjugation of similar gametes.

**isoglucosic** (*i-so-gloo-ko'-sic*). A term applied to a diabetic diet containing an amount of carbohydrates equal to the patient's tolerance.

**isogonic** (*i-so-gon'-ik*) [ ἴσος, equal; γόνος, offspring]. In biology, characterized by isogonism.

**isogonism** (*i-sog'-o-nizm*) [ἴσος, equal; γόνος, offspring]. In biology, the production by different stocks of sexual organisms of identical structure.

**isohemoagglutinin** (*i-so-hem-o-ag-gloo'-tin-in*). See *isoagglutinin*.

**isohemolysin** (*i-so-hem-ol'-is-in*) [*iso-*; αἷμα, blood; λύσιν, to loose]. In Ehrlich's lateral-chain theory, a hemolysin capable of acting against the blood of other animals of the same species as the one producing it, but capable of hemolizing the red blood-corpuscles only of such as have red blood-corpuscle receptors very similar to or identical with the receptors of the blood giving rise to the hemolysin. Cf. *isocytolysin*.

**isoidiolysin** (*i-so-id-e-ol'-is-in*). See *idioisolysin*.

**isolactose** (*i-so-lak'-tōs*). A disaccharid or true sugar formed by the action of an enzyme on lactose or milk-sugar.

**isolate** (*is'-o-lāt*) [*insula*, an island]. To separate; to place apart.

**isolation** (*is-o-la'-shun*). The act or process of isolating, or the state of being isolated; separation of those ill of contagious diseases from other persons.

**isologous** (*i-sol'-o-gus*) [ἴσος, equal; λόγος, method, ratio, or system]. Having similar relations or proportions. A qualification applied to compounds containing a like number of carbon-atoms, with a gradually decreasing number of hydrogen-atoms.

**isolophobia** (*is-o-lo-fo'-be-ah*) [*insula*, an island; φόβος, fear]. Morbid dread of being alone.

**isolysin** (*i-sol'-is-in*) [*iso-*; λύειν, to loose]. A cytolysin produced by injecting red blood-cells into an animal of the same species. An isolysin will destroy the red blood-cells of any animal of the same species except those of the immunized individual. Cf. *heterolysin*.

**isolysis** (*i-sol'-is-is*). The hemolytic action of the blood-serum of an animal of one species upon the corpuscles of another individual of the same species.

**isolytic** (*i-so-lit'-ik*). Pertaining to or caused by isolysis or an isolysin.

**isomaltose** (*i-so-mawl'-tōs*), $C_{12}H_{22}O_{11}$. A saccharose formed by the action on starch of an enzyme capable of producing maltose. It occurs in small quantity in the urine.

**isomer** (*i'-som-er*) [*iso-*; μέρος, a part]. An isomeric body. See *isomeric*.

**isomeric** (*i-so-mer'-ik*) [see *isomer*]. Of a chemical substance, composed of the same elements united in the same proportions by weight; in a restricted sense, composed of the same elements and having the same molecular weight as another substance.

**isomerism** (*i-som'-er-izm*) [see *isomer*]. The quality of being isomeric. Isomerism is of two kinds—(1) substances may have the same percentage-composition and the same molecular weights; these are termed *metameric*; (2) they may have the same percentage-composition, but different molecular weights; these are termed *polymeric*. Ammonium cyanate, $CON . NH_4$, and urea, $CON_2H_4$, are metameric; acetylene, $C_2H_2$, benzene, $C_6H_6$, and styrene, $C_8H_8$, are polymeric. i., **physical**, the form in which bodies that are isomeric and do not differ chemically present different physical properties, such as their action toward polarized light.

**isometric** (*i-so-met'-rik*) [*iso-*; μέτρον, a measure]. Of the same dimensions. i. **muscular act**, the preservation of the length of a muscle when stimulated, the muscle undergoing change in tension only.

**isometropia** (*i-so-met-ro'-pe-ah*) [*iso-*; μέτρον, a measure; ὤψ, eye]. Equality of kind and degree in the refraction of the two eyes.

**isomorphic** (*i-so-mor'-fik*) [*iso-*; μορφή, form]. Having the same form; of crystals, crystallizing in the same form.

**isomorphism** (*i-so-mor'-fizm*) [see *isomorphic*]. Similarity in crystalline form.

**isomorphous** (*i-so-mor'-fus*). See *isomorphic*.

**isonaphthol** (*i-so-naf'-thol*). A compound from naphthalene; a local antiseptic.

**isonomic** (*i-so-nom'-ik*). In chemistry, applied to isomorphism existing between two compounds of like composition.

**isopathotherapy** (*i-so-path-o-ther'-ap-e*). Same as *isopathy, q. v.*

**isopathy** (*i-sop'-ath-e*) [*iso-*; πάθος, suffering]. The treatment of disease by the administration of the causative agent or of its products, as the treatment of smallpox by the administration of variolous matter.

**isopelletierine** (*i-so-pel-et'-e-er-in*). See *pelletierine*.

**isopepsin** (*i-so-pep'-sin*) [*iso-*; *pepsin*]. 1. A body formed by heating pepsin to a point between 104° and 140° F. (40°–60° C.). It changes albumin into parapeptone. 2. Same as *parapeptone*.

**isophoria** (*i-so-fo'-re-ah*) [*iso-*; φόρος, a tending]. A condition in which the eyes lie in the same horizontal plane, the tension of the vertical muscles of each eye being equal, and the visual lines lying in the same plane.

**isopia** (*i-so'-pe-ah*) [ἴσος, equal; ὤψ, eye]. Equal acuteness of vision in the two eyes.

**isopilocarpine** (*i-so-pi-lo-kar'-pin*). An alkaloid from jaborandi isomeric with pilocarpine and similar to it in physiological effect, but weaker.

**isopleural** (*i-so-ploo'-ral*) [ἴσος, equal; πλευρά, a rib]. Bilaterally symmetrical.

**isopral** (*i'-so-pral*). Trichlorisopropyl alcohol. A hypnotic substance with odor of camphor.

**isoprecipitin** (*i-so-pre-sip'-it-in*). A precipitin which is only active against the serum of animals of the same species as that from which it is derived.

**isopters** (*i-sop'-terz*) [*iso-*; ὀπτήρ, observer]. The curves of relative visual acuity of the retina, at different distances from the macula, for form and for color.

**isoscope** (*i'-so-skōp*) [ἴσος, equal; σκοπεῖν, to see]. An instrument consisting of two sets of parallel vertical wires, one of which can be superimposed on the other; it is designed to show that the vertical lines of separation of the retina do not correspond exactly to the vertical meridians.

**isostemonous** (*i-so-stem'-o-nus*) [ἴσος, equal; στήμων, a stamen]. In biology, having the stamens of the same number as the parts of the calyx or corolla.

**isotherapeutics** (*i-so-ther-ap-ū'-tiks*). Same as *isopathotherapy*.

**isothermal** (*i-so-ther'-mal*) [*iso-*; θέρμη, heat]. Of equal or uniform temperature. i. **lines**, lines drawn through places having the same average temperature for a given period of time.

**isotonia** (*i-so-to'-ne-ah*) [ἴσος, equal; τόνος, tension]. Equality of tension.

**isotonic** (*i-so-ton'-ik*) [*iso-*; τόνος, tension]. 1. Having uniform tension or tonicity. 2. Applied to a solution of equal density, as the blood or some other fluid taken as a standard. i. **muscle**, a muscle that contracts on stimulation, its tension remaining the same. i. **salt solution**, one having the same osmotic tension as the blood-serum; a 0.9 % or "physiological salt solution."

**isotonicity** (*is-o-ton-is'-it-e*). 1. Same as *isotonia*. 2. Equality of osmotic pressure in different fluids.

**isotoxin** (*i-so-toks'-in*) [*iso-*; τοξικόν, poison]. A toxin elaborated in the blood of an animal and toxic for animals of the same species.

**isotropic, isotropous** (*i-so-trop'-ik, i-sot'-ro-pus*) [see *isotropy*]. 1. Having the same shape and appearance, from whatever point observed. 2. Being singly and uniformly refractive.

**isotropy** (*i-sot'-ro-pe*) [*iso-*; τροπή, turning]. 1. The condition of having equal or uniform properties throughout. 2. In embryology, Pflüger's term for absence of predetermined axes.

**Issaeff's period of resistance**. A temporary power of resistance to inoculation by virulent cultures of bacteria, conferred by the injection of various substances, such as salt solution, urine, serum, etc.

**issue** (*ish'-u*) [Fr., *issue*, from *exire*, to go out]. 1. An ulcer or fistulous passage made and kept up artificially for purposes of counter-irritation. 2. Offspring. 3. A discharge or flux. i. **pea**, a pea-shaped foreign body, as of ivy-wood or orris-root, inserted into an issue to keep up suppuration.

**istarin** (*is'-tar-in*). A nitrogenous, phosphorized substance of complex structure occurring in braintissue.

**isthmian, isthmic** (*is'-me-an, is'-mik*) [ἰσθμός, a narrow passage]. Pertaining to any isthmus, as that of the fauces.

**isthmitis** (*is-mi'-tis*) [ἰσθμός, a narrow passage; ιτις, inflammation]. Inflammation of the fauces.

**isthmo-** (*is'-mo-*) [ἰσθμός, a neck]. A prefix signifying the fauces.

**isthmocatarrhus** (*is-mo-kat-ar'-us*). A catarrh of the faucial isthmus.

**isthmocholosis** (*is-mo-kol-o'-sis*) [*isthmus*; χολή, bile]. Angina accompanied with bilious disorder.

**isthmodynia** (*is-mo-din'-e-ah*) [*isthmo-*; ὀδύνη, pain]. Pain in the faucial isthmus.

**isthmoid** (*is'-moid*) [ἰσθμός, a neck; εἶδος, likeness]. Resembling an isthmus.

**isthmopathy** (*is-mop'-ath-e*). A disease of the faucial isthmus.

**isthmoplegia** (*is-mo-ple'-je-ah*). Paralysis of the faucial tract.

**isthmopolypus** (*is-mo-pol'-ip-us*). A polyp of the fauces.

**isthmopyra** (*is-mo-pi'-rah*) [*isthmus;* πῦρ, fire]. Inflammation of the mucosa of the fauces.

**isthmorrhagia** (*is-mor-a'-je-ah*). Hemorrhage from the throat.

**isthmospasm** (*is'-mo-spazm*). Spasm of the isthmus of the fauces.

**isthmus** (*is'-mus*). The neck or constricted part of an organ. The part of the brain which, situated axially, serves to unite the forebrain, the cerebellum, and the spinal cord. i. **cerebri**, the midbrain. i., **gyral**, a narrow gyrus connecting two adjoining gyri; an annectant convolution or *pli de passage*. i. **of fauces**, the space between the arches of the palate. i. **rhombencephali**, the constriction between the third primary brain-vesicle and the midbrain. i. **of thyroid gland**, the narrow transverse part connecting the lobes of the thyroid body.

**istizin** (*is'-tiz-in*). A laxative preparation said to be dioxyanthiachinone. Dose 5 grains, dissolved in water.

**isutan** (*is'-ū-tan*). A proprietary compound said to consist of bismuth, resorcin, and tannic acid; used in diarrhea of children. Dose, 1–3 gr. 0.065–0.2 gm.) every 2 hours. Syn., *bismutan*.

**Italian leprosy.** See *pellagra*. I. **rhinoplasty**. See *operation, Italian*.

**Itard's catheter** (*e'-tar*) [Jean Marie Gaspard *Itard*, Parisian otologist, 1774–1838]. A Eustachian catheter.

**itch** (*ich*) [AS., *giccan*, to itch]. 1. An irritating sensation in the skin. 2. A name for various skin diseases accompanied by itching, particularly scabies. i., **barber's**. See *sycosis parasitaria*. i., **coolie**, a superficial vesicular dermatitis confined entirely to the lower extremities, caused by the larvae of *Uncinaria duodenalis*. It is endemic in Assam and other tropical regions among the laborers in tea-gardens and in damp soil. i., **Cuban**, a disease supposed to be a mild form of smallpox introduced by soldiers returning from the Cuban war. i., **dhobie**, a form of ringworm locating itself under the arms and between the legs when the skin is moist; very troublesome to soldiers in the tropics. Syn., *Manila itch*. i., **frost**, pruritus hiemalis. i.-**mite**. See *Acarus scabiei*.

**itching** (*ich'-ing*). An irritable tickling of the skin; pruritus.

**itchol** (*itch'-ol*). A proprietary ointment said to consist of lanolin, vaselin, iodoform, glycerin, phenol and oils of eucalyptus and lavender.

**-ite** (*-it*). 1. A suffix employed in mineralogy to denote a mineral or of mineral origin. 2. A suffix employed in chemistry for the salt of an acid that has the suffix *-ous*.

**iter** (*i'-ter*) [L.]. A passageway. i. **ad infundibulum**, the passage between the third ventricle of the brain and the infundibulum. i. **a tertio ad quartum ventriculum**, the aqueduct of Sylvius, extending from the third ventricle to the fourth. i. **chordae anterius**, the aperture through which the chorda tympani nerve leaves the tympanum. i. **chordae posterius**, the aperture through which the chorda tympani nerve enters the tympanum. i. **dentium**, the canal of the permanent dental sac opening behind the corresponding temporary tooth and through which the permanent tooth rises. i. **femineum**, the perineum. i. **seminarium**, the vas deferens. i. **urinae**, i. **urinarium**, the urinary passages.

**iteral** (*i'-ter-al*) [*iter*]. Relating to an iter or passage, particularly the Sylvian aqueduct.

**ithycyphes, ithycyphos** (*ith-e-si'-fēz, -fos*) [ἰθυκυφής, curved directly outward]. Having a backward angular projection of the spinal column.

**itinerarium** (*i-tin-er-a'-re-um*) [*iter*]. A lithotomy staff.

**-itis** (*-i-tis*) [ιτις, inflammation]. A suffix now used to denote inflammation, originally it had no such limited meaning, but was applied to any morbid condition.

**itrol** (*it'-rol*). Silver citrate. See under *silver*.

**itrosyl** (*it'-ro-sil*). Concentrated nitrous ether.

**IU.** Abbreviation for immunizing unit.

**ivaine** (*i'-va-ēn*) [*iva*, Latin name of *Achillea moschata*], C₁₄H₂₆O₅. An alkaloid obtained from *Achillea moschata*.

**iva-oil** (*i'-vah-oil*). A blue-green, volatile oil, of strong penetrating smell and taste of peppermint, obtained from iva, *Achillea moschata*. Its principal constituent is ivaol.

**ivaol** (*i'-vah-ol*), C₁₀H₂₀O. A pale yellow oily liquid of bitter taste and pleasant smell, the principal constituent of iva-oil (*q. v.*).

**ivory** (*i'-vor-e*) [*eboreus*, made of ivory, from *ebur*, ivory]. The hard, bone-like substance chiefly obtained from the tusks of elephants. i.-**black**, animal charcoal. i., **decalcified**, ivory treated with acid and deprived of inorganic constituents. i., **dental**, dentin.

**ivy** (*i'-ve*). An evergreen (*Hedera helix*), not used in medicine. i.-**pea**, an issue-pea made of the wood of the ivy. i., **poison**. See *rhus*.

**Iwanoff's edema of the retina**. Cystoid degeneration of the retina.

**Ixodes** (*iks-o'-dēz*) [ἰξός, bird-lime; εἶδος, form]. A genus of the order *Acarida*, including most of the parasitic ticks.

**ixodiasis** (*iks-o-di'-as-is*). Lesions and symptoms due to the presence of ticks of the genus *Ixodes;* tick fever.

**ixodic** (*iks-od'-ik*) [see *ixodes*]. Due to or derived from ticks.

**ixodin** (*iks'-od-in*) [see *ixodes*]. A ferment found in an extract of wood-ticks, obtained by means of a physiological salt solution. This substance injected intravenously in large quantities reduces bloodpressure and arrests cardiac action.

**ixyomyelitis** (*iks-e-o-mi-el-i'-tis*) [ἰξύς, waist; μυελός, marrow; ιτις, inflammation]. Inflammation of the lumbar portion of the spinal cord.

**izal** (*i'-zal*). Trade name of a proprietary disinfectant obtained in the process of coke-formation.

# J

**j.** As a Roman numeral it is used as the equivalent of i for one, or at the end of a number, as j, ij, iij, vj, vij, etc.

**J.** Symbol of *Joule's equivalent.*

**jabber** (*jab′-er*) [ME., *jaber*, to chatter]. To talk rapidly and indistinctly; to chatter.

**jaborandi** (*jab-ŏr-an′-de*). See *pilocarpus.*

**jaborandine** (*jab-o-ran′-dēn*). Synonym of *pilocarpine.*

**jaboridine** (*jab-or′-id-ēn*), C₁₀H₁₂N₂O₃. An alkaloid derived from jaborandi.

**jaborine** (*jab′-or-ēn*), C₂₂H₃₂N₄O₄. An alkaloid from jaborandi, a white amorphous powder with properties like those of atropine.

**Jaboulay's button** (*zjab-oo-la′*) [Mathieu *Jaboulay*, French surgeon, 1860-1913]. An arrangement of two cylinders which fit together and are used in lateral intestinal anastomosis. **J.'s operation**, exothyropexy.

**Jacaranda** (*jak-ar-an′-dah*). A genus of bignoniaceous plants of tropical America, several species of which are employed in syphilis in Brazil. **J. caroba** is antisyphilitic and is of service in the treatment of urethritis, rheumatism, and skin diseases. Dose of the *fluidextract* 16 min.-1 dr. (1-4 Cc.). **J. lancifoliata** is used by the natives of Brazil in urethritis. Dose of a 1 : 8 tincture 15 min. (1 Cc.); of the *fluidextract* 16-30 min. (1-2 Cc.). **J. procera** is indigenous to South America, and furnishes Caraiba bark, used in diarrhea and dysentery. The leaves are tonic, diaphoretic, and diuretic, and used in gonorrhea, gout, etc.

**Jaccoud's dissociated fever** (*zjak-kooz′*) [Sigismond *Jaccoud*, French physician, 1830-     ]. Fever with irregularity and slowness of the pulse in tuberculous meningitis of adults. **J.'s sign.** 1. A lateral displacement and rolling movement of a portion of the thoracic wall in adherent pericardium, especially when this is associated with extrapericardiac adhesions. 2. Prominence of the aorta in the region of the suprasternal notch in cases of aortic dilatation.

**jack** (*jak*). A popular term for horse-flesh salted and subsequently washed in order to deprive it of its peculiar taste. **j.-knife posture**, the patient reclines on his back with shoulders elevated, legs flexed on thighs, and thighs at right angles to abdomen.

**jacket** (*jak′-et*) [Fr., *jacque*, a coat of mail]. A short coat. **j., bark**, a jacket stuffed with powdered cinchona. **j., cotton**, a jacket lined with cotton, sometimes used in the treatment of pneumonia. **j., plaster-of-Paris**, a mould of plaster-of-Paris cast upon body or part, for keeping it rigid and fixed in a desired position in sprain or dislocation of the spine, etc. **j.-poultice**, a poultice placed between two folds of gauze or other material and applied about the whole surface of the thorax; it is sometimes used in the treatment of pneumonia. **j., Sayre's**, a plaster of Paris jacket used to support the spinal column. **j., strait**, a system of leather straps used to bind violently insane persons in order to prevent self-inflicted injury. **j., Willock's respiratory**, a jacket used in pulmonary emphysema.

**Jackson's membrane** or **veil** (*jak′-sun*) [Jabez North *Jackson*, American surgeon, 1868-     ]. A thin membrane extending from the parietal peritoneum of the right side across the front of the ascending colon to the inner side and continuous above with the transverse mesocolon.

**Jackson's pectoral syrup** [Samuel *Jackson*, of Philadelphia]. A cough medicine containing morphine hydrochloride, oil of sassafras, and syrup of acacia; one fluidrachm contains 1/32 grain of morphine hydrochloride. Dose 1 fluidrachm (4 cc.).

**Jackson's syndrome** (*jak′-sun*) [John Hughlings *Jackson*, English physician, 1834-1911]. Associated paralysis of the soft palate and larynx, accompanied by paralysis of the trapezius, the sternomastoid, and one-half of the tongue.

**Jacksonian** (*jak-so′-ne-an*). Described by John Hughlings *Jackson*, English physician, 1834-1911. **J. epilepsy**, focal, cortical, or symptomatic epilepsy; a spasm limited to a single group of muscles in the face, arm, or leg, due generally to irritative lesion on the motor area of the brain; the spasm may also involve other groups of muscles; consciousness is usually retained; there is danger of the convulsions becoming general.

**Jacob's cataract needle** (*ja′-kub*) [Arthur *Jacob*, Irish ophthalmologist, 1790-1874]. A needle with a slightly curved point, used in treating cataract. **J.'s membrane**, the layer of rods and cones of the retina. **J.'s operation**, *for trichiasis;* scalping of the edge of the lid, including the cilia and the hair-bulbs. **J.'s ulcer**, same as rodent ulcer. **J.'s wound**, chancroidal ulcer.

**Jacobson's anastomosis** [Ludwig Levin *Jacobson*, Danish anatomist, 1783-1843]. The tympanic plexus. **J.'s canal**, the tympanic canal that opens on the lower surface of the petrous portion of the temporal bone and transmits Jacobson's nerve. **J.'s cartilage**, a strip of hyaline cartilage extending from the nasal spine upward and backward between the nasal septum and vomer; it is well developed in certain animals, but rudimentary in man. **J.'s nerve**, the tympanic branch of the glossopharyngeal nerve. **J.'s organ**, a short, rudimentary canal, extending along the septum of Stenson's duct, and ending in a culdesac. **J.'s plexus**, the tympanic plexus. **J.'s retinitis**, diffuse syphilitic retinitis. **J.'s sulcus**, the vertical sulcus for the tympanic nerve on the promontory of the tympanum.

**Jacquart's angle** (*zjak-ar′*). The facial angle; that angle between the line joining the subnasal point and the glabella, and the line joining the subnasal and auricular points.

**Jacquemier's sign** (*zjak-me-a′*) [Jean Marie *Jacquemier*, French obstetrician, 1806-1879]. Blue coloration of the vaginal mucosa appearing about the twelfth week of pregnancy.

**Jacquemin's test for phenol** (*zjak-man*). Add to the solution an equal amount of anilin and then a solution of sodium hypochlorite; a blue color is produced.

**jactitation** (*jak-tit-a′-shun*) [*jactitare*, to pour forth]. A tossing about, great restlessness, a condition at times present in grave diseases. **j., periodic**, chorea.

**jaculiferous** (*jak-ū-lif′-ur-us*) [*jaculum*, a dart; *ferre*, to bear]. Prickly, bearing spines.

**Jadassohn's disease** (*yah′-das-sōn*) [Josef *Jadassohn*, Swiss dermatologist, 1863-     ]. A maculopapular erythema.

**Jadelot's lines**, **J.'s furrows** (*zjad′-lo*) [Jean François Nicolas *Jadelot*, French physician,     -1830]. Certain furrows of the face observed in conditions of disease. The lines are distinguished: The *genal* and *nasal furrows* are said to indicate disease of the gastrointestinal tract or abdominal viscera; the former runs from the mouth toward the malar bone, the latter from the nose in a semicircle about the mouth; the *labial furrow*, from the angle of the mouth outward to the lower part of the face, indicates disease of the lungs; the *oculozygomatic furrow*, beginning at the inner canthus of the eye, and passing outward below the lower lid, to be lost on the cheek; it is said to point to disorders of the nervous system.

**Jaeger's test types** (*ya′-ger*) [Edward *Jaeger* von Jastthal, Austrian ophthalmologist, 1818-1884]. A series of types of varying size, for testing the power of vision.

**Jaffé's reaction for creatinin** (*yaf-fay′*) [Max *Jaffé*, German physician, 1841-     ]. Add to the solution a solution of picric acid and a few drops of sodium hydroxide solution, and warm. The presence of creatinin is evinced by a red coloration, which changes to yellow if acid is added. Acetone and glucose give

# JAIL-FEVER 481 JAUNDICE

a similar reaction. J.'s sign, the flow of pus from a tube inserted into a subdiaphragmatic abscess is more abundant during inspiration than during expiration; if the collection is thoracic, the inverse holds true. Paralysis of the diaphragm prevents the manifestation of this sign. J.'s test for indican, add to the suspected liquid an equal amount of concentrated hydrochloric acid to, which has been added a few drops of sodium hypochlorite. A blue color denotes the presence of indican.

**jail-fever.** Typhus fever.

**(von) Jaksch's disease** (*yaksh*) [Rudolf *von Jaksch*, Austrian physician, 1855– ]. Infantile pseudoleukemia. v. J.'s test for free hydrochloric acid in gastric juice, saturate filter-paper with a solution of benzopurpurin 6-B, and dry; this gives, with dilute solutions of HCl, a beautiful violet color. If it assumes a dark-blue color, the solution contains more than 0.4 Gm. of HCl in 100 Cc. of the solution. v. J.'s test for glucose in urine, to 6–8 Cc. of urine add 2 parts of phenylhydrazin hydrochlorate and 3 parts of sodium acetate; warm, place the tube in boiling water for from 20 to 30 minutes, then in cold water. The presence of glucose is shown by a precipitate consisting of groups of yellow needles of phenylglucosazone; in doubtful cases determine the melting-point of these crystals to be 204°-205° C. v. J.'s test for melanin, treat the liquid to be tested with a few drops of a concentrated solution of ferric chloride. If melanin is present, it will turn gray, and more ferric chloride being added, the precipitate, consisting of the chloring-matter and the phosphates, is redissolved. v. J.'s test for uric acid, allow the powder to heat gently on a watch-glass with a drop or two of chlorine or bromine water. A red residue is formed which, when cold, turns a purple red when ammonia is added.

**jalap, jalapa** (*jal'-ap*, *-a*) [from *Xalapa*, a city of Mexico]. The tuberous root of *Exogonium purga*, a plant of the natural order *Convolvulaceæ*. Its active principle is a resin which contains a glucoside, *convolvulin*, $C_{31}H_{50}O_{16}$. Jalap is an active hydragogue cathartic, and is used to remove dropsical effusions by the bowel. Combined with calomel it is a favorite remedy in bilious fever. Dose of *powdered jalap* 5–30 gr. (1–2 Gm.). j., **compound powder of** (*pulvis jalapæ compositus*, U. S. P.). Dose 10 gr.–1 dr. (0.65–4.0 Gm.). j., **resin of** (*resina jalapæ*, U. S. P.) Dose 4–8 gr. (0.26–0.52 Gm.).

**jalapin** (*jal'-ap-in*) [*jalap*]. 1. A purgative glucoside, from various kinds of jalap. 2. $C_{34}H_{56}O_{16}$. The precipitate from a tincture of jalap; dose 3 gr.

**jalapinol** (*jal-ap'-in-ol*). A crystalline decomposition product of jalapin.

**jalon** (*jal'-on*). A proprietary liquid preparation of colloidal silver for internal administration, in gastric and intestinal disorders.

**Jamaica dogwood.** See *Piscidia erythrina*.

**jamaicin** (*jam-a'-is-in*). A bitter cathartic substance from the bark of the cabbage-tree, *Andira inermis*, identical with berberin.

**jambul** (*jam'-bul*) [E. Ind., *jambu*]. The dried and powdered fruit-stones of *Syzygium jambolanum*, a shrub of the order *Myrtaceæ*, growing in Western India. It is a valuable astringent in the diarrheas of children, and has also been found to lessen the amount of sugar and urine excreted in diabetes. Dose gr. ij–x; of the fluid extract of the seeds, ℥ v–x. Unof.

**James' pill** [Robert *James*, English physician, 1705–1776]. A pill containing equal parts of James' powder, ammoniac, and pill of aloes and myrrh. J.'s **powder.** See *antimony, powder of*.

**Jamestown weed.** See *Stramonium*.

**Jamieson's salve.** Lanolin, 3 parts; oil of sweet almond, ½ part; distilled water, ¼ part. A base for eye-ointments.

**Janet's method** (*zjan-a'*). The treatment of gonorrhea by irrigation with potassium permanganate.

**Janeway's pill** [Edward Gamaliel *Janeway*, American physician, 1841–1911]. A pill of aloes, podophyllum, belladonna and nux vomica.

**Janeway's sphygmomanometer** [Theodore C. *Janeway*, American physician, 1872– ]. An apparatus for determining the blood pressure.

**janiceps** (*jan'-is-eps*) [*Janus*, a two-faced divinity; *caput*, head]. A syncephalic monstrosity with two faces. j. **asymmetrus**, a j. with the two faces unequally developed.

17

**janitor** (*jan'-it-or*) [L., doorkeeper]. The pylorus. **janitrix** (*jan'-i-trix*) [L., fem. of *janitor*]. The portal vein.

**Janosik's embryo** (*yahn'-o-sik*). A human embryo described by Janosik as having two gill-pouches and three aortic arches.

**Jansen's operation** (*yahn'-sen*) [Albert *Jansen*, German otologist]. It consists of curettage of the frontal sinus after removing the lower wall and the lower part of the anterior wall of that sinus.

**janus** (*ja'-nus*). See *janiceps*.

**japaconine** (*jap-ak'-on-ēn*), $C_{25}H_{41}NO_{10}$. A decomposition-product of japaconitine.

**japaconitine** (*jap-ak-on'-it-ēn*), $C_{66}H_{88}N_2O_{21}$. The most poisonous of the known aconite alkaloids. It is obtained from *Aconitum japonicum*. On saponification it splits up into benzoic acid and japaconine.

**Japanese** (*jap-an-ēz'*) [Japan, an island on the east coast of Asia]. Pertaining to Japan or its inhabitants. J. **fanning.** See J. *method of resuscitation*. J. **hot-box,** a device for applying dry-heat to a part, as the eyes. J. **method of resuscitation,** it consists in drawing forward the tongue and making rapid passes with paper fans soaked in water and aqua ammoniæ. The object is to get as much of the vapor of ammonia into the lungs as possible. This method is also called *Japanese fanning*. J. **river fever.** See *kēdani disease*.

**jar.** A small earthen or glass vessel without handle or spout. j., **Leyden,** an electric condenser consisting of a glass jar lined, externally and internally, in its lower two thirds with tin.

**jardon** (*jar'-don*). A tumor or exostosis on the outer and lower part of the leg of a horse, below the bending of the ham.

**jargon** (*jar'-gon*). Confused, unintelligible talk, gibberish, babble, characteristic of some forms of idiocy and insanity. j. **aphasia, j. paraphasia.** See *aphasia, gibberish*.

**jargonize** (*jar'-gon-īz*). To utter unintelligible sounds.

**Jarisch's ointment** (*yah'-rish*) [O. *Jarisch*, Austrian physician, 1850– ]. An ointment containing pyrogallic acid, one dram, and lard, one ounce. J.'s **reaction.** See *Herxheimer's reaction*.

**Jarjavay's muscle** (*zjar-zjav-a'*) [Jean François *Jarjavay*, French physician, 1815–1868]. The depressor urethræ; a fasciculus of the constrictor urethræ that passes transversely over the urethra and joins the fibers of the constrictor vaginæ.

**Jarvis's adjuster** (*jar'-vis*). An appliance formerly used for reducing dislocations.

**Jarvis's snare** (*jar'-vis*) [William Chapman *Jarvis*, New York physician, 1855–1895]. A snare used for removing polypoid growths in the nose and throat.

**jasmine** (*jas'-min*). See *gelsemium*.

**jaswa** (*jas'-weh*) [Siberian]. Local name of anthrax.

**Jatropha** (*jat'-ro-fah*) [ἰατρός, a physician; τροφή, nourishment]. A genus of euphorbiaceous plants. *J. curcas* is the source of purging-nuts. *J. gossypifolia*, the tua-tua plant, indigenous to South America, West Indies, and Africa, has purgative leaves used in colic and bilious affections. It is highly extolled in Venezuela as a cure for leprosy. *J. manihot* yields tapioca.

**jaundice** (*jawn'-dis*) [Fr., *jaunisse*, from *jaune*, yellow]. A yellow discoloration of the skin, mucous membranes, and secretions, due to the presence of bile-pigments in the blood. See *icterus*. j., **acathectic, j., akathektic,** a name given by Liebermeister to the majority of cases usually classified as hematogenous icterus, but which he holds to be due to a disturbed activity of the liver-cells, which, in consequence of injury, lose their ability to secrete bile in the direction of the bile-ducts, a consequence of which is the diffusion of the bile into the blood-vessels and lymph-vessels of the liver. Syn., *diffusion icterus; functional jaundice*. j., **acute febrile.** See *Weil's disease*. j., **acute infective.** See *Weil's disease*. j., **black,** an extreme degree of jaundice. j., **black of the Tyrol,** an endemic disease, due, according to Melinkow-Raswedenkow, to *Echinococcus alveolaris*. He proposes the name *alveolar echinococcus disease*. j., **Budd's.** Rokitansky's disease. j., **catarrhal,** that due to swelling of the bile-ducts from catarrh. j., **functional.** See j., *acathectic*. j., **green,** that in which the discoloration of the skin is green or olive-colored. Syn., *icterus*

# JAVAL'S OPHTHALMOMETER 482 JEQUIRITOL

*viridis.* j., hematogenous, that form due to excessive destruction of blood-corpuscles. j., hepatogenous, that due to obstruction to the flow of bile from the liver. By some all forms of jaundice are considered hepatogenous, since bile is made only in the liver. j., lead, the earthy yellow hue of the skin in saturnine cachexia. j., malignant, acute yellow atrophy of the liver. See *icterus gravis.* j., Murphy's law of, jaundice due to gall-stones is always preceded by colic; jaundice due to malignant disease, or catarrh of the ducts accompanied by infection, is never preceded by colic. j. of the new-born. See *icterus neonatorum.* j., obstructive, that due to permanent obstruction of the common bile-duct. It is persistent and deep, and accompanied by irritability, depression, and later coma, delirium or convulsions, a slow pulse, and subcutaneous hemorrhages. j., paradoxic, Addison's disease. j., red, a non-febrile diffused redness of the skin. j., retention. See *j., obstructive.* j., saturnine, jaundice occurring in lead-poisoning. j., vernal, mild catarrhal jaundice occurring oftenest in spring and fall because of the atmospheric changes. j., white. Synonym of *chlorosis.*

**Javal's ophthalmometer** (*zjav-al'*) [Louis Emile Javal, French ophthalmologist, 1839-1907]. See *ophthalmometer.* J.'s orthoptic treatment, exercises with prisms, for use in strabismus.

**Javelle water** (*zjav-el'*) [*Javelle,* a town in France].
1. A solution of potassium or sodium hypochlorite.
2. Liquor potassæ chlorinatæ (N, F.).

**jaw** [ME.]. 1. Either of the two parts of the face (upper or lower jaw) serving the purpose of seizing or masticating the food. 2. Also the bone (jaw-bone or jaw) that forms the framework of the jaw. j., big, actinomycosis of cattle. j., bone, a maxilla, especially the superior maxilla, the inferior being called the mandible. j., disease, a term for phosphorus-necrosis. j.-fall, dislocation of lower jaw. j.-jerk, j.-clonus, a reflex contraction of the muscles of mastication produced by suddenly depressing the lower jaw. See under *reflex.* j. lever, an instrument used for opening the mouth of, and administering medicine to cattle. j., lock-, j., locked. See *lockjaw.* j., lumpy, actinomycosis of cattle. j., phossy mouth. j., pier, the os quadratum or hinge segment of the reptilian mandible; it becomes the incus or anvil bone of mammals. j., pig, abnormal prominence of the upper jaw and enlargement of the teeth in the horse. j., tooth, a molar tooth. j., wolf, cleft palate.

**Jaworski's corpuscles** (*yah-vor'-ske*) [Valery Jaworski, Polish physician, 1849- ]. Spiral bodies of mucus found in the gastric secretion in cases of pronounced hyperchlorhydria. J.'s sign. Seen in "paradoxical dilatation" of the stomach and in hourglass stomach, in which, though splashing may be elicited, no fluid can be recovered by the stomach tube.

**jecoral** (*jek'-or-al*), jecorary (*jek'-or-a-re*), jecorose (*jek'-or-ōs*) [L. *jecur, jecoris,* liver]. Hepatic, relating to the liver.

**jecorin** (*jek'-or-in*) [*jecur*]. 1. $C_{108}H_{166}N_2SP_2$. A body found in liver-substance; it resembles lecithin, but reduces Fehling's solution. It occurs also in the spleen, muscle, brain, etc. 2. A proprietary substitute for cod-liver oil.

**jecorol** (*jek'-or-ol*). A proprietary preparation said to consist of the active constituents of cod-liver oil, and offered as a substitute for it.

**jecur** (*je'-ker*) [L.]. The liver.

**Jeffersonia** (*jef-er-so'-ne-ah*) [Thomas *Jefferson,* American statesman, 1743-1826]. A genus of berberidaceous plants. J., diphylla, a N. American plant, useful in rheumatism, and a good substitute for senega. It is tonic and in large doses expectorant; it is also called *rheumatism-root* and *twin-leaf.*

**jejunal** (*jej-oo'-nal*) [*jejunum*]. Pertaining to the jejunum.

**jejunectomy** (*jej-oo-nek'-to-me*) [*jejunum;* ἐκτομή, excision]. Excision of part or all of the jejunum.

**jejunitas** (*jej-oo'-nit-as*) [L.]. Fasting.

**jejunitis** (*jej-oo-ni'-tis*) [*jejunum;* ιτις, inflammation]. Inflammation of the jejunum.

**jejunocolostomy** (*jej-oo-no-ko-los'-to-me*) [*jejunum;* κόλον, colon; στόμα, mouth]. The formation of an artificial passage between the jejunum and the colon.

**jejunoileitis** (*jej-oo-no-il-e-i'-tis*) [*jejunum; ileum;* ιτις, inflammation]. Inflammation of the jejunum and the ileum.

**jejunoileostomy** (*jej-oo-no-il-e-os'-to-me*) [*jejunum; ileum;* στόμα, mouth]. The formation of an artificial communication between the jejunum and the ileum.

**jejunoileum** -(*jej-oo-no-il'-e-um*) [*jejunum; ileum*]. That part of the small intestine extending from the duodenum to the cecum.

**jejunostomy** (*jej-oo-nos'-to-me*) [*jejunum;* στόμα, mouth]. The making of an artificial opening through the abdominal wall into the jejunum.

**jejunotomy** (*jej-oo-not'-o-me*) [*jejunum;* τέμνειν, to cut]. Incision into the jejunum.

**jejunotyphoid** (*jej-oo-no-ti'-foid*) [*jejunum; typhoid*]. Typhoid fever with involvement of the jejunum or its glands.

**jejunum** (*jej-oo'-num*) [*jejunus,* empty, because usually found empty after death]. The second division of the small intestine extending between the duodenum and the ileum, and measuring about 8 feet (2.2 meters) in length.

**jell** (*jel*) [*gelare,* to freeze]. The precipitation of colloidal solutions.

**Jellinek's sign** (*yel'-en-ek*) [Samuel *Jellinek,* Austrian physician]. The brownish pigmentation of the eyelids often found in cases of hyperthyroidism.

**jelloid** (*jel'-oid*). A form of pill coated with jujube mass.

**jelly** (*jel'-e*) [*gelare,* to freeze]. A soft, gelatinous, tremulous substance. j., bacterial, the gelatinous matrix which causes certain bacteria to adhere to one another in masses or pellicles. See *zooglea.* j., coefficient, a 2 per cent. agar solution, containing citric acid, sodium citrate, and sodium chloride, used in the *in vitro* method of studying induced cell-production. j.-glycerin, a mixture of glycerin, jelly, and zinc oxide. j., kinetic, coefficient jelly to which a dye, sodium bicarbonate, and atropine sulphate have been added to excite ameboid movement in leukocytes. j.-leaf, the mucilaginous leaf of *Sida rhombifolia,* used in making poultices. j. method of in vitro staining, the use of an agar solution containing a dye and other ingredients, spread on a glass slide, in studying living cells under the microscope. j., mineral or j., petroleum, vaseline. j., oat-, a dietetic preparation used in infant-feeding. It is prepared by soaking 4 ounces of coarse oatmeal in a quart of cold water for 12 hours. The mixture is then boiled down to 1 pint and allowed to cool. j., Wharton's. See *Wharton's jelly.*

**Jendrassik's maneuver** (*yen-dras'-sik*) [Ernst *Jendrassik,* Hungarian physician, 1858- ]. Interlocking of the fingers and forcible drawing apart of the hands, to facilitate the production of the knee-jerk.

**Jenner's stain for blood** [Louis *Jenner,* English physician]. Preparation of the neutral stain: in an open beaker mix equal parts of 1.2 or 1.25 % aqueous solution of eosin (Grubler), 1 % aqueous solution of methylene-blue (Grubler). Let stand for 24 hours. Filter. Dry the precipitate obtained. Wash the precipitate with distilled water and dry again. The staining solution: For use dissolve 0.5 Gm. of the precipitate in 100 Cc. pure methylalcohol. Method of staining: Stain in the solution for 1 to 3 minutes, covering with a watch-crystal. Pour off stain quickly and rinse in water until film is pink (5 to 10 seconds). Staining reaction: Leukocytes: nuclei, blue; granules, neutrophil, red; granules, basophil, dark violet; granules, eosinophil, brilliant crimson; malarial parasites, bacteria, and filaria, blue.

**Jennerian** (*jen-e'-re-an*). Pertaining to Edward Jenner, English physician, 1749-1823, the discoverer of vaccination against smallpox. J. va cinati n, arm-to-arm vaccination.

**jennerization.** The process of jennerizing.

**jennerize** (*jen'-er-iz*) [Edward *Jenner,* English physician, 1749-1823]. To induce immunity against a disease by repeated inoculation with attenuated cultures of the pathogenic organism producing the disease.

**Jensen's fissure** (*yen'-sen*). An inconstant fissure near the end of the fissure of Sylvius; called also the *intermedial fissure.*

**jequiritin** (*je-kwir'-it-in*). A soluble active principle obtained from the jequirity seeds.

**jequiritol** (*je-kwir'-it-ol*). An active principle of jequirity in a sterile solution of glycerin.

**jequirity** (*je-kwir'-it-e*). See *abrus*.
**jerk.** A sudden, spasmodic movement. See *chin-jerk, elbow-jerk, jaw-jerk, knee-jerk, toe-jerk*, etc.
**j.-finger**, a disease in which the flexion or extension of a finger is accomplished by a jerk.
**jerks.** Irregular and spasmodic movements of features or limbs; they are involuntary.
**jervin** (*jer'-vin*). See *veratrum*.
**jessamine** (*jes'-am-ēn*). See *gelsemium*.
**Jesuits' balsam.** Compound tincture of benzoin.
**J.s' bark.** Peruvian bark; the bark of several species of cinchona. **J.s' drops**, compound tincture of benzoin. **J.s' nut**, seed of *Trapa natans*. **J.s' powder**, powdered cinchona bark. **J.s' tea**, an infusion of leaves of *Psoralea glandulosa*.
**Jez's antityphoid extract.** An extract obtained from thymus gland, spleen, bone-marrow, brain, and spinal cord of rabbits immunized by frequent inoculations with cultures of typhoid bacilli; this is triturated in a solution of sodium chloride, alcohol, glycerol, and a minute quantity of phenol; a more recent formula contains peptone.
**jigger.** See *Pulex*.
**jimson-weed** (*jim'-son-wēd*). See *Stramonium*.
**jinked** (*jinkd*). In veterinary practice, sprained in the back.
**Jobert's fossa** (*zjo-bair'*) [Antoine Joseph Jobert de Lamballe, French surgeon, 1799–1867]. A hollow in the popliteal region, formed above, by the adductor magnus, below, by the sartorius and gracilis. It is well seen when the knee is bent and the thigh rotated strongly outward. **J.'s suture.** See *suture*.
**Joffroy's symptom** (*zjof-roy*) [Alexis *Joffroy*, French physician, 1844– ]. 1. Absence of facial contraction when the patient suddenly turns his eyes upward, seen in exophthalmic goiter. 2. Rhythmic twitching of the glutei on pressure upon the gluteal region in cases of spastic paraplegia and sciatica. Syn., *Phénomène de la hanche*.
**joha** (*yo'-hah*). A mixture of salvarsan with iodipin and sterile wool fat, used for intramuscular injections in the treatment of syphilis.
**johimbine**, johimbin (*yo-him'-bēn*). See *yohimbine*.
**Johne's bacillus** (*yo'-neh*) [Albert *Johne*, German physician]. The specific bacillus of enteritis chronica pseudotuberculosa bovis. **J.'s disease**, enteritis chronica pseudotuberculosa bovis.
**Johnson's test for albumin in urine** [Sir George *Johnson*, English physician, 1818–1896]. A concentrated solution of picric acid is poured upon the surface of the urine in a test-tube. A ring of white precipitate occurs at the junction of the two liquids; this increases on heating. Peptones and albumoses are precipitated by this reagent, but the precipitate redissolves on heating.
**Johnstoni area.** See *Celsus' area*.
**joint.** See *articulation*. **j.-bodies.** See *arthrolith*. **j., Brodie's**, hysterical arthroneuralgia. **j., Charcot's.** See *Charcot's disease* (2). **j.-mice.** See *arthrolith*. **j.-oil**, synovial fluid. **j.-water**, synovial fluid.
**Jolles's test for bile pigments in urine** (*yol'-la*) [Adolf *Jolles*, Austrian chemist, 1862– ]. Put in a stopper cylinder 50 C.c. of urine, and add a few drops of 10 per cent. hydrochloric acid and an excess of a barium chloride solution with 5 c.c. of chloroform, and shake for several minutes. Then by means of a pipet remove the chloroform and the precipitate, place in a test-tube, and heat on the water-bath to about 80 °C. When the chloroform has evaporated, carefully decant the liquid from the precipitate and let three drops of concentrated sulphuric acid, containing one-third fuming nitric acid, flow down the sides of the test-tube. If bile pigments be present, the characteristic coloration results.
**Jolly's test for muscular reaction** [Friedrich *Jolly*, German physician, 1844– ]. When the contractility of a muscle is exhausted by the faradic current, it can still be excited by the influence of the will, and, inversely, when voluntary movements are impossible, the muscle can contract itself by faradization. This phenomenon is observed in certain amyotrophies.
**Jones'** method of treating fractures of the elbow-joint [Robert *Jones*, English surgeon]. It consists in placing the arm in a position of acute flexion and retaining it in this position without passive motion until complete consolidation results.
**Jones' solution for the detection of albumin in the urine** [Henry Bence *Jones*, English physician, 1814–1873]. Corrosive sublimate, 10 parts; sodium chloride, 10 parts; succinic acid, 20 parts; distilled water, 500

parts. **J.'s test for bile-pigments in urine**, put in a stopper cylinder 50 Cc. of urine, and add a few drops of 10 % hydrochloric acid and an excess of a barium chloride solution with 5 Cc. chloroform, and shake for several minutes. Then, by means of a pipet, remove the chloroform and the precipitate, place in a test-tube, and heat on the water-bath to about 80° C. When the chloroform has evaporated, decant the liquid from the precipitate carefully and let three drops of concentrated sulphuric acid, containing one-third fuming nitric acid, flow down the sides of the test-tube. If bile-pigments are present, the characteristic coloration results.
**Jonnesco's fossa** (*yon-es'-ko*) [Thomas *Jonnesco*, Roumanian surgeon, 1861– ]. A fossa in the angle between the duodenum and jejunum. **J.'s method of anesthesia**, the producing of general anesthesia by injection of a solution of stovaine, strychnine or other alkaloid into the subarachnoid space in the dorsal or lumbar region of the spinal cord. **J.'s operation**, excision of the sympathetic ganglion on each side of the neck, for exophthalmic goiter.
**Jorissen's test for formaldehyde in milk.** To several drops of a 10 % aqueous solution of phloroglucinol add 10 Cc. of the suspected milk in a test-tube, shake well, and add a few drops of caustic soda or caustic potash. In the presence of formaldehyd a delicate red color appears.
**Jorissenne's sign** (*zjor-is-en'*) [Gustave *Jorissenne*, Belgian physician]. During the early stage of pregnancy the change of position of the woman from the horizontal to the erect does not increase the pulse-rate.
**Josseraud's sign.** A peculiar loud, metallic sound, heard over the pulmonic area, and preceding the friction-sound in acute pericarditis.
**Joule** (*jool*) [James Prescott *Joule*, English physicist, 1818–1889]. 1. A unit of electric energy, equivalent to the work expended when a current of one ampere flows for one second against a resistance of one ohm. 2. A small calory—the am<sub>t</sub> of heat required to raise the gram of water 1° C.°
**Joule's equivalent.** The mechanical equivalent of heat or the amount of work that, converted into 1° F. It is equivalent to 772 foot-pounds.
**jugal** (*joo'-gal*) [*jugum*, a yoke]. 1. Connecting or uniting, as by a yoke. 2. Pertaining to the zygoma. **j. bone**, the malar bone. **j. point**, the point situated at the angle that the posterior border of the frontal branch of the malar bone makes with the superior border of its zygomatic branch. **j. process**, the zygomatic process.
**jugate** (*joo'-gāt*) [*jugum*, a yoke]. 1. Having ridges. 2. Coupled together; yoked.
**juglandin** (*joo-glan'-din*) [*juglans*, a walnut]. A precipitate from a tincture of the root-bark of butternut, *Juglans cinerea*.
**juglans** (*joo'-glanz*) [L., "walnut".]. Butternut. The bark of the root of *J. cinerea*, of the natural order *Juglandaceæ*. It is a mild cathartic, and has also been used in intermittent and remittent fever.
**jugomaxillary** (*joo-go-maks'-il-a-re*) [*jugum*; *maxilla*, the jaw]. Relating to the jugular vein and the maxilla. **j. muscle**, the masseter.
**jugular** (*joo'-gū-lar*) [*jugulum*, the throat). Pertaining to the throat. **j. foramen.** See *foramen, jugular*. **j. fossa**, a notch in the posterior border of the petrous portion of the temporal bone, which, with a similar notch in the occipital bone, forms the foramen lacerum posterius. **j. ganglion**, the superior ganglion of the glossopharyngeal nerve. **j. notch**, the depression on the upper surface of the manubrium, between the two clavicles. **j. process**, a rough process external to the condyle of the occipital bone. **j. veins**, the *internal jugular vein* collects the blood from the brain, part of the face and neck, and unites with the subclavian vein to form the vena innominata; the *external jugular vein* carries the blood from the exterior of the cranium and parts of the face and empties into the subclavian vein.
**jugulate** (*jug'-ū-lāt*) [*jugulum*]. To check or stop any process promptly.
**jugulation** (*jug-ū-la'-shun*) [*jugulatio*, a killing]. The swift arrest of disease by therapeutic means; also the arrest of an epidemic by prompt and effective measures.
**jugulocephalic** (*jug-ū-lo-sef-al'-ik*) [*jugulum*, the throat; κεφαλή, head]. Pertaining to the throat and the head.

# JUGULUM 484 JUXTAPYLORIC

**jugulum** (*jug'-ū-lum*) [L.: *pl.*, *jugula*]. The collar-bone; also the throat.

**jugum** (*joo'-gum*) [L. *pl.*, *juga*]. 1. A yoke. 2. A ridge. **j. penis**, a compressor of the penis. **j. petrosum**, an arched elevation on the anterior surface of the petrous portion of the temporal bone over the superior semicircular canal.

**juice** (*joos*) [*jus*, broth]. 1. The liquid contained in vegetable or animal tissues. 2. Any of the secretions of the body, as the *intestinal* or *pancreatic juice*. **j.-canals**, spaces within the connective tissue forming the origin of the lymphatic vessels. **j., cancer**, a milky juice which exudes from cancerous tissue when cut. **j., gastric, j., intestinal, j., pancreatic.** See *gastric*, etc.

**jujube** (*joo'-joob*) [L., *jujuba*]. The fruit of the jujube tree, *Zizyphus jujuba*. **j.-paste**, a paste containing the pulp of jujubes and used in pulmonary disorders. It is now made of gum-arabic, of gelatin, variously flavored.

**julep** (*joo'-lep*) [Pers., *jūlāb*, a sweet drink]. A sweetened drink containing aromatic alcoholic, or medicinal substances.

**jumentous** (*joo-men'-tus*) [*jumentum*, a beast of burden]. Like a beast of burden; horse-like: applied to the odor of urine.

**jumpers** (*jum'-perz*). Persons afflicted with a peculiar neurosis by reason of which they do whatever they are told, and perform sudden leaping or jumping movements. See *lata*; *palmus*.

**junction, myoneural.** The place where a motor nerve joins the muscle which it supplies.

**junctura** (*junk-tū'-rah*) [L. a joining; *pl.*, *juncturæ*]. An articulation; a suture (of bones).

**Jungbluth's vessels** (*yoong'-bloot*) [Hermann *Jungbluth*, German physician]. Nutrient vessels lying immediately beneath the amnion and disappearing usually at an early period of embryonic life.

**jungle fever.** A severe remittent fever of India.

**juniper, juniperus** (*jū'-nip-er, jū-nip'-er-us*). 1. An evergreen shrub or tree belonging to the genus *Juniperus*. *Juniperus sabina* yields savin (*sabina*, U. S. P.). The tops of *J. virginiana*, red cedar, are used as a substitute for savin. 2. The fruit or berry of *J. communis*, containing a volatile oil and an amorphous substance, *juniperin*. The oil is a stimulant to the genitourinary mucous membrane and is used in nephritis, pyelitis, and cystitis. **j., fluidextract of** (*fluidextractum juniperi*, N. F.). Dose 60 min. (4 cc.). **j., oil of** (*oleum juniperi*, U. S. P., B. P.). Dose 1–4 min. (0.065–0.26 Cc.). **j., spirit of** (*spiritus juniperi*, U. S. P., B. P.). Dose 30 min.–1 dr. (2–4 Cc.). **j., spirit of, compound** (*spiritus juniperi compositus*, U. S. P.). Dose 1–4 dr. (4–16 Cc.). **j., tar, oil of.** See *cade*, *oil of*.

**junk.** A quilted cushion forming a sling in which to suspend a fractured limb.

**junket** (*junk'-et*). "Curds and whey," prepared by coagulating milk with rennet.

**Junod's boot** (*zjoo'-nō*) [Victor Théodore *Junod*, French physician, 1809–1881]. A boot-shaped case, usually of stiff leather, made to inclose the leg, so that, the air being exhausted, the blood rushes to the inclosed part. It has been employed to relieve inflammation and congestion of the viscera.

**jurisprudence** (*joo-ris-proo'-dens*) [*jus*, law; *prudentia*, skill]. The science of the interpretation and application of the law. **j., medical**, the application of medical knowledge to the principles of common law.

**jury** (*joo'-re*) [*jurare*, to swear]. A body of men legally appointed to determine the guilt or innocence of a prisoner, or to determine the facts in judicial inquiries. **j.-leg**, a wooden leg. **j.-of matrons**, a body of twelve matrons, formerly empaneled in

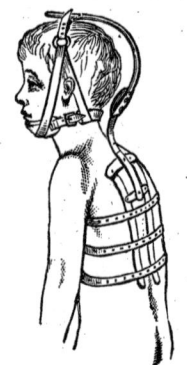

SAYRE'S JURY-MAST.

England to determine if a murderess, for whom such plea was made, were pregnant. **j.-mast**, a steel shaft with curved iron rods attached, sometimes employed to support the head in disease of the upper vertebræ; it is mainly used in connection with Sayre's treatment of spondylitis.

**jusculum** (*jus'-kū-lum*) [L. a decoction]. A soup, or broth; gruel or porridge.

**justo** (*jus'-to*). Ablative of *justum*, that which is right, or normal. **j. major**, greater than normal, larger in all dimensions than normal; applied to a pelvis. **j. minor**, abnormally small.

**Justus' test** (*jus'-tus*) [J. *Justus*, Austrian dermatologist]. Transient reduction of hemoglobin following the administration of mercury by inunction or hypodermatic injection in syphilis.

**jute** (*joot*) [Beng., *jūt*, matted hair]. The bast-fiber of several species of the genus *Corchorus*, grown chiefly in India and Ceylon. Jute is used as an absorbent dressing.

**juvantia** (*joo-van'-she-ah*) [L., "aiding"]. Adjuvant remedies or medicines.

**juxta-articular** (*juks-tah-ar-tik'-ū-lar*) [*juxta*, near; *articulus*, a joint]. Near a joint.

**juxtaposition** (*juks-tah-po-zish'-un*) [*juxta*; *positio*, position]. Situation adjacent to another; in close relationship the act of placing near; apposition.

**juxtapyloric** (*juks-tah-pi-lor'-ik*) [*juxta*, near; *pylorus*]. Near the pylorus.

# K

**K.** The chemical symbol for potassium (*kalium*).
**k., ka.** Abbreviations of *kathode* or of *kathodic*.
**Kader's method of gastrotomy** (*kah'-der*) [Bronislaw *Kader*, Polish surgeon]. Consists in the production of a funnel which projects into the stomach by suturing two sections on either side of a tube placed vertically into the stomach. Each pair of sutures increases the amount of peritoneum about the tube and further depresses its point of entrance.
**Kaes-Bechterew's layer.** See *Bechterew's layer*.
**Kahlbaum's disease** (*karl'-bowm*) [Karl Ludwig *Kahlbaum*, German physician, 1828–1899]. Katatonia; a form of insanity progressing to imbecility.
**Kahler's disease** (*kah'-ler*) [Otto *Kahler*, Austrian physician, 1849–1893]. A constitutional affection characterized by the formation of round-celled neoplasms in the skeleton, paroxysms of pain, a tendency to spontaneous fractures, especially of the ribs, enlargement of the spleen and lymphatic glands, and the presence of Bence Jones' bodies in the urine.
**K.'s law,** the ascending branches of the posterior spinal nerve-roots, after entering the cord, pass successively from the root-zone toward the mesial plane.
**Kahler-Singer's law.** See *Kahler's law*.
**kaif** (*kīf*) [Arab.]. Languor; dreamy enjoyment; sensuous tranquillity, such as follows the use of certain drugs (like opium, or hashish).
**kainogenesis** (*ki-no-jen'-e-sis*) [καινός, new, fresh; γένεσις, generation].— A renewal or improvement by infusion of fresh material.
**kainophobe** (*ki'-no-fōb*) [καινός, new; φόβος, fear]. A person that fears anything new.
**kairin** (*ki'-rin*) [καιρός, the right time], C₁₀H₁₃-NO . HCl . H₂O. Oxychinolin-ethyl hydrochloride; antipyretic, diaphoretic, emetic, and has been used as a substitute for quinine. Dose 5–15 gr. (0.32–1.0 Gm.). **k., ethyl, kairin a,** C₉H₁₀(C₂H₅)NO . HCl, recommended as antipyretic in doses of 8–25 gr. (0.5–16 Gm.).
**kairolin** (*ki'-ro-lin*) [*kairin*], C₁₀H₁₅N. Methylquinolin hydride. An antipyretic resembling kairin, but less efficient.
**kakatrophy.** See *cacotrophy*.
**kaki** (*kak'-e*). The fruit of *Diospyros kaki*, Japanese persimmon; used in vomiting of pregnancy and in diarrhea.
**kakidrosis** (*kak-id-ro'-sis*) [κακός, bad; ἱδρώς, sweat]. Fetid perspiration.
**kakké** (*kak'-ka*) [Chinese for "leg disease"]. Epidemic and endemic multiple neuritis, or beriberi.
**kako-** (*kak'-o-*). See *caco-*.
**kakodyl** (*kak'-o-dil*). See *cacodyl*.
**kakosmia** (*kak-os'-me-ah*). See *cacosmia*.
**kakotrophia** (*kak-ot-ro'-fe-ah*). See *cacotrophy*.
**kala-azar.** An obscure fatal disease, prevalent in Assam, due to a protozoan parasite, the Leishman Donovan body.
**kaladana** (*kal-ad-a'-nah*) [origin unknown]. An East Indian convolvulaceous plant, *Ipomoea* (*Pharbitis*) *nil*. Its seeds are a safe and good cathartic. Dose of the powdered drug, gr. xxx–xl; of the resin, gr. iv–viij.
**kalaf** (*kal'-af*). A medicinal fluid obtained from leaves of *Salix capensis*.
**kali** (*ka'-li*) [Ar., *qali*, potash]. Potash.
**kaligenous** (*kal-ij'-en-us*) [*kali*; *generare*, to produce]. Yielding potash.
**kaligraph** (*kal'-ig-raf*) [καλός, beautiful; γράφειν, to write]. An instrument for the use of those afflicted with writers' cramp. It is essentially a pantograph, so arranged that by making the letters very large at one point, they are reproduced of a natural size at another.
**kalimeter** (*kal-im'-et-er*). See *alkalimeter*.
**kalimetry** (*kal-im'-et-re*). See *alkalimetry*.
**kalium** (*ka'-le-um*). Potassium.
**kallak** (*kal'-ak*). A pustular dermatitis occurring among the Eskimos.

**kamala, kamela** (*kam-a'-lah, -e'-lah*) [Hind., *kamīla*]. Rottlera. The glands and hairs from the capsules of *Mallotus philippinensis* (*Rottlera tinctoria*), native to southern Asia and Abyssinia. It is purgative and anthelmintic, and is used for the expulsion of lumbricoid worms and tape-worms.
**kamalin, kamilin.** See *rottlerin*.
**kambi** (*kam'-be*) [E. Ind.]. An aromatic gum, like elemi, from *Gardenia lucida*, a plant of India.
**Kandahar sore** (*kan'-da-har*). See *furunculus orientalis*.
**kangaroo.** A marsupial mammal of Australia and the neighboring islands. **k. tendon,** a tendon derived from the tail of the kangaroo and used for surgical ligatures.
**kangri-burn** (*kang'-gre*). A squamous epithelioma frequent on the skin of the abdomen and thighs of the natives of Kashmir and attributed to the irritation caused by charcoal heaters worn beneath the clothing in cold weather.
**kaolin, kaolinum** (*ka'-o-lin, ka-o-li'-num*) [Chin., *kaoling*, "high ridge"]. Aluminum silicate. The *kaolinum* of the U. S. P. is obtained from the decomposition of feldspar. It is sometimes used as a protective application in eczema and as a coating for pills. Syn., *China-clay; white clay*.
**kaolinosis** (*ka-ol-in-o'-sis*). A pneumoconiosis occurring in workers in kaolin.
**Kaplan's test** (*kap'-lan*) [David M. *Kaplan*, American physician]. For albumin in cerebrospinal fluid: 0.5 cc. of cerebrospinal fluid is boiled in a test tube, then 2 drops of a 5 per cent. solution of butyric acid are added, the fluid is then boiled again, and 5 cc. of supersaturated solution of ammonium sulphate are underfloated; a cheesy ring in not more than 20 minutes denotes albumin.
**Kaposi's disease** (*ka-po'-se*) [Moritz Kohn *Kaposi*, Austrian dermatologist, 1837–1902]. See *xeroderma pigmentosum*.
**Karell cure** (*kar-el'*) [Philip *Karell*, Russian physician]. *In heart disease:* Rest in bed, and a light diet of milk and eggs; the milk is limited to 1½ or 2 pints a day, and the treatment is continued for one week.
**karnosin** (*kar-no'-sin*). See *carnosin*.
**karyaster** (*kar-e-as'-ter*) [*karyon*; ἀστήρ, a star]. The radiate arrangement of the chromosome during karyokinesis. Syn., *aster*.
**karyenchyma** (*kar-e-en'-ke-mah*) [*karyon*; ἐν, in; χεῖν, to pour]. The clear ground-substance occupying the meshes of the nuclear reticulum. Syn., *karyolymph; nuclear sap*.
**karyo-** (*kar-e-o-*) [*karyon*]. A prefix, signifying relating to the karyon or cell-nucleus.
**karyochromatophil** (*kar-e-o-kro-mat'-o-fil*) [*karyo-*; χρῶμα, color; φιλεῖν, to love]. 1. Having a stainable nucleus. 2. A stainable nucleus.
**karyochrome** (*kar'-e-o-krōm*) [*karyo-*; χρῶμα, color]. A nerve-cell the nucleus of which stains best.
**karyoclasis** (*kar-e-ok'-las-is*). See *karyorrhexis*.
**karyogamic** (*kar-e-o-gam'-ik*) [*karyo-*; γάμος, marriage]. Pertaining to the blending of nuclei, as in reproduction.
**karyogamy** (*kar-e-og'-am-e*) [*karyo-*; γάμος, marriage]. A conjugation of cells characterized by a fusion of the nuclei. Cf. *plastogamy*.
**karyokinesis** (*kar-e-o-kin-e'-sis*) [*karyo-*; κίνησις, movement; change]. Indirect cell-division, the common mode of reproduction of cells. It depends upon complicated changes in the mitome of the cell-nucleus that may be divided into the following steps: 1. The nucleus becomes larger; the mitome filaments thicken and form a close skein, or *spirem*. 2. The fibrils become less convoluted and more widely separated, forming the loose skein; at the same time the nuclear spindles, two cone-shaped striated bodies, appear in the achromatin. 3. The mitome fibrils split longitudinally. 4. The segments migrate

toward the poles of the new nuclei, constituting daughter-wreaths, or *asters*. 5. Transformation of asters into fully developed nuclei. 6. Division of the cell-protoplasm.

**karyokinetic** (*kar-e-o-kin-et'-ik*) [see *karyokinesis*]. Pertaining to karyokinesis, as *karyokinetic* figures, the forms assumed by the mitome in karyokinesis.

**karyoklasis** (*kar-e-ok'-las-is*) [*karyo*-; κλάσις, a breaking]. See *karyorrhexis*.

**karyolymph** (*kar'-e-o-limf*). See *karyenchyma*.

**karyolysis** (*kar-e-ol'-is-is*) [*karyo*-; λύειν, to loose]. The segmentation of the nucleus of the cell.

**karyolytic** (*kar-e-o-lit'-ik*) [*karyo*-; λύειν, to loose]. Relating to karyolysis.

**karyomicrosoma** (*kar-e-o-mik-ro-so'-mah*). See *nucleo-microsomata*, under *nucleoplasm*.

**karyomite** (*kar'-e-o-mit*). See *chromosome*.

**karyomitoic** (*kar-e-o-mit-o'-ik*) [*karyo*-; μίτος, a thread]. Relating to karyomitosis.

**karyomitoma** (*kar-e-o-mit-o'-mah*). See *cell-body*.

**karyomitome** (*kar-e-om'-it-ōm*) [*karyo*-; μίτος, thread]. The mitome threads of the nucleus.

**karyomitosis** (*kar-e-o-mit-o'-sis*) [see *karyomitome*]. Karyokinesis.

**karyomitotic** (*kar-e-o-mit-ot'-ik*) [see *karyomitome*]. Relating to karyomitosis.

**karyon** (*kar'-e-on*) [κάρυον, nucleus]. The cell-nucleus.

**karyophage** (*kar'-e-of-āj*). See *karyophagus*.

**karyophagus** (*kar-e-of'-ag-us*) [*karyo*-; φαγεῖν, to eat; pl., *karyophagi*]. A cytozoon which destroys the nucleus of the infected cell.

**karyoplasm** (*kar'-e-o-plasm*) [*karyo*-; πλάσσειν, to form]. 1. The nuclear substance of a cell. 2. The more fluid material in the meshes of the chromoplasm.

**karyorrhexis** (*kar-e-or-ek'-sis*) [*karyo*-; ῥῆξις, rupture]. Fragmentation or splitting up of a nucleus into a number of chromatin particles which become scattered in the cytoplasm; it occurs in the cells of the disappearing follicles of the ovary.

**karyosome, karyosoma** (*kar-e-o-sōm, kar-e-o-so'-mah*) [*karyo*-; σῶμα, a body; pl., *karyosomata*]. A nuclear microsoma; a round body resembling a nucleolus, contained in the segmentation-nucleus of the ovum.

**karyostasis** (*kar-e-os'-ta-sis*) [*karyo*-; στάσις, a stoppage]. The resting-stage of nuclei of cells. It is opposed to karyokinesis.

**karyostatic** (*kar-e-o-stat'-ik*). Pertaining to karyostasis.

**karyostenosis** (*kar-e-o-ste-no'-sis*) [*karyo*-; στενός, narrow]. The simple division of the nucleus of a cell. This process is called also *akinetic*, or *direct division*.

**karyota** (*kar-e-o'-tah*) [*karyon*]. Nucleated cells.

**karyotheca** (*kar-e-o-the'-ka*) [*karyo*-; θήκη, case]. Nuclear membrane.

**kasagra** (*kas-ag'-rah*). A proprietary preparation of cascara sagrada.

**kasena** (*kas'-e-nah*). An aromatic preparation of cascara and senna. Dose 1–2 dr. (4–8 Cc.).

**kasyi** (*kas'-ii*). A germicide said to consist of creosol and green soap.

**kata-** (*kat-ah-*). A prefix denoting down or intensive. For words thus beginning see *cata-*.

**katelectrotonus** (*kat-el-ek-trot'-o-nus*). See *catelectrotonus*.

**katex** (*kat'-eks*). An abbreviation of *kathode excitation*.

**katharol** (*kath'-ar-ol*). A solution of hydrogen peroxide.

**katharophore** (*kath-ar'-o-fōr*) [καθαρός, clean; φέρειν, to carry]. An instrument for cleansing the urethra.

**kathetometer** (*kath-et-om'-et-er*) [κάθετος, a plumb-line; μέτρον, measure]. 1. An instrument for ascertaining the level of fluids. 2. An apparatus for use in craniometry.

**kathion** (*kath'-e-on*). See *cation*.

**kathodal** (*kath-o'-dal*). See *cathodal*.

**kathode** (*kath'-ōd*) [*kata*-; ὁδός, way]. See *cathode*.

**kathodic** (*kath-od'-ik*). See *cathodal*.

**kation** (*kat'-e-on*). See *cation*.

**katochus** (*kat-o'-kus*) [κάτοχος, catalepsy]. An unconscious condition, resembling sleep with open eyes, observed in intermittent fever, etc.

**kava, kava-kava** (*kah'-vah*) [Hawaiian]. 1. An intoxicating beverage prepared in the Sandwich Islands from the root of *Piper methysticum*. 2. The root of *Piper methysticum*, containing a resin, *kavin*, and an alkaloid, *kavaine*. The resin is a motor depressant, locally at first an irritant, later an anesthetic; it is also a cardiac stimulant. Kava-root has been used in gonorrhea, leukorrhea, and incontinence of urine. Dose of *fluidextract* 15 min.–1 dr. (1–4 Cc.).

**kavaine** (*kah'-va-ēn*) [Hawaiian, *kava*]. An alkaloid obtained from the roots of kava-kava; also called methysticine.

**kawaine** (*kah'-wa-ēn*). See *kavaine*.

**Keating-Hart's method.** (*ke'-ting-hart'*) [Walter Valentine de *Keating-Hart*, French physician]. The treatment of external cancer by fulguration.

**kédani disease** (*ked-an'-e*). A disease common in Japan, due to inoculation with *Proteus hauseri* by the bite of a mite called kédani; also called Japanese river fever. The symptoms resemble those of abdominal typhus.

**Keeley cure** (*ke'-le*) [Leslie E. *Keeley*, American physician, 1832–1900]. A secret method of treating drunkenness. Gold was said to be administered (by the physician).

**Keen's sign** [William Williams *Keen*, American surgeon, 1837– ]. Increased diameter through the leg at the malleoli in Pott's fracture.

**kefir, kefyr** (*kef'-ir*). See *kephir*.

**Keisselbach's spot** (*ki'-sel-bakh*). A point in the anterior and lower part of the nasal septum, about ½ inch from the nostril; a favorite site for bleeding from the nose.

**keistein, keistin.** See *kyestein*.

**Keith's bundle** (*kēth*) [Arthur *Keith*, English anatomist, 1861– ]. Sinoatrial or sinoauricular bundle. K.'s node, sinoatrial or sinoauricular node.

**kelectome** (*ke'-lek-tōm*) [κήλη, a tumor; ἐκ, out; τέμνειν, to cut]. A cutting instrument introduced into a tumor, by means of a canula, in order to obtain a part of the substance for examination.

**kelene** (*kel'-ēn*). Trade name of ethyl chloride.

**kelis** (*ke'-lis*) [κηλίς, spot; pl., *kelides*]. 1. The same as morphea or scleroderma. 2. See *keloid*.

**Keller's tuberculin test plate** (*kel'-er*). A piece of adhesive plaster in the middle of which is a small circle of an ointment-like material supposed to contain tuberculin. Its action is much like that of the Moro test.

**Kelley's sign** (*kel'-e*). Of *pleural effusion in children*: a preference for lying upon the back or propped up high in bed and avoidance of bending toward or pressing upon the affected side.

**kellin** (*kel'-in*). A glucoside from the fruit of *Ammi visnaga*. It is said to affect the respiration and the pulse, and to have a paralyzant effect upon the lower extremities.

**Kelling's test for lactic acid** (*kel'-ing*) [George *Kelling*, German physician]. A weak solution of ferric chloride becomes much deeper in color when lactic acid is added to it.

**Kellock's sign.** Increased vibration of the ribs on sharply percussing them with the right hand, the left hand being placed flatly and firmly on the lower part of the thoracic wall, just below the nipple; it is elicited in pleural effusion.

**keloid** (*ke'-loid*) [from κηλίς, a scar, or κηλή, a claw; εἶδος, likeness]. A tumor-like fibrous outgrowth, usually occurring at the site of a scar. It is elevated, whitish or pink in color, and sends prolongations into the surrounding tissues resembling the claws of a crab. By many it is not considered a true tumor, but merely a hyperplastic scar. It affects the colored race more frequently than the white. Syn., Alibert's *keloid*; *cheloid*; *kelis*. k. of Addison, morphea.

**keloplasty** (*kel-o-plas'-te*). See *chiloplasty*.

**kelos** (*ke'-los*). Same as *keloid*.

**kelosomia** (*kel-o-so'-me-ah*). See *celosoma*.

**kelotomy** (*ke-lot'-o-me*). Herniotomy.

**kelp** [origin obscure]. 1. Burnt sea-weed, from which iodine is obtained. 2. The *Fucaceæ laminariæ* and other large sea-weeds.

**kelpion** (*kelp'-e-on*) [*kelp*, sea-weed yielding iodine]. An ointment containing iodine, which is volatilized when the ointment is warmed.

**kelvin** (*kel'-vin*) [William Thompson, Lord *Kelvin*, British physicist, 1824– ]. A commercial unit of electricity; one thousand watt-hours.

**Kendall's test.** See *kyestein*.

**kenencephalocele** (*ken-en-sef'-al-o-sēl*). See *cenencephalocele*.

**kenesthesia** (*ken-es-the'-ze-ah*). See *cenesthesia*.

**kenesthesis** (*ken-es-the'-sis*). See *cenesthesis*.

**Kennedy's sign** of pregnancy (*ken'-ed-e*). The umbilical or funic souffle.
**kenogenesis** (*ken-o-jen'-e-sis*). Vitiated individual development in which the phylogenetic development is not truly epitomized.
**kenophobia** (*ken-o-fo'-be-ah*) [κενός, empty; φόβος, fear]. A fear of large empty spaces.
**kenosis** (*ken-o'-sis*) [κένωσις, a draining]. 1. An evacuation. 2. Inanition.
**kenotic** (*ken-ot'-ik*) [*kenosis*]. 1. Drastic, purgative. 2. A drastic drug or agent.
**kenotoxin** (*ken'-e-toks-in*). A poisonous substance developed in the tissues during their activity and responsible for their fatigue.
**Kentmann's test for formaldehyde.** Morphine hydrochloride 0.1 Gm. is dissolved in 1 Cc. of sulphuric acid in a test-tube, and an equal volume of the solution to be examined is added without mixing; in the presence of formaldehyde the aqueous solution will be clear red violet in color after a lapse of a few minutes. The reaction is sensitive to 1 : 6000 to 1 : 5000.
**kentrokinesis** (*ken-tro-kin-e'-sis*) [κέντρον, center, spur; κίνησις, motion]. The influence of any motor nerve-center; excito-motor action.
**kephaldol** (*kef-al'-dol*). Trade name of a preparation said to be a compound of citric and salicylic acids with phenetidin, to which some quinine is added; it is an antipyretic.
**kephalin** (*kef'-al-in*). See *cephalin*.
**kephalometer** (*kef'-al-om'-et-er*). See *cephalometer*.
**kephir, kephyr** (*kef'-ir*) [Caucasian]. A nutritious substance obtained by a peculiar fermentation of cow's milk produced by certain fungi. **k., arsenical**, a combination of kephir and Fowler's solution. **k., iodo-**, a combination of kephir and sodium iodide. **k.-seed**, a substance containing the ferment (*Bacillus caucasicus*) of kephir. It is used in preparing the genuine kephir.
**keracele** (*ker'-as-ēl*) [κέρας, horn; κήλη, tumor]. A horny tumor on the hoof of horses.
**keraphyllocele** (*ker-af-il'-o-sēl*) [κέρας, horn; φύλλον, leaf; κήλη, tumor]. A horny growth between the covering of the horse's hoof and the deeper tissues.
**keraphyllous** (*ker-af'-il-us*) [κέρας, horn; φύλλον, leaf]. Composed of horny layers.
**kerasene, kerasin** (*ker'-as-ēn, ker'-as-in*) [κέρας, horn], C₁₈H₃₅NO. A nitrogenous substance; one of the cerebrins obtained from brain-substance.
**kerat-, kerato-** (*kef-at-, ker-at-o-*) [κέρας, horn; cornea]. Prefixes denoting relation to the cornea or to horn.
**keratalgia** (*ker-at-al'-je-ah*) [*kerat-*; ἄλγος, pain]. Pain in the cornea.
**keratectasia** (*ker-at-ek-ta'-se-ah*) [*kerat-*; ἔκτασις, extension]. A bulging forward of the cornea.
**keratectomy** (*ker-at-ek'-to-me*) [κέρας, horn, cornea; ἐκτομή, a cutting out]. Surgical excision of a part of the cornea.
**keratiasis** (*ker-at-i'-as-is*) [κέρας, horn]. A morbid condition characterized by the growth of horny excrescences.
**keratic** (*ker-at'-ik*) [κέρας, horn]. Horny.
**keratin** (*ker'-at-in*) [κέρας, horn]. The basis of horny tissues, hair, nails, feathers, etc. It is a mixture of various complex substances and contains sulphur. Decomposed, it yields leucin and tyrosin. It is used in pharmacy to coat pills.
**keratinization** (*ker-at-in-iz-a'-shun*) [κέροια, horn]. 1. The development of a horny quality in a tissue. 2. The coating of pills with keratin.
**keratinoid** (*ker'-at-in-oid*). Trade name of a keratin coated pill.
**keratinous** (*ker-at'-in-us*). 1. Relating to keratin. 2. Horny.
**keratitis** (*ker-at-i'-tis*) [*kerat-*; ιτις, inflammation]. Inflammation of the cornea. **k. arborescens**, **k., dendritic**, **k., furrow**, **k., mycotic**, a superficial form attributed to a specific organism and characterized by a line of infiltration of the corneal tissue near the surface and developing later into an arborescent formation. **k. bullosa**, the formation of large or small blebs upon the cornea of an eye, the seat of iridocyclitis, interstitial keratitis, or glaucoma. **k., fascicular**. See **k.**, *phlyctenular*. **k., interstitial**, a form of keratitis in which the entire cornea is invested with a diffuse haziness, almost completely hiding the iris. The surface of the cornea presents a ground-glass appearance. Later, from ciliary injection, blood-vessels form in the superficial layers of the cornea, and produce a dull-red color—the "salmon patch" of Hutchinson. The entire cornea may become cherry-red. The disease is most frequent between the ages of 5 and 15, and occurs in syphilitic individuals. **k. neuroparalytica**, keratitis following lesion of the trifacial nerve. Its cause is loss of trophic influence, aided by mechanical irritation and drying of the cornea. **k., oyster-shuckers'**, a form due to corneal traumatism from pieces of embedded oyster-shell. **k., phlyctenular**, a variety characterized by the formation of small papules or pustules, often associated with similar lesions upon the conjunctiva. It is marked by severe local congestion lacrimation, and intense photophobia. **k. punctata**, a secondary affection of the cornea in association with affections of the iris, choroid, and vitreous. It is characterized by the formation of opaque dots, generally arranged in a triangular manner upon the posterior elastic lamina of the cornea. It is sometimes designated as descemetitis. **k. purulenta**, that accompanied by the formation of pus. **k. reapers'**, that due to the irritation from grain-awns. **k., sclerosing**, an interstitial form associated with scleritis. **k., trachomatous**. See *pannus*. **k., traumatic**, that consequent upon wounds or other injury of the cornea.
**keratoangioma** (*ker-at-o-an-je-o'-mah*). See *angiokeratoma*.
**keratocele** (*ker'-at-o-sēl*) [*kerato-*; κήλη, hernia]. A hernia of Descemet's membrane through the cornea.
**keratocentesis** (*ker-at-o-sen-te'-sis*) [*kerato-*; κέντησις, a pricking]. Corneal puncture.
**keratochromatosis** (*ker-at-o-kro-mat-o'-sis*) [*kerato-*; χρῶμα, color]. Discoloration of the cornea.
**keratoconjunctivitis** (*ker-at-o-kon-junk-tiv-i'-tis*). Simultaneous inflammation of the cornea and the conjunctiva.
**keratoconometer** (*ker-at-o-ko-nom'-et-er*) [*kerato-*; κῶνος, cone; μέτρον, measure]. An instrument for estimating astigmatism by the images reflected from the cornea.
**keratoconus** (*ker-at-o-ko'-nus*) [*kerato-*; κῶνος, cone]. A conic protrusion of the cornea.
**keratocricoid** (*ker-at-o-kri'-koid*) [*kerato-*; κρίκος, ring; εἶδος, like]. The cricothyroid muscle.
**keratoderma** (*ker-at-o-der'-mah*) [*kerato-*; δέρμα, skin]. 1. The cornea. 2. A horny condition of the skin.
**keratodermatitis** (*ker-at-o-der-mat-i'-tis*) [*kerato-*; δέρμα, skin; ιτις, inflammation]. Inflammation of the keratoderma; keratitis.
**keratodermatocele** (*ker-at-o-der-mat'-o-sēl*) [*kerato-*; δέρμα, skin; κήλη, tumor]. See *keratocele*.
**keratodermatomalacia** (*ker'-at-o-der-mat-o-mal-a'-se-ah*) [*kerato-*; δέρμα, skin; μαλακία, softness]. Softening of the cornea.
**keratodermatosis** (*ker-at-o-der-mat-o'-sis*) [*kerato-*; δέρμα, skin; νόσος, disease]. A skin-affection characterized by alteration in the horny elements of the skin.
**keratoderma** (*ker-at-o-der'-me-ah*) [*kerato-*; δέρμα, skin]. See *keratoderma*. **k. erythematosa symmetrica**, tylosis of the soles and palms, in which the horny patches show a broken-up surface.
**keratodermites** (*ker-at-o-der-mi'-tēs*) [*kerato-*; δέρμα, skin; ιτις, inflammation]. A group of inflammatory scaly skin-affections.
**keratogenesis** (*ker-at-o-jen'-e-sis*) [*kerato-*; γεννᾶν, to produce]. The formation of horny material or growths.
**keratogenous** (*ker-at-oj'-en-us*) [see *keratogenesis*]. Producing a horny or horn-like substance.
**keratoglobus** (*ker-at-o-glo'-bus*) [*kerato-*; *globus*, a ball]. A globular protrusion of the cornea.
**keratoglossus** (*ker-at-o-glos'-us*). See under *muscle*.
**keratohelcosis** (*ker-at-o-hel-ko'-sis*) [*kerato-*; ἕλκωσις, ulceration]. Ulceration of the cornea.
**keratohyal** (*ker-at-o-hi'-al*) [*kerato-*; *hyoid*]. Relating to a cornu of the hyoid bone.
**keratohyalin** (*ker-at-o-hi'-al-in*) [*kerato-*; ὕαλος, glass]. A peculiar substance occurring in granules in the deeper layers of the skin.
**keratohyaline** (*ker-at-o-hi'-al-ēn*). Both horny and hyaline in structure.
**keratoid** (*ker'-at-oid*) [*kerato-*; εἶδος, like]. Hornlike.
**keratoiritis** (*ker-at-o-i-ri'-tis*) [*kerato-*; *iritis*]. Combined inflammation of the cornea and the iris.
**keratoleukoma** (*ker-at-o-lu-ko'-mah*) [*kerato-*; λευκός, white; *pl.*, *keratoleukomata*]. A leukoma or whitish opacity of the cornea.

**keratolysis** (*ker-at-ol'-is-is*) [*kerato-*; λύσις, solution]. A shedding of the skin; a rare condition in which the skin is shed periodically, that of the limbs coming off as a glove or stocking.
**keratolytic** (*ker-at-o-lit'-ik*) [*kerato-*; λύσις, solution]. Pertaining to keratolysis.
**keratoma** (*ker-at-o'-mah*) [*kerato-*; ὅμα, tumor]. 1. See *callosity*. 2. Congenital ichthyosis; the presence of horny plates upon the integument.
**keratomalacia** (*ker-at-o-mal-a'-she,-ah*) [*kerato-*; μαλακία, softness]. A softening of the cornea.
**keratome** (*ker'-at-ōm*) [*kerato-*; τομή, a cutting]. A knife with a peculiar trowel-like blade, used for making the incision into the cornea in the operation of iridectomy.
**keratometer** (*ker-at-om'-et-er*) [*kerato-*; μέτρον, a measure]. An instrument for measuring the curves of the cornea.
**keratometry** (*ker-at-om'-et-re*) [see *keratometer*]. The measurement of curves of the cornea.
**keratomycosis** (*ker-at-o-mi-ko'-sis*) [*kerato-*; mycosis]. A fungoid growth of the cornea.
**keratoncus** (*ker-at-ong'-kus*) [*kerato-*; ὅγκος, a tumor]. Any horny tumor.
**keratonosis** (*ker-at-on'-o-sis*). See *keratosis*.
**keratonosus** (*ker-at-on'-o-sus*). Any disease of the cornea.
**keratonyxis** (*ker-at-o-niks'-is*) [*kerato-*; νύξις, a pricking]. The needling of a soft cataract by puncture through the cornea; also, the old operation of couching a cataract with the needle.
**keratophagia** (*ker-at-of-a'-je-ah*). See *onychomycosis*.
**keratoplasia** (*ker-at-o-pla'-ze-ah*) [*kerato-*; πλάσσειν, to form]. The reparative renewal of the horny layer of the skin.
**keratoplastic** (*ker-at-o-plas'-tik*). Pertaining to keratoplasty.
**keratoplasty** (*ker'-at-o-plas-te*) [*kerato-*; πλάσσειν, to form]. Plastic operation upon the cornea, especially the transplantation of a portion of cornea from the eye of a lower animal to that of man.
**keratorrhexis** (*ker-at-or-eks'-is*) [*kerato-*; ῥῆξις, rupture]. Rupture of the cornea, due to ulceration or traumatism.
**keratoscleritis** (*ker-at-o-skle-ri'-tis*) [*kerato-*; σκληρός, hard; ιτις, inflammation]. Inflammation of the cornea and the sclera.
**keratoscope** (*ker'-at-o-skōp*) [*kerato-*; σκοπεῖν, to view]. An instrument for examining the cornea and testing the symmetry of its meridians of curvature.
**keratoscopy** (*ker-at-os'-ko-pe*) [see *keratoscope*]. 1. Examination of the cornea with the keratoscope. 2. Retinoscopy or skiascopy.
**keratose** (*ker'-at-ōs*) [*kerato-*]. Horny.
**keratosis** (*ker-at-o'-sis*) [κέρας, cornea]. Any disease of the skin characterized by an overgrowth of the horny epithelium. **k. follicularis**, a form of acne in which horny, prominent projections occur about the sebaceous follicles; they are firmly adherent and produce a roughness comparable to that of a nutmeg-grater. **k. pilaris**, a chronic affection of the skin marked by hard, conical elevations investing the hair-follicles, and somewhat resembling goose-flesh. **k. senilis**, a cornification of the skin of old people, often limited to certain definite regions, as the face and dorsal surfaces of the hands and feet.
**keratotome** (*ker'-at-o-tōm*). See *keratome*.
**keratotomy** (*ker-at-ot'-o-me*) [*kerato-*; τέμνειν, to cut]. Incision of the cornea.
**keraunics** (*ker-awn'-iks*) [κεραυνός, a thunderbolt]. The branch of physics treating of heat and electricity.
**keraunographic** (*ker-aw-no-graf'-ik*) [κεραυνός, a thunderbolt; γράφειν, to write]. Pertaining to the pictorial impressions of near objects sometimes seen upon the body of a person who has been struck by lightning.
**keraunoneurosis** (*ker-aw-no-nū-ro'-sis*) [κεραυνός, a thunderbolt; νεῦρον, nerve; νόσος, disease]. Nervous disease due to lightning-stroke.
**keraunophobia** (*ker-aw-no-fo'-be-ah*) [κεραυνός, a thunderbolt; φόβος, fear]. A morbid fear of lightning.
**kerectomy** (*ker-ek'-to-me*). See *keratectomy*.
**kerion** (*ke'-re-on*). See *tinea kerion*.
**keritherapy** (*ker-e-ther'-ap-e*) [κέρας, wax; *therapy*]. Treatment by means of paraffin baths.
**Kerkring's folds** (*kerk'-ring*) [Theodor *Kerkring*, Dutch anatomist, 1640–1693]. Same as **K.'s valves**.
**K.'s ossicle**, a point of ossification in the occipital bone, immediately behind the foramen magnum. **K.'s valves**, the valvulæ conniventes of the small intestine.
**kermes** (*ker'-mēz*) [Pers., *qirmis*, crimson]. A red dyestuff resembling cochineal, made from the bodies of the dried insects, *Coccus ilicis*, found on the kermes-oak. **k.-mineral**, a mixture of antimony trioxide and trisulphide.
**Kerner's reaction for creatinin**. Add to a solution of creatinin acidified with a mineral acid a solution of phosphotungstic or phosphomolybdic acid; a crystalline precipitate will be formed.
**Kernig's sign** (*ker'-nig*) [Waldemar *Kernig*, Russian physician, 1840– ]. Contracture or flexion of the knee- and hip-joint, at times also of the elbow, when the patient is made to assume the sitting posture; it is noted in meningitis.
**keroid** (*ker-oid'*) [κέρας, horn, or cornea; εἶδος, like]. 1. Horny. 2. Like the cornea.
**kerosene oil** (*ker'-o-sēn*) [κηρός, wax]. A liquid hydrocarbon, or oil extracted from bituminous coal.
**kestin** (*kes'-tin*). A proprietary antiseptic and deodorant said to contain trinitrophenol, ammonium chloride, orthoboric acid, and formic aldehyde.
**ketogenesis** (*ke-to-jen'-es-is*). The production of a ketone, or of acetone.
**ketol** (*ke'-tol*). See *indol*.
**ketols** (*ke'-tolz*). In chemistry, ketone-alcohols, containing both the ketone and alcohol groups.
**ketone** (*ke'-tōn*) [an arbitrary variation of *acetone*]. An organic compound derived by oxidation from a secondary alcohol; it contains the group =C=O.
**ketoses** (*ke-to'-ses*). In chemistry, a generic name applied to carbohydrates containing the ketone group CO.
**key** (*ke*). 1. An instrument for opening or fastening a lock. 2. A device for making and breaking an electric circuit. 3. In a system of classification a table containing the principal divisions and their distinguishing characteristics. **k., Du Bois-Reymond's**, an electric switch by means of which the circuit may be either closed or the current short-circuited. **k.-forceps, Elliot's**, two instruments are so called, the one having beaks of forceps and the handle of a key, for the extraction of teeth; the other is designed for the extraction of roots of teeth that present but one side above the alveolus. They are now but little used. **k. of Garengeot**, an instrument for the extraction of teeth, composed of a shank with a movable clasp and a cross-bar. The clasp is applied to the inner surface of the tooth, and the extraction is accomplished by turning the handle. **k., tetanizing.** See *k., DuBois-Reymond's*. **k., tooth.** See *k. of Garengeot*.
**Key and Retzius' corpuscles** (*ke', ret'-ze-us*) [Ernst Axel Henrik *Key*, Swedish physician, 1832–1901; Magnus Gustav *Retzius*, Swedish histologist, 1842– ]. Encapsulated corpuscles found in the bill of some aquatic birds and representing transition forms between Herbst's and Pacini's corpuscles. **K. and R.'s foramina.** See *Luschka's foramina*.
**Kg.** Abbreviation for *kilogram*.
**kibe** (*kīb*). A broken, or ulcerated, chilblain.
**kibisitome** (*ki-bis-it-ōm*) [κίβισις a pouch; τέμνειν, to cut]. A cystitome.
**kidney** (*kid'-ne*) [ME., *kidnere*]. One of the two large glandular organs situated in the upper and posterior portion of the abdominal cavity, and concerned in the excretion of the urine. It consists of an outer cortical substance and an inner medullary substance. The *medulla* consists of from 8 to 18 pyramids (*pyramids of Malpighi*), the apices of which, the *papillæ*, project into the calyces of the ureter. The pyramids are striated, and in places send narrow projections into the cortex—the *medullary rays*, or *pyramids of Ferrein*. Between the pyramids are extensions from the cortex—the *columns of Bertini*. The *cortex*, by the penetration into it of the medullary rays, is divided into *medullary rays* and the *labyrinth*. The secreting structure of the kidney consists of long tubes, beginning in an expanded $x_1t_nmi_1y$—the *capsule of Bowman*—which invests a tuft of blood-vessels, the *glomerulus*, and constitutes, together with this, a *Malpighian body*; extending from this is the *proximal convoluted tubule*; then comes the *spiral tubule*, then the *loop of Henle*, consisting of a descending and an ascending limb; then the *distal convoluted tubule*, which terminates in the *collecting tubule*. The blood-vessels which the kidney divide into two sets of branches, one supplying the

cortex, the other the medulla. The kidney weighs about 150 Gm. k., amyloid, a kidney the seat of amyloid degeneration. k., confluent, a single kidney formed by fusion of twin kidneys or other congenital malformation. k., fatty, one the seat of extensive fatty degeneration. k., floating. See *floating kidney*. k., gouty, k., granular, the small kidney resulting from chronic interstitial nephritis. k., horseshoe-. See *horseshoe-kidney*. k., large white, that of the advanced stage of chronic parenchymatous nephritis. k., massage, a state of uremia and renal incompetence due to improper massage of the kidney. k., movable, floating kidney. k., pigback, the large congested kidney found in alcoholic subjects. k. of pregnancy, an anemic kidney with fatty infiltration of the epithelial cells, but without any acute or chronic inflammation, occurring in pregnant women. k., red contracted. See k., gouty. k., sacculated, a condition due to hydronephrosis and absorption of the chief part of the kidney, leaving the irregularly expanded capsule. k., small white, the final stage of the large white kidney after loss of its substance from atrophy or degeneration. k., surgical, pyelonephritis. k., wandering. See k., floating. k., waxy. Same as k., amyloid.
**Kienbock's disease** (*ke'-en-bok*). Traumatic malacia of the semilunar bone of the wrist.
**Kiernan's spaces** (*kĕr'-nan*) [Francis *Kiernan*, English physician, 1800–1874]. The interlobular spaces of the liver.
**Kiesselbach's place** (*ke'-sel-bakh*) [Wilhelm *Kiesselbach*, German laryngologist, 1839– ]. The point at which the nasal septum, owing to its thinness, is especially liable to perforation.
**Kiesselbachii, locus.** See *Kiesselbach's place*.
**Kilian's line** (*kil'-e-an*) [Hermann Friedrich *Kilian*, German obstetrician, 1800–1863]. The line of the promontory of the sacrum. K.'s pelvis, the osteomalacic (halisteretic) pelvis.
**Killian's operation** (*kil'-e-an*) [Gustav *Killian*, German laryngologist, 1860– ]. Removal of the anterior wall of the frontal sinus and curettage of the frontal sinus and ethmoid cells.
**kilo** (*kil'-o*). A contraction of *kilogram*.
**kilocalory** (*kil-o-kal'-or-e*). See *calory, great*.
**kilogram** (*kil'-o-gram*) [χίλιοι, a thousand; *gram*]. One thousand grams, or 2.2 pounds avoirdupois.
**kilogrammeter** (*kil'-o-gram-ĕt'-er*) [χίλιοι, a thousand; *gram*; μέτρον, a measure]. A term denoting the energy required to raise one kilogram one meter in height; equivalent to 7.233 foot-pounds.
**kiloliter** (*kil'-o-le-ter*) [*kilo*; *liter*]. One thousand liters, or 35.31 cubic feet.
**kilometer** (*kil'-o-me-ter*) [*kilo*; *meter*]. One thousand meters, or 1093.6 yards.
**kilostere** (*kil'-o-stĕr*) [*kilo*; στερεός, solid]. One thousand cubic meters.
**kilowatt** (*kil'-o-wot*) [χίλιοι, thousand; *watt*]. One thousand watts of electricity; same as the *kelvin*.
**kilurane** (*kil'-u-rān*) [χίλιοι, a thousand; *urane*]. A thousand uranes; a unit of radioactivity.
**kinæsthesia** (*kin-es-the'-ze-ah*). See *kinesthesia*.
**kinase** (*kin'-ās*) [κίνησις, motion]. See *activator*.
**kinazyme** (*ki'-na-zim*). Trade name of a preparation of the liver, and the tryptic enzyme of the pancreas. It is said to improve digestion and nutrition, and to induce leukocytosis.
**kinematics** (*kin-em-at'-iks*) [κίνησις, to move]. The science of motion.
**kinematograph** (*kin-e-mat'-o-graf*) [κίνησις, movement; γράφειν, to write]. An apparatus used to make a continuous record of a body in movement.
**kineplasty**. A plastic amputation with the object of making a stump useful for locomotion.
**kinepock** (*kīn'-pok*). Synonym of *vaccinia, q. v.*
**kinesalgia** (*kin-es-al'-je-ah*) [*kinesis*; ἄλγος, pain]. Local pain following muscular contraction.
**kinescope** (*kin'-es-kōp*) [κινεῖν, to move; σκοπεῖν, to view]. A device to regulate with accuracy the width of an aperture through which rays of light are allowed to pass in measuring ametropia.
**kinescopy** (*kin-es'-ko-pe*). A form of retinoscopy requiring cooperation on the part of the patient; subjective retinoscopy.
**kinesialgia** (*kin-e-si-al'-je-ah*) [κίνησις, movement; ἄλγος, pain]. The condition of a muscle giving rise to pain on contraction. Cf. *kinesalgia*.
**kinesiatric** (*kin-es-e-at'-rik*) [κίνησις, movement; ἰατρικός, therapeutic]. Relating to kinesitherapy.

**kinesiatrics** (*kin-es-e-at'-riks*). Same as *kinesitherapy*.
**kinesic** (*kin-ez'-ik*). See *kinetic*.
**kinesiesthesiometer** (*kin-es-e-es-the-ze-om'-et-er*) [*kinesis*; αἴσθησις, perception; μέτρον, a measure]. An instrument for testing the muscular sense. It consists of wooden balls of the same size but of different weights.
**kinesimeter** (*kin-es-im'-et-ur*). See *kinesiometer*.
**kinesiology** (*kin-es-e-ol'-o-je*) [*kinesis*; λόγος, science]. The science of movements, considered especially as therapeutic or hygienic agencies.
**kinesiometer** (*kin-es-e-om'-et-er*) [*kinesis*; μέτρον, a measure]. An instrument for determining quantitatively the motion of a part.
**kinesiometric** (*ki-nes-e-o-met'-rik*) [κίνησις, movement; μέτρον, measure]. Relating to the measurement of motion.
**kinesioneurosis** (*kin-es-e-o-nū-ro'-sis*) [*kinesis*; *neurosis*]. A functional nervous disease associated with disorders of motion. k., external, that affecting the external muscles. k., internal, k., visceral, that affecting the muscles of the viscera.
**kinesionosos** (*kin-es-e-on'-o-sos*) [κίνησις, movement; νόσος, disease]. Any disease marked by impairment of the power of motion.
**kinesiotherapy** (*kin-es-e-o-ther'-ap-e*). See *kinetotherapy*.
**kinesipathic** (*kin-es-ip-ath'-ik*) [κίνησις, motion; πάθος, disease]. Pertaining to kinesipathy.
**kinesipathist** (*kin-es-ip'-ath-ist*) [κίνησις, motion; πάθος, disease]. One who practices the kinesistic treatment of disease.
**kinesipathy** (*kin-es-ip'-ath-e*) [*kinesis*; πάθος, disease]. Kinetotherapy.
**kinesis** (*kin-e'-sis*) [κίνησις, motion]. The general term for all physical forms of energy.
**kinesitherapy** (*kin-es-e-ther'-ap-e*) [κίνησις, motion; θεραπεία, care, cure]. See *Swedish movements*.
**kinesodic** (*kin-es-od'-ik*) [*kinesis*; ὁδός, way]. Pertaining to the motor pathways.
**kinesotherapy** (*kin-es-o-ther'-ap-e*). Same as *kinesitherapy*.
**kinesthesia, kinesthesis** (*kin-es-the'-se-ah, kin-es-the'-sis*) [κίνησις, αἴσθησις, sensation]. 1. That quality of sensations whereby we become aware of our position in space, our movements, and that gives us our impression of weight and resistance. 2. The morbid impulse that impels one looking from a height to throw himself down.
**kinesthetic** (*kin-es-thet'-ik*) [κίνησις, movement; αἴσθησις, sensation]. Relating to kinesthesia.
**kinesia** (*kin-e'-ze-ah*) [κίνησις, movement]. Movement-cure; systematic use of motion for therapeutic or hygienic purposes. 2. See *kinetia*.
**kinetia** (*ke-e'-she-ah*) [κίνησις, motion]. A term applied to all forms of motion-sickness. It includes such disorders as sea-sickness and car-sickness.
**kinetic** (*kin-et'-ik*) [*kinesis*]. Pertaining to motion; producing motion, as kinetic energy. k. jelly. See under *jelly*. k. system, Crile's term for the brain, thyroid, adrenals and muscles.
**kinetics** (*kin-et'-iks*) [κίνησις, movement]. The science of force as developing motion.
**kinetographic** (*kin-et-o-graf'-ik*) [κίνησις, movement; γράφειν, to write]. Relating to the recording of movements.
**kinetoplasm** (*kin-et'-o-plazm*). See *hyaloplasm*.
**kinetoscope** (*kin-et'-o-skōp*) [κίνησις, movement; σκοπεῖν, to view]. An apparatus for producing stereoptic pictures of objects or beings in motion; a vitascope.
**kinetotherapeutic** (*kin-es-e-o-ther-ap-ū'-tik*) [*kinetotherapy*]. Relating to the therapeutic use of systematic movements and exercises.
**kinetotherapy** (*kin-et-o-ther'-ap-e*) [*kinesis*; θεραπεία, therapy]. The treatment of disease by systematic, active or passive movements.
**king's evil**. Scrofula, on account of a belief that it could be cured by the touch of the king. k.'s yellow. See *arsenic trisulphide*.
**kink** (*kingk*). 1. The whoop in whooping-cough. 2. A synonym of whooping-cough. 3. A flexion or a twist. k.-cough, a synonym of whooping-cough. k.-host, a synonym of whooping-cough. k., Lane's. See *Lane's kink*.
**kino** (*ki'-no*) [E. Ind.]. 1. The inspissated juice of *Pterocarpus marsupium*, found in India, and similar in action to tannic acid; it is used mainly as a constituent of gargles and diarrhea mixtures. 2. A

KINOMETER 490 KNOT

general term for the astringent inspissated juice of a tree, as furnished by many species of *Eucalyptus*. **k., tincture** of (*tinctura kino*, U. S. P.). Dose 10 min.–2 dr. (0.65–8.0 Cc.).
**kinometer** (*kin-om′-et-er*) [κίνησις, movement; μέτρον, measure]. An instrument to measure the amount of displacement of the uterus in case of tumor or cellular inflammation of the pelvis.
**kinone** (*kin′-ōn*). See *quinone*.
**kinoplasm** (*kin′-o-plazm*) [κινεῖν, to move; πλάσμα, a thing molded]. Strasburger's term for the protoplasm peculiar to the centrosome; the archoplasm of Boveri.
**kinotannic acid.** A variety of tannic acid found in kino.
**kinovin** (*kin′-o-vin*). See *quinovin*.
**kionitis** (*ki-on-i′-tis*). Same as *staphylitis*.
**kionorrhaphy** (*ki-on-or′-af-e*). Same as *staphylorrhaphy*.
**kiotome** (*ki′-o-tōm*) [κίων, the uvula; τομός, cutting]. An instrument for amputating the uvula, or for dividing strictures of the bladder or rectum.
**kiotomy** (*ki-ot′-o-me*) [see *kiotome*]. Excision of the uvula.
**Kissingen salts** (*kis′-ing-en*). Effervescing salts from the mineral springs of Kissingen. **K. water,** a laxative tonic mineral water of Kissingen, in Bavaria.
**kite-tail plug.** A tampon used in controlling uterine hemorrhage. It is made by tying rolls of cotton to a string at intervals, the whole resembling a kite-tail.
**Kittel's method** (*kit′-el*) [M. J. Kittell, German physician]. Treatment of gout by massage of the effected joints.
**Kl.** Abbreviation for *kiloliter*.
**klang** [Ger.]. See *timbre*.
**Klatsch-preparation** (*klatsh-prep-ar-a′-shun*) [Ger., *Klatschpräparat*]. A cover-glass preparation made by pressing the cover-glass lightly on a bacterial colony in plate-culture.
**Klebs-Loeffler bacillus** (*klebs′-leff′-ler*) [Edwin Klebs, German bacteriologist, 1834–1913; Friedrich August Johannes Loeffler, German physician, 1852–1915]. The *bacillus diphtheriæ*.
**kleidarthrocace** (*klid-ar-throk′-as-e*) [κλείς, clavicle; ἄρθρον, joint; κάκη, evil]. Spontaneous luxation of the clavicle.
**klemmolin** (*klem′-ol-in*). A proprietary remedy for rheumatism said to be prepared from pine tops and poplar buds.
**kleptomania** (*klep-to-ma′-ne-ah*) [κλέπτειν, to steal; μανία, madness]. A form of emotional insanity manifested by a morbid desire to commit theft.
**kleptophobia** (*klep-to-fo′-be-ah*) [κλέπτειν, to steal; φόβος, fear]. 1. A morbid dread of thieves. 2. A morbid dread of becoming a kleptomaniac.
**klinocephalus.** See *clinocephalus*.
**klinostat** (*kli′-no-stat*). See *clinostat*.
**Klippel's disease** (*klip′-el*) [Maurice Klippel, French physician, 1858– ]. General paralysis occurring in arthritic patients.
**kliseometer** (*klis-e-om′-et-er*). Same as *cliseometer*.
**klopemania** (*klop-e-ma′-ne-ah*) [κλοπή, theft; μανία, madness]. Same as *kleptomania*.
**klopsophobia** (*klop-so-fo′-be-ah*) [κλοψ, thief; φόβος, fear]. Same as *kleptophobia*.
**Klumpke's paralysis** (*kloomp′-keh*) [A. Déjerine Klumpke, French neurologist]. Paralysis and atrophy of the muscles of the forearm and hand, with sensory and oculopupillary disturbances; it is due to a lesion of the seventh and eighth cervical and first dorsal nerve-roots.
**Km.** Abbreviation for *kilometer*.
**Knapp's angioid streaks** (*nap*) [Hermann Knapp, American ophthalmologist, 1832–1911]. Pigment streaks appearing occasionally in the retina after hemorrhage. **K.'s forceps.** A forceps with roller blades used in the treatment of trachoma on the palpebral conjunctiva. **K.'s operation.** For *cataract-extraction:* a broad iridectomy and peripheral opening of the capsule; the lens is expelled by gentle pressure on the lower part of the cornea.
**Knapp's test for glucose in the urine.** A solution is made of 10 Gm. of mercuric cyanid dissolved in 100 Cc. of caustic soda solution, of a specific gravity of 1.145, and diluted to one liter. When this solution is diluted with water and heated with a glucose solution, a reduction of metallic mercury takes place. Ten Cc. of this solution are reduced by 0.025 Gm. of glucose.

**kneading** (*ne′-ding*). The same as *petrissage, q. v.* See also *malaxation*.
**knee** (*nē*) [AS., *cneōw*]. The articulation between the femur and the tibia. **k.-cap,** the patella. **k., housemaid's.** See *housemaid's knee*. **k., in-.** See *genu valgum*. **k. of internal capsule,** the angle of junction of the anterior and posterior limbs of the internal capsule. **k.-jerk, k.-reflex, k.-phenomenon,** a contraction of the quadriceps extensor femoris muscle as a result of a light blow on the patellar tendon. Syn., *patellar tendon-reflex*. See also under *reflex*. **k.-joint,** a hinge-joint consisting of the articulation of the condyles of the femur with the upper extremity of the tibia and the posterior surface of the patella. **k., knock-.** See *genu valgum*. **k., out-.** See *genu varum*. **k.-pan,** the patella.
**Kneippism** (*ni′-pizm*) [E. H. Sebastian Kneipp, German priest, 1821–1897]. Hydrotherapy applied in a great variety of ways—baths, lotions, wet compresses, packs, cold affusions, and walking barefooted in the morning dew.
**knife** (*nif*) [AS., *cnif*]. An instrument for cutting. In surgery, knives are of various shapes and sizes, according to their use. **k.-needle,** a needle with a cutting edge, used in the discission of cataracts. **k.-rest crystals,** peculiarly indented crystals of triple phosphate occasionally found in urine.
**knitting** (*nit′-ing*). The union and becoming rigid of a fracture.
**knock-knee** (*nok′-ne*). See *genu valgum*.
**knock-out-drops.** A strong aqueous solution of chloral used by criminals to deprive their victims of consciousness.
**knot** (*not*) [ME., *knotte*]. An interlacement of ends or parts of one or more cords or threads so that they cannot be readily separated. **k., clove-hitch,** a knot consisting of two single, contiguous loops, the free ends toward each other. **k., double.** Same as *k., friction*. **k., false.** Same as *k., granny*. **k., friction,** one in which the ends are wound twice around each other before they are tied. **k., Gerdy's**

GRANNY, FALSE, OR DOUBLE KNOT.   REEF OR SAILOR'S KNOT.

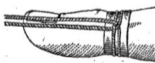

CLOVE-HITCH KNOT.   SURGICAL KNOT.

STAFFORDSHIRE OR TAIT'S KNOT.   COMBINED SURGEON'S OR REEF KNOT.

extension, resembles the clove-hitch. **k., granny,** a tie of a cord in which in the second loop the end of one cord is over, and the other under, its fellow, so that the two loops do not lie in the same line. **k., reef,** a knot so formed that the ends come out alongside of the standing parts and the knot does not jam. **k., sailor's.** Same as **k., reef. k., square.** Same as **k., reef. k., Staffordshire,** a knot used in ligating the pedicle in ovariotomy. The ligature is passed through the pedicle, and withdrawn so as to leave a loop, which is passed over the tumor, and one of the free ends is then drawn through the loop; both ends are then passed through the pedicle, tightened, and tied. k., stay, formed by two or more ligatures in the following way: on each ligature separately is made the first hitch of a reef knot, which is tightened so that the loop lies in contact with the vessel, without constricting it; then taking the two ends on one side together in one hand, and the two ends on the other side in the other hand, the vessel is constricted sufficiently to occlude it, after which the reef knot is completed. k., surgical, a double knot made by passing the thread twice through the same loop. k., Tait's. See k., *Staffordshire*.

knuckle (*nuk'-l*).  1. An articulation of the phalanges with the metacarpal bones or with each other. 2. A loop of intestine.

Kobelt's cyst (*ko'-belt*) [Georg Ludwig Kobelt, German physician, 1804–1857]. A small pedunculated cyst formed in one of Kobelt's tubes. K.'s tubes, the upper ducts of the Wolffian body which end in a culdesac.

Kobert's test for hemoglobin (*ko'-bert*) [Eduard Rudolf Kobert, German chemist, 1854– ]. Treat the solution with one of zinc sulphate or shake it with zinc powder, when a precipitate of zinc hemoglobin is formed. Alkalies color this precipitate red.

KOC. Abbreviation for *cathodal opening contraction*.

Koch's bacillus (*kōk*) [Robert Koch, German bacteriologist, 1843–1910]. 1. The *Bacillus tuberculosis*. 2. The *Spirillum cholerae asiaticae*. K.'s eruption, a morbilliform eruption following the injection of tuberculin. K.'s law, K.'s postulates, the specificity of a microorganism is conclusively demonstrated when the following conditions are fulfilled: (1) The microorganism must be present in all cases of the disease; (2) it must be cultivated in pure culture; (3) its inoculation must produce the disease in susceptible animals; (4) from such animals it must be obtained and again cultivated in pure culture. K.'s lymph. See *tuberculin*. K.'s method of sterilization, a method of interrupted heating. The culture-mediums are heated for a short time daily for from three to five successive days, usually in the steam sterilizer.

Koch-Weeks bacillus [see Koch; John Elmer Weeks, American ophthalmologist, 1853– ]. A bacillus which causes pink-eye or acute contagious conjunctivitis.

Kocher's operation (*kok'-er*) [Theodor Kocher, Swiss surgeon, 1841– ]. 1. *For excision of the ankle-joint:* the incision is made beneath the external malleolus, and is followed by division of the peroneal tendons after being secured with threads, and opening of the joint with removal of the diseased parts; the foot is replaced and the tendons sutured. 2. *For the relief of cancer of the rectum:* a long integumentary incision is made, freely exposing the bone. With a chisel each side of the sacrum is grooved along the inner side of the foramina, beginning at the third. The segment, which he calls the "*knockenspange*," is removed. This exposes the sacral canal and makes certain the protection of the nerves.

kodozonol (*kod-o-zo'-nol*). Ozonized cod-liver oil, an antiseptic dressing for wounds, burns, etc.

Koebner's multiple multiple tumors. See *mycosis fungoides*.

Koehler's disease (*ke'-ler*) [Albert Koehler, German surgeon, 1850– ]. Softening of the scaphoid bone of the foot resulting from traumatism.

von Koelliker's cells (*kel'-ik-er*) [Rudolf Albert von Koelliker, German anatomist, 1817–1905]. 1. Little cells of the seminiferous tubules which are transformed into spermatozoa. 2. See *myeloplax*. v. K.'s fibrous layer, the layer of fibrous connective tissue which forms the substantia propria of the iris. v. K.'s glands. See *Bowman's glands*. v. K.'s musclebuds. See *Kuehne's muscle-spindles*. v. K.'s nucleus, the gray matter surrounding the canal of the spinal cord. v. K.'s reticulum, the neuroglia. v. K.'s tract-cells, ganglion-cells, the axons of which pass as longitudinal fibers into the white columns of the spinal cord.

Koenig's symptom (*ker'-nig*) [Franz Koenig, German surgeon, 1832– ]. Blue-blindness in granular kidney. K.'s symptom-complex, alternation, for a long period of constipation and diarrhea, and irregular attacks of colic, which are generally of short duration and terminate suddenly. During these attacks the abdomen is distended, there exists frequently a visible peristalsis, and a loud gurgling is heard in the ileocecal region. These symptoms are characteristic of tuberculous stenosis of the cecum.

Koerte-Ballance operation (*ker'-teh bal'-ans*) [Werner Koerte, German surgeon, 1853– ; Charles Alfred Ballance, English surgeon]. 1. Anastomosis of the facial and hypoglossal nerves for the relief of facial palsy. 2. Anastomosis of the facial and spinal accessory nerves.

Kohlrausch's fold (*kōl'-rowsh*) [Otto Ludwig Bernhard Kohlrausch, German physician, 1811–1854]. A semilunar, transverse fold of the rectal mucosa, situated about 6 cm. above the anus in the anterior and right wall of the rectum. K.'s veins, the superficial veins that pass from the surface of the penis upward to empty into the dorsal vein.

koilonychia (*koi-lo-nik'-e-ah*) [κοῖλος, hollow; ὄνυξ, nail]. A condition in which the outer surface of the nail is concave; spoon-nail.

kola (*ko'-lah*). See *k.-nut*. k.-cardinette, a proprietary cordial containing from 30 to 60 gr. of kola-nut to each fluidounce. A nerve-tonic and stimulant. Dose 1–4 tablespoonfuls (16–60 Cc.) 5 or 6 times daily. k.-nut, the seed of *Cola acuminata*, used in Central Africa as a substitute for tea and coffee. It contains an alkaloid similar to caffeine and is a cerebral stimulant and cardiac tonic. k.-tannin, a compound of caffeine and tannin obtained from kola-nut.

kolabon (*ko'-lah-bon*) [*kola*; *bonbon*]. A confection prepared from undried kola-nut, containing kolanin, caffeine, and theobromine. It is recommended in treatment of sea-sickness.

kolanin (*ko'-lan-in*). The physiologically active glucoside from kola-nut; a thick extract, containing 80 to 90 % of the pure glucoside, is used in the treatment of neurasthenia and neurasthenic weakness of the heart.

Kolk's (Schroeder van der) law. A spinal nerve endows the muscles with motion through its motor branches and the parts moved with sensation through its sensory branches.

kollonema (*kol-on-e'-mah*) [κόλλα, glue; νῆμα, tissue]. Same as *myxoma*, *q. v.*

kolopexy. See *colopexy*.

koloptyphus (*ko-lo-ti'-fus*). Typhoid fever.

kolp-. For words beginning thus, see *colp-*.

kolpo- (*kol-po-*). See *colpo-*.

kolypeptic (*ko-le-pep'-tik*) [κωλύν, to hinder; πεπτικός, conductive to digestion]. Hindering or checking digestive processes.

kolyseptic (*ko-le-sep'-tik*) [κωλύν, to hinder; σῆπειν, to putrefy]. 1. Preventing putrefaction. 2. An agent that hinders a septic process.

kombe (*komb'-ba*) [African]. An African arrow-poison (*kombé inée*) extracted from *Strophanthus kombé*.

Kondoleon's operation (*kon-do'-le-on*) [Emmanuel Kondoleon, Greek surgeon]. Excision of pieces of connective tissue for the relief of elephantiasis.

koniantron (*kon-e-an'-tron*) [κόνις, dust; ἄντρον, antrum]. An instrument for spraying fluid into the tympanic cavity.

koniosis (*kon-e-o'-sis*) [κόνις, dust]. A morbid condition due to inhalation of dust.

koniscope (*kon'-is-kōp*) [κόνις, dust; σκοπεῖν, to examine]. An instrument for determining the quantity of dust in the atmosphere.

konseal (*kon'-sēl*). A form of cachet.

koosso, koosoo (*koo'-soo*). See *cusso*.

kopftetanus (*kopf'-tet-an-us*) [Ger.]. Cephalic tetanus. See *tetanus*, cephalic.

kophemia (*ko-fe'-me-ah*) [κωφᾶν, to deafen]. See *deafness*, *word*.

kopiopia (*kop-e-o'-pe-ah*) [κόπος, a straining; ὤψ, eye]. Eye-strain; weariness of the eyes. k. hysterica, a term applied to those symptoms that indicate hyperesthesia of the trigeminus and optic nerves.

Koplik's sign, K.'s spots (*kop'-lik*) [Henry Koplik, American physician, 1858– ]. Minute bluish-white spots surrounded by a reddish areola; they are seen on the mucous membrane of the cheeks and lips of the patient during the prodromal stage of measles.

Kopp's asthma (Johann Heinrich Kopp, German physician, 1777–1858]. Laryngismus stridulus. Syn., *Kopp's thymic asthma*.

kopremia, kopræmia. See *copremia*.

**koprikin** (*kop'-rik-in*) [κόπρος, dung]. Undigested animal matter in the feces.

**koprostearin** (*kop-ro-ste'-ar-in*). A modified cholesterol found in the feces.

**Koranyi's auscultation** (*ko-rahn'-ye*) [Baron F. von Koranyi, Austrian physician, 1829– ]. Auscultation with percussion upon the second joint of the forefinger applied perpendicularly to the part. **K.'s sign**. See *Grocco's sign*.

**kore-** (*kor'-e-*). See *core-*.

**koronion** (*ko-ro'-ne-on*). The apex of the coronoid process of the inferior maxilla.

**koroscopy** (*kor-os'-ko-pe*). See *retinoscopy*.

**Korsakoff's psychosis** (*kor'-sak-of*) [Sergius Korsakoff, Russian neurologist, 1853–1900]. Mental derangement, in the form of delirium, observed in cases of polyneuritis.

**kosher** (*ko'-shur*) [Heb., lawful]. Pure, lawful. **k.-meat**, the flesh of animals that have been slaughtered and inspected according to the laws of the Jewish rabbis.

**kosin** (*ko'-sin*) [Abyssinian, *cusso*], $C_{11}H_{36}O_{10}$. Same as *koussin*. See *cusso*.

**kosotoxin** (*ko-so-toks'-in*) [*koso*, the fertile flowers of *Brayera anthelmintica*; τοξικόν, poison], $C_{26}H_{34}O_{10}$. An active principle from the ethereal extract of cusso flowers. It is a strong muscle poison, but exerts little influence on the central nervous system.

**Kossel's test for hypoxanthin**. Treat the solution with hydrochloric acid and zinc and add an excess of sodium hydroxide. The presence of hypoxanthin is evinced by a ruby-red color.

**koumiss** (*koo'-mis*). See *kumiss*.

**koussin** (*koos'-in*). See under *cusso*.

**kousso** (*koos'-o*). See *cusso*.

**Kovalevski, canal of** (*kof-a-lef'-ske*) [Pavel Ivanovich *Kovalevski*, Russian embryologist, 1845– ]. The neurenteric canal; in the embryo, a passage leading from the posterior part of the medullary tube into the archenteron.

**Kowarski's test for sugar in the urine** (*ko-var'-ske*). Shake in a test-tube 5 drops of phenylhydrazin with 10 drops of strong acetic acid, add 1 Cc. of saturated sodium chloride solution and 2 Cc. or 3 Cc. of urine, and heat for 2 minutes; then cool slowly. If the amount of sugar present is as high as 0.2 %, characteristic crystals will form in a few minutes; if less, the formation of crystals will require a longer time—5 to 30 minutes.

**Koyter's muscle** (*koi'-ter*) [Volcherus *Koyter*, Dutch anatomist, 1534–1600]. The corrugator supercilii. Syn., *musculus Coiteri*.

**Kr**. Chemical symbol of *krypton*.

**Kramer's frontal band**. A head-band with appliances to hold a Eustachian catheter in place so that the surgeon's hands may be free.

**Krameria** (*kra-me'-re-ah*) [J. G. H. *Kramer*, Austrian physician]. A genus of polypetalous herbs. The *krameria* of the U. S. P. is the dried root of *K. triandra* (ratany-root) and *K. ixina*, shrubs native to South America; it possesses the same astringent qualities as tannic acid. It is used in serous diarrheas. Dose 15 gr. (1 Gm.). **k.**, **extract of** (*extractum krameriæ*, U. S. P.). Dose 5–10 gr. (0.32–0.65 Gm.). **k.**, **fluidextract of** (*fluidextractum krameriæ*, U. S. P.). Dose 5 min.–⅓ dr. (0.32–2.0 Cc.). **k.**, **infusion of** (*infusum krameriæ*, B. P.). Dose 1–2 oz. (32–64 Cc.). **k.**, **syrup of** (*syrupus krameriæ*, U. S. P.), contains of the fluidextract, 35; syrup, 65. Dose ½ dr.–¼ oz. (2–16 Cc.). **k.**, **tincture of** (*tinctura krameriæ*, U. S. P.). Dose 5 min.–1 dr. (0.32–4.0 Cc.). **k.**, **troches of** (*trochisci krameriæ*, U. S. P.), each troche contains 1 gr. (0.065 Gm.) of the extract.

**Kraske's operation** (*kras'-keh*) [Paul *Kraske*, German surgeon, 1851– ]. Removal of the coccyx and left part of the sacrum prior to resection of the rectum in case of malignant disease.

**kraurosis** (*kraw-ro'-sis*) [κραῦρος, dry]. Shriveling and dryness, especially of the vulva.

**Krause's end-bulbs** (*krow'-zer*) [1. Wilhelm *Krause*, German anatomist, 1833– ; 2. Fedor *Krause*, German surgeon, 1857– ]. [1]. Spheroid nerve-corpuscles resembling Pacinian corpuscles, but having a more delicate investment. They are found especially in the conjunctiva and the genitals. **K.'s disc**, **K.'s membrane**. [1]. The dark transverse band that divides a transparent (isotropic) sarcous element in the middle. **K.'s glands**. [1]. Acinous glands found in the conjunctiva near the fornix, especially of the upper lid. **K.'s muscle**.

[1]. The coracocervicalis muscle. **K.'s nerve**. [1]. The ulnar collateral branch of the musculospiral nerve that descends along with the ulnar nerve and enters the lower short fibers of the inner head of the triceps. **K.'s respiratory tract**. [2]. The solitary fascicle of the oblongata. **K.'s valve**. [1]. See *Béraud's valve*. **K.'s ventricle**. [1]. The terminal ventricle of the spinal cord.

**kreatin** (*kre'-at-in*). See *creatin*.

**kreatinin** (*kre-at'-in-in*). See *creatinin*.

**krelos** (*kre'-los*). Trade name of a disinfectant said to be a solution of cresols and rosin soap.

**kreolin** (*kre'-o-lin*). See *creolin*.

**kreosolid** (*kre-o-sol'-id*). See *creosote-magnesia*.

**kresamine** (*kres'-am-ēn*). See *ethylenediamine-tricresol*.

**kreotoxicon** (*kre-o-toks'-ik-on*) [κρέας, meat; τοξικόν, poison]. A general term for the active agent in poisonous meat.

**kreotoxin** (*kre-o-toks'-in*). Any basic poison generated in meat by bacteria.

**kreotoxism** (*kre-o-toks'-izm*) [see *kreotoxicon*]. Poisoning by infected meat.

**kreozonal** (*kre-o-zo'-nal*). Ozonized oil of tar for external use in skin diseases.

**kreplinum** (*krep'-li-num*). A proprietary cosmetic said to be quillaya bark in dilute (25 per cent.) alcohol to which is added oil of rosemary, lavender, or other perfuming oils.

**kresaprol** (*kres'-ap-rol*). See *cresin*.

**kresin** (*kre'-sin*). See *cresin*.

**kresoform** (*kres'-o-form*). A condensation-product of formaldehyde and creosote.

**kresofuchsin** (*kres-o-fook'-sin*). An amorphous powder of gray-blue color. It is soluble in acetic acid and acetone, less readily but quite soluble in alcohol, only very slightly so in water. The alcoholic solution appears blue, the aqueous red. It is used as a histological stain.

**Kretschmann's space** (*kretsh'-man*). A small pocket in the attic of the middle ear situated below Prussak's space.

**Kreysig's sign** (*kri'-zig*) [Friedrich Ludwig *Kreysig*, German physician, 1770–1839]. Retraction of the epigastrium and the contiguous portion of the false ribs with each systole, in adherent pericardium.

**Krishaber's disease** (*krēs-hah-bair'*) [Maurice *Krishaber*, French physician, 1836–1883]. A neurosis resembling neurasthenia, and characterized by a rapid onset, predominant cerebral symptoms (insomnia, vertigo, etc.), neuralgia, and circulatory disturbances. Syn., *neuropathie cérébrocardiaque*.

**Krisowski's sign** (*kre-sof'-ske*). Radiating lines about the mouth, and the union of the hard palate with the posterior pharyngeal wall by fibrous tissue; found in congenital syphilis.

**kristallin**. See *cristallin*.

**Kroenlein's hernia** (*kren'-līn*) [Rudolf Ulrich *Kroenlein*, Swiss surgeon, 1847–1910]. Properitoneal inguinal hernia.

**Kronecker's inhibitory center** (*krōn'-ek-er*) [Hugo *Kronecker*, Swiss physiologist, 1839–1914]. A point in the interventricular septum, puncture of which causes incoordinate fibrillary contractions of the heart. **K.'s solution**, a 5 per cent. solution of sodium chloride with sodium carbonate used in the microscopical examination of fresh tissues.

**kronethyl** (*kron-eth'-il*). An ethereal extract of Chinese cantharides. Applied in gout and neuralgia, 6–10 drops on a wet bandage.

**krymotherapy**. See *crymotherapy*.

**kryofine** (*kri'-o-fen*). Phenetidin methylglycolate, a condensation-product of paraphenetidin and methylglycolic acid. It is antipyretic and antineuralgic. Dose 5–8 gr. (0.3–0.5 Gm.).

**kryoscopy**. See *cryoscopy*.

**kryptidin** (*kript'-id-in*) [κρυπτόν, concealed], $C_{12}H_{13}N$. A base from coal-tar. Syn., *cryptidin*.

**krypto-** (*krip'-to-*). See *crypto-*.

**krypton** (*krip'-ton*) [κρυπτός, hidden]. A gaseous element found by Ramsay in liquefied air. See *elements, table of*.

**krystallose** (*kris'-tal-ōs*). Sodium saccharinate.

**kubisagari**, **kubisagri** (*koo-bis-ah-gah'-re*, *koo-bis-gah'-re*). An endemic paralytic vertigo which prevails in Japan.

**Kuehne's muscle-spindles** (*ke'-ner*) [Willy *Kuhne*, German histologist, 1837–1900]. Peculiar, fusiform enlargements occurring at the entrance of certain nerves into a muscle-bundle. K.'s muscular phenomenon. See *Porret's phenomenon*.

**Kuelz's casts.** Very short, generally hyaline, but sometimes granular, casts, occurring at the onset and during the course of diabetic coma, disappearing with the coma, and considered by Kulz as diagnostic of impending coma.

**Kuemmell's disease** (*kim'-el*) [Hermann *Kuemmell*, German surgeon, 1852– ]. Traumatic spondylitis. **K.'s kyphosis.** Kyphosis resulting from trauma, but in which the symptoms are delayed.

**Kuester's sign** (*kēs'-ter*) [Otto Ernst *Kuester*, German gynecologist, 1850– ]. The presence of a cystic tumor in the median line anterior to the uterus, disclosed by palpation and inspection; usually indicates ovarian dermoids.

**Kuestner's law.** Torsion of the pedicle of an ovarian tumor takes place toward the right if the tumor is left-sided, and toward the left if it is right-sided.

**Kuhn's tube** (*kūn*). A flexible tube containing a spiral of steel wire, which moves freely within the metal elastic tube, and terminates in a knob. It is used as a duodenal or intubation tube.

**Kuhnt's spaces** (*koont'*) [Hermann *Kuhnt*, German ophthalmologist, 1850– ]. The recesses of the posterior chamber; a series of radial spaces which communicate anteriorly with the posterior chamber of the eye and contain aqueous humor.

**kumbecephalic, kumbokephalic** (*kum-be-sef-al'-ik, kum-bo-kef-al'-ik*). See *cymbocephalic*.

**kumiss, kumyss** (*koo'-mis*) [Tartar, *kumiz*, fermented mares' milk]. An alcoholic drink originally made by the fermentation of mares' milk. At present cows' milk is used in making it.

**von Kupffer's cells** (*koop'-fer*) [Karl Wilhelm von *Kupffer*, German anatomist, 1829– ]. Stellate endothelial cells of the liver-capillaries having a large round or ovoid nucleus and frequently containing pigmentary matter.

**Kurloff's bodies** (*koor'-lof*) [Mikhail Georgiyevitch *Kurloff*, Russian physician]. Nucleoids or inclusion bodies of unknown significance, found in the large mononuclear leukocytes of the guinea pig.

**Kussmaul's aphasia** (*koos'-mowl*) [Adolf *Kussmaul*, German physician, 1822–1902]. Voluntary mutism, simulating aphasia, which sometimes affects the insane, particularly paranoiacs, with mystic ideas. **K.'s coma,** diabetic coma. **K.'s disease,** acute anterior poliomyelitis; acute atrophic spinal paralysis. **K.'s paradoxic pulse,** a pulse which becomes weaker and disappears during deep inspiration; it is observed in cases of adherent pericardium and mediastinal adhesions or tumor. **K.'s respiration,** the deep, labored respiration of diabetic coma. **K.'s symptom,** swelling of the cervical veins during inspiration in adherent pericardium and mediastinal tumor.

**Kussmaul-Landry's paralysis.** See *Landry's paralysis*.

**kusso** (*koo'-so*). See *cusso*.

**kuttarasome** (*kut-ar'-as-ōm*) [κύτταρος, any hollow cavity; σῶμα, body]. A body found by Ira van Gieson at the neck of the cone of the retina, composed of a series of parallel bars presenting a gridiron appearance. The bars had lateral anastomoses and at the top joined in a semicircular manner.

**kyanopsia** (*ki-an-op'-se-ah*). See *cyanopsia*.

**kyestein** (*ki-es'-te-in*) [κυεῖν, to be pregnant; ἐσθής, a garment]. A filmy deposit upon decomposing urine, once thought to be diagnostic of pregnancy.

**kyllopodia** (*kil-o-po'-de-ah*) [κυλλός, twisted; πούς, foot]. See *club-foot*.

**kyllosis** (*kil-o'-sis*) [κυλλός, twisted]. Same as club-foot.

**kymograph, kymographion** (*ki'-mo-graf, ki-mo-graf'-e-on*) [κύμα, a wave; γράφειν, to write]. An instrument for reproducing graphically the variations in blood-pressure.

**kymoscope** (*ki'-mo-skōp*) [κύμα, a wave; σκοπεῖν, to inspect]. A device used in the observation and study of the blood-current.

**kynocephalous** (*ki-no-sef'-a-lus*). See *cynocephalous*.

**kynophobia** (*ki-no-fo'-be-ah*). See *cynophobia*.

**kynurin** (*ki-nū'-rin*). See *cynurin*.

**kyphoscoliorachitic.** (*ki-fo-sko-le-o-rak-it'-ik*). Kyphoscoliotic.

**kyphoscoliosis** (*ki-fo-sko-le-o'-sis*) [*kyphosis; scoliosis*]. Kyphosis combined with scoliosis.

**kyphoscoliotic** (*ki-fo-sko-le-ot'-ik*) [see *kyphoscoliosis*]. Characterized by or pertaining to kyphoscoliosis.

**kyphosis** (*ki-fo'-sis*) [κύφωσις, humpbacked]. Humpback. Angular curvature of the spine, the prominence or convexity turned dorsad.

**kyphotic** (*ki-fot'-ik*) [*kyphosis*]. Relating to, of the nature of, or affected with kyphosis.

**kyphotone** (*ki'-fo-tōn*) [κυφός, a crookedness; τόνος, a brace]. An apparatus for the forcible reduction of deformity in Pott's disease.

**kyrtometric** (*kir-to-met'-rik*) [κυρτός, curved; μέτρον, a measure]. Relating to the measurements of the body-curves. Syn., *cyrtometric*.

**kysthitis** (*kis-thi'-tis*) [κύσθος, a hollow; ιτις, inflammation]. Vaginitis.

**kysthoproptosis** (*kis-tho-prop-to'-sis*) [κύσθος, a hollow; πρόπτωσις, a falling forward]. Prolapse of the vagina.

**kysthoptosis** (*kis-thop-to'-sis*). Preferred term for *kysthoproptosis*.

**kysto-** (*kis'-to-*). See *cysto-*.

**kystoma** (*kis-to'-mah*). See *cystoma*.

**kythemolytic** (*ki-them-o-lit'-ik*) [κύτος, cell; αἷμα, blood; λύσις, solution]. Pertaining to, characterized by, or causing, the destruction of blood-cells.

**kytomitome** (*ki-tom'-it-ōm*) [κύτος, cell; μίτος, thread]. The network in the body of the nucleus of the cell.

**kytoplasm** (*ki'-to-plazm*). See *cytoplasm*.

# L

**L.** Abbreviation for *Latin*, for *limes*, a boundary or threshhold (see $L_+$, and $L_0$).
**l.** Abbreviation of *left*, of *libra*, a pound, of *liter*, of *lethal*.
**L+.** Ehrlich's symbol for the quantity of toxic bouillon which is completely neutralized by one antitoxin unit.
**L₀.** Ehrlich's symbol for the minimum quantity of toxic bouillon which will kill an experimental animal.
**La.** Chemical symbol of *lanthanum*.
**lab, lab-ferment.** The ferment of rennet, producing coagulation of milk.
**Labarraque's solution** (*lab-ar-ak'*) [Antoine Germain *Labarraque*, French chemist, 1777–1850]. A solution of sodium carbonate, 10; chloride of lime, 8; water, 100; it is a disinfectant. Liquor sodæ chlorinatæ.
**Labbé's vein** (*lab-a'*) [Léon *Labbé*, French surgeon, 1832– ]. The anastomotic vein that extends from the lateral sinus to Trolard's vein or to the superior longitudinal sinus.
**labdacism** (*lab'-das-izm*). Same as *lambdacism*.
**labdanum** (*lab'-dan-um*). See *ladanum*.
**labia** (*la'-be-ah*) [L.; plural of *labium*, lip]. The lips. l. majora, two cutaneous folds from the mons Veneris to the perineum. l. minora, the *nymphae*, q. v.
**labial** (*la'-be-al*) [*labium*, a lip]. Pertaining to the lips or to a labium.
**labialism** (*la'-be-al-izm*) [*labium*, a lip]. The tendency to pronounce any articulate sounds as if they were labials; the addition of a labial or labiodental quality to an articulate sound.
**labidometer** (*lab-id-om'-et-er*) [λαβίς, forceps; μέτρον, a measure]. A forceps for measuring the fetal head in the pelvis.
**labile** (*lab'-il*) [*labi*, to glide]. 1. Gliding to and fro: applied to an electric current when the electrode is moved from place to place over the skin. 2. In chemistry, unstable. l. elements of the body, epithelial and connective tissue cells, in contradistinction to muscle and nerve cells.
**lability** (*la-bil'-i-te*) [*labile*]. 1. In electrotherapeutics, the quality of being labile. 2. Instability.
**labimeter** (*lab-im'-et-er*). Same as *labidometer*.
**labio-** (*la-be-o-*) [*labium*, lip]. A prefix meaning pertaining to the lip.
**labioalveolar** (*la-be-o-al-ve'-o-lar*) [*labio-*; *alveolus*, a small hollow]. Pertaining to the lip and to one or more dental alveoli.
**labiochorea** (*la-be-o-ko-re'-ah*) [*labio-*; χορεία, dancing]. A choreic affection of the lips and the stammering that results from it.
**labiodental** (*la-be-o-den'-tal*) [*labio-*; *dens*, a tooth]. Pertaining to the lips and the teeth.
**labioglossolaryngeal** (*la-be-o-glos-o-lar-in'-je-al*) [*labio-*; γλῶσσα, tongue; *larynx*]. Pertaining conjointly to lips, tongue, and larynx. l. paralysis. See *paralysis, bulbar*.
**labioglossopharyngeal** (*la-be-o-glos-o-far-in'-je-al*) [*labio-*; γλῶσσα, tongue; *pharynx*]. Pertaining conjointly to lips, tongue, and pharynx.
**labiograph** (*la'-be-o-graf*) [*labio-*; γράφειν, to write]. An instrument for recording the labial movements in speaking.
**labiomancy** (*la-be-o-man'-se*) [*labio-*; μαντεία, divination]. The faculty of understanding what is said by observing the motions of the lips in speech.
**labiomental** (*la-be-o-men'-tal*) [*labio-*; *mentum*, chin]. Relating to lip and chin.
**labiomycosis** (*la-be-o-mi-ko'-sis*) [*labio-*; *mycosis*]. Any affection of the lips due to fungal origin.
**labionasal** (*la-be-o-na'-sal*) [*labio-*; *nasus*, nose]. Labial and nasal; pertaining to lip and nose.
**labiopalatine** (*la-be-o-pal'-at-in*) [*labio-*; *palatum*, palate]. Relating to lip and palate in common.
**labioplastis** (*la-be-o-plas'-tik*) [*labio-*; πλάσσειν, to form]. Pertaining to an operation for restoring the lip, after injury or partial destruction of the same; cheiloplastic.
**labioplasty** (*la-be-o-plas'-te*) [*labio-*; πλάσσειν, to form]. Plastic surgery of the lips. Cheiloplasty.
**labiotenaculum** (*la-be-o-ten-ak'-u-lum*) [*labio-*; *tenaculum*, a holder]. An instrument for holding the lips or labia in a position required for examination or operation.
**labitome** (*lab'-it-ōm*) [λαβίς, forceps; τομή, cutting]. Cutting-forceps.
**labium** (*la'-be-um*) [L.; pl., *labia*]. A lip. l. cerebri, the margin of the cerebral hemisphere which overlaps the callosum. l. majus, l. pudendi majus, one of two folds of skin of the female external genital organs, arising just below the mons Veneris, surrounding the vulval entrance, and meeting at the anterior part of the perineum. l. minus, l. pudendi minus, the nympha; one of two folds of mucous membrane at the inner surfaces of the labia majora. l. tympanicum, the portion of the lamina spiralis forming the lower border of the sulcus spiralis. labia urethræ, the lateral margins of the external urinary meatus. l. vestibulare, the overhanging extremity of the lamina spiralis that forms the upper part of the sulcus spiralis.
**lablab** (*lab'-lab*). The genus *Dolichos*. l. seeds, the seeds of *Dolichos lablab*; used as food and also as a stomachic and antiperiodic.
**labor** (*la'-bor*) [L., "work"]. Parturition; the bringing forth of young. l., artificial, that effected or aided by other means than the forces of the maternal organism. l., dry, that in which there is a deficiency of the liquor amnii, or in which there has been a premature rupture of the bag of waters. l., induced, labor brought on by artificial means. l., instrumental, one requiring instrumental means to extract the child. l., mechanism of, the mechanism by which a fetus and its appendages traverse the birth-canal and are expelled. l., missed, retention of the dead fetus in the uterus beyond the period of normal gestation. l.-pains, the pains consequent upon the contractions of the uterus during labor. l., postponed, delayed beyond nine months. l., precipitate, labor in which the expulsion of the fetus and its appendages takes place with undue celerity. l., premature, labor taking place before the normal period of gestation, but when the fetus is viable. l., protracted, labor prolonged beyond the usual limit (10–20 hours in primiparæ, 2–6 hours in multiparæ). l., spontaneous, that requiring no artificial aid. l., stages of, arbitrary divisions of the period of labor—the first begins with dilatation of the os and ends with complete dilatation; the second ends with the expulsion of the child; the third (*placental*) consists in the expulsion of the placenta.
**laboratory** (*lab'-or-a-to-re*) [*laborare*, to work]. A room or place designed for experimental scientific work.
**Laborde's method of artificial respiration** (*lab-ord'*) [Jean Baptiste Vincent *Laborde*, French physician, 1830–1903]. The tongue is seized either by forceps or with thumb and finger, and rhythmical tractions are made in order to stimulate the respiratory center.
**labordin** (*la-bor'-din*). See *analgen*.
**labrum** (*la'-brum*) [L., a lip-edge, margin; pl., *labra*]. In biology, a lip-like structure, usually the upper, when two are present, the lower being the *labium*. l. cartilagineum, l. glenoideum, l. glenoidale, the cartilaginous edge or border of a cavity forming part of an articulation; as the fibrous ring bordering the glenoid cavity of the scapula.
**labyrinth** (*lab'-ir-inth*) [λαβύρινθος, a maze]. 1. A name given to the series of cavities of the internal ear, comprising the vestibule, cochlea, and the semicircular canals. 2. The parts of the cortex of the kidney between the medullary rays. See *kidney*. l., bony. See *l., osseous*. l., cortical, the tortuous

tubules and blood-vessels in the intervals of the cortex of the kidney. 1., ethmoid, 1., olfactory, the lateral portions of the ethmoid bone. 1., membranous, the membranous cavity within the osseous labyrinth, from which it is partly separated by the perilymph. l., osseous, the bony portion of the internal ear.
labyrinthal, labyrinthic, labyrinthine (*lab-er-in'-thal, lab-er-in'-thik, lab-er-in'-thin*) [*labyrinth*]. Pertaining to a labyrinth. l. vertigo. See *Ménière's disease*.
labyrinthitis (*lab-ir-in-thi'-tis*) [*labyrinth*; ιτις, inflammation]. Inflammation of the labyrinth; otitis interna. l., primary. See *Voltolini's disease*.
labyrinthus (*lab-ir-in'-thus*) [L.: *pl., labyrinthi*]. A labyrinth.
lac (*lak*) [L.]. Milk. l. sulphuris, milk of sulphur, or precipitated sulphur.
laccase (*laccol*; suffix *ase*]. An oxidizing ferment or diastase present in many plants, and capable of oxidizing laccol and other aromatic substances. It changes the colorless sap of the Japanese lac-tree by oxidation to black Japanese lacquer.
lacerable (*las'-ur-a-bl*) [*lacerare*, to tear]. Liable to become torn; capable of being torn.
lacerated (*las'-er-a-ted*) [L., *lacerare*, to tear]. Torn.
laceration (*las-er-a'-shun*) [*lacerare*, to tear]. A tear. l. of perineum, a tearing through the wall separating the lower extremity of the vagina and rectum, occurring occasionally during child-birth. l., surgical, an instrument used in effecting surgical laceration.
lacerator (*las'-er-a-tor*) [*lacerare*, to tear]. An instrument used in effecting surgical laceration.
lacerti, lacertuli cordis. See *columnæ carneæ*.
lacertus (*las-er'-tus*) [L.]. 1. The muscular part of the arm from the shoulder to the elbow. 2. A muscle or fibrous fascicle. l. fibrosus, an aponeurotic band from the biceps-tendon to the fascia of the forearm; semilunar or bicipital fascia.
lachesis (*lak'-e-sis*) [λάχεσις, destiny, fate]. 1. A genus of venomous South American reptiles. 2. The venom of L. mutus (the bushmaster snake]), and also a homeopathic preparation of the poison.
lachrymal (*lak'-rim-al*). See *lacrimal*.
laciniate (*las-in'-e-āt*) [*lacinia*, a flap]. Jagged, fringed; cut into narrow flaps.
lacmoid (*lak'-moid*). A compound of resorcin and sodium nitrite, used in alkalimetry.
lacmus (*lak'-mus*). See *litmus*.
lacrimal (*lak'-rim-al*) [*lacrima*, a tear]. Pertaining to the tears or to the organs secreting and conveying the tears. l. apparatus, the lacrimal gland, ducts, canal, sac, and nasal duct. l. artery, the first branch of the ophthalmic artery, supplying the gland. l. bone, a bone upon the nasal side of the orbit, articulating with the frontal, the ethmoid, and superior maxillary bones, in which begin the lacrimal groove and nasal duct. l. canals, l. canaliculi, superior and inferior, extend from the lacrimal punctum to the sac, and serve to convey the excess of tears from the eye to the nose. l. caruncle. See *caruncle, lacrimal*. l. ducts, 7 to 14 ducts extending obliquely from the gland to the fornix conjunctivæ, carrying the tears to the conjunctival surface of the same ball. l. fistula, a fistula communicating with a lacrimal duct. l. fossa, a depression at the upper and outer angle of the orbit. l. gland, the gland secreting the tears, situated in a depression of the frontal bone. l. lake, the inward prolongation of the palpebral fissure of the eyelids. l. papilla. See *papilla, lacrimal*. l. probe, a probe for exploring or dilating the canaliculi and nasal duct. l. puncta, the minute orifices of the canaliculi upon the eyelids near the inner canthus. l. sac, a saccular enlargement of the upper part of the nasal duct, into which the canaliculi empty. l. style, a probe used in stricture of the nasal duct.
lacrimation (*lak-rim-a'-shun*) [see *lacrimal*]. An excessive secretion of tears.
lacrimatome (*lak-rim'-at-ōm*) [*lacrima*, tear; τομή, cutting]. A cutting-instrument used in dilating the nasal duct or the canaliculi; a syringotome.
lacrimotomy (*lak-rim-ot'-o-me*) [*lacrima*, tear; τομή, cutting]. The division of strictures of the lacrimal passages.
lactaciduria (*lak-tas-id-ū'-re-ah*) [*lactic acid*; οὖρον, urine]. The presence of lactic acid in the urine.
lactagogue (*lak'-tag-og*). See *galactagogue*.
lactagol (*lak'-ta-gol*). Trade name of a galactagogue, made from cotton-seed.
lactalbumin (*lak-tal-bū'-min*) [*lac*; *albumin*]. A protein contained in milk; it resembles serum-albumin, and coagulates at a temperature of from 70° to 80° C.
lactamide (*lak'-tam-id*) [*lac*, milk; *amide*], C₃H₇NO₂. A substance formed by the union of ethyl lactate and ammonia.
lactampoule (*lak'-tam-pool*) [*lac*; *ampoule*]. A culture of the true Bulgarian bacillus especially designed for souring milk with the optimum of acidity.
lactant (*lak'-tant*) [*lactare*, to suckle]. Suckling.
lactase (*lak'-tās*). Any one of the lymphatics of the small intestine animal body which hydrolyzes lactose.
lactate (*lak'-tāt*). A salt of lactic acid.
lactation (*lak-ta'-shun*) [*lactare*, to suckle]. 1. Suckling; the period during which the child is nourished from the breast. 2. The formation or secretion of milk.
lacteal (*lak'-te-al*) [*lac*]. 1. Pertaining to milk. 2. Any one of the lymphatics of the small intestine that take up the chyle.
lactein (*lak'-te-in*) [*lac*, milk]. Same as *lactolin*.
lactescence (*lak-tes'-ens*) [*lactescere*, to turn to milk]. Milkiness (often applied to the chyle).
lactescent (*lak-tes'-ent*) [*lactescere*, to turn to milk]. Milky, or secreting a milk sap or fluid.
lactic (*lak'-tik*). Pertaining to milk or its derivatives. l. acid. See *acid, lactic*. l. acid, test for. See *Uffelmann*. l. fermentation, the souring of milk.
lactiferous (*lak-tif'-er-us*) [*lac*; *ferre*, to carry]. Conveying or secreting milk. l. ducts, the ducts of the mammary gland. l. glands, the mammary glands.
lactific (*lak-tif'-ik*) [*lac*, milk; *facere*, to make]. Producing milk.
lactiform (*lak'-tif-orm*) [*lac*, milk; *forma*, form]. Resembling milk.
lactifuge (*lak'-tif-ūj*) [*lac*; *fugare*, to drive away]. 1. Lessening the secretion of milk. 2. A drug or agent that causes a lessening in the secretion of milk.
lactigenous (*lak-tij'-en-us*) [*lac*; γενᾶν, to produce]. Milk-producing.
lactigerous (*lak-tij'-er-us*). See *lactiferous*.
lactin (*lak'-tin*). See *lactose*.
lactinated (*lak'-tin-a-ted*). Containing lactose.
lactiphagous (*lak-tif'-ag-us*) [*lac*, milk; φαγεῖν, to eat]. Consuming milk.
lactipotous (*lak-tip'-o-tus*) [*lac*, milk; *potare*, to drink]. Milk-drinking.
lactis (*lak'-tis*) [*gen. of lac*, milk]. Pertaining to milk. l. redundantia, an excessive flow of milk. l. retentio, suppression of the flow of milk.
lactivorous (*lak-tiv'-or-us*) [*lac*; *vorare*, to devour]. Subsisting on milk.
lacto- (*lak-to-*) [*lac*, milk]. A prefix denoting relation to milk.
lactobacilline (*lak-to-bas'-il-ēn*). A preparation of lactic-acid bacilli used to cause lactic-acid fermentation, or to counteract intestinal putrefaction.
lactobutyrometer (*lak-to-bū-tir-om'-et-er*) [*lac*, milk; βούτυρον, butter; μέτρον, measure]. An instrument used in estimating the proportion of butter in milk.
lactocele (*lak'-to-sēl*). See *galactocele*.
lactochloralin (*lak-to-klor'-al-in*) [*lac*, milk; χολή, bile]. A substance obtained from an aqueous solution of cholin by prolonged heating.
lactocin (*lak'-to-sin*). A sedative and hypnotic obtained from the juice of *Lactuca virosa*. Dose 1-5 gr. (0.065-0.32 Gm.).
lactocrit (*lak'-to-krit*) [*lac*, milk; κριτής, a judge]. An apparatus for testing the quantity of fatty substance in a sample of milk.
lactodensimeter (*lak-to-den-sim'-et-er*) [*lac*, milk; *densus*, dense; μέτρον, measure]. A variety of lactometer.
lactoglobulin (*lak-to-glob'-ū-lin*) [*lacto-*; *globulin*]. One of the proteins of milk.
lactoglucose (*lak-to-gloo'-kōs*) [*lac*, milk; γλυκύς, sweet]. A saccharine substance, produced in the fermentation-fluid along with galactose. It is probably the same as glucose.
lactoglycose (*lak-to-gli'-kōs*) [*lacto-*; γλυκύς, sweet]. A dry powder prepared from Mellin's food and milk, free from starch, and with the casein mechanically broken up.
lactol (*lak'-tol*), C₁₃H₇O . OC . CH(OH)CH₃, betanaphthol lactate, an intestinal antiseptic used as a substitute for benzonaphthol. It is decomposed in

the intestine into lactic acid and naphthol. Dose 3½–8 gr. (0.25–0.5 Gm.). Syn., *lactonaphthol*.
**lactola** (*lak-to'-lah*). A factitious milk made from skimmed milk, sugar, and a slight percentage of cotton-seed oil.
**lactolin** (*lak'-to-lin*) [*lac*, milk; *oleum*, oil]. Condensed milk.
**lactometer** (*lak-tom'-et-er*) [*lacto-*; μέτρον, a measure]. An instrument for determining the specific gravity of milk.
**lactonaphthol** (*lak-to-naf'-thol*). See *lactol*.
**lactone** (*lak'-tōn*), C₁₀H₈O₄. 1. An aromatic, colorless, inflammable fluid, obtained in the dry distillation of lactic acid. 2. Trade name of lactic acid bacilli tablets.
**lactopeptin** (*lak-to-pep'-tin*). The proprietary name for a mixture of pepsin, diastase, and pancreatin with lactic acid and hydrochloric acid.
**lactophenin** (*lak-to-fen'-in*) [*lacto-*; φοῖνιξ, purple red]. A derivative of phenetidin with lactic acid. It is a white powder used as an antipyretic and analgesic. Dose 8–15 gr. (0.5–1.0 Gm.).
**lactophosphate** (*lak-to-fos'-fāt*) [*lacto-*; *phosphate*]. A salt composed of a base united to lactic and phosphoric acid.
**lactoprotein** (*lak-to-pro'-te-in*) [*lacto-*; *protein*]. A protein said to exist in milk.
**lactoscope** (*lak'-to-skōp*) [*lacto-*; σκοπεῖν, to examine]. An instrument for estimating the proportions of water and fat-globules in milk.
**lactose** (*lak'-tōs*) [*lac*], C₁₂H₂₂O₁₁+H₂O. Milk-sugar (*saccharum lactis*, U. S. P.); a sugar found in the milk of mammals, and at times in the urine of nursing women. It forms white, hard, rhombic crystals, soluble in water, and has a sweetish taste. Its chief use is as a vehicle. Syn., *lactin*.
**lactoserum** (*lak-to-se'-rum*) [*lacto-*; *serum*, whey]. 1. The whey of milk. 2. The blood-serum of an animal inoculated with the milk of another animal, whereby the serum is rendered capable of precipitating casein in the milk of the variety used in the inoculation.
**lactosin** (*lak'-to-sin*), C₁₈H₂₀O₁₁. A crystallizable polysaccharide.
**lactosomatose** (*lak-to-so'-mat-ōs*). A powder, consisting of the albuminous principles of milk combined with 5 % of tannic acid; used in gastrointestinal disorders. Dose 1–3 teaspoonfuls.
**lactosuria** (*lak-tōs-u'-re-ah*) [*lactose*; οὖρον, urine]. The presence of lactose in the urine.
**lactotoxin** (*lak-to-toks'-in*) [*lac*, milk; τοξικόν, poison]. A poisonous substance found in milk.
**lactovegetarian** (*lak-to-vej-e-ta'-re-an*). Consisting of milk and vegetables.
**Lactuca** (*lak-tū'-kah*) [L., "lettuce"]. A genus of composite-flowered herbs, the lettuces. L. *sativa* is the common garden lettuce. L. *virosa* is a European species, the source of lactucarium.
**lactucarium** (*lak-tū-ka'-re-um*). The concrete milky juice of *Lactuca virosa*. It contains a substance, *laducin*, to which its properties are thought to be due; is sedative and anodyne, and has been used in cough and nervous irritability. 1., syrup of (*syrupus lactucarii*, U. S. P.). Dose 2 dr. (8 Cc.). l., tincture of (*tinctura lactucarii*, U. S. P.). Dose 1 dr. (4 Cc.).
**lactucerin** (*lak-tū'-ser-in*) [*lactuca*, lettuce; *cera*, wax]. A crystalline substance existing in lactucarium *q. v.*
**lactucerol** (*lak-tū'-ser-ol*) [*lactuca*, lettuce; *oleum*, oil], C₂₆H₄₀O₂. A crystalline substance existing in lactucerin; it occurs in two isomeric forms.
**lactucin** (*lak-tu'-sin*) [*lactuca*, lettuce]. A crystallizable extractive of lettuce and of lactucarium.
**lactucism** (*lak'-tū-sizm*). Poisoning from overdosage of hypnotic preparations from species of *Lactuca*. It is characterized by headache, dizziness, dilatation of the pupils, ataxic gait, and dyspnea.
**lactucol** (*lak'-tū-kol*) [*lactuca*, lettuce], C₁₅H₂₀O. A substance crystallizing in needles, formed when lactucerin and potassium hydroxide are melted together.
**lactumen** (*lak-tū'-men*) [*lac*, milk; *lactumina*]. A synonym of *Porrigo larvalis*. Also, applied to that form of aphthæ in which the spots have a fanciful resemblance to clots of curd.
**lactyl** (*lak'-til*), C₃H₄O. A radical found in lactic acid. l.-phenetidin. See *lactophenin*. l.-tropein, C₈H₁₄NO . CO . CH(OH) . CH₃ obtained from tropin by action of lactic acid or ethyl lactate. It is a cardiac tonic.

**lacuna** (*la-kū'-nah*) [*lacus*, a lake; *pl.*, *lacunæ*]. A little hollow space. A mucous or lymphatic follicle. l., absorption. See *Howship's l. lacunæ*. l. anatorum, the vertical groove in the center of the upper lip. l., bone. See *bone*. l. of cementum, spaces in the cement of the teeth analogous to those in bone. l. cerebri, the infundibulum of the brain. l. of cornea, spaces between the laminæ of the cornea. l. Graafianæ, Graafian follicles. l., Haversian. See *bone*. l., Howship's. See *Howship's l.* l. intervillous (of placenta), the spaces in the maternal portion of the placenta filled with blood, and in which the fetal villi hang. l. labii superioris. See *l. amatorum*. l. laterales sinus superioris, depressions along the groove of the superior longitudinal sinus. l. magna, the largest of the orifices of the glands of Littré, situated on the upper surface of the fossa navicularis. l. Morgagni. See *Morgagni's l.* l. pharyngis, a depression at the opening of the Eustachian tube in the pharynx. l. of tongue, the foramen cæcum. l. of urethra, follicular depressions in the mucous membrane of the urethra, most abundant along the floor, especially in the region of the bulb. Their mouths are directed forward.
**lacunal** (*la-kū'-nal*) [*lacuna*]. Pertaining to the lacunæ. l. spaces, the irregular fissures between the fasciculi of connective tissue, forming the beginnings of the lymphatic vessels.
**lacunar** (*la-kū'-nar*) [*lacuna*]. 1. Pertaining to the lacunæ, as *lacunar* tonsillitis. See *tonsillitis*, *lacunar*. 2. See *valve of Vieussens*.
**lacunose** (*lak-ū'-nōs*) [*lacuna*]. The condition of having pits, depressions, or spaces.
**lacunosity** (*lak-ū-nos'-it-e*) [*lacuna*, a pit]. The condition of having pits, depressions, or spaces.
**lacunosoreticulate** (*la-kū-no-so-re-tik'-ū-lāt*) [*lacuna*; *reticulum*, a network]. Both reticulate and lacunose.
**lacunosorugose** (*la-kū'-no-so-roo'-gōs*) [*lacuna*, pit, hollow; *ruga*, a wrinkle]. Deeply pitted or wrinkled, as the stone of a peach.
**lacunula** (*lak-ū'-nū-lah*) [dim. of *lacuna*, a lake: *pl.*, *lacunulæ*]. A small or minute lacuna; an airspace, such as is seen in a gray hair when magnified.
**lacus** (*la'-kus*) [L., "lake"]. A small hollow or cavity in a tissue. l. derivationis, one of the venous spaces in the tentorium cerebelli communicating with the superior longitudinal and lateral sinuses. l. lacrimalis, the space at the inner canthus of the eye, near the punctum, in which the tears collect. l. sanguineus, the uteroplacental sinus.
**ladanum** (*lad'-an-um*). The concrete gummy and resinous juice of various species of *Cistus*, growing in the Mediterranean region; as C. *ladaniferus*, C. *creticus*, C. *salvifolius*, C. *cyprius*, C. *ledon*, and C. *villosus*. It is a greenish-gray solid, of bitter taste, formerly in high esteem. It is now chiefly used in making pastils for fumigation.
**Ladendorff's test for blood** (*lad'-en-dorf*) [August *Ladendorff*, German physician]. Treat the liquid with tincture of guaiacum and then with oil of eucalyptus; in the presence of blood the upper layer becomes violet and the lower layer blue.
**ladol** (*la'-dol*). Trade name of a preparation said to be a hemostatic and uterine sedative.
**ladrèrie** (*lah-drā-re'*) [Fr.]. 1. See *measles*. 2. Leprosy. 3. A leprosarium.
**lady's bed-straw**. Cheese-rennet. The herb *Galium verum*, a refrigerant, and diuretic. Dose, fld.ext. ʒ ss–j. l.'s slipper. See *cypripedium*.
**Lady Webster's pills**. Laxative pills of aloes and mastic.
**læmoparalysis** (*le-mo-par-al'-is-is*). See *lemoparalysis*.
**Laennec's catarrh** (*len-nek'*) [René Théophile Hyacinthe *Laennec*, French physician, 1781–1826]. A form of asthmatic bronchitis with scanty, viscous, "pearly" expectoration. See *L.'s perles*. L.'s cirrhosis, L.'s disease, alcoholic cirrhosis of the liver; hobnail liver. L.'s perles, roundish, gelatinous masses forming the sputum in bronchial asthma. L.'s rale, a modified subcrepitant rale due to mucus in the bronchioles; it is heard in pulmonary emphysema. L.'s thrombus, a globular thrombus formed in the heart, especially when the heart is the seat of fatty degeneration.
**læv-**, **lævo-** (*le'-vo*). For words so commencing, see *lev-*, *levo-*.
**Lafayette mixture** (*laf-a-et'*). A mixture employed in gonorrhea. It contains copaiba, cubeb, solution

of potassium hydroxide, sweet spirit of niter, and is known also as the compound copaiba mixture.

**lagam-balsam** (*lag'-am-bawl'-sam*). A variety of gurjun-balsam brought from Sumatra.

**lagarous** (*lag'-ar-us*) [λαγαρός, lax]. Lax, loose, or soft.

**lagena** (*laj-e'-nah*) [λάγυνος, a flask, bottle; *pl.*, *lagenæ*]. The flask-like extremity of the cochlea in certain vertebrates. The third upper extremity of the scala media.

**lageniform** (*laj-en'-if-orm*) [*lagena*, a flask; *forma*, form]. Flask-shaped.

**lagentomum** (*laj-en'-to-mum*) [λαγώς, hare; ἐν, in; τομή, a cutting]. Hare-lip. Cf. *lagochilus.*

**lagmi** (*lag'-me*). A fermented wine or liquor made by the Arabs from the juice of the date-palm.

**lagnea, lagneia** (*lag-ne'-ah, lag-ni'-ah*) [λαγνεία, coition, lust]. 1. Same as *satyriasis* or *nymphomania*. 2. Coitus. 3. The semen.

**lagnesis** (*lag-ne'-sis*) [λάγνης, lewd]. Same as *satyriasis* or *nymphomania*. 1. furor, unconquerable lust.

**lagnosis** (*lag-no'-sis*). Same as *lagnesis*.

**lagocephalous** (*lag-o-sef'-al-us*) [λάγως, hare; κεφαλή, head]. Having a head like a hare.

**lagochilus, lagochilos** (*lag-o-ki'-lus, lag-o-ki'-los*) [λαγώς, hare; χεῖλος, lip]. Hare-lip. Cf. *lagentomum.*

**lagophthalmic** (*lag-off-thal'-mik*) [λαγώς, hare; ὀφθαλμός, eye]. Pertaining to or affected with lagophthalmos.

**lagophthaimos** (*lag-of-thal'-mos*) [λαγώς, hare; ὀφθαλμός, eye: from the popular notion that a hare sleeps with open eyes]. A condition in which the eyes cannot be closed.

**Lagoria's sign** (*la-gor'-e-ah*). Relaxation of the extensor muscles in intracapsular fracture of the neck of the femur.

**lagostoma** (*lag-os'-to-mah*) [λαγώς, hare; στόμα, mouth]. See hare-lip. Cf. *lagochilus* and *lagentomum.*

**Lagrange's operation** (*la-grahnj'*) [Felix *Lagrange*, French physician]. A combination of sclerectomy and iridectomy performed in cases of glaucoma.

**la grippe** (*lah-grip'*). See *influenza*.

**lag tooth**. A delayed tooth. A name for the molar or wisdom tooth.

**L. A. H.** Abbreviation for Licentiate of the Apothecaries' Hall (of Dublin).

**laibose** (*la'-bōs*). Trade name of a food said to be composed of the solids of pure whole milk and the entire digestible substance of whole wheat, in a dry granular form.

**laiose** (*la'-ōs*). A body found by Leo in diabetic urine in certain cases and regarded by him as a sugar. Syn., *Leo's sugar*.

**laity** (*la'-it-e*) [λαός, the people]. The non-professional public.

**lake-colored, laky** [Fr., *laque*, rose-colored, from Pers. *lāk*]. Applied to blood that is dark red and transparent from a solution of the hemoglobin in the serum.

**lakmoid, lacmoid** (*lak'-moid*) [*lac*, milk; εἶδος, like]. A delicate reagent in alkalimetry, made by acting on resorcin with sodium nitrate.

**lalia** (*lal'-e-ah*) [λαλία, talking]. Speech.

**lallation** (*lal-a'-shun*) [λάλος, prattle]. Any unintelligible stammering of speech, such as the prattling of a babe.

**Lallemand-Trousseau's bodies** (*lahl-mon(g)-trooso'*) [Claude François *Lallemand*, French surgeon, 1790–1853; Armand *Trousseau*, French physician, 1801–1867]. Gelatinous masses found in the secretion of the seminal vesicles.

**lalling** (*lal'-ing*) [*lallare*, to babble]. Lallation; prattle; baby-talk.

**laloneurosis** (*lal-o-nū-ro'-sis*) [λάλος, prattle; *neurosis*]. An impairment of speech arising from spasmodic action of the muscles. It includes stammering and aphthongia.

**lalopathy** (*lal-op'-ath-e*) [λάλος, prattle; πάθος, disease]. Any disorder of speech.

**lalophobia** (*lal-o-fo'-be-ah*) [λάλος, prattle; φόβος, fear]. Stutter-spasm, leading to or complicated with a dislike of speaking.

**laloplegia** (*lal-o-ple'-je-ah*) [λάλος, talking; πληγή, a stroke]. Paralysis of speech, not due, however, to paralysis of the tongue.

**Lalouette's pyramid** (*lal-oo-et'*) [Pierre *Lalouette*, French physician, 1711–1742]. A prolongation of

the upper portion of the thyroid gland, generally to the left of the median line; it is not constant.

**lambda** (*lam'-dah*) [λάμβδα, the Greek letter Λ or λ]. The angle of junction of the sagittal and lambdoid sutures.

**lambdacism** (*lam'-das-izm*) [λαμβδακισμός, a fault in pronunciation of the letter *lambda*]. 1. Difficulty in uttering the sound of the letter *l*. 2. Too frequent use of the *l* sound, or its substitution for the *r* sound.

**lambdoid, lambdoidal** -(*lam'-doid, lam-doi'-dal*) [*lambda*; εἶδος, resemblance]. Resembling the Greek letter λ. 1. suture, the suture between the occipital and the two parietal bones.

**Lamblia intestinalis** (*lam'-ble-ah in-tes-tin-a'-lis*). A flagellate protozoan parasite with a pear-shaped body, found in the intestine of man and various animals. Called also *Cercomonas intestinalis*, *Dimorphus muris* and *Megastoma entericum*.

**lame** (*lām*). Crippled; halting; limping.

**lamel** (*lam-el'*) [*lamella*, dim. of *lamina*, a plate]. A medicated disc, made with some soluble basis; it is used in the dosimetric application of drugs to the eye, etc.

**lamella** (*lam-el'-ah*) [dim. of *lamina*, a plate]. 1. Same as *lamel*. 2. A thin scale or plate. l. of bone, one of the concentric rings surrounding the Haversian canals. l., concentric, one of the plates of bone surrounding the Haversian canal. l., intermediate, one of the plates filling the spaces between the concentric layers of bone. l., periosteal, l., peripheral, a superficial lamella of bone lying under the periosteum. l., triangular, a fibrous layer connecting the choroid plexuses of the diacele. l., vascular, the endochorion. l., vitreous, the lamina basalis of the choroid.

**lamellar** (*lam-el'-ar*). Having the nature of or resembling a thin plate; composed of lamellæ or thin plates. l. cataract. See *cataract, lamellar*.

**lameness** (*lām'-nes*) [ME., *lame*]. Limping; weakness of a limb. l., intermittent. See *claudication, intermittent*.

**laminæ** (*lahm-in-ahzj'*) [Fr.]. Flattening; the compression or flatwise crushing of the fetal head to facilitate delivery.

**laminal, laminar** (*lam'-in-al, lam'-in-ar*) [*lamina*, a layer]. Composed of laminæ; having the form of a lamina. l. tissue, a synonym of *connective tissue*.

**lamine** (*la'-mēn*) [*lamium*, the dead-nettle]. An alkaloid from *Lamium album*. The sulphate is used hypodermatically as a powerful hemostatic in uterine and other internal hemorrhages.

**lamina** (*lam'-in-ah*) [L., "a plate or scale": *pl.*, *laminæ*]. A thin plate or layer. l. affixa, the line of union of the hemisphere with the thalamus. l. basalis, a structureless membrane on the inner surface of the lamina capillaris: Bruch's membrane. l. basilaris, the decidua serotina. l. choriocapillaris, the inner layer of the choroid consisting of a capillary plexus. l. cinerea, the connecting layer of gray matter between the corpus callosum and the optic chiasm. l., cribriformis, the cribriform plate of the ethmoid. '. cribrosa. 1. That portion of the choroid which is perforated for the passage of the optic nerve. 2. The fascia covering the saphenous opening. 3. The anterior or posterior perforated space of the brain. 4. The perforated plates of bone through which branches of the cochlear or auditory nerve pass. l., dental, an epithelial sheet formed by a flattening of the base of the dental band, from which the buds forming the enamel-organs of the teeth are given off. l. denticulata, a cartilaginous plate on the superior and external portion of the lamina spiral lamina. l. elastica anterior. See *Bowman's membrane*. [l. elastica posterior. See *Descemet's membrane*.] l., external elastic, the innermost layer of the adventitia, consisting of fibers of elastic tissue blending externally with the adventitia proper. l. fusca, the pigmentary tissue of the inner layer of the sclera, forming the outer layer of the perichoroid sinus. l. lateralis processus pterygoidei, the external pterygoid plate. l. medialis processus pterygoidei, the internal pterygoid plate. l., muscle. See *myocomma*. l., papyracea, the os planum of the ethmoid. l., pericalustral, a layer of white matter between the claustrum and the cortex of the insula. l. perpendicularis, the vertical plate of the ethmoid or mesethmoid. l. propria, the middle or fibrous layer of the tympanic membrane. l. quadrigemina, the part of the midbrain from which the corpora quadrigemina and the brachia are developed. l., **reticular**, the hyaline membrane of the inner ear, extending between the conjoined head of Corti's rods and the supporting cells. l. **rostralis**, the thin continuation of the rostrum of the callosum into the lamina cinerea. l. **spiralis**, a thin plate in the ear, osseous in the inner part and membranous in the outer, which divides the spiral tube of the cochlea into the scala tympani and the scala vestibuli. l. **spiralis secundaria**, a short partition projecting from the cochlear wall in its lower part only. l. **suprachoroidea**, the delicate connective-tissue membrane uniting the choroid and sclerotic coats of the eye. l. **vasculosa chorioidea**, the layers of large vessels and of capillaries of the choroid considered as one. l. **velamentosa**, the organ of Corti. l., **vitreous**, a homogeneous membrane covering the inner surface of the choroid. Syn., *membrane of Bruch*.

**laminar** (*lam'-in-ar*). See *laminal*.

**Laminaria** (*lam-in-a'-re-ah*) [*lamina*]. 1. A genus of seaweeds of the order *Laminariaceæ*. 2. The stems of *L. clouslonl* or *L. digitata*. l. **tent**, a tent made of the stem of the laminaria plant formerly used for dilatation of the cervix uteri and other canals.

**laminated** (*lam'-in-a-ted*). See *laminar*.

**lamination** (*lam-in-a'-shun*) [*lamina*]. 1. Arrangement in plates or layers. 2. An operation in embryotomy consisting in cutting the skull in slices.

**laminectomy** (*lam-in-ek'-to-me*) [*lamina; ἐκτομή,* excision]. The operation of removing the posterior vertebral arches.

**laminiform** (*lam-in'-e-form*). See *laminar*.

**laminitis** (*lam-in-i'-tis*) [*lamina,* plate; *ιτις*, inflammation]. Inflammation of the laminæ, particularly the laminæ of a horse's hoof; founder.

**laminoids** (*lam'-in-oids*). Blaud's tablets, made in two layers—one of ferrous sulphate and one of sodium bicarbonate—united by pressure.

**Lamium** (*la'-me-um*) [L., "dead-nettle"]. A genus of plants. *L. album* is a species furnishing the alkaloid lamine.

**lamnectomy** (*lam-nek'-to-me*). Same as *laminectomy*.

**Lamotte's drops** (*lam-ot'*). Ethereal tincture of ferric chloride.

**lampas** (*lam'-pas*) [Fr., *lampas*, lampas]. A congestive swelling of the fleshy lining of the roof of the mouth, in the horse.

**lampblack** (*lamp'-blak*). A fine black substance, almost pure carbon, made by burning coal-oils in an atmosphere deficient in oxygen, or by allowing a gas-flame to impinge on a cold surface.

**lamprophonia**, **lamprophony** (*lam-pro-fo'-ne-ah, lam-proff'-o-ne*) [λαμπρός, sounding; φωνή, voice]. A sonorous, ringing quality of the voice.

**lampsis** (*lamp'-sis*) [λάμψις, a shining]. Splendor, brilliancy.

**lana** (*lan'-ah*) [L.]. Wool.

**lanain** (*lan'-a-in*) [*lana,* wool]. Purified wool-fat.

**Lancaster black-drop**. Acetum opii; vinegar of opium.

**lance** (*lans*) [*lancea,* a lance or spear]. 1. A lancet. 2. To open, as with a lancet or bistoury. l., **Mauriceau's**, a lance-shaped knife for opening the fetal head in embryotomy.

**Lancereaux's interstitial nephritis** (*lan-ser-o'*) [Etienne Lancereaux, French physician, 1829-1910]. Interstitial nephritis due to rheumatism. L.'s **law**, marantic thromboses always occur at the points where there is the greatest tendency to stasis; that is, where the influence of the cardiac propulsion and of thoracic aspiration is least. L.'s **treatment** (in aneurysm of the aorta), consists in subcutaneous injections of serum gelatin.

**lancet** (*lan'-set*) [dim. of *lancea,* a lance]. A knife having a double-edged, lance-shaped blade, for incising tumors, abscesses, etc. l., **gum-**, a small lancet in which the cutting portion has a convex edge and is at right angles to the shaft; it is used for cutting the gums. l., **spring**, one in which the blade is thrust out by means of a spring controlled by a trigger. l., **thumb**, one with a double-edged, broad blade.

**lancinate** (*lan'-sin-āt*) [*lancinare,* to tear]. To lacerate, to pierce, or tear.

**lancinating** (*lan'-sin-a-ting*) [*lancinare,* to tear]. Tearing; shooting. l. **pains**, rending, tearing, or sharply cutting pains, common in posterior spinal sclerosis.

**Lancisi, nerves of** (*lan-se'-ze*) [Giovanni Maria Lancisi, Italian physician, 1654-1720]. The **striæ longitudinales**. The mesial longitudinal striæ situated on each side of the raphe of the corpus callosum. L.'s **sign**, very feeble heart-beats, amounting to a trembling of the heart, perceived by palpation in grave myocarditis.

**Landau's color test for syphilis** (*lan-do*). A modification of Wassermann's reaction. The reagent used is a 1 per cent. solution of iodine in carbon tetrachloride. L.'s **form of enteroptosis**, enteroptosis due to relaxation of the abdominal walls and pelvic floor.

**landmarks** (*land'-marks*). Superficial marks (such as eminences, lines, and depressions) that serve as guides to, or indications of, deeper-seated parts.

**Landolphi's** or **Landolfi's paste**. A caustic application composed of the chlorides of zinc, antimony, bromine and gold, which is used pure, or weakened by mixing with basilicon ointment in varying proportions. It should be spread on the surface with a spatula, or applied on charpie, and allowed to remain for 24 or 48 hours.

**Landolt's bodies** (*lan-dōlt'*) [Edmond Landolt, French oculist, 1846- ]. Small, elongated, clavate bodies lying between the rods and cones and resting upon the outer nuclear layer of the retina. **Landolt's test for phenol**. Treat the solution with bromine water; a white, crystalline precipitate of tribromphenol is produced.

**Landouzy's disease** (*lahn-doo'-ze*) [Louis Landouzy, French physician, 1845- ]. Weil's disease, *q. v.* L.'s **ischialgia**, neuralgia of the sciatic nerve, with atrophy of the muscles of all or part of the affected leg. L.'s **purpura**, a form of purpura with grave systemic symptoms.

**Landouzy-Déjérine's type of progressive muscular atrophy** (*lahn-doo'-ze-da-zjer-ēn'*) [Louis Landouzy, French physician, 1845- ; Joseph Jules Déjérine, French neurologist, 1849- ]. A form in which there is atrophy of the muscles of the face and those of the scapulohumeral group.

**Landry's disease**, L.'s **paralysis** (*lahn'-dre*) [Jean Baptiste Octave Landry, French physician, 1816-1865]. A form of paralysis characterized by loss of motor power in the lower extremities, gradually extending to the upper extremities and to the centers of circulation and respiration without sensory manifestations, trophic changes, etc. Syn., *acute ascending paralysis*.

**land-scurvy**. Purpura hæmorrhagica. See *purpura*.

**Landzert's fossa**. A fossa in the peritoneal cavity that is bounded behind by the parietal peritoneum covering the psoas, the renal vessels, the ureter, and a part of the left kidney, and below by the mesocolic fold. Syn., *paraduodenal fossa; recessus renosus*.

**Lane's kinks** (*lān*) [Sir William Arbuthnot Lane, English surgeon]. Bends or twists of the intestine at certain parts due to the upright position of the body, and the descent of the cecum. L.'s **operation**. Short-circuiting the large intestine, for chronic obstruction, constipation or colitis; the lower end of the ileum is anastomosed with the rectum; ileosigmoidostomy.

**lanesin**, **lanessin** (*lan'-es-in*) [*lana,* wool]. A proprietary preparation of wool-fat similar to lanolin.

**Lanfranc's collyrium**. A stimulant application to ulcers, containing aloes and myrrh, each 5 parts; acetate of copper, 10 parts; trisulphide of arsenic, 15 parts; rose-water, 380 parts, and white wine, 1000 parts.

**Lang's fixative and hardening fluid**. Mercuric chloride, 5 parts; sodium chloride, 6 parts; acetic acid, 5 parts; water 100 parts.

**Lang's reaction for taurin**. A white combination appearing as a precipitate on boiling a solution of taurin with freshly precipitated mercuric oxide.

**Langer's axillary arch** (*lahng'-er*) [Carl Ritter von Edenberg von Langer, German anatomist, 1819-1887]. The thickened border of fascia which forms a bridge across the bicipital groove.

**Langerhans' bodies** (*lahng'-er-hans*) [Ernst Robert Langerhans, German histologist, 1859- ]. 1. Certain modified epithelial cells forming the simplest nerve endings. 2. See *cells, centroacinar*. L.'s **granular layer**, the stratum granulosum; the layer of epidermal cells above the rete mucosum. L., **islands of**, little cellular masses in the interstitial connective tissue of the pancreas, subject to various interpreta-

tions. L.'s stellate corpuscles. Terminations of nerve fibers which have been observed in the rete mucosum of the epidermis.
**Langhans' cells** (*lähng'-hans*) [Theodor *Langhans*, German pathologist, 1839— ]. The polygonal epithelial cells, with distinct nuclei and cell-walls, constituting Langhans' layer. L.'s **giant-cell**, the giant-cell of a tuberculous granuloma. L.'s **layer**, the inner of the two layers of cells covering the chorion; it is derived from the ectoderm.
**Langier's apparatus** (*lon-je-a'*). A form of immovable splint for a limb. It is composed of strips of brown paper impregnated with starch-paste. L.'s **bandage**, a many tailed paper bandage.
**language** (*lang'-gwāj*) [*lingua*, the tongue]. The articulate sounds; signs, or symbols whereby thought is communicated. I., center for. See *aphasia*.
**languor** (*lang'-gwor*) [L., faintness]. Lassitude. Disinclination to take bodily exercise, or to exert oneself.
**lanichol** (*lan'-ik-ol*). A proprietary preparation of wool-fat.
**lanigallol** (*lan-e-gal'-ol*). A triacetate of pyrogallic acid.
**lank**. Lean, thin, attenuated.
**laniol** (*lan'-e-ol*) A proprietary wool-fat.
**lannaiol** (*lan-a-i'-ol*). An iodocresol proposed as a substitute for iodoform.
**Lannelongue's incisions**. Partial or complete interruptions of the medullary sheath of a nerve-fiber existing at irregular intervals in an interannular segment. L.'s **segments**, the cylindrical or conical segments of the medullary sheath between Lantermann's incisions.
**Lannelongue's tibia** (*lan-e-long'*) [Odilon Marc *Lannelongue*, French surgeon, 1840— ]. A deformed tibia of inherited syphilis.
**lanoform** (*lan'-o-form*). An antiseptic preparation of lanolin and 1 % of formaldehyde
**lanolin** (*lan'-o-lin*) [*lana*, wool; *oleum*, oil]. A cholesterin fat (*adeps lanæ*, U. S. P.) obtained from sheep's wool, and used as a basis for ointments. l.-**milk**, a mixture of lanolin, 10 parts; borax, 1 part; rose-water, 100 parts, and medicated soap, 2.5 parts. l. **powder**, lanolin combined with zinc oxide, magnesium carbonate, and starch. l., **sulphurated**, thilanin.
**lantanine** (*lan'-tan-en*). An alkaloid from *Lantana braziliensis*. It is a white, bitter powder, used as an antipyretic instead of quinine in intermittent fever. Dose 15–30 gr. (1–2 Gm.) daily.
**Lantermann's incisions**. Partial or complete interruptions of the medullary sheath of a nerve-fiber existing at irregular intervals in an interannular segment. L.'s **segments**, the cylindrical or conical segments of the medullary sheath between Lantermann's incisions.
**lanthanum, lanthanium** (*lan'-than-um, lan-tha'-ne-um*) [λανθάνειν, to conceal; lanthanum was a substance concealed from the knowledge of chemists]. A rare metallic element; symbol La, atomic weight, 139. Little is known of its medicinal properties. See *elements, table of*.
**lanthopine** (*lan'-tho-pēn*) [λανθάνειν, to conceal], C₂₃H₂₅NO₄. A finely crystalline alkaloid of opium. It occurs in white crystals fusible at 200° C.
**lanuginose, lanuginous** (*lan-ū'-jin-ōs, lan-ū'-jin-us*) [*lanuginosus*, wooly]. In biology, wooly, downy, lanate, lanose, lanigerous.
**lanugo** (*lan-ū'-go*) [L., "down"]. 1. The downlike hair that appears upon the fetus at about the fifth month of gestation. 2 The downy growth often seen upon the face of women and girls.
**lanulous** (*lan'-ū-lus*). Covered with short hair.
**lanum** (*la'-num*). See *lanolin*.
**lapactic** (*lap-ak'-tik*) [λαπάσσειν, to empty]. 1. Emptying; evacuant. 2. Any purgative substance.
**lapara** (*lap'-ar-ah*) [λαπάρα, the flank, loins]. 1. The loins; loosely applied to the abdomen. 2. Diarrhea.
**laparacele** (*lap'-ar-as-ēl*). See *laparocele*.
**laparectomy** (*lap-ar-ek'-to-me*). See *enterectomy*.
**laparelytrotomy** (*lap-ar-el-it-rot'-o-me*). Same as *laparo-elytrotomy*.
**laparo-** (*lap-ar-o-*) [λαπάρα, loins]. A prefix denoting pertaining to the abdomen; properly, referring to the loin or flank. See *celiotomy*.
**laparocele** (*lap'-ar-o-sēl*) [λαπάρα, loin; κήλη, tumor]. Lumbar or abdominal hernia.
**laparocholecystotomy** (*lap-ar-o-ko-les-is-tot'-o-me*) [λαπάρα, loin; χολή, bile; κύστις, bladder; τομή, cutting]. Laparotomy combined with cholecystotomy.
**laparoclysis** (*lap-ar-ok'-lis-is*) [λαπάρα, loin; κλύσις, a drenching]. An injection into the peritoneal cavity.
**laparocolectomy** (*lap-ar-o-ko-lek'-to-me*). Same as *colectomy*.

**laparocolostomy** (*lap-ar-o-ko-los'-to-me*) [*laparo-*; κόλον, colon; στόμα, mouth; τομή, a cutting]. Abdominal colostomy: the formation of a permanent opening into the colon by incision through the abdominal wall.
**laparocolotomy** (*lap-ar-o-ko-lot'-o-me*) [*laparo-*; κόλον, colon; τομή, a cutting]. Inguinal or abdominal colotomy.
**laparocolpotomy** (*lap-ar-o-kol-pot'-o-me*) See *laparo-elytrotomy*.
**laparocystectomy** (*lap-ar-o-sis-tek'-to-me*) [*laparo-*; κύστις, cyst; ἐκτομή, excision]. An operation performed in advanced extra-uterine pregnancy for removal of the fetus and the entire gestation-sac.
**laparocystidotomy** (*lap-ar-o-sist-id-ot'-o-me*). See *laparocystotomy*.
**laparocystotomy** (*lap-ar-o-sis-tot'-o-me*) [*laparo-*; κύστις, cyst; τομή, a cutting]. 1. Suprapubic cystotomy. 2. An operation in advanced extrauterine pregnancy for the removal of the fetus, the sac being allowed to remain.
**laparocystovariohysterotomy** (*lap-ar-o-sist-o-va-re-o-his-ter-ot'-o-me*). Combined ovariotomy and hysterotomy through an abdominal incision.
**laparoelytrotomy** (*lap-ar-o-el-it-rot'-o-me*) [*laparo-*; ἔλυτρον, sheath; τομή, a cutting]. An operation consisting in an incision over Poupart's ligament, dissecting up the peritoneum until the vagina is reached, incising the latter transversely, dilating the cervix, and extracting the child through the os uteri.
**laparoenterostomy** (*lap-ar-o-en-ter-os'-to-me*) [*lap-aro-*; ἔντερον, intestine; στόμα, mouth]. The formation of an artificial opening into the intestine through the abdominal wall.
**laparoenterotomy** (*lap-ar-o-en-ter-ot'-o-me*) [*lap-aro-*; ἔντερον, intestine; τομή, a cutting]. An opening of the intestine through an abdominal incision.
**laparogastrostomy** (*lap-ar-o-gas-tros'-to-me*) [*lap-aro-*; γαστήρ, stomach; στόμα, mouth]. The formation of a permanent gastric fistula through the abdominal wall.
**laparogastrotomy** (*lap-ar-o-gas-trot'-o-me*) [*laparo-*; γαστήρ, stomach; τομή, a cutting]. The opening of the stomach through an abdominal incision.
**laparohepatotomy** (*lap-ar-o-hep-at-ot'-o-me*) [*lap-aro-*; ἧπαρ, liver; τομή, a cutting]. Incision of the liver through the abdominal wall.
**laparohysterectomy** (*lap-ar-o-his-ter-ek'-to-me*) [*lap-aro-*; ὑστέρα, womb; ἐκτομή, a cutting out]. The removal of the uterus through an incision in the abdominal walls.
**laparohystero-oophorectomy** (*lap-ar-o-his-ter-o-o-of-or-ek'-to-me*) [*laparo-*; ὑστέρα, womb; ᾠοφόρος, ovary; ἐκτομή, a cutting out]. Removal of the uterus and ovaries through an incision in the abdominal wall.
**laparohysteropexy** (*lap-ar-o-his'-ter-o-peks-e*) [*lap-aro-*; ὑστέρα, womb; πῆξις, a fixing in]. Ventrofixation.
**laparohysterosalpingo-oophorectomy** (*lap-ar-o-his-ter-o-sal-pin-go-o-of-or-ek'-to-me*) [*laparo-*; ὑστέρα, womb; σάλπιγξ, tube; ᾠοφόρος, ovary; ἐκτομή, a cutting out]. Excision of the uterus, oviducts, and ovaries by the abdominal route.
**laparohysterotomy** (*lap-ar-o-his-ter-ot'-o-me*) [*lap-aro-*; ὑστέρα, uterus; τομή, a cutting]. The operation of cutting into the uterus through an abdominal incision, as for the purpose of removing a fetus.
**laparoileotomy** (*lap-ar-o-il-e-ot'-o-me*) [*laparo-*; *ileum*; τομή, a cutting]. The operation of cutting into the ileum through an abdominal incision.
**laparokelyphotomy** (*lap-ar-o-kel-if-ot'-o-me*) [*lap-aro-*; κέλυφος, egg-shell; τομή, a cutting]. Same as *laparocystotomy*.
**laparokolpotomy** (*lap-ar-o-kol-pot'-o-me*). See *lap-aroelytrotomy*.
**laparomyitis** (*lap-ar-o-mi-i'-tis*) [*laparo-*; μῦς, muscle; ιτις, inflammation]. Inflammation of the muscular portion of the abdominal wall.
**laparomyomectomy** (*lap-ar-o-mi-o-mek'-to-me*) [*lap-aro-*; μῦς, muscle; ἐκτομή, a cutting out]. Removal of a myoma through an abdominal incision.
**laparomyomotomy** (*lap-ar-o-mi-ot'-o-me*). See *laparo-myomectomy*.
**laparonephrectomy** (*lap-ar-o-nef-rek'-to-me*) [*lap-aro-*; νεφρός, kidney; ἐκτομή, a cutting out]. Nephrectomy by an incision in the loins.
**laparonephrotomy** (*lap-ar-o-nef-rot'-o-me*) [*laparo-*; νεφρός, kidney; τομή, a cutting]. Nephrotomy through an incision in the loins.

**laparorrhaphy** (*lap-ar-or'-af-e*) [*laparo-*; *ραφή*, suture]. Suture of the abdominal wall.
**laparosalpingectomy** (*lap-ar-o-sal-pin-jek'-to-me*) [*laparo-*; σάλπιγξ, tube; ἐκτομή, a cutting out]. Removal of a Fallopian tube through an abdominal incision.
**laparosalpingo-oophorectomy** (*lap-ar-o-sal-ping-go-o-of-o-rek'-to-me*). The removal of the ovaries and oviducts by an incision through the abdomen.
**laparosalpingotomy** (*lap-ar-o-sal-pin-got'-o-me*) [*laparo-*; σάλπιγξ, tube; τομή, a cutting]. 1. Cutting into an oviduct through an abdominal incision. 2. Laparosalpingectomy.
**laparoscope** (*lap'-ar-o-skōp*) [*laparo-*; σκοπεῖν, to examine]. An instrument for examining the abdomen.
**laparoscopy** (*lap-ar-os'-ko-pe*) [*laparo-*; σκοπεῖν, to examine]. Examination of the abdomen by instrumental means.
**laparosplenectomy** (*lap-ar-o-splen-ek'-to-me*) [*laparo-*; σπλήν, spleen; ἐκτομή, a cutting out]. Removal of the spleen through an abdominal incision.
**laparosplenotomy** (*lap-ar-o-splen-ot'-o-me*) [*laparo-*; σπλήν, spleen; τομή, a cutting]. Surgical entrance upon the spleen through the abdominal walls.
**laparotome** (*lap-ar-o-tōm*) [*laparo-*; τομή, a cutting]. A cutting-instrument used in laparotomy.
**laparotomist** (*lap-ar-ot'-o-mist*) [see *laparotomy*]. A surgeon who performs laparotomies.
**laparotomize** (*lap-ar-ot'-om-īz*). To make an incision in the abdominal wall; to perform laparotomy.
**laparotomy** (*lap-ar-ot'-o-me*) [*laparo-*; τομή, a cutting]. 1. An incision through the abdominal wall; celiotomy is the preferable term. 2. The operation of cutting into the abdominal cavity through the loin or flank.
**laparotyphlotomy** (*lap-ar-o-tif-lot'-o-me*). Synonym of *typhlotomy*.
**laparo-uterotomy** (*lap-ar-o-ū-ter-ot'-o-me*). Incision of the uterus through the abdomen; laparohysterotomy.
**laparovaginal** (*lap-ar-o-vaj'-in-al*) [*laparo-*; *vagina*, sheath]. Pertaining to the abdomen and the vagina.
**laparozoster** (*lap-ar-o-zos'-tur*). Zoster affecting the abdomen.
**lapathin** (*lap'-ath-in*) [*Lapathum*, the genus *Rumex*]. Chrysophanic acid.
**lapaxis** (*lap-aks'-is*) [λάπαξις, evacuation]. Evacuation.
**lapilliform** (*lap-il'-i-form*) [*lapillus*, a little stone; *forma*, form]. Presenting the appearance of little stones.
**lapis** (*la'-pis*) [L.]. A stone; an alchemic term applied to any nonvolatile substance. l. **albus**, aluminated copper. l. **imperialis**, l. **lunaris**, silver nitrate. l. **infernalis**, silver nitrate. l. **lazuli**, a beautiful blue stone of complex composition, formerly employed as a purgative and emetic and in epilepsy. l. **mitigatus**, diluted silver nitrate.
**lappa** (*lap'-ah*) [L.]. The root of the common burdock, *Arctium lappa*, containing a bitter principle, a resin, and tannin. It is aperient, diuretic, and alterative, and has been employed in gout, scorbutus, syphilis, and various skin diseases. The dose of the *root* is 1–2 dr. (4–8 Cc.); in *infusion* or *tincture* 10 min.–1 dr. (0.65–4.0 Cc.). l., **fluidextract** of (*fluidextractum lappæ*, U. S. P.). Dose 30 min.–1 dr. (2–4 Cc.).
**lapsus** (*lap'-sus*) [*labi*, to fall]. A fall; ptosis. l. **palpebræ superioris**, ptosis of the eyelid. l. **pilorum**, alopecia. l. **unguium**, falling of the nails.
**laquear** (*lak'-we-ar*) [L., a fretted ceiling]; pl., *laquearia*]. l. **vaginæ**, the vaginal vault.
**laqueus** (*lak'-we-us*) [L., a noose]. 1. A noose-shaped bandage. 2. See *fillet* (2).
**larch**. See *larix*.
**lard** (*lardum*, lard]. The fat of the interior of the abdominal cavity of the hog, constituting *adeps* (U. S. P.). Lard is much used in pharmacy as a basis for ointments. See *adeps*.
**lardacein** (*lar-da'-se-in*) [*lard*]. An amyloid substance, formed in amyloid degeneration of various organs, particularly the liver, kidney, and spleen. It is a protein, but insoluble in the ordinary solvents, is not acted upon by the gastric juice, does not readily undergo putrefaction, and gives a mahogany-brown color with iodine and a blue color with iodine and sulphuric acid.
**lardaceous** (*lar-da'-se-us*) [*lard*]. Amyloid. l. **kidney**. See *Bright's disease*.

**lardeous, lardiform** (*lar'-de-us, -form*) [*lard*; *forma*, form]. Having a fatty appearance.
**lardum, laridum** (*lard'-um, lar'-id-um*) [L.]. Lard.
**largin** (*lar'-jin*). Silver protalbin, a compound containing 11 % of silver. It is a grayish-white powder, readily soluble in water and glycerol. It is a bactericide, used in gastric ulcers and gonorrhea. Dose 5–8 gr. (0.33–0.5 Gm.) in pill; in gonorrhea a 0.25 to 1.5 % solution is employed.
**laricin** (*lar'-is-in*) [*larix*, larch]. Agaricin obtained from the larch agaric, *Polyporus officinalis*.
**larinoid** (*lar'-in-oid*) [λαρινός, fat; εἶδος, like]. Lardaceous; as larinoid carcinoma.
**larix** (*lar'-iks*) [L.]. Larch. A genus of coniferous, deciduous trees. The cortex (*laricis cortex*, B. P.) is astringent and stimulant, and has been used in purpura, hemoptysis, bronchitis, and locally in skin diseases. Dose of the *tincture* 20–30 min. (1.3–2.0 Cc.).
**larkspur** (*lark'-spur*). The plant *Delphinium consolida*, the seeds of which are diuretic and emmenagogue. Dose of the *fluidextract* 1–10 min. (0.065–0.65 Cc.).
**Larrey's amputation** (*lar'-e*) [Dominique Jean Larrey, French surgeon, 1766–1842]. Double-flap amputation at the shoulder-joint or hip-joint. L.'s **bandage**, a form of many tailed bandage, the edges being glued together. L.'s **spaces**, spaces between the parts of the diaphragm attached to the sternum and those that are attached to the ribs.
**larva** (*lar'-vah*) [*larva*, a ghost; pl., *larvæ*]. The young stage of such animals as undergo transformation. The form that insects take in emerging from the egg, commonly known as the caterpillar or "grub" stage. Also, applied to the immature form distinguishing many of the lower vertebrates before maturity.
**larvaceous** (*lar-va'-she-us*) [*larva*, a mask]. Covering the face like a mask.
**larval** (*lar'-val*) [*larva*, a ghost]. 1. Pertaining to or existing in the condition of a larva. 2. Masked; larvaceous. l. **paludism**. See *dumb ague*. l. **pneumonia**. See *pneumonia*. l. **scarlatina**, a mild case of scarlet fever in which the rash is absent.
**larvate** (*lar'-vāt*) [*larvatus*, masked]. Concealed; masked; applied to diseases and conditions that are hidden by more obvious conditions, or by some peculiarity of their symptoms.
**larvicide** (*lar'-vis-īd*) [*larva*; *cædere*, to kill]. Any agent destroying insect larvæ.
**laryngalgia** (*lar-in-gal'-je-ah*) [*larynx*; ἄλγος, pain]. Pain or neuralgia of the larynx.
**laryngeal** (*lar-in'-je-al*) [*larynx*]. Pertaining to the larynx. l. **crisis**, an acute laryngeal spasm occurring in the course of tabes dorsalis. l. **edema**, edema caused by infiltration of a fluid or semifluid into the submucous connective tissue of the larynx. l. **mirror**, a small circular mirror affixed to a long handle, used in laryngoscopy.
**laryngectomy** (*lar-in-jek'-to-me*) [*larynx*; ἐκτομή, a cutting out]. Extirpation of the larynx.
**laryngemphraxis** (*lar-in-jem-fraks'-is*) [*larynx*; ἔμφραξις, stoppage]. Closure or obstruction of the larynx.
**laryngismal** (*lar-in-jis'-mal*). Relating to laryngismus.
**laryngismus** (*lar-in-jis'-mus*) [*larynx*]. A spasm of the larynx. l. **stridulus**, a spasmodic affection of the larynx, characterized by sudden arrest of respiration, with increasing cyanosis, followed by long, loud, crowing inspirations. It is most common in rhachitic children, but may also occur as a symptom of laryngeal catarrh.
**laryngitic** (*lar-in-jit'-ik*). Pertaining to or caused by laryngitis.
**laryngitis** (*lar-in-ji'-tis*) [*larynx*; ἶτις, inflammation]. Inflammation of the larynx. It may be acute or chronic, catarrhal, suppurative, croupous (diphtheritic), tuberculous, or syphilitic. Chronic catarrhal laryngitis is divided into a hypertrophic and an atrophic stage. The symptoms of acute catarrhal laryngitis, the most common form, are hoarseness, pain, dryness of the throat, dysphagia, and cough. l., **dry**, a form characterized by heat and fatigue in the throat, persistent cough, and sometimes aphonia. Syn., *laryngitis sicca*.
**laryngo-** (*lar-in-go-*) [*larynx*]. A prefix denoting relation to the larynx.
**laryngocatarrh** (*lar-in-go-kat-ar'*). Catarrh of the larynx.

**laryngocele** (*lar-in'-go-sēl*) [*laryngo-*; κήλη, a tumor]. A saccular dilatation of the mucosa of the larynx between the hyoid bone and the cricoid cartilage.

**laryngocentesis** (*lar-in-go-sen-te'-sis*) [*laryngo-*; κέντησις, puncture]. Puncture of the larynx.

**laryngocrisis** (*lar-in-go-kri'-sis*). The paroxysmal laryngeal cramp occurring in tabes dorsalis.

**laryngofissure** (*lar-in-go-fish'-ur*) [*laryngo-*; *findere*, to cleave]. 1. Division of the larynx for the removal of tumors or foreign bodies. 2. The aperture made in the operation of laryngofissure.

**laryngograph** (*lar-in'-go-graf*) [*laryngo-*; γράφειν, to write]. An instrument for recording laryngeal movements.

**laryngography** (*lar-in-gog'-ra-fe*) [*laryngo-*; γράφειν, to write]. A description of the larynx.

**laryngologic, laryngological** (*lar-in-go-loj'-ik, lar-in-go-loj'-ik-al*) [*laryngo-*; λόγος, science]. Pertaining to laryngology.

**laryngologist** (*lar-in-gol'-o-jist*). One versed in laryngology.

**laryngology** (*lar-in-gol'-o-je*) [*laryngo-*; λόγος, science]. The science of the anatomy, physiology, and diseases of the larynx.

**laryngometry** (*lar-in-gom'-et-re*) [*laryngo-*; μέτρον, measure]. The systematic measurement of the larynx.

**laryngonecrosis** (*lar-in-go-ne-kro'-sis*) [*laryngo-*; νέκρωσις, death]. Necrosis of any portion of the larynx.

**laryngoparalysis** (*lar-in-go-par-al'-is-is*) [*laryngo-*; παράλυσις, palsy]. Paralysis of the laryngeal muscles.

**laryngopathy** (*lar-in-gop'-ath-e*) [*laryngo-*; πάθος, suffering]. Any disease of the larynx.

**laryngophantom** (*lar-in-go-fan'-tom*) [*laryngo-*; φάντασμα, an apparition]. An artificial larynx designed for illustrative purposes.

**laryngopharyngeal** (*lar-in-go-far-in'-je-al*). Pertaining conjointly to the larynx and pharynx.

**laryngopharyngeus** (*lar-in-go-far-in'-je-us*). The inferior constrictor of the pharynx.

**laryngopharyngitis** (*lar-in-go-far-in-ji'-tis*). 1. Inflammation of the larynx and the pharynx.

**laryngopharynx** (*lar-in-go-far'-inks*) [*laryngo-*; φάρυγξ, pharynx]. The inferior portion of the pharynx. It extends from the greater cornua of the hyoid bone to the inferior border of the cricoid cartilage.

**laryngophony** (*lar-in-gof'-o-ne*) [*laryngo-*; φωνή, voice]. The sound of the voice observed in auscultation of the larynx.

**laryngophthisis** (*lar-in-go-ti'-sis*) [*laryngo-*; φθίσις, wasting]. Laryngeal tuberculosis.

**laryngoplasty** (*lar-in'-go-plas-te*) [*laryngo-*; πλάσσειν, to shape]. Plastic operation upon the larynx.

**laryngoplegia** (*lar-in-go-ple'-je-ah*) [*laryngo-*; πληγή, stroke]. Paralysis of one or more muscles of the larynx.

**laryngorhinology** (*lar-in-go-ri-nol'-o-je*). Combined laryngology and rhinology.

**laryngorrhagia** (*lar-in-gor-a'-je-ah*) [*laryngo-*; ῥηγνύναι, to burst forth]. Hemorrhage from the larynx.

**laryngorrhea, laryngorrhoea** (*lar-in-gor-e'-ah*) [*laryngo-*; ῥοία, a flow]. Excessive secretion of the laryngeal mucosa, especially on attempting to use the organ.

**laryngoscleroma** (*lar-in-go-skle-ro'-mah*) [*laryngo-*; σκλήρωμα, an induration]. Scleroma affecting the larynx.

**laryngoscope** (*lar-in'-go-skōp*) [*laryngo-*; σκοπεῖν, to examine]. A mirror attached to a long handle for examining the interior of the larynx.

**laryngoscopic** (*lar-in-go-skop'-ik*). Pertaining to laryngoscopy.

**laryngoscopist** (*lar-in-gos'-ko-pist*) [see *laryngoscope*]. An expert in laryngoscopy.

**laryngoscopy** (*lar-in-gos'-ko-pe*) [see *laryngoscope*]. Examination of the interior of the larynx by means of the laryngoscope.

**laryngospasm** (*lar-in'-go-spazm*) [*laryngo-*; σπασμός, spasm]. Spasmodic closure of the glottis.

**laryngostasis** (*lar-in-gos'-tas-is*) [*laryngo-*; στάσις, stagnation]. Synonym of *croup*.

**laryngostenosis** (*lar-in-go-sten-o'-sis*) [*laryngo-*; στένωσις, contraction]. Contraction or stricture of the larynx.

**laryngostomy** (*lar-in-gos'-to-me*) [*laryngo-*; στόμα, mouth]. The establishing of a permanent opening into the larynx through the neck and trachea.

**laryngostroboscope** (*lar-in-go-stro'-bo-skōp*) [*laryngo-*; στρόβος, a twisting; σκοπεῖν, to examine]. A laryngoscope combined with an adjustable intermittent source of illumination, used in the observation of the vocal bands.

**laryngostroboscopy** (*lar-in-go-stro-bos'-ko-pe*). The inspection of the vibration of the vocal chords by means of a laryngostroboscope.

**laryngosyrinx** (*lar-in-go-si'-rinks*) [*laryngo-*; σῦριγξ, pipe; *pl.*, *laryngosyringes*]. A laryngeal tube.

**laryngotome** (*lar-in'-go-tōm*) [*laryngo-*; τομή, a cutting]. A cutting-instrument used in laryngotomy.

**laryngotomy** (*lar-in-got'-o-me*) [*laryngo-*; τέμνειν, to cut]. The operation of incising the larynx. l., complete, incision of the larynx through its whole length. l., median, incision of the larynx through the thyroid cartilage. l., subhyoid, l., superior, l., thyrohyoid, incision of the larynx through the thyrohyoid membrane.

**laryngotracheal** (*lar-in-go-tra'-ke-al*) [*laryngo-*; τραχεῖα, the windpipe]. Pertaining conjointly to the larynx and the trachea.

**laryngotracheitis** (*lar-in-go-tra-ke-i'-tis*) [*laryngo-*; *trachea*; ιτις, inflammation]. Inflammation of the larynx and the trachea.

**laryngotracheotomy** (*lar-in-go-tra-ke-ot'-o-me*) [*laryngo-*; *trachea*; τομή, a cutting]. That form of tracheotomy in which the cricoid cartilage and one or more of the upper rings of the trachea are divided.

**laryngotyphoid, laryngotyphus** (*lar-in-go-ti'-foid, lar-in-go-ti'-fus*) [*laryngo-*; *typhoid*]. Typhoid fever associated with marked laryngeal complications.

**laryngoxerosis** (*lar-in-go-zer-o'-sis*) [*laryngo-*; *xerosis*]. Dryness of the larynx.

**laryngydrops** (*lar-in'-jid-rops*) [*laryngo-*; ὕδρωψ, dropsy]. Laryngeal edema.

**larynx** (*lar'-inks*) [λάρυγξ, larynx]. The organ of the voice, situated between the trachea and the base of the tongue. It consists of a series of cartilages—the thyroid, the cricoid, and the epiglottis—and three pairs of cartilages—the arytenoids and those of Santorini and of Wrisberg— which are lined by mucous membrane and are moved by the muscles of the larynx. The mucous membrane is, on each side, thrown into two transverse folds that constitute the vocal bands, the upper being the false, the lower the true, vocal band. By the approximation or separation of the vocal bands the changes in the pitch of the voice are produced. The space between the vocal bands is termed the glottis.

**lascivia** (*las-iv'-e-ah*). Satyriasis, nymphomania.

**lascivious** (*las-iv'-e-us*) [*lascivia*, wantonness]. Libidinous. Wanton. Having an unlawful desire.

**Lasègue's law** (*las-āg'*) [Ernest Charles Lasègue, French physician, 1816–1883]. Superficial lesions or simple functional troubles of an organ increase the reflexes, while more or less pronounced organic lesions suppress them. L.'s sign. 1. Incapacity of the anesthetic hysterical individual to move the extremity which he is prevented from seeing. 2. To differentiate sciatica from hip-joint disease: in the case of the former, flexion of the thigh upon the hip is painless or easily accomplished when the knee is bent. L.'s type of mania of persecution, typical paranoia.

**lash** [LG., *lasche*, a flap]. 1. An eyelash. 2. A flagellum.

**Lassar's paste** (*las'-ar*) [Oscar Lassar, German dermatologist, 1849–1908]. A paste used in the treatment of erythema intertrigo, consisting of salicylic acid, 30 gr.; zinc oxide and powdered starch, each, 6 dr.; vaselin, 2 oz.

**lassitude** (*las'-it-ūd*) [*lassus*, tired]. A state of exhaustion or weakness, arising from causes other than fatigue.

**lata, latah** (*lah'-tah*) [Jav.]. A hysterical neurosis prevalent in Java, and nearly identical with the disorder of the jumpers in Maine and Canada, and with the miryachit of Siberia.

**latency** (*la'-tens-e*) [*latens*, to be hid]. The condition of being latent or concealed.

**latent** (*la'-tent*) [*latere*, to be hid]. Concealed; not manifest; potential. **l. heat**, that which apparently disappears when a liquid is vaporized or a solid melted. **l. period**. 1. The time required for the incubation of a disease. 2. In physiology, the time intervening between the application of a stimulus and the appearance of the resulting phenomenon.

**laterad** (*lat'-er-ad*) [*latus*, the side; *ad*, toward]. Toward the lateral aspect.

**lateral** (*lat'-er-al*) [*latus*]. 1. At, belonging to, or pertaining to the side; situated on either side of the median vertical plane. 2. Referring to structures further away from the median plane than those designated medial. l. **chain.** See *receptor*. l.-chain theory of Ehrlich, this theory presupposes that the stimulating substances introduced into the circulation have the power of combining with certain groups of molecules in the protoplasm of certain cells, which combination is succeeded by a regeneration of the lateral chains. When more of the combining substance is injected and the combining lateral chains again consumed, another still more copious regeneration occurs, and so on until the particular lateral chains are present in great excess and pass out of the cells into the blood, where they are known as antitoxin. See *immunity, theory of, Ehrlich's side-chain*. l. **column,** that column of the spinal cord between the anterior and posterior horns. l. **operation,** that form of lithotomy in which the opening is made on the right or the left side of the perineum. l. **sclerosis, amyotrophic,** a disease of the lateral columns and anterior gray matter of the cord. It is characterized by motor weakness and a spastic condition of the limbs, associated with atrophy of the muscles and final involvement of the nuclei in the medulla oblongata. l. **sclerosis, primary,** a sclerotic disease of the crossed pyramidal tracts of the cord, characterized by paralysis of the limbs, with rigidity, increased tendon-reflexes, and absence of sensory and nutritive disorders. A peculiar characteristic jerking gait is produced, and clonus of the lower limbs may be readily excited. l. **sinuses,** the two veins of the dura mater situated in the attached margin of the tentorium cerebelli. l. **ventricles.** See *ventricles, lateral*.
**lateralization** (*lat-er-al-iz-a'-shun*) [*lateralis*, lateral]. The localization of a disease upon one of the other side of the body.
**lateralized** (*lat'-er-al-īzd*) [*lateralis*, lateral]. Localized on one side; directed to one side.
**lateren** (*lat'-er-en*) [*latus*, side]. Belonging to the lateral aspect in itself.
**latericumbent** (*lat-er-ik-um'-bent*) [*latus*, side; *cumbere*, to lie]. Lying on the side.
**lateriflection, lateriflexion** (*lat-er-if-lek'-shun*). See *lateroflection*.
**laterigrade** (*lat'-er-ig-rād*) [*latus*, side; *gradus*, step]. Progressing sideways.
**lateritious, latericeous** (*lat-er-ish'-us*) [*later*, a brick]. Resembling brick-dust, as the *lateritious* sediment of the urine.
**lateriversion** (*lat-er-if-er'-shun*). See *lateroversion*.
**latero-** (*latus*, side]. A prefix signifying to one side; lateral.
**lateroabdominal** (*lat-er-o-ab-dom'-in-al*) [*latero-; abdominalis*, pertaining to the abdomen]. Pertaining both to the side and to the abdomen. l. **posture,** Sims' posture.
**laterocervical** (*lat-er-o-ser'-vik-al*) [*latero-; cervix,* the neck]. At or about the side of the neck.
**laterodeviation** (*lat-er-o-de-ve-a'-shun*) [*latero-; deviare,* to deviate]. Slight displacement or deviation to one side.
**lateroduction** (*lat-er-o-duk'-shun*) [*latero-; ducere,* to lead]. Lateral movement of the eye.
**lateroflexion** (*lat-er-o-flek'-shun*) [*latero-; flectere,* to bend]. Flexion or bending to one side.
**lateromarginal** (*lat-er-o-mar'-jin-al*) [*latero-; margo,* edge]. Placed on the lateral edge.
**lateronuchal** (*lat-er-o-nū'-kal*) [*latero-; nucha,* nape]. Situated at the side of the nape of the neck.
**lateroposition** (*lat-er-o-po-zish'-un*) [*latero-; ponere,* to place]. Displacement to one side.
**lateropulsion** (*lat-er-o-pul'-shun*) [*latero-; pellere,* to drive]. An involuntary motion to one side.
**laterotorsion** (*lat-er-o-tor'-shun*) [*latero-; torquere,* to turn]. A twisting or turning.
**lateroversion** (*lat-er-o-ver'-shun*) [*latero-; vertere,* to turn]. A turning to one side, as *lateroversion* of the uterus.
**latex** (*la'-teks*) [L., liquid: *pl., latices*]. The sap or the juice of the tubes or vessels of plants. l.-**cells,** cells giving rise to latex or milky juice.
**Latham's circle** (*la'-tham*) [Peter Mere Latham, English physician, 1789–1875]. A point midway between the left nipple and the lower end of the sternum is made the center of a circle two inches in diameter; this circle corresponds to the area of pericardial dulness.

**lathyrin** (*lath'-ir-in*) [λάθυρις, pulse]. A bitter extractive of lathyrus.
**lathyrism** (*lath'-ir-ism*) [λάθυρος, a kind of pulse]. An affection produced by the use of meal from varieties of vetches, chiefly *Lathyrus sativus* and *L. cicera*. It is a form of spastic paraplegia with tremor, involving chiefly the legs Syn., *lupinosis*.
**latibulum** (*la-tib'-ū-lum*) [*látere,* to lie hid]. A lurking-place for disease, infection, or poison.
**latissimus** (*lat-is'-im-us*) [superl. of *latus,* wide]. An adjective signifying widest. l. **colli,** the *platysma myoides*; see under *muscle*. l. **dorsi.** See under *muscle*.
**latrine** (*la-trēn'*) [Fr.]. A water-closet or privy, especially one in trough form and capable of accommodating several persons at the same time.
**lattice-work** (*lat'-is-wurk*) of **the thalamus.** The formatio reticularis, *q. v.*
**laudable** (*law'-da-bl*) [*laudare,* praise]. Praiseworthy. 1. See *pus, laudable*.
**laudanine** (*lawd'-an-en*) [*laudanum*], C₂₀H₂₅NO₄. One of the alkaloids of opium.
**laudanon** (*law'-dan-on*). A combination of various alkaloids of opium. It contains morphine, codeine, papaverine, thebaine, narceine and narcotine.
**laudanosine** (*law-dan'-o-sēn*) [λάδανον, a gum], C₂₁H₂₆NO₄. A crystallizable alkaloid of opium.
**laudanum** (*lawd'-an-um*) [Pers., *lādan,* a shrub]. Tincture of opium. l., Sydenham's, wine of opium.
**laugh** (*lahf*). 1. To make an audible expression of mirth. 2. The audible expression of mirth. l., canine, l., sardonic. Synonyms of *risus sardonicus, q. v.*
**laughing,** laughter (*lahf'-ing, lahf'-ter*) [AS., *hlehhan,* to laugh]. A succession of rhythmic, spasmodic expirations with open glottis and vibration of the vocal bands, and expressing mirth. l.-**gas,** nitrous oxide or nitrogen monoxide. See *nitrous oxide* under *anesthetic, general*.
**Laugier's hernia** (*lo-aje-a'*) [Stanislas *Laugier,* French surgeon, 1799–1872]. Femoral hernia through a gap in Gimbernat's ligament.
**Laumonier's ganglion** (*lo-mon-e-a'*) [Jean Baptiste *Laumonier,* French surgeon, 1749–1818]. The carotid ganglion.
**Laura's nucleus** (*low'-rah*). See *Deiters' nucleus*.
**laurel.** (*law'-rel*). See *cherry-laurel* and *laurus*. l.-**water,** a medicinal water distilled from leaves of the cherry-laurel. See *cherry-laurel*.
**laureol** (*law'-re-ol*). The proprietary name for a mixture of cocoanut-oil and palm-oil.
**laurocerasus** (*law-ro-ser'-as-us*). See *cherry-laurel*.
**laurotetanine** (*law-ro-tet'-an-en*), C₁₉H₂₃NO₄. An alkaloid from the bark of *Litsea citrata* and a number of other plants of the order *Laurineæ*. It is said to be a powerful poison, acting like strychnine on the spinal cord.
**laurus** (*law'-rus*) [L.]. A tree of the *Laurineæ*. L. **nobilis,** the true laurel, is indigenous in the south of Europe. Its fragrant oils (one essential; from the leaves, and the other fixed, from the berries) are chiefly used in liniments.
**Lauth's canal** (*lout*) [Thomas *Lauth,* German anatomist, 1758–1826]. See *Schlemm's canal*.
**lautissimus** (*law-tis'-im-us*) [*lautus,* washed]. Most thoroughly purified or rectified.
**lavage** (*lav-ahj*) [Fr.]. The irrigation or washing out of an organ, such as the stomach, the bowel, etc. 1. of **the blood,** washing toxic products from the blood-stream by intravenous injection of artificial serum in large doses.
**lavamentum** (*lav-am-en'-tum*) [*lavare,* to wash]. An injection.
**lavandula** (*lav-an'-dū-lah*). See *lavender*.
**lavation** (*lav-a'-shun*) [*lavare,* to wash]. Lavage.
**lave** (*lāv*) [*lavare,* to wash]. To wash; bathe.
**lavement** (*lāv'-ment*) [*lavare,* to wash]. 1. A wash; the act of washing. 2. An injection or enema.
**lavender** (*lav'-en-der*) [*lavanda,* a washing]. The flowers of *Lavandula vera,* a plant of the order *Labiatæ*. The active principle is a volatile oil. l. **flowers, oil of** (*oleum lavandulæ florum,* U. S. P.), used in the preparation of the spirit. l., **spirit of** (*spiritus lavandulæ,* U. S. P.). Dose 30 min.–1 dr. (2–4 Cc.). l., **tincture of, compound** (*tinctura lavandulæ composita,* U. S. P.). Dose 30 min.–1 dr. (2–4 Cc.). Both the spirit and tincture are stimulant and carminative.
**Laveran's crescent** (*lav-er-an'*) [Charles Louis Alphonse *Laveran,* French physician, 1845–     ].

# LAVIPEDIUM 503 LECITHIN

The sickle-shaped plasmodium found in the estivoautumnal form of intermittent fever. L.'s plasmodium, the hematozoon of malaria; see *plasmodium malariæ*.

**lavipedium** (*lav-ip-e'-de-um*) [*lavare*, to wash; *pes*, foot]. A foot-bath; a pediluvium.

**law** (*law*) [AS., *lagu*, a law]. A general rule; a constant mode of action of forces or phenomena. A rule of action prescribed by authority. l., Aran's, Bastian's, etc. See under the proper names. l. of definite proportions, when two or more chemical substances unite to form a compound, they do so in a fixed and constant proportion. l. of multiple proportions, two substances uniting to form a series of chemical compounds do so in proportions that are simple multiples of one another or of one common proportion. l. of reciprocal proportions, two elements combining with a third do so in proportions that are simple multiples or simple fractions of those in which they combine with each other. l. of refraction, rays of light entering a denser medium are deflected toward a perpendicular drawn through the point of incidence, and those entering a rarer medium are deflected away from the same perpendicular.

**lawn-tennis arm.** Displacement of the tendon, or body (or both) of the pronator radii teres. Common in lawn-tennis players. l.-t. **knee,** an affection occurring in tennis players, probably due to a contusion or laceration of the internal semilunar cartilage. l.-t. **leg,** tendinous, fascial, or muscular rupture in the calf or sura.

**lax** (*laks*) [*laxus*, loose]. Loose; not tense.

**laxaphen** (*laks'-af-en*). Trade name of a cathartic containing phenolphthalein and salicylic acid.

**laxarthrus** (*laks-ar'-thrus*) [*laxus*, loose; ἄρθρον, joint]. Luxation of a joint.

**laxative** (*laks'-a-tiv*) [*laxare*, to loosen]. · 1. Aperient; mildly cathartic. 2. An agent that loosens the bowels; a mild purgative.

**laxatol** (*laks'-at-ol*). Trade name of a preparation of phenolphthalein.

**laxator** (*laks-a'-tor*) [*laxare*]. That which loosens or relaxes. A name applied to various muscles. l. **tympani.** See under *muscle*.

**laxiquinine** (*laks-e-kwin'-ēn*). A proprietary remedy said to be quinine combined with laxatives.

**laxitas** (*laks'-it-as*) [L., looseness]. l. **alvi,** diarrhea. l. **gingivarum,** spongy gums. l. **intestinorum,** diarrhea. l. **ventriculi,** atony of the stomach.

**laxity** (*laks'-it-e*) [*laxitas*, looseness]. Lack or loss of tone or tension; a relaxed, loose, or spongy state of a tissue.

**laxol** (*laks'-ol*). Castor-oil combined with saccharin and oil of peppermint.

**layer** (*la'-er*) [ME., *leyer*]. A mass of uniform, or nearly uniform, thickness, spread over a considerable area. · l., **bacillar,** l., **bacillary,** the layer of rods and cones of the retina. l., **cellular,** the endothelial layer of the blood-vessels and lymph-vessels, composed of flattened nucleated cells. l., **cortical,** the cerebral cortex. l., **cuticular,** a striated, hyaline, refractive layer at the free end of a columnar cell. l., **ganglionic,** a layer of angular cells of the cerebral cortex, best developed in the motor area. l., **horny,** the superficial layer of the skin. l., **large pyramidcell,** the third layer of the cerebral cortex, composed of pyramidal cells larger than those of the external pyramid-cell layer, increasing in size from above downward and attaining a diameter of 40 μ. l., **molecular,** the outermost layer of the cerebral cortex, made up of neuroglia, a few small ganglion-cells, and a reticulum of medullated and nonmedullated nerve-fibers. l., **oophorous,** the outer portion of the ovary, in which the ovules are formed. l., **osteoblastic,** l., **osteogenetic,** the lower layer of periosteum, connected with the formation of bone. l. **of rods and cones.** See l., *bacillar*. · l., **serous,** the inner layer of the pericardium. l., **small pyramid-cell,** the second layer of the cerebral cortex, made up of small pyramidal cells, with a diameter of about 10 μ. l., **Waldeyer's,** internal or vascular layer of the ovary.

**layette** (*lay-et'*) [Fr.]. The full outfit of garments, bedding, etc., for a new-born child.

**layman** (*la'-man*) [λαός, the people]. A member of the laity; a person not a physician or not professionally educated.

**lazar** (*laz'-ar*) [*Lazarus*, a biblical name]. An old name for a leper, or for any person having a repulsive disease.

**lazaretto** (*laz-ar-et'-o*) [Ital., "a pest-house," from *lazar*, a leper]. A quarantine establishment; a pesthouse; also, a place for fumigation and disinfection.

**lb.** Abbreviation for Latin *libra*, a pound.

**L. D.** Abbreviation for *perception of light difference*.

**L. D. A.** Abbreviation for left dorsoanterior position of the fetus.

**L. D. P.** Abbreviation for left dorsoposterior position of the fetus.

**L. D. S.** Abbreviation of Licentiate of Dental Surgery.

**L. E.** Abbreviation for left eye.

**leaching** (*le'-ching*). The process of causing water or any fluid to percolate through some substance.

**lead** (*led*) [AS.]. See *plumbum*. l., **black-,** graphite. l.-**colic.** See *colic*, *lead*-. l.-**encephalopathy,** the cerebral manifestations of chronic lead-poisoning. They consist of epilepsy, acute delirium, and hallucinations, and may terminate in insanity. l.-**line,** the line of discoloration on the gums in cases of chronic lead-poisoning. Syn., *blue line*. l.-**pipe contraction,** the condition of the limbs in the cataleptic state, in which they maintain any position that is given them. l. **plaster,** an adhesive plaster containing lead oxide. l.-**poisoning,** a form of poisoning due to the introduction of lead into the system. The symptoms are disturbed nutrition, anemia, a blue line on the gums, lead-colic, constipation, pains in the limbs, local muscular paralysis (wrist-drop) and wasting, lead-encephalopathy, etc. Syn., *plumbism*; *saturnism*. l., **red,** red oxide of lead, $Pb_3O_4$, a poisonous red powder used in salves. l., **sugar of,** $Pb(C_2H_3O_2)_2$, lead acetate, used as an astringent and styptic. l., **test for, in system,** paint a small area of the skin with a 6 % solution of sulphite. If lead is present, the painted area will darken after a few days. [Ciccoñardi.] l., **test for, in the urine,** administer potassium iodide for four days, collecting the urine. Evaporate to a pint (500 Cc.) and filter. Pass hydrogen sulphide gas through the urine thus concentrated, when a black precipitate will form if lead is present. [White.] l., **test for, in water.** See *Blythe*. l., **white,** lead carbonate, $_2PbCO_3 \cdot Pb(OH)_2$. It is used in skin diseases.

**leader** (*le'-der*) [AS., *ladere*]. A sinew or tendon.

**leading** (*led'-ing*). Among smelters a popular term for chronic lead-poisoning.

**leaping** (*lēp'-ing*). Springing; jumping. l. **ague,** a synonym of *dancing mania*.

**leathery** (*leth'-er-e*) [AS., *lether*]. Resembling leather, as *leathery arteries*, arteries thickened and feeling like cords of leather.

**Lebbin's test, for formaldehyde in milk.** Boil a few Cc. of the suspected milk with 0.05 Gm. resorcinol, to which half, or an equal volume of a 5 % solution of sodium hydroxide is added. In the presence of formaldehyde the yellow solution changes to a fine red color, which becomes more apparent on standing.

**leben** (*leb'-en*) [Ar., *leban*]. A variety of fermented milk of the Arabs. It corresponds to the "matzoon" of the Turks.

**Leber's corpuscles** (*la'-ber*) [Theodor *Leber*, German ophthalmologist, 1840– ]. See *Gierke's corpuscles*. **L.'s disease,** hereditary optic atrophy. **L.'s plexus,** a plexus of venules in front of Schlemm's canal, with which it communicates.

**lecane** (*lek'-an-e*) [λεκάνη, a little pan]. 1. A basin. 2. The pelvis.

**lecanic** (*lek-an'-ik*). Pelvic.

**Lecat's gulf** (*lek-ah'*) [Claude Nicolas *Lecat*, French surgeon, 1700–1768]. The dilated bulbous portion of the urethra.

**lechopyra** (*lek-op'-ir-ah*) [λεχώ, a lying-in woman; πῦρ, fire, fever]. Puerperal fever.

**lecibrin** (*les'-ib-rin*). Trade name of a nucleoprotein containing lecithin.

**lecithalbumins** (*les-ith-al'-bū-mins*). More or less stable compounds of albumin and lecithin, found in the mucoza of the intestines, in the lungs, liver, kidney, and spleen, differing from nucleoproteids and nucleoalbumins in that there is no metaphosphoric acid split off and that they yield no xanthin bases.

**lecithigenous** (*les-ith-ij'-en-us*) [*lecithin*; γεννᾶν, to beget]. Producing lecithin.

**lecithin** (*les'-ith-in*) [λέκιθος, yolk of egg], $C_4H_{90}NPO_9$. A complex nitrogenous fatty substance, occurring widely spread throughout the animal body. It is found in the blood, bile, serous fluids, brain, nerves, yolk of egg, semen, pus, and white

**lecithinose** *(les-ith'-in-ōs)*. Trade name of a lecithin preparation made from the yolk of eggs.

**lecithoblast** *(les-ith'-o-blast)* [*lecithin*; βλαστός, a germ]. One of the cells of the yolk-cavity in the mammalian ovum. Syn., *lecithophore*.

**lecithoid** *(les'-ith-oid)* [*lecithin*; εἶδος, like]. Resembling lecithin.

**lecithophore** *(les-ith'-o-for)*. See *lecithoblast*.

**lecithoprotein** *(les-ith-o-pro'-te-in)*. A compound of lecithin with a protein molecule.

**lecithymen** *(les-ith-i'-men)* [λέκιθος, yolk; ὑμήν, membrane]. The vitelline membrane.

**Leclanché's battery** *(lek-lahn-sha')* [—— Leclanché, French engineer]. Positive element, zinc; negative element, carbon; exciting agent, ammonium chloride; depolarizing agent, manganese dioxide; E.M.F., 1.42 volts.

**lectual** *(lek'-tū-al)* [*lectus*, a bed]. Pertaining to a bed or couch. l. disease, a disease that confines one to bed.

**Ledoyen's disinfecting-fluid.** A solution of lead nitrate.

**Leduc current** *(led-ook')* [Stephane Armand Nicolas Leduc, French physicist]. A form of direct electric current, used in the production of electric narcosis.

**leech** *(lēch)* [AS., *læce*, physician]. 1. An old term for a physician. 2. A blood-sucking worm, the *Sanguisuga* of *Hirudo medicinalis*, found in Europe, and the *Hirudo decora*, the American leech. Leeches are used for the local abstraction of blood. A European leech draws ½–1 oz. (16–32 Cc.) of blood. l., artificial, an apparatus for cupping.

**leeches** *(lēch-ez)*. A mycotic disease of mules and cattle.

**lees** *(lēz)* [Fr., *lie*, dregs]. The dregs of vinous liquors.

**leeting** *(lēt'-ing)*. The exudation on the surface of the skin in eczema.

**left.** Sinistral; opposite of right. The left-hand side. l.-brained, having the speech-center in its normal situation, in the left third frontal, or subfrontal convolution. l.-eared, preferring the sinistral ear, as the one with which to hear sounds. l.-eyed, preferring the sinistral eye as the dominant one. l.-footed, the power is furnished and governed by the sinistral foot. l.-handed, preferring the sinistral hand for the more expert or intellectual tasks. l.-handedness, the quality of being left-handed.

**leg.** The lower extremity, especially that part from the knee to the ankle. l., badger, inequality in the length of legs. l., bakers'. See *genu valgum*. l., bandy. Same as l., bow. l., Barbados, elephantiasis of the leg. l., bayonet, uncorrected backward displacement of the leg-bones. l., black, symptomatic anthrax. l., bow, a curving outward of the legs. l., milk-, phlegmasia alba dolens. l., scissor, a crossing of the axes of the legs.

**legal** *(le'-gal)* [*lex*, *legis*, law]. Pertaining to law. l. medicine, medical jurisprudence.

**Legal's disease** *(la'-gal)* [Emmo Legal, German physician]. Paroxysmal pains and tenderness of the scalp in the region supplied by the auriculotemporal nerve, associated with pharyngotympanic catarrh. Syn., *cephalalgia pharyngotympanica*. L.'s test for acetone in urine, acidulate the urine with hydrochloric acid and distil. To the distillate add a few drops each of sodium nitroprusside and solution of potassium hydroxide. This produces a ruby-red color, which changes to purple on the addition of acetic acid. Creatinin gives a similar color, but it disappears when acetic acid is added.

**Le Gendre's nodosities** *(leh-zjandr')*. See *Bouchard's nodosities*.

**legitimacy** *(le-jit'-im-as-e)* [*legitimus*, lawful]. Born within wedlock, or within a period of time necessary to gestation, which may elapse after the death of the father.

**Legroux's remissions** *(leh-groo')* [Treves de Legroux]. Lengthy remissions which sometimes occur in the course of pulmonary tuberculosis.

**legume** *(leg'-ūm, leg-ūm')* [*legere*, to gather (so-called because it may be gathered by the hand)]. In biology, the fruit of plants of the pea or pulse family, a pod splitting along both sutures, and formed from a simple pistil.

**legumelin** *(leg-ū'-mel-in)* [*legumin*]. An albumin found in most leguminous seeds.

**legumin** *(leg-ū'-min)* [*legumen*, pulse]. A proteid found in the seeds of many plants belonging to the natural order of *Leguminosa*.

**Lehman's sign** *(la'-man)*. In the administration of chloroform, to prognosticate as to a ready or difficult anesthesia, if the eyelids closed by the anesthetizer reopen at once, wholly or in part, the anesthesia will be difficult. The eyelids will remain closed from the beginning in those who take chloroform well.

**leiocephalous** *(li-o-sef'-al-us)* [λεῖος, smooth; κεφαλή, head]. Having a smooth head.

**leiodermatous** *(li-o-der'-mat-us)* [λεῖος, smooth; δέρμα, skin]. Smooth-skinned.

**leiodermia** *(li-o-der'-me-ah)* [λεῖος, smooth; δέρμα, skin]. A disease of the skin marked by abnormal glossiness, and by atrophy.

**leiomyofibroma** *(li-o-mi-o-fi-bro'-mah)* [λεῖος, smooth; μῦς, muscle; *fibra*, fiber]. A tumor presenting the characteristics of a leioma, a myoma, and a fibroma.

**leiomyoma** *(li-o-mi-o'-mah)* [λεῖος, smooth; μῦς, muscle; ὄμα, a tumor: *pl.*, *leiomyomata*]. A tumor consisting largely of unstriped muscle-fibers.

**leiotrichous, liotrichous** *(li-ot'-rik-us)* [λεῖος, smooth; θρίξ, hair]. Having smooth or straight hair. See *lissotrichous*.

**leiphemia** *(li-fe'-me-ah)* [λείπειν, to fail; αἷμα, blood]. Failure, poverty, or a depraved state of the blood.

**leipodermia** *(li-po-der'-me-ah)* [λείπειν, to leave; δέρμα, skin]. Absence of the skin.

**leipomeria** *(li-po-me'-re-ah)* [λείπειν, to leave; μέρος, part]. A monstrosity with absence of one or more limbs.

**leipopsychia** *(li-po-psi'-ke-ah)* [λείπειν, to fail; ψυχή, spirit]. Fainting; weakness; asphyxia.

**leipothymia** *(li-po-thi'-me-ah)* [λείπειν, to fail; θυμός, mind]. A fainting or syncope.

**leipothymic** *(li-po-thi'-mik)* [λείπειν, to fail; θυμός, mind]. Faint; affected with or pertaining to leipothymia.

**leipyria** *(li-pi'-re-ah)* [λείπειν, to fail; πῦρ, fire]. Coldness, as of the extremities, during a high fever.

**Leishman-Donovan bodies** *(lish'-man-don'-o-van)* [Sir William B. Leishman, British army surgeon; C. Donovan, British army surgeon]. Small parasite-like bodies on the liver and spleen of those suffering from kala-azar. Also known as *Piroplasma donovani*; *Leishmania donovani*.

**Leishmaniosis** *(lish-man-e-o'-sis)*. A disease due to in  i n with any species of Leishman-Donovan bodyfect o

**leistungskern** *(li'-stung-skern)* [Germ.]. The central chemical nucleus or active center of a cell.

**Leiter's coil, L.'s tubes** *(li'-ter)*. Tubes of soft, flexible metal designed for application about any part of the body. Cold water is passed through the tubes, thereby reducing the temperature of the parts incased.

**lema** *(le'-mah)* [λήμη, rheum]. The collection of dried matter sometimes seen at the canthus of the eye.

**Lembert suture** *(lam-bār')* [Antoine *Lembert*, French surgeon, 1802–1851]. An intestinal suture for wounds in which the needle is passed transversely to the wound through the peritoneal and muscular coats, and out again on one side of the wound, and then carried across the wound and made to penetrate the two outer coats as before.

**lemma** *(lem'-ah)* [λέμμα, peel; *pl.*, *lemmata*]. A general name for a limiting or ensheathing membrane; neurilemma, sarcolemma, etc. It is mostly used in composition. In embryology, the outer layer of the germinal vesicle.

**lemniscus** *(lem-nis'-kus)*. See *fillet* (2).

**lemon** *(lem'-on)*. See *limo*.

**lemonade, sulphuric-acid.** Solution of sugar in water containing sulphuric acid. Used in lead manufactories to prevent lead-poisoning.

**lemoparalysis** *(le-mo-par-al'-is-is)* [λαιμός, gullet; παράλυσις, palsy]. Paralysis of the esophagus.

**lemostenosis** *(le-mo-ste-no'-sis)* [λαιμός, gullet; στένωσις, stricture]. Constriction of the pharynx or esophagus.

**Lenhossek, bundle of.** The ascending root of the vagus and glossopharyngeal nerves.

**leniceps** (*len'-is-eps*) [*lenis*, mild; *capere*, to seize]. A form of obstetric forceps, with short handles; it is so called because it was designed to be safer in use than the ordinary forms.
**lenicet** (*len'-is-et*). A preparation of aluminum acetate.
**lenient** (*le'-ne-ent*) [*leniens*, softening]. 1. Lenitive; emollient. 2. An emollient medicine or application.
**lenigallol** (*len-e-gal'-ol*). Pyrogallol triacetate; a white, nontoxic powder, soluble in aqueous alkaline solutions. It is used in psoriasis and eczema in 0.5 to 5 % ointment; it does not affect the healthy skin.
**leniment** (*len'-im-ent*) [*lenire*, to soothe]. A liniment or soothing application.
**leniol** (*len'-e-ol*). Trade name of a preparation of cod-liver oil.
**lenirenin** (*len-e-ren'-in*). A preparation containing aluminum acetate, cocaine and adrenin. It is used as a snuff in catarrhal conditions of the nose.
**lenirobin** (*len-ir'-o-bin*). Chrysarobin tetracetate. It is used in the treatment of skin diseases.
**lenitis** (*len-i'-tis*). A term for phlegmonous gastritis, or suppurative inflammation of the areolar tissue of the stomach.
**lenitive** (*len'-it-iv*) [*lenitivus*, soothing]. Emollient or demulcent; laxative or aperient. 2. An emollient remedy or application; an aperient or mildly cathartic agent. L electuary, confectio sennæ.
**Lennander's incision**. An incision to the right or left of the median line of the abdomen down to the rectus; the inner edge of this muscle is then retracted and the posterior layer of its sheath incised as well as the peritoneum.
**lennesin** (*len'-es-in*). A cholagogue, said to be a glucoside from a species of *Conyza*.
**Lennhoff's sign**. In cases of echinococcus-cyst on deep inspiration a furrow forms above the tumor between it and the edge of the ribs.
**lens** (*lenz*) [L., "a lentil"]. 1. A piece of glass or crystal for the refraction of rays of light. 2. The crystalline lens of the eye. l., achromatic. See *achromatic lens*. l., apochromatic. See *apochromatic lens*. l., biconcave (negative or minus (—) lens), a thick-edged lens having concave spherical surfaces upon its opposite sides; it is used in spectacles to correct myopia. l., biconvex (positive or plus (+) lens), a thin-edged lens; it has two convex surfaces, and is used to correct hyperopia. l., bifocal. See *bifocal lens*. l., convergent, l., converging, a double convex or planoconvex lens that focuses rays of light. l., convexoconcave, a lens having a convex and a concave surface, which would not meet if continued. Its properties are those of a convex lens of the same focal distance. l., crystalline, the lens of the eye, a biconvex transparent body lying in its capsule immediately behind the pupil of the eye and kept in place by its suspensory ligament. It serves to refract the rays of light entering the pupil and impinging on its surface so as to bring them to a focus upon the retina. l., cylindrical (either minus or plus), one with a plane surface in one axis and a concave or convex surface in the axis at right angles to the first. l., decentered, one with the optic center not opposite to the pupil of the eye. l., dispersing, a concave lens. l., orthoscopic, one which gives a flat, undistorted field of vision. l., periscopic, one with concavoconvex or convexoconcave surfaces, the opposite sides being of different curvatures; such lenses are called meniscus lenses. l., spherical, one the curved surface of which, either concave or convex, is a segment of a sphere. l., Stokes'. See *Stokes' lens*.
**lenticel** (*len'-tis-el*) [*lenticella*; dim; of *lens*, lentil]. Any one of the little mucous follicles or crypts at the base of the tongue; any lenticular gland.
**lenticonus** (*len-tik-o'-nus*) [*lens*; *conus*, a cone]. A rare, usually congenital, anomaly of the lens, in which there is a conical prominence upon its anterior or, more rarely, upon its posterior, surface.
**lenticula** (*len-tik'-u-lah*) [L., "a lentil"]. 1. The lenticular nucleus. 2. A freckle.
**lenticular** (*len-tik'-u-lar*) [*lenticula*]. 1. Pertaining to or resembling a lens. 2. Pertaining to the crystalline lens. 3. Pertaining to the lenticular nucleus of the brain. 4. Having the shape of a lentil; lentiform. l. arteries, the arteries supplying the lenticular nucleus. l. ganglion, the ophthalmic ganglion. l. nucleus, a mass of gray matter, the extraventricular portion of the corpus striatum,

situated to the outer side of the internal capsule of the brain. Syn., *lenticula*.
**lenticulate** (*len-tik'-ū-lāt*) [*lens*]. Lens-shaped; lentil-shaped.
**lenticulo-optic** (*len-tik-ū-lo-op'-tik*) [*lenticula*; *optic*]. Relating to the lenticular nucleus and the thalamus.
**lenticulostriate** (*len-tik-ū-lo-stri'-āt*) [*lenticula*; *striatus*, striated]. Pertaining to the lenticular nucleus of the corpus striatum, as *lenticulostriate artery*.
**lenticulothalamic** (*len-tik-ū-lo-thal'-am-ik*) See *lenticulo-optic*.
**lentiform** (*len'-tif-orm*) [*lens*; *forma*, form]. Lens-shaped or lentil-shaped.
**lentigines** (*len-tij'-in-ēz*) [*lens*, a lentil]. See *lentigo*. l. leprosæ, the pigmented spots of macular leprosy.
**lentiginose, lentiginous** (*len-tij'-in-ōs, len-tij'-in-us*) [*lentiginosus*, freckled]. Affected with lentigo. Speckled or freckled.
**lentigo** (*len-ti'-go*) [L., "a lentil-shaped spot"; pl., *lentigines*]. A freckle; a circumscribed patch of pigment, small in size, occurring mainly on face and hands, and due to exposure to the sun. l. æstiva, summer freckles.
**lentitis** (*len-ti'-tis*). See *phakitis*.
**lentor** (*len'-tor*) [*lentus*, adhesive]. Viscidity of a liquid; slowness of any function, or process. l. cordis, sluggishness of the heart.
**lenus** (*le'-nus*) [λῆνός, a depression]. A depression; the torcular Hierophili, *q. v.*
**Leo's sugar**. See *laiose*.
**leontiasis** (*le-on-ti'-as is*) [λίων, a lion]. A lion-like appearance of the face, seen in leprosy, elephantiasis, and leontiasis ossea. l. ossea, l. ossium, an overgrowth of the bones of the face, through which the features acquire a lion-like appearance. Syn., *megalocephaly*.
**leontodin** (*le-on'-to-din*) [λίων, a lion; ὀδούς, a tooth]. The precipitate from a tincture of the root of dandelion, *Leontodon taraxacum*; it is a tonic, diuretic, aperient, and hepatic stimulant. Dose 2–4 gr. (0.13–0.26 Gm.).
**leontodon** (*le-on'-to-don*). See *taraxacum*.
**Leopold's law** (*le'-o-pōld*) [Christian Gerhard Leopold, German physician, 1846– ]. Insertion of the placenta into the posterior uterine wall pushes the Fallopian tubes forward, so that they assume a convergent direction on the anterior wall; insertion into the anterior wall causes them to turn backward and parallel to the longitudinal axis of the recumbent woman.
**leper** (*lep'-er*) [λεπρός, scaly]. One affected with leprosy.
**lepidic** (*lep-id'-ik*) [λεπίς, a scale; a husk]. Applied to the tissues of lining membranes characterized by absence of definite stroma between the individual cells. l. tumor, lepidoma, *q. v.*
**lepido-** (*lep-id-o-*) [λεπίς, a scale]. A prefix signifying a scale or scaly.
**lepidoma** (*lep-id-o'-mah*) [λεπίς, a husk]. A term proposed by Adami for a tumor springing from the tissue of a lining membrane and distinguished as epilepidoma, hypolepidoma, mesolepidoma, and endolepidoma, according to the origin of the neoplasm from the epiblastic, hypoblastic, mesothelial, or endothelial structures.
**lepidoid** (*lep'-id-oid*) [λεπίς, scale]. Having the appearance of a scale. Squamous.
**lepidoplastic** (*lep-id-o-plas'-tik*) [λεπίς, scale; πλάσσειν, to form]. Forming scales.
**lepidoptera** (*lep-id-op'-ter-ah*) [λεπίς, scale; πτερόν, a wing]. An order of insects distinguished by featherlike scales and spirally-coiled suctorial apparatus. The order includes butterflies and moths.
**lepidosarcoma** (*lep-id-o-sar-ko'-mah*). A sarcoma covered with scales, occurring in the mouth.
**lepidosin** (*lep-id'-o-sin*) [λεπίς, scale]. A substance in the scales of fishes analogous to dentine.
**lepidosis** (*lep-id-o'-sis*). Same as *ichthyosis*. Also a synonym of *lepra* and of *pityriasis*.
**lepine** (*le'-pēn*). An antiseptic fluid said to consist of mercuric chloride 0.001 Gm.; phenol and salicylic acid, each, 0.1 Gm.; benzoic acid and calcium chloride each, 0.05 Gm.; bromine, 0.01 Gm.; quinine hydrobromide, 0.2 Gm.; chloroform, 0.2 Gm.; distilled water, 100 parts.
**lepocyte** (*lep'-o-sīt*) [λεπίς, a scale, a husk; κύτος, a hollow]. A nucleated cell possessing a cell-wall. Cf. *gymnocyte*.
**lepothrix** (*lep'-o-thriks*) [λέπος, a scale; θρίξ, a hair].

# LEPRA 506 LESION

A condition in which the hairs of the axillæ or scrotum are incased in a sheath of hardened sebaceous matter.
**lepra** (*lep'-rah*) [λέπρα, leprosy], 1. Leprosy. 2. A form of psoriasis. 1. **anæsthetica**, leprosy in which anesthesia predominates. 1. **asturiensis**, pellagra. 1. **maculosa**, the stage of true leprosy marked by the presence of pigment-spots. 1. **mutilans**, the final stage of true leprosy, marked by loss of members. 1. **tuberculosa**, a form of leprosy characterized by the presence of tubercles.
**lepraphobia.** See *leprophobia*.
**leprelcosis** (*lep-rel-ko'-sis*) [λέπρα, lepra; ἕλκωσις, ulceration]. Leprous ulceration.
**lepriasis** (*lep-ri'-as-is*) [λέπρα, leprosy]. A synonym of leprosy and of psoriasis; an old term vaguely used to designate various scaly diseases of the skin.
**lepric** (*lep'-rik*). Pertaining to lepra.
**leprid** (*lep'-rid*). A skin-lesion of leprosy.
**leprolin** (*lep'-ro-lin*). A vaccine used in the treatment of leprosy.
**leprologist** (*lep-rol'-o-jist*) [λέπρα, leprosy; λόγος, science]. An expert or specialist in leprology.
**leprology** (*lep-rol'-o-je*) [λέπρα, leprosy; λόγος, science]. The special study of leprosy.
**leproma** (*lep-ro'-mah*) [*lepra*]. The specific lesion of tubercular leprosy.
**leprophobia** (*lep-ro-fo'-be-ah*) [λέπρα, leprosy; φόβος, fear]. Morbid or insane dread of leprosy.
**leprophthalmia** (*lep-rof-thal'-me-ah*) [λέπρα, leprosy; ὀφθαλμός, the eye]. Ophthalmia of a leprous character.
**leprosarium** (*lep-ro-sa'-re-um*) [L.]. A leper-house; leprosery.
**leprosery** (*lep'-ro-ser-e*) [Fr., *léproserie*]. Same as *leprosarium*.
**leprosis** (*lep-ro'-sis*). Leprosy.
**leprosity** (*lep-ros'-it-e*). Leprousness; the state of being leprous.
**leprosy** (*lep'-ro-se*) [*lepra*]. An endemic, chronic, infectious disease, due to *Bacillus lepræ*. Two forms of leprosy are described—a tubercular and an anesthetic form. The first begins with a well-defined erythema (*macular leprosy*), which is succeeded by the formation of papules, and later of nodules, although in some cases the erythematous stage is followed by a disappearance of the pigment without nodulation (*lepra alba*). The nodules eventually break down and ulcerate. The anesthetic form begins with pains and hyperesthesia; a macular eruption appears, but later subsides, leaving spots of anesthesia; trophic lesions develop and lead to the loss of the fingers or toes, with marked deformity.
**leprotic, leprous** (*lep-rot'-ik, lep'-rus*). Affected with, or relating to, leprosy.
**leptandra** (*lep-tan'-drah*) [λεπτός, thin; ἀνήρ, male]. A former genus of plants. The *leptandra* of the U. S. P. is the rhizome and rootlets of *Veronica virginica*. Its properties are thought to be due to a glucoside *leptandrin*. It is tonic, laxative, and cholagogue, and is used in indigestion and chronic constipation. Syn., *culver's-root*. L., extract of (*extractum leptandræ*, U. S. P.). Dose 1–3 gr. (0.065–0.2 Gm.). l., fluidextract of (*fluidextractum leptandræ*, U. S. P.). Dose 20 min.–1 dr. (1.3–4.0 Cc.).
**leptandrin** (*lep-tan'-drin*). See *leptandra*.
**lepthymenia** (*lep-thi-me'-ne-ah*) [λεπτός, thin; ὑμήν, membrane]. Delicacy or thinness of membrane.
**lepto-** (*lep-to-*) [λεπτός, thin]. A prefix meaning thin.
**leptocephalia** (*lep-to-sef-a'-le-ah*) [*lepto-*; κεφαλή, head]. Abnormal smallness or narrowness of the skull.
**leptocephalic, leptocephalous** (*lep-to-sef-al'-ik, lep-to-sef'-al-us*) [*lepto-*; κεφαλή, head]. Having an abnormally small head.
**leptocephalus** (*lep-to-sef'-al-us*) [*lepto-*; κεφαλή, head]. A monster with an abnormally small head from premature union of the frontal and sphenoid bones.
**leptochasmus** (*lep-to-kas'-mus*) [*lepto-*; χάσμα, chasm]. Lissauer's term for a skull in which the angle formed by two lines drawn from the punctum alæ vomeris to the posterior nasal spine and the anterior margin of the foramen magnum respectively is between 94° and 114°.
**leptochroa** (*lep-tok'-ro-ah*) [*lepto-*; χρόα, skin]. Delicacy of the skin.
**leptochymia** (*lep-to-kim'-e-ah*) [*lepto-*; χυμός, juice]. Abnormal thinness or meagerness of the fluids of the body.

**leptodactylous** (*lep-to-dak'-til-us*) [*lepto-*; δάκτυλος, digit]. Characterized by slenderness of the fingers, or toes, or both.
**leptodermic, leptodermous** (*lep-to-der'-mik, lep-to-der'-mus*) [*lepto-*; δέρμα, skin]. Having a delicate skin.
**leptodontous** (*lep-to-don'-tus*) [*lepto-*; ὀδούς, tooth]. Having thin or slender teeth.
**leptomeninges** (*lep-to-men-in'-jes*) [*lepto-*; μῆνιγξ, a membrane]. The arachnoid and pia, or the pia alone.
**leptomeningitis** (*lep-to-men-in-ji'-tis*) [*lepto-*; μῆνιγξ, membrane; ιτις, inflammation]. Inflammation of the pia and arachnoid of the brain or the spinal cord.
**leptomeninx** (*lep-to-men'-ingks*) [*lepto-*; μῆνιγξ, membrane: *pl.*, *leptomeninges*]. The pia or the arachnoid; in the plural (*leptomeninges*), usually the arachnoid and pia taken together.
**leptophonia** (*lep-to-fo'-ne-ah*) [*lepto-*; φωνή, voice]. Delicacy, gentleness, or weakness of the voice.
**leptophonic** (*lep-to-fon'-ik*) [*lepto-*; φωνή, voice]. Having a weak voice.
**leptoprosope** (*lep-top'-ro-sōp*) [*lepto-*; πρόσωπον, face]. A person, or a head, with a long, narrow face.
**leptoprosopia** (*lep-to-pro-so'-pe-ah*) [*lepto-*; πρόσωπον, face]. Narrowness of the face.
**leptoprosopic, leptoprosopous** (*lep-to-pro-sop'-ik, lep-to-pros'-o-pus*) [*lepto-*; πρόσωπον, face]. Having a long, narrow face.
**leptorrhine** (*lep'-tor-in*)- [*lepto-*; ῥίς, nose]. Having a slender nose or proboscis.
**leptorrhinia** (*lep-tor-in'-e-ah*) [*lepto-*; ῥίς, nose]. Narrowness of the nasal bones, or smallness of the nasal index.
**Leptothrix** (*lep'-to-thriks*) [*lepto-*; θρίξ, hair]. A genus of bacteria, the elements of which form straight filaments, often of great length. L. **buccalis**. Syn., *Leptothrix gigantea*, Miller; *Leptothrix pulmonalis*; *Rasmussenia buccalis*, Saccardo. Found in the mouth of man and animals. Falsely considered the cause of dental caries. L. **epidermidis**. Syn., *Bacillus epidermidis*, Bizzozero; *Microsporon minutissimum*, Burchardt. Found on the epidermis between the toes, and held by Bizzozero to be nonpathogenic, but by Boeck to be the cause of erythrasma. L. **gigantea**, Miller. Found on the teeth of dogs, sheep, cattle, and other animals affected with pyorrhœa alveolaris. Some consider it identical with L. buccalis. L. **innominata**. See *L. buccalis*. L. **pulmonalis**. See *L. buccalis*. L. **vaginalis**, found in the vagina of animals and women. L. **variabilis**, found in saliva of healthy persons.
**leptotrichia** (*lep-to-trik'-e-ah*) [*lepto-*; θρίξ, hair]. Abnormal or excessive fineness and delicacy of the hair.
**Leptus** (*lep'-tus*) [λεπτός, thin]. A genus of beetles. L. **autumnalis**, a parasite that burrows under the skin, causing lesions like the itch. Syn., *harvest-bug; mower's mite*.
**leptynol** (*lep'-tin-ol*). A preparation said to contain colloidal palladium hydroxide in liquid paraffin; it is said to be of value in reducing obesity.
**leptystic** (*lep-tis'-tik*) [λεπτύνω, to make thin]. Relating to or affected with emaciation.
**lerema** (*ler-e'-mah*) [λήρημα, idle talk; *pl.*, *leremata*]. An idle or childish utterance, as in senility, idiocy, or dementia.
**leresis** (*ler-e'-sis*) [λήρησις, a speaking foolishly]. Garrulousness; insane or senile loquacity.
**leros** (*le'-ros*) [λῆρος, foolish speaking]. Slight delirium, with talkativeness.
**Lesbian love** (*les'-be-an*) [*Lesbos*, an island belonging to Greece]. See *tribadism*.
**lesbianism** (*les'-be-an-izm*). The doctrine and practice of Lesbian love.
**leschenia** (*les-ken-e'-mah*) [λέσχη, talk]. Insane, senile, or hysterical loquacity.
**lesion** (*le'-zhun*) [*læsio*, an injury]. An injury, wound, or morbid structural change. l., **discharging**, a brain lesion that causes sudden discharge of nervous motor impulses. l., **Ebstein's**. See *Ebstein's lesion*. l., **focal**, in the nervous system, a circumscribed lesion giving rise to distinctive and localizing symptoms. l., **functional**, l., **molecular**, a very fine lesion, not discernible by the microscope or discoverable by chemistry, but causing loss or excess of functional activity and attributed to alteration of the molecular equilibrium of that part. l., **indiscriminate**, one affecting two distinct systems. l., **initial**, of syphilis, the chancre. l., **irritative**, in the nervous system, a

lesion exciting the functions of the part wherein it is situated. l., peripheral, a lesion of the nerve-trunks or of their terminations. l., primary, of the skin, the change in the skin occurring in the developing stage of a skin disease. l., secondary. 1. In the skin, the change occurring in the primary lesion, due to irritation or other causes. It comprises erosions, ulcers, rhagades, squamæ, cicatrices or scars, crusts, and pigmentation. 2. One of the secondary manifestations of syphilis. l., structural, one working a manifest change in tissue. l., systematic, one confined to a system of organs, with a common function. l., toxic, a change in the tissues due to sepsis. l., vascular, a lesion of a blood-vessel.
**Lesser's triangle.** A triangular space bounded above by the hypoglossal nerve, its sides being formed by the bellies of the digastric.
**Lesshaft's space** (*les'-haft*) [Pyotr Frantsovich *Lesshaft*, Russian physician, 1839– ]. A locus minoris resistentiæ existing in the region of the twelfth rib in some individuals, which allows the pointing of an abscess or the protrusion of a hernia. It is bounded, in front, by the external oblique; behind, by the latissimus dorsi; above, by the serratus posticus inferior; and below, by the internal oblique. Syn., *Lesshafti rhombus*.
**lethal** (*le'-thal*) [*letum*, or *lethum*, death]. Deadly; pertaining to or producing death.
**lethality** (*le-thal'-it-e*) [*lethalitas*, from *letum*, death]. Deadliness.
**lethargic** (*leth-ar'-jik*) [λήθη, a forgetting]. Relating to, affected with, or of the nature of, lethargy.
**lethargogenic** (*leth-ur-go-jen'-ik*). Giving rise to lethargy.
**lethargus** (*leth-ar'-gus*). See *African lethargy*.
**lethargy** (*leth'-ar-je*) [λήθη, forgetfulness]. A condition of drowsiness or stupor that cannot be overcome by the will; also, a stage of hypnotism. l., **African.** See *African lethargy*.
**lethe** (*le'-the*) [λήθη, forgetfulness]. Total loss of memory; amnesia.
**letheomania** (*le-the-o-ma'-ne-ah*) [λήθη, oblivion; μανία, madness]. Morbid longing for narcotic drugs.
**letheon** (*le'-the-on*) [λήθη, forgetfulness]. An old trade name for ethylic ether, used as an anæsthetic.
**lethiferous** (*leth-if'-er-us*) [λήθη, forgetfulness; *ferre*, to bear]. Producing sleep or death.
**lethin** (*le'-thin*). The proprietary name for an alcoholic solution of camphor, acetic acid, ethereal oils, and chloroform.
**lettuce** (*let'-us*). See *lactucarium*.
**Leube Riegel test-dinner** (*loy'-beh re'-gel*) [W. O. von *Leube*, German physician, 1842– ; F. *Riegel*, German physician, 1843– ]. A dinner to ascertain the condition of the secretory function of the stomach. It consists of 400 c.c. of beef-soup, 200 grams of beefsteak, 50 grams of white bread, and 200 c.c. of water.
**leucemia** (*lu-se'-me-ah*). See *leukemia*.
**leuchemia** (*lu-ke'-me-ah*). See *leukemia*.
**leucic** (*lu'-sik*). Relating to or derived from leucin, as *leucic* acid.
**leucin** (*lu'-sin*) [λευκός, white], C₆H₁₃NO₂. A substance formed during pancreatic digestion, and also found in the urine, together with tyrosin, in acute yellow atrophy of the liver. Leucin crystallize from the urine in the form of yellowish-brown balls. See *Hofmeister, Scherer*.
**leucinosis** (*lu-sin-o'-sis*). 1. Abnormally excessive proportion or production of leucin, as in the liver. 2. Acute yellow atrophy of the liver.
**leucinuria** (*lu-sin-u'-re-ah*) [*leucin*; οὖρον, urine]. The occurrence of leucin in the urine.
**leucism, leucismus** (*lu'-sizm, lu-siz'-mus*) [λευκός, white]. In biology, whiteness resulting from bleaching or etiolation; albinism.
**leucitis** (*lu-si'-tis*). Same as *scleritis*.
**leuco-** (*lu-ko-*). For words beginning thus see *leuko-*.
**leucohæmia.** See *leukemia*.
**leucomaine** (*lu'-ko-mān*). See *leukomaine*.
**leucomma.** See *leukoma*.
**leucotoxic** (*lu-ko-toks'-ik*). Destructive to leukocytes.
**leucotoxin** (*lu-ko-toks'-in*) [*leuko-*; τοξικόν, poison]. A cytotoxin obtained by E. Metchnikoff from lymphatic ganglia. Cf. *hemotoxin, hepatotoxin, nephrotoxin, neurotoxin, spermatoxin, trichotoxin*.
**Leudet's bruit** (*loo-da'*) [Théodore Emile *Leudet*, French physician, 1825-1887]. A fine crackling

sound in the ear, audible to both the observer and the patient, in catarrhal and nervous affections of the ear. It is attributed to spasm of the external peristaphylinus muscle.
**leukæthiopia** (*lu-ke-the-o'-pe-ah*) [*leuko-*; Αἰθίοψ, Ethiopian]. Albinism in an African.
**leukæthiops** (*lu-ke'-the-ops*) [*leuko-*; Αἰθίοψ, Ethiopian]. An albino of the black race.
**leukanemia** (*lu-kan-e'-me-ah*) [*leukemia; anemia*]. A blood disease having features of leukemia and marked anemia.
**leukangeitis** (*lu-kan-je-i'-tis*). Synonym of *lymphangitis*, q. v.
**leukasmus** (*lu-kaz'-mus*). See *leukoderma*. l., acquired. See *leukoderma*. l., congenital. See *albinismus*.
**leukemia** (*lu-ke'-me-ah*) [λευκός, white; αἷμα, blood]. A disease of the blood and the blood-making organs, characterized by a permanent increase in the number of white blood-corpuscles and by enlargement of the spleen, the lymphatic glands, and the marrow of bone, together or separately. The etiology is obscure; by some the disease is considered to be infectious. l., **acute**, a generally fatal form, characterized by rapid development, high temperature, great enlargement of the spleen and lymphatics, and softening of the bone-marrow. **leukæmia cutis,** l. **of the skin,** a state of the skin characterized by formation of lymphatic enlargements of its deeper layers. **leukæmia lienalis,** splenic leukemia. l., **lienomedullary,** l., **lienomyelogenous.** See l., *splenomedullary*. l., **lymphatic,** the most common form, characterized by enlargement of the lymphatic glands, with perhaps slight changes in the spleen. l., **medullary,** l., **myelogenous,** l., **osseous,** l., **polymorphocyte,** a form in which the tissue of the bone-marrow is principally affected and obliteration of many of the small blood-vessels takes place. Syn., *myelemia*. l., **pseudo-,** l., **pseudosplenic.** See *pseudoleukemia*. l., **splenic,** that in which the blood-changes are principally due to disorders of the spleen. l., **splenomedullary,** the form characterized by excessive enlargement of the spleen and proliferation of the marrow of the bones without manifest change in the lymphatic glands.
**leukemic** (*lu-kem'-ik* or *lu-ke'-mik*) [*leukemia*]. Pertaining to leukemia.
**leukethiopia** (*lu-ke-the-o'-pe-ah*) [λευκός, white; Αἰθίοψ, Ethiopian]. Albinism in an African.
**leukine** (*lu'-kin*). An endocellular bactericidal substance found in leukocytes; an endolysin.
**leuko-** (*lu-ko-*) [λευκός, white]. A prefix meaning white.
**leukoblast** (*lu'-ko-blast*) [*leuko-*; βλαστός, a germ]. 1. The germ of a leukocyte. 2. A cell in bone-marrow, of a type which is believed to develop into a red blood-corpuscle.
**leukochroia, leukochrus** (*lu-kok'-ro-os, lu-kok'-rus*) [*leuko-*; χρώς, skin]. Having a white skin.
**leukocidin** (*lu-ko'-sid-in*) [*leukocyte; cædere*, to kill]. A cytolytic toxin capable of destroying leukocytes. See *hemotoxin* (2).
**leukocrystallin** (*lu-ko-kris'-tal-in*) [*leuko-*; κρύσταλλος, clear ice]. Peculiar crystals occasionally found in the blood of leukemic patients.
**leukocyte** (*lu'-ko-sīt*) [*leuko-*; κύτος, cell]. The colorless or white corpuscle of the blood. Leukocytes have ameboid movement and are formed in the lymphadenoid tissue of the spleen, lymphatic glands, intestinal tract, bone-marrow, etc., and probably also in the lymph and blood. Their average diameter is 0.01 mm. ($\frac{1}{2500}$ inch). In normal blood several forms are distinguished: lymphocytes, large uninuclear leukocytes, transitional forms, and multinuclear cells; according to the granules contained in their protoplasm, three varieties are described—the neutrophil, the eosinophil, and the basophil. l.s, Ehrlich's theory concerning, all varieties may be classed into two groups having separate origins, functions, and relations. The first group would comprise all lymphocytes, which are produced solely by the lymphatic tissues; and the second, the mononuclear leukocytes and transitional forms, the polynuclear neutrophils, the eosinophils, and the basophils, all of which cells are produced exclusively by the marrow. l., **polymorphonucleic,** l., **polynuclearneutrophilic,** one derived primarily from lymphocytes and secondarily from a myelocyte; originating in red bone-marrow, in the lymph-glands, or spleen. l.s, **polynuclear,** l.s, **polynuclear,** cells

in which the nucleus is either lobed or made up of several portions united by such delicate nuclear filaments as to give the impression of a multinucleated cell in distinction to mononuclear, eosinophil, and neutrophil leukocytes and lymphocytes. 1., Uskow's theory concerning, all leukocytes except the basophilic cells are but different developmental stages of the same cell. The youngest form of leukocyte, the small lymphocyte, originates in the lymph-glands, the lymphocytic bone-marrow, and the spleen.
**leukocythemia** (lū-ko-si-the'-me-ah). See *leukemia*.
**leukocythemic** (lū-ko-si-the'-mik) [*leuko-*; κύτος, cell; αἷμα, blood]. Pertaining to leukemia.
**leukocytic** (lū-ko-sit'-ik) [*leukocyte*]. Relating to or characterized by leukocytes.
**leukocytogenesis** (lū-ko-si-to-jen'-es-is) [*leukocyte*; γεννᾶν, to beget]. The formation of leukocytes.
**leukocytoid** (lū'-ko-si-toyd) [*leukocyte*; εἶδος, resemblance]. Resembling a leukocyte.
**leukocytolysin** (lū-ko-si-tol'-is-in) [see *leukocytolysis*]. A cytolysin produced by inoculation with leukocytes.
**leukocytolysis** (lū-ko-si-tol'-is-is) [*leukocyte*; λύσις, solution]. The destruction of leukocytes.
**leukocytolytic** (lū-ko-si-to-lit'-ik). Relating to the destruction of leukocytes.
**leukocytoma** (lū-ko-si-to'-mah) [*leukocyte*; ὄμα, tumor]. A tumor-like mass composed of leukocytes, as the tubercle, the gumma, etc.
**leukocytometer** (lū-ko-si-tom'-et-er) [*leukocyte*; μέτρον, a measure]. A graduated capillary tube used for counting leukocytes.
**leukocytopenia** (lū-ko-si-to-pe'-ne-ah). See *leukopenia*.
**leukocytoplania**. See *leukoplania*.
**leukocytosis** (lū-ko-si-to'-sis) [*leukocyte*]. An increase in the number of colorless blood-corpuscles in the blood. It is physiological during digestion and pregnancy; pathologic in certain anemias, especially leukemia, in some of the infectious fevers, in cachexias, and after hemorrhage. 1., **inflammatory**, that in which the lymph is concentrated in the cells while the blood is normal. 1., **pure**, a leukocytosis in which the increase of white cells affects the multinuclear form.
**leukocytotaxis** (lū-ko-si-to-tak'-sis). See *leukotaxis*.
**leukocytotic** (lū-ko-si-tot'-ik) [*leuko-*; κύτος, cell]. Pertaining to leukocytosis.
**leukocytozoa** (lū-ko-si-to-zo'-ah) [*leukocyte*; ζῷον, animal]. Infusorian parasites of the white blood-corpuscles.
**leukocyturia** (lū-ko-si-tū'-re-ah) [*leuko-*; κύτος, cell; οὖρον, urine]. The presence of colorless blood-corpuscles in the urine.
**leukoderma** (lū-ko-der'-mah) [*leuko-*; δέρμα, skin]. A condition of defective pigmentation of the skin, especially a congenital absence of pigment in patches or bands. See *achroma*, *albinism*. 1., **acquired**, vitiligo.
**leukodermic** (lū-ko-der'-mik) [*leuko-*; δέρμα, skin]. Exhibiting or pertaining to leukoderma.
**leukodiagnosis** (lū-ko-di-ag-no'-sis) [*leukocyte*; *diagnosis*]. A method of diagnosis by examining the leukocytes. Specifically used in the diagnosis of cancer.
**leukodontous** (lū-ko-don'-tus) [*leuko-*; ὀδών, tooth]. Having white teeth.
**leukoencephalitis** (lū-ko-en-sef-al-i'-tis) [*leuko-*; ἐγκέφαλος, brain]. An epizootic disease of horses characterized by drowsiness, imperfect vision, partial paralysis of the throat, twitching of the muscles of the shoulder, unsteady gait, and softening of the white substance of the frontal lobes; also called forage poisoning.
**leukogasterous** (lū'-ko-gas'-ter-us) [*leuko-*; γαστήρ, belly]. White-bellied.
**leukogene** (lū'-ko-jēn) [*leuko-*; γεννᾶν, to beget]. A substance containing sodium bisulphite, used as a bleaching agent.
**leukohemia** (lū-ko-hem'-e-ah). See *leukemia*.
**leukohemic** (lū-ko-hem'-ik). See *leukemic*.
**leukokeratosis** (lū-ko-ker-at-o'-sis) [*leuko-*]. See *leukoplakia*.
**leukol, leukolin** (lū'-kol, lu'-ko-lin). Quinolin, *q. v.*
**leukolysis** (lū-kol'-is-is). See *leukocytolysis*.
**leukolytic**. See *leukocytolytic*.
**leukoma** (lū-ko'-mah) [λευκός, white]. 1. An opacity of the cornea the result of an ulcer, wound, or inflammation, and presenting an appearance of ground glass. 2. The term has also been used for albumin. See *leukoplakia*.
**leukomaine** (lū-ko'-mah-ēn) [λευκός, white]. The name applied to any one of the nitrogenous bases or alkaloids normally developed by the metabolic activity of living organisms, as distinguished from the alkaloidal bodies developed in dead bodies, and called ptomaines. From their chemical affinities leukomaines may be divided into two groups—the *uric-acid group*, comprising adenine, carnine, guanine, heteroxanthine, hypoxanthine, paraxanthine, pseudoxanthine, spermine, xanthine; and the *creatinine group*, in which are classed amphicreatinine, crusocreatinine, xanthocreatinine, and others.
**leukomainemia** (lū-ko-ma-in-e'-me-ah) [*leukomaine*; αἷμα, blood]. The presence of leukomaines in the blood; the retention or imperfect elimination of the various excretory products of the living cells of the organism.
**leukomainic** (lū-ko-ma-in'-ik). Pertaining to, or of the nature of, a leukomaine.
**leukomatoid** (lū-ko'-mat-oid). See *leukomatous*.
**leukomatorrhea** (lū-ko-mat-or-e'-ah) [*leuko-*; *boia*, a flow]. An excessive whitish secretion. 1. **salivalis**, salivation. 1. **urinalis**, albuminuria. 1. **vaginalis**, leukorrhea.
**leukomatosis** (lū-ko-mat-o'-sis) [*leuko-*; νόσος, disease]. Abnormal increase of albumin in a part, as in the cornea, the lens, the joints; also amyloid degeneration.
**leukomatous** (lū-ko'-mat-us) [*leukoma*]. Having he nature of or affected with leukoma.
**leukomyelitis** (lū-ko-mi-el-i'-tis) [*leuko-*; μυελός, marrow; ιτις, inflammation]. Inflammation of the white substance of the spinal cord.
**leukomyelopathy** (lū-ko-mi-el-op'-ath-e) [*leuko-*; μυελός, marrow; πάθος, disease]. Any disease of the white substance of the myelon or spinal cord.
**leukonecrosis** (lū-ko-ne-kro'-sis) [*leuko-*; *necrosis*]. A rm of dry gangrene, the slough having a light color.
**Leukonostoc** (lū-ko-nos'-tok) [*leuko-*; *Nostoc*]. A genus of schizomycetes having its elements disposed in chains and enveloped in a sheath of tough jelly.
**leukonuclein** (lū-ko-nū'-kle-in). A decomposition-product of nucleohiston by action of hydrochloric acid.
**leukonychia** (lū-kon-ik'-e-ah) [*leuko-*; ὄνυξ, nail], A whitish discoloration of the nails owing ■ the presence of air beneath them.
**leukopathic** (lū-ko-path'-ik) [*leuko-*; πάθος, disease]. Relating to or affected with leukopathia.
**leukopathy, leukopathia** (lū-kop'-ath-e, lū-ko-pa'-the-ah) [*leuko-*; *pathos*, suffering]. Any deficiency of coloring-matter; albinism. See *leukoderma*.
**leukopenia** (lū-ko-pe'-ne-ah) [*leuko-*; πενία, poverty]. A decrease below the normal standard in the number of leukocytes in the peripheral blood.
**leukophlegmasia** (lū-ko-fleg-ma'-ze-ah) [*leuko-*; φλέγμα, inflammation]. 1. A condition marked by a tendency to dropsy, accompanied by a pale flabby skin and general edema of the whole body; solid edema. 2. Phlegmasia alba dolens. l. **dolens puerperarum**, phlegmasia alba dolens.
**leukophthalmous** (lū-koff-thal'-mus) [*leuko-*; ὀφθαλμός, eye]. Having unusually white eyes.
**leukopin** (lū'-ko-pin). Visual white, produced from rhodopsin by the action of light.
**leukoplakia** (lū-ko-pla'-ke-ah) [*leuko-*; πλάξ, surface]. Whitening of a surface. l. **buccalis**, l. **lingualis**, a disease characterized by the presence of pearly-white or bluish-white patches on the surface of the tongue or the mucous membrane of the cheeks, due to a hyperplasia of the epithelium.
**leukoplania** (lū-ko-pla'-ne-ah) [*leuko-*; πλάνη, a wandering]. The wandering of leukocytes or their passage through a membrane.
**leukoplasia** (lū-ko-pla'-ze-ah). See *leukoplakia*.
**leukoplast, leukoplastid** (lū-ko-plast', lū-ko-plast'-id) [*leuko-*; πλάσσειν, to form]. A starch-forming, colorless, protein bodies found in cells not exposed to light; amyloplast.
**leukopoiesis** (lū-ko-poy-e'-sis) [*leuko-*; ποιεῖν, to make]. The formation of leukocytes.
**leukoprotease** (lū-ko-pro'-te-ās). An enzyme of the polymorphonuclear leukocytes capable of proteolytic digestion in any except an acid medium.
**leukopsin** (lū-kop'-sin) [*leuko-*; ὄψις, sight]. Visual white, produced from rhodopsin by the action of light.

**leukorrhagia** (*lū-kor-a'-je-ah*) [*leuko-*; ῥηγνύναι, to burst forth]. An excessive leukorrheal flow.
**leukorrhea** (*lū-kor-e'-ah*) [*leuko-*; ῥοία, a flow]. A whitish, mucopurulent discharge from the female genital canal, popularly called "the whites."
**leukorrheal** (*lū-kor-e'-al*) [*leukorrhea*]. Of the nature of or pertaining to leukorrhea.
**leukosarcoma** (*lū-ko-sar-ko'-mah*) [*leuko-*; *sarcoma*]. A nonpigmented sarcoma.
**leukosin** (*lū'-ko-sin*) [*leuko-*]. A principle said by some to be present in asthma-crystals, and in crystals found in the blood in certain cases of leukemia.
**leukosis** (*lū-ko'-sis*) [λευκός, white]. 1. Any disease of the lymphatics. 2. Abnormal pallor of the skin. 3. The development and progress of leukoma.
**leukotactic** (*lū-ko-tak'-tik*) [*leuko-*; τάξις, arrangement]. Relating to leukotaxis.
**leukotaxis** (*lū-ko-taks'-is*). The arranging and ordering function of leukocytes.
**leukotic** (*lū-kot'-ik*) [*leuko-*]. Relating to leukoma.
**leukotin** (*lū-ko'-tin*) [*leuko-*], C₆H₃₀O₅. A crystalline substance found in *Paracoto*; it is soluble in alcohol, ether, and chloroform.
**leukotoxic** (*lū-ko-toks'-ik*). Destructive to leukocytes.
**leukotoxin** (*lū-ko-toks'-in*) [*leuko-*; τοξικόν, poison]. A cytotoxin obtained by Metchnikoff from lymphatic ganglia.
**leukotrichia** (*lū-ko-trik'-e-ah*) [*leuko-*; θρίξ, hair]. Whiteness of the hair; canities.
**leukotrichous** (*lū-kot'-rik-us*) [*leuko-*; θρίξ, hair]. White-haired.
**leukourobilin** (*lū-ko-ū-ro-bil'-in*) [*leuko-*; οὖρον, urine; *bilis*, bile]. A colorless decomposition-product of bilirubin.
**leukous** (*lū'-kus*) [*leuko-*]. White.
**leukozon** (*lū'-ko-zon*). A disinfecting powder consisting of approximately equal parts of calcium perborate and talcum.
**leukuresis** (*lū-kū-re'-sis*) [*leuko-*; οὖρον, urine]. Albuminuria.
**leusin** (*lū'-sin*). See *leucin*.
**Levaditi method** (*lev-ah-de'-te*) [Constantin *Levaditi*, French physician]. A modification of Cajal's method of staining nerve-fibers used for staining the *Treponema pallidum* in sections; a solution of silver nitrate is used, which stains the treponema a dense black.
**Levant wormseed.** See *santonica*.
**levator** (*le-va'-tor*) [L., "a lifter": *pl.*, *levatores*]. 1. That which raises or elevates, as certain muscles. See *muscles*, *table of*. 2. An instrument used for raising a depressed portion of the skull.
**lever** (*le'-ver*) [*levare*, to elevate]. 1. A vectis or one-armed tractor, used in obstetrics. 2. A dental instrument, used in lifting out decayed stumps. l., **Davy's**, an aorta-compressor. It is a wooden bar, which is introduced into the rectum.
**levicellular** (*lev-e-sel'-ū-lar*) [*levis*, smooth; *cellula*, cell]. Smooth-celled, as a levicellular myoma, or a levicellular muscle-fiber.
**levico-ochre** (*lev-ik-o-o'-ker*). A mud containing iron and arsenic, obtained from the springs at Levico in the Tyrol. It is applied in the form of a hot poultice in chronic inflammation.
**levigable** (*lev'-ig-a-bl*) [*lævigare*, to make smooth]. Susceptible of being levigated, or reduced to fine powder by a mechanical process.
**levigate** (*lev'-ig-āt*) [*lævigare*, to make smooth]. 1. To reduce to a fine powder by rubbing or grinding; to triturate. 2. Smooth. 3. Uniform: applied to the reddening in skin diseases.
**levigation** (*lev-ig-a'-shun*) [*lævigare*, to make smooth]. The trituration of a substance made into a paste with water or other liquid. When performed with a muller on a slab of porphyry it is called *porphyrization*.
**Levis's splint** (*le'-vis*). A perforated metal splint extending from below the elbow almost to the distal ends of the metacarpal bones. It is hollowed to fit the forearm and palm, and the metacarpal joints are flexed over it.
**Levisticum** (*lev-is'-tik-um*). A genus of plants containing but one species, *L. officinale*, indigenous to middle Europe. The root, plant, and fruit are diuretic and are used in dropsy and amenorrhea. Dose of *fluidextract* 15–60 min. (0.9–3.7 Cc.). See *lovage*.
**levitation** (*lev-it-a'-shun*) [*levitas*, lightness]. 1. The pretended elevation of the body into the air without support, a feat professedly performed by various modern thaumaturgists. 2. The subjective sense of being aloft, and without support; a symptom in certain cases of insanity.
**levity** (*lev'-it-e*) [*levitas*, lightness]. The antithesis of gravity.
**levoduction** (*lev-o-duk'-shun*) [*lævus*, left; *ducere*, to lead]. The movement of the eye to the left.
**levo lucosan** (*lev-o-glū'-ko-san*) [*lævus*, left; γλύκυς, sweet]g A derivative of picein obtained by heating the latter with baryta.
**levoglucose** (*le-vo-glū'-kōs*). Levulose.
**levogyrate** (*le-vo-ji'-rāt*). See *levorotatory*.
**levophoria** (*le-vo-fo'-re-ah*) [*lævus*, left; φέρειν, to bear]. A tending of the visual lines to the left.
**levorotatory** (*le-vo-ro'-tat-o-re*) [*lævus*, left; *rotare*, to turn]. Causing to turn toward the left hand: applied to substances that turn the rays of polarized light to the left.
**levorsion** (*le-vo-tor'-shun*) [*lævus*, left; *torquere*, to twist]. A turning or twisting toward the left.
**levoversion** (*le-vo-ver'-shun*). See *levotorsion*.
**Levret's law** (*lev-ra'*) [André *Levret*, French obstetrician, 1703–1780]. Marginal insertion of the umbilical cord in placenta prævia.
**levulan** (*lev'-ū-lan*) [*lævus*, left], C₆H₁₀O₅. An anhydride of levulose obtained from beet-sugar molasses.
**levulin** (*lev'-ū-lin*) [*lævus*, left]. A carbohydrate identical with synanthrose, occurring in immature grain and the tubers of certain composite flowers.
**levulinic acid** (*lev-ū-lin'-ik*). See *acid*, *levulinic*.
**levulosan** (*lev-ū la' san*) [*levulosa*], C₆H₁₀O₅. A carbohydrate prepared by heating levulose.
**levulose** (*lev'-ū-lōs*) [*lævus*, left], C₆H₁₂O₆. Fruit-sugar, the natural sugar of fruits. It is a colorless, syrupy liquid, and rotates the plane of polarized light to the left. It occurs normally in the intestine, and rarely in the urine in disease. See *invert-sugar*.
**levulosemia** (*lev-ū-lo-se'-me-ah*) [*levulose*; αἷμα, blood]. The presence of levulose in the blood.
**levulosuria** (*lev-ū-lo-sū'-re-ah*) [*levulose*; οὖρον, urine]. The presence of levulose in the urine.
**levuretin** (*lev-ū'-re-tin*) [Fr., *levure*, yeast]. A preparation of dried brewers' yeast used in skin diseases. Dose 1 teaspoonful (5 Cc.) in milk 3 times daily.
**levurin** (*lev'-ū-rin*) [see *levuretin*]. A dried extract of beer-yeast; used in cases of sepsis. Dose 1 teaspoonful (5 Cc.) 3 times daily.
**Lewin's erythema of the larynx** (*lū'-in*). Simply syphilitic catarrh of the larynx.
**lewinin** (*lū'-in-in*) [after Dr. *Lewin*]. A local anesthetic resin from kava.
**Lewisohn's method of blood transfusion** (*loo'-is-on*) [Richard *Lewisohn*, American physician]. A method of blood transfusion, in which a 10 per cent. solution of sodium citrate is used to prevent coagulation during the process.
**lexipharmac.** (*leks-if-ar'-mak*) [λῆξις, cessation; φάρμακον, poison]. An alexipharmac.
**lexipyretic** (*leks-ip-i-ret'-ik*) [λῆξις, cessation; πυρετός, fever]. 1. Febrifugal; antipyretic. 2. A febrifugal medicine.
**Leyden battery** (*li'-den*) [*Leyden*, a town in Holland]. A series of Leyden jars connected tandem. L. **jar**, a glass jar coated within and without with tin-foil, reaching nearly to the neck, and surmounted by a knobbed conductor in connection with the inner coating. It is designed for the temporary accumulation of electricity, with which the inner foil may be charged.
**Leyden's ataxia** (*li'-den*) [Ernst Victor von *Leyden*, German physician, 1832–1910]. Pseudotabes. L.'s **cells**, large, mononuclear epithelioid cells found in the anterior horns of the spinal cord in cases of anterior poliomyelitis. They are also met in other inflammatory affections of the cord. L.'s **crystals**. See *Charcot-Leyden's crystals*. L.'s **neuritis**, a variety of neuritis in which the nerve-fibers are replaced by fatty tissue. Syn., *lipomatous neuritis*. L.'s **sign**, in cases of subphrenic pyopneumothorax manometric observation shows that the pressure in the abscess-cavity rises during inspiration and falls during expiration. The reverse was held by Leyden to occur in true pneumothorax.
**Leyden-Charcot's crystals.** See *Charcot-Leyden's crystals*.
**Leyden-Moebius' type of progressive muscular atrophy.** A type commencing in the calves and often assuming the character of Duchenne's paralysis.

**Leydenia gemmipara** (*li-de'-ne-ah jem-ip'-ar-ah*) [see *Leyden's ataxia*]. The large round, or polymorphous cells with lively ameboid movement discovered by v. Leyden in the ascitic fluid in cancer patients.

**Leydig's cells** (*li'-dig*) [Franz von Leydig, German anatomist, 1821–1908]. See *Henle's cells*. L.'s duct, the Wolffian duct.

**L. F. A.** Abbreviation denoting the left frontoanterior position of the fetus in utero.

**L. F. P.** Abbreviation denoting the left frontoposterior position of the fetus in utero.

**L. F. P. S.** Abbreviation for Licentiate of the Faculty of Physicians and Surgeons (of Glasgow).

**Li.** The chemical symbol of *lithium*.

**lianthral** (*li-an'-thral*). A proprietary preparation said to be an extract of coal-tar and casein.

**liantrol** (*li-an'-trol*). A coal-tar preparation, used externally in cases of eczema.

**libanotus** (*lib-an-o'-tus*) [λίβανος, frankincense]. An old name for olibanum.

**libanus** (*lib'-an-us*) [λίβανος, frankincense]. Frankincense, or olibanum.

**liberation** (*lib-er-a'-shun*) [*liberare*, to make free]. The act of freeing. l. of the arms, in breech-presentations, the bringing down of the arms of the fetus when they have become extended along the sides of the child's head.

**liberomotor** (*lib-er-o-mo'-tor*) [*liberare*, to disengage; *motor*, mover]. Setting free or disengaging motor energy.

**libidinous** (*lib-id'-in-us*) [*libidinosus*, lustful]. Characterized by strong sexual desire.

**libido** (*lib-i'-do*) [L.]. Desire; lust. l. intestini, desire for defecation. l. sexualis, lust. l. urinæ, desire to urinate.

**libra** (*li'-brah*) [*libra*, a balance]. A pound. A weight of twelve troy ounces, or 5760 grains. Also, applied to the avoirdupois pound of sixteen ounces, or 7000 grains.

**libradol** (*lib'-rad-ol*). The proprietary name for a soft, greenish ointment recommended for relief of pain.

**lice** (*lis*). Plural of louse. See *pediculus*.

**license** (*li'-sens*) [*licentia*, license]. An official permit or authority conferring on the recipient the right and privilege of exercising his profession.

**licentiate** (*li-sen'-she-āt*) [*licentiatus*, one licensed]. A term sometimes applied to a person who practises a profession by the authority of a license.

**lichen** (*li'-ken*) [λειχήν, a lichen]. A generic term for a group of inflammatory affections of the skin in which the lesions consist of solid papules. l. acuminatus, a variety of lichen ruber in which the papules are acuminate. It is usually very acute, and is accompanied by grave constitutional symptoms (rigors, pyrexia, sweats, prostration) and by itching. l. agrius, eczema papulosum. l. disseminatus, a form with scattered lesions. l. pilaris, an inflammatory disease of the hair-follicles in which a spinous epidermic peg occupies the center of the papules. l. planus, an inflammatory skin disease, with an eruption made up of papules that are broad and angular at the base, flat and apparently glazed on the summit, slightly umbilicated, and of a dull, purplish-red color. The papules may be discrete or may coalesce, and itching may be slight or severe. l. ruber, a rare skin disease, with lesions consisting, in the beginning, of discrete, miliary, conical papules, but, as the disease advances, becoming aggregated and forming continuous red, infiltrated, and scaly patches. The whole surface may eventually become involved. The nails of the fingers and toes become affected, being of a dirty-brown color, rough, flaky, and breaking off short. The etiology and pathology are obscure. l. scrofulosus, a form occurring in strumous children. The eruption is situated on the trunk, especially upon the back, either diffusely or in patches. The papules are very small, pale, conical, and surmounted by fine scales; they cause no itching, and on fading leave a rather persistent yellowish pigmentation. l. strophulosus. See *strophulus*. l. tropicus, prickly heat.

**licheniasis** (*li-ken-i'-as-is*). 1. The formation of lichen. 2. The condition of one affected by one of the forms of lichen. l. strophulus. See *strophulus*.

**lichenification** (*li-ken-i-fi-ka'-shun*). The change of an eruption into a form resembling lichen.

**lichenin** (*li'-ken-in*), $C_6H_{10}O_5$ or $C_{18}H_{30}O_{15}$. Moss-starch; a starch-like body found in lichens. Iodine imparts a dirty-blue color to it. It is insoluble in cold water, but forms a jelly with hot water. See *cetraria*.

**lichenization** (*li-ken-i-za'-shun*). The development of lesions of lichen.

**lichenoid** (*li'-ken-oid*) [*lichen;* εἶδος, like]. Resembling lichen.

**licorice** (*lik'-or-is*). See *glycyrrhiza*.

**lid.** See *eyelid*.

**Lieben's test for acetone in urine** (modified by Ralfe) (*le'-ben*) [Adolf Lieben, Austrian chemist, 1836– ]. Dissolve 20 gr. of potassium iodide in a dram of solution of potassium hydroxide and boil; then carefully float the urine on its surface in a test-tube. A precipitation of phosphates occurs at the point of contact which, in the presence of acetone, will become yellow and studded with yellow points of iodoform.

**Lieberkuehn's ampulla** (*le'-ber-kin*) [Johann Nathaniel Lieberkuehn, German anatomist, 1711–1756]. A cavity that was supposed by Lieberkuehn to exist in an intestinal villus and to communicate at the apex with the lumen of the intestine and at the base with the lacteals. L.'s crypts, L.'s follicles, L.'s glands, minute tubular glands in the mucosa of the large and small intestine.

**Liebermann's test for proteids** (*le'-ber-man*) [Leo von Szentlorincz Liebermann, Austrian physician, 1852– ]. Wash the proteid with alcohol and ether, and treat with fuming hydrochloric acid; a beautiful violet-blue coloration is produced.

**Liebermann-Burchard's test for cholesterin** (*le'-ber-man-boork'-hart*). Allow the substance to dissolve in acetic anhydride, then add a few drops of concentrated sulphuric acid, when a beautiful violet coloration is produced, changing quickly to green if cholesterin is present.

**Liebig's extract** (*le'-big*) [Baron Justus von Liebig, German chemist, 1803–1873]. A variety of beef-extract. L.'s test for cystin, boil the substance with caustic alkali containing lead oxide. In the presence of cystin a precipitate of black lead sulphide is formed.

**lien** (*li'-en*) [L.]. The spleen. l. accessorius, accessory spleen.

**lienaden** (*li-en'-ad-en*) [*lien;* ἀδήν, gland]. The proprietary name of a preparation made from the spleen of animals.

**lienal** (*li'-en-al*) [*lien*]. Relating to the spleen.

**lienculus** (*li-en'-ku-lus*) [dim. of *lien*]. A detached part or exclave of the spleen.

**lienic** (*li-en'-ik*) [*lien*, a spleen]. Pertaining to the spleen.

**lienitis** (*li-en-i'-tis*) [*lien;* ιτις, inflammation]. Splenitis.

**lieno-** (*li-en-o-*) [*lien*]. A prefix meaning relating to the spleen.

**lienocele** (*li-en'-o-sēl*) [*lien*, a spleen; κήλη, hernia]. Hernia of some part or of all of the spleen.

**lienointestinal** (*li-en-o-in-tes'-tin-al*) [*lieno-; intestine*]. Relating to the spleen and intestine.

**lienomalacia** (*li-en-o-mal-a'-she-ah*) [*lieno-;* μαλακία, softening]. Morbid softening of the spleen.

**lienomedullary** (*li-en-o-med'-ul-ar-e*). See *lienomyelogenous*.

**lienomyelogenous** (*li-en-o-mi-el-oj'-en-us*) [*lieno-;* μυελός, marrow; γενναν, to produce]. Derived from both spleen and marrow.

**lienomyelomalacia** (*li-en-o-mi-el-o-mal-a'-she-ah*) [*lieno-;* μυελός, marrow; μαλακία, softening]. Softening of the spleen and bone-marrow.

**lienopancreatic** (*li-en-o-pan-kre-at'-ik*). Relating to the spleen and pancreas.

**lienorenal** (*li-en-o-re'-nal*) [*lieno-; ren*, kidney]. Relating to the spleen and the kidney.

**lienteric** (*li-en-ter'-ik*) [*lientery*]. Pertaining to or affected with lientery.

**lientery** (*li'-en-ter-e*) [λεῖος, smooth; ἔντερον, intestine]. A form of diarrhea in which the food passes rapidly through the bowel without undergoing digestion.

**lienunculus** (*li-en-un'-ku-lus*) [dim. of *lien*]. A detached part of the spleen.

**Lieutaud's body** (*lū-to'*) [Joseph Lieutaud, French physician, 1703–1780]. Same as L.'s triangle. L.'s sinus. The straight sinus. L.'s triangle, the trigonum vesicæ. L.'s uvula. A longitudinal mesial ridge in the trigone of the bladder.

**life** (*līf*) [AS., *lif*]. 1. The sum of properties that enables an organism to adapt itself to surrounding conditions. 2. The characteristic phenomena manifested by living beings. 3. The force or principle underlying or causing the phenomena presented by organized beings. 4. The period between birth and death. l., **animal**, the manifestations depending directly on the cerebrospinal nervous system and the voluntary muscles, as distinguished from vegetative life—that is, the functions of digestion, respiration, reproduction, etc. l., **antenatal**, the life of the fetus before birth. l., **change** of, that period in the life of a woman at which menstruation ceases. Syn., *climacteric; menopause*. l., **embryonic**, the period beginning with the differentiation of the blastoderm and ending about the end of the second month. l., **expectation** of, the average number of years which a person may expect to live, as calculated from life-tables.

**ligament, ligamentum** (*lig'-am-ent, lig-am-en'-tum*) [*ligare*, to bind]. 1. A band of flexible, compact connective tissue connecting the articular ends of the bones, and sometimes enveloping them in a capsule. 2. Certain folds and processes of the peritoneum. l., **accessory**, one that strengthens another. l., **acromioclavicular**, a ligament covering the acromioclavicular articulation and extending from the clavicle to the acromial process of the scapula. l., **adipose**, the mucous ligament of the knee-joint. l., **alar**. 1. One of the two folds of synovial membrane on each side of the mucous ligament of the knee-joint. 2. See *l., odontoid*. l., **annular** (of ankle), the broad ligament covering the anterior surface of the ankle-joint. l., annular (of wrist), a strong ligament extending from the trapezium to the unciform bone, confining the flexor tendons. l., **arcuate**, one of the arched ligaments extending from the body of the diaphragm to the last rib and to the transverse process of the first lumbar vertebra. l., **atloaxoid**, that joining the atlas and the axis. l., **atlooccipital**, that joining the atlas and the occiput. l.s, **auricular**, three ligaments uniting the external ear to the head. **ligamentum bifurcatum**, the Y-ligament. l. of Bigelow. See *l., iliofemoral*. l., **broad**. 1. A fold of peritoneum extending laterally from the uterus to the pelvic wall. 2. A ligament supporting the liver. l., **central**. See *filum terminale*. l., **check**. See *l., odontoid*. l., **ciliary**, the tissue at the junction of the cornea and sclera forming the root of the iris. **ligamenta coli**, three longitudinal bands on the surface of the large intestine, due to thickening of the longitudinal muscle-fibers. l., **conoid**, the inner portion of the coracoclavicular ligament. It is attached to the coracoid process of the scapula and the conoid tubercle of the clavicle. l., **coracoclavicular**, one extending from the coracoid process of the scapula to the clavicle. l., **coracohumeral**, that joining the coracoid process of the scapula and the upper and posterior portion of the capsule of the shoulder-joint and the upper part of the humerus. l., **coronary**, a peritoneal fold extending from the posterior edge of the liver to the diaphragm. l., **costoclavicular**. See *l., rhomboid*. l., **costocoracoid**, that joining the first rib and the coracoid process of the scapula. l., **costocolic**, a peritoneal fold joining the diaphragm and the splenic flexure of the colon. l., **cotyloid**, a ring of fibrocartilaginous tissue at the margin of the acetabulum. l., **crucial, anterior**, the smaller crucial ligament of the knee, extending from the upper surface of the tibia to the inner surface of the external condyle of the femur. l., **crucial, posterior**, one attached below to the back part of the depression behind the spine of the tibia, to the popliteal notch, and to the external semilunar fibrocartilage, and above to the inner condyle of the femur. l., **cruciform**, that formed by the transverse ligament of the atlas and a vertical ligament running from the middle of this to the body of the axis. l., **crural**. *Poupart's l.* l., deltoid, lateral internal ligament of the ankle. l., **Denucé's**. See *Denucé's ligament*. l.s, **elastic**, yellow, highly elastic ligaments lying at the back of the spinal canal, appearing in pairs between the laminæ of contiguous vertebræ from the axis to the interval between the last lumbar vertebra and the sacrum. l., **falciform**. 1. A sickle-shaped expansion of the great sacrosciatic ligament, extending along the inner margin of the tuberosity and inferior ramus of the ischium. 2. The broad ligament of the liver. l., **femoral**. See *Hey's l.*,

l., **Flood's**. See *Flood's ligament*. l.s, **funicular**, band-like ligaments accessory to capsular ligaments surrounding movable joints; they are made up of parallel bundles of flexible fibrous tissue, but without elasticity. l., **Gimbernat's**. See *Gimbernat's ligament*. l., **glenohumeral**, a portion of the coracohumeral ligament, attached to the inner and upper portion of the bicipital groove. l., **glenoid**. 1. A ring of fibrocartilaginous tissue attached to the rim of the glenoid fossa. 2. One of those joining the phalanges of the metacarpal bones. l., **glenoideobrachial**, the thickened part of the capsular ligament of the shoulder which is inserted into the lesser tuberosity of the humerus. l., **Hey's**, a sickle-shaped expansion of the fascia lata. Syn., *femoral ligament*. l., **iliofemoral**, a strong ligament extending from the anterior inferior iliac spine to the lesser trochanter and the intertrochanteric line. l., **iliotibial**. See *Maissiat's band*. l., **iliotrochanteric**, a portion of the iliofemoral ligament. l., **interclavicular**, one joining the clavicles and the sternum. l., **interfoveolar**, a thin, fibrous band extending from the posterior surface of the fascia transversalis, near the plica semilunaris, to the pubic bone and Gimbernat's ligament; it forms part of the conjoined tendon. l., **lateral**, one of the peritoneal folds between the sides of the liver and the inferior surface of the diaphragm. l., **Lockwood's**. See *Lockwood's ligament*. l., **Mauchart's**. See *Mauchart's ligament*. l., **mucous** (of the knee-joint), a fold of synovial membrane extending from the intercondyloid fossa to the lower margin of the patella. l., **nuchæ**, one at the nape of the neck, connecting the two trapezius muscles. l., **odontoid**, any one of the broad, strong ligaments arising on each side of the apex of the odontoid process and connecting the atlas with the skull. l., **orbicular** (of radius), that surrounding the head of the radius. l., **palpebral, external**, that joining the outer margin of the orbit and the tissues of the eye-lid. l., **palpebral, internal**, one extending from the nasal process of the superior maxilla to the lacrimal spine of the lacrimal bone and the inner end of the tarsal cartilage. **ligamentum patellæ**, a strong fibrous structure, extending from the tubercle of the tibia upward to become the tendon of the quadriceps extensor muscle; it embraces the patella. **ligamentum pectinatum**, the spongy tissue at the junction of the cornea and sclera in the sinus of the anterior chamber of the eye. It forms the root of the iris. l., **Poupart's**, the ligament extending from the anterior superior spine of the ilium to the spine of the pubis and the pectineal line. It is the lower portion of the aponeurosis of the external oblique muscle. l., **pterygomaxillary**, one joining the apex of the internal pterygoid plate of the sphenoid bone and the posterior extremity of the internal oblique line of the lower jaw. l., **pterygospinous**, a ligamentous band extending from the external pterygoid plate to the spine of the sphenoid. l., **pubic, inferior**, a triangular ligament extending from the symphysis pubis to the rami of the pubic bones. l., **pylorocolic**, Glénard's name for the attachment of the transverse colon to the pylorus. l., **reticular**, one holding a muscle to a bone. l., **rhomboid**, one joining the cartilage of the first rib and the tuberosity of the clavicle. l., **round** (of hip). See *ligamentum teres femoris*. l., **round** (of forearm), one joining the coronoid process of the ulna and the tuberosity of the radius. l., **round** (of liver), a fibrous cord running from the umbilicus to the notch in the anterior border of the liver. It represents the remains of the obliterated umbilical vein. l., **round** (of uterus), a ligament running from the anterior surface of the cornu of the uterus through the inguinal canal to the mons Veneris. l., **sacrosciatic, great**, a ligament extending from the sacrum, coccyx, and inferior iliac spine to the tuberosity of the ischium. l., **spinoglenoid**, one extending between the spine of the scapula and the glenoid cavity. l., **splenophrenic**, the suspensory ligament of the spleen. l., **spring**, one joining the os calcis and scaphoid bone. l., **sternoclavicular**, the capsular ligament of the articulation between the sternum and clavicle. l.s **sternopericardiac**, connecting bands between the sternum and the pericardium. l., **stylohyoid**, a fibrous cord extending from the apex of the styloid process of the temporal bone to the lesser cornu of the hyoid bone. l., **stylomaxillary**, a ligament joining the styloid process of the temporal bone and the inferior surface of the posterior margin of the ramus of the inferior maxilla. l., **subpubic**. See *l., pubic*,

# LIGAMENTAL 512 LIMBUS

*inferior.* l., **supraspinal cervical.** See *l.*, *nuchal.* l. **suspensory** (of crystalline lens), the zonule of Zinn. l., **suspensory** (of eyeball). See *Lockwood's ligament.* l., **sutural.** 1. A thin lamina of fiber occurring in the cranial sutures. 2. A thin lamina of fibrous tissue often interposed between the articulating surfaces of bones united by suture. l.s, **synovial**, synovial folds resembling ligaments. **ligamentum** teres, a rounded fibrous cord attached to the center of the articular surface of the head of the femur, and extending to the margin of the cotyloid notch of the acetabulum. l.s, **thyroarytenoid,** the vocal bands. l., **thyroarytenoid,** inferior, one of the inferior or true vocal bands. l., **thyroarytenoid, superior,** one extending between the inner surface of the upper portion of the thyroid cartilage and the anterior surfaces of the apices of the arytenoid cartilages. l., **transverse** (of atlas), one attached to two small tubercles on the inner surface of the atlas, and surrounding the odontoid process of the axis. l., **transverse** (of hip-joint), one extending across the cotyloid notch of the acetabulum. l., **transverse** (of knee-joint), one extending from the anterior margin of the external semilunar fibrocartilage to the anterior extremity of the internal fibrocartilage. l., **trapezoid,** the anterior or external portion of the coracoclavicular ligament, extending from the upper surface of the coracoid process of the scapula to the under surface of the clavicle. l., **triangular** (of the urethra), a tendinous band of triangular shape, attached by its apex to the reflected portion of Poupart's ligament, and passing inward beneath the spermatic cord and behind the inner pillar of the external abdominal ring, to join the tendon of the opposite side. l., **uterovesical,** one of the peritoneal folds connecting the bladder and the uterus. l., **vesicoumbilical.** Same as *urachus.* l., **Winslow's,** the posterior ligament of the knee-joint. l., **Y-shaped** (of Bigelow), the iliofemoral ligament.
**ligamental, li amentary** (*lig-am-en'-tal; lig-am-en'-ta-re*) [*ligare,* to bind]. Of the nature of a ligament.
**ligamentopexis** (*lig-am-en-to-peks'-is*) [*ligament;* πῆξις, fixation]. Beck's operation of suspension of the uterus on the round ligaments.
**ligamentous** (*lig-am-en'-tus*) [*ligament*]. Of the nature of, or pertaining to, a ligament.
**ligamentum** (*lig-am-en'-tum*). See *ligament.* l. **denticulatum,** a notched ligament on each side of the myelon. l. **dentis,** that portion of the gum which is attached to the neck of a tooth. l. **inguinale,** Poupart's ligament, *q. v.* l. **interfoveolare.** See *Hesselbach's ligament.* l. **mucosum,** a synovial fold. l. **nuchæ,** one at the nape of the neck, connecting the two trapezius muscles. l. **patellæ,** the ligament securing the patella to the tibia. l. **spirale,** the thick part of the cochlear basilar membrane. l. **teres,** a round ligament. See *ligament,* round. l. **teres femoris,** a fibrous cord extending from the head of the femur to the margin of the cotyloid notch of the acetabulum.
**Ligar's line.** 1. A line drawn from the posterior superior iliac spine to a point midway between the tuberosity of the ischium and greater trochanter; the upper point of trisection of this line corresponds to the point of emergence of the gluteal artery. 2. A line drawn from the posterior superior iliac spine to the inner point of trisection of a line between the tuberosity of the ischium and the greater trochanter; the middle of this line indicates the point of emergence of the sciatic artery.
**ligate** (*li'-gāt*) [*ligare,* to bind]. To apply a ligature.
**ligation** (*li-ga'-shun*) [*ligate*]. The operation of tying, especially of tying arteries.
**ligator** (*li-ga'-tor*) [*ligare,* to bind]. An instrument used in placing and fastening ligatures.
**ligature** (*lig'-at-ūr*) [*ligatura,* a band]. 1. A cord or thread used for tying about arteries or other parts. 2. Ligation. l., double, the application of two ligatures to a vessel, between which it is divided. l., **elastic,** a narrow band or thread of rubber applied tightly to a part so as to destroy the tissues and by compression to lead to separation. It is used in the treatment of hemorrhoids, anal fistula, and in the removal of pedunculated growths. l., **Erichsen's,** one consisting of a double thread, one half of which is white, the other half black; it is used in the ligation of nævi. l., **interlacing,** l., **interlocking,** one for securing a pedicle in which several loops interlace. l., **intermittent,** a tourniquet applied above a poisoned wound to interrupt the blood-current; it is occasionally relaxed to allow of renewal of the circulation. l., **lateral,** partial occlusion of the lumen of a vessel by a loose ligature. l., **provisional,** a ligature applied during an operation, with the intention of removing it before the completion of the operation. l., **Wood-ridge's,** the isolation of the ventricles by drawing a silk ligature tightly about the auricles at their junction with the ventricles.

**light.** (*līt*) [AS., *leóht*]. Wave motions of the luminiferous ether that give rise to the sensation of vision when the rays impinge upon the retina. l., **axial,** light-rays that are parallel to each other and to the optic axis. l., **central.** See *l., axial.* l.-**difference,** the difference between the two eyes in respect to their sensitiveness to light. l., **diffused,** that reflected simultaneously from an infinite number of surfaces, or that which has been scattered by means of a concave mirror or lens. l., **Finsen,** light from which the heat-rays are excluded and only the blue and violet rays remain; it is used in phototherapy. l., **oblique,** light falling obliquely on a surface. l., **polarization** of. See *polarization.* l., **reflected,** light thrown back from an illuminated object. l., **refracted,** light-rays that have passed through an object and have been bent from their original course. l., **refrigerated.** See *l., Finsen.* l.-**sense, sensibility** of the retina to luminous impressions. l.-**stroke,** narcosis or death due to exposure to light. l., **transmitted,** the light passing through an object. l.-**treatment.** See *actinotherapy, phototherapy,* and *radiotherapy.*
**lighterman's bottom.** Inflammation of the bursa over the tuberosity of the ischium, from prolonged sitting.
**lightning pains.** The lancinating pains of loco-motor ataxia, coming on and disappearing with lightning-like rapidity.
**lign aloes** (*līn al'-ōz*) [*lignum, aloes*]. Same as *Agallochum* and *eagle-wood, q. v.*
**ligneous** (*lig'-ne-us*) [*lignum,* wood]. Woody, or having a woody texture.
**lignin** (*lig'-nin*) [*lignum*], $C_{18}H_{24}O_{10}$ (?), A modification of cellulose, constituting the greater part of the weight of most dry wood.
**lignosulphin** (*lig-no-sul'-fin*). A product occurring in the manufacture of sulphocellulose, containing free sulphurous acid combined with the volatile products of wood. It is used in the disinfection of dwellings.
**lignosulphite** (*lig-no-sul'-fīt*). A liquid by-product obtained in the manufacture of cellulose from pine wood; used in laryngeal tuberculosis in inhalations of 10 to 30 % solution.
**lignum** (*lig'-num*) [L.]. Wood. l. **benedictum,** guaiac-wood. l. **cedrium,** cedar-wood. l.-**vitæ,** the tree, *Guaiacum officinale.* See *guaiacum.*
**ligroin** (*lig'-ro-in*), A product obtained from petroleum; it is used in pharmacy as a solvent and for burning in sponge-lamps.
**ligula** (*lig'-ū-lah*) [Lith. of *lingua,* a tongue]. 1. A small tongue-shaped organ. 2. The strip of white matter on the margin of the fourth ventricle. See *lingula.*
**ligule** (*lig'-ūl*) [*ligula,* a variety of *lingula,* a little tongue; a strap]. Same as *ligula.*
**ligusticum** (*li-gus'-tik-um*). See *lovage.*
**Lilienfeld's theory** of blood-coagulation. This attributes to the nucleoproteid the power of splitting the fibrinogen into globulin and thrombosin, the thrombosin uniting with calcium to form fibrin.
**lily-of-the-valley.** See under *convallaria.*
**liman** (*li'-man*) [λιμήν, a marshy lake]. A sheet of water isolated from the sea and converted into a salt lake. l. **cure,** the treatment of diseases by bathing in limans at Odessa.
**limanol** (*li'-man-ol*). An extract obtained from boiling the mud of the limans at Odessa. It is used as an application in gout.
**limatura** (*lim-at-u'-rah*) [*limare,* to file]. Filings. l. **chalybis,** l. **ferri,** iron filings.
**limb** (*lim*) [AS., *lim*]. 1. One of the extremities attached to the sides of the trunk and used for prehension or locomotion. 2. An elongated structure resembling a limb, as the *limbs* of the internal capsule.
**limbic** (*lim'-bik*) [*limbus,* a border]. Marginal; pertaining to a border. l. **fissure.** See *fissure.* l. **lobe,** that surrounding the corpus callosum.
**limbus** (*lim'-bus*) [L.]. A border; the circumferential edge of any flat organ or part. l. **acetabuli,**

LIME 513 LINE

the cotyloid ligament. l. alveolaris, the alveolar process. l. conjunctivæ, the rim of conjunctiva that overlaps the corneal epithelium. l. corneæ, the edge of the cornea at its junction with the sclerotic coat. l. fossæ ovalis, the annulus ovalis. l. laminæ spiralis, the spiral membranous cushion at the border of the osseous spiral lamina of the cochlea. l. luteus. See *macula lutea*. l. sphenoidalis, the sharp anterior edge of the groove on the sphenoid bone for the optic commissure. l. Vieussenii. The limbus fossæ ovalis.
  lime (*līm*) [Pers., līmū, a.lemon]. 1. The fruit of several species of *Citrus*, as *C. limetta*. 2. [AS., *līm*, cement.] Calcium oxide, CaO (*quicklime*). Calcium oxide has a great affinity for water and for $CO_2$. On contact with the former, slaked lime is formed, with the evolution of heat. On living tissues it acts as a caustic. See also *calcium* and *calx*. l., chlorinated (*calx chlorinata*, U. S. P., B.P.), the chloride of lime of commerce. It is not a distinct chemical compound; its chief constituent, and the one on which its disinfectant properties depend, is calcium hypochlorite, which liberates chlorine. l.-juice, the juice of the lime. l., milk of, a milky fluid consisting of calcium hydroxide suspended in water. l., slaked, a common term for lime; correctly, it is lime which has been acted on by water. l., sulphureted. See *calx sulphurata*. l., syrup of (*syrupus calcis*, U. S. P.), contains 5 % lime, 30 % sugar, 65 % water. It is the antidote to poisoning by phenol or oxalic acid. Dose ½–2 dr. (2–8 Cc.). l.-water (*liquor calcis*, U. S. P.), a solution containing about 1½ parts of lime in 1000 of water. Dose ½–2 oz. (15–60 Cc.). It is used as an antacid.
  limen insulæ (*lī'-men*) [L.; *pl.*, *limina*]. The imaginary line separating the anterior perforated substance from the island of Reil. l. nasi, the boundary-line between the osseous and cartilaginous portions of the nasal cavity.
  limes death (*lī'-mēz*) [*limes*, boundary, limit]. The smallest amount of toxin which after being mixed with an antitoxin unit, will cause the death of a guinea pig within four or five days. l. zero, the dose of toxin which is just neutralized by one antitoxin unit.
  limic (*lī'-mik*) [λιμός, hunger]. Pertaining to hunger.
  liminal (*lim'-in-al*) [*limen*, threshold]. Pertaining to the threshold, especially pertaining to the lowest limit of perception.
  limiting membrane, external. The thin layer between the outer nuclear layer of the retina and that of the rods and cones. l. membrane, internal, in the eye, the inner layer of the retina.
  limitrophes (*lim-it'-ro-fēz*) [*limes*, a boundary; τροφή, nourishment]. The sympathetic ganglia and their connections.
  limitrophic (*lim-it-rŏf'-ik*). Regulating the processes of nutrition; a qualification sometimes applied to the great ganglionic cord of the sympathetic nerve-system, or to that system at large; pertaining to the sympathetic nerves.
  limnemic (*lim-ne'-mik*) [λίμνη, marsh]. Pertaining to, or caused by the influence of, a marsh.
  limnobios (*lim-no'-be-os*) [λίμνη, a lake; βίος, life]. The organic world of fresh water.
  limnomephitis (*lim-no-me-fī'-tis*) [λίμνη, marsh; *mephitis*, noxious odor]. Any miasm or noxious odor arising from marshy ground or swamps.
  limo (*lī'-mo*) [L.]. Lemon. The fruit of *Citrus limonum*, a tree of the order *Rutaceæ*. The pulp contains a large amount of citric acid, limonis cortex (U. S. P.), the rind of lemon, yields an essential oil and a glucoside, hesperidin, $C_{21}H_{26}O_{12}$. limonis, oleum (U. S. P.), oil of lemon. Dose 1–5 min. (0.065–0.32 Cc.). limonis succus (U. S. P.), lemon-juice, is refrigerant and antiscorbutic. Locally it has been used in pruritus, sunburn, and as a gargle in diphtheria. limonis, syrupus (B. P.), is used as a refrigerant and vehicle.
  limoctonia (lim-ok-tŏ'-ne-ah) [λιμός, hunger; κτείνειν, to destroy]. Death from hunger; suicide by hunger.
  limonin (*lim'-o-nin*) [*limo*], $C_{22}H_{30}O_8$ (?). A glucoside from seeds of apples and lemons.
  limophthisis (*lim-off'-this-is*) [λιμός, hunger; φθίσις, wasting]. The wasting of the body due to privation and lack of food.
  limophoitos, limophoitosis (*lim-o-fo'-it-os, lim-o-fo-it-o'-sis*) [λιμός, hunger; φοῖτος, madness]. Insanity due to hunger or lack of nutrition.

limophoitosic (*lim-o-fo-it-o'-sik*) [*limophoitos*]. Insane from hunger or underfeeding.
  limopsora (*lim-op-so'-rah*) [λιμός, hunger; ψώρα, itch]. A kind of scabies (or pruritus?) asserted to attack man and other animals after long deprivation of food.
  limopsorus (*lim-op-so'-rus*) [λιμός, hunger; ψώρα, itch]. A disease, like scurvy, pellagra, or famine-fever, due to poor or insufficient food.
  limoseric (*lim-o-ser'-ik*) [λιμός, hunger]. Pertaining to or caused by hunger.
  limosis (*lim-o'-sis*) [λιμός, hunger]. 1. Unnatural appetite. 2. A disease distinguished by depraved appetite.
  limotherapy (*lim-o-ther'-ap-e*) [λιμός, hunger; θεραπεία, treatment]. The treatment of disease by partial or total deprivation of food. It has been used in the treatment of aneurysm.
  limp. A halting gait. See *claudicatio*. limping (*limp'-ing*). Walking with a halting gait. l., intermittent. See *claudication*, *intermittent*.
  linadin (*lin'-ad-in*). An insoluble, dark-brown powder containing 1 % of iron and 0.023 % of iodine, prepared from the spleen of animals. Dose in malarial cachexia 150–385 gr. (10–25 Gm.).
  linagogue, linagogus (*lin'-ag-og, lin-ag-o'-gus*) [*linum*, thread; ἀγωγός, leading]. An instrument used in guiding the course of a suture.
  linalool (*lin-al'-o-ol*), $C_{10}H_{18}O$. A fragrant liquid occurring in oils of lign aloe, lavender, and bergamot.
  linamarin (*lin-am-ar'-in*) [*linum*, flax; *amara*, bitter]. The toxic glucoside of common flax.
  lincture (*link'-tur*) [*lingere*, to lick]. A medicine to be taken by licking; an electuary.
  linctus (*link'-tus*) [L.]. Same as *lincture*.
  line, linea (*līn*, *lin'-e-ah*) [*linea*, a line]. 1. Extension of dimension having length, but neither breadth nor thickness. 2. The 1/12 part of an inch. 3. In anatomy, anything resembling a mathematical line in having length without breadth or thickness; a boundary or guide-mark. l., abdominal. See *abdominal*. l., linea alba. 1. A tendinous raphe extending in the median line of the abdomen from the pubes to the ensiform cartilage; it is formed by the blending of the aponeuroses of the oblique and transversalis muscles. 2. Hunter's line, the anterior peduncles of the pineal gland. lineæ albicantes, glistening white lines in either iliac region of the abdomen, seen in distention of the abdomen from pregnancy, ascites, or tumors. l., alveolobasilar, a line joining the basion and the alveolar point. l., alveonasal, a line joining the nasal and alveolar points. linea aspera, a rough longitudinal ridge on the posterior surface of the middle third of the femur, dividing below into two and above into three ridges. l., auriculobregmatic, a line passing from the auricular point to the bregma, and dividing the preauricular from the postauricular part of the cranium. l.s, axillary, anterior and posterior, vertical lines extending downward from the axilla on the side of the trunk. l., base, a line running backward from the infraorbital ridge through the middle of the external auditory meatus, and prolonged to the middle line of the head posteriorly. l., basiobregmatic, the line joining the basion and the bregma. l., Baudelocque's. See *Baudelocque's line*. l.s, Beau's. See *Beau's line*. l., biauricular, the line separating the anterior from the posterior portion of the skull; it extends from one auditory foramen over the vertex to the other. l., blue, the blue line at the dental margin of the gums in chronic lead-poisoning. l., Bryant's. See *Bryant's line*. l., Camper's. See *Camper's line*. l., Clapton's. See *Clapton's line*. l., Conradi's. See *Conradi's line*. l., Corrigan's. See *Corrigan's line*. l., costo-articular, a line drawn between the sternoclavicular articulation and the point of the eleventh rib. l., costoclavicular. See *l., parasternal*. l., curved, inferior (of the ilium), a line extending from the upper part of the anterior inferior spinous process of the ilium, and terminating at the middle of the great sciatic notch. l., curved, inferior (of the occipital bone), a ridge extending transversely across the outer surface of the occipital bone a short distance below the superior curved line. l., curved, middle (of the ilium), a line commencing about an inch or an inch and a half behind the anterior superior spine of the ilium and arching backward and downward to the upper margin of the great sciatic notch. l., curved, superior (of the ilium), a line commencing

18

about two inches in front of the posterior extremity of the crest of the ilium and curving downward and forward toward the posterior part of the great sciatic notch. l., curved, superior (of the occipital bone), a semicircular line, passing outward and forward from the external occipital protuberance. l. of demarcation, a line of division between healthy and gangrenous tissues. l.s, Eberth's. See Eberth's lines. l., Ellis'. See Ellis' line. l., embryonic, the primitive trace in the center of the germinal area of the ovum. linea eminens (of the cricoid cartilage), a mesial ridge on the dorsal half of the cricoid cartilage. linea eminens (of the patella), a ridge on the posterior surface of the patella, dividing that surface into two unequal parts, the outer of which is the larger. l., epiphyseal, the thin layer of cartilage at first separating the borders of the diaphysis and epiphysis. l., facial. 1. A straight line tangential to the glabella and some point at the lower portion of the face. 2. See Camper's-line. l. of fixation, an imaginary line drawn from the object viewed through the center of rotation of the eye. l.s, Fraunhofer's. See Fraunhofer's lines. l.s, Frommann's. See Frommann's lines. l., genal, a line seen in the faces of children, in certain diseases, running downward from the region of the malar bone to join the nasal line. See Jadelot's lines. l., gingival, Burton's, a reddish streak or margin at the reflected edge of the gums. l. of Haller. See linea splendens of Haller. l., Hilton's. See Hilton's line. l., Holden's. See Holden's line. l., Hunter's. See linea alba (2). l., iliopectineal, the bony ridge marking the brim of the true pelvis, situated partly on the ilium and partly on the pubis. l., incremental, Salter's, a curved line in dentin, supposed to indicate the laminar structure, and to correspond to the successive laminæ, or strata of dentin. l., intertrochanteric, anterior, a line upon the anterior surface of the femur, separating the neck and shaft, extending between the tubercle and a point close to and in front of the lesser trochanter. l., intertrochanteric, posterior, a ridge on the posterior surface of the femur, extending between the greater and lesser trochanters. l., intertubercular, an imaginary transverse line drawn around the abdomen at the level of the tubercles, on the iliac crests, that is about two inches behind the anterior superior iliac spines. l.s, Jadelot's. See Jadelot's lines. l.s, Kirchoff's. See Fraunhofer's lines. l., Ligar's. See Ligar's line. l., mammary. 1. A line from one nipple to the other. 2. Often, but incorrectly used, for l. mamillary, q. v. l., mammillary, a vertical line passing through the center of the nipple. l., mylohyoidean. See l., oblique, internal (of the inferior maxilla). l., nasobasilar, the line drawn through the basion and the nasal point. l., Nélaton's. See Nélaton's line. l., nigra, a dark pigmented line often present in pregnant women and extending from the pubes upward in the median line. l., nipple-. Same as l., mammillary. l., nuchal, inferior, the inferior curved line of the occiput. l., nuchal, median, the external occipital protuberance. l., nuchal, superior, the superior curved line of the occiput. l., oblique (of the fibula), a prominent ridge on the internal surface of the shaft of the fibula, commencing above at the inner side of the head, and terminating in the interosseous ridge at the lower fourth of the bone. l., oblique (of the radius), a prominent ridge running from the lower part of the bicipital tuberosity, downward and outward, to form the anterior border of the bone. l., oblique (of the thyroid cartilage), a line extending downward and outward from the tubercle of the thyroid cartilage. l., oblique (of the tibia), a rough ridge that crosses the posterior surface of the tibia obliquely downward from the back part of the articular facet for the fibula to the internal border. l., oblique, external (of the inferior maxilla), a prominent ridge on the external surface of the inferior maxilla just below the mental foramen from which it runs outward, upward, and backward to the anterior margin of the ramus. l., oblique, internal (of the inferior maxilla), a ridge on the internal surface of the lower jaw, commencing at the posterior portion of the sublingual fossa, continuing upward and outward so as to pass just below the last two molar teeth. l., oculozygomatic, one of Jadelot's lines indicative of spinal disease. l.s, Ogston's. See Ogston's line. l., parasternal, a line midway between the nipple-line and the border of the sternum. l., pectineal, the portion of the ilio-

pectineal line that is formed by the pubic bone. l., primitive, the primitive streak of the embryo. l., profile, of Camper. See Camper's line. l., quadrate, an eminence on the femur commencing about the middle of the posterior intertrochanteric line, and descending vertically for about two inches along the posterior surface of the shaft. l. of regard, in optics, the line connecting the center of rotation of the eye with the point of fixation or of regard. l., respiratory, the line connecting the bases of the upward strokes in a tracing of the pulse. l., Roser's. Same as Nélaton's line. l., Salter's. See incremental line. l., scapular, a vertical line downward from the lower angle of the scapula. l., semicircular, Douglas', the curved lower edge of the internal layer of the aponeurosis of the internal oblique muscle of the abdomen, where it ceases to cover the posterior surface of the rectus muscle. l., semilunar, of Spigelius, a curved tendinous condensation of the aponeurosis of the external oblique muscle of the abdomen, running along the outer border of the rectus abdominis. l. of sight, an imaginary line drawn from the object viewed to the center of the pupil. linea splendens of Haller, a longitudinal fibrous band extending along the middle line of the anterior surface of the spinal pia mater. l., sternal, the median line of the sternum. l., sternomastoid, a line drawn from a point between the two heads of the sternomastoid muscle to the mastoid process. l., subcostal, an imaginary transverse line drawn around the abdomen at the level of the lower border of the tenth costal cartilage. l., supraorbital, a line extending horizontally across the forehead immediately above the root of the external angular process of the frontal bone. l., test-, a line for detecting shortening of the neck of the femur. If two lines are drawn to meet at right angles, one of them backward from the anterior superior spinous process of the ilium, and the other upward from the top of the trochanter major, the latter is the testline; its length is to be compared with the same line on the uninjured side. Syn., Bryant's line. l., Thompson's. See Thompson's line. l., transverse (of the abdomen), the tendinous intersections in the course of the rectus abdominis muscle. l., trapezoid, the line of attachment of the trapezoid ligament on the inferior surface of the outer portion of the clavicle. l., Virchow's. See Virchow's line. l., visual, an imaginary line, drawn from a point looked at, through the nodal point of the eye, to the macula lutea.

**lineage** (lin'-e-āj) [linea, a line]. The line of descent from an ancestor; ancestry.

**lineal** (lin'-e-al) [linea, line]. Pertaining to lineage. See also linear.

**lineament** (lin'-e-am-ent) [lineamentum; linea, a line]. The outline of the face, or of any of its features. Also, the outline of the embryo.

**linear** (lin'-e-ar) [linea, a line]. Resembling or pertaining to a line. Applied in biology to an organ that is narrow, many times longer than broad, and that has parallel margins. l., craniectomy. See craniectomy. l., extraction. See cataract. l. fracture, one forming a line and attended with little or no displacement of the fragments.

**Ling's system** [Pier Henrik Ling, Swedish physician, 1776–1839]. A method of treatment of disease by gymnastic and other rhythmical movements of the body. Syn., kinetotherapy.

**lingam** (lin'-gam). See phallus.

**Lingism** (ling'-izm). See Ling's system.

**lingua** (ling'-gwah) [L.]. The tongue. l. exigua, the epiglottis. l. frænata, tongue-tie. l. geographica, the geographical tongue. See tongue, geographical. l. nigra. See glossophytia.

**lingual** (ling'-gwal) [lingua]. Pertaining to or shaped like the tongue. l. artery. See under artery. l. bone, the hyoid bone. l. delirium. See delirium. l. lobule. See subcalcarine convolution. l. nerve. See under nerve. l. tonsil, a quantity of lymph tissue at the base of the tongue.

**lingualis** (ling-gwa'-lis). See muscles, table of.

**linguiform** (ling'-gwif-orm) [lingua, tongue; forma, form]. Shaped like a tongue.

**lingula** (ling'-gu-lah) [dim. of lingua]. A small lobule between the valve of Vieussens and the central lobule of the cerebellum. Syn., linguetta laminosa. l. auriculæ, the cartilaginous projection toward or into the upper portion of the lobe of the ear. l. mandibularis, the prominent, thin scale of bone

# LINGULAR 515 LIPOMA

partly surrounding the inferior dental foramen of the lower jaw. l. **sphenoidalis**, a small, tongue-like process extending backward in the angle formed by the body of the sphenoid and one of its greater wings. l. **of Wrisberg**, the connecting fibers of the motor and sensory roots of the trifacial nerve.
**lingular** (*ling'-gū-lar*) [*lingula*, a little tongue]. Of or pertaining to a little tongue.
**lingulate** (*ling'-gū-lāt*) [*lingula*]. Tongue-shaped.
**liniment, linimentum** (*lin'-i-ment, lin-im-en'-tum*) [*linere*, to smear]. A liquid intended for application to the skin by gentle friction. The following are official in the U. S. P.: *linimentum ammoniæ* or volatile liniment; *l. belladonnæ; l. calcis* or Carron oil; *l. camphoræ* or camphorated oil; *l. chloroformi; l. saponis; l. saponis mollis* or tincture of green soap; *l. terebinthinæ.* l., St. **John Long's**, liniment of turpentine and acetic acid (*linimentum terebinthinæ aceticum*, B. P.).
**linin** (*li'-nin*) [*linum*, flax]. 1. A strongly purgative principle obtainable from *Linum catharticum*, or purging flax. 2. In biology, minute threads extending between the individual microsomata (*ids*) in a cell-nucleus. The achromatin of the nuclear network; *parachromatin*, less correctly called *nucleohyaloplasm.*
**linition** (*lin-ish'-un*) [see *liniment*]. The process of applying a liniment.
**linitis** (*li-ni'-tis*) [λίνον, web; ιτις, inflammation]. Inflammation of the network of filamentous areolar tissue surrounding the gastric vessels. l. **plastica,** fibrinous infiltration of the pylorus.
**linolein** (*lin-o'-le-in*) [*linum; oleum,* oil]. The neutral fat contained in linseed-oil, and to which its drying property is due.
**linseed** (*lin'-sēd*). See *linum*. l.-**oil.** See *lini, oleum,* under *linum.*
**lint** [*linum*]. A loosely woven or partly felted mass of broken linen fibers, made by scraping or picking linen cloth. It is used as a dressing for wounds. l., **common**, lint that is twilled on one side and woolly on the other. In the spreading of an ointment the twilled side is used. l., **patent,** lint that is scraped on both sides, a soft finish being thus given the two surfaces. Syn., *English charpie.*
**lintin** (*lin'-tin*). Absorbent cotton rolled or compressed into sheets.
**linum** (*li'-num*) [L.]. The seed of *L. usitatissimum,* a plant of the order *Lineæ*, containing a fatty substance, *linolein*, which is the glycerid of linoleic acid. It is a demulcent, emollient, and expectorant, useful in inflammations of mucous membranes. Syn., *flaxseed; linseed.* **lini**, **cataplasma** (B. P.), a poultice made from linseed meal. l. **catharticum**, an active purgative and vermifuge. **lini farina** (B. P.), flaxseed meal, used as a poultice. **lini, infusum** (B. P.), flaxseed tea. Dose indefinite. **lini, oleum** (U. S. P.), the fixed oil of flaxseed, a glycerid of linoleic acid. Dose ½–2 oz. (16–64 Cc.). **lini, semina** (B. P.), linseed or flaxseed.
**liodermia** (*li-o-der'-me-ah*) [λείος, smooth; δέρμα, skin]. A condition of abnormal smoothness and glossiness of the skin.
**liomyofibroma** (*li-o-mi-o-fi-bro'-mah*) [*liomyoma; fibroma*]. A tumor presenting the characteristics of a liomyoma and a fibroma.
**liomyoma** (*li-o-mi-o'-mah*) [λείος, smooth; *myoma*]. A tumor composed of unstriped muscular tissue.
**liotrichous** (*li-ot'-rik-us*). See *leiotrichous.*
**Liouville's icterus** (*le-oo-vēl'*) [Henri Liouville, French physician, 1837–1887]. Icterus neonatorum.
**lip** [AS., *lippa*]. 1. One of the two fleshy folds surrounding the orifice of the mouth. 2. One of the labia majora or labia minora. See *labium.* 3. The border of a wound.
**lipa** (*li'-pah*) [L.]. Fat.
**lipacidemia** (*lip-as-id-e'-me-ah*) [λίπος, fat; αἷμα, blood]. Presence of fatty acids in the blood.
**lipaciduria** (*lip-as-id-ū'-re-ah*) [λίπος, fat; *acid;* οὖρον, urine]. The presence of fatty acids in the urine.
**lipæmia** (*lip-e'-me-ah*). See *lipemia.*
**lipanin** (*lip'-an-in*). A substitute for cod-liver oil, consisting of pure olive-oil and 6 % of oleic acid. Dose 2 to 6 tablespoonfuls daily.
**lipara** (*lip'-ar-ah*) [λιπαρός, fatty]. An emollient plaster.
**liparia** (*lip-a'-re-ah*) [*liparia*, fatness]. Fatness; obesity.
**liparocele** (*lip'-ar-o-sēl*) [λίπος, fat; κήλη, a tumor]. A fatty tumor or cyst; a hernia containing fatty tissue.

**liparoid** (*lip'-ar-oid*) [λιπαρός, fatty; εἶδος, like]. Resembling fat.
**liparomphalos** (*lip-ar-om'-fal-os*) [λιπαρός, fat; ὀμφαλός, the navel]. A fatty tumor situated at the navel, or involving the umbilical cord.
**liparoscirrhus** (*lip-ar-o-skir'-us*) [λιπαρός, fat; σκίρρος, a carcinomatous growth]. A fatty, scirrhous tumor.
**liparotrichia** (*lip-ar-o-trik'-e-ah*) [λιπαρός, fat; θρίξ, hair]. Abnormal greasiness of the hair.
**liparous** (*lip'-ar-us*) [λιπαρός, fat]. Fat; obese.
**lipase** (*lip'-ās*) [λίπος, fat]. A fat-splitting enzyme contained in the pancreatic juice, in bloodplasma, and in many plants.
**lipectomy** (*lip-ek'-to-me*) [λίπος, fat; ἐκτομή, excision]. Excision of fatty tissue.
**lipemania** (*li-pe-ma'-ne-ah*). See *lypemania.*
**lipemia** (*lip-e'-me-ah*) [λίπος, fat; αἷμα, blood]. The presence of an emulsion of fine oil-globules in the blood, sometimes found in diabetes.
**liphemia** (*li-fe'-me-ah*). See *oligemia.*
**lipin** (*li'-pin*) [λίπος, fat]. A general term for fats, fatty acids, lipoids, soaps, etc.
**lipiodol** (*lip-i'-o-dol*). An oil containing 40 % of iodine in each cubic centimeter.
**Lipliawsky's test** (*lip-le-aw'-ske*). For diacetic acid in the urine: two solutions are needed with: (*a*) a 1 % solution of paramidoacetophenon with addition of 2 Cc. of concentrated HCl shaken thoroughly; (*b*) a 1 % aqueous solution of potassium nitrite; 6 Cc. of the first is mixed with 3 Cc. of the second, an equal volume of urine added, and a drop of ammonia. To 10 drops to 2 Cc. of this mixture add 15 to 20 Cc. of concentrated HCl, 3 Cc. of chloroform, and 2 to 4 drops of iron chloride solution. If the test-tube is corked and gently but repeatedly inverted, in the presence of diacetic acid the chloroform will show a characteristic violet color—the deepness of the color depending upon the amount present.
**lipo-** (*lip-o-*) [λίπος, fat]. A prefix meaning fat or fatty.
**lipobromol** (*lip-o-bro'-mol*). Oil of poppyseed combined with 33.3 % of bromine; a bland, almost tasteless preparation.
**lipocardiac** (*lip-o-kar'-de-ak*) [*lipo-; καρδία,* the heart]. Pertaining to a fatty heart.
**lipocele** (*lip'-o-sēl*). Synonym of *liparocele, q. v.*
**lipochondroma** (*lip-o-kon-dro'-mah*) [*lipo-; chondroma*]. A combined fatty and cartilaginous tumor.
**lipochrin** (*lip'-o-krin*) [λίπος, fat; ὠχρός, sallow]. A yellow pigment obtained from the fat-globules in the retinal epithelium.
**lipochrome** (*lip'-o-krōm*) [*lipo-; χρῶμα,* color]. Any one of a special group of fatty pigments found in animal tissues.
**lipoclastic** (*lip-o-klas'-tik*) [*lipo-; κλάειν,* to break]. Fat splitting.
**lipodermatous** (*lip-o-der'-mat-us*) [λείπειν, to leave; δέρμα, skin]. Affected with lipodermia.
**lipodermia** (*li-po-der'-me-ah*). See *leipodermia.*
**lipodystrophy** (*lip-o-dis'-tro-fe*). A disturbance of fat metabolism.
**lipoferous** (*lip-of'-er-us*) [*lipo-; ferre,* to carry]. Fat carrying.
**lipofibroma** (*lip-o-fi-bro'-mah*) [*lipo-; fibroma*]. A combined fatty and fibrous tumor.
**lipogenesis** (*lip-o-jen'-es-is*) [*lipo-; γένεσις,* birth]. The formation or deposit of fat.
**lipogenin** (*lip'-o-jen-in*). An ointment-base occurring in solid and liquid form, said to consist of a mixture of fatty acids.
**lipogenous** (*lip-oj'-en-us*) [λίπος, fat; γεννᾶν, to beget]. Fat-producing.
**lipoid** (*lip'-oid*) [λίπος, fat; εἶδος, like]. Resembling fat. A name given by Overton to a group of substances in the protoplasm of all cells, especially in the outer layer or cell membrane; they are soluble in ether or alcohol.
**linolysis** (*lin-ol'-is-is*) [*lipo-; λύειν,* to loose]. The decomposition of fat.
**lipolytic** (*lip-o-lit'-ik*). Fat-splitting.
**lipoma** (*lip-o'-mah*) [*lipo-; ὄμα,* a tumor]. A fatty tumor. l., **diffuse,** a tumor consisting of an irregular mass of fatty tissue without a capsule. l. **mixtum,** a fatty tumor, the thick capsule of which causes it to resemble fibrous growths. l., **osseous,** a fatty tumor the fibrous septa of which have become ossified.

**lipomasia** (lĭp-o-ma'-ze-ah). 1. A softened condition of bone. 2. A condition of cancellous bone, in which the spaces are widened and filled with fatty marrow, which is anemic; the bone is soft and brittle, and fractures and cuts easily.
**lipomatoid** (lĭp-o'-mat-oid). See *lipomatous*.
**lipomatosis** (lĭp-o-mat-o'-sis) [*lipoma*]. A general deposition of fat; obesity.
**lipomatous** (lĭp-o'-mat-us). Of the nature of a lipoma.
**lipomeria** (lĭp-o-me'-re-ah) [λείπειν, to leave; μέρος, a part]. A monstrosity having one limb absent.
**lipomphalus** (lĭp-om'-fa-lus) [*lipo-*; ὀμφαλός, the navel]. A fatty umbilical hernia.
**lipomyoma** (lĭp-o-mi-o'-mah) [*lipo-*; *myoma*]. A myoma with fatty elements.
**lipomyxoma** (lĭp-o-miks-o'-mah) [*lipo-*; *myxoma*]. A myxoma combined with fatty tissues.
**lipophrenia** (lĭp-o-fre'-ne-ah) [λείπειν, to fail; φρήν, mind]. Failure of mental capacity.
**lipoprotein** (lĭp-o-pro'-te-in) [*lipo-*; *protein*]. A hypothetical combination of a protein with a fatty acid.
**lipopsychia** (lĭp-o-si'-ke-ah). See *asthenia*.
**liporhodin** (lĭp-o-ro'-din) [*lipo-*; ῥόδον, a rose]. A red-colored lipochrome.
**liposarcoma** (lĭp-o-sar-ko'-mah) [*lipo-*; *sarcoma*]. Sarcoma with fatty elements.
**liposarcous** (lĭp-o-sar'-kus) [λείπειν, to leave; σάρξ, flesh]. Lean; emaciated.
**liposic** (lĭp'-o-sik). See *lipolytic*.
**liposis** (lĭp-o'-sis). See *lipomatosis*.
**lipospongosis** (lĭp-o-spun-go'-sis) [λίπος, fatty; σπόγγος, sponge]. The formation of a fatty or spongelike outgrowth.
**lipostomatous** (lĭp-o-sto'-mat-us) [λείπειν, to leave; στόμα, mouth]. Having no mouth.
**lipostomosis** (lĭp-o-sto-mo'-sis) [λείπειν, to leave; to be lacking; στόμα, mouth]. In biology, absence of the oral aperture.
**lipostomy** (lĭp-os'-to-me) [λείπειν, to leave; στόμα, mouth]. Atrophy of the mouth.
**lipothymia** (lĭp-o-thi'-me-ah) [λείπειν, to fail; θυμός, life]. Faintness.
**lipotrichia** (lĭp-o-trik'-e-ah) [λείπειν, to fail; θρίξ, hair]. Falling out of the hair.
**lipoxanthin** (lĭp-o-san'-thin) [*lipo-*; *xanthin*]. A yellow lipochrome.
**lipoxenous** (lĭp-oks'-en-us) [λείπειν, to leave; ξένος, host]. Applied to a parasite that leaves its host and completes its existence independently.
**lipoxeny** (lĭp-oks'-en-e) [λείπειν, to fail; ξένος, a host]. Desertion of a host by a parasite.
**lipoxysm** (lĭp-oks'-izm) [*lipo-*; ὀξύς, sharp]. Poisoning by means of oleic acid.
**lippa** (lĭp'-ah). Lippitudo, q. v.
**lippiol** (lĭp'-e-ol) [*lippia; oleum*, oil]. A medicinal camphor derivable from *Lippia mexicana*.
**lippitude, lippitudo** (lĭp'-e-tūd, lĭp-e-tū'-do) [*lippus*, blear-eyed]. The state of being blear-eyed, a condition marked by ulcerative marginal blepharitis.
**lipsis** (lĭp'-sis) [λεῖψις, a leaving]. Cessation; ending. l. animi, fainting.
**lipsotrychia** (lĭp-so-trik'-e-ah) [λείπειν, to leave; θρίξ, hair]. Falling out of the hair.
**lipuria** (lĭp-ū'-re-ah) [λίπος, fat; οὖρον, urine]. The presence of fat in the urine.
**liquable** (lĭk'-wah-bl) [*liquare*, to render liquid]. Capable of being liquefied.
**liquamen** (lĭk'-wam-en) [*liquare*, to render liquid]. The liquid obtained by melting solids.
**liquarium** (lĭk-wa'-re-um) [*liquarius*, pertaining to liquids]. Simple syrup of sugar.
**liquate** (lĭk'-wāt) [*liquare*, to make liquid]. To liquefy.
**liquation** (lĭk-wa'-shun) [*liquare*, to render liquid]. The process of melting.
**liquefacient** (lĭk-we-fa'-shent) [*liquefaciens*, liquefying]. 1. Having the power to liquefy or soften. 2. An agent which has the power to liquefy a hard deposit or growth.
**liquefaction** (lĭk-we-fak'-shun) [see *liquefacient*]. The process of changing or being changed into a liquid. l.-necrosis. See *necrosis, liquefactive*.
**liquefactive** (lĭk-we-fak'-tiv) [see *liquefacient*]. Pertaining to, causing, or characterized by liquefaction.
**liquescent** (lĭk-wes'-ent) [*liquescere*, to become liquid]. Becoming, or tending to become, liquid.
**liqueur** (lĭk-ār') [Fr.]. An aromatic alcoholic drink.
**liquid** (lĭk'-wid) [*liquere*, to melt]. 1. Fluid; flowing. 2. A substance that flows readily and takes the shape of the containing vessel. l. cuticle, collodion. l. smoke, pyroligneous acid.
**Liquidambar** (lĭk-wid-am'-bar) [*liquid; ambar*, from Ar., *anbar*, ambergris]. A genus of trees of the *Hamamelideæ*. L. *altingia* and L. *orientalis* are species that afford a portion of commercial styrax. L. *styraciflua*, of North America, contains a stimulant gum, and is useful in diarrheas, coughs, and colds. Syn., *bilsted; copalm; sweet-gum*.
**liquidity** (lĭk-wid'-it-e) [*liquidus*, liquid]. Fluidity; the state of being liquid.
**liquiform** (lĭk'-wif-orm) [*liquor*, liquid; *forma*, form]. Of the nature of a liquid.
**liquor** (lĭ'-kwor, lĭk'-or) [L.]. 1. Any liquid. 2. An aqueous solution of a nonvolatile substance. l. amnii, the liquid contained in the amniotic sac. l. amnii spurius, the oxidation products formed by the Wolffian bodies contained as a fluid in the sac of the allantois. Syn., *allontoic fluid*. l. Bellostii, a solution of 1 gm. of mercurous nitrate in 8 gm. of water and 2 gm. of nitric acid. It has been used as a test for helminthiasis and also for paralysis, as the substances excreted in the urine in these conditions are said to be precipitated or stained black by this reagent. l. carbonis detergens, a mixture of coal-tar and tincture of soap-bark. l. carnis ferropeptonatus. See *carniferrol*. l. Cotunnii, the perilymph of the internal ear. l. folliculi, the fluid filling the follicle or space about the developing ovum in the Graafian follicle. l. puris, the liquid portion of pus. l. sanguinis, the blood-plasma. l. Scarpæ, the endolymph. l. seminis, the fluid portion of semen.
**liquorice** (lĭk'-or-is). See *glycyrrhiza*. l., wild. See *abrus*.
**lirella** (lĭr'-el-ăt) [*lira*, a ridge]. Marked with linear ridges or furrows; ridge-like; furrow-like.
**lirelliform, lirelline, lirellous** (lĭr'-el-e-form, lĭr'-el-ēn, -us). See *lirellate*.
**Lisfranc's amputation** (lĭs-frank') [Jacques *Lisfranc*, French surgeon, 1790–1847]. A disarticulation of the metatarsal bones from the tarsus. L.'s joint, the tarsometatarsal articulation. L.'s tubercle, a rough spot on the anterior surface of the first rib near the superior border. It serves for the attachment of the scalenus anticus muscle.
**lisp**. To imperfectly pronounce the sibilant letters.
**lisping** (lĭsp'-ing) [AS., *wlispian*, to lisp]. A defect of speech in which sibilant letters are sounded like linguals, especially *s* as *th*.
**Lissauer's parietal angle** (lĭs'-ow-er) [Heinrich *Lissauer*, German neurologist, 1861–1891]. That included between lines drawn from the bregma and lambda to the most prominent point of the parietal bone. L.'s tract, the narrow bridge of white substance between the apex of the posterior horn and the periphery of the spinal cord; it is traversed by some of the root-fibers. Syn., *Lissauer's marginal zone*.
**lissencephalous** (lĭs-en-sef'-al-us) [λισσός, smooth; ἐγκέφαλος, the brain]. Having a brain with few or no convolutions.
**lissotrichous** (lĭs-ot'-rik-us) [λισσός, smooth; θρίξ, hair]. In biology, having straight, smooth hair; less correctly written leiotrichous and liotrichous.
**Lister's double salt** (lĭs'-ter) [Lord Joseph *Lister*, English surgeon, 1827–1912]. The cyanide of mercury and of zinc. L.'s dressing, gauze impregnated with phenol in some other antiseptic, or plain aseptic gauze, used as a dressing for wounds. L.'s method, Listerism, q. v.
**listerine** (lĭs'-ter-ēn) [Lord Joseph *Lister*, English surgeon, 1827–1912]. A proprietary antiseptic preparation said to contain thymol, eucalyptus, baptisia, gaultheria, mentha arvensis, benzoic and boric acids. A similar preparation is the *liquor antisepticus* of the U. S. P.
**Listerism** (lĭs'-ter-izm). A general name for the antiseptic and aseptic treatment of wounds according to the principles first enunciated by Lord Lister.
**Listing's law** (lĭs'-ting) [John Benedict *Listing*, German physicist]. When the line of sight passes from its primary position into any other position, the angle of rotation of the eyeball in this second position is the same as if the eyeball had been rotated about a fixed axis, perpendicular to both the first and the second direction of the line of sight. L.'s plane, the vertical transverse plane perpendicular

to the anteroposterior axis of the eyeball, which passes through the center of motion of the eyes and in which lie the vertical and transverse axes of normal voluntary rotation. L.'s reduced eye, a scheme for simplifying optical problems by representing the two nodal points and the two principal points of the eye by a mean nodal point and a mean principal point.

**listol** (*lis'-tōl*). An antiseptic said to consist of thymol and iodine.

**liter** (*le'-ter*) [*litra*, a pound]. The unit of capacity in the metric system, equal to 0.88036 of an imperial quart, or 1.056 U. S. quarts; it is the volume of one kilogram of water at its maximum density.

**lithagogectasia** (*lith-ag-o-jek-ta'-ze-ah*) [λίθος, stone; ἀγωγός, leading; ἔκτασις, a stretching out]. Lithectasy.

**lithagogue** (*lith'-ag-og*) [λίθος, a stone; ἀγωγός, leading]. 1. Expelling calculi. 2. Any agent tending to expel calculi from the bladder.

**lithangiuria** (*lith-an-je-ū'-re-ah*) [λίθος, a stone; ἀγγεῖον, vessel; οὖρον, urine]. A diseased condition of the urinary tract due to the presence of calculi.

**litharge** (*lith'-arj*). See *plumbi oxidum* under *plumbum*.

**lithargyrium** (*lith-ar-ji'-re-um*). Litharge.

**lithargyrius** (*lith-ar-ji'-re-us*) [L.]. A litharge of a yellowish hue; gold litharge.

**lithargyrum** (*lith-ar'-ji-rum*) [L.]. Litharge.

**lithargyrus** (*lith-ar'-ji-rus*) [L.]. Litharge, particularly silver litharge.

**lithate** (*lith'-āt*) [λίθος, a stone]. A salt of lithic (uric) acid; a urate.

**lithecboly** (*lith-ek'-bo-le*) [λίθος, a stone; ἐκβολή, a throwing out]. Expulsion of a calculus by contraction of the bladder and dilatation of its neck.

**lithectasy** (*lith-ek'-tas-e*) [λίθος, a stone; ἔκτασις, a stretching out]. Dilatation of the urethra and neck of the bladder for the removal of calculi.

**lithectomy** (*lith-ek'-to-me*) [*litho-*; ἐκτομή, a cutting out]. Same as *lithotomy*.

**lithemia, lithæmia** (*lith-e'-me-ah*) [*litho-*; αἷμα, blood]. A condition in which, owing to defective metabolism of the nitrogenous elements, the blood becomes charged with deleterious substances, principally, perhaps, of the uric-acid group, although their exact chemical nature is not determined.

**lithemic** (*lith-e'-mik*) [*lithemia*]. Pertaining to or suffering from lithemia.

**lithepsy** (*lith'-ep'-se*). See *lithodialysis*.

**lithia** (*lith'-e-ah*) [*lithium*], Li$_2$O. Lithium oxide. **l.-water**, mineral water containing lithium salts in solution.

**lithiasis** (*lith-i'-as-is*) [λίθος, a stone]. The formation of calculi in the body.

**lithiatry** (*lith-i'-al-re*). The medicinal treatment of calculus.

**lithic** (*lith'-ik*). 1. Pertaining to calculi. 2. Pertaining to lithium. **l. acid**. See *acid, uric*. **l. diathesis**, the tendency to gout.

**lithica** (*lith'-ik-ah*). Agents counteracting lithiasis.

**litholaxine** (*lith-o-laks'-ēn*). A proprietary effervescing preparation containing lithium citrate, 5 grains, and sodium phosphate, 30 grains, in each teaspoonful. It is used as an hepatic stimulant.

**lithiopiperazin** (*lith-e-o-pip'-er-az-in*). A combination of lithium and piperazin which forms a granular powder readily soluble in water. It is antiarthritic. Dose 15–45 gr. (1–3 Gm.) daily.

**lithium** (*lith'-e-um*) [λίθος, a stone]. Symbol Li; atomic weight 6.94; quantivalence 1. A soft, silverwhite metal belonging to the group of alkalies. It is the lightest solid element, having a specific gravity of 0.585. The salts of lithium are used in medicine for their solvent power of uric acid, with which they form easily soluble salts. They are, therefore, employed in rheumatic and gouty affections. **l. acetate**, LiC$_2$H$_3$O$_2$+2H$_2$O, colorless crystals soluble in water; diuretic. Dose 8–24 gr. (0.52–1.6 Gm.). **l. arsenate**, 2Li$_3$AsO$_4$+H$_2$O, an alterative. Dose $\frac{1}{75}$ gr. (0.001–0.004 Gm.). **l. benzoate** (*lithii benzoas*, U. S. P.). Dose 5–30 gr. (0.32–2.0 Gm.). **l. bitartrate**, LiC$_4$H$_5$O$_6$ · H$_2$O. It is diuretic and laxative, and is used in gout. Dose 5 gr. (0.3 Gm.). **l. bromide** (*lithii bromidum*, U. S. P.), has the action of the bromides. Dose 15–30 gr. (1–2 Gm.). **l. carbonate** (*lithii carbonas*, U. S. P.). Dose 15–15 gr. (1–1.0 Gm.). **l.-carmin**, a solution of carmin in lithium carbonate, used as a stain for tissue. **l. citrate** (*lithii citras*, U. S. P.). Dose 10–30 gr. (0.65–2.0 Gm.).

**l. citrate, effervescent** (*lithii citras effervescens*, U. S. P.). Dose 1 dr. (4 Gm.). **l. dithiosalicylate**, Li$_2$C$_{14}$H$_8$S$_2$O$_6$, used in gout and rheumatism. Dose 3–10 gr. (0.2–0.65 Gm.). **l.-diuretin**. Same as *uropherin*. **l. formate**, LiCHO$_2$+H$_2$O; used in rheumatism and gout. Dose $\frac{1}{2}$ oz. (15 Cc.) 1 % aqueous solution. **l. glycerinophosphate**, **l. glycerophosphate**, Li$_2$PO$_4$ · C$_3$H$_5$(OH)$_3$; a nerve-tonic. Dose 8–15 gr. (0.5–1.0 Gm.). **l. iodate**, LiIO$_3$, used in gout. Dose 1½–3 gr. (0.1–0.2 Gm.). **l. iodide**, LiI, used in chronic sciatica and gout. Dose 1–5 gr. (0.06–0.32 Gm.). **l. salicylate** (*lithii salicylas*, U. S. P.). Dose 20–40 gr. (1.3–2.6 Gm.). **l. sozoiodolate**, C$_6$H$_2$=I$_2$ ⟨OH / SO$_3$Li, white plates; used as an antiseptic. **l. sulphoichthyolate**, used in rheumatism. Dose 8 gr. (0.5 Gm.). **l. valerate**, **l. valerianate**, LiC$_5$H$_9$O$_2$, antispasmodic, antilithic. Dose 1–5 gr. (0.32–1.0 Gm.).

**lithiuria** (*lith-e-ū'-re-ah*) [*litho-*; οὖρον, urine]. Lithuria, *q. v.*

**litho-** (*lith-o-*) [λίθος, a stone]. A prefix denoting relation to stone or to calculi.

**lithobexis** (*lith-o-beks'-is*) [*litho-*; βῆξ, cough]. Cough with expectoration of calcareous particles.

**lithobiotic** (*lith-o-bi-ot'-ik*). See *cryptobiotic*.

**lithocenosis** (*lith-o-sen-o'-sis*) [*litho-*; κένωσις, evacuation]. The extraction of the fragments of calculi that have been crushed.

**lithoclast** (*lith'-o-klast*): See *lithotrite*.

**lithoclastic** (*lith-o-klas'-tik*) [*litho-*; κλάω, to break]. Relating to the surgical crushing of a calculus in the bladder.

**lithoclasty** (*lith'-o-klas-te*). Lithotrity, *q. v.*

**lithoclysmia** (*lith-o-kliz'-me-ah*) [*litho-*; κλύσμα, clyster]. An injection of solvent liquids into the bladder for the removal of calculi.

**lithocystotomy** (*lith-o-sis-tot'-o-me*) [*litho-*; κύστις, bladder; τομή, a cut]. Lithotomy, *q. v.*

**lithocysturia** (*lith-o-sis-tū'-re-ah*) [*litho-*; κύστις, bladder; οὖρον, urine]. Disease of the bladder caused by lithuria.

**lithodectasy** (*lith-o-dek'-tas-e*). Lithectasy, *q. v.*

**lithodialysis** (*lith-o-di-al'-is-is*) [*litho-*; διαλύειν, to dissolve]. 1. The solution of calculi in the bladder. 2. The operation of breaking a vesical calculus previous to its removal.

**lithodialytic** (*lith-o-di-al-it'-ik*) [*litho-*; διαλύειν, to dissolve]. Relating to or causing lithodialysis.

**lithofellic** (*lith-o-fel'-ik*) [*litho-*; *fel*, the gallbladder]. Relating to biliary lithiasis.

**lithogenesis** (*lith-o-jen'-es-is*) [*litho-*; γένεσις, genesis]. The formation of calculi or stones.

**lithogenous** (*lith-oj'-en-us*) [*litho-*; γεννᾶν, to beget]. Pertaining to or causing the formation of calculi or stones.

**lithogeny** (*lith-oj'-en-e*). See *lithogenesis*.

**lithoid, lithoidal** (*lith'-oid, lith-oid'-al*) [*litho-*; εἶδος, like]. Resembling a stone.

**lithokelyphopedion** (*lith-o-kel-if-o-pe'-de-on*) [*litho-*; κέλυφος, shell; παιδίον, child]. Calcification of the fetus and the fetal membranes.

**lithokelyphos** (*lith-o-kel'-if-os*) [*litho-*; κέλυφος, an egg shell]. See *lithopedion*.

**lithokonion** (*lith-o-ko'-ne-on*) [*litho-*; κονίαν, to pulverize]. An instrument formerly used in pulverizing vesical calculi.

**litholabe** (*lith'-o-lāb*) [*litho-*; λαβεῖν, to seize]. An instrument for grasping and holding a vesical calculus during an operation for its removal.

**litholaby** (*lith-ol'-a-be*). See *lithotrity*.

**litholapaxy** (*lith-ol-ap-ak'-se*) . [*litho-*; λάπαξις, removal]. An operation for crushing a stone in the bladder and removing the fragments at the same sitting.

**litholein** (*lith-o'-le-in*) [*litho-*; *oleum*, oil]. A substance similar to vaselin, and, because of its antiseptic and antiparasitic qualities, proposed as a substitute for it. It is oily, of neutral reaction, without smell or taste, and contains no fat.

**lithology** (*lith-ol'-o-je*) [*litho-*; λόγος, science]. The science of the nature and treatment of calculi.

**litholysis** (*lith-ol'-is-is*): See *lithodialysis*.

**litholyte** (*lith'-o-līt*) [*litho-*; λυεῖν, to loose]. A catheter used in the litholytic treatment of calculi.

**litholytic** (*lith-o-lit'-ik*) [*litho-*; λύειν, to loosen]. Pertaining to litholysis, or the dissolving of calculi in the bladder.

**lithomalacia** (*lith-o-mal-a'-se-ah*) [*litho-*; μαλακία

softness]. The softening of a stone in the bladder; the softening of any calculus.
**lithometer** (*lith-om'-et-er*) [*litho-*; μέτρον, measure]. An instrument for estimating the size of a vesical calculus.
**lithometra** (*lith-o-me'-trah*) [*litho-*; μήτρα, womb]. Ossification or concretion of, or within, the uterus.
**lithomyl** (*lith'-o-mil*) [*litho-*; μύλη, mill]. An instrument for pulverizing a calculus.
**lithonephria** (*lith-o-nef'-re-ah*) [*litho-*; νεφρός, kidney]. Disease due to renal calculus.
**lithonephritis** (*lith-o-nef-ri'-tis*) [*litho-*; νεφρός, kidney; ιτις, inflammation]. Inflammation of the kidney due to the presence of renal calculi.
**lithonephrosis** (*lith-o-nef-ro'-sis*). See *nephrolithiasis*.
**lithonephrotomy** (*lith-o-nef-rot'-o-me*) [*litho-*; νεφρός, kidney; τομή, a cutting]. Incision of the kidney for the removal of a renal calculus.
**lithontripsy** (*lith-on-trip'-se*) [*litho-*; τρίβειν, to rub]. Synonym of *lithotripsy*, q. v.
**lithontriptic** (*lith-on-trip'-tik*) [*litho-*; τρίβειν, to rub]. See *lithotriptic* and *antilithic*.
**lithontriptor** (*lith-on-trip'-tor*) [*litho-*; τρίβειν, to rub]. A lithotrite.
**lithopedion** (*lith-o-pe'-de-on*) [*litho-*; παιδίον, child]. A retained fetus that has undergone calcareous infiltration.
**lithophone** (*lith'-o-fōn*) [*litho-*; φωνή, sound]. An instrument for detecting by sound the presence of calculi in the bladder.
**lithoplatomy** (*lith-o-plat'-om-e*) [*litho-*; πλατύς, wide]. Removal of a vesical calculus by dilating the urethra.
**lithoplaxy** (*lith-o-plaks'-e*). See *lithotrity*.
**lithoprion** (*lith-op'-re-on*) [*litho-*; πρίων, a saw]. An instrument for sawing instead of crushing a vesical calculus.
**lithoprisy** (*lith-op'-ris-e*) [*litho-*; πρίσις, sawing]. The operation of sawing through a stone in the bladder.
**lithopthisis** (*lith-off'-this-is*) [*litho-*; φθίσις, phthisis]. Tuberculosis of the lungs with calcareous concretions.
**lithoscope** (*lith'-o-skōp*) [*litho-*; σκοπείν, to examine]. An instrument for the detection and examination of calculi in the bladder.
**lithosis** (*lith-o'-sis*) [λίθος, a stone]. A diseased condition of the lung caused by the inhalation and deposition in the lung tissue of particles of silica or aluminum silicate. Syn., *grinders' lung*.
**lithotecnon** (*lith-o-tek'-non*) [*litho-*; τέκνον, child]. Same as *lithopedion*, q. v.
**lithoterethrum**, **lithoteretron** (*lith-o-ter'-eth-rum*, *lith-o-ter'-et-ron*) [*litho-*; τέρετρον, gimlet]. A lithotrite.
**lithothlibia** (*lith-o-thlib'-e-ah*) [*litho-*; θλίβειν, to press]. The operation of crushing a vesical calculus between a sound introduced into the bladder and a finger in the rectum or vagina.
**lithothriptic** (*lith-o-thrip'-tic*) [*litho-*; θρυπτικός, breaking]. Same as *lithotritic*.
**lithothryptist** (*lith-o-thrip'-tist*) [*litho-*; θρύπτειν, to break]. One who practises lithotrity.
**lithothryptor** (*lith-o-thrip'-tor*) [*litho-*; θρύπτειν, to break]. An instrument used in lithotrity.
**lithothrypty** (*lith-o-thrip'-te*) [*litho-*; θρύπτειν, to break]. Lithotrity; the operation of crushing a calculus.
**lithotome** (*lith'-o-tōm*) [*litho-*; τομή, a cutting]. A cutting-instrument for use in lithotomy.
**lithotomist** (*lith-ot'-o-mist*) [see *lithotome*]. A surgeon who performs lithotomy.
**lithotomy** (*lith-ot'-o-me*) [see *lithotome*]. Incision into the bladder to remove a calculus. l., bilateral, a lithotomy performed by a curved transverse incision just in front of the rectum. l., high. See l., suprapubic. l., Italian, median lithotomy. l., lateral, one in which the incision is made in front of the rectum and to one side, generally the left, of the raphe. l., lithontriptic, a perineal incision following lithotrity for removal of debris of calculi. l., median, l., marian, one in which the incision is made in the median line in front of the anus. l., mediolateral, that in which the perineal incision is made in the median line, and the prostatic incision laterally. l. position, a position in which the patient rests on his back with the thighs flexed on the abdomen and the legs flexed on the thighs, the knees being widely abducted. l., quadrilateral, a modification

of bilateral lithotomy with four incisions of get prostate. l., rectal, that done by an incision throuhh the rectum. l., spontaneous, expulsion of a calculus by ulceration through the bladder and perineum. l., suprapubic, lithotomy in which the incision is made above the pubis, at a point here the bladder is not covered by peritoneum. l., vaginal, one in which the incision is made through the vaginal wall.
**lithotony** (*lith-ot'-o-ne*) [*litho-*; τείνειν, to stretch]. Removal of a vesical calculus through an artificial fistula, which is gradually dilated.
**lithotresis** (*lith-o-tre'-sis*) [*litho-*; τρῆσις, a boring]. The drilling of holes through a calculus, as a step in its destruction and removal.
**lithotripsy** (*lith'-o-trip-se*) [*litho-*; τρίβειν, to crush]. The operation of crushing calculi in the bladder.
**lithotriptic** (*lith-o-trip'-tik*) [see *lithotripsy*]. I. Relating to lithotripsy. 2. Capable of dissolving vesical calculi.
**lithotriptor** (*lith-o-trip'-tor*) [see *lithotripsy*]. An instrument for crushing calculi in the bladder.
**lithotrite** (*lith'-o-trīt*) [*litho-*; *terere*, to rub]. An instrument for crushing a vesical calculus.
**lithotritic** (*lith-o-trit'-ik*) [*litho-*; *terere*, to rub]. Pertaining to lithotrity.
**lithotrypterion** (*lith-o-trip-ter'-e-on*) [*litho-*; τρίβειν, to rub]. A small lithotrite.
**lithotrity** (*lith-ot'-rit-e*) [see *lithotrite*]. The process of crushing a stone in the bladder, with the lithotrite, into fragments small enough to pass through the urethral canal.
**lithous** (*lith'-us*). Having the nature of a stone.
**lithoxyduria** (*lith-oks-id-ū'-re-ah*) [*litho-*; ὀξύς, acid; οὖρον, urine]. The morbid presence of xanthic oxide in the urine.
**lithuresis** (*lith-ū-re'-sis*) [λίθος, a stone; οὔρησις, urination]. The voiding of small calculi with the urine.
**lithureteria** (*lith-ū-re-te'-re-ah*) [λίθος, a stone; οὐρητήρ, a ureter]. A diseased condition of the ureter due to the presence of calculi.
**lithuria** (*lith-ū'-re-ah*) [*lithium*; οὖρον, urine]. A condition marked by excess of lithic acid or its salts in the urine.
**litigation symptoms**. Various vague manifestations of nervous shock following injury, concerning which the question of malingering arises, and which may become the subject of medico-legal inquiry.
**litmus** (*lit'-mus*) [Dutch, *lack*, lac; *moes*, pulp]. A blue pigment obtained from *Roccella tinctoria*, a lichen. It is employed for determining the presence of acids and alkalies. l.-paper, blue, unsized paper steeped in a solution of litmus; it turns red on contact with acid solutions. l.-paper, red, unsized colored paper steeped in a solution of litmus colored red with acid; it turns blue on contact with alkaline solutions.
**litrameter** (*lit-ram'-et-er*) [λίτρα, pound; μέτρον, measure]. An instrument for ascertaining the specific gravity of fluids.
**litre** (*le'-ter*). See *liter*.
**Litsea** (*lit'-se-ah*) [Japanese]. A genus of plants of the order *Laurineæ*. L. *cubeba* is used as a condiment and stomachic. L. *trinervia* is a species resembling myrrh. The entire plant has stimulant and diuretic properties. The root is used as a vermifuge and emmenagogue. The berries yield an oil used as a remedy in skin diseases and on old wounds.
**Litten's sign** (*lit'-en*) [Moritz *Litten*, German physician, 1845–1907]. Retraction of the lateral portion of the thorax, where the diaphragm is inserted, the retracted portions being lowered during inspiration and arising during expiration. It is absent in pleuritic adhesions, effusion into the pleural cavity, emphysema, etc. Syn., *diaphragmatic phenomenon*.
**litter** (*lit'-er*) [*lectus*, a couch]. A stretcher or couch with handles for carrying the sick or wounded.
**Little's disease** (*lit'-el*) [William John *Little*, English surgeon, 1810–1894]. Congenital muscular rigidity; spastic cerebral diplegia of infancy.
**Littré's colotomy** (*le-tra'*) [Alexis *Littré*, French surgeon, 1658–1726]. The making of an opening into the colon through the left iliac region. L.'s **glands**, the small racemose muciparous glands in the mucous membrane of the urethra. L.'s **hernia**, a hernia in which only a diverticulum of the bowel is affected. The term is improperly applied to Richter's hernia. L.'s **sinus**, the transverse sinus.
**lituate** (*lit'-ū-āt*) [*lituus*, an augur's staff]. Forked, with the points bent slightly outward.

**live** (*liv*). Living; animate; manifesting life. **l.-birth**, birth characterized by "the manifestation of some certain sign or signs of life by the child after it is completely born." **l.-blood**, the name given to the sensation of fluttering in the eyelid, due to spasmodic action of the orbicularis palpebrarum muscle.
**live** (*liv*). To continue in being. To have life.
**livedo** (*liv-e'-do*) [*livere*, to grow black]. Same as **lividity**.
**liver** (*liv'-er*) [AS., *lifer*] The largest gland in the body, situated on the right side of the abdominal cavity, just below the diaphragm, and forming an appendage of the digestive tract. Its functions are: the secretion of bile; the formation and storage of glycogen; the production, at least at a certain period of development, of blood-corpuscles; the destruction of blood-corpuscles; the formation of a large quantity of urea; the retention and destruction of certain poisonous substances absorbed from the intestinal tract. Anatomically, it consists of five lobes—the right, the left, the lobus Spigelii, the lobus quadratus, and the lobus caudatus. These lobes are made up of lobules or acini, and these again of hepatic cells, capillaries, arteries, veins, lymphatics, and biliary channels, each lobule being surrounded by connective tissue. The weight of the liver is between 50 and 60 ounces. **l., albuminoid.** See *l., amyloid*. **l., amyloid**, one the seat of amyloid degeneration. **l., beavertail**, one the left lobe of which resembles in form a beaver's tail. **l., biliary cirrhotic**, one the seat of chronic inflammation, the result of obstruction and distention of the bile-ducts. **l., cardiac**, hepatic congestion with pulsation accompanied by disease of the right side of the heart. **l., cirrhotic**, one the seat of chronic inflammation, with overgrowth of the connective tissue and atrophy of the parenchyma. **l., degraded**, a human liver divided into an unusual number of lobes. **l.-dextrin**, a carbohydrate found in the liver. **l., fatty**, one with marked fatty infiltration and degeneration. **l., floating**, a movable condition of the liver, with displacement. **l.-fluke**, a hepatic parasitic worm. **l., foam-**, a liver containing many gas-filled cavities which give it a spongy or foamy texture. It is due to *Bacillus aerogenes capsulatus*. **l., gin-drinkers'**, the liver of atrophic cirrhosis. **l., hobnail.** See *hobnail liver*. **l., icing-**. See *icing-liver*. **l., nutmeg**, a condition of the liver occurring in heart disease, fatty infiltration, and amyloid disease. The surface of a section has a peculiar mottled appearance, the center of the lobules being dark, the periphery light, in color. **l., scrofulous**, an albuminoid liver. **l.-spot**, chloasma. **l., syphilitic**, one the seat of gummata, which, on healing, leave scars, or of a diffuse inflammation that may lead to cirrhosis, with atrophy or, especially in hereditary syphilis, enlargement. **l., tightlace**, one in which the right lobe is thickened vertically from compression, marked by the ribs, and atrophic from constant pressure by tight stays. Syn., *corset-liver*. **l., wandering**, a displaced liver. **l., waxy.** See *l., amyloid*.
**liverwort** (*liv'-er-wert*). See *hepatica*.
**livid** (*liv'-id*) [*livere*, to be dark]. Discolored from the effects of congestion or contusion; black and blue; pale lead-color.
**lividity** (*liv-id'-it-e*) [*livid*]. The state of being livid. **l., cadaveric**, **l., postmortem**, the reddish or bluish discoloration in the dependent parts of a corpse, due to the gravitation of the blood.
**livor** (*li'-vor*) [*livid*]. Lividity.
**livores mortis** (*li-vor'-ēz mor'-tis*) [L.]. Livid blotches seen on cadavers.
**lixiviation** (*liks-iv-e-a'-shun*) [see *lixivium*]. The process of leaching ashes. Also, the process of separating by solution any alkaline salt from the insoluble impurities with which it is mixed.
**lixivium** (*liks-iv'-e-um*) [*lixivia*, lye]. The filtrate obtained by leaching ashes; practically a solution of an impure potassium hydroxide.
**Lizars'** operation (*li'-zarz*) [John Lizars, Scotch surgeon, 1787–1860]. A method of exposing the upper jaw by cutting through the cheek, from the angle of the mouth to the malar bone.
**L. K. Q. C. P. I.** Abbreviation of Licentiate of the King and Queen's College of Physicians of Ireland.
**llareta** (*lar-e'-tah*). A remedy recommended in the treatment of blennorrhagia, said to be the fluid-extract of *Haplopappus slareta*, of Chili.

**LL.B.** Abbreviation of *Legum Baccalaureus*, Bachelor of Laws.
**LL.D.** Abbreviation of *Legum Doctor*, Doctor of Laws.
**L. M. A.** Abbreviation for left mentoanterior position of the fetus in utero.
**L. M. P.** Abbreviation for left mentoposterior position of the fetus in utero.
**L. M. R. C. P.** Abbreviation for Licentiate in Midwifery of the Royal College of Physicians.
**L. M. S.** Abbreviation for Licentiate in Medicine and Surgery.
**L. M. S. S. A.** Abbreviation for Licentiate in Medicine and Surgery of the Society of Apothecaries (of London).
**L. O. A.** Abbreviation for left occipitoanterior position of the fetus in utero.
**loam** (*lōm*). A mixture of clay, sand, and humus.
**lobar** (*lo'-bar*). Pertaining to a lobe. **l. pneumonia.** See *pneumonia, lobar*.
**lobate** (*lo'-bāt*). Having lobes.
**lobe** (*lōb*) [*lobus*, a lobe]. A more or less rounded part or projection of an organ, separated from neighboring parts by fissures and constrictions, as the *lobes* of the liver, of the brain, etc. **l., biventral**, the lateral continuation into the hemisphere of the pyramidal lobe. **l., cacuminal**, the superior semilunar lobe of the cerebellum. **l., caudate**, the tail-like process of the liver. **l., central.** 1. The island of Reil. 2. One of the lobes of the cerebellum. **l.s of the cerebellum**, each cerebellar hemisphere is divided into the following lobes: on the upper surface, the anterior or square lobe and the posterior or semilunar lobe; on the under surface, the flocculus or subpeduncular lobe, the amygdala or tonsil, the digastric lobe, the slender lobe, and the inferior posterior lobe. **l.s of the cerebrum**, the primary lobes into which each cerebral hemisphere is divided; they are: the frontal, parietal, occipital, temporo-sphenoidal, and central, or island of Reil. See *brain* and *convolution*. **l., clival**, the parts of the cerebellum between the preclival and postclival fissures. **l., crescentic, anterior and posterior**, two lobes on the upper surface of each cerebellar hemisphere. **l., culminal**, a part of the prevermis of the cerebellum. **l., frontal**, that part of the cerebral hemisphere in front of the central and above the Sylvian fissures. **l., gracile**, the anterior and posterior slender lobules of the cerebellum combined. **l.s of the liver.** See *liver*. **l., marginal**, the first frontal convolution running along the margin of the longitudinal fissure. **l., nodular**, the nodulus and flocculus of the cerebellum combined. **l., olfactory**, the rhinencephalon. **l.s, optic**, the corpora quadrigemina. **l., orbital**, that part of the frontal lobe which rests on the orbital plate of the frontal bone. **l., parietal**, that part of the cerebral hemisphere back of the central and above the Sylvian fissures. **l., pyramidal**, that part of the cerebellum making up the prominence of the postvermis. **l., quadrate.** 1. The anterior and posterior crescentic lobes of the cerebellum combined. 2. An oblong lobe on the inferior surface of the liver. **l., slender**, the fourth of the five lobes on the under surface of each hemisphere of the cerebellum. **l., Spigelian.** See *lobus Spigelii*. **l., temporal**, that part of the cerebral hemisphere below the Sylvian and in front of the exoccipital fissures. **l., tuberal**, the inferior semilunar and slender lobes of the cerebellum combined. **l., uvular**, the uvula and amygdala of the cerebellum combined.
**lobectomy** (*lo-bek'-to-me*) [*lobe; ἐκτομή*, excision]. Excision of a lobe of an organ or gland.
**lobelia** (*lo-be'-le-ah*) [de *Lobel*, French botanist]. The *lobelia* of the U. S. P. is the leaves and tops of *L. inflata*, of the order Lobeliaceæ. It contains a liquid alkaloid, *lobeline*; an acid *lobelic acid*; an acrid body, *lobelacrin*; and a crystalline substance, *inflatin*. Lobelia is expectorant, antispasmodic, and emetic, and has been used in asthma, whooping-cough, and other bronchial affections. A North American species, *L. syphilitica*, has been used as an antisyphilitic. **l., fluidextract** of (*fluidextractum lobeliæ*, U. S. P.), an acetic acid menstruum. Dose 8 min. (0.5 Cc.). **l., tincture** of (*tinctura lobeliæ*, U. S. P.). Dose 10–30 min. (0.65–2.0 Cc.).
**lobeline** (*lo'-be-lēn*) [*lobelia*]. 1. A precipitate from the tincture of *Lobelia inflata*; an emetic, emetico-antispasmodic, resolvent, and relaxant. Dose, as an emetic, 1–3 gr. (0.06–0.2 Gm.) in warm water, repeated in 10 minutes if necessary; as a diaphoretic

and expectorant, ¼–½ gr. (0.016–0.032 Gm.). 2. An actively poisonous alkaloid from *Lobelia inflata*. **l. hydrobromide,** used in pseudoangina pectoris and asthma. Dose, adults, ¾–6 gr. (0.05–0.4 Gm.) daily; children, ⅙–¾ gr. (0.01–0.05 Gm.) daily. **l. sulphate,** used in asthma, epilepsy, etc. Dose 1 gr. (0.06 Gm.) daily, gradually increasing to 3–8 gr. (0.2–0.52 Gm.) daily.

**lobengulism** (*lo-ben'-gū-lizm*). A condition marked by a general increase of subcutaneous fat, associated with partial or complete abeyance of sexual function. In men there is an enlargement of the mammary glands with failure of sexual appetite, and in women a cessation of menstruation.

**lobopodium** (*lo-bo-po'-de-um*) [*pl.*, *lobopodia*]. A pseudopodium which is broad and thick. Cf. *filopodium*.

**Lobstein's cancer** (*lōb'-stīn*) [Johann Georg Christian Friedrich Martin *Lobstein*, German pathologist, 1777–1835]. Retroperitoneal sarcoma. **L.'s disease,** osteopsathyrosis; fragility of the bones. **L.'s ganglion,** a small gangliform swelling of the great splanchnic nerve a short distance above the diaphragm.

**lobular** (*lob'-ū-lar*) [*lobulus*]. Pertaining to, resembling, or composed of lobules. **l. pneumonia,** bronchopneumonia.

**lobulated** (*lob'-ū-la-ted*) [*lobule*]. Consisting of lobes or lobules.

**lobule** (*lob'-ūl*) [see *lobulus*]. A small lobe. **l., cuneate.** See *cuneus*. **l., fusiform,** the subcollateral gyrus. **l., paracentral,** the superior connecting convolution of the ascending frontal and ascending parietal gyri. **l., parietal,** one of the two parts (inferior and superior) into which the parietal lobe is divided by the horizontal portion of the intraparietal fissure.

**lobulet** (*lob'-ū-let*) [dim. of *lobulus*, a lobule]. A minor lobule, especially of the lung; a group, or series of groups, of five or six air-sacs connected with the dilated end of a bronchiole.

**lobulus** (*lob'-ū-lus*) [dim. of *lobus*; pl., *lobuli*]. A lobule. **l. auriculæ,** the lobe of the ear. **l. biventer,** the biventral lobe. **l. caudatus,** the tailed lobe of the liver that separates the right extremity of the transverse fissure from the commencement of the fissure for the inferior vena cava. **lobuli cerebelli spinales,** Gordon's name for amygdalæ cerebelli. **l. cerebelli tener, l. gracilis,** the slender lobe. **l. parietalis exterior** or **superior,** the superior parietal gyrus. **l. parietalis inferior,** the subparietal gyrus. **l. posteroparietalis,** the ascending parietal gyrus. **l. quadratus,** the square lobe upon the inferior surface of the right lobe of the liver. **lobuli renales,** the Malpighian pyramids. **l. rolandicus anterior,** the ascending frontal gyrus. **l. rolandicus posterior,** the ascending parietal gyrus. **l. Spigelii,** the lobule projecting from the posterior portion of the inferior surface of the liver. **lobuli testiculi, l. testis,** pyramidal or conoid lobules varying in size and number (250–400) and converging to the mediastinum, which make up the glandular substance of the testis. Each lobule consists of from one to six seminiferous tubules.

**lobus** (*lo'-bus*) [L., a lobe]. A lobe. Any well-defined, rounded part of an organ. **l. caudatus,** the caudate lobe; see *lobulus*. **l. opertus,** the insula of the brain. **l. quadratus.** See *lobulus*. **l. Spigelii,** a prominent oblong lobe on the posterior surface of the liver.

**local** (*lo'-kal*). Limited to a part or place; not general. **l. asphyxia,** a stage of Raynaud's disease, *q. v.*

**localization** (*lo-kal-iz-a'-shun*) [*local*]. 1. The determination of the seat of a lesion. 2. The limitation of a process to a particular place; the opposite of generalization. 3. The faculty of locating sensory impressions. **l., cerebral,** the determination of the position of the centers in the brain that preside over certain physiological acts or of the seat of pathological conditions interfering with the proper function of these centers. **l., experimental,** the localization of brain-centers through experiments on animals.

**localized** (*lo'-kal-īzd*). Confined to a particular place.

**locative** (*lok'-at-iv*) [*locare*, to place]. Indicating relative position in a series. **l. name,** one that indicates the location of an organ or part; e. g., post cava.

**lochia** (*lo'-ke-ah*) [λόχιος, pertaining to child-birth]. The discharge from the genital organs during the first few weeks (from two to four) after labor. **l. alba,** the whitish flow that takes place from about the seventh day. **l. cruenta; l. rubra,** the sanguineous flow of the first few days. **l. serosa,** the serous discharge taking place about the fifth day.

**lochial** (*lo'-ke-al*) [*lochia*]. Pertaining to the lochia.

**lochiometra** (*lo-ke-o-me'-trah*) [*lochia*; μήτρα, uterus]. A collection of lochia in the uterus.

**lochiometritis** (*lo-ke-o-me-tri'-tis*) [*lochia*; *metritis*]. Puerperal metritis.

**lochiopyra** (*lo-ke-op'-ir-ah*) [*lochia-*; πῦρ, fire, fever]. Puerperal fever. Same as lochopyra.

**lochiorrhagia** (*lo-ke-or-a'-je-ah*) [*lochia*; ῥηγνύναι, to burst forth]. An excessive flow of the lochia.

**lochiorrhea** (*lo-ke-or-e'-ah*) [*lochia*; ῥοία, a flow]. An abnormal flow of the lochia.

**lochioschesis** (*lo-ke-os'-kes-is*) [*lochia*; σχέσις, retention]. Suppression or retention of the lochia.

**lochodochium** (*lo-ko-do'-ke-um*) [λόχος, child-birth; δέχεσθαι, to receive]. A lying-in hospital; a maternity.

**lochometritis** (*lo-ko-me-tri'-tis*) [*lochia*; μήτρα, womb; ιτις, inflammation]. Inflammation of the uterus consequent upon delivery; puerperal metritis.

**lochometrophlebitis** (*lo-ko-me-tro-fle-bi'-tis*) [*lochia*; μήτρα, womb; φλέψ, vein; ιτις, inflammation]. Phlebitis of the uterine veins in puerperal women.

**lochoperitonitis** (*lo-ko-per-it-on-i'-tis*). Inflammation of the peritoneum following child-birth.

**lochopyra** (*lo-kop'-ir-ah*) [*lochia*; πῦρ, fire, fever]. Puerperal fever.

**lochotyphus** (*lo-ko-ti'-fus*) [*lochia*; τῦφος, stupor]. Puerperal fever of a typhoid type.

**Lockard's treatment of typhoid.** Continuous application of ice-bags to the axilla, popliteal space, back of neck, wrist, and ankle, to reduce temperature.

**lock-finger.** A peculiar affection of the fingers in which they suddenly become fixed in a flexed position, due to the presence of a small fibrous growth in the sheath of the extensor tendon.

**lock-hospital.** An English term for a hospital for the treatment of venereal diseases; it was originally kept with locked doors, whence the name.

**lockjaw.** Tetanus; trismus.

**lock-spasm.** A spasm of the fingers in which they become firmly flexed upon the object in their grasp, as upon the pen in writing. It is sometimes seen in writer's cramp.

**Lockwood's ligament** (*lok'-wood*) [Charles Barrett *Lockwood*, English surgeon, 1858–1914]. The suspensory ligament of the eyeball, a curved fibrous band connected with Tenon's capsule, and supporting the eyeball on each side of the orbit.

**loco** (*lo'-ko*) [Sp., "crazy"]. Loco-weed, various species of leguminous plants poisonous to cattle and horses. **l.-disease,** a local epizootic disease of cattle and horses induced by eating any of the plants called *loco*.

**locoed** (*lo'-kōd*). Affected with locoism.

**locoism** (*lo'-ko-izm*). See *loco-disease*.

**locomotion** (*lo-ko-mo'-shun*) [*locus*, place; *movere*, to move]. Animal movement. **l. of an artery,** the straightening out of a curved artery under the impulse of the pulse-wave.

**locomotive** (*lo-ko-mo'-tiv*) [*locus*, a place; *movere*, to move]. Moving from place to place; able to change its place; pertaining to locomotion. **l. pulse.** See *Corrigan's pulse*.

**locomotivity** (*lo-ko-mo-tiv'-it-e*) [*locus*, place; *movere*, to move]. The power to move from one place to another.

**locomotor** (*lo-ko-mo'-tor*) [*locus*, a place; *motor*, a mover]. Pertaining to locomotion. **l. ataxia, l. ataxy.** See *ataxia*, *locomotor*, and *tabes*.

**locomotorial** (*lo-ko-mo-to'-re-al*). Relating to the locomotorium.

**locomotorium** (*lo-ko-mo-to'-re-um*) [*locus*, a place; *motor*, mover: *pl.*, *locomotoria*]. In biology, the motive apparatus of an organism; the bones, muscles, and tendons.

**locomotory** (*lo-ko-mo'-tor-e*) [*locus*, place; *movere*, to move]. Pertaining to locomotion.

**locum tenens** (*lo'-kum te'-nenz*) [L. "holding the place"]. A physician who temporarily takes charge of the patients and practice of another physician.

**locular, loculated** (*lok'-ū-lar*, *lok'-ū-la-ted*) [*loculus*]. Divided into loculi.

**loculus** (*lok'-ū-lus*) [dim. of *locus*; pl., *loculi*]. A small space or compartment.

**locus** (*lo'-kŭs*) [L.]. A place, spot, or organ. l. cinereus, l. cæruleus, l. ferrugineus, a bluish-tinted eminence on the fasciculi teretes of the fourth ventricle of the brain. l. minoris resistentiæ, a spot of diminished resistance. l. niger, a dark area in the center of a section of the crus cerebri. l. perforatus, a name given to the anterior and the posterior perforated space at the base of the brain through which blood-vessels pass. l. ruber, the red nucleus of the tegmentum.

**lodal** (*lo'-dal*). Trade name of an oxidation product of laudanosine, used to control uterine hemorrhage.

**Loebisch's formula** (*le'-bish*). The product obtained by multiplying the last two figures of the specific gravity of the urine by 2.2 indicates the number of grams of solids in 1000 Cc. of urine.

**Loeffler's alkaline solution** (*lef'-ler*) [Friedrich August Johannes *Loeffler*, German bacteriologist, 1852–1915]. A mixture of 30 parts of a concentrated alcoholic solution of methylene-blue and 100 parts of a 1 : 10,000 aqueous solution of potassium hydroxide. **L.'s bacillus**, the bacillus of diphtheria. See *Klebs-Loeffler bacillus*. **L.'s blood-serum mixture**, a mixture consisting of one part of neutral meat-infusion bouillon containing 1 % of glucose and three parts of blood-serum; it is used as a culture-medium. **L.'s toluol solution**, a solution recommended by Loeffler for the local treatment of diphtheria. It consists of menthol, 10 Gm.; toluol, sufficient to make 36 Cc.; absolute alcohol, 60 Cc.; solution of ferric chloride, 4 Cc. Another formula is: menthol, 10 Gm.; toluol, sufficient to make 36 Cc.; creolin, 2 Cc.; absolute alcohol, 65 Cc.

**loefferia** (*lef-le'-re-ah*). A disease marked by the presence of *Bacillus diphtheriæ* (Klebs and Loeffler), without diphtheritic symptoms.

**Loehlein's diameter**. The distance between the center of the subpubic ligament and the anterosuperior angle of the great sacrosciatic foramen.

**lœmia** (*lem'-e-ah*). See *loimia*.
**lœmic** (*lem'-ik*). See *loimic*.
**lœmography** (*lem-og'-ra-fe*) [λοιμός, plague; γράφειν, to write]. See *loimic*.
**lœmoid** (*lem'-oid*). See *loimic*.
**lœmology** (*lem-ol'-o-je*). See *loimology*.
**lœmophthalmia** (*lem-off-thal'-me-ah*) [λοιμός, plague; *ophthalmia*]. A contagious ophthalmia.

**Loewe's ring** (*le'-veh*). A bright circle which may appear in the visual field when the illumination is changed from blue to white. It surrounds the position of the dark ring that marks the macula lutea.

**Loewenberg's canal** (*le'-ven-berg*) [Benjamin Benno *Loewenberg*, German laryngologist, 1836– ]. That portion of the cochlear canal situated above the membrane of Corti.

**Loewenthal's reaction** (*le'-ven-tahl*) [Wilhelm *Loewenthal*, German physician, 1850–1894]. The agglutination of *Spirochæta Obermeieri*, by the blood-serum of an individual affected with relapsing fever. **L.'s test for glucose**, a glucose solution boiled with a solution of ferric chloride, dissolved in tartaric acid and sodium carbonate, becomes dark and deposits an abundant precipitate of iron oxide. This test is not applicable to urine. **L.'s tract**, the descending anterolateral tract of the spinal cord.

**Loewitt's bodies** (*le'-vit*). See *lymphogonia*.

**logadectomy** (*log-ad-ek'-to-me*) [λογάδες, whites of the eyes; ἐκτομή, excision]. Excision of a piece of the conjunctiva.

**logades** (*log'-ad-ēz*) [λογάδες]. The whites of the eyes; the sclerotic coats of the eyes.

**logaditis** (*log-ad-i'-tis*) [λογάδες, whites of the eyes; *ītis*, inflammation]. Same as *scleritis*.

**logadoblennorrhea** (*log-ad-o-blen-or-e'-ah*) [λογάδες, whites of the eyes; βλέννα, mucus; ῥοία, flow]. Conjunctival blennorrhea.

**logagnosia, logagnosis** (*log-ag-no'-ze-ah, -sis*) [λόγος, a word; ἀ, priv.; γνῶσις, a recognizing]. Aphasia; word-blindness.

**logagraphia** (*log-ag-raf'-e-ah*). Same as *agraphia*, *q. v*.
**logamnesia** (*log-am-ne'-ze-ah*) [λόγος, a word; ἀμνησία, forgetfulness]. Word-deafness; word-blindness.

**loganin** (*log'-an-in*), $C_{25}H_{34}O_{14}$. A glucoside extracted from the seeds of *Strychnos nux vomica*.

**logo-** (*lo-go-*) [λόγος, a word]. A prefix meaning relating to words or speech.

**logodiarrhea** (*log-o-di-ar-e'-ah*) [λόγος, word; διάῤῥοια, a flowing through]. Excessive or maniacal loquacity.

**logograph** (*log'-o-graf*) [λόγος, a word; γράφειν, to write]. 1. A written word. 2. Barlow's name for a device for recording spoken words.

**logographic** (*log-o-graf'-ik*) [λόγος, word; γράφειν, to write]. Pertaining to written words. **l. alalia**, that in which the thoughts cannot be expressed in writing.

**logokophosis** (*log-o-kof-o'-sis*) [*logo-*; κώφωσις, deafness]. Word-deafness; incapacity to understand spoken language.

**logomania** (*log-o-ma'-ne-ah*) [λόγος, word; μανία, madness]. 1. Insanity chracterized by talkativeness. 2. Aphasia.

**logoneurosis** (*log-o-nū-ro'-sis*) [*logo-*; *neurosis*]. 1. A neurosis marked by a speech-defect. 2. A neurosis attended with impairment of the mental powers.

**logopathy** (*log-op'-ath-e*) [*logo-*; πάθος, disease]. A disease affecting the speech.

**logoplegia** (*log-o-ple'-je-ah*) [*logo-*; πληγή, stroke]. Loss of the power of uttering articulate speech.

**logorrhea** (*log-or-e'-ah*) [*logo-*; ῥοία, a flow]. Excessive loquacity.

**logospasm** (*log'-o-spazm*) [*logo-*; σπάσμος, spasm]. Spasmodic enunciation of words.

**logwood** (*log'-wood*). See *Hematoxylon*.

**loimia** (*loi'-me-ah*) [λοιμός, plague]. A pestilence or plague.

**loimic** (*loi'-mik*) [λοιμός, plague]. Pertaining to the plague or to any pestilence.

**loimography** (*loi-mog'-ra-fe*) [λοιμός, plague; γράφειν, to write]. A description of the plague.

**loimology** (*loi-mol'-o-je*) [λοιμός, plague; λόγος, science]. The science of contagious epidemic diseases.

**loimopyra** (*loi-mop'-ir-ah*) [λοιμός, plague; πῦρ, fire]. Pestilential fever.

**loin** (*lumbus*, loin]. The lateral and posterior region of the body between the false ribs and the top of the pelvis.

**loka** (*lo'-ka*). An arrow-poison, probably of the nux vomica group.

**lolism** (*lol'-izm*). Poisoning by seeds of *Lolium temulentum*, which have found their way among grain and which contain a poisonous glucoside, loliin. It is marked by narcotic symptoms, vomiting, and diarrhea.

**lolium** (*lo'-le-um*) [L., darnel]. A genus of grasses, one of which, *L. temulentum*, a poisonous darnel, is remarkable as one of the few grasses that appear to have poisonous qualities. Its seeds in particular are said to have narcotic qualities.

**Lombardy leprosy**. Pellagra.

**loment** (*lo'-ment*) [*lomentum*, a mixture of bean-meal and rice]. A legume that when ripe breaks transversely into joints.

**lomentum** (*lo-men'-tum*). Same as *loment*.

**lomilomi** (*lo'-me-lo'-me*) [Hawaiian]. A kind of shampoo or massage, practised by the Polynesians.

**London paste**. A caustic paste containing equal parts of caustic soda and unslaked lime.

**Long's coefficient** [John Harper *Long*, American chemist, 1856– ]. The figures 2.6, by which the last two figures of the specific gravity of the urine are multiplied to obtain the number of grams of solids in 1000 Cc. of urine.

**longevity** (*lon-jev'-it-e*) [*longa*, long; *vita*, life]. Long life.

**longimanus** (*lon-je-ma'-nus*) [*longus*, long; *manus*, hand]. Long-handed.

**longing** (*long'-ing*). The earnest desire for anything; it is a condition often present in the female during pregnancy.

**longipedes** (*lon-jē-pe'-dāt*) [*longus*, long; *pes*, foot]. Long-footed.

**longissimus** (*lon-jis'-im-us*) [superl. of *longus*, long]. Longest. **l. capitis, l. cervicis**. See *trachelo-mastoid* under *muscle*. **l. dorsi**. See under *muscle*.

**longitudinal** (*lon-je-tū'-din-al*) [*longitudo*, length]. Lengthwise; in the direction of the long axis of a body. **l. fissure**. See *fissure*, *interlobular*.

**long-sightedness** (*long-sīt'-ed-nes*). See *hyperopia*.

**longus** (*lon'-gus*) [L.]. Long. **l. capitis**, the rectus capitis anticus major muscle. **l. colli**. See under muscle.

**loop** [Irish and Gael., *lub*, a bend]. 1. A bend in a cord or cord-like structure. 2. A platinum wire, in a glass handle, used with its extremity bent in a circular form; used to transfer bacterial cultures. **l. of Henle**. See *Henle's loop*.

**loose** (*loos*). Lax; wanting power of restraint; as loose bowels. 1. **ligature.** See *ligature*.
**looseness** (*loos'-nes*). Popular name for diarrhea.
**l. of the teeth**, *odontoseisis; odontosismus;* this results from disease of the gums and the gradual destruction of the alveolar processes.
**L. O. P.** Abbreviation for left occipitoposterior position of the fetus in utero.
**lophia** (*lo'-fe-ah*) [λόφος, the back of the neck]. The upper part of the back near the first dorsal vertebra.
**lophius** (*lo'-fe-us*). [λόφος, a ridge]. The ridge between two furrows or sulci of the ventricular surface of the brain.
**lophocomous** (*lo-fok'-om-us*) [λόφος, a tuft; κόμη, the hair of the head]. Having the hair in tufts.
**lophotrichea** (*lo-fo-trik'-e-ah*). Bacteria with lophotrichous ciliation.
**lophotrichous** (*lo-fot'-rik-us*) [λόφος, tuft; θρίξ, hair]. Applied to that type of ciliation in microorganisms characterized by a tuft of flagella at each pole.
**loquacity** (*lo-kwas'-it-e*) [*loquacitas; loquax*, talkative]. Volubility of speech; talkativeness; a condition that is frequently excessive in various forms of mental disorder.
**Lorain type of infantilism** (*lor-ān'*). Arrested physical development of unknown cause.
**lordoma** (*lor-do'-mah*) [λορδοῦν, to bend inward]. The anterior or forward incurvation of the spine.
**lordoscoliosis** (*lor-do-sko-le-o'-sis*). Lordosis with scoliosis.
**lordosis** (*lor-do'-sis*) [λορδοῦν, to bend back]. A curvature of the spine with a forward convexity.
**lordotic** (*lor-dot'-ik*). Bent with the convexity turned ventrad; applied to the spinal column.
**Lorenz bloodless operation** (*lor'-ens*) [Adolf *Lorenz*, Austrian surgeon, 1854– ]. *For congenital dislocation of the hip:* this consists in the reduction of the dislocation and fixation of the head of the femur against the rudimentary acetabulum until a socket is formed. L.'s **hip-redresseur**, an apparatus to correct faulty position and contraction of joints and hold the limb while fixation bandages are applied.
**Loreta's operation** (*lor-a'-tah*) [Pietro *Loreta*, Italian surgeon, 1831–1889]. 1. The forcible dilatation of the pylorus for the relief of stricture. 2. The treatment of aneurysm by the introduction of metal wire into the sac, through which an electric current is then passed.
**loretin** (*lor'-et-in*), C₉H₆IO₄SN. A nontoxic antiseptic used as a dusting-powder or in 5 to 10 % ointment or in 0.1 to 0.2 % aqueous solution.
**loripes** (*lor'-ip-ēz*) [*lorum*, a thong; *pes*, the foot]. 1. Limber-footed. 2. Crook-footed; bandylegged. 3. Talipes varus.
**losophan** (*lo'-so-fan*). Triiodometacresol, C₇H₅I₃O; a proprietary powder used in parasitic dermatoses.
**Lostorfer's corpuscles** (*los'-tor-fer*). Granular masses said to have been found in the blood of syphilitic patients.
**lotio** (*lo'-she-o*) [L.]. A lotion. 1. **hydrargyri flava** (B. P.), yellow mercurial lotion; yellow wash. 1. **hydrargyri nigra** (B. P.), black mercurial lotion; black wash.
**lotion** (*lo'-shun*) [*lotio*]. A medicinal solution for bathing a part; a wash.
**lotoflavin** (*lo-to-fla'-vin*). A yellow pigment produced by the lysis of lotusin.
**Lotus** (*lo'-tus*) [λωτός, the lotus]. A genus of leguminous plants. L. *arabicus* yields a toxic glucoside; *lotusin*, and a pigment, *lotoflavin*.
**lotusin** (*lo'-tus-in*). A toxic glucoside in *Lotus arabicus;* it is derived from maltose and gives rise to prussic acid when acted upon by a hydrolytic enzyme.
**Louis' angle** (*loo'-e*) [Antoine *Louis*, French surgeon, 1723–1792]. An angular projection existing in some individuals at the junction of the manubrium and body of the sternum. Syn., *Angulus Ludovici*.
**Louis' law** (*loo'-e*) [Pierre Charles Alexander *Louis*, French physician, 1787–1872]. The lungs always contain tubercles when tuberculosis exists elsewhere in the body.
**louse** (*lows*). See *pediculus*.
**lousiness** (*low'-ze-nes*). See *pediculosis*.
**lousy** (*low'-ze*). Affected with pediculosis.
**loutrotherapy** (*low-tro-ther'-ap-e*) [λουτρόν, a bath; θεραπεία, therapy]. The therapeutic use of artificial carbonated and Nauheim baths.
**lovage** (*luv'-aj*) [OF., *levesche*, from *ligusticum*, lovage]. The root of *Ligusticum levisticum* and *Levisticum officinale*, plants of the order *Umbelliferæ*. Lovage is stimulant, aromatic, carminative, and emmenagogue. Dose of the *fluidextract* 1–2 dr. (4–8 Cc.).
**Lower's tubercle** (*lou'-er*) [Richard *Lower*, English anatomist, 1630–1691]. A slight prominence in the right auricle between the openings of the superior and the inferior venæ cavæ.
**loxa bark** (*loks'-ah*). Pale cinchona; the bark of *Cinchona officinalis*.
**loxarthron** (*loks-ar'-thron*) [λοξός, slanting; ἄρθρον, a joint]. Any oblique or abnormal direction of a joint, not caused by spasm or luxation.
**loxarthrosis** (*loks-ar-thro'-sis*) [λοξός, awry; ἄρθρον, joint]. Distortion of a joint.
**loxia** (*loks'-e-ah*). See *torticollis*.
**loxic** (*loks'-ik*) [λοξός, oblique]. Distorted; awry; twisted.
**loxocyesis** (*loks-o-si-e'-sis*) [λοξός, awry; κύησις, pregnancy]. Oblique displacement of the gravid uterus.
**loxodont, loxodontous** (*loks'-o-dont, loks-o-don'-tus*) [λοξός, aslant; ὀδούς, tooth]. Having teeth placed at abnormal angles with the jaw.
**loxophthalmos** (*loks-off-thal'-mos*). Synonym of *strabismus, q. v.*
**Loxopterygium** (*loks-o-ter-ij'-e-um*) [λοξός, oblique; πτέρυξ, wing]. A genus of anacardiaceous trees. The bark of L. *lorentsii* and L. *sagotii*, the red quebracho, is used as a substitute for cinchona.
**loxotomy** (*loks-ot'-o-me*) [λοξός, oblique; τομή, a cutting]. Amputation by oblique section.
**lozenge** (*loz'-enj*) [OF., *losenge*]. A medicated tablet with sugar as a basis.
**L. R. C. P.** Abbreviation of Licentiate of the Royal College of Physicians.
**L. R. C. S.** Abbreviation of Licentiate of the Royal College of Surgeons.
**L. S. A.** 1. Abbreviation of Licentiate of the Society of Apothecaries. 2. Abbreviation for left sacroanterior position of the fetus in utero.
**L. S. P.** Abbreviation for left sacroposterior position of the fetus in utero.
**L. S. S.** Abbreviation for Licentiate in Sanitary Science.
**L. S. Sc.** Abbreviation for Licentiate in Sanitary Science.
**Lubarsch's crystals** (*loo'-barsh*) [Otto *Lubarsch*, German pathologist, 1860– ]. Minute crystals found postmortem in the epithelial cells of the testis and regarded as distinct from Boettcher's and Charcot's crystals.
**lubrasepic** (*lū-brah-sep'-tik*). A disinfectant containing chondrus, boric acid, and formaldehyde.
**lubricant** (*lū'-brik-ant*) [*lubricare*, to make smooth]. Making smooth, oily, or slippery.
**lubrication** (*lū-brik-a'-shun*) [*lubricare*, to make smooth]. The process of making smooth or slippery by the application of a lubricant.
**lubrichondrin** (*lū-brik-on'-drin*). A lubricant said to consist of *Chondrus crispus*, oil of eucalyptus, and formaldehyde. It is used in surgery.
**lubricity** (*lū-bris'-it-e*) [*lubricare*, to make slippery]. A synonym of lasciviousness.
**lubricous** (*lū'-brik-us*) [*lubricus, lubricare*, to make smooth]. Smooth, slippery.
**Luca's horizontal plane.** One passing through the axis of the zygomatic arches.
**Lucas' sign** (*loo'-kas*) [Richard Clement *Lucas*, English surgeon, 1846–1915]. Distention of the abdomen, an early sign of rickets.
**Lucas-Championnière's disease** (*loo-kah'-shom-pe-on-e-dr'*) [Just Marie Marcellin *Lucas-Championniere*, French physician, 1843–1913]. Chronic pseudomembranous bronchitis.
**lucent** (*lū'-sent*) [*lucere*, to shine]. Bright, shining.
**lucid** (*lū'-sid*) [*lucidus*, clear]. Clear; shining; not obscure. **l. interval**, the transitory return of the normal mental faculties in insane or delirious conditions.
**lucidification** (*lū-sid-if-ik-a'-shun*) [*lucid; facere*, to make]. A clearing-up, especially a clearing-up of the protoplasm of cells.
**lucidity** (*lū-sid'-it-e*) [*lucidus*, clear]. Clearness; lucid interval.
**lucidum** (*lū'-sid-um*) [*lucidus*, clear]. Clear. **l., septum.** See *septum*.
**luciferin** (*lu-sif'-er-in*) [*lux, lucis*, light; *ferre*, to bear]. A crystalline body obtained from the light-producing organs of certain animals.

**lucifer-match disease** or **l.-match maker's disease.** Necrosis of the jaw due to phosphorus-poisoning.
**luciform** (*lū'-sif-orm*) [*lux*, light; *forma*, form]. Resembling light.
**lucifugal** (*lū-sif'-ū-gal*) [*lux*, light; *fugere*, to flee]. Fleeing from or avoiding light.
**Lucke's operation** (*loo'-keh*). Excision of the infraorbital nerve by the pterygomaxillary route.
**lucomania** (*lū-ko-ma'-ne-ah*) [λύκος, wolf; μανία, mania]. Same as *lycanthropy, q. v.*
**lucotherapy** (*lū-ko-ther'-ap-e*) [*lux*, light; *therapy*]. Therapeutic use of light-rays.
**luctic** (*luk'-tik*) [*luctus*, sorrow]. Sorrowful, giving evidence of suffering.
**Ludovic's angle** (*lood'-ov-ik*). See *Louis' angle*.
**Ludwig's angina** (*lood'-vig*) [Wilhelm Friedrich von *Ludwig*, German surgeon, 1790–1865]. Phlegmonous cellulitis of the neck, generally secondary to specific fevers, scurvy, etc. Syn., *angina Ludovici*.
**Ludwig's ganglion** (*lood'-vig*) [Karl Friedrich Wilhelm *Ludwig*, German physiologist 1816–1895]. A collection of nerve-cells in the wall of the right auricle.
**ludyl** (*loo'-dil*). An arsenical compound, said to be of benefit in syphilis.
**Luecke's reaction for hippuric acid** (*le'-keh*) [Georg Albert *Luecke*, German surgeon, 1829–1894]. Add to the substance nitric acid at boiling temperature, and evaporate to dryness; an intense odor of nitrobenzol is produced on heating the residue.
**lues** (*lū'-es*) [L.]. Formerly a pestilential disease; at present used as a euphemism for syphilis. **l. venerea,** syphilis.
**luesan** (*lū'-es-an*). An organic compound of mercury and glidine, said to be useful in syphilis.
**luetic** (*lū-et'-ik*) [*lues*]. Affected with or relating to lues; syphilitic. **l. serum.** See *serum*.
**luetin** (*lū'-et-in*) [*lues*]. An extract of the killed cultures of several strains of the *Treponema pallidum* used in the Noguchi reaction for syphilis. **l.-reaction.** See *Noguchi reaction*.
**Luffa** (*luf'-ah*) [Arab]. A genus of cucurbitaceous plants. **L. acutangula,** the strainer vine, of India, having an edible fruit, a purgative and emetic root, and oil-bearing seeds. **L. amara** has similar uses. **L. cylindrica,** the "wash-rag," "towel gourd," or "vegetable sponge," bears a fruit which when dry contains a sponge-like network of fibers. This is sometimes used as a substitute for a sponge or towel. **L. echinata,** the irritant poisonous fruit, is used in cholera and for colic. **L. purgans,** of S. America, affords a strongly purgative resin.
**Lugol's caustic** (*loo'-gol*) [J. G. A. *Lugol*, French physician, 1786–1851]. A solution of iodine and potassium iodide, of each, one part, in water two parts. **L.'s solution,** a compound solution of iodine, containing iodine 5, potassium iodide 10, distilled water to 100; dose 3–6 ♍ (0.2–0.4 Cc.).
**lukewarm.** Tepid; about the temperature of the body.
**lumbago** (*lum-ba'-go*) [*lumbus*]. Pain in the loins.
**lumbar** (*lum'-bar*) [*lumbus*]. Pertaining to the loins. **l. colotomy,** colotomy performed in the lumbar region. **l. puncture.** See *puncture, lumbar*. **l. region.** See *under abdomen*. **l. vertebra.** See *vertebra, lumbar*.
**lumben** (*lum'-ben*) [*lumbus*, the loin]. Belonging to the loin in itself.
**lumbifragium** (*lum-bif-ra'-je-um*) [*lumbus*, loin; *frangere*, to break]. Lumbar hernia.
**lumbiplex** (*lum'-be-pleks*). The lumbar plexus.
**lumbo-** (*lum-bo-*) [*lumbar*]. A prefix meaning relating to the loins.
**lumboabdominal** (*lum-bo-ab-dom'-in-al*) [*lumbo-*; *abdomen*]. Pertaining to the loin and the abdomen.
**lumbocolostomy** (*lum-bo-ko-los'-to-me*). Colostomy after lumbar incision.
**lumbocolotomy** (*lum-bo-ko-lot'-o-me*) [*lumbo-*; κόλον, colon; τομή, a cutting]. Incision of the colon through the loins.
**lumbocostal** (*lum-bo-kos'-tal*) [*lumbo-*; *costa*, a rib]. Pertaining to the loins and ribs.
**lumbodorsal** (*lum-bo-dor'-sal*) [*lumbo-*; *dorsum*, back]. Pertaining to the lumbar and dorsal regions.
**lumbodynia** (*lum-bo-din'-e-ah*) [*lumbo-*; ὀδύνη, pain]. Same as *lumbago*.
**lumboinguinal** (*lum-bo-in'-gwin-al*) [*lumbo-*; *inguen*, groin]. Pertaining to the lumbar and inguinal regions.
**lumbosacral** (*lum-bo-sa'-kral*) [*lumbo-*; *sacrum*]. Pertaining to the lumbar vertebræ and to the sacrum.
**lumbrical** (*lum'-brik-al*). Relating to or resembling a worm of the genus *Lumbricus*.
**Lumbricales** (*lum-brik-a'-lēz*). A genus of intestinal worms.
**lumbricales** (*lum-brik-a'-lēz*). See under *muscle*.
**lumbricide** (*lum'-bris-īd*) [*lumbricus*, earth-worm; *cædere*, to kill]. A drug fatal to lumbricoid worms.
**lumbriciform** (*lum-bris'-if-orm*). See *lumbricoid*.
**lumbricoid** (*lum'-brik-oid*) [*lumbricus*; εἶδος, likeness]. Pertaining to or resembling a lumbricus.
**lumbricosis** (*lum-brik-o'-sis*). The condition of being infected with lumbricoids.
**Lumbricus** (*lum'-brik-us*) [L.]. A genus of worms, including the common earthworm and certain intestinal worms. The latter are now termed *Ascarides*.
**lumbus** [L.]. The loin.
**lumen** (*lū'-men*) [L., 'light''; pl., *lumina*]. The space inside of a tube, *e. g.*, the lumen of a thermometer, blood-vessel, etc.
**lumina** (*lū'-min-ah*). Plural of *lumen, q. v.*
**luminal** (*lū'-min-al*). 1. Pertaining to the lumen of a blood-vessel or other tubular structure. 2. Trade name for a hypnotic said to consist of phenylethylbarbituric acid.
**luminiferous** (*lū-min-if'-er-us*) [*lumen*; *ferre*, to bear]. Conveying or bearing light.
**luminosity** (*lū-min-os'-it-e*) [*luminosus*, full of light]. The property of emitting light.
**lump.** 1. A small mass; a protuberant part. 2. A dull person. 3. In the plural, a popular term for strumous enlargement of the cervical glands.
**lumpy-jaw.** See *actinomycosis*.
**lunacy** (*lū'-nas-e*) [*luna*, the moon]. Insanity, from the superstitious belief that it was influenced by the moon.
**lunar** (*lū'-nar*) [*luna*]. Pertaining to the moon or to silver (*luna* of the alchemists). **l. caustic,** silver nitrate.
**lunare,** or **os lunare** (*lū-na'-re*) [*luna*, moon]. The semilunar bone of the carpus.
**lunaria** (*lū-na'-re-ah*) [*lunaris*, pertaining to the moon]. Menstruation.
**lunate bone** (*lū'-nāt*). Semilunar bone, os lunatum.
**lunatic** (*lū'-nat-ik*) [see *lunacy*]. 1. Pertaining to or affected with insanity. 2. An insane person.
**lunella** (*lū-nel'-ah*) [dim. of *luna*, moon]. Same as *hypopyon*.
**lunet, lunette** (*loo-net'*) [Fr.]. A spectacle-lens.
**lung** [AS., *lungen*]. The organ of respiration, in which the impure venous blood is oxidized by the air drawn through the trachea and bronchi into the air-vesicles. There are two lungs, a right and a left, the former consisting of three, the latter of two, lobes. The lungs are situated in the thoracic cavity, and are enveloped by the pleura. At the root or hilum the bronchus and its arteries and the pulmonary artery and nerves enter, and the pulmonary and bronchial veins and lymphatics leave. The lung proper consists of minute air-vesicles held in place by connective-tissue trabeculæ. Capillaries traverse the walls of the air-vesicles and bring the circulating blood in close proximity to the air. The average weight of the adult right lung is 22 ounces; that of the left, 20 ounces. **l.-capacity,** breathing capacity. **l., cardiac,** proliferation of the connective tissue of a lung, producing thickening of the alveolar walls and finally obliteration of their cavity, due to organic lesions in certain heart diseases. **l., carnified,** a lung from which the blood and air have been driven out by effusion, causing it to present a slaty-gray color. **l.-fever,** croupous pneumonia. **l., saccular,** a condition of the lung marked by globular pouches at the periphery or through the whole or greater part of the lobe. **l.-stone,** a calcification of lung-tissue.
**lungwort** (*lung'-wert*). See *verbascum*.
**luniferous** (*lū-nif'-er-us*) [*luna*; *ferre*, to bear]. Crescent-shaped.
**luniform** (*lū'-ne-form*) [*luna*; *forma*, form]. Same as *luniferous*.
**lunula** (*lū'-nū-lah*) [dim. of *luna*, moon]. 1. The white semilunar area of a nail near the root. 2. A structure resembling the lunula of a nail. **l. of the cardiac valves,** the delicate edges of the leaflets of the semilunar valves. **l. of Gianuzzi.** See *Gianuzzi, crescents of*. **l. lacrymalis,** a small crest of bone separating the antrum of Highmore from the lacrymal groove. **l. scapulæ.** See *notch, suprascapular*.

# LUPANINE 524 LUTIDIN

**lupanine** ($lu'$-$pan$-$ēn$), $C_{15}H_{24}N_2O$. A bitter liquid alkaloid from the seeds of *Lupinus reticulatus*.

**luperine** ($lu'$-$per$-$ēn$). A remedy for dipsomania said to be a mixture of powdered gentian, calumba, and quassia.

**lupetazin** ($lu$-$pet'$-$az$-$in$), $HN(CH_2CH.CH_2)_2NH$. A white, crystalline powder similar to piperazin in action, application, and dosage. 1. tartrate, lycetol.

**lupia** ($lu'$-$pe$-$ah$) [*lupus*]. A name given to several kinds of malignant tumor, phagedenic ulcer, and fungoid growth.

**lupiform** ($lu'$-$pe$-$form$) [*lupus; forma*, form]. Resembling lupus.

**lupigenin** ($lu$-$pij'$-$en$-$in$) [*lupus*, a wolf], $C_{17}H_{12}O_6$. A substance resulting from the decomposition of lupiin by acids.

**lupiin** ($lu'$-$pe$-$in$) [*lupus*, a wolf], $C_{23}H_{39}O_{16} + 7H_2O$. A bitter principle found in germinating lupine seeds.

**lupine** ($lu'$-$pēn$) [*lupus*]. A plant of the genus *Lupinus*. *Lupinus albus* and other species contain a bitter glucoside, *lupinin* ($C_{25}H_{39}O_{16}$), while from *L. luteus* an alkaloid, *arginine* ($C_6H_{14}N_4O_2$), is obtained. The bruised seeds of *L. albus* have been used as an external application to ulcers.

**lupinidine** ($lu$-$pin'$-$id$-$ēn$) [*lupine*], $C_8H_{15}N$. A liquid alkaloid obtained from *Lupinus luteus*.

**lupinosis** ($lu$-$pin$-$o'$-$sis$). See *lathyrism*.

**lupinotoxin** ($lu$-$pin$-$o$-$toks'$-$in$) [*lupinus*, lupine; τοξικόν, poison]. A resinoid derived from certain poisonous species of lupine, and said to be capable of producing the symptoms of lupinosis.

**lupinus** ($lu$-$pi'$-$nus$). See *lupine*.

**lupoid** ($lu'$-$poid$). Having the nature of lupus. l. sycosis. See *ulerythema sycosiforme*. See *lupiform*.

**lupoma** ($lu$-$po'$-$mah$). The primary nodule of lupus.

**lupomania** ($lu$-$po$-$ma'$-$ne$-$ah$) [*lupus*, wolf; μανία, madness]. Rabies.

**lupotome** ($lu'$-$po$-$tōm$) [*lupus*, wolf; τομή, a cutting]. A cutting or scarifying instrument sometimes used in the treatment of lupus.

**lupous** ($lu'$-$pus$) [*lupus*, wolf]. Affected with or pertaining to lupus.

**lupulin, lupulinum** ($lu'$-$pū$-$lin$, $lu$-$pū$-$li'$-$num$) [*lupulus*, hop]. The *lupulinum* of the U. S. P, is the glandular powder obtained from the strobiles of *Humulus lupulus*. It is antispasmodic and sedative, and is used in sexual excitement, delirium tremens, renal and vesical irritation, and spermatorrhea. Dose 5–15 gr. (0.32–1.0 Gm.). l., extract of (*extractum lupulini*, B. P.). Dose 5–10 gr. (0.32–0.65 Gm.). l., fluidextract of (*fluidextractum lupulini*, U. S. P.). Dose 8 min. (0.5 Cc.). l., infusion of (*infusum lupulini*, B. P.). Dose 1–2 oz. (32–64 Cc.). l., oleoresin of (*oleoresina lupulini*, U. S. P.). Dose 3 gr. (0.2 Gm.).

**lupulus** ($lu'$-$pū$-$lus$). See *humulus*.

**lupus** ($lu'$-$pus$) [L., "a wolf"]. A chronic tuberculous disease of the skin and mucous membranes, characterized by the formation of nodules of granulation tissue. It passes through a number of phases, and terminates by ulceration or atrophy, with scar-formation. The cause of the disease is the tubercle bacillus. Syn., *lupus exedens; lupus vulgaris*. l., disseminated follicular, a variety of lupus confined to the face, especially in the situations usually occupied by acne. The papules are from a large pin-head to a pea in size, conical and deep red. l. erythematosus, a form not due to the tubercle bacillus. It occurs, as a rule, in multiple patches, with a tendency to symmetrical arrangement, chiefly about the face and head, occasionally on the extremities, and rarely on the trunk. The patches are sharply defined at the border, flat, very slightly raised, and with a tendency to the formation of crusts. The color is bright-red, and there are no nodules. It is most common in women of adult or middle age. Syn., *Cazenave's lupus; lupus erythematodes*. l. erythematosus sebaceus, a form with special involvement of the sebaceous glands. l. exedens. Synonym of *lupus*. l. hypertrophicus, that variety of lupus in which new connective-tissue formation predominates over the destructive process, and markedly raised, thick patches result. l. maculosus, a variety of lupus characterized by the eruption of very soft, smooth, brownish-red, semitranslucent miliary nodules that develop in the connective tissue of otherwise healthy skin without subjective sensations. l. nonexedens, lupus without ulceration. l. serpiginosus, that which spreads peripherally the cicatrizing centrally. l. tumidus, a form with edematous infiltration. l. vegetans, l. verrucosus, the formation in the lupus process of a warty-looking patch liable to become inflamed. l. vulgaris, lupus.

**lupuscarcinoma** ($lu$-$pus$-$kar$-$sin$-$o'$-$mah$). A carcinoma developing from lupus.

**lura** ($lū'$-$rah$) [L., the mouth of a bag]. The contracted orifice of the infundibulum after removal of the hypophysis.

**lural** ($lū'$-$ral$). Relating to the lura.

**luridity** ($lū$-$rid'$-$it$-$e$) [*luridus; luror*, yellow color]. A pale-yellow color met with in certain cachectic conditions; also in atrophied paralyzed limbs.

**Luschka's bursa** (*loosh'$-$keh*) [Herbert von *Luschka*, German anatomist, 1820–1875]. A crypt, larger and more clearly defined than the neighboring crypts, frequently located in the lower part of the pharyngeal tonsil, and regarded as a vestige of the communication existing during early fetal life between the pharynx and the hypophysis. L.'s cartilage, an inconstant, small, cartilaginous nodule, inclosed in the front part of the true vocal cord. L.'s fold, ileocolic fold; a semilunar fold of the peritoneum which is attached to the anterior layer of the mesentery, the anterior aspect of the ascending colon, and the cecum as far as the vermiform appendix. L.'s foramina, two small openings in the lateral recesses of the pia covering the fourth ventricle; they transmit the choroid plexus. L.'s fossa, a narrow fossa bounded by the ileocolic fold in front, and by the enteric mesentery, the ileum, and a small portion of the upper and inner walls of the cecum behind. Syn., *ileocolic fossa*. L.'s gland, 1. The pharyngeal tonsil. 2. The coccygeal gland. 3. The carotid gland; the intercarotid gland, a minute body of glandular structure and unknown function, situated at the bifurcation of the common carotid artery. L.'s line, an imaginary line extending from the middle of the internal palpebral ligament to the space between the first and second molars, and indicating the course of the lacrimal sac and nasal duct. L.'s subpharyngeal cartilage, a small body of hyaline cartilage situated in the areolar tissue of the lower part of the faucial tonsil. It represents a vestige of the third postoral arch of the embryo and is occasionally the seat of morbid growths. L.'s tonsil, the adenoid tissue normally existing between the orifices of the Eustachian tubes, analogous in structure to the lymphoid constituents of the tonsil.

**luscitas** (*lus'$-$it$-$as$) [*luscus*, one-eyed]. 1. The state of being blind in one eye. 2. Strabismus, especially when caused by paralysis of rheumatism.

**lusis** ($lu'$-$sis$) [λοῦσις, a washing]. A washing.

**Lusk's contraction ring**. [William Thompson *Lusk*, American obstetrician, 1838–1897]. Same as *Bandl's ring, q. v.*

**lusus naturæ** ($lu'$-$sus$ $na$-$tū'$-$re$) [*lusus*, a play; *naturæ*, nature]. A freak of nature.

**lust**. Carnal desire; sexual appetite; concupiscence.

**Lustig's plague serum** (*loos'$-$tig*) [Alessandro *Lustig*, Italian pathologist]. An antitoxic and bactericidal serum obtained by immunizing horses with injections of plague bacilli and collecting the serum from the blood of these animals.

**lustramentum** (*lus$-$tram$-$en'$-$tum*) [*lustrare*, to purify]. A purge.

**lutarious** ($lu$-$ta'$-$re$-$us$) [*lutum*, mud]. Relating to, like, or living in mud.

**lutation** ($lu$-$ta'$-$shun$) [*lutum*, mud]. The hermetic sealing of a vessel.

**Lutaud's lotion** (*loo'$-$to*). A lotion for pruritus of the vulva. It consists of eucalyptus oil, 10 parts; cocaine hydrochloride, 1 part; chloral hydrate, 10 parts; distilled water, 500 parts.

**lute** (*lūt*) [*lutum*, mud]. A composition for sealing vessels hermetically or for closing joints in apparatus. It may be made of lime and white of egg, linseed-meal, and starch, or of clay and drying oil.

**lutecium**, or **lutetium** (*loo$-$te'$-$shum*). A new chemical element, discovered in 1907; symbol Lu, atomic weight 174.

**lutein** ($lu'$-$te$-$in$) [*luteus*, yellow]. 1. A yellow pigment obtained from the corpora lutea by extraction with chloroform. 2. An internal secretion of the ovary.

**lutidin** ($lu'$-$tid$-$in$), $C_7H_9N$. A toxic liquid distilled from bituminous shale. l., beta-, $C_7H_9N$, a toxic, colorless liquid obtained by distillation of cinchonine with potassium hydroxide. It is narcotic and

# LUTREXANTHEMA 525 LYMPHANGIOMA

antispasmodic, and recommended as an antidote to strychnine.
**lutrexanthema** (*lu-treks-an'-the-mah*) [λουτρόν, a bath; *exanthema*]. An eruption due to bathing.
**Luttke's test for free hydrochloric acid in gastric juice** (*loot'-keh*). The quantitative determination successively of the total chlorine, the chlorine of the fixed chlorides, and that of the free and combined hydrochloric acid.
**luxatio erecta.** A dislocation of the shoulder-joint in which the head of the humerus is in the axilla and the shaft is directed upward against the head of the patient. l., **imperfecta**, incomplete dislocation, sprain.
**luxation** (*luks-a'-shun*). See dislocation.
**luxus** (*luks'-us*) [L.]. Excess. l.-**consumption**, a term applied to the metabolism of certain surplus protein material, which, though inside the body, does not form a component part of any of its tissues, but constitutes a kind of reservoir of force upon which the organism may draw. l., **heart**, cardiac dilatation with hypertrophy of the left ventricle.
**Luys' body, L.'s nucleus** (*lēs*) [Jules Bernard *Luys*, French physician, 1828–1897]. An almond-shaped mass of gray matter with pigmented ganglion-cells in the subthalamic region.
**Luys separator** (*lēs*) [Georges *Luys*, French physician]. An apparatus consisting of a diaphragm attached to a sound by means of which the bladder is divided into two parts, thus allowing the urine from each ureter to be collected separately.
**lycanthrope** (*li'-kan-throp*) [λύκος, wolf; ἄνθρωπος, man]. A person affected with lycanthropy.
**lycanthropic** (*li-kan-throp'-ik*) [λύκος, wolf; ἄνθρωπος, man]. Pertaining to or affected with lycanthropy.
**lycanthropy** (*li-kan'-thro-pe*) [λύκος, a wolf; ἄνθρωπος, man]. A form of mania in which the patient imagines himself a wild beast.
**lycetal, lycetol** (*lis'-et-al, -ol*), $C_6H_{14}N_2C_4H_6O_5$. A derivative of dimethylpiperazin combined with tartaric acid. It is used in gout. Dose 15–45 gr. (0.97–2.9 Gm.) well diluted in carbonated water.
**Lychnis** (*lik'-nis*) [λυχνίς, the lampflower]. A genus of plants of the order *Caryophyllaceæ*. L. *githago*, corn-cockle, is a species of Europe naturalized here. The seeds are diuretic, expectorant, and anthelmintic.
**lyciform** (*lis'-e-form*). See *lycoform*.
**lycine** (*li'-sin*). Same as *betaine*, q. v.
**lycoctonine** (*lik-ok'-ton-ēn*). An alkaloid extracted from *Aconitum lycoctonum*. It is crystallizable, very soluble in alcohol, and but slightly so in ether or water.
**lycodes** (*li-ko'-dēz*) [λύκος, wolf]. A chronic form of tonsillitis. See *lycoides*.
**lycoform** (*li'-ko-form*). A proprietary combination of alcoholic potash soap solution and formaldehyde.
**lycoid** (*li'-koid*) [λύκος, wolf; εἶδος, like]. Resembling a wolf.
**lycomania** (*li-ko-ma'-ne-ah*). See *lycanthropy*.
**Lycoperdon** (*li-ko-per'-don*) [λύκος, wolf; πέρδεσθαι, to break wind]. A genus of fungi. L. *bovista*, the fist-ball, puffball, or devil's snuff-box, has been used as a styptic, and is now employed to some extent in nervous diseases.
**lycopin** (*li'-ko-pin*) [λύκος, wolf; πούς, foot]. A precipitate from a tincture of *Lycopus virginicus*, an astringent, styptic, sedative, and tonic. Dose from 1 to 4 grains.
**lycopodium** (*li-ko-po'-de-um*). [λύκος, wolf; πούς, foot]. The moss of L. *clavatum* and other varieties, official in the U. S. P., and occurring in the form of a light, fine, yellowish powder. It is used as a desiccant and absorbent on moist and excoriated surfaces, and as an inert powder in which to embed pills to prevent their adhering to one another. Syn., *clubmoss; witch-meal; wolf's-claw*.
**lycopus** (*li'-ko-pus*) [λύκος, wolf; πούς, foot]. Bugle weed, the *Lycopus virginicus*; it is an astringent and hemostatic.
**lycorexia, lycorrhexy** (*li-kor-eks'-e-ah, li'-kor-eks-e*) [λύκος, wolf; ὄρεξις, appetite]. A wolfish or canine appetite; bulimia.
**lycostoma** (*li-kos'-to-mah*) [λύκος, wolf; στόμα, mouth]. Cleft palate.
**lycresol** (*li'-kre-sol*). A soap solution containing crude cresol.
**lye** (*li*) [AS., *leāh*]. 1. An alkaline solution obtained by leaching ashes. 2. Any alkaline solution.
**lyencephalous** (*li-en-sef'-al-us*) [λύειν, to loosen; ἐγκέφαλος, brain]. Having cerebral hemispheres that are loosely united.

**lygismus** (*li-jiz'-mus*) [λυγισμός, a bending]. A melodious, flexible voice; also, dislocation.
**lying-in.** 1. Being in confinement. 2. The puerperal state.
**lyma** (*li'-mah*) [λύμα, washings: *pl., lymata*].
1. Filth, or sordes. 2. Lochia.
**lymph** (*limf*) [*lympha*, water]. 1. The fluid in the lymphatic vessels, the product of the filtration of the liquid portion of the blood through the walls of the capillaries. 2. The coagulable exudate on an inflamed surface. 3. The liquid material used for vaccination; vaccine-lymph. l., **animal**, vaccine-lymph obtained from an animal. l.-**cell**, l.-**corpuscle**, a leukocyte occurring in the lymph. l.-**channel**. See l.-*space*. l., **fibrinous**, transparent, tenacious lymph that coagulates spontaneously. It is almost colorless and contains a large amount of fibrin and but few corpuscles. l.-**follicles**, small collections of lymphadenoid tissue occurring in mucous membranes. l.-**hearts**, certain organs found in the frog and in some fishes, which are to the lymph-stream what the blood-heart is to the blood-stream. l., **humanized**, vaccine from a human being. l., **inflammatory**, that thrown out as a product of inflammation in wounds, etc. l., **Koch's**, tuberculin. l., **plastic**, fibrinous lymph; that forming embryonic tissue. l. **reservoir**, the receptaculum chyli. l.-**scrotum**, an enlarged scrotum due to distention of the lymphatic vessels and hyperplasia of the tissues. l.-**sinus**. Same as l.-*space*. l.-**spaces**, the lacunæ occurring in connective tissue and containing lymph. l., **vaccine**. See *vaccine-lymph*.
**lymphaden** (*lim-fa'-den*) [*lympha*, water; ἀδήν, gland]. Any lymphatic gland.
**lymphadenectasis** (*lim-fad-en-ek'-tas-is*) [*lymph*; ἀδήν, gland; ἔκτασις, distention]. Dilatation of the sinuses of a lymph-gland producing a tumor-like mass.
**lymphadenhypertrophy** (*lim-fad-en-hi-per'-tro-fe*) [*lymph*; ἀδήν, gland; *hypertrophy*]. Hypertrophy of the lymphatic glands.
**lymphadenia** (*lim-fad-e'-ne-ah*) [*lympha*, lymph; ἀδήν, gland]. A general hyperplasia of the lymphatic tissue with or without leukocytosis.
**lymphadenism** (*lim-fad'-en-izm*) [*lympha*, lymph; ἀδήν, gland]. The general condition of disease that accompanies lymphadenoma.
**lymphadenitis** (*lim-fad-en-i'-tis*) [*lymph*; ἀδήν, gland; ἰτις, inflammation]. Inflammation of a lymphatic gland. l. **calculosa**, that combined with calcareous degeneration. l., **scrofulosa**, the small-celled caseous or suppurative hyperplasia of the lymph-glands. l., **tuberculous**, the formation of tubercles in the lymph-glands.
**lymphadenoid** (*lim-fad'-en-oid*) [*lymph*; *adenoid*]. Resembling, or of the nature of, a lymphatic gland or lymphatic tissue.
**lymphadenoma** (*lim-fad-en-o'-mah*) [*lymph*; *adenoma*]. Hyperplasia of the lymphatic glands. See *Hodgkin's disease*.
**lymphadenosis** (*lim-fad-en-o'-sis*) [*lympha*, lymph; ἀδήν, gland; νόσος, disease]. General lymphadenoma.
**lymphæduct** (*limf'-fe-dukt*). See *lymphduct*.
**lymphæmia** (*lim-fe'-me-ah*). See *lymphemia*.
**lymphagogue** (*limf'-fag-og*) [*lymph*; ἀγωγός, leading]. 1. Stimulating the flow of lymph. 2. An agent that stimulates the flow of lymph.
**lymphangeitis** (*lim-fan-je-i'-tis*). See *lymphangitis*.
**lymphangiectasis, lymphangiectasia** (*lim-fan-je-ek'-tas-is, -ek-ta'-se-ah*) [*lymph*; ἀγγεῖον, vessel; ἔκτασις, widening]. 1. Dilatation of the lymphatic vessels. 2. Elephantiasis.
**lymphangiectodes** (*lim-fan-je-ek-to'-dēz*). See *lymphangioma circumscriptum*.
**lymphangiectodes** (*lim-fan-je-en'-kis-is*) [*lympha*, lymph; ἀγγεῖον, vessel; ἔγχυσις, a pouring in]. Injection of the lymphatic vessels.
**lymphangioendothelioma** (*lim-fan-je-o-en-do-the-le-o'-mah*). An endothelioma originating in lymph-vessels.
**lymphangiofibroma** (*lim-fan-je-o-fi-bro'-mah*). Lymphangioma combined with fibroma.
**lymphangiology** (*lim-fan-je-ol'-o-je*) [*lympha*, water; ἀγγεῖον, vessel; γράφειν, to write]. A description of the lymphatics.
**lymphangiology** (*lim-fan-je-ol'-o-je*)[*lymph*; ἀγγεῖον, vessel; λόγος, science]. The anatomy, physiology, and pathology of the lymphatics.
**lymphangioma** (*lim-fan-je-o'-mah*) [*lymph*; *angi-*

**oma**; *pl.*, *lymphangiomata*]. A dilated or varicose condition or tumor of the lymphatic vessels. l. **capillare varicosum.** Synonym of *l. circumscriptum*. l. **cavernosum.** Synonym of *l. circumscriptum*. l. **circumscriptum**, lymphangiectodes; lupus lymphaticus, a very rare disease of the skin occurring in early life. It is marked by the formation of straw-yellow vesicles, deeply situated in the skin, with thick and tense walls, and connected with the lymphatics. Its cause is unknown. l. **tuberosum multiplex,** a very rare disease of the skin, probably congenital, characterized by the formation of large, brownish-red papules or tubercles, the size of lentils, not arranged in groups or clusters, but scattered indiscriminately over the trunk.

**lymphangiomyoma** (*lim-fan-je-o-mi-o'-mah*). See *myoma*.

**lymphangion** (*lim-fan'-je-on*) [*lympha*, lymph; ἀγγεῖον, vessel]. A lymphatic vessel.

**lymphangiophlebitis** (*lim-fan-je-o-fleb-i'-tis*) [*lymph*; ἀγγεῖον, vessel; *phlebitis*]. Inflammation of the lymphatic vessels and veins.

**lymphangioplasty** (*lim-fan-je-o-plas'-te*) [*lymph*; ἀγγεῖον, vessel; πλάσσειν, to form]. Operative formation of artificial lymphatics by means of silk threads.

**lymphangiopyra** (*lim-fan-je-op'-ir-ah*) [*lympha*, lymph; ἀγγεῖον, vessel; πῦρ, fire]. Fever due to or accompanying a diseasé of the lymphatics.

**lymphangiosarcoma** (*lim-fan-je-o-sar-ko'-mah*). Lymphangioma attended with sarcoma.

**lymphangioscopy** (*lim-fan-je-os'-ko-pe*) [*lympha*, lymph; ἀγγεῖον, vessel; σκοπεῖν, to inspect]. Inspection or observation of the lymphatics.

**lymphangiotomy** (*lim-fan-je-ot'-o-me*) [*lympha*, lymph; ἀγγεῖον, vessel; τομή, a cutting]. Dissection or anatomy of the lymphatics.

**lymphangitis** (*lim-fan-ji'-tis*) [*lymph*; ἀγγεῖον, vessel; *ιτις*, inflammation]. Inflammation of a lymphatic vessel.

**lymphangoncus** (*lim-fan-gon'-kus*) [*lympha*, lymph; ἀγγεῖον, vessel; ὄγκος, a tumor]. A firm swelling of the lymphatic vessels.

**lymphapostema** (*lim-fap-os-te'-mah*) [*lympha*, lymph; ἀπόστημα, abscess: *pl.*, *lymphapostemata*]. A lymphatic abscess.

**lymphatic** (*lim-fat'-ik*) [*lymph*]. Pertaining to lymph; containing or characterized by lymph. l. **gland.** See *gland*. l. **leukemia,** leukemia of lymphatic origin. l. **system,** a system of vessels and glands accessory to the blood-vascular system, conveying lymph. It begins as innumerable capillaries in interspaces of tissues. The form plexuses studded with lymph-glands that act as filters and finally all those below the diaphragm unite in the recaptaculum chyli on the second lumbar vertebra. From this the thoracic duct leads upward to empty into the junction of the left subclavian and internal jugular veins. The lymph from the upper right half of the body and head enters the right lymphatic duct, which empties into the junction of the right internal jugular and subclavian veins. l. **vessel,** a tube for conveying lymph.

**lymphaticosanguine** (*lim-fat-ik-o-san'-gwin*) [*lympha*, lymph; *sanguis*, blood]. Both lymphatic and sanguine. See *temperament*.

**lymphaticosplenic** (*lim-fat-ik-o-splen'-ik*). Relating to the lymphatics and the spleen.

**lymphatics** (*lim-fat'-iks*) [*lymph*]. The capillary tubes pervading the body, which convey lymph.

**lymphatism** (*lim'-fat-izm*). The lymphatic temperament; scrofula. See *status lymphaticus*.

**lymphatitis** (*lim-fat-i'-tis*). See *lymphangitis*.

**lymphatocele** (*lim-fat'-o-sēl*) [*lympha*, lymph; κήλη, tumor]. A tumor composed of dilated lymph-vessels.

**lymphatology** (*lim-fat-ol'-o-je*). See *lymphology*.

**lymphectasia** (*lim-fek-ta'-ze-ah*) [*lymph*; ἔκτασις, widening]. Dilatation with lymph.

**lymphedema** (*lim-fe-de'-mah*) [*lymph*; οἴδημα, edema]. Serous edema.

**lympheduct** (*lim'-fe-dukt*) [*lympha*, lymph; *ducere*, to lead]. A lymphatic vessel or duct.

**lymphemia** (*lim-fe'-me-ah*) [*lymph*; αἷμα, blood]. Leukemia characterized by enlargement of the lymphatic glands.

**lymphendothelioma** (*lim-fen-do-the-le-o'-mah*). A newgrowth characterized by a soft myxomatous tissue containing cysts and tubules lined with flat, scale-like cells, resembling the endothelium of the lymphatics.

**lymphenteritis** (*lim-fen-ter-i'-tis*) [*lympha*, lymph;

ἔντερον, bowels; *ιτις*, inflammation]. Inflammation of the bowels attended with serous infiltration. 2. Inflammation of the serous coat of the intestine; peritonitis.

**lymphepatitis** (*limf-hep-at-i'-tis*) [*lympha*, lymph; ἧπαρ, liver; *ιτις*, inflammation]. Inflammation of the peritoneal coat of the liver.

**lympheurysma** (*lim-fū-riz'-mah*). See *lymphaneurysma*.

**lymphexosmosis** (*lim-feks-os-mo'-sis*). Exosmosis of lymph; passage of lymph outward, through the coats of the lymph-vessels.

**lymphitis** (*lim-fi'-tis*). See *lymphangitis*.

**lymphization** (*lim-fiz-a'-shun*). The formation of lymph.

**lympho-** (*lim-fo-*) [*lymph*]. A prefix meaning relating to lymph or to the lymphatic glands.

**lymphoadenoma** (*lim-fo-ad-en-o'-mah*). A uterine neoplasm involving the interstitial lymph-tissue and the glands.

**lymphoblast** (*lim'-fo-blast*) [*lympho-*; βλαστός, germ]. Any cell with a nutritive function.

**lymphocele** (*lim'-fo-sēl*) [*lympho-*; κήλη, tumor]. A tumor containing an abnormal collection or quantity of lymph.

**lymphocyte** (*lim'-fo-sīt*) [*lympho-*; κύτος, a cell]. 1. A lymph-cell. 2. One of Ehrlich's classes of leukocytes, comprising those small cells having large nuclei and a very small amount of protoplasm. See *leukocyte*.

**lymphocythemia** (*lim-fo-si-the'-me-ah*) [*lymphocyte*; αἷμα, blood]. An excess of lymphocytes in the blood.

**lymphocytosis** (*lim-fo-si-to'-sis*). See *lymphocythemia*.

**lymphocytotoxin** (*lim-fo-si-to-toks'-in*) [*lymphocyte*; τοξικόν, poison]. A bacterial product having specific action on the lymphocytes.

**lymphoderma** (*lim-fo-der'-me-ah*) [*lympho-*; δέρμα, skin]. An affection of the lymphatics of the skin. l. **perniciosa,** leukemic enlargement of the glands.

**lymphoduct** (*lim'-fo-dukt*). See *lympheduct*.

**lymphofluxion** (*lim-fo-fluk'-shun*) [*lympho-*; *fluere*, to flow]. The increased flow of lymph induced by certain stomachics.

**lymphogenous** (*lim-foj'-en-us*) [*lympho-*; γεννᾶν, to beget]. Producing lymph.

**lymphoglandula** (*lim-fo-glan'-du-lah*) [*lympho-*; *glandula*, gland]. BNA term for a lymphatic gland.

**lymphogonia** (*lim-fo-go'-ne-ah*) [*lympho-*; γόνοι, offspring]. Large lymphocytes having a relatively large nucleus deficient in chromatin, and a faintly basic nongranular protoplasm, observed in lymphatic leukemia.

**lymphogranulomatosis** (*lim-fo-gran-ū-lo-mah-to'-sis*) [*lympho-*; *granuloma*]. Hodgkin's disease.

**lymphography** (*lim-fog'-ra-fe*) [*lympho-*; γράφειν, to write]. A description of the lymphatics.

**lymphoid** (*lim'-foid*) [*lympho-*; εἶδος, like]. Having the appearance or character of lymph.

**lymphology** (*lim-fol'-o-je*) [*lympho-*; λόγος, science]. The study of the anatomy and physiology of the lymphatic system.

**lymphoma** (*lim-fo'-mah*) [*lympho-*; ὄμα, tumor]. A tumor composed of lymphadenoid tissue. The term includes also formations not strictly tumors, as hyperplasia of the tissues proper to lymphatic glands. l., **malignant.** Synonym of *Hodgkin's disease*.

**lymphomatosis** (*lim-fo-mat-o'-sis*) [*lymphoma*]. A condition characterized by general lymphatic engorgement. l. **diffusa,** Hodgkin's disease.

**lymphomatous** (*lim-fo'-mat-us*) [*lymphoma*]. Of the nature of, or affected with, lymphoma.

**lymphomyeloma** (*lim-fo-mi-el-o'-mah*). 1. A myeloma involving the lymphatic system. 2. A sarcoma containing small round-cells.

**lymphomyxoma** (*lim-fo-miks-o'-mah*) [*lympho-*; *myxoma*]. A new-growth, usually benign, consisting of adenoid tissue.

**lymphoncus** (*lim-fong'-kus*) [*lympho-*; ὄγκος, tumor]. A hard lymphatic swelling. l. **iridis.** See *iridauxesis*.

**lymphopathy** (*lim-fop'-ath-e*) [*lympho-*; πάθος, disease]. Any disorder of the lymphatic organs.

**lymphopenia** (*lim-fo-pe'-ne-ah*) [*lympho-*; πενία, poverty]. A deficiency of lymphocytes in the circulating blood.

**lymphorrhagia** (*lim-for-a'-je-ah*) [*lympho-*; ῥηγνύναι, to burst forth]. A flow of lymph from a ruptured lymphatic vessel.

**lymphorrhea** (*lim-for-e'-ah*) [*lympho-*; ῥοία, a flow]. A discharge of lymph from a wound, internally or externally.

**lymphosarcoma** (*lim-fo-sar-ko'-mah*). A sarcoma having some of the structural elements of a lymphatic gland. 1. **malignum multiplex.** Same as *lymphadenoma*.

**lymphosarcomatosis** (*lim-fo-sar-ko-mat-o'-sis*) [*lympho-*; σάρκωμα, fleshy mass]. A condition or diathesis marked by the development of lymphosarcoma; also, the process of such development.

**lymphosis** (*lim-fo'-sis*) [*lympha*, lymph]. The elaboration of lymph.

**lymphostasis** (*lim-fos'-tas-is*) [*lympho-*; στάσις, a placing]. Stasis or stoppage of the flow of lymph.

**lymphotome** (*lim'-fo-tōm*) [*lympho-*; τέμνειν, to cut]. An instrument, on the principle of the tonsillotome, with a flexible cutting-blade for removing adenoids.

**lymphotomy** (*lim-fot'-o-me*). See *lymphangiotomy*.

**lymphotorrhea** (*lim-fot-or-e'-ah*) [*lympho-*; οὖς, ear; ῥοία, a flow]. A serous or watery discharge from the ear.

**lymphotoxemia** (*lim-fo-toks-e'-me-ah*). See *status lymphaticus*.

**lymphotoxic** (*lim-fo-toks'-ik*). Pertaining to or characteristic of a substance having toxic action on the lymphatic tissue.

**lymphotoxin** (*lim-fo-toks'-in*) [*lympho-*; τοξικόν, poison]. A cytotoxin having specific action on lymphatic tissue.

**lymphotrophy** (*lim-fot'-ro-fe*) [*lympho-*; τροφή, nourishment]. Nourishment of the cells by the lymph in regions of imperfect vascularization.

**lymphous** (*lim'-fus*). Relating to, containing, or consisting of lymph.

**lymphuria** (*lim-fū'-re-ah*) [*lympha*, lymph; οὖρον, urine]. A condition in which the urine spontaneously coagulates, but contains no fat, as it does in chyluria.

**lynx** (*links*) [λύγξ, a sobbing]. Hiccough; violent sobbing.

**lypemania** (*li-pe-ma'-ne-ah*) [λύπη, sadness; μανία, madness]. A form of dementia accompanied by profound mental depression and refusal to take food.

**lyperophrenia** (*li-per-o-fre'-ne-ah*) [λυπηρός, distressing; φρήν, mind]. Melancholia.

**lypothymia** (*li-po-thi'-me-ah*) [λύπη, sadness; θυμός, mind]. Melancholia; severe mental prostration from grief. This condition is not to be confounded with *leipothymia*, q. v.

**lyptol** (*lip'-tol*). An ointment said to consist of mercuric chloride, eucalyptus oil, formaldehyde, and benzoboric acid. It is used as a dressing for ulcers, cutaneous diseases, etc.

**lyra** (*li'-rah*). Certain longitudinal, transverse and oblique lines on the inferior surface of the fornix, the arrangement of which bears a fanciful resemblance to a lyre.

**lysargin** (*li-sar'-jin*). Colloidal silver, used as an antiseptic.

**lysatin** (*lis'-at-in*), $C_6H_{13}N_3O_2$. An alkaloid from casein.

**lysatinin** (*lis-at'-in-in*), $C_6H_{11}N_3O$ or $C_6H_{13}N_3O_2$. A mixture of equal molecules of arginin and lysin.

**lysemia** (*li-se'-me-ah*) [λύσις, solution; αἷμα, blood]. A dissolution of the blood, or a losing of the integral parts of it.

**lysidine** (*li'-sid-ēn*), $C_4H_8N_2$. A base obtained from dry distillation of sodium acetate with ethylene diamine hydrochloride. It is recommended in cases of uric-acid diathesis. Dose 15–75 gr. (1–5 Gm.) of the crystals daily in a pint of cold water. Syn., *ethylene-ethenyldiamine; methylglyoxalidin*.

**lysimeter** (*li-sim'-et-er*) [λύσις, solution; μέτρον, measure]. An apparatus for determining the solubility of a substance.

**lysin** (*li'-sin*) [λύειν, to loose]. 1. $C_6H_{14}N_2O_2$. A histon base discovered among the cleavage-products of casein and produced by the tryptic digestion of fibrin. 2. A cell-dissolving substance found in the blood-serum.

**lysinosis** (*lis-in-o'-sis*) [λύσις, λύειν, to loose (a ravelling or shred); νόσος, a disease]. A disease of the lungs due to the inhalation of cotton-fibers.

**lysis** (*li'-sis*) [ see *lysin*]. 1. The gradual decline of a disease, especially of a fever. 2. The action of a lysin.

**lysoform** (*li'-so-form*). A combination of lysol and formaldehyde, soluble in water and alcohol. It is used as an antiseptic in 1 to 3 % solutions.

**lysogen** (*li'-so-jen*) [*lysin*; γεννᾶν, to produce]. A substance or body which produces a lysin.

**lysogenic** (*li-so-jen'-ik*) [*lysis*; γεννᾶν, to produce]. Giving rise to lysins or producing lysis.

**lysol** (*li'-sol*) [λύειν, to loose]. A brown liquid substance obtained by boiling tar-oils with alkalies and fats. It is used as an antiseptic in surgery, in lupus, gonorrhea, and as a gargle in sore throat.

**lysosolveol** (*lis-o-sol'-ve-ol*). A disinfectant said to be a mixture of potassium linoleate, cresols, and water.

**lyssa** (*lis'-ah*) [λύσσα, madness]. A synonym of *hydrophobia* or *rabies*.

**lyssic** (*lis'-ik*) [λύσσα, madness]. Pertaining to rabies; due to rabies.

**lyssin** (*lis'-in*) [*lyssa*]. The specific virus of hydrophobia.

**lyssodexis** (*lis-o-deks'-is*) [λύσσα, madness; δάκνειν, to bite]. The bite of a rabid dog.

**lyssoid** (*lis'-oid*) [λύσσα, rabies; εἶδος, like]. Resembling rabies; resembling madness.

**lyssophobia** (*lis-o-fo'-be-ah*) [*lyssa*; φόβος, fear]. Morbid dread of rabies; pseudohydrophobia.

**lysulfol** (*li-sul'-fol*). Trade name of a compound of sulphur (10 %) and lysol forming a black mass. It is used in skin diseases.

**lyterian** (*li-te'-re-an*). Indicative of a lysis, or of a favorable crisis, terminating an attack of disease.

**lytic** (*lit'-ik*) [*lysis*]. 1. Relating to a lysis or to a solution. 2. Relating to a lysin.

**lytta** (*lit'-ah*). Synonym of *hydrophobia*.

**Lytta** (*lit'-ah*). A genus of vesicant coleoptera established by Fabricius. **L. vesicatoria.** See *cantharides*. **L. vittata**, the potato-fly, containing one per cent. or more of cantharidin.

# M

**M.** The abbreviation of *musculus, myopia, myopic, mille,* and *misce* (mix), as a numeral it represents the number 1000.
**m.** An abbreviation for *meter,* and *minim.*
**m-.** An abbreviation in chemistry for *meta-.*
**μ.** Greek equivalent of *m.* Used as abbreviation for a micron.
**M.A.** Abbreviation for Master of Arts.
**ma.** An abbreviation of milliampere.
**M+Am.** Abbreviation for compound myopic astigmatism.
**mabi** (*mor'-be*). A beverage, common in the West Indies, prepared from the leaves of the *Colubrina reclinata.*
**Mac., Mc., M'.,** beginning proper names will be found as if spelled Mac, and in strict alphabetical order.
**mac.** An abbreviation of *macera,* macerate.
**macaco worm** (*mah-kah'-ko wurm*) [from a Malagasy name]. The larva of a S. American fly, *Dermatobia noxialis,* which infests the skin of men and animals.
**macaja,** or **macaya butter.** The solid oil obtained from the fruit of the macaw-palm, *Acrocomia sclerocarpa.*
**macalline** (*mak-al'-ēn*). An amorphous alkaloid without taste, from macallo-bark, sparingly soluble in amyl alcohol and in chloroform, insoluble in alcohol, ether, or water.
**macaroni** (*mak-ar-o'-ne*) [Ital.]. 1. Slender tubes made of flour-paste; a favorite dish in Italy and also in the U. S. 2. An active purgative used in leadcolic. Antimony sulphide is one of the ingredients.
**McBurney's point** [Charles *McBurney,* American surgeon, 1845–1913]. The point of tenderness in appendicitis, five or six centimeters above the right anterior superior iliac spine, on a line drawn from this point to the umbilicus.
**McClintock's rule** [Alfred Henry *McClintock,* Irish physician, 1822–1881]. A pulse of 100 or more beats a minute, after parturition, indicates impending postpartum hemorrhage.
**McDonald's solution** [Ellice *McDonald,* American gynecologist]. An antiseptic solution consisting of acetone (commercial) 40, denatured alcohol 60, and pyxol 2 parts.
**Macdowel's frenum.** The intermuscular expansions given off by the posterior layer of the tendon of the pectoralis major.
**mace** (*mās*). A spice derived from the dried covering of the nutmeg, *q. v.*
**macene** (*mās'-ēn*) [*macis,* mace], $C_{10}H_{18}$. An essential oil resulting from the distillation of the flowers of nutmeg. The hydrochloride on distillation yields a camphor.
**maceration** (*mas-er-a'-shun*) [*macerare,* to make soft]. The process of softening a solid substance, or of converting into a soft mass by soaking in a liquid, as *maceration* of the fetus.
**macerator** (*mas'-er-a-tor*) [*macerare,* to make soft]. A vessel used for macerating a substance.
**Macewen's osteotomy** (*mak-ū'-en*) [Sir William *Macewen,* Scotch surgeon, 1848– ]. Supracondyloid division of the femur from the inner side for genu valgum. **M.'s space.** See *M.'s triangle.* **M.'s symptom,** increased resonance on combined percussion and auscultation of the skull in certain gross lesions of the intracranial contents—*e. g.,* in cerebral abscess or overdistended lateral ventricles. **M.'s triangle,** the suprameatal triangle; the triangular space bounded by the upper half of the posterior wall of the external auditory meatus, by the supramastoid crest, and by an imaginary line dropped from the latter at the level of the posteroinferior wall of the external meatus. It is the space selected for trephining in cases of otitic abscess of the temporosphenoidal lobe.
**Mache unit** (*mah'-keh*) [Heinrich *Mache,* Austrian physicist, 1876– ]. A term used to express the concentration of radium emanations. It is the saturation ionization current due to the radium emanation from a liter of solution or gas, expressed in electrostatic units multiplied by 1000.
**macies** (*ma'-se-ēz*) [*macies,* a wasting]. Atrophy, leanness, wasting. **m. infantum.** Synonym of *tabes mesenterica.*
**macilent** (*mas'-il-ent*) [*macilentus,* lean]. Meager; thin; lean.
**macis** (*ma'-sis*). Same as mace; see *nutmeg.*
**Mackenzie's eye-lotion.** Corrosive sublimate, 1 gr.; ammonium chloride, 6 gr.; cochineal, ½ gr.; alcohol, 1 dr.; water, 8 oz.
**mackintosh** (*mak'-in-tosh*) [Charles *Mackintosh,* the inventor]. A fabric of silk or cotton, rendered waterproof and airproof by a coating of India rubber; it is used in antiseptic surgery and in obstetrics.
**maclayin** (*mak-la'-in*), $C_{17}H_{25}O_{11}$. A powerful local irritant said to be a glucoside from *Bassia maclayana.*
**McLeod's capsular rheumatism** (*mak-lowd'*) [Roderick *McLeod,* Scotch physician, 1795–1852]. Rheumatoid arthritis attended with considerable effusion into the synovial sacs, sheaths, and bursas.
**MacMunn's test for indican in the urine.** Boil equal parts of urine and hydrochloric acid and a few drops of nitric acid; cool, and shake with chloroform. The chloroform becomes violet and shows an absorption band before D, due to indigo-blue, and one after D, due to indigo-red.
**macradenous** (*mak-rad'-en-us*) [μακρός, large; ἀδήν, gland]. Having large glands.
**macrencephalic, macrencephalia** (*mak-ren-sef-al'-ik, mak-ren-sef'-al-us*) [μακρός, long; ἐγκέφαλος, brain]. Having a large or long skull.
**macrencephalus** (*mak-ren-sef'-al-us*) [μακρός, large; ἐγκέφαλος, brain]. Lissauer's term for a skull in which the angle formed by the junction of the lines drawn from the hormion to the nasion and to the inion is between 156.5° and 170°.
**macrocranus** (*mak-ren-kra'-nus*) [μακρός, large; ἐν, in; κρανίον, skull]. Lissauer's term for a skull having a large cerebellar sector (from 20° to 27.5°).
**macritas** (*mak'-rit-as*) [*macer,* lean]. Emaciation, leanness.
**macro-** (*mak-ro-*) [μακρός, large]. A prefix meaning large, long, or great.
**macrobacteria** (*mak-ro-bak-te'-re-ah*) [*macro-; bacterium*]. Bacteria of very large size. Same as *megabacteria.*
**macrobiosis** (*mak-ro-bi-o'-sis*) [*macro-;* βίος, life]. Longevity.
**macrobiotic** (*mak-ro-bi-ot'-ik*) [*macro-;* βίος, life]. Pertaining to long life; long-lived.
**macroblast** (*mak'-ro-blast*). See *megaloblast.*
**macrobrachia** (*mak-ro-bra'-ke-ah*) [*macro-;* βραχίων, arm]. Abnormal size of the arms.
**macrocephalia, macrocephaly** (*mak-ro-sef-a'-le-ah, mak-ro-sef'-al-e*) [*macro-;* κεφαλή, head]. Abnormal largeness of the head.
**macrocephalous** (*mak-ro-sef'-al-us*) [see *macrocephalia*]. Characterized by an abnormally large head.
**macrocephalus** (*mak-ro-sef'-al-us*) [*macro-;* κεφαλή, the head]. A fetus with excessive development of the head.
**macrocephaly.** See *macrocephalia.*
**macrocheilia** (*mak-ro-ki'-le-ah*) [*macro-;* χεῖλος, lip]. Excessive development of the lips, a characteristic of certain negro tribes. It occurs also in cretinoid states, when the lips and cheeks are the seat of lymphangioma.
**macrocheiria** (*mak-ro-ki'-re-ah*) [*macro-;* χείρ, hand]. Great enlargement of the hands.
**macrochemistry** (*mak-ro-kem'-is-tre*) [*macro-; chemistry*]. Chemistry in which the reactions are observable with the naked eye. Cf. *microchemistry.*

**macrococcus** (*mak-ro-kok'-us*) [*macro-*; κόκκος, a berry: *pl.*, *macrococci*]. A term applied to cocci that are larger than the average.
**macrocoelia** (*mak-ro-ko'-le-ah*) [*macro-*; κῶλον, limb]. The possession of long limbs.
**macrocolous** (*mak-rok'-o-lus*) [*macro-*; κῶλον, limb]. Having long limbs.
**macrocoly** (*mak-ro-ko'-le*) [*macro-*; κῶλον, colon]. Simple excessive length of the colon.
**macrocomous** (*mak-ro-ko'-mus*) [*macro-*; κόμη, the hair]. Having long hairs or filaments.
**macrocornea** (*mak-ro-kor'-ne-ah*) [*macro-*; *cornea*]. Keratoglobus.
**macrocosm** (*mak'-ro-kozm*) [*macro-*; κόσμος, world]. The world or cosmos, in contradistinction to man, the microcosm, or little universe.
**macrocosmic** (*mak-ro-koz'-mik*) [*macro-*; κόσμος, world]. Pertaining to the macrocosm.
**macrocyst** (*mak'-ro-sist*) [*macro-*; κύστις, a cyst]. An abnormally large cyst.
**macrocytase** (*mak-ro-si'-tase*). An enzyme found in leukocytes and capable of digesting cells and other elements of animal origin.
**macrocyte** (*mak'-ro-sīt*) [*macro-*; κύτος, cell]. 1. A giant red blood-corpuscle found in the blood in certain anemias, especially pernicious anemia. 2. A large lymphocyte.
**macrocythemia** (*mak-ro-si-the'-me-ah*). The presence of macrocytes in the blood.
**macrocytosis** (*mak-ro-si-to'-sis*) [*macro-*; κύτος, cell]. The formation of macrocytes.
**macrodactylia** (*mak-ro-dak-til'-e-ah*) [*macro-*; δάκτυλος, finger]. An abnormally great length, or size, of fingers or toes.
**macrodactylism** (*mak-ro-dak'-til-izm*). See *macrodactylia*.
**macrodactyly** (*mak-ro-dak'-til-e*). See *macrodactylia*.
**macrodontia** (*mak-ro-don'-she-ah*) [*macro-*; ὀδούς, tooth]. Abnormally large teeth.
**macroesthesia** (*mak-ro-es-the'-ze-ah*) [*macro-*; αἴσθησις, sensation]. A disturbance of the tactile and stereognostic sense in consequence of which objects touched or handled appear much larger than they really are.
**macrogamete** (*mak-ro-gam'-ēt*) [*macro-*; γαμέτη, a wife]. The mature female cell in propagative reproduction in sporozoa.
**macrogametocyte** (*mak-ro-gam-et'-o-sīt*). The enlarged merozoite before maturation into the female cell in propagative reproduction in sporozoa.
**macrogamy** (*mak-rog'-am-e*). Conjugation of two adult protozoan cells.
**macrogastria** (*mak-ro-gas'-tre-ah*) [*macro-*; γαστήρ, belly]. Dilatation of the stomach.
**macrogastrous** (*mak-ro-gas'-trus*) [*macro-*; γαστήρ, stomach]. Having a large stomach or belly.
**macrogenesis** (*mak-ro-jen'-es-is*) [*macro-*; γεννᾶν, to beget]. Excessive development of an organ or part.
**macroglossia** (*mak-ro-glos'-e-ah*) [*macro-*; γλῶσσα, tongue]. Enlargement of the tongue, a condition seen in cretins, in whom it is probably due to lymphangioma.
**macrognathic** (*mak-rog-nāth'-ik*) [*macro-*; γνάθος, jaw]. Having long jaws; prognathous.
**macromania** (*mak-ro-ma'-ne-ah*) [*macro-*; μανία, madness]. A mania characterized by the delusion that objects are larger than they really are; or that one's own body or members are much larger than they are.
**macromastia** (*mak-ro-mas'-te-ah*) [*macro-*; μαστός, breast]. Abnormal enlargement of the breast.
**macromazia** (*mak-ro-ma'-ze-ah*). See *macromastia*.
**macromelia** (*mak-ro-me'-le-ah*) [*macro-*; μέλος, organ or member]. The excessive development of any organ or member.
**macromelus** (*mak-rom'-el-us*) [*macro-*; μέλος, organ or member]. 1. Same as *macromelia*, *q. v.* 2. One having excessively large limbs.
**macromere** (*mak'-ro-mēr*) [*macro-*; μέρος, part]. A large blastomere.
**macromerozoite** (*mak-ro-me-ro-zo'-it*) [*macro-*; *merozoite*]. A large merozoite.
**macronosia** (*mak-ro-no'-se-ah*) [*macro-*; νόσος, illness]. A protracted or chronic disease.
**macronucleus** (*mak-ro-nu'-kle-us*). A large nucleus.
**macronychia** (*mak-ro-nik'-e-ah*) [*macro-*; ὄνυξ, nail]. Excessive size of the nails.
**macropathology** (*mak-ro-path-ol'-o-je*) [*macro-*; *pathology*]. Pathology which includes no microscopic investigation.
**macrophage** (*mak'-ro-fāj*) [*macro-*; φαγεῖν, to devour]. A large phagocyte.
**macrophagocyte** (*mak-ro-fag'-o-sīt*) [*macro-*; φαγεῖν, to devour; κύτος, a cell]. A large-sized phagocyte, possessed of a single nucleus, and derived from some fixed connective-tissue element.
**macrophallus** (*mak-ro-fal'-us*) [*macro-*; φαλλός, penis]. A penis abnormally large or long.
**macropharynx** (*mak-ro-far'-ingks*) [*macro-*; *pharynx*]. A large pharynx.
**macrophonous** (*mak-rof'-o-nus*) [*macro-*; φωνή, sound]. Loud-voiced.
**macrophotograph** (*mak-ro-fo'-to-graf*) [*macro-*; *photograph*]. A large photograph, *i. e.*, a macroscopic photograph of an object, whether the object is small or large.
**macroplasia** (*mak-ro-pla'-ze-ah*) [*macro-*; πλάσις, a moulding]. Excessive development of portions of the body.
**macropodia** (*mak-ro-po'-de-ah*) [*macro-*; πούς, foot]. Excessive size of the feet.
**macropomous** (*mak-rop'-o-mus*) [*macro-*; πῶμα, a lid]. Possessing a large operculum.
**macroporous** (*mak-rop'-or-us*) [*macro-*; πόρος, a passage]. Having large pores.
**macroprosopus** (*mak-ro-pro-so'-pus*) [*macro-*; πρόσωπον, face]. A monster with abnormal development of the face.
**macropsia** (*mak-rop'-se-ah*). See *megalopsia*.
**macrorhinia** (*mak-ro-rin'-e-ah*) [*macro-*; ῥίς, nose]. Congenital hypertrophy of the nose.
**macroscelia** (*mak-ro-se'-le-ah*) [*macro-*; σκέλος, leg]. Excessive development of the legs.
**macroscopic** (*mak-ro-skop'-ik*) [*macro-*; σκοπεῖν, to see]. Large enough to be seen by the naked eye; gross; not microscopic.
**macrosis** (*mak-ro'-sis*) [μακρός, large]. A state of increase in volume.
**macrosmatic** (*mak-roz-mat'-ik*) [*macro-*; ὀσμάεσθαι, to smell]. Having well-developed olfactory organs.
**macrosomia, macrosomatia** (*mak-ro-so'-me-ah*, *mak-ro-so-ma'-she-ah*) [*macro-*; σῶμα, body]. Excessive size of the body.
**macrospore** (*mak'-ro-spōr*) [*macro-*; *spora*, seed]. In biology: (*a*) a spore of relatively large size; (*b*) one of the larger anisospores arising in the reproduction of colony-forming Radiolarians.
**macrostomia** (*mak-ro-stō'-me-ah*) [*macro-*; στόμα, mouth]. Excessive size, or width, of the mouth, or of the oral fissure.
**macrotia** (*mak-ro'-she-ah*) [*macro-*; οὖς, the ear]. Excessive length of the ears.
**macrotin** (*mak'-ro-tin*) [*macrotys*, cimicifuga]. A resin obtained by the precipitation of tincture of cimicifuga with water.
**macula** (*mak'-ū-lah*) [L.: *pl.*, *maculae*]. A macule. **macula acusticae**, the termination of the auditory nerve in the saccule and utricle. **m. arcuata**, the arcus senilis. **macula atrophicae cutis**, linear atrophy of the skin. **m. corneae**, a permanent corneal opacity from an ulcer or keratitis. **m. cribrosa**, a name for the perforations of the fossa hemisphaerica of the inner ear, for the passage of the filaments of the auditory nerve. **m. flava**, the yellow spot of the retina. **m. lutea**, the yellow spot of the retina. It is the point of clearest vision. **m. solaris**, a freckle. **m. tendineae**, thin white patches of new fibrous tissue found on the surface of the pericardium in some cases of pericarditis.
**macular** (*mak'-ū-lar*) [*macule*]. Characterized by or resembling macules.
**maculate** (*mak'-ū-lāt*). Spotted.
**maculation** (*mak-ū-la'-shun*) [*macule*]. The state or quality of being spotted; the formation of macules.
**macule** (*mak'-ūl*). A spot, especially one upon the skin, not elevated above the surrounding level. See also *macula*.
**maculopapular** (*mak-ū-lo-pap'-u-lar*). Having the characteristics of a macule and a papule.
**MacWilliam's test for albumin** [John Alexander *MacWilliam*, English physician, 1857– ]. To 20 Cc. of the liquid add a drop or two of a saturated solution of salicyl-sulphonic acid; in the presence of albumin a cloudiness or precipitate will be formed. If peptones or albumoses are present, this precipitate disappears on boiling, but reappears on cooling.
**mad.** 1. Insane. 2. Affected with rabies; rabid.

**madar** (*mad'-ar*). The plant *Calotropis gigantea*; the bark and root are used as a dye.

**madarosis** (*mad-ar-o'-sis*) [μαδάρωσις, a making bald]. Loss of the eyelashes or eyebrows.

**madarotic, madarous** (*mad-ar-ot'-ik, mad'-ar-us*) [μαδαρός, bald]. Affected with or relating to madarosis.

**madder** (*mad'-er*). See *rubia*.

**Maddox test,** or **Maddox glass-rod test** (*mad'-uks*) [Ernest Edmond *Maddox*, English ophthalmologist]. A test of heterophoria by means of a short cylinder of transparent glass about one-eighth of an inch long, fitted into a slot in an opaque disc to be set in the trial-frame before one eye, with a colored plano lens before the other eye. The rod converts the image of a distant flame into a thin line of light. The relative position of the two images thus formed permits the measure of imbalance of the muscles.

**madefaction** (*mad-e-fak'-shun*) [*madefacere*, to moisten]. The act of moistening.

**madeira** (*mad-a'-rah*).. A fine sherry wine from the island of Madeira.

**Madelung's deformity** (*mad'-el-oong*) [Otto Wilhelm *Madelung*, German surgeon, 1846-    ]. Progressive subluxation of the wrist joint, caused by relaxation of the ligaments or abnormality in the radial epiphysis. **M.'s neck,** diffuse lipoma of the neck. Syn., *Madelung's fetthals.*

**madema** (*mad-e'-mah*). See *madarosis.*

**madescent** (*mad-es'-ent*) [*madescere*, to become wet]. Becoming moist.

**madesis** (*mad-e'-sis*). Synonym of *madarosis.*

**madidans** (*mad'-id-ans*) [L.]. Weeping, oozing. See *eczema madidans.*

**madisterion, madisterium** (*mad-is-ter'-e-on, mad-is-ter'-e-um*) [μαδίζειν, to pluck bare]. Epilating-forceps.

**madness** (*mad'-nes*). See *insanity* and *mania.*

**madreporic, madreporiform** (*mad-re-por'-ik, -e-form*) [*Madrepora*, a genus of corals]. Pierced with minute openings.

**madstone** (*mad'-stōn*). A small stone, believed to have the power of absorbing poison from wounds.

**Madura-foot** (*ma-doo'-rah-foot*) [*Madura*, a district in India]. A disease occurring chiefly in India, and characterized by the formation, on the foot (sometimes on the hand), of a tender purplish swelling, which in time suppurates, the pus being evacuated through one or more sinuses. These sinuses discharge a seropurulent liquid containing peculiar bodies resembling gunpowder-grains or fish-roe. The disease is chronic, but remains local, and is believed to be due to an organism allied to the Actinomyces fungus. Syn., *mycetoma.*

**mageiric** (*maj-i'-rik*) [μαγειρικός, fit for cookery]. Relating to dietetics or the culinary art.

**Magendie's foramen** (*ma-zjon'-de*) [François *Magendie*, French physiologist, 1783-1855]. A foramen of communication between the fourth ventricle and the subarachnoid space at the tip of the calamus scriptorius. **M.'s law,** See *Bell's law.* **M.'s solution,** a solution of morphine sulphate, 16 grains to the ounce. **M.'s spaces,** imperfectly closed lymph-spaces formed by the separation of the arachnoid from the pia and corresponding to the cerebral sulci.

**magenta** (*ma-jen'-tah*) [from *Magenta*, Italy]. A coal-tar dye from which fuchsin and a large number of other dyes are prepared.

**magistery** (*maj-is'-ter-e*) [*magisterium*, masterpiece]. Formerly, a preparation considered to have especial virtue as a remedy. **m. of bismuth,** the subnitrate of bismuth. **m. of tin** (*M. Jovis*), precipitated stannous oxide.

**magistral** (*maj'-is-tral*) [*magister*, a master]. Applied to medicines prepared on prescription.

**magma** (*mag'-mah*) [μάγμα, mass]. Sediment; dregs; any pulpy mass.

**magnalium** (*mag-na'-le-um*). An alloy of magnesium and aluminum; specific gravity 2 to 3, melting-point 600° to 700° C. It is similar in quality to brass and bronze when the quantity of magnesium varies from 5 % to 30 %.

**Magnan's sign** (*man'-yan*) [Valentin *Magnan*, French neurologist, 1835-    ]. A hallucination of general sensation which takes the form of the sensation of a round foreign body beneath the skin; it is noted in chronic cocainism.

**magnesia** (*mag-ne'-ze-ah*) [from *Magnesia*, a district in Thessaly]. Magnesium oxide, MgO. **m., black,**
black oxide of manganese. **m., calcined.** See *magnesium carbonate*. **·m., white,** magnesium carbonate. See also *magnesium.*

**magnesic** (*mag-ne'-zik*). Pertaining to or containing magnesium.

**magnesite** (*mag'-nez-īt*). Native magnesium carbonate.

**magnesium** (*mag-ne'-ze-um*) [see *magnesia*]. Symbol Mg; atomic weight .24.32; quantivalence II; specific gravity 1.75. A bluish-white metal of the group to which calcium and barium belong. It is abundantly distributed throughout inorganic and organic nature; its salts are used in the arts and in medicine. The source of magnesium and its salts is chiefly the minerals dolomite and kieserite. **m. acetate,** $Mg(C_2H_3O_2)_2+4H_2O$, cathartic. Dose 5-60 gr. (0.32-4.0 Gm.). **m. benzoate,** $Mg(C_7H_5O_2)_2$; used in gout. Dose 3-20 gr. (0.2-1.3 Gm.). **m. bisulphate,** $MgH_2(SO_4)_2$; cathartic. Dose 5-20 gr. (0.32-1.3 Gm.). **m. borate,** $Mg(BO_2)_2 . 2Mg(OH)_2+7H_2O$; antiseptic. Dose 5-20 gr. (0.32-1.3 Gm.). **m. borocitrate,** a compound of citric acid, magnesium carbonate, and borax; used in lithiasis, gout, etc. Dose 15-30 gr. (1-2 Gm.). **m. bromide,** $MgBr_2+6H_2O$; sedative. Dose 10-20 gr. (0.65-1.3 Gm.). **m. carbonate** (*magnesii carbonas*, U. S. P.), $(MgCO_3)_4 . Mg(OH)_2+5H_2O$; exists in two forms—as light (*magnesii carbonas levis*) and as heavy magnesium carbonate (*magnesii carbonas ponderosa*). It is antacid, laxative, and antilithic. Dose 30-60 gr. (2-4 Gm.). **m. carbonate, solution of** (*liquor magnesii carbonatis,* B. P.). Dose 1-2 oz. (32-64 Cc.). **m. chloride,** $MgCl_2+6H_2O$; aperient and cathartic. Dose 240-465 gr. (16-30 Gm.). **m. citrate, effervescent,** is cathartic. Dose 4 dr. (4-12 Cc.). **m. citrate, solution of** (*liquor magnesii citratis,* U. S. P.). Dose 4-8 oz. (128-256 Cc.). **m. copaivate,** antiseptic, diuretic, laxative, and stimulant. Dose 10-20 gr. (0.65-1.3 Gm.). **m. creosotate.** See *creasote-magnesia.* **m. dioxide,** $MgO_2$, used in anemia. Syn., *biogen.* **m. ergotate,** used in amenorrhea and epilepsy. Dose ⅔-1 gr. (0.04-0.065 Gm.). **m. fluoride,** $MgF_2$, an antiseptic. **m. hydrate, m. hydroxide,** $Mg(OH)_2$. It is antacid, antilithic, and cathartic. Dose 60-120 gr. (4-8 Gm.). **m. hypophosphite,** $Mg(H_2PO_2)_2+6H_2O$, a nerve-stimulant. Dose 10-20 gr. (120 Cc.). **m. ichthyolate,** a combination of freshly calcined magnesia, 100 parts; ichthyol, 775 parts. Mixed with talc it is used as an antiseptic dusting-powder. **m. iodide,** $MgI_2$, alterative and sialagogue. Dose 2-10 gr. (0.13-0.65 Gm.). **m. lactate,** $Mg(C_3H_5O_3)_2+3H_2O$; a laxative. Dose 15-45 gr. (1-3 Gm.). **m. oxide** (*magnesii oxidum,* U. S. P.), MgO, is obtained by calcining magnesium carbonate, and exists in two forms—as light magnesia and as heavy magnesia (*magnesii oxidum ponderosum,* U. S. P.). It is used as an antacid and laxative, as a counter-poison, and as an antidote to arsenic. Dose 10 gr.-1 dr. (0.65-4.0 Gm.). **m. oxide, ferric hydroxide with** (*ferri hydroxidum cum magnesii oxido,* U. S. P.). Dose as arsenical antidote 4 oz. (120 Cc.). **m. oxide, heavy.** See under *m. oxide.* **m. phenolsulphonate,** an antiseptic purgative. Dose 15-30 gr. (1-2 Gm.). **m. salicylate,** $Mg(C_7H_5O_3)_2+H_2O$, an intestinal antiseptic and antirheumatic. Dose 15-120 gr. (1-8 Gm.). **m. sclerotinate.** See **m. ergotate.** **m. silicate,** $Mg_2Si_2O_5+2H_2O$, absorbent, astringent, and antiseptic. Dose 60-150 gr. (4-10 Gm.). **m. sozoiodolate,** $(C_6H_2I_2(OH)SO_3)_2Mg+8H_2O$, an antiseptic. **m. sulphate** (*magnesii sulphas,* U. S. P.), $MgSO_4+7H_2O$, Epsom salt, is an active cathartic, especially useful in inflammatory affections. Dose 1 dr.-1 oz. (4-32 Gm.). **m. sulphate, effervescent** (*magnesii sulphas effervescens,* U. S. P.), cathartic. Dose 2 dr.-1 oz. (8-32 Gm.). **m. sulphophenate.** See **m. phenolsulphonate.** **m. tartrate,** $MgC_4H_4O_6$, is cathartic and used in diseases of the urinary tract with neuralgic symptoms. Dose 8-15 gr. (0.52-1.0 Gm.).

**magnet** (*mag'-net*) [Μάγνησσα, stone of *Magnesia*, in Thessaly, where lodestone was first found]. 1. Lodestone, a magnetic iron oxide. 2. A body having the power to attract iron bodies. **m., electro-.** See *electromagnet.* **m., giant, m., Haab,** a large powerful stationary magnet for extracting particles of steel from the eye. **m., horseshoe,** an iron magnet having the shape of a horse-shoe. **m. operation,** the operation of removing foreign bodies of steel from the eye by means of a magnet. **m., permanent,** one the magnetic properties of which are permanent,

in contradistinction to a *temporary magnet*. m., **temporary**, one which derives its magnetism from another magnet or from a galvanic current.
**magnetic** (*mag-net'-ik*) [*magnet*]. Pertaining or belonging to a magnet. Possessing the property of magnetism.
**magnetism** (*mag'-net-izm*) [*magnet*]. The power possessed by a magnet to attract or repel other masses. m., animal, hypnotism.
**magnetization** (*mag-net-iz-a'-shun*) [*magnet*]. The process of rendering a substance magnetic.
**magnetoelectricity** (*mag-net-o-e-lek-tris'-it-e*) [*magnet; electricity*]. Electricity produced by means of a magnet.
**magnetograph** (*mag-net'-o-graf*) [*magnet;* γράφειν, to write]. An instrument for determining the intensity of magnetic action.
**magnetoinduction** (*mag-net-o-in-duk'-shun*) [*magnet; inductio*, induction]. The production of an induced current by the insertion of a magnet within a coil of wire.
**magnetometer** (*mag-net-om'-et-r*) [*magnet*; μέτρον, a measure]. A series of magnets suspended so as to record graphically variations in direction and intensity of magnetic force.
**magneto-optic** (*mag-net-o-op'-tik*). Relating to magnetism and light.
**magnetotherapy** (*mag-net-o-ther'-ap-e*) [*magnet;* θεραπεία, treatment]. The treatment of diseases by magnets.
**magniductor** (*mag-ne-duk'-tor*) [*magnus*, great; *ductor*, a leader]. The adductor magnus of the thigh.
**magnification** (*mag-nif-ih a' shun*) [*magnus*, large; *facere*, to make]. Enlargement, especially the enlargement of the image of an object by means of lenses.
**magnifying** (*mag'-nif-i-ing*) [see *magnification*]. Enlarging; making greater. m. power, the power of a lens to increase the diameters of the image of an object.
**magnum**, m., os. See *bones, table of*.
**mahamari** (*mah-hah-mah'-re*) [E. Ind.]. Synonym of the *plague, q. v.*
**Maher's disease** (*mah'-er*). Paracolpitis.
**Mahler's sign** (*mah'-ler*) [Richard A. *Mahler*, German obstetrician]. A gradual increase of pulserate in the puerperium, without rise of temperature, is characteristic of venous thrombosis.
**maidalakri** (*mi-dal-ak'-re*). The bark of *Litsea salicifolia*, used in the East Indies in diarrhea.
**maidenhead**. 1. Virginity. 2. The hymen.
**maidismus** (*ma-id-iz'-mus*) [*mays, maidis*, maize]. Maize-poisoning, or pellagra.
**Maier's sinus** (*mi'-er*) [Rudolf *Maier*, German physician, 1824-1888]. A small, infundibular depression in the wall of the lacrimal sac near the opening of the lacrimal ducts.
**maieusiomania** (*ma-ū-se-o-ma'-ne-ah*) [μαίευσις, delivery; μανία, madness]. Puerperal insanity.
**maieusiophobia** (*ma-ū-se-o-fo'-be-ah*) [μαίευσις, childbirth; φόβος, fear]. Morbid dread of childbirth.
**maieutics** (*ma-ū'-tiks*) [μαίευτης, an obstetrician; μαιευτική, obstetrics]. Midwifery; obstetrics.
**maim** (*mām*) [OF., *mehaigner*, to maim]. To cripple by injury or removal of a limb.
**main-en-griffe** (*mang-on(g)-grēf*). See *claw-hand*.
**m. succulente**, edema of the hands.
**maintenance** (*mān'-ten-ans*) [*manus*, hand; *tenere*, to hold]. The relationship which exists between increment and excrement, after a body has reached maturity.
**Maisonneuve's bandage** (*ma-zon-nerv'*) [Jacques Gilles Thomas *Maissonneuve*, French surgeon, 1809-1897]. A variety of plaster-of-Paris bandage formed from cloths folded, these being supported by other bandages.
**Maissiat's band** (*ma-se-ah'*) [Jacques *Maissiat*, French anatomist, 1805-1878]. The iliotibial ligament, a fibrous band in the fascia lata that extends from near the anterior superior spine of the ilium to the outer tuberosity of the tibia.
**maize** (*māz*) [W. Ind., *mahiz*]. Indian corn. See *zea mays*.
**maizole** (*ma'-zōl*). Trade name of an emulsion of cod oil, suggested as a substitute for cod-liver oil.
**maizolithium** (*ma-zo-lith'-e-um*). A diuretic and sedative said to consist of a combination of maizenic acid from cornsilk and lithium. Dose 1-2 dr. (4-8 C.c.).

**make** (*māk*) [AS., *mācian*, to make]. In electricity—(1) to establish the flow of an electric current; (2) the establishing of the flow of an electric current.
**makro-** (*mak'-ro-*). See *macro-*.
**mal** [Fr., from *malum*, evil; disease]. Disease. **m. de caderas**, a disease of horses, mules, and swine in South America, characterized by fever, emaciation, and general paresis which first appears in the hind legs. **m. de coit**. See *dourine*. **m. de mer**, seasickness. **m. des bassines**, a dermatitis affecting those engaged in winding silkworm cocoons, due to a toxic substance in the urinary product of the silkworm moths. **m., grand**. See *grand mal*. **m. perforant**, perforating ulcer of the foot; a trophic lesion of tabes. **m., petit**. See *petit mal*. **m. del pinto** or **de los pintos**, Mexican contagious psoriasis.
**mal-** [*malus*, bad]. A prefix meaning bad.
**mala** (*ma'-lah*) [L.]. The cheek-bone or the cheek.
**Malabar itch** (*mal'-ab-ar*). A cutaneous disease of the Malabar coast. **M. leprosy of**. Synonym of *elephantiasis Arabum*. **M. nut**. See *adhatoda indica*.
**M. ulcer**. See *phagedæna tropica*.
**Malacarne's pyramid** (*mal-ak-ar'-na*) [Michele Vincenzo Giacintos *Malacarne*, Italian surgeon, 1744-1816]. A crucial projection formed by the union of the vermis inferior of the cerebellum, at the junction of its posterior and middle thirds, with two transverse prolongations which pass into the corresponding hemispheres. **M.'s space**. See *Tarinus' fossa*.
**malachite-green** (*mal'-ak-it*). A salt of tetraethyl-diparaamido-triphenyl-carbinol. It is used as a stain and as a means of differentiating the colon bacillus from the typhoid bacillus; it has also been used in the treatment of trypanosomiasis.
**malacia** (*mal-a'-se-ah*) [μαλακία, a softening]. 1. A morbid softening of tissue. 2. A depraved appetite. **m. cordis**, a softening of the heart muscle. **m., vascular**, a form marked by excessive new formation of vessels.
**malacocataracta** (*mal-ak-o-kat-ar-ak'-tah*) [μαλακός, soft; καταράκτης, cataract]. A soft cataract.
**malacogaster** (*mal-ak-o-gas'-ter*) [μαλακία, a softening; γαστήρ, stomach]. Softening of the gastric walls. A synonym of *gastromalacia*.
**malacoma** (*mal-ak-o'-mah*) [μαλακός, soft; *pl., malacomata*]. The softening of any organ or part of the body.
**malacopeous, malacopœous** (*mal-ah-ko'-pe-us*) [μαλακοποιείν, to make soft]. Softening; enervating; emollient.
**malacophonous** (*mal-ak-off'-o-nus*) [μαλακός, soft; φωνή, voice]. Soft-voiced.
**malacosarcosis** (*mal-ak-o-sar-ko'-sis*) [μαλακός, soft; σάρξ, flesh]. Softness of tissues, as of muscle.
**malacosis** (*mal-ak-o'-sis*) [μαλακός, soft]. The condition distinguished by the abnormal softening of the tissues of any part of the body. Also, the process of malacoma. **m. cordis**. See *myomalacia*.
**malacosomous, malacosteosis** (*mal-ak-o-so'-mus*) [μαλακός, soft; σῶμα, body]. Soft-bodied.
**malacosteon, malacosteosis** (*mal-ah-kos'-te-on, mal-ah-kos'-te-o-sis*). See *osteomalacia*.
**malacotomy** (*mal-ak-ot'-o-me*) [μαλακός, soft; τομή, a cutting]. Incision of the abdomen; celiotomy.
**malacozoon** (*mal-ak-o-zo'-on*) [μαλακός, soft; ζῷον, animal]. A soft animal; a mollusc.
**malactic** (*mal-ak'-tik*) [μαλακός, soft]. Emollient; softening.
**malady** (*mal'-ad-e*) [*malum*, evil]. Disease.
**malagma** (*mal-ag'-mah*) [μάλαγμα, a poultice: *pl., malagmata*]. A poultice, *q. v.*
**malaise** (*mal-āz'*) [Fr.]. A general feeling of illness, accompanied by restlessness and discomfort.
**malakin** (*mal'-ak-in*). A synthetic product allied to acetphenetidin and recommended as an antirheumatic, antipyretic, and antineuralgic. Dose 60-90 gr. (4-6 Gm.) daily.
**malanders** (*mal-and'-urs*). See *malandria*.
**malandria** (*mal-an'-dre-ah*). 1. An affection related to leprosy or elephantiasis. 2. Malanders or malenders, a disease of the horse characterized by a furfuraceous eruption at the bend of the knee and on the inside of the hock. It is called *malenders* when affecting the foreleg and *salenders* when affecting the hind leg.
**malar** (*ma'-lar*) [*mala*, cheek]. Pertaining to the malar bones. **m. arch**, the zygoma. **m. bones**, the two cheek bones. **m. point**. See *point*.

**malaria** (mal-a'-re-ah). See *malarial fever*.
**malarial** (mal-a'-re-al) [*malaria*, bad air]. Pertaining to malaria. **m. cachexia**, a chronic form of malaria characterized by anemia, general failure of health, a sallow complexion, and enlargement of the spleen. **m. fever**, a disease associated with the presence in the blood of a protozoan parasite, the *Plasmodium vivax*, the *Plasmodium malariæ*, and the *Plasmodium præcox*; it is characterized by periodicity, enlargement of the spleen, and the presence in the blood, free or within the red corpuscles, of parasites (plasmodia) that exert a deleterious influence upon the red cells. The paroxysms may be intermittent, remittent, or irregular. If repeated daily, the fever is designated *quotidian*; if on alternate days, *tertian*; if with an interval of two days, *quartan*. If two paroxysms occur daily, the fever is designated a *double quotidian*. There may be a *tertian form*, a paroxysm occurring daily, but only those of alternate days being alike; a *double quartan form*, and others. A typical malarial paroxysm consists of a cold stage, a hot stage, and a sweating stage, occurring in the sequence given. Intermittent fever is characterized by the occurrence of a complete intermission of the symptoms in the interval between two paroxysms, the temperature becoming normal or subnormal. In remittent fever there is only an amelioration of the symptoms in the intervals. In certain localities in which the malarial organisms are exceedingly numerous or intensely virulent the attack displays a pernicious tendency. Of this type there may be a cerebral form, characterized either by delirium and excitement or by coma and depression; a thoracic form, in which the respiration is accelerated and there is an urgent sense of the need of air; a gastrointestinal form, attended with nausea, vomiting, jaundice, and diarrhea; or an asthenic or algid form, in which there is a condition of marked debility with a striking coldness of the surface and of the breath. To the irregular manifestations of malarial poisoning, which do not at any time present the classic association of chill, fever, and sweat, the designation of "dumb ague" is given. The enlargement of the spleen in chronic malaria is sometimes designated "ague-cake." Syn., *marsh-fever*. **m. hematuria**, the presence of blood in the urine as a result of malarial poisoning. **m. neuralgia**, neuralgia due to malarial intoxication.
**malarilabialis** (mal-a-re-la-be-a'-lis). See *zygomaticus major* under *muscle*.
**malarin** (mal'-ar-in). See *acetophenonephenetidin*.
**malarious** (mal-a'-re-us). See *malarial*.
**Malassez' disease** (mal-as-a') [Louis Charles *Malassez*, French physiologist, 1842–1910]. Cystic disease of the testis.
**malassimilation** (mal-as-im-il-a'-shun) [*mal-*; *assimilation*]. Defective assimilation.
**malate** (mal'-āt). A salt of malic acid.
**malaxation** (mal-aks-a'-shun) [μαλάσσειν, to soften]. 1. The act of kneading. 2. A form of massage.
**malcious** (mal'-shus) [μάλκιος, freezing]. Benumbing; causing to freeze; becoming congealed.
**male** (mal'-e) [μάλη]. The axilla.
**male** (māl) [*masculus*, a male]. 1. Pertaining to the male sex, or that which impregnates the female. 2. A member of the male sex. 3. Of a double-bladed instrument, the blade which is received into a hollow of the other (female) blade. **m.-fern**. See *Aspidium*. **m. organ**, the penis.
**maleic** (mal-e'-ik). Relating to or derived from malic acid.
**malemission** (mal-e-mish'-un) [*mal-*; *e*, out; *mittere*, to send]. Failure of the semen to be ejected from the penis during coitus.
**malen** (ma'-len) [*mala*, cheek]. Belonging to the malar bone in itself.
**Malerba's test for acetone** (mahl-er'-bah) [Pasquale *Malerba*, Italian physician]. Add to the acetone a solution of dimethylparaphenylendiamine; a red coloration results.
**malformation** (mal-for-ma'-shun) [*mal-*; *formatio*, a forming]. An abnormal development or formation of a part of the body.
**Malgaigne's hernia** (mahl-gān') [Joseph François *Malgaigne*, French surgeon, 1806–1865]. Hernia of infancy; descent of the intestine into the open vaginal process of the peritoneum. **M.'s hooks**, two pairs of hooks connected by a screw for approximating the fragments of a fractured patella. **M.'s triangle**, the superior carotid triangle.

**malgenic** (mal-jen'-ik) [*malum*, evil; γεννᾶν, to beget]. Producing disease.
**maliasmus** (mal-e-as'-mus). Synonym of *glanders*.
**malic acid** (ma'-lik). See *acid, malic*.
**maliform** (mal'-e-form) [*malum*, an apple; *forma*, form]. Shaped like an apple.
**malign** (mal-īn'). See *malignant*.
**malignancy** (mal-ig'-nan-se) [see *malignant*]. The quality of being malignant.
**malignant** (mal-ig'-nant) [*mal-*; *gignere*, to beget]. Virulent, compromising or threatening life. **m. cholera**, Asiatic cholera. **m. edema**. See *edema, malignant*. **m. fever**, typhus fever. **m. pustule**, anthrax. **m. tumor**, a tumor that destroys life. Malignant tumors recur and give rise to metastasis. **m. vesicle**, anthrax.
**malignin** (ma-lig'-nin). A hypothetical ferment credited with being the cause of the malignancy of cancer.
**malimali** (mah'-le-mah'-le). A convulsive tic prevalent in the Philippines.
**malingerer** (mal-in'-jer-er) [Fr. *malingre*, sickly, from *malus*, bad; *æger*, ill; sick]. One who feigns illness or defect.
**malingering, malingery** (mal-in'-jer-ing, mal-in'-jer-e) [see *malingerer*]. The feigning of disease.
**malis** (ma'-lis). A name vaguely applied to various diseases, generally of the skin, and especially to such as are due to vermin, or to parasitic worms that burrow in the skin.
**malleable** (mal'-e-a-bl) [*malleus*, hammer]. Capable of being beaten or rolled into thin sheets.
**malleability** (mal-e-ab-il'-it-e) [*malleus*, hammer]. The quality of being malleable.
**malleal, mallear** (mal'-e-al, -ar) [*malleus*]. Relating to the malleus.
**malleation** (mal-e-a'-shun) [*malleus*]. A spasmodic action of the hands, consisting in continuously striking any near object.
**malleiform** (mal'-e-if-orm) [*malleus*; *forma*, form]. Hammer-shaped.
**mallein** (mal'-e-in) [*malleus*, farcy]. A fluid obtained from cultures of *Bacillus mallei*, the microorganism of glanders. When injected into the circulation of a glanderous animal, it causes an elevation of temperature, and has been recommended for use in the early diagnosis of farcy or glanders.
**malleinization** (mal-e-in-i-za'-shun). Inoculation with mallein.
**mallenders** (mal'-en-derz). A kind of eczema or scab above the fore-foot and about the knee of the horse; see also *malandria*.
**malleoincudal** (mal-e-o-ing'-kū-dal) [*malleus*; *incus*, anvil]. Relating to the malleus and the incus.
**malleolar** (mal'-e-o-lar) [*malleolus*]. Relating to a malleolus.
**malleolus** (mal-e'-o-lus) [dim. of *malleus*; pl.; *malleoli*]. A part or process of bone having a hammer-head shape. **m., external**, the lower extremity of the fibula. **m., internal**, a process on the internal surface of the lower extremity of the tibia. **m. lateralis**. Same as *m. external*. **m. medialis**. Same as *m. internal*. **m. radialis**, the styloid process of the radius. **m. ulnaris**, the styloid process of the ulna.
**malleotomy** (mal-e-ot'-o-me) [*malleus*, or *malleolus*; τομή, incision]. 1. Incision or division of the malleus. 2. Division of the ligaments attached to the malleoli.
**mallet finger** (mal'-et). See *finger*. **m. toe**, a deformity of a toe characterized by deficient extension or undue flexion of the terminal phalanx; hammer-toe.
**malleus** (mal'-e-us) [L., "hammer"]. 1. One of the ossicles of the internal ear having the shape of a hammer. 2. Glanders.
**mallotoxin** (mal-o-toks'-in). See *rottlerin*.
**Mallotus** (mal'-o'-tus) [μαλλός, wool]. A genus of euphorbiaceous trees and shrubs. *M. philippinensis*, a species of India, yields the dyestuff kamila. The leaves and fruit are used in the treatment of snake-bites; the root, in contusions.
**mallow** (mal'-o). See *malva*. **m., marsh-**. See *althæa*.
**malnutrition** (mal-nū-trish'-un) [*mal-*; *nutrition*]. Imperfect nutrition or sustenance.
**malocclusion** (mal-ok-loo'-shun) [*mal-*; *occludere*, to shut up]. The occlusion of the teeth in positions not conformable to anatomical rule.
**malomaxillary** (ma-lo-maks'-il-a-re) [*mala*, cheek; *maxilla*]. Relating to the cheek or malar bone and the maxilla.

# MALONYL 533 MANCINISM

**malonyl** (*mal'-on-il*), $CH_2<^{CO}_{CO}$. The bivalent radical of malonic acid. **m. urea,** barbituric acid.
**maloplasty** (*mal'-o-plas-te*) [*mala*, cheek; πλάσσειν, to form]. Plastic surgery of the cheek.
**Malpighian body** (*mal-pe'-ge-an*) [Marcello *Malpighi*, Italian anatomist, 1628–1694]. The commencement of a uriniferous tubule, consisting of the glomerule of vessels (the Malpighian tuft) and the membranous envelope (Bowman's capsule). **M. capsule.** See *Bowman's capsule*. **M. cells, M. vesicles,** the pulmonary alveoli. **M. corpuscle,** any one of the minute whitish nodules of lymphadenoid tissue in the red substance of the spleen along the course of the blood-vessels. **M. pyramids,** conical masses, eight to ten in number, in the medullary portion of the kidney, having their apices directed toward the pelvis and their bases toward the cortex. **M. rete mirabile,** the network formed by the ultimate ramifications of the pulmonary artery. **M. stigmata,** the orifices of the capillary veins that join the branches of the splenic vein at right angles. **M. stratum,** the rete mucosum of the epidermis. **M. tuft,** a glomerule of the kidney.
**malposition** (*mal-po-zish'-un*) [*mal-; position*]. An abnormal position of any part or organ, especially of the fetus.
**malpractice** (*mal-prak'-tis*) [*mal-; practice*]. Improper treatment through carelessness, or ignorance, or intentionally; treatment of a disease by a method contrary to that taught by experience; also, the unlawful production of an abortion.
**malpraxis** (*mal-prax'-is*). Same as malpractice.
**malpresentation** (*mal-pre-sen-ta'-shun*) [*mal-; presentation*]. In obstetrics, such a position of the child at birth that delivery is difficult or impossible.
**malt, maltum** (*mawlt, mawl'-tum*) [AS., *mealt*]. Grain which has been soaked, made to germinate, and dried. The *maltum* of the U. S. P. is the grain of common barley, *Hordeum distichon*, made to germinate by warmth and moisture, and then baked so as to arrest the germinating process. The germinated grains contain diastase, dextrin, and maltose, as well as proteids. Malt is used as a nutrient in wasting diseases. **m. extract** (*extractum malti*, U. S. P.). Dose 1–4 dr. (¼–16 Cc.). **m. fluidextract** of. Dose ½–2 oz. (15–60 Cc.). **m.-liquors,** infusions of malt fermented so as to contain alcohol. Those in common use are beer, ale, and porter. *Beer* is made by a comparatively slow fermentation, and contains about 2.5 % of alcohol. *Ale* and *porter* are fermented more rapidly, and contain about 4.7 % of alcohol. The malt used in making porter is browned, giving the liquor a darker color. **m.-sugar,** maltose.
**Malta fever.** See *fever, Mediterranean*.
**maltase** (*mawl-tās'*). An enzyme found in the saliva and pancreatic juice which converts maltose into dextrose.
**Malthus, doctrine of** (*mal'-thus*) [Thomas Robert *Malthus*, English political economist, 1766–1834]. The doctrine that the increase of population is proportionately greater than the increase of subsistence. Syn., *Malthusianism*.
**maltine** (*mawl'-tēn*) [*malt*]. A name given to various proprietary preparations of malted wheat or barley, useful as food for invalids.
**maltobiose** (*mawl-to-bi'-ōs*). Maltose.
**maltodextrin** (*mawl-to-deks'-trin*). A form of dextrin convertible into malt.
**maltol** (*mawl'-tol*), $C_8H_8O_3$. A constituent of malt caramel, an odorless substance soluble in hot water.
**maltopepsine** (*mawl-to-pep'-sin*). Trade name of a preparation containing malt and pepsin.
**maltosazone** (*mawl-to'-saz-ōn*). An osazone formed from maltose.
**maltose** (*mawl'-tōs*) [*malt*], $C_{12}H_{22}O_{11}+H_2O$. A variety of sugar formed, together with dextrin, by the action of malt diastase upon starch.
**maltosuria** (*mawl-tōs-u'-re-ah*) [*maltose*; οὖρον, urine]. The presence of maltose in the urine.
**maltova** (*mawl-to'-vah*). A concentrated food said to be a combination of the proteids of egg.
**maltoyerbin** (*mawl-to-yer'-bin*). An expectorant said to consist of malt and yerba santa.
**malturned** (*mal-turnd'*). Term applied to a tooth so turned on its central axis as to stand in malposition.
**maltzyme** (*mawlt'-zīm*). A concentrated diastasic extract of malt; indicated in starchy indigestion.

Dose 1–2 tablespoonfuls (15–30 Cc.) during or after meals.
**malum** (*ma'-lum*) [*malum*, evil]. Disease. **m. Ægyptiacum,** diphtheria. **m. articulorum.** Synonym of *rheumatism* and of *gout*. **m. articulorum senilis.** Synonym of *arthritis deformans*. **m. caducum,** the falling sickness, or epilepsy. **m. Cotunnii,** sciatica. **m. coxæ.** Synonym of *hip disease*. **m. perforans pedis,** perforating ulcer of the foot. **m. pilare,** trichinosis. **m. primarium,** a primary or idiopathic disease. **m. Rustii,** a form of cervical Pott's disease described by Rust. **m. arteriarum senilis,** senile endarteritis deformans. **m. venereum.** Synonym of *syphilis*.
**malunion** (*mal-ū'-nyon*). Incomplete union or union in a faulty position of the fragments of a fractured bone.
**Malva** (*mal'-vah*) [L., "mallow"]. The mallow; a genus of malvaceous plants. The leaves of *M. alcea, M. rotundifolia*, and *M. sylvestris* are used as demulcents.
**Maly's test for hydrochloric acid in stomach-contents** (*mah'-le*) [Richard Leo *Maly*, Austrian chemist, 1839–1864]. Place the filtered contents of the stomach in a glass dish, and add ultramarine sufficient to make it blue. Suspend a piece of lead-paper in the upper part of the dish, and cover with a watch-glass. Warm this on the water-bath for 15 minutes, and in the presence of HCl the blue color will change to brown and the lead-paper will become dark, owing to the development of $H_2S$.
**mamanpian** (*mah-mahn-pe-ahn'*) [Fr.]. The prominent ulcer of frambesia.
**mamelon** (*mam'-el-on*). A nipple; boss; hemispherical projection.
**mamelonated** (*mam'-el-on-a-ted*). Having nipple-like elevations.
**mamma** (*mam'-ah*) [L.: *pl., mammæ*]. The breast; the milk-secreting gland of the mother. **m. aberrans,** supernumerary breast. **m. erratica,** supernumerary breast. **m. virilis,** the male breast.
**Mammalia** (*mam-a'-le-ah*) [*mamma*]. A division of the class of vertebrates including all animals that suckle their young.
**mammary** (*mam'-a-re*) [*mamma*]. Pertaining to the mammæ. **m. artery.** See under *artery*. **m. gland.** See *gland, mammary*. **m. line,** the vertical line passing through the nipple.
**mammate** (*mam'-āt*) [*mamma*, breast]. Having mammæ or breasts.
**mammiform** (*mam'-if-orm*) [*mamma*, breast; *forma*, shape]. Breast-shaped; shaped like a cone whose apex is rounded.
**mammilla** (*mam-il'-ah*) [dim. of *mamma*]. A small prominence or papilla. **m. of breast,** the nipple or teat.
**mammillaplasty** (*mam-il'-ap-las-te*) [*mammilla*; πλάσσειν, to mold]. A plastic operation for the purpose of elevating a depressed nipple.
**mammillary** (*mam'-il-a-re*) [*mammilla*]. Nipple-shaped; pertaining to a nipple.
**mammillated** (*mam'-il-al-ed*). Covered upon the surface with nipple-like protuberances.
**mammillation** (*mam-il-a'-shun*) [*mammilla*]. A granulation, especially on some mucous surface.
**mammilliplasty** (*mam-il'-ip-las-te*) [*mammilla*, nipple; πλάσσειν, to shape]. Plastic surgery of the nipple.
**mammilloid** (*mam'-il-oid*) [*mammilla*, nipple; εἶδος, like]. Nipple-shaped.
**mammillose** (*mam'-il-ōs*). Having many nipples or nipple-shaped processes.
**mammin** (*mam'-in*) [*mamma*, breast]. A preparation of mammary glands.
**mammitis** (*mam-i'-tis*). See *mastitis*.
**mammose** (*mam'-ōs*) [*mamma*]. Having full or abnormally large breasts.
**mammotomy** (*mam-ot'-o-me*). See *mastotomy*.
**mamos** (*mam'-ōs*). Trade name applied to a preparation of mammary gland substance.
**manaca** (*man'-ak-ah*) [Braz., *manacan*]. The root of *Franciscea* or *Brunfelsia uniflora*, known in Brazil as "vegetable mercury." It is used as an antisyphilitic, diuretic and cathartic.
**mancheel** (*man-chin-ēl'*). See *Hippomane mancinella*.
**mancinism** (*man'-sin-ism*) [*mancus*, imperfect, maimed]. Left-handedness.

**Mandel's test for proteins** (*man'-del*) [John A. *Mandel*, American chemist]. Add to the proteid solution a 5% solution of chromium trioxide and a precipitate will be formed.

**Mandelbaum's reaction** (*man'-del-bowm*) [M. *Mandelbaum*, German physician]. The thread-reaction.

**mandible, mandibula** (*man'-dib-l, man-dib'-ū-lah*) [*mandere*, to chew]. The inferior maxillary bone.

**mandibular** (*man-dib'-ū-lar*) [*mandible*]. Pertaining to the mandible, or lower jaw.

**mandioca** (*man-de-o'-kah*). See *manioc*.

**mandragora** (*man-drag-o'-rah*) [μανδραγόρας, the mandrake]. The mandrake. A genus of solanaceous plants. *M. officinalis* has been used as a narcotic and hypnotic. It and other species contain an alkaloid, *mandragorine*, $C_{17}H_{23}NO_3$, resembling atropine in action.

**mandrake** (*man'-drāk*). See *mandragora* and *podophyllum*.

**mandrel, mandrin** (*man'-drel, man'-drin*) [Ger.]. The firm guide or stylet (usually of metal) that gives rigidity to a flexible catheter while it is being inserted.

**manducation** (*man-dū-ka'-shun*) [*manducatio*, a chewing]. The chewing or mastication of food.

**manducatory** (*man-dū'-kat-or-e*) [*manducare*, to chew]. Pertaining to manduction.

**manganese, manganum** (*mán'-gan-ēz, man'-ganum*) [an altered form of *magnesium*]. Symbol Mn; atomic weight 54.93; quantivalence II, IV, VI. A brittle, hard, grayish-white metal, having a specific gravity of 7.2, and resembling iron in properties. It forms several oxides, the highest of which, $Mn_2O_7$, forms an acid, $HMnO_4$, from which salts, the permanganates, are produced. **m. albuminate**, used in chlorosis and anemia. **m. arsenate**, $MnHAsO_4$, alterative and tonic. Dose $\frac{1}{32}-\frac{1}{6}$ gr. (0.002–0.013 Gm.). **m. carbonate**, $MnCO_3$, used as a tonic in anemia and chlorosis. Dose 8–40 gr. (0.52–2.6 Gm.). **m. citrate**, $MnHC_6H_5O_7$, used as a tonic and astringent instead of iron citrate. Dose 1–3 gr. (0.065–0.2 Gm.). **m. dioxide, precipitated** (*mangani dioxidum præcipitatum*, U. S. P.), $MnO_2$, black oxide of manganese, is tonic and alterative, and has been used in syphilis, chlorosis, in various skin diseases, and in certain forms of dyspepsia. Dose 3–20 gr. (0.2–1.3 Gm.). It is employed in the arts, and in laboratories for the purpose of obtaining chlorine and oxygen. **m. glycerinphosphate**, $MnPO_4$·$C_3H_5(OH)_3$+$H_2O$, used in neurasthenia. Dose 2 gr. (0.13 Gm.) 3 times daily. **m. hypophosphite** (*mangani hypophosphis*, U. S. P.), $Mn(H_2PO_2)_2$+$H_2O$. Dose 10–20 gr. (0.65–1.3 Gm.). **m. lactate**, $Mn(C_3H_5O_3)_2$+$3H_2O$, tonic. Dose 5–20 gr. (0.065–0.32 Gm.). **m. oleate**, $Mn(C_{18}H_{33}O_2)_2$, used in chlorosis and anemia. **m. oxalate**, $MnC_2O_4$+$2H_2O$, a desiccant. **m. peptonate**, is used as a tonic in anemia and chlorosis. Dose 20–60 gr. (1.3–4.0 Gm.). **m. saccharate**, used as a tonic in anemia. **m. salicylate**, $Mn(C_7H_5O_3)_2$, tonic, alterative, and antirheumatic. Dose 2–10 gr. (0.13–0.65 Gm.). **m. sulphate** (*mangani sulphas*, U. S. P.), $MnSO_4$+$4H_2O$, has been used as a substitute for iron in anemia, and as a cholagogue. Dose 5–20 gr. (0.32–1.3 Gm.). **m. sulphite**, $MnSO_3$, tonic, cholagogue, and antiseptic. Dose 5–20 gr. (0.32–1.3 Gm.). **m. sulphocarbolate**, $Mn(C_6H_5SO_4)_2$+$7H_2O$, tonic and antiseptic. Dose 3–15 gr. (0.2–1.0 Gm.).

**manganicopotassic** (*man-gan-ik-o-po-tas'-ik*). Containing manganese as a bivalent radical and potassium.

**manganization** (*man-gan-iz-a'-shun*) [μάγγανον, philter]. Adulteration of drugs.

**mangasol** (*man'-gas-ol*). Magnesium chlorphenolsulphonate, used as an antiseptic powder. Dose, 10–20 gr. (0.6–1.2 gm.).

**mange** (*mānj*) [Fr., *manger*, to eat]. A parasitic skin disease of horses, cattle, and dogs, resembling scabies, and due to various species of *acarus*.

**mango** (*man'-go*) [Pl., *manga, mange*]. The fruit of *Mangifera indica*; the seeds are said to be anthelmintic; and the bark is said to be astringent and tonic to the mucous membranes.

**mangosteen** (*man'-go-stēn*) [Pg., *manga*, mango], $C_{10}H_{22}O_5$. A crystalline, bitter principle found in the pericarp of *Garcinia mangostana*.

**mania** (*ma'-ne-ah*) [μανία, madness]. A form of insanity marked by great mental and emotional excitement, by hallucinations, delusions, physical excitement, and often a tendency to violence. **m.,** alcoholic, acute mania of alcoholic origin. It differs from delirium tremens, although the term is sometimes used synonymously with it. **m. à potu.** See *delirium tremens*. **m., Bell's**, an acute delirium running a rapidly fatal course, with slight fever, and in which postmortem no lesions are found sufficient to account for the symptoms. There are the wildest hallucinations, insomnia, and intense excitement, followed by a condition called typhomania, with elevation of temperature, dry tongue, and rapid, feeble pulse. **m., dancing,** an epidemic of choreic or convulsive movements. **m., epileptic,** a maniacal outburst in an epileptic, often associated with a destructive tendency. **m., paroxysmal,** a paroxysmal neurosis in which the attacks take the form of transitory mania. **m., puerperal,** a form of mania or abnormal mental action sometimes following childbirth. **m., religious,** mania in which the central idea is religious in character, or in which a powerful religious emotion has been the exciting cause. **m., transitory,** frenzied attacks of short duration.

**maniac** (*ma'-ne-ak*) [*mania*]. An insane person; one affected with mania.

**manic** (*man'-ik*). Pertaining to mania. **m. depressive insanity**, cyclothymia; see *insanity, circular*.

**manicure** (*man'-ik-ūr*) [*manus*, the hand; *cura*, care]. 1. The processes employed in caring for one beautifying the hand. 2. One who professionally attends to the care of the hands and nails.

**manigraph** (*man'-ig-raf*) [μανία, mania; γράφειν, to write]. An alienist; one who is an expert in insanity.

**manigraphy** (*man-ig'-ra-fe*) [μανία, madness; γράφη, writing]. A treatise on, or the science of, insanity.

**Manihot** (*man'-e-hot*) [L.]. A genus of euphorbiaceous plants, yielding cassava and tapioca.

**manikin** (*man'-ik-in*) [OF., *manequin*, a puppet]. A model of the body, made of plaster, papier-mâché, or other material, and showing, by means of movable parts, the relations of the organs.

**maniluvium** (*man-il-oo'-ve-um*)[ *manus*, hand; *lavare*, to wash]. A hand-bath; a wash or lotion for the hands.

**manioc** (*man'-e-ok*) [Sp., *mandioca*]. The cassava-plant or its product, tapioca. See *manihot*.

**maniple** (*man'-ip-l*) [*manipulus*, from *manus*, hand]. A handful, or pugil.

**manipulation** (*man-ip-ū-la'-shun*) [see *maniple*]. A handling; the use of the hands for the purpose of performing some work in a skilful manner, such as reducing a dislocation, returning a hernia into its cavity, or changing the position of a fetus.

**manipulus** (*man-ip'-ū-lus*). See *maniple*.

**Mann's sign.** Diminished resistance of the scalp to the galvanic current in traumatic neuroses.

**Mann's sign** [John Dixon *Mann*, English physician, 1840–1912]. A disturbance of the normal balance of the muscles in the two orbits so that one eye appears to be on a lower level than the other; it is seen in exophthalmic goiter and other affections characterized by tachycardia.

**manna** (*man'-ah*) [μάννα, manna]. The concrete, saccharine exudation of the flowering ash, *Fraxinus ornus*, and other trees. Manna contains a sweet principle, *mannite* or *mannitol*, $C_6H_{14}O_6$, a sugar, a purgative principle, and a mucilage. Some specimens contain also a glucoside, *fraxin*. Manna is a mild laxative. Dose 1–2 oz. (32–64 Cc.).

**Mannaberg's sign** (*mah'-nah-bairg*). Accentuation of the second pulmonic sound of the heart is frequently found in abdominal disease, especially in appendicitis.

**Manning's exanthem.** A septicemic exanthem occurring as a grave complication of scarlatina and diphtheria.

**mannitan** (*man'-il-an*) [μάννα, manna], $C_6H_{12}O_5$. A sweet, syrupy substance produced by the action of sulphuric acid on mannite. **m. diacetate,** a compound of mannitan and acetic acid. Syn., *acetite*.

**mannite** (*man'-it*). See *manna* and *mannitol*.

**mannitol** (*man'-it-ol*). See under *manna*. **m. hexanitrate,** a vasodilator. **m. pentanitrate,** a body resulting from action of pyridine on mannitol hexanitrate. It reduces blood-pressure.

**mannitose** (*man'-it-ōs*) [μάννα, manna], $C_6H_{12}O_6$. An amorphous substance, isomeric with levulose, but optically inactive, obtained from mannite by oxidation.

**Mannkopff's sign, M.-Rumpf's sign** [Emil Wilhelm *Mannkopff*, German physician, 1836– ]. Acceleration of the pulse on pressure over painful points in traumatic neuroses.

**mannose** (*man'-ōs*) [*manna*], $C_6H_{12}O_6$. The aldehyde of mannitol. It exists in three forms, dextromannose, levo-mannose, and inactive mannose.

**manol** (*man'-ol*). A proprietary remedy for whooping-cough said to consist of cane-sugar, phenol, oil of anise, alcohol, and water. Syn., *Succus anisi ozonatus*.

**manola** (*man-o'-lah*). Trade name of a preparation of cod-liver oil, alcohol, coca, quinine, phosphates and other substances.

**manolymph** (*man'-o-limf*) [μανός, rare, single or separate; *lympha*, lymph]. Warthin's term for certain hemolymph glands. See *gland, hemolymph*.

**manometer** (*man-om'-et-er*) [μανός, rare; μέτρον, a measure]. An instrument for measuring the tension of liquids and gases, consisting either of a bent tube filled with mercury (*mercurial manometer*), or of a spring (*spring manometer*), connected with a writing-style.

**manometric** (*man-o-met'-rik*) [*manometer*]. Pertaining to a manometer; pertaining to tracings obtained by means of a manometer. m. flames, flames of different heights and characters seen in a rotating mirror and due to the reflection of a pulsating gas-flame when the supplying gas is set in motion by sound-waves. Syn., *Koenig's flames*.

**manoscope** (*man'-o-skōp*) [μανός, thin; σκοπεῖν, to view]. An instrument for determining the density of air.

**mansa** (*man'-sah*). The rhizome of *Houttuynia californica*; it is used in malaria and dysentery.

**mantle** (*man'-tl*). That portion of the brain substance including the convolutions, corpus callosum, and fornix; also called brain mantle, and pallium.

**manual** (*man'-ū-al*) [*manus*]. Pertaining to the hands; performed by the hands.

**manubrial** (*man-ū'-bre-al*). Pertaining to a manubrium.

**manubriate** (*man-ū'-bre-āt*) [*manubrium*]. Furnished with a handle or handle-shaped process.

**manubrium** (*man-ū'-bre-um*) [L.]. 1. A handle. 2. The first or upper piece of the sternum. m. of malleus, the handle-shaped process of the malleus of the ear. m. manus, the radius. m. of sternum, m. sterni. See *manubrium (2)*.

**manuduction** (*man-ū-duk'-shun*) [*manus*, hand; *ductio*, a leading]. The operations performed by the hands in surgical and obstetrical practice.

**manus** (*ma'-nus*) [L.]. The hand. m. curta, m. distorta, club hand.

**manustupration** (*man-ū-stū-pra'-shun*) [*manus; stuprare*, to ravish]. Masturbation.

**manyplies** (*men'-ip-līz*) [AS., *manig*, many; *plicare*, to fold]. The third compartment in the stomach of ruminants. Syn., *omasum; psalterium*.

**Manz's glands** [Wilhelm *Manz*, German ophthalmologist, 1833– ]. Utricular glands found in the orbital conjunctiva near the margin of the cornea.

**manzanita** (*man-san-e'-tah*). The *Arctostaphylos glauca*, a Californian plant whose leaves are said to be tonic and diuretic.

**M.A.O.** Abbreviation for Master of the Art of Obstetrics.

**mappy tongue** (*map'-e tung'*). See *geographical tongue*.

**Maragliano's endoglobular degeneration** (*mah-rahl-yah'-no*) [Edoardo *Maragliano*, Italian physician, 1849– ]. Vacuole-like areas seen in red blood-cells after exposure to the air. These areas are probably the result of coagulation necrosis.

**maransis** (*mar-an'-sis*). Synonym of *marasmus*.

**maranta** (*mar-an'-tah*). See *arrowroot*.

**marantic** (*mar-an'-tik*) [μαραίνειν, to make lean]. Pertaining to marasmus. m. clot, a blood-clot produced by slowing of the circulation in depressed states of the system. m. thrombosis, thrombosis due to general malnutrition.

**maraschino** (*mar-as-ke'-no*) [Sp.]. A cordial made from marasca cherries.

**marasmatic** (*mar-as-mat'-ik*). Synonym of *marasmic*.

**marasmic** (*mar-az'-mik*) [*marasmus*]. Affected with marasmus.

**marasmoid** (*mar-az'-moid*) [μαρασμός, decay; εἶδος, like]. Resembling or simulating marasmus.

**marasmopyra** (*mar-az-mop'-ir-ah*) [μαρασμός, wasting; πῦρ, fire]. Hectic fever.

**marasmus** (*mar-az'-mus*) [μαραίνειν, to grow lean].

A gradual wasting of the tissues of the body from insufficient or imperfect food-supply. There is either no organic lesion or gastrointestinal catarrh. marc (*mark*) [Fr., "dregs"]. 1. A by-product in the manufacture of wines, consisting of the stems, skins, and stones of the grapes. 2. The residue remaining after the expression of the oil from certain fruits.

**Marchi's bundle, M.'s tract** (*mar'-tshe*) [Vittorio *Marchi*, Italian physician]. See *Loewenthal's tract*.

**marcid** (*mar'-sid*) [*marcidus, marcere*, to wither]. 1. Shrunken; wasted. 2. Accompanied or characterized by wasting.

**marcor** (*mar'-kor*). See *marasmus*.

**Maréchal test for bile-pigments** (*mar-a-shal'*) [Louis Eugene *Maréchal*, French physician]. See *Smith's reaction*.

**mareo** (*mar'-e-o*) [Span.]. Mountain sickness. A malady characterized by nausea and violent headache, occasionally overcoming persons who ascend to high altitudes.

**Marey's test** (*ma'-re*) [Etienne Jules *Marey*, French physiologist, 1830–1904]. A high-tension pulse is a slow pulse.

**margarate** (*mar'-gar-āt*). The product of margaric acid and a base.

**margaric acid** (*mar-gar'-ik*). See *acid, margaric*.

**m.-acid crystals**, needle-shaped crystals consisting of compounds of the fatty acid, found in foci of fatty degeneration, in the urine, etc.

**margarin** (*mar'-gar-in*) [μάργαρος, especially glyceryl trimargarate, $C_3H_5(C_{17}H_{33}O_2)_3$, found in butter. 2. An artificial substitute for butter. See *oleomargarin*. m.-needles, fatty crystals found in putrid bronchitis and pulmonary gangrene.

**margaritoma** (*mar-gar-it-o'-mah*) [μάργαρος, the pearl-oyster; ὄμα, a tumor]. Virchow's term for a true primary cholesteatoma-formation in the auditory canal.

**margarone** (*mar'-ga-rōn*). See *palmitone*.

**Margaropus annulatus** (*mar-gar-o'-pus an-nū-la'-tus*). A cattle tick which spreads the *Babesia bigeminum*, the cause of Texas fever in cattle.

**marginal** (*mar'-jin-al*) [*margo*, margin]. Pertaining to the margin or border. m. convolution. See *convolution, marginal*.

**marginoplasty** (*mar-jin-o-plas'-te*) [*margo*, margin; πλάσσειν, to shape]. Plastic surgery of the marginal portion of the eyelid.

**margo** (*mar'-go*) [L., *pl., margines*]. A margin, edge, or border.

**Marie's disease** (*mar-ee'*) [Pierre *Marie*, French physician, 1853– ]. 1. Akromegaly. 2. Hereditary cerebellar ataxia. 3. Hypertrophic pulmonary osteoarthropathy. 4. Spondylosis rhizomelica; ankylosis of the spinal column and of the coxofemoral, less frequently also of the scapulohumeral articulations. The affection is identified with Struempell's disease (2). M.'s quadrilateral, a four-sided space bounded in front by the anterior limiting sulci of the island of Reil, behind by the posterior limiting sulci of the island of Reil, internally by the wall of the lateral ventricle, and externally by the surface of the island of Reil. M.'s symptom, tremor of the extremities or the whole body in exophthalmic goiter.

**Marie-Kahler's symptom.** See *Marie's symptom*.

**Marie-Robinson's syndrome.** A variety of diabetes with melancholia, insomnia, impotence, and the presence in the urine of a levulose that disappears rapidly on the suppression of carbohydrates.

**Marie Struempell disease.** See *Marie's disease (4)*.

**marigold** (*mar'-ig-ōld*). See *calendula*.

**Mariotte's blind spot** (*mar-e-ot'*) [Edme *Mariotte*, French physicist, 1620–1684]. The optic papilla. M.'s experiment, to demonstrate the existence of the blind spot a sheet of paper, on which a cross and a circular-spot are marked, is held a short distance in front of the eyes, the left eye being directed steadily on the cross while the right eye is closed. On moving the paper away slowly a point will be reached where the spot will no longer be visible, but it reappears when the distance is increased. M.'s law. See *Boyle's law*.

**mariscus** (*mar-is'-kus*) [*mariscA*, hemorrhoid]. Pertaining to hemorrhoids.

**maritonucleus** (*mar-it-o-nū'-kle-us*) [*maritus*, married; *nucleus*, a little nut]. The nucleus of an ovum after fecundation.

**Marjolin's ulcer** (*mar-zjo-lan'*) [René *Marjolin*, French physician, 1812-1895]. A slowly progressive, malignant ulcer with peculiar, wart-like growths commencing on a cicatrix.
**marjoram** (*mar'-jo-ram*). See *origanum*.
**mark**. Birth-mark, or mother's mark; nevus. m., portwine. See *nevus*.
**markasol** (*mar'-kas-ol*). Bismuth borophenate, used as a substitute for iodoform.
**marl**. A mixture of clay, sand, and amorphous calcium carbonate.
**Marmorek's serum** (*mar'-mo-rek*) [Alexander *Marmorek*, Austrian physician, 1865– ]. A polyvalent serum obtained by the inoculation of animals with streptococci of various origin.
**marmorekin** (*mar-mor'-e-kin*). 1. Streptococcus antitoxin. 2. Marmorek's serum.
**Marochetti's vesicles**. Small vesicles sometimes seen on the under surface of the tongue in cases of rabies.
**marrol** (*mar'-ol*). A dietetic said to contain ox-marrow and extract of hops and of malt.
**marrow** (*mar'-o*) [AS., *mearh*]. The fatty substance contained in the medullary canal of long bones and in the interstices of cancellous bone. In early life the marrow of all bones is red (*red marrow*), but later that within the shafts of long bones assumes a light color (*yellow marrow*). Red marrow is composed of a delicate reticulum of connective tissue containing blood-vessels, large connective-tissue cells, some of which in growing bone become osteoplasts (*marrow-cells*), giant-cells (*myeloplaxes*), and red corpuscles in various stages of formation. In yellow marrow most of the cells have been transformed into fat-cells. The function of bone-marrow is probably the formation of red corpuscles. In certain forms of anemia the marrow undergoes profound changes; that of the shafts of the long bones may return to its embryonic condition. Bone-marrow has been used in the treatment of pernicious anemia. **m.-space**, a cavity in the cancellous tissue of bone, containing marrow. **m., spinal**, the spinal cord.
**marrubiin** (*mar-oo'-be-in*) [*marrubium*, horehound]. The crystalline, neutral, bitter substance found in horehound.
**marrubin** (*mar-oo'-bin*). A glycerin extract of red bone marrow.
**marrubium** (*mar-oo'-be-um*) [L.]. Hoarhound. The *marrubium* of the U. S. P. is the dried leaves and tops of *M. vulgare*, of the order *Labiatæ*. It contains a volatile oil, a bitter principle, marrubiin, tannin, resin, and lignin. At present hoarhound is mainly employed in the form of candy or syrup in catarrhal affections of the respiratory tract.
**Marsden's mucilage or paste** (*marz'-den*) [Alexander Edwin *Marsden*, English surgeon, 1832-1902]. A caustic paste consisting of one part of white arsenic to two of gum arabic; formerly used in the treatment of cancer.
**Marsh's disease** [Sir Henry *Marsh*, Irish physician, 1790-1860]. Exophthalmic goiter.
**Marsh's test for arsenic** [James *Marsh*, English chemist, 1794-1846]. Introduce the substance into a flask with dilute sulphuric acid and zinc. Light a jet, and permit it to impinge on cold porcelain, or heat the delivery-tube, when a steel-white mirror of metallic arsenic is deposited. This may be distinguished from a similar deposit of antimony by the solubility of the arsenical mirror in potassium hypochlorite.
**Marshall's oblique vein** (*mar'-shal*) [Andrew *Marshall*, Scotch anatomist, 1742-1813]. A partially obliterated vein that passes along the posterior aspect of the left auricle and opens into the coronary sinus near its termination. M.'s vestigial fold, a fold of the pericardium extending from the left branch of the pulmonary artery to the left superior pulmonary vein. It contains a fibrous cord that represents the lower part of the left superior vena cava, a vessel commonly found in mammals, but rarely in man.
**Marshall-Hall's disease**. See *Hall's disease*.
**marsh-fever**. See *malarial fever*.
**marsh-gas**. See *methane*.
**marshmallow**. See *althea*.
**marsitriol** (*mar-sit'-re-ol*). A proprietary preparation of iron (ferrum glyceroarsenate). Dose ⅙ gr. (0.01 Gm.).
**marsupia patellaria** (*mar-sū'-pe-ah pat-el-a'-re-ah*). The alar ligaments of the knee-joint.
**marsupialization** (*mar-sū-pe-al-i-sa'-shun*) [μάρσι-

πος, a pouch]. The operation, recommended in certain cases of ovarian tumor, of raising the borders of the evacuated tumor-sac to the edges of the abdominal wound and stitching them there so as to form a pouch.
**marsyle** (*mar'-sil-e*). A commercial name for iron cacodylate.
**Martegiani's area** (*mar-te-zje-ah'-ne*). The slight widening of the hyaloid canal at its beginning in front of the optic disc.
**martial** (*mar'-shal*) [*mars, iron*]. Containing iron.
**Martin's bandages** (*mar'-tin*) [Henry Austin *Martin*, American surgeon, 1824-1884]. Rubber bandages, from 5 to 21 feet in length, used for making compression of a limb for the cure of ulcers, varicose veins, etc. M.'s depilatory, calcium sulphhydrate. M.'s hemostatic, surgeons' agaric impregnated with ferric chloride.
**Martinotti's cells** (*mar-tin-ot'-e*) [Giovanni *Martinotti*, Italian physician]. Ganglion-cells of the cerebral cortex, giving off a short axis-cylinder process at right angles to the surface.
**martol** (*mar'-tol*). A semifluid extract obtained from the shells of cacao-bean, consisting of carbohydrates, phosphates, iron tannate, etc.
**Maruta** (*ma-roo'-tah*) [L.]. A genus of herbs of the order *Compositæ*. *M. cotula*, may-weed, or dog's-fennel, is used as a substitute for camomile.
**maschaladenitis** (*mas-kal-ad-en-i'-tis*) [μασχάλη, axilla; ἀδήν, gland; *itis*, inflammation]. Inflammation of the glands of the axilla.
**maschale** (*mas'-kal-e*) [μασχάλη, axilla]. Axilla, armpit.
**maschaleous** (*mas-kal'-e-us*) [μασχάλη, axilla]. Pertaining to the axilla.
**maschalephidrosis** (*mas-kal-ef-id-ro'-sis*) [μασχάλη, armpit; ἐφίδρωσις, sweating]. Sweating in the axillæ.
**maschaliatria** (*mas-kal-e-a'-tre-ah*) [μασχάλη, axilla; *larpeia*, treatment]. Treatment by inunctions in the axilla.
**maschalister** (*mas-kal'-is-ter*) [μασχαλιστήρ, girth, girdle]. The second cervical vertebra, the axis.
**maschaloncus** (*mas-kal-ong'-kus*) [μασχάλη, axilla; ὄγκος, tumor]. An axillary tumor.
**maschalyperidrosis** (*mas-kal-ip-er-id-ro'-sis*) [μασχάλη, axilla; ὑπέρ, over; ἱδρωσις, sweating]. Excessive sweating in the armpits.
**(von) Maschke's reaction for creatinin** (*mash'-keh*). Add a few drops of Fehling's solution to the creatinin dissolved in a cold saturated solution of sodium carbonate. An amorphous, flocculent precipitate is formed in the cold, but better on warming to 50° to 60° C.
**masculine, masculous** (*mas'-kū-lin, mas'-kū-lus*) [dim, of *mas*, a male]. Of the male sex.
**masculonucleus** (*mas-kū-lo-nū'-kle-us*). The male pronucleus.
**mask** [Fr., *masque*, a mask]. 1. A bandage applied to the face in case of erysipelas, burns or scalds, eczema, etc. 2. Synonym of *chloasma*.
**masked** (*maskt*) [*mask*]. Covered with a. mask; concealed. m. disease, one that is concealed by concomitant symptoms.
**masochism** (*mas'-o-kizm*) [Leopold von Sacher-*Masoch*, an Austrian writer]. Sexual perversion in which the pervert takes delight in being subjected to degrading, humiliating, or cruel acts on the part of his or her associate.
**masochist** (*mas'-o-kist*). One addicted to masochism.
**masrium** (*maz'-re-um*) [Arab, *masr*, Egypt]. A metal described as a new element, found in Egypt in a mineral first called "Johnsonite," but later masrite.
**mass, massa** (*mas, mas'-ah*) [*massa*, a mass]. 1. An aggregation of particles of matter. 2. A cohesive substance that can be made into pills. m. action, chemical action as determined by the masses of the respective substances interacting. m. action, law of, chemical action is determined by the respective amounts of the substances acting in unit-volume, or by the degree of the concentrations present. m., blue- (*massa hydrargyri*, U. S. P.). See *mercury mass*. m., copaiba, copaiba, 6 parts, mixed with magnesia, 94 parts, and water; diuretic and stimulant. Dose 10-30 gr. (0.65-2.0 Gm.). m., Vallet's (*massa ferri carbonatis*, U. S. P.), mass of ferrous carbonate.
**massage** (*mas-ahzj*), massaging [Fr., from *μάσσειν*, to knead]. A method of rubbing, kneading, or stroking of the superficial parts of the body by the hand or an instrument, for the purpose of modifying

nutrition, restoring power of movement, breaking up adhesions, etc. m., **cannon-ball,** the rolling of a three-pound to five-pound cannon-ball covered smoothly with chamois skin or flannel over the course of the colon. m., **electrovibratory,** that performed by means of an electric vibrator. m., **thermic,** stroking or pressing an affected part with a heated object. m., **vapor,** treatment of a cavity by intermittent forcing of a medicated vapor into it. m., **vibratory,** light, rapid percussion either by hand or by an electric apparatus.

**massalis** (*mas-a'-lis*) [*mass*]. Mercury.
**massesis** (*mas-e'-sis*). Synonym of *mastication.*
**masseter** (*mas'-e-ter*) [μασητήρ, chewer]. One of the muscles of mastication. See *muscle.*
**masseteric** (*mas-et-er'-ik*) [*masseter*]. Pertaining to the masseter muscle.
**masseur** (*mas-ur'*) [Fr.]. A man who practises massage. Fem., *masseuse.*
**masseuse** (*mas-urs'*) [Fr.]. A woman who practises massage.
**massicot** (*mas'-ik-ot*) [Fr.], PbO. Lead oxide; litharge.
**massive** (*mas'-iv*). Heavy. m. **pneumonia,** pneumonia with absolute filling of the air-cells and bronchi with exudate.
**massol** (*mas'-ol*). The bark of *Massoia aromatica;* it is used in colic, diarrhea, and spasms.
**massotherapy** (*mas-o-ther'-ap-e*) [*massage;* θεραπεία, therapy]. Treatment by massage.
**mastaden** (*mas'-ta-den*) [μαστός, breast; ἀδήν, gland]. The mammary gland.
**mastadenitis** (*mas-tad-en-i'-tis*) [μαστός, breast; ἀδήν, gland; ιτις, inflammation]. Inflammation of the mammary gland.
**mastalgia** (*mas-tal'-je-ah*) [μαστός, breast; ἄλγος, pain]. Pain in the breast.
**mastatrophia** (*mast-at-ro'-fe-ah*) [μαστός, breast; ἀτροφία, atrophy]. Atrophy of the breast.
**mastauxe** (*mas-tawks'-e*) [μαστός, breast; αὔξη, growth]. Increase in size, or excessive size, of the breast.
**mast-cells.** See *cells, mast-.*
**mastecchymosis** (*mas-tek-im-o'-sis*) [μαστός, breast; ecchymosis]. Ecchymosis of the breast.
**mastelcosis** (*mast-el-ko'-sis*) [μαστός, breast; ἕλκωσις, ulceration]. Synonym of *masthelcosis.*
**masthelcosis** (*mas-thel-ko'-sis*) [μαστός, breast; ἕλκωσις, ulceration]. Ulceration of the breast.
**mastic, mastiche** (*mas'-tik, mas'-tik-e*) [μαστίχη, mastic]. The resin flowing from the incised bark of the *Pistacia lentiscus,* a tree of the *Terebinthaceæ.* It is used as a styptic, as a filling for teeth, and as a microscopic varnish.
**mastication** (*mas-tik-a'-shun*) [*masticare,* to chew]. The act of chewing.
**masticatory** (*mas'-tik-a-to-re*) [*mastication*]. 1. Pertaining to mastication or to the muscles of mastication. 2. A remedy to be chewed but not swallowed, used for its local action on the mouth. m. **spasm,** spasm of the muscles of mastication; trismus.
**mastiche** (*mas'-ti-ke*). See *mastic.*
**Mastigophora** (*mas-tig-of'-o-rah*) [μάστιξ, whip; φέρειν, to bear]. A class of protozoa with flagella.
**mastigophorous** (*mas-tig-off'-o-rus*) [μαστιγοφόρος, bearing a whip]. Flagellate, as certain infusoria or zoospores.
**mastigosis** (*mas-tig-o'-sis*) [μαστιγοῦν, to whip]. Flagellation as a therapeutic measure.
**mastitis** (*mas-ti'-tis*) [μαστός, breast; ιτις, inflammation]. Inflammation of the breast. m., **interstitial,** inflammation of the connective tissue of the breast. m., **parenchymatous,** inflammation of the proper glandular substance of the breast.
**masto-** (*mas-to-*) [μαστός, breast]. A prefix signifying relating to the breast.
**mastocarcinoma** (*mas-to-kar-sin-o'-mah*). Mammary carcinoma.
**mastochondroma** (*mas-to-kon-dro'-mah*) [*masto-;* χόνδρος, cartilage]. A cartilaginous tumor of the breast.
**mastodealgia** (*mas-to-de-al'-je-ah*) [μαστώδης, mastoid; ἄλγος, pain]. Pain in the mastoid process.
**mastodeocentesis** (*mas-to-de-o-sen-te'-sis*) [μαστώδης, mastoid; κέντησις, a thrust]. Surgical perforation of the mastoid process.
**mastodynia** (*mas-to-din'-e-ah*) [*masto-;* ὀδύνη, pain]. Pain in the breast.
**mastoid** (*mas'-toid*) [*masto-;* εἶδος, like]. 1. Nipple-shaped, as the *mastoid* process of the temporal bone.

2. The mastoid process. 3. Pertaining to the mastoid process, as *mastoid* foramen, *mastoid* operation. m. **abscess,** an abscess of the mastoid cells. m. **antrum,** a cavity in the mastoid portion of the temporal bone. m. **bone.** See *mastoid process.* m. **cells,** the hollow air-spaces in the mastoid process communicating with the middle ear. m. **disease,** inflammation of the mastoid cells; mastoiditis. m. **foramen.** See *foramen, mastoid.* m. **operation,** paracentesis or eradication of the mastoid cells. m. **portion,** the lower posterior portion of the mastoid bone. m. **process,** the protruding part of the temporal bone felt behind the ear. m. **sinus,** the mastoid cells.

**mastoidal, mastoideal, mastoidean** (*mas-toi'-dal, mas-toi-de'-al, mas-toi-de'-an*) [*masto-;* εἶδος, like]. Pertaining to the mastoid process.
**mastoidealgia** (*mas-toi-de-al'-je-ah*) [*masto-;* ἄλγος, pain]. Synonym of *mastodealgia.*
**mastoidectomy** (*mas-toi'-dek'-to-me*) [*mastoid;* ἐκτομή, excision]. Excision of the mastoid cells.
**mastoideocentesis** (*mas-toi-de-o-sen-te'-sis*) [*mastoid;* κέντησις, a thrust]. Surgical perforation of the mastoid process.
**mastoiditis** (*mas-toid-i'-tis*) [*mastoid;* ιτις, inflammation]. Inflammation of the mastoid cells. m., **Bezold's.** See *Bezold's mastoiditis.*
**mastoidotomy** (*mas-toid-ot'-o-me*) [*mastoid;* -τέμνειν, to cut]. Incision of the mastoid cells to relieve suppurative mastoiditis.
**mastologist** (*mas-tol'-o-jist*). A specialist in diseases of the mammary apparatus.
**mastology** (*mas-tol'-o-je*) [*masto-;* λόγος, treatise]. A treatise on the mammary apparatus, its anatomy and diseases.
**mastomenia** (*mas-to-me'-ne-ah*) [*masto-;* μήν, month]. Vicarious menstruation from the breast.
**mastoncus** (*mas-ton'-kus*) [*masto-;* ὄγκος, tumor]. Any tumor of the mammary gland or nipple.
**mastooccipital** (*mas-to-ok-sip'-it-al*) [*mastoid; occiput*]. Pertaining to the mastoid process and the occipital bone.
**mastoparietal** (*mas-to-par-i'-et-al*) [*mastoid; paries,* wall]. Pertaining to the mastoid process and the parietal bone.
**mastopathy** (*mas-top'-ath-e*) [*masto-;* πάθος, disease]. Any disease or pain of the mammary apparatus.
**mastopexy** (*mas'-to-peks-e*) [*masto-;* πῆξις, a fixing]. Surgical fixation of a pendulous breast.
**mastorrhagia** (*mas-tor-a'-je-ah*) [*masto-;* ῥηγνύναι, to break forth]. Hemorrhage from the breast.
**mastoscirrhus** (*mas-to-skir'-us*) [*masto-;* σκίρρος, hard]. A hard cancer of the breast.
**mastosis** (*mas-to'-sis*) [μαστός, breast]. Enlargement of the breast.
**mastospargosis** (*mas-to-spar-go'-sis*) [*masto-;* σπάργωσις, swelling]. Enlargement or swelling of a breast, especially that due to excess of milk.
**mastosyrinx** (*mas-to-si'-rinks*) [*masto-;* σύριγξ, pipe]. A mammary fistula.
**mastotomy** (*mas-tot'-o-me*) [*masto-;* τέμνειν, to cut]. Incision of a breast.
**mastous** (*mas'-tus*) [μαστός, breast]. Having large breasts.
**masturbation** (*mas-ter-ba'-shun*) [*masturbari,* to pollute one's self]. Production of the venereal orgasm by friction of the genitals.
**mastzellen** (*mast-tsel-en*). See *cells, mast.*
**Matas',** Rudolph (*mat'-as*) [Rudolph *Matas,* American surgeon, 1860– ]. An appliance for occluding blood-vessels while the condition of the collateral circulation is being tested. M.'s **operation.** *For the radical cure of aneurysm:* consists in arrest of the circulation in the sac, opening, evacuating, and cleansing the sac, and closing the openings by continuous fine sutures; endoaneurysmorrhaphy.
**maté** (*mah'-ta*) [Sp., *mate,* a vessel]. The leaves of *Ilex paraguayensis,* used in South America as a substitute for tea and coffee. Its properties are due to thein. Syn., *Paraguay tea.*
**materia medica** (*mat-e'-re-ah med'-ik-ah*) [L., "medical matter"]. The science that treats of the sources and preparations of the drugs and agents used in medicine.
**materies morbi** (*mat-e'-re-ēs mor'-bi*) [L., "matter of disease"]. The material that is the cause of a disease.
**maternal** (*ma-ter'-nal*) [*mater,* mother]. Pertaining to the mother. m. **impressions.** See *impressions, maternal.*

**maternity** (*ma-ter'-nit-e*) [see *maternal*]. 1. Motherhood. 2. A lying-in hospital.
**matico** (*mat-e'-ko*) [Sp.]. The leaves of *Piper angustifolium*, of the order *Piperaceæ*. It is aromatic and stimulant, and has been used as a local and general hemostatic and as an alterative stimulant to mucous membranes. Dose 60 gr. (4 Gm.). **m. fluidextract of** (*fluidextractum matico*, U. S. P.). Dose ⅓–1 dr. (2–4 Cc.). **m., tincture of.** Dose 1 dr. (4 Cc.).
**matlazahuatl** (*mat-lahs-ah-what'-l*) [Aztec word]. A form of typhus fever found in Mexico; tabardillo.
**matrass** (*mat'-ras*) [Fr., *matras*, a chemical vessel]. A glass vessel with a long neck and a round body used in various chemical manipulations.
**matricaria** (*mat-rik-a'-re-ah*) [*matrix*]. German chamomile; the flower-tops of *M. chamomilla*, of the order *Compositæ*. Matricaria contains a volatile oil and a bitter extractive principle, and is a mild tonic; and in large doses emetic and antispasmodic.
**matriculate** (*mat-rik'-u-lāt*) [*matricula*, a register]. To receive admission and to enroll one's self as a member of a college or university.
**matrix** (*ma'-triks*) [L., "a mold in which anything is cast"]. 1. A mold; the cavity in which anything is formed. 2. That part of tissue into which any organ or process is set, as the *matrix* of a tooth or of a nail. 3. The intercellular substance of a tissue, as of cartilage. 4. The uterus.
**matrixitis** (*ma-triks-i'-tis*). Same as *onychia*.
**matter** (*mat'-er*) [*materia*, matter]. 1. Physical substance. 2. Pus.
**mattoid** (*mat'-oid*) [*mattus*, drunk, stupid; εἶδος, like]. A person half-crazed; a crank, or paranoiac.
**matula** (*mat'-u-lah*) [L.]. A urinal.
**maturate** (*mat'-u-rāt*) [*mature*]. To suppurate.
**maturation** (*mat-u-ra'-shun*) [*mature*]. Ripening, as the ripening of the ovum or of a cataract.
**mature** (*ma-tūr'*) [*maturare*, to ripen]. 1. To ripen. 2. Ripe.
**maturity** (*ma-tū'-rit-e*) [*maturitas*; *maturare*, to ripen]. Full development; the quality or period of complete growth.
**matutinal** (*ma-tū'-tin-al*) [*Matuta*, goddess of the morning]. Occurring in the morning, as matutinal nausea.
**matzol** (*mat'-sol*). A mixture of cod-liver oil, 50 parts; matzoon, 45 parts; emulsifying ingredients, 5 parts.
**matzoon** (*mat'-zoon*). Milk fermented with a peculiar ferment obtained from Asiatic Turkey. It is used like kumiss in irritated states of the gastrointestinal tract.
**Mauchart's ligaments** (*mow'-shar*) [Burchard David *Mauchart*, German anatomist, 1696–1751]. The lateral or alar odontoid ligaments.
**Maumené's test for sugar** (*mōm-na'*) [Edme Jules *Maumené*, French chemist, 1818–     ]. A strip of flannel saturated with a 33⅓ % solution of stannous chloride is dipped into the liquid; on heating it to nearly 150° C. it will turn brownish-black.
**Maunoir's hydrocele** (*mo-nwar'*) [Jean Pierre *Maunoir*, French surgeon, 1768–1861]. A cystic tumor occurring in the neck between the angle of the inferior maxilla and the mastoid process. Syn., *hydrocele colli*.
**Maurer's dots or clefts** (*mow'-rer*). Large irregular formations, of uncertain significance, found in the red blood corpuscles in subtertian malaria.
**Mauriceau's lance** (*mo-ris-o'*) [François *Mauriceau*, French obstetrician, 1637–1709]. An instrument for perforating the fetal head in craniotomy.
**Mauthner's sheath** (*mowt'-ner*) [Ludwig *Mauthner*, Austrian physician, 1840–1894]. The thin, longitudinally striated, protoplasmic layer surrounding the axis-cylinder of a nerve-fiber. **M.'s test** for color vision, 33 small bottles filled with different pigments—some with one, others with two (pseudo isochromatic and anisochromatic), pigments—are employed in the manner of Holmgren's worsteds.
**mauvein** (*maw'-ve-in*) [Fr., *mauve*, mallow], C₅₄H₅₄N₄. A base derived from anilin.
**maxilla** (*maks-il'-ah*) [L.; *pl.*, *maxillæ*]. 1. The bone of the upper or lower jaw. 2. Specifically, the upper jaw-bone.
**maxillary** (*maks'-il-a-re*) [*maxilla*]. Pertaining to the maxillæ or jaws. **m. bones,** the bones of the jaws, consisting of the lower and upper jaw. **m. fissure,** the cleft in the upper maxilla for the maxillary process of the palate bone. **m. nerve, inferior.** See

under *nerve*. **m. nerve, superior.** See under *nerve*. **m. sinus,** the antrum of Highmore in the superior maxilla.
**maxillate** (*maks'-il-āt*). Furnished with jaws.
**maxillen** (*maks'-il-en*) [*maxilla*, jaw-bone]. Belonging to the maxillary bone in itself.
**maxilliferous** (*maks-il-if'-er-us*). See *maxillate*.
**maxillitis** (*maks-il-i'-tis*) [*maxilla*, jaw; ἶτις, inflammation]. 1. Inflammation of a maxilla. 2. Inflammation of a maxillary gland.
**maxillodental** (*maks-il-o-den'-tal*). Pertaining to the jaw and the teeth.
**maxillojugal** (*maks-il-o-joo'-gal*). Pertaining to the maxilla and the zygoma.
**maxillomandibular** (*maks-il-o-man-dib'-u-lar*). Pertaining to the upper jaw and the lower jaw.
**maxillomuscular** (*maks-il-o-mus'-kū-lar*). Relating to the maxillary muscles.
**maxillopalatine** (*maks-il-o-pal'-at-in*). Pertaining to the maxilla and the palatine bone.
**maxillopharyngeal** (*maks-il-o-far-in'-je-al*). Pertaining to the jaw and the pharynx.
**maxillosuprafacial** (*maks-il-o-sū-prah-fa'-shal*). Relating to the maxilla and the upper portion of the face.
**maxilloturbinal** (*maks-il-o-tur'-bin-al*). Pertaining to the maxilloturbinal bone. **m.-t. bone,** the inferior turbinate bone.
**maximal** (*maks'-im-al*) [*maximum*]. Pertaining to the maximum; highest; largest. **m. thermometer,** one registering the highest point reached by the temperature.
**maximum** (*maks'-im-um*) [L., neuter of *maximus*, the greatest]. The greatest or highest degree or amount of anything; the highest point attained or attainable by anything. **m. dose,** the largest dose of a medicament that may be given safely. **m. temperature,** the temperature above which bacterial growth does not occur.
**Maxwell's experiment** (*maks'-well*) [James Clerk *Maxwell*, English physicist, 1831–1879]. On looking through a chrome alum solution an oval purplish spot, due to the pigment of the macula lutea, is seen.
**Maxwell's ring.** See *Loewe's ring*.
**May-apple.** See *Podophyllum*.
**maydis** (*ma'-dis*). See under *ustilago*.
**Maydl's method** (*mādl*) [Karl *Maydl*, German surgeon, 1853–1903]. The transplantation of the ureters into the rectum in the treatment of exstrophy of the bladder.
**mayhem** (*ma'-hem*) [OF., *mehaigner*, to hurt]. Maiming.
**mayidism** (*ma'-i-dizm*). Pellagra.
**mayidismus** (*ma-id-iz'-mus*). Same as *pellagra*.
**mayol** (*ma'-ol*). A meat-preservative introduced by May, of Budapest, and said to be a mixture of boric acid, ammonium fluoride, glycerol, and alcohol (methyl and ethyl).
**Mayor's sign of pregnancy** (*mār*). The hearing of the fetal heart-sounds.
**Mayo-Robson's point** [Arthur William *Mayo-Robson*, English surgeon]. A spot slightly above the umbilicus where pressure causes tenderness in cases of pancreatic disease.
**mays** (*māz*) [L.]. The genus *zea*, *q. v.*
**Maytenus** (*ma'-ten-us*) [*Mayten*, Chilian name]. A genus of shrubs of the order *Celastrineæ*. *M. boariasis* is indigenous to Chili; the leaves are used on inflammatory swellings, especially in poisoning by species of *Rhus*.
**maza** (*ma'-zah*) [μᾶζα, a cake]. The placenta.
**mazalgia** (*ma-zal'-ge-ah*) [μᾶζα; ἄλγος, pain]. Mastalgia, or mastodynia.
**mazalysis** (*ma-zal'-is-is*) [μᾶζα, cake; λύσις, Retention of the placenta. Sometimes used as synonym of *mazolysis*, *q. v.*
**mazic** (*ma'-zik*) [μᾶζα, placenta]. Pertaining to the placenta.
**mazischesis** (*ma-zis'-kes-is*). Synonym of *mazalysis*.
**mazocacothesis** (*ma-zo-kak-oth'-es-is*) [μᾶζα, cake; κακός, ill; θέσις, placing]. Faulty implantation of the placenta.
**mazodynia** (*ma-zo-din'-e-ah*). See *mastodynia*.
**mazoitis** (*ma-zo-i'-tis*). See *mastitis*.
**mazology** (*ma-zol'-o-je*). Same as *mastology*.
**mazolysis** (*ma-zol'-is-is*) [μᾶζα, cake; λύσις, loosing]. Separation of the placenta.
**mazolytic** (*ma-zo-lit'-ik*). Pertaining to mazolysis.

**mazopathy** (ma-zop'-ath-e) 1. [μᾶζα, cake, placenta; πάθος, illness]. Any disease of the placenta. 2. [μαζός, breast; πάθος, illness]. Same as *mastopathy*.
**mazopexy** (ma'-zo-peks-e) [μαζός, breast; πῆξις, fixation]. Surgical fixation of a pendulous breast; mastopexy.
**Mazzoni's corpuscle** (mad-zo'-ne) [Vittorio *Mazzoni*, Italian physician]. A peripheral ending of a sensory nerve closely resembling Krause's end-bulb.
**M.B.** Abbreviation of *Medicinæ Baccalaureus*, Bachelor of Medicine.
**M.C.** Abbreviation for *Magister Chirurgiæ*, Master of Surgery.
**M.C.D.** Abbreviation for Doctor of Comparative Medicine.
**M.Ch.** Abbreviation for *Magister Chirurgiæ*, Master of Surgery.
**M.D.** Abbreviation of *Medicinæ Doctor*, Doctor of Medicine.
**M.D.S.** Abbreviation of *Master of Dental Surgery*.
**meable** (me'-a-bl) [*meabilis*, easily penetrating]. Capable of being readily traversed or passed through.
**mead** (mēd). Dilute, fermented honey or syrup flavored. See *hydromel* and *metheglin*.
**meadow-saffron.** See *colchicum*.
**measle** (mēz'-el). An individual *Cysticercus cellulosæ*. m. of pork. See *cysticercus*. m.-worm, cysticercus.
**measles** (mēz'-elz) [Du., *maselin*, measles]. 1. An acute, infectious disease, characterized by a peculiar eruption and by catarrhal inflammation of the mucosa of the conjunctiva and of the air-passages. After a period of incubation of nearly two weeks the disease begins with a chill, fever, coryza, cough, and conjunctivitis; on the third or fourth day a duskyred, papular eruption appears, arranged in the form of crescentic groups. After having reached its maximum, in three or four days, the eruption gradually fades, and is followed by a branny desquamation. The disease affects principally the young, is exceedingly contagious, and one attack of it confers almost perpetual immunity. Its cause is thought to be a bacillus. 2. A disease of hogs, cattle, and sheep, due to the presence in the body of *Cysticercus cellulosæ* and larvæ of other tape-worms. 3. The cysticerci themselves. m., black, m., hemorrhagic, a grave variety of measles in which the eruption is hemorrhagic and the constitutional symptoms profound. m., German. See *rubella*.
**measly** (mēz'-lē) [measles]. Containing measles (cysticerci).
**meat** (mēt) [AS., *mete*]. The muscular tissues of an animal, used as food.
**meatal** (me-at'-al). Pertaining to a meatus.
**meatometer** (me-at-om'-et-er) [*meatus*, meatus; μέτρον, measure]. An instrument used in measuring the caliber of any meatus, specifically of the meatus urinarius.
**meatorrhaphy** (me-at-or'-af-e) [*meatus*; ῥαφή, suture]. Suture of the cut end of a meatus, generally the urinary meatus, after a meatotomy.
**meatoscope** (me-at'-o-skōp) [*meatus*, meatus; σκοπεῖν, to inspect]. A speculum used in the examination of a meatus, specifically the distal portion of the male urethra.
**meatotome** (me-at'-o-tōm) [*meatus*, meatus; τομή, a cutting]. A cutting instrument used in performing meatotomy.
**meatotomy** (me-at-ot'-o-me) [*meatus*, meatus; τομή, section]. Surgical incision of a meatus, particularly the meatus urinarius.
**meatox** (mēt'-oks). Trade name of a preparation of beef in powder form.
**meatus** (me-a'-tus) [*meare*, to flow or pass; pl., *meatus*]. An opening or passage. m. auditorius externus, the canal extending from the concha to the membrana tympani. m. auditorius internus, the internal auditory canal. m. nasi communis, the part of the nasal cavity into which the three meatus of the nose open. m. nasopharyngeus, that part of the nasal cavity communicating with the pharynx beneath the body of the sphenoid. m. of nose, one of the three passages into which the turbinal bones divide the nasal cavity. m: urethræ, m. urinarius, the orifice of the urethra.
**Mecca balsam.** See *balm of Gilead*.
**mechanic, mechanical** (me-kan'-ik, me-kan'-ik-al) [μηχανή, a machine]. Pertaining to mechanics or to physical forces, not to chemical or vital forces. m. theory, Virchow's theory of tumor-formation,

according to which tumors are due primarily to local irritation.
**mechanics** (me-kan'-iks) [*mechanic*]. The science that treats of the influence and effects of force upon matter, and that may be divided into *statics*, the science treating of matter at rest, and *dynamics*, that treating of matter in motion.
**mechanism** (mek'-an-izm) [*mechanic*]. 1. An aggregation of parts arranged in a mechanical way to perform the functions of a machine. 2. The manner in which a mechanical act is performed, as the *mechanism* of labor.
**mechanotherapy** (mek-an'-o-ther'-ap-e) [*mechanic;* θεραπεία, treatment]. The use of mechanical agencies in the treatment of injury or disease.
**meche** (māsh) [Fr., *wick*]. A piece of gauze used as a surgical tent, or drain.
**mecism** (me'-sizm) [μῆκος, length]. A condition marked by abnormal prolongation of one or more parts of the body.
**Meckel's cartilage** (mek'-el) [1. Johann Friedrich *Meckel*, German anatomist, 1717–1774; 2. Johann Friedrich *Meckel*, German surgeon, 1781–1833]. [2]. The axis of the first branchial arch (mandibular arch) of the fetus. It disappears during the fifth or sixth month, with the exception of its posterior (tympanic) portion, which becomes the incus, malleus, and Folian process. A vestige of this cartilage (pinnal cartilage) is occasionally found in tumors of the parotid gland. M.'s cavity. [1]. A recess in the dura over the summit of the petrosa for the reception of the two roots of the fifth cranial nerve after their exit from the pons. Syn., *cavum Meckelii*. M.'s crural arch. [1]. See *Poupart's ligament*. M.'s diverticulum. [1]. The remains of the vitelline duct, frequently met as a small elongated pouch attached to the lower portion of the ileum. M.'s ganglion. [1]. The sphenopalatine ganglion. M.'s rod. Same as *M.'s cartilage*. M.'s space. [1]. A dural space lodging the Gasserian ganglion.
**meckelectomy** (mek-el-ek'-to-me) [*Meckel's ganglion*; ἐκτομή, excision]. Excision of Meckel's ganglion.
**mecometer** (me-kom'-et-er) [μῆκος, length; μέτρον, measure]. An instrument used in measuring newborn infants.
**mecon** (me'-kon) [L.]. 1. The poppy. 2. Opium.
**meconalgia** (me-kon-al'-je-ah) [*mecon;* ἄλγος, pain]. Pain or neuralgia following the disuse of opium.
**meconarceine** (mek-o-nar'-se-in) [*mecon*]. A mixture of alkaloids of opium, free from morphine, having sedative properties. Recommended in bronchial affections and neuralgia. Dose ⅛–½ gr. (0.01–0.03 Gm.).
**meconate** (mek'-on-āt) [*mecon*]. A salt of meconic acid.
**meconeuropathia** (mek-on-ū-ro-pa'-the-ah) [μῆκον, opium; νεῦρον, nerve; πάθος, illness]. Nervous disorder due to the abuse of opium, or its narcotic derivatives.
**meconic** (mek-on'-ik). Pertaining to opium. m. acid. See *acid, meconic*.
**meconidine** (mek-on'-id-ēn) [μῆκων, poppy], C₂₁H₂₃-NO₄. An amorphous alkaloid of opium.
**meconin** (mek'-on-in) [*mecon*], C₁₀H₁₀O₄. A crystalline substance that is obtained on boiling narcotine with water. It is hypnotic. Dose 1 gr. (0.06 Gm.).
**meconiorrhea** (mek-on-e-or'-e-ah) [μηκώνιον, meconium; ῥοία, flow]. A morbidly free discharge of meconium.
**meconium** (mek-o'-ne-um) [*mecon*]. The first fecal discharges of the newborn, a dark-green, viscid substance, composed of the secretion of the liver with exfoliated epithelium from the bowel.
**meconoiosin** (mek-on-oi'-o-sin) [μῆκων, poppy], C₂₁H₂₃O₃. A derivative of opium, crystalline in character and giving a dark-red color with sulphuric acid.
**meconism** (mek'-on-izm) [*meconismus;* μῆκων, poppy]. The opium-habit; opium-poisoning, especially of the chronic kind.
**meconology** (mek-on-ol'-o-je) [μῆκων, opium; λόγος, science]. The botany and pharmacology of opium, its allies and derivatives.
**meconophagism** (mek-on-off'-aj-izm) [μῆκων, opium; φαγεῖν, to eat]. The habit of opium-eating.
**meconophagist** (mek-on-off'-aj-ist) [μῆκων, opium; φαγεῖν, to eat]. An opium-eater.
**medea** (me'-de-ah) [Μήδεια, Medea, a sorceress]. 1. The genital organs. 2. Aphrodisiacs.

**media** (*me'-de-ah*) [fem. of *medius*, middle]. The middle coat of a vein, artery, or lymph-vessel.
**media** (*me'-de-ah*) [L.]. Plural of medium. **m., transparent,** of eye, the cornea, aqueous humor, lens, and vitreous humor.
**mediad** (*me'-de-ad*) [*median*]. Toward the median plane or line.
**medial** (*me'-de-al*). 1. See *median*. 2. Internal, as opposed to *lateral* (external).
**median** (*me'-de-an*) [*medius*, middle]. Situated or placed in the middle; mesal or mesial. **m. artery.** See under *artery*. **m. nerve.** See under *nerve*.
**mediastinal** (*me-de-as-ti'-nal*). Pertaining to the mediastinum.
**mediastinitis** (*me-de-as-tin-i'-tis*) [*mediastinum*; *itis*, inflammation]. Inflammation of the cellular tissue of the mediastinum.
**mediastinopericarditis** (*me-de-as-tin-o-per-ik-ar-di'-tis*) [*mediastinum*; *pericarditis*]. Combined inflammation of the mediastinum and the pericardium. **m., callous,** that attended with fibrous thickening of the pericardium.
**mediastinotomy** (*me-de-as-tin-ot'-o-me*) [*mediastinum*; τομή, an incision]. Incision into the mediastinum.
**mediastinum** (*me-de-as-ti'-num*) [*in medio stare*, to stand in the middle]. 1. A partition separating adjacent parts. 2. The space left in the middle of the chest between the two pleuræ, divided into the anterior, middle, posterior, and superior mediastinum. The *anterior mediastinum* contains the origins of the triangularis sterni muscles, the internal mammary vessels of the left side, loose areolar tissue, lymphatic vessels, and a few lymphatic glands. The *middle mediastinum* contains the heart and pericardium, the ascending aorta, the superior vena cava, the bifurcation of the trachea, the pulmonary arteries and veins, and the phrenic nerves. The *posterior mediastinum* contains a part of the aorta, the greater and lesser azygos veins, the pneumogastric and splanchnic nerves, the esophagus, the thoracic duct, and some lymphatic glands. The *superior mediastinum*, that part lying above the pericardium, contains the origins of the sternohyoid and sternothyroid muscles, and part of the longus colli muscles, the transverse portion of the aortic arch, the innominate, left carotid, and subclavian arteries, the superior vena cava and the innominate veins, the left superior intercostal vein, the pneumogastric, cardiac, phrenic, and left recurrent laryngeal nerves, the trachea, esophagus, thoracic duct, the remains of the thymus gland, and lymphatics. **m. testis,** a septum in the posterior portion of the testicle formed by a projection inward of the tunica albuginea.
**mediate** (*me'-de-āt*) [*media*]. Indirect; performed through something interposed, as mediate percussion, percussion on a pleximeter.
**medibasilic vein** (*me-de-bas-il'-ik*). The median basilic vein. See *vein*.
**medicable** (*med'-ik-a-bl*) [*medicari*, to heal]. Amenable to cure.
**medical** (*med'-ik-al*) [*medicine*]. Pertaining to medicine. **m. diseases,** diseases treated by the physician, as distinguished from surgical diseases. **m. ethics,** those principles of justice, honor, and courtesy that regulate the intercourse and conduct of physicians. **m. jurisprudence.** See *jurisprudence, medical*.
**medicament** (*med-ik'-am-ent*) [*medicine*]. A medicinal substance.
**medicamentum** (*med-ik-am-en'-tum*). See *medicament*. **m. arcanum,** a proprietary or secret remedy.
**medicaster** (*med-ik-as'-ter*). Old term for a quack.
**medicated** (*med'-ik-a-ted*). Impregnated with a medicinal substance.
**medication** (*med-ik-a'-shun*) [*medicus*]. 1. Impregnation with a medicine. 2. Treatment by medicines; the administration of medicines. **m., endermic.** See *cataphoresis.* **m., hypodermatic,** treatment by the introduction of medicines beneath the skin, usually by means of a hypodermatic syringe. **m., ionic.** See *cataphoresis*.
**medicephalic** (*me-de-sef-al'-ik*) [*medius*, middle; κεφαλή, head]. Median cephalic. See *vein*.
**medicerebellar** (*me-de-ser-e-bel'-ar*) [*medius*, middle; *cerebellum*]. Pertaining to the intermediate region of the cerebellum.
**medicerebral** (*me-de-ser'-e-bral*) [*medius*, middle; *cerebrum*, brain]. Pertaining to the central portion of the cerebrum.

**medicinal** (*med-is'-in-al*) [*medicine*]. Pertaining to, or having the nature of, a medicine. **m. rashes,** eruptions on the skin following the internal administration of certain drugs.
**medicine** (*med'-is-in*) [*medicari*, to heal]. 1. Any substance given for the cure of disease. 2. The science of the treatment of disease; the healing art. In a restricted sense, that branch of the healing art dealing with internal diseases. **m., anatomical,** that system which deals with the anatomical changes in diseased organs and their connection with symptoms manifested during life. **m., clinical,** the study of disease by the bedside of the patient. **m., experimental,** that based upon experiments on animals and the observation of pathological changes in diseases induced in them and the effect of drugs administered. **m., forensic, m., legal,** medical jurisprudence, or medicine in its relation to questions of law. **m., patent,** medicine the manufacture of which is protected by letters patent. **m., practice of,** the practical application of the principles taught by the theory of medicine. **m., preventive,** that which aims at the prevention of disease. **m., proprietary,** one the manufacture of which is limited or controlled by an owner, because of a patent, a copyright, or secrecy as regards its constitution or method of manufacture. **m., spagyric,** that of the school of Paracelsus. **m., state,** medical jurisprudence. **m., vibratory,** a method of treating nervous diseases, paralysis agitans, etc., by mechanical shaking or percussion, by means of journeys on railroad or wagon or specially devised apparatus for methodical shaking of the body.
**medicinerea** (*me-de-sin-e'-re-ah*) [*medius*, middle, and *cinereus*, ashen]. The gray matter of the claustrum and lenticula of the brain, lying between the cortex or ectocinerea and the entocinerea.
**medicisterna** (*med-e-sis-tur'-nah*) [*medius*, middle; *cisterna*, a vessel]. The cisterna venæ magnæ cerebri.
**medicochirurgical** (*med-ik-o-ki-rur'-jik-al*) [*medicine*; *chirurgicus*, a surgeon]. Pertaining conjointly to medicine and surgery.
**medicolegal** (*med-ik-o-le'-gal*) [*medicine*; *legalis*, legal]. Relating both to medicine and to the law.
**medicomechanical** (*med-ik-o-me-kan'-ik-al*). Medical and mechanical.
**medicommissure** (*me-de-kom'-is-ūr*) [*medius*, middle; *commissura*, commissure]. The middle commissure of the third ventricle. The junction of the mesal surfaces of the thalami. It is in a direct line between the porta and the aqueduct, and just dorsad of the aulix. It consists mainly of cells, and is so soft as commonly to be torn during the removal of the brain.
**medicon** (*med'-ik-on*) [μηδικόν]. A harmful or noxious drug.
**medicophysical** (*med-ik-o-fiz'-ik-al*). Both medical and physical.
**medicopsychological** (*med-ik-o-si-ko-loj'-ik-al*) [*medicus* physician; ψύχη, mind; λόγος, science]. Pertaining to medicopsychology.
**medicopsychology** (*med-ik-o-si-kol'-o-je*) [*medicus*, physician; ψυχολογια, the science of the mind]. The study of mental diseases.
**medicornu** (*me-de-kor'-nū*) [*medius*, middle; *cornu*, horn]. The middle horn of the lateral ventricle.
**medicostatistic** (*med-ik-o-stat-is'-tik*). Relating to medicine as connected with statistics.
**medicus** (*med'-ik-us*) [L.]. A physician.
**medifixed** (*me'-de-fikst*) [*medius*, middle; *fixus*, fixed]. Attached by the middle.
**medifrontal** (*me-de-fron'-tal*) [*medius*, middle; *frons*, forehead]. Middle of the forehead.
**mediglycin** (*med-ig'-lis-in*). A liquid glycerol soap used as a vehicle.
**medinal** (*med'-in-al*). Trade name of the sodium salt of veronal, used as a hypnotic in doses of 5-10 gr. (0.3-0.6 gm.).
**Medina-worm** (*me-di'-nah-wurm*). See *Filaria medinensis*.
**medio-** (*me-de-o-*) [*medius*, middle]. A prefix meaning middle.
**mediocarpal** (*me-de-o-kar'-pal*) [*medius*, middle; *carpus*]. Pertaining to the articulation between the two rows of carpal bones.
**mediocolic** (*me-de-o-kol'-ik*) [*medius*, middle; κῶλον, colon]. Pertaining to the middle portion of the colon.
**mediodorsal** (*me-de-o-dor'-sal*) [*medius*, middle; *dorsum*, back]. Both median and dorsal; on the mesial line of the back.

**mediofrontal** (*me-de-o-frun'-tal*) [*medius,* middle; *frons,* forehead]. Pertaining to the middle of the forehead.
**mediolateral** (*me-de-o-lat'-er-al*) [*medio-;* *latus,* side]. Pertaining to the middle and to a side.
**mediopalatine** (*me-de-o-pal'-at-in*). Relating to the center of the palate.
**mediopontine** (*me-de-o-pon'-tin*) [*medio-;* *pons,* bridge]. Pertaining to the central portion of the pons.
**mediotarsal** (*me-de-o-tar'-sal*) [*medio-;* *tarsus*]. Pertaining to the middle articulation of the tarsal bones.
**medipeduncle** (*me-de-pe-dung'-kl*) [*medius,* middle; *pedunculus,* peduncle]. The middle peduncle of the cerebellum; the lateral intermediate continuation of the cerebellum to the pons.
**mediscalenus** (*me-de-ska-le'-nus*) [*medius,* middle; *scalenus*]. Synonym of *scalenus medius;* see *muscles, table of.*
**medisect** (*me-de-sekt'*) [*medius,* middle; *secare,* to cut]. To make a medisection, *q. v.*
**medisection** (*me-de-sek'-shun*) [*medius,* middle; *sectio,* from *secare,* to cut]. Section of the body, or of any symmetrical part, at the median longitudinal anteroposterior plane.
**Mediterranean fever** (*med-it-er-a'-ne-an*). See *fever, Mediterranean.*
**meditrina** (*med-it-ri'-nah*). A concentrated germicidal electrozone.
**medium** (*me'-de-um*) [neuter of *medius,* middle; *pl., media*]. 1. That in which anything moves or through which it acts. 2. The soil upon which anything grows, especially a substance used for cultivating bacteria; culture or nutrient medium.
**medius** (*me'-de-us*) [*medius,* middle]. 1. The middle. 2. The middle finger.
**medoblennorrhea, medoblennorrhœa** (*me-do-blen-or-e'-ah*). Synonym of *gonorrhea* and *gleet.*
**medol** (*med'-ol*). Trade name of a preparation of creolin; chiefly used in veterinary practice.
**medorrhea, medorrhœa** (*me-dor-e'-ah*) [μήδεα, genitals; ῥοία, a flow]. A discharge from the reproductive organs. m. **urethralis,** gonorrhea. m. **virilis,** gonorrhea of the male urethra.
**medorrhoic** (*me-dor-o'-ik*) [μήδεα, genitals; ῥοία, flow]. Pertaining to medorrhea.
**medulla** (*me-dul'-ah*) [L., "marrow"]. 1. The marrow. 2. The medulla oblongata. 3. Anything resembling marrow in structure or in its relation to other parts—as a fatty substance or marrow occupying certain cavities. Also the central parts of certain organs as distinguished from the cortex. 4. The same as *corpus medullare* or *corpus dentatum.* m. of **kidney.** See under *kidney.* m. of **nerve-fiber,** the white substance of Schwann. See under *nerve-fiber.* m. **oblongata,** the upper enlarged part of the spinal cord, extending from the cord opposite the foramen magnum to the pons Varolii. m. **ossium,** bone marrow. m. **ossium rubra,** red bone-marrow; recommended in the treatment of skin diseases and in anemia. m. **spinalis,** the spinal cord or marrow.
**medulladen** (*med-ul-ad'-en*). A preparation of bone-marrow of beef; it is used in anemia, gout, etc. Dose 30–45 gr. (2–3 Gm.).
**medullar** (*me-dul'-ar*). Synonym of *medullary.*
**medullary** (*med'-ul-a-re*) [*medulla*]. 1. Pertaining to the marrow; resembling marrow. 2. Pertaining to any medulla, as that of the brain. 3. Pertaining to the medulla oblongata. m. **canal,** the hollow interior of long bones in which the marrow lies. m. **carcinoma,** a soft carcinoma very rich in cells. m. **foramen,** a nutrient foramen. m. **groove,** a longitudinal groove at the anterior part of the embryonal shield of the blastoderm. m. **rays.** See under *kidney.* m. **sheath,** the semifluid white matter between the enveloping sheath and central axis-cylinder of a nerve.
**medullated** (*med'-ul-a-ted*). Containing or covered by medulla or marrow. m. **nerve-fibers, nerve-fibers** provided with a medullary sheath, the white substance of Schwann.
**medullation** (*med-u-la'-shun*). The process of acquiring a medulla, as in the case of many nerve-fibers in the course of their development.
**medullin** (*med-ul'-in*) [*medulla,* marrow]. 1. A variety of cellulose obtained from the pith or medulla of certain plants. 2. The extract of the spinal cord of the ox; it is used in ataxia.
**medullispinal** (*med-ul-e-spi'-nal*) [*medulla; spine*]. Relating to the spinal cord.

**medullitis** (*med-ul-i'-tis*) [*medulla;* ιτις, inflammation]. 1. Inflammation of marrow. 2. Myelitis.
**medullization** (*med-ul-iz-a'-shun*) [*medulla*]. Conversion into marrow, as the softening of bone-tissue in the course of osteitis.
**medulloarthritis** (*med-ul-o-ar-thri'-tis*) [*medulla,* marrow; ἄρθρον, joint; ιτις, inflammation]. Inflammation of the marrow-elements of the cancellated articular portion of a bone.
**medullocell** (*med-ul'-o-sel*) [*medulla,* marrow; *cellula,* cell]. A marrow-cell; myelocyte.
**medulloencephalic** (*med-ul-o-en-sef-al'-ik*) [*medulla,* marrow; ἐγκέφαλος, brain]. Pertaining to the medulla and the encephalon; myeloencephalic.
**medullose, medullous** (*med-ul'-ōs, -us*). Containing much pith or marrow.
**mega-, megalo-** (*meg-ah-, meg-al-o-*) [μέγας, large]. Prefixes signifying large; also, indicating a unit 1,000,000 times greater than the unit to which it is prefixed.
**megabacteria** (*meg-ah-bak-te'-re-ah*) [*mega-;* βακτήριον,* bacterium]. The largest kind of bacteria; a group of the coccobacteria.
**megacephalia** (*meg-ah-sef-a'-le-ah*) [*mega-;* κεφαλή, head]. The megacephalic condition.
**megacephalic, megacephalous** (*meg-ah-sef-al'-ik, meg-ah-sef'-al-us*). See *megalocephalic.*
**megacheilus** (*meg-ah-ki'-lus*) [*mega-;* χεῖλος, lip]. Large-lipped.
**megacoccus** (*meg-ah-kok'-us*) [*mega-;* κόκκος, a berry]. A large-sized coccus.
**megacoly** (*meg-ak'-ol-e*). See *megalocoly.*
**megadyne** (*meg'-ah-dīn*) [*mega-;* δύναμις, power]. A unit equal to a million dynes.
**megafarad** (*meg-ah-far'-ad*) [*mega-;* Faraday]. An electric unit equal to a million farads.
**megagamete** (*meg-ag-am'-ēt*). Same as *macrogamete.*
**megagastria** (*meg-ah-gas'-tre-ah*) [*mega-;* γαστήρ, belly]. Auxesis, or abnormal enlargement of the abdomen.
**meganathus** (*meg-ah-na'-thus*) [*mega-;* γνάθος, jaw]. Large-jawed.
**megakaryocyte, megacaryocyte** (*meg-ah-kar'-e-o-sīt*). See *myeloplax.*
**megalgia** (*meg-al'-je-ah*) [*mega-;* ἄλγος, pain]. Excessively severe pain.
**megalo-.** See *mega.*
**megaloblast** (*meg-al-o-blast*) [*megalo-;* βλαστός, a germ]. A giant-corpuscle of the blood. The term is restricted to embryonic or germinal cells as distinguished from the megalocyte, which pertains to adult life.
**megalocardia** (*meg-al-o-kar'-de-ah*) [*mega-;* καρδία, heart]. Auxesis, or enlargement of the heart.
**megalocephalic** (*meg-al-o-sef-al'-ik*) [*megalocephaly*]. Large-headed; applied to a skull the capacity of which exceeds 1450 Cc.
**megalocephaly** (*meg-al-o-sef'-al-e*) [*megalo-;* κεφαλή, head]. 1. The condition of having a very large head. 2. A disease characterized by progressive enlargement of the head, face, and neck, involving both the bony and the soft tissues. Syn., *leontiasis ossea.*
**megalocerus** (*meg-al-os'-er-us*) [*mega-;* κέρας, horn]. A monstrosity with horn-like projections on the forehead.
**megalocheirous** (*meg-al-o-ki'-rus*) [*megalo-;* χείρ, hand]. Large-handed; having large antennæ.
**megalocoly** (*meg-al-ok'-ol-e*) [*megalo-;* κόλον, colon]. A uniform increase in the internal diameter of the colon, with thickening of the walls.
**megalocornea** (*meg-al-o-kor'-ne-ah*) [*megalo-;* *cornea*]. An enlarged condition of the cornea.
**megalocyte.** (*meg'-al-o-sīt*) [*megalo-;* κύτος, cell]. An abnormally large red blood-corpuscle.
**megalocytosis** (*meg-al-o-si-to'-sis*) [*megalocyte*]. The presence of large numbers of greatly enlarged erythrocytes in the blood.
**megalodactylous** (*meg-al-o-dak'-til-us*) [*megalo-;* δάκτυλος, digit]. Having abnormally large fingers and toes.
**megalogastria** (*meg-al-o-gas'-tre-ah*) [*megalo-;* γαστήρ, belly]. Abnormal enlargement of the abdomen.
**megaloglossia** (*meg-al-o-glos'-e-ah*). See *macroglossia.*
**megalokaryocyte** (*meg-al-o-kar'-e-o-sīt*). 1. A cell having a large nucleus. 2. A cell of the bone-marrow having a large, irregular, coiled nucleus.
**megalomania** (*meg-al-o-ma'-ne-ah*) [*megalo-;* μανία, madness]. 1. Mania characterized by delusions of grandeur. 2. The delirium of grandeur.

**megalomelia** (*meg-al-o-me'-le-ah*) [*megalo-*; μέλος, limb]. A monster with excessively large limbs.

**megalonychosis** (*meg-al-on-ik-o'-sis*) [*megalo-*; ὄνυξ, nail]. Universal noninflammatory enlargement of the nails.

**megalopenis** (*meg-al-o-pe'-nis*). Excessive size of the penis.

**megalophonic, megalophonous** (*meg-al-o-fo'-nik, meg-al-off'-o-nus*). Synonym of *macrophonous*.

**megalophthalmus** (*meg-al-of-thal'-mus*) [*megalo-*; ὀφθαλμός, eye]. Excessively large eyes.

**megalopia** (*meg-al-o'-pe-ah*). Synonym of *megalopsia*.

**megalopodia** (*meg-al-o-po'-de-ah*) [*megalo-*; πούς, foot]. The condition of having large feet.

**megaloporous** (*meg-al-op'-or-us*). Characterized by large pores.

**megalopsia** (*meg-al-op'-se-ah*) [*megalo-*; ὄψις, sight]. A disturbance of vision in which objects seem larger than they are.

**megaloscope** (*meg'-al-o-skōp*) [*megalo-*; σκοπεῖν, to inspect]. A magnifying endoscope or speculum.

**megaloscopy** (*meg-al-os'-ko-pe*) [*megalo-*; σκοπεῖν, to inspect]. Inspection by means of the megaloscope.

**megalosplanchnos** (*meg-al-o-splangk'-nos*) [*megalo-*; σπλάγχνον, viscus]. Possessing large viscera, especially a large liver.

**megalosplenia** (*meg-al-o-sple'-ne-ah*) [*megalo-*; *spleen*]. Enlargement of the spleen.

**megalosporon** (*meg-al-os'-po-ron*) [*megalo-*; σπόρος, seed; pl., *megalospora*]. A fungus, parasitic upon the hair; trichophyton.

**megalosyndactyly** (*meg-al-o-sin-dak'-til-e*). Syndactylism attended by hypertrophy.

**meganucleus** (*meg-an-u'-kle-us*). Same as *macronucleus*.

**megaphone** (*meg'-ah-fon*) [*mega-*; φωνή, sound]. An instrument used for assisting the hearing of the deaf, by means of large reflectors of the sound-waves.

**megarrhizin** (*meg-ar-iz'-in*). A bitter glucoside from the root of *Echinocystis fabacea*. It is said to be an active cathartic.

**megascope** (*meg-ah-skop*) [*mega-*; σκοπεῖν, to inspect]. A microscope for examining objects of comparatively large size.

**megaseme** (*meg'-as-ēm*) [*mega-*; σῆμα, sign]. With an orbital index more than 89°.

**megasoma** (*meg-ah-so'-mah*) [*mega-*; σῶμα, body]. Abnormal size and stature not reaching gigantism.

**megaspore** (*meg'-ah-spōr*) [*mega-*; σπόρος, seed]. Same as *macrospore*.

**megasthenic** (*meg-ah-sthen'-ik*) [*mega-*; σθένος, strength]. Powerful; having great bodily strength.

**Megastoma** (*meg-ah-sto'-mah*) [*mega-*; στόμα, mouth]. A genus of infusorians. **M. entericum, M. intestinale**, a species found in the intestinal canal of the cat and of certain mice, and in human feces. It is probably identical with *Cercomonas intestinalis*.

**megavolt** (*meg'-ah-volt*). A unit equal to 1,000,000 volts.

**Meglin's palatine point** (*ma-glan'*) [J. A. Meglin, French physician, 1756–1824]. The point of emergence of the large palatine nerve from the palatomaxillary canal; it constitutes at times one of the painful points in neuralgia of the superior maxillary branch of the trigeminus.

**megohm** (*meg'-ōm*) [*mega-*; *ohm*]. An electrical unit equal to one million ohms.

**megophthalmus** (*meg-of-thal'-mus*). See *keratoglobus*.

**megoxycyte** (*meg-ok'-se-sīt*) [*mega-*; ὀξύς, sharp; κύτος, cell]. A large oxyphile cell, one of the coarsely granular eosinophile cells or α-granules of Ehrlich.

**Méhu's test for albumin** (*ma'-hoo*) [Camille Jean Marie *Méhu*, French chemist, 1835–1887]. Treat the solution with 2 or 3 % of its volume of nitric acid, and add 10 volumes of a solution of 1 part phenol and 1 part acetic acid in 2 parts of 90 % alcohol and shake.

**Meibomian calculus** (*mi-bo'-me-an*) [Heinrich *Meibom*, German anatomist, 1638–1700]. The hardened secretion of the Meibomian glands that may accumulate on the inner surface of the eyelids. **M. cyst, M. tumor**, chalazion. **M. foramen**, the cecal foramen of the tongue. **M. glands**, tarsal glands; sebaceous follicles embedded in the tarsal plates of the eyelids. **M. sty**, one produced by suppuration of a Meibomian gland. Syn., *hordeolum internum*.

**Meigs' capillaries**. The capillary blood-vessels found between the muscular fibers of the heart.

**Meinert's form of enteroptosis** (*mi'-nert*). Enteroptosis occurring in chlorotic subjects.

**meio-** (*mi'-o-*). For words beginning thus, see *mio-*.

**meiostagmin reaction** (*mi-o-stag'-min*). A serum reaction based upon the lowering of the surface tension of a liquid when a specific antigen is added to a specific serum.

**Meissner's corpuscles** (*mīs'-ner*) [Georg *Meissner*, German histologist, 1829–1903]. Ovoid, laminated corpuscles connected with medullated nerve-fibers which wind around the lower pole before entering them; they are found in the papillæ of the volar surfaces of the fingers and toes. **M.'s ganglia**, the ganglionic nodes in Meissner's plexus. **M.'s plexus**, a plexus of nerves found in the submucous layer of the small intestine.

**Meissner-Billroth's plexus**. See *Meissner's plexus*.

**mel** [L.]. Honey. The product of the honeybee, *Apis mellifera*, and a few other hymenopterous insects. It contains a large amount of dextrose and levulose, and has the same properties as sugar. In medicine it is used as a vehicle, especially in gargles, and as an application to foul ulcers. **m. boracis** (B. P.), honey of borax, is used as a mouth-wash in thrush and aphthæ. **m. depuratum** (U. S. P.), clarified honey. **m. despumatum**, clarified honey. **m. rosæ** (U. S. P.), honey of rose, is used as an addition to gargles in ulcerated conditions of the mouth and throat.

**melachol** (*mel'-ak-ol*). Sodium citrophosphate, consisting of sodium phosphate, 100 parts; sodium nitrate, 2 parts; citric acid, 13 parts, rubbed together and mixed with 100 parts of water; used in liver complaints.

**melada** (*mel-a'-dah*) [Sp., fem. of *melar*, candy]. A moist brown sugar, produced like the muscovado, but not drained free of molasses.

**melæna** (*mel-e'-nah*). See *melena*.

**melagra** (*mel-a'-grah*) [μέλος, limb; ἄγρα, seizure]. Pain or gout in the limbs.

**melaleuca** (*mel-al-u-kah*). See *cajuput*.

**melalgia** (*mel-al'-je-ah*) [μέλος, limb; ἄλγος, pain]. Pain or neuralgia in the extremities.

**melamphonous** (*mel-am'-fo-nus*) [μέλας, dark; φωνή, voice]. Hoarse-voiced.

**melampyrin, melampyrit** (*mel-am'-pi-rin, -rit*). See *dulcit*.

**melanæmia** (*mel-an-e'-me-ah*). See *melanemia*.

**melanagogue** (*mel-an'-ag-og*) [*melano-*; ἀγωγός, leading]. 1. Causing an expulsion of dark feces, or of bile. 2. A remedy that causes the expulsion of dark stools or of bile; formerly, a medicine of service in the treatment of choler, or melancholy.

**melancholia, melancholy** (*mel-an-ko'-le-ah, mel'-an-kol-e*) [μέλας, black; χολή, bile]. A disorder of the mind characterized by a profound emotional depression and a tendency toward introspection, impairment of the mental and physical faculties, with or without delusions. **m., affective**, that in which the emotional nature is at fault. **m. agitata**, a form associated with excessive motor excitement. The patient rushes about, wringing his hands and lamenting loudly. **m. attonita**, a form in which the patient is perfectly motionless, lies in bed or sits up with his eyes open and fixed, and is absolutely indifferent to everything about him. Syn., *stuporous melancholia*. **m., climacteric**, that occurring at the menopause. **m., convulsive**, that associated with Jacksonian epilepsy. **m., panphobic**, that associated with the dread of everything. **m., paretic**, that preceding paresis. **m. passiva, m., passive**, a chronic form of slow development and gradual failure of the physical powers. **m., simple**, a mild form without delusions. Syn., *hypomelancholia; melancholia without delirium*. **m. simplex**, a mild form without delusions. **m. stuporosa**. Same as *m. attonita*. **m., stuporous**. See *m. attonita*.

**melancholiac** (*mel-an-ko'-le-ak*) [see *melancholia*]. 1. Suffering from melancholia. 2. A person affected with melancholia.

**melancholic** (*mel-an-kol'-ik*) [*melano-*; χολή, bile]. Sad; depressed; affected with melancholy.

**melanedema** (*mel-an-e-de'-mah*) [μέλας, black; οἴδημα, swelling]. Melanosis of the lungs.

**melanemia, melanæmia** (*mel-an-e'-me-ah*) [μέλας, black; αἷμα, blood]. The presence in the blood-plasma or the corpuscles, or in both, of dark pig-

ment-granules due to the disintegration of the hemoglobin. Syn., hemachromatosis.
**melanephidrosis** (mel-an-ef-id-ro'-sis) [μέλας, black; ἐφίδρωσις, excessive perspiration]. Black perspiration.
**melangeur** (ma-lon-zjer') [Fr.]. The graduated pipet of the hemocytometer.
**melanicterus** (mel-an-ik'-ter-us) [melano-; ἴκτερος, jaundice]. Black jaundice.
**melanidia** (mel-an-id'-e-ah). See miner's phthisis.
**melanidrosis** (mel-an-id-ro'-sis). See melanephidrosis.
**melanin** (mel'-an-in) [μέλας, black]. A black pigmentary matter occurring naturally in the choroid coat of the eye, the skin, the hair, the muscles, and, pathologically, in the skin in Addison's disease and in melanotic tumors. Melanin usually contains sulphur and rarely iron. It is a product of cell-activity and belongs to the socalled metabolic pigments.
**melanism** (mel'-an-izm) [melanin]. The abnormal deposition of dark pigment in an organ or organism.
**melano-** (mel-an-o-) [μέλας, black]. A prefix signifying black or dark-colored, or relating to melanin.
**melanoblastoma** (mel-an-o-blas-to'-mah) [melano-; blastoma]. Same as melanosarcoma.
**melanocancroid** (mel-an-o-kang'-kroid). Synonym of melanocarcinoma.
**melanocarcinoma** (mel-an-o-kar-sin-o'-mah). A carcinoma containing melanin.
**melanochlorosis** (mel-an-o-klo-ro'-sis) [melano-; χλωρός, green]. Chlorosis in which the skin has a blackish-green hue. Also the same as melanicterus.
**melanochroic, melanochrous** (mel-an-o-kru'-ik, mel-an-ok'-ro-us) [melano-; χρόα, color]. Having a dark color or complexion.
**melanocomous** (mel-an-ok'-o-mus) [melano-; κόμη, hair]. Black-haired.
**melanocyte** (mel'-an-o-sit) [melano-; κύτος, a cell]. A wandering lymph-cell which has become discolored by the absorption of dark pigment-granules.
**melanoderma, melanodermia** (mel-an-o-der'-mah, mel-an-o-der'-me-ah) [melano-; δέρμα, skin]. Black pigmentation of the skin. m., parasitic. See vagabond's disease.
**melanogen** (mel-an'-o-jen) [melano-; γεννᾶν, to produce]. A material which becomes melanin on receiving the appropriate stimulus.
**melanoid** (mel'-an-oid) [melano-; εἶδος, like]. Dark-colored; of the nature of melanosis.
**melanoma** (mel-an-o'-mah) [melano-; ὄμα, tumor: pl., melanomata]. A tumor containing melanin.
**melanomyces** (mel-an-o-mi'-sez) [melano-; μύκης, fungus]. A black fungous growth.
**melanopathy** (mel-an-op'-ath-e) [melano-; πάθος, disease]. A disease attended with a deposit of dark pigment.
**melanoplakia** (mel-an-o-pla'-ke-ah) [melano-; πλάξ, surface]. Pigmentation of the mucous membrane of the mouth.
**melanorrhagia** (mel-an-or-a'-je-ah) [melano-; ῥηγνύναι, to burst forth]. The copious discharge of blackened feces.
**melanorrhea** (mel-an-or-e'-ah) [melano-; ῥοία, a flow]. Synonym of melena.
**melanosarcoma** (mel-an-o-sar-ko'-mah) [melano-; sarcoma]. A sarcoma containing melanin.
**melanosarcomatosis** (mel-an-o-sar-ko-mat-o'-sis). The formation of melanosarcomata or the conditions favoring their formation.
**melanoscirrhus** (mel-an-o-skir'-us). A form of scirrhous scarcinoma characterized by pigmentation.
**melanosis** (mel-an-o'-sis) [melano-; νόσος, disease]. A general tendency to the formation in the blood, and the deposition in organs, of a dark granular pigment which is usually derived from the hemoglobin of the blood. m. lenticularis progressiva. See xeroderma pigmentosum.
**melanosity** (mel-an-os'-it-e) [melano-; νόσος, disease]. The condition of being melanous; darkness, as of hair, eyes, or skin.
**melanotic** (mel-an-ot'-ik) [melanosis]. Pertaining to or characterized by melanosis or by a deposit of melanin.
**melanotrichous** (mel-an-ot'-rik-us) [melano-; θρίξ, hair]. Black-haired.
**melanous** (mel'-an-us) [melanosis]. Pigmented, dark complexioned, characterized by melanosis.
**Melanthera** (mel-an'-ther-ah) [μέλας, black; ἀνθηρός, blossoming]. A genus of composite plants. M. brownei is an African plant the leaves of which in infusion are recommended as a substitute for quinine.

**melanthin** (mel-an'-thin). A glucoside, found in the seeds of Nigella sativa.
**melanuria** (mel-an-u'-re-ah) [μέλας, black; οὖρον, urine]. The presence of black pigment in the urine.
**melanurin** (mel-an-u'-rin) [melano-; οὖρον, urine]. A dark pigment found in the urine in melanuria; it is sometimes associated with the presence in the body of melanotic tumors.
**melasicterus** (mel-as-ik'-ter-us) [μέλας, black; ἴκτερος, jaundice]. Black jaundice; jaundice with great discoloration of the skin.
**melasma** (mel-az'-mah) [μέλας, black]. A deposit of dark pigment in the skin. m. suprarenale, Addison's disease.
**melatrophy** (mel-at'-ro-fe) [μέλος, limb; ἀτροφία, lack of nutrition]. Wasting of the limbs.
**melena, melæna** (mel-e'-nah) [μέλας, black]. The discharge of stools colored black by altered blood. It is quite common in the newborn. **melæna neonatorum**, an extravasation of blood into the stomach and intestines of the newborn infant, occurring most often in the first few hours of life.
**melenemesis** (mel-en-em'-es-is) [melano-; ἔμεσις, vomiting]. Black vomit.
**melenic** (mel-en'-ik) [μέλας, black]. Pertaining to melena.
**melenorrhagia** (mel-en-or-a'-je-ah). Synonym of melena.
**melezitose** (mel-ez'-it-ōs) [Fr., mélèze, larch], $C_{18}H_{30}O_{15}+2H_2O$. A sugar found in European false manna, or Briancon manna.
**Melia** (me'-le-ah) [μελία, the ash]. A genus of the order Meliaceæ. M. azadirachta is indigenous to Asia, but naturalized in southern Europe and America. The entire plant is bitter and narcotic, in small doses purgative and anthelmintic; the leaves and blossoms are vulnerary and stomachic; the bark, called margosa, is tonic and emmenagogue, the root bark is used in lepra and scrofula and as an emetic; the oil of the seeds is antiseptic. M. azedarach is indigenous to Asia and naturalized in the United States. The root bark is anthelmintic. Dose of fluidextract 10-30 min. (0.6-1.8 Cc.). The oil from the seeds is used in skin diseases and as a vulnerary.
**melicera, meliceris** (mel-is-e'-rah, mel-is-e'-ris) [mel; κηρός, wax]. A cyst containing a substance having a honey-like appearance.
**Melilotus** (mel-il-o'-tus) [mel; λωτός, lotus]. A genus of leguminous herbs. M. officinalis, the sweet clover, is official in the G. P. It contains coümarin ($C_9H_6O_2$), and melilotic acid ($C_9H_{10}O_3$), and coümaric acid ($C_9H_8O_3$), of which coümarin is the anhydride.
**melinous** (mel'-in-us) [μήλινος]. Quince-colored.
**Melissa** (mel-is'-ah) [μέλισσα, a bee]. A genus of labiate plants. M. officinalis, balm or lemon-balm, is a species growing in southern Europe. Balm is used as a drink in febrile affections and as a flavoring agent.
**melissic** (mel-is'-ik) [mel]. Obtained from honey or from beeswax.
**melitagra** (mel-it-a'-grah) [mel; ἄγρα, seizure]. Eczema associated with the formation of soft, honey-colored crusts.
**melitagra** (mel-it-a'-grah) [μέλος, limb; ἄγρα, seizure]. Any arthritic or rheumatic pain in the limbs.
**melitemia, melitæmia** (mel-it-e'-me-ah) [mel; αἷμα, blood]. The presence of an excess of sugar in the blood.
**melitis** (mel-i'-tis) [μῆλον, cheek; ιτις, inflammation]. Inflammation of a cheek.
**melitoptyalismus** (mel-it-o-ti-al-is'-mus) [μέλι, honey; πτύαλον, saliva]. The production of melitoptyalon.
**melitoptyalon** (mel-it-o-ti'-al-on) [μέλι, honey; πτύαλον, saliva]. A saliva containing glucose said to be secreted by persons suffering from hectic fever.
**melitose** (mel'-it-ōs) [mel], $C_{18}H_{32}O_{16}$. A crystalline sugar occurring in Australian manna, flour of cotton-seeds, sugar-beets, and in the molasses obtained in the manufacture of sugar.
**melituria** (mel-it-u'-re-ah) [melitose; οὖρον, urine]. Diabetes mellitus. m. inosita, the presence of inosit in the urine.
**Mellin's food**. A variety of Liebig's food for infants. Its composition is: Water 5.0, fat 0.15, grape-sugar 44.69, cane-sugar 3.51, starch none, soluble carbohydrates 85.44, albuminoids 5.95, ash 1.89.
**mellita** (mel-i'-tah) [μέλι, honey]. Pharmaceutical

**preparations known as honeys. They consist of honey, either natural, clarified, or flavored. There are three kinds official *mellita*.**
**mellite** (*mel'-īt*). See *mellitum*.
**mellithemia** (*mel-ith-e'-me-ah*). See *melitemia*.
**mellitum** (*mel-i'-tum*) [*mel*]. In pharmacy, a honey; a preparation in which honey is the menstruum. m. rosæ. See *mel rosæ*.
**melmaroba** (*mel-mar-o'-bah*). A liquid preparation said to contain *Brunfelsia uniflora*, caroba, stillingia, and potassium iodide; it is used in syphilis, chronic skin diseases, and rheumatism. Dose 1–2 dr. (3.75–7.5 Cc.).
**melocampyle** (*mel-o-kam'-pil-e*) [μέλος, limb; καμπύλη, crooked staff]. Deformity of the limbs from bending.
**melodidymus** (*mel-o-did'-im-us*) [μέλος, limb; δίδυμος, double]. A monstrosity with double limbs. Melodidymia are twins united by the limbs.
**meloil** (*mel'-oil*) [*mel*; *oleum*, oil]. Disguised castor-oil.
**melomania** (*mel-o-ma'-ne-ah*) [μέλος, song; μανία, madness]. Inordinate devotion to music.
**melomaniac** (*mel-o-ma'-ne-ak*) [μέλος, song; μανία, madness]. One who is affected with melomania.
**melomelus** (*mel-om'-el-us*) [μέλος, limb]. A monster with supernumerary limbs.
**melon** (*mel'-on*) [μῆλον, an apple]. 1. See *citrullus* and *cucumis*. 2. A proprietary cicatrizant and vulnerary. m.-root, the root of muskmelon, *Cucumis melo*. Dose of cultivated root 6 dr. (25 Gm.); of wild root 8–11 gr. (0.51–0.71 Gm.). m.-seed bodies, fibrous bodies, resembling melon-seeds in size, sometimes found in joints and cysts of tendon-sheaths.
**meloncus** (*mel-ong'-kus*) [μῆλον, cheek; ὄγκος, tumor]. A tumor of the cheek.
**melonemetin** (*mel-on-em'-et-in*). See *melonenemetin*.
**melonenemetin** (*mel-on-en-em'-et-in*). A bitter brown substance from the root of musk-melon *Cucumis melo*. Dose is used as an emetic and purgative. Dose ¼–1– gr. (0.05–0.07 Gm.).
**meloplastic** (*mel-o-plas'-tik*) [μῆλον, cheek; πλάσσειν, to form]. Pertaining to meloplasty.
**meloplasty** (*me'-lo-plas-te*) [μῆλον, cheek; πλάσσειν, to form]. A plastic operation on the cheek.
**melos** (*me'-los*) [μέλος, a limb]. Limb.
**melosalgia** (*mel-os-al'-je-ah*). See *melalgia*.
**meloschisis** (*mel-os'-kis-is*) [μῆλον, cheek; σχίσις, cleft]. A congenital cleft of the cheek.
**melosis** (*me-lo'-sis*) [μήλη, probe]. The process of probing.
**melotis** (*me-lo'-tis*) [μήλη, probe; οὖς, ear]. An aural probe.
**melotridymus** (*mel-o-trid'-im-us*) [μέλος, limb; τρίδυμος, threefold]. A fetal monstrosity with three pairs of limbs.
**melting-point.** The degree of temperature at which solids pass into the liquid state.
**Meltzer's method** (*melt'-zer*) [Samuel James *Meltzer*, American physician, 1851– ]. The introduction of an anesthetic vapor into the trachea. M.'s sign, normally, on auscultation of the heart (at the side of the xiphoid appendix) there is heard, after swallowing, a first sound produced by the flowing of fine drops, and six or seven seconds after, a "glou-glou." According to Meltzer, the second sound fails in the case of occlusion or pronounced contraction of the lower part of the esophagus.
**melubrin** (*mel'-ū-brin*). Trade name of an anti-pyrin derivative, used in acute articular rheumatism, sciatica, etc. Dose 15–30 gr. (1 to 2 Gm.).
**melulose** (*mel'-ū-lōs*). A concentrated extract of malt.
**member** (*mem'-ber*) [*membrum*, a limb]. A part of the body, especially a projecting part, as the leg or the arm.
**membra** (*mem'-brah*) [L.]. Plural of *membrum*.
**membral** (*mem'-bral*) [*membrum*, limb]. Pertaining to a limb or member.
**membrana** (*mem-bra'-nah*) [L.; *pl., membranæ*]. A membrane, as of the peritoneum. m. abdominis, the peritoneum. m. adventitia, the adventitia of blood-vessels; also, the decidua reflexa. m. agnina, the amnion. m. basilaris. See *m. propria*. m. caduca, the decidua. m. capsularis, a capsular ligament. m. decidua. See *decidua*. m. eboris, the cellular covering of tooth pulp. m. elastica laryngis, the cricothyroid membrane with the membrana quadrangularis. m. flaccida. See *Shrapnell's membrane*. m. germinativa, the blasto-

derm. m. granulosa, the layer of small polyhedral cells within the theca folliculi of the Graafian follicle. m. limitans, the limiting layer of the retina. There are two—the *internal* and the *external*. See *limiting membrane*. m. propria, the delicate membrane upon which the epithelium of mucous membranes rests; the basement-membrane. m. quadrangularis, one of the elastic membranes of the larynx. m. reuniens, the fused somatopleuric layers between the recti abdominis in the embryo. m. Ruyschiana, the middle or capillary layer of the choroid. m. sacciformis, the synovial membrane of the inferior radioulnar articulation. m. Schneideriana, the pituitary membrane. m. serotina, the part of the decidua entering into the formation of the placenta. m. tectoria, a delicate membrane of the internal ear. Syn., *Corti's membrane*. See under *ear*. m. tensa, the tympanic membrane proper, exclusive of Shrapnell's membrane. m. tenuis, the arachnoid. m. tympani. See *membrane, tympanic*. m. vestibularis. See *Reissner's membrane*. m. vibrans. Same as *m. tensa*.
**membranaceous** (*mem-bran-a'-ce-us*) [*membrum*, member]. Pertining to, consisting of, or of the nature of, a membrane.
**membrane** (*mem'-brān*) [*membrana*, from *membrum*, member]. A thin layer of tissue surrounding a part or separating adjacent cavities. m., animal, a membrane made from animal tissues, used in dialyzing. m., basement-, a delicate membrane, made up of flattened cells, underlying the epithelium of mucous surfaces. m. basilar. See *basilar*. m.-bone, any bone that originates, not in cartilage, but in membrane, as some of the cranial bones. m. of Bruch. See *Bruch's membrane*. m. Cargile's. See *Cargile's membrane*. m., cell, the cell wall. m., cloacal, the ventral wall of the cloaca of the embryo. m., compound, one made up of two distinct laminæ, as seromucous and serofibrous membranes. m. of Corti. See *membrana tectoria*. m., costocoracoid, a dense layer of fascia extending between the subclavius muscle and the pectoralis minor, and forming the anterior portion of the sheath of the axillary vessels. m., cricothyroid, the membrane connecting the thyroid and cricoid cartilages of the larynx. m., croupous, the yellowish-white membrane forming in the larynx in croup. m., Débove's. See *Débove's membrane*. m. of Descemet. See *Descemet's membrane*. m., diphtheritic, a fibrinous layer formed on a mucous membrane or cutaneous surface and extending downward for a variable depth. It is the result of coagulation-necrosis, generally brought about by the bacillus of diphtheria. m., drum-, the tympanic membrane. m., elastic, one composed of elastic fibrous tissue. m., false. See *m., diphtheritic*. m., fenestrated, the elastic membrane of the intima of arteries. Syn., *fenestrated membrane of Henle*. m., fetal, a name given to the chorion, amnion, or allantois. m., germinal, the blastoderm. m. hyaline, (1) basement membrane; (2) the membrane between the inner fibrous layer of a hair follicle and its outer root-sheath. m., hyaloid, a delicate membrane investing the vitreous humor of the eye. m., intrachoroidal, an ependymal membrane below the choroidal fissure in the embryo. m., Jacob's. See *Jacob's membrane*. m., Krause's. See *Krause's membrane*. m., limiting. See *limiting membrane*. m., meconic, a layer within the rectum of the fetus, supposed to invest the meconium. m., medullary. Same as *endosteum*. m., mucous, the membrane lining those cavities and canals communicating with the air. It is kept moist by the mucus secreted by the goblet-cells and mucous glands. m. of Nasmyth. See *Nasmyth's membrane*. m., nictitating, the winking membrane of the lower animals, represented in the human eye by the plica semilunaris. m., obturator, the fibrous membrane closing the obturator foramen. m., otolith, membrane formed of otoliths and a mesh-work of fibrous tissue in the utricle and saccule. m., palatine, the membrane covering the roof of the mouth. m., persistent pupillary. See *m., pupillary*. m., periodontal, a fibrous layer covering the cement of teeth. m., pseudoserous, one presenting the moist, glistening surface, etc., of a serous membrane, but differing from it in structure; e. g., the endothelium of the blood-vessels. m., pupillary, a delicate, transparent membrane closing the pupil in the fetus. It disappears between the seventh and eighth months; when it persists after birth it is termed persistent pupillary

**membrane. m., pyogenic,** the lining of an abscess-cavity or a fistulous tract. The term should be restricted to the lining of an abscess that is spreading and in which the membrane produces pus. **m., pyophylactic,** a protective membrane lining an abscess cavity. **m. of Reissner.** See *Reissner's membrane.* **m.,** reticular, the membrane covering the space of the outer hair-cells of the cochlea. **m. of Ruysch.** See *Ruysch's membrane.* **m., Schneiderian,** the mucosa lining the nasal fossæ. **m., secondary tympanic,** the membrane closing the fenestra rotunda. **m., serous,** a delicate membrane covered with flat 'endothelial cells lining closed cavities of the body, *e. g.,* the peritoneum and the pleura. **m., Shrapnell's.** See *Shrapnell's membrane.* **m., sutural,** fibrous tissue passing through the sutures of the cranium between the periosteum and the external layer of the dura. **m., synovial,** a membrane covering the articular extremities of bones and the inner surface of ligaments entering into the formation of a joint. **m., Tenon's.** See *Tenon's capsule.* **m., thyrohyoid,** the membrane joining the thyroid cartilage and hyoid bone. **m., tympanic,** the drum-membrane; the membrane separating the external from the middle ear. It consists of three layers: an *outer* or skin layer, a *fibrous* layer, and an *inner* mucous layer. **m., vitelline,** the true cell-membrane of the ovum, lying within the zona pellucida.

**membraniferous** (*mem-bran-if'-er-us*) [*membrane; erre,* to bear]. Having a membranous expansion.

**membraniform** (*mem-bran'-if-orm*). See *membranous.*

**membranins; membranin bodies** (*mem'-bran-ins*). A special group of proteins containing sulphur, which blackens lead; insoluble in water, salt solution, or dilute acids or alkalies, but soluble in the last two with warmth. Like mucins, they yield a reducing substance by action of dilute mineral acids with heat. They give a beautiful red coloration with Millon's reagent. Membranins constitute the substance of Descemet's membrane and of the capsule of the crystalline lens.

**membranocarneous** (*mem-bra-no-kar'-ne-us*) [*membrane; carneus,* belonging to flesh]. Both membranous and fleshy.

**membranocartilaginous** (*mem-bra-no-kar-til-aj'-in-us*). Both cartilaginous and fleshy.

**membranocranium** (*mem-bra-o-kra'-ne-um*) [*membrana,* membrane; κρανίον, skull]. The membranous skull of the fetus, prior to ossification.

**membranoid** (*mem'-bran-oid*) [*membrana,* membrane; εἶδος, like]. Resembling membrane.

**membranous** (*mem'-bran-us*). Pertaining to, having the nature of, or consisting of, a membrane. **m. labyrinth.** See *labyrinth, membranous.* **m. urethra,** the part of the urethra between the two layers of the triangular ligament.

**membrum** (*mem'-brum*) [L.: *pl., membra*]. Same as *member.* **m. muliebre,** the clitoris. **m. seminale, m. virile,** the penis.

**memory** (*mem'-o-re*) [*memor,* mindful]. That faculty of the mind by which ideas and sensations are recalled. **m. anterograde** (*an'-ter-o-grād*). Memory for events long past but amnesia in regard to recent occurrences.

**menacme** (*men-ak'-me*) [μήν, month; ἀκμή, prime]. The period of a woman's life during which menstruation persists.

**menagogue** (*men'-ag-og*). Synonym of *emmenagogue.*

**menarche** (*men-ar'-ke*) [μήν, month; ἀρχή, beginning]. The period at which menstruation is inaugurated.

**Mendel's law** (*men'-dl*) [Johann Gregor *Mendel,* Austrian naturalist, 1822-1884]. A first cross will result in offspring resembling one or the other parent, and possessing in its undeveloped form, termed "recessive," the attributes of the other. The second cross will result in fixed types possessing respectively the character of one parent, "dominant," and of both parents in varying degrees.

**Mendeléeff's law** (*men-del'-yef*) [Dimitrii Ivanovich *Mendeléeff,* Russian chemist, 1834-1907]. The properties of an element are a periodical function of its atomic weight. Also called *periodic law.*

**Mendelism** (*men'-del-izm*). The theory proposed by Mendel and comprised in his law of dichotomy in plant hybridization. See *Mendel's law.*

**mendosus** (*men-do'-sus*) [*mendax,* false]. False; incomplete.

**menelcosis** (*men-el-ko'-sis*) [μήνες, menses; ἕλκωσις, ulceration]. Ulceration of the leg, with vicarious menstruation from the sore.

**menellipsis** (*men-el-ip'-sis*) [μήνες, menses; ἔλλειψις, a falling off]. Menopause.

**menhidrosis, menidrosis** (*men-hid-ro'-sis, men-id-ro'-sis*) [μήν, month; ἱδρωσις, sweat]. The replacement of the menstrual flow by a bloody sweat.

**Ménière's disease** (*men-e-ār'*) [Prosper *Ménière,* French physician, 1799-1862]. Aural vertigo. A disease of the middle ear characterized by sudden deafness and symptoms of apoplexy. Its cause is thought to be effusion or hemorrhage into the semicircular canals of the ear.

**meningarthrocace** (*men-in-gar-throk'-as-e*) [μήνιγξ, membrane; ἄρθρον, joint; κακός, evil]. Inflammation of joint-membranes.

**meningeal** (*men-in'-je-al*) [*meninges*]. Pertaining to the meninges.

**meningematoma** (*men-in-je-mat-o'-mah*). Hematoma of the dura.

**meningeocortical** (*men-in-je-o-kor'-tik-al*). Relating to the meninges and the cortex of the brain.

**meningeorrhaphy** (*me-nin-je-or'-af-e*) [*meninges;* ῥαφή, suture]. 1. Suture of membranes. 2. Suture of the meninges of the brain or spinal cord.

**meninges** (*men-in'-jēs*) [Plural of *meninx, q. v.*]. A name applied to the membranes of the brain and spinal cord; the dura, pia, and arachnoid. **m., lepto-,** the pia and arachnoid.

**meninghematoma** (*men-ing-he-mat-o'-mah*). See *meningematoma.*

**meningina** (*men-in-ji'-nah*) [μήνιγξ, membrane]. The pia and arachnoid considered as the proper meninges, and apart from the dura; the pia-arachnoid.

**meningism** (*men'-in-jizm*) [*meninges*]. 1. Simple circulatory disturbances of the meninges, of toxic or hysterical origin. 2. Pseudomeningitis accompanied by symptoms similar to those of tuberculous meningitis.

**meningitic** (*men-in-jit'-ik*) [*meningitis*]. Pertaining to, or affected with, meningitis. **m. streak.** See *tache méningéale.*

**meningitiform** (*men-in-jit'-e-form*) [*meningitis; forma,* form]. Resembling meningitis.

**meningitis** (*men-in-ji'-tis*) [*meninges; ιτις,* inflammation]. Inflammation of the membranes of the brain or cord; that of the dura is termed *pachymeningitis;* that of the pia-arachnoid, *leptomeningitis,* or simply meningitis. Meningitis of the membranes of the brain is classified into *acute* and *chronic,* the former being subdivided into serous and purulent, the latter into fibrous, ossifying, and deep, or *en-cephalomeningitis.* According to location, two varieties are spoken of—that of the *vertex* and that of the *base.* **m., acute cerebral,** that due to traumatism, to extension of inflammation from adjacent structures, especially from the middle ear, the orbit, the nasal sinuses, or to tuberculosis (*tuberculous meningitis*); it may be secondary to acute infectious processes elsewhere in the body, as pneumonia, erysipelas, typhoid fever, influenza, smallpox, or it may be a primary disease, as in *cerebrospinal meningitis.* **m., acute spinal,** a form that may occur in tuberculosis, as a secondary process in acute infectious diseases, such as smallpox, scarlatina, pneumonia, as a part of epidemic cerebrospinal meningitis, as the result of extension of inflammation from neighboring parts, and as the result of exposure to cold and wet. The symptoms are chill, fever, pain in the back and limbs, rigidity of the muscles, dyspnea, exaggerated reflexes, later paralyses. **m., cerebrospinal,** inflammation of the membranes of the brain and spinal cord. The symptoms are fever, slow pulse, later rapid pulse, headache, delirium, rigidity and retraction of the neck, convulsions, vomiting, a scaphoid abdomen, constipation, optic neuritis; in advanced stages various palsies occur, such as ptosis, squint, and facial paralysis. In *epidemic cerebrospinal meningitis* there is usually a characteristic eruption. See under *exanthem.* **m., chronic cerebral,** a form due to syphilis or tuberculosis, or it may be associated with disease of the brain in encephalomeningitis (paretic dementia). The main symptoms are, in the first two, headache, convulsions, and rigidity of the muscles of the neck. **m., chronic spinal,** a variety due to syphilis, traumatism, and the excessive use of alcohol. It is also frequently an accompaniment of the scleroses of the

spinal cord. The symptoms are pain, hyperesthesia along the spinal nerves, increased reflexes, paralyses. m., focal, that confined to a very limited area and usually due to traumatism or syphilis. m., mechanical, that due to traumatism. m., occlusive, infantile leptomeningitis leading to the occlusion of the foramen of Magendie. m., otitic, that complicating an attack of otitis. m., septicemic, that due to an infectious process. m., tuberculous, inflammation of the pia of the brain with effusion of lymph and pus; acute hydrops.
**meningitophobia** (men-in-jit-o-fo'-be-ah) [meningitis; φόβος, fear]. A pseudomeningitis due to fear of that disease.
**meningium** (men-in'-je-um). See arachnoid.
**meningo-** (men-in-go-) [meninges]. A prefix meaning relating to the meninges.
**meningobacterin** (men-in-go-bak'-ter-in). Trade name of a vaccine for use in the prophylaxis of cerebrospinal meningitis.
**meningocele** (men-in'-go-sēl) [meningo-; κήλη, hernia]. A protrusion of the cerebral or spinal meninges through a defect in the skull or vertebral column. It forms a cyst filled with cerebrospinal fluid.
**meningocephalitis** (men-in-go-sef-al-i'-tis). See meningoencephalitis.
**meningocerebritis** (men-in-go-ser-e-bri'-tis). See meningoencephalitis.
**meningococcus** (men-in-go-kok'-us) [meningo-; coccus]. A name for the coccus of cerebro-spinal fever. See micrococcus intercellularis meningitidis.
**meningocortical** (men-in-go-kor'-tik-al). Relating to the meninges and the cortex.
**meningoencephalitis** (men-in-go-sef-al-i'-tis). Inflammation of the brain and its membranes.
**meningoencephalocele** (men-in-go-en-sef'-al-o-sēl) [meningo-; ἐγκέφαλον, brain; κήλη, hernia]. Hernia of the brain and its meninges.
**meningo-encephalomyelitis** (men-in-go-en-sef-al-o-mi-el-i'-tis). Combined inflammation of the meninges, brain, and spinal cord.
**meningomalacia** (men-in-go-mal-a'-se-ah) [meningo-; μαλακία, softness]. A softening of the cerebral or spinal meninges, or other membranes.
**meningomyelitis** (men-in-go-mi-el-i'-tis). Inflammation of the spinal cord and its meninges.
**meningomyelocele** (men-in-go-mi'-el-o-sēl). A protrusion from the spinal column of a portion of the cord and membranes.
**meningo-osteophlebitis** (men-in-go-os-te-o-fleb-i'-tis) [meningo-; ὀστέον, bone; phlebitis]. Periostitis combined with phlebitis of the veins of the bone.
**meningorrhachidian** (men-in-go-rak-id'-e-an) [meningo-; ῥάχις, spine]. Relating to the spinal meninges.
**meningorrhea, meningorrhœa** (men-in-gor-e'-ah) [meningo-; ῥοία, a flow]. Meningeal hemorrhage, or extravasation of blood.
**meningorrhagia** (men-in-gor-a'-je-ah) [meningo-; ῥηγνύναι, to burst forth]. Meningeal hemorrhage.
**meningosis** (men-in-go'-sis) [meninges]. The union of bones by a membranous attachment.
**meningotyphoid** (men-in-go-ti'-foid). Typhoid with symptoms of meningitis.
**meninguria** (men-in-gu'-re-ah) [μῆνιγξ, membrane; οὖρον, urine]. The passage or presence of membranous shreds in the urine.
**meninx** (men'-ingks) [μῆνιγξ; membrane; pl., meninges]. A membrane, especially one of the brain or spinal cord; the meninges covering the brain and spinal cord consist of the dura, pia, and arachnoid.
**menischesis** (men-is'-ke-sis). See ischomenia.
**meniscitis** (men-is-i'-tis) [μηνίσκος, crescent; ιτις, inflammation]. An inflammation of any interarticular cartilage, especially of the semilunar cartilages of the knee-joint.
**meniscoid** (men-is'-koid) [μηνίσκος, crescent; εἶδος, like]. Resembling a meniscus; crescent-shaped.
**meniscus** (men-is'-kus) [μηνίσκος, a crescent]. A crescent or crescentic body, especially an interarticular fibrocartilage. Also a concavoconvex (positive meniscus) or convexoconcave lens (negative meniscus). m. lateralis, the external semilunar fibrocartilage of the knee-joint. m. medialis, the internal semilunar fibrocartilage of the knee-joint. m., tactile, a form of nerve-ending with a concave surface turned ectad, each concavity containing a tactile cell.

**menispermin** (men-is-per'-min). A resinoid obtained from Menispermum canadense; dose about 2 gr. (0.125 Gm.).
**menispermum** (men-is-per'-mum) [μήνη, moon; σπέρμα, seed]. Yellow parilla; Canadian moonseed. The rhizome and roots of M. canadense, of the order Menispermaceæ. It is alterative and is used as a substitute for sarsaparilla.
**meno-** (men-o-) [menses]. A prefix meaning relating to the menses.
**menocelis** (men-o-se'-lis) [meno-; κηλίς, spot]. Dark erythematous or hemorrhagic spots occurring upon the skin in failure of menstruation.
**menolipsis** (men-o-lip'-sis) [meno-; λεῖψις, an omission]. The retention or absence of the menses.
**menometastasis** (men-o-met-as'-tas-is) [meno-; metastasis]. Vicarious menstruation.
**menopad** (men'-o-pad) [meno-]. A pad for catching the menstrual blood.
**menopause** (men'-o-pawz) [meno-; παῦσις, cessation]. The physiological cessation of menstruation, usually occurring between the forty-fifth and fiftieth years. Syn., climacteric.
**menophania** (men-o-fa'-ne-ah) [meno-; φαίνειν, to appear]. The first appearance of the menses.
**menoplania** (men-o-pla'-ne-ah) [meno-; πλάνη, deviation]. A discharge of blood occurring at the menstrual period, but derived from some other part of the body than the uterus.
**menorrhagia** (men-or-a'-je-ah) [meno-; ῥηγνύναι, to burst forth]. An excessive menstrual flow.
**menorrhea** (men-or-e'-ah) [meno-; ῥοία, a flow]. The normal flow of the menses; also, excessive menstruation.
**menoschesis** (men-os'-kes-is) [meno-; σχέσις, retention]. Retention of the menses.
**menosepsis** (men-o-sep'-sis) [meno-; σῆψις, putridity]. A putrid quality of the menses.
**menostasia, menostasis** (men-os-ta'-ze-ah, men-os'-tas-is) [meno-; στάσις, standing]. A suppression of the menstrual flow.
**menoxenia** (men-oks-e'-ne-ah) [meno-; ξένος, strange, foreign]. Irregularity of menstruation; vicarious menstruation.
**mens** (menz) [L.: gen., mentis]. Mind. Compos mentis, of sound mind. Non compos mentis, of unsound mind.
**mensa** (men'-sah) [L.]. A table; the upper surface of the molars.
**mensalis** (men-sa'-lis). See trapezius under muscle.
**menses** (men'-sēz) [mensis, a month]. The recurrent monthly discharge of blood from the genital canal of a woman during sexual life.
**menstrua** (men'-stroo-ah) [L.: pl., of menstruus, monthly]. 1. The menses. 2. Plural of menstruum, q. v. m. alba. Synonym of leukorrhea.
**menstrual** (men'-stroo-al) [see menstruation]. Pertaining to menstruation. m. colic, uterine colic due to menstruation.
**menstruant** (men'-stroo-ant) [menstruus, monthly]. 1. Subject to, or capable of, menstruating. 2. One who menstruates, or is capable of menstruating.
**menstruate** (men'-stroo-āt) [menstruare; from menstruus, monthly]. To discharge the menstrual flow.
**menstruation** (men-stroo-a'-shun) [menstruus, monthly, from mensis, a month]. A periodic discharge of a sanguineous fluid from the uterus, occurring during the period of a woman's sexual activity, from puberty to the menopause. m. supplementary, a menstrual flow from the uterus and also from some other site. m. suppressed, a form of amenorrhea in which the patient has formerly menstruated, but menstruation now fails to appear. m., vicarious, the discharge of blood at the time of menstruation from some organ or part other than the vagina.
**menstruous** (men'-stroo-us) [menstruus, monthly]. Having, or pertaining to, the monthly flow.
**menstruum** (men'-stroo-um) [menstruus]. A solvent.
**mensuration** (men-su-ra'-shun) [mensurare, to measure]. The act of measuring; one of the methods of physical diagnosis.
**mentagra** (men-tag'-rah). See sycosis.
**mentagrophyton** (men-tag-rof'-it-on) [mentagra, sycosis; φυτόν, a plant]. A fungus, Microsporon mentagrophytes, thought to be the cause of sycosis.
**mental** (men'-tal) 1. [mens, the mind]. Pertaining to the mind. 2. [mentum, the chin.] Pertaining to the chin.

**mentalis** (*men-ta'-lis*). The levator labii inferioris. See *muscles, table of*.
**mentality** (*men-tal'-it-e*) [*mens*, mind]. Mental activity and power; intellect.
**Mentha** (*men'-thah*) [L.]. A genus of labiate plants—the mints. **M. piperita** (U. S. P.), peppermint, the dried leaves and flowering tops of *M. piperita*. It is an aromatic stimulant, and is used to relieve nausea, flatulence, and spasmodic pain in the stomach and bowel. **menthæ piperitæ, aqua** (U. S. P.), peppermint water. Dose indefinite. **menthæ piperitæ, oleum** (U. S. P.), oil of peppermint. Dose 1–5 min. (0.065–0.32 Cc.). **menthæ piperitæ, spiritus** (U. S. P.), spirit of peppermint. Dose 10–30 min. (0.65–2.0 Cc.). **m. pulegium**, pennyroyal. **m. viridis** (U. S. P.), spearmint, the dried leaves and flowering tops of *M. spicata*. Its properties and uses are similar to those of *M. piperita*. **menthæ viridis, aqua** (U. S. P.), spearmint water. Dose indefinite. **menthæ viridis, oleum** (U. S. P.), oil of spearmint. Dose 2–5 min. (0.13–0.32 Cc.). **menthæ viridis, spiritus** (U. S. P.), spirit of spearmint. Dose 10–40 min. (0.65–2.3 Cc.).
**menthene** (*men'-thēn*) [*mentha*], $C_{10}H_{18}$. A liquid hydrocarbon produced when menthol is distilled with phosphorus pentoxide.
**menthiodol** (*men-thi'-o-dol*). A local application for neuralgia made by triturating together four parts of menthol heated in a capsule with one part of iodine.
**menthoform** (*men'-tho-form*). A combination of formaldehyde, glycerol, and menthol.
**menthol** (*men'-thol*) [*mentha*], $C_{10}H_{19}OH$. A crystalline stearopten derived from oil of peppermint, and used as an anodyne and rubefacient in neuralgia, in skin diseases associated with itching, and in rhinitis. Syn., *mint-camphor*. **m. valerate**, validol.
**mentholeate** (*men-tho'-le-āt*). A solution used in skin diseases, consisting of menthol, 200 gr., heated with oleic acid, 4 dr.
**mentholin** (*men'-thol-in*). A proprietary remedy for coryza said to consist of menthol, 1 part; pulverized coffee and boric acid, each, 10 parts.
**mentholyptine** (*men-thol-ip'-tēn*). A proprietary external antiseptic said to consist of menthol and eucalyptol.
**menthophenol** (*men-tho-fe'-nol*). An antiseptic fluid obtained by fusing together one part of phenol and three parts of menthol. It is used in the treatment of burns and wounds, and diluted (15 drops to a glass of water) as a mouth-wash; also as a local anesthetic (3 to 5 % in warm water).
**menthorol** (*men'-thor-ol*). A mixture of parachlorphenol and menthol used in tuberculosis of upper air-passages.
**menthoxol** (*men-thoks'-ol*). An antiseptic fluid mixture of 3 % solution of hydrogen dioxide, 32 to 38 % alcohol, and 1 % menthol. It is innocuous and deodorizing, and is used in suppurating wounds, ozena, etc.
**menthyl** (*men'-thil*), $C_{10}H_{19}$. The hypothetical radical of menthol. **m. acetoacetate**, $CH_3C(OH):CH . COOC_{10}H_{19}$, a bactericide.
**mentoanterior** (*men-to-an-te'-re-or*) [*mentum; anterior*, before]. Having the chin toward the front.
**mentobregmatic** (*men-to-breg-mat'-ik*) [*mentum; bregma*]. Extending from the chin to the bregma.
**mentohyoid** (*men-to-hi'-oid*) [*mentum; hyoid*]. Relating to the chin and the hyoid bone.
**mentolabial** (*men-to-la'-be-al*) [*mentum, labium,* lip]. Relating both to the chin and the lip.
**mentoposterior** (*men-to-pos-te'-re-or*) [*mentum; posterior*, after; behind]. Having the chin toward the back.
**mentula** (*men'-tū-lah*) [L.]. The penis.
**mentulagra** (*men-tū-la'-grah*) [*mentula*, penis; ἄγρα, seizure]. Painful priapism; chordee.
**mentulomania** (*men-tū-lo-ma'-ne-ah*). Synonym of *masturbation*.
**mentum** (*men'-tum*) [L.]. The chin.
**Menyanthes** (*men-e-an'-thēs*) [μήν, month; ἄνθος, flower, from its reputed emmenagogue properties]. A genus of plants of the *Gentianeæ*. *M. trifoliata*, or buckbean, contains a bitter principle, *menyanthin* ($C_{33}H_{50}O_{14}$), and has been used in malaria, scrofula, dropsy, jaundice, rheumatism, etc. Dose of the *powdered leaves* 20–30 gr. (1.3–2.0 Gm.).
**Menzer's serum** (*men'-tser*) [Arthur August Ludwig Menzer, German bacteriologist, 1871– ]. An antirheumatic serum prepared from streptococcus strains derived from rheumatic patients.

**mephitic** (*mef-it'-ik*) [*mephiticus*, pestilential]. Foul or noxious; stifling; noisome. **m. air**, carbon dioxide. **m. gangrene**, necrosis of bone associated with the evolution of offensive odors. **m. gas**, carbon dioxide.
**meralgia** (*me-ral'-je-ah*) [μηρός, thigh; ἄλγος, pain]. Neuralgic pain in the thigh. **m. paræsthetica**. See *Bernhardt's paresthesia*.
**meramaurosis** (*mer-am-aw-ro'-sis*) [μέρος, part; ἀμαυρόειν, to darken]. Partial amaurosis.
**meranesthesia** (*mer-an-es-the'-ze-ah*) [μέρος, part; *anesthesia*]. Partial or local anesthesia.
**meratrophy** (*mer-at'-ro-fe*) [μέρος, part; ἀτροφία, want of nourishment]. 1. Partial atrophy. 2. Atrophy of a limb.
**mercaptal** (*mer-kap'-tal*) [*mercury; captans*, seizing]. A thioacetal; a product of the union of a mercaptan and an aldehyde.
**mercaptan** (*mer-kap'-tan*) [*mercurius*, mercury; *captans*, seizing, on account of combining readily with mercury]. A derivative of an alcohol in which the oxygen of the latter is replaced by sulphur.
**mercauro** (*mer-kaw'-ro*). A proprietary alterative compound of an equal amount of the bromides of gold, arsenic, and mercury; it is used in syphilis and scrofula. Dose 5–15 min. (0.3–1.0 Cc.) 3 times daily after meals.
**Mercier's bar** (*mer-se-a'*) [Louis Auguste Mercier, French urologist, 1811–1882]. The transverse curved ridge joining the openings of the ureters on the inner surface of the bladder; it forms the posterior boundary of the *trigonum vesicæ*. Syn., *bar of the bladder; interureteric bar; plica ureterica*. **M.'s valve**, a valvular projection that may be formed at the vesicourethral orifice by the hypertrophied internal sphincter vesicæ.
**mercolint** (*mer'-ko-lint*). A proprietary article consisting of canton flannel impregnated with metallic mercury very finely divided.
**mercuralgam** (*mer-kū-ral'-gam*). An amalgam of mercury, aluminum, and magnesium. It is used as a substitute for mercurial ointment.
**mercurette** (*mer-kū-ret'*). Trade name of a solid preparation consisting of 30 grains of mercury in cacoa butter; used for inunction in cases of syphilis.
**mercurial** (*mer-kū'-re-al*) [*mercury*]. 1. Pertaining to or caused by mercury. 2. Any preparation of mercury or its salts. **m. ointment** (*unguentum hydrargyri*, U. S. P.), a salve containing mercury, oleate of mercury, suet, and benzoinated lard. **m. palsy, m. tremor**, paralysis or an involuntary spasmodic twitching of the voluntary muscles as a result of mercurial intoxication, and intensified on voluntary motion. **m. plaster**. See *mercury plaster*. **m. rash**, an eczema from the use of mercury.
**mercurialism** (*mer-kū'-re-al-izm*) [*mercury*]. Poisoning due to absorption of mercury.
**mercurialization** (*mer-kū-re-al-iz-a'-shun*) [*mercury*]. The act of bringing under the influence of mercury.
**mercuriate** (*mer-kū'-re-āt*). A salt of mercury.
**mercuric** (*mer-kū'-rik*) [*mercury*]. Pertaining to mercury as a bivalent element. **m. chloride**. See *mercury bichloride*.
**mercuricum** (*mer-kū'-rik-um*) [*mercury*]. Mercury when acting as a bivalent radical.
**Mercurio's position** (*mer-kū'-re-o*) [Geronimo Scipione *Mercurio*, Italian obstetrician, 1550–1595]. Same as Walcher's position, *q. v.*
**mercurioi** (*mer-kū'-re-ol*). See *mercuralgam*.
**mercurious** (*mer-kū'-re-us*) [see *mercury*]. **m. corrosivus**, corrosive sublimate. **m. dulcis**, calomel.
**mercurioiodohemol** (*mer-kū-ro-i-o-do-hem'-ol*). A combination of hemol, metallic mercury, and iodine; it is used in syphilis.
**mercurol** (*mer'-kū-rol*). A compound of nucleic acid and mercury (10 %). It is used in chronic ulcers in 2 to 5 % solution; in gonorrhea 0.5 to 2 % injection; in syphilis internally. Dose 3/4–1 1/4 gr. (0.05–0.1 Gm.) twice daily. Syn., *mercury nucleid*.
**mercuroseptol** (*mer-kū-ro-sep'-tol*). See *mercury quinosetolate*.
**mercurosum** (*mer-kū-ro'-sum*) [*mercury*]. Mercury when acting as a univalent radical.
**mercurous** (*mer'-kū-rus*) [*mercury*]. Pertaining to compounds that contain mercury as a univalent radical. **m. chloride**, calomel. See *mercury subchloride*.
**mercury** (*mer'-kū-re*) [*Mercurius*, a Latin divinity, the god of traffic]. Symbol Hg; atomic weight

200; quantivalence II. Hydrargyrum. A shining, silver-white, liquid, volatile metal, having a specific gravity of 13.55. It is insoluble in the ordinary solvents, in hydrochloric acid, and in sulphuric acid in the cold; it dissolves in the last when boiled with it, and is readily soluble in nitric acid. It boils at 357.25° C., and solidifies at —39.4° C. Mercury is found pure, but is chiefly obtained as the native sulphide, or cinnabar. It forms two classes of compounds—the *mercurous*, those in which two atoms of the metal, and the *mercuric*, those in which one atom, is combined with a bivalent radical. The mercuric salts are more soluble and more poisonous than the mercurous. The uses of mercury and its salts are as follows: as a purgative and cholagogue (calomel, blue-mass, mercury with chalk), as an alterative in chronic inflammations, as an antisyphilitic, an antiphlogistic, an intestinal antiseptic, a disinfectant, a parasiticide, a caustic, and an astringent. The absorption of mercury in sufficient quantity causes poisoning, characterized by a coppery taste in the mouth, ptyalism, loosening of the teeth, sponginess of the gums; in severer cases, ulceration of the cheeks, necrosis of the jaws, marked emaciation; at times neuritis develops, and a peculiar tremor. The soluble salts when taken in excess act as intense gastrointestinal irritants. **m. albuminate, dry,** a compound of albumin with 4 % of mercury bichloride; a white powder, soluble in water with turbidity. Triturated with milk-sugar it is used as an antiseptic dressing for wounds. **m. albuminate, liquid,** a slightly opalescent liquid containing 1 % of mercury bichloride; it is used hypodermatically in syphilis. Injection 8–15 min. (0.5–1.0 Cc.). **m., ammoniated** (*hydrargyrum ammoniatum*, U. S. P.), $NH_2HgCl$, white precipitate; mercuric ammonium chloride; it is used chiefly locally. **m., ammoniated, ointment of** (*unguentum hydrargyri ammoniati*, U. S. P.), an ointment made of ammoniated mercury, white petrolatum, and hydrous wool-fat. **m. arsenate,** $2Hg_2HAsO_4+H_2O$, used in syphilis. **m. and arsenic iodides, solution of** (*liquor arseni et hydrargyri iodidi*, U. S. P.), Donovan's solution. Dose 5 min. (0.32 Cc.). **m. asparaginate,** $Hg(C_4H_7N_2O_3)_2$, obtained from mercury oxide, and a hot aqueous solution of asparagin; it is alterative and antiseptic, and is used hypodermatically in syphilis. Dose $\frac{1}{12}$–$\frac{1}{6}$ gr. (0.005–0.01 Gm.) daily. **m. benzoate,** $Hg(C_6H_5COO)_2+H_2O$, used in injection in gonorrhea. **m. bichloride** (*hydrargyri chloridum corrosivum*, U. S. P.), $HgCl_2$, corrosive sublimate. Dose as tonic $\frac{1}{100}$–$\frac{1}{60}$ gr. (0.0006–0.0011 Gm.); as an antisyphilitic $\frac{1}{60}$–$\frac{1}{20}$ gr. (0.0011–0.003 Gm.). It is also a valuable antiseptic. **m. bichloride, peptonized.** 1. A yellowish powder containing 10 % of mercury bichloride, soluble in water. Dose $\frac{1}{2}$–$1\frac{1}{2}$ gr. (0.032–0.1 Gm.). 2. A clear yellow liquid containing 1 % of mercury bichloride. Used hypodermatically in syphilis. Dose 15 min. (1 Cc.), properly diluted, daily. **m. biniodide** (*hydrargyri iodidum rubrum*, U. S. P.), $HgI_2$, red iodide of mercury. Dose $\frac{1}{16}$–$\frac{1}{4}$ gr. (0.004–0.016 Gm.). **m. borate,** $Hg_2B_4O_7$, a brown antiseptic powder used as a dusting-powder and ointment (1 : 50) for wounds. **m. bromide,** $Hg_2Br_2$, mercurous bromide. Dose 1 gr. (0.065 Gm.) daily. **m. bromide, mercuric,** $HgBr_2$. alterative. Dose $\frac{1}{16}$–$\frac{1}{4}$ gr. (0.004–0.016 Gm.). **m, bromide, mercurous,** $HgBr_2$, a white powder; alterative and antiseptic. Dose 1 gr. (0.065 Gm.) in divided doses increasing gradually. **m. carbolate,** $Hg(C_6H_5O)_2$. Dose $\frac{1}{3}$–$\frac{1}{2}$ gr. (0.02–0.032 Gm.). **m. with chalk** (*hydrargyrum cum creta*, U. S. P.), gray powder. Dose 5 gr.–$\frac{1}{2}$ dr. (0.32–2.0 Gm.). **m.-ethylenediamine citrate,** a salt of mercury used in 3 : 1000 solution for disinfection of hands. Syn., *mercuramin*. **m., extinguished,** a trituration of metallic mercury with some fatty substance until no globules of the mercury can be discovered with a magnifying-glass of low power. **m. formamidate,** $Hg(HCONH)_2$, a solution of formamide and mercury oxide; each cubic centimeter corresponds to 0.01 Gm. of mercury bichloride. Injection in syphilis 16 min. (1 Cc.) daily. **m. gallate,** $Hg(C_7H_5O_5)_2$, a greenish-black powder containing about 37 % of mercury; alterative and antisyphilitic. Dose $1\frac{1}{2}$–3 gr. (0.1–0.2 Gm.) daily in pills. **m. iodate,** $Hg(IO_3)_2$, used subcutaneously in syphilis. Dose $\frac{1}{6}$ gr. (0.01 Gm.). **m. iodotannate,** used hypodermatically. **m. mass** (*massa hydrargyri*, U. S. P.), blue-mass; blue pill. Dose 3 gr. (0.2 Gm.). **m., metallic,** mercury in its pure state. **m. naphtholate,** $Hg(C_{10}H_7O)_2$, a lemon-colored antiseptic powder containing 30.8 % of mercury. Dose 1 gr. (0.065 Gm.). **m. nitrate, mercuric,** $Hg(NO_3)_2$, a white, deliquescent, poisonous powder, soluble in water; alterative and antiseptic. Dose $\frac{1}{60}$–$\frac{1}{8}$ gr. (0.001–0.008 Gm.). **m. nitrate, mercurous, normal,** $Hg_2(NO_3)_2+2H_2O$; antisyphilitic, antiseptic, and caustic. Dose $\frac{1}{32}$–$\frac{1}{4}$ gr. (0.002–0.016 Gm.); maximum dose $\frac{1}{4}$ gr. (0.016 Gm.), single; 1 gr. (0.065 Gm.) daily. **m. nitrate, ointment of** (*unguentum hydrargyri nitratis*, U. S. P.), citrine ointment, is used in chronic skin diseases. **m. nitrate, solution of** (*liquor hydrargyri nitratis*, U. S. P.), contains about 60 % of mercury nitrate. **m. nucleid.** See *mercurol*. **m. ointment.** See *mercurial ointment*. **m. ointment, dilute** (*unguentum hydrargyri dilutum*, U. S. P.), blue ointment. **m., oleate of** (*oleatum hydrargyri*, U. S. P.), contains yellow mercury oxide, oleic acid, and water. **m. oxide, red** (*hydrargyri oxidum rubrum*, U. S. P.), HgO, red precipitate, is used locally. **m. oxide, red, ointment of** (*unguentum hydrargyri oxidi rubri*, U. S. P.), ointment of red mercuric oxide. **m. oxide, yellow** (*hydrargyri oxidum flavum*, U. S. P.), HgO, is used locally in eye diseases. **m. oxide, yellow, ointment of** (*unguentum hydrargyri oxidi flavi*, U. S. P.), yellow mercuric oxide ointment. **m. oxycyanide,** $HgO . HgCy_2$, a white, antiseptic powder, soluble in water; application for wounds and surgical operations, 0.6 % solution. **m. paraphenylthionate,** $C_6H_4 . OH . SO_3Hg$, a stable crystalline, noncaustic compound used as a substitute for corrosive sublimate. Syn., *hydrargyrol*. **m. peptonate,** used hypodermatically. **m. plaster** (*emplastrum hydrargyri*, U. S. P.), mercurial plaster, is used to disperse indolent swellings, and is applied over the liver in chronic hepatitis. **m. protiodide** (*hydrargyri iodidum flavum*, U. S. P.), $Hg_2I_2$, yellow mercurous iodide. Dose $\frac{1}{2}$ gr. (0.033 Gm.) daily. **m. quinoseptolate,** $C_9H_5 . N . OH . SO_3Hg+2NaCl$, a compound of quinosol and mercury with sodium chloride. It is an odorless mass resembling the yolk of an egg, which in water swells into a slimy mass. It is used in syphilis. Syn., *mercuroseptol*. **m. resorcinacetate,** a yellow, crystalline powder, containing 68.9 % of mercury; it is antisyphilitic. Dose for hypodermatic use 3 min. (0.2 Cc.) of a solution of 85 gr. (5.6 Gm.) of the salt in 85 gr. (5.6 Gm.) of liquid paraffin and 30 gr. (2 Gm.) anhydrous lanolin, once a week. **m. saccharate,** a trituration of 1 part of pure mercury and 1 or 2 parts of powdered sugar; it is used as a vermifuge. **m. salicylate.** Dose $\frac{1}{6}$ gr. (0.01 Gm.). **m. silicofluoride,** $Hg_2SiF_6+2H_2O$, prismatic crystals, soluble in water; it is used as a wound antiseptic in solution of 1 : 1000 and as ointment in 1 : 2000. **m.-sozoiodol, m. sozoiodolate,** $HgC_6H_2I_2O . SO_3$, a fine powder obtained as a precipitate from mixing aqueous solutions of sodium sozoiodol and mercury nitrate; it is soluble in 500 parts of water and in solution of sodium chloride. It is alterative and antiseptic and is used locally (ointment 3 to 5 %) and subcutaneously in syphilis; injection $1\frac{1}{4}$ gr. (0.08 Gm.) a dose in solution of potassium iodide. **m. subchloride** (*hydrargyri chloridum mite*, U. S. P.), $Hg_2Cl_2$, mild mercurous chloride; calomel. Dose as alterative $\frac{1}{2}$–1 gr. (0.032–0.065 Gm.); as sedative to stomach and bowels $\frac{1}{8}$–$\frac{1}{4}$ gr. (0.008–0.016 Gm.); as purgative $\frac{1}{4}$–$\frac{1}{2}$ gr. (0.016–0.032 Gm.) every hour; or 5–15 gr. (0.32–1.0 Gm.); added to lime-water, 1 dr. to 1 pint, it forms lotio nigra, or black-wash. **m. subsulphate,** $Hg(HgO)_2SO_4$ basic mercuric sulphate; turpeth mineral. Dose as an emetic 2–5 gr. (0.13–0.32 Gm.). **m. succinimide,** $Hg(C_4H_4NO_2)_2$, soluble in 25 parts of water, slightly soluble in alcohol; it is recommended as a hypodermatic antisyphilitic (1.3 : 100). Dose $\frac{1}{5}$ gr. (0.013 Gm.). **m. sulphide, red,** cinnabar, vermilion, is used only by fumigation. **m. sulphoichthyolicum,** Unna's mixture of sodium sulphoichthyolate, 10 parts, corrosive sublimate, 3 parts. **m. suppositories** (*suppositoria hydrargyri*, B. P.), each contains 5 gr. of mercurial ointment. **m. tannate.** Dose 3 gr. (0.2 Gm.) daily. **m. thymolacetate,** $HgC_{10}H_{13}O . C_2H_3O_2+Hg(C_2H_3O_2)_2$, used in tuberculosis, syphilis, etc., by intramuscular injection. Dose $1\frac{1}{2}$ gr. (0.1 Gm.) every 3 to 5 days in liquid paraffin or glycerol. **m. thymolate,** a basic salt variable in its composition; it is used in syphilis as is mercury thymolacetate. **m. thymolsulphate,** $C_{10}H_{13}OHg_2SO_4$, a white, insoluble powder. It is used hypoder-

matically in syphilis. Dose $\frac{1}{12}-\frac{1}{6}$ gr. (0.005-0.01 Gm.). **m. tribromophenolacetate,** employed subcutaneously in syphilis, tuberculosis, etc. Dose 5 gr. (0.32 Gm.), dissolved in liquid paraffin once a week. **m. and zinc cyanide,** $Zn_4Hg(CN)_{10}$, is used locally as an antiseptic.
**merd** [*merda,* feces]. Feces. **m. diaboli,** asafetida.
**meremphraxis** (*mer-em-fraks'-is*) [μέρος, part; ἔμφραξις, stoppage]. Partial obstruction.
**mergal** (*mer'-gal*). Trade name of a mixture consisting of albumin tannate and mercuric cholate.
**meridian** (*mer-id'-e-an*) [*meridies,* midday]. A great circle surrounding a sphere and intersecting the poles. **m. of the eye,** a line drawn around the globe of the eye and passing through the poles of the vertical axis (*vertical meridian*), or through the poles of the transverse axis (*horizontal meridian*).
**meridional** (*mer-id'-e-on-al*). [*meridies,* midday]. Relating to a meridian.
**meridrosis** (*mer-id-ro'-sis*) [μέρος, part; ἱδρώς, sweat]. Local perspiration.
**merismopedia** (*mer-is-mo-pe'-de-ah*) [μερισμός, division; παῖς, child]. A genus of bacteria multiplying by two rectangular divisions, thus forming a tablet-like group of four cells in one plane.
**merispore** (*mer'-is-pōr*) [μέρος, a part; σπόρα, seed]. A spore resulting from division of another spore.
**merista** (*mer-is'-tah*). See *merismopedia.*
**meristem** (*mer'-is-tem*) [μεριστός, verbal adj. of μερίζειν, to divide]. In biology, the actively growing, undifferentiated cell-tissue of the growing tips of plants; the formative tissue of the cambium layer.
**meristiform** (*mer-is'-ti-form*). Having the shape of merismopedia; sarcinic.
**Merkel's corpuscles** (*mer'-kel*) [Karl Ludwig *Merkel,* German anatomist, 1812-1876]. See *Grandry's corpuscles.* **M.'s line,** an imaginary line extending from the middle of the internal palpebral ligament to the space between the last bicuspid and first molar teeth, and indicating the course of the lacrimal sac and nasal duct. **M.'s muscle,** the keratocricoid muscle.
**merlusan** (*mer'-lū-san*). A mercury-albumen compound, used in syphilis, and in gonorrhea.
**mero-** (*me-ro-*) [μέρος, a part]. A prefix meaning part.
**meroacrania** (*mer-o-ak-ra'-ne-ah*) [*mero-*; a, neg.; κρανίον, skull]. Congenital absence of a part of the cranium.
**meroblast** (*mer'-o-blast*) [μέρος, a part; βλαστός, a germ]. In embryology, an ovum that contains beside the formative protoplasm or yolk, more of less food-yolk or nutritive protoplasm.
**meroblastic** (*mer-o-blas'-tik*) [*mero-*; βλαστός, a germ]. Dividing only in part, as *meroblastic* ova, those in which the process of segmentation is confined to one portion of the ovum.
**merocele** (*mer'-o-sēl*) [μηρός, thigh; κήλη, hernia]. Femoral hernia.
**merocoxalgia** (*mer-o-koks-al'-je-ah*) [μηρός, thigh; *coxa,* hip; ἄλγος, pain]. Pain affecting the thigh and hip.
**merocrania** (*mer-o-kra'-ne-ah*) [*mero-*; κρανίον, the skull]. A condition of monstrosity marked by absence of part of the skull.
**merocrine** (*mer'-o-krēn*) [*mero-*; κρίνειν, to separate]. Applied to glands the cells of which, having elaborated materials of secretion, evacuate them and continue alternately to secrete and evacuate new material. Cf. *holocrine.*
**merodialysis** (*mer-o-di-al'-is-is*) [*mero-*; διάλυσις, separation]. Partial decomposition.
**merodiastolic** (*mer-o-di-as-tol'-ik*) [*mero-*; *diastole*]. Relating to a part of the diastole.
**merogastrula** (*mer-o-gas'-troo-lah*) [*mero-*; γαστήρ, belly: *pl.,* *merogastrulæ*]. In biology, the gastrula of a meroblastic ovum.
**merogenesis** (*mer-o-jen'-es-is*) [*mero-*; γένεσις, generation]. Reproduction by segmentation.
**merology** (*mer-ol'-o-je*) [*mero-*; λόγος, science]. General anatomy; the science of elementary tissues.
**meropia** (*mer-o'-pe-ah*) [*mero-*; ὤψ, sight]. Partial blindness; obscuration of vision.
**merorrhachischisis** (*mer-or-rak-is'-kis-is*) [*mero-*; ῥάχις, the spine; σχίσις, fissure]. Partial rachischisis.
**meros** (*me'-ros*) [μέρος, a part]. 1. A part. 2. [μηρός, thigh]. The thigh or femur.
**merosome** (*mer'-o-sōm*) [*mero-*; σῶμα, body]. In biology, one of the serial parts of a segmented organism; a somite or metamere.

**merosystolic** (*mer-o-sis-tol'-ik*) [*mero-*; συστολή, systole]. Relating to a part of the systole.
**merotomy** (*mer-ot'-o-me*) [*mero-*; τομή, a cutting]. The section of a living cell for the study of the ulterior transformation of the segments; by extension it is also applied to experimental division of amœbæ, etc.
**merozoite** (*mer-o-zo'-īt*) [*mero-*; ζῷον, an animal]. Any one of the segments resulting from the splitting up of the schizont in the asexual form of reproduction of protozoa.
**Merseburg triad** (*mers'-berg*) [*Merseburg,* a town in Germany]. The three classical symptoms of exophthalmic goiter: the goiter, exophthalmos, and rapid heart beat.
**Merulius** (*mer-u'-le-us*) [*merus,* bright, glistening]. A genus of fungi of the order *Basidiomycetes.* The mycelium of **M. lacrymans,** causes dry rot in timber, and diseases (sometimes fatal) of the respiratory passages are attributed to the inhaled spores.
**Méry's glands** (*ma-re'*) [Jean *Méry,* French anatomist, 1645-1722]. See *Cowper's glands.*
**merycic** (*mer-is'-ik*) [*merycism*]. Relating to merycism; ruminating.
**merycism** (*mer'-is-izm*) [μηρυκισμός, rumination]. Rumination; chewing the cud—a normal process in the ruminating animals, and sometimes occurring in man.
**merycole** (*mer'-ik-ōl*). An individual who practises merycism.
**mesad** (*mes'-ad*) [μέσος, middle; *ad,* to]. Toward the median line or plane.
**mesal** (*mes'-al*) [μέσος, middle]. Pertaining to or situated in the middle line or plane.
**mesameboid** (*mes-am-e'-boid*) [μέσος, middle; *ameboid*]. 1. A nonepithelial ameboid cell derived from the mesoderm. 2. A leukocyte.
**mesaortitis** (*mes-a-or-ti'-tis*) [μέσος, middle; *aortitis*]. Inflammation of the middle coat of the aorta.
**mesaraic** (*mes-ar-a'-ik*) [μέσος, middle; ἀραιά, belly]. Mesenteric.
**mesarteritis** (*mes-ar-ter-i'-tis*) [μέσος, middle; *arteritis*]. Inflammation of the middle coat of an artery.
**mesaticephalic** (*mes-at-e-sef-al'-ik*). With a cephalic index between 75 and 79.
**mesaticephalus** (*mes-at-e-sef'-al-us*) [μέσατος, median; κεφαλή, head]. In craniometry, a term applied to a skull having a cephalic index of between 75 and 79.
**mesatipelvic** (*mes-at-ip-el'-vik*) [μέσατος, median; *pelvis*]. A term applied to a pelvis whose index ranges between 90° and 95°.
**mescal buttons.** The dried tubercles from a species of cactus, *Anhalonium lewinii,* capable of producing inebriation and hallucinations.
**mescaline** (*mes'-kal-ēn*). An alkaloid from mescal buttons.
**mesembryo** (*mes-em'-bre-o*) [*meso-*; *embryo*]. The blastula stage of the ova of metazoans.
**mesencephal** (*mes-en'-sef-al*). Same as *mesencephalon.*
**mesencephalic** (*mes-en-sef-al'-ik*) [*meso-*; ἐγκέφαλον, the brain]. Relating to the mesencephalon.
**mesencephalon** (*mes-en-sef'-al-on*) [μέσος, middle; *encephalon*]. The midbrain; that part of the brain developed from the middle cerebral vesicle; the corpora quadrigemina, the crura cerebri, and the aqueduct of Sylvius.
**mesenchyma** (*mes-eng'-kim-ah*) [μέσος, middle; ἔγχυμα, an infusion]. The portion of the mesoderm that produces all the connective tissues of the body, the blood-vessels, and the blood, the entire lymphatic system proper, and the heart; the nonepithelial mesoblast.
**mesenna** (*mes-en'-ah*). The bark of the Abyssinian tree *Albizzia anthelmintica*; it is said to be a powerful teniafuge.
**mesenteric** (*mes-en-ter'-ik*) [*mesentery*]. Pertaining to the mesentery.
**mesentericomesocolic** (*mes-en-ter-ik-o-mes-o-kol'-ik*). Relating to the mesentery and the mesocolon.
**mesenteriolum** (*mes-en-ter-e-o'-lum*) [dim. of *mesentery*]. A little mesentery; especially the fold of peritoneum that sometimes connects the vermiform appendix with the mesentery.
**mesenteritic** (*mes-en-ter-it'-ik*) [*mesentery*; ἴτις, inflammation]. Pertaining to or affected with mesenteritis.
**mesenteritis** (*mes-en-ter-i'-tis*) [*mesentery*; ἴτις, inflammation]. Inflammation of the mesentery.

**mesenterium** (*mes-en-ter'-e-um*) [L.: *pl.*, *mesenteria*]. A mesentery.
**mesenteroid** (*mes-en'-ter-oid*) [*mesentery*; εἶδος, like]. Resembling the mesentery.
**mesenteron** (*mes-en'-ter-on*) [*mesentery*]. The middle portion of the primitive digestive tube, lined by entoderm, and giving rise to the part of the alimentary tract between the pharynx and the lower third of the rectum.
**mesentery** (*mes'-en-ter-e*) μέσος, middle; ἔντερον, bowel]. A fold of the peritoneum that connects the intestine with the posterior abdominal wall; that of the small intestine is termed *mesentery proper;* that of the colon, cecum, and rectum, *mesocolon, mesocecum, mesorectum*, respectively.
**mesentoderm, mesendoderm** (*mes-en'-to-derm, mes-en'-do-derm*) [*meso-*; ἐντός, within; δέρμα, skin]. The ental or entodermal division of the mesoderm; also, the indifferent tissue from which both entoderm and mesoderm are developed.
**mesethmoid** (*mes-eth'-moid*) [*meso-*; *ethmoid*]. The mesal element of the ethmoid bone, forming a separate bone in some of the lower animals.
**mesh.** A network, as of vessels or nerves.
**mesiad** (*mes'-e-ad*). Same as *mesad*.
**mesial** (*mes'-e-al*). Same as *mesal*.
**mesiobuccal** (*mes-e-o-buk'-al*) [*mesial*; *bucca*, cheek]. Pertaining to surfaces between the mesial and buccal aspects of the teeth.
**mesiolingual** (*mes-e-o-lin'-gwal*) [*mesial*; *lingua*, tongue]. Relating to surfaces between the mesial and lingual aspects of the teeth.
**mesion** (*mes'-e-on*). See *meson*.
**mesiris** (*mes-i'-ris*) [*meso-*; *iris*]. The middle layer of the iris, lying between the ectiris and the entiris.
**mesmeric** (*mes-mer'-ik*) [*mesmerism*]. Pertaining to or induced by mesmerism, as *mesmeric* sleep.
**Mesmerism** (*mes'-mer-izm*) [Friedrich Anton *Mesmer*, German physician, 1734-1815]. Hypnotism.
**meso-** (*mes-o-*) [μέσος; middle]. A prefix signifying middle or pertaining to the mesentery.
**mesoappendix** (*mes-o-ap-en'-diks*), [*meso-*; *appendix*]. The mesentery of the vermiform appendix.
**mesoarium** (*mes-o-a'-re-um*) [*meso-*; ᾠάριον, dim. of ᾠον, egg; *pl.*, *mesoaria*]. In biology, that fold of the peritoneum in certain animals (*e. g.*, fishes) which forms the mesentery of the ovary. See *mesovarium*.
**mesobacteria.** (*mes-o-bak-te'-re-ah*). Medium-sized bacteria.
**mesoblast** (*mes'-o-blast*) [*meso-*; βλαστός, a germ]. The middle layer of the blastoderm, probably derived from both the ectoderm and the entoderm, and giving rise to the vascular, muscular, and skeletal systems, the generative glands, and the kidneys.
**mesoblastic** (*mes-o-blas'-tik*) [*meso-*; βλαστός, sprout]. Pertaining to the mesoblast.
**mesobronchitis** (*mes-o-bron-ki'-tis*) [*meso-*; βρόγχος, bronchus; ιτις, inflammation]. An inflammation of the middle coat of the bronchial tubes.
**mesocardia** (*mes-o-kar'-de-ah*) [*meso-*; καρδία, heart]. The position of the heart in the central and anterior part of the chest, a situation that is normal at an early stage of development.
**mesocardium** (*mes-o-kar'-de-um*) [*meso-*; καρδία, heart]. A mesoblastic fold attached to the heart.
**mesocecum, mesocæcum** (*meso-se'-kum*) [*meso-*; *cecum*]. The mesentery that in some cases connects the cecum with the right iliac fossa.
**mesocele** (*mes'-o-sēl*) [*meso-*; κοιλία, a cavity]. The aqueduct of Sylvius. Syn., *iter a tertio ad quartum ventriculum*.
**mesocephalon** (*mes-o-sef'-al-on*). See *pons Varolii*.
**mesochoroidea** (*mes-o-ko-roid'-e-ah*) [*meso-*; *choroid*]. The middle coat of the choroid.
**mesococcus** (*mes-o-kok'-us*). A coccus intermediate in size between a micrococcus and a megacoccus.
**mesocolic** (*mes-o-kol'-ik*) [*mesocolon*]. Pertaining to the mesocolon. **m.** **band,** a longitudinal muscular band corresponding to the insertion of the mesocolon.
**mesocolon** (*mes-o-ko'-lon*) [*meso-*; κόλον, colon]. The mesentery connecting the colon with the posterior abdominal wall. It is divided into ascending, descending, and transverse portions.
**mesocolopexy** (*mes-o'-ko-lo-peks-e*) [*mesocolon*; πῆξις, fixation]. Same as mesocoloplication.
**mesocoloplication** (*mes-o-ko-lo-pli-ka'-shun*) [*mesocolon*; *plication*]. An operation for shortening the mesocolon, which is accomplished by folding and suturing it.
**mesocord** (*mes'-o-kord*) [*meso-*; *chorda*, cord]. An umbilical cord not inserted directly into the placenta, but received into a fold of the amnion.
**mesocornea** (*mes-o-kor'-ne-ah*) [*meso-*; *cornea*]. The proper substance of the cornea lying between the ectocornea and the entocornea.
**mesocranium** (*mes-o-kra'-ne-um*). The vertex of the skull.
**mesocuneiform** (*mes-o-kū'-ne-if-orm*) [*meso-*; *cuneiform*]. The middle cuneiform bone of the tarsus.
**mesocyst** (*mes'-o-sist*) [*meso-*; κύστις, bladder]. A double fold of peritoneum attaching the gallbladder to the liver.
**mesoderm** (*mes'-o-derm*). See *mesoblast*.
**mesodiastolic** (*mes-o-di-as-tol'-ik*) [*meso-*; *diastole*]. Occurring in the middle of the diastolic period.
**mesodme** (*mes-od'-me*). See *mediastinum*.
**mesoduodenum** (*mes-o-du-o-de'-num*) [*meso-*; *duodenum*]. That part of the mesentery that sometimes connects the duodenum with the posterior wall of the abdominal cavity. Normally, the true duodenum has no mesentery, at least in its fully developed state.
**mesoepididymis** (*mes-o-ep-id-id'-im-is*) [*meso-*; *epididymis*]. The fold of the tunica vaginalis attaching the epididymis to the upper posterior part of the testis.
**mesogaster** (*mes-o-gas'-ter*) [see *mesogastrium*]. The part of the primitive gut giving rise to the duodenum, the liver, the pancreas, the jejunum, and the ileum.
**mesogastric** (*mes-o-gas'-trik*) [see *mesogastrium*]. Pertaining to the umbilical region.
**mesogastrium** (*mes-o-gas'-tre-um*) [*meso-*; γαστήρ, stomach]. 1. The umbilical region of the abdomen. 2. A fold of mesentery that in early fetal life connects the stomach with the posterior abdominal wall.
**mesogluteus** (*mes-o-gloo-te'-us*) [*meso-*; γλουτός, the buttock]. The middle gluteal muscle.
**mesognathic** (*mes-og-na'-thik*) [see *mesognathion*]. 1. Relating to the mesognathion. 2. See under *index, alveolar*.
**mesognathion** (*mes-og-na'-the-on*) [*meso-*; γνάθος, jaw]. The intermaxillary bone; a fetal bone lying behind the fore part of the superior maxilla, with which it becomes fused.
**mesognathous** (*mes-og'-na-thus*) [*meso-*; γνάθος, jaw]. Having a gnathic index between 98° and 103°.
**Mesogonimus** (*mes-o-gon'-im-us*). A genus of flukes, same as *Paragonimus*.
**mesoileum** (*mes-o-il'-e-um*). The mesentery of the ileum.
**mesojejunum** (*mes-o-je-joo'-num*). The mesentery attached to the jejunum.
**mesolepidoma** (*mes-o-lep-id-o'-mah*). See under *lepidoma*.
**mesolobe, mesolobus** (*mes'-o-lōb, mes-ol'-o-bus*) [*meso-*; λοβός, lobe]. The corpus callosum.
**mesologic** (*mes-o-loj'-ik*) [*meso-*; λόγος, science]. Pertaining to environment in its relation to life.
**mesometritis** (*mes-o-me-tri'-tis*) [*meso-*; μήτρα, womb; *itis*, inflammation]. Inflammation of the parenchyma of the womb.
**mesometrium** (*mes-o-me'-tre-um*) [*meso-*; μήτρα, womb]. The broad ligaments.
**meson** (*mes'-on*) [μέσον, the middle]. The imaginary plane dividing the body into the right and left halves.
**mesonasal** (*mes-o-na'-sal*) [*meso-*; *nasus*, nose]. Pertaining to the median region of the nose.
**mesonephric** (*mes-o-nef'-rik*) [*meso-*; νεφρός, kidney]. Pertaining to the mesonephron.
**mesonephron, mesonephros** (*mes-o-nef'-ron, mes-o-nef'-ros*) [*meso-*; νεφρός, kidney]. 1. The Wolffian body, the middle division of the segmental organs. It precedes in the embryo the development of the permanent kidney. 2. A fold of peritoneum by which a floating kidney is attached to the abdominal wall.
**mesoneuritis** (*mes-o-nū-ri'-tis*) [*meso-*; *neuritis*]. Inflammation of the structures contained between a nerve and its sheath. **m. nodular,** a form in which there are nodular thickenings on the nerve.
**meso-omentum** (*mes-o-o-men'-tum*). The mesentery of the omentum.
**mesopexy** (*mes'-o-peks-e*) [*meso-*; πῆξις, a folding]. The operation of shortening an elongated mesentery.

**mesophilic** (*mes-o-fil'-ik*) [*meso-*; φιλεῖν, to love]. Applied to microorganisms which develop best at about body-temperature—35°–38° C.

**mesophlebion, mesophlebium** (*mes-o-fleb'-e-on, mes-o-fleb'-e-um*) [*meso-*; φλέψ, vein]. 1. The middle coat of a vein. 2. The space between two veins.

**mesophlebitis** (*mes-o-fle-bi'-tis*) [*meso-*; φλέψ, vein; ιτις, inflammation]. Inflammation of the middle coat of a vein, or mesophlebion.

**mesophryon** (*mes-of'-re-on*) [*meso-*; ὀφρύς, eyebrow]. The glabella.

**mesopleura** (*mes-o-ploo'-rah*) [*meso-*; πλευρά, rib]. An intercostal space.

**mesopneumon** (*mes-o-nū'-mon*) [*meso-*; πνεύμων, lung]. The fold of the pleura attached to the lung.

**mesopsyche** (*mes-op-si'-ke*) [*meso-*; ψυχή, soul]. Haeckel's term for the mesencephalon.

**mesorchium** (*mes-or'-ke-um*) [*meso-*; ὄρχις, testicle]. A fold of the peritoneum containing the fetal testes at about the fifth month of embryonic life.

**mesorectum** (*mes-o-rek'-tum*) [*meso-*; *rectum*]. The narrow fold of the peritoneum connecting the upper part of the rectum with the sacrum.

**mesoretina** (*mes-o-ret'-in-ah*) [*meso-*; *retina*]. The middle layer of the retina, composed of the nuclear and the rod-and-cone layer.

**mesoropter** (*mes-o-rop'-ter*) [*meso-*; ὅρος, boundary; ὀπτήρ, one who sees]. The normal position of the eyes when their muscles are at rest. m., muscular, the angle formed by the visual axes of the eyes when the external ocular muscles are at rest.

**mesorrhachischisis** (*mes-o-rak-is'-kis-is*). Partial rhachischisis; incomplete cleft of the spinal cord.

**mesorrhine** (*mes'-or-īn*) [*meso-*; ῥίς, nose]. Having a nasal index between 48° and 52°.

**mesosalpinx** (*mes-o-sal'-pingks*) [*meso-*; σάλπιγξ, a trumpet]. The upper part of the broad ligament which surrounds the Fallopian tube.

**mesoscapula** (*mes-o-skap'-ū-lah*) [*meso-*; *scapula*, shoulder-blade]. The scapular spine.

**mesoseme** (*mes'-o-sēm*) [*meso-*; σῆμα, sign]. With an orbital index of 84°–89°.

**mesosigmoid** (*mes-o-sig'-moid*) [*meso-*; *sigmoid*]. The mesentery of the sigmoid flexure of the colon.

**mesostaphyline** (*mes-o-staf'-il-īn*) [*meso-*; σταφυλή, the uvula, when swollen]. A skull with a palatal index of from 80° to 85°.

**mesostate** (*mes'-o-stāt*) [*meso-*; *statos*, placed]. A generic term for the intermediate substances formed in metabolic processes.

**mesosternum** (*mes-o-ster'-num*) [*meso-*; *sternum*]. The gladiolus, or second piece of the sternum.

**mesosystolic** (*mes-o-sis-tol'-ik*) [*meso-*; *systole*]. Relating to the middle of the systole.

**mesotan** (*mes'-o-tan*). The methyloxymethylester of salicylic acid; used in treatment of rheumatism by dermal absorption. Application 1–2 dr. (4–8 Cc.) mixed with olive-oil or castor-oil.

**mesotendon** (*mes-o-ten'-don*) [*meso-*; *tendon*]. Folds of synovial membrane extending to tendons from their fibrous sheaths.

**mesothelioma** (*mes-o-the-le-o'-mah*) [*meso-*; θηλή, nipple; ὄμα, a tumor]. A variety of epithelioma supposed to be developed from the mesoblast.

**mesothelium** (*mes-o-the'-le-um*) [*meso-*; θηλή, the nipple]. The lining of the wall of the primitive body-cavity situated between the somatopleure and splanchnopleure. It is the precursor of the endothelium.

**mesothenar** (*mes-o-the'-nar*) [*meso-*; θέναρ, palm]. The muscle drawing the thumb toward the palm of the hand; the adductor pollicis.

**mesosthenic** (*mes-o-sthen'-ik*) [*meso-*; σθένος, strength]. Having a moderate degree of muscular force.

**mesothermal** (*mes-o-ther'-mal*) [*meso-*; θέρμη, heat]. Of medium warmth.

**mesothorium** (*mes'-o-thor-e-um*). A product resulting from the disintegration of thorium, it is intermediate between radiothorium and thorium.

**mesotropic** (*mes-o-trop'-ik*) [*meso-*; τρέπειν, to turn]. Turned or situated mesad.

**mesovarium** (*mes-o-va'-re-um*) [*meso-*; *ovarium*, ovary]. A peritoneal fold connecting the ovary and the broad ligament; in the embryo with the Wolffian body.

**mesoventral** (*mes-o-ven'-tral*) [*meso-*; *venter*, belly]. Both median and ventral.

**mesoxalylurea** (*mes-oks-al-il-ū'-re-ah*). Same as *alloxan*.

**mesozoa** (*mes-o-zo'-ah*) [*meso-*; ζῷον, animal]. A class of animals intermediate between the protozoa and the metazoa.

**Mesua** (*mes'-ū-ah*) [J. Musuah, Arabian physician]. A genus of guttiferous trees. **M. ferrea**, an Indian tree of the order *Guttiferæ*. An attar is distilled from the flowers, and the oil of the seeds is used in rheumatism.

**mesuranic** (*mes-ū-ran'-ik*) [*meso-*; οὐρανίσκος, the roof of the mouth]. See *mesostaphyline*.

**meta-** (*met'-ah*) [μετά, over, among, beyond, after, or between]. 1. A prefix signifying over, beyond, among, between, change, or transformation. 2. In chemistry, a prefix denoting *unsymmetrical* derivatives of the benzene ring.

**meta-amidophenylparamethoxyquinolin** (*met-ah-am-id-o-fen-il-par-ah-meth-oks-e-kwin'-ol-in*). An antipyretic and antipyretic drug used instead of quinine. Dose 4–8 gr. (0.26–0.52 Gm.).

**metabasis** (*met-ab'-as-is*) [*meta-*; βαίνειν, to go]. Change.

**metabiosis** (*met-ah-bi-o'-sis*) [*meta-*; βίος, life]. A form of symbiosis, in which only one of the organisms is benefited; the other may remain uninfluenced or injured.

**metabolic** (*met-ah-bol'-ik*) [*metabolism*]. Pertaining to metabolism. m. equilibrium, the equality between the absorption and assimilation of food and the excretion of end-products.

**metabolism** (*met-ab'-o-lizm*) [μεταβολή, change]. A product of metabolism; a metabolite or mesostate.

**metabolism** (*met-ab'-o-lizm*) [μεταβολή, change]. The group of phenomena whereby organic beings transform food-stuffs into complex tissue-elements (*constructive metabolism*, assimilation, anabolism) and convert complex substances into simple ones in the production of energy (*destructive metabolism*, disassimilation, katabolism).

**metabolite** (*met-ab'-o-līt*) [see *metabolism*]. A product of metabolic change.

**metabolize** (*met-ab'-o-līz*) [μεταβολή, change]. To transform by means of metabolism.

**metabolon** (*met-ab'-o-lon*) [*metabolism*]. A purely transitory form of matter found in emanations from certain radioactive substances.

**metacarpal** (*met-ah-kar'-pal*) [*metacarpus*]. Pertaining to the metacarpus, or to a bone of the metacarpus.

**metacarpen** (*met-ak-ar'-pen*). Belonging to the metacarpus in itself.

**metacarpophalangeal** (*met-ah-kar-po-fa-lan'-je-al*) [*metacarpus*; *phalanges*]. Belonging to the metacarpus and the phalanges.

**metacarpus** (*met-ah-kar'-pus*) [*meta-*; καρπός, wrist]. That part of the hand between the carpus and the phalanges and consisting of five bones.

**metacasein reaction**. The coagulation of milk on boiling, after treatment with pancreatic extracts.

**metacele, metacoele, metacelia** (*met'-as-ēl, met-as-e'-le-ah*) [*meta-*; κοιλία, cavity]. The caudal or metencephalic portion of the fourth ventricle.

**metacetone** (*met-as'-et-ōn*). Diethyl-ketone.

**metacheirisis, metacheirismus** (*met-ak-i'-ris-is, met-ak-i-riz'-mus*) [*meta-*; χείρ, hand]. Manipulation in the treatment of disease.

**metachloral** (*met-ak-lo'-ral*). A tasteless, polymeric form of chloral, said to have properties not unlike those of chloral hydrate.

**metachoresis** (*met-ak-o'-res-is*) [*meta-*; χώρησις, a going]. Metastasis; dislocation.

**metachromasia** (*met-ak-ro-ma'-ze-ah*) [*meta-*; χρῶμα, color]. The chemico-chromatic changes induced in cells by the staining substances employed in histological technique.

**metachromatic** (*met-ah-kro-mat'-ik*) [*meta-*; χρῶμα, color]. Relating to a change of colors; staining with a different shade than that of the other tissues, as the mast-cell granules with basic anilin dyes. m. bodies, small granules in bacterial cells staining differently from the surrounding cytoplasm.

**metachromatism** (*met-ah-kro'-mat-izm*) [see *metachromatic*]. The quality of being different in color from other parts.

**metachrosis** (*met-ak-ro'-sis*) [*meta-*; χρώζειν, to tinge, to stain]. In biology, applied to the change or play of colors seen in the squid, chameleon, and other animals.

**metachysis** (*met-ak'-is-is*) [*meta-*; χύσις, effusion]. The transfusion of blood. Also the introduction of any substance into the blood by mechanical means.

# METACINESIS 552 METAPLASTIC

**metacinesis** (*met-ah-sin-e'-sis*). See *metakinesis*.
**metacism** (*met'-as-ism*) [μυτακισμός, from μυ, letter *m*]. Repetition of the letter *m*.
**metacondyle** (*met-ah-kon'-dīl*) [*meta-*; κόνδυλος, knuckle]. The distal phalanx of a finger, or the bone thereof.
**metacone** (*met'-ak-ōn*) [*meta-*; κῶνος, cone]. The outer posterior cusp of an upper molar tooth.
**metaconid** (*met-ak-o'-nid*) [*metacone*]. The inner-anterior cusp of a lower molar tooth.
**metaconule** (*met-ak-o'-nūl*) [*metacone*]. The posterior intermediate cusp of an upper molar tooth.
**metacresol** (*met-ah-kre'-sol*) [*meta-*; *cresol*], C₇H₈O₃. A liquid derivative of coal-tar, used as an antiseptic and disinfectant. It is stronger than phenol and less toxic. Dose 1–3 min. (0.06–0.2 Cc.). Applied in 0.5 % solution. **m.-anitol**, a 40 % solution of metacresol in anitol; recommended as an application in erysipelas. **m. bismuth**, Bi(C₇H₇O)₃, an antiseptic and astringent used in dysentery. **m.-cinnamic-ester**, an antituberculous compound of metacresol, 25 parts; cinnamic acid, 25 parts; dissolved in toluol and heated with phosphorus oxychloride, 20 parts.
**metacyesis** (*met-as-i-e'-sis*) [*meta-*; κύησις, gestation]. Extra-uterine gestation, especially that which is begun in the oviduct and continued in the abdominal cavity.
**metadermatosis** (*met-ad-er-mat-o'-sis*) [*meta-*; δέρμα, skin; νόσος, disease]. A pathological production of epidermis.
**metadiiodanilin** (*met-ah-di-i-od-an'-il-in*). See *diiodoanilin*.
**metadiphtheritic** (*met-ah-dif-ther-it'-ik*). Accompanying diphtheria.
**metadrasis** (*met-ad-ra'-sis*) [*meta-*; δρᾶσις, exertion]. Overwork of body or mind.
**metaelements** (*met-ah-el'-e-ments*). A hypothetical group of elemental substances intermediate between the elements as now known to us and protyl.
**metafacial** (*met-af-a'-shal*) [*meta-*; *facies*, face]. Posterior to the face.
**metagaster** (*met-ah-gas'-ter*) [*meta-*; γαστήρ, belly]. The permanent intestinal canal, succeeding the primitive canal, or protogaster.
**metagastrula** (*met-ah-gas'-troo-lah*) [see *metagaster*]. A modification of segmentation, producing a form of gastrula differing from the simple gastrula of the amphioxus.
**metagenesis** (*met-ah-jen'-es-is*). See *generation, alternate*.
**metagrippal** (*met-ah-grip'-al*). Occurring as a consequence of influenza; postgrippal.
**metaicteric** (*met-ah-ik'-ter-ik*). Occurring as a consequence of jaundice; posticteric.
**metakinesis** (*met-ah-kin-e'-sis*) [*meta-*; κίνησις, movement; change]. The term applied to that stage of cell-division in which the secondary threads or loops tend to pass toward the two poles of the nuclear spindle.
**metal** (*met'-al*) [μέταλλον, a metal]. An elementary substance characterized by malleability, ductility, fusibility, luster, its electric affinities, and the basic character of its oxides. **m., D'Arcet's**. See *D'Arcet's metal*.
**metalbumin** (*met-al-bū'-min*). See *paralbumin*.
**metaldehyde** (*met-al'-de-hīd*), . C₈H₁₆O₃. White needles, obtained from aldehyde by action of hydrochloric or sulphuric acid at a temperature below 0° C. It is sedative and hypnotic. Dose 2–8 gr. (0.13–0.52 Gm.).
**metalepsy** (*met'-al-ep-se*) [μετάληψις, participation]. In chemistry, change or variation produced by the displacement of an element of radical in a compound by its chemical equivalent. It is the same as *substitution, q. v.*
**metaleptic** (*met-al-ep'-tik*) [μετάληψις, participation]. Relating to metalepsy; also applied to a muscle, associated in its movement with another.
**metallesthesia** (*met-al-es-the'-ze-ah*) [*metal*; αἴσθησις, perception by the senses]. An alleged form of sensibility enabling hysterical or hypnotized subjects to distinguish between the contacts of various metals.
**metallic** (*met-al'-ik*). Similar to or resembling a metal. **m. tinkling**, peculiar metallic or bell-like sounds heard over a pneumothorax or large pulmonary cavity. The sounds are produced by coughing, speaking, or deep breathing.

**metallodynia** (*met-al-o-din'-e-ah*) [*metal*; ὀδύνη, pain]. Pain caused by metallic poisoning.
**metalloid** (*met'-al-oid*) [*metal*; εἶδος, like]. 1. Resembling a metal. 2. Any nonmetallic element.
**metallophagia, metallophagy** (*met-al-o-fa'-je-ah, met-al-off'-a-je*) [*metal*; φαγεῖν, to eat]. The insane impulse to swallow metallic objects.
**metallophobia** (*met-al-o-fo'-be-ah*). The fear of touching a metallic object.
**metalloscopy** (*met-al-os'-ko-pe*) [*metal*; σκοπεῖν, to examine]. The determination of the effects produced by the application of metals to the surface of the body. See also *metallotherapy*.
**metallotherapy** (*met-al-o-ther'-ap-e*) [*metal*; θεραπεία, therapy]. The treatment of certain nervous diseases, particularly hysteria, by the application of different metals to the affected part.
**metallotoxemia** (*met-al-o-toks-e'-me-ah*) [*metal*; τοξικόν, poison; αἷμα, blood]. Toxemia, or blood-poisoning, due to the ingestion of a metal.
**metamer** (*met'-am-ur*). A metameric substance. See *isomeric*.
**metamere** (*met'-ah-mēr*) [*meta-*; μέρος, a part]. Any one of the theoretical segments of a vertebrate animal.
**metameric** (*met-ah-mer'-ik*). Pertaining to metamerism; see *isomeric*.
**metamerid** (*met-am'-er-id*) [*metamere*]. A metameric substance; a group of metameric bodies.
**metamerism** (*met-am'-er-izm*) [*metamere*]. A variety of isomerism. See *isomerism*.
**metamorphic** (*met-am-or'-fik*) [*metamorphosis*]. Pertaining to metamorphosis; also synonymous with metamorphous.
**metamorphology** (*met-am-or-fol'-o-je*) [*metamorphosis*; λόγος, science]. In biology, the science of the changes of form passed through by individual organisms in the course of their life-histories.
**metamorphopsia** (*met-am-or-fop'-se-ah*) [μεταμορφοῦν, to change shape; ὄψις, sight]. A defect of vision in which, owing to disease of the retina or imperfection of the media, objects appear distorted.
**metamorphosing** (*met-am-or-fo'-zing*) [see *metamorphosis*]. Altering; changing. **m. breath-sound**. See *respiration, metamorphosing*.
**metamorphosis** (*met-am-or'-fo-sis*) [μεταμόρφωσις, a transformation]. A structural change or transformation. In pathology, a degeneration. **m. fatty**, fatty degeneration. **m., regressive, m., retrograde**, a disintegrating change; a degeneration. **m., viscous**, the agglutination of blood-platelets in the process of thrombosis.
**metamorphous** (*met-am-or'-fus*) [*meta-*; μορφόειν, to change]. Amorphous, but with a tendency to crystallize.
**metanephric** (*met-an-ef'-rik*) [*meta-*; νεφρός, kidney]. Pertaining to the metanephros.
**metanephros, metanephron** (*met-ah-nef'-ros, met-ah-nef'-ron*) [*meta-*; νεφρός, kidney]. The posterior of the three segmental bodies of the fetus, which is transformed into the permanent kidney and ureter.
**metanucleus** (*met-ah-nū'-kle-us*) [*meta-*; *nucleus*]. The egg-nucleus after its extrusion from the germinal vesicle.
**metapeptone** (*met-ah-pep'-tōn*) [*meta-*; πέψις, digestion]. A substance obtained by Meissner from the fluid resulting from the acid peptic digestion of any proteid after the parapeptone has been removed; it is said to be intermediate between parapeptone and dyspeptone.
**metaphases** (*met-af-a'-zēs*) [*meta-*; φάσις, a phase]. In biology, the final phase or set of phenomena of karyokinesis, from the time of division of the nuclear fibrils to the separation of the daughter-nuclei.
**metaphlogosis** (*met-af-lo-go'-sis*) [*meta-*; φλόγωσις, a burning]. Severe inflammation with much engorgement, but of short duration.
**metaphosphoric acid** (*met-ah-fos-for'-ik*). See *acid, phosphoric*.
**metaplasia** (*met-ah-pla'-ze-ah*) [*meta-*; πλάσσειν, to form]. A transformation of a tissue into another without the intervention of an embryonal tissue, as the conversion of cartilage into bone.
**metaplasis** (*met-ap'-las-is*) [*metaplasia*]. Fulfilled growth and development seen in the stage between anaplasis and cataplasis.
**metaplasm** (*met'-ah-plasm*). That portion of the protoplasm of a cell containing the products of secretion or excretion.
**metaplastic** (*met-ah-plas'-tik*) [see *metaplasia*].

# METAPLEX 553 METHEMOGLOBINEMIA

**Pertaining to** metaplasia. **m. bone,** a bone formed from periosteum.
**metaplex, metaplexus** (*met'-ah-pleks, met-ah-pleks'-us*) [*meta-; plexus*], a twining. The choroid plexus of the fourth ventricle.
**metapneumonic** (*met-ah-nū-mon'-ik*) [*meta-; pneumonia*]. Secondary to, or consequent upon, pneumonia.
**metapophysis** (*met-ah-pof'-is-is*) [*meta-; ἀπόφυσις*, a process]. A mammillary process, such as is seen upon the lumbar vertebræ.
**metapore** (*met'-ap-ōr*) [*meta-; πόρος,* passage]. Magendie's foramen (*q. v.*).
**metaprotein** (*met-ah-pro'-te-in*). A derivative by hydrolysis of a native protein.
**metapsyche** (*met-ap-si'-ke*) [*meta-; ψυχή*, soul]. The hind-brain or metencephalon.
**metaptosis** (*met-ap-to'-sis*) [*meta-; πτῶσις*, a falling]. Metastasis; sudden metabolic change.
**metapyretic** (*met-ah-pi-ret'-ik*) [*meta-; πυρετός*, fever]. 1. Occurring during fever. 2. Occurring after the decline of fever.
**metargon** (*met-ar'-gon*) [*meta-; argon*]. A gaseous element believed to exist in atmospheric air.
**metasol** (*met'-ah-sōl*). Soluble metacresol-anitol, containing 40 % of metacresol. A surgical disinfectant.
**metastasis** (*met-as'-tas-is*) [*meta-; στατός,* placed]. The transfer of a diseased process from a primary focus to a distant one by the conveyance of the causal agents through the blood-vessels or lymphchannels.
**metastasize** (*met-as'-tas-īz*). To transfer disease into a distant part by metastasis.
**metastate** (*met-as-tāt'*) [*meta-; στατός*, placed]. Any substance produced by a metabolic process; an anastate or a catastate. Same as *mesostate.*
**metastatic** (*met-ah-stat'-ik*) [see *metastasis*]. Characterized by or pertaining to metastasis. **m. abscess,** the secondary abscess in pyemia. **m. calcification,** calcareous infiltration due to an excess of lime-salts in the blood in diseases associated with rapid disintegration of bone.
**metasternum** (*met-ah-ster'-num*). The xiphoid cartilage of the sternum.
**metasyncrisis** (*met-as-in'-kris-is*). 1. An induced crisis. 2. The restoration of diseased tissues.
**metasyphilis** (*met-ah-sif'-il-is*). That form of inherited syphilis presenting only the syphilitic diathesis, *i. e.*, the degenerations and general diffuse changes in which localized lesions are absent.
**metatarsal** (*met-ah-tar'-sal*) [*metatarsus*]. Pertaining to the metatarsus.
**metatarsalgia** (*met-ah-tar-sal'-je-ah*) [*metatarsus;* ἄλγος, pain]. Morton's disease; a painful affection of the plantar digital nerves caused by pressure or pinching of them by portions of the metatarsophalangeal articulations.
**metatarsen** (*met-at-ar'-sen*). Belonging to the metatarsus in itself.
**metatarsometatarsal** (*met-ah-tar-so-met-ah-tar'-sal*). Relating to the metatarsal bones in their position to each other.
**metatarsophalangeal** (*met-ah-tar-so-fa-lan'-je-al*) [*metatarsus; phalanges*]. Pertaining to the metatarsus and the phalanges.
**metatarsus** (*met-ah-tar'-sus*) [*meta-; ταρσός*, tarsus]. The bones of the foot, five in number, situated between the tarsus and the phalanges.
**metatela** (*met-at-e'-lah*) [*meta-; tela*, tissue]. The tela of the metencephal; the velum medullare posterius, or inferior choroid tela. It constitutes the roof of the metacele.
**metathalamus** (*met-ah-thal'-am-us*) [*meta-; thalamus*]. A term including the pregeniculum and postgeniculum.
**metathesis** (*met-ath'-es-is*) [*meta-; τιθέναι,* to place]. 1. The act of changing the seat of a disease process from one part to another. 2. In chemistry, double decomposition.
**metathetic** (*met-ath-et'-ik*) [*meta-; τιθέναι*, to place]. Of the nature of a metathesis.
**metatocia** (*met-at-o'-ke-ah*) [*meta-; τόκος*, birth]. Birth by any other than the normal process, as by cesarean section.
**metatroph** (*met'-ah-trōf*). See *saprophyte.*
**metatrophia** (*met-ah-tro'-fe-ah*) [*meta-; τροφή,* nourishment]. Any morbid condition or process of nutrition.
**metatrophic** (*met-ah-tro'-fik*). Applied to sapro-

phytic organisms which cannot exist in the presence of living tissues. **m. method,** a therapeutic method of modifying the nutrition by changes in the food— with a view of administering some drug; *e. g.,* suppression of sodium chloride in food of epileptics in order to reinforce the action of bromides.
**Metazoa** (*met-ah-so'-ah*) [*meta-; ζῷον,* animal]. Animals the development of which is characterized by segmentation of the ovum. They comprise all animals except the *Protozoa.*
**Metchnikoff's larva** (*metsh'-ne-kof*) [Elie *Metchnikoff,* Russian biologist, 1845– ]. The parenchymula: the embryonic stage immediately succeeding that of the closed blastula. **M.'s phagocytic theory,** that microorganisms and other solid elements are destroyed or taken up by living cells, as by colorless blood-corpuscles.
**metecious,** metoecious (*met-e'-she-us*) [*meta-; οἶκος,* a house]. See *heterecious.*
**meteocism, metœcism** (*met'-es-izm*). See *heterecism.*
**metempiric** (*met-em-pe'-rik*) [*meta-; ἐμπειρία,* experience]. Opposed to empiric; not based on experience.
**metencephal** (*met-en'-sef-al*). Same as *metencephalon.*
**metencephalic** (*met-en-sef-al'-ik*). Pertaining to the metencephalon.
**metencephalon** (*met-en-sef'-al-on*) [*meta-; ἐγκέφαλος,* brain]. 1. The after-brain; the postoblongata, or most caudal portion of the brain. 2. Of Huxley, the cerebellum and the pons.
**meteorali** (*me-te-or'-ik*) [*μετέωρίζειν,* to elevate]. Pertaining to meteorism; also, pertaining to the atmosphere.
**meteorism** (*me'-te-or-izm*) [*μετεωρίζειν,* to elevate]. Distention of the abdomen with gas; tympanites.
**meteorograph** (*me-te-or'-o-graf*) [*μετέωρον,* a meteor; γράφειν,* to write]. An apparatus for securing a continuous record of the pressure, temperature, humidity, and velocity of the wind.
**meteorology** (*me-te-or-ol'-o-je*) [*meteor; λόγος,* a treatise]. The science of the phenomena of the atmosphere and the laws of its motions.
**metepencephalon** (*met-ep-en-sef'-al-on*) [*meta-; ἐπί,* upon; *ἐγκέφαλος,* brain]. The metencephalon and epencephalon considered together. It includes the cerebrum, the oblongata, and the pons.
**metepicele** (*met-ep'-is-ēl*) [*meta-; ἐπί,* upon; *κοιλία,* hollow]. The fourth ventricle of the brain.
**meter** (*me'-ter*). The unit of linear measure of the metric system, 39.37 inches. See *metric system.*
**m.-angle,** the angle of the visual axes, the object being one meter distant.
**-meter** (*me-ter*) [*μέτρον,* a measure]. A termination denoting an instrument for measuring.
**metergasis, metergasia** (*met-er'-gas-is, met-er-ga'-se-ah*) [*meta-; ἐργάσια,* work]. Change of function.
**metestrous** (*met-es'-trus*). Pertaining to the period in which the activity of the generative organs is diminishing in female animals.
**methacetin** (*meth-as'-et-in*) [*meta-; acetum,* vinegar], $C_9H_{11}NO_2$. An analogue of phenacetin, having the same antipyretic qualities as the latter, and more soluble. Dose from 2 to 4 grains exhibited with great caution.
**methal** (*meth'-al*) [*meta-; ὕλη,* matter], $C_{14}H_{30}O$. An alcohol, not yet isolated, occurring in spermaceti.
**methanal** (*meth'-an-al*). Same as *formaldehyde.*
**methane** (*meth'-ān*) [see *methal*], $CH_4$. Marsh-gas. The first member of the homologous series of paraffins, $C_nH_{2n+2}$. It occurs wherever decomposition of organic matter is going on, especially in marshes, and is also found at times in the stomach and intestine. It is a colorless, odorless, inflammable gas.
**methanol** (*meth'-an-ol*). Methyl alcohol.
**metheglin** (*meth-eg'-lin*). Mead; a drink made from honey by the addition of yeast and boiling water.
**methemerine** (*meth-em'-er-ēn*) [*meta-; ἡμέρα,* a day]. Quotidian.
**methemoglobin, methæmoglobin** (*met-hem-o-glo'-bin*) [*meta-; hemoglobin*]. A body similar in composition to hemoglobin, but having its oxygen more firmly united with it. It is prepared from hemoglobin by the action of potassium ferricyanide, potassium chlorate, sodium nitrite, etc. In poisoning with potassium chlorate, the nitrites, acetanilid, and other bodies, the blood contains methemoglobin.
**methemoglobinemia, methæmoglobinæmia** (*met-hem-o-glo-bin-e'-me-ah*) [*methemoglobin; αἷμα,* blood]. The presence of methemoglobin in the blood.

**methemoglobinuria, methæmoglobinuria** (*met-hem-o-glo-bin-u'-re-ah*) [*methemoglobin;* οὖρον, urine]. The presence of methemoglobin in the urine.
**methenyl** (*meth'-en-il*), CH. A hypothetical trivalent radical. **m.-orthoanisidin**, a compound of orthoanisidin and orthoformic acid ester; it is a local anesthetic. **m. tribromide**, bromoform. **m. trichloride**, chloroform. **m. triiodide**, iodoform.
**methethyl** (*meth-eth'-il*). A local anesthetic said to consist chiefly of ethyl chloride with a small quantity of methyl chloride and chloroform.
**methetic** (*meth-et'-ik*) [μέθεξις, participation]. In psychology applied to communications between the different strata of a man's intelligence.
**methogastrosis** (*meth-o-gas-tro'-sis*) [μέθυ, wine; γαστήρ, stomach; νόσος, disease]. Digestive disturbances consequent upon alcoholic excess.
**methol** (*meth'-ol*) [*meta-;* ὕλη, matter]. One of the names of methylic alcohol.
**methomania** (*meth-o-ma'-ne-ah*) [μέθυ, strong drink; μανία, madness]. 1. Same as *mania a potu*. 2. (More often) the irresistible desire for strong drink; dipsomania.
**methonal** (*meth'-on-al*), (CH₃)₂C(SO₂CH₃)₂. A hypnotic differing from sulphonal in containing methyl mercaptan instead of ethyl mercaptan. Dose 15-30 gr. (1-2 Gm.). Syn., *dimethyl sulphonedimethylmethane.*
**methoxycaffeine** (*meth-oks-e-kaf'-e-in*), C₉H₁₂N₄O₃. A white powder melting at 117° C. It is used hypodermatically as a local anesthetic and in neuralgia. Dose 4 gr. (0.26 Gm.).
**methoxyl** (*meth-oks'-il*). The characterizing group of the primary alcohols, CH₃OH.
**methozine** (*meth'-o-zēn*). Same as *antipyrine*.
**methyl** (*meth'-il*) [μέθυ, mead; ὕλη, matter], CH₃. A univalent hydrocarbon radical, the first of the univalent hydrocarbons of the marsh-gas series; the radical of methyl-alcohol. **m. acetate**, C₃H₆O₂, a fragrant liquid obtained from crude wood-vinegar. **m.-alcohol**, CH₃OH, a colorless, narcotic liquid, obtained in the destructive distillation of wood. Syn., *carbinol, wood-alcohol; wood-spirit.* **m. aldehyde**, formaldehyde. **m.-blue**, an antiseptic used as a local application in diphtheria (not to be confounded with methylene blue). **m.*chloride**, CH₃Cl, a liquid local anesthetic. **m. chloroform**, CH₃CCl₃, a volatile liquid, obtained by chlorinating ethyl chloride. It is anesthetic. **m. ether**, C₂H₆O; dimethyl oxide, an inflammable gas; also a salt of methyl. **m.-glyoxalidin**. See *lysidine*. **m.-green**, an anilin dye, used in staining tissues, also as an antiperiodic. **m.-guanidin**, C₂H₇N₃; a colorless, crystalline, strongly alkaline base, formed by the oxidation of creatin and creatinin, and also found in decomposing horseflesh and in cultures of the comma bacillus and the bacillus of anthrax. It is highly poisonous. **m.-hydantoic acid**, a crystalline substance occurring in the urine after the ingestion of sarcosin; it is also obtained by heating sarcosin and urea together for several days in baryta-water. **m. hydrate**, methyl-alcohol. **m. hydride**, CH₄; methane or marsh-gas. **m. iodide**, C₂H₃I, a reaction-product of methyl-alcohol with iodine and phosphorus; used as a vesicant instead of cantharides. **m.-phenol**, cresol. **m.-propyloxybenzol**, thymol. **m. pyridin**. See *picolin*. **m. pyridin sulphocyanate**, an energetic, noncaustic, nontoxic antiseptic employed in 1 % solution. **m. pyrocatechin**, guaiacol. **m.-quinolin**, a substance occurring with quinolin and quinalidin in coal-tar. **m. salicylate** (*methylis salicylas*, U. S. P.), C₈H₈O₃; synthetic oil of wintergreen; an oily liquid of a peculiar odor, identical with the essential constituent of the oil of wintergreen; it is used in rheumatism like the natural oil of gaultheria. **m.-salol**, CH₃(OH)(CH₃)CO₂. C₆H₅, a crystalline substance, insoluble in water, soluble in ether, chloroform, or hot alcohol; used in rheumatism. **m.-theobromin**, caffeine. **m.-uramin**. See *m.-guanidin*. **m.-violet**, an anilin dye used for staining bacteria. Under the name of *pyoktanin* it is used as an antiseptic. **m.-xanthin**. See *heteroxanthin.*
**methylacetanilid**. (*meth-il-as-et-an'-il-id*); See *exalgin*.
**methylal** (*meth'-il-al*) [*methyl; alcohol*], C₃H₈O₂. A substance prepared by distilling methyl-alcohol with sulphuric acid. It is hypnotic and antispasmodic. Dose 1 dr. (4 Cc.).
**methylamine** (*meth-il'-am-in*) [*methyl; amine*],

N(CH₃)H₂. A colorless basic gas occurring in herring-brine and in cultures of the comma bacillus.
**methylate** (*meth'-il-āt*) [*methyl*]. A compound formed from methyl-alcohol by the substitution of the hydrogen of the hydroxyl by a base.
**methylated** (*meth'-il-at-ed*). Containing methyl-alcohol. **m. spirit**. See *methyl-alcohol.*
**methylation** (*meth-il-a'-shun*). The process of mixing a substance with methyl-alcohol.
**methylbenzol** (*meth-il-ben'-zol*). Toluene.
**methylene** (*meth'-il-ēn*) [*methyl*], CH₂. A bivalent hydrocarbon radical. **m. bichloride**, CH₂Cl₂; a general anesthetic, used instead of chloroform. **m.-blue** (*methylthioninæ hydrochloridum*, U. S. P.), C₁₆H₁₈N₃SCl; a blue anilin dye used as a stain in microscopy. It has also been employed as a local application in diphtheria, tonsillitis, scarlatinal sore throat, and other inflammatory conditions, and internally in malaria and neuralgia. Dose 2-4 gr. (0.13-0.25 Gm.). **m. chloride** (of Richardson), a colorless liquid consisting of 1 volume of methyl-alcohol and 4 volumes of chloroform; it is used in inhalation-anesthesia. Syn., *methyl bichloride.* **m. creosote**, a nontoxic, yellowish powder devoid of taste or odor. It is used in the treatment of tuberculosis. Dose 8-30 gr. (0.5-2.0 Gm.). **m. oxide**, formaldehyde.
**methylenophil, methylenophilous** (*meth-il-en'-of-il, meth-il-en-of'-il-us*) [*methylene;* φιλεῖν, to love]. Having an affinity for methylene-blue.
**methylguanidine** (*meth-il-gwan'-id-ēn*). A poisonous ptomaine derived from creatinine.
**methylic** (*meth-il'-ik*). Containing methyl.
**methylil** (*meth'-il-il*). Trade name of a local anesthetic; said to be a mixture of ethyl chloride, chloroform, and methyl chloride.
**methylphenacetin** (*meth-il-fen-as'-et-in*). A hypnotic compound obtained by treating phenacetin-sodium with methyl iodide.
**methylpurin** (*meth-il-pū'-rin*). Any compound in which one or more methyl radicals have been substituted in the purin nucleus.
**methylsalol** (*meth-il-sa'-lol*). A crystalline substance used in rheumatism.
**methylthionine hydrochloride** (*meth-il-thi'-o-nin*). See *methylene-blue.*
**methysis** (*meth'-is-is*) [μέθυσις, intoxication]. Intoxication.
**methystic** (*meth-is'-tic*). 1. Intoxicant. 2. An intoxicating agent.
**methysticin** (*meth-is'-tis-in*) [μεθυστικός, intoxicating]. A glucoside, C₁₄H₁₄O₅, from *Piper methysticum*, kava-root.
**metoarion** (*met-o-a'-re-on*) [*meta-;* ὠάριον, the ovule]. The corpus luteum.
**metodontiasis** (*met-o-don-ti'-as-is*) [*meta-;* ὀδοντιᾶν, to cut teeth]. The second dentition; also, abnormality of teething.
**metol** (*met'-ol*). Trade name for methylparaamino phenol sulphate; it is used as a developer by photographers, and is capable of producing a dermatitis accompanied by ulceration and obstinate fissures of the skin.
**metopagus** (*met-op'-ag-us*) [μέτωπον, the forehead; πάγος, joined]. A twin monstrosity with united foreheads.
**metopantralgia** (*met-o-pan-tral'-je-ah*) [μέτωπον, forehead; ἄντρον, cave; ἄλγος, pain]. Pain or neuralgia of the frontal sinus.
**metopantritis** (*met-o-pan-tri'-tis*) [μέτωπον, forehead; ἄντρον, cave; *itis*, inflammation]. Inflammation of the metopantron.
**metopantron, metopantrum** (*met-o-pan'-tron, met-o-pan'-trum*) [μέτωπον, forehead; ἄντρον, cave]. The frontal sinus.
**metopic** (*met-op'-ik*) [μέτωπον, forehead]. 1. Relating to the forehead; frontal. 2. A name applied to a cranium having a medio-frontal suture. **m. points**. See *craniometric points.*
**metopion** (*met-o'-pe-on*) [μέτωπον, forehead]. See *craniometric points.*
**metopism** (*met'-o-pizm*) [μέτωπον, forehead]. Persistence of the frontal suture in adult life. See *metopon.*
**metopium** (*met-o'-pe-um*). Synonym of *metopon.*
**metopodynia** (*met-o-po-din'-e-ah*) [μέτωπον, forehead; ὀδύνη, pain]. Frontal headache.
**metopon** (*met'-o-pon*) [μέτωπον, forehead; from μετά, between; ὤψ, eye]. Forehead; also an old name for galbanum.

# METOPOPLASTY 555 METROFIBROMA

**metopoplasty** (*met-op'-o-plas-te*) [μέτωπον, the forehead; πλάσσειν, to form]. Plastic surgery of the forehead.

**metoposcopy** (*met-op-os'-ko-pe*) [μέτωπον, forehead; σκοπεῖν, to examine]. A variety of phrenology in which the character on the future of an individual is supposed to be determined by an inspection of the forehead.

**metoxenous** (*met-oks'-en-us*). See *heterecious*.

**metra** (*me'-trah*) [μήτρα, uterus]. The uterus.

**metralgia** (*met-ral'-je-ah*) [*metra*; ἄλγος, pain]. Metrodynia (*q. v.*).

**metramine** (*met'-ram-ēn*). Trade name for a brand of hexamethylenamine.

**metranastrophe** (*met-ran-as'-tro-fē*) [*metra*; ἀναστροφή, a turning upside down]. Inversion of the uterus.

**metranemia, metranæmia** (*met-ran-e'-me-ah*) [*metra*; *anemia*]. Uterine anemia.

**metraneurysm** (*met-ran'-u-rizm*) [*metra*; ἀνεύρυσμα, dilatation]. Dilatation of the uterus or vulva.

**metranoikter** (*met-rah-no-ik'-ter*). A uterine dilator with two or four branches; used when a wide, prolonged uterine dilatation is indicated.

**metratome** (*met'-rat-ōm*) [*metra*; τομή, a cutting]. An instrument for incising the uterus.

**metratonia** (*met-rat-o'-ne-ah*) [*metra*; ἀτονία, atony]. Atony of the uterus.

**metratresia** (*met-rat-re'-ze-ah*) [*metra*; ἀτρησία, atresia]. Atresia or imperforation of the womb.

**metratrophia** (*met-rat-ro'-fe-ah*) [*metra*; ἀτροφία, atrophy] of the uterus.

**metrauxe** (*met-rawks'-e*) [*metra*; αὔξη, increase]. Hypertrophy or enlargement of the uterus.

**metre** (*me'-ter*). See *meter*.

**metrechoscope** (*met-rek'-o-skōp*) [μέτρον, measure; ἠχή, sound; σκοπεῖν, to inspect]. An instrument for applying metrechoscopy.

**metrechoscopy** (*met-rek-os'-ko-pe*) [μέτρον, measure; ἠχή, sound; σκοπεῖν, to inspect]. Combined auscultation and mensuration.

**metrectasia** (*met-rek-ta'-se-ah*) [*metra*; ἔκτασις, a stretching]. Dilatation of the uterus.

**metrectatic** (*met-rek-tat'-ik*). Affected with, or pertaining to, metrectasia.

**metrectomy** (*met-rek'-to-me*) [*metra*; ἐκτομή, excision]. Excision or surgical removal of the uterus.

**metrectopia, metrectopy** (*met-rek-to'-pe-ah, met-rek'-to-pe*) [*metra*; ἔκτοπος, displaced]. Displacement of the womb.

**metrelcosis** (*met-rel-ko'-sis*) [*metra*; ἕλκος, ulcer]. Uterine ulceration.

**metremia** (*met-re'-me-ah*) [*metra*; αἷμα, blood]. Congestion of the uterus.

**metremorrhagia** (*met-rem-or-a'-je-ah*). Synonym of *metrorrhagia*.

**metremorrhoid** (*met-rem'-or-oid*) [*metra*; *hemorrhoid*]. A hemorrhoid of the uterus.

**metremphraxis** (*met-rem-fraks'-is*) [*metra*; ἐμφράσσειν, to obstruct]. Congestion, or infarction, of the uterine tissues.

**metremphysema** (*met-rem-fis-e'-mah*). Synonym of *physometra*.

**metreurynter** (*met-roo-rin'-ter*) [*metra*; εὐρύνειν to widen]. A form of colpeurynter.

**metreurysis** (*met-roo'-ris-is*). See *colpeurysis*.

**metreurysma** (*met-roo-riz'-mah*) [*metra*; εὐρύς, wide]. Morbid dilatation, or width of the uterus.

**metria** (*me'-tre-ah*) [*metra*]. Any uterine affection. The term is used also as a synonym of *puerperal fever*.

**metric** (*met'-rik*) [μέτρον, a measure]. Pertaining to the system of weights and measures, of which the meter is the basis. **m. system,** a decimal system of weights and measures employed in France, Germany, and other countries, and used generally in the sciences. The standard is the *meter*, the ten-millionth part of the distance from the equator to the north pole. The actual standard unit is the distance between two lines on a platinum-iridium rod preserved in the archives of the International Metric Commission at Paris, and is equivalent to 39.37079 inches; in the United States the length of the meter is assumed as 39.37 inches. The standard of capacity is the *liter*, a cubic volume 1/10 meter in each dimension. The standard of weight is the *gram*, the weight of 1/1000 liter (one cubic centimeter) of distilled water at its maximum density. As the unit of measurement the thousandth part of a millimeter has been adopted. It is called *micromillimeter*, or

*micron*; its symbol is μ. The multiples in the metric system are expressed by the prefixes *deca-*, *hecto-*, and *kilo-*; the subdivisions by the prefixes *deci-*, *centi-*, and *milli-*.

| 1000 | meters | = 1 kilometer. |
| 100 | meters | = 1 hectometer. |
| 10 | meters | = 1 decameter. |
| .1 | meter | = 1 decimeter. |
| .01 | meter | = 1 centimeter. |
| .001 | meter | = 1 millimeter. |
| 1000 | liters | = 1 kiloliter. |
| 100 | liters | = 1 hectoliter. |
| 10 | liters | = 1 decaliter. |
| .1 | liter | = 1 deciliter. |
| .01 | liter | = 1 centiliter. |
| .001 | liter | = 1 milliliter. |
| 1000 | grams | = 1 kilogram. |
| 100 | grams | = 1 hectogram. |
| 10 | grams | = 1 decagram. |
| .1 | gram | = 1 decigram. |
| .01 | gram | = 1 centigram. |
| .001 | gram | = 1 milligram. |

In common practice, however, the following divisions only are used, the others being expressed in figures:

| 10 millimeters | = 1 centimeter. |
| 100 centimeters | = 1 meter. |
| 1000 meters | = 1 kilometer. |
| 1000 cubic centimeters | = 1 liter. |
| 1000 milligrams | = 1 gram. |
| 1000 grams | = 1 kilogram. |

The following are the equivalent values:

| 1 meter | = 39.37 inches. |
| 1 liter | = 1 quart ½ gill, U. S. measure. |
| 1 gram | = 15.43 grains. |
| 1 minim | = 0.061 cubic centimeter. |

See also, *weights and measures*.

**metriocephalic** (*met-re-o-sef-al'-ik*) [μέτριος, moderate; κεφαλή, head]. Applied to a skull in which the arch of the vertex is moderate in height, neither akrocephalic (pointed) nor platycephalic (*q. v.*). Cf. *scaphocephalic, tapeinocephalic*.

**metritic** (*met-rit'-ik*) [*metra*; ἶτις, inflammation]. Pertaining to, or affected with, metritis.

**metritis** (*met-ri'-tis*) [*metra*; ἶτις, inflammation]. Inflammation of the uterus. **m. dissecans,** an inflammatory affection of the uterus accompanied by the sloughing away of portions of it.

**metro-** (*met-ro-*) [*metra*]. A prefix meaning relating to the uterus.

**metroblennorrhea** (*met-ro-blen-or-e'-ah*) [*metro-*; βλέννα, mucus; ῥοία, a flow]. Uterine blennorrhea.

**metrocampsis** (*met-ro-kamp'-sis*) [*metro-*; κάμψις, bending]. Obliquity or curvature of the uterus.

**metrocase** (*met-rok'-as-e*) [*metro-*; κακός, evil]. Same as *metrelcosis*.

**metrocarcinoma** (*met-ro-kar-sin-o'-mah*) [*metro-*; *carcinoma*]. Carcinoma of the uterus.

**metrocele** (*met'-ro-sēl*) [*metro-*; κήλη, hernia]. Hernia of the uterus.

**metroclyst** (*met'-ro-klist*) [*metro-*; κλύζειν, to wash out]. An instrument for giving uterine douches.

**metrocolpocele** (*met-ro-kol'-po-sēl*) [*metro-*; κόλπος, vagina; κήλη, hernia]. Protrusion of the uterus into the vagina, the wall of the latter being pushed in advance.

**metrocystosis** (*met-ro-sis-to'-sis*) [*metro-*; κύστις, a cyst]. The formation of uterine cysts or the condition giving rise to them.

**metrocyte** (*met'-ro-sīt*) [*metro-*; κύτος, cell]. A large uninuclear spheroidal cell the protoplasm of which contains hemoglobin, and which is supposed to be the source of the red corpuscles of the blood.

**metrodynamometer** (*met-ro-di-nam-om'-et-er*) [*metro-*; δύναμις, power; μέτρον, measure]. An instrument for measuring uterine contractions.

**metrodynia** (*met-ro-din'-e-ah*) [*metro-*; ὀδύνη, pain]. Pain in the uterus.

**metrodystocia** (*met-ro-dis-to'-ke-ah*) [*metro-*; δυστοκία, painful delivery]. Dystocia whose cause resides in the uterus.

**metroectasia** (*met-ro-ek-ta'-ze-ah*). Synonym of *metrectasia*.

**metroendometritis** (*met-ro-en-do-met-ri'-tis*). Combined inflammation of the uterus and endometrium.

**metrofibroma** (*met-ro-fi-bro'-mah*) [*metro-*; *fibroma*, fibrous tumor]. Uterine fibroid tumor.

**metroleukorrhea** (*met-ro-lū-kor-e'-ah*) [*metro-*; λευκός, white; ῥοία, flow]. Uterine leukorrhea.
**metrology** (*met-rol'-o-je*) [μέτρον, measure; λόγος, science]. The science of measures and of measurements.
**metroloxia** (*met-ro-loks'-e-ah*) [*metro-*; λοξός, oblique]. Obliquity of the uterus; hysteroloxia.
**metrolymphangitis** (*met-ro-lim-fan-ji'-tis*) [*metro-*; λύμφα, lymph; ἀγγεῖον, vessel; ιτις, inflammation]. Inflammation of the lymphatic vessels of the uterus. Uterine lymphangitis.
**metromalacosis** (*met-ro-mal-ak-o'-sis*) [*metro-*; μαλακός, soft]. Softening of the tissues of the uterus.
**metromania** (*met-ro-ma'-ne-ah*). 1. See *hysteromania* and *nymphomania*. 2. [μέτρον, measure; μανία, mania]. A mania for writing poetry.
**metromaniac** (*met-ro-ma'-ne-ak*) [μέτρον, measure; μανία, mania]. One insanely fond of writing verses.
**metronania** (*met-ro-na'-ne-ah*) [*metro-*; νᾶνος, dwarf]. Abnormal smallness of the uterus.
**metroncus** (*met-rong'-kus*). See *hysteroncus*.
**metroneuria** (*met-ro-nū'-re-ah*) [*metro-*; νεῦρον, nerve]. A nervous affection of the uterus.
**metroneurosis** (*met-ro-nū-ro'-sis*) [*metro-*; *neurosis*]. Any neurosis caused by uterine disease.
**metronome** (*met'-ro-nōm*) [μέτρον, measure; νόμος, law]. An instrument for measuring time in music; also one for testing the hearing.
**metroparalysis** (*met-ro-par-al'-is-is*) [*metro-*; *paralysis*]. Uterine paralysis.
**metropathic** (*met-ro-path'-ik*) [*metro-*; πάθος, disease]. Pertaining to affections of the uterus.
**metropathy** (*met-rop'-ath-e*) [*metro-*; πάθος, suffering]. Any uterine disease.
**metroperitonitis** (*met-ro-per-it-on-i'-tis*) [*metro-*; *peritonitis*]. 1. Combined inflammation of the uterus and the peritoneum. 2. Peritonitis secondary to inflammation of the uterus. 3. Inflammation of the peritoneum about the uterus.
**metropexia, metropexy** (*met-ro-peks'-e-ah, met'-ro-peks-e*). See *hysteropexy*.
**metrophlebitis** (*met-ro-fleb-i'-tis*) [*metro-*; *phlebitis*]. Inflammation of the veins of the uterus.
**metrophlogosis** (*met-ro-flo-go'-sis*). Synonym of *metritis*.
**metrophyma** (*met-ro-fi'-mah*) [*metro-*; φῦμα, a growth]. A tumor of the uterus.
**metropolypus** (*met-ro-pol'-ip-us*) [*metro-*; *polypus*]. Uterine polyp.
**metroptosis** (*met-rop-to'-sis*) [*metro-*; πτῶσις, fall]. Prolapse of the uterus.
**metrorrhagia** (*met-ror-a'-je-ah*) [*metro-*; ῥηγνύναι, to burst forth]. Uterine hemorrhage independent of the menstrual period.
**metrorrhea, metrorrhoea** (*met-ror-e'-ah*) [*metro-*; ῥοία, a flow]. Any morbid discharge from the uterus.
**metrorrhectic** (*met-ror-ek'-tik*). Pertaining to metrorrhexis.
**metrorrhexis** (*met-ror-eks'-is*) [*metro-*; ῥῆξις, rupture]. Rupture of the uterus.
**metrorthosis** (*met-ror-tho'-sis*) [*metro-*; ὀρθοῦν, to set straight]. The correction of a displaced uterus.
**metrosalpingitis** (*met-ro-sal-pin-ji'-tis*) [*metro-*; *salpingitis*]. Inflammation of the uterus and oviducts.
**metrosalpingorrhagia** (*met-ro-sal-ping-or-a'-je-ah*) [*metro-*; σάλπιγξ, tube; ῥηγνύναι, to burst forth]. Hemorrhage from the oviducts.
**metrosalpingorrhexis** (*met-ro-sal-ping-or-eks'-is*) [*metro-*; σάλπιγξ, tube; ῥῆξις, rupture]. Rupture of an oviduct.
**metrosalpinx** (*met-ro-sal'-pinks*) [*metro-*; σάλπιγξ, tube]. An oviduct or Fallopian tube.
**metroscirrhus** (*met-ro-skir'-us*) [*metro-*; σκίρρος, hard]. A scirrhous tumor of the uterus.
**metroscope** (*met'-ro-skōp*) [*metro-*; σκοπεῖν, to observe]. An instrument for examining the uterus.
**metrostaxis** (*met-ro-staks'-is*) [*metro-*; στάξις, a dropping]. Slight but persistent uterine hemorrhage.
**metrostenosis** (*met-ro-sten-o'-sis*) [*metro-*; στένωσις, contraction]. Contraction of the cavity of the uterus.
**metrosteresis** (*met-ro-ster-e'-sis*) [*metro-*; στέρησις, deprival]. Removal or absence of the uterus.
**metrotome** (*met'-ro-tōm*) [*metro-*; τομή, a cutting]. An instrument for incising the uterine neck.
**metrotomy** (*met-rot'-o-me*). See *hysterotomy*.
**metrotoxin** (*met-ro-tok'-sin*). A hypothetical hormone from the pregnant uterus which is assumed to have an inhibitory action on the ovaries.
**metrourethrotome** (*met-ro-ū-re'-thro-tōm*) [μέτρον, measure; οὐρήθρα, urethra; τομή, a cutting]. A form of urethrotome that will cut a stricture to the desired caliber; a graduated urethrotome.
**metrypercinesis** (*met-ri-per-sin-e'-sis*) [*metra*; ὑπέρ, over; κίνησις, movement]. Excessive uterine contraction.
**metryperemia, metryperæmia** (*met-ri-per-e'-me-ah*). Synonym of *metremia*.
**metryperesthesia, metryperæsthesia** (*met-ri-per-es-the'-ze-ah*) [*metra*; ὑπέρ, over; αἴσθησις, perception]. Hyperesthesia of the uterus.
**metrypertrophia** (*met-ri-per-tro'-fe-ah*). Synonym of *metrauxe*.
**Meunier's sign of measles** (*moo-ne-a'*). A daily loss of weight noticed four or five days after contagion. This may amount to 50 Gm. daily, commencing five or six days before the appearance of catarrhal or febrile symptoms.
**Mexican typhus.** See *tabardillo*.
**Meyer's disease** (*mi'-er*) [Georg Hermann *Meyer*, German anatomist, 1815–1892]. Hypertrophy of the pharyngeal tonsil; adenoid vegetations of the pharynx. **M.'s law**, mature and normal bone possesses a definite internal structure, which in every part represents the lines of greatest pressure on traction, and is so arranged as to afford the greatest resistance with the smallest amount of material. **M.'s rings**, the faint rings seen to surround a candle-flame or a similar source of light against a dark background; they appear more distinct, as Woehler has shown, when the eyes are exposed for a short time to the fumes of osmic acid. The phenomenon is due to the diffraction of light by cellular elements on the surface of the cornea. **M.'s sign**, numbness of the hands or feet associated with formication; it is observed in the eruptive stage of scarlatina.
**Meyer-Woehler's rings.** See *Meyer's rings*.
**Meynert's bundle** (*mi'-nert*) [Theodore *Meynert*, Austrian anatomist, 1833–1892]. A tract of nerve-fibers forming part of the capsula; it passes between the external geniculate body and posterior border of the putamen, and ends in the lower part of the occipital and temporosphenoid lobes. **M.'s commissure**, a tract of nerve-fibers crossing dorsally to the mesial half of the chiasm from the tuber cinereum to the opposite side; it is probably connected with Luys's body. **M.'s dorsal tegmental decussation**, one of the fountain decussations; it is situated between the red nuclei and dorsal to them. **M.'s fibers**, a tract of nerve-fibers connecting the anterior corpus quadrigeminum with the nuclei of the ocular muscles. **M.'s field**, the reticular formation of the pons. **M.'s layer**, the layer of pyramidal cells in the cerebral cortex. **M.'s radiations**, fibers in the radiary zone of the cortex cerebri. **M.'s solitary cells**, giant pyramidal cells arranged in a single row in the visual area of the cortex cerebri.
**Meynet's nodosities** (*ma-na'*) [Paul Claude Hyacinthe *Meynet*, French physician, 1831–1892]. Nodular accumulations within the capsules of joints, tendons, and tendon-sheaths, and sometimes seen in cases of rheumatism, especially in children.
**mezcal, mescal** (*mes-kahl'*) [Mex.]. An intoxicant spirit distilled from pulque, the fermented juice of various Mexican species of *Agave*.
**mezereon** (*mes-e'-re-on*). See *mezereum*.
**mezereum** (*mes-e'-re-um*) [Ar., *māsariyūn*, the camellia]. An old world shrub. The *mezereum* of the U. S. P. is the dried bark of *Daphne mezereum* and other species of *Daphne*, of the natural order *Thymelaeaceæ*. It contains a glucoside, *daphnin*, and an acrid resin. Locally applied, mezereum is an irritant and vesicant, and has been used to stimulate indolent ulcers. Internally, it has been employed in syphilis, scrofula, chronic rheumatism, and various skin diseases. Dose 10 gr. (0.65 Gm.). **m., fluidextract of** (*fluidextractum mezerei*, U. S. P.). Dose 2–5 min. (0.13–0.32 c.c.). It is used chiefly in ointments. **m. oleoresin**, ethereal extract of the bark of *Daphne mezereum* and other species. It is alterative, stimulant, and rubefacient. Dose ½–1 min. (0.03–0.06 c.c.).
**mezquit** (*mes-kēt'*) [Sp.]. The tree or shrub *Prosopis juliflora* of the S. W. United States and Mexico. Its gum resembles gum arabic.
**Mg.** Chemical symbol of *magnesium*. Also abbreviation for *milligram*.
**mho** (*mo'*) [anagram of *ohm*]. The unit of electrical conductivity; conductivity at the resistance of one ohm.

**miasm, miasma** (mi'-azm, mi-az'-mah) [μιαίνειν, to pollute: *pl.*, *miasmata*]. 1. A term loosely applied to the floating germs of any form of microbic life, especially those generating in marshy localities. 2. A noxious effluvium or emanation.
**miasmal** (mi-az'-mal) [*miasm*]. Containing, relating to, or depending upon, miasm.
**miasmatic** (mi-az-mat'-ik) [*miasm*]. Pertaining to or having the nature of miasm. m. diseases, diseases produced by miasmata.
**miasmatology** (mi-az-mat-ol'-o-je) [*miasm*; λόγος, science]. The science or study of miasmata.
**miasmifuge** (mi-az'-mif-ūj) [*miasm*; *fugere*, to put to flight]. Preventing or banishing miasmatic diseases.
**Mibelli's disease** (mib-el'-le) [Vittorio *Mibelli*, Italian physician]. Porakeratosis.
**mica** (mi'-kah) [L.]. 1. A crumb. 2. A mineral occurring in the form of thin, shining, transparent scales. m. panis, bread-crumb.
**micaceous** (mi-ka'-she-us) [*mica*, crumb]. Resembling mica; composed of crumbs; friable.
**Micajah's wafers**. A preparation said to consist of mercury bichloride, 1/78 gr.; zinc sulphate, 5 gr.; bismuth subnitrate, 15 gr.; acacia, 5 gr.; phenol, 3 gr.; water, a sufficient quantity.
**mication** (mi-ka'-shun) [*micare*, to glitter]. Quick motion; a winking; systolic contraction.
**micella** (mi-sel'-ah) [dim. of *mica*, a crumb, grain; *pl.*, *micellæ*]. One of the fundamental structural units of organized bodies; it is microscopically invisible.
**MicLallow's test for proteins**. Add ferrous sulphate to the solution, and underlay with concentrated sulphuric acid; then add carefully very little nitric acid. Besides a brown ring, a red coloration will be produced.
**micracoustic** (mi-krah-koo'-stik) [μικρός, small; ἀκουστικός, pertaining to hearing]. 1. Assisting in hearing very faint sounds. 2. An instrument possessing this property.
**micranatomy** (mi-kran-at'-o-me) [μικρός, small; *anatomy*]. Minute anatomy; histology.
**micrangiopathy** (mi-kran-je-op'-ath-e) [μικρός, small; ἀγγεῖον, a vessel; πάθος, disease]. Disease of the capillaries.
**micrangium** (mi-kran'-je-um) [μικρός, small; ἀγγεῖον, a vessel]. A capillary.
**micrencephalon** (mi-kren-sef'-al-on) [μικρός, small; ἐγκέφαλος, brain]. 1. A small brain, as in cretinism. 2. The cerebellum.
**micrencephalous** (mi-kren-sef'-al-us). Having a small brain.
**micrencephalus** (mi-kren-sef'-al-us) [μικρός, small; ἐγκέφαλος, encephalon]. Lissauer's term for a skull in which the angle formed between lines drawn from the hormion to the nasion and to the inion respectively is between 129° and 142.5°.
**micrencranus** (mi-kren-kra'-nus) [μικρός, small; ἐν, in; κρανίον, skull]. Lissauer's term for a skull with a cerebellar sector of from 8.5° to 15°.
**micro-** (mi'-kro-) [μικρός, small]. 1. A prefix signifying minute. 2. A prefix generally used to signify a unit one-thousandth, sometimes one-millionth, part of the unit to which it is prefixed.
**microaudiphone** (mi-kro-aw'-dif-ōn) [μικρός, small; *audire*, to hear; φωνή, voice]. An instrument used for rendering audible sounds that are very slight.
**microbacteria** (mi-kro-bak-te'-re-ah). Bacteria of very small size.
**microbe** (mi'-krōb) [*micro-*; βίος, life]. A living organism of very small size. The term is generally used synonymously with bacterium.
**microbemia** (mi-kro-be'-me-ah). See *microbiohemia*.
**microbian** (mi-kro'-be-an) [*microbe*]. 1. Pertaining to or of the nature of a microbe. 2. A microbe.
**microbicidal** (mi-kro-bis-i'-dal) [*microbe*; *cædere*, to kill]. Destructive to microbes.
**microbicide** (mi-kro'-bis-īd) [*microbe*; *cædere*, to kill]. 1. Destructive to microbes. 2. An agent that destroys microbes.
**microbicidin** (mi-kro-bis-'id-in). See *microcidin*.
**microbiohemia, microbiohæmia** (mi-kro-bi-o-hem'-e-ah) [*microbe*; αἷμα, blood]. A diseased condition resulting from the presence of microorganisms in the blood.
**microbiologist** (mi-kro-bi-ol'-o-jist) [*micro-*; βίος, life; λόγος, science]. An expert in the study of microbes.
**microbiology** (mi-kro-bi-ol'-o-je). The science of the nature, life, and actions of microorganisms.

**microbion, microbium** (mi-kro'-be-on, mi-kro'-be-um). Same as *microbe*.
**microbiophobia** (mi-kro-bi-o-fo'-be-ah) [*microbe*; φόβος, fear]. Morbid fear of microbes.
**microbioscope** (mi-kro-bi'-o-skōp) [*microbe*; σκοπεῖν, to view]. A microscope for the study of the changes that take place in living tissues or for the study of microorganisms.
**microbiosis** (mi-kro-bi-o'-sis). The morbid condition due to infection with pathogenic microorganisms.
**microblepharia, microblepharon** (mi-kro-blef-a'-re-ah, mi-kro-blef'-ar-on) [*micro-*; *blepharon*]. Smallness of the eyelids.
**microblepharism, microblephary** (mi-kro-blef'-ar-izm, mi-kro-blef'-ar-e). See *microblepharia*.
**microbrachia** (mi-kro-bra'-ke-ah) [*micro-*; βραχίων, arm]. Abnormal congenital smallness of the arms.
**microbrachius** (mi-kro-bra'-ke-us) [*micro-*; βραχίων, arm]. Smallness of the arms.
**microbrachycephalia** (mi-kro-bra-ke-sef-a'-le-ah) [*micro-*; βραχύς, short; κεφαλή, head]. Brachycephalia combined with microcephalia.
**microcardia** (mi'-kro-har'-de-ah) [*micro-*; καρδία, heart]. Congenital smallness of the heart.
**microcentrum** (mi-kro-sen'-trum). The dynamic center of the cell.
**microcephal** (mi-kro-sef'-al) [*micro-*; κεφαλή, head]. A person affected with microcephaly.
**microcephalia** (mi-kro-sef-a'-le-ah) [*micro-*; κεφαλή, head]. An abnormal smallness of the head.
**microcephalic** (mi-kro-sef-al'-ik) [see *microcephalia*]. Having a small head.
**microcephalism** (mi-kro-sef'-al-izm). Synonym of *microcephaly*.
**microcephalon** (mi-kro-sef'-al-on) [see *microcephalia*]. An abnormally small head.
**microcephalus** (mi-kro-sef'-al-us) [*micro-*; κεφαλή, head]. A person with a small head.
**microcephaly** (mi-kro-sef'-al-e) [*micro-*; κεφαλή, head]. Abnormal smallness of the head.
**microcheilia** (mi-kro-ki'-le-ah) [*micro-*; χεῖλος, lip]. Abnormal congenital smallness of the lips.
**microchemistry** (mi-kro-kem'-is-tre) [*micro-*; *chemistry*]. 1. The chemistry of the minute organisms and substances of nature. 2. The study of chemical reactions with the aid of the microscope.
**microcidin** (mi-kros'-id-in) [*micro-*; *cædere*, to kill]. Sodium naphtholate, an antiseptic powder.
**microclysm** (mi'-kro-klizm) [*micro-*; κλύσμα, clyster]. A small clyster; a clyster effective in small amounts.
**micrococcus** (mi-kro-kok'-us) [*micro-*; κόκκος, kernel]. A genus of bacteria the individuals of which have a spherical shape. When united in such a way as to resemble a bunch of grapes, they are called *staphylococci*; when united in couples, they are called *diplococci*; when string-like they are called *streptococci*. See table on page 558.
**microconidium** (mi-kro-kon-id'-e-um) [*micro-*; κόνις, dust; *pl.*, *microconidia*]. In biology, a relatively small-sized conidium.
**microcoria** (mi-kro-ko'-re-ah) [*micro-*; κόρη, pupil]. Same as *miosis*.
**microcornea** (mi-kro-kor'-ne-ah) [*micro-*; *cornea*]. Abnormal smallness of the cornea.
**microcosm** (mi'-kro-kozm) [*micro-*; κόσμος, world]. Man in contradistinction to the universe, or the macrocosm.
**microcosmic** (mi-kro-kos'-mik) [μικρόκοσμ]. Pertaining to the microcosm. m. salt, sodium ammonium phosphate; so-called because formerly derived from the urine of man, "the microcosm."
**microcoulomb** (mi-kro-koo'-lōm) [*micro-*; *coulomb*]. The one-millionth part of a coulomb, *q. v.*
**microcoustic** (mi-kro-koo'-stik or mik-ro-kows'-tik). Synonym of *micracoustic*.
**microcrith** (mi'-kro-krith) [*micro-*; κριθή, barley]. A unit of molecular weight, equivalent to the weight of an atom of hydrogen.
**microcrystalline** (mi-kro-kris'-tal-in) [*micro-*; *crystalline*]. Composed of crystals of microscopic size.

## TABLE OF MICROCOCCI

| Name. | Where Found. | Primary Characters. |
|---|---|---|
| M. "Coccus A" (Foutin) | Hail | Saprophytic. |
| M. (Strepto-) acidi lactici (Grotenfelt). | Milk | Zymogenic. |
| M. (Pedio-) acidi lactici (Lindner) | Milk | Zymogenic. |
| M. acidi lactici (Marpmann) | Milk | Zymogenic. |
| M. acidi lactici liquefaciens (Krüger) | Butter and cheese | Zymogenic. |
| M. acidi paralactici (Nencki) | Air, water, milk | Symbiotic-zymogenic with Bacillus chauvei. |
| M. of Adametz, Nos. I–VI | Emmerthaler cheese | Zymogenic. |
| M. aerogenes (Miller) | Water; alimentary tract | Chromogenic (yellowish). |
| M. agilis (Ali-Cohn) | Water | Chromogenic (pink-red). |
| M. agilis citreus (Menge) | Water | Chromogenic (yellow). |
| M. (Sarcina) alba (Eisenberg) | Air, water | Zymogenic. |
| M. (Diplo-) albicans amplus (Bumm). | Vaginal secretions | Saprophytic. |
| M. (Diplo-) albicans tardissimus (Eisenberg and Bumm). | Vaginal secretions | Saprophytic. |
| M. (Diplo-) albicans tardus (Unna and Tommasoli). | Skin in eczema | Pathogenesis undetermined. |
| M. albidus (Roze) | Potato scab | Saprophytic. |
| M. (Strepto-) albus (Maschek) | Water | Saprophytic. |
| M. (Staphylo-) albus | Human milk; pus in osteomyelitis. | Saprophytic. |
| M. (Pedio-) albus (Lindner) | Well-water | Saprophytic. |
| M. (Strepto-) of Aleppo boil (Nicolle and Noury). | Pus in Aleppo boil | Pathogenic. |
| M. (Strepto-) apthicola (Hallier and Schottelius). | Vesicular eruptions in sheep, cattle, and pigs. | Pathogenesis undetermined. |
| M. aquatilis (Bolton) | Water | Saprophytic. |
| M. aquatilis invisibilis (Vaughan) | Water | Saprophytic. |
| M. area celsii (Buchner and Sehlen). | Diseased hairs in Alopecia areata. | Pathogenesis undetermined. |
| M. (Strepto-) articulorum (Löffler) | Mucous membrane in diphtheria. | Pathogenic. |
| M. (Sarcina) aurantiaca (Lindner and Koch). | Air, water, "Weissbier" | Zymogenic, chromogenic (orange-yellow; linoxanthine). |
| M. (Pedio-) aurantiacus (Cohn and Schroeter). | Air, water, soil | Chromoparous (orange-yellow). |
| M. aurantiacus sorghi (Bruyning) | Blighted sorghum | Phytopathogenic. |
| M. (Sarcina) aurea (Macé) | Pulmonary exudates in pneumonia cadaver. | Zymogenic; chromogenic (golden-yellow). |
| M. (Staphylo-) aureus | Carcinoma | Saprophytic. |
| M. "Coccus B" (Foutin) | Hail | Pathogenic. |
| M. (Asco-) billrothii (Cohn) | Saccharine fluids | Zymogenic. |
| M. (Staphylo-) biskræ (Hydenreich) | Water, air, pus in Aleppo boil and Biskra button. | Pathogenic. |
| M. (Strepto-) bombycis (Bechamp) | Silkworms with "la flêcherie" or "schlafsucht." | Pathogenic. |
| M. (Strepto-) of Bonome | Meningeal exudates in cerebrospinal meningitis. | Pathogenic. |
| M. (Asco-) botryogenes (Bollinger and Rabe). | Mykodesmoids of horses. | Pathogenic. |
| M. (Sarcina) Van den Corpat | Tainted sausage | Pathogenic. |
| M. of Bovine Pneumonia (Poels and Nolen). | Lungs of cattle with infectious pleuro-pneumonia. | Pathogenic. |
| M. (Hæmato-) bovis | Blood and viscera of cattle with hemoglobinuria. | Pathogenic. |
| M. of Bronchitis (Picchini) | Sputum in bronchitis | Pathogenic. |
| M. butyri aromafaciens (Keith) | Butter | Zymogenic. |
| M. (Strepto-) cadaveris (Sternberg) | Liver of yellow-fever cadaver | Saprophytic. |
| M. candicans (Flügge) | Air, water | Saprophytic. |
| M. (Sarcina) candida (Reincke) | Water and air about breweries | Zymogenic. |
| M. candidus (Cohn) | Water | Saprophytic. |
| M. capillorum | Scalp | Chromogenic (red-yellow). |
| M. carneus (Zimmermann) | Water (Chemnitz) | Chromogenic (red). |
| M. casei amari (Freudenreich) | Bitter Swiss cheese | Zymogenic. |
| M. catarrhalis | Nasal and bronchial secretions | Pathogenic. |
| M. (Mycotetraedron) cellare (Hansgirg). | Cellar dust and soil | Saprophytic. |
| M. (Mycacantho-) cellaris (Hansgirg). | Cellar dust | Saprophytic. |
| M. cerasinus siccus (List) | Water | Chromogenic (cherry-red). |
| M. (Staphylo-) cereus albus (Passet) | Pus, water | Chromogenic (gray). |
| M. (Staphylo-) cereus aureus (Schroeter and Winkler). | Nasal secretions in coryza | Chromogenic (orange-red). |
| M. (Staphylo-) cereus flavus (Passet) | Acute abscesses | Chromogenic (lemon-yellow). |
| M. (Pedio-) cerevisiæ (Francke and Balcke). | Turbid beer | Zymogenic. |
| M. of Chicken-pox. See M. viridis flavescens. | | |
| M. chlorinus (Cohn) | Water | Chromophorous (green). |
| M. (Strepto-) cinnabareus (Flügge) | Air, water, red milk | Chromoparous (brick-red to vermilion). |
| M. citreus (List) | Water | Chromoparous (yellow). |
| M. (Diplo-) citreus conglomeratus (Bumm). | Air, dust, gonorrheal pus | Chromogenic (lemon-yellow). |

TABLE OF MICROCOCCI.—(Continued.)

| Name. | Where Found. | Primary Characters. |
|---|---|---|
| M. (Diplo-) citreus liquefaciens (Unna and Tommasoli). | Skin in eczema seborrhœicum.... | Chromogenic (lemon-yellow). |
| M. (Strepto-) coli gracilis (Escherich). | Feces........................ | Saprophytic. |
| M. concentricus (Zimmermann).... | Water (Chemnitz)............. | Chromogenic (brownish-yellow). |
| M. conglomeratus (Weichselbaum). | Water................/........ | Saprophytic. |
| M. (Strepto-) conglomeratus (Kurth) | Cases of scarlet-fever.......... | Pathogenic. |
| M. (Strepto-) coronatus (Flügge)... | Air.......................... | Saprophytic. |
| M. (Diplo-) coryzæ .(Klebs and Hajek). | Nasal secretions............... | Saprophytic. |
| M. (Strepto-) coryzæ contagiósæ equorum (Schutz). | Pus of lymphatic glands in horses having infectious pneumonia. | Pathogenic. |
| M. cremoides (Zimmermann)...... | Water (Chemnitz)............. | Chromogenic (yellow). |
| M. cumulatus tenuis (Von Besser).. | Nasal mucus (man)............ | Saprophytic. |
| M. cyaneus (Schroeter).......... | Air, water.................... | Chromogenic (blue). |
| M. decalvans (Schroeter)......... | Scalp in alopecia areata........ | Pathogenic. |
| M. delacourianus (Roze)......... | Potatoes with black gangrene.... | Phytopathogenic. |
| M. (Staphylo-) of Dengue (McLaughlin). | Blood in cases of dengue....... | Pathogenesis undetermined. |
| M. of Disse and Taguchi......... | Blood and secretions in broad condyloma. | Pathogenesis undetermined. |
| M. (Strepto-) endocarditidis rugatus (Weischselbaum). | On the valvular vegetations of ulcerative endocarditis. | Pathogenic. |
| M. (Staphylo-) epidermidis albus (Welch). | An almost constant inhabitant of the epidermis. | Pathogenic. |
| M. (Strepto-) erysipelatis (Fehleisen) | Lymph channels in cases of erysipelas. | Pathogenic. |
| M. erythromyxa (Lafar)......... | Water (Halle)................. | Chromogenic (red [liporhodinel] and yellow). |
| M. fervidosus (Adametz)......... | Water....................... | Chromogenic (faint-yellow). |
| M. finlayensis (Sternberg)........ | Viscera of yellow-fever cadaver... | Chromogenic (pale-yellow). |
| M. fioccii..................... | Conjunctival sac............... | Pathogenic. |
| M. (Sarcina) flavus (De Bary).... | Beer, cheese, etc............... | Zymogenic; chromogenic (yellow). |
| M. flavidus................... | Potato scab................... | Saprophytic. |
| M. flavus conjunctivæ........... | Human conjunctiva............ | Pathogenic. |
| M. flavus desidens (Flügge)....... | Air, water.................... | Chromogenic (yellowish-brown). |
| M. flavus liquefaciens (Flügge).... | Air, water.................... | Chromoparous (yellow). |
| M. flavus liquefaciens tardus (Unna and Tommasoli). | Skin in cases of eczema seborrhœicum. | Chromogenic (yellow). |
| M. flavus tardigradus (Flügge).... | Air, water.................... | Chromoparous (olive-green-yellow). |
| M. (Diplo-) fluorescens fœtidus; (Klamann and Rosenbach). | Human nares.................. | Chromogenic (grass-green-violet). |
| M. of Foot and Mouth Disease (Schottelius). | Vesicular eruptions in cattle, pigs, and sheep. | Pathogenic. |
| M. of Forbes.................. | Diseased larvæ of cabbage butterfly, Pieris rapæ. | Pathogenic. |
| M. freudenreichii (Guillebeau)..... | Ropy milk.................... | Zymogenic. |
| M. (Staphylo-) fulvus (Cohn)...... | Dung of horses and rabbits...... | Chromogenic (rose). |
| M. (Sarcina) fuscescens (Falkenheim). | Human stomach............... | Zymogenic. |
| M. fuscus (Maschek)............ | Water........................ | Zymogenic; chromogenic (dark-brown). |
| M. gelatinogenus (Bräutigam)..... | Ropy infusion of digitalis........ | Zymogenic. |
| M. gelatinosus................. | Ropy milk.................... | Zymogenic. |
| M. ghadialli................... | Water........................ | Saprophytic. |
| M. gingivæ pyogenes (Miller)..... | Alveolar abscess............... | Pathogenic. |
| M. gonorrhœæ (Neisser)......... | Gonorrheal pus................ | Pathogenic. |
| M. "Gray Coccus" (Maschek).... | Water........................ | Chromogenic (bluish-gray-green); zymogenic. |
| M. gummosus (Happ)............ | Ropy infusion of senega......... | Zymogenic. |
| M. hæmatodes (Babes).......... | Red sweat.................... | Chromogenic (blood-red). |
| M. (Staphylo-) hæmorrhagicus (Klein). | Vesicular eruption in sheep with "gargle." | Pathogenic. |
| M. (Strepto-) havaniensis (Sternberg). | Vomit of yellow-fever patients.... | Pathogenesis undetermined. |
| M. (Strepto-) hollandicus (Weichmann). | Ropy milk.................... | Zymogenic. |
| M. humuli lauensis (Mohl)....... | Hops......................... | Zymogenic. |
| M. (Sarcina) hyalina (Kutzing)... | Marsh water.................. | Zymogenic. |
| M. imperatoris (Roze).......... | Potato scab................... | Saprophytic. |
| M. (Strepto-) insectorum (Burrill).. | Diseased Chinch-bugs and other insects. | Pathogenic. |
| M. (Sarcina) intestinalis (Zopf).... | Intestines of fowls............. | Zymogenic. |
| M. (Diplo-) intracellularis meningitidis (Weichselbaum). | Exudates in cerebrospinal meningitis. | Pathogenic. |
| M. of Kirchner................. | Sputum in cases of influenza....* | Pathogenic. |
| M. (Diplo-) lacteus faviformis (Bumm). | Normal vaginal mucus......... | Saprophytic. |
| M. (Sphæro-) lactis acidi (Marpmann). | Milk......................... | Zymogenic. |
| M. lactis viscosus (Conn)......... | Bitter milk.................... | Zymogenic. |
| M. lardarius (Krassilochtchik).... | Grasserie of silk-worms......... | Pathogenic. |
| M. latericeus (Dobrzyniecki)...... | Mouth....................... | Chromogenic (brick-red). |

TABLE OF MICROCOCCI.—(Continued.)

| Name. | Where Found. | Primary Characters. |
|---|---|---|
| M. (Strepto-) liquefaciens (Sternberg). | Liver of yellow-fever cadaver. | Saprophytic. |
| . liquefaciens conjunctivæ. | Normal human conjunctiva. | Saprophytic. |
| . (Sarcina) litoralis (Oersted). | Sea-water. | Zymogenic; chromogenic (red). |
| M. loewenbergii. | Nose in ozena. | Pathogenic. |
| M. (Sarcina) lutea (Schroeter). | Water, potatoes, conjunctival sac. | Zymogenic; chromoparous (yellow). |
| . (Diplo-) luteus (Adametz). | Water. | Chromogenic (lemon-yellow). |
| . luteus (Schroeter). | Air, water, soil. | Chromoparous (sulphur-yellow). |
| M. lyssæ (Neisser). | Spinal cord of rabid animals. | Insufficiently studied. |
| M. (Strepto-) manfredii. | In progressive granuloma formation; pneumonia sputum. | Pathogenic. |
| . (Strepto-) of Manneberg. | Urine in acute nephritis. | Pathogenic. |
| M. (Strepto-) of Marmorek. | In erysipelas. | Pathogenic. |
| M. (Strepto-) mastobius (Nocard, Malereau, and Kitt). | Milk of sheep with gangrenous mastitis. | Pathogenic. |
| M. (Sarcina) maxima (Lindner). | Malt mashes. | Zymogenic. |
| Meningococcus. See M. intracellularis meningitidis. | | |
| M. melitensis (Bruce). | Malta fever. | Pathogenic. |
| M. (Asco-) mesenterioides (Cienkowski and Van Tieghem). | Beet-root-sap; molasses. | Zymogenic. |
| M. (Coleothrix) methystes (Veley). | Faulty rum. | Zymogenic. |
| M. (Sarcina) minuta (De Bary). | Sour milk. | Zymogenic. |
| M. (Strepto-) mirabilis (Roscoe and Lunt). | Sewage. | Saprophytic. |
| M. (Strepto-) monomorphous (Bujwid and Heryng). | Benign pharyngeal ulcers. | Saprophytic. |
| M. (Strepto-) morbillosus (Klebs and Keating). | Blood and exudates in measles. | Pathogenesis undetermined. |
| M. (Sarcina) morrhuæ. | Codfish. | Zymogenic. |
| M. (Sarcina) of Mouth and Lungs (Fischer). | Human mouth and lungs. | Zymogenic. |
| M. (Diplo-) of Mumps (Von Leyden). | Parotid saliva. | Pathogenic. |
| M. (Strepto-) of Mycosis fungoides (Rindfleisch and Auspetz). | Tissues in cases of granuloma fungoides. | Pathogenesis undetermined. |
| M. nasalis (Hack). | Nasopharynx. | Saprophytic. |
| M. neoformans. | Malta fever. | Pathogenic. |
| M. nitrificans (Van Tieghem). | Soil. | Zymogenic. |
| M. (Sarcina) nobilis (Maurea). | Old ascitic fluid. | Zymogenic; chromogenic (brick-red). |
| M. nuclei (Roze). | Potatoes. | Saprophytic. |
| M. (Strepto-) ochroleucus (Prove and Legrain). | Human urine. | Chromoparous (sulphur-yellow). |
| M. orbicularis flavus (Ravenel). | Soil. | Saprophytic. |
| M. (Diplo-) orchitidis (Hugouneng and Eraud). | Orchitis. | Pathogenic. |
| M. osteomyelitidis (Becker). | Osteomyelitis. | Pathogenic; zymogenic. |
| M. ovalis (Escherich). | Meconium and feces of infants. | Saprophytic. |
| M. (Diplo-) of Ozena (Loewenberg). | Nasal secretions. | Saprophytic. |
| M. (Sarcina) paludosa (Schroeter). | Marsh-water. | Zymogenic. |
| M. pellucidus (Roze). | Potato scab. | Phytopathogenic. |
| M. (Diplo-) of Pemphigus acutus (Gibier). | Bullas of pemphigus. | Pathogenic. |
| M. (Strepto-) peritonitidis equi (Hamburg). | Exudates in horses dead of peritonitis. | Pathogenic. |
| M. (Diplo-) of Pertussis (Ritter). | Sputum of whooping-cough. | Pathogenesis undetermined. |
| M. petrolei (Renault). | "Boghead" coal. | Saprophytic. |
| M. pfluegeri (Ludwig). | Luminous meat. | Photogenic. |
| M. phosphoreus (Cohn). | Fish. | Photogenic. |
| M. plumosus (Bräutigam). | Water. | Chromogenic (yellowish). |
| M. (Diplo-) of Pneumonia (Fraenkel). | Pulmonary exudate in acute lobar pneumonia. | Pathogenic. |
| M. porcellorum. | Swine with hepatitis. | Pathogenic. |
| M. of Progressive Abscess-formation in Rabbits (Koch). | Exudates of rabbits inoculated with putrid blood. | Pathogenic. |
| M. of Progressive Lymphoma of Animals. | Sputa of pneumonia after measles. | Pathogenic. |
| M. of Progressive Tissue Necrosis in Mice (Koch). | Exudates of mice inoculated with putrid blood. | Pathogenic. |
| M. pseudocyanus (Cohn). | Air. | Chromogenic (verdigris-green). |
| M. (Strepto-) psittaci (Eberth and Wolf). | Disease of gray parrots. | Pathogenic. |
| M. (Sarcina) pulmonum (Hauser). | Sputum. | Zymogenic. |
| M. putatus (Ravenel). | Soil. | Saprophytic. |
| M. putridus. | Water. | Zymogenic. |
| M. of Pyemia in Rabbits (Koch). | Exudates of rabbits inoculated with putrid flesh. | Pathogenic. |
| M. (Strepto-) pyogenes (Rosenbach) | Pus. | Pathogenic. |
| M. (Staphylo-) pyogenes albus (Rosenbach). | Stitch abscess. | Pathogenic. |
| M. (Staphylo-) pyogenes aureus (Rosenbach). | Air, soil, water, pus. | Pathogenic; zymogenic; chromogenic (orange-yellow). |

TABLE OF MICROCOCCI.—(Continued.)

| Name. | Where Found. | Primary Characters. |
|---|---|---|
| M. (Staphylo-) pyogenes citreus (Passet) | Pus | Pathogenic; chromogenic (lemon-yellow). |
| M. (Strepto-) pyogenes maligni (Krause and Flügge). | Leukemic spleen | Pathogenic. |
| M. pyogenes tenuis (Rosenbach) | Pus | Pathogenic. |
| M. (Diplo-) pyogenes ureæ (Rörsing). | Purulent urine | Saprophytic. |
| M. (Diplo-) pyogenes ureæ flavus (Rörsing). | Purulent urine | Saprophytic. |
| M. (Staphylo-) pyosepticus (Richet) | Carcinomatous tumor of dog | Pathogenic. |
| M. (Strepto-) radiatus (Flügge) | Air, water | Chromogenic (yellow-green). |
| M. (Sarcina) reitenbachii (Caspary) | Decaying water-plants | Saprophytic. |
| M. (Sarcina) renis (Hepworth) | Lungs in tuberculous cadavers | Zymogenic. |
| M. rheumaticus (Poynton and Payre). | Blood and synovial fluid in acute rheumatic fever. | Pathogenic. |
| M. of Rhine Water (Burri) | Water (Rhine) | Saprophytic. |
| M. rhodocrous | Stomach of goose | Chromogenic (red), liporhodine. |
| M. (Sarcina) rosea (Menge and Schroeter). | Air, red milk, beer | Chromogenic (intense red). |
| M. of Rose-red Disease of Wheat (Prillieux). | Bacteriosis of wheat | Phytopathogenic. |
| M. rosettaceus (Zimmermann) | Water (Chemnitz) | Chromogenic (grayish-yellow). |
| M. (Diplo-) roseus (Bumm) | Air | Chromogenic (pink). |
| M. roseus (Maggiora) | On the skin | Chromogenic (pink). |
| M. (Strepto-) rubiginosus (Edington). | Case of scarlatina | Saprophytic. |
| M. (Staphylo-) salivarius pyogenes (Biondi). | Saliva of child with scarlatina | Pathogenic. |
| M. (Strepto-) sanguinis canis (Pitfield). | Blood of dogs | Pathogenic. |
| M. saprogenes vini (Kramer) | Wine | Zymogenic. |
| M. scarlatinosus (Cose and Feltz) | Blood, skin, mouth, etc., of scarlet-fever patient. | Pathogenesis undertermined. |
| M. of Schmidt-Mülheim | Milk | Zymogenic. |
| M. of Schütz | Slimy milk | Zymogenic. |
| M. (Strepto-) of Septicemia in Rabbits (Koch). | Exudates of rabbits inoculated with putrid flesh. | Pathogenic. |
| M. (Strepto-) septicus (Flügge) | Soil | Pathogenic. |
| M. (Strepto-) septicus liquefaciens (Babes). | Case of septicemia | Pathogenic. |
| M. (Strepto-) septopyæmicus (Biondi). | Phlegmonous angina | Pathogenic. |
| M. sialosepticus | Saliva in case of septicemia | Pathogenic. |
| M. sordidus (Dyar) | Water | Zymogenic. |
| M. sornthalii (Adametz) | Puffy Sornthal cheese | Zymogenic. |
| M. stellatus (Maschek) | Water | Chromogenic (brownish-yellow). |
| M. (Diplo-) subflavus (Flügge) | Vaginal mucus | Pathogenic. |
| M. of Syphilis (Haberkon and Marcus). | White blood-corpuscles in case of syphilis. | Chromogenic (red). |
| M. tetragenus (Achard and Gaillard) | Variolous vesicle | Chromogenic (yellow). |
| M. tetragenus (Bosc and Galarielle) | Gangrenous pulmonary cavity | Pathogenic. |
| M. tetragenus (Gaffky) | Acute angina, "Angine sableuse" | Zymogenic; pathogenic. |
| M. tetragenus (Marotta) | Vesicles in small-pox | Zymogenic. |
| M. tetragenus febris flavæ (Finlay and Sternberg). | Case of yellow-fever | Chromogenic (lemon-yellow). |
| M. tetragenus mobilis ventriculi (Mendosa). | Stomach | Zymogenic. |
| M. tetragenus subflavus (Von Besser). | Nasal mucus | Saprophytic. |
| M. (Strepto-) toxicatus (Burrill) | On poison ivy | Pathogenic. |
| M. of Trachoma (Sattler and Michel). | Secretions and nodules in trachoma. | Saprophytic. |
| M. (Diplo-) of Trachoma (Snydaeker). | Secretions in trachoma | Saprophytic. |
| M. uberis (Dinwiddie) | Milk | Zymogenic. |
| M. unnæ (Laredde) | Vesicles in acute and chronic eczema. | Pathogenic. |
| M. ureæ (Pasteur and Cohn) | Air, water, ammoniacal urine | Zymogenic. |
| M. (Strepto-) ureæ liquefaciens (Flügge). | Urine | Zymogenic. |
| M. (Sarcina) urinæ (Welcker) | Urine | Saprophytic. |
| M. urinæ albus olearius (Doyen) | Urine of cystitis | Saprophytic. |
| M. urinæ flavus olearius (Doyen) | Urine of cystitis | Chromogenic (golden-yellow). |
| M. urinæ major (Doyen) | Urine of cystitis | Saprophytic. |
| M. urinalbus (Doyen) | Urine in cystitis and pyelonephritis. | Saprophytic. |
| M. (Strepto-) varians (Ewart) | Water | Chromophorous (green). |
| M. (Strepto-) variolæ et vaccinæ (Cohn). | Lymph of vaccine pustules | Pathogenic. |
| M. (Sarcina) ventriculi (Goodsir) | Diseased stomach | Zymogenic; chromoparous (faint-yellow). |
| M. (Strepto-) vermiformis (Maschek). | Water | Chromogenic (yellow). |
| M. versatilis | Yellow-fever cadaver | Saprophytic. |

# MICROCOCCUS 562 MICROPATHOLOGY

## TABLE OF MICROCOCCI.—(Concluded.)

| Name. | Where Found. | Primary Characters. |
|---|---|---|
| M. versicolor (Flügge) | Air, water | Chromogenic (green-yellow). |
| M. (Asco-) vibrans (Van Tieghem) | Water | Saprophytic. |
| M. (Strepto-) vini perda | Spoiled wine | Zymogenic. |
| M. (Sarcina) violaceus (Kützing) | Water | Chromogenic (violet). |
| M. violaceus (Cohn) | Water | Chromoparous (violet-blue). |
| M. violaceus (Schroeter) | Air, water | Chromoparous (violet). |
| M. (Staphylo-) viridis flavescens (Guttmann). | Lymph of varicella pustules | Chromogenic (greenish-yellow). |
| M. viscosus (Pasteur) | Ropy wort and beer | Zymogenic. |
| M. viticulosus (Katz and Flügge) | Air, water | Saprophytic. |
| M. of Weigmann | Slimy milk, "langerwei" | Zymogenic. |
| M. (Sarcina) welckerii (Rossmann) | Urine | Zymogenic. |
| M. of Whooping-cough (Letzerich) | Sputum in whooping-cough | Pathogenesis undetermined. |
| M. (Crypto-) xanthogenicus (Freire) | Yellow-fever | Pathogenic. |

**microcyst** (*mi'-kro-sist*) [*micro-*; κύστις, a cyst]. A cyst of very small size.
**microcytase** (*mi-kro-si'-tase*). An enzyme found in leukocytes and capable of digesting microorganisms.
**microcyte** (*mi'-kro-sīt*) [*micro-*; κύτος, a cell]. A small red blood-corpuscle.
**microcythemia, microcythæmia** (*mi-kro-si-the'-me-ah*) [*microcyte;* αἷμα, blood]. A condition of the blood characterized by abnormally small erythrocytes.
**microcytosis** (*mi-kro-si-to'-sis*). Same as microcythemia.
**microdactylia** (*mi-kro-dak-til'-e-ah*) [*micro-*; δάκτυλος, finger]. Abnormal smallness of the fingers.
**microdont** (*mi'-kro-dont*) [*micro-*; ὀδούς, tooth]. Having small teeth.
**microdontism** (*mi-kro-don'-tizm*) [*micro-*; ὀδούς, tooth]. Abnormal smallness of the teeth.
**microelectrometer** (*mi-kro-e-lek-trom'-e-ter*) [*micro-; electrometer*]. An apparatus for estimating minute amounts and intensities of electricity.
**microfarad** (*mi-kro-far'-ad*). The one-millionth part of a farad, *q. v.*
**microgalvanic** (*mi-kro-gal-van'-ik*). Relating to very small galvanic currents.
**microgamete** (*mi-kro-gam'-ēt*) [*micro-*; γάμος, marriage]. A male sexual cell among sporozoa.
**microgametocyte** (*mi-kro-gam-et'-o-sīt*). The cell which produces the microgametes in Protozoa.
**microgamy** (*mi-krog'-am-e*) [*micro-*; γάμος, marriage]. Conjugation between young protozoan cells.
**microgastria** (*mi-kro-gas'-tre-ah*) [*micro-*; γαστήρ, stomach]. Smallness of the stomach.
**microgenesis** (*mi-kro-jen'-es-is*) [*micro-*; γένεσις, origin]. Abnormally small development of a part.
**microgenia** (*mi-kro-je'-ne-ah*) [*micro-*; γένειον, chin]. Abnormal or congenital smallness of the chin.
**microglossia** (*mi-kro-glos'-e-ah*) [*micro-*; γλῶσσα, tongue]. Abnormal smallness of the tongue.
**micrognathia** (*mi-krog-na'-the-ah*) [*micro-*; γνάθος, jaw]. Abnormal smallness of the jaws, especially of the lower jaw.
**microgonidium** (*mi-kro-go-nid'-e-um*) [*micro-*; γονή, generation, seed: *pl., microgonidia*]. In biology, a relatively small-sized gonidium.
**microgram** (*mi'-kro-gram*) [*micro-*; gram]. A millionth part of a gram.
**micrograph** (*mi'-kro-graf*) [*micro-*; γράφειν, to delineate]. 1. A device for enabling one to draw sketches on a very small scale. 2. An instrument that magnifies the vibrations of a diaphragm and records them on a moving photographic film.
**micrography** (*mi-krog'-ra-fe*) [*micro-*; γράφειν, to write]. 1. A description of bodies that are studied under the microscope. 2. Very minute writing.
**microgyria** (*mi-kro-ji'-re-ah*) [*micro-*; γῦρος, gyrus]. Smallness of the convolutions of the brain.
**microhistology** (*mi-kro-his-tol'-o-je*). Synonym of *micristology*.
**microhm** (*mi'-krōm*) [*micro-*; *ohm*]. The millionth part of an ohm.
**microkinesis** (*mi-kro-kin-e'-sis*) [*micro-*; κίνησις, motion]. Involuntary muscular movements, especially in infants.
**microlentia** (*mi-kro-len'-te-ah*) [*micro-*; *lens*]. The state of having an abnormally small crystalline lens.

**microliter** (*mi'-kro-le-ter*) [*micro-*; *liter*]. The millionth part of a liter.
**microlith** (*mi'-kro-lith*) [*micro-*; λίθος, stone]. A microscopic calculus.
**micrology** (*mi-krol'-o-je*) [*micro-*; λόγος, a treatise]. A treatise on minute objects, especially microscopic objects.
**micromania** (*mi-kro-ma'-ne-ah*) [*micro-*; *mania*]. A form of insanity in which the patient believes himself diminutive in size and mentally inferior.
**micromazia** (*mi-kro-ma'-ze-ah*). An abnormal smallness of the breasts.
**micromelia** (*mi-kro-me'-le-ah*) [*micro-*; μέλος, limb]. Abnormal smallness of the limbs.
**micromelus** (*mi-krom'-el-us*) [*micro-*; μέλος, a limb]. A single autositic monster of the species ectromelus, characterized by the presence of abnormally small limbs.
**micromerology** (*mi-kro-me-rol'-o-je*) [*micro-*; μέρος, part; λόγος, science]. The science of anatomical segments.
**micrometer** (*mi-krom'-et-er*) [*micro-*; μέτρον, a measure]. An instrument designed for measuring minute objects seen through the microscope. **m., eyepiece, m., ocular,** a micrometer to be used with the eyepiece of a microscope. **m.-screw,** a fine screw with a scale attached showing the distance passed at each fraction of a revolution. **m., stage-,** a micrometer attached to the stage of a microscope.
**micrometry** (*mi-krom'-et-re*) [*micrometer*]. The measurement of objects by a micrometer.
**micromil** (*mi'-kro-mil*). An abbreviation of *micromillimeter*.
**micromillimeter** (*mi-kro-mil'-im-e-ter*) [*micro-*; *millimeter*]. 1. The one-millionth part of a millimeter. 2. More commonly used to denote the one-thousandth part of a millimeter or the one-millionth part of a meter. It is the unit of microscopic measurements, and is the equivalent of $\frac{1}{25400}$ of an English inch. Symbol μ. Syn., *micron*.
**micrommatous** (*mi-krom'-at-us*) [*micro-*; ὄμμα, eye]. Small-eyed.
**micromotoscope** (*mi-kro-mo'-to-skōp*) [*micro-*; *movere*, to move; σκοπεῖν, to view]. An apparatus for photographing and exhibiting motile microorganisms.
**micromyelia** (*mi-kro-mi-e'-le-ah*) [*micro-*; μυελός, marrow]. Abnormal smallness of the myel, or spinal cord.
**micron** (*mi'-kron*). See *micromillimeter* (2).
**micronemous** (*mi-kron'-em-us*) [*micro-*; νῆμα, a thread]. Furnished with short filaments.
**micronucleus** (*mi-kro-nū'-kle-us*) [*micro-*; *nucleus*]. A small or minute nucleus. In biology, Maupas' term for the paranucleus, or the nucleolus of other authorities.
**micro-organism** (*mi-kro-or'-gan-izm*) [*micro-*; *organism*]. A microscopic being of the animal or vegetable kingdom, especially the vegetable group known as bacteria.
**microparasite** (*mi-kro-par'-as-īt*) [*micro-*; παράσιτος, a parasite]. A parasitic bacterium or other microorganism.
**micropathological** (*mi-kro-path-o-loj'-ik-al*) [*micro-*; πάθος, disease; λόγος, science]. Pertaining to micropathology.
**micropathology** (*mi-kro-path-ol'-o-je*) [*micro-*; *pathology*]. 1. The study of minute pathological changes.

2. The study of microorganisms in their relation to disease.
**micropenis** (*mi-kro-pe'-nis*). Abnormal smallness of the penis.
**microphage** (*mi'-kro-fāj*) [*micro-*; φαγεῖν, to devour]. A small phagocyte.
**microphagus** (*mi-krof'-ag-us*). A microphage.
**microphobia** (*mi-kro-fo'-be-ah*). See *microbiophobia*.
**microphone** (*mi'-kro-fōn*) [*micro-*; φωνή, sound]. An instrument that amplifies feeble sounds and renders them audible.
**microphonograph** (*mi-kro-fo'-no-graf*). A combination of the microphone and the phonograph.
**microphonoscope** (*mi-kro-fo'-no-skōp*) [*micro-*; φωνή, sound; σκοπεῖν, to view]. A binaural stethoscope with a membrane in the chest-piece to accentuate the sound.
**microphotograph** (*mi-kro-fo'-to-graf*) [*micro-*; *photograph*]. 1. A photograph of microscopic size. 2. See *photomicrograph*.
**microphthalmus** (*mi-krof-thal'-mus*) [*micro-*; ὀφθαλμός, eye]. 1. The condition of having an abnormally small eye. 2. A person having such an eye.
**microphyte** (*mi-kro'-fīt*) [*micro-*; φυτόν, a plant]. Any microscopic plant, especially one that is parasitic.
**micropia** (*mi-kro'-pe-ah*). See *micropsia*.
**microplanar** (*mi-kro-pla'-nar*) [*micro-*; *planus*, flat]. The name given to anastigmatic objectives of the most perfect correction, and designed especially for use in photographing small objects, like embryos, and for microprojection.
**microplasia** (*mi-kro-pla'-ze-ah*) [*micro-*; πλάσις, a molding]. Arrested development.
**micropolariscope** (*mi-kro-po-lar'-is-kōp*) [*micro-*; *polaris*, polar; σκοπεῖν, to view]. A polariscope used in connection with a microscope.
**microprojection** (*mi-kro-pro-jek'-shun*). The projection of the image of microscopic objects on a screen.
**microprosopa** (*mi-kro-pro-so'-pah*) [*micro-*; πρόσωπον, face]. Congenital smallness of the face.
**microprotein** (*mi-kro-pro'-te-in*). See *mycoprotein*.
**micropsia** (*mi-krop'-se-ah*) [*micro-*; ὄψις, sight]. A defective state of vision in which objects appear very small.
**micropsychia** (*mi-kro-si'-ke-ah*) [*micro-*; ψυχή, mind]. Weak-mindedness.
**micropus** (*mi-kro'-pus*) [*micro-*; πούς, foot]. Abnormal smallness of the feet; a congenital defect.
**micropyle** (*mi'-kro-pīl*) [*micro-*; πύλη, gate; orifice]. The small opening in an ovum through which the spermatozoon may penetrate.
**microrrhinia** (*mi-kro-rin'-e-ah*) [*micro-*; ῥίς, nose]. Congenital atrophy or smallness of the nose.
**microscelous** (*mi-kros'-el-us*) [*micro-*; σκέλος, leg]. Short-legged.
**microscope** (*mi'-kro-skōp*) [*micro-*; σκοπεῖν, to view]. An apparatus through which minute objects are rendered visible. It consists of a lens or group of lenses by which a magnified image of the object is produced. **m., binocular**, a microscope having divergent oculars, one for each eye, so that the object is seen with both eyes. **m., compound**, one that consists of two or more lenses or lens-systems, of which one, the objective, placed near the object, gives an enlarged and inverted real image; the other, the ocular, acting like a simple microscope, gives an enlarged virtual image of the real image. **m., simple**, one consisting of one or more lenses or lens-systems acting as a single lens. The rays of light that enter the eye of the observer, after refraction through these lenses, proceed directly from the object itself.
**microscopic** (*mi-kro-skop'-ik*) [*microscope*]. 1. Pertaining to the microscope. 2. Visible only with the aid of a microscope.
**microscopist** (*mi-kros'-ko-pist*) [*microscope*]. One who is skilled in the use of the microscope.
**microscopy** (*mi-kros'-ko-pe*) [*microscope*]. The use of the microscope; examination with the microscope.
**microseme** (*mi'-kro-sēm*) [*micro-*; σῆμα, sign; index]. Having the orbital index less than 83.
**microsmatic** (*mi-kros-mat'-ik*) [*micro-*; ὀσμή, a smell]. Having ill-developed olfactory organs.
**microsol** (*mi'-kro-sol*). Trade name of an antiseptic mixture of copper sulphocarbolate, copper sulphate and diluted sulphuric acid.

**microsoma** (*mi-kro-so'-mah*) [*micro-*; σῶμα, body]. In biology, small chromatin-granules in the cell-nuclei.
**microsomia** (*mi-kro-so'-me-ah*) [see *microsoma*]. Abnormal smallness of the whole body.
**microspectroscope** (*mi-kro-spek'-tro-skōp*) [*micro-*; *spectrum*; σκοπεῖν, to view]. A spectroscope used in connection with the ocular of a microscope, and by means of which the spectra of microscopic objects can be examined.
**microsphyxia** (*mi-kro-sfiks'-e-ah*) [*micro-*; σφίξις, pulse]. Weakness or smallness of the pulse.
**Microspira** (*mi-kros'-pir-ah*) [*micro-*; σπεῖρα, a coil]. A genus of *Spirillaceæ* with rigid cells, and one, rarely two or three polar flagella.
**Micropironema** (*mi-kro-spi-ro-ne'-mah*). A genus of protozoa, same as *Spironema*, *q. v.*
**microsplanchnus** (*mi-kro-splangk'-nus*) [*micro-*; σπλάγχνον, viscus]. Having small viscera.
**Microsporidia** (*mi-kro-spor-id'-e-ah*) [*micro-*; σπόρος, seed]. A genus of the class of *Sporozoa*, occurring as parasites in the muscles of the frog, the marsh-tortoise, worms, and insects.
**microsporon** (*mi-kro-spo'-ron*) [see *microsporidia*]. A fungus to which several diseases of the skin and hair are believed to be due. **m. furfur**. See *tinea versicolor*.
**microstat** (*mi'-kro-stat*) [*micro-*; στατός, fixed]. The stage and finder of the miscrocope.
**microstethophone** (*mi-kro-steth'-o-fōn*) [*micro-*; στῆθος, chest; φωνή, sound]. A stethoscope which magnifies the sounds heard.
**microstethoscope** (*mi-kro-steth'-o-skōp*) [*micro-*; *stethoscope*]. A stethoscope which magnifies the sounds heard.
**microsthenic** (*mi-kro-sthen'-ik*) [*micro-*; σθένος, power]. Having feeble muscular power.
**microtesia** (*mi-kro-te'-ze-ah*) [μικρότης, smallness]. Congenital smallness of part of the body.
**microstomia** (*mi-kro-sto'-me-ah*) [*micro-*; στόμα, mouth]. Abnormal smallness of the mouth.
**microtherm** (*mi'-kro-therm*) [*micro-*; θέρμη, heat]. An organism in which the life-processes are carried on at a low temperature.
**microtia** (*mi-kro'-she-ah*) [*micro-*; οὖς, ear]. Abnormal smallness of the external ear.
**microtome** (*mi'-kro-tōm*) [*micro-*; τέμνειν, to cut]. An instrument for making thin sections for microscopic examination. **m., freezing**, one in which the tissue is frozen, in order to secure the hardness required for properly cutting sections.
**microtomy** (*mi-krot'-o-me*) [*microtome*]. Section-cutting.
**microtrichia** (*mi-kro-trik'-e-ah*) [*micro-*; θρίξ, hair]. Shortness or fineness of the hair.
**microunit** (*mi-kro-ū'-nit*). A unit of minute measurements; the one-millionth part of an ordinary unit.
**microvolt** (*mi'-kro-vōlt*). One-millionth of a volt.
**microxycyte** (*mi-kroks'-is-īt*) [*micro-*; ὀξύς, sharp; κύτος, a cell]. A cell containing fine oxyphil granules and a more or less pigmented nucleus, occurring in the peritoneal fluid of infected subjects.
**microxyphil** (*mi-kroks'-e-fil*). See *microxycyte*.
**microzyme** (*mi'-kro-zīm*) [*micro-*; ζύμη, leaven]. One of certain minute particles of living matter that are by some supposed to be living organisms capable of an independent existence, and which are the cause of normal and pathological fermentation; the real agents of the functions of the organism, the perversion of whose function constitutes disease.
**miction** (*mik'-shun*). Same as *micturition*.
**mictocystis** (*mik-to-sis'-tis*) [μικτός, mixed; κύστις, a bag]. An organic sac made up of different textures.
**mictopyous** (*mik-top'-e-us*) [μικτός, mixed; πύον, pus]. Mixed with pus.
**micturition** (*mik-tū-rish'-un*) [*micturire*, to urinate]. The act of passing urine. **m.-center**, the center governing the act of micturition; it is situated in the lumbar region of the spinal cord.
**mid-** [AS., *mid*, middle, with]. A prefix meaning middle; also with.
**midaxilla** (*mid-aks-il'-ah*). The center of the axilla.
**midbody** (*mid'-bod-e*). A mass of granules formed in the equator of the spindle during the anaphase of mitosis.
**midbrain**. The mesencephalon.
**midfrontal**. Pertaining to the middle of the forehead.
**midgut**. See *mesogaster*.

**midol** (*mid'-ol*). A proprietary headache remedy containing pyramidon and caffeine.
**midriff** [*mid-; hrif*, belly]. The diaphragm.
**midsternum** (*mid-ster'-num*) [*mid-; sternum*]. The mesosternum.
**midwife** [*mid; wif*, a woman]. A female nurse, or other woman, who attends women in childbirth.
**midwifery** [*midwife*]. Obstetrics.
**Miescher's tubes** (*me'-sher*) [Johann Friedrich Miescher, German pathologist, 1811–1887]. Protoplasmic masses (*Sarcosporidia*) surrounded by a distinct cuticle, and breaking up into a series of spores when mature; they are met in the muscular tissue of domestic animals.
**migraf** (*mi'-graf*). A portable microscope and camera combined, designed for the observation of microscopic objects and for making a quick and permanent record of the same.
**migrainator** (*mig'-ra-na-tor*). An apparatus for the relief of migraine consisting of two plates held by a spring for the compression of the temporal arteries and regulation of the circulation of the blood in the head.
**migraine** (*me'-grān*) [Fr., from ἡμι, half; κρανίον, skull]. A paroxysmal affection characterized by headache, usually unilateral, and by gastric, vasomotor, and visual disturbances.
**migrainin** (*mig'-ra-nin*) [*migraine*]. A name given to a mixture of antipyrine, citric acid, and caffeine. It is used for the treatment of migraine, of the headache of influenza, and of that due to alcohol, tobacco, and morphine. Dose 15-45 gr. (1-3 Gm.) in 24 hours.
**migration** (*mi-gra'-shun*) [*migrare*, to wander]. A wandering. **m., external** (of the ovum), the passage of the ovum from an ovary to the tube of the opposite side. **m., internal** (of the ovum), the passage of the ovum through the tube related to the ovary from which the ovule was discharged, into the uterus and across into the opposite tube. **m. of ovum**, the passage of the ovum from the ovary to the Fallopian tube. **m. of white corpuscles**, one of the phenomena of inflammation, consisting in the passage of the white corpuscles of the blood through the vessel-wall.
**migratory** (*mi'-grat-o-re*) [*migrare*, to wander]. Characterized by wandering, or changing locality.
**m. pneumonia.** See *pneumonia migrans*.
**migrol** (*mig'-rol*). A proprietary remedy for migraine, said to consist of caffeine, sodium bicarbonate, and guaiacetin.
**migrosine** (*mig'-ro-sin*). A mixture of menthol and acetic ether used in migraine.
**mika operation** (*mi'-kah*) [African]. A method in vogue among certain African tribes for the purpose of preventing impregnation. It consists in the formation of a permanent urethro-perineal fistula in the male.
**mikozone** (*mi'-ko-zōn*). A variety of chlorodyne, used as a hypnotic and sedative.
**mikro-** (*mi'-kro-*). See *micro-*.
**mikron** (*mi'-kron*). Same as *micron*.
**Mikulicz's cells** (*mik'-oo-lits*) [Johannes von Mikulicz, Polish surgeon, 1850–1905]. Vesicular cells found in the diseased tissue in cases of rhinoscleroma and containing *Bacillus rhinoscleromatis*. **M.'s dictum**, that it is highly dangerous to give a general anesthetic to a patient whose hemoglobin percentage is below 30. **M.'s disease**, chronic hypertrophic enlargement of the lacrimal and salivary glands. **M.'s drain**, a method of draining the abdominal cavity after operation. A piece of iodoform-gauze, with a string tied to its center, is placed in the cavity, and into this improvised sac considerably more gauze is packed. Pressure is thus induced while the capillary action of the sac secures drainage. **M.'s operation.** 1. *For the removal of tonsillar tumors:* it is done by an incision through the neck, the cut extending from the mastoid process downward and forward as far as the great cornu of the hyoid bone; the ascending ramus of the jaw is then resected and the wall of the pharynx is divided; a preliminary tracheotomy is performed. 2. *For tarsectomy:* the same as *Wladimiroff's operation*, but independently designed.
**mil** [*mille*, a thousand]. 1. The one-thousandth part of an inch. 2. The one-thousandth part of a liter; the modern equivalent of a cubic centimeter. **m., circular**, the area of a circle the one-thousandth of an inch in diameter.

**milammeter** (*mil-am'-et-ur*). Same as *milliamperemeter*.
**milchlin** (*miltsh'-lin*). A condensed skimmed milk; when diluted with water it is said to resemble ordinary milk.
**mildew** (*mil'-dū*) [AS., *meledeāw*, honey-dew]. A common name for minute fungi parasitic on plants, and als̥o found on dead vegetable substances.
**mildiol** (*mil'-di-ol*). A disinfectant said to consist of a mixture of creosote and petroleum.
**milfoil** (*mil'-foil*). See *Achillea*.
**miliaria** (*mil-e-a'-re-ah*) [*milium*]. An acute inflammatory disease of the sweat-glands, the lesions consisting of vesicles and papules, accompanied by a pricking or tingling sensation. It occurs especially in summer, is due to excessive sweating, runs an acute or subacute course, and is followed by slight desquamation. Relapses are common. **m. alba**, **m. arthritica**, a form occurring only in those affected with gouty or rheumatic cardiac disease. **m. crystallina**, a variety of miliaria in which the sweat accumulates under the superficial horny layers of the epidermis to form small, clear, transparent vesicles. Syn., *sudamina crystallina*. **m. papulosa**, the well-known "prickly heat." **m. rubra.** See *miliaria*.
**miliary** (*mil'-e-a-re*) [*milium*]. 1. Of the size of a millet-seed, as *miliary aneurysm*, *miliary tubercle*. 2. Attended or characterized by the formation of numerous lesions the size of a millet-seed, as *miliary tuberculosis*. **m. fever.** See *miliaria*. **m. tuberculosis.** See *tuberculosis, miliary*.
**milium** (*mil'-e-um*) [L., "millet-seed"]. 1. A disease of the skin characterized by the formation of small, pearly, noninflammatory elevations (*milia*) situated mainly on the face. It is due to the occlusion of the ducts of sebaceous follicles, the secretion of which accumulates and distends the follicles. 2. One of the elevations characteristic of milium. **m., amyloid.** Synonym of *molluscum contagiosum*. *q. v.* **m., colloid**, a rare skin-disease characterized by the presence, especially on the bridge of the nose, forehead, and cheeks, of minute, shining, flat, or slightly raised lesions of a pale-lemon or bright-lemon color. It is a form of colloid degeneration of the skin, affecting persons of middle or advanced age.
**milk.** The opaque white secretion of the mammary glands of the female of mammalia. Cream from which the fatty matter has been removed

| | *Fat.* | *Casein.* | *Albumin.* | *Milk-sugar.* | *Ash.* | *Total Solids.* | *Water.* |
|---|---|---|---|---|---|---|---|
| Human milk... | 2.90 | 2.40 | 0.57 | 5.87 | 0.16 | 12.00 | 88.00 |
| Cow's milk .... | 3.50 | 3.98 | 0.77 | 4.00 | 0.17 | 13.13 | 86.87 |
| Camel's milk... | 2.90 | | 3.84 | 5.66 | 0.66 | 13.06 | 86.94 |
| Goat's milk ... | 4.20 | 3.00 | 0.62 | 4.00 | 0.56 | 12.46 | 87.54 |
| Ass's milk ..... | 1.02 | 1.09 | 0.70 | 5.50 | 0.42 | 8.83 | 91.17 |
| Mare's milk ... | 2.50 | 2.19 | 0.42 | 5.50 | 0.50 | 11.20 | 88.80 |
| Sheep's milk ... | 5.30 | 6.10 | 1.00 | 4.20 | 1.00 | 17.73 | 82.27 |

is known as *buttermilk*. *Skimmed milk* is that from which the cream has been removed. **m.-catalase**, an enzyme of cow's milk capable of decomposing hydrogen dioxide and similar compounds; it is rendered inactive by heating to 80° C. **m., condensed**, cow's milk from which a large part of the water has been evaporated, a syrupy liquid remaining which is preserved with or without the addition of sugar. **m.-crust.** See *crusta lactea*. **m.-cure**, the method of treating certain diseases by an exclusive diet of milk. **m.-cyst**, a cyst of the galactophorous duct. **m., diabetic**, a prepared milk containing a small percentage of lactose. **m.-fever**, a slight rise of temperature attending the establishment of the secretion of milk. It is due to a mild degree of septic

intoxication. **m.-leg.** See *phlegmasia alba dolens.*
**m.-punch,** a preparation made by adding brandy, whisky, or rum to milk in the proportion of about one to four or six parts, and flavoring with sugar and nutmeg. **m.-sickness,** a disease of cattle communicable to persons who drink their milk or eat their flesh; it is marked by chills and trembling, vomiting, and disorder of the alimentary functions. **m.-somatose,** a food-preparation similar to somatose made from meat but containing 5 % of tannin. **m.-spot.** See *strophulus.* **m.-sugar,** the sweet principle of milk; lactose, *q. v.* **m.-teeth,** the teeth of the first dentition. **m.-tester,** a lactometer. **m.-tumor,** a tumor of the breast from retention of milk. **m.-vine,** the plant *Periploca græca.* **m., virgin's,** rose water rendered milky by the addition of tincture of benzoin.
**milkine** (*mil'-kēn*). A concentrated compound of cow's milk, 50%; malted cereals, 44 %, beef, 5 %; calcium hydroxide, 0.5 %; and sodium chloride, 0.5 %. It is a light yellow powder with sweet taste and marked odor.
**Millar's disease** (*mil'-ar*) [John *Millar,* Scottish physician, 1735–1801]. Laryngismus stridulus.
**Millard-Gubler's syndrome.** See *Gubler's hemiplegia.*
**Millard's test for albumin in urine** (*mil'-ard*) [Henry B. *Millard,* American physician, 1832–1893]. Add a mixture of carbolic acid, acetic acid, and liquor potassæ to the suspected urine; if albumin is present a white precipitate will be thrown down.
**milli-** (*mil-e-*) [*mille,* thousand]. A prefix meaning a thousand.
**milliampèrage** (*mil-e-am'-pār-ahej*). The expression of electric current-strength in milliamperes.
**milliampère** (*mil-e-am'-pār*) [*milli-, ampere*]. Onethousandth of an ampere = 1 volt divided by 1000.
**milliamperemeter** (*mil-e-am-pār'-me-ter*) [*millimpere; μέτρον,* a measure]. An instrument for measuring the strength of an electric current.
**millicurie** (*mil-e-kūr'-re*) [*milli-; curie*]. One thousandth part of a curie.
**milligram** (*mil'-e-gram*) [*milli-; gram*]. A thousandth part of a gram.
**milliliter** (*mil'-il-e-ter*) [*milli-; liter*]. A thousandth part of a liter.
**millimeter** (*mil'-im-e-ter*) [*milli-; meter*]. A thousandth part of a meter.
**millinormal** (*mil-in-or'-mal*) [*milli-; normal*]. Containing a thousandth part of what is normal.
**Millon's reagent** (*mil'-on*) [Auguste Nicolas Eugène *Millon,* French chemist, 1812–1867]. A reagent made by dissolving 10 Gm. of mercury in 20 Gm. of nitric acid, diluting the solution with an equal volume of water, and decanting in 24 hours. With proteins and with derivatives of benzene and naphthalene it gives a red color (*Millon's test*). **M.'s test.** See under **M.'s reagent.**
**millstone-maker's phthisis.** A form of pneumonokoniosis due to the inhalation of fine particles in the manufacture of millstones.
**mill-tooth.** A grinder; a molar tooth.
**milphosis** (*mil-fo'-sis*) [μίλφωσις, baldness]. Baldness of the eyebrows.
**milt.** The spleen. **m. sickness,** splenic fever, or anthrax, in cattle.
**milzbrand** (*milts'-brand*) [Ger.]. Anthrax.
**mimesis** (*mim-e'-sis*) [μίμησις, imitation]. 1. Mimicry. 2. The assumption of the symptoms of one disease by another disease.
**mimetic** (*mim-et'-ik*) [μίμος, an actor]. Imitative; mimic. **m. labor,** false labor. **m. paralysis,** paralysis of the facial muscles. **m. spasm,** spasm of the facial muscles.
**mimic** (*mim'-ik*). See *mimetic.*
**mimicry** (*mim'-ik-re*) [μίμος, an actor]. Imitation.
**mimmation** (*mim-a'-shun*) [Ar., *mim,* the name of the letter *m*]. The unduly frequent use of the sound of the letter *m* in speech.
**mimochasmesis** (*mim-o-kaz-me'-sis*) [*mimic;* χάσμησις, yawning]. Imitative yawning.
**mimography** (*mim-og'-ra-fe*) [*mimic;* γράφειν, to write]. Sign-language used by deaf-mutes.
**mimosis** (*mim-o'-sis*). Same as *mimesis.*
**min.** Abbreviation of *minimum,* or minim, the 60th part of a fluidram measure.
**mind** (*mind*). The understanding. The reasoning and intellectual faculties considered as a whole.
**mind-blindness.** A form of aphasia in which, although the patient is able to see, no intellectual impression is conveyed to his mind by the object seen. Syn., *visual amnesia.*
**mind-cure.** The alleged cure of disease through mental influence.
**mind-deafness.** A form of aphasia in which sounds, though heard and perceived as such, awaken no intelligent conception.
**Mindererus, spirit of** (*min-der-e'-rus*) [Raymond *Minderer,* German physician, 1621–    ]. See *ammonium acetate, liquor of.*
**mind-pain.** Same as *psychalgia.*
**mineral** (*min'-er-al*) [*mina,* a mine]. An inorganic chemical compound found in nature, especially one that is solid. **m., kermes,** antimony oxysulphide. **m. oil,** petroleum. **m. pitch,** bitumen. **m.-water,** water naturally or artificially impregnated with inorganic salts in sufficient quantity to give it special properties.
**mineralization** (*min-er-al-i-za'-shun*) [*mineral*]. 1. The addition of mineral substances to a body. 2. The relative amount of mineral substances dissolved in a mineral-water.
**miner's anemia** or **cachexia.** See *dochmiasis.*
**m.'s elbow,** enlargement of the bursa over the olecranon, common in miners, due to irritation while working and lying on the side. **m.'s nystagmus,** a peculiar nystagmus occurring in miners. **m.'s phthisis,** a chronic affection of the lungs due to the constant inhalation of coal-dust. Syn., *anthracosis.*
**minim** (*min'-im*) [*minimus,* least]. The onesixtieth of a fluidram. Symbol m.
**minimal** (*min'-im-al*) [see *minim*]. Least; lowest. Of doses, the least quantity that is yet effective.
**minimeter** (*min-im'-e-ter*) [*minim; μέτρον,* measure]. An apparatus for measuring liquids in minims.
**minimum** (*min'-im-um*) [see *minim*]. The least; the lowest; the lowest intensity or level. **m. lethal dose,** the quantity of a toxin which will kill a guinea pig of 250 grams weight in from 4 to 5 days. **m. temperature,** temperature below which bacterial growth does not take place. **m. thermometer.** See *thermometer, self-registering.*
**minium** (*min'-e-um*) [L., "red lead"]. Red lead oxide, Pb₃O₄, used formerly in plasters.
**minor** (*mi'-nor*) [L., "iess"]. 1. Less; lesser; smaller. 2. An individual under legal age; one under the authority of parents or guardians. **m. surgery.** See *surgery, minor.*
**mint.** See *mentha.*
**mioangioneurosis** (*mi-o-an-je-o-nū-ro'-sis*) [μείων, lesser; ἀγγείον, vessel; νεῦρον, nerve; νόσος, disease]. A nervous disorder of the smaller blood-vessels; a vaso-motor or vaso-inhibitory disturbance.
**miocardia** (*mi-o-kār'-de-ah*) [μείων, less; καρδία, heart]. The systolic diminution of the volume of the heart. See *auxocardia.*
**miodidymus, miodymus** (*mi-o-did'-im-us, mi-od'-im-us*) [μείων, less; δίδυμος, twin]. A double-headed monster joined by the occiputs.
**miopragia** (*mi-o-pra'-je-ah*) [μείων, less; πράσσειν, to do]. Diminished functional activity.
**miopus** (*mi-o'-pus*) [μείων, less; ὤψ, the face]. A double-headed monster with one face rudimentary.
**miosis** (*mi-o'-sis*) [μείων, less]. 1. Contraction or decrease in the size of an organ, especially the pupil. 2. A lessening of the intensity of existing symptoms.
**miotic** (*mi-ot'-ik*) [*miosis*]. 1. Pertaining to, or characterized by, miosis. 2. Causing contraction of the pupil. 3. An agent that contracts the pupil.
**mirbane, oil of.** A name for nitrobenzene.
**mire** (*mēr*) [*mirare,* to see]. Figures used upon the perimeter-bar of the ophthalmometer of Javal and Schiotz; by observing the variations of their images, as reflected from different meridians of the cornea, the measurement of corneal astigmatism is effected.
**mirror** (*mir'-or*) [*mirari,* to admire]. A polished surface for reflecting light or forming images of objects placed in front of it. **m., concave,** one the reflecting surface of which is concave. **m., convex,** one with a convex reflecting surface. **m., frontal,** a head-mirror. **m., head-,** a circular mirror with a central perforation, strapped to the head by a band, and used to throw light on parts to be examined. **m., laryngoscopic,** one used in examining the larynx. **m., ophthalmoscopic,** one used in ophthalmoscopy. **m., plane,** one the reflecting surface of which is flat. **m., rhinoscopic,** a mirror used in rhinoscopy. **m.-speech,** defective speech from pronouncing the words or syllables backward. **m.-writing,** a peculiar form

of writing at times observed in left-handed persons and in cases of aphasia, and characterized by a reversal of the form and arrangement of the letters, which appear as if seen in a mirror.

**miryachit** (*me-re-ash'-it*) [Russian]. A peculiar disease in which the patient mimics or imitates everything said or done by another. Cf. *jumpers; lata*.

**misanthrope** (*mis'-an-thrōp*) [μισεῖν, to hate; ἄνθρωπος, man]. A melancholy person; one who has an aversion to society.

**misanthropy** (*mis-an'-thro-pe*) [see *misanthrope*]. Aversion to human society; a symptom not rare in melancholia.

**miscarriage** (*mis-kar'-āj*). 1. The expulsion of the fetus between the fourth and the sixth month of pregnancy. 2. Abortion.

**miscarry** (*mis-kar'-e*). To give birth to a non-viable fetus.

**misce** (*mis'-e*) [L.]. Mix; a direction placed on prescriptions, and usually abbreviated M.

**miscegenation** (*mis-ej-en-a'-shun*) [*miscere*, to mix; *genus*, race]. Mixture of different races by intermarriage.

**miscible** (*mis'-ib-l*) [*misce*]. Capable of being mixed.

**miserere mei** (*miz-er-e'-re me'-i*) [L., "have mercy on me"]. An old name for volvulus or intestinal colic; also for stercoraceous vomiting.

**misocainia** (*mi-so-ki'-ne-ah*) [μισεῖν, to hate; καινός, new]. Same as misoneism.

**misogamy** (*mis-og'-am-e*) [μισεῖν, to hate; γάμος, marriage]. Aversion to marriage.

**misogyny** (*mis-oj'-in-e*) [μισεῖν, to hate; γυνή, woman]. Hatred of women.

**misologia** (*mis-o-lo'-je-ah*) [μισεῖν, to hate; λόγος, reason]. Unreasoning aversion to intellectual or literary matters.

**misoneism** (*mis-on-e'-izm*) [μισεῖν, to hate; νέος, new]. Fear or horror of novelty.

**misoneist** (*mis-on'-e-ist*) [μισεῖν, to hate; νέος, new]. One who has a morbid hatred of novelty.

**misopedia** (*mis-o-pe'-de-ah*) [μισεῖν, to hate; παῖς, child]. Morbid hatred of children, especially of one's own children.

**misopsychia** (*mis-op-si'-ke-ah*) [μισεῖν, to hate; ψυχή, life]. Morbid disgust with life.

**missed** (*mist*) [*miss*]. Passed; failed of completion. **m. abortion**, the retention of the product of conception in the uterine cavity after its death and with the appearance of some of the symptoms of abortion. **m. labor**, the retention of the product of conception in the uterus beyond term, and after the occurrence of a few ineffectual labor-pains.

**missio** (*mis'-e-o*) [L.]. A letting go. **m. sanguinis**, blood-letting.

**mist**. Abbreviation for *mistura*, mixture.

**mistletoe** (*mis'-l-tō*). See *viscum*.

**mistura** (*mis-tū'-rah*) [L.]. 1. A mixture. A preparation made by suspending an insoluble substance in watery fluids, by means of gum-arabic, sugar, yolk of egg, or other cohesive substance. When the suspended substance is of an oily nature, the preparation is termed an emulsion (*emulsum*). **m. amygdalae** (*emulsum amygdalae*, U. S. P.), emulsion of almonds. Dose 4–8 oz. (120–240 Cc.) several times daily. See also under *amygdala*. **m. creosoti** (B. P.). See *creosote mixture*. Dose (U. S. P., B. P.), chalk mixture. Dose ½ oz. (16 Cc.). **m. ferri aromatica** (B. P.), aromatic iron mixture. Dose 1–2 oz. (32–64 Cc.). **m. ferri composita** (U. S. P., B. P.), compound iron mixture; Griffith's mixture. Dose 1–2 oz. (32–64 Cc.). **m. glycyrrhizae composita** (U. S. P.), compound mixture of glycyrrhiza or brown mixture. Dose 1–2 oz. (32–64 Cc.). **m. guaiaci** (B. P.), guaiacum mixture. Dose ½–2 oz. (16–64 Cc.). **m. olei ricini** (B. P.), castor-oil mixture. Dose ½–2 oz. (16–64 Cc.). **m. rhei et sodae** (U. S. P.), mixture of rhubarb and soda. Dose, for children, ½–1 dr. (2–4 Cc.). **m. scammonii** (B. P.), mixture of scammony. Dose 2 oz. (64 Cc.). **m. sennae composita** (B. P.), compound mixture of senna. Dose 1–1½ oz. (32–48 Cc.). **m. spiritus vini gallici** (B. P.), mixture of brandy. Dose 1–2 oz. (32–64 Cc.).

**Mitchell's (Weir) disease** [Silas Weir *Mitchell*, American neurologist, 1830–1914]. Erythromelalgia. **M.'s treatment**, the rest-cure; a treatment for certain functional nervous conditions, consisting in absolute rest in bed, with massage, electricity, and the administration of abundant food, especially milk.

**mitchella** (*mit-tshel'-lah*) [John *Mitchell*, American botanist]. The *Mitchella repens*, used as a uterine tonic and as an aid to easy labor; dose 1 minim (0.6 Cc.).

**mite** (*mit*) [AS., *mite*]. A name applied to several *Acari*.

**mithridate** (*mith'-rid-āt*) [see *mithridatism*]. An old confection believed to contain an antidote to every known poison.

**mithridatism** (*mith-rid'-āt-izm*) [Μιθριδάτης, king of Pontus, who was said to have become so charged with the poisons with which he experimented that he acquired an immunity to them all]. Immunity from the effects of a poison induced by the administration of gradually increased doses.

**mitigate** (*mit'-ig-āt*) [*mitigare*, to soften]. To allay; to make milder; to moderate.

**mitigated caustic**, m. stick (*mit'-ig-a-ted*). See *argenti nitras mitigatus* under *argentum*.

**mitochondria** (*mi-to-kon'-dre-ah*). Protoplasmic granules seen in animal cells; also called *cystomicrosomes*.

**mitochysis** (*mi-tok'-is-is*) [μίτος, thread; χύσις, liquefaction]. Cell-multiplication, direct or by mitosis.

**mitoma**, **mitome** (*mi-to'-mah*, *mi'-tōm*) [μίτος, a thread]. The threads of the protoplasmic reticulum of a cell (*cytomitome*) or of the nucleus (*karyomitome*).

**mitoplasm** (*mi'-to-plazm*) [μίτος, a thread; πλάσσειν, to form]. The reticular part of the cell-nucleus, the chromatic substance or chromatin.

**mitoschisis** (*mit-os'-kis-is*). See *karyokinesis*.

**mitosis** (*mi-to'-sis*) [μίτος, a thread]. Karyokinesis. **m. heterotypic**, mitosis in which the chromosomes take the form of loops, rings, aggregations of four beads, etc., arranged longitudinally upon the spindle. **m. homeotypic**, that characterized by the reduced number of the chromosomes. **m. pathological**, irregular, atypical, asymmetric mitosis, an indication of malignancy.

**mitosome** (*mi'-to-sōm*) [μίτος, thread; σῶμα, body]. A body derived from the spindle-fibers of the secondary spermatocytes, which, according to Platner, gives rise to the middle piece and the flagellum envelope of the semen-cell.

**mitotic** (*mi-tot'-ik*) [*mitosis*]. Pertaining to mitosis.

**mitral** (*mi'-tral*) [μίτρα, a belt; a turban]. 1. Resembling a miter, as the *mitral* valve. 2. Pertaining to the auriculoventricular valve of the left side of the heart. **m. disease**, disease of the mitral valve of the heart. **m. incompetence**, **m. insufficiency**. See *m. regurgitation*. **m. murmur**. See under *murmur, cardiac*. **m. obstruction**, disease of the mitral valve causing obstruction to the flow of blood through the left auriculoventricular opening. **m. regurgitation**, imperfect closure of the mitral valve during the cardiac systole, permitting blood to be forced back into the left auricle. **m. stenosis**. See *m. obstruction*. **m. valve**, mitral.

**mixoscopia** (*miks-o-sko'-pe-ah*) [μῖξις, cohabitation; σκοπεῖν, to look]. A form of sexual perversion in which the orgasm is excited by the sight of coitus. See *voyeur*.

**mixoscopic** (*miks-o-sko'-pik*). 1. Relating to mixoscopia. 2. A sexual pervert exhibiting mixoscopia.

**mixture** (*miks'-tūr*). See *mistura*.

**M. K. O. C. P.** Abbreviation of Member of the King and Queen's College of Physicians of Ireland.

**℥m. l. d.** Abbreviation for minimum lethal dose.

**mm.** An abbreviation for millimeter and for minims.

**mmm.** Abbreviation for *micromillimeter* or *micron*.

**μ** (mu) is also used as an abbreviation for these two words.

**Mn.** Chemical symbol of *manganese*.

**mnemasthenia** (*mem-as-the'-ne-ah*) [μνήμη, memory; ἀσθένεια, weakness]. Weakness of memory not due to organic disease.

**mnemonics** (*ne-mon'-iks*) [μνήμων, mindful]. The science of cultivation of the memory by systematic methods.

**Mo.** Chemical symbol of *molybdenum*.

**moan** (*mōn*). 1. To utter a low, dull sound expressive of suffering. 2. The sound so uttered.

**mobile** (*mo'-bil*) [*movere*, to move]. Movable. **m. pain**, one that shifts from place to place. **m. spasm**, a slow, irregular movement gradually taking

place in different muscles, occurring at times in the paralyzed parts in hemiplegia.
**mobility** (*mo-bil'-it-e*) [*mobile*]. The condition of being movable.
**mobilization** (*mob-il-iz-a'-shun*) [*mobile*]. The act of rendering an ankylosed part movable.
**mochras, mochurrus,** (*mo'-kras, mo-kur'-us*). A gummy exudation from an Indian cotton tree, *Bombax malabaricum*; it is used as an astringent in diarrhea.
**modal** (*mo'-dal*) [*modus*]. A term applied to the order of response of muscles and nerves to the galvanic current.
**modality** (*mo-dal'-it-e*) [*modus*, a mode]. Any condition which modifies the action of a drug.
**moderator band.** See *Reil's band*.
**modioliform** (*mod-e-o'-le-form*) [*modiolus; forma,* form]. Having the shape of the nave of a wheel.
**modiolus** (*mo-di'-o-lus*) [L., "nave"]. 1. The central pillar or axis of the cochlea, around which the spiral canal makes two and one-half turns. 2. The crown of a trephine.
**modus** (*mo'-dus*) [L.]. A mode or method. **m. operandi,** the method of the performance of an action.
**Moebius' disease** (*me'-be-oos*) [Paul Julius *Moebius*, German physician, 1853–1907]. Periodic or recurrent paralysis of the motor oculi. **M.'s sign,** inability to retain the eyeballs in convergence in exophthalmic goiter.
**Moeller's disease** (*me'-ler*) [—*Moeller*, German surgeon, 1829–1862]. See *Barlow's disease*. **M.'s glossitis,** painful swelling of the papillæ of the tongue, associated with nervous irritability; there is imperfect covering of the filiform and fungiform papillæ.
**mogigraphia** (*mog-ig-ra'-fe-ah*) [μόγις, with difficulty; γράφειν, to write]. Writer's cramp.
**mogilalia** (*mog-il-a'-le-ah*) [μόγις, with difficulty; λαλιά, talk]. Stammering; stuttering.
**mogiphonia** (*mog-if-o'-ne-ah*) [μόγις, with difficulty; φωνή, sound]. Difficulty in speaking, excited by an effort of singing or speaking loudly.
**mogostocia** (*mog-os-to'-se-ah*) [μόγις, with difficulty; τόκος, birth]. Painful or difficult parturition.
**M.O.H.** Abbreviation for *Medical Officer of Health*.
**Mohr's test for hydrochloric acid in the contents of the stomach** [Francis *Mohr*, American chemist]. Dilute to a light yellow color a solution of iron acetate (free from alkali acetates), and treat with a few drops of a solution of potassium sulphocyanide. No change of color should take place, but if the filtered contents of the stomach are added and contain HCl, a red coloration results. This color vanishes if sodium acetate is added.
**Mohrenheim's fossa** (*mo'-ren-hīm*) [Joseph Jacob Freiherr von *Mohrenheim*, Austrian surgeon, 1799– ]. The infraclavicular fossa. It is bounded by the clavicle, pectoralis major,-deltoid, and, laterally and deeper, by the pectoralis minor. In this space the subclavian artery is found when it is to be ligated below the clavicle.
**moist** [*mustus,* sweet, like new wine]. Damp; slightly wet; characterized by the presence of fluid. **m. chamber,** a large circular glass with a lid, used in bacteriological work, especially for growing potato-cultures. **m. filter,** a filter-paper that is moistened with water. **m. gangrene,** the form of gangrene that occurs in a part filled with blood. See *gangrene, moist*. **m. rale.** See *rale.*
**mol, mole** (*mol, mōl*). See *grammolecule.*
**molar** (*mo'-lar*) [*moles* , mass]. 1. Pertaining to masses, in contradistinction to molecular. 2. Pertaining to a mole. 3. [*mola,* a millstone.] Grinding; used for grinding. 4. A grinding tooth, a grinder. **m. death,** necrosis or gangrene. **m. pregnancy,** gestation in which a mole is formed. **m. teeth,** the back, grinding teeth.
**molariform** (*mo-lar'-if-orm*) [*mola,* millstone; *forma,* form], Having the form of a molar tooth.
**molasses** (*mo-las'-ez*) [*mellaceus*, made with honey]. The syrupy liquid remaining after the refining of sugar. It contains a considerable quantity of uncrystallizable sugar, some cane-sugar, and coloring-matter. There are two kinds: *West India molasses,* from which rum is prepared, and *sugar-house molasses,* which is somewhat thicker than the first.. Molasses is used for making pills, and combined with sulphur as a domestic remedy for constipation.
**mold** (*mōld*) [AS., *molde,* dust]. 1. A variously colored deposit produced by the growth of different forms of fungi on moist surfaces. The principal molds are *Penicillium,* the *Mucorini,* and *Aspergillus.* 2. A cast; shape. 3. To make or conform to a given shape.
**mole** (*mōl*) [*moles,* a mass]. 1. A mass formed in the uterus by an ovum, the growth of which has become arrested or which has undergone degeneration. 2. Neyus. **m., blood-,** a mass of coagulated blood and retained fetal membranes and placenta, sometimes found in the uterus-after an abortion. **m., carneous.** See *m., fleshy.* **m., cystic.** Same as *m., hydatid.* **m., false,** one not containing any tissues derived from the ovum. **m., fleshy,** a bloodmole which has become more solid and has assumed a fleshy appearance; the body formed in the uterine cavity when an ovum that has died is retained within the uterus for some time. **m., hydatid, m., hydatidiform,** one formed by a proliferation and cystic degeneration of the chorionic villi; it is a form of myxoma, and has a tendency to involve the uterine wall. **m., true,** one which is the remains of an ovum. **m., vesicular.** Same as *m., hydatid.*
**molecular** (*mo-lek'-ū-lar*) [*molecule*]. Pertaining to, or composed of, molecules. **m. death,** death of a part in minute invisible particles, as ulceration, caries. **m. force,** a force acting between molecules, as cohesion. **m. heat.** See *heat.* **m. layer.** Any layer appearing to consist of minute granules without definite structure. 2. The second layer of the cerebral cortex. **m. lesion,** a very fine lesion. **m. motion,** the movements of the molecules of a substance. **m. volume,** the volume of a molecule of a substance in the gaseous state; under the same conditions of temperature and pressure the molecular volumes of all substances are equal. **m. weight,** the weight of a molecule of any substance as compared with the weight of an atom of hydrogen. It is equal to the sum of the weights of its constituent atoms.
**molecule** (*mol'-e-kūl*) [dim. of *moles,* mass]. 1. A minute portion of matter. 2. In physics and chemistry, the smallest quantity into which a substance can be divided and retain its characteristic properties; or the smallest quantity that can exist in a free state.
**molilalia** (*mol-il-a'-le-ah*) [μόλις, with toil; λαλία, speech]. Difficulty of utterance; stuttering; mogilalia.
**molimen** (*mo-li'-men*)[L.,"endeavor"; *pl., molimina*]. An effort or attempt. **m., menstrual,** any of the symptoms attendant upon the menstrual, act or function.
**Molisch's test for glucose** (*mol'-ish*) [Hans *Molisch,* Austrian chemist, 1856– ]. 1. To 5–11 Cc. of the solution add 2 drops of a 15 to 20 % alcoholic solution of naphthol. A precipitation of some of the naphthol renders the liquid cloudy, but on the addition of 1 or 2 Cc. of concentrated sulphuric acid a deep violet coloration is produced and a violet precipitate is deposited on diluting with water. 2. A 15 to 20 % solution of thymol employed instead of naphthol, applied as is the naphthol in the foregoing test. If glucose is present, it produces a ruby-red coloration, which changes to carmine on dilution with water.
**Moll's glands** [Jacobus Antonius *Moll*, Dutch physician, 1849– ]. Modified sudoriparous glands of the eyelids, opening into the follicles of the eyelashes.
**mollescence** (*mol-es'-ens*) [*mollescere,* to soften]. Softening.
**mollichthyolin** (*mol-ik-thi'-ol-in*). A compound of ichthyol and mollin.
**mollin** (*mol'-in*) [*mollis,* soft]. A soft soap of potassium hydroxide and cocoanut-oil, used as a basis for ointments.
**mollities** (*mo-lish'-e-ēz*) [*mollis,* soft]. Softness. **m. ossium.** See *osteomalacia.*
**mollosin** (*mol'-os-in*). An ointment-base consisting of yellow wax, 1 part, and liquid petrolatum, 4 parts.
**molluscous** (*mol-us'-kus*) [*molluscum*]. 1. Pertaining to the *mollusca.* 2. Pertaining to the disease molluscum.
**molluscum** (*mol-us'-kum*) [*mollusca,* shell-fish, from *mollis,* soft]. 1. A term applied to several diseases of the skin. 2. A chronic skin disease with pulpy tumors. **m.-bodies,** the products of degenerative processes occurring in the epidermic epithelial cells. **m. contagiosum,** a disease of the skin charac-

terized by the formation of pinhead-sized to pea-sized, rounded, sessile or pedunculated, pearl-like elevations of a yellowish-white or pinkish color. The lesions may be single or multiple, are usually situated upon the face, and are due to a hyperplasia of the rete mucosum, the growth probably beginning in the hair-follicles. The lesions on microscopic examination are found to contain peculiar ovoid, sharply defined bodies—*molluscum bodies*—which are by some considered as forms of epithelial degeneration, by others as protozoan parasites. **m. epitheliale,** a skin disease with, hard, round nodules containing semiliquid material. **m. fibrosum, m. simplex,** a disease of the skin characterized by the formation of multiple fibromata, which may be sessile or pedunculated, and grow from the deeper layers of the corium and the subcutaneous tissue.

**molops** (*mo'-lops*) [μώλωψ, wheal; *pl., molopes*]. A red spot on the skin, such as is seen in certain fevers. The mark of a stroke or stripe upon the skin.

**molt, moult** (*mōlt*). To shed or cast, as the skin, feathers, or hair.

**molybdamaurosis** (*mol-ib-dam-aw-ro'-sis*) [μόλυβδος, lead; ἀμαυροειν, to darken]. Retro-bulbar neuritis due to lead-poisoning.

**molybdamblyopia** (*mol-ib-dam-ble-o'-pe-ah*) [μόλυβδος, lead; ἀμβλύς, dulled; ὤψ, eye]. Impairment of vision due to lead-poisoning.

**molybdate** (*mol-ib'-dāt*). A salt of molybdic acid.

**molybdencephalia** (*mol-ib-den-sef-a'-le-ah*) [μόλυβδος, lead; ἐγκέφαλος, brain]. Brain-disease due to lead-poisoning.

**molybdencephalopathia, molybdencephalopathy** (*mol-ib-den-sef-al-o-path'-e-ah*, *mol-ib-den-sef-al-op'-ath-e*) [μόλυβδος, lead; ἐγκέφαλος, brain; πάθος, disease]. Brain-disease due to lead-poisoning.

**molybdenum** (*mol-ib-de'-num*) [μόλυβδος, lead]. A metallic element, found in nature chiefly as the sulphide—*molybdenite* (MoS₂). Atomic weight 96; symbol Mo; quantivalence II, IV, VI, VIII; specific gravity 8.6. Its principal oxide, MoO₃, forms *molybdic acid,* H₂MoO₄, the ammonium salt of which is used as a reagent in metallurgy, etc. Molybdic acid combines with phosphoric acid to form phosphomolybdic acid.

**molybdepilepsia** (*mol-ib-dep-il-ep'-se-ah*) [μόλυβδος, lead; *epilepsy*]. Epilepsy induced by lead-poisoning.

**molybdic** (*mol-ib'-dik*). Containing molybdenum as a hexad or tetrad radical. **m. anhydride,** MoO₃, a gray or bluish-white heavy powder which separates into thin scales in water. It is soluble in acids, alkalies, and solution of cream of tartar; slightly soluble in water. It is used as a reagent.

**molybdo-** (*mol-ib'-do-*) [μόλυβδος, lead]. A prefix denoting reference to or connection with lead. In chemistry, signifying composition with the element molybdenum.

**molybdocachexia** (*mol-ib-do-kak-ek'-se-ah*) [*molybdo-; cachexia*]. A depraved state of the body due to lead-poisoning; chronic lead-poisoning; plumbism.

**molybdocardialgia** (*mol-ib-do-kar-de-al'-je-ah*) [*molybdo-*; καρδία, heart; ἄλγος, pain]. Cardialgia due to lead-poisoning.

**molybdocolic** (*mol-ib-do-kol'-ik*) [*molybdo-; colic*]. Lead-colic; painters' colic.

**molybdodyspepsia** (*mol-ib-do-dis-pep'-se-ah*) [*molybdo-; dyspepsia*]. Dyspepsia caused by lead-poisoning.

**molybdonosus** (*mol-ib-don'-o-sus*) [*molybdo-*; νόσος, disease]. Synonym of *plumbism*.

**molybdoparesis** (*mol-ib-do-par'-es-is*) [*molybdo-; paresis*]. Partial paralysis, due to lead-poisoning.

**molybdosis** (*mol-ib-do'-sis*) [μόλυβδος, lead]. Lead-poisoning; plumbism.

**molybdospasmos** (*mol-ib-do-spaz'-mos*) [*molybdo-*; σπάσμος, spasm]. Spasm or cramp produced by lead-poisoning.

**molybdosynolce** (*mol-ib-do-sin-ol'-se*) [*molybdo-*; συνολκή, a drawing together]. Contraction due to lead-poisoning.

**molybdotromos** (*mol-ib-dot'-ro-mos*) [*molybdo-*; τρόμος, tremor]. Tremor due to lead-poisoning.

**molybdous** (*mol-ib'-dus*). Containing molybdenum in its lower valency.

**Momburg's belt** (*mom'-berg*) [Fritz August Momburg, German physician, 1870– ]. A band wound around the waist and then made taut; it is used to check postpartum hemorrhage.

**momentum** (*mo-men'-tum*) [*movere,* to move]. Quantity of motion. The momentum of a body depends upon its mass and velocity. Also, the quantity of potential energy possessed by a body in motion. It is usually expressed by the formula $m = vv—i.\ e.,$ the momentum equals the weight multiplied by the velocity.

**momordica** (*mo-mor'-dik-ah*). See *elaterium*.

**mon-** (*mon-*) [μόνος, one]. A prefix denoting one or single.

**monacid** (*mon-as'-id*). Applied to a base with one replaceable hydroxyl group (OH). Also, compounds uniting directly with a molecule of a monobasic acid, with half a molecule of a dibasic acid; etc.

**monad** (*mon'-ad*) [μόνος, single]. 1. A univalent element or radical. 2. Any single-celled microorganism, whether of animal or vegetable character; especially any flagellate infusorium.

**monadenoma** (*mon-ad-en-o'-mah*) [*mon-*; ἀδήν, gland; ὄμα, tumor]. A uniglandular adenoma.

**Monadina** (*mon-ad-i'-nah*) [μονάς, a unit]. The flagellate infusorians, or *monadida*; a family of animal microorganisms.

**(von) Monakow's fibers** (*mon-ah'-kow*) [Constantin von *Monakow*, Russian neurologist, 1853– ]. A tract of nerve-fibers extending from the anterior corpus quadrigeminum to the eyeball. v. **M.'s nucleus,** the lateral portion of Burdach's nucleus.

**monamide** (*mon'-am-id*). An amide formed by the replacement of the hydrogen in one molecule of ammonia by an acid radical.

**( monamine** (*mon'-am-in*) [μόνος, single; *amin*]. An amine formed by the replacement of the hydrogen in one molecule of ammonia by an alkyl radical.

**Monarda** (*mo-nar'-dah*) [after N. *Monardés,* a Spanish physician]. A genus of labiate plants, comprising *M. didyma,* the bee-balm, *M. fistulosa,* the wild bergamot, and *M. punctata,* the horsemint. The last is diaphoretic, carminative, and stimulant. It is also a source of thymol. **m. fistulosa,** wild bergamot. **m. punctata,** horse mint; it is a diaphoretic and carminative; it yields thymol.

**monargentic** (*mon-ar-jen'-tik*) [μόνος, single; *argentum,* silver]. Containing one atom of silver in a molecule.

**monarthritis** (*mon-ar-thri'-tis*) [*mon-*; ἄρθρον, a joint; *itis,* inflammation]. Arthritis affecting only a single joint.

**monarticular** (*mon-ar-tik'-u-lar*) [μόνος, single; *articulus,* a joint]. Pertaining to one joint.

**Monas** (*mon'-as*) [μονάς, unit]. A genus of infusorians.

**monaster** (*mon-as'-ter*) [μόνος, single; ἀστήρ, a star]. Mother-star. See *karyokinesis*.

**monathetosis** (*mon-ath-et-o'-sis*). Athetosis affecting one limb or side.

**monatomic** (*mon-at-om'-ik*) [μόνος, single; ἄτομος, atom]. 1. Having but one atom of replaceable hydrogen, as a *monatomic* acid. 2. Having only one atom, as a *monatomic* molecule. 3. Having the combining power of one atom of hydrogen, as a *monatomic* radical. 4. Formed by the replacement of one hydrogen atom in a compound by a radical, as a *monatomic* alcohol.

**monaxial, monaxonic** (*mon-aks'-e-al,* *mon-aks-on'-ik*) [*mon-*; ἄξων, axis]. Having a single axis.

**monaxon** (*mon-aks'-on*) [μόνος, single; *axon*]. A neuron having only one axon.

**moner, moneron** (*mo'-ner, mon'-er-on*). In biology, a non-nucleated unicellular organism of the simplest possible character.

**monerula** (*mon-er'-oo-lah*) [μονήρης, single]. The impregnated ovum at a stage when it has no nucleus.

**monesia** (*mon-e'-se-ah*) [origin unknown]. An extract from the Brazilian tree *Chrysophyllum glyciphlæum*. It is astomachic, alterative, and astringent. Dose 0.3–0.6 Gm.).

**monesin** (*mo-ne'-sin*) [*monesia*]. The acrid principle of monesia, said to be identical with saponin; it is astringent and oxytocic.

**mongumo bark** (*mon-gū'-mo*). The bark of *Ochrosia borbonica,* a tree of Madagascar; used as a tonic.

**monilethrix** (*mo-nil'-eth-riks*) [*monile,* a neck-lace; θρίξ, a hair]. An affection of the hair in which nodes are strung regularly or irregularly along the hair-shaft, giving it a beaded appearance.

**moniliform** (*mon-il'-if-orm*) [*monile,* a necklace; *forma,* form]. Shaped like a necklace; beaded or bead-like; resembling a string of beads.

**monilithrix** (*mo-nil'-ith-riks*). See *monilethrix.*

**monium** (*mo'-ne-um*) [μόνος, alone]. An element discovered spectroscopically by Sir W. Crookes in 1898; now called victorium.
**monk's-hood** (*munks'-hood*). See *aconitum*.
**Monneret's pulse** (*mon-rā'*) [Jules Edouard Auguste Monneret, French physician, 1810–1868]. The soft, full, and slow pulse of icterus.
**mono-** (*mon-o-*) [μόνος, one]. A prefix signifying one or single.
**monoanesthesia, monoanæsthesia** (*mon-o-an-es-the'-ze-ah*). Anesthesia of a single part.
**monoarticular** (*mon-o-ar-tik'-ū-lar*). Same as *monarticular, q. v.*
**monoathetosis** (*mon-o-ath-et-o'-sis*) [mono-; ἄθετος, without place]. Athetosis confined to one limb or one-half of the body.
**monobacillary** (*mon-o-bas'-il-a-re*). Due to, or characterized by, the presence of a single species of bacillus.
**monobasic** (*mon-o-ba'-sik*) [mono-; βάσις, foundation]. Of an acid, acid salt, or alcohol, having one replaceable hydrogen atom.
**monoblepsia, monoblepsis** (*mon-o-blep'-se-ah, mon-o-blep'-sis*) [mono-; βλέψις, sight]. 1. A condition in which either eye has a better visual power than both together. 2. The form of color-blindness in which but one color can be perceived.
**monobrachius** (*mon-o-bra'-ke-us*) [mono-; βραχίων, arm]. A monster having but one arm.
**monobromacetanilid** (*mon-o-brom-as-et-an'-il-id*). Same as *antisepsin*.
**monobromated** (*mon-o-bro'-ma-ted*) [mono-; bromate]. Containing one atom of bromine in the molecule.
**monobromide** (*mon-o-bro'-mid*) [mono-; bromide]. A compound having one atom of bromine in the molecule, or containing an amount of bromine which, when compared with the amount of bromine in other bromides of the same base, may be regarded as unity.
**monocalcic** (*mon-o-kal'-sik*). Containing one atom of calcium in a molecule.
**monocardian** (*mon-o-kar'-de-an*) [mono-; καρδία, heart]. Having a single heart, single-chambered or not completely divided, as in vertebrates.
**monocellular** (*mon-o-sel'-ū-lar*). Unicellular.
**monocephalus** (*mon-o-sef'-al-us*) [mono-; κεφαλή, head]. A monster consisting of a single head with two bodies more or less completely fused.
**monochloride** (*mon-o-klo'-rid*). A chlorine compound analogous to a monobromide (*q. v.*).
**monochorea** (*mon-o-ko-re'-ah*) [mono-; chorea]. Chorea confined to a single member or part of the body.
**monochroic** (*mon-o-kro'-ik*) [mono-; χρόα, color]. Having only one color. Arterial blood is monochroic.
**monochromasy** (*mon-o-kro'-mas-e*) [see *monochroic*]. The perception of one color only.
**monochromasis** (*mon-o-kro'-mat*). A person in whom all the variations of the world of color are reduced to a system of one color.
**monochromatic** (*mon-o-kro-mat'-ik*). See *monochroic*.
**monochromatophil** (*mon-o-kro-mat'-o-fil*) [mono-; χρῶμα, color; φιλεῖν, to love]. 1. A cell possessing a strong affinity for a single acid stain. 2. Exhibiting a strong affinity for a single stain.
**monocle** (*mon'-o-kl*) [mono-; oculus, eye]. 1. A lens for one eye only. 2. A bandage for one eye.
**monoclinic** (*mon-o-klin'-ik*) [mono-; κλίνη, bed]. Applied to crystals in which the vertical axis is inclined to one, but is at right angles to the other, lateral axis.
**monococcus** (*mon-o-kok'-us*) [mono-; κόκκος, grain]. A coccus occurring singly, not united in chains or pairs or in groups.
**monocranus** (*mon-ok'-ran-us*) [mono-; κρανίον, cranium]. A double monster having a single cranium.
**monocrotic** (*mon-o-krot'-ik*) [mono-; κρότος, beat]. Having but a single beat (as the normal pulse) for each cardiac systole; not dicrotic.
**monocrotism** (*mon-ok'-ro-tizm*). The condition of being monocrotic (*q. v.*).
**monocular** (*mon-ok'-ū-lar*) [mono-; oculus, eye]. 1. Pertaining to or affecting only one eye, as *monocular diplopia*; performed with one eye only, as *monocular vision*. 2. Having a single ocular or eyepiece, as a *monocular* microscope.
**monoculus** (*mon-ok'-ū-lus*) [mon-; oculus, eye].

1. A monster with but one eye. 2. In surgery, a bandage for covering one eye.
**monocyclic** (*mon-o-sik'-lik*). Arranged in a single whorl.
**monocyst** (*mon'-o-sist*) [mono-; κύστις, cyst]. A tumor made up of a single cyst.
**monocystic** (*mon-o-sis'-tik*) [mono-; κύστις, cyst]. Composed of or containing but one cyst.
**monocyte** (*mon'-o-sit*) [mono-; κυτός, a cell]. A large mononuclear leucocyte.
**monodactyiism** (*mon-o-dak'-til-izm*) [mono-; δάκτυλος, finger]. A malformation characterized by the presence of only one toe or finger on the foot or hand.
**monodactylous** (*mon-o-dak'-til-us*) [mono-; δάκτυλος, finger]. Having only one finger or toe.
**monoderic** (*mon-o-der'-ik*) [mono-; δέρος, skin]. Composed of a single layer.
**monodidymus** (*mon-o-did'-im-us*) [mono-; δίδυμος, twin]. One of twins.
**monodiplopia** (*mon-o-dip-lo'-pe-ah*) [mono-; διπλόος, double; ὄψις, sight]. Double vision with a single eye.
**monodont** (*mon'-o-dont*) [mono-; ὀδούς, tooth]. Having but one tooth.
**monogastric** (*mon-o-gas'-trik*) [mono-; γαστήρ, belly]. Having one stomach or one belly.
**monogenesis** (*mon-o-jen'-es-is*) [mono-; γένεσις, origin]. 1. Development of offspring resembling the parent, as distinguished from metagenesis. 2. Development from a single hermaphroditic parent; asexual reproduction. 3. Origin of all organisms from a single cell.
**monograph** (*mon' o graf*) [mono ; γράφειν, to write]. A treatise or memoir on a single subject.
**monohydrated** (*mon-o-hi'-dra-ted*). United with one molecule of water or of hydroxyl.
**monohydric** (*mon-o-hi'-drik*) [mono-; ὕδωρ, water]. Containing one atom of replaceable hydrogen, as *monohydric* acid, *monohydric* alcohol.
**monoideism** (*mon-o-i-de'-izm*) [mono-; ἰδέα, idea]. The domination of a single idea, as in certain cases of hypnotism and insanity.
**monoinfection** (*mon-o-in-fek'-shun*). Infection with but one kind of microorganism.
**monoiodide** (*mon-o-i'-o-did*). An iodine compound analogous to a monobromide.
**monol** (*mon'-ol*). An aqueous solution of calcium permanganate (2 : 1000) used to purify drinking-water.
**monolocular** (*mon-o-lok'-ū-lar*). See *unilocular*.
**monoma** (*mon-o'-mah*) [mono-; ὄμα, tumor]. A painful uterine tumor, always solitary, steadily progressing to a fatal termination, accompanied by severe and continuous hemorrhage.
**monomania** (*mon-o-ma'-ne-ah*) [mono-; μανία, madness]. A form of insanity characterized by a limited disturbance of the mental functions that dominates the person's thoughts and actions.
**monomaniac** (*mon-o-ma'-ne-ak*) [*monomania*]. A person affected with monomania.
**monomeric** (*mon-o-mer'-ik*) [mono-; μέρος, a part]. Consisting of a single piece.
**monometallic** (*mon-o-met-al'-ik*) [mono-; *metallic*]. 1. Containing one atom of a metal in a molecule. 2. Capable of replacing one atom of hydrogen in an acid. 3. Consisting of one metal.
**monomicrobic** (*mon-o-mi-kro'-bik*). See *monobacillary*.
**monommatous** (*mon-om'-at-us*) [mono-; ὄμμα, eye]. One-eyed.
**monomoria** (*mon-o-mo'-re-ah*) [mono-; μωρία, folly]. Melancholy.
**monomorphic** (*mon-o-mor'-fik*) [mono-; μορφή, form]. Having or existing in only one form.
**monomorphism** (*mon-o-mor'-fizm*) [mono-; μορφή, form]. The state of being monomorphic.
**monomorphous** (*mon-o-mor'-fus*) [mono-; μορφή, form]. Having but a single form; not polymorphous.
**monomphalus** (*mon-om'-fal-us*) [mono-; ὀμφαλός, navel]. A double monster united by a common umbilicus.
**monomyositis** (*mon-o-mi-o-si'-tis*) [mono-; μῦς, muscle]. Laquer's (1890) name for isolated periodic affections of the biceps muscle. Syn., *myositis acuta interstitialis*.
**monomyous** (*mon-o-mi'-us*) [mono-; μῦς, muscle]. Having only one muscle; applied to certain bivalves, the shells of which are closed by a single muscle.
**mononephrous** (*mon-o-nef'-rus*) [mono-; νεφρός, kidney]. Limited to one kidney.
**mononeuric** (*mon-o-nū'-rik*) [mono-; νεῦρον, nerve]. Applied to a nerve-cell having only one neuraxon

**mononeuritis** (*mon-o-nū-ri'-tis*) [*mono-*; *neuritis*]. Neuritis affecting a single nerve. m., **multiplex**, neuritis affecting simultaneously single nerves remote from each other.
**mononuclear** (*mon-o-nū'-kle-ar*). See *uninuclear*.
**mononym** (*mon'-o-nim*) [*mono-*; ὄνομα, name]. A name consisting of but a single word. Thus *callosum* is a mononym for *corpus callosum*; *pia*, for *pia mater*; *myel*, for *spinal cord*.
**mononymic** (*mon-o-nim'-ik*) [*mono-*; ὄνομα, name]. Having but one name.
**monopagia** (*mon-o-pa'-je-ah*) [*mono-*; πάγιος, fixed]. Fixed local pain in the head; clavus hystericus.
**monoparesis** (*mon-o-par'-e-sis*) [*mono-*; *paresis*]. Paralysis of a single part of the body, as of one limb.
**monoparesthesia** (*mon-o-par-es-the'-ze-ah*) [*mono-*; *paresthesia*]. Paresthesia confined to one limb or part.
**monopathy** (*mon-op'-ath-e*) [*mono-*; πάθος, disease]. Uncomplicated disease of a single organ.
**monophagia** (*mon-o-fa'-je-ah*) [*mono-*; φαγεῖν, to eat]. 1. Desire for a single article of food. 2. The eating of a single daily meal.
**monophasia** (*mon-o-fa'-ze-ah*) [*mono-*; *aphasia*]. A form of aphasia in which speech is limited to a single syllable, word, or phrase.
**monophobia** (*mon-o-fo'-be-ah*) [*mono-*; φόβος, fear]. Morbid dread of being alone.
**monophosphate** (*mon-o-fos'-fāt*). A phosphate with only one atom of phosphorus in the molecule.
**monophthalmia** (*mon-off-thal'-me-ah*) [*mono-*; ὀφθαλμός, eye]. Synonym of *cyclopia*.
**monophthalmous** (*mon-off-thal'-mos*) [*mono-*; ὀφθαλμός, eye]. Single-eyed; also, pertaining to a bandage for one eye.
**monophyletic** (*mon-o-fi-let'-ik*) [*mono-*; φυλή, tribe]. Derived from a single prototype. m. **hypothesis**, the doctrine of *Haeckel*, that the various organic lines of animals have descended from a common type.
**monoplasmatic** (*mon-o-plaz-mat'-ik*). See *monoplastic*.
**monoplast** (*mon'-o-plast*) [*mono-*; πλάσσειν, to form]. A simple cell.
**monoplastic** (*mon-o-plas'-tik*) [*monoplast*]. Composed of only one substance.
**monoplastid** (*mon-o-plas'-tid*) [*mono-*; πλαστός, molded]. An organism or structural element composed of only a single cell.
**monoplegia** (*mon-o-ple'-je-ah*) [*mono-*; πληγή, stroke]. Paralysis of a single limb or of a single muscle or group of muscles. It is designated as *brachial*, *crural*, or *facial*, when affecting the arm, the leg, or the face, respectively, and as *central* (*cerebral*) or *peripheral*, according to the seat of the causal lesion.
**monops** (*mon'-ops*). See *cyclops*.
**monopsia** (*mon-ops'-e-ah*). See *monophthalmia*.
**monopsychosis** (*mon-o-si-ko'-sis*) [*mono-*; ψύχη, mind: pl., *monopsychoses*]. Any kind of monomania or delusional insanity of fixed type.
**monopus** (*mon'-o-pus*) [*mono-*; πούς, foot]. 1. Congenital absence of one foot or leg. 2. A one-footed monstrosity.
**monoradicular** (*mon-o-rad-ik'-u-lar*) [*mono-*; *radix*, root]. Applied to teeth with only one root.
**monorchid, monorchis** (*mon-or'-kid, mo-nor-kis*) [*mono-*; ὄρχις, testis]. A person who has but one testicle, or in whom one testicle only has descended into the scrotum.
**monorrhinous** (*mon-or-i'-nus*) [*mono-*; ῥίς, nose]. Having a single median nasal cavity.
**monosaccharide** (*mon-o-sak'-ar-id*). Any carbohydrate whose molecule cannot be split into simpler carbohydrates; e. g., glucose, fructose.
**monoscelous** (*mon-os-el'-us*) [*mono-*; σκέλος, leg]. One-legged.
**monose** (*mon'-ōs*). Same as *monosaccharide*.
**monosodic** (*mon-o-so'-dik*). Having one atom of sodium in the molecule.
**monosomus, monosomia** (*mon-o-so'-mus, mon-o-so'-me-ah*) [*mono-*; σῶμα, body]. A double monster with a single body and two heads.
**monospasm** (*mon'-o-spazm*) [*mono-*; σπασμός, spasm]. Spasm affecting limited areas, as one side of the face, a single limb, or a single muscle or muscle-group. It is designated as *brachial*, *crural*, or *facial*, according to the part affected, and as *central* (*cerebral*) or *peripheral*, according to the seat of the causal lesion.
**monostratal** (*mon-o-stra'-tal*) [*mono-*; *stratum*, a layer]. Arranged in a single layer or stratum.

**monosymptomatic** (*mon-o-simp-tom-at'-ik*). Having but one dominant symptom.
**monotal** (*mon'-o-tal*). Trade name of guaiacol methylglycholate; analgesic and antipyretic.
**monotic** (*mon-ot'-ik*) [*mono-*; οὖς, ear]. Pertaining to but one of the ears.
**monotonia** (*mon-o-to'-ne-ah*) [*mono-*; τόνος, tone]. Uniformity of voice; in vocalization, that kind of uniformity that results from paralysis of the laryngeal tensors.
**monotrichous** (*mon-ot'-rik-us*) [*mono-*; θρίξ, hair]. Applied to that type of ciliation in bacteria which is marked by a single flagellum at one pole.
**monovalent** (*mon-ov'-al-ent*). Same as *univalent*.
**monoxenous** (*mon-oks'-en-us*) [*mono-*; ξένος, host]. Applied to parasitism confined to one host.
**monoxide** (*mon-oks'-id*). An oxide containing a single oxygen atom.
**Monro, bursa of** (*mun-ro'*) [Alexander *Monro*, Scottish anatomist, 1697-1767]. A bursa sometimes found between the subclavius muscle and the costoclavicular ligament. **M.,** **foramen of**, an opening behind the anterior pillars of the fornix, through which the lateral ventricle of the brain communicates with the third ventricle; it transmits the choroid plexus. **M.'s line**, a line drawn from the umbilicus to the anterior superior spine of the ilium. **M.'s point**, the point sometimes selected in paracentesis abdominis, midway on Monro's line. **M.'s sulcus**, a longitudinal fissure extending from the foramen of Monro to the Sylvian aqueduct, and dividing each lateral wall of the third ventricle into an upper and a lower portion.
**mons** (*monz*) [L.: pl., *montes*]. In anatomy, the *mons pubis* or the *mons Veneris*. m. **cerebelli**. See *monticulus*. m. **pubis**, the eminence in front of the body and horizontal ramus of the os pubis; it is called also, in the female, *m. Veneris*. m. **Veneris**, the mons pubis of the female.
**Monsel's salt** (*mon-sel'*). Ferric subsulphate. **M.'s solution** (*liquor ferri subsulphatis*, U. S. P.), a solution of ferric subsulphate, used as a styptic.
**monster** (*mon'-ster*) [*monstrum*, an evil omen]. An individual who, by reason of congenital faulty development, is incapable of properly performing the vital functions, or who, owing to an excess or deficiency of parts, differs in a marked degree from the normal type of the species. A teratism.
**monstricide** (*mon'-stris-īd*) [*monstrum*, monster; *cædere*, to kill]. The killing of a monster.
**monstriferous** (*mon-strif'-er-us*) [*monstrum*, monster; *ferre*, to bear]. Producing monsters.
**monstrosity** (*mon-stros'-it-e*) [*monster*]. 1. The condition of a monster. 2. A monster.

## TABLE OF MONSTROSITIES.*

According to **Geoffroy Saint-Hilaire**. Altered by **Hirst** and **Piersol**.

### HEMITERATA.

**I. ANOMALIES OF VOLUME.**

A. Of Stature.
1. *General Diminution*, as in a dwarf—delayed growth.
2. *General Increase*, as in a giant—precocious development.

B. Of Volume, strictly speaking.
1. *Local Diminution*. Affecting—
  (a) *Regions*, as a limb.
  (b) *Systems*, as undeveloped muscles.
  (c) *Organs*, as small breasts, stenosis of canals, etc.
2. *Local Increase*. Affecting—
  (a) *Regions*, as the head.
  (b) *Systems*, as the adipose tissue.
  (c) *Organs*, as large breasts in women, lactiferous breasts in men.

**II. ANOMALIES OF FORM.** *Single Order*, including—deformed heads; anomalies of shape in the stomach; deformed pelves, etc.

---

* Reproduced, with the kind consent of the publishers, from "Human Monstrosities," by Barton Cooke Hirst, M.D., and George A. Piersol, M.D., Philadelphia: Lea Brothers & Co., 1892.

## III. ANOMALIES OF COLOR.
A. *Deficiency*, complete, partial, or imperfect, as in albinism.
B. *Excess*, complete, partial, or imperfect, as in melanism.
C. *Alteration*, as in unusual color of the iris.

## IV. ANOMALIES OF STRUCTURE.
A. *Deficiency in Consistency*, as cartilaginous conditions of bones.
B. *Excess in Consistency*, as anomalous ossification.

## V. ANOMALIES OF DISPOSITION.
A. By Displacement.
  1. *Of the splanchnic organs*, as anomalous direction of heart or stomach, hernias, exstrophy of the bladder, etc.
  2. *Of the nonsplanchnic organs*; as club-foot, curvature of the spine, misplaced teeth, misplaced blood-vessels, etc.
B. By Change of Connection.
  1. *Anomalous articulations*.
  2. *Anomalous implantations*, as teeth out of line.
  3. *Anomalous attachments*, as of muscles and ligaments.
  4. *Anomalous branches*, as of arteries and nerves.
  5. *Anomalous openings*, as of veins into the left auricle, of the ductus choledochus in an unusual situation, of the vagina into the rectum, of the rectum into the male urethra, of the rectum at the umbilicus, etc.
C. In Continuity.
  1. *Anomalous imperforations*, as of rectum, vulva, vagina, mouth, esophagus.
  2. *Anomalous union of organs*, as of kidneys, testicles, digits, teeth, ribs; adhesion of the tongue to the palate.
D. By Closure, as in complete transverse septum in the vagina.
E. By Disjunction.
  1. *Anomalous perforations*, as persistence of foramen ovale, ductus arteriosus, urachus.
  2. *Anomalous divisions*, as splits, fissures in various organs, harelip, hypospadia, fissured tongue, cleft palate, fissured cheek.

## VI. ANOMALIES OF NUMBER AND EXISTENCE.
  1. *By numerical defect*, as absence of muscles, vertebræ, ribs, digits, teeth, a lung, a kidney, the uterus, the bladder, etc.
  2. *By numerical excess*, as supernumerary digits, ribs, teeth, breasts, a double uterus.

### HETEROTAXIS.
I. Splanchnic Inversion.
II. General Inversion.

### HERMAPHRODITES.
I. True Hermaphrodites.
  (a) *Bilateral hermaphrodites*.
  (b) *Unilateral hermaphrodites*.
  (c) *Lateral hermaphrodites*.
II. Pseudohermaphrodites, with double sexual formation of the external genitals, but with unisexual development of the reproductive glands (ovaries and testicles).
  (a) *Male pseudohermaphrodites*, with testicles.
    1. *Internal pseudohermaphrodites*: Development of uterus masculinus.
    2. *External pseudohermaphrodites*: External genitals approach the female type; the monstrosity presents a feminine appearance and build.
    3. *Complete pseudohermaphrodites* (internal and external): Uterus masculinus with tubes; separate efferent canals for bladder and uterus.
  (b) *Female pseudohermaphrodites*, with ovaries: Persistence of male sexual parts.
    1. *Internal hermaphrodites*: Formation of vas deferens and tubes.
    2. *External hermaphrodites*: Approach of the external genitals to the male type.
    3. *Complete hermaphrodites* (internal and external): Masculine formation of the external genitals and of a part of the sexual tract.

### MONSTERS.

#### CLASS I.—SINGLE MONSTERS.

Order I.—Autositic Monsters.

Genus I ... Species 1. *Ectromelus* ...
- Phocomelus.
- Hemimelus.
- Micromelus.
- Ectromelus.

Species 2 ......
- Symelus.
- Uromelus.
- Sirenomelus.

Genus II ... Single species. *Celosoma* ...
- Aspalasoma.
- Agenosoma.
- Cyllosoma.
- Schistosoma.
- Pleurosoma.
- Celosoma.

Genus III ... Species 1. *Exencephalus* ..
- Notencephalus.
- Proencephalus.
- Podencephalus.
- Hyperencephalus.
- Iniencephalus.
- Exencephalus.

Species 2. *Pseudencephalus* ........
- Nosencephalus.
- Thlipsencephalus.
- Pseudencephalus.

Species 3. *Anencephalus* ..
- Derencephalus.
- Anencephalus.

Genus IV .. Species 1. *Cyclocephalus* ...
- Ethmocephalus.
- Cebocephalus.
- Rhinocephalus.
- Cyclocephalus.
- Stomocephalus.

Species 2. *Otocephalus* ....
- Sphenocephalus.
- Otocephalus.
- Edocephalus.
- Opocephalus.
- Triocephalus.

Order II.—Omphalositic Monsters.

Genus I ... Species 1. *Paracephalus* ...
- Paracephalus.
- Omacephalus.
- Hemiacephalus.

Species 2. *Acephalus* ...
- Acephalus.
- Peracephalus.
- Mylacephalus.

Species 3. *Asomat.*

Genus II, Single Species, *Anideus*.

#### CLASS II.—COMPOSITE MONSTERS.

Order I.—Double Autositic Monsters.
A. *Terata katadidyma*.
  Genus I, Diprosopus.
  Genus II, Dicephalus.
  Genus III, Ischiopagus.
  Genus IV, Pygopagus.
B. *Terata anadidyma*.
  Genus I, Dipygus.
  Genus II, Syncephalus.
  Genus III, Craniopagus.
C. *Terata anakatadidyma*.
  Genus I, Prosopothoracopagus.
  Genus II, Omphalopagus.
  Genus III, Rachipagus.

Order II.—Double Parasitic Monsters.

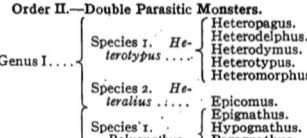

Genus I.... Species 1. Heterotypus....
- Heteropagus.
- Heterodelphus.
- Heterodymus.
- Heterotypus.
- Heteromorphus.

Species 2. Heteralius.....
- Epicomus.
- Epignathus.

Species 1. Polygnathus.
- Hypognathus.
- Paragnathus.
- Augnathus.

Genus II... Species 2. Polymelus...
- Pygomelus.
- Gastromelus.
- Notomelus.
- Cephalomelus.
- Melomelus.

Genus III.. { Endocyma..... 
- Dermocyma.
- Endocyma.

Order III.—Triple Monsters.

**monstrum** (*mon'-strum*). See *monstrosity.*
**Monteggia's dislocation** (*mon-tej'-e-ah*) [Giovanni Batista *Monteggia*, Italian surgeon, 1762–1815]. A form of dislocation of the hip-joint in which the head of the femur is near the anterior superior iliac spine, and the limb is rotated outward.
**Monteverde's sign or test.** The injection subcutaneously during life of a little ammonia solution will be followed by a port-wine congestion in surrounding parts, which does not take place in case of death.
**Montgomery's cups** (*mont-gom'-er-e*) [William Fetherston *Montgomery*, Irish physician, 1797–1859]. The enlarged epithelial depressions in the mucosa of the uterus. M.'s glands, M.'s tubercles, the sebaceous glands of the areola of the nipple appearing as small prominences, especially during pregnancy and lactation; they communicate occasionally with aberrant galactophorous glands.
**monthlies** (*munth'-lēz*). The menses.
**monthly courses, m. sickness.** The menses.
**m. nurse,** a nurse who attends a woman after childbirth.
**monticle** (*mon'-tik-l*). See *monticulus cerebelli.*
**monticulus** (*mon-tik'-ū-lus*) [L.]. A small elevation. m. cerebelli, the prominent central portion of the superior vermiform process of the cerebellum. m. Veneris. See *mons Veneris.*
**Moon's molars** [Henry *Moon*, English surgeon]. The first molar teeth, in congenital syphilis, are reduced in size and are dome shaped, owing to the dwarfing of the central tubercle of each cusp.
**moon-blindness.** Amblyopia from exposure of the eyes to moonlight during sleep.
**Moore's fracture** [Edward Mott *Moore*, American surgeon, 1814–1902]. Fracture of the lower end of the radius with dislocation of the ulna, the styloid process being tied down by the annular ligament.
**Moore's synapsis.** The tangled skein of chromatin at one side of the nucleus formed in the prophase in heterotypical division of sex-cells.
**Moore's test for glucose.** Treat the solution with one-fourth of its volume of sodium or potassium hydroxide and warm it; it will become first yellow, then orange, and finally brown, depending upon the amount of glucose present.
**Mooren's ulcer** (*moo-ren'*) [Albert *Mooren*, German oculist, 1828–1899]. A chronic serpiginous ulcer of the cornea occurring in elderly people.
**moradeine** ɪ(*mor-ad'e-in*). An alkaloid obtained from the bark of *Pogonopus febrifugus*, of South America.
**Morand's disease** (*mor-on'(g)*) [Sauveur François *Morand*, French surgeon, 1697–1773]. Paresis of the lower extremities. M.'s foot, a deformity of the foot that consists in the presence of eight toes. M.'s foramen, the *foramen, cecal (of tongue), q. v.* M.'s spur, the calcaneum.
**Morax-Axenfeld diplobacillus** (*mor'-aks-aks'-en-felt*) [Victor *Morax*, French physician; Alexander *Axenfeld*, French physician]. A bacillus causing a mild form of conjunctivitis.
**morbi** (*mor'-bi*) [L., genitive of *morbus*]. Of a disease. **agens morbi,** the cause or agent of disease. **ens morbi,** the being or essential quality of disease. **materies morbi,** the substance producing a disease.
**morbid** (*mor'-bid*) [*morbus*, disease]. Pertaining to disease or diseased parts. m. anatomy. See *anatomy, morbid.*
**morbidity** (*mor-bid'-it-e*) [*morbid*]. 1. The quality of disease or of being diseased. 2. The conditions inducing disease. 3. The ratio of the number of sick individuals to the total population of a place.
**morbidize** (*mor'-bid-īz*) [*morbid*]. To render sickly or abnormal.
**morbiferous** (*mor-bif'-er-us*) [*morbus*, disease; *ferre*, to bear]. Conveying or spreading disease.
**morbific** (*mor-bif'-ik*) [*morbus; facere,* to make]. Producing disease.
**morbigenous** (*mor-bij'-en-us*) [*morbus,* disease; γεννᾶν, to produce]. Producing disease.
**morbility** (*mor-bil'-it-e*). Same as *morbidity.*
**morbilli** (*mor-bil'-i*) [*morbus*]. Measles.
**morbilliform** (*mor-bil'-if-orm*) [*morbilli*]. Resembling measles.
**morbillous** (*mor-bil'-us*) [*morbilli,* measles]. Pertaining to measles.
**morbose** (*mor'-bōs*) [*morbus,* disease]. Diseased.
**morbus** (*mor'-bus*) [L.]. Disease. m. Addisonii, Addison's disease. m. anglicus, rickets. m. arcuatus, icterus. m. Basedowii. See *goiter, exophthalmic.* m. Brightii. See *Bright's disease.* m. caducus, epilepsy. m. cæruleus, congenital cyanosis. m. Celsi, catalepsy. m., cholera. See *cholera morbus.* m. cœliacus, chronic diarrhea in children. m. cordis, the phenomena of chronic cardiac disease. m. coxæ, m. coxarius, coxalgia. m. cucullaris, pertussis. m. divinus, epilepsy. m. gallicus, syphilis. m. maculosus neonatorum, a fatal disease occurring during the first few days of life and consisting of hemorrhages in various parts of the body. m. maculosus Werlhofii, purpura hæmorrhagica. m. magnus, epilepsy. m. major, epilepsy. m. medicorum, the mania of those who seek the advice of physicians for imaginary diseases. m. miseriæ, any disease due to poverty. m. phlyctenoides, pemphigus. m. regius, jaundice. m. sacer, epilepsy. m. vesicularis, pemphigus. m. virgineus, chlorosis. m. vulpis, alopecia. m. saltatorius, chorea. m. tuberculosis pedis, Madura foot.
**morcellation** (*mor-sel-a'-shun*) [Fr., *morceler,* to cut up or parcel out]. The art of reducing to fragments, as the fetus in embryotomy; the removal of a tumor or fetus piecemeal.
**morcellement** (*mor-sel-mon'(g)*). See *morcellation.*
**mordacious** (*mor-da'-se-us*) [*mordax,* biting]. Biting, pungent.
**mordant** (*mor'-dant*) [*mordere,* to bite]. A substance, such as alum, phenol, anilin oil, that fixes the dyes used in coloring textures or in staining tissues and bacteria.
**mordication** (*mor-dik-a'-shun*) [*mordicare,* to bite]. A burning and stinging inflammatory condition of the skin.
**Morel's ear** (*mor-el'*) [Benoit Augustin *Morel,* French alienist, 1809–1873]. A large, outstanding ear with more or less pronounced effacement of the ridges and grooves.
**Morgagni's cartilages** (*mor-gahn'-ye*) [Giovanni Battista *Morgagni,* Italian anatomist, 1682–1771]. See *Wrisberg's cartilages.* M.'s caruncle, the middle lobe of the prostate. M.'s cataract, senile cataract in which the nucleus remains hard, while the portion between it and the cortex liquefies. M.'s columns, vertical folds of the rectal mucous membrane seen at the point of union of the latter with the skin of the anus. M.'s concha, the superior turbinated bone of the ethmoid. M.'s foramen. See *Meibomian foramen.* M.'s fossa. 1. The fossa navicularis of the urethra. 2. The concave transverse depression between the upper border of the superior constrictor muscles of the pharynx and the basilar process of the occipital bone. M.'s frenum, M.'s retinaculum, the ridge formed around the cavity of the cecum by the prolongation of the folds of the ileocecal valve. M.'s glands, the small racemose muciparous glands in the mucous membrane of the urethra. M.'s globules, M.'s spheres, small, hyaline bodies found between the crystalline lens and its capsule before and after death, especially in cases of cataract. They are due to coagulation of the albuminous fluid contained in the lens. M.'s hydatid, a vesicle about the size of a pea, attached by a long thread-like stalk to the Fallopian tube in the female, and to the globus major of the epididymis in the male; it is derived from the duct of Mueller. M.'s lacunæ, small depressions in the mucosa of the urethra,

especially the bulbous portion. **M.'s liquor,** a clear fluid formed postmortem in the crystalline lens. **M.'s nodules.** See *corpora Arantii.* **M.'s prolapse,** chronic hyperplastic inflammation of the mucosa and submucosa of Morgagni's ventricle. It is not a true prolapse. **M.'s sinus.** 1. The prostatic sinus. 2. The interval between the upper border of the superior constrictor muscle of the pharynx and the basilar process of the occipital bone. **M.'s sinuses, M.'s valves,** small pouches, opening upward, formed by the rectal mucosa, just above the anus. **M.'s tubercles.** See *Montgomery's tubercles.* **M.'s ventricle,** the sacculus laryngis, the sac between the superior vocal bands and the inner surface of the thyroid cartilage.
**(de) Morgan's spots.** Bright red nevoid spots often seen on the skin in cases of cancer.
**morgue** (*morg*) [Fr.]. A place where unknown dead are exposed for identification.
**moria** (*mo'-re-ah*) [μωρία, folly]. A dementia characterized by talkativeness and silliness.
**moribund** (*mor'-ib-und*) [*moribundus,* from *moriri,* to die]. In a dying condition.
**morioplasty** (*mor'-e-o-plas-te*) [μόριον, a small piece; πλάσσειν, to mold]. Plastic surgery.
**morning-sickness.** The nausea of pregnant women, occurring chiefly in the early months of gestation; sometimes experienced by the husband during the wife's early pregnancy.
**morococcus** (*mo-ro-kok'-us*) [*morus,* mulberry; *coccus*]. A form of micrococcus found clumped in a mass.
**morocomium** (*mo-ro-ko'-me-um*) [μωρός, mad; κομεῖν, to care for]. An insane asylum.
**morodochium** (*mo-ro-do'-ke-um*) [μωρός, silly; δέχεσθαι, to receive]. An asylum for the insane.
**moron** (*mor'-on*) [μωρός, dull, stupid]. 1. A fool. 2. A child with permanently arrested mental development.
**Moro's tuberculin test or reaction** [E. *Moro,* German physician, 1874– ]. A test to determine the presence or absence of tuberculosis. A small amount of an ointment of -equal parts of "old" tuberculin and anhydrous lanolin is rubbed into the unbroken skin of the abdomen or thorax for three to five minutes. A positive reaction is indicated by the appearance, within 24 to 48 hours, of hyperemia of the area and a number of papules.
**morosis** (*mo-ro'-sis*) [μώρωσις, foolishness]. Insanity.
**morph** (*morf*) [μορφή, a blotch]. One of the dark spots that occur on the foreheads of blonde women suffering from uterine trouble, or who are pregnant.
**morphea, morphœa** (*mor-fe'-ah*) [μορφή, a blotch]. A disease of the skin characterized by the presence of rounded or oval, pinkish or ivory-white patches, due to an excess of fibrous tissue, with atrophy of the skin-structures proper. It is believed to be a trophoneurosis, and is considered a circumscribed form of scleroderma. Syn., *Addison's keloid; circumscribed scleroderma.* **m., acrotoric,** the form in which the beginning and the greatest intensity of the disease are at the extremities. **m., herpetiform,** where the lesions follow those of herpes in their distribution.
**morpheum** (*mor'-fe-um*). Morphine.
**morphia** (*mor'-fe-ah*). See *morphine.*
**morphine, morphina** (*mor'-fēn, mor-fi'-nah*) [*Morpheus,* god of sleep], $C_{17}H_{19}NO_3 + H_2O$. A colorless or white crystalline alkaloid obtained from opium, to which the chief effects of opium are due. It differs from opium in being less stimulant, less constipating, and less likely to produce disagreeable after-effects. On account of its insolubility in water morphine is used principally in the form of its salts. The dose of the salts of morphine is $\frac{1}{8}-\frac{1}{2}$ gr. (0.008–0.032 Gm.). **m. acetate** (*morphinæ acetas,* U. S. P.), $C_{17}H_{19}NO_3 \cdot C_2H_4O_2 + 3H_2O$. From it are prepared *liquor morphiæ acetatis* (B. P.) ($\frac{1}{2}$ gr. to the dram), dose 20–40 min. (1.3–2.6 Cc.), and *injectio morphinæ hypodermica* (1 gr. in 10 min.). **m. benzoate,** $C_{17}H_{19}NO_3 \cdot C_7H_6O_2$, white crystalline powder or prisms used in treatment of asthma. Dose $\frac{1}{16}-\frac{1}{8}$ gr. (0.005–0.03 Gm.). **m. borate,** a white powder containing about 33 % of morphine; recommended for hypodermatic use and for eye-lotions. **m. hydrochloride** (*morphinæ hydrochloridum,* U. S. P.), $C_{17}H_{19}NO_3 \cdot HCl + 3H_2O$. From it are prepared *liquor morphiæ hydrochloratis* (B. P.) ($\frac{1}{2}$ gr. to the dram), dose 15–30 min. (1–2 Cc.); *suppositoria morphiæ,* (B. P.) ($\frac{1}{2}$ gr. each); *tinctura chloroformi et morphinæ* ($\frac{1}{2}$ gr. to the dram); *trochisci morphiæ* (B. P.) ($\frac{1}{36}$ gr.), and *trochisci morphiæ et ipecacuanhæ* (B. P.) ($\frac{1}{36}$ gr.). **m. phthalate,** is employed hypodermatically. **m., powder of, compound** (*pulvis morphinæ compositus,* U. S. P.), Tully's powder. Dose 10 gr. (0.65 Gm.), containing $\frac{1}{4}$ gr. (0.01 Gm.) of morphine sulphate. **m. stearate,** $C_{17}H_{19}NO_3 \cdot C_{17}H_{35}COOH$, contains 25 % of morphine and is used in applications (0.5 to 50 Gm. of fixed oil of almonds), ointments (0.5 to 50 Gm. of petrolatum), and suppositories (0.02 to 2.5 Gm. of cacao-butter). **m. sulphate** (*morphinæ sulphas,* U. S. P), $(C_{17}H_{19}NO_3)_2 \cdot H_2SO_4 + 5H_2O$. From it are prepared *liquor morphiæ sulphatis* (B. P.), dose 10–40 min. (0.65–2.5 Cc.), the compound powder of morphine, and *liquor morphinæ hypodermicus* (N. F.), Magendie's solution, containing 16 gr. to the ounce. **m. tartrate,** is employed for hypodermatic use. **m. valerate, m. valerianate,** $C_{17}H_{19}NO_3 \cdot C_5H_{10}O_2$, is used as a sedative.
**morphinia** (*mor-fin'-e-ah*) [*morphine*]. Any disease due to the excessive use of morphine.
**morphinism** (*mor'-fin-izm*) [*morphine*]. 1. The condition caused by the habitual use of morphine. 2. The morphine-habit.
**morphinization** (*mor-fin-iz-a'-shun*) [*morphine*]. The production of the physiological effects of morphine.
**morphinodipsia** (*mor-fin-o-dip'-se-ah*) [*morphine;* δίψα, thirst]. Morphinomania.
**morphinomania** (*mor-fin-o-ma'-ne-ah,* *mor-fe-o-ma'-ne-ah*) [*morphine;* μανία, madness]. 1. A morbid craving for morphine. 2. Insanity due to the morphine-habit.
**morphinophagia, morphiophagy** (*mor-fin-o-fa'-je-ah,* *mor-fi-off'-aj-e*) [*morphine;* φαγεῖν, to eat]. Opiumeating.
**morphinum, morphium** (*mor-fi'-num, mor'-fe-um*). Morphine.
**morphiometry** (*mor-fe-om'-et-re*) [*morphine;* μέτρον, measure]. The determination of the quantity of morphine in a drug preparation or sample of opium.
**morphœa** (*mor-fe'-ah*). See *morphea.*
**morphogenesis** (*mor-fo-jen'-es-is*). See *morphogeny.*
**morphogeny** (*mor-foj'-en-e*) [μορφή, form; γένεια, generation]. The genesis of form; the history of the evolution of form.
**morphography** (*mor-fog'-ra-fe*) [μορφή, form; γράφειν, to write]. Systematic investigation of the structure of organisms in the most comprehensive way. See *morphology.*
**morpholecithal** (*mor-fo-les'-ith-al*) [μορφή, form; λέκιθος, yolk of an egg]. 1. Germinal; formative. 2. Pertaining to the morpholecithus.
**morpholecithus** (*mor-fo-les'-ith-us*) [μορφή, form; λέκιθος, yolk of an egg]. The formative yolk of an egg; the portion of an egg that undergoes segmentation and germination.
**morphological** (*mor-fo-loj'-ik-al*). Pertaining to morphology.
**morphology** (*mor-fol'-o-je*) [μορφή, form; λόγος, science]. The science that treats of the form and structure of organized beings.
**morpholysis** (*mor-fol'-is-is*) [μορφή, form; λύσις, a loosening]. Destruction of form.
**morphometry** (*mor-fom'-et-re*) [μορφή, form; μέτρον, a measure]. The measurement of the forms of organisms.
**morphon** (*mor'-fon*) [μορφή, form]. An individual element of an organism, characterized by a definite form, as a cell or a segment of a vertebrate.
**morphonosus** (*mor-fon'-o-sus*) [μορφή, form; νόσος, disease]. An anomalous change in the form of organs or parts.
**morphosan** (*mor'-fo-san*). Proprietary name for morphine methylbromate.
**morphosis** (*mor-fo'-sis*) [see *morphon*]. The act, mode, or order of formation of an organism.
**morphotic** (*mor-fot'-ik*) [*morphosis*]. Pertaining to morphosis; entering into the formation of the framework of an organism. **m. proteids,** those that enter into the structure of the tissues.
**morpio** (*mor'-pe-o*) [L.]. The crab-louse.
**Morrant-Baker's cysts.** See *Baker's cysts.*
**morrhua, emulsum olei** (U. S. P.), a mixture of cod-liver oil, acacia, syrup, oil of gaultheria, and water. Dose 2 dr. (8 Cc.). **morrhuæ, emulsum olei, cum hypophosphitibus** (U. S. P.), emulsion of cod-liver oil

and hypophosphites, contains of cod-liver oil, 500 Cc.; acacia, 125 Gm.; calcium hypophosphite, 10 Gm.; potassium hypophosphite, 5 Gm.; sodium hypophosphite, 5 Gm.; syrup, 10 Cc.; oil of gaultheria, 4 Cc.; water, q. s. Dose 2 dr. (8 Cc.). **morrhuæ, oleum** (U. S. P.), cod-liver oil, a fixed oil obtained from the fresh livers of *Gadus morrhua* and other species of *Gadus*. Three varieties of oil are known in commerce—a white or pale-yellow, a brownish-yellow, and a dark-brown. The oil contains *gaduin* ($C_{28}H_{48}O_3$), oleic, palmitic, stearic, myristic, and physetolic acids, glycerol, butyric and acetic acids, biliary pigments, iodine, and bromine. A crystalline substance, *morrhuol*, containing phosphorus, iodine, and bromine, has also been isolated, as well as several leukomaine and the fixed bases, *asellin* ($C_{28}H_{32}N_4$) and *morrhuin* ($C_{19}H_{27}N_3$). Cod-liver oil is used in pulmonary and other forms of tuberculosis, and in wasting conditions due to other causes. Dose 1 dr.– ½ oz. (4–16 Cc.).
**morrhuin** (*mor′-oo-in*). See under *morrhua*.
**morrhuol** (*mor′-oo-ol*). See under *morrhua*.
**Morris' test.** A rod graduated from the center and provided with sliding pointers is placed across the abdomen, so that its center corresponds to the median line of the body, and the pointers are moved along it until they reach the outer surface of the greater trochanter. In cases of fracture of the neck of the femur a discrepancy will be found on comparing the measurements on the two sides of the body.
**mors** (*morz*) [L.]. Death.
**morsal** (*mor′-sel*) [*morsus*, a bite]. Relating to the cutting or grinding portion of a tooth.
**morselling** (*mor′-sel-ing*). See *morcellation*.
**morsulus** (*mor′-sŭ-lus*) [dim. of *morsus*, a bite]. A lozenge or tablet.
**morsus** (*mor′-sus*) [L.]. A bite. **m. diaboli,** a fanciful name for the fimbriated extremity of the oviduct.
**mortal** (*mor′-tal*) [*mortalis*, from *mors*, death]. Liable to death or dissolution; terminating in death; causing death; deadly.
**mortality** (*mor-tal′-it-e*) [*mortal*]. 1. The quality of being mortal. 2. The death-rate.
**mortar** (*mor′-tar*) [*mortarium*, an urn]. An urn-shaped vessel of porcelain, iron, or glass, for pulverizing substances by means of a pestle.
**mortiferous** (*mor-tif′-er-us*) [*mors; ferre*, to bear]. Fatal.
**mortification** (*mor-tif-ik-a′-shun*). See *gangrene*.
**Morton's cough** (*mor′-tun*) [Thomas George Morton, American surgeon, 1835–1903]. A cough followed by the vomiting of food, frequently, occurring in pulmonary tuberculosis. **M.'s disease.** See *M.'s foot*. **M.'s fluid,** iodine, 10 gr.; potassium iodide, 30 gr.; and glycerin, 1 ounce; used by injection in cases of spinal meningocele. **M.'s foot,** a painful affection of the metatarsophalangeal joint of the fourth toe; metatarsalgia.
**mortuary** (*mor′-tŭ-a-re*) [*mortuarium*, a tomb]. 1. A house for temporary burial; a morgue. 2. Relating to death or burial.
**morula** (*mor′-ŭ-lah*) [dim. of *morum*, a mulberry]. The solid mass of cells resulting from the complete segmentation of the vitellus of an ovum.
**morulation** (*mor-ŭ-la′-shun*) [*morula*]. The formation of the morula during the process of the segmentation of the egg.
**moruloid** (*mor′-ŭ-loid*) [*morula*, a little mulberry; εἶδος, like]. Resembling a morula.
**morulus** (*mor′-ŭ-lus*). The lesion characteristic of frambesia. A frambesial sore.
**morum** (*mor′-rum*) [L., "a mulberry"]. 1. The fruit of the mulberry. 2. Condyloma. 3. Nevus.
**morus** (*mo′-rus*). See *mulberry*.
**Morvan's chorea** (*mor-van′*). Fibrillary contractions of the muscles of the calves and posterior portion of the thighs, often extending to the trunk and upper extremities, but leaving the face and neck intact. Syn., *chorée fibrillaire de Morvan*. **M.'s disease,** a trophic affection of the skin with pain, followed by analgesia first of one side, then of the other, and then the formation of whitlows attended with necrosis of the phalanges. Muscular atrophy, paresis, contraction of the fingers, and loss of the sensations of heat and cold are other symptoms.
¹**morve** (*morv*). See *equinia*.
**morvin** (*mor′-vin*) [Fr. *morve*, glanders, malleus]. See *mallein*.
**moschus** (*mos′-kus*). Musk; the dried secretion from the prepuce of the musk-deer, used as an antispasmodic.
**Mosler's diabetes** (*mōz′-ler*) [Karl Friedrich Mosler, German physician, 1831– ]. Polyuria due to the presence of too much inosite in the blood.
**mosquito** (*mus-ke′-to*) [Sp., "a little gnat"]. An insect the sting of which causes the formation of a wheal that itches intensely; especially the *Culex* mosquito. It is also a pathogenic agent in the transmission of disease. See *Anopheles, Culex, Stegomyia*.
**moss** [AS., *meós*]. 1. A small cryptogamic plant of the natural order *Musci*. **m., Ceylon.** See *agar*. **m., club-.** See *lycopodium*. **m., Corsican.** See *Corsican moss*. **m.-fibers,** peculiar fibers derived from the white center of the cerebellum, and characterized by having pencils of fine short branches at intervals like tufts of moss; they end partly in the granular layer, partly in the molecular layer. **m., Iceland.** See *cetraria*. **m., Irish.** See *chondrus*.
**Moszkowicz test** (*mos′-ko-vits*). *For arteriosclerosis:* the limb is elevated until the skin becomes pale; a broad, elastic bandage is applied around the thigh as high up as possible, and the bandage is allowed to remain in place for five minutes. On removing the elastic bandage a hyperemic blush spreads over the limb but is less intense as the ischemic areas of the foot or leg are approached. The contrast between the red and pale areas is marked, and varies with the extent of the arterial obstruction.
**moth, moth-patches.** Chloasma.
**mother** (*muth′-er*) [AS., *mōder*]. 1. A female parent. 2. The source of anything. 3. [allied to AS., *mud*.] A slimy film formed on the surface of fermenting liquid, as on vinegar. **m.-cell,** a cell from which other cells are formed, especially one the nucleus of which is undergoing karyokinetic changes preparatory to dividing into daughter-cells. **m.-liquor,** the liquid remaining after dissolved substances have separated by crystallization. **m.'s mark,** a birth-mark. See *nevus*.
**motile** (*mo′-til*) [*movere*, to move]. Able to move; capable of spontaneous motion, as a *motile* flagellum.
**motility** (*mo-til′-it-e*) [*motile*]. Ability to move spontaneously.
**motion** (*mo′-shun*) [*movere*, to move]. 1. The act of changing place. 2. An evacuation of the bowels; the matter evacuated.
**motive** (*mo′-tiv*) [*movere*, to move]. Causing motion; a determining impulse. **m. force, m. power,** the moving or impelling force.
**motor** (*mo′-tor*) [see *motion*]. 1. Moving or causing motion. 2. Concerned in or pertaining to motion, as *motor* cell, *motor* center, *motor* nerve. **m. aphasia.** See under *aphasia*. **m. area,** the portion of the cerebral hemisphere presiding over voluntary motion, including the precentral gyri, the posterior part of the three frontal gyri, and the paracentral lobule on the median surface of the hemisphere. **m. nerve-organs, m. nerve-plates, m.-sprays.** See *fields of innervation*. **m. oculi,** the third cranial or oculomotor nerve supplying all the muscles of the eye except the superior oblique and external rectus. **m. points,** the points on the surface of the body where the various branches of the motor nerves supplying the muscles may be stimulated by electricity.
**motorgraphic** (*mo-tor-graf′-ik*). See *kinetographic*.
**motorial** (*mo-to′-re-al*) [*motor*]. Of or pertaining to motion. **m. end-plate,** an eminence of protoplasm within the sarcolemma of a muscular fiber, representing the termination of the motor nerve-fiber.
**motorium** (*mo-to′-re-um*) [*motor*]. 1. A motor center. 2. The motor apparatus of the body, both nervous and muscular, considered as a unit.
**motorius** (*mo-to′-re-us*) [L.]. A motor nerve.
**motormeter** (*mo-tor-me′-ter*). A kinesiometer used in recording gastric movements.
**motorpathy** (*mo′-tor-path-e*) [*motor; πάθος*, disease]. Kinetotherapy.
**Mott's law of anticipation.** When children of the insane become insane they do so at a much earlier age than did their parents, and they are also liable to suffer from a much more intense form of the disease.
**mottling** (*mot′-ling*) [OF., *matellé*, clotted; curdled]. A spotted condition.
**mould** (*mōld*). See *mold*.
**mounding** (*mownd′-ding*). The rising in a lump of muscle-fibers when struck by a slight, firm blow. It is observed in the thin and feeble, and in certain

diseases, as pulmonary tuberculosis and advanced locomotor ataxia. See *myoidema*.
**mountain anemia.** Ankylostomiasis. **m.-fever, m.-sickness.** 1. A condition characterized by dyspnea, rapid pulse, headache, nausea, and vomiting, depending upon the rarefied state of the air at high altitudes. This is properly called mountain-sickness. 2. A form of typhoid fever occurring in mountainous districts.
**mounting** (*mown'-ting*) [*mount*]. The act of arranging objects, especially anatomical specimens, on a suitable support and in a proper medium for ready examination. For macroscopic specimens the medium is usually alcohol; for microscopic specimens, Canada balsam or glycerol.
**mouth** (*mowth*) [AS. *mūth*]. 1. The commencement of the alimentary canal; the cavity in which mastication takes place. In a restricted sense, the aperture between the lips. 2. The entrance to any cavity or canal. **m.-breather,** a person who habitually breathes through the mouth. **m.-breathing,** respiration through the mouth instead of, as normally, through the nose.
**movement** (*moov'-ment*) [*movere*, to move]. The act of moving. **m., ameboid,** a movement produced in certain cells, as the white corpuscles, by the protrusion of processes of the protoplasm into which the whole cell then seems to flow; so-called from the resemblance of the movement to that of the ameba. **m., angular,** the movement between two bones that may take place forward or backward, inward or outward. **m., associated,** an involuntary movement in one part when another is moved voluntarily. **m., Brownian,** a physical phenomenon, a form of communicated motion observed in aggregations of minute particles, and consisting of a rapid oscillating movement without change of the relative position of the moving particles; also called *pedesis*. **m., ciliary,** a lashing movement produced by delicate hair-like processes termed cilia, as on the epithelium of the respiratory tract and in certain microorganisms. **m., circus-,** rapid circular movements or somersaults, produced by injury of the corpus striatum, of the optic thalamus, or of the crus cerebri of one side. **m., communicated,** that produced by a force acting from without; opposed to spontaneous movement. **m.-cure,** kinesipathy. **m., fetal,** the movements of the fetus in the uterus. **m., forced,** movement of the body from injury of the motor centers or the conducting paths, as *index movement*, when the cephalic part of the body is moved about the stationary caudal part; *rolling movement*, when the animal rolls on its long axis. **m., index,** when the cephalic part of the body is moved about the fixed caudal part. **m., molecular.** Synonym of *m., Brownian*. **m., rolling,** when the animal rolls on its long axis. **m., Swedish,** kinesipathy. **m., vermicular,** peristalsis.
**mower's mite** (*mo'-er*). The harvest mite; see *Leptus autumnalis*.
**moxa** (*moks'-ah*) [Jap.]. A combustible material which is applied to the skin and ignited for the purpose of producing an eschar. It is prepared from several species of *Artemisia*; artificial moxa is made from cotton saturated with niter. **m.-bearer,** an instrument for applying the moxa. Syn., *Portemoxa*. **m., electric,** a faradic brush used as an active electrode upon the dry skin.
**moxibustion** (*moks-ib-us'-chun*) [*moxa; combustion*]. Cauterization by means of a moxa.
**moxosphyra** (*moks-os-fi'-rah*) [*moxa,* σφῦρα, hammer]. A hammer heated in boiling water and applied to the skin for purposes of counter-irritation.
**moyrapuama** (*moi-rah-poo-am'-ah*). See *muirapuama*.
**M. P. S.** Abbreviation for *Member of the Pharmaceutical Society*.
**M.R.C.P.** Abbreviation for *Member of the Royal College of Physicians*.
**M.R.C.S.** Abbreviation for *Member of the Royal College of Surgeons*.
**M. R. C. V. S.** Abbreviation for *Member of the Royal College of Veterinary Surgeons*.
**M.S.** Abbreviation 1. for *Master of Surgery*; 2. for *Master of Science*.
**Ms.** Chemical symbol of *masrium*.
**M.S.A.** Abbreviation for *Member of the Apothecaries' Society*.
**M.Sc.** Abbreviation of *Master of Science*.
**M. u.** Abbreviation of *maché unit*.
**muavine, muawine** (*moo-ah'-vin, -win*). An alkaloid from muawi-bark. The hydrobromide is used as a cardiac stimulant.
**muawi-bark** (*moo-ah'-we-bark*). The bark of a leguminous tree closely related to *Erythrophlæum coumingo*; used as an arrow-poison in Madagascar. Syn., *muawa*.
**mucago** (*mū-ka'-go*). Mucus; mucilage.
**mucedin** (*mū'-se-din*) [*mucus*]. A nitrogenous substance obtained from gluten.
**mucherus** (*mū'-ker-us*). The gum obtained from *Bombax malabaricum;* it is astringent and styptic. Dose 30–45 gr. (2–3 Gm.). Syn., *mocharas; mochras; mochurrus*.
**mucic** (*mū'-sik*) [*mucus*]. Obtained from mucus or mucilage. **m. acid,** C₆H₁₀O₈. A crystalline dibasic acid produced by the oxidation of gums and certain sugars.
**mucicarmine** (*mū-se-kar'-min*). A stain for mucin made up of carmine, 1 Gm.; aluminum chloride, 0.5 Gm.; distilled water, 2 Cc.
**muciferous** (*mū-sif'-er-us*) [*mucus; ferre,* to bear]. Producing or secreting mucus.
**muciform** (*mū'-sif-orm*) [*mucus; forma,* form]. Resembling mucus.
**mucigen** (*mū'-sij-en*) [*mucin;* γεννᾶν, to produce]. A substance producing mucin; it is contained in epithelial cells that form mucus.
**mucigenous** (*mū-sij'-en-us*) [see *mucigen*]. Producing mucus.
**mucilage** (*mū'-sil-āj*) [*mucilago,* moldy moisture]. In pharmacy, a solution of a gum in water. Mucilages (*mucilagines*) are employed as applications to irritated surfaces, particularly mucous membranes, as excipients for pills, and to suspend insoluble substances. The following are employed: *Mucilago acaciæ* (U. S. P.), *M. amyli* (B. P.), *M. sassafras medullæ* (U. S. P.), *M. tragacanthæ* (U. S. P.), *M. ulmi* (U. S. P.).
**mucilaginous** (*mū-sil-aj'-in-us*) [*mucilage*]. Pertaining to or of the nature of mucilage.
**mucilago** (*mū-sil-a'-go*). See *mucilage*.
**mucin** (*mū'-sin*) [*mucus*]. An albuminoid substance, the characteristic constituent of mucus. It is supposed to be produced by the union of an albuminous body and a colloid carbohydrate, the "animal gum" of Landwehr. Mucin occurs in saliva, bile, secretions of mucous membranes, synovial fluid, in mucous tissue, in certain cysts, etc. It is insoluble in water, and is precipitated by alcohol and acetic acid. See *levulose*.
**mucinemia** (*mū-sin-e'-me-ah*) [*mucin;* αἷμα, blood]. The presence of mucin in the blood.
**mucinoblast** (*mū-sin'-o-blast*) [*mucin;* βλαστός, a germ]. 1. A cell whose function it is to elaborate mucin. Syn., *mast-cell*. 2. A goblet-cell.
**mucinogen** (*mū-sin'-o-jen*) [*mucin;* γεννᾶν, to produce]. The antecedent principle from which mucin is derived.
**mucinoid** (*mū'-sin-oid*) [*mucin;* εἶδος, like]. Resembling mucin.
**mucinoids** (*mū'-sin-oids*). See *mucoids*.
**mucinuria** (*mū-sin-u'-re-ah*) [*mucin;* οὖρον, urine]. The presence of mucin in the urine.
**muciparous** (*mū-sip'-ar-us*) [*mucus; parere,* to bring forth]. Secreting or producing mucus.
**mucitis** (*mū-si'-tis*) [*mucus;* ιτις, inflammation]. Inflammation of a mucous membrane.
**mucivorous** (*mū-siv'-or-us*) [*mucus; vorare,* to devour]. Subsisting on mucus or gum.
**muco-** (*mū-ko-*) [*mucus*]. A prefix meaning pertaining to mucus.
**mucocele** (*mū'-ko-sel*) [*muco-;* κήλη, tumor]. 1. A mucous tumor. 2. An enlarged lacrimal sac.
**mucocolitis** (*mū-ko-ko-li'-tis*). See *colitis, mucous*.
**mucocolpos** (*mū-ko-kol'-pos*) [*muco-;* κόλπος, vagina]. A collection of mucus in the vagina.
**mucocutaneous** (*mū-ko-ku-ta'-ne-us*) [*muco-; cutaneous*]. Pertaining to a mucous membrane and the skin; pertaining to the lines where these join.
**mucoderm** (*mū'-ko-derm*) [*muco-;* δέρμα, skin]. The corium of a mucous membrane.
**mucoenteritis** (*mū-ko-en-ter-i'-tis*). Inflammation of the mucous membrane of the intestine.
**mucoid** (*mū'-koid*) [*muco-;* εἶδος, likeness]. Resembling mucus.
**mucoids** (*mū'-koids*) [see *mucoid*]. A group of glycoproteids embracing colloid, chondromucoid, and pseudomucin, and differing from true mucins in their solubilities and precipitation properties. They are found in cartilage, in the cornea and crystalline

**lens,** in white of egg, and in certain cysts and ascitic fluids.
**mucomembranous** (*mū-ko-mem'-bran-us*). See *mucosal.*
**mucoperiosteum** (*mū-ko-per-i-os'-te-um*). Periosteum possessing a mucous surface.
**mucopurulent** (*mū-ko-pū'-ru-lent*) [*muco-*; *purulent*]. Containing mucus mingled with pus.
**mucopus** (*mū'-ko-pus*) [*muco-*; *pus*]. A mixture of mucus and pus.
**Mucor** (*mū'-kor*) [*mucere,* to be moldy]. A genus of hyphomycetes. **M. corymbifer,** a species found in the cerumen of the external auditory meatus. **M. mucedo,** a species found on fecal matter and nitrogenous organic substances. **M. niger,** a parasitic fungus causing black discoloration of lingual papillæ.
**mucoriferous** (*mū-kor-if'-er-us*) [*mucor*; *ferre,* to bear]. Mold-bearing or covered with a mold-like substance.
**mucorin** (*mū'-kor-in*) [*mucor*]. An albuminoid substance from many species of the mucorinous molds.
**mucosa** (*mū-ko'-sah*) [*mucosus,* mucous]. A mucous membrane; more fully, *membrana mucosa.*
**mucosal** (*mū-ko'-sal*). Relating to mucous membranes.
**mucosanguineous** (*mū-ko-san-gwin'-e-us*). Consisting of mucus and blood.
**mucosedative** (*mū-ko-sed'-at-iv*). Soothing to mucosæ.
**mucoserous** (*mū-ko-se'-rus*). Both mucous and serous; containing mucous and serum.
**mucosin** (*mū'-ko-sin*) [*mucus*]. The form of mucin to which the nasal, uterine, and bronchial mucus owe their viscosity.
**mucosity'** (*mū-kos'-it-e*). Sliminess.
**mucous** (*mū'-kus*) [*mucus*]. Containing or having the nature of mucus; secreting mucus, as *mucous membrane*; depending on the presence of mucus, as *mucous rales.* **m. casts,** a term given to the casts found in the feces in cases of membranous enteritis. **m. catarrh,** catarrhal inflammation of a mucous membrane. **m. colitis.** See *colitis, mucous.* **m. degeneration.** See *degeneration.* **m. disease,** enterocolitis, especially of children. **m. glands,** glands containing mucous cells. **m. membrane.** See *membrane.* **m. patch,** a flattened, grayish-white exudate, occurring in secondary syphilis on mucous membranes and at mucocutaneous junctions. **m. polyp,** a soft, gelatinous outgrowth from a mucous membrane; it may be a true myxoma, but usually is a hyperplasia due to chronic inflammation. **m. tissue,** a form of connective tissue in which the intercellular substance is of a soft, gelatinous character and contains mucin. The cells from pressure assume a stellate or spindle shape. **m. tumor,** a myxoma.
**muculent** (*mū'-ku-lent*). Rich in mucus.
**Mucuna** (*mū-ku'-nah*) [Braz.]. A genus of leguminous herbs. The hairs of the pods of *M. pruriens,* cowage, were formerly used as a vermifuge and counterirritant.
**mucus** (*mū'-kus*) [L.]. The viscid liquid secreted by mucous membranes. It consists of water, mucin, and inorganic salts, together with epithelial cells, leukocytes, etc., held in suspension.
**mud-bath.** See *bath-, mud.*
**mudar** (*mū'-dar*) [E. Ind.]. The root-bark of various Asiatic species of *Calotropis* (*C. gigantea, C. procera, C. hamiltoni*); it is alterative, tonic, diuretic, sudorific, and emetic.
**mudarin** (*mū'-dar-in*) [E. Ind., *mudar*]. A bitter principle from mudar; it is said to be tonic, emetic, and alterative.
**Mueller's capsule** (*mū'-ler*) [1. Johannes *Mueller,* German physiologist, 1801-1858]. See *Bowman's capsule.* **M.'s blood-motes, M.'s dust-bodies.** See *hemokonia.* **M.'s duct or canal,** [1], a duct lying internally to the Wolffian body; it practically disappears in the male, but becomes the Fallopian tube and part of the uterus and vagina in the female. **M.'s eminence,** in the embryo, the protuberance formed by the cloaca at the point of entrance of Mueller's duct. Syn., *colliculus Muelleri.* **M.'s experiment,** [1]. 1. See *Valsalva's experiment.* 2. See *Valsalva's test.* **M.'s fibers** [2. Heinrich *Mueller,* German anatomist, 1820-1864]. Modified neuroglia cells which traverse perpendicularly the layers of the retina, and connect the internal and external limiting membranes. **M.'s fluid,** [3. Hermann Franz *Mueller,* German histologist, 1866-1898]. A fluid used for hardening tissues. Its composition is as follows: Potassium dichromate, 2 to 2.5 parts; sodium sulphate, 1 part; water, 100 parts. **M.'s ganglion.** See *Ehrenritter's ganglion.* **M.'s law.** 1. The tissue of which a tumor is composed has its type in the tissues of the animal body, either in the adult or in the embryonic condition. 2. The "law of isolated conduction." The nervous impulse, or "wave of change," passing through a neuron is not communicated to other neurons, even when these lie close alongside of it, except at the terminals. **M.'s muscle,** [2]. 1. The circular bundles of muscular fibers which form part of the ciliary muscle and are situated nearest to the iris. 2. The superior palpebral muscle. **M.'s ring,** a muscular ring formed at the internal os uteri during the later stages of pregnancy. **M.'s sarcoma,** adenofibroma of the breast. Syn., *sarcoma phyllodes.* **M.'s sign,** [4. Kolóman *Mueller,* Hungarian physician, 1849- ]. Pulsation of the tonsils and soft palate in cases of aortic insufficiency. **M.'s test for cystin,** boil the cystin with potassium hydroxide to dissolve it; when cold, dilute with water and add a solution of sodium nitroprusside. This produces a violet coloration which changes rapidly to yellow.
**Muellerian cyst.** A cyst developed from Mueller's duct.
**Muenchymeyer's disease** (*moonsh'-mi-er*). A progressive poliomyelitis with myositis ossificans.
**muguet** (*moo-gwa'*) [Fr.]. Thrush.
**muira puama** (*moo-e'-rah poo-am'-ah*) [Indian name or straight tree]. A shrub indigenous to the region of the Amazon. It is recommended in the treatment of impotence and as a nerve-tonic. Dose of *fluidextract* 15-30 min. (1-2 Cc.).
**mulberry** (*mul'-ber-e*) [*morus,* mulberry-tree]. A tree of the genus *Morus.* *Morus nigra* is the source of *mori succus* of the B. P., the latter being used to make *syrupus mori* (B. P.). Both are employed as drinks in fevers and as additions to gargles in pharyngitis. The fruit of *Morus alba* is used as food for silkworms. **m. calculus.** See *calculus, mulberry.* **m. mark,** a nevus. **m. mass.** See *morula.*
**Mulder's angle** (*mool'-der*) [Johannes *Mulder,* Dutch anatomist, 1769-1810]. In craniometry, an angle produced by the junction of Camper's line and a line joining the basi-occipital bone and the nasion.
**Mulder's test for glucose.** Alkalinize the solution with sodium carbonate and add a solution of indigocarmin. If glucose is present, the solution becomes decolorized on heating, but changes to blue again on shaking with air. **M.'s test for proteids,** proteids are colored yellow on treating with concentrated nitric acid; on the addition of ammonia or sodium or potassium hydroxide they become orange-yellow. Syn., *xanthoproteic reaction.*
**Mules' operation** (*mūls*) [Philip Henry *Mules,* English ophthalmologist, 1843-1905]. Evisceration of the globe followed by the insertion of a silver or glass ball within the sclerotic, with the view of rendering the stump better suited to an artificial eye.
**muliebria** (*mū-le-eb'-re-ah*) [L. *neut. pl.* of *muliebris,* pertaining to a woman]. The female genital organs.
**muliebris** (*mū'-li-eb-ris*). Pertaining to a woman.
**muliebrity** (*mū-le-eb'-rit-e*) [*mulier,* woman]. Womanliness; puberty in the female.
**mullen, mullein** (*mul'-en*). See *verbascum.*
**multangulum** (*mul-tan'-gū-lum*) [*multus,* many; *angulus,* an angle]. A bone with many angles. **m. majus,** the trapezium. **m. minus,** the trapezoid bone.
**multarticulate, multiarticulate** (*mul-tar-tik'-ū-lāt, mul-te-ar-tik'-ū-lāt*) [*multus,* many; *articulus,* a joint]. Furnished with many joints.
**multi-** (*mul-te-*) [*multus,* much]. A prefix signifying many.
**multicapsular** (*mul-tik-ap'-sū-lar*) [*multi-*; *capsula,* a little box]. In biology, composed of many capsules.
**multicellular** (*mul-te-sel'-ū-lar*) [*multi-*; *cellula,* cell]. Many-celled.
**multicostate** (*mul-te-kos'-tāt*) [*multi-*; *costa,* a rib]. Having many ribs.
**multicuspid, multicuspidate** (*mul-te-kus'-pid, mul-te-kus'-pid-āt*) [*multi-*; *cuspis,* a point]. Having several cusps. **m. teeth,** the molar teeth.
**multidentate** (*mul-te-den'-tāt*) [*multi-*; *dens,* a

# MULTIDIGITATE 577 MUNDIFICANT

tooth]. A term applied in biology to parts armed with many teeth or tooth-like processes.
**multidigitate** (*mul-te-dij'-it-āt*) [*multi-*; *digitus*, a finger]. Having many digits or digitate processes.
**multifetation** (*mul-tif-e-ta'-shun*) [*multi-*; *fetation*]. Pregnancy with more than two fetuses.
**multifid** (*mul'-tif-id*). Divided into many parts.
**multiflagellate** (*mul-tif-laj'-el-āt*) [*multi-*; *flagellum*, a whip]. Having many flagella.
**multiform** (*mul'-tif-orm*). Same as *polymorphous*.
**multiganglionate** (*mul-te-gang'-le-on-āt*) [*multi-*; *ganglion*]. Having many ganglia.
**multiglandular** (*mul-te-glan'-dū-lar*) [*multi-*; *glandula*, a gland]. Pertaining to several glands. m. secretions, a mixture of secretions from two or more glands, such as the saliva.
**multigravida** (*mul-ti-grav'-id-ah*) [*multi-*; *gravidus*, pregnant]. A pregnant woman who has passed through two or more pregnancies.
**multilobate** (*mul-te-lo'-bāt*) [*multi-*; λοβός, a lobe]. Composed of many lobes.
**multilobular** (*mul-ti-lob'-ū-lar*) [*multi-*; *lobule*]. Many-lobed.
**multilocular** (*mul-ti-lok'-ū-lar*) [*multi-*; *loculus*, a locule or cell]. Many-celled; polycystic.
**multinebulizer** (*mul-te-neb'-ū-li-zer*). A spraying device used in treatment of disease of the nose, throat, and ear.
**multinuclear** (*mul-ti-nū'-kle-ar*) [*multi-*; *nucleus*]. Having several or many nuclei.
**multipara** (*mul-tip'-ar-ah*) [*multi-*; *parere*, to bring forth]. 1. A pregnant woman who has already borne one or more children. Opposed to *primipara*. It has been customary to designate the number of the pregnancy of a multipara by the unpronounceable terms II-para, III-para, IV-para, etc. More commendable are the following terms: *secundipara*, *tertipara*, *quartipara*, *quintipara*, *sextipara*, *septimipara*, *octavipara*, *nonipara*, *decimipara*, etc., to designate respectively a woman in her second, third,

fourth, etc., pregnancy. 2. A woman bearing several offspring at a birth.
**multiparity** (*mul-tip-ar'-it-e*) [*multi-*; *parere*, to bring forth]. The condition, state, or fact of being multiparous.
**multiparous** (*mul-tip'-ar-us*) [see *multipara*]. Having borne several children.
**multiple** (*mul'-tip-l*) [*multi-*; *plicare*, to fold]. Manifold; affecting many parts at the same time. m. neuritis. See *neuritis, multiple*. m. pregnancy. See *pregnancy, multiple*. m. sclerosis. See *sclerosis, multiple*.
**multipolar** (*mul-te-po'-lar*) [*multi-*; *polus*, a pole]. Having more than one pole, as *multipolar* nerve-cells, those having more than one process.
**multivalent** (*mul-tiv'-al-ent*) [*multi-*; *valere*, to be worth]. In chemistry, combining with more than one atom of a univalent element.
**mummification** (*mum-if-ik-a'-shun*) [*mummy*, from Pers., *mūm*, wax; *facere*, to make]. The change of a part into a hard, dry mass; dry gangrene.
**mummified** (*mum'-if-id*) [*mummy*; *facere*, to make]. Dried, like a mummy. m. pulp, the condition of the dental pulp when it is affected by dry gangrene.
**mumps** [Dū., *mompen*, to mumble]. An acute infectious disease characterized by swelling of the parotid and at times of the other salivary glands. After a period of incubation of from two to three weeks, the disease begins with fever and pain below the ear; soon a tense, painful swelling forms in the region of the parotid gland, rendering mastication and deglutition difficult and painful. In the course of a week the swelling subsides without suppuration. The most frequent complication is orchitis; in rare cases the ovaries are affected. Syn., *parotiditis*.
**mundificant, mundificative** (*mun-dif'-ik-ant, mun-dif'-ik-at-iv*) [*mundus*, clean; *facere*, to make]. 1. Having the power to cleanse, purge, or heal. 2. A cleansing or healing agent.

## TABLE OF ENDOCARDIAL MURMURS.

| Time. | Point of Maximum Intensity. | Line of Conduction. | Lesion. | Quality. |
|---|---|---|---|---|
| Systolic. | Center of mitral area, above and to left of apex. | At sixth rib opposite apex, a line drawn from the anterior fold of axilla to lower angle of left scapula. | Mitral insufficiency or incompetence. | Variable; usually soft, blowing; may be distinctly musical. |
| Systolic. | Midsternum or to right of it, opposite third rib or second interspace. | Toward top of sternum, and along aorta and its large branches. | Aortic obstruction. | Usually loud and harsh. Harshness is one of its distinguishing characteristics. |
| Diastolic. | Midsternum opposite upper border of cartilage of third rib. | Down sternum to ensiform cartilage. | Aortic insufficiency or incompetence. | Soft, blowing, sometimes rough, frequently musical. It has the greatest area of diffusion of all the cardiac murmurs. |
| Presystolic. | Over mitral area around the apex. | Usually not transmitted. | Mitral obstruction. | Generally low-pitched, rough, churning, grinding, or blubbering. Subject to great variation of pitch and quality. |
| Systolic. | Midsternum just above the ensiform cartilage. | Toward the epigastrium. | Tricuspid insufficiency or incompetence. | Low-pitched, superficial, blowing, soft, faint. |
| Presystolic. | Midsternum opposite the cartilage of fourth rib. | Not transmitted. | Tricuspid obstruction. | Undetermined. |
| Systolic. | Second interspace to the left of sternum or at the level of third rib. | Upward a short distance and to left of sternum, stopping abruptly. | Pulmonary obstruction. | Often harsh and audible over the whole precordia; may be very faint. |
| Diastolic. | Second left interspace. | Down left edge of sternum to ensiform cartilage. | Pulmonary insufficiency or incompetence. | Soft and blowing. |

20

**mural** (*mū'-ral*) [*murus*, a wall]. Pertaining to a wall, as a *mural* fibroid. **m. gestation, m. pregnancy**, pregnancy in the uterine extremity of a Fallopian tube.

**Murat's symptom** (*mu-rah'*). In tuberculosis, vibration of the affected part of the chest, attended with a sense of discomfort while speaking.

**Murchison's pill** (*murtsh'-is-on*) [Charles *Murchison*, English physician, 1830–1879]. A pill consisting of digitalis gr. ½, squill gr. 1½, and blue massgr. 2.

**murexide** (*mū-reks'-id*) [*murex*, the purple-fish]. $C_8H_8N_6O_6+H_2O$. Ammonium purpurate, a dichroic crystalline salt obtained from guano and used as a dye. **m. test for uric acid**, cover the substance or the residue on evaporation with nitric acid; evaporate to dryness on a water-bath, and when cold, add ammonia, when it will turn purple-red.

**muriate** (*mū'-re-āt*) [*muria*, brine]. An old name for a chloride.

**muriated** (*mū'-re-a-ted*) [*muriate*]. Containing chlorine or a chloride.

**muriatic** (*mū-re-at'-ik*) [*muriate*]. Pertaining to brine. **m. acid** See *acid, hydrochloric*.

**murmur** (*mer'-mer*) [L.]. A blowing or rasping sound heard on auscultation. See also under *bruit*. **m., accidental**, a murmur dependent on an accidental circumstance, as on compression of an artery by the stethoscope. **m., anemic.** See *m., hemic*. **m., aneurysmal**, the murmur or bruit heard over an aneurysm. Syn., *aneurysmal bruit*. **m., arterial**, the sound made by the arterial current. **m., attrition**, a pericardial murmur. **m., blood-.** See *m., hemic*. **m., cardiac**, any adventitious sound heard over the region of the heart. In relation to their seat of generation, cardiac murmurs are designated as *mitral, aortic, tricuspid*, and *pulmonary*; according to the period of the heart's cycle at which they occur they are divided into *systolic*, those occurring during the systole; *diastolic*, those occurring in diastole; *presystolic* and *prediastolic*, those occurring just before systole and diastole respectively. See *table of murmurs* on page 577. **m., cardiopulmonary**, one produced by the impact of the heart against the lung. **m., diastolic**, a cardiac murmur occurring during the diastole. **m., direct**, a murmur produced by obstruction to the blood-current as it is passing in its normal direction. **m., Duroziez's**, the double murmur sometimes heard in the femoral artery in aortic regurgitation. **m., dynamic**, one resulting from tumultuous and irregular action of the heart. **m., endocardial**, a murmur produced within the cavities of the heart. (See *table* below.) **m., exocardial**, a murmur connected with the heart, but produced outside of its cavities. **m., Flint's**, a murmur sometimes heard at the apex of the heart in aortic regurgitation. It is generally presystolic in time, and is probably due to the fact that on account of the extreme ventricular dilatation the valves cannot be forced back against the walls and produce a relative narrowing of the auriculoventricular orifice. **m., friction-**, a sound produced by the rubbing of two inflamed serous surfaces upon each other. **m., functional**, a cardiac murmur occurring from excited action of the heart or anemic condition of the individual, without any structural change in the valves or orifices. **m., hemic**, a sound believed to be due to changes in the quality or amount of the blood and not to lesions of the vessels or valves. It is heard especially in anemic conditions. **m., indirect**, one produced by the blood flowing in a direction contrary to the normal course. **m., inorganic**, a murmur not due to valvular lesions; a hemic or a functional murmur. **m., mitral**, one produced at the mitral orifice. **m., muscular**. 1. The sound heard on auscultation of a contracting muscle. 2. The first sound of the heart. **m., musical**, a cardiac murmur having a musical quality. **m., organic**, a murmur due to structural changes in the heart. **m., paradox**, a systolic murmur prolonged so as to appear to be followed by a diastolic murmur. **m., presystolic**, a cardiac murmur occurring just before systole. **m., regurgitant**, one due to the blood flowing backward into the ventricle. **m., systolic**, cardiac murmur occurring during the systole. **m., vesicular**. See *vesicular*. **m., venous, m., whiffling, m., whistling.** See *bruit de diable*.

**Murphy's button** (*mur'-fe*) [John Benjamin *Murphy*, American surgeon, 1857– ]. A mechanical device for bringing together the visceral surfaces of the intestine in intestinal anastomosis. **M. drip**, proctoclysis. **M.'s law.** See under *jaundice*. **M.'s treatment.** 1. Treatment of peritonitis by continuous low-pressure proctoclysis, the patient being placed

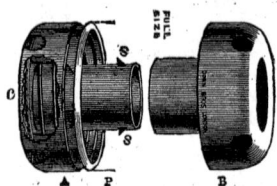

MURPHY'S BUTTON

A. Male half. B. Female half. P. Spring flange. s, s. Springs projecting through opening in hollow stem. Part of the cap of the male half has been cut away at c to show circular spring which acts as flange. The round holes in the caps are for drainage.

in Fowler's position. 2. Treatment of pulmonary tuberculosis by causing collapse of the affected lung through intrapleural injection of nitrogen.

**murrain** (*mur'-ān*) [Fr., from Lat. *mori*, to die]. 1. Any fatal disease of cattle and sheep. 2. Foot-and-mouth disease. **m., bloody**, Texas fever.

**mus** (*mus*) [L. *pl. mures*]. 1. A genus of rodents of the family *Muridæ*; it embraces rats and mice. 2. Any mouse-like formation; e. g., joint-mice. **mures articulares, mures articulorum.** See *arthrolith*.

**muscæ** (*mus'-ke*) [L.]. Plural of *musca*, a fly. **m. hispaniolæ**, cantharides: **m. volitantes**, floating specks in the field of vision due to opacities in the media of the eye.

**muscarine** (*mus'-kar-ēn*), $C_5H_{13}NO_2+H_2O$. A poisonous alkaloid obtained from *Agaricus muscarius*. It causes depression of the action of the heart and the respiration, increases the excretions of the salivary and lacrimal glands and of the intestine, and produces contraction of the pupil. **m. nitrate**, $C_5H_{13}NO_2HNO_3$. A brown mass used as an antihidrotic and antispasmodic.

**muscegenetic** (*mus-e-jen-et'-ik*) [*musca*, a fly; *generare*, to produce]. Causing muscæ volitantes or photopsia.

**muscle** (*mus'-l*) [*musculus*, a muscle]. A structure composed chiefly of muscular tissue and having the property of contracting. Muscles are of two kinds, the *striped*, or striated, and the *smooth*, or unstriated, the first being voluntary, the second involuntary, *i. e.*, not subject to the will. An alphabetical list of muscles is appended. **m., antagonistic**, one acting in opposition to another. **m., bicaudate**, one that has two distinct insertions. **m., biventer**, one with two bellies. **m.-bound**, said of muscles which have become inelastic from overuse. **m., carpophalangeus**, one extending from the carpus to a phalanx. **m.-casket**, a muscle-compartment. **m.-clot**, the clot formed in the coagulation of muscle plasma. **m.-column.** See under *muscular tissue*. **m.-compartment**, one of the divisions of a muscle-fiber produced by the extension of Krause's membrane from the sarcolemmma. **m., complex, m., compound**, a muscle possessing more than one point of origin or of insertion or of both. **m.s, congenerous**, those with related action. **m.-curve**, a *myogram, q. v.* **m., cutaneous.** 1. Having cutaneous origins and insertions. 2. Having cutaneous insertions. **m.-fiber**, the ultimate element of which muscular tissue is made up. Voluntary muscles consist of transversely striated fibers, involuntary muscles of spindle-shaped fibers or cells. **m.s, hypaxial, m.s, hyposkeletal**, those which pass below the vertebral axis; in man in front of the bodies of the vertebræ. **m.-imbalance**, lack of muscular balance, as between the muslces of the eyes. **m.s, internal**, involuntary muscles. **m., involuntary**, one not under the control of the will, as the nonstriated muscles. **m., isotonic**, one that contracts on stimulation, its tension remaining the same. **m.s, multicipital**, those having distinctly more than one origin. **m., nonstriated**, one composed of spindle-shaped muscle-fibers. See *muscular tissue.* **m.s, papillary,** the muscular eminences in the ventricles of the heart, from which the chordæ tendineæ arise. **m.s, pectinate,** the serrated muscles. **m.-plasma,** the fluid portion of the muscle tissue. It is neutral or alkaline and spontaneously coagulable, and contains myosinogen (the coagulable substance), paramyosinogen, myoglobulin, myoalbumoses, and albumin. **m.-plate,** a segment of embryonic mesoderm forming muscles. **m., polycipital.** See *m.s, multicipital.* **m.s, polygastric,** long muscles separated into parts by transverse tendinous septa. **m.s, postaxial,** muscles on the dorsal aspect of the limbs, which lie at right angles to the spine. **m.s, preaxial,** muscles on the ventral aspect of the limbs, which lie at right angles to the spine. **m.-rod,** one of the ultimate divisions of the dim band of a muscle-compartment. **m.s, serrated,** broad muscles with serrated edges by means of which they are inserted. **m.-serum,** the liquid residue after coagulation of muscle-plasma. **m.s, simple,** those having a single point of origin and one of insertion. **m., skeletal,** any one of the muscles attached to and acting on the skeleton. **m., striated, m., striped,** a muscle constituted of striped muscle-fibers. See *muscular tissue.* **m.-sugar.** See *inosit.* **m.-tonus,** the condition of slight but continued contraction in a muscle which causes the tension peculiar to waking hours. **m.s, tricipital,** muscles having three distinct origins. **m., unstriated, m., unstriped.** See *m., nonstriated.* **m., vestigial,** one that is rudimentary in man but well developed in the lower animals. **m., voluntary.** See *m., striated.* TABLE OF MUSCLES (ARRANGED ALPHABETICALLY): **abductor digiti quinti.** See *abductor minimi digiti.* **abductor hallucis,** *origin,* outer head, os calcis, plantar fascia, intermuscular septum; inner head, internal annular ligament and tendon of tibialis posticus; *insertion,* inner portion of lower surface of base of great toe and inner side of internal sesamoid bone; *innervation,* internal plantar division of posterior tibial nerve; it flexes and abducts the first phalanx of the great toe. **abductor jndicis.** See *interossei of hand, dorsal.* **abductor longus pollicis.** See *extensor ossis metacarpi pollicis* in this table. **abductor minimi digiti manus** (*A. digiti quinti*), *origin,* pisiform bone; *insertion,* first phalanx of little finger; *innervation,* ulnar; abducts little finger. **abductor minimi digiti pedis** (*A. digiti quinti*), *origin,* outer tuberosity of the os calcis and plantar fascia; *insertion,* first phalanx of little toe; *innervation,* external plantar; it abducts the little toe. **abductor pollicis brevis,** *origin.* See *abductor pollicis* in this table. **abductor pollicis longus.** See *extensor ossis metacarpi pollicis* in this table. **abductor pollicis manus,** *origin,* trapezium, scaphoid, annular ligament, palmar fascia; *insertion,* first phalanx of thumb; *innervation,* median; it abducts and flexes the first phalanx of the thumb. **abductor pollicis pedis,** *origin,* inner tuberosity of os calcis; *insertion,* first phalanx of great toe; *innervation,* internal plantar; abducts great toe. **accelerator urinæ,** *origin,* central tendon of perineum and median raphe; *insertion,* bulb, spongy and cavernous parts of the penis; *innervation,* perineal; it ejects the urine. **accessorius ad ilio-costalem,** *origin,* upper border of angles of lower six ribs; *insertion,* upper border of angles of upper six ribs and back of transverse process of seventh cervical vertebra; *innervation,* branches of dorsal nerves; it erects the spine and bends the trunk backward. **adductor brevis,** *origin,* ramus of pubes; *insertion,* upper part of the linea aspera of femur; *innervation,* obturator; it adducts, rotates externally, and flexes the thigh. **adductor hallucis,** *origin,* tarsal ends of the three middle metatarsal bones; *insertion,* base of the first phalanx of great toe; *innervation,* external plantar; it adducts the great toe. **adductor longus,** *origin,* front of pubes; *insertion,* middle of linea aspera of femur; *innervation,* obturator; it adducts, rotates outward, and flexes the thigh. **adductor magnus,** *origin,* rami of pubes and ischium; *insertion,* along the linea aspera of femur; *innervation,* obturator and great sciatic; it adducts the thigh and rotates it outward. **adductor minimus,** a name given to the upper portion of the adductor magnus. **adductor obliquus hallucis,** *origin,* tarsal end of second, third, and fourth metatarsal bones; *insertion,* base of first phalanx of great toe; *innervation,* external plantar; it adducts great toe. **adductor obliquus pollicis.** See *adductor pollicis* in this table. **adductor pollicis,** *origin,* third metacarpal; *insertion,* first phalanx of

thumb; *innervation,* ulnar; it draws the thumb to median line. **adductor transversus hallucis.** See *transversus pedis* in this table. **adductor transversus pollicis.** See *adductor pollicis* in this table. **adenoid, adenopharyngeal.** See *thyroadenoideus* in this table. **alares,** the pterygoid muscles. **anconeus,** *origin,* back of external condyle of humerus; *insertion,* olecranon process and shaft of ulna; *innervation,* musculospiral; it extends the forearm. **antitragicus,** *origin,* outer surface of antitragus; *insertion,* caudate process; *innervation,* temporal and posterior auricular. **arrectores pili,** *origin,* pars papillaris of the skin; *insertion,* hair-follicles; *innervation,* sympathetic; they elevate the hairs of the skin. **articularis genu.** See *subcrureus* in this table. **aryepiglotticus.** See *arytenoepiglottideus.* **arytenoepiglottideus inferior,** *origin,* arytenoid (anteriorly); *insertion,* epiglottis; *innervation,* recurrent laryngeal; it compresses the saccule of the larynx. **arytenoepiglottideus superior,** *origin,* apex of arytenoid cartilage; *insertion,* arytenoepiglottidean folds; *innervation,* recurrent laryngeal; it constricts the aperture of the larynx. **arytenoideus,** *origin,* posterior and outer border of one arytenoid; *insertion,* back of other arytenoid; *innervation,* superior and recurrent laryngeal; it closes the back part of the glottis. **attollens aurem,** *origin,* occipitofrontalis aponeurosis; *insertion,* pinna; *innervation,* temporal branch of facial; it elevates the pinna. **attrahens aurem,** *origin,* lateral cranial aponeurosis; *insertion,* helix; *innervation,* facial; it advances the pinna. **auricularis anterior.** See *attrahens aurem* in this table. **auricularis posterior.** See *retrahens aurem* in this table. **auricularis superior.** See *attollens aurem* in this table. **azygos uvulæ,** *origin,* posterior nasal spine of palate bone; *insertion,* uvula; *innervation,* facial through sphenopalatine ganglion; it raises the uvula. **biceps brachii,** *origin:* 1. Long head—glenoid cavity. 2. Short head—coracoid process; *insertion,* tuberosity of radius; *innervation,* musculocutaneous; it flexes and supinates the forearm. **biceps femoris,** *origin,* ischial tuberosity and linea aspera; *insertion,* head of fibula and outer tuberosity of head of fibula; *innervation,* great sciatic and external popliteal; it flexes and rotates the leg outward. **biceps flexor cruris,** the biceps femoris. **biceps flexor cubiti.** See *biceps brachii* in this table. **biventer cervicis,** *origin,* transverse processes, 2 to 4 upper dorsal; *innervation,* superior curved line of occipital bone. It is a portion of the complexus and retracts and rotates the head. **biventer mandibulæ.** See *digastric* in this table. **brachialis anticus,** *origin,* the lower half of the shaft of the humerus; *insertion,* coronoid process of ulna; *innervation,* musculocutaneous, musculospiral; it flexes the forearm. **brachiofascialis,** a part of the brachialis anticus inserted into the fascia of the arm. **brachio-radialis,** the supinator longus. **Brücke's.** See *ciliary* in this table. **buccinator,** *origin,* alveolar process of maxillary bones and pterygomaxillary ligament; *insertion,* orbicularis oris; *innervation,* buccal branch of facial; it compresses the cheeks and retracts the angle of the mouth. **bulbocavernosus.** See *accelerator urinæ* and *sphincter vaginæ* in this table. **caninus.** See *levator anguli oris* in this table. **cephalopharyngeus.** See *constrictor of pharynx, superior,* in this table. **cervicalis ascendens,** *origin,* angles of five upper ribs; *insertion,* transverse processes of fourth, fifth, and sixth cervical vertebras; *innervation,* branches of cervical; it keeps the head erect. **chondrofascialis,** a part of the pectoralis major inserted into the fascia of the arm. **chondroglossus,** *origin,* base of the lesser cornu of the hyoid b$_o$ne; *insertion,* tongue; *innervation,* hypoglossal; it depresses and draws down the side of the tongue. **ciliary** (Bowman's muscle), *origin:* 1. *Longitudinal* portion (Brücke's muscle): junction of cornea and sclera; 2. *Circular* portion (Mueller's muscle): the fibers form a circle; *insertion:* 1. Outer layers of choroid. 2. Ciliary processes; *innervation,* ciliary; it is the muscle of visual accommodation. **cleidomastoideus,** the dorsal part of the sternocleidomastoid. **cleidooccipital,** an anomalous muscle arising from the clavicle externally to the sternomastoid and inserted into the superior curved line of the occipital bone. **coccygeus,** *origin,* ischial spine; *insertion,* coccyx, sacrum, and sacrococcygeal ligament; *innervation,* sacral; it supports the coccyx and closes the pelvic outlet. **Coiter's muscle,** the corrugator supercilii. **complexus,** *origin,* transverse processes seventh cervical and six upper dorsal, and articular processes of third to sixth cervical vertebræ; *insertion*, occipital bone; *innervation*, suboccipital, great occipital, and branches of cervical; it retracts and rotates the head. **compressor narium,** *origin*, nasal aponeurosis; *insertion*, fellow muscle and canine fossa; *innervation*, facial; it compresses the nostril. **compressor narium minor,** *origin*, alar cartilage; *insertion*, skin at end of nose; *innervation*, facial; it dilates the nostril. **compressor sacculi laryngis,** *origin*, fibers of the arytenoepiglottideus; *innervation*, recurrent laryngeal; it is the compressor of the laryngeal saccule. **compressor urethræ,** *origin*, ramus of pubes; *insertion*, fellow muscle; *innervation*, perineal; it compresses the membranous urethra. **compressor vaginæ**, the analogue of the two bulbocavernosi of the male; *innervation*, perineal; it compresses the vagina. **compressor venæ dorsalis penis,** *origin*, fibers of the ischiocavernosus; *insertion*, fascial sheath of the penis, over the dorsal vessels; it is the compressor of the dorsal vein. **constrictor of pharynx (inferior),** *origin*, cricoid and thyroid cartilages; *insertion*, pharyngeal raphe; *innervation*, glossopharyngeal, pharyngeal plexus, and external laryngeal; it contracts the caliber of the pharynx. **constrictor of pharynx (middle),** *origin*, cornua of hyoid bone and stylohyoid ligament; *insertion*, pharyngeal raphe; *innervation*, glossopharyngeal and pharyngeal plexus; it contracts the caliber of the pharynx. **constrictor of pharynx (superior),** *origin*, internal pterygoid plate, pterygomaxillary ligament, jaw, and side of tongue; *insertion*, pharyngeal raphe; *innervation*, glossopharyngeal and pharyngeal plexus; it contracts the caliber of the pharynx. **constrictor urethræ.** See *compressor urethræ* in this table. **coracobrachialis,** *origin*, coracoid process of scapula; *insertion*, inner surface of shaft of humerus; *innervation*, musculocutaneous; it adducts and flexes the arm. **coracocervicalis,** an anomalous muscle arising from the coracoid process and passing upward and forward to be lost in the cervical fascia. Syn., *Krause's muscle*. **coracohumeral.** See *coracobrachialis*. **coracohyoid,** the omohyoid muscle. **coracopectoralis,** the pectoralis minor. **coracoradialis.** 1. The short head of the biceps. 2. The biceps muscle. **coracoulnaris,** the part of the biceps which has its point of insertion in the fascia of the forearm near the ulna. **corrugator cutis ani,** *origin*, submucous tissue on interior of anus; *insertion*, subcutaneous tissue on opposite side of anus; *innervation*, sympathetic; it corrugates the skin about the anus. **corrugator supercilii,** *origin*, superciliary ridge of frontal bone; *insertion*, orbicularis palpebrarum; *innervation*, facial; it draws the eyebrow downward and inward. **costoabdominal,** the obliquus externus. **costocoracoid,** the pectoralis minor. **costohyoideus,** the omohyoid muscle. **cremaster,** *origin*, upper and deep surface of middle of Poupart's ligament; *insertion*, spine and crest of pubic bone and fascia propria; *innervation*, genital branch of genitocrural; it elevates the testicle. **cricoarytenoideus lateralis,** *origin*, side of cricoid cartilage; *insertion*, angle and external surface of arytenoid; *innervation*, recurrent laryngeal; it closes the glottis. **cricoarytenoideus posticus,** *origin*, back of cricoid cartilage; *insertion*, base of arytenoid cartilage; *innervation*, recurrent laryngeal; it opens the glottis. **cricothyroid,** *origin*, cricoid cartilage; *insertion*, thyroid cartilage (lower inner border); *innervation*, superior laryngeal; it makes the vocal bands tense. **crureus.** See *vastus intermedius* in this table. **cucullaris.** See *trapezius* in his table. **deltoid,** *origin*, clavicle, acromion, and spine of scapula; *insertion*, shaft of humerus; *innervation*, circumflex; it abducts the humerus. **depressor alæ nasi,** *origin*, incisive fossa of superior maxillary bone; *insertion*, septum and ala of nose; *innervation*, facial; buccal branch; it contracts the nostril. **depressor anguli oris,** *origin*, external oblique line of inferior maxillary bone; *insertion*, angle of mouth; *innervation*, facial: supramaxillary branch; it depresses the angle of the mouth. **depressor epiglottidis,** those fibers of the thyroepiglottideus that are continued forward to the margin of the epiglottis. **depressor labii inferioris,** *origin*, external oblique line of the inferior maxillary bone; *insertion*, lower lip; *innervation*, facial: supramaxillary branch; it depresses the lip. **depressor septi.** See *depressor alæ nasi*, in this table. **depressor urethræ,** *origin*, ramus of ischium near deep transversus perinei; *insertion*, fibers of constrictor vaginæ muscle. **detrusor urinæ,** *origin*, front of pubis; *insertion*, prostate (in male), vagina (in female); *innervation*, sympathetic; it compresses bladder. **diaphragm,** *origin*, ensiform cartilage, six or seven lower ribs, ligamenta arcuata, bodies of lumbar vertebræ; *insertion*, central tendon; *innervation*, phrenic and sympathetic; *function*, respiration. **digastric** (anterior belly), *origin*, inner surface of inferior maxillary bone, near symphysis; *insertion*, hyoid bone; *innervation*, inferior dental; it elevates the hyoid bone and the tongue. **digastric** (posterior belly), *origin*, digastric groove of mastoid process; *insertion*, hyoid bone; *innervation*, facial; it elevates the hyoid bone and the tongue. **dilator naris anterioris**, *origin*, alar cartilage; *insertion*, border of ala of nose; *innervation*, facial: infraorbital branch; dilates the nostril. **dilator pupillæ,** *origin*, circumference of iris; *insertion*, margin of pupil; *innervation*, long ciliary (sympathetic); it dilates the pupil. **dilator naris posterioris,** *origin*, nasal notch of superior maxillary bone; *insertion*, skin at margin of nostril; *innervation*, facial: infraorbital branch; dilates the nostril. **dorsal interossei** (4), *origin*, sides of metacarpal bones; *insertion*, bases of corresponding phalanges; *innervation*, ulnar; abduct the fingers from the median line. **dorsal interossei** (4), *origin*, sides of metatarsal bones; *insertion*, base of first phalanx of corresponding toe; *innervation*, external plantar; abduct the toes. **dorsiscapularis,** the rhomboideus major and rhomboideus minor regarded as one. **elevator,** the levator muscles. **epicranius.** See *occipitofrontalis* in this table. **erector clitoridis,** *origin*, tuberosity of ischium; *insertion*, each side of crus of clitoris; it erects the clitoris. **erector penis,** *origin*, ischial tuberosity, crus penis, and pubic ramus; *insertion*, tunica albuginea of corpus cavernosum; *innervation*, perineal; *function*, to maintain erection. **erector pili.** See *arrectores pili* in this table. **erector spinæ,** *origin*, iliac crest, back of sacrum, lumbar and three lower dorsal spines; *insertion*, divides into sacrolumbalis, longissimus dorsi, and spinalis dorsi; *innervation*, lumbar nerves, posterior division; *function*, extension of lumbar spines on pelvis. **Eustachian,** the laxator tympani. **extensor brevis digitorum (pedis),** *origin*, os calcis, externally; *insertion*, first phalanx of great toe and tendons of extensor longus; *innervation*, anterior tibial; it extends the toes. **extensor brevis hallucis,** a name applied to that portion of the extensor brevis digitorum that goes to the great toe. **extensor brevis pollicis.** See *extensor primi internodii pollicis* in this table. **extensor carpi radialis brevior,** *origin*, external condyloid ridge of humerus; *insertion*, base of second and third metacarpal; *innervation*, posterior interosseous; it extends the wrist. **extensor carpi radialis longior,** *origin*, lower one-third of external condyloid ridge of humerus; *insertion*, base of second metacarpal; *innervation*, musculospiral; it extends the wrist. **extensor carpi ulnaris,** *origin*: *first head*, external condyle of humerus; *second head*, posterior border of ulna; *insertion*, base of fifth metacarpal; *innervation*, posterior interosseous; it extends the wrist. **extensor coccygis,** *origin*, last bone of sacrum or first of coccyx; *insertion*, lower part of coccyx; *innervation*, sacral branches; it extends the coccyx. **extensor communis digitorum,** *origin*, external condyle of humerus; *insertion*, all of the second and third phalanges; *innervation*, posterior interosseous; it extends the fingers. **extensor digitorum brevis.** See *extensor brevis digitorum* in this table. **extensor digiti quinti proprius.** See *extensor minimi digiti* in this table **extensor digitorum longus.** See *extensor longus digitorum* in this table. **extensor hallucis longus.** See *extensor proprius hallucis* in this table. **extensor indicis,** *origin*, back of ulna; *insertion*, second and third phalanges of index-finger; *innervation*, posterior interosseous; it extends the index-finger. **extensor longus digitorum pedis,** *origin*, outer tuberosity of tibia and shaft of fibula; *insertion*, second and third phalanges of toes; *innervation*, anterior tibial; it extends the toes. **extensor longus pollicis.** See *extensor secundi internodii pollicis* in this table. **extensor minimi digiti,** *origin*, external condyle of humerus; *insertion*, second and third phalanges of little finger; *innervation*, posterior interosseous; it extends the little finger. **extensor ossis metacarpi pollicis,** *origin*, back of radius and ulna and interosseous membrane; *insertion*, base of metacarpal of thumb and fascia; *innervation*, posterior interosseous; it extends the thumb. **extensor ossis metatarsi hallucis** (anomalous), *origin*, extensor proprius hallucis, or extensor communis digitorum, or tibialis anticus; *insertion*, metatarsal bone of hallux. **extensor pollicis brevis.** See *extensor primi internodii pollicis* in this table. **extensor pollicis longus.** See *extensor longus pollicis* in this table. **extensor primi internodii pollicis**, *origin*, back of radius; *insertion*, base of first phalanx of thumb; *innervation*, posterior interosseous; it extends the thumb. **extensor proprius hallucis**, *origin*, middle of fibula; *insertion*, base of last phalanx of great toe; *innervation*, anterior tibial; it extends the great toe. **extensor proprius minimi digiti**, *origin*, lower part of ulna or posterior ligament of wrist-joint; *insertion*, base of first phalanx of little finger; it extends the little finger. **extensor secundi internodii pollicis**, *origin*, back of ulna; *insertion*, base of last phalanx of thumb; *innervation*, posterior interosseous; it extends the thumb. **extensor, ulnocarpal**, the extensor carpi ulnaris. **extrarectus.** 1. The pyriformis. 2. The rectus externus muscle of the eye. **Fallopian, the pyramidalis. fascialis.** See *tensor vaginæ femoris* in this table. **flexor accessorius digitorum** (of foot) (*two heads*), *origin*, inner and outer surface of os calcis; *insertion*, tendon of flexor longus digitorum; *innervation*, external plantar; it is the accessory flexor of toes. **flexor accessorius longus digitorum** (of foot), *origin*, shaft of tibia or fibula; *insertion*, tuberosity of os calcis, and joins tendon of long flexor; *innervation*, external plantar; it assists in flexing the toes. **flexor brevis digitorum** (of foot), *origin*, inner tuberosity of os calcis and plantar fascia; *insertion*, second phalanges of the lesser toes; *innervation*, internal plantar; it flexes the lesser toes. **flexor brevis hallucis**, *origin*, under surface of cuboid, plantar ligaments, and external cuneiform; *insertion*, base of first phalanx of great toe; *innervation*, internal plantar; it flexes and slightly adducts the first phalanx of the great toe. **flexor brevis minimi digiti** (of foot), *origin*, base of fifth metatarsal; *insertion*, base of first phalanx of little toe; *innervation*, external plantar; it flexes the little toe. **flexor brevis minimi digiti (of hand)**, *origin*, unciform bone and annular ligament; *insertion*, first phalanx of little finger; *innervation*, ulnar; it flexes the little finger. **flexor brevis pollicis** (of foot). See *flexor brevis hallucis* in this table. **flexor brevis pollicis (of hand)** (*two heads*), *origin—outer*: lower border of anterior annular ligament; ridge of trapezium; *inner*: os magnum and bases of first, second, and third metacarpal bones; *insertion*, base of first phalanx of thumb; *innervation*, *outer head*—median: palmar branch; *inner head*—deep ulnar; it flexes the metacarpal bone of the thumb. **flexor carpi radialis**, *origin*, internal condyle of humerus; *insertion*, metacarpal bone of index-finger; *innervation*, median; it flexes the wrist. **flexor carpi ulnaris** (*two heads*), *origin:* 1. Internal condyle. 2. Olecranon and ulna; *insertion*, fifth metacarpal, annular ligament, and pisiform bone; *innervation*, ulnar; it flexes the wrist. **flexor digiti quinti brevis.** See *flexor brevis minimi digiti* in this table. **flexor digitorum brevis.** See *flexor brevis digitorum* in this table. **flexor digitorum longus.** See *flexor longus digitorum* in this table. **flexor digitorum profundus.** See *flexor profundus digitorum* in this table. **flexor digitorum sublimis.** See *flexor sublimis digitorum* in this table. **flexor hallucis brevis.** See *flexor brevis hallucis* in this table. **flexor hallucis longus.** See *flexor longus hallucis* in this table. **flexor indicis**, the name given to the indicial portion of the flexor profundus digitorum when it is distinct. **flexor longus digitorum** (of foot), *origin*, shaft of tibia; *insertion*, last phalanges of toes; *innervation*, posterior tibial; it flexes the phalanges and extends the ankle. **flexor longus hallucis**, *origin*, lower two-thirds of shaft of fibula; *insertion*, last phalanx of great toe; *innervation*, posterior tibial; it flexes the great toe. **flexor longus pollicis**, *origin*, shaft of radius and coronoid process of ulna; *insertion*, last phalanx of thumb; *innervation*, anterior interosseous; it flexes the thumb. **flexor ossis metacarpi pollicis.** See *opponens pollicis* in this table. **flexor pollicis brevis** See *flexor brevis pollicis* in this table. **flexor pollicis longus.** See *flexor longus pollicis* in this table. **flexor profundus digitorum**, *origin*, shaft of ulna; *insertion*, last phalanges by four tendons; *innervation*, ulnar and anterior interosseous; it flexes the phalanges; **flexor sublimis digitorum** (*three heads*), *origin:* 1. Inner condyle. 2. Coronoid process. 3. Oblique line of radius; *insertion*, second phalanges by four tendons; *innervation*, median; it flexes the second phalanges.

**gastrocnemius** (*two heads*), *origin*, condyles of femur; *insertion*, os calcis by tendo Achillis; *innervation*, internal popliteal; it extends the foot. **gemellus inferior**, *origin*, tuberosity of ischium and lesser sacrosciatic notch; *insertion*, great trochanter; *innervation*, sacral; *function*, external rotator of the thigh. **gemellus superior**, *origin*, ischial spine and lesser sacrosciatic notch; *insertion*, great trochanter; *innervation*, sacral; *function*, external rotator of the thigh. **geminous, geminus, gemelli, gemini**, the combined gemellus inferior and gemellus superior. **genioglossus.** See *geniohyoglossus* in this table. **geniohyoglossus**, *origin*, superior genial tubercle of inferior maxillary bone; *insertion*, hyoid and inferior surface of tongue; *innervation*, hypoglossal; it retracts and protrudes the tongue. **geniohyoid**, *origin*, inferior genial tubercle of inferior maxillary bone; *insertion*, body of hyoid; *innervation*, hypoglossal; it elevates and advances the hyoid. **glossopalatinus**, a small muscle constricting the action of the fauces and composing the major part of the anterior pillar. **gluteoperineal**, an anomalous branch of the superficial transverse perineal muscle which originates from the fascia covering the gluteus maximus. **gluteus maximus**, *origin*, superior curved iliac line and crest, sacrum, and coccyx; *insertion*, fascia and femur below great trochanter; *innervation*, inferior gluteal and sacral plexus; it extends, abducts, and rotates the thigh outward. **gluteus medius**, *origin*, ilium between superior and middle curved lines; *insertion*, oblique line of great trochanter; *innervation*, superior gluteal; it rotates, abducts, and advances the thigh. **gluteus minimus**, *origin*, ilium between middle and inferior curved lines; *insertion*, great trochanter; *innervation*, superior gluteal; it rotates, abducts, and draws the thigh forward. **gracilis**, *origin*, rami of pubes and ischium; *insertion*, tibia, upper and inner part; *innervation*, obturator; it flexes and abducts the leg. **gubernaculum testis.** See *cremaster* in this table. **Guthrie's.** See *compressor urethræ* in this table. **helicis major et minor**, *origin*, tubercle on helix; *insertion*, rim of helix near summit: *innervation*, auriculotemporal and posterior auricular. **Hilton's.** See *compressor sacculi laryngis* in this table. **hippicus**, the tibialis anticus. **Horner's.** See *ciliary* and *tensor tarsi* in this table. **hyoglossus**, *origin*, cornua of hyoid bone; *insertion*, side of tongue; *innervation*, hypoglossal; it depresses the side of the tongue and retracts it. **iliacus**, *origin*, iliac fossa and crest, base of sacrum; *insertion*, lesser trochanter, upper part of shaft of femur; *innervation*, anterior crural; it flexes and rotates the thigh outward. **iliacus externus**, the pyriformis. **iliocostal.** See *sacrolumbalis* in this table. **iliocostalis lumborum.** See *sacrolumbalis* in this table. **iliocostalis cervicis.** See *cervicalis ascendens* in this table. **iliocostalis dorsi.** See *accessorius ad iliocostalem* in this table. **iliocostalis lumborum.** See *sacrolumbalis* in this table. **iliolumbalis**, the quadratus lumborum. **iliopsoas**, the iliacus and psoas muscles considered as one. **infracostals** (10), *origin*, inner surface of ribs; *insertion*, inner surface of two or three ribs above; *innervation*, intercostal; *function*, expiration; by depressing the ribs. **infraoblique.** See *obliquus capitis inferior* in this table. **infraspinatus**, *origin*, infraspinous fossa; *insertion*, great tuberosity of humerus; *innervation*, suprascapular; it rotates the humerus outward. **interaccessory**, short lumbar muscles connecting the accessory processes of the vertebræ. **interarytenoid**, *origin*, one arytenoid cartilage; *insertion*, the other arytenoid cartilage; *innervation*, recurrent laryngeal; *function*, approximates arytenoid cartilages. **intercostals, external** (11), *origin*, outer lip of inferior costal border; *insertion*, superior border of ribs above; *innervation*, intercostal; they raise the ribs in inspiration. **intercostals, internal** (11), *origin*, inner lip of inferior costal border; *insertion*, superior border of rib below; *innervation*, intercostal; they depress the ribs in expiration. **internal pleuricostals.** See *intercostals, internal*, in this table. **interossei of foot, dorsal** (4), *origin*, adjacent surfaces of metatarsal bones; *insertion*, bases of first phalanges; *innervation*, external plantar; they abduct from the middle line of the second toe. **interossei of foot, plantar** (3), *origin*, inner lower surface of three outer metatarsal bones; *insertion*, bases of first phalanges of three outer toes; *innervation*, external plantar; they adduct the outer three toes. **interossei of hand, dorsal** (4), *origin*, five metacarpal bones; *insertion*, sides of aponeurosis of extensor communis and adjacent parts of first phalanges; *innervation*, ulnar; they abduct the index, middle, and ring fingers, and aid in flexing the first phalanges and extending the second and third. **interossei of hand, palmar** (3), *origin*, sides of metacarpal bones; *insertion*, aponeurosis of extensor tendons, adjacent part of first phalanges; *innervation*, ulnar; they abduct the index, ring, and little fingers, and aid in flexing the first phalanges and in extending the second and third. **interspinales**, *origin*, upper surface of spines of vertebræ, near tip; *insertion*, posterior part of lower surface of spine above; *innervation*, internal divisions of posterior branches of spinal nerves; they extend the vertebræ next above. **intertransversales or intertransversarii**, *origin*, between transverse processes of contiguous vertebræ; *innervation*, spinal nerves; they flex the spinal column laterally. **ischiocavernosus**. See *erector penis* and *erector clitoridis* in this table. **Jarjavay's**. See *depressor urethræ* in this table. **keratocricoid**. See *cricothyroid* in this table. **keratoglossus**, those fibers of the hyoglossus that arise from the greater cornu of the hyoid bone. **latissimus colli**. See *platysma myoides* in this table. **latissimus dorsi**, *origin*, spines of six lower dorsal and lumbar and sacral vertebræ, crest of ilium, and three or four lower ribs; *insertion*, bicipital groove of humerus; *innervation*, long subscapular; it draws the arm backward and downward and rotates it inward. **laxator tympani**, *origin*, spinous process of sphenoid bone and Eustachian tube; *insertion*, neck of malleus; *innervation*, facial; it relaxes the membrana tympani. **levator anguli oris**, *origin*, canine fossa of superior maxillary bone; *insertion*, angle of mouth; *innervation*, infraorbital branch of facial; it elevates the angle of the mouth. **levator anguli scapulæ**, *origin*, transverse processes of four upper cervical vertebræ; *insertion*, posterior border of scapula; *innervation*, fifth cervical and cervical plexus; it elevates the upper angle of the scapula. **levator ani**, *origin*, posterior portion of body and ramus of pubes, pelvic fascia, ischial spine; *insertion*, rectum, occcyx, and fibrous raphe; *innervation*, sacral and perineal; supports the rectum and vagina. **levator labii inferioris**, *origin*, incisive fossa of inferior maxillary bone; *insertion*, skin of lower lip; *innervation*, supramaxillary branch of facial; it elevates the lower lip. **levator labii superioris**, *origin*, lower margin of orbit; *insertion*, upper lip; *innervation*, infraorbital branch of facial; it elevates the upper lip. **levator labii superioris alæque nasi**, *origin*, nasal process of superior maxillary bone; *insertion*, alar cartilage and upper lip; *innervation*, infraorbital branch of facial; it elevates the upper lip and dilates the nostril. **levator menti**. See *levator labii inferioris* in this table. **levator palati**, *origin*, petrous portion of temporal bone; *insertion*, soft palate; *innervation*, sphenopalatine ganglion (facial); it elevates the soft palate. **levator palpebræ superioris**, *origin*, lesser wing of sphenoid; *insertion*, upper tarsal cartilage; *innervation*, third cranial; it lifts the upper lid. **levator scapulæ**. See *Levator anguli scapulæ* in this table. **levator veli palatini**. See *levator palati* in this table. **levatores costarum** (12), *origin*, transverse processes of last cervical and dorsal vertebræ; *insertion*, each to the rib below; *innervation*, intercostal; they raise the ribs. **lingualis**, *origin*, under surface of tongue; *innervation*, chorda tympani and hypoglossal; it elevates the middle of the tongue. **lingualis superior**, *origin*, a band of fibers extending from base to apex of the upper surface of the tongue. **longer straight**, the rectus capitis posticus major. **longissimus capitis**. See *trachelomastoid* in this table. **longissimus cervicis**. See *transversalis colli* in this table. **longissimus dorsi**, *origin*, erector spinæ; *insertion*, transverse processes of lumbar and dorsal vertebræ and seventh to eleventh ribs; *innervation*, branches of lumbar and dorsal; it erects the spine and bends the trunk backward. **longus capitis**. See *rectus capitis anticus major* in this table. **longus colli**. ‡. Superior oblique portion. 2. Inferior oblique portion. 3. Vertical portion. *Origin*: 1. Transverse processes third to fifth cervical vertebræ. 2. Bodies of first to third dorsal vertebræ. 3. Bodies of three second and two cervical vertebræ. *Insertion*: 1. Anterior tubercle of atlas. 2. Transverse processes fifth and sixth cervical vertebræ. 3. Bodies of second to fourth cervical vertebræ. *Innervation*, lower cervical. It flexes the cervical vertebræ. **lumbofemoralis**, the psoas magnus. **lumboiliacus**, the psoas parvus. **lumbricales** (4) (**of foot**), *origin*, tendons of flexor longus digitorum; *insertion*, first phalanges of the lesser toes; *innervation*, internal and external plantar; they are accessory flexors. **lumbricales** (4) (**of hand**), *origin*, tendons of flexor profundus digitorum; *insertion*, tendons of common extensor; *innervation*, median and ulnar; they flex the first phalanges. **malaris**, a part of the orbicularis palpebrarum originating in the inner inferior margin of the orbit and adjacent part of the dorsum of the nose and uniting partly with the zygomaticus major. **malledius**, the tensor tympani. **masseter**, *origin*, zygomatic arch; *insertion*, angle and ramus of jaw; *innervation*, inferior maxillary; it is the muscle of mastication. **mastoideus anterior, mastoideus colli**, the sternomastoid. **mastoideus lateralis**, the trachelomastoid. **mentalis**. See *levator labii inferioris* in this table. **midriff**. See *diaphragm* in this table. **Mueller's**. See *ciliary* in this table. **Mueller's** (**orbital**), *origin*, connected with the orbital periosteum; it crosses the sphenomaxillary fissure; *innervation*, sympathetic. **Mueller's** (**superior palpebral**), *origin*, connected with the levator palpebræ superioris; *insertion*, along the upper border of the tarsus; *innervation*, sympathetic; it assists in raising the upper lid. **multifidus spinæ**, *origin*, sacrum, iliac spine, articular processes of lumbar and cervical vertebræ, and transverse processes of dorsal and seventh cervical; *insertion*, laminæ and spines from last lumbar to second cervical vertebra; *innervation*, posterior spinal branches; it erects and rotates the spinal column. **musculus accessorius ad sacrolumbalem**, *origin*, angles of six lower ribs; *insertion*, angles of six upper ribs; *innervation*, branches of dorsal; it erects the spine and bends the trunk backward. **mylohyoid**, *origin*, mylohyoid ridge of inferior maxillary bone; *insertion*, body of hyoid and raphe; *innervation*, inferior dental; it elevates and advances the hyoid bone. **nasalis**, a name suggested for the pyramidalis nasi, a part of the levator labii superioris and the outer portion of the depressor alæ nasi. **nasolabialis**, *origin*, nasal septum; *insertion*, upper lip; *innervation*, facial; it connects the upper lip with the septum of the nose. **nasotransversalis**, the compressor narium. **nauticus**, the tibialis posticus. **obliquus auris or auriculæ**, *origin*, concha of ear; *insertion*, fossa of anthelix; *innervation*, temporal and posterior auricular. **obliquus capitis inferior**, *origin*, spinous process of axis; *insertion*, transverse process of atlas; *innervation*, suboccipital and great occipital; it rotates the atlas and the cranium. **obliquus capitis superior**, *origin*, transverse process of atlas; *insertion*, occipital bone; *innervation*, suboccipital and great occipital; it draws the head backward. **obliquus externus abdominis**, *origin*, eight lower ribs; *insertion*, middle line, iliac crest, Poupart's ligament; *innervation*, intercostal, iliohypogastric, ilioinguinal; it compresses the viscera and flexes the thorax. **obliquus inferior**, *origin*, orbital plate of superior maxillary bone; *insertion*, sclerotic; *innervation*, third cranial; it rotates the eyeball upward and outward. **obliquus internus abdominis**, *origin*, lumbar fascia, iliac crest, Poupart's ligament; *insertion*, three lower ribs, linea alba, pubic crest, pectineal line; *innervation*, intercostal, iliohypogastric, ilioinguinal; it compresses the viscera, flexes the thorax, and assists in expiration. **obliquus superior**, *origin*, above optic foramen, through pulley; *insertion*, sclerotic; *innervation*, fourth cranial; it rotates the eyeball downward and inward. **obturator externus**, *origin*, obturator foramen and membrane; *insertion*, digital fossa at base of great trochanter; *innervation*, obturator; *function*, external rotator of the thigh. **obturator internus**, *origin*, obturator foramen and membrane; *insertion*, great trochanter; *innervation*, sacral; *function*, external rotator of thigh. **occipitalis**. See *occipitofrontalis* in this table. **occipitofrontalis**, *origin*, superior curved line of occiput and angular process of frontal bone; *insertion*, aponeurosis; *innervation*, posterior auricular, small occipital, facial; it moves the scalp. **omohyoid**, *origin*, upper border of scapula; *insertion*, body of hyoid bone; *innervation*, descendens and communicans noni; it depresses and retracts the hyoid bone. **opisthenar**, the erector spinæ. **opponens digiti quinti**. See *opponens minimi digiti* in this table. **opponens minimi digiti**, *origin*, unciform bone; *insertion*, fifth metacarpal; *innervation*, ulnar; it flexes the little finger. **opponens pollicis**, *origin*, trapezium, anterior annular ligament; *insertion*, metacarpal bone of thumb; *innervation*, median, palmar division; it flexes the thumb. **orbicularis oculi.** See *orbicularis palpebrarum* in this table. **orbicularis oris,** *origin*, nasal septum and canine fossa of inferior maxilla, by accessory fibers; *insertion*, forms lips and sphincter of mouth; *innervation*, buccal and supramaxillary branches of facial; it closes the mouth. **orbicularis palpebrarum,** *origin*, mesal margin of orbit; *insertion*, lateral margin of orbit; *innervation*, facial; it closes the eyelids. **orbitalis.** See *Mueller's muscle* in this table. **palatoglossus,** *origin*, soft palate; *insertion*, side and dorsum of tongue; *innervation*, sphenopalatine ganglion; it constricts the fauces. **palatopharyngeus,** *origin*, soft palate; *insertion*, thyroid cartilage and pharynx; *innervation*, sphenopalatine ganglion; it closes the posterior nares. **palmaris brevis,** *origin*, annular ligament and palmar fascia; *insertion*, skin of palm of hand; *innervation*, ulnar; it corrugates the skin of the palm. **palmaris interossei** (3), *origin*, palmar surfaces second, fourth, and fifth metacarpal bones; *insertion*, bases of first phalanges of corresponding fingers; *innervation*, ulnar; adduct the fingers. **palmaris longus,** *origin*, internal condyle of humerus; *insertion*, annular ligament and palmar fascia; *innervation*, median; it renders the palmar fascia tense. **palpebral,** the orbicularis palpebrarum. **pectineus,** *origin*, iliopectineal line and pubes; *insertion*, femur below lesser trochanter; *innervation*, anterior crural, obturator; it flexes and rotates the thigh outward. **pectoralis major,** *origin*, clavicle, sternum, and costal cartilages; *insertion*, external bicipital ridge of humerus; *innervation*, anterior thoracic, external and internal; it draws the arm downward and forward. **pectoralis minor,** *origin*, third, fourth, and fifth ribs; *insertion*, coracoid process; *innervation*, anterior thoracic; it depresses the point of the shoulder. **pericardiothyroideus,** a band of fibers extending from the isthmus of the thyroid gland to the anterior surface of the fibrous layer of the pericardium. **peristaphylinus externus,** the abductor of the Eustachian tube, a muscle arising from the lower surface of the sphenoid bone and from the membranous part of the cartilaginous portion of the Eustachian tube; it is inserted into the hamular process of the pterygoid bone. **peristaphylinus internus,** the levator palati. **peroneus brevis,** *origin*, middle third of shaft of fibula, externally; *insertion*, base of fifth metatarsal bone; *innervation*, musculocutaneous; it extends the foot. **peroneus longus,** *origin*, head and shaft of fibula; *insertion*, first metatarsal of great toe; *innervation*, musculocutaneous; it extends and everts the foot. **peroneus tertius,** *origin*, lower fourth of fibula; *insertion*, fifth metatarsal bone; *innervation*, anterior tibial; it flexes the tarsus. **pharyngopalatinus.** See *palatopharyngeus* in this table. **plantaris,** *origin*, outer bifurcation of linea aspera and posterior ligament of knee-joint; *insertion*, os calcis by means of the tendo Achillis; *innervation*, internal popliteal; it extends the foot. **plantaris interossei** (3), *origin*, shafts of third, fourth, and fifth metatarsal bones; *insertion*, bases of first phalanges of corresponding toes; *innervation*, external plantar; adduct the toes. **platysma myoides,** *origin*, clavicle, acromion, and fascia; *insertion*, inferior maxillary bone, angle of mouth; *innervation*, facial and superficial cervical; it wrinkles the skin and depresses the mouth. **popliteus,** *origin*, external condyle of femur; *insertion*, shaft of tibia above oblique line; *innervation*, internal popliteal; it flexes the leg. **procerus.** See *pyramidalis nasi* in this table. **prodigus,** the supinator longus. **pronator quadratus,** *origin*, lower fourth of ulna; *insertion*, lower fourth of shaft of radius; *innervation*, anterior interosseous; it pronates the hand. **pronator radii teres,** *origin*, internal condyle of humerus and coronoid process of ulna; *insertion*, outer side of shaft of radius; *innervation*, median; it pronates the hand. **psoas magnus** or **major,** *origin*, bodies and transverse processes of last dorsal and all lumbar vertebræ; *insertion*, lesser trochanter; *innervation*, lumbar; it flexes and rotates the thigh outward, and flexes the trunk on the pelvis. **psoas parvus** or **minor,** *origin*, bodies of last dorsal and first lumbar vertebræ; *insertion*, iliopectineal eminence and iliac fascia; *innervation*, lumbar; it flexes the pelvis upon the abdomen. **pterygoid (external),** *origin*, two heads: 1, external pterygoid plate of sphenoid bone; 2, great wing of sphenoid bone; *insertion*, neck of condyle of lower jaw; *innervation*, inferior maxillary; it draws the inferior maxillary bone forward. **pterygoid (internal),** *origin*, pterygoid fossa of sphenoid bone; *insertion*, inner surface of angle of jaw; *innervation*, inferior maxillary; it raises and draws the inferior maxilla forward. **pterygopalatal, pterygopalatine,** the portion of the levator palati passing from the hamular process of the sphenoid bone to the palate bone. **pubovesical,** a bundle of muscular fibers taking origin in the back of the pubes and extending with the anterior true ligament of the bladder to lose itself in the superficial muscular layer of the bladder. **pyloric.** See *sphincter pyloricus* in this table. **pyramidalis,** *origin*, pubes; *insertion*, linea alba; *innervation*, iliohypogastric; it renders the linea alba tense. **pyramidalis nasi,** *origin*, occipitofrontalis; *insertion*, compressor naris; *innervation*, infraorbital branch of facial; it depresses the inner angle of the eyebrow. **pyriformis,** *origin*, front of sacrum, through great sciatic foramen; *insertion*, great trochanter; *innervation*, branch of sacral plexus; *function*, external rotator of thigh. **quadratus femoris,** *origin*, tuberosity of the ischium; *insertion*, quadrate line of femur; *innervation*, fifth lumbar and first sacral; *function*, external rotator of thigh. **quadratus labii inferioris.** See *depressor labii inferioris* in this table. **quadratus labii superioris.** The levator labii superioris, levator labii alæque nasi, and zygomaticus minor combined. **quadratus lumborum,** *origin*, crest of ilium, transverse processes of lower three lumbar vertebræ; *insertion*, last rib, transverse processes of upper three lumbar vertebræ; *innervation*, twelfth thoracic and upper lumbar; it flexes the thorax laterally. **quadratus plantæ.** See *flexor accessorius digitorum* in this table. **quadriceps extensor femoris,** includes the rectus, vastus internus and externus, and crureus muscles; their common tendon surrounds the patella. **rectus abdominis,** *origin*, pubic crest and fibrous tissues in front of symphysis; *insertion*, cartilages of the ribs, from the fifth to the seventh; *innervation*, intercostal, iliohypogastric, ilioinguinal; it compresses the viscera and flexes the thorax. **rectus capitis anticus** (or **anterior**) **major,** *origin*, transverse processes third to sixth cervical vertebræ; *insertion*, basilar process of occipital bone; *innervation*, first and second cervical; it flexes the head and slightly rotates it. **rectus capitis anticus** (or **anterior**) **minor,** *origin*, transverse process and lateral mass of atlas; *insertion*, basilar process of occipital bone; *innervation*, first cervical; it flexes the head. **rectus capitis lateralis,** *origin*, ventral cephalic surface of lateral mass of atlas; *insertion*, jugular process of occipital bone; *innervation*, first cervical; it flexes the head laterally. **rectus capitis posticus** (or **posterior**) **major,** *origin*, spine of axis; *insertion*, inferior curved line of occipital bone; *innervation*, suboccipital and great occipital; it rotates the head. **rectus capitis posticus** (or **posterior**) **minor,** *origin*, dorsal arch of atlas; *insertion*, below inferior curved line of occipital bone; *innervation*, suboccipital and great occipital; it draws the head backward. **rectus externus,** *origin*, two heads, outer margin of optic foramen; *insertion*, sclera; *innervation*, sixth cranial; it rotates the eyeball outward. **rectus femoris,** *origin*, anterior inferior iliac spine, brim of acetabulum; *insertion*, proximal border of patella; *innervation*, anterior crural; it extends the leg. **rectus inferior,** *origin*, lower margin of optic foramen; *insertion*, sclera; *innervation*, third cranial; it rotates the eyeball downward. **rectus internus,** *origin*, inner margin of optic foramen; *insertion*, sclera; *innervation*, third cranial; it rotates the eyeball inward. **rectus lateralis.** See *rectus externus* in this table. **rectus medialis.** See *rectus internus* in this table. **rectus superior,** *origin*, margin of optic foramen; *insertion*, sclera; *innervation*, third cranial; it rotates the eyeball upward. **retrahens aurem,** *origin*, mastoid process; *insertion*, concha; *innervation*, posterior auricular; it retracts the pinna. **rhomboatloideus,** an anomalous muscle arising from the spinous processes of the lower cervical or upper dorsal vertebræ; it is inserted into the transverse process of the atlas. **rhomboideus major,** *origin*, spines of first five thoracic vertebræ; *insertion*, root of spine of scapula; *innervation*, fifth cervical; it elevates and retracts the scapula. **rhomboideus minor,** *origin*, spines of seventh cervical and first dorsal vertebræ; *insertion*, root of spine of scapula; *innervation*, fifth cervical; it retracts and elevates the scapula. **risorius,** *origin*, fascia over masseter; *insertion*, angle of mouth; *innervation*, buccal branch of facial; it draws the angle laterally.

# MUSCLE

**rotatores spinæ,** *origin,* transverse processes of from second to twelfth thoracic vertebræ; *insertion,* lamina of next vertebra above; *innervation,* dorsal branches of spinal; they rotate the spinal column. **sacrolumbalis,** *origin,* erector spinæ; *insertion,* angle of six lower ribs; *innervation,* branches of dorsal; it erects the spine and bends the trunk backward. **sacrospinalis,** the erector spinæ. **salpingopharyngeus,** a small muscle arising from the Eustachian tube and descending to blend with the constrictors of the pharynx. **Santorini's.** 1. The risorius. 2. The involuntary muscular fibers encircling the urethra beneath the constrictor urethræ. Syn., *Santorini's circular muscle.* **sartorius,** *origin,* anterior superior spine of ilium; *insertion,* upper internal portion of shaft of tibia; *innervation,* anterior crural; it flexes the leg upon the thigh and the thigh upon the pelvis; it rotates the thigh outward. **scalenus anticus,** or *anterior, origin,* scalene tubercle on first rib; *insertion,* transverse processes third to sixth cervical vertebræ; *innervation,* lower cervical; it flexes the neck laterally. **scalenus lateralis,** an anomalous muscle originating in the dorsal tubercles of the transverse processes of the fourth, fifth, and sixth cervical vertebræ, and having its point of insertion with the scalenus medius. **scalenus medius,** *origin,* first rib; *insertion,* transverse processes of six lower cervical vertebræ; *innervation,* lower cervical; it flexes the neck laterally. **scalenus posticus,** or *posterior, origin,* second rib; *insertion,* transverse processes of three lower cervical vertebræ; *innervation,* lower cervical; it bends the neck laterally. **semimembranosus,** *origin,* tuberosity of ischium; *insertion,* inner tuberosity of tibia; *innervation,* great sciatic; it flexes the leg and rotates it inward. **semispinalis capitis.** See *complexus* in this table. **semispinalis colli,** or *cervicis, origin,* transverse processes of four upper dorsal and articular processes of four lower cervical vertebræ; *insertion,* spines of second to fifth cervical vertebræ; *innervation,* cervical branches; it erects the spinal column. **semispinalis dorsi,** *origin,* transverse processes of sixth to tenth dorsal vertebræ; *insertion,* spines of last two cervical and first four thoracic vertebræ; *innervation,* branches of dorsal; it erects the spinal column. **semitendinosus,** *origin,* tuberosity of ischium; *insertion,* upper and inner surfaces of tibia; *innervation,* great sciatic; it flexes the leg on the thigh. **serratus magnus** or *anterior, origin,* eight upper ribs; *insertion,* inner margin of dorsal border of scapula; *innervation,* posterior thoracic; it elevates the ribs in inspiration. **serratus posticus** (or *posterior*) *inferior, origin,* spines of last two thoracic and first three lumbar vertebræ; *insertion,* four lower ribs; *innervation,* tenth and eleventh intercostal; it depresses the ribs in expiration. **serratus posticus** (or *posterior*) *superior, origin,* spines of seventh cervical and first two thoracic vertebræ; *insertion,* second, third, fourth, and fifth ribs; *innervation,* second and third intercostal; it raises the ribs in inspiration. **soleus,** *origin,* shaft of fibula, oblique line of tibia; *insertion,* os calcis by tendo Achillis; *innervation,* internal popliteal and posterior tibial; it extends the foot. **sphincter ani** (external), *origin,* tip of coccyx; *insertion,* tendinous center of perineum; *innervation,* perineal, pudic, and fourth sacral; it closes the anus. **sphincter ani** (internal), a thickening of the circular fibers of the intestine an inch above the anus; *innervation,* hemorrhoidal nerves; it constricts the rectum. **sphincter antri pylorici,** a band of circular fibers distant from the pyloric orifice of the stomach about 10 cm.; also called the transverse band. **sphincter pupillæ,** involuntary fibers of the iris arranged circularly around the pupil and having a width of about 0.08 cm. **sphincter pyloricus,** the aggregation of the fibers of the circular coat at the pyloric orifice of the stomach. **sphincter urethræ membranaceæ,** the anterior fibers of the *compressor urethra, q. v.* **sphincter vaginæ,** *origin,* central tendon of perineum; *insertion,* corpora cavernosa and clitoris; the homologue of the accelerator urinæ in the man. **sphincter vesicæ,** the aggregation of the fibers of the circular layer of the muscular coat, about the neck of the bladder and the beginning of the urethra. **sphincter vesicæ internus,** *origin,* near the urethral orifice of the bladder; *innervation,* vesical nerves; it constricts the internal orifice of the urethra. **spinalis.** Same as *spinalis dorsi* in this table. **spinalis cervicis** (*normal, but inconstant*), *origin,* spines of fifth, sixth, and seventh cervical and first two thoracic vertebræ; *insertion,* spine of

# MUSCLE

axis, sometimes spines of third and fourth cervical vertebræ. **spinalis colli,** *origin,* spines of fifth and sixth cervical vertebræ; *insertion,* spine of axis, or third and fourth cervical spines; *innervation,* cervical branches; it steadies the neck. **spinalis dorsi,** *origin,* last two thoracic and first two lumbar spines; *insertion,* remaining thoracic spines; *innervation,* dorsal branches; it erects the spinal column. **splenius accessorius.** See *rhomboatloideus* in this table. **splenius capitis,** *origin,* lower two-thirds of ligamentum nuchæ, spines of seventh cervical and first two thoracic vertebræ; *insertion,* outer third of middle oblique line of occiput and outer surface of mastoid process; *innervation,* middle cervical, posterior branches; it extends the head and neck and rotates and flexes laterally. **splenius colli** or *cervicis, origin,* spines of third to sixth thoracic vertebræ; *insertion,* dorsal tubercles of transverse processes of upper three or four cervical vertebræ; *innervation,* posterior divisions of lower cervical; it extends, flexes laterally, and rotates the neck. **square.** See *quadratus lumborum* in this table. **stapedius,** *origin,* interior of pyramid; *insertion,* neck of stapes; *innervation,* facial; it depresses the base of the stapes. **sternocleidomastoid.** See *sternomastoid* in this table. **sternomastoid,** *origin,* two heads, sternum and clavicle; *insertion,* mastoid process and outer half of superior oblique line of occiput; *innervation,* spinal accessory and cervical plexus; it depresses and rotates the head. **sternocostal, sternocostalis.** See *triangularis sterni* in this table. **sterno₀humeral.** See *pectoralis major* in this table. **sternohyoid,** *origin,* sternum and clavicle; *insertion,* hyoid bone; *innervation,* descending and communicating branches of the hypoglossal; it depresses the hyoid bone. **sternomastoid.** See *sternomastoid* in this table. **sternothyroid,** *origin,* sternum and cartilage of first rib; *insertion,* side of thyroid cartilage; *innervation,* descendens and communicans hypoglossi; it depresses the larynx. **styloglossus,** *origin,* styloid process; *insertion,* side of tongue; *innervation,* hypoglossal; it elevates and retracts the tongue. **stylohyoid,** *origin,* styloid process; *insertion,* body of hyoid; *innervation,* facial; it draws the hyoid upward and backward. **stylopharyngeus,** *origin,* styloid process; *insertion,* thyroid cartilage; *innervation,* glossopharyngeal and pharyngeal plexus; it elevates the pharynx. **subanconeus,** *origin,* humerus above olecranon fossa; *insertion,* posterior ligament of elbow; *innervation,* musculospiral; *function,* tensor of ligament. **subclavius,** *origin,* cartilage of first rib; *insertion,* inferior surface of clavicle; *innervation,* fifth and sixth cervical; it draws the clavicle downward. **subcostals.** See *infracostals* in this table. **subcrureus,** *origin,* anterior distal part of femur; *insertion,* synovial sac behind patella; *innervation,* anterior crural; it draws up the synovial sac. **subscapularis,** *origin,* under surface of scapula; *insertion,* humerus, lesser tuberosity and shaft; *innervation,* subscapular; it is the chief internal rotator of the humerus. **supinator.** See *supinator radii brevis* in this table. **supinator longus,** *origin,* external condyloid ridge of humerus; *insertion,* styloid process of radius; *innervation,* musculospiral; it flexes and supinates the forearm. **supinator radii brevis,** *origin,* external condyle of humerus, oblique line of ulna; *insertion,* neck of radius and its bicipital tuberosity; *innervation,* posterior interosseous; it supinates the hand. **supraspinales,** lie on spinous processes in cervical region. **supraspinatus,** *origin,* supraspinous fossa; *insertion,* great tuberosity of humerus; *innervation,* suprascapular; supports the shoulder-joint and raises the arm. **tailors'.** See *sartorius* in this table. **temporal,** *origin,* temporal fossa and fascia; *insertion,* coronoid process of mandible; *innervation,* inferior maxillary; it elevates the mandible. **tensor fasciæ latæ.** See *tensor vaginæ femoris* in this table. **tensor palati,** *origin,* scaphoid fossa and alar spine of sphenoid; *insertion,* soft palate; winds about hamular process; *innervation,* otic ganglion; it renders the palate tense. **tensor tarsi,** *origin,* crest of lacrimal bone; *insertion,* tarsal cartilages; *innervation,* infraorbital branch of facial; it compresses the puncta and lacrimal sac. **tensor tympani,** *origin,* temporal bone, Eustachian tube and canal, sphenoid bone; *insertion,* handle of malleus; *innervation,* otic ganglion; it renders tense the membrana tympani. **tensor vaginæ femoris,** *origin,* iliac crest and anterior superior spinous process; *insertion,* fascia lata; *innervation,* superior gluteal; *function,* tensor of fascia lata. **tensor veli palatini.** See *tensor palati* in this table. **teres major,** *origin,* inferior angle of scapula; *insertion,* internal bicipital ridge of humerus; *innervation,* subscapular; it draws the arm downward and backward. **teres minor,** *origin,* axillary border of scapula;; *insertion,* great tuberosity of humerus; *innervation,* circumflex; it rotates the humerus outward and adducts it. **thenar,** the abductor and flexor muscles of the thumb. **thyroadenoideus,** a fascicle of the inferior constrictor of the pharynx having its point of insertion on the side of the thyroid gland. **thyropalatinus.** See *palatopharyngeus* in this table. **thyroarytenoideus,** *origin,* thyroid cartilage and cricothyroid membrane; *insertion,* arytenoid, inferior and anterior surface; *innervation,* recurrent laryngeal; it relaxes the vocal bands. **thyroepiglottideus,** *origin,* inner surface of thyroid cartilage; *insertion,* epiglottis and arytenoepiglottidean fold; *innervation,* recurrent laryngeal; it depresses the epiglottis. **thyrohyoid,** *origin,* side of thyroid cartilage; *insertion,* body and greater cornu of hyoid bone; *innervation,* hypoglossal; it elevates the larynx. **tibialis anticus,** or **anterior,** *origin,* outer tuberosity and upper part of shaft of tibia; *insertion,* internal cuneiform and first metatarsal bone; *innervation,* anterior tibial; it flexes the tarsus and elevates the inner border of the foot. **tibialis posticus,** or **posterior,** *origin,* shaft of fibula and tibia, interosseous membrane; *insertion,* tuberosity of scaphoid, internal cuneiform, and bases of second to fourth metatarsal; *innervation,* posterior tibial; it extends the tarsus and inverts the foot. **trachelomastoid,** *origin,* transverse processes of third to sixth thoracic, and articular processes of last three or four cervical vertebræ; *insertion,* mastoid process; *innervation,* branches of cervical; it steadies the head. **tragicus,** *origin,* tragus; *insertion,* tragus; *innervation,* temporal and posterior auricular. **transversalis** (or **transversus**) **abdominis,** *origin,* Poupart's ligament, iliac crest, six lower ribs, lumbar vertebræ; *insertion,* linea alba, pubic crest, pectineal line; *innervation,* intercostal, iliohypogastric, ilioinguinal; it compresses the viscera and flexes the thorax. **transversalis cervicis,** or **colli,** *origin,* transverse processes of third to sixth thoracic vertebræ; *insertion,* transverse processes of five lower cervical vertebræ; *innervation,* cervical branches; it keeps the neck erect. **transversus' abdominis.** See *transversalis abdominis* in this table. **transversus auris,** or **auriculæ,** *origin,* convexity of concha; *insertion,* convexity over groove of helix; *innervation,* temporal and posterior auricular; it retracts the helix. **transversus pedis,** *origin,* head of fifth metatarsal and plantar ligaments of metatarsophalangeal joints; *insertion,* first phalanx of great toe; *innervation,* external plantar; it adducts the great toe. **transversus perinei,** *origin,* ramus of ischium; *insertion,* central tendon; *innervation,* perineal; *function,* tensor of central tendon. **transversus perinei, deep.** See *compressor urethræ* in this table. **transversus' thoracis.** See *triangularis sterni* in this table. **trapezius,** *origin,* superior curved line of occipital bone, spinous processes of last cervical and all the dorsal vertebræ; *insertion,* clavicle, spine of scapula, and acromion; *innervation,* spinal accessory and cervical plexus; it draws the head backward. **triangularis.** See *depressor anguli oris* in this table. **triangularis sterni,** *origin,* ensiform cartilage, costal cartilages of three or four lower true ribs, and sternum; *insertion,* border of inner surfaces of second, third, fourth, and fifth costal cartilages; *innervation,* intercostal; *function,* expiration. **triceps brachii** (*three heads*), *origin, external* and *internal* near musculospiral groove, shaft of humerus; *middle* or *long,* lower margin of glenoid cavity; *insertion,* olecranon process of ulna; *innervation,* musculospiral; it extends the forearm. **triceps sural, triceps suræ,** the gastrocnemius and soleus considered as one. **triticeoglossus.** See *Bochdalek's muscle*. **trochlear, trochlearis.** See *obliquus superior* in this table. **of the ureters, musculi ureterum,** a bundle of thin fibers from the urinary bladder running between the openings of the ureters. **urethrobulbar,** the accelerator urinæ. **uvulæ.** See *azygos uvulæ* in this table. **varolii.** See *stapedius* in this table. **vastus externus** or **lateralis,** *origin,* anterior border of great trochanter and linea aspera of femur; *insertion,* tuberosity of tibia; *innervation,* anterior crural; it extends the leg. **vastus internus** or **medialis,** *origin,* inner lip of linea aspera of femur; *insertion,* tuberosity of tibia; *innervation,* anterior crural; it

extends the leg. **Wilson's,** a non-constant fasciculus of the compressor urethræ which is attached to the body of the pubis, near the symphysis. **Zaglas' perpendicular external,** the vertical fibers of the tongue, which, decussating with the transverse fibers and the insertions of the geniohyoglossus, curve outward in each half of the tongue. **zygomaticus major** et **minor,** *origin,* malar bone; *insertion,* angle of mouth; *innervation,* infraorbital branch of facial; draw the upper lip upward and outward. **musculamin** (*mus-kū-lā'-min*). A nitrogenous base from muscular tissue.

**muscular** (*mus'-kū-lar*) [*muscle*]. 1. Pertaining to or affecting muscles. 2. Having well-developed muscles. **m. anesthesia,** a lack of the muscular sense. **m. reflex,** a deep or tendon reflex. **m. rheumatism,** rheumatism affecting the muscles. **m. sense,** the sense of motion, weight, and position upon which the adjustment of the body to its surroundings depends. **m. system,** the muscles of the body taken together. **m. tissue,** the tissue of which muscles are composed; it is of two principal varieties—the *striped, striated,* or *voluntary,* and the *unstriped, nonstriated,* or *involuntary* muscular tissue. The striped muscular tissue is composed of muscle-fibers, the unstriped of elongated, spindle-shaped cells. A *muscle-fiber* consists of sarcolemma, muscle-nuclei, and muscle-substance. The *sarcolemma* is an elastic, homogeneous, connective-tissue sheath which lightly invests the muscle-fiber. The *nuclei* are fusiform in shape, are situated immediately beneath the sarcolemma, and run parallel with the axis of the fiber. The muscle-fiber itself consists of two substances—a dark, doubly refracting, or anisotropic contractile substance, and a lighter, semifluid, isotropic substance termed *sarcoplasm*. The *contractile substance* consists of delicate spindles, the apposition of the thicker portions of which produces the dark transverse disc. The spindles terminate in minute beads, the juxtaposition of which gives rise to the intermediate disc or *membrane of Krause*. The neutral *sarcoplasm* fills out the spaces left between the spindles, and, as ordinarily the tapering ends of the spindles are too delicate to be seen, the space between the intermediate and transverse discs—the *lateral disc*—looks homogeneous. The contractile fibrils into which the contractile substance is divided, formed by the end-to-end union of the spindles, are aggregated into bundles—the *muscle-columns*—surrounded by a layer of sarcolemma. On transverse section these muscle-columns give rise to the appearance known as *Cohnheim's fields*. The muscle-fibers are held together by delicate areolar tissue—the *endomysium*. Several grouped together form a primary bundle, which is surrounded by a sheath termed the *perimysium*. *Fasciculi* are aggregations of primary bundles, and are the units of which the complete muscle is composed, the latter being surrounded by the *epimysium*. **m.-tone.** See *muscletonus.* **m. tumor.** See *myoma*. **musculatio mucosæ** (*mus-kū-la'-ris mū-ko'-se*). The layer of unstriped muscular tissue separating the mucosa (of mucous membranes) from the submucosa.

**musculario** (*mus-kū-lar'-it-e*) [*musculus,* a muscle]. 1. The quality of being muscular. 2. The contractile power or tone of a muscle.

**musculation** (*mus-kū-la'-shun*) [*muscle*]. The muscular endowment of the body or a part; also, the action of the muscles.

**musculature** (*mus'-kū-la-tūr*) [*muscle*]. The muscular system of the body, or a part of it.

**musculi** (*mus'-kū-li*) [L.]. Plural of *musculus,* a muscle. **m. papillares,** certain muscular pillars within the cardiac ventricles. **m. pectinati** [*pecten,* a comb], small, muscular columns traversing the inner surface of the appendix auriculæ and the adjoining portion of the wall of the auricle.

**musculin** (*mus'-kū-lin*) [*muscle*]. 1. Extract of muscle tissue. See *organotherapy.* 2. See *paramyosinogen*.

**musculo-** (*mus-kū-lo-*) [*muscle*]. A prefix signifying relating to the muscles.

**musculoaponeurotic** (*mus-kū-lo-ap-on-ū-rot'-ik*). Composed of muscle and of fibrous connective tissue in the form of a membrane.

**musculocutaneous** (*mus-kū-lo-kū-tā'-ne-us*) [*musculo-; cutaneous*]. Pertaining to or supplying the muscles and skin, as the *musculocutaneous* nerve of the arm or leg.

**musculodermic** (*mus-kū-lo-derm'-ik*). See *musculocutaneous*.

**musculoelastic** (*mus-kū-lo-e-las'-tik*). Muscular and elastic; made up of muscular and elastic tissue.

**musculointestinal** (*mus-kū-lo-in-tes'-tin-al*). Relating to the muscles and the intestine.

**musculomembranous** (*mus-kū-lo-mem'-bran-us*) [*musculo-*; *membranous*]. Pertaining to or consisting of both muscles and membrane.

**musculophrenic** (*mus-kū-lo-fren'-ik*) [*musculo-*; φρήν, diaphragm]. Pertaining to or supplying the muscles and the diaphragm, as the *musculophrenic nerve*.

**musculospinal** (*mus-kū-lo-spi'-nal*). Relating to or distributed to the muscles and spine.

**musculospiral** (*mus-kū-lo-spi'-ral*). 1. Supplying muscles and having a spiral course. 2. Pertaining to the musculospiral nerve. **m. groove**, a depression of the posterior surface of the humerus, running downward and outward, and transmitting the musculospiral nerve and superior profunda artery.

**musculotegumentary** (*mus-kū-lo-teg-ū-men'-ta-re*). Affecting both muscles and integument.

**musculotonic** (*mus-kū-lo-ton'-ik*) [*musculo-*; τόνος, tone]. Relating to the tone or contractility of a muscle.

**musculous** (*mus'-kū-lus*). Composed of or containing muscular fibers.

**musculus** (*mus'-kū-lus*). See *muscle*.

**muscus** (*mus'-kus*) [L.]. Moss; lichen.

**mush.** A thick porridge, or boiled pudding, usually of maize meal, used as an article of diet, and also as a poultice.

**mushroom** (*mush'-room*). 1. See *agaric* and *fungus*. 2. To flatten-out or "upset" upon striking; said of an expansive bullet.

**musicians' cramp.** See *cramp, professional*.

**musicomania** (*mū-zik-o-ma'-ne-ah*). Monomania for, or insane devotion to, music.

**musicotherapy** (*mū-zik-o-ther'-ap-e*). The use of music in the treatment of disease, chiefly mental and nervous diseases, and in convalescence.

**musin** (*mū'-sin*). A proprietary cathartic said to be made from tamarinds.

**musk.** The dried secretions from the preputial follicles of *Moschus moschiferus*, a mammal of the order of Ruminantia. Musk (*moschus*, U. S. P.) occurs in grains or lumps, possessing a peculiar, penetrating odor. Its composition is complex, among its constituents are ammonia, stearin, olein, cholesterin, a volatile oil, gelatin, albumin, and certain salts. It is a stimulant and antispasmodic, and is used in typhoid fever and other low fevers, in adynamic pneumonia, in laryngismus stridulus, and in hiccup. Dose 10 gr. (0.65 Gm.). **m., artificial**, trinitrobutyltoluene, C₆H₁.CH₃.C₄H₉.(NO₂)₃. Dose 10 gr. (0.65 Gm.). **m., vegetable**, the seeds of *Hibiscus abelmoschus*; it is stimulant, stomachic, and antispasmodic.

**musomania** (*mū-zo-ma'-ne-ah*). See *musicomania*.

**mussanine** (*mus'-an-ēn*). An alkaloid resembling saponin, from the bark of *Albizzia anthelmintica*, a species of acacia. Syn., *moussenin*.

**Musset's sign** (*moos-sa'*) [Alfred de *Musset*, French poet, 1810–1857, who presented the phenomenon]. Rhythmic movements of the head synchronous with the radial pulse, observed in persons with an aortic affection; it is considered a pathognomonic sign of an affection of the circulatory system.

**(de) Mussey's point, (de) M.'s symptom** (*moos'-se*). A point intensely painful on pressure at the intersection of a line continuous with the left border of the sternum and of another forming a prolongation of the tenth rib. It is noted in diaphragmatic pleurisy. Syn., *Boulon diaphragmatique*.

**mussitation** (*mus-it-a'-shun*) [*mussitare*, to mutter]. Movement of the lips without the production of articulate speech.

**must** [*mustum*, new wine]. The juice freshly expressed from grapes.

**mustard** (*mus'-tard*) [ME., *mustarde*]. 1. A plant of the genus *Brassica* (*Sinapis*), of the natural order *Cruciferæ*. 2. The seed of the mustard-plant. The seeds of two species are chiefly employed, those of *Brassica nigra*, yielding *black mustard*, or *sinapis nigra* (U. S. P.), and those of *Brassica alba*, yielding *white mustard*, or *sinapis alba* (U. S. P.). The seeds contain a fixed oil consisting of the glycerol-compounds of stearic, oleic, and erucic or brassic acid, and of sinapoleic and behenic acids, sinalbin (in white mustard), and a volatile oil (in black mustard). Mustard is used in dyspepsia, as an emetic (dose 1–2 dr.—4–8 Cc.), as a rubefacient and counter-irritant, especially in the form of mustard-plaster, and in baths. **m., flour of**, black and white mustard seeds mixed and pulverized. **m., oil of, volatile** (*oleum sinapis volatile*, U. S. P.), contains allyl isosulphocyanide, the active principle, which does not exist preformed in the seeds, but is produced by the decomposition of potassium myronate, or sinigrin, under the influence of a ferment, myrosin. Dose ½ min. (0.008 Cc.). **m.-paper** (*charta sinapis*, U. S. P.), is used locally. **m.-plaster**, is made by mixing mustard and flour with water. **m.-poultice** (*cataplasma sinapis*, B. P.), is made by mixing mustard, linseed-meal, and water.

**mutacism** (*mū'-tas-izm*) [μυτακισμός, fondness for the letter μ]. The too frequent use of the "m" sound, and its substitution for other sounds.

**mutase** (*mū'-tās*). A food-preparation rich in proteins, made from leguminous plants. It is used in gastrointestinal diseases. Dose 1 dessertspoonful (10 Cc.) several times daily with food.

**mutation** (*mū-ta'-shun*) [*mutare*, to change]. 1. Change. 2. In obstetrics, a pronounced change in the presentation of the fetus. 3. A sudden variation which oversteps the limits of species and produces new species or sports.

**mute** (*mūt*) [*mutus*, dumb]. Dumb; unable to use articulate speech.

**mutilation** (*mū-til-a'-shun*) [*mutilare*, to cut]. 1. The act of maiming or disfiguring. 2. The state thereby produced. 3. The condition of the parts remaining after the excision or amputation of a member or part.

**mutism** (*mū'-tizm*) [*mutus*, dumb]. Dumbness. See *mutitas*. **m., hysterical**, obstinate and voluntary silence although the vocal organs are uninjured and there is no visible lesion of the cerebral speech-centers.

**mutitas** (*mū'-tit-as*) [*mutus*, mute]. Dumbness. **m. atonica**, dumbness arising from disorder of the nerves of the tongue. **m. organica**, that from loss of the tongue. **m. pathematica**, dumbness caused by fright or passion. **m. spasmodica**, spasmodic dumbness. **m. surdorum**, dumbness consequent upon congenital deafness.

**mutualism** (*mū'-tū-al-izm*) [*mutuus*, reciprocal]. The living together of two organisms of different species, for the advantage that each derives from the other. Syn., *symbiosis*.

**mutualist** (*mū'-tū-al-ist*) [*mutuus*, reciprocal]. An organism living with another in a state of mutualism.

**muzzle** (*mus'-l*). The projecting jaws and nose of an animal; a snout.

**muzzling** (*mus'-ling*). Same as *infibulation*.

**my.** Abbreviation for *myopia*.

**myalgia** (*mi-al'-je-ah*) [μῦς, muscle; ἄλγος, pain]. Pain in the muscles; muscular rheumatism.

**myalgic** (*mi-al'-jik*) [μῦς, muscle; ἄλγος, pain]. Pertaining to or affected with myalgia.

**myameba** (*mi-am-e'-bah*) [μῦς, muscle; *amœba*, a unicellular organism]. A muscle-cell regarded as an organism. Cf. *neuramœba, osteameba*.

**myasis** (*mi-a'-sis*). See *myiasis*.

**myasthenia** (*mi-as-the'-ne-ah*) [μῦς, muscle; *asthenia*]. Muscular debility. **m. gravis pseudoparalytica**, a disease characterized by an abnormal exhaustibility of the voluntary muscles, manifesting itself in a rapid diminution of contractility both when the muscle is innervated by the will and when stimulated by the electric current.

**myasthenic** (*mi-as-then'-ik*). Characterized by myasthenia.

**myatonia** (*mi-ah-to'-ne-ah*) [μῦς, muscle; ἀ, priv.; τόνος, tone]. Absence of muscular tone.

**mycele** (*mi-sēl'*). Same as *mycelium*.

**mycelial** (*mi-se'-le-al*) [μύκης, fungus; ἧλος, nail, wart]. Pertaining to mycelium.

**mycelioid** (*mi-se'-le-oid*) [*mycelium*; εἶδος, like]. Like molds; said of colonies of bacteria having the appearance of mold colonies.

**mycelium** (*mi-se'-le-um*) [μύκης, fungus; ἧλος, nail; wart; excrescence]. The vegetative filaments of fungi, usually forming interwoven masses.

**mycetes** (*mi-se'-tēz*) [μύκης, fungus]. The fungi.

**mycethemia** (*mi-se-the'-me-ah*). See *mycohemia*.

**mycetism** (*mi'-set-izm*) [μύκης, fungus]. Mushroom-poisoning.

**mycetogenesis** (*mi-set-o-jen'-es-is*) [μύκης, fungus; γένεσις, genesis]. Development of fungi.
**mycetogenetic** (*mi-set-o-jen-et'-ik*) [*mycetogenesis*]. Produced or caused by the growth of fungi.
**mycetogenous** (*mi-set-oj'-en-us*). Same as *mycetogenetic*.
**mycetoid** (*mi'-set-oid*) [μύκης, fungus; εἶδος, like]. Resembling a fungus.
**mycetology** (*mi-set-ol'-o-je*) [μύκης, a fungus; λόγος, science]. Same as *mycology*.
**mycetoma** (*mi-set-o'-mah*). See *Madura-foot*.
**Mycetozoa** (*mi-set-o-zo'-ah*) [μύκης, fungus; ζῷον, animal]. A group of fungus-like saprophytic organisms, the slime-fungi or slime-molds.
**mycetozoic** (*mi-set-o-zo'-ik*) [μύκης, fungus; ζῷον, animal]. Of the nature of a slime-fungus or mycetozoon.
**Mycetozoon** (*mi-set-o-zo'-on*) [μύκης, fungus; ζῷον, animal]. Any organism belonging to the class Mycetozoa.
**mychmus, mychthismus** (*mik'-mus, mik-thiz'-mus*) [μύζειν, to moan]. Sighing.
**myco-** (*mi-ko-*) [μύκης, fungus]. A prefix meaning 1. fungus or fungoid; or 2. mucus.
**mycoangioneurosis** (*mi-ko-an-je-o-nū-ro'-sis*) [μύκος, mucus; ἀγγεῖον, vessel; *neurosis*]. A neurosis accompanied by a hypersecretion of mucus producing the affection mucous colitis.
**Mycobacterium** "(*mi-ko-bak-te'-re-um*) [*myco-*; *bacterium*]. A genus of bacteria of the family *Mycobacteriaceæ*; the cells are commonly short, cylindrical, sometimes bent and irregularly swollen, clavate or cuneate; Y-shaped forms may appear or longer filaments with true branching, or short coccoid elements which may be regarded as gonidia.
**mycocyte** (*mi'-ko-sīt*). See *mucocyte*.
**mycoderm** (*mi'-ko-derm*) [*myco-*; δέρμα, skin]. The mucous membrane.
**Mycoderma** (*mi-ko-der'-mah*) [*myco-*; δέρμα, skin]. A genus of fungi forming membranes upon or in fermenting liquids. **M. aceti**, the microorganism of acetous fermentation.
**mycodermatitis** (*mi-ko-der-mat-i'-tis*). Inflammation of a mucous membrane; catarrh.
**mycogastritis** (*mi-ko-gas-tri'-tis*) [μύκος, mucus; γαστήρ, stomach; ιτις, inflammation]. Catarrhal gastritis.
**mycography** (*mi-kog'-ra-fe*). Synonym of *mycology*.
**mycohemia, mycohæmia** (*mi-ko-hem'-e-ah*) [*myco-*; αἷμα, blood]. A condition in which fungi are in the blood.
**mycoid** (*mi'-koid*) [*myco-*; εἶδος, form]. Resembling, or appearing like, a fungus; fungoid.
**mycology** (*mi-kol'-o-je*) [*myco-*; λόγος, science]. The science of fungi.
**mycomycetes** (*mi-ko-mi-se'-tēz*) [μύκης, fungus]. Fungi reproducing by oospores or zygospores.
**mycomyringitis** (*mi-ko-mi-rin-ji'-tis*) [*myco-*; *myringitis*]. Fungous inflammation of the ear-drum.
**Myconostoc** (*mi-ko-nos'-tok*) [*myco-*; *Nostoc*]. A genus of schizomycetes.
**mycophthalmia** (*mi-koff-thal'-me-ah*) [*myco-*; *ophthalmia*]. Ophthalmia due to a fungus.
**mycophylaxin** (*mi-ko-fil-aks'-in*) [μύκης, a fungus; φύλαξ, a protector]. Any phylaxin that destroys microorganisms. See *phylaxin*.
**mycoprotein** (*mi-ko-pro'-te-in*) [*myco-*; *protein*, from πρῶτος, first]. The albuminoid substance obtained from the bodies of bacteria.
**mycopus**. See *mucopus*.
**mycose** (*mi'-kōs*) [μύκης, fungus], $C_{12}H_{22}O_{11} + 2H_2O$. *Trehalose*, a substance that occurs in several species of fungi, in ergot of rye, and in the oriental *Trehala*. It is distinguished from cane-sugar by its ready solubility in alcohol, greater stability, and stronger rotatory power. It does not reduce copper-solutions.
**mycosis** (*mi-ko'-sis*) [μύκης, fungus]. 1. A growth of fungi within the body. 2. A disease caused by the presence of microorganisms in the body. **m. cutis chronica**. See *furunculus orientalis*. **m. favosa, favus,** *q. v.* **m., fungoides,** an affection of the skin characterized by the appearance, chiefly on the scalp, face, and chest, of pinkish or reddish, nodular or flattened tumors, which may go on to ulceration. Itching is often present. The disease is believed to be microorganismal in origin. **m. intestinalis.** Synonym of *anthrax*. **m. mucosina,** a form in which the fungus invades the body through the intestines and leads to abscesses in the lungs, brain, and other organs.

**mycosozin** (*mi-ko-so'-zin*) [μύκης, a fungus; σώζειν, to save]. Any sozin that acts by the destruction of microbes.
**mycothrix** (*mi'-ko-thriks*) [μύκης, fungus; θρίξ, hair]. The thread-like element in the structure of a. micrococcus.
**mycotic** (*mi-kot'-ik*) [*mycosis*]. Characterized by mycosis; due to microorganisms. **m. endocarditis,** that due to microorganismal infection. Syn., *ulcerative* or *malignant endocarditis*.
**mycoticopeptic** (*mi-kot-ik-o-pep'-tik*). Mycotic and peptic.
**mycteric** (*mik-ter'-ik*) [μυκτήρ, the nose]. Relating to the nasal cavities.
**mycterophonia** (*mik-ter-o-fo'-ne-ah*) [μυκτήρ, nose; φωνή, voice]. A nasal quality of the voice.
**mycteroxerosis** (*mik-ter-o-ze-ro'-sis*) [μυκτήρ, nose; *xerosis*]. Dryness of the nostrils.
**mydaleine** (*mid-a'-le-ēn*) [μυδαλέος, putrid]. A ptomaine obtained from putrefying cadaveric organs. It produces salivation, dilatation of the pupils, elevation of temperature, peristaltic action of the intestine, first a rise, then a fall, of temperature, and diastolic arrest of the heart.
**mydatoxin** (*mid-at-oks'-in*) [μυδᾶν, to be wet; τοξικόν, poison], $C_6H_{13}NO_2$. A ptomaine obtained from horseflesh and human flesh. It produces symptoms resembling those of curara-poisoning.
**mydesis** (*mi-de'-sis*) [μυδᾶν, to be damp]. 1. Putrefaction. 2. A discharge of pus from the eyelids.
**mydine** (*mi'-din*). [μυδᾶν, to be wet], $C_8H_{11}NO$. A nonpoisonous ptomaine produced in the putrefaction of human internal organs. The free base is strongly alkaline; has an ammoniacal odor, and is characterized by its strong reducing properties.
**mydriasia** (*mid-ri'-as-is*) [μυδρίασις, undue enlargement of the pupil of the eye]. Dilatation of the pupil of the eye. **m., alternating, m., leaping, m., springing,** mydriasis which by normal light and convergence-reaction attacks first one eye and then the other. It is due to disorder of the central nervous system. **m., paralytic,** that due to paralysis of the oculomotor nerve. **m., spasmodic, m., spastic,** that caused by overaction of the sympathetic or dilator nerve of the iris. **m., spinal,** that produced by irritation of the ciliospinal center of the spinal cord.
**mydriatic** (*mid-re-at'-ik*) [see *mydriasis*]. 1: Causing mydriasis, or dilatation of the pupil. 2. An agent causing mydriasis, or dilatation of the pupil.
**mydrin** (*mi'-drin*). A combination of ephedrine and homatropine. It is used in 10 % solution when evanescent mydriasis is required.
**mydrol** (*mi'-drol*). Iodomethylpyrazolin, a colorless, bitter powder, soluble in water and alcohol. It is used in 5 to 10 % solution as a mydriatic.
**myectomy** (*mi-ek'-to-me*) [μῦς, muscle; ἐκτομή, excision]. Excision of a portion of muscle.
**myectopy** (*mi-ek'-to-pe*) [μῦς, muscle; ἔκτοπος, displaced]. The abnormal placement of a muscle.
**myel** (*mi'-el*) [μυελός, marrow]. The spinal cord.
**myelalgia** (*mi-el-al'-je-ah*) [*myel*; ἄλγος, pain]. Pain in the spinal cord.
**myelalgic** (*mi-el-al'-jik*). Pertaining to or characterized by myelalgia.
**myelanalosis** (*mi-el-an-al-o'-sis*) [μυελός, marrow; ἀνάλωσις, wasting]. Same as *myelatrophy*; tabes dorsalis.
**myelapoplexy** (*mi-el-ap'-o-pleks-e*) [*myel*; *apoplexy*]. Hemorrhage into the spinal cord.
**myelasthenia** (*mi-el-as-the'-ne-ah*) [*myel*; *asthenia*]. Spinal exhaustion.
**myelatelia** (*mi-el-at-e'-le-ah*) [μυελός, myelon; ἀ. priv.; τέλος, completion]. Imperfect development of spinal marrow or of bone-marrow.
**myelatrophy** (*mi-el-at'-ro-fe*) [*myel*; *atrophy*]. Wasting of the spinal cord.
**myelauxe** (*mi-el-awks'-e*) [μυελός, marrow; αὔξη, increase]. Morbid enlargement of the myel.
**myelaxis** (*mi-el-aks'-is*). The neural, or cerebrospinal, axis.
**myelemia, myelæmia** (*mi-el-e'-me-ah*). Synonym of *myelogenic leukemia*.
**myelen** (*mi'-el-en*). 1. An extract made from both red and white fresh marrow. 2. One of a group of phosphorized substances found in the brain.
**myelencephal** (*mi-el-en'-sef-al*). Same as *myelencephalon*.
**myelencephalic** (*mi-el-en-sef-al'-ik*). Relating to the myelencephalon.

**myelencephalon** (*mi-el-en-sef'-al-on*) [*myel; encephalon*]. 1. The cerebrospinal axis. 2. See *metencephalon*.
**myeleterosis** (*mi-el-et-er-o'-sis*) [*myel; ἑτέρωσις, alteration*]. Any morbid alteration of the spinal cord.
**myelic** (*mi-el'-ik*) [*myel*]. Pertaining to the spinal cord.
**myelin** (*mi'-el-in*) [μυελός, marrow]. 1. The white substance of Schwann—the sheath of a medullated nerve. 2. A substance obtained from nerve tissue. See *organotherapy*.
**myelination** (*mi-el-in-a'-shun*). Same as *myelinization*.
**myelinic** (*mi-el-in'-ik*) [*myelin*]. 1. Relating to myelin, or to myelinic nerve-fibers. 2. Medullated.
**m. nerve-fibers**, nerve-fibers having the axis-cylinder inclosed in a sheath of myelin.
**myelinization** (*mi-el-in-iz-a'-shun*). The medullation of nerve-fibers.
**myelitic** (*mi-el-it'-ik*). Pertaining to or affected with myelitis.
**myelitis** (*mi-el-i'-tis*) [*myel; ιτις*, inflammation]. 1. Inflammation of the spinal cord. If it affects the gray matter, it is called *poliomyelitis*; if the white matter, *leukomyelitis*; if it extends entirely across the cord, *transverse myelitis*. The symptoms of myelitis vary with the character and the seat of the lesion. The sensory disturbances consist of hyperesthesia or anesthesia, girdle-pains, and usually a hyperesthetic zone at the level of the lesion. The reflexes are generally lost in the beginning, but later become exaggerated; sometimes they do not return. Paralysis is present in the parts below the lesion; the muscles are flaccid at first, later rigid. The sphincters are generally paralyzed. Bed-sores are common in certain forms of myelitis. 2. Inflammation, of bone-marrow; osteomyelitis. **m., acute**, that due to exposure to cold, injuries, or to acute general diseases; it is attended with sensory disturbances and motor paralysis, complete or incomplete, according to the part of the cord involved and the degree of the inflammation. **m., apoplectiform**, that in which paralysis is developed with unusual rapidity. **m., ascending**, a form in which the inflammation travels up the cord. **m., bulbar**, that affecting the medulla and manifesting itself in disturbances of the cardiorespiratory apparatus, dysphagia, vomiting, and other symptoms. **m., cavitary**, that associated with the formation of cavities. **m., central**, that limited chiefly to the gray matter of the cord. **m., chronic**, a slowly advancing form, presenting generally the same features as acute myelitis, but extended over a longer period. **m., compression-**, an inflammation of the spinal cord, secondary to compressing lesions outside of the cord (caries, carcinoma, exostosis of the vertebræ, aneurysm) or to tumors of the cord itself. **m., concussion**, that due to a spinal concussion. **m., cornual**, inflammation affecting the anterior or posterior cornua. **m., descending**, a form in which the inflammation extends downward, following the pyramidal tracts. **m., diffuse**, widely distributed inflammation of the cord involving large areas. **m., disseminated**, one in which there are several foci. **m., focal**, one in which a small area is affected. **m., hemorrhagic**, myelitis associated with or dependent upon hemorrhage. **m., parenchymatous**, that mainly limited to the proper nerve substance. **m., pressure-**. See *m., compression-*. **m., sclerotic**, a form characterized by overgrowth of the interstitial tissue, which undergoes contraction, producing an increase of hardness of the cord. **m., systemic**, a variety in which the inflammation is confined to distinct systems or tracts of the cord. **m., transverse**, that extending across the cord. **m., traumatic**, myelitis following direct injury.
**myelo-** (*mi-el-o-*) [*myelon*]. A prefix denoting reference to or connection with the spinal cord or with the bone-marrow.
**myeloblasts** (*mi'-el-o-blasts*). [*myelo-; βλαστός*, a germ]. Cells of bone-marrow from which myelocytes are formed.
**myelobrachium** (*mi-el-o-bra'-ke-um*) [*myelo-; βραχίων*; arm]. The inferior peduncle of the cerebellum.
**myelocele, myelocoele** (*mi'-el-o-sēl*) [*myelo-; κοιλία*, a cavity]. 1. The central canal of the spinal cord. 2. [κήλη, tumor]. Spina bifida.
**myelocene** (*mi'-el-o-sēn*). An ethereal extract of bone marrow to which about 1 per cent. of acetone has been added; used as an application in psoriasis, etc.
**myelocerebellar** (*mi-el-o-ser-e-bel'-ar*) [*myelo-; cerebellum*]. Relating to the spinal cord and the cerebellum.
**myelochysis** (*mi-el-ok'-is-is*) [*myelo-; χύσις*, a pouring]. Hydrorrhachis. Protrusion of the spinal marrow.
**myelocoelia** (*mi-el-o-se'-le-ah*). See *myelocele*.
**myelocyst** (*mi'-el-o-sist*) [*myelo-; κύστις*, a cyst]. A cyst springing from the medullary canal.
**myelocystic** (*mi-el-o-sis'-tik*) [see *myelocyst*]. 1. Both myeloid and cystic. 2. Pertaining to a myelocyst.
**myelocystocele** (*mi-el-o-sis'-to-sēl*) [*myelo-; cystocele*]. A cystic tumor of the spinal cord.
**myelocystomeningocele** (*mi-el-o-sis-to-men-in'-go-sēl*). Spina bifida in which the tumor contains myelic substance, membranes, and fluids.
**myelocyte** (*mi'-el-o-sīt*) [*myelo-; κύτος*, a cell]. 1. A large uninuclear leukocyte containing neutrophil granules, and supposed to be derived from the bone-marrow. 2. A free nucleus of a central or ganglionic nerve-cell. 3. Any one of the leukocytes derived from bone-marrow, as distinguished from lymphocytes found in the lymphatic glands.
**myelocythemia, myelocythæmia** (*mi-el-o-si-the'-me-ah*) [*myelocyte; αἷμα*, blood]. An excess of myelocytes in the blood; myelogenous leukemia.
**myelodiastasis** (*mi-el-o-di-as'-tas-is*) [*myelo-; διάστασις*, separation]. Severance or disintegration of the spinal cord.
**myelodiastema** (*mi-el-o-di-as-te'-mah*). Synonym of *myelodiastasis*.
**myeloencephalitis** (*mi-el-o-en-sef-al-i'-tis*) [*myelo-; ἐγκέφαλος*, brain; *ιτις*, inflammation]. Inflammation of both spinal cord and brain. **m., epidemic**, acute anterior poliomyelitis.
**myelogangliitis** (*mi-el-o-gan-gle-i'-tis*). A very severe form of cholera believed to be due to gangliitis of the solar plexus and of the hepatic plexus.
**myelogenic, myelogenous** (*mi-el-o-jen'-ik, mi-el-oj'-en-us*) [*myelo-; γεννᾶν*, to produce]. Produced in or by bone-marrow. **m. leukemia**, leukemia due to disease of the bone-marrow.
**myeloid** (*mi'-el-oid*) [*myelo-; εἶδος*, like]. 1. Resembling marrow. 2. Pertaining to the spinal cord. **m. cell**, a myeloplax. **m. sarcoma**, a form of sarcoma distinguished by the presence of multinucleated cells.
**myeloidin** (*mi-el-oid'-in*) [*myelo-; εἶδος*, like]. A nitrogenous substance containing phosphorus, found in brain-matter.
**myelolymphangioma** (*mi-el-o-lim-fan-je-o'-mah*). Same as *elephantiasis*.
**myeloma** (*mi-el-o'-mah*) [*myelo-; ὅμα*, tumor]. 1. An encephaloid tumor. 2. A giant-celled sarcoma.
**myelomalacia** (*mi-el-o-mal-a'-she-ah*) [*myelo-; μαλακία*, softening]. A softening of the spinal cord.
**myelomalacosis, myelomalaxis** (*mi-el-o-mal-ak-o'-sis, mi-el-o-mal-aks'-is*). Synonym of *myelomalacia*.
**myelomatosis** (*mi-el-o-mat-o'-sis*). Multiple myeloma.
**myelomenia** (*mi-el-o-me'-ne-ah*) [*myelo-; μῆνες*, menses]. A supposed metastasis of the menstrual blood to the spinal cord.
**myelomeningitis** (*mi-el-o-men-in-ji'-tis*). Inflammation of the membranes of the spinal cord; spinal meningitis.
**myelomeningocele** (*mi-el-o-men-in'-go-sēl*). Spina bifida.
**myelomeninx** (*mi-el-o-me'-ninks*) [*myelo-; μῆνιγξ*, membrane]. A spinal membrane.
**myelomyces** (*mi-el-om'-is-ēz*) [*myelo-; μύκης*, fungus]. An encephaloid tumor.
**myelon** (*mi'-el-on*) [μυελός, marrow]. The spinal cord.
**myelonal, myelonic** (*mi-el'-on-al, mi-el-on'-ik*). Pertaining to the myelon.
**myeloneuritis** (*mi-el-o-nū-ri'-tis*). Multiple neuritis combined with myelitis.
**myeloparalysis** (*mi-el-o-par-al'-is-is*) [*myelo-; paralysis*]. Spinal paralysis.
**myelopathic** (*mi-el-o-path'-ik*) [*myelo-; πάθος*, disease]. Relating to a myelopathy.
**myelopathy** (*mi-el-op'-ath-e*) [*myelo-; πάθος*, disease]. Any disease of the spinal cord.
**myelopetal** (*mi-el-op'-et-al*) [*myelo-; petere*, to seek]. Moving toward the myelon; said of nerve-fibers.

**myelophthisis** (*mi-el-off'-this-is*). Synonym of *tabes dorsalis*.
**myelophyma** (*mi-el-o-fi'-mah*) [*myelo-*; *φῦμα*, a growth]. A tuberculous growth in the meninges of the spinal cord.
**myeloplaque** (*mi'-el-o-plak*) [*myelo-*; πλάξ, plate]. A giant-cell of the spinal marrow; an osteoclast; a myeloplax.
**myeloplast** (*mi'-el-o-plast*) [*myelo-*; πλαστός, shaped] The peculiar cells of the bone-marrow resembling leukocytes.
**myeloplax** (*mi'-el-o-plaks*) [*myelo-*; πλάξ, a plaque]. One of the large multinucleated cells found upon the inner surface of bone, and concerned in its absorption.
**myeloplaxoma** (*mi-el-o-plaks-o'-mah*) [*myelo-*; πλάξ, plate; ὄμα, tumor]. A tumor containing myeloplaxes.
**myeloplegia** (*mi-el-o-ple'-je-ah*) [*myelo-*; πληγή, a stroke]. Paralysis of spinal origin.
**myelopore** (*mi'-el-o-pōr*) [*myelo-*; πόρος, pore]. An opening into the spinal column.
**myelorrhagia** (*mi-el-or-a'-je-ah*) [*myelo-*; ῥηγνύναι, to burst forth]. Hemorrhage into the spinal cord.
**myelorrhaphy** (*mi-el-or'-af-e*) [*myelo-*; ῥαφή, a seam]. The suturing of a severed spinal cord.
**myelosarcoma** (*mi-el-o-sar-ko'-mah*). Same as *osteosarcoma*.
**myelosclerosis** (*mi-el-o-skle-ro'-sis*) [*myelo-*; *sclerosis*]. Sclerosis of the spinal cord or of the marrow of bones.
**myelosis** (*mi-el-o'-sis*) [*myelon*]. The growth or existence of a myeloma.
**myelospasm** (*mi'-el-o-spazm*) [*myelo-*; σπασμός, spasm]. Spasm of the spinal cord.
**myelospongium** (*mi-el-o-spun'-je-um*) [*myelo-*; σπόγγος, sponge]. A network in the wall of the neural canal of the embryo, composed of processes given off by the outer extremities of the spongioblasts.
**myelosyphilis** (*mi-el-o-sif'-il-is*). Syphilis of the spine.
**myelosyphilosis** (*mi-el-o-sif-il-o'-sis*) [*myelo-*; *syphilis*]. Syphilitic disease of the spinal cord.
**myelosyringosis** (*mi-el-o-si-rin-go'-sis*). Synonym of *syringomyelia*.
**myelotherapy** (*mi-el-o-ther'-ap-e*). The therapeutic use of bone-marrow extracts.
**myelotome** (*mi'-el-o-tōm*) [*myelo-*; τέμνειν, to cut]. An apparatus for making sections of the spinal cord.
**myelotoxic** (*mi-el-o-toks'-ik*) [*myelo-*; τοξικόν, a poison]. Characteristic of or pertaining to a substance having toxic action on the cells of bone-marrow.
**myelotoxin** (*mi-el-o-toks'-in*). A cytotoxin with specific action upon bone-marrow cells.
**myelyperemia** (*mi-el-ip-er-ē'-me-ah*). See *hypermyelohemia*.
**myenergia** (*mi-en-er'-je-ah*) [μῦς, muscle; ἐνέργεια, energy]. Muscular energy.
**myentasis** (*mi-en'-tas-is*) [μῦς, muscle; ἔντασις, a stretching]. The extension or stretching of a muscle.
**myenteric** (*mi-en-ter'-ik*) [μῦς, muscle; ἔντερον, intestine]. Relating to the muscular coat of the intestine.
**myenteron** (*mi-en'-ter-on*) [μῦς, muscle; *enteron*]. The muscular coat of the intestine.
**Myers's sign** (*mi'-erz*). Numbness and formication of both hands in scarlet-fever.
**myiasis** (*mi-i'-as-is*) [μυῖα, a fly]. The presence of larvæ of flies or maggots in or on the body, as in the nose or ear, in a wound, or upon mucous membranes.
**myiocephalon** (*mi-i-o-sef'-al-on*) [μυῖα, fly; κεφαλή, head]. A minute prolapse of the iris through a corneal perforation, so-called from its resemblance to the head of a fly.
**myiodeopsia, myiodesopsia** (*mi-i-o-de-op'-se-ah, mi-i-o-des-op'-se-ah*) [μυιώδης, fly-like; ὄψις, vision]. The subjective appearance of muscæ volitantes.
**myiosis** (*mi-i-o'-sis*). See *myiasis*.
**myistos, myistus** (*mi-is'-tos, mi-is'-tus*) [μῦς, muscle; ἰστός, tissue]. Muscular tissue.
**myitis** (*mi-i'-tis*). See *myositis*.
**Mylabris** (*mil-ab'-ris*). A genus of old-world blister-flies, some of which are used like cantharides.
**mylacephalus** (*mi-las-ef'-al-us*) [μυελός, massive; ἀ, priv.; κεφαλή, head]. The lowest variety of the acephalous species of omphalositic monsters. There is such imperfect development of the fetus that the condition is but a degree above that of *fœtus amorphus* or *fœtus anideus*.
**mylacri** (*mil'-ak-ri*) [μύλη, a mill]. The molar teeth.
**mylacris** (*mil'-ak-ris*) [μύλη, a mill]. The patella.
**myle** (*mi'-le*) [μύλη, mill]. 1. Uterine mole. 2. Patella. 3. Maxilla.
**mylic** (*mil'-ik*) [μύλη, mill]. Pertaining to the molar teeth; relating to a uterine mole.
**Mylius'. modification of Pettenkofer's test** for bile-acids. Add one drop of furfurol solution and 1 Cc. of concentrated sulphuric acid to each cubic centimeter of the alcoholic solution of bile acids; cool, if necessary, so that the test does not become too warm. A red coloration is produced, which changes to bluish-violet in the course of the day.
**mylo-** (*mi'-lo-*). [μύλη, mill]. A prefix denoting connection with the lower jaw or the molar teeth.
**mylodus** (*mi-lo'-dus*) [*mylo-*; ὀδούς, tooth: *pl.*, *mylodontes*]. A molar tooth.
**myloglossus** (*mi-lo-glos'-us*) [*mylo-*; γλῶσσα, tongue]. A portion of the superior constrictor of the pharynx—that which arises from the mylo-hyoid ridge of the lower jaw. Also, an anomalous slip joining the styloglossus.
**mylohyoid, mylohyoidean** (*mi-lo-hi'-oid, mi-lo-hi-oid-ē'-an*). Pertaining to the region of the lower molar teeth and the hyoid bone. m. muscle. See under *muscle*.
**myo-** (*mi'-o-*). [μῦς, a muscle]. A prefix signifying pertaining to muscle.
**myoalbumose** (*mi-o-al'-bū-mōs*). A proteid from muscle-plasma.
**myoatrophy** (*mi-o-at'-ro-fe*) [*myo-*; *atrophy*]. Atrophy of a muscle.
**myoblast** (*mi'-o-blast*) [*myo-*; βλαστός, germ]. A cell developing into a muscle-fiber.
**myoblastic** (*mi-o-blas'-tik*) [*myo-*; βλαστός, germ]. Relating to or of the nature of a myoblast.
**myocardiograph** (*mi-o-kar'-de-o-graf*) [*myo-*; καρδία, heart; γράφειν, to write]. An apparatus for recording the movement of the heart muscles.
**myocarditis** (*mi-o-kar-di'-tis*) [*myocardium*; ιτις, inflammation]. Inflammation of the myocardium,
**myocardium** (*mi-o-kar'-de-um*) [*myo-*; καρδία, heart]. The muscular tissue of the heart.
**myocele** (*mi'-o-sēl*) [*myo-*; κήλη, hernia]. The protrusion of a muscle through its ruptured sheath.
**myocelialgia, myocœlialgia** (*mi-o-se-le-al'-je-ah*) [*myo-*; κοιλία, cavity; ἄλγος, pain]. Pain in the muscles of the abdomen.
**myocelitis, myocœlitis** (*mi-o-se-li'-tis*) [*myo-*; κοιλία, cavity; ιτις, inflammation]. Inflammation of the abdominal muscles.
**myocellulitis,** (*mi-o-sel-ū-li'-tis*). Simultaneous myositis and cellulitis.
**myocelome, myocœlome** (*mi-o-se'-lōm*) [*myo-*; κοιλία, cavity]. The cavity of a protovertebra or myotome.
**myocephalitis** (*mi-o-sef-al-i'-tis*) [*myo-*; κεφαλή, the head; ιτις, inflammation]. Inflammation of the muscles of the head.
**myochorditis** (*mi-o-kor-di'-tis*) [*myo-*; χορδή, cord; ιτις, inflammation]. Inflammation of the muscles of the vocal bands.
**myochrome** (*mi'-o-krōm*) [*myo-*; χρῶμα, color]. A reddish pigment found in muscles, and believed to be identical with hemoglobin.
**myochronoscope** (*mi-o-kro'-no-skōp*) [*myo-*; κρόνος, time; σκοπεῖν, to examine]. An instrument for measuring the rapidity of muscular contraction or the propagation of a nerve-stimulus through a muscle.
**myoclonia** (*mi-o-klo'-ne-ah*) [*myo-*; κλόνος, spasm]. Clonic spasm of the muscles.
**myoclonus** (*mi-ok'-lo-nus*) [*myo-*; κλόνος, clonus]. Clonic spasm of a muscle or of various muscles. m. multiplex. See *paramyoclonus multiplex*.
**myocolpitis** (*mi-o-kol-pi'-tis*) [*myo-*; κόλπος, vagina; ιτις, inflammation]. Inflammation of the muscular coat of the vagina.
**myocomma** (*mi-o-kom'-ah*) [*myo-*; κόμμα, segment]. Any one of the transverse segments into which embryonic muscle tissue is divided.
**myocrismus** (*mi-o-kriz'-mus*) [*myo-*; κρίζειν, to creak]. A creaking sound heard over muscles during contraction.
**myoctonine** (*mi-ok'-to-nēn*) [μῦς, mouse; κτείνειν, to kill]. 1. An alkaloid of *Aconitum lycoctonum*, a powerful poison possessing curare in action.

**myocyst** (*mi'-o-sist*) [*myo-*; *cyst*]. A cystic tumor of a muscle.
**myocyte** (*mi'-o-sīt*) [*myo-*; κύτος, cell]. A muscle-cell.
**myocytoma** (*mi-o-si-to'-mah*) [*myo-*; *cytoma*]. A tumor in which the chief cells are muscle cells.
**myodegeneration** (*mi-o-de-jen-er-a'-shun*). Muscular degeneration.
**myodemia** (*mi-o-de'-me-ah*) [*myo-*; δημός, fat]. Fatty degeneration of muscle tissue.
**myodes** (*mi-o'-dēz*) [*myo-*; εἶδος, like]. 1. Resembling muscle; muscular. 2. The platysma, or fleshy expansion of the neck. See *muscles, table of*.
**myodesopsia** (*mi-o-des-ops'-e-ah*). See *myodeopsia*.
**myodiastasis** (*mi-o-di-as'-tas-is*) [*myo-*; διά, apart; στάσις, a standing]. 1. The separation that takes place between the ends of a cut muscle. 2. The traumatic injury of a muscle by straining or stretching.
**myodynamia** (*mi-o-di-nam'-e-ah*) [*myo-*; δύναμις, force]. Muscular power or strength.
**myodynamic** (*mi-o-di-nam'-ik*) [*myo-*; δύναμις, power]. Pertaining to the force of muscular contraction.
**myodynamics** (*mi-o-di-nam'-iks*) [*myo-*; δύναμις, power]. The mechanics of muscular action.
**myodynamometer** (*mi-o-di-nam-om'-et-er*). See *dynamometer*.
**myodynia** (*mi-o-din'-e-ah*) [*myo-*; ὀδύνη, pain]. Pain in the muscles.
**myoedema, myoœdema** (*mi-o-e-de'-mah*). 1. See *myoidema*. 2. Edema of the muscles.
**myoelectric** (*mi-o-e-lek'-trik*). Pertaining to the electrical properties of muscle.
**myoendocarditis** (*mi-o-en-do-kar-di'-tis*) [*myo-*; ἔνδος, within; καρδία, heart; ιτις, inflammation]. Simultaneous inflammation of the endocardium and the myocardium.
**myoepithelial** (*mi-o-ep-e-the'-le-al*). 1. Relating to or consisting of muscle and epithelium. 2. Pertaining to myoepithelium.
**myoepithelium** (*mi-o-ep-e-the'-le-um*). Muscle-epithelium; epithelium, the cells of which possess contractile processes.
**myofibroma** (*mi-o-fi-bro'-mah*). A tumor containing muscular and fibrous tissue.
**myofibrosis** (*mi-o-fi-bro'-sis*). An increase of connective tissue between the muscle-fascicull and also between the individual fibrils; it is most likely to occur in the ventricles of the heart.
**myogaster** (*mi'-o-gas-ter*) [*myo-*; γαστήρ, belly]. The belly of a muscle.
**myogen** (*mi'-o-jen*). 1. See *myosinogen*. 2. A dietetic prepared from blood-serum of cattle.
**myogenesis** (*mi-o-jen'-es-is*) [*myo-*; γένεσις, genesis]. The development of muscular tissue.
**myogenetic, myogenic** (*mi-o-jen-et'-ik, mi-o-jen'-ik*) [*myogenesis*]. Of muscular origin.
**myogenous** (*mi-oj'-en-us*) [*myo-*; γεννᾶν, to produce]. Originating in muscle.
**myoglobulin** (*mi-o-glob'-ū-lin*) [*myo-*; *globulin*]. A substance obtained from muscles, closely resembling serum-globulin in its properties.
**myognathus** (*mi-og'-na-thus*) [*myo-*; γνάθος, jaw]. A form of double monstrosity in which the accessory head is joined to the autosite by means of muscle and integument only.
**myogram** (*mi'-o-gram*) [*myo-*; γράμμα, an inscription]. The tracing made by the myograph.
**myogramma** (*mi-o-gram'-ah*). A mark or line upon or in a muscle.
**myograph** (*mi'-o-graf*) [*myo-*; γράφειν, to write]. An instrument for recording the phases of a muscular contraction.
**myographic** (*mi-o-graf'-ik*) [see *myograph*]. Pertaining to a myograph.
**myography** (*mi-og'-ra-fe*) [see *myograph*]. 1. A description of the muscles. 2. The employment of the myograph.
**myohematin, myohæmatin** (*mi-o-hem'-at-in*). A red pigment found in muscles.
**myohysterectomy** (*mi-o-his-ter-ek'-to-me*). The incomplete removal of the uterus, more or less of the cervix uteri being left and the operation being completed without opening the vagina.
**myoid** (*mi'-oid*) [*myo-*; εἶδος, like]. Like a muscle or muscular tissue.
**myoidema** (*mi-oid-e'-mah*) [*myo-*; *edema*]. The wheal brought out by a sharp tap on a muscle in certain conditions of exhaustion.

**myoides** (*mi-oi'-dēz*) [*myo-*; εἶδος, like]. The platysma. See *muscles, table of*.
**myoideum** (*mi-oid'-e-um*) [*myo-*; εἶδος, resemblance]; Tissue resembling muscle.
**myoidism** (*mi'-oid-izm*) [see *myoideum*]. Idiomuscular contraction.
**myokymia** (*mi-o-ki'-me-ah*) [*myo-*; κῦμα, a wave]. Persistent, widespread, muscular quivering without atrophy or loss of power.
**myolemma** (*mi-o-lem'-ah*) [*myo-*; λέμμα, covering]. The sarcolemma.
**myolin** (*mi'-o-lin*) [μῦς, muscle]. The substance formerly supposed to form the contents of a muscular fibril.
**myolipoma** (*mi-o-lip-o'-mah*). A myoma containing fatty tissue.
**myologic** (*mi-o-loj'-ik*) [*myo-*; λόγος, science]. Pertaining to myology.
**myologist** (*mi-ol'-o-jist*). One versed in myology.
**myology** (*mi-ol'-o-je*) [*myo-*; λόγος, science]. The science of the nature, structure, functions, and diseases of muscles.
**myolysis** (*mi-ol'-is-is*) [*myo-*; λύσις, solution]. The degeneration of muscle tissue.
**myom** (*mi'-om*). A myoma of the uterus.
**myoma** (*mi-o'-mah*) [*myo-*; ὅμα, tumor]. A muscular tumor. If composed of nonstriped muscular tissue, it is called *liomyoma*; if of striped muscular tissue, *rhabdomyoma*. **m. telangiectodes**, an angioma surrounded by muscle-fibers; angiomyoma.
**myomalacia** (*mi-o-mal-a'-she-ah*) [*myo-*; μαλακία, softening]. Softening of muscles. **m. cordis**, softening of a portion of the heart-muscle, usually resulting from thrombosis or embolism.
**myomatous** (*mi-o'-mat-us*) [*myoma*]. Of the nature of a myoma.
**myomectomy** (*mi-o-mek'-to-me*) [*myoma*; ἐκτομή, excision]. Excision of a uterine or other myoma.
**myomelanosis** (*mi-o-mel-an-o'-sis*) [*myo-*; μελάνωσις, a becoming black]. The presence of a blackish coloration in muscular tissue.
**myomere** (*mi'-o-mēr*) [*myo-*; μέρος, a part]. A muscular flake or metamere; a myocomma or myotome.
**myometer** (*mi-om'-et-er*) [*myo-*; μέτρον, a measure]. An instrument for measuring muscle-contraction.
**myometritis** (*mi-o-mẹt-ri'-tis*). Inflammation of the uterine muscular tissue.
**myometrium** (*mi-o-me'-tre-um*) [*myo-*; μήτρα, womb]. The uterine muscular structure.
**myomohysterectomy** (*mi-o-mo-his-ter-ek'-to-me*) [*myoma*; *hysterectomy*]. The extirpation of a myomatous uterus.
**myomotomy** (*mi-o-mot'-o-me*). See *myomectomy*.
**myon** (*mi'-on*) [μυών, a group of muscles]. A unit of musculature; a group of muscles viewed as an integer.
**myonarcosis** (*mi-o-nar-ko'-sis*) [*myo-*; νάρκωσις, numbness]. Numbness of the muscles.
**myonema** (*mi-o-ne'-mah*) [*myo-*; νῆμα, a thread]. One of the long contractile fibrillæ which in the protozoa make up the layer of the cytoplasm called the myocyte. Cf. *spasmoneme*.
**myoneoplasma** (*mi-o-ne-o-plaz'-mah*) [*myo-*; νέος, new; πλάσμα, formation: *pl.*, *myoneoplasmata*]. A myoma or muscular neoplasm.
**myoneuralgia** (*mi-o-nū-ral'-je-ah*). Neuralgic pain in the muscles.
**myoneurasthenia** (*mi-o-nū-ras-the'-ne-ah*). Relaxation of the muscles occurring in neurasthenia.
**myoneure** (*mi'-o-nūr*) [*myo-*; νεῦρον, nerve]. A motor nerve-cell supplying a muscle.
**myoneuroma** (*mi-o-nū-ro'-mah*). Combined myoma and neuroma.
**myonicity** (*mi-o-nis'-it-e*) [*myo-*]. The power of living muscle to contract and to relax.
**myonitis** (*mi-on-i'-tis*) [*myo-*; ιτις, inflammation]. Synonym of *myositis*.
**myonosus** (*mi-on'-o-sus*) [*myo-*; νόσος, disease]. A disease of the muscles.
**myonymy** (*mi-on'-im-e*) [*myo-*; ὄνομα, name]. Nomenclature of the muscles.
**myopachynsis** (*mi-o-pak-in'-sis*) [*myo-*; πάχυνσις, thickening]. Muscular hypertrophy. **m. lipomatosa**. See *paralysis, pseudo-hypertrophic*.
**myopalmus** (*mi-o-pal'-mus*) [*myo-*; παλμός, a twitch]. Twitching of the muscles.
**myoparalysis** (*mi-o-par-al'-is-is*). Paralysis of a muscle or muscles.
**myopathic** (*mi-o-path'-ik*) [*myo-*; πάθος, disease].

Depending upon or relating to disease of the muscles. **m. facies,** a peculiar form of expression seen in infantile forms of myopathic muscular atrophy. It is characterized by imperfect movement of the facial muscles, sinking in of the cheeks, and drooping of the lower lip. Syn., *facies myopathique.*

**myopathy, myopathia** (*mi-op'-ath-e, mi-o-path'-e-ah*) [*myo-*; πάθος, suffering]. 1. Any disease of the muscles. 2. A group of disorders characterized by slow progressive loss of power associated with atrophy or hypertrophy of the muscles, absence of fibrillar contraction or quantitative electrical change.

**myope** (*mi'-ōp*) [see *myopia*]. A person affected with myopia.

**myopericarditis** (*mi-o-per-ik-ar-di'-tis*) [*myo-*; *pericarditis*]. A combination of pericarditis with myocarditis.

**myoperitonitis** (*mi-o-per-it-on-i'-tis*). Inflammation of the abdominal muscles combined with peritonitis.

**myophage** (*mi'-o-fāj*) [see *myophagism*]. A phagocyte which consumes muscle cells.

**myophagism** (*mi-of'-aj-izm*) [*myo-*; φαγεῖν, to eat]. The wasting away of muscular tissue observed in muscular atrophy.

**myophone** (*mi'-o-fōn*) [*myo-*; φωνή, sound]. An instrument for hearing the sounds produced during the contraction of a muscle.

**myophonia** (*mi-o-fo'-ne-ah*) [*myo-*; φωνή, sound]. Muscle-sound.

**myopia** (*mi-o'-pe-ah*) [μύων, to close; ὤψ, eye; myopes having the habit of partially closing the lids to avoid spherical aberration]. Near-sightedness; an optical defect, usually due to too great length of the antero-posterior diameter of the globe, whereby the focal image is formed in front of the retina. **m., high,** a degree of myopia greater than 6.5 diopters. **m., low,** one less than two diopters. **m., malignant,** rapidly progressing myopia. **m., progressive,** continuous increase of myopia, due to increasing elongation of the eyeball.

**myopic** (*mi-op'-ik*) [see *myopia*]. Pertaining to or having the nature of myopia; near-sighted. **m. crescent,** a yellowish-white crescentic area about the papilla, due to atrophy or breaking away of the choroid and exposure of the sclerotic.

**myopiosis** (*mi-o-pi-o'-sis*). Synonym of *myopia.*

**myoplasm** (*mi'-o-plazm*) [*myo-*; πλάσμα, something formed]. The contractile portion of a muscle fiber as opposed to the sarcoplasm or undifferentiated portion.

**myoplast** (*mi'-o-plast*) [*myo-*; πλάσσειν, to mold]. A muscle-producing cell.

**myoplastic** (*mi-o-plas'-tik*). Producing muscle; pertaining to a myoplast. **m. bodies,** the embryonic cells from which muscular fibers are developed.

**myoplasty** (*mi'-o-plas-te*). Plastic operation on muscle.

**myoplegia** (*mi-o-ple'-je-ah*) [*myo-*; πληγή, a stroke]. A condition of diminished muscular power, or of muscular paresis.

**myopolar** (*mi-o-po'-lar*) [*myo-*; *pole*]. Pertaining to muscular polarity, electric or other.

**myoporthosis** (*mi-o-por-tho'-sis*) [μύωψ, nearsighted; ὀρθός, straight]. The correction of myopia.

**myopresbytia** (*mi-o-pres-bish'-e-ah*) [μύωψ, nearsighted; πρεσβύτης, old-sighted]. Combined myopia and presbyopia.

**myoprotein** (*mi-o-pro'-te-in*). A substance found in muscle-plasma, differing apparently from the recognized albumins.

**myoproteose** (*mi-o-pro'-te-ōs*). See *myoalbumose.*

**myopsychopathy, myopsychy** (*mi-o-si-kop'-ath-e, mi-o-si'-ke*) [*myo-*; ψυχή, mind; πάθος, disease]. Myopathies associated with feebleness or defect of mind.

**myopsychoses** (*mi-o-si-ko'-sēz*) [*myo-*; ψυχή, mind]. Myopathies and neuromuscular affections associated with mental disturbances.

**myorrhaphy** (*mi-or'-af-e*) [*myo-*; ῥαφή, a seam]. The union of the abdominal recti muscles by suture when drawn apart.

**myorrheuma** (*mi-or-oo'-mah*) [*myo-*; ῥεῦμα, a flow]. A synonym of *muscular rheumatism* or *myalgia.*

**myorrhexis** (*mi-or-eks'-is*) [*myo-*; ῥῆξις, a tearing]. Laceration or rupture of a muscle.

**myosalgia** (*mi-o-sal'-je-ah*). See *myalgia.*

**myosalpingitis** (*mi-o-sal-pin-ji'-tis*). Hypertrophy of the muscular tissue of the Fallopian tube.

**myosarcoma** (*mi-o-sar-ko'-mah*). A sarcoma containing muscular tissue.

**myosclerosis** (*mi-o-skle-ro'-sis*). 1. A hardening or induration of a muscle. 2. Synonym of *pseudohypertrophic paralysis.*

**myoscope** (*mi'-o-skōp*) [*myo-*; σκοπεῖν, to inspect]. An apparatus used in observing the contraction-phenomena of muscles.

**myoseism** (*mi-o-se'-izm*) [*myo-*; σεισμός, a shake]. A symptom that consists in repeated stops in the course of muscular contractions by which the whole movement is rendered jerky.

**myoseptum** (*mi-o-sep'-tum*) [*myo-*; *septum*]. The intermuscular septum between the metameres of muscles of certain animals, as of fishes.

**myoserum** (*mi-o-se'-rum*). Muscle-juice; juice derived from meat submitted to pressure.

**myosin** (*mi'-o-sin*) [μῦς, muscle]. A protein of the globulin class, found in coagulated muscle-plasma, and formed from the antecedent globulin, myosinogen. It is also found in the cornea and in some vegetables. Myosin is soluble in strong saline solutions, and is changed into syntonin by the action of dilute hydrochloric acid and heat.

**myosinogen** (*mi-o-sin'-o-jen*) [*myosin*; γεννᾶν, to beget]. One of the proteins of muscle-plasma. It is the antecedent of myosin. Its coagulation after death is the cause of rigor mortis.

**myosinose** (*mi-o'-sin-ōs*). An albumose produced by gastric digestion of myosin.

**myosinuria** (*mi-o-sin-ū'-re-ah*). The occurrence of myosin in the urine.

**myosis** (*mi-o'-sis*). See *miosis.*

**myositic** (*mi-o-sit'-ik*). Pertaining to, or affected with myosis, or with myositis.

**myositis** (*mi-o-si'-tis*) [μῦς, muscle; ιτις, inflammation]. Inflammation of the muscles. **m. fibrosa, m., interstitial,** an inflammation of the connective tissue of muscle. **m., ossificans,** myositis due to prolonged fixation of forearm fractures by any form of bandaging which intercepts circulation through the muscles and nerves of the part. **m. ossificans,** a variety characterized by the formation of osseous deposits in the muscles. **m. ossificans progressiva,** a process of ossification attacking one muscle after another. **m. ossificans traumatica,** the formation of a mass of bone in a muscle after an injury. **m., parenchymatous,** that affecting the essential substance of a muscle. **m., specific, m., syphilitic,** that due to syphilis. **m. trichinosa,** that due to the presence of trichina spiralis in the muscles.

**myosome** (*mi'-o-sōm*) [*myo-*; σῶμα, body]. The contractile substance of muscle.

**myospasis** (*mi-os'-pas-is*) [*myo-*; σπάσις, a drawing: *pl., myospases*]. A muscular contraction.

**myospasm** (*mi-o-spazm*) [*myo-*; σπασμός, spasm]. Muscular spasm; a cramp.

**myospasmia** (*mi-o-spaz'-me-ah*) [see *myospasm*]. Diseases in which spasmodic muscular contraction is a dominant symptom.

**myosuria** (*mi-o-sū'-re-ah*). See *myosinuria.*

**myosuture** (*mi-o-sū'-tūr*). Suture of a muscle.

**myosynizesis** (*mi-o-sin-iz-e'-sis*) [*myo-*; συνίζησις, a falling together]. The adhesion of two or more muscles.

**myotactic** (*mi-o-tak'-tik*) [*myo-*; *tangere*, to touch]. Relating to muscular sense, or the sense of touch in muscles.

**myotasis** (*mi-ot'-as-is*) [*myo-*; τάσις, a stretching]. Passive tension of a muscle.

**myotatic** (*mi-ot'-at-ik*) [*myo-*; τάσις, a stretching]. Produced by or depending upon passive tension of the muscles. **m. contraction, tendon-reflex. m. irritability,** an increased irritability of muscles producing passive extension, and giving rise, when the muscle is stimulated, to *myotatic contraction,* or tendon-reflex.

**myotenotomy** (*mi-o-ten-ot'-o-me*) [*myo-*; τένων, tendon; τομή, a cutting]. Surgical division of muscles and tendons.

**myothelium** (*mi-o-the'-le-um*) [*myo-*; θηλή, nipple]. The cellular elements composing a myotome or protovertebra.

**myothermic** (*mi-o-ther'-mik*) [*myo-*; θέρμη, heat]. Pertaining to heat-development in a muscle.

**myotic** (*mi-ot'-ik*). See *miotic.*

**myotility** (*mi-o-til'-it-e*). Muscular contractility, or tonicity.

**myotome** (*mi'-o-tōm*) [*myo-*; τομή, a cutting]. 1. An instrument for performing myotomy. 2. See *myocomma.*

**myotomy** (*mi-ot'-o-me*) [see *myotome*]. 1. Division

of a muscle, particularly through its belly. 2. The dissection of muscles.
**myotone** (*mi'-o-tōn*). See *myotonia*.
**myotonia, myotonus** (*mi-o-to'-ne-ah, -ot'-o-nus*) [*myo-*; τόνος, tension] 1. Tonic muscular spasm. 2. The stretching of a muscle. 3. Muscular tone, quality, or tension. **m. acquisita,** a disease, not hereditary, characterized by tonic muscular spasm. **m. congenita, m. hereditaria,** Thomsen's disease; a hereditary disease characterized by tonic spasms in the voluntarily moved muscles.
**myotyrbe** (*mi-o-ter'-be*) [*myo-*; τύρβη, disorder]. Chorea; incoordinate muscular movements.
**Myrcia** (*mur'-se-ah*). A genus of aromatic shrubs of the natural order *Myrtaceæ*. *M. acris* is the source of oil of bay, *oleum myrciæ*, from which is prepared bay-rum, or *spiritus myrciæ*. Bay-rum is used as a local application in headache and to impart its odor to many toilet preparations.
**myriachit** (*mir-e-ah'-shit*). See *palmus*.
**myriagram** (*mir'-e-ag-ram*) [μυρίοι, ten thousand; γράμμα, gram]. Ten thousand grams.
**myrialiter** (*me-re-al-e'-ter*) [μυρίοι, ten thousand; λίτρα, liter]. Ten thousand liters.
**myriameter** (*me-re-am'-e-ter*) [μυρίοι, ten thousand; μέτρον, a measure]. Ten thousand meters.
**myricin** (*mir'-is-in*) [μυρίκη, the tamarisk]. A constituent of wax (*cera flava*), insoluble in boiling alcohol. It consists chiefly of myricyl palmitate, $C_{30}H_{61} \cdot C_{16}H_{31}O_2$, which is a compound of palmitic acid and myricyl alcohol.
**myricyl** (*mir'-is-il*). A univalent hydrocarbon, $C_{30}H_{61}$, the radical of myricyl alcohol. **m. alcohol,** $C_{30}H_{62}O$, hydrogen myricylate, is obtained by decomposing beeswax.
**myringa, myrinx** (*mir-in'-gah, mir'-inks*) [L.]. The tympanic membrane.
**myringectomy** (*mir-in-jek'-to-me*). See *myringodectomy*.
**myringitis** (*mir-in-ji'-tis*) [*myringa*; ιτις, inflammation]. Inflammation of the membrana tympani. **m. bullosa,** acute myringitis attended by small, pearly-gray blisters.
**myringodectomy** (*mir-in-go-dek'-to-me*) [*myringa*; ἐκτομή, excision]. Excision of a part or of the whole of the membrana tympani.
**myringodermatitis** (*mir-in-go-der-mat-i'-tis*) [*myringa*; *dermatitis*]. Inflammation of the external layer of the drum-membrane with the formation of blebs below or behind the malleus.
**myringomycosis** (*mir-in-go-mi-ko'-sis*) [*myringa*; *mycosis*]. Disease of the drum-membrane, due to parasitic fungi, especially *Aspergillus*.
**myringoplastic** (*mir-in-go-plas'-tik*) [*myringoplasty*]. Pertaining to myringoplasty.
**myringoplasty** (*mir-in'-go-plas-te*) [*myringa*; πλάσσειν, to shape]. A plastic operation on the membrana tympani.
**myringoscope** (*mi-ring'-go-skōp*) [*myringa*; σκοπεῖν, to examine]. An ear speculum with a magnifying lens.
**myringotome** (*mir-in'-go-tōm*) [*myringa*; τομή, a cutting]. An instrument used in incising the membrana tympani.
**myringotomy** (*mir-in-got'-o-me*) [see *myringotome*]. Incision of the tympanic membrane.
**myrinx** (*mi'-rinks*). See *myringa*.
**myrisma** (*mi-riz'-mah*) [μύρισμα]. An anointing, inunction.
**myristic acid** (*mir-is'-tik*) [*myristica*], $C_{14}H_{28}O_2$. A monobasic acid found in *Myristica fragrans*, in cocoanut-oil, in spermaceti, etc.
**myristica** (*mir-is'-tik-ah*). See *nutmeg*.
**myristicated liver** (*mir-ris'-tik-a-ted*). Nutmeg liver, *q. v.*
**myristication** (*mi-ris-tik-a'-shun*) [μυρίζειν, to anoint]. The development of a nutmeg condition of the liver.
**myristicin** (*mir-is'-tis-in*) [*myristica*.] A peculiar fatty body contained in nutmeg.
**myristicol** (*mir-is'-tik-ol*) [*myristica*; *oleum*, oil], $C_{10}H_{16}O$. An oily substance extracted from oil of nutmeg.
**myristin** (*mi-ris'-tin*) [μυρίζειν, to anoint]. A compound of glycerin and myristic acid.
**myrmecia** (*mur-me'-se-ah*) [μυρμηκία, ants' nest; μύρμηξ, ant]. A warty growth on the palm or sole.
**myrmeciasis, myrmeciasm** (*mir-me-si'-as-is, mir-me'-se-azm*) [μύρμηξ, ant]. Formication.
**myron** (*mi'-ron*) [μύρον, any sweet juice exuding from plants]. 1. An unguent. 2. A soft resin.

**myronate** (*mi'-ron-āt*). A salt of myronic acid.
**myronic acid** (*mi-ron'-ik*). See *acid, myronic*.
**myronin** (*mi'-ron-in*). An ointment-base said to be a mixture of soap, carnauba wax, and doegling oil.
**myrosin** (*mi'-ro-sin*) [*myron*]. An albuminous ferment occurring in mustard-seed, which liberates the oil of mustard from potassium myronate.
**myrrh** (*mer*) [μύρρα, myrrh]. A gum-resin (*myrrha*, U. S. P.) obtained from *Commiphora* (*Balsamodendron*) *myrrha*, a tree of the natural order *Burseraceæ*. Myrrh contains a volatile oil, a resin (*myrrhin*), and a gum, and is used as a stimulant tonic in dyspepsia, pulmonary affections, chlorosis, and amenorrhea. It is employed as a local application in various forms of stomatitis. Dose 10–30 gr. (0.65–2.0 Gm.). **m., tincture of** (*tinctura myrrhæ*, U. S. P.). Dose 10–30 min. (0.65–2.0 Cc.). Myrrh is also an ingredient of *mistura ferri composita*, *pilulæ aloes et myrrhæ*, and *tinctura aloes et myrrhæ*.
**myrrholin** (*mir'-ol-in*). A solution of equal parts of fatty oil and myrrh used as a vehicle for the administration of creosote.
**myrtaceous, myrtal** (*mir-ta'-shus, mir'-tal*). Belonging to or characteristic of the myrtle family.
**myrtiform** (*mir'-tif-orm*) [*myrtum*, a myrtle-berry-forma*, form]. Shaped like a myrtle-berry or myrtle; leaf. **m. caruncles.** See *carunculæ*. **m. fossa,** a shallow pit between the edge of the pyriform aperture of the superior maxilla above, and the sockets of the front teeth of the upper jaw below.
**myrtiformis** (*mir-te-form'-is*). See *depressor alæ nasi* and *compressor narium under muscle*.
**myrtle** (*mir'-tl*) [μύρτος, the myrtle]. A plant of the genus *Myrtus*. The leaves of *Myrtus chekan* are used in bronchitis. Dose of *fluidextract* 1–3 dr. (4–12 Cc.). The leaves of *Myrtus communis* yield *myrtol*, which distils between 160° and 180° C. The leaves are used as an antiseptic dressing for wounds. Myrtol is employed as an antiseptic in bronchitis, cystitis, and pyelitis. Dose 1–2 gr. (0.065–0.13 Gm.).
**myrtol** (*mir'-tol*) [*myrtus*; *oleum*, oil]. A constituent of the essential oil of *Myrtus communis* (see *myrtus*); it is useful in bronchitis, vaginitis, and urethritis, and as a sedative and antiseptic. Dose gtt. iv–xvj, in capsules.
**myrtus** (*mir'-tus*). See *myrtle*.
**mysophobia** (*mi-so-fo'-be-ah*) [μύσος, filth; φόβος, dread]. An abnormal dread of contact or of dirt.
**mytacism** (*mit'-as-izm*). See *mutacism*.
**mythomania** (*mith-o-ma'-ne-ah*) [μῦθος, fiction; μανία, madness]. A morbid tendency to lie or to exaggerate.
**mytilotoxicon** (*mit-il-o-toks'-ik-on*) [see *mytilotoxin*]. A general name for the active agent in mussel-poisoning.
**mytilotoxin** (*mit-il-o-toks'-in*) [μυτίλος, mussel; τοξικόν, a poison], $C_8H_{15}NO_2$. A poisonous leukomaine found in poisonous mussel; it is similar in action to curara.
**mytilotoxism** (*mit-il-o-toks'-izm*). Mussel-poisoning.
**myurous** (*mi-ū'-rus*) [μῦς, mouse; οὐρά, a tail]. Tapering like the tail of the mouse; a qualification applied to the pulse when it is progressively growing feeble.
**myxa** (*miks'-ah*) [μύξα, mucus]. Mucus.
**myxadenitis** (*miks-ad-en-i'-tis*) [μύξα, mucus; *adenitis*]. Inflammation of a mucous gland.
**myxadenoma** (*miks-ad-en-o'-mah*). Synonym of *myxoadenoma*.
**myxangitis** (*miks-an-ji'-tis*) [μύξα, mucus; ἀγγεῖον, vessel; ιτις, inflammation]. Inflammation of the duct of a mucous gland.
**myxasthenia** (*miks-as-the'-ne-ah*) [μύξα, mucus; *asthenia*]. Overdryness of the mucosa or impairment of the power to secrete mucus.
**myxedema, myxoedema** (*miks-e-de'-mah*) [μύξα, mucus; *edema*]. A disorder of nutrition in which the subcutaneous tissue, especially of the face and hands, becomes infiltrated with a mucin-like substance, giving rise to a pale, edematous swelling, which does not pit on pressure. It is associated with dulness of the intellect, slow monotonous speech, muscular weakness, tremors, and absence of sweating. It is thought to be due to atrophy, or degeneration of the thyroid gland, a view that gains strength from the facts that a similar condition to that of myxedema is produced in animals by the removal of the thyroid gland (cachexia strumipriva), and that the administration of thyroid gland in myxedema produces a marked

amelioration of symptoms. In individuals in whom the thyroid gland is congenitally diseased or absent a state resembling myxedema is likely to develop; this is known as *cretinism*.
**myxedematoid** (*miks-e-de'-mat-oid*) [μύξα, mucus; οἴδημα, edema; εἶδος, like]. Resembling myxedema.
**myxedematous** (*miks-e-dem'-at-us*) [*myxedema*]. Pertaining to, affected with, or of the nature of myxedema.
**myxemia, myxæmia** (*miks-e'-me-ah*) [μύξα, mucus; αἷμα, blood]. An accumulation of mucin in the blood.
**myxeurysma** (*miks-ū-riz'-mah*) [μύξα, mucus; εὐρύνειν, to widen]. Cavernous lymphangioma.
**myxiosis** (*miks-e-o'-sis*) [μύξα, mucus]. A mucous discharge.
**myxo-** (*miks-o-*) [μύξα, mucus]. A prefix meaning relating to mucus or mucoid.
**myxoadenoma** (*miks-o-ad-en-o'-mah*) [*myxo-*; *adenoma*]. An adenoma that has in part undergone myxomatous degeneration; an adenoma of a mucous gland.
**Myxobolus cyprini** (*miks-ob'-ol-us, sip-ri'-ni*). A pathogenic protozoon belonging to the sporozoa; it produces pox disease in carp.
**myxochondrofibrosarcoma** (*miks-o-kon-dro-fi-bro-sar-ko'-mah*). A myxochondroma containing fibrous and sarcomatous elements.
**myxochondroma** (*miks-o-kon-dro'-mah*) [*myxo-*; *chondroma*]. A tumor containing mucous and cartilaginous tissue.
**myxochondrosarcoma** (*miks-o-kon-dro-sar-ko'-mah*) [*myxo-*; χόνδρος, cartilage; σάρξ, flesh; ὄμα, tumor]. A mixed tumor containing myxomatous and cartilaginous tissue and embryonal connective tissue.
**Myxococcidium stegomyiæ** (*miks-o-kok-sid'-e-um steg-o-mi'-i-e*). A protozoon found in the body of the mosquito, *Stegomyia calopus*, and believed to be the microorganism causing yellow-fever.
**myxocylindroma** (*miks-o-sil-in-dro'-mah*) [*myxo-*; κύλινδρος, cylinder; ὄμα, tumor]. A myxomatous sarcoma in which the myxomatous tissue is disposed in the meshes of the sarcoma.
**myxocystitis** (*miks-o-sis-ti'-tis*) [*myxo-*; κύστις, bladder; ἴτις, inflammation]. Inflammation of the mucous membrane of the bladder.
**myxocystoma** (*miks-o-sist-o'-mah*). A cystoma containing mucous elements.
**myxodermia** (*miks-o-der'-me-ah*) [*myxo-*; δέρμα, skin]. Softening of the skin.
**myxodes** (*miks-o'-dēz*) [*myxo-*; εἶδος, like]. Resembling mucus.
**myxœdema** (*miks-e-de'-mah*). See *myxedema*.
**myxoendothelioma** (*miks-o-en-do-the-le-o'-mah*) [*myxo-*; ἔνδον, within; θηλή, a nipple; ὄμα, tumor; *pl., myxoendotheliomata*]. An endothelioma that contains myxomatous tissue.
**myxofibroma** (*miks-o-fi-bro'-mah*) [*myxo-*; *fibroma*]. A fibroma that has in part undergone myxomatous degeneration.
**myxofibrosarcoma** (*miks-o-fi-bro-sar-ko'-mah*). A tumor made up of myxomatous, sarcomatous, and fibromatous elements.

**myxoglioma** (*miks-o-gli-o'-mah*) [*myxo-*; *glioma*]. A glioma with myxomatous degeneration.
**myxoid** (*miks'-oid*) [*myxo-*; εἶδος, like]. Like mucus.
**myxoidedema, myxoidœdema** (*miks-oid-e-de'-mah*). A severe form of influenza.
**myxoinoma** (*miks-o-in-o'-mah*) [*myxo-*; ἴς; a fiber; ὄμα, a tumor]. A benign form of inoma with myxomatous elements.
**myxolipoma** (*miks-o-lip-o'-mah*) [*myxo-*; *lipoma*]. A fatty tumor that has in part undergone myxomatous change.
**myxoma** (*miks-o'-mah*) [*myxo-*; ὄμα, tumor]. A connective-tissue tumor after the type of the jelly of Wharton of the umbilical cord. It consists of a gelatinous, mucin-containing, intercellular substance, in which are scattered peculiar branched or stellate cells. **m., cystic, m., cystoid**, one containing parts so fluid as to resemble cysts. **m., hyaline, a** translucent form consisting almost wholly of mucous tissue. **m., medullary**, one containing many cells and presenting a white, opaque, pith-like appearance. **m., telangiectatic, m. telangiectodes, m., vascular,** a myxoma characterized by a highly vascular structure.
**myxomatous** (*miks-o'-mat-us*) [*myxoma*]. Of the nature of, or pertaining to, myxoma. **m. degeneration**, mucoid degeneration. It attacks epithelium and connective tissue, giving rise to the production of a gelatinous substance containing mucin. In epithelial tissue the cells are affected; in connective tissue, the intercellular substance.
**myxomycetes** (*miks-o-mi-se'-tēz*) [*myxo-*; μύκης, fungus]. A group of ameboid microorganisms believed at one time to be vegetable, now known to be animal in nature.
**myxoneuroma** (*miks-o-nū-ro'-mah*) [*myxo-*; *neuroma*]. 1. A glioma. 2. A neuroma with mucous elements.
**myxoneurosis** (*miks-o-nū-ro'-sis*) [*myxo-*; *neurosis*]. A neurosis which, as a functional disturbance, causes an abnormal secretion of mucus. **m. intestinalis membranacea,** intestinal catarrh combined with secretion of mucus of nervous origin.
**myxopapilloma** (*miks-o-pap-il-o'-mah*). Papilloma with mucous elements.
**myxorrhea, myxorrhœa** (*miks-or-e'-ah*) [*myxo-*; ῥοία, a flow]. A copious mucous discharge.
**myxosarcoma** (*miks-o-sar-ko'-mah*) [*myxo-*; *sarcoma*]. A sarcoma that has in part undergone myxomatous degeneration.
**myxospore** (*miks'-o-spōr*) [*myxo-*; σπόρος, seed]. A spore produced in the midst of a gelatinous mass without a distinct ascus or basidium.
**Myxosporidia** (*miks-o-spor-id'-e-ah*) [*myxo-*; σπόρος, seed]. A group or order of *Sporozoa* found as parasitic bodies in the muscles and epithelial cells of fishes; they produce a psorospermosis which is frequently widespread and destructive.
**myzesis** (*mi-ze'-sis*) [μύζειν, to suck]. Synonym of sucking.

**N**

**N.** The chemical symbol for nitrogen.
**n.** The symbol for index of refraction; also abbreviation for nasal, normal.
**Na.** Chemical symbol for sodium (natrium).
**N. A.** Abbreviation of numerical aperture.
**Nabothian cysts, N.** ovules (nah-bo'-the-an) [Martin Naboth, German anatomist, 1675-1721]. Small retention cysts formed by the Nabothian follicles. **N. follicles, N. glands,** the mucous follicles of the cervix uteri about the external os. **N. menorrhagia,** a discharge from the pregnant uterus of thin mucous that accumulates as the result of excessive secretion of the uterine glands. Syn., hydrorrhœa gravidarum.
**nacra** (na'-krah). See nakra.
**nacre** (na'-ker) [Ar., nakir, hollowed out]. Mother-of-pearl.
**nacreous** (na'-kre-us) [Ar., nakir, hollowed out]. Resembling nacre or mother-of-pearl.
**Naegele's obliquity** (na'-gel-eh) [Franz Karl Naegele, German obstetrician, 1778-1851]. Biparietal obliquity; the lateral inclination of the fetal head, at the superior pelvic strait, which brings the sagittal suture nearer to the sacral promontory. **N.'s pelvis,** the obliquely contracted pelvis; ankylosis of the sacroiliac synchondrosis of one side, with imperfect development of the sacrum on the corresponding side.
**nævoid** (ne'-void). See nevoid.
**nævose** (ne'-vōs). See nævose.
**nævus** (ne'-vus). See nevus.
**naftalan** (naf'-tal-an). An antiseptic, deodorant substance consisting of a Russian naphtha (97 %) and hard soap (3 %). It is used in skin diseases, burns, ulcers, and rheumatism.
**nagana, n'gana, nygana** (nag-ah'-nah) [African]. A disease of animals due to Trypanosoma Brucei, which is transmitted by the tsetse-fly.
**nail** (nāl) [AS., nægel]. The horny structure covering the dorsal aspect of the terminal phalanx of each finger and toe. It consists of intimately united horny epithelial cells derived from the stratum lucidum of the epidermis. **n.-bed,** a vascular tissue, corresponding to the corium and the stratum Malpighii of the skin, in which a nail rests. **n.-culture,** a term applied in bacteriology to a stab-culture showing a growth along the needle-track, and on the surface a button-like projection, giving the appearance of a nail driven into the gelatin. **n.-fold,** the portion of epidermis that covers the root and edges of the nail. **n., hang-.** See hangnail. **n., ingrowing, n., ingrown,** an overlapping of the nail by the flesh, from pressure, attended with ulceration. **n.-matrix,** the proximal end of the nail-bed; the structure from which the nail grows. **n., parrot-beak,** a nail curved like a parrot's beak. **n., reedy,** one marked with furrows. **n., turtle-back,** a nail curved in all directions; a condition seen in certain trophic disturbances.
**nailers' consumption.** See siderosis.
**Naja** (nah'-jah) [noya, the Ceylon name]. 1. A genus of serpents of the family Elapidæ. 2. A homeopathic preparation of cobra venom. **N. tripudians,** the cobra (q. v.).
**naked** (na'-ked). Unclothed; nude. **n. eye,** the eye unaided by a magnifying instrument.
**nakra** (na'-krah) [Beng.]. A Bengalese disease resembling influenza.
**namangitis** (nam-an-ji'-tis). Synonym of lymphangitis.
**nameless crime.** The name given to perversion of the genetic instinct, by which sexual gratification is secured in other than the normal way, as by buccal or anal coitus, etc.
**nanism** (na'-nism) [nanus]. Abnormal smallness of size from arrested development. Dwarfishness.
**nanocephalia** (na-no-sef-a'-le-ah) [νᾶνος, dwarf; κεφαλή, head]. The condition of being nanocephalous.

**nanocephalous** (na-no-sef'-al-us) [nanus; κεφαλή, head]. Possessing a dwarfed head.
**nanocephalus** (na-no-sef'-al-us) [νᾶνος, dwarf; κεφαλή, head]. A fetus with a dwarfed head.
**nanocormia, nanocormus** (na-no-kor'-me-ah, na-no-kor'-mus) [νᾶνος, dwarf; κορμός, trunk]. 1. A monstrosity possessing a dwarfed trunk. 2. A dwarfed condition of the trunk.
**nanoid** (na'-noid) [nanus; εἶδος, like]. Dwarf-like.
**nanomelia, nanomelus** (na-no-me'-le-ah, na-nom'-el-us) [nanus; μέλος, a limb]. A monster characterized by undersized limbs.
**nanosoma, nanosomia** (na-no-so'-mah, na-no-so'-me-ah) [νᾶνος, dwarf; σῶμα, body]. See microsomia, nanism.
**nanosomus** (na-no-so'-mus) [νᾶνος, dwarf; σῶμα, body]. One dwarfed in body.
**nanous** (nan'-us). Dwarfed, stunted.
**nanus** (na'-nus) [νᾶνος, a dwarf]. 1. A dwarf. 2. Dwarfed; stunted.
**nape** (nāp) [ME]. The back part of the neck; the nucha.
**napelline** (na-pel'-ēn) [napellus, dim. of napus, a turnip], $C_{16}H_{39}NO_{11}$. An alkaloid of Aconitum napellus. It is an anodyne and antineuralgic. Dose ⅓-½ gr. (0.01-0.03 Gm.).
**napellus** (na-pel'-us) [L., dim. of napus, a turnip]. Aconitum napellus, q. v.
**napha** (na'-fah) [L.]. Orange-blossoms.
**naphtalan** (naf'-tal-an). Same as naftalan.
**naphtha** (naf'-thah) [νάφθα, from Ar., naft, naphtha]. 1. Formerly, any strong-smelling, inflammable, volatile liquid. 2. A colorless, inflammable oil distilled from petroleum, bituminous shale, etc. **n., coal-tar,** a volatile mixture distilled from coal-tar and containing benzene, toluene, xylene, and similar hydrocarbons. **n., petroleum,** the more volatile part of petroleum collected during distillation and known as crude naphtha, or again separated by distillation into gasolene, benzene, and refined naphtha. **n.-salicili,** a disinfecting solution said to contain salicylic acid, naphthol, and borax. **n., shale,** naphtha distilled from bituminous shale. **n. vitrioli,** ethylic ether. **n., wood-,** methyl-alcohol.
**naphthalan** (naf'-thal-an). A substance obtained from the distillation of a variety of naphtha. It is used as a protective dressing.
**naphthalene, naphthalin** (naf'-thal-ēn, naf'-thal-in) [naphtha]. Naphthalinum (U. S. P.), $C_{10}H_8$. A hydrocarbon crystallizing in large, silvery, rhombic plates, slightly soluble in hot, but insoluble in cold, water, though easily soluble in methyl- and ethyl-alcohols, chloroform, ether, and benzene. It is an antiseptic, and is used in intestinal putrefaction, in typhoid fever, etc.; locally, in scabies and pruritus. Dose 5-10 gr. (0.32-0.65 Gm.).
**naphthalol** (naf'-thal-ol). See betol.
**naphthacresol** (naf-tho-kre'-sol). A brown, tarry, antiseptic liquid, insoluble in water, soluble in alcohol; used the same as creolin.
**naphthoformin** (nap-tho-form'-in). A condensation-product of naphthol, formaldehyde and ammonia. It is used as an application in skin diseases.
**naphthol, naphtol** (naf'-thol, naf'-tol), $C_{10}H_7$, OH. A substance found in coal-tar and prepared artificially from alphanaphthol. It exists in two isomeric forms, alphanaphthol and betanaphthol, and occurs in the form of pale, buff-colored crystals. Beta-naphthol (U. S. P.) is employed in dyspepsia and an intestinal antiseptic in diarrhea, typhoid fever, etc. Locally it is used in eczema, prurigo, herpes, favus, etc. Dose 5-10 gr. (0.32-0.65 Gm.). **n.-aristol, n.-diiodide.** See diiodobetanaphthol. **n.-eucalyptol,** a compound of alphanaphthol or beta-naphthol and eucalyptol; it is used as a surgical antiseptic.
**naphtholate** (naf'-thol-āt). A naphthol compound

in which a base replaces the hydrogen atom in the hydroxyl.

**naphtholism** (*naf'-thol-izm*). Poisoning from continued external application of naphthol; it is marked by nephritis, hematuria, and eclampsia.

**naphtholum** (*naf'-thol-um*). Betanaphthol.

**naphthopyrine** (*naf-tho-pi'-rin*). A molecular compound obtained by the prolonged trituration of betanaphthol with twice its weight of antipyrine.

**naphthoquinone** (*naf-tho-kwin'-ōn*), $C_{10}H_6O_2$. A crystalline substance formed by oxidation of naphthalin.

**naphthosalol** (*naf-tho-sa'-lol*). See *betol.*

**naphthoxol** (*naf-thoks'-ol*). An antiseptic fluid consisting of a 3 % solution of hydrogen dioxide, 32 to 38 % alcohol, and 2 % naphthol.

**naphthyl** (*naf'-thil*), $C_{10}H_7$. The radical of naphthalene.

**naphthylamine** (*naf-thil'-a-min*), $C_{10}H_7 . NH_2$. A crystallizable substance turning red in the air.

**naphtol** (*naf'-tol*). Same as *naphthol.*

**napiform** (*na'-pif-orm*) [*napus*, a turnip; *forma*, form]. Turnip-shaped.

**narceine** (*nar'-se-ēn*) [νάρκη, numbness], $C_{23}H_{29}NO_9$. An alkaloid contained in opium. It is sparingly soluble in water and alcohol, and forms fine, silky, inodorous, bitter crystals. It is used as a substitute for morphine. Dose $\frac{1}{4}$ gr. (0.016 Gm.). **n. hydrochloride,** $C_{23}H_{29}NO_9 . HCl + 2H_2O$, an acid substance forming colorless needles freely soluble in water and alcohol. It is used as a hypnotic in doses of $\frac{1}{6}$–3 gr. (0.01–0.2 Gm.). **n. meconate,** $C_{23}H_{29}NO_9 . C_7H_4O_7$, yellow crystals soluble in water. It is used as a sedative. Subcutaneous dose $\frac{1}{10}$–$\frac{3}{8}$ gr. (0.006–0.025 Gm.). **n., reaction for** (Arnold's), upon heating the substance containing narceine with concentrated sulphuric acid and a trace of phenol a reddish coloration results. **n. valerianate,** $C_{23}H_{29}NO_9 . C_5H_{10}O_2$, a greenish-white, unstable powder, soluble in alcohol or hot water; decomposes on exposure. It is used as a sedative in mania, hysteria, etc.

**narcism** (*nar'-sizm*) [*Narcissus*, a Greek mythological character who fell in love with his own image reflected in a fountain]. Observation of one's own naked body, with voluptuous ideas.

**narcissine** (*nar-sis'-ēn*), $C_{16}H_{17}O_4N$. An alkaloid obtained from the bulb of *Narcissus pseudonarcissus*, the common daffodil. It is a stable crystalline alkaloid, insoluble in water.

**narco-** (*nar-ko-*) [νάρκη, numbness]. A prefix meaning relating to narcosis, numbness, or stupor.

**narcohypnia** (*nar-ko-hip'-ne-ah*) [*narco-*; ὕπνος, sleep]. Waking numbness; a peculiar state in which the patient has a sense of numbness on awaking.

**narcolepsy** (*nar'-ko-lep-se*) [*narco-*; *epilepsy*]. An uncontrollable tendency to attacks of deep sleep of short duration. It has been observed in epilepsy and other affections.

**narcoleptic** (*nar-ko-lep'-tik*) [*narco-*; λαμβάνειν, to seize]. Affected with narcolepsy.

**narcoma** (*nar-ko'-mah*) [see *narcosis*]. Stupor from the use of a narcotic.

**narcomania** (*nar-ko-ma'-ne-ah*) [*narco-*; μανία, madness]. 1. Insanity characterized by stupor. 2. Insanity from use of narcotics. 3. A morbid craving for narcotics.

**narcomaniac** (*nar-ko-ma'-ne-ak*). One affected with narcomania.

**narcomatous** (*nar-ko'-mat-us*) [νάρκη, stupor]. Pertaining to, affected with, or of the nature of, narcoma.

**narcopepsia, narcopepsis** (*nar-ko-pep'-se-ah, nar-ko-pep'-sis*) [*narco-*; πέψις, digestion]. Slow or torpid digestion.

**narcophine** (*nar'-ko-fēn*). Trade name of a combination of morphine meconate and narcotine meconate; it resembles pantopon in its action, and is said to contain about 30 per cent. of morphine.

**narcose** (*nar'-kōs*) [*narcosis*]. In a condition of stupor.

**narcosis** (*nar-ko'-sis*) [ναρκοῦν, to benumb]. The state of complete unconsciousness produced by a narcotic drug or an anesthetic. **n., medullary,** anesthesia by cocainization of the spinal cord.

**narcospasm** (*nar'-ko-spazm*) [*narco-*; *spasm*]. Spasm accompanied by stupor.

**narcotic** (*nar-kot'-ik*) [see *narcosis*]. 1. Producing stupor. 2. A drug that produces narcosis.

**narcoticoacrid, narcoticoirritant** (*nar-kot-ik-o-ak'-rid, nar-kot-ik-o-ir'-it-ant*). See *acronarcotic.*

**narcotile** (*nar'-kot-īl*). A mixture of chlorides of methyl and ethyl; intended for a general anesthetic.

**narcotine** (*nar'-kot-ēn*) [see *narcosis*], $C_{22}H_{23}NO_7$. An alkaloid of opium, separated from morphine by potassium hydroxide. It crystallizes from alcohol in shining prisms, and melts at 176° C. It is sudorific and antipyretic, but has no narcotic effects. Dose 1–3 gr. (0.06–0.2 Gm.).

**narcotism** (*nar'-kot-izm*) [see *narcosis*]. The condition resulting from the use of a narcotic.

**narcotize** (*nar'-ko-tīz*) [see *narcosis*]. To put under the influence of a narcotic; to render unconscious by means of a narcotic.

**nard** [νάρδος, nard]. See *spikenard.*

**naregamia** (*nar-eg-a'-me-ah*) [E. Ind.]. Goanese ipecacuanha; the bark of *N. alata*, having properties due to an alkaloid, naregamine.

**naregamine** (*nar-eg'-am-ēn*) [*naregamia*]. An alkaloid contained in *Naregamia alata.*

**nares** (*na'-rēz*). Plural of *naris.*

**nargol** (*nar'-gol*). A preparation of silver and nucleinic acid used in the local treatment of gonorrhea and conjunctivitis.

**narinjin** (*na-rin'-jin*) [Sanskrit, *narinji*, the orange], $C_{23}H_{26}O_{12} + 4H_2O$, or $C_{23}H_{28}O_{12} + 5H_2O$. A glucoside from the blossoms of *Citrus decumana*, the grape-fruit or pomelo tree.

**naris** (*na'-ris*) [L.; pl., *nares*]. A nostril. One of a pair of openings at the anterior part (*anterior nares*) or at the posterior part (*posterior nares*) of the nasal fossæ.

**narry** (*nar'-e*) [Mongolian]. Alcoholic gastritis.

**nasal** (*na'-zal*) [*nasus*, the nose]. Pertaining to the nose. **n. artery.** See under *artery.* **n. bones,** the two small bones forming the arch of the nose. **n. capsule,** the embryonic cartilage which becomes the nose. **n. catarrh,** catarrh of the nasal mucous membrane; coryza. **n. duct.** See under *duct.* **n. eminence.** See *eminence, nasal.* **n. fossæ,** the cavities of the nose. **n. ganglion,** the sphenopalatine ganglion. **n labyrinth,** the irregular cavity formed by the turbinal bones in the nasal passages. **n. line.** See *Salle's (de) line.* **n. nerve.** See under *nerve.* **n. spine.** 1. A sharp process descending in the middle line from the inferior surface of the frontal bone between the superior maxillæ. 2. The inferior sharp edge of the nasal crest of the superior maxilla. Syn., *anterior nasal spine.* 3. The prominence formed by the junction in the median line of the elevations upon the posterior internal adjoining margins of the palatal plate of the palate bones. **n. voice,** a peculiar muffled timbre of the voice, especially marked in cases of perforation of the palate.

**nasalis.** See *muscles, table of.*

**nascent** (*nas'-ent*) [*nasci*, to be born]. A term applied to gaseous substances at the moment of their liberation from chemical combination.

**nasen** (*na'-zen*) [*nasus*, a nose]. Belonging to the nasal bone in itself.

**nasethmoid** (*na-zeth-moid*). Pertaining to the nasal and ethmoid bones.

**nasiform** (*na'-zif-orm*) [*nasus*, nose; *forma*, form]. Shaped like the nose.

**nasioalveolar** (*na-ze-o-al-ve'-o-lar*). Relating to or connecting the nasion and the alveolar point.

**nasiobregmatic** (*na-ze-o-breg-mat'-ik*). Pertaining to the nasion and the bregma.

**nasioinial** (*na-ze-o-in'-e-al*). Pertaining to the nasion and the inion.

**nasiomental** (*na-ze-o-men'-tal*). Pertaining to or connecting the nasion and the mentum.

**nasion** (*na'-ze-on*) [*nasus*, nose]. The median point of the nasofrontal suture.

**nasitis** (*na-zi'-tis*). See *rhinitis.*

**Nasmyth's membrane, N.'s cuticle** (*naz'-mith*) [Alexander Nasmyth, Scotch dentist, 1847– ]. The epithelial membrane enveloping the enamel of the tooth during its development and for a short time after birth. Syn., *cuticula dentis.*

**naso-** (*na-zo-*) [*nasus*, nose]. A prefix denoting connection with or relation to the nose.

**nasoantral** (*na-zo-an'-tral*). Relating to the nose and the maxillary antrum.

**nasoantritis** (*na-zo-an-tri'-tis*). Rhinitis combined with inflammation of the antrum of Highmore.

**nasoaural** (*na-zo-aw'-ral*). Relating to the nose and ear.

**nasobuccal** (*na-zo-buk'-al*). Relating to the nose and cheek.

**nasobuccopharyngeal** (*na-zo-buk-o-far-in'-je-al*). Relating to the nose, cheek, and pharynx.
**nasociliary** (*na-zo-sil'-e-a-re*). Applied to a nerve distributed to the nose and the ciliary body.
**nasocular** (*na-zok'-u-lar*). Pertaining to the nose and the eye; nasorbital.
**nasofrontal** (*na-zo-fron'-tal*). Pertaining to the nasal and the frontal bones.
**nasolabial** (*na-zo-la'-be-al*). Pertaining to the nose and lip.
**nasolacrimal** (*na-zo-lak'-rim-al*) [*naso-*; *lacrima*, tear]. Pertaining to the nose and the lacrimal apparatus, as the *nasolacrimal* duct.
**nasolambdoidal** (*na-zo-lam-doi'-dal*). Relating to the nasal bones and the lambdoid suture. n. line. See *Poirier's line*.
**nasology** (*na-zol'-o-je*). The study of noses.
**nasomalar** (*na-zo-ma'-lar*). Relating to the nose and the malar bone.
**nasomanometer** (*na-zo-man-om'-et-er*). A manometer supplied with tubes to introduce liquid into the nostrils in order to test the permeability of the nose.
**nasooccipital** (*na-zo-ok-sip'-it-al*). Pertaining to the nose and the occiput.
**nasopalatine** (*na-zo-pal'-at-en*). 1. Pertaining to both the nose and the palate, as the *nasopalatine* nerve. 2. Giving passage to the *nasopalatine* nerve, as the *nasopalatine* canal.
**nasopalpebral** (*na-zo-pal'-pe-bral*). Relating to the nose and the eyelids.
**nasopharyngeal** (*na-zo-far-in'-je-al*) [*nasopharynx*]. Pertaining to both the nose and the pharynx or to the nasopharynx.
**nasopharyngitis** (*na-zo-far-in-ji'-tis*) [*nasopharynx*; ιτις, inflammation]. Inflammation of the nasopharynx. Syn., *rhinopharyngitis*.
**nasopharynx** (*na-zo-far'-inks*) [*naso-*; φάρυγξ, pharynx]. The space between the posterior nares and a horizontal plane through the lower margin of the soft palate.
**nasoscope** (*na'-zo-skop*). A rhinoscope.
**nasoseptitis** (*na-zo-sep-li'-tis*). Inflammation of the nasal septum.
**nasosinuitis, nasosinusitis** (*na-zo-si-nu-i'-tis, na-zo-si-nus-i'-tis*). Inflammation of the nasal cavities and accessory sinuses.
**nasoturbinal** (*na-zo-tur'-bin-al*). Relating to the nose and the turbinal bone.
**nasrol** (*nas'-rol*). See *symphorol*.
**nasta** (*nas'-tah*) [ναστός, solid]. A fleshy tumor of the neck about the shoulders.
**nastin** (*nas'-tin*). An oily solution obtained from a streptothrix found in leprosy, which when combined with benzoyl chloride has been used as a cure for leprosy.
**nasus** (*na'-sus*) [L.: gen., *nasi*]. The nose. n. aduncus, hook nose. n. cartilagineus, the cartilaginous part of the nose. n. externus, the external nose. n. incurvus, saddle-back nose. n. osseus, the bony part of the nose. n. simus, pug nose.
**nasute** (*na'-sut*) [*nasus*, nose]. 1. Large-nosed. 2. Keen of scent.
**natal** (*na'-tl*) [*natalis*, from *nasci*, to be born]. Native; connected with one's birth.
**natal** (*na'-tl*) [*natis*, rump]. Gluteal; pertaining to the nates.
**Natal sore** (*na-tal'*) [*Natal*, a state in South Africa]. See *furunculus orientalis*.
**natality** (*na-tal'-it-e*) [*natalis*, of birth]. In State medicine and statistics, the birth-rate.
**natalom** (*na-tal'-o-in*) [*Natal*, an African state; *aloin*], C₅H₃₈O₁₁. The aloin derived from Natal aloes.
**natant** (*na'-tant*) [*natare*, to swim]. Swimming or floating on the surface of a liquid.
**nates** (*na'-tez*) [*natis*, the buttock]. The buttocks, the gluteal region of the body. n. of brain, the anterior pair of the corpora quadrigemina.
**natiform** (*nat'-if-orm*) [*nates*, buttocks; *forma*, form]. Buttock-shaped.
**National Formulary** (*nash'-on-al for'-mu-lar-e*). A collection of formulas issued by the American Pharmaceutical Association as a supplement to the United States Pharmacopœia.
**native** (*na'-tiv*) [*nativus*, born]. Of indigenous origin or growth; occurring in its natural state; not artificial. n. albumins, a class of proteins occurring ready-formed in the tissues.
**natrium** (*na'-tre-um*). See *sodium*.
**natron** (*na'-tron*) [Ar., *natrūn*, native sodium carbonate]. 1. Native sodium carbonate. 2. Soda. 3. Sodium or potassium nitrate.
**natural** (*nat'-u-ral*) [*natura*, nature]. Pertaining to nature. Not abnormal or artificial. n. history, a term including a description of all the products and phenomena of nature, but at present generally restricted to the sciences of zoology and botany. n. philosophy, the science treating of the physical properties of matter at rest and in motion; now usually called *physics*.
**Nauheim treatment** (*now'-him*) [*Nauheim*, a city in Germany]. See *Schott method* (2).
**naupathia** (*naw-pa'-the-ah*) [ναῦς, ship; πάθος, sickness]. Seasickness.
**nausea** (*naw'-she-ah*) [ναυσία, sea-sickness]. Sickness at the stomach, with inclination to vomit. n., creatic, n., kreatic, morbid aversion to eating animal food. n. gravidarum, the morning sickness of pregnancy. n. marina, n. navalis, sea-sickness.
**nauseant** (*naw'-she-ant*) [*nausea*]. 1. Nauseating; producing nausea. 2. Any agent that produces nausea.
**nauseating, nauseous** (*naw'-she-a-ting, naw'-she-us*) [*nausea*]. Producing nausea or loathing.
**nausiosis** (*naw-se-o'-sis*). 1. Nausea. 2. A venous hemorrhage in which the flow is discharged by jets.
**navel** (*na'-vel*) [AS., *nafela*]. The umbilicus. n.-string, the umbilical cord.
**navicular** (*na-vik'-u-lar*) [*navicula*, a little ship]. Boat-shaped. n. bone, the scaphoid bone of the foot. n. fossa. 1. A depression between the vaginal aperture and the fourchet. 2. A dilatation of the urethra near the glans penis.
**naviculare** (*na-vik-u-la'-re*) [*navicula*]. The scaphoid bone. See *bones, table of*.
**navicularthritis** (*na-vik-u-lar-thri'-tis*). In veterinary practice inflammation of the navicular bone and contiguous tissues resulting in incomplete extension of the joint, tumefaction of the hoof, and pain.
**naviculocuboid** (*na-vik-u-lo-ku'-boid*). Relating to the scaphoid and the cuboid bones.
**naviculocuneiform** (*na-vik-u-lo-ku-ne'-e-form*). Relating to the scaphoid and cuneiform bones.
**naviculoid** (*nav-ik'-u-loid*) [*navicula*, a small boat; εἶδος, like]. Scaphoid.
**nazeptic wool** (*naz-ep'-tik*). Trade name of a preparation consisting of strands of absorbent cotton, medicated with menthol, phenol, eucalyptol and methyl salicylate, enclosed in a glass vial; it is said to be useful for colds, catarrh, hay fever and nasal irritation due to microorganisms or dust.
**Nb.** Chemical symbol for the element *niobium*.
**Neapolitan fever** (*ne-ap-ol'-it-an*). Malta or Mediterranean fever. N. ointment, blue ointment; mercurial ointment.
**near-point.** The *punctum proximum*, the point nearest the eye at which an object can be seen distinctly. n., absolute, that near-point for either eye alone at which no effort at accommodation is made. n., relative, that near-point for both eyes at which accommodation is brought into play.
**near-sight.** See *myopia*.
**neathrosis** (*ne-ar-thro'-sis*) [νέος, new; ἄρθρον, a joint]. A new and abnormally-produced articulation, in the sequence of a fracture, dislocation, or disease of the bone.
**nebenkern** (*na'-ben-kern*) [Ger. *neben*, near; *kern*, nucleus]. See *paranucleus*.
**nebula** (*neb'-u-lah*) [L., a cloud or mist]. 1. A faint, grayish opacity of the cornea. 2. A spray, a liquid intended for use in an atomizer.
**nebulization** (*neb-u-liz-a'-shun*). Same as *atomization*.
**nebulize** (*neb'-u-liz*) [*nebula*, a mist or spray]. To convert into a spray or vapor.
**nebulizer** (*neb'-u-li-zer*). See *atomizer*.
**Necator americanus** (*ne-ka'-tor am-er-ik-a'-nus*). Same as *Uncinaria Americana*; hookworm.
**neck** (*nek*) [AS., *hnecca*]. The constricted portion of the body connecting the head with the trunk; also, the narrow portion of any structure serving to join its parts. See also *cervix*. n., anatomical, the constricted portion of the humerus, just below the articular surface, serving for the attachment of the capsular ligament. n. band, the cervical skin lesion

observed in pellagra. **n.-berg**, the vulgar name of the cervical portion of the thymus gland of animals. See *heart-berg*. **n., cephalic**, the constricted, neck-like region lying between the quadrigeminum and the thalami. Syn., *isthmus prosencephali*. **n., Derbyshire**. Synonym of *goiter*. **n., Madelung's**, diffuse lipoma of the neck. **n., Nithsdale**, goiter. **n., surgical**, the constricted part of the humerus just below the tuberosities. **n., wry-**. Synonym of *torticollis*.

**necrectomy** (nek-rek'-to-me) [*necro-*; ἐκτομή, excision]. The excision of the necrotic conductors of sound in chronic purulent otitis media; in a more general sense, any removal of necrosed material by a cutting operation.

**necremia, necræmia** (nek-re'-me-ah) [νεκρός, dead; αἷμα, blood]. Death of the blood; a condition marked by loss of vitality in the corpuscles and a tendency not to run together.

**necrencephalus** (nek-ren-sef'-al-us) [*necro-*; ἐγκέφαλος, brain]. Softening of the brain.

**necro-** (nek-ro-) [νεκρός, dead]. A prefix signifying death.

**necrobacillosis** (nek-ro-bas-il-o'-sis). A disease of animals caused by *Bacillus necrophorus*.

**necrobiosis** (nek-ro-bi-o'-sis) [*necro-*; βίος, life]. Molecular death of tissue.

**necrobiotic** (nek-ro-bi-ot'-ik) [*necro-*; βίος, life]. Pertaining to or causing necrobiosis.

**necrocedia** (nek-ro-se'-de-ah) [*necro-*; κῆδος, care]. The process of embalming.

**necrocomium** (nek-ro-ko'-me-um). See *morgue*.

**necrocytosis** (nek-ro-si-to'-sis) [*necro-*; κύτος, a cell]. Death or loss of vitality of the cells.

**necrocytotoxin** (nek-ro-si-to-toks'-in). A toxin produced by the death of cells.

**necrodermatitis** (nek-ro-der-mat-i'-tis) [*necro-*; δέρμα, skin; ιτις, inflammation]. A gangrenous inflammation of the skin.

**necrodochium** (nek-ro-do'-ke-um). See *morgue*.

**necrogenic** (nek-ro-jen'-ik) [*necro-*; γεννᾶν, to beget]. Originating from dead substances.

**necrology** (nek-rol'-o-je) [*necro-*; λόγος, science]. Tabulated mortality statistics.

**necromania** (nek-ro-ma'-ne-ah) [*necro-*; μανία, madness]. 1. A morbid desire for death or for the presence of dead bodies. 2. See *necrophilism*.

**necrometer** (nek-rom'-et-er) [*necro-*; μέτρον, a measure]. An instrument for weighing organs at an autopsy.

**necromimesis** (nek-ro-mim-e'-sis) [*necro-*; μίμησις, mimicry]. 1. The insane delusion of one who believes himself to be dead. 2. Simulation of death by a deluded person.

**necronectomy** (nek-ron-ek'-to-me) [*necro-*; ἐκτομή, excision]. The excision of a necrotic part—applied especially to the excision of the necrotic ossicles of the ear.

**necroparasite** (nek-ro-par'-as-īt) [*necro*; *parasite*]. A saprophyte, *q. v.*

**necrophagous** (nek-rof'-ag-us) [*necro-*; φαγεῖν, to eat]. A term applied to those animals that feed on dead or putrid flesh. Carrion-eaters.

**necrophile** (nek'-ro-fīl) [*necro-*; φιλεῖν, to love]. One who violates dead bodies.

**necrophilia** (nek-ro-fil'-e-ah). 1. See *necrophilism*. 2. A longing for death.

**necrophilism** (nek-rof'-il-izm) [*necro-*; φιλεῖν, to love]. A form of sexual perversion in which dead bodies are violated; insane sexual desire for a corpse.

**necrophilous** (nek-rof'-il-us). Subsisting on dead matter.

**necrophobia** (nek-ro-fo'-be-ah) [*necro-*; φόβος, fear]. 1. Insane dread of dead bodies, or of phantoms. 2. Thanatophobia; extreme dread of death.

**necropneumonia** (nek-ro-nū-mo'-ne-ah). Gangrene of the lung.

**necropsy** (nek'-rop-se) [*necro-*; ὄψις, sight]. The examination of a dead body; autopsy; postmortem examination.

**necroscopic** (nek-ro-skop'-ik) [*necro-*; σκοπεῖν, to view]. Pertaining to necroscopy.

**necroscopy** (nek-ros'-ko-pe) [*necro-*; σκοπεῖν, to inspect]. Postmortem examination of the body.

**necrose** (nek-rōs') [νεκρός, dead]. To become affected with necrosis; to cause necrosis.

**necrosemiotic** (nek-ro-sem-e-ot'-ik) [*necro-*; σημεῖον, sign]. Serving as a sign of death.

**necrosis** (nek-ro'-sis) [νεκρός, dead]. The death of cells surrounded by living tissue. *Necrosis proper* refers to death in mass; *necrobiosis* to death of individual cells. The dead tissue is called *sequestrum* in case of bone, and *sphacelus* in case of soft parts. In surgery the term necrosis is often applied specifically to the death of bone. **n., Balser's fatty**. See *Balser's fat-necrosis*. **n., central**, that in which the internal portions of a bone are involved. **n., cheesy**, necrosis characterized by the formation of a cheese-like material. **n., coagulation-**, **n., coagulative**, a variety characterized by the formation of fibrin. **n., colliquative**. See *n., liquefactive*. **n., embolic**, coagulation-necrosis in an anemic infarct following embolism. **n., endoglobular** (of Maragliano and Castellani), the degenerative decoloration of erythrocytes, giving rise to shadow corpuscles, "phantoms," or achromacytes. See *achromacyte*. **n., fat-**, a type of necrosis following fatty degeneration, reducing the entire structure to a fatty emulsion. **n. infantilis**, cancrum oris. ' **n., liquefactive**, a process analogous to coagulation-necrosis, but instead of fibrin the peculiar reaction of fluids and cells gives rise to the formation of a liquid. **n., mercurial**, a necrosis of bones due to chronic poisoning with mercury. **n., moist**, that in which the dead tissue is moist and soft. **n., phosphorus-**, a necrosis of bone, especially of the lower jaw, occurring in those exposed to the fumes of phosphorus. **n., superficial**, a necrosis in which the portion of bone just beneath the periosteum is affected. **n., total**, a form in which the bone through its entire thickness is involved. **n. ustilaginea**, dry gangrene from ergotism.

**necrosozoic** (nek-ro-so-zo'-ik) [*necro*; σῴζειν, to preserve]. Having power to preserve or to embalm.

**necrospermia** (nek-ro-sperm'-e-ah) [*necro-*; σπέρμα, seed]. Impotence due to loss of motility in the spermatozoa.

**necrosteon** (nek-ros'-te-on) [*necro-*; ὀστέον, bone]. Necrosis of bone.

**necrotic** (nek-rot'-ik) [*necrosis*]. Pertaining to or characterized by necrosis.

**necrotomic** (nek-ro-tom'-ik) [*necro-*; τομή, a cutting]. Pertaining to necrotomy.

**necrotomy** (nek-rot'-o-me) [*necro-*; τομή, a cutting]. 1. The dissection of a dead body. 2. The excision of necrotic bone or other tissue.

**nectandra** (nek-tan'-drah) [νέκταρ, nectar; ἀνήρ, a male]. A tree of the order *Laurineæ*. The bark of *Nectandra rodiæi* (*nectandræ cortex*, B. P.) contains tannic acid, resin, sugar, albumin, various salts and two alkaloids, beberine and sipirine. It is tonic, astringent, and febrifuge, and has been used in malarial fevers, but is not so good an antiperiodic as cinchona. Dose 20 gr.-1 dr. (1.3-4.0 Gm.). The root of *N. cymbarum*, of Brazil, is roborant; the balsamic oil from the bark is tonic, antispasmodic, diuretic, emmenagogue, and diaphoretic, and is also applied to ulcers. *N. pichury-major*, and *N. pichury-minor*, of Brazil, furnish seeds which are used in diarrhea and dysentery; they contain fat, an ethereal oil, and safrol.

**nectareous** (nek-ta'-re-us) [*nectar*]. Agreeable to the taste.

**nectary** (nek'-tar-e) [νέκταρ, nectar: *pl.*, *nectaries*]. In biology, that part of a flower which secretes nectar.

**nectrianin** (nek-tri'-an-in). A proposed remedy for cancer, said to be an extractive of the fungus *Nectria ditissima*, growing upon old trees.

**needle** (ne'-dl) [AS., *nædl*]. A sharp-pointed steel instrument used for sewing and for penetrating tissues for the purpose of carrying a ligature through. **n., aneurysm-**, one fixed on a handle, and with the eye at the point, especially adapted for ligating vessels. **n., cataract-**, one used for operating upon the cataractous lens or its capsule. **n., discission-**, one for insertion through the cornea, and breaking the capsule and substance of the crystalline lens. **n., exploring**, a grooved, sharp-pointed rod introduced into a cavity or a part for the purpose of determining the presence of fluid. **n., Hagedorn's**, a flat suture needle curved on its edge, with the eye perforating the side. **n.-holder**, a handle for clasping a needle. **n., hypodermic**, the fine, needle-pointed metallic tube attached to the barrel of the hypodermatic syringe. **n., knife**, one that has a sharp cutting-edge; it is used in the discission of cataracts.

**needling** (ne'-dling) [*needle*]. The process of lacerating a cataract with a needle, to afford entrance to the aqueous humor and cause absorption of the lens.

**Neef's hammer** (*năf*). An automatic arrangement for opening and breaking the current in an inductorium.
**nefrens** (*ne'-frenz*) [*ne*, not; *frendere*, to gnash the teeth]. Without teeth; edentate, whether nurslings or aged persons. Pl., **nefrendes**.
**Neftel's disease.** See *atremia* (2).
**negative** (*neg'-at-iv*) [*negare*, to deny]. 1. Denying; contradicting; opposing. 2. Of quantities, less than nothing. 3. In physics, opposed to a quality termed *positive*. **n. accommodation**, the absence of active accommodation; the state of the eye at rest, or when looking at an object at an infinite distance. **n. blood-pressure**, pressure which is less than that of the atmosphere. It exists in the large veins near the heart, owing to the aspirating action of the thorax. **n. chemotaxis**, the absence of the power of attracting leukocytes and wandering cells, or their actual repulsion. **n. electricity**, static or frictional electricity. **n. electrode**, the electrode connected with the negative pole of a battery. **n. phase**, the temporary lessening of the amount of antitoxin in the serum immediately following a second inoculation. See *opsonic index*. **n. pole**, the pole of a source of electricity to which the current returns after having passed through a circuit outside of the source. **n. variation of the muscle-current**, a diminution in the strength of muscle-current during tetanic contraction.
**negativism** (*neg'-at-iv-izm*) [*negare*, to deny]. A symptom observed in some cases of so-called *catatonia attonita*, in which the patient exhibits no spontaneous movements, although his muscles spontaneously and powerfully antagonize any passive motion.
**Negri bodies** (*na'-gre*) [Luigi *Negri*, Italian physician]. Protozoon-like bodies found in the nerve-cells of animals suffering from rabies.
**negro lethargy** (*ne'-gro*). The same as *African lethargy*, *q. v.*
**Neisseria** (*ni-se'-re-ah*) [see *Neisser's coccus*]. A genus of diplococci characterized by their coffee-bean shape, the flat sides being in apposition.
**Neisser's coccus** (*ni'-ser*) [Albert Ludwig Siegmund *Neisser*, German physician, 1855-1912]. *Micrococcus gonorrhϗœ*.
**Neisser's stain** (*ni'-ser*) [Max *Neisser*, German bacteriologist, 1869- ]. For the nuclei of *diphtheria bacilli*. It consists of 2 parts of solution (*a*) consisting of methylene blue 1 part, alcohol 20 parts, and one part of solution (*b*) consisting of crystal violet 1 part, absolute alcohol 10 parts, distilled water 300 parts. The after stain is made with chrysoidin.
**Nélaton's catheter** (*na-lah-ton(g)'*) [Auguste *Nélaton*, French surgeon, 1807-1873]. A soft-rubber catheter. **N.'s dislocation**, upward dislocation of the ankle, the astragalus being wedged in between the tibia and fibula. **N.'s fold**, a transverse fold of mucosa at the junction of the middle and lower thirds of the rectum, about 10 to 11 centimeters above the anus. **N.'s hematocele**, hematoma of the Fallopian tube. **N.'s line**, a line drawn from the anterior superior spine of the ilium to the most prominent part of the tuberosity of the ischium; in dislocation of the femur backward the trochanter is always found above this line. **N.'s operation**. For *amputation through the foot:* subastragaloid disarticulation by dorsal and plantar flaps, larger on the inner than on the outer side. **N.'s probe**, one that is capped with unglazed porcelain upon which a leaden ball makes a metallic streak. **N.'s sphincter**, a circular bundle of rectal muscular fibers situated from 8 to 10 centimeters above the anus, on a level with the prostate. It is not constant, and when present it generally occupies only a part of the circumference of the bowel. **N.'s tumor**, desmoid tumor of the abdominal wall.
**nelavan** (*nel'-av-an*). See *African lethargy*.
**nematachometer** (*nem-at-ak-om'-et-er*) [νῆμα, thread; τάχος, rapidity; μέτρον, a measure]. An instrument to measure the rapidity of transmission of impulses in peripheral nerves.
**nemathelminth** (*nem-ath-el'-minth*) [νῆμα, thread; ἔλμινς, worm]. Any nematode worm. The *nemathelminthes* (the round-worms, or thread-worms), form a class of *vermes*, many of which are endoparasitic.
**nematoblast** (*nem'-at-o-blast*) [νῆμα, thread; βλαστός, a germ]. A spermatoblast.
**Nematoda** (*nem-at-o'-dah*) [see *Nematode*]. A genus of worms, the threadworms, some of which

are parasitic in man and the lower animals. The most important of these are *Anguillula*, *Ankylostoma*, *Ascaris*, *Eustrongylus*, *Filaria*, *Oxyuris*, *Strongylus*, *Trichina*, *Trichocephalus*.
**nematode** (*nem'-at-ōd*) [νῆμα, thread; εἶδος, like]. 1. Thread-like; belonging to or resembling the *Nematoda*, or threadworms; applied to threadworms, hairworms, roundworms, pinworms. 2. The threads formed by a serial arrangement of the granules of protoplasm.
**nematoid** (*nem'-at-oid*). See *nematode* (1).
**nemomena** (*nem-o-me'-nah*) [νέμεσθαι, to devour; to spread]. Perforating ulcers.
**Nencki's test for indol** [Marcellus von *Nencki*, Polish physician, 1847-1901]. Treat with nitric acid containing nitrous acid; a red coloration results, and in concentrated solution a red precipitate may form.
**neo-** [νέος, new]. A prefix meaning new.
**neoarsycodil** (*ne-o-ar-sik'-o-dil*). Sodium methylarsenite. It is used in tuberculosis. Dose ½-1½ gr. (0.02-0.1 Gm.) for five days, then omit for five days.
**nēoarthrosis** (*ne-o-ar-thro'-sis*) [*neo-*; ἄρθρον, a joint]. A false joint.
**neoblast** (*ne'-o-blast*). See *paraplast*.
**neoblastic** (*ne-o-blas'-tik*) [*neo-*; βλαστός, a germ]. Pertaining to, or of the nature of, new tissue.
**neodermin** (*ne-o-der'-min*). An ointment containing difluordiphenyl; used on burns and ulcerated surfaces.
**neodymium** (*ne-o-dim'-e-um*) [*neo-*; δίδυμος, a twin]. According to Welsbach, a decomposition product of didymium forming red salts.
**neoferrum** (*ne-o-fer'-um*). Trade name of a preparation said to contain iron, arsenic, manganese, maltine, and sherry.
**neoformation** (*ne-o-form-a'-shun*). See *new-growth*.
**neogala** (*ne-og'-al-ah*) [*neo-*; γάλα, milk]. Same as colostrum.
**neogenesis** (*ne-o-jen'-e-sis*) [*neo-*; γενναν, to produce]. Regeneration of tissues.
**neogenetic** (*ne-o-jen-et'-ik*). Relating to neogenesis; productive of new growth. **n. zone**. See under *zone*.
**neohymen** (*ne-o-hi'-men*) [*neo-*; ὑμήν, membrane]. A new or false membrane.
**neologism** (*ne-ol'-o-jizm*) [*neo-*; λόγος, a word]. The utterance of meaningless words by the insane.
**neomembrane** (*ne-o-mem'-brān*). A new or false membrane.
**neomorphism** (*ne-o-mor'-fizm*) [*neo-*; μορφή, form]. In biology, the development of a new form.
**neon** (*ne'-on*) [νέος, new]. A gaseous element discovered by Ramsay and Travers in 1908; it is associated with liquid argon.
**neonatal** (*ne-o-na'-tal*) [*neo-*; *natus*, born]. Pertaining to the newborn.
**neonatus** (*ne-on-a'-tus*) [*neo-*; *natus*, born: *gen. pl.*, *neonatorum*]. One newly born.
**neopallium** (*ne-o-pal'-e-um*) [*neo-*; *pallium*, cloak]. The cerebral hemisphere with the exception of the rhinencephalon.
**neopathy** (*ne-op'-ath-e*) [*neo-*; πάθος, illness]. 1. A new or newly-discovered form of disease. 2. A recent complication or new condition of disease in a patient.
**neophilism** (*ne-off'-il-izm*) [*neo-*; φιλεῖν, to love]. Morbid or undue love of novelty.
**neophobia** (*ne-o-fo'-be-ah*) [*neo-*; φόβος, fear]. Insane dread of new scenes or of novelties.
**neophrenia** (*ne-o-fren'-e-ah*) [*neo-*; φρήν, mind]: Mental deterioration in early youth.
**neoplasia** (*ne-o-pla'-ze-ah*) [*neo-*; πλάσσειν, to mold]. The formation of new tissue or of a tumor.
**neoplasm** (*ne'-o-plazm*). See *newgrowth*. **n. inflammatory fungoid**, mycosis fungoides.
**neoplasmatic** (*ne-o-plaz-mat'-ik*) [*neo-*; πλάσσειν, to mold]. Of the nature of neoplasm; neoplastic.
**neoplastic** (*ne-o-plas'-tik*) [*neoplasm*]. Pertaining to, or of the nature of, a neoplasm.
**neoplasty** (*ne'-o-plas-te*) [see *neoplasm*]. The restoration of lost tissue by a plastic operation.
**neosalvarsan** (*ne-o-sal'-var-san*)' [*neo-*; *salvarsan*]. A name given by Ehrlich to a modification of salvarsan; it forms a neutral solution in distilled water without the aid of any other solvent. It is also known as 914. It is weaker than salvarsan, and so needs to be given in larger quantities.
**Neosporidia** (*ne-o-spo-rid'-e-ah*) [*neo-*; σπορίς,

# NEOSTOMY 599 NEPHROHEMIA

seed]. A class of Sporozoa in which spores are formed without terminating the existence of the individual.

**neostomy** (ne-os'-to-me) [neo-; στόμα, mouth]. The operative production of an opening into an organ or between two organs.

**nepenthe** (ne-pen'-the) [νηπενθής, banishing sorrow]. Trade name of a deodorized preparation of opium.

**nepeta** (nep'-et-ah). Catnip or catmint, from N. cateria; used for children when a carminative or mild diaphoretic is indicated.

**nephablepsia** (nef-ab-lep'-se-ah). See niphablepsia.

**nephalism** (nef'-al-izm) [νηφαλισμός, soberness]. Total abstinence from spirituous or alcoholic liquors.

**nephela** (nef'-el-ah) [νεφέλη, cloud]. Leukoma; also cloudiness of the urine.

**nephelium** (nef-e'-le-um). See nebula.

**nepheloid** (nef'-el-oid) [νεφέλη, cloud; εἶδος, like]. Cloudy or turbid, as the urine under certain conditions.

**nephelometer** (nef-el-om'-et-er) [νεφέλη, cloud; μέτρον, measure]. An apparatus for ascertaining the number of bacteria in a suspension, or the turbidity of a fluid.

**nephelometry** (nef-el-om'-et-re). The determination of the degree of turbidity of a fluid.

**nephelopia** (nef-el-o'-pe-ah) [νεφέλη, cloud; ὤψ, eye]. Cloudy or dim vision, due to some diminution of the transparency of the ocular media.

**nephradenoma** (nef-rad-en-o'-mah) [nephrus; ἀδήν, gland; ὄμα, tumor; pl., nephradenomata]. Adenoma of the kidney.

**nephralgia** (nef-ral'je-ah) [nephrus; ἄλγος, pain]. Neuralgic pain in the kidney.

**nephralgic crises** (nef-ral'-jik). Ureteral paroxysms of pain in locomotor ataxia.

**nephranuria** (nef-ran-u'-re-ah) [nephrus; ἀν, priv.; οὖρον, urine]. Suppression of the renal secretion.

**nephrapostasis** (nef-rap-os'-tas-is) [nephrus; ἀπόστασις, suppurative inflammation]. Abscess, or suppurative inflammation, of the kidneys.

**nephrapragmonia** (nef-rap-rag-mo'-ne-ah) [nephrus; ἀ, priv.; πράγμων, work]. Inactivity or torpidity of the kidneys.

**nephrarctia** (nef-rark'-te-ah) [nephrus; arctus, from arcere, to bind]. Contraction of the kidney.

**nephratonia, nephratony** (nef-rat-o'-ne-ah, nef-rat'-on-e) [nephrus; ἀ, priv.; τόνος, tone]. Atony of or paralysis of the kidneys.

**nephrauxe** (nef-rawks'-e) [nephrus; αὔξη, increase]. Enlargement of the kidney.

**nephrectasia** (nef-rek-ta'-ze-ah) [nephrus; ἔκτασις, dilatation]. Dilatation of a kidney.

**nephrectomize** (nef-rek'-to-mīz). To excise the kidney from.

**nephrectomy** (nef-rek'-to-me) [nephrus; ἐκτομή, excision]. Excision of the kidney. n., abdominal, nephrectomy performed through an abdominal incision. n., lumbar, nephrectomy through an incision in the loin.

**nephredema** (nef-re-de'-mah) [nephrus; edema]. Edema of the kidney.

**nephrelcosis** (nef-rel-ko'-sis) [nephrus; ἕλκωσις, ulceration]. Ulceration of the kidney.

**nephrelcus** (nef-rel'-kus) [nephrus; ἕλκος, an ulcer]. An ulcer of the kidney.

**nephremia, nephræmia** (nef-re'-me-ah) [nephrus; αἷμα, blood]. Renal congestion.

**nephremorrhagia** (nef-rem-or-a'-je-ah) [nephrus; αἷμα, blood; ῥηγνύναι, to burst forth]. Hemorrhage from the kidney.

**nephremphraxis** (nef-rem-fraks'-is) [nephrus; ἔμφραξις, obstruction]. Obstruction of the vessels of the kidneys.

**nephresia** (nef-re'-ze-ah) [νεφρός, kidney]. Disease of the kidney.

**nephretic** (nef-ret'-ik) [νεφρός, kidney]. Affected with nephresia.

**nephria** (nef'-re-ah). See Bright's disease.

**nephric** (nef'-rik) [nephrus]. Pertaining to the kidney; renal.

**nephridia** (nef-rid'-e-ah). Plural of nephridium.

**nephridion** (nef-rid'-e-on). Same as nephridium.

**nephridium** (nef-rid'-e-um) [dim. of νεφρός, kidney]. 1. The fat about the kidneys. 2. The suprarenal capsule. 3. A Wolffian tubule.

**nephrin, nephrina** (nef'-rin, nef-ri'-nah). See cystin.

**nephrism** (nef'-rizm) [nephrus]. The grave condition of patients suffering from pronounced or advanced disease of the kidney.

**nephritic** (nef-rit'-ik) [see nephritis]. 1. Pertaining to nephritis. 2. Improperly, pertaining to the kidney; the correct term is nephric.

**nephritides** (nef-rit'-id-ēz). The plural of nephritis; a term embracing the various forms of nephritis.

**nephritis** (nef-ri'-tis) [nephrus; ιτις, inflammation]. Inflammation of the kidney. n. caseosa, cheesy degeneration of the kidney. n., diffuse, that involving both epithelial and connective-tissue elements of the kidney. n., glomerular, glomerulonephritis. n., interstitial, that involving the connective tissue chiefly; it may be acute or chronic. n., interstitial, acute, a form due to septic infection either through the blood, as in pyemia, or through extension along the ureter or from neighboring structures. n., interstitial, chronic, a form in which the kidney is small and hard, the capsule is adherent, the cortex is granular and marked by cysts. The cortex is diminished in thickness. Syn., granular or gouty kidney. n., metastatic, that secondary to disease of another organ. n., parenchymatous, a form in which the inflammation affects the epithelium of the uriniferous tubules. Syn., catarrhal nephritis; desquamative nephritis; tubular nephritis. n., parenchymatous, acute, a form in which the kidney is enlarged, congested, its structural markings are obscured, the epithelium is in a state of cloudy swelling or fatty degeneration, and many tubules contain casts; in others the epithelium is desquamated. Syn., acute Bright's disease. n., parenchymatous, chronic, a variety in which the kidney is enlarged, pale or yellow, and soft; the epithelium presents an advanced stage of fatty degeneration. Casts are often present. Syn., large white kidney. n., saturnine, that due to chronic lead-poisoning. n., scarlatinal, an acute nephritis due to scarlatina. n., tuberculous, that due to the presence of tubercle bacilli. It presents itself either in the form of caseating masses or cavities in the substance of the kidney, or as miliary tuberculosis of the organ. n., typhoid. See nephrotyphus. n. uratica, gouty kidney, partial or more diffuse interstitial nephritis in arthritic subjects, due to deposition of urates.

**nephro-** (nef'-ro-) [nephrus]. A prefix meaning pertaining to the kidney.

**nephroabdominal** (nef-ro-ab-dom'-in-al) [nephro-; abdomen]. Pertaining to the kidneys and the abdomen.

**nephrocapsectomy, nephrocapsulectomy** (nef-ro-kap-sek'-to-me, nef-ro-kap-sū-lek'-to-me) [nephro-; capsula, capsule; ἐκτομή, excision]. Excision of the capsule of the kidney.

**nephrocapsulotomy** (nef-ro-kap-sū-lot'-om-e). Incision of the renal capsule.

**nephrocardiac** (nef-ro-kar'-de-ak) [nephro-; καρδία, heart]. Pertaining to the kidney and the heart.

**nephrocele** (nef'-ro-sēl) [nephro-; κήλη, hernia]. Hernia of the kidney.

**nephrochalæosis** (nef-ro-kal-as-o'-sis) [nephro-; χαλάζων, nodule]. Granular kidney.

**nephrocolica** (nef-ro-kol'-ik-ah) [nephro-; κωλικός, colic]. Renal colic.

**nephrocolopexy** (nef-ro-kol'-o-peks-e) [nephro-; κῶλον, colon; πῆξις, fixation]. The surgical anchoring of the kidney and colon by means of the nephrocolic ligament.

**nephrocoloptosis** (nef-ro-ko-lop-to'-sis) [nephro-; κῶλον, colon; πτῶσις, fall]. Downward displacement of the kidney and colon.

**nephrocystanastomosis** (nef-ro-sist-an-as-to-mo'-sis) [nephro-; κύστις, bladder; ἀναστόμωσις, an opening]. The surgical formation of an opening between the kidney and the urinary bladder.

**nephrocystitis** (nef-ro-sis-ti'-tis) [nephro-; κύστις, bladder; ιτις, inflammation]. Inflammation of both bladder and kidney.

**nephrocystosis** (nef-ro-sis-to'-sis) [nephro-; κύστις, cyst]. The condition of cystic kidney, or its formation.

**nephroerysipelas** (nef-ro-er-is-ip'-el-as). Simultaneous erysipelas and nephritis.

**nephrogenic, nephrogenetic** (nef-ro-jen'-ik, nef-ro-jen-et'-ik) [nephro-; γεννᾶν, to produce]. Of renal origin.

**nephrogenous** (nef-roj'-en-us) [nephro-; γεννᾶν, to beget]. Of renal origin.

**nephrography** (nef-rog'-ra-fe) [nephro-; γράφειν, to write]. A description of the kidneys.

**nephrohemia, nephrohæmia** (nef-ro-he'-me-ah). See nephremia.

**nephrohydrops, nephrydrosis** (*nef-ro-hi'-drops, nef-rid-ro'-sis*). See *hydronephrosis*.

**nephrohypertrophy** (*nef-ro-hi-pur'-tro-fe*) [*nephro-*; ὑπέρ, over; τροφή, nourishment]. Hypertrophy of the kidney.

**nephroid** (*nef'-roid*) [*nephro-*; εἶδος, form]. Kidney-shaped; reniform; resembling a kidney.

**nephrolith** (*nef'-ro-lith*) [*nephro-*; λίθος, a stone]. A calculus of the kidney.

**nephrolithiasis** (*nef-ro-lith-i'-as-is*) [*nephro-*; *lithiasis*]. The formation of renal calculi, or the diseased state that leads to their formation.

**nephrolithic** (*nef-ro-lith'-ik*) [*nephro-*; λίθος, stone]. Pertaining to, or affected with, a nephrolith.

**nephrolithocolica** (*nef-ro-lith-o-kol'-ik-ah*) [*nephro-*; λίθος, stone; κωλικός, colic]. Renal colic due to stone.

**nephrolithotomy** (*nef-ro-lith-ot'-o-me*) [*nephro-*; *lithotomy*]. An incision of the kidney for the removal of a calculus.

**nephrologist** (*nef-rol'-o-jist*) [*nephro-*; λόγος, science]. A specialist in renal diseases.

**nephrology** (*nef-rol'-o-je*) [*nephro-*; λόγος, science]. The science of the anatomy, physiology, and diseases of the kidney.

**nephrolysin** (*nef-rol'-is-in*) [*nephro-*; λύειν, to loosen]. A toxic substance capable of disintegrating kidney cells.

**nephrolysis** (*nef-rol'-is-is*) [*nephro-*; λύειν, to loosen]. 1. The disintegration of the kidney by the action of a nephrolysin. 2. The operation of loosening an inflamed kidney from surrounding adhesions.

**nephrolytic** (*nef-ro-lit'-ik*). Pertaining to nephrolysis.

**nephromalacia** (*nef-ro-mal-a'-se-ah*) [*nephro-*; μαλακία, softness]. Softening, or abnormal softness, of the kidney.

**nephromegalia** (*nef-ro-meg-a'-le-ah*) [*nephro-*; μέγας, large]. Same as *nephrauxe*.

**nephromegaly** (*nef-ro-meg'-al-e*). See *nephromegalia*.

**nephromere** (*nef'-ro-mēr*) [*nephro-*; μέρος, part]. The part of the mesoblast from which the kidney is developed.

**nephromiosis, nephromeiosis** (*nef-ro-mi-o'-sis*) [*nephro-*; μείων, less]. Contraction of the kidney.

**nephroncus** (*nef-rong'-kus*) [*nephro-*; ὄγκος, tumor]. Tumor of the kidney.

**nephroparalysis** (*nef-ro-par-al'-is-is*) [*nephro-*; *paralysis*]. Paralysis of the kidney.

**nephroparesis** (*nef-ro-par'-es-is*) [*nephro-*; *paresis*]. Same as *nephroparalysis*.

**nephropathy** (*nef-rop'-ath-e*) [*nephro-*; πάθος, disease]. Any disease of the kidney.

**nephropexy** (*nef'-ro-peks-e*) [*nephro-*; πῆξις, fixation]. Surgical fixation of a floating kidney.

**nephrophthisis** (*nef-rof'-this-is*) [*nephro-*; *phthisis*]. Cheesy degeneration of the kidney; it is due to the presence of the tubercle bacillus.

**nephroplegia** (*nef-ro-ple'-je-ah*) [*nephro-*; πληγή, stroke]. Paralysis of the kidney.

**nephrophlegmasia** (*nef-ro-fleg-ma'-ze-ah*) [*nephro-*; φλεγμασία, inflammation]. Any inflammation of the kidney.

**nephropoietin** (*nef-ro-poi'-et-in*) [*nephro-*; ποιεῖν, to make]. A substance supposed to stimulate growth of renal tissue.

**nephroptosis, nephroptosia** (*nef-rop-to'-sis, nef-rop-to'-se-ah*) [*nephro-*; πτῶσις, a falling]. Prolapse of the kidney.

**nephropyelitis** (*nef-ro-pi-el-i'-tis*) [*nephro-*; *pyelitis*]. Inflammation of the pelvis of the kidney; pyelonephritis.

**nephropyic** (*nef-ro-pi'-ik*) [*nephro-*; πύον, pus]. Relating to suppuration of the kidney.

**nephropyosis** (*nef-ro-pi-o'-sis*). Same as *pyonephrosis*.

**nephrorrhagia** (*nef-ror-a'-je-ah*) [*nephro-*; ῥηγνύναι, to burst forth]. Renal hemorrhage.

**nephrorrhaphy** (*nef-ror'-a-fe*) [*nephro-*; ῥαφή, suture]. The stitching of a floating kidney to the posterior wall of the abdomen or to the loin.

**nephros** (*nef'-ros*) [νεφρός, kidney]. The kidney.

**nephrosclerosis** (*nef-ro-skle-ro'-sis*) [*nephro-*; σκληρός, hard]. Induration of the kidney.

**nephrosis** (*nef-ro'-sis*) [*nephris*]. Any renal disease.

**nephrospasis, nephrospasia** (*nef-ro-spa'-sis, nef-ro-spa'-se-ah*) [*nephro-*; σπᾶν, to draw]. Extreme renal mobility in which the organ hangs by its pedicle, thus straining the contained vessels and nerves.

**nephrospastic** (*nef-ro-spas'-tik*) [*nephro-*; σπᾶν, to wrench]. Pertaining to spasm of the kidney.

**nephrostegnosis** (*nef-ro-steg-no'-sis*) [*nephro-*; στεγνοῦν, to cover]. A cirrhotic condition of the kidney.

**nephrostome, nephrostoma** (*nef'-ros-tōm, nef-ros-to'-mah*) [*nephro-*; στόμα, a mouth]. The internal mouth of a Wolffian tubule.

**nephrostomy** (*nef-ros'-to-me*) [see *nephrostome*]. The formation of a fistula leading to the pelvis of the kidney.

**nephrotome** (*nef'-ro-tōm*) [*nephro-*; τόμος, a slice]. An embryonic structure from which the excretory ducts of the kidneys are developed; nephromere.

**nephrotomy** (*nef-rot'-o-me*) [*nephro-*; τομή, a cutting]. Incision of the kidney. n., **abdominal**, one through an abdominal incision. n., **lumbar**, one through an incision in the loin.

**nephrotoxic** (*nef-ro-tok'-sik*). 1. Pertaining to nephrotoxin. 2. Destructive to the kidney cells; nepholytic.

**nephrotoxin** (*nef-ro-toks'-in*) [*nephro-*; τοξικόν, a poison]. A cytotoxin which has a specific action on the cells of the kidney.

**nephrotriesis** (*nef-ro-tri-e'-sis*) [*nephro-*; τρίησις, piercing]. The operation of establishing a permanent opening in the kidney and suturing the edges of the kidney incision to the edges of the external incision.

**nephrotuberculosis** (*nef-ro-tū-ber-kū-lo'-sis*). See *nephrophthisis*.

**nephrotyphoid** (*nef-ro-ti'-foid*) [*nephro-*; *typhoid*]. Enteric fever with prominent renal complications.

**nephrotyphus** (*nef-ro-ti'-fus*) [*nephro-*; *typhus*]. Typhus fever with renal hemorrhage.

**nephroureterectomy** (*nef-ro-ū-re-ter-ek'-to-me*) [*nephro-*; *ureterectomy*]. The excision of the kidney and whole ureter at one operation.

**nephrozymase** (*nef-ro-zi'-mās*) [*nephro-*; ζύμη, leaven]. A gum resembling diastase sometimes found in urine.

**nephrozymosis** (*nef-ro-zi-mo'-sis*) [*nephro-*; *zymosis*]. The condition due to or favoring zymotic disease of the kidney.

**nephrus** (*nef'-rus*) [νεφρός]. The kidney.

**nephrydrops, nephrydrosis** (*nef'-rid-rops, nef-rid-ro'-sis*). See *hydronephrosis*. n., **subcapsular**, a large collection of urine between the kidney and its capsule.

**nephrydrotic** (*nef-rid-rot'-ik*). Relating to nephrydrosis.

**nerianthin, neriantin** (*ne-re-an'-thin, -tin*). A crystalline glucoside obtained from the leaves of *Nerium oleander*.

**neriin** (*ne'-ri-in*). A glucoside from the leaves of *Nerium oleander*, apparently identical with digitalein. It is used as a heart stimulant and tonic.

**neriodorein** (*ne-re-o-do'-re-in*). Same as *neriin*.

**Nerium** (*ne'-re-um*) [νήριον, oleander]. The leaves and bark of *N. oleander*. The extractive principles exert a marked influence on the motor centers, in large doses producing paralysis and heart failure. In small doses they act as a cardiac tonic, resembling digitalis. Dose of the *extract* ⅛–⅔ gr. (0.02–0.04 Gm.). The tincture may be given cautiously in one-drop doses, gradually increased.

**Nernst lamp** (*nairnst*) [Walther *Nernst*, German physicist, 1864— ]. An incandescent electric lamp in which there are rods or filaments of metallic oxides.

**neroli** (*ner'-o-le*) [Fr.]. Oil of orange-flowers.

**nerval** (*ner'-val*) [*nervus*, a nerve]. Pertaining to a nerve or nerves.

**nerve, nervus** (*nerv, ner'-vus*) [*nervus*, nerve]. An elongated, cord-like structure made up of aggregations of nerve-fibers and having the property of transmitting nervous impulses. n., **accelerator**, a cardiac sympathetic nerve, stimulation of which causes acceleration of the heart's action. n., **afferent**, one that transmits impulses from the periphery to the central nervous system. n.-**bulb**. See *end-bud* and *motorial end-plate*. n.-**bulb**, **terminal**. See *Krause's corpuscles*. n.s, **bulbous**, amputation neuromata, round growths which form on the divided extremities of the nerves in the stumps left after amputation. n., **calorific**, a nerve stimulation of which increases the heat of the parts to which it is distributed. n.-**cavity**, the pulp cavity of a tooth. n.-**cell**, a mass of protoplasm containing a large vesicular nucleus within which lies a well-marked nucleolus. Nerve-cells have one or more elongated processes, and in accordance with the number of these are designated *unipolar*, *bipolar*, or *multipolar*. The processes are of two kinds: the axis-cylinder process and the protoplasmic processes. The *axis-cylinder* (*Deiters'*) *process* either becomes an axis-cylinder of a nerve-fiber, or divides within the gray matter into delicate filaments; it gives off minute branches termed *collaterals*; the other processes are supposed to have nutritive and conducting functions. n.-center, a group of nerve-cells acting together in the performance of a function. n., centrifugal. See n., efferent. n., centripetal. See n., afferent. n.-corpuscles. 1. Same as nerve-cells, q. v. 2. Nucleated corpuscles lying between the neurilemma and the medullary sheath of medullated nerve-fibers. n., cranial, a nerve arising directly from the brain and making its exit through a foramen in the skull. n., depressor, an afferent nerve, irritation of which depresses or inhibits the vasomotor center. n., efferent, one carrying impulses from the central nervous system to the periphery. n.-endings, the terminations of nerves at the periphery or in the nerve-centers. n., esodic, an afferent or centripetal nerve. n., exodic, an efferent or centrifugal nerve. n. of expression, the facial nerve. n.-fiber, a fiber having the property of conducting invisible or molecular waves of stimulation from one part of an organism to another, and so establishing physiological continuity between such parts without the necessary passage of waves of contraction. There are two kinds of nerve-fibers: the *medullated*, or myelinic, and the *nonmedullated*, or amyelinic. A typical medullated fiber consists of the *axis-cylinder*, which may be surrounded by a sheath, the *axilemma*; the *medullary sheath*, or white substance of Schwann; the *neurilemma*, or sheath of Schwann. The nonmedullated, pale, or Remak's fibers do not possess a medullary sheath, but consist only of axis-cylinder and neurilemma. The nerve-corpuscles are more abundant than in medullated nerve-fibers. Medullated nerve-fibers are found in the cerebrospinal nerves, while nonmedullated fibers occur in the sympathetic nerves and tend to form *plexuses*. n., frigorific, a sympathetic nerve stimulation of which causes a fall of temperature; the vasoconstrictor nerves are frigorific nerves. n.-grafting, the transplanting of a portion of healthy nerve from an animal to man, to reestablish the continuity of a divided nerve. Syn., *neuroplasty*. n.-head, the optic disc or papilla. n. hillock, a slight elevation observed where a nerve-fiber enters a muscle. n., inhibitory, one the stimulation of which inhibits or lessens the activity of an organ. n., mixed, one made up of both afferent and efferent fibers. n., motor, one containing only or chiefly motor fibers. n.-papillæ, papillæ of the skin containing tactile corpuscles, nervous plexuses, or Krause's corpuscles and sometimes blood-vessels. n.-plexus, a grouping of nerves. n., pressor, an afferent nerve, irritation of which stimulates the vasomotor center. n.-process, the axis-cylinder process of a neuron. n., secretory, an efferent nerve, stimulation of which causes increased activity of the gland to which it is distributed. n., sensory, Same as *n., afferent*. n., spinal, one of those arising from the spinal cord and making its exit through an intervertebral foramen. There are 31 pairs of spinal nerves. n.-storm, a sudden outburst or paroxysm of nervous disturbance. n.-stretching, mechanical elongation or tension of a nerve for the relief of neuralgia, spasmodic contraction, and other pathological conditions. n., sympathetic, one of a system of nerves distributed chiefly to the blood-vessels and to the viscera. See *sympathetic*. n., thermic. Same as *n., calorific*. n.-tire, neurasthenia. n.-tree, a neurodendrite. n., trisplanchnic, the system of sympathetic nerves. n., trophic, a nerve, the function of which is to preside over the nutrition of the part to which it is distributed. n. tumor, a neuroma. n.-unit, a neuron. n., vasoconstrictor, See *n., vasomotor*. n., vasodilator. See *n., vaso-motor*. n., vasomotor, any one of the nerves controlling the caliber of the blood-vessels; they are of two kinds—those stimulation of which causes contraction of the vessels—*vasoconstrictor nerves*—and those stimulation of which causes active dilation—*vasodilator nerves*. Ordinarily vasomotor is synonymous with vasoconstrictor. TABLE OF NERVES (ALPHABETICAL): abdominal, *function*, sensation and motion; *origin*, vagus; *distribution*, surface of stomach. abducens (*sixth cranial*), *function*, motion; *origin*, fourth ventricle; *distribution*, external rectus of eye. accessorius. See *spinal accessory* in this table. accessory. See *spinal accessory* (*eleventh cranial*) in this table. acusticus, the auditory nerve. ambulatorius, the vagus. ampullares, branches of the vestibular nerve distributed to the ampullæ of the semicircular canals. Arnold's. See *auricular* in this table. articular, *function*, trophic, sensory (?); *origin*, anterior crural; *distribution*, knee-joint; *branches*, capsular, synovial. articular (two), *function*, trophic, sensory (?); *origin*, ulnar; *distribution*, elbow-joint. aschianus, the first cervical nerve. auditory (*eighth cranial*; *portio mollis of seventh*), *function*, hearing; *origin*, restiform body; *distribution*, internal ear; *branches*, vestibular, cochlear. auricular, *function*, sensation; *origin*, lesser occipital; *distribution*, integument of posterior and upper portion of pinna. auricular (*Arnold's*), *function*, sensation; *origin*, vagus; *distribution*, external ear. auricular (anterior), *function*, sensation; *origin*, inferior maxillary; *distribution*, integument of external ear. auricular (posterior), *function*, motion; *origin*, facial; *distribution*, retrahens aurem, attollens aurem, occipito-frontalis; *branches*, auricular, occipital. auricularis magnus, *function*, sensation; *origin*, cervical plexus, second and third cervical; *distribution*, parotid gland, face, ear; *branches*, facial, mastoid, and auricular. auricularis profundus, the posterior auricular nerve. auriculotemporal, *function*, sensation; *origin*, inferior maxillary; *distribution*, pinna and temple; *branches*, articular, two branches to meatus, parotid, anterior auricular, superficial temporal. axillary. See *circumflex* in this table. Bell's respiratory, the long thoracic nerve. bigeminus, biradiatus, the second sacral nerve. buccal, *function*, motion; *origin*, facial; *distribution*, buccinator and orbicularis oris muscles. buccal, long, *function*, sensation, motion (?); *origin*, inferior maxillary; *distribution*, cheek; *branches*, superior and inferior buccinator and external pterygoid. buccinator. See *buccal, long*, in this table. calcanean, internal, *function*, sensation; *origin*, posterior tibial; *distribution*, fascia and integument of heel and sole. cardiac (*cervical and thoracic*), *function*, inhibition; *origin*, vagus; *distribution*, heart; *branches*, to cardiac plexuses. Casser's. See *perforating* and *musculocutaneous* in this table. cervical (eight), *function*, motion and sensation; *origin*, cord; *distribution*, trunk and the upper extremities; *branches*, anterior and posterior divisions. cervical, first (anterior division), *function*, motion and sensation; *origin*, cord; *distribution*, rectus lateralis and two anterior recti; *branches*, filaments to vagus, hypoglossal, sympathetic. cervical, first (posterior division), *function*, motion and sensation; *origin*, cord; *distribution*, recti, obliqui, complexus; *branches*, communicating and cutaneous filaments. cervical, second (anterior division), *function*, motion and sensation; *origin*, cord; *distribution*, communicating; *branches*, ascending, descending, communicating. cervical, second (posterior division), *function*, motion and sensation; *origin*, cord; *distribution*, obliquus inferior, scalp, ear, complexus, splenius, trachelomastoid; *branches*, internal or occipitalis major, and external. cervical, third (anterior division), *function*, motion and sensation; *origin*, cord; *distribution*, communicating; *branches*, ascending, descending and communicating filaments. cervical, third (posterior division), *function*, motion and sensation; *origin*, cord; *distribution*, occiput, splenius, complexus; *branches*, internal and external. cervical, fourth (anterior division), *function*, motion and sensation; *origin*, cord; *distribution*, shoulder; *branches*, communicating filaments, muscular, etc. cervicals, fifth to eighth (anterior divisions), *function*, motion and sensations; *origin*, cord; *distribution*, brachial plexus; *branches*, communicating. cervicals, fourth to eighth (posterior divisions), *function*, motion and sensation; *origin*, cord; *distribution*, muscles and skin of neck; *branches*, internal and external branches. cervicofacial, *function*, motion; *origin*, facial; *distribution*, lower part of face and part of neck; *branches*, buccal, supramaxillary, inframaxillary. chorda tympani, *function*, motion and taste; *origin*, facial; *distribution*, tongue, tympanum, submaxillary gland. ciliary, *function*, sensation, nutrition, motion; *origin*, ciliary ganglion; *distribution*, eyeball. circumflex, *function*, motion and sensation; *origin*, brachial plexus; *distribution*, teres minor, deltoid, and skin; *branches*, anterior, posterior, and articular. clunium

# NERVE 602 NERVE

**inferior medialis.** See *cutaneous, perforating* in this table. **coccygeal,** *function,* motion; *origin,* coccygeal plexus; *distribution,* coccygeus and gluteus maximus. **cochlear,** *function,* hearing; *origin,* auditory; *distribution,* cochlea. **colli superficialis,** *function,* sensation; *origin,* cervical plexus; *distribution,* platysma myoides and anterolateral parts of neck; *branches,* escending and descending branches. **communicans aervicalis.** See *communicans noni* or *hypoglossi* in this table. **communicans noni,** or **hypoglossi,** *function,* motion and sensation; *origin,* second cervical, third cervical; *distribution,* descendens noni, depressor muscles of hyoid bone; *branches,* omohyoid, ansa hypoglossi. **communicans peronei.** See *peroneal, communicating* in this table. **communicating,** *function,* motion and sensation; *origin,* cervical plexus; *distribution,* spinal accessory. **communicating,** *function,* sensation and motion; *origin,* first and second cervical; *distribution,* vagus, hypoglossal, sympathetic. of **Cotunnius.** See *nasopalatine* in this table. **crural,** *function,* sensation; *origin,* genitocrural; *distribution,* shin, upper and central part anterior aspect of thigh. **crural, anterior,** *function,* motion and sensation; *origin,* lumbar plexus, second, third, and fourth lumbar nerves; *distribution,* thigh; *branches,* middle and internal cutaneous, long saphenous, muscular, articular. **cubitalis,** the ulnar nerve. **cutaneous,** *function,* sensation; *origin,* musculospiral; *distribution,* skin of arm, radial side of forearm; *branches,* one internal, two external. **cutaneous,** *function,* sensation; *origin,* ulnar; *distribution,* wrist and palm; *branches,* first and palmar cutaneous. **cutaneous (cervical).** See *colli; superficialis* in this table. **cutaneous colli.** See *colli, superficialis* in this table. **cutaneous, dorsal,** *function,* sensation; *origin,* ulnar; *distribution,* little and ring fingers; *branches,* communicating. **cutaneous, external,** *function,* sensation; *origin,* second and third lumbar; *distribution,* skin of thigh; *branches,* anterior and posterior. **cutaneous, internal,** *function,* sensation; *origin,* brachial plexus; *distribution,* forearm; *branches,* anterior and posterior branches. **cutaneous, lesser internal** or **medial** (of Wrisberg), *function,* sensation; *origin,* brachial plexus; *distribution,* inner side of arm. **cutaneous, middle and internal,** *function,* sensation, motion (?); *origin,* anterior crural; *distribution,* skin of thigh; *branches,* communicating. **cutaneous, perforating,** *function,* sensation; *origin,* fourth sacral; *distribution,* integument covering gluteus maximus. **Cyon's.** See *depressor* in this table. **dental, inferior** or **mandibular,** *function,* sensation; *origin,* inferior maxillary; *distribution,* teeth, muscles; *branches,* mylohyoid, incisor, mental, dental. **dentals, anterior and posterior,** *function,* sensation; *origin,* superior maxillary; *distribution,* teeth. **depressor,** *function,* lowering of the blood-pressure; *origin,* in the rabbit, from the vagus; *distribution,* heart. **descendens hypoglossi,** *function,* motor; *origin,* cervical plexus; *distribution,* omohyoid, sternohyoid, sternothyroid, thyrohyoid, geniohyoid, hyoglossus, and muscles of the tongue; *branches,* muscular, lingual. **descendens noni.** See *descendens hypoglossi* in this table. **digastric,** *function,* motion; *origin,* facial; *distribution,* posterior belly of digastric. **dorsal, 12 (anterior and posterior divisions),** *function,* motion and sensation; *origin,* cord; *distribution,* muscles and skin of trunk; *branches,* external, internal, cutaneous. **dorsal (of penis),** sensation; *origin,* pudic; *distribution,* penis. **dorsospinal,** the dorsal spinal nerves. See *spinal* in this table. **erigentes,** excitor or vasodilator nerves of the penis; derived from the first and second and sometimes from the third sacral nerves. They have their origin in the sexual center of the spinal cord. **esophageal,** *function,* motion; *origin,* vagus; *distribution,* mucous and muscular coats of esophagus; *branches,* esophageal plexus. **external motor (of the eye).** See *abducens* in this table. **facial,** *function,* sensation; *origin,* great auricular; *distribution,* skin over parotid. **facial** (*seventh cranial, portio dura*), *function,* motion; *origin,* floor of fourth ventricle; *distribution,* face, ear, palate, tongue; *branches,* petrosals, tympanic, chorda tympani, communicating, posterior auricular, digastric, stylohyoid, lingual, temporal, malar, infraorbital, buccal, superior and inferior maxillary. **femoral.** See *crural (anterior)* in this table. **posterior cutaneous.** See *sciatic, small* in this table. **fibular communicating.** See *peroneal communicating* in this table. **fourth,** the trochlear nerve. **frontal,** *function,* sensation; *origin,* ophthalmic; *distribution,*

forehead and eyelids; *branches,* supraorbital, supratrochlear. **furcal,** the fourth lumbar nerve. **gastric,** *function,* motion; *origin,* vagus; *distribution,* stomach. **genital,** *function,* motion and sensation; *origin,* genitocrural; *distribution,* cremaster muscle. **genitocrural,** *function,* motion and sensation; *origin,* first and second lumbar; *distribution,* cremaster and thigh; *branches,* genital, crural, communicating. **genitofemoral.** See *genitocrural* in this table. **glossopharyngeal** (*ninth cranial*), *function,* sensation and taste; *origin,* fourth ventricle; *distribution,* tongue, middle ear, tonsils, pharynx, meninges; *branches,* tympanic, carotid, pharyngeal, muscular, tonsillar, lingual. **gluteal, inferior,** *function,* motion; *origin,* sacral plexus (second and third sacral nerves); *distribution,* gluteus maximus. **gluteal, superior,** *function,* motion; *origin,* sacral plexus; *distribution,* glutei, tensor vaginae femoris. **gustatory.** See *lingual* in this table. **hemorrhoidal, inferior,** *function,* sensation and motion; *origin,* pudic; *distribution,* external sphincter ani and adjacent integument. **hepatic,** *function* (?); *origin,* vagus; *distribution,* liver; *branches,* hepatic plexus. **Hirschfeld's,** *function,* motion; *origin,* facial; *distribution,* styloglossus and palatoglossus. **hypogastric,** *function,* sensation; *origin,* iliohypogastric; *distribution,* skin about external abdominal ring. **hypoglossal** (*twelfth cranial*), *function,* motion; *origin,* floor of fourth ventricle; *distribution,* hypoglossus and hyoid muscles; *branches,* descendens noni or hypoglossi, muscular, thyrohyoid, geniohyoid, and meningeal. **iliac,** *function,* motion; *origin,* iliohypogastric; *distribution,* integument covering fore part of gluteal region. **iliac,** *function,* sensation; *origin,* last dorsal; *distribution,* integument covering forepart of gluteal region. **iliohypogastric,** *function,* motion and sensation; *origin,* first lumbar; *distribution,* abdominal and gluteal regions; *branches,* iliac, hypogastric, communicating. **ilioinguinal,** *function,* motion and sensation; *origin,* first lumbar; *distribution,* inguinal region and scrotum; *branches,* muscular, cutaneous, and communicating. **incisive,** *function,* sensation; *origin,* inferior dental; *distribution,* canine and incisor teeth and corresponding portion of gums. **inferior medial of the buttock.** See *cutaneous, perforating,* in this table. **inframandibular.** See *inframaxillary* in this table. **inframaxillary,** *function,* motion; *origin,* facial; *distribution,* platysma myoides. **infraorbital,** *function,* sensation and motion; *origin,* facial; *distribution,* nose and lip; *branches,* palpebral, nasal, labial. **infratrochlear,** *function,* sensation; *origin,* nasal; *distribution,* skin and conjunctiva of inner part of eye, lacrimal sac. **intercostal,** *function,* motion and sensation; *origin,* spinal cord; *distribution,* muscles and integument of thorax; *branches,* muscular, anterior and lateral cutaneous. **intercostobrachialis.** See *intercostohumeral* in this table. **intercostohumeral,** *function,* sensation; *origin,* second intercostal; *distribution,* integument of upper two-thirds of inner and posterior part of arm. **interosseous, anterior,** *function,* motion; *origin,* median; *distribution,* deep muscles of forearm. **interosseous, posterior,** *function,* motion and sensation; *origin,* musculospiral; *distribution,* carpus and radial and posterior brachial regions. **ischiadic,** the great sciatic nerve. **Jacobson's.** See *tympanic* (*Jacobson's nerve*) in this table. **labial,** *function,* motion and sensation; *origin,* superior maxillary; *distribution,* muscles and mucous membrane of lips. **lacrimal,** *function,* sensation; *origin,* ophthalmic; *distribution,* lacrimal gland and conjunctiva. of **Lancisi,** longitudinal striations upon upper surface of corpus callosum. **laryngeal, recurrent** or **inferior,** *function,* motion; *origin,* vagus; *distribution,* larynx; *branches,* to all laryngeal muscles except cricothyroid. **laryngeal, superior,** *function,* sensation and motion; *origin,* vagus; *distribution,* larynx; *branches,* external-cricothyroid muscle and thyroid gland; internal-mucous membrane of larynx. **lingual,** *function,* motion and sensation; *origin,* facial; *distribution,* mucous membrane of tongue, palatoglossus and styloglossus muscles. **lingual,** *function,* sensation; *origin,* glossopharyngeal; *distribution,* circumvallate papillae and glands of tongue. **lingual,** *function,* taste and sensation; *origin,* inferior maxillary; *distribution,* tongue and mouth. **lumbar (5),** *function,* motion and sensation; *origin,* cord; *distribution,* lumbar and genital regions; *branches,* anterior and posterior divisions, lumbar plexus. **malar** (or **zygomatic**), *function,* motion; *origin,* facial; *distri-*

*bution*, lower part of orbicularis palpebrarum and eyelids. **malar** (or **zygomatico-facialis**), *function*, sensation; *origin*, orbital; *distribution*, skin over malar bone. **mandibular.** See *maxillary, inferior*, in this table. **masseteric,** *function*, motor; *origin*, inferior maxillary; *distribution*, masseter muscle (and temporal?). **masticatorius,** **masticatory,** originates chiefly in the motor nucleus in the pons, but receives an accession of fibers from a nucleus lying beneath and lateral to the mesocele. It innervates the muscles of mastication. **mastoid,** *function*, sensation; *origin*, great auricular; *distribution*, skin over mastoid process. **mastoid,** *function*, motion; *origin*, lesser occipital; *distribution*, skin over mastoid process. **maxillary.** See *maxillary (superior)* in this table. **maxillary, inferior,** *function*, sensation, motion, and taste; *origin*, trigeminus; *distribution*, muscles of mastication, ear, cheek, tongue, teeth; *branches,* masseteric, auriculotemporal, buccal, gustatory, inferior dental. **maxillary, superior,** *function*, sensation; *origin*, trigeminus; *distribution*, cheek, face, teeth; *branches*, orbital, sphenopalatine, dentals, infraorbital. **median,** *function*, motion and sensation; *origin*, brachial plexus; *distribution*, pronator radii teres, flexors, two lumbricales, fingers, palm; *branches,* muscular, anterior interosseous, palmar cutaneous. **meningeal,** *function*, sensation; *origin*, glossopharyngeal; *distribution*, pia and arachnoid. **meningeal,** *function*, sensation; *origin*, hypoglossal; *distribution,* dura mater. **meningeal,** *function*, sensation; *origin*, vagus; *distribution*, dura mater around lateral sinus. **meningeal, recurrent,** *function*, sensation; *origin*, inferior maxillary; *distribution*, dura mater and mastoid cells. **mental,** *function*, motion and sensation; *origin*, inferior maxillary; *distribution*, mucous membrane of lower lip and chin. **motor oculi** (*third cranial*), *function*, motion; *origin*, floor of aqueduct of Sylvius; *distribution*, all muscles of the eye except external rectus, superior oblique, and orbicularis palpebrarum. **muscular,** *function*, motion and sensation; *origin*, first and second cervical; *distribution*, muscles; *branches,* rectus capitis lateralis, rectus anterior major et minor. **muscular,** *function*, motion; *origin*, cervical plexus; *distribution*, sternomastoid, levator anguli scapulæ, scaleni medius, trapezius. **muscular,** *function*, motion; *origin*, brachial plexus; *distribution*, longus colli, scaleni, rhomboidei, subclavius. **muscular,** *function*, motion; *origin*, musculospiral; *distribution*, triceps, anconeus, supinator longus extensor carpi radialis longior, brachialis anticus; *branches*, internal, posterior, external. **muscular,** *function*, motion; *origin*, median; *distribution,* superficial muscles of the forearm. **muscular,** *function*, motion; *origin*, ulnar; *distribution*, flexor carpi ulnaris, flexor profundus digitorum. **muscular,** *function*, motion; *origin*, great sciatic; *distribution*, biceps, semimembranosus, semitendinosus, adductor magnus. **muscular,** *function*, motion; *origin*, sacral plexus; *distribution*, pyriformis, obturator internus, gemelli, quadratus femoris. **muscular,** *function*, motion; *origin*, anterior crural; *distribution*, pectineus and the muscles of the thigh. **musculocutaneous** (of Casser), *function*, motion and sensation; *origin*, brachial plexus; *distribution*, coracobrachialis, biceps, brachialis anticus, forearm; *branches*, anterior and posterior. **musculocutaneous,** *function*, motion and sensation; *origin*, external popliteal; *distribution*, muscles of fibular side of leg, skin of dorsum of foot; *branches*, internal, external. **musculospiral,** *function*, motion and sensation; *origin*, brachial plexus; *distribution*, back of arm and forearm, skin of back of hand; *branches*, musculocutaneous, radial, posterior interosseous. **mylohyoid,** *function*, motion; *origin*, inferior maxillary; *distribution*, mylohyoid and digastric muscles. **nasal,** *function*, sensation; *origin*, dental, anterior; *distribution*, mucous membrane of inferior meatus. **nasal,** *function*, sensation; *origin*, maxillary, superior; *distribution*, integument of lateral aspect of nose. **nasal,** *function*, sensation; *origin*, ophthalmic; *distribution*, iris, ciliary ganglion, nose; *branches*, ganglionic, ciliary, infratrochlear. **nasal, inferior,** *function*, sensation; *origin*, anterior palatine; *distribution*, mucous membrane of nose. **nasal, superior,** *function*, sensation; *origin*, Meckel's ganglion; *distribution*, mucous membrane of nose and posterior ethmoid cells. **nasopalatine,** *function*, sensation; *origin*, Meckel's ganglion; *distribution*, nasal septum. **obturator,** *function*, motion and sensation; *origin*, lumbar plexus, third and fourth nerves; *distribution*,

obturator externus, adductors, hip-joint, and skin; *branches*, anterior and posterior articular and communicating. **obturator, accessory,** *function*, motion and sensation; *origin*, lumbar plexus; *distribution*, pectineus and hip-joint. **occipital** (*smallest or third*), *function*, sensation; *origin*, third cervical; *distribution*, integument of occiput. **occipitalis magnus,** *function*, motion and sensation; *origin*, second cervical; *distribution*, complexus, trapezius, and scalp. **occipitalis major,** the internal branch of the dorsal division of the second cervical nerve. **occipitalis minimus or tertius,** a ramus from the internal branch of the dorsal division of the third cervical nerve. **occipitalis minor,** *function*, sensation; *origin*, second cervical; communicating, auricular. **oculomotor.** See *motor oculi* in this table.[1] **olfactory** (*first cranial*), *function*, smell; *origin*, frontal lobe, optic thalamus, island of Reil; *distribution,* Schneiderian membrane of nose. **ophthalmic,** *function*, sensation; *origin*, trigeminus; *distribution*, forehead, eyes, nose; *branches*, frontal, lacrimal, nasal. **optic** (*second cranial*), *function*, sight; *origin*, cortical center in occipital lobe; *distribution*, retina. **orbital,** *function*, sensation; *origin*, Meckel's ganglion; *distribution*, mucosa of posterior ethmoid cells and sphenoid sinus. **orbital or temporo-malar,** *function*, sensation; *origin*, superior maxillary; *distribution*, temple and cheek; *branches*, temporal and malar. **palatine, anterior or great,** *function*, sensation; *origin*, Meckel's ganglion; *distribution*, hard palate, gums, and nose; *branches*, two inferior. **palatine, external,** *function*, sensation; *origin*, Meckel's ganglion; *distribution*, tonsil and soft palate. **palatine, posterior or small,** *function*, motor; *origin*, Meckel's ganglion; *distribution*, levator palati and azygos uvulæ. **palmar cutaneous,** *function*, sensation; *origin*, median; *distribution*, thumb and palm. **palmar, deep,** *function*, motion; *origin*, ulnar; *distribution*, little finger, dorsal and palmar interosseous, two inner lumbricales, abductor pollicis. **palmar, superficial,** *function*, sensation and motion; *origin*, ulnar; *distribution*, palmaris brevis, inner side of hand, and little finger. **palpebral,** *function*, motor; *origin*, superior maxillary; *distribution*, integument of lower lid. **parotid,** *function*, sensation; *origin*, auriculotemporal; *distribution*, parotid gland. **parotid,** *function*, sensation; *origin*, long saphenous; *distribution*, integument over patella and plexus patellæ. **patheticus** (*fourth cranial*), *function*, motion; *origin*, valve of Vieussens; *distribution*, superior oblique of eye. **pectineus,** *function*, motion; *origin*, anterior crural; *distribution*, pectineus muscle. **perforating** (of Casser). See *musculocutaneous* in this table. **perineal,** *function*, motion and sensation; *origin*, pudic; *distribution*, perineum, genitalia, and skin of perineal region; *branches*, cutaneous and muscular. **perineal,** *function*, motion and sensation; *origin*, fourth sacral; *distribution*, external sphincter ani and integument of anus. **perineal,** the external popliteal nerve. **peroneal, common.** See *popliteal (external)* in this table. **peroneal,** *function*, communicating, a branch of the external popliteal, generally uniting with the short saphenous nerve, but at times it extends down the leg to the heel. **peroneal, deep.** See *tibial (anterior)* in this table. **peroneal, superficial.** See *musculocutaneous* in this table. **petrosals,** *function*, motion; *origin*, facial; *distribution*, ganglia and plexus; *branches*, great, small, external to Meckel's ganglion, otic ganglion, and meningeal plexus, respectively. **pharyngeal,** *function*, motion and sensation; *origin*, glossopharyngeal; *distribution*, pharynx; enters into formation of pharyngeal plexus. **pharyngeal,** *function*, sensation; *origin*, Meckel's ganglion; *distribution*, upper part of pharynx, posterior nares, and sphenoid sinus. **pharyngeal,** *function*, motion; *origin*, vagus; *distribution*, pharynx; *branches*, pharyngeal plexus, muscles, and mucosa. **pharyngeal,** *function*, sensation; *origin*, sympathetic; *distribution*, pharynx; helps to form the pharyngeal plexus. **phrenic,** *function*, motion and sensation; *origin*, third, fourth, and fifth cervical; *distribution*, diaphragm, pericardium, pleura. **plantar, external,** *function*, motion and sensation; *origin*, posterior tibial; *distribution*, little toe and deep muscles of foot; *branches*, superficial and deep. **plantar, internal,** *function*, sensation and motion; *origin*, posterior tibial; *distribution*, sole of foot, adductor pollicis, flexor brevis digitorum, toes; *branches*, cutaneous, muscular, articular, digital. **pneumogastric** (*tenth cranial, par vagum*, or *vagus*), *function*, sensation and motion; *origin*, floor of fourth ventricle; *distribution*, ear, pharynx, larynx, heart, lungs, esophagus, stomach; *branches*, auricular, pharyngeal, superior and inferior laryngeal, recurrent laryngeal, cardiac, pulmonary, esophageal, gastric, hepatic, communicating, meningeal. **popliteal, external**, *function*, sensation and motion; *origin*, great sciatic; *distribution*, extensors of foot, skin, and fascia; *branches*, anterior tibial, musculocutaneous, articular, cutaneous. **popliteal, internal**, *function*, motion and sensation; *origin*, great sciatic; *distribution*, knee, gastrocnemius, tibialis posticus, plantaris, soleus, popliteus, skin of foot; *branches*, articular, muscular, cutaneous, external saphenous, plantar. **posterior cutaneous of thigh**. See *sciatic (small)* in this table. **pterygoid, external**, *function*, motion; *origin*, inferior maxillary; *distribution*, external pterygoid muscle. **pterygoid, internal**, *function*, motion; *origin*, inferior maxillary; *distribution*, internal pterygoid muscle. **pudendal, inferior**, a branch of the small sciatic nerve distributed to the front and external part of the scrotum and perineum. **pudendal, long** (*nerve of Soemmering*), *function*, sensation; *origin*, small sciatic; *distribution*, integument of genitalia and inner and proximal part of thigh. **pudic**, *function*, motion and sensation; *origin*, sacral plexus; *distribution*, perineum, anus, genitalia; *branches*, inferior hemorrhoidal, perineal, cutaneous, dorsal of penis. **pulmonary, anterior and posterior**, *function* (?); *origin*, vagus; *distribution*, lungs; *branches* to pulmonary plexuses. **radial**. See *musculospiral* in this table. **radial (superficial ramus)**, *function*, sensation; *origin*, musculospiral; *distribution*, skin of radial side and ball of thumb; skin on posterior surface of ulnar side of thumb; skin of index-finger, middle finger, and radial side of ring-finger; *branches*, external and internal. **renal**, branches of the renal plexus following the distribution of the renal artery. **sacral (5)**, *function*, motion and sensation; *origin*, cord; *distribution*, multifidus spinæ, skin of gluteal region; *branches* to sacral plexus. **saphenous, external or short**, *function*, sensation; *origin*, internal popliteal; *distribution*, integument of foot and little toe. **saphenous, long or internal**, *function*, sensation; *origin*, anterior crural; *distribution*, knee, ankle; *branches*, cutaneous, patellar, communicating. **Sappey's**, the mylohyoid nerve. of **Scarpa**. See *nasopalatine* in this table. **sciatic, great**, *function*, motion and sensation; *origin*, sacral plexus; *distribution*, skin of leg, muscles of back of thigh, and those of leg and foot; *branches*, articular, muscular, popliteal. **sciatic, small**, *function*, sensation and motion; *origin*, sacral plexus; *distribution*, perineum, back of thigh and leg, gluteus maximus; *branches*, muscular, cutaneous, long pudendal. **second cranial**, the optic nerve. **seventh cranial**. See *facial* in this table. **sixth cranial**. See *abducens* in this table. **sixth sacral**, the coccygeal nerve. of **Soemmering**. See *pudendal, long*, in this table. **sphenopalatine**, *function*, sensation; *origin*, superior maxillary; *distribution*, Meckel's ganglion. **spinal**, *function*, motion and sensation; *origin*, spinal cord; *distribution*, trunk. There are 31 on each side: 1 coccygeal, 8 sacral, 12 dorsal, 5 lumbar, 5 sacral. **spinal accessory** (*eleventh cranial*), *function*, motion; *origin*, floor of fourth ventricle; *distribution*, sternomastoid, trapezius. **spinosus**, par *meningeal, recurrent* in this table. **spiral**. See *musculospiral* in this table. **spiralis**, the radial nerve. **splanchnic, great**, *function*, sympathetic; *origin*, thoracic ganglia; *distribution*, semilunar ganglion, renal, and suprarenal plexuses. **splanchnic, lesser**, *function*, sympathetic; *origin*, tenth and eleventh thoracic ganglia, great splanchnic; *distribution*, celiac plexus and great splanchnic. **splanchnic, renal or smallest**, *function*, sympathetic; *origin*, last thoracic ganglion; *distribution*, renal and celiac plexuses. **stapedial**, *function*, motion; *origin*, facial; *distribution*, stapedius muscle; **stapedius**. See *tympanic* in this table. **sternal**, descending cutaneous divisions of the third and fourth cervical nerves. **stylohyoid**, *function*, motion; *origin*, facial; *distribution*, stylohyoid muscle. **subclavian**, a branch of the brachial plexus which supplies the subclavius muscle. **subcostal**, the intercostal nerve. **sublingual**, a division of the lingual nerve distributed to the sublingual gland. **submaxillary**, the inframaxillary nerve. **suboccipital**, the anterior division of the first cervical nerve. **subscapular**, *function*, motion; *origin*, brachial plexus; *distribution*, subscapular, teres major, and latissimus dorsi.

**supraacromial**, *function*, sensation; *origin*, cervical plexus; *distribution*, skin over deltoid. **supraclavicular, descending**, *function*, sensation; *origin*, third and fourth cervical; *distribution*, skin of neck, breast, and shoulder; *branches*, sternal, clavicular, acromial. **supragluteal**, the superior gluteal nerve. **supramandibular**. See *maxillary, superior*, in this table. **supramaxillary**. See *maxillary, superior*, in this table. **supraorbital**, *function*, sensation; *origin*, ophthalmic; *distribution*, upper lid, forehead; *branches*, muscular, cutaneous, and pericranial. **suprapubic**, the genitocrural nerve. **suprascapular**, *function*, motion and sensation; *origin*, brachial plexus; *distribution*, scapular muscles. **suprasternal**, *function*, sensation; *origin*, cervical plexus; *distribution*, integument over upper part of sternum. **supratrochlear**, *function*, sensation; *origin*, ophthalmic; *distribution*, forehead and upper eyelid; *branches*, muscular and cutaneous. **sympathetic**. See under *sympathetic*. **temporal**, *function*, motion; *origin*, inferior maxillary; *distribution*, temporal muscle. **temporal** (or *zygomaticotemporalis*), *function*, sensation; *origin*, orbital; *distribution*, integument over temporal muscle. **temporal**, *function*, motion; *origin*, temporofacial; *distribution*, orbicularis palpebrarum, occipitofrontalis, attrahens and attollens aurem, corrugator supercilii; *branches*, muscular. **temporal, superficial**, *function*, sensation; *origin*, auriculotemporal; *distribution*, integument over temporal fascia. **temporofacial**, *function*, motion; *origin*, facial; *distribution*, upper part of face; *branches*, temporal, malar, infraorbital. **temporomalar**. See *orbital* in this table. **tenth cranial**, the vagus nerve. **third cranial**, the motor oculi. **thoracic, anterior and exterior**, *function*, motion; *origin*, brachial plexus; *distribution*, pectoralis major and minor. **thoracic, posterior or long** (*external respiratory nerve of Bell*), *function*, motion; *origin*, brachial plexus; *distribution*, serratus magnus. **thoracic, spinal**, *function*, motion and sensation; *origin*, cord; *distribution*, muscles and skin of thorax. **thyroid**, branches of the middle cervical ganglion distributed to the thyroid. **tibial**. See *popliteal* (*internal*) in this table. **tibial, anterior**, *function*, motion and sensation; *origin*, external popliteal; *distribution*, tibialis anticus, extensor longus digitorum, peroneus tertius, joints of foot; skin of great toe; *branches*, muscular, external, internal. **tibial, posterior**, *function*, motion and sensation; *origin*, internal popliteal; *distribution*, tibialis posticus, flexor longus digitorum, flexor longus pollicis, skin of heel and sole, knee-joint; *branches*, plantar, muscular, calcaneoplantar, cutaneous or internal calcanean, articular. **tonsillar**, *function*, sensation; *origin*, glossopharyngeal; *distribution*, tonsil, soft palate, and fauces. **trigeminus**, **trifacial** (*fifth cranial*), *function*, motion and sensation (taste); *origin*, floor of fourth ventricle; *distribution*, skin and structures of face, tongue, and teeth; *branches*, ophthalmic, superior and inferior maxillary. **trochlear**. See *patheticus* in this table. **twelfth cranial**, the hypoglossal nerve. **trochlear**. See *patheticus* (*fourth cranial*) in this table. **tympanic** (or *stapedius*), *function*, motion; *origin*, facial; *distribution*, stapedius, and laxator tympani muscles. **tympanic**, *function*, sensation; *origin*, glossopharyngeal; *distribution*, tympanum. **tympanic** (*Jacobson's nerve*) *function*, motion; *origin*, glossopharyngeal; *distribution*, tympanum; *branches*, tympanic plexus and communicating. **tympanichordal**. See *chorda tympani* in this table. **ulnar**, *function*, motion and sensation; *origin*, brachial plexus; *distribution*, muscles, shoulder-joint and wrist-joint, and skin of little finger; *branches*, articular, muscular, palmar cutaneous, dorsal, superior palmar, deep palmar. **vagus**. See *pneumogastric* in this table. **vestibular**, *function*, sensation; *origin*, auditory; *distribution*, utricle and ampullæ of the semicircular canals. **Vidian**, *function*, sensation; *origin*, union of large superficial and deep petrosal; *distribution*, spheno-maxillary fossa and posterior part of upper nasal meatus. of **Willis**. See *spinal accessory* in this table. **Wrisberg's**. See *cutaneous, lesser internal*, in this table. **zygomatic**. See *orbital* (*temporomalar*) in this table. **zygomatic**. See *malar* in this table. **zygomaticofacialis**. See *malar* in this table. **zygomaticotemporal** in this table.

**nervi** (*ner'-vi*) [L.]. Plural of *nervus*, a nerve. **n. erigentes**, nerve-fibers from the second and third sacral nerves to the rectum, bladder and genital organs. **n. nervorum**, the small nerves distributed

# NERVIDUCT 605 NEURAMEBIMETER

to the nerve-sheaths. n. **vasorum,** the small nerves supplying the walls of the blood-vessels.
**nerviduct** (*ner'-vid-ukt*) [*nervus,* nerve; *ductus,* duct]. The channel by which a nerve passes through a bone.
**nervimotility** (*ner-vi-mo-til'-it-e*). Capability of nerve motion.
**nervimotion** (*ner-vim-o'-shun*) [*nervus,* a nerve; *motio,* motion]. Movement caused by the stimulation of a nerve.
**nervimotor** (*ner-vim-o'-tor*) [*nervus,* a nerve; *motor,* a mover]. 1. Pertaining to or causing nervimotion. 2. That which causes nervimotion. 3. Pertaining to a motor nerve.
**nervine** (*ner'-vēn*) [*nervus,* nerve]. 1. Pertaining to the nerve. 2. Acting favorably, or decidedly, upon the nerves. 3. A remedy that calms nervous excitement or acts favorably on nervous diseases. 4. The plant *Cypripedium pubescens.* 5. An extract of the normal gray substance of sheep's brain. 6. A proprietary remedy for gout.
**nervitone** (*ner'-vit-ōn*). A proprietary remedy said to be a mixture of iron, phosphorus, asafetida, sumbul, and nux vomica.
**nervocidine** (*ner-vos'-id-ēn*) [*nervus,* nerve; *cædere,* to kill]. An alkaloid from an East Indian plant; the hydrochloride is used as a local anesthetic.
**nervosine** (*ner'-vo-sēn*). A remedy for hysteria said to consist of reduced iron 0.025 %, with valerian, orange-peel, angelica, and licorice extract.
**nervosity** (*ner-vos'-it-e*). Excessive nervousness.
**nervosism** (*ner'-vo-sizm*) [*nervus,* a nerve]. 1. Neurasthenia or nervousness. 2. The doctrine that all morbid phenomena are caused by alterations of nerve-force.
**nervotabes, peripheral** (*ner-vo-ta'-bēz*). A disturbance of an intact spinal cord, presenting clinically the appearance of tabes; due to parenchymatous neuritis of the cutaneous nerves.
**nervous** (*ner'-vus*) [*nerve*]. 1. Pertaining to or composed of nerves or nerve-structures. 2. Characterized by excessive irritability of the nervous system. **n. debility,** neurasthenia. **n. fluid,** a hypothetical fluid supposed to traverse the nerves from the nerve-centers to the periphery. **n. system,** the nervous apparatus of the body taken together; it includes the brain, spinal cord, nerves, and ganglia. **nervous exhaustion.** See *neurasthenia.*
**nervousness** (*ner'-vus-nes*) [*nervous*]. A condition of excessive excitability of the nervous system, characterized by great mental and physical unrest.
**nervule** (*ner'-vūl*) [dim. of *nervus,* nerve]. A small nerve.
**nervus** (*ner'-vus*) [*pl. nervi*]. Latin for nerve.
**nesis** (*ne'-sis*) [*νῆσις,* suture]. Suture.
**Nessler's reagent** (*nes'-ler*) [A. *Nessler,* German chemist, 1827– ]. A solution of potassium iodide, mercury bichloride, and sodium hydroxide used in estimating the amount of ammonia in water.
**nesslerizing** (*nes-ler-i'-zing*). The process of using Nessler's reagent.
**nest** [ME.]. An abode, as of eggs, insects, etc. **n.s, Brunn's epithelial.** See under *Brunn.* **n. cell-,** an aggregation or cluster of cells, as in carcinoma.
**nesteia** (*nes-ti'-ah*) [*νηστεία,* fasting]. 1. Fasting. 2. The jejunum.
**nestiatria** (*nes-te-a'-tre-ah*) [*νηστεία,* a fast; *ἰατρεία,* treatment]. Treatment by fasting; the hunger-cure.
**nestis** (*nes'-tis*) [*νῆστις,* fasting]. 1. Fasting. 2. The jejunum.
**nestitherapy** (*nes-ti-ther'-a-pe*). See *nestiatria.*
**Nestle's food** (*nesl*). A variety of milk-food for infants.
**nestoposia** (*nes-top-o'-ze-ah*) [*νῆστις,* fasting; *πόσις,* drinking]. Drinking on an empty stomach.
**net-knots.** See *neurosomes.*
**netraneurysm** (*net-ran'-ū-rizm*) [*νῆτρον,* spindle; ἀνεύρυσμα, aneurysm]. A fusiform aneurysm.
**nettle** (*net-l*). See *urtica.*
**nettlerash.** See *urticaria.*
**Nettleship's dots** [Edward *Nettleship,* English ophthalmologist, 1845– ]. Minute white dots scattered in considerable numbers between the macula and periphery of the retina; they are associated with pigment changes and night-blindness, and occur in several members of the same family.
**net-work.** The arrangement of fibers in a reticulum. **n., Gerlach's,** a network of processes of nerve-cells found in the gray matter of the spinal cord.

**n., Haller's,** the rete testis. **n., Purkinje's,** that formed by Purkinje's fibers. **n. of terminal bars,** the reticulum formed by the terminal bars on the free surface of many epithelia.
**neu** (*nū*). See *neurilemma.*
**Neubauer's artery** (*noy'-bow-er*) [Johann Ernst *Neubauer,* German anatomist, 1742–1777]. The deep thyroid artery; an occasional branch of the innominate artery, distributed to the same parts as the inferior thyroid, and often taking its place. **N.'s ganglion,** the large ganglion formed by the union of the lower cervical and first thoracic ganglion.
**Neuber's method** (*noy'-ber*). A method of treating joint and bone tuberculosis: an incision is made and all fragments and tuberculous foci are removed and the cavity filled with an emulsion of iodoform and glycerol of 10 % strength. It is then sewed up with buried sutures without drainage. **N.'s tubes,** decalcified bone drainage tubes.
**Neumann's corpuscles** (*noy'-man*). Nucleated red corpuscles, sometimes found in the blood when an active regenerative process is going on, as after hemorrhage. **N.'s crystals.** See *Charcot's crystals.* **N.'s disease,** pemphigus vegetans. **N.'s phenylhydrazin test,** consists in the use of a solution of sodium acetate in acetic acid of 50 to 75 % strength or in glacial acetic acid. A special test-tube is employed, the urine is introduced, the solution and two or three drops of pure phenylhydrazin are added, and the whole boiled down, cooled rapidly in running water, boiled one minute more, and cooled. In the presence of sugar, phenylhydrazin crystals appear. **N.'s sheaths,** the dentinal sheaths that form the walls of the dentinal tubules.
**neura** (*nū'-rah*). Synonym of *neuron.*
**neurad** (*nū'-rad*) [*neural; ad,* to]. Toward the neural aspect or axis.
**neuradynamia** (*nū-rah-din-a'-me-ah*). See *neurasthenia.*
**neuragmia** (*nū-rag'-me-ah*) [*neuron;* ἀγμός, a breaking]. The bruising or tearing of a nerve-trunk above or below its ganglion for the purpose of studying the trophic changes that follow.
**neural** (*nū'-ral*) [*neuron*]. Pertaining to nerves or nervous tissue. **n. arch,** the part of the vertebra that incloses the spinal cord, formed by the two neurapophyses. **n. axis,** the spinal cord. **n. canal.** 1. The dorsal tube of the embryo, formed by the union of the dorsal folds, and constituting the earliest traces of the nervous system. 2. The bony canal comprising the cavity of the cranium and vertebral column, which in the vertebrate animals contains the central nervous system. It is situated dorsad, the hemal canal, inclosing the heart, etc., being ventrad. See *canal, hemal.* **n. groove,** the medullary groove. **n. lamina,** the lateral portion of the neural arch of a vertebra. **n. plate,** the medullary plate. **n. spine,** the spinous process of a vertebra. **n. tube,** the closed medullary groove of the epiblast.
**neuralgia** (*nū-ral'-je-ah*) [*neuron;* ἄλγος, pain]. Severe paroxysmal pain along the course of a nerve and not associated with demonstrable structural changes in the nerve. According to their anatomical situation, the following forms of neuralgia are described: *trigeminal neuralgia,* tic douloureux, or prosopalgia; *supraorbital neuralgia; cervicooccipital neuralgia; cervicobrachial* and *brachial neuralgia; intercostal neuralgia,* sciatica or ischialgia; coccygodynia; *visceral neuralgia* (as *hepatic, gastric, intestinal, uterine, ovarian neuralgia*). According to their causes, neuralgias are classed as anemic, malarial, gouty, rheumatic, syphilitic, diabetic, toxic (*e. g.,* alcoholic, saturnine), hysterical, and reflex. The pain of neuralgia is sharp, stabbing, and paroxysmal, lasting usually but a short time; tenderness is often present at the points of exit of the nerve (*points douloureux*). Intercostal neuralgia is at times associated with herpes zoster.
**neuralgic** (*nū-ral'-jik*) [*neuralgia*]. Pertaining to, or affected with, neuralgia.
**neuralgin** (*nū-ral'-jin*). A proprietary antipyretic said to consist of a mixture of antifebrin, sodium salicylate, and caffeine. Dose 8–45 gr. (0.5–3.0 Gm.).
**neuramebimeter** (*nū-ram-eb-im'-et-er*) [*neuro-;* ἀμοιβή, return; *μέτρον,* measure]. The nerve-reply measurer; an instrument devised and used in psychophysics to obtain the reaction-time of nervous impressions.

**neuranagenesis** (*nū-ran-aj-en'-es-is*) [*neuron;* ἀναγεννάειν, to renew]. Regeneration or renewal of nerve tissue.

**neurapophysis** (*nū-rap-of'-is-is*) [*neuron;* ἀπόφυσις, offshoot]. Either one of the two apophyses on each vertebra which blend and form the neural arch, or the dorsal wall of the spinal foramen.

**neurarchy** (*nū'-rar-ke*) [*neuro-;* ἀρχη, government]. The control of the nervous system over the other systems of the body.

**neurasthenia** (*nū-ras-then-i'-ah* or *nū-ras-the'-ne-ah*) [*neuron; asthenia*]. A group of symptoms resulting from debility or exhaustion of the nerve-centers. Among the more common symptoms are a lack of energy, undue readiness of fatigue, disinclination to activity, a sense of fulness or pressure at the top of the head, pain in the back, impaired memory, and disturbed sleep; gastrointestinal symptoms, such as anorexia, constipation, fulness after eating; amenorrhea and dysmenorrhea in women, and spermatorrhea and impotence in men. Syn., *Beard's disease.* **n., cerebral,** a form marked by depression, inability to concentrate the mind, insomnia, irritability, headache, visual disturbances, etc. **n. cordis,** a neurosis in which the heart symptoms dominate. **n. gastrica,** nervous dyspepsia, a condition of disturbed functional activity of the stomach, as a rule without retardation of digestion; flatulence, pain, palpitation of the heart, and constipation are prominent symptoms. **n., sexual,** a depressed state of the nervous system associated with disturbance of the sexual function; it is characterized by pain in the back, tender points along the spine, weakness of the extremities, great prostration on slight exertion, neuralgic pains, and other nervous manifestations.

**neurastheniac** (*nū-ras-the'-ne-ak*). A person suffering from neurasthenia.

**neurasthenic** (*nū-ras-then'-ik*) [*neurasthenia*]. Relating to, or characterized by, neurasthenia.

**neurataxia, neurataxy** (*nū-rah-tak'-se-ah, nu'-rah-tak-se*) [*neuron;* ἀταξία, want of order]. 1. Ataxia of cerebrospinal origin. 2. Neurasthenia.

**neuratrophia, neuratrophy** (*nū-rat-ro'-fe-ah, nū-rat'-ro-fe*) [*neuron; atrophy*]. Atrophy, or impaired nutrition, of a nerve or nerves.

**neuraxis** (*nū-raks'-is*) [*neuron; axis*]. 1. The cerebrospinal axis. 2. An axis-cylinder.

**neuraxon** (*nū-raks'-on*) [see *neuraxis*]. The axis-cylinder process of a nerve-cell.

**neure** (*nūr*). Synonym of *neuron.*

**neurectasia, neurectasis, neurectasy** (*nū-rek-ta'-se-ah, nū-rek'-tas-is, nū-rek'-tas-e*) [*neuron;* ἔκτασις, stretching]. Nerve-stretching.

**neurectomy** (*nū-rek'-to-me*) [*neuron;* ἐκτομή, excision]. Excision of a part of a nerve.

**neurectopia, neurectopy** (*nū-rek-to'-pe-ah, nū-rek'-to-pe*) [*neuron;* ἔκτοπος, out of place]. Displacement or other abnormality of the distribution of a nerve.

**neurenteric** (*nū-ren-ter'-ik*) [*neuron; enteron*]. Pertaining to the embryonic neural canal and the intestinal tube. **n. canal,** a temporary communication existing between the neural canal and the intestinal tube of the embryo. Syn., *Kowalewsky's canal.*

**neurepithelium** (*nū-rep-ith-e'-le-um*). See *neuroepithelium.*

**neurexairesis** (*nū-reks-i-re'-sis*) [*neuron;* ἐξαίρειν, to take out]. The extraction of a nerve for relief of neuralgia.

**neuria** (*nū'-re-ah*) [*neuro-*]. Nervous tissue; a delicate layer of nerve-substance.

**neuriasis** (*nū-ri'-as-is*) [*neuro-*]. Hysterical hypochondriasis.

**neuriatry** (*nū-ri'-at-re*) [*neuro-;* ἰατρεία, therapy]. The study and treatment of nervous diseases.

**neuric** (*nū'-rik*) [*neuro-*]. Pertaining to a nerve or to nerves.

**neuricity** (*nū-ris'-it-e*). Nerve-force; nervous quality, or function.

**neuridine** (*nū'-rid-ēn*) [*neuron*], $C_5H_{14}N_2$. A ptomaine produced in the putrefaction of horseflesh, beef, human muscle, fish, cheese, etc. It has a repulsive odor and is non-poisonous.

**neurilemma** (*nū-ril-em'-ah*) [*neuron;* λέμμα, bark]. 1. The sheath incasing a nerve-fiber; the sheath of Schwann. See *nerve-fiber.* 2. See *perineurium.*

**neurilemmitis** (*nū-ril-em-i'-tis*) [*neuro-;* λέμμα, sheath; *ιτις*, inflammation]. Inflammation of the neurilemma.

**neurility** (*nū-ril'-it-e*) [*neuro-*]. The stimulus or power possessed by a nerve-fiber to cause contraction of a muscle; neuricity.

**neurimotility.** See *nervimotility.*

**neurin, neurine** (*nū'-rin*) [*neuron*]. 1. The albuminous substance forming the basis of nerve tissue. 2. $C_5H_{13}NO$; a poisonous ptomaine obtained from decomposing flesh and in the decomposition of protagon by barium hydroxide; an auxetic in cancer; it is used as a substitute for cancroin. 3. An extract of nerve tissue employed therapeutically. See *organotherapy.*

**neurinoma** (*nū-rin-o'-mah*). A neurofibroma.

**neurit** (*nū'-rit*) [*neuron*]. Synonym of *neurite.*

**neurite** (*nū'-rit*) [*neuron*]. The axis-cylinder process of a nerve-cell; a neuraxon.

**neuritic** (*nū-rit'-ik*) [*neuritis*]. Pertaining to neuritis.

**neuritis** (*nū-ri'-tis*) [*neuron; ιτις*, inflammation]. Inflammation of a nerve. **n., alcoholic.** See *n. multiple.* **n., ascending,** inflammation extending from the periphery of a nerve centrad to the spinal cord or brain. **n., atheromatous,** a form in which necrotic, inflammatory processes take place in the parts of the nerves supplied by arteries affected with atheroma, giving rise to symptoms resembling in character those of the toxic form. **n., axial, n. axialis,** optic neuritis in which the central fibers of the optic nerve, supplying the central part of the retina, are diseased. It results in central scotoma. **n., degenerative.** See *n., parenchymatous.* **n., descending,** neuritis the result of extension of disease from the spinal cord or the brain toward the periphery. **n., diabetic,** a polyneuritis sometimes seen in diabetes, and probably the result of autointoxication with the products of faulty metabolism. **n., diphtheritic,** that which follows diphtheria. **n., endemic, beriberi. n., facial,** peripheral paralysis of the facial nerve. **n. fascians,** interstitial neuritis. **n., interstitial,** inflammation of the interstitial connective tissue of a nerve-trunk. **n., leprous,** that due to the bacillus of leprosy. **n., lipomatous,** that form in which the nerve-fibers are completely destroyed and replaced by a fibrous connective tissue in which much fat is deposited. **n., lymphatic,** mesoneuritis. **n., malarial,** neuritis due to the malarial poison. **n., migrans,** a wandering neuritis. **n., multiple,** the simultaneous inflammation of several nerve-trunks, usually symmetrically situated on both sides of the body. Its most common cause is alcoholic poisoning; it may be due to arsenic, malaria, mercury, or lead; to diphtheria, pneumonia, typhoid fever, and other infectious diseases. Beriberi is a form of multiple neuritis. **n. nodosa,** neuritis with nodular formations. **n., optic,** inflammation of the optic nerve. See *papillitis.* **n., parenchymatous,** a form in which the medullary substance and the axis-cylinders are chiefly involved, the interstitial tissue being but little altered or affected only secondarily. Syn., *degenerative neuritis.* **n., postocular,** that affecting the portion of the optic nerve behind the eyeball. **n., pressure-,** inflammation of a nerve resulting from compression. **n., proliferative,** that form in which the overgrowth of the connective tissue is so extensive as to convert the whole nerve into a sclerotic cord. **n., radicular,** that in which the nerve-roots rather than the plexus are the seats of inflammation. **n., retrobulbar,** that of the optic nerve posterior to the eyeball. **n., rheumatic,** that due to rheumatism. **n., sciatic,** sciatica. **n., segmental, n., segmentary,** that affecting a segment of a nerve. **n., senile,** a form affecting the extremities of the aged. **n., toxemic,** that due to some poison or virus in the blood. **n., toxic,** that due to some poisonous substance, as lead, silver, arsenic. **n., tuberculous,** polyneuritis due to the specific action on the nerves of a poison produced by tubercle bacilli.

**neuro-** (*nū-ro-*) [*neuron*]. A prefix signifying connection with or relation to a nerve.

**neuroanatomy** (*nū-ro-an-at'-o-me*) [*neuro-; anatomy*]. The anatomy of the nervous system.

**neuroarthritism** (*nū-ro-ar'-thrit-izm*). A combined nervous and gouty diathesis.

**neuroasthenia** (*nū-ro-as-the'-ne-ah*). Same as *neurasthenia.*

**neuroblast** (*nū'-ro-blast*) [*neuro-;* βλαστός, germ]. A cell derived from the primitive ectoderm, and giving rise to nerve-fibers and nerve-cell.

**neuroblastoma** (*nū-ro-blas-to'-mah*). A tumor consisting of nerve tissue or cells.

**neurocanal** (nū-ro-kan-al') [neuro-; canalis, canal]. The central canal of the spinal axis.

**neurocardiac** (nū-ro-kar'-de-ak) [neuro-; cardia]. Pertaining to the nervous system and the heart. n. disease, exophthalmic goiter.

**neurocele** (nū'-ro-sēl) [νευρο; κοιλια, hollow]. The system of cavities and ventricles in the cerebrospinal axis.

**neurocentral** (nū-ro-sen'-tral) [neuro-; κέντρον, center]. Relating to the neural arch and the centrum of a vertebra.

**neurochitin** (nū-ro-ki'-tin) [neuro-; chitin]. The substance forming the skeletal support of nerve-fibers.

**neurochondrous** (nū-ro-kon'-drus) [νεῦρον, cord; χόνδρος, cartilage]. Fibrocartilaginous.

**neurochorioretinitis** (nū-ro-ko-re-o-ret-in-i'-tis), Chorioretinitis combined with optic neuritis.

**neurochoroiditis** (nū-ro-ko-roi-di'-tis). Combined inflammation of the choroid body and optic nerve.

**neurocranium** (nū-ro-kra'-ne-um) [neuro-; κρανίον, skull]. The brain-case, or cranial portion of the head.

**neurocyte** (nū'-ro-sīt) [neuro-; κύτος, cell]. A nerve-cell; a neuron; the essential element of nervous structures.

**neurocytoma** (nū-ro-si-to'-mah). A tumor consisting of undifferentiated nerve tissues or cells of the cerebrospinal nervous system.

**neurodealgia** (nū-ro-de-al'-je-ah) [νευρώδης, nerve-like; the retina; ἄλγος, pain]. Retinal pain.

**neurodeatrophia** (nū-ro-de-at-ro'-fe-ah) [νευρώδης, retina; ἀτροφία, atrophy]. Atrophy of the retina.

**neurodendrite** (nū-ro-den'-drīt) [neuro-; δένδρον, a tree]. A dendritic and protoplasmic extension or process of a nerve-cell, a combined neuron and dendron.

**neurodendron** (nū-ro-den'-dron) [see neurodendrite]. 1. Synonym of neuron. 2. See neurodendrite.

**neurodermatitis** (nū-ro-der-mat-i'-tis) [neuro-; dermatitis]. A neurotic affection of the skin associated with itching.

**neurodermatosis** (nū-ro-der-mat-o'-sis) [neuro-; δέρμα, skin; νόσος, disease]. A neurotic skin-affection.

**neurodermatrophia** (nū-ro-der-mat-ro'-fe-ah) [neuro-; δέρμα, skin; ἀτροφία, atrophy]. Atrophy of the skin from nervous disturbance.

**neurodes** (nū-ro'-dēz) [νευρώδης, nerve-like]. The retina, as being made up of nerve-elements.

**neurodiastasis** (nū-ro-di-as'-tas-is) [neuro-; διάστασις, separation]. Separation of nerves; neurectasis.

**neurodin** (nū'-ro-din) [neuron]. Acetylparaoxyphenylurethane, a crystalline substance used as an antineuralgic and antipyretic. Dose, antineuralgic, 15-24 gr. (1.0-1.5 Gm.); as antipyretic, 5-10 gr. (0.32-0.65 Gm.).

**neurodynamia** (nū-ro-di-nam'-e-ah) [neuro-; δύναμις, strength]. Nervous strength or energy.

**neurodynamic** (nū-ro-di-nam'-ik) [neuro-; δύναμις, strength]. Pertaining to the power of a nerve-current or of the nervous forces of the system.

**neurodynia** (nū-ro-din'-e-ah). See neuralgia.

**neuroelectrotherapeutics** (nū-ro-e-lek-tro-ther-a-pū'-tiks). The treatment of nervous affections by electricity.

**neuro₀enteric** (nū-ro-en-ter'-ik). Same as neurenteric.

**neuroepidermal** (nū-ro-ep-e-der'-mal). Relating to the nerves and the skin.

**neuroepithelial** (nū-ro-ep-ith-e'-le-al) [neuro-; ἐπί, upon; (θήλη, nipple]. Pertaining to or of the nature of neuroepithelium.

**neuroepithelioma** (nū-ro-ep-e-the-le-o'-mah) [neuro-; epithelioma]. A glioma of the retina.

**neuroepithelium** (nū-ro-ep-e-the'-le-um) [neuro-; epithelium]. The highly specialized epithelial structures constituting the terminations of the nerves of special sense, as the rod-and-cone cells of the retina, the olfactory cells of the nose, the hair-cells of the internal ear, the gustatory cells of the taste-buds.

**neurofibril** (nū-ro-fi'-bril). A conducting fibril of a nerve-cell.

**neurofibroma** (nū-ro-fi-bro'-mah) [neuro-; fibroma]. A tumor of a nerve composed of fibrous tissue.

**neurofibromatosis** (nū-ro-fi-bro-ma-to'-sis) [neurofibroma]. A disease characterized by the formation of numerous great and small tumefactions of the nerves. Syn., Recklinghausen's disease.

**neurofil** (nū'-ro-fil) [neuro-; filum, thread]. A network of protoplasmic processes arising from the commencement of the axis-cylinder and surrounding the cell.

**neuroganglion** (nū-ro-gang'-gle-on). See ganglion.

**neurogastric** (nū-ro-gas'-trik). Relating to the nerves and the stomach.

**neurogenesis** (nū-ro-jen'-es-is) [neuro-; γεννᾶν, to produce]. The formation of nerves or nerve tissue.

**neurogenetic** (nū-ro-jen-et'-ik) [neurogenesis]. Pertaining to neurogenesis.

**neurogenous** (nū-roj'-en-us). Originating in the nervous system.

**neurogeny** (nū-roj'-en-e). See neurogenesis.

**neuroglia** (nū-rog'-le-ah) [neuro-; γλία, glue]. The tissue, probably of ectodermic origin, forming the basis of the supporting framework of the nervous tissue of the cerebrospinal axis. It consists of peculiar cells, the glia-cells, having many fine branching processes.

**neurogliar** (nū-rog'-le-ar) [neuroglia]. Pertaining to or resembling neuroglia.

**neuroglioma** (nū-ro-gli-o'-mah) [neuro-; glioma]. A tumor composed of neurogliar tissue; a glioma. n., ganglionar, n. ganglionare, a glioma containing ganglion-cells.

**neurography** (nū-rog'-ra-fe) [neuro-; γράφειν, to write]. A treatise on the anatomy and physiology of the nerves and the nervous system.

**neurohistology** (nū-ro-his-tol'-o-je). The histology of the nervous system.

**neurohypnology** (nū-ro-hip-nol'-o-je) [neuro-; ὕπνος, sleep; λόγος, science]. The science or study of hypnotism.

**neuroid** (nū'-roid) [neuro-; εἶδος, like]. Resembling a nerve or nerve-substance.

**neurokeratin** (nū-ro-ker'-at-in) [neuro-; keratin]. The form of keratin found in nerve-sheaths and the white substance of Schwann.

**neurokinet** (nū-ro-kin'-et) [neuro-; κινεῖν, to move]. An apparatus for stimulating the nerves by means of mechanical percussion.

**neurokyme** (nū'-ro-kīm) [neuro-; κῦμα, a wave]. Nervous energy.

**neurolemma** (nū-ro-lem'-ah). Synonym of retina.

**neurolemmatitis** (nū-ro-lem-at-i'-tis). A synonym of retinitis.

**neurological** (nū-ro-loj'-ik-al) [neurology]. Pertaining to neurology.

**neurologist** (nū-rol'-o-jist) [neurology]. One versed in neurology.

**neurology** (nū-rol'-o-je) [neuro-; λόγος, science]. The branch of medicine dealing with the anatomy, physiology, and pathology of the nervous system.

**neurolymph** (nū'-ro-limf) [neuro-; lymph]. The cerebro-spinal fluid.

**neurolysin** (nū-rol'-is-in) [see neurolysis]. A cytolysin having specific action upon nerve-cells.

**neurolysis** (nū-rol'-is-is) [neur₀-; λύσις, solution]. 1. Exhaustion of a nerve in consequence of overstimulation. 2. Nerve stretching for the relief of excessive tension. 3. The loosening of adhesions binding a nerve. 4. The disintegration of nerve tissue.

**neurolytic** (nū-ro-lit'-ik) [neuro-; λύσις, a loosening]. Pertaining to neurolysis.

**neuroma** (nū-ro'-mah) [neuro-; ὄμα, tumor]. 1. A tumor composed of nerve tissue. 2. A fibroma on a nerve. n., amputation-, the neuroma of a stump, forming at the end of a divided nerve. n., amyelinic, a neuroma made up of nonmedullated nerve-fibers. n. cutis, a cutaneous neuroma. n. false, a fibromatous tumor forming on a nerve. n., ganglionic, n. ganglionated, a neuroma made up of nerve-cells. n., myelinic, one made up of medullated nerve-fibers. n., plexiform, the development of multiple fibromatous tumors along the course of one or more nerves, attended with hyperplasia of the nerve-fibers. n. telangiectodes, a vascular neuroma. n., traumatic, one occurring in a wound or amputation stump. n., true, a tumor containing nerve cells.

**neuromalacia** (nū-ro-mal-a'-she-ah) [νεῦρο-; μαλακία, a softening]. A softening of nerve-tissue.

**neuromatosis** (nū-ro-mat-o'-sis) [neuro-; ὄμα, tumor; νόσος, disease]. A morbid tendency to the formation of neuromata.

**neuromatous** (nū-ro'-mat-us) [neuroma]. Of the nature of a neuroma.

**neuromere** (nū'-ro-mēr) [neuro-; μέρος, a part]. A natural segment of the cerebrospinal axis.

**neuromimesis** (nū-ro-mi-me'-sis) [neuro-; μίμησις,

imitation]. Hysteric phenomena resembling true organic disease.
**neuromimetic** (*nū-ro-mi-met'-ik*). Pertaining to neuromimesis.
**neuromuscular** (*nū-ro-mus'-kū-lar*) [*neuro-; muscular*]. Pertaining conjointly to nerves and muscles.
**neuromyelitis** (*nū-ro-mi-el-i'-tis*) [*neuro-; myelitis*]. Inflammation of myelonic substance or of the medulla spinalis.
**neuromyology** (*nū-ro-mi-ol'-o-je*) [*neuro-; myology*]. The classification of muscles with regard to their innervation.
**neuromyopathic** (*nū-ro-mi-o-path'-ik*) [*neuro-; μῦς, muscle; πάθος, disease*]. Relating to disease of both muscles and nerves.
**neuromyositis** (*nū-ro-mi-o-si'-tis*) [*neuro-; myositis*]. Myositis associated with neuritis.
**neuron, neurone** (*nū'-ron, nū'-rōn*) [*νεῦρον, nerve*]. 1. The cerebrospinal axis taken as a whole. 2. One of the countless number of units of which the nervous system is composed. Each neuron consists of a cell and a series of processes. In every physiological act involving the nervous system at least two, usually more, neurons participate. The neuron at which the impulse starts is termed *archineuron*; the one at the termination, the *teleneuron*. See also *nerve-cell.* n.s, **Edinger's law** concerning. See under *Edinger.* n.s, **Goldscheider-Marinesco's law** concerning. See under *Goldscheider.* n., internuncial, one interposed between an afferent neurone and an efferent neurone. n.-threshold, the degree of excitation of a neuron which just suffices to produce a sensation in another with which it is in contact.
**neuronal, neuronic** (*nū'-ron-al, nū-ron'-ik*). 1. Relating to a neuron. 2. Trade name of a preparation said to be sedative and hypnotic.
**neuronephric** (*nū-ro-nef'-rik*) [*neuro-; νεφρός, the kidney*]. Pertaining to the nervous and renal systems.
**neuronophagia, neuronophagy** (*nū-ron-o-fa'-je-ah, nū-ron-off'-aj-e*) [*neuro-; φαγεῖν, to eat*]. The destruction of neurones by phagocytes.
**neuronosus** (*nū-ron'-o-sus*) [*neuro-; νόσος, disease*]. Synonym of *neurosis.* n. of the **skin**, neurotic skin-disease.
**neuronymy** (*nū-ron'-im-e*) [*neuro-; ὄνυμα, a name*]. Neurologic nomenclature.
**neuronyxis** (*nū-ro-niks'-is*) [*neuro-; νύσσειν, to prick*]. The puncturing of nerves.
**neuro-occipital** (*nū-ro-ok-sip'-it-al*). Relating to a neural arch and the occiput.
**neuroparalysis** (*nū-ro-par-al'-is-is*) [*neuro-; paralysis*]. Paralysis due to disease of a nerve.
**neuropathic** (*nū-ro-path'-ik*) [*neuro-; πάθος, disease*]. 1. Characterized by a diseased or imperfect nervous system. 2. Depending upon or pertaining to nervous disease. n. eschar, a bed-sore following disease of the spinal cord.
**neuropathogenesis** (*nū-ro-path-o-jen'-es-is*) [*neuro-; pathogenesis*]. The development of a disease of the nervous system.
**neuropathology** (*nū-ro-path-ol'-o-je*) [*neuro-; pathology*]. The pathology of diseases of the nervous system.
**neuropathy** (*nū-rop'-ath-e*) [*neuro-; πάθος, disease*]. Any nervous disease.
**neurophlegmon** (*nū-ro-fleg'-mon*) [*neuro-; phlegmon*]. Neuritis.
**neurophonia** (*nū-ro-fo'-ne-ah*) [*neuro-; φωνή, voice*]. A rare choreic disease of the larynx and muscles of expiration characterized by the utterance of sharp, spasmodic cries.
**neurophysiology** (*nū-ro-fiz-e-ol'-o-je*) [*neuro-; physiology*]. The physiology of the nervous system.
**neuropilem, neuropilema** (*nū-ro-pi'-lem, -pi-le'-mah*) [*neuro-; πῖλος, felt*]. The dense mat of fibrils formed in some parts by the branching nerve-processes.
**neuroplasm** (*nū'-ro-plasm*) [*neuro-; πλάσσειν, to mold*]. The protoplasm filling the interstices of the fibrils of nerve-cells.
**neuroplasty** (*nū'-ro-plas-te*) [see *neuroplasm*]. A plastic operation on the nerves; nerve-grafting.
**neuroplex, neuroplexus** (*nū'-ro-pleks, nū-ro-pleks'-us*). A plexus of nerves.
**neuroploca** (*nū-rop'-lo-kah*) [*neuro-; πλοκή, a twisting*]. A ganglion of the nerves.
**neuropodium** (*nū-ro-po'-de-um*) [*neuro-; πούς, foot; pl., neuropodia*]. A dendraxon.

**neuropore** (*nū'-ro-pōr*) [*neuro-; πόρος, pore*]. A small opening at the anterior extremity of the primary telencephalon; a pore between the neural canal and the exterior, in certain embryos.
**neuropsychology** (*nū-ro-si-kol'-o-je*) [*neuro-; psychology*]. A system of psychology based on neurology.
**neuropsychopathy** (*nū-ro-si-kop'-ath-e*) [*neuro-; ψυχή, mind; νόσος, disease*]. A mental disease based upon, or manifesting itself in, nervous disorders or symptoms.
**neuropsychosis** (*nū-ro-si-ko'-sis*) [*neuro-; psychosis*]. A combined nervous and mental disease.
**neurorelapse** (*nū-ro-re-laps'*). The manifestation of nervous symptoms in syphilis occurring after an injection of salvarsan.
**neuroretinitis** (*nū-ro-ret-in-i'-tis*) [*neuro-; retinitis*]. Inflammation of both the optic nerve and the retina.
**neurorrhaphy** (*nū-ror'-a-fe*) [*neuro-; ῥαφή, suture*]. The operation of suturing a divided nerve.
**neurorrheuma** (*nū-ror-ru'-mah*) [*neuro-; ῥεῦμα, flow*]. Nervous force.
**neurorrhexis** (*nū-ro-reks'-is*). The forcible tearing out of a nerve in the treatment of persistent neuralgia.
**Neurorrhyctes hydrophobiæ** (*nū-ro-rik'-tes hi-dro-fo'-be-e*) [*neuro-; ὀρύκτης, a digger*]. A Negri body, supposed to be the cause of rabies.
**neurosal** (*nū-ro'-sal*). Pertaining to, or of the nature of, a neurosis.
**neurosarcokleisis** (*nū-ro-sar-ko-kli'-sis*) [*neuro-; σάρξ, flesh; κλεῖσις, closure*]. An operation performed for the relief of neuralgia; pressure on the affected nerve is relieved by partial resection of the osseous canal through which it passes, and transplanting it (the nerve) in the soft tissues.
**neurosarcoma** (*nū-ro-sar-ko'-mah*). A combined neuroma and sarcoma.
**neurosclerosis** (*nū-ro-skle-ro'-sis*) [*neuro-; σκληρός, hard*]. Sclerosis of nervous tissue.
**neurosin** (*nū-ro'-sin*) [*neuron*]. A trade name for several preparations, containing calcium, glycerol, and phosphates.
**neurosis** (*nū-ro'-sis*) [*neuron*]. Any morbid nervous state. A functional disease of the nervous system—a disturbance of the nerve-centers or peripheral nerves not due to any demonstrable structural change. n., **cyclists'**, painful hyperæsthesia of the skin of the scrotum, perineum, and thighs from excess in bicycle-riding. n., **fatigue**, n., **occupation-**, n., **professional**, a functional disorder affecting groups of muscles used in the performance of special movements. n., **traumatic**, any deviation from the normal state of the nervous system caused by violence. n., **Westphal's**. See under *Westphal.*
**neurosism** (*nū'-ro-sism*). Same as *neurasthenia.*
**neuroskeleton** (*nū-ro-skel'-et-on*) [*neuro-; skeleton*]. The vertebrate endoskeleton, or true skeleton; so-called from being made up of parts that correspond with and largely serve to protect portions of the central nervous system. n., **cyclists'**, painful hyperæsthesia of the skin of the scrotum, perineum, and thighs from excess in bicycle-riding. n., **fatigue**, n., **occupation-**, n., **professional**, a functional disorder affecting groups of muscles used in the performance of special movements. n., **traumatic**, any deviation from the normal state of the nervous system caused by violence.
**neurosomes** (*nū'-ro-sōms*) [*neuro-; σῶμα, a body*]. Minute granules, variable in size, observed at the nodal points of the protoplasm of axis-cylinders.
**neurospasm** (*nū'-ro-spasm*) [*neuro-; spasm*]. Nervous spasm or twitching of a muscle.
**neurospongium** (*nū-ro-spun'-je-um*) [*neuro-; σπογγύλον, dim. of σπόγγος, a sponge*]. The inner reticular layer of the retina.
**neurostearic** (*nū-ro-ste-ar'-ik*) [*neuro-; στέαρ, fat*]. Pertaining to nervous tissue and fat. n. **acid**, C₁₈H₃₆O₂. An acid isomeric with stearic acid, occurring in the brain.
**neurosthenia** (*nū-ro-sthē'-nē-ah*) [*neuro-; σθένος, power*]. Great nervous power, or abnormal excitation of the nervous centers.
**neurosuture** (*nū-ro-su'-tūr*). The suture of a nerve.
**neurotabes** (*nū-ro-ta'-bēs*) [*neuro-; tabes, wasting*]. A form of multiple neuritis resembling posterior sclerosis.
**neurotagma** (*nū-ro-tag'-mah*) [*neuro-; τάγμα, that which has been arranged*]. A linear arrangement of the structural elements of a neuron.
**neurotension** (*nū-ro-ten'-shun*). See *neurectasis.*
**neurothele** (*nū-ro-thē'-le*) [*neuro-; θηλή, a nipple*]. A nerve-papilla.
**neurotheleitis, neurothelitis** (*nū-ro-the-le-i'-tis, nū-ro-the-li'-tis*). Inflammation of a nerve papilla.
**neurothelion, neurothelium** (*nū-ro-thē'-le-on, -um*) [*neurothele*]. A small nerve-papilla.

**neurotherapy** (nū-ro-ther'-ap-e) [neuro-; θεραπεία, treatment]. The treatment of nervous diseases.
**neurothlipsis** (nū-ro-thlip'-sis) [neuro-; θλίβειν, to press]. Pressure on a nerve.
**neurotic** (nū-rot'-ik) [neuron]. 1. Pertaining to the nerves; nervous. 2. Pertaining to neuroses. 3. Having a disordered nervous system; suffering from a neurosis.
**neurotica** (nū-rot'-ik-ah) [neuron]. Functional nervous diseases.
**neuroticism** (nū-rot'-is-ism). The condition of having a disordered nervous system or of suffering from a neurosis.
**neurotization** (nū-rot-iz-a'-shun). The regeneration of a divided nerve.
**neurotome** (nū'-ro-tōm) [neuro-; τομή, a cutting]. 1. A needle-like knife for the division of a nerve. 2. The nerve tissues of an embryonic metamere; a neural segment or neuromere.
**neurotomy** (nū-rot'-o-me) [see *neurotome*]. The division of a nerve.
**neurotonia, neurotony** (nū-ro-to'-ne-ah, nū-rot'-on-e). See *nerve-stretching*.
**neurotonic** (nū-ro-ton'-ik). 1. Pertaining to neurotony. 2. Having a tonic effect upon the nerves. n. reaction, a rare form of electric reaction exhibited in a persistent tetanic quivering of the muscles following irritation of the nerve-stems.
**neurotoxic** (nū-ro-toks'-ik) [neuro-; τοξικόν, a poison]. Having toxic action on neurons.
**neurotoxin** (nū-ro-toks'-in). A cytotoxin capable of destroying nerve cells.
**neurotripsy** (nū-ro-trip'-se) [neuro-; τρίβειν, to rub]. The crushing of a nerve.
**neurotrophasthenia** (nū-ro-trof-as-the'-ne-ah) [neuro-; τροφή, nourishment; asthenia]. Malnutrition of the nerves.
**neurotrophic** (nū-ro-tro'-fik) [neuro-; τροφή, nourishment]. Depending on or attained through the trophic influence exercised by the nerves.
**neurotrophy** (nū-rot'-ro-fe) [neuro-; τροφή, nourishment]. The nourishment of a nerve.
**neurotropic** (nū-ro-trop'-ik) [neuro-; τρόπος, a turn]. That which "turns towards" (*i. e.*, has a chemical affinity for) nervous tissue.
**neurotropism** (nū-rot'-ro-pism) [neuro-; τρόπος, a turn]. The attraction or repulsion exercised upon regenerating nerve-fibers. A substance is said to have *positive neurotropism* when these regenerating nerve-fibers have a tendency to grow toward and into it; *negative*, when they avoid it.
**neurotrosis, neurotrosmus** (nū-rot'-ro-sis, nū-ro-tros'-mus) [neuro-; τρῶσις, a wounding]. The wounding of a nerve.
**neurovaricosis** (nū-ro-var-ik-o'-sis) [neuro-; varix]. A varicosity on a nerve-fiber, or the formation of one.
**neurovascular** (nū-ro-vas'-kū-lar). Pertaining to both the nervous and vascular structures.
**neurypnology** (nū-rip-nol'-o-je). See *neurohypnology*.
**Neusser's granules** (noy'-ser) [Edmund von *Neusser*, Austrian physician, 1852– ]. Basophilic granules sometimes found in the leukocytes of the blood, near the nuclei. They are regarded by Neusser as being closely connected with the uric-acid diathesis, but their presence has been noticed also in other conditions.
**neutral** (nū'-tral) [*neuter*, neither]. Neither alkaline nor acid; bland and soothing; inactive. n. mixture, solution of potassium citrate. See *potassium citrate, solution of*.
**neutralization** (nū-tral-iz-a'-shun) [*neuter*, neither]. That process or operation that precisely counterbalances or cancels the action of an agent. In medicine, the process of checking the operation of any agent that produces a morbid effect. In chemistry, a change of reaction to that which is neither alkaline nor acid.
**neutralize** (nū-tral'-iz) [*neutral*]. To render neutral; to render inert; to counterbalance an action or influence.
**neutrolactis** (nū-tro-lak'-tis). A galactagogue said to be a liquid extract of *Galega officinalis*.
**neutrophil, neutrophile** (nū'-tro-fil) [*neuter*, neither; φίλος, loving]. 1. Stained readily by neutral dyes; applied to certain cells. 2. A leukocyte or histological element readily stainable with neutral dyes. n. leukocytes, leukocytes the protoplasm of which contains granules colored by neutral stains.
**nevoid** (ne'-void) [*nævus*, birth-mark; εἶδος, like].

Resembling a nevus. n. **elephantiasis**. See *lymph-scrotum*.
**nevolipoma** (ne-vo-lip-o'-mah). A rare form of lipoma containing a large number of blood-vessels, considered a degenerated nevus.
**nevose** (ne'-vōs) [*nævus*]. Spotted, having nævi.
**nevus, nævus** (ne'-vus) [L., *nævus*; pl., *nævi*]. 1. A circumscribed area of pigmentation; a mole. 2. An angioma of the skin, usually congenital. Syn., *mother's mark*. **nævus araneus**. See *acne rosacea*. n., **capillary**, one that involves the capillaries of the skin. n., **cutaneous**, a nevus of the skin. **nævus flammeus**, port-wine mark, a diffuse, very slightly raised, red or purplish variety of nævus maternus, involving part of the face. **nævus lipomatodes**, a large, soft mole containing a quantity of fat and loose connective tissue. **nævus maternus**. See *nevus* (2). **nævus pigmentosus**, a mole; a circumscribed, congenital pigmentary deposit in the skin, varying in color from a light fawn to a blackish tint, and often associated with hypertrophy of the hairs. **nævus vascularis**. See *nevus* (2). **nævus vascularis**, n. **vascularis tuberosus**, a cavernous angioma marked by formation of red or bluish erectile tumors.
**newgrowth** (nū'-groth). A circumscribed new formation of tissue, characterized by abnormality of structure or location. As generally used, the term includes all true tumors, as well as tumor-like growths due to microorganisms, as the gumma and tuberculous tumor. Syn., *neoplasm*.
**Newton's color-rings** (nū'-ton) [Sir Isaac *Newton*, English physicist, 1642–1726]. The colorings produced when a cover-glass is pressed upon a slide; they are the result of chromatic aberration.
**nexus** (neks'-us) [*nectere*, to bind]. A tying or binding together; an interlacing. n. **nervorum opticorum**, the chiasm. n. **staminus oculi**, the ciliary body.
**N. F.** Abbreviation of *National Formulary, q. v.*
**Ni.** The chemical symbol for *nickel*.
**nibble** (nib'-l). To gnaw; to eat in small bits.
**niccolic** (nik-ol'-ik) [*niccolum*, nickel]. Containing nickel.
**niccolum** (nik'-ol-um). Latin for *nickel*.
**nickel** (nik'-l) [G.]. Symbol Ni; atomic weight 58.68; quantivalence II, IV. A metal of silver-white luster, resembling iron in physical properties. See *elements, table of chemical*. n. **bromide**, NiBr₂+3H₂O, has been used in epilepsy. Dose 5–10 gr. (0.32–0.65 Gm.). n. **chloride**, NiCl₂, has been used as a tonic in anemia. Dose 2 gr. (0.13 Gm.). n. **sulphate**, NiSO₄+7H₂O, has been used as a tonic. Dose ¼–1 gr. (0.032–0.065 Gm.).
**nicking** (nik'-ing) [origin obscure]. The incising of a horse's tail near the root, to cause it to be carried higher.
**Nicklé's test for distinguishing glucose from cane-sugar** (ne-kla') [François Joseph Jérome *Nicklé*, French chemist, 1821–1869]. Heat the sugar for some time to 100° C. with carbon tetrachloride; cane-sugar is turned black by the process and glucose is not.
**nico** (nik'-o). Same as *symphorol, q. v.*
**Nicol's prism** (nik'-ol) [William *Nicol*, English physicist, 1768–1851]. A polished prism of Icelandspar, cut diagonally across the principal axis, the sections being joined together by means of Canada balsam. It has the property of reflecting the ordinary ray of light out of the field, while the so-called polarized ray is transmitted.
**Nicolaier's bacillus** (nik-o-li'-er) [Arthur *Nicolaier*, German physician, 1862– ]. The *Bacillus tetani*.
**nicotiana** (nik-o-she-a'-nah) [Jean *Nicot*, French diplomat, 1530–1600]. See *tobacco*.
**nicotianin** (nik-o-she-a'-nin). The volatile principle to which tobacco owes its flavor.
**nicotianomania** (nik-o-she-an-o-ma'-ne-ah) [*nicotine*; *mania*]. Insane craving for tobacco.
**nicotine** (nik'-o-tēn), C₁₀H₁₄N₂. A liquid poisonous alkaloid found in the leaves of the tobacco-plant. n. **bitartrate**, white soluble crystals used in tetanus and as an antidote in strychnine-poisoning. n. **salicylate**, hexagonal tablets containing 54% of nicotine; recommended in scabies in 1% lanolin ointment. n. **tartrate**, C₁₀H₁₄N₂S(C₄H₆O₆)+2H₂O, a solution more stable than that of the free alkaloid of the other salts.
**nicotinism** (nik'-o-tin-izm) [*nicotine*]. The morbid effects from the continued or excessive use of tobacco.

21

**nicoulin** (*nik'-oo-lin*). A drug which has been used in tetanus.
**nictation** (*nik-ta'-shun*). Same as *nictitation*.
**nictitating** (*nik'-tit-a-ting*) [see *nictitation*]. Winking. n. membrane. See *membrane, nictitating.* n. spasm, blepharospasm.
**nictitation** (*nik-tit-a'-shun*) [*nictitare*, to wink]. Abnormal frequency of winking.
**nidal** (*ni'-dal*). Pertaining to a nidus.
**nidation** (*ni-da'-shun*) [*nidus*, nest]. The development of an endometrial epithelium in an intermenstrual period.
**nidulus** (*nid'-u-lus*). The nucleus or origin of a nerve.
**nidus** (*ni'-dus*) [L., "nest"]. 1. A central point or focus of infection; a place in which an organism finds conditions suitable for growth and development. 2. A collection of ganglion-cells at the deep origin of a cranial nerve; a nucleus. n. avis, n. hirundinis, a deep fossa in the cerebellum situated between the posterior medullary velum in front and the nodule and uvula behind.
**Niemeyer's pill** (*ne'-mi-er*) [Felix von *Niemeyer*, German physician, 1820–1871]. 1. A pill of quinine, 1 gr., digitalis, ½ gr., and opium, ¼ gr. It is used in pulmonary tuberculosis, and is taken every six hours. 2. A pill of digitalis, squill and calomel or mass of mercury; used as a diuretic.
**night-blindness.** See *nyctalopia.*
**night-blooming cereus.** See *cactus grandiflorus.*
**night-cries.** A symptom of nervous or physical disorders of children, and especially of the early stage of hip-disease. The child cries out in its sleep from pain produced by reflex spasmodic twitching of the muscles already abnormally irritable.
**nightingale** (*ni'-tin-gāl*) [after Florence *Nightingale*, a nurse, 1820–1910]. A short cape used in hospitals to protect the shoulders and chest of nurses and patients.
**nightmare** [AS., *neaht*, night; *mara*, mare]. A dream characterized by great distress and a sense of oppression or suffocation.
**night-pain.** A symptom of hip-disease; pain in the hip or knee occurring during muscular relaxation of the limb in sleep.
**night palsy.** Numbness of the extremities occurring during the night, or on waking in the morning, affecting women about the period of the menopause.
**nightshade** (*nīt'-shād*). A name applied to plants of the genus *Solanum.* n., deadly, a poisonous plant, *Atropa belladonna.* See *belladonna.*
**night-soil.** The contents of privy-vaults (often removed in the night). This material is largely employed as manure.
**night-sweat.** The profuse nocturnal sweating often observed in pulmonary tuberculosis and other wasting disorders.
**night-terrors** (*nit'-ter-orz*). Distressing dreams occurring in children and causing them to wake up with cries of fear. Syn., *pavor nocturnus.*
**night-walking.** See *somnambulism.*
**nigranilin** (*ni-gran'-il-in*). Anilin-black.
**nigredo** (*ni-gre'-do*) [*niger*, black]. Same as melasma.
**nigrescent** (*ni-gres'-ent*) [*nigrescere*, to become black]. Turning black; blackish, dusky.
**nigrismus** (*ni-griz'-mus*) [L.]. Synonym of *nigredo.*
n. linguæ, black tongue.
**nigritia, nigrities** (*ni-grish'-e-ah, ni-grish'-e-ēz*). Same as *nigrismus*, and *glossophytia.*
**nigrosine** (*ni'-gro-sēn*) [*niger*, black]. A blue-black anilin dye, used in staining brain tissue.
**nihil album** (*ni'-hil*). Flowers of zinc; crude zinc oxide. n. græcum, zinc oxide.
**nihilism** (*ni'-hil-izm*) [*nihil*, nothing]. Pessimism in regard to the efficacy of drugs.
**niin** (*ni'-in*). A fatty substance allied to and probably identical with axin.
**Nikiforoff's method of fixation of blood-films** (*nik-e-for'-of*) [Mikhail Nikiforovich *Nikiforoff*, Russian physician, 1858– ]. This consists in immersion of the dried films in ether, in absolute alcohol, or in a mixture of equal parts of the two.
**Nikolsky's sign** (*nik-ol'-ske*) [Pyotr Vasilyevich *Nikolsky*, Russian dermatologist, 1855– ]. Excessive sensibility of the skin with l₀ss of the superficial layer, on receipt of a slight injury.
**ninhydrin** (*nin-hi'-drin*). Trade name of triketohydrindene-hydrate, C₆H₄(CO)₂C(OH)₂. It is soluble in water, and gives a color reaction with albumin, peptones, polypeptids and amino acids. It is also used in Abderhalden's test for pregnancy.
**ninth nerve.** The glossopharyngeal nerve. Formerly the hypoglossal nerve (the twelfth) was called the ninth nerve.
**niobium** (*ni-o'-be-um*) [Νιοβη, the daughter of Tantalus]. A rare metal, akin to bismuth and to antimony; symbol Nb; atomic weight 93.5. It is also known as *columbium.* Its medicinal properties are little known. See *Elements, Table of.*
**niphablepsia** (*nif-ah-blep'-se-ah*) [νίφα, snow; ἀβλεψία, blindness]. Snow-blindness.
**niphotyphlosis** (*nif-o-tif-lo'-sis*) [νίφα, snow; τύφλωσις, blindness]. Snow-blindness.
**niopo** [Venezuelan name]. A snuff prepared from the seeds of *Piptadenia peregrina*, which produces an intoxication approaching frenzy.
**nippers** (*nip'-erz*). An instrument for seizing small bodies. n., bone, an instrument for grasping small bits of bone.
**nipple** (*nip'-l*) [allied to *neb*, the beak of a bird, from AS., *nebb*, the face]. The conical projection in the center of the mamma, containing the outlets of the milk-ducts. n., cracked, a nipple the epidermis of which is broken in places. n., crater. See *n.*, *retracted.* n.-line, a vertical line drawn on the surface of the chest through the nipple. n.-protector, a device worn by nursing women to protect the nipple. It is called also a *nipple-shield.* n., retracted, a nipple drawn below the surrounding level. n. shield. See *n. protector.*
**nirls, nirles** (*nerls*) [origin obscure]. A variety of herpes.
**nirlus** (*nir'-lus*) [origin obscure]. An ephemeral papular eruption sometimes following measles or scarlatina.
**nirvanine** (*nir-van'-ēn*). The hydrochloride of diethylglycocoll-paraamido-o-oxybenzoic-methyl-ester. It is a local anesthetic, one-tenth as toxic as cocaine, used by Schleich's infiltration method in 0.2–0.5 % solution; in dentistry in 2 to 5 % solution.
**Nisbet's chancre** [William *Nisbet*, English physician, 1759–1822]. Nodular abscesses on the penis following acute lymphangitis from soft chancre. Syn., *Bubonuli nisbethii.*
**Nissl's bodies** (*nis'-l*) [Franz *Nissl*, German neurologist, 1860– ]. Chromophile corpuscles. The chromophilic bodies of a nerve-cell; finely granular bodies, of various sizes and shapes, brought out between the cytoreticulum by staining with Nissl's stain. Syn., *tigroid masses.* **N.'s degeneration,** the slow atrophic change which a neurone undergoes when it is prevented from functionating. **N.'s stain,** methylene-blue.
**nisus** (*ni'-sus*) [L., "effort," from *niti*, to endeavor]. 1. Any strong effort or struggle. 2. The periodic desire for procreation manifested in the spring season by certain species of animals. 3. The contraction of the diaphragm and abdominal muscles for the expulsion of the feces or the urine.
**nit** [AS., *hnitu*, a nit]. The egg or larva of a louse.
**niter** (*ni'-ter*) [Ar., *nitrûn*, natron]. Potassium nitrate or saltpeter. n., cubic, sodium nitrate. n., rough, magnesium chloride. n., sweet spirit of (*spiritus ætheris nitrosi*, U. S. P.), spirit of nitrous ether, an alcoholic solution of ethyl nitrite. Dose in fever 20–30 min. (1.3–2.0 Cc.); as a diuretic 30–60 min. (2–4 Cc.).
**Nithsdale neck** [*Nithsdale*, a valley in Dumfries, Scotland]. Goiter.
**niton** (*ni'-ton*). A name proposed by Ramsey for radium emanation considered as a new element; symbol Nt, atomic weight 222.4.
**nitragin** (*ni'-traj-in*). A nitrifying bacterial ferment obtained from the root-tubercles of leguminous plants.
**nitrate** (*ni'-trāt*). A salt of nitric acid.
**nitrated** (*ni'-tra-ted*). Combined with nitric acid.
**nitratine** (*ni'-tra-tēn*). Sodium nitrate.
**nitration** (*ni-tra'-shun*) [*nitric*]. The process of combining or treating with nitric acid.
**nitre** (*ni'-ter*). See *niter.*
**nitric** (*ni'-trik*) [*niter*]. Pertaining to or containing niter. n. acid. See *acid, nitric.* n.-acid test, a test for albumin, consisting in the addition of nitric acid to the suspected fluid—if albumin is present, a precipitate is formed. The test is usually applied by superimposing the suspected fluid upon the acid. Syn., *Heller's test.*

**nitrification** (ni-trif-ik-a'-shun) [niter; facere, to make]. The conversion of the nitrogen of ammonia and organic compounds into nitrous and nitric acids, a process constantly going on in nature under the influence of certain bacteria and other agencies.

**nitrifier** (ni'-trif-i-er). A nitrifying microorganism.

**nitrifying** (ni'-trif-i-ing). Applied to bacteria which oxidize ammonia to nitrous and nitric acids.

**nitril** (ni'-tril) [niter]. A compound of cyanogen with an alcohol radical in which the nitrogen is trivalent and the radical is united to the remaining carbon atom. The nitrils are readily converted into acids.

**nitrite** (ni'-trīt) [niter]. A salt of nitrous acid. See amyl nitrite, potassium nitrite, sodium nitrite. The nitrites produce dilatation of the blood-vessels, diminution of the blood-pressure, increased rapidity of the pulse, and depression of the motor centers in the spinal cord. They are used as antispasmodics in asthma and angina pectoris, in spasmodic dysmenorrhea, tetanus, epileptic and hysterical convulsions, and in cases of arteriosclerosis with high arterial tension. Full doses in man give rise to flushing of the face, throbbing, and headache. For test, see Griess.

**nitro-** (ni-tro-) [niter]. 1. A prefix denoting combination with the univalent radical $NO_2$. 2. A prefix denoting combination with nitrogen.

**nitro-anisol** (ni-tro-an'-is-ol), $C_7H_7NO_2$. A derivative of anisol.

**nitrobacter** (ni-tro-bak'-ter). The bacillus nitrobacter, a nitrifying bacterium.

**nitrobacteria** (ni-tro-bak-te'-re-ah) [nitro-; bacteria]. Bacteria that convert ammonia into nitric acid.

**nitrobenzol, nitrobenzene** (ni-tro-ben'-zol, ni-tro-ben'-zēn), $C_6H_5NO_2$. An oily, sweetish liquid made by the action of strong nitric acid on benzol. It is an intermediate product in the manufacture of anilin oil, and is employed as a flavoring agent under the name of artificial oil of bitter almonds or oil of mirbane. It is a powerful poison, resembling hydrocyanic acid in action. Persons engaged in its manufacture often suffer from headache and drowsiness.

**nitrobenzolism** (ni-tro-ben'-zol-izm). Poisoning by nitrobenzol through ingestion of some liquor containing it, through inhalation of its vapor, or through cutaneous absorption.

**nitrocellulose** (ni-tro-sel'-ū-lōs). See pyroxylin.

**nitroerythrol** (ni-tro-er'-ith-rol), $C_4H_6(NO_3)_4$. Butane tetranitrate, obtained by dissolving erythrol in nitric acid; large glistening plates melting at 61° C. It explodes on percussion; used in the same manner as nitroglycerin.

**nitroform** (ni'-tro-form), $CH(NO_2)_3$. An oily acid compound, chemically analogous to chloroform. It is usually obtained by treating biliary acids with nitric acid.

**nitrogen** (ni'-tro-jen) [nitro-; γεννᾶν, to produce]. Symbol N; atomic weight 14.01; quantivalence I, III, V. A nonmetallic element existing free in the atmosphere, of which it constitutes about 77 % by weight. It is a colorless, odorless gas, incapable of sustaining life. Chemically it is very inert, and combines directly with but few elements. It is an important constituent of all animal and vegetable tissues. **n.-equilibrium,** the state of an animal in which, during a definite period, the nitrogen of the excreta equals in amount the nitrogen of the food.

**nitrogenized** (ni-troj'-en-īzd). Containing nitrogen.

**nitrogenous** (ni-troj'-en-us) [nitrogen]. Containing nitrogen.

**nitrogenuric diabetes** (ni-tro-jen-ū'-rik). Same as azoturic diabetes. See under diabetes.

**nitroglucose** (ni-tro-gloo'-kōs). A substance obtained from glucose by action of nitric and sulphuric acids. It is used as an arterial stimulant. Dose of 5 % solution ½–1 min. (0.016–0.065 Cc.).

**nitroglycerin** (ni-tro-glis'-er-in) [nitro-; glycerin], $C_3H_5(NO_3)_3$. Glonoin, glyceryl trinitrate, a colorless, oily liquid produced by the action of sulphuric and nitric acids upon glycerol. It is a powerful explosive; physiologically it has the actions of the nitrites, but is more persistent than amyl nitrite, which it most resembles. Dose ₁⁄₁₅₀–¹⁄₅₀ gr. (0.0003–0.0013 Gm.). **n., spirit of** (spiritus glyceryllis nitratis, U. S. P.), spirit of glyceryl trinitrate, a 1 % alcoholic solution. Dose 1–2 min. (0.065–0.13 Cc.). **n., tablets of** (tabellæ nitroglycerini, B. P.), contain each ₁⁄₁₀₀ gr. (0.0006 Gm.) of nitroglycerin.

**nitrohydrochloric acid** (ni-tro-hi-dro-klo'-rik). See acid, nitrohydrochloric.

**nitrolevulose** (ni-tro-lev'-ū-lōs). Dextrose nitrate. It has properties similar to nitroglycerin.

**nitrolin** (ni'-tro-lin) [niter; oleum, oil]. An explosive compound consisting of a mixture of cellulose, niter, and nitrosaccharose.

**nitrometer** (ni-trom'-et-er) [nitrogen; μέτρον, measure]. An apparatus for collecting and measuring nitrogen gas, or for decomposing nitrogen oxides and estimating the resulting gases.

**nitromonas** (ni-tro-mo'-nas) [nitro-; μονάς, unit]. A group of bacteria occurring in the soil, which convert ammonium salts into nitrites and nitrites into nitrates. They will not grow in gelatin or other organic media.

**nitromuriatic acid** (ni-tro-mū-re-at'-ik). See acid, nitrohydrochloric.

**nitropropiol** (ni-tro-pro'-pe-ol). A preparation of orthonitrophenyl-propiolic acid and sodium carbonate; used for detecting sugar in the urine. **n. test for sugar in the urine,** place 10 to 15 drops of urine in test-tube, add 10 Cc. distilled water and a nitropropiol tablet, and heat two to four minutes. In the presence of sugar there is first a green coloration, followed by an intensely blue color. This will indicate 0.3 % of sugar, and only takes place if grape-sugar is actually present. If much albumin is present, first eliminate it by shaking with salt or chloroform.

**nitrosaccharose** (ni-tro-sak'-ar-ōs) [nitro-; σάκχαρον, sugar]. An unstable, resinous, explosive compound produced by treating saccharose with nitric acid.

**nitrosalol** (ni-tro-sa'-lol), $C_6H_4(OH)CO_2 . C_6H_4NO_2$. A yellowish powder melting at 148° C., soluble in alcohol or ether, insoluble in water; it is used in making salophen.

**nitroso-** (ni-tro-so-) [niter]. A prefix signifying combination with nitrosyl, the univalent radical NO.

**nitrosobacter** (ni-tro-so-bak'-ter). A rod-like form of nitrifying bacteria.

**nitrosobacteria** (ni-tro-so-bak-te'-re-ah). See nitrobacteria.

**nitrosococcus** (ni-tro-so-kok'-us). A coccus form of nitrifying bacteria. Cf. nitromonas.

**nitrosomonas** (ni-tro-so-mo'-nas). See nitromonas.

**nitrosonitric acid** (ni-tro-so-ni'-trik). Fuming nitric acid containing nitrous acid gas.

**nitrosophenyldimethylpyrazol** (ni-tro-so-fen-il-di-meth-il-pir'-az-ol), $C_{11}H_{11}(NO)N_2O$. A reaction-product of a solution of sodium nitrite with a solution of antipyrin in acidulated water; it is antipyretic, analgesic, and diuretic. Syn., isonitrosoantipyrin.

**nitro-sugars** (ni-tro-shug'-erz). A class of substances, such as nitroglucose, used as vasodilators.

**nitrosyl** (ni-tro'-sil). The univalent radical NO. **n. sulphate,** $NOHSO_4$, a nitrosyl substitution derivative of sulphuric acid; recommended as a disinfectant.

**nitrous** (ni'-trus) [niter]. 1. Containing nitrogen as a univalent or trivalent element. 2. Pertaining to or derived from nitrous acid. **n. acid,** $HNO_2$, an acid having one atom of oxygen less than nitric acid. See Griess. **n. ether,** $C_2H_5NO_2$, ethyl nitrite, a very volatile liquid having properties similar to those of amyl nitrite. **n. oxide,** $N_2O$, used as a general anesthetic in dentistry and in minor surgery. Syn., hyponitrous oxide; laughing-gas; nitrogen protoxide.

**nitroxyl** (ni-troks'-il), $NO_2$. A univalent radical found in nitric acid.

**niveau diagnosis** (ne'-vo) [Fr. niveau, level]. Localization of the level of a (spinal) tumor or other lesion.

**nizin** (ni'-zin). Trade name applied to a zinc salt of sulphanilic acid.

**N.N.R.** An abbreviation for New and Nonofficial Remedies, i. e., those that have been approved by the Council of Pharmacy and Chemistry of the American Medical Association.

**No.** An abbreviation of the Latin numero, "to the number of."

**noasthenia** (no-as-the'-ne-ah) [νοῦς, mind; ἀσθένεια, weakness]. Mental feebleness.

**Nobel's (Le) test for acetone.** A modification of Legal's test. **Le N.'s test for bile-pigments,** add to the liquid zinc chloride and a few drops of tincture of iodine. A dichroic play of colors is the result.

**noble cells.** The cells of muscles, nerves and organs, in contradistinction to epithelial and connective tissue cells. 1. The same as noble cells. 2. The same as noble metals. **n. elements.** **n. metals,** metals which do not oxidize in exposure to air;

they are gold, silver, platinum, mercury, palladium, rhodium, ruthenium, osmium and iridium.

**nocarodes** (no-kar-o'-dēz) [νῶκαρ, lethargy; εἶδος, like]. Lethargic.

**Nocht-Romanowsky stain** (nokt'-ro-man-off'-ske). This requires two solutions: I. Methylene blue, 1.0 gram; sodium carbonate, 0.5 gram; distilled water, 100.0 grams. Heat at 60° C. for two days until solution shows a slight purplish color. II. Eosin, soluble, yellowish, 1.0 gram; distilled water, 100.0 c.c. Mix a few drops of each of these solutions with about 10 c.c. of distilled water in an Esmarch dish; the smear, which has previously been fixed in absolute methyl alcohol, is then floated on this mixture for about ten minutes.

**nociassociation** (no-se-as-o-se-a'-shun) [nocere, to injure]. The release of nervous activity as manifested by shock or exhaustion, the result of trauma or surgical operation. See anociassociation.

**nociceptive** (no-se-sep'-tiv) [see nociceptor]. Capable of receiving or transmitting painful or traumatic stimuli.

**nociceptor** (no-se-sep'-tor) [nocere, to injure; capere, to take]. A peripheral nerve organ or mechanism by which stimuli of pain or trauma are received and conveyed to the cerebrum.

**noctambulation** (nok-tam-bū-la'-shun) [nox, night; ambulare, to walk]. Sleep-walking.

**noctiphobia** (nok-te-fo'-be-ah) [nox, night; φόβος, fear]. Morbid fear of night and its darkness and silence; at times a distressing accompaniment of neurasthenia.

**nocturnal** (nok-tur'-nal) [nocturnus, pertaining to the night]. Pertaining to the night. **n. emission, n. pollution,** the discharge of semen without coitus during sleep. **n. enuresis,** incontinence of urine at night during sleep. **n. epilepsy,** epilepsy in which the convulsions occur at night.

**nocuity** (nok-ū'-it-e) [nocuus, injurious]. Injuriousness; harmfulness; the quality of being noxious.

**nocuous** (nok'-ū-us) [nocuus, injurious]. Noxious; hurtful; venomous. **n. meat,** meat from animals affected with disease that may be transmitted to man.

**nod.** 1. To drop the head forward with a quick, involuntary motion. 2. The motion so made.

**nodal** (no'-dal) [nodus, a node]. Pertaining to a node. **n. point,** the point of intersection of convergent rays of light with the visual axis of the eye. The first nodal point is 6.9685 mm. behind the summit of the cornea. The second nodal point is 7.3254 mm. behind the summit of the cornea, or 0.1254 mm. behind the lens.

**nodding spasm** (nod'-ing spazm). A nodding of the head from spasm of the sternomastoid muscle.

**node** (nōd) [nodus, a node]. 1. A knob, swelling, or protuberance. 2. A point of narrowing or constriction. **n., atrioventricular, n., auriculoventricular,** a node in the right auricle which forms the starting point of the bundle of His. **n. Haygarth's.** See under Haygarth. **n., Heberden's.** See under Heberden. **n., Parrot's.** See under Parrot. **n., Ranvier's.** See under Ranvier. **n., Schmidt's.** See under Schmidt. **n. sinoatrial, n., sinoauricular,** a node at the entrance of the superior vena cava into the right auricle. **n., syphilitic,** the localized swelling on bones due to syphilitic periostitis.

**nodose** (no'-dōs). Characterized by nodes; jointed or swollen at intervals.

**nodosity** (no-dos'-it-e) [node]. 1. The state of having nodes. 2. A node. **n., Bouchard's.** See under Bouchard. **n., Féréol's.** See under Féréol. **n., Haygarth's, n., Heberden's.** See under Heberden. **n., Meynert's.** See under Meynert. **n.s, piedric,** those characteristic of piedra, a disease of the hair due to a parasitic fungus.

**nodular** (nod'-ū-lar) [nodule]. Composed of or covered by nodules; resembling a nodule.

**nodule** (nod'-ūl) [nodulus, dim. of nodus]. A small node. **n.s of Arantius.** See corpora Arantii. **n. of cerebellum,** the anterior termination of the inferior vermiform process of the cerebellum. **n.s, endolymphangeal,** small knobs formed within lymphatic vessels by localized masses of adenoid tissue. **n., lymph-,** a more correct term for lymph-follicle. **n.s lymphangeal, n.s, lymphatic, n.s, lymphoid,** localized masses of adenoid tissue consisting of branched nucleated corpuscles holding lymphoid cells in the spaces between them. **n.s of Morgagni.** Same as n.s of Arantius.

**noduli** (nod'-ū-lī). Plural of nodulus.

**nodulus** (nod'-ū-lus) [L.: pl., noduli]. 1. See nodule. 2. The nodule of the cerebellum. **n. hystericus.** Same as globus hystericus. **noduli albini,** a term given to certain small pathological knots occasionally found on the free border of the auriculoventricular valves.

**nodus** (no'-dus). See node.

**noematachograph** (no-e-mat-ak'-o-graf) [νόημα, thought; ταχύs, swift; γράφειν, to write]. An instrument for recording the time of mental operations.

**noematochometer** (no-em-at-ok-om'-et-ur) [νόημα, a thought; ταχύs, swift; μέτρον, measure]. An apparatus for estimating the time taken in recording a simple perception.

**Noguchi's luetin reaction** (no-goo'-tshe) [Hideyo Noguchi, Japanese bacteriologist]. The intracutaneous injection of a drop of luetin is followed in 24 to 48 hours by an indurated papule with a red center and a purple border, if syphilis is present. N.'s modification of Wassermann's syphilis test, the use of anti-human, instead of anti-sheep, hemolytic; also of amboceptor, complement, and antigen test-papers.

**noisome** (noi'-sum). Hurtful; noxious.

**noli-me-tangere** (no-li-me-tan'-jer-e) {L. "touch me not"}. See ulcer, rodent.

**noma** (no'-mah) [νομή, a corroding sore]. A grave usually fatal, form of stomatitis, occurring in debilitated children, generally during the convalescence from one of the exanthemata. It is characterized by the formation of a rapidly spreading ulcer involving the cheek and soon becoming gangrenous. It is a parainfectious disease due to Bacillus diphtheriticus. Syn., cancrum oris; gangrana oris; gangrenous stomatitis. **n. pudendi, n. vulvæ,** a similar ulceration occurring about the genital region of female children.

**nomadic** (no-mad'-ik) [νομάs, roving]. Spreading; said of ulcers.

**nomenclature** (no'-men-kla-tūr) [nomen, a name; calare, to call]. A systematic application and arrangement of the distinctive names employed in any science.

**non-** [non, not]. A prefix denoting negation.

**non-access** (non-ak'-ses) [non; accessus, an approach]. In medical jurisprudence, the failure to cohabit. The reverse of access, q. v. A child born under such circumstances is a bastard.

**nonadherent** (non-ad-he'-rent) [non, not; adhærere, to adhere]. Not connected to an adjacent organ or part.

**nonalbuminoid** (non-al-bū'-min-oid). A nitrogenous animal or vegetable compound of simpler composition than a proteid; nonproteid, e. g., the nitrogenous extractive of muscular and connective tissue.

**nonan** (no'-nan) [nonus, ninth]. Having an exacerbation every ninth day.

**non compos mentis** (non kom'-pos men'-tis) [L.]. Of unsound mind.

**nonconductor** (non-kon-duk'-tor). Any substance not transmitting electricity or heat.

**nonè-gravid** (non-e-grav'-id-ah) [nonus, ninth; gravida, a pregnant woman]. A woman pregnant for the ninth time.

**nonipara** (non-ip'-ar-ah) [nonus, ninth; parere, to bring forth]. A woman who has been in labor nine times.

**nonmetal** (non-met'-al). An element that is not a metal.

**non-motile** (non-mo'-til) [non; motilis, moving]. Not having the power of spontaneous motion.

**non-naturals, the six.** In the old hygiene, this term designated air, food, exercise, sleep, secretion (and excretion), and mental activity.

**nonose** (no'-nōs) [nonus, nine]. One of of a group of the glucoses, with the formula $C_9H_{18}O_9$.

**nonparous** (non-par'-us). Same as nulliparous.

**nonproteid** (non-pro'-te-id). See nonalbuminoid.

**non-restraint** (non-re-strānt') [non; Fr., restraindre, to restrain]. The treatment of insanity without any forcible means of compulsion.

**non-sexual** (non-seks'-ū-al) [non; sexus, sex]. Same as asexual.

**nonus** (no'-nus) [L., "ninth"]. The hypoglossal nerve, which was the ninth under the old classification of the cranial nerves.

**nonvalent** (non-va'-lent). Without chemical valency; incapable of entering into chemic composition.

**nonviable** (non-vi'-ab-l). Incapable of living.

**(von) Noorden treatment.** See oat treatment.

**Nordauism** (nor'-dow-ism) [Max Simon Nordau, German scientist, 1849– ]. Degeneracy.
**Nordhausen sulphuric acid** (nord'-how-zen) [Nordhausen, a town in Saxony where it was first prepared]. Fuming sulphuric acid; sulphuric acid containing more or less sulphur trioxide.
**nori** (no'-re). A Japanese gelatin obtained from *Porphyra vulgaris*, employed in cultivating protozoa.
**norm** [norma, a rule or measure]. A standard.
**norma** (nor'-mah). In anatomy, a view or aspect, essentially of the skull. **n. basilaris**, the view of the skull looking toward the inferior aspect. **n. facialis**, the aspect looking toward the face. **n. lateralis**, a profile view. **n. occipitalis** the aspect looking toward the back of the skull. **n. sagittalis**, the view of the skull seen in a mesial sagittal section. **n. verticalis**, the aspect viewed from above, or that directed toward the top of the skull.
**normal** (nor'-mal) [norma]. 1. Conforming to natural order or law. 2. Having the typical structure. **n. antitoxic serum, n. therapeutic serum,** an antitoxic blood-serum of which 0.1 Gm. is sufficient to neutralize ten times the fatal dose of toxin for a guinea-pig weighing 300 Gm. **n. salt solution, n. saline solution,** an aqueous solution of sodium chloride of a strength similar to that of the body-fluids—usually 0.6 to 0.75 %. This is the commonly accepted use of the term, but it is incorrect and should be replaced by the term *physiological solution*. **n. solution**, a solution containing in one liter a quantity of the reagent equal to the molecular weight in grams. A *decinormal* solution is one of one-tenth the strength, and a *centinormal* solution one of one-hundredth the strength, of the normal solution.
**normoblast** (nor'-mo-blast) [norma; βλαστός, a germ]. A nucleated red corpuscle of the same size as an ordinary red corpuscle.
**normocyte** (nor'-mo-sit) [norma; κύτος, a cell]. A red blood-corpuscle of normal size (7.5 μ).
**normocytosis** (nor-mo-si-to'-sis) [see *normocyte*]. A normal state of the corpuscles of the blood.
**normotonic** (nor-mo-ton'-ik) [norma; τόνος, a stretching]. Relating to normal muscular contraction; to a muscle working under normal physiological conditions.
**Norris's colorless corpuscles.** Colorless, transparent biconcave discs of the same size as the red corpuscles, invisible in the serum because their color and refractive index are the same as those of the liquor sanguinis.
**Norwegian itch.** A variety of aggravated scabies seen mainly in lepers.
**Norwood's tincture** (nor'-wood) [Wesley C. Norwood, American physician]. A tincture of veratrum viride, said to be prepared from the fresh root and to contain 240 grains of veratrum viride in each ounce of alcohol.
**nose** (nōz) [AS., nosu]. The prominent organ occupying the center of the face, the upper part (*regio olfactoria*) of which constitutes the organ of smell, the lower part (*regio respiratoria*) represents the commencement of the respiratory tract, in which the inspired air is warmed, moistened, and deprived of impurities. The nose consists of two symmetrical cavities, separated by a *septum*, and is lined internally by mucous membrane (*Schneiderian membrane*). **n., bottle,** an hypertrophied condition of the nose with a varicose condition of its veins, usually associated with alcoholism. **n. bridge of,** the prominence formed by the junction of the nasal bones. **n., saddle, n., saddleback, n., swayback,** one with a depression in the bridge due to the loss of the septum.
**nosebleed** (nōz'-blēd). A hemorrhage from the nose. Syn., *epistaxis*.
**nosegay, Riolan's.** The entire group of muscles arising from the styloid process of the temporal bone.
**nosema** (nos-e'-mah) [νόσος, disease]. 1. Illness; disease. 2. A genus of *microsporidia*.
**nosencephalus** (nos-en-sef'-al-us). Same as *notencephalus*.
**nosepiece.** A mechanical device to be attached to a microscope for holding two, three, or four objectives. It is screwed into the object-end of the tube of the microscope.
**noseresthesia** (nos-er-es-the'-ze-ah) [nosema; αἴσθησις, perception]. Perverted sensibility.
**noserous** (nos'-e-rus) [nosema]. Diseased; unhealthy.
**noso-** (nos-o-) [νόσος, disease]. A prefix signifying disease.

**nosochorologia** (nos-o-kor-o-lo'-je-ah). See *nosochthonography*.
**nosochthonography** (nos-ok-thon-og'-raf-e) [noso-; χθών, the earth; γράφειν, to write]. Geography of endemic diseases; medical geography.
**nosocomial** (nos-o-ko'-me-al) [νόσος, disease; κομεῖν, to take care of]. 1. Pertaining to a hospital, or a nosocomium. 2. Applied to disease caused or aggravated by hospital life. **n. gangrene.** Synonym of hospital gangrene.
**nosocomium** (nos-o-ko'-me-um) [νόσος, disease; κομεῖν, to take care of]. A place designed for the care of the sick. A hospital.
**nosode** (nos'-ōd) [νοσώδης, like a disease]. A homeopathic or isopathic remedy.
**nosodochium** (nos-o-do'-ke-um). Synonym of *nosocomium*.
**nosogenesis** (nos-o-jen'-es-is). Synonym of *nosogeny*.
**nosogenetic** (nos-o-jen-et'-ik) [νόσος, disease; γένεσις, genesis]. Pertaining to nosogenesis.
**nosogeny** (nos-oj'-en-e) [noso-; γεννᾶν, to beget]. The development of diseases.
**nosogeography** (nos-o-je-og'-raf-e). See *nosochthonography*.
**nosographer** (nos-og'-raf-er) [see *nosography*]. One who writes descriptions of diseases.
**nosographic** (nos-o-graf'-ik) [νόσος, disease; γράφειν, to write]. Pertaining to nosography.
**nosography** (nos-og'-raf-e) [noso-; γράφειν, to write]. A treatise on diseases.
**nosohemia, nosohæmia** (nos-o-he'-me-ah) [noso-; αἷμα, blood]. Disease of the blood.
**nosointoxication** (nos-o-in-toks-ik-a'-shun) [noso-; *intoxication*]. Autointoxication caused by pathological processes which alter the normal course of metabolism in such a way as to produce harmful products.
**nosological** (nos-o-loj'-ik-al) [νόσος, disease; λόγος, science]. Pertaining to nosology.
**nosology** (nos-ol'-o-je) [noso-; λόγος, science]. The science of the classification of diseases.
**nosomania** (nos-o-ma'-ne-ah) [noso-; μανία, madness]. 1. A morbid dread of disease. 2. A delusion that one is suffering from disease.
**nosomycosis** (nos-o-mi-ko'-sis) [νόσος, disease; μύκης, fungus]. Any disease due to the presence of a parasitic fungus, or schizomycete.
**nosonomy** (nos-on'-o-me) [noso-; ὄνομα, name]. The nomenclature of diseases.
**nosoparasites** (nos-o-par'-as-īts) [noso-; *parasite*]. Microorganisms found in conjunction with a disease process, which, while capable of modifying the course of the disease, are not its cause.
**nosophen** (nos'-o-fen). $C_6H_4C_2O_2(C_4H_2I_4OH)_2$. Tetraiodophenolphthalein, a yellowish-gray powder without odor or taste, insoluble in water or acids, slightly soluble in alcohol, more soluble in ether, chloroform, or alkalies. It is used externally as a substitute for iodoform, internally for catarrh of the stomach and intestine. Dose 5–8 gr. (0.3–0.5 Gm.). Syn., *iodophen*.
**nosophobia** (nos-o-fo'-be-ah) [νόσος, disease; φόβος, fear]. The insane, or exaggerated, fear of disease; pathophobia.
**nosophthoria** (nos-off-thor'-e-ah) [νόσος, disease; φθορά, destruction]. The eradication of diseases by prophylactic measures.
**nosophyte** (nos'-o-fīt) [noso-; φυτόν, a plant]. Any pathogenic vegetable microorganism.
**nosopoietic** (nos-o-poi-et'-ik) [noso-; ποιεῖν, to make]. Causing disease.
**nosotaxy** (nos-o-taks'-e) [νόσος, disease; τάξις, arrangement]. The classification of diseases.
**nosotoxic** (nos-o-toks'-ik). Relating to nosotoxin.
**nosotoxicity** (nos-o-toks-is'-it-e). The quality of being nosotoxic.
**nosotoxicosis** (nos-o-toks-ik-o'-sis) [see *nosotoxin*]. An abnormal condition referable to the presence of toxic basic products formed in the system in disease.
**nosotoxin** (nos-o-toks'-in) [noso-; τοξικόν, a poison]. A toxin generated in the body by a pathogenic microorganism.
**nosotrophy** (nos-ot'-ro-fe) [noso-; τροφή, nourishment]. 1. The nourishment of disease. 2. The care of the sick.
**nostalgia** (nos-tal'-je-ah) [νόστος, a return; ἄλγος, pain]. Homesickness.
**nostalgic** (nos-tal'-jik) [νόστος, return; ἄλγος, pain]. Affected with nostalgia.
**Nostoc** (nos'-tok). A genus of algæ having a gelatinous nature.

**nostology** (*nos-tol'-o-je*) [νόστος, return; λόγος, science]. In biology, the department devoted to the study of senility.

**nostomania** (*nos-to-ma'-ne-ah*) [νόστος, return; μανία, madness]. Nostalgia amounting to monomania.

**nostosite** (*nos'-to-sīt*) [νόστος, a return; σῖτος, food]. A parasite situated in or upon its permanent host.

**nostras** (*nos'-tras*) [*nostras*, of our country]. Denoting a disease belonging to the country in which it is described in contradistinction to a similar disease originating elsewhere; as *cholera nostras*, as distinguished from *Asiatic cholera*.

**nostrate** (*nos'-trāt*) [*noster*, ours]. Endemic.

**nostril** (*nos'-tril*) [AS., *nosu*, nose; *thyrl*, a hole]. One of the external orifices of the nose.

**nostrum** (*nos'-trum*) [*noster*, ours]. A quack medicine; a secret medicine.

**notal** (*no'-tal*) [νῶτον, the back]. Pertaining to the back; dorsal.

**notalgia** (*no-tal'-je-ah*) [νῶτον, back; ἄλγος, pain]. Any pain in the back.

**notanencephalia** (*no-tan-en-sef-a'-le-ah*) [νῶτον, the back; ἀνεγκέφαλος, without brain]. Congenital absence of the dorsal part of the cranium.

**notch** [O. Du., *nock*]. A deep indentation. n., **acetabular**, the cotyloid notch. n., **clavicular**, a depression at the upper end of the sternum articulating with the clavicle. n., **coracoid**. See *n., suprascapular*. n., **cotyloid**, the notch in the acetabulum near to the obturator foramen. n., **iliac, greater**, n., **ischiadic, greater**. See *n., sacrosciatic*. n., **interlobar** (of the liver), the notch in the ventral border of the liver demarcating the right and left lobes. n., **intervertebral**, one of the depressions on the vertebral pedicles, either on the upper on the lower surface. The apposition of two notches of the contiguous vertebræ forms the intervertebral foramen. n., **ischiatic**. See *n., sacrosciatic*. n., **jugular**, a notch forming the posterior boundary of the jugular foramen. n., **nasal**, an uneven interval between the internal angular processes of the frontal bone, which articulates with the nasal bone and the nasal process of the superior maxillary bone. n., **popliteal**, the depression on the posterior surface of the head of the tibia, separating the two tuberosities. n., **preoccipital**, an indentation on the inferolateral border of the cerebral hemisphere, about an inch and a half in front of the occipital pole. n. of **Rivinus**. See *Rivinian notch*. n., **sacrococcygeal**, the lateral notch at the point of union of the coccyx and sacrum. n., **sacrosciatic**, one of two notches on the posterior edge of the innominate bone. The **greater notch** is just above the spine of the ischium, and is converted into a foramen by the lesser sacrosciatic ligament; the **lesser notch** is below the spine of the ischium, and is converted into a foramen by the sacrosciatic ligaments. n., **scapular**, one at the back of the neck of the scapular through which the supraspinous and infraspinous fossæ communicate. n., **semilunar**. See *n., suprascapular*. n., **sigmoid**, a deep, semilunar depression separating the coronoid and condyloid processes of the inferior maxillary bone. n., **sphenopalatine**, the notch that separates the orbital and sphenoidal processes of the palate bone. n., **suprascapular**, a notch in the superior border of the scapula at the base of the coracoid process, for the passage of the suprascapular nerve. n., **suprasternal**, the depression at the top of the manubrium, between the two sternoclavicular articulations. n., **tympanic**. See *Rivinian notch*.

**note** (*nōt*) [*nota*, a mark]. A sound. n., **percussion**, the sound elicited on percussion.

**note-blindness** (*nōt'-blīnd-nes*). The same as *amusia, q. v.*

**notencephalia** (*no-ten-sef-a'-le-ah*). See *notencephalus*.

**notencephalocele** (*no-ten-sef-al'-o-sēl*) [νῶτον, back; ἐγκέφαλος, brain; κήλη, hernia]. Tumor of the brain in a notencephalus.

**notencephalus** (*no-ten-sef'-al-us*) [νῶτον, the back; *encephalon*]. A variety of monster in which the cranial contents are in large part outside the skull, resting upon the back of the neck.

**Nothnagel's symptom** (*nōt'-nah-gel*) [Carl Wilhelm Hermann *Nothnagel*, German physician, 1841-1905]. Paralysis of the facial muscles, which is less marked on voluntary movements than on movements connected with emotions. This symptom has been noticed in cases of tumor of the optic thalamus.

**N.'s test**, a crystal of sodium chloride placed upon the serous surface of any portion of the intestine of the rabbit causes ascending peristalsis. This test has been applied to ascertain the direction of the bowel in operations upon man, but has not been found wholly reliable. **N.'s type of facial paralysis**. See *N.'s symptom*.

**nothrous** (*no'-thrus*) [νωθρός, sluggish]. Drowsy; slow; languid; torpid.

**notifiable** (*no-tif-i'-ah-bl*) [*notificare*, to make known]. Applied to a disease which should be made known to a board of health or other authorities.

**notochord** (*no'-to-kord*) [νῶτον, the back; χορδή, a cord]. An elongated cord of cells inclosed in a structureless sheath, which in the embryo represents the vertebral column; the chorda dorsalis, or primitive backbone.

**notomelus** (*no-tom'-el-us*) [νῶτον, the back; μέλος, a limb]. A form of double monster in which the rudimentary limbs are attached to the back.

**notomyelitis** (*no-to-mi-el-i'-tis*) [νῶτον, the back; *myelitis*]. Inflammation of the spinal cord.

**novargan** (*no-var'-gan*). Trade name of a protein preparation of silver; similar to protargol.

**novaspirin** (*no-vas'-pir-in*). Trade name of anhydromethylene citric acid disalicylate; employed as a substitute for aspirin.

**novatophan** (*no-vat'-o-fan*). Trade name of a preparation of atophan (phenylquinolincarboxylic acid), said to be tasteless.

**novocaine** (*no-vo-ka'-in*). A synthetic local anesthetic compound, para-amidobenzoyl and the active principle of suprarenal extract.

**novocolchinin** (*no-vo-kol'-tshin-in*). A mixture of quinine sulphate and novocol (sodium guaiacol phosphate).

**noxa** (*noks'-ah*) [L.]. An injurious principle; especially a pathogenic microorganism or other *materies morbi*.

**noxious** (*nok'-shus*) [*noxius*, harmful]. Harmful; poisonous or deleterious.

**N-rays**. See *rays, N-*.

**Nt**. Chemical symbol for *niton*.

**nubecula** (*nū-bek'-ū-lah*) [dim. of *nubes*, a cloud]. 1. The cloudiness caused by the suspension of insoluble matter in the urine. 2. A cloudiness of the cornea.

**nubile** (*nū'-bil*) [*nubilis*; *nubere*, to marry]. Marriageable. Of an age at which there exists the possibility of procreation or child-bearing.

**nubility** (*nū-bil'-it-e*) [*nubere*, to marry]. The state of sexual development when marriage may be consummated.

**nuces** (*nū'-sēz*). Plural of *nux, q. v.*

**nucha** (*nū'-kah*) [L.]. The nape of the neck.

**nuchal** (*nū'-kal*) [*nucha*]. Pertaining to the nape of the neck.

**nucin** (*nū'-sin*). A precipitate from juglans, *q. v.*

**nucis** (*nū'-sis*). The genitive of the Latin *nux*, a nut.

**nucite** (*nū'-sīt*) [*nucis*], $C_6H_{12}O_6+4H_2O$. A carbohydrate resembling inosite, found in the leaves of *Juglans regia*.

**Nuck's canal** (*nook*) [Anton *Nuck*, Dutch anatomist, 1650-1692]. The canal formed by Nuck's diverticulum. **N.'s diverticulum**, the peritoneal covering of the round ligament of the uterus, which in the child can be traced for a short distance into the inguinal canal. **N.'s gland**. See *Blandin's gland*.

**nuclear** (*nū'-kle-ar*) [*nucleus*]. Pertaining to or resembling a nucleus. n. **cap**, a stainable mass in the form of a cone, hollowed out internally like a cap, corresponding to one pole of the nucleus upon which it sits. n. **cell**, a nucleated dendritic nerve cell. n. **figures**, the peculiar arrangement of the mitome during karyokinesis. n. **layer**, a stratum of gray matter in the cortex of the brain. n. **paralysis**, paralysis from lesions of the nuclei of origin of the nerves. n. **plate**. 1. The equatorial plate, formed by the chromosomes during the prophases of mitosis. 2. The septum which sometimes divides the nucleus in amitotic division. n. **spindle**, delicate striæ appearing in the nucleus during mitosis, arranged with the apices pointing toward the poles of the future nuclei. Its function probably is to guide the movements of the mitome threads. n. **stain**, a pigment showing a strong affinity for nuclei.

**nuclease** (*nū'-kle-ās*). 1. An immunizing enzyme found in cultures of *Bacillus pyocyaneus* and other

organisms. 2. A proposed general term to designate any of the bacteriolytic enzymes, because they digest the nucleoproteins of the bacterial cells.
**nucleated** (nū'-kle-a-ted) [*nucleus*]. Possessing a nucleus.
**nuclei** (nū'-kle-i) [L.]. Plural of *nucleus*, q. v. n. of the thalamus, the three portions into which the cinerea of the thalamus is separated. They are called the anterior, internal, and external nuclei.
**nucleide** (nū'-kle-īd). A compound of nucleol with an oxide of some metal (iron, copper, silver, mercury, etc.).
**nucleiform** (nū-kle'-if-orm) [*nucleus; forma*, form]. Resembling a nucleus.
**nuclein** (nū'-kle-in) [*nucleus*], $C_{29}H_{49}N_9P_3O_{22}$. An amorphous substance resembling the proteins, and forming the essential chemical constituent of all living cells. It is composed of nucleic acid and a base; the former seems to be the same for all nucleins, but the base varies. Nucleins are generally insoluble in dilute acids and soluble in dilute alkalies. They are supposed to represent the germicidal constituent of blood-serum. **n.-therapy**, the employment of nuclein from different glands and blood-serum in the treatment of disease.
**nucleinate** (nū'-kle-in-āt). A white, soluble powder used as a diagnostic aid in tuberculosis. Dose 30–45 gr. (2–3 Gm.).
**nucleo-** (nū-kle-o-) [*nucleus*]. A prefix meaning relating to a nucleus or to nuclein.
**nucleoalbumin** (nū-kle-o-al-bū'-min) [*nucleo-; albumin*]. A compound of a proteid and nuclein, occurring in cell-protoplasm.
**nucleoalbuminuria** (nū-kle-o-al-bū-min-ū'-re-ah). The presence in the urine of nucleoalbumin.
**nucleoalbumose** (nū-kle-o-al'-bū-mōs). A substance found in the urine in cases of osteomalacia and which is believed to be the partly hydrated albumin of nucleoalbumin.
**nucleochylema** (nū-kle-o-ki-le'-mah) [*nucleus*; χυλός, juice]. The fluid filling the interstices of the nucleohyaloplasm.
**nucleochyme** (nū'-kle-o-kīm). See *nucleochylema*.
**nucleofugal** (nū-kle-of'-ū-gal). [*nucleo-; fugere*, to flee]. Moving from a nucleus.
**nucleogen** (nū'-kle-o-jen). A proprietary preparation of nucleic acid, iron, arsenic, and phosphorus: used in cases of anemia and chlorosis.
**nucleohiston** (nū-kle-o-his'-ton) [*nucleo-; ἱστός*, tissue]. 1. A substance composed of nuclein and histon found in the leukocytes of the blood. Nuclein induces coagulation of the blood; histon prevents it. The liquid state of the blood is supposed to be dependent on the integrity of the compound formed by these two bodies. 2. An albuminoid substance obtained from lymph and thymus gland of calves; a white powder, soluble in water, alkalies, and mineral acids. It is used as a bactericide.
**nucleohyaloplasm** (nū-kle-o-hi'-al-o-plasm) [*nucleus*; ὕαλος, transparent substance; πλάσσειν, to mold]. A scarcely tangible substance, which, with chromatin, makes up the threads or mitoma of the cell-nucleus. See *linin*.
**nucleoid** (nū'-kle-oid) [*nucleo-*; εἶδος, like]. 1. Shaped like a nucleus. 2. A finely granular or fibrillar substance in the red corpuscles formed from the original nucleus.
**nucleol** (nū'-kle-ol). Trade name of a nuclein preparation obtained from yeast.
**nucleolar** (nū-kle'-o-lar) [*nucleolus*]. Pertaining to the nucleolus.
**nucleolin** (nū-kle'-o-lin). The substance of which the nucleolus is composed.
**nu le l id** (nu-kle'-ol-oid). Resembling a nucleolus. c o o
**nucleolus** (nū-kle'-o-lus) [dim. of *nucleus*]. The small spherical body within the cell-nucleus. Its true function has not as yet been established.
**nucleomicrosoma** (nū-kle-o-mi-kro-so'-mah) [*nucleus*; μικρός, small; σῶμα, body; pl., *nucleomicrosomata*]. Any one of the many minute tangible bodies that make up each fiber of the nucleoplasm.
**nucleon** (nū'-kle-on). See *paranucleon*.
**nucleonic** (nū-kle-on'-ik). Pertaining to the nucleus.
**nucleopetal** (nū-kle-op'-et-al) [*nucleo-; petere*, to seek]. Seeking the nucleus: said of the movement of the male pronucleus toward the female pronucleus.
**nucleoplasm** (nū'-kle-o-plasm) [*nucleo-; plasma*]. 1. The protoplasm of the nucleus. 2. Chromatin.

**nucleoproteins** (nū-kle-o-pro'-te-ins). Compound proteins which yield true nucleins on pepsin digestion, and also those which, on being boiled with dilute mineral acids, yield, besides proteins, xanthin bases. They occur chiefly in the cell-nuclei, and are widely diffused in the animal body.
**nucleoreticulum** (nū-kle-o-ret-ik'-ū-lum) [*nucleo-; reticulum*, a net]. Any network contained within a nucleus.
**nucleosin** (nū'-kle-o-sin). A substance isolated from spermatozoa of the salmon, identical with thymin.
**nucleotherapy** (nū-kle-o-ther'-ap-e). See *nucleintherapy*.
**nucleotoxin** (nū-kle-o-toks'-in) [*nucleo-*; τοξικόν, a poison]. A toxin derived from cell-nuclei; any toxin affecting the nuclei of cells.
**nucleus** (nū'-kle-us) [*nucleus*, from *nux*, a nut]. 1. The essential part of a typical cell, usually round in outline, and situated near the center. 2. The center around which the mass of a crystal aggregates. 3. The central element in a compound, as the carbon in hydrocarbons. 4. A collection of gray matter in the central nervous system having a distinct function. 5. The deep origin of a nerve. **n. abducens, n., abducent,** n., abducentis a gray nucleus giving origin to the abducens and facial nerves, situated within the fasciculus teres, behind the triangular nucleus, on the floor of the fourth ventricle. Syn., *nidus abducentis*. **n., accessoriovagoglossopharyngeal,** a columnar tract of nerve-cells extending from the level of the calamus scriptorius to that of the auditory stria, and from which the accessory vagus and glossopharyngeal nerves arise in succession from below upward. **n. ambiguus, n., anterior,** a collection of nerve-cells near the nucleus of the vagus. **n. amygdalae,** an irregular aggregation of gray matter situated at the apex of the temporal lobe of the brain, between it and the apex of the middle ventricular horn. **n. angularis,** the accessory auditory nucleus. **nuclei anterolateralis,** the nuclei of the lateral column. **nuclei, auditory,** the nuclei in the oblongata giving rise to the auditory nerves. **n., auditory internal,** a columnar tract of small multipolar cells embedded in the auditory eminence. **n. of Bechterew.** See under *Bechterew*. **n., Béclard's.** See under *Béclard*. **n., bony,** the center of ossification. **n., Burdach's.** See *n. funiculi cuneati.* **n., caudal.** See n., *oculomotor*. **n., caudate,** the intraventricular part of the corpus striatum. **n. centralis,** the corpus dentatum. **n. cinereus,** a term sometimes used to designate the gray substance of the restiform bodies. **n., clavate,** Burdach's nucleus. **n., cleavage-.** See n., *segmentation-*. **n., cuneate, n., cuneate, internal,** Burdach's nucleus. **n., daughter-,** one of the nuclei (usually two) produced by the division of a mother nucleus. See *karyokinesis*. **n., Deiters'.** See n. *magnocellularis*. **n. dentatus,** a folded layer of gray matter in the mesal part of the corresponding cerebellar hemisphere, and close to the roof of the fourth ventricle. **n., Edinger's.** See *Edinger's nucleus.* **n., Edinger-Westphal's.** See *Edinger-Westphal's nucleus.* **n. emboliformis,** a small mass of gray matter situated in the interval between the nucleus dentatus and nucleus fastigii, and lying nearer the former. **n., facial,** one in the reticula at the back of the pons, giving origin to the seventh or facial nerve. **n. fastigii,** a flat expanse of gray matter on each side of the inferior vermiform process of the cerebellum, directly over the roof of the fourth ventricle. **n. fimbriatus.** Synonym of *corpus dentatum*. **n. funiculi cuneati,** a mass of gray matter of the posterior column of the medulla, lying beneath the funiculus cuneatus. **n. funiculi gracilis,** an elongated, club-shaped mass of gray matter in the mesal portion of the posterior column of the medulla. **n., germinal,** the nucleus resulting from the union of the male and female pronuclei. **n., gingival,** a part of the cerebellum in the fetus (between the third and fourth months) which bears some resemblance to the gums. **n., globic, n. globosus,** the globulus, a number of small round or oval masses of gray matter situated in the interval between the nucleus dentatus and nucleus fastigii, lying near to, and probably an accessory detachment of, the latter. **n. gracilis,** a column of gray matter in the posterior pyramid of the medulla oblongata. **n., gray,** the gray matter of the spinal cord. **n., hypoglossal,** a columnar tract of large multipolar nerve-cells embedded in the cinerea of

# NUCLEUS 616 NUTRABIN

the terete funicle in the inferior triangle of the fourth ventricle, and giving origin to the nerve-fibers forming the rootlets of the hypoglossal nerve. **n. hypothalamicus**, the subthalamus. **n., insular**, one entirely separated from adjoining masses of gray matter. **n., intermediolateral**. See *column, intermediolateral*. **n., intermedullary**, great, the external and internal auditory nuclei considered as one. **n., intraventricular**, the caudate nucleus. **n. juxtaolivaris**. See *olive, accessory*. **n., Koelliker's**. See under *Koelliker*. **n., laryngeal**, the nucleus of origin of the nerve-fibers of the larynx. **n., lateral**, that part of the cornu of the cord ascending in the oblongata behind the olivary body. **n., lenticular, n. lentiformis**, the extraventricular portion of the corpus striatum, lying between the internal and external capsules. **n. of Luys**. See under *Luys*. **n. magnocellularis**, a nucleus of gray matter in the medulla oblongata, situated mesad of the restiform column at the level of entry of the auditory nerve-roots. Syn., *Deiters' nucleus*. **n., mesencephalic** (of the trigeminal nerve), a group of large nerve-cells in the gray matter surrounding the Sylvian aqueduct. **n., mesoblastic**, a nucleus of a cell belonging to the mesoblast. **n., mother-**, a cell that is in course of division into two or more parts called daughter-nuclei. See *karyokinesis*. **n., motor**, a collection of nerve-cells in the central nervous system giving origin to a motor nerve. **n., oculomotor**, the nucleus of the oculomotor nerve, lying dorsad of the posterior longitudinal bundle, under the aqueduct of Sylvius. **n., olivary**, a folded mass of gray matter in the medulla oblongata, producing a swelling on the surface—the olivary body. **n., olivary, accessory**. See *olive, accessory*. **n. of origin**, the collection of ganglion-cells in the central nervous system giving origin to a nerve. **n., ossific**. See *n., bony*. **n., peripheral**. See *n., insular*. **n., Perlia's**. See *n., Spitzka's*. **n., polymorphic**, a cell nucleus which assumes irregular forms. **nuclei, pontile, n. pontis**, scattered gray matter included in the intervals in the bundles of fibers of the ventral portion of the pons. Syn., *nidi pontis*. **n., postpyramidal**, a gray nucleus in the oblongata giving origin to the posterior pyramid. **n. pyramidalis**, the inner accessory olivary nucleus. **n. pulposus**, the remnant of the notochord appearing as a pulpy mass in the center of the intervertebral discs. **n. quintus**, the nucleus of the fifth or trigeminal nerve. **n., red**. See *n., tegmental*. **n., respiratory**, n. of respiration, Clarke's column. **n., restiform**, the gray matter of the restiform body. **n. ruber, n. tegmenti**, red nucleus, a reddish mass in the upper part of the cerebral crura, embedded among the fibers of the tegmentum. **n., sacral**, a mass of gray matter in the spinal cord at the level of the origins of the second and third spinal nerves. **n., sagittal**, the middle part of the oculomotor nucleus. **n., secondary**. See *paranucleus*. **n., segmentation-**, the nucleus that appears shortly after the fusion of the male and female pronuclei; the last step in the process of fertilization; it is so-called because within it cleavage is first established. **n., spermatic**, the male pronucleus. **n., spherical**, a gray nucleus at the junction of the hemisphere and middle cerebellar lobe. **n., Spitzka's**. See under *Spitzka*. **n., Stilling's**. Same as *n., tegmental*. **n., styloid**, a bony nodule contained in the cartilages which unite the lesser cornua with the body of the hyoid. **n., subependymal**, the internal nucleus of the eighth or auditory nerve, lying just beneath the ependyma of the fourth ventricle. **n., tegmental, n. tegmenti**, red nucleus, a mass of reticular substance in the tegmentum of the crus cerebri, to the inner side of the substantia nigra. **n., trigeminal**, several groups of nerve-cells ventrad of the facial nucleus below the lateral angle of the fourth ventricle. **n., trochlear**, the ganglionic gray substance surrounding the Sylvian aqueduct and giving origin to the fibers of the fourth nerve. **n. of the vagus**, that part of the accessorio-vagoglossopharyngeal nucleus giving origin to the pneumogastric nerve. **n., vesicular**, a rather large cell-nucleus, the membranes of which stain deeply, while the central portion remains relatively pale. **n. vestibularis**. See *Bechterew, n. of*. **n., vitelline**, a nucleus resulting from the fusion of the male and female pronuclei within the vitellus. **n., Westphal's**. See under *Westphal*. **n., white**, the white substance of the dentate body of the olive.

**Nuel's space** (*nū'-el*) [J. P. *Nuel*, Belgian oculist].

A triangular space between the outer hair-cells and the outer rods of Corti of the internal ear.
**Nuhn's gland** (*noon*) [Anton *Nuhn*, German anatomist, 1814–1884]. See *Blandin's gland*.
**nuisance** (*nū'-sans*). In medical jurisprudence, that which is noxious, offensive, or troublesome; applied to persons or things.
**nullipara** (*nul-ip'-ar-ah*) [*nullus*, none; *parere*, to bring forth]. A woman who has never borne a child.
**nulliparity** (*nul-ip-ar'-it-e*) [*nullipara*]. The condition of being nulliparous.
**nulliparous** (*nul-ip'-ar-us*) [*nullipara*]. Having never borne children.
**numb** (*num*). Having impaired sensibility.
**numbness** (*num'-nes*). Partial, or local anesthesia with torpor; deficiency of sensation; obdormition.
**n. waking.** Acroparesthesia.
**nummiform** (*num'-if-orm*) [*nummus*, a coin; *forma*, form]. Having the form of a coin; nummular.
**nummular** (*num'-ū-lar*) [*nummus*, a coin]. Resembling a coin in form, as *nummular* sputum; resembling rouleaux or rolls of coin.
**nummulation** (*num-ū-la'-shun*). The aggregation of blood-corpuscles into rolls resembling rolls of coin.
**Nunn's corpuscles**. See *Bennett's corpuscles*.
**nunnation** (*nun-na'-shun*) [Heb. *nun*, the letter *n*]. The frequent or abnormal use of the *n*-sound.
**nurito** (*nū'-rit-o*). A proprietary preparation said to contain pyramidon 6½ grains, phenolphthalein, ⅔ grain, and milk sugar, 2⅔ grains.
**nurse** (*ners*) [Fr. *nourrice*, nurse]. 1. To suckle an infant. 2. To care for the sick or for an infant. 3. The caretaker of an infant. 4. A person caring for the sick. 5. The head of a tapeworm. **n., dry**, one who does not suckle the infant. **n., hospital**, one who cares for the sick in a hospital. **n., monthly**, one who attends a woman in confinement. **n., probationer**, in hospitals a probationer is one who has entered upon her career as a nurse, and is under observation to determine her fitness for the profession. **n., professional**, one who devotes himself or herself to the care of the sick as a life-work or profession. **n., registered**, one who is licensed by the laws of the State. **n., wet**, one who suckles the infant. **n.'s contracture**, Trousseau's term for tetany found in association with debility following lactation.
**nursing** (*ners'-ing*). A term applied to the babe's taking the breast, and also to the mother's giving the breast. Also, caring for the sick. **n.-bottle**, a bottle fitted with a rubber tip or nipple for feeding infants not nursed from the breast.
**nursling** (*ners'-ling*) [Fr., *nourrice*, a nurse]. An infant that is nursed.
**Nussbaum's cell** (*noos'-bowm*) [Moritz *Nussbaum*, German histologist, 1850– ]. A granular cell, being one of the four kinds of epithelial cells forming the peptic glands; its function is unknown. **N.'s narcosis**, the condition of prolonged anesthesia induced when the administration of chloroform is preceded a few minutes by the hypodermatic injection of a full dose of morphine.
**nutation** (*nū-ta'-shun*) [*nutare*, to nod]. Nodding or oscillation. **n. of sacrum**, a partial rotation of the sacrum on its transverse axis, whereby the distance between the upper extremity or the lower extremity and the anterior pelvic wall is increased.
**nutgal** (*nut'-gawl*). An excrescence on the leaves of *Quercus lusitanica*, caused by the deposited ova of an insect.
**nutmeg** [ME., *nutmegge*]. The seed of various species of *Myristica*, of the order *Myristiceæ*. The kernel of the ripe seed of *Myristica fragrans* is the *myristica* of the U. S. P. and B. P., and is the source of a volatile oil. The covering of the nutmeg is *mace*. Nutmeg is employed as a condiment, as a corrective and mild flavoring agent, and it has also slight narcotic properties. Dose 5–20 gr.[1] (0.32–1.3 Gm.). **n. liver**, cirrhotic liver. **n., oil of** (*oleum myristicæ*, U. S. P., B. P.), contains a stearopten, *myristin*, which is the glycerid of myristic acid, $C_{14}H_{28}O_2$. Dose 2–3 min. (0.13–0.2 Cc.). **n., oil of, expressed** (*oleum myristicæ expressum*, B. P.), the oil of mace.
**nutone** (*nū'-tōn*). A nutritive tonic said to consist of cod-liver oil, malt-extract, beef-juice, and glycerol emulsion, each, 25 %; calcium hypophosphite, 1 gr.; sodium hypophosphite, 1 gr.; tincture of nux vomica, 1 drop in each teaspoonful.
**nutrabin** (*nū'-tra-bin*). A dietetic prepared from proteins of milk and beef; a brown powder flavored with vanilla.

**nutriant** (nū'-tre-ant) [*nutrire*, to nourish]. A medicine or agent that modifies nutritive processes.
**nutrient** (nū'-tre-ent) [see *nutriment*]. 1. Affording nutrition. 2. A substance that nourishes; a food.
**n. foramen**, an osseous canal for a nutrient vessel.
**n. vessel**, a vessel supplying the marrow of bones.
**nutriment** (nū'-trim-ent), [*nutrire*, to nourish]. Anything that nourishes.
**nutrin** (nū'-trin). A nutritive albuminous substance.
**nutrition** (nū-trish'-un) [*nutriment*]. The process by which tissue is built up and waste repaired.
**nutritious** (nū-trish'-us). Synonym of *nutritive*.
**nutritive** (nū'-trit-iv) [*nutrire*, to nourish]. Possessing the quality of affording nutrition.
**nutritorium** (nū-trit-o'-re-um) [*nutrire*, to nourish]. The nutritive apparatus, or that part of the organism that is directly concerned with anabolic changes.
**nutritory** (nū'-trit-o-re) [*nutrire*, to nourish]. Concerned in the processes of nutrition.
**nutrix** (nū'-triks) [L.: *pl.*, *nutrices*]. A female nurse, especially a wet-nurse.
**nutrolactis** (nū-tro-lak'-tis). A galactagogue said to be a liquid extract of *Galega officinalis*.
**nutrose** (nū'-trōs). Neutral casein sodium, a soluble powder containing 13.8 % of nitrogen, used as a food in intestinal disorders.
**nux** (*nuks*) [L.]. A nut. **n. moschata** [L., "musky nut"]. The nutmeg; myristica. **n. vomica**. The seed of *Strychnos nux-vomica*, an Indian tree of the order *Loganiaceæ*. It contains several alkaloids, the most important being *strychnine* and *brucine* (*q. v.*), which are united with a peculiar acid called *igasuric acid*. In small doses it is a bitter tonic, stimulating gastric digestion; it raises blood-pressure by stimulating the heart and the vasomotor center, and stimulates the respiratory center and the motor centers of the spinal cord. In overdoses it produces tetanic convulsions and risus sardonicus; the reflex excitability is enormously increased, and the slightest stimulus serves to bring on a convulsion. Death usually occurs from asphyxia (cramp-asphyxia), and more rarely from exhaustion. Nux vomica, or strychnine, is employed in dyspepsia, in convalescence from acute diseases, in acute infectious diseases, in shock, in poisoning by chloroform and opium, in emphysema, phthisis, and other conditions associated with dyspnea, in chronic bronchitis, in constipation, in atony of the bladder, in leadpalsy, and in amaurosis from tobacco or alcohol. Dose 1–5 gr. (0.065–0.32 Gm.). **n. vomica, extract of** (*extractum nucis vomicæ*, U. S. P.). Dose ⅛–¼ gr. (0.01–0.016 Gm.). **n. vomica, fluidextract of** (*fluidextractum nucis vomicæ*, U. S. P.). Dose 1–5 min. (0.065–0.32 Cc.). **n. vomica, tincture of** (*tinctura nucis vomicæ*, U. S. P.). Dose 5–30 min. (0.32–2.0 Cc.).
**nyctalgia** (nik-tal'-je-ah) [νύξ, night; ἄλγος, pain]. Pain which occurs chiefly during the night, *e. g.*, the osteocopic pains of syphilis.
**nyctalope** (nik'-ta-lōp) [νύξ, night; ὤψ, the eye]. One who sees better at night or in semidarkness than in a bright light.
**nyctalopia** (nik-tal-o'-pe-ah) [see *nyctalope*]. 1. Night-vision; the condition in which the sight is better by night or in semidarkness than by daylight. Dr. Greenhill and Mr. Tweedy have shown that according to the quite universal usage of modern times, the definitions of the words *nyctalopia* and *hemeralopia* have been the reverse of those of the early Greek and Latin writers. The proper derivation, therefore, of *nyctalopia* would be νύξ, night; ἀλαός, blind; ὤψ, eye, the word meaning nightblindness. *Hemeralopia* was likewise derived from

ἡμέρα, day; ἀλαός, blind; ὤψ, eye, and meant dayblindness. The attempt to reinstate the ancient usage can result only in confusion, and the words should, therefore, never be used.
**nyctamblyopia** (nik-tam-ble-o'-pe-ah) [νύξ, night; ἀμβλυωπία, dim-sightedness]. Imperfect vision at night.
**nycterine** (nik'-ter-ēn) [νυκτερινός, nightly]. 1. Occurring in the night. 2. Obscure.
**nyctophobia** (nik-to-fo'-be-ah) [νύξ, night; φόβος, fear]. Insane dread of the night.
**nyctophonia** (nik-to-fo'-ne-ah) [νύξ, night; φωνή, voice]. The hysterical loss of the voice during the day.
**nyctotyphlosis** (nik-to-tif-lo'-sis) [νύξ, night; τύφλωσις, blindness]. Night-blindness.
**nycturia** (nik-tū'-re-ah) [νύξ, night; οὖρον, urine]. Nocturnal urinary incontinence.
**nygma** (nig'-mah) [νύγμα]. A punctured wound.
**Nylander's test for glucose** (ni'-lan-der) [Claes Wilhelm Gabriel *Nylander*, Swedish chemist, 1835– ]. Dissolve 4 Gm. of Rochelle salts in 100 Cc. of a solution of caustic potash (10 %), and add 2 Gm. of bismuth subnitrate; place on the water-bath until as much of the bismuth salt is dissolved as possible; on heating 10 volumes of urine with 1 volume of the foregoing solution a black coloration or the precipitation of phosphates is produced in the presence of glucose.
**nylic standard** (ni'-lik) [Initial letters of *New York Life Insurance Company*]. A standard of weight in accordance with height and age, as adopted by the New York Life Insurance Company.
**nympha** (nim'-fah) [νύμφη, nymph; *pl.*, *nymphæ*]. A labium minus of the vulva.
**nymphectomy** (nim-fek'-to-me) [*nympha*; ἐκτομή, excision]. Surgical removal of one or both nymphæ.
**nymphitis** (nim-fi'-tis) [*nympha*; *itis*, inflammation]. Inflammation of the nymphæ.
**nymphomania** (nim-fo-ma'-ne-ah) [*nympha*; μανία, madness]. Excessive sexual desire on the part of a woman.
**nymphomaniac** (nim-fo-ma'-ne-ak) [see *nymphomania*]. One affected with nymphomania.
**nymphoncus** (nim-fong'-kus) [*nympha*; ὄγκος, tumor]. Tumor or swelling of the nympha.
**nymphotomy** (nim-fot'-o-me) [*nympha*; τομή, cutting]. 1. Incision of one or both nymphæ. 2. The surgical removal of one or both nymphæ.
**Nyssa** (nis'-ah). A genus of cornaceous trees; gum trees; tupelo or pepperidge. **N. candicans, N. capitata, N. grandidentata,** and **N. uniflora,** are species of the U. S., whose roots are used in making tupelo-tents.
**nystagmiform** (nis-tag'-me-form) [νυσταγμός, nodding of the head; *forma*, form]. Resembling nystagmus.
**nystagmus** (nis-tag'-mus) [νυστάζειν, to nod in sleep]. An oscillatory movement of the eyeballs. It may be congenital or dependent on intracranial disease, especially meningitis, or multiple sclerosis, etc. **n.,** *Cheyne's*, **n.,** *Cheyne-Stokes'*. See *Cheyne-Stokes' nystagmus*. **n., lateral**, oscillation of the eyes in the horizontal meridian. **n., rotatory**, an oscillatory, partial rolling of the eyeball around the visual axis. **n., vertical**, oscillatory movement in the vertical meridian.
**nystaxis** (nis-taks'-is). Synonym of *nystagmus*.
**Nysten's law** (ni'-sten). Rigor mortis begins in the muscles of mastication, extends to the facial and neck muscles, then to the trunk and arms, and finally to the lower extremities.
**nyxis** (niks'-is) [νύξις, puncture]. Surgical puncture or paracentesis.

# O

**O.** The chemical symbol of *oxygen;* also the abbreviation of *oculus,* eye, of *octarius,* a pint, and of *opening* of an electrical circuit.

**o-.** Abbreviation for *ortho-,* in chemical compounds.

**oak** (ōk) [AS., āc]. A genus of trees, *Quercus,* of the order *Cupuliferæ.* The dried bark of *Quercus alba,* white oak, is official in the U. S. P. It contains a peculiar tannic acid known as *quercitannic acid,* and a bitter principle, *quercin,* and is used as an astringent tonic, especially in the form of the oak-bath. It has also been employed in leukorrhea, hemorrhoids, and prolapse of the rectum. Dose of the *bark* 15 gr. (1 Gm.); of the *fluidextract (fluidextractum quercus,* U. S. P.) 15 min. (1 Cc.).

**oakum** (o'-kum) [AS., ācumba, tow]. A material made by picking old rope to pieces. It was formerly used as a dressing for wounds, and in the form of pads to absorb lochial discharges.

**oaralgia** (o-a-ral'-je-ah) [ὠάριον; ovule; ἄλγος, pain].- Ovarian neuralgia.

**oaria** (o-a'-re-ah). Plural of *oarium, q. v.*

**oarialgia** (o-ar-e-al'-je-ah). See *oaralgia.*

**oaric** (o-ar'-ik) [ὠάριον, ovule]. Ovarian; relating to the oaria.

**oariocele** (o-a'-re-o-sēl) [ὠάριον, ovule; κήλη, tumor]. Hernia involving an ovary.

**oariocyesis** (o-a-re-o-si-e'-sis) [ὠάριον, ovule; κύησις, pregnancy]. Ovarian pregnancy.

**oarioncus** (o-a-re-ong'-kus) [ὠάριον, ovule; ὄγκος, tumor]. An ovarian tumor.

**oariopathy** (o-a-re-op'-ath-e) [ὠάριον, ovule; πάθος, disease]. Any ovarian disease.

**oariophyma** (o-a-re-o-fi'-mah) [ὠάριον, ovule; φῦμα, tumor]. Same as *oarioncus.*

**oariorrhexis** (o-a-re-o-eks'-e-ah) [ὠάριον, ovule; ῥῆξις, rupture]. Rupture of the ovary.

**oarioscirrhus** (o-a-re-o-skir'-us) [ὠάριον, ovule; σκίρρος, hard]. Scirrhus of the ovary.

**oariosteresis** (o-a-re-o-ster-e'-sis) [ὠάριον, ovule; στέρησις, privation]. Same as *oariotomy.*

**oariotomy** (o-a-re-ot'-o-me) [ὠάριον, ovule; τομή, a cutting]. Surgical removal of an ovary.

**oaritis** (o-a-ri'-tis) [ὠάριον, ovule; ιτις, inflammation]. Inflammation of an ovary.

**oariule** (o-a'-re-ūl) [ὠάριον, ovary; ὑλή, scar]. A corpus luteum.

**oarium** (o-a'-re-um) [ὠάριον, ovule: *pl., oaria*]. See *ovarium.*

**oasis** (o-a'-sis) [ὄασις, a dry spot: *pl., oases*]. In surgery, an isolated spot of healthy tissue surrounded by diseased tissue.

**oat** (ōt) [AS., ātan, oats]. A cereal plant, *Avena sativa,* or other species of *Avena,* and its seed. o. treatment (of v. Noorden), in diabetes mellitus: consists in daily régime of 250 Gm. of oat-flakes per meal cooked for a long time in water, 100 Gm. of albumin, 300 Gm. of butter.

**oatmeal** (ōt'-mēl). The meal made from oats. It is used in the form of a gruel, as a food, as a demulcent and laxative, and as an emollient poultice.

**ob-** [L.]. A prefix signifying on, against, in front of, or toward.

**obdormition** (ob-dor-mish'-un) [*obdormire,* to fall asleep]. Numbness of a part due to interference with nervous function; the state of a part when it is said to be "asleep."

**obduction** (ob-duk'-shun) A necropsy.

**O'Beirne's sphincter** (o-burn') [James *O'Beirne,* Irish surgeon, 1786–1862]. A thickened circular bundle of muscular fibers, situated in the rectum, just below its junction with the colon. O'B.'s **tube,** a long, flexible tube used in making rectal injections.

**obeliac** (o-be'-le-ak) [ὀβελός, a spit]. Pertaining to, or situated near, the obelion.

**obeliad** (o-be'-le-ad). Toward the obelion.

**obelion** (o-be'-le-on). See under *craniometric point.*

**Obermayer's test for indican in the urine** (o'-ber-mi-er) [Fritz *Obermayer,* Austrian physician, 1861–]. With a lead-acetate solution (1 : 5) precipitate the urine; care must be taken not to add an excess of lead solution. Filter, and shake the filtrate for one or two minutes with an equal quantity of fuming hydrochloric acid which contains 1 or 2 parts of ferric chloride solution to 500 parts of the acid. Add chloroform, which becomes blue from the generation of indigo-blue.

**Obermeier, spirillum of** (o'-ber-mi-er) Otto Hugo Franz *Obermeier,* German physician, [1843–1873]. A spirillum found in the blood in relapsing fevre.

**Obermueller's test for cholesterin** (o'-ber-mū-ler) [Kuno *Obermueller,* German physician, 1861– ]. Place the cholesterin in a test-tube, and fuse with 2 or 3 drops of propionic acid anhydride over a small naked flame. The fused mass on cooling is violet, changing to blue, green, orange, carmin, and finally, copper-red.

**Oberst method of inducing local anesthesia** (o'-bairst) [Max *Oberst,* German surgeon, 1849– ]. Injection of a 1 per cent. solution of cocaine over the course of a nerve-trunk supplying the area to be attacked.

**obese** (o-bēs) [obesus, fat]. Extremely fat; corpulent.

**obesity** (o-bes'-it-e) [obesus, fat]. An excessive development of fat throughout the body; corpulence; polysarcia.

**obex** (o'-beks) [L., "a barrier"]. A band of white nervous matter at the point of the calamus scriptorius.

**obfuscation** (ob-fus-ka'-shun) [ob-; *fuscus,* dusky]. 1. Darkening or clouding, as *obfuscation* of the cornea. 2. Mental confusion.

**object-blindness** (ob-jekt-blind'-nes). See *apraxia.*

**object-glass** (ob'-jekt-glas). See *objective* (3).

**objective** (ob-jek'-tiv) [ob-; *jacere,* to throw]. 1. Pertaining to an object or to that which is contemplated or perceived, as distinguished from that which contemplates or perceives. 2. Pertaining to those relations and conditions of the body perceived by another, as *objective* signs of disease. 3. The lens of a microscope nearest the object.

**obligate** (ob'-lig-āt) [obligare, to bind]. Constrained; bound; not facultative. o. aerobic, of a microorganism, one that can live only as an aerobe. o. anaerobic, of a microorganism, one that can live only as an anaerobe. o. parasite, a parasite that can live only as a parasite.

**oblinition** (ob-lin-ish'-un) [oblinere, to smear]. Inunction.

**oblique** (ob-lēk' or ob-lēk') [obliquus]. Not direct; aslant; slanting. In botany, unequal-sided. In anatomy, an oblique muscle, as the external or internal oblique of the abdomen, or the superior or inferior oblique of the eye.

**obliquimeter** (ob-lik-wim'-et-er) [oblique; μέτρον, a measure]. An instrument fitted with arms employed to indicate the angle formed by comparing the plane of the pelvic brim with the perpendicular axis of the upright body.

**obliquity** (ob-lik'-wit-e) [oblique]. The state of being oblique.

**obliquus** (ob-li'-kwus) [L., "slanting"]. A term applied to various muscles. See under *muscle.*

**obliteration** (ob-lit-er-a'-shun) [obliterare, to efface, from ob-; *litera,* a letter]. Removal of a part; extirpation; complete closure of a lumen.

**oblongata** (ob-long-ga'-tah) [L.]. The medulla oblongata.

**oblongata** (ob-long-ga'-tal) [oblongata, the medulla]. Pertaining to the oblongata.

**obmutescence** (ob-mū-tes'-ens) [obmutescere, to become dumb]. Aphonia; loss of voice.

**observation** (ob-ser-va'-shun) [observatio]. The examination of a thing; a systematic study of phenomena.

# OBSESSION 619 OCOTEA

**obsession** (*ob-sesh'-un*) [*ob-; sedere*, to sit]. Possession as by evil spirits; an imperative idea; a dominant delusion. **o. dentaire**, neurasthenic neuralgia erroneously attributed to the teeth.
**obsolescence** (*ob-so-les'-ens*) [*obsolescere*, to grow old]. The state of becoming old or obsolete.
**obstetric, obstetrical** (*ob-stet'-rik, ob-stet'-rik-al*) [*obstetrics*]. Pertaining to the practice of obstetrics.
**obstetrician** (*ob-stet-rish'-an*) [*obstetrics*]. One who practises obstetrics.
**obstetrics** (*ob-stet'-riks*) [*obstare*, to stand before]. The branch of medicine that deals with the care of women during pregnancy, labor, and the puerperium.
**obstetrix** (*ob-stet'-riks*) [*obstare*, to stand before]. A midwife.
**obstipation** (*ob-stip-a'-shun*) [*obstipare*, to stop up]. Intractable constipation.
**obstruction** (*ob-struk'-shun*) [*ob-; struere*, to build]. 1. The state of being obstructed or blocked up. 2. The act of impeding or blocking up. 3. An impediment or obstacle.
**obstructive** (*ob-struk'-tiv*) [*obstruction*]. 1. Stopping or blocking up. 2. Due to an obstruction, as obstructive jaundice.
**obstruent** (*ob'-stroo-ent*) [*obstruere*, to close up]. 1. Obstructive; tending to obstruct. 2. Any remedy or agent closing the lumen or orifice of vessels or ducts.
**obstupefacient** (*ob-stū-pe-fa'-she-ent*) [*obstupefacere*, to stupefy]. Narcotic or stupefying.
**obtund** (*ob-tund'*) [*ob-; tundere*, to beat]. To blunt or dull; to lessen, as to obtund sensibility.
**obtundent** (*ob-tund'-ent*) [*obtundere*, to make dull]. Soothing, quieting; a remedy that relieves or overcomes irritation or pain.
**obturation** (*ob-tū-ra'-shun*) [*obturare*, to stop up]. The closing of an opening or passage.
**obturator** (*ob'-tū-ra-tor*) [*obturation*]. 1. Closing an opening. 2. That which closes an opening. 3. Pertaining to the *obturator* membrane, muscles, etc. **o. foramen**, a foramen in the anterior part of the os innominatum. See *foramen*, *obturator*. **o. membrane**, the membrane closing the obturator foramen. **o. muscle**. See under *muscle*. **o. nerve**. See under *nerve*.
**obtuse** (*ob-tūs'*) [*obtusus*, p.p. of *obtundere*, to blunt, dull]. Blunt.
**obtusion** (*ob-tū'-shun*) [*obtundere*, to blunt]. The blunting or weakening of normal sensation: a symptom of certain diseases.
**occalcarine** (*ok-kal'-kar-ēn*). See *occipitocalcarine*.
**occipital** (*ok-sip'-it-al*) [*occiput*]. Pertaining to or in relation with the occiput. **o. artery**. See under *artery*. **o. bone**. See *bones*, *table of*. **o. cross**, the internal occipital protuberance. **o. lobe**, one of the lobes of the cerebrum. **o. nerve**. See under *nerve*. **o. protuberance**, the prominence on the inner surface (*internal*) or on the outer surface (*external*) of the occipital bone. **o. section**, a transverse section through the middle of the occipital lobe. **o. triangle**. See under *triangle*.
**occipitalis** (*ok-sip-it-a'-lis*) [L.]. The posterior belly of the occipitofrontalis muscle.
**occipiten** (*ok-sip'-it-en*) [*occiput*]. Belonging to the occipital bone in itself.
**occipito-** (*ok-sip-it-o-*) [*occiput*]. A prefix denoting connection with or relation to the occipital bone or the occiput.
**occipitoanterior** (*ok-sip-it-o-an-te'-re-or*) [*occipito-; anterior*]. Having the occiput directed toward the front, as the *occipitoanterior* position of the fetus in the uterus.
**occipitoatloid** (*ok-sip-it-o-at'-loid*) [*occipito-; atlas; eidos*, form]. Pertaining to the occipital bone and the atlas.
**occipitoaxoid** (*ok-sip-it-o-aks'-oid*) [*occipito-; axis; eidos*, form]. Pertaining to the occipital bone and the axis.
**occipitobregmatic** (*ok-sip-it-o-breg-mat'-ik*) [*occipito-; bregma*]. Pertaining to the occiput and the bregma.
**occipitocalcarine** (*ok-sip-it-o-kal'-kar-in*). Both occipital and calcarine; referring to the posterior calcarine fissure.
**occipitocervical** (*ok-sip-it-o-ser'-vik-al*) [*occipito-; cervix*]. Pertaining to the occiput and the neck.
**occipitofacial** (*ok-sip-it-o-fa'-shal*). Pertaining to both the occiput and the face.
**occipitofrontal** (*ok-sip-it-o-fron'-tal*) [*occipito-; frontal*]. Pertaining to the occiput and forehead, or to the *occipitofrontal* muscle (*occipitofrontalis*).

**occipitofrontalis** (*ok-sip-it-o-fron-ta'-lis*) [L.]. See under *muscles*, *table of*.
**occipitomastoid** (*ok-sip-it-o-mas'-toid*) [*occipito-; mastoid*]. Pertaining to the occipital bone and the mastoid process.
**occipitomental** (*ok-sip-it-o-men'-tal*) [*occipito-; mentum*, the chin]. Pertaining to the occiput and the chin.
**occipitoparietal** (*ok-sip-it-o-par-i'-et-al*) [*occipito-; parietal*]. Pertaining to the occipital and parietal bones, or to the occipital and parietal lobes of the brain.
**occipitoposterior** (*ok-sip-it-o-pos-te'-re-or*) [*occipito-; posterior*]. Having the occiput directed backward, as the *occipitoposterior* position of the fetus in the uterus.
**occipitotemporal** (*ok-sip-it-o-tem'-por-al*) [*occipito-; temporal*]. Pertaining to the occipital and temporal bones.
**occiput** (*ok'-sip-ut*) [*ob-*, over against; *caput*, the head; *gen.*, *occipitis*]. The back part of the head.
**occluding** (*ok-lōō'-ding*) [see *occlusion*]. Closing; applied to the grinding surfaces of molars and bicuspids.
**occlusal** (*ok-lōō'-sal*). See *occluding*.
**occlusio** (*ok-lōō'-ze-o*) [L.]. Closure. **o. pupillæ**, obliteration of the pupil. **o. pupillæ lymphatica**, obliteration of the pupil by a false membrane.
**occlusion** (*ok-lōō'-zjun*) [*ob-*, against; *claudere*, to shut]. 1. A closing or shutting up. 2. The state of being closed or shut. 3. The absorption, by a metal, of gas in large quantities, as of hydrogen by platinum. 4. The full meeting or contact in a position of rest of the masticating surfaces of the upper and lower teeth; it is erroneously called articulation of the teeth. **o., buccal**, a bicuspid or molar tooth outside the line of occlusion. **o., distal**, when a tooth is more posterior than normal. **o., labial**, an incisor or cuspid tooth outside the line of occlusion. **o., lingual**, refers to a tooth inside the line of occlusion. **o., mesial**, when a tooth is nearer the median line than normal. **o., torso-**, a tooth turned on its axis.
**occlusive** (*ok-lōō'-siv*) [see *occlusion*]. Closing or shutting up, as an *occlusive* surgical dressing.
**occult** (*ok-kult'*) [*occultus*, hidden]. Hidden; concealed; not evident. **o. blood**, a concealed hemorrhage. **o. disease**, any disease the nature of which is not readily determined.
**occupation-disease**. One caused by the occupation of the patient. See *occupation-neurosis*.
**occupation-neurosis**. A functional disturbance of the part used in carrying on a certain occupation, as writer's cramp, telegrapher's cramp, etc.
**ocellus** (*o-sel'-us*) [L., a little eye: *pl.*, *ocelli*]. In biology: (*a*) one of the simple eyes or pigmentspots of invertebrate animals; (*b*) one of the elements of a compound eye; (*c*) one of the colored spots on many feathers, flowers, etc.
**ochema** (*o-ke'-mah*) [ὄχειν, to carry]. A vehicle for medicines.
**ocher, ochre** (*o'-ker*) [ὠχρός, pale]. A variety of fine clay containing iron; the common colors are yellow and red.
**ocheus** (*ok'-e-us*) [ὄχειν, to carry]. The scrotum.
**ochlesis** (*ok-le'-sis*) [ὄχλος, crowd]. Crowd-poisoning; disease due to overcrowding, and lack of ventilation.
**ochletic** (*ok-let'-ik*) [ὄχλος, crowd]. Pertaining to, or of the nature of, ochlesis.
**ochlophobia** (*ok-lo-fo'-be-ah*) [ὄχλος, crowd; φόβος, fear]. Morbid fear of crowds.
**ochriasis** (*o-kri'-as-is*) [ὠχρός, yellow]. Sallowness of complexion.
**ochrometer** (*o-krom'-et-er*) [ὠχρός, pale; μέτρον, measure]. An instrument for measuring the capillary blood-pressure.
**ochronosis** (*o-kron-o'-sis*) [ὠχρός, pale; νόσος, disease]. A brownish or blackish discoloration of cartilage and allied structures; it is probably an intensification of the pigmentation normally present in these structures, occurring in advanced life.
**ochronosus** (*o-kron'-o-sus*) [*ocher;* νόσος, disease]. Any disease marked by dark discoloration of the cartilages and allied structures.
**ochrotic** (*o-kron-ot'-ik*). Pertaining to or relating to ochronosis.
**ochropyra** (*o-kro-pi'-rah*) [ὠχρός, yellow; πῦρ, fire]. Yellow fever.
**Ocotea** (*ok-ot'-e-ah*) [native name in Guiana]. A

**genus** of laurinaceous trees. *O. cujumary* of Brazil, furnishes seeds which are used in dyspepsia. *O. opifera*, a variety the oil from the fruit and the bark of which is used as an antirheumatic. *O. pretiosa* is used in neurasthenia.
**octad** (ok'-tad) [octo, eight]. 1. An octavalent element or radical. 2. Having a valence of eight.
**octan** (ok'-tan) [octo]. Returning every eighth day, as an *octan* fever.
**octane** (ok'-tān) [ὀκτώ, eight], C₈H₁₈. The eighth member of the paraffin or marsh-gas series.
**octarius** (ok-ta'-re-us) [octo]. An eighth part of a gallon; a pint. Abbreviated O.
**octavalent** (ok-tav'-al-ent) [octo; valere, to be worth]. Having a quantivalence of eight.
**octene** (ok'-tēn). See *octylene*.
**octigravida** (ok-te-grav'-id-ah) [octo; gravida, a pregnant woman]. A woman pregnant for the eighth time.
**octipara** (ok-tip'-ar-ah) [octo; parere, to bring forth]. A woman who has been in labor eight times.
**octivalent**. See *octavalent*.
**octo-** (ok'-to-) [ὀκτώ, eight]. A prefix denoting reference to the number eight.
**octoferric** (ok-to-fer'-ik) [octo; ferrum, iron]. Containing eight atoms of iron in the molecule.
**octoroon** (ok-tor-oon') [ὀκτώ, eight]. The offspring of a white person and a quadroon; a person who has one eighth part of negro blood.
**octoses** (ok'-to-ses) [ὀκτώ, eight]. A group of the monosaccharides with the formula C₈H₁₆O₈.
**octylene** (ok'-til-ēn) [ὀκτώ, eight], C₈H₁₆. A colorless liquid derived from octane.
**ocular** (ok'-u-lar) [oculus, the eye]. 1. Pertaining to or in relation with the eye. 2. The lens of a microscope that is turned toward the eye. o., **compensating**, a lens that compensates for axial aberration of the objective. o., **Huygenian**, a. lens consisting of two planoconvex lenses, the convexities being directed toward the objective; the lower lens is the *field-lens*, the upper, the *eye-lens*.
**oculentum** (ok-u-len'-tum) [oculus, eye; pl., oculenta]. An ointment for use in the eye.
**oculin** (ok'-u-lin). An organotherapeutic preparation said to be a glycerol extract from the ciliary body of the eyes of oxen.
**oculist** (ok'-u-list). Synonym of *ophthalmologist*.
**oculo-** (ok-u-lo-) [oculus, eye]. A prefix signifying pertaining to the eye.
**oculofacial** (ok-u-lo-fa'-she-al). Relating to the eyes and the face.
**oculofrontal** (ok-u-lo-fron'-tal). Relating to the eyes and the forehead.
**oculomotor** (ok-u-lo-mo'-tor) [oculo-; movere, to move]. 1. Pertaining to the movement of the eye, as the *oculomotor* nerve. 2. Pertaining to the oculomotor nerve, as the *oculomotor* nucleus.
**oculomotorius** (ok-u-lo-mo-to'-re-us) [see *oculomotor*]. The third, or motor oculi, nerve.
**oculonasal** (ok-u-lo-na'-sal). Relating to the eye and nose.
**oculoreaction** (ok-u-lo-re-ak'-shun). Ophthalmoreaction, q. v.
**oculozygomatic** (ok-u-lo-zi-go-mat'-ik) [oculo-; zygoma]. Pertaining to the eye and the zygoma. See *Jadelot's lines*.
**oculus** (ok'-u-lus) [L., an eye: pl., oculi]. An eye. o. **bovinus**, o. **bovis**. Synonym of *hydrophthalmia*. o. **bubulus**. Synonym of o. *bovinus*. o. **cæsius**, glaucoma. o. **dexter**, the right eye. o. **duplex**, a bandage covering both eyes. o. **elephantinus**. Synonym of *hydrophthalmia*. o. **genu**, the patella. o. **lacrimans**, epiphora. o. **leporinus**, lagophthalmos. oculi **marmarygodes**. See *metamorphopsia*. o. **oviis**, cicatricula. o. **purulentus**, hypopyon. o. **scapulæ**, the glenoid cavity. o. **sinister**, the left eye. o. **simplex**. See *monoculus*.
**oculustro** (ok-u-lus'-tro). A soap consisting of oleate of potassium, glycerol, and turpentine.
**ocyodinic** (o-se-o-din'-ik) [ὠκύς, swift; ὠδίνες, labor pains]. Oxytocic; hastening the delivery o f the fetus.
**od** (od) [ὁδός, way]. The force supposed to produce the phenomena of mesmerism.
**O. D.** Abbreviation of *oculus dexter*, right eye; also of *optic disc*.
**odaxesmus** (o-daks-es'-mus) [ὀδαξησμός, a sharp biting]. The biting of the tongue, lip, or cheek, occurring during an epileptic fit.
**odaxetic** (o-daks-et'-ik) [see *odaxesmus*]. Giving rise to an itching or stinging sensation.

**odic** (o'-dik). Of or pertaining to the theoretical force or influence od. o.-**force**. See *od*.
**odinagogue** (o-din'-ag-og). Synonym of *oxytocic*.
**odol** (o'-dol). A mouth-wash, said to consist of phenyl salicylate, 2.5 parts; oil of peppermint, 0.5 part; saccharin, 0.004 part; and alcohol, 97 parts.
**odontagma** (o-don-tag'-mah) [ὀδούς, tooth; ἀγμός, a breaking]. Same as *odontoclasis*.
**odontagra** (o-don-ta'-grah) [ὀδούς, tooth; ἄγρα, seizure]. Toothache, especially a form due to gout.
**odontalgia** (o-don-tal'-je-ah) [ὀδούς, tooth; ἄλγος, pain]. Toothache. o., **phantom**, pain felt in the space from which a tooth is absent.
**odontalgic** (o-don-tal'-jik) [ὀδούς, tooth; ἄλγος, pain]. 1. Antiodontalgic. 2. Relating to toothache. 3. A remedy for toothache.
**odontatrophy** (o-don-tat'-ro-fe) [ὀδούς, tooth; ἀτροφία, atrophy]. Atrophy of the teeth.
**odonterism** (o-don'-ter-ism). Chattering of the teeth.
**odontharpagra** (o-don-thar-pa'-grah). Synonym of *dentagra*.
**odonthemodia** (o-dont-hem-o'-de-ah) [ὀδούς, tooth; αἱμωδία, a having the teeth on edge]. Hemodia; excessive sensibility of the teeth.
**odonthercos** (o-don-ther'-kos) [odonto-; ἕρκος, fence]. A set of teeth.
**odonthyalophthora** (o-dont-hi-al-of'-thor-ah) [odonto-; ὕαλος, glass; φθείρειν, to destroy]. Destruction of the enamel of the teeth.
**odonthyalus** (o-dont-hi'-al-us). Synonym of *enamel*.
**odontia** (o-don'-she-ah) [odonto-]. 1. Odontalgia. 2. Any abnormality in connection with the teeth. o. **deformis**, deformity of the teeth, arising either from error of shape, position, or malformation of the jaws or alveolar border. o. **incrustans**, tartar of the teeth.
**odontiasis** (o-don-ti'-as-is) [ὀδοντίασις, teething]. Dentition; the cutting of teeth.
**odontiater** (o-don-te-a'-ter) [odonto-; ἰατρός, physician]. A dentist.
**odontiatria** (o-don-te-a-tre'-ah) [odonto-; ἰατρεία, a healing]. Dental surgery.
**odontic** (o-don'-tik) [odonto-]. Appertaining to the teeth.
**odontinoid** (o-don'-tin-oid) [odonto-; εἶδος, resemblance]. Resembling or having the nature of teeth.
**odontitis** (o-don-ti'-tis) [ὀδούς, tooth; ιτις, inflammation]. Inflammation of the teeth.
**odonto-** (o-don-to-) [ὀδούς, tooth]. A prefix signifying pertaining to a tooth.
**odontoatlantal** (o-don-to-at-lant'-al). Same as *atloaxoid*.
**odontoblast** (o-don'-to-blast) [odonto-; βλαστός, a germ]. One of the cells covering the dental papilla and forming the dentine.
**odontoblastoma** (o-don-to-blas-to'-mah) [odontoblast; ὄμα, tumor]. A tumor composed of dentine.
**odontobothriitis** (o-don-to-both-re-i'-tis) [odonto-; βοθρίον, a little cavity; ιτις, inflammation]. Inflammation of the socket of a tooth.
**odontobothrium** (o-don-to-both'-re-um) [odonto-; βοθρίον, a little cavity]. The alveolus of a tooth.
**odontoceramic** (o-don-to-ser-am'-ik) [odonto-; κέραμος, clay]. Pertaining to porcelain teeth.
**odontochalix** (o-don-tok'-al-iks) [odonto-; χάλιξ, mortar]. Dental cement.
**odontochirurgical** (o-don-to-ki-rur'-jik-al) [odonto-; χειρουργία, surgery]. Pertaining to dental surgery.
**odontoclasis** (o-don-tok'-las-is) [odonto-; κλάειν, to break]. The breaking of a tooth.
**odontoclast** (o-don'-to-klast) [odonto-; κλάειν, to break]. A protoplasmic cell engaged in absorbing the fang of a deciduous tooth.
**odontocnesis** (o-don-tok-ne'-sis) [odonto-; κνῆσις, an itching]. A painful itching sensation in the gums, as that preceding cutting of the teeth.
**odontodol** (o-don-to'-dol). A dental anodyne said to contain cocaine hydrochloride, 1 part; oil of cherry-laurel, 1 part; tincture of arnica, 10 parts; solution of ammonium acetate, 20 parts.
**odontodynia** (o-don-to-din'-e-ah) [odonto-; ὀδύνη, pain]. Toothache.
**odontogen** (o-don'-to-jen) [odonto-; γεννᾶν, to produce]. A material producing dentine.
**odontogeny** (o-don-toj'-en-e) [odonto-; γεννᾶν, to beget]. The origin and development of teeth.
**odontoglyph** (o-don'-to-glif) [odonto-; γλύφειν, to carve]. An instrument used for scraping the teeth.

ODONTOGRAPHY 621 OHM

**odontography** (*o-don-tog'-ra-fe*) [*odonto-*; γράφειν, to write]. The descriptive anatomy of the teeth.
**odontoid** (*o-don'-toid*) [*odonto-*; εἶδος, like]. Resembling a tooth; tooth-like. o. **ligament.** See *ligament, odontoid.* o. **process,** the dentate process of the second cervical vertebra.
**odontolith** (*o-don'-to-lith*) [*odonto-*; λίθος, a stone]. The calcareous accretion on the teeth, popularly known as tartar.
**odontology** (*o-don-tol'-o-je*) [*odonto-*; λόγος, science]. The branch of science dealing with the anatomy and diseases of the teeth.
**odontoloxia, odontoloxy** (*o-don-to-loks'-e-ah, o-don'-to-loks-e*) [*odonto-*; λοξός, slanting]. Irregularity or obliquity of the teeth.
**odontoma** (*o-don-to'-mah*) [*odonto-*; ὄμα, tumor]. 1. A tumor containing dentine or tooth-like structure. 2. Any tumor in connection with the teeth.
**odontome** (*o-don'-tōm*). See *odontoma.* o', composite, a tumor made up of all the histological elements of teeth, thrown together indiscriminately.
**odontomys** (*o-don'-to-mis*) [*odonto-*; μῦς, mouse]. The dental pulp.
**odontonecrosis** (*o-don-to-nek-ro'-sis*) [*odonto-*; *necrosis*]. Necrosis or decay of the tissues of the teeth.
**odontoneuralgia** (*o-don-to-nū-ral'-je-ah*) [*odonto-*; *neuralgia*]. Neuralgia due to diseased teeth.
**odontonosology** (*o-don-to-nos-ol'-o-je*) [*odonto-*; νόσος, disease; λόγος, science]. A treatise on diseases of the teeth; also that branch of medicine that treats of diseases of the teeth.
**odontoparallaxis** (*o-don-to-par-al-aks'-is*) [*odonto-*; παράλλαξις, deviation]. Irregularity of the teeth; deviation of one or more of the teeth from the natural position.
**odontopathy** (*o-don-top'-ath-e*) [*odonto-*; πάθος, suffering]. Any disease of the teeth.
**odontoperiosteum** (*o-don'-to-per-e-os-te-um*). Synonym of *periodonteum.*
**odontoplerosis** (*o-don-to-ple-ro'-sis*) [*odonto-*; πλήρωσις, filling]. The filling of teeth.
**odontoprisis** (*o-don-to-pri'-sis*) [*odonto-*; πρίσις, a sawing]. Grinding of the teeth.
**odontorrhagia** (*o-don-tor-a'-je-ah*) [*odonto-*; ῥηγνύναι, to burst forth]. Hemorrhage from the socket of a tooth.
**odontorthosia** (*o-don-tor-tho'-ze-ah*) [*odonto-*; ὀρθός, straight]. The operation of straightening irregularly growing teeth.
**odontorthosis** (*o-don-tor-tho'-sis*). Same as *odontorthosia.*
**odontoschisis** (*o-don-tos'-kis-is*) [*odonto-*; σχίζειν, to split]. Splitting of a tooth.
**odontoschism** (*o-don'-to-skism*) [*odonto-*; σχίζειν, to split]. A fissure in a tooth.
**odontoscope** (*o-don'-to-skōp*) [*odonto-*; σκοπεῖν, to see]. A dental mirror used for inspecting the teeth.
**odontoseisis** (*o-don-to-si'-sis*) [*odonto-*; σείσις, a shaking]. Looseness of the teeth from partial or total destruction of the alveolar processes, caused most frequently by disease of the gums.
**odontosis** (*o-don-to'-sis*) [ὀδούς, tooth]. The formation and development of the teeth.
**odontosteophyte, odontosteophyton** (*o-don-tos'-te-o-fit, o-don-tos-te-off'-it-on*) [*odonto-*; ὀστέον, bone; φύειν, to grow]. A bony outgrowth from a tooth.
**odontosteresis** (*o-don-to-ster-e'-sis*) [*odonto-*; στέρησις, privation]. Loss of the teeth.
**odontotechny** (*o-don'-to-tek-ne*) [*odonto-*; τέχνη, art]. Dental surgery.
**odontotheca** (*o-don-to-the'-kah*) [*odonto-*; θήκη, case]. The follicle of a tooth.
**odontotherapy** (*o-don-to-ther'-ap-e*) [*odonto-*; θεραπεία, treatment]. The treatment of diseases of the teeth.
**odontotrimma** (*o-don-to-trim'-ah*) [*odonto-*; τρίμμα, a pulverized substance: *pl., odontotrimmata*]. A tooth-powder.
**odontotripsis** (*o-don-to-trip'-sis*) [*odonto-*; τρίβειν, to rub]. The natural abrasion or wearing away of the teeth.
**odontotrypy** (*o-don-tot'-rip-e*) [*odonto-*; τρυπᾶν, to perforate]. Perforation of a tooth to remove pus or a diseased pulp.
**odor** (*o'-dor*) [L.]. A scent, smell or perfume. Fragrance.
**odoration** (*o-dor-a'-shun*) [*odoratio,* a smelling]. 1. The act of smelling. 2. The sense of smell.
**odorator** (*o-dor-a'-tor*) [*odor,* perfume]. An atomizer for diffusing liquid perfumes.

**odoriferous** (*o-dor-if'-er-us*) [*odor; ferre,* to carry]. Fragrant.
**O'Dwyer's method of treating intubation ulcers** (*o-dwi'-er*) [Joseph P. *O'Dwyer,* American physician, 1841–1898]. This consists in the employment of tubes provided with a narrow neck and coated with a layer of gelatin and alum. The gelatinized tube is left in the larynx for five days, at the end of which time it is removed and replaced by a similar tube. This process is repeated three times, at the end of which the ulcer will usually be found to be completely healed. **O'D.'s tubes,** tubes used for intubation of the larynx.
**odyl, odyle** (*od'-il*). See *od.*
**odynacousis, odynacusis** (*o-din-ah-koo'-sis*) [ὀδύνη, pain; ἀκούειν, to hear]. Pain caused by noises.
**-odyne, -odynia** (*-o-din, -o-din'-e-ah*) [ὀδύνη, pain]. A suffix denoting pain.
**odynolysis** (*o-din-ol'-is-is*) [ὀδύνη, pain; λύειν, to loose]. Alleviation of pain.
**odynometer** (*o-din-om'-et-er*) [ὀδύνη, pain; μέτρον, measure]. An instrument for recording the amount of pain suffered by a patient.
**odynopeia** (*o-din-o-pe'-ah*) [ὀδύνη, pain; ποιεῖν, to make]. The induction of labor-pains.
**odynopeic** (*o-din-o-pe'-ik*). Oxytocic.
**odynophagia** (*o-din-o-fa'-je-ah*). See *odynphagia.*
**odynophobia** (*o-din-of-o'-be-ah*) [ὀδύνη, pain; φόβος, fear]. Morbid dread of pain; algophobia.
**odynopœia** (*o-din-o-pe'-ah*). The induction of labor pains.
**odynpphagia** (*o-din-fa'-je-ah*) [ὀδύνη, pain; φαγεῖν, to eat]. Painful deglutition.
**odynuria** (*o-din-ū'-re-ah*) [ὀδύνη, pain; οὖρον, urine]. The painful passage of urine.
**œ-.** See *e-.*
**œcology** (*e-kol'-o-je*) [οἶκος, a house, family; λόγος, science]. See *ecology.*
**Oehl's layer** (*ēl*) [Eusebio *Oehl,* Italian anatomist, 1827–1903]. The stratum lucidum of the epidermis.
**Œnanthe** (*e-nan'-the*) [οἶνος, wine; ἄνθος, a flower]. A genus of umbelliferous plants. The fruit of **Œ.** *phellandrium* is diuretic, carminative, and recommended as a specific sedative to the bronchial mucosa. Dose of *powdered fruit* 1 dr. (4 Cc.) in 24 hours.
**œnanthol** (*e-nan'-thol*), C₇H₁₄O. An aromatic liquid distilled from castor oil.
**œnanthotoxin** (*e-nan-tho-toks'-in*), C₁₇H₂₂O₅. A toxic resinoid from *Œnanthe crocata.*
**œnilism** (*e'-nil-ism*) [οἶνος, wine]. A form of alcoholism produced by abuse of wine.
**Oertel's method** (*er'-tel*) [Max Joseph *Oertel,* German physician, 1835–1897]. The treatment of circulatory disturbances of heart disease, obesity, emphysema, etc., by mechanical means. The objects aimed at are: 1. Diminution of the fatty tissue deposited, achieved by regulating the diet, etc. 2. Reduction of the body-fluids, accomplished by reducing the quantity of all fluids. 3. Strengthening of the heart-muscle and promotion of the development of compensatory hypertrophy, attained by methodical mountain-climbing and other systematic exercise. 4. Stimulation of the circulation by massage, passive movements, and sanitary gymnastics.
**œse** (*e'-zeh*) [Ger.]. An instrument consisting of a loop of platinum wire affixed to a glass handle and employed in bacteriological investigation.
**œsophagus** and allied words. See *esophagus,* etc.
**œstrum** (*es'-trum*). See *estrum.*
**œstrus** (*ēs'-trus*). See *estrus.*
**offal** (*off'-al*). Refuse of any kind.
**official** [*of-ish'-al*] [*officium,* duty; service]. Of medicines, sanctioned by the recognized authority, *i. e.,* the pharmacopeia.
**officinal** (*of-is'-in-al*) [*officina,* a workshop]. For sale in the shops; kept on sale in apothecaries' shops.
**ogmomele** (*og-mo-me'-le*) [ὄγμος, furrow; μήλη, probe]. A grooved probe.
**Ogston's line** (*og'-ston*) [Alexander *Ogston,* Scotch surgeon, 1844– ]. An imaginary line extending from the tubercle of the femur to the intercondyloid notch, and indicating the course of the tendon of the adductor magnus; it is a guide in section of the internal condyle from the shaft of the femur.
**ohm** (*ōm*) [Georg Simon *Ohm,* German physicist, 1787–1854]. The unit of electrical resistance. The ohm adopted as a standard varies: the *British Association ohm* is the resistance of a column of mercury 1 square millimeter in section and 1.049318 meters long.

The *legal ohm* is similar to that just described except that the column of mercury is 1.06 meters in length. The *international ohm*, adopted 1893, is the resistance of a column of mercury 1.063 centimeters long and weighing 14.4521 grams. O.'s law, the current strength in any conductor varies directly as the electromotive force, and inversely as the resistance.

**ohmmeter** (ōm'-e-tur). An apparatus for estimating electric resistance in ohms.

**-oid** (-oid). [εἶδος, like]. A suffix signifying likeness or resemblance.

**oidial** (o-id'-e-al). Pertaining to or due to a fungus of the genus *Oidium*.

**oidiomycetes** (o-id-e-o-mi-se'-tes). [*oidium*; μύκης, fungus]. A group of fungi which includes *Oidium*.

**oidiomycosis** (o-id-e-o-mi-ko'-sis) [ᾠόν, an egg; *mycosis*]. A disease produced by yeast-fungi of the genus *Oidium*. o. **cutis**, a cutaneous disease produced by blastomycetic fungi of the genus *Oidium*. o. **lactis**, the white mold found on milk, bread, etc.

**Oidium** (o-id'-e-um) [dim. of ᾠόν, egg]. A genus of parasitic fungi. O. **albicans**, the thrush-fungus.

**oikiomania** (oi-ke-o-ma'-ne-ah) [οἶκος, house; μανία, mania]. Domestic perversity; shrewishness manifested specially in one's own home.

**oikoid** (oi'-koid) [οἶκος, house; εἶδος, like]. The stroma of red corpuscles.

**oikologic** (oi-kol-oj'-ik) [οἶκος, house; λόγος, science]. Relating to the condition and improvement of homes.

**oikology** (oi-kol'-oj-e) [οἶκος, a house; λόγος, science]. The science of the home.

**oikophobia** (oi-ko-fo'-be-ah) [οἶκος, house; φόβος, dread]. Morbid dread, or dislike of home, or of a house.

**oikosite** (oi'-ko-sīt) [οἶκος, house; σῖτος, food]. A parasite fixed to its host.

**oil** [*oleum*, from ἔλαιον, oil]. A liquid of animal or vegetable, sometimes of mineral, origin, having a peculiar feel, and not miscible with water. *Animal and vegetable oils* are either volatile or fixed. (*For the various oils not defined here see under the qualifying word.*) o., **aleurites**, a fixed oil with nutty flavor from the seeds of the candlenut tree, *Aleurites triloba*. It is a mild cathartic, acting in the same manner as castor-oil and more promptly. Dose ½-1 oz. (15-30 Cc.). o., **almond**. See under *amygdala*. o. of **amber**. See *succinum*. o., of **anda**, a fixed oil from seeds of *Joannesia princeps*, used as a purgative. o. of **angelica**, a volatile oil from roots of *Archangelica officinalis*, used as a tonic. o., **animal**, an oil obtained from destructive distillation of bones; applied in skin diseases and used internally in hysteria. Dose 5-20 min. (0.3-1.2 Cc.). o. of **anise**, an essential oil from anise, used as a carminative. o. of **anise, star**, essential oil from fruit of *Illicium anisatum*. o. of **arachis**, fixed oil from peanuts, the fruit of *Arachis hypogæa*. o. of **arbor vitæ**. See o. of *thuja*. o. of **arnica flowers**, an essential oil from *A. montana*, used as a diuretic and an emmenagogue, and externally for rheumatism. o. of **artemesia**, an antiseptic and astringent oil from the flowers of Roman wormwood, *A. maritima*. o. of **asphalt**, a rubefacient, antiseptic oil obtained from destructive distillation of asphalt; it is applied in rheumatism and parasitic skin diseases. o. of **balm**, an essential antispasmodic oil from the leaves of *Melissa officinalis*. o., **basil-**, an essential oil from the leaves of *Ocimum basilicum*. It is antiseptic and stimulant. Dose 1-2 min. (0.06-0.12 Cc.). o. of **behen**. Same as *o. of ben*. o. of **ben**, a fixed oil from seeds of two species of *Moringa*, Asiatic trees. o. of **benne**. See *o. of sesame*. o. of **birch-bark**, a volatile oil from the bark of *Betula lenta*, almost identical with oil of wintergreen; antirheumatic and antiseptic. Dose 5-30 min. (0.3-1.8 Cc.). o. of **birch-wood**, an antiseptic black liquid from *Betula alba*. It is used in skin diseases. o. of **bitter almond**. See under *amygdala*. o. of **boldus**, a volatile oil from *Peumus fragrans*, used in genitourinary inflammations. o., **British**, a variety of petroleum. o. of **cade**, juniper tar, an empyreumatic oil distilled from the wood of *Juniperus oxycedrus*, used as an antiseptic. o. of **camomile**, (German), an essential oil from *Matricaria chamomilla*. It is used in colic, cramps, etc. Dose 1-5 min. (0.06-0.3 Cc.). o. of **camomile** (Roman), an essential oil from the flowers of *Anthemis nobilis*. It is tonic and stomachic. Dose 1-5 min. (0.06-0.3 Cc.). o., **camphorated**, a solution of camphor in olive-oil. o. of **Canada snake-root**, an antiseptic, aromatic oil from *Asarum canadense*. o., **candle-nut.** See o. *aleurites*. o. of **canella**, a volatile, stimulant oil from *Canella alba*, used as a flavor. o. of **caraway**. See under *carum*. o. of **cardamom**, a volatile oil from cardamom, used as an aromatic and stimulant. o., **Carron**, an oil consisting of equal or nearly equal parts of linseed-oil and lime-water. It is used as an application to burns, and is named after the Carron iron-works in Scotland, where it was first employed. o. of **casarilla**, a volatile oil from the bark of *Croton eluteria*; it is stimulant and aromatic and is used as an adjuvant in bitter tonics. Dose 1-2 min. (0.06-0.12 Cc.). o. of **cedar leaves**, a volatile oil from leaves of *Juniperus virginiana*, used as an antiseptic and emmenagogue. o. of **cedar** wood, volatile oil from wood of *Juniperus virginiana*. o. of **cedrat**, volatile oil from fruit-rind of *Citrus medica*, used as an aromatic. o. of **celery**, volatile oil from seeds of *Apium graveolens*, used in nervous affections. o. of **chamomile, German**, a volatile oil from the flowers of *Matricaria chamomilla*, used in cramps. o. of **chamomile, Roman**, a volatile oil from flowers of *Anthemis nobilis*, used as a stomachic. o. of **champaca**, a volatile oil from the flowers of *Michelia champaca*. o. of **cherry-laurel**, an essential oil from the leaves of *Prunus laurocerasus*; it has the odor and properties of oil of bitter almonds; used as a sedative. Dose ¼-½ min. (0.01-0.03 Cc.). o. of **citronella**, an essential oil from various species of the grass *Andropogon*. o. of **cloves**, a volatile oil from cloves, used as an antiseptic. o., **cocoanut-**, a white, semisolid fat, soluble in alcohol and ether, obtained from the nut of *Cocos nucifera*. It is used as an alterant and nutrient and as an ointment-base. Dose 2-4 dr. (8-16 Cc.). o. of **cypress**, a volatile, oily liquid obtained from the fresh leaves and shoots of *Cupressus sempervirens*. It is antiseptic and antispasmodic and is recommended in whooping-cough; used by sprinkling the clothes and room. o., **dead**, a heavy oil. o. of **dill**, a volatile oil from the fruit of *Peucedanum graveolens*, used as a carminative. Dose 3-10 min. (0.2-0.6 Cc.). o., **Dippel's**. See *o., animal*. o., **distilled**, volatile oils. o., **doegling**. See *doegling oil*. o., **dugong**, oil from the cetaceous animal *Halicare dugong*. o., **Dutch**, same as o. of *Haarlem*. o. of **ergot**, a laxative oil from ergot. o. of **erigeron**, a volatile oil from *E. canadense*, used as a hemostatic. o., **essential**, a volatile oil, so-called because it contains the essence or active principle of a plant. o., **ethereal**, a calmative, volatile liquid consisting of ether and heavy oil of wine. o. of **eulachon**, the fixed oil of candle-fish blubber. o., **fatty**, salt-like bodies composed of characteristic acids (oleic, palmitic, and stearic), known as fatty acids, and a base. In most cases the base is glyceryl, the radical of the triatomic alcohol, glycerol, so that the oils are said to be glycerids of the several fatty acids. See *o., fixed*. o. of **fir cones**, a volatile oil from the cones of *Picea excelsa*, used as an antiseptic. o. of **fir, scotch**, volatile oil from the leaves of *Pinus sylvestris*, used in chronic rheumatism. o. of **fireweed**, volatile oil from *Erechtites prælta*, used as a tonic. o., **fixed**, one not volatilizing on the application of heat. Fixed oils are also called fatty oils, because they in part constitute the animal and vegetable fats. Some are liquid, as olive-oil, cottonseed-oil, linseed-oil, castor-oil, etc.; others are solid, as tallow and beeswax, which chemically belong to the group of oils. See *o., fatty*. o., **fusel**, a fatty oily liquid obtained in rectifying brandy and whisky; it consists largely of amyl alcohol; is poisonous, and used as a solvent. o. of **garlic**, volatile oil from bulbs of *Allium sativum*, used as a diuretic and expectorant. o. of **gaultheria**. See *o. of wintergreen*. o. of **ginger**, volatile oil from ginger, used as a stomachic. o. of **gingili**, sesame oil. o., **gomenol**, a terpinol said to be obtained from *Melaleuca leucadendron*; it is recommended in pertussis. Dose 1½-3 dr. (6-12 Cc.) of 5% oil injected into the gluteal muscles. o., **Haarlem**, an oily antiseptic preparation of sulphurated linseed oil and oil of turpentine. o., **heavy, of wine**, the product obtained when alcohol is treated with an excess of sulphuric acid. See *oleum æthereum*. o. of **hemlock**, volatile oil from the bark of *Tsuga canadensis*. o., **herring**, fixed oil of herrings. o. of **hops**, a volatile oil from hops; sedative, tonic, and narcotic. Dose 1-5 min. (0.06-0.3 Cc.). o. of **horsemint**, a volatile

oil from *Monarda punctata;* used as a carminative. Dose 1–10 min. (0.06–0.6 Cc.). o. of **hyoscyamus leaves,** a green oil prepared by heating a fixed oil with fresh leaves of *Hyoscyamus niger;* used as a sedative in the cough of tuberculosis. Dose 1–5 min. (0.06–0.3 Cc.). o., **hyssop,** a volatile oil from *Hyssopus officinalis;* used in diarrhea, colic, etc. Dose 1–5 min. (0.06–0.3 Cc.). o. of **jatropha curcas,** a fixed oil from Barbados nuts, used as a purgative. o. of **juniper berries,** volatile oil from fruit of juniper, used as a diuretic and stimulant and to preserve surgical ligatures. o. of **juniper wood,** volatile oil from the fresh wood of *Juniperus communis.* Not to be used for preserving catgut; not to be confounded with oil of cade. *oil of.* o., **lard,** oil from hog's lard. o. of **lemon,** a volatile carminative oil from fresh lemon-rind. o. of **lemon balm.** Same as o. *of balm.* o. of **lemon grass,** volatile oil from several species of *Andropogon.* o. of **linaloe,** a volatile oil distilled from a Mexican wood of uncertain origin. o., **linseed-.** See *lini, oleum,* under *linum.* o. of **male-fern,** a volatile oil from *Dryopteris filix-mas* and *D. marginalis;* it is used as an anthelmintic. Dose 12–25 min. (0.7–1.5 Cc.). o., **margosa,** a bitter yellow oil from the seeds of *Melia azadirachta;* it is used as an anthelmintic and as an application in rheumatism and in sunstroke. o., **marjoram, wild,** an essential oil from the tops of *Origanum vulgare;* it is antiseptic, tonic, and emmenagogue. Dose 2–10 min. (0.12–0.6 Cc.). It is also used externally in skin diseases. o. of **matico,** a volatile antiseptic oil from *Piper angustifolium;* used in diseases of the urinary tract. Dose ½–1 min. (0.03–0.06 Cc.). o. of **menhaden,** a fixed oil from blubber of menhaden. o. of **milfoil,** a carminative oil from the flowers of *Achillea millefolium.* o., **mineral,** petroleum and certain of its derivatives. o. of **mint, curled,** volatile oil from the leaves of *Mentha aquatica.* o. of **mirbane,** nitrobenzene, *q. v.* o. of **mustard.** See under *mustard.* o. of **myrtle,** volatile oil from the leaves of *Myrtus communis,* used as an antiseptic. o. of **nagkassar** or **nahor,** oil from the seeds of *Mesua ferrea,* used locally in rheumatism. o., **neatsfoot,** fixed, lubricant oil from the feet of neat cattle. o. of **Neroli,** an essential oil from the flowers of bitter orange, used as a perfume. o., **Niaouli,** a volatile oil from the leaves of *Melaleuca leucadendron,* containing 66 % of eucalyptol; used in tuberculosis. Dose 4 min. (0.25 Cc.) 6 times daily; as an injection, 16½ min. (1 Cc.) in olive-oil. o., **orange,** volatile oils from the leaves, flowers, and fruit of various species of orange. o. of **orange-peel,** an essential oil from the fresh rind of *Citrus aurantium.* o., **origanum.** See o., *marjoram, wild;* also the common but erroneous name for oil of thyme. o. of **orris,** a volatile oil from the rhizome of several species of *Iris.* o., **paraffin,** principally hydrocarbons of the C$_n$-H$_{2n+2}$ series, distilled from petroleum. o. of **patchouli,** volatile oil from the leaves of *Pogostemon heyneanus,* used as a perfume. o. of **peanut.** Same as o. *of arachis.* o. of **peppermint,** volatile oil from leaves of *Mentha piperita,* used as a carminative and antiseptic. o. of **peppermint, Mitcham,** oil from peppermint grown at Mitcham, Surrey, England. o., **phosphorated,** one per cent. solution of phosphorus in almond oil and ether, used as a nerve stimulant. o. of **pinus pumilio,** a very fragrant volatile oil from the leaves of *Pinus pumilio;* antiseptic and expectorant, and used also in glandular enlargements and skin diseases. Dose 5–10 min. (0.3–0.6 Cc.). o. of **poho,** Japanese oil of peppermint. o. of **poppy,** a fixed oil from the seeds of various species of *Papaver.* o. of **porpoise,** fixed oil from blubber of porpoise. o. of **pumpkin seed,** thick, oily liquid from the seeds of *Curcubita pepo;* it is said to be anthelmintic. o. of **rhodium,** volatile oil from the wood of *Convolvulus scoparius,* used as a perfume. o., **rock-,** petroleum. o. of **rosewood.** See o. *of rhodium.* o. of **sassafras,** volatile oil from the root-bark of *S. officinale,* used as an aromatic and carminative. o. **seneca,** crude petroleum. o., **sesame,** a fixed oil from the seeds of *Sesamum indicum;* it is laxative and nutrient. Dose 4–8 dr. (15–30 Cc.). o., **shark,** a kind of cod-liver oil prepared on shore. o. of **spearmint,** a volatile oil from *Mentha viridis,* used as a carminative and an antiseptic. o., **sperm,** fixed oil from fat of *Physeter macrocephalus,* sperm whale. o. of **spike,** a volatile oil from the leaves and tops of *Lavandula spica;* carminative and rubefacient, and used externally in rheumatism. o., **straits,** the first oil obtained by the exposure of the livers of codfish to the sun in casks, on board ship. o. of **sumbul,** volatile oil from the root of *Ferula sumbul;* it is tonic and antispasmodic. o., **sweet-.** See *olive-oil.* o. of **sweet bay.** See o. *of laurel.* o. of **tansy,** a volatile, poisonous oil from the leaves and tops of *Tanacetum vulgare,* used as an anthelmintic and emmenagogue. o. of **teaberry.** See o. *of gaultheria.* o. of **theobroma,** cacao-butter. o., **turpentine,** a volatile oil from the concrete resin of *Pinus palustris* and other species; chiefly consisting of C$_{10}$H$_{16}$; it is anthelmintic, antiseptic, diuretic, and rubefacient. Dose 5–120 min. (0.3–7.4 Cc.), o., **valerian,** a volatile oil from the rhizome and root of *Valeriana officinalis,* used in nervous diseases. Dose 4–5 min. (0.25–0.3 Cc.). o. of **verbena.** See o. *of lemon grass.* o., **volatile,** one which vaporizes at ordinary temperatures. Volatile oils are odoriferous, and are generally obtained by distillation. o. of **wax,** a volatile oil from the distillation of wax. o. of **white cedar.** See o., *thuja.* o. of **wine, heavy,** oily liquid from the distillation of wine with excess of sulphuric acid. o. of **wintergreen,** an essential oil from the leaves of *Gaultheria procumbens;* used in rheumatism and chronic cystitis. Dose 5–10 min. (0.3–0.6 Cc.). o. of **ylang ylang,** a volatile, antiseptic oil distilled in the Philippine Islands from the flowers of *Cananga odorata.*
**oinomania** (oi-no-ma'-ne-ah) [οἶνος, wine; μανία, madness]. 1. A form of insanity characterized by an irresistible craving for, and consequent indulgence in, drink. 2. Delirium tremens.
**ointment** (oint'-ment) [L., *unguentum*]. A fatty material of the consistence of butter, generally impregnated with a medicinal substance, and used for application to the skin. o., **citrine.** See *citrine ointment.* o., **Lister's,** boric acid, 1; white wax, 1; paraffin, 2; almond oil, 2 parts. o., **Maury's,** one composed of one dram of mercuric nitrate and half a dram each of powdered opium and rhubarb to an ounce of simple ointment. It is used as an application to ulcers. o.-**muslin,** a muslin strip impregnated with ointment. o., **simple.** See *unguentum.* o., **soldiers',** o. **troopers',** mercurial ointment, *q. v.* (For other ointments not here defined see under *unguentum* or the qualifying word.)
**Oken's body** (o'-ken) [Lorenz *Oken,* German naturalist, 1779–1851]. The primitive kidneys or Wolffian bodies.
**-ol.** A termination indicating that the substance is an alcohol or a phenol.
**ol.** Abbreviation for *oleum,* Latin word for oil.
**O.L.A.** Abbreviation for *occipitolaevo anterior,* or left occipitoanterior position of the head of the fetus in labor.
**old-sight.** Presbyopia.
**Olea** (o'-le-ah) [ἐλαία, the olive-tree]. A genus of trees; see *olive.*
**olea** (o'-le-ah) [L.]. Plural of *oleum, q. v.* o. **infusa,** infused oils.
**oleaginous** (o-le-aj'-in-us) [*oleum,* oil]. Oily.
**oleamen** (o-le-a'-men) [L.: *pl.,* oleamina]. An oily, soft ointment or liniment.
**oleander** (o-le-an'-der). See *Nerium.*
**oleandrism** (o-le-an'-drizm). Poisoning by oleander; analogous to digitalism.
**olease** (o'-le-ās). An enzyme in olives which causes precipitation of the coloring-matter of olive-oil and rancidity by formation of fatty acids.
**oleate** (o'-le-āt) [*oleum,* oil]. 1. A salt of oleic acid. 2. A mixture of oleic acid with certain medicinal principles.
**olecranal** (o-lek'-ran-al) [see *olecranon*]. Pertaining to the olecranon.
**olecranarthritis** (o-lek-ran-ar-thri'-tis) [*olecranon; arthritis*]. Inflammation of the elbow-joint.
**olecranarthrocace** (o-lek-ran-arth-rok'-as-e) [*olecranon;* ἄρθρον, joint; κακός, evil]. Inflammation of the elbow-joint.
**olecranoid** (o-lek'-ran-oid) [*olecranon;* εἶδος, like]. Resembling the olecranon. o. **fossa,** the fossa at the dorsal side of the distal end of the humerus for the reception of the olecranon.
**olecranon** (o-lek'-ran-on) [ὠλένη, the ulna; κρανίον, skull]. The large concave process at the upper extremity of the ulna.

**olefiant** (o-lef'-e-ant) [see *olefin*]. Making oil. o. **gas**. See *ethylene*.
**olefin** (o'-lef-in) [*oleum*, oil; *facere*, to make]. Olefiant gas; also any one of a series of unsaturated hydrocarbons having the formula C$_n$H$_{2n}$.
**oleic** (o'-le-ik) [*oleum*]. Relating to, containing, or obtained from oil. o. **acid**. See *acid, oleic*.
**olein** (o'-le-in) [*oleum*], C$_{57}$H$_{104}$O$_6$. A neutral fat, glyceryl trioleate, occurring in olive-oil, butter, and other animal and vegetable fats. It is a colorless oil with a faint, sweetish taste, insoluble in water, readily soluble in alcohol and ether.
**oleo-** (o-le-o-) [*oleum*]. A prefix to denote connection with or relation to an oil.
**oleobalsamic mixture** (o-le-o-bawl-sam'-ik). A mixture of the oils of lavender, thyme, lemon, mace, orange-flowers, cloves, and cinnamon, with balsam of Peru and alcohol. It is used as a nervine.
**oleocreosote** (o-le-o-kre'-o-sōt) [*oleo-*; *creosote*]. A yellowish, neutral liquid composed of creosote, 33 %, and oleic acid. It is used in bronchial and pulmonary diseases. Dose 10-15 min. (0.65-1.0 Cc.).
**oleoguaiacol** (o-le-o-gwi'-ak-ol). See *guaiacol oleate*.
**oleoinfusion** (o-le-o-in-fu'-zhun). An oily solution of a drug.
**oleomargarine** (o-le-o-mar'-gar-ēn) [*oleo-*; μάργαρος, the pearl-oyster]. An artificial butter made by removing the excess of stearin from tallow or suet.
**oleometer** (o-le-om'-et-er) [*oleum*, oil; μέτρον, measure]. An instrument for ascertaining the weight and purity of oil.
**oleoresin** (o-le-o-rez'-in) [*oleo-*; *resina*, resin]. A substance consisting chiefly of a mixture of an essential oil and a resin extracted from plants with ether.
**oleosaccharose** (o-le-o-sak'-ar-ōs). A compound of saccharose with an essential oil.
**oleosaccharum** (o-le-o-sak'-ar-um) [*oleum*, oil; σάκχαρον, sugar; *pl.*, *oleosacchara*]. An oil-sugar; a preparation made by saturating thirty grains of sugar with one drop of volatile oil.
**oleum** (o'-le-um) [L.]. Oil. (*For the various oils* (*olea*) *not defined here see under* oil *or under the qualifying word.*) o. **æthereum**, a volatile, yellowish liquid consisting of equal volumes of heavy oil of wine and ether. Heavy oil of wine is produced when alcohol and sulphuric acid are distilled, and is a mixture of ethyl sulphate, ethyl sulphite, and several polymeric forms of ethylene. o. **fixum**, a fixed oil. See *oil, fixed*. o. **nigrum**, a reddish-yellow oil, obtained in the East Indies from the seeds of *Celastrus paniculatus*; it is a powerful stimulant and diaphoretic, and is used in rheumatism, gout, and various fevers. o. **phosphoratum**, a mixture of phosphorus, 1 gr.; ether, 9 gr.; almond oil, 90 Cc. It is prescribed in rhachitis.
**olfaction** (ol-fak'-shun) [*olfacere*, to smell]. The function of smelling.
**olfactive** (ol-fak'-tiv) [*olfaction*]. Synonym of *olfactory*. o. **angle**, the angle formed by the line of the olfactory fossa and the os planum of the sphenoid bone.
**olfactometer** (ol-fak-tom'-et-er) [*olfaction*; μέτρον, a measure]. An instrument for determining the power of smell.
**olfactometry** (ol-fak-tom'-et-re). The science of measuring the acuteness of the sense of smell.
**olfactory** (ol-fak'-to-re) [*olfaction*]. Pertaining to the sense of smell. o. **bulb**, the bulbous end of the olfactory nerve. o. **cells**, the cells of the nasal fossæ forming the peripheral end-organs of the olfactory nerve. o. **center**, the cerebral center for the sense of smell, supposed to be in the hippocampal gyrus. o. **glomerulus**, one of the terminations of the olfactory fibers in the olfactory bulb. o. **groove**. See o. *sulcus*. o. **islets**. See *Calleja's olfactory islets*. o. **lobe**, the olfactory tubercle, olfactory tract, and olfactory bulb considered together. o. **membrane**, the Schneiderian membrane. q. v. o. **nerve**, the first cranial nerve, the nerve of smell. o. **region**, the area of distribution of the olfactory nerve in the upper part of the nose. o. **sulcus**, the furrow for the olfactory tract and bulb on the cribriform plate of the ethmoid bone, and on the orbital surface of the cerebral hemispheres. o. **tract**, the central portion of the olfactory lobe terminating anteriorly in the olfactory bulb and posteriorly in the olfactory tubercle. o. **trigone**, the triangular mass of gray matter between the roots of the olfactory bulb. o. **tubercle**, the expanded end of the narrow olfactory lobe, commonly called olfactory nerve or first cranial nerve; the olfactory bulb. o. **vesicle**, the embryonic vesicle forming the olfactory tract and bulb.
**olibanum** (o-lib'-an-um) [Ar., *al-lubān*, frankincense]. A gum-resin produced by various species of *Boswellia*. It has been used as a substitute for the balsams of Peru and India, as an inhalation in laryngeal and bronchial inflammations, for fumigation, and in plasters. Syn., *frankincense*.
**oligemia, oligæmia** (ol-ig-e'-me-ah) [ὀλίγος, scanty; αἷμα, blood]. A state in which the total quantity of the blood is diminished. **oligæmia serosa**. Same as *hydremia*.
**olighemia, olighæmia** (ol-ig-he'-me-ah). Same as *oligemia, q. v.*
**olighidria, oligidria** (ol-ig-hi'-dre-ah, ol-ig-id'-re-ah) [*oligo-*; ἱδρώς, sweat]. Deficiency of perspiration.
**olighydria** (ol-ig-hi'-dre-ah) [*oligo-*; ὕδωρ, water]. Deficiency of the fluids of the body.
**oligo-** (ol-ig-o-) [ὀλίγος, few or scanty]. A prefix signifying want or deficiency.
**oligoblennia** (ol-ig-o-blen'-e-ah) [*oligo-*; βλέννα, mucus]. A deficient secretion of mucus.
**oligocardia** (ol-ig-o-kar'-de-ah) [*oligo-*; καρδία, heart]. Same as *bradycardia*.
**oligocholia** (ol-ig-o-ko'-le-ah) [*oligo-*; χολή, bile]. A deficiency of bile.
**oligochromemia, oligochromæmia** (ol-ig-o-kro-me'-me-ah) [*oligo-*; χρῶμα, color; αἷμα, blood]. Deficiency of hemoglobin in the blood.
**oligochrosis** (ol-ig-o-kro'-sis) [*oligo-*; χρῶσις, a coloring]. Deficiency of hemoglobin in the blood-corpuscles.
**oligochylia** (ol-ig-o-ki'-le-ah) [*oligo-*; χυλός, chyle]. A deficiency of chyle.
**oligochymia** (ol-ig-o-ki'-me-ah) [*oligo-*; χυμός, juice]. A deficiency of chyme.
**oligocopria** (ol-ig-o-kop'-re-ah) [*oligo-*; κόπρος, excrement], Deficiency of excrement.
**oligocystic** (ol-ig-o-sis'-tik) [*oligo-*; κύστις, cyst]. Having few cysts or open spaces.
**oligocythemia, oligocythæmia** (ol-ig-o-si-the'-me-ah) [*oligo-*; κύτος, cell; αἷμα, blood]. A deficiency of red corpuscles in the blood.
**oligocytosis** (ol-ig-o-si-to'-sis). See *oligocythemia*.
**oligodacrya** (ol-ig-o-dak'-re-ah) [*oligo-*; δάκρυον, tear]. Deficiency of the tears.
**oligodactylia** (ol-ig-o-dak-til'-e-ah) [*oligo-*; δάκτυλος, finger]. A condition characterized by a deficiency of fingers or toes.
**oligoerythrocythemia, oligoerythrocythæmia** (ol-ig-o-er-ith-ro-si-the'-me-ah) [*oligo-*; ἐρυθρός, red; κύτος, cell; αἷμα, blood]. Deficiency of the coloring-matter of the red corpuscles of the blood.
**oligogalactia** (ol-ig-o-gal-ak'-te-ah) [*oligo-*; *galactia*]. Deficiency in the secretion of milk.
**oligogalia** (ol-ig-o-ga'-le-ah). See *oligogalactia*.
**oligoglobulia** (ol-ig-o-glo-bū'-le-ah). See *oligocythemia*.
**oligohemia, oligohæmia** (ol-ig-o-he'-me-ah). See *oligemia*.
**oligohydramnios** (ol-ig-o-hi-dram'-ne-os) [*oligo-*; *hydramnios*]. A deficiency in the quantity of the amniotic fluid.
**oligohydria** (ol-ig-o-hi'-dre-ah). Same as *olighydria*.
**oligomania** (ol-ig-o-ma'-ne-ah) [*oligo-*; μανία, madness]. Insanity in which only a few of the mental faculties are deranged.
**oligomelus** (ol-ig-o-me'-lus). Excessive congenital thinness of the limbs, or a deficiency in their number.
**oligomenorrhea, oligomenorrhœa** (ol-ig-o-men-or-e'-ah) [*oligo-*; μήν, month; ῥοία, flow]. Insufficiency of the menstrual flow.
**oligomorphic** (ol-ig-o-mor'-fik) [*oligo-*; μορφή, form]. Applied to organisms which have but few stages of development.
**oligonitrophilous** (ol-ig-o-ni-trof'-il-us) [*oligo-*; *nitrogen*; φιλεῖν, to love]. Organisms occurring freely in nature, which develop in nutrient media containing combined nitrogen. They have the ability of assimilating and utilizing atmospheric nitrogen.
**oligopepsia** (ol-ig-o-pep'-se-ah) [*oligo-*; πέψις, digestion]. Feebleness of digestion.
**oligophosphaturia** (ol-ig-o-fos-fat-ū'-re-ah). A decrease in the amount of phosphates in the urine.
**oligophrenia** (ol-ig-o-fre'-ne-ah) [*oligo-*; φρήν, mind]. Imbecility.

**oligoplasmia** (ol-ig-o-plaz'-me-ah) [oligo-; plasma]. A deficient amount of plasma in the blood.
**oligopnœa** (ol-ig-op-ne'-ah) [oligo-; πνοή, breath]. Respiration diminished in depth or frequency.
**oligoposia, oligoposy** (ol-ig-o-po'-se-ah, ol-ig-op'-o-se) [oligo-; πόσις, drink]. Defective desire for drinking.
**oligopsychia** (ol-ig-op-si'-ke-ah) [oligo-; ψυχή, mind]. Fatuity; imbecility.
**oligosialia** (ol-ig-o-si-a'-le-ah) [oligo-; σίαλον, saliva]. Deficiency of saliva.
**oligospermatic** (ol-ig-o-sper-mat'-ik) [oligo-; σπέρμα, seed]. Pertaining to oligospermia.
**oligospermatism** (ol-ig-o-sperm'-at-izm). See *oligospermia*.
**oligospermia** (ol-ig-o-sper'-me-ah) [oligo-; σπέρμα, seed]. A deficiency in the secretion of semen.
**oligospermism** (ol-ig-o-sperm'-izm). See *oligospermia*.
**oligosteatosis** (ol-ig-o-ste-at-o'-sis) [oligo-; στέαρ, tallow]. Deficiency of the sebaceous secretion.
**oligotrichia** (ol-ig-o-trik'-e-ah) [oligo-; θρίξ, hair]. Scantiness or thinness of hair.
**oligotrophy** (ol-ig-ot'-ro-fe) [oligo-; τροφή, nourishment]. Defective or imperfect nutrition.
**oligozoospermatism** (ol-ig-o-zo-o-sper'-mat-izm). See *oligozoospermia*.
**oligozoospermia** (ol-ig-o-zo-o-sper'-me-ah) [oligo-; ζῷον, animal; σπέρμα, seed]. Deficiency of the spermatozoa in the spermatic fluid.
**oliguresia** (ol-ig-u-re'-se-ah). See *oliguria*.
**oliguresis** (ol-ig-u-re'-sis). See *oliguria*.
**oliguria** (ol-ig-u'-re-ah) [oligo-; οὖρον, urine]. A diminution in the quantity of urine excreted.
**oligydria** (ol-ig-id'-re-ah). See *olighydria*.
**olintal** (o'-lin-tal). A liquid soap containing myrrh, camphor, and menthol; used as a gargle or nasal douche in diphtheria.
**oliophen** (o-li'-o-fen). Salol and linseed oil in olive oil.
**oliva** (o-li'-vah) [L.]. The olivary body.
**olivary** (ol'-iv-a-re) [olive]. 1. Resembling an olive in shape. 2. The olivary body. o. **body**, an oval mass of gray matter situated behind the anterior pyramid of the medulla. o. **fillet**, o. *fillet*, **olivary**. o. **nucleus**. See *nucleus, olivary*. o. **peduncle**, the mass of fibers entering the hilum of the olivary body. o. **process**. See *process, olivary*.
**olive** (ol'-iv) [L., *oliva*]. 1. The olive-tree, *Olea europæa*, of the natural order *Oleaceæ*. The value of the olive lies chiefly in its fruit, from which a fixed oil is expressed—*oleum olivæ* (U. S. P., B. P.). Olive-oil consists chiefly of olein and palmitin, and is used as a nutritive food; in medicine as a laxative; in the treatment of gall-stones; as an anthelmintic; as an emollient external application to wounds, burns, etc.; and as an ingredient of liniments, ointments, and plasters. 2. The olivary body. o. *nucleus*. 1. A nucleus composed of two small masses of gray matter—an outer (the *external accessory olivary body* of Mueller) and an inner (the *internal accessory olivary body* of Mueller), within the olive, situated above and to the inner side of the dentatum. Syn., *accessory olivary nucleus; nucleus juxtaolivaris*. 2. A gray plate dorsad of the olive. 3. A similar but smaller gray plate dorsad of the pyramid. o.-**oil**. See under *olive* (1).
**Oliver's symptom** (ol'-iv-er) [Thomas *Oliver*, English physician, 1853– ]. Pulsation of the larynx, elicited by grasping the larynx between the thumb and index-finger and pressing upward, the patient being in the erect position; it is noted in aneurysm of the aortic arch and in mediastinal tumors that bring the arch of the aorta in contact with the left bronchus. O.'s **test for albumin**, float the urine on a mixture of equal parts of sodium tungstate solution (10 : 4) and a saturated solution of citric acid (10 : 6). If albumin is present, a white ring is formed at the junction of the two liquids.
**Oliver-Cardarelli's symptom**. See *Oliver's symptom*.
**olivifugal** (ol-iv-if'-u-gal) [olive; *fugere*, to flee]. In a direction away from the olivary body.
**olivipetal** (ol-iv-ip'-et-al) [olive; *petere*, to seek]. Toward the olivary body.
**Ollier's law** (ol-e-a') [Louis Xavier Edouard Léopold *Ollier*, French surgeon, 1830–1900]. When two bones are parallel and joined at their extremities by ligaments, arrest of growth in one of them entails developmental disturbances in the other. O.'s layer, the inner or osteogenetic layer of the periosteum.
**olophonia** (ol-o-fo'-ne-ah) [ὅλος, destroyed, lost; φωνή, voice]. Abnormal speech from malformation of vocal organs.
**O. L. P.** Abbreviation for *occipito lævo posterior*, or the left occipitoposterior position of the head of the fetus in labor.
**Olshausen's operation** (ōls'-how-zen) [Robert Michaelis *Olshausen*, German gynecologist, 1835– ]. For *vaginal fixation*: it consists in suturing the round ligaments and a portion of the broad ligaments instead of the uterine fundus.
**-oma** (-o'-mah). Abbreviation of ὄγκωμα, a swelling. A termination signifying a neoplasm or tumor.
**omacephalus** (o-mas-ef'-al-us) [ὦμος, shoulder; κεφαλή, head]. A variety of omphalositic monsters of the species paracephalus, in which there are present the characteristics of paracephalus except that there is more imperfect development, with absence of the upper extremities.
**omagra** (om-a'-grah) [ὦμος, shoulder; ἄγρα, seizure]. Gout in the shoulder.
**omal** (o'-mal). Trichlorphenol, a compound of chlorine and phenol, used by inhalation in bronchial diseases.
**omalgia** (o-mal'-je-ah) [ὦμος, shoulder; ἄλγος, pain]. Same as *omodynia*.
**omarthralgia** (o-mar-thral'-je-ah) [ὦμος, shoulder; ἄρθρον, joint; ἄλγος, pain]. Pain in the shoulder-joint.
**omarthritis** (o-mar-thri'-tis) [ὦμος, shoulder; ar *thritis*]. Inflammation of the shoulder-joint.
**omarthrocace** (o-mar-throk'-as-e) [ὦμος, shoulder; ἄρθρον, joint; κακός, evil]. Disease of the shoulder-joint.
**omasal** (o-ma'-sal) [*omasum*, a paunch]. Pertaining to the omasum.
**omasum** (o-ma'-sum) [*omasum*, a paunch: *pl.*, *omasa*]. The third stomach of a ruminant; it is also called the *psalterium*, and *manyplies*.
**omega melancholium** (o-meg'-ah mel-an-kol'-i-um). An omega-shaped (ω) wrinkle between the eyebrows, said to be a sign of melancholy.
**omelite** (o-mi'-ra) [African]. A vinous and acetous fermented milk used in parts of Africa.
**omental** (o-men'-tal) [*omentum*]. Pertaining to the omentum. o. **hernia**. See *epiplocele*.
**omentectomy** (o-men-tek'-to-me) [*omentum*; ἐκτομή, excision]. Excision of a portion of the omentum.
**omentitis** (o-men-ti'-tis) [*omentum*; ιτις, inflammation]. Inflammation of the omentum.
**omentocele** (o-men'-to-sēl) [*omentum*; κήλη, hernia]. Omental hernia.
**omentopexy** (o-men'-to-peks-e) [*omentum*; πῆξις, fixation]. Same as *epiplopexy*; also, *Talma's operation, q. v.*
**omentosplenopexy** (o-men-to-splen'-o-peks-e). Omentopexy followed by splenopexy to develop complementary circulation in certain hepaticosplenic lesions.
**omentulum** (o-men'-tū-lum) [dim. of *omentum*, oinentum]. The smaller omentum.
**omentum** (o-men'-tum) [L., "adipose membrane"]. A fold of the peritoneum connecting the abdominal viscera with the stomach. o., **gastrocolic**, a fold of peritoneum attached to the greater curvature of the stomach above and, after dipping down over the intestine, returning to inclose the transverse colon. Between the ascending and descending folds is the cavity of the great omentum. Syn., *great omentum*. o., **gastrohepatic**, a double fold of peritoneum passing from the lesser curvature of the stomach to the transverse fissure of the liver. On the left side it incloses the esophagus; on the right its edges are free and inclose all the structures issuing from or entering the transverse fissure of the liver; the hepatic vessels and nerves and the bile-duct. Behind it is the foramen of Winslow. Syn., *lesser omentum*. o., **gastrosplenic**, the fold of peritoneum passing from the stomach to the spleen. o. **majus**. See o., *gastrocolic*. o. **minus**. See o., *gastrohepatic*. o., **pancreaticosplenic**, a fold of peritoneum uniting the tail of the pancreas with the lower part of the inner surface of the spleen.
**omitis** (o-mi'-tis) [ὦμος, shoulder; ιτις, inflammation]. Inflammation of the shoulder.
**omnivorous** (om-niv'-o-rus) [*omnis*, all; *vorare*, to devour]. Subsisting on all kinds of food.
**omnopon** (om'-no-pon). Same as *pantopon, q. v.*

**omo-** (*o-mo-*) [ὦμος, shoulder]. A prefix denoting connection with or relation to the scapula or shoulder.
**omocace** (*o-mok'-as-e*) [*omo-*; κακός, evil]. Disease of the shoulder.
**omoclavicular** (*o-mo-kla-vik'-u-lar*) [*omo-*; *clavicula*, clavicle]. Pertaining to the shoulder and the clavicle.
**omocotyle** (*o-mo-kot'-il-e*) [*omo-*; κοτύλη, cup]. The glenoid cavity; also, the shoulder-joint.
**omodynia** (*o-mo-din'-e-ah*) [*omo-*; ὀδύνη, pain]. Pain in the shoulder.
**omohyoid** (*o-mo-hi'-oid*) [*omo-*; *hyoid*]. Pertaining conjointly to the scapula and the hyoid bone. o. muscle. See under *muscle*.
**omophagia** (*o-mo-fa'-je-ah*) [ὠμός, raw; φαγεῖν, to eat]. The practice of eating raw food.
**omoplate** (*o'-mo-plāt*) [*omo-*; πλάτη, a plate]. See *scapula*.
**omositia** (*o-mo-sit'-e-ah*) [ὠμός, raw; σιτέειν, to feed]. The eating of raw flesh.
**omosternal** (*o-mo-ster'-nal*) [*omo-*; *sternum*]. Pertaining to the shoulder and the sternum.
**omosternum** (*o-mo-ster'-num*) [*omo-*; *sternum*; *pl.*, *omosterna*]. In biology, a superior median ossification in the pectoral arch of a batrachian; the homologue of the interclavicle. The interarticular cartilage of the sternoclavicular joint.
**omphalectomy** (*om-fal-ek'-to-me*) [*omphalos-*; ἐκτομή, excision]. Excision of the navel.
**omphalelcosis** (*om-fal-el-ko'-sis*) [*omphalos*; ἕλκωσις, ulceration]. Ulceration of the navel.
**omphalexoche** (*om-fal-eks'-o-ke*) [*omphalos*; ἐξοχή, prominence]. Synonym of *exomphalos*.
**omphalic** (*om-fal'-ik*) [*omphalos*]. Pertaining to the umbilicus. o. duct, the vitelline duct: the duct connecting the umbilical vesicle with the fetal intestine during the first three months of intrauterine life.
**omphalitis** (*om-fal-i'-tis*) [*omphalos*; ιτις, inflammation]. Inflammation of the navel.
**omphalo-** (*om-fal-o-*) [*omphalos*]. A prefix denoting relation to the navel.
**omphalocele** (*om-fal'-o-sēl*) [*omphalo-*; κήλη, hernia]. Umbilical hernia.
**omphalocraniodidymus** (*om-fal-o-kra-ne-o-did'-im-us*) [*omphalos*; κρανίον, skull; δίδυμος, twin]. A form of double monstrosity in which the parasite is attached to the cranium of the autosite.
**omphalodes** (*om-fal-o'-dēz*) [*omphalos*; εἶδος, like]. Resembling the navel.
**omphaloenteric** (*om'-fal-o-en-ter'-ik*) [*omphalos*; ἔντερον, intestine]. Pertaining to the navel and the intestine.
**omphaloid** (*om'-fal-oid*) [*omphalos*; εἶδος, form]. Resembling the navel.
**omphalolysis** (*om-fal-ol'-is-is*) [*omphalos*; λύσις, loosening]. The dividing of the umbilical cord.
**omphalomesaraic, omphalomesenteric** (*om-fal-o-mez-ar-a'-ik*, *om-fal-o-mez-en-ter'-ik*) [*omphalos*; *mesentery*]. Pertaining conjointly to the umbilicus and the mesentery. o. arteries. See under *artery*. o. duct, a duct connecting the intestinal canal of the embryo with the umbilical vesicle.
**omphalomonodidymi** (*om'-fal-o-mon-o-did'-im-i*) [*omphalos*; μόνος, one; δίδυμος, twin]. A form of twin monstrosity in which the fetuses are joined at the umbilicus.
**omphaloncus** (*om-fal-ong'-kus*) [*omphalo-*; ὄγκος, tumor]. A tumor or swelling at the navel.
**omphalopagus** (*om-fal-op'-ag-us*) [*omphalo-*; πηγνύναι, to make fast]. A double monster united at the umbilicus.
**omphalophlebitis** (*om-fal-o-fleb-i'-tis*) [*omphalo-*; *phlebitis*]. Inflammation of the umbilical vein.
**omphalophyma** (*om-fal-o-fi'-mah*). Synonym of *omphaloncus*.
**omphaloproptosis** (*om-fal-o-prop-to'-sis*) [*omphalos*; πρόπτωσις, a falling forward]. Abnormal protrusion of the navel.
**omphalorrhagia** (*om-fal-or-a'-je-ah*) [*omphalo-*; ῥηγνύναι, to burst forth]. Hemorrhage from the umbilicus.
**omphalorrhea, omphalorrhœa** (*om-fal-or-e'-ah*) [*omphalo-*; ῥοία, a flow]. An effusion of lymph at the navel.
**omphalorrhexis** (*om-fal-or-eks'-is*) [*omphalos*; ῥῆξις, rupture]. Rupture of the navel, or of the navelstring.
**omphalos** (*om'-fal-os*) [ὀμφαλός, the navel]. The umbilicus.

**omphalosite** (*om-fal'-o-sīt*) [*omphalo-*; σῖτος, nourishment]. A single monster, which, lacking the heart, receives its blood-supply through the umbilical vessels, and is, therefore, incapable of extrauterine existence.
**omphalosoter** (*om-fal-o-so'-ter*) [*omphalos*; σωτήρ, preserver]. An instrument for replacing a prolapsed funis.
**omphalotaxis** (*om-fa-o-taks'-is*) [*omphalos*; τάσσειν, to arrange]. Reposition of the prolapsed funis.
**omphalotome** (*om-fal'-o-tōm*) [*omphalos*; τομή, a cutting]. An instrument for dividing the umbilical cord.
**omphalotomy** (*om-fal-ot'-o-me*) [*omphalo-*; τομή, a cutting]. The cutting of the umbilical cord.
**omphalotripsy** (*om-fal-o-trip'-se*) [*omphalo-*; τρίβειν, to rub]. Separation of the umbilical cord by a crushing instrument.
**omphalus** (*om'-fal-us*) [ὀμφαλός, navel]. See *omphalos*.
**onanism** (*o'-nan-izm*) [*Onan*, the son of Judah]. 1. Incomplete coitus. 2. Masturbation.
**onanist** (*o'-nan-ist*). One who practises onanism.
**Onchocerca** (*ong-ko-ser'-kah*). A genus of filaria.
**onchocerciasis** (*ong-ko-ser-si'-as-is*). The condition resulting from infection with *Onchocerca*.
**oncograph** (*ong'-ko-graf*) [ὄγκος, a mass; γράφειν, to record]. An instrument registering the changes of volume of an organ placed in an oncometer.
**oncography** (*ong-kog'-raf-e*) [ὄγκος, a swelling; γράφειν, to write]. The recording of the measurement of tumors by an oncometer.
**oncology** (*ong-kol'-o-je*) [ὄγκος, a mass; λόγος, science]. The branch of surgery and pathology relating to tumors.
**oncoma** (*ong-ko'-mah*) [ὄγκωμα, a swelling]. A tumor, swelling.
**oncometer** (*ong-kom'-et-er*) [ὄγκος, a mass; μέτρον, a measure]. An instrument for measuring variations in the volume of an organ, especially of the kidney or spleen.
**oncometry** (*ong-kom'-et-re*). The measurement of the size of a viscus.
**oncosis** (*ong-ko'-sis*) [ὄγκος, a mass]. The diseased state marked by the growth of tumors.
**oncosphere, oncosphæra** (*ong'-ko-sfēr*, *ong-ko-sfe'-rah*) [ὄγκος, tumor; σφαῖρα, sphere]. The embryo of tapeworms.
**oncothlipsis** (*ong-ko-thlip'-sis*) [ὄγκος, tumor; θλίψις, pressure]. Pressure caused by a tumor.
**oncotomy** (*ong-kot'-o-me*) [ὄγκος, a mass; τομή, a cutting]. The operation of incising a tumor or other swelling.
**oneiric**, **oniric** (*o-ni'-rik*) [ὄνειρος, a dream]. Relating to dreams; attended by visions.
**oneirism** (*o-ni'-rizm*) [see *oneiric*]. A condition of cerebral automatism analogous to the dream state, as a dream prolonged to the waking period.
**oneirodynia** (*o-ni-ro-din'-e-ah*) [ὄνειρος, a dream; ὀδύνη, pain]. Disquietude of the mind during sleep; painful dreaming; nightmare. o. activa, somnambulism.
**oneirogmus** (*on-i-rog'-mus*) [ὀνειρωγμός], an effusion of semen during sleep. Emission of semen during sleep.
**oneirology** (*o-ni-rol'-o-je*) [ὄνειρος, dream; λόγος, science]. The science or scientific view of dreams.
**oneironosus** (*on-i-ron'-o-sus*) [ὄνειρος, dream; νόσος, disease]. Disorder manifesting itself in dreams; morbid dreaming.
**oniomania** (*o-ne-o-ma'-ne-ah*) [ὤνιος, to be bought; μανία, madness]. A mania for buying everything.
**onion** (*un'-yun*) [L., *unio*, an onion]. The *Allium cepa* and its bulb. The latter contains a volatile oil resembling oil of garlic, and consisting principally of $C_6H_{14}S$. The onion is diuretic, expectorant, and rubefacient, and is at times used in dropsy, bronchitis, etc.; locally it has been applied as an emollient poultice. o. bodies, epithelial pearls; see under *pearl*.
**onkinocele** (*ong-kin'-o-sēl*) [ὄγκος, a mass; ἴς, a fiber; κήλη, tumor]. Inflammation of the tendon-sheaths attended by swelling.
**onobaio** (*on-o-ba'-yo*) [Nat. Obock]. An arrow-poison used by the natives of Obock.
**onomatology** (*on-o-mat-ol'-o-je*) [ὄνομα, name; λόγος, science]. The science of nomenclature; the formation of names.
**onomatomania** (*on-o-mat-o-ma'-ne-ah*) [ὄνομα, name; μανία, mania]. Functional derangement of speech, of which five varieties are described: 1. A powerful effort to recall some word. 2. An irresistible im-

pulse continually to repeat a word, by which the patient seems perplexed. 3. The patient attaches some peculiar and dreadful meaning to a commonplace word. 4. The patient attaches talismanic significance to certain words, which he repeats as a safeguard. 5. The patient is impelled to spit out some word, like a disgusting morsel.

**onomatopoiesis** (*on-o-mat-o-poi-e'-sis*) [ὄνομα, name; ποιεῖν, to make]. The extemporaneous formation of words by the insane. Words so formed are generally meaningless, or incorrect, but are sometimes quite the reverse.

**ononid** (*o-no'-nid*), C₁₈H₂₂O₈. A neutral principle contained in the root of *Ononis spinosa*.

**ononin** (*o-no'-nin*), C₃₀H₃₄O₁₃. A glucoside isolated rom the root of *Ononis spinosa*.

**Ononis** (*o-no'-nis*) [ὄνοs, an ass]. A genus of leguminous plants. The diuretic root of *O. spinosa*, rest-harrow, a shrub of Europe, is used in dropsy and gout.

**ontogenesis, ontogeny** (*on-to-jen'-es-is, on-toj'-en-e*) [ὤν; ὄντος, existing; γεννᾶν, to beget]. The development of the individual organism. See also *phylogenesis*.

**ontogenetic** (*on-to-jen-et'-ik*) [ὤν; ὄντος, existing; γένεσις, birth]. Pertaining to ontogenesis. See *evolution*.

**onychatrophia, onychatrophy** (*on-ik-at-ro'-fe-ah, on-ik-at'-ro-fe*) [onyx; atrophy]. Atrophy of the nails.

**onychauxis** (*on-ik-awks'-is*) [onyx; αὔξειν, increase]. Hypertrophy of the nail.

**onychexallaxis** (*on-ik-eks-al-aks'-is*) [ὄνυξ, nail; ἐξάλλαξις, a degeneration]. Degeneration of the nails.

**onychia** (*on-ik'-e-ah*) [onyx]. Inflammation of the matrix of the nail. o. **maligna,** a form occurring in debilitated persons, and characterized by an unhealthy ulcer in the matrix of the nail, the latter becoming discolored and thrown off. o. **simplex,** onychia without much ulceration, with loss of the nail and its replacement by a new one.

**onychitis** (*on-ik-i'-tis*). See *onychia*.

**onycho-** (*on-ik-o-*) [onyx]. A prefix meaning relating to the nails.

**onychoclasis** (*on-ik-ok'-las-is*) [onycho-; κλάσις, a breaking]. Breaking of the nail.

**onychocryptosis** (*on-ik-o-krip-to'-sis*) [onycho-; κρυπτός, hidden]. Ingrowing of the nail.

**onychogram** (*on'-ik-o-gram*) [onycho-; γράφειν, to write]. The record of the variations in blood-pressure by an onychograph.

**onychograph** (*on-ik'-o-graf*) [onycho-; γράφειν, to write]. An instrument for recording variations in blood-pressure in the capillaries of the tips of the fingers.

**onychogryphosis** (*on-ik-o-gri-fo'-sis*). See *onychogryposis*.

**onychogryposis** (*on-ik-o-gri-po'-sis*) [onycho-; γρίπωσις, curvature]. A thickened, ridged, and curved condition of the nail.

**onychohelcosis** (*on-ik-o-hel-ko'-sis*) [onycho-; ἕλκωσις, ulceration]. Ulceration of the nail.

**onychoid** (*on'-ik-oid*) [ὄνυξ, nail; εἶδος, like]. Resembling a nail. Having a texture like that of the nails.

**onycholysis** (*on-ik-ol'-is-is*) [onycho-; λύσις, a loosening]. Loosening of the nail.

**onychoma** (*on-ik-o'-mah*) [onycho-; ὄμα, tumor]. A tumor of the nail-bed.

**onychomalacia** (*on-ik-o-mal-a'-se-ah*) [onycho-; μαλακία, softness]. Abnormal softness of the nails.

**onychomycosis** (*on-ik-o-mi-ko'-sis*) [onycho-; mycosis]. A disease of the nails due to parasitic fungi, as the trichophyton, achorion, etc.

**onychonosus** (*on-ik-on'-o-sus*) [onycho-; νόσος, disease]. Any disease of the finger-nails or toe-nails.

**onychopathic** (*on-ik-o-path'-ik*) [onycho-; πάθος, disease]. Pertaining to disease of the nails.

**onychophagist** (*on-ik-off'-aj-ist*) [onycho-; φαγεῖν, to eat]. One addicted to biting the finger-nails.

**onychophagy** (*on-ik-off'-aj-e*) [onycho-; φαγεῖν, to eat]. The practice of biting the nails.

**onychophosis** (*on-ik-o-fo'-sis*) [onycho-; ὑφή, a web]. A disease of the toe-nails, consisting in an accumulation of thickened horny layers of epidermis under the nail, raising it from its bed and sometimes altering its growth.

**onychophyma** (*on-ik-o-fi'-mah*) [onycho-; φῦμα, a growth]. Morbid degeneration of the nails.

**onychoptosis** (*on-ik-op-to'-sis*) [onycho-; πτῶσις, a falling]. The falling off of the nails.

**onychorrhexis** (*on-ik-or-eks'-is*) [onycho-; ῥῆξις, rupture]. The splitting of the nails.

**onychorrhiza** (*on-ik-or-i'-zah*) [onycho-; ῥίζα, a root]. The root of the nail.

**onychosarcoma** (*on-ik-o-sar-ko'-mah*) [onycho-; σάρξ, flesh; ὄμα, tumor]. A fleshy outgrowth from a nail.

**onychosis** (*on-ik-o'-sis*) [onyx]. Any disease of the nails.

**onychostroma** (*on-ik-os-tro'-mah*) [onycho-; στρῶμα, mattress]. The matrix, or sensitive tissue of the finger, forming the bed of the nail.

**onychyphosis** (*on-ik-if-o'-sis*). See *onychophosis*.

**onym** (*on'-im*) [ὄνυμα, a name]. The technical name of an organ or of a species or other group.

**onyx** (*on'-iks*) [ὄνυξ, nail]. 1. A nail of the fingers or toes. 2. A collection of pus between the corneal lamellæ at the most dependent part.

**onyxis** (*on-iks'-is*) [ὄνυξ, nail]. An abnormal incurvation or ingrowing of the nails.

**onyxitis** (*on-iks-i'-tis*). Onychia.

**ooblast** (*o'-o-blast*) [ᾠόν, egg; βλαστός, a germ]. A cell of the germinal epithelium giving rise to an ovum.

**oocyesis** (*o-o-si-e'-sis*) [ᾠόν, egg; κύησις, pregnancy]. Ovarian pregnancy; oariocyesis.

**oocyst** (*o'-o-sist*) [ᾠόν, egg; κύστις, bladder]. 1. The encysted fertilized cell in sporozoa. 2. The envelope which surrounds the cell.

**oocytase** (*o-o-si-tās'*). A cytase which acts on the cells of the ovary.

**oocyte** (*o'-o-sīt*) [ᾠόν, egg; κύτος, a cell]. The ovarian egg-cell before the formation of the polar bodies. Syn., *ovocyte*.

**oodeocele** (*o-od'-e-o-sēl*) [ᾠόν, egg; εἶδος, like; κηλή, hernia]. Obturator hernia.

**oodocresol** (*o-od-o-kre'-sol*). See *traumatol*.

**oogamous** (*o-og'-am-us*) [ᾠόν, an egg; γάμος, marriage]. In biology, exhibiting or reproduced by the conjugation of dissimilar gametes.

**oogamy** (*o-og'-am-e*) [ᾠόν, an egg; γάμος, marriage]. In biology, the conjugation of two dissimilar gametes, as distinguished from isogamy.

**oogenesis** (*o-o-jen'-es-is*) [ᾠόν, egg; γένεσις, birth]. The process of the development of the ovum.

**oogenetic** (*o-o-jen-et'-ik*) [ᾠόν, egg; γεννᾶν, to beget]. Relating to oogenesis.

**oogonium** (*o-o-go'-ne-um*) [ᾠόν, an egg; γονή, generation: *pl.*, *oogonia*]. 1. In biology, the female sexual organ in the *Oosporeæ* before fertilization. 2. The primordial mother-cell which gives rise to the ovarian egg and its follicle. 3. The descendants of the primordial germ-cell from which ultimately arise the oocytes (Boveri); also written *ovogonium*.

**ookinesis** (*o-o-kin-e'-sis*) [ᾠόν, egg; κίνησις, movement]. The changes occurring in the egg during maturation, fertilization and segmentation.

**ookinete** (*o'-o-kin-ēt*) [ᾠόν, egg; κίνησις, movement]. The vermiform, motile, body into which the zygote develops. See *oocyst*.

**oolemma** (*o-o-lem'-ah*) [ᾠόν, an egg; λέμμα, peel, skin]. In biology, the vitelline membrane of an egg.

**oophoralgia** (*o-of-or-al'-je-ah*) [*oophoron*; ἄλγος, pain]. Pain in the ovaries.

**oophorauxe** (*o-of-or-awks'-e*) [ᾠόν, egg; φέρειν, to bear; growth]. Hypertrophy of the ovary.

**oophorectomy** (*o-o-for-ek'-to-me*) [*oophoron*; ἐκτομή, excision]. Excision of the ovary.

**oophorin** (*o-off'-or-in*). Trade name of an organotherapeutic preparation made from the ovaries of cows and hogs.

**oophoro-** (*o-of-or-i'-tis*) [*oophoron*; ιτις, inflammation]. Inflammation of the ovary.

**oophoro-** (*o-of-or-o-*) [*oophoron*]. A prefix meaning relating to the ovary.

**oophorocystosis** (*o-of-or-o-sist-o'-sis*) [*oophoro-*; κύστις, a cyst]. The formation of ovarian cysts.

**oophoroepilepsy** (*o'-off-or-o-ep'-il-ep-se*). Epileptiform disease due to an ovarian lesion.

**oophorohysterectomy** (*o-of-or-o-his-ter-ek'-to-me*) [*oophoro-*; *hysterectomy*]. Removal of the uterus and ovaries.

**oophoroma** (*o-off-or-o'-mah*) [*oophoro-*; ὄμα, a tumor: *pl.*, *oophoromata*]. A dermoid cyst of the ovary, characterized by metastasis.

**oophoromalacia** (*o-off-or-o-mal-a'-se-ah*) [*oophoro-*; μαλακία, softness]. Softening of the ovary.

**oophoromania** (*o-of-or-o-ma'-ne-ah*) [*oophoro-*; μανία, madness]. Insanity due to ovarian disorder.

**oophoromyeloma** (o-of-or-o-mi-el-o'-mah) [oophoro-; μυελός, marrow; ὄμα, tumor]. An ovarian encephaloma.

**oophoron** (o-of'-o-ron) [ᾠόν, egg; φέρειν, to bear]. The ovary.

**oophoropathia** (o-of-or-o-pa'-the-ah) [oophoro-; πάθος, disease]. Any disease of the ovary.

**oophorosalpingectomy** (o-of-or-o-sal-pin-jek'-to-me) [oophoro-; salpingectomy]. Excision of an ovary and oviduct.

**oophorosalpingitis** (o-of-or-o-sal-pin-ji'-tis) [oophoro-; salpingitis]. Inflamation of an ovary and Fallopian tube.

**oophorosalpingotomy** (o-of-or-o-sal-pin-got'-o-me) [oophoro-; salpingotomy]. Surgical removal of the ovary and oviduct.

**oophorostomy** (o-o-for-os'-to-me) [oophoro-; στόμα, mouth]. The establishment of an opening into an ovarian cyst for drainage.

**oophorrhaphy** (o-of-or'-a-fe) [oophoro-; ῥαφή, suture]. The operation of suturing an ovary to the pelvic wall.

**oophyte** (o'-o-fit) [ᾠόν, an egg; φυτόν, a plant]. Same as *oophore*.

**ooplasma** (o-o-plaz'-mah) [ᾠόν, egg; πλάσσειν, to mold]. The vitellus.

**ooplasty** (o'-o-plas-te) [ᾠόν, egg; πλάσσειν, to form]. The process of fecundation.

**ooscope** (o'-o-skōp) [ᾠόν, egg; σκοπεῖν, to view]. An apparatus for observing the developmental changes in a fertilized egg.

**oosperm** (o'-o-sperm) [ᾠόν, egg; σπέρμα, seed]. The cell formed by union of the ovum and the spermatozoon.

**oosphere** (o'-o-sfēr) [ᾠόν, an egg; σφαῖρα, a sphere]. In biology, the unfertilized germ-cell in the oogonium; the female reproductive cell.

**Oospora** (o-os'-po-rah) [ᾠόν, egg; σπόρος, seed]. A genus of fungi. O. bovis, a name proposed for the fungus of actinomycosis. O. guiguardi, a fungus that causes the deposition of carbonate of lime in the connective tissue of the walls of cystic tumors and skeletal nodosities.

**oospore** (o'-o-spor) [ᾠόν, an egg; σπόρος, seed]. In biology, a fertilized and matured oosphere.

**ootheca** (o-o-the'-kah) [ᾠόν, an egg; θήκη, a case: *pl.*, *oothecæ*]. An ovary.

**oothectomy** (o-o-thek'-to-me) [ᾠόν, egg; θήκη, a case; ἐκτομή, excision]. The surgical removal of an ovary.

**opa** (o'-pah). Trade name of a liquid dentifrice containing salol, eugenol, pinol and other substances.

**opacification** (o-pas-if-i-ka'-shun) [*opacity*]. 1. The process of becoming opaque. 2. The formation of an opacity.

**opacity** (o-pas'-it-e) [*opacus*, dull]. 1. The condition of being impervious to light. 2. An opaque spot, as *opacity* of the cornea or lens.

**opaline** (o'-pal-ēn) [ὀπάλλιος, an opal]. Having the appearance of an opal. **o. patch**, the mucous patch of syphilis, forming, in the mouth, a whitish pellicle.

**opalisin** (o-pal'-is-in). A protein found in considerable quantity in human milk, less in mares' milk, and in very small quantity in cows' milk.

**opaque** (o-pāk') [*opacus*, shaded]. Dark, obscure, not transparent, impervious to light.

**opeidoscope** (o-pi'-do-skōp) [ὄψ, a voice; εἶδος, likeness; σκοπεῖν, to view]. An instrument for studying the vibrations of the voice.

**open** [AS.]. Exposed to the air, as an *open* wound; interrupted, as an *open* circuit, one that is interrupted so that the electric current cannot pass.

**opeocele** (o'-pe-o-sēl) [ὀπή, an opening; κήλη, a tumor]. Synonym of *hernia*.

**operable** (op'-er-ab-l) [*operari*, to labor]. Admitting of an operation.

**operant** (op'-e-rant) [see *operation*]. 1. Effective; active. 2. An operator.

**operation** (op-er-a'-shun) [*operatio*, from *operari*, to labor; to do]. 1. Anything done or performed, especially anything done with instruments; a surgical procedure. 2. The mode of action of anything. Operations named after persons are entered under the proper names. **o., capital**, one involving a risk of life. **o., equilibrating**, tenotomy on the direct antagonist of a paralyzed ocular muscle. **o., high**. 1. Suprapubic lithotomy. 2. The application of the forceps to the fetal head at the superior strait. **o., Indian**, *for rhinoplasty*: a flap is taken from the forehead, with its pedicle at the root of the nose; hollow plugs are inserted into the nostrils, and the flap is secured. **o., Italian**, *for rhinoplasty*: the skin is taken from the arm over the biceps; the flap is cut on three sides, and the skin has shrunk, it is fitted to the fresh margins of the defect, the arm being bandaged in position for at least eight days. **o., major**, an important and serious operation. **o., minor**, a comparatively trivial operation. **o., radical**, one removing the cause of the disease or the diseased part itself. **o., surgical**, one performed by the surgeon by means of the hands or instruments. **o., Tagliacotian**, **o., Tagliacozzi's**. See **o., Italian**. **operative** (op'-er-a-tiv). 1. Able to act; effective. 2. Pertaining to operations.

**operator** (op'-er-a-tor) [*operari*, to labor]. A surgeon.

**opercle** (o-per'-kl) [*operculum*, a lid]. See *operculum*.

**opercular** (o-per'-kū-lar) [*operculum*]. Pertaining to an operculum. Designed for closing a cavity. Having an operculum.

**operculate** (o-per'-kū-lāt) [*operculum*]. Possessing an operculum.

**operculum** (o-per'-kū-lum) [L.: *pl.*, *opercula*]. 1. A lid or cover, as *operculum ilei*, the ileocecal valve. 2. The convolutions covering the island of Reil.

**oph**. An abbreviation sometimes employed for *ophthalmia*, *ophthalmoscope*, and *ophthalmoscopy*.

**ophelic acid** (o-fel'-ik), $C_{13}H_{20}O_{10}$. An amorphous sticky substance found in *Chiretta*. It is soluble in water, in ether, and in alcohol.

**ophiasis** (of-i'-as-is) [ὄφις, a serpent]. Alopecia areata in which the baldness progresses in a serpentine form.

**ophiosis** (of-e-o'-sis) [ὄφις, serpent]. Circumscribed baldness with scaliness.

**ophioxylin** (of-e-oks'-il-in) [ὄφις, a serpent; ξύλον, wood], $C_{15}H_{18}O_5$. A yellow crystalline body obtained from *Ophioxylon serpentinum*, a purgative and anthelmintic; its solutions stain first yellow, then brown.

**ophryitis** (of-re-i'-tis) [ὀφρύς, brow; ιτις, inflammation]. Inflammation of the eyebrow.

**ophryoalveoloauricular** (of-re-o-al-ve-o-lo-aw-rik'-u-lar). Applied to an angle formed by the ophryon, alveolar point, and auricular point.

**ophryocystis** (of-re-o-sis'-tis) [ὀφρύς, eyebrow; κύστις, bladder]. A parasitic sporozoon.

**ophryoiniac** (of-re-o-in'-e-ak) [ὀφρύς, eyebrow; ἰνίον, occiput]. Pertaining to the ophryon and the inion.

**ophryon** (of'-re-on) [ὀφρύς, eyebrow]. In craniometry, the middle of a line drawn across the forehead at the level of the upper margin of the orbits.

**ophryosis** (of-re-o'-sis) [ὀφρύς, eyebrow]. Spasm of the eyebrow.

**ophryphtheiriasis** (of-rif-thi-ri'-as-is) [ὀφρύς, eyebrow; φθειρίασις, pediculosis]. Pediculosis of the eyebrows and eyelashes.

**ophrys** (of'-ris) [ὀφρύς, eyebrow]. The eyebrow.

**ophrytic** (of-rit'-ik) [ὀφρύς, eyebrow]. Pertaining to the eyebrow.

**ophthalmagra** (of-thal-ma'-grah) [ὀφθαλμός; ἄγρα, a seizure]. Gouty or rheumatic pain in the eye.

**ophthalmalgia** (of-thal-mal'-je-ah) [*ophthalmus*; ἄλγος, pain]. Neuralgia of the eye.

**ophthalmatrophy** (of-thal-mat'-ro-fe) [*ophthalmus*; atrophy]. Atrophy of the eyeball.

**ophthalmecchymosis** (of-thal-mek-im-o'-sis) [*ophthalmus*; *ecchymosis*]. A conjunctival effusion of blood.

**ophthalmectomy** (of-thal-mek'-to-me) [*ophthalmus*; ἐκτομή, excision]. Excision of the eye.

**ophthalmemicrania** (of-thal-mem-ik-ra'-ne-ah). See *amaurosis, epileptiform*.

**ophthalmia** (of-thal'-me-ah) [*ophthalmus*]. Inflammation of the eye, especially one in which the conjunctiva is involved. **o., catarrhal**, simple conjunctivitis; a hyperemia of the conjunctiva with a mucopurulent secretion. **o., caterpillar-**, inflammation of the conjunctiva or of the cornea, the result of penetration of the tissues by the hairs of caterpillars. Syn., *ophthalmia nodosa*. **o., Egyptian**. See *trachoma*. **o., electric**, conjunctivitis due to intense electric light. **o., gonorrheal**, an acute and severe form of purulent conjunctivitis, caused by infection from urethral discharges containing the gonococcus of Neisser. **o., granular**. See *trachoma*. **o., jequirity**, that due to poisoning by jequirity. **o. neonatorum**, a gonorrheal or purulent ophthalmia

of the newborn, the eyes having been infected by the mother's vaginal discharges. o., **neuroparalytic**, disease of the eye from lesion of the Gasserian ganglion or of branches of the fifth nerve supplying the eyeball. o. **nodosa**. See o., *caterpillar-*. o., **phylctenular**, conjunctivitis characterized by phlyctenules or small vesicles situated in the epithelial layer of the conjunctiva or cornea. o., **purulent**, conjunctivitis with a purulent discharge. o., **spring**, a form common in the spring. o., **sympathetic**, a severe destructive inflammation, a form of iridocyclitis secondary to injury or disease of the fellow eye. o., **varicose**, that associated with a varicose state of the veins of the conjunctiva.
**ophthalmiater** (*off-thal-me-a'-ter*) [*ophthalmus; iatrós*, a physician]. An oculist or ophthalmologist.
**ophthalmiatric** (*off-thal-me-at'-rik*) [*ophthalmus; iatreía*, treatment]. Pertaining to the treatment of eye-diseases.
**ophthalmiatrics** (*off-thal-me-at'-riks*). The treatment of eye-diseases.
**ophthalmic** (*off-thal'-mik*). Pertaining to the eye. o. **artery**. See under *artery*. o. **ganglion**. See *ganglion, ophthalmic*. o. **nerve**. See under *nerve*.
**ophthalmitic** (*off-thal-mit'-ik*) [*ophthalmus;* ιτις, inflammation]. Pertaining to ophthalmitis.
**ophthalmitis** (*off-thal-mi'-tis*) [*ophthalmus;* ιτις, inflammation]. Inflammation of the eye. o., **sympathetic**, that following inflammation or injury of the fellow-eye.
**ophthalmo-** (*off-thal-mo-*) [ὀφθαλμός, eye]. A prefix denoting relation to the eye.
**ophthalmoblennorrhea, ophthalmoblennorrhœa**, (*off-thal-mo-blen-or-e'-ah*) [*ophthalmo-;* blennorrhea]. Blennorrhea of the conjunctiva.
**ophthalmocace** (*off-thal-mok'-as-e*) [*ophthalmo-;* κακός, evil]. Disease of the eye.
**ophthalmocarcinoma** (*off-thal'-mo-kar-sin-o'-mah*) [*ophthalmo-;* carcinoma]. Carcinoma of the eye.
**ophthalmocele** (*off-thal'-mo-sēl*). See *exophthalmos*.
**ophthalmocentesis** (*off-thal'-mo-sen-te'-sis*) [*ophthalmo-;* κέντησις, puncture]. Surgical puncture of the eye.
**ophthalmocopia** (*off-thal-mo-ko'-pe-ah*) [*ophthalmo-;* κόπος, fatigue]. Fatigue of vision because of asthenopia.
**ophthalmodesmitis** (*off-thal'-mo-dez-mi'-tis*). Synonym of *conjunctivitis*.
**ophthalmodiagnosis** (*off-thal-mo-di-ag-no'-sis*). Diagnosis by means of the ophthalmoreaction.
**ophthalmodiaphanoscope** (*off-thal-mo-di-af-an'-o-skōp*) [*ophthalmo-; diaphanoscope*]. An instrument for examining the fundus of the eye by transillumination through the mouth.
**ophthalmodiastimeter** (*off-thal-mo-di-as-tim'-et-er*) [*ophthalmo-;* διάστημα, interval; μέτρον, measure]. An instrument for use in discovering the proper adjustment of lenses to the axes of the eyes.
**ophthalmodonesis** (*off-thal-mo-don-e'-sis*) [*ophthalmo-;* δόνησις, a trembling]. A voluntary tremulous or oscillatory movement of the eye.
**ophthalmodynamometer** (*off-thal-mo-di-nam-om'-et-er*) [*ophthalmo-;* δύναμις, power; μέτρον, measure]. An instrument for measuring the power of convergence of the eyes.
**ophthalmodynia** (*off-thal-mo-din'-e-ah*) [*ophthalmo-;* ὀδύνη, pain]. Neuralgic pain in the eye.
**ophthalmography** (*off-thal-mog'-ra-fe*) [*ophthalmo-;* γράφειν, to write]. Descriptive anatomy of the eye.
**ophthalmokopia** (*off-thal-mo-kop'-e-ah*). See *ophthalmocopia*.
**ophthalmoleukoscope** (*off-thal-mo-lū'-ko-skōp*) [*ophthalmo-;* λευκός, white; σκοπεῖν, to view]. An instrument for testing color-sense by means of polarized light.
**ophthalmolith** (*off-thal'-mo-lith*) [*ophthalmo-;* λίθος, stone]. A calculus of the eye.
**ophthalmologist** (*off-thal-mol'-o-jist*) [see *ophthalmology*]. One versed in ophthalmology.
**ophthalmology** (*off-thal-mol'-o-je*) [*ophthalmo-;* λόγος, science]. The science of the anatomy, physiology, and diseases of the eye.
**ophthalmolyma** (*off-thal-mo-li'-mah*) [*ophthalmo-;* λύμη, destruction]. Destruction of the eye.
**ophthalmomacrosis** (*off-thal-mo-mak-ro'-sis*) [*ophthalmo-;* μακρός, large]. Enlargement of the eye.
**ophthalmomalacia** (*off-thal-mo-mal-a'-se-ah*) [*ophthalmo-;* μαλακία, softness]. Abnormal softness or subnormal tension of the eye.
**ophthalmomelanoma** (*off-thal-mo-mel-an-o'-mah*) [*ophthalmo-;* μέλας, black; ὄμμα, tumor]. A melanotic tumor, usually sarcoma, of the eye.
**ophthalmomelanosis** (*off-thal-mo-mel-an-o'-sis*) [*ophthalmo-;* μέλας, black; νόσος, disease]. The formation of an ophthalmomelanoma; also the growth itself.
**ophthalmometer** (*off-thal-mom'-et-er*) [*ophthalmo-;* μέτρον, a measure]. I. An instrument for measuring the capacity of the chambers of the eye. 2. An instrument for measuring refractive errors, especially astigmatism. 3. An instrument for measuring the eye as a whole.
**ophthalmometry** (*off-thal-mom'-et-re*) [*ophthalmometer*]. The determination of refractive errors by means of the ophthalmometer.
**ophthalmomyitis** (*off-thal-mo-mi-i'-tis*) [*ophthalmo-; myitis*]. Inflammation of the ocular muscles.
**ophthalmomyositis** (*off-thal-mo-mi-o-si'-tis*). See *ophthalmomyitis*.
**ophthalmomyotomy** (*off-thal-mo-mi-ot'-o-me*) [*ophthalmo-;* μῦς, muscle; τομή, a cutting]. Division of the muscles of the eye.
**ophthalmoncus** (*off-thal-mong'-kus*) [*ophthalmo-;* ὄγκος, tumor]. A tumor or swelling of the eye.
**ophthalmoneuritis** (*off-thal-mo-nū-ri'-tis*) [*ophthalmo-; neuro-*, nerve; ιτις, inflammation]. Inflammation of the ophthalmic nerve.
**ophthalmonosology** (*off-thal-mo-no-sol'-o-je*) [*ophthalmo-;* νόσος, disease; λόγος, science]. The study of the diseases of the eye.
**ophthalmopathy** (*off-thal-mop'-ath-e*) [*ophthalmo-;* πάθος, disease]. Any disease of the eye. o., **external**, an affection of the eyelids, cornea, conjunctiva, or muscles of the eye. o., **internal**, any disease affecting the deeper structures of the eye.
**ophthalmophacometer, ophthalmophakometer** (*off-thal-mo-fa-kom'-et-er*) [*ophthalmo-; phacometer*]. An instrument for measuring the curvature radius of the crystalline lens.
**ophthalmophantom** (*off-thal-mo-fan'-tom*) [*ophthalmo-; phantom*]. A model or mask for practising operations on the eye.
**ophthalmophasmatoscopy** (*off-thal-mo-faz-mat-os'-ko-pe*) [*ophthalmo-;* φαντάζειν, to make to happen; σκοπεῖν, to view]. Ophthalmoscopic and spectroscopic examination of the interior of an eye.
**ophthalmophlebotomy** (*off-thal-mo-fle-bot'-o-me*) [*ophthalmo-;* φλέψ, a vein; τέμνειν, to cut]. Bloodletting from a conjunctival vein.
**ophthalmophobia** (*off-thal-mo-fo'-be-ah*) [*ophthalmo;* φόβος, fear]. Morbid dislike of being stared at.
**ophthalmophthisis** (*off-thal-mo-ti'-sis*). See *phthisis bulbi*.
**ophthalmophyma** (*off-thal-mo-fi'-mah*) [*ophthalmo-;* φῦμα, growth]. Swelling of the eyeball.
**ophthalmoplas** (*off-thal-mo-plas'-tik*). Pertaining to ophthalmoplasty.
**ophthalmoplasty** (*off-thal'-mo-plas-te*) [*ophthalmo-;* πλάσσειν, to mold]. Plastic surgery of the eye or accessory parts.
**ophthalmoplegia** (*off-thal-mo-plē'-je-ah*) [*ophthalmo-;* πληγή, stroke]. Paralysis of the ocular muscles. o. **externa**, paralysis of the external ocular muscles. o. **interna**, paralysis of the internal muscles of the eye—those of the iris and ciliary body. o., **nuclear**, a form due to a lesion of the nuclei of origin of the motor nerves of the eyeball. o., **partial**, a form in which some of the muscles only are paralyzed. o., **progressive**, a form in which all the muscles of both eyes gradually become paralyzed. o., **total**, that form involving the iris and ciliary muscle as well as the external muscles of the eyeball.
**ophthalmoplegic** (*off-thal-mo-plē'-jik*) [*ophthalmoplegia*]. Pertaining to ophthalmoplegia.
**ophthalmoptosis** (*off-thal-mop-to'-sis*) [*ophthalmo-;* πτῶσις, a fall]. Protrusion of the eyeball; exophthalmos.
**ophthalmo-reaction** (*off-thal'-mo-re-ak'-shun*). A temporary inflammation of the conjunctiva due to the instillation of one drop of a one per cent. solution of tuberculin into the eye of a tuberculous subject.
**ophthalmorrhagia** (*off-thal-mo-ra'-je-ah*) [*ophthalmo-;* ῥηγνύναι, to burst forth]. Hemorrhage from the eye.
**ophthalmorrhea, ophthalmorrhœa** (*off-thal-mor-e'-ah*) [*ophthalmo-;* ῥοία, a flow]. A watery or sanguineous discharge from the eye.
**ophthalmorrhexis** (*off-thal-mo-reks'-is*) [*ophthalmo-;* ῥῆξις, rupture]. Rupture of the eyeball.
**ophthalmos** (*off-thal'-mos*) [ὀφθαλμός, eye]. The eye.

OPHTHALMOSCOPE 630 OPIUM

**ophthalmoscope** (*off'-thal'-mo-skōp*) [*ophthalmo-*; σκοπεῖν, to see]. An instrument for examining the interior of the eye. It consists essentially of a mirror with a hole in it, through which the observer looks, the concavity of the eye being illuminated by light reflected from the mirror into the eye and seen by means of the rays reflected from the eye-ground back through the hole in the mirror. The ophthalmoscope is fitted with lenses of different powers that may be revolved in front of the observing eye, and these neutralize the ametropia of either the patient's or the observer's eye, thus rendering the details of the fundus oculi clear.

**ophthalmoscopic** (*off-thal-mo-skop'-ik*) [*ophthalmoscope*]. Pertaining to the ophthalmoscope or its use.

**ophthalmoscopist** (*off-thal-mos'-ko-pist*) [*ophthalmo-*; σκοπεῖν, to view]. One versed in ophthalmoscopy.

**ophthalmoscopy** (*off-thal-mos'-ko-pe*) [*ophthalmoscope*]. The examination of the interior of the eye by means of the ophthalmoscope. o., **direct**, the method of the erect or upright image, the observer's eye and the ophthalmoscope being brought close to the eye of the patient. o., **indirect**, the method of the inverted image: the observer's eye is placed about 16 inches from that of the patient, and a 20 D. biconvex lens is held about two inches in front of the observed eye, thereby forming an aerial inverted image of the fundus. b., **medical**, ophthalmoscopy as an aid to internal medicine in the diagnosis of such diseases as manifest themselves in changes in the fundus of the eye. o., **metric**, that for purposes of measuring refraction.

**ophthalmospasm** (*off-thal'-mo-spasm*) [*ophthalmo-*; σπασμός, a spasm]. Ocular spasm.

**ophthalmospintherism** (*off-thal-mo-spin'-ther-ism*) [*ophthalmo-*; σπινθήρ, spark]. A condition of the eye in which luminous sparks are seen.

**ophthalmostasis** (*off-thal-mos'-tas-is*) [*ophthalmo-*; στάσις, a stopping]. Fixation of the eye during an operation upon it.

**ophthalmostat** (*off-thal'-mo-stat*) [*ophthalmo-*; ἱστάναι, to cause to stand]. An instrument used in fixing the eye in any position during an operation on it.

**ophthalmostatometer** (*off-thal-mo-stat-om'-et-er*) [*ophthalmo-*; ἱστάναι, to cause to stand; μέτρον, a measure]. An instrument for determining the position of the eyes.

**ophthalmostatometry** (*off-thal-mo-stat-om'-et-re*) [*ophthalmostatometer*]. The measurement of the position of the eyes.

**ophthalmosteresis** (*off-thal-mo-ster-e'-sis*) [*ophthalmo-*; στέρησις, deprival]. Loss, or absence of one or both eyes.

**ophthalmosynchysis** (*off-thal-mo-sin'-kis-is*) [*ophthalmo-*; σύγχυσις, a mixing together]. Effusion into the interior of the eye.

**ophthalmothermometer** (*off-thal-mo-thur-mom'-et-ur*). A device for recording local temperature in eye diseases.

**ophthalmotomy** (*off-thal-mot'-o-me*) [*ophthalmo-*; τομή, a cutting]. The dissection, or incision of the eye.

**ophthalmotonometer** (*off-thal-mo-ton-om'-et-er*) [*ophthalmo-*; *tonometer*]. An instrument for measuring intraocular tension.

**ophthalmotonometry** (*off-thal-mo-ton-om'-et-re*) [*ophthalmotonometer*]. Measurement of the ocular tension.

**ophthalmotrope** (*off-thal'-mo-trōp*) [*ophthalmo-*; τρόπος, a turn]. An instrument used for the demonstration of the direction and the position that the eye takes under the influence of each of its muscles, and the position of the false image in the case of paralysis of a given muscle.

**ophthalmotropometer** (*off-thal-mo-trop-om'-et-er*) [*ophthalmotrope*; μέτρον, measure]. An instrument for measuring the movement of the eyeballs.

**ophthalmotropometry** (*off-thal-mo-tro-pom'-et-re*) [*ophthalmotropometer*]. The measurement of the movement of the eyeballs.

**ophthalmovascular** (*off-thal-mo-vas'-kū-lar*). Pertaining to the blood vessels of the eye. o., **choke**, a condition in which pressure of the retinal vessels on each other interferes with the blood supply of the retina.

**ophthalmoxerosis** (*off-thal-mo-ze-ro'-sis*). See *xerophthalmia*.

**ophthalmoxysis** (*off-thal-mōks-i'-sis*) [*ophthalmo-*; ξύσις, a scraping]. Treatment by scraping or scarification of the conjunctiva.

**ophthalmoxyster** (*off-thal-moks-is'-ter*). The same as *ophthalmoxystrum*.

**ophthalmoxystrum** (*off-thal-moks-is'-trum*) [*ophthalmo-*; ξυστηρ, scraper]. An instrument for scraping or scarifying the conjunctiva.

**ophthalmozoa** (*off-thal-mo-zo'-ah*) [*ophthalmo-*; ζῷον, animal]. Entozoa parasitic upon the eye or its appendages.

**ophthalmula** (*off-thal'-mū-lah*) [ὀφθαλμός, eye; ὕλη, matter]. A scar of the eye.

**ophthalmus** (*off-thal'-mus*) [ὀφθαλμός, eye]. The eye.

**-opia** (-*o'-pe-ah*). See -*ops*.

**opianine** (*o-pe-an'-ēn*) [*opium*], C₂₁H₂₃N₃O₇. An alkaloid of opium.

**opianyl** (*o'-pe-an-il*) [*opium*; ὕλη, matter]. Synonym of *meconin*.

**opiate** (*o'-pe-āt*) [*opium*]. A preparation of opium.

**opiomania** (*o-pe-o-ma'-ne-ah*) [*opium*; μανία, madness]. A morbid desire for opium.

**opiophagia** (*o-pe-o-fa'-je-ah*). Synonym of *opiophagism*.

**opiophagism, opiophagy** (*o-pe-of'-aj-ism, 'o-pe-of'-aj-e*) [*opium*; φαγεῖν, to eat]. Opium-eating.

**opiophile** (*o'-pe-o-fīl*) [*opium*; φιλεῖν, to love]. A loyer, or eater, of opium; an opium-smoker.

**opisthen** (*o-pis'-then*) [ὄπισθεν, behind]. In biology, the hind part of the body of an animal.

**opisthenar** (*o-pis'-the-nar*). [*opisthen*; θέναρ, the palm]. The back of the hand. Cf. *thenar*.

**opisthiobasial** (*o-pis-the-o-ba'-se-al*). Relating to or uniting the opisthion and basion.

**opisthion** (*o-pis'-the-on*). See under *craniometric point*.

**opisthionasial** (*o-pis-the-o-na'-se-al*). Pertaining to the opisthion and nasion.

**opisthognathism** (*o-pis-thog'-nath-ism*) [ὄπισθεν, behind; γνάθος, jaw]. Recession of the lower jaw.

**opisthoporia, opisthoporeia** (*o-pis-tho-po'-re-ah, o-pis-tho-po-ri'-ah*) [ὄπισθεν, behind; πορεία, going]. Involuntary backward-walking in an attempt to go forward.

**opisthorchiasis** (*op-is-thor-ki'-as-is*). Infection of the liver with flukes of the genus *Opisthorchis*.

**Opisthorchis** (*op-is-thor'-kis*). A genus of trematodes or flukes. O. **felineus**, a parasite found in the liver and bile ducts of cats, dogs and man. O. **noverca**, the Indian liver-fluke, found in dogs and man. O. **sinensis** causes the liver-fluke disease of China and Japan which affects cats and dogs and man; called also *Distoma sinense* and *Distoma japonicum*.

**opisthotic** (*op-is-thot'-ik*) [ὄπισθεν, behind; οὖς, ear]. Relating to posterior parts of the ear-apparatus.

**opisthotonic** (*o-pis-tho-ton'-ik*). Pertaining to opisthotonos.

**opisthotonoid** (*o-pis-thot'-on-oid*) [*opisthotonos*; εἶδος, like]. Resembling opisthotonos.

**opisthotonos** (*o-pis-thot'-on-os*) [ὄπισθεν, behind; τόνος, stretching]. A condition in which, from tetanic spasm of the muscles of the back, the head and lower limbs are bent backward and the body arched forward.

**opium** (*o'-pe-um*) [ὄπιον, poppy juice]. The inspissated juice obtained by incising the unripe capsules of *Papaver somniferum*, of the order *Papaveraceæ*, occurring in commerce in the form of brownish cakes having a narcotic odor and a bitter taste. Opium contains a large number of alkaloids, of which *morphine* is the most important, since it represents the chief properties of the drug. Other alkaloids are *narcotine*, C₂₂H₂₃NO₇, *codeine*, C₁₈H₂₁NO₃, *thebaine* or *paramorphine*, C₁₉H₂₁NO₃, *papaverine*, C₂₀H₂₁NO₄, *narceine*, C₂₃H₂₉NO₉, *pseudomorphine*, C₁₇H₁₉NO₄, and *laudanine*, C₂₀H₂₅NO₄. These bases occur in opium combined with meconic and thebolactic acids. According to the U. S. P., moist opium should contain not less than 9 % of crystallized morphine. Opium acts as a narcotic, producing deep sleep, which, however, is often preceded by a stage of mental excitement and exhilaration; on awakening there may be headache, nausea, or vomiting. It slows the pulse and increases its force and raises blood-pressure; small doses do not depress the respiration, but large doses do so in a marked degree. It checks the motor activity of the stomach and intestine, and lessens all secretions except that of the skin; it produces contraction of the

pupil. When taken in poisonous doses it causes unconquerable drowsiness, passing into deep sleep, with slow, full respiration, slow pulse, and contracted pupils; later cyanosis develops, the respiration becomes exceedingly slow, and the pulse rapid and feeble; death takes place from failure of the respiration. See *poisons, table of.* There is a chronic form of opium-poisoning produced by the habitual use of opium or morphine, and characterized by mental depression, a deterioration of the moral sense, and attacks of diarrhea. The drug is used for the relief of pain of all forms except that due to cerebral inflammation; in insomnia; in inflammation of serous membranes; in spasmodic conditions; in acute colds; for cough, retention of urine, vomiting, diarrhea, certain forms of dyspnea, particularly that from heart disease; and locally as an application to sprains and inflamed surfaces. Dose 1 gr. (0.065 Gm.). o., **confection of** (*confectio opii,* B. P.). Dose 5-20 min. (0.32-1.3 Cc.). o., **deodorized** (*opium deodoratum,* U. S. P.). Dose 1 gr. (0.065 Gm.). o., **extract of** (*extractum opii,* U. S. P., B. P.). ½-½ gr. (0.016-0.032 Gm.). o., **extract of, liquid** (*extractum opii liquidum,* B. P.). Dose 5-20 min. (0.32-1.3 Cc.). o., **granulated** (*opium granulatum,* U. S. P.). Dose 1 gr. (0.065 Gm.). o., **pills of** (*pilulæ opii,* U. S. P.), pills containing 1 gr. (0.065 Gm.) of opium; those of the B. P. (*pilula saponis composita*) contain each about 20 % of opium. o. **plaster** (*emplastrum opii,* U. S. P., B. P.), contains 1½ dr. (6 Gm.) of the drug. o., **powder of, compound** (*pulvis opii compositus,* B. P.), contains 10 % of opium. o., **powder of ipecac and** (*pulvis ipecacuanhæ et opii,* U. S. P.), Dover's powder, contains 10 % each of ipecac and opium. o., **powdered** (*opii pulvis,* U. S. P.). Dose ¼-1 gr. (0.016-0.065 Gm.). o., **tincture of** (*tinctura opii,* U. S. P.), laudanum. Dose 5-15 min. (0.32-1.0 Cc.). o., **tincture of, ammoniated** (*tinctura opii ammoniata,* B. P.). Dose 30 min.-1 dr. (2-4 Cc.). o., **tincture of, camphorated** (*tinctura opii camphorata,* U. S. P.; *tinctura camphoræ composita,* B. P.), paregoric. Dose 1-4 dr. (4-16 Cc.). o., **tincture of, deodorized** (*tinctura opii deodorati,* U. S. P.). Dose 5-15 min. (0.32-1.0 Cc.). o., **troches of glycyrrhiza and** (*trochisci glycyrrhizæ et opii,* U. S. P., B. P.), each troche contains 1/12 gr. (0.005 Gm.) of opium. o., **vinegar of** (*acetum opii,* U. S. P.), black-drop. Dose 5-30 min. (0.32-2.0 Cc.). o., **wine of** (*vinum opii,* U. S. P., B. P.), Sydenham's laudanum. Dose 5-15 min. (0.32-1.0 Cc.). See also *codeine* and *morphine.*
**opiumism** (o'-pe-um-izm) [*opium*]. The condition produced by the action of opium on the system.
**opo-** [ὀπός, juice]. A prefix denoting a serum or an organic extract.
**opobalsam, opobalsamum** (o-po-bawl'-sam, o-po-bawl-sam'-um) [ὀπός, juice; βάλσαμον, balsam]. A resin from *Balsamodendron opobalsamum* and *Balsamodendron gileadense.*
**opocephalus** (o-po-sef'-al-us) [ὤψ, eye; κεφαλή, head]. A monster characterized by fusion of the ears, one orbit, and absence of mouth and nose.
**opocerebrin** (o-po-ser'-e-brin) [ὀπός, juice; *cerebrum*]. A proprietary therapeutic preparation from the gray matter of brain; used in nervous diseases, anemia, etc. Dose 4-6 gr. (0.2-0.4 Gm.) twice daily.
**opodeldoc** (op-o-del'-dok) [origin obscure]. Soap liniment; see under *soap.*
**opodidymus, opodymus** (op-o-did'-im-us, op-od'-im-us) [ὤψ, eye; δίδυμος, twin]. A monster with a single body and skull but with two distinct faces.
**opohepatoidin** (o-po-hep-at-oid'-in). A proprietary therapeutic preparation from the liver; it is used in icterus and epistaxis. Dose 8 gr. (0.5 Gm.) 3 times daily.
**opohypophysin** (o-po-hi-pof'-is-in). A proprietary preparation from the hypophysis. It is used in akromegaly. Dose ¾ gr. (0.05 Gm.).
**opolienin** (o-po-li'-en-in). A proprietary preparation from the spleen. It is used in hypertrophy of the spleen, malarial cachexia, and leukemia. Dose 20-30 gr. (1.3-2.0 Gm.) twice daily.
**opomammin** (o-po-mam'-in). A proprietary preparation from the mammary gland; it is used in uterine diseases. Dose 24 gr. (1.5 Gm.) daily.
**opomedullin** (o-po-med-ul'-in). A proprietary preparation from red bone-marrow; used in anemia,

chlorosis, and neurasthenia. Dose 3-16 gr. (0.2-1.0 Gm.) daily.
**opoorchidin** (o-po-or'-kid-in). A proprietary preparation from the testicles, used in spinal and other nervous diseases.
**opoossiin** (o-po-os'-e-in). A proprietary preparation from yellow bone-marrow; used in rhachitis and osteomalacia.
**opoovarin** (o-po-ov-ar'-e-in). A proprietary preparation from the ovaries; used in chlorosis, hysteria, and in climacteric symptoms.
**opopancreatin** (o-po-pan-kre'-at-in). A proprietary preparation from the pancreas; used in diabetes mellitus.
**opoprostatin** (o-po-pros'-tat-in). A proprietary preparation from the prostate; it is used in hypertrophy of the prostate.
**oporeniin** (o-po-ren'-e-in). A proprietary preparation from the kidneys; used in uremia, chronic nephritis, and albuminuria.
**oposuprarenalin** (o-po-su-prah-ren'-al-in). A proprietary preparation from the suprarenal capsule; used in diabetes insipidus, Addison's disease, and neurasthenia.
**opotherapy** (op-o-ther'-ap-e) [ὀπός, juice; *therapy*]. Synonym of *organotherapy.*
**opothymin** (o-po-thi'-mi-in). A proprietary preparation from the thymus; used in Graves' disease, anemia, etc.
**opothyroidin** (o-po-thi-roid'-in). A proprietary preparation from the thyroid gland; it is used in myxedema, cretinism, obesity, etc.
**Oppenheim's gait** (op'-en-him) [Hermann *Oppenheim,* German neurologist, 1858- ]. A modification of the spastic gait of disseminated sclerosis, consisting in large and irregular oscillations of the head, trunk, and extremities. O.'s **reflex,** an abnormal cutaneous reflex; slight pressure on the skin overlying the inner border of the tibia from above downward is followed by extension of the great toe or all the toes.
**Oppenheimer's test for acetone** (op'-en-hi-mer). Make a reagent by diluting 20 Cc. of concentrated sulphuric acid with a liter of water; to this add 50 Gm. of yellow oxide of mercury and set aside for 24 hours. To 3 Cc. of unfiltered urine add a few drops of the reagent. In the presence of albumin a precipitate occurs at once; in its absence the precipitate is seen some time later.
**oppilation** (op-il-a'-shun) [*oppilatio,* closure]. 1. Obstruction; closing the pores; causing constipation. 2. A constipating agent or remedy.
**oppilative** (op'-il-a-tiv) [*oppilatio,* closure]. 1. Obstruction; closing the pores; causing constipation. 2. A constipating agent or remedy.
**Oppolzer's sign.** On palpation the seat of the apex-beat is found to change with the alteration of the patient's posture in cases of serofibrinous pericarditis.
**opponens** (op-o'-nens) [*ob,* against; *ponere,* to place]. Opposing. A term applied to certain muscles that bring one part opposite another, as *opponens minimi digiti,* a muscle placing the little finger opposite the thumb. See *muscles, table of.*
**oppression** (op-resh'-un) [*oppressio; opprimere,* to bear against]. Any sensation of pressure or weight upon any part, especially the chest.
**-ops, -opsia, -optic** (-ops, -ops'-e-ah, -op'-tik) [ὤψ, eye]. Variant forms of a suffix denoting connection with or relation to the eye.
**opsialgia** (op-se-al'-je-ah) [ὤψ, face; ἄλγος, pain]. Neuralgia of the face.
**opsinogen, opsogen** (op-sin'-o-jen) (op'-so-jen). A substance producing an opsonin.
**opsinogenous** (op-sin-oj'-en-us). Capable of producing an opsonin.
**opsiometer** (op-se-om'-et-er). See *optometer.*
**opsionosis** (op-se-on'-o-sis) [ὄψις, sight; νόσος, disease]. A disease of the eye, or of vision.
**opsitocia** (op-sit-o'-ke-ah) [ὀψέ, late; τόκος, birth]. Abnormally long pregnancy.
**opsogen** (op'-so-jen). See *opsinogen.*
**opsomania** (op-so-ma'-ne-ah) [ὄψον, dainty food; μανία, mania]. Insane desire for dainty or some special food.
**opsomania** (op-so-ma'-ne-ak). One affected with opsomania.
**opsonic** (op-son'-ik). Pertaining to opsonins. o. **index,** the ratio of the number of bacteria ingested by the leukocytes of a healthy person compared with

that ingested by the leukocytes of the patient; *e. g.,* if the ratio of the healthy is 10 and that of the patient 15, then the index is 1½. **negative phase,** the decrease in opsonic power that follows the injection. **positive phase,** the subsequent increase of opsonic power.

**opsonin** (*op'-so-nin*) [ὀψώνιον, provisions]. An element in normal serum and to a greater degree in the serum of a patient successfully inoculated with dead cultures of the bacteria responsible for the disease which are thereby made susceptible to phagocytosis.

**opsonist** (*op'-son-ist*). One versed in the technique of opsonotherapy.

**opsonometry** (*op-son-om'-et-re*) [*opsonin;* μέτρον, measure]. The estimation of the opsonic index.

**opsonotherapy** (*op-son-o-ther'-ap-e*). The treatment of disease by increasing the opsonic power of the blood.

**opsophagia** (*op-so-fa'-je-ah*) [ὄψον, dainty food; φαγεῖν, to eat]. Morbid daintiness in respect of food.

**optic, optical** (*op'-tik, op'-tik-al*) [ὀπτικός, from the base ὀπ-, to see]. Pertaining to vision or to the science of optics. **o. atrophy,** atrophy of the optic nerve. **o. axis,** the axis of the eye. **o. capsule,** the embryonic structure forming the sclera. **o. center.** 1. The point in the main axis of the crystalline lens at which the rays of light meet. 2. The nervecenter concerned in the visual function. **o. chiasm, o. commissure.** See *commissure, optic.* **o. cup,** the concave area formed by the involution of the distal extremity of the primary optic vesicle. **o. disc,** the optic papilla. **o. foramen.** See *foramen, optic.* **o. groove,** the groove on the sphenoid bone for the optic chiasm. **o. lobes,** the corpora quadrigemina. **o. nerve.** See under *nerve.* **o. neuritis.** See *papillitis.* **o. papilla,** the circular prominence formed by the optic nerve after its entrance into the eyeball. **o. radiations,** a large bundle of nerve-fibers joining the optic thalamus and the occipital lobe of the cerebrum. **o. thalamus.** See *thalamus.* **o. tract,** the fibers between the visual center and the optic chiasm. **o. vesicle,** a diverticulum from each side of the primary anterior vesicle of the embryo, forming the basis of the future eye.

**optician** (*op-tish'-un*) [*optic*]. A maker of optical instruments.

**opticociliary** (*op-tik-o-sil'-e-a-re*) [*optic; ciliary*]. Pertaining to the optic and ciliary nerves.

**opticocinerea** (*op-tik-o-sin-e'-re-ah*) [*optic; cinereus,* resembling ashes]. The gray matter of the optic lobes.

**opticopupillary** (*op-tik-o-pū'-pil-a-re*) [*optic; pupillary*]. Pertaining to the optic nerve and the pupil.

**optics** (*op'-tiks*). [*optic*]. That branch of physics treating of the laws of light, its refraction and reflection, and of its relation to vision. See *dioptrics.*

**optimal** (*op'-tim-al*) [*optimus,* best]. The best; the most favorable.

**optimism** (*op'-tim-izm*) [*optimus,* best]. Delusional exaltation; delirium of grandeur; amenomania.

**optimum** (*op'-tim-um*) [*optimus,* best]. The temperature or other condition at which vital processes are carried on with the greatest activity. Midway between the minimum, or lowest endurable, and maximum, or highest endurable temperatures or other conditions.

**opto-** (*op-to-*) [*optic*]. A prefix denoting relation to the eye or to vision.

**optogram** (*op'-to-gram*) [*ὀπτο-;* γράμμα, a writing]. A faint image on the retina, for a brief period after death, of the object last seen.

**optomeninx** (*op-to-men'-inks*). Synonym of *retina.*

**optometer** (*op-tom'-et-er*) [*opto-;* μέτρον, a measure]. An instrument for determining the strength of vision, especially the degree of refractive error that is to be corrected.

**optometry** (*op-tom'-et-re*) [ὀπτός, visible; μέτρον, measure]. Measurement of the visual powers.

**optomyometer** (*op-to-mi-om'-et-er*) [*opto-; myometer*]. An instrument for measuring the strength of the muscles of the eye.

**optostriate** (*op-to-stri'-āt*) [*opto-, striatum,* striped]. Pertaining to the optic thalamus and the corpus striatum.

**optotype** (*op'-to-tīp*) [ὀπτός, visible; τύπος, type]. A test-type used in testing the acuity of vision.

**Opuntia** (*o-pun'-te-ah*) [*opuntius,* relating to Ὀποῦς, a town of Greece]. Prickly pear; a genus of cactaceous plants represented by numerous species, many of which, as *O. reticulata* and *O. tuna,* have slight medicinal properties, the former being anthelmintic, purgative, and locally sedative; the latter has been used for palpitation of the heart.

**ora** (*o'-rah*) [L.]. Margin. **o. serrata,** the jagged anterior margin of the retina.

**orad** (*o'-rad*) [*os, oris,* mouth]. Toward the mouth, or the oral region.

**oral** (*o'-ral*) [*os,* the mouth]. Pertaining to the mouth. **o. whiff,** a peculiar sound heard during expiration from the open mouth, principally in cases of thoracic aneurysm.

**orange** (*or'-anj*). See *aurantium.*

**orangeade** (*or-anj-ād'*). A drink made of orange-juice and sweetened water.

**orbicular** (*or-bik'-ū-lar*) [*orbicularis,* dim. of *orbis,* circle]. Circular. A term applied to circular muscles, as the *orbicular* muscle of the eye or of the mouth (orbicularis palpebrarum, orbicularis oris).

**orbiculare** (*or-bik-ū-la'-re*). The orbicular bone; a tubercle at the end of the long process of the incus; it is separate in early fetal life.

**orbicularis** (*or-bik-ū-la'-ris*). See under *muscle.*

**orbiculostapedial** (*or-bik-ū-lo-sta-pe'-de-al*). Relating to the orbicular process of the incus and to the stapes.

**orbiculus ciliaris** (*or-bik'-ū-lus sil-e-a'-ris*). The ciliary disc; same as *annulus ciliaris.*

**orbit** (*or'-bit*) [*orbita,* from *orbis,* a circle]. The bony pyramidal cavity containing the eye, and formed by the frontal, sphenoid, ethmoid, nasal, lacrimal, superior maxillary, and palatal bones.

**orbita** (*or'-bit-ah*) [*orbis,* a circle, orbit: *pl., orbitæ*]. The same as *orbit.*

**orbital** (*or'-bit-al*). Pertaining to the orbit. **o. height,** in craniometry, the greatest vertical width of the external opening of the orbit. **o. index,** the orbital height multiplied by 100 and divided by the orbital width. If the orbital index is above 89, it is called megaseme; if under 84, microseme; if between, mesoseme.

**orbitocele** (*or'-bit-o-sēl*) [*orbita,* orbit; κήλη, tumor]. 1. A tumor of the orbit. 2. The same as *exophthalmos.*

**orcein** (*or'-se-in*) [*Orcus,* Pluto, from its dark color], C₇H₇NO₃. A dark-red substance derived from orcin.

**orchectomy, orchiectomy** (*or-kek'-to-me, or-ki-ek'-to-me*). See *orchidectomy.*

**orcheitis** (*or-ki'-tis*). Synonym of *orchitis.*

**orchemphraxis** (*or-kem-fraks'-is*). Same as *orchidemphraxis.*

**orcheocele** (*or'-ke-o-sēl*). See *orchiocele.*

**orcheodesmosarcoma** (*or'-ke-o-dez-mo-sar-ko'-mah*) [ὄρχεα, scrotum; δεσμός, bond; σάρξ, flesh; ὄμα, tumor]. Scrotal elephantiasis.

**orcheoplasty** (*or'-ke-o-plas-te*). See *orchioplasty.*

**orcheotomy** (*or-ke-ot'-o-me*). See *orchotomy.*

**orchestromania** (*or-kes-tro-ma'-ne-ah*) [ὀρχεῖσθαι, to dance; μανία, madness]. Dancing mania; chorea, or St. Vitus' dance.

**orchi-, orchid-, orchio- (*or-ke-, or-kid-, or-ke-o-*) [ὄρχις, a testicle]. Prefixes signifying connection with or relation to the testicle.

**orchialgia** (*or-ke-al'-je-ah*) [*orchi-;* ἄλγος, pain]. Neuralgia of the testicle.

**orchic** (*or'-kik*) [ὄρχις, testicle]. Pertaining to the testicle.

**orchichorea** (*or-ke-ko-re'-ah*) [*orchi-; chorea*]. Irregular movements of the testicle due to contraction of the cremaster muscle.

**orchidalgia** (*or-kid-al'-je-ah*). Synonym of *orchialgia.*

**orchidatonia** (*or-kid-at-o'-ne-ah*) [*orchi-;* ἀτονία, atony]. Atony of the testicle; laxness of the testicle.

**orchidatrophia** (*or-kid-at-ro'-fe-ah*) [*orchi-;* ἀτροφία, atrophy]. Atrophy of the testicle.

**orchidauxe** (*or-kid-awks'-e*) [*orchi-;* αὔξη, growth]. Hypertrophy of the testicle.

**orchidectomy** (*or-kid-ek'-to-me*) [*orchi-;* ἐκτομή, excision]. Castration.

**orchidemphraxis** (*or-kid-em-fraks'-is*) [*orchi-;* ἔμφραξις, stoppage]. Obstruction of the vessels of the testis.

**orchidin** (*or'-kid-in*). A proprietary fluid from the testicle; used as a nervine.

**orchidion** (*or-kid'-e-on*) [dim. of ὄρχις, testicle]. A small testicle.

**orchiditis** (*or-kid-i'-tis*). Synonym of *orchitis.*

**orchidocatabasis** (*or-kid-o-kat-ab'-as-is*) [*orchido-;*

**katábasis**, a going down]. Descent of the testicle into the scrotum.
**orchidocele** (or-kid'-o-sēl). Synonym of orchiocele.
**orchidodynia** (or-kid-o-din'-e-ah). Synonym of orchiodynia.
**orchidoncus** (or-kid-ong'-kus) [orchido-; ὄγκος, tumor]. A tumor or tumefaction of the testicle.
**orchidopexia** (or-kid-o-peks'-e-ah). See orchidopexy.
**orchidopexy** (or'-kid-o-peks-e) [orchido-; πῆξις, fixation]. Same as orchidorrhaphy.
**orchidorrhaphy** (or-kid-or'-a-fe) [orchido-; ῥαφή, suture]. Suturing of the testicle to the surrounding tissue.
**orchidoscheocele** (or-kid-os'-ke-o-sēl) [orchido-; ὄσχη, scrotum; κήλη, tumor]. A scrotal hernia with enlargement of the testicle.
**orchidospongioma** (or-kid-o-spun-je-o'-mah) [orchido-; σπογγία, sponge; ὄμα, tumor]. A tuberculous tumor of the testicle.
**orchidotherapy** (or-kid-o-ther'-ap-e). The therapeutic use of testicular extracts.
**orchidotomy** (or-kid-ot'-o-me) [orchido-; τομή, a cutting]. Incision of the testicle.
**orchidotuberculum** (or-kid-o-tū-ber'-kū-lum). See orchidospongioma.
**orchidotyloma** (or-kid-o-ti-lo'-mah) [orchido-; τύλος, callus; ὄμα, tumor]. A callous nodule of the testicle.
**orchiectomy** (or-ke-ek'-to-me). Synonym of orchidectomy.
**orchiepididymitis** (or-ke-ep-id-id-im-i'-tis) [orchi-; epididymitis]. Inflammation of both testis and epididymis.
**orchio-**. See orchi-.
**orchiocele** (or'-ke-o-sēl) [orchio-; κήλη, tumor]. 1. A tumor of the testicle. 2. Scrotal hernia.
**orchiodynia** (or-ke-o-din'-e-ah) [orchio-; ὀδύνη, pain]. Pain in the testicles.
**orchioncus** (or-ke-ong'-kus). Synonym of orchidoncus.
**orchioneuralgia** (or-ke-o-nū-ral'-je-ah). Synonym of orchialgia.
**orchiopexy** (or-ke-o-pek'-se). Synonym of orchidopexy.
**orchioplasty** (or'-ke-o-plas-te) [orchio-; πλάσσειν, to form]. Any plastic operation on the scrotum.
**orchiorrhaphy** (or-ke-or'-af-e). Synonym of orchidorrhaphy.
**orchioscheocele** (or-ke-os'-ke-o-sēl). Synonym of orchidoscheocele.
**orchioscirrhus** (or-ke-os-kir'-us) [orchio-; σκιρρός, induration]. A hard carcinomatous tumor of the testicle.
**orchiotomy** (or-ke-ot'-o-me). Synonym of orchidotomy.
**orchis** (or'-kis) [ὄρχις, a testicle]. 1. A genus of plants furnishing salep and vanilla. 2. The testicle.
**orchitic** (or-kit'-ik). Relating to orchitis.
**orchitin** (or'-kit-in). A sterilized testicular extract.
**orchitis** (or-ki'-tis) [orchis; ιτις, inflammation]. Inflammation of the testicle.
**orchitomy** (or-kit'-o-me). Synonym of orchotomy.
**orchocele** (or'-ko-sēl). Synonym of orchiocele.
**orchos** (or'-kos) [ὄρχος, tarsus]. The tarsal cartilage.
**orchotomy** (or-kot'-o-me) 1. [ὄρχις, testis; τομή, a cutting]. Castration. 2. [ὄρχος, tarsus; τομή, a cutting]. Removal of the tarsal cartilages.
**orcin**, orcinol (or'-sin, or'-sin-ol). [Orcus, Pluto, from its dark color], C₇H₈(OH)₂. A substance found in many lichens of the genera Roccella and Lecanora. It is an antiseptic, and has been used instead of resorcinol. **o. reaction for xylose in the urine**, heat the urine with an equal volume of hydrochloric acid and a trace of orcin. As soon as a green color is apparent, cool the solution and shake with amyl-alcohol. The amyl-alcohol takes on a green color and in the spectroscope shows the characteristic band between C and D, and, in addition, a more uncertain band, more toward the red.
**ordeal bark** (or'-de-al). Casca-bark. **o. bean**, **o. nut**. See physostigma.
**order** (or'-der) [ordo, a rule]. Systematic arrangement. In biology, the taxonomic group below a class and above a family.
**orderly** (or'-der-le). A male hospital attendant.
**orectic** (or-ek'-tik) [ὄρεξις, appetite]. Stimulating appetite.
**orexin** (o-reks'-in) [ὄρεξις, appetite], C₁₄H₁₂N₃.-HCl . 2H₂O. Phenyldihydroquinazolin hydrochloride used as a stomachic. Dose 4-7 gr. (0.26-0.45 Gm.).
**o. tannate**, a light yellow powder, used as an appetizer and stomachic. Dose 4-8 gr. (0.25-0.5 Gm.).
**orexis** (o-reks'-is) [ὄρεξις, appetite]. Appetite.
**orexoids** (o-reks'-oids). Trade name of 5-grain orexin tannate tablets.
**organ** (or'-gan) [ὄργανον, an organ]. Any part of the body having a definite function to perform. **o. of Corti**. See under ear. **o., enamel-**, a club-shaped process of epithelium growing from the dental ridge and forming a cap over the dental papilla. From it the enamel of the tooth is developed. **o. of Giraldès**. See Giraldès' organ. **o., Jacobson's**. See Jacobson's organ. **o.s of reproduction**, the testicles and penis and its glands in the male, the uterus and its appendages and the vagina and its glands in the female. **o. of Rosenmueller**. See parovarium. **o., segmental**, a mesoblastic embryonic structure consisting of three parts—the pronephros, the mesonephros, or Wolffian body, and the metanephros.
**organa** (or'-gan-ah) [L.]. Plural of organum. **o. genitalia**, genital organs. **o. genitalia muliebria**, female genital organs. **o. genitalia virilia**, male genital organs. **o. oculi accessoria**, accessory organs of eye. **o. palpantia**, tactile organs. **o. sensuum**, sense organs. **o. uropoetica**, uropoietic organs. **o. urticantia**, the nematophores or nematillæ of Cœlenterates.
**organacidia** (or-gan-as-id'-e-ah). The presence of organic acids, especially in the stomach. **o. gastrica**, the presence of large quantities of organic acids in the gastric contents.
**organelle** (or'-gan-el). Any one of those parts of the protozoan protoplasm having a special function.
**organic** (or-gan'-ik) [organ]. Having, pertaining to, or characterized by organs; pertaining to the animal and vegetal worlds; affecting the structure of organs. **o. acid**, any acid containing the carboxyl group COOH. **o. chemistry**, the chemistry of the carbon compounds, carbon being the central element of compounds occurring in organized beings. **o. compound**, any chemical compound containing carbon. **o. disease**, disease of an organ attended with structural changes.
**organism** (or'-gan-ism) [organ]. A body consisting of an aggregation of organs having a definite function; any living organized being, either animal or vegetable.
**organization** (or-gan-iz-a'-shun) [organ]. 1. The orderly arrangement of organs or parts. 2. An organism. 3. The conversion into an organ, or into something resembling an organ or into living tissue.
**organo-** (or-gan-o-) [organ]. A prefix meaning relating to the organs.
**organoferric** (or-gan-o-fer'-ik). Consisting of iron and some organic substance.
**organogenesis** (or-gan-o-jen'-es-is) [organo-; γένεσις, origin]. The process of the development of an organ.
**organogenetic** (or-gan-o-jen-et'-ik) [organogenesis]. Pertaining to organogenesis.
**organography** (or-gan-og'-ra-fe) [organo-; γράφειν, to write]. A descriptive treatise of the organs of an animal or plant.
**organoid** (or'-gan-oid) [organo-; εἶδος, like]. Resembling an organ.
**organoleptic** (or-gan-o-lēp'-tik) [organo-; ληπτικός, taking]. 1. Making an impression upon some organ, chiefly of special sense. 2. Plastic; capable of receiving organization.
**organology** (or-gan-ol'-o-je) [organo-; λόγος, science]. The science that treats of the organs of plants and animals.
**organoma** (or-gan-o'-mah) [organo-; ὄμα, tumor]. A tumor containing distinct organs or parts of organs, but not so arranged as to form a body or part of one.
**organometallic** (or-gan-o-met-al'-ik). Applied to a combination of an alcoholic radical with a metal or metalloid.
**organon** (or'-gan-on) [ὄργανον, an organ; instrument]. 1. An organ, q. v. 2. A code of principles. **o. auditus**, the organ of hearing. **o. gustus**, the organ of taste. **o. olfactus**, the organ of smell. **o. spirale**, the organ of Corti. **o. tactus**, the organ of touch. **o. visus**, the organ of vision. **o. vomerona-sale**. See Jacobson's organ.
**organum** (or'-gan-um) [ὄργανον, organ; νόμος, law]. The totality of the natural laws of the conduct and functions of organic life.
**organonym** (or-gan'-o-nim) [ὄργανον, an organ; ὄνυμα, name]. The name of an organ or part.

# ORGANONYMY 634 ORTHIOPISTHIUS

**organonymy** (or-gan-on'-im-e) [ὄργανον, organ; ὄνομα, name]. A system of nomenclature of the organs.

**organopathism** (or-gan-op'-ath-izm) [ὄργανον, an organ; πάθος, disease]. The doctrine of special study and investigation of the pathology of each and every organ by itself.

**organopathy** (or-gan-op'-ath-e) [ὄργανον, organ; πάθος, disease]. 1. The disease of an organ. 2. A term used by Sharp to express the local action of drugs. 3. The same as *organopathism*.

**organopexia**, **organopexy** (or-gan-o-peks'-e-ah, or-gan'-o-peks-e) [organo-; πῆξις, a fixing]. The surgical fixation of a mixplaced organ.

**organoplastic** (or-gan-o-plas'-tik) [ὄργανον, an organ; πλάσσειν, to form, to mold]. Applied to cells or tissues from which organs are developed.

**organoplasty** (or'-gan-o-plas-te) [ὄργανον, organ; πλάσσειν, to form, to mold]. The origin or development of plant and animal organs.

**organopoiesis** (or-gan-o-poi-e'-sis) [ὄργανον, an organ; ποίησις, formation]. The same as *organoplasty*.

**organopoietic** (or-gan-o-poi-et'-ik). Relating to organopoiesis.

**organosol** (or-gan'-os-ol). A proprietary preparation of colloidal silver with an organic solvent; used in gonorrhea.

**organotherapeutic** (or'-gan-o-ther-ap-ū'-tik). Relating to the treatment of disease by means of animal extracts.

**organotherapy** (or-gan-o-ther'-ap-e) [organo-; θεραπεία; treatment]. The treatment of diseases by the administration of animal organs or extracts prepared from them.

**organotrophic** (or-gan-o-tro'-fik) [ὄργανον, organ; τροφή, nourishment]. Relating or belonging to the nourishment of organized tissue.

**organotropic** (or-gan-o-trop'-ik) [organo-; τρέπειν, to turn]. 1. Pertaining to substances which act on the organs of the body. 2. Producing degeneration of organs. See *plasmotropic*.

**organum** (or'-gan-um) [L.]. See *organ*.

**orgasm** (or'-gazm) [ὀργασμός, swelling]. Intense excitement, especially that occurring during sexual intercourse.

**orgastic** (or-gas'-tik) [ὀργάειν, to swell]. Pertaining to, or characterized by, orgasm.

**oriental** (o-re-en'-tal) [*oriens*, the east]. Pertaining to the orient or east. o. **boil**. See *furunculus orientalis*. o. **plague**. See *plague*.

**orientation** (o-re-en-ta'-shun) [see *oriental*]. The act of determining one's position in space (*subjective orientation*), or the position of surrounding objects with reference to each other (*objective orientation*). 2. The relative position of the substitution elements or radicals in the benzene ring.

**orifacial** (or-if-a'-shal) [*os*, *oris*, mouth; *facies*, face]. Pertaining to the mouth and face. o. **angle**, in craniometry, the angle formed by the junction of the plane passing through the masticating surface of the superior maxilla with the facial line.

**orifice** (or'-if-is) [*orificium*, an opening]. An opening.

**orificial** (or-if-ish'-al) [*orifice*]. Pertaining to an orifice. o. **surgery**. See *Pratt's operation*.

**orificialist** (or-if-ish'-al-ist). One who treats disease by the practice of orificial surgery.

**orificium** (or-if-ish'-e-um) [L.]. Orifice. o. **epiploicum**, a synonym of *foramen of Winslow*. o. **externum uteri**, external orifice or os of uterus. o. **infundibuli**, a synonym of *helicotrema*. o. **internum uteri**, internal orifice or os of uterus. o. **urethræ externum**, external orifice of urethra. o. **urethræ internum**, internal orifice of urethra. o. **ureteris**, orifice of ureter. o. **vaginæ**, orifice of vagina.

**oriform** (or'-if-orm) [*os*, *oris*, mouth; *forma*, form]. Having the shape of a mouth.

**Origanum** (o-rig'-an-um) [ὀρίγανον, marjoram]. A genus of plants of the order Labiatæ. *O. majorana*, sweet marjoram, is used as a condiment. *O. vulgare*, wild marjoram, contains a pungent oil (*oleum origani*), consisting chiefly of terpene, $C_{10}H_{16}$. Origanum is tonic, excitant, diaphoretic, emmenagogue, and locally anodyne. It is also employed as a clearing-agent in microscopy.

**origin** (or'-ij-in) [*oriri*, to arise]. The beginning or starting-point of anything. o. **deep**, or **ental**, of a nerve, its beginning in the cells of the nerve-center. o. **of a muscle**, the point of attachment of a muscle which remains relatively fixed during contraction of the muscle. o., **superficial**, or **ectal**, of a nerve, the point at which it emerges from the brain or cord.

**orinasal** (o-ri-na'-sal). See *oronasal*.

**orizabin** (o-riz'-ab-in). Same as *jalapin*.

**ormosine** (or'-mo-sēn). A crystalline alkaloid from the seeds of *Ormosia dasycarpa*, of South America. It is hypnotic, sedative, and narcotic. The hydrochloride is also used.

**ornus** (or'-nus) [L.]. See *fraxinus*.

**oroanal** (o-ro-a'-nal) [*os*, *oris*, mouth; *anus*]. Extending from the mouth to the anus.

**orolingual** (o-ro-lin'-gwal) [*os*; mouth; *lingua*, tongue]. Pertaining to the mouth and the tongue.

**oronasal** (o-ro-na'-sal). Pertaining to the mouth and the nose.

**oronosus** (o-ron'-o-sus) [ὄρος, mountain; νόσος, disease]. A disease prevalent in mountain regions. See *mountain sickness*.

**oropharynx** (o-ro-far'-inks) [*os*, mouth; *pharynx*]. The pharynx proper, situated below the level of the lower border of the soft palate, as distinguished from the nasopharynx.

**orotherapy** (or-o-ther'-a-pe). See *orrhotherapy*.

**oroxylon** (or-oks'-il-on) [ὄρος, mountain; ξύλον, wood]. A genus of the Bignoniaceæ. *O. indicum* is indigenous to tropical Asia. The root-bark is tonic and astringent and a powerful sudorific. The leaves are applied to ulcers.

**Oroya fever** (o-roi'-yah) [S. Amer.]. The febrile stage of *verrugas*, *q. v.*

**orphol** (or'-fol). See *bismuth betanaphtholate*.

**orpiment** (or'-pim-ent). Arsenic trisulphide.

**orrhagogus** (or-ag'-o-gus) [ὀρρός, serum; ἄγειν, to lead]. Synonym of *hydragogue*.

**orrhochezia** (or-o-ke'-ze-ah) [ὀρρός, serum; χέζειν, to relieve oneself]. Serous diarrhea.

**orrhodermitis** (o-o-der-mi'-tis). See *serodermitis*.

**orrhorrhea**, **orrhorrhœa** (or-o-re'-ah) [ὀρρός, serum; ῥοία, a flow]. An abnormally great flow of serum. Also, a watery discharge; rice-water discharges.

**orrhos** (or'-ros) [ὀρρός, serum]. Serum; whey.

**orrhosis** (or-o'-sis) [ὀρρός, serum]. The production of serum.

**orrhotherapeutic** (or-ro-ther-ap-ū'-tik). Pertaining to serum therapy.

**orrhotherapy** (or-o-ther'-ap-e) [ὀρρός, serum; θεραπεία, therapy]. 1. The treatment of disease by the use of human or animal blood-serum containing antitoxins; serum therapy. 2. Whey-cure.

**orris** (or'-is). See *iris* (2).

**orsudan** (or'-sū-dan). Trade name for sodium methylacetylaminophenylarsonate. It contains 25.4 per cent. of arsenic, and is used in the treatment of syphilis.

**ortharthragra** (orth-arth-ra'-grah) [ortho-; ἄρθρον, joint; ἄγρα, seizure]. True gout.

**orthiauchenus** (orth-e-awk'-en-us) [ὄρθιος, upright; αὐχήν, neck]. Lissauer's term for a skull in which the angle formed between the radius fixus and the line joining the basion and the inion is between 38° and 49°.

**orthin** (or'-thin), $C_7H_{18}N_7O_3$. An antiseptic derivative of phenylhydrazin.

**orthochordus** (orth-e-o-kord'-us) [ὄρθιος, upright; χορδή, cord]. Lissauer's term for a skull in which the angle formed between the radius fixus and the line joining the hormion and the basion is between 33.2° and 52°.

**orthiocoryphus** (orth-e-o-kor'-if-us) [ὄρθιος, upright; κορυφή, head]. Lissauer's term for a skull in which the angle formed between the radius fixus and the line joining the bregma and the lambda is between 29° and 41°.

**orthiodontus** (orth-e-o-don'-tus) [ὄρθιος, upright; ὀδούς, a tooth]. Lissauer's term for a skull in which the angle between the radius fixus and the line joining the alveolar and subnasal points is between 88° and 121°.

**orthiometopus** (orth-e-o-met-o'-pus) [ὄρθιος, upright; μέτωπον, forehead]. Lissauer's term for a skull in which the angle between the radius fixus and the line joining the bregma and the nasal point is between 47° and 60°.

**orthiopisthius** (orth-e-o-pis'-the-us) [ὄρθιος, upright; ὄπισθεν, behind]. Lissauer's term for a skull in which

**the angle** between the radius fixus and the line joining the lambda and the inion is between 84° and 95°.
**orthiopisthocranius** (*orth-e-o-pis-tho-kra'-ne-us*) [ὄρθιος, upright; ὄπισθεν, behind; κρανίον, skull]. Lissauer's term for a skull in which the angle formed between the radius fixus and the line joining the lambda and the opisthion is between 107° and 119°.
**orthioprosopus** (*orth-e-o-pros-o'-pus*) [ὄρθιος, upright; πρόσωπον, face]. Lissauer's term for a skull in which the angle formed between the radius fixus and the line joining the nasion and the alveolar point is between 89.4° and 100°.
**orthiopylus** (*orth-e-op'-il-us*) [ὄρθιος, upright; πύλη, gate]. Lissauer's term for a skull in which the angle formed between the radius fixus and the line joining the middle point of the anterior margin of the foramen magnum and the middle point of the posterior margin of the foramen magnum is between 15.5° and 24°.
**orthiorrhinus** (*orth-e-or-i'-nus*) [ὄρθιος, upright; ῥίς, nose]. Lissauer's term for a skull in which the angle formed between the radius fixus and the line joining the nasion and the subnasal point is between 87.5° and 98°.
**orthiuraniscus** (*orth-e-ū-ran-is'-kus*) [ὄρθιος, upright; οὐρανίσκος, canopy]. Lissauer's term for a skull in which the angle formed between the radius fixus and a line joining the posterior border of the incisor foramen and the alveolar point is between 40° and 60°.
**ortho-** (*or-tho-*) [ὀρθός, right; straight]. 1. A prefix denoting straight, normal, or true. 2. In chemistry, a prefix denoting that one among several compounds of the same elements which is considered the normal compound. Among derivatives of the benzol-ring it refers to those formed by the substitution of two adjacent hydrogen atoms.
**orthoacid** (*or'-tho-as-id*). An acid in which the hydroxyl groups are equal in number to the valence of the acidulous element; when this acid is not known, the one whose number of hydroxyl groups most nearly equals the valence of the acidulous element is improperly called an orthoacid.
**orthobiosis** (*or-tho-bi-o'-sis*) [*ortho-*; βίος, life]. Correct living; living in accordance with all the laws of hygiene.
**orthocephalic** (*or-tho-sef-al'-ik*) [*ortho-*; κεφαλή, head]. Pertaining to orthocephaly.
**orthocephalism** (*or-tho-sef'-al-izm*). Synonym of *orthocephaly*.
**orthocephalous** (*or-tho-sef'-al-us*) [*ortho-*; κεφαλή, head]. Having a skull with a vertical index of from 70.1 to 75.
**orthocephaly** (*or-tho-sef'-al-e*) [*ortho-*; κεφαλή, head]. The condition of having a skull with a vertical index of from 70.1° to 75°.
**orthochorea** (*or-tho-ko-re'-ah*) [*ortho-*; χορεία, dance]. Choreic movements in the erect posture.
**orthochromatic** (*or-tho-kro-mat'-ik*) [*ortho-*; χρῶμα, color]. A term used in photography to denote correctness in the rendering of colors.
**orthocrasia** (*or-tho-kra'-ze-ah*) [*ortho-*; κράσις, temperament]. A condition in which there is no idiosyncrasy.
**orthocresalol** (*or-tho-kres'-al-ol*). See *cresalol*.
**orthocresol** (*or-tho-kre'-sol*), C₇H₈O. One of the forms of cresol, *q. v.* It occurs in small amounts in urine.
**orthodactylous** (*or-tho-dak'-til-us*) [*ortho-*; δάκτυλος, a finger]. Having straight digits.
**orthodiagram** (*or-tho-di'-ah-gram*). The record made by an orthodiagraph.
**orthodiagraph** (*or-tho-di'-ah-graf*) [*ortho-*; diagraph]. A radiographic apparatus which records accurately and quickly the dimension, form, and position of internal organs of the body or the location of foreign bodies.
**orthodiagraphy** (*or-tho-di-ag'-raf-e*). The mode of determining by the aid of the roentgen-rays the exact dimensions of an internal organ by the shadow which it throws upon the fluorescent screen.
**orthodolichocephalous** (*or-tho-do-lik-o-sef'-al-us*) [*ortho-*; δολιχός, long; κεφαλή, head]. Having a long and straight head; having a vertical index between 70.1° and 75°, and a transverso-vertical index between 70° and 74.9°.
**orthodontia** (*or-tho-don'-she-ah*) [*ortho-*; ὀδούς, tooth] The correction of irregularities of the teeth.

**orthodontics** (*or-tho-don'-tiks*) [*orthodontia*]. "That branch of dentistry which deals with the principles and practices involved in the prevention and correction of malocclusion of the teeth, and such other malformations and abnormalities as may be associated therewith."
**orthoform** (*or'-tho-form*), C₈H₃(OH)(NH₂)COOH₃. A white, crystalline, odorless, tasteless powder, slightly soluble in water, soluble in alcohol or ether; used as a local anesthetic and antiseptic and internally in cancer of the stomach. Dose 7½–15 gr. (0.5–1.0 Gm.). Syn., *methylparaamidometaoxybenzoate*. o. **emulsion**, orthoform, 25 parts, in olive-oil, 100 parts. It is recommended in subcutaneous injection in laryngeal tuberculosis. o., **new**, metaamidoparaoxybenzoic-methylester—a cheaper product, used as is the original orthoform.
**orthognathic** (*or-thog-na'-ithik*). Same as *orthognathous*.
**orthognathism** (*or-thog'-na-thizm*) [*ortho-*; γνάθος, jaw]. The quality of being orthognathous, or of having jaws with little or no forward projection.
**orthognathous** (*or-thog'-na-thus*) [*ortho-*; γνάθος, the jaw]. Straight-jawed; having a gnathic angle of from 83° to 90°.
**orthomesocephalous** (*or-tho-mes-o-sef'-al-us*) [*ortho-*; μέσος, middle; κεφαλή, the head]. In craniometry, a term applied to a skull with a transverso-vertical index between 75.1° and 79.9°, and a vertical index between 70.1° and 75°.
**orthometer** (*or-thom'-et-er*) [*ortho-*; μέτρον, a measure]. An instrument for measuring the relative degree of protrusion of the eyes.
**orthomonochlorphenol** (*or-tho-mon-o-klor-fe'-nol*). An anesthetic drug used in rhinitis.
**orthomorphia** (*or-tho-mor'-fe-ah*) [*ortho-*; μορφή, form]. The surgical correction of deformity.
**orthonal** (*or'-tho-nal*). A local anesthetic said to contain cocaine hydrochloride, alypin, and epinephrine.
**orthopedic, orthopædic** (*or-tho-pe'-dik*) [*ortho-*; παῖς, child]. Pertaining to the correction of deformities, especially in children. o. **surgery**, the branch of surgery devoted to the correction of deformities.
**orthopedics, orthopædics** (*or-tho-pe'-diks*). See *orthopedic surgery*.
**orthopedist, orthopædist** (*or-tho-pe'-dist*). One who practises orthopedic surgery.
**orthophoria** (*or-tho-fo'-re-ah*) [*ortho-*; φόρος, a tending]. 1. A tendency of the visual lines in parallelism. 2. Normal balance of the eye muscles.
**orthophosphoric acid** (*or-tho-fos-for'-ik*). See *acid, orthophosphoric*.
**orthoplasy** (*or'-tho-plaz-e*) [*ortho-*; πλάσις, a forming]. The directive or determining influence of organic selection in evolution.
**orthopnea, orthopnœa** (*or-thop-ne'-ah*) [*ortho-*; πνεῖν, to breathe]. A condition marked by quick and labored breathing, in which the patient finds relief only by maintaining an upright position.
**orthopneic** (*or-thop-ne'-ik*) [*orthopnea*]. Characterized by orthopnea.
**orthopraxis, orthopraxy** (*or-tho-praks'-is, or'-tho-praks-e*) [*ortho-*; πρᾶξις, doing]. Correction of the deformities of the body.
**orthoptic** (*or-thop'-tik*) [*ortho-*; *optic*]. Pertaining to normal binocular vision. o. **training**, a method of correcting the defective vision of those having strabismus or muscular insufficiency, by stereoscopic and other ocular exercises of a gymnastic kind.
**orthopygium** (*or-tho-pij'-e-um*) [*ortho-*; πυγή, the rump]. Synonym of *coccyx*.
**orthoscope** (*or'-tho-skōp*) [*ortho-*; σκοπεῖν, to see]. 1. An instrument for examination of the eye through a layer of water, whereby the curvature, and hence the refraction, of the cornea is neutralized and the cornea acts as a plane medium. 2. An instrument, for use in drawing the projections of skulls.
**orthoscopic** (*or-tho-skop'-ik*) [*orthoscope*]. 1. Pertaining to an orthoscope or to orthoscopy. 2. Applied to lenses cut from the periphery of a large lens. 3. Having normal vision.
**orthoscopy** (*or-thos'-ko-pe*) [*orthoscope*]. The examination of the eye with the orthoscope.
**orthosis** (*or-tho'-sis*) [ὄρθωσις, a making straight].
**orthostatic** (*or-tho-stat'-ik*) [*ortho-*; στατός, standing]. Pertaining to or caused by standing upright. o. **albuminuria**, albuminuria which occurs when the patient stands on his feet or exercises for long periods

of time, but which disappears after a period of rest in bed.
**orthotast** (*or'-tho-tast*) [*ortho-*; τάσσειν, to arrange]. An apparatus for straightening curvatures of long bones. It has also been used as a tourniquet.
**orthoterion, orthoterium** (*or-tho-te'-re-on, or-tho-te'-re-um*) [*ortho-*]. An apparatus for straightening curved limbs.
**orthotonus** (*or-thot'-o-nus*) [*ortho-*; τόνος, tension]. Tetanic cramp in which the body lies rigid and straight.
**orthotrophy** (*or-thot'-ro-fe*) [*ortho-*; τροφή, nourishment]. Correct or normal nourishment; the normal process of nutrition.
**orthotropic** (*or-tho-trop'-ik*) [*ortho-*; τρέπειν, to turn]. Pertaining to, or exhibiting orthotropism.
**orthotropism** (*or-thot'-ro-pism*) [*ortho-*; τρέπειν, to turn]. Vertical, upward, or downward growth.
**oryza** (*o-ri'-zah*). See *rice*.
**O. S.** Abbreviation for *oculus sinister*, Latin for left eye.
**Os.** The chemical symbol of *osmium*.
**os** [L.: gen., *oris*; pl., *ora*]. The mouth. **o.** tincæ ("tench's mouth"), the os uteri, or mouth or the uterus. **o. uteri.** Same as *o. uteri internum*. **o. uteri externum,** the external opening or entrance to the uterus. **o. uteri internum,** the internal orifice of the uterus.
**os** [L.; gen., *ossis*: pl., *ossa*]. A bone. **o. acromiale,** the acromion when not united to the scapula. **o. alæforme,** the sphenoid bone. **o. alare, o. alatum,** the sphenoid bone and alisphenoid bone. **o. ballistæ,** the astragalus. **o. brachii,** the humerus. **ossa bregmatis,** the parietal bones. **o. calcis,** the bone of the heel. **o. breve,** a short bone. **o. capitatum.** See *o. magnum*. **o. convolutum,** the turbinated body. **o. coronale,** the frontal bone. **o. coxæ.** See *o. innominatum*. **o. cubitale,** the cuneiform bone of the wrist. **o. femoris,** the femur. **o. hamatum,** the unciform bone. **o. humeri,** the humerus. **o. ilii,** the ilium. **o. innominatum,** the innominate bone. **ossa intercalaria,** the Wormian bones. **o. ischii,** the ischium. **o. japonicum,** the divided malar bone, a racial characteristic of the Japanese. **o. jugale, o. jugamentum,** the malar bone. **o. juguli,** the clavicle. **o. longum,** a long bone. **o. lunatum,** the semilunar bone. **o. magnum,** the third bone of the second row of the carpus. **o. maxillaris,** the upper jaw. **o. multangulum majus,** the trapezium. **o. multangulum, minus,** the trapezoid. **o. multiforme,** the sphenoid bone. **o. naviculare,** the scaphoid bone. **o. orbiculare.** 1. The pisiform bone. 2. The lenticular process of the incus. **o. orbitale,** the upper of two portions into which the malar bone is sometimes divided by a horizontal suture. **o. pectinis,** the os pubis. **o. pectoris,** the sternum. **o. planum.** 1. A flat bone. 2. Part of the ethmoid bone. **o. pneumaticum,** a hollow bone. **o. præmaxillare,** the intermaxillary bone. **o. pubis,** the pubis. **o. sepiæ,** cuttlefish bone. **o. scutiforme,** the patella. **ossa suturarum,** the Wormian bones. Syn., *ossa triquetra; ossa Wormiana*. **o. triangulare,** the parietal bone. **o. triangulare.** See *o. cubitale*. **o. trigonum,** an ossicle due to the separation of the external tubercle of the posterior surface of the astragalus and ossification from a distinct center. **o. triquetrum.** See *o. cubitale*. **o. unguis,** the lacrimal bone. **o. ypsiloides,** the hyoid bone. **o. zygomaticum,** the malar bone.
**osazone** (*o'-saz-ōn*). A compound formed when solutions of sugar are warmed for some time with a solution of phenylhydrazin and dilute acetic acid.
**oscedo** (*os-e'-do*) [L.]. Yawning.
**oschea** (*os'-ke-ah*) [ὄσχεον, scrotum]. Synonym of *scrotum*.
**oscheal** (*os'-ke-al*). Pertaining to the scrotum.
**oscheitis** (*os-ke-i'-tis*) [*oschea*; ιτις, inflammation]. Inflammation of the scrotum.
**oscheo-** (*os-ke-o-*) [*oschea*]. A prefix meaning relating to the scrotum.
**oscheocele** (*os'-ke-o-sēl*) [*oscheo-*; κήλη, hernia]. Scrotal hernia.
**oscheohydrocele** (*os-ke-o-hi'-dro-sēl*) [*oscheo-*; hydrocele*]. A hydrocele occupying the sac of a scrotal hernia after the return of the bowel to the peritoneal cavity and the shutting off of the sac from the latter.
**oscheolith** (*os'-ke-o-lith*) [*oscheo-*; λίθος, stone]. Scrotal calculus.
**oscheoma** (*os-ke-o'-mah*) [*oscheo-*; ὅμα, tumor]. A scrotal tumor.

**oscheoncus** (*os-ke-ong'-kus*) [*oscheo-*; ὅγκος, swelling]. A swelling or tumor of the scrotum.
**oscheoplasty** (*os'-ke-o-plas-te*) [*oscheo-*; πλάσσειν, to form]. Plastic surgery of the scrotum.
**oschitis** (*os-ki'-tis*). See *oscheitis*.
**oscillation** (*os-il-a'-shun*) [*oscillatio*; *oscillare*, to sway to and fro]. A swinging or vibration; also any tremulous motion.
**oscillator** (*os'-il-a-tor*) [see *oscillation*]. An apparatus for the application of mechanical therapeutics.
**oscillometer** (*os-il-om'-et-ur*) [*oscillare*, to sway to and fro; μέτρον, measure]. An instrument for measuring oscillations, such as those seen in taking blood-pressure.
**oscitancy** (*os'-it-an-se*) [*oscitare*, to yawn]. The disposition to yawn; drowsiness.
**oscitation** (*os-it-a'-shun*) [*oscitare*, to yawn]. The act of yawning.
**osculation** (*os-ku-la'-shun*) [*osculum*]. 1. The union of vessels by their mouths. 2. The act of kissing.
**osculum** (*os'-kū-lum*) [L., "a little mouth"]. A small aperture.
**-ose** (*-ōs*) [*-osus*]. A suffix denoting a member of the carbohydrate group.
**Osiander's sign of pregnancy** (*o-ze-an'-der*). Vaginal pulsation, which may frequently be detected early in pregnancy.
**-osis** (*-o-sis*). A suffix signifying condition of, or state caused by.
**Osler's disease** (*ōs'-ler*) [Sir William Osler, English physician, 1849– ]. See *Polycythæmia cyanotica*. **O.'s phenomenon,** the agglutination of the blood-platelets observed, in blood immediately after its withdrawal from the body. **O.'s sign, O.'s spots,** small painful erythematous swellings found in the skin and subcutaneous tissues of the hands and feet, and said to be indicative of subacute or chronic malignant endocarditis.
**osmate** (*oz'-māt*) [*osmium*]. A salt of osmic acid.
**osmatic** (*oz-mat'-ik*) [ὀσμή, smell]. Characterized by a keen sense of smell; having a highly developed rhinencephalon.
**osmazome** (*oz'-maz-ōm*) [ὀσμή, smell; ζωμός, broth]. A brownish-yellow substance developed by heat in muscular fibers, and formerly supposed to give to cooked meats their peculiar flavor.
**osmesis** (*os-me'-sis*) [ὀσμησις, a smelling]. The act of smelling.
**osmic** (*oz'-mik*) [*osmium*]. Pertaining to or containing osmium. **o. acid.** See *acid, osmic*.
**osmidrosis** (*oz-mid-ro'-sis*) [ὀσμή, smell; ἱδρωσις, sweat]. The secretion of a malodorous perspiration; bromidrosis.
**osmium** (*os'-me-um*) [ὀσμή, smell]. A heavy metallic element belonging to the platinum group. Symbol Os; specific gravity 22.48; atomic weight, 190.9; quantivalence II, IV, VI, VIII. See *acid, osmic*.
**osmodysphoria** (*oz-mo-dis-fo'-re-ah*) [ὀσμή, smell; δυς, difficult; φέρειν, to bear]. Intolerance of certain odors.
**osmogen** (*os'-mo-jen*) [ὀσμή, impulse; γεννᾶν, to produce]. A substance from which a ferment or enzyme is developed.
**osmology** (*oz-mol'-o-je*). 1. The science of odors and the sense of smell. 2. That part of physical science treating of osmosis.
**osmometer** (*oz-mom'-et-er*) [ὀσμή, smell; μέτρον, a measure]. 1. An instrument for testing the sense of smell. 2. [See *osmosis*.] An apparatus for measuring osmosis.
**osmose** (*os-mōs'*). 1. Same as *osmosis*. 2. To undergo osmosis.
**osmosis** (*os-mo'-sis*) [ὠσμός, impulsion]. The passage of liquids and substances in solution through a porous septum. See *endosmosis* and *exosmosis*.
**osmotic** (*oz-mot'-ik*) [ὠσμός, impulse]. Pertaining to osmosis. **o. equivalent,** "that figure which indicates the weight of water which replaces by osmosis one part by weight of the substance subjected to the process." **o. pressure,** the pressure exerted by the particles of compounds when dissolved, directly expressed or shown by osmotic phenomena. It is equal to that which would be exerted by an equal amount of the substance if it were converted into gas and occupied the same volume at the same temperature as the solution.
**osphrasia** (*os-fra'-ze-ah*). Synonym of *osphresis*.

**osphresiology** (os-fre-ze-ol'-o-je) [ὄσφρησις, smell; λόγος, science]. The science of the sense of smell and its organs; also of odors and perfumes.
**osphresis** (os-fre'-sis) [ὄσφρησις, smell]. The sense of smell; olfaction.
**osphretic** (os-fret'-ik) [ὄσφρητικος, capable of smelling]. Same as olfactory.
**osphus** (os'-fus) [ὀσφύς, loin]. The loin.
**osphyalgia** (os-fe-al'-je-ah) [osphus; ἄλγος, pain]. Any pain in the hip or loins; sciatica.
**osphyitis** (os-fi-i'-tis) [ὀσφύς, loin; ιτις, inflammation]. Lumbar inflammation; coxitis.
**osphyomyelitis** (os-fi-o-mi-el-i'-tis) [ὀσφύς, loin; myelitis]. Myelitis of the lumbar portion of the spinal cord.
**ossa** (os'-ah) [L.; pl. of os, a bone]. Bones. See os. o. **innominata**, the irregular bones forming the sides and anterior wall of the pelvis. o. **lata**, the broad or flat bones forming the walls of cavities. o. **longa**, the long bones, e. g., those of the limbs. o. **suturarum**, o. **triquetra**, o. **wormiana**. See Wormian bones.
**ossagen** (os'-aj-en). A proprietary remedy used in rickets, said to be the calcium salt of the fatty acids of red bone-marrow. Dose 30–60 gr. (2–4 Gm.) twice daily.
**ossalin** (os'-al-in). A hygroscopic ointment-base prepared from bone-marrow. It occurs as a grayish fat with the odor of tallow. Syn., adeps ossium.
**ossalinate** (os-al'-in-āt). A proprietary substitute for cod-liver oil said to be the sodium compound of the acid ot ox-marrow.
**ossature** (os'-al-ūr) [os, a bone]. The arrangement of the bones of the body.
**ossein** (os'-e-in) [os, a bone]. The organic base of osseous tissue.
**osselet** (os'-el-et) [dim. of os, a bone]. A small bone; also a hard nodule on the inner aspect of the horse's knee.
**osseoalbumoid** (os-e-o-al'-bū-moid). A proteid substance resembling elastin, obtained from bone after hydration of the collagen.
**osseoaponeurotic** (os-e-o-ap-on-ū-rot'-ik) [os; aponeurosis]. Bounded by bone and the aponeurosis of a muscle.
**osseocartilaginous** (os-e-o-kar-til-aj'-in-us). Pertaining to or composed of both bone and cartilage.
**osseomucoid** (os-e-o-mū'-koid). A mucin discovered in bone and having the composition C₄₇.₉₇-H₈.₀₀N₁₁.₈₁S₂.₁₁O₂₁.₈₅.
**osseous** (os'-e-us) [os]. Bony; composed of or resembling bone.
**ossicle** (os'-ik-l) [ossiculum, dim. of os]. A small bone. o., **Andernach's**, the Wormian bones. o., **auditory**, one of a chain of small bones found in the tympanic cavity of the ear. o., **epactal**, a Wormian bone. o., **Kerkring's**, a point of ossification in the occipital bone, immediately behind the foramen magnum. o.s, **Riolan's**, small bones sometimes found in the suture between the inferior border of the occipital bone and the mastoid portion of the temporal bone.
**ossicula** (os-ik'-u-lah) [L.]. Plural of ossiculum. o. **auditus**. Auditory ossicles; see ossicle.
**ossiculectomy** (os-ik-ū-lek'-to-me) [ossicle; ἐκτομή, excision]. The excision of an ossicle or of the auditory ossicles.
**ossiculotomy** (os-ik-ū-lot'-o-me) [ossicle; τέμνειν, to cut]. Surgical incision of the ossicles of the ear.
**ossiculum** (os-ik'-ū-lum). [L.: pl., ossicula]. Synonym of ossicle. **ossicula calcoidea**, the cuneiform bones of the foot. **ossicula epactalia**, the Wormian bones. o. **hamuli**, a horny nodule in the trochlea near the ventral border of the lacrimal groove. o. **jugulare**, one of frequent occurrence in the posterior lacerated foramen. **ossicula triticea**, the lesser cornua of the hyoid bone.
**ossiferous** (os-if'-er-us) [os; ferre, to bear]. Containing or producing bone tissue.
**ossific** (os-if'-ik) [os; facere, to make]. Producing bone.
**ossification** (os-if-ik-a'-shun) [see ossific]. The formation of bone.
**ossifluent** (os-if'-lū-ent) [os; fluere, to flow]. Breaking down and softening bony tissue, as an ossifluent abscess.
**ossiform** (os'-if-orm) [os; forma, form]. Bone-like.
**ossifying** (os'-if-i-ing) [os; facere, to make]. Changing into bone. o. **chondroma**, a chondroma that is undergoing ossification. o. **myositis**, inflammation of muscle attended with, or followed by, deposition of bone-like masses.
**ossin** (os'-in) [os, a bone]. An extract made from bone-tissue, and used in organotherapy, q. v.
**ostagra** (os-ta'-grah) [osteo-; ἄγρα, seizure]. A bone-forceps.
**ostalgia** (os-tal'-je-ah) [osteo-; ἄλγος, pain]. Pain in a bone.
**ostalgitis** (os-tal-ji'-tis) [osteo-; ἄλγος, pain; ιτις, inflammation]. Inflammation of a bone attended by pain.
**ostarthritis** (ost-ar-thri'-tis). See osteoarthritis.
**osteal** (os'-te-al) [osteo-]. Osseous, bony; pertaining to bone.
**ostealgia** (os-te-al'-je-ah). See ostalgia.
**ostealleosis** (os-te-al-e-o'-sis) [osteo-; ἀλλοίωσις, alteration]. A metamorphosis of the substance of bone, as exemplified in osteosarcoma.
**osteameba**, **osteamœba** (os-te-am-e'-bah) [osteo-; amœba]. A bone-cell or osteoblast regarded as an organism.
**osteanabrosis** (os-te-an-ab-ro'-sis) [osteo-; ἀνάβρωσις, an eating up]. Absorption or atrophy of bone.
**osteanagenesis** (os-te-an-aj-en'-es-is) [osteo-; ἀναγεννᾶν, to regenerate]. The regeneration of bone.
**osteanaphysis** (os-te-an-af'-is-is) [osteo-; ἀναφύειν, to reproduce]. The reproduction of bone-tissue. Synonym of osteanagenesis.
**ostearthritis** (os-te-ar-thri'-tis). See osteoarthritis.
**ostearthrocace** (os-te-ar-throk'-as-e) [osteo-; ἄρθρον, joint; κακός, bad]. Malignant caries of the bones of a joint.
**ostearthrotomy** (os-te-ar-throt'-o-me). See osteoarthrotomy.
**osteauxe** (os-te-awks'-e) [osteo-; αὔξη, growth]. Abnormal enlargement of a bone.
**ostectomy** (os-tek'-to-me). See osteectomy.
**ostectopy** (os-tek'-to-pe) [osteo-; ἐκτοπός, placed]. Displacement of bone.
**osteectomy** (os-te-ek'-to-me) [osteo-; ἐκτομή, excision]. Excision of a portion of a bone.
**ostein** (os'-te-in). The same as ossein.
**osteitic** (os-te-it'-ik) [osteo-; ιτις, inflammation]. Pertaining to osteitis.
**osteitis**, **ostitis** (os-te-i'-tis, os-ti'-tis) [osteo-; ιτις, inflammation]. Inflammation of bone. o. **carnosa**, inflammation of bone, attended with the presence of an excess of fungous granulations. o., **condensing**, a form usually involving the whole of a hollow bone, and resulting in the filling of the medullary cavity with a dense bony mass; new bone usually forms on the surface, so that the bone becomes heavier and denser than normal. o. **deformans**, a rare form characterized by the production of deformity. o. **fungosa**, a simple inflammatory hyperplasia of the medulla and of the compact substance of bone characterized by fungoid granulations and leading to new ossification or destructive chronic inflammation. o., **gummatous**, a chronic form due to syphilis and characterized by the formation of gummata in the cancellous tissue of the epiphysis, in the shaft of a bone, or in the periosteum. o., **rarefying**. See osteoporosis. o., **sclerosing**. See osteosclerosis.
**ostembryon** (os-tem'-bre-on) [osteo-; ἔμβρυον, fetus]. Synonym of lithopedion.
**ostemia**, **ostæmia** (os-te'-me-ah) [osteo-; αἷμα, blood]. A morbid condition of bone distinguished by its turgescence with blood.
**ostempyesis** (os-tem-pi-e'-sis) [osteo-; ἐμπύησις, suppuration]. Suppuration of bone.
**osteo-** (os-te-o-) [ὀστέον, bone]. A prefix signifying connection with or relation to bone.
**osteoanabrosis** (os-te-o-an-ab-ro'-sis) [osteo-; ἀνά, up; βρῶσις, eating]. Absorption of bone or its destruction, as by osteoclasts.
**osteoaneurysm** (os-te-o-an'-ū-rizm) [osteo-; aneurysm]. Aneurysm of the arteries of a bone; a pulsating tumor of a bone.
**osteoarthritis** (os-te-o-ar-thri'-tis) [osteo-; arthritis]. 1. An inflammation of the bones forming a joint. 2. Chronic rheumatoid arthritis.
**osteoarthropathy** (os-te-o-ar-throp'-ath-e) [osteo-; ἄρθρον, joint; πάθος, disease]. Any disease of bony articulations. o., **hypertrophic pulmonary**, a disease characterized by a bulbous enlargement of the terminal phalanges of the fingers and toes, a thickening of the articular ends of the bones, and a peculiar curvation of the nails. The condition is usually associated with disease of the lungs or pleura (whence the name osteoarthropathie pneumonique hypertrophi-

# OSTEOARTHROTOMY 638 OSTEOPERIOSTITIS

*ante* given to it by Marie), and results from the absorption of toxic products from the diseased foci.
**osteoarthrotomy** (*os-te-o-ar-throt'-o-me*) [*osteo-*; *arthrotomy*]. Excision of the joint-end of a bone.
**osteoblast** (*os'-te-o-blast*) [*osteo-*; βλαστός, a germ]. Any one of the cells of mesoblastic origin concerned in the formation of bony tissue.
**osteoblastic** (*os-te-o-blas'-tik*). Pertaining to osteoblasts, or to the formation of bone.
**osteocachexia** (*os-te-o-kak-eks'-e-ah*). Cachexia due to disease of the bones.
**osteocampsia** (*os-te-o-kamp'-se-ah*) [*osteo-*; κάμπτειν, to bend]. Curvature of a bone without fracture, as in osteomalacia.
**osteocarcinoma** (*os-te-o-kar-sin-o'-mah*). 1. Ossifying carcinoma. 2. Carcinoma of bone.
**osteocartilaginous** (*os-te-o-kar-til-aj'-in-us*). Pertaining to or composed of both bone and cartilage.
**osteocele** (*os'-te-o-sēl*) [*osteo-*; κήλη, hernia]. A bone-like substance found in old hernial sacs; also marked hardening of the testicle.
**osteocephaloma** (*os-te-o-sef-al-o'-mah*) [*osteo-*; κεφαλή, head; ὄμα, tumor; *pl.*, *osteocephalomata*]. Encephaloma or encephaloid sarcoma of bone.
**osteochondritis** (*os-te-o-kon-dri'-tis*). Inflammation involving both bone and cartilage.
**osteochondroma** (*os-te-o-kon-dro'-mah*). A tumor that is in part bony and in part cartilaginous.
**osteochondrophyte** (*os-te-o-kon'-dro-fīt*). A bone tumor in which the proportions of bone and cartilage are nearly equal.
**osteochondrosarcoma** (*os-te-o-kon-dro-sar-ko'-mah*). An osteochondroma with sarcomatous features.
**osteoclasia** (*os-te-o-kla'-se-ah*). See *osteoclasis*.
**osteoclasis** (*os-te-ok'-la-sis*) [*osteo-*; κλάσις, a breaking]. 1. Fracture of bones for purposes of remedying deformity. 2. The destruction of bony tissue by osteoclasts.
**osteoclast** (*os'-te-o-klast*) [see *osteoclasis*]. 1. An instrument for performing osteoclasis. 2. One of the large multinuclear cells found against the surfaces of bone in little eroded depressions (Howship's lacunæ), and concerned in the removal of bone.
**osteoclastic** (*os-te-o-klas'-tik*) [*osteo-*; κλαστός, broken]. Of the nature of an osteoclast; concerned in the breaking down and absorption of bone.
**osteocomma** (*os-te-o-kom'-ah*) [*osteo-*; κόμμα, segment; *pl.*, *osteocommata*]. Any one of a series of bone-segments; in the adult skeleton, a vertebra is an example.
**osteocope** (*os'-te-o-kōp*). See *osteocopic pain*.
**osteocopic pain** (*os-te-o-kop'-ik*) [*osteo-*; κόπος, a beating]. A severe pain in a bone, usually worse at night. It is a symptom of osteitis or periostitis, especially of syphilitic origin.
**osteocranium** (*os-te-o-kra'-ne-um*). The ossified cranium as distinguished from the chondrocranium.
**osteocystoma** (*os-te-o-sis-to'-mah*). A cystic bone-tumor.
**osteodentine** (*os-te-o-den'-tēn*) [*osteo-*; *dens*, tooth]. A tissue of the nature of, and intermediate in structure between bone and dentine.
**osteodermatoplastic** (*os-te-o-der-mat-o-plas'-tik*) [*osteo-*; δέρμα, skin; πλάσσειν, to mold]. Pertaining to the formation of osseous tissue in dermal structures.
**osteodermatous** (*os-te-o-der'-mat-us*) [*osteo-*; δέρμα, skin]. Having an ossified integument. Pertaining to osteodermia.
**osteodermia** (*os-te-o-der'-me-ah*) [*osteo-*; δέρμα, skin]. Bony formations in the skin.
**osteodiastasis** (*os-te-o-di-as'-ta-sis*) [*osteo-*; διάστασις, separation]. Separation of bone (as an epiphysis) without true fracture.
**osteodiclis** (*os-te-o-dik'-lis*). See *diclidostosis*.
**osteodynia** (*os-te-o-din'-e-ah*) [*osteo-*; ὀδύνη, pain]. A chronic pain in a bone.
**osteoepiphysis** (*os-te-o-e-pif'-is-is*). A bony epiphysis.
**osteofibrolipoma** (*os-te-o-fi-bro-lip-o'-mah*). A tumor of bony, fibrous, and fatty elements.
**osteofibroma** (*os-te-o-fi-bro'-mah*). A combined osteoma and fibroma.
**osteogen** (*os'-te-o-jen*) [*osteo-*; γεννᾶν, to produce]. The substance of which osteogenic fibers are made up.
**osteogenesis** (*os-te-o-jen'-es-is*) [*osteo-*; γένεσις, origin]. The development of bony tissue.
**osteogenetic** (*os-te-o-jen-et'-ik*) [see *osteogenesis*]. Pertaining to osteogenesis. o. cell, an osteoblast. o. layer, the deep layer of periosteum from which bone is formed.

**osteogenic** (*os-te-o-jen'-ik*). Synonym of *osteogenetic*.
**osteogeny** (*os-te-oj'-en-e*). See *osteogenesis*.
**osteography** (*os-te-og'-ra-fe*) [*osteo-*; γράφειν, to write]. Descriptive anatomy of the bones and their articulations.
**osteohalisteresis** (*os-te-o-hal-is-ter-e'-sis*) [*osteo-*; ἅλς, salt; στέρησις, privation]. A loss of the mineral constituents of bone.
**osteohelcosis** (*os-te-o-hel-ko'-sis*) [*osteo-*; ἕλκωσις, ulceration]. Caries of bone.
**osteoid** (*os'-te-oid*) [*osteo-*; εἶδος, like]. 1. Resembling bone. 2. An osteoma. o. sarcoma, a sarcoma in which non-calcified bony tissue is found.
**osteolipochondroma** (*os-te-o-lip-o-kon-dro'-mah*) [*osteo-*; λίπος, fat; *chondroma*]. A chondroma with osseous and fatty elements.
**osteolith** (*os'-te-o-lith*) [*osteo-*; λίθος, stone]. A petrified bone.
**osteology** (*os-te-ol'-o-je*) [*osteo-*; λόγος, science]. Science of anatomy and structure of bones.
**osteolysis** (*os-te-ol'-is-is*) [*osteo-*; λύσις, dissolution]. 1. Absorption of bone. 2. Degeneration of bone.
**osteolytic** (*os-te-o-lit'-ik*) [*osteo-*; λύσις, dissolution]. Pertaining to, or concerned in, osteolysis.
**osteoma** (*os-te-o'-mah*) [*osteo-*; ὄμα, a tumor]. A bony tumor. o., cavalryman's, one occurring at the insertion of the long adductor muscle of the thigh. o. eburneum, a tumor consisting of hard bony tissue. o. *durum*. o. *medullare*, an osteoma containing marrow-spaces. o. *spongiosum*, an osteoma containing cancellated bony tissue.
**osteomalacia** (*os-te-o-mal-a'-she-ah*) [*osteo-*; μαλακία, softening]. Softening of bone from loss of its earthy constituents. Occurs chiefly in adults, especially in pregnancy.
**osteomalacial**, **osteomalacic** (*os-te-o-mal-a'-se-al*, *os-te-o-mal-a'-sik*) [*osteo-*; μαλακία, softening]. Pertaining to, or affected with, osteomalacia.
**osteomalacosis** (*os-te-o-mal-ak-o'-sis*). Synonym of *osteomalacia*.
**osteomalactic** (*os-te-o-mal-ak'-tik*). Pertaining to osteomalacia.
**osteomere** (*os'-te-o-mēr*). Same as *osteocomma*.
**osteometry** (*os-te-om'-et-re*) [*osteo-*; μέτρον, measure]. The study of the proportions and measurements of the skeleton.
**osteomiosis** (*os-te-o-mi-o'-sis*) [*osteo*; μείωσις, diminution]. Disintegration of bone.
**osteomyelitis** (*os-te-o-mi-el-i'-tis*) [*osteo-*; *myelitis*]. Inflammation of the marrow of bone.
**osteomyelum** (*os-te-o-mi'-el-um*) [*osteo-*; μυελός, marrow]. The marrow of bone.
**osteonabrosis** (*os-te-on-ab-ro'-sis*). See *osteoanabrosis*.
**osteonagenesis** (*os-te-on-aj-en'-es-is*). See *osteogenesis*.
**osteoncus** (*os-te-ong'-kus*) [*osteo-*; ὄγκος, tumor]. 1. A tumor of a bone. 2. An exostosis.
**osteonecrosis** (*os-te-o-ne-kro'-sis*). Necrosis of bone.
**osteoneuralgia** (*os-te-o-nū-ral'-je-ah*). Neuralgia in a bone.
**osteonosus** (*os-te-on'-o-sus*) [*osteo-*; νόσος, disease]. Disease of bone.
**osteoparectasis** (*os-te-o-par-ek'-tas-is*) [*osteo-*; παρέκτασις, a stretching out]. Abnormal lengthening of a bone; overextension in the treatment of fracture.
**osteopath**, **osteopathist** (*os-te-o-path*) [*os-te-op'-a-thist*]. One who practises osteopathy.
**osteopathic** (*os-te-o-path'-ik*). Pertaining to osteopathy.
**osteopathy** (*os-te-op'-ath-e*) [*osteo-*; πάθος, suffering]. 1. Any disease of bone. 2. A school of medicine based upon the theory that the body is a vital mechanical organism whose structural and functional integrity are coordinate and that the perversion of either is disease, while its therapeutic procedure is chiefly manipulative correction, its name indicating the fact that the bony framework of the body largely determines the structural relation of its tissues (*Committee on Osteopathic Terminology*).
**osteopedion** (*os-te-o-pe'-de-on*). See *lithopedion*.
**osteoperiosteal** (*os-te-o-per-e-os'-te-al*) [*osteo-*; *periosteum*]. Pertaining to bone and its overlying periosteum.
**osteoperiostitis** (*os-te-o-per-e-os-ti'-tis*) [*osteo-*; *peri-*

**ostitis**]. Combined inflammation of the bone and periosteum.
**osteophage** (os'-te-o-fāj) [osteo-; φαγεῖν, to eat]. A myeloplax, or osteoclastic cell.
**osteophlebitis** (os-te-o-fle-bi'-tis). Inflammation of the veins of a bone.
**osteophone** (os'-te-o-fōn) [osteo-; φωνή, sound]. An apparatus for the transmission of sounds through the bones of the face; it is miscalled *audiphone*.
**osteophony** (os-te-of'-on-e) [osteo-; φωνή, sound]. The transmission of sound through bone.
**osteophore** (os'-te-o-fōr) [osteo-; φέρειν, to bear]. A heavy tooth-forceps for crushing bone.
**osteophthisis** (os-te-off'-this-is) [osteo-; φθίσις, a wasting]. Wasting of the bones.
**osteophyma** (os-te-o-fi'-mah) [osteo-; φῦμα, swelling]. Any tumor or swelling of a bone.
**osteophyte** (os'-te-o-fīt) [osteo-; φυτόν, a plant]. A bony outgrowth of dendritic character.
**osteoplaque** (os'-te-o-plak) [osteo-; πλάξ, plate]. A layer of bone; a flat osteoma.
**osteoplast** (os'-te-o-plast) [osteo-; πλάσσειν, to form]. Same as *osteoblast*.
**osteoplastic** (os-te-o-plas'-tik) [see *osteoplasty*]. 1. Pertaining to the formation of bone. 2. Pertaining to plastic operations upon bone. o. **resection**, the Wagner-Wolff operation, in which a portion of bone, cut loose from its attachments except at one point, is laid back, the underlying diseased structure removed, and the bone replaced.
**osteoplasty** (os'-te-o-plas-te) [osteo-; πλάσσειν, to form]. Plastic operations on bone.
**osteoporoma** (os-te-o-por-o'-mah) [osteo-; πόρος, passage]. The changes produced by osteoporosis.
**osteoporosis** (os-te-o-por-o'-sis) [osteo-; πόρος, a pore]. An enlargement of the spaces of bone whereby a porous appearance is produced.
**osteopsathyrosis** (os-te-o-sath-ir-o'-sis) [osteo-; ψαθυρός, friable]. Fragility of the bones.
**osteorrhagia** (os-te-or-a'-je-ah) [osteo-; ῥηγνύναι, to burst forth]. Hemorrhage from a bone.
**osteorrhaphy** (os-te-or'-a-fe) [osteo-; ῥαφή, suture]. The suturing of bones.
**osteosarcoma** (os-te-o-sar-ko'-mah). A sarcoma containing bone.
**osteosarcomatous** (os-te-o-sar-ko'-mat-us). Of the nature of an osteosarcoma.
**osteosarcosis** (os-te-o-sar-ko'-sis) [osteo-; σάρξ, flesh]. The conversion of bone into sarcomatous tissue.
**osteoscirrhus** (os-te-o-skir'-us) [osteo-; σκίρρος, a hardening]. A scirrhous carcinoma of bone.
**osteosclerosis** (os-te-o-skle-ro'-sis). A condition in which the bone becomes hard and heavy; it is seen in sclerosing or condensing osteitis.
**osteoscope** (os'-te-o-skōp) [osteo-; σκοπεῖν, to view]. An instrument used for testing an x-ray machine by examining certain bones which are used as a standard.
**osteoseptum** (os-te-o-sep'-tum) [osteo-; *septum*]. The bony nasal septum.
**osteosis** (os-te-o'-sis) [osteo-]. Bone formation.
**osteospongioma** (os-te-o-spun-je-o'-mah). A tumor consisting of a spongy or highly cancellous growth of bony tissue.
**osteosteatoma** (os-te-o-ste-at-o'-mah). A fatty tumor of bone.
**osteostixis** (os-te-o-stiks'-is) [osteo-; στίξις, a pricking]. Surgical puncturing of a bone.
**osteosuture** (os-te-o-sū'-tūr). See *osteorrhaphy*.
**osteosynovitis** (os-te-o-sin-o-vi'-tis). Synovitis complicated with osteitis of adjacent bones.
**osteotabes** (os-te-o-ta'-bēz). Bone degeneration of infants beginning with the destruction of the cells of the lymphoid or splenoid bone-marrow, which disappears completely in parts and is replaced by soft gelatinous tissue; later the spongy bone diminishes, and lastly the compact bone.
**osteoteleangiectasis** (os'-te-o-tel-e-an-je-ek'-tas-is) [οστέον; τέλος, end; ἀγγεῖον, vessel; ἔκτασις, dilatation]. Dilatation of the blood-vessels of a bone; also a telangiectatic osteosarcoma.
**osteothrombosis** (os'-te-o-throm-bo'-sis). Thrombosis of the veins of a bone.
**osteotome** (os'-te-o-tōm) [osteo-; τομή, a cutting]. 1. An instrument for cutting bone. 2. An instrument used in cutting the bones of the fetal head in embryotomy.
**osteotomist** (os-te-ot'-o-mist) [osteo-; τομή, a cutting]. One who performs osteotomy.

**osteotomy** (os-te-ot'-o-me) [see *osteotome*]. The division of a bone. o., **cuneiform**, an osteotomy in which a wedge of bone is removed. o., **linear**, a simple division of a bone. o., **Macewen's**. See *Macewen's osteotomy*.
**osteotribe** (os'-te-o-trīb). Same as *osteotrite*.
**osteotrite** (os'-te-o-trīt) [osteo-; τρίβειν, to rub]. An instrument for scraping away carious bone.
**osteotylus** (os-te-o-ti'-lus) [osteo-; τύλος, callus]. Bone-callus.
**osteotympanic** (os-te-o-tim-pan'-ik). See *craniotympanic*.
**osteulcus** (os-te-ul'-kus) [osteo-; ἕλκειν, to draw]. A bone-forceps.
**osthelcus** (ost-hel'-kus) [osteo-; ἕλκος, an ulcer]. Caries of bone.
**osthistos** (ost-his'-tos) [osteo-; ἱστός, tissue]. Osseous tissue.
**ostia** (os'-te-ah) [L.]. Plural of *ostium*, *q. v.*
**ostial** (os'-te-al) [*ostium*]. Pertaining to an opening or orifice.
**ostiary** (os'-te-a-re) [*ostium*]. Same as *ostial*.
**ostitis** (os-ti'-tis). See *osteitis*.
**ostium** (os'-te-um) [L.: *pl., ostia*]. A mouth or aperture. o. **abdominale**, the orifice of the oviduct communicating with the peritoneal cavity o. **internum**, the uterine opening of the oviduct. o. **pharyngeum**, the pharyngeal opening of the Eustachian tube. o. **tympanicum**, the tympanic opening of the Eustachian tube. o. **vaginæ**, the external orifice of the vagina.
**ostiuli** (os'-tiuli). See *ostevli*.
**ostracosis** (os-trak-o'-sis) [ὄστρακον, oyster-shell; νόσος, disease]. The degenerative change that sometimes takes place in a portion of bone and causes it to resemble an oyster-shell.
**ostreotoxismus** (os-tre-o-toks-iz'-mus) [ὀστρεον; oyster; τοξικόν, poison]. Poisoning due to eating diseased or deteriorated oysters.
**otacoustic** (o-ta-koos'-tik) [οὖς, ear; ἀκούειν, to hear]. 1. Pertaining to or aiding hearing. 2. An ear-trumpet.
**otacousticon** (o-tak-oos'-tik-on) [oto-; ἄκουσις, hearing]. An otacoustic, or ear-trumpet.
**otagra** (o-ta'-grah). Synonym of *otalgia*.
**otalgia** (o-tal'-je-ah) [οὖς, ear; ἄλγος, pain]. Earache.
**otalgic** (o-tal'-jik) [oto-; ἄλγος, pain]. Affected with or pertaining to otalgia.
**otaphone** (o'-taf-ōn). See *olophone*.
**otectomy** (o-tek'-to-me) [oto-; ἐκτομή, excision]. Ossiculectomy.
**othelcosis** (o-thel-ko'-sis) [oto-; ἕλκωσις, ulceration]. Ulceration of the ear.
**othematoma**, **othæmatoma** (ōt-hem-at-o'-mah) [οὖς, ear; *hæmatoma*]. Hematoma of the external ear, usually the pinna; it is comparatively frequent in the insane. Syn., *Hæmatoma auris; insane ear*.
**othemorrhœa**, **othæmorrhœa** (ōt-hem-or-e'-ah) [oto-; αἷμα, blood; ῥοία, flow]. A sanguineous discharge from the ear.
**othygroma nephriticum** (ōt-hi-gro'-mah · nef-rit'-ik-um) [οὖς, ear; *hygroma; nephritic*]. Elongation of the lobule of the ear following edema from acute nephritis.
**otiatric** (o-te-at'-rik) [oto-; ἰατήρ, surgeon]. Pertaining to the treatment of diseases of the ear.
**otiatrics** (o-te-at'-riks) [οὖς, ear; ἰατρικόs, of healing]. The study of diseases of the ear and their treatment.
**otic** (o'-tik) [ὠτικός, from οὖς, ear]. Pertaining to the ear. o. **ganglion**. See *ganglion*, *otic*.
**oticodinia** (o-tik-o-din'-e-ah) [οὖς; δίνη, a whirling]. Vertigo from ear disease.
**otitic** (o-tit'-ik). Relating to otitis.
**otitis** (o-ti'-tis) [οὖς, ear; ιτις, inflammation]. Inflammation of the ear. o. **externa**, inflammation of the external ear. o. **furuncular**, the formation of furuncles in the external meatus. o. **interna**, that affecting the internal ear. o. **labyrinthica**, inflammation of the labyrinth. o. **mastoidea**, inflammation confined to the mastoid cells; mastoid disease. o. **media**, that affecting the middle ear. o. **parasitica**, that caused by a parasite. o. **sclerotica**, inflammation of the inner ear with hardening of the tissues.
**oto-** (o'-to-) [οὖς, ear]. A prefix signifying connection with or relation to the ear.
**otoblennorrhea**, **otoblennorrhœa** (o-to-blen-or-e'-ah) [oto-; βλέννα, mucus; ῥοία, flow]. Any abnormal discharge of mucus from the ear.
**otocatarrh** (o-to-kat-ar') [oto-; *catarrh*]. Catarrh of the ear.

# OTOCEPHALUS 640 OVARADEN

**otocephalus** (o-to-sef'-al-us) [oto-; κεφαλή, head]. A monster characterized by a union or close approach of the ears, by absence of the lower jaw, and an ill-developed mouth.

**otocerebritis** (o-to-ṣer-e-bri'-tis) [oto-; cerebrum, brain; itis, inflammation]. Inflammation of the brain from disease of the ear.

**otocleisis** (o-to-kli'-sis) [oto-; κλεῖσις, closure]. Occlusion of the ear.

**otoconia** (o-to-ko'-ne-ah). Plural of otoconium.

**otoconial** (o-to-ko'-ne-al) [oto-; κόνις, dust]. Pertaining to, or of the nature of, otoconia.

**otoconite** (o-tok'-o-nīt). See otoconium.

**otoconium** (o-to-ko'-ne-um) [oto-; κόνις, dust]. An otolith.

**otocrane, otocranium** (o'-to-krān, o-to-kra'-ne-um) [oto-; κρανίον, skull]. The cavity of the petrous portion of the skull holding the organ of hearing.

**otocyst** (o'-to-sist) [oto-; κύστις, bladder]. An auditory vesicle, otocell, or otidium in invertebrates, or an otolithic sac in vertebrates.

**otodynia** (o-to-din'-e-ah) [oto-; ὀδύνη, pain]. Pain in the ear.

**otoganglion** (o-to-gang'-gle-on). See ganglion, otic.

**otogenous** (o-toj'-en-us) [oto-; γεννᾶν, to produce]. Originating in the ear.

**otography** (o-tog'-ra-fe) [oto-; γράφειν, to write]. Descriptive anatomy of the ear.

**otohemineurasthenia** (o-to-hem-e-nū-ras-the'-ne-ah) [oto-; hemi, half; neurasthenia]. A condition in which hearing is limited exclusively to one ear, without the evidence of any material lesion of the auditory apparatus.

**otolith** (o'-to-lith) [oto-; λίθος, stone]. One of the calcareous concretions within the membranous labyrinth of the ear.

**otological** (o-to-loj'-ik-al). Pertaining to otology.

**otologist** (o-tol'-o-jist) [oto-; λόγος, science]. One versed in otology; an aurist.

**otology** (o-tol'-o-je) [oto-; λόγος, science]. The science of the ear, its anatomy, functions, and diseases.

**otomassage** (o-to-mas-ahzj') [oto-; massage]. The application of passive motion to the tympanic membrane and auditory ossicles.

**otomyasthenia** (o-to-mi-as-the'-ne-ah). 1. Weakness of the muscles of the ear. 2. Defective hearing due to a paretic condition of the tensor tympani and stapedius muscles.

**otomyces** (o-to-mi'-sēz) [oto-; μύκης, fungus]. A fungous growth within the ear. o. Hageni, a fungus, with green conidia, sometimes found in the external canal of the ear. o. purpureus, a dark-red fungous growth in the ear.

**otomycosis** (o-to-mi-ko'-sis). The growth of fungi within the ear, or the diseased condition caused by the same.

**otonecrectomy** (o-to-ne-krek'-to-me). See otonecronectomy.

**otonecronectomy** (o-to-nek-ro-nek'-to-me) [oto-; νεκρός, dead; ἐκτομή, excision]. Surgical removal of necrosed sound-conductors from the ear.

**otoncus** (o-tong'-kus) [oto-; ὄγκος, tumor]. A swelling or tumor of the ear.

**otoneuralgia** (o-to-nū-ral'-je-ah). Synonym of otalgia.

**otoneurasthenia** (o-to-nū-ras-the'-ne-ah). A condition of deficient tone of the auditory apparatus.

**otopathy** (o-top'-ath-e) [oto-; πάθος, disease]. Any affection of the ear.

**otopharyngeal** (o-to-far-in'-je-al). Pertaining to the ear and the pharynx; o. tube, the Eustachian tube.

**otophone** (o'-to-fōn) [oto-; φωνή, voice]. 1. An ear-trumpet or other device for gathering and intensifying sound-waves. 2. An auscultating tube used in ear diseases.

**otophthalmic** (o-tof-thal'-mik) [oto-; ὀφθαλμός, eye]. Pertaining to the ear and the eye.

**otopiesis** (o-to-pi'-es-is) [oto-; πίεσις, pressure]. Pressure on the labyrinth sufficient to cause deafness. Depression of the tympanic membrane by atmospheric pressure, owing to the rarefaction of the air within the tympanic cavity.

**otoplasty** (o'-to-plas-te) [oto-; πλάσσειν, to form]. Plastic surgery of the external ear.

**otopolypus** (o-to-pol'-ip-us). A polypus occurring in the ear.

**otopyorrhea, otopyorrhœa** (o-to-pi-or-e'-ah) [oto-; πύον, pus; ῥοία, a flow]. A purulent discharge from the ear.

**otopyosis** (o-to-pi-o'-sis) [oto-; πύον, pus]. Suppuration within the ear.

**otorhinolaryngology** (o-to-ri-no-lar-in-gol'-o-je) [oto-; ῥίς, nose; laryngology]. The anatomy, physiology, and pathology of the ear, nose, and throat.

**otorrhagia** (o-tor-a'-je-ah) [oto-; ῥηγνύναι, to burst forth]. A discharge of blood from the external auditory meatus.

**otorrhea, otorrhœa** (o-tor-e'-ah) [oto-; ῥοία, a flow]. A discharge from the external auditory meatus.

**otosalpinx** (o-to-sal'-pinks) [oto-; σάλπιγξ, tube]. The Eustachian tube.

**otoscleronectomy** (o-to-skle-ro-nek'-to-me) [oto-; σκληρός, hard; ἐκτομή, excision]. Surgical removal of sclerosed and ankylosed conductors of sound in chronic otitis media.

**otosclerosis** (o-to-skle-ro'-sis) [oto-; σκληρός, hard]. Sclerosis of the tissues of the labyrinth and middle ear.

**otoscope** (o'-to-skōp) [oto-; σκοπεῖν, to examine]. An instrument for examining the ear, especially a rubber tube, one extremity of which is inserted into the ear of the subject, and the other extremity into the ear of the examiner, a current of air being passed by means of a Politzer bag and a Eustachian catheter through the middle ear. In case of tympanic perforation the rushing sound made by the passing air is audible to the examiner.

**otoscopic** (o-to-skop'-ik) [oto-; σκοπεῖν, to view]. Pertaining to otoscopy.

**otoscopy** (o-tos'-ko-pe) [see otoscope]. Examination of the ear, especially by means of the otoscope.

**otosis** (o-to'-sis) [oto-]. A mishearing; a false impression as to sounds or words heard.

**otosteon** (o-tos'-te-on) [oto-; ὀστέον, bone]. 1. An auditory ossicle, or ear-bone. 2. An otolith.

**otostylic** (o-to-sti'-lik) [oto-; στῦλος, pillar]. Relating to the ear and to the styloid process.

**ototomy** (o-tot'-o-me) [oto-; τομή, a cutting]. Dissection of the ear.

**ottar, otto** (ot'-ar, ot'-o). See attar.

**O. U.** An abbreviation for oculus uterque, Latin for each eye.

**ouabain** (oo-ah'-ba-in), $C_{30}H_{46}O_{12}$. A poisonous glucoside from the wood of Carissa schimperi and of Acocanthera venenata. Introduced into the stomach it is nonpoisonous, but injected into the blood it is extremely virulent. It is a depressant to the heart and the respiration. As a local anesthetic it has ten times the power of cocaine. It has been recommended for whooping-cough in doses of $\frac{1}{500}$ gr. (0.000065 Gm.), repeated with caution.

**oulachon** (oo'-lak-on). See eulachon.

**oulitis** (oo-li'-tis). See ulitis.

**ouloid** (oo'-loid) [οὐλή, scar; εἶδος, like]. Resembling a scar. Also, a form of cicatrix characteristic of lupus, elephantiasis, and syphilis.

**oulorrhagia** (oo-lor-a'-je-ah) [οὖλον, gum; ῥηγνύναι, to burst forth]. Hemorrhage from the gums.

**ounce** (ouns) [uncia, a contraction of undecia, a twelfth part]. A unit of measure of weight. o., avoirdupois, the sixteenth part of the avoirdupois pound, or 437.5 gr. (31.1 Gm.). o., troy, the twelfth part of the troy pound, or 480 gr. (31.08 Gm.).

**ourari** (oo-rah'-re). Same as curare.

**ouro-** (oo-ro-). See uro-.

**ourology** (oo-rol'-o-je) [οὖρον, urine; λόγος, science]. The science of the nature and secretion of urine.

**ouroscopy** (oo-ros'-ko-pe). See uroscopy.

**-ous.** A suffix which denotes the lower of two degrees of valency assumed by an element and incidentally indicates, in many cases, a small amount of oxygen.

**outlet.** 1. The lower aperture of the pelvic canal. 2. The passage that removes the air from a room in ventilation. o. of pelvis. See outlet (1).

**outpatient** (out'-pa-shent). A hospital patient who is not treated in the wards of the institution.

**ova** (o'-vah) [L.]. Plural of ovum, an egg.

**ovadin** (o'-vad-in). A proprietary preparation of the ovaries of animals containing iodine.

**oval** (o'-val) [ovum]. 1. Egg-shaped. 2. Pertaining to an ovum.

**ovalbumin** (o-val-bū'-min) [ovum; albumen]. The albumin of the egg.

**ovaraden** (o-var-ad'-en). An organotherapeutic preparation from the ovaries of animals; used as a nervine. Dose 45–90 gr. (3–6 Gm.) daily.

**ovaralgia** (*o-var-al'-je-ah*). See *ovarialgia*.
**ovarialgia** (*o-va-re-al'-je-ah*) [*ovary;* ἄλγος, pain]. Neuralgic pain in the ovary.
**ovarian** (*o-va'-re-an*) [*ovary*]. Pertaining to the ovaries.
**ovariectomy** (*o-va-re-ek'-to-me*) [*ovary;* ἐκτομή, excision]. Excision of an ovary; oophorectomy.
**ovariin** (*o-var'-e-in*). An organotherapeutic preparation from the ovaries of cows; used in ovarian disorders. Dose 15–24 gr. (1.0–1.6 Gm.) 3 times daily in pills.
**ovarin** (*o'-var-in*). The sterilized extract of the ovaries of the pig; used in diseases of women, sterility, etc.
**ovario-** (*o-va-re-o-*) [*ovarium*, ovary]. A prefix denoting relation to the ovary.
**ovariocele** (*o-va'-re-o-sēl*) [*ovario-;* κήλη, tumor]. Tumor of the ovary; hernia of an ovary. o., **vaginal**, invasion of the vaginal wall by one or both ovaries.
**ovariocentesis** (*o-va-re-o-sen-te'-sis*) [*ovario-;* κέντησις, puncture]. Puncture of the ovary or of an ovarian cyst.
**ovariocyesis** (*o-va-re-o-si-e'-sis*) [*ovario-;* κύησις, pregnancy]. Ovarian pregnancy.
**ovariodysneuria** (*o-va-re-o-dis-nū'-re-ah*) [*ovario-;* δυς, painful; νεῦρον, nerve]. Ovarian neuralgia.
**ovarioepilepsy** (*o-va-re-o-ep'-il-ep-se*). See *hysteroepilepsy*.
**ovariohysterectomy** (*o-va-re-o-his-ter-ek'-to-me*) [*ovario-;* ὑστέρα, womb; ἐκτομή, excision]. Surgical removal of the ovaries and uterus.
**ovarioncus** (*o-va-re-ong'-kus*) [*ovario-,* ὄγκος, tumor]. An ovarian tumor.
**ovariorrhexis** (*o-va-re-or-eks'-is*) [*ovario-;* ῥῆξις, rupture]. Rupture of an ovary.
**ovariosalpingectomy** (*o-va-re-o-sal-pin-jek'-to-me*). See *oophorosalpingectomy*.
**ovariosteresis** (*o-va-re-o-ster-e'-sis*) [*ovario-;* στέρησις, deprivation]. Extirpation of an ovary.
**ovariostomy** (*o-va-re-os'-to-me*). See *oophorostomy*.
**ovariotomist** (*o-va-re-ot'-o-mist*) [see *ovariotomy*]. One who performs ovariotomy.
**ovariotomy** (*o-va-re-ot'-o-me*) [*ovario-;* τομή, a cutting]. Literally, incision of an ovary. As generally used, removal of an ovary; oophorectomy. o., **normal**, the removal of an ovary that is free from disease. Syn.: *Battey's operation*.
**ovariotubal** (*o-va-re-o-tū'-bal*) [*ovario-;* *tuba*, tube]. Pertaining to the ovary and the oviduct.
**ovaritis** (*o-var-i'-tis*). See *oophoritis*.
**ovarium** (*o-va'-re-um*) [L., ovary: *pl., ovaria*]. An ovary or oophoron.
**ovary** (*o'-var-e*) [*ovarium*, an egg-holder, from *ovum,* egg]. One of a pair of glandular organs giving rise to ova. It consists of a fibrous framework or stroma, in which are embedded the Graafian follicles, and is surrounded by a serous covering derived from the peritoneum.
**ovaserum** (*o-va-se'-rum*). The serum of an animal which has acquired specific precipitating action by the inoculation of egg-albumen; it may be used as a test for egg-albumen.
**overbite.** Lack of coaptation of upper and lower teeth.
**overcrowding.** The dwelling together of too many persons in a locality, house, or apartment, with the consequent failure in sanitary and hygienic arrangements, whence result conditions favorable to the development and spread of certain diseases.
**overeat.** To surfeit with eating.
**overextension.** Excessive extension; extension beyond the normal point or line.
**overfeed.** To feed to excess.
**overflow** (*o'-ver-flo*). A continuous escape of liquid.
**overtone.** A harmonic tone heard above the fundamental tone.
**overgrown.** Grown too large.
**overlaid.** Said of a child suffocated by the parent lying on it.
**overlying of children.** A frequent cause of death in infants from suffocation from one of the parents lying upon the child while in an intoxicated condition or intentionally.
**overmaximal.** Beyond the normal maximum, as the over-maximal contraction of a muscle.
**overpressure.** Applied chiefly to the school-system that forces too long and continuous periods of study upon the pupils, with consequent unhealthy conditions of mind and body.

**overreach.** To strike the toe of the hind-foot against the heel or shoe of the fore-foot; said of a horse.
**overriding** (*o-ver-ri'-ding*). The slipping of an end of a fractured bone over the other fragment.
**oversight.** See *hypermetropia*.
**overstrain.** To strain to excess; a condition resulting from exhausting effort.
**overtoe.** A variety of *hallux varus* in which the great toe overlies its fellows.
**overtone.** A harmonic tone heard above the fundamental tone.
**overtones.** The notes represented by the vibrating subdivisions of a string; harmonics.
**overwork, mental.** See *parathymia*.
**ovi-** [*ovum*]. A prefix denoting relating to the ovum.
**ovicapsule** (*o'-vik-ap'-sūl*) [*ovi-;* *capsula*, capsule]. An egg-case, ovisac, or Graafian follicle.
**ovicell** (*o'-vis-el*) [*ovi-;* *cella*, a cell]. An unimpregnated ovum.
**oviducal, oviducent** (*o-vid-ū'-kal, o-ve-dū'-sent*) [*ovi-;* *ducere*, to lead]. Pertaining to the oviduct, or its functions.
**oviduct** (*o'-vid-ukt*) [*ovi-;* *ductus*, a canal]. The Fallopian tube; a small tube upon each side of the uterus, through which the ovule passes to the uterus.
**oviferous** (*o-vif'-er-us*) [*ovi-;* *ferre*, to bear]. Producing or bearing ova.
**ovification** (*o-vif-ik-a'-shun*) [see *oviferous*]. The production of ova.
**oviform** (*o'-vif-orm*) [*ovi;* *forma*, form]. Egg shaped; oval.
**ovigenous** (*o-vij'-en-us*) [*ovi-;* γενής, producing]. Producing ova, as an ovary. o. **layer**, the outer layer of the ovary, in which the ovisacs containing the ova are situated.
**ovigerm** (*o'-vij-erm*) [*ovi-;* *germen*, sprout; bud]. A cell producing or developing into an ovum.
**ovigerous** (*o-vij'-er-us*) [*ovi-;* *gerere*, to carry]. Producing or carrying ova.
**ovination** (*o-vin-a'-shun*) [*ovis*, a sheep]. Inoculation with the virus of sheep-pox.
**oviparous** (*o-vip'-ar-us*) [*ovi-;* *parere*, to bring forth]. Laying eggs; bringing forth young in the egg-stage of development.
**oviposit** (*o-vip-oz'-it*) [*ovi-;* *ponere*, to place]. To lay or deposit eggs, especially with an ovipositor, as an insect.
**oviposition** (*o-vip-o-zish'-un*) [see *oviposit;* *ponere,* to place]. The act of laying or depositing eggs by the females of oviparous animals.
**ovipositor** (*o-vip-oz'-it-or*) [*ovi-;* *positor*, placer]. An organ, common among insects, composed of several modified rings of somites, forming the end of the abdomen, and employed in depositing the eggs in places fit for development.
**oviprotogen** (*o-vi-pro'-to-jen*). A proprietary dietetic, said to be a methylene compound of albumin, given in milk to infants, and used hypodermatically also.
**ovis** (*o'-vis*) [L.]. Sheep.
**ovisac** (*o'-vis-ak*) [*ovi-;* *saccus*, sac]. The capsule of an ovum; a Graafian follicle.
**oviscapt** (*o'-vis-kapt*) [*ovi-;* σκαπτείν, to dig]. Same as *ovipositor*.
**ovi vitellus** (*o'-vi vi-tel'-lus*). Latin for *yolk of egg;* used in pharmacy as an emulsifying agent.
**ovo-.** The same as *ovi-*.
**ovoblast** (*o'-vo-blast*) [*ovo-;* βλαστός, germ]. The primordial ovum.
**ovocenter** (*o-vo-sen-ter*) [*ovo-;* *center*]. The centrosome of the ovarian egg during fertilization.
**ovocyte** (*o'-vo-sīt*). The same as *oocyte*.
**ovoferrin** (*o-vo-fer'-in*). Trade name of a preparation of organic iron.
**ovogal** (*o'-vo-gal*). Trade name of a proprietary cholagogue.
**ovogenesis** (*o-vo-jen'-es-is*) [*ovo-;* γένεσις, genesis]. The process of the development or production of the ovum.
**ovoglobulin** (*o-vo-glob'-ū-lin*) [*ovo-;* *globulin*]. The globulin of white of egg.
**ovogonium.** See *oogonium*.
**ovoid** (*o'-void*) [*ovo-;* εἶδος, like]. Egg-shaped. o., **fetal**, the fetal ellipse, or the ellipse formed by the bending of the fetal body in the uterus.
**ovolecithin** (*o-vo-les'-ith-in*). See *lecithin*.
**ovomucin** (*o-vo-mū'-sin*). A glycoproteid composing about 7 % of the proteid matter of egg-white.

OVOMUCOID 642 OXYGEN

**ovomucoid** (o-vo-mū'-koid). A glycoproteid obtained from white of egg.
**ovoplasm** (o'-vo-plasm) [ovo-; plasm]. The protoplasm of the unimpregnated ovum or ovicell.
**ovos** (o'-vos). A proprietary substitute for meat-extract, prepared from yeast.
**ovovitellin** (o-vo-vit-el'-in). A protein contained in yolk of egg; a white, granular substance, soluble in dilute acids, alkalies, and a 10 % solution of common salt.
**ovoviviparous** (o-vo-vi-vip'-ar-us) [ovo-; viviparus, bringing forth alive]. Reproducing by means of eggs hatched within the body.
**ovula** (o'-vū-lah) [L.]. Plural of ovulum. o. of **Naboth**. See ovule.
**ovular** (o'-vū-lar) [ωόν, ovum, egg]. Relating to an ovule or ovum.
**ovulation** (ov-ā-la'-shun) [ovulum, dim. of ovum]. The maturation and escape of the ovum.
**ovule** (ov'-ūl) [ovum]. 1. The ovum before its escape from the Graafian vesicle. 2. A small egg; especially a small, egg-like body, as the ovule of Naboth, one of the small cysts resulting from obstruction of the ducts of the glands of the cervix uteri. o., **migration of**, the transfer of the ovule from the ovary to the oviduct.
**ovulum** (o'-vū-lum) [L.: pl., ovula]. An ovule. See ovule.
**ovum** (o'-vum) [L., "an egg": pl., ova]. The reproductive cell of an animal or vegetable; an egg. A human ovum is a cell consisting of a large amount of protoplasm (vitellus) and a large spherical nucleus, the germinal vesicle, within which is a bright spot, the nucleolus, or germinal spot. It is surrounded by an inner zone, the zona pellucida, and an outer, the vitelline membrane. o., **alecithal**, one in which the food-yolk is entirely absent or present only in very small quantity. o., **apoplectic**, one the seat of a hemorrhagic extravasation. o., **blighted**, an impregnated ovum the development of which has been arrested by disease or by hemorrhage into the chorion or amniotic cavity. o., **centrolecithal**, one in which the formative yolk is arranged in a regular layer around the whole ovum, as well as in a mass at the center in which lies the germinal vesicle. o., **holoblastic**, one in which the food-yolk is scant and more or less thoroughly intermingled with the formative yolk, and in which germination is accompanied by a practically uniform segmentation. o., **meroblastic**, an ovum with a large amount of food-yolk that takes no active part in the development of the embryo. o., **permanent**, a fully developed ovum ready for fertilization. o., **telolecithal**, one in which the food-yolk and the formative yolk divide the egg into two hemispheres, or in which the quantity of the nutritive yolk is greatly in excess.
**oxacid** (oks'-as-id). See oxyacid.
**oxalate** (oks'-al-āt) [oxalis]. A salt of oxalic acid.
**oxalemia, oxalaemia** (oks-al-e'-me-ah) [ὀξαλίς, sorrel; αἷμα, blood]. Excess of the oxalates, or of oxalic acid, in the blood.
**oxalethylin** (oks-al-eth'-il-in) C₆H₁₀N₂. An oily liquid boiling at 213°C. It is soluble in water, in chloroform, and in alcohol, is poisonous, and stimulant to the cardiac centers.
**oxalic acid** (oks-al'-ik). See acid, oxalic.
**oxalism** (oks'-al-ism). Poisoning by oxalic acid or potassium binoxalate. It is characterized by gastroenteritis with nephritis, collapse, cyanosis, mydriasis, labored breathing and dyspnea.
**oxalium** (oks-a'-le-um). Potassium binoxalate.
**oxaluria** (oks-al-ū'-re-ah) [oxalis; οὖρον, urine]. The presence of an excessive amount of calcium oxalate in the urine.
**oxalylurea** (oks-a-lil-ū-re'-ah) [oxalic; urea]. Parabanic acid, a substance produced by oxidizing uric acid or alloxan with nitric acid.
**oxaphor** (oks'-a-for). A 50 % alcoholic solution of oxycamphor; used in dyspnea. Dose 15 min. (1 Cc.) 2 or 3 times daily.
**oxatyl** (oks'-at-il). See carboxyl.
**oxhydryl** (oks-hi'-dril). Same as hydroxyl.
**oxide, oxid** (oks'-īd, or oks'-id) [ὀξύς, sharp]. A binary compound of oxygen and another element or radical. o., **acid**, an oxide which produces an acid when combined with water; an anhydride. o., **basic**, an oxide which produces a base when combined with water. o., **indifferent**, o., **neutral**, an oxide which is neither acid nor basic. o., **saline**, (1) same as

o., **indifferent** or **neutral**; (2) an oxide which is formed by the union of an acid and a basic oxide.
**oxidant** (oks'-id-ant) [oxide]. An oxidizing agent.
**oxidase** (oks'-id-ās). The inherent substance of the living cell-nucleus that possesses the power of setting free active oxygen. Columnar epithelium and glandular tissue are rich in oxidase.
**oxidation** (oks-id-a'-shun) [oxide]. The act or process of combining with oxygen.
**oxidize** (oks'-i-dīz). To combine or to cause to combine with oxygen.
**oximes** (oks'-īmz). A series of chemical compounds for the most part the product of the action of hydroxylamine upon aldehydes and ketones.
**oxols** (oks'-ols). A collective name for antiseptic fluid mixtures of a 3 % solution of hydrogen dioxide, 32 to 38 % alcohol, and 1 % naphthol, menthol, or other substance. Cf. camphoroxol; mentholoxol; naphthoxol.
**oxolyin** (oks-ol'-e-in) [ὀξύς, sharp; λύειν, to dissolve]. According to Le Conte, that one of two substances contained in globulin, casein, albumin, and fibrin which dissolves in glacial acetic acid. Cf. anoxoluin.
**oxy-** (oks-e-) [ὀξύς, sour; sharp]. 1. A prefix denoting sharp or acid. 2. A prefix denoting combined with oxygen.
**oxyacanthine** (oks-e-ak-an'-thēn) [oxy-; ἄκανθα, spine], C₁₈H₁₉NO₁₁ (?). An alkaloid of barberry, occurring in small quantities.
**oxyacid** (oks'-e-as-id). Any acid containing oxygen.
**oxyacusis** (oks-e-ak-ū'-sis). Synonym of hyperacusis.
**oxyaesthesia** (oks-e-es-the'-ze-ah). See oxyesthesia.
**oxyakoia, oxyacoa** (oks-e-a-koi'-ah, oks-e-a-ko'-ah) oxy-; ἀκοή, hearing]. Increased acuteness of hearing.
**oxyaphia** (oks-e-a'-fe-ah) [ὀξύς, acute; ἁφή, touch]. Abnormal acuteness of the sense of touch.
**oxyarteritis** (oks-e-ar-ter-i'-tis) [oxy-; arteritis]. An acute arteritis.
**oxyarthritis** (oks-e-ar-thri'-tis) [oxy-; arthritis]. An acute arthritis.
**oxyblepsia** (oks-e-blep'-se-ah) [oxy-; βλέπειν, to see]. Acuteness of vision.
**oxybolia** (oks-e-bo'-le-ah) [ὀξύς, quick; βολή, a throw]. Premature ejaculation of semen.
**oxybromide** (oks-e-bro'-mīd). A compound of an element or radical with oxygen and bromine.
**oxybutyric acid** (oks-e-bū-tir'-ik) [oxy-; butyrum, butter], C₄H₈O₃. A fatty acid found in the urine in certain fevers and in diabetes. Its presence in the body in diabetes, simultaneously with that of diacetic acid, is supposed to be the cause of diabetic coma.
**oxycamphor** (oks-e-kam'-for), C₁₀H₁₆O(OH). An oxidation-product of camphor. It is used in dyspnea. Dose 8–16 gr. (0.5–1.0 Gm.) 2 or 3 times daily.
**oxycephalia** (oks-is-ef-a'-le-ah) [oxy-; κεφαλή, head]. The character of a skull that is high and pointed; hypsicephaly.
**oxycephalus** (oks-is-ef'-al-us). See oxycephalia.
**oxycephaly** (oks-is-ef'-al-e). See oxycephalia.
**oxychinaseptol** (oks-e-kin-ah-sep'-tol). See diaphtherin.
**oxychinolin** (oks-e-kin'-o-lin). See oxyquinolin.
**oxychloride** (oks-e-klo'-rīd) [oxy-; χλωρός, green]. A compound of a basic element or radical with both oxygen and chlorine.
**oxychromatin** (oks-e-kro'-mat-in). That part of the chromatin having an affinity for acid dyes.
**oxycinesis** (oks-e-sin-e'-sis). Same as acrocinesis.
**oxydases** (oks'-e-dā-sēs). See ferment, oxidation, and oxidase.
**oxydendron** (oks-e-den'-dron) [oxy-; δένδρον, tree]. The leaves of the sorrel tree, Oxydendrum arboreum; used in ascites and disturbed portal circulation.
**oxydol** (oks'-e-dol). A solution of hydrogen dioxide.
**oxydum** (oks'-id-um). See oxide.
**oxyecoia** (oks-e-e-koi'-ah). Synonym of oxyakoia.
**oxyencephalitis** (oks-e-en-sef-al-i'-tis) [οχυ-; ἐγκέφαλος, brain; ιτις, inflammation]. Acute encephalitis.
**oxyendocarditis** (oks-e-en-do-kar-di'-tis) [oxy-; endocarditis]. Acute endocarditis.
**oxyesthesia** (oks-e-es-the'-se-ah) [oxy-; αἴσθησις, sensation]. A condition of increased acuity of sensibility.
**oxygen** (oks'-ij-en) [ὀξύς, acid; γεννᾶν, to produce]. A colorless, tasteless, odorless gas, one of the non-

metallic elements, having an atomic weight of 16.00; quantivalence II; symbol O, and constituting one-fifth of the atmosphere, eight-ninths of water, three-fourths of organized bodies, and about one-half the crust of the globe; it supports combustion, and is essential to the respiration of animals and plants. It combines with most elements, its combination with the nonmetallic substances giving rise to acids (*oxyacids*). It has been employed by inhalation in the treatment of pneumonia, pulmonary tuberculosis, grave anemias, asphyxia, and poisoning by opium.
**o.-carrier,** a katalytic substance capable of absorbing molecules of oxygen and then of splitting these to give off atomic oxygen; the nucleoproteids are the oxygen-carriers of living matter.
**oxygenated** (*oks'-e-jen-a-ted*) [*oxygen*]. Containing or impregnated with oxygen.
**oxygenation** (*oks-ij-en-a'-shun*) [*oxy-*; γεννᾶν, to produce]. The saturation of a substance with oxygen, either by chemical combination or by mixture.
**oxygeusia** (*oks-ig-ū'-se-ah*) [*oxy-*; γεῦσις, taste]. Marked acuteness of the sense of taste.
**oxyhaloid** (*oks-e-hal'-oid*) [*oxy-*; *haloid*]. A compound of an element or radical with oxygen and a halogen.
**oxyhematoporphyrin, oxyhæmatoporphyrin** (*oks-e-hem-at-o-por'-fir-in*). A peculiar substance found in urine and closely allied to urohematoporphyrin, but producing a red instead of an orange color.
**oxyhemoglobin, oxyhæmoglobin** (*oks-e-hem-o-glo'-bin*) [*oxygen*; *hemoglobin*]. Oxidized hemoglobin; that found in arterial blood.
**oxyhydrogen** (*oks-e-hi'-dro-jen*) [*oxy-*; *hydrogen*]. A mixture, in gaseous form of oxygen and hydrogen.
**o. blowpipe,** a blowpipe in which the heat is obtained by the combustion of a mixture of oxygen and hydrogen. The heat produced is intense.
**oxyiodide** (*oks-e-i'-o-did*) [*oxy-*; *iodine*]. A salt formed by the combination of an element with both oxygen and iodine.
**oxyleukotin** (*oks-il-ū'-ko-tin*) [*oxy-*; λευκός, white], C₁₄H₂₂O₁₃. A substance found in *paracotobark*.
**xymel** (*oks'-im-el*) [*oxy-*; μέλι, honey]. 1. A mixture of honey and vinegar or dilute acetic acid. 2. Any preparation containing honey and vinegar (or acetic acid) as a vehicle, as *oxymel* of squill.
**oxymethylene** (*oks-e-meth'-il-ēn*). Formaldehyde.
**oxyneurine** (*oks-in-ū'-rēn*). See *betaine*.
**oxyntic** (*oks-in'-tik*) [ὀξύς, acid]. Secreting acid.
**o. cells,** cells of the fundus glands of stomach, supposed to secrete hydrochloric acid. **o. gland,** any acid-secreting gland.
**oxyntin** (*oks-in'-tin*). Trade name of a preparation said to contain 5 per cent. of hydrochloric acid in combination with albumin.
**oxyopia** (*oks-e-o'-pe-ah*) [*oxy-*; ὤψ, eye]. Increased acuity of vision.
**oxyosphrasia** (*oks-e-os-fra'-ze-ah*). Same as *oxyosphresia*.
**oxyosphresia** (*oks-e-os-fre'-ze-ah*) [*oxy-*; ὄσφρησις, smell]. Marked or abnormal acuteness of smell.
**oxypathy** (*oks-ip'-ath-e*) [*oxy-*; πάθος, suffering]. A constitutional condition due to faulty elimination of unoxidized acids which unite with fixed alkalies of the body. Cf. *arthritism, lithemia*.
**oxyphenylethylamine** (*oks-e-fen-il-eth-il'-am-in*). A product of pancreatic digestion.
**oxyphenylsulphonic acid test for albumin.** Make a solution of 3 parts of oxyphenylsulphonic acid, 1 part salicylsulphonic acid, water 20 parts. To 1 drop of this add 1 Cc. of urine, and in the presence of albumin a white, transparent precipitate will be formed.
**oxyphil, oxyphile** (*oks'-if-il*) [*oxy-*; φίλος, loving]. Histological elements that attract acid dyes.
**oxyphonia** (*oks-if-o'-ne-ah*) [*oxy-*; φωνή, voice]. Shrillness of voice.
**oxypodia** (*oks-e-po'-de-ah*). See *talipes equinus*.
**oxypropylenediisoamylamine** (*oks-e-pro-pil-ēn-di-is-o-am-il'-am-in*). A synthetic alkaloid occurring as a clear fluid, soluble in alcohol, ether, or oils, insoluble in water, with action similar to atropine.
**oxypurin** (*oks-e-pū'-rin*). Any compound derived from purin by the addition of one or more atoms of oxygen.

**oxyquinaseptol** (*oks-e-kwin-ah-sep'-tol*). Same as *diaphtherin*.
**oxyquinolin** (*oks-e-kwin'-o-lin*), C₉H₇NO. A compound prepared by digesting quinolin with a bleaching lime-solution. Syn., *carbostyril*.
**oxyregmia** (*oks-ir-eg'-me-ah*) [*oxy-*; ἐρυγμός, eructation]. Acid eructation.
**oxyrhine** (*oks'-ir-in*) [*oxy-*; ῥίς, nose]. Possessing a sharp-pointed nose, or snout; having an acute olfactory sense.
**oxysalt** (*oks'-e-sawlt*) [*oxy-*; *salt*]. A salt of an oxyacid.
**oxysepsin** (*oks-e-sep'-sin*). An oxidized toxin prepared from cultures of *Bacillus tuberculosis*.
**oxysparteine** (*oks-e-spar'-te-in*) [*oxy-*; *sparteine*]. A derivative of sparteine; it is a cardiac stimulant. **o. hydrochloride,** C₁₆H₂₄N₂O . 2HCl, is used hypodermatically in heart disease. Dose ⅛–1½ gr. (0.05–0.1 Gm.).
**oxyspore** (*oks'-e-spōr*). See *exolospore*.
**oxytocic** (*oks-e-tos'-ik*) [*oxy-*; τόκος, labor]. 1. Hastening parturition. 2. A drug that hastens parturition.
**oxytoxin** (*oks-e-toks'-in*). An oxidized toxin.
**oxytuberculin** (*oks-e-tū-ber'-kū-lin*). An oxidized tuberculin. D₀se 5 dr. (20 Cc.) daily.
**oxyuricide** (*oks-e-ū'-ris-id*) [*oxyuris*; *cædere*, to kill]. Any anthelmintic that is destructive to worms of the genus *Oxyuris*, or pinworms.
**oxyurid** (*oks-e-ū'-rid*). A pinworm; see *Oxyuris vermicularis*.
**O.yuiais** (ȯksˌe-ū'ˑris) [*oxy-*; οὐρά, tail]. A genus of nematode worms—the pinworms. **O. vermicularis,** the common seatworm or pinworm infesting the rectum of children.
**oxyvaseline** (*oks-e-vas'-el-ēn*). Vasogen.
**oxyzymol** (*oks-e-zi'-mol*). See *carvacrol*.
**oz.** Abbreviation for *ounce*.
**ozalin** (*o'-za-lin*). A proprietary disinfectant consisting of a mixture of sulphates of calcium, magnesium, and iron with caustic soda and magnesia.
**ozena, ozœna** (*o-ze'-nah*) [ὄζειν, to smell]. Chronic disease of the nose accompanied by a fetid discharge, and depending on atrophic rhinitis, syphilitic ulceration, or caries.
**ozocerite, ozokerite** (*o-zo-se'-rīt, o-zo-ke'-rīt*) [ὄζειν, to smell; κηρός, wax]. A solid paraffin found free in Galicia and Rumania, and used in diseases of the skin.
**ozochrotia** (*o-zo-kro'-she-ah*) [ὄζειν, to smell; χρώς, skin]. An offensive odor of the skin.
**ozochrotous** (*o-zok'-ro-tus*) [ὄξη, stench; χρώς, skin]. Having a bad-smelling skin.
**ozomulsion** (*o-zo-mul'-shun*). Trade name of a preparation of ozonized cod-liver oil and guaiacol: recommended in tuberculosis and other wasting diseases.
**ozonator** (*o'-zo-na-tor*). An apparatus for generating ozone.
**ozone** (*o'-zōn*) [ὄζειν, to smell]. An allotropic form of oxygen, the molecule of which consists of three atoms. It occurs free in the atmosphere, and is a powerful oxidizing agent. In medicine it is employed as a disinfectant.
**ozonization** (*o-zo-nis-a'-shun*) [*ozone*]. The act of ozonizing, or of impregnating with ozone.
**ozonized** (*o'-zo-nīsd*) [*ozone*]. Containing ozone.
**ozonoform** (*o-zo'-no-form*). A proprietary disinfectant said to consist of ozone and a distillate of the fir tree.
**ozonometer** (*o-zo-nom'-et-er*) [*ozone*; μέτρον, measure]. A device for use in estimating the proportion of ozone in the atmosphere.
**ozonometry** (*o-zo-nom'-et-re*) [*ozone*; μέτρον, measure]. The estimation of the amount of ozone in the atmosphere.
**ozonophore** (*o-zo'-no-fōr*) [*ozone*; φέρειν, to bear]. 1. A granule of cell-protoplasm. 2. A red blood-corpuscle.
**ozonoscope** (*o-zo'-no-skōp*) [*ozone*; σκοπεῖν, to view]. A test-paper saturated with starch and iodine used in determining the amount of ozone in the air.
**ozostomia** (*o-zo-sto'-me-ah*) [ὄζειν, to smell; στόμα, mouth]. A foul odor from the mouth or on the breath.

# P

**P.** 1. The chemical symbol for *phosphorus*. 2. An abbreviation for *pharmacopœia, position* and *punctum proximum* (near-point).

**p.** An abbreviation for *para*.

**pabular** (*pab'-ū-lar*) [*pabulum,* food]. Of, pertaining to, or of the nature of, pabulum.

**pabulin** (*pab'-ū-lin*) [*pabulum,* food]. An albuminous and fatty substance present in the blood immediately after the process of digestion.

**pabulum** (*pab'-ū-lum*) [L.]. Food; anything nutritive.

**Pacchionian bodies, P. glands.** (*pak-e-o'-ne-an*) [Antonio *Pacchioni,* Italian anatomist, 1665–1726]. Hypertrophied villi of the arachnoid, occupying the convex surface of the meninges, chiefly along the superior longitudinal sinus and over the convexity of the cerebrum. **P. depressions,** the depressions produced by the Pacchionian bodies on the inner surface of the skull. **P. foramen,** the opening in the tentorium for the passage of the encephalic isthmus.

**pachemia, pachæmia** (*pak-e'-me-ah*). Synonym of *pachyemia*.

**pachismus** (*pak-iz'-mus*) [*pachy-*]. Thickening; induration.

**pachometer** (*pak-om'-et-ur*) [*pachy-*; μέτρον, a measure]. An instrument made in various forms for measuring the thickness of a body.

**pachulosis** (*pak-ū-lo'-sis*). See *pachylosis*.

**pachy-** (*pak-e-*) [παχύς, thick]. A prefix meaning thick.

**pachyacria** (*pak-e-ak'-re-ah*) [*pachy-*; ἄκρος, extremity]. Synonym of *akromegaly*.

**pachyæmia** (*pak-e-e'-me-ah*). See *pachyemia*.

**pachyblepharon** (*pak-e-blef'-a-ron*) [*pachy-*; βλέφαρον, the eyelid]. Thickening of the eyelids.

**pachyblepharosis** (*pak-e-blef-ar-o'-sis*) [*pachy-*; βλέφαρον, eyelid]. Chronic thickening and induration of the eyelids.

**pachycephalia** (*pak-e-sef-a'-le-ah*). See *pachycephaly*.

**pachycephalic** (*pak-e-sef-al'-ik*). [*pachy-*; κεφαλή, head]. Having unusual thickness of the skull.

**pachycephalous** (*pak-e-sef'-al-us*). [*pachy-*; κεφαλή, head]. One having a thick skull from union of the parietal and occipital bones.

**pachycephaly** (*pak-e-sef'-al-e*) [*pachy-*; κεφαλή, head]. Abnormal thickness of the skull.

**pachycheilia** (*pak-e-ki'-le-ah*) [*pachy-*; χεῖλος, lip]. Increased thickness of one or both lips.

**pachycholia** (*pak-e-ko'-le-ah*) [*pachy-*; χολή, bile]. An inspissated condition of the bile.

**pachychymia** (*pak-e-ki'-me-ah*) [*pachy-*; χυμός, juice]. Increased concentration of the fluids of the body.

**pachydactyl** (*pak-e-dak'-til*) [*pachy-*; δάκτυλος, finger]. A thick digit.

**pachydactylia** (*pak-e-dak-til'-e-ah*) [*pachy-*; δάκτυλος, a finger]. A condition characterized by great thickness of the fingers.

**pachydactylous** (*pak-e-dak'-til-us*) [*pachy-*; δάκτυλος, finger]. Having thick fingers.

**pachyderm** (*pak'-e-derm*) [*pachy-*; δέρμα, skin]. Thick-skinned.

**pachyderma, pachydermia** (*pak-e-der'-mah, pak-e-der'-me-ah*) [*pachy-*; δέρμα, skin]. 1. Thickening of the skin. 2. Elephantiasis. **p. laryngis,** extensive thickening of the mucous membrane of the larynx.

**pachydermatocele** (*pak-e-der-mat'-o-sēl*) [*pachy-*; *dermatocele*]. A tumor due to thickening of the skin.

**pachydermatosis** (*pak-e-der-mat-o'-sis*) [*pachy-*; *dermatosis*]. Hypertrophic rosacea.

**pachydermatous** (*pak-e-der'-mat-us*) [*pachyderma*]. Thick-skinned.

**pachyemia, pachæmia** (*pak-e-e'-me-ah*) [*pachy-*; αἷμα, blood]. Abnormal or morbid thickening of the blood.

**pachyemic, pachyemous.** Having thick blood.

**pachygastrous** (*pak-e-gas'-trus*) [*pachy-*; γαστήρ, the belly]. Having a large abdomen.

**pachyglossal, pachyglossate** (*pak-e-glos'-al, pak-e-glos'-āt*) [*pachy-*; γλῶσσα, tongue]. Having a thick tongue.

**pachygnathous** (*pak-ig'-na-thus*) [*pachy-*; γνάθος, jaw]. Having thick or heavy jaws.

**pachyhymenia** (*pak-e-hi-me'-ne-ah*). See *pachymenia*.

**pachyleptomeningitis** (*pak-e-lep-to-men-in-ji'-tis*) [*pachy-*; *leptomeningitis*]. Combined inflammation of the pia and dura.

**pachylosis** (*pak-e-lo'-sis*). [παχύς, thick]. A condition of the skin, especially of the legs, in which it is thick, dry, harsh, and scaly.

**pachymenia** (*pak-e-me'-ne-ah*) [*pachy-*; ὑμήν, a membrane]. Thickening of the skin or of a membrane.

**pachymenic** (*pak-e-men'-ik*) [*pachy-*; ὑμήν, membrane]. Affected with pachymenia.

**pachymeningitic** (*pak-e-men-in-jit'-ik*). Affected with, or pertaining to, pachymeningitis.

**pachymeningitis** (*pak-e-men-in-ji'-tis*) [*pachy-*; *meningitis*]. Inflammation of the dura of the brain (*cerebral pachymeningitis*) or the spinal cord (*spinal pachymeningitis*). **p. cervicalis hypertrophica,** a form of primary inflammation of the spinal dura producing pain and partial paralysis of one arm. **p., external,** that affecting the external layer of the dura. **p., hemorrhagic, p. hæmorrhagica interna,** an effusion of blood on the inner surface of the dura. **p., internal,** that involving the internal layer of the dura. **p., syphilitic,** that due to syphilis.

**pachymeninx** (*pak-e-me'-ninks*) [*pachy-*; μῆνιγξ, membrane]. The dura.

**pachymeter** (*pak-im'-et-er*) [*pachy-*; μέτρον, measure]. An instrument for measuring small thicknesses.

**pachynsis** (*pak-in'-sis*) [πάχυνσις, thickening]. A thickening, as of a membrane.

**pachyntic** (*pak-in'-tik*) [παχυντικός, making thick]. Pertaining to an abnormal thickening or hardening of a part; increasing the thickness.

**pachyote** (*pak'-e-ōt*) [*pachy-*; οὖς, ear]. Having thick ears.

**pachyotus** (*pak-e-o'-tus*) [*pachy-*; οὖς, the ear]. Having thick ears.

**pachypelviperitonitis** (*pak-e-pel-ve-per-it-on-i'-tis*) [*pachy-*; *pelvis*; *peritonitis*]. Pelvic peritonitis with a fibrous deposit over the uterus.

**pachyperitonitis** (*pak-e-per-it-on-i'-tis*) [*pachy-*; *peritonitis*]. An inflammation of the peritoneum characterized by thickening of the membrane.

**pachypleuritis** (*pak-e-plu-ri'-tis*) [*pachy-*; *pleura*; ιτις, inflammation]. Inflammation of the pleura, with a fibrinous deposit.

**pachypodous, pachypous** (*pak-ip'-o-dus, pak'-ip-us*) [*pachy-*; πούς, foot]. Having very thick feet.

**pachysalpingitis** (*pak-e-sal-pin-ji'-tis*). Chronic parenchymatous salpingitis.

**pachysalpingo-oothecitis** (*pak-e-sal-ping-go-o-o-thes-i'-tis*). Same as *pachysalpingo-ovaritis*.

**pachysalpingo-ovaritis** (*pak-e-sal-ping-go-o-var-i'-tis*), Inflammation of the ovary and oviduct with thickening of the parts.

**pachysomia** (*pak-e-so'-me-ah*) [*pachy-*; σῶμα, body]. Abnormal growth in thickness of the soft parts of the body; as in akromegaly.

**pachytes** (*pak'-it-ēz*) [*pachy-*]. Thickness; *pachyblepharon*.

**pachytic** (*pak-it'-ik*) [παχύτης, thickness]. Fat; thick; obese; also having the power of thickening the fluids of the body.

**pachytrichous** (*pak-it'-rik-us*) [*pachy-*; θρίξ, hair]. Furnished with thick hair.

**pachyvaginalitis** (*pak-e-vaj-in-al-i'-tis*) [*pachy-*; *vagina*; ιτις, inflammation]. Hemorrhagic inflam-

mation of the tunica vaginalis of the testicle, leading to hematocele.
**Pacini's fluid** (*pah-tshe'-nē*). A conserving and diluting fluid used in counting the red blood-corpuscles: Corrosive sublimate, 1 part; sodium chloride, 2 parts; glycerol, 13 parts; distilled water, 113 parts; allow it to stand two months. For use, mix one part of this solution with three parts of water and filter.
**Pacini's method** (*pah-tshe'-ne*). For *resuscitating asphyxiated infants:* the child lying on its back, the operator stands at its head and grasps the axillary structures, pulling the shoulders forward and upward to compress the thorax, and allowing them to fall in order to expand the chest.
**Pacinian bodies** or **corpuscles** (*pah-tshin'-e-an*) [Filippo *Pacini*, Italian anatomist, 1812–1883]. Elliptical, semitransparent bodies, that occur along the nerves supplying the skin, especially of the hands and feet, the external genitalia, the points of the extremities, the periosteum of certain bones and many other localities in man and other mammals; the so-called *corpuscles of Vater*, or *Krause's corpuscles.*
**pack** (*pak*). A blanket, either dry or soaked in hot or cold water, and wrapped about the body. p., **cold,** a blanket wrung out of cold water and wrapped about the body. p., **hot,** a blanket wrung out of hot water and wrapped about the body. p., **wet,** a blanket wrung out of warm or cold water.
**packer** (*pak'-er*). An instrument for introducing tampons or other dressings into a cavity, such as the vagina.
**packing** (*pak'-ing*). 1. The act of filling a wound or cavity with gauze or other material. 2. The material used for filling the cavity.
**Pacquelin cautery.** See *Paquelin.*
**pad.** 1. A small bag stuffed with cotton, hair, etc., used as a cushion for the support of any part of the body. See *liver-pad.* 2. Also, synonym of *compress.* p. **of corpus callosum,** the splenium of the callosum. p., **dinner,** a folded towel laid over the region of the stomach in applying a plaster-of-Paris jacket, to give space for the distention of the stomach by the food. p., **sucking,** a fatty mass situated between the masseter and the buccinator muscles; well developed in infancy.
**pæ-.** For words so beginning, see *pe-*.
**Page's disease** (*pāj*). See *Erichsen's disease.*
**pageism** (*pa'-jism*) [*pagius,* a servant]. A manifestation of masochism in which the individual affected revels in the idea of being a page to a beautiful girl.
**Pagenstecher's ointment** (*pah'-gen-stek-er*) [Alexander *Pagenstecher,* German ophthalmologist, 1828–1879]. An ointment used in ophthalmic practice. It consists of from one to three grains of the yellow oxide of mercury to the dram of vaselin. P.'s **thread,** a flax thread coated with celluloid.
**Paget's abscess** (*paj'-et*) [Sir James *Paget,* English surgeon, 1814–1899]. A residual abscess. P.'s **disease.** 1. Malignant dermatitis, attacking most often the nipple and areola. 2. Osteitis deformans; hypertrophic deforming osteitis. P.'s **recurrent fibroid,** spindle-celled sarcoma of the subcutaneous tissue.
**pagiorrheumatism** (*paj-e-or-roo'-mat-izm*) [πάγιος, firm; *rheumatism*]. Chronic rheumatism.
**pagoplexia** (*pa-go-pleks'-e-ah*) [πάγος, frost; πλῆξις, stroke]. Frost-bite; numbness due to cold. See *frost-bite.*
**paidology** (*pi-dol'-o-je*). That branch of medical science treating of childhood.
**paidonosology** (*pi-don-os-ol'-o-je*) [παῖς, child; νόσος, disease; λόγος, science]. The science of diseases of children; pediatrics.
**pain** (*pān*) [*pœna,* punishment; pain]. 1. Bodily suffering due to irritation of a sensory nerve, or possibly, in rare cases, to changes in the central nervous system. 2. One of the rhythmic contractions of the uterus during labor. p., **after-,** that following labor, and caused by the uterus contracting to expel clots, etc. p., **bearing-down,** pain with a sensation of dragging or bearing down of the pelvic organs, occurring in labor and in various inflammatory affections of the female pelvic organs. p., **boring,** severe, pain of a boring character. p., **false,** that occurring in the latter part of pregnancy and resembling labor pain, although not immediately followed by labor. p., **fulgurant,** p., **fulgurating,** the intense shooting pain affecting principally the limbs of patients suffering from locomotor ataxia. Syn., *lancinating pain.* p., **girdle-,** a painful sensation as of a cord tied about the waist; it is a symptom of organic disease of the spinal cord. p., **growing,** a popular term for the soreness about the joints in young persons at puberty. Some attribute it to increased vascularity of the epiphyses of long bones; others, to rheumatism. p.-**joy,** hysterical enjoyment of suffering. p., **labor.** See *pain* (2). p., **lancinating.** See *p., fulgurant.* p., **osteocopic,** the boneache that characterizes syphilis. p., **referred,** pain situated in a part more or less remote from the cause of the pain. p., **starting,** pain caused by a spasmodic contraction of the muscles just before the onset of sleep. It occurs in joint diseases when the cartilages are ulcerated. p., **terebrating,** p., **terebrant,** boring pain.
**painful** (*pān'-ful*). Characterized by pain. p. **heel.** See *pododynia* and *achillodynia.*
**painless** (*pān'-les*). Without pain. p. **tic.** See under *tic.*
**painter's colic.** Lead colic.
**pair** (*pār*). Two similar organs, one right and the other left, occupying the same relative position on either side of the body; as a pair of nerves.
**Pajot's hook** (*pazj'-o*) [Charles *Pajot,* French obstetrician, 1816–1896]. A hook used in decapitating the fetus; see *P.'s method.* P.'s **law,** the law governing the rotating movements of the child during labor. It is expressed as follows: When a solid body is contained within another, if the receptacle is the seat of alternations of movement and repose, and its surfaces are slippery and but slightly angular, the contained body will tend increasingly to accommodate its form and dimensions to the form and capacity of the receptacle. P.'s **maneuver.** Same as *P.'s method.* P.'s **method,** a method of decapitation of the fetus in embryotomy. It consists in passing a strong cord around a groove in a hook which is passed over the child's neck, and by a sawing movement cutting through the parts. The vagina should be protected by a speculum.
**pala** (*pa'-lah*) [L., a "spade"]. 1. A thin lamella connecting the fimbria and the tenia of the brain. It has a shape like the blade of a turf-cutter. 2. [Native Hawaiian.] The Hawaiian word for syphilis.
**Paladino's phonophore** (*pal-ah-de'-nōs fo'-no-fōr*). An instrument to facilitate hearing, consisting in a rod which connects the larynx of the speaker with the teeth of the listener.
**palæo-** (*pa'-le-o-*) [παλαιός, old]. A prefix denoting old, early, long ago, etc. (For words thus beginning see *paleo-*.
**palatal** (*pal'-at-al*) [*palate*]. Pertaining to the palate.
**palate** (*pal'-āt*) [*palatum,* palate]. The roof of the mouth. It is composed of the *hard palate,* formed by the palate and the palatal bones, and the *soft palate,* or *velum palati,* consisting of an aggregation of muscles—the tensor palati, azygos uvulæ, palatoglossus, and palatopharyngeus. p., **artificial,** a plate of hard material used as an obturator to close a fissure in the palate. p. **bone,** an L-shaped bone back of the nasal fossæ; it helps to form the floor and outer wall of the nose, roof of the mouth, and floor of the orbit; also sphenomaxillary and pterygoid fossæ and the sphenomaxillary fissure. p., **cleft,** a congenital deformity characterized by incomplete closure of the lateral halves of the palate. The soft palate and the uvula, the hard palate, or all together may be involved. p., **hard.** See *palate.* p.-**hook,** an instrument used in rhinoscopy. p.-**myograph,** an instrument for taking a tracing of the movements of the soft palate. p. **plates,** the horizontal portions of the superior maxillæ that unite to form the hard palate. p., **soft.** See *palate.*
**palatic** (*pal-at'-ik*) [*palatum,* the palate]. Palatal; palatine.
**palatiform** (*pal-at'-if-orm*) [*palatum,* palate; *forma,* form]. Resembling the palate.
**palatine** (*pal'-a-tīn*) [*palate*]. 1. Pertaining to the palate, as the *palatine* arteries. 2. Conveying palatine vessels or nerves. p. **arches,** the arches posterior and anterior, upon each side of the beginning of the pharynx. p. **bone,** the palate bone. p. **canals,** several canals in the palatal portion of the superior maxilla. p. **fossa,** a small fossa immediately behind the upper incisor teeth. p. **glands.** See *gland.*

**palatinoid** (*pal-at'-in-oid*). Trade name of a gelatin capsule used for the administration of unpalatable drugs.
**palatitis** (*pal-at-i'-tis*) [*palatum*, palate; *ιτις*, inflammation]. Inflammation of the palate.
**palato-** (*pal-a-to-*) [*palate*]. A prefix denoting relation to the palate.
**palatoglossal** (*pal-a-to-glos'-al*) [*palato-*; γλῶσσα, tongue]. Pertaining to the palate and the tongue.
**palatoglossus** (*pal-a-to-glos'-us*). See under *muscle*.
**palatognathous** (*pal-at-og'-na-thus*) [*palato-*; γνάθος, jaw]. Affected with palatognathus.
**palatognathus** (*pal-at-og'-na-thus*) [*palato-*; γνάθος, jaw]. Cleft palate.
**palatograph** (*pal-at'-o-graf*). See *palate myograph*.
**palatolabial** (*pal-at-o-la'-be-al*) [*palato-*; *labium*, lip]. Pertaining to the palate and the lips.
**palatomaxillary** (*pal-at-o-maks'-il-a-re*) [*palato-*; *maxilla*, jaw]. Pertaining to the palate and the maxilla.
**palatomyograph** (*pal-at-o-mi'-o-graf*). See *palatemyograph*.
**palatonasal** (*pal-a-to-na'-sal*) [*palato-*; *nasal*]. Pertaining to the palate and the nose.
**palatopharyngeal** (*pal-a-to-far-in'-je-al*) [*palato-*; *pharynx*]. Pertaining conjointly to the palate and the pharynx.
**palatopharyngeus** (*pal-a-to-far-in'-je-us*). See under *muscle*.
**palatoplasty** (*pal'-at-o-plas-te*) [*palato-*; πλάσσειν, to form]. Plastic surgery of the palate.
**palatoplegia** (*pal-at-o-ple'-je-ah*) [*palato-*; πληγή, a stroke]. Paralysis of the soft palate.
**palatopterygoid** (*pal-at-o-ter'-ig-oid*) [*palato-*; *pterygoid*]. Pertaining to the palate bone and the pterygoid processes of the sphenoid bone; pterygopalatine.
**palatorrhaphy** (*pal-a-tor'-a-fe*). See *staphylorrhaphy*.
**palatosalpingeus** (*pal-a-to-sal-pin'-je-us*). The tensor palati muscle.
**palatoschisis** (*pal-a-tos'-kis-is*) ‖ [*palato-*; σχίσις, cleft]. Cleft palate.
**palatostaphylinus** (*pal-a-to-staf-il-i'-nus*). See *azygos uvulæ* under *muscle*.
**palatouvularis** (*pal-a-to-ū-vū-la'-ris*). See: *azygos uvulæ* under *muscle*.
**palatum** (*pal'-a-tum*) [L.]. The palate. **p. durum**, the hard palate. **p. fissum**, cleft palate. **p. mobile**, **p. molle**, the soft palate. **p. pendulum**, the soft palate.
**paleontology** (*pa-le-on-tol'-o-je*) [παλαιός, ancient; ὤν, being; λόγος, science]. The science of the early life-forms of the earth.
**paleopathology** (*pa-le-o-path-ol'-o-je*) [παλαιός, old; *pathology*]. "The science of the diseases which can be demonstrated in human and animal remains of ancient times" (Ruffer).
**palimbolous** (*pal-im'-bo-lus*) [πάλιν, back; βάλλειν, to cast]. Changing often; applied to diseases with very inconstant symptoms.
**palimptosis** (*pal-imp-to'-sis*) [πάλιν, back; πτῶσις, a fall]. The falling back; properly, a falling back to a former position, whether after a proptosis, or abnormal forward displacement, or after the rectification of backward displacement.
**palinal** (*pal'-in-al*) [πάλιν, backward]. Moving or moved backward.
**palindromia** (*pal-in-dro'-me-ah*) [πάλιν, again; δρόμος, a course]. Recurrence or growing worse of a disease; a relapse.
**palingenesis** (*pal-in-jen'-es-is*) [πάλιν, back; γένεσις, production]. The form of development of an individual germ in which the development of its ancestors is succinctly repeated.
**palinodia** (*pal-in-o'-de-ah*) [πάλιν, back; ὀδός, a way]. A recurrence or relapse of a disease.
**palirrhea, palirrhœa** (*pal-ir-e'-ah*) [πάλιν, again; ῥοία, flow]. 1. The return of a mucous discharge. 2. Regurgitation.
**palisade-cell** (*pal-is-ād'*). A constituent cell of palisade-tissue. **p.-parenchyma**, the same as *p.-tissue*. **p.-tissue**, applied to certain cells which are elongated at right angles to the surface, occurring especially on the upper side of leaves.
**palladium** (*pal-a'-de-um*) [παλλάδιον, a statue of Pallas]. A rare metal sometimes used in making instruments of precision. **p. chloride**, a drug used in the treatment of tuberculosis. Dose 10 drops of a 3 % solution.
**palliation** (*pal-e-a'-shun*) [*palliare*, to cloak). The act of soothing or moderating, without really curing.

**palliative** (*pal'-e-a-tiv*) [*palliāre*, to cloak].' 1. Relieving or alleviating suffering. 2. A drug relieving or soothing the symptoms of a disease without curing it.
**pallidum** (*pal'-id-um*) [*pallidus*, pale]. The globus pallidus of the lenticular nucleus of the brain.
**pallium** (*pal'-e-um*) [L., "a cloak"]. Of the brain, the fissured portion of each cerebral hemisphere, exclusive of the caudatum and the rhinencephalon.
**pallor** (*pal'-or*) [L.]. Paleness, especially of the skin and mucous membranes. **p. chloroticus**, the peculiar paleness of chlorotic persons. **p. eximius**, abnormal paleness, usually due to anemia. **p. luteus**, chlorosis. **p. pathematicus**, the pallor due to terror or fright. **p. virginum**. Synonym of *chlorosis*.
**palm** (*pahm*) [*palma*, palm]. 1. The inner or flexor surface of the hand; the hollow of the hand. 2. A palm-tree. **p. oil** (*oleum palmæ*), a fixed oil obtained from the fruit of *Elæis guineensis*. It is employed in making soap.
**palma** (*pal'-mah*) [L.]. 1. The palm of the hand. 2. Palm tree. **palmæ plicatæ**, the arborescent rugæ of the anterior of the vagina.
**palmar** (*pal'-mar*). Pertaining to the palm of the hand. **p. abscess**, an abscess in the palm of the hand, usually situated beneath the palmar fascia. **p. arch**, one of the two curved arches, *superficial* and *deep*, formed by the anastomosis of the radial and ulnar arteries in the hand. **p. fascia**, the sheath investing the muscles of the hand.
**palmaris** (*pal-ma'-ris*). See under *muscle*.
**palmature** (*pal'-mat-ūr*) [*palm*]. Union of the fingers, congenital or from burns, wounds, or other trauma.
**Palmella** (*pal-mel'-ah*). A genus of fresh-water algæ.
**palmellin** (*pal-mel'-in*). A red coloring-principle of a fresh-water alga, the *Palmella cruenta*, resembling hemoglobin.
**palmetto** (*pal-met'-o*). See *saw palmetto*.
**palmiacol** (*pal-mi'-ak-ol*), C₁₆H₃₀O₃. A proprietary remedy for tuberculosis, asthma, etc., said to be a derivative of guaiacol. Dose 3 min. (o.18 Cc.) 3 or 4 times daily.
**palmic** (*pal'-mik*) [*palm*]. 1. Referring to the palm; palmitic. 2. [παλμός, throb.] Pertaining to the pulse or palpitation. 3. Relating to palmus or jumpers' disease.
**palmiped** (*pal'-mip-ed*) [*palma*, palm; *pes*, foot]. Having webbed feet.
**palmitate** (*pal'-mit-āt*) [*palm*]. A salt of palmitic acid.
**palmitic** (*pal-mit'-ik*). 1. Relating to or derived from palm-oil. 2. Relating to palmitin. **p. acid**. See *acid, palmitic*.
**palmitin** (*pal'-mit-in*) [*palm*], C₃H₅(C₁₆H₃₁O₂)₃. Glyceryl tripalmitate, a solid, crystallizable substance which, with stearin, constitutes the greater proportion of solid fats.
**palmitone** (*pal'-mit-ōn*). A ketone of palmitic acid; resembles acetone. It is distilled with slaked lime.
**palmityl** (*pal'-mit-il*) [*palm*, the palm tree; ὕλη, the stuff of which a thing is made]. The radical, C₁₆H₃₁O, of palmitic acid.
**palmodic** (*pal-mod'-ik*) [παλμῶδης, like palmus]. Pertaining to, resembling, or affected with, palmus.
**palmoplantar** (*pal'-mo-plan'-tar*) [*palma*, palm; *planta*, the sole of the foot]. Pertaining to both the palms of the hands and the soles of the feet. **p.-plantar sign**, *Filipowics' sign*; said to be diagnostic of typhoid fever. It consists of an orange or saffron coloration of the prominent parts of the palms of the hands and the soles of the feet. The change in color is attributed to feebleness of the action of the heart, causing incomplete filling of the capillaries, and dryness of the skin.
**palmoscopy** (*pal-mos'-ko-pe*) [παλμός, throb; σκοπεῖν, to observe]. The observation of the heart-beat and the pulse.
**palmus** (*pal'-mus*) [παλμός, a twitch]. 1. Jumpers' disease; lata, or miryachit; a convulsive tic, with echolalia and abulia. 2. Subsultus; palpitation; throbbing; pulsation; twitching; jerkiness. 3. The heart-beat.
**palpate** (*pal'-pāt*) [*palpare*, to feel]. 1. To examine by touch. 2. Furnished with tactile organs.
**palpation** (*pal-pa'-shun*) [*palpate*]. In physical diagnosis, the laying of the hand on a part of the body or the manipulation of a part with the hand for the purpose of ascertaining its condition or that

of underlying organs. p., bimanual, the use of the two hands in examining an organ. p., mediate, a method of physical examination performed by placing the phonendoscope on the chest after removing the tubes and resting the palmar surface of the hand upon the instrument, thus intensifying the vibrations.
**palpatometer** (*pal-pat-om'-et-er*) [*palpare*, to feel; μέτρον, a measure]. An instrument for measuring arterial tension.
**palpatometry** (*pal-pat-om'-et-re*) [*palpare*, to feel; μέτρον, measure]. A measuring of the greatest pressure that can be borne without pain.
**palpatopercussion** (*pal-pat-o-per-kush'-un*). Combined palpation and percussion.
**palpebra** (*pal'-pe-brah*) [L.: *pl.*, *palpebræ*]. The eyelid. p. **inferior**, the lower eyelid. p. **superior**, the upper eyelid.
**palpebral** (*pal'-pe-bral*) [*palpebra*]. Pertaining to the eyelid. p. **cartilage**. See *cartilage*. p. **conjunctiva**, the conjunctiva of the eyelid. p. **fascia**, the tarsal ligament of the eyelids. p. **fissure**, the opening between the upper and lower eyelids. p. **follicles**, the Meibomian glands. p. **muscle**, the orbicularis palpebrarum muscle.
**palpebralis** (*pal-pe-bra'-lis*). An old term for the orbicularis palpebrarum muscle.
**palpebrate** (*pal'-pe-brāt*) [*palpebra*]. 1. Furnished with eyelids. 2. To wink.
**palpebration** (*pal-pe-bra'-shun*) [*palpebra*]. The act of winking; nictitation.
**palpebrin** (*pal'-peb-rin*). A proprietary remedy used externally in eye diseases; it is said to consist of boric acid, mercury bichloride, zinc sulphate, and glycerol.
**palpebritis** (*pal-pe-bri'-tis*) [*palpebra*, eyelid; ιτις, inflammation]. Synonym of *blepharitis*.
**palpebrofrontal** (*pal-pe-bro-front'-al*). Relating to the eyelid and the brow.
**palpitate** (*pal'-pit-āt*) [*palpitare*; to quiver]. To flutter, to tremble or to beat abnormally fast; applied especially to the heart.
**palpitation** (*pal-pit-a'-shun*) [*palpitare*, to quiver]. A fluttering or throbbing, especially of the heart, of which the person is conscious.
**palsy** (*pawl'-se*) [from *paralysis*]. Paralysis. p., **Bell's**. See *Bell's palsy*. p., **birth-**. See *birth-palsy*. p., **bulbar**. See *bulbar paralysis*. p., **crutch-**. See *crutch paralysis*. p. **drops**, compound tincture of lavender. p., **Erb's**. See *Erb's palsy*. p., **hammer-**, that due to excessive use of the hammer. p., **lead**, paralysis of the muscles of the forearm, due to lead poisoning. p., **local**, progressive muscular atrophy. p., **night**, paresthesia of the hands occurring at night. p., **painter's**, lead paralysis. p. **scriveners'**, writers' cramp. p., **shaking**. Synonym of *paralysis agitans*. p., **wasting**, progressive muscular atrophy.
**paludal** (*pal'-ū-dal*) [*palus*, a marsh]. Pertaining to or originating in marshes; malarial.
**paludein** (*pal-ū'-de-in*) [*Paludina*, a genus of freshwater snails]. The mucus of the snail *Paludina vivipara*, which has been used to make a pectoral syrup.
**paludide** (*pal'-ū-dīd*). A cutaneous eruption supposed to be due to malaria.
**paludism** (*pal'-ū-dizm*) [*palus*, a marsh]. Malarial poisoning.
**palustral** (*pal-us'-tral*) [see *paludism*]. Pertaining to, or having the nature of, marsh-fever; paludal.
**pampiniform** (*pam-pin'-i-form*) [*pampinus*, tendril; *forma*, form]. Having the form of a tendril. p. **plexus**. See *plexus*, *pampiniform*.
**pampinocele** (*pam-pin'-o-sēl*) [*pampinus*, tendril; κήλη, hernia]. A varicocele of the veins of the pampiniform plexus.
**pamplegia** (*pam-ple'-je-ah*) [*pan-*; πληγή, stroke]. General paralysis.
**pan**. A low, flat-bottomed vessel. p., **bed**, a large, flat oval pan, usually of agate, or enameled ware or china, serving as a receptacle for the fecal discharges and urine of bed-patients. p., **brain**, p., **head**, the skull. p., **knee**, the patella.
**pan-** [πᾶς, πᾶν, all]. A prefix signifying all, every, the whole of anything.
**panacea** (*pan-a-se'-ah*) [πανάκεια, all-healing, from πᾶς, all; ἄκος, a cure]. A remedy curing all diseases; a cure-all; a quack remedy.
**panado** (*pan-a'-do*) [Sp. *panada*, from *panis*, bread]. Bread softened in water. Also a breadpoultice.
**Panama bark**. Quillaja bark. **P. fever**. 1. A pernicious form of malarial fever occurring in Panama; Chagres fever. 2. Yellow fever. **P. paralysis**, beriberi.
**panaris** (*pan'-ar-is*). See *paronychia*.
**panaritium** (*pan-ar-ish'-e-um*). See *paronychia*.
**panarthritis** (*pan-ar-thri'-tis*). Inflammation of all the structures of a joint.
**Panas' operation** (*pan-ah'*) [Photinos *Panas*, French ophthalmologist, 1832–1903]. For *ptosis*; the tarsal portion of the lid is raised by sutures and the occipito-frontalis muscle is caused to assume, to a great extent, the function of the levator palpebræ. **P.'s solution**. A mild antiseptic collyrium, serviceable in conjunctivitis, blepharitis, etc. It contains mercuric iodide 1 part, absolute alcohol 400 parts, distilled water sufficient to make 20,000 parts.
**panatrophy** (*pan-at'-ro-fe*) [*pan-*; atrophy]. 1. Atrophy affecting every part of a structure. 2. General atrophy.
**Panax** (*pan'-aks*) [πάναξ, all-healing]. A genus of araliaceous plants. Ginseng was formerly classed as *Panax*. See *aralia*, also *ginseng*.
**panbioma** (*pan-bi-o'-ma*) [*pan-*; βίος, life]. The general principle of life. Cf. *bionergy*.
**panblastic** (*pan-blas'-tik*) [*pan-*; βλαστός, a germ]. Connected with all the layers of the blastoderm.
**pancarditis** (*pan-kar-di'-tis*). General inflammation of the heart.
**panchrestous** (*pan-kres'-tus*) [*pan-*; χρηστός, useful]. Useful for everything; relating to a panacea.
**panchrestus** (*pan-kres'-tus*) [*pan-*; χρηστός, useful]. Same as *panasea*.
**pancolpohysterectomy** (*pan-kol-po-his-ter-ek'-to-me*). See *panhysterokolpectomy*.
**pancreaden** (*pan'-kre-ad-en*). The direct extract of pancreas attenuated with calcium carbonate; used in pancreatic diabetes. Dose 4–6 dr. (15–23 Gm.) daily.
**pancreas** (*pan'-kre-as*) [*pan-*; κρέας, flesh]. A compound racemose gland, from six to eight inches in length, lying transversely across the posterior wall of the abdomen; the sweetbread of animals; the abdominal salivary gland of the Germans. Its right extremity, the *head*, lies in contact with the duodenum; its left extremity, the *tail*, is in close proximity to the spleen. It secretes a limpid, colorless fluid that digests proteids, fats, and carbohydrates. The secretion is conveyed to the duodenum by the pancreatic duct, or duct of Wirsung. **p., accessory**, a small mass of glandular structure similar to the pancreas and adjacent to it. **p. Aselli**, a collection of lymph-glands in the mesentery of some mammals, resembling a pancreas. **p. glomeruli**, the islands of Langerhans. **p., lesser**, a small, partially detached portion of the gland, lying posteriorly to its head, and having occasionally a separate duct that opens into the pancreatic duct proper.
**pancreatalgia** (*pan-kre-at-al'-je-ah*) [*pancreas*; ἄλγος, pain]. Pain in the pancreas.
**pancreatectomy** (*pan-kre-at-ek'-to-me*) [*pancreas*; ἐκτομή, excision]. Excision of a portion or all of the pancreas.
**pancreatemphraxis** (*pan-kre-at-em-fraks'-is*) [*pancreas*; ἔμφραξις, stoppage]. Obstruction of the pancreatic duct.
**pancreathelcosis** (*pan-kre-ath-el-ko'-sis*) [*pancreas*; ἕλκωσις, ulceration]. Ulceration of the pancreas.
**pancreatic** (*pan-kre-at'-ik*) [*pancreas*]. Pertaining to the pancreas, as the *pancreatic duct*; depending upon disease of the pancreas, as *pancreatic diabetes*. **p. duct**, the duct of Wirsung. **p. fluid**, **p. juice**, the secretion of the pancreas, a thick, transparent, colorless, odorless fluid of a salty taste, and strongly alkaline.
**pancreaticoduodenal** (*pan-kre-at-ik-o-dū-o-de'-nal*) [*pancreas*; *duodenum*]. Pertaining to the pancreas and the duodenum, as the *pancreaticoduodenal* arteries.
**pancreaticosplenic** (*pan-kre-at'-ik-o-splen'-ik*). Pertaining to the pancreas and the spleen.
**pancreatin** (*pan-kre'-at-in*) [*pancreas*]. The active elements of the juice of the pancreas; also the commercial extract of the pancreas, supposed to possess a fermentative action similar to that of the pancreatic juice.
**pancreatinokinase** (*pan-kre-at-i-no-kin'-ās*). Trade name of a compound of pancreatin and eukinase: said to be a powerful digestant.
**pancreatitic** (*pan-kre-at-it'-ik*). Pertaining to pancreatitis.

**pancreatitis** (pan-kre-at-i'-tis) [pancreas; ιτις, inflammation]. Inflammation of the pancreas. It may be hemorrhagic, suppurative, or gangrenous. The onset of pancreatitis is usually sudden, with severe abdominal pain, vomiting, tympanites, and tenderness of the abdomen. It is generally fatal.
**pancreatolipase** (pan-kre-at-o-lip'-ās). Lipase found in the pancreatic juice.
**pancreatolith** (pan-kre-at'-o-lith) [pancreas; λίθος, a stone]. A calculus of the pancreas.
**pancreatomy** (pan-kre-at'-o-me). See. pancreatotomy.
**pancreatoncus** (pan-kre-at-ong'-kus) [pancreas; όγκος, tumor]. A tumor of the pancreas.
**pancreatopathy** (pan-kre-at-op'-a-the) [pancreas; πάθος, suffering]. Any disease of the pancreas.
**pancreatorrhagia** (pan-kre-at-or-a'-je-ah) [pancreas; ῥηγνύναι, to burst forth]. Hemorrhage from the pancreas.
**pancreatotomy** (pan-kre-at-ot'-o-me) [pancreas; τομή, a cutting]. Incision of the pancreas.
**pancreazymose** (pan-kre-az-i'-mōs) [pancreas; ζύμη, leaven]. One of the pancreatic ferments.
**pancreectomy** (pan-kre-ek'-to-me). See pancreatectomy.
**pancreobismuth** (pan-kre-o-bis'-muth). Trade name of a combination of pancreatic ferments and bismuth.
**pancreodigestin** (pan-kre-o-di-jes'-tin), Trade name of a combination of some of the digestive ferments with lactic and hydrochloric acids.
**pancreolytic** (pan-kre-o-lit'-ik). Destructive to pancreatic tissue.
**pancreon, pankreon** (pan'-kre-on). Trade name of a preparation obtained by the action of tannic acid on pancreatin. It is a grayish, odorless powder having a strong tryptolytic power. Dose 7½ gr. (0.5 Gm.) 3 times daily.
**pancreopathia** (pan-kre-o-path'-e-ah) [pancreas; πάθος, disease]. Disease of pancreas.
**pancreopathy** (pan-kre-op'-ath-e) [pancreas; πάθος, disease]. Disease of the pancreas.
**pancril** (pan'-kril). Trade name of a combination of enzymes, said to be capable of digesting nitrogenous, starchy, and fatty foods.
**pandemia** (pan-de'-me-ah) [pan-; δῆμος, people]. An epidemic that attacks all persons.
**pandemic** (pan-dem'-ik). [pan-; δῆμος, people]. Epidemic over a wide area.
**pandemy** (pan'-dem-e). Same as pandemia.
**Pander's islands** (pan'-der) [Heinrich Christian von Pander, German anatomist, 1794–1865]. The reddish-yellow patches in Pander's layer which consist of corpuscles containing hemoglobin. P.'s layer, the splanchnopleural layer of the mesoblast in which the blood-vessels are first formed.
**pandiculation** (pan-dik-u-la'-shun) [pandiculari, from pandere, to stretch out]. The act of stretching the limbs, especially on waking from sleep, accompanied by yawning.
**pandocheum, pandochium** (pan-do'-ke-um), [pan-; δέχεσθαι, to receive]. A hospital receiving all diseases.
**panelectroscope** (pan-e-lek'-tro-skōp).. An inspection apparatus for use in proctoscopy, esophagoscopy, urethroscopy, etc. It throws concentrated light through the whole tube, thus illuminating the spot that is to be inspected.
**panesthesia** (pan-es-the'-ze-ah) [pan-; αἴσθησις, perception]. 1. General or total sensation; cenesthesia. 2. The undifferentiated sensory capacity of the supposed primal germ.
**panesthetism** (pan-es'-thet-ism), Same as panesthesia.
**Paneth's cells** (pah'-nāth) [Josef Paneih, German physician, 1857— ]. Coarsely granular cells found in the crypts of Lieberkuhn in the jejunum and ileum, especially the latter.
**pang.** A momentary sharp pain. p., breast-, angina pectoris. p., brow-, hemicrania.
**pangadua** (pan-gad'-ū-in). A crystalline solid said to contain the basic principles of cod-liver oil; indicated in all affections due to faulty elimination.
**pangen** (pan'-jen) [pan-; γένεσις, birth, production]. One of the primary bearers of the individual qualities or characters of the cell, i. e., the constituent qualities of the species; one of the ultimate vital particles; a biophor.
**pangenesis** (pan-jen'-es-is) [pan-; γένεσις, origin]. Darwin's theory of heredity, which supposes the existence of gemmules or minute particles separated from the body-cells and segregated from the circulation by the reproductive glands. These preformed constituents of all parts of the fully formed animal or plant become aggregated in the germ, and give rise by a process of evolution to the new organism.
**pangenetic** (pan-jen-et'-ik). Pertaining to pangenesis.
**panglossia** (pan-glos'-e-ah) [παγγλωσσία, wordiness]. Excessive or insane garrulity.
**panhidrosis** (pan-hid-ro'-sis). See panidrosis.
**panhydrometer** (pan-hi-drom'-et-er). An instrument for determining the specific gravity of any liquid.
**panhygrous** (pan-hi'-grus) [pan-; ὑγρός, moist]. Damp as to the entire surface.
**panhyperemia, panhyperæmia** (pan-hi-per-e'-me-ah). Plethora.
**panhysterectomy** (pan-his-ter-ek'-to-me). Total extirpation of the uterus.
**panhysterokolpectomy** (pan-his-ter-o-kol-pek'-to-me) [pan-; ὑστέρα, womb; κόλπος, vagina; ἐκτομή, excision]. Complete removal of the uterus and vagina.
**panicula** (pan-ik'-ū-lah) [L.]. A swelling or tumor.
**panidrosis** (pan-id-ro'-sis) [pan-; ἱδρώς, sweat]. General perspiration.
**panis** (pan'-is) [L.]. Bread.
**panivorous** (pan-iv'-or-us) [panis, bread; vorare, to devour]. Subsisting on bread.
**Panizza's plexuses.** Two lymphatic plexuses lying in the lateral fossa of the preputial frenum; they are formed by the deeper lymphatic vessels of the integument of the glans penis.
**pankreon** (pan'-kre-on). Trade name of a preparation of pancreatin with tannic acid.
**panmeristic** (pan-mer-is'-tik) [pan-; μέρος, a part]. Relating to an ultimate protoplasmic structure composed of independent vital units.
**panmixia** (pan-miks'-e-ah) [pan-; μίξις, mingling]. Indiscriminate sexual crossing.
**panmnesia** (pan-ne'-se-ah) [pan-; μνῆσις, remembrance]. A potential remembrance of all impressions.
**pannecrotomy** (pan-nek-rot'-o-me) [pan-; νεκρός, dead; τομή, a cutting]. The dissection of all dead bodies, suggested as the best method to prevent burial of living persons.
**panneuritis** (pan-nū-ri'-tis). Multiple neuritis. p. epidemica, beriberi.
**panniculitis** (pan-ik-ū-li'-tis) [panniculus; ιτις, inflammation]. Inflammation of the abdominal panniculus adiposus.
**panniculus** (pan-ik'-ū-lus) [dim. of pannus]. A membrane or layer. p. adiposus, the layer of subcutaneous fat. p. carnosus, the layer of muscles contained in the superficial fascia. It is well developed in the lower animals, but in man is represented. mainly by the platysma. p. cordis, the pericardium. p. hymenis, p. virginis, the hymen. p. subtilis, the pia mater. p. transversus, the diaphragm.
**pannosity** (pan-os'-it-e) [pannus]. Softness of the skin.
**pannus** (pan'-us) [L., "a cloth"]. 1. Vascularization of the cornea, usually due to the irritation of trachoma granulations. 2. Chloasma. p. carnosus, p. crassus, one that has caused a considerable thickness. p. hepaticus, chloasma. p. phlyctenular, the vascularized and cloudy condition of the cornea induced by phlyctenular inflammation. p. siccus, an old pannus composed of connective tissue and poor in vessels. p. tenuis, slight pannus.
**panodic** (pan-od'-ik). See panhodic.
**panopepinone** (pan-o-pep'-tōn). Trade name of a dietetic said to consist of bread and beef, cooked, peptonised, sterilized, concentrated, and preserved in sherry.
**panophobia** (pan-o-fo'-be-ah) [pan-; φόβος, fear]. Morbid fear of everything; a symptom present in some cases of neurasthenia.
**panophthalmia, panophthalmitis** (pan-of-thal'-me-ah, pan-of-thal-mi'-tis). Inflammation of all the tissues of the eyeball. p. purulenta, a severe form with great protrusion of the eyeball and formation of pus, usually resulting in blindness.
**panosteitis** (pan-os-te-i'-tis). An inflammation of all the structures of a bone.
**panotitis** (pan-o-ti'-tis). An inflammation involving all the structures of the ear.
**panpeptin** (pan-pep'-tin). Trade name of a remedy said to contain several digestive ferments.

**panpharmacon** (*pan-far'-mak-on*) [*pan-*; φάρμακον, drug]. A panacea.
**panphobia** (*pan-fo'-be-ah*) [*pan-*; φόβος, dread]. Synonym of *panophobia*.
**panplegia** (*pan-ple'-je-ah*) [*pan-*; πληγή, stroke]. Generalized paralysis.
**pansclerosis** (*pan-skle-ro'-sis*) [*pan-*; σκληρος, hard]. Complete sclerosis or hardening of a part.
**pansinusitis** (*pan-si-nus-i'-tis*). Inflammation of all the sinuses of a part or region.
**panspermatism** (*pan-sper'-mat-izm*). The theory that germs are omnipresent.
**panspermia** (*pan-sper'-me-ah*). Same as *panspermatism*.
**pansphygmograph** (*pan-sfig'-mo-graf*). An instrument by means of which tracings can be taken simultaneously of the cardiac movements, the arterial pulse, and the respiration.
**pansymmetry** (*pan-sim'-et-re*). Entire symmetry.
**pant**. To breathe hard or quickly.
**pantachromatic** (*pan-tak-kro-mat'-ik*) [*pan-*; *achromatic*]. Colorless throughout.
**pantamorphia** (*pan-tam-or'-fe-ah*) [*pan-*; μορφή, form]. General deformity.
**pantamorphic** (*pan-tam-or'-fik*) [*pan-*; ἀμορφία, shapelessness]. Completely deformed.
**pantanencephalia** (*pan-tan-en-sef-a'-le-ah*) [*pan-*; ἀνεγκέφαλος, brainless]. Total congenital absence of the brain.
**pantanencephalic** (*pan-tan-en-sef-al'-ik*) [*pan-*; ἀ. priv.; ἐγκέφαλος, brain]. Congenitally destitute of brain.
**pantanencephalus** (*pan-tan-en-sef'-al-us*). A brainless monster.
**pantankyloblepharon** (*pan-tang-kil-o-blef'-ah-ron*) [*pan-*; ἀγκύλη, noose; βλέφαρον, eyelid]. Complete ankyloblepharon.
**pantaphobia** (*pan-taf-o'-be-ah*) [*pan-*; ἀ. priv.; φόβος, fear]. Total absence of fear.
**pantatrophia, pantatrophy** (*pan-tat-ro'-fe-ah, pantat'-ro-fe*) [*pan-*; *atrophy*]. Complete or general atrophy.
**pantatrophous** (*pant-at'-ro-fus*). Without nourishment.
**pantherapist** (*pan-ther'-ap-ist*) [*pan-*; θεραπεία, therapy]. See *eclectic*.
**panthodic** (*pan-thod'-ik*) [*pan-*; ὁδός, way]. Of nervous impulses, radiating to all parts of the body.
**panto-** (*pan'-to-*). See *pan-*.
**pantogamy** (*pan-tog'-am-e*) [*pan-*; γάμος, marriage]. Reckless indiscriminate sexual intercourse.
**pantograf** (*pan'-to-graf*) [*pan-*; γράφειν, to write]. An instrument for the mechanical copying of diagrams, etc., upon the same scale, or upon an enlarged or a reduced scale.
**pantography** (*pan-tog'-ra-fe*) [*pan-*; γράφειν, to write]. 1. General description. 2. The process of copying by a pantograph.
**pantomorph** (*pan'-to-morf*) [*pan-*; μορφή, form]. That which assumes, or exists in, all shapes.
**pantomorphia** (*pan-to-mor'-fe-ah*) [*pan-*; μορφή, form]. 1. The condition of assuming or existing in all shapes. 2. General or complete symmetry. Cf. *pantamorphia*.
**pantophobia** (*pan-to-fo'-be-ah*) [*pan-*; φόβος, fear]. Insane dread of all things.
**pantoplethora** (*pan-to-pleth'-or-ah*) [*pan-*; πληθώρη, fulness]. General hyperemia.
**pantopon** (*pan'-top-on*) [*pan-*; *opium*]. Trade name of a preparation of opium said to contain all the alkaloids of opium in the form of hydrochlorides.
**pantoscopic** (*pan-to-skop'-ik*). See *bifocal*.
**Panum's casein** (*pah'-noom*) [Peter Ludwig Panum, Danish physiologist, 1820–1885]. Serum globulin.
**panus** (*pa'-nus*) [L., "a swelling"]. An inflamed, nonsuppurating lymphatic gland. **p. faucium**, an inflamed gland in the throat. **p. inguinalis**, a bubo.
**panzyme** (*pan'-zīm*) [*pan-*; ζύμη, leaven]. Trade name of a preparation containing several enzymes.
**pap**. The nipple. 2. A soft, semiliquid food for infants. **p. pox**. Same as *cowpox*.
**papain** (*pa-pa'-in*). See under *papaya*.
**papaver** (*pa-pa'-ver*). The poppy. See *opium*.
**papaverine** (*pa-pa'-ver-ēn*) [*papaver*], C₂₁H₂₁NO₄. A crystalline alkaloid found in opium and thought to possess narcotic properties. Dose ⅙ gr. (0.016 Gm.).
**papaw** (*pa-paw'*) [a name of Malabar origin]. 1. The seed of *Asimina triloba*; it is a prompt emetic. Dose of the *fluidextract* 10–30 min. (0.6–1.9 Cc.). 2. See *papaya*.

**papaya** (*pa-pa'-yah*). Melon-tree; papaw—the *Carica papaya*, a tree of the order *Passifloraceæ*. The unripe fruit yields a milky juice containing an albuminous substance, papain or papayotin, capable of digesting fibrin and other proteid bodies. Papain in commerce occurs as a grayish powder, and has been used as a digestant in dyspepsia, as an application to false membranes, warts, epitheliomata, etc. Dose 5–10 gr. (0.32–0.65 Gm.).
**papayotin** (*pap-a'-yo-tin*). The concrete active principle of the milky juice of the papaw; it is an enzyme similar to pepsin.
**paper** (*pa'-per*) [*papyrus*]. See *charta*. **p.**, **helianthin**, **p.**, **methyl-orange**, **p.**, **tropæolin D**, paper charged with methyl-orange and used in testing for acids and alkalies. **p.**, **indigo-carmin**. See under *indigo*.
**papescent** (*pap-es'-ent*). Having the consistence of pap.
**papilla** (*pap-il'-ah*) [L., "a nipple"; *pl.*, *papillæ*]. 1. A small, nipple-like eminence. 2. Synonym of optic disc. 3. A pimple or pustule. **p.**, **acoustic**, the organ of Corti. **p.**, **bile**, the caruncula major of Santorini at the summit of which the bile and pancreatic ducts open. **p.**, **circumvallate**, one of the large papillæ at the root of the tongue, arranged like the letter V opening forward. **p.**, **clavate**. Synonym of *p.*, *fungiform*. **p.**, **conical**. See *p.*, *filiform*. **p.**, **dental**. See *organ*, *enamel-*. **p.**, **duodenal**, the elevation at the point where the common bile-duct enters the duodenum. **p.**, **filiform**, any one of the papillæ occurring on all parts of the tongue, consisting of an elevation of connective tissue covered by a layer of epithelium. **p.**, **fungiform**, any one of the low, broad papillæ found on the surface of the tongue, consisting of a connective-tissue elevation, covered by secondary papillæ. **p.**, **genital**, the primitive penis or clitoris. **p.**, **gustatory**, those papillæ of the tongue which are furnished with tastebuds. Syn., *papilla gustus*. **p.**, **lacrimal**, a small conical eminence on the eyelid at the inner canthus, pierced by the lacrimal punctum. **p.**, **lenticular**. Same as *p.*, *fungiform*. **p.**, **lingual**, one of the elevations of the mucous membrane of the dorsum of the tongue. **papillæ**, **nerve-**. See under *nerve*. **p.**, **renal**, the summit of any one of the renal pyramids projecting into the renal pelvis. **p. spiralis**, the convex spinal ridge formed by Corti's organ. **p.**, **tactile**, a little eminence of the true skin containing tactile corpuscles. **papillæ**, **vascular**, papillæ of the skin containing capillary loops.
**papillary** (*pap'-il-a-re*). 1. Pertaining to the nipple. 2. Composed of or containing papillæ; resembling a papilla. **p. body**, the papillary layer of the skin. **p. muscles**, the musculi papillares, *q. v.* **p. tumor**, a papilloma.
**papillectomy** (*pap-il-ek'-to-me*) [*papilla*; ἐκτομή, excision]. Surgical removal of papillæ.
**papilledema** (*pap-il-e-de'-mah*). Choked disc, papillitis.
**papilliferous** (*pap-il-if'-er-us*) [*papilla*; *ferre*, to bear]. Bearing or containing papillæ, as a *papilliferous cyst*.
**papilliform** (*pap-il'-if-orm*) [*papilla*; *forma*, form]. Shaped like a papilla.
**papillitis** (*pap-il-i'-tis*) [*papilla*; ιτις, inflammation]. Inflammation of the optic disc. Syn., *choked disc*; *optic neuritis*.
**papilloadenocystoma** (*pap-il-o-ad-en-o-sist-o'-mah*). Papilloma combined with adenoma and cystoma.
**papillocarcinoma** (*pap-il-o-kar-sin-o'-mah*) [*papilla*; *carcinoma*]. 1. A carcinoma in which there is the formation of papillary excrescences. 2. A papilloma which has become malignant.
**papilloma** (*pap-il-o'-mah*) [*papilla*; ὄμα, tumor]. A growth on the skin or mucous membrane resembling hypertrophied papillæ. It is a benign tumor, occurring in two forms, the *hard papilloma*, one growing from squamous epithelium, and the *soft papilloma*, one developed from columnar epithelium. **p. diffusum**, multiple papillomata occurring on the legs and buttocks. **p. neuroticum**, a painless, mostly congenital affection characterized by warty or papillomatous growths occurring on one side of the body along the course of a nerve.
**papillomatosis** (*pap-il-o-mat-o'-sis*) [*papilla*; ὄμα, tumor; νόσος, disease]. The widespread formation of papillomata; also the state of being affected with multiple papillomata.
**papillomatous** (*pap-il-o'-mat-us*) [*papilla*; ὄμα, tumor]. Pertaining to a papilloma.

**papilloretinitis** (*pap-il-o-ret-in-i'-tis*). Inflammation of the papilla and retina.
**papillose** (*pap'-il-ōs*). Bearing papillæ.
**papine** (*pap-en'*). A proprietary anodyne said to contain the pain-relieving principle of opium.
**papoid** (*pap'-oid*). A proprietary preparation resembling papain.
**pappataci fever** (*pap-at-ash'-e*). An infectious disease probably of protozoal origin, somewhat resembling dengue, but less severe and of shorter duration. It has been found in Malta, Bosnia, Herzegovina and Dalmatia, Italy and S. America.
**pappus** (*pap'-us*) [πάππος, down]. The fine down first appearing on the cheeks and chin.
**paprica, paprika** (*pap-re'-kah*). The dried and pulverized capsules of *Capsicum annuum*. Syn., *Spanish pepper; Turkish pepper*.
**papula** (*pap'-ū-lah*). See *papule*.
**papular** (*pap'-ū-lar*) [*papula*, papule]. Of the nature of a papule.
**papulation** (*pap-ū-la'-shun*) [*papula*, a pimple]. The stage, in certain eruptive diseases, marked by the formation of papules.
**papule** (*pap'-ūl*) [*papula*, a pimple: pl., *papulæ*]. A small circumscribed, solid elevation of the skin. p., moist, the syphilitic condyloma.
**papuliferous** (*pap-ū-lif'-er-us*) [*papula*, a pimple; *ferre*, to bear]. Pimply; covered with papulæ.
**papulosquamous** (*pap-ū-lo-skwa'-mus*). Characterized by both papules and scales.
**papyraceous** (*pap-ir-a'-se-us*) [*papyrus*, paper]. Resembling paper. p. bone, the ethmoid bone.
**Paquelin's cautery** (*pak-lan'*) [Claude André *Paquelin*, French surgeon, 1836— ]. A hollow platinum point kept at a uniform temperature by a current of benzene vapor; a thermocautery.
**par** [L.]. A pair. p. **vagum**, the vagus nerves.
**para-** (*par-ah-*) [παρά, beyond; beside]. 1. A prefix signifying beyond, beside, near, the opposite of, etc. 2. In chemistry, prefixed to a derivative of the benzol ring, it indicates the substitution of two atoms of hydrogen situated opposite each other.
**para-acetphenetidin** (*par-ah-as-et-fen-et'-id-in*). Phenacetin.
**para-acetophenolethyl carbonate** (*par-ah-as-et-o-fe-nol-eth'-il kar'-bon-āt*). A crystalline powder without color or taste, used as an analgesic and hypnotic. Dose 8 gr. (0.5 Gm.).
**para-amidoacetanilide** (*par-ah-am-id-o-as-et-an'-il-id*). See *paraphenylendiamine*.
**para-analgesia** (*par-ah-an-al-je'-se-ah*) [*para-; analgesia*]. Analgesia limited to the lower half of the body.
**para-anesthesia** (*par-ah-an-es-the'-ze-ah*). Anesthesia of the body below the waist.
**para-appendicitis** (*par-ah-ap-en-dis-i'-tis*). Suppurative inflammation of the connective tissue adjacent to that part of the appendix not covered with the peritoneum.
**parabanic acid** (*par-ab-an':ik*). See *oxalylurea*.
**parabiosis** (*par-ah-bi-o'-sis*) [*para-; βίωσις*, living]. 1. Union of two individuals in such a way that there is some physiological intimacy between them. 2. Temporary suppression of conductivity in a nerve.
**parabiotic** (*par-ah-bi-ot'-ik*). Pertaining to or characterized by parabiosis.
**parablast** (*par'-ah-blast*) [*para-*; βλαστός, a germ]. That part of the mesoblast from which the bloodvessels, lymphatic vessels, and other connective tissues are developed.
**parablastic** (*par-ah-blas'-tik*) [*parablast*]. Pertaining to the parablast.
**parablastoma** (*par-ah-blas-to'-mah*) [*parablast*; ὄμα, tumor]. A tumor composed of parablastic tissue.
**parablepsis** (*par-ah-blep'-sis*) [*para-*; βλέψις, vision]. False or perverted vision.
**parabulia** (*par-ab-ū'-le-ah*) [*para-*; βουλή, will]. Abnormality of volitional action.
**paracanthoma** (*par-ak-an-tho'-mah*) [*para-*; ἄκανθα, prickle; ὄμα, tumor]. A new growth affecting the prickle-cell layer of the skin.
**paracanthosis** (*par-ak-an-tho'-sis*) [*para-*; ἄκανθα, prickle; νόσος, disease: pl., *paracanthoses*]. Any skin-disease characterized by some anomaly of the prickle-cell layer.
**paracasein** (*par-ah-ka'-se-in*). A substance closely resembling casein in composition and split off from it during the coagulation of milk. Syn., *curd*.
**paracele, paracœle** (*par'-as-ēl*) [*para-*; κοιλία, a hollow]. A lateral ventricle of the brain.

**paracellulose** (*par-ah-sel'-ū-lōs*). A variety of cellulose found in pith.
**Paracelsian** (*par-as-el'-se-an*). 1. Relating to the Swiss physician and alchemist, Aurelius Phillippus Theophrastus Bombastus *Paracelsus* ab-Hohenheim, 1493–1541. 2. A follower of Paracelsus.
**paracentesis** (*par-ah-sen-te'-sis*) [*para-*; κέντησις, puncture]. Puncture; especially puncture of the wall of a cavity of the body, such as the thoracic wall, cornea, tympanic membrane.
**paracentetic** (*par-ah-sen-tet'-ik*). Pertaining to paracentesis.
**paracentral** (*par-ah-sen'-tral*) [*para-*; κέντρον, a center]. Situated near the center. p. **lobule**, convolution on the mesial surface of the cerebral hemisphere uniting the upper ends of the ascending frontal and ascending parietal convolutions.
**paracephalus** (*par-ah-sef'-al-us*) [*para-*; κεφαλή, head]. A monster characterized by a rudimentary, misshapen head and defective trunk and limbs.
**parachloralose** (*par-ak-lor'-al-ōs*) [*para-*; *chloral*]. A product of the action of chloral upon sugar. It is insoluble in water and is practically inert.
**parachlorphenol** (*par-ah-klōr-fe'-nol*), C₆H₄(Cl)OH. A substitution-product of phenol. It is antiseptic, disinfectant, and is employed in a 2 to 3 % ointment in erysipelas. p. **paste**, a paste of equal parts of lanolin, vaselin, starch, and parachlorphenol; it is used in lupus.
**parachlorsalol** (*par-ah-klōr-sa'-lol*). Parachlorphenol salicylate; used as an internal and external antiseptic like phenyl salicylate. Dose 60–90 gr. (4–6 Gm.) daily.
**paracholesterin** (*par-ak-o-les'-ter-in*) [*para-*; χολή, bile; στέαρ, fat], C₂₆H₄₄O. One of the vegetable cholesterins.
**paracholia** (*par-ak-o'-le-ah*) [*para-*; χολή, bile]. 1. Any abnormality in the secretion of bile. 2. The prodrome of disturbed liver-cell activity, in consequence of which the bile pours over the blood-vessels and lymph-vessels.
**parachordal** (*par-ak-or'-dal*) [*para-*; χορδή, a string; specifically the chorda or notochord]. 1. One of two bars of cartilage extending alongside the occipital notochord in the human fetus. 2. Adjoining the cephalochord; situated at the side of the cranial part of the notochord of the embryo. p. **Pertaining to the cartilaginous basis of the cranium in the embryo.
**parachrea, parachroia** (*par-ak-re'-ah, par-ak-roi'-ah*) [*para-*; χροιά, color]. Morbid discoloration or change of complexion.
**parachroma** (*par-ak-ro'-mah*) [*para-*; χρῶμα, color]. Change in color, especially in the natural color of the skin.
**parachromatin** (*par-ak-ro'-mat-in*) [*para-*; *chromatin*]. That part of the nucleoplasm which forms the spindle-threads during karyokinesis.
**parachromatism** (*par-ak-ro'-mat-izm*) [*para-*; χρῶμα, color]. False, or incorrect perception of color. It is not the same as true color-blindness, which it may approach more or less completely.
**parachromatoblepsia** (*par-ak-ro-mat-o-blep'-se-ah*). See *parachromatism*.
**parachromatosis** (*par-ak-ro-mat-o'-sis*) [*para-*; χρῶμα, skin; νόσος, disease]. Any one of the pigmentary skin-diseases.
**parachromophore** (*par-ak-ro'-mo-fōr*) [*para-*; χρῶμα, color; φορός, bearing]. Applied to chromogenic bacteria that produce the pigment as an excretion-product, but retain it in the organism.
**parachromophoric, parachromophorous** (*par-ah-kro-mo-fo'-ik, par-ah-kro-mof'-or-us*) [*para-*; χρῶμα, color; φορός, bearing]. Possessing color which remains within the cell as a passive metabolic product, as in some bacteria.
**parachrosis** (*par-ak-ro'-sis*) [*para-*; χρῶσις, coloring]. The existence of a pigmentary skin-disease.
**parachymosis** (*par-ak-hi'-mo-sin*) [*para-*; χυμός, juice]. The chymosin or rennin found in the human stomach and in that of the pig.
**parachymosis** (*par-ak-i-mo'-sis*) [*para-*; χυμός, juice]. A morbid state of a secretion or a secreting organ.
**paracinesis** (*par-as-in-ē'-sis*) [*para-*; κίνησις, motion]. Morbid movement of the voluntary muscles, arising from disease of the motor nerves or centers.
**paraclonus** (*par-ak'-lo-nus*). Synonym of *paramyoclonus*.
**paracmasis** (*par-ak'-mas-is*). Synonym of *paracme*.

**paracmastic** (*par-ak-mas'-tik*) [*paracme*]. Pertaining to the declining stage.
**paracme** (*par-ak'-me*) [*para-*; ἀκμή, point, prime]. 1. The degeneration or decadence of a group of organisms after they have reached their acme of development. 2. The period of decline of a disease.
**paracnemion** (*par-ak-ne'-me-on*). Synonym of *fibula*.
**paracoele** (*par'-as-ēl*). See *paracele*.
**paracolitis** (*par-ak-o-li'-tis*) [*para-*; *colon*; ιτις, inflammation]. Inflammation of the outer coat of the colon.
**paracolon** (*par-ah-ko'-lon*). A term applied to a group of bacilli intermediate between the typhoid and colon group.
**paracolpitis** (*par-ah-kol-pi'-tis*) [*para-*; κόλπος, vagina; ιτις, inflammation]. Inflammation of the connective tissue about the vagina.
**paracolpium** (*par-ah-kol'-pe-um*) [*para-*; κόλπος, vagina]. The connective tissue lying around the vagina.
**paracondylar** (*par-ah-on'-dil-ar*) [*para-*; κόνδυλος, knuckle]. Situated alongside a condyle or a condylar region.
**paracondyloid** (*par-ak-on'-dil-oid*) [*para-*; κόνδυλος, knuckle; εἶδος, form]. Adjoining the condyles.
**paracoto** (*par-ah-ko'-to*) [*para-*; Sp., *coto*, a cubit]. A South American tree the bark of which contains a neutral substance, *paracotoin*, closely resembling cotoin.
**paracotoin** (*par-ah-ko'-to-in*). See under *paracoto*.
**paracousia, paracousis** (*par-ah-koo'-se-ah, par-ah-koo'-sis*) [*para-*; ἀκούειν, to hear]. See *paracusia*.
**paracresol** (*par-ah-kre'-sol*), $C_7H_8O$. A compound of cresolsulphonate and potassium hydroxide used as a disinfectant.
**paracresotate** (*par-ah-kres'-o-tāt*), $C_6H_3(OH)(CH_3)$-CO₂Na. A crystalline powder obtained by heating sodium cresylate with carbonic acid; used in acute articular rheumatism. Dose 45–90 gr. (3–6 Gm.) daily.
**paracresylol** (*par-ah-kres'-il-ol*) [*para-*; *cresol*]. A derivative of cresol.
**paracrisis** (*par-ak'-ris-is*) [*para-*; κρίνειν, to secrete; *pl., paracrises*]. Disorder of the secretory function.
**paracrusis** (*par-ak-roo'-sis*) [*para-*; κρούειν, to strike]. 1. Insanity, delirium. 2. The checking or "driving in" of an eruption of exanthem.
**paracusia, paracusis** (*par-ah-oo'-se-ah, par-ak-oo'-sis*) [*para-*; ἀκούειν, to hear]. Any perversion of the sense of hearing. **p. acris**, excessively acute hearing, rendering the person intolerant of sounds. **p. duplicata**, a condition in which all or only certain sounds are heard double. **p. localis**, **p. loci**, difficulty in estimating the direction of sounds met with in unilateral deafness, or when the two ears hear unequally. **p. obtusa**, hardness of hearing. **p. perversa**, synonym of *p. Willisiana*. **p. imaginaria**, tinnitus aurium, *q. v.* **p. Willisiana**, deafness in quiet places with increased acuteness of hearing in the midst of noise.
**paracyclesis** (*par-ah-si-kle'-sis*) [*para-*; κύκλησις, a revolution]. A disturbance of the circulation.
**paracyesis** (*par-as-i-e'-sis*) [*para-*; κύησις, pregnancy]. Extra-uterine pregnancy.
**paracystitis** (*par-ah-sis-ti'-tis*). Inflammation of the connective tissue surrounding the bladder.
**paracystium** (*par-ah-sis'-te-um*) [*para-*; κύστις, a bladder]. The connective tissue which surrounds the bladder.
**paracytic** (*par-a-si'-tik*) [*para-*; κύτος, cell]. Lying among cells.
**paradenitis** (*par-ad-en-i'-tis*) [*para-*; ἀδήν, gland; ιτις, inflammation]. Inflammation of the areolar tissue about a gland.
**paradidymis** (*par-ah-did'-im-is*) [*para-*; δίδυμος, testicle]. The organ of Giraldēs, the atrophic remains of the tubules of the Wolffian body, lying among the convolutions of the epididymis.
**paradiphtherial, paradiphtheritic** (*par-ah-dif-thē'-re-al, par-ah-dif-ther-it'-ik*). Remotely or indirectly related to diphtheria.
**paradox** (*par'-ad-oks*). See *paradoxia*. **p., Weber's**, a muscle when so loaded as to be unable to contract, may elongate.
**paradoxia** (*par-ad-oks'-e-ah*) [παράδοξος, incredible]. An absurd or contradictory statement or proposition. **p. sexualis**, sexual excitement occurring independently of the period of the physiological maturation of the generative organs; the abnormal exhibition of sexual instincts in childhood or prior to puberty.

**paradoxical contraction**. A slow tonic contraction occurring in a muscle when suddenly relaxed or when its length is suddenly shortened. **p. pulse**. See *pulse, paradoxic*.
**paræsthesia**. See *paresthesia*.
**paraffin, paraffinum** (*par'-af-in, par-af-i'-num*) [*parum*, little; *affinis*, affinity]. 1. Any saturated hydrocarbon of the marsh-gas series, having the formula $C_nH_{2n+2}$. 2. A white, odorless, translucent hydrocarbon (*paraffinum, U. S. P.*), obtained from coal-tar or by the destructive distillation of wood. **paraffinum durum** (B. P.), hard or solid paraffin, a mixture of several of the harder members of the paraffin series of hydrocarbons. It is usually obtained by distillation from shale. **p., liquid**, a liquid hydrocarbon of the paraffin series. **paraffinum molle**, soft paraffin; the *petrolatum* of the U. S. P. See *petrolatum*.
**paraffinoma** (*par-ah-fin-o'-mah*). A tumor supposed to be due to the injection of paraffin into the tissues.
**paraflagellate** (*par-af-laj'-el-āt*) [*para-*; *flagellum*, a flagellum]. Provided with paraflagella.
**paraflagellum** (*par-af-laj-el'-um*) [*para-*; *flagellum*, a whip; *pl., paraflagella*]. A small supplementary flagellum.
**parafloccolus** (*par-af-lok'-ū-lus*). See *flocculus*.
**paraformaldehyde** (*par'-ah-form, par-ah-form-al'-de-hīd*). See *formaldehyde, para-*.
**parafuchsin** (*par-ah-fūk'-sin*). A basic triphenylmethane dyestuff.
**paraganunclinus** (*par-ah-gum-uh-sis'-imus*) [*παρ u-*; γάμμα, the Greek letter γ]. Inability to pronounce the hard "g," and also "k," other consonants being substituted, as "d" or "t."
**paraganglia cells** (*par-ah-gan'-gle-ah*). Masses or cords which originate in the embryonic sympathetic ganglia. See *chromaffin cells*.
**paraganglin** (*par-ah-gang'-lin*). A proprietary extract of the myelinic part of the suprarenal gland of the ox.
**paraganglion** (*par-ah-gang'-le-on*) [*para-*; *ganglion*; *pl., paraganglia*]. A collection of cells situated in the medullary portion of the adrenal bodies.
**parageusia, parageusis** (*par-ah-gū'-se-ah, par-ah-gū'-sis*) [*para-*; γεῦσις, taste]. Perversion of the sense of taste.
**paraglobin** (*par-ag-lo'-bin*). Same as *paraglobulin*.
**paraglobulin** (*par-ah-glob'-ū-lin*) [*para-*; *globulus*, a little ball]. A globulin found in blood-serum and other fluids of the body. Syn., *fibrinoplastin*; *fibroplastin*; *serum-globulin*.
**paraglobulinuria** (*par-ah-glob-ū-lin-ū'-re-ah*) [*paraglobulin*; οὖρον, urine]. The presence of paraglobulin in the urine.
**paraglossa** (*par-ah-glos'-ah*) [*para-*; γλῶσσα, the tongue]. 1. Swelling of the tongue; also, a hypertrophy of the tongue, usually congenital.
**paraglossia** (*par-ag-los'-e-ah*) [*para-*; γλῶσσα, tongue]. Inflammation of the muscles and connective tissues under the tongue.
**paragnathous** (*par-ag'-na-thus*) [*para-*; γνάθος, jaw]. 1. Having both mandibles of equal length, their tips falling together, as in certain birds. 2. Pertaining to paragnathus.
**paragnathus** (*par-ag'-na-thus*) [*para-*; γνάθος, jaw]. A double monster having a supernumerary mandible situated laterally.
**paragomphosis** (*par-ag-om-pho'-sis*) [*para-*; γόμφωσις, a nailing]. Impaction of the fetal head in the pelvic canal.
**paragonimiasis** (*par-ah-go-ne-mi'-a-sis*). The condition of being infected by the *Paragonimus*.
**Paragonimus** (*par-ag-on'-im-us*). A genus of nematode worms. **p. Westermanii**, the *distoma pulmonale*.
**paragonorrheal** (*par-ah-gon-o-re'-al*). Having an indirect relation to gonorrhea.
**paragraphia** (*par-ah-graf'-e-ah*) [*para-*; γράφειν, to write]. 1. A form of aphasia in which the person writes the improper word or misplaces the words. 2. Inability to express ideas in writing.
**Paraguay tea** (*par'-ah-gwi*). See *maté*.
**parahemoglobin** (*par-ah-hem-o-glo'-bin*). 1. Nencki's name for a polymeric modification of oxyhemoglobin. 2. A proprietary preparation of blood containing 5% of iron.
**parahepatic** (*par-ah-he-pat'-ik*) [*para-*; ἧπαρ, liver]. About or near the liver.
**parahydropin** (*par-ah-hi'-dro-pin*). A proprietary diuretic containing theobromine.

**parahypnosis** (*par-ah-hip-no'-sis*) [*para-;* ὕπνος, sleep]. Abnormal sleep, like that of hypnotism or of narcosis.

**parainfection** (*par-ah-in-fek'-shun*). The presence of symptoms which simulate those of an infectious disease, without the specific microorganism of that disease being present.

**parainfectious** (*par-ah-in-fek'-shus*). Pertaining to or characteristic of pathological states attributable to infection, which occur as accessory or by-conditions to some already existing disease.

**parakanthosis** (*par-ak-an-tho'-sis*). See *paracanthosis*.

**parakeratosis** (*par-ak-er-at-o'-sis*) [*para-;* κέρας, horn; νόσος, disease]. Any disease of the skin characterized by an abnormal quality of the horny layer. **p. variegata,** a rare skin-affection characterized by the presence upon the entire surface of the body of patches of red exanthem leaving small, irregular, sunken patches of normal skin, and giving to the surface a reticulated appearance.

**parakinesis** (*par-ak-in-e'-sis*). See *paracinesis*.

**paralactate** (*par-ah-lak'-tāt*). A salt of paralactic acid.

**paralactic acid** (*par-ah-lak'-tik*). See *acid, sarcolactic.*

**paralalia** (*par-ah-la'-le-ah*) [*para-;* λαλιά, speech]. Disturbance of the faculty of speech.

**paralambdacism, paralambdacismus** (*par-al-am'-das-izm, par-al-am-das-is'-mus*) [*para-;* lambda, the letter λ]. Inability to pronounce the letter *l*, or the substitution of other consonants as *t, r, s, w* for *l*.

**paralbumin** (*par-al-bū'-min*) [*para-;* albumin]. A protein substance found in ovarian cysts.

**paraldehyde** (*par-al'-de-hīd*) [*para-;* aldehyde], $C_6H_{12}O_3$. A polymeric form of aldehyde, occurring as a colorless liquid of repulsive odor and unpleasant taste, with powerful hypnotic properties. It is used in delirium tremens, mania; tetanus; and other nervous affections. Dose 30–60 min. (2–4 Cc.).

**paraldol** (*par-al'-dol*) [*para-;* aldehyde], $(C_4H_8O_2)n$. A polymer of aldol that melts at between 80° and 90° C.

**paraleipsis** (*par-al-īp'-sis*) [*para-;* ἀλείφειν, to anoint]. A disorder of the sebaceous secretion.

**paralerema** (*par-al-er-e'-mah*) [*para-;* λήρημα, utterance]. Delirium, or delirious utterance.

**paraleresis** (*par-al-er-e'-sis*) [*para-;* λήρησις, speech]. Delirium, or moderate mental disturbance.

**paralexia** (*par-al-leks'-e-ah*) [*para-;* λέξις, speech]. Disturbance of the power of reading, consisting in the transposition or substitution of words or syllables.

**paralgesia** (*par-al-je'-ze-ah*) [*para-;* ἄλγος, pain]. An abnormal painful sensation; painful paresthesia.

**paralgia** (*par-al'-je-ah*) [*para-;* ἄλγος, pain]. Any perverted and disagreeable cutaneous sensation, as of formication, cold, burning, etc.

**paralinin** (*par-al-i'-nin*) [*para-;* linin, parachromatin]. In biology, the nuclear sap or matrix, a protein of the globulin class, similar to those found in the cell-protoplasm. Cf. *paramitome, paraplasm.*

**parallactic** (*par-al-ak'-tik*). Pertaining to parallax.

**parallagma** (*par-al-ag'-ma*) [παράλλαγμα, alteration]. The overriding or overlapping of the ends of a fractured bone.

**parallax** (*par'-al-aks*) [*para-;* ἄλλος, other]. The apparent displacement of an object due to a change in the position of the observer, or by looking at it alternately with one eye and then with the other. **p., binocular,** the angle of convergence of the visual axes. **p., crossed, p., heteronymous,** that in which the object moves away from the uncovered eye. **p., homonymous,** that in which the object moves toward the uncovered eye. **p., mental,** a slight personal equation in observation due to one's standpoint. **p., stereoscopic.** See *p., binocular.* **p. test,** for locating opacities in the cornea, lens, and vitreous. It is used with the plane mirror at ten to twelve inches. A body situated anterior to the plane of the pupil will move in the direction taken by the eye, while one posterior to the plane of the lens will move against the direction taken by the eye. Bodies lying about the same plane as the pupil will show little if any movement. **p., vertical,** that in which the object moves upward or downward.

**parallelism** (*par'-al-el-izm*). See *isopathy.* **p. of disease,** the tendency in diseases to simulate others.

**paralogia** (*par-ah-lo'-je-ah*) [*para-;* λόγος, reason]. Difficulty in thinking logically. **p., thematic,** a condition in which the thought is unduly concentrated on one subject.

**paralogism** (*par-al'-o-jism*) [*para-;* λόγος, reason]. The logical error of considering effects or unrelated phenomena as the cause of a condition.

**paralysant** (*par-al-i'-zant*). See *paralysant.*

**paralysin** (*par-al'-is-in*). See *agglutinin.*

**paralysis** (*par-al'-is-is*) [*para-;* λύειν, to loosen]. A loss of motion or of sensation in a part. **p., acute amyotrophic spinal.** Same as *p., infantile.* **p., acute ascending.** See *Landry's disease.* **p., acute atrophic.** See *p., infantile.* **p., acute progressive.** See *Landry's disease.* **p. agitans,** an affection marked by tremor or alternate contraction and relaxation of the muscles of the part involved. It usually begins in one hand and seldom affects the head. The movements persist during rest and are little influenced by voluntary motion. Late in the disease there is a typical gait (festination), which consists in progressive increase of the gait until the patient breaks into a run which grows faster and faster until he falls or seizes some support. The disease is most common in males over forty and may last thirty or forty years. Syn., *Parkinson's disease; shaking palsy.* **p., alcoholic,** multiple neuritis from alcoholism. **p., amyotrophic,** that occurring as the sequel of some acute disease, and attended by atrophy of certain muscles. **p., angio-.** See *angioparalysis.* **p., ascending,** a form of paralysis marked by loss of motor power in the legs, gradually extending upward. **p., asthenic bulbar.** See *myasthenia gravis pseudoparalytica.* **p., atrophic bulbar.** Same as *p., bulbar.* **p., atrophic muscular.** See *p., atrophic.* **p. atrophic spinal.** See *p. infantile.* **p., atrophospastic.** See *amyotrophic lateral sclerosis.* **p., Bell's.** See *p., facial.* **p., bifacial,** paralysis of both sides of the face. **p., birth-.** See *paraplegia, infantile spasmodic.* **p., brachial,** palsy affecting one or both arms. **p., brachiofacial,** that affecting both arm and face. **p., Brown-Séquard's,** a motor paralysis of one side of the body with sensory paralysis of the other side. **p., bulbar,** a form due to the degeneration of the nuclei of origin of the nerves arising in the oblongata. **p., central,** a paralysis due to a lesion of the brain or spinal cord. **p., cerebral,** a paralysis due to a brain-lesion. **p., cortical,** that due to lesion of the cerebral cortex. **p., crossed,** a paralysis of the arm and leg of one side, associated with either a facial paralysis or a paralysis of the oculomotor nerve of the opposite side. **p., crural,** that chiefly affecting the thighs. **p., crutch.** See *crutch paralysis.* **p., Cruveilhier's,** progressive muscular atrophy. **p., diphtheritic,** a motor paralysis due to the action of the diphtheria toxin on the nervous system, chiefly on the peripheral nerves. **p., divers'.** See *caisson disease.* **p., Duchenne's.** See *Duchenne's paralysis.* **p., Erb's,** a partial paralysis of the brachial plexus, involving the nerves supplying the deltoid, biceps, brachialis anticus, and supinator longus, often the supinator brevis, and occasionally the infraspinatus and subscapularis muscles. **p., facial,** a paralysis of the muscles of the face, usually of one side only, due to central disease or due to a lesion of the facial nerve. **p., festinans,** a phase of paralysis agitans in which the patient walks as if hurried forward. **p., general, of the insane,** an organic disease of the brain characterized by progressive loss of power and by a deterioration of the mental faculties, ending eventually in dementia and death. The main symptoms may be divided into psychic, motor, and sensory. The psychic symptoms are principally a change of character and delusions of grandeur; the motor are weakness, tremor, disturbance of speech, apoplectiform or epileptiform seizures, and finally motor paralysis; there is often inequality of the pupils, with miosis or mydriasis; sensory symptoms are slight and consist chiefly in paresthesias. The causes are obscure—syphilis and severe nervous strain are important factors. Syn., *general paresis; paralytic dementia; paretic dementia; progressive paralysis of the insane.* **p., glossolabial.** Same as *p., bulbar.* **p., glossolabiolaryngeal,** bulbar paralysis. **p., histrionic,** a name for Bell's facial palsy, because it destroys the power of facial expression. **p., hysterical,** that associated with hysteria, but without any causative lesion. **p., incomplete,** partial loss of power. **p., infantile,** a disease peculiar to childhood, and characterized by sudden paralysis of one or more limbs or of individual muscle-groups, and

# PARALYTIC 653 PARAMUSIA

followed by rapid wasting of the affected parts, with reaction of degeneration and deformity. The paralysis is due to changes in the anterior cornua of the gray matter of the spinal cord, and is probably the result of infection. Syn. *acute anterior polio-myelitis; acute atrophic paralysis; atrophic spinal paralysis; essential paralysis.* p., **ischemic**, paralysis of a part due to stoppage of the circulation. *e. g.,* paralysis of the lower limb following embolism or thrombosis of the femoral artery. p., **Klumpke's**, a paralysis involving the lower portion of the brachial plexus, the eighth cervical and first dorsal nerves, and characterized by paralysis of the small muscles of the hand, of some of the muscles of the forearm, with anesthesia in the distribution of the ulnar and median nerves. Pupillary changes may be present. p., **Landry's**. See *Landry's disease.* p., **lead-**, a paralysis due to lead, usually of the extensors of the wrist, causing wrist-drop. It is nearly always bilateral, and is caused by a peripheral neuritis induced by the lead-poisoning. p., **Little's**, infantile spasmodic paraplegia. p., **local**, that confined to one muscle or one group of muscles. p., **mimetic**. See *Bell's paralysis*. p., **motor**, paralysis of the voluntary muscles. p., **musculospiral**, paralysis of the extensors and supinators of the wrist, due to an injury or to inflammation of the musculospiral nerve. p., **myosclerotic**. See *p., pseudohypertrophic muscular*. p., **narcosis**, pressure paralysis in the region of the brachial plexus due to prolonged narcosis, during which the arm is elevated with the head resting upon it or it is pressed against the edge of the table. p., **nuclear**, one due to a lesion of the nuclei of origin of a cranial nerve. p., **obstetrical**, any paralysis of the child resulting from injuries received during delivery. p., **oculomotor**, that attacking the oculomotor nerve. p., **peripheral**, loss of power due to a lesion of the nervous motor mechanism between the nuclei of origin and peripheral termination. p., **postdiphtheritic**. See *p., diphtheritic.* p., **pressure**, paralysis of a group of muscles supplied by a nerve which has been subjected to prolonged pressure. p., **pseudobulbar**, a symmetrical lesion of the halves of the cerebrum producing paralysis of the lips, the tongue, and the larynx or the pharynx. p., **pseudohypertrophic muscular**, a chronic disease characterized by progressive muscular weakness, associated with an apparent hypertrophy of the affected muscles. The disease usually begins in the muscles of the calf, and spreads over the body, the muscles of the hand almost always escaping. There are marked lordosis and a peculiar gait, with wide separation of the legs and swaying of the body from side to side. The characteristic symptom is the manner in which the patient arises from the floor—he "climbs" up on his legs, on account of the weakened state of the extensor muscles of the back. It is most common in young male children. No adequate nerve-lesion having as yet been discovered. The muscles are the seat of hypertrophy and atrophy of muscular fibers, hyperplasia of the connective tissue, and fatty infiltration. p., **reflex**, the paralysis sometimes following immediately upon a wound of a nerve, or the paraplegia sometimes due to irritation of an adherent prepuce. These so-called reflex palsies, as that from renal calculus, are probably due to secondary changes in the spinal cord or nerves. p., **segmental**, that of a segment of a limb produced by hypnotism. p., **sensory**, paralysis of the muscles and heightened tendon-reflexes. p., **spastic spinal**, lateral sclerosis. p., **spinalis**, paraplegia. p., **vasomotor**, paralysis of the vasomotor center or of the vasomotor nerves; it leads to dilatation of the blood-vessels. p., **wasting**, progressive muscular atrophy. p., **writers'**, writers' cramp.
**paralytic** (*par-al-it'-ik*). 1. Of the nature of paralysis; affected with paralysis. 2. A person suffering from paralysis; also one suffering from general paralysis of the insane. p. dementia, general paresis. p. **flail-joint**, flail-joint the result of paralysis.
**paralyzant** (*par-al-i'-zant*) [*paralysis*]. 1. Causing paralysis. 2. An agent or drug that induces paralysis. p., **motor**, a drug paralyzing any part of the motor apparatus: the motor cells of the spinal cord, the motor nerves, or the muscles.
**paralyzing vertigo**. See *Gerlier's disease.*
**paramagnetic** (*par-ah-mag-net'-ik*). Exhibiting a polarity in the same direction as the magnetizing force.
**paramagnetism** (*par-ah-mag'-net-izm*) [*para-; magnet*]. The phenomena exhibited by paramagnetic substances.
**paramastitis** (*par-ah-mas-ti'-tis*) [*para-; mastitis*]. Inflammation of the connective tissue about the mamma.
**paramastoid** (*par-am'-as-toid*) [*para-; mastoid*]. 1. Situated near the mastoid process. 2. The jugular process of the occipital bone.
**paramecium** (*par-am-e'-se-um*) [*para-; μῆνος*, length: *pl., paramecia*]. A longitudinal fissure.
**Paramecium** or **Paramœcium** (*par-ah-me'-se-um*). A genus of ciliate protozoa. P. **coli**, a species found in normal and diarrheal stools. Also called *Balantidium coli.*
**paramedian** (*par-am-e'-de-an*) [*para-; medius*, middle]. Situated near the median line. p. **sulcus**, a fissure present in the cervical portion of the spinal cord, not far from the posterior median fissure, and separating the column of Goll from the funiculus cuneatus.
**paramenia** (*par-ah-me'-ne-ah*) [*para-; μῆνες*, menses]. Difficult or disordered menstruation.
**paramesial** (*par-ah-me'-ze-al*) [*para-; μέσοι*, middle]. Located near the mesial line.
**parametric** (*par-ah-met'-rik*) [*parametrium*]. Pertaining to the tissues about the uterus.
**parametrism** (*par-am-met'-rizm*) [*parametrium*]. Painful spasm of the smooth muscular fibers of the broad ligament.
**parametric** (*par-am-et-rit'-ik*) [*para-; μήτρα*, uterus; *ιτις*, inflammation]. Relating to, of the nature of, or affected with, parametritis.
**parametritis** (*par-ah-met-ri'-tis*) [*parametrium; ιτις,* inflammation]. Inflammation of the cellular tissue about the uterus; pelvic cellulitis. p., **anterior**, that in which the inflammation is limited to the loose vesicouterine cellular tissue or that between the symphysis and the bladder. The swelling is anterior, and the pus generally tracks into the bladder, vagina, or inguinal region. p. **chronica atrophicans**, inflammatory hypertrophy of the connective tissue of the pelvis progressing to cicatricial atrophy. p. **chronica posterior**, chronic inflammatory processes in Douglas' folds, causing fixation of the uterus at the level of the internal os and anteflexion by shortening of the folds and torsion of the uterus when only one fold- is shortened. p., **remote**, parametritis marked by formation of abscesses in places more or less remote from the focus of the disease.
**parametrium** (*par-ah-me'-tre-um*) [*para-; μήτρα*, womb]. The connective tissue surrounding the uterus.
**paramimia** (*par-ah-mim'-e-ah*) [*para-; μιμεῖσθαι*, to mimic]. A form of aphasia characterized by the faulty use of gestures.
**paramitome** (*par-ah-mi'-tōm*) [*para-; μίτος*, a thread]. The fluid portion of the cell-substance, contained in the meshes of the mitome.
**paramnesia** (*par-am-ne'-ze-ah*) [*para-; amnesia*]. Illusion of memory, especially the illusion of feeling, as if one had already undergone the experience which may be passing.
**paramœcium**. See *paramecium.*
**paramonochlorphenol** (*par-ah-mon-o-klor-fe'-nol*), C₆H₄(Cl) . OH(1 : 4). A crystalline body obtained by the chlorination of phenol; it is antiseptic and employed in erysipelas, tuberculous diseases of throat, etc., in 5 to 20 % solution in glycerol.
**paramorphia** (*par-am-or'-fe-ah*) [*para-; μορφή*, form].
**paramorphic** (*par-am-or'-fik*) [*para-; μορφή*, form]. Pertaining to paramorphism.
**paramorphine** (*par-ah-mor'-fēn*). See *thebaine.*
**paramorphism** (*par-am-orf'-izm*) [*para-; μορφή*, form]. In chemistry, a variety of pseudomorphism in which there is a change of molecular structure without alteration of external form or chemical constitution.
**paramorphosis** (*par-am-or-fo'-sis*) [*para-; μορφή*, form]. Same as *paramorphism.*
**paramucin** (*par-ah-mū'-sin*). A colloid isolated from ovarian cysts; it differs from mucin and pseudomucin by reducing Fehling's solution before boiling with acid.
**paramusia** (*par-ah-mū'-ze-ah*) [*para-; μουσική*, music]. A form of aphasia in which there is per-

version of the musical sense, resulting in the production of improper notes and intervals.
**paramyoclonus multiplex** (*par-ah-mi-ok'-lo-nus mul'-tip-leks*). A neurosis marked by sudden, shocklike muscular contractions, which are bilateral and do not, as a rule, affect the hands or face. The etiology is unknown, and the disease is believed to be analogous to chronic adult chorea.
**paramyosinogen** (*par-ah-mi-o-sin'-o-jen*). One of the proteins of muscle-plasma, coagulating at 47° C.
**paramyotonia** (*par-ah-mi-o-to'-ne-ah*). [*para*-; μῦς, muscle; τόνος, tone]. A perversion of muscular tonicity characterized by tonic spasms. It is usually congenital. p. **congenita**, congenital paramyotonia. See *Thomsen's disease*.
**paranephrin** (*par-ah-nef'-rin*). A preparation obtained from the suprarenal gland. See also *adrenalin chloride*.
**paranephritis** (*par-ah-nef-ri'-tis*). 1. Inflammation of the paranephros. 2. Inflammation of the connective tissue about the kidney.
**paranephros** (*par-ah-nef'-ros*) [*para*-; νεφρός, kidney]. The suprarenal capsule.
**paranesthesia** (*par-an-es-the'-ze-ah*). See *paranesthesia*.
**paraneural** (*par-ah-nū'-ral*) [*para*-; νεῦρον, nerve]. Beside or near a nerve.
**paraneurismus** (*par-an-ū-riz'-mus*) [*para*-; νεῦρον, nerve]. A nervous disorder, or perversion of nervefunction.
**parangi** (*par-an'-je*). See *frambesia*.
**paranœa** (*par-an-e'-ah*). See *paranoia*.
**paranoia** (*par-ah-noi'-ah*) [*para*-; νοῦς, mind]. Mental aberration, especially a chronic disease characterized by systematized delusions.
**paranoiac** (*par-ah-noi'-ak*) [*paranoia*]. 1. Affected with paranoia. 2. A person who is affected with paranoia; a "crank."
**paranoid** (*par'-an-oid*). Resembling paranoia.
**paranomia** (*par-ah-no'-me-ah*) [*para*-; ὄνομα, a name]. See *aphasia*, *optic*, and q., *tactile*.
**paranuclear** (*par-an-ū'-kle-ar*). Pertaining to the paranucleus.
**paranucleate** (*par-an-ū'-kle-āt*). Provided with a paranucleus.
**paranuclein** (*par-ah-nū'-kle-in*). A combination of albumin with metaphosphoric acid, split off from the nucleoalbumins by action of pepsin hydrochloric acid.
**paranucleolus** (*par-ah-nū-kle'-o-lus*) [*para*-; *nucleolus*]. An irregular body sometimes found inside the nucleus of a cell prior to the division of the latter.
**paranucleon** (*par-ah-nū'-kle-on*). Phosphocarnic acid, a complex body supposed to constitute the source of muscle energy. It gives rise to lactic acid and $CO_2$ on hydrolysis.
**paranucleoprotein** (*par-ah-nū-kle-o-pro'-te-in*). A synonym of *nucleoalbumin*.
**paranucleus** (*par-ah-nū'-kle-us*) [*para*-; *nucleus*, kernel]. An irregular spherical body lying in the protoplasm of a cell near the nucleus and perhaps extruded by the latter.
**paraoxyethylacetanilide** (*par-ah-oks-e-eth-il-as-et-an'-il-id*). Acetphenetidin.
**parapancreatic** (*par-ah-pan-kre-at'-ik*). Situated beside or near the pancreas. p. **abscess**, an abscess in the tissue alongside of the pancreas.
**paraparesis** (*par-ah-par'-es-is*, or *par-ah-par-e'-sis*). Partial paralysis of the lower extremities.
**paraparetic** (*par-ap-ar-et'-ik*) [*para*-; *paresis*]. Pertaining to, or affected with, paraparesis.
**parapathia** (*par-ap-ath'-e-ah*) [*para*-; πάθος, affection]. Moral insanity.
**parapedesis** (*par-ah-ped-e'-sis*) [*para*-; πέδησις, a bending]. Passage of any secretion or excretion through other than the normal channel.
**parapeptone** (*par-ah-pep'-tōn*). See *peptone*.
**paraperitoneal** (*par-ah-per-it-o-ne'-al*). Situated near the peritoneum.
**paraphasia** (*par-ah-fa'-ze-ah*) [*para*-; φάσις, speech]. A form of aphasia in which there is inability to connect ideas with the proper words to express the ideas.
**paraphenetolcarbamide** (*par-ah-fe-net-ol-kar'-bam-id*). Sucrol.
**paraphenylendiamine** (*par-ah-fen-il-en-di'-am-in*), $C_6H_8N_2$. A crystalline substance obtained by the nitration of acetanilide and reduction with tin and hydrochloric acid. It is used in the manufacture of certain hair-dyes, and gives rise to eczema of the scalp and eyelids, or poisoning marked by vomiting, diarrhea, etc.
**paraphia** (*par-af'-e-ah*) [*para*-; ἀφή, touch]. Abnormality of the sense of touch.
**paraphimosis** (*par-ah-fi-mo'-sis*). Retraction and constriction of the prepuce behind the glans penis.
**paraphonia** (*par-ah-fo'-ne-ah*) [*para*-; φωνή, voice]. Any abnormal condition of the voice. p. **clangens**, shrillness of the voice. p. **puberum**, p. **pubescentium**, the harsh, deep, irregular voice noticed in boys at puberty.
**paraphora** (*par-af'-o-rah*) [παραφορά, wandering]. 1. Slight mental derangement or distraction. 2. Unsteadiness due to intoxication.
**paraphrasia** (*par-ah-fra'-ze-ah*) [*para*-; φράσις, utterance]. A form of aphasia characterized by incoherence of speech. p. **praceps**, precipitant utterance of incoherent speech. p. **tarda**, abnormal delay in the expression of thoughts. p. **verbalis**, the interpolation of an inappropriate word. p. **vesana**, jumbling of words and ideas.
**paraphrenitis** (*par-af-ren-e'-sis*) [*para*-; φρήν, mind]. Amentia; delirium, or insanity.
**paraphrenia, paraphrenitis** (*par-ah-fre'-ne-ah*, *par-ah-fren-i'-tis*) [*para*-; φρήν, mind; diaphragm]. 1. Delirium; a mental disease. 2. Inflammation of the diaphragm.
**paraphrenitis** (*par-ah-fre-ni'-tis*) [*para*-; φρήν, diaphragm; ιτις, inflammation]. Inflammation of the tissues adjacent to the diaphragm.
**paraphronesis** (*par-af-ro-ne'-sis*) [*para*-; φρήν, mind]. Insanity.
**paraphysis** (*par-af'-is-is*) [*para*-; φύειν, to produce: *pl.*, *paraphyses*]. 1. In biology, sterile filaments among reproductive bodies of various kinds in certain cryptogams. 2. A mesal outgrowth from the roof of the brain cephalad of the epiphysis or conarium.
**paraplasm** (*par'-ah-plazm*) [*para*-; πλάσμα, a thing formed]. 1. The fluid substance in the meshes of the cell-protoplasm. 2. A heteroplasm or false growth.
**paraplast** (*par'-ah-plast*). A proprietary plaster mass.
**paraplastic** (*par-ah-plas'-tik*) [*paraplasm*]. 1. Of the nature of paraplasm. 2. Having morbid formative powers. p. **formations**, the contractile substance of the muscular fibrils, the nervous fibers, and the red blood-corpuscles.
**paraplectic** (*par-ah-lek'-tik*) [*para*-; πληγή, a stroke]. Stricken with paraplegia.
**paraplegia** (*par-ah-ple'-je-ah*) [*para*-; πληγή, stroke]. Paralysis of the lower half of the body or of the lower extremities. p., **ataxic**, a disease characterized clinically by a combination of ataxia and exaggerated tendon-reflexes, and anatomically by sclerosis of the posterior and lateral columns of the cord. p. **diabetica**, a peripheral paralysis of the extensor muscles of the lower limbs in diabetic subjects. It differs from tabes in absence of disturbance of coordination and sensibility. p. **dolorosa**, painful pressure-paraplegia due to neoplasms in the spinal cord. p., **ideal**, reflex paraplegia due to emotion. p., **infantile spasmodic**, a spastic paralysis coming on in early childhood, and usually dependent on a cerebral lesion with failure of proper development or secondary sclerosis of the motor tracts of the spinal cord. The causes are injuries during birth, intrauterine cerebral inflammation, or anomalies of brain-development. p., **birth-palsy**; **spasmodic tabes dorsalis**; **spastic cerebral paraplegia**. p. **simplex senilis**, that dependent upon disturbed nutrition of the cortex, causing, in advanced age, paralysis of the legs without muscular atrophy. p., **spastic**, lateral sclerosis. p., **tetanoid**, lateral sclerosis.
**paraplegic** (*par-ah-ple'-jik*) [*paraplegia*]. Pertaining to, or affected with, paraplegia.
**paraplegy** (*par'-ah-ple'-je-form*). Resembling paraplegia.
**parapleuritis** (*par-ah-ploo-ri'-tis*). 1. Pleurodynia. 2. A slight degree of pleuritis. 3. Inflammation of the wall of the chest.
**paraplexus** (*par-ap-leks'-us*) [*para*-; *plexus*, a braid]. The choroid plexus of the paracele or lateral ventricle of the brain.
**parapneumonia** (*par-ah-nū-mo'-ne-ah*). A disease presenting the symptom of lobar pneumonia, but not due to the pneumococcus.
**parapophysis** (*par-ap-off'-is-is*) [*para*-; ἀπόφυσις, offshoot]. In comparative anatomy, the process homologous to the lower process of a vertebra.

**parapoplexy** (*par-ap'-o-pleks-e*) [*para-*; *apoplexy*]. A masked or slight form of apoplexy.
**paraproctitis** (*par-ah-prok-ti'-tis*). Inflammation of the connective tissue about the rectum.
**paraproctium** (*par-ap-rok'-te-um*) [*para-*; πρωκτόs, anus]. The connective tissue that surrounds the rectum.
**parapsis** (*par-ap'-sis*) [*para-*; ἅψιs, a touching]. Perversion of the sense of touch.
**parapyknomorphous** (*par-ah-pik-no-mor'-fus*) [*para-*; πυκνόs, thick; μορφή, form]. A term applied to nerve-cells in which the arrangement of the stainable portion of cell-body is intermediate between that of pyknomorphous and apyknomorphous cells.
**paraqueduct** (*par-ak'-we-duct*) [*para-*; *aqueduct*]. The lateral portion of the aquæductus cerebri.
**pararectal** (*par-ar-ek'-tal*) [*para-*; *rectum*]. Beside or near the rectum. p. pouch, a peritoneal depression behind the broad ligament and beside the rectum.
**parareducine** (*par-ah-re-dū'-sēn*) [*para-*; *reducere*; to lead back]. A leukomaine found in conjunction with reducine in the urine.
**pararhotacism** (*par-ah-ro'-tas-izm*). See *rhotacism*.
**pararhythmus** (*par-ar-ith'-mus*) [*para-*; *rhythm*]. Disturbed rhythm.
**pararthrema, pararthresis** (*par-ar-thre'-mah, par-ar-thre'-sis*) [*para-*; ἄρθρον, joint]. Subluxation.
**pararthria** (*par-ar'-thre-ah*) [*para-*; ἄρθρον, articulation]. A disorder of articulate speech.
**parasacral** (*par-ah-sa'-kral*). Beside or near the sacrum.
**parasalpingitis** (*par-ah-sal-pin-ji'-tis*). Inflammation of the tissues around an oviduct.
**parasecretion** (*par-ah-se-kre'-shun*) [*para-*; *secernere*, to secrete]. Any abnormality of secretion; any substance abnormally secreted.
**parasigmatism** (*par-ah-sig'-mat-izm*) [*para-*; σίγμα, the Greek letter 's*]. The inability to pronounce "*s*" or "*sh*," another letter, as "*f*," being substituted.
**parasinoidal** (*par-ah-sin-oi'-dal*) [*para-*; *sinus*]. Lying near or along a cerebral sinus. p. spaces, the expansion of the cerebral veins just before emptying into the superior longitudinal sinus.
**parasite** (*par'-ah-sīt*) [*para-*; σῖτοs, food]. 1. An animal or vegetable living upon or within another organism, termed the host. 2. In teratology, a fetus or fetal parts attached to or included in another fetus; an autosite. p., autochthonous, a parasite which is descended from the tissues of the host. p., endophytic, one living within the tissues of its host. p., epiphytic, one living on the surface. p., facultative, one usually parasitic, but able to live alone. p., obligate, one that dies without its host.
**parasitic** (*par-ah-sit'-ik*) [*parasite*]. 1. Of the nature of a parasite; living upon or in an animal or vegetable, as *parasitic* bacteria, *parasitic* worms. 2. Caused by parasites, as *parasitic* skin diseases.
**parasiticide** (*par-ah-sit'-is-īd*) [*parasite*; *cædere*, to kill]. 1. Destructive to parasites. 2. An agent capable of destroying parasites, especially one destroying the parasites living upon or in the skin.
**parasitifer** (*par-ah-sit'-if-er* [*parasite*; φέρειν, to bear]. The host of a parasite.
**parasitism** (*par'-ah-si-tizm*) [*parasite*]. The relation that a parasite bears to its host; infestion by parasites.
**parasitize** (*par'-as-it-īz*) [*para-*; σῖτοs, food]. To infest; the act of one organism becoming parasitic within or upon another.
**parasitogenesis** (*par-as-it-o-jen'-es-is*) [*para-*; σῖτοs, food; γένεσιs, genesis]. 1. The formation of parasites. 2. A bodily condition favoring the development of parasites.
**parasitogenetic** (*par-ah-si-to-jen-et'-ik*) [*parasite*; γεννᾶν, to beget]. Produced by parasites; depending for its origin upon parasites.
**parasitologist** (*par-ah-si-tol'-o-jist*). One versed in parasitology.
**parasitology** (*par-ah-si-tol'-o-je*) [*parasite*; λόγοs, science]. The study of parasites.
**parasitosis** (*par-as-it-o'-sis*) [*para-*; σῖτοs, food; νόσοs, disease]. Any disease dependent upon the presence of parasites. The development of a parasitic disease.
**parasitotrope, parasitotropic** (*par-ah-si'-to-trōp, par-ah-si-to-trop'-ik*). A substance in the blood with a special affinity for parasites.

**parasitotropic** (*par-ah-si-to-trop'-ik*) [*parasite*; τρόποs, a turn]. Pertaining to a substance which is attracted by a (micro-) parasite.
**parasoma** (*par-ah-so'-mah*) [*para-*; σῶμα, body]. An irregular body found in cell-protoplasm and situated near the nucleus.
**paraspadia** (*par-ah-spa'-de-ah*) [*para-*; σπᾶειν, to draw]. A condition in which the urethra opens on one side of the penis.
**paraspasm** (*par'-ah-spasm*). 1. Spasm involving both lower extremities. 2. Spastic paraplegia.
**parastata** (*par-as'-tat-ah*) [*para-*; ἱστᾶναι, to stand]. 1. The epididymis. 2. The prostate gland.
**parastatadenitis** (*par-as-tat-ad-en-i'-tis*). 1. Epididymitis. 2. Prostatitis.
**parastatitis** (*par-as-tat-i'-tis*). 1. Epididymitis. 2. Prostatitis.
**parasteatosis** (*par-as-te-at-o'-sis*) [*para-*; στέαρ, a hard fat]. An altered condition of the sebaceous secretion.
**parasternal** (*par-ah-ster'-nal*) [*para-*; *sternum*]. Beside or near the sternum. p. line, an imaginary vertical line midway between the margin of the sternum and the line passing through the nipple. p. region, the region between the sternal margin and the parasternal line.
**parastramnia, parastremma** (*par-as-tram'-ne-ah, par-as-trem'-ah*) [παραστρέφειν, to twist]. Distortion of the mouth or face.
**parasynapsis** (*par-ah-sin-ap'-sis*) [*para-*; συναπτειν, to unite]. The union of chromosomes side by side. Cf. *telosynapsis*.
**parasynovitis** (*par-ah-sin-o-vi'-tis*) Inflammation of the structures about a joint.
**parasyphilis, parasyphilosis** (*par-ah-sif'-il-is, -sif-il-o'-sis*). A series of morbid manifestations not having the anatomicopathological characteristics of syphilis, but apparently of syphilitic origin,; *e. g.* tabes, general paralysis, etc.
**parasyphilitic** (*par-as-if-il-it'-ik*) [*para-*; *syphilis*]. Not unlike syphilis, or in some way resembling syphilis.
**parasystole** (*par-as-is'-to-le*) [*para-*; *systole*]. 1. The interval between the cardiac systole and the diastole. 2. Such an interval when it is abnormally prolonged.
**parateresiomania** (*par-at-er-es-e-o-ma'-ne-ah*) [παρατήρησιs, observation; μανία, madness]. A mania for observing, or seeing new sights.
**parathelioma** (*par-ah-the-le-o'-mah*) [*para-*; θηλή, nipple; ὄμα, a tumor]. A tumor located near the nipple.
**parathenar** (*par-ah-e'-nar*) [*para-*; θέναρ, the sole of the foot]. Applied to the abductor and flexor brevis muscles of the little toe.
**parathymia** (*par-ath-i'-me-ah*) [*para-*; θυμόs, mind]. Mental strain, or overwork.
**parathyrin** (*par-ah-thi'-rēn*) [*para-*; *thyroid*]. The active principle of the parathyroid glands.
**parathyroid** (*par-ah-thi'-roid*). 1. Lying beside the thyroid gland. 2. An accessory thyroid gland.
**parathyroidectomy** (*par-ah-thi-roid-ek'-to-me*) [*para-*; *thyroid*; ἐκτομή, an excision]. Excision of a parathyroid gland.
**parathyroidin** (*par-ah-thi-roi'-din*). Trade name of a preparation made from the parathyroid glands.
**parathyroprivic** (*par-ah-thi-ro-priv'-ik*) [*parathyroid*; *privus*, deprived of]. Pertaining to the condition due to loss of function of or removal of the parathyroid glands.
**paratoloid** (*par-at'-o-loid*). A name given to the fluid used by Koch in the treatment of tuberculosis; tuberculin.
**paratonia** (*par-ah-to'-ne-ah*) [*para-*; τόνοs, tension]. Overextension; excessive tension.
**paratopia** (*par-at-o'-pe-ah*) [*para-*; τόποs, place]. Displacement.
**paratoxin** (*par-ah-toks'-in*). A bile preparation containing cholesterin, but without bile-pigment; it has been used in tuberculosis.
**paratrichosis** (*par-ah-trik-o'-sis*) [*para-*; θρίξ, hair]. A condition in which the hair is either imperfect in growth or develops in abnormal places.
**paratrimma** (*par-ah-trim'-ah*) [*para-*; τρίβειν, to rub]. Intertrigo.
**paratripsis** (*par-at-rip'-sis*) [*para-*; τρίβειν, to rub]. 1. A rubbing. 2. An increase in waste.
**paratriptic** (*par-at-rip'-tik*) [*para-*; τρίβειν, to rub]. Rubbing together; increasing waste.
**paratrope** (*par-at'-ro-pe*) [*para-*; τρέπειν, to turn]. Twisting of a limb.

**paratrophy** (*par-at'-ro-fe*) [*para-*; τροφή, nutrition]. 1. Perverted or abnormal nutrition; hypertrophy. 2. Adiposis dolorosa.

**paratuberculosis** (*par-ah-tū-ber-kū-lo'-sis*). A disease with symptoms similar to tuberculosis, but in which the tubercle bacillus cannot be found.

**paratyphlitis** (*par-ah-tif-li'-tis*). Inflammation of the connective tissue behind the cecum.

**paratyphoid** (*par-ah-ti'-foid*). An affection produced by the paracolon bacillus, presenting all the characteristic symptoms of typhoid, but in which the Widal reaction is negative; the serums, however, react promptly to other bacteria of the colon-typhoid group, which may be isolated from the blood or from the excrement.

**paratyphus** (*par-ah-ti'-fus*). Synonym of *paratyphoid*.

**paratypical, paratypicus** (*par-ah-tip'-ik-al, -us*). Irregular; not typical in character.

**paraumbilical** (*par-ah-um-bil-ik-al*) [*para-*; *umbilicus*]. Near the navel.

**paraurethral** (*par-ah-ū-re'-thral*). Beside the urethra.

**paravaginitis** (*par-av-aj-in-i'-tis*) [*para-*; *vagina*; ιτις, inflammation]. Inflammation of the connective tissue surrounding the vagina.

**paravertebral** (*par-av-er'-te-bral*) [*para-*; *vertebra*]. Situated near the spinal column.

**paravesical** (*par-av-es'-ik-al*) [*para-*; *vesica*, bladder]. Situated near the urinary bladder. **p. pouch,** the peritoneal pocket on either side of the bladder.

**paraxanthin** (*par-ah-zan'-thin*) [*para-*; *xanthin*], C₇H₈N₄O₂. Dimethylxanthin, a crystalline leukomaine occurring in normal urine and isomeric with theobromine, which it resembles in its action upon the organism, producing muscular rigidity, dyspnea, and diminution in reflex excitability.

**paraxial** (*par-aks'-e-al*) [*para-*; *axis*]. Lying near the axis of the body.

**paraxon** (*par-aks'-on*) [*para-*; *axon*]. A lateral branch of the axis-cylinder process of a nerve-cell; a collateral fiber.

**parazoon** (*par-ah-zo'-on*) [*para-*; ζῷον, an animal]. An organism parasitic upon an animal; an ectoparasite.

**parazygosis** (*par-az-i-go'-sis*) [*para-*; ζυγεῖν, to yoke]. The condition of a double monster in which there is union of the trunks above the umbilicus. It includes xiphopagus, thoracopagus, and pleuropagus.

**parchment-crackling.** The peculiar sound elicited by pressure on the cranial bones in children the subjects of rickets and congenital syphilis. It is due to a localized hypertrophy of the bones.

**parchment-induration.** A form of chancre, or primary lesion of syphilis, in which the induration is parchment-like in feel.

**parchment-skin.** See *xeroderma*.

**parecceloma** (*par-ek-se-lo'-mah*) [*para-*; ἐκ, out; κοῖλος, hollow]. A cavity produced by disease.

**pareccrisis** (*par-ek'-ris-is*) [*para-*; ἐκ, out; κρίνειν, to separate]. A disorder of a secretion.

**parecious, parœcious** (*par-e'-she-us*). [*para-*; *οἶκος*, house]. In biology, having male and female organs developed side by side.

**parectama** (*par-ek'-tam-ah*). Synonym of *parectasis*.

**parectasis** (*par-ek'-ta-sis*) [*para-*; ἔκτασις, a stretching out]. Excessive stretching or dilatation.

**paregoric** (*par-e-gor'-ik*) [παρηγορικός, soothing], 1. Soothing or assuaging. 2. A soothing remedy, as *paregoric elixir*, or paregoric, the *tinctura opii camphorata* (U. S. P.).

**pareira** (*par-a'-rah*). The root of *Chondrodendron tomentosum*, of the natural order *Menispermaceæ*. It was formerly called *pareira brava*. It contains a resin, an alkaloid, *pelosine*, identical with berberine, a bitter principle, a nitrogenous substance, calcium malate, potassium nitrate, and other salts. p., decoction of (*decoctum pareiræ*, B. P.), Dose 1–2 oz. (32–64 Cc.). p., extra t of (*extractum pareiræ*, B. P.). Dose 10–20 gr. (0.65–1.3 Gm.). p., fluidextract of (*fluidextractum pareiræ*, U. S. P.). Dose ½–1 dr (2–4 Cc.). p., liquid extract of (*extractum pareiræ liquidum*, B. P.). Dose 1 dr. (4 Cc.).

**parelectronomic** (*par-e-lek-tro-nom'-ik*) [*para-*; ἤλεκτρον, amber; νόμος, law]. Unresponsive to electromotive stimulus.

**parelectronomy** (*par-e-lek-tron'-o-me*) [*para-*; ἤλεκτρον, amber; νόμος, law]. The electric condition of a

transverse section of a muscle and its tendon, compared with that of the natural surface of the muscle. The former is negative, the latter positive.

**paremptosis** (*par-emp-to'-sis*) [*para-*; ἐμπίπτειν, to sink in]. 1. Dislocation. 2. A form of amaurosis.

**parencephalia** (*par-en-sef-a'-le-ah*) [*para-*; ἐγκέφαλος, brain]. Congenital malformation of the brain.

**parencephalis** (*par-en-sef'-al-is*). See *parencephalon*.

**parencephalitis** (*par-en-sef-al-i'-tis*). Inflammation of the cerebellum.

**parencephalocele** (*par-en-sef-al'-o-sēl*) [*para-*; ἐγκέφαλος, brain; κήλη, hernia]. Hernia of the parencephalon.

**parencephalon** (*par-en-sef'-al-on*) [*para-*; ἐγκέφαλος, brain]. The cerebellum.

**parencephalus** (*par-en-sef'-al-us*) [see *parencephalon*]. One with a congenital malformation of the brain.

**parenchyma** (*par-eng'-kim-ah*) [*para-*; ἐγχεῖν, to pour in]. The essential or specialized part of an organ as distinguished from the supporting connective tissue.

**parenchymal** (*par-eng'-kim-al*). Pertaining to, or of the nature of, parenchyma.

**parenchymatic** (*par-eng-kim-at'-ik*). Parenchymatous.

**parenchymatitis** (*par-eng-kim-at-i'-tis*) [*parenchyma*; ιτις, inflammation]. Inflammation of parenchyma.

**parenchymatous** (*par-eng-ki'-mat-us*) or *par-eng-kim'-at-us*) [*parenchyma*]. Pertaining to or affecting the parenchyma. **p. degeneration,** cloudy swelling. **p. inflammation,** inflammation of the parenchyma as distinguished from that of the interstitial tissue.

**parenchymula** (*par-eng-kim'-ū-lah*) [dim. of *parenchyma*; *pl.*, *parenchymulæ*]. The embryonic stage immediately succeeding that of the closed blastula. Synonym of *Metschnikoff's larva*.

**parenteral** (*par-en'-ter-al*) [*para-*; ἔντερον, intestine]. Outside of the intestine. **p. digestion,** digestion or dissolving of foreign-proteins or other substances by the cells of the body, in opposition to *enteral digestion* which occurs in the alimentary canal.

**parepicele** (*par-ep'-is-ēl*) [*para-*; ἐπί, upon; κοῖλος, hollow]. The lateral recess of the epicele or fourth ventricle extending latero-ventral.

**parepididymal** (*par-ep-id-id'-im-al*). Pertaining to the parepididymis.

**parepididymis** (*par-ep-id-id'-im-is*). See *paradidymis*.

**parepithymia** (*par-ep-ith-i'-me-ah*) [*para-*; ἐπιθυμία, desire]. A morbid or depraved desire or habit.

**parerethisis** (*par-er-eth'-is-is*) [*para-*; ἐριθίζειν, to excite]. Abnormal excitement or stimulus.

**parerethism** (*par-er'-eth-ism*). See *parerethisis*.

**paresis** (*par'-es-is*) [*para-*; ἱέναι, to let go]. A slight paralysis; incomplete loss of muscular power. **p., general.** See *paralysis, general, of the insane*.

**paresoanalgesia** (*par-es-o-an-al-je'-se-ah*) [*paresis*; *analgesia*]. Paresis with analgesia; a symptom of Morvan's disease.

**paresthesia** (*par-es-the'-ze-ah*) [*para-*; αἴσθησις, sensation]. 1. Morbid or perverted sensation, as numbness, formication, "pins-and-needles." 2. See *acroparesthesia*.

**paresthetic** (*par-es-thet'-ik*) [*paresthesia*]. Pertaining to, affected with, or characterized by paresthesia.

**paretic** (*par-et'-ik*) [*paresis*]. Pertaining to or affected with paresis. **p. dement,** a person suffering from paretic dementia. **p. dementia.** See *paralysis, general, of the insane*.

**pareunia** (*par-oo'-ne-ah*) [*para-*; εὐνή, a bed]. Coitus.

**parfocal** (*par-fo'-kal*). A term used to designate microscopic oculars and objectives which are so constructed or so mounted that in changing from one to another the image will remain in focus.

**parhidrosis** (*par-hid-ro'-sis*). Same as *paridrosis*.

**parhormone** (*par-hor'-mōn*). [*para-*; *hormone*]. Waste matter of cells, tissues or organs which is supposed to have an action similar to that of a hormone.

**paricine** (*par'-is-ēn*) [*par*, equal; *cinchona*], C₁₆H₁₈N₂O. An amorphous alkaloid of the cinchonas.

**paridrosis** (*par-id-ro'-sis*) [*para-*; ἵδρως, sweat]. Any abnormal condition of the secretion of sweat.

**paries** (*par'-e-ēs*) [*paries*, a wall: *pl.*, *parietes*]. An enveloping or investing structure or wall. **p. anterior,** anterior wall. **p. carotica tympani, carotid** or anterior wall of the tympanic cavity. **p. jugularis**

# PARIETAL 657 PAROPION

**tympani,** the jugular wall or floor of the tympanic cavity. **p. inferior,** inferior wall. **p. labyrinthica tympani,** labyrinthic or inner wall of the tympanic cavity. **p. lateralis,** lateral wall. **p. mastoidea tympani,** the mastoid or posterior wall of the tympanic cavity. **p. medialis,** the medial wall. **p. membranacea tympani,** the membranous or outer wall of the tympanic cavity. **p. posterior,** posterior wall. **p. superior,** superior wall. **p. tegmentalis tympani,** the tegmental wall of the tympanic cavity.
**parietal** (*par-i'-et-al*) [*paries,* wall]. 1. Forming or situated on a wall, as the *parietal* layer of the peritoneum. 2. Pertaining to or in relation with the parietal bone of the skull, as the *parietal* foramen, *parietal* lobe of the brain. **p. angle.** See under *Broca, Lissauer,* and *Quatrefages.* **p. angle,** posterior, in craniometry, that included between two lines tangent to the parietal eminence and the most prominent points of the zygomatic arch. **p. bones.** See *bones, table of.* **p. cells,** cells found in the periphery of the peptic glands of the stomach, immediately beneath the basement-membrane. Their function is supposed to be the secretion of hydrochloric acid. **p. lobe,** the cerebral lobe above the horizontal Sylvian fissure. **p. section,** a transverse vertical section through the ascending parietal convolution.
**parietale** (*par-i-et-a'-le*) [*parietalis,* belonging to walls]. One of the parietal bones.
**parietalia** (*par-i-et-a'-le-ah*) [see *parietal*]. The bones that collectively form the vault of the cranium.
**parieten** (*par-i'-et-en*) [*paries,* wall]. Belonging to the parietal bone in itself.
**parietes** (*par-i'-et-ez*) [pl. of *paries,* a wall]. The walls of a cavity.
**parieto-** (*par-i-et-o-*) [*parietal*]. A prefix meaning relating to the parietal bone.
**parietofrontal** (*par-i'-et-o-frun'-tal*) [*paries,* a wall; *frons,* front]. Of, pertaining to, or representing both the parietal and frontal bones; frontoparietal.
**parietomastoid** (*par-i-et-o-mas'-toid*) [*paries,* a wall; *mastoid*]. Pertaining to the parietal bone and the mastoid process of the temporal bone; mastoparietal.
**parieto-occipital** (*par-i-et-o-ok-sip'-it-al*) [*parieto-; occipital*]. Pertaining to the parietal and occipital bones or lobes.
**parietosphenoid** (*par-i-et-o-sfe'-noid*) [*parieto-; sphenoid*]. Pertaining to the parietal and sphenoid bones.
**parietosplanchnic** (*par-i-etvo-splangk'-nik*) [*paries,* a wall; σπλάγχνα, viscera]. Of or pertaining to the walls of the alimentary canal, as the nervous ganglia of certain molluscs.
**parietosquamosal** (*par-i-et-o-skwa-mo'-sal*) [*parieto-; squamosal*]. Of or pertaining to the parietal bone and the squamous portion of the temporal bone. **p. suture,** a suture between the squamous portion of the temporal bone and the parietal bone.
**parietotemporal** (*par-i-et-o-tem'-po-ral*) [*parieto-; temporal*]. Pertaining to the parietal and temporal bones. **p. suture,** the suture between the parietal and temporal bones.
**parietovisceral** (*par-i-et-o-vis'-er-al*) [*parieto-; visceral*]. Pertaining to the walls of a body-cavity and the contained viscera.
**parigenin** (*par-ij'-en-in*). See *parillin.*
**pariglin** (*par'-ig-lin*). See *smilacin* (*q.*).
**parillin** (*par-il'-in*) [*parilla,* dim. of *parra,* a trained vine]. A glucoside obtained from sarsaparilla; if treated by dilute mineral acids it yields *parigenin* and sugar.
**Parinaud's conjunctivitis** (*par-en-o'*) [Henri Parinaud, French ophthalmologist, 1844–1905]. A severe form of mucopurulent conjunctivitis due to infection from animals. **P.'s ophthalmoplegia,** paralysis of the external rectus of one side and spasm of the internal rectus of the other side; it is of peripheral origin.
**Paris' disease.** Acrodynia.
**Paris green** (*par'-is*), Cu(C₂H₃O₂)₂ . 3Cu(AsO₂)₂. Copper acetoarsenite, a poisonous substance used in the arts and for the destruction of the potato-bug.
**paristhmia** (*par-ist'-me-ah*) [*para-;* ἰσθμός, throat]. The tonsils.
**paristhmic** (*par-ist'-mik*) [*para-;* ἰσθμός, throat]. Relating to the tonsils.
**paristhmion** (*par-isth'-me-on*) [*para-;* ἰσθμός, throat]. A tonsil.
**paristhmitis** (*par-ist-mi'-tis*) [*para-;* ἰσθμός, throat; ιτις, inflammation]. Tonsillitis.
**Parish's camphor mixture** (*par'-ish*). Mistura

camphoræ aromaticae (N. F.). **P.'s syrup,** compound syrup of ferrous phosphate.
**parity** (*par'-it-e*) [1. *par,* equal]. Equality. 2. [*parere,* to bring forth]. The condition of being able to bear children.
**Park's aneurysm** [Henry Park, English surgeon, 1745–1831]. Arteriovenous aneurysm, the arterial dilation communicating with two contiguous veins.
**parkesin** (*park'-es-in*). A mixture of linseed-oil and chlorine sulphide in a solution of collodion in nitrobenzol. It is used as a substitute for caoutchouc.
**Parkinson's disease** (*par'-kin-sun*) [James Parkinson, English physician, 1755–1824]. Paralysis agitans. **P.'s facies or mask,** in paralysis agitans the face is expressionless, "wooden"; movements of the lips slow; eyebrows elevated.
**Parnum's test for albumin.** Add to the filtered urine one-sixth of its volume of a concentrated solution of magnesium or sodium sulphate. On acidulating with acetic acid and boiling, the albumin is precipitated.
**paroarium, paroarion** (*par-o-a'-re-um, par-o-a'-re-on*) [*para-;* ᾠάριον, dim. of ᾠόν, egg]. Same as *parovarium.*
**paroccipital** (*par-ok-sip'-it-al*). 1. Beside the occipital region. 2. The mastoid process.
**parodinia** (*par-o-din'-e-ah*). See *parodynia.*
**parodontis** (*par-o-don'-tis*). Synonym of *epulis.*
**parodontitis** (*par-o-don-ti'-tis*) [*para-;* ὀδούς, a tooth; ιτις, inflammation]. Inflammation of the tissues surrounding a tooth.
**parodynia** (*par-o-din'-e-ah*) [*parere,* to bring forth; ὀδύνη, pain]. Difficult parturition, dystocia.
**parogen** (*par'-o-jen*). A preparation used as a basis for ointments and liniments, said to contain liquid paraffin, 40 parts, oleic acid, 40 parts, and 5 per cent. ammoniated alcohol, 20 parts.
**paroleine** (*par-o'-le-in*). Trade name of a preparation of petroleum oil, used as a solvent and vehicle.
**parolivary** (*par-ol'-iv-a-re*) [*para-;* *oliva,* olive]. Situated near the olivary body. **p. body.** See *nucleus, olivary, accessory.*
**parolive** (*par-ol'-iv*). An accessory olive.
**paromphalocele** (*par-om-fal'-o-sēl*) [*para-;* ὀμφαλός, navel; κήλη, tumor]. Hernia in the region of the navel.
**paroniria** (*par-o-ni'-re-ah*) [*para-;* ὄνειρος, dream]. Depraved or morbid dreaming. **p. ambulans,** walking. **p. salax,** a restless condition attended with involuntary seminal emissions and lascivious dreams.
**paronychia** (*par-o-nik'-e-ah*) [*para-;* ὀνυχία]. An inflammation of the flexor tendons and tendinous sheaths of the fingers; whitlow.
**paronychial** (*par-o-nik'-e-al*). Having the character of paronychia.
**paronychosis** (*par-o-nik-o'-sis*) [*para-;* ὄνυξ, nail; νόσος, disease]. A diseased condition of the structures about the nails; also growth of a nail in unusual places.
**paronym** (*par'-o-nim*) [*para-;* ὄνομα, name]. A word that exactly represents a word in another language, differing from it, if at all, only in some slight modification. Thus *nerve* is a paronym of Latin *nervus;* *muscle* of *musculus;* *canal* of *canalis.* A related synonym. See *heteronymous.*
**paronymy** (*par-on'-im-e*). [*para-;* ὄνομα, name]. 1. The relation of a word in one language to its antecedent in another. 2. The principle of using in modern languages paronyms or derivations of Latin or Greek words rather than heteronyms that have no common antecedent.
**parooophoritis** (*par-o-off-or-i'-tis*) [*para-;* *oophoron;* ιτις, inflammation]. 1. Inflammation of the parovarium. 2. Inflammation of the tissues about the ovary.
**paroophoron** (*par-o-of'-o-ron*) [*para-;* *oophoron*]. The persistent tubules of the posterior part of the Wolffian body in the female, corresponding to the organ of Giraldès in the male.
**parophia** (*par-o-fo'-be-ah*) [*παρα-,* before (intensive); φόβος, fear]. Hydrophobia.
**parophthalmia** (*par-of-thal'-me-ah*) [*para-;* ὀφθαλμός, eye]. Inflammation about the eye.
**parophthalmoncus** (*par-of-thal-mong'-kus*) [*para-;* ὀφθαλμός, the eye; ὄγκος, a tumor]. A tumor near the eye.
**paropia** (*par-o'-pe-ah*) [*para-;* ὤψ, eye]. The angle of the eyelid toward the temple.
**paropion** (*par-o'-pe-on*) [*para-;* ὤψ, eye]. An eye-screen.

**paroplexia** (*par-o-pleks'-e-ah*) [*para-*; πλήσσειν, to strike]. Paraplegia.
**paropsis** (*par-op'-sis*) [*para-*; ὄψις, vision]. Disordered or false vision.
**paroptesis** (*par-op-te'-sis*) [*ptara-*; ὄπτησις, a roasting]. A hot-air bath.
**paroptic** (*par-op'-tik*) [*para-*; ὄψις, vision]. Applied to colors produced by the diffraction of light-rays.
**paroral** (*par-o'-ral*) [*para-*; *os, oris*, mouth]. In biology, alongside the mouth or oral aperture.
**parorasis** (*par-o-ra'-sis*) [*para-*; ὁράειν, to see]. Any perversion of vision or of color-perception; an hallucination.
**parorchid** (*par-or'-kid*). Same as *parorchis*.
**parorchidium** (*par-or-kid'-e-um*) [*para-*; ὄρχις, a testicle]. Abnormal position of a testicle or its non-descent.
**parorchidoenterocele** (*par-or-kid-o-en'-ter-o-sēl*) [*para-*; ὄρχις, testicle; ἔντερον, intestine; κήλη, tumor]. Inguinal hernia combined with displacement of the testis.
**parorchis** (*par-or'-kis*) [*para-*; ὄρχις, testicle]. See *epididymis*.
**parorexia** (*par-or-eks'-e-ah*) [παρά, aside;· ὄρεξις, appetite]. A perverted appetite.
**parorganum** (*par-org'-an-um*) [*para-*; ὄργανον, instrument]. A growth the tissue of which resembles that of some organ.
**parosmia** (*par-oz'-me-ah*) [*para-*; ὀσμή, smell]. A perversion of the sense of smell.
**parosphresis** (*par-os-fre'-sis*). Same as *parosmia*.
**parosteitis** (*par-os-te-i'-tis*). Synonym of *parostitis*.
**parosteosis** (*par-os-te-o'-sis*). See *parostosis*.
**parostia** (*par-os'-te-ah*) [*para-*; ὀστέον, bone]. Disorder or defect of ossification.
**parostitis** (*par-os-ti'-tis*) [*para-*; ὀστέον, bone; ιτις, inflammation]. Inflammation of the outer surface of periosteum.
**parostosis** (*par-os-to'-sis*) [*para-*; ὀστέον, bone]. The abnormal formation of bone outside of the periosteum, or in the connective tissue surrounding the periosteum.
**parotic** (*par-o'-tik*) [*para-*; οὖς, ear]. Situated near or about the ear.
**parotid** (*par-ot'-id*) [see *parotic*]. 1. Situated near the ear, as the *parotid* gland. 2. Pertaining to or affecting the parotid gland. **p. abscess**, an abscess of the parotid gland. The term is sometimes also applied to abscess of the lymphatic gland lying upon the parotid. **p. gland**, one of the salivary glands in front of and below the external ear. It is a compound racemose gland and secretes saliva containing ptyalin, a globulin-like body, potassium sulphocyanide, a trace of urea, and mineral salts. Its duct is Stenson's duct.
**parotidectomy** (*par-ot-id-ek'-to-me*) [*parotid*; ἐκτομή, excision]. Excision of the parotid gland.
**parotiditis** (*par-ot-id-i'-tis*). See *parotitis*.
**parotidoauricularis** (*par-ot-id-o-aw-rik-ū-la'-ris*) [*parotid*; *aura, ear*]. A muscle, well-developed in lower animals, arising from the surface of the parotid gland and inserted into the base of the concha. Its function is to abduct and depress the pinna.
**parotidoscirrhus** (*par-ot-id-o-skir'-us*) [*parotid*; σκιρρός, hard]. Scirrhous carcinoma of the parotid gland.
**parotis** (*par-o'-tis*) [L.]. The parotid gland. **p. accessoria**, a small lobule near the parotid gland.
**parotitic** (*par-o-tit'-ik*). Having the mumps; affected with parotitis.
**parotitis** (*par-o-ti'-tis*) [*parotid*; ιτις, inflammation]. Inflammation of the parotid gland, especially the specific infectious disease known as mumps; the name is also given to inflammation of the lymphatic gland overlying the parotid (parotid bubo). **p.**, metastatic, that secondary to disease elsewhere; it occurs in infectious fevers, as typhoid fever, and usually goes on to suppuration.
**parous** (*par'-us*) [*parere*, to bear]. Having borne one or more children.
**parovarian** (*par-o-va'-re-an*) [*para-*; *ovarium*, ovary]. 1. Situated near the ovary. 2. Pertaining to the parovarium.
**parovariotomy** (*par-o-va-re-ot'-o-me*). Excision of a parovarian cyst.
**parovaritis** (*par-o-var-i'-tis*). Inflammation of the parovarium.
**parovarium** (*par-o-va'-re-um*) [*para-*; *ovarium*, ovary]. The remnant of the Wolffian body of the female; the organ of Rosenmüller.

**paroxia** (*par-oks'-e-ah*). See *pica*.
**paroxyntic** (*par-oks-in'-tik*) [παροξύνειν, to excite]. Paroxysmal.
**paroxysm** (*par'-oks-ism*) [*para-*; ὀξύνειν, to sharpen]. 1. The periodic increase or crisis in the progress of a disease; a sudden attack, a sudden reappearance of symptoms, or a sudden increase in the intensity of existing symptoms. 2. A spasm or fit; a convulsion.
**paroxysmal** (*par-oks-iz'-mal*) [*paroxysm*]. Of the nature of or resembling a paroxysm; occurring in paroxysms.
**Parrot's atrophy of the newborn** (*par'-o*) [Joseph Marie Jules *Parrot*, French physician, 1829–1883]. Primary infantile atrophy or marasmus. Syn., *athrepsia*. **P.'s disease**, pseudoparalysis of the extremities due to epiphyseal separation which prevents spontaneous movements, in hereditary syphilis of the newborn. **P.'s nodes**, osteophytes of the frontal and parietal bones, around the anterior fontanel, in hereditary syphilis. **P.'s sign**, dilatation of the pupil when the skin is pinched; it is noted in meningitis. **P.'s ulcers**, the whitish or yellowish patches of thrush.
**parrot-disease**. See *psittacosis*.
**parrot-beak nails**. Nails that are curved strongly anteroposteriorly, like the beak of a parrot.
**Parry's disease** (*par'-e*) [Caleb Hillier *Parry*, English physician, 1755–1822]. Exophthalmic goiter.
**pars** (*pars*) [L.]. A part. **p. basilaris**, basilar process of the occipital bone. **p. calcaneocuboidea**, the internal calcaneocuboid ligament. **p. calcaneonavicularis**, the superior or external calcaneonavicular ligament. **p. cavernosa**, the cavernous or spongy portion of the male urethra. **p. centralis**, the central part or body of the lateral ventricles of the brain. **p. ciliaris retinæ**, the part of the retina in front of the ora serrata. **p. convoluta**, the convoluted part or labyrinth of the kidney. **p. flaccida**, Shrapnell's membrane. **p. horizontalis**, the horizontal plate of the palate bone. **p. intercartilaginea**, the respiratory glottis. **p. intermembranacea**, the true glottis. **p. iridica retinæ**, the uveal tract. **p. laryngea**, the laryngopharynx. **p. mastoidea**, the mastoid portion of the temporal bone. **p. membranacea**, the membranous portion of the male urethra. **p. nasalis**, the nasopharynx. **p. oralis**, the oropharynx. **p. papillaris**, the papillary layer of the skin. **p. perpendicularis**, the vertical plate of the palate bone. **p. petrosa**, the petrous portion of the temporal bone. **p. prostatica**, the prostatic portion of the male urethra. **p. pylorica**, the pyloric portion of the stomach. **p. radiata**, the pyramids of Ferrein. **p. sphincteria inferior**, the lowest portion of the esophagus. **p. spongiosa**. Same as *p. cavernosa*. **p. triangularis**, the preoperculum. **p. tympanica**, the tympanic portion of the temporal bone.
**parsley** (*pars'-le*) [πέτρον, rock; σέλινον, a kind of parsley]. The *Carum petroselinum*, a plant of the order *Umbelliferæ*, containing a volatile oil. From the seed a peculiar oily liquid, termed *apiol* (*q. v.*), is obtained. The root is used in renal diseases and dropsy; the juice of the fresh herbs and the seeds are employed as antiperiodics; apiol is an emmenagogue. **p. camphor**, apiol.
**Parsons' disease** (*par'-suns*) [James *Parsons*, English physician, 1705–1770]. Exophthalmic goiter.
**part** [*pars*, a part]. 1. A segment or section; a member or organ. 2. A portion of a cadaver allotted to a student, for dissection.
**Parthenium** (*par-ihe'-ne-um*) [παρθένος, a virgin]. A genus of herbs of the order *Compositæ*. *P. hysterophorus* contains several alkaloids, one of which, called *parthenine*, seems to be the active principle of the plant and has been used as an antipyretic and antineuralgic. *P. integrifolium*, prairie-dock, a perennial plant of the southern United States, is used as an antiperiodic.
**parthenochlorosis** (*par-then-o-klo-ro'-sis*) [παρθένος, virgin; χλωρός, green]. The chlorosis of young maidens.
**parthenogenesis** (*par-then-o-jen'-es-is*) [παρθένος, a virgin; γένεσις, production]. The development of an organism from an unfertilized ovum.
**particle** (*par'-tik-l*) [dim. of *pars*, part]. A small part. The smallest visible portion of any substance.
**particulate** (*par-tik'-ū-lāt*) [*pars*, part]. Composed of minute particles; applied to various contagia.
**Partridge's hernia**. Femoral hernia external to the femoral vessels.

**partridge-berry.** 1. A trailing plant, *Mitchella repens*, with medical uses like those of pipsissewa. 2. See *gaultheria*.

**parturiency** (*par-tū'-re-en-se*) [*parturire*, to bring forth]. The state of being parturient; parturition.

**parturient** (*par-tū'-re-ent*) [*parturition*]. 1. Being in labor; giving birth; as a *parturient* woman. 2. Traversed during birth, as the *parturient* canal.

**parturifacient** (*par-tū-re-fa'-se-ent*) [*parturition*; *facere*, to make]. 1. Promoting parturition. 2. An agent that induces parturition.

**parturiometer** (*par-tū-re-om'-et-er*) [*parturition*; μέτρον, a measure]. An instrument for determining the progress of labor by measuring the expulsive force of the uterus.

**parturition** (*par-tū-rish'-un*) [*parturitio*, from *parturire*]. The act of giving birth to young. See *labor*.

**partus** (*par'-tus*) [*parturire*, to bring forth]. The bringing forth of offspring; labor. **p. agrippinus,** labor with breech presentation. **p. cæsarius,** cesarean section. **p. difficilis,** dystocia. **p. immaturus,** premature labor. **p. maturus,** labor at term. **p. præcipitatus,** precipitate labor. **p. serotinus,** labor unduly prolonged. **p. siccus,** dry labor.

**parulis** (*par-ū'-lis*) [*para-*; οὖλον, the gum]. Abscess of the gum; a gum-boil.

**parumbilical** (*par-um-bil'-ik-al*) [παρά, beside; *umbilicus*, navel]. Situated or occurring near the umbilicus.

**paruria** (*par-ū'-re-ah*) [*para-*; οὖρον, urine]. Abnormality in the excretion of the urine.

**parurocystis** (*par-ū-ro-sis'-tis*). See *bladder*, *supplementary*.

**parvoline** (*par'-vo-lin*), C₃H₁₁N. A synthetic liquid base; also a ptomaine isomeric with it, occurring in decomposing fish and horse-flesh.

**parvule** (*par'-vūl*) [*parvus*, small]. A small pill or pellet, or granule.

**paschachurda** (*pas-kah-koor'-dah*). See *Sartian disease*.

**Paschutin's degeneration** (*pas-kū-tīn*). A special degeneration peculiar to diabetes; hydrocarbonaceous degeneration.

**pasma** (*paz'-mah*) [πασμα; πάσσειν, to sprinkle; *pl., pasmata*]. 1. A powder for sprinkling on a surface. 2. A powder mixed up into a paste.

**pass** (*pas*) [*passus*, step]. 1. To go, or to put through, or by. 2. To discharge from the intestinal canal. 3. To void. 4. To introduce an instrument into a cavity or channel.

**passage** (*pas'-āj*) [*passare*, to pass]. 1. A channel. 2. The act of passing from one place to another. 3. The introduction of an instrument into a cavity or channel. 4. An evacuation of the bowels. **p. false,** a false channel, especially one made by the unskilful introduction of an instrument into the urethra.

**Passavant's cushion** (*pahs'-af-ant*) [Gustav *Passavant*, German physician, 1815–1893]. The bulging of the posterior pharyngeal wall, produced during the act of swallowing by the upper portion of the superior constrictor pharyngis.

**passiflora** (*pas-if-lo'-rah*) [*passio*, passion; *flos*, a flower]. Passion-flower, a genus of climbing plants. *P. incarnata*, of North America, is used as a narcotic and anodyne. Dose of *fluidextract* 2–5 min. (0.13–0.3 Cc.). *P. quadrangularis*, of the West Indies; the root causes vomiting, convulsions, and paralysis, but has been prescribed as an anthelmintic.

**passion** (*pash'-un*) [*passio*, from *pati*, to suffer]. 1. Pain; suffering; as *ileac passion*, a synonym of volvulus. 2. An intense emotion of the mind; intense sexual excitement.

**passive** (*pas'-iv*) [see *passion*]. Not active; not performed or produced by active efforts, but by causes coming from without. **p. congestion,** congestion due to retention of blood in a part, and not to an active flow of blood toward the part. **p. immunity.** See *immunity*, *passive*. **p. interval,** the period of cardiac rest. **p. motion,** the movement produced by external agency and not by the person himself.

**passivism** (*pas'-iv-izm*). A form of sexual perversion in which there is a subjugation of the will of one person to that of another, with an erotic end.

**passivist** (*pas'-iv-ist*) [*pati*, to suffer]. One who is the subject of passivism, *q. v.*

**passula** (*pas'-ū-lah*) [L.]. A raisin.

**pasta** (*pas'-tah*) [L.: pl., and gen., *pastæ*]. A paste.

**paste** (*pāst*) [πάστη, mess]. Any soft, sticky substance, especially a mixture of starch or flour and water. **p., arsenical,** a caustic paste containing arsenic. **p., Canquoin's.** See *Canquoin's paste*. **p., fruit,** inspissated fruit juice. **p., London,** a mixture of equal parts of sodium hydroxide and slaked lime, moistened with alcohol. **p., phosphorus,** a rat poison made of phosphorus and flour. **p., Piffard's,** copper sulphate, 1 part; tartrated soda, 5 parts; caustic soda, 2 parts. It is used as a test for sugar in urine. **p., serum,** a sterilized mixture of serum from ox-blood with 25 % of zinc oxide; used as a film on abrasions or diseased surfaces. **p., sulphuric-acid,** a caustic mixture of equal parts of sulphuric acid and powdered saffron. **p., Vienna,** a mixture of potassium hydroxide and caustic lime moistened with water.

**paster** (*pās'-ter*). The oval or circular portion of a bifocal lens, which is used for near work.

**pastern** (*pas'-tern*). That part of a horse's foot between the fetlock-joint and the coronet of the hoof. **p.-bone,** either of the two proximal phalanges of a horse's foot. **p.-joint,** the articulation between the proximal phalanx (great pastern-bone) of the horse's foot and the cannon-bone.

**Pasteur's exhaustion theory** (*pahs'-ter*) [Louis *Pasteur*, French bacteriologist, 1822–1895]. See *immunity*, *theory of*, *exhaustion hypothesis*. P.'s **fluid,** P.'s liquid, an artificial liquid for the cultivation of bacteria, composed of water, 100 parts; crystallized sugar, 10 parts; ammonium carbonate and ashes of yeast, each, 1 part.

**Pasteur-Chamberland's filter** (*pahs'-ter-tshām'-ber-lahd*) [Louis *Pasteur*; Charles *Chamberland*, a pupil of Pasteur]. A hollow column of unglazed porcelain through which solutions are filtered by means of a vacuum exhaust or by pressure.

**pasteurella** (*pas-tur-el'-ah*). A group of polymorphic coccobacteria destitute of spores and cilia.

**pasteurellose** (*pas-tur-el'-ōs*). Hemorrhagic septicemia in animals.

**pasteurism** (*pas'-tur-izm*) [Louis *Pasteur*, French chemist and bacteriologist, 1822–1895]. Prophylactic or protective inoculation; a synonym for the word vaccination.

**pasteurization** (*pas-tur-i-za'-shun*). The process of checking fermentation in milk, wine, and other organic fluids by heating them to 60° or 70° C.

**pasteurizer** (*pas'-tū-ri-zer*). An instrument employed in pasteurization.

**pastil, pastille** (*pas'-til, pas-tēl'*) [dim. of *pasta*, paste]. 1. A small mass composed of aromatic substances and employed in fumigation. 2. A troche.

**past pointing.** A diagnostic procedure in diseases of the cerebrum, cerebellum and medulla. If the semi-circular canals of a healthy person are stimulated by injection of cold (68° F.) or warm water (112° F.) or by rotatory movements or galvanic stimulations, vertigo and nystagmus are caused and the muscles of the trunk and extremities are affected, as shown by inability of the patient, seated, with eyes closed, to touch a given point with extended arm and finger. The arm is held upright and then swung down toward the given point which it swerves past: to the right if the left canal is stimulated, and *vice versa*. The absence of these manifestations indicates possible intra-cranial disease.

**patch** [Prov. Ger., *Patschen*]. An irregular spot or area. **p., moth-,** chloasma. **p., mucous,** one of the characteristic lesions of syphilis, occurring in the so-called secondary stage, and appearing as a whitish papule or patch on mucous membranes and at mucocutaneous junctions. Syn., *condyloma latum*; *mucous papule.* **p., opaline.** See *opaline patch*.

**p.s, Peyer's.** See *Peyer's glands*.

**patchouli, patchouly** (*pat-choo'-le*). The labiate herb, *Pogostemon heyneanus*.

**pate.** The crown or top of the head.

**patefying** (*pat'-e-fi-ing*) [*patere*, to stand open]. The act of rendering patent.

**patella** (*pat-el'-ah*) [dim. of *patina*, a shallow dish]. The knee-pan, a small, round, sesamoid bone in front of the knee, developed in the tendon of the quadriceps extensor cruris muscle.

**patellar** (*pat-el'-ar*) [*patella*]. Pertaining to the patella. **p. fossa.** See *fossa*. **p. reflex, p. tendon-reflex.** See *reflex, knee-*.

**patelliform** (*pat-el'-if-orm*) [*patella*; *forma*, form]. Shaped like a patella.

**patellofemoral** (*pa-tel-o-fem'-o-ral*). Pertaining to the patella and the femur.

**patelloid, patelloidean** (*pat-el'-oid, pat-el-oid'-e-an*). Disc-like; shaped like a knee-pan; patelliform.

PATENCY 660 PEARL

**patency** (*pa'-ten-se*) [*patent*]. The state of being open; openness.
**patent** (*pat'-ent*) [*patere*, to be open]. Open; exposed. p. **medicine.** See under *medicine*.
**pathema** (*path-e'-mah*) [πάθημα; πάθος, disease]. Any disease or morbid condition.
**pathemate** (*path'-em-āt*) [πάθημα, a suffering]. Pertaining to emotional excitement.
**pathematology** (*path-em-at-ol'-o-je*). Same as *pathology*.
**pathetic** (*path-et'-ik*). [πάθος, disease]. Arousing pity; indicating sadness or sorrow; appealing; that which appeals to or stirs the passions; applied to the fourth cranial nerve (*pathetic nerve*), which innervates the *pathetic* muscle (*patheticus*, superior oblique) of the eye, by which the eye is rolled outward and downward.
**pathetism** (*path'-et-izm*) [see *pathetic*]. Hypnotism, mesmerism, animal magnetism.
**pathfinder** (*path'-fin-der*). An instrument for finding the openings of a urethral stricture.
**pathic** (*path'-ik*) [παθικός, passive]. 1. Diseased; pathological; pertaining to a morbid condition. 2. Also, one who tolerates the commission of an unnatural crime upon the person.
**patho-** (*pa-tho-*) [πάθος, disease]. A prefix denoting disease.
**pathoamine** (*path-o-am'-in*). A basic substance found in disease; a ptomaine.
**pathoanatomy** (*path-o-an-at'-o-me*) [*patho-*; *anatomy*]. Pathological anatomy.
**pathobiology** (*path-o-bi-ol'-o-je*). Same as *pathology*.
**pathogen** (*path'-o-jen*) [*patho-*; γεννᾶν, to produce]. Any microorganism or substance which produces disease.
**pathogenesis** (*path-o-jen'-es-is*) [*patho-*; γένεσις, generation]. The origin or development of disease.
**pathogenic, pathogenetic** (*path-o-jen'-ik, path-o-jen-et'-ik* [*patho-*; γεννᾶν, to produce]. Producing disease. p. **microorganism,** one that when introduced into the system causes disease.
**pathogenicity.** (*path-o-jen-is'-it-e*). The condition of being pathogenic.
**pathogeny** (*path-oj'-en-e*) [*patho-*; γενής, producing]. See *pathogenesis*.
**pathognomonic** (*path-og-no-mon'-ik*) [*patho-*; γνώμη, a sign]. Characteristic of a disease, distinguishing it from other diseases.
**pathognomy** (*path-og'-no-me*) [*patho-*; γνώμη, a sign]. The s̩ien e of the signs by which disease is recognized.c  c
**pathognostic** (*path-og-nos'-tik*). Synonym of *pathognomonic*.
**pathography** (*path-og'-ra-fe*) [*patho-*; γράφειν, to write]. A description of diseases.
**pathologic, pathological** (*path-o-loj'-ik, al*) [*pathology*]. Pertaining to pathology; pertaining to disease. p. **anatomy.** See *anatomy, morbid.* p. **histology,** the microscopic study of diseased tissues.
**pathologist** (*path-ol'-o-jist*) [*pathology*]. One versed in pathology.
**pathology** (*path-ol'-o-je*) [*patho-*; λόγος, science]. The branch of medical science that treats of the modifications of function and changes in structure caused by disease. p. **cellular,** pathology that makes the cell the basis of all vital phenomena. p. **comparative,** a study of pathological processes in lower animals, for purposes of tracing resemblances and differences among them and between them and those of the human body. p., **experimental,** the study of pathological processes artifically induced in lower animals. p., **general,** that department of pathology which takes cognizance of those morbid processes that may be observed in various diseases and in any organ, *e. g.,* inflammation, hypertrophy. p. **geographical,** pathology in its relation to climatic and geographical conditions. p., **humoral,** the old doctrine that disease is due to abnormal conditions of the blood. It has been revived in recent times in a modified form, and is now based on the theory that both immunity and susceptibility to disease reside in the juices of the body. p., **solidistic.** See *solidism.* p., **medical,** pathology of diseases not amenable to surgical treatment. p., **special,** that treating of changes in function and structure occurring in special diseases, *e. g.* pneumonia. p., **surgical,** the pathology of diseases treated by the surgeon.
**patholysis** (*path-ol'-is-is*) [*patho-*; λύειν, to dissolve]. A morbid dissolution of tissues.
**pathomaine** (*path'-o-mān*). A ptomaine.

**pathomania** (*path-o-ma'-ne-ah*) [*patho-*; μανία, madness]. Moral insanity.
**pathonomia** (*path-o-no'-me-ah*) [*patho-*; νόμος, law]. The study of the laws of pathological conditions.
**pathophobia** (*path-o-fo'-be-ah*) [*patho-*; φόβος, fear]. Exaggerated dread of disease.
**pathopoiesis** (*path-o-poi-e'-sis*) [*patho-*; ποιεῖν, to make]. The causation of disease.
**patient** (*pa'-shent*) [*pati*, to suffer]. A person under the care of a physician; a sick person.
**patten** (*pat'-en*) [Fr., *patin,* a clog]. An iron support placed under a sound foot to remove pressure from and permit extension of the diseased limb in hip-joint disease.
**Patterson's corpuscles.** (*pat'-er-sun*). The molluscum bodies; oval, shiny bodies found in the contents of the tubercles of molluscum contagiosum.
**Patterson's powder** (*pat'-er-sun*). A mixture of bismuth subnitrate and magnesia.
**patulous** (*pat'-ū-lus*) [*patere*, to lie open]. Expanded; open.
**Paul's sign** (*pawl*) [Constantin Charles Théodore *Paul,* French physician, 1833–1896]. Feeble apex-beat with forcible impulse over the body of the heart, in adherent periardium.
**Paullinia** (*paw-lin'-e-ah*). See *guarana*.
**paulocardia** (*paw-lo-kar'-de-ah*) [παῦλα, pause; καρδία, heart]. A subjective sensation of intermission or momentary stoppage of the heart-beat.
**paunch** (*pawnch*). The abdominal cavity and its contents.
**pausimenia** (*paw-sim-e'-ne-ah*). See *menopause*.
**Pauzat's disease** (*po-sah'*) Jean Eugène *Pauzat,* French physician]. Osteoplastic periostitis of the metatarsal bones.
**pavement-epithelium.** Epithelium consisting of flattened, scale-like cells fitted together by their edges like the tiles of a pavement.
**pavilion** (*pa-vil'-yon*) [*papilio,* a butterfly; a tent]. 1. The expanded extremity of a canal or tube, as the *pavilion* of the ear—the auricle; the *pavilion* of the Fallopian tube—the fimbriated extremity of the Fallopian tube. 2. In anatomy, a tent-shaped structure.
**pavimentum** (*pav-im-en'-tum*) [L.]. A floor. p. **orbitæ,** the floor of the orbit. p. **ventriculi,** the floor of a ventricle.
**pavitation** (*pav-it-a'-shun*) [*pavitatio; pavere,* to quake]. Terror, or fear, with trembling.
**pavor** (*pa'-vor*) [L.]. Fright; fear. p. **nocturnus,** night-terrors.
**Pavy's disease** (*pa'-ve*). [Frederick William *Pavy,* English physician, 1829–1911]. Cyclic albuminuria. P.'s **solution for glucose,** make a solution by mixing 120 Cc. of the ordinary Fehling's solution with 300 Cc. of strong ammonia (specific gravity, 0.88) and 400 Cc. of sodium hydroxide solution of specific gravity of 1.14; dilute with 1000 Cc. of water. This solution becomes decolorized on boiling with a glucose solution. One hundred Cc. of this solution is reduced by glucose to the same extent as 10 Cc. of Fehling's solution.
**Pawlik's folds** (*paw'-lik*) [Karl *Pawlik,* Austrian surgeon, 1849–    ]. The anterior columns of the vagina, which form the lateral boundaries of *Pawlik's* triangle and serve as landmarks in locating the opening of the ureters. P.'s **triangle,** extravesical or vaginal triangle. The triangular space formed by two divergent columns of the vagina and the transverse ridge below the external orifice of the neck of the bladder. It corresponds line for line to the trigonum vesicæ.
**pawpaw** (*paw'-paw*). The fruit of *Asina triloba.* See *Carica papaya*.
**Paxton's disease** (*paks'-tun*). Tinea nodosa.
**Pb.** Chemical symbol for *plumbum*, lead.
**P. B.** Abbreviation of *Pharmacopœia Britannica,* British Pharmacopœia.
**Pd.** The chemical symbol of *palladium*.
**P. D.** Abbreviation of *Pharmacopœia Dublinensis,* Dublin Pharmacopœia.
**P. E.** Abbreviation for *Pharmacopœia Edinensis,* Edinburgh Pharmacopœia.
**Péan's method** (*pa-an'*) [Jules *Péan,* French surgeon, 1830–1898]. Removal of a tumor in pieces when it is larger than the opening through which it is to be removed.
**peanut** (*pe'-nut*). An edible fruit of *Arachis hypogæa.* p.-**oil.** See *ground-nut oil.*
**pearl** (*perl*) [Fr., *perle,* pearl; from L., *pirula,* a little pearl]. 1. In pharmacy, a small, hollow glass body containing a dose of a volatile liquid medicine, as a

**pearl** of amyl nitrite. 2. A cataract. 3. A peculiar arrangement of the epithelial cells. **p. ash**, crude potassium carbonate. **p.-disease**, tuberculosis of serous membranes in the lower animals, especially cattle, so called on account of the most manifest lesion, the pearly nodules or tumors, which are often pendulous. **p., epidermic, p., epithelial**, one of the spheroidal concentric masses of epithelial cells often seen in hard papilloma, in squamous epithelioma and in cholesteatoma. Syn., *pearly body*. **p. tumor**. See *cholesteatoma*. **p.-white**, bismuth oxychloride.
**pearly body**. See *pearl, epidermic*.
**Pearson's solution** (*pēr'-sun*) [George Pearson, English physician, 1751–1828]. An aqueous solution of sodium arsenate, containing 1 gram of sodium arsenate in 600 c. c. of distilled water. It is one-tenth the strength of the official liquor sodii arsenatis.
**peat** (*pēt*). The product of the spontaneous decomposition of plants, especially swamp-plants, in many cases mixed with sand, loam, clay, lime, iron pyrites, ocher, etc.
**pebbles** (*peb'-lz*). Lenses for eyeglasses cut from rock crystal.
**pebeco** (*peb'-ek-o*). Trade name of a tooth-paste containing potassium chlorate.
**pebrine** (*peb'-rēn*) [Fr.]. An infectious epidemic disease of silkworms.
**peccant** (*pek'-ant*) [*peccare*, to sin]. Pathogenic; morbid; unhealthy; offensive.
**pechyagra** (*pek-i-a'-grah*) [πῆχυς, forearm; ἄγρα, seizure]. Gout in the elbow-joint.
**peciloblast** (*pes-il'-o-blast*). Ecc *poikilosyto*.
**pecilocyte** (*pes'-il-o-sīt*). Same as peciloblast.
**pecilocythemia** (*pe-sil-o-si-the'-me-ah*). The presence of pecilocytes in the blood.
**pecilocytosis** (*pe-sil-o-si-to'-sis*). See *poikilocytosis*.
**pecilonymy** (*pe-sil-on'-im-e*) [ποικίλος, various; ὄνυμα name]. The use in one publication of different names for the same part. Syn., *poikilonymy*.
**pecilothermal** (*pe-sil-o-ther'-mal*). See *poikilothermic*.
**Pecklin's glands**. See *Peyer's glands*.
**Pecquet, cistern of, P.**, reservoir of (*pek-a'*) [Jean Pecquet, French anatomist, 1622–1674]. The receptaculum chyli. **P.'s duct**, the thoracic duct.
**pectase** (*pek'-tās*) [πηκτός, fixed]. A hypothetical ferment of plants which converts pectose into pectin.
**pecten** (*pek'-ten*) [*pecten*, a comb: pl., *pectines*]. The os pubis. In biology, a comb-like structure or organ.
**pectin** (*pek'-tin*) [πηκτός, congealed]. A white, amorphous carbohydrate contained in ripe fleshy fruits and in certain roots, and believed to be formed from the pectose found in unripe fruits by the action of acids. Syn., *vegetable jelly*.
**pectinal** (*pek'-tin-al*) [*pecten*, a comb]. Comb-like.
**pectinase** (*pek'-tin-ās*). The enzyme capable of transforming pectin.
**pectinate** (*pek'-tin-āt*) [*pecten*]. Arranged like the teeth of a comb. **p. ligament**, fibers of connective tissue at the angle of the anterior chamber of the eye, between the iris and the cornea. **p. muscles**, the musculi pectinati, muscular ridges in the auricles of the heart.
**pectineal** (*pek-tin-e'-al*) [*pecten*]. 1. Comb-shaped. 2. Pertaining to the pecten or os pubis. **p. line**, that part of the iliopectineal line found on the os pubis.
**pectineus** (*pek-tin-e'-us*). See under *muscle*.
**pectiniform** (*pek-tin'-if-orm*) [*pecten*; *forma*, a form]. Comb-shaped.
**pectoral** (*pek'-tor-al*) [*pectus*, breast]. 1. Pertaining to the chest, as the *pectoral* muscles. 2. Useful in diseases of the chest. 3. A remedy useful in diseases of the chest. **p. ridge**, the external bicipital ridge of the humerus. **p. species**, a combination of pectoral herbs.
**pectoralgia** (*pek-tor-al'-je-ah*) [*pectus*, breast; ἄλγος, pain]. Neuralgic pain in the chest.
**pectoralis** (*pek-tor-a'-lis*). See under *muscle*.
**pectoriloquy** (*pek-tor-il'-o-kwe*) [*pectus*, breast; *loqui*, to speak]. The distinct transmission of articulate speech by the ear on auscultation. It may be heard over cavities in the lung, over areas of consolidation near a large bronchus, over a pneumothorax when the opening in the lung is patulous, and over some pleural effusions. **p. aphonic**. 1. The sound heard in auscultating a lung in which there is a cavity. 2. The sound heard in auscultation in pleuritic effusion when the subject speaks in a low tone. **p.,**
**whispering**, the transmission of the whispered words to the auscultating ear. The sounds seem to emanate directly from the spot auscultated.
**pectorophony** (*pek-tor-off'-o-ne*) [*pectus*, breast; φωνή, sound]. Exaggerated vocal resonance, as heard in auscultating the chest.
**pectose** (*pek'-tōs*) [πηκτός, congealed]. A compound occurring in unripe fruits, and giving rise to pectin.
**pectous** (*pek'-tus*). Relating to pectin or pectose.
**pectunculi** (*pek-tunk'-ū-li*) [*pecten*, comb]. Plural of *pectunculus*. Longitudinal striations in the walls of the Sylvian aqueduct.
**pectus** (*pek'-tus*) [L., "breast"]. The chest or breast. **p. carinatum**, a narrow chest projecting anteriorly in the region of the sternum. Syn., *keeled breast; pigeon-breast*.
**pedal** (*ped'-al*) [*pes, pedis*, foot]. 1. Pertaining to the foot. 2. Pertaining to the pes or crusta of the crus cerebri and pons. **p. system**, a ganglionic system of the brain.
**pedarthrocace** (*ped-arth-rok'-as-e*) [παῖς, child; ἄρθρον, a joint; κακός, evil]. A necrotic ulceration or caries of the joints of children.
**pedatrophia, pedatrophy** (*ped-at-ro'-fe-ah, ped-at'-ro-fe*) [παῖς, child; *atrophy*]. 1. Any wasting disease of childhood. 2. Tabes mesenterica.
**pederast** (*ped'-er-ast*). One who practices pederasty.
**pederasty** (*ped'-er-as-te*) [παῖς, boy; ἐραστής, lover]. Sexual intercourse with boys, through the anus.
**pedesis** (*pe de' sis*) [πήδησις, leaping or bounding] The dancing oscillating motion of the particles of any substance sufficiently powdered and suspended in a suitable liquid. Brownian movement.
**pedialgia** (*pe-de-al'-je-ah*) [πηδίον, foot; ἄλγος, pain]. Pain in the foot.
**pediatrician** (*ped-e-at-rish'-un*). Same as *pediatrist*.
**pediatrics, pediatry** (*pe-de-at'-riks, pe-di'-at-re*) [παῖς, child; ἰατρεία, therapeutics]. The branch of medicine dealing with the diseases of children.
**pediatrist** (*ped-e-at'-rist*). A specialist in children's diseases.
**pedicel** (*ped-i-ka'-shun*) [*paidukā*, a darling]. Sodomy with a boy.
**pedicle** (*ped'-ik-l*) [*pediculus*, dim. of *pes*, foot]. 1. A slender process acting as a foot or stem, as the *pedicle* of a tumor. 2. Of a vertebra, the portion of bone projecting backward from each side of the body and connecting the lamina with the body. **p., vertebral**. See *pedicle* (2). **p., vitelline**, the pedicle uniting the umbilical vesicle to the embryo.
**pedicterus** (*ped-ik'-ter-us*) [παῖς, a child; ἴκτερος, jaundice]. Icterus neonatorum.
**pedicular** (*ped-ik'-ū-lar*) [*pediculi*]. 1. Pertaining to a pedicle; peduncular. 2. [*pediculus*, louse]. Lousy. 3. Belonging to the genus *Pediculus*.
**pediculation** (*ped-ik-ū-la'-shun*) [*pediculus*]. The state of one suffering from pediculosis. 2. The process of developing a pedicle.
**pediculi** (*ped-ik'-ū-li*) [L.]. Plural of *pediculus*, q. v.
**pediculin** (*ped-ik'-ū-lin*) [*pediculus*]. A proprietary insecticide said to consist of limestone, 65%, and crude naphthalin, 35%.
**pediculofrontal** (*ped-ik-ū-lo-front'-al*). Relating to the pedicles of the frontal convolutions.
**pediculoides ventricosus** (*ped-ik-ū-lo'-id-ēz ven-trik-o'-sus*). A mite found in the straw of mattresses and producing straw itch.
**pediculoparietal** (*ped-ik-ū-lo-par-i'-et-al*). Relating to the pedicles of the cerebral convolutions and the parietal region.
**pediculophobia** (*ped-ik-ū-lo-fo'-be-ah*) [*pediculus*; φόβος, fear]. Morbid dread of lice.
**pediculosis** (*ped-ik-ū-lo'-sis*) [*pediculus*]. Lousiness; a skin affection characterized by the presence of pediculi or lice.
**pediculus** (*ped-ik'-ū-lus*) [L., "a louse"]. A small parasitic hemipterous insect, the louse. **p. capitis**, the head-louse. **p. corporis**, the body-louse. **p. pubis**, a species infesting the pubic region of unclean persons, occasionally spreading over other hairy parts of the body—eyebrows, axillæ, etc. Syn., *crab-louse*. **p. vestimenti**. Synonym of *p. corporis*.
**pedicure** (*ped'-ik-ūr*) [*pes*, foot; *cura*, care]. 1. Care of the feet. 2. A chiropodist.
**pediluvium** (*ped-il-oo'-ve-um*) [*pes*, foot; *lavare*, to wash]. A foot-bath.

**pediococcus** (*ped-e-o-kok'-us*) [*pes*, a foot; κόκκος, a berry]. A term formerly given to a genus of micrococci.

**pedion, pedium** (*pe'-de-on, -um*) [παῖς, child]. 1. A child; also, a fetus. 2. [πεδίον, sole]. The sole of the foot.

**pedionalgia** (*ped-e-on-al'-je-ah*) [πεδίον, sole of the foot; ἄλγος, pain]. Pain in the sole of the feet.

**peditis** (*pe-di'-tis*) [*pes*, foot; ιτις, inflammation]. A serious complication of laminitis of the horse's foot, in which not only the laminæ, but the periosteum and the coffin-bone also, are involved in the inflammatory process.

**pedobaromacrometer, pædobaromacrometer** (*pe-do-bar-o-mak-rom'-et-er*). An instrument for weighing and measuring infants.

**pedobarometer, pædobarometer** (*pe-do-bar-om'-et-ur*) [παῖς, a child; βάρος, weight; μέτρον, a measure]. An instrument for determining the weight of a child.

**pedodynamometer** (*ped-o-di-nam-om'-et-er*) [*pes*, foot; δύναμις, power; μέτρον, measure]. An instrument intended to measure the muscular strength of the leg.

**pedology** (*pe-dol'-o-je*) [παῖς, child; λόγος, science]. The science, or sum of knowledge, regarding childhood, its diseases, hygiene, etc.

**pedometer** (*pe-dom'-et-er*) 1. [παῖς, child; μέτρον, measure]. An instrument for determining the weight and height of a new-born child. 2. [*pes*, foot; μέτρον, a measure]. An instrument for automatically measuring any distance traveled. As formerly constructed, it registered the number of footsteps.

**pedometry** (*pe-dom'-et-re*) [*pedometer*]. 1. The measurement of the newborn child. 2. The use of the pedometer.

**pedonosology** (*pe-do-nos-ol'-o-je*) [παῖς, child; νόσος, disease; λόγος, science]. The nosology of disease peculiar to infancy and childhood. Pediatrics.

**pedonosos, pedonosus** (*pe-don'-o-sos, pe-don'-o-sus*) [παῖς, child; νόσος, disease]. A disease of childhood.

**pedopathy** (*pe-dop'-ath-e*) [παῖς, child; πάθος, suffering]. The science of the diseases of children, their treatment, etc.

**pedotrophy** (*pe-dot'-ro-fe*) [παῖς, child; τροφή, nourishment]. The hygiene of childhood; the care, nursing, and regimen of children.

**peduncle** (*pe'-dung-kl*) [*pedunculus*, dim. of *pes*]. A narrow part acting as a support. **p., callosal,** the anterior perforated space. **p., cerebellar, inferior,** one of two bands of white matter passing up from the medulla oblongata, connecting the medulla with the cerebellum, and forming the lower lateral wall of the fourth ventricle. **p., cerebellar, middle,** one of the bands of white matter joining the pons and the cerebellum. **p., cerebellar, posterior.** Synonym of *p., cerebellar, inferior*. **p., cerebellar, superior,** one of the two bands of white matter that pass from the cerebellum to the testes of the corpora quadrigemina. **p., cerebral,** the crus cerebri. **p. of the pineal gland,** a delicate white band passing forward from each side of the pineal gland along the edge of the third ventricle.

**peduncular** (*ped-ung'-kū-lar*) [*peduncle*]. Pertaining to a peduncle.

**pedunculate, pedunculated** (*pe-dung'-kū-lāt, pe-dung'-kū-la-ted*) [*pedunculus*, dim. of *pes*, foot]. Having a peduncle; stalked.

**pedunculus** (*pe-dung'-kū-lus*). Same as *peduncle*.

**peeling** (*pēl'-ing*) [*pellis*, skin]. A term applied to the process of desquamation, as in scarlet fever.

**peenash** (*pe'-nash*) [E. Ind.]. The Eastern name for *myiasis*, produced by *Lucilia macellaria*.

**peg.** 1. A pointed pin of wood, metal, or other material. 2. A wooden leg. **p.-leg,** a wooden leg of the simplest form. **p.-teeth,** a name given by Hutchinson to the teeth of children with hereditary syphilis, from the peg-like appearance of the crowns.

**pegmatic** (*peg-mat'-ik*) [πῆγμα, a concrement]. Pertaining to or producing coagulation.

**peinotherapy** (*pi-no-ther'-ap-e*) [πεῖνα, hunger; θεραπεία, cure]. The cure of disease by deprivation of food.

**Pekelharing's theory of blood-coagulation.** Thrombin (fibrin-ferment) is composed of nucleoalbumin and calcium; the calcium leaves the nucleoproteid and unites with fibrinogen, the compound of the two being fibrin.

**pelada, pelade** (*pel'-a-dah, pel-ahd'*) [Fr.]. 1. Alopecia areata of the scalp. 2. A disease resembling pellagra, due to eating infected maize.

**pelage** (*pel-ahj'*) [*pilus*, the hair]. The hairy system of the body.

**pelagin** (*pel'-aj-in*). A proprietary remedy for sea-sickness, said to consist of a solution of antipyrine, caffeine, and cocaine.

**pelargonic acid** (*pel-ar-gon'-ik*) [πελαργός, a stork], $C_9H_{18}O_2$. A monobasic crystalline acid obtained from the essential oil of *Pelargonium roseum* and from other oils. It is employed in the flavoring of wines.

**pelatina** (*pel-at-i'-nah*). See *pelada* (2).

**pelicochirometresis** (*pel-ik-o-ki-ro-met-re'-sis*) [πελίκη, a wooden bowl; χείρ, the hand; μέτρον, a measure]. Digital pelvimetry.

**pelicometer** (*pel-ik-om'-et-er*). Synonym of *pelvimeter*.

**pelidnoma** (*pel-id-no'-mah*). Synonym of *ecchymosis*.

**pelioma** (*pel-e-o'-mah*) [πελίωμα, a livid spot]. A livid spot, as seen in peliosis.

**peliosis** (*pel-e-o'-sis*) [see *pelioma*]. Purpura. **p. rheumatica,** purpura rheumatica; a disease characterized by a purpuric rash, with arthritis and fever.

**pellagra** (*pel-lah'-grah pel-a'-grah*) [πέλλα, skin; ἄγρα, a seizure]. A disease occurring in Italy, southern France, Spain, and in the southern States. It was formerly believed to be due to the use of diseased maize, but is now thought by some to be of protozoan origin, by others to be a deficiency disease; see *vitamine*. It is characterized in the early stages by debility, spinal pains, and digestive disturbances; later erythema develops, with drying and exfoliation of the skin. In severe cases various nervous manifestations arise, such as spasms, ataxic paraplegia, and mental disturbances. In cases presenting ataxic paraplegia the spinal cord has shown combined posterior and lateral sclerosis. Syn., *Lombardian leprosy*. **p. sine pellagra,** pellagra without the erythematous rash.

**pellagracein** (*pel-a-gra'-se-in*) [*pellagra*]. A poisonous substance found in decomposed corn-meal. Syn., *pellagrazein, pellagrocein, pellagrozein*.

**pellagraphobia** (*pel-a-graf-o'-be-ah*). Morbid dread of becoming affected with pellagra.

**pellagrin** (*pel'-a-grin*) [It., *pella*, skin; *agra*, rough; or πέλλα, skin; ἄγρα, seizure]. One who is afflicted with pellagra.

**pellagrous** (*pel-a'-grus*) [πέλλα, skin; ἄγρα, seizure]. Affected with pellagra; pertaining to pellagra.

**pellentia** (*pel-en'-she-ah*) [*pellere*, to drive]. Abortifacient drugs.

**pellet** (*pel'-et*) [*pila*, ball]. A small pill.

**pelletierine** (*pel-et'-e-er-ēn*) [Bertrand *Pelletier*, French chemist, 1761–1797], $C_8H_{13}NO$. A liquid alkaloid obtained together with an isomeric body, *isopelletierie*, also a liquid alkaloid, from pomegranate-bark. It is used as a teniafuge, chiefly in the form of the tannate. **p. hydrobromide,** used in paralysis of the eye-muscles. Dose 4–6 gr. (0.25–0.4 Gm.). **p. hydrochloride,** used as a teniafuge. Dose 4½–8 gr. (0.3–0.5 Gm.). **p. sulphate,** used as an anthelmintic. Dose 6 gr. (0.4 Gm.) with 8 gr. (0.52 Gm.) of tannin in 1 oz. (30 Cc.) of water, followed by a cathartic. **p. tannate** (*pelletierinæ tannas*, U. S. P.), a mixture of the tannates of four alkaloids obtained from *Punica granatum*. Dose 4 gr. (0.25 Gm.).

**pellicle** (*pel'-ik-l*) [*pellis*, skin]. 1. A thin membrane, or cuticle. 2. A film on the surface of a liquid.

**pellicular** (*pel-ik'-ū-lar*) [*pellicula*, dim. of *pellis*, skin]. Of the nature of, or resembling a pellicle; thin-skinned.

**pelliculate** (*pel-ik'-ū-lāt*) [*pellicula*, a small skin]. Covered with a pellicle.

**pellis** (*pel'-is*) [L.]. The skin.

**pellitory** (*pel'-it-or-e*). See *pyrethrum*.

**pellotine** (*pel'-ot-ēn*), $C_{13}H_{19}NO_3$. An alkaloid from the Mexican cactus, *Anhalonium williamsi*; it is a hypnotic, and is used chiefly in the form of the hydrochloride. Dose ⅚–1 gr. (0.056–0.06 Gm.); subcutaneously ⅓–¾ gr. (0.02–0.048 Gm.).

**pellous** (*pel'-us*) [*pellis*, skin]. Dark-skinned.

**pellucid** (*pel-ū'-sid*) [*pellucere*, to shine through]. Transparent; translucent; not opaque. **p. zone,** the zona pellucida, or inclosing membrane of the mammalian ovum. It is also called the *zona radiata*.

**pelma** (*pel'-mah*) [πέλμα, sole]. The lower surface of the toes; also the entire sole of the foot.

**pelmatic** (*pel-mat'-ik*) [πέλμα, sole]. Relating to the sole of the foot.

**pelmatogram** (*pel-mat'-o-gram*) [πέλμα, the sole of the foot; γράμμα, a writing]. An imprint of the sole of the foot.

**pelohemia, pelohæmia** (*pe-lo-he'-me-ah*) [πηλός, mud; αἷμα, blood]. Excessive thickness of the blood.

**pelopathist** (*pe-lop'-ath-ist*) [πηλός, mud; πάθος, disease]. One who practises pelopathy.

**pelopathy** (*pe-lop'-ath-e*). The treatment of diseases by the application of mud.

**pelor** (*pel'-or*) [πέλωρ, a monster]. A fetal monstrosity with some parts abnormally large.

**pelotherapy** (*pe-lo-ther'-ap-e*). See *pelopathy*.

**peltation** (*pel-ta'-shun*) [πέλτη, a shield]. The protection afforded by inoculation with a serum.

**pelveoperitonitis** (*pel'-ve-o-per-it-on-i'-tis*) [*pelvis; peritoneum;* ιτις, inflammation]. Inflammation of the pelvic peritoneum.

**pelveoscope** (*pel'-ve-o-skōp*) [*pelvis*, σκοπεῖν, to examine]. An instrument for examining the pelvis; a pelvimeter.

**pelveoscopy** (*pel-ve-os'-ko-pe*). See *pelvioscopy*.

**pelvic** (*pel'-vik*) [*pelvis*]. Pertaining to the pelvis. **p. abscess,** a suppurative inflammation of the connective tissue of the pelvic cavity, most common in women, and usually associated with puerperal or gonorrheal infection. **p. arch.** Same as *p. girdle.* **p. cellulitis,** inflammation of the connective tissue of the pelvis. **p. fascia,** the fascia lining the pelvic cavity. **p. girdle,** the arch formed by the ilium, ischium, and pubis, or in the higher vertebrates by the two innominate bones. **p. index,** the relation of the anteroposterior to the transverse diameter of the pelvis. **p. inlet,** the superior strait. **p. outlet,** the inferior strait. **p. region,** the region within the true pelvis.

**pelvicellulitis** (*pel-vis-el-ū-li'-tis*). See *pelvic cellulitis.*

**pelvicliseometer** (*pel-vik-liz-e-om'-et-er*) [*pelvis;* κλίσις, inclination; μέτρον, measure]. An instrument for determining the inclination and the diameters of the pelvis.

**pelvifixation** (*pel-ve-fiks-a'-shun*). Surgical fixation of a misplaced pelvic organ.

**pelvigraph** (*pel'-vig-raf*) [*pelvis;* γράφειν, to write]. An apparatus that automatically records the outline of the pelvic wall.

**pelvimeter** (*pel-vim'-et-er*) [*pelvis;* μέτρον, a measure]. An instrument for measuring the pelvic dimensions.

**pelvimetry** (*pel-vim'-et-re*) [see *pelvimeter*]. The measurement of the dimensions of the pelvis. **p. combined,** a combination of external and internal pelvimetry. **p. digital,** pelvimetry by means of the hand. **p., external,** measurement of the external diameters of the pelvis, by which to estimate the dimensions of the internal parts. **p., internal,** measurement of the internal dimensions of the pelvis by the hand or by the pelvimeter. **p., manual.** Same as *p., digital.* See *pelvis.*

**Table of Measurements of the Female Pelvis Covered by the Soft Parts.**

| | | |
|---|---|---|
| Between iliac spines | 26 | cm. |
| Between iliac crests | 29 | " |
| External conjugate diameter | $20\frac{1}{2}$ | " |
| Internal conjugate diagonal | $12\frac{3}{4}$ | " |
| True conjugate, estimated | 11 | " |
| Right diagonal | 22 | " |
| Left diagonal | 22 | " |
| Between trochanters | 31 | " |
| Circumference of pelvis | 90 | " |

**pelvioplasty** (*pel-ve-o-plas'-te*) [*pelvis;* πλάσσειν, to form]. 1. Pelviotomy for the purpose of enlarging the pelvic outlet. 2. Incision into the pelvis of the kidney.

**pelvioscopy** (*pel-ve-os'-ko-pe*) [*pelvis*, σκοπεῖν, to examine]. The examination of the pelvis; pelvimetry.

**pelviotomy** (*pel-ve-ot'-o-me*) [*pelvis*, τομή, a cutting]. Section or cutting of the bones of the pelvis, especially the division of the symphysis pubis in case of difficult labor.

**pelviperitonitis** (*pel-vi-per-e-ton-i'-tis*) [*pelvis; peritonitis*]. Pelvic peritonitis.

**pelvirectal** (*pel-ve-rek'-tal*). Relating to the pelvis and the rectum.

**pelvis** (*pel'-vis*) [L., "a basin"]. 1. A basin or basin-shaped cavity, as the *pelvis* of the kidney. 2. The bony ring formed by the two innominate bones and the sacrum and coccyx. 3. The cavity bounded by the bony pelvis. The pelvis consists of two parts—the *true pelvis* and the *false pelvis,* which are separated by the iliopectineal line. The entrance of the true pelvis, corresponding to this line, is known as the *inlet* or *superior strait;* the *outlet* or *inferior strait* is bounded by the symphysis pubis, the tip of the coccyx, and the two ischia. In measuring the pelvis the *cardinal points of Capuron* are used as landmarks. They are the two iliopectineal eminences and the two sacroiliac joints. **p. æquabiliter justo major,** one equally enlarged in all diameters. **p. æquabiliter justo minor,** a pelvis with all its diameters reduced below the normal. **p., axis of** (*of inlet or outlet*), a perpendicular to the middle of the anteroposterior diameter. **p., beaked,** one in which the pubic bones are compressed laterally so as to approach each other, and are pushed forward; a condition seen in osteomalacia. **p., brim of,** the entrance to the pelvic cavity, called the inlet, superior strait, margin, or isthmus. **p., cordate,** one with heart-shaped inlet. **p., diameters of,** imaginary lines drawn between certain bony points. (*a*) Of the *inlet:* the *anteroposterior* (sacropubic, or conjugate), from the upper edge of the promontory of the sacrum to a point $\frac{1}{3}$ of an inch below the upper border of the pubic symphysis; it measures 11 cm.; the *transverse,* from side to side at the widest point, measuring $13\frac{1}{2}$ cm.; the *oblique* (right and left), measuring $12\frac{3}{4}$ cm. (*b*) Of the *outlet:* the *anteroposterior,* from the tip of the coccyx to the subpubic ligament, measuring $9\frac{1}{2}$ cm.; the *transverse,* between the ischial tuberosities, measuring 11 cm.; the *oblique,* from the under surface of the sciatic ligaments to the junction of the ischiopubic rami. **p., false,** that part above the iliopectineal line. **p., floor of,** the mass of skin, connective tissue, muscles, and fascia forming the inferior boundary of the pelvis. **p., inclination of.** See *p., obliquity of.* **p., inlet of.** See *inlet.* **p., justomajor.** See *p. æquabiliter justo major.* **p., justominor.** See *p. æquabiliter justo minor.* **p., kyphotic,** one characterized by increase of the conjugate diameter of the inlet, but decrease of the transverse diameter of the outlet, through approximation of the tuberosities of the ischium. **p., malacosteon.** See *p., osteomalacic.* **p., masculine,** one narrowed progressively from above. **p., Naegele's oblique,** a pelvic deformity with ankylosis of one sacroiliac synchondrosis, lack of development of the associated lateral sacral mass, and other defects that distort the diameters and render the conjugate oblique in direction. **p., obliquity of, p., inclination of,** the angle between the axis of the pelvis and that of the body. **p., osteomalacic,** a distorted pelvis characterized by a lessening of the transverse and oblique diameters, with great increase of the anteroposterior diameter. **p., planes of,** imaginary surfaces touching all points of the circumference. The plane of pelvic expansion perforates the middle of the symphysis, the tops of the acetabula, and the sacrum between the second and third vertebræ. Its anteroposterior diameter is $12\frac{3}{4}$ cm.; its transverse diameter is $12\frac{1}{2}$ cm. The plane of pelvic contraction passes through the tip of the sacrum, the spines of the ischia, and the under surface of the symphysis. Its anteroposterior diameter is $11\frac{1}{2}$ cm.; its transverse diameter is $10\frac{1}{2}$ cm. **p., rhachitic,** one characterized by a sinking in and forward of the sacrovertebral angle, with a flaring outward of the iliac crests and increased separation of the iliac spines. **p., Robert's,** one in which there is an ankylosis of both sacroiliac joints, with a rudimentary sacrum, both lateral sacral masses being undeveloped, the oblique and transverse diameters being much narrowed. **p., rostrate.** Same as *p., beaked.* **p., simple flat,** one in which the only deformity consists in a shortening of the anteroposterior diameter. **p. spinosa,** a rhachitic pelvis in which the crest of the pubis is very sharp, and presents a spine at the insertion of the psoas parvus. **p., split,** a form in which there is congenital separation of the pubic bones at the symphysis. It is often associated with exstrophy of the bladder. **p., straits of** (*superior* and *inferior*), the planes of the inlet and outlet. **p., true,** the part below the iliopectineal line.

**pelvisacral** (*pel-ve-sa'-kral*) [*pelvis; sacrum*]. Pertaining to the pelvis and the sacrum.

**pelvisacrum** (*pĕl-ve-sa'-krum*). The pelvis and sacrum taken conjointly.
**pelvitomy** (*pel-bit'-o-me*) [*pelvis;* τομή, a cutting]. Synonym of *pelviotomy*.
**pelvitrochanterian** (*pel-ve-tro-kan-te'-re-an*). Relating to the pelvis and the great trochanter of the femur.
**pelvoscopy** (*pel-vos'-ko-pe*). See *pelvioscopy*.
**pelycalgia** (*pel-ik-al'-je-ah*) ₍[πέλυξ, pelvis; ἄλγος, pain]. Pelvic pain in general.
**pelycochirometresis** (*pel-ik-o-ki-ro-met-re'-sis*). See *pelycocheirometresis*.
**pelycotomy** (*pel-ik-ot'-o-me*) [πέλυξ, pelvis; τομή, a cutting]. Division of the os pubis; symphyseotomy. See *pelviotomy*.
**pelychirometresis** (*pel-e-o-ki-ro-met-re'-sis*) [πέλυξ, pelvis; χείρ, hand; μέτρον, measure]. Synonym of *pelvimetry, digital.*
**pelycography** (*pel-ik-og'-ra-fe*) [πέλυξ, pelvis; γράφειν, to write]. A description of the pelvis.
**pelycology** (*pel-ik-ol'-o-je*) [πέλυξ, pelvis; λόγος, science]. A treatise upon the pelvis.
**pelycometer** (*pel-ik-om'-et-er*) [πέλυξ, pelvis; μέτρον, measure]. Same as *pelvimeter*.
**pelycometresis** (*pel-ik-o-met-re'-sis*). Synonym of *pelvimetry*.
**pelyometer** (*pel-e-om'-et-er*) [πέλυξ, pelvis; μέτρον, measure]. Same as *pelvimeter*.
**pelyometresis** (*pel-e-o-met-re'-sis*) [πέλυξ, pelvis; μέτρον, measure]. Same as *pelvimetry*.
**pemmican** (*pem'-ik-an*) [Amer. Ind.]. A concentrated food consisting of a mixture of the best beef and fat dried together. Sugar is sometimes added, as well as raisins and currants.
**pemphigoid** (*pem'-fig-oid*) [see *pemphigus*]. Resembling or having the nature of pemphigus.
**pemphigus** (*pem'-fig-us*) [πέμφιξ, a blister]. An acute or chronic disease of the skin characterized by the appearance of bullæ or blebs. **p. benignus.** Same as *p. vulgaris.* **p. circinatus,** a kind with the bullæ in circles. **p. foliaceus,** a rare form characterized by crops of flaccid blebs containing a turbid fluid. The disease is usually of long duration, but eventually ends fatally. **p. hystericus.** Same as *p. pruriginosus.* **p. malignus.** Same as *p. pruriginosus.* **p. neonatorum,** an acute form occurring in infants and supposed to be due to a microorganism. **p. pruriginosus,** that associated with severe itching, purulent bullæ, and wheals. **p. solitarius,** a form with single blebs. **p. syphiliticus,** a bullous eruption due to syphilis. **p. vegetans,** an affection characterized by sore mouth, followed by some form of dermatitis attended by vesication and then by papillary growths, gradual emaciation and death. Syn., *Neumann's disease.* **p. vulgaris,** a form that is usually chronic, the blebs appearing in successive crops; on healing they leave a pigmented spot. Itching and pain may be present.
**pencil** (*pen'-sil*) [*penicillus*, pencil]. 1. In pharmacy, a medicated cylindrical stick, as a menthol pencil, which is used for local application. 2. An aggregation of rays of light meeting in a point.
**pendinski ulcer** (*pen-din'-ske*). See *furunculus orientalis.*
**pendulous** (*pen'-du-lus*) [*pendere*, to hang]. Hanging down loosely.
**pendulum-motion.** A to-and-fro movement like that of a pendulum sometimes observed in the arms in obscure nervous diseases.
**penetrating** (*pen'-e-tra-ting*) [see *penetration*]. Entering beyond the surface. **p. power.** See *focal depth.* **p. wound,** one that pierces the wall of a cavity or enters an organ.
**penetration** (*pen-e-tra'-shun*) [*penetrare*, to pierce]. 1. The act of penetrating or piercing into. 2. Of a microscope, the focal depth. 3. The entrance of the penis into the vagina.
**penetrometer** (*pen-e-trom'-et-er*). An instrument for measuring the penetrating power of the x-rays.
**pengawar, penghawar djambi.** The long soft hairs, used as a mechanical styptic, obtained from rhizomes of various ferns of Sumatra and Java, particularly from the genus *Cibotium.*
**penial** (*pe'-ne-al*) [*penis*]. Pertaining to the penis.
**penicillate** (*pen'-is-il-āt*) [*penicillus*, a painter's brush or pencil]. Shaped like a pencil of hairs.
**peniciliform** (*pen-is-il'-if-orm*) [*peniculus*, a pencil; *forma*, form]. Resembling a pencil.
**Penicillium** (*pen-is-il'-e-um*) [see *penicillate*]. A genus of fungi, of which the *Penicillium glaucum*, or common blue-mold, is a familiar example.

**penicillus** (*pen-is-il'-us*) [*penicillus*, a painter's brush; pl., *penicilli*]. One of the tufts of fine twigs into which the arteries of the spleen subdivide.
**penile** (*pe'-nil*) [*penis*]. Pertaining to the penis.
**penis** (*pe'-nis*) [*pendere*, to hang]. The male organ of copulation. It consists of the corpus spongiosum, inclosing the urethra, the two corpora cavernosa, largely composed of erectile tissue, and the glans. **p. captivus,** one held in the vagina during copulation by spasm of the perineal muscles of the female. **p. cerebri,** the pineal gland. **p. factitious,** an artificial penis. Syn., *fascinum; dildoe.* **p. feminis,** the clitoris. **p. lipodermus.** See *paraphimosis.* **p. muliebris.** Synonym of *p. feminis.* **p. palmatus,** one inclosed by the skin of the scrotum. Syn., *webbed penis.* **p. succedaneus.** See *p. factitious.*
**penischisis** (*pen-is'-kis-is*) [*penis;* σχίσις, a splitting]. A comprehensive term for epispadias and hypospadias.
**penitis** (*pe-ni'-tis*) [*penis;* ιτις, inflammation]. Inflammation of the penis.
**penjavar yambi.** See *penghawar djambi.*
**Penjdeh sore** (*penj'-deh*) [*Penjdeh,* a village of Russian Turkestan]. See *furunculus orientalis.*
**penniform** (*pen'-if-orm*) [*penna,* feather; *forma,* form]. Shaped like a feather; said of certain muscles.
**pennyroyal** (*pen-e-roi'-al*). See *hedeoma.*
**pennyweight** (*pen'-e-wāt*) [AS., *pening,* penny; *wegan,* weigh]. A weight of 24 grains.
**penologist** (*pe-nol'-o-jist*). One who makes a study of crime and its cause and prevention.
**penology, pœnology** (*pe-nol'-o-je*) [*ποινή,* penalty; λόγος, science]. The science treating of penology, its punishment and prevention; the study of the management of prisons, etc.
**penoscrotal** (*pe-no-skro'-tal*). Pertaining to the penis and the scrotum.
**pensioner, Chelsea.** See under *Chelsea.*
**pent–.** A prefix used to signify *five.*
**pentabasic** (*pen-tab-a'-sik*) [πέντε, five; *basis,* base]. Having five replaceable hydrogen atoms.
**pentad** (*pen'-tad*) [πέντε, five]. An element or radical having a valence of five.
**pental** (*pen'-tal*) [πέντε, five], $C_5H_{10}$. Trimethylethylene; it is used as an anesthetic.
**pentane** (*pen'-tān*) [πέντε, five], $C_5H_{12}$. The fifth member of the paraffin series of hydrocarbons. It is a liquid and occurs in naphtha.
**Pentastoma** (*pen-tas-to'-mah*) [πέντε, five; στόμα, mouth]. A genus of entozoa, worm-like parasites, generally referred to the class arthropoda. There are many species, several of which have been found encysted in the human liver and lungs. See *linguatula.*
**pentatomic** (*pen-tat-om'-ik*) [πέντε, five; ἄτομος, atom]. 1. Containing five atoms. 2. Having five replaceable hydrogen atoms in the molecule.
**pentavalent** (*pen-tav'-al-ent*) [πέντε, five; *valens,* having power]. Having a valence of five.
**pentose** (*pen'-tōs*) [πέντε, five], $C_5H_{10}$. Amylene, one of the olefin series of hydrocarbons.
**pentosan** (*pen'-to-san*). A complex carbohydrate capable of forming a pentose by hydrolysis.
**pentosazon** (*pen-to'-sas-ōn*). A body occurring in urine, possessing marked reducing qualities, but incapable of fermentation; it represents an abnormality in the total metabolism of the body.
**pentose** (*pen'-tōs*) [πέντε, five]. Any one of a class of carbohydrates containing five atoms of carbon. The pentoses are not fermentable, and on boiling with dilute hydrochloric acid yield furfurol, $C_5H_4O_2$.
**pentosuria** (*pen-to-su'-re-ah*) [*pentose; οὖρον,* urine]. The presence of pentose in the urine. Urine containing pentose reduces Fehling's solution, but does not ferment.
**pentoxide** (*pen-toks'-id*) [πέντε, five; *oxide*]. An oxide containing five atoms of oxygen.
**pentyl** (*pen'-til*) [πέντε, five; ὕλη, matter], $C_5H_{11}$. A univalent hydrocarbon. **p. hydride,** same as *amyl hydride.*

**Penzoldt's test for acetone** (*pen'-tsōlt*) [Franz *Penzoldt,* German physician, 1849– ]. Treat a warm saturated solution of orthonitrobenzaldehyde with the liquid to be tested for acetone, and alkalinize with sodium hydroxide. If acetone is present, the liquid becomes first yellow, then green, and lastly indigo separates, which may be dissolved with a blue color on shaking with chloroform. P.'s test for glucose in urine, to a few cubic centimeters of urine add some caustic potash, and enough of a weakly alkaline solution of diazobenzol sulphonic acid to equal the amount of urine. Shake for one-fourth to one-half of an hour to produce foam. A light Bordeaux-red or yellowish-red coloration will result, with a red foam.

**Penzoldt and Fischer's test for phenol.** Treat a strongly alkaline solution of phenol with a solution of diazobenzol sulphonic acid; a deep red coloration is produced.

**peonin** (*pe'-o-nin*), $C_{19}H_{15}O_2(NH_2)$. An indicator for alkalies.

**peotomy** (*pe-ot'-o-me*) [πέος, penis; τομή, a cutting]. Amputation of the penis.

**pepana** (*pep'-an-ah*). Trade name of a preparation of pepsin and pancreatin so arranged and coated that the pepsin is released in the stomach and the pancreatin in the intestine.

**pepo** (*pe'-po*) [πέπων, ripe; mellow]. Seed of the pumpkin, *Cucurbita pepo*; it is a teniafuge.

**peporesin** (*pe-po-rez'-in*). A hard substance in the husk of pumpkin-seeds; it is a vermicide.

**pepper** (*pep'-er*) [*piper*, from Skt., *pippala*]. The fruit of various species of *Piper*, of the order *Piperaceæ*. p., black, the *piper* of the U. S. P.; *piper nigrum* of the B. P., contains a neutral principle, *piperin* (*piperina*, U. S. P.), an acrid resin, and a volatile oil, and is used as a condiment and as a carminative stimulant, and to a slight extent is antiperiodic. Dose 5 to 30 gr. (0.32–1.3 Gm.). p., Cayenne. See *capsicum.* p., oleoresin of (*oleoresina piperis*, U. S. P.). Dose ⅓–1 min. (0.016–0.065 Cc.). p., white, similar to black pepper, but less active.

**peppermint.** See *mentha piperita*.

**peppermint test.** A method of discovering defective drain-pipes by pouring oil of peppermint down the pipes or trap connected with the drain; the odor of the peppermint enters the house if the drain pipe leaks.

**pepsencia** (*pep-sen'-she-ah*). Trade name of a preparation containing the enzymes of the gastric glands.

**pepsic** (*pep'-sik*). Same as *peptic.*

**pepsin** [πέψις, digestion]. A ferment found in the gastric juice, and capable of digesting proteids in the presence of an acid. It splits albumin into antialbumose and hemialbumose, the former of which it separates into two molecules of antipeptone, while the latter is acted upon by trypsin and split into two molecules of hemipeptone. Pepsin is used in medicine to aid digestion. The *pepsinum* of the U. S. P. and B. P. is obtained from the stomach of pigs. Dose 10–15 gr. (0.65–1.0 Gm.). **pepsinum saccharatum,** pepsin mixed with sugar of milk. **pepsini, vinum,** contains 0.3 % of hydrochloric acid. Dose ½–1 oz. (16–32 Cc.).

**pepsinate** (*pep'-sin-āt*) [πέψις, digestion]. To mix, or prepare, with pepsin.

**pepsiniferous** (*pep-sin-if'-er-us*) [*pepsin; ferre*, to bear]. Producing pepsin.

**pepsinogen** (*pep-sin'-o-jen*) [*pepsin*; γεννᾶν, to beget]. The antecedent substance or zymogen of pepsin, present in the cells of the gastric glands, and which during digestion is converted into pepsin.

**pepsinogenous** (*pep-sin-oj'-en-us*). See *pepsiniferous*.

**pepsinum** (*pep-si'-num*). See *pepsin*.

**pepsis** (*pep'-sis*) [πέψις, digestion]. Digestion.

**peptic** (*pep'-tik*) [*pepsin*]. 1. Pertaining to pepsin. 2. Pertaining to digestion. **p. glands,** the glands situated in the cardiac and middle thirds of the stomach, and secreting pepsin and hydrochloric acid. **p. ulcer,** the round ulcer of the stomach, due to erosion of the mucous membrane by the gastric juice.

**pepticity** (*pep-tis'-it-e*) [πέπτειν, to digest]. The state of being peptic; eupepsia.

**peptid** (*pep'-tid*). A compound of amino-acids intermediate between peptones and the individual amino-bodies.

**peptinotoxin** (*pep-tin-o-toks'-in*). See *peptotoxin*.

**peptogaster** (*pep-to-gas'-ter*) [πέπτειν, to digest; γαστήρ, belly]. The intestinal canal, or digestive apparatus as a whole.

**peptogastric** (*pep-to-gas'-trik*) [πέπτειν, to digest; γαστήρ, belly]. Pertaining to the peptogaster; peptic.

**peptogen** (*pep'-to-jen*) [πέψις, digestion; γεννᾶν, to beget]. A substance that favors the production of pepsin.

**peptogenic** (*pep-to-jen'-ik*) [*pepsin*; γεννᾶν, to produce]. Producing pepsin or peptones.

**peptogenous** (*pep-toj'-en-us*) [πέψις, digestion; γεννᾶν, to produce]. Producing pepsin or peptones.

**peptolysis** (*pep-tol'-is-is*) [*peptone*; λύσις, destruction]. The hydrolysis or splitting up of peptones.

**peptomangan** (*pep-to-man'-gan*). A proprietary compound said to consist of iron, manganese, and peptone; used in tuberculosis, etc.

**peptone** (*pep'-tōn*) [πέπτειν, to cook]. A protein body formed by the action of ferments on albumins during gastric and pancreatic digestion. It may be considered a hydrated albumin. Before the final formation of peptone several similar intermediate compounds are produced, as *hemipeptone* and *antipeptone*. See *pepsin*. *Amphopeptone* is a mixture of these two. *Propeptone* or hemialbumose is a mixture of several intermediate products. *Parapeptone* is also an intermediate product of digestion and is closely allied to syntonin. *Gelatin-peptone* is a peptone formed in the digestion of gelatin. Peptones are soluble, readily diffusible, are not precipitated by boiling, by nitric acid, or by potassium ferrocyanide; they are precipitated by mercury bichloride, by tannic acid, and by phosphomolybdic acid; they give Millon's test, and the xanthoproteic and biuret reactions; they are levorotatory. **p., albumin-,** a light yellow powder obtained from white of egg by action of pepsin with a little hydrochloric acid; used as a nutrient. **p. anhydride.** See *albuminate.* **p., beef,** true peptone from beef; a light brown powder, soluble in water, used as a nutrient in dyspepsia. **p., casein,** peptonized casein from milk; a yellow, hygroscopic powder used as a nutrient. **p., milk.** See *p., casein*. **p. powder,** a nutrient containing 91 % of peptone. **p.s, test for.** See *Hofmeister*.

**peptonemia, peptonæmia** (*pep-to-ne'-me-ah*) [*peptone*; αἷμα, blood]. The presence of peptone in the blood.

**peptonization** (*pep-to-ni-za'-shun*) [*peptone*]. The process of converting proteins into peptones.

**peptonize** (*pep'-to-nīz*) [*peptone*]. To digest with pepsin; to predigest; to convert into peptones.

**peptonoid** (*pep'-to-noid*) [πέπτειν, to digest; εἶδος, like]. A substance resembling or claimed to resemble peptones.

**peptonuria** (*pep-to-nū'-re-ah*) [*peptone*; οὖρον, urine]. The presence of peptones in the urine. **p. enterogenous,** peptonuria due to disease of the intestine. **p. hepatogenous,** that accompanying certain liver affections. **p. nephrogenous,** peptonuria of renal origin. **p. puerperal,** the peptonuria of the puerperal state. **p. pyogenic,** that produced by suppuration in the body.

**peptosin** (*pep-to'-sin*). A proprietary preparation of pepsin.

**peptothyroid** (*pep-to-thi'-roid*). A proprietary peptonized preparation of thyroid extract.

**peptotoxin** (*pep-to-toks'-in*) [*peptone*; τοξικόν, a poison]. A poisonous ptomaïne found in peptones and in putrefying albuminous substances, such as fibrin, casein, brain, liver, and muscle. **p., cholera,** a toxic substance generated by the cholera bacillus, and chemically allied to peptone.

**peptovarin** (*pep-to'-va-rin*). Extract of peptonized ovaries.

**pepule** (*pep'-ūl*). Trade name for a pill.

**per** [L.]. A preposition meaning through. **p. anum,** by way of the anus. **p. os,** by the mouth. **p. rectum,** by the rectum. **p. vaginam,** by the vagina. **p. viam,** by the way of.

**per-.** A prefix with an intensive meaning denoting "very"; also in chemistry denoting the highest of a series.

**peracephalus** (*per-as-ef'-al-us*) [πέρα, more than; ἀκέφαλος, without a head]. A fetal monstrosity characterized not only by want of upper extremities, but also by malformation or absence of the thorax.

**peracidity** (*per-as-id'-it-e*) [*per*, very; *acidus*, acid]. Excessive acidity.

**peracute** (*per-ak-ūt'*) [*per*, very; *acutus*, sharp]. Very acute.

**perarticulation** (*per-ar-tik-ū-la'-shun*) [*per; articulation*]. Synonym of diarthrosis.

**peratodynia** (*per-at-o-din'-e-ah*) [πέρας, end; ὀδύνη, pain]. Pain at the cardiac extremity of the stomach.

**perception** (*per-sep'-shun*) [*per, capere*, to receive].

# PERCEPTIVITY 66

1. The act of receiving impressions through the medium of the senses. 2. The faculty receiving such impressions.

**perceptivity** (*per-sep-tiv'-it-e*) [*per*, through; *capere*, to receive]. The faculty or capability of receiving impressions.

**perceptorium** (*per-sep-to'-re-um*). Same as *sensorium*.

**perchlorate** (*per-klo'-rāt*). See under *perchloric acid*.

**perchlorhydria** (*per-klor-hi'-dre-ah*). See *hyperchlorhydria*.

**perchloric acid** (*per-klo'-rik*) [*per*; χλωρός, green], HClO₄. The highest oxyacid of chlorine. It is a volatile liquid decomposing in contact with organic substances, and forming salts called *perchlorates*.

**perclusion** (*per-kloo'-shun*) [*per*, through; *claudere*, to shut up]. Inability to execute any movement.

**percolate** (*per'-ko-lāt*) [*percolare*, to strain through]. 1. To submit to the process of percolation. 2. The solution obtained by percolation.

**percolation** (*per-ko-la'-shun*) [see *percolate*]. The process of extracting the soluble constituents of a substance by allowing the solvent to trickle through a powdered mass placed in a long conical vessel—the *percolator*.

**percolator** (*per'-ko-la-tor*) [*percolare*, to strain through]. A long conical vessel with a delivery-tube at the lower extremity, employed for the purpose of extracting the soluble constituents of a substance, packed in a percolator, by means of a liquid poured over it.

**percuss** (*per-kus'*) [*percutere*, to strike]. To perform percussion upon.

**percussion** (*per-kush'-un*) [*percutere*, to strike through]. A method of physical diagnosis applied by striking upon any part of the body, with a view to ascertaining the conditions of the underlying organs by the character of the sounds elicited. **p., auscultatory**, percussion combined with auscultation. It is best performed by placing a double stethoscope at a fixed point and percussing gently all around. **p., immediate**, percussion in which the surface is struck directly, without the interposition of a pleximeter. **p., instrumental**, the use of a special hammer as a plexor, either alone or with a plate as a pleximeter. **p., mediate**, percussion in which a pleximeter is used. **p.-note**, the sound elicited on percussion. **p.-wave**, the term given to the chief ascending wave of the sphygmographic tracing.

**percussopunctator** (*per-kus-o-pungk-ta'-tor*) [*percutere*, to beat; *punctare*, to mark]. An instrument resembling a plexor or hammer, consisting principally of a group of needles by means of which multiple punctures are made into the tissues in rheumatism, lumbago, and neuralgia.

**percussor** (*per-kus'-or*) [*percutere*, to strike through]. He who or that which percusses; a percutor or plessor.

**percutaneous** (*per-ku-ta'-ne-us*) [*per*; *cutaneous*]. Performed through the skin, as *percutaneous* faradization.

**percutor** (*per-ku'-tor*) [*percutere*, to strike]. An instrument used in the percussion of massage and in therapeutic flagellation.

**Percy's operation** (*per'-se*) [J. F. *Percy*, American surgeon]. A method of destroying cancer tissue by the use of carefully regulated heat. The following agents are employed: radiant energy, hot air, hot water, steam, electro-coagulation, fulguration, and the actual cautery.

**pereirine** (*pex-i'-rēn*). An amorphous alkaloid found in the bark of *Geissospermum læve*. Its hydrochloride and valerianate are used as antipyretics.

**perencephalia, perencephaly** (*per-en-sef-a'-le-ah*, *per-en-sef'-al-e*) [πήρα, a pouch; ἐγκέφαλος, brain]. A condition marked by multiple cystic brain tumors.

**Perez's bacillus.** The supposed microorganism of ozena. **P.'s sign.** A loud friction-murmur heard over the sternum when the patient raises his arms, especially the left, over his head and lets them fall again; it is noted in cases of aneurysm of the arch of the aorta and in mediastinal tumors.

**perflation** (*per-fla'-shun*) [*perflare*, to blow through]. 1. A method of ventilation by which a current of air blowing against a dwelling is made to force its way in. 2. The act of forcing air into a cavity for the purpose of evacuating fluid.

**perforans** (*per'-for-anz*) [*perforate*]. Penetrating

**periblastic** (*per-e-blas'-tik*) [*peri-*; βλαστός, a germ]. Pertaining to the periblast. Germinating from the surface of an ovum.
**periblepsia, periblepsis** (*per-e-blep'-se-ah, -sis*) [*peri-*; βλέπειν, to look]. The wild look of a patient in delirium.
**peribronchial** (*per-e-brong'-ke-al*) [*peri-*; *bronchus*]. Surrounding a bronchus; occurring about a bronchus.
**peribronchiolitis** (*per-e-brong-ke-o-li'-tis*). Inflammation around the bronchioles.
**peribronchitis** (*per-e-brong-ki'-tis*). Inflammation of the tissue immediately surrounding the bronchi.
**peribrosis** (*per-ib-ro'-sis*) [*peri-*; βρῶσις, a feeding]. Ulceration at the canthi of the eyelids.
**pericæcal** (*per-is-e'-kal*). See *pericecal*.
**pericardiac, pericardial** (*per-e-kar'-de-ak, per-e-kar'-de-al*) [*pericardium*]. Pertaining to the pericardium.
**pericardicentesis** (*per-e-kar-de-sen-te'-sis*) [*pericardium*; κέντησις, a pricking]. Puncture of the pericardium.
**pericardiopleural** (*per-e-kar-de-o-ploo'-ral*). Relating to the pericardium and to the pleuræ.
**pericardiorrhaphy** (*per-e-kar-de-or'-af-e*) [*pericardium*; ῥαφή, suture]. The suturing of a wound in the pericardium.
**pericardiotomy** (*per-ik-ar-de-ot'-o-me*) [*pericardium*; τομή, a cutting]. Incision of the pericardium.
**pericarditic** (*per-ik-ar-dit'-ik*). Pertaining to pericarditis.
**pericarditis** (*per-e-kar-di'-tis*) [*pericardium*; ἴτις, inflammation]. Inflammation of the pericardium. The symptoms are slight fever, precordial pain and tenderness, cough, dyspnea, and rapid pulse. The physical signs vary—in the early stage there is a distinct friction-sound on auscultation, and sometimes a fremitus on palpation. In the stage of effusion there are bulging of the precordia and a triangular area of dulness, the base of which is downward; the heart-sounds are muffled. In chronic pericarditis with adhesions there is often systolic retraction of the precordia. The causes of pericarditis are rheumatism, the acute and chronic infectious diseases, Bright's disease, and extension of inflammation from neighboring parts. **p., adhesive,** that in which the two layers of pericardium tend to adhere. **p., carcinomatous,** that due to carcinoma of the pericardium. **p., dry,** a form without effusion. **p., external,** that affecting the outer layer of the pericardium. **p., fibrinous,** a form in which the membrane is covered with a fibrinous exudate, first soft and buttery in consistence, but later organizing. **p., hemorrhagic,** a form in which the fluid is hemorrhagic. This is the case most often in tuberculous pericarditis, also in scorbutus and in cachectic conditions. **p., localized,** a form giving rise to whitish areas, the so-called milk-spots. **p., moist,** that attended by an effusion. **p., obliterans,** a form leading to obliteration of the cavity by the adhesions of the layers. **p., purulent,** a variety in which the effused fluid becomes purulent. **p., serofibrinous,** a form in which there is but little lymph or fibrin, but a considerable quantity of serous fluid. **p. tuberculous,** a form due to tuberculous infection of the pericardium. **p., typhoid,** that in which there are high fever and typhoid symptoms.
**pericardium** (*per-e-kar'-de-um*) [*peri-*; καρδία, heart]. The closed membranous sac enveloping the heart. Its base is attached to the central tendon of the diaphragm; its apex surrounds for a short distance the great vessels arising from the base of the heart. It consists of an outer fibrous coat, derived from the cervical fascia; and an inner serous coat. The sac normally contains from 5 to 20 Gm. of clear serous liquid. The part in contact with the heart (*visceral pericardium*) is termed the *epicardium*; the other is the *parietal pericardium*. **p., bread-and-butter,** a peculiar appearance produced in fibrinous pericarditis by the rubbing of the two surfaces of the membrane over each other. **p., shaggy,** a pericardium upon which, as the result of fibrinous pericarditis, thick, loose, shaggy layers of fibrin are deposited.
**pericardosis** (*per-e-kar-do'-sis*). Microbic infection of the pericardium.
**pericardotomy** (*per-e-kar-dot'-o-me*) [*pericardium*; τομή, a cutting]. The operation of opening the pericardium.
**pericecal** (*per-e-se'-kal*) [*peri-*; *cecum*]. Surrounding the cecum.

**pericellular** (*per-is-el'-u-lar*) [*peri-*; *cellula*; a small cell]. Surrounding a cell.
**pericementitis** (*per-e-sem-ent-i'-tis*) [*peri-*; *cementum*, cement; ἴτις, inflammation]. Fauchard's disease; progressive necrosis of the dental alveoli.
**pericementum** (*per-e-sem-ent'-um*). A dense fibrous tissue covering the fang of a tooth.
**pericentral** (*per-is-en'-tral*). Situated around a center, or centrum.
**perichareia** (*per-ik-ar-i'-ah*) [περιχαρής, very glad]. Sudden vehement, or morbid rejoicing; a symptom in certain insanities.
**pericholangitis** (*per-e-ko-lan-ji'-tis*). See *periangiocholitis*.
**pericholecystitis** (*per-ik-ol-e-sis-ti'-tis*) [*peri-*; χολή, bile; κύστις, bladder; ἴτις, inflammation]. Inflammation near or around the gall-bladder.
**pericholous** (*per-ik'-o-lus*) [*peri-*; χολή, bile]. Excessively bilious.
**perichondral** (*per-e-kon'-dral*). Relating to the perichondrium.
**perichondrial** (*per-e-kon-dre'-al*). Pertaining to or resembling perichondrium.
**perichondritic** (*per-ik-on-drit'-ik*). Pertaining to or affected with perichondritis.
**perichondritis** (*per-ik-on-dri'-tis*) [*perichondrium*; ἴτις, inflammation]. Inflammation of the perichondrium.
**perichondrium** (*per-e-kon'-dre-um*) [*peri-*; χόνδρος, cartilage]. The fibrous connective tissue covering the surface of cartilage.
**perichondroma** (*per-ik-on-dro'-mah*) [*peri-*; χόνδρος, cartilage; ὄμα, tumor]. A tumor of the perichondrium.
**perichord** (*per'-ik-ord*) [*peri-*; χορδή, cord]. The sheath of the notochord.
**perichordal** (*per-ik-or'-dal*). Pertaining to the perichord.
**perichoroid, perichoroidal** (*per-ik-o'-roid, per-ik-o-roi'-dal*) [*peri-*; χόριον, chorion; εἶδος, like]. Surrounding the choroid.
**periclasis** (*per-ik'-las-is*) [περικλᾶν, to twist around]. A comminuted fracture.
**pericolitis** (*per-ik-o-li'-tis*) [*peri-*; *colitis*]. Inflammation of the tissues around the colon.
**pericolonitis** (*per-ik-o-lon-i'-tis*). Same as *pericolitis*.
**pericolpitis** (*per-e-kol-pi'-tis*). See *paracolpitis*.
**periconchal** (*per-ik-ong'-kal*) [*peri-*; κόγχη, a shell]. Surrounding the concha of the ear. **p. sulcus,** a sulcus separating the helix and the convex hinder surface of the concha.
**periconchitis** (*per-e-kon-ki'-tis*) [*peri-*; κόγχη, the socket of the eye]. Inflammation of the periosteum or lining membrane of the orbit.
**pericorneal** (*per-e-kor'-ne-al*) [*peri-*; *cornea*]. Surrounding the cornea.
**pericowperitis** (*per-e-kow-per-i'-tis*). Inflammation of the tissues about Cowper's glands.
**pericoxitis** (*per-e-koks-i'-tis*). Inflammation of the tissues around the hip-joint.
**pericranial** (*per-e-kra'-ne-al*) [*pericranium*]. Pertaining to the pericranium.
**pericranitis** (*per-ik-ra-ni'-tis*). Inflammation of the pericranium.
**pericranium** (*per-e-kra'-ne-um*) [*peri-*; κρανίον, skull]. The periosteum of the skull. **p. internum.** See *endocranium*.
**pericystic** (*per-e-sis'-tik*). 1. Surrounding a cyst. 2. Surrounding a bladder either gall-bladder or urinary bladder.
**pericystitis** (*per-e-sis-ti'-tis*). Inflammation of the peritoneum or the connective tissue surrounding the bladder.
**pericystium** (*per-e-sis'-te-um*) [*peri-*; κύστις, a cyst]. 1. The vascular wall of a cyst. 2. The tissues surrounding a bladder.
**pericytial** (*per-e-sit'-e-al*) [*peri-*; κύτος, a cell]. Surrounding a cell.
**peridectomy** (*per-e-dek'-to-me*). Synonym of *peritomy*.
**peridendritic** (*per-e-den-drit'-ik*). Surrounding a dentrite.
**peridental** (*per-e-den'-tal*) [*peri-*; *dens*, a tooth]. Surrounding a tooth or its root; periodontal.
**periderm** (*per'-id-erm*) [*peri-*; δέρμα, skin]. 1. The cuticle. 2. The Malpighian layer of the skin.
**peridermal, peridermic** (*per-id-er'-mal, per-id-er'-mik*). Cuticular; pertaining to the periderm.
**peridesmitis** (*per-id-es-mi'-tis*). Inflammation of the peridesmium.

**peridesmium** (*per-e-des'-me-um*) [*peri-*; δεσμός, a band]. The delicate membrane that invests a ligament.
**peridiastole** (*per-e-di-as'-to-le*) [*peri-*; *diastole*]. The pause between the systole and diastole.
**perididymis** (*per-e-did'-im-is*) [*peri-*; δίδυμοι, a testicle]. The tunica albuginea testis.
**perididymitis** (*per-e-did-im-i'-tis*) [*perididymis*; ιτις, inflammation]. Inflammation of the perididymis.
**periencephalitis** (*per-e-en-sef-al-i'-tis*). Inflammation of the pia mater.
**periencephalomeningitis** (*per-e-en-sef-al-o-men-in-ji'-tis*). See *periencephalitis*.
**periendothelioma** (*per-e-en-do-the-le-o'-mah*). A tumor originating in the endothelium of the lymphatics and the perithelium of the blood-vessels.
**periendymal** (*per-e-en'-dim-al*). Same as *periependymal*.
**perienteric** (*per-e-en-ter'-ik*) [*peri-*; ἔντερον, intestine]. Situated around the enteron; perivisceral.
**perienteritis** (*per-e-en-ter-i'-tis*) [*peri-*; ἔντερον, intestine; ιτις, inflammation]. Inflammation of the intestinal peritoneum.
**perienteron** (*per-e-en'-ter-on*) [*peri-*; ἔντερον, intestine]. The primitive perivisceral cavity; the space between the entoderm and the ectoderm; the forerunner of the schizocele or enterocele.
**periependymal** (*per-e-e-ep-en'-dim-al*) [*peri-*; *ependyma*]. Situated, or occurring, outside the ependyma.
**periepithelioma** (*per-e-e-ep-e-the-le-o'-mah*). A tumor originating in the endothelium lining the blood-vessels or lymphatics.
**Perier's operation** (*per-e-a'*). For uterine inversion; removal of the inverted uterus by the elastic ligature.
**perieresis** (*per-e-er'-es-is*) [περιαιρεῖν, to take off]. A circular incision around a tumor or abscess.
**periesophageal, periœsophageal** (*per-e-e-so-faj'-e-al*) [*peri-*; *esophagus*]. Situated, or occurring, just outside of, or around, the esophagus.
**periesophagitis, periœsophagitis** (*per-e-e-sof-aj-i'-tis*) [*peri-*; *esophagus*; ιτις, inflammation]. Inflammation of the tissues that surround the esophagus.
**perifascicular** (*per-e-fas-ik'-u-lar*) [*peri-*; *fasciculus*, a fascicle]. Surrounding a fasciculus.
**perifibral, perifibrous** (*per-if-i'-bral, per-if-i'-brus*) [*peri-*; *fibra*, a fiber]. Surrounding a fiber.
**perifistular** (*per-e-fis'-tū-lar*). Around a fistula.
**perifolliculitis** (*per-if-ol-ik-ū-li'-tis*) [*peri-*; *folliculus*, follicle; ιτις, inflammation]. Inflammation around the hair-follicles.
**perifolliculosis** (*per-if-ol-ik-ū-lo'-sis*) [*peri-*; *folliculus*, a follicle; νόσος, disease]. A follicular skin-affection.
**perigangliitis** (*per-ig-ang-gle-i'-tis*) [*peri-*; *ganglion*; ιτις, inflammation]. Inflammation of the tissues surrounding a ganglion.
**periganglionic** (*per-ig-ang-gle-on'-ik*) [*peri-*; *ganglion*]. Situated, or occurring, around a ganglion.
**perigastric** (*per-ig-as'-trik*) [*peri-*; γαστήρ, the stomach]. Surrounding or in the neighborhood of the stomach.
**perigastritis** (*per-e-gas-tri'-tis*). Inflammation of the peritoneal coat of the stomach.
**perigemmal** (*per-e-jem'-al*) [*peri-*; *gemma*, a bud]. Around a bulb or bulb-like structure; a mode of termination of certain nerve fibrils.
**periglandulitis** (*per-e-gland-ū-li'-tis*). Inflammation of the tissues about a small gland.
**periglottic** (*per-ig-lot'-ik*) [*peri-*; γλῶσσα, the tongue]. Situated around the base of the tongue and the epiglottis.
**periglottis** (*per-e-glot'-is*) [*peri-*; γλῶττις, the tongue]. The mucous membrane or villous coating of the tongue.
**perignathic** (*per-ig-na'-thik*) [*peri-*; γνάθος, jaw]. Situated about the jaw.
**perihepatic** (*per-e-he-pat'-ik*) [*peri-*; ἧπαρ, liver]. Surrounding, or occurring around, the liver.
**perihepatitis** (*per-e-hep-at-i'-tis*). Inflammation of the peritoneum surrounding the liver.
**periherniaI, perihernious** (*per-e-her'-ne-a-re, -ne-us*). Applied to tissue immediately about a hernia.
**perihysteric** (*per-e-his-ter'-ik*) [*peri-*; ὑστέρα, the womb]. Around the uterus; periuterine.
**perikaryon** (*per-e-kar'-e-on*) [*peri-*; κάρυον, a nut]. The cytoplasm of a neuron; the cell-body of a nerve.
**perikeratic** (*per-ik-er-at'-ik*) [*peri-*; κέρας, cornea]. Surrounding the cornea.
**perilabyrinthitis** (*per-e-lab-ir-inth-i'-tis*). Inflammation of the part surrounding the labyrinth.
**perilaryngeal** (*per-e-lar-in'-je-al*) [*peri-*; *larynx*]. Situated, or occurring, around the larynx.
**perilaryngitis** (*per-e-lar-in-ji'-tis*) [*peri-*; *larynx*; ιτις, inflammation]. Inflammation of the areolar tissue surrounding the larynx.
**perilymph** (*per'-e-limf*). The fluid separating the membranous from the osseous labyrinth of the ear.
**perilymphangial** (*per-il-im-fan'-je-al*) [*peri-*; *lymph*; ἀγγεῖον, vessel]. Situated, or occurring, around a lymphatic vessel.
**perilymphangitis** (*per-e-lim-fan-ji'-tis*) [*peri-*; *lymph*; ἀγγεῖον, vessel; ιτις, inflammation]. Inflammation of the tissues surrounding a lymphatic vessel.
**perilymphatic** (*per-e-lim-fat'-ik*). 1. Pertaining to the perilymph. 2. Situated or occurring about a lymphatic vessel.
**perimadarous** (*per-e-mad'-ar-us*) [*peri-*; μαδαρός, bald]. Applied to a spreading ulcer with the epiderm peeling off before its advance.
**perimastitis** (*per-e-mast-i'-tis*). Inflammation of the nne tive tissue surrounding the mammary glandʤo c
**perimeningitis** (*per-e-men-in-ji'-tis*). Inflammation of the dura mater.
**perimeter** (*per-im'-et-er*) [*peri-*; μέτρον, a measure]. 1. Circumference or border. 2. An instrument for measuring the extent of the field of vision. It consists ordinarily of a flat, narrow, metal plate bent in a semicircle, graduated in degrees, and fixed to an upright at its center by a pivot, on which it is movable. Variously colored discs are moved along the metal plate, and the point noted at which the person, looking directly in front of him, distinguishes the color.
**perimetric** (*per-im-et'-rik*) [*peri-*; μήτρα, uterus; μέτρον, measure]. 1. Situated around the uterus. 2. Pertaining to perimetry.
**perimetritic** (*per-im-et-rit'-ik*). Pertaining to perimetritis.
**perimetritis** (*per-e-met-ri'-tis*) [*perimetrium*; ιτις, inflammation]. Inflammation of the peritoneal covering of the uterus.
**perimetrium** (*per-e-me'-tre-um*) [*peri-*; μήτρα, the womb]. The serous covering of the uterus.
**perimetrosalpingitis** (*per-e-met-ro-sal-pin-ji'-tis*). A collective name for periuterine inflammations.
**perimetry** (*per-im'-et-re*) [*perimeter*]. The measuring of the field of vision.
**perimyelis** (*per-im-i'-el-is*) [*peri-*; μυελός, marrow]. The medullary membrane or endosteum; the areolar envelope of the bone-marrow; the pia mater of the spinal cord.
**perimyelitis** (*per-e-mi-el-i'-tis*). Inflammation of the pia mater of the spinal cord.
**perimyoendocarditis** (*per-e-mi-o-en-do-kar-di'-tis*). Combined pericarditis, myocarditis, and endocarditis.
**perimysial** (*per-e-mis'-e-al*) [*peri-*; μῦς, muscle]. Of the nature of, or pertaining to, perimysium; enveloping a muscle.
**perimysiitis** (*per-im-is-e-i'-tis*) [*peri-*; μῦς, muscle; ιτις, inflammation]. Inflammation of the perimysium.
**perimysium** (*per-e-mis'-e-um*) [*peri-*; μῦς, muscle]. The connective tissue enveloping the primary bundles of muscle-fibers. p. externum, the epimysium. p. internum, the endomysium.
**perinæum** (*per-in-e'-um*). See *perineum*.
**perineal** (*per-in-e'-al*) [*perineum*]. Pertaining to the perineum. p. body, the mass of tissue composed of skin, muscle, and fascia, occupying the interval between the vagina and the rectum of the woman. p. cystotomy, cystotomy performed through a perineal incision. p. fossa, the ischiorectal fossa. p. hernia, a hernia perforating the perineum by the side of the rectum or between the rectum and the bladder or the vagina. p. section, incision through the perineum for the relief of urethral stricture, the removal of calculi from the bladder, or the relief of other morbid conditions.
**perineauxesis** (*per-in-e-awks-e'-sis*) [*perineum*; αὔξησις, growth]. Any operation for the repair of a lacerated perineum.
**perineo-** (*per-in-e-o-*) [*perineum*]. A prefix meaning relating to the perineum.
**perineocele** (*per-ine'-o-sēl*) [*perineo-*; κήλη, hernia]. Perineal hernia.
**perineocolporectomyomectomy** (*per-in-e-o-kol-po-rek-to-mi-o-mek'-to-me*). Excision of a myoma by incision of the perineum, vagina, and rectum.

**perineoplasty** (*per-in-e'-o-plas-te*) [*perineo-;* πλάσ-σειν, to form]. Plastic operation upon the perineum.
**perineorrhaphy** (*per-in-e-or'-a-fe*) [*perineo-;* ῥαφή, suture]. Suture of the perineum, usually for the repair of a laceration caused during childbirth.
**perineoscrotal** (*per-in-e-o-skro'-tal*). Relating to the perineum and scrotum.
**perineosynthesis** (*per-in-e-o-sin'-thes-is*) [*perineum;* συνθλᾶν, a placing together]. A plastic operation upon the perineum in which a graft of vaginal mucosa is made to cover the wound; a variety of perineorrhaphy.
**perineotomy** (*per-in-e-ot'-o-me*) [*perineo-;* τομή, a cutting]. Incision through the perineum.
**perineovaginal** (*per-in-e-o-vaj'-in-al*). Relating to the perineum and vagina.
**perineovaginorectal** (*per-in-e-o-vaj-in-o-rek'-tal*). Relating to the perineum, vagina, and rectum.
**perinephral** (*per-in-ef'-ral*). Same as *perinephric.*
**perinephrial** (*per-in-ef'-re-al*). Pertaining to the perinephrium.
**perinephric** (*per-e-nef'-rik*) [*peri-;* νεφρός, kidney]. Situated or occurring around the kidney, as *perinephric abscess.*
**perinephritic** (*per-e-nef-rit'-ik*) [*perinephritis*]. 1. Pertaining to perinephritis. 2. Improperly used instead of perinephric.
**perinephritis** (*per-e-nef-ri'-tis*). Inflammation of the tissues surrounding the kidney.
**perinephrium** (*per-e-nef'-re-um*) [*peri-;* νεφρός, kidney]. The connective and adipose tissue surrounding the kidney.
**perinæum, perinæum** (*per-in-e'-um*) [περίναιον, the perineum]. That portion of the body included in the outlet of the pelvis, bounded in front by the pubic arch, behind by the coccyx and great sacrosciatic ligaments, and at the side by the tuberosities of the ischium. It is occupied by the terminations of the rectum, the urethra, and the root of the penis, together with their muscles, fasciæ, vessels, and nerves.
**perineurial** (*per-e-nū'-re-al*) [*peri-;* νεῦρον, nerve]. Relating to the perineurium. p. **lymph-channels,** lymph-spaces surrounding the nerve-trunks, as in the cornea.
**perineuritis** (*per-e-nū-ri'-tis*) [*perineurium;* ιτις, inflammation]. Inflammation of the perineurium.
**perineurium** (*per-e-nū'-re-um*) [*peri-;* νεῦρον, a nerve]. The connective-tissue sheath investing a funiculus or primary bundle of nerve-fibers.
**perinuclear** (*per-e-nū'-kle-ar*). Surrounding the nucleus.
**periocular** (*per-e-ok'-ū-lar*) [*peri-;* oculus, eye]. Surrounding the eye. p. **space,** the space between the globe of the eye and the orbital walls.
**period** (*pe'-re-od*) [*peri-;* ὁδός, way]. The space of time during which anything is in progress or an event takes place. p. **childbearing,** the period, from puberty to the menopause, during which the female is capable of reproducing offspring. p., **dodging,** a colloquial term for the menopause. p., **incubation-.** See *incubation* (2). p., **menstrual,** p., **monthly,** the menses. p., **reaction.** See *reaction-period.* p., **respiratory,** the interval between two successive inspirations.
**periodic** (*pe-re-od'-ik*). Recurring at more or less regular intervals.
**periodicity** (*pe-re-od-is'-it-e*) [*period*]. Recurrence at regular intervals.
**periodocasein** (*pe-re-o-do-ka'-se-in*). A proprietary compound of iodine and casein, used in myxedema.
**periodology** (*pe-re-od-ol'-o-je*) [*period;* λόγος, science]. The sum of what is known concerning the tendency of certain diseases and morbid phenomena to recur at stated periods.
**periodontal** (*per-e-o-don'-tal*) [*peri-;* ὀδούς, tooth]. Surrounding a tooth, as the *periodontal* membrane, that lining the cement of a tooth.
**periodontitis** (*per-e-o-don-ti'-tis*) [*periodontium;* ιτις, inflammation]. Inflammation of the periodontal membrane.
**periodontium** (*per-e-o-don'-she-um*) [*peri-;* ὀδούς, tooth]. The membrane surrounding a tooth; the periodontal membrane.
**periodoscope** (*pe-re-od'-o-skōp*) [*period;* σκοπεῖν, to inspect]. A calendar in the form of a movable dial, used in determining the probable date of confinement.
**periods** (*pe'-re-ods*). The menses.

**periodynia** (*per-e-o-din'-e-ah*) [*peri-;* ὀδύνη, pain]. Severe general pain throughout the body.
**periœsophagitis** (*per-e-e-sof-aj-i'-tis*). See *periesophagitis.*
**periomphacous** (*per-e-om'-fak-us*) [*peri-;* ὀμφακώδης, like unripe grapes]. Immature, unripe; applied to abscesses.
**perion** (*per'-e-on*) [*peri-;* ᾠόν, egg]. The decidua.
**perionychia** (*per-e-o-nik'-e-ah*) [*peri-;* ὄνυξ, nail]. Inflammation around the nails.
**perionychium** (*per-e-on-ik'-e-um*) [*peri-;* ὄνυξ, nail]. The border of epiderm at the root of the nail.
**perionyxis** (*per-e-o-niks'-is*). Synonym of *perionychia.*
**peri-oophoritis** (*per-e-o-of-or-i'-tis*) [*peri-;* *oophoron;* ιτις, inflammation]. Inflammation of the peritoneum and connective tissue covering the ovary.
**peri-oophorosalpingitis** (*per-e-o-of-or-o-sal-pin-ji'-tis*) [*peri-;* *oophoron;* *salpinx;* ιτις, inflammation]. Inflammation of the tissues surrounding the ovary and oviduct.
**periophthalmic** (*per-e-off-thal'-mik*). Around the eye.
**periophthalmitis** (*per-e-off-thal-mi'-tis*) [*peri-;* ὀφθαλμός, eye; ιτις, inflammation]. Inflammation of the tissues surrounding the eye.
**periople** (*per-e-op'-le*) [*peri-;* ὁπλή, a hoof]. The bands of horny matter which run obliquely inward from the heel of a horse's foot, including the frog between them. Syn., *bar; coronary frog band; perioplic band.*
**perioptic** (*per-e-op'-tik*) [*peri-;* ὀπτικός, seeing]. 1. Surrounding the orbit, or the eye. 2. Of, or pertaining to, the tissues about the eye.
**perioptometry** (*per-e-op-tom'-et-re*) [*peri-;* ὀπτός, visible; μέτρον, a measure]. The measurement of the limits of the visual field.
**perioral** (*per-e-o'-ral*) [*peri-;* *os, oris,* the mouth]. Surrounding the mouth; circumoral.
**periorbita** (*per-e-or'-bit-ah*) [*peri-;* *orbita,* orbit]. The periosteum of the eye-socket.
**periorbital** (*per-e-or'-bit-al*). 1. Surrounding the orbit. 2. Pertaining to the periorbita.
**periorbititis** (*per-e-or-bit-i'-tis*) [*peri-;* *orbita,* orbit; ιτις, inflammation]. Inflammation of the periorbita.
**periorchitis** (*per-e-or-ki'-tis*). Inflammation of the tissues surrounding the testicle. p. **adhæsiva,** a form in which adhesions are formed between the two layers of the tunica vaginalis. p. **prolifera,** periorchitis associated with proliferation of the connective-tissue elements of the tunica albuginea.
**periost** (*per'-e-ost*). Same as *periosteum.*
**periosteal** (*per-e-ost'-te-al*) [*periosteum*]. Pertaining to the periosteum.
**periosteitis** (*per-e-os-te-i'-tis*). See *periostitis.*
**periosteoma** (*per-e-os-te-o'-mah*) [*peri-;* ὀστέον, bone; ὄγκωμα, tumor]. An osteoma developed from the periosteum.
**periosteomedullitis, periosteomedullitis.** See *periosteo-osteomyelitis.*
**periosteo-osteomyelitis.** (*per-e-ost-e-o-os-te-o-mi-el-i'-tis*) [*periosteum;* *osteomyelitis*]. Inflammation of the periosteum and medulla of a bone.
**periosteophyma** (*per-e-os-te-o-fi'-mah*) [*peri-;* ὀστέον, bone; φῦμα, growth]. Swelling of the periosteum; also, a periosteophyte.
**periosteophyte** (*per-e-os'-te-o-fīt*) [*peri-;* ὀστέον, bone; φυτόν, growth]. A morbid osseous formation upon or proceeding from the periosteum.
**periosteosis** (*per-e-os-te-o'-sis*). 1. Synonym of *periostosis.* 2. See *periosteoma.* 3. The formation of a tumor of the periosteum.
**periosteotitis, periostitis** (*per-e-ost-e-ost-e-i'-tis, per-e-ost-i'-tis*). Simultaneous periostitis and osteitis.
**periosteotome** (*per-e-ost'-te-o-tōm*) [*periosteum;* τομή, a cutting]. An instrument for incising the periosteum and scraping it from the bone.
**periosteotomy** (*per-e-os-te-ot'-o-me*) [*see* *periosteotome*]. The operation of incising the periosteum.
**periosterosis** (*per-e-ost'-er-o'-sis*). Same as *periosteal.*
**periosteum** (*per-e-os'-te-um*) [*peri-;* ὀστέον, bone]. A fibrous membrane investing the surfaces of bones, except at the points of tendinous and ligamentous attachment and on the articular surfaces, where cartilage is substituted.
**periostitic** (*per-e-os-tit'-ik*) [*peri-;* *periostitis,* bone; ιτις, inflammation]. Pertaining to, resembling, or affected with, periostitis.
**periostitis** (*per-e-os-ti'-tis*) [*periosteum;* ιτις, inflammation]. Inflammation of the periosteum. It

may be acute or chronic, the latter being the more frequent form. Acute periostitis is either traumatic or the result of infection; the chronic is due to traumatism, syphilis, tuberculosis, or actinomycosis. In the acute there are swelling and diffuse suppuration, with fever and other constitutional symptoms; in the chronic, pain, which is usually worse at night, swelling, and tenderness. p. **albuminosa,** a mild form of inflammation characterized by the formation of a clear, ropy, albuminous liquid resembling synovia. It is most common in the young, and is unaccompanied by fever. p., **dental,** inflammation of the investing membrane of the roots of the teeth. p., **diffuse,** a serious inflammation usually involving the periosteum of long bones. p., **hemorrhagic,** that accompanied by bleeding between the periosteum and the bone.

**periostoma** (per-e-os-to'-mah) [peri-; ὀστέον, bone; ὄμα, tumor]. Any morbid osseous growth occurring on or surrounding a bone.

**periostosis** (per-e-os-to'-sis) [peri-; ὀστέον, bone]. An osseous formation on the exterior of a bone.

**periostotomy** (per-e-os-tot'-o-me). Synonym of *periosteotomy*.

**periotic** (per-e-o'-tik) [peri-; οὖς, ear]. 1. Situated about the ear. 2. Of or pertaining to the parts immediately about the internal ear. 3. The petrous and mastoid parts of the temporal bone.

**periovaritis** (per-e-o-var-i'-tis). See *perioophoritis*.

**periovular** (per-e-o'-vū-lar) [peri-; ovum, egg]. Surrounding the ovum.

**peripachymeningitis** (per-e-pak-e-men-in-ji'-tis) [peri-; pachymeningitis]. Inflammation of the connective tissue between the dura mater and the bone.

**peripancreatitis** (per-e-pan-kre-at-i'-tis). Inflammation of the tissues about the pancreas.

**peripapillary** (per-ip-ap'-il-a-re) [peri-; papilla, a papilla]. Occurring or situated around the circumference of a papilla, and especially of the optic disc.

**peripatetic** (per-ip-at-et'-ik) [peri-; πατεῖν, to walk]. Walking about, as in "walking typhoid."

**peripenial** (per-ip-e'-ne-al) [peri-; penis]. Surrounding the penis.

**periphacitis** (per-if-a-si'-tis). Inflammation of the periphacus.

**periphacus** (per-if-a'-kus) [peri-; φακός, crystalline lens]. The capsule surrounding the crystalline lens.

**peripharyngeal** (per-if-ar-in'-je-al) [peri-; φάρυγξ, the throat]. Surrounding the pharynx.

**peripherad** (per-if'-er-ad) [periphery; ad, toward]. Toward the periphery.

**peripheral, peripheric** (per-if'-er-al, per-if-er'-ik) [periphery]. Pertaining to or placed near the periphery.

**peripheraphose** (per-if'-er-a-fōz). See under *phose*.

**peripheric** (per-if-er'-ik). Synonym of *peripheral*.

**peripherocentral** (per-if-er-o-sen'-tral). Relating to the center and periphery.

**peripherophose** (per-if'-er-o-fōz). See under *phose*.

**periphery** (per-if'-er-e) [peri-; φέρειν, to carry]. Circumference; the external surface.

**periphlebitic** (per-if-leb-it'-ik). Pertaining to, affected with, or of the nature of, periphlebitis.

**periphlebitis** (per-if-leb-i'-tis). Inflammation of the tissues about a vein.

**periplasm** (per'-ip-lazm) [peri-; πλάσμα, anything formed]. A delicate hyaline layer around animal cells.

**periplast** (per'-ip-last) [peri-; πλάσσειν, to mold, form]. 1. The periblast or matrix of a part or organ. 2. The intercellular substance, or stroma. 3. The attraction-sphere. p., **daughter,** the centrosome.

**periplastic** (per-ip-las'-tik). 1. Of or pertaining to or resembling the matrix or periplast of a part or organ. 2. The cell-substance about the nucleus or endoplast; perinuclear protoplasm.

**peripleuritis** (per-e-ploo-ri'-tis). Inflammation of the tissues surrounding the pleura.

**Periploca** (per-ip'-lo-kah) [peri-; πλέκειν, to twine]. A genus of plants of the order *Asclepiadeæ*. P. *græca* is a European species naturalized in western New York. The leaves are used as an emmenagog; the milky juice has been used to poison animals; the bark contains a glucoside, *periplocin*.

**periplocin** (per-ip'-lo-sin), C₃₆H₅₈O₁₂. A crystalline glucoside from the bark of *Periploca græca*. It is a powerful cardiac poison used subcutaneously in heart disease. Maximum daily dose ₆₄ gr.(0.001 Gm.).

**peripneumonia** (per-e-nū-mo'-ne-ah). 1. Pneumonia. 2. Pleuropneumonia. p. **notha,** the false pneumonia of the older writers; congestion of the lungs.

**peripolar** (per-e-po'-lar) [peri-; pole]. Surrounding a pole or the poles.

**periportal** (per-e-por'-tal) [peri-; porta, door]. Surrounding the portal vein. p. **carcinoma,** a primary carcinoma developing around the portal vein, beginning at its entrance into the liver, thence, extending along the portal vessels to the remote branches.

**periproctal, periproctic** (per-ip-rok'-tal, per-ip-rok'-tik) [peri-; πρωκτός, anus]. Surrounding the anus or rectum.

**periproctitis** (per-e-prok-ti'-tis). Inflammation of the areolar tissue about the rectum or anus.

**periprostatic** (per-ip-ros-tat'-ik) [peri-; prostate]. Situated or occurring around the prostate.

**periprostatitis** (per-ip-ros-tat-i'-tis) [peri-; prostate; ιτις, inflammation]. Inflammation of the tissues situated around the prostate.

**peripyemia** (per-ip-i-e'-me-ah) [peri-; pyemia]. Suppuration about an organ or tissue.

**peripylephlebitis** (per-ip-i-le-fleb-i'-tis) [peri-; πύλη, gate, porta; phlebitis]. Inflammation of the tissues surrounding the portal vein, or of its ectal coat.

**perirectal** (per-e-rek'-tal) [peri-; rectum]. About the rectum.

**perirectitis** (per-e-rek-ti'-tis). See *periproctitis*.

**perirenal** (per-e-re'-nal) [peri-; ren, kidney]. Around the kidney.

**perirhinal** (per-i-ri'-nal) [peri-; ῥίς, nose]. Situated about the nose or nasal fossæ.

**perisalpingitis** (per-e-sal-pin-ji'-tis). Inflammation of the peritoneal covering of the Fallopian tube.

**perisalpingo-ovaritis** (per-e-sal-pin-go-o-va-ri'-tis). See *peri-oophorosalpingitis*.

**perisalpinx** (per-is-al'-pinks) [peri-; σάλπιγξ, tube]. The peritoneum covering the upper border of the Fallopian tube.

**periscelis** (per-is'-kel-is) [peri-; σκέλος, leg]. Herpes occurring around the leg where the garter binds it.

**periscleritis** (per-e-skle-ri'-tis). See *episcleritis*.

**periscopic** (per-e-skop'-ik) [peri-; σκοπεῖν, to look]. Designed for looking around, as a *periscopic* lens. See *lens, periscopic*.

**perisigmoiditis** (per-is-ig-moi-di'-tis) [peri-; sigmoid; ιτις, inflammation]. Inflammation of the tissues, especially the peritoneum, covering the sigmoid flexure of the colon.

**perisinal, perisinous, perisinuous** (per-e-si'-nal, -nus, -sin'-ū-us). Surrounding a sinus.

**perisinuitis, perisinusitis** (per-e-si-nū-i'-tis, -si'-tis). Inflammation of the tissue about a sinus, especially a cerebral sinus.

**perispermatitis** (per-is-per-mat-i'-tis). Inflammation around the spermatic cord, with an effusion of fluid; a funicular hydrocele.

**perisplenic** (per-is-plen'-ik) [peri-; spleen]. Situated or occurring near the spleen.

**perisplenitis** (per-is-plen-i'-tis). Inflammation of the peritoneal coat of the spleen.

**perispondylitis** (per-e-spon-dil-i'-tis). Inflammation of the tissues around a vertebra. p., **Gibney's,** a painful condition of the muscles of the spine.

**perissad** (per'-is-ad) [περισσός, odd]. 1. Having an uneven quantivalence, as nitrogen, the quantivalence of which is three or five. 2. An element having such a quantivalence.

**peristalsis** (per-e-stal'-sis) [peri-; στάλσις, constriction]. A peculiar wave-like movement seen in tubes provided with longitudinal and transverse muscular fibers. It consists in a narrowing and shortening of a portion of the tube, which then relaxes, while a lower portion becomes shortened and narrowed. By means of this movement the contents of this tube are forced toward the opening. p., **reversed,** peristaltic movement opposite to the normal direction.

**peristaltic** (per-e-stal'-tik) [peristalsis]. Pertaining to or resembling peristalsis. p. **unrest,** a common symptom of neurasthenia, consisting in increased peristaltic movements of the stomach coming on shortly after eating, with borborygmus and gurgling.

**peristaltin** (per-e-stal'-tin). A glucoside, readily soluble in water, derived from *Rhamnus purshiana*.

**peristaphyline** (per-e-staf'-il-in) [peri-; σταφυλή, uvula]. Situated near the uvula.

**peristaphylitis** (per-e-staf-il-i'-tis) [peri-; σταφυλή,

uvula; ιτις, inflammation]. Inflammation of the tissues surrounding the uvula.
**peristerna** (*per-e-ster'-nah*) [*peri-*; *sternum*]. A name for the lateral portions of the chest.
**peristole** (*per-is'-to-le*). Peristalsis.
**peristoma** (*per-is-to'-mah*). See *peristome*.
**peristomal** (*per-is-to'-mal*) [*peri-*; στόμα, mouth]. Surrounding the mouth.
**peristome** (*per'-is-tōm*) [*peri-*; στόμα, the mouth]. In biology: (a) the parietal region surrounding the mouth, as the oral disc of a polyp; (b) the fringe of hair-like appendages about the orifice of a moss capsule.
**peristroma** (*per-e-stro'-mah*) [*peri-*; στρῶμα, covering]. The internal layer of a tube-like covering. The villous coat of the intestine.
**perisynovial** (*per-is-i-no'-ve-al*) [*peri-*; *synovial*]. Situated or occurring around a synovial membrane.
**perisystole** (*per-e-sis'-to-le*). The slight interval between the diastole and systole.
**peritendineum** (*per-e-ten-din'-e-um*) [*peri-*; *tendo*, tendon]. The tissue surrounding the tendons like a sheath.
**perithelial** (*per-e-the'-le-al*). Relating to the perithelium.
**perithelioma** (*per-e-the-le-o'-mah*). A tumor originating in the perithelium of a vessel.
**perithelium** (*per-e-the'-le-um*) [*peri-*; θηλή, nipple]. The layer of cells surrounding the capillaries and smaller vessels.
**perithoracic** (*per-e-tho-ras'-ik*) [*peri-*; *thorax*]. Situated or occurring around the thorax.
**perithyroiditis** (*per-e-thi-roid-i'-tis*) [*peri-*; *thyroid*; ιτις, inflammation]. Inflammation of the capsule of the thyroid gland.
**peritome** (*per-it'-om-e*) [*peri-*; τομή, a cutting]. Circumcision.
**peritomy** (*per-it'-om-e*) [see *peritome*]. 1. The removal of a strip of conjunctival and subconjunctival tissue from about the cornea for the relief of pannus. 2. Circumcision.
**peritonæum** (*per-it-on-e'-um*). See *peritoneum*.
**peritoneal** (*per-it-on-e'-al*) [*peritoneum*]. Pertaining to the peritoneum.
**peritonealgia** (*per-it-on-e-al'-je-ah*) [*peritoneum*; ἄλγος, pain]. Neuralgia of the peritoneum.
**peritoneopexy** (*per-it-on-e-o-peks'-e*) [*peritoneum*; πῆξις, a fixing in]. Fixation of the uterus by the vaginal route in the treatment of retroflexions of this organ.
**peritoneorrhexis** (*per-it-on-e-or-eks'-is*) [*peritoneum*; ῥῆξις, rupture]. Rupture of the peritoneum.
**peritoneotomy** (*per-it-on-e-ot'-o-me*) [*peritoneum*; τομή, a cutting]. Incision into the peritoneum.
**peritoneum** (*per-it-on-e'-um*) [*peri-*; τείνειν, to stretch]. The serous membrane lining the interior of the abdominal cavity and surrounding the contained viscera.
**peritonism** (*per'-it-on-ism*). 1. A false peritonitis soon yielding to treatment. 2. A complex of serious phenomena complicating peritonitis or diseases of those parts covered by peritoneum.
**peritonitic** (*per-it-on-it'-ik*) [*peritoneum*; ιτις, inflammation]. Pertaining to or affected with peritonitis.
**peritonitis** (*per-it-on-i'-tis*) [*peritoneum*; ιτις, inflammation]. Inflammation of the peritoneum. It may be acute or chronic. *Acute peritonitis* may be due to exposure to cold and wet (*idiopathic peritonitis*), traumatism, perforation of an abdominal viscus, extension from neighboring parts, rheumatism, or Bright's disease. The symptoms are moderate fever, a wiry pulse, abdominal pain, tenderness, and distention; the patient lies on his back with the thighs flexed; there are vomiting and constipation. *Chronic peritonitis* is due to tuberculosis, syphilis, carcinoma, nephritis, or it may be the sequel of an acute attack. **p., adhesive**, peritonitis with adhesion between the parietal and visceral layers. **p., diffuse**, that affecting the entire peritoneum. **p., parietal**, inflammation of the serous lining of the peritoneal cavity. **p., permeation**, that produced by the penetration of the healthy intestinal wall by bacteria. **p., puerperal**, that following labor, and usually due to septic infection. **p., septic**, peritonitis due to the microorganisms of suppuration. **p., serous**, **p., serosa**, that accompanied by liquid exudation. **p., tuberculous**, that due to the deposit of miliary tubercles upon the peritoneum.

**peritonsillar** (*per-e-ton'-sil-ar*) [*peri-*; *tonsil*]. About the tonsil.
**peritonsillitis** (*per-e-ton-sil-i'-tis*) [*peri-*; *tonsilla*, tonsil; ιτις, inflammation]. Inflammation of the tissues surrounding the tonsil.
**peritracheal** (*per-it-ra'-ke-al*) [*peri-*; *trachea*]. Surrounding the trachea.
**peritracheitis** (*per-it-ra-ke-i'-tis*) [*peri-*; *trachea*; ιτις, inflammation]. Inflammation of the connective tissue about the trachea.
**Peritricha** (*per-it'-rik-ah*) [*peri-*; θρίξ, hair]. A group of bacteria having flagella projecting from the sides as well as the poles.
**peritrichous** (*per-it'-rik-us*) [*peri-*; θρίξ, a hair]. Having a band of cilia or flagella around the body.
**peritrochanteric** (*per-e-tro-kan-ter'-ik*). Situated about a trochanter.
**perityphlitic** (*per-e-tif-lit'-ik*) [*peri-*; τυφλός, cecum; ιτις, inflammation]. Of the nature of or affected with perityphilitis.
**perityphlitis** (*per-e-tif-li'-tis*). Inflammation of the peritoneum surrounding the cecum and appendix.
**periumbilical** (*per-e-um-bil'-ik-al*). Surrounding the umbilicus.
**periungual** (*per-e-ung'-wal*) [*peri-*; *unguis*, a nail]. Around the nail.
**periureteric** (*per-e-ū-re-ter'-ik*) [*peri-*; *ureter*]. Surrounding one or both ureters.
**periureteritis** (*per-e-ū-re-ter-i'-tis*). Inflammation of the tissues around a ureter.
**periurethral** (*per-e-ū-re'-thral*). Surrounding the urethra.
**periurethritis** (*per-e-ū-re-thri'-tis*) [*peri-*; *urethra*; ιτις, inflammation]. Inflammation of the connective tissue about the urethra.
**periuterine** (*per-e-ū'-ter-in*) [*peri-*; *uterus*]. About the uterus.
**perivaginal** (*per-e-vaj'-in-al*). Around or about the vagina.
**perivaginitis** (*per-e-vaj-in-i'-tis*). Synonym of *paracolpitis*.
**perivascular** (*per-e-vas'-kū-lar*). About a vessel. **p., vessel**, *peri-*; *vasculum*, vessel; ιτις, inflammation]. Inflammation of the vessel-walls, or of the perivascular sheaths.
**perivenous** (*per-iv-e'-nus*) [*peri-*; *vena*, vein]. Investing or surrounding a vein; occurring around a vein.
**perivertebral** (*per-e-ver'-te-bral*). Surrounding a vertebra.
**perivesical** (*per-iv-es'-ik-al*) [*peri-*; *vesica*, bladder]. Situated about or surrounding the bladder.
**perivisceral** (*per-iv-is'-er-al*) [*peri-*; *viscus*, viscus]. Surrounding a viscus or viscera; occurring about a viscus.
**perivitelline** (*per-iv-it'-el-in*) [*peri-*; *vitellus*, yolk]. Surrounding the vitellus or yolk. **p. space**, the space between the zona pellucida and the vitellus.
**perixenitis** (*per-e-zen-i'-tis*) [*peri-*; ξένος, a stranger; ιτις, inflammation]. Inflammation around a foreign body embedded in the tissues.
**perizoma** (*per-iz-o'-mah*) [*peri-*; ζῶμα, girdle]. 1. A girdle, *q.v.*; also, a truss. 2. Herpes zoster.
**Perkinism** (*per'-kin-ism*). A method of empirical treatment devised by Elisha *Perkins*, an American physician [1740-1810]. It consisted in drawing over the affected part the extremities of two rods (metallic tractors) of different metals. Syn., *tractoration*.
**perle** (*perl*) [Fr. and Ger., "a pearl"; pl., *perles*]. A capsule for administration of medicine. See *pearl*.
**p.s, Laennec's.** See under *Laennec*.
**perlèche** (la) (*lah pehr-lāsh*) [Fr.]. A peculiar contagious disease of the mouth occurring in children. It consists in a thickening and desquamation of the epithelium at the angles of the mouth, with occasionally the formation of small fissures, giving rise to a smarting sensation in the lips. The disease is probably microbic in origin.
**Perles' anemia-bodies** (*perlz*) [Max *Perles*, German pathologist, 1843-1881]. Small club-shaped, actively motile bodies, 3-4μ in length, found by Perles in the blood in some cases of pernicious anemia.
**Perlia's nucleus** (*per'-le-ah*) [Richard *Perlia*, German ophthalmologist]. See *Spitzka's nucleus*.
**perlsucht** (*pairl'-zoocht*). A form of tuberculosis of the pleura or peritoneum seen in cattle.
**permanent** (*per'-man-ent*) [*per*, through; *manere*, to remain]. Lasting; fixed; enduring. **p. teeth**, the teeth of the second dentition.

# PERMANGANATE 672 PERU, BALSAM OF

**permanganate** (*per-man'-gan-āt*). A salt of permanganic acid. See *manganese*.
**permanganic acid** (*per-man-gan'-ik*); HMnO₄. A monobasic acid known chiefly in its salts.
**permeable** (*per'-me-a-bl*) [*per*, through; *meare*, to pass]. Capable of affording passage. **p. stricture**, a stricture that permits the passage of an instrument.
**permeation** (*per-me-a'-shun*) [*permeare*, to pass through]. The extension of cytomata by continuous growth along natural channels.
**permixion** (*per-mik'-shun*) [*permiscere*, to mingle]. A perfect chemical mixture.
**pernambuco** wood. The wood of *Cæsalpinia echinata*. It is used as an astringent and roborant, and contains brasilin, a coloring-matter.
**pernicious** (*per-nish'-us*) [*perniciosus*, destructive]. Highly destructive; of intense severity; deadly; fatal. **p. anemia**, a disease of the blood characterized by a great diminution in the number of red corpuscles, and a relatively smaller diminution of the hemoglobin, by the presence in the blood of poikilocytes, macrocytes, microcytes, and nucleated red corpuscles. The disease most common in middle life is usually fatal, although recoveries are reported in several instances. **p. malaria**. See under *malarial fever*. **p. vomiting**, persistent, uncontrollable vomiting occurring in pregnancy.
**pernio** (*per'-ne-o*) [L.]. Synonym of *chilblain*.
**pernoctation** (*per-nok-ta'-shun*) [*pernoctatio*, wakefulness]. Wakefulness; insomnia.
**pero** (*pe'-ro*) [*pero*, boot]. The soft ectal layer of the olfactory bulb whence the olfactory nerves arise.
**perobrachius** (*pe-ro-bra'-ke-us*) [πηρός, maimed; βραχίων, arm]. A developmental defect in which the forearms and hands are malformed or wanting.
**perocephalus** (*pe-ro-sef'-al-us*) [πηρός, maimed; κεφαλή, head]. A monster with an abnormality of the conformation of the head.
**perochirus**, **perocheirus** (*pe-ro-ki'-rus*) [πηρός, maimed; χείρ, hand]. A defect in the development consisting in absence or stunted growth of the hand.
**perocormus** (*pe-ro-kor'-mus*) [πηρός, maimed; κορμός, trunk]. A monster characterized by defective development of the trunk.
**perodactylia** (*pe-ro-dak-til'-e-ah*) [πηρός, maimed; δάκτυλος, finger]. Defective development of the fingers or toes.
**perodynia** (*pe-ro-din'-e-ah*) [πήρα, pouch; ὀδύνη, pain]. Cardialgia.
**peromelia** (*pe-ro-me'-le-ah*) [πηρός, maimed; μέλος, limb]. Teratic malformation of the limbs.
**peromelus** (*pe-rom'-el-us*) [πηρός, maimed; μέλος, limb]. A monster with deficient, stunted, or misshapen limbs.
**peromoplasty** (*pe-rom'-o-plas-te*) [πήρωμα, a maiming; πλάσσειν, to form]. The formation of a new stump after an amputation when the end of a bone projects.
**peronæus** (*per-o-ne'-us*). Same as *peroneus*.
**peronarthrosis** (*per-on-ar-thro'-sis*) [περόνη, pin; ἄρθρον, joint]. A saddle-joint; a joint in which the articular surfaces are both concave and convex, as in the carpometacarpal joint of the thumb.
**perone** (*per'-o-ne*) [περόνη, pin]. The fibula.
**peroneal** (*per-o-ne'-al*) [*perone*]. Pertaining to the fibula.
**peroneo-** (*per-o-ne'-o*) [*perone*]: A prefix denoting connection with or relation to the fibula.
**peroneum** (*per-o-ne'-um*). Synonym of *perone*.
**peroneus** (*per-o-ne'-us*) [see *perone*]. Pertaining to the fibula. **p. muscles.** See under *muscle*.
**peronia** (*pe-ro'-ne-ah*) [πηροῦν, to maim]. Mutilation; malformation.
**peronin** (*per-o'-nin*), C₁₈H₂₁NO₃ . HCl. Benzylmorphine hydrochloride; a substitution-product of morphine possessing feeble narcotic properties, but useful as a somnifacient and in allaying cough. Dose ⅔–1½ gr. (0.04–0.1 Gm.); maximum daily dose 6 gr. (0.3 Gm.).
**Peronospora** (*per-o-nos'-po-rah*) [περόνη, a pin; σπόρος, spore]. A genus of fungi producing mildew. *P. ferrani* is a species that was supposed to cause cholera; *P. lutea*, one that was once held to be the cause of yellow fever.
**peroplasia** (*pe-ro-pla'-ze-ah*) [πηρός, maimed; πλάσσειν, to mold]. A malformation due to an error of development.
**peropus** (*pe'-ro-pus*) [πηρός, maimed; πούς, foot]. A developmental defect in which the feet are malformed.

**per os** [L.]. By way of, or through the mouth.
**perosis** (*pe-ro'-sis*) [πηρός, maimed]. The condition of abnormal or defective formation.
**perosomus** (*pe-ro-so'-mus*) [πηρός, maimed; σῶμα, body]. A monster presenting malformation of the entire body.
**perosplanchnica** (*pe-ro-splank'-nik-ah*) [πηρός, maimed; σπλάγχνον, viscus]. Malformation of the viscera.
**perosseous** (*per-os'-e-us*) [*per*, through; *os*, bone]. Through bone.
**peroxidate**, **peroxidize** (*per-oks'-id-āt*, *per-oks'-id-īz*) [*peroxide*]. To oxidize completely.
**peroxide**, **peroxid** (*per-oks'-id*) [*per*, through; *oxide*]. That oxide of any base which contains the most oxygen.
**peroxol** (*pur-oks'-ol*). A combination of 3% solution of hydrogen peroxide with camphor (camphoroxol), menthol (menthoxol) or naphthol (naphthoxol).
**peroxydase** (*pur-oks'-e-dās*). An enzyme found in tobacco; it is capable of producing all the phenomena of fermentation. Cf. *oxydases*.
**perpendicular** (*per-pen-dik'-ū-lar*) [*perpendicularis*, vertical]. A term applied to a line of plane, forming a right angle with another line or plane. **p. plate**, the mesal vertical plate of the ethmoid bone.
**perplication** (*per-plik-a'-shun*) [*per*, through; *plicare*, to fold]. The operation of turning an incised vessel upon itself by drawing its end through an incision in its own wall.
**perpona** (*per-po'-nah*). A proprietary analgesic said to be a coal-tar derivative. Dose 5–8 gr. (0.3–0.5 Gm.).
**per rectum** (*per rek'-tum*) [L.]. By or through the rectum.
**perseveration** (*per-sev-er-a'-shun*) [*perseverare*, to persist]. A senseless repetition of plainly spoken words or of imperfect speech.
**persimmon** (*per-sim'-on*) [Amer. Ind.]. The tree *Diospyros virginiana*, also its fruit, edible when fully ripe, but otherwise highly astringent; it is useful in diarrheas. From the fruit a beer is made, and whiskey is distilled.
**persection-time.** The period succeeding the contraction of the ventricle of the heart, but prior to the occurrence of relaxation.
**persodine** (*per'-so-dēn*). The proprietary name for a solution of 2 parts of sodium persulphate in 300 parts of water.
**personal** (*per'-son-al*) [*persona*, a person]. Pertaining to a person. **p. equation**, the peculiar difference of individuals in their reaction to various orders of stimuli.
**perspiration** (*per-spir-a'-shun*) [*perspire*]. 1. The secretion of sweat. 2. The sweat. **p., insensible**, that which takes place constantly, the fluid being evaporated as fast as excreted. **p., sensible**, that accumulating in visible drops or beads; the sweat.
**perspire** (*per-spīr'*) [*perspirare*, to breathe everywhere]. To sweat.
**perstriction** (*per-strik'-shun*) [*per*, through; *stringere*, to bind]. The arrest of hemorrhage by ligating the bleeding vessel.
**persulphate** (*per-sul'-fāt*). The sulphate which contains a greater proportion of the sulphuric acid radical than the other sulphates of the same radical.
**persulphide** (*per-sul'-fīd*). The compound of sulphur with an element or radical which contains more sulphur than the other compounds of sulphur with the same element or radical.
**Pertik's diverticulum** (*per'-tik*) [Otto Pertik, Hungarian pathologist, 1852– ]. A diverticulum of the nasopharyngeal space which may occur close to Rosenmueller's fossa and show itself as an enlargement of the latter.
**per tubam** (*per tū'-bam*) [L.]. Through a tube.
**perturbation** (*per-ter-ba'-shun*) [*perturbare*, to disturb]. Restlessness or disquietude. The employment of means that arrest or modify the development of a morbid state.
**pertussal** (*per-tus'-al*) [*per*, intensive; *tussis*, a cough]. Pertaining to or of the nature of whooping-cough.
**pertussin** (*per-tus'-in*). A proprietary remedy for whooping-cough said to consist of the fluidextract of thyme mixed with syrup to produce an infusion in the strength of 1 : 7.
**pertussis** (*per-tus'-is*). See *whooping-cough*.
**Peru, balsam of.** See *balsam*.

**perucognac** (*pe-roo-kŏn'-yak*). A preparation employed in tuberculosis said to consist of the active principle of balsam of Peru, 25 gr.; cinnamic acid, 10%, in a liter of cognac.
**peruol** (*pe'-roo-ol*). A mixture of peruscabin and castor-oil. It is used in scabies.
**peruscabin** (*pe-roo-ska'-bin*). Synthetic benzoicacid benzylester. It is used in scabies.
**Peruvian** (*pe-roo'-ve-an*) [*Peru*]. Pertaining to Peru. **P. bark.** See *cinchona*. **P. wart.** See *verruga peruana*.
**peruvin** (*pe-roo'-vin*) [*Peru*]. The name given to the cinnamic alcohol derived from balsam of Peru.
**perversion** (*per-ver'-shun*) [*per*, through; *vertere*, to turn]. The state of being turned away from the normal course, as in the modifications of function in disease. **p., sexual,** abnormality of the sexual instinct; desire for unnatural methods of sexual gratification.
**pervert** (*per'-vert*) [*per*, through; *vertere*, to turn]. One who has turned from the right way. **p., sexual,** a person whose sexual instincts are perverted.
**pervigilium** (*per-vij-il'-e-um*) [*per*, through; *vigilium*, a watch]. Insomnia; wakefulness.
**pervious** (*per'-ve-us*) [*per*, through; *via*, way]. Open; permeable.
**pes** (*pēs*) [L.]. A foot or foot-like structure. **p. accessorius,** the eminentia collateralis, a smooth white eminence in the brain, situated at the junction of the posterior and descending cornua of the lateral ventricle. **p. anserinus,** goose's foot; the radiate branching of the facial nerve after its exit at the side of the face. **p. calcaneovalgus,** talipes calcaneovalgus. **p. calcaneus,** talipes. **p. cavus, p. excavatus,** talipes cavus. **p. equinus,** talipes equinus. **pes gigas,** macropodia. **p. hippocampi,** the lower portion of the hippocampus major. **p. olfactorius,** the inner root of the olfactory lobe. **p. varus,** talipes varus.
**pessary** (*pes'-ar-e*) [πεσσόs, an oval-shaped stone]. 1. An instrument placed in the vagina to hold the uterus in position. 2. A vaginal suppository.
**pessima** (*pes'-im-ah*) [L.]. A skin-affection characterized by pustular lesions, hard and yellowish and surrounded by areolæ of inflammation, appearing over the whole surface of the body.
**pessulum, pessum** (*pes'-u-lum, pes'-sum*). A pessary.
**pest** [*pestis*, a pest]. A plague; pestilence. **p.-house,** a hospital for persons sick with pestilential diseases.
**pestiferous** (*pes-tif'-er-us*) [*pestis*, pest; *ferre*, to bear]. Causing pestilence.
**pestilence** (*pes'-til-ens*) [*pest*]. Any deadly epidemic disease, especially the plague.
**pestilential** (*pes-til-en'-shal*) [*pestilence*]. Of or having the nature of or producing a pestilence.
**pestis** (*pes'-tis*) [L.]. A plague. **p. americana,** yellow fever. **p. bubonica, p. inguinaria, p. orientalis,** the plague. **p. minor,** an oriental disease resembling the plague but not necessarily fatal. It is believed to have been a mild or modified form of bubonic plague. **p. variolosa.** Synonym of *variola*.
**pestle** (*pes'-l*) [*pistillum*, a pounder]. The instrument with which substances are rubbed in a mortar.
**petalobacteria** (*pet-al-o-bak-te'-re-ah*) [πέταλον, leaf; *bacterium*]. Bacteria in the zoogloea stage.
**petanelle** (*pet'-an-el*). A patented preparation of fibrous peat used as an absorbent.
**petechia** (*pet-e'-ke-ah*) [It., *peteche*, a flea-bite: *pl.*, *petechiæ*]. A small spot beneath the epidermis, due to an effusion of blood.
**petechial** (*pet-e'-ke-al*) [*petechia*]. Characterized by or of the nature of petechiæ. **p. fever.** (1) typhus fever, (2) cerebrospinal meningitis.
**Peter's law.** Atheromatous changes in blood-vessels are most likely to occur where there are angles and projections.
**petiolus** (*pet-i'-o-lus*) [*petiolus*, a stem or stalk of fruit; *pl.*, *petioli*]. 1. In biology, a stem, stalk, or petiole; as the petiolus of the epiglottis. 2. The manubrium of the malleus. **p. glandulæ pinealis.** The peduncle of the pineal gland.
**Petit's canal** (*pti*) [1. François Pourfour du Petit, French anatomist and surgeon, 1664–1741; 2. Jean Louis *Petit*, French surgeon, 1674–1750]. [1.]. A space, intersected by numerous fine interlacing fibers, existing between the anterior and posterior laminæ of the suspensory ligament of the crystalline lens. It extends from the periphery of the lens nearly to the apices of the ciliary processes, and transmits the secretion from the posterior chamber. **P.'s hernia** [2];
28

lumbar hernia. **P.'s ligament** [2], the concave fold formed back of the vagina by the union of Douglas' ligaments. **P.'s sinus.** [1]. See *Valsalva's sinus*. **P.'s triangle,** [2], the trigonum lumbale. The space bounded in front by the posterior border of the external oblique, behind by the anterior border of the latissimus dorsi, its base being formed by the iliac crest. Lumbar hernia usually occurs in this triangle.
**petit mal** (*pet-e mahl*) [Fr., "little illness"]. A slight epileptic seizure characterized by a momentary, scarcely recognizable loss of consciousness, often with an upward staring of the eyes and fibrillary movements of the facial muscles.
**Petri's capsules, P.'s dishes, P.'s saucers** (*pa'-tre*) [Julius *Petri*, German bacteriologist, 1852– ]. Shallow, cylindrical, covered glass vessels for bacterial culture, in which the colonies may be counted without removing the cover. **P.'s test for proteins,** a faint yellow coloration is produced by treating a protein or peptone solution with a solution of diazobenzolsulphonic acid; but if the solution is rendered alkaline by the addition of caustic alkali, the color changes to orange or brown according to concentration, and a red froth is formed on shaking.
**petrifaction** (*pet-re-fak'-shun*) [πέτρα, a stone; *facere*, to make]. Conversion into stone, as *petrifaction* of the fetus, the formation of a lithopedion.
**petrissage** (*pa-tre-sahzj'*) [Fr., kneading]. The kneading movement in massage.
**petrobasilar** (*pet-ro-baz'-il-ar*) [*petrosa-*; *basilar*]. Pertaining to the petrous portion of the temporal bone and the basilar portion of the occipital bone.
**petroccipital** (*pet-rok-sip'-it-al*). Synonym of *petro-occipital*.
**petrogen** (*pet'-ro-jen*). Trade name of a mineral oil preparation used as a vehicle and solvent for various drugs.
**petrolate** (*pet'-ro-lāt*). Same as *petrolatum*.
**petrolatum** (*pet-ro-la'-tum*) [*petroleum*]. A jellylike preparation obtained from the residuum of petroleum, soluble in ether, insoluble in water and alcohol, and known commercially as vaseline or cosmoline. It is used as a basis for ointments and as an emollient. **p. album,** white petrolatum. **p. liquidum** (U. S. P.), liquid petrolatum. **p. molle,** soft petrolatum. **p. spissum,** hard petrolatum.
**petrolene** (*pet'-ro-lēn*) [πέτρα, rock; *oleum*, oil]. A liquid hydrocarbon mixture obtained from petroleum.
**petroleum** (*pet-ro'-le-um*) [πέτρα, rock; *oleum*, oil]. An oily liquid issuing from the earth in various places, and consisting of a mixture of hydrocarbons with small amounts of oxidation-products. The hydrocarbons belong chiefly to the paraffin series. **p. ether,** a product of petroleum obtained by fractional distillation; it has a specific gravity of from 0.665 to 0.67, distilling at from 50° to 60° C.; it consists of pentane and hexane. **p.-jelly,** petrolatum.
**petrolin** (*pet'-ro-lin*) [πέτρα, a rock; *oleum*, oil]. The commercial name for a combination of hydrocarbons derived from petroleum.
**petrolization** (*pet-rol-i-za'-shun*) [*petroleum*]. The act or process of treating waters with kerosene for the extermination of mosquitoes.
**petrolize** (*pet'-rol-īz*). See *petrolization*.
**petromastoid** (*pet-ro-mas'-toid*) [*petrosa: mastoid*]. Pertaining to the petrous and mastoid portions of the temporal bone. **p. canal,** a short passage connecting the mastoid sinuses and the tympanic cavity. **p. foramen,** the tympanic orifice of the petromastoid canal.
**petromortis** (*pet-ro-mor'-tis*) [*petroleum; mors*, death]. Poisoning by automobile gas.
**petro-occipital** (*pet-ro-ok-sip'-it-al*) [*petrosa; occiput*]. Pertaining to the petrous portion of the temporal bone and to the occipital bone.
**petropharyngeus** (*pet-ro-far-in'-je-us*) [*petrosa; pharynx*]. A small muscle arising from the lower surface of the petrous portion of the temporal bone, and blending with the constrictors of the pharynx.
**petrosa** (*pet-ro'-sah*) [πέτρα, rock]. The petrous portion of the temporal bone.
**petrosal** (*pet-ro'-sal*) [*petrosa*]. 1. Pertaining to the petrosa, as the *petrosal* sinus (superior and inferior), *petrosal* nerves. 2. The petrosa.
**petrosalpingostaphylinus** (*pet-ro-sal-ping-go-staf-il-i'-nus*). Synonym of *levator palati*. See *muscles, table of*.
**petrosapol** (*pet-ro-sa'-pol*). A proprietary combination said to consist of soap and certain constituents of petroleum residue; used in scalp diseases.

**petroselinum** (*pet-ro-se-li'-num*). See *parsley*.
**petrosomastoid** (*pet-ro-so-mas'-toid*). Synonym of *petromastoid*.
**petrosphenoid** (*pet-ro-sfe'-noid*) [*petrosa*; *sphenoid*]. Pertaining to the petrous portion of the temporal bone and the sphenoid bone. p. **suture**, the suture between the temporal bone and the great wing of the sphenoid bone.
**petrosquamosal, petrosquamous** (*pet-ro-skwa-mo'-sal, pet-ro-skwa'-mus*) [*petrosa*; *squamosa*]. Pertaining to the petrous and squamous portions of the temporal bone. p. **fissure**, p. **suture**, the line of juncture of the squamous and petrous portions of the temporal bone. p. **sinus**, a venous passage formed in the dura mater at the junction of the petrous and squamous portions of the temporal bone. It opens into the lateral sinus.
**petrostaphylinus** (*pet-ro-staf-il-i'-nus*). Synonym of *levator palati muscle*. See *muscles, table of*.
**petrosulfol** (*pet-ro-sul'-fol*). A proprietary product resembling ichthyol, but with less disagreeable odor; used as is ichthyol.
**petrous** (*pet'-rus*) [πέτρα, rock]. 1. Stony; of the hardness of stone, as the *petrous* portion of the temporal bone. 2. See *petrosal* (1).
**petrox** (*pet'-roks*). A mixture of paraffin-oil, 100 parts; oleic acid, 50 parts; and alcoholic ammonia solution, 25 parts; a substitute for vasogen.
**Pettenkofer's soil-water or ground-water theory** (*pet'-en-kof-er*) [Max von *Pettenkofer*, German chemist, 1818–1901]. Cholera never prevails epidemically where the soil is impermeable to water, or where the level of the soil-water is not liable to fluctuations. P.'s **test for bile acids**, dissolve in concentrated sulphuric acid a small quantity of bile in substance in a small glass dish, or mix some of the liquid containing the bile acids with concentrated sulphuric acid and warm; in either case great care must be exercised that the temperature does not rise above 60° to 70° C. Add drop by drop a 10% solution of cane-sugar, constantly stirring with a glass rod. In the presence of bile a beautiful red coloration is produced, which becomes bluish-violet in the course of the day. This red liquid shows an absorption band at F and another near, E, between D and E.
**petuning** (*pet-ū'-ning*). A process of sprinkling tobacco with some special preparation to aid in the fermentation and flavoring.
**peucine** (*pū'-sēn*) [πεύκη, the fir]. Resin; pitch.
**peucinous** (*pū'-sin-us*) [see *peucine*[. Relating to or like the fir-tree; resinous.
**pexin** (*pek'-sin*) [πῆξις, a curdling]. Rennin or lab.
**pexinogen** (*peks-in'-o-jen*). See *renninogen*.
**Peyer's glands,** P.'s **patches** (*pi'-er*) [Johann Conrad *Peyer*, Swiss anatomist, 1653–1712]. Aggregations of lymph-follicles situated in the mucous membrane of the lower part of the small intestine, opposite the mesenteric attachment.
**Peyerian fever** (*pi-e'-re-an*). Typhoid fever.
**peyote** (*pa-yo'-ta*), Same as *mescal*. See *anhalonine* and *mescal button*.
**Pfannenstiel's incision** (*fahn'-en-stēl*) [J. *Pfannenstiel*, German gynecologist, 1862– ]. A method of entering the abdominal cavity to avoid scar and hernia by a long horizontal cut, convex downward, in the region just above the mons Veneris where pubic hair is normally present.
**Pfaundler's reaction** (*found'-ler*) [Meinhard *Pfaundler*, German physician, 1872– ]. Under certain conditions bacteria grown in an immune serum will develop in long intricated thread-like groups.
**Pfeiffer's glandular fever** (*pfi'-fer*) [Richard Friedrich Wilhelm *Pfeiffer*, German physician, 1858– ]. An acute infectious fever characterized by inflammatory swelling of the lymph-glands, anemia, and prostration. P.'s **phenomenon**. See *P.'s reaction*, and *bacteriolysis*. P.'s **reaction**, the mixing of some of the peritoneal effusion provoked in a guinea-pig by inoculating it with a mixture of blood-serum of an animal immune to cholera, and of bouillon to which a small portion of a culture of the *Spirillum choleræ asiaticæ* has been added, causes these organisms to become nonmotile and to agglutinate. The absence of this phenomenon proves that the spirillum under investigation is of a different species. The same phenomenon has been observed in the case of the typhoid bacillus and typhoid antitoxic serum, and is a valuable differential sign.
**Pflueger's law of contraction** (*pflē'-ger*) [Eduard Friedrich Wilhelm *Pflueger*, German physiologist, 1829–1910]. Galvanic stimulation of a nerve causes muscular contraction, which varies uniformly according as the kathode or the anode is applied, or as the current is closed or opened. Certain deviations from this law constitute the reaction of degeneration. The law may be briefly stated as follows:

| Strength of Current Used. | Descending Current. | | Ascending Current. | |
|---|---|---|---|---|
| | Make. | Break. | Make. | Break. |
| Very Weak | Yes. | No. | No. | No. |
| Weak | Yes. | No. | Yes. | No. |
| Moderate | Yes. | Yes. | Yes. | Yes. |
| Strong | Yes. | No. | No. | Yes. |

P.'s **law of reflex action**. 1. If stimulation of a sensory nerve is followed by a unilateral reflex movement, the latter always occurs on the side to which the sensory nerve belongs. 2. If the stimulus received by a sensory nerve extends, to motor nerves of the opposite side, contraction occurs only in the corresponding muscles. 3. If the contraction is unequal on the two sides, the stronger contraction always takes place on the side which is stimulated. 4. If the reflex excitement extends to other motor nerves, the direction of the impulse from the sensory to the motor nerve is from before backward in the brain and from below upward in the spinal cord—*i.e.*, always in the direction of the oblongata. P.'s **tubes**, ovarian tubes; sacciform or tubular ingrowths of the germ epithelium on the anterointernal surface of the Wolffian body; they ultimately form the cortex of the ovary.
**Pfuhl's sign,** P.-**Jaffe's sign** (*pfool, yah'-fa*), [Eduard *Pfuhl*, German physician, 1852– ; Max *Jaffe*, German physician, 1841–1911]. In subphrenic pyo-pneumothorax the liquid issues from the exploratory puncture or incision with considerable force during inspiration, while the contrary occurs in true pneumothorax.
P. G. Abbreviation of *Pharmacopœia Germanica*, German Pharmacopeia.
**phace, phacea** (*fa'-se, fa-se'-ah*) [φακός, a lens]. The crystalline lens.
**phacentocele** (*fa-sen'-to-sēl*) [φακός, a lens; -κήλη, hernia]. Displacement of the crystalline lens into the anterior chamber of the eye.
**phacicous** (*fa'-sik-us*) [φακός, a lens]. 1. Belonging to the crystalline lens. 2. Lentil-shaped.
**phacitis** (*fa-si'-tis*). See *phakitis*.
**phaco-** (*fa-ko-*) [φακός, a lens]. A prefix meaning pertaining to a lens or to the lens of the eye.
**phacocele** (*fa'-ko-sēl*). See *phacentocele*.
**phacocyst** (*fa'-ko-sist*) [*phaco-*; κύστις bladder]. The capsule of the crystalline lens.
**phacocystectomy** (*fa-ko-sis-tek'-to-me*) [*phaco-*; κύστις, cyst; ἐκτομή, excision]. Excision of a part of the capsule of the crystalline lens.
**phacocystitis** (*fa-ko-sis-ti'-tis*) [*phaco-*; κύστις, cyst; ιτις, inflammation]. Inflammation of the capsule of the crystalline lens.
**phacoglaucoma** (*fa-ko-glaw-ko'-mah*) [*phaco-*; glaucoma]. Structural changes in the crystalline lens induced by glaucoma.
**phacohymenitis** (*fa-ko-hi-men-i'-tis*) [*phaco-*; ὑμήν, membrane; ιτις, inflammation]. Inflammation of the capsule of the crystalline lens.
**phacoid** (*fa'-koid*) [φακός, lens; εἶδος, like]. Lens-shaped.
**phacoiditis** (*fa-koid-i'-tis*). See *phakitis*.
**phacoidoscope** (*fa-koid'-o-skōp*). Synonym of *phacoscope*.
**phacolysis** (*fa-kol'-is-is*) [*phaco-*; λύειν, to loosen]. 1. Dissolution or disintegration of the crystalline lens. 2. An operation for relief of high myopia consisting in discission of the crystalline lens followed by extraction.
**phacomalacia** (*fa-ko-mal-a'-se-ah*) [*phaco-*; μαλακία, softness]. Soft cataract.
**phacometachoresis** (*fa-ko-met-ak-or-e'-sis*) [*phaco-*; μεταχώρησις change of place]. Dislocation of the crystalline lens.
**phacometer** (*fa-kom'-et-er*) [*phaco-*; μέτρον, a measure]. An instrument for determining the refractive power of lenses.

**phacopalingenesis** (*fa-ko-pal-in-jen'-es-is*) [*phaco-;* πάλιν, again; γένεσις, genesis]. Reproduction of the crystalline lens.
**phacoplanesis** (*fa-ko-plan-e'-sis*) [*phaco-;* πλάνησις, a making to wander]. Displacement of the lens of the eye from the posterior to the anterior chamber and back again.
**phacosclerosis** (*fa-ko-skle-ro'-sis*) [*phaco-;* *sclerosis*]. Hardening of the crystalline lens.
**phacoscope** (*fa'-ko-skōp*) [*phaco-;* σκοπεῖν, to inspect]. An instrument for observing the accommodative changes of the lens.
**phacoscopy** (*fa-kos'-ko-pe*) [see *phacoscope*]. The observation and estimation of the changes in the lens of the eye caused by accommodative influences.
**phacoscotasmus** (*fa-ko-sko-taz'-mus*) [*phaco-;* σκότος, darkness]. Clouding of the crystalline lens.
**phacotherapy** (*fa-ko-ther'-ap-e*) [*phaco-;* θεραπεία, therapy], Heliotherapy.
**phaoretin** (*fe-or-et'-in*) [φαιός, dusky; ῥητίνη, resin]. C₁₄H₈O₇. A resinous extract from rhubarb-root, various species of the genus *Rheum*. It occurs as a yellowish brown powder soluble in alcohol and alkalies.
**phagedena, phagedæna** (*faj-ed-e'-nah*) [φαγέδαινα, a cancerous sore]. A rapidly spreading destructive ulceration of soft parts. **p. tropica**, tropical ulcer.
**phagedenic** (*faj-ed-en'-ik*) [*phagedena*]. Of the nature of phagedena. **p. chancroid**, a chancroid that spreads rapidly and destroys a large amount of tissue.
**phagedenism** (*faj'-ed-en-izm*). Rapidly progressive ulcerative processes of the soft parts, frequently complicated with chancroid buboes.
**phagedenoma, phagedœnoma** (*faj-ed-en-o'-mah*). A phagedenic ulcer.
**phagocytal** (*fag'-o-si-tal*). Pertaining to a phagocyte.
**phagocyte** (*fag'-o-sīt*) [φαγεῖν, to eat; κύτος, a cell]. A cell having the property of engulfing and digesting foreign or other particles harmful to the body. Phagocytes are either fixed—endothelial cells, fixed connective-tissue cells—or free—the wandering cells or leukocytes. A large phagocyte is termed a *macrophage*; a small one, a *microphage*.
**phagocytic** (*fag-o-sit'-ik*) [*phagocyte*]. Of, pertaining to, or caused by phagocytes.
**phagocytoblast** (*fag-o-si'-to-blast*) [*phagocyte*; βλαστός, a germ]. A cell giving rise to one or more phagocytes.
**phagocytolysis** (*fag-o-si-tol'-is-is*) [*phagocyte*; λύσις, solution]. 1. Destruction or dissolution of phagocytes. 2. Loss of the phagocytic action of leukocytes.
**phagocytosis** (*fag-o-si-to'-sis*) [*phagocyte*]. The ingestion of foreign or other particles, principally bacteria, by certain cells. Phagocytosis has been claimed to be the cause of immunity against infectious diseases.
**phagokaryosis** (*fag-o-kar-e-o'-sis*) [φαγεῖν, to eat; κάρυον, nucleus]. The assumption by the cell-nucleus of phagocytic action.
**phagolysis** (*fag-ol'-is-is*) [φαγεῖν, to eat; λύειν, to loosen]. Destruction or dissolution of phagocytes; phagocytolysis.
**phagomania** (*fag-o-ma'-ne-ah*) [φαγεῖν, to eat; μανία, madness]. An insatiable craving for food.
**phagotherapy** (*fag-o-ther'-ap-e*) [φαγεῖν, to eat; θεραπεία, therapy]. Treatment by superalimentation.
**phakitis** (*fa-ki'-tis*) [φακός, lens; ιτις, inflammation]. Inflammation of the crystalline lens of the eye; a condition that has, however, not been observed.
**phako-** (*fa'-ko-*). For words beginning thus, see *phaco-*.
**phalacrosis** (*fal-ak-ro'-sis*) [φαλακρός, bald]. Baldness.
**phalacrotic, phalacrous** (*fal-ak-rot'-ik, fal-ak'-rus*) [*phalacrosis*]. Bald.
**phalangeal** (*fal-an'-je-al*) [*phalanx*]. Pertaining to a phalanx.
**phalanges** (*fa-lan'-jēz*) [*phalanx*]. Plural of *phalanx*.
**phalangette, phalanget** (*fal-an-jet'*) [Fr.]. The last phalanx or terminal bone of a finger or toe.
**phalangitis** (*fal-an-ji'-tis*) [*phalanx*; ιτις, inflammation]. Inflammation of a phalanx. **p. syphilitica.** See under *dactylitis syphilitica*.
**phalangophalangeal** (*fa-lan-go-fa-lan'-je-al*) [*phalanx*]. Pertaining to the successive phalanges of the digits. **p. amputation**, removal of a finger or toe at the first or second phalangeal joints.
**phalangosis** (*fal-an-go'-sis*) [*phalanx*]. 1. A disease of the eyelids in which the lashes are arranged in rows. 2. Ptosis.

**phalanx** (*fa'-lanks*) [φάλαγξ, phalanx, *pl.*, *phalanges*]. 1. One of the bones of the fingers or toes. 2. One of the delicate processes of the headplate of the outer rod of Corti projecting beyond the inner rod.
**phallalgia** (*fal-al'-je-ah*) [*phallus*; ἄλγος, pain]. Pain in the penis.
**phallanastrophe** (*fal-an-as'-tro-fe*) [φαλλός, penis; ἀναστροφή, upturning]. Twisting or distortion of the penis.
**phallaneurysm** (*fal-an'-ū-rizm*) [*phallus*; *aneurysm*]. Aneurysm of a vessel of the penis.
**phallic** (*fal'-ik*) [*phallus*]. Pertaining to the penis.
**phallin** (*fal'-in*). A toxalbumin contained in the death cup fungus, *Amanita phalloides*.
**phallitis** (*fal-i'-tis*) [*phallus*; ιτις, inflammation]. Inflammation of the penis.
**phallocampsis** (*fal-o-kamp'-sis*) [*phallus*; κάμψις, a bending]. Chordee.
**phallocarcinoma** (*fal-o-kar-sin-o'-mah*) [*phallus*; *carcinoma*]. Carcinoma of the penis; it is usually an epithelioma.
**phallocrypsis** (*fal-o-krip'-sis*) [*phallus*; κρύψις, concealment]. Concealment of the penis by retraction.
**phallodynia** (*fal-o-din'-e-ah*) [*phallus*; ὀδύνη, pain]. Pain in the penis.
**phalloid** (*fal'-oid*) [*phallus*; εἶδος, like]. Resembling the penis.
**phalloncus** (*fal-ong'-kus*) [*phallus*; ὄγκος, a tumor]. Any tumor or swelling of the penis.
**phalloplasty** (*fal'-o-plas-te*) [*phallus*; πλάσσειν, to mold]. Plastic or restorative surgery of the penis.
**phallorrhagia** (*fal-or-a'-je-ah*) [*phallus*; ῥηγνύναι, to burst forth]. Hemorrhage from the penis.
**phallus** (*fal'-us*) [φαλλός, penis]. Penis.
**phanerogenic** (*fan-er-o-jen'-ik*) [φανερός, visible; γεννᾶν, to produce]. Noting a disease of obvious origin; the opposite of *cryptogenic*, *q. v.*
**phaneromania** (*fan-er-o-ma'-ne-ah*) [φανερός, evident; μανία, madness]. A neurotic condition in which a person pays undue attention to some external part or growth, such as a pimple, a hair or a hangnail.
**phaneroscope** (*fan'-er-o-skōp*) [φανερός, visible; σκοπεῖν, to see]. An instrument for rendering the skin transparent; it is used in examining for diseases of the skin, such as lupus.
**phaneroscopy** (*fan-er-os'-ko-pe*). The use of the phaneroscope.
**phantasia** (*fan-ta'-ze-ah*) [φαντασία, a showing]. An imaginary appearance.
**phantasm** (*fan'-tazm*) [φαντάζειν, to render visible]. An illusive perception of an object that does not exist; an optical illusion; an apparition.
**phantasmatomoria** (*fan-taz-mat-o-mo'-re-ah*) [φάντασμα, phantasm; μωρία, folly]. Childishness, or dementia, with absurd fancies.
**phantasmology** (*fan-taz-mol'-o-je*) [φάντασμα, phantasm; λόγος, science]. The science of phantasms.
**phantasmoscopia** (*fan-taz-mo-sko'-pe-ah*) [φάντασμα, phantasm; σκοπεῖν, to see]. The seeing of phantasms, in insanity or delirium.
**phantom** (*fan'-tum*) [*phantasm*]. 1. An apparition. 2. A model of a part or the whole of the human body used in practising various operations and procedures. **p. corpuscle.** See *corpuscle*. **p. tumor**, a tumor-like swelling produced artificially by the contraction of a muscle or by other causes.
**phaochrome** (*fa'-o-krōm*). Same as *chromaffin*; see *chromaffin cells*.
**phaochromoblast** (*fa-o-kro'-mo-blast*). One of the two varieties of cells into which the primary sympathetic cells become differentiated.
**pharbitin, pharbitisin** (*far-bit'-in, far-bit'-is-in*). The kalandana of the Indian pharmacy, a resinous substance isomeric with convolvulin contained in *Ipomœa hederacea*. It is used as a cathartic. Dose, 7–10 gr. (0.45–0.64 gm.).
**pharcidous** (*far'-sid-us*) [φαρκίς, a wrinkle]. Wrinkled; rugose; full of wrinkles.
**Phar.D.** Abbreviation for *Pharmaciæ Doctor*, Doctor of Pharmacy.
**pharmacal** (*far'-mak-al*) [φάρμακον, a drug]. Pertaining to pharmacy.
**pharmaceutic, pharmaceutical** (*far-ma-sū'-tik, -al*) [*pharmacy*]. Pertaining to pharmacy.
**pharmaceutics** (*far-mas-ū'-tiks*). Pharmacy.
**pharmaceutist** (*far-mas-ū'-tist*). Synonym of *pharmacist*.

**pharmacist** (*far'-ma-sist*) [*pharmacy*]. An apothecary.
**pharmaco-** (*far-ma-ko-*) [φάρμακον, a drug]. A prefix meaning pertaining to drugs.
**pharmacodynamics** (*far-ma-ko-di-nam'-iks*) [*pharmaco-*; δύναμις, force]. The science of the action of drugs.
**pharmacognosis, pharmacognosy** (*far-ma-kog-no'-sis, far-ma-kog'-no-se*) [*pharmaco-*; γνῶσις, knowledge]. The science of crude drugs.
**pharmacognostics** (*far-mak-og-nos'-tiks*). Synonym of *pharmacognosy*.
**pharmacography** (*far-ma-kog'-ra-fe*). See *pharmacognosis*.
**pharmacologist** (*far-ma-kol'-o-jist*) [*pharmacology*]. One versed in pharmacology.
**pharmacology** (*far-ma-kol'-o-je*) [*pharmaco-*; λόγος, science]. The science of the nature and properties of drugs.
**pharmacomania** (*far-mak-o-ma'-ne-ah*) [φάρμακον, drug; μανία, madness]. A morbid craving for medicines, or for self-medication.
**pharmacopeia, pharmacopœia** (*far-ma-ko-pe'-ah*) [*pharmaco-*; ποιεῖν, to make]. A collection of formulas and methods for the preparation of drugs, especially a book of such formulas recognized as a standard, as the United States or the British Pharmacopeia. The former is issued every ten years under the supervision of a national committee.
**pharmacopeial** (*far-ma-ko-pe'-al*) [*pharmacopeia*]. Contained in or sanctioned by the pharmacopeia.
**pharmacotherapeutic** (*far-mak-o-ther-ap-u'-tik*) [φάρμακον, drug; θεραπεία, treatment]. Pertaining to treatment with drugs.
**pharmacotherapy** (*far-mak-o-ther'-a-pe*) [φάρμακον, drug; θεραπεία, therapy]. The treatment of disease by means of drugs.
**pharmacy** (*far'-ma-se*) [φαρμακεία, the use of drugs]. 1. The art of preparing, compounding, and dispensing medicines. 2. A drug-store.
**pharyngalgia** (*far-in-gal'-je-ah*) [*pharynx*; ἄλγος, pain]. Pain in the pharynx.
**pharyngeal** (*far-in'-je-al*) [*pharynx*]. Pertaining to the pharynx. **p. tonsil.** See *Luschka's tonsil*. **p. spine, p. tubercle,** a small elevation near the middle of the inferior surface of the basilar process of the occipital bone, for the attachment of the pharynx.
**pharyngectomy** (*far-in-jek'-to-me*) [*pharynx*; ἐκτομή, excision]. Excision of a part of the pharynx.
**pharyngemphraxis** (*far-in-gem-fraks'-is*) [*pharyngo-*; ἔμφραξις, obstruction]. Obstruction of the pharynx.
**pharyngeus** (*far-in-je'-us*). See *constrictor* of *pharynx*, etc., under *muscles, table of*.
**pharyngismus** (*far-in-jiz'-mus*) [*pharynx*]. Spasm of the pharynx.
**pharyngitic** (*far-in-jit'-ik*). Pertaining to, affected with, or of the nature of, pharyngitis.
**pharyngitis** (*far-in-ji'-tis*) [*pharynx*; ιτις, inflammation]. Inflammation of the pharynx. **p., acute, p., catarrhal,** a form due to exposure to cold, to the action of irritant substances, or to certain infectious causes, and characterized by pain on swallowing, by dryness, later by moisture, and by congestion of the mucous membrane. **p. apostematosa,** abscess of the pharynx. **p., atrophic,** a form attended with atrophy of the mucous membrane. **p., chronic,** a form that is generally the result of repeated acute attacks, and is associated either with hypertrophy of the mucous membrane (*hypertrophic pharyngitis*) or with atrophy (*atrophic pharyngitis*). **p., croupous, p., diphtheritic,** that characterized by the presence of a false membrane, the product of the action of the diphtheria bacillus. **p., follicular,** clergyman's sore throat. **p., granular,** a form of chronic pharyngitis in which the mucous membrane has a granular appearance. **p., lithemic,** a sense of fulness in the throat with a feeling of rigidity associated with heat and dryness; it is due to the gouty diathesis. **p. sicca,** the chronic form with a dry state of the mucous membrane.
**pharyngo-** (*far-in-go-*) [*pharynx*]. A prefix signifying pertaining to the pharynx.
**pharyngocele** (*far-in'-go-sēl*) [*pharyngo-*; κήλη, hernia]. A hernia or pouch of the pharynx projecting through the pharyngeal wall.
**pharyngodynia** (*far-in-go-din'-e-ah*) [*pharyngo-*; ὀδύνη, pain]. Pain referred to the pharynx.

**pharyngodynia** (*far-in-go-din'-e-ah*) [*pharyngo-*; ὀδύνη, pain]. Pain referred to the pharynx.
**pharyngoepiglottic** (*far-in-go-ep-ig-lot'-ik*). Pertaining to the pharynx and the epiglottis.
**pharyngoepiglotticus** (*far-in-go-ep-ig-lot'-ik-us*) [*pharyngo-*; *epiglottis*]. Muscular fibers derived from the stylo-pharyngeus and inserted into the side of the epiglottis and the pharyngoepiglottic ligament.
**pharyngoesophageal** (*far-in-go-e-sof-aj'-e-al*). Pertaining to the pharynx and esophagus.
**pharyngoesophagus** (*far-in-go-e-sof'-ag-us*). The pharynx and esophagus considered as one organ.
**pharyngoglossal** (*far-in-go-glos'-al*) [*pharyngo-*; γλῶσσα, tongue]. Pertaining conjointly to the pharynx and the tongue.
**pharyngolaryngeal** (*far-in-go-lar-in'-je-al*) [*pharyngo-*; *larynx*]. Pertaining both to the pharynx and to the larynx.
**pharyngolaryngitis** (*far-in-go-lar-in-ji'-tis*) [*pharyngo-*; *laryngitis*]. Simultaneous inflammation of the pharynx and larynx.
**pharyngolith** (*far-in'-go-lith*) [*pharyngo-*; λίθος, stone]. A calcareous concretion in the walls of the pharynx.
**pharyngology** (*far-in-gol'-o-je*) [*pharyngo-*; λόγος, science]. The science of the pharyngeal mechanism, functions and diseases.
**pharyngolysis** (*far-in-gol'-is-is*) [*pharyngo-*; λύσις, a loosing]. Paralysis of the pharyngeal muscles.
**pharyngomaxillary** (*far-in-go-maks'-il-a-re*). Relating to the pharynx and the maxilla.
**pharyngomycosis** (*far-in-go-mi-ko'-sis*). Disease of the pharynx due to the action of fungi.
**pharyngonasal** (*far-in-go-na'-sal*) [*pharyngo-*; *nasus*, nose]. Pertaining to the pharynx and the nose. **p. cavity,** the upper portion of the pharynx; the nasopharynx.
**pharyngooral** (*far-in-go-o'-ral*) [*pharyngo-*; *os, oris,* mouth]. Oropharyngeal; pertaining to both pharynx and mouth. **p. cavity,** the oropharynx; the middle portion of the pharynx, communicating with the mouth.
**pharyngopalatine** (*far-in-go-pal'-a-tin*). Relating to the pharynx and the palate.
**pharyngopalatinus** (*far-in-go-pal-at-i'-nus*). Synonym of *palatopharyngeus*. See *muscles, table of*.
**pharyngoparalysis** (*far-in-go-par-al'-is-is*). See *pharyngoplegia*.
**pharyngopathy** (*far-in-gop'-ath-e*) [*pharyngo-*; πάθος, disease]. Any disease of the pharynx.
**pharyngoperistole** (*far-in-go-per-is'-to-le*). Synonym of *pharyngostenia*.
**pharyngoplasty** (*far-in'-go-plas-te*) [*pharyngo-*; πλάσσειν, to form]. Plastic surgery of the pharynx.
**pharyngoplegia** (*far-in-go-ple'-je-ah*) [*pharyngo-*; πληγή, a stroke]. Paralysis of the muscles of the pharynx.
**pharyngorhinitis** (*far-in-go-ri-ni'-tis*) [*pharyngo-*; ῥίς, nose; ιτις, inflammation]. Pharyngitis with rhinitis; inflammation of the pharynx and the nose.
**pharyngorhinoscopy** (*far-in-go-ri-nos'-ko-pe*). Synonym of *rhinoscopy, posterior*.
**pharyngorrhagia** (*far-in-go-ra'-je-ah*) [*pharyngo-*; ῥηγνύναι, to burst forth]. Hemorrhage from the pharynx.
**pharyngorrhea** (*far-in-go-re'-ah*) [*pharyngo-*; ῥοία, a flow]. A mucous discharge from the pharynx.
**pharyngoscleroma** (*far-in'-go-skle-ro'-mah*). Pharyngeal scleroma.
**pharyngoscope** (*far-in'-go-skōp*) [*pharyngo-*; σκοπεῖν, to inspect]. An instrument for use in examining the pharynx.
**pharyngoscopy** (*far-in-gos'-ko-pe*) [*pharyngoscope*]. Examination of the pharynx with the pharyngoscope.
**pharyngospasm** (*far-in'-go-spazm*) [*pharyngo-*; σπασμός, a spasm]. Spasmodic contraction of the pharynx.
**pharyngospasmodic** (*far-in-go-spas-mod'-ik*). Relating to spasmodic contraction of the pharynx.
**pharyngostaphylinus** (*far-in'-go-staf-il-i'-nus*). Synonym of *palatopharyngeus*. See *muscles, table of*.
**pharyngostenia** (*far-in-go-ste'-ne-ah*) [*pharyngo-*; στενός, narrow]. Narrowing or stricture of the pharynx.
**pharyngostenous** (*far-in-go-ste'-nus*) [*pharyngo-*; *stenosis*]. Relating to stricture of the pharynx.
**pharyngotherapy** (*far-in-go-ther'-a-pe*) [*pharyngo-*; θεραπεία, therapy]. The treatment of diseases of the pharynx by direct applications or irrigations.

# PHARYNGOTOME 677 PHENOMENON

**pharyngotome** (*far-in'-go-tōm*) [*pharyngo-*; τομή, a cutting]. An instrument for incising the pharynx.

**pharyngotomy** (*far-in-got'-o-me*) [see *pharyngotome*]. Incision into the pharynx. **p., inferior,** that in which the tissues between the hyoid bone and the cricoid cartilage are divided. **p., lateral,** incision into one side of the pharynx. **p., subhyoidean,** that through the thyrohyoid membrane.

**pharyngotonsillitis** (*far-in-go-ton-sil-i'-tis*) [*pharyngo-*; *tonsillitis*]. Inflammation of the pharynx and the tonsil.

**pharyngoxerosis** (*far-in-go-zer-o'-sis*) [*pharyngo-*; *xerosis*]. Dryness of the pharynx.

**pharynx** (*far'-ingks*) [φάρυγξ, the throat]. The musculomembranous pouch situated back of the nose, mouth, and larynx, and extending from the base of the skull to a point opposite the sixth cervical vertebra, where it becomes continuous with the esophagus. It is lined by mucous membrane, covered in its upper part with columnar ciliated epithelium, in its lower part with stratified epithelium. On the ouside of this is a layer of fibrous tissue—the *pharyngeal aponeurosis.* This in turn is surrounded by the muscular coat. The upper portion of the pharynx communicates with the nose through the posterior nares, is known as the *nasopharynx,* and functionally belongs to the respiratory tract; the lower portion is divided into the *oropharynx* and *laryngopharynx,* and is a part of the digestive tract. The pharynx communicates with the middle ear by means of the Eustachian tube.

**phase** (*fāz*) [φάσις, appearance]. The condition or stage of a disease or physiological function at a given time.

**phaselin** (*fas'-el-in*). A proprietary digestant said to be a constituent of a wild bean of Mexico.

**phatne** (*fat'-ne*) (φάτνη, socket]. Same as *alveolus.*

**phatnorrhagia** (*fat-nor-a'-je-ah*) [φάτνη, socket; ῥηγνύναι, to burst forth]. Hemorrhage from a tooth-socket.

**Ph.B.** Abbreviation for (1) *British Pharmacopœia*; (2) *Bachelor of Philosophy.*

**Ph.D.** Abbreviation for (1) *Doctor of Pharmacy*; (2) *Doctor of Philosophy.*

**phecine** (*fe'-sēn*), C₆H₄(OH)₂SO₄. Sulphometa-dihydroxybenzene, a sulphate of the dihydrate of benzene; it is said to be a nonirritant antiseptic and prophylactic, and is indicated in diseases of the skin.

**Phelps' operation** [Abel Mix *Phelps*, American surgeon, 1851–1902]. 1. For *club-foot*: a direct open incision is made through the inner and plantar surfaces of the foot. 2. For *hare-lip:* a loop is passed through the margin of the lip on either side of the cleft; the incision is curved on both sides, and a V-shaped flap is allowed to remain in the middle line, beneath the septum of the nose; the wound is closed with silk sutures.

**phenacetin** (*fe-nas'-et-in*). See *acetphenetidin.*

**phenakistoscope, phænakistoscope** (*fe-nak-is'-to-skōp*) [φενακιστής, an impostor; σκοπεῖν, to view]. That form of stroboscope in which the figures and slits revolve in the same direction. Syn., *direct stroboscope; zoetrope.*

**phenalette** (*fen-al-et'*). An effervescing headache powder, containing phenacetin.

**phenalgene** (*fen-al'-jēn*). A proprietary analgesic said to conain acetanilide, sodium bicarbonate, etc.

**phenalgin** (*fe-nal'-jin*). A proprietary antipyretic and analgesic said to be an ammoniated combination of phenyl and acetanilide. Syn., *ammoniophenylacetamide.*

**phenanthrene** (*fe-nan'-thrēn*) [*phenol*; ἄνθραξ, coal], C₁₄H₁₀. A hydrocarbon isomeric with anthracene, and found with it in the last fraction of coal-tar.

**phenate** (*fe'-nāt*) [*phenol*]. A compound of phenol and a base; a carbolate.

**phenatol** (*fe'-nat-ol*). A proprietary antipyretic and anodyne said to be a combination of sodium carbonate, bicarbonate, sulphate, and chloride, with acetanilide and caffeine.

**phenazone** (*fe'-naz-ōn*). Antipyrine.

**phene** (*fēn*). Benzene.

**phenedin** (*fe'-ned-in*). Acetphenetidin.

**phenegol** (*fe'-ne-gol*), C₆H₅ . O . NO₂ . SO₃K. Mercury potassium nitroparaphenol sulphonate. It is antiseptic and bactericidal.

**phenetidin** (*fe-net'-id-in*) [*phenol*], C₈H₁₁NO. The base from which acetphenetidin is prepared by substitution. **p. citrate,** a condensation-product of paraamidophenetol with citric acid; sedative and antipyretic.

**phenetidinuria** (*fe-net-id-in-u'-re-ah*). A condition marked by the presence of phenetidin in the urine.

**phenetol** (*fen'-et-ol*) [*phenol; oleum,* oil], C₆H₅ . O. - C₂H₅. Ethyl phenyl ether; a volatile aromatic-smelling liquid.

**phengophobia** (*fen-go-fo'-be-ah*) [φέγγος, light; φόβος, fear].· See *photophobia.*

**phenic** (*fe'-nik*) [*phenol*]. Obtained from coal-tar. **p. acid.** See *acid, carbolic,* and *phenol.*

**phenicate** (*fen'-ik-āt*). To charge with phenol or phenic acid.

**phenicism** (*fe'-nis-izm*). A synonym of *rubeola, q. v.*

**phenidin, phenin** (*fe'-nid-in, fe'-nin*). See *acetphenetidin.*

**phenigmus** (*fe-nig'-mus*) [φοῖνιξ, purple-red]. A skin disease characterized by diffuse redness, without fever.

**phenocoll** (*fe'-no-kol*) [*phenol*], C₁₀H₁₄N₂O₂. Amidophenacetin, a substance resembling acetphenetidin; the hydrochloride is used as an antipyretic. Dose 10–15 gr. (0.65–1.0 Gm.).

**phenocreosote** (*fe-no-kre'-o-sōt*). A preparation of creosote and phenol.

**phenodin** (*fe'-no-din*) [φοινώδης, blood-red]. The same as *hematin, q. v.*

**phenofax** (*fe'-no-fax*). Trade name of an antiseptic sulphonal dressing.

**phenol** (*fe'-nol*) [φοῖνιξ, purple-red]. 1. C₆H₅OH. Hydroxybenzene, obtained either from coal-tar by fractional distillation or made synthetically. More commonly known as *carbolic acid.* 2. Any member of benzene homologous with phenol. **p.-camphor,** camphorated phenol. **p. celluloid,** a protective varnish for wounds, prepared from proxylin, phenol, and camphor. **p. diiodide,** a precipitate from a combination of solutions of sodium, phenol, and potassium iodide, recommended as a wound antiseptic. Syn., *diiodophenoliodide.* **p., glycerite of** (*glyceritum phenolis,* U. S. P.), a mixture of liquefied phenol and glycerol. **p., liquefied** (*phenol liquefactum,* U. S. P.), a liquid containing not less than 86.4 % by weight of absolute phenol. Dose 1 min. (0.06 Cc.). **p., ointment of** (*unguentum phenolis,* U. S. P.), an ointment made of white petrolatum and phenol. **p., orthomonobromo-,** C₆H₅BrO, an oily, violet-colored liquid with strong odor, soluble in ether, chloroform, or 106 parts of water; used as a wound antiseptic and in erysipelas, 1 to 2 % in petrolatum, twice daily. **p., ortho-monochlor-,** C₆H₅ClO, a colorless antiseptic liquid, soluble in alcohol or ether, used with petrolatum in skin diseases, etc. **p. sulphoricinate,** a solution of phenol in sulphoricinic acid, used in 20 % solution in tuberculosis of the throat. **p., tests for.** See *Allen, Berthelot, Davy, Eijkman, Jacquemin, Landolt, Pensoldt and Fischer, Plugge.* **p. trichloride.** See *trichlorphenol.*

**phenolate** (*fe'-no-lāt*) [*phenol*]. A salt of carbolic acid.

**phenolax** (*fe'-no-laks*). Trade name of a preparation of phenolphthalein; used as a purgative.

**phenolid** (*fe'-nol-id*). A proprietary preparation said to be a mixture of acetanilide and sodium salicylate or sodium bicarbonate. Dose 5–10 gr. (0.3–0.6 gm.).

**phenolin** (*fen'-ol-in*). An antiseptic prepared from crude cresols.

**phenology, phænology** (*fe-nol'-o-je*) [φαίνειν, to bring to light; λόγος, science]. In biology, the science of the behavior of plants and animals to the periodic changes in meteorologic conditions.

**phenolphthalein** (*fe-nol-tha'-le-in*) [*phenol; phthalic acid*], C₂₀H₁₄O₄. A substance produced by the action of phenol on phthalic acid and used generally in a 1 % solution in 50 % alcohol, as a delicate test for acids and alkalies. It is turned red by alkalies and decolorized by acids. It has been recommended as a purgative. Dose 1½–4 gr. (0.09–0.2 Gm.).

**phenolsulphonic acid** (*fe-nol-sul-fon'-ik*). Sulphocarbolic acid.

**phenomenon** (*fe-nom'-en-on*) [φαινόμενον, that which is seen; from φαίνειν, to shine; *pl., phenomena*]. An event in which φαίνειν, generally of an unusual character. **p. Aubert's** etc. For this and other proper names see under the proper name. **p.,**

# PHENONE 678 PHLEBITIS

**diaphragm.** See *Litten's sign.* **p.**, **face.** See *Chvostek's sign.* **p.**, **great toe.** See *Babinski's reflex.* **p., knee.** Synonym of *patellar tendon-reflex.*
**phenone** (*fe'-nōn*) [*phenol*]. A ketone formed by the union of phenyl and a hydrocarbon of the marsh-gas series.
**phenophobia** (*fe-no-fo'-be-ah*). Synonym of *photophobia.*
**phenophthalein** (*fe-no-tha'-le-in*). See *phenolphthalein.*
**phenopyrine** (*fe-no-pi'-rēn*). A mixture of equal parts of phenol and antipyrine.
**phenoresorcin** (*fe-no-re-sor'-sin*). A mixture of 67 parts of phenol with 33 parts of resorcin.
**phenosal** (*fe'-no-sal*). Phenetidin acetosalicylate; an antipyretic compound of acetphenetidin and salicylic acid; used in acute articular rheumatism. Dose 8 gr. (0.5 Gm.) 2 to 6 times daily. Syn., *paraphenetidin salicylacetic acid.*
**phenosalyl** (*fe-no-sal'-il*). A compound of phenol, salicylic acid, lactic acid, and menthol, mixed with heat. It is an external antiseptic, used in conjunctivitis in 0.2 to 0.4 % solution; in eczema, in 1 % solution.
**phenosuccin** (*fe-no-suk'-sin*), $C_6H_4(OC_2H_5)N(COCH_2)_2$. Colorless needles derived from para-amidophenol by action of succinic acid; antipyretic and antineuralgic. Dose 15-45 gr. (1-3 Gm.) daily. Syn., *pyrantin.*
**phenosuccinate** (*fe-no-suk'-sin-āt*). The sodium salt of phenosuccin, forming a white, soluble powder. It is preferred to phenosuccin. Dose 7½-46 gr. (0.5-3.0 Gm.).
**phenoxin** (*fe-noks'-in*). A trade name for carbon tetrachloride.
**phenoxycaffeine** (*fe-noks-e-kaf'-e-in*), $C_8H_9OC_8H_9N_4O_2$. It is anesthetic and narcotic and is used subcutaneously in sciatica. Dose ¼ gr. (0.26 Gm.).
**phenyl** (*fe'-nil*) [*phenol*; ὕλη, matter]. The univalent radical, $C_6H_5$, of phenol. **p. alcohol.** Synonym of *phenol.* **p. hydrate,** phenol. **p. salicylate** (*phenylis salicylas*, U. S. P.), the salicylic ester of phenyl, a white, crystalline substance, used as an intestinal and urinary antiseptic, and as a substitute for salicylic acid. It is decomposed in the intestine into salicylic acid and phenol. Dose 5-25 gr. (0.32-1.0 Gm.). Syn., *salol.*
**phenylacetamide.** Same as *acetanilide.*
**phenylamine** (*fen-il-am'-in*). Same as *aniline.*
**phenylaniline** (*fen-il-an'-il-in*). Same as *diphenylamine.*
**phenylate** (*fe'-nil-āt*). A carbolate.
**phenylchinaldin.** See *phenylquinaldin.*
**phenylchinolin.** A derivative of chinolin by the entrance of the phenyl-group into its pyridin molecule.
**phenylene** (*fen'-il-ēn*), $C_6H_4$. A bivalent organic radical.
**phenylglucosazone** (*fe-nil-gloo-ko'-saz-ōn*), $C_{18}H_{22}N_4O_4$. A yellow, crystalline compound produced in the phenylhydrazine test for glucose.
**phenylglycuronic acid** (*fe-nil-glik-ū-ron'-ik*). A crystalline body, a compound of phenol and glycuronic acid, occurring in the urine after the ingestion of phenol.
**phenylhydrazine** (*fe-nil-hi'-dra-zēn*), $C_6H_5N_2$. A liquid base, crystallizing in plates, the hydrochloride of which is used as a test for sugar. **p. tests.** See *v. Jaksch, Neumann, Riegler.*
**phenylhydroxylamine,** $C_6H_5NHOH$, a product of nitrobenzol by reduction; very active blood-poison.
**phenylic** (*fe-nil'-ik*) [*phenyl*]. Pertaining to or containing phenyl. **p. acid,** carbolic acid, phenol.
**phenylmethane.** A crystalline analgesic and antipyretic substance.
**phenylon** (*fe'-nil-on*). Antipyrine.
**phenylone** (*fen'-il-ōn*). Antipyrine.
**phenylquinaldin** (*fe-nil-kwin-al'-din*), $C_6H_5(C_9H_5)N$, an antiperiodic prepared by the action of hydrochloric acid on a mixture of aniline, acetophenone, and aldehyde. Dose 1½-3 gr. (0.1-0.3 Gm.). It is also used externally as a local irritant.
**phenylquinolin** (*fe-nil-kwin'-o-lin*). A derivative of quinolin by the entrance of the phenyl groups into its pyridin molecule. It is more active than quinine.
**phenylurethane.** (*fe-nil-ū'-reth-ān*). See *euphorin.*
**pheochrome, pheochromoblast.** See *phaochrome, phaochromoblast.*
**phesin** (*fe'-sin*), $C_6H_3 . O . C_2H_5SO_2Na . NH . CO$ .-

$CH_3$. A proprietary antipyretic sulpho-derivative of acetphenetidin.
**Ph.G.** Abbreviation for (1) Graduate in Pharmacy; (2) German Pharmacopœia.
**phial** (*fi'-al*). See *vial.*
**philanthropist** (*fil-an'-thro-pist*) [φιλάνθρωπος, humane]. One who loves mankind.
**philanthropy** (*fil-an'-thro-pe*) [φιλάνθρωπος, humane]. The love of mankind; benevolence; charity.
**Phillips' muscle.** A small muscle extending from the styloid process of the radius and the external lateral ligament to the proximal portion of the phalanges.
**Phillyrea** (*fil-ir'-e-ah*) [φιλύρα, the linden-tree; from the similarity of its leaves]. A genus of the *Oleaceæ*. The leaves of *P. latifolia*, the stone-linden of southern Europe, are diuretic and emmenagogue and are used in ulcerations of the mouth. It contains *phillyrin.*
**phillyrin** (*fil'-i-rin*), $C_{27}H_{34}O_{11}+H_2O$. A crystalline glucoside found in the bark and leaves of *Phillyrea latifolia*, *P. angustifolia*, and *P. media.* It is antimalarial.
**philocytase** (*fil-o-si'-tās*) [φιλεῖν, to love; κύτος, a cell]. Metchnikoff's name for the intermediate body of Ehrlich; an amboceptor.
**philoneism** (*fil-o-ne'-izm*) [φιλεῖν, to love; νέος, new]. Abnormal love of novelty; the reverse of misoneism.
**philopatridomania** (*fil-o-pa-trid-o-ma'-ne-ah*) [φιλεῖν, to love; πατρίς, fatherland; μανία, madness]. An insane desire to return home; excessive nostalgia, or homesickness.
**philter, philtre** (*fil-ter*) [φίλτρον, a love-charm]. A love-potion; a preparation supposed to be efficacious in exciting sexual passion.
**philtrum** (*fil'-trum*) [φίλτρον, a love-charm]. 1. The depression on the surface of the skin of the upper lip immediately below the septum of the nose. 2. A *philter*, q. v.
**phimosientomy** (*fi-mo-si-en'-to-me*) [φιμοῦν, to constrict; ἐντομή, incision]. Incision of a constricted prepuce.
**phimosiotomy** (*fi-mo-si-ot'-o-me*). See *phimosientomy.*
**phimosis** (*fi-mo'-sis*) [φιμοῦν, to constrict]. Elongation of the prepuce and constriction of the orifice, so that the foreskin cannot be retracted to uncover the glans penis. **p. adnata,** **p. puerilis,** congenital phimosis. **p. circumligata.** See *paraphimosis.* **p. œdematodes,** phimosis with edema of the prepuce. Syn., *hydrophimosis.* **p. oris,** narrowing of the opening of the mouth. **p. palpebrarum,** Synonym of *blepharophimosis.* **p. vaginalis,** atresia of the vagina.
**phimotic** (*fi-mot'-ik*). Relating to phimosis.
**phisiotherapy** (*fiz-e-o-ther'-ap-e*) [φύσις, nature; θεραπεία, therapy]. The application of natural remedies—air, water, sunlight, etc.—in the treatment of disease.
**phleb-** (*fleb-*) [φλέψ, vein]. A prefix meaning vein.
**phlebangioma** (*fleb-an-je-o'-mah*) [*phleb-*; ἀγγεῖον, vessel]; δμα, tumor]. A venous aneurysm.
**phlebarteriectasia** (*fleb-ar-te-re-ek-ta'-ze-ah*) [*phleb-*; ἀρτηρία, artery; ἔκτασις, dilatation]. Varicose aneurysm.
**phlebarteriodialysis** (*fleb-ar-te-re-o-di-al'-is-is*) [*phleb-*; ἀρτηρία, artery; διάλυσις, separation]. Arterio-venous aneurysm.
**phlebectasia, phlebectasis** (*fleb-ek-ta'-ze-ah, fleb-ek'-tas-is*) [*phleb-*; ἔκτασις, dilatation]. Dilation of a vein; varicosity.
**phlebectomy** (*fleb-ek'-to-me*) [φλέψ, vein; ἐκτομή, excision]. Excision of a vein or a portion of a vein.
**phlebectopia** (*fleb-ek-to'-pe-ah*) [*phleb-*; ἐκ, out; τόπος, place]. The displacement or abnormal position of a vein.
**phlebemphraxis** (*fleb-em-fraks'-is*) [*phleb-*; ἔμφραξις, obstruction]. Plugging of a vein.
**phlebepatitis** (*fleb-ep-at-i'-tis*) [*phleb-*; ἧπαρ, vein; *hepatitis*]. Inflammation of the portal or hepatic veins.
**phlebexeresis** (*fleb-eks-er'-ez-mah*). Synonym of *varix.*
**phlebin** (*fleb'-in*) [*phleb-*]. A term for the venous blood-pigment as contained in the red corpuscles.
**phlebismus** (*fleb-iz'-mus*) [φλέψ, vein]. Undue prominence or swelling of a vein.
**phlebitic** (*fleb-it'-ik*). Pertaining to, of the nature of, or affected with phlebitis.
**phlebitis** (*fleb-i'-tis*) [φλέψ, vein; *ιτις*, inflammation]. Inflammation of a vein. This is generally suppura-

**tive**, (*suppurative phlebitis*), and is the result of the extension of suppuration from adjacent tissues. It leads to the formation of a thrombus within the vein (*thrombophlebitis*), which may break down and cause the distribution of septic emboli to various parts of the body. When not due to a suppurative process the phlebitis, called *plastic, adhesive*, or *proliferative*, may give rise to obliteration of the vein. The symptoms of phlebitis are pain and edema of the affected part, redness along the course of the vein, the latter appearing as a hard, tender cord. p., **sinus-**, phlebitis of the sinuses of the dura mater.
**phlebo-** (*fleb-o-*) [φλέψ, vein]. A prefix denoting pertaining to a vein.
**phlebocarcinoma** (*fleb-o-kar-sin-o'-mah*) [*phlebo-*; *carcinoma*]. Extension of carcinoma to the walls of a vein.
**phlebocholosis** (*fleb-o-ko-lo'-sis*) [*phlebo-*; χώλωσις, lameness]. Paralysis or disease of the veins.
**phlebogram** (*fleb'-o-gram*) [*phlebo-*; γράμμα, a writing]. A tracing of the movements of a vein by the sphygmograph.
**phlebograph** (*fleb'-o-graf*) [*phlebo-*; γράφειν, to write]. An instrument for recording the venous pulse.
**phlebography** (*fleb-og'-ra-fe*) [*phlebo-*; γράφειν, to write]. The anatomy and physiology of the veins; a description of the veins.
**phlebolite, phlebolith** (*fleb'-o-līt, fleb'-o-lith*) [*phlebo-*; λίθος, a stone]. Vein-stone, a hard concretion sometimes found in veins, and produced by calcareous infiltration of a vein.
**phlebolithiasis** (*fleb-o-lith-i'-as-is*) [*phlebo-*; λίθος, stone]. The formation of phleboliths.
**phlebolitic** (*fleb-o-lit'-ik*) [*phlebo-*; λίθος, a stone]. Of the nature of, containing, or characterized by, phlebolites.
**phlebology** (*fleb-ol'-o-je*) [*phlebo-*; λόγος, science]. The science of the anatomy and physiology of the veins.
**phlebopexy** (*fleb'-o-peks-e*) [*phlebo-*; πῆξις, a fixing in]. Longuet's term for the preservation of the venous reticulum which results from the extraserous transplantation of the testicle in cases of varicocele.
**phlebophthalmotomy** (*fleb-off-thal-mot'-o-me*) [*phlebo-*; ὀφθαλμός, eye; τομή, a cutting]. Scarification of the conjunctival vein.
**phleboplerosis** (*fleb-o-ple-ro'-sis*) [*phlebo-*; πλήρωσις, a filling]. Distention of the veins.
**phlebophlogosis** (*fleb-o-flo-go'-sis*). Synonym of *phlebitis*.
**phleborrhagia** (*fleb-or-a'-je-ah*) [*phlebo-*; ῥηγνύναι, to burst forth]. Venous hemorrhage.
**phleborrhaphy** (*fleb-or'-af-e*) [*phlebo-*; ῥαφή, suture]. Suture of a vein.
**phleborrhexis** (*fleb-or-eks'-is*) [*phlebo-*; ῥῆξις, rupture]. Rupture of a vein.
**phlebosclerosis** (*fleb-o-skle-ro'-sis*) [*phlebo-*; σκληρός, hard]. Sclerosis of a vein.
**phlebostasis** (*fleb-o-stas-is*) [*phlebo-*; στάσις, a standing still]. The temporary removal of some of the blood from the general circulation by means of compression in the veins in the extremities; also called "bloodless phlebotomy."
**phlebostenosis** (*fleb-o-sten-o'-sis*) [*phlebo-*; στένος, narrow]. Constriction of a vein.
**phlebostrepsis** (*fleb-o-strep'-sis*) [*phlebo-*; στρέψις, a twisting]. Torsion, or twisting, of a vein.
**phlebothrombosis** (*fleb-o-throm-bo'-sis*) [*phlebo-*; *thrombosis*]. The formation of a thrombus in a vein.
**phlebotome** (*fleb'-o-tōm*). A cutting-instrument used in phlebotomy; a fleam.
**phlebotomist** (*fleb-ot'-o-mist*) [*phlebo-*; τομή, a cutting]. One who lets blood; a bleeder.
**phlebotomus fever** (*fleb-ot'-om-us*). A fever of brief duration met with in the countries around the Mediterranean, also in India; it is apparently conveyed by sand-flies.
**phlebotomy** (*fleb-ot'-o-me*) [*phlebo-*; τομή, a cutting]. Opening of a vein for the purpose of bloodletting. The vein most often selected is the median cephalic at the bend of the elbow. p., **bloodless**. See *phlebostasis*.
**phledonia** (*fle-do'-ne-ah*) [φλεδονεία, babble]. Delirium, or delirious utterance.
**phlegm** (*flem*) [φλέγμα, phlegm]. 1. A viscid, stringy mucus, secreted by the mucosa of the upper air-passages. 2. One of the four humors of the old writers.

**phlegmasia** (*fleg-ma'-ze-ah*) [φλέγμα, a flame]. Inflammation. p. **adenosa**. See *adenitis*. p. **alba dolens**, milk-leg, a painful swelling of the leg beginning either at the ankle and ascending, or at the groin and extending down the thigh, its usual cause being septic infection after labor. p. **cellularis**, cellulitis. p. **dolens**. Same as p. *alba dolens*. p. **lactea**. See p. *alba dolens*. p. **malabarica**. Synonym of *elephantiasis arabum*. p. **membranæ mucosæ gastropulmonalis**. See *aphthæ tropicæ*. p. **myoica**, myositis.
**phlegmatic** (*fleg-mat'-ik*) [*phlegm*]. Full of phlegm; hence, indifferent; apathetic; slow, dull; lymphatic.
**phlegmon** (*fleg'-mon*) [φλεγμονή, inflammation]. An inflammation characterized by the spreading of a purulent or fibrinopurulent exudate within the tissues. p., **gas**, that in which more or less offensive gas is formed with the pus. p. **ligneux**, a peculiar form of chronic inflammation of the skin and subcutaneous tissue marked by a slow clinical course and a consistence resembling wood. It occurs most frequently on the neck.
**phlegmonodœa** (*fleg-mon-o-de'-ah*) [φλεγμονή, inflammation]. 1. See *erythematica*. 2. Peritonitis.
**phlegmonoid** (*fleg'-mon-oid*). Resembling phlegmon.
**phlegmonous** (*fleg'-mon-us*) [*phlegmon*]. Of the nature of or pertaining to phlegmon.
**phloem** (*flo'-em*) [φλοίος, bark]. In botany that portion of a fibrovascular bundle which consists of bast-tissue and sieve-tissue; leptome. Cf. *xylem*. p., **ray**, a plate of phloem-tissue between two medullary rays. p.-**sheath**, a layer of thin-walled cells surrounding the phloem-tissue; bast-sheath; periphloem; vascular bundle sheath. p.-**tissue**, phloem.
**phlogistic** (*flo-jis'-tik*) [φλογιστός, burnt]. Inflammatory.
**phlogogen, phlogogon** (*flog'-o-jen, flog'-o-gon*) [φλόγωσις, inflammation; γενναν, to produce]. Any substance having the property of exciting inflammation in a tissue with which it comes in contact; an irritant.
**phlogogenic** (*flog-o-jen'-ik*) [see *phlogogen*]. Causing inflammation.
**phlogogenous** (*flo-goj'-en-us*) [φλογός, burning; γενης, producing]. Producing inflammation.
**phlogosin** (*flog'-o-sin*) [φλόγωσις, inflammation]. A crystalline body isolated from cultures of pyogenic staphylococci, and causing suppuration when injected beneath the skin or introduced into the eye.
**phlogosis** (*flog-o'-sis*) [see *phlogosin*]. 1. Inflammation. 2. Erysipelas.
**phlogotic** (*flo-got'-ik*) [φλογός, burning]. Pertaining to or marked by phlogosis, or inflammation.
**phlogozelotism** (*flo-go-zel'-ot-izm*) [φλογός, burning; ζηλοῦν, to be eager]. A mania for ascribing to every disease an inflammatory origin.
**phloretin** (*flor-e'-tin*) [φλοιός, bark; ῥίζα, root]. A product of the treatment of phloridzin by dilute acids. Like that of phloridzin, its administration in suitable doses is followed by glycosuria or true diabetes.
**phloridzin, phlorizin, phlorrhizin** (*flor-id'-zin, flor-i'-zin*) [φλοιός, bark; ῥίζα, root], $C_{21}H_{24}O_{10} \cdot 2H_2O$. A bitter crystalline glucoside occurring in the root and trunk of apple, pear, and other fruit-trees. It is said to possess antipyretic properties. Dose 5-10 gr. (0.32-0.65 Gm.). Given to lower animals it produces glycosuria. p.-**diabetes**, the glycosuria induced in lower animals, especially dogs, by the administration of phloridzin.
**phloroglucin** (*flo-ro-gloo'-sin*) [φλοιός, bark; γλυκύς, sweet], $C_6H_3(OH)_3$. A crystalline substance found in the bark of the cherry, pear, apple, and other trees, and used as a test for woody tissue (lignin) and hydrochloric acid.
**phlorol** (*flo'-rol*) [φλοιός, bark; *oleum*, oil], $C_8H_{10}O$. A phenol found in creosote.
**phlorose** (*flo'-rōs*) [φλοιός, bark]. A glucose which is probably identical with dextrose.
**phlorrhizin** (*flor'-iz-in*). Synonym of *phloridzin*.
**phlyctena, phlyctæna** (*flik-te'-nah*) [φλύκταινα, blister]. A vesicle.
**phlyctenar** (*flik'-ten-ar*) [φλύκταινα, blister]. Affected with phlyctena; pertaining to phlyctena.
**phlyctenoid** (*flik'-ten-oid*). See *phlyctenular*.
**phlyctenosis** (*flik-ten-o'-sis*) [φλύκταινα, blister]. An eruption characterized by vesicles. p. **aggregata**, an herpetic eruption in which the vesicles are situated closely together. p. **labialis**. Synonym of *herpes*

*labialis*. **p. sparsa**, a form in which the vesicles are few and at considerable distances from each other.
**phlyctenula** (*flik-ten'-ū-lah*) [dim. of φλύκταινα, blister]. A little vesicle or blister.
**phlyctenular** (*flik-ten'-ū-lar*) [*phlyctena*]. Resembling a phlyctenule; characterized by the formation of phlyctenules, as *phlyctenular conjunctivitis*.
**phlyctenule**, **phlyctenula** (*flik-ten'-ūl, flik-ten'-ū-lah*) [*phlyctena*]. A minute phlyctena; a little vesicle or blister.
**phlysis** (*fli'-sis*) [φλύσις, eruption]. 1. A phlyctenule. 2. A whitlow.
**phlyzacion, phlyzacium** (*fli-za'-se-on, fli-za'-se-um*) [φλύζειν, to inflame]. A pustular vesicle on an indurated base. p. **acutum**. See *ecthyma*.
**phobia** (*fo'-be-ah*) [φόβος, fear]. Any obsession of fear characteristic of insanity.
**phobodipsia** (*fo-bo-dip'-se-ah*) [φόβος, fear; δίψα, thirst]. Hydrophobia.
**phobophobia** (*fo-bo-fo'-be-ah*) [φόβος, fear]. Dread of being afraid.
**Phocas' disease** (*fo-kah'*) [B. G. *Phocas*, French physician]. Chronic fibrous mastitis, characterized by the presence of multiple fibrous nodules in both breasts.
**phocomelus** (*fo-kom'-el-us*) [φώκη, a seal; μέλος, a limb]. A monster with rudimentary limbs, the hands and feet being attached almost directly to the trunk.
**phonacoscope** (*fo-nak'-o-skōp*) [φωνή, voice; σκοπεῖν, to examine]. An instrument for combined auscultation and percussion; it increases the intensity of the sounds heard.
**phonacoscopy** (*fo-nak-os'-ko-pe*). Examination of the chest with a phonoscope.
**phonal** (*fo'-nal*) [φωνή, voice]. Pertaining to the voice or to sound.
**phonation** (*fo-na'-shun*) [φωνή, voice]. The production of vocal sound or articulate speech.
**phonatory** (*fon'-a-to-re*) [*phonation*]. Pertaining to phonation. p. **band**. Same as *vocal band*.
**phonautogram** (*fo-naw'-to-gram*) [φωνή, voice; αὐτός, self; γράμμα, inscription]. The diagram of a phonautograph.
**phonautograph** (*fo-naw'-to-graf*) [φωνή, voice; *autograph*]. An apparatus for recording automatically the vibrations of the air produced by the voice.
**phone** (*fōn*) [φωνή, sound, voice]. A vocal sound.
**phonendoscope** (*fo-nen'-do-skōp*) [φωνή, voice; ἔνδον, within; σκοπεῖν, to view]. A variety of stethoscope which intensifies the auscultatory sounds.
**phonetic** (*fo-net'-ik*) [φωνή, voice]. 1. Pertaining to or representing sounds. 2. Pertaining to the voice.
**phonetics** (*fo-net'-iks*). The science dealing with the mode of production of sounds.
**phonic** (*fon'-ik*) [φωνή, voice]. Pertaining to the voice. p. **spasm**, a spasm of the laryngeal muscles occurring on attempting to speak.
**phonica** (*fon'-ik-ah*) [φωνή, voice]. Diseases affecting the vocal organs.
**phonism** (*fo'-nizm*) [φωνή, voice]. A sensation, of sound or hearing, due to the effect of sight, touch, taste, or smell, or even to the thought of some object, person, or general conception.
**phono-** (*fo-no-*) [φωνή, voice]. A prefix denoting relating to the voice or to sound.
**phonocardiogram** (*fo-no-kar'-de-o-gram*) [*phono-*; καρδία, heart; γράμμα, a writing]. An instrument for registering the sounds of the heart.
**phonocardiography** (*fo-no-kor'-de-og'-ra-fe*). Registration of the sounds of the heart.
**phonochorda** (*fo-no-kor'-dah*) [*phono-*; χορδή, cord; *pl., phonochorda*]. A vocal band.
**phonogram** (*fo'-no-gram*) [*phono-*; γράμμα, a writing]. 1. The record of a phonograph. 2. A graphic character representing a vocal sound.
**phonograph** (*fo'-no-graf*) [*phono-*; γράφειν, to record]. An instrument consisting of a wax-coated cylinder revolving under a stylus attached to a diaphragm. The vibrations of the diaphragm, set in motion by the voice, cause the cylinder to be indented by the stylus. When the cylinder is again revolved, the movement of the stylus upon the cylinder throws the diaphragm into vibration and reproduces the original sounds of the voice.
**phonology** (*fo-nol'-o-je*) [*phono-*; λόγος, science]. The science of vocal sounds; phonetics.
**phonomania** (*fo-no-ma'-ne-ah*) [φωνή, slaughter; μανία, madness]. Homicidal mania.

**phonomassage** (*fo-no-mas-ahzh'*) [*phono-*; *massage*]. Action upon the tympanum by sound vibrations conducted into the auditory canal.
**phonometer** (*fo-nom'-et-er*) [*phono-*; μέτρον, a measure]. An instrument for measuring the intensity of the voice.
**phonomyoclonus** (*fo-no-mi-ok'-lo-nus*) [*phono-*; μῦς, muscle; κλόνος, tumult]. A condition in which a sound is heard on auscultation over a muscle, denoting fibrillary contractions; these latter may be so fine as to be invisible.
**phononosus** (*fo-nom'-o-sus*). Synonym of *phonopathy*.
**phonopathy** (*fo-nop'-ath-e*) [*phono-*; πάθος, disease]. Any disorder or disease of the voice.
**phonophobia** (*fo-no-fo'-be-ah*) [*phono-*; φόβος, fear]. 1. A fear of speaking, in paresthesia of the larynx, because of the painful sensation produced during phonation. 2. Morbid dread of any sound or noise.
**phonophore** (*fo'-no-fōr*) [φωνή, sound; φορος, bearing; *pl.*, *phonophori*]. An auditory ossicle, viewed as a transmitter of sound. See *Paladino's phonophore*.
**phonopneumomassage** (*fo-no-nu-mo-mas-ahzj'*) [*phono-*; πνεῦμα, air; *massage*]. The exercise of the muscles, ligaments, and articulating surfaces of the inner ear by means of an electric apparatus.
**phonopsia** (*fo-nop'-se-ah*) [*phono-*; ὄψις, vision]. The perception of color-sensations by auditory sensations.
**phonoscope** (*fo'-no-skōp*) [*phono-*; σκοπεῖν, to examine]. A stethoscope for intensifying the tone in auscultation.
**phoria** (*fo'-re-ah*) [φορά, motion]. A colloquialism used to represent one or more of the terms *orthophoria, heterophoria, exophoria, esophoria, hyperphoria*, etc.
**phoro-** (*fo-ro-*) [φορά, motion]. A prefix meaning motion.
**phoroblast** (*for'-o-blast*) [φορεῖν, to bear; βλαστός, germ]. Connective tissue.
**phorocyte** (*for'-o-sīt*). A connective-tissue cell.
**phorocytosis** (*for-o-si-to'-sis*). Increase in the number of connective-tissue cells.
**phorometer** (*for-om'-et-er*) [*phoro-*; μέτρον, a measure]. An instrument for measuring the relative strength of the ocular muscles.
**phorone** (*fo'-rōn*), $C_9H_{14}O$. A substance prepared by saturating acetone with HCl and permitting it to stand.
**phoro-optometer** (*for-o-op-tom'-et-er*) [*phoro-*; ὀπτός, visible; μέτρον, a measure]. An apparatus for optical testing of muscular defects.
**phoroplast** (*for'-o-plast*). Connective tissue.
**phorotone** (*for'-o-tōn*) [*phoro-*; τόνος, strength]. An apparatus for exercising the eye-muscles.
**phose** (*fōz*) [φῶς, light]. A subjective sensation of light or color, as, scotoma scintillans. An *aphose* is a subjective sensation of shadow or darkness, as, muscae volitantes. *Centraphoses* are aphoses originating in the optic centers. *Centrophoses* are phoses originating in the optic centers. A *chromophose* is a subjective sensation of color. *Peripheraphoses* are peripheral aphoses. *Peripherophoses* are phoses originating in the peripheral organs of vision (the optic nerve or eyeball).
**phosgen, phosgene** (*fos'-jen, -jēn*) [φῶς, light; γεννᾶν, to produce]. Producing light. p. **gas**, $COCl_2$, carbonyl chloride; a colorless gas formed by the action of light on a mixture of carbonic oxide and chlorine.
**phosgenic** (*fos-jen'-ik*). See *photogenic*.
**phosis** (*fo'-sis'*). The formation of a phose.
**phosote** (*fo'-sōt*). A syrupy liquid, consisting of creosote, 80 %, and phosphoric anhydride, 20 %. Dose 30 min. (2 Cc.) daily.
**phosphagon** (*fos'-fag-on*). Trade name of an elixir of various glycerophosphates.
**phosphate** (*fos'-fāt*) [*phosphorus*]. A salt of phosphoric acid. The phosphates are used in medicine as tonics and alteratives in conditions associated with malnutrition of the bones (rickets, scrofula). **p., acid**, one in which one or two of the hydrogen atoms only have been replaced by metals. **p., ammoniomagnesium**, a double salt of ammonium and magnesium and phosphoric acid. **p., earthy**, a phosphate of one of the alkaline earths. **p., normal**, one in which the three hydrogen atoms, or the six of two molecules, are substituted by metals,

**e. g.**, $Na_3PO_4$, $Ca_3(PO_4)_2$. p., **triple,** ammoniomagnesium phosphate.
**phosphatic** (*fos-fat'-ik*) [*phosphate*]. Containing phosphates; characterized by the excretion of large amounts of phosphates, as *phosphatic* diathesis.
**phosphatid** (*fos'-fa-tid*) [*phosphorus*]. Any one of a large group of phosphorus-compounds found in brain-substance, and resembling the phosphates; they are esters of orthophosphoric acid.
**phosphatol** (*fos'-fat-ol*). A thick liquid obtained by action of phosphorus trichloride on creosote in an alcoholic solution of soda. It contains 90 % of creosote; used in tuberculosis.
**phosphatometer** (*fos-fa-tom'-et-er*). An instrument for estimating the amount of phosphates in the urine.
**phosphatoptosis** (*fos-fat-o-to'-sis*) [*phosphate*; πτῶσις, a falling]. Spontaneous precipitation of phosphates in the urine.
**phosphaturia** (*fos-fat-ū'-re-ah*) [*phosphate*; οὖρον, urine]. A condition in which an excess of phosphates is passed in the urine
**phosphene** (*fos'-fēn*) [φῶς, light; φαίνειν, to show]. A subjective luminous sensation caused by pressure upon the eyeball. **p. of accommodation,** a phosphene produced by the effort of accommodation. **p., pressure.** See *phosphene*.
**phosphergot** (*fos-fer'-got*). A mixture of sodium phosphate and ergot; it is indicated in general debility.
**phospherrin** (*fos-fer'-in*). A mixture said to consist of ferric chloride, phosphoric acid, and glycerol.
**phosphide** (*fos'-fīd*) [*phosphorus*]. A compound of phosphorus and another element or radical acting as a base. The phosphides are used in medicine as substitutes for phosphorus.
**phosphin** (*fos'-fin*) [*phosphorus*]. 1. Hydrogen phosphide, $PH_3$, a poisonous gas of alliaceous odor. 2. A substitution-compound of $PH_3$, bearing the same relation to it that an amine does to ammonia.
**phosphite** (*fos'-fīt*) [*phosphorus*]. A salt of phosphorous acid.
**phospho-** (*fos-fo-*) [*phosphorus*]. A prefix meaning relating to phosphorus or to its compounds.
**phosphoglyceric acid** (*fos-fo-glis-e'-rik*) [*phospho-*; γλυκύς, sweet], $C_3H_7PO_6$. A liquid body obtained from lecithin.
**phosphoglycoproteids** (*fos-fo-gli-ko-pro'-te-ids*). The same as *nucleoalbumins*.
**phosphoguaicol** (*fos-fo-gwi'-ak-ol*). See *guaiacol phosphite*.
**phosphomolybdic acid** (*fos-fo-mol-ib'-dik*) [*phospho-*; *molybdenum*]. A compound of phosphoric acid and molybdenum trioxide, used as a test for alkaloids.
**phosphonecrosis, phosphornecrosis** (*fos-fo-ne-kro'-sis, fos-for-ne-kro'-sis*). See *necrosis, phosphorus-*.
**phosphonium** (*fos-fo'-ne-um*) [*phosphorus*]. The hypothetical univalent radical $PH_4$; it is analogous to ammonium, $NH_4$.
**phosphoprotein** (*fos-fo-pro'-te-in*). A conjugated protein consisting of a compound of protein with a phosphorus-containing substance other than nucleic acid or lecithin.
**phosphorated** (*fos'-fo-ra-ted*) [*phosphorus*]. Containing phosphorus. **p.** oil, a one per cent. solution of phosphorus in expressed oil of almonds, with the addition of a small quantity of ether.
**phosphorescence** (*fos-for-es'-ens*) [*phosphorus*]. The spontaneous luminosity of phosphorus and other substances in the dark.
**phosphorescent** (*fos-for-es'-ent*) [*phosphorus*]. Possessing the quality of phosphorescence.
**phosphoreted** (*fos'-for-et-ed*) [*phosphorus*]. Combined with phosphorus.
**phosphoric acid** (*fos-for'-ik*). See *acid, phosphoric*.
**phosphoridrosis, phosphorhidrosis** (*fos-for-id-ro'-sis*) [*phospho-*; ἱδρώσις, sweat]. The secretion of phosphorescent sweat.
**phosphorism** (*fos'-for-izm*) [*phosphorus*]. Chronic phosphorus-poisoning.
**phosphorized** (*fos'-for-īzd*) [*phosphorus*]. Containing phosphorus.
**phosphorous acid** (*fos-fo'-rus*). See *acid, phosphorous*.
**phosphoruria** (*fos-for-ū'-re-ah*) [*phospho-*; οὖρον, urine]. 1. Phosphorescence of the urine. 2. Urine containing an excess of phosphates.
**phosphorus** (*fos'-for-us*) [φῶς, light; φέρειν, to bear]. A nonmetallic element having a quantivalence of III or V, and an atomic weight of 31.04. Symbol P.

In commerce it is prepared from bone-ash or from sombrerite, an impure calcium phosphate found in West Indian guano. Phosphorus may be obtained in several allotropic forms. *Ordinary phosphorus* is a yellowish-white, waxy solid, of a specific gravity of 1.837; it is exceedingly poisonous; it causes a widespread fatty degeneration, most marked in the liver. *Red* or *amorphous phosphorus* is a dark-red powder, having a specific gravity of 2.11, insoluble in carbon disulphide, noninflammable, nonluminous, nonpoisonous. *Metallic* or *rhombohedral phosphorus* is an allotropic form produced by heating phosphorus in a sealed tube with melted lead. Its specific gravity is 2.34. Medicinally, phosphorus is used as an alterative in osteomalacia and in rickets, in sexual impotence, threatened cerebral degeneration, neuralgia, chronic alcoholism, morphinomania, furunculosis, etc. Dose $\frac{1}{150}$ gr. (0.00065 Gm.). **p., pills of** (*pilulæ phosphoreted*.
**phosphori,** U. S. P.). Dose 1 pill. **p. trichloride,** $PCl_3$, a colorless liquid of unpleasant odor.
**phosphotal** (*fos'-fo-tal*). Creosote phosphite. See *phosphatol*.
**phosphotungstic acid** (*fos-fo-tung'-stik*). A crystalline compound of phosphoric and tungstic acids, used as a test for alkaloids and peptones.
**phosphuret** (*fos'-fū-ret*) [*phosphorus*]. A phosphide.
**phosphureted, phosphuretted** (*fos'-fū-ret-ed*). Synonym of *phosphoreted*.
**phosphuria** (*fos-fū'-re-ah*). Synonym of *phosphaturia*.
**phossy jaw** (*fos'-e*). See *jaw, phossy*.
**photalgia** (*fo-tal'-je-ah*) [*photo* ; ἄλγος, pain]. Pain arising from too great intensity of light.
**photaugiophobia** (*fo-taw-je-o-fo'-be-ah*) [*photo-*; αὐγή, glare; φοβός, fear]. A shrinking from the glare of light.
**phote** (*fōt*) [*photo-*]. The unit of photochemical energy employed in connection with determination of the solidity of colors to average solar light at noon.
**photesthesia, photæsthesia** (*fo-tes-the'-ze-ah*) [*photo-*; αἰσθησις, sensation]. 1. Sensitiveness to light. 2. Photophobia.
**photic** (*fo'-tik*) [*photo-*]. Relating to light.
**photism** (*fo'-tizm*) [φωτισμός, an enlightening]. An association, as of color or light, produced by hearing, taste, smell, touch, or temperature, or even by the thought of some object, person, or general conception. Cf. *phonism*.
**photo-** (*fo-to-*) [φῶς, light]. A prefix denoting relation to light.
**photoactinic** (*fo-to-ak-tin'-ik*). Emitting both luminous and actinic rays.
**photobacterium** (*fo-to-bak-te'-re-um*). A genus or form of bacteria whose cultures are phosphorescent.
**photobiotic** (*fo-to-bi-ot'-ik*). [*photo-*; βίος, life]. Living in the light exclusively.
**photocampsis** (*fo-to-kamp'-sis*) [*photo-*; κάμψις, a bending]. Refraction of light.
**photochemical** (*fo-to-kem'-ik-al*) [*photo-*; *chemical*]. Pertaining to the chemical action of light.
**photochemistry** (*fo-to-kem'-is-tre*). That branch of chemistry treating of the chemical action of light.
**photochromatic** (*fo-to-kro-mat'-ik*) [*photo-*; χρῶμα, color]. Pertaining to colored light. **p. treatment,** treatment of disease by colored light.
**photodynamic** (*fo-to-di-nam'-ik*) [*photo-*; δύναμις, power]. Pertaining to the energy of light.
**photodysphoria** (*fo-to-dis-fo'-re-ah*) [*photo-*; δυσφορία, excessive pain]. Intolerance of light; photophobia.
**photoelectricity** (*fo-to-e-lek-tris'-it-e*). Electricity produced under the influence of light.
**photoelement** (*fo-to-el'-e-ment*). The element of a galvanic battery which by decomposition gives photoelectricity.
**photofluoroscope** (*fo-to-floo'-o-ro-skōp*). See *fluoroscope*.
**photogene** (*fo'-to-jēn*) [*photo-*; γεννᾶν, to produce]. 1. A retinal impression; an after-image. 2. A liquid derived from bituminous shale.
**photogenesis** (*fo-to-jen'-e-sis*) [*photo-*; γενναν, to produce]. The production of light or of phosphorescence.
**photogenic** (*fo-to-jen'-ik*) [see *photogene*]. Light-producing.
**photogenous** (*fo-toj'-en-us*) [*photo-*; γενής, producing]. Producing light.
**photogram** (*fo'-to-gram*) [*photo-*; γράμμα, a writing]. A photographic representation of an enlargement obtained by the microscope.

**photography** (*fo-tog'-ra-fe*) [*photo-;* γράφειν, to write]. The art of producing an image of an object (*photograph*) by throwing the rays of light reflected from it upon a surface coated with a film of a substance, such as a silver salt, that is readily decomposed by light, subsequently treating the film with certain agents (developers) that bring out the image, and then dissolving the salt unacted upon by the light.

**photohemotachometer, photohæmotachometer** (*fo-to-hem-o-tak-om'-et-er*) [*photo-;* αἷμα, blood; τάχος, swiftness; μέτρον, a measure]. A hemotachometer in which the changes in level of the column of blood are photographed.

**photokinetic** (*fo-to-kin-et'-ik*) [*photo-;* κινητικός, causing movement]. Causing movement by means of light.

**photolysis** (*fo-tol'-is-is*) [*photolyte*]. Decomposition by the action of light.

**photolyte** (*fo'-to-lit*) [*photo-;* λύειν, to loosen]. A substance that is decomposed by the action of light.

**photomagnetism** (*fo-to-mag'-net-izm*) [*photo-;* magnetism]. Magnetism produced by the action of light.

**photomania** (*fo-to-ma'-ne-ah*) [*photo-;* μανία, madness]. 1. The increase of maniacal symptoms under the influence of light. 2. A morbid desire for light.

**photometer** (*fo-tom'-et-er*) [*photo-;* μέτρον, a measure]. An instrument for measuring the intensity of light.

**photometry** (*fo-tom'-et-re*) [see *photometer*]. The measurement of the intensity of light.

**photomicrograph** (*fo-to-mi'-kro-graf*). A photograph of a small or microscopic object, usually made with the aid of a microscope, and of sufficient size for observation with the naked eye. Cf. *microphotograph*.

**photomicrography** (*fo-to-mi-krog'-ra-fe*) [*photomicrograph*]. The art of producing photomicrographs.

**photonosus** (*fo-ton'-o-sus*) [*photo-;* νόσος, disease]. A diseased condition arising from continued exposure to intense or glaring light, *e. g.*, snow-blindness, etc.

**photoparesthesia** (*fo-to-par-es-the'-ze-ah*) [*photo-;* παρά, beside; αἴσθησις, sensation]. Defective, or perverted, retinal sensibility.

**photophilic** (*fo-to-fil'-ik*) [*photo-;* φιλεῖν, to love]. Seeking or loving light.

**photophobia** (*fo-to-fo'-be-ah*) [*photo-;* φόβος, fear]. Intolerance of light.

**photophobic** (*fo-to-fo'-bik*) [*photo-;* φόβος, fear]. Affected with, or pertaining to, photophobia.

**photophone** (*fo'-to-fōn*) [*photo-;* φωνή, sound]. An apparatus for the graphic representation of the character of sound-waves by means of flames.

**photophore** (*fo'-to-fōr*) [*photo-;* φέρειν, to bear]. An instrument for examination of the cavities of the body by means of the electric light.

**photopsia** (*fo-top'-se-ah*) [*photo-;* ὄψις, sight]. Subjective sensations of sparks or flashes of light occurring in certain morbid conditions of the optic nerve, the retina, or the brain.

**photoptic** (*fo-top'-tik*) [*photopsia*]. Relating to photopsia.

**photoptometer** (*fo-top-tom'-et-er*) [*photo-;* ὤψ, eye; μέτρον, a measure]. An instrument for determining visual acuity.

**photoptometry** (*fo-top-tom'-et-re*) [*photo-;* ὤψ, eye; μέτρον, measure]. The measurement of the perception of light.

**photoradiometer** (*fo-to-ra-de-om'-et-er*). An instrument for the measurement of the quantity of X-rays passing through a given surface.

**photoscope** (*fo-to'-skōp*) [*photo-;* σκοπεῖν, to view]. 1. A fluoroscope. 2. An instrument used in inspecting the antrum of Highmore as regards the translucency of its walls.

**photoscopy** (*fo-tos'-ko-pe*) [*photo-;* σκοπεῖν, to view]. The same as *skiascopy*.

**photoskioptic** (*fo-to-ski-op'-tik*) [*photo-;* σκιά, shadow; ὀπτικός, pertaining to sight]. Skiagraphic.

**photosyntaxis** (*fo-to-sin'-taks*) [*photo-;* συντασσεῖν, to arrange]. The process of the manufacture of carbohydrates by plants.

**photosynthesis** (*fo-to-sin'-the-sis*) [*photo-;* σύνθεσις, putting together]. The building up of an organic compound by the action of light through the agency of chlorophyll, considered to be due to a soluble ferment, the chlorophyll acting simply as a chemical screen or sensibilizer.

**phototachometer** (*fo-to-tak-om'-et-ur*) [*photo-;* τάχος, speed; μέτρον, measure]. An apparatus for determining the velocity of light rays.

**phototactic** (*fo-to-tak'-tik*). Pertaining to phototaxis.

**phototaxis** (*fo-to-taks'-is*) [*photo-;* τάξις, arrangement]. Same as *phototropism*. See *tropism*.

**phototherapy** (*fo-to-ther'-ap-e*) [*photo-;* θεραπεία, treatment]. 1. The treatment of disease by light. 2. Finsen's light-treatment: the treatment of skin diseases by the application of the concentrated chemic rays (blue, violet, and ultraviolet) of light. 3. The treatment of smallpox by red light.

**phototonus** (*fo-tot'-o-nus*) [*photo-;* τόνος, tension]. In biology, a condition of increased vital irritability or motility due to exposure to light, in contrast with the rigidity or quiescence produced by darkness.

**phototropism** (*fo-tot'-ro-pizm*) [*photo-;* τρόπος, a turning]. See *tropism*.

**photoxylin, photoxylon** (*fo-toks'-il-in, fo-toks'-il-on*) [*photo-;* ξύλον, wood]. A substance produced from wood-pulp by the action of sulphuric acid and potassium nitrate. It serves as a substitute for collodion in minor surgery, and as a medium for mounting microscopic specimens.

**photuria** (*fo-tū'-re-ah*) [*photo-;* οὖρον, urine]. The passage of phosphorescent urine.

**Phragmidiothrix** (*frag-mid'-i-o-thriks*) [φραγμός, a fence; θρίξ, hair]. A genus of bacteria belonging to the *Chlamidobacteriaceæ*; filaments unbranched; divisions in three directions; sheath scarcely visible. Cf. *leptothrix; cladothrix; crenothrix*.

**phren** (*fren*) [φρήν, the mind, also the diaphragm]. 1. The diaphragm. 2. Mind.

**phrenalgia** (*fren-al'-je-ah*) [*phren;* ἄλγος, pain]. 1. Melancholia; psychalgia. 2. Neuralgia of the diaphragm.

**phrenasthenia** (*fren-as-the'-ne-ah*) [*phreno-;* ἀσθένεια, weakness]. 1. Paresis of the diaphragm. 2. Congenital mental weakness.

**phrenasthenic** (*fren-as-then'-ik*). 1. Relating to phrenasthenia; idiotic, imbecile. 2. A feeble-minded person.

**phrenathesia** (*fren-as-the'-ze-ah*). Idiocy.

**phrenatrophia** (*fren-at-ro'-fe-ah*) [*phreno-;* atrophy]. Atrophy of the brain; idiocy.

**phrenauxe** (*fren-awks'e*) [*phren;* αὔξη, enlargement]. Hypertrophy of the substance of the brain.

**phrenesiac** (*fren-e'-ze-ak*) [*phreno-*]. One who is affected with phrenesis; an insane person.

**phrenesis** (*fren-e'-sis*) [φρένησις, insanity]. Frenzy; delirium; insanity.

**phrenetic** (*fren-et'-ik*) [*phren*]. Maniacal; delirious.

**phrenic** (*fren'-ik*) [*phren*]. 1. Pertaining to the diaphragm, as *phrenic* nerve, *phrenic* artery. 2. Pertaining to the mind.

**phrenicocolic** (*fren-ik-o-kol'-ik*). Same as *phrenocolic*.

**phrenicogastric** (*fren-ik-o-gas'-trik*). Same as *phrenogastric*.

**phrenicosplenic** (*fren-ik-o-splen'-ik*). Same as *phrenosplenic*.

**phrenicotomy** (*fren-ik-ot'-o-me*) [*phren;* τομή, a cutting]. Section of a phrenic nerve.

**phrenitic** (*fren-it'-ik*). Pertaining to, or affected with phrenitis. 2. Relating to the mind.

**phrenitis** (*fren-i'-tis*) [*phren;* ιτις, inflammation]. 1. Inflammation of the brain. 2. Inflammation of the diaphragm. 3. Acute delirium.

**phreno-** (*phren-o-*) [*phren*]. A prefix meaning relating either to the mind or to the diaphragm.

**phrenoblabia** (*fren-o-bla'-be-ah*) [*phreno-;* βλάβη, hurt]. Any disorder of the mind.

**phrenocolic** (*fren-o-kol'-ik*) [*phreno-;* colon]. Pertaining to the diaphragm and the colon.

**phrenocolopexy** (*fren-o-ko'-lo-peks-e*) [*phreno-;* colon; πῆξις, fixation]. The operation of suturing a prolapsed or displaced colon to the diaphragm.

**phrenocostal** (*fren-o-kos'-tal*) [*phreno-;* costa, rib]. Pertaining to the diaphragm and the ribs.

**phrenogastric** (*fren-o-gas'-trik*) [*phreno-;* γαστήρ, stomach]. Pertaining conjointly to the stomach and the diaphragm.

**phrenoglottic** (*fren-o-glot'-ik*). Pertaining to the diaphragm and the glottis.

**phrenoglottismus** (*fren-o-glot-is'-mus*). Spasm of the glottis ascribed to disease of the diaphragm.

**phrenograph** (*fren'-o-graf*) [*phreno-;* γράφειν, to write]. An instrument for registering the movements of the diaphragm.

**phrenohepatic** (*fren-o-hep-at'-ik*) [*phreno-*; ἧπαρ, liver]. Pertaining to the diaphragm and the liver.
**phrenolepsia** (*fren-o-lep'-se-ah*) [*phreno-*; λῆψις, seizure]. Insanity.
**phrenologist** (*fren-ol'-o-jist*). One versed in phrenology.
**phrenology** (*fren-ol'-o-je*) [*phreno-*; λόγος, science]. The theory that the various faculties of the mind occupy distinct and separate areas in the brain-cortex, and that the predominance of certain faculties can be predicted from modifications of the parts of the skull overlying the areas where these faculties are located.
**phrenoparalysis** (*fren-o-par-al'-is-is*). See *phrenoplegia*.
**phrenopath** (*fren'-o-path*) [*phreno-*; πάθος, disease]. One who devotes himself to phrenopathy; an alienist.
**phrenopathy** (*fren-op'-ath-e*) [*phreno-*; πάθος, disease]. Mental disease.
**phrenoplegia** (*fren-o-ple'-je-ah*) [*phreno-*; πληγή, stroke]. 1. A sudden failure of mental power. 2. Paralysis of the diaphragm.
**phrenoptosis** (*fren-op-to'-sis*) [*phreno-*; πτῶσις, falling]. Prolapse of the diaphragm.
**phrenosin** (*fren'-o-sin*) [*phren*]. A nitrogenous body obtained from brain tissue.
**phrenosplenic** (*fren-o-splen'-ik*) [*phreno-*; *spleen*]. Pertaining to the diaphragm and the spleen.
**phricasmus** (*frik-az'-mus*) [φρίκη, shivering]. Gooseskin.
**phronemophobia** (*fron-e-mo-fo'-be-ah*) [φρόνημα, a thought; φόβος, fear]. Morbid dread of thinking.
**phronesis** (*fron-e'-sis*) [φρονεῖν, to think]. Soundness of mind, or of judgment.
**phrynin** (*frin'-in*). A substance from the skin of the toad. See *bufidine*.
**phrynolysin** (*frin-ol'-is-in*) [φρύνη, a toad; λύσις, a solution]. The lysin or toxin of the fire toad, *Bombinator igneus*; it is hemolytic for the blood of various animals, and is destroyed by digestive ferments, by alkalies or by heating to 50° C.
**phthalate** (*thal'-āt*). Any salt of phthalic acid.
**phthalic acid** (*thal'-ik*) [from *naphthalene*], $C_8H_6O_4$. A crystalline substance derived from naphthalene.
**phtheiriasis, phthiriasis** (*thi-ri'-as-is*). See *pediculosis*.
**phthinoid** (*thi'-noid*) [*phthisis*; εἶδος, likeness]. Having tuberculous characteristics.
**phthiremia, phthirǣmia** (*thi-re'-me-ah*) [φθείρειν, to corrupt; αἷμα, blood]. A depraved state of the blood, with diminished plasticity.
**Phthirius** (*thi'-re-us*) [φθείρ, a louse]. A genus of *Pediculidæ* or true lice. **P. inguinalis**, pediculus pubis.
**phthisic** (*tiz'-ik*) [*phthisis*]. 1. Affected with or of the nature of phthisis. 2. A person affected with phthisis.
**phthisical** (*tiz-ik'-al*) [φθίσις, a wasting]. 1. Pertaining to or affected with phthisis or tuberculosis. 2. Popularly, same as asthmatic. **p. frame, p. habit**, a long, narrow flat chest, with depressed sternum, acute costal angle, a fair, transparent skin, light complexion, blue eyes, winged scapulæ, slender limbs. As to internal organs, the heart is relatively small, the arteries narrow, the pulmonary artery relatively wider than the aorta, and the lung-volume rather large.
**phthisin** (*tiz'-in*). A proprietary preparation of the bronchial glands of animals; used in diseases of the lungs.
**phthisiogenesis** (*tiz-e-o-jen'-es-is*) [φθίσις, wasting; γένεσις, genesis]. The production of phthisis or wasting.
**phthisiology** (*tiz-e-ol'-o-je*) [*phthisis*; λόγος, science]. The study or science of phthisis or tuberculosis; its causes, pathology, hygiene, and therapeutics.
**phthisiophobia** (*tiz-e-o-fo'-be-ah*) [*phthisis*; φόβος, dread]. Morbid dread of pulmonary consumption or tuberculosis.
**phthisiotherapy** (*tiz-e-o-ther'-ap-e*) [*phthisis*; θεραπεία, therapy]. Therapeutic measures for the cure of pulmonary tuberculosis.
**phthisis** (*ti'-sis* or *thi'-sis*) [φθίειν, to waste]. 1. A wasting away or consumption. 2. Any chronic disease characterized by emaciation and loss of strength, especially pulmonary tuberculosis. 3. Asthma. **p. bulbi**, shrinking of the eyeball. **p., fibroid**. 1. Interstitial pneumonia. 2. Chronic tuberculosis of the lungs attended with the formation of fibrous tissue, which contracts, causes shrinking of the affected part, and sometimes bronchiectasis by traction on the bronchi. **p. florida**, an acute, rapidly fatal pulmonary tuberculosis. Syn., *galloping consumption*. **p., glandular**, tuberculosis of the lymphatic glands. **p., hepatic**, tuberculosis of the liver. **p., laryngeal**, tuberculosis of the larynx. **p., nodosa**, miliary tuberculosis of the lungs. **p. pancreatica**, emaciation and cachexia from disease of the pancreas. **p., phlegmatic**, phthisis without loss of flesh. **p., pulmonary**. 1. Tuberculosis of the lung. 2. Any one of a variety of interstitial pneumonias, such as *grinder's .phthisis, miner's phthisis, stone-cutter's phthisis*, etc. **p., tuberculous**, that due to the bacillus of tuberculosis. **p. ventriculi**, atrophy of the mucous membrane and thinning of the coats of the stomach.
**phthisopyrin** (*tis-o-pi'-rin*). A proprietary remedy consisting of sodium arsenate, aspirin, and camphoric acid, used in the treatment of tuberculosis.
**phthora** (*tho'-rah*) [φθορά, decomposition]. 1. Corruption. 2. Synonym of the *plague*. 3. Abortion.
**phycochrome** (*fi'-ko-krōm*) [φῦκος, seaweed; χρῶμα, color]. The complex blue-green pigment that masks the pure green of the chlorophyl in certain Algæ, (*Cyanophyceæ*). It is composed of phycocyanin, scytonemin, etc.
**phycocyanin** (*fi-ko-si'-an-in*) [φῦκος, seaweed; κυανός, blue]. In biology, a beautiful blue pigment, characteristic of the *Cyanophyceæ* among *Algæ*.
**Phycomycetes**· (*fi-ko-mi-se'-tēz*) [φῦκος, seaweed; μύκης, fungus]. An order of fungi, with a one-celled thallus which becomes septate only during sporulation.
**phygogalactic** (*fi-go-gal-ak'-tik*) [φυγεῖν, to avoid; γάλα, milk]. 1. Stopping the secretion of milk. 2. An agent that checks the secretion of milk.
**phylacogen** (*fi-lak'-o-jen*) [φύλαξ, a guard]. Trade name of a modified vaccine.
**phylaxin** (*fi-laks'-in*) [φύλαξ, a guardian]. A defensive proteid found in animals that have acquired an artificial immunity to a given infectious disease. The phylaxins are of two varieties: one having the power to destroy pathogenic microorganisms, called *mycophylaxin*; one that counteracts the poisons of the microorganisms, called *toxophylaxin*.
**phyletic** (*fi-let'-ik*) [φῦλον, a tribe]. Pertaining to phylogenesis.
**phylogenesis, phylogeny** (*fi-lo-gen'-es-is, fi-loj'-en-e*) [φῦλον, a tribe; γεννᾶν, to beget]. The evolution of a group or species of animals or plants from the simplest form; the evolution of the species, as distinguished from *ontogeny*, the evolution of the individual.
**phylogenetic** (*fi-lo-jen-et'-ik*) [*phylogenesis*]. Pertaining to phylogenesis.
**phylum** (*fi'-lum*) [φῦλον, a tribe; *pl., phyla*]. In biology, a primary division of the animal or vegetable kingdom.
**phyma** (*fi'-mah*) [φῦμα, a growth]. 1. Formerly, any one of a variety of swellings of the skin. 2. A localized plastic exudate larger than a tubercle; a circumscribed swelling of the skin.
**phymatiasis** (*fi-mat-i'-as-is*). Same as *tuberculosis*.
**phymatoid** (*fi'-mat-oid*) [*phyma*; εἶδος, like]. Resembling a phyma or tubercle.
**phymatorhusin** (*fi-mat-or-oo'-sin*). A pigment found in the metastatic deposits of a melanotic sarcoma of the skin. It contains sulphur, is insoluble in alcohol, in water, and in ether, but dissolves readily in ammonia, and in alkaline carbonates. It is the dye of iron.
**phymatosis** (*fi-mat-o'-sis*) [*phyma*]. 1. Any disease characterized by the formation of phymata or nodules. 2. Tuberculosis.
**phymatosis** (*fi-mat-o'-sis*) [*phyma*]. 1. Any disease characterized by the formation of phymata or nodules. 2. Tuberculosis.
**phyraliphore** (*fi-ral'-if-ōr*). A cavity containing vesicles produced in endogenous cell-formation.
**physalides** (*fis-al'-id-ēz*). Plural of *physalis*.
**physaliphorous** (*fis-al-if'-or-us*) [φυσαλλίς, a bladder; φέρειν, to bear]. Furnished with vesicles or bladders; relating to or containing physaliphores.
**physalis** (*fis'-al-is*) [φυσαλλίς, a bladder]. A large giant epithelial cell of giant-cell carcinoma.
**physconia** (*fis-ko'-ne-ah*) [φύσκων, paunch]. Any abdominal enlargement, especially from tympanites. **p. adiposa**, **corpulency**. **p. aquosa**, ascites. **p. biliosa**, distention of the gall-bladder. **p. mesenterica**. Synonym of *tabes mesenterica*.
**physiatrics** (*fis-e-at'-riks*) [φύσις, nature; ἰατρεία, treatment]. The power of nature in curing disease: *vis medicatrix naturæ*.
**physic** (*fiz'-ik*) [φύσις, nature]. 1. The science of

medicine. 2. A medicine, especially a cathartic. 3. To administer medicines; also to purge.
**physical** (*fiz'-ik-al*) [see *physic*]. 1. Pertaining to nature; also pertaining to the body or material things. 2. Pertaining to physics. p. **diagnosis**, the investigation of disease by direct aid of the senses, sight, touch, and hearing. p. **examination**, examination of the patient's body to determine the condition of the various organs and parts. p. **signs**, the phenomena observed on inspection, palpation, percussion, auscultation, mensuration, or combinations of these methods.
**physician** (*fiz-ish'-an*) [φύσις, nature]. One who practises medicine.
**physicist** (*fiz'-is-ist*). 1. One skilled in physics. 2. One who holds that vital phenomena are purely physical and chemical.
**Physick's encysted rectum** (*fiz'-ik*) [Philip Syng Physick, American surgeon, 1768–1837]. Hypertrophic dilatation of the rectal pouches.
**physicochemical** (*fiz-ik-o-kem'-ik-al*) [φύσις, nature; *chemic*]. Pertaining to both physics and chemistry.
**physics** (*fiz'-iks*) [φύσις, nature]. The science of nature, especially that treating of the properties of matter and of the forces governing it.
**physinosis** (*fiz-in-o'-sis*) [φύσις, nature; νόσος, disease]. Any disease due to physical causes.
**physiobathmism** (*fiz-e-o-baih'-mism*). Inherited growth-energy which has been interfered with by physical energy.
**physiognomy** (*fiz-e-og'-no-me*) [φύσις, nature; γνώμη, knowledge]. 1. The science treating of the methods of determining character by a study of the face. 2. The countenance.
**physiologic, physiological** (*fiz-e-o-loj'-ik, -al*) [*physiology*]. 1. Pertaining to physiology. 2. Pertaining to natural or normal processes, as opposed to those that are pathological. p. **antidote**, an antidote that neutralizes a poison by effects on the system that are antagonistic to those of the poison. p. **unit**. See *unit, physiological*.
**physiologist** (*fiz-e-ol'-o-jist*) [*physiology*]. One versed in physiology.
**physiology** (*fiz-e-ol'-o-je*) [φύσις, nature; λόγος, science]. The science that treats of the functions of organic beings, as distinguished from *morphology*, etc. p., **animal**, the physiology of animals. p., **cellular**, the physiology of cells. p., **comparative**, the comparative study of the physiology of different animals and plants. p., **morbid**, the study of diseased functions or of functions modified by disease. p., **pathogenetic**, p., **pathological**, pathology. p., **special**, the physiology of special organs. p. **vegetable**, the physiology of plants.
**physiolysis** (*fiz-e-ol'-is-is*) [φύσις, nature; λύειν, to dissolve]. The disintegration of dead tissue by natural processes.
**physiomedicalism** (*fiz-e-o-med'-ik-al-izm*) [φύσις, nature; *medicari*, to heal]. The professed use of natural remedies only, poisons and minerals being rejected.
**physiopathology** (*fiz-e-o-path-ol'-o-je*). The study of function as affected by disease.
**physiotherapy** (*fiz-e-o-ther'-ap-e*). See *physiatrics*. *physiautotherapia*.
**physique** (*fiz-ēk'*) [Fr.]. Physical structure or organization.
**physocele** (*fi'-so-sēl*) [φύσα, air; κήλη, tumor]. 1. A swelling containing air or gas. 2. Emphysema of the scrotum; a hernia filled with flatus.
**physocephalus** (*fi-so-sef'-al-us*) [φύσα, air; κεφαλή, head]. Emphysematous swelling of the head.
**physohematometra** (*fi-so-hem-at-o-me'-trah*) [φύσα, air; αἷμα, blood; μήτρα, uterus]. An accumulation of gas, or air, and blood in the uterus, as in decomposition of retained menses, or placental tissue.
**physohydrometra** (*fi-so-hi-dro-me'-trah*) [φύσα, air; *hydrometra*]. An accumulation of gas and water in the uterus.
**physometra** (*fi-so-me'-trah*) [φύσα, air; μήτρα, uterus]. A distention of the uterus with gas.
**physoncus** (*fi-song'-kus*) [φύσα, air; ὄγκος, tumor]. A swelling due to the presence of air.
**physoscheocele** (*fi-sos'-ke-o-sēl*) [φύσα, air; ὄσχεον, scrotum; κήλη, tumor]. Emphysema of the scrotum.
**physospasmus** (*fi-so-spaz'-mus*) [φύσα, air, flatus; σπασμός, spasm]. Flatulent colic.
**physostigma** (*fi-so-stig'-mah*) [φύσα, air; *stigma*]. Calabar bean; ordeal-nut. The seed of *Physostigma venenosum*, of the natural order *Leguminosæ*, which is used by the natives of Africa as an ordeal poison. It contains two alkaloids—*eserine* or *physostigmine* and *calabarine*. It acts as general depressant, producing motor paralysis, and in poisonous doses causing death by paralysis of the respiration. It is a miotic, and in small doses stimulates the heart and intestinal peristalsis. In medicine it is employed as a motor depressant in tetanus and other spasms; as a stimulant in intestinal atony and dilatation, in asthma and emphysema. p., **extract** of (*extractum physostigmatis*, U. S. P., B. P.). Dose ⅛–¼ gr. (0.008–0.016 Gm.). p. **tincture** of (*tinctura physostigmatis*, U. S. P.). Dose 5–20 min. (0.32–1.3 Cc.).
**physostigmine** (*fi-so-stig'-mēn*) [*physostigma*], C₁₅H₂₁N₃O₂. An alkaloid found in the seed of *Physostigma venenosum*, Calabar bean. It is used in traumatic tetanus, tonic convulsions, strychnine poisoning, neuralgia, muscular rheumatism; chronic bronchitis, etc. Dose ₁⁄₃₀₀–₁⁄₆₀ gr. (0.0003–0.001 Gm.); maximum dose ₁⁄₁₂ gr. (0.001 Gm.), single. Syn., *eserine.* p. **salicylate** (*physostigminæ salicylas*, U. S. P.), eserine salicylate, is used internally in doses of ₁⁄₁₂ gr. (0.0008 Gm.), but its chief use is for instillation into the eye as a miotic in conditions of mydriasis, and to lessen intraocular tension in glaucoma. It is used in these conditions in solution of the strength of from one to two grains to the ounce. p. **sulphate** (*physostigminæ sulphas*, U. S. P.), eserine sulphate, is used in the same manner as the salicylate.
**physostol** (*fi-sos'-tōl*). A one per cent. sterilized solution of physostigmine in olive oil, sold in sealed tubes containing five grammes.
**phytalbumose** (*fi-tal'-bu-mōs*) [φυτόν, a plant; *albumose*]. A vegetable albumose.
**phytin** (*fi'-tin*). The potassium magnesium salt of inosit-phosphoric acid. It is found in a phosphorus compound contained in the seeds of plants.
**phyto-** (*fi-to-*) [φυτόν, a plant]. A prefix signifying relations to plants.
**phytobezoar** (*fi-to-be'-zōr*) [*phyto-*; *bezoar*]. A hair-ball or ball of vegetable fiber sometimes found in the stomach.
**phytochemistry** (*fi-to-kem'-is-tre*) [*phyto-*; *chemistry*]. Vegetable chemistry.
**phytogenesis** (*fi-to-jen'-es-is*) [*phyto-*; *genesis*]. The science of the origin and development of plants.
**phytogenetic** (*fi-to-jen-et'-ik*). Pertaining to phytogenesis.
**phytogenous** (*fi-toj'-en-us*) [see *phytogenesis*]. Produced by plants.
**phytogeny** (*fi-toj'-en-e*). Same as *phytogenesis*.
**phytoid** (*fi'-toid*) [*phyto-*; εἶδος, like]. Plant-like; *e. g.*, certain animals and organs.
**phytolacca** (*fi-to-lak'-ah*) [*phyto-*; *lacca*, lac]. The dried root of *P. decandra*, a plant of the natural order *Phytolaccaceæ*. It is emetocathartic and slightly narcotic, and has been used in rheumatism and locally in granular conjunctivitis and parasitic skin diseases. Dose 10–30 gr. (0.65–1.9 Gm.). p. **fluid-extract** of (*fluidextractum phytolaccæ*, U. S. P.). Dose 5–30 min. (0.32–1.9 Cc.).
**phytolaccin** (*fi-to-lak'-sin*) [*phyto-*; *lacca*, lac]. A resinoid, or the precipitate from a tincture of the root of *Phytolacca decandra*. It is alterative, anti-syphilitic, laxative, etc. Dose 1 to 3 grains.
**phytolin** (*fi'-tol-in*). A proprietary liquid said to be prepared from berries of *Phytolacca decandra*, used in obesity.
**phytomelin** (*fi-tom'-el-in*). See *rutin*.
**phytoparasite** (*fi-to-par'-as-īt*) [*phyto-*; *parasite*]. A vegetable parasite.
**phytopathogenic** (*fi-to-path-o-jen'-ik*) [*phyto-*; πάθος, disease; γεννᾶν, to produce]. Causing disease, in plants.
**phytopathology** (*fi-to-path-ol'-o-je*) [*phyto-*; *pathology*]. 1. The science of diseases of plants. 2. The science of diseases due to vegetable organisms.
**phytoplasm** (*fi'-to-plazm*) [*phyto-*; *plasma*]. Vegetable protoplasm.
**phytoprecipitin** (*fi-to-pre-sip'-it-in*). A precipitin produced by immunization with albumin of vegetable origin.
**phytosis** (*fi-to'-sis*) [*phyto-*; νόσος, disease: *pl., phytoses*]. 1. Any disease due to the presence of vegetable parasites. 2. The production of disease by vegetable parasites. 3. The presence of vegetable parasites.
**phytosterin** (*fi-tos'-ter-in*) [*phyto-*; στέαρ, fat]. A fat-like substance, similar to cholesterin, present in plant-seeds and sprouts.

**phytosyntax** (*fi-to-sin'-taks*) [*phyto-;* συντάσσειν, to put together]. A term designating the process of formation of complex carbon compounds out of simple ones under the influence of light. Cf. *photosynthesis*.

**phytotoxin** (*fi-to-toks'-in*) [*phyto-;* toxin]. A toxin derived from a plant, such as abrin, ricin and crotin.

**phytovitellin** (*fi-to-vit-el'-in*) [*phyto-;* vitellus, yolk]. A vegetable albumin resembling vitellin.

**phytoxylin** (*fi-toks'-il-in*) [*phyto-;* ξύλον, wood]. A substance resembling pyroxylin.

**phytozoon** (*fi-to-zo'-on*) [*phyto-;* ζῳον, animal]. A plant-like animal; a zoophyte.

**pia, pia mater** (*pi-a ma'-ter*) [L., "kind or tender mother"]. The vascular membrane enveloping the surface of the brain and spinal cord, and consisting of a plexus of blood-vessels held in a fine areolar tissue. p. m. encephali, the pia mater of the brain. p. m. spinalis, the pia mater of the spinal cord.

**pia-arachnoid** (*pi-ah-ar-ak-ni'-tis*). See *piarachnitis*.

**pia-arachnoid** (*pi-ah-ar-ak'-noid*). See *piarachnoid*.

**pial** (*pi'-al*) [*pia*]. Pertaining to the pia.

**pialyn** (*pi'-al-in*) [πίαρ, fat; λύειν, to split up, or decompose]. See *steapsin*.

**piamatral** (*pi'-ah-ma'-tral*). See *pial*.

**pian** (*pi'-an*). See *frambesia*.

**piano-player's cramp**. A painful spasm of the muscles occurring in piano-players as the result of overuse of the muscles in playing; a form of occupation neurosis.

**piarachnitis** (*pi-ah-rak-ni'-tis*) [*piarachnoid*; ιτις, inflammation]. Inflammation of the piarachnoid. Syn., *leptomeningitis*.

**piarachnoid** (*pi-ah-rak'-noid*) [*pia; arachnoid*]. The pia and arachnoid considered as one structure.

**piarolytic** (*pi-ar-o-lit'-ik*) [πίαρ, fat; λυτικός, dissolving]. Forming emulsions with fat.

**piarrhemia, piarrhœmia** (*pi-ar-e'-me-ah*) [πίαρ, fat; αἷμα, blood]. See *lipemia*.

**Piazza's fluid** (*pe-as'-ah*). Sodium chloride and ferric chloride, each, 1 Gm.; water, 4 Cc. It is used as a means of coagulating blood.

**pica** (*pi'-kah*) [L., "magpie"]. A craving for unnatural and strange articles of food; a symptom present in certain forms of insanity, hysteria, and chlorosis, and during pregnancy.

**Picea** (*pis'-e-ah*) [L., "the pitch-pine"]. A genus of coniferous trees. *P. alba* is the white spruce. *P. excelsa*, or *P. vulgaris*, the common fir or pitch-pine, yields resin and turpentine. *P. nigra* is the black spruce.

**picein** (*pis'-e-in*) [*picea*], H13O7H2O. A glucoside from the leaves of the Norway spruce, *Picea excelsa*.

**piceol** (*pi'-se-ol*). A decomposition product of picein, by action of emulsion.

**piceous** (*pis'-e-us*) [*pix*, pitch]. Resembling pitch.

**pichi** (*pe'-che*) [native Chilean]. The stems and leaves of *Fabiana imbricata*, growing in Chile. It is a terebinthin having tonic properties, and is of repute in the treatment of catarrhal inflammations of the genito-urinary tract. Dose of the fld.ext. ʒj; of the extract gr. v-x.

**Pick's bundle** (*pik*). An anomalous bundle of nerve-fibers in the oblongata connected with the pyramidal tract. **P.'s disease**, pseudocirrhosis of the liver, met occasionally as a complication of adhesive pericarditis.

**picoline** (*pik'-o-lēn*) [*picea;* oleum, oil], C6H7N. Methylpyridine, a liquid obtained by distillation from coal-tar.

**picotamour** (*pe-kōt-mon(g)* [Fr.]. A pricking sensation; tingling; formication.

**Picræna** (*pik-re'-nah*) [πικρός, bitter]. A genus of the *Simarubeæ*, indigenous to the West Indies. *P. excelsa* furnishes Jamaica quassia. *P. vellozii*, quassia, is a Brazilian species; the bark is used in dyspepsia and in intermittent fever.

**picramic acid** (*pik-ram'-ik*) [πικρός, bitter; amine], C6H5N3O5. Picric acid in which one NO2 radical has been replaced by NH2.

**picramin** (*pik-ram'-in*). A synonym of amarin.

**Picramnia** (*pik-ram'-ne-ah*) [πικρός, bitter; θάμνος, shrub]. A genus of the *Simarubeæ*. *P. pentandra*, of the West Indies, furnishes a bitter tonic. The bark of the root and stem (*Honduras bark*) is used in colic syphilis, and cholera.

**Picrasma** (*pik-ras'-mah*) [πικρός, bitter]. A genus of the *Simarubeæ*. The bark of *P. javanica*, of Java, and of *P. quassoides*, of India, is used as an antipyre-

tic, and the wood as a substitute for quassia; the wood contains quassin.

**picrate** (*pik'-rāt*) [*picric acid*]. A salt of picric acid.

**picratol** (*pik'-rat-ol*). See *silver trinitrophenolate*.

**picric acid** (*pik'-rik*). [πικρός, bitter]. See *acid, picric*. **p.-acid test**. See *Braun's reaction for glucose*.

**picrin** (*pik'-rin*) [πικρός, bitter]. A bitter substance from digitalis. Dose, ¼-½ gr. (0.016-0.033 gm).

**picroaniline** (*pik-ro-an'-il-in*). A histologic stain consisting of a mixture of saturated solutions of picric acid and aniline-blue.

**picrocarmine** (*pik-ro-kar'-min*) [πικρός, bitter; *carmine*]. A preparation for staining specimens for the microscope. Its composition is as follows: carmine, 1; ammonia, 5; distilled water, 50 parts. After solution 50 parts of a saturated watery solution of picric acid are added, and the mixture allowed to stand in a wide-mouthed bottle until the ammonia has evaporated. It is then filtered.

**picroformal** (*pik-ro-form'-al*). A fixing agent consisting of a mixture of a saturated solution of picric acid and a 6% aqueous solution of formal.

**picroglycin, picroglycion** (*pik-ro-gli'-sin, pik-ro-gli'-se-on*) [πικρός, bitter; γλυκύς, sweet]. A crystalline, bitter substance, found in *Solanum dulcamara*. It may be impure solanine.

**picrol** (*pik'-rol*), C6HI2(OH)2SO3K. A white, odorless powder containing 52% of iodine; soluble in alcohol and ether; used as a wound antiseptic. Syn., *potassium diiodoresorcin-monosulphate*.

**picromel** (*pik' ro mel*) [πικρός, bitter; μέλι, honey]. A mixture of unknown composition containing salts of glycocholic and taurocholic acids.

**picronigrosin** (*pik-ro-ni'-gro-sin*). A stain consisting of picric acid and nigrosin in alcohol.

**picropodophyllin** (*pik-ro-po-do-fil'-in*). A crystalline substance obtained from *Podophyllum peltatum*. A derivative of picric acid and antipyrine occurring in yellow inflammable needles.

**picropyrine** (*pik-ro-pi'-rēn*). A derivative of picric acid and antipyrine occurring in yellow inflammable needles.

**picrosclerotine** (*pik-ro-skle'-ro-tēn*). A poisonous alkaloid occurring in ergot.

**picrotoxin** (*pik-ro-toks'-in*) [πικρός, bitter; τοξικόν, a poison]. A bitter neutral principle prepared from *Anamirta paniculata* (*Cocculus indicus*). Picrotoxin stimulates the motor and inhibitory centers in the medulla, especially the respiratory and vagus centers; it causes epileptiform spasms by irritation of the motor centers of the cerebrum or cord. Its action is much like that of strychnine. It has been used in an ointment (10 gr. to 1 oz.) in pityriasis capitis and in pediculosis. It is useful in the night-sweats of phthisis and in the complex of symptoms known as vasomotor ataxia. Dose ⅛₀-⅒₀ gr. (0.001-0.003 Gm.).

**picrotoxinism** (*pik-ro-toks'-in-izm*). Poisoning by picrotoxin; characterized by spasms of an epileptiform nature or resembling tetanus, followed by loss of consciousness and coma.

**Pictet's chloroform** (*pik-tet'*). See *chloroform*.

**pictet liquid** (*pik'-tet*) [*pix*, pitch]. A liquid consisting of a mixture of sulphurous acid gas and carbon dioxide liquefied under pressure.

**piebald skin**. See *leukoderma* and *vitiligo*.

**piedra** (*pe-a'-drah*) [S.A.]. A disease of the hair marked by the formation of hard, pinhead-sized nodules on the shaft of the hair; it is thought to be due to a micrococcus.

**piesimeter, piesmeter** (*pi-es-im'-et-er, pi-es'-me-ter*). See *piezometer*.

**Pietrowski's reaction for proteins** (*pe-at-rof'-ske*). The biuret reaction; a violet color is produced on heating a protein with an excess of a concentrated solution of sodium hydroxide and one or two drops of a dilute solution of copper sulphate. This color is deepened by boiling.

**piezometer** (*pi-e-zom'-et-er*) [πιέζειν, to press; μέτρον, a measure]. An apparatus for measuring the degree of compression of gases or fluids. 2. An apparatus for testing the endurance of the skin to pressure.

**Piffard's paste** (*pif'-ard*) [Henry Granger *Piffard*, American dermatologist, 1842-1910]. A paste composed of 5 parts of sodium tartrate, 2 of caustic soda, and one of copper sulphate.

**pigeon-breast** (*pij'-un*). See under *breast*.

**pigment** (*pig'-ment*) [*pingere*, to paint]. 1. A dyestuff; a coloring-matter. Pigments may be in solution or in the form of granules or crystals. 2. Any organic coloring matter of the body. **p., blood-**. See *p., hematogenous*. **p., cholera-blue**, a color-base obtained by dissolving cholera-red in concentrated sulphuric acid and then neutralizing with caustic soda. **p., cholera-red,** a color-base found in cultures of cholera bacilli which, upon addition of mineral acids, gives a beautiful violet color. On rendering the solution alkaline and shaking it with benzol the cholera-red is obtained in brownish-red lamellæ. Distillation of cholera-red with zinc dust gives indol. **p.-granule,** one of the minute structureless masses of which pigment consists. **P., hematogenous,** any pigment derived from the blood. Hematogenous pigments are hemoglobin, hematoidin, hemosiderin, and the bile-pigments (*hepatogenous pigments*) which are indirectly derived from the blood-pigment. **p., metabolic,** a pigment formed by the metabolic action of cells. Melanin is the type of metabolic pigments.

**pigmentary** (*pig'-men-ta-re*) [*pigment*]. Pertaining to or containing pigment; characterized by the formation of pigment.

**pigmentation** (*pig-men-ta'-shun*) [*pigment*]. Deposition of or discoloration by pigment.

**pigmentodermia** (*pig-ment-o-der'-me-ah*). See *chromodermatosis.*

**pigmentolysin** (*pig-men-tol'-is-in*). An antibody which causes destruction of pigment.

**pigmentophage** (*pig-ment'-o-fāj*) [*pigment*; φάγειν, to eat]. A phagocyte which destroys pigment, especially that of hairs.

**pigmentum nigrum** (*pig-men'-tum ni'-grum*) [L., black pigment]. The dark coloring-matter which lines the choroid coat of the eye.

**piitis** (*pi-i'-tis*) [*pia*, pia; ιτις, inflammation]. Inflammation of the pia mater; leptomeningitis.

**pil.** Abbreviation of Latin *pilula*, pill.

**pilar, pilary** (*pi'-lar, pi'-lar-e*) [*pilaris*; *pilus*, hair]. Pertaining to the hair.

**pilastered** (*pi-las'-terd*) [*pila*, a pillar]. Flanged so as to have a fluted appearance; arranged in pilasters or columns. **p. femur,** a condition of the femur in which the backward concavity of the shaft is exaggerated and the linea aspera prominent.

**pilatio** (*pi-la'-she-o*) [*pilus*, hair]. A cranial fissure.

**pilation** (*pi-la'-shun*). See *fracture, capillary.*

**pile** (*pīl*) [*pilus*, a hair]. 1. The hair or hairs collectively of any part of the integument. 2. A hemorrhoid. 3. A battery. **p., prostatic,** a condition of enlarged prostate in which hemorrhage results. **p., thermo-electric,** a battery in which an electric current is generated on heating the bars of two kinds of metal soldered together, of which the pile consists. An index registering the exact degree of heat is moved by the current.

**pileous** (*pi'-le-us*) [*pilus*, a hair]. Pertaining to hair; hairy.

**piles** (*pīlz*). Hemorrhoids, *q.v.*

**pileum** (*pil'-e-um*) [*pileum*, a cap: *pl.*, *pilea*]. 1. In biology, the cap or whole top of the head of a bird, from bill to nape, including the forehead, vertex, and occiput. 2. A lobe of the cerebellum lying between the vermis and the paraflocculus. Its relation to the peduncle is like that of a cap.

**pileus** (*pil'-e-us*) [*pileus*, a cap: pl., *pilei*]. 1. The disc of the *Medusæ*, for which many writers have substituted the name umbrella or disc. It is also applied to the cap-like or umbrella-like summit of the stipe of many fungi. The hymenium-bearing portion is the same as *cap*. 2. A nipple shield. **p. hippocraticus,** the capeline bandage. **p. ventriculi,** the cap of the stomach; the pyloric cap, or first portion of the duodenum.

**pili** (*pi'-li*) [plural of *pilus*; a hair]. Hairs. **p., annulati,** ringed hairs, leukotrichia annularis. **p. congenital,** hair existing at birth. **p., post-genital,** that appearing some time after birth.

**piliation** (*pil-e-a'-shun*) [*pilus*, hair]. The formation and production of hair.

**piliform** (*pi'-lif-orm*) [*pilus*, hair; *forma*, a form]. Having the form or appearance of hair; filiform.

**piliganine** (*pi-lig'-an-ēn*), $C_{15}H_{24}N_2O$(?). An alkaloid obtained from *Lycopodium saururus:* it is emetic, cathartic, drastic, anthelmintic, antispasmodic. Dose $\frac{1}{6}$–$\frac{1}{3}$ gr. (0.01–0.02 Gm.). The hydrochloride is used in the same way.

**pilimiction** (*pi-lim-ik'-shun*) [*pilus*, hair; *mingere*, to urinate]. The passing of urine containing hair-like filaments.

**pilin** (*pi'-lin*). A proprietary cosmetic said to be 60 per cent. alcohol, perfumed and colored and containing benzoic acid.

**pill** [*pilula*, dim. of *pila*, a ball]. A small, round mass containing one or more medicinal substances and used for internal administration. **p., Blaud's.** See *ferrous carbonate, pills of.* **p., blue.** See *mercury mass.* **p.'s, compound cathartic.** See *compound cathartic pills.* **p., Griffith's.** Synonym of *p., Blaud's.* **p., Lady Webster's,** pill of aloes and mastic. **p.-mass,** a cohesive mass used to hold together the medicinal ingredients of a pill.

**pillar** (*pil'-ar*) [*pila*, a pillar]. A columnar structure acting as a support. **p. of the abdominal ring,** one of the columns on each side of the abdominal ring. **p. of the fauces,** one of the folds of mucous membrane on each side of the fauces. **p. of the fornix, anterior,** a band of white matter on each side passing from the anterior extremity of the fornix to the base of the brain. **p. of the fornix, posterior,** one of two bands passing from the posterior extremities of the fornix into the descending horn of the lateral ventricle.

**pillet** (*pil'-et*). A little pill, or pellet.

**pilleus, pilleum** (*pil'-e-us, pil'-e-um*) [L. a cap or caul]. The caul or membrane which sometimes covers a child's head during birth. **p. ventriculi,** the cap of the stomach; the pyloric cap, or first portion of the duodenum.

**pilo-** (*pi-lo-*) [*pilus*, a hair]. A prefix meaning relating to the hair or hairy.

**pilocarpidine** (*pi-lo-kar'-pid-ēn*). An alkaloid from jaborandi similar in physiological effect to pilocarpine but weaker.

**pilocarpine** (*pi-lo-car'-pēn*) [see *pilocarpus*]. An alkaloid isolated from pilocarpus. Used locally as a miotic. See *pilocarpus.* **p. hydrochloride** (*pilocarpinæ hydrochloridum*, U. S. P.). Dose $\frac{1}{8}$ gr. (0.008 Gm.). **p. nitrate** (*pilocarpinæ nitras*, U. S. P.). Dose $\frac{1}{8}$ gr. (0.008 Gm.).

**pilocarpus** (*pi-lo-kar'-pus*) [*pilo-*; καρπός, fruit]. The leaflets of *Pilocarpus jaborandi*, a South American shrub of the natural order *Rutaceæ:* it yields *pilocarpine*, $C_{11}H_{16}N_2O_2$, which resembles atropine in action. Jaborandi and its alkaloid pilocarpine taken internally produce salivation, perspiration, and contraction of the pupil. They are employed as diaphoretics in dropsy, Bright's disease, uremia, rheumatism, and in the early stage of cold. Dose 20–40 gr. (1.3–4.0 Gm.). **p., fluidextract** of (*fluidextractum pilocarpi*, U. S. P.). Dose 10–30 min. (0.6–2.0 Cc.).

**pilocerine** (*pi-los'-er-ēn*). A poisonous alkaloid, from a cactus, *Pilocereus.*

**pilocystic** (*pi-lo-sis'-tik*) [*pilo-*; κύστις, a sac]. Applied to encysted tumors containing hair and fat.

**pilomotor** (*pi-lo-mo'-tor*) [*pilo-*; *movere*, to move]. Causing movement of the hair. **p. nerves,** nerves causing contraction of the arrectores pili. **p. reflex,** the appearance of "goose-skin" when the skin is irritated.

**pilonidal** (*pi-lo-ni'-dal*) [*pilo-*; *nidus*, a nest]. Containing an accumulation of hairs in a cyst. **p. fistula,** a fistula in the neighborhood of the rectum depending upon the presence of a tuft of hair in the tissues.

**pilose, pilous** (*pi'-lōs, pi'-lus*) [*pilosus*, hairy]. Hairy.

**pilosebaceous** (*pi-lo-se-ba'-shus*). Pertaining to the hair follicles and sebaceous glands.

**pilosis** (*pi-lo'-sis*) [*pilus*, a hair]. The abnormal or excessive development of hair.

**pilosity** (*pi-los'-it-e*) [*pilus*, a hair]. The state of being pilose.

**Piltz's reflex.** Alteration of the size of the pupil when the attention is suddenly fixed.

**pilula** (*pil'-ū-lah*) [L., pl., *pilulæ*]. A pill.

**pilular** (*pil'-ū-lar*) [*pilula*, dim. of *pila*, a ball]. Of the nature of or pertaining to pills.

**pilule** (*pil'-ūl*) [*pilula*, a small pill]. A small pill.

**pilus** (*pi'-lus*) [*pilus*, a hair; *pl.*, *pili*]. 1. A hair. 2. In biology, a fine, slender, hair-like body. **pili gossypii,** cotton staple. **pili tactiles,** tactile hairs.

**pimeladen** (*pim-el'-ad-en*) [πιμελή, fat; ἀδήν, gland; *pl.*, *pimeladenes*]. Any sebaceous gland.

**pimelecchysis** (*pim-el-ek'-kis-is*) [πιμελή, fat; ἔκχυσις, a pouring out]. An excessive discharge of fat or of sebaceous matter.

**pimelitis** (*pim-el-i'-tis*) [πιμελή, fat; ιτις, inflammation]. Inflammation of any adipose tissue; also, of connective tissue in general.

**pimeloma** (*pim-el-o'-mah*) [πιμελή, fat; ὄμα, tumor; *pl.*, *pimelomata*]. A fatty tumor; lipoma.

**pimelopterygium** (*pim-el-o-ter-ij'-e-um*) [πιμελή, fat; πτερύγιον, a small wing]. A fatty outgrowth on the conjunctiva.

**pimelorrhea** (*pim-el-or-e'-ah*) [πιμελή, fat; ῥοία, a flow]. An excessive fatty discharge. Fecal discharge of undigested fat.
**pimelorthopnea** (*pim-el-or-thop'-ne-ah*) [πιμελή, fat; *orthopnea*]. Orthopnea due to obesity.
**pimelosis** (*pim-el-o'-sis*) [πιμελή, fat; νόσος, disease]. Conversion into fat. The fatty degeneration of any tissue; obesity, or corpulence.
**pimelotic** (*pim-el-ot'-ik*). Affected with pimelosis.
**pimeluria** (*pim-el-ū-'re-ah*) [πιμελή, fat; οὖρον, urine]. The excretion of fat in the urine; lipuria.
**pimenta** (*pi-men'-tah*) [Sp., *pimiento*]. Allspice, the nearly ripe fruit of *P. officinalis*, a tree of the natural order *Myrtaceæ*. It has a fragrant aromatic odor, due to the presence of a volatile oil. Pimenta is used as an aromatic carminative in flatulence and locally in chilblains. Dose 10–40 gr. (0.65–2.6 Gm.). **p.**, **oil of** (*oleum pimentæ*, U. S. P.). Dose 2–5 min. (0.13–1.32 Cc.).
**pimento** (*pi-men'-to*). Synonym of *pimenta*.
**Pimpinella** (*pim-pin-el'-ah*). A genus of umbelliferous plants. *P. anisum* yields anise. *P. saxifraga* is said to be diaphoretic, diuretic, and stomachic, and has been employed in asthma, dropsy, amenorrhea, etc. Dose ½ dr. (2 Cc.).
**pimple** (*pim'-pl*) [AS., *pipel*]. A small pustule or papule.
**pinapin** (*pin'-ap-in*). A fermented pineapple-juice, recommended in catarrh of the stomach, also as a spray in nasal catarrh.
**Pinard's sign** (*pe-nar'*) [Adolphe *Pinard*, French obstetrician, 1844– ]. After the sixth month of pregnancy a sharp pain upon pressure over the fundus uteri is frequently a sign of breech presentation.
**pincement** (*pans'-mo(n)g*) [Fr., "pinching"]. In massage, a pinching or nipping of the tissues.
**pincers** (*pin'-cers*) [Fr., *pince*]. Forceps.
**pincet, pincette** (*pan-set'*) [Fr.]. A small forceps.
**Pinckneya** (*pingk'-ne-ah*). A genus of the family *Rubiaceæ*. *P. pubens* is a small tree of the southern United States. The bark is astringent and tonic and is used in intermittent fevers.
**pine** (*pīn*) [*pinus*]. A genus of trees of the order *Coniferæ*, yielding turpentine, pitch, tar, and other substances.
**pineal** (*pin'-e-al*) [*pinus*, a pine-cone]. Belonging to or shaped like a pine-cone. **p. body**, **p. gland**, the epiphysis, a small, reddish-gray, vascular body situated behind the third ventricle, which is embraced by its two peduncles; it is also called the *conarium*, from its conical shape. **p. eye**, a rudimentary third, median, or unpaired eye of certain lizards, with which the pineal body of the mammalia is homologous. **p. peduncle**, a narrow white band on each side of the pineal body. **p. ventricle**, the cavity occasionally found within the pineal body.
**pinealism** (*pin-e'-al-izm*). Disturbances due to abnormality in the secretion of the pineal gland.
**Pinel's system** (*pe-nel'*) [Philippe *Pinel*, French alienist, 1745–1826]. In the treatment of the insane, suppression of all forceful proceedings.
**pinenchyma** (*pin-en'-kim-ah*) [πίναξ, tablet; ἔγχυμα, infusion]. Tissues composed of flat cells.
**pinene** (*pi'-nēn*) [*pinus*, pine], C₁₀H₁₆. A hydrocarbon, the chief constituent of many essential oils.
**pineoline** (*pin'-e-o-lēn*). A proprietary extract of the needles of *Pinus pumilio*, combined with vaseline and lanolin; it is used as an application in skin diseases.
**pinguecula, pinguicula** (*pin-gwek'-ū-lah*), *pin-gwik'-ū-lah*) [dim. of *pinguis*, fat]. A small, yellowish-white patch situated on the conjunctiva, between the cornea and the canthus of the eye; it is composed of connective tissue.
**pinguid** (*ping'-gwid*) [*pinguis*, fat]. Fat; unctuous.
**pinguoleum** (*ping-gwo'-le-um*) [*pinguis*, fat; *oleum*, oil]. A fatty or fixed oil.
**pinhole** (*pin'-hōl*). A minute perforation like that made by a pin. **p. os**, an extreme degree of atresia of the os uteri, seen in young and undeveloped women.
**p. pupil, pin-point pupil**, contraction of the iris to an extent that the pupil is scarcely larger than a pin's head. It is seen in opium-poisoning, after the use of miotics, in certain cerebral diseases, in locomotor ataxia, etc.
**piniform** (*pin'-if-orm*) [*pine; forma*, form]. Shaped like a pine-cone.
**pink-eye**. 1. A contagious, mucopurulent conjunctivitis occurring especially in horses. 2. Acute contagious conjunctivitis in man.

**pinkroot**. See *spigelia*.
**pinna** (*pin'-ah*) [L., "feather"; "wing"]. The projecting part of the external ear; the auricle. **p. nasi**. Synonym of *ala nasi*.
**pinnal** (*pin'-al*). Pertaining to the pinna.
**pinocytosis** (*pin-o-si-to'-sis*) [πίνειν, to drink; κύτος, a cell]. A name for the property exhibited by phagocytes of imbibing and absorbing liquid substances.
**pinol** (*pi'-nol*). The commercial name for the oil distilled from the needles of *Pinus pumilio*. It is recommended in tuberculosis, rheumatism, etc., and may be used externally and internally.
**pinotherapy** (*pi-no-ther'-ap-e*) [πεῖνα, hunger; θεραπεία, therapy]. Hunger-cure or nestotherapy.
**pint** (*pīnt*). The eighth part of a gallon; 16 fluidounces; an *imperial pint* contains 20 fluidounces. Symbol O (*octarius*).
**pinta disease** (*pin'-tah*) [Sp., "spot"]. Parasitic disease of the skin, confined to the tropics. Also called *pinto*, *mal de los pintos*, and *spotted sickness*.
**Pinus** (*pi'-nus*). The pine (*q. v.*). *P. pumilio* of the Alps, yields a turpentine and an oil used in medicine. The bark of *P. strobus*, white pine of the northern United States, is astringent and antiseptic, and is used in diarrhea and dysentery. Dose of the *fluidextract* 30–60 min. (1.8–3.7 Cc.). An oil distilled from the leaves of *P. sylvestris*, Scotch pine or fir, is used in medicine; the young pine-cones are used as a diuretic. An extract from the leaves is used as a diuretic and antiseptic. Dose 3–6 gr. (0.2–0.4 Gm.).
**pinworm**. See *Oxyuris* and *Ascaris*.
**piœpithelium** (*pi-o-ep-ith-e'-le-um*) [πύον, fat; *epithelium*]. Epithelium containing fat.
**pion** (*pi'-on*) [πύον, fat]. Fat.
**pionemia, pionæmia** (*pi-on-e'-me-ah*). See *lipemia*.
**pioscope** (*pi'-o-skōp*) [πύον, fat; σκοπεῖν, to see]. A variety of galactoscope.
**pip**. A contagious disease of fowls characterized by a secretion of thick mucus in the throat and mouth.
**piper** (*pi-per'*) [L.]. Pepper (*q. v.*).
**piperazidin** (*pi-per-az'-id-in*). See *piperazin*.
**piperazin** (*pi-per'-ā-zin*), C₄H₁₀N₂. Diethylendiamine, a crystalline substance produced by the action of ammonia on ethylene bromide or chloride. It is readily soluble in water. In watery solutions it acts as an excellent solvent of uric acid. It is used internally in cases of gout, lithemia, diabetes, and as a solvent for uric-acid calculi. Dose 15 gr. (1 Gm.) a day; for hypodermatic use it is best employed in a 2% solution, of which 2 c.c. may be given. Dose 75–120 gr. (5–8 Gm.) daily in broken doses. **p. water**, a combination of equal parts of piperazin and phenocoll in water.
**piperᵢc** (*pi-per'-ik*) [*piper*]. Pertaining to or containing pepper. **p. acid**, C₁₂H₁₀O₄, a monobasic acid obtained by decomposing piperin.
**piperidine** (*pi-per'-id-ēn*) [*piper*], C₅H₁₁N. A liquid base produced in the decomposition of piperine. **p. bitartratē**, is recommended in uratic diathesis. Dose 10–16 gr. (0.65–1.03 Gm.) 3 times daily. **p. guaiacolate**, C₅H₁₁N.C₇H₈O₃)₂, is used in tuberculosis. Dose 10 gr. (0.65 Gm.) twice daily.
**piperine** (*pi'-per-ēn*). A neutral principle (*piperina*, U. S. P.) obtained from pepper.
**piperism** (*pi'-per-izm*). Poisoning by pepper, marked by acute gastritis.
**piperovatine** (*pi-per-o'-va-tēn*), C₁₆H₂₁NO₂. A crystalline alkaloid isolated from *Piper ovatum*. A heart-poison and depressant of motor and sensory nerves. Acts like strychnine.
**pipet, pipette** (*pip-et'*) [Fr., dim. of *pipe*]. A glass tube open at both ends, but usually drawn out to a smaller size at one end.
**pipmenthol** (*pip-men'-thol*) [*piper*, pepper; *mentha*, mint]. A name for the menthol obtained from peppermint.
**pipsissewa** (*pip-sis'-e-wah*). See *chimaphila*.
**piptonychia** (*pip-to-nik'-e-ah*) [πίπτειν, to fall; ὄνυξ, a nail]. Shedding of the nails.
**piqûre** (*pe-kūr'*) [F.]. Puncture. **p. glycosuria**, experimental glycosuria produced by puncture of the diabetic center in the medulla.
**Piria's test for tyrosin** (*pīr'-e-ah*). Moisten the substance on a watch-glass with concentrated sulphuric acid, and warm two to ten minutes on a water-bath. Dilute with water, warm, neutralize with barium carbonate, filter while warm, and add a little solution of ferric chloride. In the presence of tyrosin a violet color results. An excess of ferric chloride destroys the color.

**piriform** (*pir'-if-orm*). Synonym of *pyriform*.
**Pirogoff's formula.** (*pir'-o-gof*) [Nikolai Vanovich *Pirogoff*, Russian surgeon, 1810–1881]. Sublimed sulphur, ½ oz.; potassium carbonate, 1 oz.; distilled water, ½ oz.; tincture of iodine, 1 oz.; lard, 3 oz. Used for the treatment of favus. **P.'s operation or amputation**, *for amputation through the foot*: a partial osteoplastic operation in which the os calcis is sawed through obliquely from above downward and forward, and the posterior portion is brought up and secured against the surface made by sawing off the lower ends of the tibia and fibula.
**Piroplasma** (*pi-ro-plaz'-mah*) [πῦρ, fire; πλάσσειν, to form]. A genus of hematozoa. **P. bigeminum**, a species found in Texas fever. **P. canis**, a species causing the bilious fever or malignant jaundice of dogs. **P. donovani**. Same as *Leishman-Donovan bodies*, *q.v.* **P. equi**, a South African species causing a bilious fever in horses. **P. hominis**, the species responsible for Rocky Mountain spotted fever in man. **P. ovis**, a species affecting sheep. **P. parvum**, a species causing the Rhodesian red-water fever of cattle in Western Africa.
**piroplasmosis** (*pi-ro-plaz-mo'-sis*). Infection with piroplasma.
**Pirquet's** (von) **reaction** (*pĕr-ka'*) [Clemens von *Pirquet*, Austrian physician 1874– ]. Apply a few drops of a 4 per cent. solution of old tuberculin to a slightly scarified area; if positive, a number of papules appear surrounded by a hyperemic area.
**piscidia erythrina** (*pis-id'-e-ah er-e-thri'-nah*) [*piscis*, fish; *cædere*, to kill]. Jamaica dogwood, a tree of the order *Leguminosæ*, the bark of which has been used for stupefying fish. It contains a neutral principle, *piscidin*. Piscidin has been used as an anodyne in neuralgia, whooping-cough, and insomnia. Dose of the *fluidextract* 1 dr. (4 Cc.).
**piscidin** (*pis'-id-in*). See under *Piscidia erythrina*.
**pisiform** (*pis'-if-orm*) [*pisum*, a pea; *forma*, form]. Pea-shaped. **p. bone**, a small bone on the inner and anterior aspect of the carpus.
**pit** [AS., *pyt*, from *puteus*, a well or pit]. 1. A depression, as the *pit* of the stomach; the armpit. 2. To indent by pressing. **p., auditory** the embryonic depression preceding the labyrinth. **p., basilar, the** depression upon the palatal surfaces of the upper incisor teeth, at the base of the cingula. **p. of the stomach:** (1) a name popularly given to that abdomen just below the sternum and between the cartilages of the false ribs; it is also termed *scrobiculus cordis*; (2) any one of the openings of gastric tubules visible on the mucous surface of the stomach. **p., tear,** the lacrimal sinus.
**pitch** [ME., *picchen*, to throw]. 1. The height of a sound; that quality which depends upon the relative rapidity of the vibrations that produce the sound. 2. [AS., *pic*, from *pix*, pitch]. A hard but viscous, shining substance, breaking with a conchoid fracture, obtained from various species of pine and from tar. **p. blende**, an oxide of uranium, the source of the radium salts known to commerce. It occurs in pitchy black masses, rarely in octahedrons. Syn., *pechurane*; *uraninite*. **p., Burgundy,** the prepared resinous exudation of *Abies excelsa*, used in the form of various plasters as a counterirritant in chronic rheumatism. **p., Canada,** a resin obtained from *Abies canadensis*, and formerly used for making plasters. **p., Jew's or mineral,** asphalt. **p., liquid** (*pix liquida*, U. S. P., B. P.), tar. See *tar*. **o. plaster,** a plaster composed of Burgundy pitch, frankincense, resin, yellow wax and olive oil.
**pith** [AS, *pitha*]. 1. The soft cellular tissue found in the center of the stalks of plants. 2. The marrow of bones. 3. The spinal marrow. 4. To cut off all connection of the brain-centers of an animal with the periphery by piercing the brain and cord.
**pithecoid** (*pith'-e-koid*) [πίθηκος, an ape; *εἶδος*, likeness]. Resembling an ape. **p. theory**, the theory of man's descent from the ape.
**pithiatic** (*pith-e-at'-ik*). Pertaining to pithiatism or hysteria.
**pithiatism** (*pith'-e-at-izm*). Same as *hysteria*.
**pithiatric** (*pith-e-at'-rik*). Capable of being relieved by suggestion or persuasion; term employed with reference to hysterical condition.
**pithing** (*pith'-ing*). The destroying of the central nervous system by piercing the brain and cord; decerebration.
**pithode** (*pith'-ōd*). See *karyokinesis*.
**Pitre's sections** (*pĕtr*) [Albert *Pitres*, French physician, 1848– ]. A series of nearly vertical sections through the brain for postmortem examinations. **P.'s sign.** 1. "Signe du cordeau"; the angle formed by the axis of the sternum and the line represented by a cord dropped from the suprasternal notch to the symphysis pubis indicates the degree of deviation of the sternum in cases of pleuritic effusion. 2. Hypesthesia of the scrotum and testis in tabes dorsalis.
**pitted** (*pit'-ed*). Marked by indentations or pits, as from smallpox.
**pitting** (*pit'-ing*). The formation of pits; also the quality of preserving, for a short time, indentations made by pressing with the finger.
**pituglandol** (*pit-ū-glan'-dol*). Trade name of a liquid preparation of the infundibulum of the pituitary gland.
**pituita** (*pit-ū'-it-ah*) [L.]. Phlegm; mucus; stringy, frothy sputum.
**pituital** (*pit-ū'-it-al*). Relating to pituita.
**pituitary** (*pit-ū'-it-a-re*) [*pituita*]. Secreting or containing mucus. **p. body, p. gland,** a small, reddish-gray vascular body, weighing about ten grains, contained within the sella turcica of the skull. It consists of two portions—the large *anterior* or *oral*, and the small *posterior* or *cerebral* division. The *anterior lobe* is derived as a diverticulum from the primitive oral cavity; the *posterior lobe* descends as an outgrowth from the brain, communicating in fetal life with the third ventricle. The stalk of this outgrowth remains as the infundibulum. The pituitary body has attracted much attention on account of pathological changes in its structure in certain obscure diseases, such as akromegaly, myxedema, and others. In some cases of akromegaly it has been much enlarged. Syn., *hypophysis cerebri*. **p. membrane,** the Schneiderian membrane.
**pituitin** (*pit-ū'-it-in*). A preparation made from the posterior lobe of the pituitary body.
**pituitous** (*pit-ū'-it-us*) [*pituita*]. Containing or resembling mucus.
**pituitrin** (*pĭt-ū'-it-rin*). Trade name of a preparation made from the posterior lobe of the pituitary gland.
**pituri** (*pit'-ū-re*). The leaves and twigs of *Duboisia hopwoodii*, used as a narcotic stimulant.
**piturine** (*pit'-ū-rēn*). A liquid alkaloid obtained from *Duboisia hopwoodii*. It is probably identical with nicotine.
**pityriasic** (*pit-ir-i-as'-ik*). Relating to or affected with pityriasis.
**pityriasis** (*pit-ir-i'-as-is*) [πίτυρον, bran]. 1. A term applied to various skin affections characterized by fine, branny desquamation. 2. Seborrhea. **p. capitis, p., capilitii,** alopecia furfuracea. **p. circinata et marginata,** a disease characterized by an eruption of rose-colored spots on the trunk, the limbs, and in the axillæ, associated with slight fever and itching. **p. furfuracea,** seborrhœa sicca. **p. gravidarum.** Same as *chloasma uterinum*. **p. pilaris.** See *keratosis pilaris.* **p. rosea.** See *p. circinata et marginata.* **p. rubra,** a chronic inflammatory skin disease, beginning in one or more localized patches, which coalesce and gradually invade the whole body. The skin is deep red in color; and covered by whitish vesicles that constantly reform. The disease lasts months or years, and generally ends fatally. Syn., *dermatitis exfoliativa.* **p. versicolor.** See *tinea versicolor.*
**pityroid** (*pit'-ir-oid*) [πίτυρον, bran; *εἶδος*, like]. Branny.
**pivot** (*piv'-ot*) [Fr., *pivot*, a pivot]. A pin on which a wheel turns. **p., clack, p., clacking,** a means devised by Magiola for attaching an artificial crown to the root of a natural tooth. **p.-joint.** See *cyclarthrosis*. **p. tooth,** an artificial crown, designed to be applied to the root of a natural tooth, by means of what is usually termed a pivot, but more properly a dowel or tenon.
**pivoting** (*piv'-ot-ing*). The fixation of an artificial crown to a tooth by means of a pivot or pin.
**pix** (*piks*). See *pitch* (2). **p. burgundica,** an exudate from *Picea* (*Abies*) *excelsa*, Norway spruce, used as a rubefacient. **P. canadensis,** that obtained from the hemlock tree, *Tsuga* (*Abies*) *canadensis*. **P. liquida,** tar, an oleoresin obtained by the destructive distillation of the pine.
**pixine** (*piks'-ēn*). A surgical dressing said to consist of Burgundy pitch with a wool-fat base.
**pixol** (*piks'-ol*). A disinfectant preparation of tar and soft soap.
**P. L.** Abbreviation for (1) *Pharmacopœia Lond-*

*inensis*, London Pharmacopœia; (2) perception of light.
**placebo** (*pla-se'-bo*) [L., "I will please," from *placere*, to please]. A medicine given for the purpose of pleasing or humoring the patient, rather than for its therapeutic effect.
**placenta** (*pla-sen'-tah*) [πλακοῦς, a cake]. The organ on the wall of the uterus to which the embryo is attached by means of the umbilical cord and from which it receives its nourishment. It is developed, about the third month of gestation, from the chorion of the embryo and the decidua serotina of the uterus. The villi of the chorion enlarge and are received into depressions of the decidua, and around them bloodsinuses form, into which, by diffusion, the wastematerials brought from the fetus by the umbilical arteries pass, and from which the blood receives oxygen and food-material being returned to the fetus by the umbilical vein. At term the placenta weighs one pound, is one inch thick at its center, and seven inches in diameter. **p., adherent,** one that is abnormally adherent to the uterine wall after childbirth. **p., annular,** one extending around the interior of the uterus in the form of a belt. **p., battledore,** one in which the insertion of the cord is at the margin of the placenta. **p., circumvallate,** a thickening or fungiform enlargement of the placenta at the point at which the decidua vera and the decidua reflexa would have united in cases in which such union has been thwarted by hypersecretion of the former or by endometritic processes. **p. cirsoides,** one in which the umbilical vessels have a cirsoid arrangement. **p., discoid,** one shaped like a disc. **p., duplex,** one divided into two parts. **p., fundal,** one attached at the fundus. **p., horse-shoe,** in twin pregnancy, a condition in which two placentæ are joined. **p., incarcerated,** one retained by irregular contraction of the uterus. **p., maternal,** the external layer developed from the decidua serotina. **p. membranacea,** one abnormally thin. **p. prævia,** a placenta that is fixed to that part of the uterine wall that becomes stretched as labor advances, so that it precedes the advance of the presenting part of the fetus. Being detached before the birth of the child, it generally causes grave hemorrhage. **p., retained,** one not expelled by the uterus after labor. **p., sanguinis,** a blood-clot. **p. student's,** a retained placenta due to improper manipulation. **p. succenturiata,** an accessory placenta.
**placental** (*pla-sen'-tal*) [*placenta*]. Pertaining to the placenta. **p. bruit, p. murmur, p. souffle,** a sound attributed to the circulation of blood in the placenta.
**p. transmission,** the conveyance of drugs and diseaseproducts through the fetoplacental circulation from m₀ther to offspring.
**placentation** (*pla-sen-ta'-shun*) [*placenta*]. The formation and mode of attachment of the placenta.
**placentitis** (*pla-sen-ti'-tis*) [*placenta; ιτις*, inflammation]. Inflammation of the placenta.
**placentoid** (*pla-sen'-toid*) [*placenta; εἶδος*, like]. Resembling a placenta.
**placentolysin** (*pla-sen-tol'-is-in*) [*placenta*; λύσις, solution]. A cytolysin formed in the blood of an animal which has received injections of placental tissue emulsions derived from some other animal.
**placentoma** (*pla-sen-to'-mah*). A neoplasm springing from a retained portion of a placenta.
**placentotherapy** (*pla-sen-to-ther'-ap-e*) [*placenta*; *therapy*]. The remedial use of preparations of the placenta of animals.
**placentula** (*pla-sen'-tū-lah*) [dim. of *placenta*]. A small placenta.
**Placido's disc** (*plas-e'-do*). A keratoscope composed of a disc with concentric circles.
**placuntitis** (*plak-un-ti'-tis*). Synonym of *placentitis*.
**placuntoma** (*plak-un-to'-mah*). Synonym of *placentoma*.
**pladaroma** (*plad-ar-o'-mah*) [πλαδάρωμα, wetness; softness]. A soft wart or tumor of the eyelid.
**pladarosis** (*plad-ar-o'-sis*) [πλαδαρός, soft]. Synonym of *pladaroma*.
**plagiobolia** (*pla-je-o-bol'-e-ah*) [πλάγιος, oblique; βάλλειν, to throw]. Imperfect or indirect emission of spermatic fluid into the vagina.
**plagiocephalic** (*pla-je-o-sef-al'-ik*) [*plagiocephaly*]. Having a skull exhibiting plagiocephaly.
**plagiocephalism.** See *plagiocephaly*.
**plagiocephalous** (*pla-je-o-sef'-al-us*). Synonym of *plagiocephalic*.

**plagiocephaly** (*pla-je-o-sef'-al-e*) [πλάγιος, oblique; κεφαλή, head]. A malformation of the head produced by the closing of half of the coronal suture, giving an oblique growth to the cranial roof.
**plague** (*plāg*) [πληγή, a stroke]. 1. Any contagious malignant, epidemic disease. 2. A contagious disease endemic in eastern Asia, and in former times occurring epidemically in Europe and Asia Minor. After a period of incubation of from three to eight days the disease begins with fever, pain, and swelling of the lymphatic glands, chiefly the femoral inguinal, axillary, and cervical. Headache, delirium, vomiting, and diarrhea may be present. When recovery is probable, the temperature falls in about a week. The cause of the disease is the *Bacillus pestis*, found by Kitasato in the blood, buboes, and internal organs of the victims of the plague. **p., black,** the plague which decimated the European nations in the 14th century. **p., bubonic,** the usual form of plague formerly prevalent in various parts of the world. **p., cold,** a fatal form of bilious pneumonia. **p., hunger,** relapsing fever. **p., levantine,** the plague of the eastern part of Europe. **p., lung,** pleuropneumonia of cattle. **p., Siberian cattle.** Synonym of *anthrax.* **p.-sore,** a sore resulting from the plague. **p.-spot,** a spot characteristic of the plague. **p., swine,** hogcholera. **p., Syrian.** Synonym of *Aleppo boil*.
**planarthragra** (*plan-ar-thra'-grah*) [πλανᾶν, to cause to wander; ἄρθρον, a joint; ἄγρα, a seizure]. Gout which wanders from one joint to another.
**plancus** (*plang'-kus*)ˋ [*planca*,ˋ a board]. 1. A person with flat feet. 2. Flat-footed.
**plane** (*plān*) [*planus*, flat]. Any flat, smooth surface, especially any assumed or conventional surface, whether tangent to the body or dividing it. **p. of section.**
**planiceps** (*pla'-ni-seps*) [*planus*, flat; *caput*, a head]. Flat-headed.
**planimeter** (*pla-nim'-et-ur*). 1. See *perimeter*. 2. An instrument which measures a plane by tracing the periphery.
**planipes** (*pla'-ne-pēz*) [*planus*, flat; *pes*, foot]. Having flat feet.
**plano-** (*pla-no-*). 1. [*planus*, flat]. A prefix signifying flat or level; also a lens having no refracting power. 2. [πλάνος, wandering.] A prefix signifying wandering.
**planocellular** (*pla-no-sel'-ū-lar*) [*plano-; cellula*, cell]. Flat-celled.
**Planococcus** (*plan-o-kok'-us*) [πλάνη, a wandering, κόκκος, a berry]. A genus of bacteria of the family *Coccaceæ* having cell division in two planes; cells separate and flagellated.
**planocompressed** (*pla-no-kom-prest'*). So compressed that the opposite sides are flat.
**planoconcave** (*pla-no-kon-kāv'*). Concave on one surface and flat on the opposite side.
**planoconic** (*pla-no-kon'-ik*). Having one side flat and the other conical.
**planoconvex** (*pla-no-kon-veks'*). Plane on one side and convex on the other.
**planoculy** (*plan'-o-sit*) [*plano-; κύτος,* a cell]. A wandering cell.
**planodia** (*plan-o'-de-ah*) [πλάνη, a wandering; ὁδός, a way]. Any false or artificial passage made by an instrument.
**planomania** (*plan-o-ma'-ne-ah*) [πλάνος, wandering; μανία, madness]. A morbid and insane desire for wandering.
**planorheumatism** (*plan-o-roo'-mat-izm*) [*plano-; rheumatism*]. Wandering or metastatic rheumatism.
**Planosarcina** (*pla-no-sar'-sin-ah*) [πλάνη, wandering; *sarcina*]. A genus of motile bacteria whose cells are flagellated and divide in three planes.
**planta** (*plan'-tah*) [L.]. The sole of the foot.
**plantar** (*plan'-tar*) [*planta*]. Pertaining to the sole of the foot. **p. arch.** See *arch, plantar*. **p. fascia,** the dense triangular shaped aponeurosis occupying the middle and sides of the sole of the foot beneath the integument. **p. reflex.** See under *reflex*.
**plantaris** (*plan-ta'-ris*). See under *muscle*.
**plantigrade** (*plan'-te-grād*) [*planta; gradi*, to walk]. Bringing the entire length of the sole of the foot to the ground in walking, as is seen in the bear.
**plantose** (*plan'-tōs*). A pale-yellow dietetic powder prepared from the oil-cake of rape-seed, containing 12 % of nitrogen.
**planum** (*pla'-num*) [L. flat]. A plane or surface. **p. nuchale,** nuchal plane. **p. occipitale,** occipital plane. **p. orbitale,** orbital plane. **p. popliteum,** popliteal plane or space. **p. sternale,** sternal plane or anterior

surface of the sternum. p. temporale, temporal plane.
**planuria** (*plan-ū'-re-ah*), [πλάνος, straying; οὖρον, urine]. The discharge of urine through abnormal passages.
**plaque** (*plak*) [Fr.]. A patch. p.s, blood-. See *blood-platelets*. p.s, opaline, scattered white spots, like those caused by silver nitrate, seen on the fauces, hard palate, cheeks, and lips; an early affection in syphilis.
**plasm** (*plazm*). Same as *plasma*.
**plasma** (*plaz'-mah*) [πλάσμα, a thing molded]. 1. The fluid part of the blood and the lymph. See *blood-plasma*. 2. Glycerite of starch. p.-cells, large, granular cells found in the connective tissue. p., lymph-, the fluid part of the lymph. p., muscle-. See *muscle-plasma*.
**plasmacules** (*plaz'-ma-kūls*). See *hemokonia*.
**plasmameba, plasmamœba** (*plaz-mam-e'-bah*) [*plasma; amœba*]. An ameba-like parasite found in the blood in dengue.
**plasmaphæresis** (*plaz-maf-e'-res-is*) [*plasma*; ἀφαίρεσις, a withdrawal]. Removal of blood plasma; a form of venesection in which blood is withdrawn but the corpuscles are returned to the circulation.
**plasmasome** (*plaz'-mas-ōm*) [πλάσμα, a. molded figure; σῶμα, body]. A protoplasmic corpuscle.
**plasmatic** (*plaz-mat'-ik*) [*plasma*]. 1. Pertaining to plasma. 2. Plastic. p. layer, the layer of plasma next to the wall of a capillary.
**plasmatorrhexis** (*plaz-mat-o-reks'-is*) [*plasma; ῥῆξις*, a bursting]. Same as *plasmorrhexis*.
**plasmatosis** (*plaz-mat-o'-sis*) [*plasma*]. The liquefaction of cell-substance as seen in the cells of the secreting milk-gland and in the cells of secreting glands of the cervix uteri.
**plasment** (*plaz'-ment*). A proprietary emollient and lubricant application made from Iceland moss.
**plasmexhidrosis** (*plaz-meks-hi-dro'-sis*) [*plasma*; ἐξ, out of; *hidrosis*]. The exudation of plasma from the blood-vessels.
**plasmic** (*plaz'-mik*) [*plasma*]. Of or pertaining to protoplasm; formative, protoplasmic; plasmatic.
**plasmin** (*plaz'-min*) [*plasma*]. A name given to the precipitate obtained from blood by treating it with a saturated solution of sodium sulphate, allowing the corpuscles to subside, then precipitating the plasma with sodium chloride, and washing the precipitate with a saturated solution of sodium chloride.
**plasmo-** (*plaz-mo-*) [*plasma*]. A prefix meaning relating to the plasma.
**plasmocyte** (*plaz'-mo-sīt*) [*plasmo-*; κύτος, a cell]. 1. Any cell, other than blood-corpuscles, free in the blood-plasma. 2. A protozoan parasite in the blood plasma.
**plasmodia** (*plaz-mo'-de-ah*). Plural of *plasmodium*.
**plasmodiblast** (*plaz-mo'-dib-last*). See *trophoblast*.
**plasmodium** (*plaz-mo'-de-um*) [*plasmo-*; εἶδος, form: *pl.*, *plasmodia*]. The mass of protoplasm formed by the fusion of two or more amebiform bodies. p. falciparum, the parasite of estivoautumnal or pernicious malaria. p. malariæ, a protozoan parasite found in the blood of persons suffering from malaria. p. præcox, the malarial parasite of birds. p. tenue, a malarial parasite, said to be found in cases of malignant malaria in India. p. vivax, the parasite of tertian malaria.
**plasmogen** (*plaz'-mo-jen*) [*plasmo-*; γεννᾶν, to produce]. Formative protoplasm; germ-plasm; bioplasm.
**plasmology** (*plaz-mol'-o-je*). [*plasmo-*; λόγος, science]. The study of cells and cell-life; the biology of histology.
**plasmolysis** (*plaz-mol'-is-is*) [*plasmo-*; λύειν, to loose]. 1. The separation of cell-protoplasm from the inclosing cell-wall. 2. The contraction of living protoplasm under the influence of reagents. 3. The escape of the soluble substances of the blood-corpuscle.
**plasmolytic** (*plaz-mo-lit'-ik*). Exhibiting or characterized by plasmolysis.
**plasmoma** (*plaz-mo'-mah*) [*plasmo-*; ὄμα, tumor]. A tumor which shows a tendency toward the formation of fibers.
**plasmon** (*plaz'-mon*) [*plasma*]. The unaltered proteid of milk. p.-butter, a mixture of plasmon (6.58 %) with butter (51.5 %), water (41.2 %), and salt (0.72 %). It resembles clotted cream in appearance and taste and serves as a substitute for cod-liver oil.

**plasmophagous** (*plaz-mof'-ag-us*) [*plasmo-*; φαγεῖν, to eat]. Living upon protoplasm; applied to organisms causing decomposition of organic matter.
**plasmoptysis**. (*plaz-mop'-tis-is*) [*plasmo-*; πτύσις, a spitting]. The escape of protoplasm from a cell due to rupture of the cell-wall.
**plasmorrhexis** (*plaz-mor-eks'-is*) [*plasmo-*; ῥῆξις, a bursting]. The rupture of a cell and the escape or loss of the plasma.
**plasmoschisis** (*plaz-mos'-kis-is*) [*plasmo-*; σχίσις, cleavage]. The splitting of a cell, as the formation of disc-shaped bodies by red blood-corpuscles.
**plasmosome** (*plaz'-mo-sōm*) [*plasmo-*; σῶμα, body]. 1. One of the granular structural elements of cells. 2. The nucleolus of a cell.
**plasmotropic** (*plaz-mo-trop'-ik*) [*plasmo-*; τροπή, a change]. Producing protoplasmic degeneration; applied to hemolytic action which leaves the red corpuscles intact in the circulation, but through the influence of poisons on the liver, spleen, and bone-marrow causes excessive destruction of them in these organs.
**plasom** (*plaz'-ōm*). A proprietary preparation from milk.
**plasome** (*plaz'-ōm*) [*plasma*; σῶμα, body, matter]. The hypothetical unit of protoplasm.
**plasson** (*plas'-son*) [πλάσσειν, to form or mold]. Primitive or undifferentiated protoplasm; the protoplasm of the cell in the nonnucleated or cytode stage.
**plastauxia** (*plas-tawks'-e-ah*) [πλάσσειν, to mold; αὔξη, increase]. An increase of plasticity.
**plaster** (*plas'-ter*) [ME., *plastre*]. 1. An adhesive, semisolid substance spread upon cloth or other flexible material for application to the surface of the body. 2. Calcined gypsum or calcium sulphate. p., adhesive (*emplastrum adhesivum*, U. S. P.), a plaster prepared by melting rubber and adding petrolatum and lead plaster. p.-bandage, a bandage stiffened with plaster of Paris. p., belladonna. See *belladonna plaster*. p., blistering, cerate of cantharides. p., capsicum. See *capsicum plaster*. p., court-, a mixture of isinglass, glycerine, and alcohol spread upon silk. p., diachylon, p., lead. See *plumbi, emplastrum*, under *plumbum*. p. jacket, a bandage of plaster of Paris for the trunk. p., mercurial. See *mercury plaster*. p.-mull, a plaster made by incorporating with mull or thin muslin a mixture of guttapercha and some medicament dissolved in benzine. It is used in skin diseases. p., mustard-, one made by spreading upon muslin powdered mustard, or a mixture of mustard and flour reduced to the consistence of paste by the addition of water. p., opium. See *opium plaster*. p.-of-Paris, a mixture of calcium sulphate (gypsum) and water, having the property of becoming hard during drying. It is used for surrounding parts, such as joints, fractured limbs, etc., with a stiff casing, to prevent mobility. p., resin, a lead-plaster with the addition of resin and wax. p., soap. See *soap plaster*. p., spice-, a plaster composed of yellow wax, suet, turpentine, oil of nutmeg, olibanum, benzoin, oil of peppermint, and oil of cloves, and used to relieve abdominal pain in children. p., strengthening, one containing iron. p., warming, a plaster of pitch and cantharides.
**plastic** (*plas'-tik*) [πλάσσειν, to mold]. 1. Formative; building up tissues; repairing defects, as *plastic* surgery, *plastic* operation. 2. Capable of being molded. p. bronchitis, pseudomembranous bronchitis. p. force, the generative force of the body. p. linitis, cirrhosis of the stomach. p. lymph, the flamed serous surfaces; and becomes organized by the development in it of blood-vessels and connective tissues.
**plasticity** (*plas-tis'-it-e*) [*plastic*]. 1. Plastic force. 2. The quality of being plastic.
**plasticule** (*plas'-tik-ūl*) [πλάσσειν, to mold]. A molecule of plastic material not yet fully organized; a plastidule.
**plastid** (*plas'-tid*) [πλάσσειν, to mold]. An elementary organism; a cell or cytode.
**plastidule** (*plas'-tid-ūl*) [dim. of *plastid*, an elementary organism]. A protoplasmic molecule; one of the physical units of which living matter is composed.
**plastin** (*plas'-tin*) [πλάσσειν, to mold]. A phosphorized protein, constituting the chief proteid of protoplasm.

**plastodynamia** (*plas-to-di-nam'-e-ah*) [*plasto-*; δύναμις, power]. Nutritive plastic power.
**plastogamy** (*plas-tog'-am-e*) [*plasma*; γάμος, marriage]. Permanent conjugation of cells which is limited to the cytoplasm.
**plate** (*plāt*) [πλατύς, broad]. 1. A flattened part, especially a flattened process of bone. 2. A thin piece of metal or some other substance to which false teeth are attached. **p., approximation-**, one of the plates of decalcified bone or other material that are used in enterectomy to bring the resected ends of intestine together. **p., auditory**, the boneplate forming the roof of the auditory meatus. **p., axial**, the primitive streak of the embryo. **p.s, blood-**. See *blood-platelets*. **p., bone-**. See *p., approximation-*. **p., cribriform**, the horizontal plate of the ethmoid bone constituting the floor of the olfactory fossa and perforated for the passage of the olfactory nerves. **p.-culture**, a method of obtaining pure cultures of bacteria by pouring the inoculated culture-medium upon sterile glass plates and allowing it to solidify. **p., dorsal**, one of the two longitudinal ridges on the dorsal surface of the embryo which subsequently join to form the neural canal. **p., end-**. See *end-plate*. **p., equatorial**, the compressed mass of chromosomes aggregated at the equator of the nuclear spindle during karyokinesis. **p.s, facial**, the frontonasal and external group of nasal and maxillary plates of the embryo. **p., foot**, the flat part of the stapes. **p., Franklin**, a glass plate partly covered on both sides with tin-foil, used as a condenser in frictional electricity. **p., frontal**, in the fetus, a cartilaginous plate interposed between the lateral parts of the ethmoid cartilage and the lesser wings and anterior portion of the sphenoid bone. **p., frontonasal**, the middle of the facial plates, which subsequently forms the external nose. **p., lateral mesoblastic**, the thick portion of the mesoblast situated one on each side of the notochord. Each plate splits into two portions, the outer divisions coalescing to form the body-wall, or somatopleure, the inner, to form the splanchnopleure, or visceral covering. **p.s, maxillary, p.s, maxillary, inferior**, the first pair of subcranial plates from which the mandible is developed. **p., medullary** or **neural**. Same as *p., dorsal*. **p., nuclear**. See *nuclear plate*. **p., palate**, the part of the palate-bone which, with its opposite fellow, forms the roof of the mouth. **p.s, pterygoid**, two plates into which the pterygoid process of the sphenoid bone divides. **p.s, subcranial**. See *arches, postoral*. **p., tympanic**, the bony sides and floor of the auditory meatus.
**platelets, blood** (*plāt'-lets*). Small discs in the blood, light gray in color, and of uncertain function.
**platetrope** (*plāt'-e-trōp*). See *platytrope*.
**platiculture** (*pla-ti-kul'-chur*). The cultivation of bacteria on plates.
**plating** (*plā'-ting*). See *platiculture*.
**platinic** (*plat-in'-ik*) [*platinum*].¹ Containing platinum as a quadrivalent element.
**platinode** (*plat'-in-ōd*) [*platinum*; ὁδός, way]. The negative or receiving plate of an electric battery, so-called because formerly often made of platinum.
**platinous** (*plat'-in-us*) [*platinum*]. Containing platinum as a bivalent element.
**platinum** (*plat'-in-um*) [Sp. *platina*, dim. of *plata*, silver]. A silver-white metal occurring native or alloyed with other metals; atomic weight 195.2; sp. gr. 21.5; quantivalence II and IV; symbol Pt. It is fusible only at very high temperatures, and is insoluble in all acids except nitrohydrochloric. On account of these properties it is extensively used for chemical apparatus—crucibles, foils, wire, etc.; it is also employed as a reagent. Platinum occurs, aside from its ordinary metallic form, as a spongy mass (*spongy platinum*) and as a fine metallic powder (*platinum-black*), which is capable of condensing a great deal of oxygen, and hence acts as a powerful oxidizing agent. Platinum forms two sets of compounds—a platinous series, in which it acts as a diad, and a platinic series, in which it acts as a tetrad. **p. chloride**, PtCl₄, is used as a reagent to detect potassium and ammonium; also in syphilis, in doses of ½–½ gr. (0.008–0.03 Gm.).
**platode, platoid** (*plat'-ōd, plat-oid'*) [πλατύς, broad; εἶδος, form]. In biology, broad or flat, as a worm.
**Platt's chlorides**. A disinfectant liquid said to be a solution of the chlorides of magnesium, potassium, sodium, zinc, and aluminum.

**platy-** (*plat-e-*) [πλατύς, broad]. A prefix signifying broad.
**platycelian, platycelous** (*plat-is-el'-e-an, plat-is-el'-us*). Concave in front and convex behind.
**platycephalic, platycephalous** (*plat-is-ef-al'-ik, plat-is-ef'-al-us*) [*platy-*; κεφαλή, head]. Having a broad skull with a vertical index of less than 70.
**platycephaly** (*plat-is-ef'-al-e*) [*platy-*; κεφαλή, head]. The quality of being platycephalous.
**platycnemia** (*plat-ik-ne'-me-ah*) [*platy-*; κνήμη, leg]. The state of being platycnemic.
**platycnemic** (*plat-ik-ne'-mik*) [see *platycnemia*]. Having a tibia which is exaggerated in breadth; broad-legged.
**platycoria, platycoriasis** (*plat-ik-o'-re-ah, plat-ik-o-ri'-as-is*) [*platy-*; κόρη, pupil]. Expansion of the pupil; mydriasis.
**platycrania** (*plat-e-kra'-ne-ah*) [*platy-*; κρανίον, skull]. The flattened condition of the skull produced artificially among savage tribes.
**platycyte** (*plat'-is-īt*) [*platy-*; κύτος, a cell]. A cell intermediate in size between a giant-cell and a leukocyte, found in tubercle nodules.
**Platyhelminthes** (*plat-e-hel-min'-thēz*) [*platy-*; ἕλμινς, a worm]. Flat-bodied, more or less elongated worms, usually containing both sexual elements at the same time. They include flat-worms, flukes, and tapeworms.
**platyhieric** (*plat-e-hi-er'-ik*) [*platy-*; ἱερός, sacrum]. Having a broad sacrum; having a sacral index of more than 100.
**platymorphia** (*plat-e-mor'-fe-ah*) [*platy-*; μορφή, form]. A flatness in the formation of the eye and shortening of the anteroposterior diameter, resulting in hyperopia.
**platymyoid** (*plat-im-i'-oid*). Applied to muscle-cells in which the contractile layer presents an even surface.
**platyopia** (*plat-e-o'-pe-ah*) [*platy-*; ὤψ, face]. Broadness of the face; the quality of being platyopic.
**platyopic** (*plat-e-op'-ik*) [*platy-*; ὤψ, face]. In biology, having a face wide across the eyes, as in the Mongolian races; having the naso-malar index below 107.5°.
**platypellic** (*plat-e-pel'-ik*) [*platy-*; πέλλα, basin]. Having a broad pelvis.
**platypodia** (*plat-e-po'-de-ah*) [*platy-*; πούς, foot]. Flat-footedness.
**platyrrhine** (*plat'-ir-in*) [*platy-*; ῥίς, nose]. Having a broad and flat nose; having a nasal index above 53.
**platyrrhiny, platyrhiny** (*plat'-ir-i-ne*) [*platy-*; ῥίς, nose]. The condition of having a platyrrhine skull.
**platysma** (*plat-is'-mah*) [πλατύς, broad]. Anything of considerable superficial dimensions; also, a plaster. **p. myoides**. See under *muscle*.
**platystencephaly, platystencephaly** (*plat-is-ten-sef-a'-le-ah, plat-is-ten-sef'-al-e*) [πλατύστατος, widest; ἐγκέφαλος, brain]. The condition of a skull very wide at the occiput and with prominent jaws.
**platytrope** (*plat'-et-rōp*) [*platy-*; τρέπειν, to turn]. In biology, one of two symmetrically related parts on opposite sides of the meson; a lateral homologue.
**Plaut's angina** (*plowt*). See *Vincent's angina*.
**Playfair's treatment** (*pla'-fār*) [William Smoult Playfair, English physician, 1836–1903]. See *Mitchell's treatment*.
**plectrum** (*plek'-trum*) [πλῆκτρον, a spur: pl., *plectra*]. The styloid process of the temporal bone; the tongue; the uvula; the malleus.
**pledget** (*plej'-et*) [origin obscure, perhaps dim. of *plug*]. A small flattened compress.
**plegaphonia** (*pleg-af-o'-ne-ah*) [πληγή, stroke; φωνή, sound]. The sound produced in auscultatory percussion of the larynx, the glottis being open.
**plegometer** (*pleg-om'-et-er*). Synonym of *pleximeter*.
**Plehn's karyochromatophilic granules** (*plān*) [Albert *Plehn*, German physician, 1861– ]. Basophile granules observed in the protozoan parasite of malaria.
**pleochroic** (*ple-o-kro'-ik*). See *pleochromatic*.
**pleochroism** (*ple-ok'-ro-izm*) [πλέων, more; χρόα, color]. The property possessed by some bodies, especially crystals, of presenting different colors when viewed in the direction of different axes.
**pleochromatic** (*ple-o-kro-mat'-ik*) [see *pleochroism*]. Pertaining to or exhibiting pleochroism.
**pleocytosis** (*ple-o-si-to'-sis*). Increase of lymphocytes in the cerebrospinal fluid.
**pleomastia, pleomazia** (*ple-o-mas'-te-ah, ple-o-ma'-*

**ze-ah)** [πλέων, more; μαστός, or μαζός, breast]. The condition of having more than two mammæ. See *polymastia*.

**pleomorphic** (*ple-o-mor'-fik*) [see *pleomorphism*]. Having more than one form.

**pleomorphism** (*ple-o-mor'-fizm*) [πλέων, more; μορφή, form]. The state of being pleomorphic, *i. e.*, of existing in widely different forms.

**pleonasm** (*ple'-o-nasm*) [πλεονασμός, an exaggeration]. Any deformity marked by superabundance of certain organs or parts.

**plerosis** (*ple-ro'-sis*) [πλήρωσις, a filling]. 1. The restoration of lost tissue. 2. Plethora.

**plesiomorphic** (*ples-e-o-mor'-fik*) [πλησίος, near; μορφή, form]. Almost identical in form.

**plesiomorphous** (*ples-e-o-mor'-fus*) [πλησίος, near; μορφή, form]. Crystallizing in similar forms but differing in chemical composition.

**plesiopia** (*ples-e-o'-pe-ah*) [πλησίος, near; ὤψ, eye]. Increased convexity of the crystalline lens, producing myopia, and due to long-continued accommodation-strain.

**plessesthesia, plessæsthesia** (*ples-es-the'-ze-ah*) [πλήσσειν, to strike; αἴσθησις, perception by the senses]. Palpatory percussion performed by placing the left middle finger firmly against the body surface and percussing with the index-finger of the right hand, allowing it to remain in contact with the left finger for a few seconds.

**plessigraph** (*ples'-e-graf*) [πλήσσειν, to strike; γράφειν, to write]. A form of pleximeter which permits close distinctions to be made in the quality of the sounds elicited, and by means of a crayon attached to the skin, organs or dull areas may be mapped on the surface of the skin.

**plessimeter** (*ples-im'-et-er*). See *pleximeter*.

**plessor** (*ples'-or*). See *plexor*.

**plethora** (*pleth'-or-ah*) [πλῆθος, fulness]. A state characterized by an excess of blood in the vessels, and marked by reddish color of the face, a full pulse, a feeling of fulness and tension in the head, drowsiness, and a tendency to nosebleed.

**plethoric** (*pleth'-or-ik*) [*plethora*]. Pertaining to or characterized by plethora.

**plethysmograph** (*pleth-is'-mo-graf*) [πληθυσμός, increasing; γράφειν, to write]. An instrument for ascertaining changes in the volume of an organ or part, dependent upon changes in the quantity of the blood.

**pleura** (*ploo'-rah*) [πλευρά, a side]. The serous membrane which envelops the lung (*p., pulmonary*), and, which being reflected back, lines the ental surface of the thorax (*p., costal*). **p. costalis.** See *p., costal.* **p., diaphragmatica,** the reflection of the pleura upon the upper surface of the diaphragm. **p. mediastinalis,** a continuation of the costal pleura covering the side of the mediastinum. **p., parietalis.** Synonym of *p., costal.* **p., pericardiaca,** the portion of the pleura contiguous to the pericardium. **p. phrenica.** Synonym of *p., diaphragmatica.* **p., pulmonalis.** Synonym of *p., pulmonary.* **p., visceralis.** Synonym of *p., pulmonary.*

**pleuracentesis** (*ploo-rah-sen-te'-sis*). Same as *pleurocentesis.*

**pleural** (*ploo'-ral*) [*pleura*]. Pertaining to the pleura.

**pleuralgia** (*ploo-ral'-je-ah*) [*pleura*; ἄλγος, pain]. Pain in the pleura or in the side; intercostal neuralgia.

**pleuralgic** (*ploo-ral'-jik*) [*pleura*; ἄλγος, pain]. Pertaining to or affected with pleuralgia.

**pleurapophyseal** (*ploo'-rap-off-iz'-e-al*) [*pleura*; ἀπόφυσις, offshoot]. Pertaining to a pleurapophysis.

**pleurapophysis** (*ploo-rap-of'-is-is*) [*pleura*; *apophysis*]. One of the lateral processes of a vertebra, having the morphologic valence of a rib.

**pleurapostema** (*ploo-rap-os-te'-mah*) [*pleura*; ἀπόστημα, abscess]. A collection of pus in the pleural cavity.

**pleurarthrocace** (*ploo-rar-throk'-as-e*) [*pleura*; ἄρθρον, joint; κακός, evil]. Disease of the costovertebral joints; also, caries of the ribs.

**pleurarthron** (*ploo-rar'-thron*) [*pleura*; ἄρθρον, joint]. The articulation of a rib.

**pleurectomy** (*ploo-rek'-to-me*) [*pleura*; ἐκτομή, a cutting out]. Excision of one or more ribs, in whole or in part.

**pleurisy** (*ploo'-ris-e*) [*pleura*]. Pleuritis; inflammation of the pleura. It may be acute or chronic. Three chief varieties are usually described, depending upon the character of the exudate: (1) Fibrinous or plastic; (2) serofibrinous; (3) purulent. In *fibrinous pleurisy* the pleura is covered with a layer of lymph of variable thickness, which, in the acute form, can be readily stripped off. *Serofibrinous pleurisy* is characterized by the presence of a considerable quantity of fluid containing flocculi of lymph, and the deposit of some fibrin on the pleural surface. *Purulent pleurisy,* or empyema, is characterized by the presence of a purulent exudate. *Acute pleurisy* is marked by sharp and stabbing pain (stitch) in the side, increased by breathing and coughing; by fever, and by a friction-fremitus felt on palpation and a to-and-fro friction-sound heard on auscultation. In the serofibrinous variety a liquid effusion takes place, the signs of which are: bulging of the intercostal spaces and chest-wall, absence of vocal fremitus, displacement of the heart, movable dulness with a curved upper line, and a tympanitic percussionnote (Skodaic resonance) beneath the clavicle and above the level of the effusion. *Chronic pleurisy* may be dry or serofibrinous. **p., diaphragmatic,** that restricted to the pleural surface of the diaphragm. **p., dry,** that attended with little or no effusion of fluid. **p., encysted,** pleurisy in which the effusion is circumscribed by adhesions. **p., false,** pleurodynia. **p., fetid,** that marked by the presence of fetid fluid. **p., hemorrhagic,** a variety in which the exudate contains blood. **p., humid, p., moist,** that accompanied by expectoration. **p., ichorous.** Same as *p., fetid.* **p., interlobar,** that affecting the pleural layers between the lobes. **p., latent,** a form without the subjective symptoms. **p., mediastinal,** inflammation of the pleural layers about the mediastinum. **p., metapneumonic,** pleurisy dependent upon a pneumonia. **p., plastic,** that marked by a deposit of a layer of semisolid exudate. **p., purulent.** Same as *empyema.* **p., serofibrinous,** a form marked by fluid exudate containing flocculi and the deposit of some fibrin. **p. s cca.** See *p., dry.*

**pleuritic** (*ploo'-rit'-ik*) [*pleurisy*]. Pertaining to, affected with, or of the nature of pleurisy.

**pleuritis** (*ploo'-ri'-tis*). See *pleurisy.*

**pleuro-** (*ploo'-ro-*) [πλευρά, side]. A prefix denoting connection with the pleura or with a side or rib.

**pleurocele** (*ploo'-ro-sēl*) [*pleuro-*; κήλη, hernia]. 1. Hernia of the lung. 2. A serous effusion into the pleural cavity.

**pleurocentesis** (*ploo-ro-sen-te'-sis*) [*pleuro-*; κέντησις, a pricking]. Surgical puncture of the pleura.

**pleurocentral** (*ploo-ro-sen'-tral*). Pertaining to a pleurocentrum.

**pleurocentrum** (*ploo-ro-sen'-trum*) [*pleuro-*; κέντρον, center: *pl., pleurocentra*]. A hemicentrum; the lateral element in a vertebral centrum.

**pleurocholecystitis** (*ploo-ro-ko-le-sist-i'-tis*) [*pleuro-*; χολή, bile; *cystitis*]. Simultaneous inflammation of the pleura and the gall-bladder.

**pleuroclysis** (*ploo-rok'-lis-is*) [*pleuro-*; κλύσις, a wash]. The injection of fluids into, or the washing out of the pleural cavity.

**pleurocolic** (*ploo-ro-kol'-ik*) [*pleuro-*; *colon*]. Costocolic; joining the side and the colon.

**pleurocollesis** (*ploo-ro-kol-e'-sis*) [*pleuro-*; κόλλησις, *culis, skin*]. Adhesion of the pleural layers.

**pleurocutaneous** (*ploo-ro-ku-ta'-ne-us*) [*pleuro-*; *culis, skin*]. In relation with the pleura and the skin, as a pleurocutaneous fistula.

**pleurodont** (*ploo'-ro-dont*) [*pleuro-*; ὀδούς, tooth]. In biology, a tooth, or an animal bearing teeth, fastened into the jaw by a lateral ankylosis; as in certain lizards.

**pleurodynia** (*ploo-ro-din'-e-ah*) [*pleuro-*; ὀδύνη, pain]. A sharp pain in the intercostal muscles, of rheumatic origin.

**pleurogenic, pleurogenous** (*ploo-ro-jen'-ik, ploo-roj'-en-us*) [*pleuro-*; γεννᾶν, to produce]. Originating in the pleura.

**pleurohepatitis** (*ploo-ro-hep-at-i'-tis*) [*pleuro-*; ἧπαρ, liver; *itis*, inflammation]. Inflammation of the pleura and the liver.

**pleurolith** (*ploo'-ro-lith*) [*pleuro-*; λίθος, a stone]. A calculus occurring in the pleura.

**pleuropericarditis** (*ploo'-ro-per-ik-ar-di'-tis*) [*pleuro-*; *pericarditis*]. Pleurisy associated with pericarditis.

**pleuroperitoneal** (*ploo'-ro-per-i-ton-e'-al*) [*pleuro-*; *peritoneum*]. Pertaining to the pleura and the peritoneum. **p. cavity,** the body-cavity.

**pleuroperitonitis** (*ploo'-ro-per-it-on-i'-tis*) [*pleuro-*; *peritonitis*]. The simultaneous existence of pleurisy and peritonitis.

**pleurophorous** (*ploo-rof'-or-us*) [*pleuro-*; : φέρειν, to bear]. Furnished with a membrane.
**pleuropneumonia** (*ploo-ro-nū-mo'-ne-ah*). Combined inflammation of the pleura and of the lung, especially a contagious variety occurring in cattle.
**pleuropulmonary** (*ploo-ro-pul'-mo-na-re*). Pertaining to the pleura and the lungs.
**pleuropyesis** (*ploo-ro-pi-e'-sis*) [*pleuro-*; πύησις, suppuration]. Purulent pleurisy.
**pleurorrhagia** (*ploo-ror-a'-je-ah*) [*pleuro-*; ῥηγνύναι, to burst forth]. Hemorrhage from the pleura.
**pleurorrhea** (*ploo-ror-e'-ah*) [*pleuro-*; ῥοία, a flow]. An effusion of fluid into the pleura.
**pleurosoma** (*ploo-ro-so'-mah*) [*pleuro-*; σῶμα, a body]. A variety of monsters of the species *Celosoma*, in which there is a lateral eventration with atrophy or imperfect development of the upper extremity on the side of the eventration.
**pleurosomus** (*ploo-ro-so'-mus*). A monster exhibiting pleurosomia.
**pleurospasm** (*ploo'-ro-spazm*) [*pleuro-*; σπασμός, spasm]. Cramp, or spasm in the side.
**pleurosthotonos** (*ploo-ros-thot'-o-nos*). See *pleurothotonos*.
**pleurothotonos** (*ploo-ro-thot'-o-nos*) [πλευρόθεν, from the side; τόνος, tension]. A form of tetanic spasm of the muscles in which the body is bent to one side.
**pleurotomy** (*ploo-rot'-o-me*) [*pleuro-*; τομή, a cutting]. Incision into the pleura.
**pleurotyphoid** (*ploo-ro-ti-foid'*). Typhoid fever with involvement of the pleura.
**pleurovisceral** (*ploo-ro vis' er al*) [*pleuro* ; *visous*]. Pertaining to the pleura or side, and to the viscera.
**plexal** (*pleks'-al*) [*plectere*, to knit]. Pertaining to or of the nature of a plexus.
**plexiform** (*pleks'-if-orm*) [*plexus*; *forma*, form]. Resembling a network or plexus.
**pleximeter** (*pleks-im'-et-er*) [πλῆξις, a stroke; μέτρον, a measure]. A disc placed on the body to receive the stroke in mediate percussion.
**pleximetric** (*pleks-im-et'-rik*) [*pleximeter*]. Pertaining to or performed with a pleximeter.
**pleximetry** (*pleks-im'-et-re*). Percussion by means of a pleximeter.
**plexor** (*pleks'-or*) [πλῆξις, stroke]. A hammer used for performing percussion.
**plexus** (*pleks'-us*) [*plectere*, to knit]. A network, especially an aggregation of vessels or nerves forming an intricate network. **p., aortic,** (1) a nerve plexus on each side and in front of the abdominal aorta; (2) one surrounding the thoracic aorta. **p., basilaris,** the basilar sinus, consisting of a number of veins connecting the two subpetrosal sinuses. **p., brachial,** a plexus formed in the neck by the union of the anterior branches of the lower four cervical and the greater part of the first dorsal nerves. Its branches are the rhomboid, subclavian, suprascapular, external anterior thoracic, musculocutaneous, subscapular, median musculospiral, posterior thoracic, internal anterior thoracic, internal cutaneous, lesser internal cutaneous, and ulnar nerves. **p., cardiac,** a plexus of nerves connected with the heart. **p., cardiac, deep or great;** the deep portion of the superficial cardiac plexus. **p., cardiac, superficial,** or **anterior,** one in the upper part of the chest, between the arch of the aorta and base of the heart. It is derived from the sympathetic nerve and is reinforced by branches of the inferior, middle, and superior cardiac, hypoglossal, and pneumogastric nerves. **p., carotid, external,** one around the external carotid artery. **p., carotid, internal,** one surrounding the internal carotid artery. **p., cavernous,** a sympathetic plexus in the cavernous sinus; it furnishes branches to the internal carotid artery and connects with the motor oculi, pathetic us, and trigeminus nerves. **p., celiac,** one close to the celiac axis. **p., cervical,** a plexus in the neck formed by the anterior branches of the upper four cervical nerves. Its branches are the *superficial,* to skin of the head and neck; and the *deep,* the phrenic, communicans noni, two musculai, and two communicating branches. **p., choroid.** See *choroid plexus.* **p., coccygeal,** one on the dorsal surface of the coccyx and caudal end of the sacrum. **p., coronary, anterior,** one between aorta and pulmonary artery. **p., coronary, gastric,** one at the lesser curvature of the stomach. **p., coronary, posterior,** one accompanying the posterior coronary artery on the dorsum of the heart. **p., crural,** one surrounding the upper portion of the femoral artery. **p., cystic,** one near the gall-bladder. **p., dental, inferior,** one around the roots of the teeth of the lower jaw. **p., diaphragmatic,** one near the phrenic artery. **p., epigastric.** Same as *p., solar.* **p., esophageal,** one around the esophagus. **p., facial,** one enveloping part of the facial artery. **p., gangliform,** one formed from roots of origin of the inferior maxillary nerve. **p., gastric,** a branch of the celiac plexus accompanying the gastric artery. **p., gastroduodenal,** a branch of the celiac plexus. **p., hemorrhoidal, inferior** and **middle,** nerve-plexus derived from the pelvic plexus near the rectum. **p., hepatic,** a branch of the celiac plexus attending the hepatic artery to the liver. **p., hypogastric,** one before the promontory of the sacrum. **p., hypogastric, inferior.** Same as *p.; pelvic.* **p., infraorbital,** one under the levator labii superioris muscle. **p., intestinal, submucous,** Meissner's, in the submucosa of the small intestine. **p., lingual,** one around the lingual artery. **p., lumbar,** one formed by the anterior divisions of the lumbar spinal nerves in the psoas muscle. **p. magnus profundus.** Same as *p., cardiac, deep.* **p., mesenteric, inferior,** one around the inferior mesenteric artery. **p., mesenteric, superior,** one around the superior mesenteric artery. **p., myenteric,** Auerbach's, one between the circular and longitudinal muscular coats of the small intestine. **p., nasopalatine,** one uniting the nasopalatine nerves in the incisor foramen. **p., obturator,** one around the obturator nerve. **p., occipital,** one around the occipital artery. **p., ophthalmic,** one around the ophthalmic artery and the optic nerve. **p., ovarian,** (1) a venous plexus in the broad ligament; (2) a nerve plexus distributed to the ovaries. **p., pampiniform,** a venous plexus of the spermatic cord. **p., pancreatic,** one that supplies the pancreas. **p., pancreaticoduodenal,** one near the head of the pancreas. **p., parotid,** the pes anserinus, *q. v.* **p., patellar,** one in front of the patella. **p., pelvic,** one at the side of the rectum and bladder, distributed to the viscera of the pelvis and plexuses of the pelvis. **p., pharyngeal,** (1) nerve-plexuses supplying the pharynx; (2) venous plexus at the side of the pharynx. **p., phrenic,** one accompanying the phrenic arteries to the diaphragm. **p., prostatic,** one occupying the sides of the prostate. **p., pterygoid,** a plexus of veins which accompanies the internal maxillary artery between the pterygoid muscles. **p., pulmonary, anterior,** one in front of the bronchus, whence branches are distributed through the lung. **p., pulmonary, posterior,** one at the back of the bronchus, whence branches are distributed through the lung. **p., pyloric,** one near the pylorus. **p., renal,** a plexus derived from the solar and abdominal aortic plexuses; it accompanies the renal artery and is distributed to the kidney. **p., sacral,** one ventrad of the sacrum. **p., semilunar, p., solar.** See *solar plexus.* **p., spermatic,** a nerve-plexus around the spermatic vessels, supplying the testes (ovaries in females). **p., sphenoid,** the upper part of the internal carotid plexus. **p., splenic,** one around the splenic artery. **p., subsartorial,** one at the posterior border of the sartorius muscle. **p., subtrapezial,** one beneath the trapezius muscle. **p., sympathetic;** a plexus formed by the branches of the sympathetic nerve. **p., thyroid, inferior,** one around the external carotid and inferior thyroid arteries, distributed to the larynx, pharynx, and thyroid gland. **p., thyroid, superior,** one around the thyroid gland. **p., tonsillar,** one in the tonsil. **p., tympanic,** the tympanic portion of the tympanic nerve. **p., uterine,** (1) a venous plexus on the walls of the uterus, extending into the broad ligament; (2) a nerve-plexus supplying the cervix and lower part of the uterus. **p., vaginal,** (1) a nerve-plexus supplying the walls of the vagina; (2) a venous plexus near the entrance of the vagina; basilar arteries. **p., vesical,** one surrounding the vesical arteries.

**pli** (*ple*) [Fr.]. 1. A gyrus or convolution. 2. A fold. 3. Plica.
**plica** (*pli'-kah*) [L.]. 1. A fold. 2. See *p. polonica.*
**p. chorioidea,** the transverse fold of invaginated roof-plate produced by the metencephalic flexure of the primitive brain. **p. epigastrica,** fold of peritoneum covering the deep epigastric artery. **p. fimbriata,** a fold of mucous membrane having a fringed free edge on either side of the frenum linguae. **p. gubernatrix,** a fold of peritoneum containing the lower part of the gubernaculum testis. **p. hypogastrica.** Same as *p. umbilicalis lateralis.* **p. lacrimalis.** See *Hasner's valves.* **p. neuropathica,**

a curling of the hair from a nervous derangement. **plicæ palmatæ**, radiating folds in the mucous membrane of the cervix. p. **polonica**, a matted, entangled condition of the hair, due to want of cleanliness in certain diseases of the scalp. Syn., *Polish plait*. p. **salpingopalatina**, a fold of mucous membrane stretching from the torus tubarius to the palate. p. **salpingopharyngea**, a vertical fold of mucous membrane stretching from the torus tubarius to the pharynx. p. **semilunaris**, a conjunctival fold in the inner canthus of the eye, the equivalent of the nictitating membrane of birds. p. **sublingualis**, a fold of mucous membrane caused by the projection of the sublingual gland. p. **triangularis**, a triangular membrane extending from the upper posterior portion of the anterior faucial pillar backward and downward until lost in the tissues at the base of the tongue. p. **umbilicalis lateralis**, fold of peritoneum covering the obliterated hypogastric artery. p. **umbilicalis media**, a fold of peritoneum covering the urachus. p. **urachi**. Same as *p. umbilicalis media*. p. **vascularis**, a fold of peritoneum containing the spermatic vessels.
**plicate** (*pli'-kāt*) [*plicare*, to fold]. Folded; plaited.
**plication** (*pli-ka'-shun*). A plica or fold.
**plicotomy** (*pli-kot'-o-mē*) [*plica*; τομή, a cutting]. Division of the posterior fold of the tympanic membrane.
**Plimmer's bodies** (*plim'-er*) [Henry George Plimmer, English protozoologist]. Intracellular bodies observed by Plimmer in cancerous tissue.
**plomb** (*plum*) [Fr., *plomber*, to plug a tooth]. A filling for a cavity. p., **iodoform** (of Mosetig-Moorhof), an antibacillary agent for filling bone-cavities after operations for tuberculosis or osteomyelitis. It consists of iodoform, spermaceti, and oil of sesame, which are sterilized, heated, and poured into the cavity, when the mixture solidifies and fills it.
**ploration** (*plo-ra'-shun*) [*plorare*, to weep]. Lacrimation.
**plug**. Something that occludes a circular opening or channel. p., **cervical**. Synonym of *p., mucous*. p., **kite-tail**, a tampon resembling a kite-tail. p., **mucous**, the mass of inspissated mucus which occludes the cervix uteri during pregnancy and is discharged at the beginning of labor. p.s, **Dittrich's**. See under *Dittrich*.
**Plugge's phenol reaction** (*plu'-geh*). A dilute phenol solution is rendered intensely red on boiling with a solution of mercuric nitrate containing a trace of nitrous acid. Metallic mercury is separated at the same time, and an odor of salicylol is given off.
**plugger** (*plug'-ur*). An instrument for the insertion and impaction of filling materials in cavities in teeth.
**plugging** (*plug'-ing*). See *tampon*. p. **instruments**, dental instruments for introducing and consolidating fillings. p. **teeth**. See *filling teeth*.
**plumbago** (*plum-ba'-go*). See *graphite*.
**plumbi** (*plum'-bi*). Genitive of Latin *plumbum*, lead.
**plumbic** (*plum'-bik*) [*plumbum*]. Pertaining to or containing lead.
**plumbism** (*plum'-bizm*) [*plumbum*]. Lead-poisoning.
**plumbite** (*plum'-bīt*). A general term for any compound formed by union of lead oxide with a base.
**plumbum** (*plum'-bum*) [L.]. Lead, a bluish-white metal occurring in nature chiefly as the sulphide, PbS, known as galena; atomic weight 207.10; sp. gr. 11.38; quantivalence II and IV; symbol Pb. The salts of lead are poisonous, producing, in sublethal doses, gastroenteritis; ingested in small quantities over a long period of time chronic lead-poisoning is produced. See *lead-poisoning*. **plumbi acetas** (U. S. P.), lead acetate, Pb(C₂H₃O₂)₂ . 3H₂O. It is used as an astringent in diarrhea and dysentery, as a hemostatic, and as an astringent and sedative in gonorrhea, leukorrhea, conjunctivitis, etc. Dose 1–3 gr. (0.065–0.2 Gm.). Syn., *sugar of lead*. **plumbi carbonas**, lead carbonate, white lead, (PbCO₃)₂ . Pb(OH)₂, is used as a local sedative in ointments and in face-powders. The prolonged use of the latter has caused poisoning. **plumbi chloridum**, lead chloride, PbCl₂, is used like the carbonate. **plumbi emplastrum** (U. S. P.), lead plaster, is made of lead acetate, soap, and water, and is used as an external application to irritated surfaces, and in the arts for glazing pottery and as an ingredient of fluid glass. Combined with olive-oil it constitutes *unguentum diachylon* (U. S. P.). **plumbi iodidum** (U. S. P.), lead iodide, PbI₂, is used as a local astringent and absorbent. **plumbi nitras** (U. S. P.), lead nitrate, Pb(NO₃)₂, is used locally as a sedative to excoriated surfaces, as sore nipples, chapped hands; in gonorrhea and leukorrhea; in onychia maligna, etc. *Ledoyen's disinfecting fluid* is a solution of lead nitrate of the strength of one dram to the ounce. **plumbi oxidum** (U. S. P.), lead oxide or litharge, PbO, is used in the making of lead plaster. **plumbi oxidum rubrum**, red lead, minium, is used extensively in the arts as a paint and in the manufacture of glass. **plumbi subacetas**, lead subacetate, is a basic salt. **plumbi subacetatis, ceratum** (U. S. P.), is made up of lead subacetate, wool-fat, paraffin, white petrolatum, and camphor. **plumbi subacetatis, liquor** (U. S. P., B. P.), Goulard's extract. **plumbi subacetatis, liquor, dilutus** (U. S. P., B. P.), lead-water, is used as a sedative and astringent in inflammations and burns. **plumbi tannas**, lead tannate, is used as a sedative astringent.
**Plummer's pill** (*plum'-er*) [Andrew *Plummer*, Scotch physician, –1756]. Compound pill of calomel and antimony. See under *antimony*.
**plumose, plumous** (*ploo'-mōs, ploo'-mus*) [*pluma*, feather]. In biology, having feathers; feathery; feathered; of bacteria, denoting a fleecy or feathery growth.
**plumper** (*plum'-per*). One of a pair of pads worn in the hollow of the cheeks to give them a rounded appearance; sometimes attached to a set of artificial teeth.
**plumula** (*ploo'-mū-lah*) [*plumula*, a little feather; *pl., plumulae*]. Minute transverse furrows on the roof of the aqueduct of Sylvius.
**Plunkett's caustic**, or **ointment** (*plunk'-et*). A caustic paste composed of the bruised plant of *Ranunculus acris* and of *R. flammula*, oak gum, arsenous acid, 3 parts; sulphur, 5 parts. These are mixed into a paste, rolled into balls, and dried in the sun. When used the ball must be reduced to a pasty consistence by rubbing with yolk of egg.
**pluriceptor** (*ploo-re-sep'-tor*) [*plus*, more; *capere*, to take]. A receptor with more than two complementophile groups.
**pluricordonal** (*ploo-ri-kord'-on-al*) [*plus*, more; *chorda*, a string]. Having several processes.
**plurifœtation** (*ploo-ri-fe-ta'-shun*) [*plus*, more; *fetus*]. The conception of twins, triplets, etc.
**pluriglandular** (*ploo-re-glan'-dū-lar*). Referring to more than one gland or to the secretions of more than one gland.
**plurilocular** (*ploo-ril-ok'-ū-lar*) [*plus*, more; *loculus*, a cell]. Having more than one cell or loculus; multilocular.
**pluripara** (*ploo-rip'-ar-ah*). See *multipara*.
**pluriparity** (*ploo-rip-ar'-it-e*) [*plus*, more; *parere*, to bring forth]. The condition of having borne several children.
**pluriseptate** (*ploo-ris-ep'-tāt*) [*plus*, more; *septum*, a partition]. Having more than one septum or partition.
**plutomania** (*ploo-to-ma'-ne-ah*) [πλοῦτος, wealth; μανία, madness]. An insane belief that one is the possessor of great wealth.
**pluviometric** (*ploo-ve-o-met'-rik*) [*pluvia*, rain; μέτρον, a measure]. Relating to the measurement of rainfalls.
**pneodynamics** (*ne-o-di-nam'-iks*) [πνεῖν, to breathe; δύναμις, power]. The dynamics of respiration.
**pneogaster** (*ne'-o-gas-ter*) [πνεῖν, to breathe; γαστήρ, stomach]. In biology, the respiratory tract.
**pneograph** (*ne'-o-graf*) [πνεῖν, to breathe; γράφειν, to write]. An instrument for recording the force and character of the current of air during respiration.
**pneometer** (*ne-om'-et-er*). Synonym of *spirometer*.
**pneophore** (*ne'-o-fōr*) [πνεῖν, to breathe; φορός, carrying]. An instrument to aid artificial respiration in the asphyxiated.
**pneoscope** (*ne'-o-skōp*) [πνεῖν, to breathe; σκοπεῖν, to examine]. An instrument for measuring respiratory movements.
**pneuma** (*nū'-mah*) [πνεῦμα, breath]. 1. Air; a breath. 2. The vital principle.
**pneumarthrosis** (*nū-mar-thro'-sis*) [πνεῦμα, air; ἄρθρον, a joint]. A collection of air or gas in a joint.

**pneumascope** (*nū'-mah-skōp*). See *pneumatoscope*.
**pneumatelectasis** (*nū-mat-el-ek'-tas-is*). Atelectasis of the lungs.
**pneumathemia** (*nū-ma-the'-me-ah*) [πνεῦμα, air; αἷμα, blood]. The presence of air or gas in the blood-vessels.
**pneumatic** (*nū-mat'-ik*) [πνεῦμα, air]. 1. Pertaining to air or gas. 2. Pertaining to respiration. 3. Pertaining to compressed or rarefied air. p. **cabinet**, a cabinet for treating a part by compressed or rarefied air. p. **speculum**. See *Siegle's speculum*. p. **trough**, a trough partly filled with water for facilitating the collection of gases.
**pneumatics** (*nū-mat'-iks*) [πνεῦμα, air]. The branch of physics treating of the physical properties of air and gases.
**pneumatinuria** (*nū-mat-in-ū'-re-ah*). See *pneumaturia*.
**pneumato-** (*nū-mat-o-*) [πνεῦμα, air]. A prefix denoting pertaining to air, gas, or breath.
**pneumatocardia** (*nū-mat-o-kar'-de-ah*) [*pneumato-*; καρδία, heart]. The presence of air or gas in the chambers of the heart.
**pneumatocele** (*nū'-mat-o-sēl*) [*pneumato-*; κήλη, tumor]. 1. A swelling containing air or gas. 2. See *pneumonocele*. 3. A swelling of the scrotum produced by the presence of gas.
**pneumatochemical** (*nū-mat-o-kem'-ik-al*). 1. Relating to the chemistry of gases. 2. Relating to the treatment of pulmonary disease by inhalation of medicated vapors. p. **apparatus of Priestley**. See *pneumatic trough*.
**pneumatodyspnea, pneumatodyspnœa** (*nū-mat-o-disp'-ne-ah*) [*pneumato-*; *dyspnea*]. Emphysematous dyspnea.
**pneumatogeny** (*nū-mat-oj'-en-e*) [*pneumato-*; γεννᾶν, to produce]. Artificial respiration.
**pneumatogram** (*nū-mat'-o-gram*) [*pneumato-*; γράμμα, inscription]. A tracing showing the frequency, duration, and depth of the respiratory movements.
**pneumatograph**. See *pneumograph*.
**pneumatology** (*nū-mat-ol'-o-je*) [*pneumato-*; λόγος, science]. 1. The science of respiration. 2. The science of gases; also their use as therapeutic agents.
**pneumatometer** (*nū-mat-om'-et-er*). An instrument for measuring the pressure of inspiration or expiration by the force exerted upon a mercuric column contained in a U-tube.
**pneumatometry** (*nū-mat-om'-et-re*) [*pneumato-*; μέτρον, measure]. 1. The measurement of the force in respiration. It is used as a means of diagnosis. 2. The treatment of pulmonary and circulatory diseases by means of a pneumatic apparatus.
**pneumatomphalocele** (*nū-mat-om-fal'-o-sēl*) [*pneumato-*; ὀμφαλός, navel; κήλη, tumor]. An umbilical hernia containing flatus.
**pneumatorrhachis** (*nū-mat-or'-a-kis*) [*pneumato-*; ῥάχις, spine]. The presence of air in the spinal canal.
**pneumatoscope** (*nū'-mat-o-skōp*) [*pneumato-*; σκοπεῖν, to examine]. 1. An apparatus for measuring the gas in expired air. 2. An instrument for internal auscultation of the thorax. 3. An instrument for determining the presence of foreign bodies in the mastoid sinuses. 4. See *pneumograph*.
**pneumatosis** (*nū-mat-o'-sis*) [πνεῦμα, air]. The presence of gas or air in abnormal places, or in an excessive quantity where a little exists normally.
**pneumatotherapy** (*nū-mat-o-ther'-ap-e*) [*pneumato-*; θεραπεία, treatment]. The treatment of diseases by means of compressed or rarefied air.
**pneumatothorax** (*nū-mat-o-tho'-raks*). See *pneumothorax*.
**pneumaturia** (*nū-mat-u'-re-ah*) [*pneumato-*; οὖρον, urine]. The evacuation of urine containing free gas.
**pneumatype** (*nū'-mat-īp*) [πνεῦμα, air; τύπος, type]. Breath-picture. The deposit formed upon a piece of glass by the moist air exhaled through the nostrils when the mouth is closed. It is employed in the diagnosis of nasal obstruction. Slate-paper may be used, pulverized sulphur or boric acid being blown upon the moistened surface to make a permanent record.
**pneumectomy** (*nū-mek'-to-me*). See *pneumonectomy*.
**pneumin** (*nū'-min*). See *methylene creosote*.
**pneumo-** (*nū-mo-*). The same as *pneumono-*.
**pneumoarctia** (*nū-mo-ark'-te-ah*) [*pneumo-*; arctare, to contract]. Contraction of the lungs; pneumonostenosis.

**pneumobacillin** (*nū-mo-bas-il'-in*). A toxic extract of pneumobacilli.
**pneumobacillus** (*nū-mo-bas-il'-us*). The *Bacillus pneumoniæ*.
**pneumobacterine** (*nū-mo-bak'-ter-ēn*). A stock vaccine obtained from cultures of the *pneumococcus*.
**pneumocace** (*nū-mok'-as-e*) [*pneumo-*; κακός, evil]. Gangrene of the lung.
**pneumocele** (*nū'-mo-sēl*). See *pneumatocele*.
**pneumocentesis** (*nū-mo-sen-te'-sis*) [*pneumo-*; κέντησις, puncture]. Paracentesis of the lung, especially for the purpose of evacuating a cavity.
**pneumocephalus** (*nū-mo-sef'-a-lus*) [*pneumo-*; κεφαλή, head]. The presence of air or gas within the cranial cavity.
**pneumochemical**. See *pneumatochemical*.
**pneumochirurgia** (*nū-mo-ki-rur'-je-ah*) [*pneumo-*; χειρουργία, surgery]. Surgery of the lungs.
**pneumochysis** (*nū-mok'-is-is*) [*pneumo-*; χύσις, a pouring]. Pulmonary edema.
**pneumococcal** (*nū-mo-kok'-al*). Pertaining to or caused by pneumococci.
**pneumococcemia, pneumococcæmia** (*nū-mo-kok-se'-me-ah*) [*pneumococcus*; αἷμα, blood]. The presence of pneumococci in the blood.
**pneumococcia** (*nū-mo-kok'-se-ah*). Generalized infection by pneumococci.
**pneumococcus** (*nū-mo-kok'-us*). Any micrococcus of the lung; especially the *micrococcus lanceolatus*.
**pneumoconiosis** (*nū-mo-kon-e-o'-sis*). See *pneumonokoniosis*.
**pneumoderma** (*nu-mo-der'-mah*) [*pneumo-*; δέρμα, skin]. Subcutaneous emphysema.
**pneumoenteritis** (*nū-mo-en-ter-i'-tis*) [*pneumo-*; *enteritis*]. Inflammation of the lungs and of the intestine. See *hog-cholera*. p., **infectious**. Synonym of *hog-cholera*.
**pneumoerysipelas** (*nū-mo-er-e-sip'-el-as*). Pneumonia associated with erysipelas.
**pneumogalactocele** (*nū-mo-gal-ak'-to-sēl*) [*pneumo-*; *galactocele*]. A galactocele containing gas.
**pneumogastric** (*nū-mo-gas'-trik*) [*pneumo-*; γαστήρ, stomach]. 1. Pertaining conjointly to the lungs and the stomach. 2. Pertaining to the pneumogastric or vagus nerve. p. **nerve**. See under *nerve*.
**pneumogram** (*nū'-mo-gram*) [*pneumo-*; γράμμα, writing]. The tracing afforded by the pneumograph.
**pneumograph** (*nū'-mo-graf*) [*pneumo-*; γράφειν, to write]. An instrument for recording the movements of the chest in respiration.
**pneumography** (*nū-mog'-ra-fe*) [see *pneumograph*]. A description of the lungs.
**pneumohemothorax** (*nū-mo-hem-o-tho'-raks*). A collection of air or gas and blood in the pleural cavity.
**pneumohydrometra** (*nū-mo-hi-dro-me'-trah*) [*pneumo-*; *hydrometra*]. Hydrometra associated with the generation of gas in the uterus.
**pneumohydropericardium** (*nū-mo-hi-dro-per-e-kar'-de-um*) [*pneumo-*; ὕδωρ, water; *pericardium*]. An accumulation of air and fluid in the pericardial sac.
**pneumohydrothorax** (*nū-mo-hi-dro-tho'-raks*). A collection of air or gas and fluid in the pleural cavity.
**pneumohypoderma** (*nū-mo-hi-po-der'-mah*) [*pneumo-*; ὑπό, under; δέρμα, skin]. Subcutaneous emphysema.
**pneumokoniosis**. See *pneumonokoniosis*.
**pneumolith** (*nū'-mo-lith*) [*pneumo-*; λίθος, a stone]. A calculus of the lung.
**pneumolithiasis** (*nū-mo-lith-i'-as-is*) [*pneumo-*; λίθος, stone]. The formation of pneumoliths.
**pneumology** (*nū-mol'-o-je*) [*pneumo-*; λόγος, science]. The sum of scientific knowledge concerning the lungs and air-passages.
**pneumolysis** (*nū-mol'-is-is*) [*pneumo-*; λύσις, a loosening]. Loosening from the intrathoracic fascia of thickened pleura which causes contraction of the lung.
**pneumomalacia** (*nū-mo-mal-a'-se-ah*) [*pneumo-*; μαλακία, softness]. Abnormal softness of the lung.
**pneumomassage** (*nū-mo-mas-sahzj'*) [*pneumo-*; μάσσειν, to knead]. The application of massage or passive motion to the tympanic membrane and auditory ossicles by pneumatic means.
**pneumomelanosis**. See *pneumonomelanosis*.
**pneumometer, pneumatometer, pneumomometer** (*nū-mom'-et-er, nū-mat-om'-et-er, nū-mo-mom'-et-er*). Synonyms of *spirometer*.
**pneumometry, pneumatometry** (*nū-mom'-et-re, nū-mat-om'-et-re*) [*pneumo-*; μέτρον, a measure]. 1. The

measurement of the force of respiration. 2. The treatment of pulmonary and circulatory diseases by means of a pneumatic apparatus.

**pneumomycosis** (*nū-mo-mi-ko'-sis*). A disease of the lungs due to fungi.

**pneumon-** (*nū'-mon*) [πνεύμων, lung]. A prefix denoting connection with or relation to the lungs.

**pneumonalgia** (*nū-mo-nal'-je-ah*) [*pneumon-*; ἄλγος, pain]. Pain in the lung.

**pneumonatelectasis** (*nū-mon-at-el-ek'-tas-is*) [*pneumon-*; *atelectasis*]. Atelectasis of the lung.

**pneumonectasia, pneumonectasis** (*nū-mon-ek-ta'-ze-ah, nū-mon-ek'-tas-is*) [*pneumono-*; ἔκτασις, distention]. Emphysema of the lung.

**pneumonectomy** (*nū-mon-ek'-to-me*) [*pneumono-*; ἐκτομή, excision]. Excision of a portion of a lung.

**pneumonedema** (*nū-mon-e-de'-mah*) [*pneumon-*; *edema*]. Edema of the lungs.

**pneumonemia, pneumonsemia** (*nū-mon-e'-me-ah*) [*pneumon-*; αἷμα, blood]. Congestion of the lungs.

**pneumonemphraxis** (*nū-mon-em-fraks'-is*) [*pneumon-*; ἔμφραξις, obstruction]. Obstruction of the lungs or the bronchi.

**pneumonemphysema** (*nū-mon-em-fis-e'-mah*) [*pneumo-*; ἐμφύσημα, inflation]. Emphysema of the lungs.

**pneumonia** (*nū-mo'-ne-ah*) [πνεύμων, lung]. Inflammation of the lung; pneumonitis. Used without qualification, the term implies lobar pneumonia (*q. v.*). p., **abortive**, acute congestion not followed by other stages. p., **acute**, lobar pneumonia, most often due to a specific microorganism. p., **alcoholic**, the croupous pneumonia of drunkards, often associated with delirium, and very fatal. p., **apex-**, p., **aspiration-**, croupous pneumonia of the apex of a lung. p., **aspiration-**, a bronchopneumonia due to the inspiration of food-particles or other irritant substances into the lung. p., **bronchial**, p., **catarrhal**. Synonym of *bronchopneumonia*. p., **central**, a croupous pneumonia beginning in the interior of a lobe of the lung. The physical signs are obscure until the inflammation reaches the surface. p., **cerebral**, a form associated with marked cerebral symptoms. It is most common in children, and in the beginning resembles meningitis. p., **cheesy**. See p., *desquamative*. p., **chronic**. See p. *interstitial*. p., **contusion**, that following contusion of the chest. p., **croupous**. See p., *lobar*. p., **deglutition-**. Synonym of p., *aspiration-*. p., **desquamative**, a form characterized chiefly by an intense desquamation of the cells lining the air-vesicles, a proliferation of the connective-tissue cells of the septa between the vesicles, and the exudation of a scanty albuminous fluid. The exudate generally undergoes caseous degeneration. p., **disseminated**, bronchopneumonia. p., **double**, lobar pneumonia of both lungs. p., **embolic**, pneumonia due to embolism of the vessels of the lung. p., **ephemeral**, congestion of the lungs. p., **fibrinous**. See p., *lobar*. p., **fibroid**, p., **fibrous**. Synonym of p., *interstitial*. p., **gangrenous**, gangrene of the lung. p., **hypostatic**, a lobular pneumonia occurring in the dependent portions of the lungs of persons debilitated by age or disease, and depending on the weakened circulation and respiration and the dorsal decubitus. p., **indurative**. See p., *desquamative*. p., **insular**. Synonym of *bronchopneumonia*. p., **interstitial**, a chronic inflammation of the lung characterized by an increase of the connective tissue. Syn., *cirrhosis of the lung; fibroid pneumonia*. p., **larval**, that presenting only initial symptoms. p., **lobar**, an acute infectious disease characterized by an inflammation of one or more lobes of the lung, the affected parts becoming consolidated, owing to the exudation of cells and fibrin into the air-vesicles. The exciting cause is usually *Diplococcus pneumoniæ* of Fränkel, but other microorganisms may produce it. Syn., *croupous pneumonia; lung-fever*. p., **lobular**. Synonym of *bronchopneumonia*. p., **massive**, lobar pneumonia in which not only the air-cells, but the bronchi of an entire lobe, or even of a lung, are filled with the fibrinous exudate. p. **migrans**, p., **migratory**, a form involving one lobe after another. p., **pleuritic**, pleuropneumonia. p., **pleurogenic**, p., **pleurogenous**, pneumonia secondary to disease of the pleura. p., **purulent**, one characterized by the formation of pus; it appears under three forms, suppuration of the minute bronchi and air-vesicles—purulent catarrh; true abscess of the lung; suppurative lymphangitis and perilymphangitis. p., **septic**, lobular pneumonia due to the inspiration of septic material or to septic emboli. p., **superficial**, that restricted to parts near the pleura. p., **syphilitic**, inflammation of the lung due to syphilis and manifesting itself as the white pneumonia of the fetus; as gumma of the lung; as interstitial pneumonia, taking its origin at the root of the lung and passing along the bronchi and vessels; and as acute syphilitic phthisis, analogous to acute pneumonic phthisis. p., **tubular**. Synonym of *bronchopneumonia*. p., **typhoid**, that attended with typhoid symptoms. p. **vera**, lobar pneumonia not complicated with other diseases or forms. p., **vesicular**, bronchopneumonia. p., **wandering**, that which affects different parts of the lung in succession and seems to be associated with erysipelas. p., **white**, a catarrhal form of pneumonia occurring in a syphilitic fetus and resulting in death. By an overgrowth of epithelium in the air-vesicles the cells die, and fatty degeneration follows, giving the lungs a white appearance, with the imprint of the ribs on their surface.

**pneumonic** (*nū-mon'-ik*) [see *pneumonia*]. Pertaining to the lungs or to pneumonia. p. **phthisis**, tuberculosis affecting a whole lobe of the lung.

**pneumonitis** (*nū-mon-i'-tis*). Pneumonia.

**pneumono-** (*nū-mon-o-*) [πνεύμων, lung]. A prefix denoting pertaining to the lungs.

**pneumonocace** (*nu-mon-ok'-as-e*) [*pneumono-*; κακός, evil]. Gangrene of the lung.

**pneumonocele** (*nū'-mon-o-sēl*) [*pneumono-*; κήλη, hernia]. Hernia of the lung.

**pneumonocentesis** (*nū-mon-o-sen-te'-sis*). Same as *pneumocentesis*.

**pneumonocirrhosis** (*nū-mon-o-sir-o'-sis*) [*pneumono-*; *cirrhosis*]. Cirrhosis of the lung, interstitial pneumonia.

**pneumonodynia** (*nū-mon-o-din'-e-ah*) [*pneumono-*; ὀδύνη, pain]. Pain referred to the lungs.

**pneumonokoniosis** (*nū-mon-o-kon-e-o'-sis*) [*pneumono-*; κόνις, dust]. A general term applied to chronic induration or fibrous inflammation of the lungs due to the inhalation of dust. Various names are given to it according to the kind of dust causing the inflammation: *anthracosis*, that due to the inhalation of coal-dust; *siderosis*, that due to inhalation of metallic dust; *chalicosis*, that due to the inhalation of mineral dust.

**pneumonolithiasis** (*nū-mon-o-lith-i'-as-is*) [*pneumono-*; λίθος, stone]. The formation of pneumoliths.

**pneumonomelanosis** (*nū-mon-o-mel-an-o'-sis*) [*pneumono-*; μέλας, black; *νόσος*, disease]. Anthracosis of the lung.

**pneumonometer** (*nū-mon-om'-et-er*). Synonym of *spirometer*.

**pneumonomycosis** (*nū-mon-o-mi-ko'-sis*). See *pneumomycosis*.

**pneumonoparalysis** (*nū-mon-o-par-al'-is-is*) [*pneumono-*; *paralysis*]. Paralysis of the lung.

**pneumonopathy** (*nū-mon-op'-ath-e*) [*pneumono-*; πάθος, disease]. Any disease of the lung.

**pneumonopexy** (*nū-mon'-o-peks-e*) [*pneumono-*; πῆξις, a fixing]. Fixation of a stump of lung tissue to the thoracic wall in connection with pneumonectomy for gangrene, hernia, or other pulmonary lesion.

**pneumonophlebitis** (*nū-mon-o-fleb-i'-tis*) [*pneumono-*; φλέψ, vein; *ιτις*, inflammation]. Inflammation of the pulmonary veins.

**pneumonophthisis** (*nū-mon-off'-this-is*) [*pneumono-*; φθίσις, a wasting]. A destructive process in the lungs.

**pneumonopleuritis** (*nū-mon-o-ploo-ri'-tis*). Synonym of *pleuropneumonia*.

**pneumonorrhagia** (*nū-mon-or-a'-je-ah*) [*pneumono-*; ῥηγνύναι, to burst forth]. Hemorrhage from the lungs.

**pneumonorrhaphy** (*nū-mon-or'-af-e*) [*pneumono-*; ῥαφή, a seam]. Suture of lacerations of the lung.

**pneumonosepsis** (*nū-mon-o-sep'-sis*) [*pneumono-*; σῆψις, putrefaction]. Septic inflammation of the lung.

**pneumonosis** (*nū-mon-o'-sis*) [*pneumo-*; *νόσος*, disease]. Any affection of the lungs.

**pneumonostenosis** (*nū-mon-o-sten-o'-sis*). Contraction of a lung.

**pneumonotomy** (*nū-mon-ot'-o-me*) [*pneumo-*; τομή, a cutting]. Surgical incision of the lung.

**pneumopaludism** (*nū-mo-pal'-ū-dizm*) [*pneumo-*; *paludism*]. A manifestation of malaria characterized by the impairment of the percussion resonance at one apex, bronchial respiratory murmurs, broncho-

**phony,** without rales, friction, or expectoration; cough occurs in paroxysms.
**pneumoparesis** (nū-mo-par'-es-is) [pneumo-; paresis]. Progressive congestion of the lungs apparently depending on vasomotor deficiency or other fault of innervation; simple respiratory failure.
**pneumopericarditis** (nū-mo-per-ik-ar-di'-tis) [pneumo-; pericarditis]. Pericarditis with the formation of gas in the pericardial sac.
**pneumopericardium** (nū-mo-per-e-kar'-de-um). The presence of air in the pericardial sac. It is due to traumatism or to communication between the pericardium and the esophagus, stomach, or lungs, and is marked by tympany over the precordial region and peculiar metallic heart-sounds.
**pneumoperitoneum** (nū-mo-per-it-on-e'-um) [pneumo-; peritoneum]. The presence of gas in the peritoneal cavity.
**pneumoperitonitis** (nū-mo-per-it-on-i'-tis) [pneumo-; peritonitis]. Peritonitis with the presence of gas in the peritoneal cavity.
**pneumopexy** (nū'-mo-peks-e). Same as pneumonopexy.
**pneumophthisis** (nū-moff'-this-is) [pneumo-; φθίσις, wasting]. A destructive process in the lung.
**pneumophyma** (nū-mo-fi'-mah) [pneumo-; φῦμα, growth; pl., pneumophymata]. A tubercle of the lung.
**pneumophymia** (nū-mo-fi'-me-ah) [pneumo-; φῦμα, growth]. Tuberculosis of the lung.
**pneumopleuritis** (nū-mo-ploo-ri'-tis) [pneumo-; pleura; ιτις, inflammation]. Conjoined inflammation of the lungs and pleura.
**pneumoprotein** (nū-mo-pro'-te-in). A protein elaborated by pneumococci.
**pneumoptysis** (nū-mop'-tis-is). Same as hemoptysis.
**pneumopyopericardium** (nū-mo-pi-o-per-e-kar'-de-um) [pneumo-; πύον, pus; pericardium]. The presence of air or gas and pus in the pericardial sac.
**pneumopyothorax** (nū-mo-pi-o-tho'-raks) [pneumo-; πύον, pus; thorax]. The presence of air and pus in the pleural cavity.
**pneumopyra** (nū-mo-pi'-rah) [pneumo-; πῦρ, fire]. Malignant bronchitis.
**pneumorrhachis.** See pneumatorrhachis.
**pneumorrhagia** (nū-mor-a'-je-ah). See hemoptysis.
**pneumosan** (nū'-mo-san). Amyl-thio-trimethylamine chloride; used for intramuscular injection in pulmonary tuberculosis.
**pneumoserothorax** (nū'-mo-se-ro-tho'-raks) [pneumo-; serum; θώραξ, chest]: The presence of air or gas and serum in the pleural cavity.
**pneumotherapy** (nū-mo-ther'-ap-e). 1. The treatment of diseases of the lung. 2. See pneumatotherapy.
**pneumothermomassage** (nū-mo-ther-mo-mas-ahzj') [pneumo-; θέρμη, heat; massage]. The application to the body of currents of air of varying degrees of pressure and temperature.
**pneumothorax** (nū-mo-tho'-raks). The presence of air or gas in the pleural cavity. It is produced by perforating wounds of the chest, by the rupture of an abscess or tuberculous cavity of the lung, by the rupture of an emphysematous vesicle, or the evacuation of an empyema into the lung or through the chest-wall. It is marked by dyspnea, shock, pain, a tympanitic (sometimes a dull) percussion-note over the affected side, displacement of the heart, bell-tympany, and diminished respiratory murmur.
**pneumotomy** (nū-mot'-o-me) [pneumo-; τομή, a cutting]. Surgical incision of the lung. Synonym of pneumonotomy.
**pneumotoxin** (nū-mo-toks'-in) [pneumo-; τοξικόν, a poison]. A toxin produced by the pneumococcus, and believed to be the cause of many of the symptoms of lobar pneumonia. Antipneumotoxin is the name given to the antitoxin supposed to exist in the blood of persons convalescent from lobar pneumonia.
**pneumotyphoid** (nū-mo-ti'-foid). Synonym of pneumotyphus.
**pneumotyphus** (nū-mo-ti'-fus) [pneumo-; typhus]. 1. Typhoid fever beginning with pneumonia dependent upon the typhoid bacillus. 2. Pneumonia occurring in the course of typhoid fever.
**pneumouria** (nū-mo-ū'-re-ah). See pneumaturia.
**pneusimeter** (nū-sim'-et-er) [πνεῦσις, a breathing; μέτρον, a measure]. An apparatus used as a spirometer to measure the vital capacity of the chest in respiration.
**pneusis** (nū'-sis) [πνεῦν, to breathe]. Respiration.
**p. pertussis.** Synonym of whooping-cough.

**pnigma** (nig'-mah) [πνίγειν, to choke]. Strangulation.
**pnigophobia** (ni-go-fo'-be-ah) [πνίγειν, to choke; φόβος, fear]. The fear of choking that sometimes accompanies angina pectoris.
**pnigos, pnix, pnixis** (ni'-gos, niks, niks'-is). Synonyms of pnigma.
**pock** (pok) [AS., poc, a pustule]. A pustule of an eruptive fever; especially of smallpox. **p.-marked,** marked with the cicatrices of the smallpox pustule. **pocked** (pokt). Pitted; marked with pustules.
**pocket** (pok'-et). In anatomy, a blind sac, or sac-shaped cavity. A diverticulum communicating with a cavity.
**pocketing** (pok'-et-ing). A mode of treating the pedicle in the operation of ovariotomy. It is accomplished by bringing the extremity of the pedicle between the inner lips of the incision, at its lower angle, thus securing its attachment to the raw surface of the abdominal wall.
**pocky** (pok'-e). Having pocks or pustules; infected with variola or syphilis.
**poculent** (pok'-ū-lent) [poculum, a goblet]. Drinkable; potable.
**poculum** (pok'-ū-lum) [potare, to drink]. 1. A drinking cup. 2. A draught or potion. **p. Diogenis** (di oj'-en-is), the palm of the hand when held so as to form a cup-like cavity.
**podagra** (pod-ag'-rah) [πούς, foot; ἄγρα, seizure]. Gout, especially of the great toe or the joints of the foot.
**podalgia** (pod-al'-je-ah) [πούς, foot; ἄλγος, pain]. Pain in the foot.
**podalic** (pod-al'-ik) [πούς, foot]. Pertaining to the feet. **p. version,** the operation of changing the position of the fetus in the uterus so as to bring the feet to the outlet.
**podarthral** (pod-ar'-thral) [πούς, foot; ἄρθρον, a joint]. Pertaining to the podarthrum.
**podarthritis** (pod-ar-thri'-tis) [πούς, foot; ἄρθρον, joint; ιτις, inflammation]. Inflammation of the joints of the feet.
**podarthrocace** (pod-ar-throk'-as-e) [πούς; foot; ἄρθρον joint; κακός, evil]. Caries of the articulations of the feet.
**podarthrum** (pod-ar'-thrum) [πούς, foot; ἄρθρον, a joint; pl., podarthra]. In biology, the foot-joint or metatarso-phalangeal articulation.
**podedema** (pod-e-de'-mah)[ πούς, foot; edema]. Edema of the foot.
**podelcoma** (pod-el-ko'-mah). See Madura foot.
**podencephalus** (pod-en-sef'-al-us) [πούς, foot; ἐγκέφαλος, brain]. A variety of monster of the species exencephalus, in which there is a protrusion of the cranial contents from the top of the head.
**podiatrist** (pod-e-at'-rist) [πούς, a foot; ιατρός, physician]. A specialist in the treatment of diseases of the feet.
**podobromidrosis** (pod-o-brom-id-ro'-sis) [πούς, foot; βρῶμος, stench; ιδρώς, sweat]. Offensive sweating of the feet.
**pododynamometer** (pod-o-di-nam-om'-et-ur) [πούς, a foot; dynamometer]. An apparatus for testing the strength of the muscles of the feet or legs.
**pododynia** (pod-o-din'-e-ah) [πούς, foot; ὀδύνη, pain], Pain in the foot, especially a neuralgic pain in the heel unattended by swelling or redness.
**podology** (pod-ol'-o-je) [πούς, foot; λόγος, science]. The anatomy and physiology, etc., of the foot.
**podometer.** See pedometer.
**podophyllin** (pod-o-fil'-in). See podophyllum.
**podophylloresin** (pod-o-fil-o-rez'-in). See podophyllum.
**podophyllotoxin** (pod-o-fil-o-toks'-in). See podophyllum.
**podophyllum** (pod-o-fil'-um) [πούς, foot; φύλλον, leaf]. The dried rhizome of P. peltatum, the Mayapple or mandrake, of the family Berberidaceæ. Its resin, commonly called podophyllin, contains podophyllotoxin (C₂₂H₂₄O₉+2H₂O), which upon treatment with ammonia yields podophyllic acid and picropodophyllin. It is used in medicine as a laxative in chronic constipation and as a cathartic in hepatic congestion and bilious fever. Dose of podophyllotoxin ½–½ gr. (0.016–0.03 Gm.). **p.,** fluidextract of (fluidextractum podophylli, U. S. P.). Dose 2–20 min. (0.13–1.3 Cc.). **p., pills of, belladonna, and capsicum** (pilulæ podophylli, belladonna et capsici, U. S. P.) Dose 1 pill. **p., resin of** (resina podophylli, U. S. P.). Dose ½–½ gr. (0.008–0.03 Gm.). **p., tincture of** (tinc-

measurement of the force of respiration. 2. The treatment of pulmonary and circulatory diseases by means of a pneumatic apparatus.
**pneumomycosis** (*nū-mo-mi-ko'-sis*). A disease of the lungs due to fungi.
**pneumon-** (*nū'-mon*) [πνεύμων, lung]. A prefix denoting connection with or relation to the lungs.
**pneumonalgia** (*nū-mo-nal'-je-ah*) [*pneumon-*; ἄλγος, pain]. Pain in the lung.
**pneumonatelectasis** (*nū-mon-at-el-ek'-tas-is*) [*pneumon-*; *atelectasis*]. Atelectasis of the lung.
**pneumonectasia, pneumonectasis** (*nū-mon-ek-ta'-ze-ah, nū-mon-ek'-tas-is*) [*pneumono-*; ἔκτασις, distention]. Emphysema of the lung.
**pneumonectomy** (*nū-mon-ek'-to-me*) [*pneumono-*; ἐκτομή, excision]. Excision of a portion of a lung.
**pneumonedema** (*nū-mon-e-de'-mah*) [*pneumon-*; *edema*]. Edema of the lungs.
**pneumonemia, pneumonæmia** (*nū-mon-e'-me-ah*) [*pneumon-*; αἷμα, blood]. Congestion of the lungs.
**pneumonemphraxis** (*nū-mon-em-fraks'-is*) [*pneumon-*; ἔμφραξις, obstruction]. Obstruction of the lungs or the bronchi.
**pneumonemphysema** (*nū-mon-em-fiz-e'-mah*) [*pneumo-*; ἐμφύσημα, inflation]. Emphysema of the lungs.
**pneumonia** (*nū-mo'-ne-ah*) [πνεύμων, lung]. Inflammation of the lung; pneumonitis. Used without qualification, the term implies lobar pneumonia (*q. v.*). p., **abortive**, acute congestion not followed by other stages. p., **acute**, lobar pneumonia, most often due to a specific microorganism. p., **alcoholic**, the croupous pneumonia of drunkards, often associated with delirium, and very fatal. p., **apex-**, p., **apical**, croupous pneumonia of the apex of a lung. p., **aspiration-**, a bronchopneumonia due to the inspiration of food-particles or other irritant substances into the lung. p., **bronchial**, p., **catarrhal**. Synonym of *bronchopneumonia*. p., **central**, a croupous pneumonia beginning in the interior of a lobe of the lung. The physical signs are obscure until the inflammation reaches the surface. p., **cerebral**, a form associated with marked cerebral symptoms. It is most common in children, and in the beginning resembles meningitis. p., **cheesy**. See *p., desquamative*. p., **chronic**. See *p. interstitial*. p., **contusion**, that following contusion of the chest. p., **croupous**. See *p., lobar*. p., **deglutition-**. Synonym of *p., aspiration-*. p., **desquamative**, a form characterized chiefly by an intense desquamation of the cells lining the air-vesicles, a proliferation of the connective-tissue cells of the septa between the vesicles, and the exudation of a scanty albuminous fluid. The exudate generally undergoes caseous degeneration. p., **disseminated**, bronchopneumonia. p., **double**, lobar pneumonia of both lungs. p., **embolic**, pneumonia due to embolism of the vessels of the lung. p., **ephemeral**, congestion of the lungs. p., **fibrinous**. See *p., lobar*. p., **fibroid**, p., **fibrous**. Synonym of *p., interstitial*. p., **gangrenous**, gangrene of the lung. p., **hypostatic**, a lobular pneumonia occurring in the dependent portions of the lungs of persons debilitated by age or disease, and depending on the weakened circulation and respiration and the dorsal decubitus. p., **indurative**. See *p., desquamative*. p., **insular**. Synonym of *bronchopneumonia*. p., **interstitial**, a chronic inflammation of the lung characterized by an increase of the connective tissue. Syn., *cirrhosis of the lung; fibroid pneumonia*. p., **larval**, that presenting only initial symptoms. p., **lobar**, an acute infectious disease characterized by an inflammation of one or more lobes of the lung, the affected parts becoming consolidated, owing to the exudation of cells and fibrin into the air-vesicles. The exciting cause is usually *Diplococcus pneumoniæ* of Fränkel, but other microorganisms may produce it. Syn., *croupous pneumonia; lung-fever*. p., **lobular**. Synonym of *bronchopneumonia*. p., **massive**, lobar pneumonia in which not only the air-cells, but the bronchi of an entire lobe, or even of a lung, are filled with the fibrinous exudate. p. **migrans**, p., **migratory**, a form involving one lobe after another. p., **pleuritic**, pleuropneumonia. p., **pleurogenic**, p., **pleurogenous**, pneumonia secondary to disease of the pleura. p., **purulent**, one characterized by the formation of pus; it appears under three forms, suppuration of the minute bronchi and air-vesicles—purulent catarrh; true abscess of the lung; suppurative lymphangitis and perilymphangitis. p., **septic**, lobular pneumonia due to the inspiration of septic material or to septic emboli. p., **superficial**, that restricted to parts near the pleura. p., **syphilitic**, inflammation of the lung due to syphilis and manifesting itself as the white pneumonia of the fetus; as gumma of the lung; as interstitial pneumonia, taking its origin at the root of the lung and passing along the bronchi and vessels; and as acute syphilitic phthisis, analogous to acute pneumonic phthisis. p., **tubular**. Synonym of *bronchopneumonia*. p., **typhoid**, that attended with typhoid symptoms. p. **vera**, lobar pneumonia not complicated with other diseases or forms. p., **vesicular**, bronchopneumonia. p., **wandering**, that which affects different parts of the lung in succession and seems to be associated with erysipelas. p., **white**, a catarrhal form of pneumonia occurring in a syphilitic fetus and resulting in death. By an overgrowth of epithelium in the air-vesicles the cells die, and fatty degeneration follows, giving the lungs a white appearance, with the imprint of the ribs on their surface.
**pneumonic** (*nū-mon'-ik*) [see *pneumonia*]. Pertaining to the lungs or to pneumonia. p. **phthisis**, tuberculosis affecting a whole lobe of the lung.
**pneumonitis** (*nū-mon-i'-tis*). Pneumonia.
**pneumono-** (*nū-mon-o-*) [πνεύμων, lung]. A prefix denoting pertaining to the lungs.
**pneumonocace** (*nu-mon-ok'-as-e*) [*pneumono-*; κακός, evil]. Gangrene of the lung.
**pneumonocele** (*nū'-mon-o-sēl*) [*pneumono-*; κήλη, hernia]. Hernia of the lung.
**pneumonocentesis** (*nū-mon-o-sen-te'-sis*). Same as *pneumocentesis*.
**pneumonocirrhosis** (*nū-mon-o-sir-o'-sis*) [*pneumono-*; *cirrhosis*]. Cirrhosis of the lung, interstitial pneumonia.
**pneumonodynia** (*nū-mon-o-din'-e-ah*) [*pneumono-*; ὀδύνη, pain]. Pain referred to the lungs.
**pneumonokoniosis** (*nū-mon-o-kon-e-o'-sis*) [*pneumono-*; κόνις, dust]. A general term applied to chronic induration or fibrous inflammation of the lungs due to the inhalation of dust. Various names are given to it according to the kind of dust causing the inflammation: *anthracosis*, that due to the inhalation of coal-dust; *siderosis*, that due to inhalation of metallic dust; *chalicosis*, that due to the inhalation of mineral dust.
**pneumonolithiasis** (*nū-mon-o-lith-i'-as-is*) [*pneumono-*; λίθος, stone]. The formation of pneumoliths.
**pneumonomelanosis** (*nū-mon-o-mel-an-o'-sis*) [*pneumono-*; μέλας, black; νόσος, disease]. Anthracosis of the lung.
**pneumonometer** (*nū-mon-om'-et-er*). Synonym of *spirometer*.
**pneumonomycosis** (*nū-mon-o-mi-ko'-sis*). See *pneumomycosis*.
**pneumonoparalysis** (*nū-mon-o-par-al'-is-is*) [*pneumono-*; *paralysis*]. Paralysis of the lung.
**pneumonopathy** (*nū-mon-op'-ath-e*) [*pneumono-*; πάθος, disease]. Any disease of the lung.
**pneumonopexy** (*nū-mon'-o-peks-e*) [*pneumono-*; πῆξις, a fixing]. Fixation of a stump of lung tissue to the thoracic wall in connection with pneumonectomy for gangrene, hernia, or other pulmonary lesion.
**pneumonophlebitis** (*nū-mon-o-fleb-i'-tis*) [*pneumono-*; φλέψ, vein; ιτις, inflammation]. Inflammation of the pulmonary veins.
**pneumonophthisis** (*nū-mon-off'-this-is*) [*pneumono-*; φθίσις, a wasting]. A destructive process in the lungs.
**pneumonopleuritis** (*nū-mon-o-ploo-ri'-tis*). Synonym of *pleuropneumonia*.
**pneumonorrhagia** (*nū-mon-or-a'-je-ah*) [*pneumono-*; ῥηγνύναι, to burst forth]. Hemorrhage from the lungs.
**pneumonorrhaphy** (*nū-mon-o'-af-e*) [*pneumono-*; ῥαφή, a seam]. Suture of lacerations of the lung.
**pneumonosepsis** (*nū-mon-o-sep'-sis*) [*pneumono-*; σῆψις, putrefaction]. Septic inflammation of the lung.
**pneumonosis** (*nū-mon-o'-sis*) [*pneumo-*; νόσος, disease]. Any affection of the lungs.
**pneumonostenosis** (*nū-mon-o-sten-o'-sis*). Contraction of a lung.
**pneumonotomy** (*nū-mon-ot'-o-me*) [*pneumo-*; τομή, a cutting]. Surgical incision of the lung.
**pneumopaludism** (*nū-mo-pal'-ū-dism*) [*pneumo-*; *paludism*]. A manifestation of malaria characterized by the impairment of the percussion resonance at one apex, bronchial respiratory murmurs, broncho-

phony, without rales, friction, or expectoration; cough occurs in paroxysms.

**pneumoparesis** (*nū-mo-par'-es-is*) [*pneumo-*; *paresis*]. Progressive congestion of the lungs apparently depending on vasomotor deficiency or other fault of innervation; simple respiratory failure.

**pneumopericarditis** (*nū-mo-per-ik-ar-di'-tis*) [*pneumo-*; *pericarditis*]. Pericarditis with the formation of gas in the pericardial sac.

**pneumopericardium** (*nū-mo-per-e-kar'-de-um*). The presence of air in the pericardial sac. It is due to traumatism or to communication between the pericardium and the esophagus, stomach, or lungs, and is marked by tympany over the precordial region and peculiar metallic heart-sounds.

**pneumoperitoneum** (*nū-mo-per-it-on-e'-um*) [*pneumo-*; *peritoneum*]. The presence of gas in the peritoneal cavity.

**pneumoperitonitis** (*nū-mo-per-it-on-i'-tis*) [*pneumo-*; *peritonitis*]. Peritonitis with the presence of gas in the peritoneal cavity.

**pneumopexy** (*nū'-mo-peks-e*). Same as *pneumonopexy*.

**pneumophthisis** (*nū-moff'-this-is*) [*pneumo-*; φθίσις, wasting]. A destructive process in the lung.

**pneumophyma** (*nū-mo-fi'-mah*) [*pneumo-*; φῦμα, growth; *pl.*, *pneumophymata*]. A tubercle of the lung.

**pneumophymia** (*nū-mo-fi'-me-ah*) [*pneumo-*; φῦμα, growth]. Tuberculosis of the lung.

**pneumopleuritis** (*nū-mo-ploo-ri'-tis*) [*pneumo-*; *pleura*; ιτις, inflammation]. Conjoined inflammation of the lungs and pleura.

**pneumoprotein** (*nū-mo-pro'-te-in*). A protein elaborated by pneumococci.

**pneumoptysis** (*nū-mop'-tis-is*). Same as *hemoptysis*.

**pneumopyopericardium** (*nū-mo-pi-o-per-e-kar'-de-um*) [*pneumo-*; πύον, pus; *pericardium*]. The presence of air or gas and pus in the pericardial sac.

**pneumopyothorax** (*nū-mo-pi-o-tho'-raks*) [*pneumo-*; πύον, pus; *thorax*]. The presence of air and pus in the pleural cavity.

**pneumopyra** (*nū-mo-pi'-rah*) [*pneumo-*; πῦρ, fire]. Malignant bronchitis.

**pneumorrhachis.** See *pneumatorrhachis*.

**pneumorrhagia** (*nū-mor-a'-je-ah*). See *hemoptysis*.

**pneumosan** (*nū'-mo-san*). Amyl-thio-trimethylamine chloride; used for intramuscular injection in pulmonary tuberculosis.

**pneumoserothorax** (*nū'-mo-se-ro-tho'-raks*) [*pneumo-*; *serum*; θώραξ, chest]. The presence of air or gas and serum in the pleural cavity.

**pneumotherapy** (*nū-mo-ther'-ap-e*). 1. The treatment of diseases of the lung. 2. See *pneumatotherapy*.

**pneumothermomassage** (*nū-mo-ther-mo-mas-ahzj'*) [*pneumo-*; θέρμη, heat; *massage*]. The application to the body of currents of air of varying degrees of pressure and temperature.

**pneumothorax** (*nū-mo-tho'-raks*). The presence of air or gas in the pleural cavity. It is produced by perforating wounds of the chest, by the rupture of an abscess or tuberculous cavity of the lung, by the rupture of an emphysematous vesicle; or the evacuation of an empyema into the lung or through the chest-wall. It is marked by dyspnea, shock, pain, a tympanitic (sometimes a dull) percussion-note over the affected side, displacement of the heart, bell-tympany, and diminished respiratory murmur.

**pneumotomy** (*nū-mot'-o-me*) [*pneumo-*; τομή, a cutting]. Surgical incision of the lung. Synonym of *pneumonotomy*.

**pneumotoxin** (*nū-mo-toks'-in*) [*pneumo-*; τοξικόν, a poison]. A toxin produced by the pneumococcus, and believed to be the cause of many of the symptoms of lobar pneumonia. *Antipneumotoxin* is the name given to the antitoxin supposed to exist in the blood of persons convalescent from lobar pneumonia.

**pneumotyphoid** (*nū-mo-ti'-foid*). Synonym of *pneumotyphus*.

**pneumotyphus** (*nū-mo-ti'-fus*) [*pneumo-*; *typhus*]. 1. Typhoid fever beginning with pneumonia dependent upon the typhoid bacillus. 2. Pneumonia occurring in the course of typhoid fever.

**pneumouria** (*nū-mo-ū'-re-ah*). See *pnuematuria*.

**pneusimeter** (*nū-sim'-et-er*) [πνεῦσις, a breathing; μέτρον, a measure]. An apparatus used as a spirometer to measure the vital capacity of the chest in respiration.

**pneusis** (*nū'-sis*) [πνεῖν, to breathe]. Respiration. **p. pertussis.** Synonym of *whooping-cough*.

**pnigma** (*nig'-mah*) [πνίγειν, to choke]. Strangulation.

**pnigophobia** (*ni-go-fo'-be-ah*) [πνίγειν, to choke; φόβος, fear]. The fear of choking that sometimes accompanies angina pectoris.

**pnigos, pnix, pnixis** (*ni'-gos, niks, niks'-is*). Synonyms of *pnigma*.

**pock** (*pok*) [AS., *poc*, a pustule]. A pustule of an eruptive fever, especially of smallpox. **p.-marked,** marked with the cicatrices of the smallpox pustule.

**pocked** (*pokt*). Pitted; marked with pustules.

**pocket** (*pok'-et*). In anatomy, a blind sac, or sac-shaped cavity. A diverticulum communicating with a cavity.

**pocketing** (*pok'-et-ing*). A mode of treating the pedicle in the operation of ovariotomy. It is accomplished by bringing the extremity of the pedicle between the inner lips of the incision, at its lower angle, thus securing its attachment to the raw surface of the abdominal wall.

**pocky** (*pok'-e*). Having pocks or pustules; infected with variola or syphilis.

**poculent** (*pok'-ū-lent*) [*poculum*, a goblet]. Drinkable; potable.

**poculum** (*pok'-ū-lum*) [*potare*, to drink]. 1. A drinking cup. 2. A draught or potion. **p. Diogenis** (*di oj'-en-is*), the palm of the hand when held so as to form a cup-like cavity.

**podagra** (*pod-ag'-rah*) [πούς, foot; ἄγρα, seizure]. Gout, especially of the great toe or the joints of the foot.

**podalgia** (*pod-al'-je-ah*) [πούς, foot; ἄλγος, pain]. Pain in the foot.

**podalic** (*pod-al'-ik*) [πούς, foot]. Pertaining to the feet. **p. version,** the operation of changing the position of the fetus in the uterus so as to bring the feet to the outlet.

**podarthral** (*pod-ar'-thral*) [πούς, foot; ἄρθρον, a joint]. Pertaining to the podarthrum.

**podarthritis** (*pod-ar-thri'-tis*) [πούς, foot; ἄρθρον, joint; ιτις, inflammation]. Inflammation of the joints of the feet.

**podarthrocace** (*pod-ar-throk'-as-e*) [πούς, foot; ἄρθρον joint; κακός, evil]. Caries of the articulations of the feet.

**podarthrum** (*pod-ar'-thrum*) [πούς, foot; ἄρθρον, a joint; *pl.*, *podarthra*]. In biology, the foot-joint or metatarso-phalangeal articulation.

**podedema** (*pod-e-de'-mah*) [πούς, foot; *edema*]. Edema of the foot.

**podelcoma** (*pod-el-ko'-mah*). See *Madura foot.*

**podencephalus** (*pod-en-sef'-al-us*) [πούς, foot; ἐγκέφαλος, brain]. A variety of monster of the species exencephalus, in which there is a protusion of the cranial contents from the top of the head.

**podiatrist** (*pod-e-at'-rist*) [πούς, a foot; ἰατρός, physician]. A specialist in the treatment of diseases of the feet.

**podobromidrosis** (*pod-o-brom-id-ro'-sis*) [πούς, foot; βρῶμος, stench; ἱδρώς, sweat]. Offensive sweating of the feet.

**pododynamometer** (*pod-o-di-nam-om'-et-ur*) [πούς, a foot; *dynamometer*]. An apparatus for testing the strength of the muscles of the feet or legs.

**pododynia** (*pod-o-din'-e-ah*) [πούς, foot; ὀδύνη, pain]. Pain in the foot, especially a neuralgic pain in the heel unattended by swelling or redness.

**podology** (*pod-ol'-o-je*) [πούς, foot; λόγος, science]. The anatomy and physiology, etc., of the foot.

**podometer.** See *pedometer.*

**podophyllin** (*pod-o-fil'-in*). See *podophyllum*.

**podophylloresin** (*pod-o-fil-o-rez'-in*). See *podophyllum, resin of.*

**podophyllotoxin** (*pod-o-fil-o-toks'-in*). See *podophyllum.*

**podophyllum** (*pod-o-fil'-um*) [πούς, foot; φύλλον, leaf]. The dried rhizome of *P. peltatum*, the May-apple or mandrake, of the family *Berberidaceæ*. Its resin, commonly called *podophyllin*, contains *podophyllotoxin* ($C_{22}H_{24}O_9+2H_2O$), which upon treatment with ammonia yields podophyllic acid and picropodophyllin. It is used in medicine as a laxative in chronic constipation and as a cathartic in hepatic congestion and bilious fever. Dose of *podophyllotoxin* $\frac{1}{6}$–$\frac{1}{2}$ gr. (0.016–0.03 Gm.). **p., fluidextract of** (*fluidextractum podophylli*, U. S. P.). Dose 2–20 min. (0.13–1.3 Cc.). **p., pills of, belladonna, and capsicum** (*pilulæ podophylli, belladonnæ et capsici*, U. S. P.) Dose 1 pill. **p., resin of** (*resina podophylli*, U. S. P.). Dose $\frac{1}{8}$–$\frac{1}{2}$ gr. (0.008–0.03 Gm.). **p., tincture of** (*tinc-*

*tura podophylli*, B. P.). Dose 15 min.–1 dr. (1–4 Cc.).
**podotrochilitis** (*pod-o-trok-il-i'-tis*) [ποῦς, foot; τροχιλία, pulley; ιτις, inflammation]. Navicular disease; an inflammatory disease of the fore-foot in the horse, involving the synovial sheath between the sesamoid or navicular bone of the third phalanx and the flexor perforans muscle over it.
**pœ-.** For words commencing thus, see *pe-*.
**Poehl's test** (*pĕl*) [Alexander Vasilyevich von *Poehl*, Russian chemist, 1850–]. For *products of bacillus choleræ*: the addition of concentrated sulphuric acid to a culture of cholera bacilli produces a rose color deepening into purple.
**pogoniasis** (*po-go-ni'-as-is*) [πώγων, beard]. Excessive growth of the beard; growth of beard in a woman.
**pogonion** (*po-go'-ne-on*) [πώγων, beard]. The most anterior point of the chin on the symphysis of the mandible.
**pogonium** (*po-go'-ne-um*) [πώγων, beard]. 1. A small beard. 2. Same as *pogonion*.
**Pohl's test for globulins.** Saturate the solution to one-half with ammonium sulphate, which precipitates the globulins. After several hours filter, and add to the filtrate a saturated solution of ammonium sulphate.
**-poietic** (*poi-et'-ik*) [ποίησις, a making]. A termination denoting making or producing, as in *hematopoietic*.
**poikiloblast** (*poi'-kil-o-blast*). A nucleated red corpuscle of irregular shape and size.
**poikilocyte** (*poi'-kil-o-sīt*) [ποικίλος, varied; κύτος, a cell]. A large red blood-corpuscle of irregular shape.
**poikilocythemia** (*poi-kil-o-si-the'-me-ah*) [ποικίλος, varied; κύτος, cell; αἷμα, blood]. The presence of poikilocytes in the blood.
**poikilocytosis** (*poi-kil-o-si-to'-sis*) [*poikilocyte*]. A condition of the blood characterized by the presence of poikilocytes; variation in the shape of the red blood corpuscles.
**poikilonymy** (*poi-kil-on'-im-e*). See *pecilonymy*.
**poikilothermal** (*poi-kil-o-ther'-mal*). Synonym of *poikilothermic*.
**poikilothermic** (*poi-kil-o-ther'-mik*) [ποικίλος, varied; θέρμη, heat]. Varying in temperature according to the surroundings; cold-blooded.
**point** [*punctum*, point, from *pungere*, to prick]. 1. The sharp end of an object, especially one used to pierce anything. 2. The limit at which anything occurs, as the melting-*point*, freezing-*point*. 3. A mark made by a sharp object; a minute spot or area; of an abscess, to come to the surface. p., **boiling**, the degree of temperature at which a liquid passes into the vaporous state with ebullition. p., **cardinal**, one of the six optical points that determine the direction of the rays entering or emerging from a series of refracting media. p., **craniometric**. See *craniometric point*. p., **critical**, *of gases*, a temperature at or above which a gas cannot be liquefied by pressure alone; *of liquids*, that temperature at which a liquid, regardless of the pressure to which it is subjected, assumes a gaseous form. p., **dew-**, the temperature at which the atmospheric moisture is deposited as dew. p., **disparate**, one of those points on the retina whence images are projected, not to the same, but to different points in space. p. of **election**, in surgery, that point at which a certain operation is done by preference. p., **far-**, the remotest point of distinct vision. p., **freezing**, the degree of temperature at which a liquid becomes solid. p., **hysteroepileptogenous**, p., **hysterogenous**. See *zone*, *hysterogenous*. p.'s, **lacrimal**, minute orifices of the lacrimal canals upon the eyelids near the inner canthus. p., **malar**, the most prominent point on the outer surface of the malar bone. p., **McBurney's**. See *McBurney's point*. p., **melting**, the degree of temperature at which fusible solids begin to melt. p., **motor**. See *motor*. p., **near-**, the nearest point at which the eyes can accommodate to see distinctly. p., **nodal**, the center of curvature of a spherical lens or refracting surface, through which rays of light pass joining conjugate points. p., **principal**, one of the two points in the optical axis of a lens that are so related that lines drawn from these points to the corresponding points in the object and its image are parallel. p. of **reflection**, the point from which a ray of light is reflected. p. of **refraction**, the point at which a ray of light is refracted. p., **spinous**, a sensitive point over a spinous process. p., **subnasal**, the middle point of the lower border of the nasal orifice. p., **supraclavicular**, the point, stimulation of which causes contraction of the arm muscles. p.s, **Valleix's**. See *Valleix's points*. p., **vital**, a spot in the oblongata corresponding to the seat of the respiratory center, puncture of which causes instant death.
**pointillage** (*pwan'-te-yahzj*) [Fr.]. Massage by means of the finger-tips.
**pointing** (*point'-ing*). The coming to a point. p. of an **abscess**, the process by which pus from the deeper structures reaches the surface.
**points douloureux** (*pwan*(g)*-doo-loo-roo*). See *Valleix's points*.
**Poirier's line.** (*pwar-e-a'*) [Paul *Poirier*, French surgeon, 1853–1907]. The nasolambdoid line used in craniocerebral topography. It begins at the nasofrontal groove, and extends outward around the base of the skull, passing 0.5 cm. above the external auditory meatus to a point 1 cm. above the lambdoid suture, or to a point 7 cm. above the inion if the suture cannot be felt. This line passes over Broca's convolution, 4 to 6 cm. of the posterior limb of the Sylvian fissure, the lower border of the supramarginal gyrus, the base of the angular gyrus, and terminates at the occipital fissure.
**Poiseuille's law** (*pwah-zoo-e'*) [Jean Léonard Marie *Poiseuille*, French physiologist, 1799–1869]. The rapidity of the current in capillary tubes is proportional to the square of their diameter. P.'s **layer**, or space the "inert" layer of the blood-current of the capillaries, in which the leukocytes roll along slowly while the red corpuscles move more rapidly in the axial stream.
**poĭs₀n** (*poi'-zn*) [Fr., from *potio*, a draught]. A substance that, being in solution in the blood or acting chemically on the blood, either destroys life or impairs seriously the functions of one or more of its organs. (See the table under this head.) p., **acrid**. See *p., irritant*. p., **acronarcotic**, one that is irritating to the part to which it is applied, but acts on the brain or myelon or both. p.s, **cellular**, cytolysins. p., **irritant**, one that causes irritation at the point of entrance or at the point of elimination. p., **muscle-**. 1. A substance that impairs or destroys the proper functions of muscles. 2. A poisonous albumin developed during muscular activity. p., **narcotic**, one affecting the cerebral centers, producing stupor. p., **ordeal-**, any one of the vegetable poisons, such as physostigma, used by savages in the trial of accused persons to determine their guilt or innocence. TABLE OF POISONS (*the antidotes are in italics*): **acid, carbolic**. See *phenol* in this table. **acid, chromic.** See *chromium trioxide* in this table. **acid, hydrochloric** (*muriatic*), *symptoms*, pain throughout digestive tract, vomiting, feeble pulse, clammy skin, collapse, eschars externally, yellow stains on clothing, but none on skin; *treatment, alkalies*; demulcent drinks; oil; stimulants (intravenous injection). **acid, hydrocyanic** (*prussic*), *symptoms*, sudden unconsciousness, slow, labored respirations, slow pulse, staring eyes, purple face, general convulsions, then relaxation and collapse, odor of peach-kernels; death may be almost instantaneous; *treatment*, stomach-tube if possible; dilute ammonia; alternate cold and warm effusions; atropine and cardiac stimulants; artificial respiration. **acid, nitric**, *symptoms*, yellow stains on skin; otherwise similar to *acid, sulphuric*; *treatment, alkalies; soap*; demulcents; stimulants. **acid, oxalic**, *symptoms*; hot, acrid taste, burning, vomiting, collapse, sometimes general paralysis, numbness, and stupor; *treatment, lime* or *chalk*. **acid, salicylic**, *symptoms*; mydriasis, quick and deep respiration, delirium, dyspnea, lessened arterial pressure, deafness, olive-green urine. **acid, sulphuric**, *symptoms*, black stains, pain throughout digestive tract, vomiting, often of tarry matter, feeble pulse, clammy skin, collapse and bloody salivation; *treatment, chalk; magnesia; soap;* demulcent drinks, **aconitum napellus** (*monkshood*), **aconite**, *symptoms*, sudden collapse, slow, feeble, and irregular pulse and respirations, tingling in the mouth and extremities, giddiness, great muscular weakness, sometimes pain in the abdomen, pupils generally dilated, but may be contracted, marked anesthesia of skin, mind clear, convulsions at times; *treatment, tannic acid* solution for washing out stomach; *digitalis*, atropine and stimulants; artificial respiration; warmth and friction; absolute quiet in recumbent position. **alcohol**, *symptoms*, confusion of thought, giddiness, tottering gait, slight cyanosis, narcosis from which patient can be aroused; full pulse; deep, stertorous

breathing; injection of eyes, dilatation of pupils, low temperature; convulsions may occur; *treatment*, evacuate stomach; coffee; battery; amyl nitrite; hot and cold douches. **ammonium and its compounds**, *symptoms*, intense gastroenteritis, often with bloody vomiting and purging, lips and tongue swollen and covered with detached epithelium, violent dyspnea, characteristic odor; *treatment, vegetable acids;* demulcents. **antimony and its compounds**, *symptoms*, metallic taste, violent vomiting, becoming bloody, feeble pulse, pain and burning in the stomach, violent serous purging, becoming bloody, cramps in extremities, thirst, great debility, sometimes prostration, collapse, unconsciousness, and convulsions without vomiting or purging; *treatment, tannic acid;* demulcent drinks; opium; alcohol; external heat. **antipyrine** (*phenazon*), *symptoms*, headache, nausea, vomiting, a rash like that of measles, vertigo, drowsiness, deafness, confusion of ideas, cyanosis, collapse; *treatment*, recumbent position; warmth; strychnine; stimulants; oxygen; artificial respiration. **apomorphine**, *symptoms*, violent vomiting, paralysis of motor and sensory nerves, delirium, depression of respiration and of heart; *treatment*, cardiac and respiratory stimulants. **arsenic and its compounds**, *symptoms*, violent burning pain in the stomach, retching, thirst, purging of blood and mucus with flakes of epithelium, tenesmus, suppression of urine; sense of constriction in throat; pulse small and frequent; *treatment, hydrated iron sesquioxide; precipitated iron carbonate;* emetics; castor-oil; demulcents. **atropa belladonna** (*deadly nightshade*) (**atropine belladonna, homatropine**), *symptoms*, heat and dryness of the mouth and throat, pupils widely dilated, scarlet rash, noisy delirium, quick pulse, at first corded, later feeble; rapid respirations, early strong, late shallow and feeble; retention of urine; sometimes convulsions, collapse, and paralysis; *treatment*, evacuate stomach; *tannic acid;* stimulants; coffee; pilocarpine; artificial respiration; physostigmine may be of benefit; evacuation of bladder. **caffeine**, *symptoms*, burning pain in the throat, giddiness, faintness, nausea, numbness, abdominal pain, great thirst, dry tongue, tremor of extremities, diuresis, weak pulse, cold skin, collapse; *treatment*, emetics; stimulants; warmth; morphine and atropine. **calabar-bean**. See *physostigma venenosum* in this table. **camphor**, *symptoms*, characteristic odor, languor, giddiness, disturbance of vision, delirium, convulsions, clammy skin, smarting in the urinary organs, pulse quick and weak, no pain, no vomiting, no purging; *treatment*, evacuate stomach; stimulants; warmth; hot and cold douches. **cannabis indica** (*Indian hemp*),'*symptoms*, pleasurable intoxication, sense of prolongation of time, anesthesia with loss of strength, especially in legs, pupils dilated, rapid pulse, heavy sleep; *treatment*, evacuate stomach; stimulants. **cantharis vesicatoria** (*Spanish fly*) (**cantharides**), *symptoms*, burning in mouth and stomach, vomiting and purging, soon becoming bloody, tenesmus, salivation, aching pains in back, strangury, priapism, unconsciousness only very late; convulsions at times; *treatment*, evacuate stomach; demulcent drinks; morphine; hot bath for the strangury; anesthetics may be necessary for the pain. **carbolic acid**. See *phenol* in this table. **chloral hydrate**, *symptoms*, deep sleep, loss of muscular power, lividity, reflexes diminished, pulse weak, respirations slowed, pupils contracted during sleep, but dilated on waking, temperature low; *treatment*, evacuate stomach; heat to the extremities; massage; coffee by the rectum; strychnine; amyl nitrite; artificial respiration. **chloroform**, *symptoms*, excitement and intoxication followed by anesthesia and unconsciousness, later profound narcosis; pain and respirations fail progressively or suddenly; *treatment*, draw tongue forward; artificial respiration; faradic current; hot and cold douches; amyl nitrite; ammonia injected into a vein; evacuation of the stomach if chloroform has been taken by mouth. **chromium trioxide**, *symptoms*, dark-yellow stains, abdominal pain, vomiting and purging, collapse; *treatment*, evacuate stomach; *chalk,* milk, or albumin; demulcent drinks. **coal-gas**, *symptoms*, headache, giddiness, loss of muscular power, unconsciousness, pupils dilated, breathing labored, coma, odor of the gas; *treatment*, fresh air; artificial respiration; ammonia; stimulants; oxygen; coffee; hot and cold douches. **cocaine**, *symptoms*, faintness, giddiness, nausea, pulse small, rapid, intermittent, dilated pupils, severe prostration, respiration slow and feeble; *treatment*, stimulants; amyl nitrite; artificial respiration. **colchicum autumnale** (*meadow-saffron*), *symptoms*, not unlike those of malignant cholera, griping pain in the stomach, vomiting and continuous purging of seromucous material, intense thirst; muscular cramps, great prostration, collapse, dilated pupils, pain in the extremities; *treatment*, evacuate stomach; *tannic* or *gallic acid;* demulcent drinks; stimulants; morphine. **conium maculatum** (*hemlock*), *symptoms*, weakness of the legs, gradual loss of all voluntary power, nausea, ptosis, dilatation of pupils; inability to speak or swallow; *treatment*, evacuate stomach; *tannic* or *gallic acid;* stimulants; warmth; artificial respiration; atropine. **croton tiglium** (*croton oil*), *symptoms*, intense pain in abdomen, vomiting, purging, watery stools, pinched face, small and thready pulse, moist skin, collapse; *treatment*, evacuate stomach; demulcent drinks; camphor; stimulants; morphine; poultices to abdomen. **cyanogen and its compounds**. Similar to *acid, hydrocyanic*, which see in this table. **datura stramonium** (*thorn-apple; Jamestown weed*), symptoms and treatment similar to those of *atropine*, which see under *atropa belladonna* in this table. **digitalis purpurea** (*Foxglove*), purging, with severe pain, violent vomiting, vertigo, feeble pulse, although heart's action is tumultuous,eyes prominent, pupilsdilated,sclera blue; delirium and .convulsions; *treatment*, evacuate stomach; *tannic* and *gallic acids;* stimulants; aconite; recumbent position. **erythroxylon coca**. See *cocaine* in this table. **fly, Spanish**. See *cantharis* in this table. **gelsemium sempervirens** (*yellow jasmine*), *symptoms* appear in about twenty minutes; great muscular weakness, diplopia, ptosis, internal squint, widely dilated pupils, dimness of vision, labored respiration, weak pulse; *treatment*, evacuate stomach; atropine; stimulants; artificial respiration; hot and cold douches. **hellebore, green and white**. See *veratrum* in this table. **hemlock**. See *conium* in this table. **iodine and its compounds**, *symptoms*, pain in throat and stomach, vomiting, purging, vomit yellow from iodine, or blue if starch is present in stomach; giddiness, faintness, convulsive movements; *treatment*, evacuate stomach; *starch;* amyl nitrite; morphine. **iodoform**, *symptoms*, slight delirium, drowsiness, high temperature, rapid pulse; symptoms resemble meningitis. **jaborandi** (*pilocarpine*), *symptoms*, copious sweating, dizziness, salivation, vomiting, diarrhea, tearing pain in eyeballs, myopia, pupils much contracted; *treatment*, evacuate stomach; stimulants; atropine. **lead acetate**, *symptoms*, sweet metallic taste, vomiting of white matter, great thirst, pain in abdomen, abdominal muscles usually rigid, constipation or diarrhea with black stools, cramps in the legs, paralysis of the extremities, convulsions; in the chronic forms, a blue line at margin of the gums; *treatment*, evacuate stomach; *dilute sulphuric acid; Epsom* or *Glauber's salts; milk;* morphine; potassium iodide to eliminate the poison. **lobelia inflata** (*Indian tobacco*), *symptoms*, severe vomiting, with intense depression and prostration, giddiness, tremors, convulsions, collapse; *treatment*, evacuate stomach; *tannic* or *gallic acid;* stimulants; strychnine; warmth; recumbent position. **mercury bichloride**, *symptoms*, acrid metallic taste, burning heat in throat and stomach, vomiting, diarrhea with bloody stools, lips and tongue white and shriveled, pulse small and frequent, death in coma or convulsions; pain may be absent; secondary *symptoms*, hectic fever, coppery taste, fetid breath, gums swollen, salivation; *treatment*, albumin in some form; raw white of egg or flour; evacuate stomach; opium; potassium iodide. **morphine**. See *opium* in this table. **nitric acid**. See *acid, nitric*, in this table. **nitroglycerin**, *symptoms*, throbbing headache, pulsation over entire body, dicrotic pulse, flushed face, mental confusion, anxiety, sudden collapse; *treatment*, recumbent position; cold to head; ergot; atropine. **nux vomica**. See *St. Ignatius bean* in this table. **opium** (**morphine, narceine, codeine, laudanum**), *symptoms*, preliminary mental excitement, acceleration of heart, soon weariness, sensation of weight in the limbs, sleepiness, diminished sensibility, pin-point pupils, pulse and respiration slow and strong; patient can be roused with difficulty, later this becomes impossible, reflexes abolished, respiration slow, irregular, and stertorous, pulse rapid and feeble; *treatment*: 1. Evacuate stomach with mustard or stomach-tube. 2. Arouse patient to maintain respiration by exercise, flagellation with wet towels, cold and hot douches alternately. 3. Stimulate by atropine, coffee, alcohol, if pulse fails; external heat; inhalations of oxygen; injection of dilute solution of *potassium permanganate*. **oxalic acid**. See *acid, oxalic*, in this table. **Paris-green**. See *arsenic* in this table. **phenol**, *symptoms*, immediate burning pain from mouth to stomach, giddiness, loss of consciousness, collapse, partial suppression of urine, which is smoky in color, characteristic odor, white, corrugated patches in mouth; *treatment*, stomach-pump, *magnesium sulphate* or *sodium sulphate*, atropine. **phosphorus**, *symptoms*, vomiting and pain, vomit may be luminous in the dark, characteristic odor; after several days deep jaundice, coffee-colored vomit, hepatic tenderness, albuminuria, marked fall in temperature, coma, failure of pulse and respiration; *treatment, sulphate of copper* as an emetic, then as an antidote in small doses with opium; purgation. **physostigma venenosum** (*calabar-bean*), *symptoms*, giddiness, prostration, loss of power in the lower limbs, muscular twitching, contracted pupils, mind clear; *treatment*, evacuate stomach; *atropine*; strychnine; stimulants; artificial respiration. **prussic acid**. See *acid, hydrocyanic*, in this table. **santonin**, *symptoms*, disturbance of color-vision—objects first assume a bluish tinge, then yellow; tinnitus, dizziness, pain in the abdomen, failure of respiration, convulsions, stupor; *treatment*, evacuate stomach; stimulants; chloral. **savin**, *symptoms*, pain, vomiting, bloody stools and tenesmus, disordered respirations, coma, convulsions, and collapse; *treatment*; evacuate stomach; castor-oil in large dose; morphine poultices to the abdomen. **silver salts**, *symptoms*, pain, vomiting, and purging; vomit white and cheesy, rapidly turning black in the sunlight; vertigo, coma, convulsions, paralysis, and marked disturbance of respiration; *treatment, salt* and water; evacuate stomach; a large amount of milk. **St. Ignatius bean**, **strychnos ignatii**, **strychnos nux-vomica** (*nux vomica*, strychnine, brucine), *symptoms*, tetanic convulsions in paroxysms at varying intervals of from five minutes to half an hour; opisthotonos during paroxysm; eyeballs prominent, pupils dilated, respiration impeded, pulse feeble and rapid, anxiety; *treatment*, evacuate stomach; *tannic acid* followed by an emetic; catheterize; keep patient quiet; bromides and chloral; amyl nitrite or chloroform to control convulsions; artificial respiration if indicated. **tobacco** (nicotine), *symptoms*, nausea, vomiting, weakness, weak pulse, cold and clammy skin, collapse; pupils contracted; then dilated; *treatment*, evacuate stomach; *tannic acid*; strychnine; stimulants; warmth; recumbent position. **veratrum album** (*white hellebore*), **veratrum viride** (*green hellebore*), *symptoms*, pain and burning in alimentary tract; vomiting and diarrhea, slow, weak pulse, labored respiration, pupils usually dilated; there may be convulsions; *treatment*, evacuate stomach; ether hypodermatically; opium; stimulants; coffee; warmth; recumbent position.

**poison-nut.** Nux vomica.
**poisonous** (*poi'-zn-us*). Having the properties of a poison; venomous.
**poitrinaire** (*pwah-tre-nār'*) [Fr. *poitrine*, chest]. A patient with pulmonary tuberculosis or other chronic disease of the chest.
**poke-root.** See *phytolacca*.
**polar** (*po'-lar*) [*pole*]. Pertaining to or situated near a pole. **p. bodies, p. cells, p. globules,** two minute cells thrown off by the unfecundated ovum during maturation. **p. method,** a method of applying electricity, in which the pole the distinctive effect of which is wanted is placed over the part to be treated and the other pole over some indifferent part. **p. ray's** the astral rays of the mitotic figure. **p. star,** a star of the diaster.
**polarimeter** (*po-lar-im'-et-er*) [*polar; μέτρον*, a measure]. An instrument for determining the degree to which an optically active substance changes the plane of polarization to the right or to the left.
**polarimetry** (*po-lar-im'-et-re*) [*polar; μέτρον*, measure]. The use of the polarimeter.
**polariscope** (*po-lar'-is-kōp*) [*polar; σκοπεῖν*, to view]. An instrument for studying the polarization of light; a polarimeter.
**polaristrobometer** (*po-lar-is-tro-bom'-et-er*) [*polar; στρόβος*, a whirling round; *μέτρον*, measure]. A form of polarimeter or saccharimeter that furnishes a delicate means of fixing the plane of polarization as rotated by the sugar solution under examination.
**polarity** (*po-lar'-it-e*). The state or quality of having poles or points of intensity with mutually opposite qualities. In electro-therapeutics, that condition of a nerve is which the part nearest the negative pole is in a state of increased, and that nearest the positive is in a state of decreased irritability.
**polarization** (*po-lar-i-za'-shun*) [*polarize*]. 1. The act of polarizing or the state of being polarized. 2. A condition produced in a ray of light by absorption, reflection, or refraction, by means of which the vibrations are restricted and take place in one plane only (*plane polarization*) or in curves (*circular or elliptic polarization*). The plane of polarization is altered or rotated when the light is passed through a quartz-crystal or solutions of certain substances (*rotatory polarization*). 3. The deposit of gas-bubbles (hydrogen) on the electronegative plate of a galvanic battery, whereby the flow of the current is impeded, and, owing to the negative plate covered with hydrogen being more electropositive than the zinc plate, the difference in potential between the two plates is reduced.
**polarize** (*po'-lar-īz*) [*polar*]. To endow with polarity; to place in a state of polarization.
**polarizer** (*po'-lar-i-zer*) [*polarize*]. An object, such as Nicol prism, by means of which light is polarized.
**pole** (*pōl*) [*πόλος*, a pole]. 1. Either extremity of the axis of a body, as of the fetus, the crystalline lens, etc. 2. One of two points at which opposite physical qualities, *e. g.,* electricity or magnetism, are concentrated; specifically, the electrode of a galvanic battery, which is positive (*positive pole*) when connected with the electronegative plate of the battery (carbon, copper, platinum), or negative (*negative pole*) when connected with the electropositive plate (zinc). **p.-changer,** a switch or key for changing or reversing the direction of a current produced by an electric battery.
**polemophthalmia** (*pol-em-off-thal'-me-ah*) [*πόλεμος*, war; *ophthalmia*]. Military ophthalmia; the ophthalmia of soldiers.
**polenta** (*po-len'-tah*) [L.]. In Italy, a maize-meal porridge.
**poleozone** (*po'-le-o-zōn*). A bactericide said to be obtained from potassium chlorate by action of sulphuric acid.
**police** (*po-lēs'*) [*πόλις*, a city]. 1. Public order. 2. An organized civil force for maintaining order. **p. sanitary,** the body of officials in the employ of a city, state or nation, whose duty it is to look after the hygienic condition as it affects the public health.
**policeman's disease.** Synonym of *tarsalgia*.
**policlinic** (*pol-ik-lin'-ik*) [*πόλις*, city; *κλίνη*, couch]. A general city hospital. Cf. *polyclinic*.
**poliencephalitis** (*pol-e-en-sef-al-i'-tis*). See *polioencephalitis*.
**polio-** (*pol-e-o*) [*πολιός*, gray]. A prefix meaning gray:
**polioencephalitis** (*pol-e-o-en-sef-al-i'-tis*) [*πολιός*, gray; *ἐγκέφαλος*, brain; *ιτις*, inflammation]. Inflammation of the gray matter of the brain. **p. acuta,** an acute inflammation of the cerebral cortex, which, when occurring in children, gives rise to infantile cerebral palsy. **p., anterior superior,** an inflammatory disease of the gray matter of the third ventricle, of the anterior portion of the brain, and of that about the Sylvian aqueduct. It is characterized by ophthalmoplegia, chiefly external, and a peculiar somnolent state. **p., inferior,** bulbar paralysis.
**poliomyelencephalitis** (*pol-e-o-mi-el-en-sef-al-i'-tis*) [*polio-; μυελός*, marrow; *ἐγκέφαλος*, brain; *ιτις*, inflammation]. Poliomyelitis and poliencephalitis existing together.
**poliomyelitis** (*pol-e-o-mi-el-i'-tis*) [*polio-; myelitis*]. Inflammation of the gray matter of the spinal cord. **p., acute anterior,** infantile paralysis, an acute inflammation of the anterior horns of the gray matter of the spinal cord. It is most common in children, coming on during the period of the first dentition, and producing a paralysis of certain muscle-groups or of an entire limb. The onset is sudden, and the paralysis is usually most extensive in the beginning, a certain amount of improvement taking place subsequently. The affected muscles atrophy rapidly, the reflexes in them are lost, and reaction of degeneration develops. From contraction of antagonistic muscles deformities occur later in life. **p., chronic anterior.** Synonym of *progressive muscular atrophy*.
**poliomyelopathy** (*pol-e-o-mi-el-op'-ath-e*)[*polio-; μυελός*, marrow; *πάθος*, disease]. Disease of the gray matter of the spinal cord and medulla oblongata.
**polioplasm** (*pol'-e-o-plazm*) [*polio-; πλάσμα*, plasm]. Granular protoplasm.

**poliosis** (*pol-e-o'-sis*) [πολιός, gray]. A condition characterized by absence of pigment in the hair.
**poliothrix** (*pol'-e-o-thriks*) [*polio-*; θρίξ, hair]. Synonym of *canities*.
**Polish plait** (*po'-lish*). See *plica polonica*.
**Politzer's bag** (*pol'-its-er*) [Adam *Politzer*, Austrian otologist, 1835– ]. A pear-shaped rubber bag used for inflating the middle ear. The tip is introduced into the nostril, and the bag compressed while the other nostril is closed and the patient performs the act of swallowing. The latter opens the Eustachian tube and allows the air to enter. P.'s luminous cone, a brightly illuminated area in the shape of an isosceles triangle which has its base near the lower circumference and its apex at the umbo of the membrana tympani. P.'s method, inflation of middle ear through the Eustachian tube. P.'s test, in cases of unilateral middle-ear disease, associated with obstruction of the Eustachian tube, the sound of a vibrating tuning fork (C₂) held before the nares during deglutition is perceived by the normal ear only; if the tube is patulous, the sound sensation is frequently stronger in the affected ear. In unilateral disease of the labyrinth the tuning-fork is heard in the normal ear whether or not deglutition occur.
**politzeration** (*pol-its-er-i-za'-shun*). The inflation of the middle ear by means of Politzer's bag.
**poll** (*pōl*). The head, especially the back portion, of an individual or of an animal. p.-evil, in farriery, an abscess behind the ears of a horse, producing a fistula.
**pollakiuria** (*pol-ak-i-ū'-re-ah*) [πολλάκις; ούρον, urine]. Abnormally frequent micturition.
**pollantin** (*pol-an'-tin*) [*pollen*; ἀντί, against]. An antitoxin used in hay-fever; it is an immune serum obtained from horses, and is used in fluid and powder form.
**pollen** (*pol'-en*) [*pollen*, fine dust]. The fecundating element produced in the anthers of flowering plants. p. disease, synonym of *hay fever*, q. v.
**pollex** (*pol'-eks*) [L. gen. *pollicis*. pl., *pollices*]. 1. The thumb. 2. The great toe. p. pedis, the great toe.
**pollicar** (*pol'-ik-ar*) [*pollex*]. Relating to the thumb.
**pollinosis** (*pol-in-o'-sis*). Same as *hay fever*, q. v.
**Pollock's operation** (*pol'-uk*) [G. *Pollock*]. For amputation at the knee-joint: a long anterior and a short posterior skin-flap, somewhat rectangular in outline, the patella being left.
**pollution** (*pol-ū'-shun*) [*polluere*, to defile]. 1. The act of defiling or rendering impure, as *pollution* of drinking-water. 2. The production of the sexual orgasm by means other than sexual intercourse. p., nocturnal, a nocturnal, involuntary, seminal discharge. p., self, masturbation.
**polonica** (*po-lon'-ik-ah*). See *plica*.
**polonium** (*pol-o'-ne-um*) [*Poland*, home of the discoverers]. A radioactive element isolated by M. and Mme. Curie from pitch-blende.
**poltophagy** (*pol-tof'-a-je*) [πολτός, porridge; φάγειν, to eat]. Complete chewing of the food before swallowing it.
**poly-** (*pol-e-*) [πολύς, many]. A prefix denoting many or much.
**polyacid** (*pol-e-as'-id*) [*poly-*; *acid*]. Applied to a base or basic radical capable of saturating several molecules of the acid radical.
**polyacoustic** (*pol-e-ak-oos'-tik*) [*poly-*; *acoustic*]. 1. Multiplying sound. 2. An instrument for intensifying sound.
**polyadenia** (*pol-e-ad-e'-ne-ah*) [*poly-*; ἀδήν, a gland]. Pseudoleukemia or Hodgkin's disease.
**polyadenitis** (*pol-e-ad-en-i'-tis*). Inflammation of many glands at once. p., malignant, bubonic plague.
**polyadenoma** (*pol-e-ad-en-o'-mah*) [*poly-*; ἀδήν, gland; ὄμα, tumor]. Multiple adenoma.
**polyadenopathy** (*pol-e-ad-en-op'-ath-e*) [*poly-*; ἀδήν, gland; πάθος, disease]. Any disease affecting many glands at once.
**polyadenous** (*pol-e-ad'-en-us*) [*poly-*; ἀδήν, gland]. Having or involving many glands.
**polyæmia** (*pol-e-e'-me-ah*). See *polyemia*.
**polyæsthesia** (*pol-e-es-the'-ze-ah*). See *polyesthesia*.
**polyanemia** (*pol-e-an-e'-me-ah*). Excessive anemia.
**polyarthritis** (*pol-e-ar-thri'-tis*). Inflammation of many joints. p. rheumatica acuta. See *rheumatism, acute articular*. p., vertebral, inflammation of the intervertebral discs with out caries of the bones of the vertebræ.
**polyarticular** (*pol-e-ar-tik'-ū-lar*) [*poly-*; *articulus*, joint]. Affecting many joints; the term multiarticular is preferable.
**polyatomic** (*pol-e-at-om'-ik*). 1. Containing several atoms. 2. Having several hydrogen atoms replaceable by bases.
**polyaxon.** (*pol-e-aks'-on*). 1. In biology, having several axes of growth. 2. A neuron having more than two axons.
**polybasic** (*pol-e-ba'-sik*) [*poly-*; *base*]. 1. Of acids, having several hydrogen atoms replaceable by bases. 2. Formed from a polybasic acid by the replacement of more than one hydrogen atom by a base.
**polyblast** (*pol'-e-blast*) [*poly-*; βλαστός, a germ]. A general term designating the various cells seen in newly developing connective tissue.
**polycardia** (*pol-e-kar'-de-ah*). See *tachycardia*.
**polycellular** (*pol-e-sel'-ū-lar*) [*poly-*; *cellula*, a cell]. Having many cells.
**polycentric** (*pol-is-en'-trik*) [*poly-*; κέντρον, center]. Having many centers or nuclear points.
**polyceptor** (*pol-e-sep'-tor*) [*poly-*; *ceptor*]. A sensitizer or "amboceptor" possessing a number of complementophile groups.
**polycholia** (*pol-e-ko'-le-ah*) [*poly-*; χολή, bile]. Excessive secretion of bile.
**polychrest, polychrestus** (*pol'-ik-rest, pol-ik-res'-tus*) [*poly-*; χρηστός, useful]. A medicine regarded as efficacious in many diseases. The word is chiefly used by homeopathic physicians.
**polychrestic** (*pol-ik-res'-tik*). Of the nature of a polychrest.
**polychroism** (*pol-ik-ro'-izm*) [*poly-*; χροιά, color]. A property possessed by certain crystals, of exhibiting two shades of color under polarized light, which vary as the polarizing instrument is rotated.
**polychromasia** (*pol-e-kro-ma'-ze-ah*). Same as *polychromatophilia*.
**polychromatic** (*pol-e-kro-mat'-ik*) [*poly-*; χρῶμα, color]. Many-colored.
**polychromatophil, polychromatophile** (*pol-e-kro-mat'-o-fil*) [*poly-*; χρῶμα, color; φιλεῖν, to love]. 1. An erythrocyte which has lost its affinity for acid stain and which with mixtures of acid and basic dyes is stained atypically by either or both elements. 2. See *polychromatophilic*.
**polychromatophilia** (*pol-e-kro-mat-o-fil'-e-ah*). The presence in the blood of polychromatophils.
**polychromatophilic** (*pol-e-kro-mat-o-fil'-ik*). Susceptible of staining with more than one dye.
**polychromemia, polychromæmia** (*pol-e-kro-me'-me-ah*) [*poly-*; χρῶμα, color; αἷμα, blood]. The increase in coloring-matter in the blood as a sequel of polycythemia.
**polychromia** (*pol-e-kro'-me-ah*) [*poly-*; χρῶμα, color]. Increased or abnormal pigmentation.
**polychylia** (*pol-ik-i'-le-ah*) [*poly-*; χυλός, chyle]. An excessive formation of chyle.
**polychylic** (*pol-ik-i'-lik*) [*poly-*; χυλός, chyle]. Relating to an excess of chyle.
**polyclinic** (*pol-e-klin'-ik*). A hospital in which many diseases are treated.
**polyclonia** (*pol-e-klo'-ne-ah*) [*poly-*; κλόνος, commotion]. An affection said to be distinct from tic and cholera but marked by clonic spasms.
**polycoria** (*pol-e-ko'-re-ah*) [*poly-*; κόρη, pupil]. The existence of more than one pupil in the iris.
**polycrotic** (*pol-e-krot'-ik*) [*poly-*; κρότος, pulse]. Of the pulse, presenting several waves for each cardiac systole.
**polycrotism** (*pol-e-kro'-tizm*). Condition of being polycrotic.
**polycyesia, polycyesis** (*pol-is-i-e'-ze-ah, pol-is-i-e'-sis*) [*poly-*; κύησις, pregnancy]. 1. The occurrence of frequent pregnancy. 2. Multiple pregnancy.
**polycystic** (*pol-e-sis'-tik*) [*poly-*; κύστις, cyst]. Containing many cysts.
**polycythemia, polycythæmia** (*pol-e-si-the'-me-ah*) [*poly-*; κύτος, cell; αἷμα, blood]. A state of the blood characterized by an excess of red corpuscles. polycythemia cyanotica, a form associated with chronic cyanosis, enlargement of the spleen, and constipation without any sign of disease of the heart, lungs, or kidneys and with no emphysema.
**polydactylism** (*pol-e-dak'-til-izm*) [*poly-*; δάκτυλος, a finger]. The existence of supernumerary fingers or toes.
**polydipsia** (*pol-e-dip'-se-ah*) [*poly-*; δίψα, thirst]. Excessive thirst.
**polyembryony** (*pol-e-em'-bre-o-ne*) [*poly-*; ἔμβρυον, an embryo]. In biology, the production of more than

POLYEMIA 702 POLYOREXIA

one embryo in a seed. Parthenogenesis occurs in most instances of polyembryony.
**polyemia, polyæmia** (*pol-e-e'-me-ah*) [*poly-*; αἷμα, blood]. Abnormal increase of the total mass of the blood; plethora. **polyæmia hyperalbuminosa,** an excess of albumin in the blood-plasma. **polyæmia polycythæmica,** an increase of the red corpuscles. **polyæmia serosa,** a condition in which the amount of blood-serum is increased.
**polyesthesia, polyæsthesia** (*pol-e-es-the'-ze-ah*) [*poly-*; αἴσθησις, sensation]. An abnormality of sensation in which a single touch is felt in two or more places at the same time.
**polyformin** (*pol-e-form'-in*). An antiseptic compound obtained by dissolving resorcinol in aqueous formaldehyde and adding an excess of ammonia; it is an insoluble, odorless powder used in the same manner as iodoform. p., soluble, $C_6H_4(OH)_2$—$(CH_2)_6N_4$ diresorcinol hexamethylenetramine, a combination of two molecules of resorcinol with one molecule of hexamethylenetetramine (urotropin); white crystals, very soluble in water and alcohol, insoluble in ether, benzol, and oils. It is used internally as an antiferment; externally in skin diseases.
**Polygala** (*pol-ig'-al-ah*) [*poly-*; γάλα, milk]. A genus of herbaceous or shrubbery plants of some 260 species. P. senega, of N. America, is therapeutically the most important. See *senega*.
**polygalactia** (*pol-e-gal-ak'-te-ah*) [*poly-*; γάλα, milk]. Excessive secretion or flow of milk.
**polygalin** (*pol-ig'-al-in*). See *senega*.
**polyganglionic** (*pol-e-gang-gle-on'-ik*) [*poly-*; ganglion]. 1. Having several ganglia. 2. Affecting several lymphatic glands at once.
**polygastria** (*pol-e-gas'-tre-ah*) [*poly-*; γαστήρ, stomach]. Excessive production of gastric juice.
**polygastric** (*pol-e-gas'-trik*) [*poly-*; γαστήρ, belly, stomach]. 1. Having several bellies (as certain muscles). 2. Having more than one stomach.
**polyglandular** (*pol-e-gland'-du-lar*) [*poly-*; *gland*]. Pluriglandular.
**polyglobulia** (*pol-e-glob-u'-le-ah*). An increase in the number of red blood corpuscles.
**polyglobulism** (*pol-e-glob'-u-lizm*). Same as *polycythemia, q. v.*
**polygnathus** (*pol-ig'-na-thus*) [*poly-*; γνάθος, jaw]. A form of monster in which the parasite is attached to the jaws of the host.
**polygonal** (*pol-ig'-o-nal*) [*poly-*; γωνία, an angle]. Having many angles.
**Polygonum** (*pol-ig'-o-num*). A genus of polygonaceous plants. P. hydropiperoides, smart weed, water pepper, is a plant common in the United States. Its active principle is thought to be polygonic acid. It stimulates the action of the heart and increases arterial tension. It is diuretic, emmenagogue and aphrodisiac. Externally, it is a rubefacient and vesicant. It is valuable in amenorrhea and impotence. Dose of the ext. gr. j–v; of the fldext. ♏ x–ʒ j.
**polygraph** (*pol'-ig-raf*) [*poly-*; γράφειν, to record]. A cylindrical recording instrument for taking simultaneous sphygmographic tracings. It is made to rotate upon its axis by clockwork.
**polygroma** (*pol-ig-ro'-mah*) [*poly-*; ὑγρός, moist; ὄμα, tumor]. A large hygroma.
**polygyria** (*pol-e-jir'-e-ah*) [*poly-*; γῦρος, gyre]. The existence of an excessive number of convolutions in the brain.
**polyhedral** (*pol-e-he'-dral*) [*poly-*; ἔδρα, a seat; a base]. Having many surfaces.
**polyhemia** (*pol-e-hem'-e-ah*). See *polyemia*.
**polyhidria** (*pol-e-hi'-dre-ah*). See *polyidrosis*.
**polyhidrosis** (*pol-e-hid-ro'-sis*) [*poly-*; ἵδρωσις, sweating]. Excessive perspiration.
**polyhydramnios** (*pol-e-hi-dram'-ne-os*). An excessive production of liquor amnii.
**polyhygroma** (*pol-e-hi-gro'-mah*). See *polygroma*.
**polyidrosis** (*pol-e-id-ro'-sis*). Same as *polyhidrosis*.
**polyinfection** (*pol-e-in-fek'-shun*) [*poly-*; *infection*]. Infection resulting from the presence of more than one organism.
**polyleptic** (*pol-e-lep'-tik*) [*poly-*; λαμβάνειν, to seize]. Characterized by numerous remissions and exacerbations.
**polymastia** (*pol-e-mas'-te-ah*) [*poly-*; μαστός, a breast]. The presence of more than two breasts.
**polymastigate** (*pol-im-as'-tig-āt*) [*poly-*; μάστιξ, a whip]. In biology, having several flagella.
**polymazia** (*pol-im-a'-ze-ah*). Synonym of *polymastia*.

**polymelia** (*pol-e-me'-le-ah*) [*poly-*; μέλος, a limb]. A malformation consisting in the presence of more than the normal number of limbs.
**polymelus, polymelius** (*pol-im'-el-us, pol-im-e'-le-us*) [see *polymelia*]. A monster having more than the normal number of limbs.
**polymenia** (*pol-e-me'-ne-ah*) [*poly-*; μῆνες, months]. Menorrhagia.
**polymenorrhea** (*pol-im-en-or-e'-ah*) [*poly-*; μῆν, month; ῥοία, a flow]. Excessive menstrual flow.
**polymer** (*pol'-im-ēr*) [*poly-*; μέρος, a part]. A polymeric substance.
**polymeric** (*pol-e-mer'-ik*) [see *polymerism*]. 1. Exhibiting polymerism. 2. Applied to muscles which are derived from two or more myotomes.
**polymerid** (*pol-im'-er-id*) [see *polymerism*]. In chemistry, a compound having the property of polymerism. Synonym of *polymer*.
**polymerism** (*pol-im'-er-izm*) [*poly-*; μέρος, a part]. 1. The existence of more than a normal number of parts. 2. A form of isomerism in which the molecular weights of the polymers are multiples of each other. See *isomeric* and *polymerization*.
**polymerization** (*pol-e-mer-i-za'-shun*) [see *polymerism*]. The apparent fusion or union of two or more molecules of a compound, forming a more complex molecule, with a higher molecular weight and somewhat different physical and chemic properties.
**polymetameric** (*pol-im-et-am-er'-ik*). Extending over or comprising two or more metameres.
**polymicrobic** (*pol-im-i-kro'-bik*) [*poly-*; μικρός, small; βίος, life]. Containing many kinds of microörganisms.
**polymicrotome** (*pol-im-i'-kro-tōm*) [*poly-*; μικρός, small; τέμνειν, to cut]. An instrument making many microscopic sections in a short time.
**polymorph** (*pol'-im-orf*) [*poly-*; μορφή, form]. A polymorphonuclear leukocyte.
**polymorphic, polymorphous** (*pol-e-mor'-fik, pol-e-mor'-fus*) [see *polymorphism*]. Having or occurring in several forms; of a crystal crystallizing in several forms.
**polymorphism** (*pol-e-mor'-fizm*) [*poly-*; μορφή, form]. The state of being polymorphous.
**polymorphocellular** (*pol-im-or-fo-sel'-u-lar*) [*poly-*; μορφή, form; *cellula*, cell]. Having cells of many forms.
**polymorphocyte** (*pol-e-mor'-fo-sīt*). A narrow cell or myelocyte.
**polymorphonuclear** (*pol-e-mor-fo-nu'-kle-ar*). Applied to multinuclear leukocytes which have nuclei exceedingly irregular in form, being twisted or knotted or presenting the appearance of being divided into distinct portions, though in reality a thin lamina of nuclear substance unites them.
**polymyoclonus** (*pol-e-mi-ok'-lo-nus*). See *paramyoclonus*.
**polymyositis** (*pol-e-mi-o-si'-tis*). Simultaneous inflammation of many muscles.
**polynesic** (*pol-e-ne'-sik*) [*poly-*; νῆσος, island]. Occurring in several foci, e. g. in *polynesic sclerosis*.
**polyneural** (*pol-in-u'-ral*) [*poly-*; νεῦρον, nerve]. Pertaining to, or supplied or innervated by several nerves.
**polyneuritis** (*pol-e-nu-ri'-tis*). See *neuritis, multiple*.
**polynuclear** (*pol-e-nu'-kle-ar*). See *multinuclear*.
**polynuclearneutrophilic** (*pol-e-nu-kle-ar-nu-tro-fil'-ik*). Relating to polynuclear leukocytes which are readily stainable with neutral dyes.
**polynucleate** (*pol-in-u'-kle-āt*). Synonym of *multinuclear*.
**polynucleosis** (*pol-e-nu-kle-o'-sis*). The condition of having many multinuclear cells in the blood or in a pathologic exudate; polymorphonuclear leukocytosis.
**polyodontia** (*pol-e-o-don'-she-ah*) [*poly-*; ὀδούς, tooth]. The presence of supernumerary teeth.
**polyonychia** (*pol-e-o-nik'-e-ah*) [*poly-*; ὄνυξ, nail]. The presence of supernumerary nails; polonychia.
**polyopia, polyopsia** (*pol-e-o'-pe-ah, pol-e-op'-se-ah*) [*poly-*; ὄψις, sight]. A condition in which more than one image of an object is formed upon the retina.
**polyophthalmica,** the phenomenon of multiple vision with a single eye.
**polyorchis** (*pol-e-or'-kis*) [*poly-*; ὄρχις, a testicle]. One who has more than two testicles.
**polyorexia** (*pol-e-or-eks'-e-ah*) [*poly-*; ὄρεξις, appetite]. Excessive hunger, or appetite; bulimia.

**polyorrhomenitis** (*pol-e-or-o-men-i'-tis*) [*poly-*; ὅρρος, serum; ὑμήν, membrane; ιτις, inflammation]. Concato's disease; a symptom-group defined by Concato as "a phthisis of serous membranes."
**polyotia** (*pol-e-o'-she-ah*) [*poly-*; οὖς, ear]. A condition in which there is more than one auricle on one or both sides of the head.
**polyp, polypus** (*pol'-ip, pol'-e-pus*) [*poly-*; πούς, foot]. A tumor having a pedicle, found especially on mucous membranes, as in the nose, bladder, rectum, uterus, etc. **p.**, **blood-**. Synonym of *p., placental.* **polypus carnosus.** Synonym of *sarcoma.* **p., fibrinous,** a polypoid mass on the uterine wall, resulting from the deposition of fibrin from retained blood. The mass may be attached to portions of an ovum or to thrombi at the placental site. **p., fibrous,** a polyp composed chiefly of fibrous tissue. **p., mucous,** a soft polyp resulting either from a localized inflammatory hyperplasia of a mucous membrane or from the formation of a true myxoma. **p., placental,** a fibrinous polyp resulting from the expulsion of fibrin upon a portion of retained placenta. **p., soft.** Synonym of *p., mucous.* **p., vascular,** a pedunculated angioma.
**polypapilloma tropicum** (*pol-e-pap-il-o'-mah*). Frambesia.
**polyparesis** (*pol-ip-ar'-es-is*) [*poly-*; πάρεσις, weakness]. General progressive paralysis of the insane, or paralytic dementia.
**polypathia** (*pol-ip-ath'-e-ah*) [*poly-*; πάθος, disease]. The presence of several diseases at one time, or the frequent recurrence of disease.
**polypeptid** (*pol-e-pep'-tid*). A complex compound of several amino-acids. See also *peptid.*
**polyphagia** (*pol-e-fa'-je-ah*) [*poly-*; φαγεῖν, to eat]. Bulimia.
**polyphalangism** (*pol-e-fal-an'-jizm*) [*poly-*; *phalanx*]. The presence of an extra phalanx on a finger or toe.
**polypharmacy** (*pol-e-far'-mas-e*) [*poly-*; φάρμακον, a drug]. The prescription of many drugs at one time; the excessive use of drugs.
**polyphobia** (*pol-if-o'-be-ah*) [*poly-*; φόβος, fear]. Morbid fear of many things.
**polyphrasia** (*pol-e-fra'-ze-ah*) [*poly-*; φράσις, speech]. A morbid state characterized by excessive speaking; morbid loquacity; verbigeration.
**polypiferous** (*pol-ip-if'-er-us*) [*polypus,* polyp; *ferre,* to bear]. Bearing or giving origin to a polypus.
**polyplasmia** (*pol-e-plaz'-me-ah*) [*poly-*; πλάσμα, plasm]. Extreme fluidity of the blood.
**polyplastic** (*pol-e-plas'-tik*) [*poly-*; πλάσσειν, to mold]. 1. Of cells, having many substances in their composition. 2. Undergoing many modifications during development.
**polypnea, polypnœa** (*pol-ip-ne'-ah*) [*poly-*; πνοία, breathing]. Great rapidity of respiration; panting respiration.
**Polypodium** (*pol-e-po'-de-um*) [*poly-*; πούς, foot]. A genus of ferns several species of which are asserted to have medical properties. The rhizome of *P. aureum,* of the West Indies, is used as a styptic; *P. calaguala,* the true calaguala, of Mexico and Peru, has a high reputation as a solvent and diaphoretic; the juice of the rhizome of *P. quercifolium,* of the East Indies, is used in inflammation of the eyes and in gonorrhea; the rhizome is used in malaria and as a tonic; *P. vulgare,* of Europe and America, has been used as an expectorant in chronic catarrh and asthma.
**polypoid** (*pol'-e-poid*) [*polyp;* εἶδος, like]. Resembling a polyp.
**Polyporus** (*pol-ip'-o-rus*) [*poly-*; πόρος, pore]. A genus of fungi. **P. amanita,** a poisonous agaric, used for killing flies, has marked excitant and narcotic properties. It has been used topically in cancerous tumors and ulcers and internally in epilepsy, skin diseases, paralysis, and tuberculosis. It contains the alkaloids muscarine and amanitine. Syn., *Amanita muscaria; bug agaric; fly agaric.* **P. officinalis,** grows on the larch and is known as purging agaric. It is used in night-sweats of tuberculosis. Dose of *fluidextract* 1–15 min. (0.06–1.0 Cc.).
**polyposis** (*pol-ip-o'-sis*) [*poly-*; πόσις, draught]. 1. Excessive thirst; polydipsia. 2. [*polyp.*] The condition of being affected with polyps. **p. ventriculi,** a plicate, warty condition of the gastric mucosa associated with hypertrophy and catarrh. Syn., *état mamelonné.*

**polypotome** (*pol-ip'-o-tōm*) [*poly-*; τομή, a cutting]. An instrument for the excision of polypi.
**polypotrite** (*pol-ip'-o-trīt*) [*polypus*; τρίβειν, to rub]. An instrument for crushing polypi.
**polypous** (*pol'-ip-us*). Of the nature of a polyp.
**polypus** (*pol'-e-pus*). See *polyp.*
**polyrrhea, polyrrhœa** (*pol-e-re'-ah*) [*poly-*; ῥοία, a flow]. An excessive secretion of fluid.
**polysaccharid, polysaccharide** (*pol-is-ak'-ar-id*). A carbohydrate which under the influence of dilute acids takes up more than two molecules of water and yields more than three sugar molecules. *Examples:* starches, gums. See *carbohydrate.*
**polysarcia** (*pol-e-sar'-se-ah*) [*poly-*; σάρξ, flesh]. Excessive corpulency; obesity.
**polysarcous** (*pol-e-sar'-kus*) [*polysarcia*]. Corpulent; exhibiting polysarcia.
**polyscelia** (*pol-e-skel'-e-ah*) [*poly-*; σκέλος, leg]. Excess in the number of legs.
**polyscelus** (*pol-is'-kel-us*) [see *polyscelia*]. A monster having supernumerary legs.
**polyscope** (*pol'-is-kōp*) [*poly-*; σκοπεῖν, to observe]. An instrument provided with an electric light which is introduced into a cavity in order to illuminate its internal surfaces.
**polyserositis** (*pol-e-se-ro-si'-tis*). Progressive malignant inflammation of the serous membranes. See *Concato's disease.*
**polysinuitis, polysinusitis** (*pol-e-sin-ū-i'-tis, -si'-nus-i-tis*). Simultaneous inflammation of several sinuses.
**polysolvol, polysolve** (*pol-e-sol'-vol, pol'-e-solv*). A thick, clear liquid which has the property of dissolving large quantities of phenol, menthol, salycilic acid, etc. It is obtained by treating castor-oil successively with strong sulphuric acid and sodium chloride and neutralizing the sulphoricinic acid obtained with sodium hydroxide. Syn., *ammonium sulphoricinate; solvin sodium.* See *solvin.*
**polysomia** (*pol-e-so'-me-ah*) [*poly-*; σῶμα, body]. A monster having more than a single body or trunk.
**polysomus** (*pol-e-so'-mus*) [*poly-*; σῶμα, body]. A monster fetus having one head and several bodies.
**polyspermia, polyspermism** (*pol-e-sper'-me-ah, pol-e-sper'-mizm*) [*poly-*; σπέρμα, seed]. The secretion and discharge of an excessive quantity of seminal fluid.
**polyspermy** (*pol-is-per'-me*) [*poly-*; σπέρμα; seed]. Impregnation of an ovum by more than one spermatozoon.
**polystichia** (*pol-e-stik'-e-ah*) [*poly-*; στίχος, row]. A condition in which the eyelashes are arranged in two or more rows.
**polystomatous** (*pol-is-to'-mat-us*) [*poly-*; στόμα, mouth]. Having many mouths or apertures.
**polythelia, polythelism** (*pol-ith-e'-le-ah, pol'-ith-el-izm*) [*poly-*; θηλή, nipple]. The presence of supernumerary nipples.
**polytrichia, polytrichosis** (*pol-e-trik'-e-ah, pol-e-trik-o'-sis*) [*poly-*; θρίξ, hair]. Excessive development of hair.
**Polytrichum** (*pol-it'-rik-um*) [πολύτριχος, having much hair]. A genus of mosses; hair-moss. *P. juniperinum* is a species indigenous to the United States and Europe, and is a powerful diuretic. Dose of the *fluidextract* 20–60 min. (1.2–3.7 Cc.).
**polytrophia, polytrophy** (*pol-it-ro'-fe-ah, pol-it'-ro-fe*) [*poly-*; τροφεία, nourishment]. Abundant or excessive nutrition.
**polyuria** (*pol-e-ū'-re-ah*) [*poly-*; οὖρον, urine]. The passage of an excessive quantity of urine. **p. spastica,** intermittent polyuria with hysterical symptoms accompanying convulsions.
**polyuric** (*pol-e-ū'-rik*) [*poly-*; οὖρον, urine]. 1. Pertaining to, or affected with polyuria. 2. One affected with polyuria.
**polyvalent** (*pol-iv'-al-ent*). Synonym of *multivalent.* **p. serum,** one obtained either by immunizing animals with different strains of the same bacterium, or a mixture of sera derived from different animals immunised with various strains.
**pomade** (*po-mād'*) [*pomum,* apple]. A perfumed ointment for applying to the scalp.
**pomander** (*po-man'-der*) [*pomum,* apple]. A ball composed of aromatics, formerly carried about the person to prevent infection; also, the globular case in which the same was kept.
**pomatum** (*po-ma'-tum*) [L.]. A pomade.
**pomegranate** (*pum-gran'-āt*). See *granatum.*
**pommel joint** (*pum'-el joint*). Condyloid joint.
**pompholyx, pompholix** (*pom'-fo-liks*) [πομφόλυξ,

a **bubble**]. A rare disease characterized by bullous eruptions on the palms of the hands and between the fingers. It occurs in depressed states of the nervous system, and is more common in women than in men. Syn., *chiropompholyx; dysidrosis.*
**pomphus** (*pom'-fus*). See *wheal.*
**pomum** (*po'-mum*) [L.]. Apple: **p. Adami,** Adam's apple, the prominence in the front of the neck caused by the projection of the thyroid cartilage.
**Pôncet's disease** (*pon'-sa*) [Antonin Poncet, French surgeon, 1846-1913]. Tuberculous rheumatism.
**Pond's extract.** A fluidextract of *Hamamelis virginiana.*
**ponderable** (*pon'-der-ah-bl*) [*pondus,* weight]. Having weight.
**Ponfick's shadows.** Achromacytes; colorless red corpuscles found in the blood in cases of hemoglobinemia.
**Pongamia** (*pon-ga'-me-ah*) [E. Ind. *pongam*]. A genus of East Indian trees. Kurung oil is the oil expressed from the seeds of *P. glabra,* native to India, China, and Australia. It is recommended in parasitic diseases of the skin, in pityriasis versicolor, herpes, rheumatism, and lepra.
**ponogen** (*pon'-o-jen*) [πόνος, work; γεννᾶν, to produce]. 1. Waste-matter of the nervous system; fatigue poison. 2. See *parhormone.*
**ponogenic** (*pon-o-jen'-ik*). Relating to *ponogen.*
**p. toxins.** See under *toxin.*
**ponograph** (*pon'-o-graf*) [πόνος, pain; γράφειν, to write]. An apparatus for determining and registering sensitiveness to pain, or to fatigue.
**ponos** (*pon'-os*) [πόνος, pain]. A chronic febrile disease endemic on the Greek islands of Spezzia and Hydra. The disease bears some resemblance to pseudoleukemia and to tuberculosis.
**pons** [L., "a bridge"]. 1. A process or bridge of tissue connecting two parts of an organ. 2. The pons Varolii. **p. basilaris,** the basilar process of the occipital bone. **p. cerebelli.** Same as *p. Varolii.* **p. hepatis,** a portion of the liver substance sometimes extending from the quadrate lobe to the left lobe. **p. Sylvii,** the quadrigeminum. **p. Varolii,** a convex white eminence situated at the base of the brain, and serving to connect the various divisions with one another. It is placed in front of the medulla oblongata, behind the cerebrum, and beneath the cerebellum, and rests upon the sphenobasilar groove. In structure the pons consists chiefly of nerve-fibers, but contains also areas of gray matter—the *pontine nuclei.* The pons is connected with the medulla, with the cerebellum (by the middle peduncles), and with the cerebrum (by the crura cerebri). **p. zygomaticus,** the zygoma.
**pontibrachium** (*pon-tib-ra'-ke-um*) [*pons,* bridge; *brachium,* arm]. The middle peduncle of the cerebellum.
**pontic** (*pon'-tik*) [*pons,* bridge]. Same as *pontile.*
**ponticinerea** (*pon-ti-sin-e'-re-ah*). A collection of gray matter in the pons.
**ponticulus** (*pon-tik'-u-lus*) [dim. of *pons,* bridge]. A small, transverse ridge between the pyramids of the oblongata and the pons. Synonym, *propons.* **p. auriculæ,** a slight prominence on the eminentia conchæ for the attachment of the retrahens aurem muscle. **p. hepaticus,** a bridge of tissue, containing a plexus of blood-vessels, and extending from the surface of the Spigelian to that of the right lobe of the liver. **p. promontorii,** a faint bony ridge on the inner wall of the tympanic cavity extending from the pyramid to the promontory and below the foramen ovale.
**pontile, pontine** (*pon'-til, pon'-tin*) [*pons*]. Pertaining to the pons Varolii. **p. hemiplegia,** a hemiplegia due to a lesion of the pons. When the lesion is situated low down in the pons, below the decussation of the fibers of the facial nerve, and above that of the pyramidal tracts, the hemiplegia is *alternate; i. e.,* the arm and leg on one side, and the face on the other, are paralyzed. **p. nuclei,** a collection of gray matter in the pons.
**pontobulbar** (*pon-to-bul'-bar*) [*pons,* bridge; *bulbus,* bulb]. Pertaining to the pons Varolii and to the bulbus spinalis (or oblongata).
**pontocrural** (*pon'-to-kroo'-ral*) [*pons,* bridge; *crus,* leg]. Pertaining to the pons Varolii and the crura cerebri.
**pontoon** (*pon-toon'*) [*ponto,* a small boat]. A loop or knuckle of the small intestine.

**pooled blood-serum.** Mixed serum from a number of persons.
**poplar** (*pop'-lar*). See *populus.*
**popliteal** (*pop-lit-e'-ad*) [*poples,* ham; *ad,* to]. Toward the popliteal aspect.
**popliteal** (*pop-lit-e'-al*) [*poples,* the ham of the knee; the hock]. Pertaining to or situated in the ham, as *popliteal* artery, *popliteal* nerve, *popliteal* space. **p. aneurysm,** aneurysm of the popliteal artery. **p. artery.** See under *artery.* **p. space.** See under *space.*
**popliteus** (*pop-lit-e'-us*) [see *popliteal*]. The ham or hinder part of the knee-joint. **p. muscle.** See under *muscle.*
**poppy** (*pop'-e*) [AS., *popig*]. *Papaver somniferum,* a plant of the order *Papaveraceæ;* the capsules yield opium. **p.-capsules** (*papaveris capsulæ,* B. P.), possess effects similar to those of opium, but far milder in degree. The decoction (*decoctum papaveris,* B. P.) is used chiefly as an anodyne application. Dose of the *extract* (*extractum papaveris,* B. P.) 5-10 gr. (0.32-0.65 Gm.); of the *syrup* (*syrupus papaveris,* B. P.) ½-1 dr. (2-4 Cc.). *Red poppy* is used in making the *syrupus rhœados* in the B. P. Dose 1 ʒ (4 Cc.).
**populin** (*pop'-u-lin*). See *Populus.*
**Populus** (*pop'-u-lus*) [L.]. Poplar, a genus of trees of the order *Salicineæ,* several species of which yield salicin (*salicinum,* U. S. P.), $C_{13}H_{18}O_7$, and *populin* or benzoyl-salicin, $C_{20}H_{22}O_8$. Poplar-buds have been used in rheumatism and Bright's disease. The leaves and buds are also employed for anodyne ointments. The wood and bark of the root of *P. monilifera,* cottonwood, of the United States, are said to be powerful antiperiodics.
**porcellaneous, porcellanous** (*por-sel-a'-ne-us, por-sel'-an-us*). Relating to or having the appearance of porcelain; applied to a condition of the skin in fever.
**porcosan** (*por'-ko-san*). A remedy for hog erysipelas prepared from weakened cultures of *Bacillus erysipelatos suis.*
**porcupine-disease.** See *ichthyosis.*
**pore** (*pōr*) [πόρος, a pore, or cavity]. A minute circular opening on a surface, as a *pore* of the skin, the opening of the duct of a sudoriparous gland. See also *porus.*
**porencephalia, porencephalus** (*por-en-sef-a'-le-ah, por-en-sef'-al-us*) [*pore;* ἐγκέφαλος, brain]. A condition characterized by the presence of depressions on the surface of the brain, due to a congenital arrest of development or to an acquired defect.
**porencephalic** (*por-en-sef-al'-ik*). Same as *porencephalous.*
**porencephalitis** (*por-en-sef-al-i'-tis*) [*porencephalia*]. Encephalitis with a tendency to the formation of cavities.
**porencephalous** (*por-en-sef'-al-us*) [*porencephalia*]. Affected with porencephalia.
**Porges reaction** or **Porges-Meier reaction** (*por'-ges-mi'-er*). A precipitation test for syphilis; "the requirements are: (1) One per cent. solution of sodium glycocholate in distilled water. (2) The patient's serum which must be absolutely clear, and heated for one-half an hour at 56° C. Two-tenths of each of the above are placed into a narrow test-tube 6 to 17 mm. in diameter, and allowed to rest for sixteen to twenty hours at room temperature. A positive reaction consists of the appearance of distinct coarse flocculi, which as a rule, collect near the surface. Mere turbidity or faint precipitates are considered as negative. The original Porges method of employing lecithin was not at all specific, the reaction being present in tuberculosis, carcinoma, and other infectious diseases."
**pornography** (*por-nog'-ra-fe*) [πόρνη, a prostitute; γράφειν, to write]. 1. A treatise on prostitution. 2. Obscene writing.
**pornotherapy** (*por-no-ther'-ap-e*) [πόρνη, a prostitute; θεραπεία, therapy]. The medical supervision of prostitutes as related to public hygiene.
**porocele** (*po'-ro-sēl*) [πόρος, callus; κήλη, hernia]. A scrotal hernia in which the coverings are indurated and thickened.
**porokeratosis** (*po-ro-ker-at'-o-sis*) [πῶρος, callus; κέρας, a horn]. A keratosis appearing in raised or smooth areas, of varying size, irregular form, circumscribed outline, at the summit of which a thin layer of horny tissue of linear arrangement is present. The affection is usually seated on the dorsal aspect of the hands and feet (never on the palmar or plantar

surface), the extensor aspect of the forearms and legs, neck, face and scalp.
**poroma** (*po-ro'-mah*) [πώρωμα], A callosity.
**poroplastic** (*po-ro-plas'-tik*) [*pore*; πλάσσειν, to mold]. Porous and plastic. p. felt, a porous felt which is readily molded; it is used in the preparation of splints and jackets.
**porosis** (*po-ro'-sis*) [πῶρος, callus]. The formation of callus.
**porosity** (*po-ros'-it-e*) [*porous*]. The condition of being porous.
**porotic** (*po-rot'-ik*) [πῶρος, callus]. Favoring the formation of callus. Of the nature of callus.
**porotomy** (*po-rot'-o-me*) [*pore*; τομή, a cutting]. Incision of the meatus of the urethra.
**porous** (*po'-rus*) [*pore*]. Having pores.
**porphyreus** (*por-fi'-re-us*) [πορφύρα, purple]. In biology, showing spots of purple upon a ground of another hue.
**porphyrin** (*por'-fir-in*) [see *porphyreus*]. A white, amorphous substance, $C_{11}H_{28}N_3O_7$, from *Alstonia constricta*, Australian fever-bark. It is antipyretic.
**porphyrization** (*por-fir-i-za'-shun*) [*porphyry*, a kind of rock]. Pulverization, so-called because generally performed on a tablet of porphyry.
**porphyruria** (*por-fir-u'-re-ah*) [πορφύρα, purple; οὖρον, urine]. The discharge of urine colored with purpurin.
**Porret's phenomenon.** When a continuous current is passed through a living muscular fiber, the sarcous substance shows an undulating movement from the positive toward the negative pole.
**porriginous** (*por-ij'-in-us*). Relating to porrigo; scurfy.
**porrigo** (*por-i'-go*) [L.]. An old term applied to several diseases of the scalp. p. **decalvans**, alopecia areata. p. **favosa**, favus. p. **larvalis**, impetigo of the scalp conjoined with eczema.
**Porro's operation**, Porro-cesarean section (*por'-o*) [Edoardo *Porro*, Italian obstetrician, 1842–1902]. Cesarean section, followed by removal of the uterus at the cervical junction, together with the ovaries and oviducts.
**Porro-Müller's operation.** *For otherwise impossible labor:* a modification of the Porro operation, in which the uterus is brought out of the abdomen before extracting the fetus.
**Poro-Veit's operation** (*por'-o-vit'*). *For otherwise impossible labor:* the stump is ligated and dropped.
**porta** (*por'-tah*) [L., "gate"]. The hilus of an organ through which the vessels enter. p. **hepatis**, p. **jecoris**, the transverse fissure of the liver through which the portal vein enters the organ. p. **labyrinthi**, the fenestra rotunda. p. **omenti**, the foramen of Winslow. p. **vestibuli**, a narrow orifice between the sinus venosus and the auricle in the embryonic heart.
**portal** (*por'-tal*) [*porta*]. Pertaining to the porta or hilum of an organ, especially to the porta hepatis or to the vein entering at the porta hepatis (*portal vein*). p. **circulation**, the passage of blood from the stomach, spleen, and intestine through the portal vein and the liver. p. **fissure**. See *fissure, transverse*. p. **vein**, the large vein entering the liver at the transverse fissure, and bringing to it the blood from the digestive tract and the spleen.
**porte-** or **port-** (*portare*, to carry). A carrier, or holder: **p.-acid**, an instrument for the local application of an acid. **p.-aiguille** = *needle-holder*. **p.-caustique**, a holder for the stick of caustic. **p.-cordon**, an instrument for replacing a prolapsed funis. **p.-crayon**. See *p.-caustique*. **p.-fil.** Synonym of *p.-ligature*. **p.-fillet**, an instrument for applying a fillet to some part of the fetal body. **p.-ligature**, an instrument for applying a ligature to a deep part. **p.-moxa**. See *moxa*. **p.-nœud**, an instrument for applying a ligature to the pedicle of a tumor. **p.-pierre**. Synonym of *p.-caustique*.
**porter** (*por'-ter*). See under *malt liquors*.
**Porter's symptom.** Tracheal tugging. See *Oliver's symptom*.
**portio** (*por'-she-o*) [L.]. Portion. Also, an abbreviated expression for portio vaginalis uteri—the vaginal portion of the uterus. p. **alba cerebri**, the white substance of the brain. p. **aryvocalis**, a short muscle attached anteriorly to the vocal band and posteriorly to the vocal process of the arytenoid cartilage. p. **axillaris**, the second part of the subclavian artery. p. **corporis striati externa**, the lenticular nucleus. p.
corporis striati interna, the caudate nucleus. p. **dura**, the facial nerve. p. **infravaginalis**, the vaginal portion of the neck of the uterus. p. **inter duram et mollem**, a small funiculus between the seventh and eighth cranial nerves. p. **intermedia Wrisbergii.** Synonym of *p. inter duram et mollem*. p. **mollis**, the auditory nerve. p. **muscularis**, the second division of the subclavian artery. p. **pectoralis**, the first division of the subclavian artery. p. **pylorica ventriculi**, the pyloric extremity of the stomach. p. **splenica ventriculi**, the cardiac extremity of the stomach. p. **thoracica**, the first part of the axillary artery. p. **ventriculi lienalis.** Synonym of *p. splenica ventriculi*.
**portiplex, portiplexus** (*por'-tip-leks*, *por-tip-leks'-us*) [*porta*, gate; *plexus*]. The plexus or vascular fringe that connects the two lateral choroid plexuses. It passes through the porta, or foramen of Monro, whence the name.
**port-wine mark, p. stain.** See *nevus* (2).
**porus** (*po'-rus*) [L.]. 1. A pore, foramen. 2. A callosity. p. **acusticus externus**, the opening of the external auditory canal. p. **acusticus internus**, the opening of the internal auditory canal into the cranial cavity. p. **opticus**, the opening in the center of the lamina cribrosa transmitting the central artery of the retina. p. **sudoriferus**, a sweat pore.
**Posadas,** protozoic disease of. See *dermatitis, blastomycetic*.
**position** (*po-zish'-un*) [*ponere*, to place]. Place; location; attitude; posture. p., **anatomical**, the person stands erect with the arms at the side and palms forward. p., **dorsal**, one in which the patient lies on the back. p., **Edebohls'.** See *Simon's position*. p., **English.** See *p., left lateral recumbent*. p. of the fetus, the relation of the presenting part of the fetus to the cardinal points of Capuron. For the *vertex*, the *face*, and the *breech* there are each four positions: a *right anterior*, a *right posterior*, a *left anterior*, and a *left posterior*. For each of the shoulders there is an *anterior* and a *posterior* position. In order to shorten and memorize these positions, the initials of the chief words are made use of, as follows: For vertex presentations the word occiput is abbreviated *O.*, and preceded by the letter *R.* or for right or left, and followed by *A.* or *P.*, according as the presenting part is anterior or posterior. We thus have the initials *L.O.A.*, left occipitoanterior, to indicate that the presenting occiput is upon the anterior left side. In the same way are obtained the terms *L.O.P.*, *R.O.A.*, *R.O.P.* For facial presentations we have in the same way *L.F.A.* (left frontoanterior), *L.F.P.*, *R.F.A.*, *R.F.P.* For breech or sacral presentations, *L.S.A.*, *L.S.P.*, *R.S.A.*, *R.S.P.*, and for shoulder (*dorsal*) presentations, *L.D.A.*, *L.D.P.*, *R.D.A.*, *R.D.P.* p., **Fowler's.** See *Fowler's position*. p., **high pelvic.** See *Trendelenburg's position*. p., **knee-chest, p., genupectoral**, one in which the patient rests upon the knees and chest, the arms being crossed above the head. p., **knee-elbow**, one in which the patient lies upon the knees and elbows, the head resting upon the hands. p., **left lateral recumbent, p., English**, one in which the patient lies on the left side with the right thigh and knee drawn up. p., **lithotomy, p., dorsosacral**, one in which the patient lies on the back with the legs flexed on the thighs and the thighs flexed on the abdomen and abducted. p., **Simon's.** See *Simon's position*. p., **semiprone**, one in which the patient lies on the left side with the right knee and thigh drawn up and the left arm placed along the back; the chest is inclined forward so that the patient rests upon it. Syn., *Sims' position*. p., **Trendelenburg's.** See *Trendelenburg's position*. p., **Walcher's.** See *Walcher's position*.
**positive** (*poz'-it-iv*) [*positivus*, from *ponere*, to place]. Real; existing; actual. In mathematics and physiology, denoting one of two quantities or conditions assumed as primary or fundamental; opposed to one assumed as negative; denoting a quantity greater than zero; to be added; additive. p. **electricity**, the kind of electricity developed by rubbing glass with silk. p. **electrode, p. pole**, the electrode or pole connected with the negative plate of a battery. p. **element, p. plate**, that plate of a battery which is acted upon by the fluid, *e. g.*, the zinc plate in the zinc-carbon battery.
positive phase. See *opsonic index*.
**positor** (*poz'-it-or*) [*ponere*, to place]. See *repositor*.
**Posner's reaction** for peptones and albumins.

The Semiprone, or Sims' Posture. Anterior View.

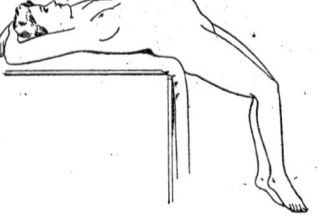

Walcher's Position.

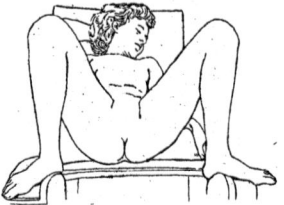

The Dorsal Elevated Posture.

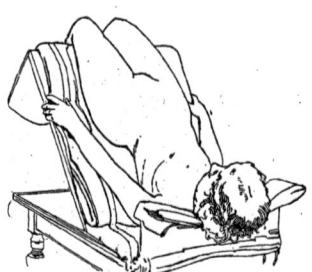

The Trendelenburg Posture.

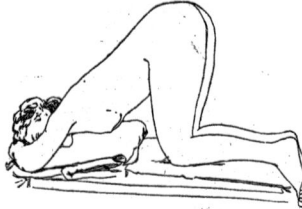

The Genu-pectoral Posture.

The Dorso-sacral Posture. Lateral View.

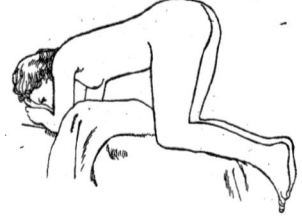

The Knee-elbow Posture.

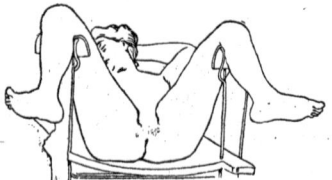

The Dorsal Recumbent Posture.

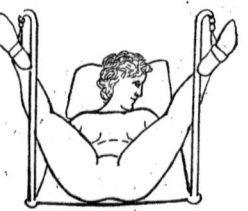

Edebohls' Posture.

Alkalinize the urine, pour it into a test-tube, and overlay it very carefully with a very dilute, almost colorless solution of copper sulphate. In the presence of peptone a violet zone will form even in the cold; the same reaction is yielded by albumin upon warming.
**posological, posologic** (*po-so-loj'-ik-al, po-so-loj'-ik*) [πόσος, how much; λόγος, science]. Pertaining to posology or quantitative dosage.
**posology** (*pos-ol'-o-je*) [πόσος, how much; λόγος, science]. That branch of medical science dealing with the dosage of medicines.
**posset** (*pos'-et*) [*posca*, sour wine and water]. A preparation of milk curdled with wine.
**possetting** (*pos'-et-ing*) [*posset*]. The regurgitation of infants.
**post-** (*pōst-*) [L.]. A prefix denoting after; behind.
**postaccessual** (*pōst-ak-ses'-ū-al*) [*post-; accessio*, a paroxysm]. Occurring after a paroxysm.
**postacetabular** (*post-as-et-ab'-ū-lar*) [*post-; acetabulum*]. Situated behind the acetabulum.
**postanal** (*pōst-a'-nal*) [*post-; anus*]. Situated behind the anus.
**postanesthetic** (*pōst-an-es-thet'-ik*). Occurring after anesthesia.
**postapoplectic** (*pōst-ap-o-plek'-tik*) [*post-; apoplexy*]. Coming on, or occurring, after a stroke of apoplexy. **p. coma**, the coma that often succeeds an apoplectic stroke.
**postauditory** (*pōst-aw'-dit-o-re*) [*post-; audire*, to hear]. Situated behind the auditory nerve, or chamber. **p. fossa**, a crescentic notch on the temporal bone separating the temporal ridge from the auditory plate.
**postaxial** (*pōst-aks'-e-al*) [*post-; axis*]. Situated behind the axis: in the arm, behind the ulnar aspect; in the leg, behind the fibular aspect.
**postbrachial** (*pōst-bra'-ke-al*) [*post-; brachium*, arm]. Situated posterior to the arm.
**postbrachium** (*pōst-bra'-ke-um*) [*post-; brachium*]. The posterior brachium of the corpus quadrigeminum, seen on the lateral slope of the mesencephal. It is between the prebrachium and the lemniscus.
**postcava** (*pōst-kav'-ah*) [*post-; cavus*, hollow]. The inferior or ascending vena cava.
**postcaval** (*pōst-kav'-al*). Pertaining to the inferior or ascending vena cava.
**postcentral** (*pōst-sen'-tral*) [*post-; center*]. 1. Situated behind a center. 2. Situated behind the fissure of Rolando, or central fissure of the brain, as the *postcentral* convolution.
**postcerebellar** (*pōst-ser-e-bel'-ar*) [*post-; cerebellum*]. Situated behind, or in the posterior portion of the cerebellum.
**postcerebral** (*pōst-ser'-e-bral*) [*post-; cerebrum*]. Situated behind, or in the posterior part of the cerebrum.
**postcibal** (*pōst-si'-bal*) [*post-; cibum*, food]. Occurring after meals.
**postcisterna** (*pōst-sis-ter'-nah*) [*post-; cisterna*, a vessel]. That portion of the spaces separating the ental layer of the arachnoid from the pia of the cerebellum, and communicating with the true encephalic cavities through the metapore, and also with the spinal subarachnoid space; the cisterna magna.
**postclavicular** (*pōst-kla-vik'-ū-lar*) [*post-; clavicle*]. Situated behind the clavicle.
**postcommissure** (*pōst-kom'-is-ūr*) [*post-; commissure*]. The posterior commissure of the brain.
**postconnubial** (*pōst-kon-nū'-be-al*) [*post-; connubium*, marriage]. Coming on, or occurring, after marriage.
**postconvulsive** (*pōst-kon-vul'-siv*) [*post-; convulsion*]. Coming on after a convulsion.
**postcordial** (*pōst-kor'-de-al*) [*post-; cor*, heart]. Situated behind the heart.
**postcornu** (*pōst-kor'-nū*) [*post-; cornu*, horn: *pl., postcornua*]. The occipital horn of the lateral ventricle of the brain.
**postcornual** (*pōst kor' nū al*) [*post ; cornu, horn*]. Pertaining to the postcornua of the ventricles of the brain or of the spinal cord).
**postcribrum** (*pōst-krib'-rum*) [*post-; cribrum*, sieve]. The posterior perforated space of the brain.
**postcubital** (*pōst-kū'-bit-al*) [*post-; cubitus*, the forearm]. Situated on the posterior aspect of the forearm.
**postdiastolic** (*pōst-di-as-tol'-ik*) [*post-; diastole*]. Occurring after the diastole.
**postdicrotic** (*pōst-di-krot'-ik*) [*post; dicrotic*]. Coming after the dicrotic wave of the pulse. **p. wave**, a second recoil-wave in the sphygmographic tracing. It is not always present.
**postdiphtheritic** (*pōst-dif-ther-it'-ik*) [*post-; diphtheric*]. Occurring after an attack of diphtheria, as *postdiphtheritic* paralysis.
**postembryonic** (*pōst-em-bri-on'-ik*) [*post-; embryo*]. Occurring after the embryonic stage.
**postepileptic** (*pōst-ep-i-lep'-tik*) [*post-; epilepsy*]. Occurring after an epileptic attack.
**posterior** (*pōs-te'-re-or*) [*posterus*, after; behind]. Placed behind or to the back of a part. **p. chamber**, the space between the iris and the lens.
**postero-** (*pōs-te-ro-*) [*posterior*]. A prefix meaning posterior.
**posteroexternal** (*pōs-ter-o-eks-ter'-nal*) [*postero-; external*]. Occupying the outer side of a back part, as the *posteroexternal* column of the spinal cord.
**posterointernal** (*pōs-te-ro-in-ter'-nal*) [*postero-; internal*]. Occupying the inner side of a back part, as the *posterointernal* column of the spinal cord.
**posterolateral** (*pōs-te-ro-lat'-er-al*) [*postero-; lateral*]. Situated behind and at the side of a part.
**posteromedian** (*pōs-te-ro-me'-de-an*). Located in the middle of a posterior aspect.
**posteroparietal** (*pōs-te-ro-par-i'-et-al*) [*postero-; parietal*]. Situated at or opposite the posterior part of the parietal bone; posterior and parietal.
**posterosuperior** (*pōs-te-ro-sū-pe'-re-or*) [*postero-; superior*]. Situated behind and above a part.
**posterotemporal** (*pōs-te-ro-tem'-por-al*) [*postero-; temporal*]. Situated at or opposite the posterior part of the temporal bone; posterior and temporal.
**posterula** (*pōs-ter'-oo-lah*) [*posterus*, posterior]. That portion of the nasopharynx between the posterior nares and the salpingo-palatal fold; a small space at the posterior ends of the turbinated bones of the nose.
**postesophageal, postœsophageal** (*pōst-e-sof-aj'-e-al*) [*post-; esophagus*]. Situated behind the esophagus.
**postfebrile** (*pōst-feb'-ril*) [*post-; febris*, a fever]. Occurring after a fever.
**postfovea** (*pōst-fo'-ve-ah*) [*post-; fovea*, pit]. The posterior fovea, a depression in the floor of the fourth ventricle of the brain.
**postgeminum** (*pōst-jem'-in-um*) [*post-; geminus*, twin]. The posterior pair of bodies of the corpora quadrigemina.
**postgeniculatum, postgeniculum** (*pōst-jen-ik-ū-la'-tum, -jen-ik'-ū-lum*) [*post-; geniculum*]. The internal geniculate body.
**postglenoid** (*pōst-gle'-noid*) [*post-; glenoid*]. Situated behind the glenoid fossa of the temporal bone, as the *postglenoid* tubercle, also called simply *postglenoid*.
**postgraduate** (*pōst-grad'-ū-āt*). 1. Belonging to or prosecuting a course of study after graduating. 2. A graduate.
**posthalgia** (*pos-thal'-je-ah*) [πόσθη, penis; ἄλγος, pain]. Pain in the penis.
**posthemiplegic** (*pōst-hem-i-ple'-jik*) [*post-; hemiplegia*]. Occurring after or following an attack of hemiplegia. **p.** chorea, choreiform movements in paralyzed limbs.
**posthemorrhagic** (*pōst-hem-o-raj'-ik*). Occurring after a hemorrhage.
**posthetomy** (*pos-thet'-o-me*) [πόσθη, prepuce; τομή, a cutting]. Circumcision.
**posthioplastic** (*pos-the-o-plas'-tik*). [*posthioplasty*]. Pertaining to, or involving, plastic surgery of the prepuce.
**posthioplasty** (*pos'-the-o-plas-te*) [πόσθη, foreskin; πλάσσειν, to mold]. Plastic surgery of the prepuce.
**posthippocampal** (*pōst-hip-o-kam'-pal*) [*post-; hippocampus*]. Pertaining to the calcar or hippocampus minor.
**posthitis** (*pos-thi'-tis*) [πόσθη, prepuce; ιτις, inflammation]. Inflammation of the prepuce.
**posthocalymma** (*pos-tho-kal-im'-ah*) [πόσθη, penis; κάλυμμα, veil]. Synonym of *condom*.
**postholith** (*pos'-tho-lith*) [πόσθη, prepuce; λίθος, a stone]. A preputial calculus.
**posthoncus** (*pos-thong'-kus*) [πόσθη, prepuce; ὄγκος, swelling]. A swelling or tumor of the prepuce.
**posthumeral** (*pōst-hū'-mer-al*) [*post-; humerus*]. Situated behind the humerus.
**posthumous** (*pos'-tū-mus*) [*postumus*, latest, last]. Occurring after death. **p. child**, one born after

the death of the father, or taken by cesarean operation from the body of its mother after her death.
**posthyoid** (*pŏst-hi'-oid*). Posterior to the hyoid bone.
**posthypnotic** (*pŏst-hip-not'-ik*) [*post-;* ὕπνος, sleep]. Succeeding the hypnotic state; acting after the hypnotic state has passed off, as *posthypnotic* suggestion.
**posthypophysis** (*pŏst-hi-pof'-is-is*) [*post-;* *hypophysis*]. The posterior and cerebral portion of the hypophysis or pituitary body.
**posticus** (*pos-ti'-kus*) [L.]. Posterior.
**postinfluenzal** (*pŏst-in-flu-en'-zal*). Occurring after influenza.
**postinsula** (*pŏst-in'-sū-lah*) [*post-;* *insula*]. 1. The posterior part of the insula. 2. Behind the insula.
**postischial** (*pŏst-is'-ke-al*). Dorsad of the ischium.
**postmalarial** (*pŏst-mal-a'-re-al*). Occurring as a sequel of malaria.
**postmastoid** (*pŏst-mas'-toid*) [*post-;* *mastoid*]. Situated behind the mastoid process of the temporal bone.
**postmedian** (*pŏst-me'-de-an*) [*post-;* *medius,* the middle]. Behind the middle transverse line of the body.
**postmediastinal** (*pŏst-me-de-as-ti'-nal*) [*post-;* *mediastinum*]. Pertaining to the postmediastinum.
**postmediastinum** (*pŏst-me-de-as-ti'-num*) [*post-;* *mediastinum*]. The posterior mediastinal space.
**postmortal** (*pŏst-mor'-tal*) [*postmortem*]. After death.
**postmortem** (*pŏst-mor'-tem*) [*post-;* *mors,* death]. 1. Occurring after death. 2. An examination of the body after death; an autopsy. **p. wart.** See *verruca necrogenica.*
**postnares** (*post-na'-rēz*) [*post-;* *naris,* nostril]. The posterior nares.
**postnarial** (*pŏst-na'-re-al*) [*post-;* *naris,* a nostril]. Pertaining to the posterior nares.
**postnasal** (*pŏst-na'-sal*) [*post-;* *nasus,* nose]. Situated behind the nose. **p. catarrh,** catarrhal inflammation of the nasopharynx.
**postnatal** (*pŏst-na'-zal*) [*post-;* *natus,* born]. Subsequent to birth, as a *postnatal disease.*
**postoblongata** (*pŏst-ob-long-ga'-tah*) [*post-;* *oblongata*]. The caudal or metencephalic portion of the oblongata, forming the floor of the metacele.
**postocular** (*pŏst-ok'-ū-lar*) [*post-;* *oculus,* the eye]. Behind the eye.
**postolivary** (*pŏst-ol'-iv-a-re*) [*post-;* *olivary*]. Behind the oliva.
**postoperative** (*pŏst-op'-er-a-tiv*) [*post-;* *operatio,* operation]. Occurring after an operation, as postoperative insanity.
**postoperculum** (*pŏst-o-per'-kū-lum*) [*post-;* *operculum*]. That one of the folds covering the insula which is formed of a part of the supertemporal gyrus; the temporal operculum.
**postopticus** (*pŏst-op'-tik-us*) [*post-;* ὤψ, eye; pl., *postoptici*]. Either one of the posterior pair of optic lobes, or corpora quadrigemina.
**postoral** (*pŏst-o'-ral*) [*post-;* *os, oris,* mouth]. Situated either behind or in the posterior part of the mouth.
**postorbital** (*pŏst-or'-bit-al*) [*post, orbita,* orbit]. Situated behind or below the orbit.
**postotic** (*pŏst-o'-tik*) [*post-;* οὖς, ear]. Behind the auditory vesicle.
**postpalatal** (*pŏst-pal'-at-al*) [*post-;* *palatum,* the palate]. Situated behind the palate bones.
**postparalytic** (*pŏst-par-ah-lit'-ik*) [*post-;* *paralysis*]. Following an attack of paralysis.
**postpartum** (*pŏst-par'-tum*) [*post-;* *partus,* birth]. Following childbirth, as *postpartum* hemorrhage.
**postpeduncle** (*pŏst-pe'-dunk-l*) [*post-;* *pedunculus,* peduncle]. The inferior cerebellar peduncle.
**postperforatus** (*pŏst-per-for-a'-tus*) [*post-;* *perforare,* to perforate]. The posterior perforated space.
**postpharyngeal** (*pŏst-far-in'-je-al*) [*post-;* *pharynx*]. Situated behind the pharynx.
**postpituitary** (*pŏst-pit-ū'-it-a-re*) [*post-;* *pituitary*]. Situated behind the pituitary body.
**postpleuritic** (*pŏst-ploo-rit'-ik*). Following pleurisy.
**postpneumonic** (*pŏst-nū-mon'-ik*). Following pneumonia.
**postpone** (*pŏst-pōn'*) [*post-;* *ponere,* to place]. Of a paroxysm, to occur after the regular time.
**postponent** (*pŏst-po'-nent*) [see *postpone*]. Delayed in recurrence.
**postpontile** (*pŏst-pon'-til*) [*post-;* *pons*]. Situated behind the pons Varolii. **p. recess,** the foramen cæcum.
**p₀stpyramidal** (*pŏst-pir-am'-id-al*) [*post-;* *pyramid*]. Situated behind the pyramidal tract. **p. nucleus,** the nucleus funiculi gracilis.
**postramus** (*pŏst-ra'-mus*) [*post-;* *ramus,* branch]. The caudal or horizontal branch of the stem of the arbor of the cerebellum.
**postrhinal** (*pŏst-ri'-nal*) [*post-;* ῥίς, nose]. Posterior and rhinal. **p. fissure,** the basirhinal fissure.
**postrolandic** (*pŏst-ro-lan'-dik*). Situated behind the fissure of Rolando.
**postsacral** (*pŏst-sa'-kral*) [*post-;* *sacrum*]. Situated behind or below the sacrum.
**postscalenus** (*pŏst-ska-le'-nus*) [*post-;* *scalenus*]. The scalenus posticus muscle.
**postscapula** (*pŏst-skap'-ū-lah*) [*post-;* *scapula*]. The part of the scapula below or posterior to the spine.
**postscapular** (*pŏst-skap'-ū-lar*). Pertaining to the postscapula.
**postscapularis** (*pŏst-skap-ū-la'-ris*) [*post-;* *scapula*]. Synonym of the infraspinatus muscle.
**postscarlatinal** (*pŏst-skar-lat-e'-nal*). Occurring after scarlatina.
**postsphenoid** (*pŏst-sfe'-noid*) [*post-;* *sphenoid*]. Situated behind the sphenoid bone; also, forming its posterior portion.
**postsylvian** (*pŏst-sil'-ve-an*). Situated behind the Sylvian fissure of the brain.
**postsyphilitic** (*pŏst-sif-il-it'-ik*). Following syphilis.
**postsystolic** (*pŏst-sis-tol'-ik*) [*post-;* *systole*]. Occurring after the systole of the heart.
**posttarsal** (*pŏst-tar'-sal*). Located behind the tarsus.
**posttibial** (*pŏst-tib'-e-al*) [*post-;* *tibia*]. Situated upon the posterior aspect of the tibia.
**posttyphoid** (*pŏst-ti'-foid*). Following typhoid.
**postulate** (*pos'-tū-lāt*) [*postulare,* to demand]. A well-known law; a basis of argument too obvious to require proof. **p.s,** Koch's. See *Koch.*
**postural** (*pos'-tūr-al*) [*ponere,* to place]. Pertaining to posture or position; performed by means of a special posture, as *postural* treatment.
**posture** (*pos'-tūr*). See *position.*
**postuterine** (*pŏst-ū'-ter-in*) [*post-;* *uterus*]. Situated behind the uterus.
**postvaccinal** (*pŏst-vaks'-sin-nal*). Following vaccination.
**postvermis** (*pŏst-ver'-mis*) [*post-;* *vermis*]. The inferior vermiform process of the cerebellum.
**potable** (*po'-ta-bl*) [*potare,* to drink]. Drinkable; fit to drink.
**Potain's solution** (*po-tan'*) [Pierre Carl Edouard Potain, French physician, 1825–1901]. *For use with the hemoglobinometer:* a mixture of a solution of gum acacia, sp. gr. 1020, 1 volume; equal parts of sodium sulphate and sodium chloride in solution of sp. gr. 1020, 3 volumes. **P.'s syndrome,** dyspepsia and dilatation of the right ventricle with accentuation of the pulmonary sound, observed during the digestive process in cases of gastrectasia.
**potamophobia** (*pot-am-o-fo'-be-ah*) [ποταμός, river; φόβος, fear]. The morbid fear of sheets of water.
**potash** (*pot'-ash*). 1. Potassium hydroxide; caustic potash. 2. Potassium carbonate.
**potassa** (*po-tas'-ah*). Potash. See *potassium hydroxide.*
**potassacol** (*po-tas'-sa-kol*). The potassium salt of guaiacol-sulphonic acid.
**potassic** (*po-tas'-ik*) [*potassa*]. Containing potassium.
**potassiocupric** (*po-tas-e-o-kū'-prik*). Containing potassium and copper.
**potassiomercuric** (*po-tas-e-o-mer-kū'-rik*). Combined with potassium and mercury.
**potassium** (*po-tas'-e-um*) [*potassa*]. A silver-white, soft, very ductile metal, belonging to the alkali group, and having a sp. gr. of 0.865; quantivalence I; atomic weight 39.10; symbol K (from the Latin *kalium*). It forms compounds with nearly all nonmetals. **p. acetate** (*potassii acetas,* U. S. P., B. P.), KC₂H₃O₂, used in rheumatism, as a diuretic in dropsy, and in cutaneous diseases. Dose 20 gr.–1 dr. (1.3–4.0 Gm.). **p. arsenate,** K₃HAsO₄, extremely poisonous crystals, soluble in water; used as an antiperiodic and alterative. Dose $\frac{1}{20}$–$\frac{1}{10}$ gr. (0.003–0.006 Gm.). **p. arsenite,** solution of (*liquor potassii arsenitis,* U. S. P.), Fowler's solution. Dose 3 min. (0.2 Cc.). **p. bicarbonate** (*potassii*

*bicarbonas*, U. S. P., B. P.), $KHCO_3$, used as an antacid. It is also highly recommended in influenza and to abort colds. Dose 20 gr.–1 dr. (1.3–4.0 Gm.). **p. binoxalate,** salt of sorrel, a constituent of wood-sorrel. **p. bisulphate,** $KHSO_4$, has been used as a laxative and refrigerant. Dose 1–2 dr. (4–8 Gm.). **p. bitartrate** (*potassii bitartras*, U. S. P., B. P.), cream of tartar, $KHC_4H_4O_6$, used as a cathartic, as a diuretic in dropsy, and for making refrigerant drinks in febrile affections. It is an ingredient of *pulvis jalapæ compositus* (U. S. P.). Dose 1 dr.–½ oz. (4–16 Cc.). **p. bromide.** See *bromidum, potassii*, under *bromine*. **p. camphorate,** $K_2C_{10}H_{14}O_4$, white, antiseptic crystals, soluble in water, used in night-sweats of tuberculosis, gonorrhea, etc. Dose 10–30 gr. (0.65–2.0 Gm.). **p. carbolate,** $C_6H_5OK$, white, antiseptic crystals, soluble in water; used in diarrhea and dysentery. Dose 1–5 gr. (0.065–0.32 Gm.). **p. carbonate** (*potassii carbonas*, U. S. P., B. P.), used as an antacid in dyspepsia and as an antilithic. **p. chlorate** (*potassii chloras*, U. S. P., B. P.), $KClO_3$, used locally and internally in various forms of stomatitis, in mercurial ptyalism, and in pharyngitis. Dose 10–20 gr. (0.65–1.3 Gm.). In overdoses it is an irritant poison. From it are prepared *trochisci potassii chloratis* (U. S. P.). **p. citrate** (*potassii citras*, U. S. P., B. P.), $K_3C_6H_5O_7+H_2O$, used as a diaphoretic in fevers, in bronchitis, and in rheumatism, the uric-acid diathesis, etc. Dose 20–30 gr. (1.3–2.0 Gm.). **p. citrate, effervescent** (*potassii citras effervescens*, U. S. P.), is used like the preceding. Dose 1–2 dr. (4–8 Cc.). **p. citrate, solution of** (*liquor potassii citratis*, U. S. P.), an aqueous liquid containing 8 % of anhydrous potassium citrate with small amounts of citric and carbonic acids. *Neutral mixture* is prepared by substituting lemon-juice for the citric acid. Dose ½–1 oz. (16–30 Cc.). **p. cobaltinitrate,** has been employed in asthma, nephritis, and valvular heart disease. Dose ¼–½ gr. (0.016–0.032 Gm.). **p. cyanide** (*potassii cyanidum*, U. S. P., B. P.), $KCN$, is similar in properties to hydrocyanic acid. It is also used locally in neuralgia. Dose ⅛ gr. (0.008 Gm.). **p. dichromate, p. bichromate** (*potassii dichromas*, U. S. P., B. P.), $K_2Cr_2O_7$, used as a caustic, and for the preparation of battery-fluids and of preserving-fluids for tissues; it was formerly employed as an alterative. **p. diiodoresorcin-monosulphate.** See *picrol*. **p. dithiocarbonate,** $K_2COS_2$, an orange-red, crystalline powder, used externally in skin diseases: eczema, 5 to 10 % ointment; psoriasis, 20 % ointment. **p. ferricyanide,** $K_6Fe_2(CN)_{12}$, is used in the arts and as a reagent. **p. ferrocyanide** (*potassii ferrocyanidum*, U. S. P., B.P.), $K_4Fe(CN)_6+3H_2O$, yellow prussiate of potash, has been used in the night-sweats of tuberculosis. Dose 10–15 gr. (0.65–1.0 Gm.). It is extensively employed in the arts, as a reagent, and in pharmacy to prepare diluted hydrocyanic acid. **p. fluoresceinate,** $K_2C_{20}H_{10}O_5$, the potassium salt of fluorescein; a yellowish-red powder, soluble in water. It is used to detect corneal ulcerations. **p.-guaiacol sulphonate,** a fine white powder, soluble in water, containing 60 % of guaiacol. Dose 5–15 gr. (0.32–0.97 Gm.). **p. hydroxide** (*potassii hydroxidum*, U. S. P., B. P.), $KOH$, caustic potash. This is a white solid, usually occurring in the form of pencils, and having powerful escharotic properties. **p. hypochlorite,** $KOCl$, a substance which in solution forms Javelle water. **p. hypophosphite** (*potassii hypophosphis*, U. S. P.), $KH_2PO_2$. Dose 10–30 gr. (0.65–2.0 Gm.). See *hypophosphite*. **p. iodate,** is used in stomatitis and pharyngitis as a substitute for potassium chlorate. Dose 4–8 gr. (0.26–0.52 Gm.). **p. iodide** (*potassii iodidum*, U. S. P., B. P.), $KI$, is used as an alterative in syphilis; in chronic poisoning by lead or mercury; as an absorbent of inflammatory exudates; in chronic rheumatism, etc. Dose 2–10 gr. (0.13–0.65 Gm.); in syphilis several drams may be administered daily. From it are prepared *unguentum iodi* and *unguentum potassii iodidi* (U. S. P.). **p. nitrate** (*potassii nitras*, U. S. P., B. P.), $KNO_3$, saltpeter, is used as a refrigerant diuretic and diaphoretic; in asthma the inhalation of fumes produced by burning niter-paper (paper dipped in a solution of the nitrate and dried) is useful. Dose 10–20 gr. (0.65–1.3 Gm.). From it is prepared *argenti nitras mitigatus* (U. S. P.). **p. nitrite,** $KNO_2$, white, amorphous sticks, soluble in water; used in asthma, epilepsy, etc. Dose ¼–2 gr. (0.016–0.13 Gm.) several times daily. **p. nitroprusside,** $K_2F_2 . NO . (CN)_5+2H_2O$, garnet-red crystals, soluble in water or alcohol, used as a test for albumin in urine. **p. perchlorate,** $KClO_4$, white crystals, slightly soluble in water; antipyretic, antiperiodic, sedative, and diuretic. Dose 5–15 gr. (0.32–1.0 Gm.). Syn., *hyperchlorate of potassium*. **p. permanganate** (*potassii permanganas*, U. S. P., B. P.), $KMnO_4$, is a salt of permanganic acid, which is derived from the manganese heptoxide, $Mn_2O_7$. It is used as an antiseptic and deodorant, in amenorrhea, and as an antidote to opium-poisoning and poisoning by snake-bites. Dose as an emmenagogue 1–2 gr. (0.065–0.13 Gm.). It is also employed as a reagent. *Condy's fluid* is an aqueous solution of potassium permanganate 2 parts in 100. **p. phosphate,** $K_2HPO_4$, has been used as an alterative in scrofula and pulmonary tuberculosis. Dose 10–30 gr. (0.65–2.0 Gm.). **p. salicylate,** $KC_7H_5O_3$, a white powder, soluble in water or alcohol; antipyretic, analgesic, antirheumatic. Dose 6–15 gr. (0.4–1.0 Gm.). **p. silicate,** $K_2SiO_3$ (approximately), soluble glass, is used as is plaster-of-Paris for making fixed dressings for fractures. **p. and sodium tartrate** (*potassii et sodii tartras*, U. S. P., B. P.), $KNaC_4H_4O_6+4H_2O$, Rochelle salt, is used as a saline cathartic; it is an ingredient of seidlitz powder. Dose ½–1 oz. (16–32 Cc.). **p. sozoiodolate,** $KHC_6H_2I_2OSO_3$, a white crystalline powder containing 58.8 % of iodine, 20 % of phenol, and 7 % of sulphur; soluble in hot water, melts at 120° C. It is bactericidal and is used externally in 3 % ointment. **p. sulphate** (*potassii sulphas*, U. S. P., B. P.), $K_2SO_4$, is used as a laxative and purgative. Dose ½–4 dr. (2–16 Gm.). **p. sulphite,** $K_2SO_3+2H_2O$, is antiseptic and laxative and used in acid fermentation of the stomach. Dose 15–60 gr. (1–4 Gm.). **p. sulphobenzoate,** $KC_7H_5SO_5+5H_2O$, soluble in water or alcohol; used as a wash, 4 to 5 % solution, in skin diseases. **p. sulphocarbolate,** $KC_6H_5SO_4+H_2O$, white crystals, soluble in water or alcohol. It is antiparasitic, germicidal, and antiseptic. **p. sulphocarbonate,** $K_2CS_3$, yellow crystals, soluble in water, used in baths in skin diseases. **p. sulphocyanate,** $KCNS$, colorless needles, soluble in water or alcohol; it is a constituent of saliva; sedative, antispasmodic, and anodyne. Dose ¾–3 gr. (0.05–0.2 Gm.); maximum dose 4½ gr. (0.3 Gm.), single; 24 gr. (1.5 Gm.) daily. **p. sulphophenate.** See *p. sulphocarbolate*. **p. tartrate** (*potassii tartras*, B. P.), $K_2C_4H_4O_6 . H_2O$, is used as a mild purgative in febrile affections. Dose 1 dr.–1 oz. (4–32 Cc.). **p. valerate,** $KC_5H_9O_2$, used as a stimulant in low fevers, hysteria, etc. Dose 2–5 gr. (0.13–0.32 Gm.) several times daily.

**potato culture.** A culture of microorganisms on potato. **p. oil,** crude amyl alcohol. **p. treatment** (of diabetes), a daily diet of 1–2½ pounds of potatoes and the exclusion of bread.

**potency** (*po'-ten-se*) [see *potential*]. 1. Power; efficacy. 2. In homeopathy, the degree of dilution of a drug.

**potentia** (*po-ten'-she-ah*) [L.]. Power, potency, ability, faculty. **p. coeundi,** capacity for copulation. **p. generandi,** procreance.

**potential** (*po-ten'-shal*) [*potens*, able]. 1. Capable of acting or doing work, as *potential* energy. 2. In electricity, a state of tension or of potential energy capable of doing work. If two bodies of different potential are brought together, a current is established between them that is capable of doing work.

**potentialization** (*po-tent-shal-i-za'-shun*). In homeopathy, the rendering of drugs potent by attenuation, dynamization, etc.

**potentiation.** See *potentialization*.

**potentize** (*po'-tent-iz*) [*posse*, to be able]. To render potent; in homeopathy, applied to drugs.

**potentor** (*po-tent'-or*) [*potentia*, power]. A device for the mechanical treatment of male impotence due to lack of penile erectility.

**potio** (*po'-she-o*) [L.]. A potion.

**potion** (*po'-shun*) [*potio*]. A drink or draught.

**potocytosis** (*po-to-si-to'-sis*) [*potio*; κύτος, cell]. The ability of cells to drink solutions.

**potomania** (*po-to-ma'-ne-ah*) [*potio*; μανία, madness]. Delirium tremens.

**Pott's aneurysm** [Percival *Pott*, English surgeon, 1714–1788]. Aneurysmal varix. **P.'s boss,** the projecting spinous process noticeable on palpation in Pott's disease or vertebral caries. **P.'s curvature, P.'s gibbus,** the deformity of Pott's disease. **P.'s disease,** caries of the vertebræ, generally of tuberculous origin. The symptoms are stiffness of

the spinal column, pain on motion, tenderness on pressure, undue prominence of one, or more of the spines; in certain cases spasmodic pain in the abdomen; formation of abscess; occasionally, in late stages, paralysis. **P.'s fracture,** fracture of the fibula about three inches above the ankle-joint, usually with a splitting-off of the tip of the inner malleolus, and at times rupture of the internal lateral ligament, with outward displacement of the foot. **P.'s gangrene,** senile gangrene. **P.'s paraplegia,** paraplegia caused by spinal caries. **P.'s puffy tumor,** circumscribed superficial swelling of the scalp connected with osteomyelitis of the skull.
**potters' asthma.** A form of pneumonokoniosis prevalent among potters. **p. bronchitis.** See *bronchitis*. **p. clay.** See *argilla*. **p. consumption,** a form of pneumonokoniosis prevalent among potters. **p. lung,** a chronic inflammation of the lungs common among potters, and due to the inhalation of dust.
**pouch** [Fr., *poche*, a pocket]. A sac or pocket. **p., Broca's.** See *Broca's pouch*. **p., pressure,** a bulge in the wall of the esophagus due to weakness. **p. of Douglas, p., rectouterine.** See *Douglas' culdesac*. **p., laryngeal,** a blind pouch of mucosa opening into the ventral part of the ventricle of the larynx.
**Poulet's disease** (*poo-la'*) [Alfred *Poulet*, French physician, 1848-1888]. Rheumatoid osteoperiostitis.
**poultice** (*pōl'-tis*) [*puls*, porridge]. A soft, semiliquid mass made of some cohesive substance mixed with water, and used for application to the skin, for the purpose of supplying heat and moisture or acting as a local stimulant. Syn., *cataplasma*.
**poultogen** (*pōl'-to-jen*). A poultice said to contain oil of wintergreen, boric acid, salicylic acid, iodine (in organic combination), and pine oils in a base of calcined siliceous earth.
**pound** (*pownd*). A unit of measure of weight. The *troy pound* contains 12 oz., or 5760 grains; the *avoirdupois pound* contains 16 oz., or 7000 grains. Symbol lb. See *weights and measures*. **p., foot-,** the force necessary to raise one pound through the height of one foot.
**poundal** (*pown'-dal*). A unit of force; that force which applied to a pound of matter for one second generates in it a velocity of one foot per second. **p., foot.** See *poundal*.
**Poupart's ligament** (*poo'-part*) [François *Poupart*, French anatomist, 1661-1709]. The ligament extending from the anterior superior spine of the ilium to the spine of the pubis and the pectineal line. It is the lower portion of the aponeurosis of the external oblique muscle.
**powder** [Fr., *poudre*, from *pulvis*, powder]. 1. A collection of fine particles. 2. In pharmacy, a collection of fine particles of one or more substances capable of passing through a sieve having meshes of a certain fineness; also a single dose of such a substance. **p., aromatic.** See *pulvis aromaticus*. **p., Dover's** (*pulvis ipecacuanhæ et opii*, U. S. P.). See *opium, powder of ipecac and*. **p., Dupuytren's.** See *Dupuytren's powder*. **p., gray,** hydrargyrum cum creta. **p., Gregory's.** See *Gregory's powder*. **p., James'.** See *antimony, powder of*. **p., Portland,** a quack remedy consisting of equal parts of the tops and leaves of *Teucrium chamædrys*, and *Erythræa centaurium*, the leaves of *Ajuga chamæpitys*, and the roots of *Gentiana lutea*, and *Aristolochia rotunda*. Dose 1 dr. (3.8 Gm.) daily before breakfast for 6 months, ½ dr. (1.2 Gm.) for 3 months; ⅓ dr. (1.9 Gm.) for 6 months, and finally ⅙ dr. (1.9 Gm.) every other day for a year. **p., Seidlitz-.** See *pulvis effervescens compositus* **p., Tully's.** See *Tully's powder*.
**pox** [*pocks*, pl. of *pock*, a vesicle or pustule]. 1. A term applied to any disease possessing a vesicular or pustular eruption. 2. Vulgarly, syphilis. **p., chicken-.** See *varicella*. **p., cow-.** See *cowpox*. **p., small-.** See *variola*.
**P.p.** Abbreviation for *punctum proximum*, Latin for near point.
**Ppt.** Abbreviation for *precipitate*.
**Pr.** Abbreviation for *presbyopia*.
**P.r.** Abbreviation for *punctum remotum*, Latin for far point.
**practice** (*prak'-tis*) [*practicare*, to practice]. The practice of medicine; the application of the principles of medicine to the diagnosis and treatment of disease.
**practician** (*prak-tish'-an*). See *practitioner*.
**practise** (*prak'-tis*) [see *practice*]. To perform the duties of a physician.

**practitioner** (*prak-tish'-on-er*) [see *practice*] One who practises medicine.
**præ-** (*pre-*) [L.]. See *pre-*.
**præcava** (*pre'-ka-vah*). See *precava*.
**præcordia** (*pre-kor'-de-ah*). See *precordia*.
**præcornu** (*pre-kor'-nū*). See *precornu*.
**præcuneus** (*pre-kū-ne'-us*). See *precuneus*.
**prægeniculatum** (*pre-jen-ik-ū-la'-tum*). See *prægeniculatum*.
**præmaxilla** (*pre-maks-il'-ah*). See *premaxilla*.
**præmedulla** (*pre-me-dul'-ah*). Synonym of *medulla oblongata*.
**præperforatus** (*pre-per-for-a'-tus*). See *preperforatus*.
**præputium** (*pre-pū'-she-um*). See *prepuce*.
**prævia** (*pre'-ve-ah*) [fem. of *prævius*, from *præ*, before; *via*, a way]. Coming before; standing before. **p., placenta.** See *placenta prævia*.
**pragmatagnosia** (*prag-mat-ag-no'-se-ah*) [πρᾶγμα, an object; ἀγνωσία, want of recognition]. Inability to recognize an object. **p., visual,** a term suggested for object-blindness.
**pragmatamnesia** (*prag-mat-am-ne'-se-ah*) [πρᾶγμα, an object; ἀμνησία, forgetfulness]. Inability to remember the appearance of an object. **p., visual,** a term suggested for that mental condition in which there is inability to call up the visual image of an object.
**Prague method** (*prāg*, or *prahg*) [*Prague*, the capital of Bohemia]. A method of delivery of the aftercoming head. The child's ankles are grasped above the internal malleoli with the right hand. The index finger of the left hand is flexed over one clavicle, and the remaining fingers of the same hand over the other clavicle. Traction directly down is now made with both hands until the perineum is well distended. The right hand then loosens its hold upon the ankles and again grasps higher up the legs, the child's feet being in contact with the back of the right hand. By a circular movement the feet are now raised toward the mother's abdomen, the obstetrician using the left hand, as originally placed, as a fulcrum around which the head moves.
**prairie-itch.** A name applied to various forms of skin diseases associated with pruritus, occurring in men engaged in work on prairies, among lumbermen, and others, and either due to scabies or produced hiemalis.
**praseodymium** (*pras-e-o-dim'-e-um*). A metallic element, one of the constituents of didymium.
**prasoid** (*pra'-soid*). 1. A solution of globularin and globulatetin; used in acute gout and rheumatism. Dose 15-20 drops 3 times daily. 2. [πράσον, the leek; εἶδος, likeness.] The color of a leek; leek-green.
**pratique** (*prah-teek'*) [Fr.]. The bill-of-health given to vessels by a health officer.
**Pratt's operation.** The practice of orificial surgery, based on the belief that many chronic diseases are due to morbid conditions of the orifices of the body, particularly the anus and the urogenital canal. The operation consists in the dilatation of either or both of these orifices and the removal of any irritating condition that may be present.
**Pravaz's syringe** (*prav-ah'*) [Charles Gabriel *Pravaz*, French physician, 1791-1853]. A hypodermic syringe.
**praxinoscope** (*praks-in'-o-skōp*) [πρᾶξις, a doing; σκοπεῖν, to examine]. A modification of the zoetrope adapted to the purposes of laryngological instruction.
**Pray's test-letters or test.** A test for astigmatism, consisting of capital letters composed of strokes that run in different directions for each letter.
**pre-** [*præ-*]. A prefix signifying before.
**preacetabular** (*pre-as-et-ab'-ū-lar*) [*pre-*; *acetabulum*]. Situated in front of the acetabulum.
**preagonal** (*pre-ag'-on-al*). Immediately preceding the death agony.
**prealbuminuric** (*pre-al-bū-min-ū'-rik*). Occurring before the appearance of albuminuria.
**preanal** (*pre-a'-nal*) [*pre-*; *anus*]. Situated in front of the anus.
**preantiseptic** (*pre-an-tis-ep'-tik*). Pertaining to the time before the adoption of antisepsis in surgical practice.
**preaortic** (*pre-a-or'-tik*) [*pre-*; *aorta*]. Situated in front of the aorta.
**preaseptic** (*pre-a-sep'-tik*) [*pre-*; *asepsis*]. Pertaining to the period before the introduction of aseptic surgery.

**preatactic** (*pre-ah-tak'-tik*). See *preataxic*.
**preataxic** (*pre-at-aks'-ik*) [*pre-; ataxia*]. Occurring before ataxia.
**preauditory** (*pre-aw'-dit-or-e*) [*præ, before; audire*, to hear]. Situated in front of the auditory nerve or canal.
**preauricular** (*pre-aw-rik'-ū-lar*). Situated in front of the auricle.
**preaxal** (*pre-aks'-al*) [*præ, before; axis*]. Prechordal; placed in front of the axon.
**preaxial** (*pre-aks'-e-al*) [*pre-; axis*]. Situated in front of the axis of the body or of a limb.
**prebacillary** (*pre-bas'-il-a-re*) [*pre-; bacillus*]. Occurring before the invasion of the system by bacilli.
**prebasilar** (*pre-bas'-il-ar*) [*præ, before; basis, base*]. Situated, or occurring, in front of any basilar structure, especially, in front of the basilar process of the occipital bone.
**prebrachial** (*pre-bra'-ke-al*) [*præ, before; brachium*, arm]. Located on the anterior aspect of the brachium or upper arm, e. g., the group of prebrachial muscles: biceps, coraco-brachialis, and brachialis anticus.
**prebrachium** (*pre-bra'-ke-um*). See *brachium quadrigeminum superius*.
**precancerous** (*pre-kan'-ser-us*) [*pre-; cancer*]. Occurring before the development of a carcinoma.
**precapillary** (*pre-cap'-il-ar-e*). An arteriole or venule.
**precardiac** (*pre-kar'-de-ak*) [*pre-; καρδία, heart*]. Anterior to the heart.
**preçava** (*pre-ka'-vah*) [*præ, before; cavus, hollow*]. The superior, or descending vena cava.
**precentral** (*pre-sen'-tral*) [*pre-; centrum, center*]. Situated in front of the central fissure of the brain. **p. convolution,** a convolution in front of the central fissure of the brain; the ascending frontal convolution.
**precerebellar** (*pre-ser-e-bel'-ar*) [*præ, before; cerebellum*]. Situated before (above) the cerebellum.
**precerebral** (*pre-ser'-e-bral*) [*præ, before; cerebrum*]. Situated, or occurring before (above), the cerebrum.
**prechordal** (*pre-kor'-dal*) [*præ, before; chorda*, chord; string]. Situated in front of the notochord.
**precipitant** (*pre-sip'-it-ant*). Any reagent causing precipitation.
**precipitate** (*pre-sip'-it-āt*) [*precipitare*, from *præ*, before; *caput*, head]. 1. The solid substance thrown down from a solution of a substance on the addition of a reagent which deprives it of its solubility. 2. The product of the reaction between precipitinogen and precipitin. 3. To throw down in an insoluble form. 4. Headlong; hasty. **p. labor.** See *labor, precipitate*. **p., red,** hydrargyrum oxidum rubrum. See *mercury oxide, red*. **p., white,** N₂(Hg₂)₃Cl₂, hydrargyrum ammoniatum. See *mercury, ammoniated*.
**precipitation** (*pre-sip-it-a'-shun*) [*precipitate*]. The process of throwing down solids from the liquids which hold them in solution. Precipitates are crystalline, amorphous, curdy, flocculent, granular, or gelatinous, according to the form assumed.
**precipitin** (*pre-sip'-it-in*). A body produced in the blood-plasma of animals by repeated injections of bacterial filtrates or foreign organic substances (serum, milk, etc.) and causing a precipitation of the bacteria or foreign substance used in the preparation.
**precipitinogen** (*pre-sip-it-in'-o-jen*). Any substance capable of causing the production of a specific precipitin.
**precipitinoid** (*pre-sip'-it-in-oid*). An inactive precipitin modified by heating to 60° C.
**preclavicular** (*pre-kla-vik'-ū-lar*). Ventrad of the clavicle.
**preclival** (*pre-kli'-val*). In front of the clivus of the cerebellum.
**precocity** (*pre-kos'-it-e*) [*præ, before; coquere*, to ripen]. Early development or maturity. It is applied especially to great development of the mental faculties at an early age.
**precommissure** (*pre-kom'-is-ūr*) [*præ, before; commissura*, commissure]. The anterior commissure of the brain.
**preconvulsant** (*pre-kon-vul'-sant*). Relating to the stage of a disease preceding convulsions.
**precordia** (*pre-kor'-de-ah*) [*pre-; cor*, the heart]. 1. A name applied by the ancients to the diaphragm, the thoracic viscera, or the epigastric region. 2. The area of the chest overlying the heart.
**precordial** (*pre-kor'-de-al*) [*precordia*]. Pertaining to the precordia.

**precordialgia** (*pre-kor-de-al'-je-ah*) [*præ, before; cor, heart; ἄλγος, pain*]. Pain in the precordial region.
**precordium** (*pre-kor'-de-um*). Syn., *precordia*.
**precornu** (*pre-kor'-nū*) [*pre-; cornu*, a horn]. Anterior horn of lateral ventricle of the brain.
**precribrum** (*pre-krib'-rum*) [*præ, before; cribrum*, sieve]. The anterior perforated space of the brain.
**precuneal** (*pre-kū'-ne-al*) [*præ, before; cuneus*, wedge]. Situated in front of the cuneus; specifically, pertaining to the precuneus.
**precuneate** (*pre-kū'-ne-āt*). See *precuneal*.
**precuneus** (*pre-kū'-ne-us*) [*pre-; cuneus*, a wedge]. The quadrate lobule of the parietal lobe situated in front of the cuneus of the occipital lobe.
**prediastolic** (*pre-di-as-tol'-ik*) [*pre-; diastole*]. Occurring before the diastole.
**predicrotic** (*pre-di-krot'-ik*). Preceding the dicrotic wave or elevation of the sphygmographic tracing.
**predigested** (*pre-dij-es'-ted*) [*præ, before; digerere*, to digest]. Partly digested by artificial means before being taken into the stomach. **p. foods,** foods that have been prepared by a process of artificial digestion through the agency of various ferments.
**predigestion** (*pre-di-jes'-chun*) [*præ, before; digerere*, to digest]. The partial digestion of food before it is eaten.
**predisposing** (*pre-dis-po'-zing*) [see *predisposition*]. Rendering susceptible or liable to attack.
**predisposition** (*pre-dis-po-zish'-un*) [*pre-; disponere*, to dispose]. The state of having a susceptibility to disease.
**predormition** (*pre-dor-mish'-un*) [*præ, before; dormition*]. Applied to the stage of unconsciousness immediately preceding actual sleep.
**preepiglottic** (*pre-ep-i-glot'-ik*). Ventrad of the epiglottis.
**preeruptive** (*pre-e-rup'-tiv*). Preceding eruption.
**prefontanel** (*pre-fon-tan-el'*) [*præ, before; fontanel*]. The anterior fontanel.
**preforceps** (*pre-for'-ceps*) [*pre-; forceps*]. Those hooked or curved anterior fibers of the callosum that reach cephalad into the frontal lobe.
**prefrontal** (*pre-fron'-tal*) [*pre-; frons*, forehead]. 1. Situated in the anterior part of the frontal lobe of the brain. 2. The middle part of the ethmoid.
**pregeminal** (*pre-jem'-in-al*) [*præ, before; geminus*, twin]. Relating to the anterior pair of the corpora quadrigemina.
**pregeminum** (*pre-jem'-in-um*) [*pre-; geminus*, twin]. The anterior pair of the corpora quadrigemina, considered as forming a single organ.
**pregeniculatum, pregeniculum** (*pre-jen-ik-ū-la'-tum, pre-jen-ik'-ū-lum*) [*pre-; geniculatum*, geniculate]. The external geniculate body.
**preglobulin** (*pre-glob'-ū-lin*) [*pre-; globulin*]. An albuminous body found in cell-protoplasm, insoluble in water, soluble in a 10 % sodium chloride solution and in dilute alkaline solution.
**pregnancy** (*preg'-nan-se*) [*prægnans*, from *præ*, before; *gignere*, to beget]. The state of being pregnant, or with child; the state of the woman from conception to the expulsion of the ovum. The normal duration of pregnancy is 280 days, or 10 lunar months, or 9 calendar months. **p., abdominal,** one in which the fetus lies in the abdominal cavity. **p., cornual,** that occurring in one of the horns of a two-horned uterus. **p., extrauterine,** ectopic gestation; the development of the ovum outside of the cavity of the uterus. **p., false,** any condition in the abdomen that simulates pregnancy. **p., hydatid,** pregnancy with the formation of a hydatid mole. **p., interstitial,** pregnancy occurring in the part of the tube that traverses the uterine wall. **p., intramural,** interstitial pregnancy. **p., mesenteric,** tubo-ligamentary pregnancy. **p., molar,** pregnancy in which the ovum is converted into a mole. **p., multiple,** that form in which the uterus contains two or more developing ova. **p., mural,** a variety of extrauterine pregnancy in which the ovum develops in the wall of the uterus. **p., ovarian,** pregnancy occurring in the ovary. **p., parietal,** interstitial pregnancy. **p., phantom,** in hysteria, abdominal enlargement simulating pregnancy. **p., sarcofetal,** that in which both a fetus and a mole are present. **p., sarcohysterical,** false pregnancy due to a mole. **p., signs of,** those clinical manifestations by which the existence of pregnancy may be demonstrated. There are three so-called absolute signs: ballottement, fetal movements, and the fetal heart-sounds.

## ELY'S TABLE OF THE DURATION OF PREGNANCY.

*Explanation.*—Find in the upper horizontal row the date of last menstruation; the figure beneath will show the expiration of 280 days or ten months of 28 days each.

| | 1 | 2 | 3 | 4 | 5 | 6 | 7 | 8 | 9 | 10 | 11 | 12 | 13 | 14 | 15 | 16 | 17 | 18 | 19 | 20 | 21 | 22 | 23 | 24 | 25 | 26 | 27 | 28 | 29 | 30 | 31 |
|---|---|---|---|---|---|---|---|---|---|---|---|---|---|---|---|---|---|---|---|---|---|---|---|---|---|---|---|---|---|---|---|
| January / *October* | 8 | 9 | 10 | 11 | 12 | 13 | 14 | 15 | 16 | 17 | 18 | 19 | 20 | 21 | 22 | 23 | 24 | 25 | 26 | 27 | 28 | 29 | 30 | 31 | 1 *Nov.* | 2 | 3 | 4 | 5 | 6 | 7 |
| February / *November* | 8 | 9 | 10 | 11 | 12 | 13 | 14 | 15 | 16 | 17 | 18 | 19 | 20 | 21 | 22 | 23 | 24 | 25 | 26 | 27 | 28 | 29 | 30 | 1 *Dec.* | 2 | 3 | 4 | 5 | | | |
| March / *December* | 6 | 7 | 8 | 9 | 10 | 11 | 12 | 13 | 14 | 15 | 16 | 17 | 18 | 19 | 20 | 21 | 22 | 23 | 24 | 25 | 26 | 27 | 28 | 29 | 30 | 31 | 1 *Jan.* | 2 | 3 | 4 | 5 |
| April / *January* | 6 | 7 | 8 | 9 | 10 | 11 | 12 | 13 | 14 | 15 | 16 | 17 | 18 | 19 | 20 | 21 | 22 | 23 | 24 | 25 | 26 | 27 | 28 | 29 | 30 | 31 | 1 *Feb.* | 2 | 3 | 4 | |
| May / *February* | 5 | 6 | 7 | 8 | 9 | 10 | 11 | 12 | 13 | 14 | 15 | 16 | 17 | 18 | 19 | 20 | 21 | 22 | 23 | 24 | 25 | 26 | 27 | 28 | 1 *Mar.* | 2 | 3 | 4 | 5 | 6 | 7 |
| June / *March* | 8 | 9 | 10 | 11 | 12 | 13 | 14 | 15 | 16 | 17 | 18 | 19 | 20 | 21 | 22 | 23 | 24 | 25 | 26 | 27 | 28 | 29 | 30 | 31 | 1 *Apr.* | 2 | 3 | 4 | 5 | 6 | |
| July / *April* | 7 | 8 | 9 | 10 | 11 | 12 | 13 | 14 | 15 | 16 | 17 | 18 | 19 | 20 | 21 | 22 | 23 | 24 | 25 | 26 | 27 | 28 | 29 | 30 | 1 *May* | 2 | 3 | 4 | 5 | 6 | 7 |
| August / *May* | 8 | 9 | 10 | 11 | 12 | 13 | 14 | 15 | 16 | 17 | 18 | 19 | 20 | 21 | 22 | 23 | 24 | 25 | 26 | 27 | 28 | 29 | 30 | 31 | 1 *June* | 2 | 3 | 4 | 5 | 6 | 7 |
| September / *June* | 8 | 9 | 10 | 11 | 12 | 13 | 14 | 15 | 16 | 17 | 18 | 19 | 20 | 21 | 22 | 23 | 24 | 25 | 26 | 27 | 28 | 29 | 30 | 1 *July* | 2 | 3 | 4 | 5 | 6 | 7 | |
| October / *July* | 8 | 9 | 10 | 11 | 12 | 13 | 14 | 15 | 16 | 17 | 18 | 19 | 20 | 21 | 22 | 23 | 24 | 25 | 26 | 27 | 28 | 29 | 30 | 31 | 1 *Aug.* | 2 | 3 | 4 | 5 | 6 | 7 |
| November / *August* | 8 | 9 | 10 | 11 | 12 | 13 | 14 | 15 | 16 | 17 | 18 | 19 | 20 | 21 | 22 | 23 | 24 | 25 | 26 | 27 | 28 | 29 | 30 | 31 | 1 *Sept.* | 2 | 3 | 4 | 5 | 6 | |
| December / *September* | 7 | 8 | 9 | 10 | 11 | 12 | 13 | 14 | 15 | 16 | 17 | 18 | 19 | 20 | 21 | 22 | 23 | 24 | 25 | 26 | 27 | 28 | 29 | 30 | 1 *Oct.* | 2 | 3 | 4 | 5 | 6 | 7 |

**p., tubal,** pregnancy which takes place within the Fallopian tube. **p., tuboabdominal,** that in which the ovum is developed in the ampulla and extends into the abdominal cavity. **p., tuboligamentary,** that in which there is secondary invasion of the broad ligament and subperitoneal tissues. **p., tuboovarian,** that in which the ovum is attached to the oviduct and ovary. **p., tubouterine.** Same as *p., interstitial.* **p., uteroabdominal,** that in which there is one fetus in the uterus and another in the abdominal cavity. **p., uteroovarian,** that in which there is one fetus in the uterus and another in the ovary. **p., uterotubal,** that in which one fetus is in the uterus and another in the oviduct.
**pregnant** (*preg'-nant*) [see *pregnancy*]. With child; gravid.
**prehemiplegic** (*pre-hem-i-ple'-jik*) [*pre-;* hemiplegia]. Occurring before an attack of hemiplegia, as *prehemiplegic* chorea.
**prehensile** (*pre-hen'-sil*) [see *prehension*]. Adapted for grasping.
**prehension** (*pre-hen'-shun*) [*prehendere*, to lay hold of]. The act of grasping or seizing.
**prehypophysis** (*pre-hi-pof'-is-is*) [*præ*, before; *hypophysis*]. The anterior and larger portion of the hypophysis, derived from the anterior mouth.
**preinsula** (*pre-in'-su-lah*) [*præ*, before; *insula*]. The cephalic region of the insula.
**prelacrymal** (*pre-lak'-rim-al*) [*præ*, before; *lacryma*, tear]. Situated in front of the lacrymal bone, or gland, or sac.
**prelimbic** (*pre-lim'-bik*) [*pre-;* limbus, a border; a band]. Lying in front of a limbus. **p. fissure,** the anterior portion of the callosomarginal fissure.
**prelum** (*pre'-lum*) [L.]. A press. **p. abdominale,** the squeezing of the abdominal viscera between the diaphragm and the rigid abdominal wall, as in the processes of defecation, micturition, and parturition.
**p. arteriale,** a tourniquet.
**prelumbar** (*pre-lum'-bar*) [*præ*, before; *lumbus*, loin]. Anterior to the lumbar vertebræ or the loins.
**premalignant** (*pre-mal-ig'-nant*). Occurring before the development of malignancy.
**premaniacal** (*pre-ma-ni'-ak-al*) [*præ*, before; μανία, mania]. Previous to insanity, or to an attack of mania.
**premature** (*pre'-ma-tūr*) [*pre-;* maturare, to ripen]. Occurring before the proper time, as *premature* labor.
**premaxilla** (*pre-maks-il'-ah*). The intermaxillary bone.
**premaxillary** (*pre-maks'-il-a-re*). In front of the maxilla. **p. bone,** the incisive or intermaxillary bone.
**premenstrual** (*pre-men'-stroo-al*). Preceding menstruation.
**premolar** (*pre-mo'-lar*) [*pre-;* molar]. 1. Situated in front of the molar teeth. 2. One of the two bicuspid teeth. 3. A molar tooth of the temporary set.
**premonitory** (*pre-mon'-it-o-re*) [*pre-;* monere, to warn]. Forewarning; foreboding, as *premonitory* symptoms, those which forebode the onset of an attack of a disease.
**prenaris** (*pre-na'-ris*) [*præ*, before; *naris*, nostril; *pl., prenares*]. The anterior naris.
**prenasal** (*pre-na'-zal*) [*præ*, before; *nasus*, nose]. In front of the nose or nasal passages.
**prenatal** (*pre-na'-tal*) [*pre-;* natus, born]. Existing or occurring before birth.
**preoblongata** (*pre-ob-lon-gah'-tah*). The cephalic portion of the oblongata; situated mainly between the pons and the fourth ventricle.
**preoccipital** (*pre-ok-sip'-it-al*) [*præ*, before; *occiput*]. Situated anterior to the occipital region. **p. fissure,** a fissure on the ventral, lateral surface of the cerebrum separating the occipital and sphenotemporal lobes. **p. notch,** a notch indicating the division between the occipital and temporal lobes of the brain.
**preoccupation** (*pre-ok-u-pa'-shun*). The condition of being so engrossed in one's own thoughts as not to answer or hear when accosted.
**preoperculum, præoperculum** (*pre-o-per'-kū-lum*) [*præ*, before; *operculum*, a lid; *pl., preopercula*]. The frontal operculum of the brain, between the presylvian and subsylvian fissures.
**preoptic** (*pre-op'-tik*). Anterior to the optic lobes; pregeminal.
**preopticus** (*pre-op'-tik-us*) [*præ*, before; ὄψ, sight]. The anterior pair of the quadrigeminal bodies.

**preoral** (*pre-o'-ral*) [*præ*, before; *os, oris,* mouth]. Situated in front of the mouth; prebuccal.
**preovarian** (*pre-o-va'-re-an*). Situated in front of the ovary.
**prepalatal** (*pre-pal'-at-al*) [*præ*, before; *palatum,* palate]. Situated in front, or in the anterior part, of the palate.
**preparation** (*prep-ar-a'ₑ-shun*) [*præparare*, to make ready]. 1. The act of making ready. 2. Anything made ready, especially, in anatomy, any part of the body prepared or preserved for illustrative or other uses. 3. In pharmacy, any compound or mixture made after a formula.
**preparative** (*pre-par'-at-iv*). An immune body; amboceptor, *q. v.*
**preparator** (*prep'-ar-a-tor*). An immune body; amboceptor.
**prepatellar** (*pre-pat-el'-ar*) [*pre-;* patella]. Situated in front of the patella, as *prepatellar* bursa.
**prepeduncle** (*pre-pe'-dung-kl*) [*præ*, before; *pedunculus*, peduncle]. The anterior of the cerebellar peduncles.
**preperforatus** (*pre-per-for-a'-tus*) [*præ*, before; *perforare*, to perforate]. The anterior perforated space.
**preperitoneal.** See *properitoneal.*
**prephthisis** (*pre-ti'-sis, pre-te'-sis*) [*præ*, before; *phthisis*]. The pretuberculous state. The predisposition to tuberculosis.
**prepituitary** (*pre-pit-ū'-it-a-re*) [*præ*, before; *pituitary*]. Situated in front of the pituitary body.
**preplacental** (*pre-plas-en'-tal*) [*præ*, before; *placenta*]. Previous to the formation of the placenta.
**prepontile** (*pre-pon'-til*) [*præ*, before; *pons,* a bridge]. Situated in front of or above the pons Varolii.
**prepotency** (*pre-po'-ten-se*) [*præ*, before; *posse*, to be powerful]. In biology, dominant "force of heredity." The power that one parent may have of impressing his or her own character upon the offspring, the peculiar characters of the other parent being less obviously transmitted.
**prepotent** (*pre-po'-tent*) [*præ*, before; *posse*, to be able]. Having a marked tendency to transmit individual characters to offspring.
**prepuberal** (*pre-pū'-ber-al*). Prior to puberty.
**prepuce** (*pre'-pūs*) [*præputium*, prepuce]. The foreskin of the penis, a fold of skin lined by mucous membrane and covering the glans penis.
**preputial** (*pre-pū'-she-al*) [*prepuce*]. Pertaining to the prepuce.
**prepyloric** (*pre-pi-lor'-ik*) [*præ*, before; *pylorus*]. Placed in front of or preceding the pylorus.
**prepyramid** (*pre-pir'-am-id*) [*præ*, before; *pyramis*, pyramid]. One of the anterior (ventral) pyramids of the oblongata.
**preramus** (*pre-ra'-mus*) [*præ*, before; *ramus*, branch]. The vertical (anterior or cephalic) branch of the stem of the arbor of the cerebellum.
**prerectal** (*pre-rek'-tal*) [*præ*, before; *rectum*]. Situated in front of the rectum.
**prerenal** (*pre-re'-nal*) [*præ*, before; *ren*, kidney]. Situated in front of the kidney.
**prereproductive** (*pre-re-pro-duk'-tiv*). Relating to the period of life preceding puberty.
**presbycusis, presbykousis** (*prez-be-koo'-sis*) [πρέσβυς, old; ἀκούειν, to hear]. The lessening of the acuteness of hearing that occurs in old age.
**presbyonosus** (*pres-be-on'-o-sus*) [πρέσβυς, old; νόσος, disease]. Any disease peculiar to old age.
**presbyope** (*prez'-be-ōp*) [see *presbyopia*]. One who is presbyopic.
**presbyophrenia** (*prez-be-o-fren'-e-ah*) [πρέσβυς, old; φρήν, mind]. Senile dementia.
**presbyopia** (*prez-be-o'-pe-ah*) [πρέσβυς, old; ὤψ, eye]. The condition of vision in the aged, due to diminished power of accommodation from impaired elasticity of the crystalline lens, whereby the near-point of distinct vision is removed farther from the eye.
**presbyopic** (*prez-be-op'-ik*). Affected with presbyopia.
**presbysphacelus** (*prez-be-o-sfas'-el-us*) [πρέσβυς, old; σφάκελος, gangrene]. Senile gangrene.
**presbytia** (*prez-bish'-e-ah*). Synonym of *presbyopia*.
**presbytic** (*prez-bit'-ik*). Suffering from Y presbyopia.
**presbytism** (*prez'-bit-izm*). Presbyopia.
**prescapularis** (*pre-skap-ū-la'-ris*) [*præ*, before; *scapula*]. The supraspinatus muscle.

**presclerosis** (*pre-skle-ro'-sis*). The vascular condition which precedes arteriosclerosis.
**presclerotic** (*pre-skle-rot'-ik*). Preceding the occurrence of sclerosis.
**prescription** (*pre-skrip'-shun*) [*pre-;* *scribere,* to write]. A formula written by the physician to the apothecary, designating the substances to be administered. A prescription consists of the heading, usually the symbol ℞ (standing for the Latin word *recipe,* imperative of *recipere,* to take), the names and quantities of the ingredients, the directions to the apothecary, the directions to the patient, the date and the signature.
**presenile** (*pre-se'-nīl*) [*pre-;* *senilis,* age]. Prematurely old.
**presenility** (*pre-sen-il'-it-e*). Premature old age. See *progeria.*
**present** (*pre-zent'*) [*præsentare,* to place before]. Of a part of the fetus, to appear first at the os uteri.
**presentation** (*pre-zen-ta'-shun*) [see *present*]. In obstetrics, that part of the fetal body which presents itself to the examining finger at the os uteri.
**preservalin** (*pre-serv'-al-in*). A mixture of borax and boric acid used by dairymen.
**presphenoid** (*pre-sfe'-noid*). The anterior part of the body of the sphenoid bone.
**presphigmic** (*pre-sfig'-mik*) [*præ,* before; σφυγμός, pulse]. Pertaining to the period preceding the pulsewave.
**prespinal** (*pre-spi'-nal*) [*præ,* before; *spina,* spine]. In anatomy, ventrad of the spine.
**presse-artère** (*pres'-ar-tăr*) [Fr.]. An instrument for the compression or temporary occlusion of an artery.
**pressor** (*pres'-or*) [*premere,* to press]. 1. Stimulating. 2. A substance found in the infundibular part of the hypophysis; it produces a rise in blood-pressure. Cf. *depressor.* **p. nerves.** Nerves which under stimulation cause the vasomotor centers to react.
**pressure** (*presh'-ur*) [*premere,* to press]. Force, weight, or tension. **p., after,** the sense of pressure that remains for a brief period after the removal of an object from the surface of the body. **p., arterial,** the tension of the blood within the arteries. **p., atmospheric,** the pressure of the atmosphere; it equals about fifteen pounds to the square inch at sea-level. **p., bipolar,** pressure on the two ends of a bone. It is used in differentiating fractures from contusions, producing pain in the case of the former. **p., endocardial,** the pressure of the blood against the walls of the vessels or of the heart. It is measured by means of the manometer. **p., endocardial,** the pressure of the blood within the heart. **p., intra-abdominal,** the pressure exerted upon the parietes by the abdominal viscera. **p., intracranial,** the pressure of the contents of the cranium upon its walls. **p., intra-ocular.** See *tension, intraocular.* **p., intrathoracic,** the pressure of the intrathoracic organs upon the walls of the chest. **p.-myelitis,** myelitis from pressure on the cord. **p., negative,** the force of suction; also absence of pressure. **p.-points** or **spots,** points of marked sensibility to pressure or weight, arranged like the temperature-spots, and showing a specific end-apparatus arranged in a punctated manner and connected with the pressure-sense. **p.-pouch.** See *pouch.* **p., pulse,** the difference between the systolic and diastolic blood-pressure. **p.-sense,** the sense by which knowledge is obtained of the amount of weight or pressure which is exerted upon a part of the body. **p. sore.** See *bedsore.*
**presternum** (*pre-ster'-num*). The manubrium or superior segment of the sternum.
**Preston's salt.** Ammonium carbonate.
**presuppurative** (*pre-sup'-ū-ra-tiv*). Pertaining to an early stage of inflammation, prior to supporation.
**presylvian fissure.** The anterior branch of the Sylvian fissure.
**presystole** (*pre-sis'-to-le*) [*præ,* before; *systole*]. The period of the heart's pause preceding the systole.
**presystolic** (*pre-sis-tol'-ik*) [*pre-;* *systole*]. Preceding the systole of the heart, as the *presystolic* murmur, *presystolic* thrill.
**pretarsal** (*pre-tar'-sal*). Located anterior to the tarsus.
**pretibial** (*pre-tib'-e-al*) [*pre-;* *tibia*]. Situated in front of the tibia.
**pretuberculous** (*pre-tū-ber'-kū-lus*) [*pre-;* *tuberculosis*]. Preceding the development of tuberculosis.
**pretympanic** (*pre-tim-pan'-ik*) [*præ,* before; *tympanum*]. Situated in front of the tympanum or tympanic region.

**preurethritis** (*pre-ū-re-thri'-tis*). Inflammation of the vestibule of the vagina, around the urethral orifice.
**preventive** (*pre-ven'-tiv*) [*prævenire,* to anticipate; to prevent]. Warding off. **p. medicine,** the department of medicine dealing with the means and methods of preventing disease.
**prevermis** (*pre-ver'-mis*) [*præ,* before; *vermis,* worm]. The superior vermiform process of the cerebellum.
**prevertebral** (*pre-ver'-te-bral*) [*pre-;* *vertebra*]. Situated in front of the vertebræ.
**prevertiginous** (*pre-ver-tij'-in-us*) [*pre-;* *vertigo*]. Dizzy, with a tendency to fall prone.
**prevesical** (*pre-ves'-ik-al*) [*præ,* before; *vesica,* bladder]. Situated in front of the bladder.
**Prévost's symptom** (*pra-vo'*) [Jean Louis *Prévost,* Swiss physician, 1838– ]. Conjugate deviation of the eyes and head, which look away from the palsied extremities and toward the affected hemisphere; it is noted in cerebral hemorrhage.
**Preyer's test for carbon monoxide in the blood.** Warm three or four drops of the suspected blood for five minutes at 30° C, with 10 Cc. of water and 5 Cc. of potassium cyanide solution (1 : 2). The spectrum of normal blood, when so treated, loses the absorption line of oxyhemoglobin and in its place shows a broad absorption band, while the spectrum of carbon monoxide blood remains unchanged.
**prezygapophysis** (*pre-zi-gap-of'-is-is*) [*præ,* before; ζυγόν, yoke; ἀπόφυσις, process]. An anterior or superior zygapophysis; a superior oblique, or articular process of a vertebra.
**priapism** (*pri'-ap-ism*) [Πρίαπος, the god of procreation]. Persistent erection of the penis, usually unaccompanied by sexual desire. It is caused by injuries to the spinal cord or to the penis, and by vesical calculus.
**priapitis** (*pri-ap-i'-tis*). Inflammation of the penis.
**prickle-cell.** A cell possessing delicate rod-shaped processes by which it is connected with neighboring cells.
**prickle-layer.** The lowest stratum of the epidermis. It is formed of prickle-cells.
**prickly heat.** See *miliaria.*
**Priessnitz bandage** (*prees'-nits*) [Vincenz *Priessnitz,* German "healer," 1799–1851]. A cold wet compress.
**Priestley's mass** (*prēst'-le*) [Joseph *Priestley,* English clergyman and scientist, 1733–1804]. A green or greenish-brown deposit sometimes seen, especially in young individuals, on the upper and lower incisor and canine teeth; it is due to a growth of chromogenic fungi in Nasmyth's membrane.
**primæ viæ** (*pri'-me vi'-e*) [L., "The primary passages"]. The alimentary canal, the lacteals being "the secondary passages."
**primary** (*pri'-ma-re*) [*primus,* first]. First in time or in importance. **p. amputation,** one that is done before the development of inflammation, usually within the first 24 hours. **p. anesthesia,** the transient anesthesia from a small amount of the anesthetic. **p. bubo,** a simple adenitis of an inguinal lymphatic gland, resulting from mechanical irritation. Syn., *bubon d'emblée.* **p. dementia,** a form of insanity occurring in young adults, and characterized by an extreme degree of apathy, the patient lying motionless, absolutely listless, without wants, and seemingly without perception of his surroundings. **p. lesion,** the original lesion which forms the starting-point for secondary lesions. **p. sore,** the initial sclerosis or chancre of syphilis.
**primigravida** (*pri-me-grav'-id-ah*) [*primus,* first; *gravidus,* pregnant]. A woman pregnant for the first time.
**primipara** (*pri-mip'-ar-ah*) [*primus,* first; *parere,* to bear]. A woman bearing or giving birth to her first child.
**primiparity** (*pri-mip-ar'-it-e*) [*primus,* first; *parere,* to bear]. The condition of being a primipara.
**primiparous** (*pri-mip'-ar-us*) [*primipara*]. Pregnant or in labor for the first time.
**primisternalis** (*pri-me-ster'-nal, pri-me-ster-na'-lis*). Relating to the manubrium of the sternum.
**primitiæ** (*pri-mish'-e-e*) [*primus,* first]. The part of the liquor amnii discharged before the extrusion of the fetus at birth.
**primitive** (*prim'-it-iv*) [*primus,* first]. First-formed; original. **p. groove,** the enlargement and deepening of the primitive streak. **p. streak, p. trace,** a

streak appearing at the end of the germinal disc, and indicating the first trace of the embryo.
**primordial** (*pri-mor'-de-al*) [*primus*, first; *ordiri*, to rise]. Existing in the beginning; first-formed; primitive; original; of the simplest character. **p. kidney**, the Wolffian body. **p. ova**, cells lying among the germ-epithelium of the surface of the embryonic ovary.
**primordium** (*pri-mor'-de-um*). An organ or structure in its earliest state.
**Primula** (*prim'-ū-lah*) [*primus*, first]. Primrose, a genus of the *Ptimulaceæ*. Many species contain saponin and other bitter principles and salicylates. *P. obconica* is a well-known irritant poison, the symptoms resembling those of saponin poisoning, and is used in the treatment of skin diseases.
**princeps** (*prin'-seps*) [L., "a leader"]. First; original. **p. cervicis**, a branch of the occipital artery supplying the trapezius muscle. **p. pollicis**, a branch of the radial artery going to the palmar surface of the thumb.
**principle** (*prin'-sipl*) [*principium*, from *princeps*]. A constituent of a compound having a definite composition and representing its essential or character istic properties. **p., proximate**. See *proximate principle*. **p., ultimate**, any one of the elements which a compound body contains.
**prinos** (*pri'-nos*) [L.]. Black alder, a shrub of the order *Aquifoliaceæ*. The bark and the berries of *P. verticillatus* have been employed as tonics and astringents in diarrhea, and locally as an application to ulcers. Dose ½-1 dr. (2-4 Cc.); of a decoction 1-2 oz. (32-64 Cc.).
**prism** (*prism*) [πρίσμα, from πρίζειν, to saw]. A solid whose bases or ends are similar plane figures and whose sides are parallelograms. In optics, a transparent solid with triangular ends and two converging sides. It breaks up white light into its component colors, and bends the rays of light toward the side opposite the angle (the base of the prism), and is used to measure or correct imbalance of the ocular muscles. **p.-diopter, p.-dioptry**, a unit of prismatic refractive power; the refractive power of a prism that deflects a ray of light one centimeter on a tangent plane situated at a distance of one meter. **p., enamel-**, one of the prismatic columns of from four to six sides composing the enamel of teeth, closely packed together and generally vertical to the surface of the underlying dentin. **p., Nicol's**. See *Nicol's prism*. **p.-optometer**, an instrument for prismatic testing of the refraction of the eye. **p.-verger**, an instrument used in the measurement and enlargement of the fusion power of the eyes.
**prismatic** (*pris-mat'-ik*) [*prism*]. Prism-shaped; produced by the action of a prism, as *prismatic* colors.
**prismoid** (*priz'-moid*) [πρίσμα, prism; εἶδος, like]. Resembling a prism.
**prismoptometer** (*priz-mop-tom'-et-er*) [*prism*; ὤψ, eye; μέτρον, a measure]. An instrument for estimating refractive defects of the eye by means of two prisms placed base to base.
**prismosphere** (*pris'-mo-sfēr*). A combination of a prism and a globular lens.
**privates** (*pri'-vets*) [*privatus*, private]. A common term for the genital organs.
**p. r. n.** Abbreviation of Latin *pro re nata*, according as circumstances may require.
**pro-** [L.]. A prefix meaning for, before, in front of.
**proagglutinoid** (*pro-ag-loo'-tin-oid*). An agglutinoid having a stronger affinity for the agglutinogen than is possessed by the agglutinin.
**proal** (*pro'-al*) [*pro*]. Having a forward direction or movement. Cf. *palinal* and *propalinal*.
**proamnion** (*pro-am'-ne-on*) [*pro-*; *amnion*]. That part of the *area embryonalis* at the sides and in front of the head of the developing embryo, which remains without mesoderm for a considerable period.
**proatlas** (*pro-at'-las*) [*pro-*; *atlas*, the first cervical vertebra]. A primitive or rudimentary atlas.
**probable duration of life**. The time, considering all circumstances, that a person of a given age may expect to live, as determined by statistics. The age at which any number of children born into the world will be reduced to one-half, so that there are equal chances of their dying before and after that age. The age at which a given population is reduced by one-half its number.
**probang** (*pro'-bang*) [*probare*, to test]. A rod of whalebone or other flexible material used for making local applications to the esophagus or larynx or for removing foreign bodies. **p., ball-**, a probang having an ivory bulb attached to one end. **p., bristle-, p., horsehair-**, one having on the end a sheath of bristles or horsehair that can be made to spread like an umbrella as the instrument is drawn out. **p., sponge-**, one provided with a small sponge at one end.
**probe** (*prōb*) [*probare*, to test]. A slender, flexible rod for exploring a channel. **p., Anel's**, see under *Anel*. **p., blunt**, one with a blunt extremity. **p., Bowman's**, see under *Bowman*. **p., drum-**, one provided with a drum or reverberator to enable the ear to detect contact with foreign bodies. **p., electric**, one having two insulated wires, so that contact with a bullet or metal completes the circuit, and thus indicates the presence of such a foreign body. **p., eyed**, a probe having a slit at one end through which a tape or ligature can be passed. **p., lacrimal**, one used for dilating the lacrimal passages. **p., meerschaum**, a probe, the end of which is tipped with meerschaum, which becomes streaked with the lead by contact with a bullet. **p., Nélaton's**, see under *Nélaton*. **p., percussion**, one made of small links so jointed together that it is very flexible.
**probilin** (*pro-bi'-lin*). Trade name of a preparation of salicylic acid, sodium oleate, sodium stearate, phenolphthalein, and camphorated menthol; said to be a cholagogue.
**procatarctic** (*pro-kat-ark'-tik*) [προκατάρχειν, to begin first]. Primary, predisposing.
**procatarxis** (*pro-kat-arks'-is*). The kindling of a disease into action by a procatarctic cause.
**procella, procuella** (*pru-se'-le-uh*) [*pi uc*, κοιλία, hollow]. Same as *paracelia*.
**procelous, procœlous** (*pro-se'-lus*) [*pro-*; κοῖλος, hollow]. Being concave in front and convex behind.
**procephalic** (*pro-sef-al'-ik*) [*pro-*; κεφαλή, head]. In biology, of or pertaining to the fore part of the head.
**procerus** (*pro-se'-rus*) [*procerus*, stretched out, long]. The pyramidalis muscle; see *muscles, table of*.
**process** (*pros'es*) [*processus*, from *procedere*, to go].
1. A course of action; a group of phenomena, as the *inflammatory process*. 2. A prominence or outgrowth, as the spinous *process* of a vertebra, the axis-cylinder *process* of a nerve. 3. In chemistry, a method of procedure, reaction, test. **p., acromion**, same as *acromion*. **p., alveolar**. See *alveolar process*. **p., anconeal**, the olecranon. **p., auditory**, the curved plate of bone surrounding the external auditory meatus for the greater part of its circumference, and serving for the attachment of the cartilage of the external ear. **p., axis-cylinder**, that process of a nerve-cell which becomes the axis-cylinder of a nerve. **p., basilar**. See *basilar process*. **p., caudate**. 1. The caudate lobe of the liver. 2. The lower end of one of the divisions of the antihelix of the external ear. **p., ciliary**. See *ciliary process*. **processes, clinoid**, projections on the body and lesser wings of the sphenoid bone. There are three: 1. The *anterior*, formed by the inner extremity of the lesser wings. 2. The *middle*, a small eminence on each side bounding the sella turcica anteriorly. 3. The *posterior*, a tubercle on each side of the posterior part of the sella turcica. **p., condyloid**, the posterior process on the upper border of the ramus of the lower jaw. **p., conoid**. See *conoid tubercle*. **p., coracoid**. See *coracoid process*. **p., coronoid**. 1. A thin, flattened process projecting from the anterior portion of the upper border of the ramus of the lower jaw, and serving for the attachment of the temporal muscle. 2. A triangular projection from the upper end of the ulna, forming the lower part of the greater sigmoid cavity. **p., cricoid**, a slight projection on the lower border of the thyroid cartilage. **p., cubital**, the lower and articular end of the humerus. **p., cuneiform**. 1. The unciform process of the ethmoid. 2. The basilar process of the occipital bone. **p., Deiters'**, the axis-cylinder process of a nerve-cell. **p., dendritic**. See *p., protoplasmic* (1). **p., dentate**, the odontoid process. **p., ensiform**, the cartilaginous tip at the lower end of the sternum. **p., ethmoid**, one of the projections from the superior border of the inferior turbinate bone. **p., external angular**, the projection on the outer extremity of the supraorbital arch of the frontal bone. **p., falciform**. 1. A process of the fascia lata of the thigh, forming the outer and upper margin of the saphenous opening. Syn., *falciform process of Burns*. 2. The falx cerebri. **p., foliaceous**, a process of the ethmoid bone projecting into the frontal cells. **p., Folian**. See *Folian process*. **p., funicular**. See *fun-*

*icular*. **p., hamate.** See *p., unciform*. **p., hamular.** 1. A hook-like process of bone on the lower extremity of the internal pterygoid plate, around which the tendon of the tensor palati turns. 2. Of the lacrimal bone, the hook-like termination of the lacrimal crest. **p., inferior vermiform,** the central projection on the lower surface of the cerebellum, between the two hemispheres. **p., internal angular,** the inner extremity of the supraorbital arch of the frontal bone. **p., jugal.** 1. The zygoma. 2. The external angular process. 3. The malar process of the superior maxilla. **p., jugular.** See *jugular process*. **p., lacrimal,** a short process of the inferior turbinate bone that articulates with the lacrimal bone. **p., lenticular,** the extremity of the long process of the incus, covered with cartilage and articulating with the stapes. **p., long (of the incus),** a slender process that descends vertically from the body of the incus and articulates, by the lenticular process, with the head of the stapes. **p., long (of the malleus),** a long, delicate process that passes from the neck of the malleus outward to the Glaserian fissure, to which it is connected by cartilaginous and ligamentous fibers. **p., malar,** a triangular eminence of the superior maxilla by which it articulates with the malar bone. **p., mammillary,** one of the tubercles on the posterior part of the superior articular processes of the lumbar vertebræ. **p., mastoid.** See *mastoid process*. **p., maxillary,** a thin plate of bone descending from the ethmoid process of the inferior turbinate bone, and hooking over the lower edge of the orifice of the antrum. **p., nasal (of the superior maxilla),** a thick, triangular process of bone that projects upward, inward, and backward by the side of the nose, forming a part of its lateral wall. **p., odontoid,** the tooth-like process of the axis which ascends and articulates with the atlas. **p., olecranon,** the olecranon. **p., olivary,** a small oval eminence situated behind the optic groove of the sphenoid bone. **p., orbicular (of the incus).** See *p., lenticular*. **p., orbital (of the palate bone),** a process directed upward and outward from the upper portion of the palate bone. **p., orbital (of the superior maxilla),** a process projecting from the orbital margin of the superior maxilla. **p., palatal,** a thick process projecting horizontally inward from the inner surface of the superior maxillary bone, and forming part of the floor of the nostril and the roof of the mouth. **p., pineal,** the epiphysis. **p., postglenoid,** a small tubercle separating the glenoid fossa from the auditory process. **p., protoplasmic.** 1. Any one of the processes of nerve-cells that are not continued as axis-cylinders. 2. A pseudopod of an amœboid cell. **p., pterygoid (of the palate bone),** a pyramidal process projecting from the posterior border of the palate bone and articulating with the sphenoid bone. **p., pterygoid (of the sphenoid bone),** one descending perpendicularly from the point of junction of the body, with the greater wing of the sphenoid bone, and consisting of an external and an internal plate. **p. of Rau.** Synonym of *p., long (of the malleus)*. **p., short (of the incus),** a conical process projecting almost horizontally backward from the body of the incus and attached by ligamentous fibers to the margin of the opening leading into the mastoid cells. **p., short (of the malleus),** a slight projection from the root of the manubrium of the malleus, lying in contact with the tympanic membrane. **p., sphenoid,** a thin plate of bone directed upward and inward from the vertical plate of the palate bone. **p., sphenomaxillary,** an inconstant downward prolongation of the greater wing of the sphenoid. **p., spinous (of the ilium),** a prominent eminence on the anterior or posterior border of the ilium. The upper process on the anterior border is the *anterior superior spinous process*; below it is the *anterior inferior spinous process*. On the posterior border there are also two processes—a *posterior superior* and a *posterior inferior*. **p., spinous (of the sphenoid bone),** a rough prominence descending from the posterior part of the greater wing of the sphenoid bone. It receives the attachment of the internal lateral ligament of the jaw and the tensor palati muscle. **p., spinous (of the tibia),** an eminence of bone on the upper surface of the tibia, between the two articular surfaces, and nearer to the posterior than the anterior border. **p., spinous (of a vertebra),** the prominent backward projection from the middle of the posterior portion of the arch of a vertebra. **p., styloid (of the fibula),** a pointed eminence projecting upward from the posterior portion of the head of the fibula. **p., styloid (of the radius),** a projection from the external border of the lower extremity of the radius. **p., styloid (of the temporal bone),** a sharp spine about an inch in length, descending downward, forward, and inward from the inferior surface of the petrous portion of the temporal bone. **p., styloid (of the ulna),** a projection from the inner and posterior portion of the lower extremity of the ulna. **p., sulcate,** an inconstant process of the palate bone connecting the orbital process with the sphenoid process. **p., superior maxillary,** an eminence on the face of the embryo which gives rise to the superior maxilla and the malar bone. **p., superior vermiform,** the upper part of the median lobe of the cerebellum, connecting the two lateral hemispheres. **p., temporal,** the posterior angle of the malar bone by which it articulates with the zygomatic process of the temporal bone. **p., transverse,** a process projecting outward from the side of a vertebra, at the junction of the pedicle and the lamina. **p., trochlear,** Hyrtl's name for a groove in a bone for the reception of a tendon. **p., unbranched.** Synonym of *p., axis-cylinder*. **p., unciform (of the ethmoid bone),** a hook-like projection from the inferior portion of each lateral mass of the ethmoid bone. It articulates with the ethmoid process of the inferior turbinated bone. **p. unciform (of the hippocampal gyrus),** a hook-like projection from the anterior end of the hippocampal convolution. **p., unciform (of the unciform bone),** a hook-like projection from the palmar surface of the unciform bone. **p., vaginal (of peritoneum),** the process of peritoneum which the testicle in its descent carries in advance, and which in the scrotum forms the tunica vaginalis testis. **p., vaginal (of the sphenoid bone),** a projection from the inferior surface of the body of the sphenoid bone, running horizontally inward from near the base of the pterygoid process. **p., vaginal (of the temporal bone),** a sheath-like plate of bone which extends from the carotid canal to the mastoid process. It separates behind into two laminæ that inclose the styloid process. **p., vermiform, inferior and superior,** the inferior and superior surfaces of the middle lobe of the cerebellum. **p., vocal,** the anterior angle of the arytenoid cartilage. **p., xiphoid,** the ensiform cartilage. **p., zygomatic (of the malar bone),** a long, serrated process which articulates with the zygomatic process of the temporal bone. **p., zygomatic (of the temporal bone),** a long projection from the lower part of the squamous portion of the temporal bone, articulating with the malar bone.

**processus** (*pro-ses′-us*) [L.]. A process. **p. accessorius,** accessory processes. **p. ad cerebrum,** the superior cerebellar peduncle. **p. ad medullam,** the inferior peduncle of the cerebellum. **p. ad pontem,** the middle peduncle of the cerebellum. **p. ad testes,** the superior cerebellar peduncles. **p. articulares inferiores,** inferior articular processes. **p. articulares superiores,** superior articular processes. **p. brevis,** the short arm of the malleus. **p. clavatus,** a thickening on the posterior pyramid of the medulla near the apex of the fourth ventricle. **p. cochleariformis,** a thin plate of bone separating the canal for the Eustachian tube from that for the tensor tympani muscle. **p. e cerebello ad cerebrum,** the superior cerebellar peduncle. **p. e cerebello ad medullam,** the inferior cerebellar peduncle. **p. e cerebello ad pontem,** the middle cerebellar peduncle. **p. e cerebello ad testes,** the superior peduncles of the cerebellum. **p. costarius,** costal process. **p. gracilis,** a long delicate process passing from below the neck of the malleus to the Glaserian fissure. **p. hamatus.** Synonym of *process, unciform*. **p. mammillaris,** mammillary process. **p. spinosus,** spinous process. **p. transversus,** transverse process.

**prochilon, procheilon** (*pro-ki′-lon*) [*pro-*; χεῖλος, lip]. The prominence in the center of the lip.

**prochondral** (*pro-kon′-dral*) [*pro-*; χόνδρος, cartilage]. Prior to the formation of cartilage.

**prochordal** (*pro-kor′-dal*) [*pro-*; χορδή, cord]. Situated in front of the chorda dorsalis.

**prochoresis** (*pro-ko-re′-sis*) [προχώρησις, a going on or advancing]. The motor activity of the stomach.

**Prochownik's diet** (*pro-kov′-nik*) [Ludwig Prochownik, German obstetrician, 1851– ]. A restricted diet for pregnant women with a narrow pelvis. **P.'s method,** resuscitation of an asphyxiated infant by compression of its chest and suspension with its head hanging backward.

**prochromatin** (*pro-kro′mat-in*). Paranuclein; the substance composing the nucleolus of a cell.

**procident** (*prōs'-id-ent*) [*procidens*, falling forward]. Affected by prolapsus.
**procidentia** (*pro-se-den'-she-ah*) [*pro-;* *cadere*, to fall]. Prolapse.
**procreate** (*pro'-kre-āt*). To beget.
**procreation** (*pro-kre-a'-shun*) [*procreare*, to bring forth]. The act of begetting offspring.
**proctagra** (*prok-ta'-grah*) [*procto-;* ἄγρα, a seizure]. Sudden pain in the anal region.
**proctalgia** (*prok-tal'-je-ah*) [*procto-;* ἄλγος, pain]. Neuralgic pain in the anus or rectum.
**proctatresia** (*prok-tat-re'-se-ah*) [*procto-;* *atresia*]. An imperforate condition of the anus or rectum.
**proctectasia, proctectasis** (*prok-tek-ta'-se-ah, prok-tek'-tas-is*) [*procto-;* ἔκτασις, dilatation]. Dilatation of the anus or rectum.
**proctectomy** (*prok-tek'-to-me*) [*procto-;* ἐκτομή, excision]. Excision of the anus or rectum.
**proctenclisis** (*prok-ten'-kli-sis* [*procto-;* ἐν, in; κλείειν, to close]. Stricture of the rectum or anus.
**procteurynter** (*prok-tū-rin'-ter*) [*procto-;* εὐρύνειν, to widen]. An instrument for dilating the anus or rectum.
**proctitis** (*prok-ti'-tis*) [*procto-;* ιτις, inflammation]. Inflammation of the anus or rectum.
**procto-** (*prok-to-*) [πρωκτός, anus]. A prefix signifying relating to the anus or rectum.
**proctocele** (*prok'-to-sēl*) [*procto-;* κήλη, hernia]. The extroversion or prolapse of the mucous coat of the rectum. p., **vaginal**, a hernia of the rectum appearing in the vagina.
**proctoclysis** (*prok-tok'-lis-is*) [*procto-;* κλύσις, a washing out]. The slow instillation of a liquid into the rectum.
**proctococcypexy, proctoccypexia** (*prok-to-kok'-se-peks-e, prok-to-kok-se-peks'-e-ah*) [*procto-;* *coccyx;* πῆξις, a fixing in]. Suturing of the rectum to the coccyx.
**proctocolitis** (*prok-to-ko-li'-tis*). Inflammation of the rectum and colon.
**proctocolonoscopy** (*prok-to-ko-lon-os'-ko-pe*) [*procto-*, colon; σκοπεῖν, to examine]. Inspection of the interior of the rectum and lower colon.
**proctocystoplasty** (*prok-to-sis'-to-plas-te*) [*procto-;* κύστις, bladder; πλάσσειν, to form]. A plastic operation on the rectum and the bladder for repair of rectovesical fistula.
**proctocystotome** (*prok-to-sist'-o-tōm*) [*procto-;* κύστις, bladder; τομή, a cutting]. An instrument designed expressly for use in proctocystotomy.
**proctocystotomy** (*prok-to-sis-tot'-o-me*). Lithotomy in which the incision is made through the walls of the rectum.
**proctodeum, proctodæum** (*prok-to-de'-um*) [*procto-;* ὀδαῖος, by the way]. An invagination of the ectoderm in the embryo that grows inward toward the cloaca until the ectoderm and endoderm come into contact; the membrane formed between the two finally breaks through, the gut then opening externally. The primitive anus.
**proctodynia** (*prok-to-din'-e-ah*) [*procto-;* ὀδύνη, pain]. Pain about the anus or in the rectum.
**proctoelytroplasty** (*prok-to-el'-it-ro-plas-te*) [*procto-;* ἔλυτρον, vagina; πλάσσειν, to form]. A plastic operation on the rectum and the vagina for repair of a rectovaginal fistula.
**proctology** (*prok-tol'-o-je*) [*procto-;* λόγος, science]. The science of the anatomy, functions, and diseases of the rectum.
**proctoparalysis** (*prok-to-par-al'-is-is*). Paralysis of the sphincter muscle of the anus.
**proctopexy, proctopexia** (*prok'-to-peks-e, -e-ah*) [*procto-;* πῆξις, a fixing in]. The fixation of the rectum to another part by sutures.
**proctoplasty** (*prok'-to-plas-te*) [*procto-;* πλάσσειν, to form]. Plastic surgery of the anus.
**proctoplegia** (*prok-to-ple'-je-ah*). Synonym of *proctoparalysis*.
**proctopolypus** (*prok-to-pol'-ip-us*) [*procto-;* *polyp*]. A polyp of the rectum.
**proctoptoma** (*prok-top-to'-mah*) [*procto-;* πτῶμα, fall]. Prolapse of the rectum.
**proctoptosis** (*prok-top-to'-sis*) [*procto-;* πτῶσις, a falling]. Prolapse of the rectum. Same as *proctoptoma*.
**proctorrhaphy** (*prok-tor'-a-fe*) [*procto-;* ῥαφή, suture]. Suture of the rectum or anus.
**proctorrhea, proctorrhœa** (*prok-tor-e'-ah*) [*procto-;* ῥοία, flow]. A discharge of mucus through the anus.
**proctoscope** (*prok'-to-skōp*) [*procto-;* σκοπεῖν, to examine]. An instrument for inspection of the rectum.
**proctoscopy** (*prok-tos'-ko-pe*). Ocular inspection of the rectum with the aid of special instruments.
**proctosigmoidectomy** (*prok-to-sig-moid-ek'-to-me*). Excision of the anus and sigmoid flexure.
**proctospasm** (*prok'-to-spasm*) [*procto-;* *spasm*]. Spasm or tenesmus of the rectum.
**proctostenosis** (*prok-to-sten-o'-sis*) [*procto-;* *stenosis*]. Stricture of the anus or rectum.
**proctostomy** (*prok-tos'-to-me*) [*procto-;* στόμα, a mouth]. The establishment of an opening into the rectum.
**proctotome** (*prok'-to-tōm*) [*procto-;* τομή, a cutting]. A cutting instrument used in proctotomy.
**proctotomy** (*prok-tot'-o-me*) [*procto-;* τομή, a cutting]. Incision into the rectum, especially for stricture. p., **external**, the operation of dividing the rectum and the anus and the tissue lying between the anus and the tip of the coccyx. p., **internal**, division of the parts as in external proctotomy, but without cutting the sphincters. p., **linear**, an incision of the dorsal rectal wall and of all the tissues lying between the anus and a subcutaneous opening made in front of the coccyx.
**proctotoreusis** (*prok-to-tor-ū'-sis*) [*procto-;* τορεύειν, to bore through]. The operation of making an opening into an imperforate anus.
**proctovalvotomy** (*prok-to-val-vot'-o-me*) [*procto-;* *valve;* τομή, a cutting]. Incision of the valves of the rectum.
**procursive** (*pro-kur'-siv*) [*pro-;* *currere*, to run]. Running forward, as *procursive* epilepsy, a form in which the patient runs during the epileptic attack.
**procurvation** (*pro-kur-va'-shun*) [*procurvare*, to bend forward]. A forward inclination of the body.
**prodigiosin** (*pro-dij-e-o'-sin*). A red pigment formed by *Bacillus prodigiosus*.
**prodigiosus toxin** (*pro-dij-e-o'-sus*). See *Coley's fluid*.
**prodromal** (*pro-dro'-mal*) [*prodrome*]. Precursory; of the nature of a prodrome.
**prodrome** (*pro'-drōm*) [πρόδρομος, a running forward] A forerunner of a disease; a symptom indicating the approach of a disease.
**prodromic** (*pro-dro'-mik*). See *prodromal*.
**product** (*prod'-ukt*) [*productive*]. Effect; result. p., **addition**, a compound resulting from the direct union of two bodies.
**productive** (*pro-duk'-tiv*) [*pro-;* *ducere*, to lead]. Forming, especially forming new tissue, as a *productive* inflammation.
**proeminent** (*pro-em'-in-ent*) [*pro-;* *eminens*, prominent]. Projecting. p. **vertebra**, the seventh cervical vertebra, so called because its spinous process projects beyond the others.
**proencephalus** (*pro-en-sef'-al-us*) [*pro-;* ἐγκέφαλος, brain]. A monster characterized by a protrusion of the brain through a fissure in the frontal region.
**proenzyme** (*pro-en'-zīm*). The substance which subsequently becomes an active ferment.
**proerythroblast** (*pro-er'-ith-ro-blast*). Same as *hematoblast*.
**profluvium** (*pro-floo'-ve-um*) [*profluere*, to flow forth]. A flux or discharge. p. **alvi**, diarrhea. p. **lactis**, excessive flow of milk. p. **muliebre**. Synonym of *leukorrhea*. p. **sanguinis**, hemorrhage. p. **seminis**. 1. Synonym of *spermatorrhea*. 2. Discharge of semen from the vagina after coitus.
**profluvium** (*pro-fuu'-ǎs*) [L., "deep"]. Deep-seated; applied to certain muscles and nerves, and in the feminine, *profunda*, to certain arteries and veins. See under *artery*.
**progaster** (*pro-gas'-ter*). Same as *archenteron*.
**progastrin** (*pro-gas'-trin*). The precursor of the gastric secretin.
**progenitor** (*pro-jen'-it-or*) [*progeny*]. Ancestor or forefather.
**progeny** (*proj'-en-e*) [*pro-;* *gignere*, to beget]. Offspring; descendants.

**progeria** (*pro-je'-re-ah*) [πρόγηρος, premature old age]. A rare form of senilism, of rapid onset, with retention of intelligence and without any perceptible cause, marked by loss of the hair, shriveling of the nails, and emaciation. Cf. *geromorphism*.
**proglossis** (*pro-glos'-is*) [*pro-*; γλῶσσα, the tongue]. The tip of the tongue.
**proglottis** (*pro-glot'-is*) [*pro-*; γλῶσσα, tongue; pl., *proglottides*]. A mature segment of a tape-worm.
**prognathic** (*prog-na'-thik*). Synonym of *prognathous*.
**prognathism** (*prog'-na-thizm*) [*pro-*; γνάθος, jaw]. The quality of having a projecting lower jaw, or of being prognathous. p., **alveolo-subnasal,** in craniometry, the prognathism measured by the angle included between the line joining the alveolar and subnasal points and the alveolo-condylean plane.
**prognathous** (*prog'-na-thus*) [*pro-*; γνάθος, jaw]. Having a projecting lower jaw.
**prognosis** (*prog-no'-sis*) [*pro-*; γνῶσις, knowledge]. An opinion or judgment in advance concerning the duration, course, and termination of a disease. p. **anceps,** a doubtful prognosis. p. **fausta,** a favorable prognosis. p. **infausta,** an unfavorable prognosis. p. **quoad vitam,** a prognosis as regards life. p. **quoad restitutio ad integrum,** a prognosis as regards recovery.
**prognostic** (*prog-nos'-tik*) [*prognosis*]. Pertaining to prognosis.
**prognosticate** (*prog-nos'-tik-āt*) [*prognosis*]. To give a prognosis.
**progression** (*pro-gresh'-un*) [*progredi,* to advance]. The act of advancing or moving forward. p., **backward,** a backward walking, a rare symptom of certain nervous lesions. p., **cross-legged,** walking with the legs almost crossing, a condition sometimes observed in bilateral hip-disease and in cerebral spastic palsy.
**progressive** (*pro-gres'-iv*) [see *progression*]. Gradually extending. p. **muscular atrophy,** chronic anterior poliomyelitis in which the large ganglion-cells of the anterior horns are gradually destroyed, which leads to atrophy of the muscles. p. **ossifying myositis,** a chronic inflammation of the muscles, associated with a tendency to bony deposits in them. p. **processes,** those which continue after the requirements of the organism have been satisfied.
**proiotia, proiotes** (*pro-i-o'-she-ah, pro-i-o'-tēz*) [πρωϊότης, earliness]. Sexual precocity.
**projectile** (*pro-jek'-til*) [*pro-*; *jacere,* to throw]. Throwing forward. p. **vomiting,** a form sometimes observed in diseases of the brain, in which the material is suddenly projected out of the mouth to some distance, generally without nausea.
**projection** (*pro-jek'-shun*) [see *projectile*]. 1. The act of throwing forward. 2. A part extending beyond the level of the surrounding surface. 3. The referring of impressions made on the organs of sense to the position of the object producing them. p.-**systems,** the name given to the pathways connecting the cerebral cortex with the periphery. The first projection system corresponds to the fibers passing through the corona radiata; the second, to the tracts proceeding downward to the gray matter from the third ventricle to the end of the spinal cord; the third, to the peripheral nerves.
**prolabium** (*pro-la'-be-um*) [*pro-*; *labium,* lip]. The red exposed part of the lip; also, the central prominence of the lip.
**prolacto bread** (*pro-lăk'-to*). A bread for the use of diabetics said to contain over 33 % of albuminoids, 17 % of fats and no starches.
**prolapse** (*pro-laps'*) [*prolabi,* to slip down]. The falling forward or downward of a part. p. **of the cord,** premature expulsion of the umbilical cord during parturition. p., **frank,** uterine displacement in which the vagina is inverted and forms a bag hanging from the vulva, at the bottom of which lies the uterus, generally retroverted, but otherwise unaltered. p. **of funis.** See *p. of the cord*. p. **of the iris,** protrusion of the iris through a corneal wound.
**prolepsis** (*pro-lep'-sis*) [*pro-*; λαμβάνειν, to seize]. 1. The return of a paroxysm before the expected time. 2. Prognosis.
**proleptic** (*pro-lep'-tik*) [*pro-*; λαμβάνειν, to seize]. 1. Prognostic. 2. Returning before the expected time.
**proliferate** (*pro-lif'-er-āt*) [*proles,* offspring; *ferre,* to bear]. To multiply; to generate.
**proliferation** (*pro-lif-er-a'-shun*) [*proliferate*]. The act of proliferating or multiplying, as *proliferation* of cells. p., **atrophic,** the multiplication of cells in atrophic tissues.

**proliferative, proliferous** (*pro-lif'-er-a-tiv, pro-lif'-er-us*) [*proliferate*]. Multiplying; characterized by the formation of new tissues or by cell-proliferation. p. **cyst,** a cyst in which the lining epithelium proliferates and produces projections from the inner surface of the cyst.
**prolific** (*pro-lif'-ik*) [*proliferate*]. Fruitful.
**proligerous** (*pro-lij'-er-us*) [*proles,* offspring; *gerere,* to bear]. Germinating; producing offspring. p. **disc.** See *discus proligerus*.
**prominence** (*prom'-in-ens*). 1. A projection on the surface of a part, especially on a bone. 2. The state of being prominent. p., **genital,** an accumulation of cells on the ventral aspect of the embryonic cloaca, from which the generative organs are developed.
**prominentia** (*prom-in-en'-she-ah*) [L. : *pl.,* prominentiæ]. A prominence, or projection. p. **albicans.** Synonym of *corpus albicans*. p. **encephali.** Synonym of *pomum Adami*. p. **lentiformis.** Synonym of *nucleus lenticular*. p. **natiformis.** Synonym of *nates of the corpora quadrigemina*. **prominentiæ orbiculares minores.** Synonym of the *testes of the corpora quadrigemina*. p. **semiovalis.** Synonym of *olivary body*. p. **sphærica,** a cerebral convolution. p. **spiralis,** a slight prominence on the outer wall of the cochlear duct, containing a small capillary vessel. **prominentiæ testiformes,** the testes of the corpora quadrigemina.
**promnesia** (*pro-ne'-se-ah*) [*pro-*; μνῆσις, remembrance]. The paradoxical sensation of remembering scenes or events which are occurring for the first time.
**promontory** (*prom'-on-to-re*) [*pro-*; *mons,* a mountain]. A projecting prominence. p. **of the sacrum,** p., **sacrovertebral,** the prominence formed by the angle between the upper extremity of the sacrum and the last lumbar vertebra.
**pronæus** (*pro-ne'-us*) [πρόναος, the first room of a temple]. The vagina; also, the vestibule of the vagina.
**pronate** (*pro'-nāt*) [see *pronation*]. To place in a prone position.
**pronation** (*pro-na'-shun*) [*pronare,* to bend forward]. 1. The condition of being prone; the act of placing in the prone position. 2. Of the hand, the turning of the palm downward.
**pronatoflexor** (*pro-na-to-fleks'-or*). Relating to the pronator and flexor muscles.
**pronator** (*pro-na'-tor*) [see *pronation*]. That which pronates, a term applied to several muscles. See under *muscle*.
**prone** (*prōn*) [*pronus,* prone]. Lying with the face downward; of the arm, having the palm directed downward; the opposite of supine.
**pronephric** (*pro-nef'-rik*) [*pro-*; νεφρός, a kidney]. In biology, of or pertaining to the peonephron, or the primitive kidney. p., **duct,** one of the four fundamental parts of the vertebrate urogenital system; the Wolffian duct.
**pronephron, pronephros** (*pro-nef'-ron, pro-nef'-ros*) [*pro-*; νεφρός, a kidney]. The primitive kidney; the anterior of the three segments of the Wolffian body, opening by means of the Muellerian duct into the cloaca. It is the first part of the urogenital system to be differentiated in the vertebrate embryo.
**pronucleus** (*pro-nū'-kle-us*) [*pro-*; *nucleus*]. One of the two nuclear bodies of a newly fecundated ovum, the *male* pronucleus and the *female* pronucleus, the fusion of which results in the formation of the germinal nucleus.
**proof-spirit.** See *spirit*.
**prootic** (*pro-o'-tik*) [*pro-*; οὖς, ear]. In front of the ear.
**proovarium.** See *parovarium*.
**propago** (*pro-pa'-go*) [L., "a slip or shoot"; pl., *propagines*]. An offset; applied to the smaller branches of vessels or nerves.
**propalinal** (*pro-pal'-in-al*) [*pro-*; πάλιν, back, backward]. In biology, applied to the forward and backward movement of the jaws of certain animals.
**propane** (*pro'-pān*) [*pro*pionic], C₃H₈. A hydrocarbon, the third member of the marsh-gas series, occurring in petroleum.
**prop-cells.** Columnar or fusiform cells placed in the intervals of the rods and hair-cells of the organ of Corti. They are also known as supporting cells and cells of Deiters.
**propedeutics, propædeutics** (*pro-pe-dū'-tiks*) [*pro-*; παιδεύειν, to teach]. Preliminary instruction.
**propenyl** (*pro'-pen-il*). See *glyceryl*.

**propepsin** (*pro-pep'-sin*). The zymogen of pepsin, found in the cells of the gastric glands.
**propeptone** (*pro-pep'-tōn*). See under *peptone*.
**propeptonuria** (*pro-pep-ton-ū'-re-ah*) [*propeptone;* οὖρον, urine]. The appearance of propeptone in the urine. It is said to occur in fevers, diphtheria, osteomalacia, and during the administration of styrax or phosphorus.
**properitoneal** (*pro-per-it-on-e'-al*) [*pro-;* peritoneum]. Situated in front of the peritoneum. p. hernia, a hernia the sac of which extends in various directions within the abdominal walls.
**prophase** (*pro'-fāz*) [*pro-;* φαίνειν, to show]. The first stage of karyokinesis or indirect cell-division.
**prophylactic** (*pro-fil-ak'-tik*) [see *prophylaxis*]. 1. Pertaining to prophylaxis. 2. A remedy or agent that prevents the development of disease.
**prophylactol** (*pro-fil-ak'-tol*). A proprietary liquid said to consist of 20% of protargol and glycerol with the addition of mercury bichloride in the proportion of one part to 2000. It is used by injection in diseases of the urinary tract.
**prophylaxis** (*pro-fil-aks'-is*) [προφυλάσσειν, to keep guard before]. Prevention of disease; measures preventing the development or spread of disease.
**propionate** (*pro'-pe-on-āt*). A salt of propionic acid.
**propione** (*pro'-pe-ōn*) [πρῶτος, first; πίων, fat], C₅H₁₀O. Diethyl ketone, a liquid obtained by the distillation of calcium propionate. Dose, as hypnotic, 8–45 gr. (0.5–3.0 Gm.); as sedative, 8 gr. (0.5 Gm.).
**propionic acid** (*pro-pe-on'-ik*). See *acid, propionic*.
**proplex** (*pro'-pleks*). See *proplexus*.
**proplexus** (*pro-pleks'-us*) [*pro-;* plexus]. The choroid plexus of the lateral ventricles of the brain.
**propons** (*pro'-pons*). The transverse white fibers in front of the pyramids and below the pons Varolii, the ponticulus.
**proprietary medicine.** A medicine that is trade-marked, patented, or secret.
**proprioceptive impulses** (*pro'-pre-o-sep-tiv*) [*proprius*, one's own; *capere*, to take]. Afferent nerve impulses which derive their stimulation from the tissues themselves.
**proprioceptor** (*pro-pre-o-sep'-tor*). A receptor which is stimulated by actions occurring within the organism. See *receptor*.
**proprius** (*pro'-pre-us*) [L.]. Individual; special, as flexor *proprius* pollicis, the special flexor of the thumb.
**proptosis** (*prop-to'-sis*) [*pro-;* πτῶσις, a falling]. A falling downward; prolapse.
**propulsion** (*pro-pul'-shun*) [*pro-*, *pellere*, to push]. 1. The act of pushing or driving forward. 2. A falling forward in walking, a condition seen in paralysis agitans. See *festination*.
**propyl** (*pro'-pil*) [*propionic*], C₃H₇. The radical of propane.
**propylamine** (*pro-pil'-am-in*) [*propyl; amine*]. 1. A liquid basic compound having the formula C₃H₉N, and existing in two forms, a *normal propylamine*, boiling at 47° C., and *isopropylamine*, boiling at 31.5° C. See under *ptomaine*. 2. A misnomer for an aqueous solution of trimethylamine, a colorless, caustic, antiseptic liquid.
**propylene** (*pro'-pil-ēn*) [*propyl*], C₃H₆. A gaseous hydrocarbon belonging to the series of the olefins.
**pro re nata** (*pro re na'-tah*) [L.]. A phrase signifying "according to the circumstances of the case."
**prorennin** (*pro-ren'-in*) [*pro-; rennin*]. The mother-substance, zymogen or preliminary material of rennin or chymosin.
**prorsad** (*pror'-sad*) [*prorsum*, forward]. Toward the anterior aspect.
**prorsal** (*pror'-sal*) [*prorsum*, forward]. Anterior; forward.
**proscolex** (*pro-sko'-leks*) [*pro-;* σκώληξ, a worm; pl., *proscolices*]. The embryo of a cestode worm immediately after leaving the egg.
**prosecretin** (*pro-se-kre'-tin*) [*pro-; secretin*]. The precursor of secretin (*q. v.*); it is secreted by the epithelium of the small intestine.
**prosector** (*pro-sek'-tor*) [*pro-; sector*, cutter]. One who prepares subjects for anatomical dissection or to illustrate didactic lectures.
**prosencephalon** (*pros-en-sef'-al-on*) [*pro-;* ἐγκέφαλον, brain]. The forebrain; that part of the anterior cerebral vesicle from which are developed the hemispheres, the corpus callosum, the anterior commissure, the fornix, the septum lucidum, the anterior perforated space, the olfactory lobes, and the corpus striatum.
**prosocele, prosocoele** (*pros'-o-sēl*) [πρόσω, forward; κοῖλος, a hollow]. The cavity of the prosencephalon. It is divided into three main divisions, a mesal cavity, the aula, and two lateral cavities, the paraceles, together with the rhinoceles (olfactory ventricles) or cavities of the olfactory bulbs.
**prosodemic** (*pros-o-dem'-ik*) [πρόσω, forward; δῆμος, people]. Of a disease which is spread by individual contact as opposed to one which is spread by general means such as the water or milk supply.
**prosogaster** (*pros-o-gas'-ter*) [πρόσω, forward; γαστήρ, belly]. The foregut.
**prosopalgia** (*pros-op-al'-je-ah*) [πρόσωπον, the face; ἄλγος, pain]. Neuralgic pain in the distribution of the trigeminal nerve. Syn., *facial neuralgia; tic douloureux.*
**prosopalgic** (*pros-o-pal'-jik*) [πρόσωπον, face; ἄλγος, pain]. Affected with prosopalgia.
**prosopantritis** (*pros-op-an-tri'-tis*) [πρόσωπον, face; ἄντρον, cavity; *itis*, inflammation]. Inflammation of the frontal sinuses.
**prosopantrum** (*pros-op-an'-trum*) [πρόσωπον, face; ἄντρον, cavity]. A frontal sinus.
**prosopectasia** (*pros-o-pek-ta'-ze-ah*) [πρόσωπον, face; ἔκτασις, enlargement]. Morbid enlargement of the face.
**prosopic** (*pros-op'-ik*) [πρόσωπον, the face]. Relating to the face; facial.
**prosopodiplegia** (*pros-o-po-di-ple'-je-ah*) [πρόσωπον, the face; *diplegia*]. Double facial paralysis.
**prosopodynia** (*pros-o-po-din'-e-ah*) [πρόσωπον, face; ὀδύνη, pain]. Facial pain, or neuralgia.
**prosoponeuralgia** (*pros-o-po-nū-ral'-je-ah*). Synonym of *prosopalgia*.
**prosoposchisis** (*pros-o-pos'-kis-is*) [πρόσωπον, face; σχίσις, fissure]. An oblique fissure or cleft of the face of a fetal monstrosity. It passes from the mouth to one of the orbits, and is usually associated with malformation of the brain.
**prosopospasm** (*pros'-o-po-spazm*) [πρόσωπον, face; *spasm*]. Risus sardonicus.
**prosoposternodymia** (*pros-o-po-ster-no-dim'-e-ah*) [πρόσωπον, face; στέρνον, breast; δίδυμος, double]. A form of double monstrosity in which the twins are united by their faces and chests.
**prosopothoracopagus** (*pros-o-po-tho-rak-op'-ag-us*) [πρόσωπον, the face; *thorax;* πάγος, that which is joined]. A double fetal monster in which the twins are united by the upper abdomen, chest, and face.
**prosopotocia** (*pros-o-po-to'-se-ah*) [πρόσωπον, face; τόκος, birth]. Face-presentation in parturition.
**prosphysectomy** (*pros-fiz-ek'-to-me*) [πρόσφυσις, a growing to; an appendage; ἐκτομή, excision]. Appendicectomy.
**prostata** (*pros'-tat-ah*). The prostate gland.
**prostatalgia** (*pros-ta-tal'-je-ah*) [*prostate;* ἄλγος, pain]. Pain in the prostate gland.
**prostatauxe** (*pros-tat-awks'-e*) [*prostate;* αὔξη, increase]. Enlargement of the prostate gland.
**prostate, p. gland** (*pros'-tāt*) [*pro-;* ἱστάναι, to stand]. The organ surrounding the neck of the bladder and beginning of the urethra in the male (*prostatic urethra*). It consists of two lateral lobes and a middle lobe, and is composed of muscular and glandular tissue; a distinct capsule surrounds it. The prostate often becomes enlarged in advanced life, and may then interfere with the emptying of the bladder.
**prostatectomy** (*pros-ta-tek'-to-me*) [*prostate;* ἐκτομή, excision]. Excision of part or all of the prostate.
**prostatelcosis, prostathelcosis** (*pros-tat-el-ko'-sis, pros-tat-hel-ko'-sis*) [*prostate;* Ἕλκωσις, ulceration]. Ulceration of the prostate.
**prostatic** (*pros-tat'-ik*) [*prostate*]. Relating to the prostate. **p. calculus**, a stone lodged in the prostate gland. **p. plexus.** 1. A collection of veins surrounding the neck and base of the bladder and the prostate gland. 2. A plexus of nerves derived from the pelvic plexus, and distributed to the prostate gland, seminal vesicles, and erectile tissue of the penis. **p. urethra**, that portion of the urethra surrounded by the prostate gland.
**prostatism** (*pros'-tat-izm*). A morbid nervous condition due to prostatic disease.

**prostatitic** (*pros-tat-it'-ik*). Affected with prostatitis.
**prostatitis** (*pros-ta-li'-tis*) [*prostate*; ιτις, inflammation]. Inflammation of the prostate gland.
**prostatocele** (*pros-tat'-o-sēl*) [*prostate*; κήλη, tumor]. An enlargement of the prostate gland, causing a tumor-like projection.
**prostatocystitis** (*pros-tat-o-sis-ti'-tis*) [*prostate*; κύστις, bladder; ιτις, inflammation]. Inflammation of the prostate and urinary bladder.
**prostatocystotomy** (*pros-tat-o-sist-ot'-o-me*) [*prostate*; *cystotomy*]. Surgical incision of the prostate and bladder.
**prostatodynia** (*pros-tat-o-din'-e-ah*) [*prostate*; ὀδύνη, pain]. Prostatalgia.
**prostatolith** (*pros-tat'-o-lith*) [*prostate*; λίθος, a stone]. A prostatic calculus.
**prostatomegaly** (*pros-tat-o-meg'-al-e*) [*prostate*; μέγαλη, great]. Prostatic hypertrophy.
**prostatometer** (*pros-tat-om'-et-er*) [*prostate*; μέτρον, measure]. An instrument for estimating the size of an enlarged prostate.
**prostatomy** (*pros-tat'-o-me*). Prostatotomy.
**prostatomyomectomy** (*pros-tat-o-mi-o-mek'-to-me*) [*prostate*; *myomectomy*]. Removal of a prostatic myoma.
**prostatoncus** (*pros-tat-ong'-kus*) [*prostate*; ὄγκος, tumor]. A tumor of the prostate.
**prostatorrhea** (*pros-tat-or-e'-ah*) [*prostate*; ῥοία, flow]. A thin urethral discharge coming from the prostate gland.
**prostatotomy** (*pros-ta-tot'-o-me*) [*prostate*; τομή, a cutting]. Incision into the prostate gland.
**prostatovesiculitis** (*pros-tat-o-ves-ik-ū-li'-tis*). Inflammation of the seminal vesicles combined with prostatitis.
**prostheon.** See *prosthion*.
**prosthesis** (*pros'-thes-is*) [πρός, to; θέσις, a placing]. Replacement of a missing part by an artificial substitute. **p., Gersuny's paraffin,** the replacement of the cartilaginous portion of the nasal septum by paraffin.
**prosthetic** (*pros-thet'-ik*) [*prosthesis*]. Pertaining to prosthesis.
**prosthetics** (*pros-thet'-iks*) [*prosthesis*]. The branch of surgery that deals with prosthesis.
**prosthion** (*pros'-the-on*) [πρόσθιον, foremost]. Alveolar point. See *craniometric points*.
**prostholytic** (*pros-tho-lit'-ik*) [πρόσθεν, before; λύειν, to loosen]. Progressive change.
**prostitution** (*pros-tit-ū'-shun*) [*prostituere*, to expose publicly]. The condition or act of a person who indiscriminately lets the body for sexual intercourse, for pay.
**prostrate** (*pros'-trāt*) [*pro-*; *sternere*, to spread out]. Lying at full length.
**prostrated** (*pros'-tra-ted*) [*pro-*; *sternere*, to spread]. Exhausted; stricken down.
**prostration** (*pros-tra'-shun*) [*prostrate*]. 1. The condition of being prostrate. 2. Extreme exhaustion of nervous or muscular force. **p., nervous,** general exhaustion from excessive expenditure of nervous energy.
**protagon** (*pro'-tag-on*) [πρῶτος, first; ἄγειν, to lead], C₁₆₀H₃₀₈N₁₀PO₃₅. A crystalline glucoside found in nervous tissue, and yielding when boiled with baryta the decomposition-products of lecithin.
**protal** (*pro'-tal*) [πρῶτος, first]. First; primary; initial; hereditary.
**protalbumose** (*pro-tal'-bū-mōs*). Same as *protoalbumose*. See under *albumose*.
**protamine** (*pro'-ta-min*) [πρῶτος, first; *amine*], C₁₆H₂₆N₆O₂. An organic base found with nuclein in the spawn of salmon.
**protan** (*pro'-tan*). A preparation of tannin and casein, insoluble in water or dilute acid solutions; employed as an intestinal astringent.
**protanopia** (*pro-tan-o'-pe-ah*) [πρῶτος, first; *anopia*]. A defect in a first constituent, essential for color-vision, as in red-blindness.
**protargol** (*pro-tar'-gol*). A silver albumose occurring as a yellow powder, containing 8 % of silver; freely soluble in water. It is recommended in gonorrhea in 0.25 to 1 % solutions.
**protean** (*pro'-te-an*) [*proteus*]. 1. Taking on many shapes, as a *protean* disease, *protean* eruption. 2. Any first hydrolytic derivative of a protein.
**protease** (*pro'-te-ās*). An enzyme which digests proteins.
**protectin** (*pro-tek'-tin*). 1. A substance which

develops in blood-serum when allowed to stand *in vitro*, and which protects the red blood corpuscles against hemolytic action. 2. Tissue paper with a coating of adhesive rubber mixture on one side.
**protective** (*pro-tek'-tiv*) [*protect*]. Covering, so as to shield against harm; guarding against harm, as a *protective dressing.* **p. protein.** See *protein, defensive.*
**proteic** (*pro-te'-ik*). Relating to protein. **p. substances, proteins.**
**proteid** (*pro'-te-id*) [πρῶτος, first]. 1. See *protein.* 2. Conjugate albumins, including nucleo-proteids and hemoglobins.
**proteiform** (*pro-te'-if-orm*). Having various forms.
**proteiin** (*pro'-te-in*) [*proteid*]. An organic substance, consisting essentially of C, H, N, O, and S, characteristic of living matter, and found, in various forms, in animals and plants; albumin. For tests, see *albumin.* **p., bacterial,** one formed by the action of a microorganism. **p., bacterial cellular,** any protein found in the bodies of bacteria. **p.s, coagulated.** See *coagulated proteins.* **p.s, compound,** a class of bodies more complex than simple proteins, which yield as nearest splitting products, simple proteins on one side and nonprotein bodies, such as coloring-matters, carbohydrates, xanthin bases, etc., on the other. They are divided into three groups—the hemoglobins, glycoproteids, and nucleoproteids. **p., defensive,** one of the proteins existing in the blood and rendering the system immune to infectious diseases. **p., pyocyaneous,** a preparation made from cultures of *Bacillus pyocyaneus* by means of potash solutions. It is recommended in the treatment of suppurating ulcers.
**proteinochromogen** (*pro-te-in-o-kro'-mo-jen*) [*protein*; χρῶμα, color; γεννᾶν, to produce]. Same as tryptophan, *q. v.*
**proteinol** (*pro'-te-in-ol*) [*proto-*; *oleum*, oil]. A form of nutritious food for infants and invalids. It contains proteids, fats, carbohydrates, and lime-salts.
**proteolysis** (*pro-te-ol-is'-is*), [πρῶτος, first; λύσις solution]. The change produced in proteins by ferments that convert them into diffusible bodies.
**proteolytic** (*pro-te-o-lit'-ik*) [*proteolysis*]. Pertaining to, characterized by, or effecting proteolysis.
**proteose** (*pro'-te-ōs*) [πρῶτος, first]. Any one of a group of bodies formed in gastric digestion, intermediate between the food proteins and the peptones, called antipeptone, hemipeptone, etc.
**Proteosoma** (*pro-te-o-so'-mah*) [Πρωτεύς, a sea god, capable of assuming many forms; σῶμα, a body]. A genus of parasitic protozoa which infest the blood of birds.
**proteosuria** (*pro-te-o-sū'-re-ah*) [*proteose*; οὖρον, urine]. The presence of proteoses in the urine.
**proteuria** (*pro-te-ū'-re-ah*). The presence of proteids in the urine.
**Proteus** (*pro'-te-us*) [L.]. A genus of *schizomycetes.*
**prothesis** (*proth'-es-is*). See *prosthesis.*
**prothetic** (*pro-thet'-ik*). See *prosthetic.*
**prothrombase** (*pro-throm'-bās*). Same as *pro-thrombin, q. v.*
**prothrombin** (*pro-throm'-bin*). Same as *thrombogen, q. v.*
**prothymia** (*pro-thi'-me-ah*) [προθυμία, zeal, readiness]. Intellectual alertness.
**protiodide** (*pro-ti'-o-dīd*) [πρῶτος, first; *iodide*]. A salt containing the least amount of iodine of the iodides of the same base. See *proto-.*
**protista** (*pro-tis'-tah*) [πρῶτος, first]. Haeckel's name for those lower organisms which are not readily referred to the plant or animal kingdom.
**proto-** (*pro-to-*) [πρῶτος, first]. 1. A prefix signifying first. 2. In chemistry, a prefix signifying the lowest of a series of compounds of the same elements, as *protoiodide, protochloride, protoxide.*
**protoalbumose** (*pro'-to-al'-bū-mōs*). See *albumose.*
**protoblast,** (*pro'-to-blast*) [*proto-*; βλαστός, germ]. A cell without a cell-wall.
**protocatechuic acid** (*pro-to-kat-e-chū'-ik*) [*proto-*; *catechu*], C₇H₆O₄. Dioxybenzoic acid, an acid sometimes occurring in the urine.
**protochloride** (*pro-to-klo'-rīd*). See *proto-* (2).
**protoelastose** (*pro-to-e-las'-tōs*) [*proto-*; ἐλαύνειν, to urge forward]. Hemielastin; one of the products of digestion of elastin.
**protogala** (*pro-tog'-al-ah*) [*proto-*; γάλα, milk]. Synonym of *colostrum.*
**protogaster** (*pro-to-gas'-ter*) [*proto-*; γαστήρ, stom-

ach]. The primitive intestinal cavity of a gastrula; the foregut.
**protogen** (*pro'-to-jen*) [*proto-*; γεννᾶν, to produce]. 1. Any albuminoid compound which does not coagulate when heated in aqueous solution. 2. A dietetic obtained by action of formaldehyde on egg-albumen.
**pr.,t.,gl.,bul,se** (*pro-to-glob'-ū-lōs*). One of the primary products of the digestion of globulin.
**protogonocyte** (*pro'-to-gon'-o-sīt*) [*proto-*; gonocyte]. One of the two cells resulting from the separation or division of the impregnated ovum.
**protoiodide** (*pro-to-i'-o-did*). See protiodide.
**protoleukocyte** (*pro-to-lū'-ko-sīt*). One of the minute lymphoid cells found in the red bone-marrow and also in the spleen.
**protometer** (*pro-tom'-et-er*) [*proto-*; μέτρον, a measure]. An instrument for measuring the forward projection of the eyeball.
**protomyosinose** (*pro-to-mi-o'-sin-ōs*) [*proto-*; myosin]. A product of the primary digestion of myosin.
**proton** (*pro'-ton*) [*proto-*]. The primitive, undifferentiated mass of cells or rudiment of a part; the equivalent of anlage.
**protonephron** (*pro-to-nef'-ron*) [*proto-*; νεφρός, kidney]. The pronephron, metanephron, and mesonephron taken together; the primitive kidney.
**protoneuron** (*pro-to-nū'-ron*) [*proto-*; νεῦρον, nerve]. One of a peculiar type of bipolar neurons or ganglion-cells of the cerebrospinal system, characterized by the entrance of the axis-cylinder of an afferent nerve-fibril through one of its prolongations, terminating in its endoplasm. It constitutes a primary independent centripetal end-organ.
**protonuclein** (*pro-to-nū'-kle-in*), C₂₉H₄₉N₁₉P₅O₇₇. A preparation from the lymphoid tissues of animals with the addition of milk-sugar and gum-benzoin. It is used as an invigorator. Dose 3–10 gr. (0.2–0.65 Gm.) several times daily. Applied also in substance to cancerous wounds.
**protopathic** (*pro-to-path'-ik*) [*proto-*; πάθος, disease]. 1. Primary; relating to the first lesion. 2. Pertaining to the first evidence of a return of function; said of the appearance of imperfect sensibility in a nerve after an injury. 3. Pertaining to nerves responding only to pain and extreme changes of temperature and in which the sensibility and localization are of a low order.
**protopathy** (*pro-top'-ath-e*) [*proto-*; πάθος, disease]. A primary disease.
**protopepsia** (*pro-to-pep'-se-ah*) [*proto-*; πέπτειν, to cook]. A primary process of digestion as that of starches by the saliva.
**protophyte** (*pro'-to-fīt*) [*proto-*; φυτόν, plant]. Any plant of the lowest and most primitive type. The *Schizomycetes*, or bacteria (*q. v.*), may be classed as protophytes, with other low vegetable forms. The protophytes have no visible reproductive organs.
**protopine** (*pro'-to-pin*) [*proto-*; ὄπιον, opium], C₂₀H₁₉NO₅. An alkaloid from opium; it is hypnotic and analgesic. Dose 40–150 gr. (2.5–10.0 Gm.).
**protoplasis** (*pro-to-pla'-sis*) [*proto-*; πλάσσειν, to form]. The primary formation of tissue.
**protoplasm** (*pro'-to-plasm*) [*proto-*; *plasma*]. 1. The viscid material constituting the essential substance of living cells, upon which all the vital functions of nutrition, secretion, growth, reproduction, irritability, and motility depend. When highly magnified, the protoplasm of most cells appears as a net-work (*spongioplasm*), containing a more fluid substance (*hyaloplasm*) in its meshes. 2. Those portions of the cell-body adjacent to the nucleus; or, the primary active portion of the cell as distinguished from the paraplasm or secondary and passive portion.
**protoplasmatic** (*pro-to-plaz-mat'-ik*). Synonym of protoplasmic.
**protoplasmic** (*pro-to-plaz'-mik*) [*protoplasm*]. 1. Pertaining to protoplasm. 2. Composed of protoplasm. p. process. See *process, protoplasmic*.
**protoplast** (*pro'-to-plast*) [*proto-*; πλαστός, formed, molded*]. 1. An embryonic cell. 2. Protoplasm.
**protoplastin** (*pro-to-plast'-in*) [*protoplasm*]. The basal substance of protoplasm.
**protoprotein** (*pro-to-pro'-te-in*). That material which, converted into organized nucleoproteids, forms living matter.
**protoproteose** (*pro-to-pro'-te-ōs*) [*proto-*; *proteose*]. Primary proteose; further digestion changes it into deuteroproteose.
**protosalt** (*pro'-to-sawlt*). In chemistry, that one of two or more compounds of a metal with an acid which contains relatively the least quantity of metal.
**protose** (*pro'-tōs*). Trade name of a food-preparation of vegetable origin.
**protospasm** (*pro'-to-spazm*) [*proto-*; σπασμός, spasm]. A spasm beginning in one part and extending to others.
**protosulphate** (*pro-to-sul'-fāt*). The one of a series of sulphates which contains relatively the smallest amount of sulphuric acid.
**protothrombin.** See *prothrombin*.
**prototoxins** (*pro-to-toks'-ins*) [*proto-*; *toxins*]. Dissociation-products of toxins differing from deuterotoxins and tritotoxins in having a stronger affinity than either for the antitoxins.
**prototoxoid** (*pro-to-toks'-oid*) [*proto-*; *toxoid*]. A transformation-product of prototoxin in which toxicity is lost, but the combining power retained.
**prototrophic** (*pro-to-tro'-fik*) [*proto-*; τροφή, nourishment]. Applied to organisms which do not require organic matter or have not the faculty of decomposing proteid stuff.
**prototype** (*pro'-to-tīp*) [*proto-*; τύπος, a type]. An original type; a type after which others are copied.
**protovertebra** (*pro-to-ver'-te-brah*). A mesoblastic segment formed on the side of the embryonic notochord.
**protoxide** (*pro-toks'-id*). See *proto-* (2).
**protoxoid** (*pro-toks'-oid*). Same as *prototoxoid*.
**protoxyl** (*pro-toks'-il*). An organic compound containing 37.69 per cent. of arsenic.
**protozoa** (*pro-to-zo'-ah*). Plural of *protozoon*.
**protozoan** (*pro-to-zo'-an*) [*proto-*; ζῶον animal]. 1. First, lowest, primitive. 2. Pertaining to protozoa.
**protozoology** (*pro-to-zo-ol'-o-je*). The study of protozoa.
**protozoon** (*pro-to-zo'-on*) [*proto-*; ζῶον, animal, pl., *protozoa*]. One of the lowest class of the animal kingdom, comprising organisms which consist of simple cells or colonies of cells, and which possess no nervous system and no circulatory organs.
**protractor** (*pro-trak'-tor*) [*pro*, before; *trahere*, to draw]. 1. A surgical instrument used for drawing extraneous bodies from wounds. 2. A muscle that draws forward.
**protrahens** (*pro-tra'-hens*) [L.]. A drawing forward. p. auriculae, a muscle drawing the pinna forward. It is known, also, as the *attrahens aurem*. See *muscles, table of*. p. lentis, the ciliary muscle of the eye.
**protropine** (*pro-tro'-pin*). See *protopine*.
**protrusor** (*pro-troo'-sor*) [*protrudere*, to thrust forward]. Anything protruded, or which protrudes. p. labii inferioris, the corrugator muscle of the lower lip. p. linguae. Synonym of *genio-glossus*. See *muscles, table of*.
**protrypsin** (*pro-trip'-sin*). See *trypsinogen*.
**protuberance** (*pro-tū'-ber-ans*) [*protuberantia*; *pro*, forward; *tuber*, a swelling]. A knob-like projecting part. p., annular. Synonym of *pons Varolii*. p., cerebral. 1. Synonym of *pons Varolii*. 2. The prominence formed by the upper or anterior of the cerebral flexures of the embryo. p., external occipital, the central prominence on the outer surface of the flat portion of the occipital bone. p., frontal. 1. The prominence of the frontal bone. 2. The prominence formed by the lower of the two flexures of the cephalic end of the embryo. p., inferior maxillary, in the embryo, a prominence formed on each side by a division of the first pharyngeal arch, and representing the rudiments of the lower jaw. p., internal occipital, a slight central prominence on the inner surface of the tabular portion of the occipital bone. p., lateral frontal, an elevation on either side of the frontal prominence of the embryo. p., mental, a synonym of *prominence, mental*. p., natiform, the nates of the corpora quadrigemina. p., occipital. See *external occipital*, and p., *internal occipital*. p., parietal, the eminence of the parietal bone, situated near the sagittal suture. p., superior maxillary, a prominence formed on each side of the embryo by a division of the first pharyngeal arch.
**protyle, protyl** (*pro-ti'-le*, *pro'-til*) [*proto-*; ὕλη, matter]. The hypothetical primal substance from which all living matter is supposed to be derived; the supposed primitive universal element.
**protylin** (*pro'-til-in*). A synthetic product con-

taining 2.7 % of phosphorus; used in neurasthenia. Dose 2-4 coffeespoonfuls.
**proud flesh.** Exuberant granulation tissue.
**provisional** (*pro-vizj'-on-al*) [*providere*, to provide]. For temporary use, as *provisional* callus.
**Prowazek's bodies** (*pro-vat'-sek*) [Stanislas Josef Mathias von *Prowazek*, German histologist]. Structures occurring in Guarnieri's vaccine bodies.
**proximad** (*proks'-im-ad*) [see *proximal*]. Toward the proximal end.
**proximal** (*proks'-im-al*) [*proximus*, nearest]. Nearest to the body or the median line of the body, or some other point considered as the center of a system.
**proximate** (*proks'-im-āt*) [see *proximal*]. Nearest; immediate, as *proximate* cause. **p. principle**, a substance existing under its own form in the animal solids or fluids, and that can be extracted by means not altering or destroying its chemical properties.
**prox. luc.** Abbreviation of *proxima luce*, the day before.
**prozonal** (*pro-zo'-nal*) [*pro-;* zona, a zone]. Applied to nerve-trunks lying in front of a sclerozone.
**prozymogen** (*pro-zi'-mo-jen*). An intranuclear substance which, after being discharged into the cytoplasm, becomes zymogen.
**prual** (*proo'-al*). A virulent toxic substance said to be obtained from the root of *Coptosapelta flavescens*, a rubiaceous plant of Malaya.
**pruinate** (*proo'-in-āt*) [*pruina*, hoar-frost]. Appearing as if covered with hoar-frost.
**prune-juice expectoration** (*proon-joos*). A peculiar bloody sputum, of a dark purple color, resembling prune-juice. It occurs in low forms of croupous pneumonia, in gangrene and carcinoma of the lung.
**prunin** (*proo'-nin*). A resinoid from *Cerasus serotina*; the dose is about two grains (0.125 gm.).
**prunum** (*proo'-num*) [L.]. Prune. The *prunum* of the U. S. P. is the partly dried fruit of *Prunus domestica*, of the order *Rosaceæ*. Prunes are laxative.
**Prunus** (*proo'-nus*) [L.]. A genus of trees of the order *Rosaceæ*. *P. domestica* is the source of the prune. The ripe seed of *P. amygdalus* is the *amygdala dulcis* of cherry U. S. P. *P. serotina* yields wild-cherry bark. **P. laurocerasus**, cherry laurel; its leaves yield hydrocyanic acid. **P. virginiana** (U. S. P.); wild-cherry, contains a volatile oil, hydrocyanic acid, tannic acid, a resin, and other substances. It is used as a tonic and sedative in gastric debility and general irritation of the system, and is a common ingredient of cough-mixtures. Dose of the *fluidextract* of wild-cherry (*fluidextractum pruni virginianæ*, U. S. P.) 1 dr. (4 Cc.); of the *infusion* (*infusum pruni virginianæ*, U. S. P.) 2-3 oz. (64-96 Cc.); of the *syrup* (*syrupus pruni virginianæ*, U. S. P.) ½ oz. (16 Cc.).
**pruriginous** (*proo-rij'-in-us*) [*prurigo*]. Pertaining to or resembling prurigo.
**prurigo** (*proo-ri'-go*) [*prurire*, to itch]. 1. A chronic inflammatory disease of the skin, characterized by small pale papules and severe itching. The papules are deeply seated, and are most prominent on the extensor surfaces of the limbs. The disease begins in early life and is usually incurable. 2. Pruritus. **p. hiemalis**, a form affecting certain persons only in winter, especially in dry climates. **p. senilis**, the pruritus of the aged, at times due to degenerative changes in the skin. **p. senilis** of the tongue, a form of glossitis occurring in the aged, characterized by itching and burning of the tongue. **p. vulvæ**, hyperesthesia of the nerves of the vulva accompanied by intense itching.
**Prussak's fibers** (*proos'-ak*) [Alexander *Prussak*, Russian otologist, 1839-1907]. The bands which, coming from the roof of the external auditory canal, pass on to the membrana tympani and form the lateral boundaries of Shrapnell's membrane; they are made up of elastic and connective-tissue fibers and contain blood-vessels and nerves. **P.'s pouch, P.'s space**, the small space formed in the attic of the middle ear by the external ligament of the malleus above, the short process of the malleus below, the neck of the malleus internally, and Shrapnell's membrane externally.
**Prussian blue.** Ferric ferrocyanide; used in the arts as a dye; it was formerly employed in medicine as a febrifuge, tonic, and alterative. Dose 3-5 gr. (0.2-0.32 Gm.).
**prussiate** (*proos'-i-āt*). 1. Any salt of prussic or hydrocyanic acid; a cyanide. 2. Particularly a ferricyanide or ferrocyanide.
**prussic acid.** See *acid, hydrocyanic*.
**psalis** (*sa'-lis*) [ψαλίς, arch]. The cerebral fornix.
**psalterial** (*sawl-te'-re-al*). Pertaining to a psalterium.
**psalterium** (*sawl-te'-re-um*) [ψαλτήριον, a psaltery: *pl.*, *psalteria*]. 1. Synonym of the *lyra*. 2. A group of longitudinal fibers on the floor of the Sylvian aqueduct. 3. In biology, the third compartment of the complicated stomach of the *pecora* or true ruminants; also called *manyplies*.
**psamma** (*sam'-ah*) [ψάμμος, sand]. Sand occurring as a urinary deposit.
**psammocarcinoma** (*sam-o-kar-sin-o'-mah*). See *carcinoma psammosum*.
**psammoma** (*sam-o'-mah*) [ψάμμος, sand; ὄμα, tumor]. A firm tumor found in the membranes of the brain, the choroid plexus, and in other parts, and characterized by the presence of peculiar calcareous particles. The tumor is generally a fibrosarcoma.
**psammotherapy** (*sam-o-ther'-ap-e*) [ψάμμος, sand; θεραπεία, treatment]. Ammotherapy; the therapeutic use of the sand-bath.
**psammous** (*sam'-us*) [ψάμμος, sand]. Sandy; sabulous.
**pselaphesis** (*sel-af-e'-sis*) [ψηλάφησις, touch]. 1. The act of groping with the fingers, seen in the low delirium of fevers; carphology. 2. Tickling, or ticklishness.
**pselaphia** (*sel-a'-fe-ah*) [ψηλαφία, touch]. 1. Digital examination or exploration. 2. The same as *pselaphesis*.
**psellism, psellismus** (*sel'-izm, sel-iz'-mus*) [ψελλισμός, a stammering]. 1. Stuttering or stammering. 2. Defective speech due to hare-lip or to cleft palate. **psellismus mercurialis**, the unintelligible, hurried, jerking speech accompanying mercurial tremor.
**pseud-, pseudo-** (*sūd-, sū-do-*) [ψευδής, false]. A prefix meaning false.
**pseudaconitine** (*sū-dak-on'-it-ēn*) [*pseud-;* *aconite*], C₃₆H₄₉NO₁₂. An extremely poisonous alkaloid from *Aconitum ferox*.
**pseudacousia, pseudacousma, pseudacusis** (*sū-dak-oos'-e-ah, -mah, -ū'-sis*) [*pseud-;* ἄκουσις, a hearing]. A disturbance of hearing in which the person's own voice sounds strange or peculiar, being altered in pitch and quality.
**pseudæsthesia** (*sū-des-the'-z-eah*). See *pseudesthesia*.
**pseudalbuminuria.** See *pseudoalbuminuria*.
**pseudamnesia** (*sū-dam-ne'-se-ah*) [*pseud-;* ἀμνησία, forgetfulness]. 1. Spurious amnesia; a condition resembling amnesia, but of a transient character. 2. An erroneous form of the word *pseudomnesia*, *q. v.*
**pseudangina** (*sū-dan'-jin-ah*). See *pseudoangina*.
**pseudankylosis** (*sū-dank-il-o'-sis*) [*pseudo-;* *ankylosis*]. A false joint; a false or fibrous ankylosis.
**pseudaphe, pseudaphia** (*sū'-da-fe, sū-da'-fe-ah*) [ψευδής, false; ἀφή, touch]. Pseudesthesia.
**pseudarthritis** (*sū-dar-thri'-tis*) [*pseud-;* *arthritis*]. Hysterical affection of a joint, simulating arthritis.
**pseudarthrosis** (*sū-dar-thro'-sis*) [*pseud-;* *arthrosis*]. A false joint.
**pseudaxis** (*sū-daks'-is*). Same as *sympodium*, *q. v.*
**pseudelminth** (*sū-del'-minth*) [*pseud-;* ἕλμινθ, a worm]. Any worm-like object mistaken for an entoparasitic worm.
**pseudencephalus** (*sū-den-sef'-al-us*) [*pseud-;* ἐγκέφαλος, brain]. A species of monster characterized by a partial development of the frontal, parietal, and occipital bones, while the brain is represented by a bunch of membranes, blood-vessels, connective and possibly nervous tissue at the base of the skull.
**pseudephedrin** (*sū-def'-ed-rin*). See *pseudoëphedrin*.
**pseudesthesia, pseudæsthesia** (*sū-des-the'-ze-ah*) [*pseud-;* αἴσθησις, feeling]. An imaginary sensation for which there is no corresponding object; a sensation referred to parts of the body that have been removed by accident or surgical operation.

**pseudinoma** (sū-din-o'-mah). Synonym of *scirrhus*.
**pseudo-** (sū'-do-). See *pseud-*.
**pseudoaconitine** (sū-do-ak-on'-it-ēn). Synonym of *pseudaconitine*.
**pseudoactinomycosis, pseudactinomycosis** (sū-do-ak-tin-o-mi-ko'-sis, sū-dak-tin-o-mi-ko'-sis). A form of pulmonary tuberculosis in which the sputum contains granular bodies resembling the grains of actinomycosis. They consist of a crystalline substance similar to leucin.
**pseudoagraphia** (sū-do-ah-gra'-fe-ah). The form of agraphia in which meaningless or disconnected words can be written.
**pseudoakromegaly** (sū-do-ak-ro-meg'-al-e). See *osteoarthropathy, hypertrophic pulmonary*.
**pseudoalbuminuria** (sū-do-al-bū-min-ū'-re-ah). False or physiological albuminuria.
**pseudoalveolar** (sū-do-al-ve'-o-lar). Simulating alveolar tissue.
**pseudoangina** (sū-do-an'-ji-nah). False angina; hysteric angina; an attack of cardiac pain somewhat resembling angina pectoris but less grave in character, longer in duration, and usually not associated with organic heart disease. It occurs in neurotic women, and is generally brought on by emotional excitement.
**pseudoangioma** (sū-do-an-je-o'-mah) [*pseudo-*; *angioma*]. The formation of a temporary angioma, as is sometimes seen in healing stumps. p., **urethral**, urethral caruncle.
**pseudoanorexia** (sū-do-an-or-eks'-e-ah). Rejection of food because of gastric distress.
**pseudoapoplexy** (sū-do-ap'-o-pleks-e) [*pseudo-*; *apoplexy*]. A diseased condition resembling apoplexy, but in which cerebral hemorrhage is not found post-mortem.
**pseudoappendicitis** (sū-do-ap-en-dis-i'-tis). 1. A condition simulating appendicitis, but with no lesion of the appendix. 2. A condition simulating appendicitis occurring in hysterical subjects or associated with secondary syphilis.
**pseudoarthrosis** (sū-do-ar-thro'-sis). See *pseudarthrosis*.
**pseudoataxia** (sū-do-ah-taks'-e-ah). See *pseudotabes*.
**pseudobacillus** (sū-do-bas-il'-us). 1. One of the very fine fat crystals sometimes found in sputum, and which stain similarly to tubercle bacilli, from which they are distinguishable by their varying size and their solubility in ether and chloroform. 2. In the plural, *pseudobacilli* (of Hayem), very minute, rod-shaped products of corpuscular fragmentation observed in poikilocytosis.
**pseudobacterium** (sū-do-bak-te'-re-um). Any object resembling a bacterium.
**pseudoblepsia, pseudoblepsis** (sū-do-blep'-se-ah, su-do-blep'-sis) [*pseudo-*; βλέψις, seeing]. A visual hallucination; a distorted visual image.
**pseudobulbar** (sū-do-bul'-bar). Not really bulbar. **p. paralysis**, symmetrical disease of both cerebral hemispheres involving the centers or paths of the nerves of speech, and thus resembling disease of the medulla oblongata.
**pseudocartilaginous** (sū-do-kar-til-aj'-in-us). Simulating cartilage in structure.
**pseudocataracta** (sū-do-kat-ar-ak'-tah) [*pseudo-*; *cataract*]. Spurious cataract.
**pseudocele** (sū'-do-sēl) [*pseudo-*; κοῖλος, a hollow]. The fifth ventricle of the brain.
**pseudoceliotomy** (sū-do-se-le-ot'-o-me). The pretended performance of abdominal section.
**pseudochalazion** (sū-do-kal-a'-ze-on). A malignant lesion of the conjunctiva resembling chalazion.
**pseudochancre** (sū-do-shang'-ker). An indurated sore simulating chancre.
**pseudochlorosis** (sū-do-klo-ro'-sis). Leache's term for a form of chlorosis in which there was no diminution of the number of erythrocytes.
**pseudochorea** (sū-do-ko-re'-ah) [*pseudo-*; χορεία, dance]. Spurious chorea, usually hysterical in origin.
**pseudochromesthesia** (sū-do-kro-mes-the'-ze-ah) [*pseudo-*; χρῶμα, color; αἴσθησις, sense]. A condition in which the vowels of a word (whether seen, heard, or remembered) each seem to have a distinct visual tint. See *photism, phonism*.
**pseudochromia** (sū-do-kro'-me-ah) [*pseudo-*; χρῶμα, color]. A false or incorrect perception of color.
**pseudocirrhosis** (sū-do-sir-o'-sis). A condition marked by ascites, dyspnea, and cyanosis, believed to be due to combined cirrhosis and peritoneal disease.
**pseudoclump** (sū'-do-klump). A clump of bacteria in broth typhoid cultures, which simulates closely the clumps formed by specific typhoid agglutinins.
**pseudocodeine** (sū-do-ko'-de-in), C₁₈H₂₁NO₃. A derivative of codeine, analogous to codeine but weaker.
**pseudocœle, pseudocœlia** (sū'-do-sēl, sū-do-se'-le-ah). Synonym of *pseudocele* and of *pseudocœlom*.
**pseudocoloboma** (sū-do-kol-o-bo'-mah) [*pseudo-*; κολόβωμα, mutilation]. A scarcely noticeable fissure of the iris, the remains of the embryonic ocular fissure, which has almost, but not perfectly, closed.
**pseudocrisis** (sū-do-kri'-sis). A false crisis; a sudden fall of temperature resembling the crisis of a disease, but subsequently followed by a rise of temperature and a continuation of the disease. It is common in pneumonia.
**pseudocroup** (sū'-do-kroop). False croup; laryngismus stridulus.
**pseudocyesis** (sū-do-si-e'-sis) [*pseudo-*; κύησις, pregnancy]. False pregnancy; the belief, on the part of a woman, in the existence of pregnancy when none exists.
**pseudocylindroid** (sū-do-sil'-in-droid) [*pseudo-*; *cylindroid*]. A band of mucus or any substance in the urine simulating a renal cast.
**pseudocyst** (sū'-do-sist) [*pseudo-*; κύστις, a bladder]. In biology, a globular body produced by the breaking up of protoplasm in a filament, in certain of the lower plants; an asexual reproductive body.
**pseudodesma** (sū-do-des'-mah) [*pseudo-*; δεσμός, bond]. A false or adventitious ligament.
**pseudodiascope** (sū-do-di'-ah-skōp) [*pseudo-*; διασκοπεῖν, to look through]. An instrument demonstrating the persistence of visual impressions made upon the retina.
**pseudodiastolic** (sū-do-di-as-tol'-ik). Apparently diastolic.
**pseudodiphtheria** (sū-do-dif-the'-re-ah) [*pseudo-*; *diphtheria*]. An inflammation characterized by the presence of a false membrane not due to the Klebs-Löffler bacillus. **p. bacillus**, a nonpathogenic bacillus resembling in form and growth the true diphtheria bacillus.
**pseudodiphtheritic** (sū-do-dif-ther-it'-ik). Simulating diphtheria; relating to pseudodiphtheria.
**pseudodyspepsia** (sū-do-dis-pep'-se-ah) [*pseudo-*; *dyspepsia*]. Nervous dyspepsia or gastric neurasthenia.
**pseudoedema** (sū-do-e-de'-mah). A condition simulating edema.
**pseudoencephalitis** (sū-do-en-sef-al-i'-tis) [*pseudo-*; *encephalitis*]. Hydrencephaloid disease.
**pseudoendometritis** (sū-do-en-do-met-ri'-tis) [*pseudo-*; *endometritis*]. A condition resembling endometritis marked by changes in the blood-vessels, hyperplasia of the glands, and atrophy.
**pseudoephedrine** (sū-do-ef'-ed-rin), C₁₀H₁₅NO. An alkaloid found in *Ephedra vulgaris* and isomeric with ephedrine.
**pseudoepilepsy** (sū-do-ep'-il-ep-se) [*pseudo-*; *epilepsy*]. Disorders simulating epilepsy, wholly or partially of rhachitic origin.
**pseudoepithelioma** (sū-do-ep-i-the-le-o'-mah) [*pseudo-*; *epithelioma*]. An affection of the skin simulating epithelioma.
**pseudoerysipelas** (sū-do-er-e-sip'-el-as) [*pseudo-*; *erysipelas*]. Inflammation of the subcutaneous cellular tissue resembling erysipelas.
**pseudoesophagism** (sū-do-e-sof'-a-jism) [*pseudo-*; *esophagism*]. A condition resembling esophagismus but without any obstruction being found in the esophagus.
**pseudoesthesia** (sū-do-es-the'-ze-ah). See *pseudesthesia*.
**pseudofever** (sū-do-fe'-ver) [*pseudo-*; *fever*]. An hysterical elevation of temperature.
**pseudofibrin** (sū-do-fi'-brin) [*pseudo-*; *fibrin*]. The coagulative material of blood in cases of "buffy coat."
**pseudofluctuation** (sū-do-fluk-tū-a'-shun) [*pseudo-*; *fluctuation*]. A tremor simulating fluctuation, sometimes observed on tapping lipomata.
**pseudofracture** (sū-do-frak-chur) [*pseudo-*; *fracture*]. A spontaneous fracture.
**pseudoganglion** (sū-do-gang'-gle-on) [*pseudo-*; *ganglion*]. A false ganglion, usually a slight thickening of a nerve.

**pseudoastralgia** (sū-do-gas-tral'-je-ah) [*pseudo-*; γαστήρ, stomach; ἄλγος, pain]. A pain resembling gastralgia, but not caused by disease of the stomach. It may be dependent upon disease of the aorta.

**pseudogelatin** (sū-do-jel'-at-in) [*pseudo-*; *gelatin*]. Any gelatinous substance obtained from vegetable tissues.

**pseudogeusesthesia, pseudogeusæsthesia** (sū-do-gū-zes-the'-ze-ah) [*pseudo-*; γεῦσις, taste; αἴσθησις, sensation]. A condition in which color sensations accompany the sense of taste.

**pseudogeusia** (sū-do-gū'-ze-ah) [*pseudo-*; γεῦσις, taste]. A false perception, or hallucination, of taste.

**pseudoglioma** (sū-do-gli-o'-mah) [*pseudo-*; *glioma*]. A name given to inflammatory changes of the vitreous humor, due to iridochoroiditis, and resembling glioma of the retina.

**pseudoglobulin** (sū-do-glob'-ū-lin). A protein found in globulin; it is not precipitated by dialysis.

**pseudoglucosazone** (sū-do-gloo-ko'-sa-zōn). A crystalline substance sometimes found in normal urine which gives the phenylhydrazine test for sugar.

**pseudogonococcus** (sū-do-gon-o-kok'-us) [*pseudo-*; *gonococcus*]. A diplococcus resembling the gonococcus found in the normal urethra.

**pseudogonorrhea** (sū-do-gon-or-e'-ah) [*pseudo-*; *gonorrhea*]. A simple nonspecific urethritis.

**pseudohematocele** (sū-do-hem-at'-o-sēl) [*pseudo-*; *hematocele*]. Hematocele occurring outside of the peritoneal cavity.

**pseudohemoglobin** (sū-do-hem-o-glo'-bin) [*pseudo-*; *hemoglobin*]. A loose combination of hemoglobin and oxygen believed to be an intermediate step between hemoglobin and oxyhemoglobin; on the reduction of the latter.

**pseudohermaphrodite** (sū-do-her-maf'-ro-dīt) [*pseudo-*; *hermaphrodite*]. An individual in which there is a double sexual formation of the external genitals, but a unisexual development of the reproductive glands (ovaries and testicles). According to the development of one or the other of the latter will the sex of the individual be determined.

**pseudohermaphroditism** (sū-do-her-maf'-ro-di-tizm) [*pseudo-*; *hermaphroditism*]. A condition simulating hermaphroditism.

**pseudohernia** (sū-do-her'-ne-ah) [*pseudo-*; *hernia*]. An empty hernia sac resembling strangulated hernia when inflamed.

**pseudoheterotopia** (sū-do-het-ur-o-to'-pe-ah) [*pseudo-*; ἕτερος, other; τόπος, place]. Misplacement of cerebral alba or cinerea by unskillful manipulation in an autopsy.

**pseudohydrarthrosis** (sū-do-hi-drar-thro'-sis) [*pseudo-*; *hydrarthrosis*]. Apparent dropsy of the knee-joint from effusion into the ligament between the patella and the tuberosity of the tibia, and especially into the fat tissue of the synovial bursa.

**pseudohydrophobia** (sū-do-hi-dro-fo'-be-ah) [*pseudo-*; *hydrophobia*]. A condition resembling hydrophobia, at times produced by dread of the disease. Syn., *lyssophobia*.

**pseudohyoscyamine.** See *hyoscyamine pseudo-*.

**pseudohypertrophic** (sū-do-hi-per-tro'-fik). Pertaining to or characterized by pseudohypertrophy. **p. muscular paralysis**, loss or diminution of the power of motion, accompanied by enlarged, and apparently hypertrophied muscles. The types are the Leyden-Moebius, Zimmerlin, and Landouzy-Déjerine. *paralysis, pseudohypertrophic.*

**pseudohypertrophy** (sū-do-hi-per'-tro-fe) [*pseudo-*; *hypertrophy*]. False hypertrophy; increase in the size of an organ on account of overgrowth of an important tissue. It is accompanied by diminution in function.

**pseudoileus** (sū-do-il'-e-us) [*pseudo-*; *ileus*]. 1. Reflex ileus due to constriction of the mesentery, or of a diverticulum, or from traction of a pedicled ovarian tumor, contusion of the testicle or abdomen, or from movable kidney. 2. Acute dilatation of the stomach. 3. Extreme constipation with paralysis of the intestinal wall.

**pseudoinfluenza** (sū-do-in-floo-en'-zah) [*pseudo-*; *influenza*]. A disease simulating influenza but not due to *Bacillus influenzæ*.

**pseudoisochromatic** (sū-do-is-o-kro-mat'-ik) [*pseudo-*; ἴσος, equal; χρῶμα, color]. Of different colors, yet apparently of the same color. Cf. *anisochromatic*.

**pseudojaundice** (sū-do-jawn'-dis). Hematogenous jaundice without hepatic disease.

**pseudoleukemia, pseudoleukæmia** (sū-do-lū-ke'-me-ah) [*pseudo-*; *leukemia*]. See *Hodgkin's disease*. **p., infantile**, von Jaksch's disease, a form of anemia occurring in young children, usually dependent on a rhachitic diathesis, and not associated with much leukocytosis.

**pseudoleukocythemia** (sū-do-lū-ko-si-the'-me-ah). Synonym of *pseudoleukemia*.

**pseudolien** (sū-do-le'-en). See *spleen, accessory*.

**pseudolipoma** (sū-do-lip-o'-mah) [*pseudo-*; *lipoma*]. A localized edema resembling an accumulation of fat, occurring above the clavicle and about the knee, especially in cases of rheumatism.

**pseudolupus** (sū-do-lū'-pus) [*pseudo-*; *lupus*]. A disease simulating lupus vulgaris, produced by a species of oidium. **p. vulgaris.** See *dermatitis, blastomycetic*.

**pseudomalady** (sū-do-mal'-ad-e) [*pseudo-*; *malum*, evil]. An imaginary or simulated illness.

**pseudomalaria** (sū-do-mal-a'-re-ah) [*pseudo-*; *malaria*]. A toxemic disease simulating malaria.

**pseudomamma** (sū-do-mam'-ah) [*pseudo-*; *mamma*]. A mamma-like structure sometimes occurring in dermoid cysts.

**pseudomania** (sū-do-ma'-ne-ah) [*pseudo-*; μανία, madness]. 1. A form of insanity in which the patient accuses himself of crimes of which he is innocent. 2. A mania characterized by lying.

**pseudomelanosis** (sū-do-mel-an-o'-sis) [*pseudo-*; μέλας, black; νόσος, disease]. The dark staining of gangrenous parts or the tissues after death, due to the deposit of ferrous sulphide, which is formed by a reaction of hydrogen sulphide and the hemoglobin of the blood.

**pseudomembrane** (sū-do-mem'-brān) [*pseudo-*; *membrane*]. A false membrane, such as is seen in diphtheria.

**pseudomembranous** (sū-do-mem'-bran-us). Characterized by, or pertaining to, false membranes. **p. inflammation**, any inflammation characterized by the formation of a false membrane.

**pseudomeningitis** (sū-do-men-in-ji'-tis) [*pseudo-*; *meningitis*]. A group of symptoms resembling that produced by meningitis, but with absence of the lesions of meningeal inflammation; meningism. **p., dental**, meningeal symptoms occurring in children during difficult dentition.

**pseudomeninx** (sū-do-men'-inks) [*pseudo-*; μῆνιγξ, membrane]. A false membrane.

**pseudomnesia** (sū-do-mne'-ze-ah) [*pseudo-*; μνῆσις, remembrance]. Perversion of the memory in which things that never occurred seem to be remembered.

**Pseudomonas** (sū-do-mo'-nas) [*pseudo-*; μόνας, monad]. A genus of microorganisms having polar flagella.

**pseudomorphine** (sū-do-mor'-fēn). A finely crystalline alkaloid of opium, insoluble in water, alcohol, ether, and chloroform; it is soluble in alkalies and lime-water, neutral, tasteless. It has the chemical but not the toxic properties of morphine.

**pseudomucin** (sū-do-mū'-sin) [*pseudo-*; *mucin*]. A substance allied to mucin, found in proliferative ovarian cysts.

**pseudomyxoma** (sū-do-miks-o'-mah) [*pseudo-*; *myxoma*]. A tumor containing colloid matter derived from a ruptured mucous cyst.

**pseudonarcotic** (sū-do-nar-kot'-ik). Sedative but not narcotic.

**pseudonarcotism** (sū-do-nar'-ko-tizm) [*pseudo-*; ναρκοῦν, to benumb]. An hysterical simulation of narcotism.

**pseudoneoplasm** (sū-do-ne'-o-plazm) [*pseudo-*; *neoplasm*]. 1. A phantom tumor. 2. A temporary swelling generally of inflammatory origin.

**pseudoneuralgia** (sū-do-nū-ral'-je-ah) [*pseudo-*; *neuralgia*]. A term given by Charcot to the pains of rickets.

**pseudoneuritis** (sū-do-nū-ri'-tis). A disease simulating a neuritis.

**pseudoneuroma** (sū-do-nū-ro'-mah) [*pseudo-*; *neuroma*]. A false neuroma. See *neuroma*.

**pseudonucleus** (sū-do-nū'-kle-in). See *paranuclein*.

**pseudoosteomalacia** (sū-do-os-te-o-mal-a'-se-ah) [*pseudo-*; *osteomalacia*]. Rhachitis in which the pelvic basin is distorted so as to resemble in form that of osteomalacia.

**pseudoparalysis** (sū-do-par-al'-is-is) [*pseudo-*; *paralysis*]. Paralysis of motion, apparently but not really due to a lesion of the nervous system. **p. agitans.** See *dystaxia agitans*. **p. myasthenica,**

bulbar paralysis without apparent anatomic lesion. **p. of rickets**, the inability to walk in severe cases of rickets, due to distortion of the bones. **p., syphilitic**, an inflammatory condition of the epiphyses of the bones in syphilis causing a marked impairment of motion.

**pseudoparanoia** (sū-do-par-an-oi'-ah) [pseudo-; paranoia]. A condition in which there is a primarily more active mentality, as well as an increased responsivity to emotional impulses, associated with a heightened desire for activity.

**pseudoparaplegia** (sū-do-par-ap-le'-je-ah) [pseudo-; paraplegia]. Spurious paraplegia. **p., tetanoid**. Synonym of paralysis, spastic.

**pseudoparasite** (sū-do-par'-ah-sīt) [pseudo-; parasite]. 1. Any object resembling a parasite. 2. A commensal.

**pseudoparesis** (sū-do-par'-es-is) [pseudo-; paresis]. An affection resembling paresis, but regarded as distinct from the ordinary forms.

**pseudopellagra** (sū-do-pel-a'-grah) [pseudo-; pellagra]. An affection regarded by some authorities as distinct from pellagra, but presenting similar skin symptoms.

**pseudopelletierine** (sū-do-pel-et'-e-er-ēn), $C_{18}H_{30}$-$N_2O_3$. An alkaloid found in the root-bark of pomegranate.

**pseudopeptone** (sū-do-pep'-tōn). Same as hemialbumose.

**pseudopericardial** (sū-do-per-e-kar'-de-al). Appearing to be connected with the pericardium.

**pseudophlegmon** (sū-do-fleg'-mon) [pseudo-; phlegmon]. A simulated furuncle due to trophic nerve lesion.

**pseudophotesthesia** (sū-do-fo-tes-the'-ze-ah) [pseudo-; φῶς, light; αἴσθησις, perception]. The seeing of photisms.

**pseudophthisis** (sū-do-ti'-sis) [pseudo-; φθίσις, wasting]. Emaciation and general wasting arising from other causes than pulmonary tuberculosis.

**pseudoplasm** (sū'-do-plazm) [pseudo-; πλάσμα, a thing molded]. Same as pseudoneoplasm.

**pseudoplegia** (sū-do-ple'-je-ah) [pseudo-; πληγή, stroke]. Simulated or hysterical paralysis.

**pseudopneumococcus** (sū-do-nū-mo-kok'-us). A diplococcus larger than the pneumococcus, but much like it in some respects.

**pseudopneumonia** (sū-do-nu-mo'-ne-ah). Any disease of the lung simulating pneumonia.

**pseudopod, pseudopodium** (sū'-do-pod, sū-do-po'-de-um) [pseudo-; πούς, foot]. A protrusion of a portion of the substance of an amoeboid cell.

**pseudopodiospore** (sū-do-po'-de-o-spōr) [pseudopodium; spore]. A spore having pseudopodia; an amæbula.

**pseudopregnancy** (sū-do-preg'-nan-se) [pseudo-; pregnancy]. Synonym of pregnancy, false.

**pseudopsia** (sū-dop'-se-ah) [pseudo-; ὤψ, eye]. Visual hallucination, or error of visual perception.

**pseudopterygium** (sū-do-ter-ij'-e-um). False, or cicatricial, pterygium.

**pseudoptosis** (sū-do-to'-sis) [pseudo-; ptosis]. A condition resembling ptosis caused by a fold of skin and fat depending below the edge of the eyelid.

**pseudopus** (sū'-do-pus) [pseudo-; pus]. A liquid that resembles pus in appearance only.

**pseudorabies** (sū-do-ra'-be-ez). See lyssophobia.

**pseudoreaction** (sū-do-re-ak'-shun) [pseudo-; reaction]. Agglutination not due to typhoid bacilli. Cf. pseudoclump.

**pseudorexia** (sū-dor-eks'-e-ah) [pseudo-; ὄρεξις, desire]. A perverted appetite.

**pseudorhachitis** (sū-do-ra-ki'-tis). Osteitis deformans.

**pseudorheumatism** (sū-do-roo'-mat-izm) [pseudo-; rheumatism]. False rheumatism. **p., infectious**, a term given by Lapersonne to certain cases of multiple synovitis or arthritis, the prime cause of which it is impossible to discover.

**pseudorhonchus** (sū-do-rong'-kus) [pseudo-; rhonchus]. A false or spurious rhonchus; a deceptive auscultatory sound.

**pseudoscarlatina** (sū-do-skar-lat-e'-nah) [pseudo-; scarlatina]. A febrile disease associated with a rash like that of scarlatina, occurring as a result of gonorrhea or after puerperal infection.

**pseudosclerosis** (sū-do-skle-ro'-sis) [pseudo-; sclerosis]. An affection similar in symptoms to multiple sclerosis of the nervous system, but without the anatomical lesions.

**pseudosmia** (sū-doz'-me-ah) [pseudo-; ὀσμή, smell]. Perversion of the sense of smell; an olfactory hallucination.

**pseudosphincter** (sū-do-sfink'-ter) [pseudo-; sphincter]. An imperfect sphincter.

**pseudospleen, pseudosplen** (sū'-do-splēn, -splen). See spleen, accessory.

**pseudosteogenesis** (sū-dos-te-o-jen'-e-sis) [pseudo-; ὀστέον, bone; γεννᾶν, to produce]. Abnormal boneformation.

**pseudosteum** (sū-dos'-te-um). An abnormal bony growth.

**pseudostoma** (sū-dos'-to-mah) [pseudo-; στόμα, a mouth]. An apparent aperture between endothelial cells that have been stained with silver nitrate.

**pseudotabes** (sū'-do-ta-bēz) [pseudo-; tabes]. 1. A disease simulating tabes dorsalis or tabes mesenterica. 2. The ataxic form of alcoholic multiple neuritis.

**pseudotetanus** (sū-do-tet'-an-us) [pseudo-; tetanus], Escherich's symptom-complex. A rare type of tetanus in which the masseters and dorsal muscles are principally involved.

**pseudotoxin** (sū-do-toks'-in). A name given to extract of belladonna containing other substances.

**pseudotrichinosis** (sū-do-trik-in-o'-sis) [pseudo-; trichinosis]. Acute polymyositis resembling trichinosis of the muscles.

**pseudotrichosis** (sū-do-trik-o'-sis) [pseudo-; trichosis]. The growth of hair in an abnormal location.

**pseudotropine** (sū-do-tro'-pēn)' [pseudo-; tropine]. $C_8H_{15}NO$. An isomeric form of tropine.

**pseudotuberculosis** (sū-do-tu-ber-kū-lo'-sis) [pseudo-; tuberculosis]. A disease resembling tuberculosis, but not caused by the tubercle bacillus.

**pseudotumor** (sū-do-tū'-mor) [pseudo-; tumor]. A phantom tumor; one that changes its location is called an abdominal tumor.

**pseudotyphoid** (sū-do-ti'-foid) [pseudo-; typhoid]. Spurious typhoid, a disease simulating typhoid fever, but in which the true lesions of this disease as well as the typhoid bacilli are absent.

**pseudovacuoles** (sū-do-vak'-ū-ōlz) [pseudo-; vacuole]. Transparent bodies containing pigment found by Laveran in blood of malarial patients.

**pseudoventricle** (sū-do-ven'-trik-l). The fifth ventricle of the brain.

**pseudovermicule** (sū-do-ver'-mik-ūl) [pseudo-; vermes, worm]. The motile stage in the development of certain plasmodia, that produced from the fertilized macrogamete in the case of the malarial parasite and allied organisms.

**pseudoxanthine** (sū'-do-zan-thēn) [pseudo-; xanthine]. 1. $C_4H_5N_5O$, a leukomaine isolated from fresh beef. 2. A body isomeric with xanthine, obtained by action of sulphuric acid upon uric acid.

**pseudoxanthoma** (sū-do-zan-tho'-mah) [pseudo-; xanthoma]. A rare chronic disease of the skin characterized by an eruption of yellowish plaques, slightly elevated, with an especial predilection for certain parts of the skin, e. g., lower abdomen, axilla, sides of neck, etc., in which it differs from true xanthoma, which affects by preference the flexures of fingers, the extensor surfaces of elbows and knees, etc. It differs histologically from true xanthoma in being a degeneration of the elastic tissue of the skin. Syn., pseudoxanthoma elastica.

**pseudozoogloea** (sū-do-zo-og-le'-ah) [pseudo-; ζῷον, animal; γλοία, glue]. A clump of bacteria something like a zooglæa mass, but not dissolving readily in water and not having the degree of compactness and gelatinization possessed by zooglææ.

**pseudulcus** (sūd-ul'-kus). A false ulcer. **p. ventriculi**, a sensory-neurosis of the stomach closely resembling gastric ulcer.

**pseudydrops** (sūd-i'-drops) [pseudo-; ὕδρωψ, dropsy]. False dropsy.

**psilosis** (si-lo'-sis) [ψίλωσις, bare]. 1. The removal of the hair from a part; depilation. 2. See aphthæ tropicæ.

**psilothin, psilothinum** (si-lo'-thin, si-lō-thi'-num). A depilatory cerate containing elemi, 40 %; benzoin, 10 %; rosin, 8 %; yellow wax, 10 % and diachylon plaster, 30 %; applied warm, and when cool, removed with the hair adhering.

**psilothric** (si-lo'-thrik) [ψίλωθρον, a depilatory]. Depilatory.

**psilothron** (si-lo'-thron) [ψίλωθρον]. A depilatory.

**psilotic** (si-lo'-tik). Relating to psilosis.

**psittacosis** (sit-ak-o'-sis) ψιττακός, a parrot]. A disease of birds, especially of parrots, transmissible to man, in whom it runs the course of a violent typhoid fever without abdominal symptoms but with pulmonary disorders resembling severe infectious pneumonia. It is due to *Bacillus psittacosis*.

**psoadic** (*so-ad'-ik*) [*psoas*]. Pertaining to a psoas muscle, or to the loin.

**psoadotomia** (*so-ad-o-to'-me-ah*) [*psoas*; τομή, a cutting]. Incision of the psoas muscle.

**psoas** (*so'-as*) [ψόα, loin]. One of two muscles—*psoas magnus* and *psoas parvus*—of the loins and pelvis. See under **muscle**. **p. abscess**, an abscess, usually dependent upon tuberculous disease of a vertebra, making its way through the sheath of the psoas muscle and pointing at the front of the thigh, below Poupart's ligament, to the outer side of the spine of the pubis.

**psodymus** (*sod'-im-us*) [*psoas*; δίδυμος, double]. A monster with two heads and chests and conjoined abdominal and pelvic cavities.

**psoitis** (*so-i'-tis*) [*psoas*; ἶτις, inflammation]. Inflammation of the psoas muscles or of the region of the loins.

**psomophagia, spomophagy** (*so-mo-fa'-je-ah, so-mof'-a-je*) [ψωμός, a bit; φαγεῖν, to eat]. Swallowing the food after imperfect and inadequate mastication; bolting the food.

**psora** (*so'-rah*) [ψώρα, the itch]. 1. Scabies. 2. Psoriasis.

**psorelcosis** (*so-rel-ko'-sis*) [*psora*; ἕλκωσις, ulceration]. Ulceration occurring during the progress of scabies.

**psorenteria** (*so-ren-te'-re-ah*) [*psora*; ἐντερον, intestine]. Inflammatory condition of solitary follicles of the intestine observed in Asiatic cholera.

**psorenteritis** (*so-ren-ter-i'-tis*) [*psora*; *enteritis*]. The intestinal condition in Asiatic cholera.

**psoriasic** (*so-ri-as'-ik*). See **psoriatic**.

**psoriasiform** (*so-ri-as'-e-form*). Resembling psoriasis.

**psoriasis** (*so-ri'-as-is*) [*psora*]. A chronic inflammatory disease of the skin characterized by the development of reddish patches covered with whitish scales. The disease affects especially the extensor surfaces of the body. **p. annularis**. Synonym of **p. circinata**. **p.**, **buccal**, **p. buccalis**. Synonym of *leukoplakia buccalis*. **p. circinata**, psoriasis in which the central part of the lesions has disappeared, leaving ring-shaped patches. **p. diffusa**, a form in which there is coalescence of large contiguous lesions. **p. guttata**. See **p. punctata**. **p. gyrata**, psoriasis with a serpentine arrangement of the patches. **p. lotricum**, a form attacking the hands and arms, particularly of washerwomen. Syn., *washerwoman's itch*. **p. osteacea**, psoriasis associated with affections of joints. **p. palmaris**, a form affecting the palms of the hands. **p. pistorum**, baker's itch. **p. punctata**, a form in which the lesions consist of minute red papules which rapidly become surmounted by pearly scales. **p. universalis**, a form in which the lesions are over all the body.

**psoriatic** (*so-ri-at'-ik*) [*psoriasis*]. Pertaining to or affected with psoriasis.

**psorocomium** (*so-ro-ko'-me-um*) [*psora*; κομεῖν, to take care of]. A hospital for patients affected with the itch.

**psoroid** (*so'-roid*) [*psora*; εἶδος, like]. Similar to psora or scabies.

**psorophthalmia** (*so-rof-thal'-me-ah*) [*psora*; ὀφθαλμός, eye]. Marginal blepharitis.

**psorosperm** (*so'-ro-sperm*) [*psora*; σπέρμα, seed]. A unicellular organism belonging to the protozoa; a coccidium; a sporozoon.

**psorospermial, psorospermic** (*so-ro-sperm'-e-al, so-ro-sperm'-ik*) [*psorosperm*]. Pertaining to, or affected with, psorosperms.

**psorospermiasis** (*so-ro-sperm-i'-as-is*) [*psorosperm*]. A state characterized by the presence of psorosperms.

**psorospermosis** (*so-ro-sperm-o'-sis*) [*psorosperm*]. A diseased condition associated with the presence of psorosperms. **p.**, **proliferative follicular**. Synonym of *keratosis follicularis* and *Darier's disease*.

**psorous** (*so'-rus*) [*psora*]. Pertaining to or affected with the itch.

**psorozoa** (*so-ro-zo'-ah*) [*psora*; ζῷον, animal]. Animal psorosperms.

**psychagogia** (*si-kag-o'-je-ah*) [ψυχή, spirit; ἀγωγός, leading]. Mental excitement or activity.

**psychagogic** (*si-kag-oj'-ik*) [ψυχή, spirit; ἀγωγός, leading]. 1. Restorative of the consciousness. 2. A remedy that restores to consciousness, as in fainting.

**psychalgia** (*si-kal'-je-ah*) [ψυχή, mind; ἄλγος, pain]. Painful cerebration in melancholia.

**psychalia** (*si-ka'-le-ah*) [ψυχή, mind]. A morbid condition attended by hallucinations.

**psychanalysis** (*si-kan-al'-is-is*) [ψυχή, mind; *analysis*]. A method of obtaining from nervous patients, against their will, a knowledge of their past experiences.

**psychasthenia** (*si-kas-the'-ne-ah*) [ψυχή, mind; *asthenia*]. Mental fatigue.

**psyche** (*si'-ke*) [ψυχή, mind]. The brain and myelon considered as one organ; the cerebrospinal axis.

**psycheism** (*si'-ke-izm*). Hypnotism.

**psychentonia** (*si-ken-to'-ne-ah*) [ψυχή, mind; ἐντονία, tension]. Mental strain or over-work.

**psychiater** (*si-ki'-at-er*) [ψυχή, mind; ἰατρός, a physician]. An alienist; one who cures mind-diseases.

**psychiatria** (*si-ki-a'-tre-ah*). Same as *psychiatry*.

**psychiatric** (*si-ke-at'-rik*) [*psychiatry*]. Pertaining to psychiatry.

**psychiatrics** (*si-ke-at'-riks*). See *psychiatry*.

**psy hiatrist** (*si-ki'-at-rist*). A specialist in psychiatry. c

**psychiatry** (*si-ki'-at-re*) [*psyche*; ἰατρεία, healing art]. The science and treatment of the diseases of the mind.

**psychic, psychical** (*si'-kik, si'-kik-al*) [*psyche*]. Pertaining to the mind. **p. blindness**. See *blindness, psychic*. **p. deafness**. See *deafness, psychic*. **p. infection**, mental infection; the development of a mental condition or disease through an influence acting upon the mind.

**psychics** (*si'-kiks*). The science of psychology.

**psychlampsia** (*si-klamp'-se-ah*) [ψυχή, mind; λάμψις, a flashing]. Mania, viewed as a discharging phenomenon of perverted cerebral activity.

**psycho-** (*si-ko-*) [ψυχή, mind]. A prefix denoting connection with the mind.

**psychoanalysis** (*si-ko-an-al'-is-is*). See *psychanalysis*.

**psychoauditory, psychauditory** (*si-ko-aw'-dit-o-re, si-kaw'-dit-o-re*) [*psycho-*; *auditory*]. Pertaining to the psychic perception of sound. **p. area**, the cortical area concerned in the conscious perception of sound.

**psychocoma** (*si-ko-ko'-mah*) [*psycho-*; *coma*]. Mental stupor.

**psychocortical** (*si-ko-kor'-tik-al*) [*psycho-*; *cortex*]. Pertaining to that part of the cerebral cortex concerned in the conscious perception of sensations.

**psychodometer** (*si-ko-dom'-et-er*). [*psycho-*; ὁδός, way; μέτρον, measure]. An instrument for measuring the rapidity of psychic processes.

**psychodynamic** (*si-ko-di-nam'-ik*). Pertaining to psychodynamics.

**psychodynamics** (*si-ko-di-nam'-iks*) [*psycho-*; δύναμις, power]. The science of the laws of mental activity.

**psychogenesis** (*si-ko-jen'-es-is*) [*psycho-*; γένεσις, generation]. The development of mental characteristics.

**psychogeny** (*si-koj'-en-e*) [*psycho-*; γενής, producing]. The development of mind.

**psychogeusic** (*si-ko-gū'-sik*) [*psycho-*; γεῦσις, sense of taste]. Pertaining to perception of taste.

**psychokinesia** (*si-ko-kin-e'-ze-ah*) [*psycho-*; κίνησις, movement]. Explosive or impulsive maniacal action, due to defective inhibition; psychlampsia.

**psychology** (*si-kol'-o-je*) [*psycho-*; λόγος, science]. The science having for its object the investigation of the mind or consciousness. **p.**, **abnormal**, the study of all irregular or unusual mental phenomena, as illusions, hallucinations, trance, hypnotism, automatism, intoxication and psychic effects of drugs, telepathy, insanity, etc. Cf. *psychopathology*; *psychiatry*. **p. experimental**. See *psychophysics*.

**psychometry** (*si-kom'-et-re*) [*psycho-*; μέτρον, a measure]. The measurement of the duration of psychic processes.

**psychomotor** (*si-ko-mo'-tor*) [*psycho-*; *movere*, to move]. Pertaining to voluntary movement, as the *psychomotor* area, disposed chiefly along each side of the central fissure.

**psychoneurology** (*si-ko-nū-rol'-o-je*). That part of neurology treating of mental action.

**psychoneurosis** (*si-ko-nū-ro'-sis*) [*psycho-*; *neurosis*]. Mental disease not dependent on any organic lesion.

**psychonomy** (*si-kon'-o-me*) [*psycho-*; ὄνομα, name]. The science of the laws of mental action.

**psychonosema** (*si-ko-no-se'-mah*) [*psycho-*; νόσημν, disease]. Any mental disease.

**psychooptic** (*si-ko-op'-tik*) [*psycho-*; ὤψ, sight]. Pertaining to the psychic perception of light. p. **area**, the cortical area concerned in conscious perception of retinal impulses.

**psychoparesis** (*si-ko-par'-es-is*) [*psycho-*; *paresis*]. Enfeeblement of the mind.

**psychopath** (*si'-ko-path*) [*psycho-*; πάθος, disease]. A morally irresponsible person.

**psychopathia** (*si-ko-pa'-the-ah*) [*psycho-*; πάθος, disease]. Psychopathy. p. **chirurgicalis**, a mania for being operated upon. p. **sexualis**, psychopathia characterized by perversion of the sexual functions.

**psychopathic** (*si-ko-path'-ik*). Pertaining to psychopathy.

**psychopathist** (*si-kop'-ath-ist*). Synonym of *psychiatrist*.

**psychopathology** (*si-ko-path-ol'-o-je*) [*psycho-*; *pathology*]. 1. The pathology of mental diseases. 2. The legal aspect of insanity.

**psychopathy** (*si-kop'-ath-e*) [*psycho-*; πάθος, disease]. Any disease of the mind.

**psychophysical** (*si-ko-fiz'-ik-al*) [*psychophysics*]. Pertaining to psychophysics. p. **law**. See *Fechner's law*.

**psychophysics** (*si-ko-fiz'-iks*) [*psycho-*; *physical*]. The study of mental processes by physical methods; the study of the relation of stimuli to the sensations which they produce, especially the determination of the differences of stimulus required to produce recognizable differences of sensation; experimental psychology.

**psychophysiology** (*si-ko-fiz-e-ol'-o-je*). Physiological psychology; mental physiology.

**psychoplasm** (*si'-ko-plazm*) [*psycho-*; πλάσμα, anything formed]. In biology: "The sentient material out of which all forms of consciousness are evolved, incessantly fluctuating, incessantly renewed." See *protyl*.

**psychoplegic** (*si-ko-ple'-jik*) [*psycho-*; πληγή, a blow]. A drug which acts by an elective affinity for the gray matter of the brain, lessening its excitability and suppressing its receptivity.

**psychorrhagia** (*si-ko-ra'-je-ah*) [*psycho-*; ῥηγνύναι, to break forth]. The death agony.

**psychosensory** (*si-ko-sen'-so-re*) [*psycho-*; *sensory*]. Pertaining to. or concerned in the conscious perception of sensory impulses.

**psychosexual** (*si-ko-seks'-u-al*). Relating to combined mental and sexual diseases.

**psychosin** (*si-ko'-sin*) [ψυχή, mind]. A cerebrosid resembling sphingosin, occurring in brain-tissue.

**psychosis** (*si-ko'-sis*) [*psyche; pl., psychoses*]. A disease of the mind, especially one without demonstrable organic lesions. Any morbid mental state. p., **Korsakoff's**. See *Korsakoff*.

**psychosomatic** (*si-ko-so-mat'-ik*) [*psycho-*; σῶμα, body]. Relating to both soul and body.

**psychotherapeutic** (*si-ko-ther-ap-ū'-tik*). Pertaining to psychotherapeutics.

**psychotherapeutics** (*si-ko-ther-ap-ū'-tiks*) [*psycho-*; θεραπεία, treatment]. The treatment of disease by mental influence, or by suggestion.

**psychotherapy** (*si-ko-ther'-ap-e*). Same as *psychotherapeutics*.

**psychotic** (*si-kot'-ik*). 1. Pertaining to psychosis. 2. Analeptic.

**psychovisual** (*si-ko-vizj'-ū-al*). Relating to subjective vision or to vision unaccompanied by stimulation of the retina.

**psychovital** (*si-ko-vi'-tal*). Psychic and vital.

**psychralgia, psychroalgia** (*si-kral'-je-ah, si-kro-al'-je-ah*) [ψυχρός, cold; ἄλγος, pain]. A morbid condition characterized by a painful subjective sense of cold.

**psychrapostema** (*si-krap-os'-te-mah*) [ψυχρόν, cold; ἀπόστημα, abscess]. Cold abscess.

**psychro-** (*si-kro-*) [ψυχρός, cold]. A prefix meaning cold.

**psychroesthesia** (*si-kro-es-the'-se-ah*) [*psychro-*; αἴσθησις, sensation]. Subjective sensation of cold.

**psychrolusia** (*si-kro-lū'-se-ah*) [ψυχρός, cold; λούειν, to wash]. Cold bathing.

**psychrometer** (*si-krom'-et-er*) [*psychro-*; μέτρον, a measure]. An instrument for determining the atmospheric moisture by estimating the amount of cold required to precipitate it.

**psychrophilic** (*si-kro-fil'-ik*) [*psychro-*; φιλεῖν, to love]. Applied to microorganisms which develop best at room-temperature from 15° to 20° C.

**psychrophobia** (*si-kro-fo'-be-ah*) [*psychro-*; φόβος, fear]. 1. Morbid dread of cold. 2. Morbid sensibility to cold.

**psychrophore** (*si'-kro-fōr*) [*psychro-*; φέρειν, to bear]. An instrument for applying cold to deeply seated parts, as a double-current catheter for applying cold to the posterior part of the urethra.

**psychrotherapy** (*si-kro-ther'-ap-e*) [*psychro-*; θεραπεία, treatment]. The treatment of disease by the use of cold.

**psydracia** (*si-dra'-se-ah*) [ψύδραξ, a blister; a pimple]. An old term for eczema.

**Pt.** Chemical symbol of platinum. Abbreviation for pint.

**ptarmic** (*tar'-mik*) [πταρμός, a sneezing]. 1. Pertaining to the act of sneezing; sternutatory. 2. A substance that produces sneezing.

**ptelein** (*te'-le-in*). An alcoholic extract from the root-bark of *Ptelea trifoliata*, used as a tonic and in dyspepsia. Dose 1–3 gr. (0.065–0.2 Gm.).

**pteleorrhine** (*tel'-e-or-in*) [πτελέο, elm; ῥίς, nose]. A term applied to the anterior nares when the aperture is asymmetric.

**ptenium, ptenum** (*te'-ne-um, te'-num*) [πτηνός, winged]. A name given to osmium because of its volatility.

**pteric** (*ter'-ik*). Pertaining to the pterion.

**pterion** (*te'-re-on*). See under *craniometric point*.

**pterna** (*ter'-nah*) [πτέρνα, the heel]. The calcaneum.

**ptero-** (*ter'-o-*) [πτερόν, wing]. A prefix to denote resemblance to a wing, or wing-shaped.

**pterygial** (*ter'-ij'-e-al*) [dim. of πτέρυξ, wing]. Pertaining to a pterygium.

**pterygium** (*ter-ij'-e-um*) [dim. of πτέρυξ, wing]. A triangular patch of mucous membrane growing on the conjunctiva, usually on the nasal side of the eye. The apex of the patch points toward the pupil, the fan-shaped base toward the canthus.

**pterygo-** (*ter'-ig-o-*) [πτέρυξ, a wing]. A prefix denoting connection with or relating to the pterygoid process.

**pterygoid** (*ter'-ig-oid*) [πτέρυξ, wing; εἶδος, like]. 1. Wing-shaped, as the *pterygoid* plate of the sphenoid bone. 2. Pertaining to the pterygoid canal, pterygoid plate, pterygoid plexus, etc. p. **fossa**, the notch separating the external and internal plates of the pterygoid process of the sphenoid. See *muscles, table of*.

**pterygoideus** (*ter-ig-o'-mah*) [*pterygo-*; ὄμα, tumor]. 1. A chronic swelling of the labia minora which interferes with coitus. 2. The lobe of the ear.

**pterygomaxillary** (*ter-ig-o-maks'-il-a-re*) [*pterygoid; maxillary*]. Pertaining to the pterygoid process and the maxilla. p. **fissure**, an elongated fissure formed by the divergence of the superior maxillary bone from the pterygoid process of the sphenoid bone. p. **ligament**. See *ligament, pterygomaxillary*.

**pterygopalatine** (*ter-ig-o-pal'-al-in*) [*pterygoid; palatine*]. Situated between the pterygoid plate of the sphenoid bone and the palate bone, as the *pterygopalatine canal*.

**pterygopharyngeus** (*ter-ig-o-far-in'-je-us*) [*pterygo-*; *pharynx*]. 1. Synonym of *palatopharyngeus*. 2. The part of the superior constrictor of the pharynx which arises from the internal pterygoid plate.

**pterygospinous** (*ter-ig-o-spi'-nus*) [*pterygo-*; *spinosus, spinous*]. Pertaining to a pterygoid process and to the spine of the sphenoid.

**pterygotemporal** (*ter-ig-o-tem'-po-ral*) [*pterygo-*; *lempus*, temple]. Pertaining to the pterygoid process and the temporal bone.

**ptiloma** (*ti-lo'-mah*) [πτίλον, down]. The part of the eyelid deprived of its cilia by ptilosis.

**ptilosis** (*ti-lo'-sis*) [πτίλον, feather]. Loss of the hair, especially that of the eyelashes.

**ptisan** (*tis'-an*) [πτισάνη, peeled barley]. 1. Barley-water. 2. A decoction of barley used as a medicinal drink. Syn., *tisane*.

**ptoma** (*to'-mah*) [πτῶμα]. Cadaver.

**ptomaine** (*to'-ma-ēn, or to'-mān*) [πτῶμα, corpse]. A basic compound resembling the alkaloids, formed by the action of bacteria on animal and vegetable tissues; a putrefactive or animal alkaloid. Some ptomaines are highly poisonous.

**ptomainemia, ptomainemia** (*to-ma-in-e'-me-ah*) [*ptomaine*; αἷμα, blood]. The presence of ptomaines in the blood.

**ptomatine** (to'-mat-ēn). Synonym of *ptomaine*.
**ptomatinuria** (to-mat-in-ū'-re-ah) [*ptomaine*; οὖρον, urine]. The presence of ptomaines in the urine.
**ptomatopsia** (to-mat-op'-se-ah) [*ptoma*; ὄψις, view]. The examination of the cadaver; necropsy.
**ptomatropine** (to-mat'-ro-pēn). A ptomaine found in decomposing meat, in the organs of persons dead of typhoid fever, etc. It resembles atropine in its physiological and chemical properties.
**ptomatropism** (to-mat'-ro-pizm). See *zootrophotoxism tropeinicus*.
**ptosis** (to'-sis) [πίπτειν, to fall]. Drooping of the upper eyelid, due to paralysis or atrophy of the levator palpebræ superioris. The term is also applied to abnormal depression of other organs; prolapse. p., **abdominal**, enteroptosis; Glénard's disease. p. **iridis**, prolapse of the iris. p. **sympathetica**, ptosis in connection with paresis of the cervical sympathetic nerve and associated with miosis and vasomotor paralysis of the side of the face affected.
**ptotic** (tot'-ik). Affected with or pertaining to ptosis.
**ptyalagogue** (ti-al'-a-gog) [πτύαλον, spittle; ἀγωγός, leading]. A medicine producing an increased flow of saliva. A sialagogue.
**ptyalin** (ti'-al-in) [πτύαλον, saliva]. A diastatic ferment found in saliva, having the property of converting starch into dextrin and sugar. The starch first becomes converted into achroodextrin and erythrodextrin; these by hydration into maltose, and the latter, by further hydration, into dextrose.
**ptyalinogen** (ti-al-in'-o-jen) [*ptyalin*; γεννᾶν, to produce]. The hypothetical antecedent of ptyalin.
**ptyalism**, p., **mercurial** (ti'-al-izm) [πτύαλον, saliva]. Salivation.
**ptyalith** (ti'-al-ith) [πτύαλον, saliva; λίθος, stone]. A salivary calculus.
**ptyalize** (ti'-al-īs) [πτύαλον, saliva]. To produce ptyalism.
**ptyalocele** (ti'-al-o-sēl) [πτύαλον, saliva; κήλη, tumor]. A cyst due to obstruction of the duct of a salivary gland.
**ptyalogogue** (ti-al'-o-gog) [πτύαλον, spittle; ἀγωγός, leading]. A medicine causing a flow of saliva.
**ptyalolith** (ti'-al-o-lith) [πτύαλον, saliva; λίθος, stone]. A salivary calculus.
**ptyalolithiasis** (ti-al-o-lith-i'-as-is). The formation or presence of a salivary calculus.
**ptyalose** (ti'-al-ōs) [πτύαλον, saliva]. A sugar found in saliva; it is identical with maltose.
**ptysis** (ti'-sis) [πτύσις]. The act of spitting.
**ptysma** (tiz'-mah) [πτύσμα]. Saliva.
**ptysmagogue** (tis'-mag-og) [πτύσμα, spittle; ἀγωγός, leading]. A drug that promotes the secretion of saliva; a ptyalagogue, or sialagogue.
**puben** (pū'-ben) [*pubes*]. Belonging to the pubes in itself.
**puber** (pū'-bur) [L., an adult]. One who has arrived at the age of puberty.
**puberal** (pū'-ber-al) [*puber*, adult]. Relating to puberty.
**pubertas** (pū-ber'-tas) [L.]. Puberty. p. **plena**, complete puberty. p. **præcox**, puberty at a very early age.
**puberty** (pū'-ber-te) [*pubertas*, from *puber*, adult]. 1. The period at which the generative organs become capable of exercising the function of reproduction, signalized in the boy by a change of voice and discharge of semen, in the girl by the appearance of the menses.
**pubes** (pū'-bēz) [L.]. 1. The pubic hair. 2. The hairy region covering the os pubis. 3. The os pubis or pubic bone; that portion of the innominate bone forming the front of the pelvis.
**pubescence** (pū-bes'-ens) [*pubescentia*]. 1. Hairiness; the presence of fine soft hairs. 2. Puberty, or the coming on of puberty.
**pubescent** (pū-bes'-ent) [*pubes*]. 1. Downy, or hairy. 2. Approaching or arriving at the age of puberty. p. **uterus**, an abnormality of the uterus in which the characters of that organ peculiar to the epoch preceding puberty persist in the adult.
**pubetrotomy** (pū-be-trot'-o-me) [*pubes*; ἦτρον, pelvis; τομή, a cutting]. Pelvic section through the pubes.
**pubic** (pū'-bik) [*pubes*]. Pertaining to the pubes. p. **bone**, the os pubis.
**pubiotomy** (pū-be-ot'-o-me) [*pubes*; τομή, a cutting]. The operation of dividing the pubic bone to facilitate delivery in cases of pelvic malformation. See also *symphyseotomy*.

**pubis** (pū'-bis) [gen. of *pubes*]. 1. Os pubis; the pubic bone. 2. One of the pubic hairs.
**pubo-** (pū'-bo-) [*pubes*]. A prefix denoting relation to the pubes.
**pubocapsular** (pū-bo-kap'-sū-lar). Pertaining to the os pubis and the capsule of the hip-joint.
**pubococcygeal** (pū-bo-kok-sij'-e-al). Pertaining, or having relation to the os pubis and the coccyx.
**pubofemoral** (pū-bo-fem'-or-al) [*pubo-*; *femur*]. Pertaining to the os pubis and the femur.
**puboprostatic** (pū-bo-pros-tat'-ik) [*pubo-*; *prostate*]. Pertaining to the os pubis and the prostate gland.
**pubotibial** (pū-bo-tib'-e-al) [*pubo-*; *tibia*]. Pertaining to the os pubis and the tibia.
**pubovesical** (pū-bo-ves'-ik-al) [*pubo-*; *vesica*, bladder]. Pertaining to the os pubis and bladder.
**puccin** (puk'-sin). A substance found in *Sanguinaria canadensis*. Its exact nature is unknown.
**pudenda** (pū-den'-dah). Plural of *pudendum*, q. v.
**pudendagra** (pū-den-da'-grah) [*pudenda*; ἄγρα, a seizure]. 1. Pain in the genital organs. 2. Primary syphilis, especially of the female genital organs.
p. **pruriens**, pruritus vulvæ.
**pudendal** (pū-den'-dal) [*pudenda*]. Pertaining to the pudenda.
**pudendum** (pū-den'-dum) [*pudere*, to be ashamed]. The external genital organs especially of the woman, generally used in the plural, *pudenda*. p. **muliebre**, the vulva.
**pudic** (pū'-dik) [*pudenda*]. Pertaining to the pudenda, as the *pudic* artery.
**puericulture** (pū-er-e-kul'-tūr) [*puer*, a child; *cultura*, culture]. That branch of hygiene which deals with the rearing of children and the care of women during pregnancy that they may bring forth healthy offspring.
**puerile** (pū'-er-il) [*puer*, a child]. Pertaining to childhood. Childish. p. **respiration**, exaggerated breath-sounds with expiration prolonged and highpitched, such as is heard in healthy children.
**puerpera** (pū-er'-pe-rah) [*puer*, child; *parere*, to bear]. A woman who is in labor or has recently been delivered.
**puerperal** (pū-er'-pe-ral) [*puerpera*]. Pertaining to, caused by, or following childbirth, as *puerperal convulsions*, *puerperal eclampsia*. p. **fever**, an acute, febrile disease of women in childbed, due to septic infection. p. **insanity**, insanity occurring during the puerperium, usually within five or ten days after delivery. It may take the form of mania (*puerperal mania*), melancholia (*puerperal melancholia*), or dementia (*puerperal dementia*).
**puerperalism** (pū-er'-per-al-ism). A comprehensive term for all the pathological conditions incident to the puerperal state. p., **infantile**, any pathologic condition incident to the newborn. p., **infectious**, puerperal disease due to infection.
**puerperant** (pū-er'-per-ant). See *puerpera*.
**puerperium** (pū-er-pe'-re-um) [*puerpera*]. 1. The state of a woman in labor or of one who has just been delivered. 2. The period from delivery to the time when the uterus has regained its normal size which is about six weeks.
**puerperous** (pū-er'-per-us). Same as *puerperal*.
**puffball**. See *Lycoperdon*.
**puffiness** (puf'-in-es). Swelling or intumescence of the tissues; an edematous condition.
**pugil, pugillus** (pū'-jil, pū-jil'-us) [L.]. A handful.
**puking** (pū'-king) [origin obscure]. Vomiting. p. **fever**. Synonym of *milk-sickness*.
**Pulex** (pū'-leks) [L., flea; *pl.*, *pulices*]. A genus of insects partly parasitic on the skin of man and animals. **P. cheopis**, the rat flea, supposed to convey the plague infection from rats to man. **P. fasciatus**, the rat flea. **P. irritans**, a species common in Europe and parasitic on the skin of man; its bite causes severe itching and localized swelling. **P. penetrans**, the chigoe, or jigger-flea, a species the female of which burrows under the skin of the feet to deposit its ova, producing a severe irritation that may proceed to serious inflammations.
**pulicaris** (pū-lik-a'-ris) [*pulex*]. Marked with little spots like flea-bites. p. **morbus**, a name for typhus, applied because of the petechiæ which occur in that disease.
**pulicatio** (pū-lik-a'-she-o) [*pulex*, flea]. The state of being infested with fleas.
**pulled elbow** (puld el'-bo). A condition in which the head of the radius has been dislodged from the orbicular ligament.

**pulling** (*pul'-ing*). One of the Swedish movements that may be either active or passive.
**pullulate** (*pul'-ū-lāt*) [*pullulare*, to put forth]. To germinate, to bud.
**pullulation** (*pul-ū-la'-shun*) [*pullulare*, to put forth; to bud; to sprout]. The act of sprouting or budding, a mode of reproduction seen, *e. g.*, in the yeast-plant.
**pulmo** (*pul'-mo*) [L.: *pl.*, *pulmones*]. Lung.
**pulmoaortic** (*pul-mo-a-or'-tik*) [*pulmo;* *aorta*]. 1. Pertaining to the lungs and the aorta. 2. Pertaining to the pulmonary artery and the aorta.
**pulmometer** (*pul-mom'-et-er*). An instrument for measuring the lung capacity; a *spirometer*.
**pulmometry** (*pul-mom'-et-re*) [*pulmo;* μέτρον, measure]. The determination of the volume of the lungs; spirometry.
**pulmonary** (*pul'-mon-a-re*) [*pulmo*, a lung]. Pertaining to or affecting the lungs, as *pulmonary arteries*, *pulmonary emphysema*.
**pulmonectomy** (*pul-mon-ek'-to-me*). See *pneumonectomy*.
**pulmonic** (*pul-mon'-ik*) [*pulmonary*]. 1. Pertaining to the lungs; pulmonary. 2. Pertaining to the pulmonary artery, as *pulmonic* valves. 3. Produced at the pulmonic valve, as *pulmonic* murmur. p. circulation, the passage of the blood from the right ventricle to the lungs and back to the left auricle. p. fever, croupous pneumonia.
**pulmonin** (*pul'-mon-in*). An organotherapeutic preparation made from calves' lungs, and used in pulmonary affections.
**pulmonitis** (*pul-mon-i'-tis*). Synonym of *pneumonia*.
**pulmotor** (*pul-mo'-tor*) [*pulmo*, lung; *motor*]. An apparatus for resuscitating persons who have been asphyxiated; it expels the gas from the lungs, introduces oxygen, and automatically establishes artificial respiration.
**pulp** [*pulp*]. 1. The soft, fleshy part of fruit. 2. The soft part in the interior of an organ, as the *pulp* of the spleen, the *pulp* of a tooth. 3. Chyme. p.-cavity, the hollow space in a tooth containing the dental pulp. p.-cells, cells found in the pulp-tissue of any organ. p., dental, a soft tissue filling the pulp-cavity of a tooth. It consists of loose connective tissue and cells, vessels, and nerves. Also the rudiment of a tooth. p., digital, the sensitive, elastic, convex prominence on the palmar or plantar surface of the terminal phalanx of a finger or toe. p. of the finger. See *p.*, *digital*. p., hair. Synonym of *papilla, hair*. p. of the intervertebral discs, the soft substance in the center of the intervertebral discs, the remains of the chorda dorsalis. p., spleen, p., splenic, the substance filling the spaces formed by the trabeculæ of the spleen. See *spleen*. p. of a tooth. See *p.*, *dental*.
**pulpar** (*pul'-par*). Pertaining to pulp.
**pulpation** (*pulp-a'-shun*). See *pulpefaction*.
**pulpefaction** (*pul-pe-fak'-shun*) [*pulp; facere*, to make]. Conversion into a pulpy substance.
**pulpitis** (*pul-pi'-tis*) [*pulp; ιτις*, inflammation]. Inflammation of the dental pulp.
**pulpy** (*pul'-pe*) [*pulp*]. Resembling pulp; characterized by the formation of a substance resembling pulp.
**pulque** (*pul'-ke*) [Sp.]. A fermented beverage prepared in Mexico from the juice of various species of *Agave*. p. brandy. See *mezcal*. p. plant. See *Agave*.
**pulsate** (*pul'-sāt*) [*pulsare*, to beat]. To beat or throb.
**pulsatile** (*pul'-sat-il*) [*pulsare*, to strike]. Pulsating; throbbing.
**pulsatilla** (*pul-sat-il'-ah*) [L.]. The herb of *Anemone pulsatilla* and of *Anemone pratensis*, of the order *Ranunculaceæ*, containing a crystalline principle, *anemonin*, C₁₅H₁₀O₃. Pulsatilla is employed in amenorrhea, dysmenorrhea, and in inflammation of mucous membranes, especially in bronchitis and asthma. Dose ¼ gr. (0.022 Gm.). Dose of pulsatilla in powder 2–3 gr. (0.13–0.2 Gm.).
**pulsating** (*pul'-sa-ting*) [see *pulsation*]. Exhibiting pulsation. p. aorta, the pulsation of the abdominal aorta seen in nervous and anemic persons. p. empyema, an accumulation of pus in the pleural cavity that transmits the pulsations of the heart.
**pulsation** (*pul-sa'-shun*) [*pulsatio*, from *pulsare*, to strike]. A beating or throbbing. p., **suprasternal**, pulsation at the suprasternal notch. It may be due to aneurysm, a dilated aortic arch, or the presence of an anomalous artery.
**pulse, pulsus** (*puls, puls'-us*) [*pulsus*, the pulse]. 1. The intermittent change in the shape of an artery due to an increase in the tension of its walls following the contraction of the heart. The pulse is usually counted at the wrist (*radial pulse*), but may be taken over any artery that is palpable, as the temporal, brachial, femoral, dorsalis pedis, etc. 2. [πόλτος, a thick pap made of meal.] Leguminous plants or their seeds, as beans, peas, etc. **p., alternating, pulsus alternans**, a variety in which a large pulsation alternates with a small one. p., **anacrotic**, one the sphygmographic tracing of which is characterized by notches in the ascending limb. p., **anatricrotic**, one with three breaks in the ascending limb. p., **angry**. Synonym of *p., wiry*. p., **ardent**, one with a quick, full wave which seems to strike the finger at a single point. p., **bigeminal, pulsus bigeminus**, one in which the beats occur in pairs, so that the longer pause follows every two beats. p.-**breath**, a peculiar audible pulsation of the breath corresponding to the heart-beats; observed in cases of dry cavities of the lungs, with thick walls not separated from the heart by permeable lung tissue. p., **capillary**, an intermittent filling and emptying of the capillaries of the skin. It is common in aortic regurgitation, and is seen under the finger-nail or on the forehead. **pulsus celer**, a quick, short pulse. p., **caprizant, goat-leap pulse, q. v.** p., **catacrotic**, one with an elevation in the line of descent in the sphygmographic tracing. **pulsus celer et altus**, a quick, full pulse, seen especially in aortic regurgitation. p., **contracted**, a small pulse with high tension. p.-**clock**, a sphygmograph. p., **cordy**, a tense pulse. p., **Corrigan's**. See *Corrigan's pulse*. p.-**curve**, the tracing of the pulse, called a sphygmogram, made by the sphygmograph. p., **decurtate**, a progressively decreasing pulse. p., **depressed**, a pulse both deep and weak. p., **dicrotic**, one in which the dicrotic wave or recoil wave is exaggerated. It is observed when the arterial tension is low, and gives to the finger the impression of two beats. p., **entoptic**, the subjective illumination of a dark visual field with each heart-beat, a condition sometimes noted after violent exercise, and due to the mechanical irritation of the rods by the pulsating retinal arteries. p., **febrile**, that characteristic of fever: full, soft, and frequent, and exhibiting a well-marked dicrotism. p., **formicant**, a small, feeble pulse likened to the movements of ants. p., **frequent, pulsus frequens**, one recurring at short intervals and differing from a quick pulse, in which the pulse-wave has a quick rise. p., **full**, one in which the artery is filled with a large volume of blood and conveys a feeling of being distended. p., **funic**, the arterial tide in the umbilical cord. p., **gaseous**, a full, compressible pulse. p., **goat-leap**, a pulse marked by a weak pulsation succeeding a strong one. p., **hard**, one characterized by high tension and rigidity. p., **high-tension**, one due to increase of the peripheral resistance, together with a corresponding increase in the force of the ventricular systole. It is gradual in its impulse, long in duration, slow in subsiding, with difficulty compressible, and the artery between the beats feels like a firm round cord. p., **hyperdicrotic, p., hyperdicrotous**, a pulse of which the aortic notch falls below the base line, indicating very low tension, a symptom of great exhaustion. p., **infrequent**, one the rhythm of which is slower than normal; *i. e.*, in which the heart-beats are fewer in a given time than normal. p., **intermittent**, one in which certain beats are dropped. p., **intricate**, an irregular, small, infrequent pulse. p., **irregular**, one in which the beats occur at irregular intervals, or in which the force, or both rhythm and force, varies. p., **jarring**. See *p., vibrating*. p., **jerky**, a pulse in which the artery is suddenly and markedly distended, as in aortic regurgitation. p., **jugular**, pulsation of the jugular veins in the neck. It is due to tricuspid regurgitation. p., **locomotive**. Synonym of *Corrigan's pulse*. p., **low-tension**, one sudden in its onset, short, and quickly declining. It is easily obliterated by pressure. p., **monocrotic**, one in which dicrotism is entirely absent. p., **paradoxic, pulsus paradoxus**, one that is weaker during inspiration, a condition sometimes observed in adherent pericardium. p., **polycrotic**, one with more than two rhythms for each heart beat. p.-**pressure**, the difference between the systolic and diastolic

pressure. **pulsus quadrigeminus**, p. **trigeminus**, a pulse in which a pause occurs after every fourth or third beat respectively. p., **quick**, one that strikes the finger rapidly, but also leaves it rapidly. p., **rate**, the number of pulsations of an artery in a minute. p., **retrosternal**, a venous pulse believed to be due to the pulsation of the left innominate vein, perceived on depressing the integument of the suprasternal notch. p., **running**, a very weak, frequent pulse with low tension in the arteries, one pulse-wave running into the next with no apparent interval; it is observed after hemorrhage. p., **senile**, one characteristic of old age. The secondary waves on the descending line of the sphygmogram are prominent and the first descending wave relatively large. p., **shabby**, an ill-defined pulse due to weak heart and relaxed arteries. p., **shuttle**, one in which the wave passes under the finger as if floating something solid with a fluid. p., **slow**, one indicating a lengthened systolic contraction of the heart and prolonged diastole—often used to signify a pulse of slow rate. p., **soft**, a pulse that is readily compressed. p., **supradicrotic**, a dicrotic pulse in which the dicrotic wave resembles the cardiac beat. **pulsus tardus**, a slow pulse. p., **thready**, one that is scarcely perceptible, feeling like a thread under the finger. p., **tricrotic**, a pulse in which the three waves normally present are abnormally distinct. p., **venous**, a pulse observed in a vein. p., **vermicular**, a pulse imitating the movement of a worm. p., **vibrating, pulsus vibrans**, a tense pulse with a wave arising quickly, giving the impression under the finger of the vibrations of a piece of tense catgut. p., **water-hammer**. See *Corrigan's pulse*. p. **wave**, the condition of expansion that begins with each cardiac systole and is propagated along the aorta and the arteries ending normally at the capillaries. p., **wiry**, a small rapid, tense pulse, feeling like a cord under the finger. It is observed in acute peritonitis.
**pulseless** (*puls'-les*). Devoid of pulse or pulsation.
**pulsellum** (*pul-sel'-um*) [*pulsellum*, dim. of *pulsus*, a beating; *pl., pulsella*]. A propulsive filament; a modified form of flagellum.
**pulsimeter** (*pul-sim'-et-er*) [*pulse*; μέτρον, a measure]. An instrument for determining the rate or force of the pulse.
**pulsometer** (*pul-som'-et-er*). Same as *pulsimeter*.
**pultus** (*pul'-sus*). A pulse. *q. v.*
**pultaceous** (*pul-ta'-shus*) [*pulp*]. Having the consistence of pulp; mushy; soft.
**pulv.** Abbreviation of Latin *pulvis*, powder.
**pulver** (*pul'-ver*). See *pulvis*.
**pulveres** (*pul'-ver-ēz*). Plural of *pulvis*.
**pulverflator** (*pul-ver-fla'-tor*) [*pulvis*, powder; *flare*, to blow]. An instrument designed for blowing or spraying impalpable powders.
**pulverization** (*pul-ver-i-za'-shun*) [*pulvis*, powder]. The act of reducing a substance to powder.
**pulverize** (*pul'-ver-īz*) [*pulvis*, powder]. To reduce to a powder.
**pulverulence** (*pul-ver'-ū-lenz*). The condition of being reduced to powder.
**pulverulent** (*pul-ver'-ū-lent*). Resembling or of the nature of a powder.
**pulvillus** (*pul-vil'-us*) [L., "a little cushion"]. An olive-shaped pad of lint used in plugging deep wounds.
**pulvinar** (*pul-vi'-nar*) [L., "couch"]. 1. The posterior tubercle of the thalamus opticus. 2. A fatty mass that occupies a part of the acetabulum. 3. A surgical pad. 4. A medicated cushion.
**pulvinate** (*pul'-vin-āt*) [*pulvinus*, a cushion]. 1. With a convex surface said of bacterial cultures. 2. Same as *pulvinar*.
**pulvis** [L.; pl., *pulveres*]. A powder. **p. acetanilidi compositus** (U. S. P.), a mixture of acetanilid, caffeine, and sodium bicarbonate. **p. aromaticus** (U. S. P.), a mixture of cinnamon, ginger, nutmeg, and cardamom seeds; used as a carminative. Dose 10–30 gr. (0.65–2.0 Gm.). **p. cretæ compositus** (U. S. P.), consists of prepared chalk, powdered acacia, and sugar, and is used as a mild astringent. Dose 10 gr.– 1 dr. (0.65–4.0 Gm.). **p. effervescens compositus** (U. S. P.), Seidlitz powder, a preparation consisting of two powders: the white paper contains 35 gr. of tartaric acid, the blue paper, 40 gr. of sodium bicarbonate and 2 dr. of Rochelle salt. **p. glycyrrhizæ compositus** (U. S. P.), consists of sehna, licorice, oil of fennel, washed sulphur, sugar; it is used as a laxative. Dose ½–2 dr. (2–8 Gm.). **p. ipecacuanhæ et opii** (U. S. P.). See *opium, powder of ipecac and*. **p. jalapæ compositus** (U. S. P.), consists of jalap, 35 parts; potassium bitartrate, 65 parts; it is used as a hydragogue cathartic. Dose ½–1 dr. (2–4 Gm.). **p. morphinæ compositus** (U. S. P.), consists of morphine sulphate, camphor, glycyrrhiza, precipitated calcium carbonate, and alcohol. Dose 7½ gr. (0.5 Gm.). **p. rhei compositus** (U. S. P.), consists of rhubarb, magnesia, and ginger; it is used as a mild laxative. Dose ½–1 dr. (2–4 Gm.).
**pumex** (*pū'-meks*) [*spumex*, foam]. See *pumice*.
**pumice** (*pum'-is*) [*pumex*]. Pumice-stone, used as a detergent for the skin and an ingredient in some dentifrices.
**pumiline** (*pū'-mil-ēn*). The oil from the young branches of *Pinus pumilio*.
**pump.** An apparatus either drawing up a liquid into its hollow chamber, or, after sucking up the liquid, forcibly ejecting it from one end. p., **air-**, one used to exhaust the air from a chamber or to force more air into a chamber already filled with air. p., **Alvegniat's.** See under *Alvegniat*. p., **breast-**, a pump for removing milk from the breast. p., **dental**, one for removing saliva during dental operations. p., **stomach-**, one for removing the contents of the stomach in cases of poisoning.
**pumpkin-seed.** The seed of *Cucurbita pepo* of the order *Cucurbitaceæ*. The seed *pepo* (U. S. P.) is used against tape-worm. Dose 4 oz. (130 Cc.).
**puncta** (*punk'-tah*) [pl. of *punctum*, a point]. See *punctum.* p. **dolorosa**, tender or painful points at the exit or in the course of nerves the seat of neuralgia; also called Valeix's points. p. **lacrimalia**, the orifices of the lacrimal canaliculi in the eyelids near the inner canthus. p. **vasculosa**, minute red spots studding the cut surface of the white central mass of the brain. They are produced by the blood escaping from divided blood-vessels.
**punctate, punctated** (*punk'-tāt, punk'-ta-ted*) [*punctum*, point]. Dotted; full of minute punctures.
**puncticulum** (*punk-tik'-ū-lum*) [dim. of *punctum*, a point; pl., *puncticula*]. A small point; petechia.
**punctiform** (*punk'-tif-orm*) [*punctum*, point; *forma*, form]. 1. Having the nature or qualities of a point; seeming to be located at a point; as a *punctiform* sensation. 2. Denoting very minute colonies, in bacteriology.
**punctum** (*punk'-tum*) [L.]. A point. p. **cæcum**. See *blind spot*. p. **lacrimale**. See *puncta lacrimalia*. **p. proximum.** See *near-point*. **p. remotum.** See *far-point*. **p. saliens**, the first trace of the embryonic heart.
**puncture** (*punk'-chur*) [*pungere*, to prick]. 1. A hole made by a pointed instrument. p., **diabetic**, puncture of the fourth ventricle, which produces glycosuria. p., **exploratory**, the puncture of a cyst or cavity for removal of a portion of its contents for examination. p., **lumbar**, puncture of the spinal canal for the withdrawal of cerebrospinal fluid for examination or for the relief of abnormal tension; first suggested by Quincke. p., **spinal**. See *p.*, *lumbar*.
**punctured** (*punk'-tūrd*) [*puncture*]. Produced by a prick, as a *punctured* wound.
**pungent** (*pun'-jent*) [*pungere*, to prick]. Acrid; penetrating; producing a pricking or painful sensation.
**Punica** (*pū'-nik-ah*) [*punicum*, the pomegranate]. A genus of polypelatous plants. **P. granatum**. See *pomegranate*.
**puniceous** (*pū-nish'-us*) [*puniceus*, red]. Bright-carmin color.
**punicin** (*pū'-nis-in*) [*puniceus*, reddish]. 1. A crystalline coloring-matter obtained from the colorless juices of certain kinds of shell-fish (*Purpura lapellus, P. patulo*); on exposure to the sunlight it becomes of a purple color.
**punicine** (*pū'-nis-ēn*). Synonym of *pelletierine*.
**punktograph** (*punk'-to-graf*) [*punctum*, point; γράφειν, to record]. A radiographic instrument for the surgical localization of foreign bodies, as bullets embedded in the tissues.
**pupa** (*pū'-pah*) [*pupa*, a girl, doll, puppet; pl., *pupæ*]. In biology, the second stage of development from the egg, of such insects as undergo complete metamorphosis.
**pupal** (*pū'-pal*) [*pupa*, a doll]. Pertaining to a pupa.
**pupil** (*pū'-pil*) [*pupilla*, a little girl; the name is believed to be derived from the small images seen

in the pupil]. The aperture in the iris of the eye for the passage of light. **p.**, **Argyll Robertson.** See *Argyll Robertson pupil*. **p.**, **artificial,** an aperture made by iridectomy when the normal pupil is occluded. **p.**, **cat's-eye,** an elongated, slit-like pupil. **p.**, **multiple,** the presence of bands dividing the pupil into several portions due to persistence of portions of the fetal pupillary membrane. **p.**, **pinhole,** extreme miosis.
**pupilla** (*pū-pil'-ah*) [L.]. The pupil of the eye.
**pupillary** (*pū'-pil-a-re*) [*pupil*]. Pertaining to the pupil. **p. membrane.** See *membrane, pupillary*. **p. membrane, persistent.** See under *membrane, pupillary*. **p. reflex.** See under *reflex*.
**pupillometer** (*pū-pil-om'-et-er*) [*pupil*]. An instrument for measuring the pupil of the eye.
**pupilloscopy** (*pū-pil-os'-ko-pe*) [*pupil*; σκοπεῖν, to inspect]. 1. Examination of the pupil. 2. Retinoscopy.
**pupillostatometer** (*pū-pil-o-stāt-om'-e-ter*) [*pupilla*, pupil; στάτος, placed; μέτρον, a measure]. An instrument for measuring the exact distance between the centers of the two pupils.
**pural** (*pū'-ral*). Trade name of a disinfecting agent consisting of powdered wood charcoal, saturated with a mixture of menthol, carbolic and benzoic acids, and compressed into cylinders, which are ignited for disinfection.
**Purdy's solution** (*pur'-de*) [Charles Wesley *Purdy*, American physician, 1846-1901]. A modification of Fehling's solution for the quantitative estimation of glucose. Solution I: pure crystallized copper sulphate, 4.158 gm.; distilled water, q. s. ad 500 c.c. Solution II: Rochelle salt, 20.4 gm.; pure potassium hydroxide, 20.4 gm.; ammonium hydrate (sp. gr. 0.88), 300 c.c.; distilled water, q. s. ad 500 c.c. Mix 5 c.c. of each solution and this mixture will indicate 0.005 gm. of glucose.
**pure** (*pūr*). Unstained; unalloyed.
**purgament, purgamentum** (*per'-gam-ent, per-gamen'-tum*) [*purgare*, to purge; *pl.*, *purgamenta*]. 1. A purge. 2. In the plural, the lochia; also, excrement.
**purgatin, purgatol** (*pur'-ga-tin, -ol*). See *anthrapurpurin diacetate*.
**purgation** (*pur-ga'-shun*) [*purge*]. 1. The evacuation of the bowels by means of purgatives. 2. Cleansing.
**purgative** (*pur'-ga-tiv*) [*purge*]. 1. Producing purgation. 2. A drug producing copious evacuations of the bowel.
**purge** (*purj*) [*purgare*, to purge]. 1. To cause free evacuation of the bowel. 2. A drug that causes free evacuation of the bowel.
**purgen** (*pur'-jen*). See *phenolphthalein*.
**purgerine** (*pur'-jer-ēn*). A proprietary syrup of senna; laxative.
**purging nut.** Curcas, the seed of *Jatropha purgans* or *J. curcas*, having cathartic properties. Syn., *Barbados nut*.
**purified** (*pū'-rif-īd*) [*purus*, pure; *facere*, to make]. Cleansed; freed from extraneous matter.
**puriform** (*pū'-re-form*) [*pus*; *forma*, form]. Resembling pus.
**purin** (*pū'-rin*). A synthesized substance ($C_5H_4N_4$) from which may be derived a series of compounds known as xanthin bases, and uric acid. **p.-bases, p.-bodies,** those derived from purin by simple substitution of the various hydrogen atoms by hydroxyl, amide, or alkyl groups; *e. g.*, adenin, hypoxanthin, guanin, xanthin, and the methylxanthins, theobromine, caffeine, etc., together with the nuclein. **p.s, endogenous,** those originating from nuclein cleavage during metabolic processes. **p.s, exogenous,** those derived from the purin-bodies of food-stuffs.
**purinemia, purinæmia** (*pū-rin-e'-me-ah*) [*purin; αἷμα*, blood]. The presence of purin bodies in the blood.
**purinometer** (*pū-rin-om'-et-er*) [*purin*; μέτρον, a measure]. An apparatus consisting of a graduated separator with a uniform bore for the clinical estimation of urinary purins.
**Purkinje's cells** (*poor-kin'-ye*) [Johannes Evangelista *Purkinje*, Bohemian anatomist and physiologist, 1787-1869]. Large ganglion-cells of the cerebellar cortex, disposed as a single row at the junction of the nuclear and the molecular layer, and presenting pyriform or flask-shaped bodies, 60-70 μ in their longest diameter. **P.'s corpuscles,** the lacunæ of bone. **P.'s fibers,** anastomosing muscular fibers found in the subendocardial tissue of some animals. They are made up of polyhedral nucleated cells, the margins of which consist of fine, transversely striated fibers. **P.'s figures,** the dark lines which are seen on a yellow back-ground when a candle is held a short distance from the eye in a darkened room. They are produced by the retinal vessels. **P.'s granular layer.** See *Czermak's interglobular spaces*. **P.'s images,** three pairs of images of one object seen in an observed pupil: the first, erect, reflected from the anterior surface of the cornea; the second, erect, reflected from the anterior surface of the lens; the third, inverted, reflected from the posterior capsule of the lens. **P.'s network,** the network of beaded fibers (*Purkinje's fibers*) visible to the naked eye in the subendocardial tissue of the ventricles. **P.'s vesicle,** the germinal vesicle.
**Purkinje-Sanson's images** (*poor-kin'-ye san'-sun*). See *Purkinje's images*.
**puro** (*pū'-ro*). A meat-juice, made by expression of the meat. It contains 21 % of unchanged albumin.
**puroform** (*pū'-ro-form*). A proprietary antiseptic and disinfectant said to be a combination of zinc and formaldehyde, thymol, menthol, and eucalyptol.
**purohepatitis** (*pū-ro-hep-at-i'-tis*) [*pus*; ἧπαρ, liver; ιτις, inflammation]. Purulent hepatitis.
**puromucous** (*pū-ro-mū'-kus*) [*pus*; *mucus*]. Purulent and mucous.
**puronal** (*pū'-ron-al*). A proprietary antiseptic compound said to contain acetanilid, 97.22 %, and bismuth subiodide, 2.35 %.
**purple, visual.** See *rhodopsin*.
**purples** (*pur'-pls*) [*purpureus*, purple]. 1. A popular name for purpura; also, 2. for petechial spots. 3. Swine fever.
**purposive** (*pur'-po-siv*). Functional; not vestigial, and not rudimentary; regarded as fulfilling an end or purpose in the economy. **p. acts,** those acts performed with the consent of the will.
**purpura** (*pur'-pūr-ah*) [L.]. A disease characterized by hemorrhages into the skin, taking the form of petechiæ, macules, or large patches. It may occur as an independent affection, or be symptomatic of other diseases. **p. fulminans,** a grave form of purpura developing in young children as a sequel to acute infectious diseases. It is of short duration, is marked by extensive extravasations, grave constitutional symptoms, and usually ends fatally. **p. hæmorrhagica,** a systemic disease with marked constitutional symptoms, followed by an eruption of hemorrhagic petechiæ upon the legs, and extending in successive crops over the whole body-surface, coalescing to form extensive irregular ecchymotic patches or even raised bloody tumors. Hemorrhages may take place from the mucous surfaces or into the serous cavities. Syn., *land-scurvy; morbus maculosus Werlhofii; Werlhof's disease.* **p. rheumatica,** a form with fever and rheumatic pains. **p. simplex,** the mildest degree of purpura. **p. urticans.** See *urticaria hæmorrhagica*.
**purpuraceous** (*pur-pū-ra'-se-us*) [*purpura*, purple]. Of a purple color.
**purpuric** (*pur-pū'-rik*) [*purpura*]. Pertaining to or resembling purpura.
**purpuriferous** (*pur-pū-rif'-ur-us*) [*purpura*, purple; *ferre*, to bear]. Producing a purple pigment.
**purpurin** (*pur'-pū-rin*) [*purpura*]. $C_{14}H_8O_5$. 1. A dye present with alizarin in madder-root, but also prepared artificially. 2. Uroerythrin, a red coloring-matter sometimes present in urinary deposits.
**purpurinuria** (*pur-pū-rin-ū'-re-ah*). The excretion of purpurin in the urine; porphyruria.
**purr** (*pur*). A low murmur.
**purring thrill.** A fine trembling vibration like the purring of a cat, perceived by palpation over the precordium. It may be due to aneurysm or to valvular heart lesion, especially mitral stenosis.
**purulence** (*pūr'-ū-lens*) [*pus*]. The state of being purulent; suppuration.
**purulent** (*pūr'-ū-lent*) [*pus*]. Having the character of or containing pus; characterized by the formation of pus. **p. catarrh,** an inflammation of a mucous membrane accompanied by the production of pus. **p. edema,** a severe infiltration of pus together with much fluid.
**puruloid** (*pūr'-ū-loid*) [*pus*; εἶδος, like]. Resembling pus; puriform.
**pus** [L.]. A liquid substance consisting of cells and an albuminous fluid (liquor puris), formed in certain kinds of inflammation. See *Donné's test*.

PUSTULA MALIGNA 732 PYGODIDYMUS

**p., blue,** pus colored blue by *Bacillus pyocyaneus.* **p.-corpuscles,** the corpuscles found in pus. **p., curdy,** pus containing cheesy-looking flakes. **p., ichorous,** pus that is thin and acrid. **p., laudable,** a whitish, inodorous pus, formerly thought to be essential to the healing of wounds. **p., orange,** pus colored by the presence of hematoidin crystals. **p., sanious,** pus mixed with blood. **p.-tube.** See *pyosalpinx.*
**pustula maligna** (*pus-tū'-lah ma-lig'-nah*). Anthrax.
**pustulant** (*pus'-tū-lant*) [*pustule*]. 1. Causing the formation of pustules. 2. An irritant substance giving rise to the formation of pustules.
**pustular** (*pus'-tū-lar*) [*pustule*]. Characterized by the presence of pustules.
**pustulation** (*pus-tū-la'-shun*) [*pustule*]. The formation of pustules.
**pustule** (*pus'-tūl*) [*pustula*, a pustule]. A small circumscribed elevation of the skin containing pus. **p., malignant;** anthrax.
**pustuliform** (*pus'-tū-lif-orm*) [*pustula, forma,* form]. Resembling a pustule.
**pustulocrustaceous** (*pus-tū-lo-krus-tā'-shus*) [*pustule; crusta,* crust]. Characterized by the formation of pustules and crusts.
**pustuloderma** (*pus-tū-lo-der'-mah*) [*pustule;* δέρμα, the skin]. Any skin disease characterized by the formation of pustules.
**pustulose, pustulous** (*pus'-tū-lōs, pus'-tū-los*) [*pustula,* pustule]. Characterized by pustules.
**pusula** (*pus'-ū-lah*) [L.]. 1. Pustule. 2. Erysipelas.
**putamen** (*pū-ta'-men*) [L., "a husk"]. The outer darker part of the lenticular nucleus of the brain.
**Putnam-Dana's symptom-complex** [James Jackson *Putnam,* American neurologist, 1846– ; Charles Loomis *Dana,* American neurologist, 1852– ]. Combined sclerosis of the lateral and posterior columns of the spinal cord.
**Putnam's sign** (*put'-nam*) [James Jackson *Putnam,* American neurologist, 1846– ]. Absolute increase of measurements from the anterior superior iliac spine to the internal malleolus; it is observed in hysterical hip disease.
**putrefaction** (*pū-tre-fak'-shun*) [*putrid; facere,* to make]. The decomposition of nitrogenous organic matter under the influence of microorganisms, accompanied by the development of disagreeable odors, due to the evolution of ammonia, hydrogen sulphide, and other gases, and the production of aromatic bodies. In addition, many other compounds are formed, among which ptomaines are the most important. The end-products are water, nitrogen, methane, and carbon dioxide.
**putrefactive** (*pū-tre-fak'-tiv*) [see *putrefaction*]. Pertaining to or causing putrefaction. **p. alkaloid,** a ptomaine.
**putrescence** (*pū-tres'-ens*) [*putrescere,* to become rotten]. The state or process of putrefaction.
**putrescent** (*pū-tres'-ent*) [see *putrescence*]. Undergoing putrefaction.
**putrescentia** (*pū-tres-en'-she-ah*). See *putrescence.* **p. uteri,** the severest form of puerperal endometritis, with sanious ulceration of the wall of the uterus extending to the peritoneum.
**putrescibility** (*pū-tres-e-bil'-it-e*) [*putrescere,* to grow rotten]. Capacity for undergoing putrefaction.
**putrescine** (*pū-tres'-in*) [see *putrescence*], C₄H₁₂N₂. Tetramethylenediamine. A poisonous ptomaine, a clear, rather thin liquid, of a disagreeable odor, boiling at 156°–157° C.
**putrid** (*pū'-trid*) [*putridus,* rotten]. Rotten; characterized by putrefaction. **p. fever.** Synonym of *typhus.*
**putrify** (*pū'-tre-fi*) [*putrefacere,* to putrefy]. To render putrid.
**putrilage** (*pū'-tril-āj*) [*putrid*]. Putrescent material.
**putrilaginous** (*pū-tril-aj'-in-us*) [*putrid*]. Gangrenous.
**putromaine** (*pū-tro-ma'-in*). A ptomaine developed in putrefactive processes.
**pyæmia** (*pi-e'-me-ah*). See *pyemia.*
**pyapostasis** (*pi-ap-os'-tas-is*) [πύον, pus; ἀπόστασις, a standing off]. Metastasis of pus.
**pyarthrosis** (*pi-ar-thro'-sis*) [πύον, pus; ἄρθρον, a joint]. Suppuration of a joint.
**pycnocardia** (*pik-no-kar'-de-ah*) [πυκνός, frequent; καρδία, heart]. See *tachycardia.*

**pycnometer** (*pik-nom'-et-er*) [πυκνός, thick; μέτρον, measure]. 1. An instrument for the determination of the specific gravity of fluids. 2. An instrument for the measurement of the thickness of objects.
**pycnomorphous, pyknomorphous** (*pik-no-morf'-us*) [πυκνός, thick; μορφή, form]. Applied to nerve-cells in which the stained parts of the cell-body are compactly arranged.
**pycnophrasia** (*pik-no-fra'-ze-ah*) [πυκνός, thick; φράσις, speech]. Thickness of speech.
**pycnosis** (*pik-no'-sis*) [πυκνός, thick]. 1. Thickening; inspissation. 2. A degenerative change in cells whereby the protoplasm is condensed and the cells shrink in volume.
**pycnosphygmia** (*pik-no-sfig'-me-ah*). See *tachycardia.*
**pycnotic** (*pik-not'-ik*). Pertaining to or characterized by pycnosis.
**pyecchysis** (*pi-ek'-is-is*) [πύον, pus; ἔκχυσις, effusion]. Effusion of pus.
**pyedema, pyœdema** (*pi-e-de'-mah*) [πύον, pus; edema]. Edema due to purulent infiltration.
**pyelitic** (*pi-el-it'-ik*). Relating to or affected with pyelitis.
**pyelitis** (*pi-el-i'-tis*) [πύελος, a trough; ιτις, inflammation]. Inflammation of the pelvis of the kidney. **p., calculous,** that due to calculi.
**pyelo-** (*pi-e-lo-*) [πύελος, a trough]. A prefix denoting relation to the kidney or to the pelvis of the kidney.
**pyelocystitis** (*pi-el-o-sis-ti'-tis*) [*pyelo-; cystitis*]. Pyelitis with cystitis.
**pyelocystostomosis** (*pi-e-lo-sist-o-sto-mo'-sis*) [*pyelo-;* κύστις, a bladder; στόμα, a mouth]. The establishment of direct communication between the kidney and the bladder.
**pyelography** (*pi-el-og'-ra-fe*). Skiagraphy of a renal pelvis and ureter which have been filled with a silver salt solution.
**pyelolithotomy** (*pi-el-o-lith-ot'-o-me*) [*pyelo-; lithotomy*]. Removal of a renal calculus through an incision into the pelvis of the kidney.
**pyelometer** (*pi-el-om'-et-er*). Synonym of *pelvimeter.*
**pyelonephritic** (*pi-el-o-nef-rit'-ik*). Pertaining to pyelonephritis.
**pyelonephritis** (*pi-el-o-nef-ri'-tis*) [*pyelo-; nephritis*]. Inflammation of the kidney and its pelvis.
**pyelonephrosis** (*pi-el-o-nef-ro'-sis*). 1. Synonym of *pyelonephritis.* 2. Any disease of the pelvis of the kidney.
**pyeloplication** (*pi-el-o-pli-ka'-shun*) [*pyelo-; plica*]. Operative infolding of a dilated renal pelvis.
**pyelotomy** (*pi-el-ot'-o-me*) [*pyelo-;* τομή, a cutting]. Incision of the renal pelvis.
**pyemesis** (*pi-em'-is-is*) [πύον, pus; ἔμεσις, vomiting]. Vomiting of pus.
**pyemia, pyæmia** (*pi-e'-me-ah*) [πύον, pus; αἷμα, blood]. A disease due to the presence of pyogenic microorganisms in the blood and the formation, wherever these organisms lodge, of embolic or metastatic abscesses. The disease is generally fatal. **p., arterialization** of a cardiac thrombus and the dissemination of emboli through the arterial circulation. **p., cryptogenic,** a condition in which the primary suppuration occurs in the deeper tissues of the body. **p. otogenous,** pyemia originating in the ear.
**pyemic** (*pi-em'-ik* or *pi-e'-mik*) [*pyemia*]. Pertaining to or affected with pyemia.
**pyemide** (*pi'-em-ēd*) [πύον, pus; pl., *pyemides*]. A cutaneous manifestation the result of metastases in pyemia.
**pyencephalus** (*pi-en-sef'-al-us*) [πύον, pus; ἐγκέφαλος, brain]. Suppuration within the cranium.
**pyenin** (*pi'-en-in*). Paranuclein.
**pyesis** (*pi-e'-sis*). Synonym of *suppuration.*
**pygal** (*pi'-gal*) [πυγή, buttock]. Pertaining to the buttocks.
**pygalgia** (*pi-gal'-je-ah*) [πυγή, rump; ἄλγος, pain]. Pain in the buttocks.
**pygalopubic** (*pi-gal-o-pū'-bik*) [πυγή, buttock; *pubes*]. Relating to the buttocks and the pubes.
**pygmalionism** (*pig-ma-le-on-izm*) [*Pygmalion,* king of Cyprus, who fell in love with an ivory image of a maiden]. Erotic love with a statue, a form of erotomania.
**pygodidymus** (*pi-go-did'-im-us*) [πυγή, buttock; δίδυμος, twin]. A double monster united by the buttocks.

**pygomelus** (*pi-gom'-el-us*) [πυγή, buttock; μέλος, a member]. A monster with a parasite attached to the hypogastric region or to the buttock.
**pygopagus** (*pi-gop'-ag-us*) [πυγή, buttock; πάγος, joined]. A monster with conjoined buttocks or backs.
**pyic** (*pi'-ik*). Synonym of *purulent*.
**pyin** (*pi'-in*) [πῦον, pus]. An albuminous substance of complex constitution occurring in pus. It may be separated by adding sodium chloride and filtering.
**pykno-**. For words beginning thus, see under *pycno-*.
**pyla** (*pi'-lah*) [πύλη, gate]. Opening between the third ventricle and Sylvian aqueduct.
**pylar** (*pi'-lar*). Relating to the pyla.
**pyle** (*pi'-le*) [πύλη, a gate]. The portal vein.
**pylema** (*pi-le'-mah*) [*pyle*; αἷμα, blood]. The blood of the portal vein.
**pylemphraxis** (*pi-lem-fraks'-is*) [πύλη, gate; ἔμφραξις, obstruction]. Obstruction of the portal circulation.
**pylephlebectasis, pylephlebectasia** (*pi-le-fleb-ek'-tas-is, pi-le-fleb-ek-ta'-ze-ah*). [πύλη, gate; φλέψ, vein]. Dilatation of the portal vein, which is usually caused by some obstruction in the liver, or it may be due to relaxation of the vessel-walls from some disturbance of innervation.
**pylephlebitis** (*pi-le-fleb-i'-tis*) [*pyle*; *phlebitis*]. Inflammation of the portal vein. The condition is usually secondary to disease of the intestine, is generally suppurative in character, and gives rise to the symptoms of pyemia.
**pylethrombophlebitis** (*pi-le-throm-bo-fleb-i'-tis*) [*pyle*, *thrombosis*; *phlebitis*]. Inflammation and thrombosis of the portal vein.
**pylethrombosis** (*pi-le-throm-bo'-sis*) [*pyle*; *thrombosis*]. Thrombosis of the portal vein.
**pylic** (*pi'-lik*) [*pyle*]. Pertaining to the portal vein.
**pyloralgia** (*pi-lor-al'-je-ah*) [*pylorus*; ἄλγος, pain]. Pain in the region of the pylorus.
**pylorectomy** (*pi-lor-ek'-to-me*) [*pylorus*; ἐκτομή, excision]. Excision of the pylorus.
**pyloric** (*pi-lor'-ik*) [*pylorus*]. Pertaining to the pylorus. **p. glands**, glands situated in the region of the pylorus and secreting the gastric juice. **p. orifice**. See *pylorus* (1). **p. valve**. See *pylorus* (2).
**pyloristenosis** (*pi-lor-is-ten-o'-sis*) [*pyloro-*; στενός, narrow]. Contraction of the pylorus.
**pyloritis** (*pi-lor-i'-tis*) [*pyloro-*; ιτις, inflammation]. Inflammation of the pylorus.
**pyloro-** (*pi-lo-ro-*) [*pylorus*]. A prefix meaning relating to the pylorus.
**pylorochesis** (*pi-lor-o-ke'-sis*) [*pyloro-*; ὄχησις, a holding]. Obstruction of the pylorus.
**pylorocolic** (*pi-lor-o-kol'-ik*). Pertaining to or connecting the pyloric end of the stomach with the transverse colon.
**pylorodiosis** (*pi-lor-o-di-o'-sis*) [*pyloro-*; δίωσις, a pushing through]. Loreta's operation: digital divulsion of the pyloric orifice following gastrotomy.
**pyloroplasty** (*pi-lor'-o-plas-te*) [*pyloro-*; πλάσσειν, to form]. Plastic operation upon the pylorus.
**pyloroptosis, pyloroptosia** (*pi-lor-op-to'-sis, pi-lor-op-to'-se-ah*) [*pyloro-*; πτῶσις, falling]. Downward displacement of the pylorus.
**pyloroscirrhus** (*pi-lor-o-skir'-us*) [*pyloro-*; σκίρρος, induration]. Scirrhus of the pylorus.
**pylorospasm** (*pi-lor'-o-spazm*). Spasm of the pylorus.
**pylorostenosis** (*pi-lor-o-ste-no'-sis*) [*pyloro-*; *stenosis*]. Stenosis, or stricture, of the pylorus.
**pylorostomy** (*pi-lor-os'-to-me*) [*pyloro-*; στόμα, mouth]. Making an opening through the abdominal wall into the pylorus.
**pylorus** (*pi-lo'-rus*) [πυλωρός, a gate-keeper]. 1. The circular opening of the stomach into the duodenum. 2. The fold of mucous membrane and muscular tissue surrounding the aperture between the stomach and the duodenum. **p., antrum of**, the portion of the stomach between the pyloric orifice and the sphincter antri pylorici or transverse band.
**pyo-** (*pi-o-*) [πῦον, pus]. A prefix denoting pertaining to pus.
**pyoblenna** (*pi-o-blen'-ah*) [*pyo-*; βλέννα, mucus]. Muco-pus.
**pyoblennorrhea** (*pi-o-blen-or-e'-ah*) [*pyo-*; βλέννα, mucus; ῥοία, a flow]. A muco-purulent discharge.
**py**₀**ele** (*pi'-o-sēl*) [*pyo-*; κήλη, hernia]. Hernia with pus in its sac.
**pyocelia, pyocœlia** (*pi-o-se'-le-ah*) [*pyo-*; κοιλία, a hollow]. Pus in the abdominal cavity.

**pyocenosis** (*pi-o-sen-o'-sis*) [*pyo-*; κένωσις, emptying]. The evacuation of a pus-cavity.
**pyochezia** (*pi-o-ke'-ze-ah*) [*pyo-*; χέζειν, to defecate]. Discharge of pus from the intestines.
**pyococcus** (*pi-o-kok'-us*). Any pus-producing coccus.
**pyocolpocele** (*pi-o-kol'-po-sēl*). A suppurating tumor of the vagina.
**pyocolpos** (*pi-o-kol'-pos*) [*pyo-*; κόλπος, vagina]. An accumulation of pus within the vagina.
**pyoctanin** (*pi-ok'-tan-in*). See *pyoktanin*.
**pyocyanase** (*pi-o-si'-an-ās*) [*pyo-*; κύανος, blue]. The specific bacteriolytic enzyme of *Bacillus pyocyaneus*. It is said to digest the bacilli of typhoid, diphtheria, anthrax and cholera, and also fibrin. It is a yellowish-green, alkaline, amorphous substance, soluble in water, to which it imparts a greenish tint.
**pyocyanin** (*pi-o-si'-an-in*) [see *pyocyanase*]. C₁₄H₁₄NO₂. A colored substance derived from blue pus and from cultures of *Bacillus pyocyaneus*.
**pyocyanogenic** (*pi-o-si-an-o-jen'-ik*). Producing pyocyanin.
**pyocyanolysin** (*pi-o-si-an-ol'-is-in*) [*pyo-*; κύανος, blue; λύειν, to loosen]. A hemolysin produced in broth cultures by *Bacillus pyocyaneus*.
**pyocyst** (*pi'-o-sist*) [*pyo-*; *cyst*]. A cyst containing pus.
**pyocyte** (*pi'-o-sīt*) [*pyo-*; κύτος, a cell]. A pus-corpuscle.
**pyodermatitis** (*pi-o-der-mat-i'-tis*). A skin-affection produced by inoculation with pyogenic material.
**pyoderma** (*pi-o-der'-mah*) [*pyo-*; δέρμα, skin]. Any cutaneous lesion due to pus-producing microorganisms.
**pyodermitis** (*pi-o-der-mi'-tis*) [*pyo-*; δέρμα, skin; ιτις, inflammation]. An inflammatory skin-affection attended by pus-formation.
**pyogenes** (*pi-oj'-en-ēs*). Synonym of *pyogenic*.
**pyogenesis** (*pi-o-jen'-es-is*) [*pyo-*; γενᾶν, to produce]. The formation of pus.
**pyogenic, pyogenetic** (*pi-o-jen'-ik, pi-o-jen-et'-ik*) [*pyogenesis*]. Producing pus. **p. membrane**. See *membrane*, *pyogenic*. **p. microorganisms**, the microorganisms producing pus. The ordinary pyogenic microorganisms are staphylococci and streptococci. Under certain circumstances pus may be produced by the pneumococcus of Fraenkel, *Bacillus coli communis*, the bacillus of typhoid, the gonococcus, and others.
**pyogenin** (*pi-oj'-en-in*). C₈H₁₃N₂O₁₉. A substance obtained from the cell-body of pus-cells.
**pyohemia, pyohæmia** (*pi-o-he'-me-ah*). See *pyemia*.
**pyohemothorax, pyohæmothorax** (*pi-o-hem-o-tho'-raks*). The presence of pus and blood in the pleural cavity.
**pyoid** (*pi'-oid*) [*pyo-*; εἶδος, like]. Resembling pus.
**pyoktanin** (*pi-ok'-tan-in*) [*pyo-*; κτείνειν, to kill]. A name given to methyl-violet and methylene-blue on account of their germicidal properties. Pyoktanin has been used in diphtheria, cystitis, gonorrhea, ulcers, and inflammations of the conjunctiva, and as an injection in carcinoma. **p., blue**, C₂₄H₂₈N₃Cl, antiseptic, disinfectant, and analgesic. Dose 1–5 gr. (0.065–0.32 Gm.). Dusting-powder, 1 : 1000–1 : 100; aqueous solution, 1–4 : 10,000. **p.-mercury**, a compound of pyoktanin and mercury. Applied in 1 : 200 solution or with equal parts of starch. **p., yellow**, C₁₇H₂₄N₃ClO, antiseptic and disinfectant, but weaker than blue pyoktanin. Dose 1–8 gr. (0.065–0.52 Gm); aqueous solution, 1–4 : 10,000; dusting-powder, 1 to 2%; ointment, 2 to 10%.
**pyolymph** (*pi'-o-limf*) [*pyo-*; *lymph*]. Lymph containing pus-corpuscles.
**pyometra** (*pi-o-me'-trah*) [*pyo-*; μήτρα, womb]. A collection of pus in the uterus.
**pyonephritis** (*pi-o-nef-ri'-tis*) [*pyo-*; *nephritis*]. Suppurative inflammation of the kidney.
**pyonephrosis** (*pi-o-nef-ro'-sis*) [*pyo-*; νεφρός, kidney]. An accumulation of pus in the pelvis of the kidney.
**pyonephrotic** (*pi-o-nef-rot'-ik*). Pertaining to pyonephrosis.
**pyo-ovarium** (*pi-o-o-va'-re-um*) [*pyo-*; *ovarium*, ovary]. Ovarian abscess.
**pyopericarditis** (*pi-o-per-e-kar-di'-tis*). Suppurative pericarditis.
**pyopericardium** (*pi-o-per-e-kar'-de-um*). The presence of pus in the pericardium.
**pyoperitonitis** (*pi-o-per-it-on-i'-tis*). Synonym of *peritonitis, purulent*.

**pyophthalmia** (*pi-of-thal'-me-ah*) [*pyo-*; ὀφθαλμός, eye]. Purulent ophthalmia.
**pyophylactic** (*pi-o-fil-ak'-tik*) [*pyo-*; φυλάσσειν, to guard]. Protecting against pus. **p. membrane**, the lining membrane of an abscess cavity.
**pyophysometra** (*pi-o-fi-so-me'-trah*) [*pyo-*; φῦσα, wind; μήτρα, womb]. The presence of pus and gas in the uterus.
**pyoplania** (*pi-o-pla'-ne-ah*) [*pyo-*; πλανάειν, to wander]. Infiltration of tissues with pus.
**pyopneumopericarditis** (*pi-o-nū-mo-per-e-kar-di'-tis*) [*pyo-*; πνεῦμα, air; *pericarditis*]. Pericarditis complicated by the presence of pus and air in the pericardium.
**pyopneumopericardium** (*pi-o-nū-mo-per-ik-ar'-de-um*) [*pyo-*; πνεῦμα, air; *pericardium*]. Pus and air or gas in the pericardium.
**pyopneumoperitonitis** (*pi-o-nū-mo-per-it-on-i'-tis*) [*pyo-*; πνεῦμα, air; *peritonitis*]. Peritonitis complicated by the presence of pus and air in the peritoneal cavity.
**pyopneumothorax** (*pi-o-nū-mo-tho'-raks*). An accumulation of air or gas and pus in the pleural cavity. **p., subphrenic**, a collection of air and pus beneath the diaphragm. See *Pfuhl's sign*.
**pyopoiesis** (*pi-o-poi-e'-sis*). Synonym of *suppuration*.
**pyopoietic** (*pi-o-poi-et'-ik*) [*pyo-*; ποιεῖν, to make]. Secreting pus; suppurative.
**pyoptysis** (*pi-op'-tis-is*) [*pyo-*; πτύειν, to spit]. The expectoration of pus.
**pyorrhagia** (*pi-o-ra'-je-ah*) [*pyo-*; ῥηγνύναι, to burst forth]. A profuse discharge of pus.
**pyorrhea, pyorrhœa** (*pi-or-e'-ah*) [*pyo-*; ῥοία, a flow]. A purulent discharge. **p., alveolaris.** Progressive necrosis of the dental alveoli; Riggs' disease, Fauchard's disease.
**pyosalpingitis** (*pi-o-sal-pin-ji-tis*) [*pyo-*; σάλπιγξ, tube; ιτις, inflammation]. Purulent inflammation of the Fallopian or Eustachian tube.
**pyosalpingo-oophoritis** (*pi-o-sal-ping-go-o-of-or-i'-tis*) [*pyo-*; σάλπιγξ, a tube; *oophoritis*]. Combined suppurative inflammation of the ovary and oviduct.
**pyosalpinx** (*pi-o-sal'-pinks*) [*pyo-*; σάλπιγξ, tube]. An accumulation of pus in the oviduct.
**pyosapremia, pyosapræmia** (*pi-o-sap-re'-me-ah*) [*pyo-*; *sapremia*]. Same as pyemia.
**pyoscheocele** (*pi-os'-ke-o-sēl*) [*pyo-*; ὄσχεον, scrotum; κήλη, tumor]. A suppurative swelling of the scrotum.
**pyoscope** (*pi'-o-skōp*) [*πύος*, colostrum; σκοπεῖν, to examine]. An instrument for determining the richness of milk by its color.
**pyosepticemia, pyosepticæmia** (*pi-o-sep-tis-e'-me-ah*) [*pyo-*; *septicemia*]. The association of pyemia and septicemia.
**pyosin** (*pi'-o-sin*). $C_{57}H_{110}N_2O_{15}$. A substance obtained from the body plasma of pus-cells. Cf. *pyogenin*.
**pyosis** (*pi-o'-sis*) [*πύον*, pus]. 1. Suppuration. 2. Suppuration of the eye.
**pyostatic** (*pi-o-stat'-ik*) [*pyo-*; στατικός, causing to stand]. 1. Preventing the formation of pus. 2. An agent arresting the secretion of pus.
**pyothorax** (*pi-o-tho'-raks*). An accumulation of pus in the pleural cavity; empyema. **p., subphrenic,** an abscess beneath the diaphragm.
**pyotorrhea, pyotorrhœa** (*pi-o-tor-e'-ah*) [*pyo-*; οὖς, ear; ῥοία, aflow]. Purulent otorrhea.
**pyoturia** (*pi-o-tū'-re-ah*). See *pyuria*.
**pyoureter** (*pi-o-ū-re'-ter*) An accumulation of pus in a ureter.
**pyoxanthin, pyoxanthose** (*pi-o-zan'-thin, pi-o-zan'-thōs*) [*πύον*, pus; ξανθός, yellow]. A yellow substance sometimes found in pus, and resulting from the oxidation of pyocyanin.
**pyra** (*pi'-rah*). Synonym of *anthrax*.
**pyracetosalyl** (*pi-ras-e-to-sal'-il*). A preparation of antipyrine and aspirin: used as an antipyretic and antineuralgic; dose 8 grains (0.5 gm.).
**pyraloxin** (*pi-ral-oks'-in*). An oxidation-product of pyrogallol, used in skin-diseases.
**pyramid** (*pe'-ram-id*) [πυραμίς]. Any conical eminence of an organ; especially a body of longitudinal nerve-fibers on each side of the anterior median fissure of the oblongata. **p., anterior,** one of the two pyramidal bundles of white matter on each side of the anterior median fissure of the medulla. **p. of the cerebellum,** a conical projection forming the central portion of the inferior vermiform process. **p.s of Ferrein.** See *Ferrein's pyramids*. **p., lateral.** Same as *restiform body*. **p., Malpighian,** one of the conical masses composing the medullary substance of the kidney. **p., posterior,** one of the two narrow bundles of white matter placed on each side of the posterior median fissure of the medulla oblongata. They are continuous with the posterior median columns of the spinal cord. **p., renal.** See *p., Malpighian,* and *Ferrein's pyramids.* **p., temporal,** the petrosa. **p., thyroid.** See *Lalouette's pyramid.* **p. of the tympanum,** a hollow conical process on the inner wall of the tympanum; the stapedius muscle passes through an aperture at its apex.
**pyramidal** (*pe-ram'-id-al*) [*pyramid*]. Shaped like a pyramid. **p. bone,** the carpal cuneiform. **p. tract.** See *tract pyramidal.*
**pyramidale** (*pir-am-id-a'-le*) [πυραμίς, pyramid]. The cuneiform bone of the carpus; the os pyramidale.
**pyramidalis** (*pe-ram-id-a'-lis*). Pyramidal, as *pyramidalis* muscle. See under *muscle*.
**pyramidon** (*pe-ram'-id-on*). Dimethylamidophenyldimethylpyrazolon, a yellowish-white powder, recommended as an antipyretic. Dose 5–8 gr. (0.32–0.51 Gm.) every 2 hours.
**pyramis** (*pir'-am-is*). 1. Synonym of *pyramid*. 2. The modiolus. 3. The pyramid of the thyroid. 4. The petrosa. 5. The anterior pyramid of the oblongata. 6. The posterior pyramid. See *pyramid of the cerebellum.* **p. cochleæ,** the modiolus. **p. laminosa.** See *pyramid of the cerebellum.* **p. ossis temporis, p. trigona,** the petrosa. **p. vermis,** the pyramid of the cerebellum. See *crista vestibuli.*
**pyrantin** (*pi-ran'-tin*). See *phenosuccin.* **p., soluble,** sodium salt of paraethoxylphenylsuccinamic acid. A useful antipyretic. Dose 15–45 gr. (1–3 Gm.) daily.
**pyranum** (*pi-ra'-num*). The sodium salt of a combination of benzoic acid, thymol, and salicylic acid; used as an antirheumatic and analgesic. Dose 7–30 gr. (0.45–2.0 Gm.) 2 or 3 times daily.
**pyrazine** (*pi-raz'-ine*). Antipyrine. **p. hexahydride.** Piperazine.
**pyrazol** (*pi'-raz-ōl*) [πῦρ, fire; *azotum*, nitrogen], $C_3H_4N_2$, a derivative of pyrrol. It is used as a diuretic. Dose 15–30 gr. (1–2 Gm.).
**pyrenemia, pyrenæmia** (*pi-ren-e'-me-ah*) [πυρήν, a fruit-stone; αἷμα, blood]. The existence of nucleated red cells in the blood.
**pyrenoid** (*pi'-re-noid*) [πυρήν, the stone of a fruit; εἶδος, form]. One of the small, bright globules found imbedded in the chromatophores of green algæ and of certain invertebrates, and having the reactions of nuclein and the function of forming starch and similar carbohydrates.
**pyrethrum** (*pi-re'-thrum*). Pellitory; the root of *Anacyclus pyrethrum,* a plant of the order *Compositæ.* Pyrethrum is used as a sialagogue and masticatory in headache, toothache, and neuralgic affections of the face. It is employed either in powder or in the form of the tincture. **p., tincture of** (*tinctura pyrethri,* U. S. P.), is never given internally.
**pyretic** (*pi-ret'-ik*) [πυρετός, fever]. Pertaining to or affected with fever.
**pyreticosis** (*pi-ret-ik-o'-sis*) [πυρετός, fever]. Feverishness.
**pyretin** (*pi'-re-tin*) [see *pyretic*]. An ant'pyretic said to consist of acetanilide, caffeine, sodium bicarbonate, and calcium carbonate. Dose 3–8 gr. (0.19–0.52 Gm.).
**pyreto-** (*pi-ret-o-*) [πυρετός, fever]. A prefix meaning fever.
**pyretogenesia, pyretogenesis** (*pi-ret-o-jen-e'-se-ah, pi-ret-o-jen'-es-is*) [*pyreto-*; γένεσις, origin]. The origin and process of fever.
**pyretogenic, pyretogenous** (*pi-ret-oj-en'-ik, pi-ret-oj'-en-us*) [*pyreto-*; γεννᾶν, to produce]. Causing or producing fever.
**pyretogenin** (*pi-ret-oj'-en-in*) [see *pyretogenic*]. A substance formed by microorganisms, and said to have the property of producing fever when inoculated into animals.
**pyretography** (*pi-ret-og'-ra-fe*) [*pyreto-*; γράφειν, to write]. A treatise on fevers.
**pyretologist** (*pi-ret-ol'-o-jist*) [*pyretology*]. A specialist in fevers.
**pyretology** (*pi-ret-ol'-o-je*) [*pyreto-*; λόγος, a science]. The science of the nature of fevers.
**pyretometer** (*pi-ret-om'-et-ur*) [*pyreto-*; μέτρον, measure]. A clinical thermometer.

**pyretotyphosis** (*pi-ret-o-ti-fo'-sis*) [*pyreto-*; τύφωσις, delirium]. The stupor or delirium of fever.
**pyretotyposis** (*pi-ret-o-ti-po'-sis*) [*pireto-*; τύπωσις, a forming]. Intermittent fever.
**pyrexia** (*pi-reks'-e-ah*) [πύρεξις, fever]. Elevation of temperature above the normal; fever.
**pyrexial** (*pi-reks'-e-al*). Pertaining to pyrexia.
**pyrheliometer** (*pir-he-li-om'-et-ur*) [πῦρ, fire; ἥλιος, sun; μέτρον, measure]. An instrument for measuring the heating and chemical effects of light.
**pyridine** (*pir'-id-ēn*) [πῦρ, fire], C₅H₅N. A liquid base obtained as a distillation-product from tobacco, coal-tar, and other organic matter, and forming the first of a long and important series of homologous bases. Pyridine has been used in asthma by inhalation. p. **tricarboxylic acid**, C₅H₂(CO₂H)₃N, is antiseptic, antipyretic, antiperiodic, and antispasmodic. It has been used in malaria, asthma, and typhoid fever. Dose 2–10 gr. (0.13–0.65 Gm.).
**pyriform** (*pi'-rif-orm*) [*pyrus*, pear; *forma*, a form]. Pear-shaped.
**pyriformis** (*pe-rif-orm'-is*). Pyriform, as *pyriformis* muscle. See under *muscle.*
**pyro-** (*pi-ro-*) [πῦρ, fire]. A prefix signifying fire or heat.
**pyroacetic** (*pi-ro-as-e'-tik*). Pertaining to or obtained from acetic acid by the action of heat. p. **spirit**. Synonym of *acetone*.
**pyroacid** (*pi-ro-as'-id*). A product obtained by subjecting certain organic acids to heat.
**pyroarsenic acid** (*pi-ro-ar-sen'-ik*) [*pyro-*; *arsenic*], H₄As₂O₇. A tetrabasic acid produced when arsenic is heated to 180° C.
**pyroborate** (*pi-ro-bo'-rāt*). A salt of pyroboric acid. Syn., *biborate.*
**pyroboric acid** (*pi-ro-bo'-rik*) [*pyro-*; *boron*], H₂B₄O₇. Tetraboric acid, a dibasic acid produced by heating boric acid.
**pyrocatechin** (*pi-ro-kat'-e-kin*) [*pyro-*; *catechu*], C₆H₆O₂ = C₆H₄(OH)₂. Catechol; a crystalline substance formed by the dry distillation of catechu and sometimes occurring in the urine. It has been used as an antipyretic. Dose 1–2 gr. (0.065–0.13 Gm.).
**pyrocatechinuria** (*pi-ro-kat-e-kin-ū'-re-ah*) [*pyrocatechin*; οὖρον, urine]. The presence of pyrocatechin in the urine.
**pyroctin** (*pi-rok'-tin*). A proprietary febrifuge.
**pyrodextrin** (*pi-ro-deks'-trin*) [*pyro-*; *dextrin*]. C₆H₁₀O₅. A brownish solid resulting from the action of heat upon dextrin.
**pyrodin** (*pi-ro'-din*) [πῦρ, fire], C₆H₅. C₂H₃O.N₂H₃. Acetylphenylhydrazine; a crystalline, poisonous substance, used as a substitute for chrysarobin in psoriasis and in other cutaneous affections; also as an antipyretic. Dose ½–3 gr. (0.03–0.2 Gm.).
**pyroform** (*pi'-ro-form*). Bismuth oxyiodopyrogallol; used in skin diseases and said to be less toxic than pyrogallol.
**pyrogallic acid** (*pi-ro-gal'-ik*). See *pyrogallol.*
**pyrogallol** (*pi-ro-gal'-ol*) [*pyro-*; *galla*, galls], C₆H₃(OH)₃. Pyrogallic acid; a phenol derivative produced by the action of heat on gallic acid. It is used locally in diseases of the skin. p.-**bismuth**, a darkgreen powder containing equal parts of bismuth and pyrogallol; used as an intestinal disinfectant and wound antiseptic. p., **oxidized**, a stable brown or black powder, slightly soluble in water, insoluble in alcohol or ether, used as a substitute for pyrogallol as less irritating and only slightly toxic. Dose ¼–15 gr. (0.05–1.0 Gm.) daily. Ointment in skin diseases 75 gr. (5 Gm.) to 375 gr. (25 Gm.) each of vaseline and lanolin.
**pyrogenic** (*pi-ro-jen'-ik*) [*pyro-*; γεννᾶν, to produce]. Producing fever.
**pyroleum** (*pi-ro'-le-um*) [*pyro-*; *oleum*, *oil*]. 1. Petroleum. 2. An oil produced by dry distillation.
**pyroligneous** (*pi-ro-lig'-ne-us*) [*pyro-*; *lignum*, wood]. Pertaining to the destructive distillation of wood. p. **acid**, wood-vinegar. See *acid*, *pyroligneous.*
**pyrolusite** (*pi-ro-lū'-sīt*) [*pyro-*; λοῦσις, a washing]. Native manganese dioxide.
**pyrolysis** (*pi-rol'-is-is*) [*pyro-*; λύσις, solution]. Decomposition by means of heat.
**pyrolytic** (*pi-ro-lit'-ik*). Pertaining to pyrolysis.
**pyromania** (*pi-ro-ma'-ne-ah*) [*pyro-*; μανία, madness]. A monomania for incendiarism.
**pyromaniac** (*pi-ro-ma'-ne-ak*) [*pyro-*]. One affected with pyromania.
**pyrometer** (*pi-rom'-et-er*) [*pyro-*; μέτρον, a measure]. An instrument for measuring the intensity of heat of too high a degree to be estimated by the ordinary thermometer.
**pyronin** (*pi'-ro-nin*). A basic triphenylmethane dyestuff.
**pyronyxis** (*pi-ro-niks'-is*) [*pyro-*; νύξις, a pricking]. Ignipuncture.
**pyrophobia** (*pi-ro-fo'-be-ah*) [*pyro-*; φόβος, dread]. Morbid dread of fire.
**pyrophosphate** (*pi-ro-fos'-fāt*). A salt of pyrophosphoric acid.
**pyrophosphoric acid** (*pi-ro-fos-for'-ik*). See *acid*, *pyrophosphoric.*
**pyroptothymia** (*pi-rop-to-thi'-me-ah*) [*pyro-*; πτοεῖν, to terrify; θυμός, mind]. A form of insanity in which the person imagines himself enveloped in flame. Puncturing with hot needles.
**pyropuncture** (*pi-ro-pungk'-tūr*) [*pyro-*; *puncture*].
**pyrosal** (*pi'-ro-sal*). Antipyrine salicylacetate, C₆H₃O₃C₁₁H₁₁N₂O; antipyretic and antineuralgic. Dose 8 gr. (0.5 Gm.) 2 to 6 times daily.
**pyroscope** (*pi'-ro-skōp*) [*pyro-*; σκόπειν, to examine]. An instrument employed in determining the intensity of thermal radiation.
**pyrosis** (*pi-ro'-sis*) [πῦρ, fire]. An affection of the stomach characterized by a burning sensation, accompanied by eructations of an acrid, irritating fluid; heartburn.
**Pyrosoma** (*pi-ro-so'-mah*). See *Piroplasma.* P. **bigeminum** (*pi-ro-so'-mah bi-jem'-in-um*) [*pyrus*, pear; σῶμα, a body]. The parasite which is the cause of Texas fever in cattle.
**pyrotic** (*pi-rot'-ik*) [*pyrosis*]. 1. Inflammable. 2. Caustic.
**pyrotoxic** (*pi-ro-toks'-ik*) [*pyro-*; τόξικον, poison]. A caustic poison.
**pyrotoxin** (*pi-ro-toks'-in*) [*pyro-*; τοξικόν, a poison]. A toxic agent generated in the course of the febrile process.
**pyrotoxina bacterica** (*pi-ro-toks'-in-ah bak-ter'-ik-ah*) [see *pyrotoxin*]. A pyogenic substance believed to be produced by many forms of bacteria.
**pyroxylin** (*pi-roks'-il-in*) [*pyro-*; ξύλον, wood]. Guncotton; cotton-fiber treated with a mixture of nitric and sulphuric acids, by which the cellulose is changed into various nitro-compounds. Soluble gun-cotton (*pyroxylinum*, U. S. P.) is used in the preparation of collodion. The explosive gun-cotton is the hexanitrate of cellulose.
**pyrozol** (*pi'-ro-sol*). A proprietary antiseptic said to be a coal-tar derivative.
**pyrozone** (*pi'-ro-zōn*). A proprietary preparation of hydrogen dioxide, an external antiseptic. It is also used externally in a 5% and a 25% ethereal solution.
**pyrrhol, pyrrol** (*pir'-ol*) [*pyro-*; *oleum*, oil], C₄H₄-(NH). A liquid base obtained in the distillation of Dippel's oil and other organic substances. p. **tetraiodide**. See *iodol.*
**pythogenesis** (*pi-tho-jen'-es-is*) [πύθειν, to rot; γένεσις, genesis]. Production from decaying matter.
**pythogenic** (*pi-tho-jen'-ik*) [πύθειν, to rot; γεννᾶν, to produce]. Producing or arising from decomposition. p. **fever**. Synonym of *typhoid fever.*
**pyuria** (*pi-ū-re-ah*) [πῦον, pus; οὖρον, urine]. The passage of urine containing pus.

# Q

**q. h.** Abbreviation of *quaque hora*—every hour.
**q. 2 h.**, abbreviation of *quaque secunda hora*—every second hour. **q. 3 h.**, abbreviation of *quaque tertia hora*—every third hour.
**q. l.** Abbreviation of *quantum libet*—as much as is desired.
**q. p.** Abbreviation of *quantum placet*—as much as you please.
**q. s.** Abbreviation of *quantum sufficit*—as much as suffices.
**quack** (*kwak*). A pretender of medical skill; a vender of nostrums; a medical charlatan.
**quackery** (*kwak'-er-e*). The practice of medicine by a quack; medical charlatanism.
**quack-salver**. A quack, or mountebank; a peddler of his own medicines and salves.
**quader** (*kwa'-der*) [Ger., square]. The precuneus, or quadrate lobe of the cerebrum.
**quadrangular** (*kwod-rang'-gū-lar*) [*quadrangulum*, a four-cornered figure]. Having four angles, as the *quadrangular* lobe, the square lobe of the cerebellum.
**quadrant** (*kwod'-rant*) [see *quadratus*]. 1. The fourth part of a circle, subtending an angle of 90 degrees. 2. One of the four regions into which the abdomen may be divided for purposes of physical diagnosis. **q. of Wilder**, such an area of the ventral aspect of the crus cerebri in the cat.
**quadrate** (*kwod'-rāt*) [*quadrant*]. Square; four-sided. **q. bone**, the bone which in birds and reptiles articulates with the squamosal above, the mandible below, the pterygoid internally, and the quadrojugal externally. **q. cartilages**, small quadrangular cartilaginous plates often found in the alæ of the nose. **q. lobe**, (1. A small lobe of the liver. 2. A lobe of the cerebellum. **q. lobule**. See *precuneus*.
**quadratipronator** (*kwod-ra-ti-pro-na'-tor*). Same as *pronator quadratus*. See *muscles, table of*.
**quadratus** (*kwod-ra'-tus*) [L.]. Squared; having four sides. **q. muscle**. See under *muscle*.
**quadri-** (*kwod-re-*) [L.]. A prefix denoting four or four times.
**quadribasic** (*kwod-re-ba'-sik*) [*quadri-; basis*, base]. In chemistry, applied to an acid having four replaceable hydrogen atoms.
**quadriceps** (*kwod'-re-seps*) [*quadri-; caput*, head]. Four-headed, as a *quadriceps* muscle. A large muscle of the thigh. See under *muscle*. **q. suræ** [L., the quadriceps muscle of the calf]. The muscle mass comprising the gastrocnemius, soleus, and plantaris.
**quadrigeminal** (*kwod-re-jem'-in-al*) [see *quadrigeminum*]. Fourfold; consisting of four parts, as the *quadrigeminal* bodies. See *corpora quadrigemina*.
**quadrigeminum** (*kwod-re-jem'-in-um*) [*quadrigeminus*, fourfold]. One of the corpora quadrigemina.
**quadrilateral** (*kwod-re-lat'-er-al*) [*quadri; latus*, a side]. Having four sides. **q., of Marie**. See under *Marie*.
**quadrille** (*kwad-ril'*) [Fr.]. An embryological term designating the complex movement undergone by the dividing centrosomes previous to the formation of the cleavage centrosomes.
**quadripara** (*kwod-rip'-ar-ah*) [*quadri*, four; *parere*, to bear]. A woman who is bearing or has borne, her fourth child, or has had her fourth confinement.
**quadriparity** (*kwod-re-par'-it-e*). The state of having borne four children.
**quadriparous** (*kwod-rip'-a-rous*). Pertaining to a quadripara, or to a fourth confinement.
**quadriplegia** (*kwod-ri-ple'-je-ah*) [*quadri*, four; πληγή, stroke]. Paralysis of all four limbs.
**quadrisect** (*kwod'-ri-sekt*) [*quadri*, four; *secare*, to cut]. To divide into four parts.
**quadriurate** (*kwod-re-u'-rāt*) [*quadri-; urate*]. A term applied to the hyperacid urate of human urine and the urine of birds and reptiles. The quadriurates are mixtures of biurates and uric acid, and have the general formula of $MH(C_5H_2N_4O_3)H_2C_5H_2N_4O_3$.

**quadrivalent** (*kwod-riv'-al-ent*) [*quadri-; valere*, to be worth]. In chemistry, having a combining power equivalent to that of four hydrogen atoms. See *quantivalence*.
**quadroon** (*kwod-roon'*) [*quartus*, fourth]. Offspring of a white person and a mulatto.
**quadruplet** (*kwod-roo'-plet*) [*quadruplare*, to make fourfold]. Any one of four children born at one birth.
**Quain's fatty heart** (*kwān*). Fatty degeneration of the cardiac muscular fibers.
**Quain's method of removing a foreign body** (*kwān*). An incision is made at some little distance from the foreign body, the latter is then grasped with forceps at right angles to its long axis, and then pushed out through another incision at the point of entrance.
**quaker's black drop**. Vinegar of opium, acetum opii.
**quaker-button**. A popular name for nux vomica.
**qualitative** (*kwol'-it-a-tiv*) [*qualitas*, quality]. Pertaining to quality. **q. analysis**. See *analysis, qualitative*.
**quantimeter** (*kwon-tim'-ei-er*). An instrument for measuring the dosage of the Roentgen rays.
**quanti-Pirquet's reaction** (*kwon'-te-pēr-ka*). A quantitative Pirquet's reaction undertaken with the idea of estimating the degree of tuberculous infection.
**quantitative** (*kwon'-tit-a-tiv*) [*quantus*, how much]. Pertaining to quantity. **q. analysis**. See *analysis, quantitative*.
**quantivalence** (*kwon-tiv'-al-ens*) [*quantus*, how much; *valere*, to be worth]. The combining power of an element or radical expressed in terms of the number of atoms of hydrogen with which it will unite. *Univalent* or monad atoms, as chlorine, are saturated with one atom; *bivalent* or diad atoms require two; *trivalent* or triad, as boron, take three; *quadrivalent* or tetrad, *quinquivalent* or pentad, *sextalent* or hexad, require four, five, and six atoms of hydrogen respectively.
**quantum** (*kwon'-tum*) [L.]. 1. As much as. 2. A certain prescribed amount. **q. libet**, as much as you please. **q. sufficit**, as much as suffices. **q., normal**, a constant quantity or standard. **q. vis**, as much as you wish.
**quarantine** (*kwor'-an-tēn*) [It., *quaranta*, forty]. 1. The time (formerly forty days) during which vessels or travelers from ports infected with contagious or epidemic dieases are required by law to remain outside the port of their destination, as a safeguard against the spreading of such diseases. 2. The place of detention. 3. The act of detaining vessels or travelers from suspected ports or places for purposes of inspection of disinfection. **q., land-**, the isolation of a person or district on land for purposes similar to those of detention of persons arriving at a place by sea. **q. period**, the length of time required to insure immunity after exposure, or the length of time necessary after an attack, to render the disease innocuous.
**quart** (*kwort*) [*quartus*, fourth]. The fourth part of a gallon. **imperial q.**, contains about 20 per cent. more than the ordinary quart.
**quartan** (*kwor'-tan*) [*quartus*]. 1. Recurring on the fourth day. 2. A form of intermittent fever the paroxysms of which occur every fourth day. **p., double, quartan** fever characterized by milder and severer paroxysms, each occurring every fourth day. **q. fever**. See *quartan* (2). **q. parasite**, the *Plasmodium malariæ*.
**quarter-crack**. In farriery, a fissure of the hoof on the inner side of the fore-foot of a horse.
**quarter-evil** (*kwor'-ter-e-vil*). A synonym of *blackleg*.
**quartipara** (*kwor-tip'-ar-ah*) [*quartus; parere*, to bring forth]. A woman in her fourth pregnancy. See *multipara*.
**quartiparous** (*kwor-tip'-ar-us*) [*quartipara*]. Pregnant four times.
**quartisternum** (*kwor-te-ster'-num*) [*quartus*, fourth;

## QUARTONOL 737 QUINCOCA

*sternum*]. A part of the sternum having a special center of ossification corresponding with the fourth intercostal space.

**quartonol** (*kwor'-ton-ol*). A proprietary mixture of calcium, sodium, quinine, strychnine, and glycerophosphates.

**quartz** (*kworts*). See *silica*.

**quassation** (*kwas-a'-shun*) [*quassatio*, a shaking or shattering]. The reduction of barks, roots, and other drugs to morsels, in preparation for further pharmaceutical treatment. Syn., *cassation*.

**quassia** (*kwosh'-e-ah*) [after *Quassi*, a negro slave in Surinam who first used it]. The wood of several trees of the order *Simarubaceæ*. The quassia of the U. S. P. and B. P. is the wood of *Picrasma excelsa*, known as *Jamaica quassia*, or of *quassia amara*, known as *Surinam quassia*. It is a simple bitter, and is used in dyspepsia and constipation; in the form of an enema it is employed against seatworms. q., extract of (*extractum quassiæ*, U. S. P.). Dose 1 gr. (0.065 Gm.). q., fluidextract of (*fluidextractum quassiæ*, U. S. P.) Dose 8 minims (0.5 Cc.). q., infusion of (*infusum quassiæ*, B. P.). Dose ½ to 1 ounce (15 to 30 Cc.). q., tincture of (*tinctura quassiæ*, U. S. P., B. P.). Dose ½ to 1 dram (2 to 4 Cc.).

**quassia cup**. A cup made of quassia wood, which is filled with water and allowed to stand; the water acquires the bitter taste of the quassia.

**quassiin** (*kwos'-se-in*). Same as *quassin*.

**quassin** (*kwos'-in*). The active principle of quassia; very bitter, white crystals, soluble in alcohol and chloroform; used as a tonic. Dose $\frac{1}{32}-\frac{1}{8}$ gr. (0.002-0.02 Gm.).

**quaternary** (*kwa-ter'-na-re*) [*quaterni*, four each]. 1. Consisting of four elements. 2. Fourth in order.

**q. syphilis**, parasyphilis.

**Quatrefages, parietal angle of** (*katr-fahzj*) [Jean Louis Armand de *Quatrefages* de Bréau, French naturalist, 1810–1892]. In craniometry, that formed by the lines drawn through the extremities of the transverse maximum or bizygomatic diameter and the maximum transverse frontal diameter (called *positive* when it opens downward, and *negative* when it opens upward).

**quatuor** (*kwat'-u-or*) [L.]. Four. **q. pills**, pills made up of iron sulphate, quinine, aloes, nux vomica and gentian.

**queasy** (*kwe'-ze*). Nauseated; inclined to vomit. (Colloquial.)

**quebrabunda** (*ke-brah-bun'-dah*) [Port.]. Straddling disease, a tropical disease similar to beriberi which attacks horses and pigs.

**quebrachamine** (*ke-brah'-kam-ēn*). See under *quebracho*.

**quebrachine** (*ke-brah'-kēn*). An alkaloid of quebracho (q. r.). It is used internally and hypodermatically in dyspnea. Dose $\frac{2}{3}-1\frac{1}{2}$ gr. (0.04–0.1 Gm.).

**quebracho** (*ke-brah'-ko*) [from Pg. *quebra-hacho*, ax-breaker]. The name of several hard-wooded trees of South America. The white quebracho (*quebracho blanco*) is *Aspidosperma quebracho*, of the order Apocynaceæ. It contains the following alkaloids: *Aspidospermine* C$_{22}$H$_{30}$N$_2$O$_2$; *aspidospermatine* C$_{22}$H$_{26}$-N$_2$O$_2$, *aspidosamine*, C$_{22}$H$_{26}$N$_2$O$_2$; *quebrachine* C$_{21}$H$_{26}$-N$_2$O$_2$, and *quebrachamine* C$_{22}$H$_{28}$N$_2$O$_2$. It is used in emphysema, bronchitis, and in asthma. See *aspidospermine*.

**quebrachol** (*ke-brah'-kol*). A levorotary crystalline substance found in the bark of *Aspidosperma quebracho blanco*.

**queen of the meadow**. *Spiræa ulmaria*; *eupatorium*

**queen's-delight, queen's-root**. See *stillingia*.

**queen's metal**. An alloy of antimony, tin, etc.

**Quénu's operation of thoracoplasty** (*ka-noo'*) [E. *Quénu*, French surgeon, 1852– ]. An operation for empyema, consisting in simple section of the ribs, without resection, to favor retraction of the chest walls.

**Quénu-Mayo operation** (*ka-noo'-mu'-o*) [see *Quénu*; William J. *Mayo*, American surgeon, 1861– ]. An operation for cancer of the rectum, consisting of excision of the rectum with removal of neighboring lymph-glands.

**quercetin** (*kwur'-se-tin*). A neutral principle derived from quercitrin.

**quercin** (*kwer'-sin*) [*quercus*], C$_6$H$_6$(OH)$_6$. A bitter, crystallizable carbohydrate extracted from acorns and oak-bark.

**quercitannic acid** (*kwer-si-tan'-ik*) [*quercus; tannin*],

C$_{17}$H$_{16}$O$_9$. A variety of tannic acid found in oak-bark.

**quercitannin** (*kwer-sit-an'-in*). Quercitannic acid.

**quercite** (*kwer'-sīt*) [*quercus*], C$_6$H$_7$(OH)$_5$. A sweet principle found in acorns.

**quercitol** (*kwer'-sit-ol*). See *quercite*.

**quercitrin** (*kwer'-sit-rin*) [*quercus; citrus*, lemon], C$_{36}$H$_{38}$O$_{20}$. A glucoside found in the bark of *Quercus tinctoria* and in many other plants. It is tonic and astringent.

**quercus** (*kwer'-kus*) [L.,'oak]. The *quercus* of the U. S. P. is the dried bark of *Quercus alba*. Dose 15 grains (1 gm.). q., fluidextract of (*fluidextractum quercus*, U. S. P.). Dose 15 min. (1 Cc.). See *oak*.

**quercynol** (*kwur'-sin-ol*). A proprietary remedy said to consist of extract of *Quercus alba* with cyanol and extract of hyoscyamus. It is used in vaginal wafers.

**Quevenne's iron** (*ke-ven'*) [Theodore Auguste *Quevenne*, French physician, 1805–1855]. Ferrum reductum.

**quick** (*kwik*) [AS., *cwic*, alive]. 1. A sensitive, vital, tender part, as the flesh under a nail. 2. Pregnant, and able to feel the movements of the fetus.

**quicken** (*kwik'-en*). To experience the sensation of quickening, q. v.

**quickening** (*kwik'-en-ing*) [see *quick*]. The first feeling on the part of the pregnant woman of fetal movements, occurring between the fourth and fifth months of pregnancy.

**quicklime** [*quick; lime*]. Calcium oxid; unslacked lime. Seen under *lime*.

**quicksilver** [*quick; silver*]. The popular name for mercury.

**quickwater**. Solution of mercuric nitrate.

**quillaia**, **quillaja** (*kwil-a'-yah*) [Chilian, *quillean*, to wash]. A genus of trees of the order Rosaceæ. The *quillaja* of the U. S. P. is the dried bark of *Quillaja saponaria*. It contains saponin and produces a froth when agitated in water. It is used in pulmonary affections and as a sternutatory, and in the arts as a substitute for soap. Syn., *soap-bark*. Q., fluidextract of (*fluidextractum quillajæ*, U. S. P.). Dose 3 min. (0.2 Cc.). Q., tincture of (*tinctura quillajæ*, U. S. P.), *tinctura quillajæ*, B. P.). Dose 1 dr. (4 Cc.).

**quill-suture**. See *suture*, *quill-*.

**quina** (*kwin'-ah*). Same as *cinchona*. q. calisaya, yellow cinchona bark. q. colorada, red cinchona bark.

**quinacetine sulphate** (*kwin-as'-et-ēn*) (C$_{17}$H$_{16}$NO$_2$)$_2$-H$_2$SO$_4$H$_2$O. An antipyretic and anodyne. Dose 5–15 gr. (0.32–0.97 Gm.).

**quinaldin** (*kwin-al'-din*), C$_{10}$H$_9$N. Methylquinolin; formed by digesting anilin with paraldehyde and hydrochloric acid.

**quinalgen** (*kwin-al'-jen*). See *analgen*.

**quinamicine** *kwin-am'-is-ēn*). An artificial alkaloid, C$_{19}$H$_{24}$N$_2$O$_2$, obtained from quinamine.

**quinamidine** (*kwin-am'-id-ēn*). An isomere of quinamicine.

**quinamine** (*kwin'-am-ēn*) [Sp. *quina*, bark; *amine*], C$_{19}$H$_{24}$N$_2$O$_2$. An alkaloid of the cinchonas.

**quinaphenin** (*kwin-a-fen'-in*). A white, tasteless powder, obtained by action of quinine on the hydrochlorate of eloxyphenylcarbamic acid. Used in whooping-cough. Dose for young children 1–2½ gr. (0.065–0.16 Gm.) daily; older children 3–5 gr. (0.2–0.3 Gm.).

**quinaphthol** (*kwin-af'-thol*). See *chinaphthol*.

**quinaquina** (*kwin-ah-kwin'-ah*). Cinchona.

**quinaseptol** (*kwin-ah-sep'-tol*). See *diaphtol*. q., argentic, an odorless, harmless antiseptic and hemostatic which promotes granulation.

**quinate** (*kwin'-āt*). A salt of quinic acid.

**quince-seed** (*kwins-sēd*). See *cydonium*.

**Quincke's disease** (*kving'-keh*) [Heinrich Irenaeus *Quincke*, German physician, 1842– ]. Angioneurotic edema; acute circumscribed edema. Q.'s pulse. Q.'s disease. Q.'s pulse, rhythmic reddening and blanching of the finger-nails at each diastole of the heart, depending upon oscillations of blood-pressure which are propagated into the capillaries; it is found in aortic insufficiency. Q.'s puncture, lumbar puncture to examine or remove cerebrospinal fluid. Q.'s sign. See *Q.'s pulse*. Q.'s space, the space between the third and fourth lumbar vertebræ. Q.'s suture. See *puncture, lumbar*.

**quincoca** (*kwin-ko'-kah*). A tonic said to be a combination of quinine, coca-leaves, gentian, wild

25

cherry, orange peel, and aromatics exhausted with port wine.
**quinetine** (*kwin'-et-ēn*). A mixture of cinchona alkaloids, similar to *febrifuge*.
**quinetum** (*kwin-e'-tum*). The mixed alkaloids from red cinchona bark used as a cheap febrifuge in India. Dose gr. j–v. It is an antiperiodic.
**quinhydrone** (*kwin-hi'-drōn*). A reaction product of an aqueous solution of quinone and hydroquinone; green prisms with pungent taste, soluble in hot water, alcohol, ether, or ammonium.
**quinia** (*kwin'-e-ah*). See *quinine*.
**quinic** (*kwin'-ik*) [Peruvian, *kina*, bark]. Pertaining to quinine. **q. acid**, $C_7H_{12}O_6$, an acid occurring in cinchona bark, in the ivy, oak, elm, ash, coffee-plant, etc. **q. fever**, febrile symptoms, with an eruption; it occurs among workmen making quinine.
**quinicine** (*kwin'-is-ēn*). A cinchona alkaloid, isomeric with quinine and quinidine.
**quinidamine** (*kwin-id'-am-ēn*). An alkaloid obtained from cinchona.
**quinidine** (*kwin'-id-ēn*) [*quinine*], $C_{20}H_{24}N_2O_2$. An alkaloid of cinchona bark isomeric with quinine, which it resembles in action, differing only in being less powerful. **q. sulphate**, is used as an antiperiodic in doses of 20–60 gr. (1.3–4.0 Gm.). **q. tannate**, is used in diarrhea, nephritis and malaria. Dose 2–12 gr. (0.1–0.8 Gm.) twice daily.
**quinimetry** (*kwin-im'-et-re*) [*quinia*; μετρον, meas-ure]. See *quiniometry*.
**quinina** (*kwin-i'-nah*). See *quinine*.
**quininæ** (*kwin-i'-ne*). Genitive of quinina.
**quinine** (*kwin-een*, *kin'-een*) [Peruvian, *kina*, bark], $C_{20}H_{24}N_2O_2+3H_2O$. Quinine (*quinina*, U. S. P.) is a bitter amorphous or crystalline alkaloid obtained from the bark of various species of cinchona. It is soluble in 900 parts of water, readily soluble in alcohol, ether, and chloroform, and gives a beautiful emerald-green color when it or its salts are treated with a solution of chlorine and then with ammonia. Quinine acts as a stimulant to the nervous system, causing in large doses cerebral congestion and lessening of the reflexes; it is a slight respiratory stimulant and a depressant to the circulation; it lessens the ameboid movement of the white corpuscles, and during fever is strongly antipyretic; it also possesses antiseptic properties. In large doses it causes ringing in the ears, a feeling of fulness in the head, dizziness, slight deafness and at times disturbances of vision; occasionally also a rise of temperature (*quinine fever*). It is used as an anti-periodic in malaria, in which disease it has a specific action; it is also employed as an antipyretic in other febrile affections, as a tonic in convalescence, as a stimulant to the uterus during parturition, in whooping-cough, coryza, and hay-fever. **q. acetate**, $C_{20}H_{24}N_2O_2 \cdot C_2H_4O_2$. Dose 1–15 gr. (0.065–1.0 Gm.). **q. albuminate**. Dose 1–15 gr. (0.065–1.0 Gm.). **q. bisulphate** (*quininæ bisulphas*, U. S. P.). Dose same as that of the sulphate. **q. bromate**, $C_{20}H_{24}N_2O_3 \cdot HBrO_3$, antiseptic and antipyretic. Dose 1–30 gr. (0.065–2.0 Gm.). **q. camphorate**, $(C_{20}H_{24}N_2O_2)_2 \cdot C_{10}H_{16}O_4$, antiseptic and antipyretic. Dose 1–30 gr. (0.065–2.0 Gm.). **q. carbolate**, $C_{20}H_{24}N_2O_2 \cdot C_6H_6O$, antiseptic and antipyretic. Dose 1–30 gr. (0.065–2.0 Gm.). **q. chlorate**, $C_{20}H_{24}N_2O_2 \cdot HClO_3 + 1\frac{1}{2}H_2O$, explosive white crystals, soluble in water and alcohol; used in fevers with symptoms of angina. **q. chlorophosphate**, $C_{20}H_{24}N_2O_2 \cdot HCl \cdot 2PO_4H_3 + 3H_2O$; used in obstinate cases of malaria. **q. cinnamate**, $C_{20}H_{24}N_2O_2 \cdot C_9H_8O_2$; antipyretic and antiseptic. Dose 1–30 gr. (0.065–2.0 Gm.). **q. citrate**, $(C_{20}H_{24}N_2O_2)_2 \cdot C_6H_8O_7 + 7H_2O$. Dose 2–20 gr. (0.13–1.3 Gm.). **q. ethylsulphate**, $C_{20}H_{24}N_2O_2$, obtained from a hot alcoholic solution of sodium sulphovinate and quinine sulphate; recommended for subcutaneous use. Dose 3–8 gr. (0.10–0.52 Gm.). **q. ferrocyanide**, $C_{20}H_{24}N_2O_2 \cdot H_4Fe(CN)_6 + 3H_2O$, used in night-sweats of tuberculosis. Dose 5–10 gr. (0.32–0.65 Gm.). **q. formate**, $C_{20}H_{24}N_2O_2 \cdot CH_2O_2$, used as is the sulphate. **q. glycerophosphate**, $(C_{20}H_{24}N_2O_2)_2 \cdot C_3H_7O_3 \cdot PO_3$; used in malaria, neuralgia, etc. Dose 2 gr. (0.1 Gm.). **q. hydrobromide** (*quininæ hydro-bromidum*, U. S. P.), $C_{20}H_{24}N_2O_2 \cdot HBr + H_2O$. Used in the same doses as the sulphate. **q. hydrochloride** (*quininæ hydrochloridum*, U. S. P., B. P.), given as is the sulphate. **q. hydrochlorosulphate**, $C_{20}H_{24}N_2O_2 \cdot HCl \cdot H_2SO_4 + 3H_2O$; for hypodermatic use as causing less pain than any other salt of quinine. **q. hydro-**
**iodate**, $C_{20}H_{24}N_2O_2 \cdot HI$, used as a nervous sedative and in neuralgia internally or subcutaneously. Dose 1–1½ gr. (0.06–0.1 Gm.). **q. lygosinate**, a combination of quinine and lygosine, a fine, orange-yellow powder. It is bactericide and antiseptic and is used as a dusting-powder. **q. muriate**. See *q. hydrochloride*. **q. oleate** (*oleatum quininæ*), a mixture of exsiccated quinine, 1 part, and oleic acid, 3 parts; used in the administration of quinine by inunction. **q. peptonate**, a brown powder containing 80 % of peptone and 20 % of quinine; nutrient and tonic. Dose 5–60 gr. (0.32–4.0 Gm.). **q. phenate**, **q. phenolate**. See *q. carbolate*. **q. phosphate**, $(C_{20}H_{24}N_2O_2)_2H_3PO_4 + 8H_2O$, antiperiodic. Dose 1–30 gr. (0.065–2.0 Gm.). **q. phthalate**, $(C_{20}H_{24}N_2O_2)_2C_8H_6O_4$; used as is the sulphate. Dose 1–30 gr. (0.065–2.0 Gm.). **q. quinate**, $C_{20}H_{24}N_2O_2 \cdot C_7H_{12}O_6 + 2H_2O$; used subcutaneously. Dose, as the sulphate. **q. quino-vate**, $C_{20}H_{24}N_2O_2 \cdot C_{30}H_{48}O_4$ (?), antiperiodic. Dose 1–30 gr. (0.065–2.0 Gm.). **q. saccharate**, **q. saccharinate**, $C_{20}H_{24}N_2O_2 \cdot C_7H_5NO_3$, antipyretic and antiseptic. Dose 1–30 gr. (0.065–2.0 Gm.). **q. salicylate** (*quininæ salicylas*, U. S. P.), $2C_{20}H_{24}N_2O_2 \cdot C_7H_6O_3 + H_2O$; antiperiodic, and used to relieve the pains of rheumatism and gout. Dose 4 gr. (0.25 Gm.). **q. stearate**, $C_{20}H_{24}N_2O_2 \cdot C_{18}H_{36}O_2$; used as is the sulphate, but by inunction. **q. sulphate** (*quininæ sulphas*, U. S. P., B. P.), is the salt most commonly employed. Dose in malaria 5–24 gr. (0.32–1.6 Gm.) before the paroxysms; as a prophylactic 2–4 gr. (0.13–0.26 Gm.); as a tonic 1–2 gr. (0.065–0.13 Gm.); in whooping-cough 1½ gr. (0.1 Gm.) for each year of the child's age, or locally in solution of 1–2 gr. (0.065–0.13 Gm.) to the ounce (32 Cc.) by the atomizer. **q. sulphochlorhydrate**, used by injection in carcinoma. Dose 0.50–0.60 cg. every other day. **q. sulphocresotate**, an intestinal antiseptic. **q. sulphoethylate**. See *q. ethylsulphate*. **q. sulpho-muriate**. See *q. hydrochlorosulphate*. **q. sulpho-tartrate**, a compound of quinine sulphate and tartaric acid; antipyretic, antiseptic. Dose 1–30 gr. (0.065–2.0 Gm.). **q. sulphovinate**. See *q. ethylsulphate*. **q. thymate**, used as is the sulphate. Dose 1–30 gr. (0.065–2.0 Gm.). **q. and urea hydrochloride**, employed chiefly for hypodermatic use. **q.-urethane**, a nonirritant compound made by heating 3 parts of quinine hydrochloride with 15 parts of urethane and 3 parts of water; used for intravenous injection. **q. valerate**, used in doses of 1–2 gr. (0.065–0.13 Gm.) in nervous debility and hemicrania.
**Cinchonism**, **quinism**. (*kwin-ēn'-izm*, *kwin'-izm*).
**quiniometry** (*kwin-e-om'-et-re*) [*quinia*; μετρον, measure]. The determination of the amount of alkaloids contained in samples of cinchona bark.
**Quinlan's test for bile** (*kwin'-lan*). On examination through a spectroscope absorption lines appear in the violet end of the spectrum, in the presence of bile.
**quinochloral** (*kwin-o-klo'-ral*). See *chinoral*.
**quirofoform** (*kwin'-o-form*). See *chinoform*.
**quinoidine** (*kwin-oi'-dēn*). See *chinoidine*. **q.**, *animal-*, a basic substance obtained from animal tissues and having the property of fluorescence like quinine.
**quinol** (*kwin'-ol*). See *hydroquinone*.
**quinoline** (*kwin'-o-lēn*) [*quinine*], $C_9H_7N$. A liquid alkaloid obtained in the destructive distillation of quinine, or cinchonine, with potassium hydroxide; it occurs also in coal-tar. It is antipyretic and antiseptic. Dose 4–10 min. (0.2–0.6 Cc.). **q. bismuth sulphocyanate**, $(CHN \cdot HSCN)_3Bi(SCN)_3$, a granular, orange-red powder, insoluble in water, alcohol, or ether; melts at 76° C. It is used in the treatment of gonorrhea, skin diseases, and ulcers in 0.5 to 1 % solution. **q. monohypochlorite**. See *chinol*. **q. salicylate**, $C_9H_7N \cdot C_7H_6O_3$, antiseptic and antirheumatic. Dose 5 gr. (0.5–1.0 Gm.). Application, 0.7 % aqueous solution. **q. sulphate**, $C_9H_7N \cdot H_2SO_4$; antiseptic and used as is quinolin. **q. tartrate**, $(C_9H_7N)_2(C_4H_6O_6)$, antipyretic and antiseptic. Dose 5–15 gr. (0.3–1.0 Gm.). Injection in gonorrhea, 0.7 % aqueous solution.
**quinology** (*kwin-ol'-o-je*) [Sp., *quina*, bark; λογος, discourse]. The scientific study of the cinchona trees and of their alkaloids.
**quinone** (*kwin'-ōn*) [*quinine*], $C_6H_4O_2$. 1. A yellow, crystalline substance obtained by heating quinic acid with manganese dioxide and sulphuric acid,

2. A general name for certain derivatives of the benzene series.
**quinopyrine** (*kwin-o-pi'-rēn*). A concentrated aqueous solution of quinine hydrochloride and antipyrin, used subcutaneously in malaria.
**quinosol** (*kwin'-o-sol*). 1. $C_9H_6N \cdot OSO_3K + H_2O$. Oxyquinoline potassium sulphate, a yellow powder, soluble in water; antipyretic, antiseptic, styptic, and deodorant. Syn., *chinosol*. 2. The proprietary name for a neutral combination of tricresyl sulphonate and quinoline, with tricresol. It is not caustic, and is soluble in water to the extent of 1 in 25. A disinfectant for surgical instruments, and bactericide. Application, 0.1 to 2 % solutions.
**quinotannic acid** (*kwin-o-tan'-ik*) [*quinine; tannin*]. A form of tannic acid found in cinchona bark.
**quinotropine** (*kwin-o-tro'-pēn*). Urotropine quinate.
**quinovin** (*kwin'-o-vin*) [Peruvian, *kina*, bark], $C_{38}H_{52}O_{11}$. Kinovin, a bitter glucoside found in cinchona bark.
**quinoxim** (*kwin-oks'-im*) [Sp., *quina*, bark; ὀξύς, acid]. Nitrosophenol; prepared by the action of nitrous acid upon phenols.
**Quinquaud's disease** (*kang-ko'*) [Charles Eugene *Quinquaud*, French physician, 1841–1894]. A disease of the hair-follicles attended with cicatrization of the skin. Syn., *acne decalvans; folliculitis decalvans*. Q.'s **panaris**, phlegmonous inflammation of the fingers and toes of neuropathic origin, differing from Morvan's disease in that it is painful and never accompanies paretic phenomena, and ordinarily does not entail necrosis of the phalanges. Q.'s **phenomenon** or **sign**, an involuntary crepitus of hand and fingers when extended, often found in alcoholics. Q.'s **sign of chronic alcoholism**, the subject for examination is directed to hold the tips of the outstretched fingers of one hand perpendicularly to the outspread palm of the examiner and to press upon it with only moderate firmness. In the course of two or three seconds, if the person is addicted to alcohol, crepitation of the phalanges will be perceptible, as if the bones of each finger impinged roughly upon each other. The sound ranges in intensity from a slight grating to crashing.
**quinquevalent** (*kwin-kwev'-al-ent*) [*quinque*, five; *valere*, to be worth]. Having a valence of five; capable of combining with or replacing five atoms of hydrogen or their equivalent.
**quinquina** (*kwin-kwi'-na*). Cinchona.
**quinquinina** (*kwin-kwe-ni'-nah*). A preparation containing alkaloids of cinchona bark, extracted by macerating in acidulated water, and precipitated by a soluble alkali.
**quinquivalent** (*kwin-kwiv'-al-ent*). See *quinquevalent*.

**quinsy** (*kwin'-ze*) [σύν, with; ἄγχειν, to choke]. Acute inflammation of the tonsils, usually tending to suppuration. **q., lingual,** quinsy originating in the lingual tonsil and involving the tongue.
**quintan** (*kwin'-tan*) [*quintus*, fifth]. An intermittent fever, the paroxysms of which recur every four days, *i. e.*, on the fifth, ninth, thirteenth, etc.
**quintes'-ens** (*kwin-tes'-ens*) [*quintus*, fifth; *essentia*, essence]. The active principle of any substance, concentrated to the utmost degree.
**quintipara** (*kwin-tip'-ar-ah*) [*quintus*, fifth; *parere*, to bring forth]. A woman who has been in labor five times, or who is in labor for the fifth time.
**quintisternum** (*kwin-te-ster'-num*) [*quintus*, fifth, *sternum*]. A part of the sternum having a special center of ossification corresponding with the fifth intercostal space.
**quintuplet** (*kwin-tū'-plet*) [*quintuplex*, five-fold]. One of five children born at one time.
**quionine** (*kwi'-o-nēn*). "Tasteless quinine." A mixture of cinchona alkaloids, principally cinchonidine.
**quitenidine** (*kwit-en'-i-dēn*). An alkaloid formed by the oxidation of quinidine.
**quittor, quitter** (*kwit'-or, kwit'-er*). In farriery, a fistulous wound upon the quarters or the heel of the coronet, caused by treads, pricks in shoeing, or other injuries which produce suppuration at the coronet or within the foot.
**quiz** (*kwiz*) [*quæso*, I ask]. 1. A recitation, conducted by questions and answers, in which the student familiarizes himself with his studies. 2. To teach by this method of questions and answers.
**quizzer** (*kwiz'-er*) [*quiz*]. One who conducts a quiz.
**quotidian** (*kwo-tid'-e-an*) [*quot*, as many as; *dies*, day]. 1. Recurring every day. 2. An intermittent fever, the paroxysms of which recur daily. **q., double,** a fever having two paroxysms a day, usually differing in character.
**quotient** (*kwo'-shent*) [*quoties*, how often]. The result of the process of division. **q., blood,** the result obtained by dividing the quantity of hemoglobin in the blood by the number of erythrocytes, expressed in each case as a percentage of the normal amount. **q., protein,** the result of dividing the amount of globulin in the blood-plasma by the amount of albumin in it. **q. respiratory,** the result obtained by dividing the carbon dioxide expired by the oxygen absorbed. This is normally $\frac{4 \cdot 5}{5} = 0 \cdot 9$.

**q. v.** Abbreviation for (1) *quantum vis,*—as much as you wish. (2) For *quod vide*—which see.

# R

**R.** The abbreviation of *Réaumur*, of *resistance* (electric), of *residuum*, of *right*, and of *recipe*, take (generally written ℞.).
— **R.** Abbreviation for Rinne's test negative.
**+ R.** Abbreviation for Rinne's test positive.
**Raabe's test for albumin** (*rah'-beh*) [Gustav Raabe, German physician, 1875–  ]. Place in a test-tube 1 Cc. of the liquid to be tested; on the addition of a small piece of trichloracetic acid a white zone or ring will be formed in the presence of albumin. The ring produced by uric acid is diffused and not sharply defined.
**rabbeting** (*rab'-et-ing*) [OF., *rabouter*, to push back]. The interlocking of the broken serrated edges of a fractured bone.
**rabelaisin** (*rab-el-a'-is-in*). A glucoside from the bark of *Lunasia amara*, of the Philippine Islands; the bark is used in inflammation of the eye and as an arrow-poison by the Negritos.
**rabiate** (*ra'-be-āt*) [*rabies*, rage]. Rabid.
**rabic** (*rab'-ik*) [*rabies*]. Pertaining to rabies, as *rabic* virus.
**rabid** (*rab'-id*) [*rabies*]. Affected with rabies or hydrophobia; pertaining to rabies, as *rabid* virus.
**rabies** (*rab'-e-ēz*) [L.]. Lyssa or hydrophobia. The latter term is generally applied to the human disease consequent upon the bite of a rabid dog or other animal. Rabies is an acute infectious disease of animals dependent upon a specific agent, *Bacillus lyssæ*, and communicable to man by inoculation. All animals are liable to the disease, but it occurs most frequently in the wolf, the cat, and the dog, and is chiefly propagated by the latter, which is specially susceptible. The toxin has a special affinity for the nervous system, and is found in the secretions, particularly in the saliva. See *hydrophobia*. **r., dumb**, rabies in rodents, in which the preliminary and second periods are absent, and the paralytic stage is pronounced from the onset (Osler). **r. canina, r. felina**, rabies in or acquired from the dog or cat respectively. **r., false**. See *r., pseudo*. **r. paralytic.** 1. Of Gamaleia, rabies in which the third stage is the only manifestation of the infection. 2. An acute ascending spinal paralysis due to infection, probably rabietic. **r., pseudo.** 1. A neurotic or hysterical manifestation closely simulating rabies, but of longer duration and amenable to treatment (Osler). 2. A morbid condition resembling rabies induced experimentally in animals, and occurring in dogs infested with the *Strongylus gigas*. **r., street**, Pasteur's term for the rabies of dogs infected naturally. Fr. *rage des rues*. **r. tanacetic**, a morbid condition resembling rabies induced in rabbits by the intravenous injection of oil of tanacetum.
**rabietic** (*ra-be-et'-ik*) [*rabies*, rage]. Pertaining to affected with, or of the nature of, rabies.
**rabific** (*ra-bif'-ik*) [*rabies*, rage]. Causing rabies; communicating hydrophobia.
**Rabuteau's test for hydrochloric acid in the contents of the stomach** (*rab-oo-to'*). Make a solution containing 50 Cc. of starch mucilage, 1 Gm. of potassium iodate, and 0.5 Gm. of potassium iodide; add to it the filtered contents of the stomach. The solution will become blue in the presence of free HCl.
**race** (*rās*). 1. A genealogic, ethnic, or tribal stock; a breed or variety of plants or animals made permanent by constant transmission of its characters through the offspring. 2. A root, especially of ginger. **r.-ginger**, ginger in the race or root.
**raceme** (*ra-sēm'*) [*racemus*, a cluster of grapes]. In biology, an indeterminate inflorescence having a common peduncle with one-flowered pedicels arranged along its sides. **r., compound**, a raceme in which the pedicels branch and form secondary racemes. **r., false**, a circinate, or scorpioid, cyme.
**racemose** (*ras'-e-mōs*) [*racemus*, a bunch of grapes]. Resembling a bunch of grapes, as a *racemose* gland. **r. aneurysm**, aneurysm by anastomosis. **r. cells**, clusters of cells arranged around a central duct **r. varix**, anastomotic varix.
**rachi-** (*ra-ke-*) [*rhachis*]. A prefix meaning relating to the spine. For words beginning thus, see *rhach-*.
**racial** (*ra'-se-al*) [origin obscure]. Pertaining or due to one's race.
**raclage** (*rak-lahzj'*) [Fr.]. The destruction of a soft growth by rubbing, as with a brush, or harsh sponge; grattage.
**raclement.** See *raclage*.
**rad.** Abbreviation of Latin *radix*, root.
**radal** (*ra'-dal*). A 20 % solution of protargol; used as a prophylactic in gonorrhea.
**Radcliffe's elixir** (*rad'-klif*). Compound tincture of aloes.
**radesyge** (*rah-da-sū'-geh*) [Norwegian]. A disease also known as Scandinavian syphilis, or Norwegian leprosy, and characterized by ulceration and other cutaneous lesions. It is probable that under this name are included syphilitic and leprous lesions.
**radiad** (*ra'-de-ad*) [*radius; ad*, toward]. Toward the radial side.
**radial** (*ra'-de-al*) [*radius*]. 1. Radiating; diverging from a common center. 2. Pertaining to or in relation with the radius or bone of the forearm, as the *radial* artery.
**radiale** (*ra-de-a'-le*). The scaphoid bone of the carpus.
**radialis** (*ra-de-a'-lis*) [L.]. Pertaining to the radius. Various muscles are so called. See *extensor* and *flexor*, under *muscles, table of*.
**radian** (*ra'-de-an*). An arc whose length is equal to the radius of the circle of which it is a part.
**radiant** (*ra'-de-ant*) [*radius*]. 1. Radiate. 2. Emitting rays. **r. energy**, a form of energy emitted by all bodies in proportion to their temperature, and propagated by undulations in the luminiferous ether. When the body reaches 600° C., it begins to radiate light as well as heat. That portion of radiant energy which does not produce the sensation of light is generally spoken of as *radiant heat*, in distinction from *radiant light*. **r. matter**, matter in the ultragaseous state, as in a Crookes tube.
**radiate** (*ra'-de-āt*) [see *radiation*]. Diverging from a central point.
**radiated substance of kidney.** The medullary portion of the kidney.
**radiatio** (*ra-de-a'-she-o*) [L.]. See *radiation*.
**radiation** (*ra-de-a'-shun*) [*radiare*, to radiate]. 1. The act of radiating or diverging from a central point, as *radiation* of light; divergence from a center, having the appearance of rays. 2. In cerebral anatomy, certain groups of fibers that diverge after leaving their place of origin. **r., acoustic**, a tract of fibers extending from the medial geniculate body to the superior and transverse temporal gyri. **r., cortico-striate**, fibers running between the corpus striatum and the equatorial zone of the cortex. **r., occipitothalamic**, same as *optic radiation*. **r., optic**, a large strand of fibers continuous with those of the corona radiata, derived mainly from the pulvinar, the external and internal geniculate bodies, and the optic tract, and radiating into the occipital lobes. **r., striothalamic**, a system of fibers connecting the corpus striatum with the optic thalamus and the subthalamic region. **r., tegmental**, the radiating fibers of the hind portion of the internal capsule. **r., temporothalamic**, same as *acoustic radiation*. **r., thalamic**, certain tracts of fibers from the optic thalami that radiate into the hemispheres.
**radical** (*rad'-ik-al*) [*radix*, a root]. 1. Belonging to the root; going to the root, or attacking the cause of a disease; the opposite of *conservative*. 2. A group of atoms that acts in combination as a simple element, but is incapable of existence in the free state, as $NH_4$, ammonium, $C_6H_5$, phenyl. **r. operation**, an operation for a complete cure of a morbid condition.

**radices** (ra-di'-sēz). Plural of radix.
**radicle** (rad'-ik-l) [dim. of radix]. 1. A little root, as the radicle of a nerve, one of the ultimate fibrils of which a nerve is composed; radicle of a vein, one of the minute vessels uniting to form a vein. r., ascending (of the fornix), the anterior crura or fibers, extending upward from the corpora albicantia. r., descending (of the fornix), the posterior crura or those fibers of the fornix extending from the optic thalami to the corpora albicantia. r., electro-negative, the nonmetallic constituent of a compound which, in electrolysis, is evolved at the anode. r., electro-positive, that constituent of a salt which, in electrolysis, appears at the kathode, and which is either a base or a group of atoms having basic properties. r., vascular, vessels uniting to form a larger vessel. r.s, venous, the capillaries forming the smallest veins. Syn., capillaries, venous. 2. See radical.
**radicotomy** (rad-ik-ot'-o-me) [radix, root; τόμη, incision]. Same as rhizotomy, q. v.
**radicula, radicule** (rad-ik'-ū-lah, rad'-ik-ūl) [radicula, little root]. Same as radicle.
**radiculalgia** (rad-ik-ū-lal'-je-ah) [radicula, a little root; ἄλγος, pain]. Neuralgia affecting the nerve-roots.
**radicular** (rad-ik'-ū-lar). Pertaining to a root or to a radicle; specifically, pertaining to the roots of the spinal nerves. r. arteries, arteries which accompany nerve roots into the spinal cord.
**radiculectomy** (rad-ik-ū-lek'-to-me) [radicula; ἐκτομή, excision]. Excision of a nerve rootlet; resection of the posterior spinal nerve-roots.
**radiculitis** (rad-ik-ū-li'-tis) [radicula; ιτις, inflammation]. Inflammation of a nerve root.
**radien** (ra'-de-en) [radius]. Belonging to the radius in itself.
**radii** (ra'-de-i) [Plural of radius, a ray]. r. auriculares, lines projected on the cranium at right angles to a line passing through the auricular points. r. ciliares, the ciliary processes. r. frontis, wrinkles of the forehead. r. lentis, lines radiating from the poles of the crystalline lens. r. medullares, bundles of receiving tubules of the kidney, beginning in one tubule at the apices of the papillæ, dividing dichotomously, and extending nearly to the cortical surface.
**radio-** (ra-de-o-) [radiare, to emit rays]. 1. A prefix meaning pertaining to radiant energy or to radium. 2. A prefix meaning relating to the radius.
**radioactive** (ra-de-o-ak'-tiv) [radio-; active]. Exhibiting radiant energy.
**radioactivity** (ra-de-o-ak-tiv'-it-e). A property possessed by certain substances of spontaneously emitting radiations which are capable of penetrating substances which are opaque to ordinary rays of light.
**radiobe** (ra'-de-ōb) [radius, a ray; βίος, life]. A peculiar, microscopic, radium formation, thought to be intermediate between a crystal and a living microorganism.
**radiobicipital** (ra-de-o-bi-sip'-it-al) [radio-; biceps]. Pertaining to the radius and the biceps.
**radiocarpal** (ra-de-o-kar'-pal) [radio-; carpus]. Pertaining to the radius and the carpus.
**radiochemistry** (ra'-de-o-kem'-is-tre). That branch of chemistry which deals with radioactive phenomena.
**radiochronometer** (ra-de-o-kro-nom'-et-er) [radio-; chronometer]. An instrument for testing the character of Roentgen-tubes, and the penetrating quality of the X-rays.
**radiode** (ra'-de-ōd) [radio-; ὁδός, a way]. An electric attachment for the application of radium.
**radiodermatitis** (ra-de-o-der-mat-i'-tis). See actino-dermatitis.
**radiodiagnosis** (ra-de-o-di-ag-no'-sis) [radio-; diagnosis]. The diagnosis of a lesion by means of radiography or radioscopy.
**radiodigital** (ra-de-o-dij'-it-al) [radio-; digital]. 1. Pertaining to the radius and the fingers. 2. Pertaining to the fingers on the radial side of the hand.
**radio-element** (ra'-de-o-el'-em-ent). An element which possesses radioactivity.
**radiogen** (ra'-de-o-jen). A trade name for certain radioactive products.
**radiogram** (ra'-de-o-gram). See skiagram.
**radiograph** (ra'-de-o-graf) [radio-; γράφειν, to write]. 1. To obtain a picture by the action of radiant energy upon a sensitive plate. 2. Apparatus for obtaining such a picture. 3. A picture so produced.

**radiographer** (ra-de-og'-raf-ur). One skilled in radiography.
**radiography** (ra-de-og'-ra-fe). See skiagraphy.
**radiohumeral** (ra-de-o-hū'-mer-al) [radio-; humerus]. Pertaining to the radius and the humerus.
**radiology** (ra-de-ol'-o-je) [radio-; λόγος, science]. The science of radiant energy.
**radiolus** (ra-di'-o-lus) [dim. of radius, a ray]. A probe or sound.
**radiometacarpalis** (ra-de-o-met-ak-ar-pa'-lis) [radio-; metacarpus]. The flexor carpi radialis brevis when the insertion is at a metacarpal bone.
**radiometer** (ra-de-om'-et-er) [radio-; μέτρον, a measure]. An instrument for testing the penetration in radiography; a skiameter.
**radiomuscular** (ra-de-o-mus'-kū-lar) [radius; muscular]. Relating to the radius and its muscles. The name of branches of the radial artery distributed to the forearm, and of filaments of the radial nerve going to the same muscles.
**radion** (ra'-de-on). A particle thrown off by a radioactive substance.
**radioneuritis** (ra-de-o-nū-ri'-tis). A form of neuritis observed in persons who have worked for a long time with x-rays.
**radiopalmar** (ra-de-o-pal'-mar) [radio-; palm]. 1. Pertaining to the radius and the palm. 2. Pertaining to the outer side of the palm.
**radiopraxis** (ra-de-o-praks'-is) [radio-; πρᾶξις, action; practice]. The art of applying radiant energy either in therapeutics or for other purposes.
**radioscopy** (ra-de-os'-ko-pe) [radio-; σκοπεῖν, to view]. The process of securing an image of an object upon a fluorescent screen by means of radiant energy.
**radiostereoscopy** (ra-de-o-ster-e-os'-ko-pe) [radius, ray; στερεός, solid; σκοπεῖν, to view]. The application of the principle of the stereoscope, obtaining a viewpoint for the left eye and one for the right by lateral displacement of the tube along the plane of the plate, determining this displacement by the formula of Marie and Ribault for the purpose of demonstrating the different planes in which various objects shown by radioscopy are situated.
**radiotherapeutic** (ra-de-o-ther-ap-ū'-tik) [radiotherapy]. Having reference to the therapeutic use of radiant energy.
**radiotherapeutics** (ra-de-o-ther-ap-ū'-tiks). See radiotherapy.
**radiotherapy** (ra-de-o-ther'-a-pe). The treatment of disease by means of X-rays, radium, and other radioactive substances.
**radiothorium** (ra-de-o-thor'-e-um). A radioactive substance which is neither radium nor thorium, but has properties like those of thorium.
**radioulnar** (ra-de-o-ul'-nar) [radio-; ulna]. Pertaining to the radius and ulna.
**radium** (ra'-de-um) [radiare, to emit rays]. An elementary body from pitch-blende, characterized by the phenomenon radioactivity. It is obtained by the fractional reprecipitation or recrystallization of the barium chloride prepared from pitch-blende. See elements, table of.
**radius** (ra'-de-us) [L., "a spoke of a wheel"]. 1. A ray. 2. The outer of the two bones of the forearm. r. fixus, an imaginary line connecting the inion and the hormion.
**radix** (ra'-diks) [L.: gen., radicis; pl., radices]. A root. Any one of the spinal nerve roots. r. arcus vertebræ, a root or pedicle of the vertebral arch.
**raffinase** (raf'-in-ās). The enzyme which decomposes raffinose; it is found in the seed of the cotton plant, in the root of the sugar beet, in certain yeasts, and in barley and wheat during germination.
**raffinose** (raf'-in-ōs), $C_{18}H_{32}O_{16}+5H_2O$. A trisaccharid derived from beets.
**rafle** (ra'-fl) [Fr.]. A pustular disease of cattle.
**rag-picker's disease.** An acute febrile disease occurring in workmen engaged in sorting rags in paper-factories. It is supposed to be due to the inhalation of anthrax bacilli or spores, and is characterized by an exudation into the pulmonary tissue, bronchial glands, and pleural cavity. Syn., hadernkrankheit.
**rage** (rāj) [ME.]. 1. Violent passion or anger. 2. Any intensely painful affection. 3. (rahzj) [Fr.]. Hydrophobia; rabies.
**ragle** (rahgl) [Fr.]. An hallucination due to isolation and insomnia, observed in French troops

while in the desert, in which they imagined they saw prairie and water.
**railway sickness.** See *car-sickness*. **r. kidney,** a renal affection said to be due to the constant jar of railway journeys. **r.-spine,** a term given by Erichsen to a varied group of spinal symptoms consequent on slight injuries or concussions received in railway accidents. The condition is classed with the traumatic neuroses and is a form of neurasthenia. It is frequently a cause for litigation. See *Erichsen's disease.*
**Rainey's capsules, corpuscles, or tubes** (*ra'-ne*) [George Rainey, English anatomist, 1801-1884]. See *Miescher's tubes.*
**raise** (*rās*) [ME. *raisen,* to raise]. To expectorate.
**raised** (*rāzd*) [ME., *raisen,* to raise]. Elevated. **r. base** (*for artificial teeth*), a term applied in mechanical dentistry to a metallic base, surmounted by a box or chamber soldered to it, and designed to compensate for the loss of substance which the parts have sustained. A base thus constructed is named termed by dentists a raised plate.
**raising** (*ra'-zing*). 1. Expectoration. 2. One of the Swedish movements, either active or passive. It is used for deformities of the back, to relieve constipation, to act upon the abdomen, etc.
**raisins** (*ra'-zins*). Dried grapes; passulæ. **r., Corinth,** currants.
**rake teeth.** A term applied to teeth separated by intervals, like those of a rake.
**rale** (*rahl*) [Fr., *râler,* to rattle]. An adventitious sound heard over the chest during respiration and indicating some local disturbance. Rales are either *dry* or *moist* (produced by the bubbling of air through liquid), *sonorous* or *sibilant.* They are also classified according to their place of production into *laryngeal, tracheal, bronchial, vesicular, cavernous, pleural, pericardial.* **r., amphoric,** a large, musical, tinkling rale, heard in inspiration and expiration, in tuberculous and abscess cavities; produced by movement of air in a tense-walled cavity containing air and communicating with a bronchus. **r., bubbling, large,** a moist rale, larger than the medium bubbling, heard in inspiration and expiration in bronchitis and pulmonary engorgement; produced by passage of air through frothy mucus in the trachea and larger bronchi. **r., bubbling, medium,** a moist rale larger than the small bubbling, heard in inspiration and expiration in capillary bronchitis, especially in children; produced by the passage of air through mucus in the larger tubes. **r., bubbling, small,** a small, moist rale, sounding like the bursting of small bubbles, heard in inspiration and expiration in capillary bronchitis, especially in children; produced by the passage of air through mucus in the bronchioles. **r., cavernous,** a hollow, metallic rale, heard in inspiration and expiration in the third stage of pulmonary tuberculosis; produced by the passage of air through a small cavity with flaccid walls that collapse with expiration. **r., clicking,** a small, sticky rale heard in inspiration in the early stage of pulmonary tuberculosis; caused by passage of air through softening material in the smaller bronchi. **r., consonating,** a bright, clear, ringing rale, heard in inspiration and expiration in tuberculous pneumonia; produced when the bronchial tubes are surrounded by a consolidated tissue. **r., crackling, dry,** a sharp, short, clicking rale, heard in inspiration in the second or softening stage of pulmonary tuberculosis and in pulmonary gangrene; produced by the breaking down of lung tissue. **r., crackling, large,** a dry rale larger than the medium crackling, heard in inspiration and expiration, in pulmonary tuberculosis and pneumonia, after the formation of small cavities; produced by fluid in very small cavities. **r., crackling, medium,** a dry rale, larger than the small crackling, heard chiefly in inspiration, in softening of tuberculous deposit or pneumonic exudation; caused by fluid in the finer bronchi. **r., crackling, small,** a small, dry rale, sounding like the breaking of small shells, heard chiefly in inspiration, in softening of tuberculous deposit or pneumonic exudation; produced by fluid in the finer bronchi. **r., crepitant,** a small rale, sounding like the rubbing of hair between the fingers, heard at the end of inspiration in pneumonia, early stage, edema of the lungs, hypostatic pneumonia; localized in pulmonary tuberculosis. Produced by the passage of air into vesicles, collapsed or containing fibrinous exudation; usually at the base of the lungs. **r. de retour.** Same as *rale redux.* **r., dry,** a large and sonorous, or small and hissing or whistling rale heard in inspiration and expiration in bronchitis and asthma, localized in beginning pulmonary tuberculosis; produced by narrowing of the bronchial tubes from thickening of the mucous lining, from spasmodic contraction of the muscular coat, viscid mucus within, or pressure from without. **r., extrathoracic,** one produced in the trachea or larynx. **r., friction,** a grazing, rubbing, grating, creaking, or crackling rale heard in inspiration and expiration, most distinct at the end of inspiration, in pleurisy and pericarditis; produced by the rubbing together of serous surfaces, roughened by inflammation or deprived of their natural secretion. **r., gurgling,** a moist rale, larger than the large bubbling, sounding like the bursting of large bubbles, heard in inspiration and expiration in pulmonary tuberculosis after the formation of large cavities. **r., guttural,** one produced in the throat. **r., moist,** one produced by the passage of air through bronchi containing fluid. **r., mucous** (of Laennec), a modification of the subcrepitant rale, heard in inspiration and expiration in pulmonary emphysema; produced by viscid bubbles bursting in the bronchial tubes. **r. redux,** return of the crepitant rale heard in the resolution stage of pneumonia; produced by the passage of air through fluid in a bronchial tube. Syn., *rale de retour.* **r., sibilant,** a high-pitched and even hissing or piping rale, heard in inspiration and expiration in bronchitis, asthma, and localized in beginning pulmonary tuberculosis; produced by narrowing of the smaller bronchi from viscid mucus adhering to the walls, from thickening of the lining membrane, or spasmodic contraction. **r., sonorous,** a low-pitched, snoring rale, heard in inspiration and expiration, most frequently in bronchitis and spasmodic asthma; produced by lessened caliber of the larger bronchi, from spasm, tumefaction of mucous lining, or external pressure. **r., subcrepitant,** a small, moist rale heard in inspiration and expiration in capillary bronchitis; produced by the passage of air through mucus in the capillary bronchial tubes. **r., subcrepitant, Hirtz's,** a moist, metallic rale, pathognomonic of tuberculous softening. **r. vesicular.** Same as *r. crepitant.*
**Ralfe's test** [Charles Henry Ralfe, English physician, 1842-1896]. 1. *For acetone in urine:* Boil 4 Cc. of liquor potassæ with 1.5 gm. of potassium iodide; overlay it with 4 Cc. of urine; a yellow ring studded with specks of iodoform appears at the line of contact. 2. *For peptones in urine:* Place 4 Cc. of Fehling's solution in a test-tube, and overlay it with an equal amount of urine; a rose-colored halo appears above the zone of phosphates.
**ramal** (*ra'-mal*) [*ramus,* a branch]. Pertaining to a ramus; branching. **ramalis vena,** the portal vein and its branches.
**ramaninjana** (*ram-an-in-yah'-nah*). A nervous disease of Madagascar.
**Ramdohr's operation** (*ram'-dōr*) [C. A. von Ramdohr, American surgeon, 1855-1912]. *For enterorrhaphy;* the insertion of the proximal within the distal end of the intestine, and suturing. R.'s suture, invagination of the upper portion of the intestine into the lower, followed by suture.
**ramenta** (*ra-men'-tah*) [L.; *pl., filings, scrapings*]. Shreds, filings, or shavings. **r. ferri,** iron filings. **r. intestinorum,** shreds of intestinal mucus discharged with the evacuations in severe dysentery.
**ramex** (*ra'-meks*) [*gen., ramicis: pl., ramices*]. A hernia, or hernial or scrotal tumor. **r. varicosus,** variocele.
**rami** (*ra'-mi*) [L.]. Plural of ramus, *q. v.* **r. accelerantes,** accelerator nerves. **r. alares,** branches of the lateral nasal artery supplying the nasal pinnæ. **r. anteriores nervorum spinalium,** the anterior divisions of the spinal nerves. **r. cardiaci** (**nervi vagi**), the cardiac branches of the pneumogastric nerve. **r. communicantes noni,** the branch of the descendens noni which join the communicating branches of the second and third cervical nerves. **r. emissaria,** branches of the anterior spinal plexuses which emerge through the intervertebral and anterior sacral foramina. **r. intestinales,** branches of the cerebrospinal nerves supplying the abdominal viscera. **r., ischio-pubic,** the descending rami of the ischium and the pubes taken as one. **r. linguales** (**nervi glossopharyngei**), the terminal branches of the ninth nerve. **r. marginales,** the branches of the palpebral arteries which aid in forming the arcus arteriosus palpebræ. **r. musculares,** unnamed

branches of nerves or blood-vessels distributed to the muscles.  r. olfactorii, the olfactory nerve. r. pharyngei (nervi vagi), the branches of the vagus going to the pharynx.  r. ventrales, the branches of the intercostal arteries distributed to the intercostal muscles and to the ribs.
ramie (ram'-e) [Malay]. See r. fiber.  r. fiber, China-grass. The bast fiber from two varieties of *Boehmeria nivea*, known in India as *Rhea*, and in the Malay Archipelago as *Ramie*. The properly prepared fiber is of fine, silky luster, soft, and extraordinarily strong. It is the most perfect of all the vegetable fibers, and is composed of pure cellulose.
ramification (ram-if-ik-a'-shun) [*ramus; facere*, to make]. 1. The act or state of branching. 2. A branch.
ramify (ram'-e-fi) [see *ramification*]. To form branches; to branch.
ramolescence (ram-o-les'-ens) [Fr. *ramollir*, to soften]. A softening; mollification.
ramolissement (rah-mo-lēs-mon(g)) [Fr.]. Morbid softening of any tissue or part.
Ramón y Cajal's cells (rah-mōn'-e-kah-hahl'). See *Cajal*.
ramose (ra'-mōs) [*ramus*]. Having many branches; branching.
Ramsden's eye-piece [Jesse *Ramsden*, English optician, 1735-1800]. An eye-piece having two plano-convex lenses, used with a micrometer.
ramulus (ram'-ū-lus) [L.: *pl., ramuli*]. A small branch, or ramus.
ramus (ra'-mus) [L.; pl., *rami*]. 1. A branch, especially of a vein, artery, or nerve. 2. A slender process of bone projecting like a branch or twig from a large bone, as the ramus of the lower jaw; ascending ramus of the ischium; ascending or horizontal ramus of the pubes.  r. abdominalis, the hypogastric nerve. r. acetabuli, a branch of the internal circumflex artery supplying the hip-joint.  r. anastomoticus, the branch of an artery by which an anastomosis is established. r. anterior ascendens, r. anterior nervi acustici. See *cochlear nerve*.  r. ascendens, the anterior branch of the fissure of Sylvius.  r. ascendens glabellaris, the branch of the angular artery going to the inner angle of the orbit..  r. ascendens nervi vagi, a branch of the superior laryngeal nerve going to the epiglottis. r. ascendens (inferior) ossis ischii, the ascending branch of the ischium.  r., ascending (*of the ischium*), the portion between its tuberosity and the acetabulum.  r. ascending (*of the pubic bone*), the portion between its body and the acetabulum.  r. auricularis nervi vagi. See *nerve of Arnold*. r. bulbocavernosus. See *artery of the bulb of the urethra*.  r. canalis spinalis, the branch of the intercostal artery supplying the walls of the spinal canal..  r. cardiacus nervi vagi inferior, inferior cardiac nerve.  r. cardiacus nervi vagi superior, superior cardiac nerve.  r. cervicofacialis (*nervi facialis*), cervicofacial nerve.  r. cochleæ, r. cochlearis, the cochlear nerve.  r. communicans, a branch of a spinal nerve connecting it with the sympathetic ganglia.  r. communicans anterior, anterior communicating artery of the brain.  r. communicans medullæ spinalis, a branch of a spinal nerve uniting it with the sympathetic. r. communicans posterior, the posterior communicating artery of the brain.  r. cruralis, lumbo-inguinal nerve.  r. cutaneus nervi radialis. See *radial nerve*.  r. cutaneus palmaris longus. See *r. palmaris longus nervi mediani*.  r. descendens. 1. The descendens noni nerve.  2. The inferior division of the inferior maxillary nerve.  r. descendens nervi hypoglossi, the descendens noni nerve.  r. descendens (superior) ossis ischii, the descending branch of the ischium.  r. descendens ossis pubis, the descending branch of the pubic bone.  r., descending (*of the ischium*), the portion between its body and tuberosity.  r., descending (*of the pubic bone*), the portion included between its body and its junction with the ischium.  r. dexter arteriæ pulmonalis, the right pulmonary artery.  r. dorsalis nasi, the dorsal artery of the nose.  r. dorsalis nervi radialis. See *radial nerve*.  r. dorsalis nervi ulnaris. See *ulnar nerve*.  r. dorsalis pollicis radialis, that part of the radial artery which winds round the outer side of the carpus and crosses the thumb beneath the extensor tendons.  r. duræ matris vagi, a branch from the jugular ganglion going to the meninges.  r. externus, r. femoralis, lumbo-inguinal nerve. r. hepaticus dexter, the right hepatic artery.  r. hepaticus sinister, the left hepatic artery.  r., horizontal (*of the pubic bone*). See *r., ascending* (of the pubic bone).  r. horizontalis fissuræ Sylvii, the posterior limb of the fissure of Sylvius.  r. horizontalis mandibulæ, the body of the inferior maxilla. r. horizontalis (superior) ossis pubis (pectinis), the horizontal ramus of the pubic bone.  r. of the inferior maxilla, the portion ascending from the angle, and terminating in the condyle and coronoid process. r. intermedius. See *r. anastomoticus*.  r. lingualis recurrens. See *r. ascendens nervi vagi*.  r. magnus nervi mediani, the musculocutaneous nerve.  r. major nervi maxillaris inferioris, the greater of the two primary branches of the inferior maxillary nerve. r. malaris, the inferior branch of the orbital nerve. r. marginalis, the dorsal branch of the radial nerve supplying the thumb.  r. mastoideus.  1. The small occipital nerve.  2. The occipital branch of the posterior auricular artery.  3. The posterior division of the great auricular nerve.  r. maxillaris, inferior nervi trigemini. See *inferior maxillary nerve*. r. maxillaris superior (medius) nervi trigemini. See *superior maxillary nerve*.  r. medullæ spinalis, the spinal branch of the intercostal artery supplying the spinal cord.  r. meningeus posterior. See *r. duræ matris vagi*.  r. minor nervi vidiani, the great superficial petrosal nerve.  r. muscularis nervi radialis, the radial nerve.  r. nasalis nervi ophthalmici.  See *naso-ciliary nerve*.  r. nutriens, the nutrient artery of a bone.  r. ophthalmicus (quinti), the ophthalmic nerve.  r. ossis maxillæ inferioris. See *r. of the inferior maxilla*.  r. ovarii. See *artery, ovarian*.  r. palmaris longus nervi mediani, a branch of the median nerve distributed to the integument of palm of the hand.  r. perpendiculares (mandibulæ). See *r. of the ilio-lumbar inferior maxilla*.  r. posterior. 1. The ilio-lumbar artery.  2. The posterior limb of the fissure of Sylvius.  r. primus nervi trigemini (primus quinti). See *ophthalmic nerve*.  r. profundus nervi radialis. See *radial nerve*.  r. recurrens vagi, a branch of the jugular ganglion which goes to the transverse sinus through the jugular foramen.  r. secundus (ganglii Gasserii), r. secundus nervi trigemini (quinti paris), the superior maxillary nerve.  r. sinister arteriæ pulmonalis, the left pulmonary artery.  r. sinualis, the recurrent branch of the ophthalmic nerve going to the tentorium.  r. sublimis (volaris superficialis) arteriæ radialis, the superficial volar artery.  r. superficialis nervi radialis, the radial nerve.  r. superior ossis ischii, the descending ramus of the ischium.  r. superioris nervi trigemini, r. superioris quinti, the ophthalmic nerve.  r. supraspinatus, the supraspinous artery.  r. tertius nervi trigemini, the inferior maxillary nerve.  r. vestibularis, the vestibular nerve.  r. volaris nervi ulnaris, the volar distribution of the ulnar nerve.
ramuscule (ra-mus'-kūl) [*ramusculus*]. A little branch, especially of the pial arteries.
rancid (ran'-sid) [*rancidus*, sour]. Having a rank or musty smell or taste; a term applied to fats and oils that have undergone decomposition with the development of volatile principles.
rancidity (ran-sid'-it-e) [*rancid*]. The state of being rancid.
Randia (ran'-de-ah) [Isaac *Rand*, an English botanist of the eighteenth century]. A genus of cinchonaceous shrubs. R. aculeata, of West India; ink-berry, indigo plant. The juice of the fruit is astringent. R. dumetorum, of India; has a poisonous and strongly emetic fruit. R. longiflora, of Bengal; the cortex is used in intermittent fever.
Randolph's test for peptones in urine [Nathaniel Archer *Randolph*, American physician, 1858-1887]. To 5 c.c. of faintly acid urine add 2 drops of saturated solution of potassium iodide and 3 or 4 drops of Millon's reagent; a yellow precipitate indicates the presence of peptones.
range (rānj) [Fr., *ranger*, to dispose]. Scope; extent.  r. of accommodation. See *accommodation, range of relative*.
ranine (ra'-nīn) [*rana*, a frog]. 1. Pertaining to a frog.  2. Pertaining to a ranula or to the region in which a ranula occurs, as *ranine artery*.
Ranke's angle (ran'-keh) [Hans Randolph *Ranke*, Dutch anatomist, 1849-1887]. The angle between the horizontal plane of the head and a line from the center of the alveolar border to the center of the frontonasal suture.
Ransohoff's operation (ran'-so-hof) [Joseph *Ransohoff*, American surgeon, 1853-    ]. Discission of the pulmonary pleura, employed as a substitute for decortication.

**ranula** (*ran'-û-lah*) [*rana*, a frog]. A cystic tumor beneath the tongue, due to the occlusion of the duct of the sublingual or submaxillary gland, or of a mucous gland of the floor of the mouth. Syn. *frog-tongue*. r., **apidea**, salivary calculi. r. **pancreatica**, a dilated saccular condition of the larger pancreatic ducts due to calculous obstruction. r., **suprahyoid**, a cystic tumor situated above the hyoid bone.

**ranunculaceous** (*ra-nung-kû-la'-se-us*) [*ranunculus*, a medicinal plant]. Noting, or relating to plants of the order *Ranunculaceæ*.

**ranunculus** (*ra-nung'-kû-lus*) [L.]. A genus of acrid herbs. Many of the species are poisonous and have been used as a counterirritants and vesicants. *R. acris* is very irritant and causes erythema æstivum.

**Ranvier's accessory plexus** (*ron(g)-ve-a*) [Louis Antoine *Ranvier*, French histologist, 1835– ]. The superficial stroma plexus of the cornea. R.'s **cells**, connective-tissue corpuscles occurring in tendon. R.'s **crosses**, black, crucial figures seen at Ranvier's nodes on staining with silver nitrate. The transverse branch of the cross is represented by the line of constriction, and the longitudinal branch by the axis-cylinder. R.'s **nodes**, annular constrictions of the neurilemma, with discontinuity of the medullary sheath of the nerve-fiber. R.'s **tactile discs**, nerve-endings consisting of small, cup-shaped bodies, the concave side of which is directed toward the free surface of the epidermis.

**rape** (*râp*) [*rapere*, to seize]. Sexual intercourse with a woman without her free consent. r. **seed**, the seed of wild turnip, *Brassica campestris*.

**raphania** (*raf-a'-ne-ah*) [ῥαφανίς, radish]. A nervous affection attended with spasmodic disorder of the joints and the limbs. It has been attributed to a poisonous principle in the seeds of the wild radish, which become mixed with grain. The affection is allied to ergotism and pellagra.

**raphe** (*raf'-a*) [ῥαφή, a seam]. A seam or ridge, especially one indicating the line of junction of two symmetrical halves. r. **of the ampulla**, longitudinal ridge on the roof of the ampulla of the semicircular canal. r. **exterior**, the stria longitudinalis medialis. r. **inferior corporis callosi**, the raphe on the inferior surface of the corpus callosum. r. **palati duri**. Same as r. *palatine*. r., **palatine**, the narrow ridge of mucosa in the mesial line of the palate. r. **of the penis**, a continuation of the raphe of the scrotum upon the penis. r., **perineal**, the ridge of skin in the median line of the perineum. r. **of the pharynx**, a fibrous band in the median line of the posterior wall of the pharynx. r. **of the pons**, the intersection of the fibers at the meson as seen in transection. r. **postoblongata**, the posterior median fissure of the medulla oblongata. r. **of the scrotum**, a median ridge dividing the scrotum into two lateral halves; it is continuous posteriorly, with the raphe of the perineum, anteriorly with the raphe of the penis. r. **Stilling's**, a narrow band connecting the pyramids of the oblongata. r. **superior corporis callosi**, the longitudinal raphe in the middle of the superior surface of the corpus callosum. r. **of the tongue**, a median furrow on the dorsal surface of the tongue corresponding to the fibrous septum which divides it into symmetrical halves.

**raphidiospore** (*ra-fid'-e-o-spôr*). See *exotospore*.

**raptus** (*rap'-tus*) [*rapere*, to seize]. Any sudden attack or seizure; rape. r. **hæmorrhagicus**, a sudden hemorrhage. r. **maniacus**, transient frenzy. r. **melancholicus**, sudden and vehement melancholy. r. **nervorum**, cramp or spasm.

**rarefaction** (*râr-e-fak'-shun*) [*rarus*, thin; rare; *facere*, to make]. The act of rarefying or of decreasing the density of a substance, especially the air. r. **of bone**, the process of rendering bone more porous.

**rarefy** (*râr'-e-fi*) [see *rarefaction*]. To make less dense or more porous.

**rarefying osteitis.** See *osteoporosis*.

**raritas** (*râr'-it-as*) [L.]. Rarity. r. **dentium**, fewness of teeth; less than the usual number of teeth, with or without interspaces between them.

**rasceta** (*ras-e'-tah*) [L.]. The transverse lines or creases on the inner side of the wrist.

**Rasch's sign** [Hermann *Rasch*, German obstetrician, 1873– ]. Fluctuation obtained by applying two fingers of the right hand to the cervix, as in ballottement, and steadying the uterus through the abdomen with the left hand. It depends upon the presence of the liquor amnii, and is an early sign of pregnancy.

**rash** [OF., *rasche*, from *radere*, to scrape]. A superficial eruption of the skin or mucous membrane. r., **amygdalotomy**, one that generally appears on the neck, chest, or abdomen- two or three days after an operation on hypertrophied tonsils. r., **caterpillar-**, a localized eruption attributed to the irritant action of the hairs of certain caterpillars. r., **drug-**, one produced by drugs. r., **medicinal.** See r., *drug-*. r., **mulberry-**, an eruption resembling an exanthem of measles, sometimes occurring in typhus. r., **nettle-**. See *nettlerash*. r., **rose-**. See *roseola*. r., **scarlet**. See *scarlatina*. r., **tonsillotomy**. See r., *amygdalotomy*. r., **tooth-**, any rash attributed to dentition.

**rasion** (*ra'-zjun*) [*radere*, to scrape]. The scraping of drugs with a file.

**Rasmussen's aneurysm.** Dilatation of an artery in a tuberculous cavity; its rupture is a frequent cause of hemorrhage. R.'s **test for urobilin**, shake together thoroughly equal parts of urine and ether to which has been added 6 or 7 drops of tincture of iodine. Allow it to stand until the solution separates into an upper layer of ether and iodine and a lower one of urine. In the presence of bile the lower layer turns green if biliverdin also exists.

**Raspail's reaction for albumins.** These are colored red by sugar and concentrated sulphuric acid. R.'s **sedative water**, a lotion containing camphor and ammonia.

**raspatory** (*ras'-pa-to-re*) [*raspatorium*, from *radere*, to scrape]. A rasp or file for trimming the rough surfaces of bones or for removing the periosteum.

**raspberry** (*raz'-ber-e*). The fruit of *Rubus idæus*, a plant of the order *Rosaceæ*. A syrup is used as a vehicle and as a drink in fevers.

**rasura** (*ra-zû'-rah*) [L.]. 1. The process of rasping, shaving, or scraping. 2. Scrapings; filings.

**rat** [ME., *ratte*, rat]. A rodent of the family *muridæ*. r.-**tail sutures**, fibers from the rat's tail, used instead of silk or gut, for surgical sutures. r.-**tooth forceps.** See *forceps*.

**ratafia** (*rat-a-fe'-ah*) [Malay, *arag*, arrack; *tafia*, a spirit distilled from molasses]. A name for various liqueurs, or aromatized and sweetened cordials.

**ratany**, **ratanhia** (*rat'-an-e*, *rat-an'-he-ah*). See *krameria*.

**Rathke's duct** (*rath'-keh*) [Martin Heinrich *Rathke*, German anatomist, 1793–1860]. That portion of Mueller's duct which intervenes between the latter and the sinus pocularis; it may persist after birth as a patulous duct. R.'s **folds**, two projecting folds of the fetal mesoderm which are placed between the orifice of the intestine and the allantois and unite in the median line to form Douglas' septum. R,'s **glands.** See *Jacobson's organ.* R., **investing mass of**, the membranous capsule covering the end of the chorda dorsalis in the developing embryo and forming the rudiment of the base of the skull. It molds itself on the cerebral vesicles, so as to constitute the membrane in which the vault of the skull is developed. The membranous capsule at the base of the skull presents two thickenings, the lateral trabeculæ of Rathke, directed forward and enclosing the pituitary opening. R., **lateral trabeculæ of.** See *R., investing mass of.* R.'s **pouch**, in the embryo, the diverticulum of the pharyngeal membrane which is connected with the midbrain, and ultimately forms the anterior lobe of the hypophysis.

**ratio** (*ra'-she-o*) [L.]. 1. The mind or reasoning faculties. 2. In chemistry and pharmacy, the proportion of ingredients or of atomic composition. r. **medendi**, the theory or scheme of a course of medical treatment. r. **ocular micrometer**, the number obtained by finding the number of divisions on the ocular micrometer required to include the image of an entire millimeter of the stage micrometer.

**ration** (*ra'-shun*) [*ratio*, proportion]. The daily allowance of food or drink. r., **emergency**, one with high force-value and with sufficient available nitrogen for the needs of hard labor, prepared in compact form and designed for occasions when the use of the regular ration is impracticable.

**rational** (*rash'-un-al*) [*ratio*, reason]. Based upon reason; reasonable. In therapeutics, opposed to empirical. r. **formula**, a chemical formula which shows, either partly or completely, the constitution of a compound. r. **symptoms**, the symptoms elicited by questioning the patient, as opposed to those ascertained by physical examination.

**ratsbane** (*ratz'-bān*). 1. Arsenic trioxide. 2. A name given to any rat-poison containing arsenic.

**rattle** (*rat'-l*) A rale. **r., death-**, a gurgling sound observed in dying persons, due to the passage of the air through mucus in the trachea.

**Rau's process** (*row*) [Johannes Jacobus *Rau*, Dutch anatomist, 1668–1719]. The longer process at the junction of the handle with the neck of the malleus. It is also called the *Folian process*.

**Rauber's layer** (*row'-ber*) [August Antinous *Rauber*, German anatomist, 1845– ]. A superficial stratum of flat cells occurring in the center of the embryonal spot at an early stage in the development of the blastodermic membranes.

**raucedo** (*raw-se'-do*) [*raucus*, hoarse]. Hoarseness arising from inflammation of the mucosa of the larynx and throat. **r. catarrhalis**, hoarseness, resulting from laryngitis. **r. potatorum**, hoarseness caused by drinking whisky or other distilled liquors. **r. syphilitica**, chronic hoarseness due to secondary syphilitic affections of the larynx.

**Rauchfuss's triangle** (*rowk'-foos*) [Charles Andreyevich *Rauchfuss*, Russian physician, 1835– ]. Same as *Grocco's triangle*.

**raucitas** (*raw'-sit-as*). See *raucedo*.

**rauschbrand**. (*rowsh'-brant*) The German name for black-leg.

**ray** (*rā*) [*radius*, a ray]. 1. A beam of light or heat; one of the component elements of light or heat. 2. One of a number of lines diverging from a common center. **r.s, actinic**. See *r.s, chemical*. **r.s, alpha-**, rays discovered by Rutherford, emanating from uranium, thorium, and radium, and differing from kathode rays in having much less penetrating power and in not being deviated ordinarily either by a magnet or an electrically charged body. **r.s, Becquerel**, invisible radiations of electrified particles or ions projected from radioactive bodies, such as uranium, radium, polonium, or their salts, without evident cause, and persisting over long periods. **r.s, beta-**, Rutherford's name for the kathode rays emitted by radioactive substances. They differ from the alpha-rays in greater penetrating power, weaker electric power, and in carrying a negative charge. **r.s, chemical**, solar rays that produce chemical change; see *phototherapy*. **r.s, diakathodic**, bluish rays obtained by directing the ordinary kathode rays upon a piece of wire gauze or upon a spiral of wire which is itself negatively electrified. They are not directly affected by a magnet. **r.-fungus**. See *actinomyces*. **r.s, gamma-**, a type of Becquerel rays more highly penetrating than the alpha-rays and beta-rays, but insignificant in energy compared with them. **r.s, Goldstein**, kathode rays which have been altered by being passed through a perforated metallic plate. **r.s, hard**, Roentgen rays coming from a tube the exhaustion of which is sufficient to cause a considerable difference in the potential between the kathode and the anode and in the velocity of the kathode rays. They have high penetrating powers. Cf. *r.s, soft*. **r.s, Hertzian**, radiant energy having the greatest wave length of any yet discovered in the spectrum, supposed to be several miles in length. These rays have the peculiar property of converting poor electric contacts into good ones when they fall upon them. **r.s, kathode**, the stream of negatively electrified particles emanating from the kathode of a Crookes tube and passing in straight lines regardless of the anode. They are capable of deflection with a magnet and produce fluorescence and heat wherever they impinge. **r.s, Lenard**, cathode rays outside the vacuum tube as described by Philipp Lenard (1894) and secured by him by means of an aluminum window. **r., medullary, of the kidney**, any one of the bundles of tubules that are the continuation into the cortex of the malpighian pyramids. **r.s, N-**, a form of etherwaves discovered by Blondlot (1903) and named after the initial letter of Nancy, in the university of which his researches were conducted. They increase the brightness of an electric spark or the luminosity of phosphorescent bodies; they are emitted by the Roentgen-ray tube, by an Auer-Welsbach incandescent gas-burner, by the ordinary gas-flame, but not by a Bunsen burner; the sun emits these in abundance, as does the Nernst lamp. Compression, torsion, and strain of many solids will cause the emission of the rays; living bodies, plants, and animals emit them. **r.s, Niewenglowski's**, certain luminous rays emitted from phosphorescent substances which may pass through opaque screens and affect sensitive plates. Niewengloski was probably the first to establish the existence of such rays. **r.s, photographic**. See *r.s, ultraviolet*. **r.s, positive**. See *r.s, Goldstein*. **r.s, Roentgen-**, the ether-rays or waves discovered by Roentgen, of Würzburg, and named by him *x-rays*. A vacuum-tube of glass (called a *Geissler tube*, a *Hittorf* or a *Crookes tube*) is used with two wires sealed through the glass. These wires are connected with the two poles of a battery, and Roentgen found that the rays from the kathode (*kathode rays*) had peculiar penetrative powers through matter opaque to other ether-rays, and that by means of these rays photographs ("*shadowgrams*") may be taken of bones, metallic substances, etc., situated in the tissues; they readily traverse living tissues and influence the nutrition of the deeper ones; they have no appreciable effect on the vitality of bacteria. Hertz, and especially Lenard, prior to Roentgen, had discovered this penetrating power of the kathode rays, but failed to make the application suggested by Roentgen. **r.s, S of Sagnac**, secondary rays emanating from metals on which Roentgen rays fall and distinguish from the primary rays irregularly refracted by difference in character, not being nearly so penetrating; the lighter the metal struck by the primary rays, the more penetrating the secondary rays. **r.s, soft**, rays coming from a tube the pressure in which is fairly low; they are readily absorbed. Cf. *r.s, hard*. **r.s, ultraviolet**, waves of the luminiferous ether which do not affect the retina. They can be reflected, refracted, and polarized; they will not traverse many bodies that are pervious to the rays of the visible spectrum; they produce photographic and photochemical effects, and destroy rapidly the vitality of bacteria. Syn., *actinic rays; photographic rays*. **r.s, uranium**. See *r.s, Becquerel*. **r.s, X-**. See *r.s, Roentgen*.

**Ray's mania**. Moral insanity, regarded by Ray as a distinct form of mental disorder.

**Raygat's test of live birth** (*ra'-gat*). Place the lungs in water and note their specific gravity. If inflation has occurred they will float. Also called *hydrostatic test*.

**Raynaud's disease** (*ra-nō'*) [A. G. Maurice *Raynaud*, French physician, 1834–1881]. 1. A trophoneurosis characterized by three grades of intensity: (*a*) Local syncope, observed most frequently in the extremities, and producing the condition known as dead fingers or dead toes. (*b*) Local asphyxia, which usually follows local syncope, but may develop independently. The fingers, toes, and ears are the parts usually affected. In the most extreme degree the parts are swollen, stiff, and livid, and the capillary circulation is almost stagnant. (*c*) Local or symmetrical gangrene. Small areas of necrosis appear on the pads of the fingers and of the toes, also at the edges of the ears and tip of the nose. Occasionally symmetrical patches are seen on the limbs or trunk, and in severe cases terminate in extensive gangrene. Some cases are attended by hemoglobinuria. The pathology of the disease is obscure. 2. Paralysis of the muscles of the throat following parotitis. See *R.'s disease* (1). **R.'s phenomenon**, a white and cold condition of the fingers, alternating with burning heat and redness, occurring in Raynaud's disease (*q. v.*), and showing vasomotor disturbance.

**Rb**. Chemical symbol of *rubidium*.

**R. C. P**. Abbreviation for *Royal College of Physicians*.

**R. C. S**. Abbreviation for *Royal College of Surgeons*.

**R. D**. Abbreviation for *reaction of degeneration*.

**R. D. A**. Abbreviation denoting the *right dorsoanterior* position of the fetus.

**R. D. P**. Abbreviation denoting the *right dorsoposterior* position of the fetus.

**R. E**. Abbreviation for *right eye*

**Ré**. Abbreviation for *Réaumur*, or the degree of Réaumur's thermometer scale.

**re-**. A Latin prefix signifying back or again.

**reabsorption** (*re-ab-sorp'-shun*). See *resorption*.

**reaching** (*rēch'-ing*) [ME., *rechen*, to reach]. To attempt to vomit; to retch.

**reacquired** (*re-ak-wīrd'*) [*re*, again; *acquirere*, to acquire]. Acquired a second time. **r. movements**. See *acquired movements*.

**reaction** (*re-ak'-shun*) [*re*, again; *agere*, to act].

1. Counteraction; opposite action; interaction. 2. The response of an organ or part to a stimulus. 3. In chemistry—(a) interaction of two or more substances when brought in contact; (b) the response to a certain test, as *acid reaction, alkaline reaction*, responding to the test for acid and alkali respectively. r., **addition**, the direct union of two or more molecules to form a new molecule. r., **amphigenous**, r., **amphoteric**, a double reaction occurring occasionally in the urine; owing to the presence of substances by which the liquid responds to both the acid and alkaline tests. r., **chameleon**, the peculiar change of color observed in cultures of *Pseudomonas pyocyanea*, from green to brown and back again. r., **chemical**. See *reaction* (3). r., **clump**, the agglutination of bacteria or of leukocytes as the result of the action of certain enzymes, lysins, or toxins. r., **consensual**, reaction which is independent of the will. r. of **degeneration**, the reaction obtained when an electric stimulus is applied to a muscle deprived of its trophic nerve influence. It is characterized by the following conditions: diminution or abolition of the excitability of the muscles for the faradic current, with a temporary increase in excitability for the galvanic current. In the nerves there is a diminution or abolition of both faradic and galvanic excitability. The reactions of the muscle to the galvanic current (the true reaction of degeneration) may be formulated as follows:

AnClC < KaClC } Muscle normal.
AnOC > KaOC
AnClC = KaClC } Muscle in first stage of de-
AnOC = KaOC } generation.
AnClC > KaClC } Muscle in more advanced
AnOC < KaOC } stage of degeneration.
AnClC = anodal closing contraction; AnOC = anodal opening contraction; KaClC = kathodal closing contraction; KaOC = kathodal opening contraction.

See *Pflueger's law of contraction*. r., **electric**, a response in a muscle or a part produced by electric stimulation. r. **of exhaustion**, a variety of reaction to electric excitation seen in states of exhaustion, in which a certain reaction produced by a given current-strength cannot be reproduced without an increase of current-strength. r., **Franklinic, of degeneration** a rare form of reaction of degeneration produced by static electricity and similar to that obtained by the faradic current. r., **hemianopic pupillary**, of Wernicke, a reaction obtained in some cases of hemianopia in which a pencil of light thrown on the blind side of the retina gives rise to no movement in the iris, but thrown upon the normal side, produces contraction of the iris. It indicates that the lesion producing the hemianopia is situated at or anterior to the geniculate bodies. r., **mixed**, a reaction normal in the nerve and altered in the muscle. It is called by Erb the middle form of degenerative reaction. r., **myasthenic**, that in which the normal tetanic contraction of a muscle under faradic stimulation becomes less intense and of shorter duration with every consecutive stimulus and finally ceases, the muscle being exhausted. Syn., *Faradic exhaustibility*. r., **myotonic**, a reaction seen in Thomsen's disease, in which there is quantitative increase in the faradic excitability. See *Erb's waves*. r., **neurotonic**, in electrotherapy, a tonic persistence of contraction, after the current has been broken, upon galvanic and faradic stimulation of the nerve alone, in contradistinction to the myotonic persistency that follows faradic stimulation of the muscle. r., **neutral**, a reaction indicating the absence of both acid and alkaline properties. r.-**period**, the period following the incident shock. r.-**time**, the interval between the application of a stimulus and the beginning of the corresponding motor act. r., **thread**, a peculiar reaction, consisting in the formation of long interlacing threads, produced in certain bacteria, *Bacillus coli communis, Bacillus proteus*, etc., when brought in contact with blood-serum, especially that of the individual from whom the bacteria were obtained.

**reactivate** (*re-ak'-tiv-āt*). To render active again, as by the addition of fresh normal serum to an immune serum which has lost its activity.

**reactivation** (*re-ak-tiv-a'-shun*). The rendering active again a serum which has become inactivated; it is accomplished by the addition of complement.

**reader's cramp.** A spasm of the ocular muscles following prolonged reading.

**readjustment** (*re-ad-just'-ment*). See *advancement*.
**reagent** (*re-a'-jent*) [*re*, again; *agere*, to act]. In chemistry, anything used to produce a reaction; a test. r., **general**, a reagent that indicates the group of substances to which a body belongs, without determining which one of the group it is. r., **special**, a reagent which indicates the presence of an individual substance, and not only the group of which it is a member.
**reagin** (*re'-a-jin*). An antibody, *q. v.*
**real focus.** See *focus* (2). r. **image**. See *image, real*.
**realgar** (*re-al'-gar*) [Ar., *rahj al-ghar*, powder of the mine]. Arsenic disulphide, $As_2S_2$.
**reamer** (*re'-mer*) [ME., *remen*, to widen]. An instrument for gouging out holes.
**reamputation** (*re-am-pū-ta'-shun*) [*re-; amputation*]. An amputation upon a member on which the operation has been performed before.
**reanimate** (*re-an'-im-āt*) [*re*, again; *animare*, to animate]. To revive; to resuscitate; to restore to life, as a person apparently dead.
**Réaumur's thermometer** (*ra-o-mūr'*) [René Antoine Ferchault de Réaumur, French physicist, 1683–1757]. See *thermometer, Réaumur*.
**Récamier's operation** (*ra-kam-e-a'*) [Joseph Claude Anselme Récamier, French gynecologist, 1774–1852]. Curettage of the uterus.
**receiver** (*re-se'-ver*) [*recipire*, to receive]. 1. The vessel receiving the products of distillation. 2. In an air-pump, the jar in which the vacuum is produced.
**receptacula** (*re-sep-tak'-ū-la*). Plural of Latin *receptaculum, q. v.* r. **durae matris**, the sinuses of the dura. r. **lactis**, the ampullae of the galactophorous ducts.
**receptacular** (*re-sep-tak'-ū-lar*) [*receptaculum*, a receptacle]. Pertaining to a receptaculum.
**receptaculum** (*re-sep-tak'-ū-lum*) [L., "a receptacle"]. A receptacle. r. **chyli**, the sac-like beginning of the thoracic duct opposite the last dorsal vertebra. r. **Cotunni**, a triangular space near the middle of the posterior surface of the petrous portion of the temporal bone, at the termination of the aquae-ductus vestibuli. It is formed by the separation of the laminae of the dura mater. r. **Pecqueti**, same as r. *chyli*. r. **seminis**, see *cistern, seminal*.
**receptive** (*re-sep'-tiv*) [*recipere*, to receive]. Having the quality of or capacity for receiving. r. **centers**. In physiology and psycho-physics, nerve-centers to which influences arrive that may excite sensations or some kind of activity not associated with consciousness.
**receptor** (*re-sep'-tor*) [*recipere*, to receive]. 1. A name given by Ehrlich to the atomic lateral chain or haptophorous group, which, existing in each cell in addition to its nucleus, combines with intermediary bodies such as toxins, food molecules, foreign substances. 2. Peripheral nerve endings in the skin and special sense organs. r., **free**, an antibody. r. **of the first order**, one with a single anchoring or haptophore group. r. **of the second order**, one containing a haptophore and a functional, fermentive, or zymophore group. r. **of the third order**, one possessing two haptophore and a zymophore group.
**recess** (*re-ses', re-ses'-sus*) [*recessus*, a recess]. A fossa, ventricle, or ampulla; an anatomical depression. r. **acetabuli**, the cotyloid cavity. r., **ampullar**, the ampulla of the semicircular canal of the inner ear. r., **auditory**, a depression of the ectoderm on each side of the cephalic extremity of the embryo, constituting the first foreshadowing of the internal ear. It is the precursor of the auditory vesicle. r. **aulae**. See r., **aulic**. r., **aulic**, Bergmann's name for a triangular recess between the columns of the fornix immediately dorsad of the anterior commissure. Syn., r., *triangular; recessus aulae; foveola triangularis seriata*. r. **chiasmal**, a pit in front of the infundibulum bounded by the optic chiasm and the cinereous lamina. r. **chiasmatica**. See *lamina cinerea*. r. **cochlear**, an elliptic pit below the oval window of the vestibule, forming part of the cochlea. r. **cochlearis**, a shallow depression between the diverging portions of the crista vestibuli. r. **conarii**, the cavity at the base of the pineal body, situated between the supracommissure and postcommissure. r., **duodenojejunal**, a pouch of the mesentery on the right side of the jejunum and near its union with the duodenum. Syn., *duodenojejunal fossa*. r. **ellipticus**. See *fovea hemielliptica*. r., **epiphyseal**, the preferred name for pineal recess. r., **epitympanic**,

synonym of *attic*. r. hemielliptĭcus. See *fovea hemielliptica*. r. hemisphæricus, a tiny perforated cavity in the inner wall of the vestibule, anterior to the crista vestibuli; it transmits the branches of the auditory nerve. r. hepaticorenal, that formed by the hepaticocolic ligament. r. ileocecal, the ileocecal fossa. r. incisive, a depression on the nasal septum immediately above the anterior palatine canal. r. infrapineal. See *recessus conarii*. r. infundibuliform. See *fossa of Rosenmueller*. r. intercruralis, the interpeduncular space. r. jugular, see *jugular fossa*. r. abyrinthi, a cavity formed in the base of the fetal skull, developing into the primitive auditory vesicle, and finally into the internal ear. r., laryngopharyngeal, the lower pyramidal part of the pharynx from which the esophagus and larynx open. r. lateral, the lateral extension of the fourth ventricle in the angle between the cerebellum and the oblongata. r. naso-palatinus, the nasal orifice of the naso-palatine canal. r. occipitalis, the posterior horn of the lateral ventricle. r. opticus, the conical depression at the beginning of each optic nerve where it leaves the chiasma, the remnant of the cavity of the stalk of the optic vesicle (Wilder). See *lamina cinerea*. r., palatal. See *fossa, supratonsillar*. r. peritonæi, the pocket-like processes formed by the peritoneum. r. pharyngeus, a pouch-like process of the mucosa of the pharynx situated below the opening of the Eustachian tube; Rosenmueller's fossa. r. pineal. See *recessus conarii*. r., postpontile, the foramen cæcum of the brain. r. sacci lachrymalis, an inconstant anterior pouch of the lacrimal sac. r., salpingopharyngeal, an inconstant pharyngeal diverticulum adjacent to the pharyngeal opening of the Eustachian tube. r. sphæricus. See *fovea hemisphærica*. r. spheno-ethmoidalis, a small depression or groove between the sphenoid bone and the superior turbinated bones. r., Stensonian. See *r., incisive*. r. subsigmoid, the pouch made in the peritoneum by the mesentericomesocolic ligament. r. sulciform. See *fossa, sulciform*. r. superior sacci omenti, the cavity of the lesser omentum. r. suprapinealis, a space between the habenal commissures and the post-commissure. r. tecti, a space beneath the valvula and velum of the cerebellum. r., triangulare. See *r., aulic*. r. utriculi, a recess at the upper part of the utricle of the inner ear. r. venosus. See *fossa, Landzerti's*. r. vesicæ urinariæ, the lower portion of the urinary bladder.

recession (*re-sesh'-un*) [*recedere*, to recede]. The gradual withdrawal of a part from its normal position, as the recession of the gums from the necks of the teeth.

recidivation (*re-sid-iv-a'-shun*) [*recidivus*, a falling back]. The relapsing of a disease. In criminology, a relapsing into crime.

recidivist (*re-sid'-iv-ist*) [*recidivus*, falling back]. 1. A patient who returns to a hospital for treatment, especially an insane person who so returns. 2. In criminology, a confirmed or relapsed criminal; (*a*) one who for the most part has no mental or bodily signs of degeneration, caused by bad bringing up, society, poverty, sexual disorders, and who makes crime a trade or a vengeance; (*b*) one with inborn criminal inclinations and a positive tendency to insanity or epilepsy; and (*c*) one whose antecedents and environment lead him to crime by blunting his sense of honor and morality. The latter classes are inclined to coarseness, boldness, resistance, and wilful spoiling of their clothes; but are not legally regarded as insane.

recidivity (*re-sid-iv'-it-e*) [*recidivus*, a falling back]. Tendency to return or to relapse.

recipe (*res'-ip-e*). 1. The imperative of *recipere*, used as the heading of a physician's prescription, and signifying take. Symbol ℞. 2. Also the prescription itself.

recipiomotor (*re-sip-e-o-mo'-tor*) [*recipere*, to receive; *motor*]. Receiving motor impulses.

reciprocal (*re-sip'-ro-kal*) [*reciprocus*, to receive]. In psychology, applied to those instances in which there is both agency and percipience at each end of the telepathic chain. r. proportions, law of, two elements combining with a third do so in proportions that are simple multiples or simple fractions of those in which they combine with each other. r. reception, a mode of articulation in which the articular surface is convex on one side and concave on the other.

Recklinghausen's canals (*rek'-ling-how-sen*) [Friedrich Daniel von *Recklinghausen*, German pathologist, 1833–1910]. Minute channels supposed to exist in all connective tissue, which are directly continuous with the lymphatic vessels, and hence may be said to form their origin. R.'s disease. 1. Neuro-fibromatosis. 2. Hemachromatosis; an affection characterized by bronzing of the skin, hypertrophic cirrhosis of the liver, enlargement of the spleen, and, in later stages, diabetes from pancreatic sclerosis.

reclinatio (*rek-lin-a'-she-o*). See *reclination*. r. palpebrarum, ectropion.

reclination (*rek-lin-a'-shun*) [*reclinare*, to recline]. 1. An old operation for cataract, called also "couching," in which the lens was pushed back into the vitreous chamber. 2. The act of lying down.

recoil-atom. See *rest-atom*.

Reclus' disease (*rek-loo'*) [Paul *Reclus*, French surgeon, 1847– ]. Cystic disease of the mammary gland; chronic cystic or interstitial mastitis.

recomposition (*re-kom-po-zish'-un*) [*recomponere*, to reunite]. Reunion of parts or constituents after sorbed after fulfilling its function.

recompression (*re-kom-presh'-un*). Subjection again to the action of compressed air; said of persons working in air under high pressure.

reconstituent (*re-kon-stit'-u-ent*) [*re-; constituere*, to constitute]. A medicine which promotes continuous repair of tissue-waste or makes compensation for its loss.

reconstitution (*re-kon-stit-u'-shun*) [*re*, again; *constituere*, to constitute]. Continuous repair of decaying tissue, or restoration to compensate loss by tissue-waste.

recrement (*rek'-re-ment*) [*recrementum*, from *re*, again; *crescere*, to grow]. A secretion that is reabsorbed after fulfilling its function.

recrementitious (*rek-re-men-tish'-us*) [*recrement*]. Pertaining to or of the nature of a recrement.

recrudescence (*re-kroo-des'-ens*) [*re-; crudescere*, to become raw]. An increase in the symptoms of a disease after a remission or a short intermission.

rectal (*rek'-tal*). Pertaining to the rectum or performed through the rectum. r. alimentation, see *alimentation, rectal*. r. crises, attacks of rectal pain and tenesmus occurring in locomotor ataxia.

rectalgia (*rek-tal'-je-ah*) [*rectum*; ἄλγος, pain]. Pain in the rectum; proctalgia.

rectectomy (*rek-tek'-to-me*). See *proctectomy*.

rectification (*rek-tif-ik-a'-shun*) [*rectus*, straight; *facere*, to make]. 1. A straightening, as *rectification* of a crooked limb. 2. The redistillation of weak spirit in order to strengthen it.

rectified spirit. Alcohol containing 94.9 % of ethyl-alcohol.

rectify (*rek'-tif-i*) [see *rectification*]. To make right or straight; to refine.

rectitis (*rek-ti'-tis*). See *proctitis*.

recto- (*rek-to-*) [*rectum*]. A prefix meaning relating to the rectum.

rectoabdominal (*rek-to-ab-dom'-in-al*). Relating to the rectum and the abdomen.

rectocele (*rek'-to-sēl*) [*recto-*; κήλη, hernia]. Prolapse of the rectum into the vagina.

rectococcygeal (*rek-to-kok-sij'-e-al*) [*recto-*; *coccygeal*]. Pertaining to the rectum and the coccyx.

rectococcypexia, rectococcypexy (*rek-to-koks-e-peks'-e-ah, rek-to-koks-e-peks'-e*) [*recto-*; κόκκυξ, coccyx; πῆξις, a fastening]. Suturing of the rectum to the coccyx.

rectocolitis (*rek-to-kol-i'-tis*). Inflammation of the mucosa of the rectum and colon combined.

rectocolonic (*rek-to-ko-lon'-ik*) [*recto-*; *colon*]. Pertaining to the rectum and the colon.

rectocystotomy (*rek-to-sist-ot'-o-me*) [*recto-*; *cystotomy*]. Incision of the bladder through the rectum.

rectogenital (*rek-to-jen'-it-al*). Pertaining to the rectum and the genital organs.

rectolabial (*rek-to-la'-be-al*). Relating to the rectum and the labia pudendi.

rectopexia, rectopexy (*rek-to-peks'-e-ah, rek-to-peks'-e*) [*recto-*; πῆξις, a fastening]. Surgical fixation of a prolapsed rectum.

rectophobia (*rek-to-fo'-be-ah*) [*rectum*, φόβος, fear]. 1. A presentiment or sense of impending ill experienced by patients having rectal disease. 2. A morbid dread of rectal disease.

rectoplasty (*rek'-to-plas-te*) [*recto-*; πλάσσειν, to form]. See *proctoplasty*.

rectoromanoscope (*rek-to-ro-man'-o-skōp*) [*recto-*; *S romanum*, the sigmoid flexure; σκοπεῖν, to view]. A speculum used in examining the rectum and the sigmoid flexure.

**rectoscope** (*rek'-to-skōp*) [*recto-*; σκοπεῖν, to inspect]. A rectal speculum.

**rectoscopy** (*rek-tos'-ko-pe*) [*rectum*; σκοπεῖν, to inspect]. An examination of the rectum.

**rectosigmoidoscopy** (*rek-to-sig-moy-dos'-ko-pe*) [*recto-*; *sigmoid*; σκοπεῖν, to inspect]. Ocular inspection of the rectum and sigmoid flexure of the colon with the aid of special instruments.

**rectostenosis** (*rek-to-sten-o'-sis*) [*recto-*; *stenosis*]. Stenosis of the rectum.

**rectostomy** (*rek-tos'-to-me*). See *proctostomy*.

**rectotome** (*rek'-to-tōm*) [*recto-*; τομή, a cutting]. A cutting instrument used in rectotomy.

**rectotomy** (*rek-tot'-o-me*). See *proctotomy*.

**rectourethral** (*rek-to-u-re'-thral*) [*recto-*; *urethra*]. Pertaining to the rectum and the urethra.

**rectouterine** (*rek-to-u'-ter-in*) [*recto-*; *uterus*]. Pertaining to the rectum and the uterus.

**rectovaginal** (*rek-to-vaj'-in-al*) [*recto-*; *vagina*]. Pertaining to the rectum and the vagina. r. **fistula**, an opening between the vagina and the rectum.

**rectovesical** (*rek-to-ves'-ik-al*) [*recto-*; *vesica*, the bladder]. Pertaining to the rectum and the bladder.

**rectum** (*rek'-tum*) [*rectus*, straight]. The lower part of the large intestine, extending from the sigmoid flexure to the anus. It begins opposite the left sacroiliac synchondrosis, passes obliquely downward to the middle of the sacrum, and thence descends in the median line to terminate at the anus.

**rectus** (*rek'-tus*) [L.]. Straight; applied to anything having a straight course. r. **muscle**, see under *muscle*.

**recumbent** (*re-kum'-bent*) [*recumbere*, to recline]. Leaning back; reclining.

**recuperate** (*re-kū'-per-āt*) [*recuperare*, to regain]. To regain strength or health.

**recuperation** (*re-kū-per-a'-shun*) [*recuperatio*; *recuperare*, to recover]. Convalescence. Restoration to health.

**recuperative** (*re-kū'-per-a-tiv*) [*recuperativus*]. Pertaining to, or tending to, recovery of health or strength.

**recurrence** (*re-kur'-ens*) [*recurrere*, to run back]. The return, as of a disease.

**recurrens** (*re-kur'-enz*) [see *recurrent*]. Relapsing fever.

**recurrent** (*re-kur'-ent*) [*re*, back; *currere*, to run]. 1. Returning. 2. In anatomy, turning back in its course, as recurrent laryngeal nerve. r. **fever**, relapsing fever.

**recurring** (*re-kur'-ing*) [*recurrere*, to run back]. Returning; occurring again. r. **disease**, one that returns or relapses. r. **utterance**, the involuntary utterance of certain words, usually a symptom of motor aphasia.

**recurvation** (*re-kur-va'-shun*) [*recurvatus*, curved back]. The act or process of recurving or of bending backward.

**red. in pulv.** Abbreviation of the Latin *redactus in pulverem*, reduced to powder.

**red** [AS., *reád*]. The least refrangible of the spectral colors; of a color resembling that of the blood. r. **bark**, see *cinchona*. r.**-blindness**, see under *blindness*, *color-*. r. **blister**, unguentum hydrargyri iodidi rubri. r. **cerate**, calamine ointment. r. **chalk**, red-dle; hydrated aluminum silicate containing a quantity of ferric oxide. r., **Chinese**, mercuric sulphide. r. **corpuscles**, see *corpuscles*. R.**-Cross Society**, an international society founded by Clara Barton, and intended to act upon the principles laid down in the Geneva convention of 1864. It furnishes nurses and supplies for service in wars, and relieves the distress, needs, or wants of those who suffer in floods, pestilences, and public calamities. r. **gum**. 1, A red, papular eruption of infants. Syn., *strophulus*. 2. Eucalyptus. r. **lead**, red lead oxide, formerly used in plasters. r. **lotion**. See *r. wash*. r. **mixture**, a combination of rock salt, potassium nitrate, sodium carbonate and molasses, used for injecting bodies for the dissecting room. It imparts a beautiful red color to the muscles. It is also called *Horner's mixture*. r. **nucleus**, see *nucleus*, *tegmental*. r., **oil**, oleic acid as a by-product in the manufacture of stearic acid candles. r. **pepper**, see *capsicum*. r. **plague**, a form of the plague characterized by a red spot, boil, or bubo. r. **precipitate**, see *mercury oxide, red*. r. **softening**, a form of acute softening of the brain or spinal cord, characterized by a red, punctiform appearance due to the presence of blood. r. **tartar**, argols, impure cream of tartar. r. **wash**, lotio rubra, a solution of zinc sulphate in compound tincture of lavender and water. r.**-water**, a common name for hemoglobinuria in cattle.

**reddle** (*red'-del*). Red chalk.

**redia** (*re'-de-ah*) [*Redi*, an Italian naturalist, 1626–1698]. *pl.*, *redia*. In biology; the larval stage of a trematode, which results from the development of a parthenogenetic egg of the first larval stage (sporocyst). The rediæ have at the anterior extremity of their body a sucker-like formation, a pharynx, a simple intestinal tube, and a birth-aperture. The first rediæ give rise to a second and these to a third parthenogenetic generation and these finally to larvæ called *cercariæ*, *q. v.*

**redintegration** (*red-in-te-gra'-shun*) [*redintegrare*, to renew]. The complete restitution of a part that has been injured or destroyed.

**redresser** (*re-dres'-er*) [Fr. *re*, *dresseur*]. An instrument used to replace a displaced organ or part.

**redressment** (*re-dres'-ment*) [Fr., *redressement*]. Correction of a deformity or replacement of a dislocated part.

**redressement forcé** (*ra-dres-mon(g) for-sa*) [Fr.]. The forcible correction of a deformity or restoration of a displaced part.

**reduce** (*re-dūs'*) [*re*, back; *ducere*, to lead]. 1. To restore a part to its normal relations, as to *reduce* a hernia or fracture. 2. In chemistry, to bring back to the metallic form; to deprive of oxygen.

**reduced** (*re-dūsd'*) [*reduce*]. 1. Restored to the proper place. 2. In chemistry, brought back into the metallic form, as *reduced* iron. 3. Diminished in size. r. **eye**, see *eye, reduced, of Donders*. r. **hematin**, the product of the production of hematin in alkaline solution. r. **hemoglobin**, the result of deoxidation of oxyhemoglobin. r. **iron**, iron by hydrogen; ferrum reductum.

**reducible** (*re-dū'-si-bl*) [*reduce*]. Capable of being reduced.

**reducin** (*re-dū'-sin*). A leukomaine, $C_{11}H_{24}N_6O_9$, found in urine.

**reduction** (*re-duk'-shun*) [*reduce*]. The act of reducing. r. **en bloc**; r. **en masse**, the reduction of a strangulated hernia still surrounded by its sac, thus failing to relieve the strangulation.

**reductor** (*re-dukt'-or*) [L.]. 1. An instrument for effecting reduction. 2. A retractor muscle.

**reduplicated** (*re-dū'-plik-a-ted*) [see *reduplication*]. Doubled; as *reduplicated* heart-sounds. See *reduplication*.

**reduplication** (*re-dū-plik-a'-shun*) [*re*, again; *duplicare*, to double]. A doubling. r. **of the heart-sounds**, a doubling of either the first or the second sound of the heart.

**redux** (*re'-duks*) [L.]. Returning. r., **crepitus**, the small mucous rales heard in the early stage of the resolution of lobar pneumonia.

**reed** (*rēd*). See *Abomasum*.

**reef-knot** (*rēf-not*). A sailor's knot used in the ligature of arteries. It is not likely to slip or loosen. See *knot*.

**reel, cerebellar**. The peculiar staggering gait in diseases of the cerebellum, particularly in tumor.

**Rees' test for albumin** [George Owen Rees, English physician, 1813–1889]. Small amounts of albumin are precipitated by an alcoholic solution of tannic acid.

**reevolution** (*re-ev-ol-ū'-shun*). Hughlings Jackson's term for a symptom following an epileptic attack, which consists of three stages: (1) Suspension of power to understand speech (word-deafness); (2) perception of words and echolalia without comprehension; (3) return to conscious perception of speech with continued lack of comprehension.

**refine** (*re-fīn'*) [*re*, again; *finire*, to finish]. To purify; to separate a substance from foreign matter.

**reflected** (*re-flek'-ted*) [*reflectere*, to bend backward]. Cast or thrown back. In anatomy, turned back upon itself; r. **light**, see *light*.

**reflection** (*re-flek'-shun*) [*reflex*]. 1. A bending or turning back; specifically, the turning back of a ray of light from a surface upon which it impinges without penetrating. 2. When used in speaking of membranes such as the peritoneum it refers to the folds which it makes in passing from the wall of the cavity over an organ and back again to the wall which bounds such cavity.

**reflector** (*re-flek'-tor*). A polished surface by which light is reflected.

**reflex** (*re'-fleks*) [*re*, back; *flectere*, to bend]. 1. Anything reflected or thrown back. 2. A reflex act.

r., abdominal, contraction of the muscles about the umbilicus, on sharp, sudden stroking of the abdominal wall from the margin of the ribs downward; it shows integrity of the spinal cord from the eighth to the twelfth dorsal nerve. r. act, an act following immediately upon a stimulus without the intervention of the will. r., anal, a contraction of the sphincter ani on anal irritation. r., ankle, clonic contractions of the tendo Achillis, dependent upon alternate contraction and relaxation of the anterior, tibial, and calf-muscles; obtained by sudden complete flexion of the foot, by pressing the hand against the sole. Syn., ankle-clonus. r. arc, the mechanism necessary for a reflex action; it consists of an afferent or sensory nerve; a nerve-center to change this sensory impulse into a motor one; and an efferent or motor nerve to carry a motor impulse to the muscle or group of muscles. r., biceps, contraction of the biceps muscle on tapping the tendon of the biceps; a normal reflex, but increased by the causes which increase the knee-jerk. r., bone, a reflex muscular contraction evoked by blows over a bone. r., bulbocavernous, see r., virile. r. center, the nerve center in a reflex arc. r., chin, see r., jaw. r., ciliospinal, pupillary dilatation on irritation of the skin of the neck. r., contralateral, a flexion or extension of the leg on one side when the other leg is flexed passively; it occurs in children in meningitis. r., corneal. Same as r., eyelid-closure. r., cranial, any brain reflex. r., cremasteric, retraction of the testicle on the corresponding side, obtained on stimulation of the skin on the front and inner aspect of the thigh; it shows integrity of the cord between the first and second pairs of lumbar nerves. r., crossed, one in which stimulation of one side of the body produces a reflex on the opposite side. r., deep, reflexes developed by percussion of tendons or bones. r., dorsal, same as r., erector spinæ. r., elbow-jerk, same as r., biceps. r., epigastric, dimpling in the epigastrium, due to contraction of the highest fibers of the rectus abdominis muscle, on stimulation of the skin in the fifth or sixth intercostal space near the axilla; it shows integrity of the cord from the fourth to the seventh dorsal nerves. r., erector spinæ, local contraction of erector spinæ muscle on stimulation of the skin along the border; it shows integrity of the dorsal region of the cord. r., eyelid-closure, closure of the lid on irritation of the conjunctiva. r., faucial, vomiting on irritation of the fauces. r., femoral, plantar flexion of the first three toes and of the foot, and extension of the knee-joint upon irritation of the skin on the upper anterior aspect of the thigh; it occurs in disease of the spinal cord, e. g., in some cases of transverse myelitis. r., front-tap, see r., tendo Achillis. r., gluteal, contraction of the glutei upon firm, sudden stroking of the skin over the buttock; it shows integrity of the cord at the fourth and fifth lumbar nerves. r., guttural, a reflex observed in cases of diseased genitalia in women, in which the patient is desirous of spitting but cannot. r., Haab's pupil-, see Haab's pupil-reflex. r., interscapular, see r., scapular. r., iris-contraction, see r., pupillary. r., jaw, clonic movements of the inferior maxilla, obtained on a downward stroke with a hammer on the lower jaw hanging passively or gently supported by the hand; it is rarely present in health; increased in sclerosis of the lateral columns of cord. Syn., jaw-clonus; jaw-jerk. r., knee, contraction of the quadriceps muscle, the foot being jerked forward on striking the patellar tendon after rendering it tense by flexing the knee at a right angle; it is normal in health; absent in locomotor ataxia, destructive lesions of the lower part of the cord, multiple neuritis, affections of the anterior gray cornua, infantile paralysis, meningitis, diphtheric paralysis, atrophic palsy, pseudohypertrophic muscular paralysis, diabetes, etc.; increased in diseases of the pyramidal tracts, in spinal irritability, tumors of the brain, cerebrospinal sclerosis, lateral sclerosis, after epileptic seizures or unilateral convulsions. Syn., knee-jerk. r., laryngeal, coughing, produced by irritation of the fauces, larynx, etc. r., lumbar, same as r., erector spinæ. r. multiplicator, an apparatus for the registration of tendon-reflexes. r., nasal, sneezing, on irritation of the Schneiderian membrane. r., obliquus, contraction of the fibers of the obliquus externus in females (corresponds to cremasteric in males, although it can also be caused in males) on irritation of skin below Poupart's ligament. r., ophthalmic, see r., supraorbital. r., palatal, swallowing produced by irritation of the palate. r., palmar, contraction of the digital flexors upon tickling the palm; it shows that the cervical region of the cord is normal. r., patellar. Same as r., knee. r., patellar, paradoxical, contraction of the adductor but not of the quadriceps muscle on percussing the patellar tendon, with the patient in the dorsal decubitus. If the patient is in the sitting posture, the normal reflex is elicited; it shows spinal concussion. r.s, pathic, movements resulting from stimulation of a sensory nerve. r., penile, see r., virile. r., periosteal, sharp contractions of the muscles upon tapping the bones of the forearm or leg; it indicates disease of the lateral columns of the spinal cord. r., peroneal, reflex movements caused by a stroke on the peroneus muscles when tense or when the foot is turned inward. r., pharyngeal, swallowing produced by irritation of the pharynx. r., plantar, contraction of the toes upon stroking the sole of the foot. r., platysmal, dilatation of the pupil upon pinching the platysma myoides muscle. r., pupillary, contraction of the iris on exposure of the retina to light; it is absent in basal meningitis, etc. r., pupillary, paradoxical, dilatation of the pupil on stimulation of the retina by light. r., rectal, the reflex by which the aggregation of feces in the rectum induces defecation. r., scapular, contraction of the scapular muscles on irritation of the interscapular region; it shows integrity of the cord between the upper two or three dorsal and lower two or three cervical nerves. r., skin, see r., platysma. r., sole, same as r., plantar. r., spinal, those reflex actions emanating from centers in the spinal cord. r., superficial, such as are developed from irritation of the skin. r., supraorbital, a slight contraction of the orbicularis palpebrarum muscle on striking the supraorbital nerve or one of its branches a slight blow. r., tendo Achillis, reflex contraction of the gastrocnemius muscle, produced by striking the muscles on the anterior part of leg while in extension, the foot being extended by the hand upon the sole; it is considered a delicate test of heightened spinal irritability. Syn., front-tap contraction. r., tendon, muscle reflex action; myostatic reaction; deep reflex. r., toe, involuntary flexion of the foot, then flexion of the leg, and, lastly, flexion of the thigh on the pelvis upon strong flexion of the great toe; it is seen in cases in which the knee-jerk and other tendon-reflexes are strongly developed. r., triceps, extension of the forearm on tapping the tendon of the triceps muscle. r., virile, retraction of the bulbocavernous portion upon sharp percussion of the back of the penis, the sheath having been made tense; it occurs in health. r., visceral, one of a group of reflexes, as, blinking, from touching the cornea; penile, erection on slight contact or produced by passing a catheter; rectal, constriction of the bowel following introduction of a foreign body, as a suppository; sneezing, that produced by a draft of cold air or a brilliant light; vesical, contraction of the bladder following irritation of the urethral orifice, e. g., incontinence of urine in children, by reason of a long prepuce; vomiting, from tickling the fauces. r., wrist, a series of jerking movements of the hand produced by pressing the hand backward to extreme extension; observed in the late rigidity of hemiplegia. Syn., wrist-clonus.

reflexa (re-fleks'-ah) [L.]. See decidua reflexa.
reflexio (re-fleks'-e-o) [L.]. See reflection. r. palpebrarum, see ectropion.
reflexogenic (re-fleks-o-jen'-ik) [reflexus, reflex; γεννᾶν, to produce]. Causing or increasing a tendency to reflex action; producing reflexes.
reflexograph (re-fleks'-o-graf) [reflex; γράφειν, to write]. An instrument for measuring, timing, and charting automatically knee-jerks and other tendon-reflexes.
reflexometer (re-fleks-om'-et-er) [reflexus, reflex; μέτρον, a measure]. An instrument used to measure the force required to produce myotatic movement.
reflexophile (re-fleks'-o-fil) [reflex; φιλεῖν, to love]. Attended by reflex activity.
reflux (re'-fluks) [re, back; fluere, to flow]. A return flow.
refoulement (ra-fool'-mon(g)) [Fr.]. A forcing back. r. du sacrum en arrière, a backward inclination of the sacrum that makes one of the changes in the pelvis of girls at puberty.
refract (re-frakt')-[re, back; frangere, to break]. 1. To bend. 2. To change direction by refraction. 3. To estimate the degree of ametropia, heterophoria, and heterotropia present in an eye.
refracta dosi (re-frak'-tah do'-si) [L.]. In broken or divided doses.

**refraction** (*re-frak'-shun*) [*refract*]. 1. The act of refracting or bending back. 2. The deviation of a ray of light from a straight line in passing obliquely from one transparent medium to another of different density. 3. The state or refractive power, especially of the eye; the ametropia, emmetropia, or muscle-imbalance present. 4. The act or process of correcting errors of ocular refraction. **r., angle of,** the angle formed by a refracted ray of light with the perpendicular at the point of refraction. **r.**, atomic, the product of the refractive index of the constituent elements of a compound and their atomic weights. **r., coefficient of,** the quotient of the sine of the angle of refraction into the sine of the angle of incidence. **r., double,** the power possessed by certain substances, as Iceland spar, of dividing a ray of light and thus producing a double image of an object. **r., dynamic,** the static refraction of the eye, plus that secured by the action of the accommodative apparatus. **r. equivalent, r., specific.** See *refractive power, specific*. **r., errors of,** departures from the power of producing a normal or well-defined image upon the retina, because of ametropia. **r. of the eye,** the influence of the ocular media upon a cone or beam of light, whereby a normal or emmetropic eye produces a proper image of the object upon the retina. **r.-image,** see *image*. **r., index of,** the refractive power of any substance as compared with air. It is the quotient of the angle of incidence divided by the angle of refraction of a ray passing through the substance. **r., molecular,** the molecular weight of a compound. The molecular refraction of a liquid carbon compound is equal to the sum of the atomic refractions. **r., static,** that of the eye when accommodation is at rest.

**refractionist** (*re-frak'-shun-ist*) [*refrangere*, to break up]. One who corrects errors of ocular refraction, or ametropia.

**refractive** (*re-frak'-tiv*) [*refract*]. Refracting; capable of refracting or bending back; pertaining to refraction. **r., equivalent.** See *refractive power, specific*. **r. index,** same as refraction, index of, *q. v.* **r. power,** the measure of influence which a transparent body exercises on the light which passes through it. **r. power, specific,** an almost constant quantity representative of the relation between the coefficient of refraction, the temperature, and the specific gravity of a given fluid.

**refractivity** (*re-frak-tiv'-it-e*) [*refract*]. Power of refraction; ability to refract.

**refractometer** (*re-frak-tom'-et-er*) [*refract*; μέτρον, a measure]. 1. An instrument for measuring the refraction of the eye. 2. An instrument for the determination of the refractive indexes of liquids.

**refractory** (*re-frak'-tor-e*) [*refractarius*, stubborn]. 1. Resisting treatment. 2. Resisting the action of heat; slow to melt.

**refracture** (*re-frak'-tūr*) [*re*, back; *frangere*, to break]. The breaking again of fractured bones that have joined by faulty or improper union.

**refrangibility** (*re-fran-jib-il'-it-e*) [see *refract*]. Capability of undergoing refraction.

**refresh** (*re-fresh'*) [*re*, again; *friscus*, new]. In surgery, to give to an old lesion the character of a fresh wound.

**refrigerant** (*re-frij'-er-ant*) [see *refrigeration*]. 1. Cooling; lessening fever. 2. A medicine or agent having cooling properties or lowering body-temperature.

**refrigeration** (*re-frij-er-a'-shun*) [*re*, again; *frigus*, cold]. The act of lowering the temperature of a body by conducting away its heat to a surrounding cooler substance.

**refringent** (*re-frin'-jent*) [*refringere*, to break]. See *refractive*.

**refuse** (*ref'-ūs*). Waste from manufacturing or other establishments, and all inorganic waste.

**refusion** (*re-fu'-zjun*) [*refusio*, an overflowing]. The act of withdrawing blood from the vessels, exposing it to the oxygen of the air, and passing it back again.

**regenerate** (*re-jen'-er-āt*) [*regenerare*, to generate again]. To generate anew; to reproduce.

**regeneration** (*re-jen-er-a'-shun*) [*re*, again; *generare*, to beget]. 1. The new growth or repair of structures or tissues lost by disease or by injury. 2. In chemistry, the process of obtaining from the byproducts or end-products of an operation a substance which was employed in the earlier part of the operation. **r., cell-processes in,** the process consist in either simple hypertrophy, (increase in the size of existing cells), or numerical hypertrophy, hyperplasia (increase in the number of cells in the tissue). **r. after inflammation,** repair by multiplication of the tissue cells. **r. after necrosis,** repair by absorption of dead tissue and its replacement by newly formed normal tissue. **r., pathological,** the renewal of destroyed tissue by a pathological rather than a physiological process.

**regenerin** (*re-jen'-er-in*). Trade name of a lecithin and iron preparation; used as a tonic.

**regime** (*ra-zjēm*) [Fr.]. See *regimen*.

**regimen** (*rej'-im-en*) [*regere*, to rule]. The regulated use of food and the sanitary arrangement of surroundings to suit existing conditions of health or disease.

**regio** (*re'-je-o*). Latin for *region*.

**region** (*re'-jun*) [*regio*, a region]. One of the divisions of the body possessing either natural or arbitrary boundaries. **r.s of the abdomen,** see under *abdomen*. **r., acromial,** the parts near the acromion. **r., anal,** pertaining to the anus. **r., aulic,** the area about the aula. **r., auricular,** the parts near the ear. **r., axillary,** a region upon the lateral aspect of the thorax, extending from the axilla to a line drawn from the lower border of the mammary region to that of the scapular region. **r., basilar,** the region at the base of the skull. **r., brachial,** the region of the arms. **r. of Broca,** the third left frontal convolution of the brain. **r., central gray,** the medullary substance of the cerebellar hemispheres. **r., cervical,** the parts around the neck. **r., ciliary.** 1. The zone of the eyeball in which the ciliary body is situated. 2. The part of the eyelid containing the cilia. **r., clavicular,** the area about the clavicle. **r., costal,** the lateral chest area. **r., diaphragmatic,** the region of the diaphragm. **r., epicranial,** the region above the diaphragm. **r., epigastric,** the region over the stomach bounded laterally by two vertical lines passing through the middle of Poupart's ligament; above by a horizontal line touching the lower margin of the sternum, and below by a horizontal line touching the lowest part of the thorax, and include the pyloric end and middle of stomach, the left lobe of the liver, the lobulus Spigelii, the pancreas, the duodenum, parts of the kidneys, the aorta, vena cava, thoracic duct, semilunar ganglia. **r., femoral,** the parts about the femur. **r., fibular,** the parts about the fibula. **r., gastric,** the region over the stomach. **r., gluteal,** the region of the gluteus muscle. **r., gustatory,** the tips, margins, and root of the tongue in the neighborhood of the circumvallate papillæ; also the lateral parts of the soft palate and the anterior surface of the anterior pillars of the fauces. **r., humeral,** the parts about the humerus. **r., hyo-mental,** see **r.,, supra-hyoid. r., hyo-sternal,** see **r., infra-hyoid. r., hypochondriac,** the region that joins the epigastric region laterally; The right hypochondriac region includes the surface of the abdomen covering the right lobe of the liver, the gall-bladder, the hepatic flexure of the colon, and part of the right kidney; the left that covering the spleen, the splenic end of the stomach, the extremity of the pancreas, the splenic flexure of the colon, and part of the left kidney. **r., hypogastric,** that part of the abdominal surface between a horizontal line drawn through the anterior superior crests of the ilia above and on either side by vertical lines drawn through the center of Poupart's ligament. It overlies the small intestines, the bladder in children and in adults when distended, the uterus during pregnancy, sometimes the vermiform appendix, the cecum, and the sigmoid flexure of the colon. **r., iliac,** the region of the ilium; see also **r., inguinal. r., ilioinguinal,** the iliac region and the groin conjointly. **r., inferior sternal,** the space corresponding to the part of the sternum below the lower margin of the third costal cartilages. **r., infra-axillary,** the space between the anterior and posterior axillary lines. **r., infraclavicular,** the area circumscribed superiorly by the lower border of the clavicle, inferiorly by the lower border of the third rib, on one side by a line extending from the acromion to the pubic spine, and on the other side by the edge of the sternum. **r., infra-hyoid,** the space below the hyoid bone, between the sterno-cleido-mastoidei and the sternum. **r., inframammary,** the space between a line drawn along the upper border of the xiphoid cartilage and the margin of the false ribs, and between the middle line of the xiphoid cartilage and a vertical line passing through the pubic spine. **r., infrascapular,** the region on either side of the vertebral column below a horizontal line drawn through the inferior angle of each scapula. It is called also the subscapular region. r., **infraspinous**, that included between the spine of the scapula and a line passing through the angle of the scapula. r., **inguinal**, r., **iliac**, the right and left inguinal or iliac regions are two of the nine abdominal regions. The right includes the abdominal surface covering the cecum and the cecal appendix, the ureter, and the spermatic vessels; the left that covering the sigmoid flexure of the colon, the ureter, and the spermatic vessels. r., **interscapular**, the space between the scapulæ. r., **ischio-rectal**, the region corresponding to the posterior part of the pelvic outlet, between the ischium and the rectum. r., **jugal**, the space over the zygoma. r., **laryngo-tracheal**, the parts about the larynx and the trachea. r., **lenticulo-striate**, the anterior parts of the lenticular and caudate nuclei, and the intervening portion of the internal capsule. r., **lenticulo-thalamic**, the posterior part of the lenticular nucleus, the optic thalamus, and the part of the internal capsule which intervenes. r., **lingual**, the region of the tongue. r., **lumbar**, the surface of the abdomen between a curved line drawn parallel with the ca. tilage of each ninth rib above and a curved line parallel with the iliac crests below, and a vertical line through the center of Poupart's ligament anteriorly, and the lumbar vertebræ posteriorly. r., **mammary**, the space on the anterior surface of the chest between a line drawn through the lower border of the third rib, and one drawn through the upper border of the xiphoid cartilage. r., **maxillary**, the parts about the jaws. r., **mesogastric**, the umbilical and the right and left lumbar regions together. r., **middle cervical**, the area between the lower jaw, the sternum, and the anterior edges of the sterno-cleido-mastoidei. r., **motor**, see *area, motor*. r., **mylohyoid**, see *triangle, mylo-hyoid*. r., **nasal**, the parts around the nose. r., **olfactory**, the region of the nasal mucous membrane including the ramifications of the olfactory nerve. r., **orbital**, the region of the orbits. r., **palatal**, the parts about the palate. r., **palpebral**, the region of the eyebrows. r., **parasternal**, the space between the mid-axillary line and the edge of the sternum. r., **parotid**, see *r., retromaxillary*. r., **perineal**, the region of the perineum. r., **pharyngeal**, the parts about the pharynx. r., **popliteal**, see *popliteal space*. r., **precordial**, the surface of the chest covering the heart. r., **prefrontal**, the part of the frontal lobe anterior to the precentral fissure; it is also called prefrontal lobe. r., **prevertebral**, the ventral surface of the vertebral column. r., **psycho-motor**, the cerebral cortex. r., **pterygo-maxillary**, the parts connecting or lying between the pterygoid process of the sphenoid and the maxillary bone. r., **pulmo-cardiac**, the region of the left thorax in which the left lung overlaps the heart. r., **pulmo-gastric**, the portion of the left thorax in which the lung overlaps the stomach. r., **pulmo-hepatic**, the portion of the right thorax in which the lung overlaps the liver. r., **pulmovascular**, the part of the thorax in which the lung overlaps the origins of the large vessels. r., **respiratory** (*of the nose*), the portion of the nasal passages having to do with the act of respiration. r., **retromaxillary**, the area dorsad of the superior maxilla. Syn., *r., parotid*. r., **sacral**, the area above the sacrum. r., **sacrococcygeal**, that part of the dorsal wall of the pelvis corresponding to the ventral surface of the sacrum and coccyx. r., **scapular**, the space over either scapula. r., **sensory**, see *sensorium*. r., **sternal**, the region overlying the sternum. r., **sternal, inferior**, the part of the sternal region lying below the margins of the third costal cartilages. r., **sternal, superior**, that portion of the sternal region lying above the lower margins of the third costal cartilages. r., **sub-auricular**, the space immediately below the ear. r., **subclavicular**, same as *r., infraclavicular*. r., **sub-mammary**, same as *r., inframammary*. r., **sub-maxillary**, same as *r., supra-hyoid*. r., **sub-mental**, the region just beneath the chin. r., **sub-ocular**, the anterior extremity of the temporo-sphenoidal lobe. r., **subscapular**, see *r., infrascapular*. r., **subthalamic**, the extension of the tegmentum beneath the posterior portion of the optic thalamus. r., **superior** (*of the skull*), the space between the superior curved line of the occipital bone behind, the supra-orbital ridge in front and, laterally, between the temporal lines. r., **supra-clavicular**, the space between the upper margin of the hyoid bone, the lower border of the inferior maxilla and the sternocleido-mastoid muscles. r., **supra-inguinal**, that bounded by the rectus abdominis muscle, Poupart's ligament, and a line through

the iliac crest. r., **supra-mammary**, same as *r., infra-clavicular, q. v.* r., **supra-scapular**, the area above the spine of the scapula. r., **supraspinous**, the region corresponding to the supraspinous fossa of the scapula. r.; **supra-sternal**, see *notch, supra-sternal*. r. **tegmental**, the tegmentum and corresponding parts of the pons and oblongata to the decussation of the pyramids. r., **temporo-maxillary**, the area over the junction of the temporal and maxillary bones. r., **thoracic**, relating to the entire surface of the thorax. r., **thyrohyoid**, the region around the thyroid cartilage and the hyoid bone. r., **umbilical**, the surface of the abdomen immediately about the umbilicus, bounded as follows: above, by a horizontal line connecting the cartilages of the ninth ribs; below, by a line joining the crests of the ilia, and laterally, by lines passing vertically through the center of Poupart's ligament. r., **vertebral**, relating to the region over the vertebral column.

**regional** (*re'-jun-al*) [*region*]. Pertaining to a region. r. **anatomy**, the branch of anatomy that treats of the relations of the structures in a region of the body to each other and to the body-surface.

**register** (*rej'-ist-er*) [*registrum*]. The compass of a voice; also a subdivision of its compass, consisting in a series of tones produced in the same way and of a like character.

**registration** (*rej-is-tra'-shun*) [*registratio*, a registering]. The act of recording, as of deaths, births, etc.

**reglementation** (*reg-le-men-ta'-shun*) [Fr., *reglementer*, to make regulations]. The legal restriction of prostitution.

**regression** (*re-gresh'-un*). Retrogression.

**regressive** (*re-gres'-iv*) [*re*, back; *gradi*, to go]. Going back; returning; subsiding.

**regular** (*reg'-ū-lar*) [*regere*, to rule]. According to rule or custom. r. **physician**, one belonging to the regular school. r. **school of medicine**, the great mass of the profession, whose practice is based on the results of experience and experimental research without adherence to any exclusive theory of therapeutics.

**regulin** (*reg'-ū-lin*). Trade name of a preparation said to consist of agar-agar and cascara sagrada; used in the treatment of constipation.

**regulus** (*reg'-ū-lus*). A metal reduced from its ore to the metallic state.

**regurgitant** (*re-gur'-jit-ant*). Flowing backward.

**regurgitation** (*re-gur-jit-a'-shun*) [*re*, again; *gurgitare*, to engulf]. 1. A back-flow of blood through a heart-valve that is defective. 2. The return of food from the stomach to the mouth soon after eating, without the ordinary efforts at vomiting. r., **aortic**, that of the blood-stream through the aorta from incompetence of the valves. r., **functional**, a form of mitral regurgitation due to contraction of the chordæ tendineæ and papillary muscles. r., **mitral**, see *mitral regurgitation*.

**Reichardt's test for arsenic in the urine.** Concentrate 200 Cc. of urine with about 2 Gm. of caustic soda; dissolve the residue in a little water acidulated with hydrochloric acid, and then test in a Marsh's apparatus.

**Reichel's cloacal duct** (*ri'-kel*) [Friedrich Paul *Reichel*, German obstetrician, 1858– ]. In the embryo the narrow cleft separating Douglas's septum from the cloaca.

**Reichert's canal** (*ri'-kert*) [Karl Bogislaus *Reichert*, German anatomist, 1811–1884]. See *Hensen's canal*. **R.'s cartilages**, the hyoid bars which constitute the skeletal elements of the hyoid branchial arch of the embryo and ultimately become the styloid processes, the stylohyal ligaments, and the lesser cornua of the hyoid bone. R.'s **membrane**, see *Bowman's membrane*. R.'s **scar**, an area over the embedded ovum consisting of a fibrinous lamella instead of decidual tissue as over the rest of the ovum.

**Reichl's test for proteids**. To the proteid solution add 2 or 3 drops of an alcoholic solution of benzaldehyde, and then some sulphuric acid, previously diluted with an equal bulk of water. Finally, add a few drops of a ferric sulphate solution, and a deep blue coloration will be produced in the cold after some time, or at once on warming. Solid proteids are stained blue by this reaction.

**Reichl-Mikosch's reagent for albumins**. Benzaldehyde and sulphuric acid containing ferric sulphate.

**Reichmann's disease** (*rik'-mahn*). A chronic disease of the stomach characterized by permanent gastric hypersecretion, associated with marked dilatation of the stomach, with thickening of its walls, and hypertrophy of the glands. It is accompanied by violent attacks of pain with vomiting, and may be followed by the formation of a round ulcer on the wall of the stomach; gastrosuccorrhea. **R.'s sign,** the presence in the stomach, before eating in the morning, of an acid liquid mixed with alimentary residues; it is indicative of gastrosuccorrhea and pyloric stenosis.

**Reid's lines** [Robert William *Reid*, Scotch anatomist, 1851– ]. Three imaginary lines serving for measurements in craniocerebral topography; one of them, the base line, is drawn from the lower margin of the orbit through the center of the external auditory meatus to just below the external occipital protuberance. The two others are perpendicular to it, one corresponding to the small depression in front of the external auditory meatus, the other to the posterior border of the mastoid process. The fissure of Rolando extends from the upper limit of the posterior vertical line to the point of intersection of the anterior line and the fissure of Sylvius.

**Reil's ansa** (*ril*) [Johann Christian *Reil*, Dutch anatomist, 1759–1813]. A tract of fibers passing from the optic thalamus downward and outward toward the white substance of the hemisphere. Syn., *ansa peduncularis; goose's foot*. R.'s covered band. 1. The lateral longitudinal stria; the longitudinal fibers which cross the transverse striæ beneath the fornicate convolution. 2. A fibromuscular fillet that frequently extends across the right ventricle of the heart. Syn., *moderator band*. **R.'s island,** the insula; a group of three to five small convolutions (gyri operti) situated at the bottom of the fissure of Sylvius. **R.'s line,** a ridge descending posteriorly from the summit of the pyramid of the cerebellum. **R.'s sulcus,** the sulcus in the bottom of the Sylvian fissure, separating the insula from the remainder of the hemisphere.

**reimplantation** (*re-im-plan-ta'-shun*) [*re*, again; *plantare*, to plant]. In dental surgery, the replacing of a drawn tooth into its socket.

**reinfection** (*re-in-fek'-shun*) [*re*, again; *infection*]. Infection a second time with the same kind of virus.

**reinforcement** (*re-in-fors'-ment*) [OF., *reinforcer*, to strengthen]. The act of reinforcing. Any augmentation of force. **r. of reflexes,** increased myotatic irritability (or reflex response) when muscular or mental actions are synchronously carried out, or other stimuli are coincidently brought to bear upon other parts of the body than that concerned in the reflex arc.

**reinoculation** (*re-in-ok-u-la'-shun*) [*re*, again; *inoculare*, to inoculate]. Inoculation a second time with the same kind of virus.

**Reinsch's test.** A test for arsenic. The suspected fluid is strongly acidulated with hydrochloric acid and boiled, some slips of bright copper being added; a grayish coating on the copper may be shown to be arsenic by heating in a glass tube held obliquely, when, if it is arsenic a crystalline coating will be sublimated on the glass above the copper. A similar test, but with different end reactions may be used for antimony, bismuth, and mercury.

**reinversion** (*re-in-ver'-shun*) [*re*, again; *invert*]. The act of reducing an inverted uterus by the application of pressure to the fundus.

**Reisseisen's muscles** (*ris'-is-en*) [Francis Daniel *Reisseisen*, German anatomist, 1773–1828]. The muscular fibers of the bronchi.

**Reissner's Canal** (*ris'-ner*) [Ernst *Reissner*, German anatomist, 1824–1878]. See *canal, membranous, of the cochlea*. **R.'s corpuscles,** the epithelial cells covering Reissner's membrane. **R.'s membrane,** the membrana vestibularis; a delicate membrane which separates the membranous cochlea (scala media) from the scala vestibuli.

**rejuvenescence** (*re-joo-ven-es'-ens*) [*re*, again; *juvenescere*, to grow young]. A renewal of youth; a renewal of strength and vigor.

**relapse** (*re-laps'*) [*re*, again; *labi*, to fall]. A return of an attack of a disease shortly after the beginning of convalescence.

**relapsing fever.** An acute infectious disease due to *Spirochæta Obermeieri*. After a period of incubation of from 5 to 7 days, the disease sets in with chill, fever, and pains in the back and limbs. The spleen enlarges, sweats and delirium occur, and the symptoms continue for 5 or 6 days, then suddenly cease by crisis. After a variable interval, usually in about a week, a second paroxysm occurs, which may be followed by a third and fourth. The disease prevails where conditions of overcrowding and defective food-supply obtain; hence the name sometimes given it, *famine-fever*.

**relation** (*re-la'-shun*) [*relatio*]. 1. Inter-dependence; mutual influence or connection between organs or parts. 2. Connection by consanguinity; kinship. 3. In anatomy, the position of parts of the body as regards each other.

**relax** (*re-laks'*) [see *relaxation*]. To loosen, or make less tense. To cause a movement of the bowels.

**relaxant** (*re-laks'-ant*) [see *relaxation*]. 1. Loosening; causing relaxation. 2. An agent that diminishes tension.

**relaxation** (*re-laks-a'-shun*) [*re*, again; *laxare*, to loosen]. A diminution of tension in a part; a diminution in functional activity, as *relaxation* of the skin.

**relief** (*re-lēf*) [OF., *relef*, a raising]. 1. The partial removal of anything distressing; alleviation. r. incision, one to relieve tension, as in an abscess.

**religiosus** (*re-lij-e-o'-sus*) [L. "religious"]. A name given to the superior rectus muscle of the eye.

**Remak's band** (*rem'-ak*) [Robert *Remak*, German anatomist, 1815–1865; Ernst *Remak*, German neurologist, 1849– ]. The axis-cylinder of a nerve-fiber. **R.'s contractions,** so-called diplegic contractions occasionally seen in progressive muscular atrophy when an electric current is applied. The positive electrode is placed above and the negative below the fifth cervical vertebra, the contractions occurring on the side opposite to the anode. **R.'s fibers,** the non-medullated nerve fibers. **R.'s fibrils,** the fibrils composing a nonmedullated nerve-fiber. **R.'s ganglion,** a ganglion of nerve cells located in the tissue of the heart near the superior vena cava. **R.'s layer,** the inner longitudinal fibrous layer of the tunica intima of large arteries. **R.'s sign,** the production by the pricking of a needle, of a double sensation, the second being painful; it is noted in tabes dorsalis. **R.'s type of palsy,** paralysis affecting the muscles of the arm— the deltoid, biceps, brachialis anticus, and supinator longus.

**remedial** (*re-me'-de-al*) [*remedy*]. Having the nature of a remedy; relieving; curative.

**remedy** (*rem'-ed-e*) [*re*, again; *mederi*, to heal]. Anything used in the treatment of disease.

**Remijia** (*re-mij'-e-ah*) [*Remijo*, a Spanish surgeon]. A genus of rubiaceous shrubs and trees closely related to cinchona.

**remission** (*re-mish'-un*) [*re*, back; *mittere*, to send]. 1. Abatement or subsidence of the symptoms of a disease. 2. The period of diminution of the symptoms of a disease.

**remittent** (*re-mit'-ent*) [*remission*]. Characterized by remissions. r. fever, a malarial fever characterized by periods of remission without complete apyrexia.

**remulus** (*rem'-u-lus*) [L., "a small oar"]. The narrow dorsal portion of a rib.

**ren** [L.: *gen., renis; pl., renes*]. The kidney. r. amyloideus, amyloid degeneration of the kidneys. r. mobilis, *movable kidney, q. v.* r. unguiformis, see *horseshoe kidney*.

**renaden** (*ren'-ad-en*). A proprietary preparation from kidneys; used in chronic nephritis. Dose 1½–2 dr. (6–8 Gm.).

**renal** (*re'-nal*) [*ren*]. Pertaining to the kidney. r. apoplexy, ischuria, or suppression of urine from hemorrhage into the substance of the kidney or other renal lesion. r. calculus, a concretion in the kidney. r. glands, the suprarenal capsules. r. inadequacy, the condition in which the amount of urinary solids, and often the quantity of urine itself, is considerably diminished. It is probably due to an exhausted condition of the epithelial cells of the kidney. **r. plexus,** see *plexus, renal*. r. storm, Murchison's term for a peculiar form of neurosal attack referred to the kidney, frequently seen in patients suffering from aortic regurgitation. There is sudden excruciating pain over the region of the kidney, like renal colic, but without nausea or retraction of the testicle, and with the passage of normal urine.

**renalina** (*ren-al-e'-na*). Trade name for a preparation of suprarenal gland.

**renculin** (*ren'-ku-lin*) [*ren*]. An albuminoid said to exist in the suprarenal capsules.

**renculus** (*ren'-ku-lus*). See *reniculus*.

**Rendu's type of tremor** (*ron'-doo*) [Henri Jules Louis Marie *Rendu*, French physician, 1844–1902].

A hysterical tremor provoked or increased by volitional movements.
**renicapsule** (*ren-e-kap'-sūl*) [*ren; capsula,* a capsule]. A suprarenal capsule.
**reniculus** (*ren-ik'-ū-lus*) [L., dim. of *ren, kidney*]. A lobule of the kidney; renculus.
**reniform** (*ren'-e-form*) [*ren; forma,* form]. Kidney-shaped.
**renin** (*ren'-in*) [*ren*]. A renal substance used in organotherapy.
**reniportal** (*ren-ip-or'-tal*) [*ren,* kidney; *porta,* gate]. Relating to the venous capillary circulation of the kidney.
**renipuncture** (*ren-e-punk'-chur*) [*ren; pungere,* to prick]. Puncture of the capsule of the kidney.
**renitent** (*ren'-it-ent*) [*reniti,* to resist]. Resistant to pressure.
**rennet** (*ren'-et*) [AS., *rinnan,* to run]. 1. The prepared inner membrane of the fourth stomach of the calf, or an infusion of this membrane. It contains a milk-curdling ferment that decomposes casein. 2. Rennin, *q. v.*
**rennin** (*ren'-in*) [see *rennet*]. The milk-curdling ferment of the gastric juice.
**rennihogen, rennogen** (*ren-in'-o-jen, ren'-o-jen*) [*rennet*]. The zymogen whence rennin is formed; it exists in the cells of the mucous membrane of the stomach.
**renocutaneous** (*ren-o-kū-ta'-ne-us*) [*ren; cutaneous*]. Relating to the kidneys and the skin.
**renoform** (*ren'-o-form*). Trade name for a preparation of suprarenal gland.
**renogastric** (*ren-o-gas'-trik*) [*ren;* γαστήρ, stomach]. Relating to the kidney and the stomach.
**renointestinal** (*ren-o-in-tes'-tin-al*). Relating to the kidney and the intestines.
**renopulmonary** (*ren-o-pul'-mon-a-re*). Relating to the kidney and the lungs.
**renostypticin** (*ren-o-stip'-lis-in*). Trade name for a preparation of suprarenal gland.
**renostyptin** (*ren-o-stip'-tin*). Trade name for a preparation of suprarenal gland.
**renovation** (*ren-o-va'-shun*) [*renovatio; renovare,* to render new]. The repair or renewal of that which has been impaired.
**renuent** (*ren'-ū-ent*) [*renuens,* nodding back the head]. In anatomy, throwing back the head; applied to certain muscles.
**renule** (*ren'-ūl*) [*ren,* the kidney]. A small kidney. Reoch's test for albumin. See *Macwilliam.* R.'s test for hydrochloric acid in the contents of the stomach, on the addition of a mixture of citrate of iron and quinine and potassium sulphocyanide to the gastric juice or contents of the stomach, containing free hydrochloric acid, a red coloration will be produced.
**reorganization** (*re-or-gan-iz-a'-shun*) [*re,* again; *organization*]. Healing by the development of tissue elements similar to those lost through some morbid process.
**rep.** Abbreviation for *repetatur* [L.]. Let it be repeated.
**repand** (*re-pand'*) [*re,* back; *pandus,* bent, crooked]. In biology, applied to a leaf-margin which is toothed like the margin of an umbrella. A bacterial culture with a wrinkled or wavy edge.
**repatency** (*re-pa'-len-se*) [*re,* again; *patens,* open]. The reopening of a part or vessel. r. of a vessel, after ligation, the reopening of the lumen of a ligated vessel from too rapid absorption of the ligature, or from slipping of the knot.
**repellent** (*re-pel'-ent*) [*re,* back; *pellere,* to push]. 1. Driving back. 2. Causing resolution of morbid processes.
**repercolation** (*re-per-ko-la'-shun*) [*re,* again; *percolare,* to percolate]. Repeated percolation; the passage of a percolate for a second time, or oftener, through the percolator.
**repercussion** (*re-per-kush'-un*) [*re,* again; *percussion*]. 1. Ballottement. 2. A driving in or dispersion of a tumor or eruption.
**repercussive** (*re-per-kus'-iv*) [see *repercussion*]. 1. Repellent. 2. A repellent drug.
**repercutient** (*re-per-kū'-she-ent*) [*re,* again; *percutere,* to percuss]. Effecting a repercussion; pertaining to a process or function of rebound, or reaction.
**replantation** (*re-plan-ta'-shun*) [*replantare,* to plant again]. The act of planting again. r. of the teeth, the replacement of teeth which have been extracted or otherwise removed from their cavities; when

diseased, the thickened periosteum is scraped off before returning such teeth to their sockets.
**repletion** (*re-ple'-shun*) [*re,* again; *plere,* to fill]. The condition of being full.
**replication** (*rep-lik-a'-shun*) [*re,* back; *plica,* a fold]. A refolding or turning back of a part so as to form a duplication.
**reposing** (*re-po'-zing*) [see *reposition*]. Returning an abnormally placed part to its proper position. reposing the features, in dentistry, a term including everything necessary to bring each and all of the visible parts of the face and mouth into harmony of relation to each other; this necessarily includes the teeth, the relation of the lower to the upper jaw, the lips, cheeks, and soft parts of the face that have assumed a wrong position by reason of the loss of the natural organs.
**reposition** (*re-po-zish'-un*) [*re,* back; *ponere,* to place]. Return of an abnormally placed part to its proper position. Reduction of hernia, dislocation, uterus, etc.
**repositor** (*re-pos'-it-or*) [see *reposition*]. An instrument for replacing parts that have become displaced, especially for replacing a prolapsed umbilical cord; an instrument used in the replacement of a displaced uterus.
**repoussoir** (*ra-poo-swahr'*) [Fr.]. An instrument for extracting the roots of teeth.
**réprise** (*ra-prēz'*) [Fr. "recovery"]. That part of the cry of a child which is heard during the act of inspiration. The loud inspiration in pertussis. The whoop.
**reproduction** (*re-pro-duk'-shun*) [*re,* again; *producе*]. 1. The conscious repetition of perceived sensations. 2. The act of producing again; the procreation of one's kind; the producing of something like that lost. r., asexual, that without sexual intercourse. r., endogenous, internal cell formation. r., sexual, that by the union of sexually distinct cells.
**reproductive** (*re-pro-duk'-tiv*) [*reproduction*]. Pertaining to reproduction, as the *reproductive* organs.
**repullulation** (*re-pul-ū-la'-shun*) [*re,* again; *pullulare,* to sprout]. The return of a morbid growth.
**repulsion** (*re-pul'-shun*) [*re,* back; *pellere,* to drive; to push]. 1. The act of repelling or driving back or apart. 2. The influence tending to drive two bodies apart; the opposite of attraction. r., capillary, repulsion due to the forces causing movements of liquids in small tubes.
**resacetin** (*res-as'-et-in*). A salt of oxyphenyl-acetic acid.
**resalicin** (*res-al'-dol*). An acetyl derivative of saliformin and resorcinol; an intestinal astringent and antiseptic. Dose 8 dr.–2½ oz. (30–75 Gm.) daily.
**resalgin** (*rez-al'-jin*). A compound of resorcin and antipyrin.
**resection** (*re-sek'-shun*) [*re,* again; *secare,* to cut]. The operation of cutting out. r. of a joint, the cutting away of the ends of the bones forming a joint, or a portion of bone, nerve, or other structure.
**reserve air.** See *respiration.*
**reservoir** of Pecquet (*pek-a'*). See *receptaculum chyli.*
**residual air** (*re-zid'-ū-al*). See under *respiration.*
**r. ear,** a middle ear which has been the seat of a suppurative process, which process has ceased and left the tympanic structures in a permanently damaged condition.
**residue** (*rez'-id-ū*) [*residere,* to remain]. That remaining after a part has been removed; balance or remainder.
**residuum** (*re-zid'-ū-um*). 1. See *residue.* 2. Behring's term for the mass of tubercle bacilli used in the manufacture of tuberculase.
**resilience** (*re-zil'-e-ens*) [*resilient*]. 1. The quality of being elastic or resilient. 2. Healthy reaction.
**resilient** (*re-zil'-e-ent*) [*re,* back; *salire,* to leap]. Rebounding; elastic. r. stricture, one that contracts again immediately after being dilated.
**resin** (*rez'-in*) [*resina*]. 1. One of a class of vegetable substances exuding from various plants, and characterized by being soluble in alcohol, in ether, and in the volatile oils, and insoluble in water; they are readily fusible and inflammable. They are obtained in pharmacy by treating the substances containing them with alcohol, and then precipitating the alcoholic solution with water. 2. See *rosin.*
r., gum-, one differing from a true resin only in containing some gum capable of softening in water.

# RESINA 754 RESPIRATION

**r. of jalap** (*resina jalapæ*, U. S. P., B. P.). Dose 2–5 gr. (0.13–0.32 Gm.). **r.-plaster.** See *plaster*. **r. of podophyllum** (*resina podophylli*, U. S. P.). Dose ⅛–¼ gr. (0.008–0.032 Gm.). **r. of scammony** (*resina scammonii*, U. S. P., B. P.). Dose 4–8 gr. (0.26–0.52 Gm.).

**resina** (*res-i'-nah*). Colophony. The residue left after distilling off the volatile oil of turpentine. See *rosin*.

**resinate** (*rez'-in-āt*). A compound of a resin with a base.

**resineon** (*rez-in'-e-on*). A volatile oil distilled from resin with potash and freed from phenol. It is used as a wound antiseptic and in the treatment of skin diseases.

**resinoid** (*rez'-in-oid*) [*resina*, a resin; είδος, like]. 1. Resembling a resin. 2. A substance which has some of the properties of a resin. Most of the so-called resinoids are of indefinite chemical composition; others are impure resins.

**resinol** (*rez'-in-ol*). See *retinol*.

**resinous** (*res'-in-us*) [*resin*]. Having the nature of a resin.

**resistance** (*re-sis'-tans*) [*resistare*, to withstand]. 1. Opposition to force or external impression. 2. In electricity, the opposition offered by a conductor to the passage of the current: **r.-coil,** a coil of wire for increasing the resistance in a circuit. **r., essential, r., internal,** the resistance to conduction within the battery itself. **r., extraordinary, r., external,** the resistance to conduction outside of the battery. **r., Issaeff's period of,** a temporary power of resistance to inoculation by virulent cultures of bacteria, conferred by the injection of various substances, such as salt solution, urine, serum, etc.

**resistivity** (*re-zis-tiv'-it-e*) [*resistance*]. The amount or hara ter of electrical resistance exhibited by a body.

**resol** (*res'-ol*). A disinfectant mixture of saponified wood-tar and methyl-alcohol.

**resolution** (*rez-o-lū'-shun*) [*resolvere*, to resolve]. The return of a part to the normal state after a pathological process.

**resolve** (*re-zolv'*) [*resolvere*, to resolve]. 1. To return to the normal state after some pathological process. 2. To separate anything into its component parts.

**resolvent** (*re-zol'-vent*) [*resolve*]. 1. Causing solution or dissipation of tissue. 2. An agent causing resolution.

**resolving power.** The capability of a lens of making clear the finest details of an object.

**resonance** (*rez'-o-nans*) [*re*, again; *sonare*, to sound]. 1. The sound obtained on striking a hollow object, especially the note obtained on percussing the chest or abdomen. 2. The sound of the voice as transmitted to the ear applied to the chest. **r., amphoric,** a sound resembling that produced by blowing across the mouth of a bottle. **r., bell-metal,** a bell-like sound heard on auscultation in pneumothorax when the chest is percussed with two coins used as plexor and pleximeter. **r., cracked-pot,** a sound elicited by percussing over a pulmonary cavity communicating with a bronchus. **r., hydatid,** a peculiar sound heard in combined auscultation and percussion of hydatid cysts. It is not heard in other cystic conditions. **r., Skodaic,** the increased percussion resonance over the upper part of a lung when the lower part is compressed by a pleural effusion. **r., tympanitic,** a hollow sound elicited on percussion over the intestine and over large pulmonary cavities with thin yielding walls. **r., vesicular,** the normal pulmonary resonance. **r., vesiculotympanitic,** an admixture of vesicular and tympanitic resonance. **r., vocal,** the sound heard on auscultation of the chest during ordinary speech. **r., whispering,** the sound heard on auscultation of the chest during the act of whispering.

**resonant** (*rez'-o-nant*) [*resonans*, resounding]. 1. Sounding or ringing in the nasal passages. 2. A resonant or nasal sound; see *consonant*.

**resonator** (*rez'-o-na-tor*) [see *resonance*]. An instrument used to intensify sounds.

**resopyrine** (*res-o-pi'-rēn*). A compound of resorcinol and antipyrine. It is used in any condition in which antipyrine and resorcinol are indicated. Dose 5–10 gr. (0.32–0.65 Gm.).

**resorbent** (*re-sorb'-ent*) [*resorbere*, to draw to itself]. 1. Favoring resorption. 2. A drug which aids in the process of resorption.

**resorbin** (*re-sorb'-in*). A penetrating ointment-base consisting of an emulsion of sweet almond oil, wax, and a dilute aqueous solution of gelatin or soap.

**resorcin** (*re-zor'-sin*). See *resorcinol*.

**resorcinism** (*re-zor'-sin-izm*) [*resorcin*]. A toxic condition caused by injudicious or excessive use of resorcin.

**resorcinol** (*re-zor'-sin-ol*) [*resin*; *orcin*]. 1. C₆H₆O₂. A crystalline substance isomeric with pyrocatechin and hydroquinone, and usually prepared by fusing sodium benzene disulphonate and sodium hydroxide. It is an antipyretic and antiseptic, but is chiefly used in ointments for chronic skin diseases. Dose 2–5 gr. (0.13–0.32 Gm.). 2. Equal parts of resorcinol and iodoform fused together; it is used as a surgical dusting-powder, 20 to 50 % with starch, or 7 to 15 % ointment.

**resorcinopyrin** (*re-zor-sin-o-pi'-rin*). See *resopyrin*.

**resorcinum** (*re-zor'-sin-um*) [L.; *gen.*, *resorcini*]. Resorcinol.

**resorcylalgin** (*re-zor-sil-al'-jin*). A crystalline derivative of β-resorcylic acid and antipyrin, soluble in alcohol, ether, or hl r rm, or 150 parts of water; melts at 115°C. Oft is antipyretic and anodyne. Syn., *resalgin*.

**resorption** (*re-sorp'-shun*) [*re*, again; *sorbere*, to absorb]. 1. The absorption of morbid deposits, as of the products of inflammation. 2. The process through which the roots of temporary teeth disappear. Occasionally the roots of permanent teeth suffer resorption. **r., cutaneous.** See *absorption, cutaneous*. **r.-infection,** a mode of infection marked by the development of bacteria at a distance from the point of introduction. **r., lacunar** (*of bone*), resorption of bone by osteoclasts forming and occupying Howship's lacunæ.

**resosalyl** (*re-so-sal'-il*). A proprietary antiseptic containing the salicylate ester of ethyl-resorcin with boric acid, benzoic acid, camphor, chloral, and other substances.

**respirable** (*res-pi'-rah-bl*) [see *respiration*]. Capable of being inspired and expired; capable of furnishing the gaseous interchange in the lungs necessary for life.

**respiration** (*res-pi-ra'-shun*) [*re*, again; *spirare*, to breathe]. 1. The interchange between the gases of living organisms and the gases of the medium in which they live, through any channel, as *cutaneous respiration*. 2. The act of breathing with the lungs; the taking into and the expelling from the lungs of air. It consists of two acts—*inspiration*, or the taking in of the atmospheric air, and *expiration*, the expelling of the modified air. Expired air contains less oxygen and more carbon dioxide than inspired air. The volume of air taken into the lungs and given out during an ordinary respiration (*tidal air*) is 500 Cc.; the volume that can be inspired in addition by a forcible inspiration (*complemental air*) is 1500 Cc.; that which remains in the chest after a normal expiration (*reserve* or *supplemental air*) is 1500 Cc.; the amount remaining in the chest after the most complete expiration (*residual air*) is from 1200–1600 Cc. The volume of air that can be forcibly expelled after the most forcible inspiration is termed *vital* or *respiratory capacity* and is equal to the tidal air, complemental air, and reserve air, or about 3500 Cc. See *breath* and *breathing*. **r., abdominal,** a type of respiration caused by the contraction of the diaphragm and the elasticity of the abdominal walls and viscera. It is more common in men than in women. **r., absent,** suppression of respiratory sounds. **r., accelerated,** when exceeding 25 respirations a minute. **r., aerial,** respiration in which the respiration membrane receives oxygen and is relieved of carbon dioxide by means of atmospheric air. **r., amphoric,** a blowing respiration engendered in large cavities with firm walls. Its peculiar character is due to an echo from the walls of the cavity. **r., aquatic,** respiration in which the respiratory membrane, the branchial mucosa (gills), the skin, etc., receive oxygen and are relieved of carbon dioxide by means of water. **r., artificial,** artificial production of the normal respiratory movements; see *artificial respiration*. **r., blowing.** See **r., bronchial**. **r., branchial,** respiration by means of gills or branchiæ as in aquatic animals. **r., bronchial,** respiration as heard over the trachea or bronchial tubes in health; it is high in pitch, equal in inspiration and expiration, blowing in character, especially the expiratory element, and is marked by a brief pause be-

tween inspiration and expiration. It is well defined only in case of pulmonary consolidation. r., broncho-cavernous, a form intermediate in character between bronchial and cavernous respiration. r., broncho-vesicular, respiration having the characters of both bronchial and vesicular respiration. It is heard over areas of consolidation surrounded by patches of healthy lung-tissue. r., buccal. See *mouth-breathing*. r., cavernous, a blowing respiration of low pitch, circumscribed, alternating with gurgling, and deriving its chief character from the nature of the cavity in which it is generated. r., center of, the nervous center regulating the act of respiration is situated in the floor of the fourth ventricle near the point of the calamus. It is automatic in its action. r., cerebral, respiration in which the lips are closed, the cheeks distended, the nostrils dilate with each expiration, which is attended with a puffing sound; the respirations are irregular. It is observed especially in typhus fever and in apoplexy. r., clavicular, a form resorted to by singers and in which the clavicle is brought into play in the respiratory movements, the shoulders being elevated. r., cogged or cogwheel. See r., *interrupted*. r., costal, respiration in which the chest-movement predominates over the diaphragmatic movement. It is seen especially in women, and is supposed to be related to gestation or perhaps partially to the mode of dress. r., costo-inferior, respiration in which the elevation and depression (respiratory movements) are confined chiefly to the lower ribs. It is best seen in dogs. r., costo-superior, respiration in which the respiratory movements involve chiefly the upper ribs. It is most common in women. r., cutaneous, the giving off of carbon dioxide and taking up of oxygen through the skin. r., diaphragmatic. See r., *abdominal*. r., direct, respiration in which the living substance of an organism, as an ameba, takes oxygen directly from the surrounding medium and returns carbon dioxide directly to it, no respiratory blood being present. r., divided, respiration in which there exists a distinct interval between inspiration and expiration. It is seen in emphysema as a result of the distention of the air-vesicles and consequent reduction in expelling force. r., exaggerated, an increase in intensity, without alteration in character or rhythm of the respiratory movements. r., external. See r., *blood*. r., extrinsic (*of F. Hewitt*), the inspiration and immediate expiration of a gas, so that a portion that has once been inspired is not inspired again. r., facial, a term applied to all the movements of the face during inspiration and expiration. r., feeble, diminution in the intensity without alteration in the character or rhythm of the respiratory movements. r., fetal, the interchange of gases between the fetal and the maternal blood through the medium of the placenta. r., forced, respiration induced by blowing air into the lungs by means of a bellows, or in some other way, as in physiological experiments. r., harsh. See r., *broncho-vesicular*. r., hissing, an increased vesicular murmur causing a hissing sound. r., hollow. See r., *amphoric*. r., indeterminate, the most pronounced vesicular grade of broncho-vesicular respiration. r., indirect, respiration in which the living substance of the organism, as in all the higher animals, gets rid of carbon dioxide and obtains oxygen by means of a circulating respiratory blood. r., inner or internal, the taking up of oxygen and giving off of carbon dioxide by the body-elements for their own requirements. It occurs in man in the capillary system. r., interrupted, respiration in which either inspiration or expiration is divided into two or more parts. It is most often heard at the apex of the right lung, anteriorly. r., intestinal, the passage of gases of respiratory gases in the mucous membrane of the intestines. r., intra-uterine, respiration by the fetus before delivery. r., intrinsic (*of F. Hewitt*), the breathing over and over again of a limited volume of gas. r., jerking. See r., *interrupted*. r., labored, respiration in which, owing to lack of ability on the part of the ordinary muscles of respiration to sufficiently aerate the blood, the auxiliary muscles of respiration are called into play. r., laryngeal, the widening of the glottis during inspiration and its narrowing during expiration. r., lung. See r., *pulmonary*. r., metamorphosing (*of Seitz*), respiration in which the first part of the inspiratory sound is tubular and the last part cavernous; a cavernous element is also heard during expiration. It is a certain sign of a cavity (Vierordt). r., muscle, respiration by a muscle when in action. r., nasal, nose-breathing. r., nervous. See r., *cerebral*. r., normal, respiration as it occurs in a normal individual in a state of rest or moderate action. r., oral. See *mouth-breathing*. r., ordinary. See r., *normal*. r., organs of, any parts of the body by means of which certain constituents of the blood are exchanged for those of the surrounding air or water. r., pharyngeal (*of Garland*), rhythmic expansions and contractions of the pharynx in connection with other movements of respiration. The expansion is pre-inspiratory and the contraction inspiratory. r., placental. See r., *fetal*. r., puerile. See r., *exaggerated*. r., postural, r., prone. See *artificial respiration, Hall's method*. r., puerile. See r., *exaggerated*. r., pulmonary, respiration in which the interchange of gas between the blood and air occurs in the lungs. r., rough, a variety of broncho-vesicular respiration. r., rude. See r., *rough*. r., senile, the feeble respiration of old age. r., sighing, deep respiration accompanied with sighing. It is seen in pulmonary congestion and dyspepsia. r., stertorous, the sound produced by breathing through the nose and mouth at the same time, causing vibration of the velum pendulum palati between the two currents of air. r. subsibilant (*of Laennec*), a dull, whistling sound heard over the bronchi, and due to an obstruction by mucus. r., superficial. See r., *blood*. r., supplementary. See r., *exaggerated*. r., thoracic. See r., *costal*. r., tissue. See r., *internal*. r., to-and-fro. See r., *intrinsic*. r., tracheal, the respiratory murmur heard in a normal individual by placing a stethoscope over the supra-sternal fossa. r., tranquil. See r., *normal*. r., tubular. See r., *bronchial*. r., uremic. See *Cheyne-Stokes' respiration*. r., vaginal, the movements of the vagina caused by the movements of the diaphragm in respiration. r., ventral. See r., *abdominal*. r., vesicular, a soft, gradual, low-pitched inspiration immediately followed by a shorter and less distinct expiration-sound heard over the normal lung during respiration. r., vesiculo-bronchial. See r., *bronchovesicular*. r., vesiculocavernous, respiration that is both vesicular and cavernous. r., wavy. See r., *interrupted*. respirator (*res'-pi-ra-tor*) [*respiration*]. An appliance by which the inspired air, in passing through it, is warmed, purified, or medicated.
respiratory (*res-pi'-ra-to-re*) [*respiration*]. Pertaining to respiration. r. blood, Huxley's name for the fluid present in the pseudohemal system of vessels of certain invertebrates (Annelida). It contains a dissolved red substance allied to hemoglobin. r. bundle, the ascending root of the glossopharyngeal nerve, probably arising in the posterior horns of the cord. r. capacity, the capacity of the blood for taking up oxygen in the respiratory organs and depositing it in the tissues, and of taking up carbon dioxide from the tissues and giving it off in the respiratory organs. r. cavity, the same as the thoracic cavity; also used as a general term to describe the air-passages. r. center. See *center, respiratory*. r. chamber, a respiratory cavity. r. excursion, the entire movement of the chest during the complete act of respiration. r. filaments, thread-like organs arranged in tufts near the head of the larva of the gnat. r. glottis, that part of the glottis between the arytenoid cartilages. r. murmur, the sound produced by the air entering and escaping from the lungs during respiration. r. nerve, one of two nerves supplying important muscles of respiration: the *external* is the posterior thoracic nerve; the *internal*, the phrenic nerve. r. percussion, Da Costa's term for the method of physical examination by noting the sound elicited by percussion of the chest while the breath is held after a full inspiration, and also after a prolonged expiration. r. periods, the time elapsing between the beginning of one inspiration and that of the next. r. pulse, the modifications in the pulse produced by respiration. r. quotient, the quotient resulting from dividing the quantity of carbon dioxide exhaled, by the amount of oxygen inhaled. r. sound. See r., *murmur*. r. surface, the entire surface of pulmonary tissue coming in contact with the respired air. r. tract, all the air-passages and air-cells concerned in respiration. r. tubes, a term applied to all tubular organs of respiration. r. vesicular murmur, the normal respiratory murmur; see *respiration, normal*.

**respire** (*re-spīr'*) [*re*, back; *spirare*, to breathe]. To breathe.

**respirometer** (*res-pi-rom'-et-er*) [*respiration*; μέτρον, a measure]. A device to determine the character of the respiration.

**response** (*re-spons'*) [*respondere*, to answer]. The reaction or movement of a muscle or other part due to the application of a stimulus.

**responsibility** (*re-spon-sib-il'-it-e*) [*respondere*, to answer]. In medical jurisprudence, the accountability of a person for an act committed. It usually turns upon the question as to whether or not the person was of sound mind and capable of controlling his actions and thoughts.

**rest** [ME., *resten*, to rest]. 1. Cessation of labor or action; to sleep; to lie dormant. 2. A mass of embryonic cells which, having been misplaced during organic evolution, remain quiescent and fail to reach their normal evolution. They at times act as foci for the development of new growths or other pathologic phenomena. **r.s, adrenal, r.s, suprarenal,** masses of aberrant adrenal tissue occasionally observed beneath the capsule of the kidney. **r.-atom,** the part of an atom which remains after an alpha-ray has been discharged from it; also called recoil-atom. **r.-cure.** See *Mitchell's treatment*.

**restibrachium** (*res-te-bra'-ke-um*) [*restis; brachium*]. The inferior peduncles of the cerebellum.

**restiform** (*res'-te-form*) [*restis; forma,* form]. Corded or cord-like. **r. body,** a part of the medulla oblongata, which as the inferior cerebellar peduncle connects the medulla with the cerebellum. It contains fibers from the lateral column of the spinal cord (the lateral cerebellar tract), from the posterior column, and from the inferior olivary nucleus.

**resting** (*rest'-ing*). Ceasing from motion; at rest. **r.-cell.** Same as *r.-spore*. **r.-sporangium,** in biology, Pringsheim's term for peculiar resting-cells formed by the mycele of a few fungi (*e. g., Saprolegnia*), in which zoöspores are produced. **r.-spore,** in biology, a spore invested with a firm cell-wall, which remains dormant for a period, often during the whole winter, before it germinates. **r.-stage,** in biology, the period of dormancy in the history of a plant or germ. **r.-state,** in biology, a state of suspended activity, the condition of perennial plants, bulbs, seeds, and spores during their period of dormancy.

**restis** (*res'-tis*) [L., "a rope"]. The restiform body.

**restitutio ad integrum** [L.]. Complete restoration to a healthy condition.

**restitution** (*res-tit-ū'-shun*) [*re*, again; *statuere*, to set up]. 1. The act of restoring. 2. In obstetrics, a rotation of the fetal head immediately after its birth.

**restoration** (*res-tor-a'-shun*) [*restaurare*, to restore]. The renewal of or return to a state of health.

**restorative** (*re-sto'-ra-tiv*) [*restore*]. A remedy that is efficacious in restoring health and strength.

**restraint** (*re-strānt'*) [*restringere*, to draw back]. 1. Hindrance of any action, physical, moral, or mental. 2. The state of being controlled; specifically, abridgment of liberty in the care of the insane. **r.-bed and r.-chair.** See *r., mechanical*. **r., mechanical,** restraining the insane by mechanical means. **r., medicinal,** the use of narcotics and sedatives in quieting the insane.

**restringent** (*re-strin'-jent*) [*restringere*, to restrain]. An astringent or styptic.

**resublimation** (*re-sub-lim-a'-shun*) [*re*, again; *sublimare*, to raise on high]. The process of subliming a drug for the second time.

**resudation** (*re-sū-da'-shun*) [*re*, again; *sudor*, sweat]. The return of sweating as a symptom.

**resupinate** (*re-sū'-pin-āt*) [*re*, again; *supinare*, to bend backward]. Turned in a direction opposite to normal; as an ovary with its apex downward.

**resurrectionist** (*res-ur-ek'-shun-ist*) [*resurgere*, to rise again]. Colloquially, one who steals dead bodies from the grave as subjects for dissection.

**resuscitate** (*re-sus'-it-āt*) [*resuscitare*, to revive]. To revive; to recover from apparent death.

**resuscitation** (*re-sus-it-a'-shun*) [*re*, again; *suscitare*, to raise up]. The bringing back to life of one apparently dead.

**resuscitator** (*re-sus'-it-a-tor*) [see *resuscitation*]. One who or that which resuscitates. **r., intragastric,** an apparatus devised by Fenton B. Türck, for the purpose of reducing surgical shock and collapse. It consists simply of a double stomach-tube, at one end of which is attached a soft-rubber bag. By this means heat is applied in a uniform and diffuse manner, up to 135° F.

**retainer** (*re-ta'-nur*) [*retinere*, to keep back]. A dental appliance for holding in position teeth which have been moved.

**retamine** (*ret-am'-ēn*) [*retama,* the Spanish name for genista], $C_{15}H_{26}N_2O$. An alkaloid from the bark of *Genista sphærocarpa*.

**retardation** (*re-tar-da'-shun*). [*retardatio; retardare,* to delay]. Any hindering or delaying of a function. In obstetrics, delay in expelling the fetus. In biology, the change of structure during growth accomplished by the subtraction of parts. The opposite of *acceleration, q. v.*

**retarding** (*re-tar'-ding*) [*retardare,* to delay]. Hindering; delaying. **r. ague,** a variety of ague in which the paroxysm is postponed to a later hour each day.

**retch** [AS., *hræcan,* to clear the throat]. To strain at vomiting.

**rete** (*re'-tē*) [L., a net; *pl., retia*]. Any network or decussation and interlacing, especially of capillary blood-vessels. **r., acromiale,** a plexus of arteries on the surface of the acromial process, formed by anastomoses between the acromial branch of the acromiothoracic, the suprascapular, and the anterior and posterior circumflex. **r. arteriosum capitis,** a network over the upper part of the cranium formed by the anastomosis of the frontal, temporal, and occipital arteries. **r. arteriosum faciei,** a network formed by the terminal branches of the facial, infraorbital, ophthalmic, and internal maxillary arteries. **r. articulare cubiti,** an arterial anastomosis over the elbow. **r. articulare genu,** one formed by the anastomosis of the arteries over the anterior and lateral surfaces of the knee. **r., bipolare,** applied to blood-vessels that unite into larger stems and again divide and end in capillaries. **r. calcaneum,** formed by the terminal branches of the posterior tibial. **r. carpi,** anterior, and posterior, two plexuses of arteries formed by the carpal branches of the radius and ulna, one in front and the other at the back of the wrist. **r. carpi dorsale,** the posterior carpal arch. **r. carpi volare,** an arterial meshwork made up of branches from the radial and ulnar arteries and deep palmar arch upon the anterior surface of the carpus. **r. choroideum,** vascular prolongations of the pia. **r., cubitale.** See *r. articulare cubiti*. **r. dorsalis pedis,** an arterial-network on the dorsum of the foot formed by branches of the tarsal and metatarsal arteries joined by perforating plantar branches. **r. epidermal.** Same as *r. mucosum*. **r. Halleri,** the upper part of the Wolffian body by which the communication between the seminiferous tubules and the Wolffian duct is established and maintained. **r. majus,** the great omentum. **r. malleolare internum** and **externum,** the network surrounding the inner and the outer ankle. **r. Malpighii,** the layers of epithelial cells above the corium. **r. mirabile** (*pl., retia mirabilia*), is seen when an artery splits into branches and reunites in a trunk, without forming capillaries. **r. mirabile duplex.** See *r. mirabile geminum* or *conjugatum*. **r. mirabile** (*of Galen*), a network of vessels formed by the intracranial portion of the internal carotid artery in some animals. **r. mirabile geminum** or **conjugatum,** a plexus in which arteries and veins are combined. **r. mirabile of Malpighi,** the network formed by the ultimate ramifications of the pulmonary artery. **r. mirabile simplex,** a network involving only veins or arteries. **r. mirabile unipolar.** See *r., unipolar*. **r. mucosum,** the three lower layers of living cells of the epidermis. **r. olecrani,** the network of vessels around the olecranon and at the back of the elbow, formed by the divisions of the profunda and other arteries. **r. patellare,** the plexus of vessels surrounding the patella. **r. tarseum dorsale,** an arterial network upon the dorsal surface of the tarsus. **r. testis,** the network of seminal tubules in the corpus Highmori of the testicle. **r., unipolar,** the capillary divisions of blood-vessels which do not reunite. **r. vasculosum testis,** a network of blood-vessels. See *r. Halleri*. **r. venosum dorsale manus,** a venous network on the back of the hand. **r. venosum dorsale pedis,** a venous network on the dorsum of the foot. **r. venosum volare manus,** a palmar network of the hand.

**retene** (*re'-tēn*), $C_{18}H_{18}$. A hydrocarbon occurring in the highest fractions of coal-tar and also a deriva-

tive of phenanthrene. It occurs in the tar of highly resinous pines and in some mineral resins.
**retentio mensium** (*re-ten'-she-o men'-se-um*) [L.]. Retention of the menses; a condition in which menstruation occurs but its products are retained in consequence of atresia of the genital canal.
**retention** (*re-ten'-shun*) [*re*, back; *tenere*, to hold]. The act of retaining or holding back. r. **cyst.** See *cyst, retention*. r.-**hypothesis** (of Chauveau). See *immunity, theory of, Chauveau's retention*. r. **of urine,** the holding of the urine in the bladder on account of some hindrance to urination.
**retia** (*re'-te-ah*) [L., plural of *rete*]. See *rete*.
**retial** (*re'-te-al*) [*rete*]. Relating to, or of the nature of, a rete.
**reticula** (*ret-ik'-u-lah*) [pl. of *reticulum*, a network]. The preferred name for formatio reticularis.
**reticular** (*ret-ik'-u-lar*) [*reticulum*]. Resembling a net; formed by a network. r. **formation.** See *formatio reticularis*. r. **lamina,** the membrane covering the organ of Corti. r. **layer of the skin,** the deep layer of the skin, consisting of interlacing bands of white and yellow fibrous tissue. r. **tissue,** the stroma of adenoid tissue; adenoid tissue; cellular tissue in general.
**reticulated** (*ret-ik-u-la'-ted*). Having net-like meshes.
**reticulin** (*re-tik'-u-lin*). A body found by Siegfried in the fibers of reticular tissue with a percentage composition: C, 52.88; H, 6.97; N, 15.63; S, 1.88; P, 0.34; ash, 2.27; but believed by other authorities to be simply collagen coagulated by reagents combined with protein and nuclein residues of cells.
**reticulose** (*ret-ik'-u-lōs*) [*reticulum*, a net]. Minutely or finely reticulate.
**reticulum** (*ret-ik'-u-lum*) [*reticulum*, dim. of *rete*, net]. A network.
**retiform** (*ret'-if-orm*). Net-shaped; reticular. r. **tissue.** See *reticular tissue*.
**retina** (*ret'-in-ah*) [*rete*]. The delicate membrane of the eye representing the terminal expansion of the optic nerve, and extending from the point of entrance of the nerve forward to its termination in the ora serrata. It consists of the following layers, named from behind forward: (*a*) the pigment-layer; (*b*) the neuroepithelial layer, comprising the layer of rods and cones (Jacob's membrane; bacillary layer), the outer limiting membrane, and the outer nuclear layer; (*c*) the cerebral layer, comprising the outer reticular layer (outer granular layer), the inner nuclear layer, the inner reticular layer (inner granular layer), the ganglion-cell layer, the nerve-fiber layer. These layers are cemented together by a supporting framework of connective tissue, the fibers of Mueller, or radiating fibers. r., **central artery of,** a branch of the ophthalmic artery that pierces the optic nerve in the orbit, branching within the globe, and supplying the retina. r., **coarctate,** a term used to describe the morbid condition caused by an effusion of liquid between the retina and the choroid; it gives the retina a funnel shape. r., **detachment of,** disconnection from the choroid. r., **epilepsy of,** a symptom of migraine or of epilepsy, characterized by transient loss of sight. r., **fovea centralis of.** See *fovea*. r. **leopard or tiger,** the appearance of the retina in chronic retinitis pigmentosa. r., **limbus luteus of.** See *macula lutea*. r., **membrana limitans of.** See *membrana limitans*. r., **physiological,** middle point of. See *fovea centralis*. r. **pulsation of.** See *pulsation*. r., **shot-silk appearance of.** See *reflex, watered silk*. r., **sustentacular fibers of.** See *Mueller, fibers of*. r. **watered-silk appearance of.** See *reflex, watered-silk*.
**retinacula** (*ret-in-ak'-u-lah*) [L.]. Plural of *retinaculum, q. v*.
**retinaculum** (*ret-in-ak'-u-lum*) [L., "a band"]. A band or membrane holding back an organ or part. r. **costæ ultimatæ.** Same as lumbo-costal ligament, *q. v*. r. **retinacula cutis,** fibrous bands connecting the corium with the underlying fascia. r. **ligamenti arcuati,** the short external lateral ligament of the knee-joint. r. **Morgagni,** r. of the **ileocecal valve,** the ridge formed by the coming together of the valve-segments at each end of the opening between the cecum and the ileum. r. **muscularis tendinis subscapularis majoris,** a name for the inconstant brachio-capsularis muscle originating in the shaft of the humerus and inserted into the capsular ligament of the shoulder joint. **retinacula ossis brachii,** fibrous bands inserted into the neck of the humerus and having their origin in the capsule of the humero-scapular articulation. r. **patellæ externum,** the lateral patellar ligaments. r. **patellæ internum,** the ligamentum patellæ mediale. r. **peroneorum inferius,** a fibrous band running over the peroneal tendons as they pass through the grooves on the outer side of the calcaneum. r. **peroneorum superius,** the external annular ligament of the ankle-joint. r. **tendineum,** the annular ligament of the wrist or ankle. **retinacula valvulæ.** See *retinaculum Morgagni*.

**retinal** (*ret'-in-al*) [*retina*]. Pertaining to or affecting the retina. r. **apoplexy,** hemorrhage into the retina. r. **horizon,** a term used by Helmholtz to describe the horizontal plane passing through the transverse axis of the eyeball. r. **image,** the image of external objects as reflected on the retina. r. **ischemia,** anemia of the retina. r. **melanin.** See *fuscin*. r. **purple.** Same as *rhodopsin*.

**retinitis** (*ret-in-i'-tis*) [*retina*; *ιτις*, inflammation]. Inflammation of the retina. r. **albuminurica,** the form due to nephritis, usually chronic. r. **apoplectica,** retinal apoplexy. r., **central punctate,** a form of syphilitic retinitis in which a small number of striæ or small yellowish-white spots and pigment-dots are most seen in the aged. A great number of striæ or white spots are visible in the fundus. r., **central recurrent,** a rare form of syphilitic retinitis characterized by a central dark scotoma which disappears in a few days to return in a few weeks; the attacks becoming more frequent. r., **central relapsing,** a form of syphilitic retinitis in which there is a gray or yellow area in the muscular region, or numerous small yellowish-white spots and pigment-dots. r. **cerebralis,** retinitis due to intracranial inflammation. r., **choroido-,** a form of syphilitic retinitis with cellular infiltration, exudation, atrophy, and proliferation of the pigment-epithelium in the choroid, between the choroid and retina, and in the retinal layers. r. **circumpapillaris,** a form in which there is proliferation of the outer layers of the retina around the disc. r., **diabetic,** the form of retinitis occurring in diabetes. r., **diffuse.** See *r. serosa*. r., **diffuse parenchymatous,** the parenchymatous and small yellowish-white spots and pigment-dots. r., **glycosuric.** Same as *r., diabetic, q. v*. r. **gravidarum,** a form occurring in pregnant women and which is similar to retinitis albuminurica, and is of grave prognostic import. r. **hæmorrhagica,** a form in which there is swelling of the papilla and opaque infiltration of the surrounding retina; there are distended, dark, and tortuous veins, and the arteries are small; there are hemorrhages, linear or irregular and round in appearance. r. **hepatica,** a rare form whih sometimes occurs in cases of parenchymatous hepatitis. r. **leukæmica,** a form characterized by pallor of the retinal vessels and optic disc, the boundary of the latter being indistinct. Hemorrhages appear at various points of the membrane, while numerous white patches and round bodies are visible about the disc in the retina. r. **macularis.** Same as *r., central relapsing, q. v*. r. **nephritica.** See *r. albuminurica*. r. **nyctalopia,** a diffuse, streaked opacity of the retina and swelling of the disc, with central scotoma or color-scotoma, and more or less marked amblyopia. It indicates retro-bulbar neuritis. r. **paralytica** (of Klein), retinitis caused by paralysis affecting the optic nerve. r., **parenchymatous,** a simple chronic retinitis affecting the connective tissue of the retina. r. **pigmentosa,** an affection involving all the layers of the retina, and consisting in a slowly-progressing connective-tissue and pigment-cell proliferation of the entire membrane, with wasting of its nerve-elements. r. **postica,** inflammation of the ectal retinal layer. r. **proliferating,** a development of connective tissue with the formation of dense bluish white masses within the retina, and extending into the vitreous humor. r. **punctata albescens.** Same as *r., central punctate*. r., **purulent,** a form in which there are small circumscribed white spots near the papilla and in the macular region. r., **renal.** See *r. albuminurica*. r., **septic.** Same as *r., purulent*. r. **serosa,** a form characterized by an infiltration, most marked in the nerve-fiber and ganglionic layer of the retina, creating opacity, edema, and hyperemia, most marked in the fovea. r., **simple syphilitic,** a form of syphilitic retinitis in which the ophthalmoscope shows a gray opacity surrounding the papilla, which is discolored and cloudy, and the veins darker than normal. r. **simplex.** Same as

r. *serosa*. r., **solar**, retinal change from the effect of sunlight. r. **sympathetica**, retinitis of sympathetic origin, and attended with retinal hyperemia, redness of the disc, engorgement of the veins, and great disturbance of vision. r., **syphilitic**, the form occurring in syphilis; it is chronic, diffuse, and a late manifestation of the systemic disease.

**retinochoroiditis** (*ret-in-o-ko-roi-di'-tis*) [*retina; choroiditis*]. Inflammation of the retina and choroid.

**retinoid** (*ret'-in-oid*) [ῥητίνη, resin; εἶδος, form]. Resin-like, or in the form of a resin.

**retinol** (*ret'-in-ol*) [*resin*], C₁₀H₁₆. A liquid hydrocarbon obtained in the destructive distillation of resin. It is used as a solvent and has also been employed in gonorrhea.

**retinoscopy** (*ret-in-os'-ko-pe*) [*retina;* σκοπεῖν, to view]. A method of determining the refraction of the eye by observation of the movements of the retinal images and shadows through the ophthalmoscopic mirror. Syn., *skiascopy*.

**retort** (*re-tort'*) [*re*, back; *torquere*, to twist]. A vessel employed in distillation, consisting of an expanded globular portion and a long neck, and containing the liquid to be distilled.

**retract** (*re-trakt'*) [*re*, back; *trahere*, to draw]. To draw back; to contract; to shorten.

**retractile** (*re-trak'-til*) [*retrahere*, to draw back]. That which may be drawn back. r. **carcinoma**, mammary carcinoma with retraction of the nipple.

**retractility** *re-trak-til'-it-e*) [*retract*]. The power of retracting or drawing back.

**retraction** (*re-trak'-shun*) [*retract*]. The act of retracting or drawing back, as a retraction of the muscles after amputation. Shortening.

**retractor** (*re-trak'-tor*) [*retract*]. An instrument for drawing back the lips of a wound so as to give a better view of the deeper parts.

**retrad** (*re'-trad*) [*retro*, backward]. In or toward the rear.

**retrahens aurem** (*re-tra'-hens aw'-rem*). Drawing back the ear. See under *muscle*.

**retrahent** (*re'-tra-hent*) [*retrahens*, drawing back]. Drawing backward; retracting.

**retrenchment** (*re-trench'-ment*) [Fr., *retrenchement*]. A plastic operation the object of which is to obtain cicatricial contraction by the removal of superfluous tissue.

**retro-** (*re-tro-*). A prefix meaning back, backward, or behind.

**retroaction** (*re-tro-ak'-shun*) [*retro-; agere*, to do]. Reverse action.

**retroanteroamnesia** (*re-tro-an-ter-o-am-ne'-ze-ah*). See *amnesia*, *retroanterograde*.

**retroanterograde** (*re-tro-ant'-er-o-grād*) [*retro-; anterius*, before; *gradi*, to go]. Reversing the order of succession. r. **amnesia.** See *amnesia, retroanterograde*.

**retroauricular** (*re-tro-aw-rik'-u-lar*). Dorsad of the auricle of the ear or of the heart.

**retrobuccal** (*re-tro-buk'-al*) [*retro-; bucca*, the cheek]. Pertaining to the back part of the mouth or of the cheek.

**retrobulbar** (*re-tro-bul'-bar*) [*retro-; bulbar*]. 1. Situated or occurring behind the eyeball. 2. Behind the medulla oblongata. r. **neuritis,** inflammation in the orbital part of the optic nerve. r. **perineuritis,** inflammation of the sheath of the orbital part of the optic nerve.

**retrocecal** (*re-tro-se'-kal*). Pertaining to the back of the cecum.

**retrocedent** (*re-tro-se'-dent*) [*retro-; cedere*, to go]. Going back; disappearing from the surface. r. **gout**, a form of gout in which the joint-inflammation suddenly disappears and is replaced by affections of the internal organs.

**retroceps** (*re'-tro-seps*) [Fr.]. A variety of obstetrical forceps used to grasp the fetal head from behind.

**retrocervical** (*re-tro-ser'-vik-al*) [*retro-; cervix*, neck]. Situated behind the cervix uteri.

**retrocession** (*re-tro-sesh'-un*) [*retrocede*]. The act of going back.

**retroclusion** (*re-tro-kloo'-zhun*) [*retro-; claudere*, to shut]. A form of acupressure in which the pin is passed first above the artery into the tissues on the other side, then below the artery into the tissues upon the side first entered.

**retrocolic** (*re-tro-kol'-ik*). Behind the colon.

**retrocollic** (*re-tro-kol'-ik*) [see *retrocollis*]. Pertaining to the muscles at the back of the neck.

r. **spasm**, spasm of the muscles at the back of the neck, causing retraction of the head.

**retrocollis** (*re-tro-kol'-is*) [*retro-; collis*, the nape of the neck]. Torticollis.

**retrocopulation** (*re-tro-kop-ū-la'-shun*) [*retro-; copulare*, to copulate]. The act of copulating from behind or aversely.

**retrodeviation** (*re-tro-de-ve-a'-shun*) [*retro-; deviation*]. Any backward displacement; a retroflexion or retroversion.

**retrodisplacement** (*re-tro-dis-plās'-ment*) [*retro-; displacement*]. Backward displacement of a part or organ.

**retroesophageal** (*re-tro-e-sof-aj'-e-al*) [*retro-; esophagus*]. Located behind the esophagus.

**retroflected** (*re-tro-flek'-ted*). Same as *retroflexed*.

**retroflection** (*re-tro-flek'-shun*). See *retroflexion*.

**retroflex** (*re'-tro-fleks*) [*retro-; flectere*, to turn]. Turning back abruptly.

**retroflexed** (*re-tro-flekst'*) [*retro-; flectere*, bend]. Bent backward.

**retroflexion** (*re-tro-flek'-shun*) [*retro-; flexion*]. The state of being bent backward. r. **of the uterus**, a condition in which the uterus is bent backward upon itself, producing a sharp angle in its axis.

**retrograde** (*ret'-ro-grād* or *re'-tro-grād*) [*retro-; gradi*, to go]. Going backward; undoing. r. **carcinoma**, a carcinoma which grows firmer and less in size and remains so. r. **embolism**, embolism in which the embolus has gone against the normal direction of the blood-stream. r. **metamorphosis**, katabolic change.

**retrography** (*re-trog'-ra-fe*) [*retro-;* γράφειν, to write]. Backward writing; mirror-writing.

**retroinsular** (*re-tro-in'-sū-lar*) [*retro-; insula*, island]. Situated behind the island of Reil, as the *retroinsular convolutions*.

**retroiridian** (*re-tro-i-rid'-e-an*) [*retro-; iris*]. Behind the iris.

**retrojection** (*re-tro-jek'-shun*) [*retro-; jectio*, a throwing]. The washing out of a cavity from within outward.

**retrojector** (*re'-tro-jek-tor*) [*retro-; jacere*, to throw]. An instrument for washing out the uterus.

**retrolingual** (*re-tro-ling'-gwal*) [*retro-; lingua*, the tongue]. Relating to that part of the throat back of the tongue.

**retromalleolar** (*re-tro-mal-e'-o-lar*) [*retro-; malleolus*]. Located back of a malleolus.

**retromammary** (*re-tro-mam'-ar-e*) [*retro-; mamma*, breast]. Situated or occurring behind a mammary gland.

**retromastoid** (*re-tro-mas'-toid*). Behind the mastoid.

**retromaxillary** (*re-tro-maks'-il-a-re*) [*retro-; maxilla*]. Situated behind the maxilla.

**retromorphosis** (*re-tro-mor-fo'-sis*) [*retro-;* μορφή, form]. Katabolism; retrograde metamorphosis; katabolic change.

**retronasal** (*re-tro-na'-zal*) [*retro-; nasus*, nose]. Situated behind the nose or nasal cavities.

**retro-ocular** (*re-tro-ok'-ū-lar*). See *retrobulbar* (1).

**retroperitoneal** (*re-tro-per-it-on-e'-al*) [see *retroperitoneum*]. Situated behind the peritoneum.

**retroperitoneum** (*re-tro-per-it-on-e'-um*) [*retro-; peritoneum*]. The space lying behind the peritoneum and in front of the spinal column and lumbar muscles.

**retroperitonitis** (*re-tro-per-it-on-i'-tis*) [*retro-; peritonitis*]. Inflammation of the retroperitoneal structures.

**retropharyngeal** (*re-tro-far-in'-je-al*) [*retro-; pharynx*]. Situated behind the pharynx, as *retropharyngeal abscess*.

**retropharyngitis** (*re-tro-far-in-ji'-tis*) [*retro-; pharyngitis*]. Inflammation of the retropharyngeal tissues.

**retropharynx** (*re-tro-far'-inks*). The posterior portion of the pharynx.

**retroplacental** (*re-tro-pla-sent'-al*). Behind the placenta.

**retroposed** (*re'-tro-pōzd*) [*retro-; ponere*, to place]. Displaced backward.

**retroposition** (*re-tro-po-zish'-un*) [*retro-; position*]. Backward displacement of the uterus without flexion or version.

**retropulsion** (*re-tro-pul'-shun*) [*retro-; pellere*, to drive]. 1. A driving or turning back, as of the fetal head. 2. A running backward; a form of walking sometimes seen in paralysis agitans.

**retrostalsis** (*re-tro-stal'-sis*) [*retro-;* στάλσις, com-

pression]. Reversed peristalsis; peristaltic action that tends to drive the intestinal contents cephalad instead of caudad.
**retrosternal** (*re-tro-ster'-nal*) [*retro-*; *sternum*]. Situated behind the sternum.
**retrotarsal** (*re-tro-tar'-sal*) [*retro-*; *tarsus*]. Situated behind the tarsus, as the *retrotarsal* fold of the conjunctiva. r. fold. See *fornix conjunctivæ*.
**retrotracheal** (*re-tro-tra'-ke-al*) [*retro-*; *trachea*]. Situated or occurring behind the trachea.
**retrouterine** (*re-tro-ū'-ter-in*) [*retro-*; *uterus*]. Behind the uterus. r. **hematocele**, a blood-tumor behind the uterus in the pouch of Douglas.
**retrovaccination** (*re-tro-vak-sin-a'-shun*) [*retro-*; *vaccination*]. Vaccination with virus from a cow that had been inoculated with the virus of smallpox from a human subject.
**retrovaccine** (*re-tro-vak'-sēn*) [*retro-*; *vaccine*]. The virus obtained after inoculating a cow with human virus.
**retroversioflexion** (*re-tro-ver-se-o-flek'-shun*). Combined retroversion and retroflexion.
**retroversion** (*re-tro-ver'-shun*) [*retro-*; *version*]. A turning back. r. of uterus, a condition in which the uterus is tilted backward without curvature of its axis.
**retroverted** (*re'-tro-ver-ted*). Tilted or turned backward, as a *retroverted* uterus.
**Retzius' brown striæ** (*ret'-ze-us*) [1. Anders Adolf *Retzius*, Swedish anatomist, 1796–1860; and 2. Magnus Gustav *Retzius*, Swedish histologist, 1842–   ]. [2] Brownish concentric lines in the enamel of the teeth, running nearly parallel to the surface. **R.'s capsule**, [1] the fascial formation investing the intrapelvic and bulbous portions of the urethra and Cowper's glands. **R.'s fibers**, [1] the rigid filaments of Deiters' cells in the organ of Corti. **R.'s ligament**, [1] the outer portion of the anterior annular ligament of the ankle which forms a loop around the peroneus tertius and the extensor longus digitorum. **R.'s space**, [1] a triangular space the basis of which lies between the spines of the pubes, the apex being from 5 to 7 centimeters above. In this space, which is filled with connective tissue, the bladder is not covered by the peritoneum. Syn., *cavum Retzii*. **R.'s veins**, [1] the veins forming anastomoses between the mesenteric veins and the inferior vena cava.
**reunient** (*re-ūn'-yent*) [*re*, again; *unire*, to unite]. Uniting divided parts.
**reunion** (*re-ūn'-e-ol*) [*Reunion*, an island in the Indian Ocean; *oleum*, oil]. A proprietary substitute for attar of rose, said to be derived from Algerine, French and Reunion geranium oil. It resists oxidation and has the perfume of the tea rose. It is soluble in alcohol, fats, and fixed oils.
**reunion** (*re-ūn'-yun*) [*re*, again; *unio*, to become one]. The joining of parts whose continuity has been destroyed. r. of wound. See *healing*.
**Reusner's sign of early pregnancy** (*roys'-ner*). An increase in the volume of the pulsation of the uterine arteries may be perceived through the vagina in the posterior culdesac as early as the fourth week.
**Reuss' formula** (*roys*). The formula by means of which the amount of albumin contained in pathological exudates and transudates can be approximately calculated when the specific gravity, that depends upon the amount of albumin present, is known: $E = \frac{3}{4}(S - 1000) - 2.8$; $E$ = percentage of albumin contained in the fluid; $S$ = specific gravity of the fluid. **R.'s test for atropine**, heat the substance to be tested with sulphuric acid and an oxidizing agent; in the presence of atropine a fragrance as of roses and orange-flowers is given off.
**revaccination** (*re-vak-sin-a'-shun*) [*revaccinatio*]. Renewed or repeated vaccination.
**revalenta** (*rev-al-en'-tah*). A commercial and proprietary food-preparation for invalids, said to be composed principally of lentil meal.
**reveilleur** (*ra-va'-yur*) [Fr.]. The instrument used in Baunscheidtism.
**revellent** (*re-vel'-ent*). See *revulsive*.
**Reverdin's method, or operation** (*re-vēr-dan'*) [Auguste *Reverdin*, Swiss surgeon, 1849–1908]. 1. *For blepharoplasty:* removal of the cicatricial tissue, suturing of the lid to the opposite one in its normal position, and skin-grafting of the raw surface. 2. *For skin-grafting:* a point of skin is raised on an ordinary sewing-needle, and shaved off with a scalpel or scissor; the graft is then transferred to the fresh surface next to the healthy granulations. 3. *For* *symblepharon:* detachment of the lid and transplantation of a small flap from the cheek.
**reverie** (*rev'-er-e*) [Fr., *reverie*]. A state of dreamy abstraction; visionary mental or ideational movement, the mind itself; at least so far as volition is concerned, being passive.
**reverse** (*re-vers'*) [*revertere*, to turn back]. In bandaging, a half-turn employed to change the direction of a bandage.
**reversible reaction, or equation.** One in which the displacement may occur in either direction.
**reversion** (*re-ver'-shun*) [*revertere*, to turn back]. In theology: 1. The appearance of characteristics which existed in remote ancestors. 2. The backward development of plant-organs, as stamens into petals, etc. 3. Becoming wild after having been domesticated or cultivated. 4. The chemical action opposed to inversion (the hydrolytic cleavage of compound sugars into monosaccharids) whereby monosaccharids are condensed into complicated carbohydrates. r., **neogenetic**, the anomalous adult development of an embryonic rudiment. r., **paleogenic**, reversion to an atavus so remote that the rudiment is not even represented in the embryo.
**Révilliod's sign** (*ra-ve-yo'*) [Henri *Révilliod*, Swiss physician]. Inability of the patient to close the eye of the affected side only; it is observed in paralysis of the facial nerve. Syn., *signe de l'orbiculaire*.
**revitalization** (*re-vi'-tal-i-za'-shun*) [*re*, again; *vita*, life]. The act or process of refreshing or revitalizing.
**revive** (*re-viv'*) [*re*, again; *vivere*, to live]. To return to life after seeming death.
**revivification** (*re-viv-if-ik-a'-shun*) [*revivificatio*]. 1. Restoration to consciousness. 2. The refreshening of surfaces by paring before placing them in apposition.
**reviviscence** (*re-viv-is'-ens*) [*reviviscere*, inceptive of *revivere*, to revive]. The awaking from a period of dormancy; state of insects after hibernation.
**revulsant** (*re-vul'-sant*) [*revellere*, to push away]. 1. Revulsive. 2. A medicine or agent that, by irritation, draws the blood from a distant part of the body.
**revulsion** (*re-vul'-shun*) [*re*, back; *vellere*, to pluck]. A plucking or driving backward; specifically, the diverting of disease from one part to another by the sudden withdrawal of the blood from the part.
**revulsive** (*re-vul'-siv*) [see *revulsion*]. 1. Causing revulsion. 2. An agent that causes revulsion.
**revulsor** (*re-vul'-sor*). 1. An apparatus for effecting revulsion by the alternate application of heat and cold. 2. A plate or cylinder set with needles, used in producing counterirritation.
**Reybard's suture** (*ra'-bar*). An interrupted loop-suture for wounds of the intestine.
**Reynold's test for acetone.** To the liquid to be tested add freshly precipitated mercuric oxide, shake and filter. If acetone is present, the filtrate will contain mercury, owing to the acetone dissolving freshly precipitated mercuric oxide. The mercury may be detected by overlaying the filtrate with ammonium sulphide, which turns black.
**R. F. A.** Abbreviation for *right frontoanterior* position of the fetus.
**R. F. P.** An abbreviation for *right frontoposterior* position of the fetus.
**Rh.** Chemical symbol of rhodium.
**Rhabditis** (*rab-di'-tis*) [$\dot{\rho}\alpha\beta\delta os$, a rod]. A genus of the nematode worms a few species of which are parasitic in man.
**rhabdium** (*rab'-de-um*) [dim. of $\dot{\rho}\alpha\beta\delta os$, a rod]. A fiber of striped or voluntary muscle.
**rhabdoid** (*rab'-doid*) [$\dot{\rho}\alpha\beta\delta os$, a rod; $\epsilon l\delta os$, like]. Rod-like. r. **suture**, the sagittal suture.
**rhabdomyoma** (*rab-do-mi-o'-mah*) [$\dot{\rho}\alpha\beta\delta os$, a rod; *myoma*]. A form of myoma characterized by the presence of striated muscular fibers.
**Rhabdonema** (*rab-do-ne'-mah*) [$\dot{\rho}\alpha\beta\delta os$, a rod; $\nu\hat{\eta}\mu\alpha$, a thread]. A genus of parasitic round-worms.
**rhachi-** (*ra'-ke*) [*rhachis*]. A prefix meaning relating to the spine.
**rhachiagra, rachiagra** (*ra-ke-a'-grah*) [*rhachi-*; $\ddot{\alpha}\gamma\rho\alpha$, a seizure]. Gouty or rheumatic pain in the muscles of the spine.
**rhachial, rachial** (*ra'-ke-al*) [$\dot{\rho}\dot{\alpha}\chi\iota s$, spine]. Pertaining to the spine.
**rhachialgia, rachialgia** (*ra-ke-al'-je-ah*) [$\dot{\rho}\dot{\alpha}\chi\iota s$, spine; $\ddot{\alpha}\lambda\gamma os$, a pain]. Any pain in the spine. Spinal irritation. r. **mesenterica**, see *tabes mesenterica*, *q. v.*

**rhachialgitis, rachialgitis** (rak-e-al-ji'-tis) [ῥάχις, spine; ἄλγος, pain; ιτις, inflammation]. Inflammatory rhachialgia.

**rhachiasmus, rachiasmus** (ra-ke-as'-mus) [ῥάχις, spine]. Spasm of the muscles at the back of the neck, as seen in the early part of many epileptic attacks.

**rhachicentesis, rachicentesis** (ra-kis-en-te'-sis) [rhachi-; κέντησις, puncture]. Puncture into the spinal canal.

**rhachicocainization, rachicocainization** (ra-ke-ko-ka-in-i-za'-shun) [rhachi-; cocainization]. The induction of anesthesia by the injection of a solution of cocaine hydrochloride into the subarachnoid space by means of a lumbar puncture.

**rhachidial, rachidial** (ra-kid'-e-al) [ῥάχις, spine]. Pertaining to a rhachis, or spine.

**rhachidian, rachidian** (ra-kid'-e-an) [see rhachidial]. Spinal; vertebral.

**rhachilysis, rachilysis** (ra-kil'-is-is) [ῥάχις; spine; λύειν, to loose]. A method of treating lateral curvature of the spine by mechanical counteraction on the abnormal curves.

**rhachio-** or **rachio-** (ra'-ke-o) [ῥάχις, spine]. A prefix denoting connection with or relation to the spine.

**rhachiocampsis, rachiocampsis** (ra-ke-o-kamp'-sis) [ῥάχις, spine; κάμψις, a bending]. Curvature of the spine.

**rhachiochysis, rachiochysis** (ra-ke-o-ki'-sis) [ῥάχις, spine; χύσις, a pouring]. An accumulation of water or watery substance within the spinal canal.

**rhachiodynia, rachiodynia** (ra-ke-o-din'-e-ah) [rhachi-; ὀδύνη, pain]. Spasmodic pain in the vertebral column.

**rhachiokyphosis, rachiokyphosis** (ra-ke-o-ki-fo'-sis) [ῥάχις, spine; κύφωσις, a bending]. Gibbosity, or hunch of the back.

**rhachiometer, rachiometer** (ra-ke-om'-et-er) [ῥάχις, spine; μέτρον, a measure]. An instrument used to measure the degree of spinal deformities.

**rhachiomyelitis, rachiomyelitis** (ra-ke-o-mi-el-i'-tis) [ῥάχις, spine; μυελός, marrow; ιτις, inflammation]. Inflammation of the spinal cord. Myelitis.

**rhachiomyelophthisis, rachiomyelophthisis** (ra-ke-o-mi-el-off'-this-is) [ῥάχις, spine; μυελός, marrow; φθίσις, a wasting]. See tabes dorsalis.

**rhachiomyelos, rachiomyelos** (ra-ke-o-mi'-el-os) [ῥάχις, spine; μυελός, marrow]. See spinal cord.

**rhachioparalysis, rachioparalysis** (ra-ke-o-par-al'-is-is) [ῥάχις, spine; paralysis]. Spinal paralysis.

**rhachiophyma, rachiophyma** (ra-ke-o-fi'-mah) [ῥάχις, spine; φῦμα, a growth]. A spinal tumor.

**rhachioplegia, rachioplegia** (ra-ke-o-ple'-je-ah) [ῥάχις, spine; πληγή, stroke]. Spinal paralysis.

**rhachiorrheuma, rachiorrheuma** (ra-ke-or-roo'-mah) [ῥάχις, spine; rheuma]. Spinal rheumatism.

**rhachioscoliosma, rachioscolioma** (ra-ke-o-sko-le-o'-mah) [ῥάχις, spine; σκολίωμα, a curve]. Lateral distortion and curvature of the spine.

**rhachioscoliosis, rachioscoliosis** (ra-ke-o-sko-le-o'-sis) [ῥάχις, spine; scoliosis]. The condition and progress of curvature of the spine.

**rhachiostrophosis, rachiostrophosis** (ra-ke-o-stro-fo'-sis) [ῥάχις, spine; στρόφος, twisted]. Curvature of the spine.

**rhachiotome, rachiotome** (ra'-ke-o-tōm) [ῥάχις, spine; τομή, cutting]. A cutting instrument used in rhachiotomy.

**rhachiotomy, rachiotomy** (ra-ke-ot'-o-me) [rhachi-; τομή, a cutting]. 1. The operation of cutting into or through the vertebral column. 2. The operation of cutting through the spine of the fetus to facilitate delivery.

**rhachipagus, rachipagus** (ra-kip'-a-gus) [ῥάχις, spine; πάγος, anything fixed]. A double fetal monstrosity in which the twins are joined back to back by any portion of the spinal column.

**rhachis, rachis** (ra'-kis) [ῥάχις]. The spinal column. In biology, (a) the main petiole of a compound leaf; (b) the axis of inflorescence; (c) the shaft of a feather; (d) the arched middle area of the dorsal surface of a trilobite. r. nasi, the line extending from the tip to the root of the nose.

**rhachischisis, rachischisis** (ra-kis'-kis-is) [ῥάχις, spine; σχίζειν, to cleave]. A cleft in the vertebral column. Same as spina bifida.

**rhachistovainization, rachistovainization** (ra-ke-sto-va-in-i-za'-shun) [rhachi-; stovaine]. The induction of anesthesia by the injection of a solution of stovaine into the subarachnoid space by means of a lumbar puncture.

**rhachitæ, rachitæ** (ra-ki'-te) [ῥαχῖται]. The muscles attached to the vertebral column.

**rhachitic, rachitic** (ra-kit'-ik) [rhachitis]. Affected with, resembling, or produced by rhachitis; rickety. r. rosary, the row of nodules appearing on the ribs, at their junction with the cartilages, in rhachitis.

**rhachitis, rachitis** (ra-ki'-tis) [rhachi-; ιτις, inflammation]. Rickets, a constitutional disease of infancy, characterized by impaired nutrition and changes in the bones, the symptoms being a diffuse soreness of the body, slight fever, and profuse sweating about the head and neck, and changes in the osseous system, consisting in a thickening of the epiphyseal cartilages and periosteum and a softening of the bones. Through the action of the muscles on the soft bones various deformities are produced, while the periosteal hyperplasia leads to nodular hyperostoses, especially about the head, giving the latter a square appearance (caput quadratum). Dentition and closure of the fontanels are delayed. Nervous symptoms are often present, as feverishness, laryngismus stridulus and convulsions. The liver and spleen are usually enlarged. The etiology is obscure—it has been ascribed to deficiency in the earthy salts, to defect in the osteoblasts, and to microorganismal infection. r. adultorum, osteomalacia; mollities ossium. r. annularis, congenital rhachitis characterized by the production after birth, of furrows of the bones and fractures (Winckler). r. micromelica, intrauterine rhachitis, characterized by shortening of the limbs and thickening of the diaphyses (Winckler). r. senilis. See r. adultorum.

**rhachitism, rachitism** (ra'-kit-ism). Rhachitis.

**rhachitol** (ra'-kit-ol). An extract of suprarenal glands; used in the treatment of rhachitis.

**rhachitome, rachitome** (ra'-kit-ōm) [rhachi-; τέμνειν, to cut]. An instrument for opening the spinal canal.

**rhachitomy, rachitomy** (ra-kit'-o-me) [rhachi-; τέμνειν, to cut]. 1. Section of the spine. 2. Decollation of the fetus.

**rhachitropocainization, rachitropocainization** (ra-ke-tro-pa-ko-ka-in-i-za'-shun). Subarachnoid cocainization by means of tropacocaine.

**rhachus** (rā'-kus) [ῥάχος]. A ragged wound.

**rhaciodynia, rachiodynia** (ra-ke-o-din'-e-ah) [ῥάχις, spine; ὀδύνη, pain]. Pain in the spinal cord.

**rhacoma** (ra-ko'-mah) [ῥάκωμα, to rend]. Excoriation, rent, or chapping. Also, a pendulous condition of the scrotum.

**rhacosis** (ra-ko'-sis) [ῥάκος, a rag]. The condition of one affected with rhacoma.

**rhacous** (ra'-kus) [ῥάκος, a rag]. Wrinkled; lacerated.

**rhæbocrania** or **rhebocrania** (re-bo-kra'-ne-ah) [ῥαιβός, crooked; κρανίον, the skull]. The condition of wry-neck.

**rhæboscelia, rhæbosis** (re-bo-se'-le-ah, re-bo'-sis) [ῥαιβός, crooked; σκέλος, leg]. Crooked-legged.

**rhæstocythemia** (rēs-to-si-the'-me-ah). See rhestocythemia.

**rhagades** (rag'-ad-ēz) [ῥαγάς, fissure]. Linear cracks or fissures, especially in the skin.

**rhagadia** (rag-a'-de-ah). See rhagades.

**rhagoid** (ra'-goid) [ῥάξ, a grape; εἶδος, likeness]. Resembling a grape.

**rhamma** (ram'-ah). Suture, q. v.

**rhamnegin** (ram'-ne-jin), C₁₅H₁₀O₅. A glucoside derived from buckthorn-berries.

**rhamnetin** (ram-ne'-tin), C₁₂H₁₂O₅. See rhamnin.

**rhamnin** (ram'-nin) [ῥάμνος, the buckthorn]. 1. A yellow, neutral, crystalline substance found in buckthorn. It contains rhamnetin, a valuable yellow coloring-matter. 2. A proprietary fluidextract of cascara sagrada; it is recommended in the treatment of obstinate constipation.

**rhamnocathartin** (ram-no-kath-art'-in) [ῥάμνος, buckthorn; καθαρτικός, purging]. A yellow, amorphous, translucent substance; a bitter principle contained in the berries of rhamnus cathartica.

**rhamnose** (ram'-nōs) [rhamnus], C₆H₁₂O₅. One of the glucoses. It results upon decomposing various glucosides with dilute sulphuric acid.

**rhamnoxanthin** (ram-no-san'-thin) [ῥάμνος, buckthorn; ξανθός, yellow]. See frangulin.

**Rhamnus** (ram'-nus) [ῥάμνος, buckthorn]. A genus of trees and shrubs; buckthorns. **R. purshiana** (U. S. P.); cascara sagrada. The dried bark of R. purshiana, the California buckthorn. It is used

as a laxative in habitual constipation. Dose 15 gr. (1 Gm.). Dose of the *extract* (*extractum rhamni purshianæ*, U. S. P.) 4 gr. (0.25 Gm.); of the *fluidextract* (*fluidextractum rhamni purshianæ*, U. S. P.) 15 min. (1 Cc.); of the *aromatic fluidextract* (*fluidextractum rhamni purshianæ aromaticum*, U. S. P.) 15 min. (1 Cc.). See also *Cascara sagrada*.
**rhanter** (*rant'-er*) [ῥαντήρ, a sprinkler]. The inner canthus, *q. v.*
**raphagra** (*raf-a'-grah*) [ῥαφή, a seam; ἄγρα, a seizure]. Pain in the cranial sutures.
**rhaphanedon** (*raf-an'-ed-on*) [ῥαφανηδόν]. A transverse fracture.
**rhaphania** (*raf-a'-ne-ah*). See *raphania*.
**rhaphe** (*raf'-e*). See *raphe*.
**rhatany** (*rat'-an-e*). See *krameria*.
**rhebosis** (*re-bo'-sis*) [ῥαιβός, bent; σκέλος, leg]. Curvature of the legs.
**rhegma** (*reg'-mah*) [ῥῆγμα, a rent]. A rupture of the walls of a vessel or of the containing membrane of a tissue, as, for example, the coats of the eye, the walls of the peritoneum. Also, the bursting of an abscess.
**rhein** (*re'-in*) [*rheum*]. 1. The precipitate from a tincture of *Rheum palmatum;* it is cathartic, tonic, cholagogue, and antiseptic. Dose 1-4 gr. (0.065-0.25 Gm.). 2. Same as *chrysarobin*.
**rhembasmus** (*rem-bas'-mus*) [ῥέμβειν, to wander]. Mental distraction, or wandering.
**rheo-** (*re-o-*) [ῥέος, current]. A prefix denoting pertaining to a current.
**rheochord** (*re'-o-kord*). An instrument serving to graduate the strength of the galvanic current. See *rheostat*.
**rheometer** (*re-om'-et-er*) [*rheo-*; μέτρον, a measure]. 1. A galvanometer. 2. An apparatus for measuring the velocity of the blood-current.
**rheophore** (*re'-o-fōr*) [*rheo-*; φέρειν, to bear]. An electrode.
**rheoscope** (*re'-o-skōp*) [*rheo-*; σκοπεῖν, to see]. An instrument for demonstrating the existence of an electric current; a galvanoscope.
**rheostat** (*re'-o-stat*) [*rheo-*; ἱστάναι, to stand]. An instrument introduced into an electric current and offering a known resistance, for the purpose of regulating the strength of the current.
**rheotachygraphy** (*re-o-tak-ig'-raf-e*) [*rheo-*; ταχύς, swift; γράφειν, to write]. The registration of the curve of variation in electromotive action of muscles.
**rheotaxis** (*re-o-taks'-is*) [*rheo-*; τάξις, orderly arrangement]. The reaction of a body to a current of fluid, whereby that body is induced to move either with or against the current of the fluid.
**rheotome** (*re'-o-tōm*) [*rheo-*; τέμνειν, to cut]. An instrument for breaking and making a galvanic circuit; an interrupter. r., **differential**, one for indicating the negative variation in muscle-currents.
**rheotrope** (*re'-o-trōp*) [*rheo-*; τρέπειν, to turn]. An apparatus for reversing the direction of an electric current.
**rheotropism**. Rheotaxis.
**rhestocythemia, rhestocythæmia** (*res-to-si-the'-me-ah*) [ῥαιστός, destroyed; κύτος, cell; αἷμα, blood]. The presence of broken-down erythrocytes in the blood.
**rhēum** (*re'-um*). See *rhubarb*.
**rheum** (*room*) [ῥεῦμα, from ῥεῖν, to flow]. Any watery or catarrhal discharge. r., **salt-**, eczema.
**rheuma** (*roo'-mah*). Same as *rheum*. r. **epidemicum**. Synonym of *influenza*. r. **ventris**. Synonym of *dysentery*.
**rheumagon** (*roo'-ma-gon*) [*rheum*; ἄγειν, to carry off]. A proprietary preparation of sodium iodide and sodium phosphate for use in gout and syphilis.
**rheumarthritis, rheumarthrosis** (*roo-mar-thri'-tis, roo-mar-thro'-sis*) [*rheum*; ἄρθρον, a joint; ἴτις, inflammation]. Acute articular rheumatism.
**rheumatalgia** (*roo-mat-al'-je-ah*) [*rheum*; ἄλγος, pain]. Rheumatic pain.
**rheumatic** (*roo-mat'-ik*) [*rheum*]. Pertaining to, of the nature of, or affected with rheumatism. r. **diathesis**, the condition of body tending to the development of rheumatism. r. **fever**, acute articular rheumatism. r. **gout**. Synonym of *rheumatoid arthritis*.
**rheumatin** (*roo'-mat-in*). See *saloquinine salicylate*.
**rheumatisant** (*roo-mat'-is-ant*). One affected with rheumatism.
**rheumatism** (*roo'-mat-izm*) [*rheum*]. A constitutional disease characterized by pain in the joints and muscles, tending to recur, and associated with exposure to cold and wet. r., **acute articular**, a form characterized by fever, by swelling of various joints, beginning usually in one and rapidly spreading to others, by acid sweats, and by a marked tendency to involve the endocardium, less frequently the pericardium, pleura, and peritoneum. The iris and conjunctiva may also become affected. r., **chronic**, a chronic form in which the symptoms are milder and in which the disease attacks either the muscles (*muscular rheumatism*) or the joints (*chronic articular rheumatism*). r., **gonorrheal**, joint-inflammation occurring in association with gonorrheal urethritis. It generally involves but one. joint; if several are affected, it is usually the smaller joints. The course is chronic. r., **inflammatory**, acute articular rheumatism. r., **muscular**, muscular pain with or without fever and other rheumatic symptoms. r., **synovial**, a rheumatic disorder of the synovial membranes with serous accumulation. r., **tuberculous**, arthritis due to the toxins of tuberculosis; Poncet's disease.
**rheumatismal** (*roo-ma-tiz'-mal*) [*rheumatism*]. Pertaining to rheumatism. r. **edema**, rheumatism with painful subcutaneous edema.
**rheumatismoid** (*roo-mat-iz'-moid*). See *rheumatoid*.
**rheumatismus** (*roo-ma-tiz'-mus*). Rheumatism.
**rheumato-, rheumo-** (*roo-mat-o-, roo-mo-*) [*rheumatism*]. Prefixes meaning relating to rheumatism.
**rheumatocolica** (*roo-mat-o-kol'-ik-ah*) [*rheumato-*; *colic*]. Rheumatic colic.
**rheumatodynia** (*roo-mat-o-din'-e-ah*) [*rheumato-*; ὀδύνη, pain]. A dull rheumatic pain.
**rheumatoid** (*roo'-ma-toid*) [*rheumato-*; εἶδος, like]. Resembling rheumatism. r. **arthritis**. See *arthritis, rheumatoid*.
**rheumatokelis** (*roo-mat-o-ke'-lis*) [*rheumato-*; κηλίς, a spot]. Purpura occurring in conjunction with rheumatism.
**rheumatophthisis** (*roo-mat-off'-this-is*) [*rheumato-*; *phthisis*]. Atrophy the result of rheumatism.
**rheumatopyra** (*roo-mat-o-pi'-rah*) [*rheumato-*; πῦρ, fire]. Rheumatic fever.
**rheumatosis** (*roo-mat-o'-sis*) [*rheum*]. The condition due to the action of poisons in the blood affecting the articular and endocardial parts.
**rheumatospasm** (*roo-mat'-o-spazm*) [*rheumato-*; σπασμός, a spasm]. Spasms due to rheumatism.
**rheumic** (*roo'-mik*) [*rheum*]. Pertaining to rheum.
**rheumodontalgia** (*roo-mo-don-tal'-je-ah*) [*rheumo-*; *odontalgia*]. Toothache of rheumatic origin.
**rheumoparotiditis** (*roo-mo-par-o-tid-i'-tis*). Rheumatic parotiditis.
**rheumophthalmia** (*roo-moff-thal'-me-ah*) [*rheumo-*; *ophthalmia*]. Ophthalmia due to rheumatism.
**rheumorchitis** (*roo-mor-ki'-tis*). Orchitis of rheumatic origin.
**rheumorrhea, rheumatorrhœa** (*roo-mo-tor-e'-ah*). Rheumatic otorrhea.
**rheumotylus** (*roo-mo-til'-us*) [*rheumo-*; τύλος, a knob]. A callus the result of rheumatism.
**rhexis** (*reks'-is*) [ῥῆξις, rupture]. Rupture of a vessel or of an organ.
**rhicnosis** (*rik-no'-sis*) [ῥικνός, shriveled]. A wrinkling of the skin, the result of muscular atrophy.
**rhigolene** (*rig'-o-lēn*) [ῥῖγος, cold]. A very volatile liquid obtained from petroleum by distillation, and used as a local anesthetic. Its rapid evaporation freezes and benumbs the part upon which it is sprayed.
**rhigos** (*ri'-gos*) [ῥῖγος, cold]. Synonym of *rigor*.
**rhin-, rhino-** (*rin-, ri-no-*) [ῥίς, nose]. A prefix signifying pertaining to the nose.
**rhinæsthesia** (*ri-nes'-the'-ze-ah*) [*rhin-*]. See *rhinesthesia*.
**rhinæus** (*ri-ne'-us*) [*rhin-*]. Synonym of *compressor naris*. See *muscles, table of*.
**rhinal** (*ri'-nal*) [*rhin-*]. Pertaining to the nose.
**rhinalgia** (*ri-nal'-je-ah*) [*rhin-*; ἄλγος, pain]. Pain in the nose.
**rhinalgin** (*ri-nal'-jin*). A nasal suppository, recommended in coryza, said to contain cacao-butter, 1 Gm.; alumnol, 0.01 Gm.; menthol, 0.025 Gm.; and oil of valerian, 0.025 Gm.
**rhinanchone** (*ri-nan'-ko-nē*) [*rhin-*; ἀγχόνη, strangulation]. Painful constriction of the nasal passages.
**rhinatralgia** (*ri-nat-ral'-je-ah*) [*rhino-*; ἄντρον, a cavity; ἄλγος, pain]. Pain in the cavities of the nose.
**rhinedema, rhinœdema** (*ri-ne-de'-mah*) [*rhin-*; *edema*]. Edema affecting the nose.

**rhinelcos** (*ri-nel'-kos*) [*rhin-*; ἕλκος, an ulcer] A nasal ulcer.

**rhinencephalia** (*ri-nen-sef-a'-le-ah*) [*rhino-*; ἐγκέφαλος, the brain]. A monstrosity with an extreme elongation of the nose.

**rhinencephalic** (*ri-nen-sef-al'-ik*) [*rhino-*; ἐγκέφαλος, brain]. 1. Pertaining to or of the nature of a rhinencephalus. 2. Pertaining to the rhinencephalon.

**rhinencephalon** (*ri-nen-sef'-al-on*) [*rhin-*; ἐγκέφαλος, brain]. The olfactory lobe of the brain.

**rhinencephalus** (*ri-nen-sef'-al-us*). See *rhinencephalus*.

**rhinenchysia** (*ri-nen-ki'-ze-ah*) [*rhino-*; ἔγχυσις, a pouring in]. Douching of the nasal passages.

**rhinenchysis** (*ri-nen'-ki-sis*) [*rhin-*; ἐγχεῖν, to pour in]. The injection of liquid into the nasal cavities.

**rhinenchyta** (*ri-nen'-kit-ah*) [*rhino-*; ἔγχυτος, poured in]. A nasal syringe.

**rhinenchytous** (*ri-nen'-kit-us*) [*rhino-*; ἔγχυτος, poured in]. Pertaining to nasal injections.

**rhinenchytum** (*ri-nen'-kit-um*) [*rhino-*; ἐγχεῖν, to pour in]. A liquid used in nasal douching.

**rhinesthesia, rhinæsthesia** (*ri-nes-the'-ze-ah*) [*rhin-*; αἴσθησις, sensation]. The sense of smell.

**rhineurynter** (*ri-nū-rin'-ter*) [*rhin-*; εὐρύνειν, to dilate]. A distensible bag or sac which is inflated after insertion into the nostril.

**rhinhematoma, rhinhæmatoma** (*rin-hem-at-o'-mah*) [*rhin-*; hematoma]. An effusion of blood into the nasal cartilage.

**rhiniatry** (*ri-ni'-at-re*) [*ῥίς*,¹ the nose; ἰατρεία, a medical treatment]. Synonym of *rhinology*.

**rhinic** (*rin'-ik*) [*rhino-*]. Pertaining to the nose.

**rhinion** (*rin'-e-on*) [*ῥίς*, nose]. The lower point of the suture between the nasal bones. See under *craniometric point*.

**rhinismus** (*ri-nis'-mus*) [*rhino-*]. A nasal quality of voice.

**rhinitis** (*ri-ni'-tis*) [*rhin-*; ιτις, inflammation]. 1. Inflammation of the nasal mucous membrane. 2. A medicinal preparation of belladonna, camphor, and quinine. **r.**, acute, coryza; cold in the head. **r.**, atrophic, that followed by atrophy of the mucous membrane. **r. caseosa**, that marked by gelatinous fetid discharge. **r.**, chronic, a form usually due to repeated attacks of acute rhinitis, and producing in the early stages hypertrophy of the mucous membrane (*hypertrophic rhinitis*) and in the later stages atrophy (*atrophic rhinitis*), and the presence of dark, offensively smelling crusts. **r.**, fibrinous, a rare form of rhinitis characterized by the development of a false membrane in the nose. **r.**, hypertrophic, that marked by hypertrophy of the nasal mucous membrane. **r.**, pseudomembranous. See **r.**, *fibrinous*. **r.**, syphilitic, a chronic form due to syphilis, and usually attended by ulceration and caries of the bone and an offensive discharge (ozena). **r.**, tuberculous, that due to the tubercle bacillus; it is usually associated with ulceration and caries of the bone. **r.**, vasomotor, hay-fever.

**rhino-** (*ri'-no-*) [*ῥίς, ῥινός*, nose]. A prefix denoting relation to or connection with the nose.

**rhinoantritis** (*ri-no-an-tri'-tis*) [*rhino-*; antritis]. Inflammation of the nasal mucous membrane and of the antrum of Highmore.

**rhinoblennorrhœa, rhinoblennorrhea** (*ri-no-blen-or-e'-ah*) [*rhino-*; blennorrhea]. Synonym of *rhinorrhea*.

**rhinobyon** (*ri-no'-be-on*) [*rhino-*; βύειν, to stop]. A nasal plug or tampon.

**rhinocace** (*ri-nok'-as-e*) [*rhino-*; κακός, evil]. Fetid ulceration of the nose.

**rhinocanthectomy** (*ri-no-kan-thek'-to-me*). See *rhinommectomy*.

**rhinocarcinoma** (*ri-no-kar-sin-o'-mah*) [*rhino-*; carcinoma]. Nasal carcinoma.

**rhinocatarrhus** (*ri-no-kat-ar'-rus*) [*rhino-*; catarrh]. Synonym of *coryza*.

**rhinocaul** (*ri'-no-kawl*) [*rhino-*; καυλός, a stalk]. The crus, peduncle, or support of the olfactory bulb.

**rhinocœle, rhinocœle** (*ri'-no-sēl*), or **rhinocœlia** (*ri-no-se'-le-a*) [*rhino-*; κοιλία, hollow]. The hollow, or ventricle, of the rhinencephalon; in man it is very small or quite obliterated.

**rhinocephalus** (*ri-no-sef'-al-us*) [*rhino-*; κεφαλή, head]. A monster in which the nose resembles a tube and the eyes are fused below the nose.

**rhinocheiloplasty** (*ri-no-ki'-lo-plas-te*) [*rhino-*; cheiloplasty]. Plastic surgery of the nose and lip.

**rhinocleisis** (*ri-no-kli'-sis*) [*rhino-*; κλεῖσις, fastening]. Nasal obstruction.

**rhinocnesmus** (*ri-nok-nes'-mus*) [*rhino-*; κνησμός, an itching]. Itching of the nose.

**rhinodacryolith** (*ri-no-dak'-re-o-lith*) [*rhino-*; *dacryolith*]. A lacrimal stone in the nasal duct.

**rhinoderma** (*ri-no-der'-mah*). See *keratosis pilaris*.

**rhinodynia** (*ri-no-din'-e-ah*) [*rhino-*; ὀδύνη, pain]. Any pain in the nose.

**rhinogramma** (*ri-no-gram'-ah*) [*rhino-*; γράμμα, a line]. The nasal line.

**rhinolalia** (*ri-no-la'-le-ah*) [*rhino-*; λαλιά, speech]. A nasal tone in the voice due to nasal defect. The imperfect articulation may be due to undue closure (*rhinolalia clausa*) or to undue patulousness (*rhinolalia aperta*) of the posterior nares.

**rhinolaryngitis** (*ri-no-lar-in-ji'-tis*). Simultaneous inflammation of the mucosa of the nose and larynx.

**rhinolaryngology** (*ri-no-lar-in-gol'-o-je*). The science of the anatomy, physiology and pathology of the nose and larynx.

**rhinolerema** (*ri-no-ler-e'-mah*) [*rhino-*; λήρημα, silly talk]. Same as *rhinoleresis*.

**rhinoleresis** (*ri-no-ler-e'-sis*) [*rhino-*; λήρησις, folly]. Perverted olfactory sense.

**rhinolethrum** (*ri-no-leth'-rum*) [*rhino-*; ὄλεθρος, destruction]. Destruction of the nose.

**rhinolin** (*ri'-no-lin*). A proprietary antiseptic and analgesic substance.

**rhinolith** (*ri'-no-lith*) [*rhino-*; λίθος, a stone]. A nasal calculus.

**rhinolithiasis** (*ri-no-lith-i'-as-is*) [*rhino-*; λίθος, stone]. The formation and presence of nasal calculi.

**rhinolite** (*ri'-no-līt*). See *rhinolith*.

**rhinologic** (*ri-no-loj'-ik*) [*rhinology*]. Pertaining to *rhinology*.

**rhinologist** (*ri-nol'-o-jist*) [*rhinology*]. A specialist in the treatment of diseases of the nose.

**rhinology** (*ri-nol'-o-je*) [*rhino-*; λόγος, science]. The science of the anatomy, functions, and diseases of the nose.

**rhinomanometer** (*ri-no-man-om'-et-er*) [*rhino-*; manometer]. A manometer used for measuring the amount of nasal obstruction.

**rhinometer** (*ri-nom'-et-er*). [*rhino-*; μέτρον, a measure]. An instrument for measuring the nose.

**rhinomiosis** (*ri-no-mi-o'-sis*) [*rhino-*; μείωσις, a lessening]. Operative shortening of the nose.

**rhinommectomy** (*ri-nom-ek'-to-me*) [*rhino-*; ὄμμα, the eye; ἐκτομή, a cutting out]. Excision of the inner canthus of the eye.

**rhinonecrosis** (*ri-no-ne-kro'-sis*). Necrosis of the nasal bones.

**rhinopharyngeal** (*ri-no-far-in'-je-al*). Pertaining to the nose and pharynx, or to the nasopharynx.

**rhinopharyngitis** (*ri-no-far-in-ji'-tis*) [*rhino-*; pharyngitis]. Inflammation of the nose and pharynx, or of the nasopharynx.

**rhinopharyngolith** (*ri-no-far-ing'-go-lith*) [*rhino-*; pharynx; λίθος, a stone]. A nasopharyngeal calculus.

**rhinopharynx** (*ri-no-far'-ingks*). See *nasopharynx*.

**rhinophonia** (*ri-no-fo'-ne-ah*) [*rhino-*; φωνή, sound]. A nasal tone in the voice.

**rhinophyma** (*ri-no-fi'-mah*) [*rhino-*; φῦμα, tumor]. A form of acne rosacea of the nose characterized by a marked hypertrophy of the blood-vessels and the connective tissue, producing a lobulated appearance of the nose.

**rhinoplastic** (*ri-no-plas'-tik*) [*rhino-*; πλάσσειν, to mold]. Pertaining to or having the character of rhinoplasty. **r. operation**, a surgical operation for creating an artificial nose or reconstructing a nose partially destroyed.

**rhinoplasty** (*ri'-no-plas-te*) [*rhino-*; πλάσσειν, to mold]. A plastic operation upon the nose, to replace lost tissue. **r.**, English, Syme's operation, in which flaps are taken from the cheek. **r.**, German, v. Graefe's modification of the Tagliacotian rhinoplasty. The entire operation is done at a single sitting. **r. of v. Graefe**. Same as **r.**, *German*. **r.**, heteroplastic, rhinoplasty in which the tissues are removed from some person other than the one operated upon. **r.**, Indian, an operation originating in India, in which the flap is taken from the forehead. **r.**, Italian. Synonym of **r.**, *Tagliacotian*. **r.**, Langenbeck's, a modification of the Indian method, in which the periosteum is included in the frontal flap. **r.**, osteoplastic, rhinoplasty with transplantation of a cartilaginous flap to replace the septum nasi. **r.**, periosteal. See **r.**, *Langenbeck's*. **r. of Post**, a

**modified Tagliacotian rhinoplasty,** in which the flap is taken from the finger of the patient. **r.** of Syme, English rhinoplasty. **r., Tagliacotian,** rhinoplasty as performed by Tagliacozzi. The flap is taken from the skin of the arm. **r.** of Wood. See *Wood*.

**rhinopolyp, rhinopolypus** (*ri-no-pol'-ip, -us*) [*rhino-*; *polyp*]. Polyp of the nose.

**rhinoptia** (*ri-nop'-she-ah*) [*rhino-*; ὄπτόs, seen]. Internal strabismus.

**rhinorrhagia** (*ri-nor-a'-je-ah*) [*rhino-*; ῥηγνύναι, to burst forth]. Hemorrhage from the nose.

**rhinorrhaphy** (*ri-nor'-a-fe*) [*rhino-*; ῥαφή, suture]. Reduction of the tissue of the nose by section, and by suturing the edges of the wound.

**rhinorrhea, rhinorrhœa** (*ri-nor-e'-ah*) [*rhino-*; ῥοία, a flow]. A mucous discharge from the nose.

**rhinosalpingitis** (*ri-no-sal-pin-ji'-tis*). Simultaneous inflammation of the nasal mucosa and the Eustachian tube.

**rhinosclerin** (*ri-no-skle'-rin*) [*rhinoscleroma*]. A preparation from cultures of *Bacillus rhinoscleromatis;* used in the treatment of rhinoscleroma.

**rhinoscleroma** (*ri-no-skle-ro'-mah*) [*rhino-*; σκληρόs, hard; ὄμα, tumor]. A new growth of almost stony hardness, affecting the anterior nares and adjacent parts. The disease commences in the mucous membrane of the anterior nares and adjoining skin, the lesions consisting of flat, isolated, or coalescent nodules. It is thought to be due to *Bacillus rhinoscleromatis.*

**rhinoscope** (*ri'-no-skōp*) [*rhino-*; σκοπεῖν, to examine]. An instrument for examination of the cavities of the nose.

**rhinoscopic** (*ri-no-skop'-ik*) [*rhino-*; σκοπεῖν, to view]. Pertaining to the rhinoscope, or to rhinoscopy.

**rhinoscopy** (*ri-nos'-ko-pe*) [see *rhinoscope*]. Examination of the nasal fossæ by means of the rhinoscope; that of the anterior nares is termed *anterior rhinoscopy;* that of the posterior nares, *posterior rhinoscopy.*

**rhinosis** (*ri-no'-sis*). Synonym of *rhicnosis*.

**Rhinosporidium kinealyi.** A neosporidium found in India in tumors of the septum nasi.

**rhinostegnosis** (*ri-no-steg-no'-sis*) [*rhino-*; στέγνωσις, obstruction]. Nasal obstruction.

**rhinothrix** (*ri'-no-thriks*) [*rhino-*; θρίξ, a hair; pl., *rhinotriches*]. A hair growing in the nostril; a vibrissa.

**rhinotomy** (*ri-not'-o-me*) [*rhino-*; τομή, á cutting]. Incision of the nose.

**Rhipicephalus** (*ri-pis-ef'-al-us*) [ῥιπίs, a fan; κεφαλή, head]. A genus of ticks. **R. shipleyi,** the brown tick of South Africa, the agent of transmission of Rhodesian cattle disease.

**rhiptasmus** (*rip-taz'-mus*). Synonym of *ballismus*.

**rhizagra** (*riz-a'-grah*) [*rhizo-*; ἄγρα, seizure]. An instrument for extracting the roots of teeth.

**rhizo-** (*ri-zo-*) [*rhizome*]. A prefix meaning root.

**rhizodontropy** (*ri-so-don'-tro-pe*) [*rhizo-*; ὀδούς, tooth; τροπή, turn, pivot]. The pivoting of an artificial crown upon the root of a tooth.

**rhizodontrypy** (*ri-so-don'-trip-e*) [*rhizo-*; ὀδούς, tooth; τρύπη, hole]. Surgical puncture of the root of a tooth.

**rhizoid** (*ri'-zoid*) [*rhizo-*; εἶδος, form]. 1. Like a root. 2. Slender, root-like filaments, the organs of attachment in many cryptogams. 3. A bacterial plate culture of an irregular branched or root-like character.

**rhizoma** (*ri-zo'-mah*) [*pl., rhizomata*]. Same as *rhizome*.

**rhizome** (*ri'-zōm*) [ῥίζα, root]. A subterranean stem having roots at its nodes and a bud or shoot at its apex.

**rhizomelic** (*ri-zo-mel'-ik*) [*rhizo-*; μέλοs, a limb]. Affecting or relating to the roots of members.

**rhizomorphoid** (*ri-zo-mor'-foid*) [*rhizo-*; μορφή, form; εἶδος, like]. Having the form of a root.

**rhizoneure** (*ri'-zo-nūr*) [*rhizo-*; νεῦρον, a nerve]. One of those cells that form nerve roots.

**rhizoneuron** (*ri-zo-nū'-ron*) [see *rhizoneure*]. A neuron the nerve-processes of which leave the spinal cord through the anterior horn; a motor nerve-cell.

**rhizonychia** (*ri-zo-nik'-e-ah*) [*rhizo-*; ὄνυξ, the nail]. The root of the nail.

**rhizonychium** I(*ri-zo-nik'-e-um*) [*rhizo-*; ὄνυξ, the nail]. The root of the nail.

**rhizopod** (*ri'-zo-pod*) [*rhizo-*; πούs, foot]. A member of the *Rhizopoda*, a subclass of protozoa or animalcules.

**rhizotomy** (*ri-zot'-om-e*) [*rhizo-*; τομή, a cutting]. Section of the posterior spinal nerve roots.

**rhodalline** (*ro-dal'-ēn*). See *thiosinamine*.

**rhodanate** (*ro'-dan-āt*). A sulphocyanate.

**rhodeorrhetin** (*ro-de-or-re'-tin*). Synonym of *convolvulin*.

**Rhodesian cattle disease.** An African disease of cattle transmitted by the brown tick, *Rhipicephalus shipleyi.*

**rhodium** (*ro'-de-um*) [ῥόδιοs, rosy]. A rare metal (symbol, Rh; at. wt., 102.9) of the platinum group. Its medicinal qualities are little known. See *elements, table of.*

**rhodogenesis** (*ro-do-jen'-es-is*) [ῥόδον, rose; γέννἀν, to produce]. The regeneration of visual purple which has been bleached by light.

**rhodophane** (*ro'-do-fān*) [ῥόδον, rose; φανήs, appearing]. A red pigment found in the retinal cones.

**rhodophylaxis** (*ro-do-fil-aks'-is*) [ῥόδον, rose; φύλαξις, a guarding]. The property possessed by the retinal epithelium of producing rhodogenesis.

**rhodopsin** (*ro-dop'-sin*) [ῥόδον, rose; ὤψ, eye]. Visual purple; a retinal substance the color of which is preserved by darkness, but bleached by daylight; it is contained in the retinal rods.

**rhœadine** (*re'-ad-in*) [ῥοίαs, a kind of poppy], CHNO₄. A crystallizable alkaloid obtained from *Papaver rhœas.*

**rhœbdesis** (*reb-de'-sis*) [ῥοίβδησις]. Absorption; resorption.

**rhois** (*ro-is'*) [L.]. Genitive of *rhus, q. v.*

**rhombencephalon** (*rom-ben-sef'-al-on*) [ῥόμβοs, a lozenge shaped figure; ἐγκέφαλοs, brain]. The metencephalon or hind-brain together with the myelencephalon or after-brain.

**rhomboatloideus** (*rom-bo-at-loid'-e-us*). See under *muscle*.

**rhombocele, rhombocœle** (*rom'-bo-sēl*). Same as *rhombocœlia.*

**rhombocœlia** (*rom-bo-se'-le-ah*) [ῥόμβοs, rhomb; κοιλία, hollow]. The *sinus rhomboidalis;* a dilatation of the cavity of the spinal cord in the sacral region.

**rhomboid** (*rom'-boid*) [ῥόμβοs, a rhomb; εἶδος, resemblance]. Having a shape similar to that of a rhomb, a quadrilateral figure with opposite sides equal and parallel and oblique angles. **r. body.** See **r.** *fossa*. **r. fossa,** the fourth ventricle of the brain. **r. ligament.** See *ligament, rhomboid*. **r. muscle.** See under *muscle*. **r. sinus.** See **r.** *fossa*.

**rhomboideus** (*rom-boid'-e-us*). See under *muscle*.

**rhoncal** (*rong'-kal*). Same as *rhonchal*.

**rhonchal, rhonchial** (*rong'-kal, rong'-ke-al*) [*rhonchus*]. Relating to or produced by a rhonchus, as *rhonchal* fremitus.

**rhonchus** (*rong'-kus*) [ῥόγχοs, snore]. A rattling sound produced in the throat or bronchial tubes during respiration. See *rale*.

**rhotacism** (*ro'-tas-izm*) [ῥῶ, the Greek ρ, r]. The use of the *r* sound in place of other speech-sounds; the too strong utterance of the letter *r*.

**rhubarb** (*roo'-barb*) [ῥῆον, rhubarb]. The general name for plants of the genus *Rheum*, of the order *Polygonaceæ*. The official drug (*rheum*, U. S. P.; *rhei radix*, B. P.) is the bark of *Rheum officinale* or *Rheum palmatum;* it contains chrysophanic acid, tannic acid (rheotannic acid), and several coloring principles, and is used as a laxative, stomachic, and astringent. Its chief uses are in dyspepsia with constipation, in the diarrhea of children, and in the beginning of bilious fevers. Dose 5–30 gr. (0.32–2.0 Gm.). **r., extract of** (*extractum rhei*, U. S. P., B. P.). Dose 10–15 gr. (0.65–1.0 Gm.). **r., fluidextract of** (*fluidextractum rhei*, U. S. P.). Dose 10–30 min. (0.65–2.0 Cc.). **r., infusion of** (*infusum rhei*, B. P.). Dose 32–64 Cc.). **r., pills of, compound** (*pilulæ rhei composita*, U. S. P.), pills of rhubarb and aloes. Dose 2–4 pills. **r., powder of, compound** (*pulvis rhei compositus*, U. S. P., B. P.), Gregory's powder. Dose ½–1 dr. (2–4 Gm.). **r. and soda, mixture of** (*mistura rhei et sodæ*, U. S. P.), Dose ½–1 dr. (2–4 Gm.). **r., syrup of** (*syrupus rhei*, U. S. P., B. P.). Dose 1 dr. (4 Cc.). **r., syrup of, aromatic** (*syrupus rhei aromaticus*, U. S. P.). Dose 4 Cc.). Both the syrup and the aromatic syrup are used chiefly for children, in the doses given. **r., tincture of** (*tinctura rhei*, U. S. P., B. P.). Dose 1–2 dr. (4–8 Cc.). **r., tincture of, aromatic** (*tinctura rhei aromatica*, U. S. P.). Dose ½–1 dr. (2–4 Cc.). **r., wine of** (*vinum rhei*, B. P.). Dose 1–4 dr. (4–16 Cc.).

**Rhus** (*rus*) [gen., *rhois*]. [ῥοῦς, sumac]. A genus of shrubs or small trees of the order *Anacardiaceæ*. The dried fruit of *R. glabra*, sumac, constitutes the *Rhus glabra* of the U. S. P., and is used as an astringent in inflammations of the mouth and throat, in the form of a decoction or the official *fluidextractum rhois glabræ*. *R. toxicodendron*, the poison-ivy, is a powerful irritant and produces in susceptible persons a violent dermatitis with vesicles and intense itching (ivy-poisoning). The active agent seems to be an acid called toxicodendric acid. In overdoses taken internally it acts as a narcotic poison. It has been employed in chronic rheumatism and in incontinence of urine. *R. venenata*, swamp-sumac, is also poisonous.

**rhusin** (*roo'-sin*). A precipitate from a tincture of the root-bark of sumach, *Rhus glabra*; it is tonic, astringent, and antiseptic. Dose 1 to 2 grains.

**Rhynchota** (*rin-ko'-tah*). An order of sucking insects, including the *Pediculidæ* and the *Acanthiidæ*.

**rhypophobia** (*ri-po-fo'-be-ah*) [ῥύπος, filth; φοβεῖν, to fear]. A morbid dread of filth.

**rhyptic** (*rip'-tik*) [ῥύπτειν, to cleanse]. Detergent; cleansing; cathartic.

**rhypus** (*rip'-us*) [ῥύπος]. Dirt; sordes.

**rhysema** (*ri-se'-mah*) [ῥύσημα]. Wrinkle.

**rhythm** (*rithm*) [ῥυθμός, rhythm]. Action or function recurring at regular intervals. **r., gallop**, a form of heart action in which the cardiac sounds occur in groups of three.

**rhythmic** (*rith'-mik*). Pertaining to or having the quality of rhythm. **r. segmentations**, a term suggested by Cannon for rhythmic localized contractions occurring in the small intestine during digestion.

**rhythmophone** (*rith'-mo-fōn*) [ῥυθμ; φωνή, sound]. A form of microphone for studying the heart-beat and pulse-beat.

**rhytidosis** (*rit-id-o'-sis*). A wrinkling. See also *rutidosis*.

**rib** [AS., *ribb*]. One of the 24 long, flat, curved bones forming the wall of the thorax. **r.s, abdominal**, 1. The floating ribs. 2. Ossifications of the inscriptiones tendineæ. **r.s, asternal**, the false ribs. **r.s, cervical**, rib-like processes extending ventrally from the cervical vertebræ. **r., false**, one of the five lower ribs not attached to the sternum directly. **r., floating**, one of the last two ribs which have one end free. **r.s, short**, the false ribs. **r.s, sternal**, the true ribs. **r., true**, one of the seven upper ribs that are attached to the sternum. **r.s, vertebrochondral**, the highest three false ribs; they are united in front by their costal cartilages.

**Ribble's bandage** (*ribl*). The spica bandage for the instep.

**Ribes' bag** (*rēb*) [Camille Champetier de *Ribes*, French obstetrician, 1848–]. A rubber bag used to dilate the cervix uteri.

**Ribes' ganglion** (*rēb*) [François *Ribes*, French physician, 1800–1864]. A small ganglion of the sympathetic system situated on the anterior communicating artery.

**ribesin** (*ri-be'-zin*) [*ribesium*, currant]. The juice of the black currant, *Ribes nigrum*, used for staining microscopic sections.

**rice** (*ris*). A plant, *Oryza sativa*, of the *Gramineæ*; also its seed. Rice is used as a food, as a demulcent, and, in the form of rice-water, as a drink in fevers. **r.-seed bodies**, peculiar small, white bodies resembling grains of rice, found in the so-called ganglia occurring on tendons. **r.-water evacuations**, the name given to the bowel discharges in cholera.

**Richardson's method of auscultation** (*ritsh'-ard-son*) [Sir Benjamin Ward *Richardson*, English physician, 1828–1896]. The introduction into the esophagus of an elastic bougie or tube connected with the ear-pieces of a stethoscope. **R.'s sign**, a fillet applied to the veins of the arm will not cause filling of the veins on the distal side of the fillet if death be present.

**Richet's bandage** (*re'-sha*) [Didier Dominique Alfred *Richet*, French surgeon, 1816–1891]. A form of plaster-of-Paris bandage to which a small amount of gelatin has been added.

**Richter's hernia** (*rik'-ter*) [August Gottlieb *Richter*, German surgeon, 1742–1812]. Partial enterocele; strangulated enterocele in which only part of the circumference of the gut is constricted; called also *Littré's hernia*.

**Richter-Monro's line**. See *Monro's line*.

**ricin** (*ris'-in*) [*ricinus*, castor oil]. A poisonous proteid found in the castor-oil bean.

**ricini oleum**. Castor oil; see *ricinus*.

**ricinin** (*ris'-in-in*) [*ricinus*]. A poisonous crystalline substance obtained from castor-oil.

**ricinism** (*ris'-in-izm*). Poisoning from the seeds of *Ricinus communis*; it is marked by hemorrhagic gastroenteritis and icterus.

**ricinus** (*ris'-in-us*) [L., "a tick," from the resemblance of the seed to that insect]. A plant or tree, *R. communis*, or castor-oil plant, of the order *Euphorbiaceæ*. **ricini, oleum** (U. S. P.), castor-oil, the fixed oil expressed from the seeds of *R. communis*; it is used as a cathartic in constipation, colic, and irritative diarrheas. Dose ½ oz. (16 Cc.).

**rickets** (*rik'-ets*). See *rhachitis*.

**rickety** (*rik'-et-e*) [*rickets*]. Affected with or distorted by rickets.

**Ricord's chancre** (*re-kor'*) [Philippe *Ricord*, French surgeon, 1800–1889]. The parchment-like initial lesion of syphilis. Syn., *chancre parcheminé*. **r. lupinus**, cleft palate.

**rider's bone**. An osseous formation in the adductor muscles of the leg, from long-continued pressure of the leg against the saddle. **r.'s bursa**, an enlarged bursa, produced in the same way as the rider's bone. **r.'s leg**, strain of the adductor muscles of the thigh.

**ridge** (*rij*) [ME., *rigge*, the back of a man or beast]. An extended elevation or crest. **r., genital**, the germ-ridge, in front of and internal to the Wolffian body, from which the internal reproductive organs are developed. **r., intervertebral**, that on the vertebral end of a rib dividing the articular surface into two portions. **r., maxillary**, the dental crest; a ridge of muscular fibrous tissue along the alveolar processes of the fetus. **r. oblique**, a ridge on the grinding surface of an upper molar tooth. **r.s, occipital**, the superior and inferior curved lines of the occipital bone. **r.s, palatine**, the central ridge together with the lateral corrugations of the mucosa of the hard palate; they are especially noticeable in the human fetus. **r., temporal**, that extending from the external angular process of the frontal bone, across the frontal and parietal bones, and terminating in the posterior root of the zygomatic process.

**ridgel** (*rij'-el*) [origin uncertain]. A male animal having one testicle removed or wanting.

**ridgeling** (*rij'-ling*). See *ridgel*.

**riding of bones**. In surgery, the displacement of the fractured ends of bones which are forced past each other by muscular contraction, instead of remaining end to end.

**Ridley's sinus** [Humphrey *Ridley*, English anatomist, 1653–1708]. The circular sinus.

**Riedel's process, or lobe** (*re'-del*) [Bernhard Moritz Carl Ludwig *Riedel*, German surgeon, 1846– ]. A tongue-shaped process of the liver extending downward, and frequently felt over the enlarged gall-bladder in cases of cholelithiasis.

**Riegel's pulse** (*re'-gel*) [Franz *Riegel*, German physician, 1843–1904]. A pulse which becomes smaller during expiration. **R.'s syndrome**, Riegel's disease; the association of tachycardia with troubles simulating asthma.

**Riegler's test for albumin**. 1. Calcium naphtholsulphonate, 8; citric acid, 8; dissolve in distilled water, 200; 10 Cc. of urine is mixed with 10 to 20 drops of the reagent. Traces of albumin are indicated by a turbidity; larger quantities by a precipitate. Quantitative determination may be made with an albuminometer. 2. Ten Gm. betanaphthalinsulphonic acid are well shaken with 200 Cc. water and filtered. A turbidity or precipitate on adding 20 to 30 drops of reagent to 5 to 6 Cc. of fluid indicates albumin. Sensitiveness, 1 : 40,000. Albumoses and peptones react in a similar manner, but the precipitate disappears on warming and reappears on cooling. **R.'s test for albumoses and peptones**, dissolve 5 Gm. parasitranillin in 25 Cc. water and 6 Cc. concentrated sulphuric acid; add 100 Cc. water, then a solution of sodium nitrite 3 Gm. in 25 Cc. water, and make up to 500 Cc. with water. Filter and preserve in the dark. Mix 10 Cc. reagent with 10 Cc. fluid to be tested, then add 30 drops 10% solution NaOH—if very small quantities of albumoses or peptones are present, a yellowish orange color develops; with notable quantities a blood-red, even the froth on shaking being red. On now adding excess of $H_2SO_4$ the color becomes an orange or brownish precipitate forms. **R.'s test for aldehydes and glucose**, heat 0.1 Gm. phenylhydrazin hydrochloride,

0.5 Gm. crystal sodium acetate, and 1 Cc. sugar solution until dissolved. When near boiling-point add 20 to 30 drops 10% NaOH without shaking—in a few seconds to 5 minutes liquid becomes violet-red, even if there is but 0.005% sugar present. If no sugar present, color will be a slight pink. For sugar in urine, color must develop within one minute to afford physiological significance. Reaction also occurs with aldehydes, hence absence of these must be assured. According to Jolles, absence of albumin must also be assured. Reaction uninfluenced by uric acid or creatinin. **R.'s test for bile-pigments,** on adding an excess of paradiazonitranilin solution to an alkaline solution of bilirubin or biliverdin, intensely colored reddish-violet flocks are precipitated, soluble in chloroform, alcohol, or benzine, and affording reddish-violet or violet solution. **R.'s test for nitrites,** 15 Cc. of the fluid to be examined is mixed in a test-tube with 0.02 to 0.03 Gm. of the naphthol reagent (equal parts naphthionic acid and pure betanaphthol) and 2 or 3 drops concentrated HCl, shaken, and 1 Cc. strong NH₃ poured down the side of the tube, while held in a slanting position; presence of nitrites is indicated by appearance of a red zone, and on shaking the whole solution turns red. **R.'s test for uric acid,** paranitranilin, 0.5 Gm.; water, 10 Cc.; pure concentrated H₂SO₄, 15 drops. Put into a glass flask of 150 Cc. capacity, and heat with agitation until dissolved. Water 20 Cc. is now added, and the mixture cooled quickly, 2.5% NaNO₂ solution to 10 Cc. is added, and diluted, after 15 minutes, with water 60 Cc. The mixture is shaken up repeatedly and filtered. The formation of a blue or green color on adding the reagent and 10% NaOH solution indicates presence of uric acid.
**Rieux's hernia** (re-oo). Retrocecal hernia.
**Riga's disease** (re′-gah). Papillomatous ulceration of the frenum of the tongue, covered with a whitish, diphtheroid exudate.
**Rigal's suture** (re-gal′) [Joseph Jean Antoine Rigal, French surgeon, 1797-1865]. Twisted rubber suture for harelip operations.
**Riggs's disease** [John M. Riggs, American dentist, 1810-1885]. Pyorrhœa alveolaris. See *Fauchard's disease.*
**right** (rīt). Belonging to or located upon that side which, with mammals contains less of the heart and is on the east when the face is toward the north; dextral. **r.-brained,** having the speech-center in the right instead of the left hemisphere. **r.-eared,** preferring the dextral ear as the one with which to hear sounds. **r.-eyed,** preferring the dextral eye as the dominant one. **r.-eyedness,** dextrocularity, the condition of using the right eye with more expertness and correctness than the left. **r.-footed,** choosing the dextral foot as the one to guide and base action, from which to spring in beginning to march, in spading, etc. **r. hand,** see *dexter*. **r.-handed,** using the right hand with more freedom and effect than the left; preferring the right hand for the more expert or intellectual tasks. **r.-handedness,** the condition of being right-handed.
**rigid** (rij′-id) [rigidus, stiff]. Stiff, hard. **r. os,** see *rigidity, anatomical.*
**rigiditas** (rij-id′-it-as) [L.]. Stiffness; rigidity. **r. articulorum,** spurious ankylosis. **r. cadaverica,** rigor mortis.
**rigidity** (rij-id′-it-e) [rigidus, stiff]. Stiffness; inflexibility; immobility; tonic contraction of muscles. **r., anatomical** (*of the cervix uteri*), rigidity in which the cervix, though neither edematous nor tender, is not wholly effaced in labor, but retains its length and dilates only to a certain extent, beyond which the contractions of the uterus are without effect. **r., cadaveric,** rigor mortis. **r., cerebellar,** rigidity of the spinal muscles due to tumor of the middle lobe of the cerebellum. The head is drawn backward, the spine curved, and the arms and legs made rigid. **r., hemiplegic,** spastic rigidity of the paralyzed limbs in hemiplegia. **r., muscular,** see *Thomsen's disease.* **r., pathological** (*of the cervix uteri*), rigidity due to organic disease or cicatricial contraction. **r., postmortem,** rigor mortis. **r., spasmodic** (*of the cervix uteri*), rigidity due to spasmodic contraction of the cervix.
**rigor** (ri′-gor) [rigor, from rigere, to be cold]. Chill. **r. mortis,** the muscular rigidity that occurs a short time after death, due to chemical changes resulting in coagulation of the muscle-plasma and the development of an acid reaction. **r. nervorum,** tetanus.
**rima** (ri′-mah) (L., pl., rimæ). A chink or cleft.

**r. glottidis,** the cleft between the true vocal bands; the glottis. **r. laryngis,** see *r. glottidis.* **r. oris,** the line formed by the junction of the lips. **r. palpebrarum,** the palpebral fissure. **r. pudendi,** the fissure between the labia majora. **r. respiratoria,** the space back of the arytenoid cartilages. **r. vocalis,** see *r. glottidis.*
**Rimini's test for formaldehyde** (re′-min-e). Add to the suspected fluid 3 drops of a dilute solution of phenolhydrazine hydrochloride and then 3 drops of a five per cent. aqueous solution of sodium nitroprusside; then an excess of a saturated aqueous solution of sodium hydroxide; then warm. An intense blue color, gradually changing to green and then to ashy gray follows in the presence of formaldehyde even in minute quantity.
**rimose** (ri′-mōs) [rimosus, full of chinks]. In biology, full of crevices or furrows.
**rimous** (ri′-mus) [rima, a cleft]. Having cracks, clefts, or fissures.
**rimula** (rim′-ū-lah) [dim. of rima, a chink]. A small cleft or fissure, especially of the spinal cord or the cerebellum.
**rinderpest** (rin′-der-pest) [Ger., "cattle-pest"]. An acute infectious disease of cattle, appearing occasionally among sheep and other ruminants.
**Rindfleisch's granule-cells** (rĭnt′-flīsh) [Georg Eduard *Rindfleisch,* German physician 1836-1908]. Eosinophile leukocytes with granulations.
**ring** [ME.]. A circular opening or the structure surrounding it. See *annulus*. **r., abdominal,** see *abdominal.* **r.-bodies,** peculiar ring-shaped bodies in the erythrocytes in pernicious anemia, leukemia, and lead-poisoning. **r., contraction,** see *Bandl.*
**ringed hair,** a very rare form of canities, in which the hairs are white or colored in rings or bands.
**Ringer's solution** (ring′-er). An artificial blood serum, in two strengths: 1. sodium chloride, 7.500; calcium chloride, 0.125; potassium chloride, 0.075; sodium bicarbonate, 0.125; distilled water, 1000. 2. Sodium chloride 9.00; calcium chloride, 0.24; potassium chloride, 0.42; sodium bicarbonate, 0.30; distilled water, 1000. Each of these solutions is to be sterilized.
**ringworm.** Tinea trichophytina, a contagious disease of the skin due to a vegetable parasite, the trichophyton. See *tinea.*
**Rinmann's sign of early pregnancy.** Slender cords radiating from the nipple; they are considered to be hypertrophic acini of the glands.
**Rinné's test** (rin′-nā) [Friedrich Heinrich *Rinné,* German otologist]. A test to determine the condition of the various parts of the ear, performed by applying a vibrating tuning-fork first over the mastoid process, leaving it there until the patient seems no longer to hear the sound, and then as quickly as possible bringing it immediately in front of the external meatus, avoiding all contact with the head or ear. If the patient is then able to hear the sound of the tuning-fork once more, it indicates that the conduction through the air is better than through the bone.
**rinolite** (ri′-no-līt). See *rhinolith.*
**Riolan's arch** (re-ol-on(g)) [Jean *Riolan,* French physician, 1580-1657]. The arch of the mesentery which is attached to the transverse mesocolon. **R.'s bouquet,** the muscular bundle attached to the styloid process and composed of the stylloglossus, stylohyoid, and stylopharyngeus. **R.'s muscle,** the ciliary portion of the orbicularis palpebrarum. **R.'s ossicles,** small bones sometimes found in the suture between the inferior border of the occipital bone and the mastoid portion of the temporal bone.
**ripa** (ri′-pah) [ripa, a bank]. The line formed by the reflection of the endyma upon any plexus or tela of the brain.
**Ripault's sign** (re-po′) [Louis Henri Antonin *Ripault,* French physician, 1807-1856]. A change in the shape of the pupil on pressure upon the eye, transitory during life, but permanent after death.
**ripe** (rīp). Mature, competent.
**ripples** (rip′-els). Scotch vernacular term for locomotor ataxia.
**risiccol** (ris′-ik-ol). A preparation containing chiefly castor oil and magnesia.
**risidontrophy** (ris-id-on′-tro-fe) [ῥίζα, a root]. The operation of drilling the root of a tooth.
**risipola lombarda** (ris-ip-o′-lah lom-bar′-dah). Synonym of *pellagra.*
**risorius** (ri-so′-re-us) [ridere, to laugh]. Laughing. **r. muscle.** See under *muscle.*

**ristin.** The monobenzoic acid ester of ethylene glycoll; it has been recommended in scabies.
**risus** (*ri'-sus*) [L.]. A grin or laugh. r. **caninus,** see *r. sardonicus.* r. **sardonicus,** the sardonic grin, a peculiar grinning distortion of the face produced by spasm of the muscles about the mouth, seen in tetanus.
**Ritgen's method** [Ferdinand August Marie Franz *Ritgen,* German physician, 1787–1867]. A method of manual delivery of the fetal head. It consists in lifting the head upward and forward through the vulva, between the pains, by pressure made with the tip of the fingers upon the perineum behind the anus close to the extremity of the coccyx.
**Ritter's disease** (*rit'-er*) [Gottfried *Ritter* von Rittersheim, German physician, 1820–1883]. Dermatitis exfoliativa of the newborn. **R.'s fiber,** a delicate fiber regarded as a nerve-fiber, seen in the axis of a retinal rod, near the peripheral end of which it forms a small enlargement. **R.'s law** of **contraction,** stimulation of a nerve occurs both at the moment of closing and of opening of the electric current. **R.'s tetanus,** tetanic contractions occurring on the opening of the constant current which has been made to pass for some time through a long section of a nerve. In man the phenomenon does not occur under physiologic conditions, but it is seen in tetany.
**Ritter-Rollet's phenomenon.** Flexion of the foot following the application of a mild galvanic current, and extension following that of a strong current.
**Ritter-Valli's law.** Section of a living nerve is followed by a gradual loss of irritability, preceded by a slight increase, the phenomenon taking place centrifugally from the divided end.
**Rivalle's paste.** A caustic made by adding concentrated nitric acid to lint.
**rivalry** (*ri'-val-re*) [*rivales,* near neighbors who used the same brook]. A struggle for supremacy. r. **of colors,** a rivalry of the visual fields of the two eyes, a different color being presented to each. r. **of contours,** a rivalry of the contours of two objects, one of which is presented to each eye, when they overlap in the binocular field of vision. **r.,** **retinal,** see *r., strife.* r., **strife,** the alternate mastery of one or the other sensation, color, contour, etc., in the eyes when the fields of vision of the two eyes are incapable of being combined into one image. r. **of visual fields,** see *r., strife.*
**Rivalta's test** (*re-val'-tah*) [Sebastiano *Rivalta,* Italian veterinary surgeon]. For *differentiating exudate from transudate:* A drop of the fluid is allowed to fall into a solution of acetic acid (2 drops of glacial acetic acid to 100 Cc. of distilled water). If the drop sinks and leaves a turbidity the fluid is an exudate.
**Riverius' draft** (*rev-e'-re-us*) [Lazarus *Riverius,* French physician, 1589–1655]. A solution of sodium citrate.
**Rivinian canals,** R. **ducts** [Augustus Quirinus *Rivinus,* German anatomist, 1652–1723]. The ducts of the sublingual gland. R. **foramen.** See *Bochdalek's Canal.* R. **glands,** the sublingual glands. R. **ligament.** See *Shrapnell's membrane.* R. **notch,** R. **segment,** a notch of irregular outline at the upper border of the sulcus tympanicus; it is marked at each end by a small spine.
**Rivolta's disease.** Actinomycosis.
**rivulose** (*riv'-u-lōs*) [*rivulus,* a small stream]. In biology, marked with small sinuate lines.
**rixolin** (*riks'-ol-in*). A mixture of petroleum and light oil of camphor.
**riziform** (*ris'-if-orm*). Resembling grains of rice.
**rizine** (*ri'-zēn*). Rice that has been acted upon by superheated steam.
**R. M. A.** An abbreviation for *right mentoanterior position* of the fetus.
**R. M. P.** An abbreviation for *right mentoposterior position* of the fetus.
**R. N.** Abbreviation for *Registered Nurse.*
**R. O. A.** An abbreviation for *right occipito-anterior* position of the fetus.
**roaring** (*rōr'-ing*). A disease of horses that causes them to make a singular noise in breathing under exertion. The disease is due to paralysis and wasting of certain laryngeal muscles, usually of the left side, resulting in a narrowing of the glottis.
**rob.** A confection made of fruit-juice, especially of that of the mulberry.
**Robert's pelvis** (*ro-bair'*) [César Alphonse *Robert,* French surgeon, 1801–1862]. The transversely contracted or doubly synostotic pelvis; ankylosis of both sacroiliac synchondroses, the sacrum being absent or undeveloped.
**Roberts' test for albumin** [Sir William *Roberts,* English physician, 1830–1899]. Float the urine on the surface of a saturated common salt solution containing 5% of hydrochloric acid, of specific gravity 1.052. A white ring or zone formed between the two liquids indicates albumin. Roberts suggests that a mixture of 1 part strong nitric acid and 5 parts saturated magnesium sulphate solution may be employed also. R.'s **test for glucose in urine,** find the specific gravity of the urine at a known temperature by means of a urinometer supplied with a thermometer. Acidify slightly with tartaric acid, and add a piece of yeast the size of a pea, and shake. Let it stand in a warm place (20°–25° C.) for 24 hours. Filter through a dry filter and cool to the same temperature at which the specific gravity was previously taken. Take the specific gravity again. Every degree of density lost represents 1 grain of glucose to the ounce of urine.
**Robertson's pupil.** See *Argyll Robertson pupil.*
**robin** (*ro'-bin*). A toxic albuminoid from the bark of the locust tree, *Robinia pseudacacia;* its action is similar to that of abrin and ricin.
**Robin's myelopaxes** (*ro-ban'*) [Charles Philippe *Roban,* French physician, 1821–1885]. Osteoclasts.
**Robinson's circle** (*rob'-in-sun*) [Byron *Robinson,* American surgeon]. An arterial anastomosis consisting of the following arteries: uterine, ovarian, abdominal aorta, common iliacs and internal iliacs.
**Robinson's disease** (*rob'-in-sun*) [Andrew Rose *Robinson,* American dermatologist, 1845– ]. Hydrocystoma.
**Robiquet's paste** (*rob-e-ka'*) [Pierre Jean *Robiquet,* French physician, 1780–1840]. A caustic paste consisting of equal parts of zinc chloride and flour with gutta-percha. It is firm and tenacious.
**robor** (*ro'-bor*) [L.]. Strength.
**roborant** (*ro'-bor-ant*) [*robor,* strength]. 1. Tonic, strengthening. 2. A tonic or strengthening remedy.
**roborat** (*ro'-bor-at*) [see *roborant*]. An albuminous dietetic prepared from maize, containing lecithin and glycerinophosphoric acid.
**roborin** (*ro'-bor-in*). A grayish-green powder or brown mass, obtained from blood, and said to consist of water, 7.6 %, calcium carbonate, 10.23%, common salt, 1.7%, iron oxide, 0.49%, other mineral substances, 1.28%, albuminoids, 78.63%; the last are principally calcic albuminates.
**Robson's point** (*rob'-sun*) [A. W. Mayo *Robson,* English surgeon]. A point one-third of the way on a line from the umbilicus to the right nipple; it is the point of greatest tenderness in inflammation of the gall bladder.
**Roccella** (*rok-sel'-ah*). A genus of plants of the *Roccelleæ.* R. *tinctoria* is the litmus-plant.
**Rochelle salt** (*ro-shel'*). Potassium and sodium tartrate.
**rock-oil.** See *petroleum.*
**Rocky mountain fever.** A form of fever occurring at high altitudes; mountain fever; and see *Texas fever.*
**rod** [ME.]. One of numerous slender bacillary structures, as in the retina. r.-**and-cone layer,** r. **and cones,** see under *retina.* r.-**granules,** cells of the outer nuclear layer of the retina; they are characterized by transverse striæ and give off processes connected with the nerve-fiber layer of the retina. **r.'s, retinal,** cylindrical bodies found in the rods and cones of the retina.
**rodagen** (*rod'-ah-jen*). A proprietary preparation of the milk of thyroidectomized goats, for use in exophthalmic goiter.
**rodent ulcer.** See *ulcer, rodent.*
**rodostrophone** (*ro-dos'-tro-fōn*). An instrument for transmitting articular sounds from the skull of one person directly to that of another.
**Rodriguez' aneurysm** (*rod-re'-ga*). Varicose aneurysm in which the sac is immediately contiguous to the artery.
**Roederer's ecchymoses** [Johann Georg *Roederer,* German obstetrician, 1727–1763]. See *Bayard's ecchymoses.* R.'s **obliquity,** flexion of the chin when the child is engaged at the superior pelvic strait during labor.
**roentgenism** (*rent'-gen-izm*) [Wilhelm Konrad *Roentgen,* German physicist, 1845– ]. 1. The application of the Roentgen-rays in therapeutics. 2. Disease or disability from misuse of the Roentgen rays.
**roentgenization** (*rent-gen-iz-a'-shun*). Exposure or subjection to the action of Roentgen rays.

**roentgenogram** (*rent-gen'-o-gram*). A Roentgen-ray photograph; a skiagram.
**roentgenograph** (*rent-gen'-o-graf*). To make a roentgenogram.
**roentgenography** (*rent-gen-og'-ra-fe*). Same as skiagraphy, *q. v.*
**roetgenologist** (*rent-gen-ol'-o-jist*). One who is expert in the diagnosis and treatment by the Roentgen rays.
**roentgenology** (*rent-gen-ol'-o-je*) [*Roentgen rays;* λόγος, treatise]. The study of the roentgen rays.
**roentgenometry** (*rent-gen-om'-et-re*). Measurement of the penetrating power or of the quantity employed of the Roentgen rays.
**roentgenoscope** (*rent-gen'-o-skōp*). Same as fluoroscope, *q. v.*
**roentgenoscopy** (*rent-gen-os'-ko-pe*) [*Roentgen rays;* σκοπέω, to view]. Examination of solid bodies by means of Roentgen rays.
**roentgenotherapy** [*Roentgen rays;* θεραπεία, treatment]. The treatment of disease by means of the Roentgen rays.
**Roentgen-rays.** See *rays, Roentgen-.*
**roetheln.** See *rubella.*
**Roger's disease** [Henri Louis Roger, French physician, 1811–1892]. The presence of a congenital abnormal communication between the ventricles of the heart. **R.'s symptom,** subnormal temperature during the third stage of tuberculous meningitis, regarded by Roger as pathognomonic of the disease.
**Rokitansky's disease** (*roh it an' she*) [Carl Freiherr von Rokitansky, Austrian pathologist, 1804–1878]. Acute yellow atrophy of the liver. **R.'s pelvis,** pelvic deformity due to spondylolisthesis. **R.'s tumor,** an ovarian tumor made up of a large number of cysts.
**Rolandic** (*ro-lan'-dik*) [Louis Rolando, Italian anatomist, 1773–1831]. Described by Rolando, as the *rolandic* fissure. **R. angle,** the angle formed by the fissure of Rolando with the superior border of the cerebral hemisphere. **R. area,** the excitomotor area of the cerebral hemispheres, comprising the ascending frontal and ascending parietal convolutions.
**Rolando's arciform fibers.** [see *Rolandic*]. The external arcuate fibers of the oblongata. **R.'s cells,** the ganglion-cells found in Rolando's gelatinous substance. **R.'s fissure,** a fissure on the lateral aspect of the cerebrum extending downward from near the longitudinal fissure at about its middle point. It separates the frontal from the parietal lobe. The central fissure. **R.'s funiculus,** the lateral cuneate funiculus, a longitudinal prominence caused by Rolando's gelatinous substance on the surface of the oblongata, between the cuneate funiculus and the line of roots of the spinal accessory nerve. **R.'s gelatinous substance,** the elongated column which forms a continuation of the apices of the posterior horns of the spinal cord, extending from the lumbar portion of the cord upward into the pons. It consists of neuroglia and a number of ganglion-cells. **R.'s tubercle,** a mass of gray matter forming the upper termination of Rolando's funiculus. The fibers given off from its cells go to make up the sensory root of the trigeminus.
**rolandometer** (*ro-land-om'-et-er*). A device for locating on the head the place of the fissure of Rolando.
**Roller's nucleus** (*rol'-er*) [Christian Friedrich Wilhelm Roller, German physician, 1802–1878]. 1. A nucleus situated near the hilum of the olivary body of the oblongata; it is connected with the fibers of the anterolateral fundamental tract of the spinal cord. 2. An aggregation of small ganglion-cells situated anteriorly to the nucleus of the hypoglossal nerve.
**roller-bandage.** A bandage made into a cylindrical roll.
**Rollet's chancre.** One partaking of the characteristics of both simple and true chancre. **R.'s nerve-corpuscles,** see *Golgi's corpuscles.* **R.'s secondary substance,** see *Engelmann's lateral disc.*
**Rollet's delomorphous-cells** [Alexander Rollet, Austrian physiologist, 1834–1903]. Large, well defined cells between the membrana propria and the chief cells of the fundus glands of the gastric mucous membrane. They are supposed to secrete the hydrochloric acid. **R.'s stroma,** an insoluble, spongy network forming the structure of an erythrocyte, within the interstices of which is embedded the hemoglobin.
**Roman-Delluc's test for urobilin in urine.** Shake 100 Cc. urine with 20 Cc. chloroform, after acidulating with 8 to 10 drops acetic acid. Overlay 2 Cc. of clear chloroformic solution with 4 Cc. of 1 : 1000 solution zinc acetate in 95% alcohol. At the line of separation a characteristic green fluorescence will appear if urobilin is present, more easily recognized against a black background. On shaking, fluorescence is more marked and the mixture acquires a pink tint.

**romanoscope** (*ro-man'-o-skōp*). A speculum for examining the sigmoid flexure.
**Romberg's disease** [Moritz Heinrich Romberg, German physician, 1795–1873]. Progressive facial hemiatrophy. **R.'s sign.** 1. Swaying of the body and inability to stand when the eyes are closed and the feet placed together; it is seen in tabes dorsalis, hereditary cerebellar ataxia, etc. It is also called the Brauch-Romberg symptom. 2. Neuralgic pain in the course and distribution of the obturator nerve, pathognomonic of obturator hernia. **R.'s spasm,** masticatory spasm, a spasm affecting the muscles supplied by the motor fibers of the fifth nerve. **R.'s trophoneurosis,** see *Romberg's disease.*
**Romershausen's eye-water** (*ro'-merz-how-zen*). A wash employed in chronic ophthalmic catarrh. It is a mixture of fennel water and tincture of fennel.
**Rommelaere's law** [Guillaume A. V. Rommelaere, Belgian physician, 1836– ]. Constant diminution of the nitrogen in the urine in cases of carcinoma. **R.'s sign,** diminution of the normal phosphates and sodium chloride in the urine is pathognomonic of cancerous cachexia.
**rongeur forceps** (*rŏng-zjur*). A strong pair of forceps for breaking off pieces of bone, especially in enlarging a trephine opening.
**roof-cell.** A nerve-cell of the roof-nucleus.
**roof-nucleus.** A nucleus in the roof of the fourth ventricle.
**root.** 1. The descending axis of a plant. 2. The part of an organ embedded in the tissues, as the *root* of a tooth. **r.-arteries,** the radicular vessels. **r. of a nerve,** one of two bundles of nerve-fibers, the anterior and posterior roots, joining to form a nerve-trunk. **r.-sheath,** the epithelium of the hair-follicle. **r.-zone,** a name given to the column of Burdach of the spinal cord.
**R. O. P.** An abbreviation for *right occipitoposterior position* of the fetus.
**rophetic** (*ro-fet'-ik*) [ῥοφητικός, given to sopping up]. A mechanical absorbent agent, as a dusting-powder, sponge, etc.
**rosa** (*ro'-zah*) [L.]. A rose; see *rose.*
**rosacea** (*ro-za'-se-ah*). See *acne rosacea.*
**rosalia** (*ro-sa'-le-ah*). 1. Scarlatina. 2. Measles. 3. Erythema.
**rosanilin, rosaniline** (*ro-zan'-il-in*) [*rose; anilin*], $C_{20}H_{19}N_3O$. A colorless, crystalline derivative of aniline. It is used as the basis of various dyes. **r. acetate, r. hydrochloride,** the red dye fuchsin.
**rosary, rhachitic.** See *rhachitic rosary.*
**rose** (*rōz*) [ῥόδον, from Ar., *ward,* a rose]. A genus of plants of the order Rosaceæ. **r.s, attar of,** see *r., oil of.* **r. catarrh,** see *hay-fever.* **r.-cold,** see *hay-fever.* **r., confection of** (*confectio rosæ,* U. S. P.), a confection prepared from the petals of the red rose (*rosa gallica,* U. S. P.). **r., dog-,** the common wild rose of Europe. The fruit (*rosæ caninæ fructus,* B. P.) is used in Europe as a vehicle. **r., fluidextract of** (*fluidextractum rosæ,* U. S. P.), used as an astringent and vehicle. Dose 1–2 dr. (4–8 Cc.). **r., honey of** (*mel rosæ,* U. S. P.), a syrup made of fluidextract of rose and clarified honey. Dose 1 dr. (4 Cc). **r., oil of** (*oleum rosæ,* U. S. P.), a volatile oil distilled from the petals of *Rosa damascena,* and employed as a perfume and flavoring agent. Syn., *attar of roses.* **r. red,** see *roseola.* **r., red** (*rosa gallica,* U. S. P.; *rosæ gallicæ petala,* B. P.), the dried petals are slightly astringent and tonic, but are chiefly employed as a vehicle. **r., syrup of** (*syrupus rosæ,* U. S. P.), a syrup made of fluidextract of rose, diluted sulphuric acid, sugar, and water. **r.-water** (*aqua rosæ,* U. S. P.). Dose 4 dr. (16 Cc.). **r.-water, ointment of** (*unguentum aquæ rosæ,* U. S. P.), cold cream. **r.-water, triple** (*aqua rosæ fortior,* U. S. P.), stronger rose-water.
**Rose's biuret reaction for albumins.** Alkalinize the albumin solution with soda-lye and add, drop by drop, with constant shaking, a dilute copper sulphate solution (17 or 18 Gm. crystallized cupric sulphate in 1 liter of water). The solution will become rose-red, then violet, and finally blue; the blue appears of a reddish tint when compared with a normal alkaline copper solution.
**Rose's operation** (*rōz*) [William Rose, English

surgeon, 1847]. Removal of the Gasserian ganglion, for the relief of trifacial neuralgia.
**rosella** (*ro-sel'-ah*). See *rubella*.
**rosemary** (*rŏs'-ma-re*) [*rosmarinus*, marine dew; from *ros*, dew; *marinus*, marine]. The *Rosmarinus officinalis*, a plant of the order *Labiatæ*. r., oil of (*oleum rosmarini*, U. S. P.), a volatile oil used as a stimulant and in rubefacient liniments. Dose 3–6 min. (0.2–0.3 Cc.). r., spirit of (*spiritus rosmarini*, B. P.), prepared from the oil and used as a perfume and in liniments.
**Rosen's liniment.** A liniment composed of oil of nutmeg, spirit of juniper, and oil of cloves.
**Rosenbach's disease** (*ro'-zen-bakh*) [Ottomar *Rosenbach*, German physician, 1851–1907]. A nodular enlargement, painful to the touch, of the dorsal aspect of the proximal ends of the last phalanges. The affection is regarded as identical with Heberden's nodes. r.'s **modification of Gmelin's test for bile-pigments**, when the liquid has all been filtered through a very small filter, apply to the inside of the filter a drop of nitric acid containing only a very little nitrous acid, when a pale yellow spot will form, surrounded by colored rings, which are yellowish-red, violet, blue, and green. r.'s **sign.** 1. Loss of the abdominal reflex in inflammatory intestinal diseases. 2. Tremor of the eyelids when the patient is asked to close them, often associated with insufficient closure of the lids. It is seen in neurasthenia. 3. Tremor of the upper lids in exophthalmic goitre when the eyes are gently closed. R.'s **syndrome**, a variety of paroxysmal tachycardia consisting in the association of cardiac, respiratory, and gastric troubles. R.'s **test for indirubin**, boil the liquid with nitric acid, and indigo-blue will be formed from indirubin.
**Rosenbach-Semon's law.** See *Semon's law*.
**Rosenberg's method.** By requiring the patient to read aloud a difficult passage, the production of the knee-jerk is facilitated.
**Rosenheim's sign** (*ro'-zen-him*). A friction-sound heard on auscultation over the left hypochondrium in fibrous perigastritis.
**Rosenmueller's fossa** (*ro'-zen-mü-ler*) [Johann Christian *Rosenmueller*, German anatomist, 1771–1820]. A depression behind the pharyngeal orifice of the Eustachian tube, frequently the seat of morbid growths. R.'s **gland**. 1. The palpebral portion of the lacrimal gland. 2. The largest of the group of deep subinguinal glands in the crural ring. R.'s **organ**, the parovarium, a vestige of the Wolffian body and duct. R.'s **valve**, a semilunar fold of the mucous membrane seen occasionally in the lacrimal duct above its junction with the lacrimal sac.
**Rosenthal's canal** (*ro'-zen-tahl*) [Friedrich Christian *Rosenthal*, German anatomist, 1780–1829; Isidor *Rosenthal*, German physiologist, 1836– ]. The spiral canal of the modiolus. r.'s **hyperacid vomiting**, the vomiting of very acid material, indicative of exaggerated secretion of HCl in the gastric juice; Rossbach's disease. R.'s **sign**, the application of a strong faradic current to the sides of the vertebral column causes burning and stabbing pains in cases of spondylitis. R.'s **vein**, the basilar vein, a branch of Galen's vein.
**roseola** (*ro-ze'-o-lah*) [*roseus*, rosy]. 1. Rose-rash, a name given to any rose-colored eruption. 2. Synonym of *rubella*. r. **cholerica**, an eruption sometimes appearing in cholera. r., **syphilitica**, an eruption of rose-colored spots appearing early in secondary syphilis. r. **typhosa**, the eruption of typhoid or typhus fever. r. **vaccinia**, a general rose-colored eruption sometimes occurring during vaccinia.
**roséoles à verre bleu.** Faint syphilides discovered by means of cobalt-blue glasses worn close to the eyes, before they are revealed to the naked eye.
**roseolous** (*ro-ze'-o-lus*) [*roseus*, rosy]. Having the character of roseola.
**Roser's position.** With head dependent over the end of the table.
**Roser-Braun's sign.** Absence of pulsations of the dura in cases of cerebral abscess, tumors, etc.
**Roser-Nélaton's line.** See *Nélaton's line*.
**roset, rosette** (*ro-set'*). 1. See *karyokinesis*. 2. A congery of cells from the neuroepithelial layer of the retina described by Wintersteiner as a characteristic of glioma of the retina. They correspond to the external limiting membrane of the retina, with rudimentary rods and cones projecting into the central cavity.
**rosin** (*roz'-in*). The residue left after distilling off the volatile oil from turpentine. See also *resina*. r. **cerate** (*ceratum resinæ*, U. S. P.), a mixture of rosin, yellow wax, and lard. r. **cerate, compound** (*ceratum resinæ compositum*, U. S. P.), a mixture of rosin, yellow wax, prepared suet, turpentine, and linseed-oil.
**Rosin's test for indigo-red.** Render the liquid alkaline with sodium carbonate and extract with ether, which is colored red by the indirubin.
**rosinol** (*roz'-in-ol*). See *retinol*.
**rosin-weed**, Compass plant. The plant *Silphium laciniatum* secretes an oleo-resin commonly used as a chewing-gum. Tonic, alterative and emetic. Dose of fld.ext. ʒ ss–j.
**rosmarinus** (*roz-ma-ri'-nus*). See *rosemary*.
**rosolene** (*roz'-o-lēn*) [*rosin*, a variation of *resin*; *oleum*, oil]. The oily distillate of colophony.
**rosolic acid** (*ro-zol'-ik*), $C_{20}H_{16}O_3$. A substance used as a test for acids and alkalies: acids decolorize it; with alkalies it gives a red color.
**Ross**, cycle of. That phase of development of *plasmodium malariæ* which occurs in the mosquito. See *Golgi, cycle of*.
**Ross's in vitr, method.** A method of studying, under the microscope, reproduction and other phenomena in living cells on glass slides covered with nutrient jelly.
**Rossbach's disease** (*ros'-bahk*) [Michael Joseph *Rossbach*, German physician, 1842–1894]. Gastroxynsis; a neurosis of the stomach attended with paroxysmal hypersecretion.
**Rossell's test for blood in the stools** (*ros-el'*) [Otto *Rossell*, Swiss physician]. To an ethereal extract of the feces are added oil of turpentine and solution of aloin. In the presence of blood the mixture assumes a red color.
**rostellum** (*ros-tel'-um*) [dim. of *rostrum*]. A little beak, especially the hook-bearing portion of the head of certain worms.
**rostral** (*ros'-tral*) [*rostrum*, beak]. 1. Pertaining to or resembling a rostrum. 2. See *cephalic*.
**rostrate** (*ros'-trāt*) [*rostrum*]. Furnished with a beak or beak-like process.
**rostriform** (*ros'-trif-orm*) [*rostrum*, beak; *forma*, form]. Shaped like a rostrum.
**rostrum** (*ros'-trum*) [L.]. A beak; a projection or ridge. r. **corporis callosi**, the anterior tapering portion of the corpus callosum. r. **sphenoidale**, the vertical ridge on the inferior aspect of the body of the sphenoid bone, which is received in the upper grooved border of the vomer. Syn., *beak of the sphenoid*.
**rot** 1. To suffer putrefactive fermentation. 2. Decay; decomposition. 3. A disease of sheep. r. **potato**-. See *mildew*.
**rotary** (*ro'-ta-re*) [*rotation*]. Producing or characterized by rotation. r. **joint**, a pivot joint.
**rotate** (*ro'-tāt*) [*rotare*, to revolve]. Wheel-shaped. In dentistry, the term implies the turning of a tooth on its axis. r.-**plane**, in biology, wheel-shaped and flat.
**rotating** (*ro-ta'-ting*) [*rotare*, to revolve]. Revolving. r. **devices**, appliances, either single or double, for correcting torsion of single-rooted teeth.
**rotation** (*ro-ta'-shun*) [*rotare*, to turn, from *rota*, a wheel]. 1. The act of turning about an axis passing through the center of a body, as *rotation* of the eye, *rotation* of the arm. 2. In dentistry, the operation by which a tooth is turned or twisted into its normal position. r. **joint**, a lateral ginglymus. r.-**stage of labor**, one of the stages of labor consisting in a rotatory movement of the fetal head or other presenting part, whereby it is accommodated to the birth-canal. It may be internal, occurring before the birth of the presenting part, or external occurring afterward. r., **wheel of**, Helmholtz, the tilting of the vertical meridians of the eye.
**rotator** (*ro-ta'-tor*) [see *rotation*]. Anything, especially a muscle, that produces rotation.
**Rotch's sign** [Thomas Morgan *Rotch*, American physician, 1848– ]. Dulness on percussion in the right fifth intercostal space in pericardial effusion.
**Roth's disease.** R.'s **symptom-complex.** "Meralgia paræsthetica." See *Bernhardt's paresthesia*. R.'s **spots**, white spots resembling those of albuminuric retinitis, seen in the region of the optic disc and the macula in cases of septic retinitis. R.'s **vas aberrans**, an inconstant diverticulum of the head portion of the rete testis.
**rötheln** (*ret'-eln*). See *rubella*.
**rottlera** (*rot-le'-rah*). See *kamala*.

**rottlerin** (*rot'-ler-in*), $C_{22}H_{30}O_6$. A bitter principle from kamala; used as an anthelmintic.
**rotula** (*rot'-ū-lah*) [dim. of *rota*, a wheel]. 1. The patella. 2. A troche or lozenge.
**rotulad** (*rot'-ū-lad*) [*rotula*; *ad*, towards]. Toward the patella.
**rotular** (*rot'-ū-lar*) [*rotula*, a little wheel]. Of or pertaining to the patella, e. g., the patellar aspect of a limb; opposed to *popliteal*.
**Rouge's operation** (*roozj*) (*for access to the nasal cavities*). The upper lip is freed from the jaw by an incision through the mucous membrane; the cartilaginous séptum and lower lateral cartilages are then detached so that the nose and lips can be raised to the necessary extent.
**Rouget's bulb.** (*roo-zja'*) [Antoine D. *Rouget*, French physiologist]. The bulb of the ovary, a plexus of veins lying on the surface of the ovary and communicating with the uterine and pampiniform plexuses. **R.'s motorial end-plates,** small cellular elements connected, within the sarcolemma, with the endings of motor nerves. **R.'s muscle,** see *Mueller's muscle* (1).
**Roughton's band.** Collapse, from atrophy of the tissues, of the zone corresponding to the junction of the alæ nasi with the lateral cartilages. The resulting contact of this zone with the septum causes obstruction during inspiration.
**Rougnon-Heberden's disease** (*roon-yong'-heb'-er-den*) [Nicholas François *Rougnon*, French physician, 1727–1799; William *Heberden*, English physician, 1710–1801]. Angina pectoris.
**rouleau** (*roo-lō*) [Fr.; pl., *rouleaux*]. A roll, especially a roll of red blood-corpuscles, resembling a roll of coins.
**round ligament.** 1. One of the ligaments of the uterus passing through the inguinal canal. 2. One of the ligaments of the liver lying in the longitudinal fissure. 3. One of the ligaments of the hip-joint—ligamentum teres.
**rounding** (*rown'-ding*). A term given to that propensity manifested by certain hypochondriac individuals to run the round of all the free dispensaries in a vicinity. Such patients are termed "rounders."
**roundworm.** The ascaris.
**roup** (*roop*). An infectious respiratory disease of fowls.
**Roussel's sign** (*roo'-sel*). A sharp pain caused on light percussion, in the subclavicular region between the clavicle and the third or fourth rib, originating 3 to 4 cm. from the median line and extending to and beyond the shoulder and the suprapinal fossa; it is observed in incipient tuberculosis.
**routinist** (*roo-te'-nist*) [*route*, a beaten path]. A physician who does not deviate in his treatment from an unvarying routine.
**Roux's serum** (*roo*) [Pierre Paul Emile *Roux*, French bacteriologist, 1853– ]. An antitetanic serum. **R.'s unit,** one mil of an antitoxic (antitetanic) serum should be sufficient to protect 1,000,000 grammes of guinea-pig against the minimum lethal dose of tetanus toxin.
**Roux's sign of suppurative appendicitis** (*roo*). On palpation the empty cecum presents a special soft resistance comparable to that of a wet pasteboard tube.
**Rovighi's sign.** Hydatid fremitus; a thrill observed on combined palpation and percussion in cases of superficial hydatid cyst of the liver.
**Rovsing's sign** (Niels Thorkild *Rovsing*, Danish surgeon, 1862– ]. Pressure on the descending colon at a point corresponding to McBurney's point will, in case of appendicitis, cause pain at McBurney's point.
**R. S. A.** An abbreviation for *right sacroanterior* position of the fetus.
**R. S. P.** An abbreviation for *right sacroposterior* position of the fetus.
**Ru.** Chemical symbol for ruthenium.
**rubber** (*rub'-er*). The *elastica* of the U. S. P. The prepared milk juice of several species of *Hevea*. Syn., *caoutchouc; India-rubber; Para rubber.* **r.-dam,** a sheet of rubber used to confine the flow of secretions or of discharges from a wound. **r. tissue,** gutta-percha in sheets.
**rubedo** (*roo-be'-do*) [*ruber*, red]. Any diffused redness of the skin.
**rubefacient** (*roo-be-fa'-she-ent*) [*ruber; facere,* to make]. 1. Causing redness of the skin. 2. An agent that causes redness of the skin.
**rubella** (*roo-bel'-ah*) [dim. of *rubeola*]. An acute contagious eruptive disease, of short duration and mild character. After a period of incubation varying from one to three weeks, the disease sets in abruptly with pains in the limbs, sore throat, and slight fever. The eruptions appear at the end of the first day, and consists of red papules, and disappears usually without desquamation in about three days. The disease is associated with enlargement of the superficial cervical and posterior auricular glands. Syn., *epidemic roseola; French measles; German measles; rötheln.*
**rubeola** (*roo-be'-o-lah*). See *measles*. **r. notha,** same as *rubella*.
**rubeolin** (*roo-be'-ol-in*) [*rubeola*]. A name given to the specific toxin of measles.
**rubescence** (*roo-bes'-ens*) [*rubescere*, to become red]. Blushing; redness of countenance or complexion.
**rubescent** (*roo-bes'-ent*) [*rubescere*, to become red]. Growing red.
**rubia** (*roo'-be-ah*). The *Rubia tinctorum* or dyers' madder, containing the coloring principles alizarin ($C_{14}H_8O_4$) and purpurin ($C_{14}H_9O_5$). It is used as a dye.
**rubidium** (*roo-bid'-e-um*) [*rubidus*, red]. A rare alkaline metal, resembling potassium in physical and chemical properties; its salts are used in medicine. See *elements, table of*.
**rubiginous** (*roo-bij'-in-us*) [*rubiginosus,* rusty]. Rust-colored.
**rubigo** (*roo-bi'-go*) [L.]. Rust.
**rubijervine** (*roo-bij-er'-vin*) [*rubeus,* red; *jerva,* green hellebore root], $C_{26}H_{48}NO_2$. An alkaloid of veratrum album.
**rubin** (*roo'-bin*). Synonym of *fuchsin*.
**Rubner's test for carbon monoxide in the blood** (*roob'ner*) [Max *Rubner*, German physiologist, 1854– ]. Agitate the blood with 4 or 5 volumes of solution of lead acetate for one minute. If the blood contains CO, it will retain its bright color; if it does not, it will turn chocolate-brown. **R.'s test for glucose,** add to the liquid an excess of lead acetate; filter, and add to the filtrate ammonium hydrate until no further precipitate is produced. Warm gently, when the precipitate formed will gradually become pink; this color decreases on standing.
**rubor** (*roo'-bor*) [L.]. Redness or discoloration due to inflammation. **r., regional,** isolated spots which become red, with elevation of temperature, before and after local cyanosis.
**rubrescin** (*roo-bres'-in*). A combination of resorcinol, 50 Gm., and chloral hydrate, 25 Gm. It is used in 1% solution as an indicator for alkalimetry and acidimetry.
**rubrin** (*roo'-brin*). See *hematin*.
**rubrol** (*roo'-brol*). A solution used by injection in gonorrhea, and said to consist of boric acid, thymol, and a coal-tar derivative in water.
**rubrospinal tract.** Monakow's bundle.
**rubrum** (*roo'-brum*) [*ruber,* red]. The preferred name for the nucleus ruber.
**rubus** (*roo'-bus*) [L.]. Blackberry. A genus of plants of the order *Rosaceæ*. The *rubus* of the U. S. P. is the dried bark of the rhizome of *R. villosus, R. nigrobaccus,* and *R. cuneifolius*. It is used as an astringent tonic in diarrhea. Dose 20–30 gr. (1.3–2.0 Gm.). **r., fluidextract of** (*fluidextractum rubi* U. S. P.). Dose ½–1 dr. (2–4 Cc.). **r., syrup of** (*syrupus rubi,* U. S. P.). Dose 1–2 dr. (4–8 Cc.). The fruit of *R. idæus,* the raspberry, is used to prepare *syrupus rubi idæi,* which is used as a tonic.
**ructation** (*ruk-ta'-shun*). An eructation or belching of wind.
**ructus** (*ruk'-tus*) [L.]. A belching of gas from the stomach. **r. hystericus,** hysteric belching, the gas escaping with a loud, sobbing, gurgling noise.
**rudimentary** (*roo-dim-en'-ta-re*) [*rudimentum,* a rudiment]. Undeveloped; unfinished.
**rudimentum** [L.; *pl., rudimenta*]. A rudiment.
**rue** (*roo*) [*burh,* rue]. A plant, *Ruta graveolens,* of the order *Rutaceæ,* yielding an oil (*oleum rutæ,* B. P.) which is a local irritant and has been employed in amenorrhea and menorrhagia. Dose 2–5 min. (0.13–0.32 Cc.).
**Ruffini's end organs** (*roof-fe'-ne*) [Angelo *Ruffini,* Italian anatomist]. Small bodies found in the skin where Pacinian corpuscles exist; they are made up of the terminal arborizations of a nerve and a fibrous framework.
**Rufus's pills.** Pills of aloes and myrrh.
**ruga** (*roo'-gah*) [L.; *pl., rugæ*]. 1. A wrinkle, furrow, crease, or ridge, as e. g., in the mucosa of the stomach,

**vagina**, etc. 2. A fold of pia on the ental surface of the piarachnoid. **rugæ, palatal,** the elevations upon the mucous covering of the hard palate; they assist in speech and deglutition.
**rugitus** (*roo-gi'-tus*). See *bombus*.
**rugose** (*roo'-gōs*) [*ruga*]. Characterized by folds.
**rugosity** (*roo-gos'-it-e*) [*ruga*]. A condition of being in folds.
**rugous** (*roo'-gus*). See *rugose*.
**Ruhmkorff's coil** (*room'-korf*) [Heinrich Daniel *Ruhmkorff*, German physicist, 1823–1887]. An induction coil.
**rum** [abb. from ME. *rumbooze*, alcoholic liquor]. A spirit obtained from the molasses of the sugar-cane by fermentation and distillation.
**rumbling** (*rum'-bling*). See *borborygmus*.
**rum-blossom,** a pimple on the nose caused by excessive drinking; rum-bud; acne rosacea. **r.-bud,** see **r.-blossom.**
**rumen** (*roo'-men*) [L.]. The first stomach of ruminants, also called the paunch, from which the food is returned to the mouth for remastication.
**rumenotomy** (*roo-men-ot'-o-me*) [*rumen*, the gullet; τέμνειν, to cut]. Incision of the rumen or paunch of an animal.
**rumex** (*roo'-meks*) [L.]. Yellow dock, a genus of plants of the order *Polygonaceæ*. The root of *R. crispus* is astringent and tonic, and has been employed externally and internally in various diseases of the skin. Dose of the *extract* 1 dr. (4 Cc.).
**rumicin** (*roo'-mis-in*). Chrysophanic acid.
**rumin** (*roo'-min*) [*rumex*]. A precipitate from a tincture of the root of *Rumex crispus*. It is antiscorbutic, alterative and astringent.
**rumination** (*roo-min-a'-shun*). See *merycism*.
**Rummo's disease** (*room'-mo*) [Gaetano *Rummo*, Italian physician]. Cardioptosis.
**rump** [Icel., *rumpr*]. The region near the end of the backbone; the buttocks or nates.
**Rumpel-Leede sign** (*room'-pel-la'-deh*) [O. *Rumpel*, C. *Leede*, German physicians]. A bandage is placed half way up the arm, drawn tight enough to produce a decided blue discoloration of the forearm, and left in place from three to eight minutes; petechiæ then appear on the anterior surface of the elbow joint. It is observed in scarlet fever and other exanthemata.
**Rumpf's sign** [Theodor *Rumpf*, German physician, 1851– ]. Fibrillary twitching of muscles in traumatic neurosis.
**run.** In pathology, to discharge pus or purulent matter from a diseased part. **r.-around,** see *paronychia*.
**Runeberg's type of pernicious anemia** (*roo'-na-berg*). [Johan Wilhelm *Runeberg*, Finnish physician, 1843– ]. A form of pernicious anemia with remissions.
**Runge's method** (*roon'-geh*). A method of dressing the umbilical cord. The stump is powdered with a mixture of boric acid and starch, one part to three.
**rupia** (*roo'-pe-ah*) [ῥύπος, filth]. A form of eruption occurring especially in tertiary syphilis, and characterized by the formation of large, dirty-brown, stratified, conic crusts.
**rupial** (*roo'-pe-al*) [*rupia*]. Resembling or characterized by rupia.
**rupophobia** (*roo-po-fo'-be-ah*). See *rhypophobia*.
**ruptio** (*rup'-she-o*) [*rumpere*, to break]. Rupture of a vessel or organ.
**rupture** (*rup'-tūr*) [*rumpere*, to break]. 1. A forcible tearing of a part, as *rupture* of the uterus, *rupture* of the bladder. 2. Hernia.

**ruptured** (*rup'-tūrd*) [*rumpere*, to break]. Burst; affected with hernia.
**Rusconi's anus** (*roos-ko'-ne*) [Mauro *Rusconi*, Italian biologist, 1776–1849]. The blastopore.
**Russell's bodies** [William *Russell*, Scotch physician]. Fuchsin bodies. Roundish colloid or hyaline bodies, of varying size, found in a variety of conditions, notably in carcinomatous growths and certain morbid changes of the mucosa of the nose and stomach.
**Russian oil.** A pure petroleum, odorless and tasteless, and said to have been refined in Russia.
**Russo's reaction** (*roos'-so*) [Mario *Russo*, Italian physician]. To 4 or 5 Cc. of the patient's urine add four drops of a 0.1 per cent. aqueous solution of methylene blue; mix well and examine against the light; a mint or emerald green coloration is positive, but a bluish tinge renders the test negative. The reaction is said to be positive in typhoid, smallpox, measles, and advanced tuberculosis.
**rust** [AS.]. 1. The oxide and hydroxide of iron formed on the surface of iron exposed to the air. 2. A disease common on cereals, causing rust-like masses to break out on the tissues of the plant.
**Rust's disease** [Johann Nepomuk *Rust*, German physician, 1775–1840]. Tuberculous spondylitis affecting the first and second cervical vertebræ. **R.'s symptoms,** at every change of position of the body a patient suffering from caries or carcinoma of the upper cervical vertebræ supports his head with the hand.
**rusty** (*rus'-te*). Of the nature or appearance of rust.
**r. expectoration,** the common name for the usual form of expectoration in croupous pneumonia, due to the presence of a small amount of blood in the sputa.
**rut** (*rut*) [OF., "a roaring"]. 1. The state of concomitant menstruation and ovulation in the lower animals. 2. The condition of a male animal in which it is capable of inseminating.
**ruta** (*roo'-tah*). See *rue*.
**ruta graveolens.** Rue; the leaves are used as an emmenagog.
**ruthenium** (*roo-the'-ne-um*) [*Ruthenia*, a province of Russia]. A rare metal of the platinum group. Symbol, Ru; atomic weight, 101.7. Little is known of its medicainl properties. See *elements, table of*.
**Rutherford's solution.** A decalcifying and hardening solution for tissue-specimens: chromic acid 1 grm., water 200 Cc.; then add 2 Cc. nitric acid.
**rutidosis** (*roo-tid-o'-sis*) [ῥυτίς, a wrinkle]. A wrinkling; the contraction or puckering of the cornea that just precedes death.
**rutin** (*roo'-tin*) [*ruta*, rue]. A crystalline neutral substance obtained from the leaves of rue.
**Ruysch's glomerulus** (*rīsh*) [Fredericus *Ruysch*, Dutch anatomist, 1638–1731]. See *Malpighian tuft*. **R.'s membrane,** the middle or capillary layer of the choroid. **R.'s tube,** a minute tubular cavity in the nasal septum, opening by a small, round orifice a little below and in front of the nasopalatine foramen. It is best seen in the fetus, and represents the rudimentary homologue of Jacobson's organ. **R.'s uterine muscle,** the muscular tissue of the fundus uteri; it was believed by Ruysch to act independently of the rest of the uterine muscle.
**rye** (*rī*) [AS., *ryge*]. The plant *Secale cereale* and its grain. The grain is used in the manufacture of bread.
**rye asthma,** a form of hay-fever occurring at the time of the flowering rye. **r., ergot of,** see *ergot*. **r., spurred,** same as *ergot*.
**rypia** (*ri'-pe-ah*). See *rupia*.
**rytidosis corneæ.** See *rutidosis*.

# S

**S.** 1. The chemical symbol of *sulphur.* 2. An abbreviation in prescriptions, of *signa,* sign or label. **S. romanum,** the sigmoid flexure of the colon.
**s.** Abbreviation of *sinister,* left; also of *semis,* half (usually *ss*).
**Sa.** The chemical symbol of *samarium.*
**sabadilla** (*sab-ad-il'-ah*). Cevadilla; *Schœnocaulon officinale* (*Asagræa officinalis*), a plant of the order *Melanthaceæ,* containing the alkaloids veratrine, sabadine, and sabadinine. It is an emetocathartic, and was formerly used as a teniacide and to destroy vermin in the hair. Its chief value is as a source of veratrine. It is official in the B. P.
**sabadine** (*sab'-ad-ēn*), C₂₉H₅₁NO₈ (Merck). An alkaloid from the seeds of *Schœnocaulon officinale,* occurring in white, acicular crystals, soluble in water, alcohol, and ether; melts at 240°C. It is sternutatory.
**sabal** (*sab'-al*). Saw-palmetto. The *sabal* of the U. S. P. is the partly dried ripe fruit of *Serenoa serrulata.* A fluidextract has been recommended in inflammations of the genitourinary tract, atonic impotence, and in bronchitis and pulmonary tuberculosis. Dose 1-2 dr. (4-8 Cc.).
**sabalol** (*sab'-al-ol*). A substance prepared from the active principles of the saw-palmetto (*Serenoa serrulata*).
**Sabbatia** (*sab-a'-she-ah*) [after Liberatus *Sabati,* an Italian botanist]. A genus of the order *Gentianeæ.* *S. angularis* is the American centaury.
**sabbattin** (*sab'-at-in*). A glucoside obtained from *Sabbatia elliottii,* quinine flower; it is antiperiodic and antipyretic.
**saber shin.** Term applied to the anterior border of the tibia, which has a sharp convex edge; found in hereditary syphilis.
**sabina** (*sa-bi'-nah*). See *savin.*
**sabromine** (*sab'-ro-mēn*). Trade name of a preparation of calcium, bromine and behenic acid; used as bromides in general.
**sabulous** (*sab'-ū-lus*) [*sabulum,* sand]. Gritty; sandy.
**sabulum conarii** (*sab'-ū-lum ko-nar'-e-i*) [L., sand of the conarium]. A sandy substance contained in the pineal gland.
**saburra** (*sab-ur'-ah*) [L., "coarse sand"]. Foulness of the stomach or of the tongue or teeth; sordes.
**saburral** (*sab-ur'-al*) [*saburra*]. 1. Pertaining to or affected with saburra. 2. Resembling or pertaining to coarse sand.
**sac** (*sak*) [*saccus,* a bag]. The bag-like bulging or covering of a natural cavity, hernia, cyst, or tumor. **s.s, air-,** the air-cells of the lung. **s., allantoid,** synonym of *allantois.* **s., amniotic,** the amnion. **s., auditory,** the rudimentary organs of hearing of the embryo of certain vertebrates. **s., conjunctival,** that formed by reflection of the palpebral conjunctiva. **s., dental,** see *dental sac.* **s., dorsal,** a recess between the epiphysis cerebri and the roof of the third ventricle. Syn., *suprapineal recess.* **s., embryonic,** the sac-like stage of the embryo, which it presents early in its development, just after the abdominal plates have closed. **s., endolymphatic,** a sac of the dura included in the aqueduct of the vestibule. See under *duct, endolymphatic.* **s. of the epididymis,** the visceral layer of the tunica vaginalis covering the epididymis. **s., epiploic,** see *sac, omental.* **s., fetal.** See *gestation-.* **s., gestation-,** the sac inclosing the embryo in ectopic pregnancy. **s., hernial,** the peritoneal covering of a hernia. **s., lacrimal,** the dilated upper portion of the lacrimal duct. **s., omental,** the sac formed between the ascending and descending portions of the great omentum. **s., pericardial,** the pericardium. **s., peritoneal,** the cavity formed by the peritoneal serous membrane. **s., pleural,** the cavity formed by the pleura. **s. of the pulmonary veins,** the left auricle of the heart. **s., serous,** the closed cavity formed by any serous membrane. **s., tubotympanic,** the diverticulum of the primitive gut forming the tympanic cavity and the Eustachian tube. **s., umbilical,** the umbilical vesicle. **s. of the venæ cavæ,** the right auricle of the heart. **s., vitelline,** the sac inclosing the vitellus or yolk in the embryo.
**saccàde** (*sak-ahd'*) [F., *saccade,* pull, draw]. The involuntary jerk of deglutition.
**saccaneurysma** (*sak-an-ū-riz'-mah*) [σάκκος, bag; εὑρύνειν, to dilate]. A sacculated aneurysm.
**saccate, saccated** (*sak'-āt, sak'-a-ted*) [*saccus,* a sac]. Sac-shaped; contained in a sac; encysted.
**saccharated** (*sak'-ar-a-ted*) [*saccharin*]. Containing sugar. **s. ferrous carbonate** (*ferri carbonas saccharatus,* U. S. P.). Dose 4 gr. (0.25 Gm.).
**saccharide** (*sak'-ar-id*) [σάκχαρον, sugar]. A compound of a base with sugar. A sucrate. See *casein saccharide.*
**saccharephidrosis** (*sak-ar-ef-id-ro'-sis*) [σάκχαρον, sugar; ἐφιδρῶσις, ephidrosis]. A form of hyperidrosis characterized by the excretion of sugar in sweat.
**saccharic** (*sak-ar'-ik*). Pertaining to or obtained from sugar.
**sacchariferous** (*sak-ar-if'-er-us*) [*saccharum;* sugar; *ferre,* to bear]. Containing or producing sugar.
**saccharification** (*sak-ar-if-ik-a'-shun*) [*saccharin; facere,* to make]. The act of converting into sugar.
**saccharimeter** (*sak-ar-im'-et-er*) [*saccharin;* μέτρον, a measure]. An apparatus for determining the amount of sugar in solutions, either in the form of a hydrometer, which indicates the strength in sugar by the specific gravity of the solution; or of a polarimeter, which indicates the strength in sugar by the number of degrees of rotation of the plane of polarization of light to the right. **-s. test,** a solution of dextrose rotates the plane of polarized light to the right.
**saccharimetry** (*sak-ar-im'-et-re*) [*saccharum;* μέτρον, measure]. The operation or art of ascertaining the amount or proportion of sugar in solution in any liquid.
**saccharin** (*sak'-ar-in*) [σάκχαρον, sugar], C₇H₅SO₃N. A crystalline substance nearly 280 times sweeter than cane-sugar, and used as a substitute for the latter in diabetes. It is also employed as an antiseptic. Syn., *benzoyl sulphonicimide; glucusimide; gluside; orthosulphaminbenzoic anhydride; saccharinol; saccharinose; sycose; zuckerin.* **s.-sodium,** a soluble powder containing 90 % of saccharin; used as an intestinal antiseptic. Dose 15 gr. (1 Gm.) once or twice daily. **s. soluble.** Same as *sodium saccharin.*
**saccharine** (*sak'-ar-in*) [*saccharum*]. Containing sugar; sugary; as sweet as sugar.
**saccharins** (*sak'-ar-ins*). A name given to the lactones of the saccharic acids.
**saccharobacillus** (*sak-ar-o-bas-il'-us*). See *bacillus pasteurianus,* in table of *bacteria.*
**saccharobiose** (*sak-ar-o-bi'-ōs*). A disaccharid, with the formula C₁₂H₂₂O₁₁.
**saccharogalactorrhea, saccharogalactorrhœa** (*sak-ar-o-gal-ak-tor-e'-ah*) [*saccharum;* γάλα, milk; ῥοία, a flow]. The excretion of an excess of sugar with the milk.
**saccharogen** (*sak'-ar-o-jen*) [σάκχαρον, sugar; γεννᾶν, to produce]. A material found in milk and convertible into lactose. A glucoside.
**Saccharomyces** (*sak-ar-o-mi'-sēz*) [*saccharum;* μύκης, a fungus]. A genus of unicellular vegetable organisms, of which the yeast-plant is a common example. *S. albicans,* same as *Oidium albicans,* the fungus of thrush. **S. cerevisiæ,** the ferment of beer-yeast. **S. granulomatosus,** Sanfelice (1898) obtained from granulomatous nodule of a pig. Inoculated in swine, it produced similar lesions, but was not pathogenic to other animals. **S. lithogenes,** Sanfelice (1895), from a carcinomatous metastasis in an ox, the primary tumor occurring in the liver. It killed white

mice in 8 days after subcutaneous inoculation. S. neoformans, Sanfelice (1895), isolated from fermenting grape-juice. It produced nodules in all organs of guinea-pigs except brain, heart, and suprarenals; death occurred in 20 to 30 days after inoculation. He emphasized the similarity of the organisms to the so-called coccidia of cancers. S. niger, isolated from the tissue of a guinea-pig which died of marasmus. It produced enlargement of lymph-glands and suppuration at the point of inoculation in guinea-pigs, rabbits, chickens, and dogs. S. ruber, Demme (1891), a red, budding fungus found in milk, which produced gastroenteritis in children. Shown by Casagrandi (1897) to be pathogenic for guinea-pigs, dogs, and mice when inoculated subcutaneously or into the abdomen. S. septicus, de Galtano, found in urinary sediment. An exceptionally virulent species producing fatal fibrinous peritonitis and septicemia in guinea-pigs in 12 hours. S. theobromæ, the yeast causing the fermentation in the curing of cacao. S. tumefaciens albus, Foulerton, isolated from patients in cases of pharyngitis.

saccharomycosis (sak-ar-o-mi-ko'-sis). A pathological condition due to yeasts or saccharomyces. s. hominis, a name given by Busse (1894) to pyemia produced by a pathogenic yeast. Syn., saccharomycosis subcutaneous tumefaciens, Curtis.

saccharorrhea, saccharorrhœa (sak-ar-or-e'-ah) [σάκχαρον, sugar; ῥοία, flow]. The secretion of saccharine fluid. Glycosuria. s. cutanea. See saccharephidrosis. s. lactea. See saccharogalactorrhea. s. pulmonalis, the exudation of sweetish sputa. s. urinosa. See diabetes mellitus.

saccharoscope (sak'-ar-o-skōp) [saccharum; σκοπεῖν, to view]. An instrument for determining and registering the amount of sugar in the urine.

saccharose (sak'-ar-ōs) [saccharum]. 1. C₁₂H₂₂O₁₁. A crystalline carbohydrate, cane-sugar, occurring in the juice of many plants, chiefly in sugar-cane, in some varieties of maple, and in beet-root; it melts at 160° C.; at 190°–200° C. it changes into a brown, noncrystallizable mass called caramel, used in coloring liquids. It is not directly fermentable, and does not reduce alkaline copper solutions. 2. Any one of a group of carbohydrates isomeric with cane-sugar.

saccharosuria (sak-ar-o-sū'-re-ah) [saccharose; οὖρον, urine]. The presence of saccharose in the urine.

saccharum (sak'-ar-um) [σάκχαρον, sugar], C₁₂H₂₂O₁₁, sugar. The sugar of the pharmacopeia (saccharum, U. S. P.; saccharum purificatum, B. P.) is the refined sugar obtained from s. officinarum and from various species of sorghum. See saccharose (1). s. album, white or pure crystallized sugar. s. canadense, maple-sugar, obtained from Acer saccharinum. s. candidum, rock-candy. s. lactis, sugar of milk. s. purificatum, pure white sugar.

saccharure (sak'-ar-ūr) [saccharum]. A preparation obtained by saturating sugar with a tincture, then drying, and pulverizing.

sacchorrhea, sacchorrhœa (sak-or-e'-ah) [saccharum; ῥεῖν, to flow]. Glycosuria.

sacchulose (sak'-ū-lōs). A product resulting from the treatment of sawdust with a weak solution of sulphurous acid under a pressure of about 100 lb. to the square inch (Classen Process). It has been proposed as a food stuff.

sacciform (sak'-sif-orm) [sac; forma, form]. Resembling a sac. s. disease of the anus, distention and inflammation of the pouches of the rectum.

saccular (sak'-ū-lar) [sac]. Sac-shaped, as a saccular aneurysm.

sacculated (sak'-ū-la-ted) [sac]. Divided into small sacs.

sacculation (sak-ū-la'-shun) [sac]. 1. The state of being sacculated. 2. The formation of small sacs.

saccule (sak'-ūl) [sacculus, dim. of sac, a sac]. 1. A small sac. 2. The smaller of two vestibular sacs of the membranous labyrinth of the ear. See sacculus labyrinthi.

sacculocochlear (sak-ū-lo-kok'-le-ar). Relating to the saccule of the vestibule and the cochlea.

sacculus (sak'-ū-lus). A saccule. s. alveolaris, an air cell. s. buccalis, hanging cheek or pouch, in animals. s. cæcalis, s. laryngis, the laryngeal pouch between the superior vocal bands and the inner surface of the thyroid cartilage. sacculi chalicophori, the lacunæ of bone. s. chylifer, s. rorifer, the receptaculum chyli. s. cordis, the pericardium. s. ellipticus, s. hemiellipticus. See utricle. s., Horner's, the anal pocket; a saccular fold of the rectal mucosa. s. labyrinthi, s. proprius, s. rotundus, s. sphæricus, the saccule of the vestibule.

saccus (sak'-us) [L.]. A sac. s. endolymphaticus, a small sac contained in the aqueduct of the vestibule and serving to establish a communication between the endolymph and the subdural space. s. lacrimalis, the lacrimal sac. s. reuniens, the sinus venosus. s. vitellinus, the vitelline sac.

sacer (sa'-ser) [L.]. Sacred. s. ignis, erysipelas. s. morbus, epilepsy.

sachet (sash-a') [saccus, a sac]. A small bag of perfumed or medicated substances. s. resoluti (Fr.), a sachet of equal parts of sal ammoniac, iron sulphate, and calcium sulphate.

Sachsse's solution (sak'-seh) [Georg Robert Sachsse, German chemist, 1840–1895]. See under S.'s test. S.'s test, a quantitative test for the determination of sugar in urine, consisting in the reduction of the test solution, a solution of red iodide of mercury 18 Gm., potassium iodide 25 Gm., potassium hydroxide 80 Gm., water to make a liter. The end of the reaction is ascertained by means of a solution of stannous chloride, supersaturated with sodium hydroxide.

sack (sak) [siccus, dry]. 1. An old name for dry Spanish and Canary wine; s herry. 2. Synonym of sac.

sacrache (sāk'-rāk) [sacrum; ache]. Sacral pain in the gravid woman.

sacrad (sa'-krad) [sacrum, the sacrum]. Toward the sacral aspect.

sacral (sa'-kral) [sacrum]. Pertaining to the sacrum. s. bone. See sacrum. s. canal. See canal. s. cornua. See cornu. s. flexure, the curve of the rectum in front of the sacrum. s. foramen. See foramen. s. groove. See groove. s. index, the sacral breadth multiplied by 100, and divided by the sacral length. s. nerves. See nerves. s. plexus. See plexus.

sacralgia (sa-kral'-je-ah) [sacrum, sacrum; ἄλγος, pain]. Pain in the sacrum; hieralgia.

sacra media (sa'-krah me'-de-ah) [arteria, understood]. The artery running down the middle of the anterior surface of the sacrum and representing the termination of the aorta.

sacrectomy (sa-krek'-to-me) [sacrum; ἐκτομή, excision]. Excision of part of the sacrum. See operation, Kraske's.

sacred (sa'-kred). Hallowed; holy. s. bark. See cascara sagrada. s. malady. Synonym of epilepsy.

sacren (sa'-kren) [sacrum, sacrum]. Belonging to the sacrum in itself.

sacriplex (sa'-krip-leks) [sacrum, plexus]. The sacral plexus of nerves.

sacro- (sa-kro-) [sacrum]. A prefix denoting relating to the sacrum.

sacroanterior (sa-kro-an-te'-re-or). Applied to a fetus having the sacrum directed forward.

sacrococcygeal (sa-kro-kok-sij'-e-al) [sacro-; coccyx]. Pertaining to the sacrum and the coccyx.

sacrocoxalgia (sa-kro-koks-al'-je-ah). See sacroiliac disease.

sacrocoxitis (sa-kro-koks-i'-tis). See sacroiliac disease.

sacrodynia (sa-kro-din'-e-ah) [sacro-; ὀδύνη, pain]. Pain referred to the region of the sacrum in cases of hysteria or neurasthenia.

sacroiliac (sa-kro-il'-e-ak) [sacro-; ilium]. Pertaining to the sacrum and the ilium. s. disease, an inflammation, usually tuberculous, of the sacroiliac joint, characterized by pain, tenderness, and swelling and elongation of the limb. s. synchondrosis, the junction of the sacrum and ilium.

sacrolumbalis (sa-kro-lum-ba'-lis). See under muscle.

sacrolumbar (sa-kro-lum'-bar) [sacro-; lumbus, loin]. Pertaining to the sacrum and the loins. s. angle, the angle formed by the articulation of the sacrum and the last lumbar vertebra.

sacroposterior (sa-kro-pōs-te'-re-or) [sacro-; posterior]. Of the fetus, having the sacrum directed backward.

sacropromontory (sa-kro-prom'-on-to-re). The promontory of the sacrum.

sacrosciatic (sa-kro-si-at'-ik) [sacro-; sciatic]. Pertaining to the sacrum and the ischium, as the sacrosciatic notch, sacrosciatic ligaments.

sacrospinal (sa-kro-spi'-nal) [sacrum; spina, spine]. Pertaining to the sacrum and the spine.

**sacrospinalis** (*sa-kro-spi-na'-lis*) [*sacro-*; *spine*]. The erector spinæ muscle.
**sacrotomy** (*sa-krot'-o-me*) [*sacro-*; τέμνειν, to cut]. Excision of the lower portion of the sacrum.
**sacrouterine** (*sa-kro-ū'-ter-in*). Pertaining to the sacrum and the uterus.
**sacrovertebral** (*sa-kro-ver'-te-bral*) [*sacro-*; *vertebra*]. Pertaining to the sacrum and the vertebræ. **s. angle**, the promontory of the sacrum.
**sacrum** (*sa'-krum*) [*sacer*, sacred; *os*, bone, understood]. A curved triangular bone composed of five united vertebræ, situated between the last lumbar vertebra above, the coccyx below, and the ossa innominata on each side, and forming the posterior boundary of the pelvis.
**sactosalpinx** (*sak-to-sal'-pinks*) [σακτός, crammed; σάλπιγξ, tube]. The obstruction of a Fallopian tube and consequent distention from retained secretion. **s. hæmorrhagica**. See *hematosalpinx*.
**saddle** (*sad'-l*) [ME., *sadel*]. A contrivance secured on the back of a horse or other animal to serve as a seat for a rider. **s.-arch**, that form of dental vault the section of which represents the shape of a saddle. **s.-back**, lordosis. **s.-bags**, a pair of leathern cases, formerly, and still locally, carried by physicians upon the saddle, and containing their medicines and instruments. **s.-head**. See *clinocephalus*. **s.-joint**, an articulation in which each surface is concave in one direction and convex in the other. **s.-nose**, a nose of which the bridge is sunken in. **s.**, Turkish. See *sella turcica*.
**sadism** (*sa'-dizm*) [Donatien Alphonse François; Marquis de Sade, 1740–1814]. Sexual perversion in which pleasure is derived from inflicting cruelty upon another.
**sadist** (*sa'-dist*). One affected with sadism.
**sadistic** (*sa-dis'-tik*). Pertaining to sadism.
**Saemisch's operation** (*sa'-mish*) [Edwin Theodor *Saemisch*, Austrian ophthalmologist, 1833–1909]. *For hypopyon ulcer:* the cornea is transfixed, and the intervening tissue, including the base of the ulcer, is divided by cutting outward. **S.'s ulcer**, a serpiginous infecting ulcer of the cornea.
**Saenger's macula** (*seng'-er*) [Max *Saenger*, Austrian gynecologist and obstetrician, 1853–1903]. A bright red spot marking the orifice of the duct of Bartholin's gland in cases of gonorrheal vulvitis. Syn., *macula gonorrhæica*.
**Saenger's pupil reaction.** For the differential diagnosis of cerebral syphilis and tabes; in amaurosis and optic atrophy of cerebral syphilis the pupil reflex to light may be preserved and even increased after a protracted stay in the dark, which is never the case in tabes dorsalis.
**sæpimentum** (*se-pe-men'-tum*) [*sæpire*, to fence]. 1. The tissue enclosing the three umbilical vessels. 2. Pons Varolii.
**sæptum** (*sep'-tum*). See *septum*.
**safflower** (*saf'-low-er*). See *carthamus*. **s. carmin**. Same as *carthamin*.
**saffron** (*saf'-ron*) [Ar., *zafarān*, saffron]. The *Crocus sativus*, a plant of the order *Irideæ*. Its stigma (*crocus*, B. P.) contains a glucoside, coloring-matter (crocin), and a bitter principle. Saffron is used as a coloring and flavoring agent, and in the form of a tea to bring out the eruption of the exanthematous diseases. **s.**, American. See *carthamus*. **s., meadow**. See *colchicum*. **s. substitute**. Same as *Victoria yellow*. **s.-tea**. See *carthamus*. **s.-yellow**. Same as Martius's yellow.
**safranine, safranin** (*saf'-ra-nin*) [*saffron*], C₁₈H₁₈N₄. A coal-tar dye used in microscopy, especially in studying karyokinesis. It is a powerful cardiac and respiratory poison.
**safranophile** (*sa-fran'-of-fil*) [Fr., *safran*, saffron; φιλεῖν, to love]. In bacteriology, or histology, applied to microbes or histological elements that show a peculiar affinity for safranine.
**safrene** (*saf'-rēn*) [*saffron*], C₁₀H₁₆. A hydrocarbon obtained from sassafras.
**safrol** (*saf'-rol*) [*saffron*; *oleum*, oil]. *Safrolum* (U. S. P.), C₁₀H₁₀O₂. The stearoptene of sassafras oil, used in headache, neuralgia, and rheumatism. Dose 10–20 min. (0.65–1.3 Cc.).
**sagapenum** (*sag-a-pe'-num*). A fetid gum-resin believed to be the concrete juice of *Ferula persica*. Its properties resemble those of asafetida and galbanum.
**sage** (*sāj*) [*salvia*]. *Salvia officinalis*, a plant of the order *Labiatæ*. Its leaves (*saliva*, U. S. P.) contain several terpenes, an oil, salviol, C₁₀H₁₈O, and camphor. Sage is tonic, astringent and aromatic, is used in dyspepsia. Was formerly employed in colliquative sweats. Dose 20–30 gr. (1.3–2.0 Gm.).
**sage-femme** (*sahzj-fam'*) [Fr., literally, a wise woman]. A midwife.
**sagittal** (*sōj'-it-al*) [*sagitta*, an arrow]. 1. Arrow-like, as the *sagittal* suture of the skull. 2. Pertaining to the anteroposterior median plane of the body. **s. furrow**, a channel extending along the median line of the inner surface of the vault of the cranium. **s. nucleus**, that of the oculo-motor nerve. **s. plane**, the median plane of the body. **s. section**. See *section*. **s. sinus**, the longitudinal sinus. **s. suture**, the suture uniting the parietal bones.
**sago** (*sa'-go*) [Malay, *sāgu*]. A food and demulcent. **s.-spleen**, a spleen presenting on section the appearance of sago-grains, as a result of amyloid degeneration of the Malpighian bodies. **s.-grain**, a vesicular granulation of the eyelid, seen in granular ophthalmia.
**Sagotia racemosa** (*sa-go'-she-ah ras-e-mo'-sah*). A South American species of the order *Euphorbiaceæ*, used as a tonic and aphrodisiac.
**sagrada** (*sag'-rah-din*). The proprietary name for a 20 % solution of extract of cascara sagrada with spirit of peppermint.
**Sahli's desmoid test** (*sah'-le*) [Hermann *Sahli*, German physician, 1856– ]. For estimating the functional activity of the stomach. A pill of 0.05 Gm. of methylene-blue and 0.1 Gm. of iodoform is inclosed in a bag of rubber-dam and tied with dry catgut. The gut is digested by gastric juices and not by pancreatic juices. The pill is then absorbed and in about 6 hours the urine is green. The iodine will be found in the saliva in two hours.
**Saigon cinnamon.** A variety of cinnamon (*cinnamomum saigonicum*, U. S. P.) obtained from Saigon, the capital of French Cochin-China. See *cinnamon*.
**St. Anthony's fire.** Erysipelas; anthrax.
**St. Gothard's disease.** Ankylostomiasis.
**St. Hubert's disease.** Hydrophobia.
**St. Ignatius' bean.** See *ignatia*.
**St. Roch's disease.** Bubo.
**St. Sement's disease.** Syphilis.
**St. Vitus' dance.** See *chorea*. **St. Vitus' dance of the voice**, stammering.
**saiodine** (*sa-i'-o-den*). Trade name of a preparation containing calcium, iodine and behenic acid.
**sajodin** (*sah-yo'-din*). See *saiodine*.
**saké** (*sāh'-ka*) [Jap.]. Japanese rice-beer or other alcoholic beverage.
**sal** [L.]. 1. Salt. 2. Any substance resembling salt. **s. acetosellæ**, potassium binoxalate. **s. aeratus**. 1. Sodium bicarbonate. 2. Potassium bicarbonate. **s. alembroth**. See *alembroth*. **s. amarum**, magnesium sulphate. **s. ammoniac**, ammonium chloride. **s. communis**. See *salt, common*. **s. de duobus**, potassium sulphate. **s. enixum**, potassium bisulphate. **s. Glauberi**. See *salt, Glauber's*. **s. kissingense**, a salt obtained from the mineral springs of Kissingen, in Bavaria. **s. polychrest**, potassium sulphate with sulphur. **s. prunellæ**, **s. prunelle**. 1. A fused mixture of potassium nitrate, 128 parts, and sulphur, 1 part. 2. Fused potassium nitrate. **s. rupium**, rock-salt. **s. seignette**, potassium and sodium tartrate. **s. sodæ**. See *salt of soda*. **s. volatile**, ammonium carbonate, or aromatic spirit of ammonia.
**Sala's cells.** Stellate connective-tissue cells found in the network of fibers forming the sensory nerve-endings in the pericardium.
**salaam convulsion** (*sa-lahm'*) [Ar., *salām*, saluting]. A clonic spasm of the muscles of the trunk, producing a bowing movement; it is usually due to hysteria.
**salacetin** (*sal-as'-et-in*). Trade name of phenyl-aminoacetosalicylate; said to be antiseptic and analgesic.
**salacetol** (*sal-as'-et-ol*) [*salix*, willow; *acetum*, vinegar]. The salicylic acid ester of acetone-alcohol. It is proposed as a substitute for sodium salicylate and salol. It has been recommended as an intestinal and genito-urinary antiseptic, and for the treatment of acute or chronic rheumatism. Dose, 30–45 grains.
**salacious** (*sa-la'-se-us*) [*salax*, lustful]. Lustful.
**salacity** (*sa-las'-it-e*) [*salax*, lustful]. Lustful or venereal desire.
**salactol** (*sal-ak'-tol*). A combination of the sodium salts of salicyclic and lactic acids dissolved in a 1 %

# SALAMIDE 774 SALIVA

solution of hydrogen dioxide; it is recommended in diphtheria. The solution is applied as a spray or with a brush, and given internally in doses of a tablespoonful (15 Cc.).

**salamide** (*sal'-am-id*). An amidogen derivative of salicylic acid, which it closely resembles in therapeutic properties, but acts more promptly and in smaller doses.

**salantol** (*sal-an'-tol*). See *salacetol*.

**salborol** (*sal-bo'-rol*). A compound of phenyl salicylate and boric acid, used in rheumatism.

**salbromalide** (*sal-bro'-mal-id*). See *antinervin*.

**saldanin** (*sal'-dan-in*). A local anesthetic said to be prepared from *Datura arborea*.

**salen** (*sal'-en*). Trade name of a mixture of methyl and ethyl glycolic esters; soluble in ether, alcohol and castor-oil.

**salenal** (*sal'-en-al*). Trade name of an ointment containing 33.3 per cent of salen.

**salenders** (*sal'-en-derz*). See *malandri*.

**salep** (*sal'-ep*) [Ar., *sahleb*]. The dried tubers of various species of the genus *Orchis* and the genus *Eulophia*. It is used as a food, like sago and tapioca.

**saleratus** (*sal-er-a'-tus*) [*sal, aeratus*, aerated salt]. Properly, sal-aeratus. Potassium bicarbonate; also, sodium bicarbonate.

**salethyl** (*sal-eth'-il*). A proprietary preparation said to be pure ethyl salicylate.

**salhypnone** (*sal-hip'-nōn*), C₆H₄O(COC₆H₄)COO-CH₃. A benzoylmethylsalicylic ester; long colorless needles, insoluble in water, sparingly soluble in alcohol and ether; melts at 113°-114°. It is used as an antiseptic.

**salicamar** (*sal-ik'-am-ar*), CH₂OH . CHOH . CH₂-O . C₆H₄CO . CH₂ . CHOH . CH₂OH. A glycerol ether of glycerolsalicylic acid; recommended as a stomachic and antirheumatic.

**salicin** (*sal'-is-in*) [*salix*, willow], C₁₃H₁₈O₇. A crystalline glucoside found in the bark and leaves of the willow. Salicin (*salicinum*, U. S. P., B. P.) is used as a substitute for salicylic acid in doses of 5-30 gr. (0.3-2.0 Gm.); maximum daily dosage 150 gr. (9.7 Gm.).

**salicol** (*sal'-ik-ol*). A proprietary solution said to consist of methyl-alcohol, salicylic acid, and oil of wintergreen in water; used as an antiseptic and cosmetic.

**salicyl** (*sal'-is-il*) [*salicylic acid*], C₇H₅O₂. The hypothetical radical of salicylic acid. **s. acetate**, C₆H₁₁ . O . C₇H₅O₃O, acetosalicylic anhydride. **s. acetol**, salactol. **s.-anilide**, salifebrine. **s. bromanilide**, same as *antinervin*. **s.-creosote**, a paste prescribed by Unna in skin diseases, consisting of a mixture of salicylic acid, creosote, wax, and cerate. **s.-p-phenetidin**, See *malakin*. **s.-quinine salicylate**. See *saloquinine salicylate*. **s.-resorcinol**, C₁₃H₁₀O₄, obtained from salicylic acid and resorcinol with heat. It occurs in plates slightly soluble in water; melts at 133° C. It is antiseptic, antipyretic, and analgesic, and used in typhoid, diarrhea, etc. Dose 5-15 gr. (0.32-1.0 Gm.); maximum dose 15 gr. (1 Gm.) single; 60 gr. (4 Gm.) daily. **s. urate**, a salt of salicyluric acid.

**salicylage** (*sal'-is-il-āj*) [*salicylic acid*]. The addition of salicylic acid to foods for their preservation.

**salicylamide** (*sal-is-il'-am-id*) [*salicylic acid*], C₆H₄(OH)CONH₂. A tasteless compound produced by treating methyl salicylate with an alcoholic solution of ammonia, and used as a substitute for salicylic acid.

**salicylate** (*sal-is'-il-āt*) [*salicylic acid*], A salt of salicylic acid. The salicylates of lithium, methyl, and sodium, which are official, and those of ammonium and strontium, which are unofficial, are used in rheumatism, in doses of 10-15 gr. (0.65-1.0 Gm.). *Bismuth salicylate* is employed as an intestinal antiseptic; *naphthol salicylate* is betol; *phenyl salicylate* is salol.

**salicylated** (*sal-is'-il-a-ted*). Impregnated with salicylic acid.

**salicylic acid** (*sal-i-sil'-ik*). See *acid, salicylic*.

**s.-acid glycerolester**, recommended as a valuable antirheumatic remedy. **s. alcohol**. See *saligenin*. **s. aldehyde**, salicylous acid; used as an internal antiseptic. Dose 2-8 gr. (0.1-0.5 Gm.) daily. **s.-amide**. See *salicylamide*. **s. amylester**. See *amyl salicylate*. **s. anhydride**. See *s. aldehyde*. **s. cream**, an antiseptic mixture of powdered salicylic acid, 2 dr.; phenol, 1 dr.; glycerol, 10 dr. **s. naphthylic ester**. See *betol*. **s. phenylester**. See *salol*.

**s. silk**, a dressing made of silk waste impregnated with 10 % salicylic acid and a little glycerol. **s. suet**, one part of salicylic acid in 49 parts of mutton-suet; used as a dressing for sores. **s. thymolester**, thymol acetate. **s. wool**, cotton impregnated with 4 to 10 % of salicylic acid and an equal amount of glycerol.

**salicylid** (*sal-is'-il-id*). An anhydride of salicylic acid. **s.-phenetidin**. See *malakin*.

**salicylism** (*sal'-is-il-ism*) [*salix*, willow]. A toxic condition, produced by the injudicious or excessive use of salicylic acid or its salts.

**salicylize** (*sal'-is-il-īz*). To treat with salicylic acid.

**salicylol** (*sal'-is-il-ol*). See *salicylic aldehyde*.

**salicyluric acid** (*sal-is-il-ū'-rik*) [*salicylic, uric*]. A compound of glycol and salicylic acid found in the urine after the administration of salicylic acid.

**salifebrin** (*sal-e-feb'-rin*). Salicylanilide; C₁₃H₁₁-NO₂, a white, permanent powder, insoluble in water, freely soluble in alcohol; recommended as an antipyretic and antineuralgic. It colors blue litmuspaper red.

**saliferous** (*sal-if'-er-us*) [*sal*, salt; *ferre*, to bear]. Producing salt.

**salifiable** (*sal-if-i'-a-bl*) [*sal; fieri*, to become]. Forming a salt by union with an acid.

**saliformin** (*sal-if-orm'-in*), (CH₂)₆N₄ . C₆H₄(OH)-COOH. A white, crystalline powder, of sour taste, soluble in water and alcohol. It is an antiseptic and uric-acid solvent. Dose 15-30 gr. (1-2 Gm.). Syn., *Formin salicylate*; *hexamethylenetetraminesalicylate*; *urotropin salicylate*.

**saligallol** (*sal-e-gal'-ol*). Pyrogallol disalicylate, a resinous solid, soluble in acetone or chloroform. Used as a vehicle for cutaneous applications and as a varnish.

**saligenin** (*sal-ij'-en-in*) [*salicin*; γεννᾶν, to produce], C₇H₈O₂. Orthooxybenzylalcohol, a substance obtained from salicin by boiling with dilute hydrochloric or sulphuric acid.

**salimeter** (*sal-im'-et-er*) [*sal*, salt; μέτρον, a measure]. A hydrometer for ascertaining the strength of saline solutions.

**salinaphtol** (*sal-in-af'-tol*). See *betol*.

**saline** (*sa'-līn*) [*sal*, salt]. 1. Salty; containing salt or substance resembling salt. 2. A salt of an alkali or alkaline earth. **s. solution**, a 0.6 % solution of sodium chloride; physiological (wrongly called normal) salt solution.

**salines** (*sa'-līnz*) [*sal*, salt]. Salts of the alkalies or of magnesium, used as hydragogue cathartics. Magnesium sulphate and citrate; sodium sulphate and Rochelle salts are examples.

**salinigrin** (*sal-in-i'-grin*). A substance said to be a glucoside from the bark of *Salix nigra*.

**salinometer** (*sal-in-om'-et-er*). Synonym of *salimeter*.

**saliodine** (*sal-i-o'-dīn*). A proprietary preparation, said to be "an iodated, acetosalicylate, with adjuvants"; recommended in rheumatism, malaria, influenza, syphilis, etc. Dose 10 to 30 grains.

**salipen** (*sal'-if-en*). Salicylphenetidin; a compound of salicylic acid and phenetidin.

**salipyrine** (*sal-e-pī'-rin*) [*salicylate; πῦρ*, fire], C₁₁H₁₂N₂O . C₇H₆O₃. Antipyrine salicylate, consisting of 57.7 parts of salicylic acid and 42.3 parts of antipyrine; it is soluble in water, and is used in rheumatism, neuralgia, and as an antipyretic. Dose 15-30 gr. (1-2 Gm.).

**saliretin** (*sal-e-ret'-in*) [*saligenin; ῥητίνη*, resin], C₁₄H₁₄O₄. An amorphous resinous body, produced by treating saligenin with acids.

**Salisbury treatment** (*sawls'-ber-e*). The treatment of obesity by meat diet and hot water.

**salitannol** (*sal-e-tan'-ol*), C₁₄H₁₆O₇. A condensation-product of salicylic and gallic acids by action of phosphorus oxychloride; a white, amorphous powder, soluble in solutions of caustic alkalies, slightly soluble in alcohol. Recommended as a surgical antiseptic.

**salithymol** (*sal-e-thī'-mol*). Thymol salicylate.

**salitonia** (*sal-i-to'-ne-ah*). A saline tonic.

**saliva** (*sa-lī'-vah*) [L.]. The mixed secretion of the parotid, submaxillary, sublingual, and mucous glands of the mouth. It is opalescent, tasteless, alkaline, and has a specific gravity of from 1004 to 1009, and contains serum-albumin, globulin, mucin, urea, an amylolytic ferment called ptyalin, and a proteolytic and a lipolytic ferment; also salts, among which is potassium sulphocyanate, derived especially

# SALIVANT 775 SALPINGO-OOPHORITIS

from the parotid gland. Among formed elements are epithelial cells, salivary corpuscles, and bacteria. The functions of saliva are to moisten the food and lubricate the bolus, to dissolve certain substances, to facilitate tasting, to aid in deglutition and articulation, and to digest starches, which it converts into maltose, dextrin, and glucose. **s., chorda,** that produced by stimulation of the chorda tympani nerve. **s., ganglionic,** that produced by irritating the submaxillary glands. **s., sympathetic,** that produced by stimulation of the sympathetic nerve.
**salivant** (*sal'-iv-ant*) [*saliva,* saliva]. 1. Stimulating the secretion of saliva. 2. A drug which increases the flow of saliva.
**salivary** (*sal'-iv-a-re*) [*saliva*]. Pertaining to or producing saliva; formed from saliva. **s. calculus,** a calcareous concretion found in the salivary ducts. **s., corpuscles,** pale, spherical, nucleated bodies found in saliva. **s. diastase.** Same as *ptyalin.* **s. digestion,** the conversion of starches into dextrin and sugar by the action of saliva. **s. fistula,** an abnormal opening communicating with a salivary duct. **s. glands,** the glands, six in number, situated three on each side of the mouth, which secrete the saliva. See *parotid, submaxillary,* and *sublingual.*
**salivate** (*sal'-iv-āt*) [*salivare,* to spit out]. To cause an excessive discharge of saliva.
**salivation** (*sal-iv-a'-shun*) [*salivate*]. An excessive secretion of saliva; a condition produced by mercury, pilocarpin, and by nervous disturbances.¹
**salivator** (*sal'-iv-a-tor*) [*salivate*]. An agent causing salivation.
**salivatory** (*sal-iv-a'-to-re*). Salivant; stimulating the secretion of saliva.
**salivin** (*sal'-iv-in*). Same as *ptyalin.*
**salivolithiasis** (*sal-iv-o-lith-i'-as-is*) [*saliva;* λίθος, stone]. Formation of a salivary calculus.
**salix** (*sa'-liks*) [L.]. The bark of the common white willow, *S. alba.* Its properties are due to a constituent, salicin, which is tonic and antiseptic. It is useful as an antipyretic in rheumatic fever. **s. nigra,** the bark of the black willow, recommended in nocturnal emissions and ovarian neuralgia.
**Salkowski's modification of Hoppe-Seyler's test for CO in the blood** (*sal-kow'-ske*) [Ernst Leopold Salkowski, German physician, 1844– ]. Add to the blood to be tested 20 volumes of water and an equal quantity of a sodium hydroxide solution of specific gravity 1.34. In the presence of carbon monoxide the mixture will soon become milky, changing to bright red. On standing, red flakes collect on the surface. Normal blood treated in this way gives a dirty brown coloration. **S.'s reaction for cholesterin,** dissolve the substance in chloroform and add an equal volume of concentrated sulphuric acid. The cholesterin solution becomes bluish-red, changing gradually to violet red, while the sulphuric acid appears red with a green fluorescence. **S.'s test for indol,** to the indol solution add a few drops of nitric acid, and then, drop by drop, a 2 % solution of potassium nitrite. The presence of indol is evinced by a red color, and finally by a red precipitate of nitrosoindol nitrate.
**(de) Salle's line.** A line beginning at the upper margin of the ala nasi, encircling the angle of the mouth, and ending at the edge of the orbicularis oris. For significance see *Jadelot's lines.*
**sallenders** (*sal'-en-durz*). See *malandria.*
**salmiac** (*sal'-me-ak*). Ammonium chloride.
**salmin** (*sal'-min*), $C_{30}H_{57}N_{17}O_6 + 4H_2O$. A protamine from the spermatozoa of salmon, identical or isomeric with clupein.
**Salmon's back-cut.** An incision along the track of an anal fistula.
**salmon patch** (*sam'-un*). See *Hutchinson's patch.*
**salochinin.** See *saloquinine.*
**salocoll** (*sal'-o-kol*). Phenocoll salicylate. A white powder, odorless and tasteless, only slightly soluble in cold water. It is said to be antipyretic, antineuralgic, and antirheumatic. Dose 15 to 30 grains.
**salol** (*sal'-ol*) [*salix,* willow]. See *pheny salicylate.* **s., camphorated,** a mixture of 75 % of phenyl salicylate with 25 % of camphor; an oily liquid, used in alcohol, ether, chloroform, or oils, a local anesthetic, antiseptic, and analgesic. Dose 3–10 gr. (0.2–0.65 Gm.). Syn., *camphor salol.* **s. tribromide,** $C_6H_4$.-$OH$.$COO$.$C_6H_2Br_3$, a white, odorless, tasteless powder, freely soluble in chloroform and glacial acetic acid, insoluble in ether, or alcohol. It is a combined hypnotic and hemostatic. Dose 32 gr. (2 Gm.). Syn., *cordol.*
**salolism** (*sa'-lol-ism*). Poisoning by phenyl salicylate, a mixture of salicylism and carbolism in which the symptoms of the latter predominate.
**Salomon-Saxl's reaction** (*sal'-om-on-saksl'*) [Hugo Salomon, Austrian physician]. Excess of neutral sulphates in the urine of patients with cancer. It is present in many cases of cancer, but it is not specific, and is even found in healthy individuals.
**saloop** (*sa-loop'*) [see *salep*]. A drink prepared from salep; also from sassafras bark and herbs. It is regarded as a cure for drunkenness.
**salophen** (*sal'-o-fen*) [*salix,* willow], $C_6H_4$.$OH$.-$CO_2$.$C_6H_4$.$NH(C_2H_3O) = C_{15}H_{13}NO_4$. Acetylparamidophenyl salicylate, a crystalline substance containing 50 % of salicylic acid, and used as a substitute for the latter, and as an intestinal antiseptic. Dose 15 gr. (1 Gm.).
**saloquinine** (*sal-o-kwin'-ēn*), $C_6H_4$.$OH$.$CO$.$O$.-$C_{20}H_{23}N_2O$. The quinine ester of salicylic acid; a crystalline, absolutely tasteless substance, insoluble in water, readily soluble in alcohol or ether; melts at 130° C. It is used as is quinine. Dose 10–30 gr. (0.65–2.0 Gm.) several times daily. **s. salicylate,** $C_6H_4$.$OH$.$COO$.$C_{20}H_{23}N_2O$.$C_6H_4$.$OH$.$COOH$, crystallizes in white needles, soluble with difficulty in water; melts at 179° C. It is tasteless and recommended in rheumatism. Dose 15 gr. (1 Gm.) 3 times daily. Syn., *rheumatin.*
**salosantal** (*sal-o-san'-tal*). A 33 % solution of phenyl salicylate in sandalwood oil with the addition of a little oil of peppermint. It is indicated in cystitis, prostatitis, etc. Dose 10–20 drops 3 times daily after meals.
**salpingectomy** (*sal-pin-jek'-to-me*) [*salpinx;* ἐκτομή, excision]. Excision of a Fallopian tube.
**salpingemphraxis** (*sal-pin-jem-fraks'-is*) [*salpinx;* ἔμφραξις, obstruction]. Closure of the Eustachian or Fallopian tube.
**salpingian, or salpingic** (*sal-pin'-je-an, sal-pin'-jik*) [σάλπιγξ, tube]. Pertaining to a Eustachian or Fallopian tube.
**salpingion** (*sal-pin'-je-on*). The point at the inferior surface of the apex of the petrosa.
**salpingitic** (*sal-pin-jit'-ik*) [*salpinx;* ιτις, inflammation]. Pertaining to or affected with salpingitis.
**salpingitis** (*sal-pin-ji'-tis*) [*salpinx;* ιτις, inflammation]. 1. Inflammation of the Fallopian tube. 2. Inflammation of the Eustachian tube. **s., chronic parenchymatous,** pachysalpingitis, chronic interstitial inflammation and thickening of the muscular coat of the Fallopian tube. **s., chronic vegetating,** excessive hypertrophy of the mucosa of the Fallopian tube. **s., gonorrheal,** that due to infection with gonococci. **s., hemorrhagic,** hematosalpinx. **s., interstitial,** that marked by excessive formation of connective tissue. **s., isthmic nodular,** follicular inflammation of the small constricted portion (isthmus) of the oviduct, with formation of small nodules of muscular and connective tissue. **s., mural.** See **s., *chronic parenchymatous.* **s., parenchymatous.** See **s., *chronic parenchymatous.* **s., pneumococcus,** that due to infection with pneumococci. **s., pseudofollicular,** adenomyoma originating in the tubal epithelium. **s., purulent,** salpingitis with secretion of pus instead of mucus or serum. **s., tuberculous,** that marked by the infiltration of the lining membrane and walls of the tube with tuberculous nodules.
**salpingo-** (*sal-ping'-go-*) [σάλπιγξ, tube]: A prefix denoting relation to the Fallopian or the Eustachian tube.
**salpingocatheterism** (*sal-ping-go-kath'-et-er-izm*). Catheterization of the Eustachian tube.
**salpingocele** (*sal-ping'-go-sēl*) [*salpingo-;* κήλη, a hernia]. Hernia of the oviduct.
**salpingocyesis** (*sal-ping-go-si-e'-sis*) [*salpingo-;* κύησις, pregnancy]. Tubal pregnancy.
**salpingolysis** (*sal-ping-gol'-is-is*) [*salpingo-;* λύειν, to loosen]. The breaking down of adhesions of the Fallopian tube.
**salpingomallearis, salpingomalleus** (*sal-ping-go-mal-e-a'-ris, sal-ping-go-mal'-e-us*). The tensor tympani muscle. See *muscles, table of.*
**salpingo-oophorectomy** (*sal-ping-go-o-o-for-ek'-to-me*) [*salpingo-; oophoron;* ἐκτομή, excision]. Excision of the Fallopian tube and the ovary.
**salpingo-oophoritis** (*sal-ping-go-o-of-or-i'-tis*) [*sal-*

# SALPINGO–OOPHOROCELE 776 SALVE

**pingo-; oophoron; ιτις, inflammation].** Inflammation of the Fallopian tube and the ovary.
**salpingo-oophorocele** (sal-ping-go-o-of'-or-o-sēl) [salpingo-; oophorocele]. Hernial protrusion of the ovary and oviduct.
**salpingo-oothecectomy** (sal-ping-go-o-o-the-sek'-to-me). Same as salpingo-oophorectomy.
**salpingo-oothecitis** (sal-ping-go-o-o-the-si'-tis). Same as salpingo-oophoritis.
**salpingo-oothecocele** (sal-ping-go-o-o-the'-ko-sēl). Same as salpingo-oophorocele.
**salpingo-ovariectomy** (sal-ping-go-o-va-re-ek'-to-me). Same as salpingo-oophorectomy.
**salpingo-ovariotomy** (sal-ping-go-o-var-e-ot'-om-e). Same as salpingo-oophorectomy.
**salpingo-ovaritis** (sal-ping-go-o-var-i'-tis). See salpingo-oophoritis.
**salpingopalatal** (sal-ping-go-pal'-at-al) [salpingo-; palatum, palate]. Pertaining to the Eustachian tube and the palate. **s.** fold, a fold of mucosa covering the levator palati muscle.
**salpingoperitonitis** (sal-ping-go-per-it-on-i'-tis). Inflammation of the peritoneum lining the oviduct.
**salpingopharyngeal** (sal-ping-go-far-in'-je-al) [salpingo-; pharynx]. Pertaining to the Eustachian tube and the pharynx.
**salpingopharyngeus** (sal-ping-go-far-in-je'-us) [salpingo-; pharynx]. A muscular bundle passing from the Eustachian tube downward to the constrictors of the pharynx.
**salpingorrhaphy** (sal-ping-gor'-a-fe) [σάλπιγξ, tube; ῥαφή, suture]. Suture of the Fallopian tube.
**salpingosalpingostomy** (sal-ping-go-sal-ping-gos'-to-me). The operation of uniting the two Fallopian tubes.
**salpingoscope** (sal-ping'-go-skōp) [salpingo-; σκοπεῖν, to look]. A modified cystoscope provided with an electric lamp of low voltage for exploration of the nasopharynx.
**salpingostaphylinus** (sal-ping-go-staf-il-i'-nus) [salpingo-; σταφυλή, uvula]. The abductor muscle of the Eustachian tube. **s. internus.** Synonym of levator palati. See under muscle.
**salpingostenochoria** (sal-ping-go-ste-no-ko'-re-ah) [salpingo-; στενός, narrow; χώρα, space]. Stenosis or stricture of the Eustachian tube.
**salpingostomatomy** (sal-ping-go-sto-mat'-om-e). Salpingostomy.
**salpingostomy** (sal-ping-gos'-to-me) [salpingo-; στόμα, mouth]. The operation of making an artificial fistula between a Fallopian tube and the body-surface.
**salpingotomy** (sal-ping-got'-o-me) [salpingo-; τομή, a cutting]. The operation of cutting into a Fallopian tube.
**salpingsterocyesis** (sal-ping-kis-ter-o-si-e'-sis) [σάλπιγξ, tube; ὑστέρα, womb; κύησις, gestation]. Interstitial pregnancy.
**salpinx** (sal'-pinks) [σάλπιγξ, tube]. A tube, especially the Eustachian or the Fallopian tube.
**salpyrine** (sal-pi'-rēn). See salipyrine.
**salt** [sal, salt]. 1. Sodium chloride. 2. Any compound of a base and an acid. **s., acid,** a salt formed from a dibasic or polybasic acid in which only a part of the replaceable hydrogen atoms has been replaced by the base. **s., alkaline.** See **s., basic. s., aperient, of Frederick,** sodium sulphate. **s. of barilla,** sodium carbonate. **s., basic,** a salt containing an excess of the basic element, and formed by the union of a normal salt with a basic oxide or hydroxide. **s., bay-,** sodium chloride; also the sea-salt obtained by the evaporation of sea-water by solar heat. **s. of bones,** ammonium carbonate. **s., Carlsbad,** a salt prepared from one of the springs at Carlsbad or made in imitation of it. Each spring contains in varying degrees carbonates of magnesium, iron, manganese, calcium, strontium, lithium, and sodium, sulphates of sodium and potassium, sodium chloride, sodium fluoride, sodium borate, and calcium phosphate. **s. of colcothar,** sulphate of iron. **s., common,** sodium chloride. **s., crab orchard,** a mild saline purgative produced from the evaporated water of springs at Crab Orchard, Kentucky. It contains magnesium, sodium, and potassium sulphates and a little iron and lithium. **s., diuretic,** potassium acetate. **s., double,** one in which the hydrogen atoms of an acid are replaced by two metals. **s., Epsom-,** magnesium sulphate. **s. fever,** fever caused by giving salt solution intravenously or otherwise. **s.-frog,** a frog from whose vascular system all blood has been artificially removed and replaced by physiological salt solution. Syn., *Cohnheim's frog.* **s., Glauber's.** sodium sulphate. **s., halogen,** s., **haloid,** any salt of the halogen elements, fluorine, chlorine, bromine, and iodine. **s.s of lemon,** potassium binoxalate. **s., Monsel's,** subsulphate of iron, used chiefly in solution as a styptic. **s., neutral.** 1. A salt which has a neutral action towards litmus. 2. Often used as the equivalent of *s. normal, q. v.* **s., normal,** a salt in which all of the available hydrogen has been replaced by a metal or its equivalent. **s. oxy-,** a salt of an oxyacid, one containing oxygen. **s.s, Preston's,** English smelling-salts. **s., purging, tasteless,** sodium phosphate. **s.-rheum,** chronic eczema. **s., Rochelle,** sodium and potassium tartrate. **s., rock-,** native sodium chloride, occurring in crystalline masses. **s., sea-,** the sodium chloride obtained by the evaporation of sea-water. **s., secondary.** Same as **s. neutral. s.-sickness.** See *sickness, salt-.* **s., smelling-,** any pungent, irritant salt which when inhaled usually acts reflexly as a respiratory or circulatory stimulant. Ammonium carbonate is generally used. **s. of soda,** sodium carbonate. **s. solution,** a solution of sodium chloride in distilled water. One containing from 0.6 to 0.75 % of sodium chloride is known as a *physiological* or (incorrectly) *normal salt solution,* and is used in physiological experiments on living tissues. In medicine it has been employed to restore to the system the fluids lost by severe hemorrhage or profuse diarrheal discharges. The solution is introduced into the subcutaneous tissues or into a vein; sometimes also into the rectum. **s. of sorrel,** potassium binoxalate derived from species of *Oxalis* and *Rumex.* **s. spirit of,** hydrochloric acid. **s.-starvation.** See *hypochlorization.* **s., table,** sodium chloride. **s. of tartar,** pure potassium carbonate. **s. of urine,** ammonium carbonate. **s. of vitriol,** zinc sulphate. **s. of wisdom,** sal ammbroth. **s. of wormwood,** potash prepared from wormwood.
**saltans rosa.** Urticaria.
**saltation** (sal-ta'-shun) [saltare, to dance]. The dancing or leaping sometimes noticed in chorea.
**saltatory, saltatoric** (sal'-tat-o-re, sal-tat-or'-ik) [saltare, to dance]. Dancing or leaping. **s. spasm,** a clonic spasm that causes the patient to leap or jump when he attempts to stand.
**salted.** A term applied to animals that have recovered from South African horse-sickness.
**Salter's incremental lines** [Sir James A. *Salter,* English dentist]. Dentinal lines more or less parallel to the surface of the tooth. Caused by imperfectly calcified dentin.
**saltpeter** (sawlt-pe'-tre) [salt; petra, a rock]. Potassium nitrate. **s., Chili,** sodium nitrate. **s., wall,** calcium nitrate.
**salts.** A saline cathartic, especially magnesium sulphate, sodium sulphate, or Rochelle salt.
**salubrin** (sal-ū'-brin) [salubritas, healthfulness]. A compound said to contain 2 % of anhydrous acetic acid, 25 % of acetic ether, 50 % of alcohol, and the remainder distilled water. It is antiseptic, astringent, and hemostatic, and is used diluted with water as a gargle and on compresses.
**salubrious** (sa-lū'-bre-us) [salus, health]. Healthful.
**salubrity** (sa-lū'-brit-e) [salubritas, healthfulness]. The state or character of being wholesome.
**salubrol** (sal-ū'-brol). Tetrabromomethylenediantipyrine. An inodorous, antiseptic powder used in the same way as iodoform.
**salufer** (sal'-ū-fer). Sodium silicofluoride.
**salumin** (sal'-ū-min). See *aluminum salicylate.* **s., soluble,** ammoniated aluminum salicylate.
**salutarium** (sal-ū-ta'-re-um) [salus, health]. A sanitarium.
**salutary** (sal'-ū-ta-re) [salus, health]. Promotive of health.
**salvarsan** (sal'-var-san). Ehrlich's "606." Dioxydiamidoarsenobenzol, $C_{12}H_{12}O_2N_2As_2$. A sulphuryellow powder furnished by Ehrlich as a remedy for syphilis. **s. milk,** milk from a goat that has been subjected to injections of salvarsan; used for syphilitic children.
**salvatella** (sal-vat-el'-ah) or. **vena salvatella** [salvatus, from salvare, to save]. The vein on the back of the little finger. See *vein.*
**salve** (sahv) [AS., sealf]. Ointment. **s., Deshler's,** compound rosin cerate. **s. pencil,** ointment in the form of a pencil or stick.

**salveol** (*sal'-ve-ol*). Trade name of an antiseptic solution of sodium creosotate in cresol.
**salvia** (*sal'-ve-ah*) [L.]. Official name for the dried leaves of *Salvia officinalis*. See *sage*.
**salviol** (*sal'-ve-ol*) [*salvia*, sage; *oleum*, oil], C₁₉H₁₆O. A liquid substance obtained from oil of sage.
**Salzer's operation** [Fritz Adolf *Salzer*, Austrian surgeon, 1858– ]. Excision of the whole of the third division of the fifth nerve.
**samadera** or **samandura** (*sam-ad-e'-rah* or *sa-man'-du-rah*) [E. Ind.]. A genus of old-world trees of the simarubaceous type. **s. indica** produces a bitter, febrifugal bark.
**samarium** (*sam-a'-re-um*) [L.]. A metallic element belonging to the didymium group. Symbol Sa; atomic weight 150.4.
**sambucin** (*sam-bū'-sin*) [*Sámbucus*, the elder]. An alcoholic fluidextract of the bark of *Sambucus nigra*. It is a diuretic.
**sambucus** (*sam-bū'-kus*) [L.]. Elder; a shrub or tree of the order *Caprifoliaceæ*. The flowers of *S. canadensis* and the berries are sudorific; the latter have been used as an alterative in rheumatism and syphilis. The inner bark has been employed in epilepsy, dropsy, and various chronic diseases. *Aqua sambuci* (B. P.) is used as a vehicle.
**samol** (*sam'-ol*). Trade name of an ointment containing 25 per cent. of salimenthol.
**samshu** (*sam'-shoo*) [Chinese]. An alcoholic drink distilled in China from rice or millet, or both.
**sanative** (*san'-a-tiv*) [*sanare*, to heal]. Promoting health; healing.
**sanatogen** (*san-at'-o-jen*). A proprietary food said to contain 90 % of casein and 5 % of sodium glycerophosphate. Dose 1 teaspoonful (5 Cc.) added to soup, cocoa, etc.
**sanatol** (*san'-at-ol*). The trade name of a disinfectant said to consist of sulphuric acid, esters of phenol, and its homologues.
**sanatolyn** (*san-at'-ol-in*). A disinfectant said to consist of phenol and sulphuric acid with a percentage of ferrous sulphate.
**sanatorium** (*san-at-o'-re-um*). [*sanare*, to heal]. An establishment for the treatment of the sick; especially a private hospital. See *sanitarium*.
**sanatory** (*san'-at-o-re*). See *sanative*.
**sanatose** (*san'-at-ōs*). A proprietary preparation said to consist of sodium glycerophosphate and casein.
**sand** [AS.]. An aggregation of fine grains of silicic oxide. **s., auditory**, otoliths. **s.-bath**. 1. A vessel containing dry sand in which a substance requiring a slowly rising or uniform temperature may be heated. **s.-blind**. See *metamorphopsia*. **s.-bodies**. See *corpora aranaceæ*. **s., brain**. See *acervulus*. **s.-crack**, a crack or fissure in the hoof of a horse, extending from the coronet toward the sole, and due to a diseased condition of the horn-secreting membrane. **s.-flea**. See *chigoe*. **s., intestinal**, gritty material passed with the stools. **s., pineal**. See *acervulus*. **s.-tumor**. See *psammoma*.
**sandalwood**. 1. Red sanders (*Santalum rubrum*, U. S. P.; *Pterocarpi ligni*, B. P.), the wood of *Pterocarpus santalinus*, of the order *Leguminosæ*. It is used as a coloring agent. 2. The wood of *Santalum album*, of the order *Santalaceæ*, containing a volatile oil. **s. oil** (*oleum santali*, U. S. P., B. P.), used in bronchitis and gonorrhea. Dose 15–20 min. (1.0–1.3 Cc.).
**sandarac** (*san'-dar-ak*). A white, transparent resin produced by *Callitris quadrivalvis*, a tree of North Africa. It is now little used except as a varnish and incense. **s. varnish**, in dentistry a solution of sandarac in alcohol used as a separating medium in making plaster casts.
**sandaracin** (*san-dar'-as-in*). Giese's name for sandarac which has been exposed to the action of alcohol. It is a mixture of two of the three resins of which sandarac is said to be composed.
**sanders** (*san'-durz*), *Sandalwood*.
**Sanders' sign** [James *Sanders*, English physician, 1777–1843]. Undulatory character of the cardiac impulse, most marked in the epigastric region, in adherent pericardium. **S.'s type of paranoia**, paranoia appearing in youth. Syn., *paranoia originaria*.
**Sanderson's method of attenuation**. The passing of virus through the system of another animal (*e. g.*, the guinea-pig, in anthrax) so that it becomes modified in virulency.
**Sandstroem's bodies** or **glands** (*sant'-strēm*) [Ivar *Sandstroem*, Norwegian physician]. The parathyroid glands; also called Gley's glands.
**sane** (*sān*) [*sanus*, whole]. Of sound mind.
**sangaree** (*sang-ga-re'*) [Sp., *sangria*]. A sweetened and flavored drink, consisting essentially of diluted wine or porter.
**Sanger's macula**. A bright red spot marking the orifice of the duct of Bartholin's gland in cases of gonorrheal vulvitis. Syn., *macula gonorrhœica*.
**Sanger's operation**. A method of performing cesarean section: a modification of the usual operation in which the uterus is brought out through a long abdominal incision before extraction of the fetus.
**Sanger's pupil-reaction**. For the differential diagnosis of *cerebral syphilis* and *tabes*: in amaurosis and optic atrophy of cerebral syphilis the pupil-reflex to light may be preserved and even increased after a protracted stay in the dark, which is never the case in tabes dorsalis.
**S.-angle**. See *angle, sigma*.
**sangrenal** (*sang'-gre-nal*). A preparation made from adrenal glands; used as an astringent, hemostatic, and cardiac stimulant.
**sanguicolous** (*sang-gwik'-o-lus*) [*sanguis*, blood; *colere*, to inhabit]. Living in the blood, as a parasite.
**sanguiferous** (*san-gwif'-er-us*) [*sanguis*, blood; *ferre*, to carry]. Carrying, or conveying, blood.
**sanguiferrin** (*sang-gwif'-er-in*). Trade name of a preparation said to contain hemoglobin and manganese.
**sanguification** (*sang-gwif-ik-a'-shun*) [*sanguis*, blood; *facere*, to make]. 1. The formation of blood. 2. Conversion into blood, as the *sanguification* of substances absorbed from the intestinal tract.
**sanguimotion** (*sang-gwi-mo'-shun*) [*sanguis*; *motion*]. The circulation of the blood.
**sanguimotory** (*sang-gwi-mo'-tor-e*). Relating to the circulation of the blood.
**sanguinal** (*sang'-gwin-al*). A hematinic consisting of evaporated blood and hemoglobin in liquid form, and free from the intermediate products of the degeneration of albuminous bodies. It consists of natural blood-salts 46 parts, oxyhemoglobin 10 parts, and peptonized muscle-albumin 44 parts.
**sanguinaria** (*sang-gwin-a'-re-ah*) [*sanguis*]. Bloodroot, a genus of plants of the order *Papaveraceæ*. The rhizome of *S. canadensis* (*sanguinaria*, U. S. P.) is emetic and narcotic; in large doses it is an expectorant in bronchitis. Dose, as an expectorant, 1–5 gr. (0.065–0.32 Gm.); as an emetic, 5–10 gr. (0.32–0.65 Gm.). **s., fluidextract of** (*fluidextractum sanguinariæ*, U. S. P.). Dose 1–10 min. (0.065–0.65 Cc.). **s., tincture of** (*tinctura sanguinariæ*, U. S. P.). Dose 30–60 min. (2–4 Cc.).
**sanguinarine** (*sang-gwin'-ar-ēn*), C₂₀H₁₅NO₄, the most important alkaloid derived from the rhizome of *Sanguinaria canadensis*. Dose $\frac{1}{16}$–$\frac{1}{2}$ gr. (0.005–0.011–0.05 Gm.) in solution. Small doses expectorant, large doses emetic. **s. nitrate**, C₁₇H₁₅NO₄. HNO₃, a red, crystalline powder, soluble in water and alcohol. Dosage and uses the same as the alkaloid. **s. sulphate**, (C₁₇H₁₅NO₄)₂ . H₂SO₄, red crystalline powder, soluble in water and alcohol. Dosage and uses the same as the alkaloid.
**sanguine** (*sang'-gwin*) [*sanguis*]. 1. Resembling blood; bloody. 2. Hopeful; active, as *sanguine* temperament.
**sanguineous** (*sang-gwin'-e-us*) [*sanguis*]. 1. Pertaining to the blood; containing blood. 2. Sanguine.
**s. cyst**, a cyst containing blood-stained fluid.
**sanguino** (*sang'-gwin-o*). A proprietary preparation said to contain all iron salts, albumins, fats, and carbohydrates formed in the animal organism.
**sanguinoform** (*sang-gwin'-o-form*). A therapeutic preparation of blood said to be obtained from the embryonic blood-forming organs of animals.
**sanguinolent** (*sang-gwin'-o-lent*) [*sanguis*]. Tinged with blood.
**sanguis** (*sang'-gwis*) [L.]. Blood.
**sanguisuction** (*sang-gwis-uk'-shun*) [*sanguis*, blood; *suctus*, p. p. of *sugere*, to suck]. The abstraction of blood by suction, as by a leech or other parasite.
**sanguisuga** (*sang-gwi-sū'-gah*). See *leech*.
**sanies** (*sa'-ne-ēz*) [L.]. A thin, fetid, greenish, seropurulent fluid discharged from an ulcer, wound, or fistula.
**sanious** (*sa'-ne-us*) [*sanies*]. Pertaining to or resembling sanies, as *sanious* pus.

**sanitarian** (san-it-a'-re-an) [sanitas, health]. One skilled in sanitary science and matters of public health.

**sanitarium** (san-it-a'-re-um) [sanitas, health]. A place where the conditions are such as especially to promote health; a resort for convalescents. s., **ocean**, a ship so constructed as to be specially adapted to the requirements of invalids or convalescents and to making long cruises.

**sanitary** (san'-it-a-re) [sanitas]. Pertaining to health. **s. cordon**, a line of guards to control ingress to or egress from an infected locality. **s. police**. See *police, sanitary*. **s. science**, the science that includes a consideration of all that can be done for the prevention of disease and the promotion of the public health.

**sanitas** (san'-it-as). 1. Health. 2. A class of proprietary antiseptic solutions, made from turpentine.

**sanitation** (san-it-a'-shun) [sanitary]. The act of securing a healthful condition; the application of sanitary measures.

**sanity** (san'-it-e) [sanitas, from sanus, sound]. Soundness of mind.

**sanmethyl** (san-meth'-il). A proprietary preparation said to consist of methylene-blue, copaiba, phenyl salicylate, oils of sandalwood and cinnamon, and the oleoresins of cubebs and matico. It is used in gonorrhea. Dose in capsules 10 min. (0.66 Cc.).

**sanmetra** (san-met'-rah) [sanus,. sound; μήτρα, womb]. A combination of zinc sulphate, 1 gr.; antipyrine, 2 gr.; ichthyol, 5 gr.; fluidextract of hydrastis, 5 gr.; creosote, 1-2 gr.; extract of hyoscyamus, 1-2 gr.; menthol and thymol, each, 1-2 5 gr.; oil of eucalyptus, 1 gr. It is indicated in vaginal, uterine, and pelvic diseases, and is used in suppositories.

**sanmetto** (san-met'-o). A proprietary preparation recommended in genitourinary diseases and said to consist of sandalwood and saw-palmetto. Dose 1 teaspoonful (5 Cc.) 4 times daily.

**sano** (san'-o). A proprietary dietetic remedy said to consist of dextrinised barley flour with a high percentage of proteids; according to analysis, it consists of water, 13.7 %; proteids, 12.5 %; fat, 1.6%; mineral matter, 1.85 %; soluble carbohydrates, 4.1 %; cellulose, 1.4 %; and starch, 64.9 %.

**sanoderma** (san-o-der'-mah). A sterilized muslin bandage saturated with bismuth subnitrate.

**sanoform** (san'-o-form), C₆H₂I₂OHCOOCH₃. A methyl ether of diiodosalicylic acid. It is a white, tasteless, odorless, permanent powder, containing 60.7 % iodine; melts at 110° C.; soluble in alcohol, ether, or vaseline. It is used as a surgical dressing in powder or 10 % ointment.

**sanose** (san'-ōs). A proprietary dietetic said to contain 80 % of casein and 20 % of albumose; a white, odorless, tasteless powder forming an emulsion when stirred with water or milk. Dose 5 dr.–1½ oz. (20–50 Gm.) in a pint of milk.

**sanosin** (san'-o-sin). A mixture of sulphur, charcoal, and eucalyptus leaves. The fumes of this when ignited are used by inhalation in the treatment of pulmonary tuberculosis.

**Sansom's sign** (san'-sum) [Arthur Ernest Sansom, English physician, 1839–1907]. 1. Considerable extension of dulness in the second and third intercostal spaces in pericardial effusion. 2. A rhythmic murmur transmitted through the air in the mouth when the lips of the patient are applied to the chestpiece of the stethoscope; it is heard in cases of aortic aneurysm.

**Sanson's images** [Louis Joseph Sanson, French physician, 1790–1841]. See *Purkinje's images*.

**santal** (san'-tal). Santalum, white sandalwood. See *sandalwood* (2).

**santalal** (san'-tal-al), C₁₅H₂₄O. A constituent of santal oil found by Chapoteau.

**santalol** (san'-tal-ol), C₁₅H₂₆O. A constituent of oil of santal found by Chapoteau.

**Santa Lucia bark.** The bark of *Exostemma floribundum*, a rubiaceous tree of the Antilles.

**santalin** (san'-tal-in) [santalum, sandalwood]. The coloring-matter of red sandalwood, obtained by evaporating the alcoholic infusion to dryness. It is a red resin, fusible at 212° F., and is very soluble in acetic acid as well as in alcohol, essential oils, and alkaline lyes.

**santalum** (san'-tal-um) [L.]. *Pterocarpi lignum* (B. P.). White sandalwood. The wood of a species of *S. album* and *S. citrinum*, or yellow sandalwood.

It yields oil of santal, an astringent oil, useful in chronic bronchitis and gonorrhea. It is often adulterated with oil of cedar. Dose of the volatile oil dr. xxxx, in emulsion or capsule. **s. rubrum**, red-sanders, the wood of *Pterocarpus santolinus*, imparts a brilliant-red color to ether and alcohol.

**santol** (san'-tol). 1. A crystalline substance C₆H₆O₃, found by H. Weidel (1870) in white sandalwood. 2. A proprietary preparation of sandalwood, used for gonorrhea, etc.

**santonica** (san-ton'-ik-ah) [σαντονικόν, wormwood]. Levant wormseed, the unexpanded flower-heads of *Artemisia pauciflora* (U. S. P.) or *A. maritima*, var., *Stechmanniana* (B. P.), of the order *Compositæ*, the essential constituent of which is santonin, C₁₅H₁₈O₃ (santoninum, U. S. P., B. P.). Santonin is a neutral crystalline principle, producing, in overdoses, xanthopsia, giddiness, stupor, at times convulsions, and death from failure of respiration. The urine is colored yellow. Santonica and santonin are used as vermicides against the lumbricoid worm. Dose of santonica 10–30 gr. (0.65–2.0 Gm.); of santonin 1–2 gr. (0.065–0.13 Gm.). Sodium santoninate was formerly used as a substitute for santonin, but has produced poisoning.

**santonin** (san'-to-nin). See under *santonica*. **s., troches of** (*trochisci santonini*, U. S. P., B. P.), those of the U. S. P. contain each about ½ gr. (0.033 Gm.) of santonin; those of the B. P. contain 1 gr. (0.065 Gm.) of the drug.

**santoninoxime** (san-ton-in-oks'-ēm), C₁₅H₁₉O₂(N-OH). A derivative of santonin by action of an alcoholic solution of hydroxylamine hydrochloride with sodium; a white, crystalline powder, less toxic than santonin, and used as a vermicide. Dose for adults ⅕ gr. (0.32 Gm.) divided into two doses and taken at intervals of one to two hours, followed by a cathartic. Repeat for 2 or 3 days.

**santonism** (san'-ton-izm). Poisoning from overdosage of santonin.

**Santorini's canal** (san-to-re'-ne) [Giovanni Domenico Santorini, Italian anatomist, 1681–1737]. See *Bernard's canal*. **S.'s cartilages**, cornicula laryngis; the cartilaginous nodules on the tips of the arytenoid cartilages. **S.'s circular muscle**, involuntary muscular fibers encircling the urethra beneath the constrictor urethræ. **S.'s concha**, a small, supernumerary, spongy bone sometimes found above the superior turbinated bone of the ethmoid. **S.'s fissures**, two fissures separating the cartilaginous portions of the external auditory canal into three incomplete rings. Syn., *incisura Santorini*. **S.'s muscle**. 1. The risorius q. v., under *muscles, table of*. 2. Same as Santorini's circular muscle. **S.'s papilla**, the papilla of the duodenum. **S.'s plexus**. 1. The vesicoprostatic plexus of veins in the male; the venous plexus surrounding the front and sides of the urethra in the female. 2. An anastomotic network formed at the foramen ovale by the filaments of the two roots of the inferior maxillary nerve. **S.'s tubercle**, the cornicula laryngis. **S.'s veins**, the emissary veins forming a communication between the cerebral sinuses and the veins of the scalp; especially the small veins passing through the parietal foramen and connecting the parietal with the superior longitudinal sinus.

**santozea** (san-to-ze'-ah). Trade name of a preparation said to contain santal, saw-palmetto, etc.; used for cystitis and other genitourinary disorders.

**santyl** (san'-til). Santalyl salicylate, a proprietary gonorrhea remedy.

**sap**. The nutritive fluid which circulates by endosmosis in plants.

**saphena** (sa-fe'-nah) [σαφηνής; manifest: pl., *saphenæ*]. A name given two large veins of the leg—the internal or long, and the external, or short saphena. **saphenous** (sa-fe'-nus) [σαφηνής, manifest]. Apparent; superficial; manifest; applied to two veins of the lower limb, the internal or long saphenous vein and external or short saphenous vein, situated just beneath the surface; also applied to the nerves accompanying these veins. **s. nerves**. See *nerves*. **s. opening**, an opening in the fascia lata at the upper part of the thigh through which the long saphenous vein and nerve pass. **s. veins**. See above and also *veins*.

**sapid** (sap'-id) [*sapere*, to taste]. Capable of being tasted.

**sapidity** (sap-id'-it-e) [*sapid*]. The property or quality of a substance which gives it taste.

**sapientia** (*sa-pe-en'-she-ah*) [L.]. Wisdom. **sapientiæ dentes**, the posterior or third molar teeth.
**sapo** (*sa'-po*) [L.]. Soap. See *soap*.
**sapocarbol** (*sa-po-kar'-bol*). A disinfectant solution of cresol and soft soap.
**sapodermin** (*sap-o-der'-min*). An antiseptic soap containing albuminate of mercury; used in the treatment of parasitic and fungoid diseases.
**sapogenin** (*sap-oj'-en-in*), $C_{24}H_{36}O_3$ (Hesse). A derivative of saponin by action of dilute acids with heat. It occurs in needles grouped in stars, soluble in alcohol or ether. Syn., *saporetin*.
**sapolan** (*sap'-ol-an*). A compound said to consist of a naphtha product, 2.5 parts; soap, 3 to 4 %; lanolin, 1.5 parts; it is used in skin diseases.
**sapolanolin** (*sa-po-lan'-o-lin*). A preparation of soft soap and lanolin; used in eczematous conditions.
**saponaceous** (*sap-o-na'-se-us*) [*sapo*, soap]. Having the nature of soap.
**saponal** (*sap'-o-nal*). A cleansing compound said to consist of soap, 20 %; sodium carbonate, 60%; sodium chloride, 2.2 %; and water, 11 %.
**Saponaria** (*sap-o-na'-re-ah*) [*sapo*, soap]. A genus of plants of the order *Caryophylleæ*. *S. officinalis*, or soapwort, bouncing-bet, is a species growing wild abundantly in the United States and Europe in the vicinity of houses. The root, rhizome, and stolons are used in gout, syphilis, and as an expectorant. It contains saponin, sapotoxin, sapogenin, etc.
**saponarius** (*sap-o-na'-re-us*). Of a soapy character.
**saponatus** (*sap-o-na'-tus*) [L.]. Mixed with soap.
**saponetin** (*sap-on-et'-in*). A microcrystalline body, $C_{60}H_{66}O_{18}$, obtained by heating saponin with dilute acids.
**saponification** (*sa-pon-if-ik-a'-shun*) [*sapo*; *facere*, to make]. The act of converting into soap; the process of treating a neutral fat with an alkali, which combines with the fatty acid, forming a soap. **s. equivalent**, a term used to indicate the number of grams of an oil saponified by one equivalent in grams of an alkali. **s., fermentation**, saponification brought about by the action of a ferment.
**saponiform** (*sap-on'-e-form*). Soap-like in appearance and consistence.
**saponify** (*sa-pon'-e-fi*) [see *saponification*]. To convert into soap; to convert a neutral fat by the action of an alkali into free glycerol and a salt of the alkali, the latter forming a soap.
**saponiment** (*sap-on'-im-ent*) [*sapo*, soap]. A term denoting a medicinal compound of soap.
**saponin** (*sap'-o-nin*) [*sapo*], $C_{32}H_{54}O_{18}$. A glucoside contained in the roots of soapwort and other plants, and in aqueous solution forming a strong lather. **s., coal-tar**. See *liquor carbonis detergens*.
**saponule, saponulus** (*sap'-on-ûl*, *sap-on'-û-lus*). Imperfect soaps formed by combination of essential oils with bases.
**saporetin** (*sap-or-e'-tin*). See *sapogenin*.
**saporific** (*sap-o-rif'-ik*) [*sapor*, savor; *facere*, to make]. Producing taste, flavor, or relish.
**saporosity** (*sap-or-os'-it-e*) [*sapid*]. Sapidity.
**sapotin** (*sap'-o-tin*). A glucoside, $C_{15}H_{22}O_{20}$, extracted from the seed of the sapodilla-plum, the fruit of *Achras sapota*, occurring in minute crystals which melt at 240° C. It is readily soluble in water, less so in alcohol, and insoluble in ether, benzine, or chloroform.
**sapotiretin** (*sap-o-tir-et'-in*), $C_{11}H_{20}O_{10}$. A product obtained from sapotin by boiling it with dilute sulphuric acid; insoluble in water, readily soluble in alcohol.
**sapotoxin** (*sap-o-toks'-in*) [*sapo*; τοξικόν, poison], $C_{17}H_{26}O_{10}$. A poisonous glucoside obtained from saponin.
**sappan-wood** (*sap-an'-wood*). The wood of *Cæsalpinia sappan*; used as a dye as a substitute for hematoxylon.
**sappanin** (*sap'-an-in*). A substance obtained by the fusion of an extract of the wood of *Cæsalpinia sappan* with caustic soda.
**Sappey's accessory portal veins** (*sap'-e*) [Marie Philibert Constant *Sappey*, French anatomist, 1810–1896]. A system of venules uniting to form small trunks, which redivide in the liver and empty into the sublobular veins. They consist of the minute nutrient veins of the portal vein, hepatic artery, and bile-ducts; of venules lying in the gastrohepatic omentum, the suspensory ligament of the liver, and about the fundus of the gall-bladder; and of the group of small veins in the umbilical region. Through the branches lying in the suspensory ligament of the liver and through the parumbilical group the portal vein communicates with the venæ cavæ. **S.'s fibers**, smooth muscular fibers found in the check ligaments of the eyeball close to their orbital attachment.

**sapphism** (*saf'-izm*) [from Σαπφώ, Sappho, a Greek poetess]. Tribadism. Unnatural passion of one woman for another.
**sapremia, sapræmia** (*sap-rem'-e-ah*) [σαπρός, putrid; αἷμα, blood]. The intoxication produced by absorption of the results of putrefaction.
**sapremic** (*sap-re'-mik*) [*sapremia*]. Affected with, of the nature of, or pertaining to, sapremia.
**saprine** (*sap'-rin*) [σαπρός, putrid]. A nonpoisonous ptomaine formed in the putrefaction of animal tissues.
**sapro-** (*sap'-ro-*) [σαπρός, putrid]. A prefix signifying decay, putridity, etc.
**saprodontia** (*sap-ro-don'-she-ah*) [*sapro-*; ὀδούς, tooth]. Caries or rottenness of the teeth.
**saprogenic, saprogenous** (*sap-ro-jen'-ik, sap-roj'-en-us*) [*sapro-*; γεννᾶν, to beget]. 1. Causing putrefaction. 2. Produced by putrefaction.
**saprol** (*sap'-rol*) [σαπρός, putrid]. A mixture of crude cresols with hydrocarbons; used as a disinfectant.
**Saprolegnia** (*sap-ro-leg'-ne-ah*) [*sapro-*; λέγνον, an edge]. Fly-fungus. A genus of fungi of the order *Saprolegniaceæ*. Four species are known: *S. monoïca*, *S. dioeca*, *S. asterophora*, and *S. ferox*. They are all saprophytes on dead plants and animals, especially flies, in water, with the exception of the last-named species, which is both saprophyte and facultative parasite. It is the cause of fish or salmon disease.
**saprophagous** (*sap-rof'-a-gus*) [*sapro-*; φαγεῖν, to eat]. Subsisting on decaying matter.
**saprophilous** (*sap-rof'-il-us*) [*sapro-*; φιλεῖν, to love]. Infesting decaying matter. Saprophytic.
**saprophyte** (*sap'-ro-fīt*) [*sapro-*; φυτόν, a plant]. A vegetable organism living on dead organic matter.
**saprophytic** (*sap-ro-fit'-ik*) [*saprophyte*]. Growing in dead organic matter, as *saprophytic* bacteria.
**sapropyra** (*sap-ro-pi'-rah*) [*sapro-*; πῦρ, fire]. 1. Malignant typhus, or putrid fever. 2. Any fever due to putrid infection.
**saprostomous** (*sap-ros'-to-mus*) [*sapro-*; στόμα, mouth]. Having offensive breath.
**saprotyphus** (*sap-ro-ti'-fus*) [*sapro-*; *typhus*]. Malignant or putrid typhus fever.
**saprozoic** (*sap-ro-zo'-ik*) [*sapro-*; ζῷον, an animal]. Living in decaying organic matter.
**sar, sara** [E. Ind., "rotten"]. Vernacular for trypanosomiasis (surra) (*q. v.*).
**sarapus** (*sar'-ap-us*) [σαίρειν, to sweep; ποῦς, foot]. A flat-footed person.
**Sarbo's sign**. Analgesia of the peroneal nerve, occasionally observed in tabes dorsalis.
**sarc** (*sark*) [σάρξ, flesh]. The belly, body, or fleshy portions of a muscle.
**sarcepiplocele** (*sar-sep-ip'-lo-sēl*) [σάρξ, flesh; ἐπίπλοον, omentum; κήλη, tumor]. An omental hernia with sarcocele, or with great thickening of the omentum.
**Sarcina** (*sar-si'-nah*) [L., "a bundle"; pl., *sarcinæ*]. A genus of schizomycetes consisting of cocci dividing in three directions, thus producing cubic masses. See *micrococci, table of*. 1. See *hypoxanthine*. 2. Sarcina, *q. v.*
**sarcinic** (*sar-si'-nik*) [*sarcina*]. Pertaining to or caused by sarcinæ.
**sarcinuria** (*sar-sin-ū'-re-ah*) [*sarcin*; οὖρον, urine]. The discharge of sarcin with the urine.
**sarcitis** (*sar-si'-tis*) [σάρξ, flesh; ιτις, inflammation]. Inflammation of fleshy tissue; especially inflammation of muscle.
**sarco-** (*sar ko-*) [σάρξ, flesh]. A prefix denoting composed of or pertaining to flesh.
**sarcoadenoma** (*sar-ko-ad-en-o'-mah*) [*sarco-*; ἀδήν, gland; ὄμα, tumor]. A fleshy glandular tumor. See *adenosarcoma*.
**sarcoblast** (*sar'-ko-blast*) [*sarco-*; βλαστός, a germ]. 1. In biology, a protoplasmic germinal mass. 2. Marchesini's term for *sarcoplast*.
**sarcocarcinoma** (*sar-ko-kar-sin-o'-mah*). A tumor composed of malignant growth of both carcinomatous and sarcomatous types.

# SARCOCARP 780 SARRACENIA

**sarcocarp** (sar'-ko-karp) [sarco-; καρπόs, fruit]. In biology, a fleshy, succulent mesocarp.
**sarcocele** (sar'-ko-sēl) [sarco-; κήλη, a tumor]. Any fleshy swelling of the testicle. **s. malleosa**, that due to *Bacillus mallei*. **s., syphilitic**, syphilitic orchitis.
**Sarcocephalus** (sar-ko-sef'-al-us) [sarco-; κεφαλή, head]. A genus of the *Rubiaceæ*. *S. esculentus*, a shrub of western Africa, the Guinea or Sierra Leone peach, yields an astringent antipyretic bark, doundaki or doundaké (q. v.); it is the quinquina africané or kina du Rio Nuñez of the French. The wood, called nijmo, is tonic and astringent. It contains the alkaloid doundakine.
**sarcocol, sarcocolla** (sar'-ko-kol, sar-ko-kol'-ah) [σάρξ, flesh; κόλλα, glue: named from its vulnerary power]. 1. A gum-like drug, much used in India and Arabia, supposed to be the product of some species of *Astragalus*. 2. An African resin with purgative qualities, the product of various plants of the genera *Pænæa* and *Sarcocolla*. It is acrid and nauseous.
**Sarcocystis** (sar-ko-sis'-tis) [sarco-; κύστιs, a cyst]. A group of the sporozoa. **S. miescheri**, a parasite found in pork and beef.
**sarcocyte** (sar'-ko-sīt). See *ectoplasm*.
**sarcode** (sar'-kōd) [σάρξ, flesh]. Animal protoplasm.
**Sarcodina** (sar-ko-di'-nah) [sarco-; δίνη, a whirling]. A class of protozoa moving and feeding by means of pseudopodia, e. g., ameba.
**sarcoenchondroma** (sar-ko-en-kon-dro'-mah). A combined sarcoma and enchondroma.
**sarcoepiplomphalus** (sar-ko-ep-e-plom'-fal-us) [sarco-; ἐπίπλοον, caul; ὀμφαλός, navel]. An umbilical hernia forming a fleshy mass, from great thickening of the omentum.
**sarcogenic** (sar-ko-jen'-ik) [sarco-; γεννᾶν, to beget], Producing flesh or muscle.
**sarcoglia** (sar-kŏg'-le-ah). [sarco-; γλία, glue]. Sarcoplasm; a protoplasmic substance containing the granules and nuclei composing the eminence of Doyen, or the point of entrance of a motor nerve into muscular fiber.
**sarcohydrocele** (sar-ko-hi'-dro-sēl) [sarco-; ὕδωρ, water; κήλη, tumor]. A sarcocele complicated with hydrocele of the tunica vaginalis.
**sarcoid** (sar'-koid) [sarco-; εἶδος, shape]. Resembling or having the nature of flesh.
**sarcolactic acid** (sar-ko-lak'-tik). See *acid, sarcolactic*.
**sarcolemma** (sar-ko-lem'-ah). [sarco-; λέμμα, husk]. The delicate membrane enveloping a muscle-fiber.
**sarcolemmic, sarcolemmous** (sar-ko-lem'-ik, sar-ko-lem'-us) [sarco-; λέμμα, covering]. Pertaining to or of the nature of sarcolemma.
**sarcology** (sar-kol'-o-je) [sarco-; λόγος, science]. 1. The anatomy treating of the soft tissues, as distinguished from osteology. 2. Myology.
**sarcolyte** (sar'-ko-līt) [σάρξ, flesh; λυεῖν, to dissolve]. A cell which is actively concerned in effecting the retrograde metamorphosis of soft tissues.
**sarcoma** (sar-ko'-mah) [sarco-; ὅμα, tumor]. A tumor made up of embryonal connective tissue. It is characterized by a great preponderance of cells and very little homogeneous or fibrillar intercellular substance. **s. of Abernethy**. See *s., adipose*. **s., adipose**, one in which groups of sarcoma-cells are contained in alveolar spaces. **s., angiolithic**. Synonym of *psammoma*. **s., angioplastic**, a tumor of the testicle first described by Malazzez and Monod as composed of a protoplasmic network with irregular spaces and trabeculæ, the latter made up of anastomosing giant-cells. The name *epithelioma syncytiomatodes testiculi* is proposed for it. **s. botryoides**, a grape-like variety of sarcoma found in the cervix uteri. **s. carcinomatodes**, a scirrhous cancer. **s. deciducellulare**. See *deciduoma malignum*. **s., encephaloid**, a soft, rapidly growing sarcoma, usually of the round-celled variety. **s. epulis**. See *epulis, malignant*. **s., giant-celled**, one containing giant-cells as a prominent feature. **s., glandular**, Hodgkin's disease. **s., granulation**. See *s., mammary*. **s. lipomatodes, s., lipomatous**, one characterized by infiltration of fat. **s. lymphadenoides**. See *lymphosarcoma*. **s., mastoid**, a sarcoma of the mammary gland. **s., melanotic**, a sarcoma, usually spindle-celled, in which the cells contain melanin. **s. molle**. See *lymphosarcoma*. **s. molluscum**, multiple connective-tissue tumors of the skin containing few spindle-cells. **s., mucous**. See *myxosarcoma*. **s., Mueller's**, "sarcoma phyllodes"; adenofibroma of the breast. **s., myeloid**. See *s., giant-celled*. **s. myxomatodes**, a myxosarcoma (q. v.). **s. phyllodes**. See *s., Mueller's*. **s., round-celled**, one made up of round-cells. There are two varieties, the small round-celled and the large round-celled. **s. scroti**, a sarcocele. **s., spindle-celled**, one made up of spindle-cells. Syn., *recurrent fibroid*.
**sarcomatoid** (sar-ko'-mat-oid) [sarcoma; εἶδος, resemblance]. Resembling a sarcoma.
**sarcomatosis** (sar-ko-mat-o'-sis) [sarcoma]. The formation of multiple sarcomatous growths in various parts of the body. **s. generalis**. Synonym of *granuloma fungoides*.
**sarcomatous** (sar-ko'-mat-us) [sarcoma]. Of the nature of or resembling sarcoma.
**sarcomere** (sar'-ko-mēr) [sarco-; μερόs, a part]. One of the segments into which a muscle-fibril appears to be divided by transverse septa.
**sarcomoscheocele** (sar-ko-mos'-ke-o-sēl) [sarco-; ὄσχεον, scrotum; κήλη, tumor]. A fleshy scrotal tumor.
**sarcomphalocele, sarcomphalon** (sar-kom-fal'-o-sēl, sar-kom'-fal-on) [sarco-; ὀμφαλός, navel; κήλη, tumor]. A fleshy tumor at the umbilicus.
**sarcophyma** (sar-ko-fi'-mah) [sarco-; φῦμα, a tumor]. A fleshy tumor; sarcoma.
**sarcoplasm** (sar'-ko-plazm) [sarco-; πλάσσειν, to mold]. The hyaline or finely granular interfibrillar material of muscle tissue; the term is opposed to the *myeloplasm* or contractile substance.
**sarcoplasmic** (sar-ko-plaz'-mik). Containing or relating to sarcoplasm.
**sarcoplast** (sar'-ko-plast) [sarco-; πλάσσειν, to mold]. A cell lying between muscular fibrils and capable of developing into a muscular fiber.
**sarcoplastic** (sar-ko-plas'-tik) [sarcoplast]. Forming flesh.
**sarcopoietic** (sar-ko-poi-et'-ik) [σάρξ, flesh; ποιεῖν, to make]. Producing flesh or muscle.
**Sarcopsylla** (sar-kop-sil'-ah) [σάρξ, flesh; ψύλλα, flea]. A genus of siphonapterous or aphanipterous insects. **S. penetrans**, the chigoe.
**Sarcoptes** (sar-kop'-tēs) [sarco-; κόπτειν, to cut]. A genus of mites. **S. hominis**, the itch-mite. **S. scabiei**. See *acarus scabiei*.
**sarcosepsis** (sar-ko-sep'-sis) [sarco-; sepsis]. The presence of bacteria directly in the tissues.
**sarcosin** (sar'-ko-sin) [sarcin], C₄H₉NO₂. Methylglycocoll, a crystalline substance produced when creatin and caffeine are heated with baryta.
**Sarcosporidia** (sar-ko-spor-id'-e-ah) [sarco-; *Sporidia*]. A variety of psorosperms found in the muscles of cattle, sheep, swine, and other mammals.
**sarcosporidiasis** (sar-ko-spo-rid-i'-a-sis) [*Sarcosporidia*, a genus of psorosperms]. A disease produced by sporozoa of the order *Sarcosporideæ*.
**sarcostosis** (sar-kos-to'-sis). 1. Bone formation in muscular tissues. 2. See *osteosarcoma*.
**sarcostroma** (sar-ko-stro'-mah) [sarco-; στρῶμα, a covering]. A thick, fleshy, false membrane.
**sarcostyle** (sar'-ko-stīl) [sarco-; στῦλος, a pillar]. One of the fine longitudinal fibrils of which a striated muscle-fiber is composed and into which it can be split up.
**sarcotherapeutics** (sar-ko-ther-ap-ū'-tiks) [sarco-; θεραπεία, treatment]. The treatment of disease by means of animal extracts or substances. See *organotherapy*.
**sarcotic** (sar-kot'-ik) [σάρξ, flesh]. Pertaining to, or causing fleshy formation or sarcosis.
**sarcotome** (sar'-ko-tōm) [sarco-; τομή, a cutting]. A surgical instrument for the division of soft tissues.
**sarcotripsy** (sar'-ko-trip-se). See *écrasement*.
**sarcous** (sar'-kus) [σάρξ, flesh]. Pertaining to flesh or muscle. **s. element**, Bowman's name for one of the ultimate fibrils of striped muscle-fibers. **s. substance**, the substance of a sarcous element.
**sardonic grin** (sar-don'-ik). See *risus sardonicus*.
**sarkine** (sar'-kin). Same as *sarcine* (1).
**Sarracenia** (sar-as-e'-ne-ah) [Dr. Sarrazin, of Quebec]. A genus of American insectivorous plants, e. g., side-saddle flower, or pitcher-plant, remarkable for their trumpet shaped leaves. **S. purpurea, S. flava**, and **S. variolaris**, are said to afford roots serviceable in dyspepsia and gout. **S. purpurea** and **S. violaris** have been vaunted as a cure for smallpox. They are diuretic, diaphoretic, and stimulant. Dose of the fldext. gtt. xxv.

**sarsa** (sar'-sah). Same as *sarsaparilla*.
**sarsaparilla** (sar-sap-ar-il'-ah) [Sp., zarza, a bramble]. The *Smilax officinalis* and other species of *Smilax*, of the order *Liliaceæ*. The dried root (*sarsaparilla*, U. S. P.; *sarsæ radix*, B. P.) contains a crystalline glucoside, *parillin*, $C_{26}H_{44}O_{10}$. Sarsaparilla has been employed as an alterative in syphilis, rheumatism, and scrofulous affections. **s., decoction of** (*decoctum sarsæ*, B. P.). Dose 4–6 oz. (128–192 Cc.). **s., decoction of, compound** (*decoctum sarsæ compositum*, B. P.). Dose 4–6 oz. (128–192 Cc.). **s., fluidextract of** (*fluidextractum sarsaparillæ*, U. S. P.). Dose 30–60 min. (2–4 Cc.). **s., fluidextract, compound** (*fluidextractum sarsaparillæ compositum*, U. S. P.). Dose 30–60 min. (2–4 Cc.). **s., liquid extract of** (*extractum sarsæ liquidum*, B. P.). Dose 2–4 dr. (8–16 Cc.). **s., syrup of, compound** (*syrupus sarsaparillæ compositus*, U. S. P.). Dose 1–4 dr. (4–16 Cc.).
**sarsasaponin** (sar-sah-sap'-on-in), $12(C_{22}H_{36}O_{10})$ + 2H₂O. A glucoside found by Kobert (1892) in sarsaparilla. It is the most poisonous of its constituents.
**sartian disease** (sar'-shun). An endemic affection of the tropics, characterized by red indurated spots that finally ulcerate; probably *furunculus orientalis*, q. v.
**sartorius** (sar-to'-re-us) [*sartor*, tailor]. The tailor's muscle, so called from being concerned in crossing the one leg over the other. See under *muscle*.
**sassafras** (sas-a-fras) [Sp., from *saxifraga*, from *saxum*, rock; *frangere*, to break]. The *S. variifolium*, a tree of the order *Laurineæ*. The root-bark (*sassafras*, U. S. P.; *sassafras radix*, B. P.) is employed as an aromatic stimulant. The pith (*sassafras medulla*, U. S. P.) yields a mucilage (*mucilago sassafras medullæ*, U. S. P.) that is used as an application to inflamed eyes, and as a demulcent drink in inflammation of the mucous membranes and kidneys. **s. nuts,** pichurim beans; the seeds of *Nectandra pichury-major* and *N. pichury-minor*.
**sassafrid** (sas'-ah-frid). A peculiar principle of *Sassafras officinale*, isolated by Reinach.
**sassafrol** (sas'-af-rol). See *safrol*.
**sassolin** [It.]. Boric acid extracted from the deposits in lagoons of Tuscany.
**Sassy bark** (sas'-e). The bark of *Erythrophlæum*.
**sat.** Abbreviation of saturated. **sat. sol.,** abbreviation of saturated solution.
**satamuli** (sat-ah-moo'-le). The native name in India for *Asparagus racemosus*. It is used as a diuretic and as a sedative in nervous pain.
**satellite** (sat'-el-it) [*satelles*, an attendant]. In anatomy, the vein accompanying an artery.
**satellitism** (sat'-el-it-ism). Mutualism; symbiosis.
**satellitosis** (sat-el-i-to'-sis). A condition in which there is an accumulation of free nuclei around the ganglion cells of the cortex of the brain; it is found in general paralysis and other affections.
**sathe, sathon** (sa'-the, sa'-thon) [σάθη]. The penis.
**satiety** (sa-ti'-e-te) [*satis*, enough]. Fulness beyond desire.
**satisfied hydracarbon.** One that has no free valences; and see *saturated*.
**Satterthwaite's method of artificial respiration** (sat'-er-thwāt). Pressure upon the abdomen alternating with relaxation to allow descent of the diaphragm.
**Sattler's vascular layer.** The layer of blood-vessels of the choroid lying internally to Haller's tunica vasculosa.
**saturated** (sat'-u-ra-ted) [*saturare*, to fill]. 1. Of a liquid, containing in solution all of a substance that it can dissolve. 2. Of a chemical compound, having all the affinities of its component atoms satisfied, with the maximum number of hydrogen atoms or their equivalents; a term especially applied to the hydrocarbons. And see *satisfied hydrocarbon*.
**saturation** (sat-u-ra'-shun) [*saturare*, to fill]. 1. A state in which a liquid holds in solution all of a substance that it can dissolve; the state of being or becoming saturated. 2. Of a chemical compound, a state in which the affinities of all its atoms are saturated. **s. of the atmosphere,** that condition in which any reduction of temperature will be followed by a precipitation of the aqueous vapor mingled with the atmosphere. **s-. points,** the temperature at which the atmosphere contains as much moisture as it can possibly hold, in the form of vapor.
**satureia** (sa-tur-e'-ya). A plant of the order *Labiatæ*. *S. hortensis* resembles thyme and is used as a culinary herb.

**saturnine** (sat'-ur-nīn) [*Saturnus*, a Roman deity; the alchemists' name for lead]. 1. Pertaining to or produced by lead. 2. Of gloomy nature. **s. breath,** the peculiar sweet breath characteristic of lead-poisoning. **s. encephalopathy.** See *lead encephalopathy*.
**saturnism** (sat'-ur nism). Lead-poisoning; plumbism.
**satyria** (sat-i'-re-ah). A genitourinary tonic said to consist of saw palmetto (*Serenoa serrulata*), false bittersweet (*Celastrus scandens*), muria-puama (*Liriosoma ovata*), couch-grass (*Agropyron repens*), and phosphorus, administered in an aromatic vehicle. Dose 1 teaspoonful (5 Cc.) 4 times daily after meals.
**satyriasis** (sat-ir-i'-as-is) [σάτυρος, a satyr]. Excessive venereal desire in the man. Erotic insanity. See *priapism*. 2. Leprosy.
**satyromania** (sat-ir-o-ma'-ne-ah). Same as *satyriasis* (1).
**saunders** (sawn'-derz). See *sandalwood*.
**sauriasis** (saw-ri'-as-is). Ichthyosis.
**sauriderma** (saw-re-der'-mah) [σαύρα, lizard; δέρμα, skin]. Ichthyosis.
**sauriosis** (saw-re-o'-sis) [σαύρα, lizard]. Ichthyosis.
**sausage-poisoning.** A state of gastroenteritis produced by the ingestion of decomposed sausage. Syn., *allantiasis*; *botulism*.
**sausarism** (saw'-sar-izm) [σαυσαρισμός]. 1. Paralysis of the tongue. 2. Dryness of the tongue.
**sauterne** (so'-tern) [*Sauterne*, a place in France]. A certain white wine.
**Sauvineau's ophthalmoplegia** (so-vin-o). [Charles Sauvineau, French ophthalmologist, 1862— ]. Paralysis of the internal rectus muscle of one side and spasm of the external rectus of the opposite side. This affection is the reverse of Parinaud's ophthalmoplegia.
**Savill's disease** [Thomas Dixon *Savill*, English physician, 1856–1910]. An epidemic skin disease characterized by the appearance of a papular rash, followed by a branny desquamation and by marked constitutional symptoms. A fatal result may follow. Syn., *dermatitis exfoliativa epidemica*; *epidemic eczema*.
**savin, savine** (sav'-in). A shrub, *Juniperus sabina*; of the order *Coniferæ*. The tops (*sabina*, U. S. P.; *sabinæ cacumina*, B. P.) contain a volatile oil (*oleum sabinæ*, U. S. P.) and possess marked irritant properties. Savin is employed in amenorrhea, chronic rheumatism, gout, and as a local application to warts, ulcers, and parasitic affections of the skin. **s., fluidextract of** (*fluidextractum sabinæ*, U. S. P.). Dose 3–8 min. (0.2–0.5 Cc.). **s., oil of** (*oleum sabinæ*, U. S. P.). Dose 2–5 min. (0.13–0.32 Cc.). **s., ointment of** (*unguentum sabinæ*, B. P.). Dose 20 min.–1 dr. (1.3–4.0 Cc.).
**Saviotti's canals** (sah-ve-ot'-e). Fine artificial passages formed between the secreting cells of the pancreas by the forcible injection of a colored fluid into the ducts of that organ.
**savonal** (sav'-on-al). Trade name of a soap mixture, used as a base for ointments.
**savory** (sa'-vo-re) [*savor*, odor, or flavor]. Having a pleasant odor or flavor. See *summer savory*.
**saw.** An instrument having a thin blade with sharp teeth on one edge, and used for dividing bones and other hard substances. **s., Adams',** a small straight saw with a long handle. **s., Butcher's,** one in which the blade can be fixed at any angle. **s., chain-,** one in which the teeth are set in links movable upon each other, the saw being moved by pulling alternately upon one and the other handle. **s., crown-.** See *trephine*. **s., Gigli's,** a wire with a serrated edge, used in cranial operations. **s., Hey's,** a serrated disc attached to a handle, and used for enlarging an opening in a bone.
**saw-palmetto.** The fruit of *Serenoa serrulata*; sedative; nutritive and tonic. Dose of the fluidextract, ʒ ss ij.
**saxifragant** (saks-if'-rag-ant) [*saxum*, a stone; *frangere*, to break]. Having the power of dissolving or breaking up calculi.
**saxifrage** (saks'-if-raj) [*saxum*, a rock; *frangere*, to break]. Any plant of the genus *Saxifraga*, including many species of herbs, to some of which doubtful medicinal properties are ascribed.
**saxin** (saks'-in). Trade name of a sweetening agent more powerful than saccharin, and about 600 times sweeter than sugar.

**saxoline** (*saks'-ol-ēn*). A proprietary soft petrolatum.

**Sayre's apparatus, S.'s jacket** (*sair*) [Lewis Albert *Sayre*, American surgeon, 1820–1900]. A jacket of plaster-of-paris molded to support the spine in diseases of the vertebral column.

**Sb.** Chemical symbol of antimony (*stibium*).

**Sc.** Chemical symbol of scandium.

**scab** [*scabere*, to scratch]. 1. The crust formed by the desiccation of the secretion of an ulcer. 2. Scabies.

**scabbard** (*skab'-ard*) [ME., *scauberd*, a sheath]. A veterinary term for the prepuce of the horse.

**scabbed** (*skabd*) [*scabere*, to scratch]. Mangy, affected with scabies.

**scabby** (*skab'-e*). Same as *scabbed*.

**scabies** (*ska'-be-ēz*) [*scabere*, to scratch]. Itch; a disease of the skin caused by an animal parasite, *Sarcoptes scabiei*, or itch-mite. The insect forms burrows or cuniculi beneath the skin, and causes irritation, with vesicles, papules, or pustules, which are frequently modified by scratching. **s. agria**, lichen s., B., e., k's, scabies crustosa; Norwegian itch. **s. capitis**, a disease of the hairy scalp marked by exudation and formation of crusts. Syn., *achores capitis*. **s. capitis favosa**, favus. **s. crustosa**, an extreme form of general scabies of the body resulting in fish-scale-like desquamation. Syn., *Norway itch*. Cf. *radesyge*. **s. fera**. See *ecthyma*. **s. ferina**, mange. **s. humida, s. miliaris**, eczema. **s. lymphatica**, that accompanied by vesicular eruption. Syn., *watery itch*. **s. papuliformis, s. papulosa**, a form marked by papular efflorescence. Syn., *rank itch*. **s. pecorina**, a form affecting sheep. Syn., *sheep-itch*. **s. purulenta, s. pustulosa**, that in which there is formation of large pustules resembling those of smallpox, occurring on the wrists of children. Syn., *Rocky itch*.

**scabiophobia** (*ska-be-o-fo'-be-ah*) [*scabies*, itch; φόβο, fear]. Morbid or insane fear of scabies.

**scabious** (*ska'-be-us*) [*scabiosus*, rough, scabby]. 1. Scabby or scaly. 2. As a noun, a plant of the genus *Scabiosa*; popularly regarded as useful in skin diseases and gout, and as a vulnerary.

**scabrities** (*ska-brish'-e-ēz*) [*scaber*, rough]. Roughness; scabbiness. **s. unguium**, abnormal thickening and roughness of the nails.

**scala** (*ska'-lah*) [L.]. A staircase or ladder. **s. anterior cochleæ, s. externa cochleæ**. See *s. vestibuli*. **s. clausa, s. inferior cochleæ, s. interna cochleæ**. See *s. tympani*. **s. media**, the space between the membrane of Reissner and the basilar membrane, containing the essential peripheral organs of hearing. **s. rhythmica**. See *nucleus, hypoglossal*. **s. tympani**, the canal lying below the osseous lamina and the basilar membrane of the internal ear. **s. vestibuli**, the canal bounded by the osseous lamina and the membrane of Reissner. See under *ear*.

**scald** (*skawld*) [*excaldere*, to wash in hot water]. 1. The burn caused by hot liquids or vapors. 2. [Icel., *skalli*, a bare head]. A disease of the skin accompanied by the formation of scabs. **s.-head**, see *favus*.

**scale** (*skāl*) [AS., *scealu*, a husk; a scale]. 1. The dry, semiopaque lamina of horny epidermis, shed from the skin in health and in various diseases. 2. [*scala*, a ladder]. Anything bearing marks placed at regular intervals and used as a standard in measuring, as *barometric scale*. 3. To remove the tartar from the teeth.

**scalene** (*ska'-lēn*) [σκαληνός, uneven]. Having unequal sides. **s. muscle**. See *scalenus* under *muscle*. **s. tubercle**, a tubercle on the upper surface of the first rib for the insertion of the scalenus anticus muscle.

**scalenus** (*ska-le'-nus*). See *muscles, table of*.

**scaler** (*ska'-ler*) [ME., *scale*, scale]. In dentistry, an instrument for removing the tartar from the teeth.

**scaling** (*ska'-ling*) [*scale*]. 1. Desquamating; producing scales. 2. A pharmaceutical method consisting of drying concentrated solutions of drugs on glass plates. **s. the teeth**, an old name for the operation, in dentistry, which consists in the removal of salivary calculus, commonly called tartar, from the teeth.

**scall** (*skawl*) [ME., *skalle*, a scab]. Favus, impetigo, psoriasis, eczema, or other skin-diseases. **s., dry**, psoriasis, scabies. **s., milk**, crusta lactea. **s., moist**, eczema.

**scallard** (*skal'-lard*). Porrigo.

**scalled** (*skawld*) [ME., *skalle*, a scab]. Affected with scall.

**scalp** [ME., *scalp*, the top of the head]. The hairy integument covering the cranium.

**scalpel** (*skal'-pel*) [*scalpere*, to cut]. A small knife having a convex edge.

**scalprum** (*skal'-prum*) [*scalpere*, to scrape]. 1. A toothed raspatory used in trephining and in removing carious bone. 2. A strong and large scalpel.

**scaly** (*ska'-le*) [*scale*]. 1. Resembling scales; characterized by scales, as *scaly* desquamation. 2. Covered with or having scales. **s.-skin**, a contagious disease common in the Louisiade, Marshall, and Gilbert groups of South Sea Islands. It is an eruption of small, dry, horny scales, giving the sufferers a repulsive appearance. It is probably mycetogenic in origin. **s. tetter**, see *psoriasis*.

**scamma** (*skam'-ah*) [σκάμμα, a trench]. Same as *fossa*.

**scammonin** (*skam'-o-min*) [*scammony*]. A glucoside found in scammony. It may be identical with jalapin.

**scammonium** (*skam-o'-ne-um*). See *scammony*.

**scammony** (*skam'-o-ne*) [σκαμμωνία, scammony]. The dried juice of the root of *Convolvulus scammonium*. It is a drastic cathartic, and is generally given in combination with other drugs. **s., resin of** (*resina scammonii*, U. S. P.). Dose as an active purge for adults 5–15 gr. (0.3–1.0 Gm.).

**scandium** (*skan-de-um*) [*Scandia, Scandinavia*]. A rare metal belonging to the aluminum group. Symbol Sc; atomic weight 44.1. See *elements, table of*.

**scanning, s. speech** [*scandere*, to climb]. A peculiar slow and measured form of speech, occurring in various nervous affections, especially in multiple sclerosis.

**scansorius** (*skan-so'-re-us*). See *muscles, table of*.

**Scanzoni's operation** (*skan-zo'-ne*) [Friedrich Wilhelm *Scanzoni*, German obstetrician, 1821–1891]. A method of rotating the fetal head with the forceps in order to hasten delivery.

**scapha** (*ska'-fah*) [σκάφη, trough]. 1. A trough. 2. The scaphoid fossa.

**scaphocephalic, scaphocephalous** (*skaf-o-sef-al'-ik, skaf-o-sef'-al-us*) [see *scaphocephaly*]. Having a boat-shaped head, from early ossification of the sagittal suture, which projects like the keel of a boat.

**scaphocephalus** (*skaf-o-sef'-al-us*) [σκάφη, a skiff; κεφαλή, head]. A boat-shaped appearance of the cranium, due to a premature union of the sagittal suture, or abnormal development.

**scaphocephaly** (*skaf-o-sef'-al-e*) [σκάφη, boat; κεφαλή, head]. The condition of having a skull characterized by a projecting, keel-like sagittal suture, due to its premature ossification.

**scaphocuboid** (*skaf-o-kū'-boid*) [σκάφη, boat; *cuboid*]. Pertaining to the scaphoid and cuboid bones.

**scaphohydrocephalus, scaphohydrocephaly** (*skaf-o-hi-dro-sef'-al-us, skaf-o-hi-dro-sef'-al-e*) [σκάφη, boat; *hydrocephalus*]. Scaphocephaly due to hydrocephalus.

**scaphoid** (*skaf'-oid*) [σκάφη, boat; εἶδος, like]. Boat-shaped. **s. abdomen**, the sunken abdomen seen in meningitis and in great emaciation. **s. bone**, a name given to a boat-shaped bone of the tarsus and of the carpus. **s. fossa**. See *fossa, scaphoid*.

**scaphoideum** (*skaf-oid'-e-um*) [L.]. The scaphoid bone of the wrist or ankle.

**scaphoidoastragalan** (*skaf-oid-o-as-trag'-al-an*). Relating to the scaphoid bone and the astragalus.

**scaphoidocuboid** (*skaf-oid-o-kū'-boid*). Relating to the scaphoid and cuboid bones.

**scapholunar** (*skaf-o-loo'-nar*) [σκάφη, boat; *luna*, moon]. Pertaining to the scaphoid and semilunar bones.

**scaphula** (*skaf'-ū-lah*) [σκάφη, boat]. The fossa navicularis.

**scapula** (*skap'-ū-lah*) [L.]. The shoulder-blade, the large, flat, triangular bone forming the back of the shoulder. See *bones, table of*. **scapulæ alatæ**, of Galen and Aristotle, a wing-like appearance of the shoulder-blade in thin persons of weak musculature, especially in paralysis of the serratus magnus.

**scapulocromial** (*skap-ū-lak-ro'-me-al*) [*scapula*, shoulder-blade; *acromion*]. Pertaining to the acromion process of the scapula.

**scapulalgia** (*skap-ū-lal'-je-ah*) [*scapula*, shoulder-blade; ἄλγος, pain]. Pain in the neighborhood of the shoulder-blade.

**scapular** (*skap'-ū-lar*) [*scapula*]. Pertaining to the shoulder-blade. **s. line**, a vertical line drawn on the back through the inferior angle of the scapula. **s. point**, a tender point developed in neuralgia of the brachial plexus and situated at the inferior angle of

# SCAPULARY 783 SCENT

the scapula. **s. reflex.** See under *reflex*. **s. region,** the region of the back corresponding to the position of the scapula, the spine of which divides it into a supraspinous and an infraspinous region.
**scapulary** (*skap'-ū-la-re*). A bifurcated bandage, the two ends of which pass over the shoulders, while the single end passes down the back, all three being fastened to a body-bandage.
**scapulectomy** (*skap-ū-lek'-to-me*) [*scapula;* ἐκτομή, excision]. Surgical removal of the scapula.
**scapulen** (*skap-ū-len*) [*scapula*, the shoulder-blade]. Belonging to the scapula in itself.
**scapulo-** (*skap'-ū-lo-*) [*scapula*, scapula]. A prefix denoting relation to the shoulder or scapula.
**scapuloclavicular** (*skap-ū-lo-kla-vik'-ū-lar*) [*scapula; clavicle*]. Pertaining to the scapula and the clavicle.
**scapulocoracoid** (*skap-ū-lo-ko'-rak-oid*) [*scapula, coracoid*]. Pertaining to the scapula and the coracoid process.
**scapulodynia** (*skap-ū-lo-din'-e-ah*). Synonym of *scapulalgia*.
**scapulohumeral** (*skap-ū-lo-hū'-mer-al*) [*scapula; humerus*]. Pertaining to the scapula and the humerus. **s. amputation,** removal of the arm at the shoulder-joint.
**scapulopexy** (*scap-ū-lo-pek'-se*) [*scapulo-;* πῆξις, fixation]. The operation of fixing the scapula to the ribs.
**scapulothoracic** (*skap-ū-lo-tho-ras'-ik*) [*scapulo-; thoracic*]. Pertaining to the scapula and the thorax.
**scapulovertebral** (*skap-ū-lo-ver'-te-bral*) [*scapulo; vertebra*]. Pertaining to the scapula and the spine.
**scapus** (*ska'-pus*) [L.: pl., *scapi*]. A stem, shaft. **s. penis,** the body of the penis. **s. pili,** the hairshaft.
**scar** (*skar*). See *cicatrix*.
**scarfskin, scurfskin** (*skarf'-skin, skerf'-skin*) [AS., *scearfe*, a fragment; *skin*]. The epidermis or cuticle.
**scarification** (*skar-if-ik-a'-shun*) [*scarify*]. The operation of making numerous small, superficial incisions.
**scarificator** (*skar'-if-ik-ā-tor*) [*scarify*]. An instrument used in scarification, consisting of a number of small lancets operated by a spring.
**scarify** (*skar'-if-i*) [*scarificare*, to scratch]. To make a number of small, superficial incisions.
**scarlatina** (*skar-lat-e'-nah*) [*scarlatinus*, scarlet]. Scarlet fever. An acute, contagious, febrile disease, having a period of incubation varying from several hours to a week, setting in with vomiting or a chill, which is followed by high fever, rapid pulse, sore throat, and the appearance, at the end of the first or the second day of the disease, of a punctiform, scarlet-red eruption. The tongue, at first heavily coated and red at the tip and edges, soon shows prominence of the papillæ, which are red and swollen (strawberry tongue). The eruption, at the appearance of which all the symptoms become intensified, gradually fades after five or six days, and is followed by a scaly desquamation. A peculiarity of scarlatina is the tendency it has to involve the kidneys. **s. anginosa,** scarlatina with marked inflammation of the throat.
**s. cynanchica,** see *s. anginosa*. **s. gastrica,** scarlet fever complicated with gastro-enteritis. **s. gravior,** malignant scarlet fever. **s. hæmorrhagica,** scarlet fever, or more usually septic fever with hemorrhagic spots. **s. lævis,** mild scarlet fever. **s. latens,** scarlet fever without eruptions. **s. papulosa,** scarlet fever in which there are prominent papules, due to involvement of the hair follicles. **s. pruriginosa,** synonym of *urticaria*. **s. puerperalis,** see *s. puerperal*. **s. pustulosa,** scarlet fever with a pustular eruption. **s. rheumatica,** synonym of *dengue*. **s. septica,** a grave form of scarlet fever characterized by symptoms of septic intoxication. **s. simplex,** mild scarlet fever. **s. sine angina,** scarlet fever without throat symptoms. **s. sine eruptione, s. sine exanthemate,** scarlet fever without the rash. **s. traumatica,** the eruption similar to that of scarlet fever, accompanied by febrile symptoms, which sometimes follow wounds or surgical operations. **s. typhosa,** malignant scarlet fever, with grave nervous symptoms. **s. urticata,** urticaria. **s. maligna, s., malignant,** a form characterized by an abrupt onset, high fever, convulsions, coma, and death, usually before the appearance of the eruption. **s. puerperal,** Littre's name for a rash resembling scarlatina sometimes followed by vesication and pustulation of the affected parts, but

without fever; observed in puerperants. Syn., *erythema diffusum* (Braun); *porphyra* (Retzius).
**scarlatinal, scarlatinoid, scarlatinous** (*skar-lat-e'-nal, skar-lat'-in-oid, skar-lat'-in-us*) [*scarlatina*]. 1. Pertaining to or caused by scarlatina. 2. A disease simulating scarlatina. **s. nephritis,** the acute catarrhal nephritis arising in the course of or during the convalescence from scarlatina.
**scarlatiniform** (*skar-lat-in'-if-orm*). Synonym of *scarlatinoid*.
**scarlatinosis** (*skar-lat-in-o'-sis*). The toxic state due to the specific toxin of scarlatina.
**scarlet fever.** See *scarlatina*.
**scarlet red.** A synthetic dye, the sodium salt of amidoazobenzeneazobetanaphtholdisulphonic acid. It has been used to heal wounds, in addition to its uses as a staining reagent.
**Scarpa's fascia** [Antonio *Scarpa*, Italian anatomist, 1747–1832]. The deep layer of the superficial abdominal fascia. **S.'s foramina,** the nasopalatine foramina, bony canals opening into the incisor canal, transmitting the nasopalatine nerves. **S.'s ganglion,** a ganglion near the internal auditory meatus, at the point of junction of the facial nerve and the vestibular branch of the auditory nerve; *roots,* facial and auditory nerves; *distribution,* internal ear. **S.'s habenula.** See *Haller's habenula*. **S.'s hiatus.** 1. See *Breschet's helicotrema.* 2. See *foramen of Winslow*. **S.'s liquor,** the endolymph of the labyrinth. **S.'s membrane,** the membrane which closes the fenestra rotunda of the tympanic cavity. **S.'s nerve,** the nasopalatine nerve. **S.'s staphyloma,** posterior staphyloma; staphyloma of the posterior segment of the sclera. **S.'s triangle,** a triangular space having for its base Poupart's ligament, and for its apex the point of intersection of the sartorius and adductor longus muscles.
**scat** (*skat*). A hermaphrodite.
**scatacratia** (*skat-ak-ra'-she-ah*). See *scoracratia*.
**scatemia, scatæmia** (*skat-e'-me-ah*) [σκῶρ, σκατός, dung; αἷμα, blood]. Autointoxication from retained fecal matter.
**scatiatria** (*skat-i-at'-re-ah*) [σκῶρ, σκατός, dung; ἰατρεία, a healing]. Medical treatment directed to the condition of the feces.
**scatocyanin** (*skat-o-si'-an-in*) [σκῶρ, σκατός, dung; κύανος, dark-blue]. A derivative of chlorophyll, resembling but not identical with phyllocyanin, discovered by E. Schunck (1901). It crystallizes in rhombic plates, pale brown by transmitted light, purplish-blue with brilliant metallic luster by reflected light, decomposed by heat, insoluble in ether, alcohol, or benzol; soluble in chloroform.
**scatol** (*ska'-tol*). See *skatol*.
**scatologia, scatology** (*skat-o-lo'-je-ah, skat-ol'-o-je*) [σκῶρ, σκατός, dung; λόγος, science]. The science or study of excreta.
**scatologic** (*skat-ol-oj'-ik*) [σκῶρ, σκατός, dung; λόγος, science]. Pertaining to scatologia.
**scatophagous** (*skat-of'-ag-us*) [σκῶρ, σκατός, dung; φαγεῖν, to eat]. Coprophagous; excrement-eating.
**scatoscopy** (*skat-os'-ko-pe*) [σκῶρ, σκατός, dung; σκοπεῖν, to inspect]. Inspection of the excreta.
**scatosin** (*skat'-o-sin*) [σκῶρ, σκατός, dung], $C_{18}H_{18}N_2O_2$. A base isolated by F. Baum, 1893, from the products of pancreatic autodigestion; it is probably related to skatol.
**scatt** (*skat*). See *anthrax*.
**scatula** (*skat'-ū-lah*) [L.]. An oblong, flat box for powders or pills.
**scatulation** (*skat-ū-la'-shun*) [ML., *scedūvian*, to show]. The state or condition of incasement.
**scavenger** (*skav'-en-jer*) [AS., *sceāwian*, to show]. One who cleans; a remover of waste and filth. **s.-cells,** wandering cells that take up debris; they are present in the nervous system.
**Sc. D.** Abbreviation for *Scientiæ Doctor*, Doctor of Science.
**scelalgia** (*se-lal'-je-ah*) [σκέλος, leg; ἄλγος, pain]. Pain in a leg. **s. puerperarum,** synonym of *phleg masia alba dolens*.
**scelotyrbe** (*sel-o-ter'-be*) [σκέλος, leg; τύρβη, vacillation]. Weakness or indecision in stepping, often due to a palsied condition. **s. agitans, s. festinans** (Sauvages), paralysis agitans. **s. fibrilis.** See *subsultus tendinum*. **s. pituitosa.** See *enteritis, pseudomembranous*. **s. spastica,** chronic spasms affecting the lower limbs and causing lameness. **s. tarantismus,** chorea.
**scent** (*sent*) [ME., *senten*, to smell]. An effluvium

from any body capable of affecting the olfactory sense; odor, fragrance. **s.-bag**, same as *s.-organ*. **s. gland**, an odoriferous gland, or one secreting an odoriferous substance. **s.-pore**, the orifice of a scent-gland. **s.-test**, for plumbing. It is made by putting into the pipes a quantity of some pungent chemical, such as peppermint oil, the odor of which will escape from the defects in the pipes if there are any [Price]. **s.-vesicle**, a vesicle containing odoriferous matter.

**Schacher's ganglion** (*shah'-ker*) [Polycarp Gottlieb *Schacher*, German physician, 1674-1751]. The ophthalmic ganglion.

**Schachowa's spiral tube** (*shak-ko'-vah*) [Seraphina *Schachowa*, Russian histologist]. The section of a uriniferous tubule that lies between a convoluted and a looped tubule.

**Schaefer's dumb-bells**. The dumb-bell shaped elements regarded by Schaefer as constituting the primitive fibrils of striped muscular tissue. **S.'s reflex**, pinching of the Achilles tendon at its middle or upper third causes slight flexion of the foot and toes in cases of organic hemiplegia. The significance of this reflex is the same as that of Babinski's toe phenomenon. **S.'s** [sign. 1. *Of pregnancy:* a characteristic discoloration in stripes, reddish on a livid background, which appears in the neighborhood of the urethra or on the vestibule of the vagina. The stripes run for the most part crosswise or oblique. The condition is regarded as due to a vasomotor reaction dependent upon the life of the child as the stripes disappear as soon as the child within the womb is dead. 2. *Of hemiplegia:* See *Schacher's reflex*.

**Schaefer's method** (*sha'-fer*) [Edward Albert *Schaefer*, British physiologist, 1850-, ]. *Method of resuscitation in asphyxia or drowning:* the patient is placed face downward, and pressure is made intermittently over the lower part of the thorax to induce natural breathing.

**Schede's method** (*sha'-deh*) [Max *Schede*, German surgeon, 1844-1902]. A method of treating caries of bone. The diseased tissue is scraped away and the cavity allowed to fill with a blood-clot. The latter is kept moist and aseptic by a covering of gauze and protective. **S.'s operation**, a radical thoracoplasty in which the ribs from the second down and from their tubercles to the costal cartilages are excised with intercostal structures and parietal pleura. The skin and muscle flap is then sutured and in contact with the collapsed lung.

**Scheele's acid** (*sheel*) [Karl Wilhelm *Scheele*, Swedish chemist, 1742-1783]. A 4 per cent. solution of hydrocyanic acid. **S.'s green**, cupric arsenite, CuHAsO₃.

**Scheiner's experiment** (*shi'-ner*) [Christopher *Scheiner*, German physicist, 1575-1650]. An experiment illustrating refraction and accommodation of the eye. The person looks through two pinholes made in a card and placed at a less distance than the diameter of the pupil. If the eye is emmetropic, or if accurately focused, the two sets of rays, passing through the pinholes, unite and form a single image. In a myopic or a hyperopic eye the object appears double.

**schema** (*ske'-mah*) [σχῆμα, form]. 1. A simple design to illustrate a complex mechanism. 2. An outline of a subject.

**schematic** (*ske-mat'-ik*) [*schema*]. Pertaining to or of the nature of a schema. **s. eye**, one showing the proportions of a normal or typical eye.

**schemograph** (*ske'-mo-graf*) [*schema;* γράφειν, to write]. An apparatus for tracing the outline of the field of vision; the measurement of the field is made with the perimeter.

**Schenk's method** (*shenk*) [Leopold *Schenk*, Austrian physiologist, 1842-1902]. The determination of sex of infants by regulation of the mother's diet before and during pregnancy; that it is possible to govern the process of gestation so as to determine the sex of human offspring: "When no sugar is secreted, not even the smallest quantity, then the ovum will be developed which is qualified to become a male child."

**Scherer's test for inosite** (*sha'-rer*) [Johan Joseph *Scherer*, 1814-1869]. Evaporate the substance to dryness on a platinum foil with nitric acid, add ammonia and one drop of calcium chloride solution, and carefully reevaporate to dryness. In the presence of inosite a rose-red residue is obtained. **S.'s test for leucin**, carefully evaporate the leucin to dryness on platinum foil with nitric acid. Add a few drops of sodium hydroxide and warm, and the colorless residue changes to a color varying from pale yellow to brown, according to the purity of the leucin; and further evaporation agglomerates it into an oily drop, which rolls about on the foil. **S.'s test for tyrosin**, carefully evaporate the substance to dryness on platinum foil with nitric acid. A yellow residue is formed (nitrotyrosin), which becomes a deep reddish-yellow color on the application of caustic soda.

**scherlievo** (*skair-le-a'-vo*) [Ital.]. A form of ulcerative syphilis prevalent in the Austrian seaports during the last century.

**scheroma** (*ske-ro'-mah*) [σχερος, dry]. Xerophthalmia.

**Scheurlen's bacillus** (*shoir'-lenz*). A bacillus at one time thought to be the cause of carcinoma.

**Schick's reaction** (*shik*) [——— *Schick*, Austrian physician]. Schick uses a diphtheria toxin in a dilution of such strength that 0.1 c.c. equals 1/50 of the lethal dose of a 250 gram guinea pig. Of this toxin he injects 0.1 c.c. of a 1/1000 dilution. In those who react positively there is a reddening and infiltration developing in twenty-four hours and reaching a maximum in forty-eight hours; this indicates susceptibility to diphtheria; a negative reaction indicating immunity.

**Schiefferdecker's intermediate discs** (*she'-fer-dek-er*). The substance which is assumed to fill in the space existing at Ranvier's nodes between Schwann's sheath and the axis-cylinder. It appears as a black line on staining with silver nitrate and forms the horizontal branch of Ranvier's Latin cross.

**Schiff's reaction for cholesterin** [J. Moritz *Schiff*, German physiologist, 1823-1896]. Evaporate the substance over a small flame in a porcelain dish with a few drops of a mixture consisting of 1 part of a medium solution of ferric chloride and 2 or 3 parts of concentrated hydrochloric or sulphuric acid. In the presence of cholesterin a reddish-violet residue is first obtained and then a bluish-violet. **S.'s test for carbohydrates in urine**, dip strips of paper in a mixture of equal parts of glacial acetic acid and xylidin, with a very little alcohol, and dry. Warm the urine with sulphuric acid, and expose the paper to the fumes. In the presence of carbohydrates the paper will be stained red. **S.'s test for urea**, add to the urea a drop of a concentrated watery solution of furfurol, and next a drop of hydrochloric acid of specific gravity 1.10. A play of color is produced, changing from yellow, green, and blue to purple. The same reaction is given by allantoin, but it is less intense. **S.'s test for uric acid**, allow the substance to dissolve in sodium carbonate, and on the addition of a solution of silver nitrate a reduction of black silver oxide is obtained. If a piece of filter-paper previously treated with silver nitrate solution is treated with a drop of the solution of the substance in sodium carbonate, a reduction of black silver oxide will also be obtained on the paper.

**schindylesis** (*skin-dil-e'-sis*) [σχινδύλησις, a cleavage]. A form of articulation in which a plate of one bone is received into a fissure of another bone.

**schirrus** (*skir'-us*). Synonym of *scirrhus*.

**schisto-, schiz-, schizto-** (*skis-to-, skiz-, skiz-to-*) [σχιστός, cleft]. Prefixes meaning split or fissured.

**schistocelia, schistocœlia** (*skis-to-se'-le-ah*) [*schisto-;* κοιλία, cavity]. Abdominal fissure.

**schistocephalus** (*skis-to-sef'-al-us*) [*schisto-;* κεφαλή, head]. 1. Having a fissured head. 2. A monster with a fissured skull.

**schistocormus** (*skis-to-kor'-mus*) [*schisto-;* κορμός, trunk]. A monstrosity having a cleft thorax, neck, or abdominal wall.

**schistocystis** (*skis-to-sis'-tis*) [*schisto-;* κύστις, bladder]. Fissure of the bladder.

**schistocyte** (*skis'-to-sīt*) [*schisto-;* κύτος, a cell]. 1. A blood-corpuscle in process of segmentation. 2. Ehrlich's name for a poikilocyte.

**schistocytosis** (*skis-to-si-to'-sis*) [*schistocyte*]. 1. An aggregation of schistocytes in the blood. 2. The splitting process of blood-corpuscles.

**schistoglossia** (*skis-to-glos'-e-ah*) [*schisto-;* γλῶσσα, tongue]. Cleft tongue.

**schistomelus** (*skis-tom'-el-us*) [*schisto-;* μέλος, limb]. A monstrosity with a cleft lower extremity.

**schistometer** (*skis-tom'-et-er*) [*schisto-;* μέτρον, a measure]. A device for measuring the distance between the vocal cords.

**schistoprosopia** (*skis-to-pro-so'-pe-ah*) [*schisto-;* πρόσωπον, face]. Congenital fissure of the face.

**schistoprosopus** (*skis-to-pros-o'-pus*) [see *schistoprosopia*]. 1. Having a cleft or fissured face. 2. A monster having a fissure of the face.

**schistorrhachis, schistorrachis** (*skis-tor'-a-kis*) [*schisto-*; ῥάχις, spine]. Spina bifida.
**Schistosoma** (*skis-to-so'-mah*) A genus of trematode worms of flukes. **S. hæmatobium**, a blood-fluke causing Egyptian hematuria. **S. japonicum**, an Asiatic blood-fluke the cause of a disease endemic in certain parts of China and Japan; there are enlargement of the liver and spleen, increased appetite, diarrhea, and frequently mucous, bloody stools. Syn., *Schistosoma cattoi*.
**Schistosomum** (*skis-to-so'-mum*). See *Schistosoma*.
**schistosomus** (*skis-to-so'-mus*) [*schisto-*; σῶμα, a body]. A variety of monster in which there is a lateral or median eventration extending the whole length of the abdomen, the lower extremities being absent or rudimentary.
**schistosternia** (*skis-to-ster'-ne-ah*) [*schisto-*; στέρνον, sternum]. Sternal fissure. Synonym of *schistothorax*.
**schistothorax** (*skis-to-tho'-raks*) [*schisto-*; *thorax*]. Fissure of the thorax.
**schistotrachelus** (*skis-to-tra'-kel-us*) [*schisto-*; τράχηλος, neck]. Fissured neck or cervix.
**schizaxon** (*skiz-aks'-on*) [*schiz-*; *axis*]. An axon which divides in its course into equal or nearly equal branches.
**schizo-** (*ski-zo*) [σχίζειν, to split]. A prefix denoting split or cleft.
**schizoblepharia** (*skiz-o-blef-a'-re-ah*) [σχίζειν, split; βλέφαρον, eyelid]. Fissure of the eyelid.
**schizocyte.** See *schistocyte*.
**schizocytosis.** See *schistocytosis*.
**schizogenesis** (*skiz-o-jen'-es-is*) [*schizo-*; γένεσις, production]. Reproduction by fission.
**schizognathism** (*skiz-og'-na-thism*) [*schizo-*; γνάθος, jaw]. Cleavage of the jaw.
**schizogonic** (*skiz-o-gon'-ik*). Relating to schizogony.
**schizogony** (*skiz-og'-o-ne*) [*schizo-*; γονία, generation]. 1. Same as *schizogenesis*. 2. A form of multiple division in which the contents of the oocyst eventually split up into swarm spores. Cf. *sporogony*.
**Schizomycetes** (*skiz-o-mi-se'-tēs*) [*schizo-*; μύκης, a fungus]. The cleft fungi or bacteria, so called because multiplying by fission.
**schizomycosis** (*skiz-o-mi-ko'-sis*) [*schizo-*; μύκης, fungus]. A disease due to schizomycetes.
**schizont** (*skiz'-ont*) [σχίζειν, to divide]. Schaudinn's term for the mother-cell in coccidia which, by multiple division, gives rise to the crescentic swarm spores called merozoites. Syn., *oudelerospore* (E. R. Lankester, 1900); *sporocyte* (Ron, 1899). Cf. *sporont*.
**schizophrenia** (*skiz-o-fre'-ne-ah*) [*schizo-*; φρήν, mind]. Dementia præcox.
**schizophyta** (*skiz-o-fi'-tah*) [*schizo-*; φυτόν, plant]. Dried but viable schizomycetes. Fission-plants.
**schizothorax** (*skiz-o-tho'-raks*). Synonym of *schistothorax*.
**schizotrichia** (*skiz-o-trik'-e-ah*) [*schizo-*; θρίξ, hair]. Splitting of the hair.
**Schlange's sign.** In cases of intestinal obstruction the intestine is dilated above the seat of obstruction and peristaltic movements are absent below that point.
**Schlatter's disease** (*shlaht'-er*) [Carl *Schlatter*, Swiss surgeon, 1864– ]. A condition characterized by pain in the tubercle of the tibia, increased by extension and pressure; it occurs in athletes and is said to be due to separation of the tubercle of the tibia.
**Schleich infiltration anesthesia** (*shlīkh*) [Carl Ludwig *Schleich*, German surgeon, 1859– ]. A local anesthesia produced by the hypodermatic injection of cocaine combined with a weak salt solution; by the addition of a little morphine the anesthetic action is prolonged. S. method of producing general anesthesia, the administration of small doses of chloroform, petroleum ether, and sulphuric ether. S.'s solution, 1½ gr. of cocaine hydrochloride ⅛ gr. of morphine hydrochloride, 3 gr. common salt, dissolved in 3 oz. and 3 dr. of sterilized water.
**Schlemm's canal** (*shlem*) [Friedrich *Schlemm*, German anatomist, 1795–1858]. An irregular space or plexiform series of spaces occupying the sclero-corneal region of the eye; it is regarded by some as a venous sinus, by others as a lymph-channel. **S.'s ligament,** one of two ligaments connected with the shoulder-joints; the glenoideobrachial ligament.
**Schlesinger's type of syringomyelia.** The dorsolumbar type.

**Schmalz's operation** (*shmolts*). For stricture of the lacrimal duct: the introduction of a thread through the sac, and as far into the duct as possible.
**schmerzfreude** (*schmärts'-froy'-deh*) [Ger., pain-joy]. A rare symptom of hysteria, in which pain or normally painful operations seem to the patient pleasant.
**Schmidel's anastomoses** (*shme'-del*) [Casimir Christopher *Schmidel*, German anatomist, 1716–1792]. An abnormal anastomosis between the vena cava and one of the veins of the portal system.
**Schmidt's blood-coagulation theory.** Para-globulin under the influence of fibrin-ferment enters into combination with fibrinogen, the result being fibrin. **S.'s incisions.** See *Lantermann's incisions*. **S.'s (Ad.) method for demonstrating disturbances in the functions of the intestine,** it is formed upon the amount of the fermentation of the feces. The patient is given daily 1560 Gm. milk, 4 eggs, 3 pieces (100 Gm.) of zwieback, a plate of oatmeal-soup (40 Gm.), with 10 Gm. of sugar, a plate of flour soup made with 25 Gm. of wheat flour and 10 Gm. of sugar, and a cup of bouillon; 120 Gm. of potatoes are also given. A small amount (0.3 Gm.) of carmin is given to color and designate the first stool to be examined. A small portion of the stool is dried to constant weight and weighed. It is then mixed with water and placed in a fermentation-tube and kept at 37° C. Fermentation with the evolution of gases sets in and is divided into an early and a late fermentation. Early fermentation occurs during the first 24 or 48 hours. Later fermentation begins slowly on the second or third day. In the early fermentation it is the starch that is acted upon, while in the late it is the albuminous cellulose materials. Early fermentation can be considered as present only when in the first 24 hours an evident amount of gas is formed. Normally after the diet described there should be no such fermentation. Its occurrence indicates faulty starch digestion and an abnormal condition of the bowels, especially of the small intestine. **S.'s nodes,** a term for the medullated interannular segments of a nerve-fiber.
**Schmiedel's ganglion.** The inferior carotid ganglion.
**Schneiderian membrane** (*shni-de'-re-an*) [Conrad Victor *Schneider*, German anatomist, 1614–1680]. The nasal mucous membrane.
**Schoen's theory of accommodation.** See under *accommodation*.
**Schoenbein's reaction for copper** (*shen'-bīn*) [Christian Friedrich *Schoenbein*, German chemist, 1799–1868]. On the addition of potassium cyanide and tincture of guaiac to a solution of a copper salt a blue coloration is produced.
**Schoenlein's disease** (*shen'-līn*) [Johann Lucas *Schoenlein*, German physician, 1793–1864]. Peliosis rheumatica. **S.'s triad,** purpuric exanthem, rheumatic phenomena, and gastrointestinal disorders in peliosis rheumatica.
**Schott's method** (*shot*) [Theodor *Schott*, German physician, 1852– ]. 1. A method of treating heart disease by resisted exercise and special forms of baths. 2. A system of gymnastic movements, accompanied by baths containing Nauheim salts, for the treatment of heart disease, anemia, and chronic rheumatism.
**Schreger's lines** (*shra'-ger*) [Christian Heinrich Theodor *Schreger*, Danish anatomist, 1768–1833]. Curved lines in the enamel of the teeth, parallel to the surface; they are due to the optical effect produced by the simultaneous curvature of the dentinal fibers.
**Schreiber's maneuver** (*shri'-ber*) [Julius *Schreiber*, German physician, 1849– ]. Friction of the skin of the thigh and leg to reinforce the patellar and Achilles tendon-reflexes.
**Schreiber's base** (*shri'-ner*). See *spermin*.
**Schroeder's contraction ring** (*shro'-der*). See *Bandl's ring*. **S.'s test for urea,** when added to a solution of bromine in chloroform the urea will decompose, with the formation of gas.
**Schroeder's method** (*shro'-der*). For resuscitation of asphyxiated infants: the babe while in a bath is supported by the operator on the back, its head, arms, and pelvis being allowed to fall backward; a forceful expiration is then effected by bending up the body over the bath, thereby compressing the thorax.
**Schroetter's catheters** (*shret'-er*) [Leopold von Kristelli *Schroetter*, Austrian laryngologist, 1837–1908]. Instruments of hard rubber and of varying caliber, somewhat triangular on section, used for the dilata-

tion of laryngeal strictures. S.'s chorea, laryngeal chorea.

Schroth's cure (*shrōt*) [Johann *Schroth*, German physician, 1800–1856]. Dipsotherapy; a method of treating certain diseases by reducing to a minimum the liquid ingested by the patient.

Schueffner's dots (*shoof'-ner*) [Wilhelm *Schueffner*, German pathologist]. Red granules seen in erythrocytes, after Romanowski staining, in benign tertian malarial infections.

Schuele's sign (*shoo'-leh*) [Heinrich *Schuele*, German neurologist]. Vertical folds between the eyebrows, forming the Greek letter omega (*omega melancholicum*) frequently seen in subjects of melancholia.

Schueller's ducts. The ducts of Skene's glands.

Schultze's cells. The olfactory cells. S.'s comma-shaped tract, a small tract of descending fibers in the posteroexternal column of the spinal cord near the gray commissure. S.'s fold, a fold formed by the amnion near the insertion of the umbilical cord when the cephalic end of the fetus encroaches upon the latter. S.'s granules, finely granular masses in the blood formed by the breaking-up of the blood-plaques. S.'s method of resuscitation, the child is seized from behind with both hands, by the shoulders, in such a way that the right index finger of the operator is in the right axilla of the child from behind forward, and the left index finger in the left axilla, the thumbs hanging loosely over the clavicles. The other three fingers hang diagonally downward along the back of the thorax. The operator stands with his feet apart and holds the child as above, practically hanging on the index fingers in the first position, with the feet downward, the whole weight resting on the index fingers in the axillæ, the head being supported by the ulnar borders of the hands. At once the operator swings the child gently forward and upward. When the operator's hands are somewhat above the horizontal, the child is moved gently, so that the lower end of the body falls forward toward its head. The body is not flung over, but moved gently until the lower end rests on the chest. In this position the chest and upper end of the abdomen are compressed tightly. The child's thorax rests on the tips of the thumbs of the operator. As a result of this forcible expiration the fluids usually pour out of the nose and mouth of the infant. The child is allowed to rest in this position one or two seconds. The operator gradually lowers his arms, the child's body bends back, and he again holds the infant hanging on his index fingers with its feet downward. These movements are repeated 15 or 20 times in the minute. S.'s position of the placenta, the position assumed by the placenta when its central portion bulges downward and is expelled in advance of the periphery. S.'s reagent for cellulose, iodine dissolved to saturation in a zinc chloride solution of specific gravity 1.8, and the addition of 6 parts of potassium iodide. This reagent turns cellulose blue. S.'s test for cholesterin, evaporate to dryness with nitric acid, using a porcelain dish on the water-bath. In the presence of cholesterin a yellow residue is obtained, which changes to yellowish-red on the addition of ammonia. S.'s test for proteids, to a solution of the proteid add a few drops of a dilute cane-sugar solution and then concentrated sulphuric acid. On warming and keeping the temperature at 60° C. a bluish-red color is produced.

Schultze-Chvostek's sign. See *Chvostek's symptom*.

Schwabach's test (*shvah'-bak*) [Dagobert *Schwabach*, German otologist, 1846– ]. The duration of the perception of a vibrating tuning-fork placed upon the cranium is prolonged beyond the normal in cases of middle-ear disease, but shortened when the deafness is due to a central cause.

Schwalbe's convolution (*shval'-beh*) [Gustav Albert *Schwalbe*, German anatomist, 1844– ]. The first occipital convolution. S.'s fissure, one between the lower portion of the temporosphenoidal and the occipital lobes. S.'s nucleus, the principal vestibular nucleus. S.'s sheath, the delicate sheath which covers elastic fibers. S.'s space, the subvaginal space of the optic nerve.

Schwann, primitive bundle of [Theodor *Schwann*, German anatomist, 1810–1882]. A muscular fiber. S., sheath of, the neurilemma of a nerve-fiber. S., white substance of, the myelin of a medullated nerve-fiber.

Schwarz's reaction for sulphonal. Upon heating sulphonal with charcoal the odor of mercaptan is evolved.

Schwediauer's disease (*shva'-de-ow-er*) [François Xavier *Schwediauer*, Austrian physician, 1748–1824]. Same as Albert's disease.

Schweinfurth green (*shvīn-foort*). Synonym of *Paris-green*.

Schweitzer's reagent for cellulose (*shvi'-tser*). Copper sulphate, 10 parts; water, 100 parts. Add potassium hydroxide, 5 parts, in water 50 parts. Wash the precipitate and dissolve in 20% ammonia solution. This reagent dissolves cellulose.

schwelle (*shvel'-eh*) [Ger., "threshold"]. The threshold, or limen, of any sensation; nerve-excitation which just fails of producing a sensation.

sciage (*se'-ahsj*) [Fr., "sawing"]. A to-and-fro sawing movement in massage, practised with the ulnar border, or with the dorsum of the hand.

sciagram. See *skiagram*.

sciagraphy. See *skiagraphy*.

sciameter. See *skigmeter*.

sciascopia, sciascopy (*ski-as-ko'-pe-ah, ski-as'-ko-pe*). See *retinoscopy*.

sciatic (*si-at'-ik*) [ἰσχίον, ischium]. 1. Pertaining to the ischium, as the *sciatic* notch. 2. Pertaining to the sciatic nerve. In addition to pain there are numbness and tingling, tenderness along the course of the nerve, and eventually wasting of the muscles. See *Felt treatment of sciatica*.

sciatica (*si-at'-ik-ah*) [*ischiatica*, from *ischium*]. A disease characterized by neuralgic pain along the course of the sciatic nerve. It usually follows exposure to cold and wet, and is dependent upon inflammation of the nerve. In addition to pain there are numbness and tingling, tenderness along the course of the nerve, and eventually wasting of the muscles. See *Felt treatment of sciatica*.

science \(*si'-ens*) [*scire*, to know]. Systematized and classified knowledge. S., Christian, a method of treating disease upon principles similar to those upon which faith-cure rests.

scientific (*si-en-tif'-ik*) [*scientia*, knowledge; *facere*, to make]. Relating to science. That which is based upon science.

scientist (*si'-en-tist*) [*scientia*, science]. A savant; one versed in science.

scieropia (*si-er-o'-pe-ah*) [σκιερός, shady; ὤψ, eye]. Defective vision in which all objects appear dark.

scilla (*sil'-ah*). See *squill*.

scillain (*sil'-a-in*). See *scillitoxin*.

scillin (*sil'-in*). An inactive substance obtained from squills.

scillipicrin (*sil-ip-ik'-rin*) [*scilla*; πικρός, bitter]. A yellowish-white, amorphous, hygroscopic powder obtained from squill. It is used as a diuretic in doses of 8–45 gr. (0.5–3.0 Gm.) daily.

scillism (*sil'-izm*) [*scilla*]. Poisoning from extracts or tinctures of squill due to the contained glucoside scillitoxin. It is marked by vomiting, retarded pulse, and stupor.

scillitic (*sil-it'-ik*). Pertaining to or containing squill.

scillitin (*sil'-it-in*). A white or yellowish resinous substance, the bitter principle of squill.

scillitoxin (*sil-it-oks'-in*) [*scilla*; τοξικόν, poison]. An amorphous, light-brown, bitter, active principle of squill. It is soluble in alcohol, insoluble in ether and water, and a cardiac poison somewhat resembling digitalis. It is used as a diuretic in doses of $\frac{1}{40}$–$\frac{1}{6}$ gr. (0.001–0.002 Gm.) several times daily; maximum daily dose $\frac{1}{4}$ gr. (0.05 Gm.).

scillocephalus (*sil-o-sef'-al-us*) [σκίλλα, squill; κεφαλή, head]. 1. Congenital deformity of the head, in which it is small and conically pointed, or squill-shaped. 2. A person with a squill-shaped head, usually an idiot.

scillopicrin (*sil-o-pik'-rin*). See *scillipicrin*.

scillotoxin (*sil-o-toks'-in*). See *scillitoxin*.

scintillascope (*sin-til-ah-skōp*) [*scintilla*, a spark; σκοπεῖν, to observe]. Same as *spinthariscope*, *q. v.*

scintillation (*sin-til-a'-shun*) [*scintillare*, to sparkle]. An emission of sparks. Also a subjective visual sensation as of sparks.

scirrhencanthis (*skir-en-kan'-this*) [σκιρρός, hard; ἐν, in; κάνθος, canthus]. Scirrhus of the lacrymal gland.

scirrhoblepharoncus (*skir-o-blef-ar-ong'-kus*) [σκιρρός, hard; βλέφαρον, eyelid; ὄγκος, tumor]. A hard tumor of the eyelid.

**scirrhocele** (*skir'-o-sēl*) [σκιρρός, hard; κήλη, tumor]. Scirrhous tumor of the testicle.
**scirrhoid** (*skir'-oid*) [*scirrhus*; εἶδος, like]. Resembling a scirrhus.
**scirrhoma** (*skir-o'-mah*). See *scirrhus*. s. **caminariorum**, chimney-sweep's carcinoma.
**scirrhophthalmia** (*skir-off-thal'-me-ah*) [σκιρρός, hard; ὀφθαλμός, eye]. Scirrhus of the eyeball.
**scirrhosarca** (*skir-o-sar'-kah*) [*scirrhus*; σάρξ, flesh]. Hardening of the flesh, especially of new-born infants; sclerema neonatorum.
**scirrhosis** (*skir-o'-sis*) [σκιρρός, hard]. The formation of a scirrhous carcinoma.
**scirrhous** (*skir'-us*) [*scirrhus*]. Hard.
**scirrhus** (*skir'-us*) [σκίρρος, hard]. A hard carcinoma.
**scissile** (*sis'-l*) [*scindere*, to divide]. Capable of being divided.
**scission** (*sish'-un*) [*scindere*, to cut]. A cutting or splitting of anything; fission.
**scissiparity** (*sis-ip-ar'-it-e*) [*scissus*, p. p. of *scindere*, cut, divide; *parere*, to bring forth]. In biology, generation by fission; schizogenesis.
**scissor-leg** (*siz'-or-leg*). A deformity that sometimes follows double hip-joint disease; the legs are crossed in walking.
**scissors** (*siz'-orz*) [*scindere*, to cut]. An instrument consisting of two blades held together by a rivet, and crossing each other so that in closing they cut the object placed between them. The blades may be straight, angular, or curved. s., **artery**, a scissors, one blade of which is probe-pointed, for introduction into a duct or canal. s., **canalicular**, delicate scissors, one blade of which is probe-pointed, used in slitting the lacrymal canal. s., **cannula**, scissors for slitting any canal or tube longitudinally. s., **craniotomy**, a strong S-shaped instrument used in craniotomy for perforating the skull and cutting away portions of bone. s., **iris**, one having flat blades which are bent in such a manner that they may be applied to the eyeball. Also, scissors used in iridectomy. s., **perforator**, see s., *craniotomy*. s., **skingrafting**, an instrument consisting of a forceps and a scissors, the former for seizing a small piece of skin, and the latter for cutting it off. s., **uvula**, one designed for removal of the uvula. s., **de Wecker's**, a peculiar modification of iris-scissors.
**scissura** (*sis-ū'-rah*) [*scindere*, to cut]. A fissure; a splitting. s. **pilorum**, a splitting of the ends of the hairs.
**Sclavo's serum** (*sklah'-vo*). A serum used in the treatment of anthrax; it may be given hypodermically, intravenously, or by mouth; the average dose is 40 Cc.
**sclera** (*skle'-rah*) [σκληρός, hard]. The sclerotic coat of the eye; the firm, fibrous, outer membrane of the eyeball, continuous with the sheath of the optic nerve behind and with the cornea in front. s. **testis**, the tunica albuginea of the testis.
**scleracne** (*skle-rak'-ne*) [σκληρός, hard; *acne*]. Acne indurata.
**scleradenitis** (*skle-rad-en-i'-tis*) [σκληρός, hard; ἀδήν, a gland; ιτις, inflammation]. See *adenosclerosis*.
**scleral** (*skle-ral'*) [*sclera*]. Pertaining to the sclera.
**sclerangia** (*skle-ran'-je-ah*) [*sclera*; ἀγγεῖον, a vessel]. 1. A sense of hardness yielded by a vessel. 2. See *angiosclerosis*.
**scleratitis** (*skle-rat-i'-tis*). Same as *scleritis*.
**sclerectasia** (*skle-rek-ta'-ze-ah*) [*sclera*; ἔκτασις, extension]. Localized bulging of the sclera.
**sclerectoiridectomy** (*skle-rek-to-ir-id-ek'-to-me*). Excision of a portion of the sclera and of the iris, for glaucoma.
**sclerectomy** (*skle-rek'-to-me*) [*sclero-*; ἐκτομή, excision]. 1. Excision of a portion of the sclera. 2. The excision of the sclerosed and ankylosed conductors of sound in chronic catarrhal otitis media.
**scleredema** (*skle-re-de'-mah*). See *sclerema œdematosum*.
**sclerema** (*skle-re'-mah*) [*sclera*]. Sclerosis, or hardening, especially of the skin. s. **adiposum**, a grave form of sclerema neonatorum marked by extreme hardness of the skin, atrophy, and adherence to the subcutaneous tissues. s. **adultorum**, see *morphea*. s. **cutis**, scleroderma. s. **neonatorum**, a disease of the newborn characterized by a hardening of the subcutaneous tissue, especially of the legs and feet, and probably dependent on a coagulation of the fat. s. **œdematosum**, a generally fatal form of sclerema neonatorum marked by edema of the skin with induration, impairment of muscular action, and subnormal temperature. Syn., *compact edema of infants*. s. **partial**, Schwimmer's name for scleroderma occurring in limited areas. Syn., *Sclérème en plaques*. s. **universale**, Schwimmer's name for scleroderma affecting at once the whole surface of the body, or from single areas of sclerosis of the skin gradually diffusing itself over the entire body. Syn., *Carcinus eburneus* (Alibert); *Cutis tensa chronica* (Fuchs); *Elephantiasis sclerosa* (Rasmussen); *Sclerosis corii* (Wilson).
**scleremia, scleremus** (*skle-re'-me-ah, -mus*). Same as *sclerema*.
**sclerencephalia** (*skle-ren-sef-a'-le-ah*) [σκληρός, hard; ἐγκέφαλος, brain]. Sclerosis of brain-tissue.
**sclererythrin** (*skle-rer'-ith-rin*) [σκληρός, hard; ἐρύθρος, red]. A red substance obtained from ergot.
**scleriasis** (*skle-ri'-as-is*) [*sclera*]. Scleroderma.
**scleritic** (*skle-rit'-ik*) [σκληρός, hard]. Sclerous.
**scleriritomy** (*skle-rir-it'-o-me*) [σκληρός, hard, *sclera*; τομή, a cutting]. Incision of the conjunctiva, sclera, and iris, followed by excision of a piece of the iris and anterior capsule, in staphyloma of the cornea and secondary glaucoma.
**scleritis** (*skle-ri'-tis*) [*sclera*; ιτις, inflammation]. Inflammation of the sclerotic coat of the eye. It may exist alone (simple scleritis or episcleritis) or may be combined with inflammation of the cornea, iris, or choroid.
**sclero-** (*skle-ro-*) [σκληρός, hard]. 1. A prefix meaning hard. 2. A prefix denoting connection with the sclera.
**sclerocataracta** (*skle-ro-kat-ar-ak'-tah*) [*sclero-*; *καταράκτης*, cataract]. A hard cataract.
**sclerochoroiditis** (*skle-ro-ko-roid-i'-tis*) [*sclero-*; *choroiditis*]. Inflammation of the sclerotic coat and the choroid of the eye.
**scleroconjunctival** (*skle-ro-kon-jungk-ti'-val*) [*sclero-*; *conjunctiva*, conjunctival]. Pertaining conjointly to the sclerotic coat of the eye and the conjunctiva.
**scleroconjunctivitis** (*skle-ro-kon-junk-ti-vi'-tis*). Simultaneous conjunctivitis and scleritis.
**sclerocornea** (*skle-ro-kor'-ne-ah*) [*sclero-*; *cornea*]. The sclera and cornea regarded as one.
**sclerocorneal** (*skle-ro-kor'-ne-al*) [*sclero-*; *cornea*]. Pertaining conjointly to the sclerotic coat and the cornea of the eye.
**sclerocyclotomy** (*skle-ro-si-klot'-o-me*) [*sclero-*; *κύκλος*, a circle; τομή, a cutting]. Hancock's operation of division of the ciliary muscle.
**sclerodactylia, sclerodactyly** (*skle-ro-dak-til'-e-ah, skle-ro-dak'-til-e*) [*sclero-*; *δάκτυλος*, finger]. A disease of the fingers (or toes) allied to scleroderma. It is usually symmetrical, occurs chiefly in women, and leads to marked deformity.
**scleroderma** (*skle-ro-der'-mah*) [*sclero-*; *δέρμα*, skin]. A disease characterized by a progressive induration of the skin, occurring either in circumscribed patches (see *morphea*) or diffusely. The skin becomes hard, pigmented, and firmly attached to the underlying tissues; destructive changes may also occur, and joints may become immobile from adhesions of the skin. The cause of scleroderma is not known. s. **circumscribed**. See *morphea*. s. **neonatorum**. See *sclerema neonatorum*. s. **œdematosum**. See *sclerema œdematosum*.
**sclerodermatitis** (*skle-ro-der-mat-i'-tis*) [*sclero-*; *δέρμα*, skin; ιτις, inflammation]. Inflammatory thickening and hardening of the skin.
**sclerodermatous** (*skle-ro-der'-mat-us*) [*sclero-*; *δέρμα*, skin]. Having a hard outer covering.
**sclerodermitis** (*skle-ro-der-mi'-tis*). Sclerodermatitis.
**sclerogenous** (*skle-roj'-en-us*) [*sclero-*; *γεννᾶν*, to beget]. Producing a hard substance.
**sclerogeny** (*skle-roj'-en-e*) [see *sclerogenous*]. The formation of sclerous tissue.
**scleroid** (*skle'-roid*) [σκληρός, hard; εἶδος, form]. Hard or bony in texture.
**scleroiritis** (*skle-ro-i-ri'-tis*) [*sclero-*; *ιρις*, iris; ιτις, inflammation]. Inflammation of the sclera and the iris.
**sclerokeratitis** (*skle-ro-ker-at-i'-tis*) [*sclero-*; *bern*, *tisis*]. Inflammation of the sclera and the cornea.
**sclerokeratoiritis** (*skle-ro-ker-at-o-i-ri'-tis*). Combined inflammation of the sclera, cornea, and iris.
**scleroma** (*skle-ro'-mah*) [*sclero-*; *ὄμα*, tumor]. Abnormal hardness or induration of a part. s. **adultorum**. Synonym of *scleroderma*. s., **respiratory**, rhinoscleroma.
**scleromatocystis** (*skle-rom'-at-o-sist-is*) [*sclero-*; *κύστις*, a bladder]. Induration of a cyst, but especially of the gall-bladder or urinary bladder.

**scleromeninx** (skle-ro-me'-ninks) [sclero-; μῆνιγξ, membrane]. The dura mater.
**scleromere** (skle'-ro-mēr) [sclero-; μέρος, a part]. Any metamere or segment of the skeleton, such as a primitive vertebra.
**sclerometer** (skle-rom'-et-er) [sclero-; μέτρον, a measure]. An apparatus for determining the hardness of substances.
**scleromucin** (skle-ro-mū'-sin) [sclero-; mucus]. A gummy substance obtained from ergot, and considered one of its active principles.
**scleronychia** (skle-ro-nik'-e-ah) [sclero-; ὄνυξ, nail]. Induration and thickening of the nails.
**scleronyxis** (skle-ron-ik'-sis) [sclero-; νύξις, a pricking]. Puncture of the sclera.
**sclero-ophoritis** (skle-ro-o-for-i'-tis) [sclero-; ᾠόν, egg; φορός, bearing; ιτις, inflammation]. Sclerosis of the ovary.
**sclerophthalmia** (skle-rof-thal'-me-ah) [sclero-; ὀφθαλμος, eye]. Xerophthalmia.
**sclerosal** (skle-ro'-sal) [σκληρός, hard]. Of the nature of sclerosis.
**sclerosarcoma** (skle-ro-sar-ko'-mah) [sclero-; sarcoma]. A hard, fleshy tumor, especially of the gums.
**sclerose** (skle'-rōz) [σκληρός, hard]. To affect with sclerosis; to become affected with sclerosis.
**sclerosed** (skle'-rōzd) [sclerosis]. Affected with sclerosis; hardened.
**sclérose en plaques** (skla-ros'-on(g) plahk). Synonym of *sclerosis, multiple.*
**sclerosis** (skle-ro'-sis) [σκληρός, hard]. Hardening, especially a hardening of a part from an overgrowth of fibrous tissue; applied particularly to hardening of the nervous system from atrophy or degeneration of the nerve-elements and hyperplasia of the interstitial tissue; also to a chronic inflammation of the arteries characterized by thickening of their coats. s., **amyotrophic lateral**, a combination of chronic anterior poliomyelitis with lateral sclerosis. s., **annular**, a chronic myelitis, in which the sclerosis extends about the cord like a ring. s., **arterio-**. See *endarteritis*. s., **atrophic**, sclerosis with atrophy. s., **cerebrospinal**. See *s. disseminated*. s., **combined**, simultaneous sclerosis of the posterior and the lateral columns of the spinal cord. s., **corii**. Synonym of *scleroderma*. s. **dermatis**. Synonym of *scleroderma*. s., **diffuse**, one extending through a large part of the brain and cord. s., **disseminated**, a form in which numerous sclerotic patches are scattered through the brain and cord. s., **focal**, one confined to a particular region of the brain or cord. s., **general**, a connective tissue hyperplasia affecting an entire organ. s., **initial**, the syphilitic chancre. s., **insular**. See *s., multiple*. s., **lateral**. See *lateral sclerosis*. s. **lobar**, sclerosis of a lobe of the brain. s. of the **lung**. Synonym of *pneumonia, interstitial*. s. of **middle ear**, v. Troeltsch's name for otitis media hypertrophica. s., **miliary**, small sclerotic patches such as have been observed in the spinal cord in some cases of pernicious anemia. s. **multilocular**. See *sclerosis, disseminated*. s., **multiple**, chronic induration occurring in patches in different parts of the nervous system. The principal symptoms are muscular weakness and tremor upon essaying voluntary action. s., **multiple cerebral**, multiple sclerosis affecting only the brain. s., **multiple cerebrospinal**, multiple sclerosis affecting both the brain and the spinal cord. s., **neural**, sclerosis attended by chronic neuritis. s. **ossium**. Synonym of *ostitis, condensing*. s., **posterior spinal**, locomotor ataxia; tabes dorsalis, q. v. s., **postero-lateral**. See *Friedreich's ataxia*. s., **progressive muscular**. Synonym of *pseudohypertrophic muscular paralysis*. s., **renal**. Synonym of *nephritis, interstitial*. s., **syphilitic**, **arterio-**, the arterial sclerosis due to syphilis. It affects chiefly the intima, but also the adventitia. s. **telæ cellularis et adiposæ**, scleroderma. s. **testis**, sarcocele. s., **tuberous**, a form marked by hypertrophy and increased density of the involved areas. s., **ulcerating**, the primary lesion of syphilis; Hunterian or indurated chancre. s., **vascular**, sclerosis of the walls of blood-vessels; arteriosclerosis.
**scleroskeletal** (skle-ro-skel'-et-al) [sclero-; σκελετόν, a dry body]. Pertaining to a scleroskeleton.
**scleroskeleton** (skle-ro-skel'-et-on) [sclero-; skeleton]. In biology, ossifications other than the bones of the main endoskeleton.
**sclerostenosis** (skle-ro-sten-o'-sis) [sclero-; stenosis]. 1. Sclerosis with stenosis. 2. *Scleroderma*. s. **cutanea**, scleroderma.

**sclerosteous** (skle-ros'-te-us) [sclero-; ὀστέον, bone]. A bony formation resulting from osseous deposit in a tendon.
**Sclerostoma** (skle-ros'-to-mah) [σκληρός, hard; στόμα, mouth]. A genus of nematoid worms. S. **duodenale**. Same as *Ankylostoma duodenale*.
**Scleroth's cure** (skla'-rōt). The treatment of pleuritic effusions by diet, i. e., the withdrawal of fluids for the purpose of causing absorption of the effusion.
**sclerothrix** (skle'-ro-thriks) [sclero-; θρίξ, hair]. 1. Abnormal hardness of the hair. 2. Of Metchnikoff, a genus of *Mycobacteriaceæ* included in *Mycobacterium*, Lehmann and Neumann.
**sclerotic** (skle-rot'-ik) [sclera]. 1. Hard; indurated. 2. Pertaining to the outer coat of the eye. 3. Related to or derived from ergot. s. **coat**. See *sclera*.
**scleroticectomy** (skle-rot-ik-ek'-to-me). [sclero-; τομή, excision]. The removal of a part of the sclera.
**scleroticochoroiditis** (skle-rot-ik-o-ko-roid-i'-tis). See *sclerochoroiditis*.
**scleroticonyxis** (skle-rot-ik-on-ik'-sis). See *scleronyxis*.
**scleroticopuncture** (skle-rot-ik-o-punk'-tūr). Same as *scleronyxis*, q. v.
**scleroticotomy** (skle-rot-ik-ot'-o-me) [sclero-; τομή, a cutting]. Incision of the sclerotic.
**sclerotidectomy** (skle-rot-id-ek'-to-me). See *scleronyxis*.
**sclerotis** (skle-ro'-tis) [σκληρός, hard]. The ergot of rye, q. v.
**sclerotitic** (skle-ro-tit'-ik) [sclero-; ιτις, inflammation]. Affected with sclerotitis.
**sclerotitis** (skle-ro-ti'-tis). See *scleritis*.
**sclerotium** (skle-ro'-she-um) [sclera]. A thick mass of hyphæ constituting a resting-stage in the development of some fungi, as the ergot.
**sclerotome** (skle'-ro-tōm) [sclero-; τέμνειν, to cut]. 1. A knife used in sclerotomy. 2. A hard tissue separating successive myotomes in certain of the lower vertebrates. 3. The skeletal tissue of an embryonic metamere.
**sclerotomy** (skle-rot'-o-me) [see *sclerotome*]. The operation of incising the sclera. s., **anterior**, the making of an incision through the sclera anterior to the ciliary body, and entering the anterior chamber, as is done in glaucoma. s., **posterior**, sclerotomy by an incision through the sclera behind the ciliary body, and entering the vitreous chamber.
**sclerotonyxis** (skle-ro-to-niks'-is) [sclero-; νύξις, a pricking]. An operation for cataract formerly practised, in which a broad needle was introduced into the sclera, behind the ciliary region, passed between the iris and the lens, and the lather depressed into the vitreous.
**sclerotrichia** (skle-ro-trik'-e-ah) [sclero-; θρίξ, hair]. A harsh and dry state of the hair.
**sclerous** (skle'-rus) [σκληρός, hard]. Hard; indurated.
**sclerozone** (skle'-ro-zōn) [sclero-; zone]. That portion of the surface of a bone giving attachment to the muscle derived from a given myotome.
**sclopetarius** (sklo-pet-a'-re-us) [L.]. Relating to a gun. **sclopetaria vulnera**, gunshot wounds. Syn., *sclopetica vulnera*.
**scobinate** (sko'-bin-āt) [scobus, a file]. Having a rough surface.
**scoleciform** (sko-les'-if-orm) [σκώληξ, a worm; forma, form]. Having the form or character of a scolex.
**scolecitis** (sko-le-si'-tis) [scolex; ιτις, inflammation]. Appendicitis.
**scolecoid** (sko'-le-koid) [σκωληκοειδής, worm-like]. 1. Vermiform. 2. Resembling a scolex. Removal of the vermiform appendix.
**scolecoiditis** (sko-le-koid-i'-tis) [scolex; εἶδος, likeness; ιτις, inflammation]. Appendicitis.
**scolecology** (sko-le-kol'-o-je). See *helminthology*.
**scolectomy** (sko-lek'-to-me). Appendicectomy.
**scolex** (sko'-leks) [σκώληξ, a worm]. The head of a tape-worm, giving rise to the chain of proglottides.
**scolices** (sko'-lis-ēz). Plural of *scolex*.
**scolicoiditis** (sko-le-koid-i'-tis). Gerster's name for appendicitis.
**scoliocoiditis** (sko-le-o-koid-i'-tis). Nothnagel's term for appendicitis.
**scoliolordosis** (sko-le-o-lor-do'-sis). Combined scoliosis and lordosis.

**scolioma** (*sko-le-o'-mah*). Curvature of the spine. See *scoliosis*.
**scoliometer, scoliosometer** (*sko-le-om'-et-er, sko-le-o-som'-et-er*) [σκολιόs, bent; μέτρον, measure]. An instrument for measuring the extent of a scoliosis.
**scolioneirosis** (*sko-le-o-ni-ro'-sis*) [σκολιόs, bent; ὄνειροs, a dream]. Oppressive, disagreeable dreaming.
**scoliorrhachitic, scoliorachitic** (*sko-le-o-ra-kit'-ik*) [*scoliosis; rachitis*]. Pertaining to or produced by scoliosis and rickets.
**scoliosiometry** (*sko-le-o-se-om'-et-re*) [*scoliosis;* μέτρον, a measure]. The estimation of the degree of deformity in scoliosis.
**scoliosis** (*sko-le-o'-sis*) [σκολιόs, curved]. A morbid lateral curvature of the spine. s., **cicatricial**, scoliosis due to cicatricial contraction, such as occurs after costal necrosis. s., **empyematic**, that due to empyema. s., **habit**, scoliosis as a result of faulty posture. s., **inflammatory**, scoliosis due to caries of the vertebræ. s., **myopathic**, a form due to paresis of the muscles of the spine. s., **osteopathic**, spinal curvature caused by disease of the vertebræ. s., **paralytic**, the same as s., *myopathic*. s., **rhachitic**, spinal curvature due to rhachitis. s., **rheumatic**, temporary scoliosis caused by rheumatism of the muscles of the spine. s., **sciatic**, scoliosis in sciatica with the convexity toward the affected side. Frequently there is compensatory curvature higher up, and the leg is slightly flexed and supported on the toe. s., **static**, scoliosis as a result of inequality in the length of the lower limbs.
**scoliosometer** (*sko-le-os-om'-e-ter*) [σκολιόs, curved; μέτρον, measure]. An instrument for measuring the amount of deformity in scoliosis.
**scoliotic** (*sko-le-ot'-ik*) [*scoliosis*]. Pertaining to or marked by scoliosis.
**scoliotone** (*sko'-le-o-tōn*) [*scoliosis*; τόνοs, a stretching]. An apparatus for elongating the spine and lessening the rotation in lateral curvature.
**scolopsia** (*sko-lop'-se-ah*) [σκόλοψ, anything pointed]. A suture between two bones having reciprocal movement.
**scombrin** (*skom'-brin*) [*Scomber*, a genus of fishes]. A protamine obtained from mature spermatozoa of mackerel.
**scombron** (*skom'-bron*). Bang's name for a histon obtained from immature spermatozoa of mackerel.
**scoop** [AS., *skopa*]. An instrument resembling a spoon, for the extraction of bodies from cavities, as an *ear-scoop, lithotomy-scoop*.
**scooper's pneumonia**. The chronic form of pneumonia occurring in grain-scoopers from exposure to cold and dust.
**scoparin** (*sko'-par-in*). See under *scoparius*.
**scoparius** (*sko-pa'-re-us*) [*scopa*, a broom]. The *Cytisus scoparius*, a shrub of the order *Leguminosæ*. The dried tops constitute the *scoparius* of the U. S. P. (*scoparii cacumina*, B. P.); they contain the alkaloid *sparteine*, C₁₅H₂₆N₂, and a neutral principle, *scoparin*, C₂₁H₂₂O₁₀. Scoparius is diuretic and cathartic, these actions probably depending upon scoparin. Dose of the *fluidextract* 20–40 min. (1.3–2.6 Cc.). For properties of sparteine, see *sparteine*.
**-scope** (*skōp*) [σκοπεῖν, to examine]. A suffix, signifying to see or examine; usually forming a part of the name of some instrument.
**scopola** (*sko'-po-lah*) [after Giovanni Antonio *Scopoli*, Italian naturalist and physician, 1723–1787]. The dried rhizome of *Scopola carniolica*. Dose 1–3 gr. (0.05–0.15 Gm.). The rhizomes of *S. japonica* and *S. carniolica* contain the alkaloid scopolamine or scopoleine, used as a mydriatic.
**scopolamine, scopoleine** (*sko-pol'-am-ēn, sko-pol'-e-in*). The active principle of *Scopola carniolica*, C₁₇H₂₁NO₄, an alkaloid apparently identical with hyoscine, used with morphine in producing anesthesia by Schneiderlin's and Korff's method (see under *anesthetic, local*). s. **hydrobromide** (*scopolaminæ hydrobromidum*, U. S. P.), C₁₇H₂₁NO₄HBr, hygroscopic crystals, used as a mydriatic and sedative. Externally in ophthalmology, $\tfrac{1}{10}-\tfrac{1}{3}$ % solution; subcutaneously for the insane, $\tfrac{1}{216}-\tfrac{1}{64}$ gr. as **narcophine anesthesia**. Same as twilight sleep, *q. v.*
**scopolia** (*sko-po'-le-ah*). See *scopola*.
**scopomorphinism** (*sko-po-mor'-fin-izm*). Associated chronic addiction to scopolamine and morphine.
**scopophobia** (*sko-po-fo'-be-ah*) [σκοπεῖν, to examine; φόβοs, fear]. A morbid dread of being seen.
**-scopy** [σκοπεῖν, to examine]. A suffix denoting inspection or examination.

**scoracratia** (*sko-rak-ra'-she-ah*) [σκῶρ, feces; ἀκρατία, want of conrtol]. Involuntary evacuation of the bowels.
**scorbutic** (*skor-bū'-tik*) [*scorbutus*]. Pertaining to, affected with, or caused by scorbutus or scurvy.
**s. cancer**. Synonym for *cancrum oris*.
**scorbutus** (*skor-bū'-tus*) [L.]. See *scurvy*. **s. alpinus**. See *pellagra*. **s. nauticus**. See *scurvy*. **s. oris**. See *cancrum oris*.
**scordinema** (*skor-din-e'-mah*). See *pandiculation*.
**scoretemia, scoretæmia** (*skor-e-te'-me-ah*). See *scatemia*.
**scotodinia** (*skot-o-din'-e-ah*) [σκότοs, darkness; δῖνοs, a whirl]. Vertigo associated with the appearance of black spots before the eyes.
**scotogram** (*skot'-o-gram*) [σκότοs, darkness; γράφειν, to write]. See *skiagram*.
**scotograph** (*skot'-o-graf*) [σκότοs, darkness; γράφειν, to write]. 1. An instrument for aiding the blind to write. 2. A name given to the picture produced by means of the so-called Roentgen-rays. See *rays, Roentgen-*.
**scotography** (*skot-og'-raf-e*). Skiagraphy.
**scotoma** (*sko-to'-mah*) [σκότωμα, darkness: *pl., scotomata*]. A dark spot in the visual field. **s., absolute**, scotoma with perception of light entirely absent. **s., annular**. See s., *ring*. **s., central**, one limited to the region of the visual field corresponding to the macula lutea. **s., color-**, color-blindness limited to a part of the visual field, and which may exist without interruption of the field for white light. **s., flittering**. See s. *scintillans*. **s., negative**, a defect due to the destruction of the retinal center, and which is not noticeable to the patient. **s., positive**, a scotoma perceptible to the patient as a dark spot before his eyes. **s., relative**, a scotoma within which perception of light is only partially impaired. **s., ring-**, a zone of scotoma surrounding the center of the visual fie ld. **s. scintillans, s., scintillating**, a scotoma with serrated margins extending peripherally and producing a large defect in the visual field. Syn., *fortification-spectrum*.
**scotomatous** (*skot-o'-mat-us*). Pertaining to or affected with scotoma.
**scotometer** (*skot-om'-et-ur*) [σκότοs, darkness; μέτρον, a measure]. 1. An instrument for detecting, locating, and measuring scotomata. 2. An instrument used in the detection of central scotomata.
**scotoscopy** (*skot-os'-ko-pe*) [σκοτία, darkness; σκοπεῖν, to inspect]. See *retinoscopy*.
**scotosis** (*skot-o'-sis*). See *scotoma*.
**scototherapy** (*skot-o-ther'-ap-e*) [σκότοs, darkness; θεραπεία, therapy]. The treatment of malaria and other diseases by keeping the patient in a dark room and in the intervals between the attacks of the disease clothing him in garments impenetrable by light.
**Scott's dressing, S.'s ointment**. Compound mercury ointment.
**scourge** (*skerj*) [ME., *scourge*, scourge]. 1. Any severe epidemic disease of a fatal character. 2. To strike the skin with light withs or with knotted cords in order to produce counter-irritation.
**scouring** (*skowr'-ing*) [ME., *scouren*, to scour]. Purging; also, diarrhea. **s. rush**, the stalks of *Equisetum hyemale*. Diuretic and astringent. Dose of fldext. ♏ xx–ʒj. See *equisetum*.
**scr. scruple; s. brassage**.
**scraper** (*skra'-per*) [ME., *scrapien*, to scrape]. An instrument used to produce an abrasion. **s., tongue**, an instrument used to remove accumulations of exfoliated epithelium and other foreign material from the tongue.
**scrattage** (*skrat-ahzj*) [Fr.]. Ophthalmoxysis, the oldest method of mechanical treatment of trachoma, the scratching out of the granules; revived in 1890. Syn., *brassage*.
**screatus** (*skre-a'-tus*) [L.]. 1. A hawking. 2. A neurosis characterized by paroxysms of hawking.
**screw-worm**. The larva of the fly *Chrysomyia macellaria*. It is found in tropical America where it may cause fatal results in man by burrowing into the nasal or aural cavities.
**scrivener's palsy**. See *writer's cramp*.
**scrobiculate** (*skro-bik'-ū-lat*) [*scrobiculus*, a little ditch or trench]. Pitted or grooved. Possessing minute or shallow depressions.
**scrobiculus** (*skro-bik'-ū-lus*) [L.]. A small pit. **s. cordis**, the depression at the epigastrium; the pit of the stomach. **s. variolæ**, a scar caused by a small-pox pustule.

**scrofula** (*skrof'-ū-lah*) [*scrofa*, a sow]. A term formerly applied to a peculiar condition characterized by enlargement of the lymphatic glands and necrosis of the bones; it is at present considered a form of tuberculosis.
**scrofuleicosis** (*skrof-ū-lel-ko'-sis*) [*scrofula*; ἕλκωσις, ulceration]. Scrofulous ulceration.
**scrofulide** (*skrof'-ū-līd*). See *scrofuloderm*.
**scrofulism** (*skrof'-ū-lism*). The scrofulous diathesis or condition.
**scrofuloderm** (*skrof'-ū-lo-derm*) [*scrofula*; δέρμα, the skin]. A disease of the skin due to scrofula, and generally characterized by superficial irregular ulcers with undermined edges. The cause is probably the tubercle bacillus.
**scrofulome** (*skrof'-ū-lōm*) [*scrofula*; ὅμα, tumor]. A tumor of a supposed scrofulous nature or origin.
**scrofulonychia** (*skrof-ū-lo-nik'-e-ah*) [*scrofula*; ὄνυξ, nail]. Onychia maligna.
**scrofulophyma** (*skrof-ū-lo-fi'-mah*) [*scrofula*; φυμά, growth]. Scrofuloderma tuberculosum. **s. diffusum**, elephantiasis scrofulosa.
**scrofulosis** (*skrof-ū-lo'-sis*) [*scrofula*]. The state characterized by the presence of scrofula; a scrofulous diathesis.
**scrofulotuberculosis** (*skrof-ū-lo-tū-ber-kū-lo'-sis*). Attenuated tuberculosis.
**scrofulous** (*skrof'-ū-lus*) [*scrofula*]. Having the nature of, affected with, or produced by scrofula.
**scroll** (*skrōl*) [ME., *scrolle*]. A roll of paper, or anything folded so as to resemble a roll. **s.-bone**, a turbinate bone. **s.s. olfactory**, the turbinate bones.
**scrophularin** (*skrof-ū-la'-rin*). A principle obtained by Walz from *Scrophularia nodosa*.
**scrotal** (*skro'-tal*) [*scrotum*]. Pertaining to or contained in the scrotum, as *scrotal hernia*.
**scrotitis** (*skro-ti'-tis*) [*scrotum*; ἰτις, inflammation]. Inflammation of the scrotum.
**scrotocele** (*skro'-to-sēl*) [*scrotum*; κήλη, tumor]. Same as *scrotal hernia*.
**scrotopexy** (*skro'-to-peks-e*) [*scrotum*; πῆξις, a fixing in]. Longuet's term for the preservation of the scrotum which results from the extraserous transplantation of the testicle in cases of varicocele. Cf. *orchidopexy; vaginopexy*.
**scrotum** (*skro'-tum*) [L.]. The pouch containing the testicles, consisting of skin, dartos, spermatic fascia, cremasteric fascia, infundibuliform fascia, and parietal tunica vaginalis. **s. cardis**, the pericardium. **s., lymph**, dilatation of the scrotal lymphatics; elephantiasis of the scrotum.
**scruff** (*skruf*) [origin obscure]. A popular name for the nape, or back of the neck.
**scrumpox** (*skrum'-poks*). A name used in England among school-children for impetigo contagiosa.
**scruple** (*skroo'-pl*) [*scrupulus*, dim. of *scrupus*, a sharp stone]. In apothecaries' weight, 20 grains; represented by the sign ℈.
**scrupulosity** (*skroo-pū-los'-it-e*) [*scrupulosus*, exact]. An over-precision, or morbid conscientiousness as to one's thoughts, words, and deeds. It is somewhat common among insane persons of a certain type.
**sculcopin** (*skul'-ko-pin*). The proprietary name for a preparation of hydrastis and skull-cap, used as a local astringent.
**Scultetus' bandage** (*skul-te'-tus*) [Johann *Scultet*, German surgeon, 1595-1645]. A bandage used in compound fractures, so arranged that the short pieces of which it is composed may be removed without motion of the limb.
**scurf** (*skerf*) [AS.]. A bran-like desquamation of the epidermis, especially from the scalp; dandruff.
**scurvy** (*sker'-ve*) [*scurf*]. A disease observed among persons who have been deprived of proper food for a length of time; it is characterized by spongy gums, extravasations of blood beneath the skin, hemorrhages from the mucous membranes, fetor of the breath, and painful contractions of the muscles. It is most common among sailors living on salt meats. **s. of the Alps**, pellagra. **s.-grass**, *Cochlearia officinalis*, a plant of the order *Cruciferæ*, the properties of which reside in a volatile oil resembling oil of mustard. It is used in scurvy and in chronic rheumatism. **s., land-**. See *purpura hæmorrhagica*.
**scute** (*skūt*) [*scutum*, a shield]. A crescentic plate forming the outer wall of the attic.
**scutellaria** (*skū-tel-a'-re-ah*) [*scutellum*, a little shield]. A genus of the *Labiatæ*. The dried plant of *S. lateriflora*, skullcap, is the *scutellaria* of the

U. S. P., and is employed in neuralgia, chorea, delirium tremens, and other nervous affections. **s., fluidextract of** (*fluidextractum scutellariæ*, U. S. P.). Dose 1-2 dr. (4-8 Cc.).
**scutellarin** (*skū-tel-ar'-in*). 1. An impure precipitate from an alcoholic tincture of scutellaria. Dose 3-4 gr. (0.2-0.26 Gm.). 2. $C_{21}H_{20}O_{12}$. A nontoxic principle derived from the root of *Scutellaria lateriflora*, forming flat yellow needles, soluble in alcohol, ether, or alkalies, melting at 190° C. It is used as a tonic and sedative in nervous diseases. Dose ⅛-4 gr. (0.05-0.26 Gm.).
**scutiform** (*skū'-tif-orm*) [*scutum*, a shield]. Shield-shaped. **s. leaf**, the first-formed leaf or cotyledon in *Salvinia*, so named from its peculiar shape.
**scutulate** (*skū'-tū-lāt*) [*scutulum*]. Shaped like a lozenge.
**scutulum** (*skū'-tū-lum*) [dim. of *scutum*, a shield]. Any one of the thin plates of the eruption of favus.
**scutum** (*skū'-tum*) [*scutum*, a shield; pl., *scuta*]. 1. A shield-like plate of bone. 2. The thyroid cartilage. 3. The patella. **s. cordis**, the sternum. **s. genu**, the patella. **s. pectoris**, the thorax. **s. thoracis**, the sternum. **s. tympanicum**, the semilunar plate or bone separating the attic of the tympanum from the outer mastoid cells.
**scybala** (*sib'-al-ah*). Plural of *scybalum* (*q. v.*).
**scybalous** (*sib'-al-us*) [*scybalum*]. Of the nature of a scybalum.
**scybalum** (*sib'-al-um*) [σκύβαλον, fecal matter]. A mass of abnormally hard fecal matter.
**scymnol** (*sim'-nol*) [*Scymnus*, a genus of sharks], $C_{27}H_{46}O_5$ or $C_{25}H_{40}O_4$. An organic base obtained by Hammarsten from the bile of sharks.
**scyphoid** (*si'-foid* or *ski'-foid*) [σκύφος, a drinking-cup]. Cup-shaped.
**Scythian disease** (*sith'-e-an*). Atrophy of the male genital organs with loss of strength.
**scythropasmus** (*si-thro-spaz'-mus*) [σκυθρός, angry; σπασμός, spasm]. A heavy or fatigued expression, regarded as an evil symptom in grave disease.
**scytitis** (*si-ti'-tis*) [σκῦτος, skin; ἰτις, inflammation]. Inflammation of the skin; dermatitis.
**scytoblasta** (*si-to-blas'-tah*). See *scytoblastema*.
**scytoblastema** (*si-to-blas'-te-mah*) [σκῦτος, skin; βλάστημα, germ]. The primitive or embryonic stage of the development of the skin.
**scytoblastesis** (*si-to-blas-te'-sis*) [σκῦτος, skin; βλάστημα, germ]. The condition and progress of scytoblastema.
**scytodephic**, **scytodepsic** (*si-to-de'-fik, si-to-dep'-sik*) [σκῦτος, a hide; δέψειν, to soften]. Relating to tannin; tannic.
**scytomorphosis** (*si-to-mor-fo'-sis*) [σκῦτος, skin; μόρφωσις, shaping]. An abnormal development of the skin.
**Se**. Chemical symbol of selenium.
**seal** (*sēl*) [ME., *seel*, seal]. A body of water, or other material, placed in the trap of a house-drain for the purpose of preventing the ingress of sewer air.
**seam** (*sēm*). See *suture* and *raphe*.
**seamstress's cramp**. A painful cramp affecting the fingers of seamstresses; an occupation-neurosis analogous to writer's cramp.
**searcher** (*serch'-er*). A sound used for the detection of stone in the bladder.
**sea-onion**. See *squill*.
**sea-sickness**. A condition occurring in persons aboard ship, produced by the rolling of the ship, and characterized by vertigo, nausea, retching, and prostration. A similar state may be induced by riding in cars, elevators, etc.
**sea-tangle**. See *laminaria*.
**seat-worm**. See *oxyuris*.
**sebaceofollicular** (*se-ba-se-o-fol-ik'-ū-lar*). Relating to a sebaceous follicle.
**sebaceous** (*se-ba'-shus*) [*sebum*, fat]. Pertaining to sebum; secreting sebum. **s. crypt**, *sebaceous gland*. **s. cyst**, a cystic tumor formed by occlusion of the duct of a sebaceous gland, with retention of the secretion, dilatation, and thickening of the wall of the gland. It contains a grayish-white cheesy material. **s. glands**, **s. follicles**, compound saccular glands associated with the hair-follicles, and secreting a semifluid substance, the sebum, composed of oil-droplets and broken-down epithelial cells.
**sebadilla** (*seb-ad-il'-ah*). See *sabadilla*.
**sebastomania** (*se-bas-to-ma'-ne-ah*) [σεβαστός, revered; μανία, madness]. Religious insanity.

**sebiferous** (*se-bif'-er-us*) [*sebum*, fat; *ferre*, to bear]. Same as *sebiparous*.

**Sebileau's sublingual hollow** (*seb-il-o*). A pyramidal area with its base upward, extending along beneath the tongue, and formed by the oral mucosa and the sublingual glands, the apex below at the point where the mylohyoid muscle covers the geniohyoid.

**sebiparous** (*seb-ip'-ar-us*) [*sebum*; *parere*, to produce]. Secreting sebum.

**sebolith** (*seb'-o-lith*) [*sebum*; λίθος, a stone]. A concretion in a sebaceous gland.

**seborrhagia** (*seb-or-a'-je-ah*). See *seborrhea*.

**seborrhea, seborrhœa** (*seb-or-e'-ah*) [*sebum*; *pola*, a flow]. A functional disease of the sebaceous glands, characterized by an excessive secretion of sebum, which collects upon the skin in the form of an oily coating or of crusts or scales. **s. congestiva**, lupus erythematosus. **s. capillitii, s. capitis**, seborrhea of the scalp. **s. corporis**, seborrhea of the trunk. **s. faciei**, seborrhea of the face. **s. flavescens**. See *s. nasi*. **s. ichthyosis**, a variety characterized by the formation of large, plate-like crusts. **s. nasi**, seborrhea of the sebaceous glands of the nose. **s. nigra, s. nigricans**, seborrhea with the formation of darkcolored crusts, the coloration being usually from dirt. **s. oleosa**, a form characterized by an excessive oiliness of the skin, especially about the forehead and nose. **s. sicca**, the commonest form of seborrhea, characterized by greasy, brownish-gray scales. **seborrheic, seborrhoic** (*seb-or-e'-ik, seb-or-o'-ik*). 1. Affected with seborrhea. 2. One suffering with seborrhea.

**sebum** (*se'-bum*) [L.]. 1. The secretion of the sebaceous glands. 2. Suet; see also *sevum*. **s. palpebrale**, the dried glandular secretion of the eyelids. **s. præputiale, s. præputii**, smegma præputii.

**sec** (*sek*) [Fr.]. Dry; said of bloodless surgical operations.

**secacornin** (*sek-ak-or'-nin*). A solution of the active principles of ergot in water, glycerin, and alcohol.

**secale** (*se-ka'-le*) [L.]. Rye. **s. cereale**, common rye. **s. cornutum**. See *ergot*.

**secalose** (*sek'-al-ōs*). A carbohydrate from green rye, soluble in water.

**secernent, secerning** (*se-sern'-ment, se-sern'-ing*) [*secernere*, to separate]. Secreting; applied to the function of a gland or a follicle.

**secohm** (*sek'-ōm*) [*secundus*, following; *ohm*]. A unit of electric self-induction.

**second intention**. See under *healing*. **s. nerve**, the optic nerve. **s. sight**. See *clairvoyance*.

**secondaries** (*sek'-un-da-rēz*) [*secundus*, second). A name sometimes applied to the secondary symptoms of syphilis, in contradistinction from the primaries.

**secondary** (*sek'-un-da-re*) [*secundarius*, from *secundus*, second]. 1. Second in the order of time or development, as the *secondary* lesions of syphilis. 2. Second in relation; subordinate; produced by a cause considered primary. **s. amputation**, an amputation done after the subsidence of inflammatory symptoms. **s. cataract**. See *cataract*, *recurrent capsular*. **s. coil**, the coil of wire in which the induced current is generated. **s. degeneration** (os nerve-fibers), a degeneration following injury or disease of the trophic centers. **s. hemorrhage**. See *hemorrhage, secondary*.

**secreta** (*se-kre'-tah*) [*secernere*, to separate]. The substances secreted by a gland, follicle, or other organ; products of secretion.

**secretagogue, secretagog** (*se-kre'-tag-og*) [*secretion*; ἀγωγός, leading]. 1. Stimulating the secretory function. 2. An agent which stimulates secretion.

**secrete** (*se-krēt'*) [*secernere*, to separate]. To separate; specifically, to separate from the blood, or from out of materials furnished by the blood a certain substance termed secretion.

**secretin** (*se-kre'-tin*) [see *secrete*]. A hormone produced in the epithelial cells of the duodenum by the contact of acid. It is absorbed from the cells by the blood and excites the pancreas to activity; it also stimulates the secretion of bile.

**secreting** (*se-kre'-ting*) [*secernere*, to separate]. Effecting secretion. **s. fringes**, synovial fringes.

**secretion** (*se-kre'-shun*) [*secrete*]. 1. The act of secreting or forming from materials furnished by the blood a certain substance which is either eliminated from the body or used in carrying on special functions. 2. The substance secreted. **s. antilytic**, the saliva secreted by a submaxillary gland with intact nerves, as distinguished from that which flows from a gland which has had its nerves divided. **s., external**, a secretion thrown out upon the external or internal surface of the body. **s., internal**, a secretion that is not thrown out upon a surface, but is absorbed into the blood. **s., menstrual**, menstrual blood. **s., paralytic**, the abnormal discharge from a gland after section of its motor nerve. **s., sebaceous**, sebum.

**secretodermatosis** (*se-kre-to-der-mat-o'-sis*) [*secretio*, a secretion; δέρμα, skin; νόσος, disease]. An affection of the secretory apparatus of the skin.

**secretomotor** (*se-kre'-to-mo'-tor*) [*secretio*, a secretion; *motor*, a mover]. Applied to nerves intermediating the function of secretion.

**secretory** (*se'-kre-to-re*) [*secretion*]. Pertaining to secretion; performing secretion. **s. capillaries**, minute canaliculi into which gland-cells discharge their secretion; they are simple or branched, sometimes anastomose, forming a network enveloping the gland-cell, and open individually or united in a single trunk into the lumen of the gland. They occur in the fundus glands of the stomach, where the capillary networks envelop the parietal cells, in the liver, and in other glands. **s. fibres**, centrifugal nerve-fibers exciting secretion.

**sectile** (*sek'-til*) [*secare*, to cut]. Capable of being cut.

**sectio** (*sek'-she-o*) [L.]. See *section*. **s. abdominis**. See *celiotomy*. **s. agrippina**, cesarean section. **s. alta**, suprapubic cystotomy. **s. cadaveris**, an autopsy. **s. cæsarea**, cesarean section. **s. franconiana**, suprapubic cystotomy. **s. lateralis**, lateral lithotomy. **s. mariana, s. mediana**, median lithotomy. **s. nympharum**, nymphotomy.

**section** (*sek'-shun*) [*secare*, to cut]. 1. The act of cutting or dividing. 2. A cut; a cut surface. **s., abdominal**. See *celiotomy*. **s., cesarean**. See *cesarean section*. **s.-cutter**, a microtome. **s., frontal**, a section dividing the body into dorsal and ventral parts. **s., occipital**, a transverse section through the middle of the occipital lobe. **s., parietal**, a transverse vertical section through the ascending parietal convolution. **s., perineal**, external urethrotomy without a guide. **s., Pitres'**, a series of sections through the brain for postmortem examination. **s., sagittal**, a section parallel with the sagittal suture, and hence with the median plane of the body, and serving to divide the body into equal parts.

**sector** (*sek'-tor*) [*secare*, to cut]. An area of a circle included between two radii and an arc.

**secundigravida** (*se-kun-de-grav'-id-ah*) [*secundus*, second; *gravidus*, pregnant]. A woman pregnant the second time.

**secundina** (*se-kun-di'-nah*) [*secundinus*, from *secundus*, second]. 1. Something following. 2. The afterbirth, generally used in plural *secundinæ*; see *secundines*. **s. cerebri**, the arachnoid and pia. **s. oculi**, the middle coat of the choroid. **s. uteri**, the chorion.

**secundines** (*sek'-un-dēnz*) [*secundus*, second]. The placenta, part of the umbilicus, and the membranes discharged from the uterus after the birth of the child.

**secundipara** (*se-kun-dip'-ar-ah*) [*secundus*, second; *parere*, to bring forth]. A woman who has borne two children (not twins).

**secundiparity** (*se-kun-dip-ar'-it-e*). The state of being a secundipara.

**secundiparous** (*se-kun-dip'-ar-us*). Having borne two children.

**secundum artem** (*se-kun'-dum ar'-tem*) [L., "according to art"]. In the approved, professional, or official manner.

**sedans** (*se'-danz*) [L.; pl., *sedantia*]. Sedative; a sedative medicine.

**sedatin** (*sed'-at-in*). See *valeryl-phenetidin*.

**sedatine** (*sed'-at-ēn*). See *antipyrine*.

**sedation** (*se-da'-shun*) [*sedare*, to soothe]. 1. A state of lessened functional activity. 2. The production of a state of lessened functional activity.

**sedative** (*sed'-at-iv*) [see *sedation*]. 1. Quieting or lessening functional activity. 2. An agent lessening functional activity. **s. salt**, boric acid.

**sedentaria** (*sed-en-ta'-re-ah*) [L.; plural (neuter) of *sedentarius*, sedentary]. **s. ossa**, the ischia and coccyx, the bones on which the body rests while in a sitting posture.

**sedentary** (*sed'-en-ta-re*) [*sedentarius*; *sedere*, to sit]. Occupied in sitting; sitting at one's work. Pertaining to the habit of sitting.

**sediment** (*sed'-im-ent*) [*sedimentum*, from *sedere*, to sit]. The material settling to the bottom of a liquid.

**sedimentation** (*sed-im-en-ta'-shun*) [*sediment*]. The process of producing the deposition of a sediment, especially the rapid deposition by means of a centrifugal machine. **s. test**, Widal's reaction.

**sedimentator** (*sed-i-ment'-at-or*). A centrifugal apparatus for producing a rapid deposit of the sediment of urine.

**sedimentum** (*sed-im-en'-tum*) [L.]. Sediment, deposit. **s. lateritium,** brickdust deposit.

**Sedlitz** (*sed'-litz*). See *Seidlitz*.

**sedox** (*se'-doks*). A proprietary preparation used as a dressing for wounds, etc.

**Seebeck-Holmgren's test**. See *Holmgren's test*.

**seed** (*sēd*). 1. A fertilized ovule or ovum, as the egg of the silkworm-moth. 2. Seminal fluid; sperm or milt. 3. Offspring. **s.-coat,** the testa or exterior coat of the seed. **s.-lac.** See *lac*.

**Seegen's dietetic regimen**. A regimen for diabetics, consisting of meats of all kinds, eggs, corn, vegetables, cheese, and gluten bread.

**seehear** (*se'-hēr*). Of W. Rollins, a stethoscope fitted with a sound chamber and fluorescent screen by means of which the heart and lungs are rendered both visible and audible.

**Seeligmueller's sign** (*za'-lik-mü-ler*) [Otto Ludwig Adolf *Seeligmueller*, German neurologist, 1837– ]. Mydriasis on the affected side in cases of neuralgia.

**see-saw eczema.** A form of eczema alternating with some other disease.

**Seessel's pocket** (*za'-sel*) [A. *Seessel*, German embryologist]. A slight depression in the epithelial lining of the pharyngeal membrane of the embryo, behind Rathke's hypophyseal pouch.

**segestor** (*se-jes'-tor*) [*se*, self; *gerere*, to carry]. A proprietary embalming fluid, introduced into the vessels of the cadaver by a syringe; so-called because it has the alleged property of finding its way to all parts of the dead organism.

**Séglas' type of paranoia** (*sa-glah'*) [Jules *Séglas*, French physician, 1856– ]. **Psychomotor type** of paranoia.

**segment** (*seg'-ment*) [*segmentum*, from *secare*, to cut]. 1. A small piece cut from the periphery of anything; a part bounded by a natural or imaginary line. 2. A natural division, resulting from segmentation; one of a series of homologous parts, as a myotome; the part of a limb between two consecutive joints. A subdivision, ring, lobe, somite, or metamere of any cleft or articulated body. **s. of Bandl.** See *Bandl's ring*. **s., interannular,** the portion of a nerve included between two consecutive nodes of Ranvier. **s., intermediate** (*of a cilium*), the isotropous, delicately striated portion of a cilium between the cilium proper and its pedicle. **s., lower** (*of the uterus*), all that portion of the uterus situated below the ring of Bandl. **s.s, medullary,** the incisures of Schmidt and Lantermann, or oblique markings in the medullary sheath of a nerve-fiber. **s., primitive,** Minot's word for a primitive division of the vertebrate celom. **s., pubic** (*of the pelvic floor*), this "consists of what extends from the symphysis pubis to the anterior vaginal wall, inclusive of the latter, and is chiefly made up of bladder." (D. B. Hart.) **s., Rivinian** (*of the tympanic ring*), that portion of the temporal bone between the two points of attachment of its tympanic portion to its squamous portion. **s., sacral** (*of the pelvic floor*), that portion which "extends from the sacrum to the posterior vaginal wall." (D. B. Hart.) **s., Schmidt-Lantermann's,** the elongated pieces making up the medullary substance of nerve-fibers, several pieces being included within each internode. **s. vertebra.** 1. See *somatome*. 2. The cusps of the heart-valves.

**segmental** (*seg-men'-tal*) [*segment*]. 1. Pertaining to a segment; made up of segments. 2. Undergoing or resulting from segmentation. **s. duct,** the duct of the pronephron. **s. organs,** a tubular structure found in the embryos of amniotic animals, and comprising the pronephron, the mesonephron, and the metanephron.

**segmentation** (*seg-men-ta'-shun*) [*segmentum*, a piece cut off]. The process of cleavage or division. In embryology, the term is restricted by usage "to the production of cells up to the period of development when the two primitive germ-layers are clearly differentiated and the first trace of organs is beginning to appear." (Minot.) Merogenesis. **s.-cavity,** the central space in the blastula stage of the segmentation of an ovum. **s.-cells,** homogeneous indifferent cells formed by the repeated division of the fecundated egg-cell, and which compose first of all the solid mulberry germ. (Haeckel.) **s., centro-lecithal,** a form of segmentation in which the spheres enclose a central nutritive yolk. **s., complete,** holoblastic segmentation. **s. direct,** amitosis, or direct cell-division. **s., discoidal,** a form of segmentation in which the germinal disc alone is involved. **s., duplicative,** segmentation peculiar to the gonococcus, marked by an interval between the two segments. **s., free,** cleavage of gymnoplasts. **s., germ,** segmentation of the impregnated ovum, or of the first embryonic segmentation-sphere, or blastosphere. **s., holoblastic,** segmentation in which all the contents of the ovum undergo cleavage. **s., incomplete, s., meroblastic,** segmentation in which only a portion of the contents of the ovum, the formative yolk, undergoes cleavage, the other portion, or food-yolk, being a reserve store of food for the developing embryo. **s., metameric,** division of the embryo into metameres; **s.-nucleus.** See under *nucleus*. **s., partial.** See **s., incomplete. s., protovertebral,** division of the mesoblast on each side of the notochord into somites, or protovertebrae. **s., regular,** segmentation in which the spheres are equal in size and symmetrically arranged; **s.-sphere,** one of the cells of an ovum during the early stages of segmentation. **s., total.** See **s., holoblastic. s., unequal,** a variety of segmentation, in which, after cleavage of an ovum has four equal segments, the spheres of one pole are smaller and more numerous than those of the other.

**segregator** (*seg'-re-ga-tor*) [*segregare,* to separate]. An instrument by means of which urine from each kidney may be secured without danger of admixture.

**Seguin's signal symptom** [Edouard *Seguin*, French alienist, 1812–1880]. The initial convulsion of an attack of Jacksonian epilepsy, which indicates the seat of the cortical lesion.

**Seidel's reaction for inosite** (*si'-del*). Evaporate to dryness a little of the substance in a platinum crucible with nitric acid of specific gravity 1.1–1.2, and treat the residue with ammonia and a few drops of a solution of strontium acetate. If inosite is present, a green color and a violet precipitate are obtained.

**Seidlitz powder** (*sid'-litz*). Pulvis effervescens compositus.

**Seiler's cartilage** (*si'-ler*). A small cartilaginous rod attached to the vocal process of the arytenoid cartilage. It is more developed in the female than in the male.

**seisesthesia** (*si-zes-the'-ze-ah*) [σεῖσις, a concussion; αἴσθησις, sensation]. Perception of concussion.

**seismotherapy** (*sīs-mo-ther'-ap-e*) [σεισμός, a shaking; θεραπεία, therapy]: The therapeutic use of mechanical vibration; vibrotherapeutics. Syn., *shaking cure*.

**seizure** (*sēz'-ūr*) [ME.; *seisen*, to seize]. The sudden onset of a disease or an attack. In surgery, the grasping of a part to be operated upon.

**sejunction** (*se-junk'-shun*) [*sejungere*, to disunite]. In psychology the interruption of the continuity of association-complexes, tending to break up personality.

**sel** [Fr.]. Salt. **s. alembroth,** a solution of mercuric chloride and ammonium chloride, each gr. x, in one pint of distilled water. **s. amarum; s. amer,** magnesium sulphate. **s. ammonia, s. ammoniac, s. ammoniacum,** ammonium chloride. **s. ammoniac martial,** ammonia-chloride of iron. **s. de Chrestien,** gold and sodium chloride. **s. commune, s. culinare,** sodium chloride. **s. digestif,** potassium chloride. **s. digestif de Vichy,** sodium bicarbonate. **s. d'Epsom,** magnesium sulphate. **s. de Figuier.** See *s. de Chrestien*. **s. de Glauber,** sodium sulphate. **s. de Perse,** sodium borate. **s. de saturne,** lead acetate. **s. secret de Glauber,** ammonium sulphate. **s. de Seidlitz,** magium sulphate. **s. de Seignette,** potassium and sodium tartrate. **s. de soude,** sodium carbonate. **s. vegetale,** potassium tartrate.

**selection** (*se-lek'-shun*) [*seligere*, to choose]. The act of choosing. **s., artificial,** the artificial choice, definitely planned, of such forms of animals or plants as will by differentiation develop and reproduce given or desired characteristics. **s., natural,** the selective action of external conditions, whereby characters favorable to the species of animal or plant

SELECTOR 793 SEMINIFEROUS

are preserved. s., **physiological,** the selection of those varieties, the individuals of which are fertile among themselves, but sterile or less fertile with other varieties and with the parent stock. s., **sexual,** the selection produced by preferences of the one sex for a member of the other sex in some way specially endowed.
\ **selector** (se-lek'-tor) [selection]. A device for selecting or separating. s., **cell-,** an appliance for regulating the current strength in galvanic electricity. A good selector must admit of an increase or a decrease of electromotive force through the introduction of one cell at a time; it must permit of such increase or decrease without producing any interruption in the flow of the current. All selectors are constructed upon one of three principles: the crank, the rider, or the plug system.
**selene** (se-le'-ne) [σελήνη, moon]. The white spot sometimes occurring on the finger-nails. Cf. lunula.
**seleniasis** (sel-en-i'-as-is) [σελήνη, moon]. Lunacy; epilepsy; somnambulism.
**seleniate** (sel-en'-e-āt). A salt of selenic acid.
**selenic** (se-len'-ik) [selenium]. A compound containing selenium combined directly with three atoms of oxygen. s. **acid,** H₂SeO₄, a dibasic acid resembling sulphuric acid in its properties.
**seleniferous** (sel-en-if'-er-us) [selenium; ferre, to bear]. Containing selenium.
**selenin B** (sel'-en-in) [selenium]. The active toxic element in cultures of Diplococcus semilunaris.
**selenite** (se'-len-īt) [selenium]. 1. A salt of selenous acid. 2. A translucent form of calcium sulphate.
**selenitic** (se-len-it'-ik). Containing selenite.
**selenium** (se-le'-ne-um) [σελήνη, the moon], Se = 79.2: usually bivalent, sometimes quadrivalent or hexivalent. A rare element resembling sulphur in its properties.
**selenogamia** (sel-en-o-gam'-e-ah) [σελήνη, the moon; γαμός, marriage]. Somnambulism.
**selenoplegia** (sel-e-no-ple'-je-ah) [σελήνη, moon; πληγή, stroke]. A kind of apoplexy said to be caused by exposure to the moon's rays.
**selenoplexia** (sel-e-no-pleks'-e-ah). See selenoplegia.
**selenopyrine** (sel-en-o-pi'-rin). A reaction product of potassium selenide with a so-called antipyrine chloride.
**self.** Same; identical; own; personal. s.**-abuse.** See masturbation. s.**-differentiation,** the theory that cells control themselves; that is to say, the fate of the cells is determined by forces situated within them, and not by external influences. s.**-digestion.** See autodigestion. s.**-fertilization,** fertilization of a flower by its own pollen. s.**-heal,** Prunella vulgaris; heal all; a perennial herb growing in North America, Europe, and Asia. s.**-incasement,** a condition in which the small intestine is inclosed, as in a pouch, between the layers of the mesentery. s.**-infection,** the spread of infectious material from a circumscribed area to others or to the entire organism. s.**-inflation,** a process by which a person in danger of drowning may render himself buoyant. After having made a puncture in the mucous membrane of the mouth, at the reflection of the cheek from the lower jaw, air is forced into the subcutaneous tissue of the neck by vigorous blowing efforts with the mouth and nose closed. s.**-limited,** a-term applied to certain diseases, which even without treatment run a definite course within a given time. s.**-pollution.** See masturbation. s.**-repositor,** pneumatic, a curved and bulbous glass tube used at bed-time for the reposition of the displaced uterus, the instrument being used by the patient, and operated by airpressure. s.**-suggestion.** See autosuggestion. s.**-suspension,** suspension of the body for the purpose of stretching or making extension on the vertebral column: see suspension. s.s, **axillocephalic,** suspension by the axillæ and the head. s.s, **cephalic,** suspension by the head.
**sella** (sel'-ah) [L.]. A saddle. s. **turcica,** the pituitary fossa of the sphenoid bone lodging the pituitary body.
**sellanders, sellenders** (sel'-an-derz, sel'-en-derz) [origin obscure]. A kind of eczema occurring on the tarsus of the horse. See mallenders.
**Selters, Seltzer water** (sel'-ters, selts'-er). An effervescent mineral water obtained at Selters in Prussia.
**semeiography** (sem-i-og'-ra-fe) [σημεῖον, sign; γράφειν, to write]. Symptomatology.

**semeiology** (sem-i-ol'-o-je) [σημεῖον, sign; λόγος, discourse]. Symptomatology.
**semeiotic** (sem-i-ot'-ik) [σημεῖον, sign]. Pertaining to symptoms.
**semeiotics** (sem-i-ot'-iks) [see semeiotic]. Symptomatology.
**semeiincident** (sem-el-in'-sid-ent) [semel, once; incidere, to happen]. Happening only once in the same person, as a semelincident disease.
**semen** (se'-men). 1. A seed. 2. The fecundating fluid of the male, chiefly secreted by the testicles, composed of liquor seminis; seminal granules, oilglobules, and spermatozoa. s. **contra,** wormseed.
**semenuria.** See seminuria.
**semester** (se-mes'-ter) [semestris, half yearly; sex, six; mensis, month]. A period of six months.
**semi-** (sem-i-) [L.]. A prefix denoting half.
**semiarticulate** (sem-e-ar-tik'-ū-lāt) [semi-; articulus, a joint]. Loose-jointed.
**semicanal** (sem-e-kan-al') [semi-; canal]. A canal open on one side; a sulcus or groove.
**semicanalis** (sem-e-kan-a'-lis). See semicanal. s. **humeri,** the bicipital groove. s. **nervi vidiani,** the groove on the temporal bone for the passage of the Vidian nerve. s. **tensor tympani,** a depression situated close to the hiatus of Fallopius in the anterior wall of the tympanum. The tendon of the tensor tympani is transmitted through an aperture át its apex. s. **tubæ Eustachii.** See sulcus tubæ Eustachii.
s. **tympanicus,** the tympanic canal.
**semicartilaginous** (sem-ik-ar-til-aj'-in-us) [semi-; cartilago, gristle]. Partially cartilaginous.
**semicircular** (sem-e-sir'-kū-lar) [semi-; circulus, a circle]. Having the form of a half-circle. s. **canals.** See under ear.
**semiconscious** (sem-ik-on'-shus) [semi-; conscius, knowing]. Half-conscious; partially conscious.
**semicordate** (sem-e-kor'-dāt) [semi-; cor, the heart]. Shaped like the half of a heart that has been divided longitudinally.
**semicretin** (sem-e-kre'-tin) [semi-; cretin]. A person having a form of cretinism in which the rudiments of language have been developed. Intellection reaches only to the most ordinary bodily wants.
**semicretinism** (sem-e-kre'-tin-ism). The condition of being a semicretin.
**semidecussation** (sem-e-de-kus-a'-shun). Partial decussation.
**semiflexion** (sem-e-flek'-shun) [semi-; flexion, a flexion]. A posture half-way between flexion and extension.
**semiglutin** (sem-e-gloo'-tin), C₁₈H₁₈N₁₇O₂₂. A derivative of gelatin resembling a peptone.
**semilunar** (sem-e-lū'-nar) [semi-; luna, moon]. Resembling a half-moon in shape. s. **bone,** one of the carpal bones. s. **cartilages,** two interarticulating cartilages of the knee. s. **fold,** the conjunctival folding at the inner canthus. s. **ganglia.** See ganglia. s. **lobe,** a lobe on the upper surface of the cerebellum. s. **notch,** a notch in the scapula through which the suprascapular nerve passes. s. **space of** Traube, that portion of the left inferior anterior thoracic region corresponding to the tympanitic resonance of the stomach. s. **valves.** See valves.
**semilunare** (sem-e-loo-na'-re) [L.]. Semilunar. The semilunar bone of the carpus.
**semiluxation** (sem-e-luks-a'-shun) [semi-; luxus, a luxation]. Subluxation.
**semimembranosus** (sem-e-mem-bra-no'-sus). See muscles, table of.
**semimembranous** (sem-e-mem'-bra-nus). Partly membranous, as the semimembranous muscle (semimembranosus). See under muscle.
**seminal** (sem'-in-al) [semen]. Pertaining to the semen. s. **cyst,** a cyst of the spermatic cord or testicle containing semen. s. **fluid,** semen (2). s. **vesicle.** See vesicle, seminal.
**seminalism** (sem'-in-al-izm) [seminalis, relating to seed; primary]. A vitalistic theory proposed by Bouchet, of Paris, which teaches that the vital forces of man and beasts are totally distinct and that beasts have an intelligence of instinct and man one of abstraction.
**seminatio** (sem-in-a'-shun) [seminatio, a sowing]. The intromission of semen into the uterus or vagina.
**seminervosus** (sem-in-er-vo'-sus). See seminendinosus.
**seminiferous** (sem-in-if'-er-us) [semen; ferre, to carry]. Producing or carrying semen, as the seminiferous tubules of the testicle.

# SEMINIFIC 794 SENSATION

**seminific** (*sem-in-if'-ik*) [*semen,* semen; *facere,* to make]. Producing semen.
**seminormal** (*sem-i-nor'-mal*) [*semi-*; *norma,* rule]. Half-normal. s. solution, one containing in solution half the quantity of the substance contained in the normal solution.
**seminuria** (*sem-in-ū'-re-ah*) [*semen,* seed; οὖρον, urine]. The discharge of semen in the urine.
**semiography.** See *semeiography.*
**semiology** (*se-me-ol'-o-je*). See *semeiology.*
**semiotic** (*se-me-ot'-ik*). See *semeiotic.*
**semiplegia** (*sem-ip-le'-je-ah*). See *hemiplegia.*
**semipronation** (*sem-ip-ro-na'-shun*) [*semi-*; *pronatus,* prone]. The assumption of a semiprone, or partly prone position; an attitude of semisupination.
**semiprone** (*sem'-ip-rōn*) [*semi-*; *pronus,* bent]. Half prone. s. **posture.** See *positions, table of.*
**semiptosis** (*sem-ip-to'-sis*) [*semi-*; πτῶσις, a falling]. Partial ptosis.
**semis** (*se'-mis*) [L..]. Half; abbreviated in prescriptions to *ss,* which is placed after the sign indicating the measure.
**semisideratio** (*sem-is-id-er-a'-she-o*). Synonym of *hemiplegia.*
**semisomnis** (*sem-e-som'-nis*) [*semi-*; *somnus,* sleep]. Coma.
**semisomnous** (*sem-e-som'-nus*). Relating to a comatose condition.
**semisoporus** (*sem-e-so'-por-us*) [*semi-*; *sopor,* sleep]. Coma.
**semispinalis** (*sem-e-spi-na'-lis*). See under *muscle.*
**semissis** (*sem-is'-is*) [L.]. One-half. See *semis.*
**semisulcus** (*sem-e-sul'-kus*). A half sulcus which, uniting with another sulcus, forms a complete sulcus.
**semisupination** (*sem-is-ū-pin-a'-shun*) [*semi-*; *supinare,* to bend backward]. The assumption of a position half-way between supination and pronation.
**semitendinosus** (*sem-it-en-din-o'-sus*). See *muscles, table of.*
**semitendinous** (*sem-e-ten'-din-us*). Partly tendinous, as a *semitendinous* muscle (*semitendinosus*). See under *muscle.*
**semitertian** (*sem-it-er'-shan*) [*semi-*; *tertius,* third]. Partly tertian and partly quotidian (applied to intermittent fevers).
**semivalent** (*sem-iv'-al-ent*) [*semi-*; *valere,* to be able]. Of one-half the normal valency.
**Semon's law** [Sir Felix Semon, English laryngologist, 1849— ]. In progressive organic lesions of the motor laryngeal nerves, the cricoarytenoidei postici—the abductors of the vocal cords—are the first, and sometimes the only, muscles affected. S.'s **symptom,** impaired mobility of the vocal cords in carcinoma of the larynx.
**Semon-Rosenbach's law.** See *Semon's law.*
**sempules** (*sem'-pūls*). Suppositories shaped like a dumb-bell. It is said that this shape renders them more easy of introduction, and also more liable to remain *in situ* until they are absorbed.
**senalbin** (*sen-al'-bin*), C₃₀H₄₄N₂S₂O₁₆. A glucoside found in white mustard, *Brassica alba.*
**senecin** (*sen'-es-in*). 1. An oleoresin from *Senecio gracilis* and *S. vulgaris*; it is emmenagogue, emetic, and astringent. 2. A proprietary elixir of *Senecio jacobœa,* recommended as an emmenagogue; it must not be confounded with the oleoresin of senecio.
**Senecio** (*se-ne'-se-o*) [*senex,* an old man]. Groundsel, a genus of composite-flowered plants, said to contain 960 species, many of them medicinal. *S. aureus* is the common liferoot. *S. canicula,* yerba del Puebla, a Mexican species, is diuretic and is recommended in treatment of epilepsy. *S. cineraria* is a species of South America; the fresh juice of the leaves, stems, and flowers is recommended in treatment of capsular and lenticular cataracts and other diseases of the eye. *S. gracilis* is a slender species, generally regarded as a variety of *S. aureus. S. jacobœa,* ragwort or ragweed, is tonic and astringent.
**senectus** (*se-nek'-tus*) [*senex,* old]. Old age.
**senega** (*sen'-e-ga*) [L.]. The *Polygala senega,* a plant of the *Polygaleæ.* Its root is official (*senega,* U. S. P.; *senegœ radix,* B. P.); it contains a bitter principle, senegin or polygalic acid (or polygalin), which is probably identical with saponin. It is used as a stimulant, expectorant, and diuretic; in large doses it is emetocathartic. It is chiefly employed in bronchitis and laryngitis, as a diuretic in dropsy, and in amenorrhea. Dose 10–20 gr. (0.65–1.3 Gm.); of the *infusion* 1 oz. (30 Cc.). s., fluid-extract of (*fluidextractum senegœ,* U. S. P.). Dose 10–20 min. (0.65–1.3 Cc.). s., syrup of (*syrupus senegœ,* U. S. P.). Dose 1–2 dr. (4–8 Cc.). s., tincture of (*tinctura senegœ,* B. P.). Dose 1 dr. (4 Cc.).
**senegin** (*sen'-e-jin*), C₃₂H₅₄O₁₇ (Hesse). Polygallic acid, a saponin-like glucoside from senega; it is a yellowish powder, soluble in water; used as an expectorant and diuretic. Dose ⅓–2 gr. (0.032–0.13 Gm.).
**senescence** (*se-nes'-ens*) [*senex,* old]. The condition or time of growing old. Senility.
**seng.** A proprietary digestant said to be derived from ginseng, *Aralia quinquefolia.*
**senile** (*se'-nīl*) [*senilis*; from *senex,* old]. Pertaining to or caused by old age. s. **gangrene.** See *gangrene.*
**senilis** (*se-ni'-lis*) [L.]. Old; pertaining to old age.
**senilism** (*se'-ni-lizm*) [*senile*]. A condition of prematurity. See *progeria.* Cf. *ateleiosis*; *infantilism.*
**senility** (*sen-il'-it-e*) [*senile*]. The state of being senile; the weakness of body and mind characteristic of old age.
**seniocine** (*sen-i'-o-sin*). An alkaloid obtained from *Senecio vulgaris* and *S. jacobœa.*
**senki.** A disease resembling lepra and associate with colic, described by Kömpfer in 1713 as peculiar to Japan.
**Senn's bone-plates** [Nicholas Senn, American surgeon, 1844–1908]. Plates of decalcified bone used in intestinal anastomosis. S.'s **test,** the introduction of hydrogen gas into the bowel through the rectum, for the detection and localization of an abnormal opening.
**senna** (*sen'-ah*) [Ar., *sena*]. The leaflets of various species of *Cassia,* a genus of the order *Leguminosæ. Senna* of the U. S. P. is derived from *Cassia acutifolia;* that of the B. P. is of two varieties—*Alexandrian senna,* from *Cassia acutifolia,* and *East India* or *Tinnevelly senna,* from *Cassia angustifolia. Deresinate senna* is that from which the resin has been removed by maceration in alcohol to prevent griping. Senna contains cathartic acid, a glucoside representing the purgative properties of senna, the bitter principles *sennapicrin* and *sennacrol,* and a coloring-matter, *chrysophan.* Senna is used as a purgative, generally in combination with an aromatic to prevent griping. Dose 1–2 dr. (2–8 Cc.). s., **compound infusion of** (*infusum sennæ compositum,* U. S. P.), black draught, contains senna, manna, and magnesium sulphate. Dose 4 oz. (128 Cc.). s., **confection of** (*confectio sennæ,* U. S. P., B. P.). Dose 2 dr. (8 Gm.). s., **fluidextract of** (*fluidextractum sennæ,* U. S. P.). Dose 1–4 dr. (4–16 Cc.). s., **syrup of** (*syrupus sennæ,* U. S. P., B. P. Dose 1–4 dr. (4–16 Cc.). s., **tincture of** (*tinctura sennæ,* B. P.). Dose 1–4 dr. (4–16 Cc.).
**sennacrol** (*sen'-ak-rol*). See under *senna.*
**sennapicrin** (*sen-ap-ik'-rin*). See under *senna.*
**sennatin** (*sen'-at-in*). A preparation of senna leaves used as a cathartic, but administered by subcutaneous or intramuscular injection.
**sennin** (*sen'-in*) [Nicholas Senn, American surgeon, 1844–1908]. A proprietary antiseptic preparation described as a chemically pure product of boric acid, iodine and phenol. It is a fine, white powder, odorless, slightly astringent, and of sweetish taste.
**sennit** (*sen'-it*), C₁₂H₂₄N₁₂. A nonfermentable sugar found in senna, occurring in soluble warty crystals. Syn., *cathartomannite.*
**sensation** (*sen-sa'-shun*)— [*sensatio,* from *sentire,* to feel]. A feeling or impression produced by the stimulation of an afferent nerve. s., **correlative,** stimulation of the cerebrum by a sensation carried by a single sensory nerve. s., **cutaneous,** a sensation produced through the medium of the skin. s., **eccentric,** the conception of locality. s., **external,** a sensation transmitted from a peripheral sense-organ. s., **general.** See *subjective sensation.* s., **girdle, girdle-pain.** s., **internal.** See *subjective sensation.* s., **objective,** an external sensation due to some objective agency. s.s, **psychovisual,** sensations of sight without the stimulation of the retina; visions. s., **radiating.** See s., *secondary* (1). s., **secondary.** 1. Mueller's name for the excitement of one sensation by another or the extension of morbid sensations in disease to unaffected parts. 2. A sensation of one type attending a sensation of another type. Cf. *audition colorée.* s., **special,**

any sensation produced by the special senses. **s., subjective.** See *subjective sensation*. **s., tactile,** one produced through the sense of touch. **s., transference of,** clairvoyance.

**sense** (*sens*) [*sensus*, from *sentire*, to feel]. 1. Any one of the faculties by which stimuli from the external world or from within the body are received and transformed into sensations. The faculties receiving impulses from the external world are the senses of sight, hearing, touch, smell, and taste, which are the special senses, and the muscular and temperature-sense. Those receiving impulses from the internal organs (visceral senses) are the hunger-sense, thirst-sense, and others. 2. A sensation. **s.-body,** a peripheral sense-organ. **s.-capsule,** the hollow cup-like receptacle of a peripheral sense-organ. **s.-club.** See *rhopalium*. **s.-epithelium,** a tract of epithelium having some specialized function of sensation. **s.-filament,** the thread-like peripheral termination of a sensory nerve. **s.-scale.** See *squama rhopalaris*. **s.-seta,** the bristle-like termination of a peripheral sensory nerve-fiber. **s.-shock,** a condition observed in hysterical women and overworked men, and occurring at the moment of waking from sleep. A sensation like an aura rises from the feet or hands, and, passing upward to the head, disappears in the sense of a blow or shock, or of a bursting in the head. It is of no serious significance.

**sensibilin** (*sen'-sib-il-in*). A specific antibody derived from sensibilisinogen; toxogenin.

**sensibilisinogen** (*sen-sib-il-is-in'-o-jen*). One of the substances in an antigen; it produces a specific antibody called sensibilisin.

**sensibility** (*sen-sib-il'-it-e*) [see *sense*]. 1. The ability to receive and feel impressions. 2. The ability of a nerve or end-organ to receive and transmit impulses. **s., organic,** the capability of transmitting and receiving impressions without being conscious of them (Bichat). **s., range of.** See *Fechner's law*. **s., recurrent,** Longet's and Magendie's term for the sensibility observed in the anterior roots of the spinal nerves, which appeared to be dependent on the posterior root, and not inherent, like the sensibility of the posterior root itself. The existence of recurrent sensibility was denied by Longet. **s., transference or externalization of.** See *sensitivisation*.

**sensibilizer** (*sen'-sib-il-i-zer*) [see *sense*]. A substance which, acting as a chemical screen, conduces to synthesis or other chemical processes.

**sensible** (*sen'-si-bl*) [*sense*, to feel]. Perceptible by the senses, as *sensible* perspiration; capable of receiving an impression through the senses; endowed with sensation.

**sensiferous** (*sen-sif'-er-us*) [*sensus*, sense; *ferre*, to bear]. Conveying a sensation, or sense-impression.

**sensigenous** (*sen-sij'-en-us*) [*sensus*, sense; *gignere*, to produce]. Giving rise to a sensory impulse.

**sensitive** (*sen'-sit-iv*) [*sensitivus*]. 1. Capable of feeling; capable of transmitting sensation. 2. Reacting to a stimulus. **s. soul** (of Stahl), the immortal principle.

**sensitization** (*sen-sit-i-za'-shun*). The rendering of a cell liable to destruction by a complement, through the action of a specific amboceptor.

**sensitized** (*sen'-sit-izd*). Rendered sensitive.

**sensitizer** (*sen-sit-i'-zer*). Bordet's name (1899) for the intermediary body of Ehrlich. Syn., *substance sensibilisatrice*. See *amboceptor*.

**sensomobile** (*sen-so-mo'-bil*) [*sensus*, feeling; *mobilis*, movable]. Moving in response to stimulation.

**sensomobility** (*sen-so-mo-bil'-it-e*). The capacity for movement in response to a sensory stimulus.

**sensomotor** (*sen-so-mo'-tor*). Sensorimotor.

**sensorial** (*sen-so'-re-al*) [*sensorium*]. Pertaining to the sensorium.

**sensoriglandular** (*sen-so-re-gland'-ū-lar*). Causing glandular action by stimulation of the sensory nerves.

**sensorimetabolism** (*sen-so-re-met-ab'-ol-izm*). Metabolism resulting from stimulation of the sensory nerves.

**sensorimotor** (*sen-so-re-mo'-tor*) [*sensus*, feeling; *motor*]. Both sensory and motor; concerned with the perception of sensory impulses and with motor impulses. **s. centers,** centers that are concerned both with the perception of sensation and with motor impulses.

**sensorimuscular** (*sen-so-re-mus'-kū-lar*). Producing muscular action in response to stimulation of the sensory nerves.

**sensorium** (*sen-so'-re-um*) [L.]. A center for sensations, especially the part of the brain concerned in receiving and combining the impressions conveyed to the individual sensory centers. **s. commune,** a portion of the cerebral cortex dominating the sensory impulses.

**sensorivolitional** (*sen-so-riv-o-lish'-un-al*) [*sensus*, sense; *volitio*, willing]. Pertaining to or concerned in sensation and volition.

**sensory** (*sen'-so-re*) [*sensus*, feeling]. Pertaining to or conveying sensation. **s. aphasia.** See under *aphasia*. **s. aura,** an aura affecting the special senses. **s. crossway,** the posterior third of the posterior limb of the internal capsule, where the afferent fibers conveying sensory impulses cross to the opposite side. **s. decussation,** the superior pyramidal decussation. **s. epilepsy,** various disturbances of sensation occurring in paroxysms that replace the epileptic convulsion. **s. nerve,** one that conveys sensations from the periphery to the centers.

**sensualism** (*sen'-shoo-al-izm*) [*sensus*, sense]. The condition or character of one who is controlled by the animal passions.

**sensus** (*sen'-sus*) [L.]. Sense; feeling. **s. communis,** the state of the consciousness or sense of normal sensations at any one time.

**sentient** (*sen'-she-ent*) [*sentire*, to feel]. Having sensation; capable of feeling.

**sentina** (*sen-ti'-nah*) [L., "the hold of a ship"]. The epiphysis cerebri.

**sentinel-pile.** The thickened wall of the anal pocket at the lower end of an anal fissure.

**sentisection** (*sen-tis-ek'-shun*) [*sentire*, to feel; *sectio*, section]. Painful vivisection; vivisection of an animal not under the influence of anesthetics.

**separator** (*sep'-ar-a-tor*) [*separare*, to separate]. 1. Anything that separates, especially an instrument for separating the teeth. 2. An instrument for detaching the pericranium or periosteum. 3. An appliance for preventing the urine from the two ureters from mixing in the bladder.

**separatorium** (*sep-ar-a-to'-re-um*) [*separare*, to separate]. In pharmacy, a strainer. In surgery, an instrument for separating the pericranium from the skull.

**sepedogenesis** (*se-ped-o-jen'-e-sis*) [*sepedon*; γενν᾽ᾶν, to produce]. Putrescence.

**sepedon** (*se'-ped-on*) [σῆπια, to be rotten]. Putridity.

**sepia** (*se'-pe-ah*) [σηπία, the cuttle fish]. 1. The ink or black secretion of the common cuttle-fish; used as a pigment. 2. See *sepiost*.

**sepiost** (*se'-pe-ost*) [σηπία, the cuttle-fish; ὀστέον, bone]. The endoskeleton of the cuttle-fish (*sepia*); cuttle-fish bone, sepium, is sometimes prescribed as an antacid and used in dentifrices.

**sepium** (*se'-pe-um*) [σήπιον, the bone of the cuttle-fish]. Same as *sepiost*.

**sepsine** (*sep'-sēn*) [*sepsis*]. A poisonous ptomaine obtained from decomposed yeast and blood.

**sepsis** (*sep'-sis*) [σῆψις, putrefaction]. A state of poisoning produced by the absorption of putrefactive substances. **s., gas,** a septic condition due to the gas bacillus, *Bacillus aerogenes capsulatus*. **s., puerperal,** sepsis occurring after childbirth, from absorption of putrefactive products from the parturient canal.

**sepsometer** (*sep-som'-et-ur*). See *septometer* (2).

**septa** (*sep'-tah*). Plural of *septum*.

**septal** (*sep'-tal*) [*septum*, septum]. Pertaining to a septum. **s. gland.** See under *gland*.

**septan** (*sep'-tan*) [*septem*, seven]. Recurring every seventh day, as *septan* fever.

**septate** (*sep'-tāt*) [*septum*, a fence]. Possessing septa or partitions.

**septectomy** (*sep-tek'-to-me*) [*septum*; ἐκτομή, excision]. Excision of part of the nasal septum.

**septemia, septæmia, septhemia** (*sep-te'-me-ah, sep-the'-me-ah*). See *septicemia*.

**septentrionaline** (*sep-ten-tre-on'-al-ēn*). An alkaloid obtained from *Aconitum lycoctonum*. It is a sensory paralyzant, resembling curara; it has been suggested as a local and general anesthetic, and is used as an antidote to strychnine and in treatment of tetanus and hydrophobia.

**septic** (*sep'-tik*) [*sepsis*]. Relating to sepsis. Pertaining to or produced by putrefaction. **s. fever, septicemia. s. intoxication,** a form of poisoning resulting from the absorption of products of putrefaction. **s. tank,** in sewage treatment a large closed

chamber through which the sewage is allowed to pass slowly. Cf. *contact-bed*.
**septicemia, septicæmia** (*sep-te-se'-me-ah*) [*sepsis;* αἷμα, blood]. An infection characterized by the presence in the blood of bacteria; clinically the term is also used to include toxemia, whether or not there is invasion of the blood by bacteria. **s., bacillar, of chickens,** a disease of chickens described by Fuhrmann as due to a specific bacillus belonging to the colon group. **s., goose,** a rapidly fatal disease of geese due to a specific microbe. The infection takes place by way of the mucosa of the head. **s., mouse,** a form of septicemia occurring in mice and produced by *Bacillus murisepticus*. It is usually fatal in from 40 to 60 hours, the animal early becoming apathetic. **s., phlebitic.** See *pyemia*. **s., rabbit,** a form of septicemia occurring in rabbits and due to *Bacillus septicæmiæ hæmorrhagicæ* or bacillus of chicken cholera. **s., sputum,** a form of septicemia produced by inoculation with microorganisms found in sputum, especially the pneumococcus.
**septicemic, septicæmic** (*sep-te-sem'-ik* or *sep-te-se'-mik*) [*septicemia*]. Pertaining to or affected with septicemia.
**septicine** (*sep'-tis-in*) [*septic*]. A ptomaine obtained from decaying flesh.
**septicogenic** (*sep-tik-o-jen'-ik*) [*septic;* γεννᾶν, to produce]. Applied to a group of microorganisms established by Cohn producing ordinary putrefaction as distinguished from that which produces disease. Cf. *pathogenic*.
**septicophlebitis** (*sep-tik-o-fleb-i'-tis*). Phlebitis due to septic poisoning.
**septicopyemia, septicopyæmia** (*sep-tik-o-pi-e'-me-ah*) [*septicemia; pyemia*]. Combined septicemia and pyemia. **s., primary,** that in which the general infection is produced by the same bacteria as those causing the primary lesion. **s., secondary,** that in which the general infection is due to other bacteria than those causing the pimary lesion.
**septicopyemic** (*sep-tik-o-pi-e'-mik*). Pertaining to septicopyemia.
**septigravida** (*sep-te-grav'-id-ah*) [*septem*, seven; *gravida*, pregnant]. A woman who is pregnant for the seventh time.
**septimetritis** (*sep-ti-met-ri'-tis*). Metritis due to septic poisoning.
**septipara** (*sep-tip'-ar-ah*) [*septem*, seven; *parere*, to bear]. A woman who has been in labor for the seventh time.
**septivalent** (*sep-tiv'-al-ent*) [*septem*, seven; *valere*, to be worth]. Having an atomicity of seven.
**septoforma** (*sep-to-form'-ah*). A condensation-product of formaldehyde dissolved in an alcoholic solution of linseed-oil potassium soap. It is used as an antiseptic and antiparasitic in veterinary practice.
**septomarginal** (*sep-to-mar'-jin-al*). Relating to the margin of a septum.
**septometer** (*sep-tom'-et-er*) [*septum;* μέτρον, a measure]. 1. An instrument for determining the thickness of the nasal septum. 2. [*sepsis*.] An apparatus for determining organic impurities in the air.
**septonasal** (*sep-to-na'-zal*) [*septum*, septum; *nasus*, nose]. Pertaining to the nasal septum.
**septopyemia, septopyæmia** (*sep-to-pi-e'-me-ah*). See *septicopyemia*.
**septotome** (*sep'-to-tōm*) [*septum;* τομή, a cutting]. An instrument for cutting the nasal septum.
**septotomy** (*sep-tot'-o-me*) [*septum;* τομή, a cutting]. The operation of cutting the nasal septum.
**-septula** (*sep'-tu-lah*). Plural of *septulum* (*q. v.*).
**s. fibrosa,** fibrous trabeculæ extending from the deep fascia of the penis into the corpus cavernosum. **s. interalveolaria,** the septa dividing the alveoli of the lungs. **s. medullaria,** processes radiating from the periphery of the gray substance of the spinal cord into the white substance. **s. renum.** See *columna Bertini*. **s. testis,** septules of the testis.
**septulum** (*sep'-tu-lum*) [L.; *pl., septula*]. A small septum. **s. testis.** See *septula testis*.
**septum** (*sep'-tum*) [*sepire*, to hem in]. A partition; a division-wall. **s. atriorum, s. atrium, s. auricularum,** the septum between the right and left auricles of the heart. **s., Bigelow's,** the calcar femorale, an early vertical spur of compact tissue in the neck of the femur, a little in front of the lesser trochanter. **s. cordis,** the wall between the two sides of the heart. **s. crurale,** the layer of areolar tissue closing the femoral ring. **s., Douglas',** in the fetus the septum formed by the union of Rathke's folds transforming the rectum into a complete canal. **s. intermusculare,** septum between muscles. **s. linguæ,** the vertical mesal partition of the tongue, which divides the muscular tissue into two halves. **s. lucidum,** a thin, translucent septum forming the internal boundary of the lateral ventricles of the brain and inclosing between its two laminæ the fifth ventricle. **s., nasal,** the septum between the two nasal cavities. **s., pectiniform,** that between the corpora cavernosa of the penis. **s. pellucidum.** See *s. lucidum*. **s. of the pons,** the median raphe of the pons formed by the decussation of nerve-fibers. **s., rectovaginal,** the tissue forming the partition between the rectum and the vagina. **s. scalæ,** lamina spiralis. **s. scroti,** that dividing the scrotum into two cavities. **s. subarachnoid,** a partition formed by bands of fibro-elastic tissue passing from the arachnoid to the pia along the posterior median line of the spinal cord. **s. thoracis.** See *mediastinum*. **s. transversum.** 1. The diaphragm. 2. The tentorium cerebelli. **s., triangular medullary.** See *s. lucidum*. **s., ventricular,** 1. Same as *s. lucidum*. 2. Same as *s. ventriculorum*. **s. ventriculorum,** the septum between the two ventricles of the heart.
**septuplet** (*sep'-tū-plet*) [*septem*, seven]. One of seven offspring born from a single gestation.
**sepulture** (*sep'-ul-tūr*) [*sepultura; sepelire*, to entomb]. The disposal of the dead by burial.
**séquardin** (*sa-kwar'-din*) [Charles Édouard Brown-Séquard, French physiologist, 1817–1894]. A sterilized testicular extract.
**sequel, sequela** (*se'-kwel, se-kwel'-ah*) [*sequi*, to follow; *pl., sequels, sequelæ*]. A disease or abnormal condition following an attack of a disease, and directly or indirectly dependent upon it.
**sequence** (*se'-kwens*) [*sequentia*]. 1. The order of occurrence, as of symptoms. 2. A sequela.
**sequential** (*se-kwen'-shal*) [*sequentia*]. Occurring as a sequence, as seasonal insanity.
**sequester** (*se-kwes'-ter*). Sequestrum.
**sequestral** (*se-kwes'-tral*) [*sequestrum*, sequestrum]. Pertaining to, or of the nature of, a sequestrum.
**sequestration** (*se-kwes-tra'-shun*) [*sequestrum*]. 1. The formation of a sequestrum. 2. The isolation of persons suffering from disease for purposes of treatment or of protecting others.
**sequestrectomy** (*se-kwes-trek'-to-me*). See *sequestrotomy*.
**sequestrotomy** (*se-kwes-trot'-o-me*) [*sequestrum;* τομή, a cutting]. The operation of removing a sequestrum.
**sequestrum** (*se-kwes'-trum*) [*sequestrare*, to separate; *pl., sequestra*]. A detached or dead piece of bone within a cavity, abscess, or wound. **s., primary,** that entirely detached and demanding removal. **s., secondary,** one that is partially detached, and that unless very loose may be pushed into place. **s., tertiary,** cracked or partially detached and remaining firmly in place.
**sera** (*se'-rah*) [L.]. Plural of serum, *q. v.*
**seralbumin** (*se-ral-bū'-min*) [*serum; albumin*]. Serum-albumin, the albumin found in the blood.
**serempion** (*se-rem'-pe-on*). A form of epidemic measles encountered in the West Indies, and causing great mortality, especially among children.
**Serenoa** (*ser-e-no'-ah*) [Sereno Watson, American botanist]. A genus of palms of one species, *S. serrulata*, the saw-palmetto of North and South America. The fruit is diuretic and sedative and used in diseases of the genitourinary tract. Dose of extract 8–20 gr. (0.52–1.3 Gm.); of the *fluidextract* 57–114 min. (3.7–7.4 Cc.).
**serial** (*se-re'-al*) [*series*, a succession]. Following in regular order; occurring in rows. **s. sections,** microscopic sections made in consecutive order and arranged in the same manner.
**séribèle** (*sa-re-bāl*). A teniafuge said to consist of the seeds and root bark of *Connarus guianensis*. Dose 2 oz. (60 Gm.) in decoction.
**sericeps** (*ser'-is-eps*) [*sericum*, silken; *caput*, head]. A device made of loops of ribbon, used in place of the forceps in making traction upon the fetal head.
**sericum** (*ser'-ik-um*) [L.]. Silken, silk (*q. v.*). Sericum was formerly much prescribed as a cordial, tonic, nervine, and as a restorative of the memory, reason, and reproductive power. It was an ingredient of various electuaries.
**series** (*se'-rēz*) [L.]. A succession or chain of similar parts or activities. **s., aliphatic,** the open

chain series of organic compounds, derived from methane. s., aromatic, the organic compounds derived from benzene. s., fatty. Same as s. *aliphatic*. s., homologous, a series of organic compounds the consecutive members of which differ by a common ratio (generally $CH_2$). s. dentium, a row of teeth. s., numbering parts in, the rule almost universally followed is to commence with the part at the proximal, or at the cephalic aspect. *e. g.*, the most cephalic vertebra (atlas) is number one. The shoulder-girdle is the proximal segment of the pectoral limb.
**seriflux** (*se'-rif-luks*) [*serum; fluxus*, flow]. Any serous or watery discharge, or a disease characterized by such a discharge.
**serin** (*se'-rin*), $CH_2(OH)$ . $CH(NH_2)$ . $CO_2H$. 1. amidoglycerol, obtained by boiling serecin with dilute sulphuric acid. It forms hard crystals, soluble in water, but insoluble in alcohol and ether.* 2. Serum albumin.
**serious** (*se'-re-us*) [*serius*, grave]. Applied to such morbid conditions or symptoms as indicate a grave prognosis.
**seriscission** (*ser-is-ish'-un*) [*sericum*, silken; *scissio*, a cutting]. Division of soft tissues by a silken ligature.
**sero-** (*se-ro-*) [*serum*]. A prefix meaning relating to serum or serous.
**serobacterins** (*se-ro-bak'-ter-ins*). Emulsions of killed bacteria which have been sensitized by treatment with a specific immune serum.
**serochrome** (*se'-ro-krōm*) [*sero-*; χρῶμα, color]. Gilbert's name for the pigments (lipochrome, lutein) which serve to give color to normal serum.
**serocolitis** (*se-ro-ko-li'-tis*). Inflammation of the serous covering of the colon.
**serocyst** (*se'-ro-sist*). A tumor containing cysts filled with serum.
**serocystic** (*se-ro-sis'-tik*) [*sero-*; κύστις, a bladder]. Composed of cysts filled with a serous fluid.
**serodermatosis** (*se-ro-der-mat-o'-sis*) [*serum*; δέρμα, skin; νόσος, disease]. A skin-disease characterized by serous effusion into the tissue of the skin.
**serodermitis** (*se-ro-der-mi'-tis*) [*sero-*; *dermitis*]. An inflammatory skin affection attended with serous effusion.
**serodiagnosis** (*se-ro-di-ag-no'-sis*). Diagnosis based upon the reaction of blood-serum of patients. See *Widal's reaction*.
**seroenteritis** (*se-ro-en-ter-i'-tis*). Inflammation of the serous covering of the small intestine.
**serofibrinous** (*se-ro-fi'-brin-us*) [*sero-*; *fibrin*]. 1. Composed of serum and fibrin, *e. g.*, a *serofibrinous* exudate. 2. Characterized by the production of a serofibrinous exudate, as a *serofibrinous* inflammation.
**serofibrous** (*se-ro-fi'-brus*). Pertaining to a serous membrane and a fibrous tissue.
**seroformalin** (*se-ro-form'-al-in*). An antiseptic dusting-powder of dried coagulated blood-serum and formalin.
**seroglobulin** (*se-ro-glob'-ū-lin*). See *paraglobulin*.
**serohepatitis** (*se-ro-hep-at-i'-tis*) [*sero-*; *hepatitis*]. Inflammation of the hepatic peritoneum.
**seroid** (*se'-roid*) [*sero-*; εἶδος likeness]. Resembling a serous membrane.
**seroimmunity** (*se-ro-im-ū'-nit-e*). Passive immunity; see under *immunity*.
**serolactescent** (*se-ro-lak-tes'-ent*) [*serum*; *lac, lactis*, milk]. Having the characters of both serum and milk. The secretion of Montgomery's glands is said to be serolactescent.
**serolemma** (*se-ro-lem'-ah*) [*sero-*; λέμμα, a¹ husk; a peel]. The embryonic external layer of the amnion.
**serolin** (*se'-ro-lin*) [*sero-*; *oleum*, oil]. A neutral fatty constituent of blood, occurring in small amount; its nature is undetermined.
**serolipase** (*se-ro-lip'-ās*). Lipase as found in blood-serum.
**serological** (*se-ro-loj'-ik-al*). Pertaining to serology.
**serologist** (*se-rol'-o-jist*). One versed in serology.
**serology** (*se-rol'-o-je*). That branch of science which deals with serum; especially immune and hemolytic sera.
**seromembranous** (*se-ro-mem'-bran-us*). Serous and membranous.
**seromucous** (*se-ro-mū'-kus*). Having the nature of or containing both serum and mucus.
**seropneumothorax** (*se-ro-nū-mo-tho'-raks*) [*serum*; πνεῦμα, air; *thorax*]. Pleurisy with serous effusion, associated with pneumothorax.

**seropurulent** (*se-ro-pū'-roo-lent*) [*sero-*; *purulent*]. Composed of serum and pus, as a *seropurulent* exudate.
**seropus** (*se'-ro-pus*) [*sero-*; *pus*]. A fluid consisting of serum and pus.
**seroreaction** (*se-ro-re-ak'-shun*). 1. Any reaction occurring in a serum, such as complement fixation. 2. Serum disease.
**serosa** (*se-ro'-sah*) [*serous; membrana*, understood]. A serous membrane.
**serosanguineous** (*se-ro-san-gwin'-e-us*) [*sero-*; *sanguis*, blood]. Having the nature of, or containing, both serum and blood.
**seroserous** (*se-ro-se'-rus*) [*serous*]. Pertaining jointly to two serous surfaces.
**serosine** (*se'-ro-sēn*). A proprietary remedy said to be antipyretic and antiseptic; same as *bromaniline*.
**serositis** (*se-ro-si'-tis*). Inflammation of a serous membrane.
**serosity** (*se-ros'-it-e*) [*serum*]. The quality of being serous; a serous fluid not the true secretion of serous membranes.
**serosynovial** (*se-ro-si-no'-ve-al*) [*sero-*; *synovia*]. Having the characters of both serum and synovia; pertaining to both a serous and a synovial membrane.
**serosynovitis** (*se-ro-si-no-vi'-tis*) [*sero-*; *synovitis*]. A synovitis with increase of synovial fluid.
**serotaxis** (*se-ro-taks'-is*) [*sero-*; τάξις, arrangement]. In diagnosis the determination of the blood to the skin by application of a solution of caustic potash.
**serotherapy** (*se-ro-ther'-ap-e*) [*sero-*; *therapy*]. 1. The treatment of disease by means of human or animal blood-serum containing antitoxins. 2. Whey cure.
**serothorax** (*se-ro-tho'-raks*). Hydrothorax.
**serotina** (*ser-o-ti'-nah*). See *decidua serotina*.
**serous** (*se'-rus*) [*serum*]. 1. Pertaining to, characterized by, or resembling serum. 2. Producing serum, as a *serous* gland; containing serum, as a *serous* cyst. **s. cavity**, a large lymph-space. **s. effusion**, an effusion of serum. **s. exudate**, an exudate consisting largely of serum. **s. fluid**, normal lymphatic fluid. **s. inflammation**, an inflammation characterized by the formation of a serous exudate. **s. membrane**. See *membrane*.
**serovaccination** (*se-ro-vak-sin-a'-shun*). A method of obtaining mixed immunity by injecting a serum (to secure passive immunity) and also vaccinating (to secure active immunity).
**serozyme** (*se'-ro-zīm*) [*serum*; ζύμη, leaven]. Same as thrombogen, *q. v.*
**serpedo** (*ser-pē'-do*) [*serpere*, to creep]. Same as *psoriasis*.
**serpens** (*ser'-penz*) [L.]. Serpentine, sinuous; creeping. **s., ulcus**, a fistulous ulcer; a sinuous ulcer of the cornea.
**serpentaria** (*ser-pen-ta'-re-ah*) [L.]. Virginia snake-root, the root of several species of *Aristolochia*, of the order *Aristolochieæ*. The rhizome and rootlets of *Aristolochia serpentaria* and *Aristolochia reticulata* constitute the *serpentaria* of the U. S. P. (*serpentariæ rhizoma*, B. P.). Serpentaria contains a volatile oil, a bitter principle, and a nitrogenous principle called *aristolochin*. It is a stimulant, tonic diaphoretic, and diuretic, and is used in intermittent fever and in dyspepsia. **s., fluidextract** of (*fluidextractum serpentariæ*, U. S. P.). Dose 20–30 min. (1.3–2.0 Cc.). **s., infusion** of (*infusum serpentariæ*, B. P.). Dose 1 oz. (32 Cc.). **s., tincture** of (*tinctura serpentariæ*, U. S. P., B. P.). Dose 1 dr. (4 Cc.).
**serpentine** (*ser'-pen-tīn*) [*serpens*, serpent]. Sinuous; snake-like.
**serpes** (*ser'-pēz*) [*serpere*, to creep]. Herpes.
**serpiginous** (*ser-pij'-in-us*) [*serpiginosus*, from *serpere*, to creep]. Creeping. **s. ulcer**, one that extends in one direction while healing in another.
**serpigo** (*ser-pi'-go*) [L.]. Ringworm; herpes.
**serra** (*ser'-ah*) [L., a saw]. In biology, a saw or sawlike structure. **s. salvia**, mountain sage, an herb of the U. S., introduced as a substitute for quinine in the treatment of periodic fevers. Also of service in rheumatism, scarlet fever, and diphtheria. Dose of the fluidextract 3 j–ij.
**serrago** (*ser-a'-go*) [L., *gen.*, *serraginis*]. Sawdust.
**serrate, serrated** (*ser'-āt, ser'-a-ted*) [*serra*, a saw]. Provided with sharp projections like the teeth of a saw.
**serratiform** (*ser-at'-if-orm*) [*serra*, a saw; *forma*, a form]. Same as *serrated*.
**serration** (*ser-a'-shun*) [see *serrate*]. The state or condition of being serrate.

**serratus** (*ser-a'-tus*) [L.]. Serrated; applied to muscles arising or inserted by a series of processes resembling the teeth of a saw. See under *muscle*.

**serre-fine** (*săr-fēn*) [Fr.]. A small spring-forceps for seizing and compressing bleeding vessels.

**serre-nœud** (*săr-neh'*) [Fr.]. An instrument used for drawing tight a ligature thrown around a part, as around the pedicle of a tumor.

**Serres' glands** (*sair*) [Etienne Rénaud Auguste Serres, French physician, 1787–1868]. Pearl-like masses frequently seen in the infant near the gum and resulting from the fragmentation of the dental epithelium. They may give rise to cysts or other abnormal growths.

**serrulate** (*ser'-roo-lāt*) [*serrula*, dim. of *serra*, a saw]. Minutely notched or serrated.

**Sertoli's cells, S.'s columns** (*ser'-to-le*) [Enrico Sertoli, Italian histologist]. The supporting cells of the seminiferous tubules, arranged radially on the membrana propria, and forming long columns between the spermatoblasts.

**serum** (*se'-rum*) [L.; *pl., sera*]. 1. The clear, yellowish fluid separating from the blood after the coagulation of the fibrin. 2. Any clear fluid resembling the serum of the blood. 3. An antitoxin for therapeutic use. **s., Adamkiewicz's.** See *cancroin*. **s. adapted**, a serum produced by immunization. **s.-albumin**, the albumin found in the blood-serum and other animal fluids. **s., allergic**, one which produces hypersensitiveness to injections of serum. **s., allergic**, one which does not produce hypersensitiveness to injections of serum. **s., anthrax**, one used in cattle in the form of protective inoculations. **s., anticancerous**, a serum provided by the inoculation of an ass previously infected with the filtered juice of a neoplastic tumor. **s., antidiphtheritic** (*serum antidiphthericum*, U. S. P.), one prepared by (1) the production of diphtheric toxin by means of bouillon cultures; (2) the immunization of horses, and (3) the collection and separation of the blood-serum from the immunized animals. If 1 Cc. of this serum suffices to protect perfectly a guineapig against a fatal dose of the toxin, and without even the occurrence of a localized reaction at the site of the injection, the serum is said to contain one immunizing unit in the cubic centimeter; if 0.1 Cc. suffices, it has 10 units; if 0.01 Cc., it has 100 units per cubic centimeter. It is the least allowed by law is 100 units. **s., antiepitheliomatous**, one obtained from animals by inoculation with cultures of pathogenic yeast isolated from cancerous tumors. Syn., *serum of Hoffmann* and of *Villiers and Wlaeff*. **s., antimorphine**, a resistance substance conferring immunity to the action of morphine poison, obtained by L. Hirschlaff from animals treated with increasing doses of morphine. **s., antiscarlatinal.** See *s., v. Leyden's*, and *s., Moser's*. **s., antistaphylococcic, s., antistaphylococcous**, a serum produced in the goat by Proscher that will immunize rabbits against from 5 to 7 times the lethal dose of culture of staphylococcus when used in doses of 1–5 Cc. **s., antistreptococcic, s., antistreptococcous.** 1. A specific serum obtained by Piorkowski against the streptococcus, which causes the disease called *pferdedruse*, a contagious, catarrhal affection of the nasal and pharyngeal membranes of horses. The serum has protective and curative properties. It agglutinates the specific streptococcus in dilutions of 1 : 100 the streptococcus of angina but slightly or not at all, and other pathogenic species in dilutions of less than 1 : 25. Piorkowski concludes that there are specific races of streptococci, and that success in the use of an antiserum for the specific group must depend on the particular group of cases is used. 2. See *s., Moser's*. **s., antitoxic**, that which acts upon the bacterial toxins and is not bactericidal. **s., antityphoid**, a sterilized culture of typhoid bacilli used by vaccination as a prophylaxis against typhoid. Cf. *Jess's antityphoid extract*. **s., antivenomous;** see *antivenin*. **s., bactericidal**, that which destroys bacteria but has no effect upon toxins. **s. bacteriolytic**, one which contains a lysin capable of destroying certain bacteria. **s., Bardel's**, sodium chloride, 1 Gm.; phenol, 0.5 Gm.; sodium phosphate, 3 Gm.; sodium sulphate, 2 Gm.; water, to 100 Cc. **s., Behring's, s. of Behring-Roux**, see *s., antidiphtheric*. **s., bichlorureted, of Chéron**, mercury bichloride, 0.5 cg.; sodium chloride, 2 Gm.; distilled water, 200 Cc.; crystallized phenol added when the serum is quite cold, 2 Gm. Dose 300 gr. (20 Gm.) injected into the gluteal region every eight days for syphilis. **s., Calmette's**, see *antivenin*. **s., cancer.** 1. Cancer-juice. 2. See *cancroin*. **s., Cantani's**, sodium chloride, 4 Gm.; sodium carbonate, 3 Gm.; water, 1000 Cc. **s.-casein, Panum's** name for paraglobulin. **s., cerebrospinal**, cerebrospinal fluid. **s., Chéron's**, phenol, 1 Cc.; sodium chloride, 2 Gm.; sodium phosphate, 4 Gm.; sodium sulphate, 8 Gm.; boiled distilled water, 100 Cc. **s., clumping**, a serum capable of producing agglutination of bacteria. **s., Crocq's**, sodium phosphate, 2 Gm.; distilled water, 100 Cc. **s.-disease**, name given to various symptoms which appear some days after the injection of a serum; urticaria, fever. Swollen glands, edema, albuminuria, and arthralgia may thus be present. **s.-diagnosis**, see *serodiagnosis*. **s., Flexner's**, serum used in the treatment of diplococcic cerebrospinal meningitis. **s., globulicidal**, a hemolytic serum. **s.-globulin**, see *paraglobulin*. **s., Haffkine's**, a sterilized culture of cholera bacilli for conferring immunity against cholera. See *Haffkine's method* under *immunization*. **s., Haffkine's** prophylactic, a serum obtained by heating a virulent culture of plague bacilli to 70° C. It is used as a prophylactic against cholera. **s., Hayem's.** 1. Sodium chloride, 5 Gm.; sodium sulphate, 10 Gm.; sterilized water, 1 liter. 2. Sodium chloride, 7.5 Gm.; sterilized water, 1000 Cc. Syn., *physiological serum*. **s., hemolytic**, any blood-serum which produces hemolysis. **s. of Hoffmann**, see *s., antiepitheliomatous*. **s., Huchard's**, sodium phosphate, 10 Gm.; sodium chloride, 5 Gm.; sodium sulphate, 2.5 Gm.; distilled water, to 100 Cc. **s., Huchard's** concentrated, sodium chloride, 5 Gm.; sodium phosphate, 10 Gm.; sodium sulphate, 2.5 Gm.; phenol, 1.5 Gm.; water, 100 Cc. **s., immune**, the serum of an immunized animal, containing a specific antibody. **s., inactivated**, see *inactivate*. **s., jequiritol**, an antitoxin prepared on the principle of Beyring's method, which has the power of rapidly and surely paralyzing the effects of jequiritol in the human system when applied locally in the conjunctival sac and when injected subcutaneously. **s., Kronecker and Lichtenstein's**, sodium chloride from 6 to 7.5 Gm.; sodium carbonate, 0.1 Gm.; water, 1000 Cc. **s. lactis**, whey. **s., Latta's**, sodium chloride, from 3 to 5 Gm.; sodium carbonate, 1.7 Gm.; water, 3400 Cc. **s., Leclerc's** (very strong), sodium chloride, 4 Gm.; sodium phosphate, sodium sulphate, of each, 0.5 Gm.; boiled distilled water, 100 Cc. **s., leukotoxic**, one which destroys the leukocytes. **s., v. Leyden's**, antistreptococcus serum taken from convalescent scarlatina patients. Cf. *s., Moser's*. **s., luetic**, emulsion of liver or kidney of a syphilitic fetus, cleared by filtration. **s.-lutein**, the pigment contained in the serum from the blood of most animals. **S., Luton's**, crystallized sodium phosphate, 5 Gm.; sodium sulphate, 10 Gm.; boiled distilled water, 100 Cc. **s., Maragliano's**, an antituberculous serum obtained from an ass or horse treated with repeated injections of tuberculous toxin. **s., Marmorek's**, a polyvalent serum obtained by the inoculation of animals with streptococci of various origin. **s., Mathieu's**, sodium sulphate, 6 Gm.; sodium phosphate, 4 Gm.; sodium chloride, 1 Gm.; glycerol, 20 Gm.; distilled water, to 100 Cc. **s., meningococcic**, serum used in the treatment of diplococcic cerebrospinal meningitis. **s., Moser's** (Paull), an antistreptococcus serum obtained by simultaneous inoculation of horses with several varieties of streptococci taken from the blood of scarlatina patients. **s., neurotoxic**, one which acts directly upon the nerve-tissues. **s., normal**, that of which 0.1 Cc. neutralizes 10 times the minimal lethal dose of a specific bacterial poison. **s., Paquin's**, an antitoxic serum of tuberculosis produced by successive inoculation of horses. It is injected in daily doses of from 10 to 150 drops. **s., Parascandolo's**, an immunizing serum produced by inoculation of animals with mixed cultures of streptococci and staphylococci. **s., physiological, s., pooled**, mixed serum from different individuals. **s., pneumococcic**, serum used in the treatment of pneumonia. See *s., Hayem's* (2). **s.s., polyvalent**, serums derived from animals infected by a number of different streptococci. **s., protective**, any immunizing serum. Cf. *s., Haffkine's*. **S., Renzi's**, iodine, 1 Gm.; potassium iodide, 3 Gm.; sodium chloride, 6 Gm.; water, 1000 Cc. **S., Richet and Héricourt's**. See *s., anticancerous*. **s., Roussel's**, sodium phosphate, 50 Gm.; water, 1000 Cc. **s., Roux's**. See *s., antidiphtheric*. **s., Sapellier's**, sodium chloride, 60 Gm.; potassium chloride, 5 Gm.; sodium carbonate, 31 Gm.; sodium phosphate, 4.5 Gm.; potassium sulphate, 3.5 Gm.; boiled

water, 100 Cc. s., **Schiess's,** sodium chloride, 75 Gm.; sodium bicarbonate, 50 Gm.; water, 1000 Ccf s., **Schwartz's,** sodium chloride, 6 Gm.; solution o. caustic soda, 2 drops; water, 1000 Gm. s., **Sclavo's,** cultures of pneumococci in egg-albumen. s., **seraphthin,** a proprietary prophylactic against foot-and-mouth disease. **s.-sickness,** see *s.-disease.* s. **specific.** See *s., immune.* s., **streptococcic,** serum used in the treatment of streptococcic septicemia and pyemia. s. **sublimatum,** one part of corrosive sublimate to from 50 to 100 parts of serum. It is used subcutaneously as an antiseptic and for impregnating bandages. s., **Syndmann's,** sodium chloride, 6 Gm.; sodium bicarbonate, 1 Gm.; water, 1000 Cc. **s.-therapy.** See *serotherapy.* **s.-thyroid,** serum used in the treatment of exophthalmic goiter. s. of **Tizzoni** and **Cattani,** obtained by evaporating in a vacuum the serum of an immunized horse. Each gram of the powdered residue corresponds to 10 Cc. of the serum. s., **Trunecek's,** for the treatment of symptoms caused by arteriosclerosis; sodium chloride, 4.92 Gm.; sodium sulphate, 0.44 Gm.; sodium carbonate, 0.21 Gm.; potassium sulphate, 0.4 Gm.; sodium phosphate, 0.15 Gm. This is given in hypodermatic injections of 1 Cc. every 3 or 4 days, increasing to 5 to 7 Cc., or in rectal injections of 35 Cc. **s.-unit.** See *unit.* s., **Vandervelde's.** 1. Sodium glycerophosphate, sodium chloride, of each, 3 Gm.; water, 1000 Cc. 2. Sodium chloride, potassium chloride, of each, 3 Gm.; sodium carbonate, 2.5 Gm.; sodium phosphate, 3 Gm.; potassium sulphate, 2 Gm.; water, to 100 Cc. s. **of Villiers and Wlaeff.** See *s., antiepitheliomatous.* s. **of Wlaeff,** see *s., antiepitheliomatous.* s., **Yersin's,** serum of a horse immunized by intravenous injection of a virulent culture of the plague bacillus.
**serumal** (*se'-roo-mal*). Relating to or derived from serum. s. **calculus,** a calculus formed about the teeth by exudation from diseased gums.
**serumuria** (*se-rum-ū'-re-ah*). Same as *albuminuria.*
**Servetus's circulation** (*ser-ve'-tus*) [Michael *Servetus,* Spanish physician, 1509-1553]. The pulmonary circulation.
**servol** (*ser'-vol*). An alcohol soap solution containing 12 per cent. of formaldehyde; it is used as a disinfectant.
**sesame** (*ses'-am-e*). See *sesamum.*
**sesamoid** (*ses'-am-oid*) [*sesame; eἶδος,* like]. Resembling a sesame-seed. s. **bone,** a small bone developed in a tendon subjected to much pressure. s. **cartilage,** small cartilages in the ale of the nose.
**sesamoiditis** (*ses-am-oi-di'-tis*) [*sesamoid; ιτις,* inflammation]. Inflammatory disease of the sesamoid bones of the fetlock of the horse.
**sesamum** (*ses'-am-um*) [*σήσαμον,* sesame]. A genus of plants of the order *Pedalineæ. S. indicum* and *S. orientale* yield a bland, sweetish oil. s., **oil of** (*oleum sesami*), sesame oil, employed like olive-oil. Syn., *benne oil; teel oil.*
**sesqui-** (*ses-kwe-*) [L.]. A prefix denoting one and one-half.
**sesquibasic** (*ses-kwe-ba'-sik*) [*sesqui,* one-half more; *base*]. Applied to salts formed from a tribasic acid by the replacement of three atoms of hydrogen by two of a basic element or radical.
**sesquih.** Abbreviation of *sesquihora,* an hour and a half.
**sesquioxide** (*ses-kwe-oks'-īd*) [*sesqui-;* ὀξύς, acid]. A compound of oxygen and another element, containing three parts of oxygen to two of the other element.
**sesquisalt** (*ses'-kwe-sawlt*) [*sesqui-; salt*]. A salt containing one and one-half times as much of the acid as of the radical or base.
**sessile** (*ses'-il*) [*sessilis,* from *sedere,* to sit]. Attached by a broad base; not pedunculated, as a *sessile* tumor.
**sesunc.** Abbreviation of *sesuncia,* an ounce and a half.
**set** [ME., *setten*]. 1. To reduce the displacement in a fracture and apply suitable bandages. 2. To harden; to solidify—as a cement or amalgam.
**seta** (*se'-tah*) [*seta,* a bristle: *pl., setæ*]. A stiff, stout, bristle-like appendage; a chæta, vibrissa.
**setaceous** (*se-ta'-se-us*) [*seta,* a hair, a bristle]. Bristly, bristling, bristle-shaped.
**setaria** (*se-ta'-re-ah*) [*seta,* a bristle]. A genus of grasses including millet, *S. italica.*
**setarin** (*se-ta'-rin*). A toxic glucoside isolated by E. F. Ladd, 1899, from millet, *Setaria italica.*

**Setchenow's inhibitory center** (*setsh'-en-of*). See *Setschenow.*
**setiform** (*se'-tif-orm*) [*seta,* a bristle; *forma,* form]. Bristle-like in shape.
**setigerous,** or **setiferous** (*se-tij'-er-us, se-tif'-er-us*) [*seta,* bristle; *gerere* or *ferre,* to bear]. Bearing bristles or stiff hairs.
**seton** (*se'-ton*) [*seta,* a bristle]. 1. A thread or skein of threads drawn through a fold of the skin, so as to produce a fistulous tract; it is used as a counterirritant. 2. The tract thus produced.
**setose** (*se'-tōs*). Beset with bristle-like appendages.
**Setschenow's inhibitory center** (*setch'-en-of*) [Ivan *Setschenow,* Russian physician]. A cerebral center for the inhibition of reflex movements, situated in the corpora quadrigemina and the medulla oblongata.
**setula** (*set'-ū-lah*) [dim. of *seta,* a bristle: *pl., setulæ*]. A diminutive bristle.
**sevadilla,** see *cebadilla.*
**seven-day fever.** Relapsing fever.
**seventh nerve.** The facial nerve. See *nerves.*
**sevetol** (*sev'-et-ol*). Trade name of a preparation of predigested animal and vegetable fats.
**seviparous** (*se-vip'-ar-us*) [*sevum; parere,* to produce]. Sebiferous; fat-producing.
**sevum** (*se'-vum*) [L.]. Suet. s. **præparatum,** the prepared suet of the U. S. P.
**sewage** (*sū'-aj*). The heterogeneous substances constituting the excreta and waste matter of domestic economy and the contents of drains. It consists mainly of putrescent animal and vegetable tissues, fecal matter, and urine—the latter in a state of ammoniacal fermentation—mixed with water or dissolved in it. **s.-farming,** the use of sewage as a manure. **sewer** (*sū'-er*). A canal for the removal of sewage. **s.-air throat,** acute tonsillitis. **s.-gas,** the mixture of air, vapors, and gases, which emanates from sewers. It varies greatly in respect to its pathogenic qualities. **s.-g. pneumonia.** See *pneumonia.*
**sewerage** (*sū'-er-aj*). 1. The collection and removal of sewage. 2. The system of pipes, etc., for the removal of sewage.
**sewing spasm.** See *seamstress's cramp.*
**sex-** (*seks*) [L.]. A numeral used as a prefix, meaning six.
**sex** (*seks*) [*sexus,* also *secus,* sex]. The state or condition of being either male or female.
**sexidigital, sexidigitate** (*seks-e-dij'-it-al, seks-e-dij'-it-āt*) [*sex,* six; *digitus,* a finger]. Having six fingers or six toes.
**sexivalent** (*seks-iv'-al-ent*) [*sex,* six; *valere,* to be worth]. Having an atomicity of six as compared with that of hydrogen.
**sextan** (*seks'-tan*) [*sex,* six]. Occurring every sixth day, as a *sextan* fever.
**sextigravida** (*seks-te-grav'-id-ah*) [*sextus,* sixth; *gravida,* pregnant]. A woman pregnant for the sixth time.
**sextipara** (*seks-tip'-ar-ah*) [*sextus,* sixth; *parere,* to produce]. A woman in labor for the sixth time.
**sextonol** (*seks'-to-nol*). Trade name of a mixture of the glycerophosphates of calcium, iron, manganese, quinine, sodium, and strychnine.
**sextuplet** (*seks-tū'-plet*) [*sex,* six]. One of six offspring of a single gestation.
**sexual** (*seks'-ū-al*) [*sexus,* sex]. Pertaining to or characteristic of sex, as the *sexual* organs. s. **bondage,** the dependence of one person upon another of the opposite sex that is abnormal but not perverse. s. **diseases,** diseases of the sexual organs. s. **intercourse,** copulation. s. **inversion,** a variety of sexual perversion in which there is an abnormal liking for a person of the same, instead of for one of the opposite sex. s. **involution,** the menopause. s. **metamorphosis,** a variety of sexual perversion in which the individual has the tastes and feelings and assumes the dress and habits of the opposite sex. s. **selection.** See *evolution.*
**sexuality** (*seks-ū-al'-it-e*) [*sexus,* sex]. The collective differences which in an individual make one male or female.
**sexually** (*seks'-ū-al-e*) [*sexus,* sex]. In a sexual manner.
**sexvalent** (*seks'-va-lent*). Sexivalent.
**shackle** (*shak-l*) [ME., *schakkyl,* shackle]. Something that hinders or confines. **s.-joint,** a variety of articulation formed by passing a bony ring of one part through a perforation of another part. It is seen in the exoskeleton of some fishes. **s.-vein,** a vein of

the horse, probably the median antebrachial, from which blood was formerly abstracted.

**shaddock** (*shad'-ok*) [Captain *Shaddock*, who introduced the tree into the West Indes from Java in the early part of the eighteenth century]. The fruit of *Citrus decumana*, grape fruit.

**shadow** (*shad'-o*) [ME., *schadowe*, shadow]. A phantom cell, or skeletonized blood-cell, formed by the removal of the hemoglobin from a red corpuscle. **s.-test.** See *retinoscopy*.

**shadowgram.** See *skiagram*.

**shadowgraph** (*shad'-o-graf*). See *skiagraph*.

**shaft.** The trunk of any columnar mass, especially the diaphysis of a long bone.

**shakes** (*shāks*). See *ague*.

**shaking** (*sha'-king*) [ME., *shaken*, to shake]. A passive Swedish movement used in the treatment of nervous affections. **s. cure,** the treatment of disease by a shaking or vibratory movement, advocated by Charcot in paralysis agitans, by means of a vibrating arm-chair. **s. palsy.** See *paralysis agitans*.

**shampoo** (*sham-poo'*) [Hind., *tshanpna*, shampoo]. 1. Synonym of *massage*. 2. To lather, rub, or wash the head.

**shampooing** (*sham-poo'-ing*). The performance of massage with the application of a liniment or other medicinal substance, and also in connection with the Turkish bath.

**shank** (ME., *shanke*, the chief bone of the leg]. The leg from the knee to the ankle; the tibia or shin-bone.

**share-bone.** The os pubis.

**Sharpey's intercrossing fibers** (*shar'-pe*) [William *Sharpey*, English anatomist, 1802-1880]. The collagenous fibers forming the lamellæ which constitute the walls of the Haversian canals in bone; same as osteogenic fibers. **S.'s perforating fibers,** calcified white or elastic fibers which connect the lamellæ to the walls of the Haversian canals.

**shaven-beard appearance.** A peculiar appearance of the agminated glands of the intestine in typhoid fever, resembling that of a recently shaven beard.

**shawl-muscle,** the trapezius.

**shears** (*shērs*) [ME., *sherēs*, shears]. A large pair of scissors. **s. bandage,** strong shears for cutting bandages, usually bent at an angle.

**sheath** (*shēth*). An envelope; a covering. In anatomy, applied to the coverings of arteries, muscles, nerves, fascia, etc. **s. arachnoidean,** a delicate partition lying between the pial sheath and the dural sheath of the optic nerve. **s., axis-cylinder.** See *Huxley's layer*. **s., capillary,** or **s., circumvascular,** a wide lymphatic tube surrounding some of the smallest blood-vessels. **s. cellular.** See *epineurium*. **s., cortical,** the bast-bundles. **s., crural,** the femoral sheath; see under *femoral*. **s., dentinal,** the structure lining the dentinal canals. **s., dural,** a strong fibrous membrane forming the external investment of the optic nerve; see *dura*. **s., femoral.** See *femoral*. **s., fibril,** a sheath formed by connective-tissue fibrils and surrounding individual nerve-fibers. **s. of Henle.** 1. An attenuated extension of the perineurium investing the fibers composing funiculi of a nerve-trunk; it consists of a delicate fibrous envelope lined with endothelial plates, which in some cases alone represent the entire sheath. 2. The cellular layer forming the outer portion of the inner root-sheath of the hair. **s., Huxley's.** See *Huxley's layer*. **s., lamellar.** See *perineurium*. **s. of Mauthner,** a protoplasmic investing membrane beneath the neurilemma and the nodes of Ranvier, passing inward to separate the myelin from the axis-cylinder. **s., medullary,** the myelin-sheath surrounding the axis-cylinder. **s., myelin,** *medullary sheath* and *neurilemma*. **s., nerve.** See *perineurium*. **s., Neumann's.** See *dentinal*. **s., neural.** See *s., medullary*. **s., perivascular.** See *s., capillary*. **s. of the optic nerve.** See *s., dural*. **s., pial,** the extension of the pia which closely invests the surface of the optic nerve. See *pia*. **s., primitive.** See *neurilemma*. **s. of rectus,** that formed by the aponeurosis of the external and internal oblique muscles and the transversalis. **s., root.** See *root sheath*. **s. of Schwann.** See *neurilemma*. **s., Schwalbe's,** the delicate sheath which covers elastic fibers. **s., synovial,** a synovial membrane which lines the cavity attached to a bone and through which a tendon glides. **s., tangential,** the fibro-cellular sheath surrounding the carotids.

**shed.** To throw off.

**shedding** (*shed'-ing*). Throwing off. **s. teeth,** the teeth of the first dentition; the term is also applied to the loss of the first or temporary set of teeth.

**sheep-pox.** A contagious pustular disease of sheep similar to cow-pox.

**sheet** (*shēt*). A large piece of linen or cotton used as bed-clothing. **s.-bath.** See *bath*. **s., draw, a sheet** so folded as to be placed, or removed, from beneath the patient with the least inconvenience.

**Sheldon's method of hemostasis in disarticulation of the hip-joint.** Consists in a preliminary disarticulation of the head of the femur, followed by the introduction of the artery forceps into the wound behind the femur and clamping of the femoral vessels.

**shellac** (*shel'-ak*). See *lac*. **s. cement,** see *cement*.

**shells** (*shelz*) [ME., *schelle*, shell]. Tinted spectacles, for protection of the eyes. Coquilles.

**Shepherd's fracture.** A fracture of the outer portion of the astragalus.

**sherbet** (*shur'-bet*) [Pers. *sharbat*]. An oriental, cooling drink made from fruit juices and water, sweetened, flavored, and iced with mountain snow.

**Sherrington's law** [Charles Scott *Sherrington*, English physiologist]. The peripheral branches of the spinal nerve-roots—anterior and posterior—form anastomoses in such a manner as to supply any given region of the integument with the branches of three roots—a middle one and the ones next above and below. **S.'s solution,** *for use with the hemocytometer:* methylene blue, 0.1 Gm.; sodium chloride, 1.2 Gm.; neutral potassium oxalate, 1.2 Gm.; distilled water, 300 Cc.

**sherry wine** (*sher'-e*). See *vinum xericum*.

**shield** (*shēld*) [ME., *sheeld*]. 1. A protective stiucture or apparatus. 2. In biology, a protective plate, scute, lorica, or carapace. **s., antithermic,** a protective covering of the cautery to prevent destruction of the tissues about the field of operation. **s. bone,** the scapula. **s., nipple-,** a protective covering for sore nipples. **s.-shaped,** shaped like a buckler or shield. **s., Sims',** an instrument used in the application of wire sutures.

**Shiga's bacillus** (*she'-ga*) [K. *Shiga*, Japanese bacteriologist]. The *bacillus dysenteriæ*; also called the *Shiga-Krause bacillus*.

**shikimi, shikimia.** See *sikimin*.

**shima-mushi** (*shi-mah-mush'-e*) [Jap.]. A Japanese febrile disease supposed to be due to the bite of an insect.

**shimu-mushi.** See *shima-mushi*.

**shin** [AS., *scina*]. The sharp anterior margin of the tibia. **s.-bone,** the tibia.

**shingles** (*shing'-glz*). Herpes zoster.

**ship-fever.** Typhus fever.

**shirt-stud abscess.** See *abscess*.

**shiver** (*shiv'-er*) [ME., *chiveren*, to shiver]. A slight tremor or shaking of the body due to cold, etc.

**shock** [Fr., *shoc*]. 1. A sudden grave depression of the system produced by operations, accidents, or strong emotion. It is due to a profound influence on the nervous system. If not fatal, it is followed by a stage of reaction. 2. The agent causing a general or local depression, as an *electric shock*. **s., deferred,** that curious condition in which the manifestations of shock, due not to severe bodily injury but to purely mental causes develop after the lapse of some time from the occurrence. This variety of shock may be even more profound than that produced by bodily injury. **s., discharging,** a shock produced by a discharge of electricity. **s., electric,** the physiological effect produced upon an organism by the opening or closing of an electric circuit in which it is included. **s., epigastric,** the result of a blow upon the epigastrium. **s., erethismic,** a form of shock attended with symptoms of excitement. **s., fetal,** the sensation produced by movements of the fetus in utero. **s., railway,** the mental impression produced by a railway accident. **s., secondary,** or **insidious,** a second attack occurring after the first. **s., sexual,** shock caused by rape or coitus. **s., torpid,** shock in which marked depression is a prominent symptom. **s., traumatic,** shock due to traumatism.

**shoddy fever.** A diseased condition caused by the inhalation of the dust in shoddy factories; it is characterized by feverishness, headache, nausea, dryness of the mouth, dyspnea, cough, and expectoration.

**shoemaker's spasm.** An occupation-neurosis, analogous to writer's cramp, occurring in shoemakers.

**short circuit.** One in which an electric current encounters an abnormally small resistance. **s.-circuiting,** a modification of Nélaton's operation for in-

testinal obstruction consisting either in lateral approximation and union or lateral implantation. s.-sight, s.-sightedness, myopia. s.-windedness, dyspnea.
**shot-gun prescription**, one with many ingredients, written with the expectation that some one may prove curative. s.-gun quarantine, the extemporized and unauthorized establishment of a cordon against a place suspected of being the seat of an epidemic of a communicable disease. s.-silk. See *retina*.
**shoulder** (*shōl-'der*) [AS., *sculder*]. The region where the arm joins the trunk, formed by the meeting of the clavicle and the scapula and the overlying soft parts. s.-blade, the scapula. s.-girdle. See *girdle, shoulder-*. s., **noisy,** of R. H. Sayre, a grating of the muscles over the scapula on moving the shoulder up and down, believed to be due to a snapping tendon between or a bursa beneath the scapula. s., **slipped,** s., **splayed,** a dislocated shoulder. s.-wrench, a sprain or dislocation of the shoulder.
**show** (*shō*). 1. A bloody discharge from the birth-canal prior to labor. 2. The first appearance of a menstrual flow.
**shower** (*shour*) [ME., *shour*, shower]. A light fall of rain. s.-bath, see *bath*. s.-bath, electric, see *bath, electric* s., **uric acid,** a temporary increase in the amount of uric a.id in the urine; it occurs in gouty patients.
**Shrapnell's membrane** (*shrap'-nel*) [Henry Jones *Shrapnell*, English anatomist]. A small portion of the drum-membrane filling the notch of Rivinus.
**shreds.** Patches of filmy material passed with the fecal discharges in some cases of enteritis and diarrhea. They may be composed of false membrane, or actual sloughs from the intestinal mucosa, or of flakes of hardened mucus.
**shrivel** (*shriv'-l*). To shrink in bulk and become wrinkled.
**shucks.** A strong tea of corn-shucks, used as a remedy for chronic malaria in the southern United States.
**shudder** (*shud'-er*). A convulsive but momentary tremor, caused usually by fright, disgust, or nervous shock.
**shunt.** In electricity, a conductor of low resistance, joining two points in an electric current, and completing a path through which the current will pass.
**shuttle-bone.** The scaphoid bone.
**Si.** The chemical symbol of *silicon*.
**siagantritis** (*si-ag-an-tri'-tis*). See *siagonantritis*.
**siagon** (*si'-ag-on*) [σιαγών, jaw-bone]. In biology, the mandible of a crustacean.
**siagonagra** (*si-ag-on-a'-grah*) [σιαγών, jaw-bone; ἄγρα, seizure]. Gouty pain in the maxilla.
**siagonantritis** (*si-ag-on-an-tri'-tis*) [σιαγών, the jaw-bone; ἄντρον, autrum]. Inflammation within the antrum of Highmore.
**sialaden** (*si-al'-ad-en*) [σίαλον, saliva; ἀδήν, a gland]. A salivary gland.
**sialadenitis** (*si-al-ad-en-i'-tis*) [σίαλον, saliva; ἀδήν, gland; ιτις, inflammation]. Inflammation of a salivary gland.
**sialadenoncus** (*si-al-ad-en-ong'-kus*) [σίαλον, saliva; ἀδήν, gland; ὄγκος, a tumor]. A tumor of a salivary gland.
**sialagogue, sialagog** (*si-al'-a-gog*). See *sialogogue*.
**sialaporia** (*si-al-ap-o'-re-ah*) [σίαλον, spittle]. Deficiency in the amount of saliva.
**sialemesis** (*si-al-em-e'-sis*) [σίαλον, saliva; *emesis*]. The hysterical vomiting of saliva.
**sialic, sialine** (*si-al'-ik, si-al-ēn*) [σίαλον]. Having the nature of saliva.
**sialism, sialismus** (*si'-al-izm, si-al-iz'-mus*). See *ptyalism*.
**sialodochitis** (*si-al-o-do-ki'-tis*) [σίαλον, spittle; δοχεῖον, receptacle; ιτις, inflammation]. Inflammation of the salivary ducts. s. **fibrinosa,** inflammation of a salivary duct obstructed by a fibrinous exudate.
**sialodochium** (*si-al-o-do'-ke-um*) [σίαλον, saliva; δοχεῖον, receptacle]. A salivary duct.
**sialoductitis** (*si-al-o-duk-ti'-tis*). Inflammation of tenson's duct.
**sialogenous** (*si-al-oj'-en-us*) [σίαλον, spittle; γεννᾶν, to produce]. Generating saliva.
**sialogogic** (*si-al-o-goj'-ik*) [σίαλον, spittle; ἀγωγός, leading]. 1. A sialogogue. 2. Promoting a flow of saliva.
**sialogogue, sialogog** (*si-al'-o-gog*) [σίαλον, spittle; ἀγωγός, leading]. 1. Producing a flow of saliva. 2. A drug producing a flow of saliva.

**sialoid** (*si'-al-oid*) [σίαλον, spittle; εἶδος, like]. Pertaining to, or like saliva.
**sialolith** (*si'-al-o-lith*) [σίαλον, spittle; λίθος, stone]. A salivary calculus.
**sialolithiasis** (*si-al-o-lith-i'-as-is*) [σίαλον, spittle; λίθος, stone]. The presence of salivary calculi.
**sialon** (*si'-al-on*). Saliva.
**sialoncus** (*si-al-ong'-kus*) [σίαλον, spittle; ὄγκος, a tumor]. A tumor under the tongue, arising from the obstruction of the duct of a salivary gland by calculus or other cause.
**sialorrhea, sialorrhoea** (*si-al-or-e'-ah*) [*sialon*; ῥοία, a flow]. Salivation. s., **pancreatic,** a flow of pancreatic juice.
**sialoschesis** (*si-al-os'-kes-is*) [σίαλον, spittle; σχέσις, holding]. Suppression of the secretion of saliva.
**sialosemeiology** (*si-al-o-se-mi-ol'-o-je*) [*sialon*; *semeiology*]. Diagnosis based upon examination of the saliva.
**sialostenosis** (*si-al-o-ste-no'-sis*) [σίαλον, spittle; στένος, narrow]. Occlusion of a salivary duct.
**sialosyrinx** (*si-al-o-si'-ringks*) [σίαλον, saliva; σῦριγξ, tube]. 1. A salivary fistula. 2. A syringe for washing out the salivary ducts. 3. A drainage-tube for the salivary ducts.
**sialozemia** (*si-al-o-ze'-me-ah*) [σίαλον, spittle; ζημία, loss]. Loss of saliva; salivation.
**sibbens** (*sib'-enz*) [Gael., *subhan*, raspberries]. A disease formerly endemic in the Scotch highlands, and by some identified with syphilis, by others with yaws.
**sibilant** (*sib'-il-ant*) [*sibilare*, to hiss]. Hissing or whistling, as a *sibilant* rale.
**sibilation** (*sib-il-a'-shun*) [*sibilare*, to hiss]. Pronounciatin in which the *s* sound predominates.
**sibilismus** (*sib-il-is'-mus*) [*sibilare*, to hiss]. 1. A hissing sound. 2. A sibilant rale. s. **aurium,** tinnitus aurium.
**sibilus** (*sib'-il-us*) [*sibilare*, to hiss]. A sibilant rale.
**Sibson's aortic vestibule** (*sib'-sun*) [Francis *Sibson,* English physician, 1814–1876]. The chamber formed by the left ventricle just below the aortic orifice for the reception of the semilunar valves during diastole. S.'s **groove,** a furrow formed in some individuals by a prominence of the lower border of the pectoralis major. S.'s **notch,** the inward curve of the upper left border of precordial dulness in acute pericardial effusion.
**siccant, siccative** (*sik'-ant, sik'-at-iv*) [*siccare*, to dry]. 1. Drying; tending to make dry. 2. A drying agent or medicine.
**sicchasia** (*sik-a'-se-ah*) [σικχαίνειν, to feel disgust]. 1. Morbid loathing of food. 2. Nausea. 3. Nausea of pregnancy.
**sicco** (*sik'-o*). Dried hematogen (*q. v.*); a blackbrown, tasteless powder, soluble in water, indicated in anemia, chlorosis, etc. Dose 75–105 gr. (5–7 Gm.) daily; children 4 gr. (0.25 Gm.).
**siccolabile** (*sik-o-lab'-il*) [*siccus,* dry; *labile*]. Liable to be altered or destroyed by drying.
**siccostabile** (*sik-o-sta'-bil*) [*siccus*, dry; *stabile*]. Not altered by drying.
**siccus** (*sik'-us*) [L.]. Dry.
**sick** (*sik*) [ME., *sik*, sick]. 1. Ill; not well. 2. Nauseated, or "sick at the stomach." 3. Menstruating. s. **time,** popularly used for the period of menstruation. s.-**headache,** headache with anorexia, nausea, vomiting, etc.; migraine. s.-**list,** a list of persons, especially in military or naval service, who are disabled by sickness. s.-**report,** a sick-list. s.-**room,** a room occupied by one who is sick. s.-**stomach,** synonym of *nausea*, and of *milk-sickness*.
**Sickingia** (*sik-in'-je-ah*) [Count v. *Sickingen,* of Vienna]. A genus of rubiaceous plants. *S. rubra*, casca de araribá, is a species found in Brazil and Japan furnishes arariba bark, used in intermittent fever. It contains the alkaloid *aribine* and a red coloring-matter. *S. viridiflora,* casca de araribá branca, of Brazil, furnishes a bark used in malaria.
**sickle-germs** (*sik'-l-jermz*). A falciform stage in the development of *Coccidia*.
**sickliness** (*sik'-le-nes*) [ME., *sik,* sick]. Predisposition to easily contract disease; insalubrity of climate.
**sickly** (*sik'-le*) [ME., *sik,* sick]. Predisposed to disease. Unhealthy.
**sickness** (*sik'-nes*) [*sick*]. 1. The state of being unwell. 2. Nausea. 3. Menstruation. s., **African horse-**. See *edemamycosis*. s., **African sleeping-**. See

*African lethargy.* s., **bleeding,** hemophilia. s., **Ceylon,** beriberi. s., **country,** nostalgia. s., **creeping,** chronic ergotism. s., **falling,** epilepsy. s., **green,** chlorosis. s., **jumping,** a form of choromania. See **jumpers.** s., **leaguer,** typhus. ' s., **milk,** a form of poisoning due to the ingestion of diseased milk or meat. s., **miners'.** See ankylostomiasis. s., **monthly,** the menstrual epoch. s., **mountain-,** a sensation of nausea, with impeded respiration and irregular heart's action due to the rarefied air of high altitudes. s., **painted.** See *pinta disease.* s., **railway.** See *car-sickness.* s., **salt,** a condition of starvation due to animals being confined on poor pastures consisting of dry wire grass and other inferior vegetation. s., **sea-.** See *sea-sickness.* s., **serum.** See *serum disease.* s., **spotted.** See *pinta disease.* s., **sweating.** See *sweating-sickness.* s., **theater,** Paul's name for malaise with dyspnea and oppression followed by weak pulse and syncope, usually observed in women who have dined hurriedly and reached a crowded theater in a heated condition; frequently a result of eye-strain. See *vertigo, stomachal.*

**sicopirin, sicopyrin** (*sik-o-pi'-rin*). C₁₈H₁₃O₅. A glucosidal body found by Peckolt in the root-bark of *Bowditchia virgiloides.*

**Sida** (*si'-dah*) [σίδη, a malvaceous plant]. A genus of plants of the order *Malvaceæ.* S. *paniculata,* a species of Peru, is an active vermifuge. Its action is believed to be due to the very minute but resisting bristles which cover its leaves. *S. rhombifolia,* Queensland hemp, containing a great amount of mucilage, is used in Australia for snake-bite, pulmonary complaints, and in making poultices.

**side** (*sid*) [ME.]. A lateral half of the body or of any bilateral organ. **s.-bone.** 1. The hip-bone. 2. The diseased or disordered condition in horses which causes the lateral cartilages above the heels to ossify. 3. An abnormal ossification of the lateral elastic cartilage in a horse's foot. **s.-chain,** see *receptor.* **s.-chain theory,** see under *lateral* and under *immunity.*

**siderante, siderante** (*sid'-er-ant, sid-er-an'-te*) [*siderari,* to be blasted or planet struck]. Characterized by sudden and abrupt onset as though the result of malign astral influences.

**sideration** (*sid-er-a'-shun*) [*sideratio,* blight produced by the stars]. 1. Apoplexy. 2. Gangrene. 3. Lightning-stroke. 4. Therapeutic application of electric sparks.

**siderism** (*sid'-er-izm*) [*siderites,* the lodestone]. The curative influences long supposed to be exerted over the body by the lodestone; metallotherapy.

**siderodromophobia** (*sid-er-o-dro-mo-fo'-be-ah*) [σίδηρος, iron; δρόμος, way; φόβος, fear]. Morbid dread of traveling by railway.

**siderophilous** (*sid-er-off'-il-us*) [σίδηρος, iron; φιλεῖν, to love]. Applied to cells that show a tendency to take up iron, e. g., the red blood-corpuscles.

**siderophone** (*sid'-er-o-fōn*) [σίδηρος, iron; φωνή, sound]. An electric appliance devised by Martin Jannson (1902) as an improvement upon Asmus' sideroscope for detecting the presence of small splinters of iron.

**sideroscope** (*sid-er-o-skōp'*) [σίδηρος, iron; σκοπεῖν, to examine]. An instrument for the detection of particles of iron or steel in the eye.

**siderosis** (*sid-er-o'-sis*) [σίδηρος, iron]. 1. A pigmentation by a deposit of particles of iron; specifically, a chronic interstitial pneumonia caused by the inhalation of particles of iron. 2. A recognized type of lung disease (pneumokoniosis) due to the inhalation of metallic dust. Cf. *anthracosis; chalicosis;* **silicosis.** 3. An excess of iron in the system.

**sidonal** (*si'-don-al*). See *piperazin quinate.* s., **new,** quinic acid anhydride, a white, tasteless powder, soluble in water, used as a uric-acid solvent. Dose 7½–120 gr. (.5–8 Gm.) daily, given in 4 or more doses.

**Siebold's operation** (*se'-bolt*) [Eduard Caspar Jacob von *Siebold,* German surgeon, 1801–1861]. Hebotomy, *q. v.*

**Siegle's otoscope,** S.'s **speculum** (*se'-gleh*) [Emil *Siegle,* German otologist, 1833– ]. An instrument consisting of a glass-covered box with a conical projection and a rubber tube attached laterally. When the conical projection is inserted firmly into the external auditory canal, and the air is compressed or rarefied, the movements of the drum-membrane may be observed. It is also used for the purpose of rendering the articulations of the ossicles mobile.

**Siemerling's nucleus** (*se'-mer-ling*). The anteroventral nucleus of the anterior group of oculomotor nuclei in the gray matter below the Sylvian aqueduct.

**Sieur's sign.** "Signe du sou." A clear, metallic sound sometimes heard in cases of pleural effusion on percussing the chest in front with two coins and auscultating behind.

**sieve** (*siv*). A vessel with a reticulated bottom, used for the separation of pulverized from coarse substances. The gauge of the sieve is usually expressed in the number of meshes per square inch. **s., bone,** the ethmoid bone. **s.-cells,** long cells of tubular or prismatic form constituting an essential element in fibro-vascular bundles of the inner bark of exogenous stems. They are peculiar in the possession of circumscribed panels, with fine perforations, which allow of communication between contiguous cells. **s.-disc.** See *s.-plate.* **s.-hypha,** a hypha in which sieve-plates occur. **s.-plates,** the perforated panels of sieve-cells occurring at the points of contact of sieve-cells. **s.-pores,** the perforation in the panels of sieve-cells. **s.-tissue,** a cellular tissue made up of thin-walled cells which possess areas with sieve-like markings. The tissue is characteristic of the phloem. **s.-tubes.** See *s.-cells.* **s.-vessel.** See *s.-cells.*

**sig.** (*sig*). Abbreviation for *signa,* "label it," or for "*signetur,*" "let it be labeled."

**Sigault's, Sigaultian Operation** (*se-go, se-go'-shun*) [Jean Réné *Sigault,* French obstetrician]. Symphyseotomy.

**sigh** (*si*) [AS., *sican,* to sigh]. A prolonged and deep inspiration followed by a shorter expiration. Syn., *suspirium.*

**sighing** (*si'-ing*) [ME., *sighen,* to sigh]. 1: The act of giving forth a sigh. 2. A deep respiration accompanied by sighs. 3. Characterized by sighs.

**sight** (*sit*) [AS., *sihti*]. The act of seeing; the special sense concerned in seeing. **s., day-,** hemeralopia. **s., far-,** s., **long,** hyperopia. **s., night-,** nyctalopia. **s., old,** presbyopia. **s., short-,** myopia. **s., weak,** asthenopia.

**sigillation** (*sij-il-a'-shun*) [*sigillum,* a seal]. The mark of a cicatrix.

**sigmatism** (*sig'-mat-izm*) [*sigmoid*]. 1. Defective utterance of the sound of *s.* 2. The too frequent use of the *s* sound in speech.

**sigmoid** (*sig'-moid*) [σίγμα, the Greek σ; εἶδος, likeness]. 1. Shaped like the letter S. 2. Pertaining to the sigmoid flexure of the colon, as the *sigmoid artery,* the *sigmoid* mesocolon. **s. catheter,** one shaped like an S, for passing into the female bladder. **s. cavities,** two depressions on the head of the ulna; the *greater* is for articulation with the humerus; the *lesser,* on the outer side of the coronoid process, is for articulation with the radius. **s. flexure,** an S-shaped bend in the colon between the descending portion and the rectum, usually occupying the left iliac fossa. **s. fossa,** an S-shaped groove on the mastoid process. **s. gyrus,** the S-shaped cerebral fold about and behind the cruciate fissure in carnivora. **s. mesocolon,** the fold of the peritoneum attaching the sigmoid flexure of the colon to the left iliac fossa. **s. n t h,** see *n. valves,* the cardiac semilunar valves.

**sigmoidectomy** (*sig-moi-dek'-to-me*) [*sigmoid* (2); ἐκτομή, excision]. Excision of a part of the sigmoid flexure of the colon.

**sigmoiditis** (*sig-moi-di'-tis*) [*sigmoid;* ιτις, inflammation]. Inflammation of the sigmoid flexure of the colon.

**sigmoido-** (*sig-moi-do-*) [*sigmoid*]. A prefix denoting relation to the sigmoid flexure.

**sigmoidopexy** (*sig-moid'-o-peks-e*) [see *sigmoido-;* πῆξις, a fixing]. An operation for prolapse of the rectum by fixation of the sigmoid flexure.

**sigmoidoproctostomy** (*sig-moid-o-prok-tos'-to-me*) [*sigmoido-; proctostomy*]. Anastomosis of the sigmoid flexure with the rectum.

**sigmoidoscope** (*sig-moid'-o-skōp*) [*sigmoido-;* σκοπεῖν, to view]. An appliance for the inspection of the sigmoid flexure; it differs from the proctoscope in its greater length and diameter.

**sigmoidoscopy** (*sig-moid-os'-ko-pe*) [see *sigmoidoscope*]. Visual inspection of the sigmoid flexure with the aid of special instruments.

**sigmoidostomy** (*sig-moid-os'-to-me*) [*sigmoido-;* στόμα, mouth]. The formation of an artificial anus in the sigmoid flexure of the colon.

**sign** (*sin*) [*signum,* a mark]. 1. A mark or evidence; in a restricted sense, a physical sign. 2. A conventional character used in pharmacy or other-

**wise.** s., **accessory**, a non-pathognomonic sign. s., **antecedent**, a sign which precedes an attack of a disease. s., **assident**, same as *sign, accessory*. s., **cling.** See *Gersuny's symptom*. s., **coin.** See *bell sound*. s., **commemorative**, a sign of some previous disease. s., **echo**, the involuntary repetition of the last syllable, word, or clause of a sentence. s.-**language**, the method of intercommunication employed by deaf-mutes, in which ideas are communicated by means of signs. s., **objective**, one apparent to the observer. s., **palmoplantar.** See *Filipowitch's sign*. s.s, **physical**, the symptoms derived from auscultation, percussion, etc. s., **subjective**, one recognized only by the patient.

**signa** (*sig'-nah*) [*sign*]. Mark. In prescription-writing, a term placed before the physician's directions to the patient concerning the medicine prescribed; abbreviated to S. or Sig.

**signal** (*sig'-nal*) [*signum*, a sign]. A sign. s., **Marcel Duprez'**, the interruption of an electric current produced by a tuning-fork of 100 vibrations per second.

**signaletic** (*sig-nal-et'-ik*). Relating to signalization.

**signalization, signalment** (*sig-nal-i-za'-shun, sig'-nal-ment*). See *Bertillonage* and *identification*.

**signatura** (*sig-nat-ū'-rah*) [L.]. 1. Signature. 2. A characteristic mark. 3. The directions showing how medicines are to be taken.

**signature** (*sig'-nat-ūr*) [*sign*]. 1. The part of the prescription that is to be placed on the label. 2. A distinguishing character. **signatures, doctrine of**, a theory that the medicinal uses of plants or other objects can be determined from the signatures or peculiar characters.

**Signorelli's sign** (*sēn-yor-el'-e*) [Angelo *Signorelli*, Italian physician]. Pressure on the glenoid fossa, in front of the mastoid process, causes pain in cases of meningitis.

**signum** (*sig'-num*) [L., *pl., signa*]. A mark, sign, or indication.

**siguatera** (*sig-wah-te'-rah*) [Sp., "fish-poisoning"]. 1. The name given by Spanish colonists to a complex of symptoms that resulted from eating poisonous fishes indigenous to certain hot countries. 2. Poisoning from the ingestion of fresh food uninfected by bacteria, but in which the toxin is a leukomaine formed by the physiological activity of the tissues.

**sikimin** (*sik'-im-in*) [*sikkim*, a region of the Himalaya]. A poisonous principle derived from *Illicium religiosum*.

**silbamine** (*sil'-bam-ēn*). Fluoride of silver, used for irrigating the urethra and bladder.

**silberol** (*sil'-ber-ol*). See *silver paraphenolsulphonate*.

**silex** (*si'-leks*) [L.]. See *silica*.

**Silex's sign** (*si'-leks*). Radial furrows about the mouth, and coincidently in other parts of the face; a pathognomonic sign of congenital syphilis. S.'s **test for glucose in urine**. Add ammonia in excess to a strong solution of silver nitrate; add the urine, and boil. In the presence of glucose a metallic silver mirror is deposited at the bottom of the tube. Aldehyde and tartaric acid give the same reaction.

**silica** (*sil'-ik-ah*) [*silex*, flint]. Silicon dioxide, SiO₂, occurring in nature in the form of quartz, flint, and other minerals.

**silicate** (*sil'-ik-āt*) [*silica*]. A salt of silicic acid.

**silicic acid** (*sil-is'-ik*) [*silica*), H₄SiO₄. A tetrabasic acid forming the silicates. See *sodium silicate*.

**silicide** (*sil'-is-id*) [*silex*, flint]. A combination of silicon with another element.

**silicious, siliceous** (*sil-ish'-us*) [*silex*, flint]. Having the nature of or containing silicon.

**silicium** (*sil-ish'-e-um*). See *silicon*.

**silicofluoride** (*sil-ik-o-flu'-o-rid*). A compound of silicon and fluorine with some other element.

**silicol** (*sil'-ik-ol*) [*silex*, flint]. An alcohol from a silicon or silicon-carbon radical.

**silicon** (*sil'-ik-on*) [*silica*]. A nonmetallic element occurring widely distributed in nature as silica, SiO₂, and in the form of silicates. Atomic weight 28.3; symbol Si; valence IV. It resembles carbon in its chemical behavior. s. **carbide**, a compound prepared by heating in an electric furnace silica and carbon in the presence of salt. Next to the diamond it is the hardest substance known. The pure salt forms colorless, transparent laminæ of diamond-like luster. Its specific gravity is 3.22 and its index of hardness 9.5. s. **tetracetate**, Si(O.C₂H₃O)₄, aceto-orthosilicic anhydrid, a substance occurring in prismatic crystals.

**silicosis** (*sil-ik-o'-sis*) [*silica*]. A deposit of particles of silica in the tissues; specifically, a chronic fibroid condition of the lung or the bronchial lymphatic glands, produced by the inhalation of particles of silica.

**siliqua** (*sil-ik'-wah*) [*siliqua*, a husk or pod; pl., *siliquæ*]. Same as *silique*. s. **olivæ**, s. **olivæ externa**, the nerve-fibers encircling the olive. s. **olivæ interna, dentoliva.**

**silique** (*sil-ēk'*) [*siliqua*, a husk or pod]. In biology, the slender, two-valved capsule of some *Cruciferæ*.

**siliquose** (*sil'-ik-wōs*). Resembling a silique. s. **cataract**, see *cataract*.

**silk** [ME.]. The simplest and most perfect of the textile fibers. It differs from all other fibers in that it is found in nature as a continuous fine thread. Silk is the product of the silkworm (*Bombyx mori*), and is simply the fiber that the worm spins around itself for protection when entering the pupa or cyrysalis state. The silk-fiber consists, to the extent of rather more than half its weight, of *fibroin*, C₁₅H₂₃N₅O₆, a nitrogenous principle. Covering this is the silk-glue, or *sericin*, C₁₅H₂₅N₅O₈. The most important physical properties of the silk-fiber are its luster, strength, and avidity for moisture. Besides the true silk, we have several socalled "wild silks," the most important of which is the *tusser silk*, the product of the larva of the moth, *Antherœa mylitta*, found in India. The cocoons are much larger than those of the true silkworm, are egg-shaped, and of a silvery drab color. The cocoon is very firm and hard, and the silk is of a drab color. It is used for the buff-colored Indian silks, and latterly largely in the manufacture of silk plush. Other wild silks are the *eria silk* of India, the *muga silk* of Assam, the *atlas* or *fagara silk* of China, and the *yama-mai silk* of Japan. Silk has been used as a hemostatic. s., **dentists'**, silk containing some vesicant. s. **floss, dentists'**, untwisted filaments of fine silk prepared expressly for the purpose of cleaning the surfaces of the teeth, and used by some dentists for finishing the surfaces of fillings in the sides of teeth. s. **gelatin**, a glutinous mass formed by boiling certain kinds of raw silk in water. It is used in bacteriology as a culture-medium for the majority of bacilli of water and air. s.-**grass**, pineapple fiber. s., **saddler's**, a heavy silk used by saddlers and to some extent in surgery. s., **Tait's**, cable twist; it differs from ordinary silk in containing the gums or animal matter imparted by the worm in the spinning process.

**silkworm-gut.** The thread drawn from the silkworm killed when ready to spin the cocoon.

**sillonneur** (*sil-on-ur'*) [Fr.]. A three-bladed scalpel used by ophthalmologists.

**silphologic** (*sil-fol-oj'-ik*) [σιλφη, an insect; λόγος, science]. Larval.

**silphology** (*sil-fol'-o-je*) [σιλφη, an insect; λόγος, science]. The morphology and development of larvæ.

**silver.** See *argentum*. s. **arsenite**, Ag₃AsO₃, an alterative and antiseptic; used in skin diseases. Dose $\frac{1}{160}$ to $\frac{1}{60}$ gr. (0.0006–0.0011 Gm.). s.-**casein**, a fine white powder, soluble in hot water, obtained from sodium casein by action of silver nitrate and alcohol. It is used in gonorrhea in 2 to 10% solutions. s. **chloride**, AgCl, a white powder, soluble in ammonium, potassium thiosulphate, or potassium cyanide. It is used as an antiseptic and a nerve-sedative. Dose $\frac{1}{8}$–$1\frac{1}{2}$ gr. (0.02–0.5–0.1 Gm.). Syn., *horn-silver*; *Luna cornea*. s. **citrate**, Ag₃C₆H₅O₇, a fine dry powder soluble in 3800 parts of water, used as a surgical antiseptic and disinfectant. Application 1 to 2% ointment or 1 : 4000 solution. Syn., *itrol*. s., **colloidal**, a form of metallic silver consisting of heavy greenish-black particles of metallic luster which, when triturated with water, form a greenish-black fluid. It is used in the treatment of septic diseases, applied in the form of an ointment. Syn., *argentum colloidale*; *argentum Credé*; *collargol*. s. **cyanide.** See *argenti cyanidum* under *argentum*. s. **fluoride**, AgFl, a brown, glassy, elastic solid, very soluble in water, discovered by Paterno in 1901. It is used as an antiseptic. Syn., *tachiol*. s.-**fork deformity**, a peculiar deformity of the wrist and hand in Colles' fracture, resembling the curve on the back of a fork. s. **gelatose**, albargin. s. **ichthyolate**, see *ichthargan*. s. **iodide**, see *argenti iodidum* under *argentum*. s. **lactate**, AgC₃H₅O₃ + H₂O, a white, soluble powder, recommended as a surgical antiseptic. Injection in erysipelas 5 gr. (0.3 Gm.) to 3 fl. oz. (100 Cc.) of water; as a wash, 1 teaspoonful of solution 1 : 50 in a glass of water. Syn., *actol*. s. **nitrate**, see *argenti* *nitras* under *argentum*. **s. nucleate, s. nucleide,** see *nargol*. **s. oxide,** see *argenti oxidum* under *argentum*. **s. paraphenol-sulphonate,** an external antiseptic. Syn., *silberol*. **s. and potassium cyanide,** AgK(CN)₂, very poisonous white crystals, soluble in 4 parts of water at 20° C. or 25 parts of 85% alcohol. It is antiseptic and bactericide. One part in 50,000 destroys anthrax bacilli. **s.-protalbin,** see *largin*. **s. sulphocarbolate, s. sulphophenate,** a fine, crystalline powder containing about 28% of metallic silver; it is a noncorrosive antiseptic, used in eye diseases and wounds. **s. test for glucose in urine,** add ammonia in excess to a strong solution of silver nitrate; add the urine and boil. In the presence of glucose a metallic silver mirror is deposited at the bottom of the tube. Aldehyde and tartaric acid give the same reaction. **s. thiohydrocarburosulphonate,** see *ichthargan*. **s. trinitrophenolate,** a compound containing 30% of silver, used as an antiseptic on inflamed mucous surfaces. Syn., *picratol*. **s. vitelline,** see *argyrol*.

**Silvester's method of artificial respiration** (*sil-ves'-ter*) [Henry Robert *Silvester*, English physician, 1828–1908]. It consists chiefly of movements of the arms; this method is valueless in asphyxia neonatorum, owing to nondevelopment of the pectoral muscles.

**Simaba** (*sim-a'-bah*) [native name in Guiana]. A genus of simarubaceous tropical trees. The seeds of *S. cedron* are antiperiodic and tonic. Dose of fluidextract 1–8 min. (0.06–0.5 Cc.). It contains, according to Tanret, the alkaloid *cedronine* and also *cedrin*. The bark also has tonic and febrifuge properties.

**Simaruba** (*sim-ar-oo'-bah*). A genus of trees of the order *Simarubaceæ*. The bark of the root of *S. officinalis* has been used as a simple bitter.

**similia similibus curantur, doctrine of.** A sophism formulated by Hippocrates, later by Paracelsus ("simile similis cura, non contrarium"), and later, as one of the results of the reaction against the heroic measures of venesection and drastic medication, by Samuel Christian Friedrich Hahnemann, the founder of homeopathy, whose doctrine that *like is to be cured by like* led naturally to the practice of *isopathy* (*q. v.*), according to which smallpox is to be treated by variolous pus, tapeworm by ingestion of proglottides, etc.

**similimum** (*sim-il'-im-um*) [L., "most like"]. The homeopathic remedy which will produce the symptom complex "most like" that of a given disease.

**Simon's operation.** 1. Perineorraphy. 2. Colpocleisis. **S.'s posture,** the dorsal posture with the legs and thighs flexed, the hips elevated, and the thighs abducted. **S.'s symptom,** immobility or retraction of the umbilicus during inspiration; sometimes seen in tuberculous meningitis. **S.'s triangles,** two roughly triangular areas covering— (1) the lower portion of the abdomen, the inner surface of the thigh to a point 10 to 12 centimeters below the groins, and the inguinal region as far outward as the trochanter (abdominocrural or femoral triangle); and (2) the axillary and pectoral regions and the inner surface of the arm (brachial triangle). They are frequently the seat of petechial or petechio-erythematous rashes during the first three days of smallpox.

**Simon's symptom-complex** (*se'-mon*). In primary cancer of the female breast, metastasis may involve the hypophysis and produce polyuria.

**Simonart's bands, S.'s threads** (*se-mo-nar'*) [Pierre Joseph Cécilien *Simonart*, Belgian obstetrician, 1817–1847]. Amniotic bands formed by drawn-out adhesions between the fetus and the amnion where the cavity has become distended through the accumulation of fluid.

**Simonelli's test for renal inadequacy** (*se-mo-nel'-le*) [Francesco *Simonelli*, Italian physician]. If the kidneys are healthy, iodine administered appears at the same time in the urine and the saliva.

**simple** (*sim'-pl*) [*simplex*, simple]. 1. Not complex; consisting of but one substance, or containing only one active substance; not compound. 2. Wanting in intellect. 3. A medicinal plant. See *simples*. **simpler, simplist** (*sim'-pler, sim'-plist*). A herb-doctor.

**simples** (*sim'-plz*) [*simple*]. A term for herbs having a medicinal value.

**Simpson's plug or splint** [William Kelly *Simpson*, American laryngologist, 1855–1914]. A tampon or splint, cut to fit the nares, and inserted to stop epistaxis or to retain the parts in apposition after operation on the nasal septum.

**Sims' depressor** [James Marion *Sims*, American gynecologist, 1813–1883]. An instrument for depressing the anterior vaginal wall. **S.'s posture,** the semiprone position for vaginal operations. The patient lies on the left side with the right knee and thigh drawn up and the left arm placed along the back; the chest is inclined forward so that the patient rests upon it. **S.'s speculum,** the duckbill vaginal speculum.

**simul** (*si'-mul*) [L.]. At once; at the same time.

**simulation** (*sim-u-la'-shun*) [*simulatio; simulare,* to feign]. In medicine, the feigning or counterfeiting of disease. The pretence of a malingerer.

**simulium reptans** (*sim-u'-le-um*). A biting insect believed to convey the infective agent of pellagra.

**simulo** (*sim'-u-lo*). The fruit of certain species of *Capparis*, especially *C. coriacea*, of Peru; it is recommended as a cure for epilepsy, and possesses antiscorbutic and stimulant properties. Dose of the tincture ʒ i–iij; of the fluidextract ʒ ss–iij.

**sinal** (*si'-nal*). Relating to or situated within a sinus.

**sinalbin** (*sin-al'-bin*). A white crystalline substance, found in mustard, *q. v.*

**sinamine** (*sin'-am-ēn*), C₃H₆CN. Allyl cyanamide, a substance obtained from crude oil of mustard.

**sinapeleum** (*sin-ap-el'-e-um*) [σίναπι, mustard; ἔλαιον, oil]. Mustard-oil.

**sinapin** (*sin'-ap-in*) [σίναπι, mustard], C₁₆H₂₃O₅. A substance occurring as a sulphocyanate in white mustard.

**sinapis** (*sin-a'-pis*). Mustard. See *mustard*.

**sinapiscopy** (*sin-ap-is'-ko-pe*) [σίναπι, mustard; σκοπεῖν, to view]. The use of mustard as a test of sensory disturbances, analogous to a similar use of *metalloscopy*.

**sinapism** (*sin'-ap-izm*) [*sinapis*]. A mustard-plaster.

**sinapized** (*sin'-ap-īzd*) [*sinapis*]. Containing mustard.

**sinapol** (*sin'-ap-ol*). A mixture recommended as an application for neuralgia, rheumatism, etc., said to consist of spirit of rosemary (1 : 15), 780 Gm.; castor-oil, 120 Gm.; menthol, 30 Gm.; essence of mustard, 30 Gm.; aconitine, 0.4 Gm.

**sinapolin** (*sin-ap'-ol-in*). Diallylurea, a substance obtained from mustard oil by heating with water and lead oxide.

**sincalin** (*sing'-ka-lin*). A base found in mustard and identified with cholin.

**sinciputa** (*sin-sip'-it-al*) [*sinciput*]. Pertaining to the sinciput.

**sinciput** (*sin'-sip-ut*) [*semi*, half; *caput,* head]. The superior and anterior part of the head. Also, the top of the head; the bregma.

**sinew** (*sin'-ū*). A tendon (*q. v.*).

**singers' nodes or nodules.** *Chorditis nodosa* or *tuberosa, q. v.*

**singult** (*sin'-gult*) [*singultus,* a sobbing]. A sob.

**singultation** (*sin-gul-ta'-shun*) [*singultus,* hiccup]. Hiccupping.

**singultient** (*sin-gul'-she-ent*). Sobbing; sighing.

**singultous** (*sin-gult'-us*). Relating to or affected with hiccup.

**singultus** (*sin-gul'-tus*). See *hiccup*.

**sinigrin** (*sin'-ig-rin*). A glucoside found in black mustard.

**sinister, sinistra, sinistrum** [L.]. Left.

**sinistrad** (*sin'-is-trad*) [*sinister*, left; *ad*, toward]. Toward the left.

**sinistral** (*sin'-is-tral*) [*sinister*]. 1. On the left side. 2. Showing preference for the left hand, eye, foot, etc., for certain acts or functions.

**sinistrality** (*sin-is-tral'-it-e*) [*sinister*]. The preference generally for the left hand, eye, foot, etc., in performing certain acts.

**sinistration** (*sin-is-tra'-shun*). 1. A turning to the left. 2. Sinistrality.

**sinistraural** (*sin-is-traw'-ral*) [*sinister; auris,* ear]. Left-eared; the reverse is *dextraural*.

**sinister** (*sin'-is-tren*) [*sinister*, left]. Belonging to the sinistral side in itself.

**sinistrin** (*sin'-is-trin*) [*sinister*]. A substance resembling dextrin, found in squill. See *animal.* See under *helicoprotein*.

**sinistro-** (*sin-is-tro-*) [*sinister*, left]. A prefix meaning left or toward the left side.

**sinistrocardial** (*sin-is-tro-kar'-de-al*) [*sinistro-*; καρδία, heart]. Having the heart to the left of the median line; the reverse is *dextrocardial*.

**sinistrocerebral** (*sin-is-tro-ser'-e-bral*). 1. Located in the left cerebral hemisphere. 2. Functionating preferentially with the left side of the brain; the reverse is *dextrocerebral*.

**sinistrocular** (*sin-is-trok'-ū-lar*). Left-eyed; the reverse is *dextrocular*.

**sinistrogyric** (*sin-is-tro-ji'-rik*). See *sinistrorse*.

**sinistrohepatal** (*sin-is-tro-hep'-at-al*) [*sinistro-;* ἧπαρ, liver]. Having the liver to the left of the median line; the reverse is *dextrohepatal*.

**sinistromanual** (*sin-is-tro-man'-ū-al*) [*sinistro-; manus*, hand]. Left-handed; the reverse is *dextromanual*.

**sinistropedal** (*sin-is-trop'-ed-al*) [*sinistro-; pes*, foot]. Left-footed; the reverse is *dextropedal*.

**sinistrophoria** (*sin-is-tro-fo'-re-ah*). See *levophoria*.

**sinistrorse** (*sin'-is-trors*) [*sinistro-; vertere*, to turn]. In biology, turning from right to left.

**sinistrose** (*sin'-is-trōs*). A levorotatory sugar; levulose.

**sinistrosplenic** (*sin-is-tro-splen'-ik*). Having the spleen to the left of the median line; the reverse is *dextrosplenic*.

**sinistrotorsion** (*sin-is-tro-tor'-shun*) [*sinistro-; torquere*, to turn]. A twisting or turning toward the left; the reverse is *dextrotorsion*.

**sinistrous** (*sin'-is-trus*). Awkward; unskilled; the reverse is dextrous, skilled, expert.

**sinkaline,¹ sinkoline** (*sink'-al-ēn, sink'-ol-ēn*). An alkaloid found in mustard, identical with *choline*, *q. v.*

**sinual** (*sin'-ū-al*) [*sinus*, a curve]. Possessing the characteristics of a sinus.

**sinuation** (*sin-ū-a'-shun*) [*sinuatus*, from *sinuare*, to bend]. 1. The state of being sinuate or sinuous. 2. A cerebral gyre.

**sinuatrial.** Same as *sinuauricular*.

**sinuauricular** (*si-nū-aw-rik'-ū-lar*) [*sinus; auricula*, auricle]. Pertaining to the sinus venosus and the right auricle of the heart.

**sinuitis.** See *sinusitis*.

**sinuose** (*sin'-ū-ōs*). Same as *sinuous*.

**sinuosity** (*sin-ū-os'-it-e*) [*sinuare*, to bend]. Anfractuosity; the state of being sinuous or bent.

**sinuous** (*sin'-ū-us*) [*sinuosus; sinus*, a curve]. Wavy; applied especially to tortuous fistulæ and sinuses.

**sinus** (*si'-nus*) [L., "a gulf or hollow"]. 1. A hollow or cavity; a recess or pocket. 2. The space between the breasts. 3. A large channel containing blood, especially one containing venous blood. 4. A suppurating tract. 5. A cavity within a bone.

**sinuses, accessory, of the nose,** the maxillary, frontal, ethmoid, and sphenoid sinuses. **s., air-,** a cavity within bones containing air, especially one communicating with the nasal passages. **s. alæ parvæ,** the sphenoparietal sinus situated along the posterior border of the lesser wing of the sphenoid bone. **s., aortic,** one of the pouch-like dilatations of the aorta opposite the segments of the semilunar valves. **s., cavernous,** a large venous sinus extending from the sphenoidal fissure to the apex of the petrous portion of the temporal bone, communicating behind with the inferior and superior petrosal sinuses and receiving the ophthalmic vein in front. **s., circular,** a venous sinus surrounding the pituitary body, and communicating on each side with the cavernous sinus. **s. circularis iridis.** See *Schlemm's canal*. **s., common, of the vestibule.** See *utricle* (1). **s., coronary** (of the heart), a large venous sinus in the transverse groove between the left auricle and left ventricle of the heart. **s. ensiformis,** the sinus of Eternod, a vascular loop connecting the vessels of the chorion with the vessels on the under aspect of the yolk-sac. **s., ethmoid,** the ethmoid cells. **s., frontal,** one of the two irregular cavities in the frontal bone containing air and communicating with the nose by the infundibulum. **s. of the heart,** the chief cavity of either of the auricles. **s., inferior longitudinal,** a venous sinus which extends along the posterior half of the lower border of the falx cerebri and terminates in the straight sinus. **s., inferior petrosal,** a large venous sinus arising from the cavernous sinus running along the lower margin of the petrous portion of the temporal bone, and joining the lateral sinus to form the internal jugular vein. **s. intercavernosus, anterior** and **posterior,** sinuses extending across the hypophyseal fossa and connecting the cavernous sinuses of both sides. **s. of kidney,** the prolongation inward of the hilum of the kidney. **s. of the larynx,** the ventricle of the larynx. **s., lateral,** a venous sinus which begins at the torcular Herophili and runs horizontally on the inner surface of the occipital bone to the base of the petrous portion of the temporal bone, where it unites with the inferior petrosal sinus to form the internal jugular vein. **s., lymph,** spaces in the parenchyma of a lymphatic gland between the pulp of the gland and the dilatations of lymphatic vessels. **sinuses, mastoid,** the mastoid cells. **s., maxillary,** the antrum of Highmore. **s., occipital,** a small venous sinus in the attached margin of the falx cerebelli, opening into the torcular Herophili. **s., petrosquamosal.** See *petrosquamosal sinus*. **s.-phlebitis,** inflammation of one of the sinuses of the cranial cavity. **s., placental,** slanting venous channels issuing from the placenta at its uterine surface by piercing the decidua serotina. **s. pocularis,** a large lacuna in the center of the prostatic portion of the urethra. **sinuses, precaval.** See *ducts of Cuvier*. **s., precervical,** a recess between the lowermost branchial arch and the trunk of the embryo. **s., prostatic,** a fossa on each side of the verumontanum. **s. rectus.** Same as *s. straight*. **s., rhomboid, s. rhomboideus,** the fourth ventricle of the brain. **s., sagittal.** See *s., inferior* and *superior longitudinal*. **s., sphenoid,** the air-space in the body of the sphenoid bone, communicating with the nasal cavity. **s., sphenoparietal.** Same as *s. alæ parvæ*. **s., squamosopetrosal.** See *petrosquamosal sinus*. **s., straight,** a venous sinus running from the inferior longitudinal sinus along the junction of the falx cerebri and tentorium to the lateral sinus. **s., superior longitudinal,** a venous sinus which runs along the upper edge of the falx cerebri, beginning in front of the crista galli and terminating at the torcular Herophili. **s., superior petrosal,** a venous sinus running in a groove in the petrous portion of the temporal bone, extending from the posterior part of the cavernous sinus to the lateral sinus. **s., terminal, s. terminalis,** a vein that encircles the vascular area of the blastoderm, and empties either by one trunk, the anterior vitelline vein, into the left vitelline vein, or by two trunks into both vitelline veins. **s.-thrombosis,** thrombosis of the sinuses of the dura mater. It is usually septic in character, and is likely to lead to pyemia. The most frequent cause is disease of the middle ear. **s., transverse,** a sinus uniting the inferior petrosal sinuses. **s., urogenital,** the canal or duct into which, in the embryo, the Wolffian ducts and the bladder empty, and which opens into the cloaca. **s. venosus.** 1. The chamber of the lower vertebrate heart into which empty the veins returning the blood from the body. 2. The vessel in the septum transversum of the embryonic mammalian heart into which open the vitelline, and allantoic veins, and the ducts of Cuvier. 3. That portion of the adult right auricle back of the crista terminalis.

**sinusitis** (*si-nus-i'-tis*). Inflammation of a sinus. **s., serous, s.,** chronic catarrhal. See *mucocele*.

**sinusoid** (*si'-nus-oid*) [*sinus; εἶδος*, likeness]. 1. Resembling a sinus. 2. One of the relatively large spaces or tubes constituting the embryonic circulatory system in the suprarenal gland, liver, and other viscera.

**sinusoidal** (*si-nus-oid'-al*). Pertaining to or derived from a sinusoid. **s. current,** an alternating induced electrical current with equal current strokes.

**sinusoidalization** (*sin-us-oi-dal-i-za'-shun*). The application of a sinusoidal current.

**sionagra** (*si-on-a'-grah*) [ἰσχιάων, the jaw-bone; ἄγρα, a seizure]. Gout in the jaw-bone.

**siphon** (*si'-fon*) [σίφων, a tube]. A tube bent at an angle, one arm of which is longer than the other, for the purpose of removing liquids from a cavity or vessel.

**siphonage** (*si'-fon-aj*) [σίφων, a siphon]. The action of a siphon, such as in washing out the stomach, in drainage of wounds, or in house-plumbing.

**siphonoma** (*si-fon-o'-mah*) [*siphon; ὄμα*, tumor]. A tumor composed of fine tubes. Syn., *Henle's tubular tumor*.

**siren** (*si'-ren*). Same as *sirenomelus*.

**sirenomelus** (*si-ren-om'-el-us*) [σειρήν, mermaid; μέλος, a limb]. A form of monster in which the lower extremities are infernally fused, the feet being absent.

**siriasis** (*sir-i'-as-is*). Sunstroke.

**sirolin** (*sir'-ol-in*). Thiocol, 10 %, in a syrup of orange bark. It is used in tuberculosis, bronchitis,

and intestinal catarrh. Dose 3 or 4 teaspoonfuls (15-20 Cc.) daily.
**sirup** (*sir'-up*). See *syrup*.
**Sisymbrium** (*sis-im'-bre-um*) [σισύμβριον, from σῦς, a pig; ὄμβριος, rainy, wet; a plant growing in wet places where swine wallowed]. A genus of cruciferous plants. *S. officinale*, wild mustard, singer's herb, is a European species, laxative, diuretic, and expectorant, and is employed in laryngeal catarrh and laryngitis. Dose, 3 cupfuls a day of a decoction of 30 Gm. of the leaves, sweetened with 60 Gm. of a syrup of the drug made in the usual way.
**site** (*sīt*) [*situs*, place]. Situation. **s., placental,** the area to which the placenta is attached.
**sitfast.** In farriery, a piece of dead tissue in the skin which would be thrown off but that it has formed firm connections with the fibrous skin beneath, or with the deeper tissues, and is thus bound in its place as a persistent source of irritation.
**sitieirgia** (*sit-e-ir'-je-ah*) [σιτίον, food; εἴργειν, to shut out]: Sollier's term for hysterical anorexia.
**sitiology** (*si-te-ol'-o-je*). See *sitology*.
**sitiomania** (*sit-e-o-ma'-ne-ah*). See *sitomania*.
**sitiophobia** (*sit-e-o-fo'-be-ah*). See *sitophobia*.
**sitogen** (*si'-to-jen*). A vegetable food-product intended to replace meat-extracts.
**sitology** (*si-tol'-o-je*) [σιτιόν, nourishment; λόγος, a treatise]. The science of nourishment or dietetics.
**sitomania** (*si-to-ma'-ne-ah*) [σῖτος, food; μανία, madness]. 1. A periodic craving for food; periodic bulimia. 2. Sitophobia.
**sitophobia** (*si-to-fo'-be-ah*) [σῖτος, food; φόβος, fear]. Morbid aversion to food.
**sitotoxicon** (*si-to-toks'-ik-on*) [σῖτος, food; τοξικόν, poison]. The active poisonous agent in sitotoxism; all sitotoxicons are not of bacterial origin.
**sitotoxin** (*si-to-toks'-in*) [see *sitotoxicon*]. Any basic poison generated in vegetable food by growth of bacteria or fungi.
**sitotoxism** (*si-to-toks'-izm*) [see *sitotoxicon*]. Poisoning with vegetable food infected with molds and bacteria.
**situs** (*si'-tus*) [L., "site"]. A position. **s. perversus,** malposition of one or more of the viscera. **s. transversus.** Same as *s. inversus*. **s. viscerum inversus,** an anomaly in which the viscera of the body are changed from the normal to the opposite side of the body.
**sitz-bath** (*sits'-bath*) [Ger., *Sitz*, a seat; *bath*]. A hip-bath; a bath taken in a sitting posture.
**six hundred and six.** See *salvarsan*.
**sixth nerve.** The abducens nerve. See *nerves, table of*.
**Sjoeqvist's test for the quantitative estimation of free HCl in the gastric juice** (*syo'-kvist*) [John August Sjoeqvist, Swedish physician, 1863– ]. It depends upon the action of barium carbonate on the acid of the secretion, the hydrochloric acid being estimated as barium chloride by means of titration with a solution of potassium dichromate.
**skatol** (*skat'-ol*) [σκατός, gen. of σκώρ, dung], C₉H₉N, methyl indol; it is a nitrogenous compound produced by the decomposition of proteids in the intestinal canal.
**skatophagia** (*skat-o-fa'-je-ah*). See *scatophagia*.
**skatoxyl** (*skat-oks'-il*) [σκῶρ σκατ-), dung; ὀξύς, acid]. A product of the oxidation of skatol. It is obtained from the urine in cases of disease of the large intestine.
**Skeer's sign.** A yellowish-brown ring near the pupillary margin of the iris, observed in the early stage of some cases of tuberculous meningitis.
**skein** (*skān*) [ME., *skeyne*, skein]. 1. A fixed length of any thread or yarn of silk or other material; doubled again and again and knotted. 2. A synonym of *spirem*. **s., close.** See *spirem*. **s., loose,** the thickened chromatin fibrils resulting from a loosening of the spirem or close skein in mitotic cell-division. **skeins, test.** See *Holmgren's test*.
**skeletal** (*skel'-et-al*) [*skeleton*]. Pertaining to or connected with the skeleton or supporting structure of a body. **s.-muscle,** a muscle attached to the skeleton. **s.-tissue,** the tissue of the framework of the body.
**skeletins** (*skel'-et-inz*) [*skeleton*]. A name given to a number of insoluble epithelial products found chiefly in invertebrates.
**skeletization** (*skel-et-i-za'-shun*) [*skeleton*]. 1. The process of converting into a skeleton; gradual wasting of the soft parts, leaving only the skeleton.

**skeleto-** (*skel-et-o-*) [*skeleton*]. A prefix meaning relating to the skeleton.
**skeletogenous** (*skel-et-oj'-en-us*) [*skeleto-*; γεννᾶν, to produce]. Producing a skeleton or skeletal tissues.
**skeletography** (*skel-et-og'-ra-fe*) [*skeleto-*; γράφειν, to write]. A description of the skeleton.
**skeletology** (*skel-et-ol'-o-je*) [*skeleto-*; λόγος, science]. The branch of anatomy treating of the skeleton.
**skeleton** (*skel'-et-on*) [σκελετόν, a dried body, from σκέλλειν, to dry up]. A supporting structure, especially the bony framework (*osseous skeleton*) supporting and protecting the soft parts of an organism. **s., appendicular,** the skeleton of the limbs. **s., axial,** the skeleton of the head and trunk. **s., cartilaginous,** the cartilaginous structure from which the bony skeleton is formed through ossification.
**skeletonize** (*skel'-e-ton-īz*). To reduce to a skeleton.
**skeletopy** (*skel-et'-op-e*). See *skeletotopy*.
**skeletotopic** (*skel-et-o-top'-ik*). Applied by Waldeyer to such topographic description as refers a part or organ to its relation to the skeleton.
**skeletotopy** (*skel-et-ot'-o-pe*) [*skeleto-*; τόπος, a place]. Waldeyer's term for the relation of an organ or part to the osseous skeleton of the whole organism. Cf. *holotopy*; *idiotopy*; *syntopy*.
**Skene's glands** (*skēn*) [Alexander Johnston Chalmers Skene, American gynecologist, 1838–1900]. Two complex tubular glands in the mucosa of the female urethra opening by small ducts just within the meatus urinarius.
**skérljivo.** See *scherlievo*.
**skew muscles.** Triangular-shaped or quadrilateral shaped muscles, the plane of whose line of origin intersects that of the insertion.
**skiagram** (*ski'-ag-ram*) [σκία, shadow; γράμμα, a writing]. The finished, printed Roentgen-ray picture. Syn., *inductogram*; *shadowgram*.
**skiagraph** (*ski'-ag-raf*). See *Skiagram*.
**skiagrapher** (*ski-ag'-raf-er*). An adept in skiagraphy.
**skiagraphy** (*ski-ag'-ra-fe*) [σκία, shadow; γράφειν, to write]. Photography by the Roentgen-rays; skotography, skiography, radiography, electrography, electroskiography, Roentgography, and the new photography are names that have been used to designate the method.
**skiameter** (*ski-am'-et-er*) [σκία, shadow; μέτρον, a measure]. An apparatus devised by Biesalski for measuring the intensity of the Roentgen-rays and for the recognition of fine differences in the density of Roentgen-ray shadows.
**skiametry** (*ski-am'-et-re*). Shadow mensuration applied to a method of determining the density of Roentgen-ray shadows.
**skiaporescopy** (*ski-ap-or-es'-ko-pe*). See *retinoscopy*.
**skiascope** (*ski'-as-kōp*) [σκία, shadow; σκοπεῖν, to view]. An instrument employed in retinoscopy. **s.-optometer,** an optometer designed for the determination of the refraction of the eye by retinoscopy.
**skiascopy, skiascopia** (*ski-as'-ko-pe, ski-as-ko'-pe-ah*). 1. See *retinoscopy*. 2. Examination by either skiagraph or fluoroscope.
**skiatherapy** (*ski-ah-ther'-ap-e*) [σκία, shadow; θεραπεία, therapy]. The therapeutic application of Roentgen-rays.
**skimmetin** (*skim'-et-in*), C₉H₆O₃. A dissociation product of skimmin by action of dilute mineral acid with heat; it is perhaps identical with umbelliferone.
**Skimmia** (*skim'-e-ah*) [*mijama-skimmi*, Japanese name]. A genus of the *Rutaceæ*. *S. japonica* is a species of Japan; the flowers are used to flavor tea; the leaves contain an ethereal oil; the bark contains the glucoside *skimmin*.
**Skimmin** (*skim'-in*), C₁₅H₁₆O₈. A glucoside similar to scopolein and escullin isolated from the bark of *Skimmia japonica*, occurring in long, colorless needles, soluble in hot water, alcohol, or alkalies, insoluble in chloroform or ether; melts at 210° C.
**skin** [ME.]. The protective covering of the body, composed of the epidermis, *scarf-skin*, or cuticle, and the corium, or *true skin*. The epidermis consists of a deep layer, the *stratum Malpighii*, and three superficial layers—the *stratum granulosum*, the *stratum lucidum*, and the *stratum corneum*. The corium, derma, or *true skin* consists of a papillary and reticular layer (*stratum papillare* and *stratum reticulare*), the former projecting upward in the form of papillæ. The true skin is made up of elastic

tissue, white fibrous tissue, and nonstriped muscular tissue (the arrectores pili). The subcutaneous tissue consists of fibroelastic and adipose tissue. The appendages of the skin are the nails, hairs, and sweat- and sebaceous glands, which are derivatives of the epithelial layer of the skin. In the skin are also placed terminal nerve-organs subserving the sense of touch. s., atrophy of the, a wasting-away or retrogressive change in the skin. Syn., *dermatatrophia.* s.-bound. See *scleroderma.* s.-bound disease. See *scleroderma neonatorum.* s., bronzed. See *Addison's disease.* s., congestion of the, engorgement of the blood-vessels of the skin. Syn., *dermathemia.* s., edema of the, effusion of serum into the areolar tissue of the skin. Syn., *dermatochysis.* s., fish. See *ichthyosis.* s., glossy, a peculiar shiny, glazed skin seen in conditions in which the trophic nerve-supply to the skin is cut off, as after injury to a nerve. s., goldbeaters', a thin tenacious sheet from the cecum of cattle, occasionally used as a surgical dressing. s., goose-. See *goose-flesh.* s.-grafting, the application of pieces of the outer layers of healthy skin to a granulating surface for the purpose of hastening its cicatrization. (1) *Autoepidermic-*When the epithelial cells are taken from the patient, it includes—(a) scrapings from healthy skin; (b) corn shavings; (c) pellicles from blisters; (d) the Ollier-Thiersch method, in which one-half the skin thickness (epidermis, rete, and part of the cutis proper) is required; and (e) the Krause method, in which the whole thickness of the skin is used in grafting. (2) *Heteroepidermic:* When the epithelial cells are furnished by another person. (3) *Zoodermic:* When the skin is removed from lower species, as the use of—(a) small pieces of sponge; (b) frog skin; (c) inner membrane of hens' eggs; (d) inner surface of pullets' wings; (e) skin of pups; (f) skin of guinea-pigs; (g) skin of rabbits. s., hypertrophy of the, excessive growth of the skin. Syn., *dermahypertrophia.* s., neuralgia of the. See *dermatalgia.* s., pigmentation of the. See *dermatodyschroia.* s.-shedding. See *keratolysis.*

skinny (*skin'-e*) [ME.]. 1. Cutaneous. 2. Emaciated.

skirt. The diaphragm.

skleriasis (*skle-ri'-as-is*). See *scleroderma.*

sklerodactylia. See *sclerodactylia.*

Skoda, consonating rales of (*sko'-dah*) [Joseph Skoda, Austrian physician, 1805–1881]. Bronchial rales heard through the consolidated pulmonary tissue of pneumonia. S.'s resonance sign, S.'s tympany, a tympanitic note heard above the line of fluid in a pericardial effusion, or above the line of consolidation in pneumonia. It is almost as tympanitic as the abdomen.

Skodaic resonance. See *Skoda's sign.*

skolikoiditis (*sko-le-koid-i'-tis*) [σκωληκοειδής, worm-like]. Synonym of *appendicitis.*

skoliosis. See *scoliosis.*

skoliosometer (*sko-le-o-som'-et-er*) [σκολιός, curved; μέτρον, measure]. See *scoliosometer.*

skookum chuck [Amer. Ind., "good water"]. A homeopathic remedy for skin diseases consisting of a trituration of the salts of the spring of this name. Dose 2 gr. (1.03 Gm.) in one-half glass of water; teaspoonful every 2 or 3 hours.

skopophobia (*sko-po-fo'-be-ah*) [σκοπος, a spy; φόβος, fear]. Insane dread of spies.

skotograph (*skot'-o-graf*). See *skiagraph.*

skotography (*skot-og'-raf-e*). Synonym of *skiagraphy.*

skull (*skul*) [Icel., *skål*, a bowl]. The bony framework of the head, consisting of the cranium and the face. The cranium is made up of the occipital, frontal, sphenoid, and ethmoid bones, and the two parietal and two temporal bones. The face is composed of two nasal, two superior maxillary, two lacrimal, two malar, two palate, and two inferior turbinated bones, and the vomer and inferior maxillary bone. *Modes of measuring the capacity of the skull. Drocu's method.* 1. The skull, made impermeable, is filled with water, which can be weighed or measured. 2. The skull is packed with shot, which is then measured; but both the filling and measuring are aided by certain implements, and especially by a funnel of certain dimensions, which controls the flow of the shot, and every step of the procedure follows definite rules. *Method of Busk, Flower, or Tiedemann:* the skull is filled with small, rounded seeds, beads, shot, or other substance, and the contents are then measured. The filling or the measuring (or both) is aided by certain manipulations (tilting, tapping, etc.). *Method of Schmidt or Matthews.* See *Broca's method* above. *Welcker's method:* the mode of filling the skull, so long as efficient and uniform, is immaterial; all that is required is that each worker should, with the aid of a standard skull, find the exact size of the funnel necessary to give him, in measuring the skull, the correct result with his particular method and substance used for the filling of the skull. s.-cap. 1. The top of the skull. 2. See *scutellaria.* s., natiform, a skull covered with osteophytes. s.-roof, the roof of the skull; skull-cap. s., tower. See *oxycephalia.*

skunk-cabbage. *Dracontium fœtidum*, the rhizome of which is stimulant, antispasmodic, and narcotic, and has been used in asthma, rheumatism, hysteria, and dropsy.

slabber (*slab'-ur*). See *slaver.*

slag [Sw., *slagg*, dross]. The earthy matter separated, in a more or less completely fused and vitrified condition, during the reduction of a metal from its ore. slag-wool, a product of blowing a jet of steam into melted slag; it is noninflammable and a nonconductor of heat. Syn., *mineral wool.*

slake (*slāk*) [AS., *sleccan*, to quench; extinguish]. 1. To quench or appease. 2. To disintegrate by the action of water.

slaver (*slav'-er*) [ME.]. Drivel; saliva, especially such as is discharged involuntarily.

sleep. The periodic state of rest in which voluntary consciousness and activity cease. s.-drunkenness. See *somnolentia.* s.-epilepsy. See *narcolepsy.* s., hypnotic, s., magnetic, s., mesmeric. See *hypnotism.* s.-paralysis, paralysis produced by pressure during sleep. s., paroxysmal. See *narcolepsy.* s., twilight. See *twilight.* s.-walking. See *somnambulance.*

sleeping dropsy or sickness. A peculiar disease of West Africa characterized by increasing somnolence. See *African lethargy.*

sleeplessness (*slēp'-les-nes*). See *insomnia.*

slender column. See *funiculus gracilis.* s. lobe of cerebellum, a small lobe in the inferior surface of the cerebellum.

slide (*slīd*). A small, rectangular plate of glass upon which objects intended for examination with the microscope are placed.

sling. A swinging bandage for supporting an arm or other part.

slit [ME.]. A narrow opening; a visceral cleft; the separation between the labia; the vulvar cleft. s., genitourinary, s., urinogenital, s., urogenital, the urogenital opening. s. of the microspectroscope, the spectral ocular, in place of an ordinary diaphragm, has two movable knife edges so arranged that a slitlike opening of greater or less width and length may be obtained by the use of screws for that purpose.

slobber (*slob'-ur*). See *slaver.*

sloid, sloyd (*sloid*) [Sw., *slojd*, slight skill]. A system of manual training taught in elementary school; it is of Swedish origin.

slough (*sluf*) [ME., *slouh*, the skin of a snake]. The separated dead matter in an ulceration.

sloughing (*sluf'-ing*) [*slough*]. Pertaining to or characterized by sloughs.

slows (*slōz*). Synonym of *milk-sickness.*

Sluder's method (*sloo'-der*) [Greenfield *Sluder*, American laryngologist, 1865– ]. *Of tonsillectomy:* Removal of the tonsil with capsule complete.

sludge (*sluj*) [AS., *slog*, mud]. Sewage-deposit.

slumber (*slum'-ber*) [ME., *slumberen*, to slumber]. 1. To sleep lightly. 2. Light sleep.

smallpox. See *variola.*

smear-cultures (*smēr*). See *culture.*

Smee's battery [Alfred *Smee*, English surgeon, 1818–1877]. Positive element, zinc; negative element, platinized silver; exciting agent, sulphuric acid, dilute; depolarizing agent, none; E. M. F., 0.5 to 1.0 volt.

smegma (*smeg'-mah*) [σμήγμα, a cleansing substance]. 1. Sebum. 2. See *s. præputii.* s. clitoridis, the substance secreted by the sebaceous glands of the clitoris and labia minora. s. embryonum. See *vernix caseosa.* s. præputii, or simply smegma, the substance secreted by the sebaceous glands of the prepuce.

smegmatic (*smeg-mat'-ik*) [σμήγμα, a cleansing substance]. Pertaining to, or of the nature of, smegma.

**smell.** 1. The perception of odor. 2. Odor.
**smelling-salts.** A name applied to various preparations of ammonium carbonate scented with aromatic substances.
**smelting** (*smelt'-ing*) [ME., *smelten*, to smelt]. The treatment of ore by which it is subjected to intense heat for the purpose of separating the contained metal.
**smesches** (*sme'-chez*). Puffs of arsenic trioxide gas which occasionally escape from the doors of the calcining furnaces in Cornish arsenic works, and which give rise to pulmonary irritation among the workmen.
**smilacin** (*smi'-las-in*) [*smilax*]. 1. The precipitate from a tincture of the root of sarsaparilla, *Smilax officinalis*, alterant, detergent, diaphoretic, and stimulant. Dose 2–5 gr. (0.13–0.32 Gm.). 2. $C_{20}H_{30}O_8$ (Flückiger) or $C_{14}H_{22}O_6$ (Poggiale) or $C_{18}H_{30}O_9$ (Peterson). Folchi's name for a saponin-like glucoside found by Palotta, in 1824, in sarsaparilla-root (various species of *Smilax*), and named by him *pariglin*. It forms a yellowish-white powder, soluble in water and alcohol; alterative, expectorant, and emetic, and used in syphilis and colds. Dose 1–3 gr. (0.065–0.2 Gm.). Syn., *parillin* (Batha); *salseparin* (Thubeuf).
**smilax** (*smi'-laks*). See *sarsaparilla*.
**smile** (*smil*) [ME., *smil*]. A joyful expression.
**s., levator, s., nasal,** W. R. Gowers' name for a peculiarity of expression in some patients affected with myasthenia, consisting in absence of normal movement at the corners of the mouth.
**smith's cramp.** An occupation-neurosis occurring in smiths, and characterized by painful cramps in the arm or hand.
**Smith's disease.** [Eustace *Smith*, English physician]. Mucous colitis. **S.'s sign,** a murmur audible over the sternum when the chin is drawn up.
**Smith's dislocation of the foot.** [Robert William *Smith,* Irish surgeon]. Dislocation upward and backward of all the metatarsal bones, together with the internal cuneiform. **S.'s fracture,** transverse fracture about 5 cm. above the lower extremity of the radius.
**Smith's operation** [Henry *Smith*, English surgeon]. For hemorrhoids: crushing by means of a clamp, and applying the Paquelin cautery to the stump after cutting away the projecting part.
**Smith's phenomenon** [Theobald *Smith*, American scientist, 1859– ]. Animals injected with a foreign serum or inert protein often die or show severe symptoms after a second injection, even in minute quantity.
**Smith's reaction for bile-pigments.** Pour tincture of iodine carefully over the liquid to be tested. A green ring appears between the two liquids.
**smoker** (*smo'-ker*). One who uses tobacco. **s.s' cancer.** See *cancer*. **s.s' dyspepsia.** See *dyspepsia*. **s.s' patch,** a chronic inflammation of a small spot of the mucous membrane of the mouth arising from an irritation produced by the pipe. It varies in size from a quarter to a half of an inch in diameter, and is smooth and red in appearance. **s.s' sore-throat,** the condition of catarrh of the pharynx and larynx, with hoarseness, common in habitual smokers. **s.s' vertigo.** See *vertigo*.
**smoke test for plumbing.** "By means of bellows, or some smoke-producing rocket, smoke is forced into the system of pipes, the ends plugged up, and the escape of the smoke watched for, as wherever there are defects in the pipes the smoke will appear" [Price].
**smudging** (*smuj'-ing*). A form of defective speech in which the difficult consonants are dropped.
**Sn.** Chemical symbol of tin [L., *stannum*].
**snaggle-teeth** (*snag'-l*). Irregular and oblique dentition.
**snakeroot.** See *cimicifuga*, *senega*, and *serpentaria*.
**snap-finger.** See *spring-finger*.
**snare** (*snār*). A light or small écraseur, or wire loop, used in removing polpi and small excrescences. **s. cold,** the ordinary snare. **s. galvanocaustic, s., hot,** a snare in which the wire is heated by a galvanic current.
**sneeze** (*snēz*) [AS., *fneósan*, to sneeze]. A sudden, noisy, spasmodic expiration through the nose.
**sneezing.** The act of expelling air violently through the nose. **s., pregnancy,** spasmodic fits of sneezing from hyperemia of the nasal mucosa, following a circulatory disturbance due to pregnancy.

**Snell's laws** [Simeon *Snell,* English ophthalmologist, 1851–1909]. The two laws which govern single refraction: (1) The sine of the incident angle bears a fixed ratio to the sine of the angle of refraction for the same two mediums, the ratio varying with different mediums. (2) The incident and the refracted ray are in the same plane, which is perpendicular to the surface separating the two mediums.
**Snellen's types** [Hermann *Snellen*, Dutch ophthalmologist, 1834–1908]. See *test-types*.
**snore, snoring** [ME., *snoren,* to snore]. 1. To breathe through the nose in such manner as to cause a vibration of the soft palate, thereby producing a rough, audible sound. 2. The sound so produced.
**Snow, external symptoms of** (*sno*). Bulging of the sternum when the thymus gland and its lymphatics are involved secondary to cancer of the breast.
**snow-blindness.** See *blindness, snow-*.
**snuff** (*snuf*) [ME., *snuffen,* to snuff]. 1. Powdered tobacco, variously perfumed and mixed, used for inhalation into the nostrils. 2. A medicated powder to be insufflated into the nostrils. 3. To inhale; to smell. **s.-box, anatomist's,** the *foveola radialis*.
**snuffles.** Coryza, especially of infants, which is frequently due to inherited syphilis.
**soamin** (*so-am'-in*). Trade name of sodium paraaminophenylarsonate, an arylarsonate. It is similar to atoxyl, and is used in syphilis, trypanosomiasis and pellagra.
**soap** (*sapo,* soap]. A chemical compound made by the union of certain fatty acids with an alkali or other metal. According to the alkali used, the soap formed is a potash-soap, soda-soap, ammonia-soap, lead-soap, lime-soap, etc. **s.-bark.** See *quillaia*. **s., Castile,** soap made from olive-oil. **s., gray,** soap to which mercury and benzoinated fat are added. Syn., *sapo cinereus*. **s., green** (*sapo mollis*, U. S. P.), soft soap, made from linseed-oil and potash. The *sapo mollis* of the B. P. is made from olive-oil and potash. Syn., *potash-soap.* **s., green, tincture of** (*linimentum saponis mollis,* U. S. P., B. P.), liniment of soft soap, used as an anodyne. **s., hard.** See **s., soda-.** **s. liniment** (*linimentum saponis,* U. S. P., B. P.), liquid opodeldoc, used as a sedative liniment in rheumatic affections and sprains. **s. plaster** (*emplastrum saponis,* U. S. P., B. P.), used as a local sedative. **s., potash-.** See **s., green.** **s., soda-** (*sapo,* U. S. P.; *sapo durus,* B. P.), hard soap, made from sodium hydroxide and olive-oil. It enters into the composition of various pills, and from it are also prepared soap plaster and soap liniment. **s., soft.** See **s., green.** **s., Spanish,** castile soap.
**sob.** A convulsive inspiration due to contraction of the diaphragm and spasmodic closure of the glottis.
**socaloin** (*so-kal'-o-in*), $C_{16}H_{16}O_7$. Aloin obtained from Socotrine aloes.
**socia parotidis** (*so'-se-ah par-ot'-id-is*). A small separate lobe of the parotid gland.
**social** (*so'-shal*) [*socius,* a companion]. Gregarious, growing near, or together. **s. evil,** prostitution.
**society screw.** The screw at the lower end of the drawtube or body-tube of a microscope for receiving the objective.
**sociology** (*so-se-ol'-o-je*) [*socius,* a fellow-being; λόγος, a treatise]. A treatise on the mutual relations of people and of social organization.
**sock** (*sok*) [ME., *socke*]. 1. A short-legged stocking. 2. An insole. **s. instep arch,** a device to be worn inside the shoe in cases of flat-foot. **s.s, Neapolitan,** socks containing mercurial ointment, which are to be worn continuously for the purposes of inunction.
**socket** (*sok'-et*) [ME., *soket*]. The concavity into which a movable part is inserted.
**socordia** (*so-kor'-de-ah*) [*socors,* silly]. Hallucination.
**soda** (*so'-dah*) [Ital., from L., *solidus,* solid]. 1. Sodium oxide, $Na_2O$. 2. Sodium carbonate or sodium bicarbonate. See *sodium*. 3. Sodium hydroxide, $NaOH$. **s., baking-,** sodium bicarbonate. **s., caustic,** sodium hydroxide. **s., chlorinated,** a mixture of sodium chloride and sodium hypochlorite. **s.-soap.** See *soap, soda-*. **s., washing,** sodium carbonate. **s.-water,** water impregnated with carbon dioxide.
**sodacol** (*so'-dak-ol*). The sodium salt of guaiacol sulphonic acid.

**sodic** (*so'-dik*) [*soda*]. Derived from or containing soda or sodium.
**sodii** (*so'-de-i*) [L.; genitive of sodium]. Of sodium.
**sodium** (*so'-de-um*) [*soda*]. A metallic element of the alkaline group of metals, melting at 95.6° C., and having a specific gravity of 0.97, an atomic weight of 23, and a valence of one. Symbol Na, from the Latin *natrium*. Sodium occurs widely distributed in nature, and forms an important constituent of animal tissues. It has a strong affinity for oxygen and other nonmetallic elements. It is also a constituent of many medicinal preparations. **s.-acetanilid sulphonate,** a white, crystalline mass, readily soluble in water, used as a substitute for antipyrin. **s. acetate** (*sodii acetas*, U. S. P.), NaC₂H₃O₂+3H₂O, is diuretic. Dose 2o gr.-2 dr. (1.3-8.0 Gm.). **s. acid sulphosalicylate.** See *s. sulphosalicylate, acid*. **s. anhydromethylenecitrate,** an antilithemic remedy depending for its action upon the liberation of formaldehyde in the blood. Syn., *citarin*. **s. anisate,** 2NaC₈H₇O₃+H₂O, small colorless scales, soluble in water; antipyretic and antirheumatic. Dose 5-15 gr. (0.32-1.0 Gm.). **s. arsenate** (*sodii arsenas*, U. S. P.), NaH₂AsO₄+7H₂O, clear, colorless, poisonous prisms, with mild alkaline taste, soluble in 4 parts of water; alterative, tonic, antiseptic. From it is prepared *liquor sodii arsenatis* (U. S. P.). Dose 1/64-1/20-1/2 gr. (0.001-0.003-0.008 Gm.). Antidotes—emetics, stomach siphon, fresh ferric hydrate, dialyzed iron, ferric hydrate and magnesia, demulcents, stimulants, warmth. **s. arsenate, exsiccated** (*sodii arsenas exsiccatus*, U. S. P.), Na₂HAsO₄. Dose 1/12 gr. (0.003 Gm.). **s. seniate.** See *s. arsenate*. **s. arsenotrioxate,** a soluble arsenic salt recommended as a substitute for potassium arsenite and arsenic trioxide. **s. aurochloride,** AuCl₃NaCl+2H₂O, a golden-yellow powder said to contain 30 % of gold and freely soluble in water, sparingly so in alcohol.—It is used in syphilis. Dose 1/4-1 gr. (0.01-0.06 Gm.). **s. benzoate** (*sodii benzoas*, U. S. P., B. P.), NaC₇H₅O₂, is used in gout, rheumatism, lithemia, influenza, etc. Dose 1-2 dr. (4-8 Gm.). **s.-benzoyl-sulphonicimide.** See *saccharin*. **s. biborate.** See *s. borate*. **s. bicarbonate** (*sodii bicarbonas*, U. S. P., B. P.), NaHCO₃, is used as an antacid in dyspepsia, gout, rheumatism, lithemia, and diabetes. Dose 10 gr.-1 dr. (0.65-4.0 Gm.). **s. biiodosalicylate.** See *s. diiodosalicylate*. **s. bismuth citropyroborate,** lustrous leaflets, soluble in water, insoluble in alcohol; used in gastralgia. **s. bisulphate,** NaHSO₄+H₂O; it is used as a means of rendering water infected by typhoid bacilli drinkable and harmless for troops in the field. **s. bisulphite** (*sodii bisulphis*, U. S. P.), NaHSO₃+H₂O, opaque prisms or granular powder of disagreeable taste, soluble in 4 parts of water, 72 parts of alcohol, or 2 parts of boiling water. It is antipyretic and antiseptic, used in gastric fermentation and as a parasiticide in skin diseases. Dose 10-30 gr. (0.65-2.0 Gm.). Syn., *leucogen*. **s. biurate,** the deposit of this salt in or upon the tissues of the joints is held to be the etiological factor in arthritic manifestations of gout. **s. borate** (*sodii boras*, U. S. P., B. P.), borax, Na₂B₄O₇. 10H₂O, is used in dysmenorrhea, in uric-acid diathesis, in stomatitis, and as an antiseptic. In over-doses it is a depressant poison. See *boron*. **s. borobenzoate,** a compound of borax, 3 parts, and sodium benzoate, 4 parts, dissolved in water and evaporated. **s. borosalicylate,** a hard mass obtained by triturating 32 parts of sodium salicylate and 25 parts of boric acid with a little water; it is a soluble antiseptic. Syn., *borsalicylate; borsalyl; borosalicyl*. **s. borosulphate,** SO₂. OBO. Na, odorless, vitreous masses of faint, harsh, acidulous taste, soluble in 5 parts of water. It is an internal and external antiseptic. Dose 5-10 gr. (0.3-0.6 Gm.) 5 or 6 times daily, in water. Application, 1 to 2 % solution. Syn., *borol*. **s. bromide** (*sodii bromidum*, U. S. P., B. P.), NaBr, is used like the other bromides. Dose 1/2-2 dr. (2-8 Gm.), **s. cacodylate, s. methylarsenate.** See *arrhenal* and *neoarsycodil*. **s. caffeine sulphate, s. caffeine sulphonate, s. and caffeine sulphonate,** C₈H₉N₄O₃. SO₃Na, soluble in 50 parts of water, in 7 parts of boiling water; used as a diuretic in obesity and dropsy. Syn., *symphorol-sodium; symphorol N*. **s. cantharidinate,** a compound of cantharidin, 0.2 Gm., and sodium hydroxide, 0.3 Gm., dissolved in 20 Cc. of water with heat; the solution is made up to 1000 Cc. It is used subcutaneously in tuberculosis of the throat. Dose 8 gr. (0.5 Gm.). **s. carbolate.** See *s. phenate*. **s. carbonate,** Na₂CO₃.-10H₂O, is used as an antacid, and locally in diseases of the skin and in superficial burns. Dose 10 gr.-1/2 dr. (0.65-2.0 Gm.). *Dried sodium carbonate* is used like the carbonate. Dose 5-15 gr. (0.32-1.0 Gm.). **s. carbonate, monohydrated** (*sodii carbonas monohydratus*, U. S. P.), Na₂CO₃+H₂O. Dose 5 gr. (0.25 Gm.). **s. cetrarate,** Na₂C₁₈H₁₄O₈, a microacicular powder, soluble in water and used as a tonic. Dose 2-15 gr. (0.13-1.0 Gm.). **s. chlorate** (*sodii chloras*, U. S. P.), NaClO₃, has medicinal properties similar to those of potassium chlorate, but is more soluble. Dose 5-15 gr. (0.32-1.0 Gm.). **s. chloride** (*sodii chloridum*, U. S. P., B. P.), NaCl, common salt, is a constituent of animal fluids and tissues and of food. In medicine it is used as a stomachic; in hemoptysis; as an application to sprains and bruises; as a tonic and stimulant in the form of salt-water baths; as a cathartic; and in the form of a 0.6-0.75 % solution to replace the loss of fluids from hemorrhage or profuse diarrhea, being used as an intravenous, subcutaneous, or rectal injection. **s. chloroborate,** a combination of boric acid, borax, sodium chloride, and sodium sulphate used as an antiseptic in typhoid, etc. **s. choleate;** dried purified oxgall, a yellow powder, soluble in water, and used as a tonic and laxative in chronic constipation. Dose 5-10 gr. (0.32-0.65 Gm.). **s. cinnamate,** NaC₉H₇O₂, a white powder, soluble in water; used intravenously in tuberculosis. Dose 1/4-1 gr. (0.005-0.009 Gm.) thrice weekly. Syn., *hetol*. **s. citrate** (*sodii citras*, U. S. P.), 2Na₃C₆H₅O₇+11H₂O. Dose 15 gr. (1 Gm.). **s. citrate, neutral,** 2Na₃C₆H₅O₇+11H₂O, white crystals, soluble in water; used as a purgative in diseases of genitourinary origin, fever, etc. Dose 10-60 gr. (0.65-4.0 Gm.). **s. citrobenzoate,** a white, bulky powder, soluble in water; diuretic, antiseptic, and antilithic. **s. citrophosphate.** See *melachol*. **s. citrotartrate** (*sodii citrotartras effervescens*, B. P.), is refrigerant and laxative. Dose 1-2 dr. (4-8 Gm.). **s. copaivate,** NaC₂₀H₃₀O₂, a yellow, powdery mass, soluble in water, used as antiseptic and diuretic. Dose 10-30 gr. (0.65-2.0 Gm.). **s. corallinate.** See *s. rosolate*. **s. cresylate,** used as in cresol. **s.-diiodoparaphenol sulphonate.** See *s. sozoiodolate*. **s. diiodosalicylate,** 2NaC₇H₃I₂O₃. +5H₂O, white needles or leaflets, soluble in 50 parts of water at 20° C. It is analgesic and antiseptic and used externally on parasitic and syphilitic sores. **s. dioxide,** Na₂O₂, a white powder, soluble in water, used as a bleaching agent and disinfectant in dentistry. Syn., *sodium peroxide*. **s. dithionate.** See *s. thiosulphate*. **s. α-dithiosalicylate,** Na₂C₁₄H₈O₄S₂, a yellowish powder, soluble in water, used in foot-and-mouth disease. **s. β-dithiosalicylate,** Na₂C₁₄H₉O₆S₂, a grayish powder, more soluble in water than the alkali salt. It is used internally in rheumatism, gonorrhea, etc. Dose 1-10 gr. (0.065-0.65 Gm.). Wash in foot-and-mouth disease 2.5 to 5 % solution. **s.-ethoxyphenyl succinamide.** See *s. phenosuccinate*. **s. ethylate,** NaOC₂H₅, is used as a caustic. **s. ethylsulphate,** NaC₂H₅SO₄+H₂O, flat, aromatic crystals, soluble in water or alcohol; used in constipation. Dose 60-300 gr. (4-20 Gm.). **s. fluoride,** NaF, shining crystals, soluble in water, used internally in malaria, epilepsy, etc. Dose 1/16-1/4 gr. (0.005-0.01 Gm.) in solution with sodium bicarbonate; externally as an antiseptic dressing for wounds. Syn., *fluorol*. **s. fluosilicate.** See *s. silicofluoride*. **s. glycerinoborate, s. glyceroborate,** glycerite of borax obtained by heating 40 parts of borax with 60 parts of glycerol and forming a translucent, brittle mass, soluble in water or alcohol; antiseptic. **s. glycerinophosphate, s. glycerophosphate,** Na₂PO₄C₃H₅(OH)₂+H₂O, soluble in water; used in neurasthenia, Addison's disease, phosphaturia, etc. Injections, 3-4 gr. (0.2-0.26 Gm.) daily in solution of sodium chloride. **s. glycholate,** NaC₂₆H₄₂NO₆, a white powder, soluble in water or alcohol; used in chronic constipation and tuberculosis, and as a remedy for gall-stones. Dose 5 gr. (0.32 Gm.) thrice daily. **s. guaiacol carbonate,** a white powder, soluble in water, similar to but milder than sodium salicylate in action. **s. gynocardate,** NaC₁₄H₂₂O₂, a yellow-white powder, soluble in water and alcohol; an antiseptic and alterative, used in leprosy. Dose 5-15 gr. (0.32-1.0 Gm.) twice daily in capsules containing 3 grains each. **s. hippurate,** NaC₉H₈NO₃, a white powder, soluble in boiling water; a solvent for uric acid. **s. hydrate,**

# SODIUM

See *s. hydroxide*. **s. hydroxide** (*sodii hydroxidum*, U. S. P.; *soda caustica*, B. P.), NaOH, is an extremely corrosive substance, occurring in the form of white pencils; it is used as a caustic and as an antacid, like the bicarbonate. From it is prepared *liquor sodii hydroxidi* (U. S. P.), Dose 5–30 min. (0.32–2.0 Cc.). **s. hypochlorite**, NaOCl, is a constituent of *liquor soda chlorinata* (U. S. P., B. P.), Labarraque's solution. **s. hypophosphite** (*sodii hypophosphis*, U. S. P., B. P.), NaPH$_2$O$_2$H$_2$O, is used like the other hypophosphites in pulmonary tuberculosis, scrofula, rickets, etc. Dose 10–30 gr. (0.65–2.0 Gm.). **s. hyposulphite**. See *s. thiosulphate*. **s. ichthyol, s. ichthyolsulphonate**, a dark-brown mass prepared by neutralizing ichthyol-sulphonic acid with an aqueous solution of sodium hydroxide. **s. indigosulphate**, indigo-carmin. **s. iodide** (*sodii iodidum*, U. S. P., B. P.), NaI, is used like potassium iodide. **s. kussinate**, NaC$_{11}$H$_{27}$O$_{10}$, an intensely bitter, yellowish, amorphous mass, soluble in hot water and alcohol; used as a vermifuge. **s. lactate**, NaC$_3$H$_5$O$_3$, a thick syrup used as a hypnotic. Dose 2–4 dr. (8–16 Gm.) in sweetened water. **s. and magnesium borocitrate**, a white antiseptic powder used in lithiasis. Dose 5–30 gr. (0.32–2.0 Gm.). **s. mercurophenyl disulphonate**. See *hermophenol*. **s. metavanadate**, an alterative and succedaneum for arsenic. Dose $\frac{1}{60}-\frac{1}{8}$ gr. (0.001–0.008 Gm.). **s.-methoxysalicylate**. See *s.-guaiacol carbonate*. **s. methylarsenite**. See *neoarsycodil*. Cf. *arrhenal*. **s. β-naphtholate**, **s. betanaphthol**, **s.-naphthol**, microcidin. **s. nitrate** (*sodii nitras*, U. S. P., B. P.), NaNO$_3$, has been used in dysentery. Dose $\frac{1}{2}-1$ oz. (16–32 Cc.). **s. nitrite** (*sodii nitris*, U. S. P., B. P.), NaNO$_2$, forms white crystals, soluble in water, used as a diuretic and antispasmodic like the other nitrites, but its effects are more slowly produced and more permanent. Dose $\frac{1}{2}-2$ gr. (0.032–0.13 Gm.). Recommended for lowering blood-pressure in doses of 2–3 gr. (0.13–0.2 Gm.) every 2 to 4 hours. See *nitrite*. **s. nitroprusside**, Na$_2$Fe(CN)$_5$NO, is used as a reagent. **s. nucleinate**, a white powder, soluble in water, used in puerperal affections and pneumonia and in diagnosing tuberculosis. Dose 30–46 gr. (2–3 Gm.). **s. oleate**, a compound of NaC$_{18}$H$_{33}$O$_2$, with excess of oleic acid, a yellowish mass, soluble in water. Syn., *eunatrol*. **s. orthodinitrocresylate**. See *antinonnin*. **s. orthophosphate**. See *s. phosphate*. **s. ossalinate**, used to prevent the action of cod-liver oil; the sodium compound of the acid of ox-marrow. **s.-oxynaphtholate**, C$_{10}$H$_5$- (OH). COONa, a white, odorless powder, antiseptic and antithermic in action. **s. paracresotate**, NaC$_8$H$_7$O$_3$Na, a fine, microcrystalline, bitter powder, soluble in 24 parts of warm water, used as an antipyretic and antiseptic. Dose 1–20 gr. (0.065–1.3 Gm.). **s. parafluorobenzoate**, is used in tuberculous processes. Dose 8 gr. (0.5 Gm.) 3 times daily. **s. peroxide**. See *s. dioxide*. **s. persulphate**, Na$_2$S$_2$O$_8$- a bactericide and vulnerary, used in 3 to 10 % solution. It is also used as an aperient, 30 gr. (2 Gm.) in 10 oz. (295 Cc.) of water, 1 tablespoonful daily before the principal meals. Syn., *persodine*. **s.- phenacetinsulphonate**, a soluble succedaneum for acetphenetidin. **s. phenate**, NaC$_8$H$_5$O, white crystals, soluble in water, used as an antiseptic. Dose 2–10 gr. (0.13–0.65 Gm.). **s.-phenolphthaleinate**. See *s. phenolsulphonate*. **s. phenolsulphonate** (*sodii phenolsulphonas*, U. S. P.), NaC$_6$H$_5$O$_4$S+2H$_2$O, is used locally as an antiseptic and internally as an antiseptic in intestinal fermentation. Dose 10–30 gr. (0.65–2.0 Gm.). **s.-phenolsulphoricinate**, synthetic phenol, 20 %, and sodium sulphoricinate, 80 %; used in 20 % aqueous solution in skin diseases and in painting false diphtheric membranes, etc. **s. phenosuccinate**, the sodium salt of phenosuccin, a white powder, soluble in water. It is antipyretic and antineuralgic. Dose 7$\frac{1}{2}$–16 gr. (0.5–1.0 Gm.). **s. phosphate** (*sodii phosphas*, U. S. P., B. P.), disodium orthophosphate, Na$_2$HPO$_4$+12H$_2$O, occurs in colorless, translucent, monoclinic prisms, of a saline taste. It is soluble in boiling water and melts at 35° C. It is a mild cathartic and antilithic. Dose 5–40 gr. (0.32–2.6 Gm.). **s. phosphate, effervescing** (*sodii phosphas effervescens*, U. S. P., B. P.), is used like the phosphate. Dose 2–4 dr. (8–16 Gm.). **s. phosphate, exsiccated** (*sodii phosphas exsiccatus*, U. S. P.). Dose 15 gr. (1 Gm.). **s. polyborate**. See *s. tetraborate*. **s. and potassium tartrate**, Rochelle salt; see *potassium and sodium tartrate*. **s. pyrophosphate** (*sodii pyrophosphas*, U. S. P.), Na$_4$P$_2$O$_7$+ 10H$_2$O, is used for preparing ferric pyrophosphate; soluble in boiling water; used in lithiasis. Dose 2–20 gr. (0.13–1.3 Gm.). **s.-rosanilinsulphonate**, NaC$_{20}$H$_{18}$N$_3$O$_6$S$_2$, crystals with green luster obtained from fuchsin by action of fuming sulphuric acid. Syn., *acid fuchsin*; *fuchsin-sodium*; *magenta*; *rubin*. **s. rosolate**, NaC$_{20}$H$_{14}$O$_3$, red masses with green luster used as a dye. Syn., *sodium corallinate*. **s. saccharinate**, the sodium salt of soluble saccharin; a white, crystalline powder containing 90 % of saccharin, soluble in water and having a sweetening capacity 450 times greater than cane-sugar. It is recommended as a valuable intestinal antiseptic. Dose 1 gr. (0.065 Gm.) once or twice daily. Syn., *crystallose*. **s. salicylate** (*sodii salicylas*, U. S. P., B. P.), NaC$_7$H$_5$O$_3$, has the properties and uses of salicylic acid. It occurs in shining white scales, soluble in 0.9 part of water or 6 parts of alcohol. It is antiseptic, antirheumatic, and antipyretic. Dose 2–30 gr. (0.13–2.0 Gm.); maximum dose 60 gr. (4 Gm.) single. **s. santonínate**, 2NaC$_{15}$H$_{19}$O$_4$+7H$_2$O, bitter acicular crystals in stellate groups, soluble in 3 parts of water, 12 parts of alcohol, 0.5 of boiling water, 3.4 of boiling alcohol. It is given for intestinal worms. Dose for adults $\frac{1}{4}-1$ gr. (0.016–0.065 Gm.). **s. silicate**, Na$_2$SiO$_3$, whitish crystals occurring in flat pieces, used in preparing *liquor sodii silicatis*, which is employed as a surgical dressing. Syn., *soluble glass*. **s. silicofluoride**, Na$_2$SiF$_6$, white crystals or granular powder soluble in 200 parts of water. It is used as a styptic, antiseptic, and germicide in aqueous solution of 2 : 1000. Its solution is known as *salufer*. **s. sozoiodolate**, NaOC$_6$H$_3$I$_2$OHSO$_3$+2H$_2$O, long crystals, soluble in 14 parts of water, alcohol, or 20 parts of glycerol. It is alterative and antiseptic. Dose 5–30 gr. (0.32–2.0 Gm.) daily. In whooping-cough 3 gr. (0.2 Gm.) blown into the nose. Externally in skin diseases, syphilis, etc., ointment 10 %, or 1 % aqueous solution. **s. stearate**, **s. stearinate**, NaC$_{18}$H$_{35}$O$_2$, soapy, acicular crystals or scales, soluble in water; it is used in treatment of parasitic skin diseases. **s. succinate**, Na$_2$C$_4$H$_4$O$_4$+6H$_2$O, white crystals, freely soluble in water, recommended in catarrhal icterus. Dose 45 gr. (3 Gm.) daily. **s. sulphanilate**, C$_6$H$_4$- NH$_2$SO$_2$ONa . 2H$_2$O, white plates, soluble in water, recommended in coryza. **s. sulphantimonate**, Na$_3$SbS$_4$+9H$_2$O, large yellow or colorless crystals of the acid of *Schlippe's salt*. **s. sulphate** (*sodii sulphas*, U. S. P., B. P.), Na$_2$SO$_4$ . 10H$_2$O, Glauber's salt, is a hydragogue cathartic and diuretic. Dose $\frac{1}{2}-1$ oz. (16–32 Gm.). **s. sulphate, effervescing** (*sodii sulphas effervescens*, B. P.), is used for the same purposes as the sulphate. Dose 2–4 dr. (8–16 Gm.). **s. sulphite** (*sodii sulphis*, U. S. P., B. P.), Na$_2$SO$_3$+7H$_2$O, is used as to sodium thiosulphate. **s. sulphite benzoate**, a white powder, soluble in water; it is used as a wound antiseptic in the form of a dusting-powder. **s. sulphocaffeate**, bitter crystals, slightly soluble in water; a nontoxic, nonirritating, powerful diuretic. Dose 15 gr. (1 Gm.) in capsules. Syn., *nasrol*; *symphorol*. **s. sulphocarbolate**. See *s. phenolsulphonate*. **s. sulphoricinate**, a compound of sulphoricinic acid and sodium hydroxide. A brown, syrupy liquid, soluble in alcohol and water; used as a solvent for iodine, iodoform, etc. Syn., *polysolve*; *solvin*. **s. sulphoricinate, phenolized**. See *s.-phenolsulphoricinate*. **s. sulphosalicylate, acid**, NaC$_7$H$_5$O$_6$SO$_3$, white, crystalline powder, soluble in water, used as an antiseptic and antipyretic. Dose 10–30 gr. (0.65–2.0 Gm.). **s. sulphovinate**. See *s. ethylsulphate*. **s. sulphurosobenzoate**, a clear, colorless liquid, said to be a harmless antiseptic for wounds. **s. tartrate**, Na$_2$C$_4$H$_4$O$_6$+2H$_2$O, white needles or prisms, soluble in water. Cathartic and diuretic. Used as an antacid and refrigerant in fevers. Dose 4–8 dr. (15–30 Gm.) once a day. **s. taurocholate**, NaC$_{26}$H$_{44}$NSO$_7$, a white powder, obtained from bile of herbivora, soluble in water and alcohol. It is used in deficient biliary secretion. Dose 2–6 gr. (0.13–0.4 Gm.). **s. tellurate**, NaTeO$_4$+2H$_2$O, a white powder, used as an antipyretic, antiseptic, and antihidrotic. Dose $\frac{1}{4}-1$ gr. (0.016–0.05 Gm.) in elixirs. **s. tetraborate**, a compound of equal parts of boric acid and sodium biborate, forming an unctuous, insipid powder, neither toxic nor caustic. It is used in conjunctivitis and keratitis. Syn., *antipyonin*. **s. thioantimonate**. See *s. sulphantimonate*. **s. thiophenate**, C$_6$H$_5$S . SO$_2$Na, a white powder, slightly soluble in water; used on prurigo in 0.5 to 1 % ointment. **s. thiophenesulphonate,** NaC₄H₃S₂O₃+H₂O, a white, scaly powder, containing 33 % of sulphur; used as an antiseptic on prurigo and skin diseases in 5 to 10 % ointment. **s. thiosulphate** (*sodii thiosulphas*, U. S. P., Na₂S₂O₃+5H₂O, is used to check fermentation, and locally in parasitic diseases of the skin and mouth. Dose. 10–20 gr. (0.65–1.3 Gm.). **s. trichlorocarbolate, s. trichlorophenol,** C₆H₃Cl₃. ONa, white needles or crystalline powder, soluble in hot water; antiseptic. **s.-tumenol sulphonate,** a compound of sodium and sulphotumenolic acid. A dark-colored, dry powder, soluble in water. Syn.; *tumenol.* **s. tungstate,** Na₂WO₄+2H₂O, colorless, rhombic, bitter crystals, soluble in 4 parts of water and 2 parts of boiling water. Syn., *sòdium wolframite.* **s. valerate, s. valerianate** (*sodii valerianas,* B. P.), is used as a nervous stimulant. Dose 1–5 gr. (0.065–0.32 Gm.). **s. xanthogenate,** NaC₂H₅OS₂, a compound obtained by adding a saturated alcoholic solution of soda to carbon disulphide. It is antiseptic and germicide.

**sodomist, sodomite** (*sod′-om-ist, sod′m-īt-o*) [*sodom*]. One guilty of sodomy.

**sodomy** (*sod′-om-e*) [*Sodom*, a city of ancient Palestine]. Sexual connection by the anus.

**sodor** (*so′-dor*). The proprietary name for capsules of liquid carbonic acid for preparation of carbonated beverages.

**Soemmering's bone** (*sem′-er-ing*) [Samuel Thomas von Soemmering, German anatomist, 1755–1830]. The marginal process of the malar bone. **S.'s crystalline swelling,** an annular swelling formed in the lower part of the capsule, behind the iris, after extraction of the crystalline lens. **S.'s foramen.** See *S.'s yellow spot.* **S.'s ganglion, S.'s gray substance,** the substantia nigra (locus niger) of the cerebral peduncles. **S.'s ligament,** the suspensory ligament of the lacrimal gland. **S.'s nerve,** the long pudendal nerve. **S.'s yellow spot,** the macula lutea of the retina.

**soft.** Yielding readily to pressure; not hard. **s. palate.** See *palate.* **s. parts,** the tissues of the body other than bone and cartilage. **s. soap.** See *soap, green.* **s. water,** one containing but little mineral matter and forming free lather with soap.

**softening** (*sof′-en-ing*) [ME.]. The act of becoming less cohesive, firm, or resistant. **s., acute gastric,** a disease of childhood in which the stomach and intestines are said to undergo softening. It is probably a post-mortem phenomenon. **s., anemic,** disintegration and liquefaction of the brain-substance from lack of blood-supply. **s. of the bones,** osteomalacia. **s. of the brain,** a disease of the cerebral tissue dependent upon inflammation or blood failure, the symptoms varying according to the part affected, but consisting in loss of function, partial or complete. According to the appearances presenting the softening has been distinguished as red, yellow, or white. See *general paralysis of the insane.* **s. colliquative,** the name applied to that condition in which the affected tissues liquefy. **s., esophageal,** softening of the lower portion of the esophagus due to the solvent action of the gastric juice. **s., gray,** an inflammatory softening of the brain or cord with a gray discoloration. **s., green,** a purulent softening of nervous matter. **s. of the heart,** myomalacia cordis, a softening of the cardiac muscle consequent on arterial anemia. **s., hemorrhagic,** the softening of parts involved in a hemorrhage. **s., mucoid,** myxomatous degeneration. **s., red or yellow** (of the brain), when hemorrhage accompanies the ischemic softening, and the products of disintegration of the blood mingle with the nerve substance, giving it a red or yellow hue. **s. of the spinal cord,** various stages in myelitis known by the terms gray, green, red, white, and yellow softening. **s. of the stomach,** gastromalacia, consequent upon highly acid contents with a feeble circulation in the walls, but usually a post mortem phenomenon. See *auto-digestion.* **s. of a thrombus,** may be *simple* or *red, puriform* or *yellow,* the latter resulting in the extremely unfavorable condition of *thrombophlebitis.* **s., white** (*of the brain*), when the ischemia is unaccompanied with hemorrhage.

**soil** [ME.]. The ground; earth. **s.-diseases,** those diseases supposed to be produced by emanations from a decomposing organic soil, or arising from imperfect drainage of decaying animal matter. **s. pipe,** the main discharge-pipe of a system of house-plumbing; usually an upright, hollow cylinder of iron. **s. water.** See *subsoil-water.*

**soja-beans, soy-beans** (*so′-yah*). The edible seeds of *Glycine soja,* a leguminous plant of the East Indies. The meal of the soja-beans is used in diabetes. They contain a diastatic ferment, casein, cholesterin, lecithin, asparagin, leucin, cholin, hypoxanthin bases, phenylamidopropionic acid, oil (18 %), sugar (12 %).

**sokodu** (*so′-ko-doo*). A disease which follows the bite of rats; observed in Japan and China.

**sokra** [E. Ind., "without flesh or blood; skeleton"]. Vernacular for trypanosomiasis (*q. v.*).

**sol** (*sol*). A colloid in solution.

**sol.** Abbreviation of *solution.*

**solanidine** (*so-lan′-id-ēn*). An alkaloid obtained by decomposing solanine.

**solanine** (*so′-lan-ēn*) [*solanum,* the nightshade]. An alkaloid found in various species of solanum.

**solanism** (*so′-lan-izm*). Nightshade poisoning from ingestion of berries of *Solanum dulcamara* or *S. nigrum,* or rarely through eating unripe potatoes; due partly to the contained glucoside causing vomiting, pain, and diarrhea, partly to tropeine, marked by symptoms of belladonna poisoning.

**solanoid** (*sol′-an-oid*) [*solanum,* nightshade; *eldos,* like]. Of a potato-like texture, as a solanoid carcinoma.

**solanoma** (*so-lan-o′-mah*). A solanoid tumor.

**Solanum** (*so-la′-num*) [L.]. A genus of the *Solaneæ,* including the tomato, potato, bitter-sweet, and black nightshade. *S. carolinense,* horse-nettle, is indigenous to the United States; a fluidextract from the fresh berries is recommended in epilepsy, tetanus, and convulsions of pregnancy. Dose 10–30 min. (0.6–1.8 Cc.). It is also used as an abortifacient. *S. crispum, S. gayanum,* and *S. tomatilo* are indigenous to Chili and Peru, and are used under the name of *natrix* in inflammatory fevers, in typhus, etc. *S. dulcamara,* bittersweet, is indigenous to Europe and Asia, and contains dulcamarine and solanine. An extract from the young branches is employed as an alterant and diuretic in dropsy, cutaneous diseases, and rheumatism. Dose 5–20 gr. (0.32–1.3 Gm.). *S. insidiosum* and *S. paniculatum,* jurubeba, are species of Brazil, and are alterative and antiblennorrheic. Dose of *fluidextract* in gonorrhea and syphilis 15–30 min. (0.9–1.8 Cc.). *S. nigrum,* nightshade, is found in Europe, Asia, and America, is used as a diuretic and emetic, and externally as a cataplasm. *S. tuberosum* is the potato, indigenous to Chili; it contains solanine, solanidine, solaneine, and a small amount of tropeine. The tubers contain the proteid tuberin.

**solar ganglion** (*so′-lar*) [*solaris,* from *sol,* the sun, so-called because of the radiating nerves]. See under *s. plexus.* **s. plexus,** a plexus consisting of a network of nerves and ganglia (*solar ganglia*), and situated behind the stomach and in front of the aorta and crura of the diaphragm. It receives the great splanchnic nerves and filaments from the right pneumogastric nerve, and supplies branches to all the abdominal viscera.

**solarium** (*so-la′-re-um*) [*solaris;* solar; *sol,* sun]. A room enclosed with glass, and arranged for the administration of sun-baths.

**solarization** (*so-lar-i-za′-shun*) [*sol,* the sun]. The application of solar or electric light for therapeutic purposes.

**solaro** (*so-la′-ro*) [*sol,* the sun]. A cloth, shot with an orange colored material, and recommended for use in tropical countries as a protection from the rays of the sun.

**Solayrès' obliquity** (*so-lār-a*) [François Louis Joseph *Solayrès de Renhac,* French obstetrician, 1737–1772]. Lateral obliquity. Descent of the child's head by its occipitomental diameter into the oblique diameter of the pelvis.

**Soldaini's solution for glucose** (*sol-dah-e′-ne*) [Arturo *Soldaini,* Italian chemist]. Fifteen Gm. of copper carbonate dissolved in 1500 Cc. of water, to which is added 416 Gm. of potassium bicarbonate. A reduction of copper suboxide is obtained by heating the foregoing solution with a glucose solution.

**sole** (*sōl*) [ME.]. The plantar surface of the foot. **s.-leather.** See *leather.* **s.-plate,** (*a*) the name given by Boas to the parmar side of claws and hoofs, as distinguished from the volar side (*Sohlenhorn*); (*b*) the flattened nucleated mass of soft, faintly granular protoplasm closely applied to the surface of a voluntary muscle to receive the ultimate

fibrillæ of the medullated nerve-fibers composing its motor supply. It forms part of the motor disc or endplate. **s.-reflex.** See *reflex, plantar*.
**solen** (*so'-len*) [σωλήν, a channel]. 1. A channel. 2. The central canal of the spinal cord.
**solenochalasis** (*so-len-o-kal-a'-sis*) [σωλήν, channel; χαλαστικός, making supple]. Dilatation of a tubular organ.
**solenoid** (*so'-len-oid*) [σωλήν, a pipe; εἶδος, likeness]. A spiral of conducting wire wound into a cylindrical shape so that it is almost equivalent to a number of equal and parallel circuits arranged upon a common axis; in therapeutics the name is applied to a large cage used for holding the patients in teslaization in such manner that they are not in direct communication with the current.
**solenostegnosis** (*so-len-o-steg-no'-sis*) [σωλήν, a channel; στέγνωσις, stenosis]. Constriction of a tubular organ.
**soleus** (*so-le'-us*). A flat muscle of the calf. See under *muscle*.
**solferino** (*sol-fer-e'-no*) [an Italian city, the scene of a battle]. A synonym of *fuchsin*.
**solicictus** (*so-lis-ik'-tus*). Synonym of *heat-stroke*.
**solid** (*sol'-id*) [*solidus*, solid]. 1. Firm; dense; not fluid or gaseous. 2. Not hollow. 3. A firm body; a body the molecules of which are in a condition of strong mutual attraction.
**Solidago** (*sol-id-a'-go*) [*solidus*, solid: *gen.*, *solidaginis*]. Golden-rod, a genus of some 100 species of composite flowered plants, mostly American. **S. odora**, is carminative, diaphoretic, stimulant, diuretic, and antemetic. **S. rigida**, is tonic and astringent. **S. virgaurea**, of both continents, is astringent, tonic, and vulnerary.
**solidarity** (*sol-id-ar'-it-e*) [*solidus*, solid]. The unitary nature of the relations of the various parts of an organism, whereby all individual parts are subordinated to the welfare of the whole.
**solidification** (*sol-id-if-ik-a'-shun*) [*solidus*, solid; *facere*, to make]. The act of becoming solid, or of possessing molecular attraction.
**solidism** (*sol'-id-izm*) [*solid*]. The theory that diseases depend upon alterations in the solids of the body.
**solidist** (*sol'-id-ist*) [*solidus*, solid]. The name given to one opposed to the doctrines of the humoralists.
**solitary** (*sol'-it-a-re*) [*solitarius*, solitary]. Single; existing separately; not collected together. **s. bundle, s. fasciculus**, a strand of nerve-fibers in the medulla. **s. follicles**, **s. glands**, minute lymphatic nodules in the mucous membrane of the intestine.
**solium** (*so'-le-um*). A variety of tape worm. See *tænia solium*.
**sollunar** (*sol-lū'-nar*) [*sol*, sun; *luna*, moon]. Influenced by or relating to the sun and the moon.
**Solly's arciform band.** See *Rolando's arciform fibers*.
**soloid** (*sol'-oid*). Trade name applied to chemical and other substances which are compressed.
**Solomon's seal** (*sol'-o-monz sēl*). The root of *Convallaria polygonatum*, a tonic, mucilaginous and slightly astringent. It was formerly a popular domestic remedy for rheumatism and gout, and is externally employed in contusions. Dose of the fldext. ʒ j–ij.
**solphinol** (*sol'-fe-nol*). A mixture of borax, boric acid, and sulphurous alkalies; a white, crystalline, odorless powder, soluble in 10 parts of water or in 20 parts of glycerol. It is used as an antiseptic.
**Solpugidæ** (*sol-pū'-je-de*) [*solpuga*, a venomous spider]. A group of spider-like arachnids having closer relationship to the scorpions than to the true spiders. Their bite is poisonous.
**solubility** (*sol-ū-bil'-it-e*) [*solubilis*, from *solvere*, to dissolve]. The state of being soluble.
**soluble** (*sol'-ū-bl*). Capable of being dissolved.
**solurol** (*sol'-ū-rol*). Trade name of a preparation of thyminic acid; said to be a uric acid solvent.
**solute** (*so-lūt'*). The substance dissolved in a solution.
**solutio** (*so-lū'-she-o*). See *solution*.
**solution** (*so-lū'-shun*) [*solutio*, from *solvere*, to loosen]. 1. A separation or break, as *solution* of continuity. 2. The process of dissolving a solid or of being dissolved. 3. A liquid in which a substance has been dissolved. 4. **colloidal**, one obtained by dipping bars of metal into pure water and passing a heavy electric current from one bar to the other through the water. The metal under these conditions is torn off in in a state of such fine division that it remains suspended in the water in the form of a solution. Syn., *pseudosolution*. **s. of contiguity**, a dislocation. **s. of continuity**, the division of a tissue. **s. of cresol, compound** (*liquor cresolis compositus*, U. S. P.), cresol, 500 Gm.; linseed-oil, 350 Gm.; potassium hydroxide, 80 Gm.; water, to make 1000 Gm. **s. grammolecular**, one in which each liter contains the weight of one molecule of the active chemical expressed in grams. **s.s, isotonic**, such as are equal in osmotic pressure. **solutio lithantracis acetonica, a. solution** of coal-tar 10 parts, in benzol 20 parts, and acetone 77 parts. It is employed in skin diseases. See *s., normal*. *normal solution*. **s., normal saline.** See *saline solution*. **s., potassium silicate**, a colorless, slightly turbid, syrupy liquid with alkaline reaction, consisting of 10 % of potassium silicate in water, K₂SiO₃-H₂O. **solutio retinæ**, detachment of the retina. **s., saturated.** See *saturated* (1). **s., standard**, a solution containing a definite quantity of a reagent. **s., test-**, a standard solution. **s., volumetric**, a standard solution. See **s., water-glass.** See *s., potassium silicate*. (For solutions not defined here see the qualifying word.)
**solutol** (*sol-ū'-tol*) [*solution*]. An alkaline solution of cresol in sodium cresylate, used as a disinfectant.
**solv.** Abbreviation of Latin *solve*, dissolve.
**solvella** (*sol-vel'-ah*) [pl., *solvellæ*]. A soluble tablet.
**solvent** (*sŏl'-vent*) [*solvere*, to dissolve]. 1. Capable of dissolving. 2. A liquid capable of dissolving.
**solveol** (*sol'-ve-ol*). A neutral solution of cresol in sodium cresylate, used as a disinfectant.
**solvin** (*sol'-vin*). Sodium sulphoricinate; one of a series of liquids obtained from certain oils by the action of concentrated sulphuric acid; it is a powerful solvent, and also possesses the property of dissolving the red corpuscles.
**soma** (*so'-mah*) [σῶμα, the body: *pl.*, *somata*]. 1. The body alone, considered without the limbs. 2. The entire body with the exclusion of the germ-cells.
**somacule** (*so'-mak-ūl*) [dim. of σῶμα, the body]. A physiological unit corresponding to, but greatly more conplex than, the chemical molecule; the smallest possible division of protoplasm.
**somæsthesia.** *Somatesthesia*.
**somal** (*so'-mal*). Pertaining to the body.
**somascesis** (*so-mas-se'-sis*). See *gymnastics*.
**somatesthesia, somatæsthesia** [σῶμα, body; αἰσθησις, sensation]. Bodily sensation, the consciousness of the body.
**somatic** (*so-mat'-ik*) [σωματικός, from σῶμα, body]. 1. Pertaining to the body. 2. Pertaining to the framework of the body and not to the viscera. **s. cavity**, the body-cavity or perivisceral cavity. **s. cells**, undifferentiated body-cells or parenchymatous cells. **s. death**, the final cessation of all vital activities in the body at large; see *death*. **s. mesoderm**, the upper or outer leaf of the mesoderm separated by the cœlomic fissure from the lower or inner leaf. **s. musculature**, the muscles of the outer wall of the body somatopleure, as distinguished from those of the splanchnopleure, the splanchnic musculature.
**somaticosplanchnic** (*so-mat-ik-o-splank'-nik*) [σῶμα, body; σπλάγχνα, viscera]. Same as *somaticovisceral*. **somaticovisceral** (*so-mat-ik-o-vis'-er-al*) [σῶμα, body; *viscera*]. Relating to the body and the viscera.
**somatoblast** (*so-mat'-o-blast*) [σῶμα, body; βλαστός, a germ]. Any plastidule from which cell-material (in contradistinction to nuclear material) is built up or developed.
**somatochrome** (*so-mat'-o-krōm*) [σῶμα, body; χρῶμα, color]. Applied by Nissl to a group of nerve-cells possessing a well-defined cell-body completely surrounding the nucleus on all sides, the protoplasm having a distinct contour, and readily taking a stain. This group is divided into arkyochrome, stichochrome, arkyostichochrome, and gyrrochrome nerve-cells.
**somatodidymus** (*so-mat-o-did'-im-us*) [σῶμα, body; δίδυμος, twin]. A double monster having the trunks united.
**somatodymia** (*so-mat-o-dim'-e-ah*) [σῶμα, body; δύειν, to enter]. A twin monstrosity in which the trunks are united. There are several varieties: *ischiodymia*, union by the hips; *infraomphalodymia*, union in the inferior umbilical region; *omphalo-*

# SOMATOGENIC 813 SOPHOMANIA

**dymia**, union in the umbilical region; *supraomphalodymia*, union in the superior umbilical region; *sternodymia*, by the sternum; *sternoomphalodymia*, union by the sternal and the umbilical regions; *vertebrodymia*, union by the vertebræ.

**somatogenic** (*so-mat-o-jen'-ik*) [σῶμα, the body; γεννᾶν, to produce]. Pertaining to somatogeny.

**somatogeny** (*so-mat-oj'-en-e*) [σῶμα, body; γεννᾶν, to produce]. The acquirement of bodily characters, especially the acquirement of characters due to the environment.

**somatologic** (*so-mat-o-loj'-ik*) [σῶμα, body; λόγος, science]. Pertaining to somatology.

**somatology** (*so-mat-ol'-o-je*) [σῶμα, body; λόγος, science]. The study of anatomy and physiology of organized bodies; biology apart from psychology.

**somatome** (*so'-mat-ōm*) [σῶμα, body; τομή, a cutting]. 1. A transverse segment of an organized body; a somite. 2. An embryotome.

**somatomegaly** (*so-mat-o-meg'-al-e*) [σῶμα, body; μέγα, large]. Gigantism.

**somatomic** (*so-mat-om'-ik*) [σῶμα, body; τεμνεῖν, to cut]. Pertaining to a somatome.

**somatopagus** (*so-mat-op'-ag-us*) [σῶμα, body; πάγος, fixed]. A double monstrosity having two trunks.

**somatoplasm** (*so'-mat-o-plazm*) [σῶμα, the body; πλάσμα, anything formed]. The protoplasm of the body-cells; Weismann's term for that form of living matter which composes the mass of the body, and which is the subject of death, as distinguished from germ-plasm, which composes the reproductive cells and is possessed of potential immortality.

**somatopleural** (*so-mat-o-ploo'-ral*) [σῶμα, body; πλευρά, side]. Pertaining to a somatopleure.

**somatopleure** (*so-mat'-o-ploor*) [σῶμα, body; πλευρά, the side]. The body-wall; the somatic mesoblast.

**somatose** (*so'-mat-ōs*) [σῶμα, body]. A proprietary albumose food-product. s., ferro-, s., iron, a preparation of somatose containing 2 % of iron. Dose 75-150 gr. (5-10 Gm.) daily. s., milk, a tasteless, inodorous food in the form of a powder prepared from milk with 5 % of tannic acid; used in chronic diseases of the digestive tract. Dose for adults 2 or 3 tablespoonfuls (30-45 Cc.) daily.

**somatosplanchnopleuric** (*so-mat-o-splank-no-ploo'-rik*). Relating to the somatopleure and the splanchnopleure.

**somatotomy** (*so-mat-ot'-o-me*) [σῶμα, body; τομή, section]. Anatomy; dissection.

**somatotridymus** (*so-mat-o-trid'-im-us*) [σῶμα, body; τρίδυμος, triple]. A monster with three trunks or bodies.

**somesthetic, somæsthetic** (*so-mes-thet'-ik*) [σῶμα, body; αἴσθησις, sensation]. Pertaining to general sensory structures. s. area, Munk's *Körperfühlsphäre*, the region of these cortex in which the axons of the general sensory conduction-path terminate. s. path, the general sensory conduction-path leading to the cortex.

**somiology** (*so-mi-ol'-o-je*) [σῶμα, body; λόγος, science]. A term proposed by Rafinesque, 1814, as a common name under which to consider the phenomena of organic nature, now covered by the term biology. Syn., *organology*; *organomy*.

**somite** (*so'-mīt*) [σῶμα]. 1. A segment of the body of an embryo. 2. One of a series of segments of the mesoblast on each side of the dorsal ridge of the embryo; a protovertebra; a protovertebral or mesoblastic somite.

**somitic** (*so-mit'-ik*) [σῶμα, the body]. Resembling or pertaining to a somite.

**somnal** (*som'-nal*) [*somnus*, sleep]. A crystalline substance, a compound of chloral hydrate and urethane. It is diuretic and hypnotic. Dose 30 gr. (2 Gm.).

**somnambulance** (*som-nam'-bū-lans*). Same as somnambulism.

**somnambulation** (*som-nam-bū-la'-shun*). Same as somnambulism.

**somnambulator** (*som-nam'-bū-la-tor*). Same as *somnambulist*.

**somnambulism** (*som-nam'-bū-lizm*) [*somnus*, sleep; *ambulare*, to walk]. 1. The condition of half-sleep, in which the senses are but partially suspended; also sleep-walking, a condition in which the individual walks during sleep. 2. The type of hypnotic sleep in which the subject is possessed of all his senses, often having the appearance of one awake, but whose will and consciousness are under the control of the hypnotizer. **somnambulisme, provoque**, sleep-walking induced by mesmerism, hypnotism, or "electrobiology."

**somnambulist** (*som-nam'-bū-list*). One who walks in his sleep.

**somnial** (*som'-ne-al*) [*somniatio*, dreaming]. Relating to dreams.

**somniation** (*som-ne-a'-shun*) [*somniatio*]. Dreaming.

**somniative, somniatory** (*som'-ne-at-iv, som'-ne-at-o-re*). Relating to dreaming; producing dreams.

**somniculous** (*som-nik'-ū-lus*) [*somnus*, sleep]. Drowsy; sleepy.

**somnifacient** (*som-ne-fa'-shent*) [*somnus*, sleep; *facere*, to make]. 1. Producing sleep. 2. A medicine producing sleep; a hypnotic.

**somniferin** (*som-nif'-er-in*) [*somnus*, sleep; *ferre*, to bear]. A morphine ether discovered by Bombelon, said to be stronger than morphine and without bad effects or influence upon the heart.

**somniferine** (*som-nif'-er-ēn*) [*somnus*, sleep; *ferre*, to bear]. An alkaloid derived from *Withania somnifera*, a solanaceous plant of Asia and the Mediterranean region. It is said to be narcotic.

**somniferous** (*som-nif'-er-us*) [*somnus*, sleep; *ferre*, to bear]. Producing sleep.

**somnific** (*som-nif'-ik*) [*somnus*, sleep]. Causing sleep.

**somnifugous** (*som-nif'-ū-gus*) [*somnus*, sleep; *fugere*, to flee]. Driving away sleep.

**somniloquence, somniloquism, somniloquy** (*som-nil'-o-kwens, som-nil'-o-kwizm, som-nil'-o-kwe*) [*somnus*, sleep; *loqui*, to talk]. The act of talking during sleep.

**somniloquist** (*som-nil'-o-kwist*). One given to talking during sleep.

**somnipathist, somnipathy.** See *somnopathist*, *somnopathy*.

**somnoform** (*som'-no-form*). An anesthetic consisting of ethyl chloride, 60 %; methyl chloride, 35 %; ethyl bromide, 5 %.

**somnol** (*som'-nol*). A synthetic product of chloralurethane with a polyatomic alcohol radical; used as a hypnotic and cerebral sedative. Dose 2-4 dr. (7.7-15.5 Cc.).

**somnolence** (*som'-no-lens*) [*somnolentia*; *somnus*, sleep]. A condition of drowsiness or sleep.

**somnolent** (*som'-no-lent*) [*somnolentus*]. Inclined to sleep.

**somnolentia** (*som-no-len'-she-ah*) [L.]. Sleep-drunkenness, a condition of incomplete sleep in which a part of the faculties are abnormally excited, while the others are in repose.

**somnolescent** (*som-no-les'-ent*) [*somnus*, sleep]. 1. Drowsy. 2. Inducing drowsiness.

**somnolism** (*som'-no-lizm*). Hypnotism.

**somnone** (*som'-nōn*). A proprietary hypnotic said to contain opium, lupulin, and lactucarium. Dose 16-32 min. (1-2 Cc.).

**somnopathist** (*som-nop'-ath-ist*) [*somnus*, sleep; πάθος, disease]. One subject to hypnotic trance.

**somnopathy** (*som-nop'-ath-e*). Hypnotic somnambulism.

**somnos** (*som'-nos*) [*somnus*, sleep]. A proprietary hypnotic formed by the chemical reaction between chloral hydrate and glycerol in certain proportions.

**somnovigil** (*som-no-vij'-il*). See *coma-vigil*.

**somnus** (*som'-nus*) [L.]. Sleep; see *hypnosis*.

**sonifer** (*son'-if-er*) [*sonus*, sound; *ferre*, to carry]. A variety of ear-trumpet.

**sonitus** (*son'-it-us*). See *tinnitus*.

**sonometer** (*so-nom'-et-er*) [*sonus*, a sound; μέτρον, a measure]. 1. An instrument for determining the pitch of sounds and their relation to the musical scale. 2. An instrument for testing hearing.

**sonorous** (*so-no'-rus*) [*sōnus*, sound]. Capable of producing a musical sound, resonant; of rales, low-pitched.

**sonus** (*so'-nus*). See *sound*.

**soor.** See *thrush*.

**soot-cancer, soot-wart.** Epithelioma of the scrotum; so-called from its frequency in chimney-sweeps.

**sophistication** (*so-fis-tik-a'-shun*) [σοφιστικός, deceitful]. The adulteration or imitation of a substance.

**sophol** (*so'-fol*). Proprietary name of a compound of formaldehyde, nuclein and silver; silver methylenenucleinate.

**sophomania** (*sof-o-ma'-ne-ah*) [σοφός, wise; μανία, madness]. Insanity in which the patient believes himself to excel in wisdom.

**Sophora** (*so-fo'-rah*) [Arab.]. A genus of leguminous trees, shrubs, and herbs, mostly growing in warm regions. **S. sericea** (see *loco*) is a poisonous plant of the U. S.; its seeds contain sophorine. **S. speciosa**, a tree of Texas, also yields sophorine.
**sophorine** (*so-fo'-rēn*) [Arab.]. A paralyzant, poisonous alkaloid which exists in the seeds of some species of *Sophora*.
**sophronistæ dentes** (*sof-ro-nis'-te den'-tēz*). Wisdom teeth, or dentes sapientiæ.
**sopor** (*so'-por*) [L.]. Sleep, especially the profound sleep symptomatic of a morbid condition.
**soporate** (*so'-por-āt*) [*sopor*]. To stupefy; to render drowsy.
**soporifacient** (*so-por-if-a'-se-ent*) [*sopor*, sleep; *facere*, to make]. A drug producing sleep; a hypnotic.
**soporiferous** (*so-por-if'-er-us*). See *soporific*.
**soporific** (*so-por-if'-ik*). [*sopor*; *facere*, to make]. 1. Producing sleep. 2. A sleep producer. 3. Narcotic.
**soporose, soporous** (*so'-por-ōs, so'-por-us*) [*sopor*, sleep]. Sleepy; partaking of the nature of sound sleep.
**sora** (*so'-rah*). Synonym of *urticaria*.
**sorbefacient** (*sor-be-fa'-shent*) [*sorbere*, to suck; *facere*, to make]. 1. Promoting absorption. 2. A medicine or agent that induces absorption.
**sorbic** (*sor'-bik*) [*sorbus*, the sorb-tree]. Pertaining to or derived from the mountain ash. **s. acid.** See *acid*.
**sorbin** (*sor'-bin*). See *sorbinose*.
**sorbinose** (*sor'-bin-ōs*) [*sorbus*, the sorb-tree]. *Sorbine*, a ketone alcohol, found in mountain-ash berries, and consisting of large crystals, which possess a very sweet taste. It reduces alkaline copper-solutions, but is incapable of fermentation under the influence of yeast.
**Sorbite** (*sor'-bīt*) [*sorbus*, the sorb-tree], C₆H₁₄O₆+H₂O. A hexahydric alcohol occurring in mountain ash berries, forming small crystals which dissolve readily in water. They melt at 110° C. Sorbite corresponds, in all probability, to grape-sugar.
**sorbose** (*sor'-bōs*). Same as *sorbinose*.
**Sorby's cells** (*sor'-be*). For *spectroscopic examination of blood:* a narrow-lumen glass receptacle made of barometer tubing, both ends of which are accurately ground to parallel surfaces, one end being cemented to a small polished glass plate.
**sordes** (*sor'-dēz*) [*sordere*, to be foul]. Filth, dirt, especially the crusts that accumulate on the teeth and lips in continued fevers. **s. aurium**, cerumen. **s. gastricæ**, undigested gastric debris. Syn., *saburra gastrica*.
**sordid** (*sor'-did*) [*sordidus*, dirty; filthy]. In biology, of a dull or dirty color.
**sordidin** (*sor'-did-in*), C₁₄H₁₈O₇ or C₁₅H₂₀O₈. A substance isolated from the lichen, *Zeora sordida*.
**sore.** 1. Painful; tender. 2. An ulcer or wound. **s., bed-.** See *bed-sores*. **s., Delhi, s., Penjdeh, s., natal.** See *furunculus orientalis*. **s. feet of coolies, s.s, water-.** See *itch*, *coolie*. **s., hard,** chancre. **s. mouth**, Ceylon. See *aphthæ tropicæ*. **s. soft**, chancroid. **s. throat of Fothergill**, ulcerative angina of severe scarlatina (scarlatina anginosa). **s., Veld.** See under *Veld*. **s. venereal**, chancroid.
**Soret's band.** An absorption band in the extreme violet end of the spectrum of blood; it is characteristic of hemoglobin.
**sorghum** (*sor'-gum*). A variety of sugar-cane. **s. saccharatum** of the family Gramineæ. Also, a syrup made from the expressed inspissated juice of the same.
**soroche** (*so-ro'-cha*) [Sp.]. Mountain sickness.
**sororiation** (*sor-or-e-a'-shun*) [*soror*, a sister]. The development which takes place in the female breasts at puberty.
**sorrocco** (*sor-ok'-o*): Puna. An affection resembling sea-sickness, common in the high regions of South America. See *soroche*.
**sorts** (*sorts*). In the drug-trade, refuse or culls; the poorest grade of any drug.
**soson** (*so'-son*). Unaltered meat-albumin, 98.5 %, in powder. It is odorless and palatable.
**sostrum** (*sos'-trum*) [*sostron*, a reward for saving life]. A physician's fee.
**soterocyte** (*so'-ter-o-sīt*) [σωτήρ, a preserver; κύτος, a cell]. A blood platelet.
**sotopan** (*so'-to-pan*). A proprietary remedy said to contain iron, quinine, bromine, calcium and phosphoric acid.

**souffle** (*soo'-fl*) [Fr.]. A blowing sound; an auscultatory murmur; a bruit. **s., cardiac.** See *heart-murmur*. **s., fetal**, an inconstant murmur heard over the uterus during pregnancy, and supposed to be due to the compression of the umbilical cord. **s., funic, s., funicular,** a hissing sound, synchronous with the fetal heart-sounds, heard over the abdomen of a pregnant woman, and supposed to be produced in the umbilical cord. **s., placental, s., uterine,** a sound heard in the latter months of pregnancy, and caused by the entrance of blood into the dilated arteries of the uterus. **s., splenic,** a sound said to be audible over the spleen in cases of malaria and leukemia. **s., umbilical.** See *s., funic*.
**soul** (*sōl*) [ME., *soule*]. The moral and emotional part of man's nature. **s.-blindness.** See *blindness, psychic*, and *apraxia*. **s.-deafness**, deprivation of all sensation of sound or reminiscence of it. **s., Stahl's**, according to the doctrine of George Ernst Stahl (1660–1734), the supreme, life-giving, life-preserving principle, distinct from the spirit; when hindered in its operation, disease resulted; it governed the organism chiefly by way of the circulation. His doctrine was called *animism*, and was a reaction against the chemical and mechanical theories of the seventeenth century.
**sound** (*sonus*). 1. The sensation produced by stimulation of the auditory nerve by aerial vibrations. 2. [Fr., *sonder*, to probe]. An instrument for introduction into a channel or cavity, for determining the presence of constriction, foreign bodies, or other morbid conditions, and for the purpose of treatment. **s., anasarcous,** a moist bubbling sometimes heard on auscultation when the skin is edematous. **s., bandbox**, the resonant percussion note sometimes heard in emphysema. **s., Bellocq's.** See *Bellocq's cannula*. **s., bellows**, an endocardial murmur which sounds like a bellows; see *bellows*. **s., blowing, s., bottle.** See *amphoric breathing*. **s., bronchial**, the large harsh sound of bronchial respiration. **s.s, cardiac.** See *s., heart*. **s., cracked-pot**, a form of tympanitic resonance indicative of a cavity. **s., esophageal**, a long flexible sound for examination of the esophagus. **s.s, fetal heart-,** the sounds produced by the beating of the fetal heart, best heard near the umbilicus of the mother. **s., flapping,** the clap made by the closure of the cardiac valves. **s.s, friction-,** the sounds produced by the rubbing of one rough surface upon another. **s., funicular bellows.** See *souffle*. **s.s, heart-,** the two sounds heard over the cardiac area. The first dull and prolonged, is said to sound like *lubb*, and is isochronous with the systole of the ventricles. The second, sharp and short, is said to sound like *dup*, and is isochronous with the closure of the semilunar valves. **s., kettle-singing**, a chest-sound sometimes heard in incipient pulmonary tuberculosis. It resembles water boiling in a kettle. **s., lacrimal**, a fine sound for exploring or dilating the lacrimal canal. **s., metallic heart-.** See *metallic tinkling*. **s., metamorphosing breath-,** a sound due to the passage of air through a narrow opening into and out of a pulmonary cavity. **s., muscle-,** the sound heard through the stethoscope when placed over a muscle in the state of contraction; sussurus. **s., osseous,** a high-pitched intense auscultatory sound having a slightly metallic timbre. **s., pulmonary,** the respiratory murmur. **s., respiratory,** respiratory murmur. **s., sawing,** a cardiac murmur resembling the sound produced by sawing. **s.-shadow,** the interference with a sound-wave caused by an object being placed between the ear and the source of sound. **s., subjective.** See *phantasm*. **s., to-and-fro,** the friction-sound of pericarditis and pleuritis. **s., tubular,** the sound of tracheal respiration. **s., urethral,** an elongated steel instrument, usually slightly conical, for examination and dilatation of the urethra. **s., uterine,** a graduated probe for measurement of the uterine cavity.
**sour** (*sowr*). Having an acid taste; fermented. **s. dough.** See *leaven*. **s.-wood,** sorrel tree; the leaves of *Oxydendron arboreum;* they are tonic, refrigerant and diuretic, and of reputed value in dropsy. Dose of the fldext. ℞ xxx–℥ ij.
**Southern fever.** Synonym of *Texas fever*.
**southernwood.** See *artemisia abrotanum*.
**Southey's drainage-tubes** [Reginald *Southey*, English surgeon, 1835–1899]. Tubes of small caliber, employed for draining away the fluid from limbs that are the seat of extensive anasarca.

**Soxhlet's apparatus** (*soks'-let*) [Franz *Soxhlet*, German chemist, 1848-]. 1. An apparatus for sterilizing milk. 2. An apparatus for the determination of the fat in milk.

**soy bean**, or **soya bean** (*soi* or *soi'-yah*) [Jap.]. A kind of bean, the seed of *Glycine soja* (also referred to as *Soja hispida* and *Dolichos sinensis*), a plant of Japan and China and India. Diabetic bread, biscuits, and cakes are prepared from its flour, which contains no starch. The sauce called *soy* is also made from this bean.

**Soyka's plates** (*soi'-keh*). Dishes employed in the cultivation of bacteria. They are similar to Petri's capsules, but differ from them in having from eight to ten depressions ground in the lower plate, which resemble the "wells" in hollow slides.

**Soymida** (*soi'-mid-ah*) [Telugu name]. A genus of the *Meliaceæ*. *S. febrifuga* is an East Indian tree that furnishes rotun bark, introduced as a medicine in 1807 and used as a tonic and antiperiodic.

**sozal** (*so'-zal*). See *aluminum sulphocarbolate*.

**sozin** (*so'-zin*) [σώζειν, to save; keep]. A defensive proteid occurring naturally in the animal body. One capable of destroying microorganisms is termed a *mycosozin*, one antagonizing bacterial poisons, a *toxosozin*.

**sozoborol** (*so-zo-bo'-rol*). A mixture used in coryza said to consist of aristol, sozoiodol, and borates.

**sozodont** (*so'-zo-dont*) [σώζειν, to protect; ὀδούς, tooth]. A dentifrice supposed to be prepared mainly of Castile soap and alcohol.

**sozoiodol, sozoiodolic acid** (*so-zo-i'-o-dol, so-zo-i-o-dol'-ik*) [σώζειν, to save; *iodol*], C₆H₂I₂(SO₃H)OH. A crystalline, odorless powder used as an antiseptic, disinfectant, and parasiticide, chiefly in the form of its salts, of which the following have been employed: sodium sozoiodol, potassium sozoiodol, zinc sozoiodol, and mercury sozoiodol. **s., lead**, fine acicular crystals, sparingly soluble in water. **s., sodium**. See *sodium sozoiodolate*.

**sozolic acid** (*so-zo'-lik*). See *aseptol*.

**sp.** Abbreviation of Latin *spiritus*, spirit.

**space** (*spās*) [ME., from L., *spatium*, space]. A name given for purposes of description to sundry inclosed or semiinclosed spaces within or about the body. **s., anterior perforated**, a triangular space at the mesal side of the Sylvian fissure. **s., arachnoid.** See *s., subarachnoid*, and *s., subdural*. **s., axillary**, the axilla. **s., bregmatic**, the anterior fontanel. **s., circumdental**, the interspace between the ciliary body and the equator of the lens. **s.s, circumvascular lymph-**, channels surrounding the bloodvessels and communicating with lymphatic vessels. **s., complemental** (of pleura), the portion of the pleural cavity just above the attachments of the diaphragm which is not filled with lung during inspiration. **s., corneal**, that between the corneal layers. **s., epidural** (of the spinal canal), a lymph-space between the spinal dura and the periosteum lining the canal. **s.-feelings**, Hering's term for the perceptions or inferences of space-relations resulting from the retinal image. **s., Haversian**, the anterior fontanel. *Haver's spaces*. **s., hypoprostatic**, the space between the rectum and the prostate. **s.s, intercellular**, cavities formed by the splitting or separation of the walls of adjoining cells. **s., intercostal**, the space between two contiguous ribs. **s., intercrural**. See *s., interpeduncular*. **s.s, interfascicular**, spaces between the bundles in fibrous tissue. **s., interglobular**, an apparent, irregular space in the interglobular substance of the dentine. **s.s, interlamellar**, the spaces between the lamellæ of the cornea. **s., intermesoblastic**, the cavity between the visceral and parietal laminæ of the mesoblastic plates of the embryo. **s.s, intermetarsal**, spaces between the metatarsal bones. **s.s, intermuscular**, in the popliteal region, the spaces between the quadriceps extensor and the posterior muscles of the thigh. **s., interosseous**, the space between two parallel bones. **s., interparietal.** See *Virchow-Robin's*. **s., interpeduncular**, a diamond-shaped depression at the base of the brain, lying between the optic tracts and the crura cerebri. **s., interpleural**, the mediastinum. **s., interproximate**, in dentistry, the V-shaped space between the proximate surfaces of the teeth and the alveolar septum which is filled by the gum. **s., intertunical.** See *Virchow-Robin's*. **s., intervaginal** (of the optic nerve). **s., subvaginal. s.s, investing.** See *s., lymph-*. **s., ischiorectal.** See *fossa, ischiorectal*. **s., lacunar.** See

*lacuna.* **s., lymph-**, a sinus or space through which lymph passes. **s., marrow-.** See *canal, medullary*. **s., mediastinal**, the mediastinum. **s.-nerves**, the fibers of the auditory nerve in the semicircular canals. **s., pelvirectal.** See *fossa, ischiorectal*. **s., perforated.** See *perforated space*. **s.s, peri-cellular**, lymph-spaces in the brain. **s., perichoroid**, a lymph-space between the sclera and the choroid. **s., perigastric**, the cavity surrounding the stomach and other viscera. **s., periienticular**, the space surrounding the crystalline lens holding the zonule of Zinn. **s.s, perineural**, lymph-spaces between the lamellæ of the perineurium. **s., perivascular.** See *s., circumvascular lymph-*. **s., perivitelline.** See *s., yolk-*. **s., pituitary**, the space between the two cranial trabeculæ wherein the hypophysis appears. **s.s, placental blood-**, the intervillous lacunæ of the placenta. **s.s, pleuroperitoneal.** See *s., intermesoblastic*. **s., pneumatic**, an accessory sinus of the nose. **s., popliteal**, a lozenge-shaped space at the back of the knee and thigh. **s., posterior perforated**, the depression just behind the albicantia at the base of the brain. **s., posterior triangular**, the space lying above the clavicle and between the sternomastoid and the trapezius muscle and the occiput. **s., prevesical**, a space lying immediately above the pubis and between the transversalis fascia and the posterior surface of the rectus abdominis. **s., quadrilateral**, the anterior and posterior triangles of the neck taken together. **s., rectovesical**, the space between the bladder and the rectum. **s., retroperitoneal**, that behind the peritoneum, but in front of the spinal column and lumbar muscles. **s., retropharyngeal**, that behind the pharynx; it contains loose areolar tissue. **s., semilunar** (of Traube). See *semilunar space of Traube*. **s.-sense.** 1. The faculty by which the form of objects is recognized. 2. A sense by which we judge of the relation of objects in space; it is a part of the sense of sight. **s., subarachnoid**, the space between the arachnoid and the pia proper. It contains the cerebrospinal fluid. **s., subdural**, the space between the dura and the arachnoid. Normally it contains only a capillary layer of fluid. **s., subumbilical**, a triangular space in the bodycavity having its base at the umbilicus. **s., subvaginal**, a lymph-space within the sheath of the optic nerve. **s., suprachoroid**, the space between the velum interpositum and the fornix. **s., uterorectal**. See *Douglas' culdesac*. **s., visual**, the visual field. **s., yolk-**, the space formed by the retraction of the vitellus from the zona pellucida.

**spagiric** (*spaj-ir'-ik*) [σπᾶν, to stretch or rend; ἀγείρειν, to collect]. Pertaining to the obsolete chemical, alchemistic or Paracelsian, school of medicine.

**spagirism** (*spaj'-e-rizm*) [σπᾶειν, to stretch; ἀγείρειν, to collect]. The Paracelsian, or spagiric school, or doctrine, of medicine.

**spagirist** (*spaj'-e-rist*) [σπᾶειν, to stretch; ἀγείρειν, to collect]. A Paracelsian; a physician of the obsolete alchemistic school.

**Spallanzani's law** (*spal-lan-tsah'-ne*) [Lazaro *Spallanzani*, Italian physiologist, 1729–1799]. The regenerative power of the cells depends on the age of the individual; it decreases with age.

**spamenorrhea** (*spa-men-or-e'-ah*) [σπάνις, scarcity; μήν, month; ῥοία, flow]. Scantiness of menstruation.

**spanamenorrhea** (*span-ah-men-or-e'-ah*). See *spamenorrhea*.

**spanemia, spanæmia** (*span-e'-me-ah*) [σπάνις, scarcity; αἷμα, blood]. Poverty of the blood; anemia.

**spanemic, spanæmic** (*span-e'-mik*). See *anemic*.

**spaniocardia** (*span-e-o-kar'-de-ah*) [σπάνεως, seldom; καρδία, heart]. Landois' name for bradycardia.

**Spanish fever**. Synonym of *Texas fever*. **S. fly.** See *cantharides*. **S. white**, bismuth subnitrate.

**S. windlass.** See *windlass*.

**spanopnea, spanopnœa** (*span-op'-ne-ah*) [σπάνις, scarcity; πνεῖν, to breathe]. Infrequency of respiratory actions.

**sparadrap** (*spar'-a-drap*) [*sparadrapum*]. A plaster spread upon cotton, linen, silk, leather, or paper; adhesive plaster.

**spargosis** (*spar-go'-sis*) [σπαργᾶν, to be distended]. 1. Enlargement of a part. 2. Enlargement of the breasts from accumulation of milk. 3. Elephantiasis.

**sparteine** (*spar'-te-in*). An alkaloid found in scoparius. **s. hydrochloride**, C₁₅H₂₆N₂ · 2HCl, colorless crystals, soluble in water or alcohol, used as is the sulphate. s. hydroiodide, $C_{15}H_{24}N_2 \cdot HI$, white needles, soluble in water or alcohol; usage and dose the same as the sulphate. s. sulphate, $C_{15}H_{24}N_2 \cdot H_2SO_4 + 5H_2O$ (*sparteinæ sulphas*, U. S. P.), bitter, colorless prisms, soluble in water or alcohol, boiling at 136° C. It is a heart stimulant and diuretic. Dose $\frac{1}{6}-\frac{1}{3}$ gr. (0.011–0.022 Gm.); maximum dose $\frac{1}{3}$ gr. (0.032 Gm.) single; $\frac{1}{4}-2$ gr. (0.05–0.13 Gm.) daily. s. triiodide, $C_{15}H_{26}N_2I_3$, a black powder obtained from an ethereal solution of iodine and sparteine; soluble in alcohol; usage and dosage the same as the sulphate.

**spartism** (*spar'-tizm*). Poisoning from sparteine; characterized by vomiting, somnolence, paralytic-like weakness, and accelerated pulse.

**spartium** (*spar'-te-um*). Same as *scoparius*.

**spasm** (*spazm*) [σπασμός, spasm]. A sudden muscular contraction. s. of accommodation, spasm of the ciliary muscles, producing accommodation for objects near by. s., Bell's, convulsive facial tic. s., bronchial, asthma. s., carpopedal, a contraction causing flexion of the fingers and wrist or ankles and toes. s., clonic, a spasm broken by relaxations of the muscles. s., clonic, in the area of the nervus accessorius. Synonym of *torticollis*, *spasmodic*. s., clonic, in the area of the portio dura. Synonym of *tic, painless*. s., cynic. See *cynic spasm*. s., deglutition, a paroxysm of rapid swallowing, noted by Young (1901) as a symptom of whooping-cough. s., drivers', one of the so-called professional neuroses, it consists of cramp-like pains in the arms of drivers upon taking the reins in the hands. s., facial, a peculiar clonic contraction of the muscles supplied by facial nerve, at times confined to the muscles surrounding the eye, or else involving one entire side of the face. s.s, fatigue, Poore's term for a group of affections characterized by spasmodic contractions, either clonic or tonic, brought about by voluntary movement, the exciting cause being limited to some particular action. Syn., *business spasms*; *coordinated business neuroses*; *functional spasms*; *handicraft spasms*; *movement spasms*; *occupation spasms*; *professional spasms*. s., fixed, permanent or continuous tetanic rigidity of one or more muscles. s.s, Friedreich's. Synonym of *paramyoclonus multiplex*. s.s, functional, s s, handicraft. See *s.s, fatigue*, and *occupation-neurosis*. s. of the glottis. See *laryngismus stridulus*. s., habit-, half voluntary spasmodic movements, the result of habit, sometimes called habit-chorea. s., hammer. See *palsy, hammer*. s., histrionic, a condition in which focal involuntary twitchings of the face, acquired in childhood, persist during adult life, and are increased by emotional causes. s., idiopathic muscular. See *tetany*. s., inspiratory, a spasmodic contraction of nearly all the inspiratory muscles. s., laryngeal congenital, a peculiar stridor developing at birth, and disappearing after one or two years. s., lingual. See *aphthongia*. s., lock-, a form of writer's cramp in which the fingers become locked on the pen. s., masticatory (of the face). See *trismus*. s., mimic, facial neuralgia. s., mobile, slow, irregular movements depending upon hemiplegia. s.s, movement. See *s.s, fatigue*. s., muscular, idiopathic. See *tetany*. s., myopathic, one attending a disease of the muscles. s., nictitating, s., nodding. See *eclampsia nutans*. s., occupation, s., professional. See *s.s, fatigue*. s., pantomimic. Synonym of *tic, painless*. s., penman's. Synonym of *writers' cramp*. s., perineal. See *vaginodynia*. s., phonetic (*of the glottis*), spastic aphonia consisting of a spasm of the glottis, with elevation, resulting in interference with respiration. s., retrocollic, clonus of the deeper muscles of the back of the neck. s., Romberg's, masticatory spasm affecting the muscles supplied by the motor fibers of the fifth nerve. s., salaam, clonic spasm of the muscles of the leg, causing jumping movements. s. saltatoric, s., saltatory, s., static reflex. See under *saltatory*. s., sewing, an affection of tailors, seamstresses, and shoemakers, in which clonic and tonic spasms attack the muscles of the hands on attempting to use them in the regular work. s., smiths', a spasm that occurs in those engaged in penblade manufacturing, saw straightening, razor-blade striking, scissors-making, file-forging, etc. It consists in spasmodic movements of the arm used, and finally paralysis;· see *hemiplegia, hephestic*. s., spinal accessory. See *tic rotatoire*. s., synclonic, tremulous agitation. s., telegraphist's, an affection described first by Onimus, 1875. See *s.s, fatigue*. s., tetanic.

See *s., tonic*. s., tonic, a spasm that persists without relaxation for some time. s., toxic, one due to poison. s., winking. See *spasmus nictitans*. s., writer's, writers' cramp, *q. v.*

**spasmo-** (*spas-mo-*) [σπασμός, spasm]. A prefix denoting pertaining to a spasm.

**spasmodermia** (*spaz-mo-der'-me-ah*) [*spasmo-*; δέρμα, skin]. A spasmodic skin-affection.

**spasmodermic** (*spaz-mo-der'-mik*). Relating to a spasmodic affection of the skin.

**spasmodic** (*spaz-mod'-ik*) [*spasm*]. Pertaining to or characterized by spasm. s. spinal paralysis. See *lateral sclerosis*.

**spasmodism** (*spaz'-mod-izm*) [*spasm*]. Fleury's term for those nervous states that originate in medullary excitation.

**spasmodyspnea** (*spaz'-mo-disp-ne-ah*) [*spasmo-*; *dyspnea*]. Spasmodic difficulty of breathing.

**spasmology** (*spaz-mol'-o-je*) [*spasmo-*; λόγος, a treatise]. The sum of scientific knowledge of the nature and causes of convulsions.

**spasmolygmus** (*spaz-mo-lig'-mus*). See *hiccough*.

**spasmoneme** (*spaz'-mo-nēm*) [*spasmo-*; νῆμα, a thread]. The central reticulum or undulating bundle of fibrils eccentrically located in the peduncle of a stalked infusorian (*Vorticella*) and derived from the myonemes of the body.

**spasmophilia** (*spaz-mo-fil'-e-ah*) [*spasmo-*; φιλεῖν, to love]. A morbid tendency to convulsions.

**spasmorthopnea** (*spaz-mor-thop-ne'-ah*) [*spasmo-*; ὀρθός, straight; πνεῖν, to breathe]. Spasmodic orthopnea.

**spasmotin** (*spaz-mo'-tin*). See *sphacelotoxin*.

**spasmotoxin** (*spas-mo-toks'-in*) [*spasmo-*; τοξικόν, poison]. I. A ptomaine base of composition yet undetermined, obtained by Brieger from cultures of the tetanus germ, together with other unnamed toxins, one of which induced complete tetanus, with salivation and lacrimation. Spasmotoxin induces in animals violent clonic and tonic convulsions. 2. See *sphacelotoxin*.

**spasmous** (*spaz'-mus*) [σπασμός, spasm]. Having the nature of a spasm.

**spasmus** (*spaz'-mus*) [σπαξμός, spasm]. A spasm. s. bronchialis. Synonym of *bronchial asthma*. s. cynicus, spasmodic contraction of muscles on both sides of the mouth, giving a grinning expression. s. glottidis, spasm of the glottis or larynx, laryngismus stridulus. s. intestinorum. Synonym of *enteralgia*. s. muscularis. Synonym of *cramp*. s. nictitans, spasmodic action of the orbicularis palpebrarum muscle, causing a winking-like movement of the lid. s. nutans, salaam convulsions, nodding spasm. s. oculi. Synonym of *nystagmus*. s. ventriculi. Synonym of *enteralgia* and of *gastrodynia*.

**spastic** (*spas'-tik*) [σπαστικός, spastic]. Pertaining to or characterized by spasm; produced by spasm. s. diplegia. See *paraplegia, infantile spasmodic*. s. paralysis. See *paralysis, spastic*.

**spasticity** (*spas-tis'-it-e*) [*spastic*]. The state of being spastic.

**spathologic** (*spath-o-loj'-ik*) [σπαθᾶν, to go fast; λόγος, science]. Relating to rapid proliferation of leukocytes.

**spatial** (*spa'-shal*) [*spatium*, space]. Relating to space.

**spatium** (*spa'-she-um*). Latin for *space*.

**spatula** (*spat'-u-lah*) [L.; dim. of *spatha*, a ladle]. A flexible blunt blade used for spreading ointments.

**spatule** (*spat'-ūl*) [*spatula*, a blade]. A structure having a spatulate shape.

**spavin** (*spav'-in*). A disease of horses affecting the hock-joint, or joint of the hind leg between the knee and the fetlock. s., blood, a dilatation of the vein that runs along the inside of the hock of a horse, forming a soft swelling. s., bog, an encysted tumor on the inside of the hock of a horse, containing gelatinous matter. s., bone, a disease of the bones upper and inner part of the hock.

**spay** [Gael., *spoth*]. To remove the ovaries.

**spearmint** (*spēr'-mint*). See *mentha viridis*.

**specialism** (*spesh'-al-izm*). See *specialty*.

**specialist** (*spesh'-al-ist*) [*specialis*, particular, special]. One, especially a physician or surgeon, who limits his practice to certain specified diseases, or to the diseases of a single organ or class.

**speciality** (*spesh'-al-te*) [*species*]. The particular branch of medicine or surgery pursued by a specialist.

**species** (*spe'-shēz*) [L.]. 1. A subdivision of a

genus of animals or plants the individuals of which are either identical in character or differ only in unimportant and inconstant details. 2. A name in German and French pharmacy, and in the National Formulary, for certain mixtures of herbs, used in making decoctions and infusions. **s.-cycle,** the entire series of forms exhibiting or illustrating all the phases in the life-history of a species. **s. emollientes,** a mixture of the leaves of althea and mallow, of the leaves and branches of the melilot, of matricaria and flaxseed, in equal parts; used as an emollient cataplasm. **s. laxantes.** See *tea, Saint Germain.* **s., morphological,** one of "such living beings as constantly resemble one another so closely that it is impossible to draw any line of demarcation between them while they differ only in such characters as are associated with sex." (Huxley.) **s., nascent,** an incipient species; a form undergoing modification. **s., origin of,** a term employed by naturalists to denote the evolution of differentiated groups or species from groups of individuals characterized by general similarity or by homogeneity of structure.

**specific** (*spe-sif'-ik*) [*species; facere,* to make]. 1. Of or pertaining to a species, or to that which distinguishes a thing or makes it of the species of which it is. 2. A medicine which has a distinct curative influence on an individual disease. 3. Produced by a single microorganism, as a *specific* disease; in a restricted sense, syphilitic. **s. gravity.** See *gravity, specific.* **s. heat.** See *heat, specific.* **s. remedy,** a remedy peculiarly curative of a certain disease.

**specificity** (*spes-if-is'-it-e*) [*specific*]. The quality of being specific.

**specillum** (*spe-sil'-um*) [L.: *pl., specilla*]. A probe, especially one of silver, armed with a button-shaped head, for exploring wounds, fistulæ, etc. 2. A lens.

**specimen** (*spes'-im-en*) [L.]. An example; a sample. **s.-cooler,** a small water-cell immediately under the specimen in microprojection, to prevent injury from the heat of the radiant.

**spectacles** (*spek'-tak-ls*) [*spectare,* to view]. Framed or mounted lenses for the correction of optical or muscular defects of the eye; see *lens.* **s., bifocal.** See *bifocal.* **s., orthoscopic.** See *orthoscopic.* **s., pantoscopic.** A synonym of *s., bifocal, q. v.* **s., periscopic.** See *periscopic.* **s., prismatic,** spectacles with prismatic lenses, either alone or combined with spherical or cylindrical lenses. **s., protective,** lenses, usually tinted, to shield the eyes from light, dust, heat, etc.

**spectral** (*spek'-tral*) [*spectrum*]. Pertaining to a spectrum.

**spectro-** (*spek-tro-*) [*spectrum*]. A prefix meaning relating to the spectrum.

**spectrocolorimeter** (*spek-tro-kul-or-im'-et-er*) [*spectro-; color; μέτρον,* measure]. An apparatus for the isolation of a single spectral color. It is used for the detection of color-blindness.

**spectrometer** (*spek-trom'-et-er*) [*spectro-; μέτρον,* a measure]. An instrument for determining the deviation of a ray of light produced by a prism or diffraction-grating, or for ascertaining the wave-length of a ray of light.

**spectrometry** (*spek-trom'-et-re*) [*spectrometer*]. The use of the spectrometer.

**spectromicroscope** (*spek'-tro-mi'-kro-skōp*). See *microspectroscope.*

**spectrophone** (*spek'-tro-fōn*) [*spectro-; φωνή,* sound]. An apparatus devised by Painter and Bell (1881) for the production of sound by the rays of the spectrum.

**spectrophotometer** (*spek-tro-fo-tom'-et-er*) [*spectro-; φῶς,* light; *μέτρον,* a measure]. 1. An apparatus for determining the amount of color in spectrum-analysis. 2. Helmholtz's apparatus for mixing colors.

**spectrophotometry** (*spek-tró-fo-tom'-et-re*). The quantitative estimation of the coloring-matter in a substance by means of the spectroscope.

**spectropolarimeter** (*spek-tro-po-lar-im'-et-er*) [*spectro-; polus,* pole; *μέτρον,* measure]. An instrument in which a spectroscope and polarizing apparatus are combined for the purpose of determinating the concentration of solutions of substances that rotate the plane of polarized light.

**spectroscope** (*spek'-tro-skōp*) [*spectro-; σκοπεῖν,* to see]. An instrument for the production and examination of the spectrum.

**spectroscopic** (*spek-tro-skop'-ik*) [*spectroscope*]. Pertaining to the spectroscope.

**spectrotherapy** (*spek-tro-ther'-ap-e*) [*spectro-; therapy*]. Apéry's term for the therapeutic employment of prismatically decomposed rays.

**spectrum** (*spek'-trum*) [L., pl.; *spectra:* "an image"]. 1. The band of rainbow colors produced by decomposing light by means of a prism or a diffraction-grating. 2. An after-image or ocular spectrum. **s., absorption,** a spectrum which contains dark lines or bands. These are produced in a continuous spectrum by the absorption of adventurous vapors, through which the light has passed, as in the solar atmosphere. **s.-analysis,** determination of the nature of bodies by the character of their spectra. **s., auditory.** See *phonism, photism.* **s., comparison,** the arrangement side by side of the spectra of two different substances. **s., complementary,** a spectrum derived from bodies which change in chemical or molecular constitution before reaching a sufficiently high temperature to become luminous. **s., continuous,** a spectrum without sudden variations of hue, in which the various rainbow or spectral colors merge gradually into one another. **s., double,** see *s., comparison.* **s., line,** the spectrum resulting from incandescent gas. It consists not of the various rainbow colors, but of sharp, narrow, bright lines, the color depending on the substance; all the rest of the spectrum is dark. **s., normal,** a spectrum in which the red color occupies about the same space as the blue and the violet. **s., solar,** the spectrum afforded by the refraction of a ray of sunlight.

**speculum** (*spek'-u-lum*) [L.: *pl., specula*]. 1. A mirror. 2. An instrument for dilating the opening of a cavity of the body in order that the interior may be more easily visible, as *vaginal speculum, rectal speculum, nasal speculum,* etc. 3. A tendinous structure. **s. citrinum,** ointment. **s. helmontii,** the central tendinous part of the diaphragm, the centrum nerveum. **s. indicum,** iron filings. **s. lucidum.** Same as *septum lucidum.* **s. matricis,** womb-mirror; a vaginal speculum. **s.-metal,** an alloy of copper and tin. **s. oris,** an oral speculum for "mouth mirror." **s. rhomboideum,** a rhomboid area formed by the tendon of the trapezius muscles at the level of the upper dorsal and lower cervical spines. **s., Sims',** a vaginal speculum invented by J. Marion Sims.

**spedalskhed** (*sped-als'-ked*). A Scandinavian term for leprosy.

**speech** [AS., *sprecan,* to speak]. 1. The faculty of expressing thought by spoken words; the act of speaking. 2. The words spoken. **s. center,** the cerebral center for speech. See *center, speech.* **s., staccato,** see *scanning.*

**spell-bone.** The fibula.

**spelter** (*spel'-ter*). Crude zinc; an alloy of zinc and copper.

**Spence's test.** A tumor of the mammary gland can be distinguished from an inflammatory enlargement by the absence, in the latter case, of any tumefaction, there being only the lumpy and wormy sensation of the swollen acini and ducts.

**Spencer's area.** A cortical area in the frontal lobe just outside of the olfactory tract and anterior to the point where it joins the temporosphenoid lobe, as indicated by the crossing of the Sylvian artery. Faradic stimulation of this area influences the respiratory movements, causing stoppage of the respiration when sufficiently intense.

**spend** [*dispendere,* to lay out, to expend]. To ejaculate the semen.

**Spengler's bodies** (*speng'-gler*) [*Carl Spengler,* Swiss physician, 1861– ]. Small particles resembling fragments of bacilli, found in tuberculous sputum, and having the same staining reaction as tubercle bacilli. **S.'s method of examining sputum,** five Cc. of 0.4 per cent. of soda are added to 5 Cc. of sputum; 0.1 Cc. trypsin or pancreatin, and two or three drops of chloroform are then added. This is corked and incubated at body temperature. The tube must be agitated occasionally during the first few hours. Next day the supernatant fluid is poured off, and the residue used for making smears which are then examined for the tubercle bacilli.

**Spens' syndrome** [Thomas *Spens,* Scotch physician, 1764–1842]. The same as *Adams Stokes'* disease, *q. v.*

**spent** (*spendere,* to spend). Exhausted; impotent. **s. acid,** a battery-acid that has become too weak for efficient action.

**sperm, sperma** (*sperm, sper'-mah*) [*σπέρμα,* seed]. The semen. **s.-ball,** a spherical cluster of spermatozoa. **s.-blastoderm,** a blastodermic layer of formative spermatozoa. **s. blastophore,** the residual mass of the sperm-mother cell. **s.-blastula,** a spherical blastula whose surface is a sperm-blastoderm. **s.-cell,** a spermatoblast. **s.-morula,** a spermatic morula. **s.-mother cell,** Lankester's term for the spherical male germs of the malaria parasite as found in the mosquito. **s.-nucleus,** the nucleus of a spermatozoon. **s.-oil,** an oil procured from the deposits in the head of the sperm-whale. **s.-rope,** a string of spermatozoa. **sperm-cell.** Spermatozoon.

**spermaceti** (*sper-mạs-e'-te*) [σπέρμα, seed; κῆτος, whale], $C_{15}H_{31}O_2C_{16}H_{33}$. A white, semitransparent substance (*cetaceum,* U. S. P., B. P.), consisting of a mixture of various fats of which cetyl palmitate, $C_{16}H_{33}(C_{16}H_{31}O_2)$, is the most important. It is obtained from the head of the sperm whale, and is used internally as an emollient and as an ingredient of various ointments. *Ceratum cetacei* and *unguentum cetacei* B. P.) are prepared from it. See *cetaceum.*

**spermacrasia** (*sper-mak-ra'-ze-ah*) [σπέρμα, seed; *acrasia*]. 1. Imperfection of the semen. 2: Spermatorrhea.

**spermaduct** (*sper'-ma-dukt*) [*sperm; ductus,* a duct]. A sperm-duct, the vas deferens.

**spermagone** (*sper'-mag-ōn*) [σπέρμα, seed; γονεία, generation]. Same as *spermatogonium.*

**spermagonium** (*sper-mag-o'-ne-um*) [σπέρμα, seed; γονεία, generation: *pl., spermagonia*]. Same as *spermogonium.*

**spermalist** (*sper'-mal-ist*). Same as *spermist.*

**spermary** (*sper'-ma-re*) [σπέρμα, seed]. The analogue in the male of the ovary; *i. e.,* the organ generating the sperm-cells; in the higher animals, called the testis, or testicle.

**spermatanergia** (*sper'-mat-an-ur'-je-ah*) [σπέρμα, seed; *anergia*]. Sterility in the male.

**spermatemphraxis** (*sper-mat-em-fraks'-is*) [σπέρμα, seed; ἔμφραξις, obstruction]. An obstruction to the discharge of semen.

**spermatic** (*sper-mat'-ik*) [*sperm*]. 1. Pertaining to the semen. 2. Conveying the semen, as the *spermatic cord.* 3. Pertaining to the spermatic cord, as the *spermatic* fascia. **s. artery,** a branch of the aorta supplying the testicle. **s. canal,** see inguinal canal. **s. cones.** See *cone.* **s. cord,** the cord of arteries, veins, lymphatics, nerves, and the excretory duct of the testicle passing from the testicle to the internal abdominal ring. **s. crystals,** a variety of crystals formed in seminal fluid after prolonged standing; see under *spermin.* **s. fascia,** a thin fascia attached to the internal abdominal ring, and prolonged down over the outer surface of the spermatic cord. **s. gelatin,** a gelatinous substance found in the spermogonia of certain cryptogams. **s. plexus,** the pampiniform plexus. **s. rete.** See *rete testis.*

**spermatid** (*sper'-mat-id*) [σπέρμα, seed]. A seminal cell. A cell produced by fission of a secondary spermatocyte.

**spermatin** (*sper'-mat-in*) [*sperm*]. An odorless, mucilaginous substance found in semen.

**spermatism** (*sper'-mat-izm*) [σπέρμα, seed]. A discharge of semen.

**spermatismus** (*sper-mat-iz'-mus*) [σπέρμα, seed]. The emission of semen.

**spermatitis** (*sper-mat-i'-tis*) [σπέρμα, seed; *ιτις*, inflammation]. Same as *funiculitis.*

**spermatize** (*sper'-mat-īz*). To discharge semen.

**spermato-** (*sper-mat-o-*) [σπέρμα, seed]. A prefix meaning pertaining to the semen.

**spermatoal** (*sper-mat-o'-al*) [σπέρμα, seed; ᾠόν, an egg]. Pertaining to a spermatoon.

**spermatoblast, spermoblast** (*sper'-mat-o-blast, sper'-mo-blast*) [*spermato-*; βλαστός, a germ]. A cell resulting from the division of the spermatogenic cell and developing into a spermatozoon.

**spermatoblastic** (*sper-mat-o-blas'-tik*). Pertaining to spermatoblasts.

**spermatocele** (*sper'-mat-o-sēl*) [*spermato-*; κήλη, tumor]. A spermatic cyst or encysted hydrocele containing spermatozoa.

**spermatocidal** (*sper-mat-o-si'-dal*) [*spermato-*; *cadere,* to kill]. Destructive to spermatozoa.

**spermatoclemma** (*sper-mat-o-klem'-ah*) [*spermato-*; κλέμμα, a stealing: *pl., spermatoclemmata*]. Involuntary emission of semen. A synonym of *pollution.*

**spermatocratia** (*sper-mat-o-kra'-she-ah*). Synonym of *spermatorrhea.*

**spermatocyst** (*sper'-mat-o-sist*) [*spermato-*; κύστις, cyst]. A seminal vesicle; a pathological cyst containing spermatozoa.

**spermatocystectomy** (*sper-mat-o-sist-ek'-to-me*) [*spermato-*; *cystectomy*]. Excision of a spermatic cyst.

**spermatocystic** (*sper-mat-o-sis'-tik*). Pertaining to a spermatocyst.

**spermatocystitis** (*sper-mat-o-sis-ti'-tis*). Inflammation of the seminal vesicles.

**spermatocystotomy** (*sper-mat-o-sis-tot'-o-me*) [*spermato-*; *cystotomy*]. Surgical incision of a seminal vesicle.

**spermatocytal** (*sper-mat-o-si'-tal*). Pertaining to a spermatocyte.

**spermatocyte** (*sper'-mat-o-sīt*) [*spermato-*; κύτος, cell]. The germinal cell from which the spermatozoon develops.

**spermatogenesis, spermatogeny** (*sper-mat-o-jen'-es-is, sper-mat-oj'-en-e*) [*spermato-*; γένεσις, origin]. The formation of spermatozoa.

**spermatogenic** (*sper-mat-o-jen'-ik*) [see *spermatogenesis*]. Producing spermatozoa, as the *spermatogenic* cells of the testicle.

**spermatogenous** (*sper-mat-oj'-en-us*) [*spermato-*; γένης, producing]. Producing spermatozoa.

**spermatogeny** (*sper-mat-oj'-en-e*). The same as *spermatogenesis.*

**spermatogonium** (*sper-mat-o-go'-ne-um*) [*spermato-*; γονή, generation]. A formative seminal cell or mass of spermatoblasts.

**spermatoid** (*sper'-mat-oid*). See *spermatozoon.*

**spermatology** (*sper-mat-ol'-o-je*) [*spermato-*; λόγος, a treatise]. The sum of what is known regarding the origin, nature, qualities, and characteristics of the seminal fluid.

**spermatolysin** (*sper-mat-ol'-is-in*). A substance causing spermatolysis.

**spermatolysis** (*sper-mat-ol'-is-is*) [*spermato-*; λύσις, solution]. Destruction or solution of spermatozoa.

**spermatomere** (*sper'-mat-o-mēr*) [*spermato-*; μέρος, share]. Any one of the portions into which a pronucleus of the fertilized ovum may divide.

**spermatoon** (*sper-mat-o'-on*) [*spermato-*; ᾠόν, egg]. The nucleus of a sperm-cell or spermatozoon.

**spermatopathy** (*sper-mat-op'-ath-e*) [*spermato-*; πάθος, disease]. Disease of the sperm-cells or of their secreting mechanism.

**spermatophobia** (*sper-mat-o-fo'-be-ah*) [*spermato-*; φόβος, fear]. False spermatorrhea; morbid dread of spermatorrhea.

**spermatophore** (*sper'-mat-o-fōr*) [*spermato-*; φέρειν, to bear]. 1. The part of the spermatospore that is not converted into a spermatoblast. 2. A semitransparent capsule surrounding a group of spermatozoa.

**spermatoplania** (*sper-mat-o-pla'-ne-ah*) [*spermato-*; πλάνη, a wandering]. A supposed metastasis of the semen.

**spermatopoietic** (*sper-mat-o-poi-et'-ik*) [*spermato-*; ποιεῖν, to make]. Pertaining to the production or secretion of semen.

**spermatorrhea, spermatorrhoea** (*sper-mat-or-e'-ah*) [*spermato-*; ῥοία, a flow]. Involuntary discharge of semen without sexual excitement. **s. dormientum,** a nocturnal emission of semen. **s., false,** when spermatozoids are not in the fluid; called also prostatorrhea. **s., true,** when spermatozoids are present.

**spermatoschesis** (*sper-mat-os'-kes-is*) [*spermato-*; σχέσις, suppression]. Suppression of the seminal fluid.

**spermatospore** (*sper'/mat-o-spōr*) [*spermato-*; σπόρος, seed]. A primitive cell giving rise by division to spermatoblasts.

**spermatotoxin, spermatoxin** (*sper-mat-o-toks'-in, sper-ma-toks'-in*). See *spermolysin.*

**spermatovum** (*sper-mat-o'-vum*) [*spermato-*; *ovum,* egg: *pl., spermatova*]. An impregnated ovum.

**spermatozemia** (*sper-mat-o-ze'-me-ah*). See *spermatorrhea.*

**spermatozoa** (*sper-ma-to-zo'-ah*). Plural of *spermatozoon, q. v.*

**spermatozoal, spermatozoan, spermatozoic** (*sper-mat-o-zo'-al, sper-mat-o-zo'-an, sper-mat-o-zo'-ik*). Relating to a spermatozoon.

**spermatozoicide** (*sper-mat-o-zo'-is-īd*) [*spermatozoon; cadere,* to kill]. 1. Destructive to spermatozoa. 2. An agent destructive to spermatozoa.

**spermatozoid or spermatozooid** (*sper-mat-o-zo'-id, sper-mat-o-zo'-oid*). Same as *spermatozoon.*

**spermatozoon** (*sper-mat-o-zo'-on*) [*spermato-*; ζῷον, animal]. The male element capable of fecundating the ovum. It consists of an oval head and a long, mobile cilium or tail. It is the essential element of the semen.

**spermaturia** (*sper-mat-ū'-re-ah*) [*spermato-*; *ouron*, urine]. The presence of semen in the urine.
**spermic** (*sper'-mik*). Same as *spermatic*.
**spermiduct** (*sper'-mid-ukt*) [*σπέρμα*, seed; *ductus*, a duct]. A duct for the passage of semen; the vas deferens.
**spermin** (*sper'-min*) [*sperm*], C₅H₅N. A non-poisonous base obtained from sputum, human semen, the organs of leukemic patients, etc. It has been used in neurasthenia, senile debility, diabetes mellitus, and pulmonary tuberculosis. 2. A preparation of the testicles of animals. **s. phosphate**, constitutes the Charcot-Leyden crystals.
**spermism** (*sper'-mizm*) [*σπέρμα*, seed]. The theory that the animal is the result of the development of a spermatozoon, the ovum acting only as an accessory matrix.
**spermist** (*sper'-mist*). A believer in spermism.
**spermoblast** (*sper'-mo-blast*), see *spermatoblast*.
**spermocenter** (*sper-mo-sen'-ter*). The sperm-centrosomes during fertilization of the egg.
**spermolith** (*sper'-mo-lith*) [*σπέρμα*, semen; *λίθος*, a stone]. A calculus in the spermatic duct or seminal vesicles.
**spermolysin** (*sper-mol'-is-in*) [*sperm*; *λύειν*, to loosen]. Metchnikoff's name for a cytolysin produced by inoculation with spermatozoa. Syn., *spermatoxin*.
**spermolysis** (*sper-mol'-is-is*). Dissolution of spermatozoa.
**spermoneuralgia** (*sper-mo-nū-ral'-je-ah*) [*σπέρμα*, seed; *νεῦρον*, nerve; *ἄλγος*, pain]. Neuralgia of the testicles and spermatic cord.
**spermophlebectasia** (*sper-mo-fleb-ek-ta'-ze-ah*) [*σπέρμα*, seed; *φλέψ*, vein; *ἔκτασις*, distention]. Varicosity of the spermatic vein.
**spermoplasm** (*sper'-mo-plazm*) [*sperm*; *πλάσσειν*, to mold]. The protoplasm of a spermatozoon.
**spermorrhagia** (*sper-mor-a'-je-ah*). See *spermatorrhea*.
**spermorrhea** (*sper-mor-e'-ah*). See *spermatorrhea*.
**spermosphere** (*sper'-mo-sfēr*) [*sperm*; *σφαῖρα*, sphere]. A mass of spermatoblasts.
**spermospore** (*sper'-mo-spōr*). See *spermatospore*.
**spermotoxin** (*sper-mo-toks'-in*). See *spermolysin*.
**sp. gr.** Abbreviation of *specific gravity*.
**sph.** Abbreviation for *spherical*; also for *spherical lens*.
**sphacelate, sphacelated** (*sfas'-el-āt, sfas'-el-a-ted*) [*sphacelus*]. Necrosed; gangrenous; mortified.
**sphacelation** (*sfas-el-a'-shun*). The formation of a sphacelus; moist gangrene; necrosis.
**sphacele** (*sfas'-ēl*). The uncorticated apical cell of the branches of certain marine algæ.
**sphacelism** (*sfas'-el-ism*) [*σφάκελος*, gangrene]. 1. The condition of being affected with sphacelus. 2. Necrosis. 3. Inflammation of the brain.
**sphaceloderma** (*sfas-el-o-der'-mah*) [*sphacelus*; *δέρμα*, skin]. Gangrene of the skin, especially symmetrical gangrene, or Raynaud's disease.
**sphaceloid** (*sfas'-el-oid*) [*σφάκελος*, gangrene; *εἶδος*, like]. Resembling a sphacelus or gangrenous part.
**sphacelotoxin** (*sfas-el-o-toks'-in*) [*sphacelia*, a stage in the growth of ergot; *toxicon*, a poison], C₂₆O₄H₄. A yellowish, pulverulent body obtained from ergot, insoluble in water, soluble in ether, chloroform, alcohol, and alkaline solutions. It is used as a tonic, astringent, and emmenagogue. Dose ⅓–1½ gr. (0.032–0.1 Gm.). Syn., *spasmotin*; *spasmotoxin*.
**sphacelous** (*sfas'-el-us*) [*σφάκελος*, gangrene]. Pertaining to sphacelus; gangrenous; necrosed.
**sphacelus** (*sfas'-el-us*) [*σφάκελος*, gangrene]. A slough.
**sphæræsthesia** (*sfe-res-the'-ze-ah*). See *spheresthesia*.
**sphærobacteria** (*sfe-ro-bak-te'-rah*). See *spherobacteria*.
**sphærobacterium** (*sfe-ro-bak-te'-re-um*). See *spherobacterium*.
**Sphærococcus** (*sfe-ro-kok-us*) [*σφαῖρα*, a ball; *κόκκος*, a berry]. A genus of marine algæ of the order *Sphærococcoideæ*. **S. compressus**, said to furnish in part the Japanese isinglass or agar of commerce.
**sphæroma** (*sfe-ro'-mah*). See *spheroma*.
**sphage** (*sfāj*) [*σφαγή*, the throat]. The throat; the anterior portion of the neck.
**sphagiasmus** (*sfa-je-az'-mus*) [*σφαγή*, throat]. Epileptic spasm of the muscles of the neck.
**sphagitis** (*sfa-ji'-tis*) [*σφαγή*, the throat; *ιτις*, inflammation]. 1. Inflammation of the jugular vein. 2. Sore-throat.

**sphenencephalus** (*sfe-nen-sef'-al-us*). See *sphenocephalus*.
**sphenethmoid** (*sfe-neth'-moid*). Same as *sphenoethmoid*.
**sphenic** (*sfe'-nik*) [*σφήν*, wedge]. Wedge-like.
**sphenion** (*sfe'-ne-on*) [*σφήν*, wedge]. The apex of the sphenoid angle of the parietal bone on the surface of the skull. See craniometric points.
**spheno-** (*sfe-no-*) [*σφήν*, a-wedge]. A prefix denoting pertaining to the sphenoid bone.
**sphenobasilar** (*sfe-no-bas'-il-ar*) [*spheno-*; *βάσις*, base]. Pertaining conjointly to the sphenoid bone and the basilar portion of the occipital bone. **s. groove**, the depression on the body of the sphenoid bone and the basilar portion of the occipital bone, upon which the pons rests.
**sphenoccipital** (*sfe-nok-sip'-it-al*) [*spheno-*; *occiput*]. Pertaining to the sphenoid and the occipital bones; sphenobasilar.
**sphenocephalus** (*sfe-no-sef'-al-us*) [*spheno-*; *κεφαλή*, head]. A variety of monster in which the two eyes are well separated, the ears united under the head, the jaws and mouth distinct, and the sphenoid bone altered in shape, so that it is analogous in form to what is found normally in birds.
**sphenoethmoid** (*sfe-no-eth'-moid*) [*spheno-*; *ethmoid*]. Relating to both the sphenoid and the ethmoid bones. **s. recess**, the groove at the back of the roof of the nasal fossa.
**sphenofrontal** (*sfe-no-frun'-tal*). Belonging or relating to both the sphenoid and frontal bones.
**sphenoid** (*sfe'-noid*) [*σφήν*, wedge; *εἶδος*, like]. 1. Wedge-shaped, as the *sphenoid bone*. 2. The sphenoid bone. **s. bone**, see *sphenoidale* under *bones, table of*. **s. sinus**, see sinus. **sphenoidal**, sphenoid.
**sphenoidale** (*sfe-noid-a'-le*) [neuter of *sphenoidalis*, sphenoid]. The sphenoid bone. **s. basilare anterius**, the anterior portion of the body of the sphenoid. **s. basioposticum**, the lower portion of the body of the sphenoid. **s. laterale posterius**, the lateral portion of the sphenoid. **sphenoidalia lateralia**, the greater wings of the sphenoid.
**sphenoides** (*sfe-noi'-dez*) [*σφήν*, wedge; *εἶδος*, like]. The sphenoid bone.
**sphenoiditis** (*sfe-noid-i'-tis*). Inflammation of the sphenoid sinus.
**sphenoido-** (*sfe-noi-do-*). The same as *spheno-*.
**sphenoidoauricular** (*sfe-noi-do-aw-rik'-u-lar*) [*sphenoid*; *auricula*, auricle]. Pertaining to the sphenoid and binauricular diameters of the skull. **s.-a. index**, the ratio of the minimum sphenoid diameter of the skull with the binauricular diameter, the latter being taken as 100.
**sphenoidofrontal** (*sfe-noi-do-fron'-tal*) [*sphenoido-*; *frontal*]. 1. Pertaining to the sphenoid and frontal bones. 2. Pertaining to the sphenoid and frontal diameters of the skull. **s.-f. index**, the relation between the minimum sphenoid diameter of the skull and the minimum frontal diameter taken as 100.
**sphenoidoparietal** (*sfe-noi-do-par-i'-et-al*) [*sphenoido-*; *parietal*]. 1. Pertaining to the sphenoid and parietal bones. 2. Belonging or relating to the sphenoid and parietal diameters of the skull. **s.-p. index**, the relation between the minimum sphenoid diameter of the skull and the maximum frontal diameter taken as 100.
**sphenomalar** (*sfe-no-ma'-lar*). Pertaining to the sphenoid and malar bones.
**sphenomandibular** (*sfe-no-man-dib'-ū-lar*). Pertaining to the sphenoid and inferior maxillary bones. **s. ligament**. See *sphenomaxillary ligament*.
**sphenomaxillary** (*sfe-no-maks'-il-a-re*) [*spheno-*; *maxilla*]. Pertaining to the sphenoid and maxillary bones, as the sphenomaxillary fossa. **s. fissure**. See *fissure, sphenomaxillary*. **s. fossa**, a triangular space at the angle of the sphenomaxillary and pterygomaxillary fissure. **s. ligament**, a ligament extending from the inferior maxilla, near the interdental dental foramen, to the spinous process of the sphenoid bone.
**sphenometer** (*sfe-nom'-et-er*) [*σφήν*, wedge; *μέτρον*, measure]. An instrument for measuring the wedge to be removed in osteotomy for curvature.
**spheno-occipital** (*sfe-no-ok-sip'-it-al*), see *sphenoccipital*.
**spheno-orbital** (*sfe-no-or'-bit-al*) [*spheno-*; *orbit*]. Pertaining to the sphenoid bone and the orbit.
**sphenopalatine** (*sfe-no-pal'-a-tīn*) [*spheno-*; *palatum*, palate]. Pertaining to the sphenoid bone and the palate, as the *sphenopalatine foramen*. **s.-foramen**, the spheno-palatine notch converted into a foramen

by articulation with the sphenoidal turbinated bone. s.-p. ganglion. See *ganglion.* s.-p. notch, a deep notch separating the orbital and sphenoid process of the palate bone; see notch.

**sphenoparietal** (*sfe-no-par-i'-et-al*) [*spheno-*; *par-ietal*]. Pertaining to the sphenoid and parietal bones.

**sphenopetrosal** (*sfe-no-pe-tro'-sal*) [*spheno-*; πέτρα, rock]. Pertaining to the sphenoid bone and the petrous portion of the temporal bone.

**sphenopterygoid** (*sfe-no-ter'-ig-oid*). Pertaining to the body of the sphenoid bone and to the pterygoid process.

**sphenorbital** (*sfe-nor'-bit-al*). See *sphenoorbital*.

**sphenosis** (*sfe-no'-sis*) [σφήν, wedge]. The wedging of the fetus in the pelvis.

**sphenosquamosal, sphenosquamous** (*sfe-no-skwa-mo'-sal, sfe-no-skwa'-mus*). Belonging or relating to both the sphenoid bone and the squamous portion of the temporal bone.

**sphenotemporal** (*sfe-no-tem'-po-ral*). Pertaining conjointly to the sphenoid and temporal bones.

**sphenotic** (*sfe-no'-tik*) [*spheno-*; οὖς, ear]. A part of the sphenoid bone, existing as a distinct bone in the fetus, and forming the parts adjacent to the carotid groove.

**sphenotresia** (*sfe-no-tre'-ze-ah*) [σφήν, wedge; τρῆσις, perforation]. A variety of craniotomy in which the basal portion of the fetal skull is perforated.

**sphenotribe** (*sfe'-no-trīb*) [*spheno-*; τρίβειν, to rub]. An instrument for crushing the basal portion of the fetal skull.

**sphenotripsy** (*sfe'-no-trip-se*) [see *sphenotribe*]. Crushing of the fetal skull.

**sphenoturbinal, sphenoturbinate** (*sfe-no-ter'-bin-al, sfe-no-ter'-bin-āt*). 1. Pertaining to the sphenoid and turbinate bones. 2. One of the sphenoidal spongy bones situated cephalad of the body of the sphenoid.

**sphenovomerine** (*sfe-no-vo'-mer-in*). Pertaining to the sphenoid bone and the vomer.

**spheral** (*sfe'-ral*) [σφαῖρα, sphere]. Like a sphere.

**sphere** (*sfēr*) [σφαῖρα, a sphere]. 1. A ball or globe. 2. A space. s. of attraction, a clear spot in the cellplasma, outside and close to the nucleus of an ovum undergoing mitosis. It contains the centrosoma of Boveri, and is the center of the formation of the amphiasters in karyokinesis (*q. v.*). s., embryonic. s., segmentation-. s.- granule, a large granular corpuscle found in serous exudations. s., hearing, the area in the brain which is supposed to be the seat of hearing. It is in the temporal lobe. s., motor, a region in the central nervous system which, when stimulated, gives rise to motion. s., protoplasmic primordial. See s., segmentation-. s., segmentation-, a nucleated cell derived from division of the vitellus in the process of segmentation. s., sensory, a sensory area of the central nervous system. s., vitelline, s., yolk-, the mulberry-like mass of cells that results from the fission of the substance of the ovum after fertilization. s., yeast-, in biology, an aggregation of certain sprouting forms of the genus *Mucor*.

**spheresthesia, spheræsthesia** (*sfe-res-the'-ze-ah*) [σφαῖρα, globe; αἴσθησις, sensation]. Perverted feeling, as of the contact of a ball or globe-shaped body.

**spheric, spherical** (*sfer'-ik, sfer'-ik-al*). Having the shape of or pertaining to a sphere. s. aberration. See *aberration, spherical*.

**spherobacteria** (*sfe-ro-bak-te'-re-ah*) [σφαῖρα, sphere; *bacteria*]. The micrococci.

**spheroid** (*sfe'-roid*) [σφαῖρα, sphere; εἶδος, like]. Having the form of a sphere. A solid resembling a sphere. s., oblate, one in which the polar axis is less than the equatorial diameter. s., prolate, one in which the polar axis exceeds the equatorial diameter.

**spheroma** (*sfe-ro'-mah*) [σφαῖρα, sphere; ὄμα, tumor]. Any spherical shaped tumor or protuberance.

**spherometer** (*sfe-rom'-et-er*) [σφαῖρα, sphere; μέτρον, a measure]. An instrument for determining the degree of curvature of a sphere or part of a sphere, especially of optic lenses, or of the tools used for grinding them.

**sphincter** (*sfingk'-ter*) [σφίγγειν, to bind]. A muscle surrounding and closing an orifice, as the *anal sphincter*, the *pyloric sphincter*, etc. s. ani. See under *muscle*. s. antripylorici. See under *muscle*. s. gulæ, the constrictor of the pharynx. s. ilei, the ileocecal valve. s. intestinalis. See *s. ani (internal)* undermuscle. s. laborium, the orbicularis oris muscle. s. laryngis, the arytenoepiglottideus muscles of both sides surrounding the laryngeal opening. s. oculi, s. palpebrarum, the orbicularis palpebrarum muscle.

s. œsophageus. See *foramen, esophageal.* s., oral, the orbicularis oris. s. pharyngolaryngeus, the inferior constrictor of the pharynx, the anterior cricothyroid, and the thyroid muscles considered as one. s. pyloricus. See under *muscle*. s., third, of the rectum, a duplicature of the mucosa projecting well into the lumen of the gut from the right side, forming rather more than a semicircle, and involving more of the ventral than of the dorsal wall. It is also called *Kohlrausch's fold*. s. vaginæ. See under *muscle*.

**sphincteralgia** (*sfingk-ter-al'-je-ah*) [*sphincter;* ἄλγος, pain]. Pain in the sphincter ani muscle, or about the anus.

**sphincterectomy** (*sfingk-ter-ek'-to-me*) [*sphincter;* ἐκτομή, excision]. 1. Oblique blepharotomy; Stellwag's operation for the dilatation of the palpebral fissure, or for blepharospasm. 2. The surgical removal of the pyloric sphincter.

**sphincterial, sphincteric** (*sfingk-te'-re-al, sfingk-ter'-ik*). Pertaining to a sphincter or to its function.

**sphincterismus** (*sfingk-ter-iz'-mus*). A spasmodic contraction of the sphincter ani muscle, usually attendant upon fissure or ulcer of the anus, but occasionally occurring independently of such lesion.

**sphincterolysis** (*sfingk-ter-ol'-is-is*) [*sphincter*; λύσις, solution]. The operation of freeing the iris in anterior synechia.

**sphincteroplasty** (*sfingk'-ter-o-plas-te*) [*sphincter;* πλάσσειν, to form]. The formation of an artificial sphincter by plastic operation.

**sphincteroscope** (*sfingk-ter'-o-skōp*) [*sphincter;* σκοπεῖν, to examine]. An instrument for making visual inspection of a sphincter.

**sphincteroscopy** (*sfingk-ter-os'-ko-pe*) [see *sphincteroscope*]. Visual inspection of a sphincter by means of special instruments.

**sphincterotomy** (*sfingk-ter-ot'-o-me*) [*sphincter*; τομή, a cutting]. The operation of incising a sphincter.

**sphinctrate** (*sfingk'-trāt*). Contracted or constricted as if by a sphincter.

**sphingoine** (*sfin'-go-in*) [σφίγγειν, to bind]. A leukomaine derived from cerebral tissue.

**sphingomyelin** (*sfin-go-mi'-el-in*) [σφίγγειν, to bind; μυελός, marrow]. A brain-phosphatide allied to myelin. It is capable of being decomposed into neurin and a substance which is convertible into sphingosin.

**sphingosine** (*sfin'-go-sēn*) [σφίγγειν, to bind]. An alkaloidal cerebroside occurring in brain-tissue.

**sphygmic, sphygmical** (*sfig'-mik, sfig'-mik-al*) [σφυγμός, pulse]. Pertaining to the pulse.

**sphygmo-** (*sfig-mo-*) [σφυγμός, pulse]. A prefix signifying pertaining to the pulse.

**sphygmobolometer** (*sfig-mo-bo-lom'-et-er*) [*sphygmo-;* βόλος, a throw; μέτρον, a measure]. An instrument for measuring and recording the force of the pulse.

**sphygmocardiograph** (*sfig-mo-kar'-de-o-graf*) [*sphygmo-;* καρδία, heart; γραφεῖν, to record]. An instrument for the recording of the movements of the pulse and the heart.

**sphygmocardioscope** (*sfig-mo-kar'-de-o-skōp*) [*sphygmo-;* καρδία, heart; σκοπεῖν, to examine]. Same as sphygmocardiograph.

**sphygmochronograph** (*sfig-mo-kro'-no-graf*) [*sphygmo-;* *chronograph*]. A registering sphygmograph.

**sphygmochronography** (*sfig-mo-kro-nog'-raf-e*). The registration of the extent and oscillations of the pulsewave.

**sphygmodic** (*sfig-mo'-dik*) [σφυγμός, pulse]. Like the pulse; throbbing.

**sphygmodynamometer** (*sfig-mo-di-nam-om'-et-er*) [*sphygmo-;* δύναμις, power; μέτρον, measure]. An instrument for measuring the force of the pulse.

**sphygmogenin** (*sfig-moj'-en-in*) [*sphygmo-;* γεννᾶν, to produce]. A substance isolated by Fränkel from the suprarenal capsule, which causes increase of bloodpressure. It is used as an antidote in nicotine poisoning.

**sphygmogram** (*sfig'-mo-gram*) [*sphygmo-;* γράμμα, a writing]. The tracing made by the sphygmograph.

**sphygmograph** (*sfig'-mo-graf*) [*sphygmo-;* γράφειν, to write]. An instrument for recording graphically the features of the pulse and the variations in blood-pressure.

**sphygmographic** (*sfig-mo-graf'-ik*). Pertaining to the sphygmograph.

**sphygmography** (*sfig-mog'-ra-fe*) [σφυγμός, pulse; γράφειν, to write]. A description of the pulse, its pathological variations and their significance.

**sphygmoid** (sfĭg'-moid) [σφυγμός, pulse;˙ εἶδος, resemblance]. Resembling or having the nature of continuous pulsation.
**sphygmology** (sfĭg-mol'-o-je) [sphygmo-; λόγος, treatise]. The branch of medicine dealing with the characters of the pulse.
**sphygmomanometer** (sfĭg-mo-man-om'-et-er) [sphygmo-; manometer]. An instrument for measuring the tension of the blood-current or arterial pressure.
**sphygmometer** (sfĭg-mom'-et-er), see sphygmograph.
**sphygmometroscope** (sfĭg-mo-met'-ro-skōp) [sphygmo-; μέτρον, measure; σκοπεῖν, to inspect]. An instrument used for listening to the pulse while the blood pressure is being estimated.
**sphygmo-oscillometer** (sfĭg-mo-os-il-om'-et-er). A form of sphygmomanometer in which the systolic and diastolic blood pressure are indicated by an oscillating needle.
**sphygmopalpation** (sfĭg-mo-pal-pa'-shun). The palpation of the pulse.
**sphygmophone** (sfĭg'-mo-fōn) [sphygmo-; φωνή, sound]. A sphygmograph in which the vibrations of the pulse produce a sound.
**sphygmoscope** (sfĭg'-mo-skōp) [sphygmo-; σκοπεῖν, to examine]. An instrument for showing the movements of the heart or the pulsations of a blood-vessel.
**sphygmoscopy** (sfĭg-mos'-ko-pe) [sphygmo-; σκοπεῖν, to observe]. 1. The art of tracing the pulse-curve by the sphygmoscope. 2. Examination of the pulse.
**sphygmosystole** (sfĭg-mo-sis'-to-le) [sphygmo-; systole]. That part of the sphygmogram produced under the influence of the cardiac systole upon the pulse.
**sphygmotechny** (sfĭg-mo-tek'-ne) [sphygmo-; τέχνη, art]. The art of diagnosis and prognosis by means of the pulse.
**sphygmotonograph** (sfĭg-mo-to'-no-graf) [sphygmo-;˙ τόνος, tension; γράφειν, to write]. An instrument which records simultaneously the blood pressure, the apex beat and the pulse.
**sphygmotonometer** (sfĭg-mo-to-nom'-et-er) [sphygmo-; τόνος, tone; μέτρον, measure]. An instrument for use in estimating the elasticity of the arterial walls.
**sphygmus** (sfĭg'-mus) [σφυγμός, pulse]. Pertaining to or having the nature of a pulse.
**sphygmus** (sfĭg'-mus) [σφυγμός, pulse]. The pulse; a pulsation.
**sphyra** (sfi'-rah) [σφύρα, a hammer]. The malleus.
**sphyrectomy** (sfi-rek'-to-me) [sphyra; ἐκτομή, excision]. Excision of the malleus.
**sphyrotomy** (sfi-rot'-o-me) [sphyra; τομή, a cutting]. Surgical removal of part of the handle of the malleus, or of the malleus or its handle together with a portion of the membrana tympani.
**sphyxis** (sfīks'-is). See pulsation.
**spica** (spi'-kah) [L.]. 1. A spike or spur. 2. A spiral bandage with reversed turns. s.-bandage. See under bandage.
**spice** (spīs) [ME., spice, spice]. An aromatic vegetable substance used for flavoring; a condiment. s.-berry, a popular name for Gaultheria procumbens. s.-plaster. See under plaster. s.-poultice, a poultice made from the mixture of a variety of spices. s.-wood. See fever bush.
**spicula** (spĭk'-ū-lah) [dim. of spica, a spike: pl., spiculæ]. A small spike-shaped bone or fragment of bone.
**spicular** (spĭk'-ū-lar) [spicula, a spicule]. Having the form of a spicule.
**spicule** (spĭk'-ūl) [dim. of spica, a spike]. A needle-shaped body; a spike. s., bony, a needle-shaped bone or fragment of bone. s.-sheath, the investment of a sponge-spicule.
**spiculum** (spĭk'-ū-lum) [L.]. See spicula, and spicule.
**spider** (spi'-der) [ME., spither]. An arthropod of the class Arachnida. s. cancer, see acne rosacea. s.-cells, in biology—(a) Bacilli the flagella of which are present in such numbers as to give the microbes the appearance of minute spiders. (b) The characteristic cells of the neuroglia. They have numerous long and delicate prolongations. s., Menarody, a poisonous species of Latrodectus found in Madagascar. s. nevus, see acne rosacea. s.-web, the web spun by the spider, formerly much used as a hemostatic, and also in some systemic diseases; used also as a moxa with the blow-pipe.
**Spiegelberg's sign** (spe'-gel-berg). A sensation like that of passing over wet rubber, imparted to the finger which presses on, and moves along, the affected part; it is noted in cancer of the cervix uteri.
**Spiegel's line, lobe.** See Spigelius' line, lobe.
**Spiegler's test for albumin** (spe'-gler) [Edward Spiegler, Austrian dermatologist, 1863–1908]. Acidulate the solution by the addition of acetic acid to remove the mucin; filter, and overlay the filtrate with a solution prepared by dissolving 8 Gm. of mercuric chloride and 4 Gm. of tartaric acid in 200 Cc. of water, and adding 20 Gm. of glycerol to it. In the presence of albumin a white ring will form between the two liquids.
**Spigelia** (spi-je'-le-ah) [after Adrian van der Spiegel; see Spigelius]. Pinkroot, a genus of plants of the order Loganiaceæ. The rhizome and rootlets of S. marilandica constitute the spigelia of the U. S. P.; they contain a volatile alkaloid, spigeline, and are used as an anthelmintic against the roundworm. Dose 10–20 gr. (0.65–1.3 Gm.) for a child; 1–2 dr. (4–8 Gm.) for an adult. s., fluidextract of (fluidextractum spigeliæ, U. S. P.). Dose 10–20 min. (0.65–1.3 Cc.) for a child; 1–2 dr. (4–8 Cc.) for an adult.
**spigeline** (spi-je'-lēn). An alkaloid said to exist in the anthelmintic species of Spigelia, of which it appears to be an active principle.
**Spigelian line, lobe** (spi-je'-le-an). See Spigelius' line, lobe.
**Spigelius' line** [Adrian van der Spiegel, Belgian physician and anatomist, 1578–1625]. The semilunar line marking the insertion of the muscular fibers of the transversalis abdominis into its tendon. S.'s lobe, a small triangular lobe on the under surface of the right lobe of the liver.
**spike** (spīk) [spica, a spike, ear of corn]. 1. That form of indeterminate anthotaxy in which the flowers are sessile, or nearly so, and arranged on a lengthened axis. 2. A sharp point. s.-lavender, a plant, Lavandula spica; it yields oil of spike.
**spikenard** (spīk'-nard). A name given to the rhizome of various species of Valeriana.
**spilania** (spi-lo'-mah), see nevus (2).
**spilomania** (spi-lo-pla'-ne-ah) [σπίλος, a stain; πλανή, wandering]. A condition characterized by transient or wandering maculæ of the skin. Also, a synonym of elephantiasis græcorum.
**spiloplaxia** (spi-lo-plaks'-e-ah) [σπίλος, spot; πλάξ, a broad surface]. 1. A condition marked by the spots symptomatic of elephantiasis. 2. A synonym of leprosy.
**spilus** (spi'-lus) [σπίλος, a spot]. A mole or colored mark on the skin; nevus.
**spina** (spi'-nah) [L.]. 1. A thorn. 2. The spine. s. accessoria ischii, an inconstant projection into the great sciatic notch at the junction of the ischium and ilium. s. angularis, the spine of the sphenoid bone. s. bifida, a protrusion of the spinal membranes through a congenital cleft of the lower part of the vertebral column. s. bifida occulta, spina bifida in which there is no protrusion of the spinal membranes. s. dorsalis, the spinal column. s. frontalis, the nasal spine. s. helicis. See crista helicis. s. iliaca, the iliac spine (anterior superior, anterior inferior, posterior superior and posterior inferior). s. ischiadica, s. ischiatici, s. ischii. See spine of ischium. s. mentalis, the mental spine; genial tubercle. s. nasalis, the nasal spine. s. nodosa, rhachitis. s. scapulæ, the spine of the scapula. s. supra meatum, an elevation just above the superior angle of the mastoid process of the temporal bone. It appears to be the posterior part of the zygomatic line. s. temginis, a bony process in the tympanum, just above the entrance to the mastoid antrum. s. ventosa, a rarefying form of osteitis in which the bone is eroded or destroyed, and the subperiosteal tissue and osseous marrow contain numerious small cells with transuded red blood-corpuscles. It is frequently a result of syphilis.
**spinal** (spi'-nal). 1. Pertaining to the spine. 2. Pertaining to the spinal cord. s. accessory nerve. See under nerve. s. canal. See canal, vertebral. s. column, the vertebral column, composed of vertebræ, intervertebral cartilages, and ligaments. s. cord, the neural structure occupying the vertebral canal and extending from the atlas to the first lumbar vertebra, and terminating in the filum terminale. It is covered by the spinal membranes (the pia mater, arachnoid, and dura mater) and is divided into symmetrical halves by the anterior and posterior median fissures. These halves are joined together by the anterior white commissure and the gray commissure. In the middle of the latter is the central canal, a con-

tinuation of the ventricular cavities of the brain. Each half of the spinal cord consists of an internal mass of gray matter and an outer covering of white matter. The former is subdivided into the anterior and posterior horns, which are made up of ganglion-cells, nerve-fibers, and delicate fibrilla, and a modified neuroglia, the substantia gelatinosa. The white matter is divided by the two gray horns into three columns: the anterior, lateral, and posterior. These are again subdivided into distinct physiological tracts. Thus the anterior column includes the direct pyramidal tract (Türck's column) and the anterior groundbundle, or anterior radicular zone, which is continuous with the adjacent part of the lateral column. In the latter the following tracts are distinguished: the crossed pyramidal, direct cerebellar, anterolateral (Gowers' tract), and mixed lateral tract. The posterior column contains the posteromedian tract (Goll's column) and the posterolateral or posteroexternal tract (Burdach's column). The spinal cord is the conductor of impulses from and to the brain, as well as a center for reflex acts. s. curvature. See *lordosis*, *kyphosis*, and *scoliosis*. s. epilepsy. See *epilepsy*, *spinal*. s. irritation, a form of neurasthenia characterized by pain in the back, tenderness along the spines of the vertebræ, fatigue on slight exertion, and occasionally numbness and tingling in the limbs. s. marrow, the spinal cord. s. nerves, the 31 pairs of nerves arising from the spinal cord, and grouped into 8 *cervical*, 12 *dorsal*, 5 *lumbar*, 5 *sacral*, 1 *coccygeal*. Each arises by two roots, a dorsal and a ventral. On the dorsal root is the spinal ganglion. Beyond the ganglion the two roots unite to form, in the spinal canal, the mixed trunk of a spinal nerve. The anterior roots supply efferent fibers to all the voluntary muscles of the trunk and extremities, to the smooth muscular fibers of the bladder, ureter, uterus, etc., vasomotor, inhibitory, secretory, and trophic fibers. The posterior roots carry afferent impulses.
**spinalgia** (*spi-nal'-je-ah*) [*spine*,ἄλγος, pain]. Tenderness of a vertebral spine to pressure.
**spinalis** (*spi-na'-lis*) [*spine*]. 1. Spinal. 2. A muscle attached to the spinous processes of the vertebræ; see under *muscle*.
**spinant** (*spi'-nant*) [*spine*]. A drug or other agent increasing the reflex excitability of the spinal cord; strychnine is a spinant.
**spinate** (*spi'-nāt*) [*spinatus*, having spines]. Armed with spines or thorn-shaped processes.
**spindle** (*spin'-dl*) [ME.]. A tapering rod or pin. A body having a fusiform shape. s., achromatic, s., cleavage, s., karyokinetic, s., nuclear, s., segmentation, the double cone-like appearance of the nucleus during certain stages of karyokinesis. s.-cataract, a form of cataract characterized by a spindle-shaped opacity extending from the posterior surface of the anterior portion of the capsule to the anterior surface of the posterior portion of the capsule, with a central dilatation. s.-cell, a fusiform cell. s.-celled, having fusiform cells, a form of cell typical of certain morbid growths, especially sarcoma; fusocellular. s., central, the lining filaments spanning the interval between the centrosome at the completion of the prophase. s.-legged, having long, thin legs. s., neuro-muscular, small fusiform end-organs found in almost all the muscles of the body. s., nuclear, the cone-like appearance of the nucleus during certain stages of karyokinesis. s.-oils, lubricating oils. s.-shanked, same as *s.-legged*. s.-shaped, shaped like a spindle; fusiform. s.-tree, see *euonymus*.
**spine** (*spīn*) [*spina*, a thorn]. 1. A sharp process of bone. 2. The backbone or spinal column. s.-ache, pain in or about the spine. s., angular, curvature of the spine. s., cauda equina of. See *cauda equina*. s., cleft, or cloven. See *spina bifida*. s., cruciate. See *s. of tibia*. s., ethmoid, the spine on top of the sphenoid bone. s., filum terminale of. See *filum terminale*. s., frontal. See *frontal crest*. s., hemal, the part that closes in the hemal arch of a typical vertebra. s., irritable. See *spinal irritation*. s., ischiatic. See *s. of ischium*. s. of ischium, a pointed eminence on the posterior border of the body of the ischium. It forms the lower border of the great sciatic notch. s., mental. See *genial tubercles*. s., navicular, a pointed projection on the inner edge of the navicular bone. s., neural, the part that closes in the neural arch of the typical vertebra. s., occipital, external, the external occipital crest. s., palatine. See *s., nasal*. s., pharyngeal, the ridge on the under surface of the basilar process of the occipital bone. s. of the pubes, the prominent tubercle on the upper border of the body of the pubes. s., pubic. See *s. of the pubes*. s., railway, the designation given to a series of nervous symptoms developed from shock produced by a railway accident or from the concussion produced by constant travel. s. of the scapula, the plate of bone crossing the dorsum of the scapula and dividing it into two unequal parts. s., sciatic. See *s. of ischium*. s., sphenoid, the spinous process of the greater wing of the sphenoid bone. s. of the sphenoid. See *s., sphenoid*, and *s., ethmoid*. s. of the tibia, the elevation upon the upper surface of the tibia between its two articulating surfaces. s., trochlear, a small projection on the upper ventral part of the inner wall of the orbit for the trochlea. s., typhoid, acute inflammation of one or more vertebræ following typhoid fever. s., zygomatic, a projection from the zygomatic process.
**spinicerebrate** (*spi-ni-ser'-e-brāt*). Furnished with a brain and spinal cord.
**spinideltoid** (*spi-ni-del'-toid*). The part of the deltoid muscle arising from the spine of the scapula.
**spinifugal** (*spi-nif'-u-gal*) [*spine*; *fugere*, to flee]. Moving from the spinal cord.
**spinipetal** (*spi-nip'-et-al*) [*spine*; *petere*, to seek]. Moving toward the spinal cord.
**spinitis** (*spi-ni'-tis*). See *myelitis*.
**spinitrapezius** (*spi-ni-tra-pe'-ze-us*). The spinal part of the trapezius as distinguished from the cranial part.
**spinobulbar** (*spi-no-bul'-bar*) [*spine*; *bulbus*, the medulla oblongata]. Pertaining to the spinal cord and the medulla oblongata.
**spinogalvanization** (*spi-no-gal-van-i-za'-shun*). Galvanization of the spinal cord.
**spinoglenoid** (*spi-no-glen'-oid*). Relating to the spine of the scapula and the glenoid cavity.
**spinol** (*spin'-ol*). An extract of young, fresh spinach (*Atriplex hortensis*) leaves, containing about 2.6% of iron and occurring both as a liquid and as a powder. It is used in the spinach cure for children. Dose 1–8 gr. (0.65–0.52 Gm.) several times daily. s. siccum, spinol in the form of a green powder.
**spinomuscular** (*spi-no-mus'-ku-lar*). Relating to the spinal cord and the muscles.
**spinoneural** (*spi-no-nū'-ral*). Pertaining to the spinal cord and the peripheral nerves.
**spinose** (*spi-nōs*) [*spinosus*, full of thorns]. Possessing thorns; or shaped like a thorn.
**spinous** (*spi'-nus*) [*spine*]. Pertaining to the spine; spiny or spiniform. s. process, the apophysis or prominence at the posterior part of each vertebra.
**spinthariscope** (*spin-thar'-is-kōp*) [σπινθήρ, a spark; σκοπεῖν, to view]. An instrument devised by Sir William Crookes for demonstrating the physical properties of radium. It consists of a fluorescent screen in front of which is placed a small quantity of radium bromide, with or without a lens for examining the scintillations.
**spintherism** (*spin'-ther-ism*) [σπινθερίζειν, to emit sparks]. The sensation of sparks dancing before the eyes.
**spintheropia** (*spin-ther-o'-pe-ah*) [σπινθήρ, spark; ὤψ, sight]. Same as *spintherism*.
**spintometer** (*spin-tom'-et-er*) [σπινθήρ, spark; μέτρον, a measure]. An apparatus for measuring the length of sparks in the Roentgen tube.
**spiradenitis** (*spi-rad-en-i'-tis*) [σπεῖρα, a coil; *adenitis*]. Unna's name for phlegmonous hidrosadenitis.
**spiradenoma** (*spi-rad-en-o'-mah*) [σπεῖρα, a coil; ἀδήν, a gland; ὄμα, a tumor]. Adenoma of the sweat-glands.
**spiral** (*spi'-ral*) [σπεῖρα, a coil]. 1. Winding like the threads of a screw, as a *spiral* bandage. 2. A curve having a *spiral* course. s. bandage. See *bandage*. s. canal. See *canal*. s. lamina. See *lamina spiralis*.
**Spirasoma** (*spi-rah-so'-mah*) [σπεῖρα, a coil; σῶμα, a body]. A genus of bacteria of the *Spirillaceæ* having rigid cells without flagella.
**spirem, spirema, spireme** (*spi'-rem*, *spi-re'-mah*, *spi-rēm'*) [σπεῖρα, a twist]. The close skein, or mother skein, or wreath, of chromatin-fibrils in a cell undergoing mitotic divisions.
**spirillicidal** (*spir-il-e-sīd'-al*) [*Spirillum*; *cadere*, to kill]. Said of an agent which is capable of destroying spirilla or spirochætes.
**spirillosis** (*spir-il-o'-sis*). 1. Any affection due to

# SPIRILLUM 823 SPIROBACTERIA

**Spirillum.** 2. A disease of cattle in the Transvaal. s. of fowls, a disease of geese, ducks, guinea-fowls, turtle-doves, pigeons, and sparrows. The affected fowls exhibit diarrhea, loss of appetite, pale combs, and in acute cases die suddenly of convulsions. It is due to a spirillum which is transmitted by *Argus persicus*. The serum of animals which have recovered from a first attack possesses strong immunizing properties.
**Spirillum** (*spi-ril'-um*) [*spirillum*, dim. of *spira*, a coil]. A genus of bacteria having a-spiral shape. See following *table of spirilla*. **s.-feve**. See *relapsing fever*.

**rectified** (*spiritus rectificatus*, B. P.), contains 16 % of water. **s. of salt**, hydrochloric acid. **s. of wine**, alcohol.
**spirituous** (*spir'-it-ū-us*). Alcoholic; pertaining to alcoholic liquors.
**spiritus** (*spir'-it-us*) [L.]. See *spirit*. **s. ætheris nitrosi**. See *niter, sweet spirit of*. **s. chloroformi**, is used as a carminative. Dose 10–60 min. (0.65–4.0 Cc.). See also *chloroform, spirit of*. **s. frumenti**, whisky, a spirit obtained by the distillation of fermented grain. **s. juniperi**, gin or whisky with which juniper-berries and hops have been distilled.
**s. myrciæ**, bay-rum; a hydroalcoholic solution of

## TABLE OF SPIRILLA.

| Name. | Where Found. | Character. |
|---|---|---|
| S. amyliferum (Van Tieghem) | Water | Saprophytic. |
| S. anserum (Sakharoff) | Blood of septicemic geese | Pathogenic. |
| S. aquatilis (Günther) | Water (Spree) | Saprophytic. |
| S. attenuatum (Warming) | Sea-water | Saprophytic. |
| S. aureum (Weibel) | Air, sewage | Chromogenic (golden-yellow). |
| S. beroliniensis (Neisser) | Water (Berlin) | Saprophytic. |
| S. bonhoffii | Water | Saprophytic. |
| S. choleræ asiaticæ (Koch) | Dejecta of cholera patients; water. | Pathogenic, zymogenic. |
| S. concentricum (Kitasato) | Putrid blood | Saprophytic. |
| S. danubicus (Heiden) | Water (Danube) | Saprophytic. |
| S. denticola (Miller) | Mouth | Saprophytic. |
| S. desulfuricans (Beyerinck) | Pit-water | Zymogenic. |
| S. dunbarii (Dunbar and Oergel) | Water (Elbe) | Saprophytic. |
| S. endoparagogicum (Sorokin) | Exudate of poplar tree | Saprophytic. |
| S. flavescens (Weibel) | Sewage | Chromogenic (yellowish-green). |
| S. flavum (Weibel) | Sewage | Chromogenic (ocher-yellow). |
| S. of hospital gangrene (Vincent) | Membranous pulp covering the ulcers. | Pathogenic. |
| S. jenensis (Ehrenberg) | Water | Saprophytic. |
| S. leucomelænum (Perty) | Water | Saprophytic. |
| S. linguæ (Weibel) | Tongue of mouse | Saprophytic. |
| S. litorale (Warming) | Bog-water | Saprophytic. |
| S. luteum (Jumella) | Bog-water | Chromogenic (citron-yellow). |
| S. maasei (Van't Hoff) | Water (Rotterdam) | Pathogenic. |
| S. marinum (Russell) | Sea-water | Saprophytic. |
| S. (*Vibrio*) metchnikovi (Gamaleia) | Intestines of fowls | Pathogenic. |
| S. nasale (Weibel) | Nasal mucus | Saprophytic. |
| S. obermeieri (Cohn) | Blood in cases of relapsing fever | Pathogenic. |
| S. plicatile (Dujardin) | Water | Saprophytic. |
| S. (*Vibrio*) proteus (Finkler-Prior) | Feces in cases of cholera nostras | Saprophytic. |
| S. of pseudocholera (Renon) | Well-water (Billancourt) | Pathogenic. |
| S. recti physeteris (Beauregard) | Ambergris | Zymogenic. |
| S. rosenbergii (Warming) | Brackish water | Saprophytic. |
| S. roseum | Feces | Chromogenic (red). |
| S. roseum (Macé) | Blennorrhagic pus | Chromogenic (rose-red). |
| S. rubrum (Esmarch) | Water | Chromoparous (wine-red). |
| S. rufum (Perty) | Well-water | Chromophorous (rose- to blood-red). |
| S. rugula (Müller) | Water, mouth | Zymogenic (fecal odor). |
| S. saprophiles (Weibel) | Sewage | Saprophytic. |
| S. (*Vibrio*) schuylkillensis (Abbot) | Water (Schuylkill) | Pathogenic. |
| S. serpens (Müller) | Water | Saprophytic. |
| S. smithii | Intestines of swine | Saprophytic. |
| S. sputigenum (Müller) | Healthy mouth. | |
| S. tenue (Ehrenberg) | Water | Saprophytic. |
| S. terrigenus (Günther) | Soil | Saprophytic. |
| S. tyrogenum (Denecke) | Milk | Zymogenic. |
| S. undula (Müller) | Water | Saprophytic. |
| S. violaceum (Warming) | Brackish water | Chromophorous (violet). |
| S. volutans (Ehrenberg) | Marsh-water | Saprophytic. |
| S. of Wernicke | Water | Pathogenic. |

**spirit** (*spir'-it*) [*spiritus*, breath, from *spirare*, to breathe]. 1. The soul. 2. An alcoholic solution of a volatile substance. 3. Alcohol. **s., adiaphorous**, a liquid obtained by the distillation of cream of tartar e., **ammonia**. See *ammoniæ, spiritus*, under *ammonia*. **s., anise**. See *anisi, spiritus*, under *anisum*. **s., Columbian**, deodorized methyl alcohol. **s., corn–**, whisky obtained by the distillation of corn. **s., methylated**, denatured alcohol, ethyl alcohol with one-ninth its volume of methyl alcohol. **s. of Mindererus**, a solution of ammonium acetate, used as a diuretic. **s., potato–**, whisky obtained by the distillation of potatoes. **s., proof–**, diluted alcohol (*alcohol dilutum*, U. S. P.; *spiritus tenuior*, B. P.), containing about 41 % by weight of absolute ethyl-alcohol. **s.,**

various essential oils, and containing 0.8 per cent. of oil of myrcia. **s. odoratus**, Cologne-water. **s. vini gallici** (U. S. P.), brandy; a liquor obtained by the distillation of wine. (*For other spirits see the different drugs.*)
**Spiro's test** (*spe'-ro*) [Karl Spiro, German chemist, 1867– ]. A test for the determination of ammonia and urea in urine by the use of barium oxide and petroleum; it is based on the tests of Folin and Mörner-Sjoqvist.
**spiro–** (*spi-ro-*). 1. [σπεῖρα, a coil]. A prefix meaning spiral. 2. [*spirare*, to breathe.] A prefix meaning relating to respiration.
**spirobacteria** (*spi-ro-bak-te'-re-ah*) [*spiro-*; *bacteria*].

# SPIROCHETE 824 SPLANCHNOSKELETON

**Spiral bacteria,** including spirilla, spirochetes, and vibrios.
**Spirochete, Spirochæta** (*spi'-ro-kēt, spi-ro-ke'-tah*) [*spiro-;* χαίτη, a bristle]. A genus of bacteria characterized by flexible spiral filaments. See under *spirillum*. **S. duttoni,** the cause of African tick fever. **S. novyi,** found in relapsing fever in South America. **S. pallida,** same as *Treponema pallidum, q. v.* **S. pallidula,** same as *s. pertenuis*. **S. pertenuis,** believed to be the cause of yaws. **S. phagedenis,** an anaerobe obtained from phagedenic ulcers on the external genitals. **S. plicatilis,** occurs in stagnant water and is of large size, being about 0.75μ thick and 20 to 500μ long. **S. recurrentis,** the spirillum of Obermeier found in the blood in cases of relapsing fever. **S. refringens,** occurs in primary syphilitic lesions along with *s. pallida*. **S. vincenti,** found in Vincent's angina or ulcerative disease of the tonsils.
**spirochetosis** (*spi-ro-ke-to'-sis*). An infection caused by a spirochete.
**spirofibrillæ** (*spi-ro-fi-bril'-ē*) [*spiro-;* *fibrilla,* a small fiber]. The term applied by Fayod in his theory of the structure of protoplasm to supposed long, twisted, hollow fibrils constituting the protoplasm and nuclei of vegetable cells and uniting to form the spirospartas (*q. v.*). Fayod asserts also that the blood-plasma consists of spirofibrillæ and that they penetrate here and there into the hematoblasts. In this case Bütschli holds that Fayod mistakes coagulation of fibrin for spirofibrillæ.
**spirograph** (*spi'-ro-graf*) [*spiro-;* γράφειν, to write]. An instrument for registering the movements of respiration.
**spirographidin** (*spi-ro-graf'-id-in*) [σπείρα, a coil; γράφειν, to write]. The hyalin obtained from spirographin.
**spirographin** (*spi-ro-graf'-in*) [σπείρα, coil; γράφειν, to write]. A substance obtained from the cartilage and skeletal tissues of the worm, *Spirographis*.
**spiroid** (*spi'-roid*) [*spira,* spire]. Resembling a screw; having spiral convolutions.
**spirometer** (*spi-rom'-et-er*) [*spiro-;* μέτρον, a measure]. An instrument for measuring the quantity of air taken in and given out in forcible respiration.
**spirometric** (*spi-ro-met'-rik*) [*spiro-;* μέτρον, measure]. Pertaining to the spirometer or to spirometry.
**spirometry** (*spi-rom'-et-re*) [see *spirometer*]. Pertains to the measurement of respiration.
**Spiromonas** (*spi-ro-mo'-nas*) [*spiro-;* μονάς, a unit]. A genus of biflagellate monads or free-swimming animalcules established by Perty (1852), now referred to *Bodo* (Ehrenberg), Stein.
**spironema** (*spi-ro-ne'-mah*) [*spiro-;* νῆμα, a thread]. Treponema.
**spirophore** (*spi'-ro-fōr*) [*spiro-;* φέρειν, to bear]. An instrument for performing artificial respiration.
**spirosal** (*spi'-ro-sal*). Trade name of a monoglycolic ester of salicylic acid, used externally in rheumatism and similar conditions.
**spirospartæ** (*spi-ro-spar'-te*) [*spiro-;* σπάρτη, a rope]. The term applied by Fayod in his theory to twisted hollow strings the walls of which are formed by the twisting together of the fibrils or spirofibrillæ. The cavities of the spirospartæ and spirofibrillæ are said to be filled in the normal condition by "granular plasma"; spirospartæ pass from the protoplasm into the nucleus and *vice versa,* and also may be traced frequently from one cell into a neighboring one, so that the cell loses its value as a morphological and physiological unit. These results were obtained in vegetable cells, chiefly by injection with quicksilver, by which method Fayod believes he filled the cavities of the spirospartæ and spirofibrillæ with metal. Cf. *spirofibrillæ.*
**spirulina** (*spi-ru-li'-nah*) [*spirula,* from *spira,* a coil]. A spiral microorganism of spindle shape.
**spissated** (*spis'-a-ted*). Inspissated.
**spissitude** (*spis'-it-ūd*) [*spissare,* to thicken]. The state of being inspissated.
**spit** [ME., *spitten,* to spit]. 1. To eject sputum from the mouth. 2. Saliva. 3. A frothy secretion produced by certain insects as a means of protection.
**spittle** (*spit'-l*). See *saliva*.
**Spitzka's bundle** (*spitz'-kah*) [Edward Charles *Spitzka,* American neurologist, 1852-1914]. A tract of nerve-fibers which passes from the cerebral cortex through the pyramidal region of the crus cerebri to the oculomotor nuclei of the opposite side.

**S.'s nucleus,** the central nucleus of the oculomotor group in the gray matter below the aqueduct of Sylvius. **S.'s postorbital limbus,** a welt-like projection of the orbital surface of the frontal lobe into the middle cranial fossa.
**Spitzka-Lissauer's tract.** See *Lissauer's tract.*
**Spix, angles of** (*spiks*) [Joannes Baptist *Spix,* German anatomist, 1781-1826]. In craniometry, those angles formed: (1) between the alveolo-nasal line and the coronal line; (2) between the alveolo-nasal line and the nasobasilar line. **S., horizontal plane of,** in craniometry, the alveolocondylean plane. **S.'s spine,** the bony spine at the inner border of the inferior dental foramen, giving attachment to the sphenomaxillary ligament; the lingula of the inferior maxillary bone.
**splanchna** (*splangk'-nah*) [σπλάγχνα, viscera]. 1. The intestines. 2. The viscera.
**splanchnapophyseal** (*splangk-nap-off-iz'-e-al*) [*splanchna; apophysis*]. Pertaining to a splanchnapophysis.
**splanchnapophysis** (*splangk-nap-off'-is-is*) [σπλάγχνα, viscera; *apophysis*]. An apophysis or outgrowth of a vertebra on the opposite side of a vertebral axis from a neurapophysis, and inclosing some viscus.
**splanchnectopia** (*splank-nek-to'-pe-ah*) [σπλάγχνα, viscera; ἔκτοπος, displaced]. The abnormal position or dislocation of a viscus.
**splanchnemphraxis** (*splangk-nem-fraks'-is*) [σπλάγχνα, viscera; ἔμφραξις, obstruction]. Obstruction of the intestine.
**splanchneurysma** (*splangk-nū-riz'-mah*) [*splanchno-; aneurysm*]. Distention of the intestines.
**splanchnic** (*splangk'-nik*) [σπλάγχνα, viscera]. 1. Pertaining to or supplying the viscera. 2. A remedy efficient in diseases of the bowels. **s. nerves,** three nerves, the great, lesser, and least, or renal splanchnic, derived from the sympathetic system.
**splanchno-** (*splangk-no-*) [σπλάγχνα, viscera]. A prefix denoting pertaining to the viscera.
**splanchnoblast** (*splangk'-no-blast*) [*splanchno-;* βλαστός, a germ]. An anlage, proton, or incipient rudiment destined to take part in the formation of one or more of the viscera.
**splanchnocele** (*splangk'-no-sēl*) [*splanchno-;* κοῖλος, hollow]. 1. A protrusion of any abdominal viscus. 2. Splanchnocœle.
**splanchnocœle** (*splangk'-no-sēl*) [*splanchno-;* κοῖλος, hollow]. That part of the cœlom which persists in the adult, and gives rise to the pericardial, pleural, and abdominal cavities; the ventral cœlom, or pleuroperitoneal space. It appears as a narrow fissure in the parietal zone of the mesoblast.
**splanchnodiastasis** (*splangk-no-di-as'-tas-is*) [*splanchno-; diastasis*]. Displacement or separation of the viscera.
**splanchnography** (*splank-nog'-raf-e*) [*splanchno-;* γράφειν, to write]. The descriptive anatomy of viscera.
**splanchnolith** (*splangk-no'-lith*) [*splanchno-;* λίθος, a stone]. Calculus of a viscus.
**splanchnolithiasis** (*splangk-no-lith-i'-as-is*) [*splanchno-;* λίθος, stone]. The condition of calculus of the intestine.
**splanchnology** (*splangk-nol'-o-je*) [*splanchno-;* λόγος, science]. The branch of medical science treating of the viscera.
**splanchnomegaly** (*splangk-no-meg'-al-e*) [*splanchno-;* μέγας, large]. Giant growth of the viscera.
**splanchnopathy** (*splangk-nop'-ath-e*) [*splanchno-;* πάθος, disease]. Disease of viscera.
**splanchnopleural** (*splangk-no-ploo'-ral*) [*splanchnopleure*]. Relating to the splanchnopleure.
**splanchnopleure** (*splangk'-no-ploor*) [*splanchno-;* πλευρά, the side]. The visceral layer of mesoderm forming the covering of the digestive tube.
**splanchnoptosia, splanchnoptosis** (*splangk-nop-to'-she-ah, -sis*) [*splanchno-;* πτῶσις, a falling]. A condition of relaxation of the abdominal viscera; it includes gastroptosis, enteroptosis, nephroptosis, less commonly hepatoptosis and splenoptosis.
**splanchnosclerosis** (*splangk-no-skle-ro'-sis*) [*splanchno-;* σκληρός, hard]. Visceral induration.
**splanchnoscopy** (*splangk-nos'-ko-pe*) [*splanchno-;* σκοπεῖν, to examine]. Visual examination of the viscera.
**splanchnoskeleton** (*splangk-no-skel'-et-on*) [*splanchno-; skeleton*]. That portion of the skeleton related to the viscera.

# SPLANCHNOTOMY 825 SPLENOPARECTAMA

**splanchnotomy** (*splangk-not'-o-me*) [*splanchno-*; τέμνειν, to cut]. Dissection of the viscera.

**splanchnotribe** (*splangk'-no-trīb*) [*splanchno-*; τρίβειν, to crush]. An instrument for crushing the intestine and so occluding its lumen, previous to resecting the intestine.

**splashing** (*splash'-ing*) [origin obscure]. Making a splashing sound. **s. fremitus**, a noise heard in succession in some cases of pleural effusion; it may be simulated by the presence of fluid in a distended stomach. **s. in the stomach**, a sign of atony of that organ.

**splay-foot.** See *talipes*.

**spleen** (*splēn*) [σπλήν, spleen]. One of the abdominal viscera, situated just below the diaphragm on the left side, and connected with the hematopoietic system. It is covered by a fibroelastic capsule from which trabeculæ radiate into the organ. In the spaces formed by these are found collections of lymphoid tissue (the Malpighian corpuscles) and the splenic pulp. The Malpighian corpuscles surround the small branches of the splenic artery. The splenic pulp consists of a delicate reticulum containing large connective-tissue cells, lymphoid cells, and red corpuscles. The spleen receives a large amount of blood, which in passing from the termination of the splenic artery to the beginning of the splenic vein is probably not held within walls, but comes in direct contact with the lymphoid tissue. The spleen normally weighs about 200 Gm. **s., accessory**, a detached portion of splenic tissue in the neighborhood of the spleen. Syn., *splenculus*. **s., bacon,** a uniformly lardaceous spleen. **s., floating.** See *s., wandering*. **s., Indian,** an indurated spleen sometimes found in Anglo-Indians. **s., lardaceous,** an enlargement of the spleen due to waxy degeneration. **s.-pulp,** the proper substance of the spleen. **s., sago-,** one of which the Malpighian follicles are the seat of amyloid change. **s., wandering,** one that, owing to relaxation of its attachments, is movable. **s., waxy.** See *s., lardaceous*.

**splen-** (*splen-*) [σπλήν, spleen]. A prefix denoting pertaining to the spleen.

**splenadenoma** (*splen-ad-en-o'-mah*) [*splen-*; *adenoma*]. Hyperplasia of the lymphoid tissue of the spleen.

**splenæmia.** See *splenemia*.

**splenalgia** (*splen-al'-je-ah*) [*splen-*; ἄλγος, pain]. Neuralgic pain in the spleen.

**splenauxe** (*splen-awks'-e*) [*spleen;* αὔξη, increase]. Enlargement of the spleen.

**splenculus** (*splen'-kū-lus*) [*spleen*]. An accessory spleen.

**splenectasis** (*splen-ek'-tas-is*) [*spleen;* ἔκτασις, enlargement]. Enlargement of the spleen.

**splenectomize** (*splen-ek'-tom-īz*) [*splenectomy*]. To excise the spleen.

**splenectomy** (*splen-ek'-to-me*) [*spleen-;* ἐκτομή, excision]. Excision of the spleen.

**splenectopia, splenectopy** (*splen-ek-to'-pe-ah, splen-ek'-to-pe*) [*splen-;* ἔκτοπος, dislocated]. Displacement of the spleen.

**splenelcosis** (*splen-el-ko'-sis*) [*splen-;* ἕλκωσις, ulceration]. Ulceration of the spleen.

**splenelcus** (*splen-el'-kus*) [*splen-;* ἕλκος, ulcer]. An ulcer upon the spleen.

**splenemia, splenæmia** (*splen-e'-me-ah*) [*splen-;* αἷμα, blood]. Splenic leukemia.

**splenemphraxis** (*splen-em-fraks'-is*) [*splen-;* ἔμφραξις, obstruction]. Congestion of the spleen from any cause.

**splenepatitis** (*splen-ep-at-i'-tis*) [*splen-;* ἧπαρ, liver; ιτις, inflammation]. Inflammation involving both liver and spleen.

**splenetic** (*splen-et'-ik*). Splenic. Pertaining to the spleen; ill-humored; fretful; hypochondriacal.

**splenial** (*sple'-ne-al*) [σπλήνιον, bandage]. 1. Serving as a bandage or splint. 2. Pertaining to the splenium or to the splenius.

**splenic** (*splen'* *ik*) [*spleen*]. 1. Pertaining to or affecting the spleen. 2. A remedy efficient in disorders of the spleen. 3. Affected with splenitis. **s. apoplexy, s. fever.** See *anthrax*.

**splenicogastric** (*splen-ik-o-gas'-trik*) [σπληνικός, splenic; γαστήρ, stomach]. Belonging or pertaining to both the spleen and the stomach.

**splenicopancreatic** (*splen-ik-o-pan-kre-at'-ik*) [σπληνικός, splenic; *pancreas*]. Belonging or pertaining to both the spleen and the pancreas.

**splenicterus** (*splen-ik'-ter-us*) [*spleen;* *icterus*]. Inflammation of the spleen associated with jaundice.

**spleniculus** (*splen-ik'-ū-lus*). See *splenculus*.

**splenicus** (*splen'-ik-us*). [*spleen*]. 1. Splenic. 2. A drug acting upon the spleen.

**spleniferrin** (*splen-i-fer'-in*). An organic iron preparation said to be obtained from the spleen.

**splenification** (*splen-if-ik-a'-shun*). See *splenization*.

**splenified** (*splen'-if-īdy*). Of a tissue, resembling the tissue of the spleen, as *splenified* bone-marrow.

**splenin** (*splen'-in*). An organotherapeutic preparation made from the spleen of animals.

**spleniserrate** (*splen-is-er'-āt*) [σπλήνιον, bandage; *serra*, saw]. Pertaining to the splenius and serrate muscles.

**splenitic** (*splen-it'-ik*). See *splenic*.

**splenitis** (*splen-i'-tis*) [*splen-;* ιτις, inflammation]. Inflammation of the spleen. **s., spodogenous,** that due to accumulation of waste-matter.

**splenitive** (*splen'-it-iv*). Capable of acting upon the spleen.

**splenium** (*sple'-ne-um*) [σπλήνιον, a bandage]. 1. A bandage. 2. The rounded posterior extremity of the corpus callosum.

**splenius** (*sple'-ne-us*) [*splenium*]. Shaped like a splenium, as the *splenius* muscle or simply *splenius*. See under *muscle*.

**splenization** (*splen-i-za'-shun*) [*spleen*]. The change in an organ, especially the lung, produced by congestion, whereby it comes to resemble the tissue of the spleen.

**spleno-** (*splen'-o-*) [σπλήν, spleen]. A prefix denoting pertaining to the spleen.

**splenoblast** (*splen'-o-blast*) [*spleno-;* βλαστός, a germ]. A cell from which a splenocyte is derived.

**splenocele** (*splen'-o-sēl*) [*spleno-;* κήλη, hernia]. 1. Hernia of the spleen. 2. A tumor of the spleen.

**splenocleisis** (*splen-o-klī'-sis*) [*spleno-;* κλείειν, to shut in]. Causing the production of new fibrous tissue on the spleen, as by friction with gauze.

**splenocolic** (*splen-o-kol'-ik*) [*spleno-;* κόλον, colon]. Pertaining to the spleen and the colon.

**splenocyte** (*splen'-o-sīt*) [*spleno-;* κύτος, a cell]. The cell peculiar to splenic tissue.

**splenodynia** (*splen-o-din'-e-ah*) [*spleno-;* ὀδύνη, pain]. Pain in the spleen.

**splenography** (*splen-og'-ra-fe*) [*spleno-;* γράφειν, to write]. The descriptive anatomy of the spleen.

**splenohemia, splenohæmia** (*splen-o-he'-me-ah*) [*spleno-;* αἷμα, the blood]. Congestion of the spleen. Hyperemia of the spleen.

**splenohepatomegaly** (*splen-o-hep-at-o-meg'-al-e*) [*spleno-;* ἧπαρ, liver; μέγας, great]. Enlargement of the liver and spleen.

**splenoid** (*splen'-oid*) [*spleno-;* εἶδος, resemblance]. Resembling the spleen.

**splenokeratosis** (*splen-o-ker-at-o'-sis*) [*spleno-;* κέρας, horn]. Splenic induration.

**splenolaparotomy** (*splen-o-lap-ar-ot'-o-me*). See *laparosplenotomy*.

**splenology** (*splen-ol'-o-je*) [*spleno-;* λόγος, science]. The sum of what is known of the splenic structure, function, and diseases.

**splenolymph** (*splen'-o-limf*). Intermediate in character between the spleen and a lymph-gland. See *glands, splenolymph*.

**splenolymphatic** (*splen-o-lim-fat'-ik*). Relating to the spleen and the lymph-glands.

**splenolymphoma** (*splen-o-lim-fo'-mah*). See *splenadenoma*.

**splenolysin** (*splen-ol'-is-in*) [*spleno-;* *lysin*]. An antibody destructive to splenic tissue or cells.

**splenoma** (*splen-o'-mah*) [*spleno-;* ὄμα, tumor]. Tumor of the spleen.

**splenomalacia** (*splen-o-mal-a'-se-ah*) [*spleno-;* μαλακία, softness]. Softening of the spleen.

**splenomedullary** (*splen-o-med'-ul-a-re*). Relating to the spleen and the marrow of bones.

**splenomegalia, splenomegaly** (*splen-o-meg-a'-le-ah, splen-o-meg'-al-e*) [*spleno-;* μέγας, great]. Enlargement of the spleen, especially simple enlargement of the spleen without leukemia; by some it is considered merely as Hodgkin's disease of splenic type. **s., tropical,** kala azar.

**splenomyelogenous** (*splen-o-mi-el-oj'-en-us*). Referring to the spleen and bone marrow; splenomedullary.

**splenoncus** (*splen-ong'-kus*). See *splenoma*.

**splenoparectama, splenoparectasis** (*splen-o-par-ek'-ta-mah, splen-o-par-ek'-ta-sis*) [*spleno-;* ἔκτασις, distention]. Enlargement of the spleen.

**splenopathia** (*splen-o-path'-e-ah*). See *splenopathy*.
**s. leukocythæmica**, splenic leukemia.
**splenopathy** (*splen-op'-ath-e*) [*spleno-*; πάθος, suffering]. Any disease of the spleen.
**splenopexia, splenopexis, splenopexy** (*splen-o-peks'-e-ah, splen'-o-peks-is, splen'-o-peks-e*) [*spleno-*; πῆξις, a fixing in]. Fixation of a wandering spleen to the abdominal wall by means of sutures.
**splenophlegmone** (*splen-of-fleg'-mon-e*) [*spleno-*; φλεγμόνη, inflammation]. Phlegmonous inflammation of the spleen.
**splenophrenic** (*splen-o-fren'-ik*) [*spleno-*; φρήν, diaphragm]. Pertaining to the spleen and the diaphragm.
**splenophthisis** (*splen-off'-this-is*) [*spleno-*; φθίσις, wasting]. Atrophy of the spleen.
**splenopneumonia** (*splen-o-nū-mo'-ne-ah*) [*spleno-*; *pneum'nia*]. Pneumonia with splenization of the lung. o
**splenoptosis** (*splen-op-to'-sis*) [*spleno-*; πτῶσις, a falling]. Downward displacement of the spleen.
**splenorrhagia** (*splen-or-a'-je-ah*) [*spleno-*; ῥηγνύναι, to burst forth]. Hemorrhage from the spleen.
**splenorrhaphy** (*splen-or'-af-e*) [*spleno-*; ῥαφή, suture]. Suture of the spleen.
**splenoscirrhus** (*splen-o-skir'-us*) [*spleno-*; σκίρρος, hardness]. Cancer of the spleen.
**splenotomy** (*splen-ot'-o-me*) [*spleno-*; τέμνειν, to cut]. 1. The operation of incising the spleen. 2. Dissection of the spleen.
**splenotyphoid** (*splen-o-ti'-foid*) [*spleno-*; *typhoid*]. Typhoid fever with splenic complication.
**splenule** (*splen'-ūl*). An accessory or rudimentary spleen.
**splenunculus** (*splen-ung'-kū-lus*). Accessory spleen; lienunculus.
**splint** [Swedish, *splint*, a kind of spike]. A piece of wood, metal, or other material for keeping the ends of a fractured bone or other movable parts in a state of rest. **s., anchor**, a splint used for fracture of the jaw. Metal loops fit over the teeth, and are held in contact by a rod and nut. **s. bandage**, an immovable bandage. **s. Bavarian**, coarse flannel is cut to fit the part, and stitched over the limb. A thick paste of plaster of Paris is rubbed upon the cloth to secure immobility. **s. bone**, the fibula. **s.-box**. See *fracture-box*. **s., bracketed**, a splint consisting of two pieces of wood or metal joined by brackets. **s., interdental**, an appliance used in the treatment of fractured jaws. **s., poroplastic**, a splint which can be softened with hot water and molded upon the limb, to harden and retain the shape when dried.
**splintage** (*splint'-āj*). The application of splints.
**splinter** (*splin'-ter*) [ME., *splinteren*, to split]. See *sequestrum*. Applied, also, popularly to a bit of wood or other material that pierces the skin. **s.-bone**. 1. The fibula. 2. A term applied to one of the two small bones extending from the knee to the fetlock of the horse, behind the shank-bone.
**split**. A longitudinal fissure. **s. cloth**, a bandage for the head with six or eight tails attached to a central part. **s. pelvis**, congenital non-union of the bones of the pubes at the symphysis.
**spodiomyelitis** (*spo-de-o-mīel-i'-tis*) [σποδός, gray; μυελός, marrow; ιτις, inflammation]. An acute inflammation in the anterior cornua of the spinal cord, in which the larger multipolar ganglion-cells are destroyed. Poliomyelitis.
**spodium** (*spo'-de-um*) [σποδός, ashes]. An old term for animal charcoal.
**spodogenous** (*spo-doj'-en-us*) [σποδός, ashes; γεννᾶν, to produce]. Pertaining to or produced by wastematerial, as *spodogenous* enlargement of the spleen, a swelling of the spleen produced by the accumulation of the detritus of red corpuscles.
**spodophagous** (*spo-dof'-ag-us*) [σποδός, ashes; φαγεῖν, to eat]. Destroying the waste-material of the body.
**spodophorous** (*spo-dof'-or-us*) [σποδός, ashes; φέρειν, to bear]. Carrying or conveying waste-material.
**Spoendel's foramen**. A small opening in the cartilaginous base of the skull between the ethmoid and the lesser wings of the sphenoid and the anterior ethmoid.
**spokebone** (*spōk'-bōn*). See *radius*.
**spoke-shave** (*spōk'-shāv*). A ring-knife, devised by Carmalt Jones, for use in operations on the nasal cavities.
**spondyl-, spondylo-** (*spon-dil-, spon-dil-o-*) [σπόν-

δυλος, vertebra]. A prefix denoting pertaining to a vertebra.
**spondylalgia** (*spon-dil-al'-je-ah*) [*spondyl-*; ἄλγος, pain]. Pain referred to a vertebra.
**spondylarthritis** (*spon-dil-ar-thri'-tis*) [*spondyl-*; ἄρθρον, joint; ιτις, inflammation]. Inflammation of a vertebral articulation. **s. synovialis**, inflammation of the synovial membranes of the articular process of the vertebræ (Huter).
**spondylarthrocace** (*spon-dil-ar-throk'-as-e*) [*spondyl-*; ἄρθρον, joint; κακή, evil]. Caries of a vertebra.
**spondyle** (*spon'-dil*) [σπόνδυλος, a vertebra]. A vertebra.
**spondylexarthrosis** (*spon-dil-eks-ar-thro'-sis*) [*spondyl-*; ἐξ, out; ἄρθρον, joint]. Dislocation of a vertebra.
**spondylitis$_c$** (*spon-dil-it'-ik*). Relating to spondyl* itis.
**spondylitis** (*spon-dil-i'-tis*) [*spondyl-*; ιτις, inflammation]. Inflammation of one or more vertebræ; Pott's disease. **s. cervicalis**, arthritis of one or more cervical vertebræ. **s. deformans**, chronic inflammation of the vertebræ, of a gouty or rheumatic nature, terminating in ankylosis and deformity. **s. tuberculosa**, tuberculous spondylitis; Pott's disease.
**spondylizema** (*spon-dil-i-ze'-mah*) [*spondyle*; ἵζημα, a subsiding]. The settling of a vertebra into the place of a subjacent one that has been destroyed.
**spondylocace** (*spon-dil-ok'-as-e*). See *spondylar-. throcace*.
**spondylodidymia** (*spon-dil-o-did-im'-e-ah*) [*spondyle*; δίδυμος, twin]. A form of somatodymia in which the union is in the vertebræ. Syn., *vertebradymia*.
**spondylodymus** (*spon-dil-od'-im-us*) [*spondyle*; δύειν, to enter]. A twin monster united by the vertebræ.
**spondylodynia** (*spon-dil-o-din'-e-ah*) [*spondyle*; ὀδύνη, pain]. Pain in a vertebra.
**spondylisthesis** (*spon-dil-o-lis-the'-sis*) [*spondyle*; ὀλίσθησις, a slipping]. Deformity of the spinal column produced by the gliding forward of the lumbar vertebræ in such a manner that they overhang the brim and obstruct the inlet of the pelvis; especially the separation of the last lumbar vertebra from, and its slipping forward on, the sacrum.
**spondylolisthetic** (*spon-dil-o-lis-thet'-ik*). Pertaining to or caused by spondylolisthesis.
**spondylolizema** (*spon-dil-o-liz-e'-mah*). Same as *spondylizema*.
**spondylomyelitis** (*spon-dil-o-mi-el-i'-tis*). See *spondylitis*.
**spondylopathia** (*spon-dil-o-pa'-the-ah*). See *spondylopathy*.
**spondylopathy** (*spon-dil-op'-ath-e*) [*spondyle*; πάθος, a suffering]. Any disease of the vertebræ.
**spondylopyosis** (*spon-dil-o-pi'-o-sis*) [*spondyle*; πύον, pus]. Suppurative inflammation of one or more vertebræ.
**spondyloptosis** (*spon-dil-op-to'-sis*). See *spondylolisthesis*.
**spondyloschisis** (*spon-dil-os'-kis₁is*). Deficient ossification in the arch of the fifth lumbar vertebra; this is said to be one of the causes of spondylolisthesis. The condition may affect one or both sides of the vertebra.
**spondylosis** (*spon-dil-o'-sis*) [*spondyle*]. Vertebral ankylosis. **s., rhizomelic**, spondylose rhizomelique; Marie's term for a variety of arthritis deformans with ankylosis of the vertebræ and arthritis of the hips and shoulders.
**spondylotherapy** (*spon-dil-o-ther'-ap-e*). Spinal therapeutics; the treatment of diseased conditions by various manipulations applied to the spinal column.
**spondylotomy** (*spon-dil-ot'-o-me*) [*spondyle*; τομή, section]. Section of a vertebra in embryotomy; section of a vertebra in correcting a deformity. Cf. *rhachiotomy*.
**spondylous** (*spon'-dil-us*) [*spondyle*]. Vertebral; like a vertebra.
**spondylus** (*spon'-dil-us*) [*spondyle*]. A vertebra.
**sponge** (*spunj*) [σπόγγος, a sponge]. A marine animal of the class *Porifera*, having a porous, horny skeleton; also the skeleton itself, used as an absorbent. **s.-bath**, the application of water to the surface of the body by means of a sponge. **s., burnt**, sponge-charcoal made from fine sponges cleansed and burned, then powdered and sifted through a No. 100 silk sieve. **s., compressed**, a fine sponge cleansed, exposed to pressure, and dried. **s.-gatherer's disease**,

a disease of divers due to a secretion of a species of *Actinia* found in waters where sponges grow. This viscid excretion causes at the point of contact upon the body a swelling and intense itching, followed by a papule surrounded by a zone of redness which later becomes black and gangrenous and forms a deep ulcer. **s.-graft.** See *graft, sponge-*. **s. holder,** an instrument consisting of a rod which serves as a handle, furnished at the distal end with a device for clasping a sponge. **s. prepared,** a sponge rendered soft and elastic and suitable for surgical uses by soaking in cold water and separation of calcareous matter. **s.-tent.** See *tent, sponge-*. **s. test,** a hot sponge is passed up and down the spine; in the presence of caries, pain is felt as the sponge passes over the seat of the lesion.
**spongework** (*spunj-werk*). Synonym of *spongioplasm.*
**spongia** (*spun'-je-ah*). See *sponge*. **s. cerata.** See *sponge, waxed*. **s. compressa.** See *sponge, compressed*. **s. fluviatilis,** small sponges found on stones and on water-plants in streams, ponds, and marshy places. **s. lacustris,** a Russian variety used by homeopaths in the preparation of a tincture. **s. officinalis, s. præparata.** See *sponge, compressed,* and *s. prepared*. **s. usitatissima,** sponges with fine pores cleansed and pressed. **s. usta.** See *sponge, burnt.*
**spongiform** (*spun'-je-form*) [*sponge; forma,* a form]. Resembling a sponge.
**spongin** (*spun'-jin*) [*sponge*]. The horny substance forming the skeletal fibers of the sponge.
**spongioblast** (*spun'-je-o-blast*) [*sponge;* βλαστός, a germ]. A variety of cell derived from the ectoderm of the embryonic neural tube, and forming later the neuroglia. **s.** of **inner molecular layer of retina.** See *cells amacrine*.
**spongiocyte** (*spun'-je-o-sīt*) [*sponge;* κύτος, a cell]. Fish's term for the glia or neuroglia cell.
**spongioid** (*spun'-je-oid*) [*sponge;* εἶδος, resemblance]. Spongiform.
**spongiopilin** (*spun-je-o-pī'-lin*) [*sponge;* πῖλος, felt]. Felted or woven cloth into which tufts of sponge are incorporated and one side of which is coated with rubber; it is used as a poultice.
**spongioplasm** (*spun'-je-o-plazm*) [*sponge;* πλάσσειν, to mold]. The fine, elastic protoplasmic threads forming the reticulum of cells.
**spongiose** (*spun'-je-ōs*) [*σπόγγος,* sponge]. Full of pores, like a sponge.
**spongiositis** (*spun-je-o-si'-tis*). Inflammation of the corpus spongiosum.
**spongy** (*spun'-je*) [*σπόγγος,* sponge]. Having the texture of sponge; very porous. **s. body,** the corpus spongiosum. **s. bones,** bones having a porous, reticulated structure, especially the turbinated bones of the nose, and the sphenoid and ethmoid bones. **s. portion of the urethra,** that contained in the corpus spongiosum of the penis.
**spontaneous** (*spon-ta'-ne-us*) [*spons,* will]. Voluntary; occurring without extraneous impulse; automatic.
**spoon.** An instrument consisting of an oval or circular bowl fixed to a handle; it is used in surgery to scrape away dead tissue, granulations, etc. **s.-nail,** a nail with a concave outer surface.
**spoonful.** A spoon is full when the contained liquid comes up to but does not show a curve above the upper edge or rim of the bowl. A teaspoonful equals 5 Cc.; a dessertspoonful, 10 Cc.; a tablespoonful, 15 Cc.
**spora** (*spo'-rah*) [L.]. See *spore*.
**sporadic** (*spor-ad'-ik*) [*σποραδικός,* scattered]. Scattered; occurring in an isolated manner. **s. cholera,** cholera morbus.
**sporadoneure** (*spor-ad'-o-nūr*) [*σπορás,* scattered; νεῦρον, a nerve]. An isolated nerve-cell.
**sporangia** (*spor-an'-je-ah*). Plural of *sporangium*.
**sporangial** (*spor-an'-je-al*). Relating to a sporangium.
**sporangium** (*spor-an'-je-um*) [*spore;* ἀγγεῖον, a vessel; pl., *sporangia*]. In biology, a capsule producing or inclosing spores.
**sporation** (*spor-a'-shun*) [*σπόρος,* seed]. See *sporulation.*
**spore** (*spōr*) [*σπόρος,* seed]. 1. A reproductive body of a cryptogam. 2. Any germ or reproductive element less organized than a true cell; also any spermatic or ovulary cell. **s.-capsule,** a spore-case. **s.-case,** the sporangium or covering of a spore.

**s.-cell,** a spore. **s., compound,** a spore that produced secondary spores. **s., daughter,** a spore produced in a mother-cell. **s.-formation,** the origination of spores. **s.-group.** Same as *sporidesm*. **s., inactive,** a non-motile fertile cell. **s., mother-,** a mother-cell. **s., naked,** a gymnospore. **s.-plasm,** the protoplasm of a sporangium. **s., primary,** a spore the germination of which produces a prothallium; a protospore. **s.-sac,** the sac lining the cavity of the sporangium of mosses; see *sporangium*. **s., secondary,** a merispore; cf. *s. compound*. **s., swarm,** a spore endowed with the power of locomotion.
**sporicidal** (*spor-is'-i-dal*) [*spore; cædere,* to kill]. Destructive to spores.
**sporicide** (*spor'-e-sīd*) [*spore; cædere,* to kill]. Any agent which destroys spores.
**sporidesm** (*spor'-id-ezm*) [*spore;* δέσμη, a bundle]. In biology, a septate or compound spore.
**sporidium** (*spor-id'-e-um*) [*spore;* ἰδίον, a dim.; pl., *sporidia*]. 1. In biology, a spore borne upon a promycelium. 2. A provisional genus of *Sporozoa*. S. vaccinale, Funck, a species of sporozoa occurring as: (1) small, spheric, highly refractive bodies (2-10 μ), of green color and slow movement; (2) small refracting spheres inclosed in capsules; (3) morula masses or spore-casts. They can be cultivated and the culture produces typical vaccinia when inoculated in calves.
**sporiferous** (*spor-if'-er-us*) [*spore; ferre,* to bear]. Spore-bearing.
**sporification** (*spor if ih a' shun*). The formation of spores.
**sporiparous** (*spor-ip'-ar-us*) [*spore; parere,* to produce]. In biology, reproducing by means of spores.
**sporo-** (*spor-o-*) [*spore*]. A prefix meaning relating to a spore or seed.
**sporoblast** (*spor'-o-blast*) [*sporo-;* βλαστός, a germ]. One of the four round bodies produced by the process of endogenous cell-formation in a coccidium.
**sporocyst** (*spor'-o-sist*) [*sporo-;* κύστις, a bag]. 1. The mother-cell of a spore. 2. That stage of a sporozoon resulting from the development of a sporoblast and in its turn giving rise to two sporozoites.
**sporocyte** (*spor'-o-sīt*) [*spore;* κύτος, a hollow]. In biology, the mother-cell of a spore; a sporocyst.
**sporoderm** (*spor'-o-derm*) [*spore;* δέρμα, skin]. In biology, the coat of a spore, including exospore and endospore.
**sporoduct** (*spor'-o-dukt*) [*spore; ducere,* to lead]. A passage through which spores are conducted.
**sporogenesis** (*spor-o-jen'-es-is*) [*sporo-;* γένεσις, generation]. The development of spores; reproduction by spores.
**sporogenous** (*spor-oj'-en-us*) [*spore;* γενής, producing]. In biology, spore-producing.
**sporogeny** (*spor-oj'-en-e*). Same as *sporogenesis, q. v.*
**sporogone** (*spor'-o-gōn*). Same as *sporogonium.*
**sporogonium** (*spor-o-go'-ne-um*) [*spore;* γόνη, generation; pl., *sporogonia*]. In biology, the nonsexual generation of a moss, proceeding from the fertilized oösphere; also called *sporogone*.
**sporogony** (*spor-og'-o-ne*). 1. See *sporogenesis*. 2. A form of exogenous sporulation; an oöcyst containing a sporont divides into four sporoblasts, which ripen into sporocysts and in turn divide into a crescentic nucleated body, the sporozoite. It occurs among coccidia. Cf. *schizogony*.
**sporont** (*spor'-ont*) [*sporo-; ὤν,* being]. 1. In biology a gregarine without an epimerite, as distinguished from a cephalont. 2. Schaudinn's term for the single-celled contents of the coccidial oöcyst. Cf. *schizont*.
**sporophore** (*spor'-o-fōr*) [*sporo-; φέρειν,* to bear]. That portion of a fungus bearing the spores.
**sporophyl, sporophyll, sporophyllum** (*spor'-o-fil, spor-o-fil'-um*) [*spore; φύλλον,* leaf]. In biology, the modified leaf which bears the spores, or receptacles holding the spores, in many of the vascular cryptogams; the fertile leaf.
**sporophyte** (*spor'-o-fīt*) [*spore; φυτόν,* plant]. In biology, the nonsexual generation of one of the vascular cryptogams and higher cellular cryptogams. It is often of great size and extended length of life, and is that which is commonly known as the fern, clubmoss, etc. On it are produced, without any process of fertilization, the *spores*.
**sporoplasm** (*spor'-o-plazm*) [*sporo-;* πλάσσειν, to form]. The cytoplasm of the asexual reproductive cell.

## SPOROTHECA 828 SPUTUM

**sporotheca** (*spor-o-the'-kah*) [*sporo-*; θήκη, a case]. 1. See *sporangium*. 2. The envelope of the sporulating cell.
**sporothrix.** See *sporotrichum*.
**sporotrichosis** (*spor-o-trik-o'-sis*). Infection by sporothrix, producing indolent subcutaneous abscesses.
**Sporotrichum** (*spo-rot'-rik-um*) [*spore;* θρίξ, hair]. A genus of fungi, some of whose members such as *S. beurmanni* and *S. schenkii* give rise to sporotrichosis.
**Sporozoa** (*spor-o-zo'-ah*) [*sporo-*; ζῷον, an animal]. A class of parasitic *Protozoa* subdivided into the *Gregarinidea*, parasitic in various worms and arthropods; the *Coccidiea* or oviform psorosperms, parasitic in the hepatic and intestinal epithelium of various mammals, including man; the *Sarcosporidia*, or tubuliform psorosperms, parasitic in the muscles of various animals; the *Myxosporidia*, the psorosperms of fishes; the *Microsporidia*, the psorosperms of articulates.
**sporozoite** (*spor-o-zo'-it*) [see *Sporozoa*]. The sickle-shaped, nucleated organism which results from the division of a sporocyst among the sporozoa. Syn., *germinal rod; zygotoblast; gametoblast.*
**sporozooid** (*spor-o-zo'-oid*) [*spore; zooid,* animal-like]. 1. One of the two "sickle-cells" or "falciform bodies" produced by every spore of the true coccidia. 2. Any oöspore.
**sporozoon** (*spor-o-zo'-on*) [see *Sporozoa*]. In biology, a member of the *sporozoa*.
**sport.** An animal or plant that exhibits decided variation from the normal type.
**sporular** (*spor'-u-lar*). Having the character of a sporule.
**sporulation** (*spor-u-la'-shun*) [*spore*]. The production of spores. s., **arthrogenous**, the change of bacteria into resistant forms, which are capable of germinating again under favorable conditions.
**sporule** (*spor'-ul*) [σπόρος, seed]. A term applied to a minute spore; also sometimes to minute granules within a spore.
**sporuliferous** (*spor-u-lif'-er-us*) [*sporule; ferre,* to bear]. Bearing sporules.
**spot** [ME.]. See *macule*. **s.s, acoustic.** See *maculæ acusticæ* under *macula.* **s.s, Bitot's**, xerosis conjunctivæ; silver-gray, shiny, triangular spots on both sides of the cornea, within the region of the palpebral aperture, consisting of dried epithelium, flaky masses, and microorganisms. They are observed in some cases of hemeralopia. **s., blind** (of Mariotte), the entrance of the optic nerve where the rods and the cones are absent. **s., blue** (of the integument), a tegumentary spot over the sacral region characteristic of the Mongolian race; due to aggregations of long, spindle-shaped, and stellate cells in the cutis containing pigment. It appears in the fourth month of fetal life and persists sometimes to the seventh year. **s., corneal,** an opacity of the cornea; leukoma. **s.s, cribriform,** the perforations of the fovea hemisphærica for the passage of the filaments of the auditory nerve. Syn., *macula cribrosa.* **s.-disease.** See *Pébrine.* **s., embryonic,** the nucleolus of the ovum. **s.s, genital,** nasal parts which show increased sensitiveness during menstruation. **s., germinal, s., germ-.** See *s.; embryonic.* **s., hectic,** the bright flush on the cheeks of a person suffering from hectic fever. **s.s, Koplik's.** See *Koplik's spots.* **s.s, lenticular.** See *s.s., rose.* **s., light** (on the membrana tympani), a cone of light on the anterior and inferior part of the tympanic membrane, with its apex directed inward. **s., Mariotte's,** the optic disc. **s., milk-,** a spot found postmortem on the external surface of the visceral layer of the pericardium, usually over the right ventricle; it varies from one-half to one inch in diameter, and is of common occurrence in persons who have passed middle life. **s., mother's,** nevus. **s.s, rose,** a red papulous eruption forming spots the size of a small lentil, effaced by pressure of the finger and occurring mostly on the abdomen and loins during the first seven days of typhoid fever. They are due to inflammation of the papillary layer of the skin from invasion of typhoid bacilli. Syn., *typhoid roseola; typhoid spots;* Fr., *tache rosées lenticulaires.* Cf. *tache bleuâtre.* **s.s, soldiers'.** Same as *macula lutea.* **s., Soemmering's.** See *macula lutea.* **s.s, sun.** See *lentigo.* **s.s, typhoid.** See *s.s, rose.* **s. of Wagner,** the embryonic spot. **s.s, white,** grayish or yellowish-white elevated spots from the size of a pin-head to that of a one-cent piece, of varying shape and distinctness of outline, often occurring on the ventricular surface of the anterior leaflet of the mitral valve. **s., wine-,** port-wine mark; strawberry mark. **s., yellow.** See *macula lutea.* **spotted fever.** 1. Cerebrospinal fever. 2. Typhus. 3. Tick fever. **s. sickness.** See *pinta disease.*
**sprain** (*sprān*) [OF., *espreindre,* from L., *exprimere,* to press out]. A wrenching of a joint, producing a stretching or laceration of the ligaments. **s.-fracture,** an injury in which a tendon together with a shell of bone is torn from its attachment. **s., riders',** a condition of the adductor longus muscle of the thigh, resulting from a sudden effort on the part of the horseman to maintain his seat owing to some unexpected movement of his horse.
**spray** (*sprā*). A liquid blown into minute particles by a strong current of air. **s.-cure,** a form of douche applied by means of a spraying apparatus.
**Sprengel's deformity** [Otto Gerhard Carl *Sprengel,* German surgeon, 1852– ]. Congenital upward displacement of one scapula.
**sprew** (*sproo*). See *sprue.*
**spring** [ME.]. The first of the four seasons of the year; also, a device having resiliency. **s., conjunctivitis.** See *vernal conjunctivitis.* **s., fever,** lassitude. **s.-finger,** a condition in which there is an obstruction to flexion and extension of one or more fingers at a certain stage of these movements. It is due to injuries or may result from inflammation of the tendinous sheaths. **s.-halt,** an involuntary convulsive movement of the muscles of either hind leg in the horse, by which the leg is suddenly and unduly raised from the ground and lowered again with unnatural force. **s.-knee,** a condition of the knee similar in general features to the condition known as spring-finger. Just before full extension of the joint is reached there is a slight hitch, and then the limb straightens itself with a sharp, rather painful jerk. **s.-ligament,** the inferior, calcaneoscaphoid ligament of the sole of the foot. **s.-nail,** a hangnail. **s. ophthalmia.** See *vernal conjunctivitis.* **s.-worm.** See *oxyuris vermicularis.*
**sprue.** 1. Thrush. 2. The name given by the Dutch in Java to aphthæ tropicæ (*q. v.*); a chronic catarrhal inflammation of the entire alimentary tract, especially prevalent in Malaya. Syn., Ceylon *sore mouth; diarrhœa alba; sprouw; tropical sprue.*
**sprung knee.** In the horse an alteration in the direction and articulation of the bones which form the various carpal joints, so that instead of forming a vertical line from the distal end of the forearm to the cannon-bone, the knee (wrist) is more or less bent forward.
**spud** [Dan., *spyd,* a spear]. 1. An instrument used in the detachment of the mucosa in flaps in operations necessitating the removal of bone. 2. A short flattened blade used to dislodge a foreign substance.
**spunk** (*spungk*). Surgeon's agaric saturated with potassium nitrate. See under *agaric.*
**spur** (*sper*) [ME., *spure*]. 1. A sharp point or projection. 2. The angle made by any branch with the main blood-vessel. 3. In biology, a pointed, spine-like outgrowth, either of the integument or a projecting appendage. 4. Ergot. **s.-blind,** myopic. **s.-gall,** a callous and hairless place on the side of a horse, caused by the use of a spur. **s. of the septum,** an outgrowth of the nasal septum.
**spurge** (*sperj*) [ME., *sporgeon,* spurge]. A general name for plants of the genus *Euphorbia.*
**spurious** (*spū'-re-us*) [*spurius,* false]. Not legitimate; bastard. **s. false pains. s. melanosis,** see *miners' phthisis.* **s. pregnancy,** see *pseudocyesis.*
**spurred** (*sperd*) [ME., *spure*]. Having spurs. **s. rye,** see *ergot.*
**sputa** (*spu'-tah*). Plural of *sputum, q. v.*
**sputum** (*spu'-tum*) [*spuere,* to spit; pl., *sputa*]: The secretion ejected from the mouth in spitting. It consists of saliva and mucus from the nasal fossæ and the fauces. In diseased conditions of the air-passages or lungs it may be purulent, mucopurulent, fibrinous, or bloody. **s., æruginous,** sputum of a green color. **s., black-pigmented,** having a black color from inhaled particles of carbon. **s., cavernous,** nummular sputum, from a pulmonary cavity. **s. coctum,** opaque, yellowish or greenish, viscid, generally partially confluent, through occasionally nummulated, sputum of the later stages of acute bron-

chitis. s. crudum, the scanty, viscid expectoration of the early stages of acute bronchitis. s., egg-yolk, sputum having a yellow color. s., globular, spherical masses of sputum of the later stages of bronchitis; yellow in color, and consisting of epithelium, pus-corpuscles, mucus, etc. s., green, bloody sputum in which oxidation of the hemoglobin has taken place; it is seen in pneumonia. s., hailstone, spherical masses of sputum of the later stages of bronchitis. s., icteric, sputum tinged green or yellow, due to the presence of bile pigment; it is observed in icterus. sputa margaritacea, see *s., pearly*. s., mucopurulent, small lumps or pellets in a viscid, mucoserous fluid, seen in bronchitis. s., nummular, a sputum characterized by round, coin-like masses; it is seen in pulmonary tuberculosis. s., pearly, sputum consisting of small translucent pellets. s., prune-juice, s., rusty, the typical, dark-colored sputum of the third stage of pneumonia; the color is due to the admixture of blood. s. puriforme, sputum having the appearance of pus. s., rusty, the dark-colored sputum of lobar pneumonia, the color of which is due to the admixture of blood. s. septicemia, see *septicemia, sputum*. s. tuberculosum, a purulent or mucopurulent sputum containing tubercle-bacilli, occurring in pulmonary tuberculosis and in caseous pneumonia. s., yellow, sputum having a yellow color, due to the presence of fungi; the term is also applied to sputum rendered yellow by oxidation of the contained hemoglobin.
squalor (*skwol'-or*, or *skwa'-lor*) [L.]. Filth. Disorder and uncleanliness.
squama (*skwa'-mah*) [L: pl., *squamæ*. A scale or scale-like mass, as the *squama* of the temporal bone. s. frontalis, the vertical portion of the frontal bone. s. occipitalis, the supraoccipital bone. s. temporalis, the squamosa.
squamate (*skwa'-māt*) [*squama*, a scale]. Scaly, or scale-like.
squamo- (*skwa-mo-*) [*squama*]. A prefix denoting relating to the squamous portion of the parietal or temporal bone.
squamocellular (*skwa-mo-sel'-ū-lar*) [*squamo-*; *cellula*, a small cell]. Flat-celled.
squamoid (*skwa'-moid*) [*squama*, scale; είδος, like]. Resembling a squama.
squamomandibular (*skwa-mo-man-dib'-ū-lar*). Relating the squamosa and maxillary bone.
squamomastoid (*skwa-mo-mas'-toid*) [*squamo-*; *mastoid*]. Pertaining to the squamous and mastoid portions of the temporal bone.
squamoparietal (*skwa-mo-par-i'-et-al*) [*squamo-*; *paries*, wall]. Pertaining to the squamous portion of the parietal bone.
squamopetrosal (*skwa-mo-pe-tro'-sal*) [*squamo-*; πετρά, rock]. Pertaining to the squamous and petrous portions of the temporal bone.
squamosa, squamosal (*skwa-mo'-sah, skwa-mo'-sal*). The squamous portion of the temporal or occipital or frontal bone.
squamosphenoid (*skwa-mo-sfe'-noid*) [*squamo-*; *sphenoid*]. Pertaining to the squamous portion of the temporal bone and to the sphenoid bone.
squamotemporal (*skwa-mo-tem'-po-ral*) [*squamo-*; *temporal*]. Pertaining to the squamous portion of the temporal bone.
squamotympanic (*skwa-mo-tim-pan'-ik*) [*squamo-*; *tympanic*]. Pertaining to the squamosal and tympanic bones.
squamous (*skwa'-mus*) [*squamosus*, scaly]. 1. Of the shape of a scale, as the *squamous* portion of the temporal bone. 2. Scaly. s. bone, the circular plate forming the upper anterior portion of the temporal bone. s. suture, the suture between the squamous portion of the temporal bone and the parietal and parietal bones.
squamozygomatic (*skwa-mo-zi-go-mat'-ik*) [*squamo-*; *zygomatic*]. Pertaining to the squamous and zygomatic portions of the temporal bone.
square lobe. 1. The lobus quadratus of the liver. 2. A lobe on the upper surface of the cerebellar hemisphere.
squarious, squarrous (*skwa'-re-us, skwar'-us*) [*squarrosus*, scurfy]. Scurfy.
squarra (*skwar'-ah*) [ἐσχάρα, a scab]. A rough crust of *tinea*. s. tondens, alopecia areata.
Squibb's diarrhea mixture [Edward Robinson *Squibb*, American manufacturing chemist, 1819–1900]. A mixture containing tincture of opium 25 Cc., tincture of capsicum 12.5 Cc., spirit of camphor 25 Cc., chloroform 10 Cc., and alcohol to make 125 Cc. Average dose 30 minims (2 Cc.).
squill (*skwil*). The bulb of *Urginea maritima* (U. S. P.) or *Urginea scilla* (B. P.), of the order *Liliaceæ*. Squill (*scilla*, U. S. P., B. P.) contains several bitter principles, *scillitin, scillipicrin, scillitoxin*, and *scillin*, a carbohydrate *sinistrin*, and other substances. It is expectorant, diuretic, and emetocathartic, and is used in dropsy and in croup. Dose 1–2 gr. (0.065–0.13 Gm.). s., fluidextract of (*fluidextractum scillæ*, U. S. P.). Dose 2–3 min. (0.13–0.2 Cc.). s., oxymel of (*oxymel scillæ*, B. P.). Dose 1–2 dr. (4–8 Gm.). s., pills of, compound (*pilula scillæ compositæ*, B. P.). Dose 5–10 gr. (0.32–0.65 Gm.). s., syrup of (*syrupus scillæ*, U. S. P., B. P.). Dose ½–1 dr. (2–4 Cc.). s., syrup of, compound (*syrupus scillæ compositus*, U. S. P.), hive-syrup. Dose 20–30 min. (1.3–2.0 Cc.). s., tincture of (*tinctura scillæ*, U. S. P., B. P.). Dose 5–30 min. (0.32–2.0 Cc.). s., vinegar of (*acetum scillæ*, U. S. P., B. P.). Dose 10–30 min. (0.65–2.0 Cc.). squillitic (*skwil-it'-ik*). Pertaining to or containing squill.
squint (*skwint*). See *strabismus*.
Squire's catheter [Truman Hoffman *Squire*, American surgeon, 1823–1889]. A vertebrated catheter.
Squire's sign [G. W. *Squires*]. A rhythmic dilatation and contraction of the pupil in basilar meningitis.
squirting cucumber. See *elaterium*.
Sr. Chemical symbol of *strontium*.
S romanum (*ro-ma'-num*). The sigmoid flexure.
ss. Abbreviation for Latin *semis*, one-half.
S. S. paste. A paste made of brown sugar changed to the consistency of condensed milk by the addition in water of nitrate of silver a 1 to 3,000 solution. It is applied to septic and sloughing or gangrenous wounds.
stab-culture, a culture in which the inoculating point is thrust into a tube of agar, or other suitable solid culture material; it is used for the propagation of anaerobic bacteria; it is also called *stick-culture* or *thrust-culture*.
stabile (*sta'-bil*) [*stabilis*, from *stare*, to stand]. Not moving; fixed. s. current, an electric current produced by holding both the electrodes in a fixed position.
staccato speech. See *scanning*.
stachydrine (*sta-kid'-rēn*), C₇H₁₃NO₂. / An alkaloid from the bulb of *Stachys palustris*, forming colorless crystals which liquefy on exposure; soluble in water and alcohol; melt at 210° C.
Stacke's operation (*stah'-keh*) [Ludwig *Stacke*, German otologist, 1859– ]. Removal of the posterior and superior wall of the auditory meatus, so that the tympanum, attic, antrum and meatus make one cavity; this affords free exit for pus in suppurative disease of the middle ear.
stactometer (*stak-tom'-et-er*) [σтактós, a dropping; μέτρον, a measure]. An instrument for measuring drops.
Staderini's nucleus. The nucleus intercalatus, an aggregation of ganglion-cells situated between the dorsal nucleus of the pneumogastric and the nucleus of the hypoglossal nerve.
stadium (*sta'-de-um*) [L.]. Stage. s. acmes, the height of a disease. s. amphiboles, see *stage amphibolic*. s. annihilationis, the convalescent stage. s. augmenti, the period in which there is increase in the intensity of the disease. s. caloris, the period during which there is fever; the hot stage. s. contagii, the prodromal stage of an infectious disease. s. convalescentiæ, the period of recovery from disease. s. decrementi, defervescence of a febrile disease; the period in which there is a decrease in the severity of the disease. s. decrustationis, the stage of an exanthematous disease in which the lesions form crusts. s. desquamationis, the period of desquamation in an exanthematous fever. s. eruptionis, that period of an exanthematous fever in which the exanthem appears. s. exsiccationis. See *s. decrustationis*. s. floritionis, the stage of an eruptive disease during which the exanthem is at its height. s. frigoris, the cold stage of a fever; see *stage, algid*. s. incrementi, the stage of increase of a fever or of a disease. s. incubationis, see *stage, latent*. s. maniacale, the last stage of excitement in mania, after which the nervous manifestations gradually subside. s. nervosum, the paroxysmal stage of a disease. s. prodromorum, in eruptive fevers, the stage prior to the appearance of the eruption. s. staseos, see *s. acmes*. s. sudoris, the sweat-

ing stage. **s. suppurationis,** the period in the course of variola in which suppuration occurs. **s. ultimum,** the final stage of a febrile affection.
**staff.** An instrument for passing into the bladder through the urethra and used as a guide in oper tions on the bladder or for stricture.
**staffa** (*staf'-ah*). 1. The stapes. 2. A figure-of-8 bandage.
**Staffordshire knot.** See *knot.*
**stage** (*stāj*) [*stare*, to stand]. 1. A definite period of a disease characterized by certain symptoms; a condition in the course of a disease. 2. The horizontal plate projecting from the pillar of a microscope for supporting the slide or object. **s., algid,** a condition characterized by subnormal temperature, feeble, flickering pulse, various nervous symptoms, etc. It occurs in cholera and other diseases marked by exhausting intestinal discharges. **s., amphibolic,** the stage of a disease intervening between its height and its decline. **s., asphyxial,** the preliminary stage of Asiatic cholera, marked by extreme thirst, muscular cramps, etc., due to loss of water from the blood. **s., cold,** the rigor or chill of an attack of a malarial paroxysm. **s., eruptive,** that in which an exanthem makes its appearance. **s., expulsive** (of labor), the stage which begins when dilatation of the cervix uteri is complete and during which the child is expelled from the uterus. See *labor, stages of.* **s., first** (of labor), that stage in which the molding of the fetal head and the dilatation of the cervix are effected. **s., hot,** the febrile stage of a malarial paroxysm. **s. of invasion,** the period in the course of a disease in which the system comes under the influence of the morbific agent. **s. of latency,** the incubation-period of an infectious disease, or that period intervening between the entrance of the virus and the manifestations of the symptoms to which it gives rise. **s., placental** (of labor), the period occupied by the expulsion of the placenta and fetal membranes. **s., preeruptive,** the period of an eruptive fever following infection and prior to the appearance of the eruption. **s., pyrogenetic,** the stage of invasion in febrile diseases. **s., second** (of labor), see **s., expulsive. s., sweating,** the third or terminal stage of a malarial paroxysm, during which there is sweating. **s., third** (of labor), see **s., placental.**

**staggers** (*stag'-erz*). One of the various forms of functional and organic disease of the brain and spinal cord in domestic animals, especially horses and cattle. Enzootic cerebritis of horses, sheep, etc. **s. blind,** staggers due to cerebral disease. **s.-bush,** *Andromeda mariana;* a plant allied to the mountain laurel, growing in the seaboard States of North America. **s.-grass,** *loco,* or *loco disease, q. v.,* produced by eating various so-called loco-weeds. **s., mad, s., sleepy,** staggers due to inflammation of the cerebral envelopes. **s., stomach,** staggers due to cerebral disturbance dependent on gastric disorder.

**stagnation** (*stag-na'-shun*) [*stagnare*, to settle]. A cessation of motion. In pathology, a cessation of motion in any fluid; stasis.

**stagnum chyli.** Same as *receptaculum chyli.*

**Stahl's ear** [Friedrich Carl *Stahl*, German physician, 1811–1873]. A congenital deformity of the ear which consists in a broadening of the helix, the fossa ovalis and upper part of the scaphoid fossa being covered.

**Stahlian** (*stah'-le-an*). An animist, a follower of the doctrine of George Ernst Stahl, German chemist, 1660–1734.

**stain** (*stān*) [from *distain,* from *dis,* priv.; *tingere,* to color]. 1. A discoloration. 2. A pigment employed in microscopy to color the tissues or to produce certain reactions. The common microscopic stains are hematoxylin, carmin, osmic acid, and the anilin dyes. **s., intra vitam,** one that will act upon living material. **s., inversion, of Rawitz,** a process in which under the influence of a mordant, a basic anilin dye behaves as a plasma or acid dye. TABLE OF STAINS, REAGENTS, REACTIONS, STAINING METHODS, etc.: Appended are those most important to students of medicine. For a full definition of all the stains, etc., in use in special and general branches, see the Illustrated Dictionary and the Dictionary of New Medical Terms. **acid fuchsin,** a diffuse stain, having a special affinity for axis-cylinders. A solution of 2 Gm. in 40 Cc. of 90% alcohol and 160 Cc. of distilled water is employed. Wash out in 90% alcohol. Weigert stains sections of tissue hardened in Mueller's fluid in a saturated aqueous solution of acid fuchsin for from 1 to 24 hours, then rinses them quickly in water, immerses for a few minutes in a saturated solution of potassium hydroxide, 1 part, alcohol, 10 parts. Wash thoroughly to remove the alkali, dehydrate, clear, and mount. This process differentiates the finer nervefibers in the spinal cord. **alcohol,** an excellent fixing medium, suitable for all tissues except those of the central nervous system and those undergoing fatty infiltration or degeneration. It is also used to harden and preserve objects that have been fixed in other fluids. As a preservative it is not without defects, as it alters the structure of tissues by continuously dehydrating their albuminoids. Toluol, ether, and xylol are recommended as substitutes (Kultschitzky). As a fixing agent alcohol is usually employed in gradually increasing strengths, beginning with 50 or 70%. **alcohol, absolute,** one of the most penetrating fixing agents, which has the advantage of preserving the structure of glands and of nuclei. It should be employed in large quantities. Hydration may be prevented by suspending in the alcohol strips of gelatin (Lowrie). After fixation, preserve the object in 90% alcohol. **amyloid reaction** (*in tissues having undergone amyloid degeneration*). 1. With iodine: Dilute Lugol's solution with distilled water until it has the color of port-wine, and add 25% of glycerol; in this mount the sections for 3 minutes, wash in water, and mount in glycerol. The amyloid substance is brown-red; the remaining tissues are light-yellow. (For permanent preparations see the method of Langhans for glycogen.) 2. With iodine-green: Stain for 24 hours in iodine-green (0.5 Gm. dissolved in 150 Cc. of distilled water) and wash in water. The amyloid masses are red-violet, the remaining tissues green. 3. With iodine and sulphuric acid: Place sections that have been treated with Lugol's solution (see *Iodine Reaction* in this table) in 1% sulphuric acid. The brown of the amyloid substance becomes intensified or it changes to a violet or blue to green color. 4. With methyl-green: Stain for from 3 to 5 minutes in 1% solution of the dye and wash in distilled water containing 1% of hydrochloric acid. Amyloid substance violet, nuclei, green. 5. With methyl-violet: The process of staining is the same as with methyl-green. The amyloid is purple-red, the remaining tissue blue. **anilin oil,** an important medium because of its ability to clear watery objects; it will even clear aqueous media without the intervention of alcohol, which sometimes renders it valuable as a penetrating medium prior to paraffin embedding. It is also used for clearing celloidin sections. **anilin water** (*Ehrlich*), shake up 3 Cc. of anilin oil with 97 Cc. of distilled water and filter. The filtrate should be clear. Used as a mordant for anilin dyes. It does not keep well, and should be freshly prepared. **Apathy cement,** heat together, in a porcelain capsule, equal parts of hard paraffin (60° C. 140° F.) (melting-point) and Canada balsam until the mixture assumes a golden tint and no longer emits vapors of turpentine. On cooling, this forms a firm mass, which for use is warmed and applied with a glass rod. This cement is suitable for closing glycerol mounts. **balsam-paraffin for cells** (*Julien*), this substance consists of paraffin saturated with balsam-cement, and is prepared as follows: Reduce commercial Canada balsam to a wax-like consistence by slow evaporation in a shallow tin can over a low flame. Test by cooling a few drops from time to time. Melt slowly one-fourth pound of paraffin, with a melting-point above 45° C. (113° F.); add a lump of balsam-cement about the size of a marble, and then digest at gentle heat, stirring frequently, for about an hour. The appearance of a slight yellow tinge indicates the saturation of the paraffin by the balsam. When it is desired to prepare a cell, the balsam-paraffin is cautiously heated to the melting point in a shallow porcelain capsule. These paraffin cells are suitable for dry or liquid mounts, excepting for the latter when Canada balsam, dammar, or oils are used as preservatives. The great advantage of the balsam-paraffin is its chemic indifference to the reagents employed in the preservation mediums. Boston's mixture (for the preservation of casts in urine), liquor acidi arsenosi (U. S. P.), 1 oz.; salicylic acid, ½ gr.; glycerol, 2 dr. Dissolve by warming gently and add "whole tears" of acacia to saturation. Let the mixture settle, decant the supernatant liquid, and add a drop of formalin. Place a drop of urine containing casts on a slide, evaporate nearly to dryness, add a drop of the preservative, mix the two with a delicate needle, apply a cover-glass, and when the mount has hardened, seal with cement.

# STAIN 831 STAIN

**Bremer's method** (*for diabetic blood*), fix the films for 6 minutes in the oven at 135° C. (275° F.). Stain for 3 minutes with 1% solution of methyl-blue or with the Ehrlich-Biondi mixture. The yellow-green reaction of the erythrocytes may also be obtained by using eosin, congo red, or biebrich scarlet in 1% solution. **Canada balsam**, evaporate the balsam in a water-bath to dryness, and dissolve in an equal volume of xylol, benzol, toluol, chloroform, or turpentine. Filter through paper and keep in a "capped" bottle. If it gets too thick, dilute by adding more of the solvent employed. **carbolfuchsin**, fuchsin, 1 Gm.; phenol, 5 Gm.; alcohol, 10 Gm.; distilled water, 100 Gm. **celloidin**, stated to be a preparation of pure pyroxylin. It is nonexplosive and is soluble in ether and alcohol. Celloidin should be used in thin (2%) and in thick (6%) solutions. The object is thoroughly dehydrated in absolute alcohol, placed in a mixture of equal parts of ether and alcohol for from 12 to 24 hours, or longer if the object is large. It is then placed for 24 hours in a thin solution of celloidin (8 Gm. in 100 Cc. each of alcohol and ether), and transferred from this to a thick solution of celloidin (8 Gm. in 50 Cc. each of alcohol and ether). Select a cork or a piece of soft, dry wood, and dip it in the thick celloidin solution; when dry, place the prepared tissue upon it and drop the thick celloidin solution upon this with a pipet until it is embedded in a jelly-like mass. Blocks or cylinders of glass or vulcanized fiber serve better than cork, as they sink in the liquids used. Harden in 90% alcohol, and preserve in 70% alcohol. When making sections, keep the knife and section wet with alcohol. Apathy advises previous smearing of the knife with vaselin, as it cuts better and is protected from the alcohol. Use bergamot oil to clear. Sections may be kept in from 60 to 80% alcohol. **Chenzinsky's stain** (*for blood*), concentrated aqueous solution of methylene-blue and distilled water equal parts. To this is added an equal quantity of 0.5% solution of eosin in 60% alcohol. Stain blood-films 4 to 5 minutes. Red blood-corpuscles stain a rose-red, nuclei of leukocytes blue, and malarial parasites blue. **Claudius' method** (*for bacteria*), stain in gentian or methyl-violet, after Gram (see *Gram's Method* in this table), differentiate in a saturated aqueous solution of picric acid diluted with an equal volume of water, decolorize in chloroform. **collodion**, prepared by dissolving gun-cotton or soluble cotton in equal parts of 95% alcohol and sulphuric ether; it is in every way as good as celloidin, and considerably cheaper. **Conn's method** (*for preserving cultures of bacteria as museum specimens*). Inoculate 2% agar slants and seal the tubes with paraffin and plaster-of-paris. In a few days the cultures cease growing and remain indefinitely unaltered. **copper acetate**, a solution of 1 Gm. of copper acetate and 4 Gm. of mercuric chloride in 250 Cc. of glycerol and 1 Cc. of glacial acetic acid is used in preserving and mounting green algæ. **corrosive sublimate**, a most excellent reagent, generally applicable, useful particularly for fixing glands and glandular structures. Saturated aqueous and alcoholic solutions are usually employed. The tissues turn white when exposed. All the corrosive sublimate must be washed out in iodine or the sections will be sprinkled with crystals of the salt. The hardening is completed in alcohol. Glass, wood, or platinum should be used in manipulating objects immersed in this reagent. Sections may be stained with any of the usual reagents. **cover-glass preparations**, such preparations are usually made in examining blood, sputum, or other fluid or semifluid substance. In the case of sputum, a tiny mass is placed on a cover-glass, another is pressed gently down upon this, and the two glasses are separated by sliding one over the other, the object being to secure a thin, even film on each glass. The film may also be spread with the edge of a cover-glass or with a platinum spatula. The preparations are then left to dry in air, or they may be dried by exposing them to a temperature of 120° for twenty minutes, or by passing them quickly thrice through the flame of a spirit-lamp or Bunsen burner. When dry, they are ready to stain. To obtain a cover-glass preparation of blood, cleanse the finger, prick the pad, wipe off the first drop of blood that exudes, touch the apex of the second drop with a cover-glass, spread in the manner described, and dry in air. **Craig's method** (*for obtaining the flagellated malarial plasmodium*), cleanse the ear or finger, also the slide and cover-glass, with alcohol. Make a puncture with a sterile needle and wipe away the first drops of blood. Gently breathe upon the slide and take up on it the blood from the summit of the second drop and immediately apply the cover-glass. The brief exposure to air and the moisture on the slide are said to hasten flagellation. **creosote**, the properties of this agent are similar to those of phenol. Beechwood creosote is a good clearing medium for celloidin sections. **Czenzynke's double stain**, concentrated aqueous solution of methylene-blue, 40 Cc.; 0.5% solution of eosin in 70% alcohol, 20 Cc.; distilled water, 40 Cc. This is used to stain the blood, and colors the red corpuscles red, the leukocytes blue; also for *Plasmodium malariæ*, the gonococcus, and the influenza bacillus of Pfeiffer and Canon. **dahliaviolet**, a nuclear stain, recommended for demonstrating the granules in Ehrlich's mastzellen. Tissues hardened in alcohol are stained for several hours in a solution of dahlia-violet 2 Gm., in 90% alcohol 25 Cc. Wash in alcohol until nearly colorless. **Delafield's hematoxylin**, see under *Delafield*. **eau de Labarraque** (*sodium hypochlorite*), rub up 20 Gm. of chlorinated lime in 100 Cc. of distilled water and mix with 40 Gm. of crystallized sodium carbonate dissolved in the same quantity of water. Let the mixture stand for an hour and filter. This is used in the same way as Javelle water. With the aid of heat, chitin is dissolved in either of the solutions in a short time (Loos). Chitinous structures, macerated for 24 hours or more in these solutions diluted with 4 to 6 volumes of water, become soft and transparent, and permeable to staining fluids, aqueous or alcoholic. This method is especially applicable to nematodes and their ova. **Ehrlich's acid hematoxylin**, used for staining sections and in the mass. Dissolve 1 Gm. of hematoxylin in 30 Cc. of alcohol and add 50 Cc. each of glycerol and water, alum in excess, and 4 Cc. of glacial acetic acid. Let the mixture ripen in the light until it acquires a deep-red color. Objects stained in it should be washed in undistilled water. **Ehrlich's anilin gentian-violet**, a mixture of 5 Cc. of a saturated alcoholic solution of gentian-violet and 100 Cc. of anilin water. **Ehrlich-Biondi-Heidenhain triple stain**, to 100 Cc. of a saturated aqueous solution of orange add, with continual agitation, 20 Cc. of a saturated, aqueous solution of acid fuchsin and 50 Cc. of a like solution of methyl-green; dilute with from 60 to 100 volumes of water. A drop on blotting-paper should form a spot bluish-green in the center, orange at the periphery; a red zone outside the orange indicates that the mixture contains too much fuchsin. From 6 to 24 hours is required to stain. Wash out in alcohol and clear in xylol. Chromatic elements are colored blue; cytoplasm, violet or orange-red; karyoplasm, the same but in lighter tones, and all the denser protoplasmic elements the same, but darker (Gilson). This is by far the best stain for photomicrography, except for connective tissue (Lindsay Johnson). A slightly acid reaction of the alcohol used for washing out will produce a relatively strong coloration by the methyl-green, while that by the fuchsin will be relatively pale; the opposite result will be obtained if the alcohol contains a trace of alkali. The addition of very dilute acetic acid, until the red tint is markedly intensified, will restore the energy of the fuchsin, which is likely to decline after a time (Heidenhain). **Ehrlich's iodine method**, stain the fixed film in a syrupy solution of gum-arabic containing 1% of Lugol's solution: leukocytes stained brown indicate a suppurative process. **Ehrlich's stains** (*for the granules of leukocytes*). 1. *Acidophilous* or *eosinophilous mixture*: Two parts each of indulin, aurantia, and eosin; glycerol, 30 parts. Suitable for staining sections and cover-glass preparations. This is also known as "Mixture C." 2. "*Triacid*" *mixture*: Dissolve—(a) 1 Gm. of orange-yellow (extra) in 50 Cc. of distilled water; (b) 1 Gm. of acid fuchsin extra in 50 Cc. of distilled water; (c) 1 Gm. of crystalline methyl-green in 50 Cc. of distilled water. Let the solutions settle. Then mix 11 Cc. of solution a with 10 Cc. of solution b, and 18 Cc. of distilled water and 10 Cc. of absolute alcohol; to this mixture add a mixture of 13 Cc. of solution c, 10 Cc. of distilled water, and 3 Cc. of absolute alcohol. Let the stain stand for one or two weeks before using. **Farrant's solution.** See under *Farrant*. **Flemming's fluid.** 1. Chromium trioxide 0.2 Gm.; glacial acetic acid, 0.1 Cc.; water, 100 Cc. This is especially recommended for fixing the achromatic spindle-fibers in nuclei. (2) Chromium trioxide (1%), 45 Cc.; 2% osmic acid, 12 Cc.; glacial acetic acid, 3 Cc. This fixes small pieces (2-3 mm. thick) in from a few to 24 hours, and is useful for fixing the figures in cell-division and for many other purposes. A weaker solution is also used: 1% osmic and glacial acetic acids, each, 100 Cc. The second formula is the one generally known as *Flemming's fluid*. **Frænkel-Gabbet method** (*for tubercle bacilli*). See under *Gabbet*. **Futcher-Lazear method** (*for the malarial parasite*), fix the film for one minute in a mixture of 10 Cc. of 95% alcohol and 2 drops of formalin; wash, dry, and stain for 15 seconds in carbolthionin, prepared by mixing 20 Cc. of a saturated solution of the dye in 50% alcohol and 100 Cc. of 2% aqueous solution of phenol. **gentian-violet. 1.** (*Concentrated alcoholic*). Gentian-violet, 25 Gm.; absolute alcohol, 100 Cc. **2.** (*Aqueous*). Gentian-violet, 1 Gm.; 90% alcohol, 20 Cc.; distilled water, 80 Cc. A nuclear stain, prepared by dissolving 0.5 Gm. of the dye in 80 Cc. of distilled water, 20 Cc. of 90% alcohol, and 1 Cc. of glacial acetic acid. Stains in 5 minutes. Dissolved in indifferent media it may be used for staining *intra vitam*, and in acid solutions colors the nuclei of fresh tissues. It may be used according to Gram's method. **glycerin-jelly**, soak in 150 Cc. of distilled water 25 Gm. of gelatin for two hours, and add 3 Cc. of phenol and 175 Cc. of glycerol; heat for 15 minutes and filter through spun glass. Wrap the cork of the bottle in which the jelly is preserved in linen dipped in dilute phenol. For use, melt it in hot water, place a drop on the section, upon which gently press a cover-glass. **glychemalum**, hematein, 0.4 Gm. (rubbed with a few drops of glycerol until it dissolves); alum, 5 Gm.; glycerol, 30 Cc.; distilled water, 70 Cc. (*Mayer*). **gold chloride**, recommended for tracing nerve-endings in fresh tissues and for staining connective tissue and cartilage-cells. Place small pieces of tissue, ½ inch square, in from 0.5 to 1% solution of commercial gold chloride in distilled water. Keep in the dark, and when the tissue has become yellow, wash in distilled water. Then expose to the light in 50 Cc. of water containing 2 drops of acetic acid for 48 hours, or until the tissue acquires a purple tint. Mount in glycerol. **Goldhorn's stain** (*for blood*), preparation of the solution of polychrome methylene-blue. Solution A (Merck's medicinal methylene-blue: Grübler's methylene-blue rectified, and methylene-blue—Koch): Dissolve 2 Gm. methylene-blue in 300 Cc. warm water. Add to this 4 Gm. lithium carbonate, shaking constantly. Heat in an evaporating dish on a water-bath, the water touching the dish. Stir the solution occasionally. Remove in 15 to 20 minutes. Do not filter. Set aside for several days. Then add dilute acetic acid (5%) until the solution is only *faintly* alkaline. Solution B: A 0.1% *aqueous* solution of eosin. Fix blood-films in *methyl-alcohol* for 15 seconds. Wash in running water. Stain in solution B for 7 to 30 seconds. Wash. Stain in solution A for 30 seconds to 2 minutes. Wash *thoroughly* in running water. Dry by agitating in air, *not* between filter-paper. The eosin may be added to the methyl-alcohol (enough to make a 0.1% solution); or solution B may be added to solution A (1 : 4), but this easily produces a precipitate (the neutral stain). They give good results. Mixtures of methyl-alcohol, eosin, and polychrome methylene-blue give poor results. The depth of the chromatin stain depends on the length of staining. To stain the chromatin of half-grown malarial parasites 1½ to 2 minutes is necessary while the chromatin of the hyaline forms stains in 10 seconds. Repeated staining may improve the chromatin violet. To do this the blood-film may be stained with solution B for 5 seconds with solution A for 10 seconds. **Golgi's method** (*for the restoration of overhardened tissue*), wash in a half-saturated solution of copper acetate until it yields no precipitate, and return for 5 or 6 days to Golgi's mixture (see below). The tissue will then take the silver and the sections can be mounted in thickened cedar oil under a cover-glass. **Golgi's mixture**, potassium dichromate (3.5% solution), 54 Cc.; osmic acid (2%), 6 Cc. gonococcus, the gonococci are seen in the pus-cells grouped around the nucleus. Watery solutions of anilin dyes, preferably methylene-blue, stain the cocci intensely. **Gram's method**, heat for from 2 to 5 minutes, or stain cold for from 20 to 30 minutes (tubercle bacilli, 12 to 24 hours), in saturated solution of gentian-violet anilin water; rinse quickly in absolute alcohol; transfer to Gram's solution (1 to 1½ minutes), in which the specimen turns black; wash in alcohol until the black color vanishes and a pale-gray color appears; dry and mount in Canada balsam. The decolorization may be hastened by adding 3% nitric acid to the alcohol and then washing in pure alcohol. All the tissue-cells are decolorized by this method, while the bacteria are stained a deep blue. The cells may be subsequently stained with a watery or alcoholic solution of Bismarck brown for from 2 to 5 minutes, then washed in absolute alcohol until the section is yellowish-brown. This method is of diagnostic value, as certain bacteria are stained, others decolorized, by it. The bacteria that are stained by Gram's method are: tubercle bacillus; Fraenkel-Weichselbaum pneumococcus; *Streptococcus pyogenes*; streptococcus of erysipelas; *Staphylococcus pyogenes aureus*, *albus*, *citreus*, and *flavus*; anthrax bacillus; bacillus of hog erysipelas. The bacteria that are decolorized by Gram's method are: Typhoid bacillus; gonococcus; Friedländer's capsule bacillus; Koch's comma bacillus; glanders bacillus, and the spirillum of relapsing fever. Botkin advises washing the preparation in plain anilin water before decolorizing in the iodin solution. *Modifications of Gram's Method:* 1. *Guenther's modification:* Transfer from the iodine-potassium-iodide solution to alcohol, then to a mixture of alcohol, 1 volume, and nitric acid, 3 volumes, and from this again into alcohol. 2. *Nicolle's modification:* Decolorize in a mixture of alcohol, 2 volumes, and acetone, 1 volume. 3. *Ribbert's modification:* Decolorize in alcohol containing 10% of acetic acid. 4. *Weigert's modification:* The sections, stained with gentian-violet or methyl-violet, are not transferred to alcohol from the iodine solution, but are laid upon slides and covered with anilin oil. This is removed with blotting-paper, and followed by xylol and xylol-balsam. The anilin oil dehydrates and differentiates. **Grenacher's alcoholic borax-carmin**, dissolve 4 Gm. of borax in 100 Cc. of distilled water; add 3 Gm. of carmin, warm, and dilute with 100 Cc. of 70% alcohol. Filter before using, and transfer the tissue from the stain directly into alcohol acidulated with from 4 to 6 drops of hydrochloric acid, in which it should remain until it acquires a bright, transparent appearance. This solution is used for staining in bulk and gives a splendid color. **Gruber and Durham's method** (*for the agglutination of typhus and cholera bacilli*), place a drop of immunization serum on a cover-glass and beside it a drop of equal size of the culture, as finely divided as possible. Mix and examine on a slide with a ground cell. In doubtful cases put the preparation in the oven for from 15 to 30 minutes. **Haffkine's bouillon** (*for the culture of the bacilli of bubonic plague*), chop 1 kilo of goat's flesh and heat it at a pressure of 3 atmospheres for 6 hours in dilute hydrochloric acid. Filter, neutralize, dilute with water to 3 liters, and sterilize. **Haffkine's prophylactic**, inoculate a flask containing 3 liters of Haffkine's bouillon with a pure culture of pest bacilli; when the stalactite growth develops, shake the flask until the colony sinks to the bottom, and when the growth reappears, shake again; when the stalactite culture forms the third time, heat to 60° C. (140° F.) for 3 hours. Decant the clear fluid and preserve in hermetically sealed tubes. Dose 16-32 min. (1-2 Cc.) injected beneath the skin. **Harris' carboltoluidin**, dissolve 1 or 2 Gm. of toluidin-blue in a saturated solution of phenol. Before staining treat the sections with water; stain for from 5 minutes to 24 hours, wash, and differentiate in glycerol-ether (Grübler) diluted 15 times with water or in acidulated alcohol; after from 5 to 15 minutes wash in alcohol. Eosin in alcohol may be used as a counterstain. In this case omit the differentiation and stain for from ¼ to 2 minutes and wash in alcohol. **Harris' hematoxylin**, dissolve 1 Gm. of hematoxylin in 10 Cc. of alcohol and add to 200 Cc. of a saturated aqueous solution of alum; heat to boiling and add 0.5 Gm. of mercuric oxid; when the solution turns a dark purple, remove from the flame and cool quickly. For use dilute to the color of port-wine with aqueous solution of alum. **Haug's phloroglucin fluid**. One of the most rapid decalcifying agents, and without injurious action on the tissue-elements, with the exception of blood. It is prepared as follows: Warm slowly and carefully 1 Gm. of phoroglucin in 10 Cc. of pure nitric acid, and to the resulting ruby-colored solution add 50 Cc. of distilled water. If a larger quantity is desired, add nitric acid and water to the foregoing proportion until the volume measures 300 Cc., the limit of the protective influence of the phloroglucin. Previously to being brought into this fluid the tissues should be well fixed. Fetal bones and those of lower vertebrates are decalcified in half an hour. Older and harder bones require several hours. When decalcification is completed, wash in running water for two days. The sections stain well. Another formula, useful for teeth when rapid action is necessary, consists of phloroglucin, 1 Gm.; nitric acid, 5 Cc.; 95% alcohol, 70 Cc.; distilled water, 30 Cc. The function of the phloroglucin is to protect the organic tissue-elements against the action of the acid. **Heidenhain's fluid**, saturate hot 0.5% sodium chloride solution with mercuric chloride. **Heller's method**. 1. *For the osmication of medullated nerve-fibers:* Harden the tissue in Mueller's fluid. Stain the sections in 1% osmic acid—in the oven for 10 minutes, at room-temperature for a half-hour; wash in water; reduce in 5% pyrogallic acid for a half-hour, oxidize in 2.5% potassium permanganate for from 3 to 5 minutes, decolorize in 2% oxalic acid for from 3 to 5 minutes. 2. *For mounting objects for sectioning:* Pin a piece of paper about the cork or block so that it projects and forms a trough into which the celloidin can be poured around the object. Harden in the vapor of alcohol by suspension in a closed cylinder containing a few centimeters of alcohol. **hemosiderin**, amorphous yellow to black-brown iron-containing fragments occurring in thrombi or hemorrhagic infarcts. In sections of material hardened in alcohol or formalin, treated for a few minutes with a 2% aqueous solution of potassium ferrocyanide and examined in glycerol containing 0.5% of hydrochloric acid, the pigment appears in the form of dark-blue granules. **Hermann's fluid**, a modification of Flemming's fluid. Platinum chloride is used instead of chromium trioxide; in other respects the formulas are alike. **His' medium** (*for the differential culture of the typhoid bacillus*). 1. The tube culture-medium: triturate 5 Gm. of agar, 80 Gm. of gelatin, 5 Gm. of beef-extract, and 5 Gm. of salt; add a liter of water and enough hydrochloric acid or soda solution to produce a reaction of 1.5% of normal acid, using phenolphthalein as the indicator. Clear with 1 or 2 eggs beaten in 25 Cc. of water; add 10 Gm. of glucose; boil for 25 minutes, and filter through absorbent cotton. 2. For the plate-culture use 10 Gm. of agar, 25 Gm. of gelatin, 5 Gm. each of beef-extract and salt, and 10 Gm. of glucose. The medium must contain not less than 2% of normal acid. The typhoid bacillus alone has the power of clouding these media. **Hofbauer's method**. (*for staining the iodinophil granules of leukocytes*), dry the film and stain 1 minute in a solution of iodine, 1 part, potassium iodide, 3 parts, and water, 100 parts, brought to a syrupy consistence by the addition of gum-arabic. Remove the excess of the stain with filter-paper to prevent diffuse coloring. **iodine-alcohol**, alcohol, 90%, to which enough tincture of iodine is added to impart the color of port-wine. See also *Zenker's fluid* in this table. **iron hematoxylin**, sections are treated with a weak aqueous solution of ferric acetate, washed in water, and stained in 0.5% aqueous solution of hematoxylin. A blue-black or black-brown stain is obtained. This process is recommended by Bütschli for staining sections of protozoa 1μ thick. Another method is as follows: Treat sections for from ½ hour to 2 or 3 hours with a 1.5 to 4% solution of ferric ammonium sulphate; wash in water, and stain for from 1 to 12 hours in an aqueous solution of hematoxylin, about 0.5%. Rinse with water and treat again with the iron solution. As soon as differentiation is complete, wash for 15 minutes in running water and mount. The results vary according to the duration of the treatment with the iron and hematoxylin solutions; short baths give a blue preparation, in which the nuclear structures are highly differentiated; prolonged baths give black preparations, showing connective-tissue fibers and red blood-corpuscles black, central and polar bodies intensely black, cytoplasm sometimes colorless, sometimes gray, in which case cell plates and achromatic spindle-fibers are stained. Microorganisms are sharply stained. **Jenner's stain** (*for blood*). See under *Jenner*. **karyokinesis**. Place small pieces of tissue hardened in strong Flemming's solution in an alcoholic solution of safranin (2 Gm. to 60 Cc.) for from 24 to 48 hours. Wash for a few minutes in water, and carry to acidulated absolute alcohol (10 drops of acetic acid to 100 Cc.) for from ½ to 1 minute. When thick clouds of color are no longer given off, carry to absolute alcohol. After

1 or 2 minutes clear and mount. **Leishman's stain** (*for blood*). Perparation of the neutral stain. Solution A: The solution of polychrome methylene-blue. A 1% aqueous solution of methylene-blue med. (Grübler) is made alkaline with 0.5% $Na_2CO_3$. This is heated for 12 hours at 65° C., and then allowed to stand for 10 days before use. Solution B: A 0.1% aqueous solution of eosin (extra BA Grübler). Equal parts of solutions A and B are mixed in an open vessel and allowed to stand for 5 or 6 hours, with occasional stirring. The precipitate formed is collected on a filter, washed with water, dried, and powdered. The staining solution: Dissolve 0.1 Gm. of the dry precipitate in 100 Cc. pure methyl-alcohol (Merck "for analysis"). To stain: Four drops of the solution are poured on the blood-film, and allowed to stain for ½ minute. Without pouring off the stain, 6 to 8 drops of distilled water are added and the mixture is allowed to stain for 5 minutes. Wash gently. Put a few drops of water on the blood-film for 1 minute. Then dry and mount. Staining reactions: Red blood-corpuscles stain pale pink or greenish; lymphocytes: nuclei, dark ruby red; protoplasm, pale blue; mononuclears: nuclei, ruby red; protoplasm, pale blue; polymorphonuclear neutrophils: nuclei, ruby red; granules, red; "coarse-grained eosinophils": nuclei, ruby red; granules, purplish black; blood-platelets stain deep ruby red; malarial parasites: nuclei, chromatin portion, ruby red; cytoplasm, blue. **lithia-water**. Saturated aqueous solution of lithium carbonate 1 Cc., and distilled water 30 Cc., used as an intermediate agent in staining microorganisms. **living cells**. Young larvæ of *Amphibia* are the best objects for the study of cells *intra vitam*. Place the larvæ of *Salamandra* in a watch-glassful of water containing 5 to 10 drops of a solution of 1 part curarn in 100 parts each of water and glycerol. From ½ to 1 hour's immersion is required for curarization. It is not necessary to wait until the larvæ are motionless; they may be removed as soon as their movements have become slow. The gills and the caudal fin may then be studied. The tail may be excised from the living animal and studied for some time in 1% salt solution or other indifferent medium. The adult animal offers for study the thin, transparent bladder. Larvæ may be bred from adults, if well fed with aquatic worms, and supplied with a vessel of water. The larvæ will be deposited in the water. The cytoplasm of living cells may be stained with methylene-blue; dahlia, or gentian-violet dissolved in the water or in an indifferent liquid. **Loeb's method** (*for producing artificial parthenogenesis*). Place the unfertilized eggs of sea-urchins in sea-water containing magnesium chloride in the proportion of 5000 ($\frac{1}{2}$n MgCl) to 5000 Cc. of water. After 2 hours restore them to normal sea-water. The eggs from normal gastrulæ and plutei. **Loeffler's methylene-blue**. Add 30 Cc. of a concentrated alcoholic methylene-blue solution to 100 Cc. of a solution of caustic potash (0.01 : 100). Filter before using. **Loeffler's stain** (*for flagella*). Mix 10 Cc. of 20% solution of tannin, 5 Cc. of saturated solution of ferrous sulphate, and 1 Cc. of aqueous or alcoholic solution of fuchsin, methyl-violet, or "Wollschwarz." For typhoid bacilli add 1 Cc. of 1% solution of soda; for *Bacillus subtilis* add 30 drops; for bacilli of malignant edema, 36 drops; for cholera bacilli add 1 drop of sulphuric acid to the soda solution; for *Spirillum rubrum*, 9 drops. **McCrorie's method** (*for flagella*). Stain the cover-glass preparation in warmed mixture of equal parts of a saturated solution of night blue, a 10% solution of tannin, and a 10% solution of alum. **Mallory's method**. 1. *For neuroglia:* Fix for 4 days in 10% formalin, then for 4 days in a saturated solution of picric acid; after this mordant for 4 days in 5% solution of ammonium bichromate at 37° C. Stain the sections for 2 minutes in 1% aqueous solution of acid fuchsin, rinse, and treat for 2 minutes with 1% aqueous solution of phosphomolybdic acid; wash in two changes of water and stain for 5 minutes in a mixture of water-soluble anilin blue, 0.5 Gm.; orange G, 2 Gm.; oxalic acid, 2 Gm.; and water, 100 Cc.; wash in water and dehydrate in alcohol. Result: connective tissue, blue; neuroglia, deep red; ganglion-cells and axis-cylinders, light red. 2. *For neuroglia:* Fix the tissues after the method given in No. 1, and treat the sections for 15 minutes with a 0.5 aqueous solution of potassium permanganate and after washing for the same time with 1% solution of oxalic acid, wash, and stain in hematoxylin prepared by dis-

28

solving 0.1 Gm. of the dye in a little hot water, and when cool adding water up to 80 Cc., 20 Cc. of 10% aqueous solution of phosphotungstic acid, and last 0.2% of hydrogen dioxid. Wash in water, dehydrate in alcohol, clear in oil of origanum, and mount in balsam. Nuclei, neuroglia, and fibrin blue; axiscylinders and ganglion-cells pale pink; connective tissue deep pink. 3. *For connective tissue:* Fix in Zenker's fluid or sublimate and stain the sections for 2 minutes in 0.1% aqueous solution of acid fuchsin. For further treatment see No. 1. Result: fibrous tissue, mucus, amyloid and hyaline substances, blue; nuclei, cytoplasm, elastin, fibrin, neuroglia, and axiscylinders, red; erythrocytes and myelin sheaths, yellow. 4. *For nuclei and fibrin:* Stain sections of tissue fixed in any medium except formaldehyde for 3 minutes in 10% aqueous solution of ferric chloride; drain and dry and stain for 3 minutes in a 1% aqueous solution of hematoxylin; wash and differentiate in a 0.25% solution of ferric chloride. Result: nuclei, dark blue; fibrin, gray to dark blue. In sublimate preparations the erythrocytes are greenish gray; connective tissue, pale yellow. 5. *For staining Amœba coli in tissues:* Use alcohol material and treat the sections for from 5 to 20 minutes with saturated aqueous solution of thionin; wash, and differentiate for from 30 to 60 seconds in 2% aqueous solution of oxalic acid; wash, dehydrate, clear, and mount in the usual way. **Mallory's phosphomolybdic-acid hematoxylin.** Mix 10% solution of phosphomolybdic acid, 1 part; hematoxylin, 1 part; water, 100 parts; chloral, from 6 to 10 parts. Expose to sunlight for a week. Filter before using and save the used portions. Stain sections for from 10 minutes to an hour; wash in 40 to 50% alcohol, changing it 2 or 3 times. Dehydrate and mount. If the solution does not stain readily, add a little hematoxylin. The stain is blue, and in its general effect similar to nigrosin. It is recommended for preparations of the central nervous system. **Mallory-Wright method** (*for staining tubercle bacilli*). Stain lightly in alum-hematoxylin, then for 2 or 3 minutes in steaming hot carbolfuchsin; decolorize for 30 seconds in acid alcohol. **Marchi's method.** Used to demonstrate early degeneration of nerves, prior to sclerosis. After hardening in Mueller's fluid place the tissue in a large quantity of a mixture of Mueller's fluid, 2 parts; 1% osmic acid, 1 part. The degenerated fibers are stained black; the normal are yellow or uncolored. **Mayer's carmalum.** Take 1 Gm. of carminic acid, 10 Gm. of alum, and 200 Cc. of distilled water; heat the mixture and filter, adding an antiseptic to keep it clear. The fluid is light red in color, shading toward violet, and is said to have good penetrating powers, even in osmium preparations, and to be better than alum-carmin for staining *in toto.* **Mayer's carmin and indigo-carmin.** Dissolve 0.1 Gm. of indigo-carmin in 50 Cc. of distilled water or of 5% alum solution; add 1 volume of indigo-carmin solution to 4 volumes of carmalum. **Mayer's hemalum.** An excellent stain for large objects. It consists of two solutions—one of hematein, or ammonium hematein, 1 Gm., dissolved by the aid of heat in 50 Cc. of 90% alcohol; the other of alum 50 Gm. and distilled water 1 liter. The solutions are mixed, left to cool, and then filtered. A crystal of thymol may be added to prevent the formation of mold. For most purposes it is advisable to dilute this stain with water or alum solution. Hemalum plus 2% glacial acetic acid gives a more precise nuclear stain. **Mayer's hemalum and indigo-carmin.** Add 1 volume of a 0.05% aqueous solution of indigo-carmin to 4 volumes of hemalum. **Mayer's paracarmin.** Dissolve carminic acid, 1 Gm., aluminium chloride, 0.5 Gm., calcium chloride, 4 Gm., in 100 Cc. of 70% alcohol, with or without heat. Filter, after precipitation, and the solution will have a clear red color. Suitable for staining bulky objects with large cavities, such as *Salpa.* **methylene-blue.** An important reagent, which gives a specific stain for lymph-spaces and intercellular cement, closely resembling gold and silver impregnation, for medullated nerves, and for plasma-cells. It also stains *intra vitam*, and is a specific reagent for the axis-cylinders of sensory nerves in living animals (Ehrlich). Small and permeable aquatic organisms may be stained during life by adding to the water containing them enough of the dye to give it a very pale tint. Nerve-tissue may be stained by injecting the dye into the vascular system of a living, narcotized animal, or by removing the organ and immersing it in the solution. From 0.5 to 1% solutions in physiological salt solution are employed for this purpose. The color is not permanent, but may be fixed by ammonium picrate. Parker fixes the color by dehydrating in a solution of mercury bichloride, 1 Gm., in methylol 5 Cc.; washing in a mixture of 2 parts of the methylol and sublimate solution, 1 part pure methylol; 3 parts xylol. The object is then placed in xylol for 4 or 5 days, when it is ready to mount or embed. Mayer's albumin should not be used to fix sections to the slide, as it discharges the color. A solution of 0.25 Gm. in 90% alcohol, 20 Cc., and distilled water, 80 Cc., is used for tissue-staining. A 1% and a saturated alcoholic (15 Gm. to 100 Cc.) solution are used for staining microorganisms. **methyl-green.** This is chiefly used as a nuclear stain for fresh or recently fixed tissues; it is also a reagent for amyloid degeneration (Heschl), giving a violet color. Use 0.5 Gm. of methyl-green in 20 Cc. of 90% alcohol, 80 Cc. of distilled water, and 1 Cc. of acetic acid. Stain the tissue for 5 minutes, wash in acidulated water, differentiate in 90% alcohol, and dehydrate. The nuclein reaction depends on the presence of acetic acid. Arnold recommends a dilute solution of methyl-green containing 0.6% sodium chloride for staining cells and nuclei. Bizzozero has observed that the elements of blood and pus, also ciliated epithelium and spermatozoa, do not stain with methyl-green if the cells are highly alkaline: if the alkalinity is diminished, they are dyed violet; if the cells are acid, they are colored green. Carnoy regards methyl-green as the best stain for nucleoli. **methyl-violet.** A good chromatin stain. Dissolve 0.5 Gm. in 200 Cc. of distilled water and 5 Cc. of glacial acetic acid. Stain sections for 20 minutes, wash in distilled water, and then in equal parts of glycerol and water. Mount in Farrant's medium. This is also a reagent for tissues undergoing amyloid degeneration. The amyloid substance stains pink. 1. (*Alcoholic*). Methyl-violet, 25 Gm.; absolute alcohol, 100 Cc. 2. (*Aqueous*). Methyl-violet, 1 Gm.; alcohol, 20 Cc.; distilled water, 80 Cc. **muchmateïn.** A specific stain for mucin. 1. Pulverize 0.2 Gm. of hematin with a few drops of glycerol and then add 0.1 Gm. of aluminum chloride, 40 Cc. of glycerol, and 60 Cc. of water. 2. Dissolve 0.2 Gm. of hematin and 0.1 Gm. of aluminum chloride in 100 Cc. of 70% alcohol. Two drops of nitric acid may be added. **mucicarmin.** A specific stain for mucin. Rub 1 Gm. of carmin in a mortar with 0.5 Gm. of aluminum chloride and 2 Cc. of distilled water; heat for 2 minutes, until the light-red color changes to dark; stir and add a little 50% alcohol; when dissolved, make up to 100 Cc. with 50% alcohol, and after 24 hours filter. For use dilute tenfold with water or with 50% alcohol. **Mueller's Berlin blue.** Precipitate a strong solution of Berlin blue with 90% alcohol. The fluid is neutral and the precipitate finely divided. **Mueller's fluid.** This agent is very extensively used, as it penetrates well and hardens evenly. It has the following composition: potassium dichromate, 2.5 parts; sodium suphate, 1 part; water, 100 parts. The addition of a little camphor, chloral, thymol, or naphthalene will prevent the formation of mold. The time required for hardening depends on the size of the object. This fluid diluted to 0.2% is used as a macerating agent. **Nissl's method** (*for ganglion-cells*). 1. Stain sections of tissue hardened in 10% formalin or in graded alcohols in hot concentrated aqueous fuchsin solution. 2. Stain in hot 0.5% methylene-blue; when cool, transfer to a mixture of anilin (20 parts) and 90% alcohol (200 parts); then treat with origanum oil, then with benzine, and mount in solution of colophonium in out embedding), cut sections and stain them in hot Nissl's methylene-blue; treat with the anilin-alcohol mixture, then with cajeput oil, then as in 2. **Nissl's methylene-blue.** Methylene-blue (B patent), 3.75 parts; Venice soap, 1.75 parts; distilled water, 1000 parts. nitric acid. An efficacious agent, which causes no swelling and does not attack the tissue-elements. It is used in 1 and in 10% solution, the latter for large, hard bones, the former for young bones. The specimens should previously have been fixed in absolute alcohol, and the decalcifying fluid changed daily. They must be removed as soon as decalcification is complete or they will become discolored. They are then washed in running water for two hours and preserved in alcohol, which should be renewed in a few days. **Nocht's stain** (*for blood*). Original method: Unna's polychrome methylene-blue is neutralized with dilute acetic acid. Solution A: 1 Cc. of this neutralized polychrome methylene-blue is mixed in a watch-crystal with a saturated aqueous solution of ordinary methylene-blue until its red color disappears and the solution becomes blue. Solution B: Dilute 3 or 4 drops of 1% aqueous solution of eosin with 1 or 2 Cc. water. Add solution A drop by drop to solution B until B is dark blue; a precipitate has then been formed. In this mixture blood-films are to be stained for several hours up to 24 hours. Fix films in alcohol or by heat. Subsequent modification: Solution A: The polychrome methylene-blue solution. To a 1% aqueous solution of methylene-blue add 1 or 0.5% Na₂CO₃. Heat at 50° C. to 60° C. (122°–140° F.) for several days. Solution B: Dilute 2 or 3 drops of 1% aqueous solution of eosin with 1 or 2 Cc. water. To solution B add solution A drop by drop until the mixture is dark blue and has lost its eosin tint. To stain, float blood-films face down on this mixture for from 5 to 10 minutes. **normal salt solution.** Sodium chloride, 6–7.5 Gm.; distilled water, 1000 Cc. Used in the study of living structures. **paraffin infiltration and embedding.** The initial step in this process consists in the infiltration of the object with a clearing agent; that is, by some substance which is a solvent of paraffin. It is then immersed in melted paraffin until it is thoroughly saturated. The paraffin should be kept just at the melting-point and should be renewed if the object is large. The duration of the bath depends on the size of the object. When this second step in the process is completed, embed in paraffin, as in simple embedding. To prevent crystallization of the paraffin the embedded object should be quickly cooled, which may be done by floating it in the containing receptacle on cold water. **phenol and xylol.** A mixture of 1 part of phenol and 3 parts of xylol is used to clarify celloidin sections, which may be taken from 70% alcohol, and do not require further dehydration. A layer of previously heated copper sulphate in the bottom of the bottle will keep the mixture free from water. **pianese double stain.** Prepare a saturated solution of nigrosin in a saturated alcoholic solution of picric acid; mix 2 volumes of this with 1 volume of anilin water and evaporate in open air. The crystals deposited are dissolved in absolute alcohol, and from this solution green crystals are obtained soluble in alcohol, ether, and water. For tissues, make a 2% solution in alcohol; for microorganisms, in water. Stain sections first in lithium-carmin, treat with acid alcohol, wash, and immerse in an alcoholic solution of picronigrosin until they assume a brown hue. Decolorize in oxalic acid. Nuclei are stained red; plasma, dark-yellow; cartilage, yellow; connective tissue, pale green; elastic fibers, violet. **picric acid.** A fixing agent of great penetration, and, therefore, especially suitable for the preparation of chitinous structures. A saturated solution is employed. The time required for fixation varies from a minute to a day, and depends on the size of the object. Wash out in alcohol and stain in alcoholic solutions. **picric alcohol.** A saturated solution of picric acid in 50% alcohol. **picronigrosin.** A solution of 1 Gm. of picric acid in 100 Cc. of distilled water with the addition of 1 Gm. of nigrosin. **Plehn's method.** 1. *For the study of the living malarial parasite:* Place a drop of fluid paraffin on a slide and a drop on a cover-glass; take up the drop of blood on the latter and so place it on the slide that the blood is between the drops of paraffin. Examine on a warm stage. The addition of a drop of methylene-blue will stain the living organisms. 2. *For malarial films fixed in absolute alcohol:* Stain for 5 minutes in a mixture of concentrated aqueous solution of methylene-blue, 60 Cc., 0.5% solution of eosin in 75% alcohol, 20 Cc., distilled water, 20 Cc., and 20% potash lye, 12 drops. **polychrome methylene-blue.** A reddish-violet dye sometimes present as an impurity in commercial methylene-blue, or that develops in old, ripened, or alkaline solutions of methylene-blue. It is used for staining cell-granules. See *Unna's Method.* **potassium permanganate** (*Du Pleiss*). Useful for the study of isolated and very contractile cells, as spermatozoa. It is said to kill more rapidly than any other agent, 2% osmic acid not excepted. A saturated aqueous solution is used. It is also used for washing out overstaining with carmin, and in 1% solution as a mordant for anilin dyes (Henneguy) and for reducing silver impregnations. **Rosenberger's method.** 1. *For staining blood:* Fix the films by heat or in absolute alcohol or alcohol and ether and stain in a mixture of 10 Cc. of a saturated aqueous solution of methylene-blue, 4 Cc. of a saturated aqueous solution of phloxin, 6 Cc. of 95% alcohol, and 12 Cc. of distilled water. 2. *For staining the tubercle bacillus:* The essential point in this process is the use of sweet spirit of niter for bleaching; it is also mixed with alcoholic solutions of methylene-blue, malachite green, Bismarck brown, and gentian-violet. **safranin.** *Pfitzner's formula:* Safranin (Grübler's), 1 part; absolute alcohol, 100 parts; water, 200 parts. *Flemming* uses a concentrated alcoholic solution diluted one-half with water. *Babes' formula:* (*a*) Equal parts of a concentrated alcoholic and a concentrated aqueous solution; (*b*) water, 100 parts; anilin oil, 2 parts; safranin, in excess. The latter may be used according to the method of Gram, and is recommended for the demonstration of mitotic figures. **Scheele's green mass.** (*a*) Mix 80 Cc. of a saturated solution of potassium arseniate and 50 Cc. of glycerol. (*b*) Take 40 Cc. of a saturated solution of copper sulphate and 50 Cc. of glycerol. Combine the two solutions with three volumes of the vehicle. **substantive staining.** A histological stain obtained by direct absorption of the pigment from the solution in which the tissue is immersed. Dyes that combine directly with the substance acted on are called substantive dyes. **subtractive staining.** A socalled theory of Heidenhain's, based on the hypothesis that a general stain satisfies the affinities of some cell-structures, that hold it in subsequent treatment with specific dyes, while the other structures give up the general stain and then take the specific stain. **sudan III.** A selective stain for fat. Prepare a saturated solution in 95% alcohol, dilute two-thirds with 50% alcohol, and filter. Stain sections for from 5 to 10 minutes, wash for about the same time in 60 or 70% alcohol, and mount in glycerol. Small oil-drops yellow, large ones orange. For staining the fat-granules in the elements of tissues undergoing fatty degeneration use the undiluted stain. The tissue may be fixed in Mueller's fluid or cut fresh on the freezing microtome. **thionin.** The uses and technique are the same as for methylene-blue. A saturated solution in 50% alcohol diluted with 5 volumes of water is used for staining. **Thoma's method** (*for the numeration of blood-cytes*). Dilute the blood in the proportion of 1 : 10 with water containing 0.3% anhydrous acetic acid. This dissolves the colored blood-cells. **Unna's hematoxylin.** A constant half-ripe stock solution. Hematoxylin, 1. Gm; alum, 10 Gm.; alcohol, 100 Cc.; water, 200 Cc.; sublimed sulphur, 2 Cc. If the sulphur is added 2 or 3 days after preparing the hematoxylin solution, it will arrest oxidation and the stain will be ready for use at this stage. The oxidation of alum-hematoxylin solutions can be instantaneously accomplished by adding a little neutralized hydrogen dioxid. **Unna's method.** *For collagen:* 1. Stain sections of alcohol material for 5 minutes in strong solution of polychrome methylene-blue, then for 15 minutes in neutral 1% solution of orcein in absolute alcohol; wash in alcohol; bergamot; balsam. Collagen, dark red; nuclei, blue; granules of mast-cells, carmin red; cytoplasm of plasma-cells, blue. 2. Stain sections for 20 seconds in 1% solution of water-blue (Wasserblau); wash and stain for 5 minutes in neutral aqueous 1% solution of safranin; wash in water and then treat with absolute alcohol until the blue color reappears; collagen, sky-blue; nuclei, red; cytoplasm, violet. 3. *For collagen, elastin, and smooth muscle:* Stain with hot orcein for 10 minutes, wash in dilute alcohol, stain with hematein for 10 minutes, and treat for a few seconds with acid alcohol; wash, and place in a 2% solution of acid fuchsin for 5 minutes, in saturated aqueous solution of picric acid for 2 minutes, then in saturated alcoholic solution of picric acid for 2 minutes; absolute alcohol; oil; balsam. Elastin, brown-red; collagen, red; muscle-fibers, yellow with gray-violet nuclei. 4. *For elastin and smooth muscle:* Stain as in 3, substituting polychrome methylene-blue for hematein and 1% potassium permanganate for the acid alcohol. Elastin, brown-red; collagen, decolored; muscle-fibers, violet. 5. *For smooth muscle:* Stain sections for 10 minutes in polychrome methylene-blue; wash, and fix in 1 % red prussiate of potash; differentiate in acid alcohol for 10 minutes; absolute alcohol; oil; balsam. The collagen is decolored. 6. *For keratohyalin:* Overstain in hematoxylin, treat for 10 seconds with 0.5 % solution of potassium permanganate, and wash in alcohol; or place the stained sections in 33 % solution of iron sulphate for 10 seconds, or in 10 % solution of iron chloride. 7. *For epithelia:* Stain sections for 10 minutes in neutral aqueous 1 % solution of water blue; wash, and stain for 10 minutes in 1 % solution of orcein. Or overstain sections of alcohol material in polychrome methylene-blue and differentiate in Unna's glycerol-ether mixture (Grübler); or in a mixture of alcohol, 10 parts; xylol, 15 parts; anilin, 25 parts; and transfer to xylol; or in a mixture of xylol, 30 parts; alcohol, 20 parts; then transfer to xylol and then to anilin containing alum to saturation (agitated and filtered before using). 8. *For plasma-cells and mast-cells:* Apply the methods for epithelia. 9. *For overcoming the decoloration of bacteria in the process of dehydrating in alcohol:* Transfer the section from the decolorizing fluid to the slide, remove as much as possible of the water by means of filter-paper, and then heat the slide over flame until the section is dry; when cold, mount in balsam. **Waldeyer's method** (*for the fixing and decalcification of bone*). Fix the fresh object in chromium trioxide (1 : 600); decalcify in a mixture of chromium trioxide (1 : 200) 100 Cc. and nitric acid 2 Cc. Wash thoroughly and harden in alcohol. **Weigert's differentiating fluid.** Borax, 2 Gm.; potassium ferricyanide, 2.5 Gm.; distilled water, 200 Cc. Used after hematoxylin. **Weigert's method.** 1. *For fibrin:* Make celloidin sections and stain one minute in Weigert's fibrin stain (5 % solution of gentian-violet, 4.4 Cc.; 96 % alcohol, 6 Cc.; anilin oil, 1 Cc.). Dry with unsized printing paper and add a drop of Gram's solution saturated with iodine. Most of the stained parts are decolorized. Remove the iodine with printing paper; clear in equal parts of anilin oil and xylol, renewing it until all the water is removed. The water gives the water a white appearance. Dry with filter-paper, wash well with xylol, and mount in xylol balsam. 2. *For neuroglia:* Fix for 8 days in the following mixture: dissolve 2.5 Gm. chrome alum in 100 Cc. water, by heat, and while hot add 5 parts each of acetic acid and pulverized copper acetate, when cold, 10 parts of formalin. Embed in celloidin. Treat the sections for 10 minutes with 0.3 % solution of potassium permanganate, wash in water, and reduce in the following: 5 parts each of chromogen and formic acid in 100 parts of water, to which, after filtering, add 10 parts of a 10 % solution of sodium sulphate. After 3 hours transfer to 5 % chromogen and after 24 hours stain in the following: saturate hot 75 %, alcohol with methyl-violet, decant when cold, and to each 100 Cc. add 5 Cc. of 5 % aqueous solution of oxalic acid; differentiate in a saturated solution of iodine in 5 % solution of potassium iodide; decolorize in a mixture of equal volumes of anilin and xylol, wash in xylol, and mount in balsam. 3. *Without decolorizing:* Tissues hardened in Mueller's fluid and alcohol are embedded in celloidin, and then put into a mixture of equal parts of a 10 % solution of sodium-potassium tartrate and a cold saturated solution of copper acetate, which is kept at from 38° to 40° C. (86°–104° F.). They are next placed in a half-saturated solution of copper acetate at the same temperature for 48 hours. The fluid is then rinsed in water, may be kept in 80 % alcohol and cut at any time. The staining fluid is composed of 1 part of an alcoholic hematoxylin solution (1 : 10) and 9 parts of a saturated solution of lithium carbonate; this fluid is to be freshly made. Stain for from 4 to 12 hours; wash, dehydrate in 90 % alcohol, and clear in anilin xylol (2 : 1), then in pure xylol, and mount in xylol balsam. The advantage of the method is the clearness with which the fine medullated fibers are distinguished from the cells and other parts, and it is less tedious than the old method. **Wright's stain** (*for blood*). Preparation of the neutral stain. Solution A: Make a 0.5 % aqueous solution of the NaHCO₃, being careful to bring all the salt into solution before going on to the next step. Then add 1 % of methylene-blue (Grübler's methylene-blue, "Bx," "Koch," or "Ehrlich's rectified"). Steam this in an Arnold sterilizer for 1 hour after steam is up. Cool. Solution B: 0.1 % aqueous solution of eosin (Grübler, "yellowish, soluble in water"). Add solution B to solution A until the mixture becomes purple, a metallic scum precipitate appears in suspension. (About 500 Cc. of solution B to 100 Cc. of solution A.) Filter off the precipitate. Do not wash it. Dry. Preparation of the staining solution.

Make a saturated solution of the precipitate in pure methyl-alcohol (0.3 Gm. in 100 Cc. methyl-alcohol). Filter, and add an additional 25 % of the original volume of methyl-alcohol used. This prevents precipitation of the stain on the film. Cover the film with the stain for 1 minute. Without pouring off the stain add water drop by drop until the mixture is translucent at the edges and a yellowish metallic scum forms on the surface. Stain in this diluted stain for 2 to 3 minutes. Wash in distilled water until the film becomes pink. Dry between filter-papers. Staining reactions: Lymphocytes: nuclei, dark purplish-blue; cytoplasm, robin's-egg blue; large mononuclears: nuclei, blue; cytoplasm, pale blue; polymorphonuclear neutrophils: nuclei, blue; granules, reddish-lilac; eosinophils: nuclei, blue; granules, blue; mastzellen: nuclei, blue to purplish; granules, dark blue or purple; myelocytes: nuclei, dark blue or lilac; granules, dark or reddish-lilac; blood-platelets stain blue or purplish; malarial parasites: nuclei, chromatin portion, lilac-red to black; cytoplasm, blue. Used for paraffin and celloidin sections. It causes shrinkage if the sections are not thoroughly dehydrated. **Zenker's fluid.** Dissolve 25 Gm. potassium dichromate, 10 Gm. sodium sulphate, and 50 Gm. mercuric chloride in 1000 Cc. warm distilled water. At the time of using add to each 20 Cc. 1 Cc. of glacial acetic acid. Fix the tissue for from 24 to 48 hours; wash for 24 hours in running water. Harden in the dark in the ascending series of alcohols. For the removal of the precipitate add to the 90 % alcohol enough tincture of iodine to impart the color of port-wine, and repeat the addition daily until the color does not fade. Preserve in 90 % alcohol. **Ziehl-Neelsen method.** Float the corner-glass preparation upon Ziehl's carbolfuchsin; heat until vapor arises (about 3 to 5 minutes), wash in water, and decolorize in 15 % nitric or 5 % sulphuric acid, then in 60 to 80 % alcohol to remove the remnant of color. Wash well, dry, and mount in balsam. In the case of tissue-sections, stain cold for 15 minutes and decolorize as detailed; upon removal from the alcohol counterstain with methylene-blue; wash, dehydrate, clear, and mount.

**staining, in vitro method of.** A method of studying, under the miscrocope, the diffusion into living cells of dyes contained in agar jelly spread on glass slides.

**staircase** (*stăr'-kās*) [ME., *staire*, stair; *case*]. A continuous series of responses to nerve-stimuli, varying from a minimal intensity to a maximum intensity. (Romanes.) See *summation*.

**stairs sign.** Difficulty in descending stairs; one of the early symptoms of locomotor ataxia.

**stalagmometer** (*stal-ag-mom'-et-er*) [σταλαγμός, a dropping; μέτρον, a measure]. An instrument for measuring the size of drops, or the number of drops in a given volume of liquid. It is used to measure the surface tension of liquids.

**stalagmometry** (*stal-ag-mom'-et-re*) [*stalagmometer*]. A method of diagnosis based upon the determination of the relative degree of surface tension or capillarity possessed by the body-fluids.

**stalk** (*stawk*) [ME., *stalken*]. Any lengthened support to an organ.

**stamen** (*sta'-men*). The pollen-bearing organ of the flower, when complete consisting of a stalk of filament and a pollen-sac or anther.

**stamina** (*stam'-in-ah*). Natural strength of constitution. Vigor. Inherent force.

**staminode** (*stam'-in-ōd*). Same as *staminodium*.

**staminodium** (*stam-in-o'-de-um*) [στήμων, a thread; εἶδος, form]. In biology, a stamen-like organ; a rudimentary or aborted stamen; a parastemon. See *lepal*.

**stammer** (*stam'-er*) [AS., *stamur*, stammering]. To speak interruptedly or with hesitation.

**stammering** (*stam'-er-ing*). Interrupted or hesitating speech. **s. bladder.** See *bladder, stammering*.

**stamp-licker's tongue.** An inflammatory condition of the mouth occurring in those who moisten stamps or other labels with the tongue.

**stamper** (*stamp'-er*). A name for one affected with locomotor ataxia, from the stamping gait incident to it.

**stanch** (*stanch*, or *stawnch*) [ME., *staunche*, stanch]. To check or stop (a flow); as to stanch a hemorrhage or a wound.

**stand.** To have an upright posture. Also, a

frame or a table to place things upon. s., microscope, the tripod or base of the microscope with the tube, but without eye-pieces and objectives.
**standard** (stan'-dard) [extendere, to spread out]. 1. An established rule or model. 2. Something used for comparison. s. candle, a spermaceti candle used as a standard of light; it burns at the rate of two grains a minute. s. solution, a solution containing a definite quantity of a reagent.
**standardization** (stan-dar-diz-a'-shun). Regulation by a standard; conformity to or use as a standard of comparison; the bringing of a preparation up to a definite standard.
**standstill** (stand'-stil). A state of quiescence dependent upon suspended action. s., expiratory, suspension of action at the end of expiration. s., inspiratory, halt in the respiratory cycle at the end of inspiration when the lungs are filled with air. The condition can be produced by stimulating the central end of the cut vagus. s., respiratory, suspended respiration.
**stannate** (stan'-āt). A salt of stannic acid.
**stannic** (stan'-ik) [stannum]. 1. Pertaining to stannum, or tin. 2. Containing tin as a tetrad element. s. acid, H₂SnO₃. A gelatinous white precipitate which, on drying, forms a translucent vitreous mass. It is dibasic. s. chloride, SnCl₄, a thin white liquid.
**stanniferous** (stan-if'-er-us) [stannum; ferre, to bear]. Yielding or containing tin.
**stannite** (stan'-tī). Tin sulphide; bell-metal.
**Stannius' ligature,** or **experiments** (stan'-e-us) [Herman Friedrich Stannius, German physiologist, 1808–1883]. 1. Separation by a ligature of the sinus venosus from the remainder of the frog's heart causes the latter to remain distended in diastole, while the former continues its rhythmic pulsations. Mechanical excitation of the auricle or ventricle produces a single contraction, which is repeated only when a new stimulus is applied. 2. If a ligature is placed around the groove dividing the auricles from the ventricle, there occurs a rhythmic contraction of the ventricle, while the auricles remain quiescent.
**stannous** (stan'-us). Containing tin as a bivalent element.
**stannum** (stan'-um) [L.]. Tin; see tin. s. cinereum, bismuth. s. glaciale, bismuth. s. indicum, zinc.
**stapedectomy** (sta-pe-dek'-to-me) [stapes; ἐκτομή, excision]. Excision of the stapes.
**stapedial** (sta-pe'-de-al) [stapes]. 1. Shaped like a stirrup. 2. Relating to the stapes.
**stapediotenotomy** (sta-pe-di-o-ten-ot'-om-e) [stapes; tenotomy]. Cutting of the tendon of the stapedius muscle.
**stapediovestibular** (sta-pe-de-o-ves-tib'-ū-lar) [stapes; vestibulum, vestibule]. Relating to the stapes and the vestibule.
**stapedius** (sta-pe'-de-us) [stapes]. See under muscle.
**stapes** (sta'-pēz) [L., "a stirrup"]. The stirrup-shaped bone of the middle ear, articulating with the incus and the fenestra ovalis.
**staphisagria** (staf-is-ag'-re-ah) [σταφίς, a dried grape; ἄγριος, wild]. Stavesacre. The staphisagria of the U. S. P. is the ripe seed of Delphinium staphisagria; of the order Ranunculaceæ. It contains the alkaloids delphinine, C₂₂H₃₇NO₆, delphinoidine, C₄₂H₆₈-N₂O₇, delphisine, C₂₇H₄₅N₂O₄, and staphisagrine, C₂₂H₃₃NO₅. It has been used locally as an application in rheumatism, and as an ointment to destroy lice and itch-mites. s., fluidextract of (fluidextractum staphisagria, U. S. P.). Dose 1 min. (0.06 Cc.).
**staphisagrine** (sta-fis-a'-grēn) [σταφίς, dried grape; ἄγριος, wild]. An amorphous alkaloid obtained from staphisagria.
**staphylagrum, staphylagra** (staf-il-a'-grum, staf-il-a'-grah) [σταφυλή, uvula; ἀγρεῖν, to take hold of]. An instrument formerly used to hold the uvula during amputation of the same.
**staphyle** (staf'-i-le) [σταφυλή, a bunch of grapes]. The uvula.
**•staphyledema** (staf-il-e-de'-mah) [staphylo-; edema]. Edema of the uvula. Any morbid enlargement of the uvula.
**staphyleus** (staf-il-e'-us) [σταφυλή, the uvula]. Pertaining to the uvula.
**staphylhematoma, staphylhæmatoma** (staf-il-hem-at-o'-ma) [σταφυλή, uvula; αἷμα, blood; ὄμα, tumor]. An extravasation of blood into the uvula.

**staphyline** (staf'-il-in) [σταφυλή, uvula]. Pertaining to the uvula or to the entire palate. s. glands. Synonym of palatine glands.
**staphylinopharyngeus** (staf-il-i-no-far-in'-je-us) [staphyle; pharynx]. 1. Relating to the palate and pharynx. 2. The palatopharyngeus. See under muscle.
**staphylinus** (staf-il-i'-nus) [staphyle]. 1. Palatal. 2. See s. medius. s. externus, the tensor palati. s. internus, the levator palati. s. medius, the azygos uvulæ muscle. See under muscle.
**staphylion** (sta-fil'-e-on) [σταφυλή, uvula]. The middle point of the posterior nasal spine.
**staphylitis** (staf-il-i'-tis) [staphyle; ιτις, inflammation]. Inflammation of the uvula.
**staphylo-** (staf-il-o-) [σταφυλή, uvula]. A prefix denoting pertaining to the uvula.
**staphyloangina** (staf-il-o-an'-jin-ah). Walsh's term for pseudomembranous inflammations of the throat due to infection by staphylococci.
**staphylocausticum** (staf-il-o-kaws'-tik-um) [staphylo-; caustic]. A caustic used for application to the uvula.
**staphylococcemia, staphylococcæmia** (staf-il-o-kok-se'-me-ah) [staphylococcus; αἷμα, blood]. A morbid condition due to the presence of staphylococci in the blood.
**staphylococcia** (staf-il-o-kok'-se-ah). General infection with staphylococci.
**Staphylococcus** (staf-il-o-kok'-us) [σταφυλή, grape; κόκκος, berry]. A micrococcus; a genus of Schizomycetes in which the cocci are irregularly clustered like a bunch of grapes. See Micrococci, table of.
**staphylocosis** (staf-il-o-ko'-sis). Infection by staphylococcia.
**staphylodialysis** (staf-il-o-di-al'-is-is) [staphylo-; διάλυσις, relaxation]. Relaxation of the uvula.
**staphylohemia, staphylohæmia** (staf-il-o-he'-me-ah). See staphylococcemia.
**staphylolysin** (staf-il-ol'-is-in) [staphylococcus; λύειν, to loosen]. Neisser and Wechsberg's name for a hemolysin produced by Staphylococcus aureus and S. albus.
**staphyloma** (staf-il-o'-mah) [σταφύλωμα, a defect in the eye]. A bulging of the cornea or sclera of the eye. s. æquatoriale, s. æquatoris, s., equatorial, staphyloma of the sclera in the equatorial region. s., annular, one surrounded on all sides by atrophic choroid. s., anterior. See keratoglobus. s., ciliary, one in the region of the ciliary body. s. corneæ, a bulging of the cornea due to a thinning of the membrane with or without previous ulceration. s., intercalary, one developing in that region of the iris which is united with the periphery of the iris. s., posterior, s. posticum, a backward bulging of the sclerotic coat at the posterior pole of the eye. s., Scarpa's, posterior staphyloma; staphyloma of the posterior segment of the sclera. s. uveale, thickening of the iris. Syn., iridoncosis.
**staphylomatic, staphylomatous** (staf-il-o-mat'-ik, staf-il-o'-mat-us). Pertaining to, of the nature of, or affected with, staphyloma.
**staphylomycosis** (staf-il-o-mi-ko'-sis) [staphylococcus; mycosis]. A morbid condition due to staphylococci.
**staphyloncus** (staf-il-ong'-kus) [σταφυλή, uvula; ὄγκος, tumor]. Swelling of the uvula.
**staphylopharyngorrhaphy** (staf-il-o-far-in-gor'-af-e) [staphylo-; pharynx; ῥαφή, a seam]. See Passavant's operation.
**staphyloplasmin** (staf-il-o-plaz'-min) [staphylococcus; plasmin]. Staphylococcus toxin.
**staphyloplasty** (staf-il-o-plas-te) [staphylo-; πλάσσειν, to mold]. A plastic operation on the soft palate or uvula.
**staphyloptosis** (staf-il-op-to'-sis) [staphylo-; πτῶσις, falling]. Abnormal elongation of the uvula.
**staphylorrhaphy** (staf-il-or'-a-fe) [staphylo-; ῥαφή, suture]. Suture of a cleft soft palate.
**staphylostreptococcia** (staf-il-o-strep-to-kok'-se-ah). Infection by both staphylococci and streptococci.
**staphylotome** (staf'-il-o-tōm) [staphylo-; τομή, a cutting]. A cutting instrument used in staphylotomy.
**staphylotomy** (staf'-il-ot-o-me) [staphylo-; τομή, a cutting]. 1. The operation of incising or removing the uvula. 2. The operation of incising a staphyloma.
**staphylotoxin** (staf-il-o-toks'-in). See staphylolysin.
**staphylygroma** (staf-il-ig-ro'-mah) [σταφυλή, uvula; ὑγρός, wet]. Synonym of staphyledema.

**staphysina** (*staf-is-i'-nah*) [σταφίς, a dried grape]. A product obtained by Thompson from the seeds of *Delphinium staphisagria*.

**star** [ME., *siarre*]. In biology applied to various radiate structures, granules, cells, groups of cells, or organisms. **s.-anise.** See *illicium*. **s.-cells,** endothelial cells of vessels, first described by Kupffer, and regarded as nervous elements on account of their shape and thin, elongated processes, but afterward shown to belong to the endothelial tissues; they have the power of inclosing various granules. Syn., *Kupffer's cells.* **s., daughter-.** See *diaster.* **s.-grass.** See *aletris.* **s.s of Verheyn,** the star-shaped figures formed by the stellate veins of the kidney, beneath the capsule. **s.s, Winslow's,** capillary whorls which form the beginning of the vorticose veins of the choroid. Syn., *Stellæ vasculosæ winslowii.*

**starblind** [AS., *stærblind*]. Half blind; blinking.

**starch** [AS., *stearc*, stiff]. A carbohydrate ($C_6H_{10}O_5$)n, widely distributed in the vegetable kingdom, occurring in peculiar concentrically marked granules or grains. When heated with water, the granules swell up, burst, partially dissolve, and form starch paste. The soluble portion is called granulose, the insoluble is cellulose. Iodine produces a characteristic blue coloration with starch. The most important varieties of starch are: potato-starch, leguminous starch, wheat-starch (*amylum*, U. S. P., B. P.), sago-starch, rice-starch. **s., animal.** 1. See *glycogen.* 2. See *bodies, amylaceous.* **s., corn-.** See *corn-starch.* **s.-enema,** an enema consisting of starch-water. **s., glycerite of.** See *amyli, glyceritum,* under *amylum.* **s., iodized,** iodide of starch, a dark powder containing 2 % of iodine; a disinfectant and internal and external antiseptic. Dose 3–10 gr. (0.2–0.66 Gm.). **s., soluble,** a white powder obtained by heating starch and glycerol and adding strong alcohol during the cooling; it is used as an emulsifier. Syn., *amylodextrin.* **s.-water,** a mixture of wheatstarch and water, used chiefly as an emollient enema.

**starter** (*start'-er*) [ME., *starten*, to start]. A pure culture of bacteria employed to start some particular fermentation, as in the ripening of cream.

**Startin's bandage** (*star'-tin*). A bandage impregnated with a mixture of paraffin and stearin.

**starvation** (*star-va'-shun*) [AS., *steorfan*, to die]. Deprivation of food; the state produced by deprivation of food.

**stasibasiphobia** (*sta-se-ba-se-fo'-be-ah*). [στάσις, standing; βάσις, a step, walk; φόβος, fear]. A peculiar fear in consequence of which the act of walking or of standing becomes impossible.

**Stas's process** (*stahs*). See *Stas-Otto method.*

**stasimetry** (*stas-im'-et-re*) [*stasis*; μέτρον, a measure]. Bitot's term for the estimation of the consistence of soft organic bodies.

**stasimorphy** (*stas'-e-mor-fe*) [*stasis*; μορφή; form]. Deviation from the normal from arrest of development.

**stasiphobia** (*stas-i-fo'-be-ah*) [*stasis*; φόβος, fear]. Fear of standing upright.

**stasis** (*sta'-sis*) [στάσις, from ιστάναι, to stand]. A standstill of the current of any of the fluids of the body, especially of the blood. **s., diffusion,** stasis in which there occurs diffusion of serum or lymph. **s., intestinal,** an undue delay in the passage of fecal material along the intestines. **s., venous,** stasis due to venous congestion.

**Stas-Otto method** (*stahs'-ot'-o*) [Jean Servais *Stas*, Belgian chemist, 1813–1891]. A method of extracting the putrefactive alkaloids from tissues. It depends upon the fact that the salts of the alkaloids are soluble in water and in alcohol, and generally insoluble in ether, and may be removed from alkaline fluids by agitation with ether. The method is applied as follows: "Treat the mass with twice its weight of pure 90 per cent. alcohol, and from 10 to 30 grains of tartaric or oxalic acid; digest the whole for some time at about 70° C. and filter. Evaporate the filtrate at a temperature not exceeding 35° C., filter in a strong current of air or in vacuo over sulphuric acid. Take up the residue with absolute alcohol, filter, and evaporate again at a low temperature. Dissolve the residue in water, alkalinize with sodium bicarbonate, and agitate with ether. After separation, remove the ether and allow it to evaporate spontaneously. The residue may be further purified by redissolving in water and again extracting with ether. The method has been modified in some of its details, especially by Selmi and Marino-Zuco.

**state** (*stāt*) [*status; stare*, to stand]. A condition. **s. medicine,** that department of medical study that concerns public health, and is in part occupied with the statistics of disease.

**statement** (*stāt'-ment*) [*stare*, to stand]. A declaration. **s., ante-mortem,** a declaration made immediately before death, and which if made with the consciousness of impending death is legally held as binding as a statement sworn to.

**static** (*stat'-ik*) [στατικός, causing to stand]. At rest. In equilibrium. **s. ataxia.** See *ataxia.* **s. breeze,** a method of administration of static electricity, consisting in the withdrawal of a static charge from a patient by means of a pointed electrode. **s. electricity.** See *electricity.* **s. pelvis,** the bony pelvis. **s. shock,** a mode of applying Franklinic electricity, placing the patient on an insulated stool, and applying one pole of a static machine to this platform, while the other pole is applied to the body of the patient by the operator. **s. test,** this consists in ascertaining the absolute weight of the lungs and comparing this weight with the average lung-weights of still born children, and of children who have died soon after birth. Foderé fixes the weight of the lungs of still born children born at term at 480 grains, and 960 grains as the weight of the lungs soon after breathing has been established. This test is of but slight value. **s. theory** (*of Golts*), every position of the head causes the endolymph of the semicircular canals to exert the greatest pressure upon some part of the canals, thus in varying degree exciting the nerve terminations of the ampullæ.

**Statice** (*stat'-is-e*) [στατική, an astringent herb]. A genus of plants of the order *Plumbagineæ*. *S. antarctica* and *S. brasiliensis*, baycuru or guaycura, South American species, are used to produce uterine contractions. *S. gmelini*, a species indigenous to southern Russia, is used as a gargle and in diarrhœa. *S. limonium* grows upon the coasts of Europe and North America; the plant, seed, and root are used as astringents.

**statics** (*stat'-iks*) [see *static*]. The science relating to forces in a condition of equilibrium.

**statim** (*stat'-im*) [L.]. Immediately, at once.

**station** (*sta'-shun*) [*statio*, from *stare*, to stand]. 1. Standing position or attitude. 2. In obstetrics, the location of the head or presenting part; *e. g.*, it may be at the outlet of the pelvis, or above the inlet. **s.-test,** the patient is made to stand with his eyes shut and feet together; an unusual swaying of the body denotes ataxia.

**stationary** (*sta'-shun-a-re*) [see *station*]. Standing still; not moving. **s. air,** the amount of air which is constantly in the lungs during normal respiration.

**statistics** (*sta-tis'-tiks*) [*status*, a state]. A numerical collection of facts relating to any subject. **s., medical,** that part of medicine pertaining to details of mortality, climate, and the geographical distribution of diseases. **s., vital.** Same as *s., medical.*

**statocyst** (*stat'-o-sist*) [στατός, standing; κύστις, cyst]. One of the vestibular sacs of the labyrinth which is supposed to act as the nervous mechanism on which static equilibrium depends.

**statometer** (*stat-om'-et-er*) [στατός, standing; μέτρον, a measure]. An instrument for measuring the degree of exophthalmos.

**stature** (*stat'-ūr*) [*statura*, stature]. The height of any animal when standing. In quadrupeds, it is measured at a point over the shoulders. In man, it is the measured distance from the heel to the top of the head.

**status** (*sta'-tus*) [L.]. A state. **s. arthriticus,** the nervous manifestations preceding an attack of gout. **s. cribrosus,** a scarcely macroscopic sieve-like condition of the brain or nerve-substance, due to absorption of minute vessels; observed in autopsies. Fr., *état criblé.* **s. epilepticus,** a condition in which epileptic attacks occur in rapid succession, the patient not regaining consciousness during the interval. **s. gastricus,** gastritis. **s. lymphaticus,** a condition in which all the lymphatic tissues, the thymus, the spleen, and the bone marrow are hyperplastic; sudden death is liable to occur especially in surgical anesthesia. Syn., *lymphaism; lymphotoxemia; status thymicus.* **s. parathyreoprivus,** a pathological state caused by complete loss of parathyroid tissue. **s. præsens,** the state of a patient at the time of examination. **s. thymicolymphaticus, s. thymicus**

See *s. lymphaticus.* **s. typhosus.** See *typhoid state.*
**s. verminosus.** See *helminthiasis.* **s. vertiginosus,** persistent vertigo.
**statuvolence** (*sta-tū-vo'-lens*) [*status,* state; *volens,* willing]. Autohypnotism; voluntary somnambulism or clairvoyance; a trance into which one voluntarily enters without aid from another.
**staurion** (*staw'-re-on*) [σταυρός, cross]. The craniometric point where the transverse palatine suture crosses the median suture.
**stauroplegia** (*staw-ro-ple'-je-ah*) [σταυρός, crossed; πληγή, a stroke]. Crossed hemiplegia.
**stave of the thumb.** See *Bennett's fracture.*
**stavesacre** (*stāvz-a'-ker*). See *staphisagria.*
**staxis** (*staks'-is*). See *stillicidium.*
**stay knot.** See *knot.*
**steam** (*stēm*). The vapor of water; water in a gaseous state. **s.-atomizer.** See *atomizer.* **s.-doctor,** an old name for a Thompsonian physician, from the extensive use of steaming and sweating made by that school. **s.-tug murmur,** the double murmur of aortic obstruction and insufficiency. It may be expressed by the word *hoo-chee, hoo* representing the obstructive murmur, and *chee* the regurgitant murmur.
**steapsin** (*ste-ap'-sin*) [στέαρ, fat]. A ferment of the pancreatic juice which causes fats to combine with an additional molecule of water and then split into glycerin and their corresponding acids; lipase.
**stear** (*ste-ar*) [στέαρ, fat]. See *adeps.*
**stearate** (*ste'-ar-āt*) [στεατίν]. A salt of stearic acid. Glycerol stearate is called stearin (*q. v.*).
**stearerin** (*ste-ar'-er-in*) [στέαρ, fat]. A fatty substance found in the oil of sheeps' wool and which is analogous to stearin.
**stearic acid** (*ste-ar'-ik*). See *acid, stearic,* and *stearin.*
**steariform** (*ste-ar'-if-orm*) [στέαρ, fat; *forma,* form]. Having the appearance of or resembling fat.
**stearin** (*ste'-ar-in*) [στέαρ, fat], C₃H₅O₃(C₁₈H₃₅O₂)₃. 1. A compound of stearic acid and glyceryl occurring in the harder animal fats, especially in tallow. It crystallizes in white, pearly scales. 2. Stearic acid.
**stearoconotum** (*ste-ar-ok-on-o'-tum*) [στέαρ, fat; κόνις, dust]. An insoluble but fusible solid yellowish fat occurring in brain tissue; it contains sulphur and phosphorus.
**stearodermia** (*ste-ar-o-der'-me-ah*) [στέαρ, fat; δέρμα, the skin]. An affection of the sebaceous glands of the skin.
**stearol** (*ste'-ar-ol*) [στέαρ, fat]. A medicament having fat as an excipient.
**stearone** (*ste'-ar-ōn*) [στέαρ, fat], C₃₅H₇₀O. A volatile liquid obtained by partial decomposition of stearic acid.
**stearoptene** (*ste-ar-op'-tēn*) [στέαρ, fat; πτηνός, winged; volatile]. The crystalline substance occurring naturally in solution in a volatile oil.
**stearrhea, stearrhœa** (*ste-ar-e'-ah*) [στέαρ, fat; ῥοία, a flow]. See *seborrhea.* **s. congestiva.** Synonym of *seborrhœa congestiva.* **s. flavescens,** a seborrhea in which the sebaceous matter turns yellow after being deposited upon the skin. **s. nigricans.** See *chromidrosis.* **s. simplex,** ordinary seborrhea.
**stearyl** (*ste'-ar-il*) [στέαρ, fat; ὕλη, matter], C₁₈H₃₅O. The radical of stearic acid.
**steatin** (*ste'-at-in*) [στεάτινον, pertaining to suet]. 1. Same as *stearin.* 2. Any cerate containing a considerable proportion of tallow.
**steatinum** (*ste-at-i'-num*) [στεάτινον, pertaining to suet]. A name given to certain pharmaceutical preparations similar to cerates.
**steatite** (*ste'-at-īt*). See *talc.*
**steatitis** (*ste-at-i'-tis*) [στέαρ, fat]. Inflammation of the fatty tissues.
**steato-** (*ste-at-o-*) [στέαρ, fat]. A prefix meaning fatty.
**steatocele** (*ste'-at-o-sēl*) [*steato-*; κήλη, tumor]. A swelling formed by a collection of fatty matter in the scrotum.
**steatocryptosis** (*ste-at-o-krip-to'-sis*) [*steato-*; κρύπτη, a crypt, or sac]. Abnormality of function of the sebaceous glands.
**steatodes** (*ste-at-o'-dēz*) [στέαρ, fat; εἶδος, resemblance]. Fatty.
**steatogenous** (*ste-at-oj'-en-us*) [*steato-*; γενής, produced]. Producing steatosis.
**steatolysis** (*ste-at-ol'-is-is*) [*steato-*; λύσις, solution]. The emulsifying process by which fats are prepared for absorption and assimilation.

**steatolytic** (*ste-at-o-lit'-ik*). Accomplishing a steatolysis.
**steatoma** (*ste-at-o'-mah*) [*steato-*; ὄμα, tumor]. 1. A sebaceous cyst. 2. A lipoma. **s., Mueller's,** a lipofibroma.
**steatopathic** (*ste-at-o-path'-ik*) [*steato-*; πάθος, disease]. Pertaining to diseases of the sebaceous glands.
**steatopygia** (*ste-at-o-pi'-je-ah*) [*steato-*; πυγή, buttock]. Enormous fatness of the buttock, common among the women of some African tribes.
**steatopygous** (*ste-at-op'-ig-us*) [see *steatopygia*]. Characterized by excessive development of the buttocks.
**steatorrhea, steatorrhœa** (*ste-at-or-e'-ah*) [*steato-*; ῥοία, flow]. 1. An increased flow of the secretion of the sebaceous follicles; see *seborrhea.* 2. Fatty stools. **s. amianthaca,** a form of seborrhea in which the excess of solid constituents gives the appearance of scaliness of the skin; see, also, *seborrhea.* **s. nigricans.** Same as *seborrhœa nigricans.* **s. simplex,** excess of sebaceous excretion of the face.
**steatosis** (*ste-at-o'-sis*). 1. Fatty degeneration. 2. An abnormal accumulation of fat. **s. cordis,** fatty heart.
**steatozoon** (*ste-at-o-zo'-on*) [*steato-*; ζῶον, an animal]. The parasite, *Demodex folliculorum,* contained in comedones.
**stechiology, stœchiology, stoicheiology** (*stek-e-ol'-o-je*) [στοιχεῖον, a first principle; λόγος, science]. The doctrine of elements and of elementary principles.
**stechiometry, stœchiometry, stoichiometry** (*stek-e-om'-et-re, stoi-ke-om'-et-re*) [στοιχεῖον, a first principle; μέτρον, a measure]. 1. The mathematical side of chemistry. 2. The estimation of the proportions in which elements combine to form compounds.
**steel** (*stēl*). Iron chemically combined with a certain proportion of carbon. It holds an intermediate position between white cast iron and wrought iron, partaking of the most valuable qualities of both. Steel of good quality is fine-grained, elastic, and tough; also, *ferrum.* **s. drops,** tincture of chloride of iron. **s.-grinders' phthisis.** See *pneumonokoniosis.* **s. mixture,** *mistura ferri composita.* **s.-pen palsy.** See *writers' cramp.* **s. tincture.** See *ferric chloride, tincture of.* **s. wine.** Synonym of *vinum ferri.*
**Steele's sign** (*stēl*). Exaggerated pulsation over the whole area of the cardiac region; it is noted in intrathoracic tumor.
**steep.** 1. A name for rennet. 2. To stand in water for making an infusion.
**stege** (*ste'-je*) [στέγος, roof]. The inner layer of the rods of Corti.
**stegmonth** (*steg'-munth*) [στέγειν, to cover]. The period (about a month) between childbirth and complete return to health; the puerperium.
**stegnosis** (*steg-no'-sis*) [στέγνωσις, a checking of a discharge; a soldering]. Constipation, or costiveness; the checking of a discharge; the closing of a passage; stenosis.
**stegnotic** (*steg-not'-ik*). Effecting stegnosis. Astringent.
**Stegomyia** (*steg-o-mi'-e-ah*) [στεγανός, covered; μυῖα, a fly]. A genus of mosquitos or *Culicidæ,* represented in most tropical and subtropical countries. The adults are usually very vicious biters, both by day and night. According to the experiments of the American Commission on Yellow Fever *S. calopus* or *fasciata* is the agent which spreads the germs of this disease. S. calopus, S. fasciata, a very distinct and common species, easily distinguished by the thoracic ornamentation and by the last hind tarsal joint being white. It is one of the most troublesome and annoying of mosquitos; the bite is very irritating. It is the intermediate host of the germ of yellow fever and of the hematozoon *Filaria Bancrofti,* which also occurs in *Culex fatigans,* and in *Anopheles.* Syn., *Brindled* or *Tiger mosquito.*
**stella** (*stel'-ah*) [L., "star"]. A star-shaped bandage; stellate bandage. **stellæ vasculosæ Wlusslowii.** See *Winslow's stars.*
**stellate** (*stel'-āt*) [*stella,* a star]. Star-shaped, or with parts radiating from a center. **s. bandage,** one that is wound crosswise on the back. **s. cells,** small polyaxonic nerve-cells in the molecular layer of the cortex cerebri. **s. fracture,** a fracture in which there are numerous fissures radiating from the central point of injury. It usually occurs in flat bones. **s. hair,** a hair which divides at the end in a star-shaped fashion. **s. laceration,** one involving the

tissues in several directions, as a stellate laceration of the cervix uteri. **s. ligament**, the anterior costovertebral ligament. **s. veins**, minute venous radicles arranged in stellate fashion and located just beneath the capsule of the kidney.

**stellula** (*stel'-ū-lah*) [dim. of *stella*, a star]. In anatomy, a plexus of veins in the cortex of the kidney.

**stellulæ Verheynii**, a stellate network of veins in the outer part of the cortex of the kidney. **s. Winslowii.** See *Winslow's stars*.

**stem.** The pedicle of a tumor; the shaft of a hair; the supporting stalk of a leaf or plant. **s., brain**, the brain, less the fissured portion of the cerebrum. **s.-eelworm.** See *s. sickness*. **s., gland**, a glandduct. **s. of hair.** Synonym of *hair-shaft*. **s.-pessary**, a pessary having a stem or rod which enters the os uteri. **s.-sickness**, a parasitic disease of clover, due to the presence of the stem-eelworm. (*Tylenchus devastatrix*).

**Stellwag's sign** (*stel'-vahg*) [Carl *Stellwag* von Carion, Austrian ophthalmologist, 1823–1904]. Absence or diminution in frequency of the winking movements of the eyelids and abnormal width of the palpebral aperture; it is seen in exophthalmic goiter.

**stenagma** (*sten-ag'-mah*) [στενάζειν, to sigh]. Synonym of *sigh*.

**stenagmus** (*sten-ag'-mus*) [στενάζειν, to sigh]. Sighing.

**stench** [ME., *stench*, a smell]. An ill smell; an offensive odor. **s.-pipe**, an upright pipe that reaches above the roof of a house; it is intended to give vent to foul vapors that accumulate in waste-pipes, and water-closets. **s.-trap**, in sewerage and plumbing, a device for preventing a reflux of foul vapors and gases.

**Stender dish** (*sten'-der*) [Wilhelm P. *Stender*, German manufacturer of scientific apparatus]. A vessel used in staining sections of tissues.

**stenion** (*sten'-e-on*) [στενός, narrow]. A craniometric point at the extremity of the smallest transverse diameter in the temporal fossa.

**Steno's duct.** See *Stensen's duct*.

**steno-** (*sten-o-*) [στενός, narrow]. A prefix meaning narrow or constricted.

**stenobregmate** (*sten-o-breg'-māt*) [*steno-*; βρέγμα, the bregma]. The condition in which the upper and fore-part of the head is narrow.

**stenocardia** (*sten-o-kar'-de-ah*) [*steno-*; καρδία, heart]. Angina pectoris.

**stenocephalous** (*sten-o-sef'-al-us*) [*steno-*; κεφαλή, head]. Having a head narrow in one or more of its diameters.

**stenocephaly** (*sten-o-sef'-al-e*) [see *stenocephalous*]. Narrowing of the head in one or more of its diameters.

**stenochasmus** (*sten-o-kaz'-mus*) [*steno-*; χάσμα, a chasm]. Lissauer's term applied to a skull in which a line drawn from the point upon the rostrum of the sphenoid where it is included between the alæ vomeris, to the center of the posterior nasal spine and to the basion, intersects with an angle of 74° to 94°.

**stenochoria** (*sten-o-ko'-re-ah*) [*steno-*; χῶρος, space]. Narrowing; stenosis; partial obstruction, particularly of the lacrymal duct. **s. saccilacrimalis**, stenosis of the lacrymo-nasal duct.

**stenocompressor** (*ste-no-kom-pres'-or*) [*Steno's duct*; *compressor*]. An instrument used to compress Stenson's ducts during dental operations.

**stenocoriasis** (*sten-o-ko-ri'-as-is*) [*steno-*; κόρη, pupil]. Narrowing of the pupil.

**stenocrotaphia**, or **stenocrotaphy** (*sten-ok-ro-ta'-fe-ah*, or *sten-o-kro'-ta-fe*) [*steno-*; κρόταφος, the temple]. A narrowing of the temporal region of the skull.

**stenodont** (*sten'-o-dont*) [*steno-*; ὀδούς, tooth]. Provided with narrow teeth.

**stenomycteria** (*sten-o-mik-te'-re-ah*) [*steno-*; μυκτήρ, the nose]. Nasal stenosis.

**Stenon's duct.** See *Stensen's duct*.

**Stenonian, Stenonine** (*sten-o'-ne-an, sten'-o-nēn*). Named for Nicholas Stenson (latinized Stenonianus), a Danish anatomist, 1638–1686.

**stenopeic** (*sten-o-pe'-ik*) [*steno-*; ὀπή, an opening]. Pertaining to or having a narrow slit; applied to lenses that allow the passage of rays only through a narrow slit.

**stenosin** (*sten'-o-sin*), AsCH₃O₃Na₂₂H₂O, disodic methylarsenate, discovered by Baeyer; said to be a nontoxic arsenical salt. Dose 1 cg. 1 to 5 times daily.

**stenosis** (*sten-o'-sis*) [στενός, narrow]. Constriction or narrowing, especially of a channel or aperture. **s., aortic**, a narrowing of the aortic orifice at the base of the heart or a narrowing of the aorta itself. **s., cardiac**, as a consequence of inflammation of the connective tissue in the myocardium, the conus arteriosus upon either side of the heart may become diminished in diameter, with consequent hindrance to the free passage of blood from the ventricle into its corresponding artery. This constitutes what is called stenosis of the heart. The second sound is fully formed and sharply defined, thus distinguishing the condition from valvular stenosis. **s., cicatricial**, stenosis due to a contracted cicatrix. **s., granulation**, narrowing caused by encroachment of contraction of granulations. **s., mitral**, stenosis of left auriculoventricular orifice. **s., post-tracheotomy**, stenosis after tracheotomy. **s., Dittrich's**, stenosis of the conus arteriosus.

**stenostegnosis, stenostenosis** (*sten-o-steg-no'-sis, sten-o-sten-o'-sis*) [*stenononianus*, or *Stensen*; στένωσις, constriction]. Stenosis of Stensen's duct.

**stenostomatous** (*sten-o-sto'-mat-us*) [*steno-*; στόμα, mouth]. Having a small mouth.

**stenostomia** (*sten-o-sto'-me-ah*) [*steno-*; στόμα, mouth]. A narrowing or closure of the mouth.

**stenostomy** (*sten-os'-to-me*) [*steno-*; στόμα, a mouth]. The contraction of any mouth or aperture.

**stenothermal** (*sten-o-ther'-mal*) [*steno-*; narrow; θέρμος, heat]. Capable of sustaining a small range of temperature.

**stenothorax** (*sten-o-tho'-raks*) [*steno-*; *thorax*]. Having a straight, short thorax.

**stenotic** (*sten-ot'-ik*) [*stenosis*]. Characterized by stenosis; produced by stenosis.

**Stensen's** (*Steno's*) **duct** [see *Stenonian*]. The duct of the parotid gland. **S.'s experiment**, temporary ligation of the aorta of the rabbit immediately below the point at which the renal arteries are given off, for the purpose of cutting off the blood-supply of the lower portion of the spinal cord. **S.'s foramina**, the incisive foramina which transmit the anterior branches of the descending palatine vessels. **S.'s plexus**, the venous plexus surrounding Stensen's duct. **S.'s veins**, the venæ vorticosæ of the choroid.

**stentorin** (*sten'-to-rin*). A blue pigment obtained from infusorians of the genus *Stentor*.

**stentorophonous** (*sten-tor-of'-on-us*) [Στέντωρ, a loud-voiced Greek in the Trojan war; φωνή, sound]. Having a loud voice.

**stephanial, stephanic** (*stef-an'-e-al, stef'-an-ik*) [στέφανος, a wreath]. Pertaining to the stephanion.

**stephanion** (*stef-an'-e-on*) [στέφανος, a wreath; crown]. The point of intersection of the temporal ridge and coronal suture. See under *craniometric point*.

**steppage-gait** (*step'-āj-gāt*). The peculiar highstepping gait seen in tabes dorsalis and certain forms of multiple neuritis.

**stercobilin** (*ster-ko-bi'-lin*) [*stercus*, dung; *bilis*, bile]. A brown coloring-matter found in feces, and identical with hydrobilirubin.

**stercoraceous** (*ster-ko-ra'-shus*) [*stercus*]. Fecal; having the nature of feces; containing feces as, *stercoraceous* vomiting.

**stercoral** (*ster'-ko-ral*) [*stercus*]. See *stercoraceous*.

**stercorary** (*ster'-ko-ra-re*) [*stercus*]. Fecal.

**stercoremia, stercoræmia** (*ster-ko-re'-me-ah*) [*stercus*; αἷμα, blood]. A condition of the blood resulting from arrest of intestinal excretion and the absorption of toxic matters from the feces.

**stercorin** (*ster'-ko-rin*) [*stercus*]. An extractive from the feces resembling cholesterin.

**Sterculia** (*ster-ku'-le-ah*) [*stercus*, dung]. A genus of some 85 species of tropical trees. **S. urens**, of India, and **S. tragacantha** of Africa afford some part of the gums known as tragacanth. **S. acuminata** produces the kola-nut; see *kola*.

**stercus** (*ster'-kus*) [L., "dung"]. Feces.

**stere** (*stēr*) [στερεός, solid]. A measure of 1000 liters; a kiloliter.

**stereo-** (*ster-e-o-*) [στερεός, solid]. A prefix meaning solid or relating to solidity.

**stereoagnosis** (*ster-e-o-ag-no'-sis*). See *astereognosis*.

**stereochemistry** (*ster-e-o-kem'-is-tre*) [στερεός, solid; *chemistry*]. Stereo-isomerism; theoretical explanations of close isomerisms, by which it is assumed that the differences between the various isomers are due to the different positions of the same atoms or radicals in tri-dimensional representations of the molecules.

Ordinary structural formulæ involve only two dimensions, length and breadth, but these are not sufficient to explain numerous cases of isomerism now known, and a "spatial" or "solid" conception of the molecule is necessary. The term allo-isomerism has been proposed for these cases.
**stereocyst** (*ster'-e-o-sist*) [*stereo-*; κύστις, cyst]. A hard cyst, or cystic growth.
**stereognosis** (*ster-e-og-no'-sis*) [*stereo-*; γνῶσις, knowledge]. The faculty of recognizing the nature and use of objects by contact and handling them. Cf. *astereoagnosis*.
**stereognostic** (*ster-e-og-nos'-tik*) [see *stereognosis*]. 1. Pertaining to the cognition of solidity, or tridimensional forms. 2. Recognizing by sense of touch.
**stereogram** (*ster'-e-o-gram*) [*stereo-*; γράμμα, a writing]. A stereoscopic picture.
**stereograph** (*ster'-e-o-graf*) [*stereo-*; γράφειν, to write]. 1. Of Broca, an instrument used to make outline drawings of parts of the cranium. 2. Same as *stereogram*.
**stereography** (*ster-e-og'-ra-fe*) [*stereo-*; γράφειν, to write]. Graphic representation of the skull; a branch of craniometry.
**stereoisomerism** (*ster-e-o-i-som'-er-izm*) [*stereo-*; *isomerism*]. The condition in which two or more substances having the same molecular formulæ have different properties; these differences are due to the different relative positions of the atoms in the molecule.
**stereometer** (*ster-e-om'-et-er*) [*stereo-*; μέτρον, a measure]. An apparatus for the determination of the specific gravity of liquids, porous substances, powders, etc., as well as solids.
**stereometry** (*ster-e-om'-et-re*) [see *stereometer*]. 1. The determination of the specific gravity of substances. 2. The measurement of volume.
**stereomonoscope** (*ster-e-o-mon'-o-skōp*) [*stereo-*; μόνος, single; σκοπεῖν, to view]. An instrument with two lenses for producing a single picture giving the effect of solidity.
**stereoneura** (*ster-e-o-nū'-rah*) [*stereo-*; νεῦρον, nerve]. A term proposed by Wilder for the invertebrates whose nervous axis, when it exists, presents no cavity as in the vertebrates or celoneura.
**stereophantoscope** (*ster-e-o-fan'-to-skōp*). A panorama stereoscope using rotating discs in place of pictures.
**stereophoroscope** (*ster-e-o-for'-o-skōp*) [*stereo-*; φέρειν, to carry; σκοπεῖν, to see]. A stereoscopic zoetrope, an instrument for producing a series of images apparently in motion.
**stereoplasm** (*ster'-e-o-plazm*) [*stereo-*; πλάσσειν, to mold]. The solid part of the protoplasm of cells.
**stereopsis** (*ster-e-op'-sis*) [*stereo-*; ὄψις, vision]. Stereoscopic vision.
**stereoscope** (*ster'-e-o-skōp*) [*stereo-*; σκοπεῖν, to see]. An instrument by which two similar pictures of the same object are made to overlap so that the reflected images are seen as one, thereby giving the appearance of solidity and relief.
**stereoscopic** (*ster-e-o-skop'-ik*) [*stereoscope*]. Pertaining to stereoscopy. **s. vision**, binocular vision. See *stereoscope*.
**stereoscopy** (*ster-e-os'-ko-pe*) [*stereoscope*]. The use of the stereoscope.
**stereoskiagraphy** (*ster-e-o-ski-ag'-ra-fe*) [*stereo-*; *skiagraphy*]. The use of the stereoscope in the study of skiagrams.
**stereostroboscope** (*ster-e-o-stro'-bo-skōp*) [*stereo-*; στρόβος, a twisting; σκοπεῖν, to view]. An apparatus for the experimental study of points moving in three dimensions.
**stereotics** (*ster-e-ot'-iks*). Lesions or deformities affecting the harder portions of the body.
**stereotypy** (*ster-e-ot'-o-pe*) [*stereo-*; τύπος, a type]. Morbid persistence of a volitional impulse when once started.
**stereosol** (*ster'-e-sol*). A liquid said to be an alcoholic solution of gum lac, benzoin, tolu balsam, phenol, oil of ginger, and saccharin. It is used in diphtheria and skin diseases.
**steriform** (*ster'-e-form*). An almost tasteless and odorless powder consisting essentially of sugar of milk and 5 % of formaldehyde. **s. chloride**, a mixture of formaldehyde, 5 parts; ammonium chloride, 10 parts; pepsin, 20 parts; and milk-sugar, 65 parts. **s. iodide**, formaldehyde, 5 parts; ammonium iodide, 10 parts; pepsin, 20 parts; and milk-sugar, 65 parts.

**sterigma** (*ster-ig'-mah*) [στήριγμα, a prop, support; pl., *sterigmata*]. In biology, a stalk or support.
**sterile** (*ster'-il*) [*sterilis*, barren]. 1. Not fertile; not capable of reproducing. 2. Free from microorganisms or spores.
**sterility** (*ster-il'-it-e*) [*sterile*]. The condition of being sterile, infertile, or incapable of reproducing. **s., facultative**, sterility caused by the prevention of conception. **s., idiopathic**. See *azoospermia*. **s., one-child**, sterility occurring in a woman after she has given birth to one child. **s., relative**, sterility due to other causes than abnormality of the sexual organs.
**sterilization** (*ster-il-i-za'-shun*) [*sterile*]. The act of rendering anything sterile; the destruction of microorganisms, particularly by means of heat. **s. fractional**, **s., intermittent**, a method of sterilization in which an interval of time is allowed to elapse between the several heatings, giving an opportunity for any spores present to develop into adult microorganisms, in which form they readily succumb to the action of heat.
**sterilized** (*ster'-il-īzd*). Rendered sterile.
**sterilizer** (*ster'-il-i-zer*) [*sterile*]. An apparatus for destroying the microorganisms attached to an object, especially by means of heat.
**sterisol** (*ster'-is-ol*). A preparation containing sugar of milk, 2.98 parts; sodium chloride, 0.672 part; potassium phosphate, 0.322 part; formaldehyde, 0.520 part; water, 95.506 parts. Used as an antiseptic in infectious diseases.
**Stern's position in heart examination** [Heinrich Stern, American physician, 1868– ]. The murmur is heard more clearly in cases of tricuspid regurgitation if the patient is placed on his back with his neck extended and head lowered.
**sternad** (*ster'-nad*) [*sternum*]. Toward the sternal aspect.
**sternal** (*ster'-nal*) [*sternum*]. Pertaining to the sternum.
**sternalgia** (*ster-nal'-je-ah*) [*sternum*; ἄλγος, pain]. Pain in the sternum.
**sternalgic** (*ster-nal'-jik*) [*sternum*; ἄλγος, pain]. Affected with sternalgia.
**sternalis** (*ster-na'-lis*) [*sternum*]. Connected with the sternum; sternal.
**Sternberg's disease**. The tuberculous form of pseudoleukemia.
**sternebra** (*ster'-ne-brah*) [*sternum*; *vertebra*; pl., *sternebra*]. Any one of the serial segments of the sternum.
**sternebral** (*ster'-ne-bral*) [*sternum*; *vertebra*]. Pertaining to or of the nature of a sternebra.
**sternen** (*ster'-nen*) [*sternum*]. Belonging to the sternum in itself.
**sterniform** (*ster'-nif-orm*) [*sternum*; *forma*, form]. Shaped like a sternum.
**sterno-** (*ster-no-*) [*sternum*]. A prefix denoting connection with the sternum.
**sternoabdominalis** (*ster-no-ab-dom-in-a'-lis*) [*sterno-*; *abdomen*]. The triangularis sterni and the transversus abdominis considered as a single muscle.
**sternochondroscapularis** (*ster-no-kon-dro-skap-u-la'-ris*) [*sterno-*; χόνδρος, cartilage; *scapula*]. An inconstant muscle arising from the sternum and the first costal cartilage and extending to the upper border of the scapula.
**sternoclavicular** (*ster-no-kla-vik'-u-lar*) [*sterno-*; *clavicle*]. Pertaining to the sternum and the clavicle.
**sternocleidal** (*ster-no-kli'-dal*). Same as *sternoclavicular*.
**sternocleidomastoid** (*ster-no-kli-do-mas'-toid*) [*sterno-*; κλείς, key; *mastoid*]. Pertaining to the sternum, the clavicle, and the mastoid process, as the *sternocleidomastoid muscle*. See under *muscle*.
**sternocoracoid** (*ster-no-kor'-ak-oid*). Relating to the sternum and the coracoid.
**sternocostal** (*ster-no-kos'-tal*) [*sterno-*; *costa*, a rib]. Pertaining to the sternum and the ribs.
**sternodymia** (*ster-no-dim'-e-ah*) [*sterno-*; δύειν, to enter]. A form of somatodymia in which the union is in the sternum.
**sternodynia** (*ster-no-din'-e-ah*) [*sterno-*; ὀδύνη, pain]. Sternalgia; pain in the sternum.
**sternofacial** (*ster-no-fa'-shal*) [*sterno-*; *facies*, face]. Pertaining to the sternum and the face.
**sternoglossal** (*ster-no-glos'-al*) [*sterno-*; γλῶσσα, tongue]. Pertaining to the sternum and the tongue.
**sternohyoid** (*ster-no-hi'-oid*) [*sterno-*; *hyoid*].

**Pertaining** to the sternum and the hyoid bone, as the *sternohyoid* muscle. See under *muscle*.
**sternoid** (*ster'-noid*) [*sternum*; εἶδος, resemblance]. Resembling the sternum.
**sternomastoid** (*ster-no-mas'-toid*). Relating to the sternum and the mastoid process of the temporal bone.
**sternomaxillary** (*ster-no-maks-il'-ar-re*) [*sterno-; maxilla*, jaw]. Pertaining to the sternum and the mandible.
**sterno-omphalodymia** (*ster-no-om-fal-o-dim'-e-ah*) [*sterno-;* ὀμφαλός, a navel; δύειν, to enter]. A form of somatodymia in which the union is in both the sternal and umbilical regions.
**sternopagia** (*ster-no-pa'-je-ah*) [*sterno-;* πάγος, fixed]. The condition of a sternopagus.
**sternopagus** (*ster-nop'-ag-us*) [*sterno-;* πάγος, fastened]. A double monster the parts of which are united at the sternum.
**sternopericardiac** (*ster-no-per-e-kar'-de-ah*). Relating to the sternum and the pericardium.
**sternoscapular** (*ster-no-skap'-ū-lar*) [*sterno-; scapula*]. Pertaining to the sternum and the scapula.
**sternothyroid** (*ster-no-thi'-roid*) [*sterno-; thyroid*]. Pertaining to the sternum and the thyroid cartilage, as the *sternothyroid* muscle. See under *muscle*.
**sternotracheal** (*ster-no-tra'-ke-al*) [*sterno-; trachea*]. Pertaining to the sternum and the trachea.
**sternotrypesis** (*ster-no-tri-pe'-sis*) [*sterno-;* τρύπησις, a boring]. Perforation of the sternum.
**sternoxiphoid** (*ster-no-zif'-oid*). Relating to or connecting the sternum and the xiphoid process.
**sternum** (*ster'-num*) [στέρνον, breast-bone]. The flat, narrow bone in the median line in the front of the chest, composed of three portions—the manubrium, the gladiolus, and the ensiform or xiphoid appendix.
**sternutament** (*ster-nū'-tam-ent*) [*sternutamentum; sternutare*, to sneeze]. A substance causing sneezing.
**sternutatio** (*ster-nū-ta'-she-o*) [L.]. Sneezing. **s. convulsiva,** paroxysmal sneezing, as in hay fever.
**sternutation** (*ster-nū-ta'-shun*) [*sternutatio*, a sneezing]. The act of sneezing.
**sternutatory** (*ster-nu'-tat-o-re*) [see *sternutation*]. 1. Producing sneezing. 2. An agent that causes sneezing.
**sterochemistry** (*ste-ro-kem'-is-tre*). See *stereochemistry*.
**sterol** (*ster'-ol*). A class of compounds which are non-saponifiable, but are soluble in ether; they are derived from plants and animals; cholesterol is an example.
**stertor** (*ster'-tor*) [L., "a snoring"]. Sonorous breathing or snoring; the rasping, rattling sound produced when the larynx and the air-passages are obstructed by mucus.
**stertorous** (*ster'-to-rus*) [*stertor*]. Characterized by stertor, as *stertorous* breathing.
**sterule** (*ster'-ūl*). Trade name for a glass capsule containing a sterile solution.
**stetharteritis** (*steth-ar-ter-i'-tis*) [στῆθος, .chest; ἀρτηρία, artery; ιτις, inflammation]. Inflammation of the arteries of the thorax.
**stethemia, stethæmia** (*steth-e'-me-ah*) [στῆθος, chest; αἷμα, blood]. An accumulation of blood in the pulmonary vessels.
**stethendoscope** (*steth-en'-do-skōp*) [*stetho-;* ἔνδον, within; σκοπεῖν, to view]. A variety of fluoroscope used for examining the chest.
**stetho-** (*steth-o-*) [στῆθος, chest]. A prefix denoting pertaining to the chest.
**stethocatharsis** (*steth-o-kath-ar'-sis*). Synonym of *expectoration*.
**stethocele** (*steth'-o-sēl*). See *pneumonocele*.
**stethocelodyspnea** (*steth-o-se-lo-disp'-ne-ah*) [*stetho-;* κήλη, hernia; *dyspnea*]. Dyspnea due to hernia of the lung.
**stethochysis** (*steth-ok'-is-is*). See *hydrothorax*.
**stethocyrtograph.** See *stethokyrtograph*.
**stethogoniometer** (*steth-o-go-ne-om'-et-er*) [*stetho-;* γωνία, angle; μέτρον, measure]. An instrument for measuring the curvature of the chest.
**stethograph** (*steth'-o-graf*) [*stetho-;* γράφειν, to write]. An instrument recording the respiratory movements of the chest.
**stethokyrtograph** (*steth-o-kir'-to-graf*) [*stetho-;* κυρτός, curved; γράφειν, to write]. An apparatus designed for measuring and recording the dimensions of the chest.

**stethomenia** (*steth-o-me'-ne-ah*)[*stetho-;* μήν, month]. Vicarious menstruation by way of the bronchial tubes.
**stethometer** (*steth-om'-et-er*) [*stetho-;* μέτρον, a measure]. An instrument for measuring the degree of expansion of the chest.
**stethomyitis** (*steth-o-mi-i'-tis*) [*stetho-;* μῦς, muscle; ιτις, inflammation]. Inflammation of the muscles of the chest.
**stethonoscope** (*steth-on'-o-skōp*) [*stetho-;* σκοπεῖν, to view]. An apparatus for use in auscultation which may be attached to a binaural stethoscope.
**stethoparalysis** (*steth-o-par-al'-is-is*). Paralysis of the muscles of the chest.
**stethophone** (*steth'-o-fōn*) [*stetho-;* φωνή, sound]. Stethoscope.
**stethophonometer** (*steth-o-fo-nom'-et-er*) · [*stetho-;* φωνή, sound; μέτρον, a measure]. An instrument for measuring the phenomena elicited by auscultation.
**stethophonometry** (*steth-o-fo-nom'-et-re*) [see *stethophonometer*]. The determination of the intensity of the acoustic phenomena associated with the lungs and heart.
**stethopolyscope** (*steth-o-pol'-is-kōp*) [*stetho-;* πολύς, many; σκοπεῖν, to view]. A stethoscope having several tubes for the simultaneous use of several observers.
**stethoscope** (*steth'-o-skōp*) [*stetho-;* σκοπεῖν, to view]. An instrument for ascertaining the condition of the organs of circulation and respiration by the sounds made by these organs. It consists of a hollow tube, one end being placed over the locality to be examined, the other at the ear of the examiner. The **binaural** stethoscope consists of a Y-shaped tube, the flexible branches being applied each to an ear of the listener. **s.**, differential, one determining the time rather than the quality of the sounds heard, so that murmurs at two localities may be compared.
**stethoscopic** (*steth-o-skop'-ik*). Pertaining to or detected by means of the stethoscope.
**stethoscopy** (*steth-os'-ko-pe*), [see *stethoscope*]. Examination with the aid of the stethoscope.
**stethospasm** (*steth'-o-spazm*) [*stetho-;* *spasm*]. Spasm of the pectoral muscles.
**sthenia** (*sthen'-e-ah*) [σθένος, strength]. Normal or excessive force or vigor (opposed to asthenia).
**sthenic** (*sthen'-ik*) [σθένος, strength]. Strong; active. **s. fever,** a form of fever marked by high temperature, quick and tense pulse, and highly colored urine.
**sthenopyra** (*sthen-o-pi'-rah*) [σθένος, strength; πῦρ, fever]. Sthenic fever.
**stibiacne** (*stib-e-ak'-ne*) [*stibium; acne*]. Acne caused by the use of antimony.
**stibial** (*stib'-e-al*) [*stibium*]. Pertaining to stibium, or antimony.
**stibialism** (*stib'-e-al-izm*) [*stibium*]. Antimonial poisoning.
**stibiated** (*stib'-e-a-ted*) [*stibium*, antimony]. Containing antimony.
**stibiation** (*stib-e-a'-shun*). Excessive use of antimonials.
**stibine** (*stib'-ēn*). Antimony trihydride; antimoniureted hydrogen, SbH₃.
**stibium** (*stib'-e-um*). Antimony.
**stibogram** (*stib'-o-gram*) [στίβος, a beaten path; γράμμα, a writing]. A record of footsteps.
**stibonium** (*stib-o'-ne-um*). The radical SbH₄; similar in constitution to ammonium, NH₄.
**stichochrome** (*stik'-o-krōm*) [στίχος, a row; χρῶμα, color]. Applied by Nissl to a somatochrome nerve-cell in which the chromophilic substance is arranged in striæ running in the same direction and usually parallel with the contour of the cell-body, partly also with the surface of the nucleus. Cf. *arkyostichochrome*.
**Sticker's disease** [G. *Sticker*, German physician, 1860— ]. Erythema infectiosum.
**sticking plaster.** Adhesive plaster.
**stictacne** (*stik-tak'-ne*) [στικτός, punctated; *acne*]. Acne punctata; acne in which the pustules have a red, raised base, with a central black point.
**stiff** (*stif*). Inflexible, unyielding, immovable in continuity; applied especially to normally movable parts. **s. joint.** See *ankylosis*. **s. neck.** See *torticollis*.
**stiff-neck fever.** Epidemic cerebrospinal meningitis.
**stifle** (*sti'-fl*) [ME., *stifl*, to choke]. 1. To choke; to kill by impeding respiration. 2. The **stifle-joint,** *q. v.* 3. Disease or other affection of the

stifle-bone, q. v. s.-bone, the patella of the horse.
s.-joint, the knee-joint of the horse.
stigma (stig'-mah) [στίγμα, a point; pl., stigmata].
1. A small spot or mark, especially a spot of hemorrhage in the palm or sole, occurring in hysterical persons. 2. Any one of the marks or signs characteristic of a condition; generally used in the plural, as *hysterical stigmata*. 3. That part of a pistil which receives the pollen. stigmata, bakers', nodules on the backs of the fingers caused by kneading dough. stigmata, Cohn's, minute gaps in the interalveolar walls of the normal lung. s., Giuffrida-Rugieri's, of degeneration, the absence or incompleteness of the glenoid fossa. s. of Graafian follicle, the point where the blood-vessels of the walls are absent and where it finally ruptures. stigmata, hereditary, psychical stigmata resembling those of an ancestor and supposed to be inherited. stigmata, hysterical, the specific, peculiar phenomena or symptoms of hysteria as the anesthesia, hyperesthesia, hysterogenic zones, reversal of the color field, contraction of the visual field, the phenomena of transport, amblyopia, impairment of the sense of hearing, of taste, and of muscular sense, etc. stigmata, Malpighi's, the orifices of the capillary veins that join the branches of the splenic vein at right angles. stigmata maydis, zea mays. See under *zea*. stigmata, neurasthenic. See *stigmata, hysterical*. stigmata nigra, the black spots caused by the presence of grains of gunpowder in the skin. stigmata ovariorum, small cicatrices seen in the ovaries after the escape of the ova. stigmata, psychical, certain mental states characterized by susceptibility to particular suggestions. stigmata rubra, petechiæ due to various causes. stigmata, somatic, the objective signs of certain nervous affections. stigmata, venous, varicose veins.
stigmal (stig'-mal) [στίγμα, stigma]. Pertaining to a stigma.
stigmatic (stig-mat'-ik) [stigma]. Pertaining to a stigma.
● stigmatism (stig'-mat-ism) [στίγμα, point]. 1. A condition of the refractive media of the eye in which rays of light from a point are accurately brought to a focus on the retina. Synonymous with emmetropia. See, also, *astigmatism*. 2. The condition of having stigmata.
stigmatization (stig-mat-i-za'-shun) [stigma]. The formation of stigmata.
stigmatodermia (stig-mat-o-der'-me-ah) [στίγμα, a prick; δέρμα, skin]. Disease of the prickle-cell layer of the skin.
stigmatose (stig'-mat-ōs) [στίγμα, stigma]. Marked with stigmata.
stilet, stilette (stil-et') [Fr., dim. of *stilus*, a point]. 1. A small, sharp-pointed instrument inclosed in a cannula. 2. A wire passed into a flexible catheter.
still-birth. The birth of a dead child.
still-born. Born lifeless.
Still-Chauffard symptom-complex (stil'-sho-far') [see *Still's disease*; A. Chauffard, French physician]. The symptoms of Still's disease, q. v., observed in pseudotuberculosis.
Still's disease [George Frederic Still, English physician, 1868—    ]. A form of polyarthritis with enlargement of spleen and lymph-glands; it occurs in infancy and childhood.
Stiller's sign [Berthold Stiller, Austrian physician, 1837—    ]. Marked mobility or fluctuation of the tenth rib in neurasthenia and enteroptosis.
stillicidium (stil-is-id'-e-um) [stilla, a drop; cadere, to fall down]. The flow of a liquid drop by drop. s. lacrimarum, overflow of tears from obstruction of the canaliculus or nasal duct; epiphora. s. narium, coryza. s. urinæ, dribbling of urine.
Stilling's bundle [Benedict Stilling, German anatomist, 1810–1879]. See *Krause's respiratory tract*. S.'s canal. 1. The central canal of the spinal cord. 2. See *canal, hyaloid*. S.'s cells, S.'s columns, groups of multipolar cells near the gray commissure in the posterior cornua of the cervical and lumbar spinal cord. They correspond to Clarke's vesicular columns. S.'s fibers, the association fibers of the cerebellum. S.'s fleece, the meshwork of fibers formed around the dentate nucleus of the cerebellum. S.'s gelatinous substance, the gelatinous substance surrounding the central canal of the spinal cord. S.'s nucleus. 1. The nucleus ruber of the subthalamic region. 2. The nucleus of the hypoglossal nerve in the fourth ventricle. S.'s raphe, a narrow band connecting the pyramids of the oblongata. S.'s sacral nucleus, an island of ganglion-cells in the region of the spinal cord. S.'s scissors of the brain, the supposed resemblance to the outline of a pair of scissors seen in a horizontal section of the brain through the thalamus, nucleus ruber, and the nucleus dentatus cerebelli.
Stilling-Clarke's cells or dorsal nucleus. See *Clarke's column*.
Stillingia (stil-in'-je-ah) [Benjamin Stillingfleet, English botanist, 1702–1771]. A genus of plants of the order Euphorbiaceæ. The *stillingia* of the U. S. P. is the root of *S. sylvatica*, queen's root or, queen's delight, and is used as an alterative in syphilis, scrofula, diseases of the skin, etc. s., fluidextract of (*fluidextractum stillingiæ*, U. S. P.). Dose ½–1 dr. (2–4 Cc.).
stillingin (stil-in'-jin) [Benjamin Stillingfleet, English botanist, 1702–1771]. A precipitate from a tincture of the root of *Stillingia sylvatica*; resolvent, stimulant, diuretic, antisyphilitic. Dose 1 to 3 grains.
stilus (sti'-lus) [stilus, a point]. 1. A more correct form of the word *stylus*, used as an anatomical term.
2. A small tube or a bit of wire sometimes retained in the obstructed lacrymal duct, with a view to the restoration of its function. 3. An ointment or other medicament in the shape of a pencil or stick.
stimulant (stim'-ū-lant) [stimulus, a goad]. 1. s., cardiac, one that increases the heart's action. s., cerebral, one that exalts the action of the cerebrum. s., cutaneous, one that increases the activity of the skin, producing diaphoresis. s., diffusible or diffusive, one that has a prompt but transient effect. s., hepatic, one that excites the activity of the liver. s., intestinal, one that acts upon the intestinal tract. s., local, one acting directly on the end organs of the sensory nerves of the skin. s., renal, one producing diuresis. s., spinal, one exciting the spinal cord. s., stomachic, one giving tone to the stomach, aiding digestion, etc. s., vasomotor, one exciting the vasomotor apparatus.
● stimulate (stim'-ū-lāt) [see *stimulant*]. To quicken; to stir up; to excite; to increase functional activity.
stimulation (stim-u-la'-shun) [see *stimulant*]. 1. The act of stimulating. 2. The effect of a stimulant.
stimulator (stim'-ū-la-tor). A stimulating drug or agent.
stimulin (stim'-ū-lin). Metchnikoff's name for substances supposed to stimulate the phagocytes to destroy germs.
stimulus (stim'-ū-lus) [L.; pl., *stimuli*]. A goad; an impulse; anything capable of causing stimulation. s., adequate. See s., *homologous*. s., chemical, one due to or produced by chemical means. s., difference, the difference in activity between two stimuli. s., heterologous, one acting upon the nervous elements of the sensory apparatus along their entire course. s., homologous, one acting only upon the end-organ. s., maximal, a stimulus, increase above which cannot be appreciated. s., mechanical, one acting by mechanical means, as pinching or striking. s., minimal, the smallest stimulus which can be appreciated. s., subminimal, one too weak to produce any obvious effect. s., summation of. See *summation*. s., thermal, the application of heat.
Stipa (sti'-pah) [στύπη, tow]. A genus of grasses. *S. vaseyi*, sleepy grass, is a species found in New Mexico in the Sacramento Mountains, the ingestion of which causes in horses a stupor which endures for several days.
stipate (sti'-pāt) [stipare, to press together].
stipatio (sti-pa'-she-o) [L.]. An aggregation forming an obstruction. s. telæ cellulosæ infantum, sclerema neonatorum.
stirp (sturp) [stirps, a stock, root, race]. 1. The sum total of hereditary organic units contained in the fertilized ovum. 2. A race, lineage, or family.
stirpiculture (stur-pik-ul'-tū-ral) [stirps, a race; cultura, culture]. Pertaining to stirpiculture.
stirpiculture (stur-pik-ul'-tūr) [stirps, stock, race; cultura, culture]. The proposed improvement of the human species by attention to the laws of breeding.
stirrup, stirrup-bone (stir'-up). The stapes.
stitch. 1. A sudden, sharp, lancinating pain.
2. See *suture* (2). s.-abscess, an abscess forming in

a suture. s., **Marcy's cobbler.** See *suture, cobbler's.*
s., **sclerocorneal,** a peculiar stitch devised by Kalt to secure rapid union of the wound and to prevent prolapse of the iris after simple extraction of cataract. Syn., *Kalt stitch.* s., **in the side,** intercostal neuralgia.
**stith, stithe** (*stith, stīth*). The incus.
**stock** (*stok*). A quantity of solution, or other material, kept on hand for use as occasion requires.
**stocking, elastic.** A stocking of elastic fibers for the compression of a limb affected with varicose veins and other diseases.
**stœchiology** (*stek-e-ol'-o-je*) [στοιχεῖον, an element; λόγος, a treatise]. The study of the chemical elements of the gases, fluids, and solids of the body; see *stechiology.*
**Stoerk's blennorrhea** [Karl *Stoerk,* Austrian laryngologist, 1832–1899]. Profuse chronic suppuration and consequent hypertrophy of the mucosa of the nose, pharynx, and larynx.
**stoichiometry.** See *stechiometry.*
**Stokes, astigmatic lens of,** an apparatus consisting of two plano-cylindrical lenses, one concave, the other convex, the two of equal focal distance; it is used in the diagnosis of astigmatism.
**Stokes' disease** [William *Stokes,* Irish physician, 1804–1878]. See *goiter, exophthalmic.* S.'s **expectorant,** a preparation used in the treatment of bronchitis. It consists of pulverized carbonate of ammonium 16 grains, fluidextract of senega and squills each ½ dram, paregoric 3 drams, syrup of Tolu sufficient to make two ounces. A dram of this is given p. r. n. S.'s **law,** inflammation of serous or mucous membranes leads to paralysis of subjacent muscles. S.'s **liniment,** a liniment containing turpentine, acetic acid, oil of lemon, egg, and rose water. S.'s **pulse.** See *Corrigan's pulse.* S.'s **sign.** 1. A violent abdominal throbbing felt on palpation to the right of the umbilicus in acute enteritis. 2. Marked feebleness of the first heart-sound, when occurring during fevers, calls for alcoholic stimulation. S.'s **syndrome.** See *Adams-Stokes' disease.*
**Stokes-Adams' symptom-complex** or **disease.** See *Adams-Stokes' disease.*
**Stokes' operation** [Sir William *Stokes,* Irish surgeon, 1839–1900]. 1. *For amputation above the knee:* the same as Gritti's operation, except that section of the femur is made above the condyles. 2. *For excision of the tongue:* a modification of Jaeger's operation. 3. *For flat-foot:* by removing a wedge-shaped piece of bone from the head and neck of the astragalus. 4. *For single hare-lip:* the prolabium is formed by tissue from both sides of the cleft by means of incisions skirting the red margin; the upper part of the cleft is incompletely pared and the partially dissected flaps turned back, while the edges of the skin are brought together and the prolabial flaps drawn downward and outward.
**Stokes's reagent for reducing hemoglobin** [William Royal *Stokes,* American pathologist, 1870– ]. Add some citric or tartaric acid to a solution of ferrous sulphate and ammonia enough to make it alkaline.
**Stokvis' test for bile-pigments** [Barend Joseph *Stokvis,* Dutch physician, 1834–1902]. To 20 to 30 Cc. of urine add 5 to 10 Cc. of a zinc acetate solution (1 : 5). Wash the precipitate on a small filter with water, and dissolve in a little ammonia. When filtered, the filtrate will give, after standing in the air, a brownish green color, and show the absorption bands of bilicyanin, one between C and D, the second at D, and the third between D and E.
**stolidity** (*stol-id'-it-e*) [*stoliditas*]. A term designating stupidity of various degrees, even to amentia, or complete imbecility—oftener, however, signifying merely a phlegmatic or immobile temperament.
**Stoll's pneumonia.** Bilious pneumonia; a variety of pneumonia with gastrohepatic symptoms.
**stolon** (*sto'-lon*). In biology: (*a*) a slender, prostrate branch, taking root, or bearing a bulb at the tip, where it forms one or more new plants; (*b*) an analogous budding stock in certain compound animals.
**stolonization** (*sto-lon-iz-a'-shun*) [*stolo,* a shoot]. The process of transforming, in certain organisms, one organ into another through external influences, such as gravitation, contact, light, etc.
**Stoltz's operation** (*stōlts*) [Joseph Alexis *Stoltz,* French gynecologist, 1803–1896]. Pubiotomy.
**stoma** (*sto'-mah*) [στόμα, mouth; pl., *stomata*]. 1. A mouth. 2. A pore, as that between endothelial cells, establishing direct communication between adjacent lymph-channels.
**stomacace** (*sto-mak'-as-e*) [στόμα, mouth; κάκος, evil]. Canker of the mouth. Fetor of the mouth with ulcerated gums, also scorbutic sore-mouth.
**stomach** (*stum'-ak*) [στόμαχος, the stomach]. The most dilated part of the alimentary canal, situated below the diaphragm in the left hypochondriac, the epigastric, and part of the right hypochondriac regions. It is connected at one end (cardiac end) with the esophagus, at the other (pyloric end) with the duodenum. Its wall consists of four coats—the serous, muscular, submucous, and mucous. The mucous coat contains the gastric glands (cardiac and pyloric glands), which secrete the gastric juice and mucus. s.-**bed,** the shelf-like support upon which the stomach rests, formed by the portion of the pancreas situated to the left of the median line. This is quite thick anteroposteriorly, and its upper surface (anterior surface of His) makes a large portion of the shelf. s.-**bucket,** a small bucket for extracting some of the gastric contents. s.-**cough,** a reflex cough excited by irritation of the stomach. s.-**pump,** a pump for withdrawing the contents of the stomach. s.-**reefing.** Synonym of *gastrorrhaphy.* s.-**tooth,** a lower canine tooth, especially one of the first dentition. s.-**tube,** a flexible tube for irrigation or evacuation of the stomach. s.-**worm disease,** a disease of cattle due to species of *Strongylus*— *S. contortus, S. osterfagi, S. curticei, S. parkeri, S. retortæformis, S. fillicollis, S. oncophorus.*
**stomachal** (*stum'-ak-al*) [*stomach*]. Pertaining to the stomach.
**stomachalgia** (*stum-ak-al'-je-ah*) [*stomach;* ἄλγος, pain]. Pain in the stomach.
**stomachic** (*stum-ak'-ik*) [*stomach*]. 1. Pertaining to the stomach. 2. Stimulating the secretory activity of the stomach. 3. One of a class of substances which have an influence upon the work of the digestive organs.
**stomachoscopy** (*stum-ak-os'-ko-pe*) [*stomach;* σκοπεῖν, to view]. Examination of the stomach; gastroscopy.
**stomata** (*sto'-mat-ah*). Plural of *stoma, q. v.*
**stomatal** (*sto'-mat-al*) [*stomach*]. Relating to stomata.
**stomatalgia** (*sto-mat-al'-je-ah*) [στόμα, mouth; ἄλγος, pain]. Pain in the mouth.
**stomatic** (*sto-mat'-ik*) [*stoma,* mouth]. Relating or belonging to the mouth.
**stomatitis** (*stom-at-i'-tis*) [*stoma;* ιτις, inflammation]. Inflammation of the mouth. s. **aphthosa,** s., aphthous. See *aphtha.* s., **catarrhal,** a simple form characterized by swelling of the mucous membrane, pain, and salivation. s., **epidemic,** an acute infectious stomatitis, which occurs in epidemic. s., **gangrenous.** See *cancrum oris.* s., **mercurial,** that arising from poisoning by mercury. s., **mycotic.** See *thrush.* s., **parasitic.** See *thrush.* s., **scorbutic,** that due to scurvy. s., **ulcerative,** a form characterized by the formation of small ulcers on the cheeks, lips, and tongue, with copious salivation, pain, fetid breath, slight fever, and at times great prostration. s., **vesicular.** Same as s., *aphthous.*
**stomato-** (*sto-mat-o-*) [στόμα, mouth]. A prefix meaning pertaining to the mouth.
**stomatocace** (*sto-mat-ok'-as-e*) [*stomato-;* κάκος, evil]. Fetid ulceration of the mouth.
**stomatocatharsis** (*sto-mat-o-kath-ar'-sis*). Synonym of *salivation.*
**stomatodynia** (*sto-mat-o-din'-e-ah*) [*stomato-;* ὀδύνη, pain]. Pain in the mouth.
**stomatodysodia** (*sto-mat-o-dis-o'-de-ah*) [*stomato-;* δυσωδία, foul odor]. A foul odor of the breath; ill smelling breath.
**stomatogastric** (*sto-mat-o-gas'-trik*) [*stomato-;* γαστήρ, stomach]. In biology, applied to the nerves, pertaining to the mouth and the stomach.
**stomatol** (*sto'-mat-ol*). An antiseptic compound said to consist of terpineol, 4 parts; soap, 2 parts; alcohol, 45 parts; aromatics, 2 parts; glycerol, 5 parts; water, 42 parts.
**stomatologic, stomatological** (*sto-mat-o-loj'-ik, sto-mat-o-loj'-ik-al*) [*stomato-;* λόγος, science]. Pertaining to stomatology.
**stomatologist** (*sto-mat-ol'-o-jist*) [*stomato-;* λόγος, science]. One versed in stomatology.
**stomatology** (*sto-mat-ol'-o-je*) [*stomato-;* λόγος, science]. The sum of what is known about the mouth.

**stomatomalacia** (*sto-mat-o-mal'-a-se-ah*) [*stomato-*; μαλακία, softening]. Sloughing or softening of parts of the mouth.
**stomatomenia** (*sto-mat-o-me'-ne-ah*) [*stomato-*; μήν, month]. Vicarious menstruation by way of the mouth.
**stomatomia** (*sto-mat-o'-me-ah*) [*stomato-*; τέμνειν, to cut]. A general term for the incision of a mouth, as of the uterus.
**stomatomy** (*stom-at'-o-me*) [*stoma*; τομή, a cutting]. Incision of the os uteri.
**stomatomycosis** (*stom-at-o-mi-ko'-sis*) [*stomato-*; *mycosis*]. A disease of the mouth due to fungi, especially *Oidium albicans*.
**stomatonecrosis, stomatonoma** (*stom-at-o-ne-kro'-sis; stom-at-on'-o-mah*). See *cancrum oris*.
**stomatopathy** (*stom-at-op'-ath-e*) [*stomato-*; πάθος, disease]. Any disease of the mouth.
**stomatoplasty** (*sto'-mat-o-plas-te*) [*stomato-*; πλάσσειν, to form]. A plastic operation upon the mouth.
**stomatopoiesis** (*sto-mat-o-poi-e'-sis*). See *stomatoplasty*.
**stomatoplastic** (*sto-mat-o-plas'-tik*). Pertaining to stomatoplasty.
**stomatorrhagia** (*sto-mat-or-a'-je-ah*) [*stomato-*; ῥηγνύναι, to burst forth]. Copious hemorrhage from the mouth.
**stomatoscope** (*stom'-at-o-skōp*) [*stomato-*; σκοπεῖν, to inspect]. An instrument for inspecting the cavity of the mouth.
**stomatosis** (*sto mat o' sis*) [στόμα, mouth]. Disease of the mouth.
**stomatosyrinx** (*sto-mat-o-sir'-ingks*) [*stomato-*; σῦριγξ, a tube]. The Eustachian tube.
**stomatotomy** (*sto-mat-ot'-o-me*) [*stomato-*; τομή, a cutting]. Incision of the os uteri.
**stomatotyphus** (*sto-mat-o-ti'-fus*) [*stomato-*; *typhus*]. A form of typhus in which the beginning lesions are found in the mouth.
**stomenorrhagia** (*sto-men-or-a'-je-ah*) [στόμα, mouth; *menorrhagia*]. Vicarious menstruation from the mouth.
**stomocephalus** (*sto-mo-sef'-al-us*) [*stoma*; κεφαλή, head]. A variety of monster in which there is the same deformity as in rhinocephalus or in cyclocephalus, associated with a defect of the maxillary bones, so that the skin hangs in folds around the mouth.
**stomodæal** (*sto-mo-de'-al*) [στόμα, mouth; ὁδαῖος, by the way]. Having the character of a stomodæum.
**stomodæum** (*stom-o-de'-um*) [*stoma*; ὁδαῖος, by the way]. The primitive oral cavity of the embryo, formed by a depression of the ectoderm and afterward forming the mouth and upper part of the pharynx.
**stomoschisis** (*sto-mos'-kis-is*) [στόμα, mouth; σχίσις, fissure]. Fissure of the mouth, particularly of the soft palate.
**Stomoxys calcitrans** (*sto-moks'-is kal'-sit-ranz*). The common stable fly which, by its bite, is believed to spread trypanosomes.
**stone.** A hardened mass of mineral matter. See *calculus*. **s.**, blue, copper sulphate crystals. **s.**, **gall-**, a biliary calculus; see *gall-stone*.
**stool.** The evacuation of the bowels. **s.s**, **acholic.** 1. Light gray or clay-colored stools having the consistence of putty, which follow stoppage of the flow of bile into the duodenum. The color is due to the presence of the normal urobilin. The stools show, under the microscope, an abnormal amount of fat. This form of acholic stool is accompanied by icterus and choluria. 2. Stools of the same color may occur in the absence of interference with the flow of bile, but when the stool contains an excessively large amount of fat and fatty acids. **ss.**, **bilious**, the discharge is bilious diarrhea, as after large doses of calomel. **ss.**, **caddy**, yellow-fever stools which resemble fine, dark, sandy mud. **ss.**, **fatty**, stools in which fat is present; due to pancreatic disease. **s.**, **insulated**, in electricity, a stool provided with insulated legs. **s.s**, **lead-pencil**, fecal discharges of a very small caliber. They occur independent of any general nervousness or local intestinal spasm, and cannot be regarded as evidence of stricture or stenosis of the colon. **ss.**, **mucous**, stools containing mucus. They indicate the existence of intestinal inflammation. **ss.**, **pea-soup**, the peculiar liquid evacuation of typhoid fever. **ss.**, **rice-water**, the stools of cholera, in which there is a copious serous exudation containing detached epithelium. **s.s**, **Schafkoth**,

see *s.s*, *sheep-dung*. **s'.s**, **sheep-dung**, the small round fecal masses (similar to the dung of sheep) due to atony of the intestine; this form of passage may occur in the socalled "starvation" or "hunger" evacuation which is found in cases of inanition; *e. g.*, after carcinomatous cachexia when the intestine becomes very much contracted. **s.-sieve of Boas**, an apparatus by means of which feces may be thoroughly and conveniently washed, so that undigested remains of food, bits of mucus, concretions, and parasites are readily seen and isolated.
**stop** [ME., *stoppen*, to stop]. To plug up; to hold back; to hinder. **s.-cock**, a turning cock, connected with a pipe, for regulating the flow of gases or liquids. **s.-needle**, a lance-pointed needle used in the operation of discission, having an enlargement or shoulder upon the shank to prevent too deep penetration.
**stoppage** (*stop'-aj*) [ME., *stoppen*, to stop]. Cessation of flow or action; closure or stegnosis.
**stopper, stopple** (*stop'-er*, *stop'-l*) [ME., *stoppen*, to stop]. A plug or other closure for a bottle, commonly made of cork, rubber, or glass. **s.-dropper**, a combination of stopper and medicine pipet in one piece.
**stopping** (*stop'-ing*). See *filling*.
**storax** (*sto'-raks*). See *styrax*.
**storesin** (*sto-rez'-in*) [*storax*]. An amorphous resin forming the largest ingredient of storax.
**Stoughton's elixir** (*stou'-ton*). Tinctura absinthii composita; a tincture of wormwood, germander, gentian, rhubarb, orange-peel, cascarilla, and aloes; used as a flavor in alcoholic drinks and as a general tonic.
**stout** (*stowt*). 1. Hardy, sturdy, corpulent. 2. A heavy beer or porter.
**stovaine** (*sto-vān'*), C₁₄H₂₁NO₂HCl. Amylene hydrochloride. A local anesthetic, also used in spinal anesthesia.
**stovainization** (*sto-va-ni-za'-shun*). The production of local anesthesia by the subarachnoid injection of stovaine.
**strabilismus** (*strab-il-iz'-mus*). See *strabismus*.
**strabism** (*strab'-izm*). See *strabismus*.
**strabismal**, **strabismic** (*strab-iz'-mal*, *strab-iz'-mik*) [*strabismus*]. Relating to strabismus.
**strabismometer** (*strab-iz-mom'-et-er*). See *strabometer*.
**strabismometry** (*strab-iz-mom'-et-re*) [*strabismus*; μέτρον, measure]. The measurement of the degree of strabismus.
**strabismus** (*strab-iz'-mus*) [στραβισμός, from στραβός, crooked]. Squint; that abnormality of the eyes in which the visual axes do not meet at the desired objective point, in consequence of incoordinate action of the external ocular muscles. **s.**, **alternating**, one in which either eye fixes alternately. **s.**, **bilateral**, same as *s.*, *alternating*. **s.**, **concomitant**, one in which the squinting eye has full range of movement. **s.**, **convergent**, one in which the squinting eye is turned to the nasal side. **s.**, **divergent**, one in which the squinting eye is turned to the temporal side. **s.**, **external**, see *s.*, *divergent*. **s.**, **Hirschberg's test for**, a rough estimate of the amount of strabismus is made by observing the position of the corneal reflection of a candle-flame held one foot in front of the eye to be tested, the examiner placing his own eye near the candle and looking just over it. **s.**, **internal**, see *s.*, *convergent*. **s.**, **paralytic**, due to paralysis of one or more muscles. **s.**, **spastic**, due to a spastic contraction of an ocular muscle. **s.**, **sursumvergens**, one in which the visual axis is directed upwards.
**strabometer** (*strab-om'-et-er*) [*strabismus*; μέτρον, measure]. An instrument for the measurement of the deviation of the eyes in strabismus.
**strabometry** (*strab-om'-et-re*) [*strabismus*; μέτρον, measure]. The determination of the degree of ocular deviation in strabismus.
**straboscope** (*strab'-o-skop'-ik*) [*strabismus*; crooked; σκοπεῖν, to see]. Pertaining to the appearance of objects as seen by one with strabismus. **s. disc**, an instrument producing distortion of objects.
**strabotome** (*strab'-o-tōm*) [*strabismus*; τέμνειν, to cut]. A knife used in strabotomy.
**strabotomy** (*strab-ot'-o-me*) [see *strabotome*]. An operation for the correction of strabismus.
**Strachan's disease** (*strorn*) [William Henry Williams Strachan, English physician]. Pellagra.
**strain** (*strān*) [OF., *estraindre*, from L., *stringere*, to draw tight]. 1. Excessive stretching; overuse of a part. 2. The condition produced in a part by overuse or wrong use, as eyestrain. 3. To overexert; to

use to excess; to make violent efforts. 4. A sub-variety of any domestic animal, often locally called breed. 5. In pharmacy, to separate insoluble substances from the liquid in which they occur; to filter.
**strainer** (*stra'-ner*). In pharmacy, a sieve for filtration.
**strait** (*strāt*) [Fr., *étroit*, from *strictus*, drawn tight]. A narrow or constricted passage, as the inferior or superior *strait* of the pelvis. **s., inferior** (of the pelvis), see under *pelvis* (3), **s.-jacket**, a strong jacket placed on the insane or delirious to prevent injury to themselves or to others. **s., superior** (of the pelvis), see under *pelvis* (3). **s.-waistcoat**, see *s.-jacket*.
**stramonium** (*stra-mo'-ne-um*). The thorn-apple. The *stramonium* of the U. S. P. is the dried leaves of *Datura stramonium*, Jamestown weed or jimson-weed, a plant of the order *Solanaceæ*. It contains two alkaloids, *daturine*, identical with atropine, and *hyoscyamine*. The action of stramonium resembles that of belladonna. It is used in asthma, dysmenorrhea, neuralgia, rheumatism, and pains of syphilitic origin. In asthma the leaves may be smoked in a tobacco pipe. Locally stramonium is employed as an ointment or cataplasm in irritable ulcers and inflamed surfaces. **s., extract of** (*extractum stramonii*, U. S. P.). Dose ½ gr. (0.01 Gm.). **s., fluidextract of** (*fluidextractum stramonii*, U. S. P.). Dose 1 min. (0.05 Cc.). **s. ointment** (*unguentum stramonii*, U. S. P.), an ointment made of stramonium extract, diluted alcohol, hydrous wool-fat, and benzoinated lard. **s., tincture of** (*tinctura stramonii*, U. S. P.). Dose 8 min. (0.5 Cc.).
**strangalesthesia, strangalæsthesia,** (*stran-gal-es-the'-ze-ah*). See *zonesthesia*.
**strangles** (*strang'-ls*) [στραγγάλη, a halter]. An infectious catarrh of the upper air-passages especially of the nasal cavity, of the horse, ass, and mule, associated with suppuration of the submaxillary and other lymphatic glands.
**strangling.** See *strangulation*.
**strangulated** (*strang'-gū-la-tĕd*). 1. Choked. 2. Compressed so that the circulation is arrested, as *strangulated* hernia.
**strangulation** (*strang-gū-la'-shun*) [*strangulare*, to choke]. 1. The act of choking. 2. Constriction of a part producing arrest of the circulation; as *strangulation* of a hernia.
**strangury** (*strang'-gū-re*) [στράγξ, a drop; οὖρον, urine]. Painful urination, the urine being voided drop by drop.
**strap.** 1. A long band, as of adhesive plaster. 2. To compress a part by means of bands, especially bands of adhesive plaster.
**Strasburger's cell-plate** [Edward *Strasburger*, German histologist, 1844– ]. The equatorial plate in which division of the nucleus occurs during katyokinesis.
**Strassburg's test for bile-acids** (*strahs'-boorg*) [Gustav Adolf *Strassburg*, German physiologist, 1848– ]. Dip filter-paper into urine to which cane-sugar has been added; dry it, and apply a drop of sulphuric acid. In the presence of bile-acids a red coloration will be shown on the paper. For this test the liquid must be free from albumin.
**stratification** (*strat-e-fik-a'-shun*) [*stratum; facere*, to make]. Arrangement in layers.
**stratified** (*strat'-e-fīd*) [see *stratification*]. Arranged in layers.
**stratiform** (*strat'-e-form*) [*stratum; forma*, form]. Formed into a layer. **s. fibrocartilage**, fibrocartilage lining bony grooves through which the tendons of muscles pass.
**stratum** (*stra'-tum*) [L., from *sternere*, to strew]. A layer. **s. albocinereum**, the alternate white and gray matter of the corpus striatum. **s. bacillatum, s. bacillosum, s. bacillorum**, the bacillary layer, the layer of rods and cones of the retina. **s. choriocapillare**, see *tunica ruyschiana*. **s. cinereum**, the most superficial layer of the cortex of the cerebellum, also of the anterior lobes of the corpora quadrigemina and of the floor of the fourth ventricle. **s. corneum, s. granulosum, s. lucidum, s. Malpighii**, see under *skin*. **s. corticale**, see *cortex* (3). **s. cutaneum**, the outer dermic layer of the tympanic membrane. **s. cylindrorum**, the bacillary layer of the retina. **s. episclerata**, the part of Tenon's capsule on the sclerotic coat. **s. epitrichiale**, see *epitrichium*. **s. gelatinosum**, the fourth layer in the olfactory bulb, composed of large ganglion-cells with branched processes. Syn., **ganglion-cell layer. s. glomerulorum**, the layer of the olfactory lobe (the second from the ventral side) containing the olfactory glomerules. **s. granulosum**, a layer of minute cells or one of cells containing many granules. 1. The external granular layer of the retina. 2. Meynert's name for the layer of small, irregular cells composing the fourth stratum of the cortex in the five-stratum type. 3. The layer of the olfactory lobe lying between the medullary ring and the stratum gelatinosum. 4. The layer of the epidermis covering the rete mucosum. 5. A histological appearance in that portion of the dentin immediately underlying the enamel and cementum of a tooth. **s. griseum centrale**, see *entocinerea*. **s. lacunosum**, the inner portion of the fifth or outer layer of the hippocampus. **s. lucidum**, a translucent layer of the epidermis consisting of irregular transparent cells with traces of a nucleus. **s. moleculare, s. molecular. s. mucosum**, see *rete mucosum*. **s. nerveum** of Henle, the layers of the retina exclusive of the rods and cones. Syn., *Bruecke's tunica nervea*. **s. oriens**, the third layer, counting from within outward, of the hippocampus. **s. nucleare**, that part of the gray matter of the medulla forming the floor of the fourth ventricle. **s. proligerum**, the discus proligerus and cumulus proligerus regarded as one. Syn., *membrana cumuli*. **s. reticulatum**, Arnold's, the network formed by the fibers connecting the occipital lobe with the thalamus before they enter the latter. **s. spinosum**, see *prickle-layer*. Consecutive hypertrophy of this layer constitutes acanthosis. **s. vasculosum**, see *tunica vasculosa*. **s. vasculosum cutis**, the subpapillary layer of the derma; the part of the corium immediately below the papillæ. **s. zonale**, the superficial portion of the fifth or outer layer of the hippocampus.

**Straus' sign** (*strows*) [Isidore *Straus*, French physician, 1845–1906]. In facial paralysis from a central cause the hypodermatic injection of pilocarpine causes no appreciable difference in the perspiration of the two sides, either as to time or quantity, whereas there is a marked retardation of the secretion on the affected side in severe peripheral paralysis. **s.'s réaction**. The injection of material containing the bacillus of glanders into the abdominal cavity of a male guineapig is followed in a few days by a characteristic, generally purulent, inflammation of the testes.

**Strauss's sign** (*strows*) [Hermann *Strauss*, German physician, 1868– ]. The administration of fatty food by the mouth causes an increase in the amount of fatty constituents in the effusion of chylous ascites.
**strawberry-marks.** Same as mother's-marks, *q. v.*
**strawberry-tongue.** The characteristic tongue of scarlatina, in which the vessels of the fungiform papillæ become turgid, causing the papillæ to stand out as red points, in marked contrast with the thick coating of fur on the filiform papillæ.
**streak** (*strēk*). A furrow, line, or stripe. **s., cultūre**, a bacterial culture in streaks. **s.s, Knapp's angioid**, pigment streaks appearing occasionally in the retina after hemorrhage. **s'., medullary**, see *medullary groove*. **s., meningitic**, see *tache cérébrale*. **s., primitive**, an opaque band extending some distance forward from the posterior margin of the area pellucida, and forming the first noticeable sign of the development of the blastoderm. **s., reflex**, a shining white streak running along the center of the vessels in the retina. It is due to the reflection of the light from the anterior surface of the column of blood.
**stream** (*strēm*) [ME., *streem*]. To flow; applied to movement in protoplasm and in blood corpuscles.
**stremma** (*strem'-ah*) [στρέμμα, a sprain twist]. A sprain.
**strengthening plaster.** Emplastrum roborans or iron-plaster.
**strephotome** (*stref'-o-tōm*) [στρέφειν, to twist; τέμνειν, to cut]. An instrument shaped like a corkscrew, formerly used to secure union in the operation for the radical cure of hernia.
**strepitus** (*strep'-it-us*) [L., noise]. A sound, a noise. **s. aurium**, see *tinnitus aurium*. **s. coriaceous**, auscultatory sound resembling the creaking of leather. **s. uteri**, see *souffle, uterine*. **s. uterinus**, the *uterine bruit*, *q. v.*
**strepto-** (*strep-to-*) [στρεπτός, twisted]. A prefix signifying twisted.
**streptoangina** (*strep-to-an'-jin-ah*). A pseudomembranous deposit in the throat due to streptococci (J. E. Walsh). Cf. *diphtheroid* (2).

**streptobacillus** (*strep-to-bas-il'-us*). A bacillus forming twisted chains.

**streptobacteria** (*strep-to-bak-tē'-re-ah*) [*strepto-;* βακτήριον, bacterium]. Short, rod-shaped bacteria forming chains.

**streptococcal, streptococcic, streptococcous** (*strep-to-kok'-al, -ik, -us*). Relating to or due to streptococci.

**streptococcemia, streptococcæmia** (*strep-to-kok-se'-me-ah*) [*streptococcus;* αἷμα, blood]. The presence of streptococci in the blood.

**streptococcolysin, streptocolysin** (*strep-to-kok-ol'-is-in, strep-to-kol'-is-in*) [*streptococcus;* λύειν, to loosen]. A hemolysin produced in cultures of streptococci.

**Streptococcus** (*strep-to-kok'-us*) [*strepto-;* κόκκος, a kernel]. A genus of schizomycetes of which the cocci are arranged in strings. See *micrococci, table of.* **s. angina,** angina due to streptococci. **s.-curve,** the remitting temperature-curve in hectic fever, supposed to depend upon the streptococcus (Petruschky).

**streptocolysin** (*strep-to-ko-li'-is-in*) [*streptococcus;* λύειν, to loose]. A hemolysin produced in cultures of streptococci.

**streptocosis** (*strep-to-ko'-sis*). Infection by streptococci.

**streptocyte** (*strep'-to-sīt*) [*strepto-;* κύτος, cell]. A cell presenting a twisted appearance or occurring with others in twisted chains; a streptococcus.

**streptomycosis** (*strep-to-mi-ko'-sis*) [*streptococcus;* μύκης, fungus]. Infection with streptococci.

**Streptopus** (*strep'-to-pus*) [*strepto-;* πούς, foot]. Twisted stalk, a genus of liliaceous plants. *S. distortus* is indigenous to Europe and America, and is used in infusion as a gargle.

**streptosepticemia, streptosepticæmia** (*strep-to-sep-tis-e'-me-ah*). Septicemia due to invasion of streptococci.

**streptothrical** (*strep-to-thrik'-al*). Relating to or due to members of the genus *Streptothrix*.

**streptothricosis** (*strep-to-thrik-o'-sis*). Infection with streptothrix.

**Streptothrix** (*strep'-to-thriks*). A class of schizomycetes.

**stretch.** To draw out to full length. **s.-walk,** a position in walking for physical development, with the arms stretched upward.

**stretcher** (*stretsh'-er*). A cot litter for carrying the sick.

**stria** (*stri'-ah*) [L., a streak: *pl., striæ*]. **s. cornea,** a narrow white streak interpolated between the thalamus and the caudatum on the ventricular floor. A streak or white line. **s. medullaris,** a band of white matter adjacent to the tænia thalami (Barker). **s. medullaris thalami,** an oblique furrow on the superior aspect of the thalamus. **s., pineal,** the habena or habenula. **s. terminalis,** tænia semicircularis. **s. vascularis,** the vascular upper part of the spiral ligament of the scala media.

**striæ** (*stri'-e*) [L., *pl.* of *stria*]. **s. acusticæ** transverse white lines on the lower part of the floor of the fourth ventricle, which unite with the auditory nerve-roots. **s. atrophicæ,** whitish, cicatricial lines of the skin caused by the contractions of skin that have been stretched by fat, pregnancy, etc. **s. gravidarum,** the atrophic striæ observed upon the abdomen in pregnant women. **s. longitudinales,** long, slightly elevated lines on the upper surface of the corpus callosum. **s. medullares,** see *s. acusticæ.* **s. musculares,** the transverse markings of striated muscles. **s. Schreger's,** Schreger's lines, *q. v.*

**striate, striated** (*stri'-āt, stri'-ā-ted*) [*stria*]. Striped. **s. body,** the corpus striatum. **s. muscle,** see under *muscle,* and *muscular tissue.*

**striation** (*stri-a'-shun*) [*stria*]. 1. The state of being striated. 2. A striated structure. **s., tabby-cat,** see *tabby-cat striation.*

**striatum.** 1. See *stratum.* 2. The corpus striatum. **s. oriens,** see *stratum oriens.*

**stricture** (*strik'-tūr*) [*strictura,* from *stringere,* to draw tight]. A narrowing of a canal from external pressure, or as a result of inflammatory or other changes in its walls. **s., annular,** a ring-like obstruction produced by a fold of mucous membrane and constriction all around the urethra, gut, etc. **s., bridle,** a fold of mucous membrane forming a crescentic obstruction, or perforated in its center; called, also, *s., linear, s., pack-thread, s., valvular,* and *s., hour-glass,* according to the peculiar appearances. **s., cicatricial,** a stricture due to cicatricial tissue. **s., congestive,** a temporary obstruction of the urethra from subacute prostatitis or other passing inflammation. **s.-cutter,** an instrument for dividing a stricture! **s.-fever,** the constitutional disturbances sometimes the result of acute stricture. **s., functional,** see *s., spasmodic.* **s., impermeable,** or **s., impassable,** one not permitting the passage of a bougie or catheter. **s., irregular,** or **s., tortuous,** so named from the complications or peculiarities. **s., irritable,** one in which the passage of the instrument causes great pain. **s., organic,** narrowing of a canal due to tissue-change, to deposits, or to pressure from without. **s., permeable** or **passable,** one permitting the passage of an instrument. **s., recurrent** or **contractile,** one in which the constriction returns after dilatation. **s., simple,** one that produces no interruption of function, pain, etc. **s., spasmodic,** a stricture due to muscular spasm and not to organic change.

**stricturotome** (*strik'-tū-ro-tōm*) [*stricture; τέμνειν,* to cut]. An instrument for dividing a stricture.

**stricturotomy** (*strik-tū-rot'-o-me*) [see *stricturotome*]. The operation of incising a stricture.

**stridor** (*stri'-dor*) [*stridere,* to make a creaking sound]. A peculiar, harsh, vibrating sound produced during expiration. **s. dentium,** grinding of the teeth. **s., inspiratory,** the sound heard in inspiration through a spasmodically closed glottis. **s., laryngeal,** stridor due to laryngeal stenosis. **s., laryngeal, congenital.** Respiratory croaking (in babies). **s. serraticus,** a sound like that of sharpening a saw, sometimes produced by expiration through a tracheotomy-tube.

**stridulous** (*strid'-ū-lus*) [*stridor*]. Characterized by stridor. **s. laryngismus,** see *laryngismus stridulus.*

**stringent** (*strin'-jent*) [*stringere,* to bind]. Binding.

**stringo-galvanometer.** Same as electrocardiograph, *q. v.*

**stringhalt** (*string'-hawlt*) [a corruption of *springhalt*]. A popular name for a nervous affection manifested in involuntary, convulsive movements of one or both hind legs of a horse. See *springhalt.*

**striocellular** (*stri-o-sel'-ū-lar*). Relating to or composed of striated muscle-fiber and cells.

**stripe** (*strīp*) [ME., *stripe,* a stripe]. A streak; a discolored mark.

**stripping** (*strip'-ing*) [ME., *stripen,* to rob]. Uncovering; unsheathing. In the plural, the last and richest milk given at any one milking; so called because it is slowly removed by the milker, who strips the teats between the fingers. **s. of the pleura,** removal of the lining membrane of the thorax of an animal used for food, to remove the traces of pleurisy and of tuberculosis.

**strobic** (*strob'-ik*) [στρόβος, a top]. Resembling or pertaining to a top. **s. discs,** discs drawn with concentric circles, so as to produce an illusory impression as if they were revolving.

**strobila** (*strob-i'-lah*) [στρόβιλος, a pine-cone: *pl., strobilæ*]. 1. A form of development occurring in the *cnidaria* and *cestoda,* in which the products of asexual generation by a sort of fission remain attached to the proliferating organism or to each other. 2. A sexual form of strobilation. **s., monodisc,** the simplest form of strobilation, in which the disc (*scyphistoma*) separates from its peduncle. **s., polydisc,** that form in which successively formed discs remain attached.

**strobilation** (*strob-il-a'-shun*) [στρόβιλος, a pine-cone]. The formation of zooids, discs or joints by metameric division, gemmation, or fission.

**strobile** (*strob'-il*) [στρόβιλος, a pine-cone]. 1. A multiple fruit in which the seeds are enclosed by prominent scales, as a pine-cone. 2. A strobila.

**strobiloid** (*strob'-bil-oid*) [στρόβιλος, a pine-cone; εἶδος, like]. Like a strobile.

**strobilus** (*stro-bi'-lus*) [*strobilus,* a pine-cone]. The tape-worm.

**stroboscope** (*stro'-bo-skōp*) [στρόβος, a twisting; σκοπεῖν, to view]. An instrument by which a series of slightly different pictures presented rapidly in succession is made to appear as a continuous object in motion; a zoetrope. **s., direct,** one in which the figures and slits revolve in the same direction; a phenakistoscope. **s., reverse,** one in which the figures and slits revolve in opposite directions; a dedalum.

**stroboscopic** (*strob-o-skop'-ik*) [*stroboscope*]. Pertaining to the stroboscope.

**strobostereoscope** (*stro-bo-ster'-e-o-skōp*). See *stereostroboscope.*

**stroke** (*strōk*). 1. In pathology, a sudden and severe seizure or fit of disease. 2. A popular term for

**apoplexy.** 3. To pass the hands gently over the body. **s., apoplectic,** see *apoplexy.* **s., back,** of the heart, the supposed "reaction-impulse," or recoil of the ventricles at the moment the blood is discharged into the aorta. **s., heat,** see *hyperpyrexia.* **s., paralytic,** sudden loss of muscular power from lesion of the brain or spinal cord.
**stroma** (*stro'-mah*) [στρῶμα, a bed]. The tissue forming the framework for the essential part of an organ. **s., cancer,** the fibrous-tissue element of a cancer.
**stroma fibrin.** Landois' term for fibrin formed directly from stroma, as distinguished from plasma-fibrin or that formed in the usual way. **s.-plexus,** a plexus of axis-cylinders formed by the corneal nerves.
**stromatic** (*stro-mat'-ik*). Resembling a stroma.
**stromatolysis** (*stro-mat-ol'-is-is*) [*stroma*; λύειν, to loosen]. A dissolution of the stroma or surrounding membrane of a cell, without the cell body being affected.
**strombodes jenneri** (*strom-bo'-dēs jen'-er-i*). A name proposed by Sjöbring for the microorganism of vaccinia.
**Stromeyer's cephalhematocele** (*stro'-mi-er*) [Georg Friedrich Louis *Stromeyer*, German surgeon, 1804–1876]. Subperiosteal cephalhematoma communicating with veins and becoming tensely filled during strong expiratory efforts. **S.'s splint,** one used to prevent stiffness of the joints in case of fracture. It consists of two hinged parts that can be fixed at any angle.
**stromuhr** (*stro'-moor*) [Ger.]. An instrument for measuring the velocity of blood-flow.
**strongylosis** (*stron-jil-o'-sis*). Infection with worms of the genus *Strongylus.*
**Strongylus** (*stron'-jil-us*) [στρογγύλος, round]. A genus of nematode worms found in the lower animals, and occasionally in man.
**strontia** (*stron'-she-ah*). Strontium oxide.
**strontium** (*stron'-she-um*) [*Strontian*, a town in Scotland]. A metallic element belonging to the group of alkaline earths. It has a specific gravity of 2.5, an atomic weight of 87.63, and a valence of two. Symbol Sr. **s. acetate,** $2Sr(C_2H_3O_2) + H_2O$, a white crystalline powder, soluble in water, used as an anthelmintic. Dose ¼–⅔ gr. (0.016–0.05 Gm.). **s. arsenite,** Sr $(AsO_2)_2 + 4H_2O$, a white powder, soluble in water, used as an alterative and tonic in skin diseases and malarial conditions. Dose $\frac{1}{60}-\frac{1}{15}$ gr. (0.002–0.004 Gm.). **s. bromide** (*strontii bromidum,* U. S. P.), $SrBr_2.6H_2O$, has been used in epilepsy, diabetes, gastrectasis, rheumatoid arthritis, and lithemia. Dose 15–30 gr. (1–2 Gm.). **s. and caffeine sulphonate,** $(C_8H_9N_4O_2.SO_3)_2Sr$, soluble in water, used as a diuretic. Syn., *symphorol, strontium.* **s. glycerinophosphate,** a white powder, soluble in water, containing 26–27% of phosphoric acid. **s. iodide** (*strontii iodidum,* U. S. P.), $SrI_2.6H_2O$, is used like the other oxides. **s. lactate,** $Sr(C_3H_5O_3)_2.3H_2O$, is used in nephritis, albuminuria, rheumatism, and gout. Dose 10–30 gr. (0.65–2.0 Gm.). **s. loretinate** (basic), Sr. I.-O.C₉H₄N.SO₃, fine, bright needles, slightly soluble in water, decomposed at 300° C. **s. loretinate** (normal), Sr(I.OH.C₉H₄N.SO₃)₂.H₂O, orange-red, prismatic crystals, sparingly soluble in water. **s. oxide,** strontia, SrO, strontium combined with oxygen. **s. phosphate,** $Sr_3(PO_4)_2$, a white powder, devoid of taste, soluble in acids; used as a nutritive and tonic. Dose 10–30 gr. (0.65–2.0 Gm.). **s. salicylate** (*strontii salicylas,* U. S. P.), Sr(C₇H₅O₃)₂.2H₂O, white crystals, soluble in water; used in gout, chorea, etc. Dose 10–40 gr. (0.65–2.6 Gm.).
**strophanthin** (*strof-an'-thin*) [*strophanthus*] *Strophanthinum* (U. S. P.), $C_{30}H_{40}O_{10}$. A toxic glucoside, soluble in water and alcohol, derived from strophanthus. Dose $\frac{1}{360}-\frac{1}{240}$ gr. (0.0002–0.0003 Gm.). **s.** represents a yellowish, amorphous powder containing 59% of strophanthin, soluble in water; used as a heart tonic. Dose $\frac{1}{60}-\frac{1}{30}$ gr. (0.0004–0.001) Gm.
**Strophanthus** (*strof-an'-thus*) [στρόφος, a twisted band; ἄνθος, flower]. A genus of plants of the order *Apocynaceæ,* some of the species of which are used for the preparation of arrow-poison in Africa. The strophanthus of the U. S. P. is the ripe seed of *S. kombé*; it contains a crystalline glucoside, *strophanthin,* and an alkaloid, *ineine.* Strophanthus is a muscle-poison, but in small doses is a cardiac and perhaps a vascular stimulant. It is used in the same cases as digitalis. **s., tincture of** (*tinctura strophanthi,* U. S. P., B. P.). Dose 5–15 min. (0.32–1.0 Cc.).

**strophantism** (*stro-fan'-tizm*). Poisoning from strophanthin; the symptoms resemble those of digitalism.
**strophium** (*strof'-e-um*) [στρόφος, a cord]. A bandage.
**strophocephalus** (*strof-o-sef'-al-us*) [στρόφος, twisted; κεφαλή, head]. A monster having displacement of the parts forming the head and face.
**strophocephaly** (*strof-o-sef'-al-e*) [*strophocephalus*]. Distortion of the head; the condition of having a distorted head.
**strophulus** (*strof'-ū-lus*) [στρόφος, a twisted band]. A form of miliaria occurring in infants. Syn., *red gum; tooth-rash.* **s. albidus,** same as *milium.* **s. confertus,** see *s. intertinctus.* **s. infantum,** an urticarial disease of infants. **s. intertinctus,** a popular dermatitis of more or less acute form, a variety of eczema common in infants. **s. puriginosus,** an eruption occurring in children, and characterized by disseminated, intensely itching papules. **s. volaticus,** an acute skin disease, a typical erythema papulatum, characterized by slight maculæ.
**structural** (*struk'-tū-ral*) [*structura,* structure]. Pertaining to or affecting the structure.
**structure** (*struk'-tūr*) [*structura; struere,* to build]. The manner or method of the building up, arrangement, and formation of the different tissues and organs of the body or of a complete organism. Also, an organ, a part, or a complete organic body.
**Struempell's disease** (*strūm'-pel*) [Adolf von *Struempell,* German physician, 1853– ]. 1. Poliencephalitis. 2. Chronic ankylosing inflammation of the vertebral column. **S.'s type of spastic paralysis,** the hereditary, familiar form of spastic spinal paralysis.
**Struempell-Leichtenstern's disease** (*strum'-pel-lik'-ten-stern*). Acute encephalitis of infancy.
**struma** (*stroo'-mah*) [L.]. 1. Scrofula. 2. Goiter. **s. aberrate,** a goiter of an accessory thyroid gland. **s. maligna,** carcinoma of the thyroid gland. **s. suprarenalis,** a peculiar fatty tumor of the suprarenal bodies. **strumæ lipomatodes aberratæ renis,** Grawitz's term for a group of new-growths of kidney, usually benign, but, at times, serving, as the foci of origin of malignant tumors. Regarded by some pathologists as endotheliomata, by others as adrenal rests (cf. under *rest*).
**strumectomy** (*stroo-mek'-to-me*) [*struma;* ἐκτομή, excision]. Excision of an enlarged or strumous gland, or of a goiter.
**strumiform** (*stroo'-mif-orm*) [*struma; forma,* form]. Having the appearance of struma.
**strumiprival, strumiprivous** (*stroo-mi-pri'-val, -vus*) [*struma; privare,* to deprive]. Deprived of the thyroid; due to removal of the thyroid; thyroprival.
**strumitis** (*stroo-mi'-tis*) [*struma; ῖτις,* inflammation]. Inflammation of a goitrous thyroid gland.
**strumoderma, strumoderm** (*stroo-mo-der'-mah, stroo'-mo-derm*). See *scrofuloderma.*
**strumose** (*stroo'-mōs*) [*struma*]. Swollen on one side; possessing a wen-like protuberance.
**strumosis** (*stroo-mo'-sis*). See *strumositas.*
**strumositas** (*stroo-mos'-it-as*) [*struma*]. The tendency toward, or diathesis of, goiter or of scrofula.
**strumous** (*stroo'-mus*) [*struma*]. 1 Scrofulous. 2. Goitrous.
**Struve's test for blood in urine** (*stroo'-veh*) [Heinrich *Struve,* German physician]. To the urine, previously treated with ammonia or caustic potash, add tannin and acetic acid until the mixture has an acid reaction. In the presence of blood a dark precipitate is formed. When this is filtered and dried, the hemin crystals may be obtained from the dry residue by adding ammonium chloride and glacial acetic acid.
**strychnia** (*strik'-ne-ah*). Same as *strychnine.*
[see *strychnos*]. *Strychnina* (U. S. P.), $C_{21}H_{22}N_2O_2$. Dose $\frac{1}{60}-\frac{1}{12}$ gr. (0.001–0.0033 Gm.); hypodermatically in chronic alcoholism, $\frac{1}{15}-\frac{1}{30}$ gr. (0.003–0.006 Gm.). **s. acetate,** $C_{21}H_{22}N_2O_2.C_2H_4O_2$, small white crystals, soluble in 96 parts of water. Use and doses the same as the alkaloid. **s. arsenate,** $(C_{21}H_{22}N_2O_2)_3.As_2O_5$, a white, crystalline powder with bitter taste, soluble in 14 parts of cold water, 5 parts of hot water. It is used as a tonic and alterative in tuberculosis, malaria, etc. Dose $\frac{1}{60}-\frac{1}{15}$ gr. (0.001–0.004 Gm.); hypodermatically, 0.5% in liquid paraffin; of this $\frac{1}{4}-15$ min. (0.25–0.9 Cc.) daily. **s. arsenite,** $(C_{21}H_{22}N_2O_2)_3As_2O_3$, a white crystalline powder, soluble in 10 parts of boiling water. It is tonic, alterative, and antiperiodic. Dose

# STRYCHNINISM 849 STYLOMAXILLARY

$\frac{1}{10}-\frac{1}{18}$ gr. (0.001–0.004 Gm.); subcutaneous dose 4–15 drops of a 0.5% solution in liquid paraffin. **s. bisaccharinate, s.**-diorthosulphamin-benzoate, used as is the arsenite. **s. camphorate,** $C_{21}H_{22}N_2O_2 \cdot C_{10}H_{16}O_4$, small white crystals or crystalline powder soluble in water, used as is the alkaloid. **s. citrate,** $C_{21}H_{22}N_2O_2 \cdot C_6H_8O_7$, white crystals, soluble in water; usage and dosage the same as the alkaloid. **s. ferricitrate,** iron and strychnine citrate. **s. hydride,** obtained by the action of metallic sodium on strychnine in a boiling alcoholic solution and differing in physiological action from strychnine, it may, therefore, be used as a physiological antidote in strychnine poisoning. **s. hydrobromide,** $C_{21}H_{22}N_2O_2 \cdot HBr$, white acicular crystals, soluble in 32 parts of water, used as a tonic and sedative. Dose $\frac{1}{32}-\frac{1}{12}$ gr. (0.002–0.005 Gm.). **s. hydrochloride,** $C_{21}H_{22}N_2O_2 \cdot HCl + 3H_2O$, white needles, soluble in 50 parts of water at 22°C. Usage and doses the same as the alkaloid. **s. hydroiodide,** $C_{21}H_{22}N_2O_2 \cdot HIO_3$, white crystals, soluble in water. Used as is the alkaloid. **s. hypophosphite,** a white powder, used as a tonic in tuberculosis. Dose $\frac{1}{32}-\frac{1}{12}$ gr. (0.002–0.005 Gm.). **s. with iron and quinine citrate,** iron and quinine citrate with strychnine; greenish-brown, transparent scales, soluble in water, and containing 3.4% of pure strychnine. It is tonic and antiperiodic. Dose 3–7 gr. (0.2–0.45 Gm.). **s. lactate,** $C_{21}H_{22}N_2O_2 \cdot C_3H_6O_3$, a white, crystalline powder, soluble in water. Usage and dosage the same as the alkaloid. **s. nitrate** (*strychninæ nitras*, U. S. P.), $C_{21}H_{22}N_2O_2 \cdot NHO_3$, silky needles, soluble in 50 parts of water, 60 parts of alcohol, or 2 parts of boiling water or alcohol. Usage and dosage the same as the alkaloid. **s. oleate,** a mixture of strychnine in oleic acid, soluble in ether and oleic acid; it is used in the external administration of strychnine. **s.-orthosulphamin-benzoate,** see *s. saccharinate*. **s. phenolsulphate,** see *s. sulphocarbolate*. **s. phosphate,** $(C_{21}H_{22}N_2O_2)_3H_3PO_4 + 9H_2O$, a white, crystalline powder, soluble in water. Usage and dosage the same as the alkaloid. **s. saccharinate,** $C_{21}H_{22}N_2O_2 \cdot C_6H_4(SO_2)(CO)NH$, a true salt of strychnine and saccharin; a white sweet powder, used in all cases where the alkaloid is indicated in doses one-third larger. **s. salicylate,** $C_{21}H_{22}N_2O_2 \cdot C_7H_6O_3$, a white powder, soluble in water, recommended in rheumatism and chorea. Dose about the same as the alkaloid. **s. sulphate** (*strychninæ sulphas*, U. S. P.), $(C_{21}H_{22}N_2O_2)_2H_2SO_4 + 5H_2O$, white, odorless, very bitter prisms, which effloresce in dry air; soluble in 50 parts of water, 109 parts of alcohol, 2 parts of boiling water, or 8.5 parts of boiling alcohol; melt at 200°C. The action differs but slightly from the alkaloid. Dose $\frac{1}{32}-\frac{1}{12}$ gr. (0.002–0.005 Gm.). **s. sulphocarbolate, s. sulphophenate,** a white crystalline powder, soluble in water or alcohol. **s. and zinc hydroiodide,** $C_{21}H_{22}N_2O_2 \cdot HI \cdot ZnI_2$, small white crystals, soluble in water.
**strychninism** (*strik'-nin-ism*) [*strychnine*]. The state of being under the influence of strychnine.
**strychninization** (*strik-nin-is-a'-shun*) [*strychnine*]. The condition produced by large doses of strychnine or nux vomica.
**strychninomania** (*strik-nin-o-ma'-ne-ah*) [*strychnine;* μανία, madness]. Delirium from the use of strychnine or nux vomica.
**strychnism** (*strik'-nism*). Same as *strychninism*.
**strychninize, strychnize** (*strik'-nin-īz,' strik'-nīz*). To bring under the influence of strychnine.
**Strychnos** (*strik'-nos*) [στρύχνος, the nightshade]. A genus of the *Loganiaceæ*. *S. icaja* is found in the Gaboon region; it contains strychnine in the bark, leaves and root. From the stem the arrow-poison, tarfa, toomba, M'boundou, n'caza, icaja, or akanga, is prepared. The seeds of *S. ignatii* (St. Ignatius' beans) of the Philippines act in the same manner as nux vomica, but contain more strychnine and less brucine than it does. *S. potatorum* is indigenous to the East Indies. The seeds, nirmali, chilliji, chilbing, are used largely to clear muddy water. They contain no strychnine or brucine, and are used as a remedy in diabetes and gonorrhea. The fruit is employed in dysentery. The bark of *S. pseudoquina*, of South America, contains no poisonous alkaloid, but a bitter substance, and is used as a substitute for quinine. *S. tieuté* is a species of Java; from the root-bark the Javanese arrow-poison, upas radju or tachetsik, containing 1.5% strychnine and a little brucine, is prepared. The seed and leaves contain 1.4% of strychnine and only traces of brucine. *S. toxifera,* of Guiana, furnishes curara.

**stub-thumb.** Abbreviation and clubbing of the phalanx of the thumb.
**student's placenta.** A retained placenta from improper manipulation.
**Stuetz's test** (*stūts*). See *Fuerbringer's test*.
**stultitia** (*stul-tish'-e-ah*) [*stultus*, a fool]. Foolishness; dulness of intellect.
**stump.** The extremity, pedicle, or basis of the part left after surgical amputation, excision, or ablation. **s. of eyeball,** the remainder of the globe after excision of an anterior staphyloma or after other capital operation on the globe that deprives it of vision. **s.-foot,** synonym of *club-foot*. **s., sugar-loaf,** a conical stump due to undue retraction of the muscles; called, also, *conical stump*. **s. of tooth,** that part remaining after removal or destruction of the corona.
**stun** [AS., *stunian*, to make a din]. To render temporarily insensible, as by a blow.
**stupe** (*stūp*) [*stupa*, tow]. A cloth used for applying heat or counterirritation; especially a cloth wrung out of hot water and sprinkled with a counter-irritant as *turpentine-stupe*.
**stupefacient, stupefactive** (*stū-pe-fa'-shent, stū-pe-fak'-tiv*) [*stupor; facere,* to make]. Narcotic.
**stupefaction** (*stū-pe-fak'-shun*) [see *stupefacient*]. Stupor, and the process of reaching it.
**stupemania** (*stū-pe-ma'-ne-ah*) [*stupor; mania*]. Mental stupor with insanity.
**stupor** (*stū'-por*) [L.]. The condition of being but partly conscious or sensible. Also a condition of insensibility. **s., anergic, acute dementia. s., delusional,** melancholic dulness of mind, with delusions; it is sometimes a kind of auto-hypnotism. **s., epileptic, s., post-convulsive,** the stupor following an epileptic convulsion. **s. formicans,** formication. **s., lethargic,** see *trance*. **ø. melancholicus,** the stupor found in association with melancholia. **s. miliaris,** paresthesia of the fingers and toes in connection with miliary fever. **s. vigilans,** catalepsy.
**stuporous** (*stū'-por-us*) [*stupor*]. In a condition of, or attended with stupor. **s., insanity,** see *insanity, confusional*.
**stupration, stuprum** (*stū-pra'-shun, stū'-prum*) [*stuprum*, defilement]. Rape.
**sturdy** (*stur'-de*). 1. Vigorous; hardy. 2. See *gid* and *staggers*.
**sturin** (*stu'-rin*) [*sturio*, sturgeon]. A protamine obtained from the sperm of the sturgeon.
**Sturm's focal interval** (*stoorm*) [Johann Christoph Sturm, 1635–1703]. The interval between the principal focal lines of a cylindrical lens.
**stutter** (*stut'-er*) [Ger., *stottern*, to stutter]. To hesitate or make repeated efforts to articulate a syllable. Stuttering is a variety of stammering; see *stammering*.
**s.-spasm,** see *lalophobia*.
**stuttering** (*stut'-er-ing*) [Icel., *stauta*, to stutter]. A hesitation in speech due to an inability to enunciate the syllables without repeated efforts.
**sty, stye** (*sti*). See *hordeolum*. **s., Meibomian,** abscess of a Meibomian gland. **s., Zeissian,** abscess of one of Zeiss's glands.
**style, stylet** (*stīl, stī-let'*) [στυλός, pillar]. 1. A probe. 2. A wire inserted into a catheter or cannula in order to stiffen the instrument or to perforate the tissues.
**styliform** (*sti'-lif-orm*) [*stilus*, stake; *forma*, form]. Shaped like a style.
**stylo-** (*sti-lo-*) [στυλός, pillar]. A prefix denoting pertaining to the styloid process of the temporal bone.
**styliscus** (*sti-lis'-kus*) [στυλός, a pillar: *pl.*, *stylisci*], 1. A slender cylindrical talent. 2. In biology, the passage leading from the stigma to the ovary through the style.
**styloglossal** (*sti-lo-glos'-al*) [*stylo-; glossal*]. Connected with or relating to the styloid process of the temporal bone and the tongue.
**styloglossus** (*sti-lo-glos'-us*). See under *muscle*.
**stylohyal** (*sti-lo-hi'-al*) [*styloe; hyoid*]. One of the bones of the hyoid arch of vertebrates.
**stylohyoid** (*sti lo hi' oid*) [*stylo ; hyoid*], 1. Pertaining to the styloid process of the temporal bone and the hyoid bone, as the *stylohyoid muscle*. See under *muscle*. 2. Pertaining to the stylohyoid muscle.
**styloid** (*sti'-loid*) [στυλός, pillar; εἶδος, like]. Resembling a stylus.
**stylomastoid** (*sti-lo-mas'-toid*) [*stylo-; mastoid*]. Pertaining to the styloid and mastoid processes.
**stylomaxillary** (*sti-lo-maks'-il-a-re*) [*stylo-; maxilla*]. Pertaining to the styloid process and the maxilla.

**stylopharyngeus** (*sti-lo-făr-in-je'-us*). See under *muscle*.

**styloslaphyline** (*sti-lo-staf'-il-ĭn*) [*stylo-*; *staphyle*]. Connected with or relating to the styloid process of the temporal bone and the velum palati.

**stylosteophyte** (*sti-los'-te-o-fīt*) [*stylo-*; φύτον, plant]. A style-shaped exostosis.

**stylostixis** (*sti-lo-stiks'-is*). See *acupuncture*.

**stylus** (*sti'-lus*) [L., "a stake"]. A pointed instrument for making applications. A stylet.

**styma** (*sti'-mah*). See *priapism*.

**stymatosis** (*sti-mat-o'-sis*) [στύμα, stiffness]. A violent erection of the penis attended with hemorrhage.

**stype** (*stīp* or *sti-pe*) [στύπη, tow]. A tampon or pledget, especially such as is used in producing local anesthesia.

**styphage, stypage** (*ste-fahzj*, *ste-pahzj*). The production of local anesthesia by an application made with a stype. s., Bailly's, a revulsive, the application of cotton pledgets wet with methyl chloride.

**stypsis** (*stĭp'-sis*) [στυπτικός, astringent]. 1. Constipation. 2. The use of a styptic.

**styptase** (*stĭp'-tās*). Trade name of a styptic said to contain tannin, hamamelis, calcium chlorate and fluorides.

**styptic** (*stĭp'-tik*) [στυπτικός, astringent]. 1. Checking hemorrhage by contracting the blood-vessels. 2. An agent that checks hemorrhage by causing contraction of the blood-vessels.

**stypticin** (*stĭp'-tis-in*). See *cotarnine hydrochloride*.

**stypticity** (*stĭp-tis'-it-e*) [στυπτικός, astringent]. The quality of being styptic.

**styptol** (*stĭp'-tol*). Cotarnine phthalate; an internal styptic.

**styracin** (*sti'-ra-sin*). See under *styrax*.

**styracol** (*sti'-ra-kol*). Guaiacol cinnamate, C₆H₄(OCH₃)C₆H₇O₃. It is given internally in catarrhal affections of the digestive tract and in pulmonary tuberculosis.

**styrax** (*sti'-raks*) [στύραξ, storax]. Storax; a balsam obtained from the inner bark of *Liquidambar orientalis*, or oriental sweet-gum. It contains a volatile oil, styrol, several resins, an amorphous substance called storesin, cinnamic acid, and styracin (the cinnamate of cinnamyl). It is stimulant, expectorant, and antiseptic, acting like benzoin and tolu, and is used in bronchial affections and catarrh of the urinary passages. Externally it is an antiseptic and parasiticide. It is a constituent of friars' balsam. Dose 5–20 gr. (0.32–1.3 Gm.).

**styrene** (*sti'-rēn*) [στύραξ, storax]. C₈H₁₀O. Styryl alcohol, cinnamyl-alcohol; a substance obtained by saponifying styracin, its cinnamic ester, with potassium hydroxide. It crystallizes in shining needles, is sparingly soluble in water, possesses a hyacinth-like odor, melts at 33° C., and distils at 250° C.

**styrol** (*sti'-rol*), C₈H₈. Cinnamene; phenylethylene, a colorless, highly refractive liquid hydrocarbon, obtained by heating styracin with calcium hydrate.

**styrolene** (*sti'-ro-lēn*). Same as *styrol*, q. v.

**styrone** (*sti'-rōn*) [*sturax*]. Cinnamic alcohol, C₉H₁₀O. s. crystals, s., crystallized, cinnamic alcohol.

**styryl alcohol** (*sti'-ril*). See *styrone*.

**sub-** [L.]. A prefix denoting under or beneath; in chemistry, a prefix denoting—(1) the lower of two compounds of the same elements; (2) a basic salt.

**subabdominal** (*sub-ab-dom'-in-al*) [*sub-*; *abdomen*]. Beneath the abdomen.

**subacetabular** (*sub-as-et-ab'-u-lar*). Below the acetabulum.

**subacetate** (*sub-as'-et-āt*) [*sub-*; *acetum*, vinegar]. A basic acetate.

**subacid** (*sub-as'-id*) [*sub-*; *acidum*, acid]. Moderately acid or sour.

**subacidity** (*sub-as-id'-it-e*) [*subacid*]. A condition of moderate acidity.

**subacromial** (*sub-ak-ro'-me-al*) [*sub-*; *acromial*]. Below the acromion.

**subacute** (*sub-ak-ūt*) [*sub-*; *acutus*, sharp]. 1. Moderately acute. 2. The stage of a disease when it is intermediate between an acute and a chronic form.

**subagitatrix** (*sub-aj-it-a'-triks*) [L.]. One who practises tribadism.

**subanal** (*sub-a'-nal*) [*sub-*; *anus*]. Situated below the anus.

**subancestral** (*sub-an-sest'-ral*). Not in the direct line of descent.

**subanconeal** (*sub-an-ko-ne'-al*)[*sub-*; *anconeus*]. Beneath the anconeus muscle.

**subanconeus** (*sub-an-ko-ne'-us*). See under *muscle*.

**subapical** (*sub-a'-pik-al*) [*sub-*; *apex*]. Beneath the apex.

**subaponeurotic** (*sub-ap-on-ū-rot'-ik*) [*sub-*; *aponeurosis*]. Beneath an aponeurosis.

**subaqueous** (*sub-a'-kwe-us*) [*sub-*; *aqua*, water]. Living beneath the water.

**subarachnoid** (*sub-ar-ak'-noid*) [*sub-*; *arachnoid*]. Beneath the arachnoid membrane, as the *subarachnoid* space.

**subarcuate** (*sub-ar'-kū-āt*) [*sub-*; *arcus*, an arc]. Slightly arcuate.

**subareolar** (*sub-ar-e'-o-lar*) [*sub-*; *areola*]. Situated, or occurring beneath the mammary areola.

**subastragalar, subastragaloid** (*sub-as-trag'-al-ar*, *sub-as-trag'-al-oid*) [*sub-*; *astragalus*]. Below the astragalus. s. amputation, a partial removal of the foot, in which only the astragalus is left.

**subastringent** (*sub-as-trin'-gent*) [*sub-*; *astrinjens*, astringent]. Only slightly astringent.

**subatloidean** (*sub-at-loid'-e-an*). Located beneath the atlas.

**subatomic** (*sub-at-om'-ik*). Underlying atoms.

**subaudition** (*sub-aw-dish'-un*) [*sub-*; *audire*, to hear]. The act or ability of comprehending what is not expressed.

**subaural** (*sub-aw'-ral*) [*sub-*; *aura*, ear]. Beneath the ear.

**subauricular** (*sub-aw-rik'-u-lar*). Below the auricle of the ear.

**subaxial** (*sub-aks'-e-al*) [*sub-*; *axis*]. Lying below the axis.

**subaxillary** (*sub-aks'-il-a-re*) [*sub-*; *axilla*]. Situated below the axilla.

**subbasal** (*sub-ba'-sal*) [*sub-*; *base*]. Situated below or near the base or basal membrane.

**subbrachial, subbrachiate** (*sub-bra'-ke-al*, *sub-bra'-ke-āt*) *sub-*; *βραχίον*, arm]. 1. Under the pectoral muscles. 2. Beneath the brachium.

**subbrachycephalic** (*sub-bra-ke-sef-al'-ik*). Having a cephalic index from 80° to 84°.

**subcalcareous** (*sub-kal-ka'-re-us*) [*sub-*; under; *calx*, lime]. Somewhat calcareous.

**subcalcarine** (*sub-kal'-ka-rēn*) [*sub-*; *calcarine*]. Situated beneath the calcarine fissure, as the *subcalcarine* convolution. s. convolution, a narrow convolution ventrad of the cuneus and lying between the collateral and calcarine fissures.

**subcallosal** (*sub-kal-lo'-sal*) [*sub-*; *callosum*]. Below the corpus callosum.

**subcapsular** (*sub-kap'-su-lar*) [*sub-*; *capsula*, capsule]. Beneath a capsule.

**subcarbonate** (*sub-kar'-bon-āt*) [*sub-*; *carbonate*]. A basic carbonate.

**subcartilaginous** (*sub-kar-til-aj'-in-us*) [*sub-*; *cartilago*, cartilage]. 1. Situated beneath cartilage. 2. Partly cartilaginous.

**subcecal** (*sub-se'-kal*) [*sub-*; *cecum*]. Lying below the cecum.

**subcentral** (*sub-sen'-tral*) [*sub-*; *center*]. 1. Situated near the center. 2. Ventrad of the central fissure of the brain.

**subcerebellar** (*sub-ser-e-bel'-ar*) [*sub-*; *cerebellum*]. Situated beneath the cerebellum.

**subcerebral** (*sub-ser-e'-bral*) [*sub-*; *cerebrum*]. Situated beneath the cerebrum.

**subchloride** (*sub-klor'-īd*). That chloride of a series which contains relatively the least chlorine.

**subchondral** (*sub-kon'-dral*) [*sub-*; χόνδρος, cartilage]. Lying beneath cartilage.

**subchordal** (*sub-kor'-dal*) [*sub-*; *chorda*, cord]. Beneath the notochord.

**subchorionic** (*sub-ko-re-on'-ik*) [*sub-*; *chorion*]. Lying beneath the chorion.

**subchoroidal** (*sub-ko-roid'-al*) [*sub-*; *choroidal*]. Situated or occurring under the choroid.

**subchronic** (*sub-kron'-ik*) [*sub-*; *chronic*]. More nearly chronic than is indicated by the term subacute.

**subclavian** (*sub-kla'-ve-an*) [*sub-*; *clavis*, key]. Lying under the clavicle, as the *subclavian* artery.

**subclavicula** (*sub-kla-vik'-ū-lah*). The first rib.

**subclavicular** (*sub-kla-vik'-ū-lar*). Beneath the clavicle.

**subclavius** (*sub-kla'-ve-us*). See under *muscle*.

**subcollateral** (*sub-kol-at'-er-al*) [*sub-*; *collateral*]. Ventrad of the collateral fissure of the brain. s. gyrus, a convolution connecting the occipital and temporal lobes.

**subconjunctival** (*sub-kon-jŭngk-ti'-val*) [*sub-*; *conjunctiva*]. Situated beneath the conjunctiva.

**subconscious** (*sŭb-kon'-shus*). Imperfectly conscious.

**subconsciousness** (*sub-kon'-shŭs-nes*) [*subconscious*]. Imperfect consciousness; that state in which mental processes take place without the mind being distinctly conscious of its own activity.

**subcontinuous** (*sub-kon-tin'-ū-us*) [*sub-*; *continuous*]. Almost continuous.

**subcoracoid** (*sub-kor'-ak-oid*) [*sub-*; *coracoid*]. Situated below the coracoid process.

**subcordate** (*sub-kor'-dāt*) [*sub-*; *cor*, heart]. Having nearly the shape of a heart.

**subcorneous** (*sub-kor'-ne-us*) [*sub-*; *corneus*, horny]. Somewhat horny.

**subcortex** (*sub-kor'-teks*). That part of the brain substance which immediately underlies the cortex.

**subcortical** (*sub-kor'-tik-al*) [*sub-*; *cortex*]. Beneath the cortex.

**subcostal** (*sub-kos'-tal*) [*sub-*; *costa*, rib]. Lying beneath a rib or the ribs.

**subcostales** (*sub-kos-ta'-lēz*). The infracostal muscles. See *muscles*.

**subcranial** (*sub-kra'-ne-al*) [*sub-*; κρανίον, cranium]. Situated beneath the cranium.

**subcrepitant** (*sub-krep'-it-ant*) [*sub-*; *crepitare*, to make a crackling noise]. Almost crepitant, as *subcrepitant* rale. See *rale*, *subcrepitant*.

**subcrepitation** (*sub-krep-it-a'-shun*). An indistinctly crepitant sound.

**subcrureus** (*sub-kroo-re'-us*). See under *muscle*.

**subculoyd** (*sub-kū'-loid*). A trade name to designate certain preparations designed for hypodermic injection.

**subculture** (*sub-kul'-tūr*) [*sub-*; *culture*]. In bacteriology, a secondary culture made from a primary culture.

**subcuneus** (*sub-kū-ne'-us*) [*sub-*; *cuneus*, a wedge]. An area of the occipital lobe ventrad of the cuneus, and caudad of the collateral fissure.

**subcutaneous** (*sub-kū-ta'-ne-us*) [*sub-*; *cutaneous*]. Beneath the skin; hypodermatic.

**subcutaneus colli** (*sub-kū-ta'-tik'-ū-lar*) [*sub-*; *cutis*, skin]. The platysma myoides muscle.

**subcuticular** (*sub-kū-tik'-ū-lar*) [*sub-*; *cutis*, skin]. Beneath the epidermis, as a *subcuticular* suture.

**subcutin** (*sub-kū'-tin*). Paraphenolsulphonate of paraamidobenzoic ethyl ester; small acicular crystals melting at 195.6° C.; soluble in 100 times its weight in water. Its solutions can be sterilized. It is recommended as a local anesthetic.

**subcutis** (*sub-kū'-tis*) [*sub-*; *cutis*, skin]. The deeper portion or layer of the true skin.

**subdelirium** (*sub-de-lir'-e-um*) [*sub-*; *delirium*]. A slight or muttering delirium, with lucid intervals.

**subdeltoid** (*sub-del'-toid*). Beneath the deltoid muscle.

**subdental** (*sub-den'-tal*) [*sub-*; *dens*, a tooth]. Situated beneath the teeth.

**subdermal** (*sub-der'-mal*). See *subcutaneous*.

**subdiaphragmatic** (*sub-di-a-frag-mat'-ik*) [*sub-*; *diaphragm*]. Under the diaphragm.

**subdicrotic** (*sub-di-krot'-ik*). Obscurely dicrotic.

**subdivided** (*sub-div-i'-ded*) [*sub-*; *dividere*, to divide]. Re-divided; making secondary or smaller divisions.

**sobdolichocephalic** (*sub-dol-ik-o-sef-al'-ik*) [*sub-*; dolicocephalic]. Somewhat dolichodephalic; having the cephalic index above 75° and below 77°.

**subdorsal** (*sub-dor'-sal*) [*sub-*; *dorsum*, back]. Situated on the side of or below the dorsal surface of the body.

**subduction** (*sub-duk'-shun*) [*sub-*; *ducere*, to lead]. Maddox's term for deorsumduction.

**subdural** (*sub-dū'-ral*) [*sub-*; *dura*]. Beneath the dura.

**subectodermal** (*sub-ek-to-der'-mal*) [*sub-*; *ectodermal*]. Beneath the ectoderm.

**subencephalon** (*sub-en-sef'-al-on*) [*sub-*; ἐγκέφαλος, brain]. The medulla oblongata, pons, and corpora quadrigemina taken together.

**subendocardial** (*sub-en-do-kar'-de-al*) [*sub-*; *endocardium*]. Beneath the endocardium.

**subendothelial** (*sub-en-do-the'-le-al*) [*sub-*; *endothelial*]. Situated or occurring under an endothelial structure.

**subendothelium** (*sub-en-do-the'-le-um*). The layer of connective-tissue cells between the mucosa and the epithelium of the bladder, intestine, and bronchi.

**subendymal** (*sub-en'-dim-al*). Beneath the ependyma.

**subepidermal, subepidermatic, subepidermic** (*sub-ep-e-der'-mal*, *sub-ep-e-der-mat'-ik*, *sub-ep-e-der'-mik*) [*sub-*; *epidermis*]. Situated beneath the epidermis.

**subepithelial** (*sub-ep-e-the'-le-al*) [*sub-*; *epithelium*]. Situated under an epithelial surface.

**suberin** (*sū'-ber-in*) [*suber*, cork]. 1. Pulverized cork; used as a dressing for wounds. 2. The impure cellulose forming the cellular tissue of cork.

**subese** (*sub-ēs*) [*sub-*; *edere*, to eat]. Underfed; thin; the opposite of obese.

**subesophageal** (*sub-e-so-faj'-e-al*) [*sub-*; *esophagus*]. Beneath the esophagus.

**subfalcial** (*sub-fal'-se-al*) [*sub-*; *falx*]. At the free edge of the falx cerebri.

**subfalciform** (*sub-fal'-se-form*) [*sub-*; *falx*]. Somewhat sickle-shaped.

**subfascial** (*sub-fash'-e-al*) [*sub-*; *fascia*]. Beneath the fascia.

**subfebrile** (*sub-feb'-ril*) [*sub-*; *febris*, fever]. Slightly febrile.

**subfemoralis** (*sub-fem-o-ra'-lis*). Same as *subcruræus*.

**subfissure** (*sub-fish'-ūr*) [*sub-*; *fissura*, fissure]. A fissure of the brain which is concealed by a supergyre, and invisible until the lips of the superfissure are divaricated.

**subflavor** (*sub-fla'-vor*). A secondary or subordinate flavor.

**subflavous** (*sub-fla'-vus*) [*sub-*; *flavus*, yellow]. Somewhat yellow. **s. ligament**, the ligament of yellowish elastic material found between the laminæ of adjacent vertebræ.

**subfoliar** (*sub-fo'-le-ar*) [*sub-*; *folium*, leaf]. Having the character of sobfolium.

**subfolium** (*sub-fo'-le-um*) [*sub-*; *folium*, leaf]. A leaflet going to make up a part of any folium of the cerebellum.

**subfornical** (*sub-for'-nik-al*) [*sub-*; *fornix*]. Beneath the fornix of the brain.

**subfrontal** (*sub-frun'-tal*) [*sub-*; *frons*, forehead]. Applied to a fissure and gyre (Broca's) in the ventral region of the frontal lobe of the brain.

**subgallate** (*sub-gal'-lāt*). A basic salt of gallic acid. **s., bismuth.** See *dermatol*.

**subgelatinous** (*sub-jel-at'-in-us*) [*sub-*; *gelatin*]. Partly gelatinous.

**subgemmal** (*sib-jem'-al*). Beneath a taste-bud.

**subgeneric** (*sib-jen-er'-ik*). Relating to a subgenus.

**subgeniculate** (*sub-jen-ik'-ū-lāt*). Incompletely geniculate.

**subgenus** (*sub-je'-nus*). A subordinate genus, a subdivision of a genus higher than a species.

**subgerminal** (*sub-jer'-min-al*) [*sub-*; *germ*]. Situated beneath a germinal structure.

**subglenoid** (*sub-gle'-noid*) [*sub-*; *glenoid*]. Beneath the glenoid fossa, as *subglenoid* dislocation of the humerus.

**subglossal** (*sub-glos'-al*). See *hypoglossal*; *sublingual*.

**subglossitis** (*sub-glos-i'-tis*) [*sub-*; γλῶσσα, tongue; ιτις, inflammation]. Inflammation of the tissues under the tongue. See *ranula*.

**subglottic** (*sub-glot'-ik*). See *infraglottic*.

**subgrundation** (*sub-grun-da'-shun*) [Fr., *subgrondation*]. The intrusion of one part of a cranial bone beneath another.

**subgyre** (*sub-jīr*) [*sub-*; *gyrus*]. A gyre that is encroached upon or covered by another or *supergyre* (covering-gyre).

**subhepatic** (*sub-he-pat'-ik*) [*sub-*; ἧπαρ, liver]. Situated beneath or on the under surface of the liver.

**subhumeral** (*sub-hū'-mer-al*) [*sub-*; *humerus*]. Below the humerus.

**subhyaloid** (*sub-hi'-al-oid*). Beneath the hyaloid membrane of the eye.

**subhyoid** (*sub-hi'-oid*) [*sub-*; *hyoid*]. Beneath the hyoid bone. **s. bursa**, a bursa lying between the thyrohyoid membrane and hyoid bone and the sen joint insertion of the onohyoid, sternohyoid, amd stylohyoid muscles. Syn., *Boyer's bursa*.

**subicteric** (*sub-ik-ter'-ik*) [*sub-*; ἴκτερος, jaundice]. Moderately or slightly icteric.

**subiculum** (*sub-ik'-ū-lum*) [*subex*, a layer]. The uncinate gyrus. **s. promontorii**, support of the promontory; the posterior boundary of the fenestra vestibuli.

**subiliac** (*sub-il'-e-ak*) [*sub-*; *ilium*]. Pertaining to the subilium.

**subilium** (*sub-il'-e-um*) [*sub-*; *ilium*]. The lowest portion of the ilium.
**subimaginal** (*sub-im-aj'-in-al*) [*sub-*; *imago*]. Having the character of a subimago.
**subinfection** (*sub-in-fek'-shun*) [*sub-*; *infection*]. 1. A slight degree of infection. 2. A chronic intoxication due to frequent small doses of a toxic agent introduced from without or produced within the body.
**subinflammation** (*sub-in-flam-a'-shun*) [*sub-*; *inflammation*]. A slight degree of inflammation.
**subinflammatory** (*sub-in-flam'-at-or-e*). Of the nature of a slight inflammation.
**subintegumentary** (*sub-in-teg-u-men'-tar-e*) [*sub-*; *integumentum*, integument]. Situated beneath the integument.
**subintestinal** (*sub-in-tes'-tin-al*) [*sub-*; *intestinum*, intestine]. Situated beneath the intestines.
**subintrance** (*sub-in'-trans*) [*subintrare*, to enter secretly]. Anticipation of recurrence.
**subintrant** (*sub-in'-trant*). Entering secretly; applied to malarial fevers in which a new paroxysm begins before the termination of the preceding one.
**subinvolution** (*sub-in-vo-lū'-shun*) [*sub-*; *involutio*, a rolling up]. Imperfect involution. **s. of the uterus**, the imperfect contraction of the uterus after delivery.
**subiodide** (*sub-i'-o-dīd*). That iodide of a series having the least iodine.
**subjacent** (*sub-ja'-sent*) [*sub-*; *jacere*, to lie]. Lying beneath.
**subject** (*sub'-jekt*) [*sub-*; *jacere*, to throw]. 1. An individual that serves for purposes of experiment or study, or that is under observation or treatment. 2. A cadaver. 3. The matter of a discourse.
**subjective** (*sub-jek'-tiv*) [*subject*]. 1. Pertaining to the individual himself. 2. Of symptoms, experienced by the patient himself, and not amenable to physical exploration. **s. sensation**, one not caused by external stimuli.
**subjectivity** (*sub-jek-tiv'-it-e*) [*subjicere*, to throw under]. Illusiveness.
**subjectoscope** (*sub-jek'-to-skōp*). An instrument for examining subjective visual sensations.
**subjugal** (*sub-joo'-gal*) [*sub-*; *jugum*, yoke]. Below the malar bone.
**subkatabolism** (*sub-kat-ab'-ol-izm*). Katabolic stasis, a condition marked by inactivity, devitalization, and premature senility of the cells due to suboxidation, excessive strain, fatigue, etc.
**sublamine** (*sub'-lam-in*). A soluble compound of mercury sulphate and ethylenediamine containing 43% of mercury. It is used as a disinfectant, and intramuscularly in syphilis. Dose 2-6 dr. (7.7-23.3 Cc.) of 1% solution in normal salt solution.
**sublaryngeal** (*sub-lar-in'-je-al*) [*sub-*; *laryngeal*]. Situated below the larynx.
**sublatio** (*suo-la'-she-o*) [*sublatio*, removal]. 1. Removal; ablation. 2. Depression, or couching, of the lens in cataract. **s., retinal**, detachment of the retina.
**sublation** (*sub-la'-shun*). See *sublatio*.
**subligamen** (*sub-li-ga'-men*) [*sub-*; *ligare*, to bind]. A form of truss used in hernia.
**subliminal** (*sub-lim'-in-al*) [*sub-*; *limen*, threshold]. Below the threshold of consciousness or of sensation. See *threshold*.
**sublimate** (*sub'-lim-āt*) [*sublimare*, to lift up high]. A substance obtained by sublimation. **s., corrosive**, mercuric chloride; see *mercury bichloride*.
**sublimation** (*sub-lim-a'-shun*). The vaporization and condensation of a volatile solid.
**sublime** (*sub-līm*) [see *sublimate*]. 1. To subject to sublimation. 2. To undergo sublimation.
**sublimis** (*sub-lī'-mis*) [L.]. Elevated; superficial; a qualification applied to certain muscles.
**sublingual** (*sub-ling'-gwal*) [*sub-*; *lingua*, tongue]. 1. Lying beneath the tongue. 2. Pertaining to the parts lying beneath the tongue.
**sublinguitis** (*sub-ling-gwi'-tis*) [*sub-*; *lingua*, tongue; *ιτις*, inflammation]. Inflammation of the sublingual gland.
**sublobular** (*sub-lob'-ū-lar*) [*sub-*; *lobule*]. Situated beneath a lobule. **s. veins**, the radicles of the hepatic veins, situated at the base of a cluster of lobules.
**sublumbar** (*sub-lum'-bar*) [*sub-*; *lumbus*, loin]. Situated under the loins.
**subluxation** (*sub-luks-a'-shun*) [*sub-*; *luxation*]. Incomplete luxation; sprain.
**submalleolar** (*sub-mal-e'-o-lar*) [*sub-*; *malleolus*]. Under the malleoli. **s. amputation**, removal of the foot at the ankle-joint.

**submammary** (*sub-mam'-a-re*) [*sub-*; *mamma*, breast]. Situated beneath the breast.
**submarginal** (*sub-mar'-jin-al*) [*sub-*; *margin*]. Situated near the border or margin.
**submarine** (*sub-mar-ēn*) [*sub-*; *marine*]. A dental term applied to conditions and materials in the treatment and management of which the parts are filled with the fluids of the mouth.
**submaksillaritis** (*sub-maks-il-ar-i'-tis*) [*sub-*; *maxilla*, jaw; *ιτις*, inflammation]. Inflammation of the submaxillary gland.
**submaxillary** (*sub-maks'-il-a-re*) [*sub-*; *maxilla*]. 1. Lying beneath the lower maxilla, as the *submaxillary gland*. 2. Pertaining to the submaxillary gland.
**submaxillitis** (*sub-maks-il-i'-tis*). Inflammation of the submaxillary gland.
**submedial** (*sub-me'-de-al*) [*sub-*; *medius*, middle]. Situated beneath or near the middle.
**submembranous** (*sub-mem'-bra-nus*) [*sub-*; *membrana*, a membrane]. Somewhat membranous.
**submeningeal** (*sub-men-in'-je-al*). Beneath the meninges.
**submental** (*sub-men'-tal*) [*sub-*; *mentum*, chin]. Situated under the chin.
**submerge** (*sub-merj'*) [*sub-*; *mergere*, to dip]. To place under the surface of a liquid.
**submersion** (*sub-mer'-shun*) [*submerge*]. The act of submerging; the condition of being under the surface of a liquid.
**submesaticephalic** (*sub-mes-at-e-sef-al'-ik*). Having a cephalic index of 75° to 76°.
**submetallic** (*sub-met-al'-ik*) [*sub-*; *metallum*, metal]. To a certain extent metallic.
**submicroscopic** (*sub-mi-kro-skop'-ik*). Pertaining to a particle which is visible by the aid of the ultramicroscope.
**submissio** (*sub-mis'-e-o*) [L.]. A lowering. **s. cordis**, the systole of the heart.
**submorphous** (*sub-mor'-fus*) [*sub-*; *μορφή*, form]. Having the characters both of a crystalline and of an amorphous body; applied to calculi.
**submucosa** (*sub-mū-ko'-sah*) [*sub-*; *mucosus*, mucous]. The layer of fibrous connective tissue that attaches the mucous membrane to the subjacent parts.
**submucous** (*sub-mū'-kus*) [*sub-*; *mucous*]. Situated beneath a mucous membrane.
**submuscular** (*sub-mus'-kū-lar*) [*sub-*; *muscular*]. Beneath a muscle.
**subnarcotic** (*sub-nar-kot'-ik*). Moderately narcotic.
**subnasal** (*sub-na'-sal*) [*sub-*; *nasus*, nose]. Situated below the nose. **s. point**. See under *craniometric point*.
**subneural** (*sub-nū'-ral*) [*sub-*; *νεῦρον*, nerve]. Situated under the neuron or under a nerve. **s. gland**, the homologue in the amphioxus of the hypophysis of higher vertebrates.
**subnitrate** (*sub-ni'-trāt*) [*sub-*; *nitrate*]. A basic nitrate.
**subnodal** (*sub-no'-dal*) [*sub-*; *nodus*, node]. Behind or under a node.
**subnormal** (*sub-nor'-mal*) [*sub-*; *norma*, rule]. Below normal.
**subnotochordal** (*sub-no-to-kord'-al*). Below the notochord.
**subnucleus** (*sub-nū'-kle-us*) [*sub-*; *nucleus*]. Any one of the smaller groups of cells into which a large nerve-nucleus is divided by the passage through it of nerve-bundles.
**subnutrition** (*sub-nū-trish'-un*). Defective nutrition.
**suboccipital** (*sub-ok-sip'-it-al*) [*sub-*; *occiput*]. Situated beneath the occiput.
**suboccipitobregmatic** (*sub-ok-sip-it-o-breg-mat'-ik*) [*sub-*; *occiput*; *bregma*]. Situated in the region extending from the bregma to beneath the occiput.
**subocular** (*sub-ok'-ū-lar*) [*sub-*; *oculus*, eye]. Beneath the eye.
**subopercular** (*sub-o-per'-kū-lar*) [*sub-*; *operculum*, lid]. Pertaining to a suboperculum.
**suboperculum** (*sub-o-per'-kū-lum*) [*sub-*; *operculum*, lid]. A gyrus of the brain between the presylvian and subsylvian fissures; the orbital operculum.
**suboptic** (*sub-op'-tik*). Same as *suborbital*.
**suboral** (*sub-o'-ral*) [*sub-*; *os*, *oris*, mouth]. Beneath the mouth.
**suborbicular, suborbiculate** (*sub-or-bik'-ū-lar*, *sub-or-bik'-ū-lāt*) [*sub-*; *orbicular*]. Almost orbicular.

# SUBORBITAL 853 SUBSTANTIA

**suborbital** (*sub-or'-bit-al*) [*sub-*; *orbit*]. Beneath the órbit. Synonym of infraorbital.

**subordination** (*sub-or-din-a'-shun*) [*sub-*; *ordo*, order]. The condition of being under subjection or control; the condition of organs that depend upon or are controlled by other organs.

**suboxidation** (*sub-oks-id-a'-shun*). Deficient oxidation.

**suboxide** (*sub-oks'-ĭd*) [*sub-*; ὀξύs, acid]. One of two oxides containing the less oxygen.

**subpallial** (*sub-pal'-e-al*) [*sub-*; *pallium*, a mantle]. Beneath the pallium.

**subpapular** (*sub-pap'-ū-lar*). Indistinctly papular.

**subparalytic** (*sub-par-al-it'-ik*) [*sub-*; *paralytic*]. Slightly paralytic.

**subparietal** (*sub-par-i'-et-al*) [*sub-*; *paries*, wall]. Situated beneath the parietal bone, convolution, or fissure.

**subpatellar** (*sub-pat-el'-ar*) [*sub-*; *patella*, kneecap]. Situated beneath the patella.

**subpectoral** (*sub-pek'-tor-al*) [*sub-*; *pectus*, chest]. Situated beneath the chest.

**subpeduncular** (*sub-pe-dung'-kū-lar*) [*sub-*; *pedunculus*, peduncle]. Situated beneath a peduncle. s. lobe, the *flocculus*, q. v.

**subpericardial** (*sub-per-e-kar'-de-al*) [*sub-*; *pericardium*]. Situated beneath the pericardium.

**subpericranial** (*sub-per-e-kra'-ne-al*). Beneath the pericranium.

**subperiosteal** (*sub-per-e-os'-te-al*) [*sub-*; *periosteum*]. Beneath the periosteum. s. operation, excision of bone without removing the periosteum.

**subperitoneal** (*sub-per-e-ton-e'-al*) [*sub-*; *peritoneum*]. Beneath the peritoneum.

**subperitoneoabdominal** (*sub-per-it-on-e-o-ab-dom'-in-al*). Beneath the abdominal peritoneum.

**subperitoneopelvic** (*sub-per-it-on-e-o-pel'-vik*). Beneath the peritoneum of the pelvis.

**subpersonal** (*sub-per'-son-al*). Having individuality in a very slight degree.

**subpetrosal** (*sub-pet-ro'-sal*). Below the petrosa.

**subpharyngeal** (*sub-far-in'-je-al*) [*sub-*; *pharynx*]. Beneath the pharynx.

**subphrenic** (*sub-fren'-ik*). Synonym of subdiaphragmatic.

**subpial** (*sub-pi'-al*) [*sub*, under; *pia*]. Situated or occurring beneath the pia.

**subplacenta** (*sub-pla-sent'-ah*). The decidua vera.

**subplantigrade** (*sub-plant'-e-grād*). Incompletely plantigrade, walking with the heel slightly elevated.

**subpleural** (*sub-ploo'-ral*) [*sub-*; *pleura*]. Beneath the pleura.

**subplexal** (*sub-pleks'-al*) [*sub*, under; *plexus*]. Lying under a plexus of the brain.

**subpontile** (*sub-pon'-tīl*) [*sub*, under; *pons*]. Situated or occurring beneath the pons.

**subpontine** (*sub-pon'-tīn*) [*sub-*; *pons*]. Beneath the pons.

**subpreputial** (*sub-pre-pū'-she-al*) [*sub-*; *preputium*, prepuce]. Beneath the prepuce.

**subprostatic** (*sub-pros-tat'-ik*) [*sub*, under; *prostate*]. Beneath the prostate gland.

**subpubic** (*sub-pū'-bik*) [*sub-*; *pubes*, pubis]. Situated beneath the pubic arch or symphysis.

**subpulmonary** (*sub-pul'-mon-a-re*) [*sub-*; *pulmo*, the lung]. On the ventral side of the lungs.

**subpyramidal** (*sub-pir-am'-id-al*) [*sub*, under; *pyramis*, pyramid]. 1. Beneath a pyramid. 2. Approximately pyramidal.

**subreniform** (*sub-ren'-if-orm*) [*sub*, under; *ren*, kidney; *forma*, form]. Shaped somewhat like a kidney.

**subresin** (*sub-rez'-in*) [*sub*, under; *resina*, resin]. That ingredient of a resin which is soluble in boiling alcohol, but is precipitated on cooling.

**subretinal** (*sub-ret'-in-al*) [*sub-*; *retina*]. Beneath the retina.

**subsacral** (*sub-sa'-kral*) [*sub*, under; *sacrum*]. Situated or occurring ventrad of the sacrum.

**subsalt**. A basic salt.

**subsaturation** (*sub sat ū ra' shun*). Incomplete saturation.

**subscapular** (*sub-skap'-ū-lar*) [*sub-*; *scapula*]. 1. Beneath the scapula, as the *subscapular* muscle, or subscapularis. 2. Pertaining to the subscapular muscle.

**subscapularis** (*sub-skap-ū-la'-ris*). See under *muscle*.

**subscleral** (*sub-skle'-ral*). Beneath the sclera.

**subsclerotic** (*sub-skle-rot'-ik*) [*sub*, under; σκληρός, hard]. Beneath the sclerotic.

**subscription** (*sub-skrip'-shun*) [*sub-*; *scribere*, to write]. That part of a prescription containing the directions to the pharmacist, indicating how the ingredients are to be mixed and prepared.

**subsensation** (*sub-sen-sa'-shun*). A subordinate sensation.

**subseptal** (*sub-sep'-tal*). Situated below a septum.

**subseptate** (*sub-sep'-tāt*) [*sub-*; *septum*, a hedge]. Partially divided.

**subserous** (*sub-se'-rus*) [*sub*; *serous*]. Beneath a serous membrane.

**subserrate** (*sub-ser'-āt*). Slightly serrate.

**subsibilant** (*sub-sib'-il-ant*). Having a sound like muffled whistling.

**subsidence** (*sub'-sid-ens*, or *sub-si'-dens*) [*sub*, under; *sedere*, to sit]. The gradual cessation and disappearance of an attack of disease.

**subsigmoid** (*sub-sig'-moid*) [*sub*, under; *sigmoid*]. Under the sigmoid cavity or flexure. s. fossa, a fossa bounded in the median line by the attached mesentery of the sigmoid flexure, and above by the limit of the attachment of the mesentery of the descending colon. It may be seen by lifting the sigmoid flexure of the large intestine, and varies much in size in different individuals.

**subsistence** (*sub-sis'-tens*) [*sub*, under; *sistere*, to stand]. That which nourishes or gives support. Food.

**subsoil** (*sub'-soil*). The under-soil. s. water, water which has penetrated the soil, and is found immediately above the first impervious stratum.

**subspinous** (*sub-spi'-nus*) [*sub-*; *spine*]. 1. Beneath a spine. 2. Beneath the spinal column. s. dislocation, luxation of the head of the humerus below the spine of the scapula.

**subspiral** (*sub-spi'-ral*) [*sub*, under; *spira*, coil]. Somewhat spiral.

**subsplenial** (*sub-sple'-ne-al*) [*sub*, under; σπληνίον, a bandage]. Beneath the splenium.

**substage** (*sub'-stāj*). The parts beneath the stage of a microscope, including the diaphragm, condenser, illuminator, and other accessories.

**substance** (*sub'-stans*) [*substantia*, substance]. 1. The material of which anything is composed. 2. A tissue. s., alible, the portion of the chyme which is utilized for nourishing the body. s., alimentary, an article of food. s., basis, the intercellular or ground-substance. s., cell-, cell-protoplasm. s., contractile. 1. Living protoplasm which has the property of contracting. 2. The contractile portion of a muscular fiber. s., cortical, the peripheral portion of an organ, situated just beneath the capsule. s., gray. See *substantia cinerea*. s., ground-, the homogeneous matrix or intracellular substance of a tissue in which the cellular elements and fibers are embedded. s., haptophorous, a toxoid. s., immune, the immune body. s., interfilar. See *enchylema*. s., interstitial. 1. The connective tissue of an organ. 2. Achromatin. s., intertubular, the matrix of dentine in which the dentinal canals are placed. s., intervertebral, the intervertebral discs. s., living, protoplasm. s., medullary. 1. The part of an organ constituting its central in contradistinction to its peripheral or cortical portion. 2. The tissue forming the medulla, as in bone. s., parietal, the matrix of the cartilage. s. of Rolando. See *substantia gelatinosa*. s., Rollet's secondary. See *disc, interstitial*. s., Rovidas' hyaline, a nucleoproteid, insoluble in water, forming a large proportion of the constituents of pus-corpuscles. It expands into a tough, slimy mass when treated with a 10% common salt solution. It is soluble in alkalies, but quickly changed by them. s., sarcous, the substance of a sarcous element. s. of Schwann, white, the medullary sheath of a nerve-fiber. s. sensibilisatrice, Bordet's name (1899) for the later mediate body. s., Stilling's gelatinous, the gelatinous substance surrounding the central canal of the spinal cord. s., supporting, a supporting tissue, as the neuroglia and connective tissue. s., white reticular, the reticulated layer of white tissue on the anterior half of the uncinate convolution.

**substandard** (*sub-stand'-ard*). Below the standard requirements.

**substantia** (*sub-stan'-she-ah*) [L.]. Substance. s. alba, the white fibrous tissue of the brain and nerves. Syn., *alba*. s. cinerea, the gray matter of the nervous system. s. corticalis. See *substance, cortical*. s. eburnea, dentine. s. ferruginea. Synonym of *locus cinereus*. s. filamentosa dentium; a

name given by Malpighi to the enamel of the teeth. **s. fusca.** See *locus niger*. **s. gelatinosa,** that part of the gray matter of the cord which caps the head of the posterior horns and surrounds the central canal. **s. gelatinosa centralis,** the light zone surrounding the central canal of the developing spinal cord. **s. glomerulosa,** the cortical substance of the kidney. **s. grisea,** the gray matter of the spinal cord. **s. grisea centralis.** See *entocinerea*. **s. hyalina,** Leydig's term for the interreticular portion of protoplasm. **s. intermedia,** the portion of the cerebellar substance situated between the cortical and the central gray matter. **s. medullaris,** the medullary substance of the kidney; also, of a hair. **s. nigra,** the locus niger. **s. opaca,** Leydig's term for the reticulum of protoplasm. **s. ossea,** cement. **s. ossea dentium,** a name given by Malpighi to dentine. **s. perforata anterior,** one of two perforated spaces at the base of the brain bounded by the olfactory trigone and the optic chiasm and tract. **s. perforata posterior,** a perforated area between the peduncles of the brain. **s. primaria,** the medullary portion of the central nervous system. **s. propria,** the essential tissue of an organ; especially the modified connective-tissue lamellæ of the cornea; also the middle or fibrous tissue layer of the tympanic membrane. **s. reticularis,** the network of nerve-fibers and gray matter found in the deep parts of the medulla and in the pons. **s. rubra** (*lienis*), the splenic pulp. **s. spongiosa,** the entire gray matter of the cord except those parts occupied by the s. gelatinosa. **s. striata,** synonym of enamel. **s. vasculosa,** the pulp of the spleen.
**substernal** (*sub-stur'-nal*) [*sub-*; *sternum*]. Beneath the sternum.
**substitution** (*sub-stit-ū'-shun*) [*sub-*; *statuere*, to place]. The replacement of one thing by another. In chemistry, the replacing of one or more elements or radicals in a compound by other elements or radicals.
**substratum, substrate** (*sub-stra'-tum, sub'-strāt*) [*sub-*; *stratum*, a layer]. 1. An under layer or stratum. 2. A substance upon which an enzyme acts.
**substriate** (*sub-stri'-āt*) [*sub-*; *stria*]. Having imperfect striæ.
**subsulphate** (*sub-sul'-fāt*) [*sub-*; *sulphur*]. A basic sulphate.
**subsultory** (*sub-sul'-tor-e*) [*sub-*; *saltire*, to leap]. Leaping; twitching.
**subsultus** (*sub-sul'-tus*) [see *subsultory*]. A morbid jerking or twitching. **s. clonus.** See *s. tendinum*. **s. tendinum,** involuntary twitching of the muscles, especially of the hands and feet, seen in low fevers.
**subsylvian** (*sub-sil'-ve-an*). Beneath the Sylvian fissure.
**subsynovial** (*sub-si-no'-ve-al*) [*sub-*; *synovia*, synovia]. Situated within a synovial sac.
**subtarsal** (*sub-tar'-sal*). Below the tarsus.
**subtegmen** (*sub-teg'-men*) [*subtexere*, to weave under]. Weft. **s. fornicis,** in the fornix the layer of nerve-fibers situated beneath the superficial longitudinal bundles.
**subtegumental** (*sub-teg-ū-men'-tal*) [*sub-*; *tegumentum*, a cover]. Subcutaneous.
**subtemporal** (*sub-tem'-por-al*) [*sub-*; *tempus*; temple]. Situated beneath the temporal bone or muscle, or below the temple.
**subtenial** (*sub-te'-ne-al*). Situated beneath the tenia.
**subternatural** (*sub-ter-nat'-ū-ral*) [*subter*, below; *natura*, nature]. Below what is natural.
**subthalamic** (*sub-thal-am'-ik*) [*sub-*; θάλαμος, thalamus]. Beneath the optic thalamus.
**subthalamus** (*sub-thal'-am-us*). See *hypothalamus*.
**subthoracic** (*sub-tho-ras'-ik*) [*sub-*; *thorax*]. Situated below the thorax.
**subthyroideus** (*sub-thi-roid'-e-us*). An anomalous bundle of fibers uniting the inferior and lateral thyroarytenoid muscles.
**subtrapezial** (*sub-tra-pe'-ze-al*). Beneath the trapezium.
**subtrochanteric** (*sub-tro-kan-ter'-ik*) [*sub-*; *trochanter*]. Below the trochanter.
**subtrochlear** (*sub-trok'-le-ar*) [*sub-*; *trochlear*]. Beneath the trochlea.
**subtropical** (*sub-trop'-ik-al*). Pertaining to regions almost tropical in climate.
**subtuberal** (*sub-tū'-ber-al*). Situated beneath a tuber.

**subtympanitic** (*sub-tim-pan-it'-ik*). See *hypotympanic*.
**sububeres** (*sub-ū'-ber-ēz*) [*sub-*; *ubera*, the breasts]. Children at the breast. Suckling children.
**subumbilical** (*sub-um-bil'-ik-al*). Situated below the umbilicus.
**subungual, subunguial** (*sub-un'-gwal, sub-ung'-gwe-al*) [*sub-*; *unguis*, nail]. Beneath the nail.
**suburethral** (*sub-ū-re'-thral*) [*sub-*; *urethra*]. Beneath the urethra.
**subvaginal** (*sub-vaj'-in-al*) [*sub-*; *vagina*, sheath]. Beneath a sheath. **s. space,** the space beneath the sheath of dura mater surrounding the optic nerve.
**subvertebral** (*sub-ver'-te-bral*) [*sub-*; *vertebra*]. Beneath a vertebra.
**subvirile** (*sub-vir'-il*). Deficient in virility.
**subvitrinal** (*sub-vit'-rin-al*). Beneath the vitreous humor.
**subvola** (*sub-vo'-lah*) [*sub-*; *vola*, the palm of the hand]. 1. The space between the second and fifth fingers. 2. See *hypothenar*.
**subvolution** (*sub-vo-lū'-shun*) [*sub-*; *volvere*, to roll]. A method of operating (as is done for pterygium) in which a flap is turned over so that an outer or cutaneous surface comes in contact with a raw, dissected surface. Adhesions are thus prevented.
**subzonal** (*sub-zo'-nal*) [*sub-*; ζώνη, zone]. Beneath the zona pellucida.
**subzygomatic** (*sub-zi-go-mat'-ik*) [*sub-*; *zygoma*]. Below the zygoma.
**succagogue, succagog** (*suk'-ag-og*) [*succus*, juice; ἀγωγός, a leading]. 1. A drug which stimulates the secretory function. 2. An agent which stimulates the flow of a digestive juice, particularly the gastric juice.
**succedaneous** (*suk-se-da'-ne-us*) [*succedere*, to take the place of]. Relating to or acting as a succedaneum.
**succedaneum** (*suk-se-da'-ne-um*) [see *succedaneous*]. A substitute. **s.,** caput. See *caput succedaneum*.
**succenturiate** (*suk-sen-tū'-re-āt*) [*succenturiare*, to receive as a substitute]. Accessory. **s. kidney,** the suprarenal body.
**succi** (*suk'-i*) [L.]. Genitive and plural of *succus*, *q. v.*
**succiferous** (*suk-sif'-er-us*) [*succus*, juice; *ferre* to bear]. Producing sap.
**succinamic acid** (*suk-sin-am'-ik*) [*succinum*, amber], $C_4H_7NO_3$. A crystalline monobasic acid, of pleasant acid taste.
**succinamide** (*suk-sin'-am-id*) [*succinum*, amber; *amide*], $C_4H_8N_2O_2$. A substance produced by shaking succinic ester with aqueous ammonia. It is a white powder, insoluble in water and in alcohol; is crystallized from hot water in needles.
**succinate** (*suk'-sin-āt*) [*succinum*, amber]. A salt of succinic acid.
**succinctum** (*suk-sink'-tum*). Synonym of *diaphragm*.
**succinic acid** (*suk-sin'-ik*). See *acid, succinic*.
**succinimide** (*suk-sin'-im-id*) [*succinum*, amber], $C_4H_5NO_2$. A crystalline substance produced by gentle ignition of the anhydride in a current by dry ammonia. It crystallizes from acetone in rhombic octahedra without any water; when anhydrous, it melts at 126° C. and boils at 288° C.
**succinin** (*suk'-sin-in*) [*succinum*, amber], $C_7H_{10}O_2$. 1. A dark-brown substance produced when equal parts of succinic acid and glycerin are heated together to about 230° C. 2. The insoluble portion of amber.
**succinone** (*suk'-sin-ōn*) [*succinum*, amber]. An oily liquid obtained in the dry distillation of calcium succinate. It has a decided empyreumatic odor.
**succinonitril** (*suk-sin-o-ni'-tril*). Ethylene cyanide, $C_4H_4N_2$, a crystalline body.
**succinum** (*suk-si'-num*) [L.]. Amber, a fossil resin found in the alluvial deposits of Central Europe, and thought to be derived from an extinct species of pine. It contains a volatile oil, *oleum succini*, used in hysteria, whooping-cough, amenorrhea, and locally as a rubefacient in chronic rheumatism, whooping-cough, and infantile convulsions. Dose 5-15 min. (0.32-1.0 Cc.).
**succorrhœa, succorrhoea** (*suk-or-e'-ah*) [*succus*; ῥεῖν, to flow]. An excessive flow of a secretion. **s., pancreatic,** a pathological increase of the pancreatic juice when the secretory innervation of the gland is exaggerated.
**succory** (*suk'-or-e*). The chicory, *Cichorium intybus*. See *chicory*.

succuba (suk'-ū-bah) [sub-; cumbere, to lie; pl., succubæ]. A female demon formerly believed to consort with men in their sleep. Cf. incubus (2).
succubate (suk'-ū-bāt) [succubare, to lie under]. To have carnal knowledge of a man.
succubus (suk'-ū-bus) [L.: pl., succubi]. A male demon, once considered to be the counterpart of the succuba.
succulent (suk'-ū-lent) [succus]. Juicy.
succursal (suk-ur'-sal) [succursalis, subsidiary]. Subsidiary. s. hospital, or asylum, a branch provincial hospital, usually for mild cases.
succus (suk'-us) [L.]. 1. A vegetable juice. 2. An animal secretion. s. anisi ozonatus. See manol. s. entericus, the intestinal juice, secreted by the glands of the intestinal mucous membrane. It is thin, opalescent, alkaline, and has a specific gravity of 1011. Its chief function is probably to act as a diluent. It contains an amylolytic and a proteolytic ferment. s. gastricus, the gastric juice. s. glandulæ suprarenalis, liquid extract of suprarenal extract. s. intestinalis. Same as s. entericus. s. pancreaticus, the pancreatic juice. s. prostaticus, the prostatic fluid, a constituent of the semen. s. spissatus, any extract prepared by evaporation of the natural juice of a plant.
succussion (suk-ush'-un) [succutere, to shake up]. A shaking, especially of the individual from side to side, for the purpose of determining the presence of fluid in a cavity or hollow organ of the body. s.-sound, s.-splash, the peculiar splashing sound heard when the patient is shaken in hydropneumothorax or pyopneumothorax, or in cases of dilated stomach containing fluid.
sucholoalbumin (sū-ko-lo-al'-bū-min) [σύς, swine; χολή, bile; albumin]. A poisonous proteid classed among the albumoses obtained from cultures of the bacillus of hog cholera.
sucholotoxin (sū-ko-lo-toks'-in) [σύς, swine; χολή, bile; τοξικόν, poison]. A feebly toxic base obtained by de Schweinitz from cultures of swine-plague bacillus. Cf. susotoxin.
suck (suk) [sugere, to draw in]. To take nourishment, as a babe, at the breast; to draw in with the aid of the mouth.
sucking (suk'-ing) [sugere, to suck]. Giving suck; nursing; drawing with the mouth. s.-bottle, a nursing-bottle. s.-pad, a fatty mass on the outer side of the buccinator muscle, well developed in infants.
suckle (suk'-l) [sugere, to suck]. To give suck. To nurse at the breast.
suckling (suk'-ling) [sugere, to suck]. A suckling child; a nursling.
sucramin (sū-kram'-in). The ammoniacal salt of saccharin. A sweetening agent differing from saccharin in its insolubility in the solvents of that substance.
sucrate (sū'-krāt). A chemical compound containing sucrose.
sucrol (sū'-krol). See dulcin.
sucrose (sū'-krōs). See saccharum.
suction (suk'-shun) [suctio, a sucking]. The act of sucking. s.-plate, in dentistry, a plate constructed so as to be held in place by atmospheric pressure.
suctorial (suk-to'-re-al) [sugere, to suck]. Pertaining to, or suitable for sucking.
sudamen (sū-da'-men) [sudor, sweat]. An eruption of translucent, whitish vesicles, due to a noninflammatory disturbance of the sweat-glands, consisting in a collection of sweat in the ducts of the sweat-glands or beneath the epidermis, and occurring in fevers and profuse sweating.
sudamina (sū-dam'-in-ah). Plural of sudamen.
sudaminal (sū-dam'-in-al) [sudare, to sweat]. Of the nature of sudamina.
sudan (sū-dan'), C₂₂H₁₄N₂O. A diazo-compound from alphanaphthalamine with naphthol, a brown powder used as a stain, soluble in alcohol, ether, fats, and oils. Syn., pigment brown. S. III, C₂₂H₁₆-N₄O, a diazo-compound from amidoazobenzene and betanaphthol; a brown powder, soluble in alcohol, ether, benzene, petroleum ether, oils, and fats. s. yellow g, C₁₂H₁₀N₂O₂, a diazo-compound from anilin and resorcinol; a brown powder used as a stain, soluble in alcohol, fats, and oils.
sudanophile (sū-dan'-o-fil). A leukocyte which, owing to fatty degeneration, is stained readily by sudan III.

sudation (sū-da'-shun) [sudor]. The act of sweating.
sudatoria (sū-dat-o'-re-ah). See ephidrosis and miliaria.
sudatorium (sū-dat-o'-re-um) [sudor]. 1. A hot-air bath. 2. A room for the administration of a hot-air bath.
sudol (sū'-dol). A preparation used to check excessive sweating, said to consist of wool-fat and glycerol with 30 % of formaldehyde and oil of wintergreen.
sudolorrhea (sū-do-lor-e'-ah) [sudare, to sweat; oleum, oil; ῥοία, a flow). Synonym of eczema seborrhæicum.
sudor (sū'-dōr) [L.]. Sweat. s. anglicus. See miliaria. s. cruentus. Synonym of hematidrosis. s. nocturnus, night-sweat. s. sanguinosus. See hematidrosis. s. urinosus. See uridrosis.
sudoral (sū'-dor-al) [sudor]. Pertaining to or characterized by sweating.
sudoresis (sū-dor-e'-sis). Excessive sweating.
sudoriferous (sū-dor-if'-er-us) [sudor; ferre, to bear]. Producing sweat.
sudorific (sū-dor-if'-ik) [sudor; facere, to make]. 1. Inducing sweating. 2. An agent inducing sweating.
sudorikeratosis (sū-dor-e-ker-a-to'-sis). Keratosis of the sudoriferous ducts.
sudoriparous (sū-dor-ip'-ar-us) [sudor; parere, to beget]. Secreting sweat.
suet (su'-et) [sebum, suet]. The internal fat of the abdomen of the sheep or cattle. s., mutton- (sevum præparatum, U. S. P., B. P.), consists of stearin, palmitin, and olein, and is used as an emollient and in the preparation of ointments. s. prepared, sevum preparatum, see suet, mutton.
suffocation (suf-o-ka'-shun). Interference with the entrance of air into the lungs by means other than external pressure on the trachea.
suffocative catarrh. Capillary bronchitis.
suffraginis (suf-raj'-in-is) [suffrago, hock]. The large pastern-bone, a very compact bone in the foot of a horse, set in an oblique direction downward and forward, and extending from the cannon-bone to the coronet.
suffragination (suf-raj'-in-us) [suffrago, hock]. Pertaining to the suffrago of the horse.
suffrago (suf-ra'-go) [L.: gen., suffraginis; pl., suffragines]. The hock of a horse's hind leg, whose convexity is backward. It corresponds to the human heel.
suffumigation (suf-ū-mig-a'-shun) [suffumigatio]. 1. Fumigation. 2. A substance used for fumigation.
suffumigium (suf-ū-mij'-e-um) [L.: pl., suffumigia]. A medicinal smoke, vapor, or fumigation.
suffusion (suf-ū'-zhun) [sub-; fundere, to pour]. 1. A spreading or flow of any fluid of the body into surrounding tissue; an extensive superficial extravasation of blood. 2. The pouring of water upon a patient as a remedial measure.
sugar (shoog'-ar) [ME., suger, sugar]. The generic name of a class of sweet carbohydrates. See saccharum. Chemically, sugars are divided as follows: cane-sugar, C₁₂H₂₂O₁₁; glucose (grape-sugar or starch-sugar), C₆H₁₂O₆; lactose, sugar of milk; and inosit, a variety found in certain muscular tissues and in the juice of asparagus. s., acid of, oxalic acid. s., acorn, quercite. s., beet-, saccharose obtained from species of Beta, especially the common beet, Beta vulgaris. s., brown, an impure cane sugar. s., cellulose, sugar derived from cellulose; it has the same formula and properties as glucose. s. chestnut, glucose. s.-coated, coated with sugar, as some pills. s., date-, sugar from the fruit of Phœnix dactylifera. s., diabetic, glucose. s., fruit-, levulose. s., grape-, glucose in the solid state. s., gum, arabinose. s., honey, glucose. s.-house eczema, an eczema sometimes observed in laborers employed in sugar refineries. s., invert. See invert-sugar. s. of lead, plumbi acetas. s., left rotating, levulose. s., Leo's. See laiose. s., liver, another name for glucose which is derived from the liver; glycogen. s. of malt, maltose. s., manna. Synonym of mannite. s., maple, saccharose obtained from the sugar-maple. s., meat, inosit. s. of milk. See lactose. s.-mite, an acarid of the genus Glyciphagus that infests certain unrefined commercial sugar, and is said to be a cause of grocers' itch. s., mucin, levulose. s., muscle, inosite. s., refined, purified cane-sugar. s.-teat, a nipple-shaped linen rag containing a lump of sugar.

It is given (by those who know no better) to an infant to quiet it.
**sugarine** (*shoog'-ar-ēn*). Methylbenzol-sulphinide, a compound said to have 500 times greater sweetening power than sugar.
**sugent** (*sū'-jent*) [*sugere*, to suck]. Sucking; absorbent.
**suggescent** (*suj-es'-ent*). Fitted for sucking.
**suggestible** (*suj-es'-tib-l*) [*suggestion*]. Amenable to suggestion.
**suggestion** (*suj-es'-chun*) [*suggerere*, to suggest, from *sub*, under; *gerere*, to bring]. 1. The artificial production of a certain psychic state in which the individual experiences such sensations as are suggested to him or ceases to experience those which he is instructed not to feel. 2. The thing suggested. **s., hypnotic.** See *hypnotism*. **s., posthypnotic,** the command to do certain acts given the subject while in the hypnotic stage, and causing him to execute these acts after his return to his normal condition. **s., self-,** a suggestion conveyed by the subject from one stratum of his personality to another without external intervention. **s.-therapy,** treatment of disordered states by means of suggestion.
**suggestionize** (*sug-jes'-chun-īz*). To treat a person by suggestion.
**suggestotherapist** (*suj-est-o-ther'-a-pist*). One who treats disease by means of suggestion.
**suggillation, sugillation** (*suj-il-a'-shun*) [*suggillare*, to beat black and blue]. An ecchymosis or bruise.
**suicidal** (*sū-is-ī'-dal*) [*sui*, of himself; *cædere*, to kill]. Self-destroying; having a tendency to suicide.
**suicide** (*sū'-is-īd*) [*sui*, of himself; *cædere*, to kill]. 1. The intentional taking of one's own life; self-murder. 2. One who takes his own life.
**suint** (*swint*) [Fr.]. A soapy substance rich in potash and cholesterin, derivable from sheeps' wool. Lanolin, agnin, and potash salts are obtained from it.
**sulcate** (*sul'-kāt*) [*sulcus*]. Furrowed; grooved.
**sulciform** (*sul'-sif-orm*) [*sulcus*, a furrow; *forma*, form]. Like a groove or sulcus.
**sulcus** (*sul'-kus*) [L.; *pl. sulci*]. A furrow or groove; applied especially to the fissures of the brain. See under *fissure*. 2. A furrow on the mesal or ventricular surface of the brain. **s. ad aquæductum vestibuli.** See *fossa, sulciform*. **s., cacuminal,** in comparative neurology, one beginning at the laterocephalic angle of the flocculus and curving around in a direction caudodorsomesad usually fuses with its opposite from the other pileum. **s. centralis,** the central fissure or *fissure of Rolando*, *q. v.* **s. chiasmatis,** the optic groove of the sphenoid bone. **s. cinguli,** the callosomarginal fissure. **s. costæ,** the subcostal groove. **s. circularis Reilii,** the circuminsular fissure bounding the insula. **s., culminal,** in comparative neurology, one dividing the culmen from the central lobe and extending laterad to the mesal border of the middle cerebellar peduncle. **s., Ecker's,** the anterior or transverse occipital sulcus, usually joined to the horizontal part of the interparietal sulcus. **s., floccular,** in comparative neurology, a sulcus separating the flocculus from the pileum and from the surface of the middle cerebellar peduncle. It arises just dorsad of the auditory nerve and extends dorsocaudoventrad in the form of a loop to the caudal limit of the flocculus. **s. frontalis superior, medius,** and **inferior,** the superfrontal, medifrontal, and subfrontal fissures, respectively. **s., furcal,** in comparative neurology, a sulcus just caudad of the culmen, forming a landmark of division between the horizontal and vertical branches of the stem of the arbor of the cerebellum. **s. habenæ,** the furrow along the dorsomesal angle of the thalamus just dorsad of the habena. **s., hippocampal,** a constant fissure of the cerebrum extending from the splenium to near the tip of the temporal lobe; it is collocated with the hippocampus major. Syn., *hippocampal fissure*. **s. horizontalis. cerebelli,** one between the upper and lower surfaces of the cerebellum. **s. hypothalamicus.** See *aulix*. **s., interfloccular,** in lower mammals a deep sulcus dividing the paraflocculus into two lobes, the supraflocculus dorsad and the mediflocculus ventrad. **s. intertubercularis,** the bicipital groove. **s. intraparietalis,** a more or less confluent group of fissures in the parietal lobe. **s. limitans,** the fissure between the striatum and the thalamus. **s. longitudinalis,** interventricular groove. **s. lunatus,** the lateral occipital fissure. **s. midgracilis,** a fissure in the slender lobe of the cerebellum. **s., Monro's.** See *Monro's sulcus*. **s. nervi radialis,** the musculospiral groove. **s., nodular,** in comparative neurology, one apparently representing the central fissure and separating the nodule of the cerebellum from the uvula. **s. olfactorius,** the fissure occupied by the olfactory tract and bulb. **s., paracentral** (of Wilder), a fissure surrounding the paracentral lobule. **s. paramedialis,** one between the superfrontal fissure and the dorsimesal border of the hemisphere. **s. parolfactorius anterior** and **posterior,** fissures limiting the parolfactory area. **s., peduncular,** of Wilder, a groove on the inner edge of the crus cerebri lodging the third nerve. Syn., *oculomotor furrow*. **s. postcentralis,** the mesodorsal segment of the postcentral fissural complex, back of the fissure of Rolando. **s. postdeclivis,** a fissure separating the declivil lobe from the folium vermis. **s. postnodularis,** a fissure between the nodule and uvula of the cerebellum. **s. postpyramidalis,** one situated between the pyramid and the tuber vermis. **s. præauricularis,** part of the sulcus around the auricular surface of the ilium. **s. præclivalis.** See *s., furcal*. **s., precentral,** one situated in front of the fissure of Rolando and running nearly parallel with it. **s. predeclivis,** a fissure bounding the declive and posterior part of the quadrangular lobule. **s. prepyramidalis,** one situated between the uvula and pyramid. **s. pulmonalis,** the vertical groove in the back between the ribs and spine. **s., pyramidal,** in comparative neurology, one arising just caudad of the peduncular sulcus and extending caudodorsomesad in the form of a crescent; it divides the tuberal and pyramidal lobes. **s., Reil's,** the sulcus in the bottom of the Sylvian fissure, separating the insula from the remainder of the hemisphere. **s. rostralis,** any one of the fissures on the mesal surface of the hemisphere and parallel to the mesorbital border. **s. spiralis,** the grooved extremity of the lamina spiralis of the cochlea. **s. subcentralis,** the laterovental segment of the postcentral fissural complex back of the fissure of Rolando. **s. temporalis superior, medius,** and **inferior,** the supertemporal, meditemporal, and subtemporal fissures. **s., triradiate,** the orbital fissure. **s. tubæ eustachii. s. tubarius,** a depression on the petrosa for the cartilaginous part of the Eustachian tube. Syn., *groove for the Eustachian tube; semicanalis tubæ Eustachii*. **s. tympanicus,** a furrow on the concave surface of the tympanic plate for attachment of the membrana tympani. **s., uvular,** in comparative neurology, one marking the boundary of the pyramidal lobe, arising at the caudal angle of the flocculus and extending caudodorsomesad. **s., vertical.** Same as *s., precentral*. **s., Waldeyer's,** the sulcus spiralis of the cochlea.
**sulf-.** For words beginning thus, see *sulph-*.
**sulfur.** See *sulphur* and its derivatives.
**sulph-** (*sulf-*). See *sulpho-*.
**sulphaldehyde** (*sulf-al'-de-hīd*) [*sulphur; aldehyde*]. A substance produced by the action of hydrogen sulphide on ethylic aldehyde. It occurs in the form of an oleaginous liquid of a repulsive odor, solidifying at a temperature slightly below the freezing point. It is a hypnotic, and is said to produce tranquil sleep without any phenomena of excitation.
**sulphamide** (*sul-fam'-id*) [*sulphur; amide*]. One of several compounds formed by the action of sulphuryl chloride upon the free secondary amines.
**sulphaminol** (*sul-fam'-in-ol*) [*sulphur; amine*], $C_{13}H_9S_2NO$. Thioxydiphenylamine. An antiseptic substance obtained by the action of sulphur on the salts of methoxydiphenylamine. It is used by insufflation in diseases of the antrum and frontal sinuses. It has been used with success by insufflation in the treatment of laryngeal tuberculosis. Dose in cystitis gr. ij–v.
**sulphanilic acid** . (*sul-fan-il'-ik*). See *acid, sulphanilic*.
**sulphas** (*sul'-fas*) [L.: gen., *sulphatis*]. A sulphate.
**sulphate** (*sul'-fāt*) [*sulphur*]. A salt of sulphuric acid.
**sulphemoglobin, sulphæmoglobin** (*sulf-hem-o-glo'-bin*) [*sulphur; hemoglobin*]. A substance formed by the interaction of hemoglobin and hydrogen sulphide.
**sulphemoglobinemia, sulphæmoglobinæmia** (*sulf-hem-o-glo-bin-e'-me-ah*). The condition and symptoms, due to the presence of sulphemoglobin in the blood.
**sulphhydrate** (*sulf-hī'-drāt*) [*sulphur*; ὕδωρ, water]. A compound of a base with the univalent radical, *sulphhydryl*, SH.
**sulphhydric acid.** Used improperly as a synonym of sulphureted hydrogen.

# SULPHIDE 857 SUMAC

**sulphide** (*sul'-fid*) [*sulphur*]. A compound of sulphur with an element or basic radical.
**sulphin.** See *aureolin*.
**sulphinide** (*sul'-fin-id*). Saccharin.
**sulphite** (*sul'-fit*) [*sulphur*]. A salt of sulphurous acid.
**sulpho-** (*sul-fo-*) [*sulphur*]. An prefix denoting containing sulphur, or $SO_2$.
**sulphoazotized** (*sul-fo-az'-o-tīzd*). Containing sulphur and nitrogen.
**sulphobenzide** (*sul-fo-ben'-zīd*), $C_{12}H_{10}SO_2$. A crystalline substance obtained from benzene by action of fuming sulphuric acid; soluble in ether; melts at 129° C., boils at 376° C. Syn., *diphenylsulphone*.
**sulphocalcine** (*sul-fo-kal'-sēn*). A proprietary antiseptic and solvent said to contain calcium oxide, washed sulphur, benzoboric acid, extract of pancreas, and oils of wintergreen and eucalyptus; used as a gargle or spray in diphtheria.
**sulphocarbol** (*sul-fo-kar'-bol*). See *acid, sulphocarbolic*.
**sulphocarbolate** (*sul-fo-kar'-bo-lāt*) [*sulphur*; *carbolic*]. A salt of sulphocarbolic acid.
**sulphocarbolic acid** (*sul-fo-kar-bol'-ik*). See *acid, sulphocarbolic*.
**sulphocarbonated** (*sul-fo-kar'-bon-a-ted*). Containing sulphur and carbonic acid.
**sulphocarbonilid** (*sul-fo-kar-bon-il'-id*), $C_{13}H_{12}N_2S$. A crystalline substance obtained from anilin by action of alcohol and carbon disulphide with heat; soluble in alcohol and ether, melts at 153° C. Syn., *thiocarbonilid*.
**sulphocarbonism** (*sul-fo-kar'-bon-izm*). Poisoning by carbon disulphide through ingestion of some substance containing it or through inhalation of the fumes in manufactures (caoutchouc, etc.); marked by narcosis, with fall of temperature, convulsive chills, odor of radish on the breath, and in severe cases with peripheral paralysis, general anesthesia, and muscular atrophy.
**sulphoform** (*sul'-fo-form*). Trade name for a triphenylstibine sulphide.
**sulphogen** (*sul'-fo-jen*). A proprietary antiferment said to consist of sulphur, magnesia, aromatics, and the active principle of *Genista*. It is indicated in gastritis, dyspepsia, etc.
**sulphonal** (*sul'-fo-nal*) [*sulphur*], $C_7H_{16}S_2O_4$. Diethylsulphone-dimethylmethane, a crystalline substance soluble in 15 parts of boiling water and about 450 parts of cold water. It is used as an hypnotic in insomnia from functional causes. Dose 10-40 gr. (0.65-2.6 Gm.).
**sulphonalism** (*sul-fon'-al-izm*). A group of symptoms said to be occasioned by the prolonged administration of sulphonal.
**sulphonaphthol** (*sul-fo-naf'-thol*). A proprietary antiseptic.
**sulphonate** (*sul'-fon-āt*). A salt of sulphonic acid.
**sulphonation** (*sul-fon-a'-shun*). In chemistry the introduction of a sulpho-group in place of aromatic hydrogen atoms.
**sulphonethylmethane** (*sul-fon-eth-il-meth'-ān*). *Sulphonethylmethanum* (U. S. P.), $C_8H_{18}S_2O_4$. Trional, q. v. A product of the oxidation of mercaptol. Dose 15 gr. (1 Gm.).
**sulphonic acids** (*sul-fon'-ik*). Organic acids which contain the group $SO_2OH$ instead of the carboxyl group, $COOH$.
**sulphonmethane** (*sul-fon-meth'-ān*). *Sulphonmethanum* (U. S. P.), $C_7H_{16}S_2O_4$. Sulphonal, q. v. Dose 15 gr. (1 Gm.).
**sulphoparaldehyde** (*sul-fo-par-al'-de-hīd*), $(C_4H_8S_2)_3$. A crystalline substance, soluble in alcohol, insoluble in water; recommended as a hypnotic. Syn., *trithialdehyde*.
**sulphophenate, sulphophenylate** (*sul-fo-fe'-nāt, sul-fo-fen'-il-āt*). See *sulphocarbolate*.
**sulphophenol** (*sul-fo-fe'-nol*). See *acid, sulphocarbolic*.
**sulphophnon** (*sul-fo-fon*). A mixture of zinc sulphide and calcium sulphate.
**sulphosalicylic acid.** See *acid, sulphosalicylic*.
**sulphosote** (*sul'-fo-sōt*). Potassium creosote sulphonate. It is antituberculous. Dose 5-20 gr. (0.3-1.3 Gm.) several times daily.
**sulphourea** (*sul-fo-ū'-re-ah*). See *thiourea*.
**sulphovinic acid** (*sul-fo-vi'-nik*), $C_2H_5.HSO_4$, ethylsulphuric acid, a monobasic acid formed by the action of sulphuric acid on alcohol.

**sulphoxism** (*sul-foks'-izm*). Poisoning with sulphuric acid.
**sulphur** (*sul'-fur*) [L.: *gen., sulphuris*]. A nonmetallic element found native in volcanic regions (*volcanic sulphur*), and occurring combined with several metals, especially iron and copper, in the form of sulphides, called iron and copper pyrites. Sulphur can exist in various allotropic forms. The ordinary sulphur is a yellow, brittle solid, having a specific gravity of 2.07 and an atomic weight of 32.07. Symbol S. Its valence is two or six. Sulphur combines with oxygen to form sulphurous oxide (*sulphur dioxide*), $SO_2$, and sulphuric oxide (*sulphur trioxide*), $SO_3$, which by uniting with water form corresponding acids—sulphurous acid, $H_2SO_3$, and sulphuric acid, $H_2SO_4$. Other acids are also formed: hyposulphurous acid, $H_2SO_2$, thiosulphuric acid, $H_2S_2O_3$, and a series of acids termed thionic acids, viz., $H_2S_2O_6$, $H_2S_3O_6$, $H_2S_4O_6$, and $H_2S_5O_6$. Sulphurous oxide, $SO_2$, is employed as a disinfectant by fumigation. With hydrogen sulphur forms the offensively smelling gas, hydrogen sulphide (hydrosulphuric acid or sulphureted hydrogen), $H_2S$. With metals and other bases it forms sulphides. Sulphur is laxative and diaphoretic. It has been used in hemorrhoids, chronic rheumatism, gout, and locally in diphtheria and in various diseases of the skin, especially acne and scabies. **s.-alcohol,** mercaptan. **s., balsam of,** a solution of sulphur in linseed-oil. **s. dioxide.** See *sulphur*. **s. iodide** (*sulphuris iodidum*, U. S. P., B. P.), $S_2I_2$, employed in various skin diseases. From it is prepared *unguentum iodidi* (B. P.). **s., liver of,** potassium sulphide. **s., milk of.** See *s., precipitated*. **s. ointment** (*unguentum sulphuris*, U. S. P., B. P.), an ointment prepared from washed sulphur. **s., precipitated** (*sulphur præcipitatum*, U. S. P., B. P.). Dose 1-3 dr. (4-12 Gm.). **s., ruby.** See *arsenic disulphide*. **s., spirit of,** sulphuric acid. **s., sublimed** (*sulphur sublimatum*, U. S. P., B. P.), a fine yellow powder, having a slight characteristic odor, and a faintly acid taste. Dose 1-3 dr. (4-12 Gm.). Syn., *flowers of sulphur*. **s. trioxide.** See *sulphur*. **s., vegetable.** See *lycopodium*. **s., washed** (*sulphur lotum*, U. S. P.), a fine yellow powder without odor or taste.—1 dr.-½ oz. (2-16 Gm.).
**sulphuraria** (*sul-fu-ra'-re-ah*). A sediment of the San Filippo Springs, used in skin diseases. A yellow powder, containing sulphur, 32.96 %; calcium sulphide, 36.55 %; organic substances, 13.44 % silica and strontium sulphate, 1.07 %.
**sulphurated** (*sul'-fū-ra-ted*) [*sulphur*]. Combined with sulphur.
**sulphuration** (*sul-fu-ra'-shun*) [*sulphur*]. The act of dressing, anointing, or impregnating with sulphur.
**sulphuret** (*sul'-fū-ret*). A sulphide. **s., golden,** a sulphuret of antimony obtained by precipitating antimonic acid by sulphureted hydrogen.
**sulphureted** (*sul'-fū-ret'-ed*) [*sulphur*]. Combined with sulphur. **s. hydrogen.** See under *sulphur*.
**sulphureus** (*sul-fū'-re-us*). 1. Used by Mayou (1679) and early chemists in the sense of combustible, as those substances capable of burning were supposed to contain a "sulphur" which gave them that property. 2. See *sulphurous*.
**sulphuric** (*sul-fū'-rik*) [*sulphur*]. Combined with sulphur; derived from sulphur trioxide, $SO_3$. **s. acid.** See *acid, sulphuric*, under *sulphur*.
**sulphuricity** (*sul-fū-ris'-it-e*). The state of being sulphurous.
**sulphurize** (*sul'-fū-rīz*). To impregnate with sulphur.
**sulphurous** (*sul-fū'-rus* or *sul'-fū-rus*) [*sulphur*]. 1. Of the nature of sulphur. 2. Combined with sulphur; derived from sulphur dioxide, $SO_2$. **s. acid.** See *acid, sulphurous*, and *sulphur*.
**sulphume** (*sul'-fūm*). A proprietary preparation said to be "liquid sulphur." Also said to be similar to sulphurine, or to Vleminckx's solution.
**sulphurine** (*sul'-fū-rēn*). A preparation of some of the higher sulphides of sodium and potassium with sulphur.
**sulphydryl** (*sulf-hi'-dril*). The univalent radical SH.
**sum.** Abbreviation of Latin *sume, take,* or *sumendus, a um,* to be taken; used as a direction in prescriptions.
**sumac, sumach** (*sū'-mak*) [Ar., *summoq, sumac*]. The powdered leaves, peduncles, and young branches of *Rhus coriaria*, *R. cortinus*, and other species of *Rhus*, used in the manufacture of leather. Sumac

**contains** from 16 to 24 per cent. of a tannin that seems to be identical with gallotannic acid. See *rhus*.
**sumbul** (*sum'-bul*). Musk-root. The *sumbul* of the U. S. P. is the dried rhizome and root of an undetermined plant, probably of the family *Umbellifera*. It contains angelic acid, $C_5H_8O_2$, and a little valerianic acid, $C_5H_{10}O_2$. It is used as a nervine in neurasthenia, hysteria, and in anemia, chronic bronchitis, etc. Dose ½–2 dr. (2–8 Gm.). **s., extract of** (*extractum sumbul*, U. S. P.). Dose 4 gr. (0.25 Gm.). **s., fluidextract of** (*fluidextractum sumbul*, U. S. P.). Dose 30 min. (2 Cc.). **s., tincture of** (*tinctura sumbul*, B. P.). Dose 20 min.–1 dr. (1.3–4.0 Cc.).
**summation** (*sum-a'-shun*) [*summatio*]. The accumulation of effects, especially of those of muscular, sensory, or mental stimuli. **s. of stimuli,** if a stimulus in itself insufficient to cause contraction of a muscle be repeatedly applied in proper tempo and strength, contraction will finally be produced. Similar summation occurs in nervous tissue, and the cardiac contractions exhibit a rhythm of increased force, called *staircase* or *treppe rhythm*.
**summational** (*sum-a'-shun-al*) [*summatio*, a summing up]. Produced by summation. **s. tones,** supposed production of new tones by the summation or addition of the number of vibrations of existing tones.
**summer catarrh.** See *hay-fever*. **s. complaint.** See *cholera infantum*. **s. granulations.** See *trachoma*. **s. rash.** Same as *lichen tropicus*. **s. savory,** the leaves of *Satureia hortensis*, stimulant, carminative, and emmenagogue. Dose of the Fld. ext. ʒj–iv.
**summer-rash.** Lichen tropicus.
**sunburn.** Superficial inflammation of the skin caused by exposure to the sun.
**Sun cholera mixture** [New York "Sun," in which the formula was originally published]. It consists of tincture of opium 25, tincture of capsicum 12.5, tincture of rhubarb 12.5, spirit of camphor 25, spirit of peppermint 25, and alcohol 25. Dose 30 minims (2 Cc.).
**sunstroke.** Insolation.
**super-** (*sū-per-*) [L., "above" or "upon"]. A prefix denoting above, upon, or excessive.
**superabduction** (*sū-per-ab-duk'-shun*) [*super-; abduction*]. Excessive abduction.
**superacidity** (*sū-per-as-id'-it-e*). See *hyperacidity*.
**superacromial** (*sū-per-ak-ro'-me-al*) [*super-; acromion*]. Situated or occurring above or upon the acromion.
**superacute** (*sū'-per-ak-ūt*) [*super-; acutus*, pointed]. Extremely acute.
**superalbal** (*sū-per-al'-bal*). Situated in the upper part of the substantia alba.
**superalbuminosis** (*sū-per-al-bū-min-o'-sis*) [*super-; albumin*]. The over-production of albumin.
**superalimentation** (*sū-per-al-im-en-ta'-shun*) [*super-; alimentation*]. Overfeeding.
**superanal** (*sū-per-a'-nal*). Same as *supraanal*.
**superatrophy** (*sū-per-at'-ro-fe*). Excessive atrophy.
**supercallosal** (*sū-per-kal-o'-sal*) [*super-; callosum*]. Situated above or occurring above the callosum.
**supercarbonate** (*sū-per-kar'-bon-āt*). A bicarbonate.
**supercentral** (*sū-per-sen'-tral*). Lying above the center.
**supercerebral** (*sū-per-ser'-e-bral*). In the superior part of the cerebrum.
**supercilia** (*sū-per-sil'-e-ah*). Plural of *supercilium*.
**superciliary** (*sū-per-sil'-e-a-re*) [*super-; cilium*, eyelash]. Pertaining to the eyebrow. **s. entropion,** incurvation of hairs of the eyebrow against the conjunctiva. **s. ridges,** the projecting apophyses at the anterior surface of the frontal bone.
**supercilium** (*sū-per-sil'-e-um*) [L.]. The eyebrow.
**superconception** (*sū-per-kon-sep'-shun*). Same as *superfetation*.
**superdentate** (*sū-per-den'-tāt*) [*super-; dens*, tooth]. Having teeth only in the upper jaw.
**superdistention** (*sū-per-dis-ten'-shun*) [*super-; distendere*, to distend]. Excessive distention.
**superduct** (*sū'-per-dukt*). To elevate; to lead upward.
**superduction** (*sū-per-duk'-shun*). Maddox's term for sursumduction.
**superdural** (*sū-per-du'-ral*). Lying in the upper part of the dura.

**superevacuation** (*sū-per-e-vak-ū-a'-shun*). Excessive evacuation.
**superexcitation** (*sū-per-ek-si-ta'-shun*). Excessive excitement.
**superextension** (*sū-per-eks-ten'-shun*) [*super-; extendere*, to extend]. Excessive extension.
**superfecundation** (*sū-per-fe-kun-da'-shun*) [*super-; fecundus*, fertile]. The fertilization of more than one ovum of the same ovulation resulting from separate acts of coitus.
**superfecundity** (*sū-per-fe-kun'-dit-e*) [*super-; fecundus*, fertile]. Superabundant fecundity.
**superfetation, superfœtation** (*sū-per-fe-ta'-shun*) [*super-; fetus*]. A fertilization of an ovum when there is another from a previous ovulation in the uterus. Conception by a pregnant woman.
**superfibrination** (*sū-per-fib-rin-a'-shun*) [*super-; fibrin*]. Excessive formation of fibrin in the blood.
**superficial** (*sū-per-fish'-al*) [*super-; facies*, face]. Confined to or pertaining to the surface. **s. fascia,** a sheet of fatty areolar tissue under the skin.
**superficialis** (*sū-per-fish-e-a'-lis*) [L.]. 1. Superficial 2. A superficial artery, or muscle, or other part, as *superficialis volæ*, a superficial branch of the radial artery.
**superficies** (*sū'-per-fish-ēz*) [L.]. The surface or outside.
**superfissure** (*sū'-per-fish-er*) [*super-; fissure*]. The lines of overlapping of a supergyre. Also, the lines of two supergyres meeting from opposite directions.
**superflexion** (*sū-per-flek'-shun*) [*super-; flexion*]. Excessive flexion.
**superfrontal** (*sū-per-frun'-tal*) [*super-; frons*, forehead]. Superior or upper, as a fissure in the upper part of the frontal lobe of the brain.
**superfunction** (*sū-per-funk'-shun*). Excessive action of an organ or structure.
**supergenual** (*sū-per-jen'-ū-al*) [*super-; genu*, knee]. Situated above the knee.
**supergyre** (*sū'-per-jīr*). See *subgyre*.
**superhumeral** (*sū-per-hū'-mer-al*). Borne upon the shoulders; situated above the shoulders.
**superhumerale** (*sū-per-hū-mer-a'-le*). The acromion.
**superimposed** (*sū-per-im-pōzd'*) [*super-; imposed*]. Placed one upon another.
**superimpregnation** (*sū-per-im-preg-na'-shun*). 1. See *superfetation* and *superfecundation*. 2. Polyspermy, the piercing of the ovum by several spermcells.
**superincumbent** (*sū-per-in-kum'-bent*) [*super-; incumbere*, to lie upon]. Lying or resting upon something else.
**superinduce** (*sū-per-in-dūs'*) [*superinducere*, to bring upon]. To bring on as a complication of a condition already existing.
**superinvolution** (*sū-per-in-vo-lū'-shun*) [*super-; involutere*, to involute]. Hyperinvolution; excessive rolling up.
**superior** (*sū-pe'-re-or*) [comparative of *superus*, high]. Higher; denoting the upper of two parts.
**superlabia** (*sū-per-la'-be-ah*) [*super-; labium*, a lip]. The clitoris.
**superlactation** (*sū-per-lak-ta'-shun*) [*super-; lac*, milk]. 1. Excess of the secretion of milk. 2. Excessive continuance of lactation.
**superligamen** (*sū-per-li-ga'-men*) [*super-; ligamen*, a bandage]. An outer bandage to hold a surgical dressing in place.
**supermedial** (*sū-per-me'-de-al*). Above the middle.
**supermotility** (*sū-per-mo-til'-it-e*). Excessive motility.
**supernatant** (*sū-per-na'-tant*) [*super-; natans*, swimming]. Floating upon the surface of a liquid.
**supernidation** (*sū-per-nid-a'-shun*) [*super-; nidus*, nest]. Excessive proliferation of the menstrual decidua, resulting sometimes in membranous dysmenorrhea.
**supernormal** (*sū-per-nor'-mal*). Pertaining to a faculty or phenomenon which is beyond the level of ordinary experience; pertaining to a transcendental world.
**supernumerary** (*sū-per-nū'-mer-a-re*) [*super-; numerus*, a number]. Existing in more than the usual number.
**supernutrition** (*sū-per-nū-trish'-un*) [*super-; nutrire*, to nourish]. Excessive nourishment. See *hypertrophy*.
**superoccipital** (*sū-per-ok-sip'-it-al*) [*super-; occi-*

# SUPEROLATERAL 859 SUPRAMAMMARY

**put]**. Situated at or near the upper part of the occiput.

**superolateral** (sū-per-o-lat'-er-al). Located in the upper part of the side of a structure.

**superoxidized** (sū-per-oks'-id-īzd) [super-; ὀξύς, acid]. Having an excess of oxygen above the usual amount which satisfies the combining capacities of the other elements of a body.

**superoxygenation** (sū-per-oks-e-jen-a'-shun). Excessive oxygenation.

**superparasite** (sū-per-par'-as-īt) [super-; parasite]. In biology, a parasite of parasites.

**superparasitic** (sū-per-par-as-it'-ik). Pertaining to superparasitism.

**superparasitism** (sū-per-par'-as-i-tizm) [super-; parasite]. The infestation of parasites by other parasites.

**superpetrosal** (sū-per-pet-ro'-sal). Situated on the upper part of the petrosa.

**superphosphate** (sū-per-fos'-fāt). An acid phosphate.

**superpigmentation** (sū-per-pig-men-ta'-shun). Excessive pigmentation.

**supersacral** (sū-per-sa'-kral) [super-; sacrum]. Situated over the sacrum.

**supersalt** (sū'-per-sawlt). An acid salt.

**supersaturate** (sū-per-sat'-ū-rāt) [super-; saturare, to saturate]. To saturate to excess; to add more of a substance than a liquid can normally and permanently dissolve.

**superscapular** (sū-per-skap'-ū-lar). Same as suprascapular.

**superscription** (sū-per-skrip'-shun). The sign ℞ abbreviation of Latin recipe, take), at the beginning of a prescription.

**supersecretion** (sū-per-se-kre'-shun) [super-; secernere, to secrete]. Excessive secretion.

**supersensitive** (sū-per-sen'-sit-iv). Abnormally sensitive.

**supersphenoid** (sū-per-sfen'-oid) [super-; sphenoid]. Situated cephalad or dorsad of the sphenoid bone.'

**superspinatus** (sū-per-spi-na'-tus). In veterinary anatomy an extensor of the humerus which has no exact analogue in man.

**supersquamosal** (sū-per-skwa-mo'-sal) [super-; squama, scale]. A bone of the skull of ichthyosaurus, behind the postfrontal and postorbital.

**supertemporal** (sū-per-tem'-po-ral) [super-; temporal]. Situated high up in the temporal region.

**supertension** (sū-per-ten'-shun). See hypertension.

**supervenosity** (sū-per-ve-nos'-it-e) [super-; venosus, venous].· The condition in which the blood has become venous to a high degree.

**supervention** (sū-per-ven'-shun) [super-; venire, to come]. That which is added; an extraneous, or unexpected condition added to another, as the supervention of septicemia, or other complication in disease.

**superversion** ℞ (sū-per-vur'-shun). See sursumversion.

**supination** (sū-pin-a'-shun) [supinus, on the back]. 1. The turning of the palm of the hand upward. 2. The condition of being supine; lying on the back.

**supinator** (sū'-pin-a-tor). See under muscle.

**supine** (sū-pīn') [supinus, on the back]. Lying on the back face upward or palm upward.

**suplagalbumin** (sū-plag-al'-bū-min). See sucholoalbumin.

**suplagotoxin** (sū-plag-o-toks'-in). See sucholotoxin.

**suppedaneous** (sup-ed-a'-ne-us) [sub, under; pes, foot]. Pertaining to the sole of the foot.

**supplemental** (sup-le-men'-tal) [supplere, to complete]. Additional. s. air. See under respiration.

**support** (sup-ort') [supportare, to carry]. 1. The act of holding, anything in its position. 2. Any appliance acting as a supporter.

**supporter** (sup-or'-ter) [supportare, to carry]. An apparatus intended to aid in supporting a prolapsed organ (as the uterus), or a pendulous abdomen.

**suppositorium** (sup-oz-it-o'-re-um). [supponere, to place under: gen., suppositorii; pl., suppositoria]. See suppository.

**suppository** (sup-oz'-it-o-re) [suppositorium, from sub-, under; ponere, to place]. A solid medicated compound designed to be introduced into the rectum, urethra, or vagina. Its consistence is such that while retaining its shape at ordinary temperatures, it readily melts at the temperature of the body. The basis of most suppositories is oil of theobroma. For urethral suppositories a mixture of gelatin and glycerol is used. The only suppositories that are official are the glycerol suppositories (suppositoria glycerini, U. S. P.). See under glycerin. s., tannic-acid, one part of tannin to five parts of cacao-butter.

**suppression** (sup-resh'-un) [suppressio, a keeping back]. A sudden cessation of secretion, as suppression of the urine or of the menses.

**suppurant** (sup'-ū-rant) [suppuration]. 1. Promoting suppuration. 2. An agent promoting suppuration.

**suppuration** (sup-ū-ra'-shun) [subpurare, to form pus]. The formation of pus.

**suppurative** (sup'-ū-ra-tiv) [suppuration]. 1. Producing pus. 2. An agent that favors suppuration. s. fever, pyemia, q. v.

**supra-** (sū-prah-) [L., "above"]. A prefix signifying upon or above.

**supra-acromial** (sū-prah-ak-ro'-me-al) [supra-; acromion]. Situated above the acromion.

**supra-anal** (sū-prah-a'-nal) [supra-; anus]. Situated above the anus.

**supra-auricular** (sū-prah-aw-rik'-ū-lar) [supra-; auricle]. Above the external ear. s. point. See under craniometric point.

**supra-axillary** [supra-; axilla]. Above the axilla.

**suprabuccal** (sū-prah-buk'-al) [supra-; bucca, mouth]. Above the buccal region.

**supracapsulin** (sū-prah-kap'-sū-lin). Trade name of a preparation of the suprarenal capsule.

**supracephalic** (sū-prah-sef-al'-ik) [supra-; κεφαλή, head]. Placed on the head.

**suprachoroid** (sū-prah-ko'-roid) [supra-; choroid]. Above the choroid of the choroid plexus.

**suprachoroidea** (sū-prah-ko-roid'-e-ah). The choroid layer next to the sclera.

**supraciliary** (sū-prah-sil'-e-a-re). Same as superciliary.

**supraclavicular** (sū-prah-kla-vik'-ū-lar) [supra-; clavicle]. Above the clavicle.

**supracommissure** (sū-prah-kom'-ish-ūr) [supra-; commissure]. The commissure of the brain just in front of the stalk of the epiphysis.

**supracondylar, supracondyloid** (sū-prah-kon'-dil-ar, sū-prah-kon'-dil-oid) [supra-; condyle]. Above a condyle.

**supracostal** (sū-prah-kos'-tal) [supra-; costa, a rib]. Above the ribs.

**supracotyloid** (sū-prah-kot'-il-oid) [supra-; cotyloid]. Above the cotyloid cavity.

**supradiaphragmatic** (sū-prah-di-af-rag-mat'-ik) [supra-; diaphragm]. Situated above the diaphragm.

**supradin** (sū-prad'-in). A powdered preparation of the suprarenal capsules, containing 0.015 per cent. of iodine.

**supradorsal** (sū-prah-dor'-sal) [supra-; dorsum, back]. Dorsal; placed dorsally.

**supraepicondylar** (sū-prah-ep-e-kon'-dil-ar). Situated above an epicondyle.

**supraesophageal** (sū-prah-e-so-faj'-e-al) [supra-; esophagus]. Situated above the gullet.

**supraflocculus** (sū-prah-flok'-ū-lus). The dorsal lobe of the paraflocculus in the lower mammals.

**supragenual**. See supergenual.

**supraglenoid** (sū-prah-glen'-oid). Above the glenoid cavity.

**supraglottic** (sū-prah-glot'-ik) [supra-; γλωττίς, glottis]. Above the glottis.

**suprahepatic** (sū-prah-hep-at'-ik). Above the liver. s. veins, the hepatic veins.

**suprahyoid** (sū-prah-hi'-oid) [supra-; hyoid]. Above the hyoid bone.

**supra iliac** (sū-prah-il'-e-ak) [supra-; ilium]. Above or at the upper end of the ilium.

**suprainguinal** (sū-prah-in'-gwin-al) [supra-; ilium]. Above the groin.

**supraintestinal** (sū-prah-in-test'-in-al) [supra-; intestine]. Above the intestine.

**supralabial** (sū-prah-la'-be-al) [supra-; labium, lip]. Pertaining to or situated above the upper lip.

**suprabialis** (sū-prah-la-be-a'-lis). See levator labii superioris under muscle.

**supraliminal** (sū-prah-lim'-in-al) [supra-; limen, threshold]. Lying above the threshold. s. consciousness, the empirical self of common experience.

**supralumar** (sū-prah-lum'-bar). Above the loin.

**supramalleolar** (sū-prah-mal-e'-o-lar) [supra-; malleolus]. Above a malleolus.

**supramammary** (sū-prah-mam'-a-re) [supra-; mamma, breast]. Above the mammary gland.

# SUPRAMANDIBULAR  860  SURGEONSHIP

**supramandibular** (sū-prah-man-dib'-ū-lar). Situated above the mandible.
**supramarginal** (sū-prah-mar'-jin-al) [supra-; margin]. Above an edge or margin, as the supramarginal convolution of the brain.
**supramastoid** (sū-prah-mas'-toid) [supra-; mastoid]. Above the mastoid process of the temporal bone.
**supramaxilla** (sū-prah-maks-il'-ah) [supra-; maxilla]. The supramaxillary bone.
**supramaxillary** (sū-prah-maks'-il-a-re) [supra-; maxilla]. Pertaining to the superior maxilla.
**supramental** (sū-prah-men'-tal) [supra-; mentum, chin]. Above the chin.
**supranasal** (sū-prah-na'-sal) [supra-; nasus, nose]. Above the nose. s. point. See craniometric points.
**supraneural** (sū-prah-nū'-ral) [supra-; νεῦρον, nerve] Over or above the neural axis.
**supranuclear** (sū-prah-nū'-kle-ar). Above the nucleus.
**supraobliquus** (sū-prah-ob-li'-kwus) [supra-; obliquus, slanting]. Coues' name for the obliquus superior muscle of the eye.
**supraoccipital** (sū-prah-ok-sip'-it-al) [supra-; occiput]. 1. Above the occipital bone. 2. The upper part of the occipital bone.
**supraocclusion** (sū-prah-ok-lū'-zjun). The condition of a tooth which has erupted further from its socket than normal.
**supraomphalodymia** (sū-prah-om-fal-o-dim'-e-ah) [supra-; ὀμφαλός, navel; δύειν, to enter]. A form of somatodymia in which the union is in the superior umbilical region.
**supraorbital** (sū-prah-or'-bit-al) [supra-; orbit]. 1. Above the orbit, as of the supraorbital nerve. 2. Pertaining to the supraorbital nerve. s. ridge. The curved prominent margin forming the upper boundary of the orbit. s. foramen, a foramen at the inner third of the orbit; it transmits the supraorbital artery, vein, and nerve. Sometimes it is incomplete, being but a notch or groove, and then is called the s. notch. s. point. See craniometric points.
**suprapatellar** (sū-prah-pat-el'-ar) [supra-; patella]. Above the patella.
**suprapedal** (sū-prah-ped'-al) [supra-; pes, foot]. Above the foot.
**suprapelvic** (sū-prah-pel'-vik) [supra-; pelvis]. Above the pelvis.
**suprapharyngeal** (sū-prah-far-in'-je-al) [supra-; pharynx]. Above the pharynx.
**suprapineal** (sū-prah-pin'-e-al) [supra-; pineal]. Above the pineal gland.
**suprapontine** (sū-prah-pon'-tin) [supra-; pons, a bridge]. Above or in the superior part of the pons.
**suprapubic** (sū-prah-pū'-bik) [supra-; pubis]. Above the pubes.
**suprarenaden** (sū-prah-ren'-ad-en). A preparation made from the suprarenal capsules; used in Addison's disease, neurasthenia, etc. Dose 15–23 gr. (1.0–1.5 Gm.) daily.
**suprarenal** (sū-prah-re'-nal) [supra-; ren, the kidney]. 1. Above the kidney, as the suprarenal capsule. 2. Pertaining to the suprarenal capsule. s. body, s. capsule, a small triangular organ situated above the kidney, and consisting of an external or cortical and an internal or medullary portion. The cortex consists of polygonal cells disposed in three layers—the zona glomerulosa, zona fasciculata, and zona reticularis. Fibrous septa, derived from the capsule, extend into the organ and separate the groups of cells. The medulla contains cords and networks of polygonal cells, and in its center ganglion-cells and nonmedullated nerve-fibers. The function of the suprarenal body is not definitely known—it is believed to bear some relation to pigment production. s. epithelioma. See Grawitz's tumor. s. rests. See rests, adrenal.
**suprarenalin** (sū-prah-ren'-al-in). A preparation of suprarenal glands; used as a vasoconstrictor and hemostatic.
**suprarene** (sū'-prah-rēn). A suprarenal capsule.
**suprarenin** (sū-prah-ren'-in). A synonym of epinephrin.
**suprascapular** (sū-prah-skap'-ū-lar) [supra-; scapula]. Above or in the upper part of the scapula.
**supraseptal** (sū-prah-sep'-tal) [supra-; septum]. Situated above a septum.
**supraspinal** (sū-prah-spi'-nal) [supra-; spine]. Above a spine.
**supraspinales** (sū-prah-spi-na'-lēz). See under muscle.

**supraspinatus** (sū-prah-spi-na'-tus) [supra-; spine]. Above the spine, as the supraspinatus muscle. See under muscle.
**supraspinous** (sū-prah-spi'-nus) [see supraspinatus]. Above the spinous process of the scapula or of a vertebra. s. fossa, the triangular depression above the spine of the scapula.
**suprastapedial** (sū-prah-sta-pe'-de-al) [supra-; stapes]. Above the stapes.
**suprasternal** (sū-prah-ster'-nal) [supra-; sternum]. Above the sternum.
**suprasylvian** (sū-prah-sil'-ve-an). Above the Sylvian fissure.
**suprasymphyseal** (sū-prah-sim-fiz'-e-al). Above the symphysis pubis.
**supratemporal** (sū-prah-tem'-po-ral) [supra-; temporal]. Above the temporal region.
**suprathoracic** (sū-prah-tho-ras'-ik) [supra-; thorax]. Above the thorax.
**supratrochlear** (sū-prah-trok'-le-ar) [supra-; trochlea]. Above the trochlea or pulley of the superior oblique muscle.
**supraturbinal** (sū-prah-tur'-bin-al). The superior turbinate bone.
**supratympanic** (sū-prah-tim-pan'-ik) [supra-; tympanum]. Above the tympanum.
**supravaginal** (sū-prah-vaj'-in-al) [supra-; vagina]. 1. Above a sheath; on the outside of a sheath. 2. Above the vagina.
**supraverge** (sū'-prah-verj) [supra-; vergere, to incline]. To diverge in a vertical plane.
**supravergence** (sū-prah-ver'-jens). The ability of the two eyes to diverge in a vertical plane; an ability measured by a prism of 2°–3°. Syn., sursumvergence. s., right, the ability to overcome prisms, base down, before the right eye, or base up before the left eye.
**sura** (sū'-rah) [L.]. 1. The calf of the leg. 2. A form of toddy made in Western Africa from the sap of the oil-palm.
**sural** (sū'-ral) [sura]. Pertaining to the calf of the leg.
**suralimentation** (sur-al-im-en-ta'-shun) [super-; alimentation]. The method of forced feeding or overalimentation sometimes employed in pulmonary tuberculosis and other diseases.
**suranal** (sur-a'-nal). Same as supra-anal.
**surcingle** (sur'-sin-gl) [super-; cingulum, a belt]. Above the thorax.
**surculus** (sur'-kū-lus) [L., "a twig, shoot"; pl., surculi]. In biology, a sucker. surculi fellei, the ductules conveying the bile to the hepatic ducts.
**surditas** (sur'-dit-as). Synonym of deafness. s. verbalis, see aphasia.
**surdity** (sur'-dit-e) [surdus, deaf]. Deafness.
**surdomute** (sur'-do-mūt) [surdus, deaf; mutus, mute]. A deaf and dumb person.
**surdomutitas** (sur-do-mū'-tit-as) [surdus; mutus, mute]. Deaf-mutism.
**surdus** (sur'-dus). See deaf.
**surexcitation** (sur-eks-i-ta'-shun) [super-; excitatio, a rousing]. Excessive excitement.
**surface** (sur'-fas) [Fr., surface]. 1. The exterior of a body. 2. The face or surface of a body; a term frequently used in anatomy in the description of bones. s., fixation, a curved surface the points of which occupy in the two monocular fields positions which are identical horizontally, regardless of vertical disparity. s., labial, the surface of a tooth-crown which is toward the lips. s.-markings, (in anatomy), marks made upon the skin to indicate the size, shape, and position of underlying structures. s.-wells, those which obtain their supply from the subsoil water.
**surgeon** (sur'-jun) [see surgery]. One who practises surgery. s.-apothecary, in England, one who is licensed to practise by the Royal College of Surgeons and by the Apothecaries' Society. s.-aurist, an otologist. s.-dentist, a dentist who practises the surgical as well as the mechanical parts of his profession. s.-general, the title of certain surgeons of high rank, chiefly in the military and naval services. s.-generalship, the office of a surgeon-general. s.-ship, the office of a surgeon. s., veterinary, one who treats disease of the domestic animals.
**surgeon** (sur'-jun-se) [surgeon]. The office of surgeon, military or naval.
**surgeonry** (sur'-jun-re). The practice of a surgeon.
**surgeonship** (sur'-jun-ship). The office of a surgeon.

**surgery** (*sur'-jer-e*) [χειρ, hand; εργειν, to work]. The branch of medicine dealing with diseases requiring operative procedure. **s., antiseptic**, the application of antiseptic methods in the treatment of wounds. **s., aseptic**, operative procedure in the absence of germs, everything coming in contact with the wound being sterile. **s., conservative**, measures directed to the preservation rather than to the removal of a part. **s., major**, that in which the operations are important and involve risks to life. **s., military**, that pertaining to gunshot wounds and other injuries peculiar to military life. **s., minor**, that part of surgery including procedures not involving danger to life, as bandaging, the application of splints, dressings, sutures, counterirritation, cauterization, and bloodletting. **s., operative**, that which refers to the performance of operations. **s., orthopedic**, the remedy of deformities by manual and instrumental measures. **s., plastic**, repair of absent or defective tissue by transference of tissue from another part or person. **s., railway**, deals with injuries received on railways. **s., veterinary**, the surgery of domestic animals.

**surgical** (*sur'-jik-al*) [*surgery*]. 1. Pertaining to surgery. 2. Produced by *surgical* operations. 3. A name applied in some hospitals to a piece of cotton or other material, used by the physician to remove the lubricant from his fingers after vaginal or rectal examination, and before washing his hands. **s. fever**, fever following operation or injury. **s. kidney**, suppuration of the kidney due to disease of the genitourinary tract. **s. neck** (of the humerus), the constricted part of the shaft below the tuberosities, so called because it is a common seat of fracture. **s. sore-throat**, sore-throat due to absorption of septic matters in hospitals; it sometimes attacks internes and nurses. **s. tuberculosis**, tuberculous disease that may be reached by operative treatment, *e. g.*, that involving glands, joints, bone, and the like.

**Surinam bark** (*sū'-rin-am*). The bark of *Andira retusa*, used as an anthelmintic.

**surinamine** (*sū-rin-am'-ēn*). An alkaloid found by Hüttenschmid, 1824, in the bark of *Andira retusa*. It forms fine, gleaming needles, without taste or odor and of neutral reaction, soluble in water, and with anthelmintic action. Syn., *andirine; geoffroyine.*

**surons** (*sū'-rons*). Skins which have served the purpose of carrying drugs, especially from South America.

**surra** (*soor'-rah*) [native Indian name]. An epizootic pernicious anemia in horses, mules, and camels due to *Trypanosoma evansi*.

**surrenal** (*sur-re'-nal*) [*supra-; ren*, the kidney]. 1. Suprarenal. 2. A suprarenal gland.

**surrogate** (*sur'-o-gāt*) [*surrogatus*, substituted]. Any medicine or ingredient used as a substitute for another and more expensive ingredient, or one toi which there is a special objection in any particular case.

**sursumduction** (*sur-sum-duk'-shun*) [*sursum*, up; *ducere*, to lead]. 1. The power of the two eyes of fusing two images when one eye has a prism placed vertically before it. 2. See *supravergence*. 3. A movement of either eye alone upward. **s., right**, the absolute power that the right eye has to rotate upward.

**sursumvergence** (*sur-sum-vur'-jenz*) [*sursum; vergere*, to bend]. The turning of the eyes upward; supravergence.

**sursumversion** (*sur-sum-ver'-shun*) [*sursum; vertere*, to turn]. The movement of both eyes up.

**surumpe.** The name in the Andes for hyperesthesia of the retina observed at great altitudes.

**survival** (*sur-vi'-vl*). The persistence of an individual or race after the general extinction of related forms.

**survivorship** (*sur-vi'-vor-ship*) [*super; vivere*, to live]. In medical jurisprudence the probability of a certain individual having survived others when all concerned were in the same accident and all were killed.

**susceptible** (*sus-sep'-tib-l*) [*suscipere*, to undertake]. Sensitive to an influence. In pathology, liable to become affected with a disease.

**susceptivity** (*sus-sep-tiv'-it-e*) [*suscipere*, to undertake]. The state or quality of being susceptible.

**suscitability** (*sus-si-ta-bil'-it-e*) [*suscitare*, to lift up]. The quality of being easily roused or excited.

**suscitation** (*sus-si-ta'-shun*). The act of exciting.

**susotoxin** (*sū-so-tok'-sin*) [*sus*, pig; τοξικόν, poison], $C_{10}H_{26}N_2$. A toxin found in cultures of the bacillus of hog cholera.

**suspended** (*sus-pen'-ded*) [*suspendere*, to hang up]. 1. Hanging; applied to an ovule hanging from the ovarian wall, or a seed from the summit of a cell. 2. Interrupted. **s. animation**, a term sometimes applied to the temporary cessation of the vital functions. It may be due to asphyxia, to syncope, or to the trance-like condition that closely simulates death, in which the patient may remain for some hours or even days. **s. matter**, undissolved particles diffused throughout a liquid.

**suspension** (*sus-pen'-shun*) [*sub-; pendere*, to hang]. 1. Hanging; a mode of treatment of tabes dorsalis and other nervous diseases, in which the patient hangs by the neck, chin, and shoulders. 2. Temporary cessation of a function or process.

**suspensoid** (*sus-pen'-soid*). An apparent solution which is seen, by the microscope, to consist of small particles of the solute in active Brownian movement.

**suspensorium** (*sus-pen-so'-re-um*) [*sub*, under; *pendere*, to hang]. That upon which anything hangs for support. **s. hepatis**, the suspensory ligament of the liver. **s. testis**, the cremaster muscle. **s. vesicæ**, the superior false ligament of the urinary bladder.

**suspensory** (*sus-pen'-so-re*) [see *suspension*]. 1. Serving for suspension or support, as *suspensory* ligament, *suspensory* bandage. 2. A device for suspending a part.

**suspiration** (*sus-pi-ra'-shun*) [*suspiratio*]. A sigh, *q. v.*; the act of sighing.

**sustentacular** (*sus-ten-tak'-ū-lar*) [*sustentaculum, sustentare*, to support]. Pertaining to or serving as a sustentaculum. **s. cells**, a name given to certain supporting cells in the testicle. **s. tissue**, supporting tissue.

**sustentaculum** (*sus-ten-tak'-ū-lum*) [*sustentare*, to support]. A support. **s. lienis**, the suspensory ligament of the spleen. **s. tali**, a process of the os calcis supporting the astragalus.

**sustoxin.** See *susotoxin.*

**susurration** (*sū-sur-ra'-shun*) [*susurratio*]. A murmur, or susurrus.

**susurrus** (*sū-sur'-rus*) [L.]. A soft murmur in aneurysm, cardiac diseases, contracting muscle, etc. **s. aurium**, see *tinnitus aurium.*

**sutura** (*sū-tū'-rah*) [L.]. See *suture*

**sutural** (*sū'-tū-ral*) [*suture*]. Pertaining to or having the nature of a suture.

**suture** (*sū'-tūr*) [*sutura*, a seam]. 1. A line of joining or closure, as a *cranial suture*. 2. A stitch or series of stitches used in closing the lips of a wound. **s., arcuate, s., basilar**, the junction between the basilar surface of the occipital bone and the posterior surface of the body of the sphenoid. **s., biparietal.** See **s., sagittal**. **s., buried**, one completely covered by and not involving the skin. **s., catgut**, one in which the material employed is catgut. **s., chainstitch**, the sewing machine stitch. **s., circular**, one that is applied to the entire circumference of a divided part, as the intestine. **s., cobbler's**, one made by arming a needle with two threads. **s., continuous, s., glover's**, one in which the thread passes across the wound continually in the same direction, and is tied only at the beginning and end. **s., coronal**, the union of the frontal with the parietal bones transversely across the vertex of the skull. **s., cranial**, the line of union of two or more cranial bones. **s., cross**, the application of two single stitches to a T-wound. **s., dentate**, an irregular notched suture, as that between the parietal bones. **s., dry**, one carried through adhesive-plaster strips applied to the lips of the wound. **s., ethmofrontal**, the union between the frontal and ethmoid bones. **s., ethmolacrimal**, the union between the lacrimal and ethmoid bones. **s., ethmosphenoid**, the union between the sphenoid and ethmoid bones. **s. false**, sutura notha, any suture in which there is interlocking of the bones without serration. **s., the four masters'**, a suture of the intestine used in the thirteenth century in which the trachea of a goose was used as a means of support and the ends of the severed intestine brought into position on it and sutured with four interrupted stitches which did not include the trachea. **s., frontal**, a suture which at birth joins the two frontal bones from the vertex to the root of the nose, but which afterward becomes obliterated. **s., frontomalar**, the union between the malar and frontal bones. **s., frontomaxillary**, the union between the superior maxillary and frontal bones. **s.,**

Superficial and deep interrupted sutures.

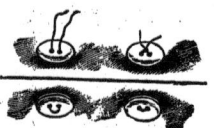

Button suture. (Moullin.)

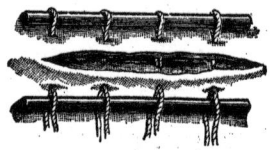

Quilled suture. (Stewart.)

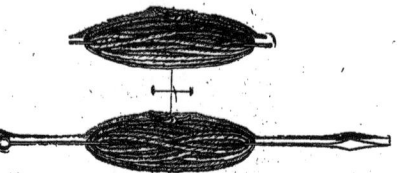

Twisted suture. (Esmarch and Kowalzig.)

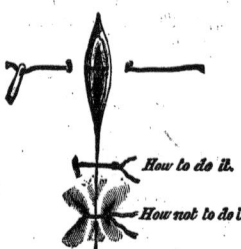

Tension in suturing. (Moullin.)

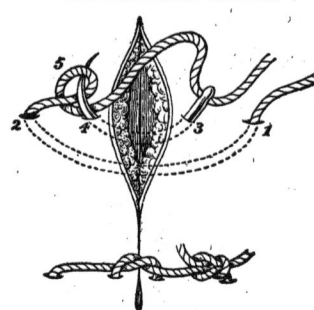

Combined retention and coaptation suture. The needle is inserted at 1, brought out at 2, reinserted at 3, and emerges at 4, passing through the loop at 5. When drawn tight it holds the wound edges firmly together and prevents inversion of the skin, as shown in the lower part of the illustration. (Stewart.)

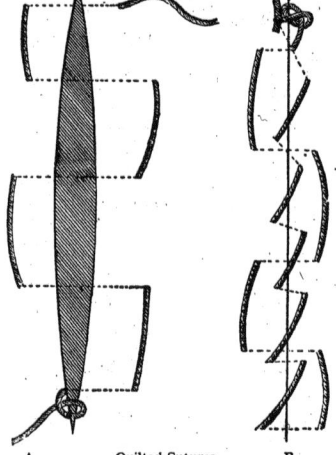

A   Quilted Sutures.   B

knot. s., harelip, s., twisted, one in which the edges of the wound are transfixed with pins and approximation secured by twisting or wrapping the ends of the pins with thread. s., harmonic, same as *harmonia*. s., intermaxillary, the union between the superior maxillary bones. s., internasal, the union between the nasal bones. s., interparietal. See s., *sagittal*. s., interrupted, one of a series of sutures passed through the margins of the wound, and each of which is tied separately. s., jugal. See s., *sagittal*. s., lambdoid, the union between the two superior borders of the occipital bone and the parietal bones. s., longitudinal. See s., *sagittal*. s., masto-occipital. See s., *occipitomastoid*. s., mastoparietal. See s., *parietomastoid*. s., mattress, a continuous suture which is made back and forth through both lips of a wound. s., maxillolacrimal, the union between the lacrimal and superior maxillary bones. s., mediofrontal. See s., *frontal*. s., metopic, See s., *frontal*. s., nasofrontal, the frontonasal suture. s., naso-

frontonasal, the union between the nasal and frontal bones. s., frontoparietal. See s., *coronal*. s., frontosphenoid, the union between the wings of the sphenoid bone and the frontal bone. s., frontotemporal, the union between the frontal and temporal bones. s., granny-knot, a single-knot stitch is formed and the needle is passed in the opposite direction from which it was inserted under the thread in forming a square

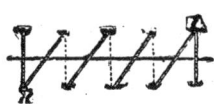

Continuous or Glover's suture. (Esmarch and Kowalzig.)

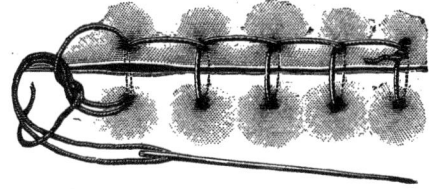

Continuous button-hole suture. (Walsham.)

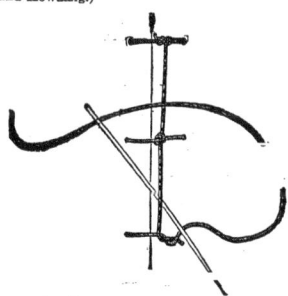

Ford's suture: showing two square knots, a single knot, and the method of completing a square knot. (DaCosta.)

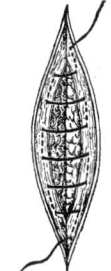

Halsted's subcuticular suture. (Stewart)

maxillary, the union between the superior maxillary and nasal bones. s., occipital. See s., lambdoid. s., occipitomastoid, the union between the mastoid portion of the temporal bone and the occipital bone. s., occipitoparietal. See s., lambdoid. s., palatine, the union between the palate bones. s., palatine transverse, See s., palatomaxillary. s., palatomaxillary, a suture between the palatal processes of the os palatinum and the superior maxilla. s., parallel, a continuous subcuticular suture. s., parietomastoid, the union between the mastoid portion of the temporal bone and the parietal bone. s., petro-occipital, the union between the occipital bone and the petrous portion of the temporal. s., petrosphenoid, the union between the great wing of the sphenoid bone and the petrous portion of the temporal. s., pin-. See s., harelip. s., quill-, s., quilled, one in which a doubled thread is passed and tied over quills or pieces of a soft catheter. s., quilted, one similar to a mattress suture. s., relaxation-, one introduced some distance from the wound-margin, carried through its depths, and made to emerge at some distance on the opposite side, to relieve the tension of the wound-sutures proper. s., sagittal, the union between the superior borders of the parietal bones. s., sclerocorneal, of Kalt. See stitch, sclerocorneal. s., shotted, one in which both ends of the suture are passed through a perforated shot, which is then tightly compressed. s., sphenomalar, the union between the malar bone and the great wing of the sphenoid. s., sphenopalatine, a cranial suture between the sphenoid and palatine bones. s., sphenoparietal, the union between the greater wing of the sphenoid bone and the parietal bone. s., sphenotemporal, the union between the temporal and the sphenoid bone. s., squamoparietal, s., squamosal, the union between the squamous portion of the temporal bone and the parietal bone. s., squamosphenoid, the union between the great wing of the sphenoid and the squamous portion of the temporal bone. s., subcuticular, a buried, continuous suture in which the needle is passed horizontally into the true skin back and forth until the wound is closed. s., tension, same as s., relaxation. s., twisted, one made by winding a thread around a needle that pierces the wound transversely. s., Wormian, anterior, the upper part of the lambdoid suture.

Suzanne's gland (soo-zan') [Jean Georges Suzanne, French physician, 1859– ]. A mucous gland found in the floor of the mouth close to the median line.
svapnia (svap'-ne-ah). Denarcotized opium.
s. v. r. Abbreviation of spiritus vini rectificatus, rectified spirit of wine.
swab (swob). A piece of cloth or sponge upon the end of a stick, used in feeding the sick, making applications to the throat, cleansing the mouth and teeth, etc. s.-stick, a rod or shaft, one extremity of which is to be wound with cotton.
swallow (swol'-o). 1. To take into the stomach through the throat. The cavity of the throat and gullet; the esophagus. 2. A fissirostral oscine passerine bird, of the genus Hirundo. s.-'s nest, the nidus hirundinis; a deep fossa of the cerebellum between the commissure of the flocculus and uvula. s.-tails, notches on the teeth of a horse.
swallowing. Deglutition.
swamp (swomp) [origin obscure]. A piece of low, wet, marshy land. s.-apple, a parasitic growth found on swamp-azalea. s.-dogwood. 1. Buttonbush. 2. A plant allied to Cornus florida; Cornus sericea. s.-fever, a malarial fever. s.-itch, same as army itch. s.-laurel, kalmia glauca. s.-milkweed, Asclepias incarnata. s.-pine, Pinus australis; broom-pine. s. sassafras. The Magnolia glauca, the bark of which is aromatic and diaphoretic.
swarming. 1. Moving in a swarm; 2. Breeding multitudes. A form of endogenous cell-formation noted in certain algæ.
sweat (swet) [AS., swāt]. The secretion of the sudoriferous glands, consisting of a transparent, colorless, aqueous fluid, holding in solution neutral fats, volatile fatty acids, -cholesterin, traces of albumin and urea, free lactic acid, sodium lactate, sodium chloride, potassium chloride, and traces of alkaline phosphates. s., bloody, see hæmatidrosis. s., blue, perspiration which has a blue color, it is thought by some to be due to oxidation of the colorless indican secreted in the sweat. s., English, see sweating-sickness. s.-gland, one of the small glands secreting the sweat, situated in the true skin and subcutaneous areolar tissue, consisting of a convoluted tube from which the excretory duct passes outward through the skin. In its passage through the epidermis the duct

is more or less spiral. **s., green,** sweat having a bluish or greenish color, seen mainly in copper-workers, and due to copper taken into the system by the inhalation of particles or fumes, or with food and drink. **s.-house,** a room or inclosure furnished with apparatus for subjecting the occupant to the sudorific effects of steam. **s., night,** drenching perspiration occurring at night or whenever the patient sleeps, in the course of pulmonary tuberculosis. **s., phosphorescent,** a very rare peculiarity of the sweat, in which it becomes phosphorescent; it has been observed in some cases of miliaria and after eating phosphorescent fish. The phosphorescence is thought to be due to bacilli. **s., Picardy,** see *sweating-sickness*. **s., red,** a peculiar, red perspiration noted in the axillæ and genital region, and due to microorganisms which have developed in the hairs of these warm, moist parts.
**sweating-sickness.** An infective, febrile epidemic disorder, characterized by a rapid course and profuse perspiration. It was prevalent in England at the end of the fifteenth and first half of the sixteenth century. Syn., *English sweat; miliaria; Picardy sweat; sudor anglicus;* Fr., *suette miliaire;* Ger., *schweissfreisel.*
**Swediaur's disease** (*sva'-de-our*) [François Xavier *Swediaur*, Austrian physician, 1748–1824]. See *Albert's disease.*
**Swedish green.** See *Scheele's* green. **S. movements.** Certain systematic gymnastic exercises intended to exercise and develop the human body, and affect function, nutrition, etc.
**sweeny** (*swe'-ne*) [origin obscure]. A wasting, or shrinkage, of the shoulder-muscles of the horse, generally due to some lameness of the foot or foreleg; it is also called *swinney.*
**sweet.** Having a taste like that of sugar or honey. **s. precipitate,** mercuric chloride. **s. principle of fats,** glycerin. **s. spirit of niter,** spiritus ætheris nitrosi.
**sweetbread.** 1. The pancreas. 2. The thymus.
**swell** (*swel*). To grow in bulk; to puff out. **s.-bodies,** Zuckerkandl's term for the venous plexuses found over the turbinated bodies.
**swelling** (*swel'-ing*) [ME., *swellen,* to swell]. Any morbid enlargement, inflation, tumor, or protuberance. **s., albuminous,** see *s., cloudy*. **s., blennorrhagic** (of the knee), the enlargement of the knee which occurs in gonorrheal synovitis. **s., cloudy,** a swelling of cells due to disturbed nutrition. **s., glassy,** amyloid degeneration. **s., lacteal,** a swelling of the breast due to obstruction of the lacteal ducts. **s., Sœmmering's crystalline,** an annular swelling formed in the lower part of the capsule, behind the iris, after extraction of the crystalline lens. **s., white.** 1. A disease of the bones which causes them to dilate as if distended by pressure from within. 2. A tumor, usually of the wrist or ankle, due to caries. Syn., *hydrarthrosis.*
**swine-erysipelas.** A contagious disease of swine marked by fever and a red eruption on the neck and belly.
**swine-fever.** Hog cholera.
**swine-plague.** An infectious disease of swine, due to the bacillus of swine plague.
**swoon.** Syncope.
**sycephalus** (*si-sef'-al-us*) [*syn; κεφαλή,* head]. A double monster having two incomplete heads joined together.
**sycoceryl** (*si-ko-ser'-il*). Applied to an alcohol the acetate of which is a constituent of sycoretin. **s. hydrate,** sycoceryl aldehyde.
**sycoma** (*si-ko'-mah*) [*σῦκον,* a fig: *pl., sycomata*]. A condyloma, or wart; a condition characterized by warty or fig-like excrescences on the soft tissues of the body, as the eyelids, tongue, anus, or genital.
**sycoretin** (*si-ko-re'-tin*) [*σῦκον,* fig; *κηρός,* wax]. A resin found in 1855 in *Ficus rubiginosa,* of New South Wales.
**sycose** (*si'-kōs*). Saccharin.
**sycosiform** (*si-ko'-se-form*) [*sycosis; forma,* form]. Resembling sycosis.
**sycosis** (*si-ko'-sis*) [*σύκωσις,* a fig-like excrescence on the flesh]. An inflammatory disease affecting the hair-follicles, particularly of the beard, and characterized by papules, pustules, and tubercles, perforated by hairs, together with infiltration of the skin and crusting. **s., bacillogenic,** a variety of so-called "non-parasitic" sycosis, ascribed to the *Bacillus sycosiferus fœtidus.* **s. barbæ,** sycosis of the beard. **s. capillitii.** 1. Dermatitis papillaris capillitii, of Kaposi; s. frambœsiformis, of Hebra; acne keloid, of Bazin. 2. Kerion. 3. Pustular eczema of the scalp. **s., coccogenic,** the so-called "non-parasitic" sycosis which is caused by organism belonging to the schizomycetes (Unna). **s. contagiosa,** see *s. parasitaria*. **s. frambœsiformis.** See *dermatitis papillaris capillitii.* **s., hypertrophic.** See *s., keloid*. **s., hypogenic, s., hyphomycetous, s. parasitaria,** of Unna; tinea sycosis, the inflammation excited by the *Trichophyton tonsurans.* **s., keloid,** sycosis in which keloid degeneration occurs in the cicatrices resulting from the follicular inflammation (Milton). Also called *ulerythema sycosiforme* (Unna). **s., lupoid.** See *s., keloid*. **s. mentagra.** See *s. barbæ*. **s., non-parasitic,** sycosis due to the presence of coccogenic organisms. See *s., coccogenic*. **s. palpebræ marginalis,** sycosis affecting the edge of the eyelid. **s. parasitiria, s. parasitica,** barbers' itch; a disease of the hair follicles, usually affecting the region covered by the beard, and due to the presence of the *Trichophyton tonsurans.* See *s., hyphogenic*. **s., parasitic.** See *s. parasitaria*. **s., schizomycetic.** See *s., coccogenic.*
**Sydenham's chorea** (*sid'-en-ham*) [Thomas *Sydenham,* English physician, 1624–1689]. Chorea minor; infectious chorea. **S.'s cough,** spasm of the respiratory muscles in hysteria. **S.'s laudanum,** wine of opium.
**syl.** A flavoring agent in syrup.
**syllable-stumbling** (*sil-ab-l-stum'-bling*). A form of dysphasia wherein each sound and syllable can be distinctly uttered, but the word as a whole is spoken with difficulty. It occurs in paretic dementia.
**syllabic utterance.** Scanning speech, observed in insular sclerosis; the words are enunciated slowly and separately and there may be a staccato accentuation of individual syllables.
**syllabus** (*sil'-ab-us*) [*σύλλαβος,* a collection]. A compendium containing the heads of a discourse; the main propositions of a course of lectures; an abstract.
**syllepsiology** (*sil-lep-se-ol'-o-je*) [*σύλληψις,* pregnancy; *λόγος,* science]. The physiology of conception and pregnancy.
**syllepsis** (*sil-ep'-sis*) [*σύλληψις*]. Conception, or impregnation.
**Sylvester's method.** See *Silvester*.
**Sylvian** (*sil'-ve-an*). Described by the anatomist Jacobus *Sylvius* (1478–1555), or Franciscus *Sylvius* (1614–1672). **S. angle,** the angle formed by the posterior limb of the Sylvian fissure with a line perpendicular to the superior border of the hemisphere. **S. aqueduct,** a narrow canal extending from the third to the fourth ventricle. **S. artery,** the middle cerebral artery, lying in the fissure of Sylvius. **S. fissure,** a deep fissure of the brain beginning on the outer side of the anterior perforated space, and extending outward to the lateral surface of the hemisphere. It has two branches—a short vertical and a long horizontal, the latter separating the parietal from the temporosphenoid lobe. Between the branches lies the island of Reil. **S. fossa,/S. valley,** the depression which appears on the surface of the brain about the end of the second month of fetal life and afterward becomes the Sylvian fissure. **S. vein,** one of the veins of the convexity of the brain, which courses at first along the fissure of Sylvius and then ascends across the hemisphere. **S. ventricle,** the fifth ventricle.
**Sylviduct** (*sil-ve-dukt*). The aqueduct of Sylvius.
**sym-** (*sim-*). The same as *syn-*.
**symbion, symbiont** (*sim'-be-on, sim'-be-ont*) [*syn;* *βίος,* a life]. In biology, either of two organisms living in intimate association; a commensal.
**symbiosis** (*sim-bi-o'-sis*) [*syn; βίος,* a life]. In biology, the intimate association of two living organisms, not parent and offspring, male and female, or parasite and host. Commensalism.
**symblepharon** (*sim-blef'-ar-on*) [*σύν,* together; *βλέφαρον,* the eyelid]. Adhesion of the eyelids to the eyeball. **s., anterior,** when the edge of the lid is adherent. **s., posterior,** when the adhesion is at the conjunctival fold. **s., total,** when the entire lid is adherent.
**symblepharopterygium** (*sim-blef-ar-o-ter-ij'-e-um*). A variety of symblepharon in which a cicatricial band resembling a pterygium connects the lid and the eyeball.
**symblepharosis** (*sim-blef-ar-o'-sis*)- [*syn; βλέφαρον,* eyelid]. Adhesion of the eyelids to the globe of the eye or to each other.

SYMBOL 865 SYMPTOM

**symbol** (sim'-bol) [σύμβολον, a pledge]. 1. A sign or character denoting an idea. The following are commonly employed in medicine; ℞, Recipe, take of; ℈, scruple; ℨ, dram; f℥, fluidram; ℥, ounce; f℥, fluidounce; ♏, Minim. 2. In chemistry, a conventional abbreviation of the name of an element, generally consisting of the initial letter or letters of the name in Latin or English. It denotes one atom of the element.

**symbolism** (sim'-bol-izm) [symbol]. The delusional or hallucinational interpretation of all events or objects as having a mystic significance; a habit not uncommon in certain forms of insanity.

**Syme's operation** (sim) [James Syme, Scotch surgeon, 1799-1870]. 1. Amputation at the ankle-joint, the malleoli being sawn through, and a flap made with the skin of the heel. 2. External urethrotomy.

**symmelic** (sim-el'-ik) [syn; μέλος, limb]. Characterized by a coalition of the limbs.

**symmelus, symelus** (sim'-el-us) [σύν, together; μέλος, a limb]. A species of monster characterized by imperfect development of the pelvis and lower extremities with more or less intimate fusion of the latter.

**symmetric, symmetrical** (sim-et'-rik, sim-et'-rik-al) [συμμετρία, proportion]. Pertaining to or exhibiting symmetry. s. **gangrene.** See *sphaceloderma*.

**symmetry** (sim'-et-re) [syn; μέτρον, measure]. In anatomy, a harmonious correspondence of parts; also the relation of homologous parts at opposite sides or ends of the body. In pathology, the theory that constitutional diseases affect both lateral halves of the body impartially.

**symparalysis** (sim-par-al'-is-is). Conjugate paralysis, a term given to the abolition of certain synkineses of the eye.

**sympathectomy, sympathicectomy** (sim-path-ek'-to-me, sim-path-is-ek'-to-me). Excision of part of the sympathetic nerve.

**sympatheoneuritis** (sim-path-e-o-nū-ri'-tis). Inflammation of the sympathetic nerve.

**sympathetic** (sim-path-et'-ik) [see *sympathy*]. 1. Pertaining to or produced by sympathy. 2. Conveying sympathy or sympathetic impulses, as the sympathetic system. s. **ganglia.** See under *s. system*. s. **irritation**, irritation of an organ arising from irritation of another related organ, as *sympathetic irritation* of one eye from irritation of the other. s. **nerve**, see *s. nervous system*. s. **ophthalmia**, inflammation of one eye arising subsequent to inflammation of the other eye. s. **plexuses**, see under *s. nervous system*. s. **nervous system**, a system of ganglia (*sympathetic ganglia*) forming a chain from the cranium to the end of the spinal column, connected together by nerve-fibers, and supplying the viscera and blood-vessels. At intervals the sympathetic nerves and ganglia form plexuses (*sympathetic plexuses*).

**sympatheticectomy** (sim-pa-thet-e-sek'-to-me). Excision of parts of the sympathetic nerve.

**sympatheticism** (sim-path-et'-is-izm). See *sympathism*.

**sympatheticoparalytic** (sim-path-et-ik-o-par-al-it'-ik). Due to paralysis of the sympathetic nerve.

**sympatheticotonic** (sim-path-et-ik-o-ton'-ik). Applied to migraine caused by tonic contraction of the arteries due to overaction of the sympathetic.

**sympatheticus** (sim-path-et'-ik-us). See *sympathetic nerve*.

**sympathetiplex** (sim-path-et'-ip-leks) [*sympathetic*; *plexus*]. A sympathetic plexus.

**sympathic** (sim-path'-ik). Synonym of *sympathetic*.

**sympathiconeuritis** (sim-path-ik-o-nū-ri'-tis). Inflammation of the sympathetic nerve.

**sympathicotripsy** (sim-path-ik-o-trip'-se) [*sympathy*; τρίβειν, to crush]. In treatment of mental diseases, crushing of the superior cervical ganglion.

**sympathism** (sim'-path-izm). Susceptibility to hypnotic suggestion.

**sympathist** (sim'-path-ist). One who is susceptible to hypnotic suggestion.

**sympathizer** (sim-path-i'-zer). An eye with *sympathetic ophthalmia*, q. v.

**sympathoblast** (sim-path'-o-blast). A primitive sympathetic nerve cell.

**sympathy** (sim'-path-e) [σύν, together; πάθος, suffering]. The mutual relation between parts more or less distant, whereby a change in the one has an effect upon the other.

29

**symperitoneal** (sim-per-it-on-e'-al) [syn-; *peritoneum*]. Connecting artificially two parts of the peritoneum.

**sympexia** (sim-peks'-e-ah). See *sympexis*.

**sympexion** (sim-peks'-e-on) [augmented form of *sympexis*]. A sympexis.

**sympexis** (sim-peks'-is) [σύμπηξις]. A concretion found in the vessels or crypts of certain glands, especially the thyroid and lymphatic, the prostate, and in the seminal vesicles.

**symphiocephalus** (sim-fi-o-sef'-al-us) [syn-; φύειν, to grow; κεφαλή, the head]. A twin monster with the union in the head.

**symphora** (sim'-for-ah). Synonym of *congestion*.

**symphorema** (sim-for-e'-mah) [syn-; φερείν, to bear]. The state of being congested.

**symphoresis** (sim-fo-re'-sis) [syn-; φερείν, to bear]. Congestion, or a congestive process.

**symphoricarpus** (sim-for-ik-ar'-pus) [syn-; φερείν, to bear; καρπός, fruit]. In biology, bearing clustered fruits.

**symphorol** (sim'-for-ol) [syn-; φορά, that which is brought forth]. A generic name for the caffeine sulphates or sulfocaffeinates. s. **L**, lithium and caffeine sulphate. s. **N**, sodium and caffeine sulphate. s. **S**, strontium and caffeine sulphonate.

**symphyseal** (sim-fiz'-e-al) [*symphysis*]. Pertaining to a symphysis.

**symphyseorrhaphy** (sim-fiz-e-or'-af-e) [*symphysis*; ῥαφή, a seam]. Suture of a divided symphysis.

**symphyseotome** (sim-fiz-e-ot'-om). An instrument used in performing symphyseotomy.

**symphyseotomy** (sim-fiz-e-ot'-o-me) [*symphysis*; τομή, a cutting]. The operation of dividing the symphysis pubis for the purpose of increasing the diameters of the pelvic canal and facilitating labor.

**symphysic** (sim-fiz'-ik). Same as *symphyseal*. See also *teratism*.

**symphysion** (sim-fiz'-e-on). The middle point of the outer border of the alveolar process of the mandible.

**symphysis** (sim'-fiz-is) [σύν, together; φύειν, to grow]. The line of junction of two bones. s. **cartilaginosa**, synchondrosis. s. **ligamentosa**, syndesmosis. s. **mandibulæ**, s. **menti**, the central vertical ridge upon the outer aspect of the lower jaw, showing the line of union of the two halves. s. **ossium muscularis**. See *syssorcosis*. s. **pubis**, the line of junction of the pubic bones. s., sacrococcygeal, the sacrococcygeal articulation. s., sacroiliac, the union between the sacrum and the ilium.

**symphysodactylia** (sim-fiz-o-dak-til'-e-ah). Synonym of *syndactylism*.

**symphysopsia** (sim-fiz-ops'-e-ah). Synonym of *cyclopia*.

**symphysoskelia** (sim-fiz-o-ske'-le-ah) [*symphysis*; σκελός, leg]. The condition in which the lower extremities are united.

**sympiesis** (sim-pi-e'-sis) [syn-; πίεσις, a squeezing]. A pressing together of parts.

**sympiesometer** (sim-pe-es-om'-et-ur) [σύν, together; πίεσις, a squeezing; μέτρον, a measure]. An apparatus for estimating pressure.

**symplocarpus fœtidus** (sim-plo-kar'-pus fet'-id-us). *Dracontium fœtidum*, or skunk-cabbage, the rhizome of which is stimulant, antispasmodic, and narcotic, and has been used in asthma, chronic rheumatism, chorea, hysteria, etc. Dose 10-20 gr. (0.65-1.3 Gm.).

**Symplocos** (sim'-plo-kos) [σύμπλοκος, braided]. A genus of the *Styraceæ*. The leaves of *S. alstonia*, a tree of South America, are used as mate and also as a digestive and diaphoretic. *S. platyphylla*, sweetleaf, is indigenous to the southern United States and South America; the root bark is used in intermittent fever. *S. racemosa* is indigenous to India, and furnishes a red coloring-matter and lodh-bark, used in plasters. It contains lotourin, colloturin, and lotouridin. *S. tinctoria*, sweetleaf, of South Carolina, contains in its sweet leaves a yellow coloring-matter. The root is used as a digestive.

**sympodia** (sim-po'-de-ah) [syn-; πούς, foot]. The condition in which the lower extremities are united.

**sympodial** (sim-po'-de-al) [syn-; πούς, foot]. Pertaining to a sympodium.

**sympodium** (sim-po'-de-um) [syn-; πούς, foot: pl., *sympodia*]. In biology, a stem which consists of a series of secondary stems or axes which have arisen as branches one from another, as in the grape-vine, the tomato, the linden, etc. See *pseudaxis*.

**symptom** (simp'-tom) [σύν, together; πτώμα, a falling]. The change in a patient occurring during disease and serving to point out its nature and location. See *sign*. s., **accessory**, s., **assident**, a minor symptom. s., **accidental**, one intervening in the course of a disease without having any connection with it. s.-**complex**, the ensemble of symptoms of a disease. See also *syndrome*. s.-**complex, Avellis'**, paralysis of one-half of the soft palate, associated with a recurrent paralysis on the same side. s.-**complex, Benedikt's**, tremor of one side of the body, in part or wholly, and oculomotor paralysis of the other side. s.-**complex, Bernhardt-Roth's**. See *Bernhardt's paresthesia*. s.-**complex, Erb-Goldflam's**. See *Erb's disease*. s.-**complex, Escherich's**. See *pseudotetanus*. s.-**complex, Friedmann's vasomotor**, a train of symptoms following injury to the head, consisting of headache, vertigo, nausea and intolerance of mental and physical exertions and of galvanic excitation; it is occasionally associated with ophthalmoplegia and mydriasis. These phenomena may subside and recur with greater intensity, with fever, unconsciousness, and paralysis of the cranial nerves, ending in fatal coma. They are probably due to an encephalitis of slow development with acute exacerbations. s.-**complex, Hoppe-Goldflam's**, see *Erb's disease*. s.-**complex, Putnam-Dana's**, combined sclerosis of the lateral and posterior columns of the spinal cord. s.-**complex, Roth's**, see *Bernhardt's paresthesia*. s.-**complex, Wilk's**, see *Erb's disease*. s.s, **concomitant**, accessory phenomena occurring in connection with the essential phenomena of a disease. s., **constitutional**, s., **general**, one produced by the effect of the disease on the whole body. s. **delayed**, see *deferred shock*. s. **direct**, one depending directly upon disease. s., **dissociation-**, see *dissociation-symptom*. s., **equivocal**, one of doubtful significance. s. **grouping**, the same as symptom-complex, *q. v.* s., **indirect**, one only indirectly due to disease. ss., **labyrinthine**, a group of symptoms due to lesion or disease of the internal ear. s., **local**, one indicating the concentration of a disease in a certain part of the body. s., **negatively pathognomonic**, one which never occurs in a certain disease and therefore by its presence shows the absence of that disease. s., **objective**, one observed by the physician. s., **passive**. See *s. static*. s., **pathognomonic**, a symptom which exhibits itself only in a certain disease and therefore undeniably proves its presence. ss., **physical**, the physical signs of morbid conditions. s., **rational**, a subjective symptom. s., **signal**, the first disturbance of sensation prededing a more extensive convulsion, as the aura heralding an attack of epilepsy. s., **static**, a symptom which indicates the condition in a single organ without reference to the rest of the body. s. **subjective**, that observed only by the patient. ss., **sympathetic**, symptoms for which no adequate cause can be given other than so-called sympathy. s. **turpitudinis**, nymphomania.

**symptomatiatria** (*simp-to-mat-e-a'-tre-ah*) [*symptom*; ιατρεία, treatment]. Treatment directed solely to the symptoms.

**symptomatic** (*simp-tom-at'-ik*) [*symptom*]. 1. Pertaining to or of the nature of a symptom. 2. Affecting symptoms, as *symptomatic* treatment. s. **anthrax**, see *black-leg*.

**symptomatography** (*simp-to-mat-og'-ra-fe*) [*symptom*; γράφειν, to write]. A written or printed description of symptoms.

**symptomatologic** (*simp-to-mat-o-loj'-ik*) [*symptom*; λόγος, science]. Pertaining to symptomatology.

**symptomatology** (*simp-tom-at-ol'-o-je*) [*symptom*; λόγος, science]. The science of symptoms; the symptoms of disease taken together as a whole.

**symptosis** (*simp-to'-sis*) [συμπίπτειν, to fall together]. Wasting; emaciation; collapse.

**sympus** (*sim'-pus*) [*syn-*; πούς, foot]. A vice of development consisting in coalescence of the lower limbs; a siren monster. There may be but one foot (*s. monopus*) or two (*s. dipus*), or the feet may be represented only by toes (*s. apus*).

**syn-** (*sin-*) [σύν, together]. A prefix signifying with or together.

**synadelphus** (*sin-ad-el'-fus*) [*syn-*; ἀδελφός, brother]. A monster having eight limbs with but one head and trunk.

**synæsthesia** (*sin-es-the'-ze-ah*). See *synesthesia*.

**synalgia** (*sin-al'-je-ah*) [*syn-*; ἄλγος, pain]. Pain felt in a distant part from an injury or stimulation of another part.

**synanastomosis** (*sin-an-as-to-mo'-sis*) [*syn-*; *anatomosis*]. The joining of several blood-vessels.

**synanche** (*si-nang'-ke*). Synonym of *diphtheria*.

**synangium** (*sin-an'-je-um*) [*syn-*; ἀγγεῖον, vessel]: *pl., synangia*]. An arterial axis, or trunk.

**synanthema** (*sin-an'-them-ah*) [συναυθεῖν, to blossom together]. A group of efflorescences on the skin.

**synantherin** (*sin-an'-ther-in*). See *inulin*.

**synanthrose** (*sin-an'-throse*). See *levulose*.

**synaphymenitis** (*sin-af-i-men-i'-tis*) [συναφή, connection; ὑμήν, a membrane; ιτις, inflammation]. A former synonym for conjunctivitis.

**synapse** (*sin-aps'*). See *synapsis*.

**synapsis** (*sin-ap'-sis*) [*syn-*; ἄπτειν, to clasp]. 1. The anatomical relation of one neuron with another. The intertwining of the terminal arborizations of the neurons by means of which the nerve-impulses may pass from one to another. Syn., *synapse*. 2. The joining together of chromosomes.

**synaptase** (*sin-ap'-tās*). See *emulsin*.

**synarthrodia** (*sin-ar-thro'-de-ah*). See *synarthrosis*.

**synarthrodial** (*sin-ar-thro'-de-al*) [*synarthrosis*]. Pertaining to or of the nature of a synarthrosis.

**synarthrophysis** (*sin-ar-thro-fi'-sis*) [*syn-*; *arthrosis*; φύειν, to grow]. Progressive ankylosis of the joints.

**synarthrosis** (*sin-ar-thro'-sis*) [*syn-*; ἄρθρον, a joint]. A form of articulation in which the bones are immovably bound together without any intervening synovial cavity. The forms are *sutura*, in which processes are interlocked; *schindylesis*, in which a thin plate of one bone is inserted into a cleft of another bone, and *gomphosis*, in which a conical process is held by a socket.

**syncaryosis** (*sin-kar-e-o'-sis*) [*syn-*; κάρυον, a nut]. A term proposed by His for syncytial formation or the growth of the multinuclear giant-cells.

**syncephalus** (*sin-sef'-al-us*) [*syn-*; κεφαλή, head]. A monster with two heads fused into one.

**synchilia, syncheilia** (*sin-ki'-le-ah*) [*syn-*; χεῖλος, lip]. Atresia of the lips.

**synchondrosial** (*sin-kon-dro'-ze-al*) [*syn-*; χόνδρος, cartilage]. Pertaining to a synchondrosis.

**synchondrosis** (*sin-kon-dro'-sis*) [*syn-*; χόνδρος, a cartilage]. A joint in which the surfaces are connected by a growth of cartilage. s., **sacroiliac**, the union between the sacrum and the ilium.

**synchondrotomy** (*sin-kon-drot'-o-me*) [*syn-*; χόνδρος, cartilage; τέμνειν, to cut]. A division of the cartilage uniting bones, especially of that of the symphysis pubis.

**synchopexia** (*sin-ko-peks'-e-ah*). Same as *tachycardia*.

**synchronism** (*sin'-kro-nizm*) [*syn-*; χρόνος, time]. Concurrence in time of two or more events.

**synchronous** (*sin'-kro-nus*) [*syn-*; χρόνος, time]. Occurring at the same time.

**synchysis scintillans** (*sin'-kis-is sin'-til-lans*) [σύγχυσις, a mixing together]. The presence of bright, shining particles in the vitreous humor of the eye.

**syncleisis** (*sin-kli'-sis*). Synonym of *occlusion*.

**synclinal** (*sin'-klin-al*). Bending or inclining in the same direction.

**synclisis** (*sin'-klis-is*). Same as *synclitism*.

**synclitic** (*sin-klit'-ik*) [συγκλίτης, leaning together]. Exhibiting or characterized by synclitism.

**synclitism** (*sin'-klit-izm*) [συγκλίτης, bending together]. A condition marked by parallelism or similarity of inclination; parallelism between the pelvic planes and those of the fetal head.

**synclonus** (*sin'-klo-nus*) [*syn-*; κλόνος, clonus]. 1. Clonic movements occurring simultaneously in several muscles. 2. A disease thus characterized, as chorea. s. **ballismus**, paralysis agitans. s. **tremens**, general tremor.

**syncopal** (*sin'-ko-pal*) [*syncope*]. Pertaining to or characterized by syncope.

**syncope** (*sin'-ko-pe*) [*syn-*; κόπτειν, to strike or cut]. Swooning or fainting, a partial or complete temporary suspension of the functions of respiration and circulation from cerebral anemia. s. **anginosa**, synonym of *angina pectoris*. s., **laryngeal**, laryngeal vertigo. s., **local**, sudden pallor and insensibility of a part.

**syncopexia, syncopexy** (*sin-ko-peks'-e-ah, sin-ko-peks'-e*). See *synesthesia*.

**syncopic** (*sin-kop'-ik*). Of the nature of syncope; syncopal.

**syncyanin** (*sin-si'-an-in*). A blue pigment elaborated by *Bacillus cyanogenus*.

**syncytial** (sin-sit'-e-al) [see syncytium]. Pertaining to a syncytium.
**syncytiolysin** (sin-sit-o-ol'-is-in) [syncytium; λύειν, to loosen]. A cytolisin produced by injections of an emulsion made from placental tissue.
**syncytioma** (sin-sit-e-o'-mah) [syncytium; ὅμα, a tumor]. A tumor composed of syncytial tissue. s. **malignum**, malignant degeneration of the villi of the chorion. See *deciduoma malignum*.
**syncytium** (sin-sit'-e-um) [syn-; κύτος, a cell]. 1. A mass of protoplasm with numerous nuclei. 2. The collection of epithelial cells forming the outermost covering of the chorionic villi.
**syndactyl** (sin-dak'-til) [syn-; δάκτυλος, a digit]. Having the adjoining fingers or toes bound together.
**syndactylia, syndactylism, syndactyly** (sin-dak-til'-e-ah, sin-dak'-til-izm, sin-dak'-til-e) [syn-; δάκτυλος, finger]; webbed toes.
**syndactylous** (sin-dak'-til-us). Same as *syndactyl*.
**syndectomy** (sin-dek'-to-me). See *peritomy*.
**syndelphus** (sin-del'-fus) [syn-; ἀδελφός, brother]. A monocephalic double monstrosity with a single pelvis, united thoraces, four upper and four lower extremities.
**syndesis** (sin-de'-sis) [συνδεῖν, to bind together]. The state of being bound together.
**syndesmectopia** (sin-dez-mek-to'-pe-ah) [σύνδεσμος, a bond; ἔκτοπος, out of place]. Ligamentous displacement.
**syndesmitis** (sin-dez-mi'-tis) [σύνδεσμος, a ligament; ιτις, inflammation]. 1. Inflammation of a ligament. 2. Conjunctivitis.
**syndesmodiastasis** (sin-dez-mo-di-as'-tas-is) [σύνδεσμος, ligament; διάστασις, separation]. Separation of the ligaments.
**syndesmography** (sin-dez-mog'-ra-fe) [syndesmus; γράφειν, to write]. The branch of anatomy treating of ligaments.
**syndesmology** (sin-dez-mol'-o-je). See *syndesmography*.
**syndesmoma** (sin-dez-mo'-mah) [σύνδεσμος, ligament; ἤμα, tumor]. A new growth containing a large amount of connective tissue.
**syndesmoplasty** (sin-dez-mo-plas'-te) [σύνδεσμος, ligament; πλάσσειν, to form]. A plastic operation on a ligament.
**syndesmorrhaphy** (sin-dez-mor'-af-e) [σύνδεσμος, ligament; ῥαφή, a suture]. Suture or repair of ligaments.
**syndesmosis** (sin-dez-mo'-sis) [syndesmos]. A form of articulation in which the bones are connected by ligaments.
**syndesmotic** (sin-dez-mot'-ik) [σύνδεσμος, ligament]. Bound together.
**syndesmotomy** (sin-dez-mot'-o-me) [syndesmus; τέμνειν, to cut]. 1. Dissection of the ligaments. 2. The division of a ligament.
**syndesmus** (sin-dez'-mus) [σύνδεσμος, ligament]. A ligament.
**syndrome** (sin'-drōm) [σύν, together; δρόμος, a running]. The aggregate symptoms of a disease; a complex of symptoms. See *symptom-complex*. s., **levulosuric**, a variety of diabetes with melancholia, insomnia, impotence, and the presence in the urine of a levulose that disappears rapidly on the suppression of carbohydrates. s. **temporanea di Gubler**, see *Gubler's hemiplegia*.
**synechia** (si-nek'-e-ah or sin-e-ki'-ah) [σύν, together; ἔχειν, to hold; pl., *synechiæ*]. A morbid union of parts; especially, adhesion of the iris to a neighboring part of the eye. s., **annular**, s., **circular**, exclusion of the pupil. s., **anterior**, adhesion between the iris and transparent cornea. s., **posterior**, adhesion between the iris and crystalline lens. Syn., *ptosis diplopia*. s., **total**, adhesion of the entire surface of the iris to the lens.
**synechiæ pericardii** (sin-ek'-e-e per-e-kar'-de-i) [synechia]. Adhesions of the pericardium.
**synechiotomy** (sin-ek-e-ot'-o-me). See *synechotomy*.
**synechotome** (sin-ek'-o-tōm) [synechia; ἔχειν, to hold together; τέμνειν, to cut]. An instrument for the division of adhesions, particularly of the membrana tympani.
**synechotomy** (sin-ek-ot'-o-me) [synechia; τέμνειν, to cut]. The division of a synechia.
**synectenterotomy** (sin-ek-ten-ter-ot'-o-me) [synechia; *enterotomy*]. The division of an intestinal adhesion.

**synencephalia** (sin-en-sef-al'-e-ah) [syn-; ἐγκέφαλος, the brain]. The condition of synencephalus.
**synencephalocele** (sin-en-sef'-al-o-sēl) [syn-; *encephalocele*]. An encephalocele arising from abnormal adhesions, probably the result of some intrauterine inflammation.
**synencephalus** (sin-en-sef'-al-us) [syn-; ἐγκέφαλος, brain]. A monster having two bodies with but a single head.
**synergetic** (sin-er-jet'-ik). Exhibiting synergy; working together.
**synergic** (sin-er'-jik) [synergy]. Pertaining to synergy.
**synergist** (sin'-er-jist) [synergy]. An agent cooperating with another.
**synergistic**. See *synergetic*.
**synergy** (sin'-er-je) [syn-; ἔργον, work]. The cooperative action of two or more agents (*synergists*) or organs.
**synesis** (sin'-e-sis) [σύνεσις, a coming together]. Faculty of comprehension, intelligence, sagacity.
**synesthesia, synæsthesia** (sin-es-the'-ze-ah) [syn-; αἴσθησις, sensation]. A secondary sensation or subjective impression accompanying an actual perception.
**synetion, synætion** (sin-e'-shun) [syn-; αἰτία, cause]. A cause which cooperates with another to produce disease.
**syngenesis** (sin-jen'-es-is) [syn-; γένεσις, generation]. 1. The theory that the embryo is the product of the union of the male and female elements; also the theory that the embryo contains within itself the germs of all future generations developed from it. 2. Reproduction by union of male and female elements. **sygenetic** (sin-jen-et'-ik) [see *syngenesis*]. Propagated by means of both parents.
**syngenic** (sin-jen'-ik). Synonym of *congenital*.
**syngignoscism** (sin-gig'-no-sizm) [syn-; γιγνώσκειν, to know]. Hypnotism, so termed from the agreeing of one mind with another.
**synidrosis** (sin-id-ro'-sis) [syn-; ἱδρώς, sweat]. Concurrent sweating. The association of perspiration with another condition.
**synizesis** (sin-iz-e'-sis) [syn-; ἴζειν, to sit]. Closure. s. **pupillæ**, closure of the pupil.
**synkaryon** (sin-kar'-e-on). A nucleus resulting from the fusion of two pronuclei.
**synkinesis, synkinesia** (sin-kin-e'-sis, sin-kin-e'-she-a) [syn-; κίνησις, movement]. Involuntary movement taking place in one part of the body synchronously with or in consequence of a voluntary or reflex movement in another part.
**synneurosis** (sin-nū-ro'-sis). See *syndesmosis*.
**synocha, synochus** (sin'-o-kah, sin'-o-kus) [syn-; ἔχειν, to hold on]. Any continued fever.
**synochal** (sin'-o-kal) [σύνοχος, continued]. Pertaining to synocha.
**synonym** (sin'-o-nim) [syn-; ὄνομα, a name]. A word which can replace another word without alteration of meaning. In medicine, any variant name by which a disease, an organ, or a part of the body may be known.
**synophrys** (sin-off'-ris) [σύν, together; ὀφρύς, the eyebrow]. The growing together of the eyebrows.
**synophthalmia** (sin-off-thal'-me-ah) [syn-; *ophthalmia*]. A malformation in which the orbits form a single, continuous cavity. This condition is called, also, *cyclopia*.
**synophthalmus** (sin-off-thal'-mus). See *cyclops*.
**synopsia** (sin-ops'-e-ah) [syn-; ὤψ, eye]. Congenital union of the eyes.
**synopsis** (sin-op'-sis) [syn-; ὄψις, a seeing]. A classified collation. A general view.
**syn‌orchidism, synorchism** (sin-or'-kid-izm, sin-or'-kizm) [syn-; ὄρχις, testicle]. Partial or complete fusion of the two testicles.
**synoscheos** (sin-os'-ke-os) [syn-; ὄσχεος, scrotum]. A condition of adherence between the skin of the penis and that of the scrotum.
**synosteography** (sin-os-te-og'-ra-fe) [syn-; ὀστέον, bone; γράφειν, to write]. The descriptive anatomy of the joints.
**synosteology** (sin-os-te-ol'-o-je) [syn-; ὀστέον, bone; λόγος, knowledge]. The sum of what is known regarding the joints.
**synosteophyte** (sin-os'-te-o-fīt) [syn-; ὀστέον, a growth]. Congenital bony ankylosis. Syn., *synostosis congenita*.
**synostosis** (sin-os-to'-sis) [syn-; ὀστέον]. See *synostosis*.
**synosteotome** (sin-os'-te-o-tōm). A knife for the dissection of joints.

**synosteotomy** (*sin-os-te-ot′-o-me*) [*syn-*; ὀστέον, bone; τομή, a cutting]. The dissection of the articulations of bones; anatomy of the joints.
**synostology** (*sin-os-tol′-o-je*). See *synosteology*.
**synostosed** (*sin-os-tōsd′*) [*syn-*; ὀστέον, bone]. Joined in bony union.
**synostosis** (*sin-os-to′-sis*) [*syn-*; ὀστέον, bone]. A union of normally separate bones by osseous material. **s. congenita**. See *synosteophyte*. **s., tribasilar**, shortening of the base of the skull and consequent curvature of the basal parts of the brain; a cause of imbecility.
**synostotic** (*sin-os-tot′-ik*) [*synostosis*]. Pertaining to or of the nature of synostosis.
**synotia** (*sin-o′-she-ah*) [*syn-*; οὖς, ear]. Union of the ears, as in certain monsters.
**synotus** (*sin-o′-tus*) [*syn-*; οὖς, ear]. A monster characterized by fused ears.
**synovectomy** (*sin-o-vek′-to-me*). 1. Excision of synovial membrane. 2. Arthrectomy.
**synovia** (*sin-o′-ve-ah*) [*syn-*; ᾠόν, an egg]. The clear, alkaline, lubricating fluid secreted within synovial membranes.
**synovial** (*sin-o′-ve-al*) [*synovia*]. Pertaining to the synovia. **s. bursa**. See *bursa, synovial*. **s. membrane**. See *membrane, synovial*. **s. sheath**, a synovial membrane which lines the cavity attached to a bone and through which a tendon glides.
**synovin** (*sin′-o-vin*) [*syn-*; ᾠόν, egg]. The form of mucin found in synovia.
**synoviparous** (*sin-o-vip′-ar-us*) [*synovia*; *parere*, to produce]. Producing or secreting synovia.
**synovitis** (*sin-o-vi′-tis*) [*synovia*; ιτις, inflammation]. Inflammation of a synovial membrane. **s., acute suppurative**, a very acute purulent form, of rheumatic or traumatic origin, leading to ankylosis. Syn., *anthropyosis empyema articuli; pyarthrosis*. **s. chronic purulent**, synonym of *fungous arthritis*. **s. chronic serous**, synonym of *hydrarthrosis*. **s., dry**, synovitis with little if any exudate. **s., exanthematous**, synovitis produced by the exanthemata. **s., fibrinous**. See *s., dry*. **s., fungous**, synonym of *fungous arthritis*. **s., gonorrheal**, synonym of *rheumatism, gonorrheal*. **s. hyperplastica**, **s. hyperplastica granulosa**, **s. hyperplastica lævis**, **s. hyperplastica pannosa**, arthritis fungosa. **s., lipomatous**, synovitis in which the new formation undergoes fatty degeneration. **s., metritic**, a synovitis secondary to uterine infection. **s., puerperal**, synovitis occurring after childbirth, and due to septic infection. **s., purulent**, synovitis with suppuration. **s., scarlatinal**, synovitis occurring in an attack of scarlet fever. **s., syphilitic**, synovitis due to syphilitic inflammation. **s., tendinous**, inflammation of the synovial sheath surrounding a tendon. **s., tuberculosis**, synovitis with deposits of tubercle. **s., urethral**, synonym of gonorrheal synovitis.
**syntasis** (*sin′-tas-is*) [συντείνειν, to stretch together]. A stretching, or tension.
**syntaxis** (*sin-taks′-is*) [συντάσσειν, to arrange]. 1. Articulation. 2. Reduction, taxis. 3. A suture.
**syntectic** (*sin-tek′-tik*). Pertaining to or characterized by syntexis; wasting.
**syntenosis** (*sin-ten-o′-sis*) [*syn-*; τένων, tendon]. Articulation by means of tendons, as in the human digits.
**synteresis** (*sin-ter-e′-sis*) [συντηρεῖν, to keep safe]. Preventive treatment, or hygiene; prophylaxis.
**synteretic** (*sin-ter-et′-ik*) [συντηρεῖν, to keep safe]. Pertaining to prophylaxis, or synteresis; hygienic; preventive.
**synteretics** (*sin-ter-et′-iks*) [συντηρεῖν, to watch closely]. Hygiene.
**syntexis** (*sin-teks′-is*) [συντήκειν, to melt together]. A wasting; tabes; phthisis.
**synthermal** (*sin-ther′-mal*). Same as *isothermal*.
**synthesis** (*sin′-thes-is*) [σύν, with; τιθέναι, to place]. In chemistry, the artificial formation of a compound by combining its constituents.
**synthetic** (*sin-thet′-ik*) [*synthesis*]. Pertaining to or produced by synthesis.
**synthetism** (*sin′-thet-izm*) [*synthesis*]. The sum of operations and means necessary for reducing a fracture and holding the parts in position.
**synthol** (*sin′-thol*). A chemically pure synthetic substitute for alcohol. It is colorless and non-irritant.
**synthorax** (*sin-tho′-raks*). Synonym of *thoracopagus*.
**syntonin** (*sin′-to-nin*) [σύντονος, contracted]. An acidalbumin obtained by the action of dilute hydrochloric acid upon the myosin of muscle.
**syntopic** (*sin-top′-ik*) [σύν, together; τόπος, place]. Applied by Waldeyer to a topographic description which points out the relation of a part or organ to the viscera of soft parts of the organism.
**syntopy** (*sin′-to-pe*). The relation of an organ or part to the viscera of the organism. Cf. *holotopy*; *idiotopy*; *skeletopy*.
**syntoxoid**, (*sin-toks′-oid*) [*syn-*; τοξικόν, poison; εἶδος, likeness]. A toxoid having the same affinity as toxin for antitoxin.
**syntrimma, syntripsis** (*sin-trim′-ah, sin-trip′-sis*) Synonym of *comminution*.
**syntrophus** (*sin′-tro-fus*) [*syn-*; τροφός, a nurse]. A congenital disease.
**syntropic** (*sin-trop′-ik*) [*syn-*; τρέπειν, to turn]. Similar, and turned in the same direction (thus the ribs of either side are *syntropic*; those of opposite sides are *antitropic*).
**synulodynia** (*sin-ū-lo-din′-e-ah*) [συνούλωσις, a scarring over; ὀδύνη, pain]. Pain in a cicatrix.
**synulosis** (*sin-ū-lo′-sis*) [*syn-*; ὀυλή, scar]. Cicatrization; cicatrix.
**synulotic** (*sin-ū-lot′-ik*) [συνυλωτικός, healing]. Promoting cicatrization.
**synzygia** (*sin-zij′-e-ah*). See *syzygy*.
**syphilelcos, syphilelcus** (*sif-il-el′-kos, sif-il-el′-kus*) [*syphilis*; ἕλκος, ulcer]. Syphilitic ulcer. Chancre.
**syphilelcosis** (*sif-il-el-ko′-sis*) [*syphilis*; ἕλκος, ulcer]. The condition or progress of syphilitic ulceration; the condition of having a chancre.
**syphilelcus** (*sif-il-el′-kus*) [*syphilis*; ἕλκος, ulcer]. A syphilitic ulcer.
**syphilicoma** (*sif-il-ik-o′-mah*) [*syphilis*; κομεῖν, to take care of]. A hospital for syphilitics.
**syphilide, syphilid** (*sif′-il-id*) [*syphilis*]. Any disease of the skin due to syphilis. Syphilides may be erythematous, macular, acneiform, lenticular, squamous, vesicular, pustular, bullous, tubercular, rupial, etc. **s., secondary**, any syphilide occurring during the secondary stage of syphilis. **s., tertiary**, any syphilide occurring during the tertiary stage of syphilis.
**syphilidiatria** (*sif-il-id-e-a′-tre-ah*) [*syphilis*; ἰατρεία, treatment]. The medicinal treatment of syphilis.
**syphilidocolpitis** (*sif-il-id-o-kol-pi′-tis*) [*syphilis*; κόλπος, vagina; ιτις, inflammation]. Syphilitic inflammation of the vagina.
**syphilidography** (*sif-il-id-og′-ra-fe*). See *syphilography*.
**syphilidologist**. See *syphilologist*.
**syphilidology** (*sif-il-id-ol′-o-je*). See *syphilology*.
**syphilidomania** (*sif-il-id-o-ma′-ne-ah*). See *syphilomania*.
**syphilidophobia** (*sif-il-id-o-fo′-be-ah*). See *syphilophobia*.
**syphilidophthalmia** (*sif-il-id-off-thal′-me-ah*) [*syphilis*; *ophthalmia*]. Syphilitic ophthalmia.
**syphilin** (*sif′-il-in*). See *syphilitoxin*.
**syphilinthus** (*sif-il-e-on′-thus*) [*syphilis*; ἴονθος, eruption]. Any copper-colored scaly eruption in syphilis.
**syphilphobia** (*sif-il-fo′-be-ah*). See *syphilophobia*.
**syphilis** (*sif′-il-is*) [origin obscure]. A chronic infectious, contagious, venereal disease, characterized by a variety of structural lesions of which the chancre, the mucous patch, and the gumma are the most distinctive. A spirochete (*Spirochæta pallida*, or *Treponema pallidum*) is the cause. The disease is generally acquired in sexual congress, hence its earliest manifestations appear upon the genital organs, but any abraded surface of the body, if brought in contact with the syphilitic poison, may give entrance to the infection. The earliest lesion of acquired syphilis is the *chancre*, *initial sclerosis*, or *primary sore*, which appears after a period of incubation varying from two to three weeks. It is usually a reddish-brown papule with an ulcerated central spot, and has a slight serous or purulent discharge. Taken between the fingers it is found to have a peculiar cartilaginous hardness. Microscopically it consists of an accumulation of round cells, epithelioid cells, with, perhaps, a giant-cell here and there. The blood-vessels present a hyperplasia of the intima, to which in part the induration of the chancre is due. Very soon after the appearance of the chancre the nearest lymphatic glands become enlarged and indurated—the *indolent buboes* of syphilis. The *mucous patch, condyloma latum*,

**moist papule**, or **mucous tubercle** is located upon mucous membranes, at mucocutaneous junctions, or where two skin surfaces are in habitual contact, and is a flat, scarcely elevated patch, generally covered by a whitish pellicle. The **gumma** or **gummy tumor** is a rounded nodule, varying in size from the dimensions of a pea to those of a small apple. Its favorite seats are the periosteum of flat bones, the membranes of the brain, the liver, spleen, and testicle. It is usually soft, and contains in its interior a gelatinous "gummy" material. Another important though not distinctive lesion produced by syphilis is a diffuse sclerosis of the blood-vessels, especially of the parenchymatous organs. The clinical course of syphilis is generally divided into three stages: the *primary* (*primary syphilis*), characterized by the presence of the chancre and of the indolent bubo; the *secondary* (*secondary syphilis*), by the mucous patch, cutaneous eruptions, sore throat, and general enlargement of the lymphatic glands; the *tertiary* (*tertiary syphilis*), by the gumma and by severe skin-lesions. Between the appearance of the chancre and the secondary manifestations a period of six weeks usually elapses. The tertiary phenomena follow the secondary after a stage of quiescence of variable length. Syphilis also bears an important, but as yet obscure, relation to certain diseases of the nervous system, such as locomotor ataxia and paretic dementia. **s., congenital.** See *s., hereditary.* **s. d'emblée**, the invasion of syphilis without a local lesion. **s., extragenital**, that in which the first lesion is situated elsewhere than on the genital organs. **s., hereditary**, syphilis transmitted from parent to offspring. See *Colles' law; Demarquay's, Hutchinson's, Krisowski's, Silex's, Wegner's sign.* **s. ingenita.** See *s., hereditary.* **s. insontium**, syphilis of the innocent, *i. e.*, syphilis acquired in an innocent manner, or nonvenereal syphilis. **s., marital**, syphilis acquired in lawful wedlock. **s. neonatorum**, syphilis of the newborn. **s., nonvenereal.** Synonym of *s. insontium.* **s. œconomica**, a form of syphilis insontium in which the disease is acquired through eating and drinking or household utensils, or by incidental contact with syphilitic persons. **s., pulmonary**, a rare disease which is either hereditary or follows the initial attack after from 10 to 20 years or longer. Two forms may be differentiated—a chronic interstitial indurative process and a growth of gummata. **s. technica**, syphilis acquired in following one's occupation, as by physicians, midwives, nurses. **s., venereal**, syphilis acquired in illegitimate sexual intercourse. **s., visceral**, syphilis of the viscera—the lesions are either inflammatory or gummatous.
**syphilitic** (sif-il-it'-ik) [syphilis]. Pertaining to or affected with syphilis.
**syphilitoxin** (sif-il-e-toks'-in) [syphilis; τοξικόν, poison]. A term formerly used for the supposed specific virus of syphilis, before the discovery of the *Treponema pallidum.*
**syphilization** (sif-il-i-za'-shun) [syphilis]. 1. Inoculation with syphilis, especially inoculation for the purpose of conferring immunity to future attacks. 2. The state produced by inoculation with syphilis.
**syphilized** (sif'-il-īzd) [syphilis]. Affected with hereditary syphilis.
**syphilocerebrosis** (sif-il-o-ser-e-bro'-sis). Syphilis affecting the brain.
**syphiloderm, syphiloderma** (sif'-il-o-derm, sif-il-o-der'-mah). See *syphilide.*
**syphilogenesis, syphilogeny** (sif-il-o-jen'-e-sis, sif-il-oj'-en-e) [syphilis; γεννᾶν, to produce]. The origin or development of syphilis.
**syphilographer** (sif-il-og'-ra-fer) [see *syphilography*]. One who writes on syphilis.
**syphilography** (sif-il-og'-ra-fe) [syphilis; γράφειν, to write]. A treatise on syphilis.
**syphiloid** (sif'-il-oid) [syphilis; εἶδος, like]. 1. Resembling syphilis. 2. A disease resembling syphilis.
**syphilolepis** (sif-il-ol'-ep-is) [syphilis; λεπίς, scale]. A scaly or furfuraceous eruption of syphilitic origin.
**syphilologist** (sif-il-ol'-o-jist) [syphilis; λόγος, science]. A specialist in the treatment of syphilis.
**syphilology** (sif-il-ol'-o-je) [syphilis; λόγος, science]. The sum of knowledge regarding the origin, nature, and treatment of syphilis.
**syphiloma** (sif-il-o'-mah) [syphilis; ὄμα, tumor]. 1. A syphilitic gumma. 2. A term introduced by Ernst Wagner as a substitute for gumma.
**syphilomania** (sif-il-o-ma'-ne-ah) [syphilis; mania].

The inclination to attribute diseases to syphilis. The morbid belief in the presence of syphilis.
**syphilomatous** (sif-il-o'-mat-us) [syphilis; ὄμα, tumor]. Pertaining to syphiloma.
**syphilonychia** (sif-il-o-nik'-e-ah) [syphilis; ὄνυξ, nail]. An onychia of syphilitic origin. **s. exulcerans**, syphilitic onychia with ulceration. **s. sicca**, syphilitic onychia without ulceration.
**syphilopathy** (sif-il-op'-ath-e) [syphilis; πάθος, disease]. Any syphilitic disease.
**syphilophobe** (sif'-il-o-fōb) [syphilis; φόβος, fear]. One affected with syphilophobia.
**syphilophobia** (sif-il-o-fo'-be-ah) [syphilis; φόβος, dread]. 1. A condition in which the patient imagines himself to be infected with syphilis. 2. A morbid dread of syphilitic infection.
**syphilopyra** (sif-il-o-fi'-mah) [syphilis; φῦμα, growth]. 1. Syphiloma of the skin. 2. Any growth due to syphilis.
**syphilosis** (sif-il-o'-sis) [syphilis]. Syphilitic disease.
**syphionthus** (sif-e-on'-thus) [syphilis; ἴονθος, an eruption on the face]. The copper-colored eruptions or fawn-colored, furfuraceous patches of syphilitic origin.
**syr.** Abbreviation of Latin *syrupus*, syrup.
**syrgol** (sir'-gol). An organic silver compound, said to contain 20 per cent. of colloidal silver oxide.
**Syriac ulcer.** Synonym of *diphtheria.*
**syrigmophonia** (sir-ig-mo-fo'-ne-ah) [συρίσσειν, to hiss; φωνή, voice]. A piping or whistling state of the voice.
**syrigmus** (sir-ig'-mus) [συρίσσειν, to hiss]. Any subjective hissing, murmuring or tinkling sound heard in the ear.
**syringe** (sir'-inj) [σύριγξ, a pipe]. An apparatus for injecting a liquid into a cavity.
**syringeal** (sir-in'-je-al) [σύριγξ, a pipe]. Relating or belonging to a fistula or to the Eustachian tube.
**syringenin** (sir-in'-jen-in) [*Syringa*, a genus of shrubs]. $C_{15}H_{22}O_8 + H_2O$. A dissociation product of syringin by action of dilute acids; a clear, rose-red, amorphous mass, soluble in alcohol, insoluble in water and ether.
**syringin** (sir-in'-jin) [*Syringa*]. A crystalline glucoside obtained from *Syringa vulgaris*, $C_{17}H_{24}O_9$ + $H_2O$, white, tasteless, acicular crystals, soluble in alcohol and hot water, boils at 191° C. It is antipyretic and antiperiodic; used in malaria. Syn., *lilacin; ligustrin.*
**syringing** (sir-in-je'-us) [σύριγξ, a tube]. Fistulous.
**syringitis** (sir-in-ji'-tis) [*syrinx*; ιτις, inflammation]. Inflammation of the Eustachian tube.
**syringobulbia** (si-rin-go-bul'-be-ah) [σύριγξ, tube; bulb]. The presence of cavities in the medulla oblongata similar to syringomyelia.
**syringocele, syringocœlia** (sir-ing'-go-sēl, sir-ing-go-se'-le-ah) [σύριγξ, tube; κοιλία, hollow]. The cavity or central canal of the myelon or spinal cord.
**syringocystadenoma** (sir-ing-go-sis-tad-en-o'-mah) [*syrinx; cystadenoma*]. A peculiar disease of the skin that probably begins in embryonic sweat-glands.
**syringomeningocele** (sir-ing'-go-men-in'-go-sēl) [σύριγξ, pipe; *meningocele*]. A meningocele resembling a syringomyelocele.
**syringomyelia** (sir-in-go-mi-e'-le-ah) [*syrinx*; μυελός, marrow]. A condition characterized by the presence of cavities in the substance of the spinal cord. Syn., *myelosyringosis.* **s.**, Grasset-Rauzier's type of, a form with marked sudoral and vasomotor symptoms. **s., Schlesinger's type of**, the dorsolumbar type.
**syringomyelitis** (sir-ing-go-mi-el-i'-tis) [σύριγξ, tube; μυελός, marrow; ιτις, inflammation]. The inflammation coincident with or preceding syringomyelus.
**syringomyelocele** (sir-in-go-mi'-el-o-sēl) [*syrinx*; μυελός, marrow; κοιλία, cavity]. A form of spina bifida in which the protruding mass consists of membranes and nerve substance, and the cavity of which communicates with the central canal of the spinal cord.
**syringomyelus** (sir-ing-go-mi'-el-us) [σύριγξ, tube; μυελός, marrow]. An abnormal dilatation of the central canal of the spinal cord in which the central gray column is converted into connective tissue, the interior softening and forming a cavity. A disease similar to this in children is called *hydromyelia.*
**syringotome** (sir-in'-go-tōm) [*syrinx*; τέμνειν, to cut]. An instrument for incising a fistula.

**syringotomy** (sir-in-got'-o-me) [syrinx; τομή, a cutting]. The operation of cutting a fistula, especially a fisula in ano.
**syrinx** (sir'-ingks) [σύριγξ, a tube]. 1. A fistula. 2. The Eustachian tube.
**syrup** (sir'-up) [syrupus, syrup]. 1. A concentrated solution of sugar in water (syrupus, U. S. P., B. P.). 2. A preparation composed of a solution of a medicinal substance in syrup. **s., hive**, compound syrup of squills. **s., simple**, the aqueous solution of sugar without other ingredients.
**syrupy** (sir'-up-e) [syrup]. Resembling a syrup.
**sysoma** (si-so'-mah) [syn-; σῶμα, body: pl., sysomata]. A double monstrosity with two separate heads, but with the bodies fused in more or less intimate union.
**sysomic** (si-so'-mik). Of the nature of a sysoma.
**syssarcosis** (sis-ar-ko'-sis) [σύν, together; σάρξ, flesh]. The union of bone by the interposition of muscular tissue.
**syssomus** (sis-so'-mus) [σύν, together; σῶμα, body]. A double monster joined by the trunks.
**systaltic** (sis-tal'-tik) [systole]. Pulsatory; contracting; having a systole.
**systasis** (sis'-tas-is) [σύστασις]. Consistency, density.
**system** (sis'-tem) [σύστημα, from σύν, together; ἱστάναι, to stand]. 1. A methodical arrangement. 2. A combination of parts into a whole, as the digestive system, the nervous system. 3. The body as a whole. **s., Bertillon.** See under identification; also Bertillonage. **s., centimeter-gram-second**, the system based upon the use of the centimeter, gram, and second as units of length, mass, and time respectively. **s., Galton.** See under identification. **s.-disease, s.-lesion**, a disease of the cerebrospinal axis affecting a tract of nerve-fibers or nerve-cells having common anatomic relations and physiological properties. **s., kinetic.** See under kinetic. **s., pedal**, a ganglionic system of the brain. **s., portal**, the system of veins collecting the venous blood from the digestive tract.
**systema** (sis-te'-mah). See system.
**systematic** (sis-tem-at'-ik) [system]. Pertaining to or affecting a system.

**systematology** (sis-tem-at-ol'-o-je) [system; λόγος, science]. The science of arrangement and classification.
**systemic** (sis-tem'-ik) [system]. 1. Of or pertaining to a system. 2. Pertaining to the whole organism.
**systemoid** (sis'-tem-oid) [system; εἶδος, form]. A term applied to tumors composed of a number of tissues resembling a system of organs; teratoid.
**systole** (sis'-to-le) [συστολή, contraction]. The contraction of the heart and arteries. **s., aborted**, a cardiac systole which on account of insufficient energy or mitral regurgitation, does not increase the arterial pressure. **s., anticipated**, an aborted systole due to an imperfectly filled ventricle. **s., arterial**, the arterial retraction following cardiac systole. **s., auricular**, auricular contraction. **s., ventricular**, the contraction of the ventricles.
**systolic** (sis-tol'-ik) [systole]. Pertaining to the systole; occurring during systole.
**systolometer** (sis-to-lom"-et-er) [systole; μέτρον, measure]. An instrument for estimating the intensity and quality of cardiac sounds and murmurs, and the length of the pauses.
**systremma** (sis-trem'-ah) [σύστρεμμα, a swelling; pl., systremmata]. Cramp in the muscles of the leg.
**syzygial** (siz-ij'-e-al) [σύζυγος, yoked gtoteher]. Pertaining to syzygy.
**Syzygium** (sis-ij'-e-um) [σύζυγος, yoked]. A genus of East Indian trees, of which S. jambolanum is used in diabetes.
**syzygy** (sis'-ij-e) [σύζυγος, yoked together; pl., syzygies]. In biology: 1. A fusion of two bodies, without loss of identity. 2. A zygote or conjugate body, formed by the union or conjugation of two similar gametes, and usually followed by encapsulation and later by sporulation; a syzygium. **s. bone**, one shaped like the letter s, e. g., the episternum.
**Szabo's test for hydrochloric acid in the contents of stomach** (tsah'-bo). Mix together equal parts of 0.5 % solutions of ammonium sulphocyanide and sodioferric tartrate. This makes a pale yellow liquid, which changes to brownish-red on the addition of a solution containing HCl.

# T

**T.** An abbreviation of *tension, temperature,* and *absolute temperature*.
**T+.** Abbreviation for *increased tension*.
**T−.** Abbreviation for *diminished tension*.
**t.** Abbreviation for *temporal*.
**TA.** Abbreviation for *tuberculin A*.
**Ta.** Chemical symbol of *tantalum*.
**tabacosis** (*tab-ak-o'-sis*) [*tabacum*]. A state of poisoning produced by the excessive use of tobacco.
**tabacum** (*tab-ak'-um*). See *tobacco*.
**tabanid** (*tab'-an-id*) [*tabanus*, a gad-fly]. Any horse-fly or gad-fly.
**Tabanus** (*tab-an'-us*) [see *tabanid*]. A genus of horse-flies or gad-flies. More than 1300 species are known, the females of many of them being capable of inflicting a severe and painful bite.
**tabardillo** (*tab-ar-dĕl'-yo*) [Spanish]. 1. Mexican typhus. 2. An infectious disease endemic in certain parts of Mexico.
**tabasheer** (*tab-a-shĕr'*) [Hindu, *tabasher*]. An opal-like substance found in the joints of certain species of bamboos. It is used as a tonic, aphrodisiac, pectoral, astringent, and antispasmodic.
**tabatière anatomique** (*tahb-aht-ē-air', ahn-aht-ōm-ēk*) [Fr., "anatomical snuff-box"]. The depression at the base of the thumb between the tendons of the extensor primi and extensor secundi internodii pollicis.
**tabby-cat striation.** Peculiar markings occurring on muscles that have undergone extreme fatty degeneration, especially seen in the heart muscle.
**tabefaction** (*ta-be-fak'-shun*) [*tabefacere*, to melt]. Wasting; emaciation.
**tabella** (*ta-bel'-ah*) [L.: *pl.*, *tabellæ*]. A troche. A tablet.
**tabes** (*ta'-bēz*) [L.]. A wasting or consumption. The word is generally used as a synonym of *tabes dorsalis*. **t. coxaria,** wasting from hip disease. **t., diabetic,** a peripheral neuritis affecting diabetics. **t. diuretica.** Same as *diabetes mellitus*. **t. dolorosa,** a form in which pain is the dominating feature. **t. dorsalis,** locomotor ataxia, a disease dependent upon sclerosis of the posterior columns of the spinal cord. The symptoms are lightning-pains; unsteadiness and incoordination of voluntary movements, extending to the upper extremities; disorders of vision, among others the Argyll Robertson pupil; cutaneous anesthesia; girdle-sense; abolition of the patellar reflex; diminution of sexual desire; disturbance of the sphincters. **t. ergotica,** a toxemia resulting from the use of ergot; its symptomatology closely simulates that of locomotor ataxia. **t., hereditary.** See *Friedreich's disease*. **t. mesenterica,** tuberculous disease of the mesenteric glands in children, with progressive wasting. **t., spasmodic,** lateral sclerosis of the spinal cord.
**tabescence** (*tab-es'-ens*) [*tabes,* wasting]. Wasting; marasmus; emaciation.
**tabescent** (*tab-es'-ent*) [*tabescere,* to waste away]. Wasting, or becoming wasted or emaciated.
**tabetic** (*tab-et'-ik*). 1. Affected with tabes; of or pertaining to tabes. 2. Pertaining to or affected with tabes dorsalis.
**tabetiform** (*tab-et'-if-orm*). Resembling tabes.
**tabic** (*tab'-ik*). See *tabetic*.
**tabid** (*tab'-id*). See *tabetic*.
**tablature** (*tab'-lat-ūr*) [*tabula,* a table]. Separation into tables, as exemplified in the frontal, parietal, and occipital bones.
**table** (*ta'-bl*) [*tabula*]. 1. A flat-topped piece of furniture, as an *operation table, examining table*. 2. A flat plate, especially one of bone, as a *table* of the skull. 3. **vitreous,** the inner cranial table.
**tablespoon.** A large spoon, holding about 15 Cc. or 4 fluidrams.
**tablet** (*tab'-let*) [*table*]. A lozenge; a troche. **t. triturate,** a small troche containing a triturated medicine.

**tablogestin** (*tab-lo-jes'-tin*). Chologestin in tablet form.
**tabloid** (*tab'-loid*) [*table;* εἶδος, like]. 1. A flat troche. 2. A trade name for a compressed or other tablet.
**taboparalysis** (*ta'-bo-par-al-is-is*). A condition in which tabes is associated with general paralysis.
**taboparesis** (*ta'-bo-par'-es-is*). Same as *taboparalysis*.
**tabophobia** (*ta-bo-fo'-be-ah*) [*tabes;* φόβος, fear]. A morbid fear of becoming affected with tabes; it is a frequent symptom of neurasthenia.
**tabula** (*tab'-ū-lah*). See *table*.
**tabular** (*tab'-ū-lar*) [*tabula,* table]. Having the form of a table. **t. bone,** a flat bone or one composed of two tables of compact bone with cancellous tissue or diploe between them.
**tabule** (*tab'-ūl*). A tablet.
**tac** (*tak*). 1. Synonym of *influenza*. 2. Rot; scabies in the sheep.
**tacahout** (*tak'-a-howt*). A kind of gall produced upon the tamarisk; it is an astringent.
**tacamahac, tacamahaca** (*tak'-am-a-hak, tak-am-a-hak'-ah*) [Mexican]. A resin produced by various trees.
**tache** (*tahsh*) [Fr.]. A spot. **taches blanches,** certain white spots described by Hanot as occurring on the liver, especially on its convex surface, in infectious diseases. Microscopically, they present a leukocytic infiltration and bacteria. **t. bleuâtre,** a spot of a delicate blue tint, sometimes observed on the skin of typhoid-fever patients. **t. cérébrale, t. méningéale,** the red line made when the fingernail is drawn over the skin; due to vasomotor paresis and occurring especially in meningeal irritation. **t. motrice,** an eminence of protoplasm within the sarcolemma where the nerve-fiber pierces the latter; a motorial end-plate. **t. spinale,** a bulla-like spot seen in certain diseases of the spinal cord.
**tachometer** (*tak-om'-et-er*). Same as *tachometer*.
**tachetic** (*tak-et'-ik*) [*tache*]. Relating to the formation of reddish-blue or purple patches (taches).
**Tachia** (*tak'-e-ah*) [*tachi,* an ant, so-called by the Galibis because they harbor ants]. A genus of shrubs and trees of the *Gentianaceæ*. The root of *T. guianensis,* a species of Brazil and Guiana, is used as is gentian and also as an antipyretic and prophylactic against malaria. Dose of *tincture* 1 or 2 drops.
**tachiol** (*tak'-e-ol*). A modification of silver fluoride; employed as a surgical antiseptic in solution of 1 : 1000 to 1 : 100 and in ophthalmic practice.
**tachistoscope** (*tak-is'-to-skōp*) [ταχύς, swift; σκοπεῖν, to view]. A form of stereoscope giving rapid impressions by means of a movable diaphragm.
**tachography** (*tak-og'-raf-e*) [ταχύς, swift; γραφεῖν, to write]. The estimation of the rate of flow of arterial blood by means of the tachygraph.
**tachometer** (*tak-om'-et-er*). See *hemotachometer*.
**tachy-** (*tak-e-*) [ταχύς, swift]. A prefix meaning swift.
**tachycardia** (*tak-e-kar'-de-ah*) [*tachy-;* καρδία, heart]. Excessive rapidity of the heart's action. **t., essential,** that occurring in paroxysms, and due to functional disturbance of the cardiac nerves. **t., paroxysmal,** tachycardia occurring periodically in paroxysms. **t. reflex,** tachycardia due to other causes than those producing essential t. **t. strumosa exophthalmic,** the tachycardia occurring in exophthalmic goiter.
**tachycardiac** (*tak-e-kar'-de-ak*). Pertaining to or suffering from tachycardia.
**tachygraf** (*tak'-ig-raf*). See *hemotachometer*.
**tachygraphy** (*tak-ig'-raf-e*) [ταχύς, swift; γραφεῖν, to write]. The estimation of the rate of flow of arterial blood by means of the tachygraph.
**tachyiatria** (*tak-e-i-at'-re-ah*) [*tachy-;* ἰατρεία, healing]. The art of curing quickly.

**tachymeter** (*tak-im'-et-er*). Same as *hemotachometer*.
**tachyphagia** (*tak-e-fa'-je-ah*) [*tachy-*; φαγεῖν, to eat]. Rapid eating.
**tachyphrasia** (*tak-e-fra'-ze-ah*) [*tachy-*; φράσις, speech]. Morbid rapidity or volubility of speech.
**tachyphrenia** (*tak-e-fre'-ne-ah*) [*tachy-*; φρήν, mind]. Morbid mental activity.
**tachypnea, tachypnœa** (*tak-ip-ne'-ah*) [*tachy-*; πνοή, breath]. Abnormal frequency of respiration.
t., **nervous**, respiration of 40 or more to the minute accompanying neurotic disorders, particularly hysteria and neurasthenia.
**tachytomy** (*tak-it'-o-me*) [*tachy-*; τομή, a cutting]. The art of operating quickly.
**tactile** (*tak'-til*) [*tactus*]. Pertaining to the sense of touch. t. **cells**, cells representing special sensory nerve-endings, found in the deeper layers of the epidermis or the adjacent stratum of corium. t. **corpuscles**, special sensory nerve-endings exhibiting more complexity of structure than the tactile cells. t. **disc**, the flattened terminal expansion of the axis cylinder in a special sensory nerve ending, or tactile corpuscle. t. **irritability**, the property of cellular repulsion. t. **meniscus**, a peculiar crescentic expansion of a nerve-fiber over the ental surface of a tactile cell. t. **papilla**. See *papilla*. t. **reflexes**, reflex movements from stimulation of the tactile corpuscles.
**taction** (*tak'-shun*) [*tactio*, a touch]. A touch, a touching, the tactile sense.
**tactometer** (*tak-tom'-et-er*) [*tactus*; μέτρον, a measure]. An instrument for estimating tactile sensibility; an esthesiometer.
**tactor** (*tak'-tor*) [*tactus*, touch]. A tactile organ.
**tactual** (*tak'-tu-al*) [*tactus*, touch]. Relating to the sense of touch; tactile.
**tactus** (*tak'-tus*) [*tangere*, to touch]. Touch. t. **eruditus**, t. **expertus**, special sensitiveness of touch acquired by long experience. ■
**tædium vitæ** (*te'-de-um vi'-te*) [L.]. Weariness of life, a symptom witnessed in many cases of insanity; it is sometimes a precursor of suicide.
**Tænia, Tenia** (*te'-ne-ah*) [L., "a band"]. 1. A band or band-like structure. 2. Tenia, see *tapeworm*. T. **cœnurus**, a parasite found in the intestine of the dog. T. **cucurbitana**, long tapeworm (pork-worm). T. **echinococcus**. See *tapeworm, dog-*. t. **fornicis**, one of the peduncles of the pineal gland. t. **hippocampi**, the corpus fimbriatum of the hippocampus major. T. **mediocanellata**, T. **saginata**. See *tapeworm, beef-*. t. **semicircularis**, a narrow band on the floor of the lateral ventricle, between the caudate nucleus and the optic thalamus. T. **solium**. See *tapeworm, pork-*. t. **thalami**, the habenula. t. **tubæ**, an occasional thickening of the upper border of the perisalpinx. t. **ventriculi quarti**, the tenia of the fourth ventricle, the ligula. t. **ventriculi tertii**, the tenia of the third ventricle, the stria medullaris. t. **violacea**, a bluish, longitudinal band on the floor of the fourth ventricle. T. **vulgaris**, broad tapeworm.
**tæniacide** (*te'-ne-as-id*). See *teniacide*.
**tæniafuge** (*te'-ne-af-ūj*). See *teniafuge*.
**Tagetes** (*ta-je'-tēs*). A genus of plants of the order *Compositæ*. T. **erecta**, African marigold, and T. **patula**, French marigold, are used as substitutes for calendula.
**Tagliacotian operation** (*tah-le-ah-ko'-she-an*) [Gaspard *Tagliacozzi*, Italian surgeon, 1546–1599]. *Tagliacotian operation, Italian.*
**tagma** (*tag'-mah*) [τάγμα, that which has been arranged; *pl., tagmata*]. An aggregate of molecules.
**tagulawaya**. See *balsam, tagulawaya*.
**tail** (*tāl*). 1. The caudal extremity of an animal. 2. Anything resembling a tail. t. **bone**, the coccyx. t. **fold**, an embryonic infolding or hollow, enclosing the hind-gut. t. **gut**, that part of the archenteron which is in the tail of the embryo. t. of **pancreas**, the splenic end of the pancreas.
**Taillefer's valve**. A valvular fold of mucous membrane about the middle of the nasal duct.
**tailor's cramp, or spasm**. An occupation-neurosis occurring in tailors, and characterized by spasm of the muscles of the arm and head.
**tailor's muscle**. Sartorius.
**taint** (*tānt*). An infection, or pathogenic influence; as a syphilitic *taint*. A spot or blemish.
**Tait's knot** (*tāt*) [Lawson *Tait*, English gynecologist, 1845–1909]. A peculiar method of ligating the pedicle in the operation of ovariotomy. See

**knot, Staffordshire**. T.'s **law**, in every disease of the abdomen or pelvis in which the health is destroyed or the life threatened, and in which the condition is evidently not due to malignant disease, an exploration of the cavity by celiotomy should be made. T.'s **method**, perineorrhaphy. T.'s **operation**, perineorrhaphy.
**takadiastase** (*ta-kah-di'-as-tās*) [Jokichi *Takamine*, Japanese chemist, 1853– ]. A diastatic ferment obtained from wheat-bran by action of the spores of the fungus *Eurotium oryzæ* (Taka-moyash). It is used in digestive disorders, especially those resulting from deficient secretion of saliva and hyperacidity of the stomach. Dose 2–5 gr. (0.1–0.3 Gm.).
**take** (*tāk*). To become infected, as by vaccine virus.
**takosis** (*ta-ko'-sis*) [τῆκειν, to waste]. A highly contagious fatal disease of goats.
**talalgia** (*tal-al'-je-ah*) [*talus*, heel; ἄλγος, pain]. Pain in the heel.
**Talbot's law**. When the visual stimuli proceeding from a revolving disc are completely fused and the sensation is uniform, the intensity is the same as that which would occur if the same amount of light were spread uniformly over the disc.
**Talbot-Plateau's law**. See *Talbot's law*.
**talc, talcum** (*talk, tal'-kum*) [Ar., *talq*, talc]. 4MgO.5SiO₂.H₂O. The talcum of the U. S. P. is a native hydrous magnesium silicate. It is a white, greasy powder, used as a dusting-powder. Syn., *soapstone; steatite*. t., **purified** (*talcum purificatum*, U. S. P.), talc, hydrochloric acid, and water.
**Taliacotian** (*tal-e-ak-o'-she-an*). See *Tagliacotian*.
**taliped** (*tal'-ip-ed*) [*talus*, ankle; *pes*, foot]. A person affected with talipes; club-footed.
**talipedic** (*tal-ip-e'-dik*) [*talipes*]. Belonging or relating to talipes; club-footed.
**talipes** (*tal'-ip-ēs*) [*talus*, ankle; *pes*, foot]. Club-foot, a deformity depending upon contraction of one or more muscles or tendons about the foot, either congenital or acquired. t. **arcuatus**. See t. *cavus*. t. **calcaneus**, talipes in which the patient walks upon the heel alone. t. **cavus**, an increased curvature of the arch of the foot. t. **equinus**, talipes in which the heel is elevated and the weight thrown upon the anterior portion of the foot. t. **percavus**, excessive plantar curvature. t. **planus**, flat-foot; splay-foot. t. **spasmodic**, non-congenital talipes due to muscular spasm. t. **valgus**, talipes in which the foot is everted. t. **varus**, a variety, the reverse of the last, in which the foot is bent inward. Combinations of these occur, as t. *equinovalgus*, t. *equinovarus*, t. *calcaneovalgus*, t. *calcaneovarus*, etc.
**talipomanus** (*tal-ip-o-ma'-nus*) [*talipes; manus*, hand]. Deformity of the hand, analogous to club-foot; club-hand.
**Tallerman treatment**. The local application of superheated dry air, the affected part being introduced into a cylinder.
**tallow** (*tal'-o*) [O. D. *talgh*]. The fat extracted from suet, the solid fat of cattle, sheep, and other ruminants.
**Tallqvist's method** (*tal'-kvist*) [Theodor Waldemar *Tallqvist*, Finnish physician]. To determine approximately hemoglobin percentages allow a drop of blood to soak into a bit of filter-paper and compare with the naked eye the color strength of the stain with a series of printed standard tints of known value.
**Talma's disease** [Sapé *Talma*, Dutch physician, 1847– ]. Myotonia acquisita. T.'s **operation**, suture of the omentum to the abdominal wall for relief of ascites due to cirrhosis of the liver.
**talo-** (*ta-lo-*) [*talus*, ankle]. A prefix denoting pertaining to the ankle or to the astragalus.
**talocalcanean** (*ta-lo-kal-ka'-ne-an*). See *astragalocalcanean*.
**talocrural** (*ta-lo-kroo'-ral*) [*talo-*; *crus*, leg]. Relating to the astragalus and the bones of the leg.
**talofibular** (*ta-lo-fib'-ū-lar*). Relating to the astragalus and the fibula.
**talonavicular** (*ta-lo-nav-ik'-ū-lar*). See *astragaloscaphoid*.
**taloscaphoid** (*ta-lo-skaf'-oid*). See *astragaloscaphoid*.
**talotibial** (*ta-lo-tib'-e-al*). See *astragalotibial*.
**talpa** (*tal'-pah*) [L.]. A mole or wen.
**talpiform** (*tal'-pe-form*) [*talpa*; *forma*, form]. Wen-shaped.
**talus** (*ta'-lus*) [L.]. 1. The astragalus. 2. The ankle.

**tama** (*tam'-ah*) [L.]. Swelling of the feet and legs.
**tamar indien.** An aromatic confection of senna.
**tamarac** (*tam'-ar-ak*) [Am. Ind.]. The bark of *Larix americana*, a tonic and mild astringent acting on mucous membranes. Dose of the fluidextract ʒss–j.
**tamarind, tamarindus** (*tam'-ar-ind, tam-ar-in'-dus*) [Ar., *tamr*, a ripe date; *Hind*, India]. *Tamarindus indica*, a tree of the order *Leguminosæ*. The preserved pulp of the fruit (*tamarindus*, U. S. P., B. P.) is laxative and refrigerant. Dose 1 dr.–1 oz. (4–32 Gm.).
**tambour** (*tam'-boor*) [Fr.]. A drum; a drum-like instrument used in physiological experiments, and consisting of a metal cylinder over which is stretched an elastic membrane, from which or to which passes a tube for transmitting a current of air. It is connected with another apparatus upon which changes in pressure in the tambour are recorded.
**tampicin** (*tam'-pis-in*). A purgative resin, $C_{14}H_{24}O_{14}$, from the root of Tampico jalap, *Ipomœa simulans*.
**tampol** (*tam'-pol*). A medicated tampon, for gynecological use.
**tampon** (*tam'-pon*) [Fr.]. 1. A plug of cotton, sponge, or other material inserted into the vagina, nose, or other cavity. 2. To plug with a tampon.
**tamponade** (*tam-pon-ād'*) [Fr.]. The act of plugging with a tampon.
**Tamus** (*ta'-mus*). A genus of dioscoreaceous oldworld plants. The pulp of the bulb of *T. communis* (black bryony) is discutient, rubefacient, diuretic, and laxative.
**tanacetin** (*tan-as'-et-in*). See under *tansy*.
**tanacetum** (*tan-as-e'-tum*). See *tansy*.
**tanalum** (*tan-al'-um*) [*tannin*; *alum*]. Aluminum tannotartrate; used in diseases of the nose and throat.
**tanargan** (*tan-ar'-gan*). A tannin-silver-albumin preparation.
**tanargentan** (*tan-ar-jen'-tan*). A trade name applied to a compound or mixture of silver with tannin and albumin.
**tanghin** (*tang'-gin*). A poisonous extractive obtained from *tanghinia* (*q. v.*).
**tanghinia** (*tan-gin'-e-ah*). *T. venenifera*, the ordeal-bean of Madagascar, a cardiac and respiratory poison. Its active principle is *tanghinin*.
**tanghinin** (*tan-gin'-in*). See under *tanghinia*.
**tangle** (*tang'-gl*). See *laminaria*. t.-**tent**. See *sea-tangle*.
**tango foot** (*tang'-go*). Tenosynovitis of the dorsal flexors of the foot, particularly of the tibialis anticus, found in those addicted to modern dances.
**tannagen** (*tan'-a-jen*). See *tannigen*.
**tannal** (*tan'-al*) [*tannin*; *alum*]. Aluminum tannate; it is used in diseases of the nose and throat. t., **insoluble**, aluminum tannate. t., **soluble**, aluminum tannotartrate.
**tannalbin** (*tan-al'-bin*). A compound of tannin and albumin; a brown, tasteless powder, insoluble in water, and containing 50 % of tannin. An intestinal astringent. Dose 15 gr. (1 Gm.) 2 to 4 times daily. t., **veterinary**, a tannalbin specially prepared for a veterinary intestinal astringent.
**tannas** (*tan'-as*). Latin form of *tannate*.
**tannase** (*tan'-ās*). A zymase occurring in certain plants containing tannin, and produced in cultures of *Penicilium glaucum*.
**tannate** (*tan'-āt*) [*tannin*]. A salt of tannic acid.
**tannic acid.** See *acid, tannic*.
**tannichthol** (*tan-ik'-thol*). A trade name for suppositories containing tannic acid, phenol, ichthyol, belladonna, stramonium, witch-hazel and sometimes opium.
**tannigen** (*tan'-ij-en*). See *acetyl tannin*.
**tannin** (*tan'-in*). See *acid, tannic*. t., **formaldehyde**, tannoform.
**tannismuth** (*tan-is'-muth*). Trade name of bismuth bitannate.
**tannisol** (*tan'-is-ol*). Methylene ditannin, said to be a condensation product of tannin and formaldehyde.
**tannipyrine** (*tan-ni-pi'-rēn*). A condensation product of antipyrine and tannic acid; it is used as a styptic.
**tannobromine** (*tan-o-bro'-mēn*). A product formed from formaldehyde and dibromtannin. It is said to be a nerve sedative.
**tannocasum** (*tan-o-ka'-sum*). A compound of tannin and casein; it is used as an intestinal astringent.
**tannochloral** (*tan-o-klo'-ral*). See *captol*.
**tannochrome** (*tan'-o-krōm*). Trade name of a preparation containing resorcinol and chromium bitannate.
**tannocol** (*tan'-o-kol*). A combination of equal parts of gelatin and tannic acid.
**tannocreosoform** (*tan-o-kre-so'-so-form*). A compound of tannin, creosote, and formaldehyde.
**tannoform** (*tan'-o-form*), $CH_2(C_{14}H_9O_9)_2$. A condensation-product of tannin and formaldehyde. Used internally in chronic intestinal catarrh. Dose 4–8 gr. (0.25–0.5 Gm.); externally in skin diseases, burns, etc., in 10 % ointment, or dusting-powder with 2 to 4 parts of starch. Syn., *methylene ditannin; tannin-formaldehyde*.
**tannogelatin** (*tan-o-jel'-at-in*). Same as *tannocol*.
**tannoguaiaform** (*tan-no-gwi'-a-form*). A compound of tannic acid, guaiacol and formaldehyde, employed as an intestinal antiseptic and astringent.
**tannon** (*tan'-ōn*), $(CH_2)_6N_4(C_{14}H_{10}O_9)_3$. A condensation-product of tannin and urotropin; used in acute catarrh and subacute and chronic enteritis. Dose 15 gr. (1 Gm.) 3 or 4 times daily.
**tannopin** (*tan'-o-pin*). See *tannon*.
**tannopumilin** (*tan-o-pū'-mil-in*). A proprietary remedy for skin diseases, said to consist of tannic acid and oil of *Pinus pumilio*.
**tannosal** (*tan'-o-sal*). Tannic acid ester of creosote, containing 60 per cent. of creosote; antitubercular. Same as *creosal*.
**tannothymal** (*tan-o-thi'-mal*). Trade name of a condensation product of formaldehyde, thymol and tannin.
**tannyl** (*tan'-il*). A compound of tannin and oxychlorcasein.
**Tanret's reagent for albumin** (*tahn-ra*) [Charles Tanret, French physician, 18 – ]. Potassium iodide, 3.32 Gm.; mercuric chloride, 1.35 Gm.; acetic acid, 20 Cc., diluted with distilled water to 60 Cc. This reagent, added to an albumin solution, gives a white precipitate.
**Tansini's operation** (*tan-se'-ne*) [Iginio *Tansini*, Italian surgeon, 1855– ]. 1. An operation for the removal of the breast, including the skin covering it, followed by the covering of the bare area with a flap of skin taken from the back. 2. An operation for the removal of a hepatic cyst.
**tansy** (*tan'-ze*) [O. Fr., *tanasie*, from Low L., *tanacetum*, from *ἀθανασία*, immortality]. A perennial herb, *Tanacetum vulgare*, of the order *Compositæ*. The leaves and tops contain a bitter principle, *tanacetin*, $C_{11}H_{16}O_4$, tannic acid, and an essential oil (*oleum tanaceti*). Tansy is an aromatic bitter and irritant narcotic, and has been used in malaria, in hysteria, and as an emmenagogue and anthelmintic. In overdoses it produces abdominal pain, vomiting, epileptiform convulsions, and death from failure of respiration. Dose 30 gr.–1 dr. (2–4 Gm.); of the oil 1–4 min. (0.065–0.26 Cc.).
**tantalum** (*tan'-tal-um*). A rare metal, allied in properties to antimony and bismuth; symbol Ta, atomic weight 181.5. See *elements, table of*.
**tap.** 1. A sudden slight blow. 2. To empty of fluid, as to *tap* a hydrocele. 3. An East Indian term for trypanosomiasis.
**tapeinocephalic, tapeinocephaly.** See *tapinocephalic, tapinocephaly*.
**tapetal** (*tap'-e-tal*). Pertaining to the tapetum.
**tapetum** (*ta-pe'-tum*) [*τάπης*, a mat or rug]. 1. The layer forming the roof of the posterior and middle cornua of the lateral ventricles of the brain; it is composed of fibers from the corpus callosum. 2. The brilliant greenish layer of the eyes of nocturnal animals, which are by it visible in the dark. Syn., *tapetum lucidum*. t. **alveoli**, the alveolar periosteum. t. **cellulosum**. See *t. fibrosum*. t. **fibrosum**, or shining structure in the choroid of the eye. It takes the place of the *t. cellulosum* of the carnivora, the iridescent portion of the choroid in these animals. t. **lucidum**, the brilliant, greenish, reflecting layer of the membrana versicolor of the eyes of many of the lower animals. t. **nigrum**, the pigmentary layer of the retina. t. **ventriculi**, a bundle of white fibers of the brain uniting the cortex of the frontal with that of the occipital lobe.
**tape-worm.** One of the *Cestoda*, a class of worms parasitic in man and the lower animals. The adult worm (*strobilus*) consists of a head (*scolex*) and

## TAPHOPHOBIA 874 TARSOPHALANGEAL

**numerous segments** (*proglottides*), which are capable of leading for some time a separate existence, are hermaphroditic, and contain numerous ova. If the ova are swallowed by the proper host, they develop into embryos (*proscolices*), which are transformed into the *cysticerci*, containing the *scolices*. If the meat of animals containing living scolices is eaten, the latter develop into the mature tapeworm, or strobilus. t., armed. See *t., pork*. t., beef- (*Tænia mediocanellata* or *saginata*), also termed the *unarmed tape-worm*, the cysticercus of which occurs in beef. t., dog-. (*Tænia echinococcus*), also called *hydatid tape-worm*. The mature parasite lives in the intestine of the dog; the scolices occur in the internal organs of man and give rise to the echinococcus or hydatid cysts. t., fish-, t., broad, t., Swiss (*Bothriocephalus latus*), the cysticercus of which occurs in fish. t., hydatid. See *t., dog-*. t., pork- (*Tænia solium*), also known as the *armed tape-worm*, from the presence of several hooklets on the head, is derived from pork which contains the cysticerci. Other tape-worms occasionally found in man are: *Tænia cucumerina* or *elliptica*, most frequent in the dog and cat; *Tænia nana* has been found in man in Italy; *Tænia leptocephala*, common in the mouse, has also been observed in man.
**taphophobia** (*taf-o-fo'-be-ah*) [τάφος, burial; φόβος, fear]. Morbid fear of being buried alive.
**taphosote** (*taf'-o-sōt*). Creosote tannophosphate.
**tapinocephalic** (*tap-in-o-sef-al'-ik*) [see *tapinocephaly*]. Affected with tapinocephaly.
**tapinocephaly** (*tap-in-o-sef'-al-e*) [ταπεινός, low; κεφαλή, head]. Flatness of the top of the cranium.
**tapioca** (*tap-e-o'-kah*) [Sp.]. A variety of starch obtained from the cassava or manioc plant, *Jatropha manihot*. It is used as a food.
**tapir mouth** (*ta'-per*). A separation and thickening of the lips, with disease of the orbicularis oris muscle, causing the lips to resemble those of the tapir. It is sometimes seen in facial muscular atrophy of the Landouzy-Déjérine type.
**tapotement** (*tap-ōt-mon*(g)) [Fr.]. In massage, the operation of percussing or tapping.
**tapping** (*lap'-ing*). See 1. *tapotement*; 2. *paracen-tesis*.
**taproot** (*tap'-rūt*). The main root, or downward continuation of the plant axis.
**tar** (*tahr*) [AS., *teoru*, tar]. An empyreumatic liquid resin obtained by the destructive distillation of the wood of various species of *Pinus*, of the order *Coniferæ*. Tar (*Pix liquida*, U. S. P., B. P.) contains a great variety of compounds, among which are pyroligneous acid, toluene, xylene, pseudocumene, cresol, phenol, guaiacol, resol, paraffin, naphthalene, pyrocatechin, etc. It is employed in chronic bronchitis and in diseases of the urinary tract; externally, in tinea capitis, psoriasis, chronic eczema, and other affections of the skin. **t.-acne**. See *acne picealis*. **t. balls**, coal tar camphor, naphthalene. **t., Barbados**, a black petroleum of Barbados of the consistency of molasses and with bituminous taste. **t., birch**, crude oil of birch. **t.-camphor**, naphthalene. **t., coal**, a dark, highly complex, semi-liquid substance obtained by the destructive distillation of coal. **t., gas**. See *t., coal*. **t., juniper**, oil of (*oleum cadinum*, U. S. P.). See *cade, oil of*. **t., oil of** (*oleum picis liquidæ*, U. S. P.), a volatile oil distilled from tar. Dose 3 min. (0.2 Cc.). **t. ointment** (*unguentum picis liquidæ*, U. S. P., B. P.), a mixture of tar, yellow wax, and lard. **t. spirit**, benzol. **t., syrup of** (*syrupus picis liquidæ*, U. S. P.). Dose 1-2 dr. (4-8 Cc.). **t.-water**, an infusion containing one part of tar to four of water. **t., wood**, a thick, shining, black liquid obtained by the distillation of the wood of various species of conifers.
**tara** (*lah'-rah*). A nervous disease occurring in Siberia.
**taracanin** (*tar-ak'-an-in*). The same as *antihydropin*.
**tarantism, tarantismus** (*tar'-an-tizm, tar-an-tis'-mus*). A choreic affection, ascribed to the bite of a tarantula, and supposed to be cured by dancing.
**tarantula** (*tar-an'-tū-lah*) [*Tarentum*]. 1. A species of spider, *Lycosa tarantula*, closely resembling the trap-door spider, *Mygale hensii*, with which it is often confounded. Its bite is poisonous. See *larantism*. 2. The Italian form of dancing mania; tarantism.
**tarantulism** (*tar-an'-tū-lizm*). Same as *tarantism*.
**tarassis** (*tar-as'-is*) [ταράσσειν, to trouble]. Hysteria in the male.

**taraxacerin** (*tar-aks-as'-er-in*). A waxy substance found in dandelion.
**taraxacin** (*tar-aks'-as-in*). A crystallizable material derivable from the common dandelion; said to be tonic and diuretic.
**taraxacum** (*tar-aks'-ak-um*). Dandelion; the *T. officinale* (*T. dens-leonis*), a plant of the order *Compositæ*. Its root (*taraxacum*, U. S. P.; *taraxaci radix*, B. P.) contains two crystalline principles, *taraxacin* and *taraxacerin*, and is used in chronic congestion of the liver and spleen. **t., decoction of** (*decoctum taraxaci*, B. P.). Dose 2 oz. (64 Cc.). **t., extract of** (*extractum taraxaci*, U. S. P., B. P.). Dose 10 gr. (0.65 Gm.). **t., fluidextract of** (*fluid-extractum taraxaci*, U. S. P., B. P.). Dose 1 dr. (4 Cc.). **t., juice of** (*succus taraxaci*, B. P.). Dose 2-4 dr. (8-16 Cc.).
**taraxis** (*tar-ak'-sis*) [τάραξις, trouble]. A slight conjunctivitis, or eye trouble.
**Tardieu's ecchymoses** or **spots** [Auguste Ambroise *Tardieu*, French physician, 1818-1879]. Ecchymotic spots found beneath the pleura and the pericardium after death from strangling. They have also been observed in death from asphyxia due to other causes.
**tared** (*tārd*). Allowed for as a tare or deduction; having the weight previously ascertained, as a *tared* filter. The term is used in pharmacy and chemistry.
**Tarin's, Tarinus' fascia** (*ta-ran', ta-ri'-nus*) [Pierre *Tarin*, French anatomist, 1725-1761]. The fascia dentata Tarini; the gyrus dentatus. See *fascia dentata*. T.'s foramen. See *Fallopian hiatus*. T.'s fossa, T.'s pons, the posterior perforated space which forms part of the floor of the third ventricle. T.'s space. See *T.'s fossa*. T.'s tenia, tænia semicircularis; a white band lying below the vena corporis striati, and extending from near the anterior extremity of the thalamus, along the inner border of the inferior cornu of the lateral ventricle, into the gray substance of the hippocampus major. T.'s valve, the posterior medullary velum.
**Tarnier's sign** (*tar-ne-a'*) [Etienne Stéphane *Tarnier*, French obstetrician, 1828-1897]. Effacement of the angle between the upper and lower segments of the uterus; it is an indication of inevitable abortion.
**tarropetrolin** (*tar-o-pet'-ro-lin*). A compound of wood tar and petroleum, used as a salve in various skin diseases.
**tarsadenitis (meibomica)** (*tar-sad-en-i'-tis*) [*tarsus*; ἀδήν, a gland; ιτις, inflammation]. Inflammation of the Meibomian glands and tarsal cartilage.
**tarsal** (*tar'-sal*) [*tarsus*]. 1. Pertaining to the tarsus of the foot. 2. Pertaining to the tarsus of the eye. **t. cartilage**, the cartilaginous layers in the free edge of each eyelid. **t. cyst**. See *chalazion*. **t. glands**, the Meibomian glands.
**tarsale** (*tar-sa'-le*). Any bone of the tarsus, but especially one in the distal row.
**tarsalgia** (*tar-sal'-je-ah*) [*tarsus*; ἄλγος, a pain]. Pain, especially one of neuralgic character, in the tarsus.
**tarsalia** (*tar-sa'-le-ah*) [pl. of *tarsale*]. The tarsal bones.
**tarsalis** (*tar-sa'-lis*). A tarsal muscle.
**tarsectomy** (*tar-sek'-to-me*) [*tarsus*; ἐκτομή, excision]. 1. Excision of tarsal bones. 2. Excision of part of a tarsal cartilage.
**tarsectopia** (*tar-sek-to'-pe-ah*) [*tarsus*; ἔκτοπος, out of place]. Tarsal displacement.
**tarsen** (*tar'-sen*) [*tarsus*]. Belonging to the tarsus in itself.
**tarsitis** (*tar-si'-tis*) [*tarsus*; ιτις, inflammation]. Inflammation of the tarsus; and see *blepharitis*.
**tarso-** (*tar-so-*) [ταρσός, *tarsus*]. A prefix denoting pertaining to the tarsus.
**tarsocheiloplasty** (*tar-so-ki'-lo-plas-te*) [*tarso-*; χεῖλος, lip; πλάσσειν, to form]. Plastic surgery of the edge of the eyelid.
**tarsochasis** (*tar-so-klā'-sis*) [*tarsus*; κλάσις, rupture]. 1. Rupture of the tarsal cartilage. 2. Intentional fracture of the tarsus, for the correction of club-foot.
**tarsomalacia** (*tar-so-mal-a'-she-ah*) [*tarso-*; μαλακία, softening]. Softening of the tarsus of the eyelid.
**tarsometatarsal** (*tar-so-met-ah-tar'-sal*) [*tarso-*; *metatarsus*]. Relating to the tarsus and the metatarsus.
**tarso-orbital** (*tar-so-or'-bit-al*). Relating to the framework of the eyelids and the walls of the orbit.
**tarsophalangeal** (*tar-so-fa-lan'-je-al*) [*tarso-*; φάλαγξ, phalanx]. Pertaining to the tarsus and the phalanges.

**tarsophyma** (tar-so-fi'-mah) [tarso-; φῦμα, a growth]. Any morbid growth or tumor of the tarsus.
**tarsoplasia** (tar-so-plā'-ze-ah). Same as tarsoplasty.
**tarsoplasty** (tar'-so-plas-te) [tarso-; πλάσσειν, to form]. Plastic surgery of the eyelid; blepharoplasty.
**tarsorrhaphy** (tar-sor'-a-fe) [tarso-; ῥαφή, suture]. The operation of sewing the eyelids together for a part or the whole of their extent.
**tarsotarsal** (tar-so-tar'-sal) [tarsus]. Between the tarsal bones; midtarsal.
**tarsotibial** (tar-so-tib'-e-al). Same as tibiotarsal.
**tarsotomy** (tar-sot'-o-me) [tarso-; τομή, a cutting]. 1. Operation upon the tarsal cartilage. 2. Operation upon the tarsus of the foot. t. **cuneiform,** removal of a wedge-shaped piece of any of the tarsal bones.
**tarsus** (tar'-sus) [ταρσός, tarsus]. 1. The instep, consisting of the calcaneus, astragalus, cuboid, scaphoid, internal, middle, and external cuneiform bones. 2. The cartilage of the eyelid, called the tarsal cartilage, a dense connective tissue forming the support of the lid.
**tartar** (tar'-tar) [Low L., tartarum, from Ar., durd, dregs]. 1. A hard mineral deposited on the inside of wine-casks, and consisting mainly of acid potassium tartrate (cream of tartar). 2. A hard incrustation on the teeth, consisting of mineral and organic matter. t., **alkali of,** potassium carbonate. t., **borated,** potassium and sodium borotartrate. t., **cream of** (potassii bitartras, U. S. P.). See potassium bitartrate. t. **emetic,** antimony and potassium tartrate. See under antimony. t., **soluble,** potassium tartrate. t., **vitriolated,** potassium sulphate.
**tartarated** (tar'-tar-a-ted) [tartar]. Containing tartar. t. **antimony,** tartar emetic. t. **soda,** sodium and potassium tartrate.
**tartaric acid.** See acid, tartaric.
**tartarization** (tar-tar-i-za'-shun). The treatment of syphilis with tartar emetic.
**tartarized** (tar'-tar-īzd). See tartarated.
**tartarlithin** (tar-tar-lith'-in). See lithium bitartrate.
**tartarus** (tar'-tar-us) [L.]. 1. Tartar. 2. Certain salts of potassium. t. **boraxatus,** potassium and sodium borotartrate. t. **natronatus,** potassium and sodium tartrate. t. **tartarisatus,** potassium tartrate. t. **vitriolatus,** potassium sulphate.
**tartrate** (tar'-trāt). A salt of tartaric acid.
**tartrated** (tar'-tra-ted). Containing tartar; combined with tartaric acid.
**tartrophen** (tar'-tro-fen). A combination of phenetidin and tartaric acid.
**Tashkend ulcer** (tash-kend'). See Sartian disease.
**taste** (tāst). 1. The sensation produced by stimulation of special organs in the tongue (taste-organs) by soluble bodies. 2. The faculty by which these sensations are appreciated. t., **after-,** a secondary taste perceived after the immediate taste has ceased. t.-**bud,** an oval, flask-shaped body, embedded in the epithelium of the tongue, and serving the sense of taste. It is also called t.-bulb. t.-**bulb.** See t.-bud. t.-**cell,** one of a number of peculiarly shaped, flask-like bodies found between the epithelial cells covering the slopes of the circumvallate papillæ. They are the terminal end-organs of the gustatory nerve. t.-**center,** the gustatory nervous center. Its position is not determined. t.-**end.** See t.-cell. t.-**goblets,** flask-like bodies on the sides and base of the tongue enclosing the gustatory cells; see t.-cell. t.-**pore,** the minute canal connecting the interior of a taste-bud with the surface of the mucous membrane.
**tattóoing** (tat-too'-ing) [Tahitian]. The production of permanent colors in the skin by the introduction of foreign substances, such as carbon, india-ink, etc., a common practice among sailors. t. **of the cornea,** a method of hiding leukomatous spots. t., **electrolytic,** the electrolytic treatment of angioma or nevus by means of a negative electrode carrying from 10 to 20 needles.
**taurin** (taw'-rin) [taurus, bull], $C_2H_7NSO_3$. Amidoethylsulphonic acid, a crystalline decomposition-product of bile. See Lang.
**taurocholate** (taw-ro-ko'-lāt). Any salt of taurocholic acid.
**taurocholic acid** (taw-ro-kol'-ik). See acid, taurocholic.
**taurocol** (taw'-ro-kol). A preparation containing sodium glycocholate, sodium taurocholate, cascara sagrada, phenolphthalein, and aromatics. It is a cholagogue.

**tautomenial** (taw-to-me'-ne-al) [ταὐτό, the same; μήν, month]. Relating to the same menstrual period.
**tautomeral, tautomeric** (taw-tom'-er-al, taw-to-mer'-ik) [ταὐτό, the same; μέρος, part]. 1. Exhibiting tautomerism; a qualification applied to compounds to which two different structural formulæ may be rightly attributed. 2. Applied to neurons of the cinerea of the spinal cord, the axons of which pass into the white matter of the cord on the same side in which they are located.
**tautomerism** (taw-tom'-er-ism) [ταὐτό, same; μέρος, a share]. 1. The attribution of two different formulæ to one compound. 2. The quality exhibited by those cases in which two structural formulæ are possible, while but one compound appears to be obtainable. It is assumed that in such bodies the formulæ are susceptible of change from one arrangement to the other. The phenomenon has also been called desmotropy. t., **virtual,** term for phasotropy.
**Tawara's node** (tah-vah'-rah) [S. Tawara, Japanese physician]. A node of interlacing muscle fibers in the auricular septum at the beginning of the muscle bundle of His.
**taxine** (taks'-ēn) [τάξος, yew-tree]. A poisonous alkaloid from the leaves and seeds of the Taxus baccata, or yew-tree. It is used in epilepsy.
**taxis** (taks'-is) [τάξις, from τάσσειν, to arrange]. 1. An arranging. 2. A manipulation, especially manipulation for the reduction of hernia. 3. The reaction of protoplasm to a stimulus; tropism, chemotaxis, q. v. t., **bipolar,** the replacement of a retroverted uterus by upward pressure through the rectum and drawing the cervix down in the vagina. t., **positive,** t., **negative.** See chemotaxis.
**taxodium** (taks-o'-de-um) [τάξος, yew-tree; εἶδος, form]. The common bald or black cypress of the southern United States and Mexico; said to be useful in hepatic diseases, in rheumatism, and as a diuretic.
**taxonomic** (taks-o-nom'-ik) [τάξις, arrangement; νόμος, law]. Pertaining to systematic classification.
**taxonomy** (taks-on'-o-me) [τάξις, arrangement; νόμος, law]. The principles of classification.
**Taxus** (taks'-us) [τάξος, yew-tree]. A genus of cone-bearing trees, the yews. T. **baccata,** the common European yew-tree. Its leaves and seeds are poisonous and have sedative qualities.
**Tay's choroiditis.** Choroidal degeneration, characterized by irregular yellowish spots visible around the macula lutea, and thought to be due to an atheromatous condition of the arteries. Syn., choroiditis guttata senilis.
**Taylor's test for acetone.** A few drops of a freshly prepared aqueous solution of sodium nitroprusside are added to 10 c.c. of urine or distillate; concentrated ammonium hydroxide is then stratified upon the mixture. A magenta color at the point of contact indicates the presence of acetone.
**Tay-Sach's disease** [Warren Tay, English physician; Bernard Sachs, American neurologist, 1858- ]. Amaurotic family idiocy.
**tayuya** (ta-ū'-yah). The roots of various plants, Dermophylla pendulina, Cayaponia martiana, etc., used in the treatment of syphilis.
**tayuyin** (ta-u'-yin). A bitter principle from tayuya.
**Tb.** Chemical symbol of terbium.
**T-bandage.** See under bandage.
**TC.** See under tuberculin.
**Te.** 1. Chemical symbol for tellurium. 2. Abbreviation for tetanic contraction.
**tea** (te) [Chinese]. 1. The dried leaves of Thea chinensis, of the order Ternstromiaceæ, used for preparing a beverage, also called tea. 2. Any vegetable infusion used as a beverage. t., **James',** t., **Labrador,** the leaves of Ledum latifolium used as a substitute for tea. t., **teamsters'.** See tepopote.
**teaberry.** See gaultheria.
**teacher's' nodes** of nodules. See chorditis tuberosa.
**Teale's amputation** (tēls) [Thomas Pridgin Teale, English surgeon, 1801-1868]. 1. For amputation of the arm: the long flap is placed upon the anteroexternal aspect of the arm; the brachial artery and the median and ulnar nerves are divided with the posterior flap. 2. For amputation of the leg: a rectangular flap operation, in which a long anterior and a short posterior flap are made, each consisting of both integument and muscle; the length of the anterior flap is equal to half the circumference of the

limb, and the posterior flap is one-quarter of the length.

**tears.** 1. The secretion of the lacrimal gland. 2. Hardened lumps, or drops, of any resinous or gummy drug.

**tease** (*tēs*). To tear a tissue into its component parts with needles.

**teaspoon.** A small spoon holding about 4 Cc. or 1 dr.

**teat** (*tēt*). A nipple.

**technic, technique** (*tek'-nik, tek-nēk'*) [τέχνη, art]. The method of procedure in operations or manipulations of any kind.

**technocausis** (*tek-no-kaw'-sis*) [τέχνη, art; καῦσις, a burning]. Mechanical cauterization, in counterdistinction to that produced by chemicals.

**tecnology** (*tek-nol'-o-je*) [τέκνον, a child; λόγος, study]. The study or scientific knowledge of childhood, its hygiene, diseases, etc.

**tecnotonia** (*tek-no-to'-ne-ah*) [τέκνον, a child; κτείνειν, to kill]. Child-murder; infanticide.

**tecosis.** See *takosis*.

**tectiform** (*tek'-ti-form*) [*tectum*, a roof; *forma*, form]. Roof-shaped.

**tectocephalic** (*tek-to-sef-al'-ik*) [*tectum*, a roof; κεφαλή, head]. Pertaining to a roof-shaped skull.

**tectocephaly** (*tek-to-sef'-al-e*) [*tectum*, a roof; κεφαλή, head]. The condition of having a roof-shaped skull.

**tectology** (*tek-tol'-o-je*) [τέκτων, a builder; λόγος, science]. Structural morphology.

**tectorial** (*tek-to'-re-al*) [*tectorium*]. Serving as a roof or covering. t. **membrane.** See *membrana tectoria*.

**tectorium** (*tek-to'-re-um*) [L.: *pl.*, *tectoria*]. 1. A covering. 2. See *membrana tectoria*.

**tectum** (*tek'-tum*) [L.; *gen.*, *tecti*]. A roof or covering. t. **ventriculi quarti,** *Vieussen's valve, q. v.*

**tedious** (*te'-de-us*) [*tædium*, weariness]. Unduly protracted, as *t. labor*.

**tedium vitæ.** See *tædium*.

**teel oil.** See *sesamum, oil of*.

**teeth** (*tēth*) [plural of *tooth, q. v.*]. t., **auditory,** the tooth-like projections on the edge of the limbus laminæ spiralis of the ear. They extend between the epithelial cells and give the limbus an uneven, highly refracting surface. They are composed of the osteogenous tissue of the crista. t., **Chiaie** [Prof. Stephano *Chiaie*, of Naples]. A peculiar deterioration of the dental enamel among the inhabitants of the Italian littoral; characterized by the teeth becoming black and destitute of enamel (*denti neri*), though apparently strong and serviceable; or the teeth remain white and finely formed but marked by a line of fine black, script-like marks (*denti scritti*). t., Corti's. See *t., auditory*. t., **Horner's,** incisor teeth presenting horizontal grooves that are due to a deficiency of enamel. t., **Huschke's.** See under *Huschke*. t., **Hutchinson's.** See *Hutchinson's teeth*. t., **master,** a name given by early writers to the venom fangs of serpents. t., **notched.** See *Hutchinson's teeth*. t., **numbering of the,** in numbering the teeth, the incisor next the symphysis menti is first, the wisdom-tooth last, or eighth. The first incisor is also said to be central, mesal, or proximal, and the last or wisdom-tooth, distal. In numbering the groups of teeth, as incisors, bicuspids or premolars, molars, the one nearest the symphysis is number one of the particular group. t., **pegged.** See *Hutchinson's teeth*. t., **permanent**; those of the second dentition. t., **pivot.** See under *pivot*. t., **sectorial,** the cutting teeth of the carnivora. t., **springing,** a name given by early writers to the venom fangs of serpents. t., **succedaneous,** the permanent teeth which take the places of the temporary teeth. t., **temporary,** the teeth of the first dentition; milkteeth; deciduous teeth; also, a provisional set of artificial teeth. t., **test-,** the central upper incisors of the permanent teeth, which are observed as a test, being "notched" or "pegged" in cases of congenital lues. t., **wall,** molars.

**teething** [AS., *tōth*, tooth]. The eruption of the first teeth in an infant; dentition.

**Teevan's law** (*te'-van*). [William Frederick *Teevan*, English surgeon, 1834–1887]. Fracture of a bone occurs in the line of extension, not in that of compression.

**tegmen** (*teg'-men*) [*tegere*, to cover]. A cover. t. **mastoideum,** the roof of the mastoid cells. t. **tympani,** the roof of the tympanic cavity. t. **ventriculi quarti,** the roof of the fourth ventricle.

**tegment** (*teg'-ment*). The tegmentum.

**tegmental** (*teg-men'-tal*) [*tegmen*]. Pertaining to the tegmentum. t. **nucleus,** the red nucleus. See *nucleus, tegmental*.

**tegmentum** (*teg-men'-ium*) [*tegmen*]. A covering; specifically, the dorsal portion of the crus cerebri and pons Varolii. t. **auris,** the membrana tympani. t., **hypothalamic,** or **subthalamic,** the continuation of the tegmentum under the thalamus. t. **tympani,** the tegmen tympani. t. **ventriculi, lateralis,** the centrum ovale majus. t. **ventriculorum,** the centrum ovale majus.

**tegmin** (*teg'-min*). A white, aseptic substance used as in collodion, in sealing small wounds that do not require drainage; it is said to consist of an emulsion of wax, acacia, water, zinc oxide, and lanolin.

**tegone** (*teg'-ōn*). A proprietary medicated plaster similar to *gelone*, *q. v.*

**tegumen** (*teg'-ū-men*). See *tegmen*.

**tegument** (*teg'-ū-ment*) [*tegmen*]. The integument.

**tegumental, tegumentary** (*teg-ū-men'-tal, teg-ū-men'-a-re*) [*tegmen*]. Relating to the skin or tegument.

**Teichmann's crystals** (*tīk'-man*) [Ludwig *Teichmann*, German histologist, 1825–1895]. Hemin crystals. T.'s test for hemin; to the dry residue placed on a slide a small crystal of sodium chloride is added and a cover-glass laid over it. A few drops of glacial acetic acid are allowed to flow in under the cover-glass, and the whole is heated gently so as not to boil the liquid. On cooling, rhombic crystals of hemin (*Teichmann's crystals*) will be found. If no crystals appear after the first warming, warm again, and, if necessary, add more acetic acid.

**teichopsia** (*ti-kop'-se-ah*) [τεῖχος, wall; ὄψις, vision]. A temporary amblyopia, with subjective visual images like fortification-angles; it is probably due to vasomotor disturbances of the visual center.

**teinodynia** (*ti-no-din'-e-ah*). See *tenodynia*.

**teinophlogosis** (*ti-no-flo-go'-sis*) [τείνειν, to stretch; φλόγωσις, inflammation]. Inflammation of the tendons.

**tela** (*te'-lah*) [L.]. A web or tissue. t. **adiposa,** adipose tissue. t. **aranea,** spiders' web, cobweb; it is used as a styptic. t. **cellulosa,** connective or areolar tissue. t. **choroidea,** the membranous roof of the third and fourth ventricles of the brain. t. **choroidea ventriculi quarti,** a fold of pia forming a part of the roof of the fourth ventricle. t. **choroidea ventriculi tertii,** the velum interpositum, *q. v.* t. **epithelialis,** epithelial tissue. t. **erectilis,** erectile tissue. t. **flava,** elastic tissue. t. **hæmalis,** lymph cells and blood-corpuscles, and also, splenic tissue. t. **vasculosa,** the choroid plexus.

**telæsthesia** (*tel-es-the'-ze-ah*). Telesthesia, telepathy, *q. v.*

**telangiectasia** (*tel-an-je-ek-ta'-ze-ah*). See *telangiectasis*.

**telangiectasis** (*tel-an-je-ek'-ta-sis*) [τέλος, end; ἀγγεῖον, a vessel; ἔκτασις, a stretching]. Dilatation of groups of capillaries or smaller blood-vessels. t. **faciei,** acne rosacea. t. **lymphatica,** lymphangiectasis.

**telangiectatic** (*tel-an-je-ek-tat'-ik*). Pertaining to or characterized by telangiectasis.

**telangiectoma** (*tel-an-je-ek-to'-mah*) [see *telangioma*], Birthmark; simple nevus.

**telangioma** (*tel-an-je-o'-mah*) [τέλος, end; ἀγγεῖον, a vessel; ὄγμα, tumor]. A tumor composed of dilated capillaries.

**telangiosis** (*tel-an-je-o'-sis*) [τέλος, end; ἀγγεῖον, a vessel]. Disease of the capillaries or minute blood-vessels.

**telar** (*te'-lar*) [*tela*, a web]. Pertaining to a tela; of the nature of a tela.

**teledactyl** (*tel-e-dak'-til*) [τῆλε, afar; δάκτυλος, finger]. A device to avoid stooping when wishing to pick up things from the floor (in disease of the spine, injuries, etc.). It consists of six spring-forceps at the end of a cane, operated by a cord passing to the handle and provided with a ring for the index finger.

**teledendrite** (*tel-e-den'-drīt*). See *telodendron*.

**telediastolic** (*tel-e-di-as-tol'-ik*) [τῆλε, end; *diastole*]. Relating to the last phase of a diastole.

**telegony** (*tel-eg'-on-e*) [τῆλε, far away; γονή, offspring]. The influence of a previous husband on the children of a subsequent one through the same woman.

**telegrapher's cramp.** See *occupation-neurosis*.
**telekinesis** (*tel-ek-in-e'-sis*) [τῆλε, afar; κίνησις, movement]. The power claimed by certain persons of causing objects to move without touching them.
**telelectrotherapeutics** (*tel-e-lek-tro-ther-ap-ū'-tiks*) [τῆλε, afar; *electrotherapeutics*]. The treatment of hysterical paralysis by a series of electric discharges near the patient without actual contact.
**telencephal** (*tel-en'-se-fal*). Telencephalon.
**telencephalon** (*tel-en-sef'-al-on*). The end-brain, a part of the prosencephalon, *q. v.*
**teleneurite** (*tel-e-nū'-rīt*) [τῆλος, end; νεῦρον, nerve]. One of the terminal filaments of the main stem of an axis-cylinder process.
**teleneuron** (*tel-e-nū'-ron*) [see *teleneurite*]. The neuron forming the terminus of an impulse in a physiological act involving the nervous system.
**teleologic** (*tel-e-o-loj'-ik*) [τέλεος, complete; λόγος, treatise]. Relating to the final cause of things.
**teleology** (*tel-e-ol'-o-je*) [τέλεος, complete; λόγος, treatise]. The doctrine of final causes.
**teleorganic** (*tel-e-or-gan'-ik*) [τέλεος, complete; *organic*]. Necessary to organic life.
**teleotherapeutics** (*tel-e-o-ther-ap-ū'-tiks*) [τῆλε, afar; *therapeutics*]. Suggestive therapeutics.
**telepathist** (*tel-ep'-ath-ist*) [τῆλε, afar; πάθος, disease]. One who is versed in telepathy.
**telepathy** (*te-lep'-ath-e*) [τῆλε, afar; πάθος, disease]. The action, real or supposed, of one mind upon another when the two persons are separated by a considerable distance; thought transference.
**telephic** (*tel-ef'-ik*). Malignant; incurable; relating to a telephium.
**telephium** (*tel-ef'-e-um*) [*Telephus*, son of Hercules, whose wound received from Achilles did not heal]. An old inveterate ulcer.
**teleradiography** (*tel-e-ra-di-og'-ra-fe*) [τῆλε, afar; *radiography*]. Radiography with the tube held at a distance of about six feet from the body.
**telesthesia, telæsthesia** (*tel-es-the'-ze-ah*) [τῆλε, afar; αἴσθησις, sensibility]. Distant perception; a perception of objects or conditions independently of the recognized channels of sense.
**telesystolic** (*tel-e-sis-tol'-ik*) [τέλος, end; *systole*]. Pertaining to the last phase of systole.
**teletherapy** (*tel-e-ther'-ap-e*) [τῆλε, afar; *therapy*]. Absent treatment.
**tellicherry bark.** The bark of *Wrightia zeylanica*; it is used in dysentery.
**tellurate** (*tel'-ū-rāt*). A salt of telluric acid.
**telluric** (*tel-ū'-rik*) [*tellus*, earth]. 1. Derived from the earth. 2. Relating to tellurium. t. acid, an acid, H₂TeO₄, whose salts are known as tellurates.
**tellurism** (*tel'-ū-rizm*) [see *telluric*]. Telluric miasm; influence of the soil as a cause of disease.
**tellurium** (*tel-ū'-re-um*) [see *telluric*]. A nonmetallic element of bluish-white color, having a specific gravity of 6.23, a quantivalence of two or six, an atomic weight of 127.5. Symbol Te.
**Tellyesniczky's fluid** (*tel-yes-nits'-ke*) [Kálmán *Tellyesniczky*, Hungarian histologist]. A 3 per cent. solution of potassium bichromate in water; to which 5 per cent. glacial acetic acid is added just before use. It is used as a hardening fluid.
**teloblast** (*tel'-o-blast*) [τέλος, end; βλαστός, germ]. A segmentation-sphere at the extremity of the germ-band, which becomes elongated by cells arising from the mesoblast.
**teloblastic** (*tel-o-blas'-tik*) [*teloblast*]. Pertaining to a teloblast.
**telodendron** (*tel-o-den'-dron*) [τέλος, end; δένδρον, tree]. The terminal arborization of an axis-cylinder process.
**telokinesis** (*tel-o-kin-e'-sis*). See *telophase*.
**tololecithal** (*tel-o-les'-ith-al*) [τέλος, end; λέκιθος, yolk]. Of an ovum, having a relatively large mass of food-yolk placed eccentrically.
**tolemma** (*tel-o-lem'-ah*) [τέλος, end; λέμμα, husk]. The membrane covering the eminence of Doyère, or the point of entrance of a motor nerve into a muscular fiber.
**telophase** (*tel'-o-fāz*) [τέλος, end; φάσις, an appearance]. The final phase of any process, as that of karyokinesis.
**telosporidia** (*tel-o-spo-rid'-e-ah*) [τέλος, end; *Sporidia*]. A class of sporozoa which end their individual existence at the stage of spore formation.
**telosynapsis** (*tel-o-sin'-ap-sis*) [τῆλε, afar; *synapsis*]. The union of chromosomes end to end.
**temper** (*tem'-per*). To make metals hard and elastic by heating them and then suddenly cooling them.
**temperament** (*tem'-per-am-ent*) [L., *temperamentum*]. A term applied to mental disposition and physical constitution of an individual, as the bilious, lymphatic, nervous, and sanguine temperaments.
**temperance** (*tem'-per-ans*) [*temperantia*]. Moderation in satisfying desire; especially as regards the use of alcoholic beverages.
**temperate** (*tem'-per-āt*) [*temperatus*, moderated]. Moderate, without excess. t. zone, the zone of climate situated between 30° and 70°.
**temperature** (*tem'-per-a-tūr*) [L., *temperatura*]. The degree of intensity of heat of a body, especially as measured by a scale termed a thermometer. t., absolute, that reckoned from the absolute zero of temperature, estimated at −273° C. t., critical, the temperature at which a gas can, by pressure, be reduced to a liquid. t., mean, the average temperature of a place for a given period of time. t., normal, the temperature of the body in a state of health, *i. e.,* 98.6° F. t., optimum, the temperature most favorable for the cultivation of microorganisms. t.-sense, the sense by which differences in temperature are appreciated, consisting of a sense for cold (*cryesthesia*) and a heat-sense (*thermoesthesia*). These are represented on the surface by different nerve-endings, the so-called cold and hot points.
**temple** (*tem'-pl*) [*tempus*, time]. The portion of the head behind the eye and above the ear.
**Templin oil** (*tem'-plin*) [*Templin*, a town of Prussia]. Oil of pine-cones, from *Pinus pumilio;* it resembles ordinary oil of turpentine.
**tempolabile** (*tem-po-la'-bil*) [*tempus*, time; *labilis*, unstable]. Becoming changed in the course of time.
**temporal** (*tem'-po-ral*) [*temple*]. 1. Pertaining to the temple, as the *temporal* bone, the *temporal* artery. 2. Pertaining to time. t. artery. See *artery*. t. bone. See *bone*. t. crest, a ridge on the frontal bone. t. diplopia. Same as *homonymous diplopia*. t. fossa. See *fossa*. t. muscle. See *muscle*. t. operculum. See *postoperculum*. t. ridge. See *ridge*.
**temporalis** (*tem-po-ra'-lis*) [*temporal*]. The temporal muscle; see *muscles, table of.*
**temporary** (*tem'-po-ra-re*) [*temporarius*, lasting but for a time]. Not permanent. t. stopping, a preparation consisting principally of bleached gutta-percha, carbonate of calcium, and quartz, for filling teeth. t. teeth. See *teeth*.
**tempora** (*tem'-po-ren*) [*tempora*, the temples]. Belonging to the temporal bone in itself.
**temporization** (*tem-po-ri-za'-shun*) [*tempus*, time]. The expectant treatment of disease.
**temporo-** (*tem-po-ro-*) [*temple*]. A prefix denoting pertaining to the temple.
**temporoauricular** (*tem-po-ro-aw-rik'-ū-lar*) [*temporo-; auricular*]. Pertaining to the temporal and auricular regions of the head.
**temporofacial** (*tem-po-ro-fa'-shal*) [*temporo-; facies, face*]. 1. Pertaining to the temple and the face. 2. The larger of the main branches of the facial nerve.
**temporohyoid** (*tem-po-ro-hi'-oid*) [*temporo-; hyoid*]. Pertaining to the temporal and hyoid bones or regions.
**temporomalar** (*tem-po-ro-ma'-lar*) [*temporo-; mala, cheek*]. Pertaining to the temporal and malar bones, or to the temple and cheek.
**temporomandibular** (*tem-po-ro-man-dib'-ū-lar*) [*temporo-; mandible*]. Pertaining to the temporal bone and the mandible.
**temporomastoid** (*tem-po-ro-mas'-toid*) [*temporo-; mastoid*]. Pertaining to the temporal and mastoid regions of the skull.
**temporomaxillary** (*tem-po-ro-maks'-il-a-re*) [*temporo-; maxilla*]. 1. Pertaining to the temporal region and the upper jaw. 2. Temporomandibular.
**temporo-occipital** (*tem-po-ro-ok-sip'-it-al*) [*temporo-; occiput*]. Pertaining to the temporal and occipital bones or regions.
**temporoparietal** (*tem-po-ro-par-i'-et-al*) [*temporo-; paries, wall*]. 1. Pertaining to the temporal and parietal bones. 2. Pertaining to the temporal and parietal lobes of the brain.
**temporosphenoid** (*tem-po-ro-sfe'-noid*) [*temporo-; sphenoid*]. Pertaining to, or in relation with, the temporal and sphenoid bones.
**temporozygomatic** (*tem-po-ro-zi-go-mat'-ik*). Relating to the temporal and zygomatic bones or regions.
**tempostabile** (*tem-po-sta'-bil*) [*tempus*, time; *stabilis,*

**stable].** Not undergoing spontaneous change in the course of time.
**temulence** (*tem'-ū-lens*) [*tēmulentia,* inebriety]. Inebriety; drunkenness.
**temulentia** (*tem-ū-len'-she-ah*). See *temulence.*
**temulin** (*tem'-ū-lin*). The narcotic principle of *Lolium temulentum.*
**tenacious** (*te-na'-shus*) [*tenax,* tough]. Tough; cohesive.
**tenacity** (*te-nas'-it-e*) [*tenacitas*]. Toughness.
**tenaculum** (*ten-ak'-u-lum*) [*tenere,* to hold]. A hook-shaped instrument for seizing and holding parts.
**tenalgia** (*ten-al'-je-ah*) [τένων, tendon; ἄλγος, pain]. See *tenodynia.* t. **crepitans.** See *tendosynovitis.*
**tenalgin** (*ten-al'-jin*). A proprietary teniafuge prepared from the areca-nut.
**tenalin** (*ten'-al-in*). A teniafuge from areca-nut, with the toxic principle arecolin eliminated as far as possible.
**tenax** (*te'-naks*) [L.]. Trade name of oakum especially prepared for surgeons' use.
**tenderness** (*ten'-der-nes*). The condition of abnormal sensitiveness to touch; soreness.
**tendinitis** (*ten-din-i'-tis*). See *tenonitis.*
**tendinoplasty** (*ten'-din-o-plas-te*) [*tendo-,* a tendon; πλάσσειν, to form]. Plastic surgery of tendons.
**tendinosus** (*ten-din-o'-sus*). The semitendinosus muscle; see *muscles.*
**tendinosuture** (*ten-din-o-sū'-tūr*). See *tenorrhaphy.*
**tendinotrochanteric ligament** (*ten'-din-o-tro-kan-ter'-ik*). A ligament extending from the capsular ligament of the hip-joint to the great trochanter of the femur.
**tendinous** (*ten'-din-us*) [*tendon*]. Pertaining to or having the nature of tendon. t. **spot,** a deposit of fibrin on a serous membrane.
**tendo** (*ten'-do*) [*tendo, tendinis,* a tendon; pl., *tendines*]. A tendon. t. **Achillis,** the Achilles tendon or common tendon of the gastrocnemius and soleus muscles inserted into the heel. t. **Achillis reflex.** See *reflexes.* t. **calcaneus.** See *t. Achillis.* t. **oculi,** t. **palpebræ.** See *ligament, palpebral, internal.*
**tendomucoid** (*ten-do-mū'-koid*). A mucin found in tendons.
**tendon** [*tendere,* to stretch]. A band of dense fibrous tissue forming the termination of a muscle and attaching the latter to a bone. t., **central,** the aponeurosis in the center of the diaphragm. t., **cordiform.** See *t., central.* t., **hamstring.** See *hamstring.* t.-**reflex,** a reflex produced by stimulating the tendon of a muscle. t., **reindeer,** tendon from the neck of the reindeer; used as ligatures. t.-**spindles.** See *corpuscles, Golgi's.* t. **of Zinn,** the ligament of Zinn.
**tendophone** (*ten'-do-fōn*). See *dermatophone.*
**tendophony** (*ten-dof'-o-ne*). See *tenophony.*
**tendoplasty** (*ten'-do-plas-te*). See *tenoplasty.*
**tendosynovitis** (*ten-do-si-no-vi'-tis*). See *tenosynovitis.*
**tendotome** (*ten'-do-tōm*). See *tenotome.*
**tendotomy** (*ten-dot'-o-me*). See *tenotomy.*
**tendovaginal** (*ten-do-vaj'-in-al*) [*tendon; vagina,* a sheath]. Relating to a tendon and its sheath.
**tendovaginitis** (*ten-do-vaj-in-i'-tis*) [*tendon; vagina,* sheath; ιτις, inflammation]. Inflammation of a tendon and its sheath; tenosynovitis. t. **crepitans.** See *tenalgia crepitans.* t. **granulosa,** tuberculosis of tendon sheaths, the sheaths being filled with granulation or fungous tissue.
**tenectomy** (*ten-ek'-to-me*). See *tenonectomy.*
**tenesmic** (*ten-ez'-mik*) [*tenesmus*]. Of the nature of, or affected with tenesmus.
**tenesmus** (*ten-ez'-mus*) [τείνειν, to strain]. A straining, especially the painful straining to empty the bowels or bladder without the evacuation of feces or urine.
**tenia** (*te'-ne-ah*) [*tænia,* tape-worm]. A tapeworm. See *tape-worm; tænia.* t.-**toxin,** the toxin produced by tape-worms and to which the pathological changes wrought in the intestine are partly due.
**teniacide** (*te'-ne-as-īd*) [*tenia; cædere,* to kill]. 1. Destructive of tape-worms. 2. An agent that destroys tape-worms.
**teniafuge** (*te'-ne-af-ūj*) [*tenia; fugare,* to drive]. 1. Expelling tape-worms. 2. An agent that expels tape-worms.
**tenial** (*te'-ne-al*). Pertaining to a tenia or tænia.
**teniasis** (*te-nī'-as-is*) [*tenia*]. The ensemble of symptoms resulting from the presence of tenia in the body.

**tenicide** (*te'-nis-īd*). See *teniacide.*
**tenide** (*ten'-īd*). A remedy for diabetes.
**tenifuge** (*ten'-if-ūj*). See *teniafuge.*
**tenioid** (*te'-ne-oid*) [*ταινία,* a band, ribbon; εἶδος, form]. Ribbon-like, or resembling a tape-worm.
**teniola** (*ten-i'-o-lah*) [L.]. A small ribbon. t. **cinerea,** a thin grayish ridge separating the striæ of the fourth ventricle from the cochlear division of the acoustic nerve.
**teniophobia** (*te-ne-o-fo'-be-ah*) [*tænia,* tape-worm; φόβος, dread]. Morbid dread of becoming the host of a tape-worm.
**Tennesson's acne.** A disseminate variety of acne cornea.
**tennis-arm, tennis-elbow.** A strain of the elbow, said to be frequent in tennis-players.
**tennysine** (*ten'-is-ēn*). An alkaloid occurring in brain tissue.
**teno-** (*ten-o-*) [τένων, tendon]. A prefix meaning pertaining to a tendon.
**tenodynia** (*ten-o-din'-e-ah*). [*teno-;* ὀδύνη, pain]. Pain in a tendon.
**tenography** (*ten-og'-ra-fe*). See *tenontography.*
**tenology** (*ten-ol'-o-je*). See *tenontology.*
**tenomyotomy** (*ten-o-mī-ot'-o-me*) [*teno-; myotomy*]. Abadie's operation to enfeeble one of the recti muscles, consisting of incising the lateral parts of its tendon near its sclerotic insertion and removing a small portion of the muscle on each side.
**Tenon's capsule** (Jacques René *Tenon,* French anatomist and surgeon, 1724–1816). A fibroelastic membrane surrounding the eyeball. It is covered by a continuous layer of endothelial plates, and corresponds to a synovial sac. **T.'s fascia.** See *Tenon's capsule.* **T.'s membrane.** See *Tenon's capsule.*
**T.'s space,** the lymph-space existing between the sclerotic and Tenon's capsule.
**tenonectomy** (*ten-on-ek'-to-me*) [*teno-;* ἐκτομή, excision]. Excision of a portion of a tendon.
**tenonitis** (*ten-on-i'-tis*). 1. Inflammation of Tenon's capsule. 2. Tenontitis.
**tenonometer** (*ten-on-om'-et-er*) [*teno-;* μέτρον, a measure]. An instrument for measuring the tension of the eyeball.
**tenonostosis** (*ten-on-os-to'-sis*). See *tenostosis.*
**tenontagra** (*ten-on-ta'-grah*) [*teno-;* ἄγρα, seizure]. Gout in the tendons.
**tenontitis** (*ten-on-ti'-tis*) [*teno-;* ιτις, inflammation]. Inflammation of a tendon.
**tenontodynia** (*ten-on-to-din'-e-ah*). See *tenodynia.*
**tenontography** (*ten-on-tog'-ra-fe*) [*teno-;* γράφειν, to write]. The descriptive anatomy of the tendons.
**tenontolemmitis** (*ten-on-to-lem-i'-tis*) [*teno-;* λέμμα, a limiting membrane; ιτις, inflammation]. See *tenosynovitis.*
**tenontology** (*ten-on-tol'-o-je*). See *tenontography.*
**tenontophyma** (*ten-on-to-fī'-mah*) [*teno-;* φῦμα, growth]. A tumor growing on a tendon.
**tenontophyte** (*ten-on'-to-fīt*) [*teno-;* φυτόν, plant]. A new formation upon a tendon.
**tenontoplasty.** See *tenoplasty.*
**tenontothecitis** (*ten-on-to-the-si'-tis*). See *tenosynovitis.* t. **prolifera calcarea,** necrobiosis of the tendons in their sheaths accompanied by calcareous deposit.
**tenophony** (*ten-of'-on-e*) [*teno-;* φωνή, sound]. A sound elicited by auscultation supposed to be produced by the chordæ tendinæ.
**tenophyte** (*ten'-o-fīt*) [*teno-;* φυτόν, a growth]. A bony or cartilaginous growth on a tendon.
**tenoplasty** (*ten'-o-plas-te*) [*teno-;* πλάσσειν, to form]. Plastic surgery of a tendon.
**tenorrhaphy** (*ten-or'-a-fe*) [*teno-;* ῥαφή, suture]. The uniting of a divided tendon by sutures.
**tenositis** (*ten-o-si'-tis*). See *tenonitis.*
**tenostosis** (*ten-os-to'-sis*) [*teno-;* ὀστέον, a bone]. Ossification of a tendon.
**tenosuture** (*ten-o-sū'-tūr*). Same as *tenorrhaphy.*
**tenosynitis** (*ten-o-sin-i'-tis*). Same as *tenosynovitis.*
**tenosynovitis** (*ten-o-sin-o-vi'-tis*) [*teno-;* synovia; ιτις, inflammation]. Inflammation of a tendon and its sheath.
**tenotomania** (*ten-ot-o-ma'-ne-ah*) [*teno-;* τομή, cutting; μανία, name]. A morbid desire to perform tenotomy.
**tenotome** (*ten'-o-tōm*) [*teno-;* τέμνειν, to cut]. A knife for performing tenotomy.
**tenotomist** (*ten-ot'-o-mist*). One skilled in tenotomy.

**tenotomize** (*ten-ot'-o-mīz*) [*teno-*; τέμνειν, to cut]. To perform tenotomy.

**tenotomy** (*ten-ot'-o-me*) [see *tenotome*]. The operation of cutting a tendon. t., **graduated**, cutting a part of the fibers of the tendon of an ocular muscle for heterophoria or slight degrees of strabismus. t., **tarsal**, division of the peroneal tendon for the relief of spavin.

**tenovaginitis** (*ten-o-vaj-in-i'-tis*) [*teno-*; *vagina*, sheath; ιτις, inflammation]. Inflammation of the sheath of a tendon.

**tension** (*ten'-shun*) [*tendere*, to stretch]. 1. The act of stretching; the state of being stretched. 2. In electricity, the power of overcoming resistance. t., **arterial**, the strain in the arterial walls at the height of the pulse wave. t., **elastic**, stretching by means of an elastic material. t. of **gases**, the tendency of a gas to expand on account of the mutual repulsion of its molecules. t., **intravenous**, the strain of the blood current upon the walls of the veins. t., **intraocular**, the pressure of the ocular contents upon the sclerotic coat. It may be estimated by means of an instrument called a tonometer, or by palpation with the fingers; and is recorded by symbols as follows: Tn = normal tension; T + 1, T + 2, T + 3. indicate various degrees of increased tension, and T — 1, T — 2, T — 3, corresponding degrees of decreased tension. t., **muscular**, the state of muscular contraction which occurs when muscles are passively stretched.

**tensity** (*ten'-sit-e*) [see *tension*]. Tenseness, the condition of being stretched.

**tensive** (*ten'-siv*). Giving the sensation of stretching or contraction.

**tensor** (*ten'-sor*) [see *tension*]. A stretcher; a muscle that serves to make a part tense. See under *muscle*.

**tensure** (*ten'-shur*). Tension, a stretching or straining.

**tent** (L., *tenta*). A plug of soft material, as lint, gauze, or other material that increases in volume by the absorption of water; it is used chiefly for dilating an orifice and for keeping a wound open. t., **laminaria**, a tent made of sea-tangle. t., **sponge**, a tent made of compressed sponge, used for dilating the os uteri. t., **tupelo**, one made of the wood of the root of the water-tupelo.

**tentacle** (*ten'-tak-l*) [*tentare*, to handle, touch, feel]. In biology, loosely applied to any slender, tactile or prehensile organ, as a feeler, horn, proboscis, antenna, vibrissa, ray, or arm.

**tentacula** (*ten-tak'-ū-lah*). Same as *tentacle*.

**tentaculate** (*ten-tak'-ū-lāt*). Having tentacles.

**tentative** (*ten'-ta-tiv*) [*tentare*, to try; to prove]. Empirical; experimental.

**tenth cranial nerve.** The pneumogastric or vagus nerve.

**tentiginous** (*ten-tij'-in-us*) [*tentigo*, lust]. Characterized by insane lust.

**tentigo** (*ten-ti'-go*) [L.]. Lust, satyriasis. t. **prava**. Synonym of *lupus*. t. **venerea**. Synonym of *nymphomania*.

**tentorial** (*ten-to'-re-al*) [*tentorium*]. Pertaining to the tentorium. t. **sinus**, the straight sinus.

**tentorium**, t. **cerebelli** (*ten-to'-re-um*) [L., a tent]. The partition between the cerebrum and the cerebellum formed by an extension of the dura mater. t. of the **hypophysis**, the process of the dura mater covering the hypophysis cerebri; the diaphragma sellæ.

**tentum** (*tent'-tum*) [*tendere*, to stretch]. The penis.

**tenuate** (*ten'-ū-āt*) [*tenuis*, thin]. To make thin.

**tenuis** (*ten'-ū-is*) [L.]. Slender, thin. t. **mater**, same as pia mater.

**tenuity** (*ten-ū'-it-e*). Thinness; the condition of being thin.

**tenuous** (*ten'-ū-us*). Thin; minute.

**tephromyelitis** (*tef-ro-mi-el-i'-tis*) [τέφρος, ash-colored; *myelitis*]. See *poliomyelitis*.

**tephrosis** (*def-ro'-sis*) [τέφρα, ashes]. Incineration; cremation.

**tephrylometer** (*tef-ril-om'-et-er*) [τέφρα, ashes; μέτρον, measure]. A graduated glass tube for measuring the thickness of the gray matter of the brain by means of the segment or core removed.

**tepid** (*tep'-id*) [*tepidus*, warm]. About blood-heat.

**tepidarium** (*tep-id-a'-re-um*) [*lepid*]. A warm bath.

**tepopote** (*tep-o-po'-ta*). The twigs of *Ephedra antisyphilitica*; used in venereal diseases.

**tepor** (*tep'-or*) [L.]. Warmth; moderate heat.

**ter** [L., "three times"]. A common prefix meaning three, or threefold. t. **in die**, three times a day; abbreviated to *t. i. d.*

**terabdella** (*ter-ab-del'-ah*) [τερέειν, to bore, βδέλλα, a leech]. An artificial leech.

**teramorphous** (*ter-ah-mor'-fus*) [*teras*; μορφή, form]. Of the nature of a monstrosity.

**teras** (*te'-ras*) [τέρας, a monster; fl., *terata*]. A monster.

**teratic** (*ter-at'-ik*) [*teras*]. Monstrous.

**teratism** (*ter'-at-izm*) [*teras*]. An anomaly of conformation, congenital or acquired. t., **acquired**, deformity which is the result of disease, violence, nor operation. t., **atresic**, deformity, in which the natural openings are occluded. t., **casemic**, deformity in which parts which should be united remain in their primitive, fissured state. t., **ectogenic**, one in which certain parts of the body are absent or defective. t., **ectopic**, one in which there is displacement of one or more parts. t., **hermaphroditic**, one in which the organs of both sexes exist. t., **hypergenetic**, one in which certain organs are disproportionately large. t., **symphysic**, one in which certain organs or parts are abnormally fused.

**teratoblastoma** (*ter-at-o-blas-to'-mah*) [*teras*; βλάστος, germ; -ωμα, tumor]. Same as *teratoma*.

**teratogenesis** (*ter-at-o-jen'-es-is*). Same as *teratogeny*.

**teratogeny** (*ter-at-oj'-en-e*) [*teras*; γεννᾶν, to beget]. The formation or bringing forth of monsters.

**teratoid** (*ter'-at-oid*) [*teras*; εἶδος, like]. Resembling a monster. t. **tumor**, a complex tumor due to the growth of tissue embryologically misplaced.

**teratological** (*ter-at-o-loj'-ik-al*). Pertaining to teratology.

**teratology** (*ter-at-ol'-o-je*) [*teras*; λόγος, science]. The science of malformations and monstrosities.

**teratoma** (*ter-at-o'-mah*) [*teras*; ὅμα, tumor; pl., *teratomata*]. A tumor containing teeth, hair, and other material not found in the part wherein it grows, and resulting from an embryonic misplacement of tissue or from the inclosure of parts of a rudimentary fetus.

**teratomatous** (*ter-at-o'-mat-us*) [*teratoma*]. Of the nature of, or resembling a teratoma.

**teratophobia** (*ter-at-o-fo'-be-ah*) [*teras*; φόβος, dread]. 1. Morbid fear of monsters or of deformed or peculiar individuals. 2. Morbid dread, on the part of a pregnant woman, of giving birth to a teratism.

**teratosis** (*ter-at-o'-sis*) [*teras*]. 1. A congenital deformity. 2. Also used as a synonym of teratism.

**terbasic** (*ter-ba'-sik*). Synonym of *tribasic*.

**terbium** (*tur'-be-um*) [*Ytterby*, in Sweden]. A rare metallic element, symbol Tb. See *elements, table of*.

**terchloride** (*ter-klo'-rīd*) [*ter*, three; *chloride*]. Synonym of *trichloride*.

**tere** (*te'-re*). Latin for *rub*.

**terebene** (*ter'-eb-ēn*) [τερέβινθος, terebinth tree]. *Terebenum* (U. S. P.), C₁₀H₁₆. A hydrocarbon obtained by the oxidation of oil of turpentine by means of sulphuric acid. It is soluble in alcohol, and is used in bronchitis, dyspepsia, and diseases of the genitourinary tract. Dose 5–10 min. (0.32–0.65 Cc.). t. **glycerol**, a mixture of terebene, 4 parts; glycerol, 7 parts and water, 1 part, shaken together and exposed until the separated glycerol remains turbid when allowed to stand. It is used as an application to purulent wounds.

**terebenthene** (*ter-e-ben'-thēn*). Oil of turpentine.

**terebinth** (*ter'-e-binth*) [see *terebene*]. 1. The turpentine-tree, *Pistacia terebinthus*, which yields Chian turpentine. 2. Turpentine.

**terebinthina** (*ter-eb-in'-thin-ah*). See *turpentine*.

**terebinthinate** (*ter-eb-in'-thin-āt*) [see *terebene*]. 1. Containing turpentine. 2. A member or derivative of the turpentine group.

**terebinthinism** (*ter-e-bin'-thin-izm*). Poisoning with oil of turpentine.

**terebinthinize** (*ter-e-bin'-thin-īz*). To charge with turpentine.

**terebrachesia** (*tı re br̥sh o' sis*) [*teras*, round; βράχος, vs, short]. The operation of shortening the round ligament of the uterus.

**terebrant, terebrating** (*ter'-e-brant, ter'-e-bra-ting*) [*terebrare*, to bore]. Piercing, boring, said of pain.

**terebration** (*ter-eb-ra'-shun*) [see *terebrant*]. The operation of boring.

**teremorrhu** (*ter-e-mor'-ū*) [*terebene; morrhua*, the cod]. A proprietary preparation of pure terebene and cod-liver oil.

**teres** (*te'-rēz*) [L., round; *gen.*, *teretis*; *pl.*, *ters*-

# TERETE 880 TESTICLE

**tes].** 1. Round, as the *ligamentum teres*. 2. A muscle having a cylindrical shape, as *teres major, teres minor*. See under *muscle*.

**terete** (*ter-ēt'*) [*teres*]. Cylindrical. **t. eminence,** a slight thickening of the funiculus teres on the floor of the fourth ventricle. **t. funicle.** See *funiculus teres*.

**teretipronator** (*te-re-te-pro-na'-tor*). The pronator radii teres muscle. See *muscles*.

**teretiscapularis** (*te-re-te-skap'-ū-la-ris*). The teres major muscle. See *muscles*.

**ter in die** [L.]. Three times daily; generally abbreviated to *t. i. d.*

**tergal** (*ter'-gal*) [*tergum*, back]. Pertaining to the back, or dorsal surface or aspect.

**tergolateral** (*ter-go-lat'-er-al*) [*tergum*, back; *latus*, side]. Pertaining to the back and the side.

**term** [τέρμα, a limit]. A limit; the time during which anything lasts. The time of expected delivery. The menses.

**terma** (*ter'-mah*) [see *term*]. The layer of gray matter between the corpus callosum and the optic commissure; the lamina terminalis or lamina cinerea of the brain.

**termatic** (*ter-mat'-ik*) [τέρμα, limit]. Pertaining to the terma; as the *termatic artery*.

**terminad** (*ter'-min-ad*) [see *term*]. Situated in or toward the terminus.

**terminal** (*ter'-min-al*) [see *term*]. Pertaining to the end; placed at or forming the end. In the plural, a name sometimes applied to the poles of a battery or other electric source, or to the ends of the conductors or wires connected thereto. **t. artery.** See *artery*. **t. carbon atoms,** those combined with three hydrogen atoms. **t. genital corpuscles,** the round dilatations terminating the nerves in the dermis covering the extremity of the penis in horses. **t. infection,** an infection occurring late in the course of another disease and often causing the death of the patient. **t. neuritis.** See *erythromelalgia*.

**terminology** (*ter-min-ol'-o-je*) [*terminus*, a name, term; λόγος, science]. Nomenclature; a system of technical names or terms.

**terms** (*terms*). The menses.

**ternary** (*ter'-na-re*) [*ter*, three times]. Of chemical compounds, made up of three elements or radicals.

**ternitrate** (*ter'-ni-trāt*) [*ter*]. Trinitrate.

**teroxide** (*ter-oks'-id*) [*ter*, three; *oxide*]. A trioxide.

**terpene** (*ter'-pēn*) [a modified form of *terebene*]. One of a number of hydrocarbons having the formula C₁₀H₁₆, and contained in many volatile oils. **t. hydrochloride,** artificial camphor, C₁₀H₁₆·HCl; obtained from dry pinene by the action of dry chlorine in the cold. It is used as an internal antiseptic in tuberculosis and to check the flow of saliva; externally, it is used with phenol in skin diseases. **t. iodide.** See *iodoterpin*.

**terpenism** (*ter'-pen-izm*). Poisoning by terpene from internal use or inhalation; marked by abdominal pain, vomiting, inflammation of bladder and kidneys, bronchitis, paroxysms of asphyxia, and collapse. The urine has the odor of violets.

**terpin** (*ter'-pin*), C₁₀H₁₆(H₂O)₂H₂O. A diatomic alcohol obtained from turpentine; used in bronchial and pulmonary diseases to facilitate expectoration. **t. hydrate** (*terpini hydras*, U. S. P.), C₁₀H₂₀O₂+H₂O, a colorless, crystalline substance used as an expectorant and diuretic. Dose as expectorant 3–6 gr. (0.2–0.4 Gm.); diuretic, 10–15 gr. (0.65–1.0 Gm.) several times daily.

**terpineol** (*ter-pin'-e-ol*) [*terpin; oleum*, oil], C₁₀H₁₇OH. A viscous liquid obtained by heating terpin hydrate with phosphoric acid; used as is terpin.

**terpini hydras** (*ter-pi'-ni*). Terpin hydrate.

**terpinol** (*ter'-pin-ol*). An oily liquid obtained by the action of dilute mineral acids on terpin hydrate with heat; soluble in alcohol or ether. It is used as a bronchial stimulant, antiseptic, and diuretic. Dose 8–15 m.

**terra** (*ter'-ah*) [L.: gen., and pl., *terræ*]. Earth. **t. adamica,** any red bole, as Armenian bole. **t. alba,** white clay. **t. cariosa,** rotten-stone. **t. foliata,** sulphur. **t. foliata mineralis,** impure sodium carbonate. **t. foliata tartari,** potassium acetate. **t. fullonica,** fuller's earth. **t. fullonum,** fuller's earth. **t. japonica,** catechu. **t. livonica,** a very astringent sealed earth from Livonia. **t. miraculosa Saxoniæ,** Saxony earth, a mottled and variegated lithomarge or marrow stone. **t. oriana, t. orleana,** annotto. **t. ponderosa,** baryta or barium sulphate. **t. disienna,** same as

ocher. **t. sigillata,** a sealed earth; any bole so highly valued as to be formed into a small mass and stamped with a seal. **t. sigillata alba,** white bole. **t. sigillata rubra,** red bole. **t. umbra,** umber.

**terracing a suture.** A term indicating the closure of a wound by means of the insertion of successive tiers of sutures.

**terrain-cure** (*ter-an(g)-kūr*) [Fr.]. A method of treatment consisting in mountain-climbing, dietetics, etc., for plethora, corpulence, neurasthenia, chlorosis, incipient pulmonary tuberculosis, etc.

**terralin** (*ter'-al-in*). An ointment-vehicle, consisting of calcined magnesia, kaolin, silica, glycerol, and an antiseptic. It can be readily removed from the skin by water.

**terraline** (*ter'-al-in*) [*terra*, earth]. A proprietary preparation of petroleum, recommended as a substitute for cod-liver oil.

**terrol** (*ter'-ol*). A mixture of hydrocarbons of the paraffin series, offered as a substitute for cod-liver oil.

**terroline** (*ter'-ol-ēn*). A variety of petroleum-jelly.

**tersulphate** (*ter-sul'-fāt*) [*ter*, thrice; *sulphate*]. A salt in which the base is united with three sulphuric acid radicals.

**tersulphide** (*ter-sul'-fīd*). See *trisulphide*.

**tertian** (*ter'-she-an*) [*tertius*, third]. Recurring every other day, as *tertian fever*, a form of intermittent fever. **t., double,** quotidian.

**tertiarism** (*ter'-she-ar-izm*). Tertiary syphilis.

**tertiary** (*ter'-she-a-re*) [*tertian*]. Third in order. **t. alcohol,** an alcohol which contains the trivalent group COH. **t. syphilis.** See under *syphilis*.

**tertipara** (*ter-tip'-a-rah*) [*ter*, three times; *parere*, to bear]. A woman who has been in labor three times.

**tervalence** (*ter-va'-lens*). Synonym of *trivalence*.

**tescalama** (*tes-kal-am'-ah*). The milky juice of *Ficus nymphæifolia*, a tree of Mexico and South America. It is used in plasters.

**Tesla currents** (*tes'-lah*) [Nikola Tesla, electrician, 1857– ]. Rapidly alternating electric currents of high tension; they were applied therapeutically by d'Arsonval.

**teslaization** (*tes-lah-iz-a'-shun*). The therapeutic application of Tesla currents. Syn., *arsonvalization*.

**tessellated** (*tes'-el-a-ted*) [*tessellatus*, from *tessella*, a small square stone]. **t. epithelium,** flattened epithelial cells joined at their edges.

**test** [*testum*, a crucible]. 1. A trial. In chemistry, a characteristic reaction which distinguishes one body from others. 2. The reagent for producing a tested reaction. **t. breakfast.** See *test-meal*. **t. glass,** a small glass vessel, used in the chemical laboratory for purposes of experimentation and investigation. **t.-meal,** one given for the purpose of studying the secretory power of the stomach. **t.-paper,** paper impregnated with a chemical reagent, and used for detecting the presence of certain substances or conditions which cause a change in the color of the paper. **t. solution.** See *standard solution*. **t.-spoon,** a small spoon with a spatula-shaped handle used in chemical experiments. **t.-tube,** a cylinder of thin glass closed at one end, used in various chemical procedures. **t.-types,** letters or figures of different sizes to test acuteness of vision. Those most commonly employed are *Snellen's test-types*, a series of letters which at proper distances subtend an angle of five minutes.

**testa** (*tes'-tah*) [L.]. A shell. **t. ovi,** egg-shell. **t. præparata,** crushed and powdered oyster-shell.

**testaceous** (*tes-ta'-shus*) [*testa*, a shell]. Pertaining to a shell.

**testaden** (*tes'-tad-en*). A preparation from the testes of the bull. Used in affections of the spinal cord and in nervous troubles. Dose 92–123 gr. (6–8 Gm.) daily.

**testectomy** (*tes-tek'-to-me*) [*testis*, testicle; ἐκτομή, excision]. Orchidectomy, castration.

**testes** (*tes'-ēz*) [L.] [*testis*]. See *testicle*, and *testis*. **t. of brain,** see *corpora quadrigemina*. **t., female,** the ovaries. **t. muliebres,** the ovaries.

**testibrachial** (*tes-te-bra'-ke-al*) [*testis; brachium*, arm]. Pertaining to the testibrachium.

**testibrachium** (*tes-te-bra'-ke-um*) [*testis; brachium*, arm]. The process connecting the cerebellum with the testes of the brain; the superior peduncle of the cerebellum.

**testicle** (*tes'-tik-l*) [*testiculus*, dim. of *testis*]. See *testis* (1). **t., displaced,** a testicle in an abnormal situation, as in the pelvic cavity. **t., inverted,** a testicle which is so placed in the scrotum that the epididymis is attached to the anterior part of the

**tetanus** (*tet'-an-us*) [τέτανος, from τείνειν, to stretch].
1. An infectious disease characterized by tonic spasm of the voluntary muscles, an intense exaggeration of reflex activity, and peculiar convulsions. It is due to the bacillus of tetanus. The poison may enter through a wound (*traumatic tetanus*); at times no point of entrance is discoverable (*idiopathic tetanus*). 2. A tense, contracted state of a muscle, especially when caused experimentally. **t., acoustic.** See *acoustic.* **t. antitoxin.** See under *antitoxin.* **t., artificial,** that produced by a drug. **t., cephalic, t., cerebral, t., kopf-,** a special form of tetanus that has sometimes been observed to follow injuries of the head, especially those in the neighborhood of the eyebrow, trismus and facial paralysis occur upon the side of the injury, there is dysphagia, and death frequently results. **t. dolorificus,** synonym of *cramp.* **t., extensor,** a form of tetanus in which the extensors act more powerfully than the flexors. **t., hydrophobic,** tetanus characterized by violent spasm of the muscles of the throat. **t., imitative,** hysteria which simulates tetanus. **t. infantum.** See *t. neonatorum.* **t., idiopathic,** tetanus in which there is no history of injury. **t., localized,** tetanic spasm of a part. **t. neonatorum,** that due to infection of the umbilicus or the circumcision-wound. **t., postoperative,** that following operation. **t., puerperal,** that following labor. **t., remittent.** See *tetanilla.* **t., Ritter's,** the series of contractions, or apparent tetanus, observed on the opening or interrupting of an electric current which has been passing through the nerve for some time; opening tetanus. **t., toxic,** tetanus produced by an overdose of nux vomica or its alkaloids. **t., traumatic,** tetanus following an injury. **t., Wundt's,** a prolonged tetanic contraction induced in a frog's muscle by injury or the passage of a strong current.

**tetany** (*tet'-an-e*) [*tetanus*]. A disease characterized by intermittent, bilateral, painful, tonic spasms of the muscles, especially of the upper extremities. It is most common in young adults, but may occur in others. The cause appears to be a toxic agent. It occurs in connection with typhoid fever, gastrointestinal inflammation, in rickets, dilatation of the stomach, and after extirpation of the thyroid or parathyroid glands. **t.; duration,** a continuous tetanic spasm occurring in degenerated muscles when a strong continuous current is applied. **t., epidemic, t., rheumatic,** a form occurring over large portions of Europe, especially in the winter season. It is acute, lasting only two or three weeks, and rarely proving fatal. **t., gutturo-,** a stammering due to tetanoid spasm of the laryngeal muscles. **t., parathyreoprival,** tetany following removal of the parathyroid glands. **t. rheumatic.** See *t., epidemic.* **t., thyreoprival,** a form following removal of or suspension of the function of the thyroid gland.

**tetarelle** (*ta-tar-el'*) [Fr., *téter*, to suck]. An appliance for enabling a weakly infant to obtain milk from its mother. It consists of a nipple shield and two tubes; the mother sucks one of the latter, and the milk flows to the infant's mouth through the other.

**tetartocone** (*tet-ar'-to-kōn*) [τέταρτος, fourth; κῶνος, cone]. Posterointernal cone; the fourth or posterointernal cusp of an upper molar tooth.

**tetartoconid** (*tet-ar-to-kon'-id*) [*tetartocone*]. Posterointernal or fourth cone of the lower molar teeth.

**tetmil** (*tet'-mil*). Ten millimeters.

**tetra-** (*tet-rah-*) [τέτρα, four]. A prefix meaning four.

**tetra-allylammonium-alum** (*tet-rah-al-il-am-o-ne-um-al'-um*), $N(C_3H_5)_4.Al_2(SO_4)_3 + 12H_2O$. A uric-acid solvent.

**tetrabasic** (*tet-rah-ba'-sik*) [*tetra-*; βάσις, base]. Having four atoms of replaceable hydrogen.

**tetrablastic** (*tet-rah-blas'-tik*) [*tetra-*; βλαστός, a germ]. Having four germ-layers, namely an ectoderm, entoderm, somatopleure, and a splanchnopleure.

**tetraboric acid** (*tet-rah-bo'-rik*). See *acid, tetraboric.*

**tetrabrachius** (*tet-rah-bra'-ke-us*) [*tetra-*; βραχίων, arm]. A monster having four arms.

**tetracetate** (*tet-ras'-et-āt*). A combination of a base with four molecules of acetic acid.

**tetracheirus** (*tet-rah-ki'-rus*) [*tetra-*; χείρ, hand]. A monster with four hands.

**tetrachloride, tetrachlorid** (*tet-rah-klo'-rid*) [*tetra-*; *chloride*]. A binary compound consisting of an element or radical and four chlorine atoms.

**tetrachlormethane** (*tet-rah-klor-meth'-ān*). Carbon tetrachloride.

**tetracid** (*let-ras'-id*) [*tetra-*; *acid*]. Having four atoms of hydrogen that are replaceable by acid radicals.
**tetracoccus** (*tet-rah-kok'-us*) [*tetra-*; κόκκος, berry]. A micrococcus occurring in groups of four.
**tetracrotic** (*tet-rah-krot'-ik*) [*tetra-*; κρότος, a beat]. Same as *katatricrotic*.
**tetrad** (*tet'-rad*) [τέτρα, four]. 1. An element having an atomicity of four. 2. A group of four.
**tetradactyl** (*tet-rah-dak'-til*) [*tetra-*; δάκτυλος, a finger, or toe]. Having four digits on each limb.
**tetraethylene iodide** (*tet-rah-eth'-il-ēn*). See *diiodoform*.
**tetragenic** (*tet-raj-en'-ik*). Pertaining to or produced by the *Micrococcus tetragenus*.
**tetragenous** (*tet-raj'-en-us*) [*tetra-*; γίγνεσθαι, to be born]. In biology, applied to bacteria and other organisms which produce square groups of four as the result of fission.
**tetragon, tetragonum** (*tet'-rag-on, tet-rah-go'-num*) [*tetra-*; γωνία, angle, corner]. A four-sided figure.
**tetragonum lumbale**, an irregular, rhomboid space in the lumbar region beneath the aponeurosis of the latissimus dorsi, bounded externally by the dorsal margin of origin of the obliquus externus muscle, internally by the margin of the sacrospinalis, above by the serratus posticus inferior, and below by the upper margin of the obliquus internus.
**tetragonus** (*tet-rah-go'-nus*) [*tetra-*; γωνία, angle, corner]. The platysma muscle.
**tetrahydric** (*tet-rah-hi'-drik*) [*tetra-*; *hydrogen*]. Containing four replaceable atoms of hydrogen.
**tetrahydrobetanaphthylamine** (*tet-rah-hi-dro-ba-tan-naff-thil-am'-ēn*). See *thermin*.
**tetraiodoethylene** (*tet-rah-i-o-do-eth'-il-ēn*). See *diiodoform*.
**tetraiodophenolphthalein** (*tet-rah-i-o-do-fe-nol-thal'-e-in*). Nosophen.
**tetraiodopyrrol** (*tet-trah-i-o-do-pir'-ol*). See *iodol*.
**tetramastia** (*tet-rah-mas'-te-ah*) [*tetra-*; μαστός, breast]. *Tetramazia.*
**tetramastigote** (*tet-ram-as'-tig-ōt*) [*tetra-*; μάστιξ, a whip]. In biology, applied to microorganisms having four flagella.
**tetramazia** (*tet-rah-ma'-ze-ah*) [*tetra-*; μαζός, breast]. The presence of four breasts or mammary glands.
**tetramerism** (*tet-ram'-er-izm*) [*tetra-*; μέρος, part]. In biology, division into four parts.
**tetramethylenediamine** (*tet-rah-meth-il-ēn-di-am'-in*), C₄H₈(NH₂)₂. Putrescine.
**tetramethylputrescine** (*tet-rah-meth-il-pū-tres'-in*), C₄H₂₀N₂. A crystalline base derived from putrescine, having very poisonous properties.
**tetranitrin, tetranitrol** (*tet-rah-ni'-trin, tet-rah-ni'-trol*). See *erythrol tetranitrate*.
**tetranopsia** (*tet-ran-op'-se-ah*) [*tetra-*; ὄψις, vision]. A contraction of the field of vision limited to one quadrant.
**tetra-ophthalmus, tetrophthalmus** (*tet-rah-off-thal'-mus, tet-roff-thal'-mus*) [*tetra-*; ὀφθάλμος, eye]. A form of monster having four eyes. See *diprosopus*.
**tetraotus, tetrotus** (*tet-rah-o'-tus; tet-ro'-tus*) [*tetra-*; οὖς, the ear]. A form of monster having four ears. See *diprosopus*.
**tetraplegia** (*tet-rah-ple'-je-ah*) [*tetra-*; πληγή, stroke]. Paralysis of all four extremities.
**tetrapus** (*tet'-rah-pus*) [*tetra-*; πούς, foot]. 1. Having four feet. 2. A monster having four feet.
**tetrascelus** (*tet-ras'-el-us*) [*tetra-*; σκέλος, leg]. A monster having four legs.
**tetraschistic** (*tet-rah-skis'-tik*) [*tetra-*; σχίσις, division]. Dividing into four similar parts; tetragenous.
**tetraster** (*tet-ras'-ter*) [*tetra-*; ἀστήρ, a star]. A karyokinetic figure characterized by an arrangement of four stars, due to a fourfold division of the nucleus.
**tetrastichiasis** (*tet-rah-stik-i'-as-is*) [*tetra-*; στίχος, row]. Anomalous arrangement of the eyelashes in four rows.
**tetrastoma** (*tet-ras'-to-mah*) [*tetra-*; στόμα, a mouth]. A genus of entozoa.
**tetratomic** (*tet-rat-om'-ik*) [*tetra-*; *atom*]. 1. Containing four atoms. 2. Having four atoms of replaceable hydrogen.
**tetravalent** (*tet-rav'-al-ent*). See *quadrivalent*.
**tetrelle** (*tet-rel'*). See *tetarelle*.
**tetronal** (*tet'-ron-al*) [τέτρα, four], C₉H₂₀S₂O₄. Diethylsulphondethylmethane, a hypnotic resembling sulphonal. Dose 10–20 gr. (0.65–1.13 Gm.).
**tetronerythrin** (*tet-ron-er'-ith-rin*) [*tetra-*; ἐρυθρός, red]. A pigment found in some animals.

**tetrophthalmus** (*tet-roff-thal'-mus*). See *tetraophthalmus*.
**tetrotus** (*tet-ro'-tus*). See *tetraotus*.
**tetroxide, tetroxid** (*tet-roks'-id*) [*tetra-*; *oxid*]. A binary compound composed of a base and four atoms of oxygen.
**tetryl** (*tet'-ril*). See *butyl*.
**tetter** (*tet'-er*). A name for various skin eruptions, particularly herpes, eczema, and psoriasis. t., brawny, seborrhœa capitis. t., dry, dry or squamous eczema. t., humid, t., moist. See *eczema*. t., milky. See *crusta lactea*. t., moist. See *eczema*. t., running, includes various forms of eczema. t., scaly, psoriasis and squamous eczema.
**tety** (*tet'-e*). A skin disease found in Madagascar, and characterized by a pustular or squamous eruption in the neighborhood of the mouth and nostrils.
**teucrin** (*tū'-krin*), 1.C₂₁H₃₄O₁₁ or C₂₁H₃₆O₁₁. A crystalline glucoside from *Teucrium fruticans*. 2. A purified, sterilized aqueous extract of *Teucrium scordium*; a pungent brown liquid used in the treatment of tuberculous abscesses to arrest development. Hypodermic dose 50 min. (3 Cc.); locally 10 gr. (0.65 Gm.) with lanolin once daily.
**Teucrium** (*tū-kre-um*) [τεύκριον, germander]. A genus of labiate plants, germander or spleenwort. T. chamædrys, is used as an alterative. T. maritimum, cat-thyme, has errhine and antispasmodic properties, and was formerly used in coughs and nervous affections.
**teutlose** (*tūt'-lōs*) [τεῦτλον, beet]. A sugar found in beetroot.
**tewfikose** (*tū-fik'-ōs*). A sugar obtained from the milk of the Egyptian buffalo.
**Texas fever.** An infectious disease of cattle characterized by high fever, hemoglobinuria, and enlargement of the spleen. The disease is due to the parasite *Pyrosoma bigeminum*, which invades the red bloodcorpuscle and is transmitted by the cattle-tick, *Boophilus bovis*.
**texis** (*teks'-is*) [τέξις]. Child-bearing.
**text-blindness** (*tekst'-blind-ness*). See *word-blindness under aphasia*.
**textiform** (*teks'-te-form*) [*textum*, a web; *forma*, form]. Reticular, forming a mesh.
**textural** (*teks'-tū-ral*) [*textum*]. Pertaining to the tissues.
**texture** (*teks'-tūr*) [*texere*, to weave]. 1. Any organized substance or tissue of which the body is composed. 2. The arrangement of the elementary parts of tissue.
**textus** (*teks'-tus*) [*texere*, to weave]. A tissue.
**T-fiber.** A fiber given off at right angles from an axis-cylinder process.
**tfol.** Arabian soapstone, a natural product, closely related to steatite, proposed as a vehicle for antiseptic emulsion: 20 parts of tfol in 100 parts of heavy tar oil.
**TGl** [*tuberculin; Fr. globulineuse*, globulinous]. A symbol for a globulin contained in tubercle bacilli, soluble in 10 per cent. salt solution.
**thalamencephal** (*thal-am-en'-se-fal*). See *thalamencephalon*.
**thalamencephalic** (*thal-am-en-sef-al'-ik*). Pertaining to the thalamencephalon.
**thalamencephalon** (*thal-am-en-sef'-al-on*) [*thalamus; encephalon*]. The posterior portion of the anterior brain-vesicle; the interbrain. Syn., *diencephalon*.
**thalami** (*thal'-am-i*). Plural of *thalamus*.
**thalamic** (*thal-am'-ik*) [*thalamus*]. Pertaining to the thalamus. t. epilepsy, epilepsy from disease of the optic thalamus.
**thalamo-** (*thal-am-o-*) [*thalamus*]. A prefix denoting relation to the thalamus.
**thalamocœle, thalamocœle** (*thal'-am-o-sēl*) [*thalamo-*; κοιλία, a hollow]. The third ventricle.
**thalamocortical** (*thal-am-o-kor'-tik-al*) [*thalamo-; cortex*]. Pertaining to the thalamus and the cortex of the brain.
**thalamocrural** (*thal-am-o-kroo'-ral*) [*thalamo-; crus*, leg]. Pertaining to the thalamus and a crus cerebri.
**thalamolenticular** (*thal-am-o-len-tik'-ū-lar*) [*thalamo-; lenticular*]. Pertaining to the thalamus and the lenticular nucleus.
**thalamomammillary** (*thal-am-o-mam'-il-la-re*) [*thalamo-; mammillary*]. Pertaining to the thalamus and the mammillary bodies. t. fasciculus, the bundle of Vicq d'Azyr.
**thalamotegmental** (*thal-am-o-teg-ment'-al*). Relating to the thalamus and tegmentum.

**thalamus** (*thal'-am-us*) [θάλαμος, couch; pl., *thalami*]. A mass of gray matter at the base of the brain, developed from the wall of the vesicle of the third ventricle, and forming part of the wall of the latter cavity. The posterior part is called the *pulvinar*. The thalamus receives fibers from all parts of the cortex, and is also connected with the tegmentum and with fibers of the optic tract. t., optic, t. opticus. The same as *thalamus*.

**thalassophobia** (*thal-as-o-fo'-be-ah*) [θάλασσα, sea; φόβος, fear]. A morbid fear of the sea.

**thalassotherapy** (*thal-as-o-ther'-ap-e*) [θάλασσα, sea; θεραπεία, treatment]. Treatment of disease by sea-voyages, sea-bathing, sea-air, etc.

**thaletts** (*thal'-lets*). Pieces of chocolate containing phenolphthalein, and used as a laxative.

**thalictrine** (*thal-ik'-trēn*). A poisonous alkaloid obtained from *Thalictrum*.

**thalleine** (*thal'-e-ēn*). A combination of thalline with a compound ether or alcoholic radical.

**thallic** (*thal'-ik*). Pertaining to the metal thallium.

**thalline** (*thal'-in*) [θαλλός, a green shoot], C₁₀H₁₃-QN. A liquid basic substance tetrahydroparam-ethyloxychinolin. **t. acetate**, used for night-sweats of tuberculosis. Daily dose, 1½ gr. (0.1 Gm.). It has the peculiar property of causing rapid falling of the hair. **t.-alopecia**, falling of the hair following the ingestion of thalline acetate. **t. periodate**, used as the sulphate. **t. salicylate**, an antiseptic, antipyretic, and antirheumatic. Dose 3–8 gr. (0.2–.52 Gm.). **t. sulphate**, white needles or crystalline powder turning brown on exposure; soluble in water, 5 parts; alcohol, 100 parts; boiling water, 0.5 parts. It is antiseptic, antipyretic, and hemostatic; dose, 3–8 gr. (0.2–0.52 Gm.). Injection in chronic gonorrhea, 5% solution in oil. **t. tannate**, used as the sulphate. **t. tartrate**, crystalline powder soluble in 10 parts of water or 300 parts of alcohol; used as the sulphate.

**thallinization** (*thal-in-iz-a'-shun*). Continuous influence of thalline (or its salts) by frequent repetition of the dose.

**thallium** (*thal'-e-um*) [see *thalline*]. A metallic element having an atomic weight of 204, a specific gravity of 11.19; symbol, Tl. The salts are poisonous. See *elements, table of chemic*.

**thallophyte** (*thal'-o-fīt*) [θαλλός, a green shoot; φυτόν, a plant]. One of a class of very low cryptogams.

**thalocol** (*thal'-o-kol*). Tablets containing phenolphthalein and calomel.

**thalosen** (*thal'-o-sen*). Tablets containing phenolphthalein, sulphur, senna, and aromatics; used as a laxative.

**thallus** (*thal'-us*) [θαλλός, a young shoot]. In biology, applied to a plant-body in which there is no differentiation into root, stem and leaves.

**thamuria** (*tham-ū'-re-ah*) [θαμά, often; οὖρον, urine]. Frequent urination.

**thanato-** (*than-at-o-*) [θάνατος, death]. A prefix denoting pertaining to death.

**thanatobiologic** (*than-at-o-bi-o-loj'-ik*) [*thanato-*; βίος, life]. Pertaining to life and death.

**thanatognomonic** (*than-at-og-no-mon'-ik*) [*thanato-*; γνώμων, sign]. Indicative of death.

**thanatoid** (*than'-at-oid*) [θάνατος, death; εἶδος, like]. Resembling death.

**thanatol** (*than'-at-ol*). Same as *guaethol*.

**thanatology** (*than-at-ol'-o-je*) [*thanato-*; λόγος, science]. The sum of scientific knowledge regarding death.

**thanatomania** (*than-at-o-ma'-ne-ah*) [*thanato-*; μανία, madness]. Suicidal mania.

**thanatometer** (*than-at-om'-et-er*) [*thanato-*; μέτρον, a measure]. A thermometer introduced into a body-cavity to determine if the depression of temperature is so great as to be a sign of death.

**thanatophidia** (*than-at-o-fid'-e-ah*) [*thamato-*; ὄφις, a serpent]. Those serpents whose bite produces toxic symptoms or death.

**thanatophobia** (*than-at-o-fo'-be-ah*) [*thanato-*, φόβος, fear]. A morbid fear of death.

**thanatopsy** (*than'-at-op-se*) [*thanato-*; ὄψις, view]. Autopsy, necropsy.

**thanatosis** (*than-at-o'-sis*). See *gangrene*.

**Thane's method**) (*thān*) [George Dancer Thane, English anatomist]. To find the fissure of Rolando in operations upon the brain, the middle point of a line passing from the root of the nose to the occipital protuberance is determined. The upper extremity of the fissure lies half an inch behind this point.

**Thapsia** (*thap'-se-ah*) [θαψία]. A genus of old-world umbelliferous plants. Thapsia resin (*resina thapsiæ*) is the product of *T. garganica, T. silphium*, and *T. villosa*. The resin is strongly counter-irritant. The root of *T. garganica* is used by the natives of North America as a counter-irritant. Internally a tonic; dose of the fluidextract ♏x–xxx. The root of *T. villosa* (deadly carrot) is purgative.

**thaumatrope** (*thaw'-mah-trōp*) [θαῦμα, wonder; τροπή, a turning]. A device containing figures, on opposite sides of a rotating board, which blend when in motion. It shows the duration of visual impressions.

**thaumatropy** (*thaw-mat'-ro-pe*) [θαῦμα, wonder; τροπή, change]. The transformation of one kind of tissue into another.

**thea** (*the'-ah*). Tea; the dried leaves of *Thea sinensis*, a shrub of the order *Ternstrœmiaceæ*, containing the alkaloid *theine*, C₈H₁₀N₄O₂, identical with caffeine. Thea is astringent and gently stimulant to the nervous system; its infusion is used as a beverage. **t., nigra**, black tea, is less pungent and less fragrant than green tea and is made from leaves that have undergone fermentation and are then slowly dried. **t. viridis**, green tea, is prepared from leaves that have been dried quickly, having undergone no fermentation.

**theaism** (*the'-ah-izm*). See *theism*.

**theatrin** (*the'-at-rin*). An ointment-vehicle consisting of wax, oil, and water.

**thebaic** (*the-ba'-ik*) [Thebes, where opium was once prepared]. Pertaining to or derived from opium.

**thebaine** (*the'-ba-ēn*) [Θῆβαι, Thebes], C₁₉H₂₁NO₃. An alkaloid found in opium, analogous to strychnine in its physiological effects. It is also called paramorphine.

**thebaism** (*the'-ba-izm*) [Θῆβαι, Thebes]. The condition induced by thebaine or paramorphine; opiumism.

**Thebesian foramina,** (*the-be'-se-an*). T. **valve,** T. **veins**. See under *Thebesius*.

**Thebesius' foramina** [Adam Christianus Thebesius, German physician, 1686–1732]. The orifices of Thebesius' veins. T.'s **valve**, an endocardial fold at the orifice of the coronary vein in the right auricle. T.'s **veins**, venæ minimæ cordis; the venules which convey the blood directly from the myocardium into the auricles.

**thebolactic acid** (*theb-o-lak'-ik*). A variety of lactic acid occurring in opium.

**theca** (*the'-kah*) [θήκη, a sheath, pl., *thecæ*]. A sheath, especially one of a tendon. **t. cerebri**, the cranium. **t. cordis**, the pericardium. **t. folliculi**, a membranous formation around a Graafian vesicle caused by fibrillation of a layer of young connective tissue subsequent to the increased vascularity accompanying the process of maturation. **t. tendinis**, the synovial sheath of a tendon. **t. vertebralis**, the membranes of the spinal cord.

**thecal** (*the'-kal*) [*theca*]. Pertaining to a sheath or theca. **t. abscess**, tenosynovitis, paronychia, or whitlow.

**thecate** (*the'-kāt*) [*theca*]. Contained within a sheath.

**thecitis** (*the-si'-tis*) [*theca*; ιτις, inflammation]. Inflammation of the sheath of a tendon.

**thecodont** (*the'-ko-dont*) [*theca*; ὀδούς, tooth]. Having the teeth covered or sheathed in alveoli.

**thecosoma, thecosomum** (*the-ko-so'-mah, -mum*) [θήκη, a sheath; σῶμα, body]. Same as *schistosomum*.

**thecostegnosis** (*the-ko-steg-no'-sis*) [*theca*; στεγνοῦν, to contract]. The shrinking or contraction of the sheath of a tendon.

**Theden's bandage** (*ta'-den*) [Johann Christian Anton *Theden*, German surgeon, 1714–1797]. A form of roller bandage applied from below upward over a graduated compress. to control hemorrhage from a limb.

**theic** (*the'-ik*) [*thea*, tea]. A tea-drunkard; an immoderate user of tea.

**theine** (*the'-in*). Same as *caffeine*. See under *thea*.

**theinism** (*the'-in-izm*). Same as *theism*.

**theism** (*the'-izm*) [*thea*]. The morbid condition due to the excessive use of tea; it is characterized by headache, palpitation, tremor, insomnia, cachexia, etc.

**thelalgia** (*the-lal'-je-ah*) [θηλή, a nipple; ἄλγος, pain]. Pain in the nipple.

**thelasis, thelasmus** (*thel-as'-is, thel-as'-mus*) [θηλάζειν, to suckle]. The act of sucking.

**thele** (*the'-le*) [θηλή, a nipple]. The nipple of the female breast.

## THELEPLASTY 884 THERMESTHESIA

**theleplasty** (*the'-le-plas-te*) [*thele;* πλάσσειν, to form]. Plastic surgery of the nipple.

**thelerethism** (*the-ler'-eth-izm*) [*thele;* ἐρεθισμός, irritation]. Erection of the nipple.

**thelitis** (*the-li'-tis*) [*thele;* ιτις, inflammation]. Inflammation of the nipple.

**thelium** (*the'-le-um*) [*thele*]. 1. A papilla. 2. A layer of cells. 3. The nipple.

**theloncus** (*the-long'-kus*) [*thele;* ὄγκος, a tumor]. Tumor of the nipple.

**thelorrhagia** (*the-lor-aj'-e-ah*) [*thele;* ῥεῖν, to flow], Hemorrhage from the nipple.

**thelothism** (*the'-lo-thizm*) [*thele;* ὠθέειν, to push]. Projection of the nipple, caused by contraction of the transverse muscular fibers.

**thelyblast** (*thel'-e-blast*) [θῆλυς, female; βλαστός, a germ]. The female element of the bisexual nucleus; the ovum after the polar globules have been extruded.

**thelygonia** (*thel-ig-o'-ne-ah*) [θῆλυς, female; 'γονή, birth]. 1. The procreation of female offspring. 2. Nymphomania.

**thelymania** (*thel-im-a'-ne-ah*) [θῆλυς, female; μανία, madness]. Satyriasis.

**thenad** (*the'-nad*) [*thenar;* ad, toward]. Toward the thenar eminence.

**thenal** (*the'-nal*) [*thenar*]. Pertaining to the palm, or the thenar eminence.

**thenar** (*the'-nar*) [θέναρ, palm]. 1. The palm of the hand. 2. The fleshy prominence of the palm corresponding to the base of the thumb. t. eminence. See *thenar* (2). t. muscles, the abductor and flexor muscles of the thumb.

**thenen** (*the'-nen*) [θέναρ, palm]. Belonging to the thenar aspect in itself.

**theobroma** (*the-o-bro'-mah*) [θεός, a god; βρῶμα, food]. A genus of trees of the *Sterculiaceæ*. The seeds of *T. cacao* yield a fixed oil (*oleum throbromatis*, U. S. P.), and contain the alkaloid *theobromine*, C₇H₈N₄O₂, which is closely related to caffeine and xanthin. The seeds are used in the preparation of chocolate and cocoa; the oil (cacao-butter) is employed as an ingredient of cosmetic ointments and for making pills and suppositories. Theobromine acts similarly to caffeine.

**theobromine** (*the-o-bro'-mēn*). See under *theobroma*. t. sodiosalicylate. See *diuretin*.

**theocin** (*the'-o-sin*). Synthetically prepared theophylline isomeric with theobromine, and used as a diuretic. Dose 4 gr. (0.25 Gm.) 2 or 3 times daily.

**theolactin** (*the-o-lak'-tin*). A proprietary diuretic containing sodium lactate and theobromine-sodium.

**theolin** (*the'-o-lin*). Same as *heptane*.

**theomania** (*the-o-ma'-ne-ah*) [θεός, a god; μανία, madness]. 1. Religious mania. 2. Insanity in which the individual believes himself to be a divine being.

**theomaniac** (*the-o-ma'-ne-ak*). One who is affected with theomania.

**theopathy** (*the-op'-ath-e*) [θεός, a god; πάθος, disease]. Cure by prayer.

**theophobia** (*the-o-fo'-be-ah*) [θεός, a god; φόβος, fear]. Morbid fear of the deity.

**theophorin** (*the-off'-or-in*). Proprietary name of a double salt of theobromine-sodium and sodium formate. It is said to be diuretic.

**theophylline** (*the-off'-il-ēn*) [*thea*, tea; φύλλον, leaf], C₇H₈N₄O₂. An alkaloid occurring in tea and isomeric with theobromine and with paraxanthin.

**theoplegia** (*the-o-ple'-je-ah*) [θεός, a god; πληγή, a stroke]. Apoplexy.

**theoretical** (*the-o-ret'-ik-al*). Based on theory; speculative.

**theory** (*the'-o-re*) [θεωρία, a view]. The abstract principles of a science. Also a reasonable supposition or assumption, generally one that is better developed and more probable than a mere hypothesis.

**theotherapy** (*the-o-ther'-ap-e*) [θεός, a god; *therapy*]. The treatment of disease by prayer and religious exercises.

**therapeusis** (*ther-ap-ū'-sis*). See *therapeutics*.

**therapeutic** (*ther-ap-ū'-tik*) [see *therapeutics*]. Pertaining to therapeutics; curative. t: test, a method of diagnosis by administering certain remedies known to influence a given disease, *e. g.*, quinine in malaria, potassium iodide and mercury in syphilis.

**therapeutics** (*ther-ap-ū'-tiks*) [θεραπευτική, the art of medicine]. The branch of medical science dealing with the treatment of disease. .t., empirical, treatment based upon experience. t., mediate, medicating a child through its mother's milk. t., rational, treatment based upon a knowledge of the symptoms

of the disease and the physiological action of the remedy. t., specific, treatment of a disease by a specific remedy. t., suggestive, hypnotic suggestion in the treatment of disease.

**therapeutist** (*ther-ap-ū'-tist*) [see *therapeutics*]. One skilled in therapeutics.

**therapia sterilisans magna.** Ehrlich's mode of treatment by destruction of the parasites in the body of a patient without doing serious harm to the patient; it is accomplished by the administration, in one large dose, of a sufficient quantity of a drug having a special affinity for the parasite causing the disease.

**therapic** (*ther-ap'-ik*) [*therapy*]. Pertaining to therapy; therapeutic.

**therapist** (*ther'-ap-ist*). Same as *therapeutist*.

**therapol** (*ther'-ap-ol*) [*therapy;* oleum, oil]. A vegetable oil containing ozone; it has been used in diphtheria.

**therapy** (*ther'-ap-e*). See *therapeutics*. t., bacterial. Same as *opsonic therapy*. t., opson,c, treatment by the use of bacterial vaccines which increase the opsonic index. t., psychic, treatment of disease by influence of the mind. t., serum. See *serotherapy*. t., vaccine. Same as *therapy, opsonic*.

**therencephalous** (*ther-en-sef'-al-us*) [θήρ, a wild beast; ἐγκέφαλος, the brain]. A term applied to a skull in which the lines from the inion and nasion to the hormion make an angle of from 116° to 129°.

**theriaca** (*the-ri'-ak-ah*) [θηριακή, from θηρίον, a wild beast, because believed to be an antidote against the poison of animals]. Treacle; molasses. t. Andromachi, Venice treacle, a compound containing nearly 70 ingredients, and used as an antidote against poisons.

**theriatrics** (*the-re-at'-riks*) [θήρ, a wild beast; ἰατρική, the art of healing]. The medical treatment of animals.

**theriodic** (*the-re-od'-ik*) [*θηρίον*, a wild beast]. Malignant.

**therioma** (*the-re-o'-mah*) [θήρ, a wild beast; *pl., theriomata*]. A malignant ulcer or tumor.

**theriomimicry** (*the-re-o-mim'-ik-re*) [θηρίον, a wild beast; μῖμος, an imitator]. Imitation of the acts of animals.

**theriotherapy** (*the-re-o-ther'-ap-e*) [θηρίον, a wild beast; *therapy*]. Veterinary therapy.

**theriotomy** (*the-re-ot'-o-me*) [θηρίον, a wild beast; τομή, an incision]. Zoötomy; the anatomy or dissection of animals.

**therm** [θέρμη, heat]. 1. The amount of heat required to raise the temperature of one gram of water from 0° C. to 1° C.; it is also called a calorie. 2. Also a heat unit equivalent to one thousand kilogram (large) calories; it is designated T.

**thermacogenesis** (*ther-mak-o-jen'-es-is*) [θερμή, heat; φαρμάκον, drug; γένεσις, production]. The raising of the body temperature by the action of drugs.

**thermæ** (*ther'-me*) [θέρμη, heat]. Hot baths; hot springs.

**thermaerotherapy** (*ther-mah-e-ro-ther'-ap-e*). The therapeutic application of hot air.

**thermæsthesia** (*thur'-mes-the'-ze-ah*). See *thermesthesia*.

**thermal** (*ther'-mal*) [see *therm*]. 1. Pertaining to heat. 2. Hot, as *thermal spring*. t. capacity, the amount of heat required to raise the temperature of a body from 0° to 1° C. t. death-point, the degree of a heat required to kill a fluid culture in ten minutes. t. unit, the amount of heat required to raise the temperature of a pound of water one degree F. or C.

**thermalgesia** (*ther-mal-je'-ze-ah*) [*therm;* ἄλγος, pain]. The condition in which heat causes pain.

**thermalgesia** (*ther-man-al-je'-ze-ah*). See *thermoanalgesia*.

**thermanesthesia, thermanæsthesia** (*ther-mān-es-the'-ze-ah*). See *thermoanesthesia*.

**thermantidote** (*thur-man'-tid-ōt*) [θέρμη, heat; *antidote*]. An apparatus for cooling the air, much used in some hot regions.

**thermasma** (*thur-mas'-mah*). A warm fomentation.

**thermatology** (*thur-mat-ol'-o-je*) [θέρμη, heat; λόγος, science]. The scientific use or understanding of heat or of the waters of thermal springs in the cure of disease.

**thermesthesia, thermæsthesia** (*ther-mes-the'-ze-ah*) [*therm;* αἴσθησις, sensation]. 1. The heat-sense. 2. Sensitiveness to heat.

**thermesthiometer, thermæsthesiometer** (*thur-mes-the-ze-om'-et-er*) [θέρμη, heat; αἴσθησις, sensation; μέτρον, measure]. An instrument for measuring the sensibility to heat of different regions of the skin.
**thermic** (*ther'-mik*) [θέρμη, heat]. Pertaining to heat. t. **fever**, sunstroke; heat-fever. t. **sense**, thermesthesia.
**thermifugin** (*thur-mif'-ū-jin*) [θέρμη, heat; *fugare*, to expel]. Trade name of sodium carbonate, recommended as an antipyretic.
**thermin** (*ther'-min*), C₁₀H₁₁ . NH₂HCl. A colorless liquid obtained from a solution of β-naphthylamine in amyl-alcohol by action of metallic sodium. It is used as a mydriatic. Syn., *tetrahydro-β-naphthylamine*. t. **hydrochloride**, C₁₀H₁₁NH₂ . HCl, used to increase body-temperature.
**thermo-** (*ther-mo-*) [θέρμη, heat]. A prefix meaning heat.
**thermoaerophore** (*ther-mo-a-e'-ro-fōr*) [*thermo-*; ἀήρ, air; φέρειν, to bear]. An apparatus for the therapeutic local application of hot air.
**thermoæsthesia** (*thur-mo-es-the'-ze-ah*). See *thermesthesia*.
**thermoalgesia** (*ther-mo-al-je'-sia*). Same as *thermalgesia*.
**thermoanalgesia** (*ther-mo-an-al-je'-ze-ah*) [*thermo-*; *analgesia*]. Insensibility to heat or to contact with heated objects; due to cerebral lesion.
**thermoanesthesia, thermoanæsthesia** (*ther-mo-an-es-the'-ze-ah*) [*thermo-*; *anesthesia*]. Loss of the perception of thermal impressions, a condition sometimes present in syringomyelia.
**thermocauterectomy** (*ther-mo-kaw-ter-ek'-to-me*). See *igniextirpation*.
**thermocautery** (*ther-mo-kaw'-ter-e*). See *Paquelin's cautery*.
**thermochemistry** (*ther-mo-kem'-is-tre*) [*thermo-*; *chemistry*]. That branch of chemical science embracing the mutual relations of heat and chemical changes.
**thermochroic** (*ther-mō-krō'-ik*) [*thermo-*; χρόα, color]. Transmitting some thermal rays and absorbing others.
**thermochroism** (*thur-mo-kro'-izm*) [*thermo-*; χρῶσις, a coloring]. The property possessed by certain substances of transmitting some thermal radiations while they absorb or change others.
**thermochrosis** (*thur-mo-kro'-sis*). See *thermochroism*.
**thermocurrent** (*thur-mo-kur'-ent*) [*thermo-*; *current*]. An electric current produced by heat.
**thermodiffusion** (*thur-mo-dif-ū'-zjun*) [*thermo-*; *diffusion*]. Diffusion of a gas by inequalities in temperature.
**thermodin** (*ther'-mo-din*) [θέρμη, heat], C₁₃H₁₇NO₄. An antipyretic derivative of ethyl carbamate.
**thermoelectricity** (*ther-mo-e-lek-tris'-it-e*) [*thermo-*; *electricity*]. Electricity generated by heat.
**thermoesthesia, thermoæsthesia** (*thur-mo-es-the'-ze-ah*). See *thermesthesia*.
**thermoesthesiometer**. See *thermesthesiometer*.
**thermoexcitory** (*thur-mo-ek-si'-to-re*) [*thermo-*; *excitor*, *excitor*]. Having the function of exciting the production of heat (opposed to *thermoinhibitory*).
**thermofuge** (*ther'-mo-fūj*). An external antiseptic, emollient, and detergent, said to be a compound of aluminum silicate, glycerol, boric acid, menthol, thymol, oil of eucalyptus, and ammonium iodide.
**thermogen** (*ther'-mo-jen*) [see *thermogenesis*]. An appliance for keeping up the temperature of patients during an operation. It consists of a quilted cushion through which pass wires the temperature of which can be raised by the passage of an electric current.
**thermogenesis** (*ther-mo-jen'-es-is*) [*thermo-*; γεννᾶν, to produce]. The production of heat.
**thermogenetic, thermogenic, thermogenous** (*ther-mo-jen-et'-ik, ther-mo-jen'-ik, ther-moj'-en-us*) [see *thermogenesis*]. Pertaining to thermogenesis; producing heat.
**thermogenics** (*ther-mo-jen'-iks*) [see *thermogenesis*]. The science of the production of heat.
**thermogram** (*ther'-mo-gram*) [*thermo-*; γράμμα, a written character]. The record of a thermograph.
**thermograph** (*ther'-mo-graf*) [*thermo-*; γράφειν, to write]. A device for registering variations of temperature automatically.
**thermohyperalgesia** (*ther-mo-hi-per-al-je'-ze-ah*) [*thermo-*; *hyperalgesia*]. Painful sensation felt on contact with a hot body.
**thermohyperesthesia** (*ther-mo-hi-per-es-the'-ze-ah*) [*thermo-*; *hyperesthesia*]. Abnormal sensitiveness to the application of hot bodies.
**thermohypesthesia** (*ther-mo-hi-pes-the'-ze-ah*) [*thermo-*; *hypesthesia*]. Abnormal indifference or insensibility to heat or to contact with heated objects.
**thermohypoesthesia** (*ther-mo-hi-po-es-the'-ze-ah*). See *thermohypesthesia*.
**thermoinhibitory** (*ther-mo-in-hib'-it-o-re*) [*thermo-*; *inhibitory*]. Inhibiting the production of heat.
**thermol** (*ther'-mol*), C₁₄H₁₄NO₃. A coal-tar derivative forming as white crystals, soluble in water and alcohol. It is analgesic, antipyretic, and antiseptic. Dose 2 gr. (0.2 Gm.) every 3 hours.
**thermolabile** (*ther-mo-la'-bil*) [*thermo-*; *lapsus*, a gliding or falling]. Destroyed or changed by heat.
**thermology** (*ther-mol'-o-je*) [*thermo-*; λόγος, science]. The science of heat.
**thermolusia** (*ther-mo-lū'-se-ah*) [*thermo-*; λούειν, to wash]. A hot bath.
**thermolysis** (*ther-mol'-is-is*) [*thermo-*; λύσις, a loosening]. 1. Dissipation of animal heat. 2. Chemical decomposition by means of heat.
**thermolytic** (*ther-mo-lit'-ik*) [see *thermolysis*]. Pertaining to thermolysis.
**thermomagnetism** (*ther-mo-mag'-net-izm*) [*thermo-*; *magnetism*]. Magnetism produced by heat.
**thermomassage** (*ther-mo-mas-ahzj'*). Massage with application of heat.
**thermometer** (*ther-mom'-et-er*) [*thermo-*; μέτρον, a measure]. An instrument for determining the intensity of heat, consisting of a substance capable of

## COMPARISON OF THERMOMETERS.

| Fahr. | Cent. | Reau. | Fahr. | Cent. | Reau. | Fahr. | Cent. | Reau. |
|---|---|---|---|---|---|---|---|---|
| 212 | 100 | 80 | 122 | 50 | 40 | 32 | 0 | 0 |
| 210 | 98.9 | 79.1 | 120 | 48.9 | 39.1 | 30 | − 1.1 | − 0.9 |
| 208 | 97.8 | 78.2 | 118 | 47.8 | 38.2 | 28 | − 2.2 | − 1.8 |
| 206 | 96.7 | 77.3 | 116 | 46.7 | 37.3 | 26 | − 3.3 | − 2.7 |
| 204 | 95.6 | 76.4 | 114 | 45.6 | 36.4 | 24 | − 4.4 | − 3.6 |
| 202 | 94.4 | 75.6 | 112 | 44.4 | 35.6 | 22 | − 5.6 | − 4.4 |
| 200 | 93.3 | 74.7 | 110 | 43.3 | 34.7 | 20 | − 6.7 | − 5.3 |
| 198 | 92.2 | 73.8 | 108 | 42.2 | 33.8 | 18 | − 7.8 | − 6.2 |
| 196 | 91.1 | 72.9 | 106 | 41.1 | 32.9 | 16 | − 8.9 | − 7.1 |
| 194 | 90 | 72 | 104 | 40 | 32 | 14 | −10 | − 8 |
| 192 | 88.9 | 71.1 | 102 | 38.9 | 31.1 | 12 | −11.1 | − 8.9 |
| 190 | 87.8 | 70.2 | 100 | 37.8 | 30.2 | 10 | −12.2 | − 9.8 |
| 188 | 86.7 | 69.3 | 98 | 36.7 | 29.3 | 8 | −13.3 | −10.7 |
| 186 | 85.6 | 68.4 | 96 | 35.6 | 28.4 | 6 | −14.4 | −11.6 |
| 184 | 84.4 | 67.6 | 94 | 34.4 | 27.6 | 4 | −15.6 | −12.4 |
| 182 | 83.3 | 66.7 | 92 | 33.3 | 26.7 | 2 | −16.7 | −13.3 |
| 180 | 82.2 | 65.8 | 90 | 32.2 | 25.8 | 0 | −17.8 | −14.2 |
| 178 | 81.1 | 64.9 | 88 | 31.1 | 24.9 | − 2 | −18.9 | −15.1 |
| 176 | 80 | 64 | 86 | 30 | 24 | − 4 | −20 | −16 |
| 174 | 78.9 | 63.1 | 84 | 28.9 | 23.1 | − 6 | −21.1 | −16.9 |
| 172 | 77.8 | 62.2 | 82 | 27.8 | 22.2 | − 8 | −22.2 | −17.8 |
| 170 | 76.7 | 61.3 | 80 | 26.7 | 21.3 | −10 | −23.3 | −18.7 |
| 168 | 75.6 | 60.4 | 78 | 25.6 | 20.4 | −12 | −24.4 | −19.6 |
| 166 | 74.4 | 59.6 | 76 | 24.4 | 19.6 | −14 | −25.6 | −20.4 |
| 164 | 73.3 | 58.7 | 74 | 23.3 | 18.7 | −16 | −26.7 | −21.3 |
| 162 | 72.2 | 57.8 | 72 | 22.2 | 17.8 | −18 | −27.8 | −22.2 |
| 160 | 71.1 | 56.9 | 70 | 21.1 | 16.9 | −20 | −28.9 | −23.1 |
| 158 | 70 | 56 | 68 | 20 | 16 | −22 | −30 | −24 |
| 156 | 68.9 | 55.1 | 66 | 18.9 | 15.1 | −24 | −31.1 | −24.9 |
| 154 | 67.8 | 54.2 | 64 | 17.8 | 14.2 | −26 | −32.2 | −25.8 |
| 152 | 66.7 | 53.3 | 62 | 16.7 | 13.3 | −28 | −33.3 | −26.7 |
| 150 | 65.6 | 52.4 | 60 | 15.6 | 12.4 | −30 | −34.4 | −27.6 |
| 148 | 64.4 | 51.6 | 58 | 14.4 | 11.6 | −32 | −35.6 | −28.4 |
| 146 | 63.3 | 50.7 | 56 | 13.3 | 10.7 | −34 | −36.7 | −29.3 |
| 144 | 62.2 | 49.8 | 54 | 12.2 | 9.8 | −36 | −37.8 | −30.2 |
| 142 | 61.1 | 48.9 | 52 | 11.1 | 8.9 | −38 | −38.9 | −31.1 |
| 140 | 60 | 48 | 50 | 10 | 8 | −40 | −40 | −32 |
| 138 | 58.9 | 47.1 | 48 | 8.9 | 7.1 | −42 | −41.1 | −32.9 |
| 136 | 57.8 | 46.2 | 46 | 7.8 | 6.2 | −44 | −42.2 | −33.8 |
| 134 | 56.7 | 45.3 | 44 | 6.7 | 5.3 | −46 | −43.3 | −34.7 |
| 132 | 55.6 | 44.4 | 42 | 5.6 | 4.4 | −48 | −44.4 | −35.6 |
| 130 | 54.4 | 43.6 | 40 | 4.4 | 3.6 | −50 | −45.6 | −36.4* |
| 128 | 53.3 | 42.7 | 38 | 3.3 | 2.7 | −52 | −46.7 | −37.3 |
| 126 | 52.2 | 41.8 | 36 | 2.2 | 1.8 | −54 | −47.8 | −38.3 |
| 124 | 51.1 | 40.9 | 34 | 1.1 | 0.9 | −56 | −48.9 | −39.1 |

expanding and contracting, and a graduated scale by means of which variations in the volume of the substance can be determined. In the ordinary thermometer the expansive substance is mercury

# THERMOMETRIC 886 THIOACETALDEHYDE

(*mercurial thermometer*), expanding into a vacuous capillary tube, the degree of heat being measured by the length of the column of mercury. t., **air**, one in which the expansive substance is air. t., **alcohol**, one in which the expansive substance is alcohol. t., **Centigrade**, t., **Celsius**, one in which the freezing-point is at 0° and the boiling-point at 100°. t., **clinical**, a self-registering thermometer for ascertaining the bodily temperature. t., **differential**, one for determining slight variations of temperature. t., **Fahrenheit**, one in which the interval between the freezing-point and the boiling-point is divided into 180 equal parts, each called a degree, the zero-point being 32° or divisions below the freezing-point of water. t., **fever**, a *clinical thermometer*. t., **maximum**, one which registers the maximum heat to which it has been exposed. t., **mercurial**, one in which the expansive substance is mercury. t., **minimum**, one that registers the lowest temperature to which it has been exposed. t., **Réaumur**, one in which the freezing-point of water is 0° and the boiling-point 80°. t., **self-registering**, one that by means of an index shows the highest (*maximum*) or lowest (*minimum*) temperature to which it has been exposed. t., **spirit**, one in which alcohol or ether is used. t., **surface-**, one for registering the surface-temperature of any portion of the body.

**thermometric** (*ther-mo-met'-rik*) [*thermometer*]. Pertaining to a thermometer or to thermometry.

**thermometry** (*ther-mom'-et-re*) [*thermometer*]. The measuring of temperature by means of the thermometer.

**thermoneurosis** (*ther-mo-nū-ro'-sis*) [*thermo-*; *neurosis*]. Pyrexia of vasomotor origin.

**thermonosus** (*ther-mon-o'-sus*) [*thermo-*; νόσος, disease]. Disease caused by heat.

**thermopalpation** (*ther-mo-pal-pa'-shun*) [*thermo-*; *palpation*]. Palpation of the surface of the body with a view to the determination of variations of temperature.

**thermophagy** (*ther-moff'-aj-e*) [*thermo-*; φάγειν, to eat]. The habit of swallowing very hot food.

**thermophile** (*ther'-mo-fil*) [*thermo-*; φιλεῖν, to love]. 1. A microorganism which develops best at relatively high temperatures. 2. A thermoelectric battery.

**thermophilic** (*ther-mo-fil'-ik*) [*thermo-*; φιλεῖν, to love]. Applied to microorganisms which develop best at relatively high temperatures from 50° to 55° C. or above.

**thermophobia** (*ther-mo-fo'-be-ah*), [*thermo-*; φόβος, fear]. Morbid dread of heat.

**thermophore** (*ther'-mo-fōr*) [*thermo-*; φερεῖν, to bear]. 1. Any appliance adapted to hold heat; as used in local treatment, a receptacle for hot water, a water-bag. 2. A receptacle containing chemicals which absorb a large amount of heat in the process of fusing and which give it off gradually as recrystallization takes place. Used as hand or foot warmers and in local treatment.

**thermophylic** (*ther-mo-fil'-ik*) [*thermo-*; φυλάσσειν, to guard]. Resistant to the effect of heat, said of certain microorganisms.

**thermopile** (*ther'-mo-pīl*) [*thermo-*; *pile*]. A contrivance consisting of a series of connected metallic plates, in which, under the influence of heat, a current of electricity is produced which acts upon a registering index. By means of it very minute amounts of heat can be measured.

**thermoplegia** (*ther-mo-ple'-je-ah*) [*thermo-*; πληγή, a stroke]. Heat-stroke.

**thermopolypnea** (*ther-mo-pol-ip-ne'-ah*) [*thermo-*; πολύς, many; πνεῖν, to breathe]. Rapid respiration due to high temperature.

**thermoregulator** (*ther-mo-reg'-ū-la-tor*). See *thermostat*.

**thermoscope** (*ther'-mo-skōp*) [*thermo-*; σκοπεῖν, to view]. An instrument for measuring minute differences of temperature without registering the degree or amount of heat.

**thermostabile** (*ther-mo-sta'-bil*) [*thermo-*; *stabilis*, firm; steadfast]. Not destroyed or changed by heat.

**thermostat** (*ther'-mo-stat*) [*thermo-*; στατός, standing]. A device for automatically regulating and maintaining a constant temperature.

**thermosteresis** (*ther-mo-ster-e'-sis*) [*thermo-*; στέρησις, deprivation]. Deprivation of heat.

**thermosystaltic** (*ther-mo-sis-tal'-tik*) [*thermo-*; συστέλλειν, to contract]. Contracting under the influence of heat; pertaining to muscular contraction due to heat.

**thermosystaltism** (*ther-mo-sis-tal'-tizm*) [*thermo-*; συστέλλειν, to contract]. Muscular, or other, contraction caused by heat.

**thermotactic, thermotaxic** (*ther-mo-tak'-tik, ther-mo-taks'-ik*) [see *thermotaxis*]. Regulating the heat of the body, as a *thermotactic* center.

**thermotaxis** (*ther-mo-taks'-is*) [*thermo-*; τάξις, from τάσσειν, to arrange]. 1. The regulation and correlation of heat production and heat dissipation. 2. Thermotropism.

**thermoterion** (*ther-mo-te'-re-ōn*). An apparatus for keeping food warm, consisting of a glass case surrounded by a hot water chamber and an air space to prevent the radiation of heat.

**thermotherapy** (*ther-mo-ther'-ap-e*) [*thermo-*; θεραπεία, cure]. Treatment of disease by heat.

**thermotics** (*ther-mot'-iks*) [*thermo-*]. The science of heat; thermology.

**thermotonometer** (*ther-mo-ton-om'-et-er*) [*thermo-*; τόνος, a stretching; μέτρον, a measure]. An apparatus for determining the amount of muscular contraction induced by thermic stimuli.

**thermotoxin** (*ther-mo-toks'-in*) [*thermo-*; *toxin*]. A poison produced by heat in the body.

**thermotracheotomy** (*ther-mo-tra-ke-ot'-o-me*). Tracheotomy by means of the actual cautery.

**thermotropism** (*ther-mot'-ro-pizm*) [*thermo-*; τρόπος, a turn]. That property possessed by some cells and organisms of bending towards or away from a source of heat.

**theroid** (*the'-roid*) [θήρ, a wild beast; εἶδος, resemblance]. Like a beast, bestial.

**theromorph** (*ther'-o-morf*) [θήρ, a wild beast; μορφή, form]. A monstrosity resembling an animal.

**theromorphism** (*ther-o-mor'-phism*) [θήρ, a wild beast; μορφή, form]. Apparent reversion, in a human subject, to an animal form of lower type.

**thesis** (*the'-sis*) [θέσις, a proposition]. A dissertation. Usually, the essay presented by an undergraduate at the time of his candidature for a degree.

**thevetin** (*thev'-et-in*) [Andre *Thevet*, a French traveler]. A poisonous glucoside from certain species of *Thevetia*.

**thew** (*thū*). A muscle, a sinew.

**thial** (*thi'-al*). Trade name of hexamethylenetetramineoxymethylsulphonate. It is used as a disinfectant. t. liquid, a 50 per cent. solution of thial.

**thialdin** (*thi-al'-din*), $C_6H_{13}NS_2$, a crystalline substance obtained by the action of sulphureted hydrogen on aldehyde-ammonium; it is used as a heart stimulant.

**thialion** (*thi-al'-e-on*). A proprietary preparation containing lithium; used in gouty conditions.

**thick wind**. A colloquial term for impaired respiration in the horse, somewhat louder and less free than normal breathing.

**Thielmann's diarrhea mixture** (*tēl'-man*) [Karl Heinrich *Thielmann*, German physician, 1802–1872]. Wine of opium 32, tincture of valerian 50, ether 16, oil of peppermint 4, fluidextract of ipecac 1, and alcohol enough to make 125. Dose 30 minims (2 Cc.).

**Thiersch's method of skin-grafting** (*tērsh*) [Karl *Thiersch*, German surgeon, 1822–1895]. Long, broad strips of skin are removed from the arm or leg and placed on a wound previously deprived of its granulations by means of a sharp curet. T.'s **solution**, a valuable antiseptic wash for the nose, throat, or stomach, consisting of salicylic acid, 2 parts; boric acid, 12 parts; water, 1000 parts.

**thigenol** (*thi'-jen-ol*). The sodium salt of the sulphonic acid extracted from a synthetic sulphur oil. It is used in the treatment of skin diseases in the form of pomades containing 20 parts in 100.

**thigh** (*thī*). The part of the lower limb extending from pelvis to knee. t.-**bone**, the femur. t.-**friction**, a form of masturbation. t.-**joint**, the hip-joint.

**thigmotaxis** (*thig-mo-taks'-is*). See *thigmotropism*.

**thigmotropism** (*thig-mot'-ro-pizm*) [θίγμα, touch; τρόπος, a turn]. That property possessed by some cells and organisms of being attracted by mechanical stimuli.

**thilanin** (*thil'-lan-in*) [θεῖον, sulphur; *lanolin*]. A brownish-yellow substance derived from and resembling lanolin. It contains 3 % of sulphur and is used in the treatment of eczema and other diseases of the skin.

**thio-** (*thi-o-*) [θεῖον, sulphur]. A prefix denoting containing sulphur in the place of oxygen.

**thioacetaldehyde** (*thi-o-as-et-al'-de-hīd*). See *sulphaldehyde*.

**thioacid** (*thi-o-as'-id*). One of a group of acids produced by the substitution of sulphur for the oxygen in an oxygen acid.
**thioalcohol** (*thi-o-al'-ko-hol*). See *mercaptan*.
**thiocamph** (*thi'-o-kamf*) [*thio-*; *camphor*]. A fluid disinfectant, used for fumigation. It is formed by the action of sulphur dioxide on camphor.
**thiocarbamide** (*thi-o-kar'-bam-id*). See *thiourea*.
**thiochromogen** (*thi-o-kro'-mo-jen*). See *aureolin*.
**thiocol** (*thi'-o-kol*). See *potassium-guaiacol sulphonate*.
**thiocyanate** (*thi-o-si'-an-āt*) [*thio-*; *cyanate*]. A salt of thiocyanic acid.
**thiocyanic acid** (*thi-o-si-an'-ik*), CNHS. Sulphocyanic acid, a monobasic acid forming the thiocyanates or sulphocyanates. Potassium thiocyanate, CNKS, occurs in saliva.
**thiodinaphthyloxide** (*thi-o-di-naf-thil-oks'-id*). An orange-colored powder used in the treatment of skin diseases.
**thioether** (*thi-o-e'-ther*). An ether in which sulphur replaces the oxygen; a sulphur ether.
**thioform** (*thi'-o-form*). See *bismuth dithiosalicylate*.
**thiogenic** (*thi-o-jen'-ik*) [*thio-*; γεννᾶν, to produce]. Applied to bacteria able to convert hydrogen sulphide into higher sulphur compounds.
**thiol** (*thi'-ol*) [θεῖον, sulphur]. A substance prepared from gas-oil by heating with sulphur, and occurring in a dry and a liquid form. It has been used as an application to ulcers and in diseases of the skin.
**thiolin** (*thi'-ol-in*). See *acid, thiolinic*.
**thionic** (*thi-on'-ik*) [θεῖον, sulphur]. Pertaining to sulphur. t. acid, thioacid.
**thionin** (*thi'-o-nin*) [θεῖον, sulphur], $C_{12}H_9N_3S$. A sulphur compound of the aromatic group, used as a stain in microscopy. Its solutions are of a dark-blue color.
**thiophene** (*thi'-o-fēn*) [*thio-*; *phenol*], $C_4H_4S$. A hydrocarbon of the aromatic series; a colorless, oily liquid, miscible with water. The iodide, $C_4H_2I_2S$, has been used as a substitute for iodoform. t. **sodium sulphonate**, $C_4H_3S \cdot NaSO_3$, a white powder used in prurigo.
**thiophil** (*thi'-o-fil*) [*thio-*; φιλεῖν, to love]. Loving sulphur; applied to microorganisms.
**thiopyrine** (*thi-o-pi'-rin*). A derivative of antipyrine, formed by the substitution of sulphur for oxygen.
**thioresorcinol** (*thi-o-rez-or'-sin-ol*) [*thio-*; *resorcinol*], $C_6H_4(SH)_2$. A compound of sulphur and resorcinol used as a powder or ointment as a substitute for iodoform.
**thiosapol** (*thi-o-sa'-pol*). A sulphureted soda soap containing 10% of sulphur.
**thiosavonals** (*thi-o-sav'-on-als*). Potash sulphur soaps that contain sulphur in a chemically combined state.
**thiosinamine** (*thi-o-sin'-am-ēn*) [*thio-*; *sinapis*, mustard], $C_4H_8N_2S$. A crystalline substance prepared from mustard-oil and ammonia. It is used in lupus, glandular enlargements, and night-sweats.
**thiosulphate** (*thi-o-sul'-fāt*). A salt of thiosulphuric acid.
**thiosulphuric acid** (*thi-o-sul-fū'-rik*) [*thio-*; *sulphuric*], $H_2S_2O_3$. An acid derived from sulphuric acid by the substitution of one atom of sulphur for one of oxygen.
**Thiothrix** (*thi'-o-thriks*) [θεῖον, sulphur; θρίξ, hair]. A genus of the family *Beggiatoaceæ*; filaments non-motile; surrounded by a delicate sheath; sulphur granules in cell contents; at ends of filaments rod-shaped gonidia; filaments unequal in diameter.
**thiourea** (*thi-o-ū-re'-ah*) [*thio-*; *urea*], $CS(NH_2)_2$. Sulphocarbamide; a derivative of urea in which sulphur replaces the oxygen of the latter.
**thiourethane** (*thi-o-ū'-re-thān*) [*thio-*; οὖρον, urine]. Anyone of the crystalline esters of sulphocarbamic acid.
**third corpuscle. Platelet.** t. **cranial nerve**, the oculomotor nerve; see *motor oculi* under *nerve*. t. **intention**. See *healing*. t. **tonsil**. See *Luschka's tonsil*. t. **ventricle**. See *ventricle*.
**thirst** (*thurst*) [AS., *thurst*]. A desire for drink.
**thirst-cure**. See *Schroth's cure*.
**Thiry's fistula** (*te'-re*) [Jean Hubert Thiry, Belgian physician, 1817–1879]. A fistula for obtaining the intestinal juice. A piece of intestine about four inches long is separated from the bowel without dividing the mesentery and its blood-vessels. One end of the tube is closed, and the other is stitched to the abdominal wound. The two ends of intestine from which the piece was cut out are then united by sutures. From the excised piece a pure intestinal juice is obtained.
**Thiry-Vella fistula** (*te'-re-vel'-lah*) [Jean Hubert Thiry, Belgian physician, 1817–1879; Luigi Vella, Italian physiologist, 1825–1886]. See *Vella's fistula*.
**thiuret** (*thi'-ū-ret*) [θεῖον, sulphur], $C_2H_7N_3S_2$. A crystalline antiseptic which readily yields its sulphur in a nascent condition. On this property depends its antiseptic action.
**thlipsencephalus** (*thlip-sen-sef'-al-us*) [θλίψις, pressure; ἐγκέφαλος, brain]. A monster in which there is extensive exposure of the base of the brain from non-development of the occipital bone and even of the upper vertebræ.
**thliptol** (*thlip'-tol*). A proprietary antiseptic and deodorant liquid, said to consist of benzoboric acid combined with oil of eucalyptus, thyme, etc.
**Thomas's splints** (*tom'-as*) [Hugh Owen Thomas, English surgeon, 1834–1891]. Rigid splints, made of curved iron rods, adapted to the shape of the limb, and kept in place by plaster of Paris bandages: They are employed in hip-joint disease, and are designed to secure rest, avoid friction and to allow the weight of the limb gradually to remedy the deformity in place of more active extension.
**Thompson's line**. A red line along the border of the gums, frequently seen in pulmonary tuberculosis.
**Thompson's test** [Sir Henry Thompson, English surgeon, 1820–1904]. The collection of the morning urine in two glasses to determine whether a gonorrheal process is localized in the anterior portion of the urethra or whether it has extended into the posterior portion.
**Thomsen's disease** [Asmus Julius Thomsen, Danish physician, 1815– ]. Myotonia congenita, a disease commonly congenital and occurring in families, and characterized by tonic spasm or rigidity of the muscles, coming on when they are first put in action after a period of rest. As the muscles are used the stiffness gradually wears off.
**Thomsonianism** (*tom-so'-ne-an-izm*). A system of medicine introduced by Samuel Thomson (1769–1843), of Massachusetts. It insisted on the use of vegetable remedies only.
**thoracal** (*tho'-rak-al*). Pertaining to the thorax or chest.
**thoracalgia** (*tho-rak-al'-je-ah*) [*thorax*; ἄλγος, pain]. Pain in the thorax.
**thoracaorta** (*tho-rak-a-or'-ta*) [*thorax*; *aorta*]. The thoracic aorta.
**thoracectomy** (*tho-rak-sek'-to-me*) [*thorax*; ἐκτομή, excision]. Thoracotomy with resection of a part of one or more ribs.
**thoracentesis** (*tho-ras-en-te'-sis*) [*thorax*; κέντησις, a piercing]. Puncture of the thorax for the removal of fluid.
**thoracic** (*tho-ras'-ik*) [*thorax*]. Pertaining to or situated in the chest or thorax. t. **aorta**. See *arteries, table of*. t. **axis**, the acromiothoracic artery; see *arteries, table of*. t. **choke**, in the horse, the lodgment of a foreign body in the thoracic portion of the esophagus. t. **duct**. See *duct, thoracic*. t. **index**. See *index, thoracic*. t. **nerve**. See *nerves, table of*.
**thoracicoabdominal** (*tho-ras-ik-o-ab-dom'-in-al*). Pertaining to the thorax and the abdomen.
**thoracicoacromial** (*tho-ras-ik-o-ak-ro'-me-al*). Acromiothoracic, relating to the chest and the shoulder; applied to a group of muscles.
**thoracicoacromialis** (*tho-ras-ik-o-ak-ro-me-a'-lis*). See *artery, acromiothoracic*.
**thoracicohumeral** (*tho-ras-ik-o-hū'-mer-al*). Relating to the chest and upper arm.
**thoracicolumbar** (*tho-ras-ik-o-lum'-bar*). Pertaining to the thoracic and lumbar regions.
**thoracispinal** (*tho-ras-is-pi'-nal*). Relating to the thoracic portion of the spinal column.
**thoraco-** (*tho-rak-o-*) [θώραξ, thorax]. A prefix denoting pertaining to the thorax.
**thoracoacromial** (*tho-rak-o-ak-ro'-me-al*). Acromiothoracic, relating to the chest and the shoulder; applied to a group of muscles.
**thoracoceloschisis** (*tho-rak-o-se-los'-kis-is*) [*thoraco-*; κοιλία, belly; σχίσις, a cleaving]. Congenital fissure of the chest and abdomen.
**thoracocentesis** (*tho-rak-o-sen-te'-sis*). See *thoracentesis*.

**thoracocyllosis** (*tho-rak-o-sil-o'-sis*) [*thoraco-*; κύλλωσις, curvation]. Deformity of the thorax.
**thoracocyrtosis** (*tho-rak-o-sur-to'-sis*) [*thoraco-*; κυρτός, curved]. Excessive curvature of the thorax.
**thoracodelphus** (*tho-rak-ad-el'-fus*). See *thoradelphus*.
**thoracodidymus** (*tho-rak-o-did'-im-us*) [*thoraco-*; δίδυμος, double]. A double monster joined at the thorax.
**thoracodynia** (*tho-rak-o-din'-e-ah*) [*thoraco-*; ὀδύνη, pain]. Pain in the chest.
**thoracogastrodidymus** (*tho-rak-o-gas-tro-did'-im-us*) [*thoraco-*; γαστήρ, belly; δίδυμος, double]. A twin monstrosity united by the thorax and abdomen.
**thoracogastroschisis** (*tho-rak-o-gas-tros'-kis-is*). See *thoracoceloschisis*.
**thoracograph** (*tho-rak'-o-graf*) [*thoraco-*; γράφειν, to write]. An instrument for recording the movements and the outline of the chest wall.
**thoracometer** (*tho-rak-om'-et-er*) [*thoraco-*; μέτρον, a measure]. A stethometer.
**thoracometry** (*tho-rak-om'-et-re*) [see *thoracometer*]. Measurement of the movement of the walls of the chest.
**thoracomyodynia** (*tho-rak-o-mi-o-din'-e-ah*) [*thoraco-*; μῦς, muscle; ὀδύνη, pain]. Pain in the muscles of the chest.
**thoracopagus** (*tho-rak-op'-ag-us*) [*thoraco-*; πάγος, that which is firmly set]. A double monster with portions of the thorax or abdomen coalescent. t. **tribrachius**, with two of the upper limbs coalescent. t. **tripus**, with two of the lower limbs coalescent.
**thoracopathia** (*tho-rak-o-pa'-the-ah*) [*thoraco-*; πάθος, disease]. A disease of the thorax.
**thoracoplasty** (*tho-rak'-o-plas-te*) [*thoraco-*; πλάσσειν, to form]. Plastic operation upon the thorax.
**thoracopneumoplasty** (*tho-rak-o-nū'-mo-plas-te*) [*thoraco-*; πνεύμων, lung; πλάσσειν, to form]. Plastic operation upon the lung and chest.
**thoracoschisis** (*tho-rak-os'-kis-is*) [*thoraco-*; σχίσις, a cleaving]. Congenital fissure of the thorax.
**thoracoscope** (*tho-rak'-o-skōp*) [*thoraco-*; σκοπεῖν, to view]. A stethoscope.
**thoracoscopy** (*tho-rak-os'-ko-pe*) [see *thoracoscope*]. Examination of the chest, especially by the stethoscope.
**thoracostenosis** (*tho-rak-o-sten-o'-sis*) [*thoraco-*; stenosis]. Contraction or compression of the walls of the chest.
**thoracostomy** (*tho-rak-os'-to-me*) [*thoraco-*; στόμα, mouth]. The operation of making an opening in the thorax.
**thoracotomy** (*tho-rak-ot'-o-me*) [*thoraco-*; τέμνειν, to cut]. Incision of the thorax or chest-wall.
**thoradelphus** (*tho-rad-el'-fus*) [*thorax*; ἀδελφός, brother]. A double monster united above the umbilicus, with one head, four lower and two upper extremities.
**thorax** (*tho'-raks*) [θώραξ, a breastplate]. The chest; the framework of bones and soft tissues bounded by the diaphragm below, the ribs and sternum in front, the ribs and dorsal portion of the vertebral column behind, and above by the structures in the lower part of the neck, and containing the heart inclosed in the pericardium, the lungs invested by the pleura, and the mediastinal structures. t., **region of**. See *region, thoracic*.
**Thorel's bundle** (*tor'-el*) [—— *Thorel*, German physician]. A structure in the heart wall connecting the sinoauricular and auriculoventricular nodes.
**thorium** (*tho'-re-um*) [Icel., *Thōrr*]. A rare metal related chemically to tin. Symbol Th; atomic weight, 234.4. It is a radioactive substance and gives off several emanations, indicated by various letters such as A, B, X, etc. t. **paste**, a preparation made of thorium protoxide, lead sulphate, sulphuric acid, and hydrochloric acid. It is used for the treatment of lupus, and epithelioma.
**Thormaehlen's test for melanin in urine** (*tōr'-ma-len*) [Johann *Thormaehlen*, German physician]. To the urine to be tested add sodium nitroprusside, caustic potash, and acetic acid, and in the presence of melanin a deep-blue coloration will be produced.
**thorn-apple**. See *stramonium*.
**Thorn's maneuver** (*torn*) [Wilhelm *Thorn*, German obstetrician, 1859—  ]. Changing of a face presentation into a vertex presentation by combined external and internal version.
**Thornton's sign** [J. Knowsley *Thornton*, English physician, 1845—  ]. Violent pain in the flanks in nephrolithiasis.
**Thornwaldt's disease** (*torn'-volt*). Nasolaryngeal stenosis associated with the formation of a cyst-like cavity in the midst of the racemose glands in the pharyngeal mucosa, and containing pus or mucopus.
**thoroughjoint**. Diarthrosis, or arthrodia.
**thoroughwort**. See *eupatorium*.
**thought-reading**. See *telepathy*.
**thought-transference**. See *telepathy*.
**thoxos** (*thok'-sos*). Trade name for a preparation containing lithium and strontium salicylate, colchicum, ash-bark, etc.
**thread**. The spun and twisted fibers of cotton, linen, or silk. t.s, **mycelial**, the hyphæ of the mycelium. t.s, **nuclear**, chromatin fibrils of the cell-nucleus. t.s, **Simonart's**. See *bands, amniotic*. t.-**fungus**, a general term for any kind of Trichophyton. t.-**granules**. See *mitochondria*. t.-**reaction**. See *Pfaundler's reaction*.
**threadworm**. See *Oxyuris*.
**thready** (*thred'-e*). Like a thread. See *pulse*.
**three-cornered bone**. The cuneiform bone of the carpus.
**three-day fever**. Synonym of *dengue*.
**thremmatology** (*threm-at-ol'-o-je*) [θρέμμα, a nurseling; λόγος, science]. Experimental or artificial evolution. It includes the science of breeding, and the laws of heredity and variation.
**threpsology** (*threp-sol'-o-je*) [θρέψις, nutrition; λόγος, treatise]. The science of nutrition.
**threshold** (*thresh'-old*). 1. The lower limit of stimulus capable of producing an impression upon consciousness. 2. The entrance of a canal. t., **absolute**, the lowest limit of perception of a sensation. t., **auditory**, the minimum perceptible sound. t., **differential**, the lowest limit at which two stimuli can be discriminated. t., **double-point**, the smallest distance apart at which two points can be felt as one. t., **neuron**. See *neuron*. t., **relational**, the ratio of two stimuli when their difference is just perceptible. t., **stimulus**. Same as *t. absolute*.
**thridacium** (*thri-da'-se-um*) [θρίδαξ, lettuce]. The expressed and inspissated juice of the lettuce; a variety of lactucarium.
**thrill**. A fine vibration felt by the hand. A thrill may be felt on palpation over an aneurysm, over a heart the seat of valvular disease, and over hydatid cysts. t., **presystolic**, a thrill which can sometimes be felt before the systole when the hand is placed over the apex-beat. t., **purring**, a thrill resembling that felt when the hand is placed on the back of a cat.
**throat** (*thrōt*) [AS., *throte*, throat]. 1. The anterior part of the neck. 2. The pharynx and larynx; the fauces. t.-**cough**, a cough due to irritation of the pharynx apart from diseases of the respiratory tract, as from an elongated uvula. t.-**mirror**. See *laryngeal mirror*. t., **sore**, pharyngitis. t., **sore**, **clergyman's**, laryngitis caused by overuse through public speaking.
**throb**. A pulsation or beating.
**throbbing** (*throb'-ing*). A rhythmic beating. t. **aorta**, exaggerated pulsation of the abdominal aorta perceptible to the patient.
**throe** (*thro*). A violent pang, or pain, as in parturition.
**thromballosis** (*throm-bal-o'-sis*) [*thrombus*; ἀλλοίωσις, a change]. The changed condition caused by coagulation of the venous blood.
**thrombase** (*throm'-bās*). Same as *thrombin*.
**thrombectomy** (*throm-bek'-to-me*) [*thrombus*; ἐκτομή, excision]. Excision of a thrombus.
**thrombin** (*throm'-bin*) [*thrombus*]. The fibrin-ferment, the enzyme that causes coagulation of shed blood.
**thrombo-** (*throm-bo-*) [θρόμβος, a thrombus]. A prefix denoting pertaining to a thrombus.
**thromboangiitis** (*throm-bo-an-je-i'-tis*) [*thrombo-*; ἀγγεῖον, a vessel; ἰτις, inflammation]. Thrombosis with inflammation of the intima of a vessel.
**thromboarteritis** (*throm-bo-ar-ter-i'-tis*) [*thrombo-*; *arteritis*]. Inflammation of an artery associated with thrombosis.
**thrombocystis** (*throm-bo-sis'-tis*) [*thrombo-*; κύστις, a bladder]. Sac sometimes enveloping a thrombus.
**thrombocyte** (*throm'-bo-sīt*) [*thrombo-*; κύτος, a cell]. Same as *blood-platelet*.
**thrombogen** (*throm'-bo-jen*) *[thrombo-*; γεννᾶν,

to produce]. The substance which, when activated by thrombokinase, becomes the fibrin-ferment, thrombin.
**thrombogenic** (*throm-bo-jen'-ik*) [*thrombo-*; γεννᾶν, to produce]. 1. Producing thrombi. 2. Relating to thrombogen.
**thromboid** (*throm'-boid*) [*thrombo-*; εἶδος, like]. Resembling or having the nature of a thrombus.
**thrombokinase** (*throm-bo-kin'-ase*). An activating substance capable of transforming thrombogen into thrombin.
**thrombolymphangitis** (*throm-bo-lim-fan-ji'-tis*) [*thrombo-*; *lymphangitis*]. Lymphangitis, with thrombosis.
**thrombopenia** (*throm-bo-pe'-ne-ah*) [*thrombo-*; πενία, poverty]. Same as purpura hæmorrhagica, q. v.
**thrombophlebitis** (*throm-bo-fleb-i'-tis*) [*thrombo-*; *phlebitis*]. Inflammation of a vein associated with thrombosis.
**thrombosed** (*throm'-bōzd*). 1. Affected with thrombosis. 2. Clotted.
**thrombosin** (*throm'-bo-sin*). One of the products of the cleavage of fibrinogen by acetic acid; it is a proteid body which passes into fibrin in the presence of soluble calcium salts.
**thrombosis** (*throm-bo'-sis*) [*thrombus*]. The formation of a thrombus. t., **atrophic**, that due to general malnutrition. t., **cardiac**, thrombosis of the heart. t., **coagulation**, that caused by fibrin coagulation. t., **compression**, that due to compression of a vessel, as by a tumor. t., **dilatation**, that which results from the slowing of the blood current next to the vessel-walls as the result of dilatation of a vessel (as in aneurysms, a curve of the heart. t., **Lancereaux's law** of, marantic thromboses always occur at the points where there is the greatest tendency to stasis; that is where the influence of the cardiac propulsion and of thoracic aspiration is least. t., **marantic**. Same as *t.*, *atrophic*. t., **marasmic**. Same as *t.*, *atrophic*. t. **placental**, that of the uterine veins of the site of the placenta. t., **plate**. See *thrombus, autochthonous*. t., **puerperal venous**, puerperal thrombosis of the uterine veins.
**thrombostasis** (*throm-bo-sta'-sis*) [*thrombo-*; *stasis*]. Stasis of blood leading to formation of a thrombus.
**thrombotic** (*throm-bot'-ik*) [*thrombosis*]. Pertaining to or produced by thrombosis.
**thrombus** (*throm'-bus*) [θρόμβος, a clot: pl., *thrombi*]. A clot of blood formed within the heart or blood-vessels due usually to a slowing of the circulation or to alteration of the blood or vessel-walls. t., **annular**, one that involves the whole circumference of the vessel but does not entirely occlude it. t., **antemortem**, the white thrombi in the heart and large vessels formed before death. t., **autochthonous**. Same as *t. blood-plate*. t., **ball**, a small or large, rounded, antemortem clot found in the heart, especially in the auricles. t., **blood-plate**, that ascribed by Eberth to agglutination of blood-plates. t., **currant-jelly**, a soft, reddish, postmortem clot. t., **Laennec's**, a globular thrombus formed in the heart, especially its latter is the seat of fatty degeneration. t., **lateral**, a clot attached to the vessel-wall, and not obstructing the lumen completely. t. **neonatorum**. Same as *cephalhematoma*. t., **obstructing**, one completely obstructing the lumen of the vessel. t., **parietal**, or **valvular**, one adherent to the wall of a vessel or the heart and not entirely occluding the vessel. t., **progressive**, one that grows into the lumen of the vessel. t., **stratified**, one in which there are successive layers of fibrinous deposit and of varying color. t. **vulvæ, t. vaginæ**, hematoma of the labium pudendi majus. t. **white**. See *t.*, *antemortem*.
**throttle** (*throt'-l*). 1. The throat. 2. To choke; to suffocate.
**through-drainage**. A method of drainage in which a perforated tube is carried through the cavity to be drained, so that the latter can be flushed through and through by the injection of fluid into one end of the tube.
**through-illumination**. Transillumination.
**throwback** (*thro'-bak*). 1. To show reversion in characters to those of the offspring of a known sire or to those of the first sire. See *infection* (2), *telegony*, *reversion*. 2. To reduce in class or rank.
**thrush**. A form of stomatitis due to a specific fungus, *Oidium albicans* or *Saccharomyces albicans*, and characterized by the presence of diffuse white patches. It occurs especially in weakly children,

but may affect adults depressed by wasting diseases. 2. A diseased condition of the frog of the horse's foot, with a foul-smelling discharge.
**thrypsis** (*thrip'-sis*) [θρύψις, a crushing]. A comminuted fracture.
**Thudichum's test** (*too'-de-koom*) [Johann Ludwig Wilhelm *Thudichum*, German physician, 1829–1901]. *For creatinine:* a dilute solution of ferric chloride is added to the suspected fluid; the presence of creatinine is shown by a dark red color which is increased by warming.
**Thuja** (*thū'-jah*). A genus of trees of the order *Coniferæ*. *T. occidentalis* or arbor vitæ has been used in intermittent fever, rheumatism, scurvy, and as an emmenagogue. t., **oleum**, a volatile oil of camphoraceous odor, composed of thujol and terpene. Dose, ♏ j–v.
**thujetin** (*thū'-jet-in*). A dissociation product of thujin by prolonged heating with sulphuric acid. A yellow crystalline powder similar to quercitrin, soluble in alcohol and ether.
**thujin** (*thū'-jin*). A glucoside similar to quercitrin found in *Thuja occidentalis*.
**thujol** (*thū'-jol*). An oily liquid, the chief constituent of *Thuja occidentalis*. It increases the blood-pressure and has antipyretic properties.
**thujone** (*thū'-jōn*). A colorless oily ketone, which causes the symptoms found in absinthism.
**thulium** (*thū'-le-um*) [*Thule*, northland]. An element occurring in some rare metals; symbol Tm; atomic weight, 168.5. See *elements, table of*.
**thumb** (*thum*). The digit on the extreme radial side of the hand, differing from the other digits in having but two phalanges, and in that its metacarpal bone is separately movable. t.-**exercise**. See under *exercise*. t.-**lancet**, a lancet with a broad pointed extremity and a double cutting edge. t.-**marks**, an impression made by the thumb. t., **stub**-. See *stub-thumb*.
**thumps**. An affection in the horse, identical with hiccough in man, due to spasmodic action of the diaphragm.
**thunder-struck disease**. Synonym of *apoplexy*.
**thus** [L., *gen.*, *thuris*, "incense"]. 1. True frankincense or olibanum. 2. Turpentine of pinetrees.
**thuya** (*thū'-yah*). See *thuja*.
**thylacitis** (*thi-las-i'-tis*) [θυλάκιον, a little bag]. Inflammation of the sebaceous glands. Acne rosacea.
**thyma** (*thi'-mah*). A corruption of ecthyma and also of thymion.
**thymacetin** (*thi-mas'-et-in*) [*thyme*; *acetum*, vinegar], C₆H₃(CH₃)(C₃H₇)(OC₂H₅)NH(C₂H₃O). A derivative of thymol used as an antineuralgic. Dose 3–15 gr. (0.2–1.0 Gm.).
**thymasthma** (*thi-mas'-mah*). See *thymic asthma*.
**thyme** (*tīm*) [θύμον, thyme]. The genus *Thymus*, of the order *Labiatæ*. *Thymus vulgaris* yields a volatile oil in which are found *cymene*, *thymene*, and *thymol*. t., **oil of** (*oleum thymi*, U. S. P.), a volatile oil distilled from the leaves and tops of *Thymus vulgaris*; often misnamed oil of origanum. Dose 3 min. (0.2 Cc.).
**thymectomize** (*thi-mek'-to-mīz*) [*thymectomy*]. To excise the thymus gland.
**thymectomy** (*thi-mek'-to-me*) [*thymus*; ἐκτομή, excision]. Excision of the thymus.
**thymegol** (*thi-me'-gol*). An antiseptic and emetic forming a red-brown powder; said to be a parasulphonic derivative of potassium, thymol, and mercury.
**thymelcosis** (*thi-mel-ko'-sis*) [*thymus*; ἕλκωσις, ulceration]. Ulceration of the thymus gland.
**thymene** (*ti'-mēn* or *thi'-mēn*) [*thyme*]. A hydrocarbon existing in oil of thyme.
**thymhydroquinone** (*thīm-hi-dro-kwin'-ōn*). A reduction product of thymoquinone.
**thymic** (*thi'-mik*, also for first definition, *ti'-mik*) [*thyme*; *thymus*]. 1. Pertaining to or contained in thyme. 2. Pertaining to the thymus gland. t. **acid**, thymol. t. **asthma**, a form of laryngismus stridulus consisting in a temporary suspension of respiration, attributed to enlargement of the thymus. t. **death**, sudden death, occurring in status lymphaticus.
**thymicolymphatic** (*thi-mik-o-lim-fat'-ik*). Pertaining to the status lymphaticus.
**thymin** (*thi'-min*), C₅H₆N₂O₂. A crystalline body obtained by boiling nucleic acid from the thymus gland of the calf with dilute sulphuric acid.
**thyminol** (*thi'-min-ol*). An antiseptic liquid said to be a solution of thymol, eucalyptol, menthol, *Baptisia*, benzoic, boric, and salicylic acids.

**thymiodide** (*thi-mi'-o-did*). Thymol iodide.
**thymi oleum.** The volatile oil of garden thyme; it is important as a source of thymol. It has the general properties of the terpenes and mints; it is often sold as oil of marjoram, which it resembles.
**thymion** (*thi'-me-on*) [θύμιον]. A wart; a condyloma.
**thymiosis** (*thi-me'-o-sis*) [*thymion*]. 1. Yaws. 2. A condition associated with the formation of warty growths.
**thymitis** (*thi-mi'-tis*) [*thymus; ιτις*, inflammation]. Inflammation of the thymus gland.
**thymoform** (*thi'-mo-form*). A reaction-product of thymol and formaldehyde; used as are iodoform and dermatol.
**thymohydroquinone** (*thi-mo-hi-dro-kwin'-ōn*). A substance occurring in the urine after the ingestion of thymol.
**thymokesis** (*thi-mo-ke'-sis*). Persistence or enlargement of the thymus gland in an adult.
**thymol** (*ti'-mol*, or *thi'-mol*) [*thyme; oleum*, oil], $C_{10}H_{13} \cdot HO$. A phenol derived from the volatile oils of *Thymus vulgaris*, *Monarda punctata*, and *Carum ajowan*. It is a crystalline solid, melting at 44° C., very slightly soluble in water, and is used as a local antiseptic and deodorant in ulcers, leukorrhea, and stomatitis, as an intestinal antiseptic, and as an anthelmintic. Dose 1–2 gr. (0.065–0.13 Gm.). t. **camphor.** See *camphor*. t. **carbonate,** recommended as preferable to thymol in uncinariasis. Dose 30 gr. (2 Gm.). Syn., *thymotol*. t. **chlormethylsalicylate,** a condensation product of thymol and chlormethyl salicylic acid. Antiseptic. t. **gauze,** contains 1 per cent of thymol. t. **inhalation,** thymol gr. xx, alcohol ℥ iij, magnesium carbonate gr. x, water ad ℥ iij; add a teaspoonful to a pint of water. t. **iodide.** Same as *aristol*. t. **salicylate,** t. **solution,** C7H8O3, an intestinal antiseptic. t. **solution,** for spraying, 1 : 1000. thymol solution (**Volkmann's**), thymol 1, alcohol 20, glycerin 20; dissolve and add to water 1000. It is used as a spray and antiseptic lotion; it does not produce eczema, as carbolic lotions do. t. **urethane,** a compound of thymol and thymol carbonic ester forming colorless crystals insoluble in water; used as an anthelmintic.
**thymolize** (*thi'-mol-īz*). To treat with thymol.
**thymoloform** (*thi-mol'-o-form*). Thymoform.
**thymolol** (*thi'-mol-ol*). Thymol iodide; aristol.
**thymopathy** (*thi-mop'-ath-e*) 1. [*thymus; πάθος*, suffering]. Any disease of the thymus gland. 2. Also [θυμός, the mind; πάθος, disease]. Mental disorder.
**thymoprivous** (*thi-mop'-riv-us*) [*thymus; privus*, bereft of]. Pertaining to or caused by removal or premature atrophy of the thymus.
**thymotol** (*thi'-mo-tol*). See thymol carbonate.
**thymoxalme** (*thi-moks-al'-me*) [*thyme; ὀξύς*, sharp; ἅλς, salt]. A mixture of thyme, vinegar, and salt.
**thymozone** (*thi'-mo-zōn*). A combination of *Eucalyptus globulus*, *Thymus vulgaris*, and *Pinus sylvestris*, with benzoic, boric, and salicylic acids. A nonirritating compound used as an internal antiseptic and externally as is phenol.
**thymuin** (*thi'-mū-in*). A trade name of a preparation containing thymus adrenals, steapsin, sodium cacodylate, and nascent ferrous carbonate.
**thymus** (*thi'-mus*) [θύμος, the thymus]. 1. An organ situated in the anterior superior mediastinum. It continues to develop until the second year of life, afterward remains stationary until about the fourteenth, and then undergoes fatty metamorphosis and atrophy. The thymus consists of lobules largely composed of lymphadenoid tissue in which minute concentric bodies, the corpuscles of Hassal, are found. The latter are remnants of epithelial structures. 2. A genus of labiate plants. See *thyme*. t. **death,** sudden death assumed to be due to enlargement of the thymus gland. t. **gland.** Same as *thymus* (1).
**thyraden** (*thi'-ra-den*). See *thyroidin*.
**thyrein** (*thi'-re-in*). See *iodothyrin*.
**thyremphraxis** (*thi-rem-fraks'-is*) [*thyroid; ἔμφραξις*, stoppage]. Lessened or abolished function of the thyroid gland.
**thyreo-** (*thi-re-o-*). See *thyro-*.
**thyreoadenitis.** See *thyroadenitis*.
**thyreoantitoxin.** See *thyroantitoxin*.
**thyreocele.** See *thyrocele*.
**thyreochondrotomy.** See *thyrochondrotomy*.
**thyreocricotomy.** See *thyrocricotomy*.
**thyreoepiglottideus.** Thyroepiglottideus muscle. See under *muscles*.
**thyreohyoideus.** See *thyrohyoid*.
**thyreoid** (*thi'-re-oid*). See *thyroid*.
**thyreoidectomy.** See *thyroidectomy*.
**thyreoidin.** See *thyroidin*.
**thyreoiditis.** See *thyroiditis*.
**thyreoidotomy.** See *thyroidotomy*.
**thyreojitis.** See *thyroiditis*.
**thyreoncus.** Same as *thyrocele*.
**thyreophyma.** See *thyrophyma*.
**thyreoprivus.** See *thyroprivus*.
**thyreoprotein.** See *thyroprotein*, and *thyroantitoxin*.
**thyreotomy.** See *thyrotomy*.
**thyreotoxin.** See *thyrotoxin*.
**thyrine** (*thi'-rēn*). Proposed name for the active principle of the thyroid gland.
**thyro-** (*thi-ro-*) [θυρεός, a shield, and hence, from similarity of shape and function, the thyroid gland or cartilage]. A prefix signifying relationship to the thyroid gland.
**thyroadenitis** (*thi-ro-ad-en-i'-tis*) [*thyroid; ἀδήν*, gland; *ιτις*, inflammation]. Inflammation of the thyroid gland.
**thyroantitoxin** (*thi-ro-an-te-toks'-in*) [*thyro-; antitoxin*]. 1. C8H11N2O5. A proteid constituent of the thyroid gland. 2. A preparation of the thyroid gland used in exophthalmic goiter, bronchocele, etc. Dose 2 gr. (0.13 Gm.) daily.
**thyroarytenoid** (*thi-ro-ar-it'-en-oid*) [*thyro-; arytenoid*]. Pertaining to the thyroid and arytenoid cartilages, as the thyroarytenoid ligaments, thyroarytenoid muscle (*thyroarytenoideus*).
**thyrocele** (*thi'-ro-sēl*) [*thyro-; κήλη*, a tumor]. A tumor affecting the thyroid gland; goiter.
**thyrochondrotomy** (*thi-ro-kon-drot'-o-me*) [*thyro-; χόνδρος*, cartilage; *τέμνειν*, to cut]. Incision of the thyroid cartilage.
**thyrochrom** (*thi'-ro-krōm*). An alcoholic extract of the thyroid gland of the calf.
**thyrocolloid** (*thi-ro-kol'-oid*). A proprietary preparation of the thyroid glands of sheep.
**thyrocricotomy** (*thi-ro-kri-kot'-o-me*) [*thyroid; cricotomy*]. Tracheotomy performed through the cricothyroid membrane.
**thyrodyl** (*thi'-ro-dil*). Trade name of a preparation containing thyroid, pituitary, and adrenal glands.
**thyroepiglottic** (*thi-ro-ep-e-glot'-ik*) [*thyro-; epiglottis*]. Pertaining to the thyroid cartilage and the epiglottis, as the thyroepiglottic muscle (*thyroepiglottideus*).
**thyrogenous** (*thi-roj'-en-us*) [*thyro-; γεννᾶν*, to produce]. Originating in the thyroid gland.
**thyroglandin** (*thi-ro-gland'-in*). A compound of iodoglobulin and thyroidin, in the form and proportion in which they exist in the thyroid gland. Dose 3–5 gr. (0.19–0.32 Gm.) for myxedema and obesity.
**thyroglobulin** (*thi-ro-glob'-ū-lin*). The iodineproteid of the thyroid secreted by it and lodged in the colloid substance.
**thyroglossal** (*thi-ro-glos'-al*) [*thyro-; γλῶσσα*, tongue]. Pertaining to the thyroid and the tongue. t. **duct,** a fetal passage between the thyroid gland and the tongue.
**thyrohyal** (*thi-ro-hi'-al*). See *thyrohyoid*.
**thyrohyoid** (*thi-ro-hi'-oid*) [*thyro-; hyoid*]. Pertaining to the thyroid cartilage and hyoid bone, as the *thyrohyoid* membrane. See under *muscles*.
**thyroid** (*thi'-roid*) [θυρεός, shield; εἶδος, like]. 1. Shield-shaped. 2. Pertaining to the thyroid gland. 3. Pertaining to the thyroid cartilage. 4. Pertaining to the thyroid foramen. 5. The thyroid gland. t. **accessory,** an outlying portion of the thyroid gland. t. **axis.** See *arteries, table of*. t. **body.** See *t. gland*. t. **cartilage,** the largest of the laryngeal cartilages, united at an angle in front called the pomum adami. t. **extract.** See under *t. gland*. t. **foramen.** See *foramen, thyroid*. t. **gland,** one of the so-called ductless glands, lying in front of the trachea, and consisting of two lateral lobes, connected centrally by an isthmus. The organ is composed of follicles lined by epithelium, producing a peculiar colloid material. The function of the organ is not definitely known, but it is supposed to be the production of some substance necessary to the body—an internal secretion that may counteract poisons produced in the system. Hypertrophy of the gland (goiter) is sometimes associated with a peculiar disease known as exophthalmic goiter; absence of the gland leads to

**cretinism** or **myxedema**. An extract prepared from the thyroid gland of animals (*thyroid extract*) and other preparations of the gland are used medicinally. See *organotherapy*. **t. therapy,** the treatment of disease by the administration of thyroid-extract.
**thyroidectin** (*thi-roi-dek'-tin*). Trade name of a substance prepared from the blood of thyroidectomized animals; it has been used in exophthalmic goiter in 5 grain doses.
**thyroidectomized** (*thi-roid'-ek-tom-īzd*). See *thyroprival*.
**thyroidectomy** (*thi-roi-dek'-to-me*) [*thyroid*; ἐκτομή, excision]. Excision of the thyroid gland.
**thyroidin** (*thi-roi'-din*) [*thyroid*; εἶδος, like]. A proprietary lactose trituration of dried extract of thyroid gland; one part represents two parts of fresh gland. It is an alterative used in myxedema, struma, and psoriasis. Dose 15-24 gr. (1.0-1.5 Gm.) daily.
**thyroidism** (*thi'-roid-izm*). 1. Disturbances produced by hypertrophy of the thyroid gland. 2. A series of phenomena due to continued use of thyroid preparations. 3. Disturbances due to removal of the thyroid.
**thyroiditis** (*thi-roi-di'-tis*) [*thyroid*; ιτις, inflammation]. Inflammation of the thyroid gland.
**thyroidization** (*thi-roid-i-za'-shun*). Treatment with thyroid gland preparations.
**thyroidotomy** (*thi-roi-dot'-o-me*) [*thyroid*; τομή, a cutting]. Incision of the thyroid gland.
**thyroidotoxin** (*thi-roid-o-toks'-in*). A substance specifically toxic for the cells of the thyroid gland.
**thyroigenous** (*thi-roi'-jen-us*). Originating in disturbances of the thyroid gland.
**thyroiodine** (*thi-ro-i'-od-in*). A substance found principally combined with a thyroid, but also free in the thyroid gland. Syn., *iodothyrin*.
**thyrolaryngeal** (*thi-ro-lar-in'-je-al*). Relating to the larynx and the thyroid body.
**thyrolingual** (*thi-ro-ling'-gwal*). Relating to the thyroid and the tongue: thyroglossal.
**thyrolytic** (*thi-ro-lit'-ik*) [*thyroid*; λύσις, dissolution]. Destruction of thyroid tissue.
**thyron** (*thi'-ron*). Trade name of a substance prepared from pigs' thyroids.
**thyroncus** (*thi-ronk'-us*) [*thyroid*; ὄγκος, tumor]. Same as *thyrocele*.
**thyrophyma** (*thi-ro-fi'-mah*) [*thyro-*; φῦμα, a tumor]. Enlargement of the thyroid gland.
**thyroprival** (*thi-ro-pri'-val*) [*thyro-*; *privare*, to deprive]. Due to loss of function or removal of the thyroid gland.
**thyroprivus** (*thi-ro-pri'-vus*) [*thyroprival*]. 1. Deprived of the thyroid gland. 2. A morbid condition due to loss of the thyroid gland.
**thyroprotein** (*thi-ro-pro'-te-in*). 1. An albumin from the thyroid gland. 2. A toxic protein from the thyroid gland.
**thyroptosis** (*thi-rop-to'-sis*) [*thyro-*; πτῶσις, a falling]. Displacement of a goitrous thyroid so as to be concealed in the thorax.
**thyrotherapy** (*thi-ro-ther'-ap-e*) [*thyroid*; *therapy*]. Treatment of disease by thyroid gland preparations.
**thyrotomy** (*thi-rot'-o-me*) [*thyro-*; τομή, a cutting]. Incision or splitting of the thyroid cartilage.
**thyrotoxicosis** (*thi-ro-toks-ik-o'-sis*) [*thyroid*; τόξικον, a poison]. 1. Poisoning by thyroid secretion. 2. Exophthalmic goiter.
**thyrotoxin** (*thi-ro-toks'-in*). A cytotoxin obtained by injections of emulsion of thyroid glands.
**thyrsus** (*thur'-sus*) [θύρσος, a stalk]. The penis.
**Ti.** Chemical symbol of titanium.
**tibia** (*tib'-e-ah*) [L., "shin"]. The larger of the two bones of the leg, commonly called the shinbone, articulating with the femur, fibula, and astragalus.
**tibiad** (*tib'-e-ad*) [*tibia*; *ad*, to]. Toward the tibial aspect.
**tibial** (*tib'-e-al*) [*tibia*]. 1. Pertaining to or in relation with the tibia, as the *tibial* muscle (*tibialis*), *tibial* artery, *tibial* nerve. 2. Referring to the inner or medial border of the foot.
**tibialis** (*tib-e-a'-lis*). 1. Tibial; pertaining to the tibia. 2. A muscle connected with the tibia. See *muscles, table of*.
**tibien** (*tib'-e-en*) [*tibia*]. Belonging to the tibia in itself.
**tibio-** (*tib-e-o-*) [*tibia*]. A prefix meaning pertaining to the tibia.
**tibiocalcanean** (*tib-e-o-kal-ka'-ne-an*) [*tibio-*; *calcaneus*]. Pertaining to the tibia and the calcaneus.

**tibiofemoral** (*tib-e-o-fem'-or-al*) [*tibio-*; *femur*]. Pertaining to the tibia and the femur.
**tibiofibular** (*tib-e-o-fib'-u-lar*) [*tibio-*; *fibula*]. Pertaining to the tibia and the fibula.
**tibionavicular** (*tib-e-o-nav-ik'-u-lar*) [*tibio-*; *navicula*, a boat]. Relating to the tibia and the navicular or scaphoid bone of the tarsus.
**tibioperoneal** (*tib-e-o-per-o-ne'-al*) [*tibio-*; *peroneus*]. Same as *tibiofibular*.
**tibioscaphoid** (*tib-e-o-skaf'-oid*). Same as *tibionavicular*.
**tibiotarsal** (*tib-e-o-tar'-sal*) [*tibio-*; *tarsus*]. Pertaining to the tibia and the tarsus.
**tic** (*tik*) [Fr.]. A twitching, especially of the facial muscles: **t. convulsif, t., convulsive,** spasm of the facial muscles. **t. douloureux,** neuralgia of the trifacial nerve. **t. impulsive.** Same as *Gilles de la Tourette's disease*. **t., painless,** the occurrence, at intervals, of sudden rapid involuntary contraction in a muscle or group of muscles. **t. rotatoire,** or **t. giratoire,** spasmodic torticollis, spinal accessory spasm: a spasm of certain muscles by which the head and neck are forcibly rotated to one side or from one side to the other.
**tick** (*tik*). A name applied to several species of *Acarus*. **t. fever.** 1. Texas fever. 2. Rocky Mountain spotted fever. 3. African relapsing fever.
**tickle** (*tik'-l*). To touch so as to cause a peculiar sensation (tickling or titillation), usually associated with laughing and reflex muscular movements.
**Ticorea** (*ti-ko'-re-ah*). A genus of rutaceous plants of S. America. The bark of *T. febrifuga* is used in fevers; other species also are medicinal.
**ticpolonga** (*tik-po-long'-gah*). The cobra manil; a venomous serpent of Ceylon.
**ticuna** (*ti-ku'-nah*). A powerfully convulsant arrow poison of S. American origin and of unknown derivation.
**t. i. d.** An abbreviation for the Latin *ter in die*, three times a day.
**tidal air.** See under *respiration*. **t. breathing,** Cheyne-Stokes respiration. **t. wave.** See *wave, tidal.*
**tide** (*tīd*). A definite period of time. **t., acid,** a transient condition of increased acidity of the urine, sometimes seen after fasting. **t., alkaline,** the transient condition of alkalinity of the urine, occurring during digestion, when by reason of the determination of acid to the stomach there is a diminution of the acid salts secreted by the kidney.
**Tidy's test for albumin in the urine** (*ti'-de*) [Charles Meymott Tidy, English physician, 1843–1892]. Phenol and acetic acid, or phenol and alcohol will cause a white precipitate if albumin is present.
**Tiedemann's glands** (*tē'-de-man*) [Friedrich Tiedemann, German anatomist, 1781–1861]. See *Bartholin, glands of*. **T.'s nerve,** a plexus of delicate nervefibers derived from the ciliary nerves, and surrounding the central artery of the retina.
**tiglic aldehyde** (*tig'-lik al'-de-hīd*). See *croton aldehyde*.
**tiglium** (*tig'-le-um*). A plant, *Croton tiglium*, of the order *Euphorbiacea*. It contains a fixed oil.
**tigli, oleum** (U. S. P.), croton oil (*oleum crotonis*, B. P.); a powerful local irritant, and used, locally, as a counterirritant in gout, rheumatism, neuralgia, glandular swellings, etc. It is an active purgative, especially useful when a prompt effect is desired, as in mania, coma, etc. Dose 1 or 2 drops (0.065–0.13 Cc.).
**tigretier** (*te-gra-te-a'*) [Fr.]. A form of tarentism due to the bite of a poisonous spider.
**tigroid** (*ti-groid'*) [τιγροειδής, spotted]. A term applied to chromophil corpuscles. **t. bodies, t. masses.** See *Nissl's bodies*.
**tigrolysis** (*ti-grol'-is-is*). Disintegration of the tigroid masses in a cell.
**Tilia** (*til'-e-ah*) [L.]. A genus of exogenous trees—the linden or basswood.
**tiliacin** (*til-i'-as-in*). A glucoside found in the leaves of the linden tree (*tilia*).
**Tillaux-Phocas' disease** (*te-lo'*) [Paul Jules Tillaux, French surgeon, 1834– ]. See *Phocas' disease*.
**tilletia** (*til-e-she'-ah*) [L.]. A genus of ustilagineous fungi.
**tilmus** (*til'-mus*) [τίλμος, a pulling]. Carphology.
**timbre** (*tam'-br*) [Fr.]. The peculiar quality of a tone, other than pitch and intensity, that makes it distinctive. It depends upon the overtones of the vibrating body.

**time.** The duration of an event or phenomenon. **t., inertia,** in the stimulation of a muscle or sense-organ the latent time required to overcome the inertia of the muscle or organ after the reception of the stimulus through the nerve. **t., persistence.** See under *persistence*. **t., reaction,** that required for the conduction of a sensory impulse of the center, combined with that of the duration of the perception, of the direction of attention (apperception), of the voluntary impulse, and of the return of a motor impulse to the muscles, with their consequent activity. **t., recognition,** the time required for the recognition of the kind of stimulus after its application. **t.-sense,** the perception of the lapse of time.

**tin.** A silvery-white, metallic, malleable element, having a specific gravity of 7.25, an atomic weight of 119, an atomicity of two or four. Symbol Sn, from the Latin *stannum*. **t. chloride,** stannous chloride, $SnCl_2 + 2H_2O$, is used as a reagent. **t., precipitated** (galvanically), recommended as a vermifuge against tape-worm. Dose 9 gr. (0.5 Gm.) every fifteen minutes until five or six doses have been taken.

**tincæ, os** (*ting'-ke*) [L. "the tench's mouth"]. An old name for the os uteri.

**tincal** (*ting'-kal*). Crude or native borax.

**tinct.** An abbreviation of *tinctura*, tincture.

**tinctable** (*tink'-tab-l*). Tingible; stainable.

**tinction** (*ting'-shun*) [*tingere*, to dye]. A staining material. A tint. The process of staining.

**tinctorial** (*ting-to'-re-al*) [*tingere*, to dye]. Pertaining to staining or dyeing.

**tinctura** (*ting-tū'-rah*) [L.; gen. and pl. *tincturæ*]. Tincture.

**tincturation** (*ting-tū-ra'-shun*) [*tinctura*, tincture]. The preparation of a tincture; the treatment of a substance in such a way as to make a tincture from it.

**tincture** (*tingk'-tūr*) [*tinctura*, from *tingere*, to tinge]. 1. A solution of the medicinal principles of a substance in a fluid other than water or glycerol. 2. Specifically, an alcoholic solution of a medicinal substance. **t., ammoniated,** one made with ammoniated alcohol. **t., ethereal,** one made with ether. **t.s of fresh herbs** (*tincturæ herbarum recentium*, U. S. P.), prepared by macerating fresh herbs with alcohol.

**tinea** (*tin'-e-ah*) [L., "a moth; a worm"]. Ringworm; a generic term applied to a class of skin diseases caused by parasitic fungi, formerly applied to many spreading cutaneous diseases. **t. amiantacea, t. asbestina,** seborrhea of the scalp in which the crusts resemble asbestos. **t. axillaris,** ringworm of the axilla. **t. barbæ.** Same as *t. sycosis*. **t. capitis,** ringworm of the scalp. **t. circinata.** See under *t. trichophytina*. **t. corporis.** Same as *t. trichophytina*. **t. cruris.** See under *t. trichophytina*. **t. decalvans,** alopecia areata. **t. favosa,** favus. **t. furfuracea,** seborrhœa sicca. **t. imbricata,** a disease occurring in the East Indies, and characterized by the formation of concentric scaly patches and intense itching. Syn., *Bowditch Island ringworm; Tokelau ringworm*. **t. kerion,** a markedly inflammatory form of ringworm of the scalp (*tinea tonsurans*), giving rise to the formation of an edematous, boggy swelling discharging a mucoid secretion. **t. lupinosa.** Same as *t. favosa*. **t. nodosa,** a nodose condition of the hair of the mustache, accompanied by thickening, roughness, and fragility. **t. sycosis.** See under *t. trichophytina*. **t. tarsi,** blepharitis ulcerosa. **t. tonsurans.** See under *t. trichophytina*. **t. trichophytina,** ringworm, a spreading, contagious disease of the skin due to a vegetable fungus, the *Trichophyton*. On the non-hairy parts of the body (*tinea trichophytina corporis, tinea circinata*) it presents itself by spreading, scaly patches, tending to clear in the center. On the thighs and scrotum (*tinea trichophytina cruris*) it is apt to assume the appearance of eczema, hence it is also called *eczema marginatum*. The nails may be affected (*tinea trichophytina unguium*), becoming grayish, opaque, and brittle. On the scalp (*tinea trichophytina capitis, tinea tonsurans*) it forms roundeded, grayish, slightly elevated, scaly patches, with brittleness and loss of the hair. Ringworm of the bearded region (*tinea trichophytina barbæ, tinea sycosis*, parasitic sycosis, barber's itch) forms at first rounded, scaly patches, which soon become nodular and lumpy and tend to break down. **t. vera.** Synonym of *favus*. **t. versicolor,** a disease of the skin due to a vegetable parasite, *Microsporon furfur*. It is characterized by brownish-yellow macules that coalesce to form extensive areas of eruption. There is usually slight itching. C. W. Allen's iodine test is of value for the recognition of suspected areas; it consists in the application of iodine solution, preferably Lugol's (iodine, 5; potassium iodide, 10; water, 100). The diseased portion will stain deep brown or mahogany color, in contrast to the light-yellow coloration of healthy tissue.

**tingible** (*tinj'-ib-l*) [*tingere*, to tinge]. Capable of being stained; stainable.

**tingle** (*ting'-gl*). A pricking or stinging sensation; the feeling of a slight, sharp, and sudden thrill, as of pain.

**tinkling** (*tink'-ling*). A chinking sound, heard over a pneumothorax or a large pulmonary cavity. Syn., *metallic tinkling*.

**tinnitus** (*tin-i'-tus* or *tin'-it-us*) [*tinnire*, to tinkle]. A subjective ringing, roaring, or hissing sound heard in the ears. Syn., *tinnitus aurium*. **t., telephone,** a professional neurosis or abnormal nervous condition of the auditory apparatus, believed to be caused by the continual use of the telephone.

**tintometer** (*tint-om'-et-ur*) [*tint*; μέτρον, measure]. An instrument to measure the amount of coloring-matter in a liquid.

**tip.** The point or summit of anything. **t.-foot,** talipes equinus; a variety of club-foot. **t., Woolner's,** the apex of the helix of the ear.

**tiqueur** (*te-ker'*) [Fr.]. One exhibiting the clonic or tonic movements designated as tics.

**tire** (*tīr*). 1. A sense of weariness and exhaustion; fag. 2. To pass a wire (as a tire around a wheel) around a fractured patella.

**tireball** (*tīr-bal'*) [Fr. *tirer*, to draw; *balle*, ball]. An instrument for extracting bullets from a part. It resembles a corkscrew.

**tirefond** (*tēr-fon(g)'*) [Fr. *tirer*, to draw; *fond*, bottom]. An instrument for penetrating a cavity or tissue, transfixing and withdrawing foreign bodies, and usually made in the form of a gimlet.

**tisane** (*te-zan'*). Any decoction or beverage having slight curative or restorative qualities. See *ptisan*.

**tissue** (*tish'-oo*) [Fr., *tissu*, from L., *texere*, to weave]. An aggregation of similar cells and fibers, forming a distinct structure, and entering as such into the formation of an organ or organism. **t., adenoid.** See *t., lymphadenoid*. **t., adipose,** fatty tissue, a form of connective tissue consisting of fat-cells lodged in the meshes of areolar tissue. **t., animal,** a general name for any of the textures which form the elementary structures of the body, and of which there are four classes: epithelial, connective, muscular and nervous. **t., areolar,** a form of connective tissue consisting of cells and delicate, elastic fibers interlacing in every direction. **t., basement,** the tissue of the basement membrane. **t., cancellous,** the spongy tissue of bones. **t., cartilaginous.** See *cartilage*. **t., connective,** a general term for all those tissues of the body that support the essential elements or parenchyma. The most important varieties are adipose tissue, areolar tissue, osseous tissue, cartilaginous tissue, elastic tissue, fibrous tissue, lymphoid tissue. **t., corneous,** tissue found in the nails, hair, epidermis, etc. **t., dental.** See *dentine*. **t., elastic,** connective tissue composed of yellow elastic fibers. **t., embryonal connective.** See *t., mucoid*. **t., epithelial.** See *epithelium*. **t., erectile,** a spongy tissue that becomes expanded and hard when filled with blood. **t., fibrous connective,** there are three varieties; white fibrous, yellow elastic, and areolar, the variety depending upon the character of the fibers. The fibers are imbedded in a matrix or a soft, homogeneous material that contains mucin. **t. gelatinous,** mucous tissue. **t. glandular,** a form of epithelial tissue. **t., granulation.** See *granulation tissue*. **t., interstitial,** tissue formed during inflammation. **t., interstitial connective.** See *t., areolar*. **t., intertubular,** the dense tissue of dentine. **t., lepidic.** See *lepidic*. **t., lymphadenoid, t., lymphoid,** a form of connective tissue in which reticular meshes contain lymphoid cells; it composes the greater part of the lymphatic glands and is found in the spleen, tonsils, and the alimentary mucosa. **t., mesenchymal,** the embryonic tissue from which the connective tissues are derived. **t., mucoid,** mucous, or gelatinous, connective tissue such as is present in the umbilical cord of the fetus. **t., mucous,** a connective tissue such as is present in the umbilical cord of the fetus. **t., muscu-**

lar. See *muscular tissue*. t., **nervous**, the intrinsic substance of a nerve or nerve-fiber. t., **osseous**. See *bone*. t., **parenchymal**, the areolar tissues that accompany vessels and nerves into the interior of organs and glands, giving them protection. t. retiform, adenoid tissue. t., **simple**, that having but one or two structural elements, e. g. blood, lymph, epithelium, connective tissue of cartilage and bone, and nervous and muscular tissues. t., **white fibrous**, a form of connective tissue consisting of exceedingly fine, inelastic, transparent filaments. This tissue forms the greater part of ligaments, tendons, fascia, sheaths of muscles, periosteum, etc. t., **yellow elastic**, a very elastic yellowish tissue predominating in the subflavous ligament, vocal bands, inner coats of blood-vessels, and the longitudinal coats of the trachea and bronchi.
**tit**. The nipple. See *teat*.
**titanium** (*tit-a'-ne-um*) [Τιτάν, Titan]. A metal having a certain relationship to iron, chromium, and tin. It is extremely infusible and will scratch glass. Its medicinal properties are little understood. See *elements, table of chemical*.
**titer, titre** (*te'-ter*) [Fr.]. A standard of fineness or strength.
**titillation** (*tit-il-a'-shun*) [*titillare*, to tickle]. The act of tickling; the sensation produced by tickling.
**titration** (*tit-ra'-shun*) [Fr., *titre*, standard of fineness]. Volumetric analysis by the aid of standard solutions.
**titubation** (*tit-u-ba'-shun*) [*titubare*, to stagger]. A staggering gait seen especially in diseases of the cerebellum. t., **lingual**, stammering, stuttering.
**tixol** (*tiks'-ol*). A preparation of arsenic into which animals are dipped, to exterminate ticks.
**Tizzoni's test** (*tid-so'-ne*) [Guido *Tizzoni*, Italian physician, 1853– ]. For iron in the tissues: A section of the tissue is treated with a 2 per cent. solution of potassium ferrocyanide and a one-half per cent. solution of hydrochloric acid. A blue color indicates the presence of iron.
**Tl**. Chemical symbol of *thallium*.
**Tm**. Chemical symbol of *thulium*.
**Tn**. Abbreviation of *normal intraocular tension*.
**TO**. Abbreviation for *original* or *old tuberculin*. See *tuberculin*.
**toadhead** (*tōd-hed*). A kind of head sometimes found in certain so-called acephalous monsters.
**toast** (*tōst*). Bread browned by the fire. t.-**water**, water in which toasted bread has been steeped; it is used as a beverage by invalids.
**tobacco** (*to-bak'-o*) [Sp., *tabaco*, tobacco]. A plant, *Nicotiana tabacum*, of the order *Solanaceæ*, the dried leaves of which (*tabaci folia*, B. P.) contain a liquid alkaloid, *nicotine*, C₁₀H₁₄N₂, which is also present in the seeds and root. Nicotine is one of the most active poisons known. Tobacco-smoke contains a series of bases, among which are pyridin, picolin, lutidin, parvolin, and others. Tobacco is used as a sedative in nearly all parts of the world, being smoked, chewed, or used as snuff. Its physiological action is that of a nauseant, antispasmodic, and depressant; it is also a local irritant. In medicine it has been employed as a relaxant in intestinal obstruction, being given in the form of an enema, but it is now rarely used except in asthma and locally in hemorrhoids. t.-**amblyopia**, amblyopia produced by the prolonged and excessive use of tobacco. t.-**heart**, an irritable state of the heart, characterized by irregular action and palpitation, produced by excessive indulgence in tobacco. t., **Indian**. See *lobelia*.
**tobaccoism** (*to-bak'-o-izm*). A morbid condition due to the use of tobacco.
**Tobin's tubes**. A method of ventilation of rooms by the introduction of air through tubes placed in the wall.
**Tobold's apparatus** (*to'-bōld*) [Adelbert August Oskar *Tobold*, German laryngologist, 1827– ]. An illuminating apparatus with a movable reflector for use with the laryngoscope.
**tocanalgine** (*tok-an-al'-jēn*) [τόκος, birth; ἀν, priv.; ἄλγος, pain]. A morphine derivative used for producing analgesia in childbirth.
**tocodynamometer**. See *tokodynamometer*.
**tocograph**. See *tokograph*.
**tocology**. See *tokology*.
**tocomania**. See *tokomania*.
**tocometer** (*tok-om'-et-er*). See *tokodynamometer*.
**tocus**. See *tokus*.
**Todd's ascending process** [Robert Bentley *Todd*,

English physician, 1809–1860]. See *Scarpa's fascia*.
T.'s **cirrhosis**, hypertrophic cirrhosis of the liver.
**toddalia** (*tod-a'-le-ah*). A genus of rutaceous plants. t. **aculeata**, of S. Asia, is a useful aromatic stimulant and tonic.
**toddy** (*tod'-e*) [Hindu, *tadi*, a palm tree]. 1. The fermented juice of the cocōa-nut palm, obtained by incision of the palm, and collected in pots hung to the trees under the cuts. It is then fermented and distilled. 2. A drink composed of sweetened spirits and water.
**toe** [AS., *tá*]. A digit of the foot. t.-**brace**, an appliance for correction of flat-foot and deformed toes. t.-**clonus**, contraction of the great toe on sudden extension of the first phalanx. t.**drop**, inability to raise or extend the toes owing to paralysis of the muscles which dorsally flex the foot. t., **flexed**, t., **hammer-**, a claw-like permanent distortion of a toe in which it is abnormally flexed at the last joint, allowing the tip to rest on the ground while the first joint is raised above the proper level. t., **Morton's**. See *Morton's foot*. t. **reflex**. See *reflex*, *toe*.
**Toepfer's test for free HCl in gastric contents** (*tep'-fer*) [Alfred Eduard Franz *Toepfer*, German physician, 1858– ]. A few drops of a 0.5 alcoholic solution of dimethylamidoazobenzol gives a cherry red color to a fluid containing free hydrochloric acid.
**toilet** (*toi'-let*) [OF., *toilette*, a cloth]. In surgery, the cleansing, washing, and dressing of an operative wound. Also the cleansing of the parts after parturition.
**Toison's solution** (*twah-zorn'*) [J. *Toison*, French histologist, 1858– ]. A solution containing methyl violet 0.025, sodium sulphate 8, sodium chloride 1, glycerine 30, water to 200. It is used as a diluting fluid and stain for white corpuscles.
**Tokelau ringworm**. See *linea imbricata*.
**tokodynamometer** (*tok-o-di-nam-om'-et-er*) [τόκος, birth; *dynamometer*]. An instrument for measuring the force of the expulsive efforts of the uterus in childbirth.
**tokograph** (*tok'-o-graf*) [τόκος, birth; γράφειν, to record]. A recording tokodynamometer.
**tokology** (*tok-ol'-o-je*) [τόκος, birth; λόγος, science]. The science of obstetrics.
**tokomania** (*tok-o-ma'-ne-ah*) [τόκος, birth; μανία, madness]. Puerperal insanity.
**tokus** (*to'-kus*) [τόκος, birth]. Childbirth.
**tolerance, toleration** (*tol'-er-ans, tol-er-a'-shun*) [*tolerare*, to bear]. The ability of enduring the influence of a drug or poison, particularly when acquired by a continued use of the substance.
**tolerant** (*tol'-er-ant*) [*tolerare*, to bear]. Withstanding the action of a medicine without injury.
**Tollen's reagent for glucose**. An ammoniacal silver solution obtained by precipitating silver nitrate solution with caustic potash and adding just enough ammonia to dissolve the precipitate yielded. This solution is reduced by glucose.
**tolokno** (*to-lok'-no*). A food prepared chiefly from oats. Used in Russia for superalimentation in tuberculosis.
**tolphite** (*tol'-fīt*). A dusting-powder containing talcum.
**tolu** (*to-loo'*) [Santiago de *Tolu*, in the United States of Colombia, where it was first obtained]. Short for *balsam of tolu*. t., **balsam of** (*balsamum tolutanum*, U. S. P., B. P.), a balsam obtained from *Toluifera balsamum*, an evergreen tree of the order *Leguminosæ*. It is used as a stimulant expectorant. Dose 10-30 gr. (0.65-2.0 Gm.). t., **syrup of** (*syrupus tolutanus*, U. S. P., B. P.). Dose 4 dr. (16 Cc.). t., **tincture of** (*tinctura tolutana*, U. S. P., B. P.). Dose 30 min. (2 Cc.).
**toluene** (*tol'-u-ēn*) [*tolu*], C₇H₈. Methylbenzene; a hydrocarbon obtained from coal-tar and also produced in the dry distillation of tolu balsam and many resins.
**toluidine** (*tol-u'-id-ēn*) [*tolu*], C₇H₇.NH₂. A homologue of aniline, prepared from toluene. t. **blue**, C₁₅H₁₆N₃SClZnCl, the double salt of zinc chloride and dimethyltoluthionin. It occurs as a black powder dissolving in water and alcohol with a fine blue coloration. It acts upon lower organisms as a powerful poison and may be employed as is methylene-blue in infectious conjunctivitis, and also as a substitute for fluorescein in fixing the limits of corneal lesion.

**toluol** (tol'-ŭ-ol). Same as *toluene*.
**tolylacetamide** (tol-il-as-et-am'-id), C₉H₁₁(CH₃NH).- (C₇H₇O). A derivative of coal-tar; used as an antiseptic. Dose 2–10 gr. (0.32–0.65 Gm.).
**tolyliantipyrine.** See *tolypyrine*.
**tolypyrine** (tol-e-pi'-rēn), C₁₁H₁₄N₂O. Colorless crystals of an intensely bitter taste, soluble in water and alcohol; used as is antipyrine. t. **salicylate**, C₁₈H₁₄N₂O . C₇H₆O₃. Dose, in rheumatism, etc., 15–30 gr. (1–2 Gm.).
**tolysal** (tol'-is-al). Tolypyrine salicylate.
**Tomaselli's disease** (to-mah-sel'-le). Quinine fever, produced by large doses of quinine; it is attended with hematuria, dysuria, dyspnea, threatened collapse.
**tomato** (to-ma'-to). The ripe fruit of the common tomato, *Lycopersicum esculentum*. It is said to be useful in canker of the mouth, sore mouth, etc.
**tomentum** (to-men'-tum) [L.]. A lock of wool. t. **cerebri**, the network of small blood-vessels of the pia penetrating the cortex of the brain.
**Tomes' fibers** (tōms) [Sir John Tomes, English dentist, 1836–1895]. Elongated, branched processes of the odontoblasts of the plup filling the dentinal tubules of teeth.
**tomomania** (tom-o-ma'-ne-ah) [τομή, a cutting; μανία, madness]. 1. An excessive desire to perform operations. 2. An excessive desire to submit to surgical operations.
**tomotokia** (tom-o-to'-ke-ah) [τομή, a cutting; τόκος, birth].- Cesarean section.
**tone** (tōn) [τόνος, from τείνειν, to stretch]. 1. A distinct sound. 2. The normal state of tension of a part or of the body. t.-**deafness**, sensory amusia.
**tonga** (tong'-gah). A mixture of various barks, probably of *Premna taitensis* and *Raphidophora vitiensis*, brought from the Fiji islands, and used in neuralgia. Dose of a *fluidextract* ½ dr. (2 Cc.).
**tongaline** (tong'-gal-ēn). A proprietary preparation said to contain tonga (bark of *Premna taitensis*), extract of *Cimicifuga racemosa*, sodium salicylate, pilocarpine salicylate, and colchicine salicylate. It is recommended in gouty diathesis.
**tongine** (ton'-jin). An alkaloid obtained from tonga.
**tongue** (tung). The movable muscular organ attached to the floor of the mouth, and concerned in tasting, masticating, swallowing, and speaking. It consists of a number of muscles, and-is covered by mucous membrane from which project numerous papillæ, and in which are placed the terminal organs of taste. t., **bifid**, a tongue the anterior portion of which is cleft in the median line. t., **black**, a condition-in which the dorsal surface of the tongue is covered with a black coating. t.-**bone**. See *hyoid*. t.-**depressor**, a spatula for pushing down the tongue during the examination of the mouth and throat. t., **fern-leaf pattern**, a name given to a tongue presenting a well-marked central furrow (mid-rib) with lateral branches. t., **filmy**, one with whitish, symmetrical patches on both sides. t., **furred**, a coated tongue the papillæ of which are prominent, giving the mucous membrane the appearance of a whitish fur. t., **geographical**, one with localized thickening of the epithelium, giving to the surface the appearance of a geographical chart. t., **hairy**, one with a hyperplasia of the papillæ; giving rise to hair-like projections. t., **parrot**, a shriveled dry tongue that cannot be protruded, found in typhus fever. t., **stamp-licker's**, an infectious process in those employed in industries where small packets are labeled. It gives rise to ulcers of the tongue and mouth. t., **strawberry**, a hyperemic tongue, the fungiform papillæ of which are very prominent; it is seen especially in scarlatina. t.-**swallowing**, a condition in which there is an abnormal mobility of the organ, so that it falls backward, giving rise to danger of suffocation. t.-**tie**, a congenital shortening of the frenum of the tongue, interfering with its mobility. Syn., *ankyloglossia*. t., **wooden**, one the seat of actinomycosis.
**tonic** (ton'-ik) [tone]. 1. Pertaining to tone; producing normal tone or tension. 2. Characterized by continuous tension or contraction, as a *tonic spasm, tonic convulsion*. 3. An agent or drug producing normal tone of an organ or part. t., **cardiac**, strengthening the heart-muscle. t., **intestinal**, one strengthening the tone of the intestine. t., **nervine**, one increasing the tone of the nervous system. t. **spasm**, the continued, rigid, contraction of a muscle or muscles. t., **stomachic**, one increasing the tone of the stomach. t. **treatment**. 1. Treatment of disease by tonics. 2. The continuous treatment of syphilis by the use of the protiodide of mercury for two or three years. t., **vascular**, one increasing the tone or tension of the blood-vessels.
**tonicity** (ton-is'-it-e) [tone]. The condition of normal tone or tension of organs; a state of tone.
**tonicize** (ton'-is-īz) [tone]. To give tone or tension to anything.
**toninervin** (ton-e-ner'-vin). A water-soluble salt of quinine said to contain 4.5 % of iron. Dose as antipyretic 2–5 gr. (0.1–0.3 Gm.) every three hours; as tonic ¾ gr. (0.05 Gm.) twice daily.
**tonitruphobia** (ton-it-roo-fo'-be-ah) [tonitrus, thunder; φόβος, fear]. Morbid dread of thunder.
**tonka-bean** (tong'-kah-bēn). The seed of *Dipteryx odorata*, a tree of South America; it contains coumarin, and is used as a flavoring agent.
**tonogram** (ton'-o-gram) [τόνος, tone; γράφειν, to write]. A record made by a tonograph.
**tonograph** (ton'-o-graf) [τόνος, tension; γράφειν, to write]. A device for recording the tension of the arterial blood-current.
**tonol** (to'-nol). Trade name for a preparation of glycerophosphates.
**tonometer** (ton-om'-et-er) [τόνος, tone; tension; μέτρον, a measure]. An instrument to measure tension, as that of the eyeball. t., **Gaertner's**, one for estimating blood-pressure. t., **Musken's**, one for measuring the tonicity of the Achilles tendon.
**tonometry** (ton-om'-et-re). The measurement of tonicity.
**tonophant** (ton'-of-ant) [τόνος, tone; φαίνειν, to make apparent]. An apparatus to render visible the vibrations of sound.
**tonoplasts** (ton'-o-plasts) [τόνος, tension; πλάσμα, a thing molded]. Small intracellular bodies which build up strongly osmotic substances within themselves and in this way swell to small vacuoles.
**tonoscope** (ton'-o-skōp) [τόνος, tone; σκοπείν, to view]. An instrument for examination of the interior of the cranium by means of sound.
**tonquinol** (ton'-kwin-ol). Trinitroisobutyltoluol. A substitute for musk.
**tonsil** (ton'-sil) [tonsilla]. 1. A small, almond-shaped body, situated on each side of the fauces, between the anterior and posterior pillars of the soft palate. It consists of an aggregation of from 10 to 18 lymph-follicles, and is covered by mucous membrane, which dips into certain depressions called crypts. 2. A small lobe of the cerebellar hemisphere, situated on the inferior mesial aspect. t., **cerebellar**. See *tonsilla* and *amygdala*. t., **epipharyngeal**. See t., *pharyngeal*. t., **faucial**. See *tonsil* (1). t., **Gerlach's tubal**, a mass of adenoid tissue in the lower part of the Eustachian tube, particularly along its median wall and about the pharyngeal orifice. t.-**guillotine**. See *guillotine*. t., **lingual**, an accumulation of lymphadenoid tissue at the base of the tongue. t., **Luschka's**. See *Luschka*. t., **palatine**. See *tonsil*. t., **pharyngeal**, a mass of lymphadenoid tissue in the pharynx, between the Eustachian tubes; Luschka's tonsil. t., **third**. See t., *pharyngeal*.
**tonsilla** (ton-sil'-lah) [L.]. 1. Tonsil. 2. One of the five lobes of the hemisphere of the cerebellum, situated in the mesal side of the hemisphere, by the vallecula. t. **cerebelli**, tonsil of the cerebellum. See *tonsil* (2), and *tonsilla* (2). t. **intestinalis**, Peyer's patches. t. **lingualis**, lingual tonsil. -t. **palatina**, palatine tonsil. t. **pharyngea**, pharyngeal tonsil.
**tonsillar** (ton'-sil-ar) [tonsil]. 1. Pertaining to the tonsil, as the *tonsillar artery*. 2. Affecting the tonsil, as *tonsillar abscess*.
**tonsillectomy** (ton-sil-ek'-to-me) -[tonsilla, tonsil; ἐκτομή, excision]. Removal of the tonsil.
**tonsillith**. See *tonsillolith*.
**tonsillitis** (ton-sil-i'-tis) [tonsil; ιτις, inflammation]. Inflammation of the tonsil. t., **follicular**, a form in which the follicles are especially involved and project as whitish points from the surface of the tonsil. t., **herpetic**, a form characterized by an eruption of herpetic vesicles, which soon rupture, leaving small, circular ulcers that coalesce and become covered with a fibrinous exudation. The disease has an acute onset, a continuous fever, and a critical decline, affects those subject to herpes elsewhere, and tends to recur. t., **lacunar**. Same as t., *follicular*. t., **mycotic**, tonsillitis due to fungi. t., **phlegmonous**.

Same as t., *suppurative*. t., **pustular**, a form characterized by the formation of pustules, as in smallpox. t., **suppurative**. Synonym of *quinsy*.
**tonsillitic** (*ton-sil-it'-ik*) [*tonsilla*, tonsil; ιτις, inflammation]. Pertaining to or affected with tonsillitis.
**tonsillolith** (*ton-sil'-o-lith*) [*tonsil*; λίθος, a stone]. A concretion within the tonsil.
**tonsillotome** (*ton'-sil-o-tōm*) [*tonsil*; τομή, a cutting]. An instrument for removing or cutting off the tonsil.
**tonsillotomy** (*ton-sil-ot'-o-me*) [see *tonsillotome*]. The operation of cutting away the whole or a part of the tonsil. t. **rash**. See *rash, amygdalotomy*.
**tonsilsector** (*ton-sil-sek'-tor*). A tonsillotome consisting of a pair of circular scissor-blades moving inside a circular guarding ring.
**tonsure** (*ton'-shur*) [*tondere*, to clip]. The shaving or removal of the hair from any part.
**tontine** (*ton-tēn'*) [Lorenzo *Tonti*, Italian banker, 17th century]. A species of life insurance in which the policy holders receive no dividend or return premiums, till the end of a fixed period, when the profits are divided among the survivors who have kept their policies in force.
**tonus** (*to'-nus*). See *tone*. t., **chemical**, the condition of the muscles when at rest and undergoing no mechanical exertion.
**tooth** (*tooth*) [AS., *tōth*; plural, *teeth, q. v.*]. One of the small, bone-like organs occupying the alveolar processes of the upper and lower jaws, and serving for tearing and comminuting the food. The teeth begin to appear in the human being about the seventh month; by the end of the third year, the eruption of the so-called *temporary, deciduous*, or *milk-teeth*, numbering 20, is completed. The *permanent teeth* begin to replace the deciduous teeth about the seventh year. In the adult, the permanent teeth number 32, or 16 in each jaw, and are divided as follows: 2 incisors, 1 canine, 2 bicuspids, and 3 molars in each lateral half of the jaw. Each tooth is composed of a *crown*, the exposed part, a constricted part, called the *neck*, and a part within the alveolus, called the *fang* or *root*. In structure a tooth consists of an outer hard substance, the *enamel*, incasing the crown; the *dentine*, within the enamel; and the *pulp*, a soft, vascular tissue filling the pulp-cavity. The dentine of the root is surrounded by the *cement* or *crusta petrosa*. t.-**ache**, any pain in or about the teeth; see *odontalgia*. t.-**ache tree**. See *prickly ash* and *Xanthoxylum fraxineum*. t.-**cough**, reflex cough due to dental irritation. t., **eye**-, the upper canine tooth. t.-**key**, an instrument formerly used for the extraction of teeth. t.-**paste**. See *dentrifrice*. t.-**plugger**, a dental instrument for filling teeth. t.-**pulp**. See *pulp*. t.-**rash**, a skin eruption sometimes occurring during dentition; strophulus. t.-**sac**, in the fetus the connective tissue surrounding the germ of a tooth. t., **stomach**-, the lower canine tooth. t., **wisdom**-, the third molar tooth.
**Tooth's** type of progressive muscular atrophy [Howard Henry *Tooth*, English physician]. See *Charcot-Marie's type*.
**toothed** (*tootht*). Provided with teeth or indentations; dentate. t. **vertebra**, the axis.
**topalgia** (*top-al'-je-ah*) [τόπος, place; άλγος, pain]. Pain in a circumscribed area not referable to the distribution of any nerve.
**topasol** (*to'-pas-ol*). See *anticornutin, anticoroin, antimucorin, antiperonosporin*.
**topesthesia, topæsthesia** (*top-es-the'-ze-ah*) [τόπος, place; αἴσθησις, sensation]. Local tactile sensibility.
**tophaceous** (*to-fa'-shus*) [*tophus*]. Of the nature of tophi; sandy, or gritty.
**tophi** (*to'-fi*). Plural of *tophus*.
**tophus** (*to'-fus*) [τόφος, stone]. 1. The hard, stonelike deposits occurring in gout, especially about the knuckles and the cartilages of the ear, and consisting of sodium urate. 2. The tartar of the teeth. 3. A syphilitic node.
**topic, topical** (*top'-ik, top'-ik-al*) [τόπος, place]. Local.
**topica** (*top'-ik-a*). See *topicum*.
**topicum** (*top-ik'-um*) [τόπος, place: *pl., topica*]. Any remedy for local and external application.
**Topinard's angle** (*top-en-ar'*) [Paul *Topinard*, French anthropologist, 1830–1912]. In craniometry, that included between two lines from the nasal spine to the ophryon and auricular point; also called ophryospinal-facial angle. T.'s **profile line**, a line joining the intersuperciliary point and the most prominent point of the chin.
**topo**- (*top-o-*) [τόπος, place]. A prefix meaning relating to a locality; localized.
**topoalgia** (*top-o-al'-je-ah*) [*topo-*; άλγος, pain]. Localized pain, common in neurasthenia, and often appearing suddenly after emotional disturbances.
**topognosis** (*top-og-no'-sis*) [*topo-*; γνῶσις, knowledge]. Same as *topesthesia*.
**topographical** (*top-o-graf'-ik-al*) [see *topography*]. Pertaining to a locality. t. **anatomy**, the study of the regions occupied by a part, or in which anything occurs.
**topography** (*top-og'-ra-fe*) [*topo-*; γράφειν, to write]. A study of the regions of the body or its parts, as *cerebral topography*.
**topology** (*top-ol'-o-je*) [*topo-*; λόγος, science]. 1. Topographical anatomy. 2. The relation of the presenting part of the fetus to the pelvic canal.
**toponarcosis** (*top-o-nar-ko'-sis*) [*topo-*; νάρκωσις, a benumbing]. Local insensibility or anesthesia.
**toponeurosis** (*top-o-nū-ro'-sis*) [*topo-*; *neurosis*]. A local neurosis.
**toponomy** (*top-on'-o-me*). See *toponymy*.
**toponym** (*top'-on-im*) [τόπος, place; ὄνυμα, a name]. A term relating to position and direction.
**toponymy** (*top-on'-im-e*) [τόπος, place; ὄνυμα, name]. Topical terminology; the system of anatomical terms indicating the direction and position of parts. It is either *intrinsic*, having reference only to the **organism**; or *extrinsic*, based upon the relation of the organism toward the earth's surface.
**topophone** (*top'-o-fōn*) [τόπος, place; φωνή, voice]. An instrument to determine the direction of a source of sound.
**topophobia** (*top-o-fo'-be-ah*) [*topo-*; φόβος, fear]. Morbid dread of certain places.
**topothermesthesiometer** (*top-o-ther-mes-the-ze-om'-et-er*) [*topo-*; θέρμη, heat; αἴσθησις, sensation; μέτρον, measure]. An instrument for estimating local sensitiveness to impressions of heat.
**torcular Herophili** (*tor'-kū-lar her-of'-il-i*) [L., "the wine-press of Herophilus"]. The expanded extremity of the superior longitudinal sinus, placed in a depression on the inner surface of the occipital bone. It receives the blood from the occipital sinus, and from it is derived the straight sinus and the lateral sinus of the side to which it is deflected.
**tori** (*to'-ri*). Genitive and plural of *torus*.
**toric** (*to'-rik*) [*torus*]. Having the properties of a torus. t. **lens**. See *lens*.
**toril** (*tor'-il*). An extract of meat containing its albuminoids prepared with the addition of savory herbs.
**tormen** (*tor'-men*). See *tormina*.
**tormentil, tormentilla** (*tor'-men-til, tor-men-til'-ah*). The root of *Potentilla tormentilla*, a mild tonic and astringent.
**tormentum** (*tor-men'-tum*) [L., a "rack"]. An old name for various obstructive disorders of the intestine. t. **intestinorum**. Synonym of *dysentery*.
**tormina** (*tor'-min-ah*) [plural of *tormen*, a racking pain]. Griping pains in the bowel. t. **alvi**, colic. t. **Celsi**, dysentery. t. **intestinorum**, dysentery. t., **post-partum**, the pains of parturition. t. **ventriculi nervosa**. See *hyperperistalsis*.
**torminal, torminous** (*tor'-min-al, tor'-min-us*). Affected with tormina.
**Tornwaldt's disease**. See *Thornwaldt's disease*.
**torosity** (*to-ros'-it-e*) [*torosus*, brawny; muscular]. Muscular strength.
**torpent** (*tor'-pent*) [*torpor*]. 1. Incapable of the active performance of a function. 2. A medicine or agent that reduces or subdues irritative action.
**torpescence** (*tor-pes'-ens*) [*torpor*]. Numbness; torpidity.
**torpid** (*tor'-pid*) [*torpor*]. Affected with torpor.
**torpidity** (*tor-pid'-it-e*). See *torpor*.
**torpify** (*tor'-pe-fi*). To make numb or torpid; to stupefy.
**torpitude** (*tor'-pe-tūd*). Torpidity; numbness.
**torpor** (*tor'-por*) [L.]. Sluggishness; inactivity. t. **intestinorum**, constipation. t. **retinæ**, dulled perceptive power of the retina.
**torrefaction, torrefication** (*tor-e-fak'-shun, tor-e-fik-a'-shun*) [*torrefacere*, to dry by heat]. Roasting; drying by means of high heat.
**torrefy** (*tor'-e-fi*) [*torrefacere*, to dry by heat]. To parch; to dry by heat.
**Torricellian vacuum** (*to-re-tshel'-e-an*) [Evangelista

**Torricelli,** Italian physicist, 1608–1647]. The vacuum above the mercury-column of a barometer.
**torsiometer** (*tor-se-om'-et-er*) [*torsion;* μέτρον, a measure]. An instrument for measuring ocular torsion.
**torsion** (*tor'-shun*) [*torquere*, to twist]. 1. A twisting; also, the rotation of the eye about the visual axis. 2. The tilting of the vertical meridian of the eye. t. of an artery, twisting of the free end of an. artery to check hemorrhage. t.-balance, an instrument for measuring horizontal forces. t. of teeth, the forcible turning of teeth in their cavities for the purpose of correcting irregularity in position. t. of the umbilical cord, the spontaneous twisting of the umbilical cord. From eight to ten twists are normal; great torsion usually occurs after the death of the fetus.
**torsoclusion** (*tor-sok-lū'-zjun*) [*torsion; occlusion*]. A form of acupressure in which the point of the pin is pushed through a portion of the tissue parallel with the course of the vessel to be secured, then carried over its anterior surface, and at the same time swept around until brought to a right angle with the artery, when the point is thrust into the soft parts beyond.
**torso-occlusion** (*tor-so-ok-lū'-zjun*) [*torsion; occlusion*]. Said of a tooth turned on its axis.
**tort** [*torquere,* to turn]. To tilt the vertical meridian of the eye. **Extort,** to tilt the vertical meridian outward. **Intort,** to tilt the vertical meridian inward. **Intorter,** the muscle tilting the vertical meridian of the eye inward.
**torticollar** (*tor-tik-ol'-ar*) [*torticollis*]. Affected with wry-neck, or torticollis.
**torticollis** (*tor-te-kol'-is*) [*tort; collum,* neck]. Wry-neck, a contraction of one or more of the cervical muscles, usually of one side, resulting in an abnormal position of the head. t., **intermittent.** See t., spasmodic. t., **rheumatic,** stiff-neck, a form due to rheumatism of the sternomastoid or other muscle of the neck. t., **spasmodic,** t. **spastica,** spasmodic contraction of the muscles of the neck of one side, especially the sternomastoid, causing a drawing of the head toward the opposite side.
**tortuous** (*tor'-tū-us*) [*tortus,* twisted]. Twisted, sinuous.
**Torula** (*tor'-oo-lah*) [*torulus,* a small tuft]. 1. A genus of fungi reproducing by budding, many species of which are alcoholic ferments. 2. A chain of spherical bacteria. T. **cerevisiæ.** See *Saccharomyces cerevisiæ.*
**toruliform** (*tor-oo'-li-form*) [*torula; forma,* form]. Resembling an organism of the genus *Torula.*
**toruli tactiles** (*tor'-oo-li tak'-ti-lēs*) [*torulus*]. Tactile elevations.
**toruloid** (*tor'-oo-loid*). Toruliform.
**torulose** (*tor'-oo-lōs*) [*torula*]. Knobbed.
**torulus** (*tor'-oo-lus*) [*torus,* a swelling, protuberance: pl., *toruli*]. An elevation.
**torus** (*tor'-us*) [L.]. 1. A surface having a regular curvature, with two principal meridians of dissimilar curvature at right angles to each other. 2. An elevation or prominence. 3. The tuber cinereum of the brain. t. **frontalis,** a protuberance in the region of the frontal sinuses, at the root of the nose. t. **manus,** the metacarpus. t. **occipitalis,** one sometimes found on the occipital bone about the superior curved line. t. **palatinus,** a protuberance on the surface of the hard palate, marking the point of junction of the intermaxillary and palatomaxillary sutures. t. **spiralis,** the stria acustica. t. **tubarius,** a rounded eminence of mucous membrane in the naso-pharynx near the opening of the Eustachian tube; also called Eustachian cushion. t. **uretericus,** a ridge in the bladder-wall connecting the ureteral orifices. t. **uteri,** a rounded ridge on the posterior wall of the uterus due to the reflection of the peritoneum upon the posterior wall of the vagina.
**touch** (*tutsh*) [Fr., *toucher*]. 1. The tactile sense. The act of judging by the tactile sense; palpation. 2. In obstetrics, digital examination of the female genital organs and adjacent parts through the vagina. t., **abdominal,** application of the hands to the abdomen for the diagnosis of intra-abdominal conditions. t., **after,** the sensation which persists for a short time after contact with an object has ceased. t.-**corpuscle,** a touch-body; a tactile corpuscle. See under *tactile.* t. **double,** combined vaginal and abdominal vaginal and rectal palpation. t.-**me-not.** See *noli me tangere.* t., **rectal,** examination made by the finger in the rectum. t., **royal,** the laying on of the hands by a king, formerly believed to be efficacious in scrofula or kings' evil. t., **vaginal.** See *touch.* t., **vesical,** examination through the bladder, the urethra having been dilated to admit the finger. The latter can only be done in the female. t.-**wood.** See *amadou.*
**tour de maitre** (*toor-d'-mātr*) [Fr., "the master's turn"]. A method of passing a catheter into the bladder in which it is introduced into the urethra with the convexity upward, the shaft lying obliquely across the left thigh of the patient, and as the point enters the bulb, the handle is swept around toward the abdomen, when the beak passes into the membranous urethra, and is carried into the bladder by depressing the shaft between the patient's thighs. A sound may be introduced into the uterus in an analogous manner, by entering the instrument with the convexity upward, and then sweeping the shaft around.
**Tourette's disease** (*too-ret'*) [Georges Gilles de la *Tourette,* French physician]. A convulsive form of tic characterized by motor incoordination with echolalia and coprolalia. T.'s **sign,** inversion of the ratio existing normally between the earthy phosphates and alkaline phosphates of the urine; it is found in paroxysms of hysteria.
**tourniquet** (*toor-nik-et*) [Fr., from *tourner,* to turn]. An instrument for controlling the circulation by means of compression, usually consisting of two metallic plates united by a thumb-screw, and a strap provided with a pad. The strap is fastened about the part, the pad being made to lie over the arrtery to be occluded. The screw is placed diametrically opposite the pad, and the strap tightened by separating the metallic plates of the screw. t., **Dupuytren's,** one for compressing the abdominal aorta, consisting of a semicircle of metal with a pad at one extremity. t., **Esmarch's,** one consisting of a stout, elastic rubber band applied above the proximal turn of an elastic bandage passing around the part to be rendered bloodless. t., **field-,** one consisting of a strap and buckle with a pad to be placed over the artery. t., **horseshoe,** one shaped like a horseshoe, to compress only two points, and thus permit venous return. t., **lip,** one consisting of a U-shaped piece of steel, the arms being provided with plates which are approximated by a central screw. t., **provisional,** one applied loosely, so that it may be tightened at once upon the recurrence of hemorrhage. t., **screw,** that invented by the French surgeon, John Lewis Petit (1674–1750). See *tourniquet* (1). t., **Signorini's.** See *t., horseshoe.* t., **Skey's,** also a modification of the horseshoe or Signorini's. t., **Spanish windlass,** a knotted bandage or handkerchief twisted by a stick and used as a tourniquet. t., **torcular.** See *t., Spanish windlass.*
**tous les mois** (*too-la-mwah'*) [Fr., "every month"]. A variety of arrow-root starch prepared from *Canna edulis;* canna starch.
**tow.** The coarse part of flax or hemp, used as an absorbent.
**towelling** (*tow'-el-ing*). Rubbing with a towel.
**tower-skull.** See *oxycephalia.*
**tox-, toxico-, toxo-** (*toks-, toks-ik-o-, toks-o-*) [τοξικόν, poison]. Prefixes signifying poisonous or caused by a poison.
**toxæmia** (*toks-e'-me-ah*). See *toxemia.*
**toxalbumin** (*toks-al'-bū-min*) [*tox-; albumin*]. A poisonous proteid. Toxalbumins have been obtained from cultures of bacteria and from certain plants. See *toxin* (2).
**toxalbumose** (*toks-al'-bū-mōs*). A toxic albumose.
**toxanemia, toxanæmia** (*toks-an-e'-me-ah*) [*tox-; anemia*]. Anemia produced by poison.
**toxemia, toxæmia** (*toks-e'-me-ah*) [*tox-;* αἷμα, blood]. Blood-poisoning, a condition in which the blood contains poisonous products, either those produced by the bacteria, or due to the growth of micro-organisms.
**toxemic** (*toks-e'-mik*) [see *toxemia*]. Pertaining to, affected with, or caused by toxemia.
**toxenzyme** (*toks'-en-zīm*). A toxic enzyme.
**toxic** (*toks'-ik*) [*toxin*]. 1. Poisonous; produced by a poison. 2. Pertaining to a toxin. t. **unit.** See *unit.*
**toxicant** (*toks'-ik-ant*) [*toxic*]. 1. Poisonous or toxic. 2. A poisonous agent.
**toxicemia** (*toks-is-e'-me-ah*). See *toxemia.*
**toxichemia** (*toks-ik-e'-me-ah*). See *toxemia.*
**toxichemitosis** (*toks-e-ke-mit-o'-sis*) [*toxic;* αἷμα, blood]. Blood-poisoning.

**toxicide** (*toks'-i-id*) [*tox-; cedere*, to kill]. A remedy or principle that destroys toxic agents.
**toxicity** (*toks-is'-it-e*) [*toxic*]. 1. The quality of being toxic. 2. The kind and amount of poison or toxin produced by a microorganism.
**toxicodendrol** (*toks-ik-o-den'-drol*). A toxic nonvolatile oil from the poison ivy, *Rhus toxicodendron*, and from poison sumach, *Rhus venenata*.
**toxicodendron** (*toks-ik-o-den'-dron*) [*toxico*; δένδρον, tree]. See *rhus*.
**toxicoderma** (*toks-ik-o-der'-mah*) [*toxico-*; δέρμα, skin]. Disease of the skin due to poison.
**toxicodermatitis** (*toks-ik-o-der-mat-i'-tis*) [*toxico-; dermatitis*]. Inflammation of the skin due to poison.
**toxicogenic** (*toks-ik-o-jen'-ik*) [*toxico-*; γεννᾶν, to produce]. Producing poisons.
**toxicohemia** (*toks-ik-o-he'-me-ah*). See *toxemia*.
**toxicoid** (*toks'-ik-oid*) [τοξικόν, poison; εἶδος, like]. Resembling a poison.
**toxicologist** (*toks-ik-ol'-o-jist*) [*toxico-*; λόγος, science]. One versed in toxicology.
**toxicology** (*toks-ik-ol'-o-je*) [*toxico-*; λόγος, science]. The science of the nature and effects of poisons, their detection, and the treatment of their effects.
**toxicomania** (*toks-ik-o-ma'-ne-ah*) [*toxico-*; μανία, madness]. 1. Morbid desire to consume poison. 2. Toxiphobia.
**toxicomucin** (*toks-ik-o-mū'-sin*). See *toxomucin*.
**toxicopathy** (*toks-ik-op'-a-the*) [*toxico-*; πάθος, disease]. Disease of toxic origin.
**toxicophobia** (*toks-ik-o-fo'-be-ah*) [*toxico-*; φόβος, fear]. Morbid dread of being poisoned.
**toxicophylaxin** (*toks-ik-o-f-laks'-in*) [*toxico-*; *phylaxin*]. A phylaxin which destroys or counteracts the toxic products of pathogenic bacteria. See *phylaxin*.
**toxicosis** (*toks-ik-o'-sis*) [*toxin*]. A state of poisoning. t., auto-, one with clinical symptoms that are caused by the formation of toxic basic products from morbid matter, such as pathological fluids lodged in certain parts of the system. t., **exogenic**, one with clinical symptoms induced by the action of toxic bases taken into the system with the food, such as the poison of sausages and cheese. t., **noso-**, one with clinical symptoms referable to the presence of basic products which are formed in the system (blood, etc.) in disease and eliminated with the urine. t., **retention**, one with clinical symptoms depending upon the retention of the physiological bases (e. g. uremia).
**toxicosozin** (*toks-ik-o-so'-zin*) [*toxico-*; *sozin*]. Same as *toxosozin;* and see *sozin*.
**toxidermitis** (*toks-e-der-mi'-tis*). See *toxicodermatilis*.
**toxiferous** (*toks-if'-er-us*) [*toxin; ferre*, to bear]. Producing or conveying poison.
**toximucin** (*toks-e-mū'-sin*). See *toxomucin*.
**toxin** [τοξικόν, poison]. 1. Any poisonous nitrogenous compound produced by animal or vegetable cells. 2. Any poisonous substance, protein in nature, produced by animal or vegetable cells, by immunization with which specific antitoxins may be obtained. Syn., *toxalbumin*. t., **animal**, one produced by the metabolic activity of animal cells, as snake-venom. t., **bacterial**, one produced by the metabolic activity of bacteria, as diphtheria toxin. t., **extracellular**, a bacterial toxin elaborated by a microorganism and thrown off into the surrounding medium. The majority of the best known toxins are extracellular. t., **fatigue**, see *t., ponogenic*. t., **intracellular**, a bacterial toxin contained in the bodies of the bacteria themselves. t.s, **ponogenic**, toxins such as are characteristic of nerve-tissue waste. t.-**unit**, consists of two parts, a haptophore complex which unites it with the cell receptor (or lateral chain), and the toxophore complex, which is the poisonous element. t., **vegetable**. 1. Any toxin produced by vegetable cells. 2. Specifically, one produced by higher plants, as ricin (produced by the castor-oil plant), abrin (produced by the jequirity plant).
**toxinemia** (*toks-in-e'-me-ah*). See *toxemia*.
**toxinfection** (*toks-in-fek-shun*) [*toxico-; infection*].
Infection by means of a toxin.
**toxinic** (*toks-in'-ik*) [*toxin*]. Pertaining to a toxin.
**toxinicide** (*toks-in'-is-id*) [*toxin; cædere*, to kill]. Any substance that destroys a toxin.
**toxinosis** (*toks-in-o'-sis*). See *toxicosis*.
**toxipeptone**. See *toxopeptone*.
**toxiphobia** (*toks-e-fo'-be-ah*) [*toxin*; φόβος, fear]. Morbid dread of being poisoned.
**toxiphoric** (*toks-if-or'-ik*). See *toxiferous*. t. side-

**chain**, applied by Ehrlich to atom groups which combine with the toxin of any particular disease-germ.
**toxiresin** (*toks-ir-ez'-in*) [*toxin; resina*, a resin]. A poisonous decomposition product of digitalis, resulting when the latter is treated with dilute acids or heated to 240° C.
**toxis** (*toks'-is*). See *toxicosis*.
**toxitherapy** (*toks-e-ther'-ap-e*). The therapeutic use of antitoxins.
**toxituberculide** (*toks-e-tū-ber'-kū-lid*). A skin lesion to be due to the action of tuberculous toxin.
**toxoalexin** (*toks-o-al-eks'-in*). See *toxophylaxin*.
**toxoid** (*toks'-oid*) [*toxin; είδος*, likeness]. A toxin transformation-product destitute of toxic effect.
**toxolipoid** (*toks-ō-lip'-oid*). An antigen formed by the combination of a lipoid with a toxin.
**toxolysin** (*toks-ol'-is-in*). Same as *antitoxin*.
**toxomucin** (*toks-o-mū'-sin*) [*toxo-; mucus*]. A toxic substance obtained from cultures of tubercle bacilli.
**toxon** (*toks'-on*) [*toxin*]. Ehrlich's name for any class of several substances which appear in fresh toxins; they neutralize antitoxin and are feebly poisonous.
**toxonosis** (*toks-on-o'-sis*) [*toxo-*; νόσος, disease]. An affection resulting from the action of a poison.
**toxopeptone** (*toks-o-pep'-tōn*). A poisonous proteid, resembling peptone in its behavior to heat and reagents, produced in peptone cultures by the comma bacillus.
**toxophile** (*toks'-o-fil*) [*toxo-*; φιλεῖν, to love]. Having an affinity for toxins or poisons.
**toxophore** (*toks'-o-fōr*) [*toxo-*; φέρειν, to bear]. That complex of atoms of a toxin-unit which is the poisonous element of a toxin. t. **group**, that part of the toxin molecule which exerts the poisonous effects.
**toxophorous** (*toks-off'-or-us*). Pertaining to the toxophore.
**toxophylaxin**. See *toxicophylaxin*.
**toxosis** (*toks-o'-sis*). See *toxonosis*.
**toxosozin** (*toks-o-so'-zin*). See under *sozin*.
**Toynbee's corpuscles** [Joseph *Toynbee*, English otologist, 1815–1866]. The corneal corpuscles. T.'s **experiment**, rarefaction of the air contained in the tympanic cavity by swallowing while the mouth and nose are closed. **T.'s law**. See *Gull-Toynbee's law*.
**T.'s ligament**, the tensor ligament; the fibrous sheath of the tendon of the tensor tympani. **T.'s otoscope**, an otoscope by means of which the physician can listen to the sounds in the patient's ear during politzerization.
**T. P.** Abbreviation for *tuberculin precipitation;* Calmette's tuberculin, or purified tuberculin. See *tuberculin*, and *Calmette's reaction*.
**T. R.** Abbreviation for *tuberculin residuum;* new tuberculin. See *tuberculin*.
**tr.** Abbreviation for *tinctura* or *tincture*.
**trabal** (*tra'-bal*) [*trabs*, beam]. Pertaining to the trabs cerebri; callosal.
**trabecula** (*tra-bek'-ū-lah*) [L., "a small beam"]. Any one of the fibrous bands extending from the capsule into the interior of an organ. t. **cerebri**, the corpus callosum. t. **cinerea**, the middle or gray commissure of the cerebrum. t. **cranii**, a structure in the embryo from which the sella turcica is developed. t., **Rathke's**. See *t. cranii*.
**trabeculæ** (*tra-bek'-ū-le*) [pl. of *trabecula*]. t. **carneæ**, the columnæ carneæ of the heart. t. **corporum cavernosorum**, the trabeculæ of the corpora cavernosa. t. **lienis**, the trabeculæ of the spleen.
**trabecular** (*tra-bek'-ū-lar*) [*trabecula*]. Of the nature of a trabecula. t. **duct**, a duct whose cavity or lumen is crossed by ligneous threads or bands. t. **region**, that part of the skull in the embryo where the sella turcica is later developed; trabecula cranii.
**trabecularism** (*tra-bek'-ū-lar-izm*) [*trabecula*, a little beam]. Arrangement like the beams of a framed building; support by a trabecular structure.
**trabeculum** (*tra-bek'-ū-lāt*) [*trabecula*, a little beam]. Having trabeculæ.
**trabs** (*trabs*) [L., "a beam"]. The corpus callosum, called also *trabs cerebri*.
**trace** (*trās*). 1. A mark. 2. A barely recognizable quantity. t., **primitive**. See *primitive streak*.
**tracer** (*tra'-ser*). An instrument used in dissection for isolating nerves and vessels by tearing the connective tissue.
**trachea** (*tra'-ke-ah*) [τραχεῖα, a windpipe]. The windpipe; the cartilaginous and membranous tube extending from the lower part of the larynx to its division into the two bronchi.

**tracheaectasy** (tra-ke-ah-ek'-tas-e) [trachea; ἔκτασις, dilatation]. Dilatation of the trachea.
**tracheal** (tra'-ke-al) [trachea]. Pertaining to or produced in the trachea. **t. catarrh.** See tracheitis. **t. triangle,** the inferior carotid triangle; see triangle. **t. tugging,** the downward tugging movement of the larynx, sometimes observed in aneurysm of the aortic arch.
**trachealgia** (tra-ke-al'-je-ah) [trachea; ἄλγος, pain]. 1. Pain in the trachea. 2. Croup.
**trachealis muscle** (tra-ke-a'-lis) [trachea]. The intrinsic transverse muscle-fibers found in the trachea.
**tracheitis** (tra-ke-i'-tis) [trachea; ιτις, inflammation]. Inflammation of the trachea.
**trachelagra** (tra-kel-ag'-rah) [τράχηλος, neck; ἄγρα, seizure]. Rheumatic or gouty pain in the neck.
**trachelalis** (trak-el-a'-lis) [τράχηλος, neck]. The trachelomastoid muscle. See under muscle.
**trachelectomopexy** (tra-kel-ek-to-mo-peks'-e) [τράχηλος, neck; ἐκτομή, excision; πῆξις, a fixing]. Partial excision with fixation of the neck of the uterus.
**trachelectomy** (tra-kel-ek'-to-me) [trachelo-; ἐκτομή, excision]. Excision of the neck of the uterus.
**trachelematoma** (tra-kel-e-ma-to'-mah) [trachelo: hematoma]. A hematoma of the neck, or in the sternomastoid muscle.
**trachelian** (tra-ke'-le-an) [τράχηλος, neck]. Pertaining to the neck, particularly its dorsal part; cervical.
**trachelismus** (tra-kel-is'-mus) [τράχηλος, neck]. Spasmodic contraction of the muscles of the neck.
**trachelitis** (tra-kel-i'-tis) [trachelo-; ιτις, inflammation]. Inflammation of the neck of the uterus.
**trachelo-** (trak-el-o-) [τράχηλος, neck]. A prefix denoting pertaining to the neck.
**tracheloacromial** (tra-kel-o-a-kro'-me-al) [trachelo-; acromion]. Connecting the shoulder-blade and vertebræ.
**tracheloacromialis** (tra-kel-o-a-kro-me-a'-lis) [trachelo-; acromion]. An inconstant muscle, arising from the occipital bone and inserted into the acromion process.
**trachelobregmatic** (tra-kel-o-breg-mat'-ik) [trachelo-; bregma]. Relating to the neck and the bregma.
**trachelocele** (tra-kel'-o-sēl). Same as tracheocele.
**tracheloclavicular** (tra-kel-o-kla-vik'-u-lar). Relating to the neck and the collar-bone.
**trachelocyllosis** (tra-kel-o-sil-lo'-sis) [trachelo-; κύλλωσις, a bending]. Torticollis.
**trachelocyrtosis** (tra-kel-o-sir-to'-sis) [trachelo-; κυρτός, curved]. Same as trachelokyphosis.
**trachelocystitis** (tra-kel-o-sis-ti'-tis) [trachelo-; cystitis]. Inflammation of the neck of the bladder.
**trachelodynia** (tra-kel-o-din'-e-ah) [trachelo-; ὀδύνη, pain]. Pain in the neck.
**trachelokyphosis** (tra-kel-o-ki-fo'-sis) [trachelo-; κύφωσις, kyphosis]. An anterior curvature of the cervical portion of the spinal column.
**trachelologist** (tra-kel-ol'-o-jist). An expert in diseases of the neck.
**trachelology** (tra-kel-ol'-o-je) [trachelo-; λόγος, science]. The science of the neck and its diseases.
**trachelomastoid** (trak-el-o-mas'-toid) [trachelo-; mastoid]. Pertaining to the neck and the mastoid process. See muscles, table of.
**trachelomyitis** (tra-kel-o-mi-i'-tis) [trachelo-; μῦς, muscle; ιτις, inflammation]. Inflammation of the muscles of the neck.
**trachelo-occipital** (tra-kel-o-ok-sip'-it-al). Relating to the nape of the neck and the occiput.
**trachelopanus** (tra-kel-o-pan'-us) [trachelo-; panus, swelling]. Tumefaction of the cervical lymphatic glands.
**trachelopexia** (tra-kel-o-peks'-e-ah) [trachelo-; πῆξις, a fixing]. Fixation of the neck of the uterus.
**trachelophyma** (tra-kel-o-fi'-mah) [trachelo-; φῦμα, growth]. Swelling of the neck.
**tracheloplasty** (trak'-el-o-plas-te) [trachelo-; πλάσσειν, to mold]. Plastic operation on the neck of the uterus.
**trachelorrhaphy** (trak-el-or'-a-fe) [trachelo-; ῥαφή, suture]. Repair of a laceration of the cervix uteri.
**tracheloschisis** (tra-kel-os'-kis-is) [trachelo-; σχίσις, fissure]. A congenital fissure of the neck.
**trachelosyringorrhaphy** (tra-kel-o-sir-in-gor'-af-e) [trachelo-; σύριγξ, a pipe; ῥαφή, a seam]. An operation for vaginal fistula with stitching of the cervix uteri.
**trachelotomy** (trak-el-ot'-o-me) [trachelo-; τομή, a cutting]. Incision into the cervix uteri.

**tracheo-** (tra-ke-o-) [τραχεία, trachea]. A prefix denoting connection with or relation to the trachea.
**tracheoaerocele** (tra-ke-o-a'-er-o-sēl) [tracheo-; ἀήρ, air; κήλη, tumor]. A diverticulum of the trachea.
**tracheoblenorrhea, tracheoblenorrhœa** (tra-ke-blen-or-e'-ah) [tracheo-; βλέννα, mucus; ῥοία, a flow]. A profuse discharge of mucus from the trachea.
**tracheobronchial** (tra-ke-o-brong'-ke-al) [tracheo-; bronchial]. Pertaining to the trachea and a bronchus or the bronchi.
**tracheobronchitis** (tra-ke-o-brong-ki'-tis) [tracheo-; bronchitis]. Inflammation of the trachea and bronchi.
**tracheobronchoscopy** (tra-ke-o-brong-kos'-co-pe) [tracheo-; βρόγχος, bronchus; σκοπέω, I view]. Inspection of the interior of the trachea and bronchi.
**tracheocele** (tra'-ke-o-sēl) [tracheo-; κήλη, tumor]. 1. Protrusion of the mucous membrane of the trachea. 2. Goiter.
**tracheoesophageal** (tra-ke-o-es-of-aj'-e-al) [tracheo-; esophagus]. Pertaining to the trachea and the esophagus.
**tracheolaryngeal** (tra-ke-o-lar-in'-je-al) [tracheo-; larynx]. Pertaining to the trachea and the larynx.
**tracheolaryngotomy** (tra-ke-o-lar-ing-got'-o-me) [tracheo-; laryngotomy]. Incision into the larynx and trachea; combined tracheotomy and laryngotomy.
**tracheopathia osteoplastica** (tra-ke-o-path'-e-ah os-te-o-plas'-tik-ah). A deposit of cartilage and bone in the mucosa of the trachea.
**tracheophony** (tra-ke-off'-o-ne) [tracheo-; φωνή, voice]. The sound heard over the trachea on auscultation.
**tracheophyma** (tra-ke-o-fi'-mah) [tracheo-; φῦμα, tumor]. A goiter, bronchocele.
**tracheoplasty** (tra'-ke-o-plas-te) [tracheo-; πλάσσειν, to form]. Plastic surgery of the trachea.
**tracheopyosis** (tra-ke-o-pi-o'-sis) [tracheo-; πύον, pus]. Purulent tracheitis.
**tracheorrhagia** (tra-ke-or-a'-je-ah) [tracheo-; ῥηγνύναι, to burst forth]. Hemorrhage from the trachea.
**tracheoschisis** (tra-ke-os'-kis-is) [tracheo-; σχίσειν, to split]. Fissure of the trachea.
**tracheoscopic** (tra-ke-o-skop'-ik) [tracheo-; σκοπεῖν, to view]. Pertaining to tracheoscopy.
**tracheoscopy** (tra-ke-os'-ko-pe) [tracheo-; σκοπεῖν, to inspect]. Inspection of the interior of the trachea by means of a laryngoscopic mirror and reflected light.
**tracheostenosis** (tra-ke-o-sten-o'-sis) [tracheo-; stenosis]. Abnormal constriction or narrowing of the trachea.
**tracheotome** (tra'-ke-o-tōm) [tracheo-; τομή, a cutting]. A cutting instrument used in tracheotomy.
**tracheotomist** (tra-ke-ot'-om-ist). One skilled in tracheotomy.
**tracheotomize** (tra-ke-ot'-om-īz). To perform tracheotomy upon.
**tracheotomy** (tra-ke-ot'-o-me) [see tracheotome]. The operation of cutting into the trachea through the cricothyroid membrane, or through the cricoid cartilage and the upper part of the trachea. **t.** inferior, one performed below the isthmus of the thyroid gland. **t., superior,** one performed above the isthmus of the thyroid gland. **t.-tube,** a metal tube placed in the opening made in tracheotomy, and through which breathing is carried on.
**trachielcosis** (tra-ke-el-ko'-sis) [tracheo-; ἕλκος, an ulcer]. Ulceration of the trachea.
**trachielcus** (tra-ke-el'-kus). An ulcer of the trachea.
**trachitis** (tra-ki'-tis). See tracheitis.
**trachoma** (tra-ko'-mah) [τραχύς, rough]. A contagious disease of the eyelids characterized by small, sago-like elevations on the conjunctiva, and later by cicatricial contraction and deformity of the lids. The friction of the elevations (trachoma-granulations) against the cornea often produces ulcer or panus. Syn., granular conjunctivitis; granular lids. **t., Arlt's,** the granular form. **t., brawny,** a late stage of mixed trachoma, in which the surface of the conjunctiva is rather smooth although lymphoid infiltration persists. **t. deformans,** a name given to a form of vulvitis at the stage when it results in diffuse scar-tissue. **t., diffuse,** a high degree of mixed trachoma in which large growths cover the tarsal conjunctiva. **t., follicular, t., mixed,** the usual form of trachoma; see **t., Arlt's. t., laryngis.** See **t. of vocal cords. t., papillary,** in which the granulations are red and papillary. **t., Tuerck's.** See Tuerck's trachoma. **t.** of

vocal bands, nodular swellings on the vocal cords; singers' nodes.
trachomatous (tra-ko'-mat-us) [trachoma]. Affected with or pertaining to trachoma.
trachychromatic (trak-e-kro-mat'-ik) [τραχύs, rough; χρῶμα, color]. Said of a nucleus with a deeply staining chromatin.
trachyphonia (trak-if-o'-ne-ah) [τραχύs, rough; φωνή, voice]. Roughness or hoarseness of the voice.
tract [tractus]. 1. A distinct, more or less defined region having considerable length. 2. Any one of the columns of white matter of the spinal cord. 3. A track or course. t., acusticocerebellar, a tract of fibers arising in the cerebellar nuclei and terminating in the nucleus of Deiters. t., alimentary, the alimentary canal, extending from the mouth to the anus. t., anterior ascending cerebellospinal. Same as Gowers's tract. t., cerebellar, an ascending tract of fibers at the periphery of the posterior portion of the lateral column of the spinal cord. t., crossed or lateral pyramidal, that part of the pyramidal tract which decussates in the medulla. t., descending anterolateral, a few long fibers scattered in the anterior and lateral ground-bundles of the spinal cord. t., digestive, the alimentary tract. t., direct or anterior pyramidal, that part of the pyramidal tract which does not decussate in the medulla. t., direct cerebellar. See t., cerebellar. t., frontopontal, a tract of nerve-fibers from the frontal lobe of the brain to the nucleus pontis. t., genitourinary, the genitourinary organs in continuity. t., habenular, a tract of fibers passing from the habenula to the mesal side of the red nucleus. t., intermediate, a tract of nerve-fibers from the corpus striatum to the motor cerebral nuclei, the nucleus pontis, and the opposite cerebellar hemisphere. t., intermediolateral, a tract of nerve-fibers in the lateral column of the spinal cord, midway between the anterior and posterior gray horns. t., motor, the path for motor impulses from the brain to a muscle. t., olfactory. See olfactory tract. t., optic. See optic tract. t., oval, a part of the descending posteromedial tract of the spinal cord. t., pontospinal, medial and lateral, tracts of nerve-fibers arising in the nuclei of the pontile reticular substance and terminating at various levels in the spinal cord. t., prepyramidal, the ventral pyramids of the cord. t., pupillodilator, the bulbar and spinal portions of the anterior longitudinal bundle. t., pyramidal, the continuation in the spinal cord of the ventral pyramids of the oblongata. t., respiratory, the respiratory organs in continuity. t., rubrospinal, a tract of nerve-fibers from the red nucleus to the gray matter of the spinal cord. t., semilunar, a band of fibers in the outer portion of the cerebellum. t., sensory, any tract of fibers conducting sensation to the brain. t., septomarginal, a narrow strip of fibers in the posterior column close to the septum as high as the eleventh dorsal segment. t., spinothalamic, that part of the fibers in the anterior ascending cerebellospinal tract which goes to the lateral nucleus of the thalamus. t., spinovestibular, a tract of fibers in the posterior portion of the direct cerebellar tract going to the vestibular nucleus. t., temporopontal, a tract of nerve-fibers from the temporal lobe to the substantia nigra and nucleus pontis.
tractellum (trak-tel'-lum) [dim. of tractus, a tract; pl., tractella]. That flagellum of a protozoan which precedes in locomotion.
traction (trak'-shun) [trahere, to draw]. The act of drawing or pulling. t.-aneurysm, an aneurysm due to traction on the aorta by an incompletely atrophied ductus arteriosus. t., axis, traction in the axis or direction of a channel, as of the pelvis, through which a body is to be drawn. t., axis-, forceps, an obstetric forceps for performing axis-traction in the delivery of fetus. t.-diverticulum, a circumscribed sacculation of the esophagus from the traction of adhesions. t., elastic, traction by an elastic force.
tractograph (trak'-to-graf) [trahere, to draw: γράφειν, to write]. An apparatus used to make traction tests.
tractor (trak'-tor) [tractio]. 1. An instrument for making traction. 2. See Perkinism.
tractoration (trak-tor-a'-shun) [tractor]. Treatment by metallic tractors. See Perkinism.
tractus (trak'-tus) [L.]. See tract. t. centralis, a. central tract. t. iliotibialis, the iliotibial band. t. olfactorius, the olfactory tract. t. opticus, the optic tract. t. solitarius, the respiratory bundle. t. spinalis nervi trigemini, the spinal tract, or ascending root, of the trigeminal nerve.

tragacanth (trag'-a-kanth) [τραγάκανθα, "goatthorn"]. A gummy exudation from various species of Astragalus, of the order Leguminosæ, constituting the tragacantha of the U. S. P. and B. P. It resembles gum-arabic, and is used as a demulcent, and is added to water to suspend insoluble powders and for making troches. t., glycerin of (glycerinum tragacanthæ, B. P.). t., mucilage of (mucilago tragacanthæ, U. S. P.). Dose 4 dr. (16 Cc.). t., powder of, compound (pulvis tragacanthæ compositus, B. P.). Dose 30 gr.– 1 dr. (2–4 Gm.).
tragal (tra'-gal) [tragus]. Pertaining to the tragus.
tragalism (trag'-al-izm) [τράγοs, goat]. Salaciousness; sensuality.
tragi (tra'-gi) [pl. of tragus]. Hairs of the external auditory meatus.
tragicus (traj'-ik-us). See under muscle.
tragophonia, tragophony (trag-off-'o'-ne-ah, trag-off'-on-e) [τράγοs, goat; φωνή, voice]. Synonym of egophony.
tragopodia (trag-op-o'-de-ah) [τράγοs, goat; πούs, foot]. Knock-knee.
tragus (tra'-gus) [τράγοs, goat]. 1. The small prominence of cartilage projecting over the meatus of the external ear. 2. One of the hairs at the external auditory meatus.
trailer. See hand, trailing.
training (trān-ing). Systematic exercise for physical development or for some special attainment. t.-school, an institution where persons are instructed in nursing.
trait (trāt, or trā) [Fr., trait, a line]. Any natural characteristic or feature, that is peculiar to an individual.
trajector (tra-jek'-tor) [L., "a piercer"]. An instrument used to determine the approximate location of a bullet in the cranium or elsewhere.
trance (trans) [transitus, a passing or passage]. 1. A form of catalepsy, characterized by a prolonged condition of abnormal sleep, in which the vital functions are reduced very low, and from which the patients ordinarily cannot be aroused. The breathing is almost imperceptible, and sensation abolished. The onset and awakening are both very sudden. 2. The state of syncope much protracted. t.-doctor, a mesmerist. t., ecstatic, catalepsy. t., hysterical, the trance-like condition sometimes met with in hysteria.
trans- [trans, across]. A prefix denoting through or across.
transanimation (trans-an-im-a'-shun) [trans-; anima, life]. The performing of artificial respiration on a stillborn infant.
transaudient (trans-aw'-de-ent) [trans-; audire, to hear]. Allowing the transmission of sound.
transcendental (tran-sen-den'-tal) [trans-; scandere, to climb]. Beyond the bounds of experience. t. anatomy, philosophical anatomy.
transcortical (trans-kor'-tik-al) [trans-; cortex]. Across or through the cortex. From one part of the cortex to another.
transect (tran-sekt') [trans-; secāre, to cut]. To make a transection.
transection (trans-ek'-shun) [trans-; section]. A section made across the long axis of a part.
transfer (trans'-fer) [trans-; ferre, to carry]. The change of anesthesia or hyperesthesia or other symptom from one part of the body to another, a phenomenon present in some cases of hysteria.
transference (trans-fer'-ens). 1. See transfer. 2. See telepathy.
transfix (trans-fiks') [trans-; figere, to fix]. To pierce through and through.
transfixion (trans-fik'-shun) [transfix]. 1. The act of piercing through and through. 2. A method of amputation in which the knife is passed directly through the soft parts, the cutting being done from within outward.
transforation (trans-for-a'-shun) [trans-; forare, to pierce]. The act of perforating, as transforation of the fetal skull.
transforator (trans'-for-a-tor) [trans-; forare, to pierce]. An instrument for transforation of the fetal head.
transformation (trans-for-ma'-shun) [trans-; formare, to form]. 1. A change of form or constitution. 2. A change of one form of connective tissue into another. 3. Degeneration.
transformism (trans-for'-mizm) [trans-; formare, to form, shape]. In biology, the doctrine of descent with modification; the transmutation of species.

**transfrontal** (*trans-fron'-tal*) [*trans-*; *frons*, forehead]. Crossing the frontal lobe of the brain.

**transfuse** (*trans-fūs'*). To perform transfusion.

**transfuser, transfusionist** (*trans-fū'-sur, trans-fū'-zjun-ist*). One skilled in the transfusion of blood.

**transfusion** (*trans-fū'-zjun*) [*trans-*; *fundere*, to pour]. 1. A transfer of blood into the veins. 2. The introduction into a vessel of the body of blood, saline solution, or other liquid. 3. The pouring of liquid from one vessel to another. **t., arterial**, transfusion of blood into an artery. **t., direct, t., immediate,** the transfusion of blood from one person to another without exposure of the blood to the air. **t., indirect, t., mediate,** the introduction of blood that was first drawn into a vessel. **t. peritoneal,** transfusion into the peritoneal cavity. **t., reciprocal,** the exchange of equal volumes of blood between a patient suffering from a febrile disease and one who is convalescent from that disease. **t., venous,** transfusion into a vein.

**transic** (*trans'-ik*) [*trance*]. Relating to a trance.

**transiliac** (*trans-il'-e-ak*) [*trans-*; *ilium,* ilium]. Passing across from one ilium to the other, as the transiliac diameter or axis.

**transilient** (*trans-il'-e-ent*) [*trans-*; *salire*, to leap]. Extending across. **t. fiber,** a nerve-fiber passing from one convolution of the brain to another not immediately adjacent.

**transillumination** (*trans-il-ū-min-a'-shun*) [*trans-*; *illumination*]. Illumination of the walls of a cavity by a light passed through them, or, the throwing of an intense light through the substance of a hollow organ as a means of diagnosis.

**transinsular** (*trans-in'-su-lar*) [*trans-*; *insula*; island]. Traversing the insula of the brain or the island of Reil.

**transischiac** (*trans-is'-ke-ak*) [*trans-*; *ischium*]. Extending transversely from one ischium to the other.

**transition** (*trans-ish'-un*) [*transire*, to go over]. Change; passage from one state to another. **t. resistance,** the resistance introduced into an electric current by the accumulation of decomposition-products upon the electrodes.

**translation** (*trans-la'-shun*) [L., *translatus*]. A change of location.

**translucent** (*trans-lū'-sent*) [*trans-*; *lucere*, to shine]. Permitting a partial transmission of light; somewhat transparent.

**translucid** (*trans-lū'-sid*). Semitransparent.

**translumination** (*trans-lū-min-a'-shun*). Synonym of *transillumination*.

**transmigration** (*trans-mi-gra'-shun*) [*trans-*; *migrare*, to wander]. 1. A wandering across or through; as *transmigration of the ovum, transmigration* of the white corpuscles. 2. Diapedesis. **t., external,** the passage of an ovum from one ovary to the opposite oviduct without traversing the uterus. **t., internal,** the passage of the ovum through its proper oviduct into the uterus and across to the opposite oviduct.

**transmissibility** (*trans-mis-ib-il'-it-e*) [*trans-*; *mittere*, to send]. The capability of being transmitted or communicated from one person to another.

**transmission** (*trans-mish'-un*) [*trans-*; *mittere*, to send]. 1. The communication or transfer of anything, especially disease, from one person or place to another. 2. See *heredity*. **t., duplex,** the property of nerves of transmitting impulses in two directions. **t., placental,** the conveyance of certain drugs and bacteria and their products through the fetoplacental circulation.

**transmitting power.** The faculty which an individual organism has of transmitting its individual peculiarities to its progeny.

**transmutation** (*trans-mū-ta'-shun*) [*transmutare*, to change]. The process of changing; the conversion of one substance or one form into another substance or form.

**transocular** (*trans-ok'-ū-lar*) [*trans-*; *oculus*, eye]. Extending across the eye.

**transonance** (*trans'-o-nans*) [*trans-*; *sonare*, to sound]. Transmitted resonance; the transmission of sounds through an organ, as of the cardiac sounds through the lungs and chest-wall.

**transpalatine** (*trans-pal'-at-in*) [*trans-*; *palatine*]. 1. Transverse, as a palatine bone, which extends on either side from the median line. 2. A bone of certain sauropsidan vertebrates.

**transpalmar** (*trans-pal'-mar*) [*trans-*; *palma*, palm]. Situated across the palm.

**transpalmaris** (*trans-pal-ma'-ris*) [*trans-*; *palma*, palm]. The palmaris brevis muscle. See *muscles*.

**transparent** (*trans-pa'-rent*) [*trans-*; *parere*, to appear]. Having the property of permitting the passage of light-rays without material obstruction, so that objects beyond the transparent body can be seen.

**transperinæus** (*trans-per-in-e'-us*) [*trans-*; *perinæum*, perineum]. The transversus perinæi muscle.

**transperitoneal** (*trans-per-it-on-e'-al*). Across the peritoneal cavity; through the peritoneum.

**transpinalis** (*trans-pi-na'-lis*) [*trans-*; *spinalis*, spinal]. Any intertransverse muscle of the vertebral column.

**transpiration** (*trans-pir-a'-shun*) [*trans-*; *spirare*, to breathe]. 1. The act of exhaling fluid or gas through the skin. 2. The material exhaled. **t., pulmonary,** the exhalation of watery vapor from the lungs.

**transplantar** (*trans-plan'-tar*) [*trans-*; *planta*, sole]. Lying across the sole.

**transplantation** (*trans-plan-ta'-shun*) [*trans-*; *plantare*, to plant]. The operation of transplanting or of applying to a part of the body tissues taken from another body or from another part of the same body. See *graft*. **t. of cornea,** see *keratoplasty*. **t. of teeth,** the insertion of a natural tooth from a foreign source in a natural alveolus.

**transpleural** (*trans-ploo'-ral*) [*trans-*; πλευρά, side]. Crossing the pleural sac.

**transposition** (*trans-po-zish'-un*) [*trans-*; *position*]. A change of position. **t. of the viscera,** a change in the position of the viscera whereby they are placed on the side opposite to that normally occupied.

**transprocess** (*trans-pros'-es*) [*trans-*; *processus*, process]. A transverse process.

**transsection** (*trans-sek'-shun*). A cross-section.

**transsegmental** (*trans-seg-ment'-al*). Across a segment of a limb or organ.

**transtemporal** (*trans-tem'-po-ral*) [*trans-*; *tempus*, temple]. Crossing the temporal lobe.

**transthalamic** (*trans-thal-am'-ik*). Across the thalamus.

**transthoracic** (*trans-tho-ras'-ik*) [*trans-*; *thorax*]. Extending across the thorax.

**transthoracotomy** (*trans-tho-rak-ot'-o-me*) [*trans-*; *thorax*; ῥοή, a cutting]. The operation of cutting across the thorax.

**transubstantiation** (*trans-sub-stan-she-a'-shun*) [*trans-*; *substantia*, substance]. The replacement of one tissue by another.

**transudate** (*trans'-ū-dāt*) [*trans-*; *sudare, to perspire*]. A liquid or other substance produced by transudation.

**transudation** (*trans-ū-da'-shun*) [*transudate*]. 1. The passing of fluid through a membrane, especially of blood-serum through the vessel-walls. 2. Transudate.

**transudatory** (*trans-ū'-da-to-re*) [*trans-*; *sudare*, to perspire]. Passing by or pertaining to transudation.

**transvaginal** (*trans-vaj'-in-al*). Across or through the vagina.

**transversal** (*trans-ver'-sal*) [*trans-*; *vertere*, to turn]. Transverse; running across.

**transversalis** (*trans-ver-sa'-lis*) [*trans-*; *vertere*, to turn]. Transverse; an artery (*transversalis colli*) or a muscle (*transversalis abdominis*) running transversely. See under *artery* and under *muscle*. **t. fascia,** the fascia on the inner surface of the transversalis abdominis between the latter and the peritoneum.

**transverse** (*trans-vers'*) [*trans-*; *vertere*, to turn]. Crosswise; at right angles to the longitudinal axis of the body. **t. presentation,** a presentation of the fetus at right angles to the longitudinal axis of the uterus.

**transversectomy** (*trans-ver-sek'-to-me*) [*transverse*; ἐκτομή, excision]. Removal of the transverse process of a vertebra.

**transversus** (*trans-ver'-sus*) [L.]. Transverse, as *transversus* muscle. See under *muscle*.

**transvestism** (*trans-vest'-izm*) [*trans-*; *vestis*, a garment]. Cross dressing; a man wearing woman's clothes, or *vice versa*.

**trap** (*trap*). A device intended to prevent the escape of foul vapors from sewers and waste-pipes into a house. It generally consists of one or more S-shaped pipes, filled with water; some are provided also with valves. **t.-door flap,** a semicircular or horseshoe flap made in trephining the skull.

**trapezate** (*trap'-es-āt*). Same as *trapeziform*.

**trapezial** (*tra-pe'-ze-al*). Pertaining to the trapezium, or to the trapezius.

**trapeziform** (*tra-pez'-if-orm*) [τράπεζα, table; *forma*, form]. Having the shape of a trapezium.

**trapeziometacarpal** (*tra-pe-ze-o-met-ah-kar'-pal*) [*trapezium; metacarpal*]. Pertaining to the trapezium and the metacarpus.
**trapezium** (*tra-pe'-ze-um*) [τραπεζά, a table]. 1. The multangulum majus, the first bone of the second row of carpal bones. 2. A tract of transverse fibers situated in the lower part of the pons, inclosing the superior olivary nucleus, and connected with the accessory auditory nucleus.
**trapezius** (*tra-pe'-ze-us*). See under *muscle*.
**trapezoid** (*trap'-ez-oid*) [τράπεζά, a table; εἶδος, like]. A geometrical four-sided figure having two parallel and two diverging sides. t. bone, or simply *trapezoid*, the multangulum minus, the second bone of the second row of the carpus. t. ligament. See *ligament*, *trapezoid*. t. line, a rough line on the clavicle to which the t. ligament is attached.
**Trapp's formula** [Hermann *Trapp*, German physician]. The product obtained by doubling the last two figures of the specific gravity of the urine roughly indicates the number of grams of solids per 1000 Cc. of urine.
**Traube's corpuscles** (*trow'-beh*) [Ludwig *Traube*, German physician, 1818–1876]. Normal red bloodcorpuscles appearing as pale yellowish rings. Syn., *phantom corpuscles*. T.'s curves, large rhythmic undulations seen in a sphygmographic tracing soon after respiration has ceased; they are attributed to stimulation of the vasomotor center in the oblongata. T.'s dyspnea, dyspnea with slow respiratory movements, marked expansion of the thorax during inspiration, and collapse during expiration; it is noted in diabetes mellitus. T.'s phenomenon, a double sound, systolic and diastolic, heard over peripheral arteries, especially the femoral, in aortic insufficiency, occasionally also in mitral stenosis, lead-poisoning, etc. T.'s plugs. See *Dittrich's plugs*. T.'s semilunar space, the space in which the tympanitic sound of the stomach can be heard within the thorax under normal conditions; it is bounded by the liver, the lower border of the left lung, the spleen, and the arch of the free ribs.
**Traube-Hering's curves.** (*trow'-beh-ha'-ring*) [*Traube; Ewald Hering*, German physician, 1834– ]. See *Traube's curves*.
**traulism, traulismus** (*traw'-lizm, traw-liz'-mus*) [τραυλισμός, a lisping]. A lisping; a stammer; drawling, or imperfect utterance.
**trauma** (*traw'-mah*) [τραῦμα, a wound: *pl.*, *traumata*]. A wound; an injury.
**traumatic** (*traw-mat'-ik*) [*trauma*]. Pertaining to or caused by a wound or injury. t. degeneration, the degeneration of the ends of nerves at the point of section, extending to the nearest node of Ranvier, after which fatty degeneration begins. t. fever, fever following within from eight to thirty-six hours of an operation or injury. It is due to absorption of poisonous material from the seat of injury. t. hysteria. See *fright-neuroses*. t. infective diseases, a class of diseases characterized by definite symptoms following wounds or abrasions. t. suggestion. See *autosuggestion*.
**traumaticin** (*traw-mat'-is-in*) [*trauma*]. A solution of guttapercha in chloroform used for closing superficial wounds.
**traumatism** (*traw'-mat-ism*) [*trauma*]. 1. The condition produced by trauma. 2. Improperly, trauma.
**traumatol** (*traw'-mat-ol*) [*trauma*]. Iodocresol, C₇H₇IO, obtained by the action of iodine on cresol. It is an odorless, reddish-violet precipitate containing 54% of iodine, soluble in chloroform and carbon disulphide, insoluble in water, acid, and alcohol. It is a surgical antiseptic, used pure as a dusting-powder and in 5–10% pastes and ointments.
**traumatology** (*traw-mat-ol'-o-je*) [*trauma*; λόγος, science]. The science or description of wounds.
**traumatonesis** (*traw-mat-o-ne'-sis*) [τραῦμα, a wound; νῆσις, a suture]. Suture of a wound.
**traumatopnea** (*traw-mat-op-ne'-ah*) [*trauma*; πνοή, breath]. The passage of respiratory air through a wound in the chest-wall.
**traumatopyra** (*traw-mat-o-pi'-rah*) [τραῦμα, wound; πῦρ, fever]. Synonym of *traumatic fever*.
**traumatosepsis** (*traw-mat-o-sep'-sis*) [τραῦμα, a wound; σῆψις, putrefaction]. Synonym of *hospital gangrene*.
**traumatosis** (*traw-mat-o'-sis*) [*trauma*]. Traumatism.
**travail** (*trav'-il*). Labor in childbed.
**travel-sickness.** Car-sickness.

**tray** (*tra*). A flat, shallow vessel of glass, hard rubber, or metal, for holding instruments during a surgical operation.
**treacle** (*tre'-kl*). The uncrystallized residue remaining after the refining of cane-sugar. See *theriaca*, and *molasses*.
**treat** (*trēt*). To manage disease by the application of remedies.
**treatment** (*trēt'-ment*) [*tractare*, to treat]. The means employed in effecting the cure of disease; the management of disease or of diseased patients. t., active, that which is vigorously applied to the disease. t., causal, that which is directed to the removal of the cause of a disease. t., conservative, that which abstains from any interference until absolutely indicated; in surgical cases it aims at preservation rather than mutilation. t., empirical, see *empiric*. t., expectant. See *expectant*. t., mixed, treatment of syphilis with mercury and potassium iodide. t., palliative, that which is directed towards relief of symptoms rather than to cure of the disease. t., preventive, t., prophylactic. See *preventive*, *prophylactic*. t., rational. See *rational*. t., specific. See *specific*. t., supporting, that which is directed to keeping up the strength of the patient. t., symptomatic, See *symptomatic*. t., terrain. See *terrain-cure*.
**trefoil tendon.** The central tendon of the diaphragm.
**trefusia** (*tre-fū'-ze-ah*). A red-brown, soluble powder, obtained by drying defibrinated blood. It is used in chlorosis.
**trehala** (*tre-hah'-lah*) [Turkish]. Turkish manna; a variety of manna derived from the cocoons of *Larinus maculatus*, an insect that feeds upon an Asiatic thistle, *Echinops persica*.
**trehalose** (*tre-hal'-ōs*), C₁₂H₂₂O₁₁. A carbohydrate resembling sugar, derived from ergot and from trehala manna.
**Treitz's fossa** [Wenzel *Treitz*, Austrian physician, 1819–1872]. The inferior duodenal fossa; a fossa in the peritoneum on the left side of the ascending duodenum. T.'s hernia. Retroperitoneal hernia; duodenojejunal hernia. T.'s ligament, a fold of the peritoneum extending from the duodenojejunal junction to the left crus of the diaphragm. T.'s muscle, the suspensory muscle of the duodenum; a thin, triangular muscle that arises from the left crus of the diaphragm and the connective tissue surrounding the celiac axis, and is inserted into the duodenojejunal flexure.
**trema** (*tre'-mah*) [τρῆμα, a hole]. 1. A synonym of *foramen*. 2. The vulva. 3. A genus of the *Ulmaceæ*. t. orientalis, Indian nettle-tree. The bark, leaves, and root are used as a nervine for epilepsy.
**trematode** (*trem'-at-ōd*) [τρῆμα, hole; εἶδος, like]. A member of the *Trematoda*, a class of worms, some of which are parasitic in man and the lower animals.
**tremble** (*trem'-bl*) [*tremere*, to tremble]. To be affected with slight, quick, and continued vibratory movements; to quiver.
**trembles** (*trem'-blz*) [*tremere*, to tremble]. Synonym of *milk-sickness*. Also used as a synonym of *paralysis agitans*.
**trembling** (*trem'-bling*) [*tremere*, to tremble]. A tremor; quivering; affected with involuntary muscular agitation. t. chair, a chair used in the treatment of paralysis agitans for giving vibratory motion to the body of the patient seated in it. t. palsy. Synonym of *paralysis agitans*.
**tremellose** (*trem'-el-os*) [*tremere*, to tremble]. In biology, jelly-like.
**tremogram** (*trem'-o-gram*) [*tremere*, to tremble; γράμμα, a writing]. The tracing of tremor made by means of the tremograph.
**tremograph** (*trem'-o-graf*) [*tremor*; γραφή, a writing]. A device for recording tremor.
**tremolo** (*trem'-o-lo*). An apparatus for performing massage; a vibrator.
**tremor** (*trem'-or*) [*tremere*, to shake]. A trembling of the voluntary muscles. t. arsenical, a tremor the result of arsenical intoxication. t. artuum, paralysis agitans. t. capitis, tremor affecting the muscles of the neck and head. t. coactus. Synonym of *t., forced*. t., forced, a form of tremor which resembles that of paralysis agitans; it is, however, likely to be remittent, and may be diminished or arrested by voluntary effort. t., convulsive. See *paramyoclonus*. t. cordis, a sudden rapid fluttering of the heart, and the ordinary full pulse of health suddenly drops to a mere tremulous thread. A symptom often

met in neurotic persons. **t., epileptoid,** intermittent clonus with tremor. **t., fibrillary,** tremor caused by consecutive contractions of separate muscle-fibrillæ. **t., forced,** the convulsive movements persisting during repose after voluntary motion, due to an intermittent and rhythmic irritation of the nervous centers. **t., hysterical,** the tremor observed in hysteria, and due to the uncertainty of nervous impulse. **t., intention,** one appearing on voluntary movement. **t., intermittent,** the tremor commonly observed in hemiplegics on any attempt at voluntary motion. **t., mercurial,** a peculiar form of tremor observed among smelters and others exposed to the fumes of mercury. It is sudden or gradual in onset, and is usually unaccompanied by salivation. The arms are first involved, and then the entire muscular system. If allowed to go on, paralysis, mania and idiocy may result. **t. metallicus.** See *t., mercurial.* **t., muscular,** slight, oscillating, rhythmical muscular contractions. **t. potatorum,** delirium tremens. **t., purring.** Synonym of *purring thrill.* **t., Rendu's type of,** a hysterical tremor provoked or increased by volitional movements. **t. saturninus,** the tremor of lead-poisoning. **t. tendinum.** Synonym of *subsultus tendinum.* **t., vibratile.** Synonym of *fremitus.* **t., volitional,** a trembling of the entire body during voluntary effort as observed in multiple sclerosis. See *t., intentional.* **t.s from zinc-poisoning.** Synonym of *brass-founder's ague* (*q. v.* under *ague*).

**tremorless** (*trem'-or-less*) [*tremere,* to tremble]. Free from tremor.

**tremulation** (*trem-u-la'-shun*) [*tremulare,* to tremble]. A tremulous condition.

**tremulor** (*trem'-u-lor*). An appliance for the administration of vibratory massage.

**tremulous** (*trem'-u-lus*) [*tremor*]. Trembling, quivering, as *tremulous iris.*

**Trendelenburg position** [Friedrich *Trendelenburg,* German surgeon, 1844– ]. One in which the patient lies on the back on a plane inclined at about 45°, the pelvis higher than the head.

**trepan** (*tre-pan'*) [τρυπᾶν, bore]. An old form of the word trephine.

**trepanatio** (*trep-an-a'-she-o*). See *trephining.* **t. corneæ,** an operation for conical cornea, by means of the conical trephine.

**trepanation** (*trep-an-a'-shun*) [*trepan*]. The operation of trephining.

**trepanize** (*trep'-an-iz*) [*trepan*]. To trepan.

**trepanning** (*tre-pan'-ing*). Boring; using the trephine. **t.-elevator,** a lever used to raise the piece of bone detached by the trepan.

**Trepanosoma** (*trep-an-o-so'-mah*). See *Trypanosoma.*

**trepanosomiasis** (*trep-an-o-so-mi'-as-is*). See *trypanosomiasis.*

**trephination** (*tref-in-a'-shun*). See *trephining.*

**trephine** (*tre-fin*) [Fr., *trephine*]. 1. An instrument for cutting out a circular piece of bone, usually from the skull. 2. To operate with the trephine. **t. brace,** a trephine with an ordinary carpenters' brace. **t., conical,** a trephine with a truncated cone-shaped crown and provided with oblique ridges on its outer surface to stop its progress as soon as the bone is penetrated. **t., corneal,** a small cutting trephine used to remove a circular section from the summit of a conical cornea. It is manipulated with the thumb and finger. **t. of Gault,** a form of conical trephine. **t., nasal,** an instrument made of a steel shaft ending in a small, fenestrated tube, having a knife or saw edge. **t., tympanic,** an instrument made of a small steel shaft ending in a small, polished tube, 2 mm. in diameter, with a cutting edge.

**trephining** (*tre-fi'-ning*) [*trephine*]. The operation of cutting bone with a trephine.

**trepidatio** (*trep-id-a'-she-o*) [*trepidare,* to be agitated]. The state of agitation. **t. cordis,** palpitation of the heart.

**trepidation** (*trep-id-a'-shun*) [*trepidare,* to tremble]. 1. Trembling. 2. A peculiar oscillatory movement at times seen in the muscles after hemiplegia.

**Treponema pallidum** (*tre-po-ne'-mah pal'-id-um*). The pathogenic parasite of syphilis. Syn., *Spirochæta pallida.* **T., pertenue,** the supposed pathogenic parasite of yaws. Syn., *Spirochæta pertenuis.*

**treppe** (*trep'-eh*). See *summation.*

**Tresilian's sign, of mumps** (*tres-il'-e-an*) [Frederick James *Tresilian,* British physician]. The opening of Stenson's duct on the inner surface of the cheek opposite the second upper molar becomes a bright red papilla.

**tresis** (*tre'-sis*) [τερπαίνειν, to pierce]. Wound, perforation. **t. causis,** see *burn.* **t. punctura,** a puncture. **t. vulnus,** a wound.

**Tretop's test for albumin in urine.** Four or 5 Cc. of fresh urine are heated in a test-tube nearly to boiling-point, and a few drops of 40% formalin added after it is removed from the flame. Any albumin in the urine is coagulated like the white of an egg, and accumulates on the surface and walls of the tube.

**Treves' bloodless fold** [Sir Frederick *Treves,* English surgeon, 1853– ]. The ileoappendicular fold. A quadrilateral fold of the peritoneum attached by its upper border to the ileum, opposite the mesenteric attachment, and by its lower border to the mesoappendix or to the appendix itself. The outer or right border is attached to the inner aspect of the cecum as far down as the appendix, the left or inner concave margin being free.

**tri-** [τρεῖς, or *tres,* three]. A prefix denoting three.

**triacetate** (*tri-as'-e-tāt*). An acetate containing three molecules of the acetic-acid radical.

**triacetin** (*tri-as'-et-in*) [*tres,* three; *acetum,* vinegar], $C_3H_5(C_2H_3O_2)_3$. An oily liquid found in cod-liver oil, in some of the fats, in the oil of *Euonymus europæus* and in a mixture of glycerin and glacial acetic acid.

**triacid** (*tri-as'-id*) [*tri-; acidum,* acid]. Of an alcohol, containing three atoms of hydrogen replaceable by a base.

**triacol** (*tri'-ak-ol*). Trade name of a preparation containing sodium, potassium and ethyl-morphine salts of guaiacol-sulphonic acid; used in phthisis and other chronic lung affections.

**triad** (*tri'-ad*). See under *quantivalence.* **t., Hutchinson's,** the combination of notched teeth, interstitial keratitis and otitis, found in subjects of hereditary syphilis. **t.-jar,** a jar in which mixed liquids are allowed to stand in order that they may separate by gravity.

**triallylamine** (*tri-al-il-am'-in*). A volatile base having the formula $(C_3H_5)_3N_3$.

**triamine** (*tri-am'-in*). A compound derived from three molecules of ammonia in which the hydrogen has in part or wholly been replaced by bases.

**triangle** (*tri'-ang-gl*) [*tri-; angulus,* an angle]. A figure having three sides and three angles. **t. of Bryant.** See *t., iliofemoral.* **t., carotid, inferior,** a triangle located in the neck; it is bounded in front by the median line of the neck, behind by the anterior margin of the sternomastoid, and above by the anterior belly of the omohyoid. Its floor is formed by the longus colli below and the scalenus anticus above. The common carotid artery, internal jugular vein, vagus nerve, superficialis colli nerve, a branch of the communicans noni, the inferior thyroid artery, the recurrent laryngeal nerve, the sympathetic nerve, the trachea, thyroid gland, and larynx are the important structures within it. Syn., *triangle of necessity.* **t., carotid, superior,** a triangle located in the neck; it is bounded behind by the sternomastoid, in front by the anterior belly of the omohyoid, and above by the posterior belly of the digastric. Its floor is formed by the thyrohoid, hyoglossus, and inferior and middle constrictors of the pharynx. The most important structures contained within it are the common carotid artery and its bifurcation into the external and internal carotids, the superior thyroid artery, the lingual artery, the facial artery, the occipital and ascending laryngeal arteries, the internal jugular vein, and the veins corresponding to the arteries mentioned, the descendens noni, hypoglossal, pneumogastric, sympathetic, spinal accessory, superior laryngeal, and external laryngeal nerves. Syn., *triangle of election.* **t., cephalic,** a triangle on the anteroposterior plane of the skull, bounded by lines joining the occiput with the forehead and with the chin and a line joining the latter two. **t., digastric.** See *t., submaxillary.* **t. of elbow,** a triangle lying in front of the elbow with the base directed upward toward the humerus, and bounded externally by the supinator longus and internally by the pronator radii teres. Its floor is formed by the brachialis anticus and supinator brevis. Its contents are the brachial artery and veins, the radial and ulnar arteries, the median and musculospiral nerves, and the tendon of the biceps. **t. of election.** See *t., carotid, superior*. **t., extravesical.** See *Pawlik's triangle*. **t., fascial,** a triangle formed by lines uniting the basion with the alveolar and nasal points and a line joining the latter two. **t., frontal,** a triangle bounded by the maximum frontal diameter and lines joining its extremities and the glabella. **t. of Hesselbach.** See *Hesselbach's triangle*. **t., hypoglossohyoid,** a triangular space in the lateral subhyoid region, limited above by the hypoglossal nerve, in front by the posterior border of the mylohyoid muscle, behind and below by the tendon of the digastric muscle. The area is occupied by the hyoglossal muscle, which covers the lingual artery. **t., iliofemoral,** a triangle located at the hip. Its hypothenuse is formed by Nélaton's line, a second side by the continuation outward of a line drawn through the two superior iliac spines, and the third by a line drawn at right angles to this form the summit of the greater trochanter. **t., inferior occipital,** a triangle having the bimastoid diameter for its base and the inion for its apex. **t., infraclavicular,** a triangle situated below the clavicle; it is bounded above by the clavicle, below and to the inner side by the upper border of the great pectoral muscle, and to the outer side by the anterior border of the deltoid. It contains the axillary artery. **t., inguinal.** Same as *t., Scarpa's*. **t., interdeferential.** See *trigone*. **t. of Lesser,** a triangle located in the neck. Its boundaries are as follows: at its upper border, the hypoglossal nerve; the two sides are formed by the anterior and posterior bellies of the digastric muscle. It is covered by the skin, superficial and deep fascia, and apex of the submaxillary gland. The floor is formed by the hyoglossus muscle. It contains the ranine vein and hypoglossal nerve. **t., lumbocostoabdominal,** a triangle bounded anteriorly by the external oblique, superiorly by the lower border of the serratus posticus inferior and the point of the twelfth rib, posteriorly by the outer edge of the erector spinae, and inferiorly by the internal oblique. **t., Malgaigne's.** See *t., carotid, superior*. **t., mylohyoid,** the space bounded by the mylohyoid and the two bellies of the digastric. **t. of necessity.** Same as *t., carotid, inferior*. **t. of the neck, anterior,** a triangle bounded anteriorly by a line extending from the chin to the sternum, posteriorly by the anterior margin of the sternomastoid, the base being formed by the lower border of the body of the inferior maxilla and a continuation of this line to the mastoid process of the temporal bone. It is subdivided into three smaller triangles by the digastric muscle above and the anterior belly of the omohyoid below. These are named from below upward the *inferior carotid*, the *superior carotid*, and the *submaxillary*. **t. of the neck, posterior,** a triangle bounded anteriorly by the sternomastoid muscle, posteriorly by the anterior margin of the trapezius; the base is formed by the upper border of the clavicle; the apex corresponds to the occiput. It is divided by the posterior belly of the omohyoid muscle into two triangles, the *occipital* or *upper*, and the *subclavian* or *lower*. **t., occipital,** a triangle with the following boundaries: anteriorly, the sternomastoid muscle; posteriorly, the trapezius; and below, the omohyoid muscle. Its important contents are the spinal accessory nerve, the ascending and descending branches of the cervical plexus, and the transversalis colli artery and vein. **t., omoclavicular.** Same as *t., subclavian, q. v.* **t., omohyoid.** Same as *t., superior carotid*. **t., omotracheal.** See *t., carotid inferior*. **t., palatal,** a triangle having the width of the palate as its base and the alveolar point as its apex. **t. of Petit.** See *Petit's triangle*. **t., pubourethral,** a triangle situated in the perineum. Its boundaries are externally the sartorius muscle; internally, the adductor longus muscle, and above, Poupart's ligament. Its important contents are the femoral artery and vein, the anterior crural nerve, and the crural branch of the genitocrural nerve. **ts., Simon's,** the groin, the internal face of the thighs, and the hypogastric region form the *femoral triangle of Simon*;

the surface of the axilla, the pectoral region, and the inner aspect of the arm, the *brachial triangle of Simon*. **t., subclavian,** a triangle bounded above by the posterior belly of the omohyoid muscle, below by the upper border of the clavicle. Its base is formed by the sternomastoid muscle. It contains the subclavian artery and occasionally the vein, the brachial plexus of nerves, the suprascapular vessels, the transversalis colli artery and vein, and the external jugular vein. **t., submaxillary,** a triangle formed above by the lower border of the body of the inferior maxilla and a continuation of this line to the mastoid process of the temporal bone, below by the posterior belly of the digastric and the stylohyoid muscle, and anteriorly by the middle line of the neck. It contains the submaxillary gland, the facial artery and vein, the submental artery, the mylohyoid and nerve, and the stylomaxillary ligament, behind which is the external carotid artery. **t., suboccipital,** a triangle in the posterior part of the neck, formed by the rectus capitis posticus major and superior and inferior oblique muscles, and containing the vertebral artery. **t., suprameatal.** See *Macewen's triangle*. **t., surgical,** a triangular space containing important vessels and nerves which may require to be operated upon. **t., vesical,** a triangle at the base of the bladder—the trigonum of the urinary bladder. The apex is at the beginning of the urethra, and the other two angles at the orifices of the ureters.

**triangular** (*tri-ang'-gū-lar*) [*triangle*]. Having three sides or angles, as the *triangular ligament*.

**triangularis** (*tri-ang-gū-lā'-ris*). A triangular muscle. See under *muscle*.

**triatomic** (*tri-at-om'-ik*) [*tri-; atom*]. 1. Consisting of three atoms. 2. Having three atoms of replaceable hydrogen.

**tribade** (*trib'-ād*) [*tribadism*]. 1. One who indulges in tribadism. 2. The active agent in tribadism.

**tribadism** (*trib'-ad-izm*) [τρίβειν, to rub]. Unnatural sexual relations between women, produced by friction of the genitals.

**tribasic** (*tri-bā'-sik*) [*tri-; basis*, a base]. Having three hydrogen atoms replaceable by bases.

**tribrachius** (*tri-brā'-ke-us*) [*tri-;* βραχίων, arm]. A monster with three arms.

**tribromaniline** (*tri-bro-man'-il-in*), C₆H₄Br₃N. Colorless needles obtained from aniline by action of bromine. **t. hydrobromide.** See *bromamide*.

**tribromide** (*tri-brōm'-īd*) [*tri-; bromide*]. A compound of bromine containing three atoms of bromine in the molecule.

**tribrommethane** (*tri-brōm-meth'-ān*) [*tri-; bromine; methane*]. Bromoform.

**tribromobenzol** (*tri-bro-mo-ben'-zol*) C₆H₃Br₃. A bromine substitution-product of benzene.

**tribromosalol, tribromsalol** (*tri-bro-mo-sa'-lol, tri-brōm-sa'-lol*), C₆H₅ . C₇H₂Br₃O₃. A crystalline substance used as an intestinal antiseptic and hypnotic. It is produced by the reaction of phenol with bromine. It is antiseptic, especially for the intestinal tract. Dose 1-4 gr. (0.06-0.26 Gm.). t.-bismuth, xeroform.

**tribromphenol** (*tri-brōm-fe'-nol*) [*tri-;* βρώμος, stench; *phenol*], C₆H₂Br₃OH. A substance produced by the reaction of phenol with bromine. It is antiseptic, especially for the intestinal tract. Dose 1-4 gr. (0.06-0.26 Gm.). t.-bismuth, xeroform.

**tribromphenyl salicylate** (*tri-brom-fen'-il*). Tribromsalol.

**tributum** (*trib-ū'-tum*) [*tribuere*, to render]. A tribute. t. lunare. Synonym of *menstruation*.

**tricalcic** (*tri-kal'-sik*) [*tri-; calcium*]. Containing three atoms of calcium.

**tricaudalis** (*tri-kaw-da'-lis*) [*tri-; cauda*, a tail]. The retrahens aurem muscle; so-called because it is composed of three slips.

**tricellular** (*tri-sel'-ū-lar*) [*tri-; cellula*; a cell]. Having three cells.

**tricephalus** (*tri-sef'-al-us*) [*tri-;* κεφαλή, head]. A monster with three heads.

**triceps** (*tri'-seps*) [*tri-; caput*, head]. Three-headed; a muscle having three heads. See under *muscle*.

**trich-, tricho-** (*trik-, trik-o-*) [θρίξ, a hair]. A prefix signifying pertaining to a hair.

**trichangeia** (*trik-an-ji'-ah*) [*trich-;* ἀγγεῖον, vessel]. The capillary blood-vessels.

**trichangiectasis** (*trik-an-je-ek-tā'-sis*) [*trichangeia;* ἔκτασις, extension]. Dilatation of the capillaries.

**trichatrophia** (*trik-at-ro'-fe-ah*) [*trich-; atrophy*]. A brittle state of the hair from atrophy of the hair-bulbs.

TRICHAUXIS 904 TRICHOPHAGY

**trichauxis** (trik-awks'-is) [trich-; αὔξησις, increase]. Hypertrichiasis.

**trichesthesia, trichæsthesia** (trik-es-the'-ze-ah) [trich-; αἴσθησις, sensibility]. 1. A peculiar form of tactile sensibility in regions covered with hairs. 2. See *trichoesthesia*.

**trichiasis** (trik-i'-as-is) [θρίξ, a hair]. A state of abnormal position of the eyelashes, so that they produce irritation by friction upon the globe. t. of the anus, an incurvation of the hairs about the anus, so that they irritate the mucous membrane.

**Trichina** (trik-i'-nah) [θρίξ, a hair]. A genus of nematode worms, of which one species, *T. spiralis*, is parasitic in the hog and at times in man. See *trichinosis*.

**Trichinella spiralis** (trik-in-el'-ah spi-ra'-lis). Same as *Trichina spiralis*.

**trichiniasis** (trik-in-i'-as-is). See *trichinosis*.

**trichiniferous** (trik-in-if'-er-us) [trich-; ferre, to bear]. Containing trichinæ.

**trichinization** (trik-in-iz-a'-shun). Infestation with trichinæ.

**trichinophobia** (trik-i-no-fo'-be-ah) [trichina; φόβος, fear]. Morbid fear of trichinosis.

**trichinoscope** (trik-i'-no-skōp). A microscope for the detection of *Trichina spiralis*.

**trichinosis** (trik-in-o'-sis) [trichina]. A disease produced by the ingestion of pork containing *Trichina spiralis*. It is characterized by nausea, vertigo, fever, diarrhea, prostration, stiffness and painful swelling of the muscles, edema of the face, and in some cases perspiration, insomnia, and delirium.

**trichinotic** (trik-in-ot'-ik) [trich-; νοσός, disease]. Pertaining to or affected with trichinosis.

**trichinous** (trik'-in-us) [trichina]. Infested with or containing trichinæ.

**trichismus** (trik-is'-mus) [θρίξ, a hair]. 1. A scarcely perceptible fracture. 2. A capillary fissure or crack.

**trichitis** (trik-i'-tis) [trich-; ιτις, inflammation]. Inflammation of the hair-bulbs.

**trichiurus** (trik-e-u'-rus). The trichocephalus or threadworm.

**trichloracetic acid** (tri-klor-as-e'-tik). See *acid, trichloracetic*.

**trichloraldehyde** (tri-klor-al'-de-hīd). Chloral.

**trichlorhydrin** (tri-klor-hi'-drin), $C_3H_5Cl_3$. A colorless oily liquid with odor of alcohol.

**trichloride** (tri-klor'-īd) [tri-; chloride]. A compound containing chlorine in the proportion of three atoms to one of the base.

**trichlormethane** (tri-klor-meth'-ān). Chloroform.

**trichloropropane** (tri-klor-o-pro'-pān). Same as *trichlorhydrin*.

**trichloroquinone** (tri-klo-ro-kwin'-ōn), $C_6HCl_3O_2$. A crystalline substance obtained from a sulphuric-acid solution of phenol by action of potassium chlorate with HCl.

**trichlorphenol** (tri-klōr-fe'-nol) [tri-; χλωρός, green; phenol], $C_6H_2Cl_3(OH)$. A derivative of phenol used as a disinfectant.

**tricho-.** See *trich-*.

**trichoesthesia.** See *trichesthesia*.

**trichobacteria** (trik-o-bak-te'-re-ah) [tricho-; bacteria]. 1. Flagellate bacteria. 2. Filamentous bacteria.

**trichobezoar** (trik-o-be'-zo-ar) [tricho-; bezoar]. A hair ball or concretion in the stomach or intestine. See *egagropilus*.

**trichocardia** (trik-o-kar'-de-ah) [tricho-; καρδία, heart]. Inflammation of the pericardium with pseudomembranous elevations.

**trichocephaliasis** (trik-o-sef-al-i'-as-is) [trichocephalus]. The diseased condition produced by threadworms.

**Trichocephalus** (trik-o-sef'-al-us) [tricho-; κεφαλή, head]. A genus of nematode worms, the threadworms. T. dispar, a variety parasitic in the intestine, especially the large intestine. T. trichiuris. Same as *T. dispar*.

**trichocirsus** (trik-o-sir'-sus) [tricho-; κιρσός, a varix]. Abnormal capillary dilatation.

**trichoclasis, trichoclasia** (trik-ok'-las-is, trik-o-kla'-ze-ah). See *trichorrhexis nodosa*.

**trichocryptosis** (trik-o-trip-to'-sis) [trich-; κρυπτός, hidden]. Any disease of the hair-follicles.

**trichocyst** (trik'-o-sist) [trich-; κύστις, bladder]. In biology, a small vesicle containing a thread, which can be shot out rapidly, like the nematocyst of a cœlenterate, and found in the ectoplasm of the Infusoria and in some of the Flagellata.

**trichodangeia** (trik-od-an'-je-ah) [τριχώδης, hair-like; ἀγγεῖον, vessel; pl. of *trichodangeium*]. A term synonymous with capillaries.

**trichodangeitis** (trik-od-an-je-i'-tis) [τριχώδης, hair-like; ἀγγεῖον, vessel; ιτις, inflammation]. Capillary inflammation.

**trichodarteria** (trik-od-ar-te'-re-ah) [τριχώδης, hair-like; ἀρτηρία, an artery]. An arteriole.

**trichodarteriitis** (trik-od-ar-ter-e-i'-tis) [tricho-; ἀρτηρία, artery; ιτις, inflammation]. Inflammation of the arterioles.

**trichodophlebitis** (trik-od-o-fleb-i'-tis) [τριχώδης, hair-like; φλέψ, a vein; ιτις, inflammation]. Inflammation of the venules.

**trichoepithelioma** (trik-o-ep-e-the-le-o'-mah) [tricho-; epithelioma]. A skin-tumor originating in the hair-follicles.

**trichoesthesia, trichoæsthesia** (trik-o-es-the'-ze-ah) [tricho-; αἴσθησις, sensibility]. The sensation perceived when a hair is touched.

**trichoesthesiometer** (trik-o-es-the-ze-om'-et-ur) [tricho-; αἴσθησις, sensibility; μέτρον, measure]. An electrical appliance for determining the sensibility of the hair.

**trichogen** (trik'-o-jen) [tricho-; γεννᾶν, to produce]. A substance that stimulates the growth of the hair.

**trichogenous** (trik-oj'-en-us) [tricho-; γεννᾶν, to produce]. Encouraging the growth of hair.

**trichoglossia** (trik-o-glos'-e-ah) [tricho-; γλῶσσα, tongue]. Hairy tongue, a thickening of the papillæ, producing an appearance as if the tongue were covered with hair.

**trichohyaline** (trik-o-hi'-al-in) [tricho-; hyaline]. The hyaline of the hair; it is like keratohyaline.

**trichoid** (trik'-oid) [tricho-; εἶδος, like]. Resembling hair.

**tricholabis, tricholabium** (trik-ol'-ab-is, trik-o-la'-be-um) [tricho-; λάβη, a handle]. Tweezers for pulling out hairs.

**tricholith** (trik'-o-lith) [tricho-; λίθος, a stone]. A hairy concretion.

**trichologia** (trik-o-lo'-je-ah) [tricho-; λέγειν, to pick out]. 1. Carphologia; floccitation. 2. The plucking out of one's hair.

**trichology** (trik-ol'-o-je) [tricho-; λόγος, science]. 1. The science of the hair and its diseases. 2. Trichologia.

**trichoma** (trik-o'-mah) [τρίχωμα, a growth of hair]. 1. Trichomatosis. See *trichotillomania*.

**trichomania.** See *trichotillomania*.

**trichomaphyte** (trik-o'-maf-īt) [τρίχωμα, a growth of hair; φυτόν, a plant]. A cryptogamic growth which was formerly thought to be the cause of trichomatosis.

**trichomatose** (trik-o'-mat-ōs) [τρίχωμα, a growth of hair]. Matted together.

**trichomatosis** (trik-o-mat-o'-sis) [see *trichomatose*]. An affection of the hair characterized by a matted condition due to fungoid growths. See *plica polonica*.

**Trichomonas** (trik-om'-o-nas) [tricho-; μονάς, a monad]. A genus of infusorians. T. intestinalis, is found in the feces in some cases of diarrhea, enteritis, and typhoid. T. vaginalis, a species occasionally found in the vagina.

**trichomyces** (trik-om'-is-ēs) [tricho-; μύκης, a mushroom]. Synonym of *trichophyton*.

**trichomycosis** (trik-o-mi-ko'-sis) [tricho-; mycosis]. A disease of the hair produced by a vegetable parasite. t. barbæ. Synonym of *sycosis parasitaria*, t. capillitii. Synonym of *t. circinata*. t. circinata, ringworm of the scalp, produced by the *Trichophyton tonsurans*. t. favosa. See *favus*. t. nodosa, a peculiar condition, generally nodose in character, affecting the hairs of the axilla and scrotum, and due to the growth and encapsulation in the cortical layers of the shaft of a small rod-shaped bacterium. t., palmellina, a disease affecting the hairy parts of the trunk; t. nodosa. t. pustulosa, a pustular, parasitic disease affecting hairy regions.

**trichonosis, trichonosus** (trik-on'-o-sis, trik-on'-o-sus) [tricho-; νόσος, disease]. Any disease of the hair. t. cana. See *canities*. t. discolor. See *canities*. t. furfuracea. Synonym of *tinea tonsurans*. t. versicolor. See *ringed hair*.

**trichopathia** (trik-o-path'-ik) [tricho-; πάθος, disease]. Relating to disease of the hair.

**trichopathy** (trik-op'-ath-e) [tricho-; πάθος, disease]. Any disease of the hair.

**trichophagy, trichophagia** (trik-of'-aj-e, trik-o-fa'-je-ah). The eating of hair.

**trichophobia** (*trik-o-fo'-be-ah*) [*tricho-*; φόβος, fear]. Morbid fear of hair.

**trichophytic** (*trik-of-it'-ik*). 1. Relating to the genus *Trichophyton*. 2. [φύειν, to grow.] Promoting the growth of hair. 3. An agent promoting the growth of hair.

**trichophytinous** (*trik-off-it-i'-nus*) [*tricho-*; φυτόν, a plant]. Pertaining to the presence of *Trichophyton tonsurans*.

**Trichophyton** (*tri-kof'-it-on*) [*tricho-*; φυτόν, a plant]. A fungus parasitic upon the hair, and causing tinea trichophytina, or ring-worm. T. **tonsurans**, the cause of tinea tonsurans.

**trichophytosis** (*trik-off-it-o'-sis*) [*tricho-*; φυτόν, a plant]. A contagious disease of the skin and hair, occurring most often in children, due to the invasion of the epidermis by the trichophyton-fungus, and characterized by the formation of circular or annular, scaly patches and partial loss of hair. See *tinea*. t. **barbæ**. Synonym of *dermatomycosis maculo-vesiculosa*. t. **cruris**. Synonym of *marginal eczema*.

**trichopoliosis** (*trik-o-pol-e-o'-sis*) [*tricho-*; πολιούσθαι, to become gray]. Synonym of *canities*.

**trichoptilosis** (*trik-op-til-o'-sis*). Synonym of *trichorrhexis nodosa*.

**trichorrhea trichorrhœa** (*trik-or-e'-ah*) [*tricho-*; ῥοία, a flow]. Rapid loss of the hair.

**trichorrhexis** (*trik-or-eks'-is*) [*tricho-*; ῥῆξις, a breaking]. Brittleness of the hair. t. **nodosa**, an atrophic condition of the hair, affecting more often the male beard, and characterized by irregular thickenings resembling nodes on the hair-shaft, the hairs often breaking with a "green-stick fracture" immediately through a node.

**trichoschisis** (*trik-os'-kis-is*) [*tricho-*; σχίσις, a splitting]. The splitting of the hair.

**trichoscopy** (*trik-os'-ko-pe*) = [*tricho-*; σκοπεῖν, to examine]. The examination of the hair.

**trichosis** (*trik-o'-sis*) [θρίξ, hair; νόσος, disease]. Any morbid affection of the hair. t. **athrix**. Synonym of *alopecia*. t. **decolor**, morbid discoloration of the hair. t. **distrix**. Synonym of *trichoptilosis*. t. **hirsuties**. Same as *hirsuties*. t. **plica**. See *plica polonica* and *trichomatosis*. t. **poliosis**. See *canities*. t. **sensitiva**, a sensitive state of the scalp; any manipulation causing pain. t. **setosa**, a disease in which the hair grows thick, rigid, and bristly.

**trichostereticus** (*trik-o-ster-et'-ik-us*) [*tricho-*; στερητικός, depriving]. Causing loss of hair.

**trichosyphilis** (*trik-o-sif'-il-is*) [*tricho-*; *syphilis*]. Any syphilitic disease, or affection of the hair.

**trichosyphilosis** (*trik-o-sif-il-o'-sis*). Synonym of *trichosyphilis*.

**Trichothecium** (*trik-o-the'-se-um*) [*tricho-*; θήκη, a chest]. A vegetable parasite of the hair. T. **roseum**, a fungous growth found in the ear.

**trichotillomania** (*trik-o-til-o-ma'-ne-ah*) [*tricho-*; τίλλειν, to pluck out; μανία, madness]. An uncontrollable impulse to pull out one's hair.

**trichotomy** (*tri-kot'-o-me*) [τρίχα, in three; τομή, a cutting off]. Division into three parts.

**trichotoxicon** (*trik-o-toks'-i-kon*) [*tricho-*; τοξικόν, a poison]. A supposed toxin, existing in respired air, which, when introduced into the blood, exerts a poisonous action upon the hair, thus causing alopecia.

**trichotoxin** (*trik-o-toks'-in*). A cytotoxin obtained by E. Metchnikoff from the ciliated epithelia.

**trichroic** (*tri-kro'-ik*) [*trichroism*]. Possessing trichroism.

**trichroism** (*tri'-kro-izm*) [*tri-*; χρόα, color]. The property of exhibiting three different colors when viewed under three different aspects.

**trichromat** (*tri-kro'-mat*) [τρεῖς, three; χρῶμα, color]. Persons for whom the end regions of the spectrum are of constant hue and differ only in intensity. Just inside of each end region there is an intermediate region in which any color can be produced by mixtures of the end color with the color of the intermediate region. Between these intermediate regions lies the middle region, which requires the presence of some third color in addition to colors from the end regions. Cf. *dichromat*; *monochromat*.

**trichromatic** (*tri-kro-mat'-ik*) [τρεῖς, three; χρῶμα, color]. Having three colors.

**Trichuris** (*tri-kū'-ris*) [*trich-*; οὐρα, tail]. A genus of trematodes. T. **trichiura**, the *Trichocephalus dispar*, q. v.

**tricipital** (*tri-sip'-it-al*) [*triceps*, three-headed]. 1. Three-headed. 2. Pertaining to the triceps.

**tricorn** (*tri'-korn*) [*tri-*; *cornu*, horn]. A lateral ventricle of the brain.

**tricornis** (*tri-kor'-nis*) [*tri-*; *cornu*, horn]. Having three horns or processes or prominences; a name applied to each of the lateral ventricles of the brain.

**tricornute** (*tri-kor'-nūt*) [*tres*, three; *cornutus*, horned]. In biology, having three horn-like appendages.

**tricresol, trikresol** (*tri-kre'-sol*). A refined mixture of metacresol, 40 %; paracresol, 33 %; orthocresol, 27 %; soluble in 40 parts of water. It has three times the germicidal value of phenol.

**tricresolamine** (*tri-kres-ol-am'-in*). A solution containing 2 % each of ethylenediamine and tricresol; it is a clear, colorless, alkaline liquid turning yellow on exposure. It is stronger and less irritating than tricresol.

**tricrotic** (*tri-krot'-ik*) [*tri-*; κρότος, stroke]. Having three waves corresponding to one pulse-beat; exhibiting tricrotism.

**tricrotism** (*tri'-krot-izm*) [see *tricrotic*]. The quality of being tricrotic.

**tricrotous** (*tri'-kro-tus*) [τρεῖς, three; κρότος, stroke]. Same as *tricrotic*.

**tricuspid** (*tri-kus'-pid*) [*tri-*; *cuspis*, a point]. 1. Having three cusps, as the tricuspid valve. 2. Affecting or produced at the tricuspid valve.

**tridactyl** (*tri-dak'-til*) [τρεῖς, three; δάκτυλος, finger]. Having three digits.

**tridymus** (*trid'-im-us*) [τρίδυμος]. Synonym of *triplet*.

**trielcon** (*tri-el'-kon*) [*tri-*; ἕλκειν, to draw]. A three-pronged instrument for extracting bullets or other foreign bodies from the body.

**triencephalus** (*tri-en-sef'-al-us*) [τρεῖς, three; ἐγκέφαλος, brain]. A fetal monster without smell, hearing, or sight.

**triethylamine** (*tri-eth-il-am'-in*) [*tri-*; *ethyl*; *amine*], C₆H₁₅N. A ptomaine obtained from putrid haddock.

**trifacial nerve** (*tri-fa'-shal*) [*tri-*; *facies*, face]. The fifth cranial nerve, so-called because it divides into three main branches that supply the face.

**triferrin** (*tri-fer'-in*). See *iron paranucleinate*.

**trifid** (*tri'-fid*) [*tres*, three; *findere*, to cleave]. Threecleft.

**triflagellate** (*tri-flaj'-el-āt*) [*tres*, three; *flagellum*, a whip]. Having three flagella; trimastigate.

**trifolium** (*tri-fo'-le-um*) [*tri-*; *folium*, leaf]. Clover. T. **pratense** (common red clover) is vaunted in the treatment of whooping-cough, syphilis, and carcinoma.

**triformal** (*tri-form'-al*). See *formalin*.

**triformol** (*tri-form'-mol*). Same as *paraform*.

**trigastric** (*tri-gas'-trik*) [τρεῖς, three; γαστήρ, belly]. Having three fleshy bellies (as certain muscles).

**trigemin** (*tri-jem'-in*). A substance obtained from pyramidon by action of butyl-chloral hydrate, forming white needles soluble in water; antineuralgic. Dose 8–20 gr. (0.5–1.3 Gm.).

**trigeminal** (*tri-jem'-in-al*) [*tri-*; *geminus*, twinborn]. 1. Triple; dividing into three parts, as the trigeminal nerve. 2. Pertaining to the trigeminal nerve. See *trifacial*.

**trigeminus** (*tri-jem'-in-us*) [see *trigeminal*]. The trifacial nerve.

**trigger** (*trig'-er*). A device by means of which a catch or spring is released. t.-**area**, a sensitive region of the body, irritation of which may give rise to certain peculiar phenomena, either physiological or pathological, in some part of the body. t.-**finger**, a condition in which flexion or extension of a finger is at first obstructed, but finally accomplished with a jerk or sweep. t. **knee**, a condition characterized by a sudden arrest of the movement of the knee during flexion or extension; this arrest is followed by a sudden jerking and lateral movement of the leg and the production of a clicking sound. It is apparently due to laxity of the joint capsule. t.-**material**, an apheter; any theoretical catastatic substance whose sudden breaking up communicates an explosive decomposition to the protoplasm directly concerned in any function. The trigger-material itself must be acted upon by another trigger-material; and thus every nerve-impulse and every functional act must be accompanied by the destruction of a fuse-like train of protoplasm. See *apheter*.

**trigocephalus** (*tri-go-sef'-al-us*). See *trigonocephalus*.

**trigonal** (*trig'-o-nal*) [τρεῖς, three; γωνία, angle]. Same as *trigonous*.

**trigone** (*tri'-gōn*) [*tri-*; γωνία, angle]. Triangle.

See *trigonum*. **t. of the bladder,** a smooth triangular space on the inside of the bladder, immediately behind the orifice of the urethra. **t., olfactory,** the gray root of origin of the olfactory tract.
**Trigonella** (tri-go-nel'-ah) [τρίγωνος, three-cornered]. A genus of *Leguminosæ*. **T. elatior,** a variety, the seeds of which have been used in affections of the bladder and as poultices. **T. fœnum græcum,** a variety used in plasters and salves and in veterinary medicine. **T. monspeliaca,** a variety a decoction of the seeds of which is used by the Italians in various forms of diarrhea.
**trigonitis** (tri-go-ni'-tis) [trigonum; ιτις, inflammation]. Inflammation of the trigonum vesicæ.
**trigonocephalic** (trig-o-no-sef-al'-ik) [τρίγωνος, three cornered; κεφαλή, head]. Pertaining to trigonocephaly.
**trigonocephalus** (trig-o-no-sef'-al-us) [τρίγωνος, three cornered; κεφαλή, head]. A triangular shaped skull with the small end anterior, due to a premature union of the coronal suture.
**trigonocephaly** (trig-o-no-sef'-al-e) [τρίγωνος, three cornered; κεφαλή, head]. A deformity of the skull produced by a premature union of the medio-frontal or metopic suture. See *trigonocephalus*.
**trigonum** (tri-go'-num) [τρίγωνος, three cornered]. A triangle; also the interpeduncular space; and see *trigone*. **t. acustici,** a three-cornered space on the dorsal surface of the medulla. **t. cerebrale.** Synonym of *fornix cerebri*. **t. cervicale,** the base of the dorsal gray cornu of the spinal cord. **t. clavipectorale,** a triangle of the chest. Its boundaries are the clavicle, the pectoralis minor muscle, and the thorax. **t. collaterale,** a triangular area at the junction of the posterior and inferior horns of the lateral ventricles. **t. colli medianum,** relating to the space occupied by the two anterior triangles of the neck. **t. coracoacromiale,** a triangular space whose boundaries are the coracoid process, the apex of the acromion, and the concave border of the clavicle. **t. deltoideopectorale,** the infraclavicular fossa. **t. dorsale,** the space between the anterior pair of the corpora quadrigemina. **t. femorale,** Scarpa's triangle. **t. fluctuans,** the posterior cerebral commissure. **t. habenulæ,** the triangular space behind the upper surface of the optic thalamus; in front of the lamina quadrigemina, and between the sulcus habenulæ and the sulcus subpinealis. **t. hypoglossi,** a triangular space on the dorsal surface of the oblongata. Its boundaries are, above, the striæ medullares acusticæ, internally, the posterior longitudinal fissure, and, externally, the ala cinerea. **t. inferius commissuræ posterioris,** the lower triangular half of the posterior commissure of the brain. **t. lemnisci,** the fillet. **t. lumbale.** See *Petit's triangle*. **t. olfactorium.** See *Broca's olfactory area*. **t. pensile,** the posterior cerebral commissure. **t. vagi,** a small, triangular space on the medulla oblongata, marking the origin of the vagus nerve. **t. ventriculi lateralis,** a triangular projection located between the entrances to the posterior and descending horns of the lateral ventricle. **t. vesicæ,** the triangular surface of the bladder immediately behind the urethral orifice.
**trihydrate** (tri-hi'-drāt) [tri-; *hydrate*]. A compound containing the hydroxyl-radical in the proportion of three to one atom of the base.
**trihydric** (tri-hi'-drik) [tri-; *hydric*]. Containing three atoms of hydrogen replaceable by bases.
**trihydroxide** (tri-hi-droks'-īd). See *trihydrate*.
**triiniodymus** (tri-in-e-od'-im-us) [τρεῖς, three; ἰνίον, the nape of the neck; δίδυμος, double]. A monster having three heads united posteriorly and attached to a single body.
**triiodide** (tri-i'-o-dīd) [tri-; *iodide*]. A compound containing iodine in the proportion of three atoms to one of the base.
**triiodocresol** (tri-i-o-do-kre'-sol). See *losophan*.
**triiodomethane** (tri-i-o-do-meth'-ān). Iodoform.
**triketohydrindenhydrate** (tri-ke-to-hi-drin-den-hi'-drāt). Same as ninhydrin, q. v.
**trikresol** (tri-kre'-sol). See *tricresol*.
**trilabe** (tri'-lāb) [tri-; λαμβάνειν, to grasp]. A three-pronged instrument for withdrawing small calculi or other foreign bodies from the bladder, through the urethral passage.
**trilaminar** (tri-lam'-in-ar) [tri-; *lamina*, plate]. In biology, three-layered.
**trilateral** (tri-lat'-er-al) [tri-; *latus*, a side]. Having three sides.

**trilaurin** (tri-law'-rin). A crystalline glyceride found in cocoanut oil and some other oils.
**trilinolein** (tri-lin-o'-le-in). A glyceride contained in linseed oil, hempseed oil, sunflower oil, etc.
**trillin** (tril'-e-in) [*trillium*]. A precipitate from a tincture of the root of *Trillium pendulum*, styptic, tonic, expectorant, antiseptic, and emmenagogue. Dose, 2 to 4 grains. See *beth-root*.
**trillin** (tril'-in). An alcoholic extract of *Trillium erectum*; it is astringent, tonic and expectorant.
**trilobate** (tri-lo'-bāt, or tri'-lo-bāt) [tri-; *lobatus*, lobed]. In biology, three-lobed.
**trilobed** (tri'-lōbd). Same as *trilobate*.
**trilocular** (tri-lok'-ū-lar) [tres, three; *loculus*, cell]. In biology, having three chambers or cells.
**trimanual** [tri-; *manus*, a hand]. Pertaining to a maneuver accomplished by the aid of three hands.
**trimastigate** (tri-mas'-tig-āt) [τρεῖς, three; μάστιξ, whip, scourge]. In biology, having three flagella; triflagellate.
**trimercuric** (tri-mer-kū'-rik) [tres, three; *mercury*]. Containing three atoms of bivalent mercury.
**trimester** (tri-mes'-ter) [*trimestris*, of three months]. A stage or period of three months.
**trimethyl** (tri-meth'-il) [tres, three; *methyl*]. The chemical group (CH₃)₃.
**trimethylamine** (tri-meth-il-am'in) [tri-; *methyl*; *amine*]. (CH₃)₃N. A colorless liquid ptomaine obtained from herring-brine and various animal and vegetable substances.
**trimethylenediamine** (tri-meth-il-ēn-di-am'-in) [tri-; *methylene*; *diamine*]. A ptomaine obtained from cultures of the comma bacillus on beef-broth. It causes convulsions and muscle-tremor.
**trimethylxanthine** (tri-meth-il-zan'-thin). See *caffeine*.
**trimorphic** (tri-mor'-fik). Same as *trimorphous*.
**trimorphism** (tri-mor'-fizm) [τρεῖς, three; μορφή, form]. 1. In biology, a term used to indicate the fact that hermaphrodite flowers of three different kinds, short-styled, mid-styled, and long-styled, are produced on the same species of plant. 2. Existence of three distinct forms, as certain insects.
**trimorphous** (tri-morf'-us) [τρεῖς, three; μορφή, form]. Pertaining to trimorphism.
**trineuric** (tri-nū'-rik) [tri-; νεῦρον, nerve]. Applied to a nerve-cell provided with three neuraxons.
**trinitrate** (tri-ni'-trāt) [tri-; *nitrate*]. A nitrate containing three nitric-acid radicals.
**trinitrin** (tri-ni'-trin). See *nitroglycerin*.
**trinitrocellulose** (tri-ni-tro-ū-lōs). See *pyroxylin*.
**trinitrocresol** (tri-ni-tro-kre'-sol), C₇H₅N₃O₇. Antiseptic crystals, obtained from nitration of coal-tar cresol; antiseptic.
**trinitroglycerin** (tri-ni-tro-glis'-er-in). Nitroglycerin.
**trinitrol** (tri-ni'-trol). Erythrol nitrate, similar to nitroglycerin.
**trinitrophenol** (tri-ni-tro-fe'-nol). Picric acid.
**trinophenon** (tri-no-fe'-non). A remedy for burns said to be an aqueous solution of picric acid.
**triocephalus** (tri-o-sef'-al-us) [tri-; κεφαλή, head]. A monster characterized by an absence of the ocular, nasal, and buccal apparatus, the head being merely a small spheroidal mass.
**triolein** (tri-o'-le-in). See *olein*.
**trional** (tri'-on-al), C₂H₅CH₃—C—(SO₂C₂H₅)₂. Sulphonethyl-methane, a hypnotic. Dose 15 gr. (1 Gm.).
**trionym** (tri'-o-nim) [tri-; ὄνομα, name]. A name consisting of three terms.
**triophthalmos** (tri-off-thal'-mos) [tri-; ὀφθαλμός, eye]. A diprosopic monster with three eyes and other deformities of the face and head.
**triopodymus** (tri-op-od'-im-us) [tri-; ὄψ, the face; δίδυμος, double]. A monster with three faces and but a single head.
**triorchid** (tri-o'-kid) [tri-; ὄρχις, a testicle]. 1. Having three testicles. 2. An individual having three testicles.
**triorchis** (tri-or'-kis) [tri-; ὄρχις, testicle]. An individual that has three testicles.
**triose** (tri'-ōs). A monosaccharid containing three carbon atoms in the molecule.
**triotonol** (tri-o-to'-nol). Trade name of a mixture containing the glycerophosphates of sodium, calcium and strychnine.
**triotus** (tri-o'-tus) [tri-; οὖς, ear]. A diprosopic monster with three ears, and generally with four eyes.

**trioxide** (*tri-oks'-īd*) [*tri-*; *oxide*]. A compound containing oxygen in the proportion of three atoms to one of the base.

**tripalmitin** (*tri-pal'-mit-in*). See *palmitin*.

**tripara** (*trip'-ar-ah*) [*tri-*; *parere*, to bear]. A woman who has borne three children.

**tripes** (*tri'-pēz*) [L.]. 1. Three-footed. 2. A monster having three feet.

**tripharmacon, tripharmacum** (*tri-far'-mak-on, tri-far'-mak-um*) [*tri-*; φάρμακον, a drug]. A medicine made up of three ingredients.

**triphasic** (*tri-fa'-sik*). Having three phases or variations.

**triphenamine** (*tri-fen'-am-in*). A mixture of phenocoll, phenocoll salicylate, and phenocoll acetate; recommended in rheumatic complaints.

**triphenetolguanidin hydrochloride** (*tri-fen-et-ol-gwan'-id-in*). A local anesthetic used in 0.1 % solution in treatment of eyes.

**triphenin** (*tri-fen'-in*). Propionyl-phenetidin, $C_6H_4$-$OC_2H_5NHC_2H_5CO$, obtained by boiling paraphenetidin with propionic acid. It is used as an antipyretic and sedative. Daily dose 46 gr. (3 Gm.); single dose 8–15 gr. (0.5–1.0 Gm.). Syn., *methylphenacetin*.

**triphenyl albumin** (*tri-fen'-il*). A culture-medium made by heating dry egg-albumen with phenol. It is odorless, tasteless, insoluble in water, alcohol, and potassa solution, but soluble in phenol.

**triphenylstibine sulphide** (*tri-fen-il-stib'-in*). A preparation used as a substitute for sulphur in skin diseases. It releases sulphur in nascent condition.

**triphthemia, triphthæmia** (*trif-the'-me-ah*) [τριπτός, rubbed, pounded; αἷμα, blood]. The retention of waste material in the blood. t. **carbonifera**, that due to excessive ingestion of carbohydrates.

**Tripier's amputation** (*trip-e-a'*). [Léon Tripier, French surgeon, 1842–1801]. One differing from Chopart's only in that the portion of the os calcis below the sustentaculum tali is removed.

**triple** (*trip'-l*) [L., *triplex*]. Threefold. t. **phosphate**, ammoniomagnesium phosphate, a phosphate occurring in urine and in phosphatic calculi.

**triplegia** (*tri-ple'-je-ah*) [*tri-*; πληγή, stroke]. Hemiplegia with the additional paralysis of one limb on the opposite side.

**triplet** (*trip'-let*) [*triple*]. 1. One of three children born at one birth. 2. In optics, a system consisting of three lenses.

**triplex** (*trip'-leks*) [L.]. Triple. t. **pills**, *pilulæ triplices*, pills containing three principal ingredients; pills of aloes, podophyllin and blue mass.

**triploblastic** (*trip-lo-blas'-tik*) [*triple;* βλαστός, germ]. Possessing three blastodermic membranes.

**triplokoria** (*trip-lo-ko'-re-ah*) [τριπλόος, threefold; κόρη, pupil]. An iris having three pupils.

**triplopia** (*trip-lo'-pe-ah*) [*triple;* ὤψ, eye]. A disturbance of vision in which three images of a single object are seen.

**tripod** (*tri'-pod*) [τρίπους, three-footed]. An object having three legs or supports. t., **anatomical**, the three piers on which the foot rests when a person stands erect; these piers are (1) the heel, (2) the three inner metatarsal bones, and (3) the two outer metatarsals. t., **Haller's**, the celiac axis. t., **vital**, the brain, heart, and lungs, viewed as the triple support of life.

**tripper-faden** (*trip'-er-fah'-den*). [Germ.]. Gonorrheal threads. Thread-like structures seen in the urine in gonorrhea. t.-**kokken**, gonococci.

**triprosopus** (*trip-ro-so'-pus*) [*tri-*; πρόσωπον, face]. A form of fetal monstrosity in which there is a fusion of three faces in one.

**tripsis** (*trip'-sis*) [τρίβειν, to rub]. 1. Same as *trituration*. 2. Massage.

**triptokoria** (*trip-to-ko'-re-ah*) [*tri-*; πίπτειν, to fall; κόρη, the pupil]. A condition of the iris in which there are three distinct pupils.

**tripus** (*tri'-pus*) [*tri-*; πούς, foot]. Same as *tripod*. t. **cellatus**, the three branches of the celiac artery.

**triquetrous** (*tri-kwet'-rus*) [*triquetrum*]. Three-cornered, as the *triquetrous* bone (*os triquetrum*), a Wormian bone. See *triquetrum*.

**triquetrum** (*tri-kwet'-rum*) [*triquetrus*, three-cornered]. 1. Any one of the Wormian bones. 2. The cuneiform bone of the carpus.

**triradial, triradiate** (*tri-ra'-de-al, tri-ra'-de-āt*) [see *triradius*]. Radiating in three directions. t. **pelvis**, one in which the promontory is pushed forward and the acetabula pressed inward.

**triradius** (*tri-ra'-de-us*) [*tres*, three; *radius*, ray; pl., *triradii*]. In the impression of the palmar surface in the Galton system a triangluar area composed of transverse ridges at the base of each of the four fingers; used in the classification of palmar impressions.

**trisaccharid** (*tri-sak'-ar-id*) [*tri-*; *saccharum*]. A carbohydrate which under the influence of a dilute acid yields three other sugar molecules and takes up two molecules of water.

**trismic** (*triz'-mik*). Relating to trismus.

**trismoid** (*triz'-moid*) [*trismus*]. A form of trismus neonatorum thought to be due to pressure on the occipital bone during labor.

**trismus** (*triz'-mus*) [τριγμός, from τρίζειν, to gnash]. Lockjaw, a tonic spasm of the muscles of mastication. t. **capistratus**, a condition in which the jaws cannot be separated because of adhesions between the cheeks and the gums, following ulceration of the parts. t. **catarrhalis maxillaris**, neuralgia of the jaw. t. of **cerebral origin**, persistent spasm of the muscles of the lower jaw, due to cerebral disease. t. **cynicus**, risus sardonicus. t. **dolorificus**, tic douloureux. t. **maxillaris**. See t. *catarrhalis maxillaris*. t. **nascentium**, t. **neonatorum**, a form of trismus occurring in newborn infants, and supposed to be due to septic infection of the umbilical stump. t. **sardonicus**. See t. *cynicus*. t., **traumatic**, trismus following a wound or injury. t. **uteri**, trismus occurring during and as a result of the puerperium.

**trisplanchnic** (*tri-splangk'-nik*) [*tri-*; σπλάγχνον, viscus]. Distributed to the viscera of the three largest cavities of the body, as the *trisplanchnic* nerve (the sympathetic nerve).

**tristearin** (*tri-ste'-ar-in*) [*tri-*; στέαρ, fat]. $C_3H_5$-$(C_{18}H_{35}O_2)_3$. See *stearin*.

**tristichiasis** (*tris-tik-i'-as-is*) [τριστιχία, a triple row]. A form of congenital distichiasis in which there are three rows of cilia.

**tristimania** (*tris-tim-a'-ne-ah*) [*tristis*, sad; μανία, frenzy]. Melancholia.

**tristis** (*tris'-tis*) [L.]. Sad; gloomy; having a dull color.

**trisubstituted** (*tri-sub'-sti-tū-ted*) [*tri-*; *substituere*, to substitute]. Having three atoms or radicals substituted by other atoms or radicals.

**trisulphide** (*tri-sul'-fīd*) [*tri-*; *sulphur*]. A compound containing sulphur in the proportion of three atoms to one of the base.

**tritanopia** (*trit-an-o'-pe-ah*) [*tri-*; *anopsia*]. A defect in a third principal element essential for color vision, as in violet-blindness.

**triticeoglossus** (*trit-is-e-o-glos'-us*) [*triticum;* γλῶσσα, tongue]. An anomalous muscle having its origin from the arytenoid cartilage and its insertion in the side of the tongue.

**triticeous** (*trit-ish'-us*) [*triticum*]. Having the shape of a grain of wheat. t. **cartilage**, t. **nodule**, corpus triticeum, a small cartilaginous nodule in the thyrohyoid ligament.

**triticeum** (*trit-is'-e-um*) [*triticum*, wheat]. The triticeous nodule.

**triticin** (*trit'-is-in*) [*triticum;* wheat]. 1. A gum-like substance found in *Triticum repens*. 2. A proprietary food preparation.

**triticum** (*trit'-ik-um*) [L.]. A genus of the *Gramineæ*. *T. sativum* (*T. vulgare*) is wheat. Triticum is official in the U. S. P. in the form of the rhizome of *Agropyron repens*, and is used in cystitis and irritable bladder. t., **fluidextract of** (*fluidextractum tritici*, U. S. P.). Dose 3–6 dr. (12–24 Cc.). t., **repens**, triticum.

**tritipalm** (*trit'-e-pahm*). A proprietary genitourinary tonic said to consist of the fluidextract of saw palmetto, *Serenoa serrulata*, and couch-grass, *Agropyron repens*.

**tritol** (*tri'-tol*). Any emulsion of oil, 4 parts, and diastasic extract of malt, 1 part.

**tritopine** (*tri'-to-pin*). An alkaloid from opium.

**tritorium, triturium** (*tri-to'-re-um, tri-tū'-re-um*) [*tritus,* a rubbing]. A vessel used in separating liquids of different density.

**tritotoxin** (*tri-to-toks'-in*) [τρίτος, third; τοξικόν, poison]. One of the third group into which Ehrlich classifies toxins, according to the avidity with which they combine with antitoxins, tritotoxin combining least readily.

**tritoxide** (*tri-toks'-īd*) [τρίτος, third; ὀξύς, acid]. Same as *trioxide*.

**tritubercular** (*tri-tū-bur'-kū-lar*) [*tri-;* *tuberculum*, tubercle]. Having three tubercles or cusps; tricuspid.

**triturable** (*trit'-ū-rabl*). Capable of being powdered.

**triturate** (*trit'-ū-rāt*) [*triturare*, from *terere*, to rub]. 1. To reduce to a fine powder. 2. A finely divided powder. In the U. S. P. a medicinal substance rubbed up with milk-sugar. t., tablet-, a triturate compressed into tablet form.

**trituration** (*trit-ū-ra'-shun*) [*triturate*]. The process of reducing a solid substance to a powder by rubbing.

**triturium** (*trit-ū'-re-um*). See *tritorium*.

**trivalence** (*tri'-va-lens*, or *triv'-al-ens*) [*tri-; valere*, to be worth]. The quality of being trivalent.

**trivalent** (*tri'-va-lent*, or *triv'-al-ent*) [*tri-; valere*, to be worth]. Combining with or equivalent to three atoms of hydrogen.

**trivalve** (*tri'-valv*) [*tri-; valva*, door]. Having three valves or blades (as a speculum).

**trivalvular** (*tri-val'-vū-lar*) [*tri-; valvula*, a small valve]. Having three valves.

**trizonal** (*tri-zo'-nal*) [*tri-; zona*, a belt or girdle]. Possessing, or arranged in, three layers or zones.

**trocar** (*tro'-kar*) [Fr., *trois-quarts*, from its triangular point]. An instrument used in paracentesis, or tapping a cavity, as in hydrocele. It consists of a perforator and a metallic tube. t., lancet, a trocar having a lancet-shaped perforator. t., piloting, Durham's trocar, used for introducing the articulated tracheotomy-tube. t., rectal, a curved trocar used in tapping the bladder through the rectum.

**troch.** Abbreviation of *trochischus*, troche.

**trochanter** (*tro-kan'-ter*) [τροχαντήρ, from τροχός, a wheel or pulley]. One of two processes on the upper extremity of the femur below the neck. The *greater trochanter* is situated on the outer, and the *lesser trochanter* on the inner, side of the bone. t., major, the greater trochanter. t. minor, the lesser trochanter. t. tertius, an anomalous process at the upper portion of the popliteal space of the femur. t., third. Same as *t. tertius*.

**trochanteric** (*tro-kan-ter'-ik*) [*trochanter*]. Pertaining to a trochanter.

**trochantin** (*tro-kan'-tin*) [*trochanter*]. The lesser trochanter.

**trochantinian** (*tro-kan-tin'-e-an*) [τροχαντήρ, trochanter]. Pertaining to the trochantin.

**troche** (*tro'-ke*) [τροχός]. A lozenge.

**trochin, trochinus** (*tro'-kin, tro-ki'-nus*). [τροχός, a wheel]. The lesser tuberosity of the head of the humerus.

**trochinian** (*tro-kin'-e-an*) [τροχός, wheel]. Pertaining to the trochin.

**trochischi** (*tro-kis'-ki*) [L.]. Plural of *trochischus*.

**trochiscus** (*tro-kis'-kus*). See *troche*.

**trochiter** (*trok'-it-er*) [τροχός, a wheel]. The greater tuberosity of the proximal end of the humerus.

**trochiterian** (*trok-it-e'-re-an*) [τροχός, a wheel]. Pertaining to the trochiter.

**trochlea** (*trok'-le-ah*) [L. a pulley]. A part or process having the nature of a pulley. t. of the astragalus, the surface of the astragalus articulating with the tibia. t. of the femur, the intercondyloid fossa of the femur. t. of the humerus, an articulation at the extremity of the humerus, over which a band of cartilage passes. t. labyrinthi. See *cochlea*. t. of the obliquus oculi superior, t. of the orbit, the ligamentous ring or pulley, attached to the upper margin of the orbit, which transmits the tendon of the superior oblique muscle of the eye. t. tali. See *t. of the astragalus*.

**trochlear** (*trok'-le-ar*) [see *trochlea*]. 1. Pertaining to or of the nature of a pulley. 2. Pertaining to the trochlear muscle. 3. Pertaining to the trochlear nerve.

**trochlearis** (*trok-le-a'-ris*) [see *trochlea*]. Pulley-shaped, as the trochlearis muscle or simply *trochlearis*, the superior oblique muscle of the eye.

**trochocardia** (*trok-o-kar'-de-ah*) [τροχός, wheel; καρδία, heart]. A rotary displacement of the heart on its long axis.

**trochocephalus** (*trok-o-sef'-al-us*) [τροχός, wheel; κεφαλή, head]. A rounded appearance of the head, due to partial synostosis of the frontal and parietal bones.

**trochoginglymus** (*trok-o-ging'-lim-us*) [τροχός, wheel; γίγγλυμος, ginglymus]. A combination of a hinge-joint and a pivot-joint, as in the humero-radial articulation.

**trochoid** (*tro'-koid*) [τροχός, wheel]. Serving as a pulley or pivot; involving a pivotal action.

**trochoides** (*tro-ko'-id-ēz*) [τροχός, a wheel]. A pivot-joint or pulley-joint, such as the atloaxoid joint.

**(von) Troeltsch's corpuscles** (*treltsh*) [Anton Friedrich *von Troeltsch*, German otologist, 1829–1890]. Spindle-shaped connective-tissue corpuscles, stellate on transverse section, found between the middle fibrous and inner circular layers of the membrana tympani. v. **T.'s spaces**, two small pockets formed in the upper part of the attic of the middle ear by folds of mucous membrane.

**Troisier's ganglion**, T.'s sign (*tro-se-ā'*) [Émile Troisier, French physician, 1844– ]. Enlargement of the left supraclavicular lymph-glands, an indication of malignant disease of the intraabdominal region.

**Trolard's vein** (*tro-lar'*) [Paulin *Trolard*, French anatomist]. An anastomotic vein that extends from the superior longitudinal sinus to the superior petrosal or the cavernous sinus.

**trolley-buzz.** A buzzing sound constantly heard by people who ride much on noisy trolley-cars.

**trolley-eye.** See *chalcitis*.

**Trombidium** (*trom-bid'-e-um*). A genus of mites which includes the harvest mite. By some, the chigoe is considered as belonging to this group.

**Trommer's test for glucose** (*trom'-er*) [—— *Trommer*, German chemist, 1806–1879]. To the liquid rendered alkaline by caustic soda a fairly strong solution of cupric sulphate is added drop by drop until a little of the copper hydrate formed remains undissolved on shaking. On warming in the presence of glucose, a yellow reduction of hydrate suboxide of copper is first formed, and then red suboxide separates, even below the boiling-point. If not enough copper salt has been used, the reaction will be yellowish-brown in color; but if the copper salt is in excess, the excess of hydrate is changed by boiling into a dark-brown hydrate, which interferes with the test.

**tromomania** (*trom-o-ma'-ne-ah*) [τρόμος, tremor; μανία, madness]. Delirium tremens.

**trona** (*tro'-nah*). Native sodium carbonate, Na$_2$CO$_3$.

**tropacocaine** (*tro-pa-ko'-ka-ēn*) [*atropine; cocaine*]. An alkaloid obtained from a small-leaved coca-plant of Java. t., **hydrochloride**, C$_{15}$H$_{19}$NO$_2$. C$_5$H$_5$CO. HCl, in 2 to 3% solutions, is preferred to cocaine hydrochloride, as a local anesthetic, as being less toxic and more certain.

**tropæolin.** See *tropeolin*.

**tropate** (*tro'-pāt*). A salt of tropic acid.

**tropein** (*tro'-pe-in*). A salt of tropin and an organic acid.

**tropeinism** (*tro'-pe-in-ism*). Poisoning by any of the tropeines or by plants (*Solanaceæ*) containing tropeins. It is characterized in light cases by dryness of the mouth, dysphagia, and acceleration of the pulse; in severer cases by dilatation of the pupils, ataxia, clonic spasms, psychic disturbances with excessive excitement; the severest cases are marked by loss of consciousness, anesthesia, paralysis of the sphincters, and cardiac and respiratory paralysis.

**tropeolin** (*tro-pe'-o-lin*). One of a group of orange anilin dyes, so-called from the resemblance of their colors to those of the flowers of *Tropæolum*, the garden nasturtium. Its solutions are turned brown by free acids, and are used as a test for such acids.

**tropesis** (*tro-pe'-sis*) [τροπή, a turn]. Inclination.

**trophe** (*trof'-e*) [τροφή, nourishment]. Aliment.

**trophedema.** See *trophoedema*.

**trophesial, trophesic** (*tro-fe'-ze-al, tro-fe'-sik*) [τροφή, nourishment]. Pertaining to or of the nature of a trophesy.

**trophesy** (*trof'-es-e*) [τροφή, nourishment]. Defective nutrition of a part resulting from disorder of the nerves regulating nutrition; trophoneurosis.

**trophic** (*trof'-ik*) [τροφή, nourishment]. Pertaining to the functions concerned in nutrition, digestion, and assimilation. t. **centers**, centers regulating the nutrition of nerves, or through them, of organs.

**tropho-** (*trof-o-*) [τροφή, nourishment]. A prefix denoting relation to nutrition or to nourishment.

**trophoblast** (*trof'-o-blast*) [τροφή, nourishment; βλαστός, a germ]. In biology, the outer epiblastic layer of the extra-embryonic somatopleure.

**trophoblastic** (*trof-o-blas'-tik*) [τροφή, nourishment; βλαστός, germ]. Pertaining to a trophoblast.

**trophoedema** (*trof-o-e-de'-mah*) [*tropho-; edema*].

A condition marked by localized permanent edema. t., chronic, frequently hereditary, marked by hard, white, painless swellings on the legs, lasting through life without material injury to health.

**tropholecithal** (*trof-o-les′-ith-al*) [τροφή, nourishment; λέκιθος, the yolk of an egg]. Pertaining to a tropholecithus.

**tropholecithus** (*trof-o-les′-ith-us*) [τροφή, nourishment; λέκιθος, the yolk of an egg]. In biology, the food yolk of a meroblastic egg. Cf. *morpholecithus*.

**trophology** (*trof-ol′-o-je*) [*tropho-*; λόγος, science]. The science of nutrition.

**trophoneurosis** (*trof-o-nū-ro′-sis*) [*tropho-*; *neurosis*]. Any disease of a part due to disturbance of the nerves or nerve-centers with which it is connected. t., disseminated. Synonym of *scleroderma*. t., facial, progressive facial atrophy; facial hemiatrophy. t., muscular, trophic changes in the muscles in connection with disease of the nervous system. t. of Romberg, unilateral atrophy of the face; hemiatrophy.

**trophoneurotic** (*trof-o-nū-rot′-ik*). Pertaining to or caused by a trophoneurosis.

**trophonine** (*trof′-on-ēn*). A proprietary food said to consist of beef, nuclealbumin, gluten of wheat, and enzymes of the digestive gland.

**trophonosis, trophonosus** (*trof-on′-o-sis, trof-on′-o-sus*). See *trophopathy*.

**trophonucleus** (*trof-o-nū′-kle-us*) [*tropho-*; *nucleus*]. The nucleus which is concerned with the nutrition of a cell and not with its reproduction.

**trophopathy** (*trof op′ ath-e*) [*tropho-*; πάθος, disease]. A disorder of nutrition.

**trophoplasm** (*trof′-o-plazm*) [*tropho-*; πλάσσειν, to mold]. The vital substance of the cell; the formative plasm.

**trophoplast** (*trof′-o-plast*) [see *trophoplasm*]. A mass of formative plasm.

**trophospongia** (*tro-fo-spun′-je-ah*) [τροφή, nourishment; σπογγία, a sponge]. In biology, the outer or maternal layer of the *trophoblast, trophodisc* or *trophocalyx*.

**trophotonos** (*trof-ot′-on-os*) [*tropho-*; τόνος, tension]. Rigidity of contractile tissue due to trophic disturbances.

**trophotropic** (*trof-o-trop′-ik*) [τροφή, nourishment; τρέπειν, to turn]. In biology, exhibiting trophotropism.

**trophotropism** (*trof-ot′-ro-pizm*) [*tropho-*; τρέπειν, to turn]. The attraction and repulsion exhibited by certain organic cells to various nutritive solutions.

**tropic** (*trop′-ik*) [τροπή, a turning]. An affix used by Ehrlich and Wright. See *bacteriotropic*. t. acid [*atropine*], C₉H₁₀O₃. An acid produced by treating atropine with baryta-water, alkalies or acids.

**tropidine** (*trop′-id-ēn*) [*atropine*], C₈H₁₃N. A substance resulting from the decomposition of atropine in the presence of hydrochloric and glacial acetic acids; it is an oily fluid having an odor like that of coniine.

**tropine** (*trop′-ēn*) [*atropine*], C₈H₁₅NO. A crystalline base obtained in the decomposition of atropine.

**tropism** (*tro′-pizm*) [τροπή, a turn]. The striving of living cells after light and darkness, heat or cold, etc. t., chemo-, the directing influence of chemical agents. t., photo-, that exerted by light. t., galvano-, that due to galvanic electricity.

**tropococaine.** See *tropacocaine*.

**tropometer** (*trop-om′-et-er*) [τροπή, turn; μέτρον, a measure]. 1. An instrument for measuring the various rotations of the eyeball. 2. An apparatus for estimating the amount of torsion in long bones.

**tropon** (*tro′-pon*) [τροφή, nourishment]. An albuminous substance obtained from animal and vegetable sources, containing 90 % of albumin. It is a light brown, nonhygroscopic powder, intended as a nutrient for convalescents. One teaspoonful to one tablespoonful is given with each meal in cocoa, soup, etc.

**Trousseau's disease** (*troo-so′*) [Armand Trousseau, French physician, 1801–1867]. Stomachal vertigo. **T.'s marks,** "taches cérébrales," circumscribed spots produced by mechanical irritation in tuberculous meningitis and other diseases seriously affecting the nutrition of the nervous system. **T.'s phenomenon,** muscular spasm, which continues as long as pressure is applied on the large arteries or on the nerve trunk in tetany, showing heightened neuromuscular irritability. **T.'s points apophysaires,** points sensitive to pressure over the dorsal and lumbar vertebræ in intercostal and lumboabdominal neuralgias. See *Valleix's points douloureux*. **T.'s roseola,** rubeola; rötheln. **T.'s symptom,** the production of paroxysms of tetany by pressure upon the principal nerve-trunks or blood-vessels of the parts affected; it is observed in tetany. **T.'s test for bile-pigments.** See *Smith's reaction*.

**troy ounce.** A unit in troy weight, equal to 480 grains. **t. weight.** See *weights and measures*.

**true.** Real; not false. **t. aneurysm.** See *aneurysm, true*. **t. conjugate.** See under *conjugate*. **t. corpus luteum,** the corpus luteum of pregnancy. **t. pelvis,** that part of the pelvic cavity situated below the iliopectineal line. **t. rib.** See *rib, true*. **t. skin,** the corium. **t. vocal bands,** the inferior bands, or those concerned in the production of the voice.

**truncal** (*trung′-kal*) [*trunk*]. Pertaining to a trunk.

**truncated** (*trung′-ka-ted*) [*trunk*]. Deprived of limbs or accessory parts.

**trunci** (*trung′-ki*) [L.]. Plural of *truncus*. **t. lumbales,** lumbar trunks.

**truncus** (*trung′-kus*) [L.: pl., *trunci*]. A trunk. **t. bronchomediastinalis dexter,** right bronchomediastinal trunk. **t. corporis callosi,** trunk or body of corpus callosum. **t. intestinalis,** intestinal trunk. **t. jugularis,** jugular trunk. **t. lumbosacralis,** lumbosacral trunk or cord. **t. costocervicalis,** costocervical trunk or cord. **t. subclavius,** subclavian trunk. **t. sympathicus,** sympathetic trunk. **t. thyreocervicalis,** the thyroid axis.

**Trunecek's method** (*troo′-nek-sek*). See under *serum, Trunecek's*.

**trunk** [*truncus*, a trunk]. 1. The body except the head and limbs. 2. The main stem of a nerve or vessel.

**truss** (*trus*) [Fr., *trousse*]. An apparatus for maintaining a hernia in place after reduction. Also an appliance for making pressure. **t., carotid,** a truss for compressing the carotid artery. **t., French,** a truss for inguinal hernia, in which pressure is exerted by an elastic, steel spring that supports the pad. **t., Hainsby's,** a truss for approximating the edges of a wound; it is used in the operation for harelip. **t., suspensory,** a suspensory bandage.

**trypan-blue.** A dye of the benzopurpurin series, used as a trypanocide.

**trypanocidal** (*tri-pan-o-si′-dal*) [*trypanosoma*; *cædere*, to kill]. An agent that destroys trypanosomes.

**trypanocide** (*tri′-pan′-o-san*). A dye-stuff of trypanocidal properties when combined with arseno-phenylglycin.

**Trypanosoma** (*tri-pan-o-so′-mah*) [τρύπανον, a borer; σῶμα, body]. A genus of protozoan parasitic organisms. **T. brucei,** the organism causing the tsetse fly disease of horses. **T. castellanii,** probably the same as *T. gambiense*. **T. equiperdum,** the exciting cause of dourine. *q. v.* **T. equinum,** the exciting cause of *mal de Cáderas* in the horse. **T. evansi,** the organism found in surra. **T. gambiense,** the organism causing sleeping-sickness. **T. lewisi,** one found in rats. **T. theileri,** one found in galziekte, a disease of cattle.

**trypanosome** (*tri′-pan-o-sōm*). One of any species of *Trypanosoma*.

**trypanosomiasis** (*tri-pan-o-so-mi′-a-sis*) [*Trypanosoma*]. Any of the several diseases due to infection with the various species of *Trypanosoma*.

**trypan-red.** A reddish-brown powder recommended in the treatment of trypanosomiasis.

**trypanroth.** Same as *trypan-red*.

**trypanrosan** (*tri-pan′-o-san*). A preparation of chlorinated parafuchsin, used in the treatment of trypanosomiasis.

**trypesis** (*trip-e′-sis*) [τρυπᾶν, to bore]. The operation of trephining.

**trypsalin** (*trip′-sal-in*). Trade name of a powder of trypsin said to be capable of dissolving dead tissue, it is designed for use by insufflation in nose and throat diseases.

**trypsase** (*trip′-sās*). See *trypsin*.

**trypsin** (*trip′-sin*) [τρίψις, a rubbing]. The proteolytic ferment of the pancreatic juice, which in an alkaline medium converts proteids into peptones. It has lately been advocated for curative use in cancer.

**trypsinogen** (*trip-sin′-o-jen*) [*trypsin*; γεννᾶν, to produce]. The zymogen from which trypsin is formed.

**tryptic** (trĭp'-tĭk) [*trypsin*]. Pertaining to or caused by trypsin.
**tryptolytic** (trĭp-tŏl-ĭt'-ĭk) [*trypsin*; λύειν, to loosen]. Of or pertaining to the peculiar cleavage properties of trypsin.
**tryptone** (trĭp'-tōn) [*trypsin*]. Peptone formed by the action of trypsin.
**tryptonemia, tryptonæmia** (trĭp-to-nē'-me-ah). See *peptonemia*.
**tryptophan** (trĭp'-to-fan). One of the end products of tryptic digestion. With a solution of chlorine or bromine it gives a violet color. Synonym, *proteinochromogen*. t. test, tryptophan is present in the stomach, as a result of pepsin digestion in cases of cancer of the stomach; the test is made by the addition of bromine water, as above.
**T. S.** Abbreviation of *test solution*.
**tsetse-fly** (tsĕt'-se). *Glossina morsitans* and *G. palpalis*, dipterous insects of South Africa, which carry the *Trypanosoma gambiense*. t. disease. See *disease, tsetse-fly*.
**Tsuga** (tsoo'-gah). A genus of Coniferæ, a species of which, *T. canadensis*, yields Canada pitch.
**tsutsugamushi disease** (tsoo-tsoo-ga-moo'-she). Japanese river fever.
**T. U.** Abbreviation of *toxic unit*.
**tua-tua.** See *Jatropha gossypifolia*.
**tub.** To treat by means of a cold bath.
**tuba** (tu'-bah) [L.]. A tube. t. acustica. Same as t. auditiva. t. auditiva, the auditory or Eustachian tube. t. Eustachii, Eustachian tube. t. fallopiana, t. Fallopii, t. uterina, Fallopian tube.
**tubage** (tu'-bāj) [*tuba*, a tube]. The introduction of a tube or catheter. t. of the glottis. See *intubation*.
**tubal** (tu'-bal) [*tube*]. Pertaining to a tube, especially the Fallopian tube or the renal tubules. t. abortion, internal rupture of the ovum in extra-uterine gestation, with a pouring out of blood through the fimbriated extremity of the tube into the abdominal cavity. t. mole, a tubal ovum that has been destroyed by hemorrhage. t. pregnancy, pregnancy in one or the other Fallopian tube.
**tubba, tubboe** (tub'-ah, tub'-o). Yaws attacking the palms of the hands and the soles of the feet; crab-yaws.
**tubbing** (tub'-ing). The employment of the cold bath in the treatment of fever.
**tube** (tūb) [*tuba*, a tube]. A hollow, cylindrical structure, especially the Fallopian tube or the Eustachian tube. t., air-, a bronchial tube. t., alimentary, the alimentary canal. t., auditory, the external auditory canal. t., auricular, the external auditory meatus. t., auscultation, one used to test the acuteness of hearing. t., capillary, a tube with minute lumen. t., cardiac, the embryonic heart. t.-casts, casts of the renal tubules; they indicate disease of the kidneys. t., Crookes', t., Geissler's, t., Hittorf's. See under *rays, Roentgen-*. t., drainage-, a hollow tube of glass, rubber, or other material inserted into a wound or cavity to allow of the escape of fluids. t., Eustachian. See *Eustachian*. t.s, Fallopian. See *oviducts*. t., feeding, one for introducing food into the stomach. t.s, fusion, Priestley Smith's name for a miniature stereoscope by which the two images formed by a straight and a squinting eye may be fused together and seen simultaneously. Cf. *heteroscope*. t., intubation, a tube for insertion into the larynx through the mouth in laryngeal diphtheria, etc. t., sediment, a glass cylinder constricted to a fine point at one end and both ends open; it is used in precipitating urine. t., stomach, a flexible tube used for lavage. t., tracheotomy. See *tracheotomy*. t., vacuum, a sealed glass tube out of which the air has been pumped-and which has at each end a piece of platinum wire passed through the glass and entering the tube.
**tuber** (tu'-ber) [L., "a bump or swelling"]. 1. A thickened portion of an underground stem. 2. Any rounded swelling. t. anatomica, a protuberance, tumor, or swelling. t. annulare, the anterior surface of the pons; see *pons Varolii*. t. anterius. See *t. cinereum*. t. calcanei, the tuberosity of the calcaneum. t. cinereum, tubera candicantia. See *corpora albicantia*. t. cinereum, a tract of gray matter extending from the optic chiasma to the corpora albicantia and forming part of the floor of the third ventricle. t. cochleæ, the promontory of the tympanum. t. corporis callosi, the splenium. t. Eustachii, a slight protuberance below the fenestra ovalis on the inner wall of the tympanic cavity. t. frontale, the frontal eminence. tubera geniculata, the internal and external geniculate bodies. t. gutturosum. See *goiter*. t. ischiadicum, the tuberosity of the ischium. t. ischii, the tuberosity of the ischium. t. maxillæ, the tuberosity of the superior maxilla. t. maxillare, the maxillary tuber. t. omentale hepatis, a prominence on the left lobe of the liver, corresponding to the lesser curvature of the stomach. t. omentale pancreatis, a prominence of the middle part of the pancreas, corresponding to the lesser omentum. t. parietale, the parietal eminence. t. posticum. Same as t. vermis. t. supracondyloideum, an eminence opposite the distal end of the internal border of the linea aspera of the fibula. t. supraorbitale, the superciliary ridge. t. syphiliticum. See *syphiloma*. t. tympani. See *t. Eustachii*. t. valvulæ (cerebelli), a small prominence at its anterior extremity in front of the uvula. t. vermis, the posterior end of the inferior worm of the cerebellum; also called *t. valvulæ*. t. verrucosum, a callosity often found on the great toe at the metatarso-phalangeal joint. t. zygomaticum, a prominence of the zygoma on its lower border near the union of the superior maxilla and the zygomatic process.

**tubercle** (tu'-ber-kl) [*tuberculum*, a tubercle]. 1. A small nodule. 2. A rounded prominence on a bone. 3. The specific lesion produced by the tubercle bacillus, consisting of a collection of round-cells and epithelioid cells, with at times giant-cells. t., acoustic, the nucleus of the dorsal cochlear nerve, a leaf-like mass of cinerea wrapped about the dorsolateral surface of the restis. t., adductor, a slight protuberance at the lower end of the internal supracondylar line of the femur, giving attachment to the tendon of the adductor magnus. t., amygdaloid, a prominence on the roof of the descending cornu of the lateral ventricle. t., anatomical, a wart-like tuberculous growth sometimes appearing on the hands of dissectors. t., anterior, a tubercle at the anterior part of the extremity of the transverse process of certain vertebræ. t. bacillus. See under *bacteria*. t., carotid, a prominence of the sixth cervical vertebra on the anterior part of its transverse process. t., conoid, a broad projection of the clavicle on its posterior border at the union of its middle and outer thirds, to which the conoid ligament is attached. t., deltoid, a projection on the anterior border of the clavicle, giving origin to a part of the deltoideus. t., dissection, anatomical tubercle. t., fibrous, a tubercle which has been modified by the formation of connective tissue within its structure. t., genial, one of the tubercles on each side of the middle line on the inner surface of the lower maxilla. t., genital, the rudimentary penis or clitoris in the urogenital region of the embryo in front of the cloaca. t., gray. See *tubercle* (3). t., hepatic, in the embryo, the bile-tubules. t., hyaline. See *tube-cast, hyaline*. t., lacrimal, one of the small papillary prominences at the margin of the eyelids, in the center of which is the punctum lacrimale. t. of Lower, a small eminence on the wall of the right auricle, between the orifices of the venæ cavæ. t., lymphoid, a tubercle consisting chiefly of round or lymphoid cells. t., mammillary. See *corpora albicantia*. t., miliary. See *tubercle* (3). t., migratory, the tube of ectodermal tissue from which the nerve-system is developed. t., olfactory. See *bulb, olfactory*. t., otopharyngeal, the Eustachian tube. t., painful, a painful nodule in the subcutaneous tissue in the region of the joints. t., posterior, a tubercle at the posterior end of the lumbar and several of the thoracic vertebræ. t., postglenoid, a process of the temporal bone that descends behind the condyle of the jaw and prevents backward displacement during mastication. t., prostatic, a middle lobe of the prostate. t., pterygoid, a tubercle on the inner surface of the inferior maxilla; it gives attachment to the internal pterygoid muscle. t. of Rolando, one of the rounded masses close under the surface of the lateral columns of the medulla oblongata, formed by the enlarged dorsal horns of the gray matter. t., scalene-, a tubercle on the first rib, giving attachment to the anterior scalene muscle. t., supraglenoid, one above the superglenoid fossa of the scapula, giving attachment to the long head of the biceps. t. of the vagina, a prominence on the anterior wall of the vagina. t., zygomatic,

# TUBERCULA 911 TUBERCULUM

one at the junction of the zygoma with its anterior root.
**tubercula** (tū-ber'-kū-lah) [plural of *tuberculum*, a tubercle]. **t. coronæ dentis,** tubercles of the crown of a tooth. **t. dolorosa.** See *tubercle, painful.* **t. quadrigemina.** See *corpora quadrigemina.*
**tubercular** (tū-ber'-kū-lar) [*tubercle*]. 1. Presenting the appearance of a tubercle. 2. Provided with tubercles.
**tuberculase** (tū-ber'-kū-lās). See *Behring's tulase.*
**tuberculate, tubercled** (tū-bur'-kū-lat, tū'-bur-kld) [*tuberculum*, a tubercle]. Warty; bearing tubercles.
**tuberculated** (tū-bur'-kū-la-ted) [*tuberculum*, a tubercle]. Furnished with tubercles; tuberculous.
**tuberculation** (tū-bur-kū-la'-shun) [*tuberculum*, a tubercle]. The formation, development, or arrangement of tubercles; the process of affecting a part with tubercles.
**tuberculid, tuberculide** (tū-ber'-kū-lid). Any cutaneous manifestation due to the toxins of the tubercle bacilli.
**tuberculin** (tū-ber'-kū-lin) [*tubercle*]. A glycerol extract of cultures of the bacillus of tuberculosis. It is a brownish, neutral liquid, soluble in water, and is used as a means of diagnosing tuberculosis, and treating; when injected into tuberculous individuals, a reaction is produced which differs from that given by healthy individuals. Syn., *paratoloid.* T. A, the result of extracting the bacilli with a 10% normal caustic soda solution and filtering and neutralizing the product. **t. filtrate,** the bouillon from cultures on which tubercle bacilli of the human type have been grown to maturity and freed from germs by filtration through porcelain; no heat is used in its manufacture. Syn., *tuberculin Denys, B. F.* **t.,** Koch's. See *t., new;* and *t., old.* **t., new** (T. R.), an unsterilized, unfiltered, glycerol-water semisolution of living, dried, pulverized, and washed bacilli. **t., original,** or **old** (T. O.). See *tuberculin.* **t., purified,** the resultant redissolved precipitate of the tuberculin original with 60% of alcohol. T. R. See *t., new.* **t. test,** the injection of a small amount of tuberculin will produce fever and local swelling in a person or animal who has tuberculosis; but there is no reaction in one free from tuberculosis.
**tuberculine** (tū-ber'-kū-len). A ptomaine produced from the tubercle bacillus.
**tuberculinization** (tū-ber-kū-lin-iz-a'-shun). Treatment of tuberculosis by the use of tuberculin.
**tuberculinose** (tū-ber'-kū-lin-ōs). Dialyzed tuberculin.
**tuberculitis** (tū-ber-kū-li'-tis) [*tubercle;* ιτις, inflammation]. Inflammation in the tissues surrounding a tuberculous node.
**tuberculization** (tū-ber-kū-liz-a'-shun) [*tuberculum*, tubercle]. 1. The formation of tubercles, or the condition of being charged with tubercles. 2. Treatment with tuberculin.
**tuberculoalbumin** (tū-ber-kū-lo-al-bū'-min). A tuberculin preparation similar to tuberculase.
**tuberculocele** (tū-ber'-kū-lo-sēl) [*tuberculosis;* κήλη, a tumor]. Tuberculous disease of the testicle.
**tuberculocidin** (tū-ber-kū-lo-si'-din) [*tubercle; cædere*, to kill]. An albumose obtained from tuberculin by precipitation with platinum chloride. It is said to possess the beneficial effects of tuberculin without producing an injurious reaction.
**tuberculoderma** (tū-ber-kū-lo-der'-mah). A cutaneous manifestation of the action of tubercle bacilli, a tuberculid.
**tuberculofibroid** (tū-ber-kū-lo-fi'-broid). Relating to a tubercle that has undergone fibroid degeneration.
**tuberculoid** (tū-ber'-kū-loid) [*tuberculum*, tubercle; εἶδος, like]. Resembling tubercle or tuberculosis.
**tuberculoidin** (tū-ber-kū-loi'-din). Tuberculin which has been treated with alcohol and so cleared of its bacilli.
**tuberculol** (tū-ber'-kū-lol). Tuberculin which has been freed from secondary products.
**tuberculoma** (tū-ber-kū-lo'-mah). A tuberculous tumor.
**tuberculomyces** (tū-ber-kū-lo-mi'-sēz). A group of bacilli containing the different varieties of tubercle bacilli.
**tuberculophobia** (tū-ber-kū-lo-fo'-be-ah) [*tuberculosis;* φόβος, fear]. Morbid fear of tuberculosis.
**tuberculoplasmin** (tū-ber-kū-lo-plaz'-min). The filtered watery solution of the protoplasm of moist living bacilli, extracted by crushing with hydraulic pressure.

**tuberculosamine** (tū-ber-kū-lo'-sam-ēn). An amine isolated from tubercle bacilli.
**tuberculose** (tū-ber'-kū-lōs). See *tuberculated.*
**tuberculosis** (tū-ber-kū-lo'-sis) [*tubercle*]. An infectious disease due to *Bacillus tuberculosis*, discovered by Koch. The lesion produced by the growth of the bacillus is the tubercle (miliary or gray tubercle or nodule), a small, grayish, translucent nodule, from $\frac{1}{10}$ to 2 mm. in diameter, firmly embedded in the surrounding tissues. By the coalescence of neighboring tubercles large masses, the so-called tuberculous infiltrations, are produced. The tendency of tuberculous lesions is to undergo cheesy necrosis. For this degeneration two factors are responsible: the absence of blood-vessels and the action of peculiar poisons elaborated by the bacillus. The breaking down of tuberculous areas in the interior of organs gives rise to cavities, which may be seen in muscles, bones, brain, lymphatic glands, and elsewhere, but are most pronounced in the lungs. On surfaces—skin and mucous membranes—tuberculosis often leads to the formation of ulcers. The most frequent seats of tuberculosis are the lung, the intestinal tract, the lymphatic glands, the serous membranes, the bones, the skin, the testicle, the epididymis, the brain, the Fallopian tubes, the uterus, the spleen. The symptoms of tuberculosis vary with the localization of the disease. A few general phenomena are common to nearly all forms, viz,. emaciation, loss of strength, anemia, fever, and sweats. **t., acute miliary,** an acute febrile disease, characterized by the formation of minute tubercles in great numbers in various parts of the body. It is due to the discharge into the circulatory stream of tubercle bacilli. Three forms are usually described: (1) a general or typhoid form; (2) one with marked pulmonary symptoms; (3) one in which cerebral symptoms predominate. **t., attenuated,** tuberculosis with tendency to cold abscesses and various skin complications. **t., avian,** tuberculosis affecting birds. **t., bovine,** tuberculosis occurring in cattle. Syn., *pearl disease.* **t., cestodic,** a disease resembling tuberculosis, due to infestation with cestodes. **t., disseminated,** acute miliary tuberculosis. **t., general miliary.** See *t., acute miliary.* **t., laryngeal,** tuberculosis of the larynx, usually secondary to tuberculosis of the lungs, but in rare cases primary. **t., miliary,** tuberculosis characterized by the formation of miliary tubercles. **t., pulmonary,** phthisis, pulmonary. **t., surgical,** tuberculosis of parts amenable to surgical treatment, as the bones and joints.
**tuberculotoxin** (tū-ber-kū-lo-toks'-in). A toxin generated by the tubercle bacillus.
**tuberculotoxoidin** (tū-ber-kū-lo-toks-oid'-in). A solution of tubercle bacilli in sulphuric acid, said to have some immunizing value and to raise the opsonic index.
**tuberculous** (tū-ber'-kū-lus) [*tubercle*]. Affected with or caused by tuberculosis.
**tuberculum** (tū-ber'-kū-lum). See *tubercle.* **t. acusticum,** a group of nerve-cells connected with the auditory fibers. **t. anterius.** 1. The conical prominence on the anterior arch of the atlas. 2. The frontal extremity of the thalamus. **t. articulare,** articular tubercle. **t. auriculæ** (*Darwinii*), Darwinian tubercle of auricle. **t. caudatum,** the caudate lobe of the liver. **t. cinereum,** gray or ashen tubercle: (1) the cuneate tubercle of the oblongata; (2) the tuberculum Rolandi, found below the clava. **t. corniculatum** (*Santorini*), corniculate tubercle of Santorini. **t. costæ,** tubercle of the rib. **t. cuneiforme** (*Wrisbergi*), cuneiform cartilage. **t. epiglotticum,** epiglottic tubercle or cushion of epiglottis. **t. impar,** a rounded elevation between the ventral ends of the mandibular and hyoid arches and from which the papillary portion of the tongue is developed. **t. intercondyloideum laterale,** lateral intercondyloid tubercle. **t. intercondyloideum mediale,** medial intercondyloid tubercle. **t. intervenosum** (*Loweri*), intervenous tubercle of Lower. **t. jugulare,** jugular tubercle. **t. majus,** larger tubercle or greater tuberosity. **t. mentale,** mental tubercle. **t. minus,** smaller tubercle or lesser tuberosity. **t. obturatorium anterius,** anterior obturator tubercle. **t. obturatorium posterius,** posterior obturator tubercle. **t. pharyngeum,** pharyngeal tubercle. **t. posterius,** the rudimentary spinous process of the atlas. **t. pubicum,** pubic tubercle or spine of os pubis. **t. scaleni** (*Lisfranci*), scalene tubercle of Lisfranc. **t. sebaceum.** See *milium.* **t. supratragicum,** supra-

tragic tubercle. **t. thyreoideum inferius,** inferior thyroid tubercle. **t. thyreoideum superius,** superior thyroid tubercle. **t. vestibularis.** Same as *t. acusticum.*

**tuberiferous** (*tū-ber-if'-er-us*) [*tuber,* a tuber; *erre,* to bear]. Producing tubers.

**tuberose** (*tū'-ber-ōs*) [*tuber*]. Resembling a tuber.

**tuberositas** (*tū-ber-os'-it-as*) [L.]. A tuberosity. **t. coracoidea,** the coracoid tuberosity, an impression for the conoid ligament. **t. costalis,** costal tuberosity or impression for rhomboid ligament. **t. deltoidea,** deltoid tuberosity. **t. glutæa,** gluteal tuberosity. **t. iliaca,** iliac tuberosity. **t. infraglenoidalis,** infraglenoidal tuberosity. **t. masseterica,** masseteric tuberosity. **t. ossis cuboidei,** tuberosity of cuboid bone. **t. ossis navicularis,** tuberosity of scaphoid bone of tarsus. **t. pterygoidea,** pterygoid tuberosity. **t. radii,** tuberosity of radius, or bicipital tuberosity. **t. sacralis,** sacral tuberosity. **t. supraglenoidalis,** supraglenoidal tuberosity or tubercle. **t. tibiæ,** tuberosity or tubercle of the tibia. **t. ulnæ,** tuberosity of the ulna. **t. unguicularis,** ungual tuberosity.

**tuberosity** (*tū-ber-os'-it-e*) [*tuber*]. A protuberance on a bone. **t., greater,** a rough projection on the outer side of the head of the humerus. **t. of the ischium,** a thick, downward projection of the ischium, on which the body rests in sitting. **t., lesser,** a small tuberosity in front of the head and on the inner side of the bicipital groove of the humerus. **t. maxillary,** a rounded eminence at the lower part of the zygomatic surface of the superior maxillary bone, especially prominent after the growth of the wisdom-teeth. **ts. of the os calcis,** two prominences, an external and an internal, on the posterior inferior aspect of the os calcis. **t. of the palate bone,** a pyramidal process at the lower part of the posterior border of the external surface of the palate bone. **t. of the radius,** a rough eminence at the inner and interior aspect of the bone just beneath the neck.

**taberous** (*tu'-ber-us*) [*tuber*]. Like a tuber, as *tuberous* angioma.

**Tuebingen heart.** A disease of the heart, first observed at Tuebingen, marked by cardiac dilatation and hypertrophy and believed to be due to overindulgence in alcoholic drinks.

**tubo-** (*tū-bo'-*) [*tube*]. A prefix meaning relating to a tube.

**tuboabdominal** (*tū-bo-ab-dom'-in-al*) [*tubo-; abdomen*]. Pertaining to a Fallopian tube and to the abdomen. **t. pregnancy,** one that begins in the tube, but that finally becomes abdominal.

**tuboadnexopexy** (*tū-bo-ad-neks'ʻo-peks-ē*). Surgical fixation of the uterine adnexa.

**tuboligamentus** (*tū-bo-lig-am-enī'-us*). Relating to the oviduct and the broad ligament.

**tubo-ovarian** (*tū-bo-o-va'-re-an*) [*tubo-; ovary*]. Pertaining to the Fallopian tube and the ovary. **t. pregnancy,** an extrauterine pregnancy in which the ovum develops between the fimbriæ of the oviduct and the ovary.

**tubo-ovariotomy** (*tū-bo-o-va-re-ot'-om-ē*). Excision of a Fallopian tube and ovary.

**tuboperitoneal** (*tū-bo-per-it-on-e'-al*). Relating to the oviduct and the peritoneum.

**tubotympanal** (*tū-bo-tim'-pan-al*). Pertaining to a Eustachian tube and the tympanum of the ear.

**tubouterine** (*tū-bo-ū'-ter-in*) [*tubo-; uterus*]. Pertaining to the Fallopian tube and the uterus. **t. pregnancy,** a form of tubal pregnancy in which the ovule develops in the uterine wall, a portion of the sac often projecting into the uterus, and having on the outer side the round ligament and the greater portion of the tube. Also known as interstitial pregnancy.

**tubovaginal** (*tū-bo-vaj'-in-al*). Pertaining to a Fallopian tube and the vagina.

**tubular** (*tū'-bū-lar*) [*tubulus,* a small tube]. 1. Shaped like a tube. 2. Pertaining to or affecting tubules, as *tubular* nephritis. 3. Produced in a tube, as *tubular* breathing. **t. adenoma,** an adenoma after the type of tubular glands. **t. breathing.** Synonym of *breathing, bronchial.* **t. epithelioma,** a carcinoma found in the salivary glands composed of irregular cells, the cell-masses extending in tubes or cylindrical plugs in various directions. Pearly bodies are also present. **t. gland,** a secreting gland tubelike or cylindrical in shape; also the enteric glands or follicles of Lieberkuehn. **t. gestation,** extrauterine fetation in the oviduct. **t. membrane.** See *perineurium, neurilemma.* **t. pneumonia.** Synonym of *pneumonia, lobular.* **t. rale,** one produced in a bronchial tube.

**tubulature** (*tū'-bū-la-tūr*) [*tubule*]. The short tube of a retort or receiver.

**tubule** (*tū'-būl*) [*tubulus,* dim. of *tubus,* a tube]. A small tube. In anatomy, any minute, tube-shaped structure; see also *tubulus.* **t., communicating,** or junctional, that part of a uriniferous tubule between the distal convoluted, and the straight collecting tubule. **t., dentinal,** the tubular structure of the teeth. **t. of Ferrein.** See *Ferrein, tube of.* **t.s, segmental, t.s, Wolffian.** See under *Wolffian body.* **t., seminiferous,** any one of the tubules of the testicles. **t., uriniferous,** one of the numerous winding tubules of the kidney.

**tubuli** (*tū'-bū-lī*) [L.]. Plural of *tubulus.* **t. renales,** renal tubules. **t. renales contorti,** convoluted renal tubules. **t. lactiferi,** the excretory ducts of the mammæ. **t. renales recti,** straight renal tubules. **t. seminiferi contorti,** convoluted seminiferous tubules. **t. seminiferi recti,** straight seminiferous tubules.

**tubuliform** (*tū'-bū-lif-orm*) [*tubulus,* tubule; *forma,* form]. Shaped like a tubule.

**tubulization** (*tū-bū-li-za'-shun*). Protection of the ends of nerves, after neurorrhaphy, by a paraffin tube.

**tubulocyst** (*tū'-bū-lo-sist*). A cystic dilatation occurring in an occluded canal or duct.

**tubulodermoid** (*tū-bū-lo-der'-moid*). A dermoid tumor in fetal tubular structure, which should have become occluded.

**tuboluracemose** (*tū-bū-lo-ras'-em-ōs*). Denoting a gland that is both tubular and racemose.

**tubulus** (*tū'-bū-lus*) [*tubulus,* a small tube: pl., *tubuli*]. A small tube-like organ; a tubule.

**tubus** (*tū'-bus*) [L.]. A tube, canal. **t. acusticus,** an ear trumpet. **t. digestorius,** the digestive canal. **t. medullaris,** the vertebral canal. **t. respiratorius,** the respiratory canal. **t. vertebralis,** the spinal or vertebral canal.

**Tuerck's bundle** [Ludwig *Tuerck,* Austrian neurologist, 1810–1878]. A tract of nerve-fibers passing from the cortex of the temporosphenoidal lobe through the outer portion of the crusta of the cerebral peduncle and the pons into the internal geniculate body. **T.'s column,** the anterior or direct pyramidal tract. **T.'s degeneration,** secondary parenchymatous degeneration of the spinal nerve tracts. **T.'s hemianesthesia,** anesthesia affecting the functions of the posterior spinal roots of one side, at times also those of the nerves of special sense. It is caused by lesions of the posterior portion of the capsula and the contiguous region of the corona radiata. **T.'s trachoma,** granular laryngitis affecting the postero-internal wall of the larynx. Syn., *laryngitis sicca.*

**Tuffier's inferior ligament** (*toof-e-a'*) [Théodore *Tuffier,* French surgeon]. That portion of the enteric mesentery which is inserted into the iliac fossa. **T.'s syndrome,** a congenital state of general tissue debility, resulting in relaxation and displacement of various organs, such as splanchnoptosis, varicocele, uterine displacements, etc.

**Tuffnell's bandage** (*tuf'-nel*) [Thomas Joliffe *Tuffnell,* English surgeon, 1819–1885]. An immovable bandage stiffened with a paste of white of egg and flour. It is also called *egg-and-flour bandage.* **T.'s method,** T.'s treatment. A treatment for aneurysm, consisting in absolute rest, dry diet, and the administration of potassium iodide.

**tuft, Malpighian.** See *Malpighian body.*

**tugging, tracheal.** See *tracheal tugging.*

**tulase** (*tu'-lās*). See *Behring's tulase.*

**tulipine** (*tu'-lip-ēn*). A poisonous alkaloid from the tulip.

**Tully's powder** [William *Tully,* American physician, 1785–1859]. A powder containing morphine sulphate, 1 part; camphor, licorice, and calcium carbonate, each, 20 parts.

**Tulpius' valve** [Nicholas *Tulp,* Dutch physician, 1593–1674]. The ileocecal valve.

**tumefaction** (*tū-me-fa'-shent*) [*tumefaction*]. Swelling; swollen.

**tumefaction** (*tū-me-fak'-shun*) [*tumefacere,* to cause to swell]. A swelling.

**tumenol** (*tū'-men-ol*) [*bitumen; oleum,* oil]. A sulphonated preparation of certain hydrocarbons. It may be used in three forms: 1. *tumenol* itself, a dark-brown or brownish-black liquid; 2. *tumenol sulphone,* an aromatic, syrupy liquid; 3. *tumenol*

# TUMESCENCE 913 TURNING

*sulphonic acid*, a black powder, soluble in water, 10 per cent. tincture is used in the itching dermatoses.
**tumescence** (*tū-mes'-ens*) [*tumescere*, to swell]. The condition of growing tumid; a swelling.
**tumescent** (*tū-mes'-ent*) [*tumescere*, to become swollen]. Swelling or enlargement.
**tumid** (*tū'-mid*) [see *tumescence*]. Swollen.
**tumidity** (*tū-mid'-it-e*) [*tumidus*, swollen]. The state of being swollen.
**tumor** (*tū'-mor*) [*tumere*, to swell]. 1. A swelling. 2. A new growth not the result of inflammation. The appended classification is based, at least as regards classes A and B, on the blastodermic origin of the dominant tissue of the tumor. 3. A mass of cells, tissues, or organs, resembling those normally present in the body, but arranged atypically, growing at the expense of the body, but subserving no useful purpose therein. **t. albus**, white swelling; tuberculous enlargement of a joint. **t., benign,** one which does not give rise to metastasis or recur after removal. **t., cystic,** one made up of cysts. **t., dentinoid,** a dental osteoma arising from the crown of a tooth. **t., fibroid,** a fibroma. **t., Gubler's,** a prominence on the back of the wrist seen in wrist-drop. **t., gummous** or **gummy,** a syphilitic gumma. **t., heterologous,** one composed of tissue differing from that in which it grows. **t., histoid,** one composed of a single tissue. **t., homologous,** one composed of tissue resembling that from which it grows. **t., malignant,** one which gives metastasis or recurs, or does both, and eventually destroys life. **t., mucous, a myxoma. t., muscular,** a myoma. **t., phantom,** an apparent tumor due to flatus or contraction of a section of an abdominal muscle; it is seen in hysterical patients. **t., potato,** an endothelioma derived from the carotid body. **t., sebaceous,** one of a sebaceous gland; an atheroma. **t., splenic,** a term sometimes applied to an enlarged spleen. **t., teratoid, a teratoma.**

### A. Mesodermic Tumors.

1. Sarcoma.
   - Round-cell... { Large. Small. Lymphosarcoma.
   - Spindle-cell, { Large. Small.
   - Giant-cell.
   - Melanotic.
   - Alveolar.
   - Endothelioma.
   - Angiosarcoma.
   - Cylindroma.
   - Chloroma.
   - Psammoma.
2. Fibroma. { Hard. Soft.
3. Myxoma.
4. Lipoma.
5. Chondroma. { Hyaline. Fibrous.
6. Osteoma. { O. durum or O. eburneum. O. spongiosum, O. medullare.
7. Hemangioma. { Telangiectatic. Cavernous.
8. Lymphangioma. { Simple. Cystic.
9. Myoma. { Liomyoma. Rhabdomyoma.

### B. Ectodermic and Entodermic Tumors.

1. Glioma.
2. Neuroma. { N. myelinicum. N. amyelinicum.
3. Epithelioma.
   - Adenoma. { Tubular. Racemose.
   - Carcinoma. { Squamous. Cylindrical. Glandular.
   - Epithelial cystoma. { C. simplex. C. papilliferum.

### C. Teratoid Tumors or Teratomata.

1. Dermoid cyst.
2. Cholesteatoma.

**tumoraffin** (*tū'-mor-af-fin*) [*tumor*; *affinity*]. Said of substances (drugs, radiant energy, etc.) which are supposed to have some special affinity for tumor cells.

**tumultus** (*tū-mul'-tus*) [L.]. Tumult. **t. cordis,** irregular heart-action. **t. sermonis,** a stuttering manner of reading, from pathologic cause.
**tungstate** (*tung'-stāt*). See under *tungsten*.
**tungsten** (*tung'-sten*) [Swed., "heavy stone"]. A metallic element having a specific gravity of 19.26, an atomic weight of 184. Symbol W (from the German name *Wolfram*). It forms *tungstic acid*, H₂WO₄, the latter combining with bases to form *tungstates*, which are used as reagents. See *elements*, *table of chemical*.
**tungstic acid** (*tung'-stik*). See under *tungsten*.
**tunic** (*tū'-nik*) [*tunica*]. A coat or membrane; see *tunica*.
**tunica** (*tū'-nik-ah*) [L.]. A tunic. **t. adnata,** the conjunctiva covering the eyeball. **t. adventitia,** the outer coat of an artery. **t. albuginea corporum cavernosorum,** the fibrous covering of the corpora cavernosa. **t. albuginea oculi,** the sclerotic coat of the eye. **t. albuginea ovarii,** the compact connective tissue immediately under the epithelium of the cortex of the ovary. **t. albuginea testis,** the fibrous covering of the testis. **t. extima,** see *t. adventitia*. **t. intima,** the inner coat of an artery. **t. media,** the middle coat of an artery. **t. ruyschiana,** the layer of capillary vessels of the choroid coat of the eye. **t. vaginalis,** the serous covering of the testis derived from the peritoneum. **t. vasculosa,** the vascular layer of the testis, called also the pia mater of the testis; also the lamina vasculosa of the choroid.
**tunicin** (*tū' nis-in*), (C₆H₁₀O₅). A substance obtained from the mantles of ascidians; considered by some as identical with vegetable cellulose, by some as identical with animal cellulose, and by others as a distinct body convertible into sugar.
**tuning-fork** (*tū'-ning-fork*). A pronged, metallic instrument capable of vibrating so as to form a certain definite note.
**tunnel-anemia.** See *ankylostomiasis*.
**tunnel-disease.** See (1) *caisson-disease*; (2) *ankylostomiasis*.
**tupelo** (*tū'-pel-o*). The *Nyssa grandidentata*, of the order *Cornaceæ*. Its root has been used for making tents (*tupelo-tent*).
**turacin** (*tū'-ras-in*) [African, *turakoo*]. A crimson coloring-matter obtained from the feathers of the turakoo. It is slowly soluble in water, but easily soluble in alkaline fluids, and contains about 6 per cent. of copper.
**turbid pneumonia.** A term applied to the indistinct pneumonic symptoms following injections of tuberculin; it is also called *injection pneumonia*.
**turbinal** (*tur'-bin-al*) [*turbo*, top]. 1. Turbinated. 2, A turbinated bone.
**turbinated** (*tur'-bin-a-ted*) [*turbinal*]. Top-shaped; scroll-shaped. **t. bodies,** the turbinated bones with their covering of vascular and mucous membrane. **t. bone,** one of the three (superior, middle, and inferior) bony projections upon the outer wall of each nasal fossa. They are covered by an erectile vascular mucous membrane.
**turbinectomy** (*tur-bin-ek'-to-me*) [*turbinal*; ἐκτομή, a cutting out]. Excision of a turbinated bone.
**turbinotome** (*tur'-bin-ot-ōm*). An instrument used in turbinotomy.
**turbinotomy** (*tur-bin-ot'-o-me*) [*turbinal*; τομή, a cutting]. Incision into a turbinated bone.
**turbo cerebri** [L., "the top-shaped-body of the brain"]. The pineal body.
**turgescence** (*tur-jes'-ens*) [*turgid*]. Swelling.
**turgid** (*tur'-jid*) [*turgidus*, swollen]. Swollen; congested.
**turgometer** (*tur-gom'-et-er*) [*turgor*, swelling; μέτρον, measure]. An apparatus to determine the degree or amount of turgescence.
**turgor** (*tur'-gor*) [L., "a swelling"]. Active hyperemia; turgescence. **t. vitalis,** the normal fulness of the blood-vessels.
**Turlington's balsam** (*tur'-ling-tun*). The compound tincture of benzoin.
**turmeric** (*tur'-mer-ik*). See *curcuma*.
**turmerol** (*tur'-mer-ol*). An oily substance derived from turmeric.
**turn.** 1. To cause to revolve about an axis. 2. To change the position of the fetus so as to facilitate delivery. **t.** of life, see *menopause*.
**Turner's cerate.** The ceratum calaminæ or ointment of calamin (20 per cent.). **T.'s yellow,** same as *Cassel yellow*.
**turning** (*turn'-ing*). See *version*.

# TURPENTINE 914 TYMPANY

**turpentine** (*tur'-pen-tīn*) [τερέβινθος, terebinth]. A concrete or liquid oleoresin obtained from various species of *Coniferæ*. The ordinary or *white turpentine* (*terebinthina*, U. S. P.; *thus americanum*, B. P.), derived from *Pinus palustris* and other species of *Pinus*, contains a volatile oil, oil or spirits of turpentine. t., **camphor**, terpene hydrochlorate. t., **Canada** (*terebinthina canadensis*, U. S. P., B. P.), is obtained from *Pinus balsamea*, and under the name of Canada balsam is used as a mounting medium in microscopy. t., **Chian**, collected on the island of Chios, from *Pistacia terebinthus*, was formerly used in cancer. t., common European, t., **Bordeaux**, is obtained from several species of pine; chiefly *Pinus sylvestris* and *Pinus maritima*. It yields large quantities of oil of turpentine. t., **confection of** (*confectio terebinthinæ*, B. P.). Dose ½–1 dr. (2–4 Gm.). t., **enema of** (*enema terebinthinæ*, B. P.), oil of turpentine, 1 oz.; mucilage of starch, 15 oz. t. **liniment** (*linimentum terebinthinæ*, U. S. P., B. P.), resin cerate and oil of turpentine. t., **liniment of, acetic acid** (*linimentum terebinthinæ aceticum*, B. P.), oil of turpentine, acetic acid, liniment of camphor, of each, 1 oz. t., **oil of** (*oleum terebinthinæ*, U. S. P., B. P.), a volatile oil, recently distilled from turpentine. When pure, it consists only of carbon and hydrogen, but on exposure absorbs oxygen. Oil of turpentine is stimulant, diuretic, and anthelmintic; in large doses it acts as a cathartic; locally it is a rubefacient. In overdoses it acts as an irritant, especially to the kidneys, producing bloody urine and strangury. It is used as a stimulant in typhoid and other low fevers; in tympanites; as a hemostatic; in chronic renal diseases, dysentery, and whooping-cough; as an inhalation in bronchitis; as a cathartic in the form of enema; and as a teniafuge. Dose 5–30 min. (0.32–2.0 Cc.). t., **oil of, emulsion of** (*emulsum olei terebinthinæ*, U. S. P.). Dose 1 dr. (4 Cc.). t., **oil of, rectified** (*oleum terebinthinæ rectificatum*, U. S. P.). Dose 5–30 min. (0.32–2.0 Cc.). t., **ointment of** (*unguentum terebinthinæ*; B. P.), used on burns. t., Strasburg, a variety derived from *Abies picea*. t., Venice, a variety obtained from *Larix europæa*; it yields oil of turpentine.
**turpentole** (*tur'-pen-tōl*). A purified petroleum.
**turpeth** (*tur'-peth*) [Pers., *turbad*, a purgative root]. The *Ipomœa turpethum*, a purgative plant resembling jalap, found in Asia. t. **mineral**, the yellow, or subsulphate of mercury, used as an emetic. See *mercury subsulphate*.
**turpethin** (*tur'-peth-in*), C₃₄H₅₆O₁₈. A resin obtained from the root of *Ipomœa turpethum*.
**turps**. The trade-name for oil of turpentine.
**turtle** (*tur'-tl*) [A corruption of *tortoise*]. A tortoise; a member of the *testudinata*. t.-**back nail**. See *nail*. t.-**lung**. See *bronchiectasis*.
**turunda, turundula** (*tū-run'-dah, tū-run'-dū-lah*) [L.; *pl., turunda, turundulæ*]. A surgical tent.
**tussal** (*tus'-al*) [*tussis*, cough]. Pertaining to of the nature of a cough.
**tussedo** (*tus-e'-do*). Tussis.
**tussicular** (*tus-ik'-ū-lar*) [*tussicula*, a slight cough]. Characterized by a slight cough.
**tussiculation** (*tus-ik-ū-la'-shun*). A hacking cough.
**tussilago** (*tus-il-a'-go*). A genus of plants of the order *Compositæ*. The leaves of *T. farfara*, coltsfoot, and also other parts of the plant are used as a demulcent in pulmonary affections associated with cough.
**tussis** (*tus'-is*) [L.]. A cough. t. **convulsiva**, whooping-cough.
**tussive** (*tus'-iv*) [*tussis*]. Pertaining to or caused by cough.
**tussol** (*tus'-ol*). Antipyrine mandelate.
**tutamen** (*tū-ta'-men*) [L., defense: *pl., tutamina*]. A defense or protection. **tutamina cerebri**, the skull and meninges of the brain. t. **oculi**, the appendages of the eyes—the lids, brows, lashes, etc.
**tutty** (*tut'-e*) [Tamul word, *tutum*]. Impure oxide of zinc deposited as an incrustation on the chimneys of furnaces during the smelting of lead ores containing zinc; used as an external desiccant when pulverized.
**T. V.** Abbreviation of *tuberculin volution*, a principle said to exist in the tubercle bacillus.
**twang**. A personal quality of the voice, usually nasal.
**tween-brain** (*twēn'-brān*). See *diencephalon*.
**tweezers** (*twe'-zers*). See *volsella*.
**twelfth cranial nerve**. The hypoglossal nerve. See under *nerve*.

**twilight sleep**. A method of childbirth popularly supposed to be painless; but it is the memory of the pain that is abolished, and not the pain itself. The method is not without danger, and its value is still undecided. "The patient is delivered in a delirium" (Lequeux).
**twin**. One of two individuals born at the same birth.
**twitch** (*twich*). To give a short, sudden pull or jerk; see also, *uvular twitch*. t.-**grass**, triticum repens; see *triticum*.
**twitching** (*twich'-ing*). An irregular spasm of a minor extent.
**twixt-brain** (*twikst'-brān*). See *diencephalon*.
**T.X.** Symbol of a derivative of tuberculin prepared by Behring.
**tyle** (*ti'-le*) [τύλος, a knob]. A callus.
**tylion** (*til'-e-on*) [*tyle*]. A craniometric point on the anterior border of the optic groove in the mesal line.
**tyloma** (*ti-lo'-mah*) [*tyle*]. A callus.
**tylosis** (*ti-lo'-sis*) [*tyle*]. 1. A state characterized by the formation of callus. 2. A form of blepharitis with thickening and hardening of the edge of the lid. t. **lingua**, same as *leukoplakia buccalis*.
**tylosteresis** (*ti-lo-ster-e'-sis*) [τύλος, knot; στέρεσιν, to deprive]. Extirpation or removal of a callosity.
**tylotic** (*ti-lot'-ik*) [τύλος, a knot]. Pertaining to, affected with, or of the nautre of tylosis.
**tympanal** (*tim'-pan-al*). See *tympanic ring*.
**tympanectomy** (*tim-pan-ek'-to-me*) [*tympanum*; ἐκτομή, a cutting out]. Excision of the tympanic membrane.
**tympania** (*tim-pan'-e-ah*). Same as *tympanites*.
**tympanic** (*tim-pan'-ik*) [*tympanum*]. Pertaining to the tympanum. t. **bone**, t. **plate**, the thin plate of bone separating the tympanum from the cranial cavity. t. **membrane**. See *membrane, tympanic*. t. **ring**, an osseous ring forming part of the temporal bone at the time of birth and which develops into the tympanic plate. t. **tegmen**, the bony plate forming the roof of the tympanum.
**tympanichord** (*tim-pan'-ik-ord*) [*tympanum*; χορδή, a string]. The chorda tympani, a branch of the facial nerve.
**tympanichordal** (*tim-pan-ik-or'-dal*). Pertaining to the tympanichord.
**tympanicity** (*tim-pan-is'-it-e*). The quality of being tympanic.
**tympaniform** (*tim-pan'-if-orm*). Shaped like a tympanum.
**tympanism** (*tim'-pan-izm*) [*tympanum*]. Distention with gas; tympanites.
**tympanites** (*tim-pan-i'-tēz*) [*tympanum*]. A distention of the abdominal walls from accumulation of gas in the intestine or peritoneal cavity.
**tympanitic** (*tim-pan-it'-ik*) [*tympanites*]. Caused by or of the nature of tympanites. t.: **abscess**, an abscess containing air. t. **resonance**, the note obtained on percussing a cavity distended with gas.
**tympanitis** (*tim-pan-i'-tis*) [*tympanum*; ιτις, inflammation]. Inflammation of the tympanum; otitis media.
**tympanoeustachian** (*tim-pan-o-ū-sta'-ke-an*) [*tympanum*; *Eustachian*]. Pertaining to the tympanum and the Eustachian tube.
**tympanohyal** (*tim-pan-o-hi'-al*) [*tympanum*; *hyoid*]. A small cartilage of the human fetus subsequently fusing with the styloid process of the temporal bone.
**tympanomalleal** (*tim-pan-o-mal'-e-al*) [*tympanum*; *malleus*]. Pertaining to the tympanic bone and the malleus.
**tympanomandibular** (*tim-pan-o-man-dib'-ū-lar*) [*tympanum*; *mandible*]. Pertaining to the tympanum and the mandible.
**tympanomastoiditis** (*tim-pan-o-mas-toid-i'-tis*). Inflammation of the tympanum and mastoid cells.
**tympanophony** (*tim-pan-of'-o-ne*). See *autophony* 2).
**tympanosis** (*tim-pan-o'-sis*). Tympanites.
**tympanosquamosal** (*tim-pan-o-skwa-mo'-sal*) [*tympanum*; *squama*, scale]. Common to the tympanic and the squamosal bone.
**tympanotomy** (*tim-pan-ot'-o-me*) [*tympanum*; τομή, a cutting]. Incision of the membrana tympani.
**tympanous** (*tim'-pan-us*). Distended with gas; relating to tympanum.
**tympanum** (*tim'-pan-um*) [τύμπανον, drum]. The middle ear.
**tympany** (*tim'-pan-e*). 1. Tympanites. 2. A tympanitic percussion-note.

**tyndallization** (tin-dal-iz-a'-shun) [John Tyndall, English physicist, 1820–1893]. See *sterilization, intermittent*.

**type** (tīp) [τύπος, a stamp]. Imprint; emblem; symbol; character. A normal average example. In pathology, the distinguishing features of a fever, disease, etc., whereby it is referred to its proper class. t., **test**, see *test-types*.

**typembryo** (ti-pem'-bre-o) [*type*; *embryo*]. That stage or period in the development of an embryo when the characteristics of the main type to which it belongs are first discoverable.

**typewriters' backache.** An occupation-neurosis; it is said to be best guarded against by the use of a high seat and a footstool.

**typewriter's cramp.** See under *spasms, fatigue*.

**typhase** (ti'-fās). The special bacteriolytic enzyme of *Bacillus typhi abdominalis*.

**typhemia, typhæmia** (ti-fe'-me-ah) [τῦφος, smoke; αἷμα, blood]. The presence of typhoid bacilli in the blood.

**typhfever** (tif-fe'-ver). Typhoid or typhus fever.

**typhia** (ti'-fe-ah). See *fever, typhoid*.

**typhinia** (ti-fin'-e-ah). Synonym of *relapsing fever*.

**typhization** (tif-iz-a'-shun). 1. Infection with typhoid or typhus fever. 2. Preventive inoculation with typhoid vaccine.

**typhlatony, typhlatonia** (tif-lat'-on-e, tif-lat-o'-ne-ah) [*typhlo-*; *atony*]. An atonic condition of the wall of the cecum, generally due to a catarrhal condition.

**typhlectomy** (tif-lek'-to-me) [*typhlo-*; ἐκτομή, excision]. Excision of the cecum.

**typhlenteritis** (tif-len-ter-i'-tis) [τυφλός, blind; ἔντερον, bowel; ιτις, inflammation]. Typhlitis.

**typhlitis** (tif-li'-tis) [τυφλόν, cecum; ιτις, inflammation]. Inflammation of the cecum.

**typhlo-** (tif'-lo-) [τυφλόν, cecum]. A prefix signifying relating to the cecum.

**typhlocele** (tif'-lo-sēl). See *cecocele*.

**typhlodiclidiitis** (tif-lo-di-kli-di'-tis) [*typhlo-*; δυκλίς, a folding door; ιτις, inflammation]. Inflammation of the ileocecal valve.

**typhloempyema** (tif-lo-em-pi-e'-mah) [*typhlo-*; *empyema*]. Abscess attending typhlitis or appendicitis.

**typhloenteritis** (tif-lo-en-ter-i'-tis). See *typhlitis*.

**typhloid** (tif'-loid) [τυφλός, blind]. Having defective vision.

**typhlolithiasis** (tif-lo-lith-i'-as-is) [*typhlo-*; *lithiasis*]. The formation of calculi in the cecum.

**typhlology** (tif-lol'-o-je) [τυφλός, blind; λόγος, science]. The science of blindness.

**typhlomyxorrhea** (tif-lo-miks-or-e'-ah) [*typhlo-*; μύξα, mucus; ῥείν, to flow]. Evacuation of mucus derived from the cecum.

**typhlopexy** (tif'-lo-pek-se) [*typhlo-*; πῆξις, fixation]. Operation of fixing the cecum to the abdominal wall.

**typhlosis** (tif-lo'-sis) [τυφλός, blind]. Blindness.

**typhlosole** (tif'-lo-sōl) [*typhlo-*; σωλήν, tube]. A tube lying in the dorsal middle line of the intestine of certain worms (*Lumbricidæ*).

**typhlospasm** (tif'-lo-spazm) [*typhlo-*; *spasm*]. Spasm of the cecum.

**typhlostenosis** (tif-lo-sten-o'-sis) [*typhlo-*; *stenosis*]. Stenosis of the cecum.

**typhlostomy** (tif-los'-to-me) [*typhlo-*; στόμα, mouth]. A form of colostomy in which the opening is made in the cecum.

**typhlotomy** (tif-lot'-o-me) [*typhlo-*; τομή, a cutting]. Division or section of the cecum.

**typho-** (ti-fo-) [*typhoid*]. A prefix meaning relating to typhoid or of a typhoid character.

**typhobacillosis** (ti-fo-bas-il-o'-sis) [*typho-*; *bacillus*]. The systemic poisoning produced by the toxins formed by the typhoid bacillus.

**typhobacterin** (ti-fo-bak'-ter-in). A vaccine prepared from the typhoid bacillus.

**typhogenic** (ti-fo-jen'-ik) [τῦφος, stupor; γεννᾶν, to produce]. Producing typhus or typhoid fever.

**typhoid** (ti'-foid) [τῦφος, stupor; εἶδος, like]. Resembling typhus. t., **abenteric**, typhoid fever involving other organs than those of the intestinal tract. t., **abortive**, is characterized by abrupt onset of symptoms, which subside quickly, convalescence following in a few days. t., **afebrile**, typhoid fever with the usual symptoms, positive diazo and Widal reaction, presence of rose-spots, but absence of increased temperature. t., **ambulatory**, typhoid fever in which the patient does not, or will not, take to his bed. t. **carrier.** See *carriers*. t. **condition**, see *typhoid state*. t. **fever**, enteric fever, abdominal typhus, ileotyphus. An infectious disease caused by *Bacillus typhosus* discovered by Eberth. It is introduced into the body with the food and drinking-water, and is found in the intestine, the spleen, and the fecal discharges, but may also occur in the various complicating lesions. The principal lesions of typhoid fever are an enlargement and necrosis of Peyer's patches, and enlargement of the spleen and the mesenteric glands. The mucous membrane of the intestine is also the seat of a catarrhal inflammation. After a period of incubation of from two to three weeks the disease sets in with weakness, headache, vague pains, a tendency to diarrhea, and nose-bleed. The temperature gradually rises, being higher each evening than the previous evening, and reaches its maximum (104°–105° F.) in from one to two weeks. It then remains at this level for from one to two weeks, and finally sinks by lysis. The pulse is soft and dicrotic, but often not so rapid as would be expected from the high temperature. The tongue is at first coated on the dorsum and red at the tip and edges, but soon becomes dry, brown, and tremulous, and, like the teeth and lips, covered with sordes. There is usually complete anorexia, the bowels are loose, and the stools have a peculiar "pea-soup" color. At times constipation exists. Slight congestion of the lungs with cough is usually present. On the seventh, eighth, or ninth day the peculiar eruption appears—it consists of small, slightly elevated, rose-colored spots, disappearing on pressure, and coming out in successive crops. Nervous symptoms are prominent in typhoid fever, and are headache, slight deafness, stupor, muttering delirium, carphology, subsultus tendinum; and coma vigil. Complications are frequent, the most important being intestinal hemorrhage, perforation of the bowel, peritonitis, pneumonia, and nephritis. Relapses are fairly common, although second attacks are rare. t. **state**, the condition of stupor and hebetude, with dry, brown tongue, sordes on the teeth, rapid, feeble pulse, incontinence of feces and urine, and rapid wasting, seen in typhoid fever and other continued fevers. t. **vaccination**, vaccination against typhoid; see *vaccination* (2). t., **walking**, see *t., ambulatory*.

**typhoidal** (ti-foi'-dal) [τῦφος, stupor; εἶδος, like]. Resembling typhoid.

**typhoidet, typhoidette** (ti-foi-det') [Fr.]. A mild or benign type of typhoid fever.

**typhoid spine**, a neurosis sometimes following typhoid fever, characterized by the production of acute pains in the vertebral column on the slightest movement.

**typhoin** (ti'-fo-in). A preparation of dead typhoid bacilli used by injection in the treatment of typhoid fever.

**typholysin** (ti-fol'-is-in) [*typho-*; *lysin*]. A hemolysin formed by the *Bacillus typhosus*.

**typhomalarial** (ti-fo-mal-a'-re-al) [*typho-*; *malaria*]. Exhibiting symptoms of both typhoid and malarial fevers. t. **fever**, a fever exhibiting symptoms both of typhoid and of malarial fever, but probably malarial in nature.

**typhomania** (ti-fo-ma'-ne-ah) [*typho-*; μανία, madness]. The lethargic state, with delirium, sometimes observed in typhus, typhoid, and other low fevers.

**typhonia** (ti-fo'-ne-ah). Same as *typhomania*.

**typhopaludism** (ti-fo-pal'-ū-dizm) [*typho-*; *paluda*, a marsh]. Fever of malarial origin accompanied by symptoms of typhoid.

**typhopneumonia** (ti-fo-nū-mo'-ne-ah) [*typho-*; *pneumonia*]. Pneumonia occurring in the course of typhoid fever.

**typhosepsis** (ti-fo-sep'-sis). The systemic poisoning of typhoid fever.

**typhosis** (ti-fo'-sis). See *typhoid state*. t., **syphilitic**, a form of intestinal neuralgia associated with secondary syphilis.

**typhotoxin** (ti-fo-toks'-in) [*typho-*; τοξικόν, poison]. A poisonous ptomaine produced by the typhoid bacillus. It is isomeric with the base $C_7H_{17}NO_2$, obtained from *putrefying horseflesh*, and induces lethargy, paralysis, and death.

**typhous** (ti'-fus) [*typhus*]. Pertaining to or having the nature of typhus.

**typhus** (ti'-fus) [τῦφος, stupor]. An acute infectious and contagious disease chiefly characterized by a petechial rash, marked nervous symptoms, and a high fever, ending by crisis in from 10 to 14 days. The only peculiar lesions noted postmortem are a dark fluid state of the blood and a staining of the endocardium and intima of the blood-vessels. The disease is caused by the *Bacillus typhi exanthemataci* (Plotz). After a period of incubation of from a few hours to two weeks, the disease sets in abruptly with pains in the head, back, and limbs, the fever rising rapidly to 104° or 105° F. The nervous symptoms resemble those of typhoid fever. The eruption appears on the fourth or fifth day as rose-colored spots scattered over all the body, and quickly becoming hemorrhagic. It does not disappear on pressure. The chief complications are hyperpyrexia, pneumonia and nephritis. Syn., *jail-fever; ship-fever.* t., abdominal, typhoid fever. t. biliosus, same as *Weil's disease.* t. icterodes. Synonym of *yellow fever.* t. levissimus, a mild form of typhus. t. Mexican, Same as tabardillo, *q. v.* t. petechialis, cerebrospinal fever. t. recurrens, relapsing fever. t. siderans, a malignant form of typhus fever ending fatally in two or three days.

typical (*tip'-ik-al*) [τύπος, a stamp], Constituting a type or form for comparison; illustrative; complete.

typoscope (*ti'-po-skōp*) [τύπος, a stamp; σκοπεῖν, to look]. A small device to exclude extraneous light, for the use of cataract patients and amblyopes in reading.

tyramine (*ti'-ram-ēn*). A trade name applied to parahydroxyphenylethylamine.

tyrannism (*tir'-an-ism*) [τύραννος, a tyrant]. Cruelty of morbid inception, of which sadism is an erotic variety.

Tyree's antiseptic powder. A proprietary preparation said to contain alum, sodium biborate, eucalyptus, phenol, thymol, wintergreen, and peppermint; it is recommended for leukorrheal and purulent discharges.

tyrein (*ti'-re-in*) [τυρός, cheese]. Coagulated casein.

tyremesis (*ti-rem'-es-is*) [τυρός, cheese; ἔμεσις, a vomiting]. The vomiting of caseous matter; an ailment common among nursing infants.

tyriasis (*tir-i'-as-is*). 1. Elephantiasis. 2. Alopecia.

tyro- (*ti-ro-*) [τυρός, cheese]. A prefix meaning cheese or cheese-like.

tyroid (*ti'-roid*) [τυρο-; εἶδος, like]. Cheese-like.

tyroleucin (*ti-ro-lū'-sin*) [*tyro-*; leucin. A substance obtained from decomposing albumin.

tyroma (*ti-ro'-mah*). [*tyro-*; ὄμα, tumor]. 1. A caseous mass. 2. A tuberculous tumor.

tyromatosis (*ti-ro-ma-to'-sis*). Caseation.

tyrosal (*ti'-ro-sal*). See *salipyrine.*

tyrosin (*ti'-ro-sin*) [τυρός, cheese]. C₉H₁₁NO₃. A crystalline amidoacid, a decomposition product of proteids. t., tests for. See *Hoffmann, Piria, Scherer,* (*von*) *Udransky, Wurster.*

tyrosinase (*ti-ro'-sin-ās*) [*tyrosin*]. An oxidizing enzyme found in many fungi, and in dahlia and beetroot. It acts upon all the cresols.

tyrosinuria (*ti-ro-sin-ū'-re-ah*) [*tyrosin*; *urine*]. The presence of tyrosin in the urine.

tyrosis (*ti-ro'-sis*) [τυρός, cheese]. Caseation.

Tyrothrix (*ti-ro'-thriks*) [τυρός, cheese; θρίξ, hair]. A genus of *Schizomycetes.*

tyrotoxicon (*ti-ro-toks'-ik-on*) [*tyro-*; τοξικόν, a poison]. A ptomaine obtained from poisonous cheese, milk, ice-cream, etc. It induces vertigo, nausea, vomiting, chills, rigors, severe pains in the epigastric region, dilatation of the pupils, griping and purging, a sensation of numbness or of pins and needles, especially in the limbs, and marked prostration or even death. The poison is thought to be the cause of many cases of summer diarrhea of infants.

tyrotoxin (*ti-ro-toks'-in*) [see *tyrotoxicon*]. A curara-like poison from poisonous cheese; it is not identical with tyrotoxicon.

tyrotoxism (*ti-ro-toks'-izm*). Cheese-poisoning.

Tyrrell's fascia (*tir'-el*) [Frederick Tyrrell, English physician, 1797–1843]. See *Dénonvillier's fascia.* T.'s hook, a blunt, slender hook used in certain operations upon the eye, as in iridectomy.

Tyson's glands (*ti'-sun*) [Edward Tyson, English anatomist, 1649–1708]. The sebaceous glands of the corona glandis and the inner layer of the prepuce secreting the smegma.

T.Z. Symbol of *tuberculin symoplastiche;* the dried residue of tubercle bacilli, soluble in alcohol.

tzetze. Same as *tsetse.*

# U

**U.** 1. The chemical symbol of *uranium*. 2. Abbreviation for *unit*. 3. Symbol for *kilurane q. v.*
**uabain.** See *ouabain*.
**uarthritis** (ū-ar-thri'-tis). See *Arthritis urica*.
**uber** (ū'-ber) [L., *udder*; *gen., uberis*; *pl., ubera*]. The mamma; also the nipple.
**uberous** (ū'-ber-us) [*uber*, fruitful]. Fruitful, prolific.
**uberty** (ū'-ber-te) [*uber*, udder]. Fertility; productiveness.
**ucambin, ukambin** (ū-kam'-bin). An African arrow-poison with the effects of strophanthin, but more powerful.
**udder** (ud'-er). The mammary apparatus, especially of the cow.
**(von) Udransky's test** for bile acids (oo-dran'-ske) [Lasǎlo von *Udránsky*, Austrian physiologist]. To 1 Cc. of a watery or alcoholic solution of the substance add one drop of a 0.1% watery solution of furfurol, and underlay with 1 Cc. of concentrated sulphuric acid; then cool. In the presence of bile acids a red color with a shade of blue will be produced. v. U.'s test for tyrosin, to 1 Cc. of a solution of the substance add one drop of a 0.5% watery solution of furfurol, and underlay with 1 Cc. of concentrated sulphuric acid. The mixture becomes pink. The mixture should not rise above 50° C.
**Uffelmann's test** for hydrochloric acid in the contents of the stomach (oo'-len-hoot) [Julius *Uffelmann*, German physician, 1837–1894]. Strips of filter-paper saturated in an extract of bilberries in amylic alcohol and dried, when dipped into the contents of a stomach containing HCl, will be turned pink. U.'s test for lactic acid in the contents of the stomach, make a mixture of 10 Cc. of a 4% solution of phenol, 20 Cc. of water, and a few drops of ferric chloride solution; this will have a blue coloration. Add the liquid to be tested, and in the presence of lactic acid a yellow coloration will result.
**Uhlenhuth's test** (oo'-len-hoot) [Paul *Uhlenhuth*, German bacteriologist, 1870– ]. A method of examination of tubercle bacilli in sputum, by adding antiformin.
**Uhthoff's sign** (oot'-hof) [Wilhelm *Uhthoff*, German ophthalmologist, 1853– ]. The nystagmus of multiple cerebrospinal sclerosis.
**ukambin.** See *ucambin*.
**ula** (ū'-lah) [οὖλον, gum]. The gums.
**ulæmorrhagia.** See *ulemorrhagia*.
**ulaganectesis** (ū-lag-an-ek'-te-sis) [οὖλον, gum; ἀγανάκτησις, irritation]. Irritation or uneasy sensations in the gums.
**ulatrophia** (ū-lat-ro'-fe-ah) [οὖλον, gum; ἀτροφία, atrophy]. A shrinkage of the gums.
**ulcer** (ul'-ser) [*ulcus*, ulcer]. A loss of substance occurring on the skin or mucous membranes, and due to a gradual necrosis of the tissues. u., Aden. See *phagedæna tropica*. u., **adherent**, an ulcer of the skin, the base of which becomes adherent to the underlying fascia. u., **Anamite**, a phagedenic sore of hot countries; it is very unyielding to treatment. u., **amputating**, an ulcerating process encircling a part and destroying the tissues to the bone. u., **arterial**, a superficial ulcer due to arterial disease. u., **atheromatous**, a loss of substance in the wall of an·artery or the endocardium, due to the breaking down of an atheromatous patch. u., **atonic**, an ulcer which has unhealthy granulations, with little or no tendency to cicatrization. u., **autochthonous**, chancre. u., **carious**, an ulcer producing gangrene. u., **catarrhal**, a form of intestinal ulcer due to a superficial loss of epithelial cells. u., **chancroidal**, a chancroid. u., **creeping**, a serpiginous ulcer. u., **Curling's**, an ulcer of the duodenum observed after severe burns of the body. u., **endemic**, an ulcer more or less local as regards countries in which it is found, as Aleppo boil. u., **erethistic**, a name given to an extremely sensitive ulcer, such as about the anus or the matrix of the nails. u., **fissurated**, laceration of the cervix uteri. u., **fistulous**, an ulceration communicating with a fistula. u., **follicular**, a small ulcer on a mucous membrane having its origin in a lymph-follicle. u., **fungous**, one covered by fungous granulations. u., **gastric**, perforating ulcer of the lining membrane of the stomach. u., **hard**, a chancre. u., **indolent**, one with an indurated, elevated edge and a nongranulating floor, usually occurring on the leg. u., **inflamed**, one surrounded by marked inflammation. u., **Jacob's**. See *u., rodent*. u., **lipoid**, an ulceration resembling lupus. u., **menstrual**, an ulcer from which vicarious menstruation takes place. u., **Marjolin's**, an ulcer having for its seat an old cicatrix. u., **peptic**. See *peptic ulcer*. u., **perforating**, an ulcer that perforates the tissues of a part, particularly the foot or the stomach. u., **phagedenic**, one which rapidly eats away the tissues. u., **phlegmonous**. Synonym of *u., inflamed*. u., **kissing**, an ulcer occurring on two parts which are frequently or constantly in apposition. u., **rodent**, a form of ulcer, probably epitheliomatous, which gradually involves and eats away soft tissues and bones. u., **round**; the peptic ulcer of the stomach. u., **Saemisch's**, an infectious ulcer of the cornea. u., **serpiginous**, one healing in one place while spreading in another. u., **simple**, a mild form of ulceration, not due to a poison or systemic disease. u., **symptomatic**, an ulcer indicative of general disease. u., **tuberculous**, one due to the tuberde bacillus. u., **varikose**, an ulcer due to varicose veins. u., **venereal**, chancre or chancroid. u., **weak**, one with exuberant and flabby granulations.
**ulcera** (ul'-ser-a) [L., *pl.* of *ulcus*]. Ulcers.
**ulcerate** (ul'-ser-āt) [*ulcus*]. To become converted into or affected with an ulcer.
**ulcerated** (ul'-ser-a-ted) [*ulcus*, ulcer]. Affected with ulceration.
**ulceration** (ul-ser-a'-shun) [*ulcer*]. The formation of an ulcer; a process of liquefaction-necrosis or molecular death on a free surface.
**ulcerative** (ul'-ser-a-tiv) [*ulcer*]. Pertaining to ulceration; characterized by ulceration.
**ulcerine** (ul-ser-ēn'). An ointment used for x-ray burns. It contains belladonna, poppy, henbane, balsam of Peru and lard.
**ulcerous** (ul'-ser-us) [*ulcer*]. Exhibiting ulceration; having the character of an ulcer.
**ulcus** (ul'-kus) [L.: *pl., ulcera*]. An ulcer. u. **cancrosum**, (1) cancer; (2) rodent ulcer; (3) chancre. u. **cruris**, indolent ulcer of the leg. u., **exedens**, rodent ulcer. u. **grave**, Maduro foot. u. **induratum**, chancre. u. **molle**, chancroid. u. **rodens**, rodent ulcer. u. **phagedænicum**, eating or phagedenic ulcer. u. **tuberculosum**, lupus. u. **venereum**, (1) chancre; (2) chancroid. u. **venereum molle**, chancroid. u. **ventriculi**, gastric ulcer.
**ulcuscle, ulcuscule, ulcusculum** (ul-kus'-kl, ul-kus'-kūl, ul-kus'-kū-lum) [*ulcusculum: pl., ulcuscula*]. A small ulcer.
**ule** (ū'-le) [οὐλή, a scar]. A cicatrix.
**ulectomy** (ū-lek'-to-me) [οὐλή, a scar; ἐκτομή, excision]. Excision of scar tissue.
**ulegyria** (ū-le-ji'-re-ah) [οὐλή, a scar; γῦρος, a circle]. Convolutions in the cortex of the brain, made irregular by scar-formation.
**ulemorrhagia** (ū-lem-or-aj'-e-ah) [οὐλή, a scar; αἷμα, blood; ῥηγνύναι, to break forth]. 1. Hemorrhage from a cicatrix. 2. [οὖλον, the gum]. Bleeding of the gums.
**ulerythema** (ū-ler-ith-e'-mah) [*ule*; ἐρίθημα, erythema]. An erythematous disease marked by the formation of cicatrices. u. **centrifugum**, lupus erythematosus. u. **ophryogenes**, ulerythema of the eyebrows with loss of hair. u. **sycosiforme**, Unna's name for lupoid sycosis, a form in which keloid degeneration occurs in cicatrices resulting from follicular inflammation.

**uletic** (ū-let'-ik) [οὖλον, gum; οὐλή, scar]. 1. Pertaining to the gums. 2. Pertaining to scars, cicatricial.
**uletomy** (ū-let'-o-me) [οὐλή, a scar; τομή, incision]. Incision of a cicatrix.
**ulexine** (ū-leks'-en) [ulex, a shrub], $C_{13}H_{14}N_2O$. An alkaloid from the seed of *Ulex europæus*, the common gorse of Europe. It is a local anesthetic and powerful diuretic.
**ulitis** (ū-li'-tis) [ulon; ιτις, inflammation]. Inflammation of the gums.
**ulmarene** (ul'-mar-ēn). A mixture of definite quantities of salicylic ether and aliphatic alcohols; recommended as an external application in gout, rheumatism, etc.
**ulmus** (ul'-mus) [L.]. Slippery elm. The *ulmus* of the U. S. P. is the dried bark of *Ulmus fulva*, or elm, of the order *Urticaceæ*. It is used as a demulcent in diarrhea, dysentery, and diseases of the urinary tract; as a poultice in inflammations, and in the form of tents for dilating the os uteri. ulmi, mucilago (U. S. P.), mucilage of elm.
**ulna** (ul'-nah) [L., "a cubit"]. The bone on the inner side of the forearm, articulating with the humerus and the head of the radius above and with the radius below. See *bones, table of.*
**ulnad** (ul'-nad) [ulna]. Toward the ulnar aspect.
**ulnar** (ul'-nar) [ulna]. 1. Pertaining to or in relation with the ulna, as the *ulnar* artery, *ulnar* nerve. 2. Pertaining to the ulnar artery or ulnar nerve. u. phenomenon, a condition of analgesia of the trunk of the ulnar nerve on one side, mostly absent in general paralytics but generally found in other insane patients.
**ulnare** (ul-na'-re). The cuneiform bone of the carpus.
**ulnaris** (ul-na'-ris) [L.]. 1. Ulnar. 2. The ulnar muscle, a muscle on the ulnar side of the forearm; see *muscles, table of.*
**ulnen** (ul'-nen) [ulna, ulna]. Belonging to the ulna in itself.
**ulnocarpal** (ul-no-kar'-pal) [ulna; carpus]. Pertaining to the ulna and the carpus.
**ulnoradial** (ul-no-ra'-de-al) [ulna; radius]. Pertaining to the ulna and the radius.
**ulocace** (ū-lok'-as-e) [οὖλον, gum; κακός, evil]. Ulcerative inflammation of the gums.
**ulocarcinoma** (ū-lo-kar-sin-o'-mah) [ulon; carcinoma]. Carcinoma of the gums.
**ulodermatitis** (ū-lo-der-ma-ti'-tis) [οὐλή, scar; δέρμα, skin; ιτις, inflammation]. Inflammation of the skin with formation of cicatrices.
**uloglossitis** (ū-lo-glos-i'-tis) [οὖλον, gum; γλῶσσα, tongue; ιτις, inflammation]. Inflammation of gums and tongue.
**uloid** (ū'-loid) [ule; εἶδος, like]. Scar-like. u. cicatrix. A scar-like lesion due to subcutaneous degeneration.
**ulon** (ū'-lon) [οὖλον, gum]. The gums.
**uloncus** (ū-long'-kus) [ulon; ὄγκος, a tumor]. A tumor or swelling of the gums.
**ulorrhagia** (ū-lor-a'-je-ah) [ulon; ῥηγνύναι, to burst forth]. Bleeding from the gums.
**ulorrhea, ulorrhœa** (ū-lor-e'-ah) [ulon; ῥοία, flow]. Bleeding from the gums.
**ulosis** (ū-lo'-sis) [ule]. Cicatrization.
**ulotic** (ū-lot'-ik) [ulosis]. Pertaining to or tending toward cicatrization.
**ulotrichous** (ū-lot'-rik-us) [οὖλος, woolly; θρίξ, hair]. Having woolly hair.
**ulsanin** (ul'-san-in) [ulcus, ulcer; sanare, to heal]. Trade name of a preparation containing iodine and boric acid; used in treatment of ulcers, particularly of the larynx.
**ultimate** (ul'-tim-āt) [ultimus, last]. Farthest, or most remote, final. u. analysis. See *analysis.* u. principle. See *principle.*
**ultimisternal** (ul-tim-is-tur'-nal) [ultimus; last; sternum]. Pertaining to the last segment of the sternum.
**ultimum** (ul-tim-um) [L.]. Last. u. moriens (last dying). 1. The right auricle; so called from the belief that it is the last part of the heart to cease its contractions. 2. The upper part of the trapezius muscle which usually escapes in progressive muscular atrophy.
**ultra-** (ul-trah-) [L. beyond]. A prefix denoting excess.
**ultrabrachycephaly** (ul'-trah-brak-is-ef'-al-e) [ultra-; brachycephaly]. Brachycephaly in which the cephalic index exceeds 90°.

**ultradolicocephaly** (ul'-trah-dol-ik-o-sef'-al-e) [ultra-; dolicocephaly]. Dolicocephaly in which the cephalic index is less that 64°.
**ultrafiltration** (ul-trah-fil-tra'-shun) [ultra-; filtration]. Filtration by forcing under pressure a liquid through a filter which has been reinforced with some colloidal material.
**ultragaseous state** (ul-trah-gas'-e-us). The state in which matter is supposed to be less ponderable than gas, or in which gas is rarefied to such an extent that its molecules do not collide; also called radiant matter.
**ultramicroscope** (ul-trah-mi'-kro-skōp) [ultra, beyond; microscope]. A microscope for the examination, by powerful side illumination, of objects beyond the power of ordinary microscopes.
**ultramicroscopic** (ul-trah-mi-kro-skop'-ik). Too small to be seen by the aid of the microscope.
**ultramicroscopy** (ul-trah-mi-kros'-ko-pe). The scientific use of the ultramicroscope.
**ultraquinine** (ul-trah-kwin-ēn'). Homoquinine.
**ultrared** (ul-trah-red'). Infra-red.
**ultratoxon** (ul-trah-toks'-on) [ultra-; toxon]. A toxin of a low degree of avidity.
**ultraviolet rays.** See *rays, ultraviolet.*
**ultromotivity** (ul-tro-mo-tiv'-it-e) [ultro, spontaneously; motivity]. Power or capability of moving spontaneously.
**Ultzmann's reaction for bile-pigments** (oolts'-mahn) [Robert *Ultzmann*, German chemist, 1842–1889]. To 10 Cc. of the liquid add 3 or 4 Cc. of a caustic potash solution (1 : 3) and then an excess of hydrochloric acid. In the presence of bile-pigments the solution will become emerald green.
**ululation** (ū-lū-la'-shun) [ululare, to howl]. A hysterical howling.
**ulyptol** (ū-lip'-tol). See *eulyptol.*
**umbel** (um'-bel) [umbella, diminutive of *umbra*, shade]. In botany, that form of indeterminate inflorescence in which the axis is very short and the pedicels radiate from it like the ribs of an umbrella.
**umbelliferon** (um-bel-if'-er-on) [umbella, umbel; ferre, to bear], $C_9H_6O_3$. Oxycoumarin. Fine needles, sparingly soluble in hot water and ether. Found in the bark of *Daphne mezereum*; it is obtained by distilling different resins, such as galbanum, asafetida, etc.
**Umbellularia** (um-bel-ū-la'-re-ah) [umbellula, a little umbel]. A genus of the *Laurineæ*. The principal species, *U. californica*, California laurel or spice tree, contains in its leaves a pungent volatile oil recommended for inhalation in nasal catarrh; the leaves are used in neuralgic headache, colic, and atonic diarrhea. Dose of *fluidextract* 10–30 min. (0.65–2 Cc.).
**umber** (um'-ber) [umbra, shade]. A dark-brown pigment somewhat resembling ocher.
**Umber's test** for scarlet fever. The solution consists of concentrated hydrochloric acid, 30 gm.; paradimethylamidobenzaldehyde, 2 gm.; water 70 Cc. Two drops of this solution are added to a small amount of urine, and a red coloration is said to denote scarlet fever.
**umbilical** (um-bil'-ik-al) [umbilicus]. 1. Pertaining to the umbilicus, as the *umbilical* cord, *umbilical* vessels. 2. Pertaining to the umbilical cord or umbilical vessels. u. arteries, the arteries of the umbilical cord. u. cord. See *cord* (2). u. duct. See *duct.* u. fissure. See *fissure.* u. region, the central of the regions into which the abdomen is divided for purposes of physical diagnosis. u. ring, the aperture, closed in the adult, through which the umbilical vessels pass in fetal life. u. souffle, the peculiar sound heard occasionally over the umbilical cord of the fetus. u. stalk, u. duct. u. vesicle, the part of the yolk-sac remaining outside of the embryo and supplying nutriment to it through the omphalomesaraic duct. u. vessels, the umbilical arteries and veins.
**umbilicate, umbilicated** (um-bil'-ik-āt, um-bil'-ik-a-ted) [umbilicus]. Having a depression like that of the navel.
**umbilication** (um-bil-ik-a'-shun) [umbilicus]. 1. A depression like that of the navel. 2. The state of being umbilicated.
**umbilicus** (um-bil-i'-kus) [L.]. The navel; the round, depressed cicatrix in the median line of the abdomen, marking the site of the aperture which in fetal life gave passage to the umbilical vessels. u., posterior, a depression in the spinal region due to imperfect closure of the vertebral groove.

**umbo** (*um'-bo*) [L.: *pl.*, *umbones*]. A boss or bosselation; any central convex eminence, as, the *umbo* of the membrana tympani.
**umbonate** (*um'-bo-nāt*) [*umbo*]. Bossed; furnished with a low, rounded projection, like a boss (umbo).
**umbonation** (*um-bon-a'-shun*) [*umbo*]. The formation of a low, rounded projection.
**umbrascopy** (*um-bras'-ko-pe*) [*umbra*, shadow; σκοπεῖν, to view]. See *retinoscopy*.
**unavoidable hemorrhage.** See *hemorrhage, unavoidable.*
**unazotized** (*un-as'-o-tīzd*) [*un*, not; *azote*]. Deprived of nitrogen.
**unbalance.** See *imbalance.*
**unc,** Abbreviation for *uncia*, an ounce.
**uncia** (*un'-se-ah*) [L.]. (1) An ounce. (2) An inch.
**unciform** (*un'-se-form*) [*uncus; forma,* form]. Hook-shaped. **u. bone,** a hook-shaped bone in the second row of the carpus. **u. eminence,** the hippocampus minor. **u. process,** a hook-shaped process on the ethmoid and other bones.
**unciforme** (*un-se-form'-e*) [L.]. The unciform bone.
**uncinal** (*un'-sin-al*). 1. Uncinate. 2. Furnished with hooks.
**Uncinaria** (*un-sin-a'-re-ah*) [*uncinus*, a hook]. A genus of parasitic nematode worms. And see *Ankylostomum.* **U. americana,** a species of hook worm found in the Southern States and the West Indies. **U. duodenalis.** See *Ankylostomum.*
**uncinariasis** (*un sin a ri' a sis*). Disease produced by parasites of the genus *Uncinaria.* Infection with hook-worm; hook-worm disease. Syn., ankylostomiasis; dochmiasis; hook-worm disease.
**uncinate** (*un'-sin-āt*) [*uncus*]. Hooked. **u. convolution, u. gyrus,** the continuation of the hippocampal convolution, or fornicate convolution, ending in a hook-like process near the end of the temporal lobe.
**uncinatum** (*un-sin-a'-tum*) [*uncinatus*, hooked]. The unciform bone.
**uncipressure** (*un'-sip-resh-ur*) [*uncus; pressure*]. A method of arresting hemorrhage by the use of two hooks dug into the sides of the wound so as to compress the vessel.
**uncomplemented** (*un-kom'-ple-men-ted*) [*un,* not; *complement*]. Not joined with complement, and therefore inactive.
**unconscious** (*un-kon'-shus*) [*un*, not; *consciens,* knowing]. Not conscious. **u. cerebration,** see *cerebration.*
**unconsciousness** (*un-kon'-shus-nes*) [*un*, not; *consciens,* knowing]. The state of being without sensibility, and having abolished reflexes.
**unction** (*unk'-shun*) [*unguere;* to anoint]. 1. The act of anointing. 2. An ointment. 3. Calomel ointment.
**unctuous** (*unk'-tū-us*) [*unctus,* an anointing]. Greasy; oily.
**uncture** (*unk'-tūr*) [*unctus,* an anointing]. An unguent.
**uncus** (*ung'-kus*) [L.]. 1. A hook. 2. The hook-like anterior extremity of the uncinate gyrus of the brain. **u. gyri hippocampi,** hook of the hippocampal gyrus.
**under.** Below; beneath. **u.-cut,** in dentistry, a depression made beyond a general surface for the purpose of retaining a filling. **u.-hung,** applied to a projecting lower jaw. **u.-jawed,** same as *underhung.* **u.-toe,** a variety of hallux varus in which the great toe underlies its neighbors.
**undulant** (*un'-dū-lant*) [*unda*, a wave]. Characterized by fluctuations. **u. fever.** See *fever, Mediterranean.*
**undulation** (*un-dū-la'-shun*) [see *undulant*]. A wave-like motion; fluctuation. **u. jugular,** the venous pulse. **u.,** respiratory, the variations in the blood pressure due to respiration.
**undulatory** (*un'-dū-lat-o-re*) [see *undulant*]. Moving like waves; vibratory.
**ung.** Abbreviation for *unguentum* [L]., ointment.
**ungual** (*un'-gwal*) [*unguis;* nail]. 1. Pertaining to a nail. 2. Resembling a nail in size, as the *ungual* bone (the lacrimal bone). **u. phalanx,** the terminal phalanx of the fingers and toes.
**unguent** (*un'-gwent*). See *unguentum.*
**unguentine** (*un'-gwen-tēn*). An alum and petroleum ointment containing 2% of phenol and 5% of ichthyol.
**unguentum** (*un-gwen'-tum*) [L., ointment]. 1. An ointment. 2. See *u. simplex.* **u. acidi tannici,** ointment of tannic acid; tannic acid, 20 Gm.; glycerol, 20 Gm.; unguentum, 60 Gm. **u. Credé,** soluble silver 15%, incorporated in lard and 10% of wax added. It is scented with benzoinated ether; used in treatment of septic diseases. **u. durum,** an ointment base consisting of paraffin (solid), 4 parts; wool-fat, 1 part; liquid paraffin, 5 parts; used for ointments containing liquid antiseptics. **u. hydrargyri.** See *mercurial ointment.* **u. hydrargyri dilutum,** blue ointment; mercurial ointment, 670 Gm., with petrolatum, 330 Gm. **u. iodi,** iodine ointment; iodine, 4 Gm.; potassium iodide, 4 Gm.; glycerol, 12 Gm.; benzoinated lard, 80 Gm. **u. molle,** an ointment base consisting of solid paraffin, 11 parts; lanolin, 5 parts; and liquid paraffin, 34 parts. **u. potassii iodidi,** ointment of potassium iodide; potassium iodide, 10 Gm.; potassium carbonate, 0.6 Gm.; water, 10 Gm.; benzoinated lard, 80 Gm. **u. simplex** (*unguentum,* U. S. P.) consists of lard, 80 parts, and wax 20 parts. **u. stramonii.** See *stramonium ointment.*
**ungues** (*un'-gwēs*) [pl. of *unguis*]. Nails. **u. adunci,** hooked nails; see *onychogryposis.*
**unguiculate** (*un-gwik'-ū-lāt*) [*unguis*]. Having nails or claws.
**unguinal** (*un'-gwin-al*) [*unguis*]. Pertaining to a nail or to the nails.
**unguis** (*un'-gwis*) [L.]. 1. A nail. 2. The lacrimal bone.
**ungula** (*un'-gū-lah*) [L., "a claw"]. 1. An instrument for extracting a dead fetus from the uterus. 2. A hoof; a claw.
**ungulate** (*ung'-gū-lāt*) [*ungula,* hoof]. Having hoofs; applied to certain orders of mammalia.
**uni-** (*ū-ne-*) [*unus,* one]. A prefix denoting one.
**uniarticulate** (*ū-ne-ar-tik'-ū-lāt*) [*uni-; articulus,* joint]. Having but one joint.
**uniaxial** (*ū-ne-aks'-e-al*) [*uni-; axis*]. Having but one axis.
**unibasal** (*ū-ne-ba'-sal*) [*uni-; basis,* base]. Having but one base.
**unicamerate** (*ū-ne-kam'-er-āt*) [*uni-; camera,* chamber]. Having but one cavity; unilocular.
**unicellular** (*ū-ne-sel'-ū-lar*) [*uni-; cellula,* dim. of *cella,* a cell]. Composed of but one cell.
**unicentral** (*ū-nis-en'-tral*) [*uni-; centrum,* center]. Having a single center of growth.
**uniceptor** (*ū'-ne-sep-tor*) [*uni-; capere,* to take]. An antitoxin or receptor which has only one uniting arm (viz., the haptophore group).
**unicism** (*ū'-nis-izm*) [*unicus,* single]. The belief that there is but a single venereal virus.
**unicorn** (*ū'-ne-korn*) [*uni-; cornu,* horn]. Having a single horn. **u. root.** See *Aletris.* **u. uterus,** a uterus with but a single cornu.
**unicuspid** (*ū-nik-us'-pid*) [*uni-; cuspis,* point]. 1. Having but a single cusp (as a tooth). 2. A tooth with but a single cusp or point.
**unifilar** (*ū-ne-fi'-lar*) [*uni-; filum,* a thread]. Connected by one thread; furnished with one filament.
**uniflagellate** (*ū-nif-laj'-el-āt*) [*uni-; flagellum,* a tail]. Having a single flagellum.
**uniforate** (*ū-nif'-o-rāt*) [*uni-; foratus,* pierced]. Having one opening.
**unigravida** (*ū-ne-grav'-id-ah*) [*uni-; gravida,* pregnant]. A woman who is pregnant for the first time.
**unilaminar, unilaminate** (*ū-ne-lam'-in-ar, -āt*) [*uni-; lamina,* a layer]. Occurring in a single layer.
**unilateral** (*ū-ne-lat'-er-al*) [*uni-; latus,* side]. Pertaining to or affecting but one side.
**unilobar, unilobed** (*ū-ne-lo'-bar, ū'-ne-lōbd*). Furnished with one lobe.
**unilocular** (*ū-ne-lok'-ū-lar*) [*uni-; loculus,* dim. of *locus,* a place]. Having but one loculus or cavity.
**uninterrupted** (*un-in-ter-up'-ted*) [*un,* not; *interrupted*]. Continuous; not broken.
**uninuclear, uninucleated** (*ū-ne-nū'-kle-ar, ū-ne-nū'-kle-a-ted*) [*uni-; nucleus*]. Having but a single nucleus.
**uniocular** (*ū-ne-ok'-ū-lar*) [*uni-; oculus,* eye]. 1. Pertaining to or performed with one eye. 2. Having only one eye.
**union** (*ūn'-yun*). Joining. See under *healing.*
**unioval** (*ū-ne-o'-val*) [*uni-; ovum,* egg]. Formed from one ovum.
**unipara** (*ū-nip'-ar-ah*) [*uni-; parere,* to bear]. A woman who has borne but one child.
**uniparous** (*ū-nip'-ar-us*) [*unipara*]. 1. Having borne but one child. 2. Producing one at a birth.
**unipolar** (*ū-ne-po'-lar*) [*uni-; polus,* a pole]. 1.

# UNISEXUAL 920 URANYL

Having but one pole or process. 2. Pertaining to one pole.
**unisexual** (ū-ne-seks'-ū-al) [uni-; sexus, sex]. Provided with the sexual organs of one sex only.
**unit** (ū'-nit) [unus, one]. 1. A single thing; a group considered as a whole or as forming one of many similar groups composing a more complex body. 2. A quantity with which others are compared. u., **antitoxic**, see u., immunizing. u., C. G. S., a unit in the centimeter-gram-second system. u.s, **electric**, unit of capacity, the farad; unit of current, the ampere; unit of electromotive force, the volt; unit of power, the watt; unit of quantity, the coulomb; unit of resistance, the ohm. u. **of force**, the dyne. u. **of heat**, the calory. u. **of length**, u. **of volume**, u. **of weight**. See under weights and measures. u. **of light**, the light of a standard candle, i. e., a spermaceti candle burning 120 grains an hour. u., **physiological**, a term used by Herbert Spencer to express a unit between the chemical and the morphological units in complexity, and of an aggregation of which units the body is composed, and which represents the character of the species. u., **serum-**, u., **immunizing**, according to Behring, 1 Cc. of an antitoxic blood-serum, of which 0.1 Cc. protects a guinea-pig of 500 grams against ten times the fatal dose of diphtheria toxin. u., **toxic**, the smallest dose of a toxin which is capable of proving fatal to a guinea-pig of about 250 grams weight, in three or four days. u. **of work**, the erg.
**unitary** (ū'-nit-a-re) [unit]. 1. Pertaining to or having the qualities of a unit. 2. Pertaining to monsters having the organs of a single individual.
**u. theory**. 1. The theory that all disease is single in its nature. 2. The theory that the serum of each animal contains only one alexin or complement.
**univalence** (ū-niv'-al-ens) [uni-; valere, to be worth]. The state of being univalent.
**univalent** (ū-niv'-al-ent) [uni-; palere, to be worth]. Having a valence of one; capable of replacing a single hydrogen atom in combination.
**universal** (ū-niv-ur'-sal) [universalis]. General. u. **joint**, a ball-and-socket joint, movable in any direction.
**Unna's dermatosis** (oōn'-ah) [Paul Gerson Unna, German dermatologist, 1850- ]. Seborrhea. U.'s **layer**. See Langerhans' granular layer. U.'s **papillary hair**, a complete hair and hair-follicle. U.'s **paste**, a salve of zinc oxide in glycerine and mucilage of acacia; used in skin lesions. U.'s **plasma- cells**, cubic or rhombic cells, found especially in granulomatous inflammations, the protoplasm of which stains deeply with methylene blue, while the nucleus is readily decolorized (by creosote or styrone).
**unof.** Abbreviation of unofficial.
**unofficial** (un-of-ish'-al). Not included in the pharmacopeia; not sanctioned by recognized authority.
**unorganized** (un-or'-gan-izd). Without organs; not arranged in the form of an organ or organs.
**unpolarized** (un-po'-lar-izd). Not polarized.
**unrest, peristaltic**. A condition characterized by spasmodic and irregular movements of the stomach or intestine.
**unsatisfied** (un-sat'-is-fid). A term applied to a hydrocarbon which has one or more free valences.
**unsaturated** (un-sat'-ū-ra-ted) [un, not; saturated]. 1. Not saturated. 2. A term applied to hydrocarbons of the methane or paraffin series when their greatest possible valence is not satisfied.
**Unschuld's sign** (oon'-shoolt) [Paulus Unschuld, German physician, 1835- ]. A tendency to cramps in the calf of the leg; it is an early sign in diabetes.
**unsex** (un-seks'). To spay or castrate.
**unsound** (un-sownd'). Not healthy; diseased.
**unsoundness** (un-sownd'-nes). The state of being unsound. u. **of mind**, incapacity to govern one's affairs.
**unstriated** (un-stri'-a-ted). Not striated, as unstriated muscle.
**ununited** (un-ū-ni'-ted). Not united, as an ununited fracture.
**unwell**. 1. Ill; sick. 2. Menstruating.
**upas** (ū'-pas) [Malay, upas, poison]. A name applied to several trees found in the East Indies and containing a poisonous principle. It is used as an arrow-poison. u. **antiar**, Javanese arrow-poison; it acts directly on the vasomotor centers. u. **radju**, or u. **tienté**. See under strychnos tienté.

**urachal** (ū'-ra-kal) [οὐραχός, urachus]. Pertaining to the urachus.
**urachus** (ū'-ra-kus) [οὐρον, urine; ἔχειν, to hold]. The allantoic stalk connecting in the fetus the bladder with the allantois, in after-life represented by a fibrous cord passing from the apex of the bladder to the umbilicus. u., **patent**, a condition in which the urachus of the embryo does not become obliterated, but persists to adult life.
**uracil** (ū'-ra-sil) [οὐρον, urine], $C_4H_4N_2O_2$. The ureid of β-oxyacrylic acid; also known in the form of its derivatives.
**uracrasia, uracratia** (ū-rak-ra'-ze-ah, ū-rak-ra'-she-ah) [οὐρον, urine; ἀκρασία, incontinence]. Incontinence of urine; enuresis, q. v.
**uræmia** (ū-re'-me-ah). See uremia.
**uræmic** (ū-re'-mik). See uremic.
**uragogue** (ū-rag-og') [οὐρον, urine; ἀγωγός, drawing forth]. Increasing urinary secretion; a diuretic.
**ural, uralin, uralium** (ū'-ral, ū'-ral-in, ū-ra'-le-um) [οὐρον, urine]. Chloral-urethane, $CCl_3CH:OH.NH-CO_2C_2H_5$, a hypnotic. Dose 10–20 gr. (0.65–1.3 Gm.).
**urali** (ū-ra'-le). Synonym of curare.
**uramil** (ū'-ram-il), $C_4H_5N_3O_3$. Amidobarbituric acid, obtained by boiling alloxanthin with an ammonium chloride solution.
**uramine** (ū'-ran-ēn). See guanidine.
**uranalysis** (ū-ran-al'-is-is). Analysis of the urine.
**urane** (ū'-rān). 1. Uranium oxide; it is used to give a yellow fluorescence to glass. 2. A unit of radio activity. See kilurane.
**uranic** (ū-ran'-ik). Containing uranium as a hexad radical. u. **acid**, uranium trioxide.
**uranin** (ū'-ran-in). See fluorescein-sodium.
**uraninite** (ū-ran'-in-it). Pitch-blende.
**uranischasma** (ū-ran-is-ko-kaz'-mah) [οὐρανίσκος, the roof of the mouth; χάσμα, chasm]. Cleft palate.
**uranisconitis** (ū-ran-is-ko-ni'-tis) [uraniscus; itis, inflammation]. Inflammation of the uraniscus, or palate.
**uraniscoplasty** (ū-ran-is'-ko-plas-te) [uraniscus; πλάσσειν, to form]. A plastic operation for the repair of cleft palate.
**uraniscorrhaphy** (ū-ran-is-kor'-a-fe) [uraniscus; ραφή, suture]. Suture of a palatal cleft; staphylorrhaphy.
**uraniscus** (ū-ran-is'-kus) [οὐρανίσκος, the roof of the mouth]. The palate.
**uranism** (ū'-ran-izm) [Οὐρανός, the Greek personification of heaven]. Sexual perversion in which the desire is for individuals of the same sex.
**uranist** (ū'-ran-ist). A sexual pervert having a passion for one of his own sex.
**uranium** (ū-ra'-ne-um) [Uranus]. A heavy white metal. See oxide of chemical. Its phosphate and nitrate are used as tests for phosphoric acid. The salts are very poisonous. u. **acetate**, $(UO_2)(C_2H_3O_2)_2+2H_2O$; recommended in coryza in solution as nasal douche. u. **ammonium fluoride**, $UO_3.F_2(NH_4).H_2O$, a greenish-yellow, crystalline powder, used for the detection of Roentgen-rays. u. **nitrate**, $(HNO_3)_2UO_2.6H_2O$, used in diabetes. Dose ¼–1 gr. (0.014–0.026 Gm.) twice daily in aqueous solution with saccharin. u. **oxide**, red, u. **trioxide**, $UO_3$, a reddish powder. Syn., uranic acid. u. **x**, a radioactive precipitate obtained from uranium nitrate by means of ammonium carbonate. u. **yellow**, sodium uranate.
**uranomania** (ū-ran-o-ma'-ne-ah) [οὐρανός, palate; μανία, madness]. Religious mania with exaltation.
**uranoplastic** (ū-ran-o-plas'-tik) [οὐρανός, palate; πλάσσειν, to form]. Belonging or pertaining to uranoplasty.
**uranoplasty** (ū'-ran-o-plas-te). Same as uranisco-plasty.
**uranoplegia** (ū-ran-o-ple'-jah) [οὐρανός, palate; πληγή, stroke]. Paralysis of the muscles of the soft palate.
**uranorrhaphy** (ū-ran-or'-af-e) [οὐρανός, palate; ραφή, seam]. See uraniscorrhaphy.
**uranoschisis** (ū-ran-os'-kis-is) [οὐρανός, palate; σχίσις, a cleft]. Cleft palate.
**uranoschism** (ū'-ran-o-skizm) [οὐρανός, palate; σχίσμα, a cleft]. Cleft palate.
**uranostaphyloplasty** (ū-ran-o-staf'-il-o-plas-te). See uraniscoplasty.
**uranostaphylorrhaphy** (ū-ran-o-staf-il-or'-af-e) [οὐρανός, palate; staphylorrhaphy]. Same as uraniscorrhaphy.
**uranyl** (ū'-ran-il). Uranium dioxide. See uranium.

**u.-ammonium fluoride.** See *uranium-ammonium fluoride.*
**urapostema** (ū-ra-pos-tē'-mah) [οὖρον, urine; ἀπόστημα, abscess]. An abscess containing urine.
**urare, urari** (ū-rah'-re). See *curare.*
**urarize** (ū'-ra-rīz). To bring under the influence of *curare;* curarize.
**urarthritis** (ū-rar-thri'-tis). See *arthritis urica.*
**urase** (ū'-rās). An insoluble enzyme associated with the bacteria which ferment urea; it is very plentiful in the urine in catarrh of the bladder.
**urasol** (ū'-rah-sol). Acetylmethylene-disalicylic acid; an antiseptic, diaphoretic, and uric-acid solvent.
**urate** (ū'-rāt) [οὖρον, urine]. A salt of uric acid.
**uratemia, uratæmia** (ū-ra-tē'-me-ah) [urate; αἷμα, blood]. An abnormally large quantity of urates in the blood.
**uratic** (ū-rat'-ik) [urate]. Pertaining to or characterized by urates. **u. diathesis,** a condition in which there is a tendency to the deposition of urates in the joints and elsewhere; a tendency to gout. **u. inspissation,** uric acid infarct.
**uratolysis** (ū-ra-tol'-is-is) [urate; λύειν, to loosen]. The decomposition or solution of urates.
**uratolytic** (ū-rat-o-lit'-ik) [urate; λύειν, to loosen]. Capable of dissolving urates.
**uratoma** (ū-ra-to'-mah) [urate; ὄμα, tumor; *pl.*, *uratomata*]. A concretion composed of urates, and occurring chiefly about the joints; a tophus.
**uratosis** (ū-rat-o'-sis) [urate]. A morbid condition marked by the deposit of urates.
**uraturia** (ū-rat-ū'-re-ah) [urate; οὖρον, urine]. A condition marked by an excess of urates in the urine.
**urbanization** (ur-ban-iz-a'-shun) [urbs, a city]. A term devised to express the tendency of modern society to develop into cities at the expense of the country population, with a consequent influence upon disease, the death-rate, etc.
**urceolate** (ur'-se-o-lāt) [urceolus, a little pitcher or urn]. Pitcher-shaped, urn-shaped.
**urea** (ū-re'-ah) [οὖρον, urine]. CO(NH₂)₂. The chief nitrogenous constituent of urine, and principal end-product of tissue metamorphosis; it occurs also in the blood, the lymph, and the liver. See *biuret, Schiff, Schroeder.* **u.-bromine,** calcium bromocarbamide. **u. enzyme, u. ferment.** See *urase.* **u. quinate,** a combination of urea and quinic acid; used in the treatment of gout and uric concrements in the kidneys. Daily dose 30–80 gr. (2–5 Gm.) in 400 Cc. of hot water. **u. salicylate,** recommended as a substitute for sodium salicylate. Dose 7 gr. (0.45 Gm.) one to four times daily.
**ureal** (ū'-re-al) [urea]. Pertaining to or containing urea.
**ureameter** (ū-re-am'-e-ter) [urea; μέτρον, a measure]. An apparatus for determining the amount of urea contained in a liquid.
**ureametry** (ū-re-am'-et-re) [ureameter]. The determination of the amount of urea in a liquid.
**urease** (ū'-re-ās). See *urase.*
**Urechites suberecta** (ū-rek-i'-tēz sub-e-rek'-tah). Savannah flower; yellow nightshade, a poisonous West Indian plant.
**urechitin** (ū-rek'-it-in), C₂₁H₃₀O₉. A poisonous glucoside from *Urechites suberecta.*
**urechitoxin** (ū-rek-it-oks'-in) [urechites; τόξικον, poison]. A highly poisonous principle from the leaves of *Urechites suberecta.*
**urechysis** (ū-rek'-is-is) [οὖρον, urine; ἔκχυσις, an effusion]. An effusion of urine into areolar tissue.
**urecidin** (ū-re-si'-din). A proprietary preparation of lemon-juice and lithium citrate for use in gout and uric-acid diathesis.
**uredema, uroedema** (u-re-de'-mah) [οὖρον, urine; *edema*]. Distention of tissues from extravasation of urine.
**uredo** (ū-re'-do) [uredo, a blight]. 1. A genus of fungi. 2. Urticaria. 3. A sensation of burning in the skin.
**ureid, urelde** (ū'-re-id) [urea]. A compound of urea and an acid radical.
**ureine** (ū'-re-in). A yellow, oily liquid isolated from the urine, and said to be the cause of the symptoms observed in uremia.
**urelcosis** (ū-rel-ko'-sis) [οὖρον, urine; ἕλκωσις, ulceration]. Ulceration of the urethra or urinary organs.
**uremia** (ū-re'-me-ah) [οὖρον, urine; αἷμα, blood]. The symptoms due to the retention in the blood of excrementitious substances normally excreted by the kidneys; it is characterized by headache, vertigo, vomiting, amaurosis, convulsions, coma, sometimes hemiplegia, and a urinous odor of the breath.
**uremic** (ū-rem'-ik, ū-re'-mik) [uremia]. Due to or characterized by uremia.
**uremide** (u'-rem-īd). A skin eruption found in cases of uremic poisoning.
**ureometer** (ū-re-om'-et-er). See *ureameter.*
**ureometry** (ū-re-om'-et-re). See *ureametry.*
**ureorrhea** (ū-re-or-e'-ah) [οὖρον, urine; ῥοία, a flow]. Polyuria; an increased flow of urine.
**urerythrin.** See *uroerythrin.*
**uresiesthesis** (ū-res-e-es'-the-sis) [uresis; αἴσθησις, sensation]. Constant desire to urinate.
**uresin** (ū'-re-sin). A citrourotropin dilithic salt, useful as a uric-acid solvent.
**uresis** (ū-re'-sis) [οὔρησις]. Same as *urination.*
**-uret** (ū-ret). A suffix denoting a binary compound of carbon, sulphur, etc., with another element; in modern chemistry the suffix *-ide* is used.
**-uret** (ū'-ret). The group CH₂NO; it replaces a hydrogen atom in ammonia to form urea.
**uretal** (ū'-ret-al). Same as *ureteric.*
**ureter** (ū-re'-ter) [οὐρητήρ, ureter]. The long, narrow tube conveying the urine from the pelvis of the kidney to the bladder.
**ureteral** (ū-re'-ter-al) [ureter]. Pertaining to the ureter.
**ureteralgia** (ū-re-ter-al'-je-ah) [ureter; ἄλγος, pain]. Pain in the ureter.
**ureterocystoscope** (ū-re-tur-sist'-o-skōp) [ureter; *cystoscope*]. Same as *ureterocystoscope.*
**ureterectasis** (ū-rel-er-ek'-tas-is) [ureter; ἔκτασις, a stretching]. Dilatation of a ureter.
**ureterectomy** (ū-re-ter-ek'-to-mē) [ureter; ἐκτομή, excision]. Excision of a ureter.
**ureteric** (ū-re-ter'-ik) [ureter]. Pertaining to the ureters or to a ureter.
**ureteritis** (ū-re-ter-i'-tis) [ureter; ιτις, inflammation]. Inflammation of a ureter.
**uretero-** (ū-re-ter-o-) [ureter]. A prefix denoting relating to the ureter.
**ureterocele** (ū-re'-ter-o-sēl) [uretero-; κήλη, hernia]. A hernia containing a ureter.
**ureterocervical** (ū-re-ter-o-ser'-vik-al). Relating to or connecting the ureter and the cervix uteri.
**ureterocystoneostomy** (ū-re-ter-o-sist-o-ne-os'-to-me). See *ureterocystostomy.*
**ureterocystoscope** (ū-re-ter-o-sist'-o-skōp). An electric cystoscope holding in its grooved wall a catheter for insertion into the ureter.
**ureterocystostomy** (ū-re-ter-o-sis-tos'-to-me) [uretero-; κύστις, bladder; στόμα, mouth]. The surgical formation of a communication between a ureter and the bladder.
**ureterodialysis** (ū-re-ter-o-di-al'-is-is) [uretero-; διάλυσις, a breaking]. Rupture of the ureter.
**ureteroenterostomy** (ū-re-ter-o-en-ter-os'-to-me) [uretero-; ἔντερον, bowel; στόμα, mouth]. Surgical formation of a passage from a ureter to the intestine.
**ureterography** (ū-ret-er-og'-raf-e) [uretero-; γράφειν, to write]. Radiography of the ureter after the injection of some opaque substance.
**ureterolith** (ū-re'-ter-o-lith) [uretero-; λίθος, stone]. Calculus in the ureter.
**ureterolithiasis** (ū-re-ter-o-lith-i'-as-is) [uretero-; λίθος, stone]. The presence or formation of a calculus in the ureter.
**ureterolithotomy** (ū-re-ter-o-lith-ot'-o-me). Incision of the ureter for removal of a calculus.
**ureterolysis** (ū-re-ter-ol'-is-is) [uretero-; λύσις, a loosening]. Rupture of the ureter.
**ureteroneocystostomy** (ū-re-ter-o-ne-o-sist-os'-to-me). See *ureterocystostomy.*
**ureteroneopyelostomy** (ū-re-ter-o-ne-o-pi-el-os'-to-me) [uretero-; νέος, new; πύελος, trough; στόμα, mouth]. Excision of part of a ureter and implantation into a new aperture made into the pelvis of the kidney, on the corresponding orifice of the ureteral end.
**ureteronephrectomy** (ū re ter o nef reh' to me) [*uretero-*; νεφρός, kidney; ἐκτομή, excision]. Removal of the kidney and its ureter.
**ureterophlegma** (ū-re-ter-o-fleg'-mah) [uretero-; φλέγμα, phlegm]. Accumulation of mucus in the ureter.
**ureterophlegmasia** (ū-re-ter-o-fleg-ma'-ze-ah). Same as *ureteritis.*
**ureteroplasty** (ū-re-ter-o-plas'-te) [uretero-; πλάσσειν, to form]. A plastic operation on a ureter.
**ureteroproctostomy** (ū-re-ter-o-prok-tos'-to-me)

URETEROPYELITIS 922 URIC ACID

[*uretero-*; προκτός, anus; στόμα, mouth]. The surgical formation of a passage from the ureter to the anus.
**ureteropyelitis** (ū-re-ter-o-pi-el-i'-tis) [*uretero-*; *pyelitis*]. Inflammation of a ureter and the pelvis of a kidney.
**ureteropyeloneostomy** (ū-re-ter-o-pi-el-o-ne-os'-to-me). See *ureteroneopyelostomy*.
**ureteropyelonephritis** (ū-re-ter-o-pi-el-o-nef-ri'-tis). Inflammation of the ureter and of the kidney and its pelvis.
**ureteropyosis** (ū-re-ter-o-pi-o'-sis). [*uretero-*; *pyosis*]. Purulent inflammation of the ureter.
**ureterorectostomy** (ū-re-ter-o-rek-tos'-to-me) [*uretero-*; *rectum*; στόμα, mouth]. Ureteroproctostomy.
**ureterorrhagia** (ū-re-ter-or-a'-je-ah) [*uretero-*; ῥηγνύναι, to burst forth]. Hemorrhage from the ureter.
**ureterorrhaphy** (ū-re-ter-or'-af-e) [*uretero-*; ῥαφή, suture]. Suture of the ureter.
**ureterostegnosis** (ū-re-ter-o-steg-no'-sis) [*uretero-*; στέγνωσις, stenosis]. Stenosis or constriction of the ureter.
**ureterostenoma** (ū-re-ter-o-sten-o'-mah) [*uretero-*; στένωμα, a narrow place]. Narrowing of the ureter.
**ureterostenosis** (ū-re-ter-o-sten-o'-sis). See *ureterostegnosis*.
**ureterostoma** (ū-re-ter-o-sto'-mah) [*uretero-*; στόμα, a mouth]. 1. The renal or the cystic opening or mouth of the ureter. 2. A ureteral fistula.
**ureterostomatic** (ū-re-ter-o-sto-mat'-ik). Relating to the ureteral orifice, or to a ureteral fistula.
**ureterostomy** (ū-re-ter-os'-to-me) [see *ureterostoma*]. The formation of a ureteral fistula.
**ureterotomy** (ū-re-ter-ot'-o-me) [*uretero-*; τομή, a cutting]. Incision of the ureter.
**ureteroureteral** (ū-re-ter-o-ū-re'-ter-al) [*ureter*]. Pertaining to both ureters, or to two parts of one ureter. u. anastomosis; See *ureteroureterostomy*.
**ureteroureterostomy** (ū-re-ter-o-ū-re-ter-os'-to-me) [*uretero-*; *ureterostomy*]. Surgical formation of a passage between the ureters or between different parts of the same ureter.
**ureterouterine** (ū-re-ter-o-ū'-ter-in) [*uretero-*; *uterus*]. Pertaining to the ureter and the uterus.
**ureterovaginal** (ū-re-ter-o-vaj'-in-al) [*uretero-*; *vagina*]. Pertaining to the ureter and the vagina.
**urethane** (ū'-reth-ān) [*urea*; *ether*]. 1. C₂H₅CH₂·NO₂, ethyl carbamate. A hypnotic. Dose 20-40 gr. (1.3-2.6 Gm.). 2. In a wider sense, any ester of carbamic acid is called a urethane. u. chloral. See *uralium*. u., ethyl. See *urethane* (1).
**urethra** (ū-re'-thrah) [οὐρήθρα, urethra]. The canal through which the urine is discharged, extending from the neck of the bladder to the meatus urinarius. It is divided in the man into the *prostatic portion*, the *membranous portion*, and the *spongy* or *penile portion*, and is from 8 to 9 inches long. In the woman it is about 1½ inches in length. u. muliebris, female urethra. u. virilis, male urethra.
**urethral** (ū-re'-thral) [*urethra*]. Pertaining to the urethra; produced in or arising from the urethra, as *urethral* fever. u. arthritis, gonorrheal rheumatism.
**urethralgia** (ū-re-thral'-je-ah) [*urethra*; ἄλγος, pain]. Pain in the urethra.
**urethrameter** (ū-re-thram'-e-ter). See *urethrometer*.
**urethrascope** (ū-re'-thra-skōp). See *urethroscope*.
**urethratresia** (ū-re-thrat-re'-ze-ah) [*urethra*; *atresia*]. 1. Occlusion of the urethra. 2. Imperforate urethra.
**urethrectomy** (ū-re-threk'-to-me) [*urethra*; ἐκτομή, excision]. Excision of a urethra or a portion of it.
**urethremorrhage** (ū-re-threm'-or-āj) [*urethra*; αἷμα, blood; ῥηγνύναι, to burst forth]. Hemorrhage from the urethra.
**urethremphraxis** (ū-re-threm-frak'-sis). See *urethrophraxis*.
**urethreurynter** (ū-re-throo-rin'-ter) [*urethra*; εὐρύνειν, to dilate]. An apparatus for dilating the urethra.
**urethrism, urethrismus** (ū'-re-thrizm, ū-re-thris'-mus) [*urethra*]. Urethral irritability.
**urethritis** (ū-re-thri'-tis) [*urethra*; ιτις, inflammation]. Inflammation of the urethra. u., anterior, inflammation of the part situated anterior to the anterior layer of the triangular ligament. u., posterior, inflammation of the prostatic and membranous portions. u., simple, a nonspecific inflammation of the urethra. u., specific, that due to the gonococcus; gonorrhea. u.venerea, gonorrhea.
**urethro-** (ū-re-thro-) [οὐρήθρα, *urethra*]. A prefix denoting pertaining to the urethra.
**urethrobulbar** (ū-re-thro-bul'-bar). Relating to the urethra and the bulb of the corpus spongiosum.

**urethrocele** (ū-re'-thro-sēl) [*urethro-*; κήλη, a hernia]. A protrusion or thickening of the wall of the female urethra.
**urethrocystitis** (ū-re-thro-sis-ti'-tis) [*urethro-*; *cystitis*]. Inflammation of the urethra and bladder.
**urethrograph** (ū-re'-thro-graf) [*urethro-*; γράφειν, to record]. A recording urethrometer.
**urethrometer** (ū-re-throm'-et-er). [*urethro-*; μέτρον, a measure]. An instrument for determining the caliber of the urethra or for measuring the lumen of a stricture.
**urethropenile** (ū-re-thro*pe'-nīl*). Relating to the urethra and the penis.
**urethroperineal** (ū-re'-thro-per-in-e'-al). Relating to the urethra and the perineum.
**urethroperineoscrotal** (ū-re-thro-per-in-e-o-skro'-tal). Relating to the urethra, perineum, and scrotum.
**urethrophraxis** (ū-re-thro-fraks'-is) [*urethro-*; φράξις, a blocking]. Urethral obstruction.
**urethroplasty** (ū-re'-thro-plas-te) [*urethro-*; πλάσσειν, to form]. Plastic operation upon the urethra.
**urethroprostatic** (ū-re-thro-pros-tat'-ik). Relating to the urethra and the prostate.
**urethrorectal** (ū-re-thro-rek'-tal). Relating to the urethra and the rectum.
**urethrorrhagia** (ū-re-thror-a'-je-ah) [*urethro-*; ῥηγνύναι, to burst forth]. Hemorrhage from the urethra.
**urethrorrhaphy** (ū-re-thror'-af-e) [*urethro-*; ῥαφή, suture]. Suturing of an abnormal opening into the urethra.
**urethrorrhea, urethrorrhœa** (ū-re-thro-re'-ah) [*urethro-*; ῥοία, a flow]. A morbid discharge from the urethra.
**urethroscope** (ū-re'-thro-skōp) [*urethro-*; σκοπεῖν, to view]. An instrument for inspecting the interior of the urethra.
**urethroscopic** (ū-re-thro-skop'-ik). Relating to the urethroscope.
**urethroscopy** (ū-re-thros'-ko-pe) [see *urethroscope*]. Inspection of the urethra with the aid of the urethroscope.
**urethrospasm** (ū-re'-thro-spazm) [*urethro-*; σπασμός, spasm]. A spasmodic stricture of the urethra.
**urethrostenosis** (ū-re-thro-sten-o'-sis) [*urethro-*; στένωσις, a constriction]. Stricture of the urethra.
**urethrostomy** (ū-re-thros'-to-me) [*urethro-*; στόμα, a mouth]. Perineal section with permanent fixation of the membranous urethra in the perineum.
**urethrotome** (ū-re'-thro-tōm) [*urethro-*; τομή, a cutting]. An instrument used for performing urethrotomy. u., dilating, a combined urethrotome and dilator.
**urethrotomy** (ū-re-throt'-o-me) [see *urethrotome*]. The operation of cutting a stricture of the urethra. u., external, division of a stricture by an incision from without. u., internal, division of a urethral stricture from within the urethra.
**urethroureteral** (ū-re-thro-ū-re'-ter-al) [*urethro-*; *ureter*]. Relating to the urethra and the ureter.
**urethrovaginal** (ū-re-thro-vaj'-in-al) [*urethro-*; *vagina*]. Pertaining to the urethra and the vagina.
**urethrovesical** (ū-re-thro-ves'-ik-al) [*urethro-*; *vesica*, bladder]. Pertaining to the urethra and the bladder.
**urethylane** (ū-reth'-il-ān), C₂H₅NO₂. A colorless crystalline substance soluble in water and alcohol; used as a hypnotic and a diuretic.
**uretic** (ū-ret'-ik) [οὐρητικός]. 1. Pertaining to urine; stimulating the flow of urine. 2. An agent or medicine that stimulates the flow of urine.
**Urgens'** reaction for sulphocyanates in saliva. Add to the saliva a saturated solution of hydriodic acid 1 part and starch-paste 5 parts; if sulphocyanates are present, a blue tint appears. The sulphocyanates of potassium and sodium normally present in saliva are usually absent in chronic suppurative conditions of the middle ear, and the progress of the disease can be estimated by the presence and extent of the reaction.
**Urginea** (ur-jin'-e-ah) [*urgere*, to press]. A genus of liliaceous plants. **U. scilla**, the plant that produces the official squill. See *scilla*.
**urgosan** (ur'-go-san). Trade name of a preparation containing hexamethylenamine and gonosan.
**urhidrosis** (ū-ri-dro'-sis). Uridrosis.
**urian** (ū'-re-an). Urochrome.
**uriasis** (ū-ri'-as-is). See *lithiasis*.
**uric** (ū'-rik) [οὖρον, urine]. Pertaining to the urine.
**uric acid** (ū'-rik) [οὖρον, urine], C₅H₄N₄O₃. A dibasic acid; one of the nitrogenous end-products of metabolism. It is found in the urine and in the

magnesium, potassium, and sodium. The most important abnormal constituents present in disease are albumin, sugar, blood, pus, acetone, diacetic acid, fat, chyle, tube-casts, various cells, and bacteria. u., **incontinence of,** inability to retain the urine. See *enuresis*. u., **residual,** urine that remains in the bladder after urination in prostatic hypertrophy and in cystic disease. u., **retention of,** inability to pass the urine. See *ischuria*. u., **suppression of.** See *ischuria*.
**urinemia, urinæmia** (*ū-rin-e'-me-ah*) [*urine; αἷμα*, blood]. The presence of urinary constituents in the blood; uremia.
**uriniferous** (*ū-rin-if'-er-us*) [*urine; ferre*, to bear]. Carrying or conveying urine, as *uriniferous* tubule.
**urinific** (*ū-rin-if'-ik*) [*urine; facere*, to make]. Excreting or producing urine.
**uriniparous** (*ū-rin-ip'-ar-us*) [*urine; parere*, to produce]. Secreting urine.
**urino-** (*ū-rin-o-*) [*urine*]. A prefix denoting relation to the urine.
**urinocryoscopy** (*ū-rin-o-kri-os'-ko-pe*) [*urino-; κρύος*, cold; *σκοπεῖν*, to examine]. Cryoscopy applied to urine.
**urinogenital** (*ū-rin-o-jen'-it-al*) [*urino-; γεννᾶν*, to produce]. Urogenital.
**urinoglucosometer** (*ū-rin-o-gloo-ko-som'-et-er*). An apparatus for quantitative estimation of glucose in the urine.
**urinologist** (*ū-rin-ol'-o-jist*). One skilled in urinology.
**urinology** (*ū-rin-ol'-o-je*). See *urology*.
**urinoma** (*ū-rin-o'-mah*) [*urino-; ὄγκος*, a tumor]. A cyst containing urine.
**urinometer** (*ū-rin-om'-et-er*) [*urino-; μέτρον*, a measure]. A hydrometer for ascertaining the specific gravity of urine.
**urinometry** (*ū-rin-om'-et-re*) [see *urinometer*]. The determination of the specific gravity of the urine by means of the urinometer.
**urinoscopic** (*ū-rin-o-skop'-ik*) [*urino-; σκοπεῖν*, to examine]. Pertaining to examination of the urine.
**urinoscopy** (*ū-rin-os'-ko-pe*). See *uroscopy*.
**urinose, urinous** (*ū'-rin-ōs, ū'-rin-us*) [*urine*]. Having the characters of urine, as a *urinose* odor.
**uriseptin** (*ū-ris-ep'-tin*). Trade name of a preparation containing formaldehyde and lithia; it is said to be a diuretic and genito-urinary antiseptic.
**urisolvent** (*ū-ris-ol'-vent*). Dissolving uric acid.
**urisolvin** (*ū-re-sol'-vin*). A compound of urea and lithium citrate, a uric-acid solvent and diuretic.
**uristamine** (*ū-ris'-tam-ēn*). A trade name for a brand of hexamethylenetetramine.
**uritis** (*ū-ri'-tis*) [*urere*, to burn; *ιτις*, inflammation]. Inflammation following a burn.
**uritone** (*ū'-rit-ōn*). A trade name for a brand of hexamethylenetetramine.
**urning** (*urn'-ing*). [*Οὐρανός*, the Greek personification of heaven]. A homosexual individual; a pervert in whom the desire is only for individuals of the same sex.
**urnism** (*urn'-izm*). See *uranism*.
**uro-** (*ū-ro-*) [*οὖρον*, urine]. A prefix denoting pertaining to urine or uric acid.
**uroacidimeter** (*ū-ro-as-id-im'-et-er*). An instrument for measuring the acidity of the urine.
**uroammoniac** (*ū-ro-am-o'-ne-ak*). Relating to or containing uric acid and ammonia.
**uroazotometer** (*ū-ro-az-ot-om'-et-er*). An apparatus for quantitative estimation of the nitrogenous substances in urine.
**urobacillus** (*ū-ro-bas-il'-us*) [*uro-; bacillus*]. A bacillus occurring in urine, particularly in decomposing urine.
**urobenzoic acid** (*ū-ro-ben-zo'-ik*). Hippuric acid.
**urobilin** (*ū-ro-bil'-in*) [*uro-; bile*], $C_{32}H_{40}N_4O_7$. A yellowish-brown, amorphous pigment derived from bilirubin. It is the principal pigment of the urine, and is increased in febrile and other conditions. See *Gerhardt, Grünbart*. u., **jaundice,** a jaundice supposed to be due to the presence of urobilin in the blood. u., **pathogenic,** the excessive coloring matter of certain dark urines occurring in various diseases, as pernicious anemia, febrile diseases, etc.
**urobilinemia, urobilinæmia** (*ū-ro-bil-in-e'-me-ah*) [*urobilin; αἷμα*, blood]. The presence of urobilin in the blood.
**urobilinicterus** (*ū-ro-bil-in-ik'-ter-us*). Pigmentation of the skin, cornea, etc., from absorption of extravasated blood and contained urobilin.

**urobilinogen** (ū-ro-bil-in'-o-jen). A chromogen from which urobilin is formed by oxidation when urine is allowed to stand.

**urobilinoidin** (ū-ro-bil-in-oid'-in). A form of urinary pigment derived from hematin and resembling urobilin, though not identical with it. It occurs in certain pathological conditions.

**urobilinuria** (ū-ro-bil-in-ū'-re-ah) [urobilin; urine]. The presence of an excess of urobilin in the urine.

**urobromohematin** (ū-ro-bro-mo-he'-mat-in), $C_{68}H_{84}N_8Fe_3O_{24}$. A coloring matter found in the urine of leprous patients; it is closely allied to the coloring-matter of the blood.

**urocanin** (ū-ro-kan'-in). See under *urocaninic acid*.

**urocaninic acid** (ū-ro-kan-in'-ik) [uro-; canis, a dog], $C_8H_{14}N_2O_2+2H_2O$. An acid found in the urine of dogs when there is a diminution in the amount of urea. By heating it is decomposed into carbonic acid, water, and a base, *urocanin*, $C_{11}H_{10}N_4O$.

**urocele** (ū'-ro-sēl) [uro-; κήλη, a tumor]. A swelling of the scrotum from extravasation of urine.

**urocheras** (ū-rok'-er-as) [uro-; χερás, sand]. The sandy substance deposited from standing urine.

**urochesia** (ū-ro-ke'-ze-ah) [uro-; χέζειν, to defecate]. Discharge of urine through the anus.

**urochrome** (ū'-ro-krōm) [uro-; χρώμα, color]. A yellow coloring-matter found in urine, supposed to be impure urobilin.

**urocinetic.** See *urokinetic*.

**uroclepsia** (ū-ro-klep'-se-ah) [uro-; κλεψία, theft]. Unconscious discharge of urine.

**urocol** (ū'-ro-kol). A trade name for a preparation containing urea quinate (urol) and colchicine; used in gout.

**urocrisis, urocrisia** (ū-rok'-ris-is, ū-ro-kris'-e-ah) [uro-; crisis]. 1. A disease crisis attended with excessive urination. 2. Diagnosis by examination of the urine. 3. A vesical crisis; see *vesical*.

**urocriterion** (ū-ro-kri-te'-re-on). In diagnosis by inspection of urine, the indication which determines the diagnosis.

**urocyanin** (ū-ro-si'-an-in). See *uroglaucin*.

**urocyanogen** (ū-ro-si-an'-o-jen) [uro-; κύανος, blue]. A blue pigment found in urine, particularly in cases of cholera.

**urocyanose** (ū-ro-si'-an-ōs). See *urocyanogen*.

**urocyanosis** (ū-ro-si-an-o'-sis) [uro-; κύανος, blue]. Blue discoloration of the urine from the presence of indican.

**urocyst, urocystis** (ū'-ro-sist, ū-ro-sist'-is) [uro-; κύστις, bladder]. The urinary bladder.

**urocystic** (ū-ro-sis'-tik). Pertaining to the urinary bladder.

**urocystitis** (ū-ro-sis-ti'-tis) [urocyst; ιτις, inflammation]. Inflammation of the urinary bladder; cystitis.

**urodialysis** (ū-ro-di-al'-is-is) [uro-; διάλυσις, a cessation]. Partial and temporary cessation of the secretion of urine.

**urodochium** (ū-ro-do-ki'-um) [uro-; δοχείον, receptacle]. A urinal.

**urodynia** (ū-ro-din'-e-ah) [uro-; ὀδύνη, pain]. Painful micturition.

**uroedema** (ū-re-de'-mah). See *uredema*.

**uroerythrin** (ū-ro-er'-ith-rin) [uro-; ἐρυθρόs, red]. An amorphous, reddish pigment with an acid reaction, occurring in the urine in rheumatic and other diseases.

**urofuscohematin** (ū-ro-fus-ko-hem'-at-in) [uro-; fuscus, dark; hematin]. A red pigment derived from hematin, occurring in the urine.

**urogaster** (ū'-ro-gas-ter) [uro-; γαστήρ, stomach]. The urinary intestine or urinary passages collectively.

**urogenin** (ū-roj'-en-in). Trade name of a diuretic said to contain lithium, hippuric acid and theobromine.

**urogenital** (ū-ro-jen'-it-al) [uro-; genital]. Pertaining to the urinary and genital organs. u. ducts. See *ducts of Mueller*. u. sinus, the anterior part of the cloaca, into which the urogenital ducts open.

**urogenous** (ū-roj'-en-us) [uro-; γεννᾶν, to produce]. Producing urine.

**uroglaucin** (ū-ro-glaw'-sin) [uro-; γλαυκόs, bluish-green]. A blue pigment, at times occurring in urine, as in scarlatina, and supposed to result from the oxidation of a chromogen.

**uroglycosis** (ū-ro-gli-ko'-sis) [uro-; γλυκύs, sweet]. Diabetes mellitus.

**urogravimeter** (ū-ro-grav-im'-et-er). See *urinometer*.

**urohematin** (ū-ro-hem'-at-in) [uro-; hematin]. Altered hematin in the urine.

**urohematoporphyrin** (ū-ro-hem-at-o-por'-fir-in) [uro-hematin; πόρφυρος, purple]. Urohematin; a urinary pigment occasionally occurring in the urine in certain pathological states.

**urokinetic, urocinetic** (ū-ro-kin-et'-ik) [uro-; κίνησις, movement]. Due to a reflex from the urinary apparatus; generally used of a form of dyspepsia due to irritation or disease of the urinary tract.

**urol** (ū'-rol). 1. See *urea quinate*. 2. A trade name for a preparation containing urea quinate.

**urolagnia** (ū-ro-lag'-ne-ah) [uro-; λαγνεία, lust]. A form of sexual perversion in which sexual excitement is produced by the sight of urine or of a person urinating.

**urolite** (ū'-ro-līt). Same as *urolith*.

**urolith** (ū'-ro-lith) [uro-; λίθος, a stone]. A calculus occurring in the urine.

**urolithiasis** (ū-ro-lith-i'-a-sis) [uro-; lithiasis]. 1. The presence of, or a condition associated with urinary calculi. 2. The formation of urinary calculi.

**urolithic** (ū-ro-lith'-ik) [uro-; λίθος, a stone]. Pertaining to, or having the nature of urinary calculi.

**urolithology** (ū-ro-lith-ol'-o-je) [urolith; λόγος, science]. The science of urinary calculi.

**urologic** (ū-ro-loj'-ik) [uro-; λόγος, science]. Pertaining to urology.

**urologist** (ū-rol'-o-jist). One versed in urology.

**urology** (ū-rol'-o-je) [uro-; λόγος, science]. The scientific study of the urine.

**urolutein** (ū-ro-lū'-te-in) [uro-; luteus, yellow]. A yellow pigment sometimes found in urine.

**urolytic** (ū-ro-lit'-ik) [uro-; λύειν, to loosen]. Capable of dissolving urinary calculi.

**uromancy** (ū'-ro-man-se) [uro-; μαντεία, divination]. Diagnosis or prognosis by observation of the urine.

**uromelanin** (ū-ro-mel'-an-in) [uro-; melanin]. A black pigment sometimes found in the urine, derived from the decomposition of urochrome.

**uromelus** (ū-ro-me'-lus) [οὐρά, tail; μέλος, a limb]. A monster in which there is more or less complete fusion of the limbs, with but a single foot.

**urometer** (ū-rom'-et-er). See *urinometer*.

**uromphalus** (ū-rom'-fal-us) [urachus; ὀμφαλός, navel]. A monstrosity with the urachus protruding at the navel.

**uroncus** (ū-rong'-kus) [uro-; ὄγκος, tumor]. A tumor or swelling containing urine.

**uronephrosis** (ū-ro-nef-ro'-sis) [uro-; *nephrosis*]. See *hydronephrosis*.

**urology** (ū-ron-ol'-o-je). See *urology*.

**urophan** (ū'-ro-fan) [uro-; φαίνειν, to appear]. A generic name for substances which, taken into the body, appear again unchanged chemically in the urine.

**urophanic** (ū-ro-fan'-ik) [see *urophan*]. Appearing in the urine.

**urophein** (ū-ro-fe'-in) [uro-; φαιός, gray]. A pigment body to which the characteristic odor of the urine has been ascribed.

**uropherin** (ū-rof'-er-in). Lithiotheobromine salicylate; a white powder, soluble in water, used as a diuretic. Dose 15 gr. (1 Gm.). Syn., *lithium diuretin*. u. benzoate, $LiC_7H_7N_4O_2+LiC_6H_5CO_2$, theobromine and lithium benzoate, a fine white powder containing 50% of theobromine; it is a diuretic and nerve stimulant. u. salicylate, $LiC_7H_7N_4O_2+LiC_7H_5O_3$, theobromine and lithium salicylate, a white powder used as the benzoate.

**urophthisis** (ū-rof-thi'-sis) [uro-; *phthisis*]. Synonym of *diabetes mellitus*.

**uropittin** (ū-ro-pit'-in) [uro-; πίττα, pitch], $C_9H_{10}N_2O_3$. A nitrogenous derivative of urochrome.

**uroplania** (ū-ro-pla'-ne-ah) [uro-; πλάνη, a wandering]. The presence of urine in other localities than the urinary organs; the discharge of urine from an abnormal orifice.

**uropoiesis** (ū-ro-poi-e'-sis) [uro-; ποιεῖν, to make]. Secretion of the urine by the kidneys.

**uropoietic** (ū-ro-poi-et'-ik) [see *uropoiesis*]. Concerned in uropoiesis.

**uropsammus** (ū-rop-sam'-us) [uro-; ψάμμος, sand]. Urinary gravel.

**uropyoureter** (ū-ro-pi-o-ū-re'-ter). An infected uroureter.

**urorhodin** (ū-ro-ro'-din) [uro-; ῥόδον, rose]. A red pigment found in urine and derived from uroxanthin.

**urorhodinogen** (ū-ro-ro-din'-o-jen) [urorhodin;' γεννᾶν, to produce]. The chromogen which by decomposition produces urorhodin.

**urorosein** (ū-ro-ro'-ze-in) [uro-; rosa, rose]. A rose-colored pigment found in the urine in various diseases.

**urorrhagia** (ū-ror-a'-je-ah) [uro-; ῥηγνύναι, to burst forth]. Excessive secretion and discharge of urine.

**urorrhea, urorrhoea** (ū-ror-e'-ah) [uro-; ῥοία, a flow]. 1. The normal flow of urine. 2. Involuntary passage of urine.

**urorubin** (ū-ro-roo'-bin) [uro-; ruber, red]. A red pigment obtained by treating urine with hydrochloric acid, and also in the preparation of uropittin.

**urorubrohematin** (ū-ro-rbo-bro-hem'-at-in). See urobromohematin.

**urosacin** (ū-ro'-sas-in). See urorhodin.

**urosanol** (ū-ro-sa'-nol). The trade name of a solution of protargol in gelatin, used in urethritis.

**urosceocele** (ū-ros'-ke-o-sēl). See urocele.

**uroschesis** (ū-ros'-kes-is) [uro-; σχέσις, retention]. Suppression of urine.

**uroscopic** (ū-ro-skop'-ik). Same as urinoscopic.

**uroscopist** (ū-ros'-ko-pist) [see uroscopy]. One who makes a specialty of urinary examinations.

**uroscopy** (ū-ros'-ko-pe) [uro-; σκοπεῖν, to view]. Examination of the urine.

**urosemiology** (ū-ro-se-mi-ol'-o-je) [uro-; semiology]. Examination of the urine as an aid to diagnosis.

**urosepsin** (ū-ro-sep'-sin). The toxin concerned in urosepsis.

**urosepsis** (ū-ro-sep'-sis) [uro-; σῆψις, sepsis]. The condition of intoxication due to the extravasation of urine.

**uroseptic** (ū-ro-sep'-tik) [see urosepsis]. Relating to or characterized by urosepsis.

**uroses** (ū-ro'-sēs). See urosis.

**urosin** (ū'-ro-sin). Lithium quinate, a uric-acid solvent. Dose 7½ gr. 6 to 10 times daily.

**urosis** (ū-ro'-sis) [uro-; νόσος, disease; pl., uroses]. Any disease of the urinary organs.

**urospasm** (ū'-ro-spazm) [uro-; σπασμός, spasm]. Spasm of some part of the urinary tract.

**urospectrin** (ū-ro-spek'-trin). A pigment similar to hematoporphyrin obtained from normal urine by shaking the urine with acetic ether.

**urostealith** (ū-ro-stē'-al-ith) [uro-; στέαρ, fat; λίθος, stone]. A fat-like substance occurring in some urinary calculi.

**urosteatoma** (ū-ro-ste-at-o'-mah). See urostealith.

**urotheobromine** (ū-ro-the-o-bro'-min). See paraxanthin.

**urotoxia** (ū-ro-toks'-e-ah). Same as urotoxy.

**urotoxic** (ū-ro-toks'-ik) [uro-; τοξικόν, poison]. 1. Pertaining to poisonous substances eliminated in the urine. 2. Pertaining to poisoning by urine or some of its constituents. u. coefficient, the number of urotoxies formed in 24 hours by one kilogram of an individual. The normal urotoxic coefficient in man is about 0.4, i. e., a man produces for each kilogram of body-weight 0.4 urotoxies, or sufficient poison to kill 400 Gm. of animal. u. unit, a urotoxy.

**urotoxicity** (ū-ro-toks-is'-it-e). The toxic quality of urine.

**urotoxin** (ū-ro-toks'-in). The poison of urine.

**urotoxy** (ū'-ro-toks-e) [see urotoxic; pl., urotoxies]. The unit of toxicity of urine—the amount necessary to kill a kilogram of living substance.

**urotropin** (ū-ro-tro'-pin), (CH₂)₆N₄, hexamethylentetramine; obtained by action of formaldehyde on ammonia. A uric-acid solvent. Daily dose 8–30 gr. (0.5–2.0 Gm.). u. quinate, used in gout and as a uric-acid solvent.

**uroureter** (ū-ro-ū-re'-ter). A partial or complete non-evacuation of the urine from the ureter.

**urous** (ū'-rus) [οὖρον, urine]. Having the nature of urine. u. acid, uric acid.

**uroxanthin** (ū-ro-zan'-thin) [uro-; ξανθός, yellow]. A yellow pigment occurring in human urine and yielding indigo-blue on oxidation. Indigogen.

**uroxin** (ū-roks'-in). Same as alloxantin.

**urozemia** (ū-ro-ze'-me-ah) [uro-; ζημία, loss]. Diabetes. u. albuminosa, Bright's disease. u. mellita, diabetes mellitus.

**urrhodin** (ū'-rod-in). See urorhodin.

**ursal** (ur'-sal). See urea salicylate.

**ursin** (ur'-sin). See arbutin.

**ursone** (ur'-sōn). See ericolin.

**urtica** (ur'-tik-ah) [L.]. Nettle. 1. A genus of plants of the order Urticaceæ. U. dioica, the common nettle, and U. urens, the dwarf nettle, are used as diuretics, local irritants, and hemostatics. 2. A wheal.

**urticaria** (ur-tik-a'-re-ah) [urtica]. A disease of the skin characterized by the development of wheals, which give rise to sensations of burning and itching. They appear suddenly in large or small numbers, remain for from a few minutes to several hours, and disappear suddenly. The disease may be acute or chronic, and is due to agencies bringing about vasomotor system, such as gastrointestinal disorders, the ingestion of certain foods, as shell-fish, strawberries, etc. Syn., hives; nettlerash. u. bullosa, a form characterized by the formation of bullæ. u. conferta, a form in which the lesions are grouped. u. evanida, a form marked by sudden vanishing and reappearance of the symptoms. u. factitia, u., facticious, the form produced in individuals with an irritable skin by any slight external irritation. Syn., dermographia; dermographism. u., giant, u. gigans. See u. œdematosa. u. hæmorrhagica, purpura urticans, a variety characterized by hemorrhage into the wheals from rupture of the extremely congested capillaries; it is regarded as a variety of erythema multiforme. u. medicamentosa, a variety due to the use of certain drugs. u. œdematosa, giant urticaria, a variety characterized by the sudden appearance of large, soft, edematous swellings of the skin and subcutaneous tissue, which may measure several inches in diameter. u. papulosa, a form occurring in children, in which, as a result of the inflammatory effusion, a small, solid papule remains after the subsidence of the wheal. Syn., Lichen urticatus. u. pigmentosa, a rare type which begins within the first few months of life and consists of large, reddish, wheal-like tubercles that eventually change to a brownish-red or yellowish color. u. tuberosa, a form in which the wheals assume a tuberous form and become very large—as big as a walnut, hen's egg, or even larger. u. vesiculosa, urticaria characterized by the presence of vesicles.

**urticarial, urticarious** (ur-tik-a'-re-al, ur-tik-a'-re-us). Pertaining to urticaria.

**urticate** (ur'-tik-āt) [urticare, to sting]. 1. To sting like a nettle. 2. To flagellate with nettles.

**urtication** (ur-tik-a'-shun) [urtica]. 1. Flagellation with nettles, a method of treatment formerly employed in paralysis and to produce local irritation. 2. A sensation as if one had been stung by nettles.

**usanæ** (ū'-sān). A local anesthetic used in dentistry.

**Uskow's pillars** (oos'-kof). In the embryo, two folds or ridges which grow from the dorsolateral region of the body-wall and unite with the septum transversum to form the diaphragm.

**Usnea** (us'-ne-ah). A genus of lichen or tree moss.

**U. S. P., U. S. Phar.** Abbreviation for United States Pharmacopœia.

**Ustilago** (us-til-a'-go) [L.]. A genus of parasitic fungi—the smuts. U. maydis, corn-smut, is a fungus parasitic upon maize or Indian corn. In properties it resembles ergot of rye.

**ustion** (us'-chun) [ustio; urere, to burn]. A burning. In chemistry, incineration. In surgery, cauterization.

**ustulation** (us-tū-la'-shun) [L., ustulatio]. The act of roasting, drying, or parching.

**ustus** (us'-tus) [urere, to burn]. Calcined; burned.

**usure** (ū'-zjur) [uti, to use]. Circumscribed atrophy of a part or organ through pressure of neoplasms which have developed from it, of aneurysms, etc. u. des cartilages articulaires, osteoarthritis.

**uta** (oo'-tah). Peruvian vernacular term for a dermatophytic process analogous to Biskra button.

**utend.** (ū'-tend). Abbreviation of utendus, to be used.

**uteralgia** (ū-ter-al'-je-ah) [uterus; ἄλγος, pain]. Pain in the uterus.

**uterectomy** (ū ter at'-to me). Same as hysterectomy.

**uterine** (ū'-ter-ēn) [uterus]. Pertaining to the uterus. u. appendages, the ovaries and oviducts. u. milk. 1. A fluid between the villi of the placenta in the cow. 2. The small quantity of albuminous fluid contained in the small spaces between the epithelial covering of the villi of the chorion and the crypts or depressions in the decidua. u. pregnancy, normal pregnancy. u. souffle, a vascular sound heard on auscultation through the abdominal wall between the fifth and sixth months of pregnancy.

**uterism** (ū'-ter-izm). Uteralgia.
**uteritis** (ū-ter-ī'-tis). Inflammation of the uterus. See *metritis*.
**utero-** (ū-ter-o-) [*uterus*]. A prefix denoting pertaining to the uterus.
**uteroabdominal** (ū-ter-o-ab-dom'-in-al) [*utero-; abdomen*]. Pertaining to the uterus and the abdomen.
**uterocervical** (ū-ter-o-ser'-vik-al). Relating to the uterus and the cervix of the uterus.
**uterocolic** (ū-ter-o-kol'-ik). Relating to the uterus and the colon.
**uterofixation** (ū-ter-o-fiks-a'-shun). See *hysteropexy*.
**uterogastric** (ū-ter-o-gas'-trik). Relating to the uterus and the stomach.
**uterogestation** (ū-ter-o-jes-ta'-shun) [*utero-; gestatio, gestation*]. Gestation within the cavity of the uterus; normal pregnancy.
**uterointestinal** (ū-ter-o-in-tes'-tin-al). Relating to the uterus and the intestine.
**uteromania** (ū-ter-o-ma'-ne-ah). See *nymphomania*.
**uterometer** (ū-ter-om'-et-er) [*utero-; μέτρον, measure*]. An instrument used to accurately measure the uterus and determine its position.
**utero-ovarian** (ū-ter-o-o-va'-re-an) [*utero-; ovary*]. Pertaining to the uterus and the ovary.
**uteroparietal** (ū-ter-o-pa-ri'-et-al) [*utero-; paries, wall*]. Pertaining to the uterus and the abdominal wall, applied to a form of hysteropexy.
**uteropelvic** (ū-ter-o-pel'-vik). Pertaining to the uterus and the pelvis.
**uteropexia, uteropexy** (ū-ter-o-peks'-e-ah; ū'-ter-o-peks-e). See *hysteropexy*.
**uteroplacental** (ū-ter-o-pla-sen'-tal) [*utero-; placenta*]. Pertaining to the uterus and the placenta.
**u. vacuum,** the vacuum caused by the traction upon the funis of a detached placenta, causing it to cling to the uterine wall.
**uterorectal** (ū-ter-o-rek'-tal). Relating to the uterus and the rectum.
**uterosacral** (ū-ter-o-sa'-kral) [*utero-; sacrum*]. Pertaining to the uterus and the sacrum.
**uteroscope** (ū'-ter-o-skōp) [*utero-; σκοπεῖν, to inspect*]. A uterine speculum.
**uterotome** (ū'-ter-o-tōm) [*utero-; τομή, a cutting*]. A cutting instrument used in uterotomy. See *hysterotome*.
**uterotomy** (ū-ter-ot'-o-me). See *hysterotomy*.
**uterotonic** (ū-ter-o-ton'-ik). Supplying muscular tone to the uterus.
**uterotractor** (ū-ter-o-trak'-tor) [*utero-; trahere, to draw*]. A variety of forceps having several teeth on each blade employed in making traction on the cervix uteri.
**uterotubal** (ū-ter-o-tū'-bal). Relating to the uterus and the oviducts.
**uterovaginal** (ū-ter-o-vaj'-in-al). Relating to the uterus and vagina.
**uteroventral** (ū-ter-o-vent'-ral) [*utero-; venter, the belly*]. Relating to the uterus and the abdomen.
**uterovesical** (ū-ter-o-ves'-ik-al). Relating to the uterus and the bladder.
**uterus** (ū'-ter-us) [L.]. The womb; the organ of gestation, receiving the ovum in its cavity, retaining and supporting it during the development of the fetus, and becoming the principal agent in its expulsion during parturition. It is a pear-shaped, muscular organ, three inches long, two inches wide, and one inch thick, and is divided into three portions—the *fundus*, the *body*, and the *cervix*. The *fundus* is the upper and broad portion; the *body* gradually narrows to the *neck*, which is the contracted portion. The orifice, os *uteri*, communicates with the vagina. The inner surface is covered with mucous membrane continuous with that of the vagina. The outer surface of the fundus and body is covered with peritoneum. The whole organ is suspended in the pelvis by means of the broad ligaments. The Fallopian tubes enter, one on each side of the fundus, at the cornua of the organ. **u. acollis,** a uterus in which the vaginal part is absent. **u. arcuatus,** a subvariety of uterus bicornis in which there is merely a vertical depression in the middle of the fundus uteri. **u. bicornis,** a uterus divided into two horns or compartments on account of an arrest of development. **u. bicornis unicollis,** a variety of double uterus in which the cervix is large and single. **u., bifid.** See *u. septus*. **u. biforis,** one in which the external os is divided anteroposteriorly by a septum. **u. bilocularis, u., bipartite.** See *u. septus*.

**u. biparititus unicollis,** one in which the cervix is simple and only the body of the uterus is double. **u. cordiformis,** a heart-shaped uterus, a form due to faulty development. **u. didelphys.** See *u. duplex*. **u. duplex,** a uterus that is double from failure of the Muellerian ducts to unite. **u., fetal,** one of defective development, in which the length of the cervical canal exceeds the length of the cavity of the body. **u., gravid,** a pregnant uterus. **u., infantile,** a uterus normally formed, but arrested in development. **u. masculinus,** a small culdesac situated at the middle of the highest portion of the crest of the urethra. It is the analogue of the uterus of the female. Syn., *prostatic vesicle; sinus pocularis; utricle*. **u. parvicollis,** a malformation described by Herman in which the vaginal portion is small but the body normal; also called *uterus acollis*. **u., sacciform, u., sacculated,** a sacculation of the retroverted pregnant uterus at term. **u., semiduplex,** one in which the two horns join at the os internum, and below the joint of junction there is no division at all, or a division not reaching to the os externum. **u. semipartitus.** See *u. subseptus*. **u. septus,** a uterus divided internally by a septum into two halves, more or less complete, anteroposteriorly. **u. subseptus,** one divided internally by an incomplete septum; it may start from the fundus and reach all the way, or be present in the cervix only. Also called *uterus semipartitus*. **u. unicornis,** a uterus having but a single lateral half with usually only one Fallopian tube; it is the result of faulty development.
**utricle** (ū'-trik-l) [*utriculus*, dim. of *uter*; a small bag]. 1. A delicate membranous sac communicating with the semicircular canals of the ear. 2. The uterus masculinus.
**utricular** (ū-trik'-ū-lar) [*utricle*]. 1. Pertaining to the utricle. 2. Shaped like a bladder.
**utriculitis** (ū-trik-ū-lī'-tis). Inflammation of the utricle.
**utriculosaccular** (ū-trik-ū-lo-sak'-ū-lar). Pertaining to the utricle and saccule of the ear.
**utriculus** (ū-trik'-ū-lus). See *utricle*. **u. hominis.** See *uterus masculinus*. **u. lachrymalis,** the lacrymal sac. **u. masculinus, u. prostaticus, u. urethræ, u. virilis.** See *uterus masculinus*.
**utriform** (ū'-tre-form) [*uter*; bag; *forma*, form]. Bladder-shaped.
**uva** (ū'-vah) [L.; *pl., uvæ*]. A grape. **u. passa,** a raisin. **u. ursi,** the *Arctostaphylos uva-ursi*, or bearberry of the order *Ericaceæ*. Its leaves (*uva ursi*, U. S. P.; *uvæ ursi folia*, B. P.) contain a bitter, crystalline glucoside, arbutin, $C_{12}H_{16}O_7$, splitting up into glucose and hydroquinone, $C_6H_6O_2$. Uva ursi is astringent and tonic, and is used in chronic nephritis, pyelitis, cystitis, incontinence of urine, gleet, leukorrhea, etc. Dose 20 gr.-1 dr. (1.3-4.0 Gm.). **u. ursi, extract of.** Unof. Dose 15-30 gr. (1-2 Gm.). **u. ursi, fluidextract of** (*fluidextractum uvæ ursi*, U. S. P.). Dose 1 dr. (4 Cc.). **u. ursi, infusion of** (*infusum uvæ ursi*, B. P.). Dose 1-2 oz. (32-64 Cc.).
**uvæ** (ū'-ve) [*uva*, a grape]. Raisins. The ripe fruit of *Vitis vinifera*, imported from Spain. **u. passæ majores,** ordinary raisins. **u. passæ minores,** Corinth raisins, or true currants.
**uvæformis** (ū-ve-for'-mis) [*uva*, a grape; *forma*, form]. Like a grape or bunch of grapes.
**uvea** (ū'-ve-ah) [*uva*]. The pigmented layer of the eye, comprising the iris, ciliary body, and choroid. The middle layer of the choroid coat.
**uveal** (ū'-ve-al) [*uvea*]. Pertaining to the uvea.
**uveitic** (ū-ve-it'-ik). Pertaining to, or resembling uveitis.
**uveitis** (ū-ve-ī'-tis) [*uvea*; ιτις, inflammation]. Inflammation of the uvea.
**uviform** (ū'-vi-form) [*uva*, a grape; *forma*, form]. Like a grape or bunch of grapes.
**uviol** (ū'-ve-ol) [*u*(*ltra*)-*viol*(*et*)]. A kind of glass which allows the ultraviolet rays to pass through the glass, for supplying the ultraviolet ray.
**uvula** (ū'-vū-lah) [L.]. The conical appendix hanging from the free edge of the soft palate and formed by muscles (azygos uvulæ, levator and tensor palati), mucous membrane, and connective tissue. **u. cerebelli,** a small lobule of the inferior vermis of the cerebellum, forming the posterior boundary of the fourth ventricle. **u. palatina,** the uvula. **u.-twitch,** an expedient for keeping the uvula forward in posterior rhinoscopy. **u. vermis,** uvula of the vermis, *u. cerebelli*. Lieutaud's **u.,** a ridge along

the middle of the trigone of the bladder. **u. vesicæ,** a prominence at the internal orifice of the urethra.
**uvulæ** (ū'-vū-le). The azygos uvulæ muscle. See *muscles*.
**uvular** (ū'-vū-lar) [*uvula*]. Pertaining to the uvula.
**uvularis** (ū-vū-la'-ris) [*uvula*]. The azygos uvulæ muscle. See under *muscle*.
**uvuloptosis, uvulaptosis** (ū-vū-lop-to'-sis, ū-vū-lap-to'-sis) [*uvula*; πτῶσις, falling]. A related and pendulous condition of the uvula.
**uvulotome, uvulatome** (ū'-vū-lot-ōm, ū'-vū-lat-ōm) [see *uvulotomy*]. An instrument used in performing uvulatomy.
**uvulotomy, uvulatomy** (ū-vū-lot'-o-me, ū-vū-lat'-o-me) [*uvula*; τομή, a cutting]. The operation of cutting off the uvula.
**uvulitis** (ū-vū-li'-tis) [*uvula*; ιτις, inflammation]. Inflammation of the uvula.
**uzara** (ū-zah'-räh). A preparation made from the root of an African plant belonging to the *Asclepiadaceæ*. It is used in bacillary dysentery and in diarrhea, and is said to contain no tannin.

# V

**V.** 1. Abbreviation of *vision* or *acuity of vision*, also of *volt*. 2. The chemical symbol of *vanadium*.
**vaccigenous** (*vak-sij'-en-us*) [*vaccine*; γενναν, to produce]. Producing or cultivating vaccine virus.
**vaccin** (*vak'-sin*) [*vacca*, a cow]. See *vaccine*.
**vaccina** (*vak-si'-nah*). See *vaccinia*.
**vaccinable** (*vak'-sin-a-bl*) [*vaccine*]. Susceptible of successful vaccination.
**vaccinal** (*vak'-sin-al*) [*vaccine*]. Pertaining to vaccination or to vaccine. **v. fever**, a mild fever after vaccination.
**vaccinate** (*vak'-sin-āt*) [*vaccine*]. 1. To inoculate with the virus of vaccinia. 2. To inoculate with any virus in order to produce immunity against an infectious disease.
**vaccination** (*vak-sin-a'-shun*) [*vaccinate*]. 1. Inoculation with the virus of cowpox in order to protect against smallpox. 2. Inoculation with any virus to produce immunity against an infectious disease. **v., animal**. See *v., bovine*. **v., arm-to-arm**, that method of vaccination in which the virus is carried from the arm of one patient to that of another. **v., bovine**, that practised by the aid of vaccine-lymph cultivated in bovine animals. **v., compulsory**, the law compelling the vaccination of infants within a certain period after birth. **v., Jennerian**, vaccination (1). **v.-rash**, a rash sometimes following vaccination; it is usually transitory but sometimes assumes an eczematous or erythematous form. It may also be syphilitic. **v.-syphilis**. See *vaccino-syphilis*.
**vaccinationist** (*vak-sin-a'-shun-ist*). An advocate of Jennerian vaccination.
**vaccinator** (*vak'-sin-a-tor*) [*vaccinate*]. 1. One who vaccinates. 2. An instrument used for vaccinating.
**vaccine** (*vak'-sēn*) [*vacca*, a cow]. 1. Lymph from a cowpox vesicle. 2. Any substance used for preventive inoculation. **v., autogenous**, a vaccine made from a culture obtained from the patient himself. **v., bacterial**, an emulsion of dead bacteria in normal salt solution used hypodermically for the purpose of raising the opsonic index of a patient suffering from infection by the same bacteria; and see *bacterine*. **v., body**, cytorrhyctes. **v., bovine**, that derived from the cow. **v., corresponding**, a vaccine prepared from vaccine of the same species as those causing an infection, but not derived from the patient himself. **v.-farm**, a farm upon which vaccine virus is systematically produced and collected. **v., Haffkine's, v., Wright's**. See *Haffkine, Wright*. **v., heterogen₀us**, one prepared from organisms derived from some source other than the patient in whose treatment they are to be used; the source is usually a "stock" culture. **v., humanized**, that from vaccinal vesicles of man. **v., lymph**, the virus of vaccine. **v., mixed**, a vaccine prepared from more than one species of bacteria. **v., multivalent**. Same as **v., polyvalent**. **v., point**, a slip of quill or bone coated at one end with vaccine lymph. **v., polyvalent**, a bacterial vaccine made from cultures of two or more strains of the same species of bacteria. **v. rash**, an erythema after vaccination. **v., stock**. Same as *vaccine, corresponding*. **v., virus**, the virus of vaccinia.
**vaccinella** (*vak-sin-el'-ah*) [*vaccinia*]. Spurious vaccinia. A secondary eruption sometimes following cowpox.
**vaccinia** (*vak-sin'-e-ah*) [*vacca*, a cow]. Cowpox, a contagious disease of cows transmissible to man by vaccination and conferring immunity against smallpox. In the human subject inoculated with cowpox a small papule appears at the site of inoculation in from one to three days, which becomes a vesicle about the fifth day, and at the end of the first week is pustular, umbilicated, and surrounded by a red areola. Desiccation begins in the second week and a scab forms, which soon falls off, leaving a white, pitted cicatrix.

**vaccinifer** (*vak-sin'-if-er*) [*vaccine; ferre*, to bear]. A person or animal from whom vaccine-virus is taken; a vaccine-point.
**vacciniform** (*vak-sin'-if-orm*) [*vaccine; forma*, form]. Resembling vaccinia.
**vacciniola** (*vak-sin-i'-o-lah*) [dim. of *vaccinia*]. A secondary eruption, sometimes following vaccinia, and resembling the eruption of smallpox.
**vaccinin** (*vak-sin'-e-in*). The same as *arbutin*.
**vaccinism** (*vak'-sin-izm*). The theory of the efficacy of vaccination.
**vaccinist** (*vak'-sin-ist*). A practiser or defender of vaccination; one who believes in the efficacy of vaccination.
**Vaccinium** (*vak-sin'-e-um*) [L., "blueberry"]. A genus of plants to which belong the cranberry, blueberry, bilberry, etc., of many species. *V. crassifolium* is used in catarrhal inflammations of the urinary tract. *V. myrtillus*, the bilberry, is indigenous to Europe, and yields fruits which are dried for use in decoction for diarrhea and leukoplakia; the leaves are used in diabetes. An extract, *extractum myrtilli winternitzi*, is a specific for stomatitis and a prominent remedy for affections of the mouth and tongue. Paint the affected parts every hour.
**vaccinization** (*vak-sin-i-za'-shun*). Thorough vaccination by repeated inoculations.
**vaccinogen** (*vak-sin-o-jen*) [*vaccine*; γενναν, to produce]. The person or animal from which or from whom vaccine virus is taken.
**vaccinogenous** (*vak-sin-oj'-en-us*). See *vaccigenous*.
**vaccinoid** (*vak'-sin-oid*) [*vaccine*; ειδος, Resembling vaccinia; vaccinella.
**vaccinophobia** (*vak-sin-o-fo'-be-ah*). [*vaccine*; φόβος, fear]. Morbid dread of vaccination.
**vaccinostyle** (*vak-sin'-o-stile*) [*vaccine*; *stylus*, a pointed instrument]. A small metallic lance for use in vaccinating.
**vaccinosyphilis** (*vak-sin-o-sif'-il-is*) [*vaccine*; *syphilis*]. Syphilis conveyed by vaccination with contaminated virus, or by a contaminated instrument.
**vaccinotherapy** (*vak-sin-o-ther'-ap-e*). The therapeutic use of bacterial vaccines.
**vacillatio** (*vas-sil-a'-she-o*) [*vacillare*, to stagger]. Staggering, reeling. **v. dentium**, looseness of the teeth.
**vacuolar** (*vak'-u-o-lar*) [*vacuole*]. Pertaining to or of the nature of a vacuole.
**vacuolate, vacuolated** (*vak'-u-o-late, vak'-u-o-la-ted*). Having or pertaining to vacuoles.
**vacuolated** (*vak'-u-o-la-ted*). Of a cell, containing one or more vacuoles.
**vacuolation** (*vak-u-o-la'-shun*) [*vacuole*]. The formation of vacuoles; the state of being vacuolated.
**vacuole** (*vak'-u-ōl*) [*vacuus*, empty]. A clear space in a cell. **v., contractile**, a vacuole in the protoplasm of certain protozoa, which gradually increases in size and then collapses. **v., diffusion**, in the *in vitro* method of examining living cells, minute droplets of the surrounding colored liquid which have been absorbed by the cell.
**vacuolization** (*vak-u-o-li-za'-shun*). Same as *vacuolation*.
**vacuum** (*vak'-u-um*) [L.]. A space from which the air has been exhausted. **v., high**, a vacuum in which the exhaustion of air has been very great. **v., plate**, in dentistry, a term applied to a plate on which artificial teeth are mounted, having an air chamber to assist in its retention in the mouth. **v., Toricellian**, the vacuum above the mercury in the tube of a barometer.
**vadum** (*va'-dum*) [L., a shallow]. Plural, *vada*. A shallow in the depths of any fissure of the brain.
**vagabond's disease**. Parasitic melanoderma, a pigmentation of the skin from chronic irritation by pediculi.
**vagal** (*va'-gal*) [*vagare*, to wander]. Pertaining to the vagus nerve.

**vagina** (*va-ji'-nah*) [L.]. 1. A sheath. 2. The musculomembranous canal extending from the vulvar opening to the cervix uteri, insheathing the latter and the penis during copulation. **v. bulbi.** See *v. oculi*. **v., bulb of,** bulbus vestibuli, a small body of erectile tissue on each side of the vestibule of the vagina. **v. cordis,** the pericardium. **v. femoris,** the fascia lata of the thigh. **v. oculi,** Tenon's capsule.
**vaginal** (*vaj'-in-al*) [*vagina*]. 1. Pertaining to or of the nature of a sheath, as the *vaginal* tunic (tunica vaginalis of the testicle). 2. Pertaining to the vagina.
**vaginalectomy** (*vaj-in-al-ek'-to-me*). See *vaginectomy* (2).
**vaginalis** (*vaj-in-a'-lis*) [*vagina*]. Vaginal.
**vaginalitis** (*vaj-in-al-i'-tis*) [*vaginalis*, or a sheath; ιτις, inflammation]. Inflammation of the tunica vaginalis of the testicle.
**vaginant** (*vaj'-in-ant*) [*vaginare*, to sheath]. Sheathing; vaginal.
**vaginapexy** (*vaj-in-a-pek'-se*). 1. See *vaginopexy*. 2. See *colpopexy*.
**vaginate** (*vaj'-in-āt*) [see *vaginant*]. Sheathed.
**vaginectomy** (*vaj-in-ek'-to-me*) [*vagina*; ἐκτομή, excision]. 1. Excision of the vagina. 2. Excision of the tunica vaginalis.
**vaginicoline** (*vaj-in-ik'-o-lēn*) [*vagina*, vagina; *colere*, to inhabit]. Living in the vagina, as an animalcule.
**vaginiferous** (*vaj-in-if'-er-us*) [*vagina*, vagina; *ferre*, to bear]. Producing or bearing a vagina.
**vaginigluteus, vaginigluteæus** (*vaj-in-i-gloo-te'-us*) [*vagina*; *gluteus*]. The tensor vaginæ femoris. See under *muscle*.
**vaginismus** (*vaj-in-is'-mus*) [*vagina*]. Painful spasm of the vagina. **v., mental,** that due to extreme aversion to the sexual act. **v., perineal,** that due to spasm of the perineal muscles. **v., posterior,** that due to spasm of the levator ani muscle. **v., vulvar,** that due to spasm of the levator ani.
**vaginitis** (*vaj-in-i'-tis*) [*vagina*; ιτις, inflammation]. 1. Inflammation of the vagina. 2. Inflammation of a sheath.
**vagino-** (*vaj-in-o-*) [*vagina*]. A prefix denoting pertaining to the vagina.
**vaginoabdominal** (*vaj-in-o-ab-dom'-in-al*). Relating to the vagina and abdomen.
**vaginocele** (*vaj'-in-o-sēl*) [*vagino-*; κηλή, a hernia, or tumor]. Colpocele.
**vaginodynia** (*vaj-in-o-din'-e-ah*) [*vagino-*; ὀδύνη, pain]. Neuralgic pain of the vagina.
**vaginofixation** (*vaj-in-o-fiks-a'-shun*) [*vagino-*; *fixation*]. 1. An operation whereby the vagina is rendered immovable. 2. Vaginal hysteropexy.
**vaginomycosis** (*vaj-in-o-mi-ko'-sis*). Mycosis affecting the vagina.
**vaginoperitoneal** (*vaj-in-o-per-it-o-ne'-al*). Relating to the vagina and the peritoneum.
**vaginopexy** (*vaj'-in-o-peks-e*) [*vagino-*; πῆξις, a fixing]. 1. The preservation of the tunica vaginalis which results from extraseous transplantation of the testicle in cases of varicocele. 2. Vaginofixation.
**vaginoscope** (*vaj'-in-o-skōp*) [*vagino-*; σκοπεῖν, to view]. A vaginal speculum.
**vaginoscopy** (*vaj-in-os'-ko-pe*) [*vagino-*; σκοπεῖν, to view]. Inspection of the vagina.
**vaginotomy** (*vaj-in-ot'-o-me*) [*vagino-*; τομή, section]. Incision of the vagina; colpotomy.
**vaginovesical** (*vaj-in-o-ves'-ik-al*). See *vesicovaginal*.
**vaginovulvar** (*vaj-in-o-vul'-var*). See *vulvovaginal*.
**vagitus** (*va-ji'-tus*) [*vagire*, to cry]. The cry of in infant. **v. uterinus,** the cry of a child while till in the uterus. **v., vaginalis,** the cry of a child while the head is still in the vagina.
**vagoaccessorius** (*va-go-ak-ses-o'-re-us*). The vagus .nd accessorius nerves considered as one.
**vagotomized** (*va-got'-om-īzd*). Applied to an animal n which the vagi nerves have been severed intentionally.
**vagotomy** (*va-got'-o-me*) [*vagus*; τομή, a cutting]. Division of the vagus nerve.
**vagotonia, vagotony** (*va-go-to'-ne-ah, va-got'-on-e*; *vagus*; τόνος, tension). Irritability of the vagus nerve.
**vagotonic** (*va-go-ton'-ik*). Pertaining to or characterized by vagotonia.
**vagrant** (*va'-grant*) [*vagare*, to wander]. Wandering, as a vagrant cell. **v.'s disease,** a discoloration of the skin occurring especially in elderly persons who are of uncleanly habits and infested with vermin.
**vagus** (*va'-gus*) [*vagare*, to wander]. The pneumogastric nerve. See under *nerve*. **v.-pneumonia,** pneumonia following section of the vagi in the lower animals, and due to the aspiration of food into the air-passages. **v.-pulse,** a slow pulse due to the inhibitory action of the vagus on the heart.
**Valangin's solution** (*va-lan'-jin*) [Francis Joseph Pahud de *Valangin*, English physician, 1725–1805]. A solution of arsenic trioxide in dilute hydrochloric acid; the liquor acidi arsenosi of the U. S. P.
**valdivin** (*val'-div-in*). An emetic principle derived from the fruit of *Simaba valdivia*.
**valence, valency** (*va'-lens, va'-len-se*) [*valere*, to be worth]. The relative combining capacity of an atom compared with that of the atom of hydrogen.
**Valenta's test for fats** (*val-en'-tah*). Mix thoroughly in a test-tube equal volumes of fat and glacial acetic acid, sp. gr. 1.0562; apply heat if the oil does not dissolve in the cold. Three classes of oils are distinguished, according as solution takes place at ordinary temperatures, at temperatures up to the boiling-point of glacial acetic acid, or whether even then solution is incomplete. In the case of oils dissolving upon application of heat, the temperature is observed at which upon cooling turbidity appears.
**Valentin's corpuscles** (*val'-en-tin*) [Michael Bernard *Valentin*, German anatomist, 1657–1729]. Small bodies, said to be amyloid, occasionally found in nerve tissue. **V.'s ganglion,** a ganglion enlargement found occasionally above the root of the second bicuspid, at the junction of the middle and posterior dental nerves; *root,* posterior and middle dental nerves; *distribution,* posterior teeth. **V.'s limiting membrane.** See *Schwann, sheath of*.
**Valentine's reaction for fuchsin.** Upon shaking ether with a solution containing fuchsin the ether does not dissolve the coloring-matter, but upon adding ferrous iodide the ether is colored violet.
**valeral** (*val'-ur-al*), $C_4H_9COH$. Isovaleric aldehyde, or amyl aldehyde, an oxidation product of amyl alcohol.
**valeraldehyde** (*val-ur-al'-de-hīd*). Amyl aldehyde, $C_5H_9$. COH.
**valeraldine** (*val-ur-al'-dēn*). A synthetic alkaloid formed from valeral ammonia by the action of hydrogen sulphide.
**valerate** (*val'-er-āt*). Any salt of valeric acid; same as valerianate.
**valerene** (*val'-ur-ēn*). Amylene.
**valerian** (*val-e'-re-an*). A plant of the genus *Valeriana*. The root of *Valeriana officinalis* (*valeriana*, U. S. P.; *valerianæ rhizoma*, B. P.) contains a volatile oil, from which valeric acid is obtained. Valerian is employed as a mild nervous stimulant in hysteria, migraine, low fevers, etc. Dose 30 gr. (2 Gm.). **v., fluidextract of** (*fluidextractum valerianæ*, U. S. P.). Dose 1 dr. (4 Cc.). **v., infusion of** (*infusum valerianæ*, B. P.). Dose 1–2 oz. (32–64 Cc.). **v., oil of** (*oleum valerianæ*, B. P.). Dose 4–5 min. (0.26–0.32 Cc.). **v., tincture of** (*tinctura valerianæ*, U. S. P., B. P.). Dose 1–3 dr. (4–12 Cc.). **v., tincture of, ammoniated** (*tinctura valerianæ ammoniata,* U. S. P., B. P.). Dose 1–3 dr. (4–12 Cc.).
**Valeriana** (*va-le-re-a'-nah*). A genus of plants; also the rhizome and rootlets of *V. officinalis*; it is an antispasmodic and stimulant. See *valerian*.
**valerianate** (*val-e'-re-an-āt*). A salt of valerianic acid; those of ammonium, iron, quinine, and zinc are official.
**valeric acid, valerianic acid** (*va-le'-rik, val-e-re-an'-ik*). See *acid, valeric*.
**valeridin** (*val-er'-id-in*). See *valeryl-phenetidin*.
**valerol** (*val'-er-ol*). A clear oily liquid of unpleasant odor, obtained from valerian; like valerian camphor.
**valerophen** (*va-ler'-o-fen*). A phenolphthalein methyl derivative of valeric acid.
**valeryl** (*val'-er-il*). The radical $C_5H_9O$. **v.-phenetidin,** $C_6H_4(OC_2H_5)NH.C_5H_9O$; it is sedative and antineuralgic. Dose 8–15 gr. (0.5–1.0 Gm.).
**valetudinarius; valetudo, health]. An invalid.
**valeur globulaire** (*val'-er glob-u-lair'*) [Fr. "*globular value*"]. The proportion of hemoglobin to the number of red corpuscles, expressed in terms of the amount of hemoglobin in an individual corpuscle. The color index.

**valgoid** (*val'-goid*) [*valgus;* εἶδος, likeness]. Resembling valgus.

**valgus** (*val'-gus*) [L., bow-legged]. 1. Bow-legged. 2. A condition in which the arch of the foot is depressed so that the inner side of the sole rests upon the ground. Syn., *genu varum; splay-foot; talipes valgus*.

**validol** (*val'-id-ol*) [*valerian; menthol*]. The chemically pure combination of menthol and valeric acid with the addition of 30% free menthol. It is a colorless, somewhat viscous fluid, with a pleasant odor and cooling taste. It is claimed to have powerful analeptic and carminative properties, and is an excellent solvent and vehicle for menthol. It is also employed as an antispasmodic; in migraine; as a specific in alcoholic intoxication; as a prophylactic against sea-sickness, etc. Dose 10–15 drops daily on sugar. **v., camphorated,** validol containing 10% of camphor, used in scotoma scintillans. Dose 10–15 drops.

**valin** (*val'-in*). (CH₃)₂ . CH . CHNH₂ . COOH. Alpha aminoisovaleric acid.

**vallate** (*val'-āt*) [*vallum,* rampart]. Surrounded with a walled depression; cupped. And see *circumvallate*.

**vallecula** (*val-ek'-ū-lah*) [*vallis,* a valley; *pl., vallecula*]. A shallow groove or depression. **v. cerebelli,** the depression between the cerebellar hemispheres. **v. epiglottica,** a depression between the lateral and median glosso-epiglottic folds on each side. **valleculæ linguæ,** the glosso-epiglottic fossæ. **v. ovata,** the fissure of the liver which contains the gallbladder. **v. Sylvii,** a cerebral depression which develops into the fissure of Sylvius. **v. unguis,** the depression in the skin for the root of the nail.

**Valleix's points douloureux** (*val-lay'*) [François Louis Isidore *Valleix,* French physician, 1807–1855]. Painful points found in peripheral neuralgias where the nerves pass through openings in fascia or issue from bony canals.

**Vallet's mass** (*val-la*). Massa ferri carbonatis. Ferrous sulphate, 100; sodium carbonate, 110; honey, 38; sugar, 25; syrup and distilled water, of each, enough to make 100 parts.

**valley of the cerebellum.** See *vallecula cerebelli*.

**Valli-Ritter's law.** See *Ritter-Valli's law*.

**vallis** (*val'-is*). See *vallecula cerebelli*. **v. alarum,** valley of the arm-pits; the axilla. **v. femorum,** the vulva.

**vallum** (*val'-um*) [L.]. The supercilium or eyebrow. **v. unguis,** the nail wall.

**valoid** (*val'-oid*) [*valere,* to be equal]. 1. Trade name applied to certain galenical preparations. 2. A name for certain fluidextracts, equal weights of which and of the drugs from which they are prepared, have the same strength.

**valonia** (*val-o'-ne-ah*) [βάλανος, an acorn]. The acorn cups of *Quercus ægilops;* it is used as an astringent in diarrhea.

**Valsalva's experiment** (*val-sal'-vah*) [Antoine Marie *Valsalva,* Italian anatomist, 1666–1723]. Strong expiratory efforts made while the mouth and nose are closed cause at first an increase, and when continued, finally a diminution, of blood-pressure. The phenomenon is due to reflex actions of the vasomotor center through the pulmonary nerves. **V.'s ligaments,** the extrinsic ligaments of the pinna of the ear. **V.'s liquor.** See *Scarpa's liquor*. **V.'s method of** treating internal aneurysm, by general depletion, such as purging, bleeding, and restricted diet. **V.'s sinus,** one of the pouch like dilatations of the aorta or pulmonary artery opposite the segments of the semilunar valves. **V.'s test,** inflation of the tympanic cavity with air by means of forcible expiratory efforts made while the nose and mouth are tightly closed. Perforation of the tympanic membrane may be detected by this test.

**valsol** (*val'-sol*). An ointment-vehicle consisting of a mixture of oxygenized hydrocarbons, which forms an emulsionized mass with water and readily dissolves iodine, iodoform, ichthyol, etc.

**Valsuani's disease.** Pernicious progressive anemia occurring in pregnancy.

**value, globular.** A fraction of which the numerator is the percentage of hemoglobin and the denominator the percentage of red corpuscles. It indicates the percentage of hemoglobin in a corpuscle. Syn., *valeur globulaire*.

**valufe** (*val'-āl*). Trade name for capsules containing divided doses of a substance.

**valva** (*val'-vah*) [*valva,* the leaf of a door; *pl., valvæ*]. A valve. **v. Tulpii,** the ileocecal valve.

**valval, valvar** (*val'-val, val'-var*) [*valva,* valve]. Pertaining to a valve.

**valvate** (*val'-vate*) [*valva,* the leaf of a door]. Resembling or functioning as a valve; provided with a valve.

**valve** (*valv*) [*valva,* a door]. 1. A device placed in a tube or canal so as to permit free passage one way, but not in the opposite direction. 2. A fold of membrane acting as a valve, as *valve* of the heart. **v. Amussat's.** See *Heister's valve*. **v., aortic,** the valve consisting of three semilunar segments, situated at the junction of the aorta with the heart. **v.s, auriculoventricular,** the mitral and tricuspid valves. **v., Bauhin's,** the ileocecal valve. **v., bicuspid.** See *v., mitral*. **v., coronary,** the valve protecting the orifice of the coronary sinus and preventing regurgitation of blood during the contraction of the right auricle. **v., Eustachian,** that between the inferior vena cava and the right auricle of the fetus. **v. of Hasner,** an imperfect valve at the inferior meatus of the nose. **v., Heister's,** a fold of mucous membrane at the neck of the gall-bladder. **v. of Houston,** three oblique folds in the mucous membrane of the rectum at about the level of the prostate. **v., ileocecal,** the folds of mucous membrane at the junction of the ileum and cecum. **v. of Kerkring,** any one of the *valvulæ conniventes* (*q. v.*). **v.s, laryngeal,** a term applied to the superior or false vocal bands, because of their supposed use in holding the breath. **v., mitral,** the valve that controls the opening from the left auricle to the left ventricle; it is constituted of two leaflets. **v., pulmonary, v., pulmonic,** the valve composed of three semilunar leaflets, and situated at the junction of the pulmonary artery and the right ventricle. **v.'s, rectal,** semilunar folds fixed to the rectum by their convex borders, occupying in their attachments from one-third to one-half the circumference of the gut. They are composed of a duplicature of the mucous membrane inclosing some cellular tissue and a few circular muscular fibers. The margins and diameters of these pass each other when the rectum is empty and present an additional barrier to the involuntary evacuation of the feces, retarding downward movement. **v.s semilunar,** the three valves guarding the orifice of the pulmonary artery and aorta. **v.s, sigmoid.** See *v.s, semilunar*. **v.-test,** Azoulay's, auscultation of the heart while the patient is lying with the arms raised perpendicularly and the legs lifted obliquely. **v. of Thebesius,** the coronary valve or fold of the endocardium of the right auricle **v., tricuspid,** that which controls the opening from the right auricle to the right ventricle; it consists of three segments. **v. of Varolius,** the ileocecal valve. **v. of Vieussens,** a thin leaf of medullary substance forming the roof of the anterior portion of the fourth ventricle of the brain.

**valviform** (*val'-vif-orm*) [*valva,* valve; *forma,* form]. Valvular.

**valvotomy** (*val-vot'-o-me*) [*valva,* valve; τομή, incision]. Cutting a valve; especially the valves of the rectum.

**valvula** (*val'-vū-lah*) [dim. of *valva,* a valve; *pl., valvulæ*]. 1. A small valve. 2. The superior medullary velum. **v. bicuspidalis,** the mitral valve. **v. cæci or coli,** the ileocecal valve. **v. cerebelli,** valve of Vieussens. **v. Eustachii,** the Eustachian valve. **v. fossæ navicularis,** the valve of Guerin. **v. lacrimalis inferior,** the plica lacrimalis or valve of Hasner. **v. processus vermiformis,** a fold of mucous membrane at the opening of the appendiceal canal. **v. pylori,** circular fold of mucous membrane at the pyloric orifice. **v. semilunaris,** one of the semilunar valve leaflets of the heart. **v. sinus coronarii,** the coronary valve or valve of Thebesius. **v. sinus sinistri,** the interauricular valve. **v. spiralis,** the valve of Heister. **v. tricuspidalis,** the tricuspid valve. **v. vaginæ,** the hymen. **v. venæ cavæ inferioris,** the Eustachian valve. **v. vestibuli sinistra,** the right venous valve of the embryonic heart.

**valvulæ** (*val'-vū-le*) [*pl.* of *valvula*]. **valvulæ conniventes,** the transverse folds of mucous membrane of the small intestine. Syn., *valves of Kerkring*. **valvulæ cuspidales,** the mitral and tricuspid valves.

**valvular** (*val'-vū-lar*) [*valve*]. Pertaining to or originating at a valve.

**valvule** (*val'-vūl*) [*valvula*]. A small valve.

**valvulitis** (*val-vū-li'-tis*) [*valve;* ιτις, inflammation].

Inflammation of a valve, especially of a cardiac valve.

**valyl** (*val'-il*), CH₃.CH₂.CH₂N(C₂H₅)₂, valerianic-acid diethylamide; a colorless, limpid fluid, used in nervous diseases. Dose 2 gr. (0.125 Gm.) in capsule three times daily.

**valylene** (*val'-il-ēn*) C₅H₆. A hydrocarbon with an alliaceous odor, a homologue of vinyl acetylene; pentone.

**valzin** (*val'-zin*). See *sucrol*.

**vampirism** (*vam'-pi-rizm*). The insane belief that one's blood is being sucked by another person at night.

**vanadate** (*van'-ad-āt*). A salt of vanadic acid.

**vanadic acid** (*van-ad'-ik*), H₃VO₄. An acid derived from vanadium; it forms salts called vanadates.

**vanadin** (*van'-ad-in*). A remedy recommended in pulmonary tuberculosis, said to consist of a solution of a vanadium salt with sodium chloride.

**vanadium** (*van-a'-de-um*) [*Vanadis*, a goddess of Scandinavian mythology]. A rare metallic element. Symbol, V; atomic weight, 51. See *elements, table of chemical*.

**Van Buren's disease** (*van-bū'-ren*) [William Holme Van Buren, American surgeon, 1819–1883]. Chronic circumscribed infiltration of the corpus cavernosum, one of the erectile bodies of the penis. **V. B.'s operation,** *for prolapse of the anus;* a linear cauterization of the mucosa with the Paquelin's cautery.

**Van Deen's test for blood in the urine** [Izaak Van Deen, Dutch physician, 1804–1869]. The addition of 2 Cc. of tincture of guaiac and 2 Cc. of old oil of turpentine produces a blue color in the presence of blood or pus.

**Vandellia** (*van-del'-e-ah*) [— *Vandelli*, Italian botanist]. A genus of scrophulariaceous plants. V. diffusa, of S. America, is emetic and purgative, and said to be useful in hepatic and intestinal diseases.

**Van den Velden's test** [Reinhardt *Van den Velden*, German physician, 1851– ]. For free hydrochloric acid in the gastric juice; methylene blue solution is turned from violet to blue or green in presence of the force acid. Also called Maly's test.

**Van der Kolk's law.** See *Kolk's law*.

**Van Gehuchten's fixative and hardening fluid.** Consists of glacial acetic acid, 10 parts; chloroform, 30 parts; absolute alcohol, 60 parts.

**Van Gieson's stain** (*van-ge'-son*) [Ira *Van Gieson*, American histologist]. Satured aqueous solution of picric acid 100 Cc., with 5 Cc. of a one per cent. solution of acid fuchsin.

**Van Harlingen's formula.** It consists of 1 dram of precipitated sulphur, with five grains of powdered camphor, 10 of powdered gum tragacanth, and one ounce each of rose-water and lime-water. Used in treatment of acne rosacea.

**Van Helmont's mirror** (*van-hel'-mont*) [Jean Baptiste *Van Helmont*, Belgian physician, 1577–1644]. The central tendon of the diaphragm.

**Van Hook's operation** [Weller *Van Hook*, American surgeon]. Ureteroureterostomy.

**Van Hoorne's canal** [John *Van Hoorne*, Dutch anatomist, 1621–1670]. Thoracic duct.

**vanilla** (*van-il'-ah*) [L.]. A genus of plants of the order *Orchideæ*. The fruit of *V. planifolia* is the vanilla of the U. S. P. It contains from 1 to 3% of vanillin (*q. v.*). Vanilla is used as a flavoring agent, and as an ingredient of a test-solution for hydrochloric acid. **v., tincture of** (*tinctura vanillæ*, U. S. P.), vanilla, sugar, alcohol, and water.

**vanillin** (*van-il'-in*). *Vanillinum* (U. S. P.), C₈H₈O₃, an aromatic crystalline principle, the methyl ether of protocatechuic aldehyde. Dose ½ gr. (0.03 Gm.). **v.-paraphenetidin,** a crystalline condensation-product of vanillin with paraphenetidin; it is hypnotic, antineuralgic, and styptic. Dose 24–30 gr. (1.5–2.0 Gm.).

**vanillism** (*van-il'-izm*) [*vanilla*]. A form of dermatitis characterized by marked itching, occurring among vanilla workers.

**Van Swieten's liquor, Van S.'s solution** [*van-sve'-tens*] [Gérard *Van Swieten*, Dutch physician, 1700–1772]. A solution of mercuric chloride 2 gr., alcohol 3 gr., distilled water sufficient to make 4 oz.

**Van't Hoff's law** [Jacobus Henricus *Van't Hoff*, Dutch chemist, 1852–1911]. The osmotic pressure exerted by any substance in solution is the same as it would exert if present as a gas in the same volume as that occupied by the solution, provided that the solution is so dilute that the volume occupied by the solute is negligible in comparison with that occupied by the solvent.

**vapoaural massage** (*va-po-aw'-ral*). Massage of the tympanum by medicated vapors.

**vapocauterization** (*va-po-kaw-ter-iz-a'-shun*). See *atmocausia*.

**vapocresolin** (*va-po-kres'-ō-lin*) [*vapor; cresolin*]. A popular remedy in the treatment of laryngeal diphtheria.

**vapor** (*va'-por*) [L.]. A gas, especially the gaseous form of a substance which at ordinary temperatures is liquid or solid. **v. bath,** the therapeutic application of steam or of some other vapor to the body, in a suitable apparatus or apartment. **v. douche,** a jet of vapor impinging upon some part of the surface of the body.

**vaporarium** (*va-por-a'-re-um*) [L.]. A vapor-bath; an establishment for giving vapor-baths.

**vapores uterini.** Synonym of *hysteria*.

**vaporimeter** (*va-po-rim'-et-ur*) [*vapor; μέτρον*, measure]. An apparatus for determining the tension of vapor.

**vaporish** (*va'-por-ish*). Hysterical, splenetic.

**vaporium** (*va-por'-e-um*). An apparatus for giving vapor baths or douches.

**Vaquez's disease** (*vak-kay'*) [H. *Vaquez*, French physician]. Polycythemia with cyanosis, enlarged spleen, and disease of the bone-marrow.

**vaporization** (*va-por-i-za'-shun*) [*vapor*]. The conversion of a solid or liquid into a vapor.

**vaporize** (*va'-por-īz*) [*vapor*]. To convert into vapor.

**vaporizer** (*va'-por-i-zer*) [*vapor*]. An atomizer, a nebulizer.

**vaporole** (*va'-por-ōl*) [*vapor*]. Trade name of a glass capsule containing a drug for inhalation, or for hypodermic injection.

**vapors** (*va'-porz*). Lowness of spirits; hysteria.

**varalette** (*var-al-et'*). Trade name of a compressed effervescent tablet.

**variability** (*va-re-a-bil'-it-e*) [*variare*, to change]. Ability of the organism or race to adapt itself to its environment.

**variation** (*va-re-a'-shun*) [*variare*, to change]. Deviation from a given type as the result of environment, natural selection, or cultivation and domestication. **v., double,** the double current produced in a muscle by the passage of a single induction shock. **v., negative,** the diminution of the muscle current caused by stimulation of the motor nerve.

**varicated** (*var'-ik-a-ted*) [*varix*, varix]. Having varices.

**varication** (*var-ik-a'-shun*) [*varix*, varix]. The formation of a varix; a system of varices.

**varicella** (*var-is-el'-ah*) [dim. of *variola*, smallpox]. Chickenpox; an acute, contagious disease of childhood, characterized by an eruption of transparent vesicles which appear in successive crops on different parts of the body. **v. gangrænosa,** varicella in which the eruption leads to a gangrenous ulceration. **v., pustular.** Same as *varioloid*.

**varicelliform** (*var-is-el'-e-form*). Characterized by vesicles resembling those of varicella.

**varicelloid,** (*var-is-el'-oid*) [*varicella; εἶδος,* like]. Resembling varicella.

**varices** (*var'-is-ēs*) [L.]. Plural of *varix*.

**variciform** (*var-is'-i-form*) [*varix; forma,* a form]. Having the form of a varix.

**varicoblepharon** (*var-ik-o-blef'-ar-on*) [*varix; βλέφαρον,* eyelid]. A varicosity of the eyelid.

**varicocele** (*var'-ik-o-sēl*) [*varix; κήλη,* a tumor]. Dilatation of the veins of the spermatic cord, forming a soft, elastic swelling. **v., ovarian,** varicosity of the broad ligament. **v., utero-ovarian,** a varicose condition of the veins of the pampiniform plexus in the broad ligament.

**varicocelectomy** (*var-ik-o-se-lek'-to-me*) [*varicocele; ἐκτομή,* excision]. Excision of a varicocele.

**varicoid** (*var'-ik-oid*) [*varix; εἶδος,* resemblance]. Same as *varicose*.

**varicola** (*var'-ik-ōl*). Same as *varicocele*.

**varicomphalus** (*var-ik-om'-fal-us*) [*varix; ὀμφαλός,* navel]. A varicosity at the navel.

**varicose** (*var'-ik-ōs*) [*varix*]. 1. Of blood-vessels, swollen, knotted, and tortuous. 2. Due to varicose veins, as *varicose ulcer*. **v. aneurysm.** See under *aneurysm, arteriovenous*.

**varicosis** (*var-ik-o'-sis*) [*varicose*]. An abnormal dilatation of the veins.

**varicosity** (*var-ik-os'-it-e*) [*varicose*]. The condition of being varicose; a varicose portion of a vein.

**varicotomy** (*var-ik-ot'-o-me*) [*varix; τόμη*, a cutting]. Excision of a varicose vein. See *cirsotomy*.
**varicula** (*var-ik'-u-lah*) [dim. of *varix*]. A varix of the conjunctiva.
**variety** (*va-ri'-et-e*) [*varietas*, difference]. A subdivision of a species; a stock, strain, breed.
**variform** (*var'-e-form*) [*varius*, various; *forma*, form]. Having diversity of form.
**variola** (*va-re-o-lah*) [*varius*, variegated; spotted]. Smallpox, a contagious infectious disease ushered in with severe febrile symptoms, which, in the course of two or three days, are followed by a papular eruption spreading over all parts of the body. During the succeeding two weeks the eruption passes through the stage of vesicles and pustules, the latter going on to the formation of crusts. The falling off of the crusts leaves a pitted appearance of the skin (pock-marks). The period of incubation is about thirteen days. **v., black.** See *v., hemorrhagic*. **v., coherent,** a form in which the pustules coalescence but retain their individuality. **v., confluent, v. confluens,** a severe form in which the pustules spread and run together. **v., discrete,** a form in which the pustules preserve their distinct individuality. **v., hemorrhagic,** smallpox in which hemorrhage occurs into the vesicles, which gives them a blackish appearance. **v., malignant,** black smallpox, a severe and very fatal form of the hemorrhagic type. **v., mitigated, v., modified.** See *varioloid*. **v. notha, varicella. v. vera,** true smallpox as distinguished from varioloid.
**variolar** (*va-ri'-o-lar*) [*variola*]. Pertaining to smallpox.
**Variolaria amara** (*var-e-o-la'-re-ah am-a'-ra*). A lichen used as a febrifuge and anthelmintic.
**variolate** (*var'-e-o-lāt*) [*variola*]. 1. Having small pustules like those of variola. 2. To inoculate with smallpox.
**variolated** (*var'-e-o-la-ted*) [*variola*]. Having, or having had smallpox.
**variolation, variolization** (*var-e-o-la'-shun, var-e-o-li-za'-shun*) [*variola*]. The inoculation of smallpox.
**variolic** (*var-e-ol'-ik*) [*variola*]. Pocky, variolous.
**varioliform** (*var-i-o'-lif-orm*) [*variola; forma*, form]. Resembling variola.
**variolin** (*var-i'-o-lin*) [*variola*]. The specific virus of smallpox.
**variolinum** (*var-e-o-li'-num*). A homeopathic preparation from the virus of variola.
**variolization.** See *variolation*.
**varioloid** (*var'-e-o-loid*) [*variola; εἶδος*, like]. A mild form of variola occurring in persons that have been vaccinated or inoculated with smallpox virus.
**variolous** (*var-i'-o-lus*) [*variola*]. Pertaining to or having the nature of variola.
**variolovaccine** (*var-e-o-lo-vak'-sēn*). A vaccine lymph or crust obtained from a heifer which has been inoculated with smallpox virus.
**variolovaccinia** (*var-e-o-lo-vak-sin'-e-ah*) [*variola; vacca*, cow]. A form of vaccinia or cowpox induced in the heifer by inoculating it with smallpox virus.
**varisse** (*va-rēs'*). A lump on the inner side of the hind leg of a horse.
**varium** (*var'-e-um*), Trade name of an ovarian extract.
**varix** (*var'-iks*) [*varus*, crooked; *pl.*, *varices*]. A dilated and tortuous vein. **v., aneurysmal.** See *aneurysmal varix* under *aneurysm, arteriovenous*. **v., lymphaticus,** dilatation of the lymphatic vessels, especially that due to the *Filaria sanguinis hominis*; and see *lymph-scrotum*.
**varnish** (*var'-nish*). A quickly-drying solution of some resin.
**Varolian** (*var-o'-le-an*) [*Constanzio Varioli*, Italian anatomist, 1543-1575]. Relating to the pons Varolii.
**Varolii, pons.** The mesencephalon; that part of the brain which connects the oblongata with the cerebral peduncles and the cerebellum. See under *pons*.
**V. valvula,** the ileocecal valve.
**varus** (*va'-rus*). A condition in which the foot is turned inward. See *talipes varus*, and *acne*.
**vas** (*vas*) [L.; *pl.*, *vasa*]. A vessel. **v. aberrans,** a blind tube projecting from the lower part of the epididymis. **v. deferens,** the excretory duct of the testis.
**vasa** [*pl.* of *vas*]. **vasa afferentia,** the branches of a lymphatic or lacteal vessel entering a lymphatic gland. **vasa brevia,** the gastric branches of the splenic artery. **vasa centralia retinae,** the central artery and veins of the retina. **vasa ciliaria,** the ciliary arteries and veins. **vasa efferentia.** 1. The terminal ducts of the rete testis. 2. The efferent vessels of lymphatic glands. **vasa intestini tenuis,** small vessels arising from the superior mesenteric artery and distributed to the jejunum and ileum. **vasa recta,** the tubules of the rete testis. **vasa vasorum,** the vessels supplying the arteries and veins with blood. **vasa vorticosa.** See *vena vorticosa*.
**vasal** (*va'-sal*) [*vas*, a vessel]. Pertaining to a vessel or to vessels; vascular.
**vasalium** (*vas-a'-le-um*) [*vas*; *pl.*, *vasalia*]. Tissue peculiar to vascular or closed cavities.
**vascula** (*vas'-ku-lah*). Plural of *vasculum* q. v.
**vascular** (*vas'-ku-lar*) [*vasculum*]. Consisting of, pertaining to, or provided with vessels.
**vascularity** (*vas-ku-lar'-it-e*) [*vascular*]. The quality of being vascular.
**vascularization** (*vas-ku-lar-i-za'-shun*) [*vascular*]. The process of becoming vascular. The formation and extension of vascular capillaries.
**vasculin** (*vas'-ku-lin*). Extract of vascular tissue.
**vasculitis** (*vas-ku-li'-tis*). See *angiitis*.
**vasculomotor** (*vas-ku-lo-mo'-tor*) [*vasculum; motor*]. Acting as a vasomotor upon the capillaries.
**vasculum** (*vas'-ku-lum*) [L.; *pl.*, *vascula*]. A small vessel. **v. aberrans.** See *vas aberrans*.
**vasectomy** (*vas-ek'-to-me*) [*vas; ἐκτομή*, a cutting out]. Resection of the vas deferens.
**vaseline** (*vas'-el-ēn*). See *petrolatum*.
**vaselon** (*vas'-el-on*). An ointment-base consisting of a mixture of margerin and stearin dissolved in mineral oil.
**vasicine** (*vas'-is-in*). An alkaloid from *Adhatoda vasica*; it is used in bronchial affections and as an insecticide.
**vasifactive** (*vas-if-ak'-tiv*) [*vas; facere*, to make]. Giving rise to new blood-vessels.
**vasiform** (*vas'-if-orm*) [*vas; forma*, form]. Resembling a vessel or duct.
**vaso-** (*va-zo-*) [*vas*, a vessel]. A prefix denoting pertaining to a vessel.
**vasoconstrictine** (*va-zo-kon-strik'-tēn*). Trade name of a preparation of the active principle of the medulla of the suprarenal bodies.
**vasoconstriction** (*va-zo-kon-strik'-shun*) [*vaso-; con stringere*, to bind]. The constriction of blood-vessels.
**vasoconstrictive** (*va-zo-kon-strik'-tiv*) [see *vasocon striction*]. Promoting or stimulating constriction of blood-vessels.
**vasoconstrictor** (*va-zo-kon-strik'-tor*). 1. Causing a constriction of the blood-vessels. 2. A nerve or a drug that causes constriction of blood-vessels.
**vasocorona** (*va-zo-ko-ro'-nah*) [*vaso-; corona*, crown]. The system of arterioles that supply the periphery of the spinal cord.
**vasodentine** (*va-zo-den'-tēn*) [*vaso-; dentin*]. Dentine possessing blood-vessels.
**vasodilatation** (*va-zo-dil-a-ta'-shun*). Dilatation of the blood-vessels.
**vasodilator** (*va-zo-di-la'-tor*). 1. Pertaining to the dilating motility of the nonstriped muscles of the vascular system. 2. A nerve-element or a drug that causes dilatation of blood-vessels.
**vasofactive** (*va-zo-fak'-tiv*). See *vasifactive*.
**vasoformative** (*va-zo-for'-mat-iv*) [*vaso-; formare*, to form]. Forming or producing vessels. **v. cells,** those engaged in the production of vascular tissue.
**vasoganglion** (*va-zo-gang'-gle-on*) [*vaso-; γάγγλιον*, ganglion]. A knot or rete of blood-vessels.
**vasogen** (*vas'-o-jen*). A proprietary oxygenated vaseline.
**vasohypertonic** (*va-zo-hi-per-ton'-ik*). See *vasocon strictor*.
**vasohypotonic** (*va-zo-hi-po-ton'-ik*). See *vasodi lator* (1).
**vasoinhibitor** (*va-zo-in-hib'-it-or*) [*vaso-; inhibēre*, to inhibit]. A drug or agent tending to inhibit the action of the vasomotor nerves.
**vasoinhibitory** (*va-zo-in-hib'-it-o-re*) [see *vasoin hibitor*]. Inhibiting vasomotor action, especially vasoconstrictor action.
**vasol** (*va'-sol*). A mixture of liquid petrolatum with ammonium oleate. **v., iodized,** vasol containing 7% of iodine.
**vasoligation** (*va-zo-li-ga'-shun*) [*vas; ligation*]. Ligation of the vas deferens.
**vasolimentum** (*vā-so-lin'-im-ent*). Parogen.
**vasomotion** (*va-zo-mo'-shun*) [*vaso-; motio, motion*]. Increase or decrease of the caliber of a blood-vessel.
**vasomotor** (*va-zo-mo'-tor*) [*vaso-; motor*, from *mov ere*, to move]. Regulating the tension of blood-ves-

šels. **v. ataxia,** instability of the circulatory mechanism characterized by abnormal readiness of disturbance of the equilibrium of the cardiovascular ap, paratus, with tardiness of restoration. **v. catarrh** or **rhinitis,** hay fever. **v. centers,** centers situated in the medulla oblongata and spinal cord, and governing the caliber of the blood-vessels. **v. nerves,** the nerves passing to the blood-vessels; they are of two kinds, the *vasoconstrictor* (*vasohypertonic*) nerves, or those stimulation of which causes contraction of the blood-vessels, and the *vasodilator* (*vasohypotonic*) nerves, stimulation of which causes dilation of the vessels.
**vasomotorial, vasomotory** (*va-zo-mo-to'-re-al, va-zo-mo'-tor-e*) [*vas,* vessel; *motor,* motor]. Relating to the vasomotor function.
**vasomotricity** (*va-zo-mo-tris'-it-e*) [*vas,* vessel; *motor,* motor]. The quality of having a vasomotor action.
**vasoneurosis** (*va-zo-nū-ro'-sis*) [*vas; neurosis*]. Angioneurosis.
**vasoparesis** (*va-zo-par'-e-sis*) [*vaso-; paresis*]. Paresis affecting the vasomotor nerves.
**vasosection** (*va-zo-sek'-shun*) [*vas,* the vas deferens; *sectio,* a cutting]. Severing of the vas deferens.
**vasosensory** (*va-zo-sen'-so-re*) [*vaso-; sensory*]. Serving as a sensory apparatus for the vessels.
**vasospasm** (*va'-zo-spasm*) [*vaso-; σπασμός,* tension]. Vasoconstriction, angiospasm.
**vasospastic** (*va-zo-spas'-tik*). Angiospastic.
**vasostimulant** (*va-zo-stim'-ū-lant*). Inducing or exciting vasomotor action.
**vasostomy** (*va-zos'-to-me*). [*vas; στόμα,* mouth]. The making of an artificial opening into the vas deferens. Syn., *Belfield's operation.*
**vasothion** (*va-zo-thi'-on*). A compound of vasogen and sulphur, 10%; it is used in chronic skin diseases.
**vasotomy** (*va-zot'-om-e*) [*vas,* the vas deferens; *τεμνειν,* to cut]. Incision of the vas deferens.
**vasotonic** (*va-zo-ton'-ik*) [*vaso-; τόνος,* tone]. 1. Pertaining to the normal tone or tension of the blood-vessels. 2. A vasostimulant.
**vasotonin** (*va-zo-to'-nin*). A mixture of urethane and yohimbine, used for lowering the blood-pressure.
**vasotribe** (*va'-zo-trīb*) [*vaso-; τρίβειν,* to grind]. An instrument for controlling hemorrhages; an angiotribe.
**vasotrophic** (*va-zo-trof'-ik*) [*vaso-; τροφή,* nourishment]. Concerned in the nutrition of vessels.
**vasovagal** (*va-zo-va'-gal*) [*vaso-; vagus,* the vagus nerve]. Pertaining to the vasomotor action of the vagus.
**vasovesiculectomy** (*va-zo-ves-ik-ū-lek'-to-me*). Excision of the vas deferens and seminal vesicles.
**vastus** (*vas'-tus*) [L.]. 1. Large; extensive. 2. A large muscle of the thigh. See under *muscle.*
**Vater's ampulla** (*fah'-ter*) [Abraham *Vater,* German anatomist, 1684-1751]. V., **ampulla of,** a depression in the internal and posterior wall of the descending portion of the duodenum, into which the ductus communis choledochus and the pancreatic duct open. **V.'s corpuscles.** See *Pacinian corpuscles.* **V.'s fold,** a vertical fold of mucous membrane at the lower angle of Vater's ampulla.
**Vater-Pacini's corpuscles** (*fah'-ter-pa-chē-ne*). See *Pacinian corpuscles.*
**Vaughan-Novy test** (*vorn'no'-ve*) [Victor Clarence *Vaughan,* American physician, 1851- ; Frederick George *Novy,* American bacteriologist, 1864- ]. For *tyrotoxicon*: a few drops each of phenol and sulphuric acid are added to the suspected substance in solution; a yellow or orange-red color denotes the presence of tyrotoxicon.
**V. C.** Abbreviation for color vision.
**vecordia** (*ve-kor'-de-ah*) [*vecors,* destitute of reason]. Insanity; especially dementia or idiocy.
**vectis** (*vek'-tis*) [*vehere,* to carry]. An instrument similar to the single blade of a forceps, used in hastening the delivery of the fetal head in labor.
**vector** (*vek'-tor*) [*vector,* a carrier, from *vehere,* to carry]. An insect which carries microorganisms from a sick person to some other person; the process is purely mechanical.
**vegetable** (*vej'-et-ab-l*) [*vegetare,* to quicken]. 1. A plant, especially one used as food. 2. See *vegetal.* **v. albumin.** See *phytalbumose.* **v. proteids.** See under *proteid.* **v. sulphur.** See *lycopodium.*
**vegetal** (*vej'-e-tal*) [*vegetus,* lively; *vegere,* to move, quicken]. Of or pertaining to plants, characteristic of plants, plant-like in habit. **v. functions,** the vital phenomena common to plants and animals, viz.,

irritability, digestion, assimilation, growth, secretion, excretion, circulation, respiration, generation.
**vegetality** (*vej-e-tal'-it-e*) [*vegetare,* to quicken]. The possession of *vegetal functions* (*q. v.*); the opposite of *animality.*
**vegetarian** (*vej-et-a'-re-an*) [see *vegetable*]. One who lives on vegetable food alone.
**vegetarianism** (*vej-et-a'-re-an-izm*) [see *vegetable*]. 1. The doctrine that vegetable food is the only kind proper for man. 2. The practice of living only on vegetable food.
**vegetation** (*vej-et-a'-shun*) [see *vegetable*]. An outgrowth resembling a plant in outline, as the fibrous projections on the cardiac valves in endocarditis, papillomata, polypoid growths, etc. **v.,** aural growths of lymphoid tissue in the nasopharyngeal cavity.
**vegetative** (*vej'-et-a-tiv*) [see *vegetable*]. Having the power of growth, like a plant.
**vegeto-** (*vej'-et-o-*) [*vegere,* to grow]. A prefix employed to denote connection with or relation to the vegetable kingdom.
**vegetoalkali** (*vej-et-o-al'-ka-li*). An alkaloid.
**vegetoanimal** (*vej-et-o-an'-i-mal*). Common to plants and animals.
**vehicle** (*ve'-hik-l*) [*vehiculum,* from *vehere,* to carry]. An excipient or substance serving as a medium of administration of medicines.
**Veiel's paste** (*vīl*) [Theodor *Veiel,* German dermatologist, 1848- ]. A paste used in the treatment of furuncles. It consists of equal parts of zinc oxide and vaseline, with 4 per cent. of boric acid. It is to be well rubbed into the skin around the boil three times a day.
**veil** (*vāl*) [*velum,* veil]. See *velum, velamen.* A caul or piece of the amniotic sac covering the face of a new-born infant. **v., acquired,** an obscuration or imperfection of voice from exposure to cold, catarrhal conditions, or overuse, or from bad training. **v., uterine,** a cap fitted over the cervix uteri, to prevent the entrance of the semen.
**vein** (*vān*) [*vena*]. A blood-vessel carrying blood from the tissues to the heart. Veins, like arteries, have three coats, but less well developed; many also possess valves. **v.,** angular, a continuation of the frontal vein downward to become the facial at the lower margin of the orbit. **v.,** anterior internal maxillary. Same as **v.,** *facial, deep.* **v.,** auricular (*anterior* and *posterior*). 1. The vein of the ear. 2. A vein from the cardiac auricles. **v.,** axillary, a large vein formed by the junction of the brachial veins. **v.s,** azygos, three veins situated in front of the bodies of the thoracic vertebræ; they are a means of communication between the superior and inferior venæ cavæ. **v.,** basilar, a large vein passing back over the crus cerebri to unite with the veins of Galen. **v.,** basilic, a vein on the inner side of the arm. **v.s, brachial,** the veins accompanying the brachial artery. **v.,** brachiocephalic. See **v.s,** *innominate.* **v.s of Breschet,** the veins of the diploe. **v., brooch,** an instrument for compressing veins. **v.,** cardiac, great. See **v.,** *coronary* (1). **v.,** cephalic, a large vein of the arm, formed by the union of the median cephalic and superficial radial, and opening into the axillary vein. **v.s, cerebral,** veins coming from the cerebrum; they are cortical and central. **v.,** coronary. 1. The great cardiac vein, a vein opening into the coronary sinus of the heart. 2. See **v.,** *gastric.* **v.,** dorsispinal, one of the veins forming a reticulum around the vertebræ. **v.s, emissary,** small veins passing through the cranial foramina and connecting the cerebral sinuses with external veins. **v.s, emulgent,** the renal veins. **v.,** facial, a continuation of the angular vein; it joins the internal jugular at the level of the hyoid bone. **v., facial, deep,** one joining the facial vein below the malar bone; it receives the blood from the pterygoid plexus. **v., femoral,** common, a short thick trunk, corresponding to the femoral artery; it becomes the external iliac at Poupart's ligament. **v., femoral, deep,** a vein accompanying the femoral artery; it empties into the superficial femoral. **v., femoral, superficial,** a name given to the femoral vein before it is joined by the deep femoral vein to form the common femoral vein. **v., frontal,** the anterior vein of the scalp as it crosses the frontal bone. **v.s of Galen,** two large veins of the brain, continuations of the internal cerebral veins, and opening into the straight sinus. **v., gastric,** a vein accompanying the artery of the same name. **v.s, hemiazygos,** small, accessory

veins of the azygos veins. **v.s, hemorrhoidal,** a plexus of veins surrounding the rectum. **v., iliac, common,** a vein formed opposite the sacroiliac synchondrosis by the confluence of the external and internal iliac veins. **v., iliac, external,** a continuation upward of the common femoral; it extends from the lower border of Poupart's ligament to the lower border of the sacroiliac synchondrosis. **v., iliac, internal,** a short trunk extending from the top of the great sciatic notch to the great sacroiliac synchondrosis. **v.s, innominate,** two large valveless veins returning the blood from the head, neck, and upper extremity. **v., jugular, anterior,** a vein beginning at the level of the chin and ending at the clavicle in the external jugular vein. **v., jugular, external,** a vein formed at the angle of the lower jaw by the union of the posterior auricular and temporomaxillary veins; it empties into the subclavian. **v., jugular, internal,** a continuation of the lateral sinus, beginning at the jugular fossa, accompanying the internal and common-carotid arteries, and joining the subclavian vein to form the innominate. **v.** of Marshall. See **v., oblique. v., maxillary anterior,** a small vein of the anterior portion of the face. **v., maxillary, internal,** one accompanying the first part of the internal maxillary artery. **v., median basilic,** a vein uniting with the ulnar to form the basilic. **v., median cephalic,** a vein uniting with the superficial radial to form the cephalic. **v., median, deep,** a vein formed by the union of the outer vena comes of the ulnar artery and the muscular and radial recurrent veins. **v., median, superficial,** one starting at the anterior plexus of the wrist and uniting with the deep median to form the median. **v., mesenteric, inferior,** one that accompanies the inferior mesenteric artery and joins the splenic vein behind the pancreas. **v., mesenteric, superior,** one that accompanies the superior mesenteric artery and joins the splenic vein to form the portal. **v., oblique, of Marshall,** a vein crossing the dorsal portion of the left auricle of the heart. It is the remnant of the left duct of Cuvier. **v.s, omphalomesenteric,** several venous trunks of the primitive embryonic circulation which carry the blood from the terminal sinus to the sinus venosus, a short vitelline vein. **v., ophthalmic,** a short trunk carrying the blood from the eye and emptying into the cavernous sinus. **v.s, plantar,** veins accompanying the plantar arteries. **v., popliteal,** one formed by the union of the venæ comites of the anterior and posterior tibial arteries; it accompanies the popliteal artery, and becomes the femoral vein at the junction of the lower with the middle third of the thigh. **v., portal,** a short trunk entering the liver at the transverse fissure and formed by the junction of the superior mesenteric and splenic veins. **v.s, pulmonary,** four veins, two from each lung, returning the aerated blood from the lungs to the heart. **v., radial, superficial,** a vein accompanying the musculocutaneous nerve up the radial side of the forearm. **v., ranine,** the chief vein conveying blood from the tongue. It originates near the tip beneath the mucosa, accompanies the hypoglossal nerve across the hypoglossus muscle, and empties into the internal jugular vein. **v., renal,** a vein accompanying the renal artery. **v., saphenous, long or internal,** a long superficial vein running up the inner aspect of the leg and thigh, terminating in the femoral vein below Poupart's ligament. **v., saphenous, short or external,** a superficial vein running up the outer aspect of the foot, leg, and back of the calf, and emptying into the popliteal vein. **v., spermatic,** one returning the blood from the testicle; on the right side it terminates in the inferior vena cava and on the left in the left renal vein. **v., splenic,** one returning the blood from the spleen, and forming the portal vein by its union with the superior mesenteric vein. **v., stellate.** See *Verheyen's stars*. **v.-stone.** See *phlebolith*. **v., subclavian,** a continuation of the axillary vein, uniting with the internal jugular vein to form the innominate vein at the sternoclavicular articulation. **v.s, temporal,** veins returning the blood from the temporomaxillary region of the head; they join the internal maxillary vein to form the temporomaxillary vein. **v., temporomaxillary,** one formed by the union of the temporal and internal maxillary veins in the parotid gland; it terminates in the external jugular. **v.** of Trolard, a vein of the cerebrum passing along the posterior branch of the fissure of Sylvius, and emptying into the superior petrosal sinus. **v., ulnar,** one running from the wrist up the anterior and inner surface of the forearm. **v., umbilical,** a vein conveying the blood from the placenta to the fetus. **v., Vesalius'.** See *Vesalius's vein*. **v.s, vitelline.** See *vitelline veins*.

**velamen** (*ve-la'-men*) [L.]. A veil or covering membrane. **v. nativum,** the skin. **v. vulvæ,** the Hottentot apron; see *apron*.

**velamentous** (*vel-am-en'-tus*) [*velamen*]. Resembling a veil.

**velamentum** (*ve-la-men'-tum*) [L.; *pl. velamenta*]. A veil, or covering membrane. **v. abdominale,** peritoneum. **v. cerebrale,** one of the meninges. **v. cerebri,** *v. cerebrale*. **v. corporis commune,** the skin. **v. infantis,** one of the fetal membranes. **v. linguæ,** the glosso-epiglottic ligament.

**velar** (*ve'-lar*) [*velum*]. Pertaining to a velum; especially the velum palati.

**Velden's** (von den) test for hydrochloric acid in the contents of the stomach. Filter paper dipped into a watery or alcoholic solution of tropeolin 00 turns ruby red or brownish red on the application of free hydrochloric acid.

**Veld sore.** [Dutch, *veld*, field]. A lesion common among troops during the Boer war, a running sore probably due to the sting of a fly.

**veliform** (*vel'-if-orm*) [*velum*, veil; *forma*, form]. Forming a velum.

**Vella's fistula** (*vel'-lah*) [Luigi *Vella*, Italian physiologist, 1825–1886]. An intestinal fistula for obtaining gastric juice.

**vellication** (*vel-ik-a'-shun*) [*vellicare*, to twitch]. Spasmodic twitching of muscular fibers.

**vellolin** (*vel'-o-lin*). A purified wool fat, lanolin.

**vellosine** (*vel-o'-sēn*), $C_{22}H_{26}N_2O_4$. An alkaloid contained in Paopereira bark, *Geissospermum vellosii*; it resembles brucine in physiological action.

**Velpeau's bandage** (*vel'-po*) [Alfred Armand Louis Marie *Velpeau*, French surgeon, 1795–1867]. A bandage for the shoulder. **V.'s deformity,** the "silver fork," deformity in Colles' fracture. **V.'s hernia,** femoral hernia anterior to the blood-vessels.

**velum** (*ve'-lum*) [L., a veil; a sail]. A veil or veil-like structure. **v., anterior medullary.** See *valve of Vieussens*. **v. interpositum,** the membranous roof of the third ventricle. **v. palati,** the soft palate. **v. pendulum palati,** the soft palate, especially the uvula. **v., posterior medullary,** the commissure of the flocculus of the cerebellum. **v. staphylinum,** soft palate. **v. Tarini, v. posterior medullary. v. terminale,** lamina terminalis. **v. triangulare, v. interpositum.**

**vena** (*ve'-nah*) [L.; *pl. venæ*]. A vein. See *vein*. **v. azygos, v. azygos major, v. azygos dextra,** a vein connecting the right lumbar, right renal vein, or postcava with the precava. **v. azygos minor.** See *v. hemiazygos.* **v. cava anterior.** See *v. cava superior.* **v. cava inferior,** a vein formed by the junction of the two common iliac veins and emptying into the right auricle of the heart. It receives the lumbar, right spermatic, renal, suprarenal, phrenic, and hepatic veins. **v. cava superior,** a vein formed by the union of the innominate veins, and conveying the blood from the upper half of the body to the right auricle. **v. comes,** a vein accompanying an artery in its course. **v. corporis striati,** a vein which helps to form the internal cerebral vein and returns the blood from the corpus striatum. **v. hemiazygos, v. azygos minor,** a vein from the left lumbar or left renal vein to the v. azygos major. **v. hemiazygos accessoria,** an inconstant vein which may take the place of the left superior intercostal vein.

**venæ** (*ve'-ne*) [*pl.* of *vena*]. **venæ advehentes.** The vessels passing from the vitelline veins to the liver. **v. comites,** the two veins accompanying an artery. **v. Galeni,** two venous trunks in the brain. **v. minimæ cordis,** the smallest of the cardiac vessels, entering into the cavities of the heart. **v. revehentes,** the vessels passing from the liver to the sinus of the embryo. **v. Thebesii,** the small veins by which the blood passes from the walls of the heart to the right auricle. **v. vorticosæ,** the stellate veins of the choroid coat of the eyeball.

**venenation** (*ven-en-a'-shun*) [*venenum*, a poison]. The condition due to poisoning.

**venenatus** (*ven-en-a'-tus*) [see *venenation*]. Poisonous.

**venenific** (*ven-en-if'-ik*) [*venenum*, poison]. Poison-forming.

**venenosalivary** (ven-en-o-sal'-iv-a-re). See *venomosalivary*.
**venenose, venenous** (ven'-en-ōs, ven'-en-us) [*venenosus*]. Toxic; poisonous.
**venenosity** (ven-en-os'-it-e). The condition of being toxic.
**venereal** (ven-e'-re-al) [*Venus*, the goddess of love]. Pertaining to or produced by sexual intercourse. **v. diseases**, gonorrhea, syphilis, and chancroid. **v. sore**. See *chancre*. **v. wart**, see *verruca acuminata*.
**venerismus pyorrhoicus**. Gonorrhea.
**venery** (ven'-er-e) [see *venereal*]. Sexual intercourse.
**venesection** (ven-e-sek'-shun). See *blood-letting*.
**venesuture** (ven-e-sū'-tūr) [*vena; sutura*, suture]. The suturing of a vein.
**Venetian red**. An ochre, whose color is due to ferric oxide.
**veniplex** (ven'-ip-leks) [*vena; plexus*]. A plexus of veins.
**venipuncture** (ven'-e-punk-chur). Puncture of a vein.
**venisuture** (ven-i-sū'-tūr). See *venesuture*.
**venom** (ven'-om) [*venenum*, poison]. Poison, especially a poison secreted by certain reptiles and insects. **v. albumin**, the albumin of the venom of a snake. **v. globulin**, a globulin found in snake poisons. **v. hemolysis**, dissolution of red blood corpuscles by snake venom. **v. leukolysis**, destruction of leukocytes by the action of venom. **v. peptone**, a peptone found in the venom of certain serpents.
**venomosalivary** (ven-om-o-sal'-iv-a-re). Secreting a toxic saliva.
**venomotor** (ven-o-mo'-tor) [*vena; movere*, to move]. Causing the veins to contract or dilate.
**venomous** (ven'-om-us) [*venom*]. Poisonous; secreting venom.
**venosclerosis** (ven-o-skle-ro'-sis) [*vena; σκληρός*, hard]. Induration of the veins.
**venosity** (ven-os'-it-e) [*venosus*]. A condition in which the arterial blood shows venous qualities.
**venous** (ve'-nus) [*venosus*]. Pertaining to or produced in a vein. **v. blood**, the dark blood in the veins. **v. hum**, the murmur or rushing sound heard in auscultation of a vein. **v. sinus**, a cerebral sinus.
**venovenostomy** (ven-o-ven-os'-to-me) [*vena; στόμα*, mouth]. The making of an anastomosis between two veins.
**vent** (vent). An outlet, especially the anal opening.
**venter** (ven'-ter) [L.]. 1. The belly or abdomen. 2. The belly of a muscle. 3. The cavity of the abdomen. 4. The concavity of any expanded part, as the *venter* of the scapula, *venter* of the ilium. **v. imus**, the hypogastrium. **v. medius**, the thorax. **v. renum**, the pelvis of the kidneys. **v. supremus**, the skull.
**ventilation** (ven-til-a'-shun) [*ventilare*, to fan]. The act or process of supplying fresh air; the act or process of purifying the air of a place.
**ventose** (ven'-tōs) [*ventosa*, a cupping-glass]. 1. A cupping-glass. 2. [*ventosus*, windy]. Flatulent.
**ventosity** (ven-tos'-it-e). Flatulence.
**ventrad** (ven'-trad) [*venter; ad*, toward]. Toward the ventral aspect.
**ventral** (ven'-tral) [*venter*]. 1. Pertaining to the belly 2. Referring to the anterior aspect of the body or to the flexor aspect of the limbs. **v. decubitus**, lying down on the abdomen; prone.
**ventricle** (ven'-trik-l) [*ventriculus*, dim. of *venter*, a belly]. A small cavity or pouch. **v., aortic**, the left ventricle of the heart. **v. of Arantius**, a culdesac at the lower end of the fourth ventricle. **v.s of the brain**, cavities in the interior of the brain, comprising the two lateral ventricles, the third, fourth, and fifth ventricles. **v. of cord**, the central canal of the spinal cord. **v., fifth**, the cavity vetween the laminæ of the septum lucidum. **v. of corpus callosum**, the space between the labium cerebri and the corpus callosum; the callosal fissure. **v., fourth**, the space between the oblongata and pons in front, and the cerebellum behind. **v. of larynx**, a depression between the true and false vocal bands. **v.s lateral**, serous cavities, one in each cerebral hemisphere, and communicating with the third ventricle through the foramen of Monro. Each ventricle consists of a triangular central cavity or body and three smaller cavities or cornua. The corpus callosum forms the roof of the body, the septum lucidum the mesal boundary and the floor is formed by the corpus striatum, tænia semicircularis, thalamus, choroid plexus, corpus fimbriatum, and fornix. **v., left, of heart**, that upon the dorsal and left side of the heart, and which, through the aorta, forces the blood throughout the body. **v. of myelon**, the central canal of the spinal cord. **v., pineal**, one found occasionally within the pineal body; it is thepersistence of a fetal condition. **v., right, of heart**, that forcing the blood through the pulmonary artery into the lungs. **v., terminal**, the dilated portion of the central canal of the spinal cord in the filum terminale internum. **v., third**, an open space between the optic thalami and extending to the base of the brain. **v. Verga's**, a space occasionally found between the corpus callosum and the fornix.
**ventricornu** (ven-tri-kor'-nū) [*venter; cornū*]. The anterior horn of the gray matter of the myelon.
**ventricose** (ven'-trik-ōs) [*venter*, abdomen]. Inflated or swollen on one side, resembling an abdomen.
**ventricular** (ven-trik'-ū-lar) [*ventricle*]. Pertaining to a ventricle. **v. aqueduct**. See *aquæductus Sylvii*. **v. bands**, the longitudinal folds of mucous membrane above and parallel to the vocal bands. The false vocal cords. **v. ligament**, a false vocal band. **v. muscle**, the thyroepiglottideus. **v. septum**, (1) the septum between the ventricles of the heart; (2) the septum pellucidum.
**ventricularis** (ven-trik-ū-la'-ris). The thyroepiglottideus muscle. See under *muscle*.
**ventriculi** (ven-trik'-ū-li) [pl. of *ventriculus*]. Ventricles.
**ventriculus** (ven-trik'-a-lus). 1. See *ventricle*, (2) the stomach. **v. cerebri**, ventricle of the brain. **v. cordis**, ventricle of the heart. **v. dexter**, right ventricle. **v. lateralis**, lateral ventricle. **v. medius**, middle (third) ventricle. **v. quartus**, fourth ventricle. **v. sinister**, left ventricle. **v. tertius**, third ventricle. **v. tricornis cerebri**, lateral ventricle of the brain.
**ventricumbent** (ven-tre-kum'-bent) [*venter; cumbere*, to lie]. Lying with the ventral surface down.
**ventriduction** (ven-tre-duk'-shun) [*venter; ducere*, to lead]. The act of drawing a part toward the belly.
**ventrifixation**. See *ventrofixation*.
**ventriloquism** (ven-tril'-o-kwizm) [*venter*, belly; *loqui*, to speak]. Peculiar vocal utterance without the usual modifications of the resonance-organs, so that the voice seems to come from a closed space or from a distance.
**ventrimeson** (ven-trim-e'-son) [*venter; μέσον*, middle]. The mesial line on the ventral aspect of the body.
**ventripyramid** (ven-trip-ir'-am-id) [*venter; pyramid*]. An anterior pyramid of the oblongata.
**ventro-** (ven-tro-) [*venter*, the belly]. A prefix signifying relation to the belly.
**ventrocystorrhaphy** (ven-tro-sis-tor'-a-fe) [*κύστις*, cyst; *ῥαφή*, suture]. Suture of an opened cystwall to the wall of the abdomen so as to provide a free discharge of its contents.
**ventrofixation** (ven-tro-fiks-a'-shun) [*ventro-; fixation*]. The stitching of a displaced viscus to the abdominal wall.
**ventrohysteropexy** (ven-tro-his'-ter-o-peks-e) [*ventro-; ὑστέρα*, womb; *πῆξις*, a fixing]. Ventrofixation of a uterus.
**ventroinguinal** (ven-tro-in'-gwin-al) [*ventro-; inguinal*]. Pertaining to the abdomen and the groin.
**ventrolateral** (ven-tro-lat'-er-al) [*ventro-; latus*, side]. Toward the ventral and lateral aspects.
**ventrolateral** (ven-tro-lat'-er-al). Relating to the ventral and lateral aspects of a part.
**ventromyel** (ven-tro-mi'-el) [*ventro-; μυελός*, marrow]. The anterior portion of the spinal cord.
**ventroptosis** (ven-tro-to'-sis). See *gastroptosis*.
**ventroscopy** (ven-tros'-ko-pe) [*ventro-; σκοπεῖν*, to view]. Direct examination of the abdominal and pelvic cavities by means of an apparatus resembling the cyscoscope.
**ventrose** (ven'-trōs) [*ventrosus*]. Having a belly, or a swelling like a belly (pot belly).
**ventrosuspension** (ven-tro-sus-pen'-shun). See *ventrofixation*.
**ventrotomy** (ven-trot'-o-me) [*ventro-; τομή*, a cut]. Celiotomy.
**ventrovesicofixation** (ven-tro-ves-ik-o-fiks-a'-shun) [*ventro-; vesica*, bladder; *fixation*]. The suturing of the uterus to the bladder and abdominal wall.
**venule, venula** (ven'-ūl, ven'-ū-lah) [*venula*, dim. of *vena*, a vein]. A small vein.
**venus** (ve'-nus) [*Venus*, goddess of love]. 1.

Sexual intercourse. 2. Alchemic name for copper. **v., crystals of,** copper acetate.
**veratralbine** (*ver-at-ral'-bēn*) [*veratrum; albus*, white]. An alkaloid obtained from white hellebore.
**veratrina.** See *veratrine*.
**veratrine** (*ver-at'-rēn*). *Veratrina* (U. S. P.); a mixture of alkaloids obtained from the seeds of *Asagræa officinalis* (sabadilla), of the order *Liliaceæ*. It is a local irritant, and produces tetanic convulsions followed by paralysis; it first stimulates, then paralyzes, the vasomotor center. It is used externally in the form of an ointment in rheumatism, gout, and neuralgia. **v. ointment** (*unguentum veratrinæ*, U. S. P.), an ointment composed of veratrine, expressed oil of almond, and benzoinated lard. **v., oleate of** (*oleatum veratrinæ*, U. S. P.), veratrine, oleic acid, and olive-oil.
**veratrinize** (*ver-at'-rin-īz*). To bring under the influence of veratrine.
**veratrize.** See *veratrinize*.
**veratroidine** (*ver-at-roi'-din*). See under *Veratrum*.
**veratrol** (*ver-at'-rol*), C₈H₁₀O₂. A colorless oil with aromatic odor obtained from veratric acid by action of baryta with heat; it is used as an antiseptic by inhalation and by application in 1% solution; less poisonous than guaiacol but more caustic.
**Veratrum** (*ver-at'-rum*). A genus of plants of the order *Liliaceæ*. The rhizome and roots of *V. viride*, American hellebore, or *V. album*, white hellebore (*veratrum*, U. S. P.; *veratri viridis rhizoma*, B. P.), contain the alkaloids jervine and veratroidine. The former is a depressant to the vasomotor centers and the motor centers of the spinal cord. In toxic doses it produces slowness of the pulse, fall in blood-pressure, relaxation, epileptiform convulsions, paralysis, and death from failure of the respiration. *Veratroidine* is irritant and produces vomiting and purging; it is also a depressant to the motor centers of the spinal cord and to the pulse. When veratrum is administered the combined action of the alkaloids is obtained, and consists chiefly in slowing of the pulse and lessening of blood-pressure, with vomiting in the case of large doses. It is employed in sthenic inflammations, as penumonia, peritonitis, in puerperal eclampsia, and in excessive cardiac hypertrophy. **v., fluidextract of** (*fluidextractum veratri*, U. S. P.). Dose 1–3 min. (0.065–0.2 Cc.). **v., tincture of** (*tinctura veratri*, U. S. P.). Dose 1–3 min. (0.065–0.2 Cc.).
**Verbascum** (*ver-bas'-kum*) [L.]. Mullein, a genus of plants of the order *Scrophularineæ*. The leaves and flowers of *V. thapsus* have been used as demulcent in catarrhal inflammation of mucous membranes and as an application to hemorrhoids.
**Verbena** (*ver-be'-nah*) [L.]. A genus of flowering plants of some 80 species once highly esteemed in medicine, but now little used.
**verbigeration** (*ver-bij-er-a'-shun*) [*verbigerare*, to carry words about]. The frequent and uncontrollable repetition of the same word, sentence, or sound without reference to its meaning.
**verdigris** (*ver'-dig-ris*) [Fr., *verd de gris*, probably from L., *viridis*, green; *æs*, copper]. 1. A mixture of copper acetates. 2. A deposit upon copper vessels, from the formation of cupric salts.
**Verga's lacrimal groove** (*vair'-gah*) [*Verga*, Italian anatomist, 1811–1895]. A more or less pronounced groove extending downward from the lower orifice of the nasal duct. **V.'s ventricle,** a cleft-like space between the fornix and the callosum.
**vergences** (*ver'-jen-sēz*) [*vergere*, to bend]. A term applied to associated disjunctive movements of the eyes, *e. g.*, convergence, divergence.
**vergens** (*ver'-jens*) [L.]. Inclining. **v. deorsum,** inclining downward, as of the axis of vision in one eye in strabismus. **v. sursum,** upward inclination.
**Verheijn's stars.** See *Verheyen, stars of*.
**Verheyen, stars of** (*fer-hi'-en*) [Philippus *Verheyen*, Flemish anatomist, 1648–1710]. Venous plexuses of stellate form situated on the surface of the kidney, beneath its capsule.
**verjuice** (*ver'-joos*) [Fr., *verd*, green; *jus*, juice]. The acid juice of unripe fruits.
**Vermale's amputation** (*ver-māl'*) [Raymond de *Vermale*, French surgeon]. An amputation with a double flap.
**vermiceous** (*ver-mish'-us*) [*vermis*, a worm]. Relating to worms.
**vermicidal** (*ver-mis-i'-dal*) [*vermis*, worm; *cædere*, to kill]. Destroying worms.

**vermicide** (*ver'-mis-īd*) [*vermis*, a worm; *cædere*, to kill]. An agent that destroys intestinal worms.
**vermicular** (*ver-mik'-u-lar*) [*vermis*]. Wormlike. **v. motion,** peristalsis. **v. sulci,** grooves between the vermis and the lateral hemispheres of the cerebellum.
**vermiculate** (*ver-mik'-u-lāt*) [*vermiculatus*]. Resembling or shaped like a worm.
**vermiculation** (*ver-mik-u-la'-shun*) [*vermis*]. A worm-like motion; peristaltic motion.
**vermicule** (*ver'-mik-ūl*) [*vermiculus*, a little worm]. 1. A small worm. 2. A sexually produced embryo of the malarial parasite.
**vermiculus** (*ver-mik'-u-lus*) [L.]. A little worm or grub.
**vermiform** (*ver'-mif-orm*) [*vermis; forma*, a form]. Worm-shaped. **v. appendix.** See *appendix, vermiform*. **v. process, inferior and superior,** the inferior and superior surfaces of the middle lobe of the cerebellum.
**vermifugal** (*ver-mif'-u-gal*) [*vermifuge*]. Having the qualities of a vermifuge; expelling worms.
**vermifuge** (*ver'-mif-ūj*) [*vermis; fugare*, to expel]. An agent that expels intestinal worms.
**vermilingual, vermilinguid** (*ver-me-lin'-gwal, -gwe-al*) [*vermis; lingua*, tongue]. Having a worm-shaped tongue.
**vermilion** (*ver-mil'-yun*). Red mercuric sulphide. **v. border,** the margin of the lips where skin and mucous membrane meet.
**vermin** (*ver'-min*) [*vermis*, worm]. A general (and mainly collective) name for parasitic animals and for semi-parasites, such as fleas and bed-bugs.
**verminal** (*ver'-min-al*). Relating to or due to worms.
**vermination** (*ver-min-a'-shun*) [*vermis*]. 1. Infestation with worms. 2. The generation of worms.
**verminous** (*ver'-min-us*) [*vermis*]. Infested with, or pertaining to worms.
**vermis** (*ver'-mis*) [L.]. 1. A worm. 2. The middle lobe of the cerebellum. **v., inferior,** of the cerebellum. See *process, vermiform*. **v., superior,** of the cerebellum. See *process, superior vermiform*.
**vermix** (*ver'-mix*). A contraction of the term *vermiform appendix*.
**vermouth, vermuth** (*ver'-mooth*) [Ger. *wermuth*, wormwood]. A cordial prepared from white wine and flavored with wormwood; used as an appetizer.
**vernal** (*ver'-nal*) [*vernalis*, of the spring]. Pertaining to the spring. **v. catarrh** or **conjunctivitis**, a form of conjunctivitis recurring each spring or summer, and disappearing with frost. **v. fever,** malarial fever.
**Verneuil's neuroma** [Aristide Auguste Stanislas *Verneuil*, French surgeon, 1823–1895]. A plexiform neuroma or neuroma cirsoideum. **V.'s operation,** a form of iliac colotomy.
**vernier** (*ver'-ne-er*) [after the inventor, Pierre *Vernier*, French physicist, 1580–1637]. In physics, a contrivance attached to various instruments of precision for the estimation of minute fractions of any unit of distance.
**vernine** (*ver'-nēn*), C₁₀H₁₀N₅O₃. A leukomaine base found in young vetch, clover, ergot, etc., and yielding guanine on heating with hydrochloric acid.
**vernix caseosa** [L., "cheesy varnish"]. A sebaceous deposit covering the surface of the fetus.
**vernonin** (*ver-no'-nin*) C₁₈H₂₈O₇. A glucoside from the root of *Vernonia nigritiana*. Its action is similar to that of digitalin, and it is used as a cardiac tonic.
**veronal** (*ver'-on-al*). Diethylmalonylurea, a white crystalline substance used as a hypnotic. Dose 7–20 gr. (0.5–1.3 Gm.).
**Veronica** (*ve-ron'-ik-ah*). A genus of scrophulariaceous herbs and shrubs. **V. virginica,** leptandra, Culver's physic; the root is a purgative and cholagogue.
**verrucose** (*ver-oo'-kah*) [L.: pl., *verrucæ*]. Wart. **v. acuminata,** a venereal wart. **v. necrogenica,** a warty excrescence found on the fingers of those who frequently handle the tissues of tuberculous subjects. Syn., *anatomic tubercle; dissection tubercle*.
**verruciform** (*ver-oo'-sif-orm*) [*verruca; forma*, form]. Wart-like.
**verrucose, verrucous** (*ver'-oo-kōs, ver'-oo-kus*) [*verruca*]. Warty; covered with or having warts.
**verruga** (*ver-oo'-gah*) [Sp.]. See *verruca*. 2. **verruga peruana.** **v. peruana,** an endemic specific disease of the skin, occurring in the western Andes in Peru. Syn., *Carrion's disease; Peruvian wart*.
**verruges** (*ver'-gahs*). See *verruga peruana*.
**version** (*ver'-zjun*) [*vertere*, to turn]. Turning; an

operation whereby one part of the fetus is made to replace another at the mouth of the uterus. **v., abdominal**, same as *v., external*. **v., bipolar**, version by acting upon both poles of the fetus. **v., cephalic**, turning of the fetus so as to bring the head to present. **v., combined**, bipolar version consisting of a combination of external and internal version. **v., external**, that effected by external manipulation. **v., internal**, that performed by entering the hand within the uterus. **v., mixed**, same as *v., combined*. **v., pelvic**, turning the fetus to bring about a breech presentation. **v., podalic**, that in which one or both feet are brought to the mouth of the uterus. **v., spontaneous**, the process whereby without external influence, a transverse position is changed into a longitudinal one.
**Verstraetin's bruit**. A bruit heard over the lower border of the liver in some cachectic individuals.
**vertebra** (*ver'-teb-rah*) [L., "a joint; a bone of the spine"; *pl. vertebræ*]. One of the bones forming the spinal or vertebral column. There are 33 vertebræ, divided into 7 cervical, 12 thoracic or dorsal, 5 lumbar, 5 sacral (the sacrum), 4 coccygeal (the coccyx). A typical vertebra consists of a body and an arch, the latter being formed by 2 pedicles and 2 laminæ. The arch supports 7 processes: 4 articular, 2 transverse, and 1 spinous. **v., basilar**, the last lumbar vertebra. **v. dentata**, the axis. **v., false**, one of the sacral or coccygeal vertebræ. **vertebræ, flexion**, all except the first two cervical vertebræ. **v. magna**, the sacrum. **v. prominens**, the seventh cervical vertebra. **vertebræ, rotation**, the first and second cervical vertebræ. **v., tricuspid**, the sixth cervical vertebra in the lower animals. **v., true**, one of the cervical dorsal or lumbar vertebræ.
**vertebradymia** (*ver-te-brah-dim'-e-ah*). See *spondylodidymia*.
**vertebral** (*ver'-teb-ral*) [*vertebra*]. 1. Pertaining to or characteristic of a vertebra; made up of or possessing vertebræ. 2. Pertaining to the vertebral artery. **v. artery**. See *artery*. **vertebral.** **v. column**, the spinal column; the backbone. **v. groove**, the groove between the spinous and transverse processes of the spinal column, the floor being formed by the laminæ. **v. ribs**, the last two ribs.
**vertebralis** (*ver-te-bra'-lis*): Vertebral, pertaining to one or more of the vertebræ.
**vertebrarium** (*ver-te-bra'-re-um*) [L.]. The spinal column.
**vertebrarterial** (*ver-teb-rar-te'-re-al*) [*vertebra; artery*]. Giving passage to the vertebral artery, as the *vertebrarterial* foramina in the transverse processes of the cervical vertebræ.
**Vertebrata** (*ver-te-bra'-tah*) [*vertebra*, a vertebra]. A great division of the animal kingdom, including all animals having a spinal column, or its equivalent body axis.
**vertebrate, vertebrated** (*ver'-teb-rāt, ver'-teb-ra-ted*) [*vertebra*]. 1. Having a vertebral column. 2. Resembling a vertebral column in flexibility, as a *vertebrate* catheter.
**vertebrectomy** (*ver-te-brek'-to-me*) [*vertebra-*; ἐκτομή, excision]. Excision of a portion of a vertebra.
**vertebro-** (*ver-teb-ro-*) [*vertebra*]. A prefix denoting pertaining to a vertebra.
**vertebroarterial**. See *vertebrarterial*.
**vertebrobasilar** (*ver-te-bro-bas'-il-ar*) [*vertebro-;* βάσις, base]. Belonging to the vertebræ and the base of the skull. **v. plexus**, the vertebrobasilar plexus.
**vertebrochondral** (*ver-teb-ro-kon'-dral*) [*vertebro-;* χονδρός, cartilage]. Connecting the costal cartilages with the vertebræ.
**vertebrocostal** (*ver-teb-ro-kos'-tal*) [*vertebro-; costa*, a rib]. Pertaining to the vertebræ and the ribs.
**vertebrodidymia** (*ver-te-bro-did-im'-e-ah*) [*vertebro-;* δίδυμος, twin]. A monstrosity formed by two individuals united by the vertebræ.
**vertebrofemoral** (*ver-te-bro-fem'-or-al*) [*vertebro-; femur*]. Pertaining to the vertebral column and the femur.
**vertebroiliac** (*ver-teb-ro-il'-e-ak*) [*vertebro-; ilium*]. Pertaining to the vertebræ and the ilium.
**vertebromammary** (*ver-te-bro-mam'-ar-e*) [*vertebro-; mamma*, breast]. Relating to the vertebræ and the mammary region of the thorax.
**vertebrosacral** (*ver-teb-ro-sa'-kral*) [*vertebro-; sacrum*]. Pertaining to the vertebræ and the sacrum.
**vertebrosternal** (*ver-teb-ro-ster'-nal*) [*vertebro-; sternum*]. Extending from the spinal column to the sternum. **v. ribs**, the true ribs.

**vertex** (*ver'-teks*) [L.]. The crown or top of the head; calvaria, **v. cordis**, the apex of the heart. **v. cubiti**, the olecranon. **v. presentation**, a presentation of the vertex of the fetal skull.
**vertical** (*ver'-tik-al*) [*vertex*]. 1. Pertaining to the vertex. 2. Perpendicular; referring to the position of the long axis of the body in the erect posture. **v. diameter of cranium**, an imaginary line from the basion to the bregma.
**verticil** (*ver'-tis-il*) [*verticillus*, the whirl of a spindle]. In biology, a whorl; a circle of leaves, tentacles, hairs, organs, or processes radiating from an axis on the same horizontal plane.
**verticillate** (*ver-tis-il'-āt*) [*verticillus*, a whirl]. Whorled.
**verticomental** (*ver-tik-o-men'-tal*) [*vertex; mentum*, the chin]. Pertaining to the vertex and the chin.
**vertiginous** (*ver-tij'-in-us*) [*vertigo*]. Resembling or affected with vertigo.
**vertigo** (*ver'-tig-o;* also *ver-ti'-go*) [L., from *vertere*, to turn]. Giddiness, dizziness; a sensation of lack of equilibrium. It may be due to disease of the ears (*auditory* or *aural vertigo*), the eyes (*ocular vertigo*), the brain (*cerebral vertigo*), the stomach (*gastric vertigo*), the blood, etc. **v., auditory or aural**. See *Ménière's disease*. **v., cerebral**, that due to cerebral disorder. **v., epileptic**, vertigo associated with or preceding an attack of epilepsy. **v., essential**, one not due to any discoverable cause. **v., gastric**, that arising from dyspepsia. **v., intestinal**. 1. That caused by intestinal disorder. 2. That caused by pressure on the terminal portions of the intestine by gas or feces, or even when the finger is introduced into the rectum and irritates the intestinal wall. It is thought to be due to pressure on the hemorrhoidal plexus of the sympathetic system. **v., labyrinthine**. See *Ménière's disease*. **v., lithemic**, a form associated with gout and lithemia. **v., neurasthenic**, subjective vertigo found in neurasthenia. **v., objective**, one in which objects seem to the patient to move. **v., ocular**, that due to eye-disease. **v., organic**, that due to brain lesion. **v., paralyzing**. See *Gerlier's disease*. **v., peripheral**, that due to irritation that is not central. **v., stomachal**, gastric vertigo, caused by disorder of the stomach. **v., subjective**, one in which the patient has a sensation as if he himself were moving. **v. tenebricosa**, that accompanied by dimness of vision and headache. **v., toxemic**, that due to some poison in the blood. **v., vertical**, that caused by looking downward from or upward to a height.
**verumontanum** (*ver-oo-mon-ta'-num*) [*veru*, a spit; *mons*, a mountain]. The caput gallinaginis, a longitudinal ridge on the floor of the prostatic urethra.
**vervain** (*ver'-vān*). See *verbena*.
**Vesalius' foramen** (*ves-a'-le-us*) [Andreas *Vesalius*, Italian anatomist, born in Belgium, 1514-1564]. An inconstant foramen in the base of the skull, anterointernal to the foramen ovale; it transmits an emissary vein. **V.'s glands**, the bronchial and pulmonary glands. **V.'s ligament**. See *ligament, Poupart's*. **V.'s sesamoid bones**, fibrocartilaginous or osseous bodies often found in the tendons of the gastrocnemius. **V.'s vein**, a small vein through which the pterygoid plexus communicates with the cavernous sinus.
**vesania** (*ves-a'-ne-ah*) [L.]. Unsoundness of mind.
**vesanic** (*ves-an'-ik*) [*vesania*, unsoundness of mind]. Relating to insanity.
**vesica** (*ves-i'-kah*) [L.; *gen*. and *pl*. *vesicæ*]. The bladder. **v. fellea**, the gall-bladder. **v. urinaria**, the urinary bladder.
**vesical** (*ves'-ik-al*) [*vesica*]. Pertaining to the bladder. **v., calculus**, a stone in the bladder. **v. crisis**, severe paroxysmal pain in the bladder occurring in locomotor ataxia. **v. triangle**, the trigone.
**vesicant** (*ves'-ik-ant*) [*vesicāre*, to blister]. 1. Blistering. 2. A blistering agent.
**vesication** (*ves-ik-a'-shun*) [see *vesicant*]. The formation of a blister; a blister.
**vesicatory** (*ves'ik-at-or-e*) [see *vesicant*]. 1. Blistering. 2. A blistering agent.
**vesicle** (*ves'-ik-l*) [*vesicula*, dim. of *vesica*, bladder]. 1. A small bladder; especially a small sac containing fluid. 2. A small blister on the skin, as a *herpetic* or *smallpox vesicle*. **v., allantoic**, the internal hollow portion of the allantois. **v., auditory**, an ectodermic sac, a part of the cerebral vesicle, from which the internal ear is formed. **v., blastodermic**. See *blastoderm*. **v.s, cerebral** or **encephalic**, divisions of the anterior extremity of the neural tube of the

embryo, subsequently forming the segments of the brain. **v., germinal**, the nucleus of the ovum. **v., Graafian**. See *follicle, Graafian*. **v., ocular**, a protrusion of the anterior cerebral vesicle, the first indication of the eye. **v., olfactory**, the primitive vesicle that develops into the olfactory lobe. **v., optic**, a hollow process of the cerebral vesicle forming the essential part of the eye. **v., otic**. See *v., auditory*. **v., prostatic**. See *uterus masculinus*. **v., seminal**, one of the two little sacs situated at the base of the bladder and serving as reservoirs for the semen. **v., umbilical**. See *yolk-sac*.

**vesico-** (*ves-ik-o-*) [*vesica*]. A prefix denoting pertaining to the bladder.

**vesicoabdominal** (*ves-ik-o-ab-dom'-in-al*) [*vesico-; abdomen*]. Pertaining to the abdomen and the urinary bladder.

**vesicocele** (*ves'-ik-o-sēl*) [*vesico-;* κήλη, hernia]. Hernia of the bladder; cystocele.

**vesicocervical** (*ves-ik-o-ser'-vik-al*) [*vesico-; cervix*]. Pertaining to the cervix uteri and the urinary bladder.

**vesicoclysis** (*ves-ik-ok'-lis-is*) [*vesico-;* κλυσις, a washing out]. The injection of fluid into the bladder.

**vesicofixation** (*ves-ik-o-fiks-a'-shun*) [*vesico-; fixation*]. 1. The operation of suturing the bladder to the abdominal wall. 2. The surgical attachment of the uterus to the bladder.

**vesicoprostatic** (*ves-ik-o-pros-tat'-ik*) [*vesico-; prostate*]. Pertaining to the prostate gland and the urinary bladder.

**vesicopubic** (*ves-ik-o-pu'-bik*) [*vesico-; pubis*]. Pertaining to the urinary bladder and to the pubes.

**vesicorectal** (*ves-ik-o-rek'-tal*) [*vesico-; rectum*]. Pertaining to the bladder and the rectum.

**vesicosigmoid** (*ves-ik-o-sig'-moid*) [*vesico-; sigmoid*]. Pertaining to the urinary bladder and the sigmoid flexure.

**vesicosigmoidostomy** (*ves-ik-o-sig-moid-os'-tom-e*) [*vesico-; sigmoid;* στόμα, mouth]. The operation of forming a communication between the urinary bladder and the sigmoid flexure.

**vesicospinal** (*ves-ik-o-spi'-nal*) [*vesico-; spina*, spine]. Pertaining to the urinary bladder and the spinal cord.

**vesicotomy** (*ves-ik-ot'-o-me*) [*vesico-;* τέμνειν, to cut]. Incision of the bladder; cystotomy.

**vesicoumbilical** (*ves-ik-o-um-bil'-ik-al*) [*vesico-; umbilicus*]. Pertaining to the umbilicus and the urinary bladder.

**vesicourachal** (*ves-ik-o-u'-rak-al*). Relating to the bladder and the urachus.

**vesicoureteral** (*ves-ik-o-u-re'-ter-al*) [*vesico-; ureter*]. Pertaining to the urinary bladder and the ureter.

**vesicourethral** (*ves-ik-o-u-re'-thral*) [*vesico-; urethra*]. Pertaining to the bladder and the urethra.

**vesicouterine** (*ves-ik-o-u'-ter-in*) [*vesico-; uterus*]. Pertaining to the urinary bladder and the uterus.

**vesicouterovaginal** (*ves-ik-o-u-ter-o-vaj'-in-al*). Relating to the bladder, uterus, and vagina.

**vesicovaginal** (*ves-ik-o-vaj'-in-al*) [*vesico-; vagina*]. Pertaining to the bladder and the vagina.

**vesicovaginorectal** (*ves-ik-o-vaj-in-o-rek'-tal*) [*vesico-; vagina; rectum*]. Pertaining to the bladder, vagina, and rectum.

**vesicula** (*ves-ik'-u-lah*) [dim. of *vesica*, a bladder; *pl., vesiculæ*]. A vesicle. **v. fellis**, the gall-bladder. **v. Graafiana**. See *follicle, Graafian*. **vesiculæ Nabothii**. See *ovule* (2). **v. prostatica**, the sinus pocularis. **vesiculæ seminales**. See *vesicle, seminal*.

**vesicular** (*ves-ik'-u-lar*) [*vesicle*]. 1. Pertaining to or composed of vesicles. 2. Produced in vesicles, as *vesicular* breathing, *vesicular* murmur. **v. column**, a column of ganglion-cells at the base of the posterior horn of the spinal cord. **v. column, posterior**. See *column of Clarke*. **v. eczema**, eczema attended with the formation of vesicles. **v. murmur**, a fine, normal, inspiratory, auscultatory sound heard over the chest. **v. rale**, the crepitant rale.

**vesiculate** (*ves-ik'-u-lāt*). 1. Having a vesicle. 2. To become vesicular.

**vesiculated** (*ves-ik'-u-la-ted*) [*vesicle*]. Composed of vesicles.

**vesiculation** (*ves-ik-u-la'-shun*) [*vesicle*]. The formation of vesicles; the state of becoming vesiculated.

**vesiculectomy** (*ves-ik-u-lek'-to-me*) [*vesicula;* ἐκτομή, excision]. Resection, complete or partial, of the seminal vesicles.

**vesiculiferous** (*ves-ik-u-lif'-er-us*) [*vesicle; ferre*, to bear]. Bearing or having vesicles.

**vesiculiform** (*ves-ik'-u-li-form*) [*vesicula; forma*, form]. Having the form of a vesicle.

**vesiculitis** (*ves-ik-u-li'-tis*) [*vesicle;* ιτις, inflammation]. Inflammation of the seminal vesicles.

**vesiculobronchial** (*ves-ik-u-lo-brong'-ke-al*) [*vesicle; bronchus*]. Both vesicular and bronchial.

**vesiculocavernous** (*ves-ik-u-lo-kav'-er-nus*) [*vesicle; cavernous*]. Both vesicular and cavernous.

**vesiculopapular** (*ves-ik-u-lo-pap'-u-lar*) [*vesicle; papule*]. Consisting of vesicles and papules.

**vesiculopustular** (*ves-ik-u-lo-pus'-tu-lar*) [*vesicle; pustule*]. Consisting of vesicles and pustules.

**vesiculose** (*ves-ik'-u-lōs*). Vesiculiform.

**vesiculotomy** (*ves-ik-u-lot'-om-e*). [*vesicle;* τομή, a cutting]. Division of a seminal vesicle.

**vesiculotubular** (*ves-ik'-u-lo-tu'-bu-lar*) [*vesiculo-; tubulus*, a tubule]. Both vesicular and tubular (a qualification for certain respiratory sounds).

**vesiculotympanitic** (*ves-ik-u-lo-tim-pan-it'-ik*) [*vesicle; tympanum*]. Both vesicular and tympanitic.

**vesipyrine** (*ves-ip-i'-rin*). Acetyl saloi, used like saloi in influenza, rheumatism, and neuralgia.

**vespajus** (*ves-pa'-jus*) [*vespa*, a wasp]. A follicular, suppurative inflammation of the hairy part of the scalp.

**vessel** (*ves'-el*) [Fr., from *vasculum*, a vessel]. A receptacle for fluids, especially a tube or canal for conveying blood or lymph. **v.s, absorbent**, the lymphatics and lacteals. **v.s, chyliferous**, absorbent vessels extending from the intestinal walls to the thoracic duct. **v.s, hemorrhoidal**, varicose veins of the rectum. **v.s, Jungbluth's**, nutrient vessels lying immediately beneath the amnion and disappearing usually at an early period of embryonic life. **v.s, lacteal**. Same as *v.s, chyliferous*. **v.s, radicular**, branches of vertebral arteries supplying cerebral nerve-roots. **v.s, umbilical**, the umbilical arteries and veins.

**vestibular** (*ves-tib'-u-lar*) [*vestibule*]. Pertaining to a vestibule.

**vestibulate** (*ves-tib'-u-lāt*) [*vestibulum*]. Having a vestibule; vestibular.

**vestibule** (*ves'-tib-ūl*) [*vestibulum*, a porch]. An approach; an antechamber. **v., aortic**, the space formed by the left ventricle adjoining the root of the aorta. **v. of the ear**, the oval cavity of the internal ear, which forms the entrance to the cochlea. **v. of the mouth**, that portion of the mouth outside of the teeth. **v. of the nose**, the anterior part of the nose. **v. of the vagina, v. of the vulva**, a triangular space below the clitoris and between the nymphæ.

**vestibulotomy** (*ves-tib-u-lot'-o-me*) [*vestibule;* τομή, a cutting]. Surgical operation, making an opening into the vestibule of the labyrinth.

**vestibulourethral** (*ves-tib-u-lo-u-re'-thral*). Relating to the bulbi vestibuli and to the urethra.

**vestibulum** (*ves-tib'-u-lum*). See *vestibule*. Generally applied to the vestibule of the ear.

**vestige** (*ves'-tij*) [*vestigium*, footprint]. A trace or remnant of something formerly present or more fully developed.

**vestigial** (*ves-tij'-e-al*) [*vestige*]. Of the nature of a vestige or trace; rudimentary. **v. fold**, a fibrous band of the pericardium representing the obliterated left innominate vein.

**vestigium** (*ves-tij'-e-um*) [L., a foot-print; *pl., vestigia*]. An anatomical relic of fetal or embryonic life. Thus, the thymus gland becomes in adults a *vestigium*.

**vestosol** (*ves'-to-sol*). An ointment said to contain formaldehyde, boric acid and zinc oxide.

**vesuvin** (*ves-u'-vin*) [*Vesuvius*, a volcano near Naples]. Brownish-brown, triamidobenzol; a stain used in microscopy.

**veta** (*ve'-tah*) [Sp.]. Mountain sickness.

**veterinarian** (*vet-er-in-a'-re-an*) [see, *veterinary*]. One who practises veterinary medicine.

**veterinary** (*vet'-er-in-a-re*) [*veterinarius*, of, or belonging to beasts of burden]. Pertaining to domestic animals. **v. medicine**, medicine as applied to the domestic animals.

**vetrinol** (*vet'-rin-ol*). An unguentine for veterinary use.

**vetol** (*vet'-ol*). A yohimbine for veterinary use.

**V. F.** Abbreviation for visual field.

**via** (*vi'-ah*) [L.; *pl., viæ*]. A way.. **viæ naturales**, the natural passages. See *prima viæ*.

**viability** (*vi-ab-il'-it-e*) [*viable*]. The state of being viable.

**viable** (*vi'-ab-l*) [Fr. *vie*, from L.*vita*, life]. Capa-

ble of living; likely to live; applied to a fetus capable of living outside of the uterus.
**vial** (*vi'-al*) [φιάλη, a shallow cup]. A small glass bottle.
**vibex** (*vi'-beks*) [L.; *pl., vibices*]. A linear ecchymosis.
**vibrate** (*vi'-brāt*) [*vibrare*, to shake]. To move to and fro.
**vibratile** (*vi'-bra-til*) [see *vibrate*]. Moving to and from; vibrating.
**vibration** (*vi-brā'-shun*) [see *vibrate*]. The act of moving to and fro.
**vibrator** (*vi'-bra-tor*) [see *vibrate*]. A device for conveying mechanical vibration to a part.
**Vibrio** (*vib'-re-o*) [see *vibrate*]. A genus of *Schizomycetes*. See *Spirillum*. V. choleræ, the spirillum of Asiatic cholera.
**vibrissa** (*vib-ris'-ah*) [L.; *gen*. and *pl., vibrissæ*]. One of the hairs near the opening of the anterior nares.
**vibromassage** (*vi-bro-mas-ahzj'*). 1. See *massage, vibratory*. 2. A form of pneumomassage for the ear.
**vibrometer** (*vi-brom'-et-er*) [*vibrate*; μέτρον, a measure]. A device for the treatment of deafness, by which rapid vibrations of the membrana tympani are induced.
**vibrophone** (*vi'-bro-fōn*) [*vibrate*; φωνή, sound]. A device for applying sound massage to the membrana tympani in treatment of deafness.
**vibrotherapeutics** (*vi-bro-ther-ap-ū'-tiks*). The therapeutic application of vibration.
**viburnin** (*vi-bur'-nin*) [*Viburnum*]. A precipitate from a tincture of *Viburnum opulus*; antispasmodic, antiperiodic, expectorant, tonic. Dose 1 to 3 grains.
**Viburnum** (*vi-bur'-num*). A genus of the *Caprifoliaceæ*. The dried bark of V. opulus, cranberrytree, cramp-bark, is official in the U. S. P,. and is used in dysmenorrhea, scurvy, asthma, etc. The dried bark of the root of V. prunifolium or of V. lentago is official in the U. S. P., and is used in dysmenorrhea, threatened abortion, menorrhagia, etc. V. opulus, fluidextract of (*fluidextractum viburni opuli*, U. S. P.). Dose 1–2 dr. (4–8 Cc.). V. prunifolium, fluidextract of (*fluidextractum viburni prunifolii*, U. S. P.). Dose 1–2 dr. (4–8 Cc.).
**vicarious** (*vi-ka'-re-us*) [*vices*, changes]. Taking the place of something else; of a habitual discharge occurring in an abnormal situation, as *vicarious* menstruation.
**vice** (*vis*) [L., *vitium*]. 1. A physical defect, as a *vice* of conformation. 2. A moral defect; a bad habit.
**Vichy water** (*ve-she*). A mildly laxative and antacid mineral water obtained from Vichy, in France, and used in rheumatic and gouty conditions and in disorders of the liver.
**vicious** (*vish'-us*) [from *vice*]. Defective, faulty. v. union, the union of the ends of a fractured bone with deformity.
**vicocoa** (*vi-ko'-ko*). A combination of malt, kola, and cocoa.
**Vicq d'Azyr's band, V. d'A.'s stripe** (*vik-dah-zēr*) [Félix Vicq d'Azyr, French anatomist, 1748–1794]. See *Baillarger's layer*. V. d'A.'s bundle, a tract of nerve-fibers passing from the corpus albicans to the anterior nucleus of the optic thalamus. V. d'A.'s foramen, the foramen cæcum at the upper end of the median groove of the anterior surface of the oblongata. V. d'A.'s line, V. d'A.'s band. V. d'A.'s operation, rapid tracheotomy through the cricothyroid membrane.
**Victoria blue** (*vik-to'-re-ah*) [after Queen *Victoria* of England, 1819–1901]. A blue stain used in histology. It is phenyltetramethyl-amidoalphanaphthyldiphenylcarbinol hydrochloride. V. orange, a yellow stain used in histology: it is a salt of dinitrocresol.
**victorium** (*vik-to'-re-um*) [after Queen *Victoria* of England, 1819–1901]. A supposed element of the yttrium-cerium group discovered by Sir William Crookes; its existence is not proved.
**Vidal's operation** (*ve'-dal*) [Auguste Théodore *Vidal* de Cassis, French surgeon, 1803–1856]. Subcutaneous ligation of the veins involved, in treatment of varicocele.
**Vidian artery** (*vid'-e-an*) [relating to, described by, or named after Vidus *Vidius* (Guido Guidi), Italian anatomist, 1545–1569]. A branch of the internal maxillary artery; it passes through the Vidian canal and is distributed to the pharynx and Eustachian tube. V. canal, a canal of the sphenoid bone at the base of the internal pterygoid plate, opening anteriorly into the sphenomaxillary fossa, and posteriorly into the lacerated foramen. It transmits the Vidian nerve and vessels. V. nerve, a branch given off from the sphenopalatine ganglion.
**vieirin** (*vi-e'-ir-in*). A principle from the bark of *Remijia vellosii*, one of the cuprea-barks. It is an amorphous white substance with an aromatic odor and bitter taste. It is soluble in alcohol and chloroform, and is used as a febrifuge instead of quinine. Dose 1–4 gr. (0.065–0.25 Gm.) several times daily.
**Vienna paste** (*ve-en'-ah*). See *paste, Vienna*.
**Vienna powder**. Potassa cum calce.
**Vierordt's hemotachometer** (*fēr'-ort*) [Karl *Vierordt*, German physiologist, 1818–1884]. An instrument for measuring the rate of flow of the blood.
**Vieussens' annulus** (*ve-oo-son'*) [Raymond *Vieussens*, French anatomist, 1641–1715]. A small nerve passing between the middle and lower cervical, or first dorsal, ganglia and forming a loop around the subclavian artery. Syn., *ansa subclavia*. V.'s centrum ovale. See *centrum ovale Vieussenii*. V.'s ganglion, the solar plexus. V.'s isthmus, V.'s ring, the annulus ovalis. V.'s valve. See *valve of Vieussens*. V.'s ventricle, the fifth ventricle.
**vigil** (*vij'-il*) [L.]. Watchful wakefulness. v., coma. See *coma vigil*.
**vigilambulism** (*vij-il-am'-bū-lizm*) [*vigil; ambulare*, to walk]. Ambulatory automatism in the waking state.
**vigintinormal** (*vij-in-te-nor'-mal*) [*viginti*, twenty; *norma*, rule]. Possessing one-twentieth of what is normal.
**Vignal's cells** (*vēn'-yal*) [Guillaume *Vignal*, French physiologist, contemporary]. Embryonic connective-tissue (mesenchymatous) cells lying upon the axis-cylinders of which the fetal nerve-fibers are made up. At first globular, these cells elongate and gradually fuse until they form a complete sheath around the axis-cylinder.
**Vigo plaster** (*ve'-go*) [Giovanni da *Vigo*, Italian surgeon, circ. 1500]. A plaster containing mercury, turpentine, wax, lead-plaster, and other substances. V.'s powder, red oxide of mercury.
**vigoral** (*vig'-o-ral*). A proprietary preparation of pulverized beef and beef extract.
**Vigouroux's sign** (*ve-goo-roo'*) [Auguste *Vigouroux*, French neurologist]. Diminished resistance of the skin to electric stimulation in exophthalmic goiter.
**Villard's button** (*ve-lar'*) [—— *Villard*, French surgeon]. A modification of Murphy's button.
**Villatte's liquor, or solution**. A preparation used for injecting into carious bones, consisting of zinc sulphate and copper sulphate, each, 15 gr.; lead subacetate solution, ½ dr., and dilute acetic acid, 3½ dr.
**villi** (*vil'-li*). Plural of *villus*, a tuft. Tufts of hair, or hair-like processes or projections of a mucous membrane giving it a velvety appearance. v., arachnoid, Pacchionian bodies. v. of the chorion, fringes growing from the external surface of the vitelline membrane, finally covering the entire chorion. v., intestinal, minute, highly vascular tongue-like processes projecting from the free surface of the mucous membrane of the small intestine throughout its whole extent. They are larger and more numerous in the duodenum and jejunum and are fewer and smaller in the ileum. They constitute the chief organs of absorption of fatty emulsions. v. pericardiaci, villi upon the ental surface of the pericardium. v. peritoneales, villi upon the free surface of the peritoneum. v. pleurales, villi on the parietal pleura. v., synovial, small, tongue-like processes projecting from the fringes of synovial membranes.
**villiferous** (*vil-if'-er-us*) [*villus; ferre*, to bear]. Furnished with tufts of hairs or villi.
**villiform** (*vil'-if-orm*) [*villus; forma*, form]. Villose in form.
**villiplacental** (*vil-ip-las-en'-tal*) [*villus; placenta*]. Having a tufted or villous placenta.
**villitis** (*vil-i'-tis*) [*villus; ιτις*, inflammation]. Inflammation of the cushion or soft part of the wall of a horse's hoof. See *coronitis*.
**villoid** (*vil'-oid*) [*villus; εἶδος*, like]. Villiform.
**villoma** (*vil-o'-mah*) [*villus; ὄμα*, tumor]. A villous tumor.
**villose, villous** (*vil'-ōs, vil'-us*) [*villus*]. Pertaining to a villus; covered with villi; characterized by the formation of villus-like projections.

**villositis** (vil-os-i'-tis) [villus; -ιτις, inflammation]. Inflammation of the villous surface of the placenta.
**villosity** (vil-os'-it-e) [villus]. 1. The state of being villous. 2. A proliferation of a membranous surface.
**villus** (vil'-us) [L., "a tuft of hair"; pl., villi]. 1. One of the minute club-shaped projections from the mucous membrane of the intestine, consisting of a lacteal vessel, an arteriole, and a vein, inclosed in a layer of epithelium. 2. One of the vascular tufts of the chorion.
**vin.** Abbreviation of Latin vinum, wine.
**vin** (van) [Fr.]. Wine.
**vina** (vi'-nah) [L., pl. of vinum; wine]. Wines.
**v. medicata**, medicated wines.
**vinasse** (ve-nas) [Fr.]. Potash obtained from the residue of the wine-press.
**Vincent's angina** [H. Vincent, French physician, 1862– ]. Diphtheroid angina due to the bacillus of pseudodiphtheria. Syn., ulceromembranous angina.
**V.'s sign.** See Argyll-Robertson pupil.
**Vincetoxicum** (vin-se-toks'-ik-um) [vincere, to subdue; toxicum, poison]. A genus of the order Asclepiadaceæ. The root of V. officinale, swallowwort, indigenous to Europe, is used as an emetic and in menstrual disorders.
**vincula accessoria tendinum** [L., pl. of vinculum, a band]. The slender tendinous filaments which connect the phalanges with the flexor tendons.
**vinculum** (vin'-ku-lum) [vinculum, a band, fetter]. A ligament, frenum. **v. linguæ**, frenum of the tongue. **v. linguæ**, the lateral prolongation of the lingula of the cerebellum. **v. præputii**, the frenum of the prepuce. **v. umbilicale**, the umbilical cord.
**vinegar** (vin'-e-gar) [Fr., vin, wine; aigre, sour]. 1. An impure solution of acetic acid, obtained by acetous fermentation of wine, beer, cider, etc., or by the dry distillation of wood. It is used as a condiment. 2. A solution of a medicinal substance in vinegar or acetic acid. Only two vinegars are official in the U. S. P,: vinegar of opium (acetum opii, U. S. P.) and vinegar of squill (acetum scillæ, U. S. P.). **v. of lead**, solution of lead subacetate. **v., radical**, glacial acetic acid.
**vinic** (vi'-nik) [vinum]. Pertaining to wine; obtained from wine.
**vinometer** (vi-nom'-e-ter) [vinum, wine; μέτρον, measure]. An instrument for measuring the percentage of alcohol in a liquor.
**vinopyrine** (vi-no-pi'-rin). Trade name of an antipyretic, said to be paraphenetidin bitartrate.
**vinous** (vi'-nus) [vinum]. Having the nature of wine; containing wine.
**vinum** (vi'-num) [L.: gen., vini; pl., vina]. Wine. The fermented juice of fruits, especially that of grapes. See wine. There are 10 official vina, of which 8 are medicated. **v. absinthiatum**, wormwood-wine; made by macerating artemisia absinthium in white wine. **v. album**, white wine, an alcoholic liquid made by fermenting the juice of the fresh grape; it contains from 10 to 14 per cent. by weight of absolute alcohol. **v. antimonii**, wine of antimony. Dose ♏ x-xxx. **v. aromaticum** consists of strong white wine 94 per cent., with one per cent. each of lavender, origanum, peppermint, rosemary, sage, and wormwood. **v. cocæ**, wine of coca, contains 6.5 per cent. of fluidextract of coca. Dose ʒ iv. **v. colchici seminis**, wine of colchicum-seed. Dose ♏ x-xxx. **v. ergotæ**, wine of ergot. Dose ʒ j-iij. **v. ferri**, wine of iron. Dose ʒ j-ij. **v. ferri amărum**, bitter wine of iron. Dose ʒ j-iij. **v. ipecacuanhæ**, wine of ipecac. Dose ♏ v-x. **v. opii**, 1.3 to 1.5 gm. morphine in 100 cc. Dose ♏ v-xv. **v. portense**, port wine, is fortified with 25-30 per cent.; and **v. xericum**, sherry wine, until it contains about 25 per cent. of alcohol. **v. rubrum**, red wine, an alcoholic liquid made by fermenting the juice of the fresh, colored grapes, the fruit of Vitis vinifera, in the presence of their skins. It contains from 10-14 per cent. by weight of absolute alcohol. **vini gallici spiritus**, brandy.
**vioform** (vi'-o-form). See iodochloroxyquinolin.
**Viola** (vi'-o-lah) [L., "violet"]. A genus of plants of the Violarieæ, including V. tricolor, heart's-ease, V. odorata, V. cucullata. V. odorata, as well as other species, is used in bronchitis.
**violation** (vi-o-la'-shun) [violare, to ravish]. Rape. Sometimes used to express the fact of coitus without force, but by deception, with the weak-minded, etc.
**violet** (vi'-o-let) [viola]. 1. One of the colors of the spectrum, very closely resembling the purple of violets and possessing the greatest refrangibility of the spectral colors. 2. A violet dyestuff. **v. blindness**, retinal insensibility to violet tints. **v. gentian-**, a violet anilin dye used for staining in histological and bacteriological work. **v., methyl-.** See methyl-violet.
**violine** (vi'-o-lĕn) [viola]. An emetocathartic alkaloid from Viola tricolor.
**violinist's cramp, violin-player's cramp.** An occupation-neurosis occurring in violin-players, and characterized by spasm of the fingers used in playing.
**viperine** (vi'-pur-in) [viperinus]. 1. Pertaining to a viper. 2. Virginia snake-root. 3. A toxalbumin extracted from the venom of vipers.
**viraginity** (vir-aj-in'-it-e) [virago, a bold man-like woman]. A form of sexual perversion in which the female individual is essentially male in her feelings and tastes.
**Virchow's angle** (fĕr'-ko) [Rudolf Virchow, German pathologist, 1821–1902]. In craniometry, the angle formed by the union of a line joining the naso-frontal suture and the most prominent point of the lower edge of the superior alveolar processes, and a line joining the superior border of the external auditory meatus and the lower border of the orbit. **V.'s axiom**, "omnis cellula e cellula," every cell (is derived) from a cell. **V.'s bone-cells**. The cells found in lacunæ of bone. **V.'s corpuscles**. See Toynbee's corpuscles. **V.'s crystals**, bright yellow or orange-colored crystals of hematoidin sometimes found in extravasated blood. **V.'s degeneration**, amyloid degeneration. **V.'s disease**, leontiasis ossea. **V.'s gland**, the jugular gland; a lymphatic gland situated behind the clavicular insertion of the sternomastoid. **V.'s granulations**, granulations consisting principally of ependymal and neuroglia fibers, commonly found in the walls of the ventricles of the brain in progressive general paralysis. **V.'s law**, the cellular elements of a tumor are derived from preexisting tissue-cells. **V.'s line**, the line extending from the root of the nose to the lambda.
**Virchow-Hassall's bodies.** See Hassall's bodies.
**Virchow-Holder angle.** Virchow's angle.
**Virchow-Robin's space.** An adventitious lymph-space found between the adventitia and media of the blood-vessels of the brain and communicating with the subarachnoid space.
**virgin** (vur'-jin) [virgo, a maid]. A person who has never had sexual intercourse.
**virginal** (vur'-jin-al) [virgin]. Pertaining to virginity. **v. membrane**, the hymen.
**Virginia creeper.** Vitis hederacea (Ampelopsis quinquefolia); the leaves and twigs are alterative, tonic, astringent, and expectorant. **V. snake-root.** See Serpentaria.
**virginity** (vur'-jin'-it-e) [virgin]. The condition of being a virgin.
**viridin** (vir'-id-in) [viridis, green]. An oily substance, C11H18N, derived from coal-tar.
**viridine** (vir-id-in) [viridis, green]. An alkaloid obtained from Veratrum viride, and supposed to be identical with jervine.
**virile** (vir'-il) [virilis, from vir, a man]. Pertaining to or characteristic of the man. **v. member**, the penis.
**virilescence** (vir-il-es'-ens) [virile]. The assumption of male characters by an aged woman; the growth of a beard, the development of a manly voice on the part of a woman after the menopause.
**virilia** (vir-il'-e-ah) [plural of virilis, manly]. The male generative organs.
**virilin** (vir-il'-in). An aphrodisiac preparation composed of yohimbine, strychnine, and glycerophosphates.
**virility** (vir-il'-it-e) [virile]. The condition of being virile; procreative power.
**viripotent** (vir-ip'-o-tent) [vir, a man; potens, able; hence, ripe for a man]. Marriageable. The term should be used of the female only.
**virogen** (vi'-ro-jen). A preparation said to be composed of glycerophosphates and soluble protein of milk.
**virol** (vi'-rol). A proprietary substitute for cod-liver oil.
**virola-tallow.** An oil or fat from the seeds of Myristica sebifera; a remedy for rheumatism.
**virose, virous** (vi'-rōs, vi'-rus) [virosus, poisonous]. Poisonous; having a poisonous taste or smell.
**virtual cautery** (vur'-tū-al kor'-ter-e). Cautery by

the application of caustics; term used in opposition to *actual cautery.*
**virtual focus** (*vur'-tū-al fo'-kus*). See *focus, negative.*
**virtual image** (*vur'-tū-al im'-āj*). The image formed by rays prolonged after reflection.
**virulence** (*vir'-oo-lens*) [*virus*]. Malignity; noxiousness; infectiousness. The disease-producing power of a microorganism.
**virulent** (*vir'-oo-lent*) [*virus*]. Having the nature of a poison.
**viruliferous** (*vir-oo-lif'-er-us*) [*virus; ferre,* to carry]. Containing or conveying a virus.
**virulin** (*vir'-oo-lin*). Antiphagin. A constituent of virulent bacteria which enables them to resist the action of phagocytes.
**virus** (*vi'-rus*) [L.]. 1. The poison of an infectious disease, especially one found in the secretion or tissues of an individual or animal suffering from an infectious disease. 2. Vaccine-lymph. **v., attenuated,** a virus whose pathogenicity has been lessened by unfavorable conditions of cultivation. **v. fixé,** or **v., fixed, v.** of rabies which has been rendered as virulent, as possible. **v., humanized,** vaccine-lymph taken from the vaccine pustule of a human subject. **v., organized,** a pathogenic microorganism. **v., street, v.** of rabies ordinarily found in rabid dogs. **v., unorganized,** a poisonous chemical substance developed in the body by the action of the body-cells or of microorganisms.
**vis** [L.: *pl., vires*]. Force; energy; power. **v. a fronte,** a force that attracts. **v. a tergo,** a force that pushes something before it. **v. conservatrix,** the healing power of nature. **v. formativa,** energy manifesting itself in the formation of new tissue to replace that which has been destroyed. **v. inertiæ,** that force by virtue of which a body at rest remains at rest. **v. medicatrix naturæ,** the healing power of nature apart from medicinal treatment. **v. vitæ,** vital force.
**viscera** (*vis'-er-ah*). Plural of *viscus.*
**viscerad** (*vis'-er-ad*) [*viscera; ad,* to]. Toward the viscera.
**visceral** (*vis'-er-al*) [*viscera*]. Pertaining to a viscus or to viscera. **v. arches and clefts,** four slit-like depressions with intermediate thickenings of the lateral wall of the cervical region of the embryo. **v. skeleton,** that part of the bony skeleton which encloses viscera, such as the pelvis, ribs, and sternum.
**visceralgia** (*vis-er-al'-je-ah*) [*viscera;* ἄλγος, pain] Pain in a viscus.
**visceralism** (*vis'-ur-al-izm*). The doctrine that all disease has its origin in the viscera.
**viscerimotor** (*vis-er-im-o'-tor*) [*viscera; motor*]. Conveying motor impulses to a viscus.
**visceripericardial** (*vis-er-ip-er-ik-ar'-de-al*) [*viscera; pericardium*]. Relating to the pericardium and the viscera.
**viscero-** (*vis-er-o-*) [*viscera*]. A prefix denoting pertaining to the viscera.
**visceroinhibitory** (*vis-er-o-in-hib'-it-o-re*). Inhibiting the movements of viscera.
**visceromotor** (*vis-er-o-mo'-tor*). Visceromotor.
**visceroparietal** (*vis-er-o-pa-ri'-et-al*) [*viscero-; paries,* wall]. Pertaining to the viscera and the abdominal wall.
**visceropericardial** (*vis-er-o-per-ik-ar'-de-al*). See *visceripericardial.*
**visceroperitoneal** (*vis-er-o-per-it-on-e'-al*) [*viscero-; peritoneum*]. Relating to the abdominal viscera and the peritoneum.
**visceropleural** (*vis-er-o-ploo'-ral*) [*viscero-;* πλευρά, side]. Pertaining to the thoracic viscera and the pleura; pleuroviscéral.
**visceroptosis** (*vis-er-op-to'-sis*) [*viscero-;* πτῶσις, a falling]. Abdominal ptosis; Glénard's disease.
**viscerosensory** (*vis-er-o-sen'-so-re*) [*viscero-; sensory*]. Relating to sensation in the viscera.
**visceroskeletal** (*vis-er-o-skel'-et-al*) [*viscero-; skeleton*]. Pertaining to the visceral skeleton.
**viscerosomatic** (*vis ᴇ s o mat' ih*) [*viscero-;* σῶμα, body]. Relating to the viscera and the body.
**viscid** (*vis'-id*) [*viscidus,* sticky]. Sticky; adhesive; glutinous.
**viscidity** (*vis-id'-it-e*) [*viscid*]. The state of being viscid. Same as *viscosity.*
**viscin** (*vis'-in*) [*viscum*]. A mucilaginous extract of mistletoe.
**viscometer** (*vis-kom'-et-er*). See *viscosimeter.*
**viscose** (*vis'-kōs*) [*viscum*]. 1. See *viscous.* 2. A gummy product of viscous fermentation.

**viscosimeter** (*vis-cos-im'-et-ur*) [*viscosity;* μέτρον, a measure]. An apparatus for determining the degree of viscosity of a fluid, especially blood.
**viscosity** (*vis-kos'-it-e*). The state of being viscous.
**viscous** (*vis'-kus*). 1. Viscid. 2. Pertaining to a viscus or internal organ.
**Viscum** (*vis'-kum*) [L.]. A genus of plants, including the mistletoe, of the order *Loranthaceæ,* growing as parasites upon trees. *V. album,* European mistletoe, and *V. flavescens,* or *Phoradendron flavescens,* American mistletoe, contain a viscid principle, *viscin,* which is the chief constituent.
**viscus** (*vis'-kus*) [L.: *pl., viscera*]. Any one of the organs inclosed within one of the four great cavities, the cranium, thorax, abdominal cavity, or pelvis; especially one within the abdominal cavity.
**visibility** (*vis-ib-il'-it-e*). The state of being visible.
**visible** (*vis'-ib-l*) [*vision*]. Capable of being seen.
**vision** (*vizh'-un*) [*videre,* to see]. The act of seeing; sight. **v., binocular.** See *binocular vision.* **v., central,** vision with the macula lutea. **v., chromatic,** pertaining to the color sense. **v., direct.** See *v., central.* **v., double.** See *diplopia.* **v., field of.** See *field.* **v., indirect,** vision with other parts of the retina than the macula. **v., multiple,** a condition of the eye wherein more than one image of an object is formed upon the retina. **v., qualitative,** vision in which there is ability to distinguish objects. **v., quantitative,** mere perception of light. **v., solid, v., stereoscopic,** the perception of relief or depth of objects obtained by binocular vision.
**visit** (*vis'-it*) [*videre,* to see]. A professional call upon a patient.
**viskolein** (*vis-ko'-le-in*). A proprietary preparation said to be antiseptic and antipyretic.
**visual** (*vis'-ū-al*) [*vision*]. Pertaining to vision. **v. angle.** See *angle.* **v. axis.** See *axis.* **v. cells,** the rods and cones and external nuclear layer of the retina. **v. field,** the area within which objects may be seen. **v. purple,** a pigmentary substance in the retina reacting to light in a peculiar manner, and thought to be intimately connected with vision. See *rhodopsin.*
**visualization** (*vis-ū-al-iz-a'-shun*). The act of rendering a mental perception visible to the eye; the recalling of a mental image with such distinctness that it seems reality.
**visuoauditory** (*vis-ū-o-aw'-dit-o-re*) [*vision; audire,* to hear]. Pertaining to hearing and seeing; of nerve-fibers, connecting the visual and auditory centers.
**visuometer** (*vis-ū-om'-et-er*) [*visus,* vision; μέτρον, a measure]. An apparatus for determining range of vision.
**visus** (*vi'-sus*) [*videre,* to see]. Vision. **v. acrior,** nyctalopia. **v. acris,** acuteness of vision. **v. brevior,** myopia. **v. coloratus,** chromatopsia. **v. debilitas,** asthenopia. **v. decoloratus,** achromatopsia. **v. dimidiatus,** hemiopia. **v. diurnus,** hemeralopia. **v. duplicatus,** diplopia. **v. habetudo,** amblyopia. **v. juvenum,** myopia. **v. lucidus,** photopsia. **v. muscarum,** specks before eyes. **v. senilis,** presbyopia.
**vita** (*vi'-tah*) [L.]. Life.
**vitafer** (*vi'-ta-fer*). A proprietary preparation, containing casein and glycerophosphates, and used as a tonic and nutrient.
**vital** (*vi'-tal*) [*vita,* life]. Pertaining to life. **v. capacity,** the volume of air that can be expelled from the lungs after a full inspiration. **v. center,** the respiratory center in the medulla. **v. knot,** the respiratory center in the medulla. **v. principle,** the energizing principle on which individual life depends. **v. signs,** respiration, pulse, and temperature. **v. statistics,** statistics of births, deaths, marriages, and diseases in a community.
Vitali's test (*ve-tah'-le*) [Dioscoride *Vitali,* Italian physician]. 1. *For alkaloids:* The addition of sulphuric acid, potassium chlorate, and an alkaline sulphide will give various color reactions with an alkaloid. 2. *For atropine:* After evaporation with fuming nitric acid and moistening with alcoholic solution of potassium hydroxide, atropine causes a violet color which changes to red. 3. *For bile pigments in the urine:* Add to the liquid a few drops of a potassium nitrite solution and then some dilute sulphuric acid. A beautiful green color will be produced, changing to red or blue, and finally to yellow. 4. *For pus in the urine:* The urine is acidified with acetic acid, then filtered; to the filtrate a small

quantity of guaiacum is added; in the presence of pus a dark blue color results.
**vitalism** (*vi'-tal-ism*) [see *vital*]. The doctrine that ascribes the phenomena exhibited by living organisms to the action of a vital force distinct from mechanical or chemical force.
**vitalist** (*vi'-tal-ist*) [see *vital*]. A believer in vitalism.
**vitality** (*vi-tal'-it-e*) [*vita*, life]. The vital force, or principle of life; also the condition of having life; vigor; activity.
**vitalize** (*vi'-tal-iz*) [*vita*, life]. To endow with life.
**vitals** (*vi'-talz*) [see *vital*]. The organs essential to life.
**vitamine** (*vi'-tam-ēn*) [*vita*, life; *amine*]. A substance, belonging to a group of organic bases of unknown composition, which is present in small quantities in food, and is necessary for the normal processes of metabolism; the absence or insufficiency of these substances is supposed to be the cause of beriberi, pellagra, rickets, and scurvy.
**vitaminosis** (*vi-tam-in-o'-sis*) [*vitamine; νόσος*, disease]. An indefinite term used to include the diseases supposed to be due to deficiency, scurvy, pellagra, beriberi, and rickets.
**vitellary** (*vit'-el-a,re*) [*vitellus*, yolk]. Pertaining to the vitellus.
**vitellicle** (*vit-el'-ik-l*) [*vitellus*, yolk]. The yolksac; umbilical vesicle.
**vitellin** (*vit-el'-in*) [*vitellus*, yolk]. A globulin found in egg-yolk.
**vitelline** (*vit-el'-īn*) [*vitellus*, yolk]. Pertaining to the vitellus or yolk. **v. artery**, an artery passing from the yolk-sac to the primitive aorta of the embryo. **v. duct**, the omphalomesaraic duct. **v. membrane**, the true membrane of the ovum, lying inside of the zona pellucida. **v. veins**, veins returning the blood from the yolk-sac to the primitive heart of the embryo.
**vitellolutein** (*vit-el-o-lū'-te-in*) [*vitellus, luteus*, golden yellow]. A yellow dye from the lutein of eggs.
**vitellomesenteric** (*vit-el-o-mes-en-ṯer'-ik*) [*vitellus; mesentery*]. Omphalomesenteric.
**vitellorubin** (*vit-el-o-roo'-bin*) [*vitellus; ruber,* red]. A reddish pigment obtained from the yolk of egg.
**vitellose** (*vit-el-ōs'*) [*vitellus*]. A protease obtained from vitellin.
**vitellus** (*vit-el'-us*) [L.]. A yolk; specifically, the yolk of the egg of the common fowl, *Gallus domesticus*. **v. ovi**, the yolk of an egg.
**vitiation** (*vish-e-a'-shun*) [*vitiare*, to corrupt]. 1. The contamination of any substance. 2. Lessening of efficiency or utility.
**vitiligines** (*vit-i-lij'-in-ēz*) [pl. of *vitiligo*]. The lineæ albicantes.
**vitiligo** (*vit-il-i'-go*) [L.]. Piebald skin, a disease of the skin characterized by a disappearance of the natural pigment, occurring in patches and leaving whitish areas.
**vitiligoid** (*vit-il'-ig-oid*). Resembling vitiligo.
**-vitiligoidea** (*vit-il-ig-oi'-de-ah*) [*vitiligo; εἶδος*, like]. Xanthoma.
**vitium** (*vish'-e-um*) [L.: pl., *vitia*]. A vice, defect, disease, or fault. **v. caducum**, epilepsy. **v. cordis**, organic heart disease. **v. primæ conformationis**, a malformation.
**vitodynamic** (*vi-to-di-nam'-ik*) [*vita*, life; δύναμις, energy]. Relating to vital forces.
**vitreocapsulitis** (*vit-re-o-kap-sū-li'-tis*). See *hyalitis*.
**vitreodentine** (*vit-re-o-den'-tēn*) [*vitreus*, glassy; *dens*, a tooth]. A variety of dentine of particularly hard texture.
**vitreous** (*vit'-re-us*) [*vitrum*, glass]. Glassy; hyaline. The vitreous humor (*q. v.*). **v. body**. See *v. humor*. **v. chamber**, the portion of the eye posterior to the crystalline lens. **v. degeneration**, hyaline degeneration. **v. humor**, the transparent, jelly-like substance filling the posterior chamber of the eye. **v. membrane**, the inner membrane of the choroid. **v. table**, the hard, brittle, inner table of the skull.
**vitrescence** (*vit-res'-ens*) [*vitrum*, glass]. The condition of becoming hard and transparent like glass.
**vitreum** (*vit'-re-um*) [*vitreus*, glassy]. The vitreous body of the eye; same as vitreous humor.
**vitric** (*vit'-rik*). Relating to glass or any vitreous substance.
**vitrina** (*vit-ri'-na*) [*vitrum*, glass]. The vitreous

body. **v. auditoria**, or **v. auris**, the endolymph. **v. oculi**, the vitreous body.
**vitriol** (*vit'-re-ol*) [*vitriolum; vitrum*, glass]. A term formerly used to denote any substance having a glassy fracture or appearance. 1. Sulphuric acid, more commonly called oil of vitriol. 2. Any crystalline salt of sulphuric acid. **v., blue**, copper sulphate. **v., elixir of**, aromatic sulphuric acid. **v., green**, ferrous sulphate or copperas. **v., oil of**, sulphuric acid. **v., white**, zinc sulphate.
**vitriolated** (*vit'-re-ol-a-ted*). Containing vitriol; containing sulphur or sulphuric acid. **v. soda**, sodium sulphate. **v. tartar**, potassium sulphate.
**vitriolation** (*vit-re-o-la'-shun*). Conversion into glass or into a hyaloid structure.
**vitriolum cupri**. Blue vitriol.
**vitrum** (*vit'-rum*) [L.]. Glass.
**vitular** (*vit'-ū-lar*) [*vitulus*, calf]. Relating to a calf or to calving. **v. apoplexy**, apoplexy of cows occurring at parturition. **v. fever**. 1. Vitular apoplexy. 2. A fever following parturition in the cow.
**vividiffusion** (*viv-e-dif-ū'-zjun*) [*vivus*, living; *diffusion*]. The temporary flow of some of the arterial blood of a living animal through an extra-circuit of collodion tubes surrounded with physiological saltsolution, which circuit is inserted into one of the peripheral arteries. The blood thus circulates outside of the body through a dialyser, and back again into a vein.
**vivification** (*viv-if-ik-a'-shun*) [*vivus*, living; *facere*, to make]. The act of making alive or of converting into living tissue.
**viviparity** (*viv-ip-ar'-it-e*) [*vivus*, living; *parere*, to bring forth]. The bringing forth of living offspring; the state of being viviparous.
**viviparous** (*viv-ip'-ar-us*) [see *viviparity*]. Bringing forth the young alive—distinguished from *oviparous*.
**vivipation** (*viv-ip-a'-shun*) [*vivus*, alive; *parere*, to bring forth]. A form of generation in which the ovum matures in the uterus.
**viviperception** (*viv-ip-er-sep'-shun*) [*vivus*, living; *percipere*, to perceive]. The study of physiological processes without dissection or vivisection.
**vivisect** (*viv'-is-ekt*) [*vivus*, living; *secare*, to cut]. To practise or perform vivisection.
**vivisection** (*viv-is-ek'-shun*) [*vivus*, living; *secare*, to cut]. The dissection of a living animal; experimentation upon an animal while still alive.
**vivisectionist** (*viv-is-ek'-shun-ist*) [see *vivisection*]. A practiser or defender of vivisection; a vivisector.
**vivisector** (*viv-is-ek'-tor*) [see *vivisection*]. One who practises vivisection.
**vivisectorium** (*viv-is-ek-to'-re-um*) [L.]. A place or laboratory where vivisection is performed.
**Vleminckx's solution** (*flem'-inx*) [Jean François *Vleminckx*, Belgian physician, 1800–1876]. An application used in Austria and Germany for treating acne. It consists of lime, 1; sulphur, 2; water, 20. Slake the lime, add the sulphur, and boil to 12 parts.
**v. vocal** (*vo'-kal*) [*vox*, voice]. Pertaining to the voice; pertaining to the organs producing the voice. **v. area**, the portion of the glottis lying between the vocal cords. **v. bands**, **v. cords**. See under *larynx*. **v. fremitus**, the thrill conveyed to the hand when applied to the chest during speaking. **v. ligaments**, the true vocal cords. **v. resonance**, the resonance produced by the voice as heard on auscultating the lung.
**vocalis** (*vo-ka'-lis*). See *muscles, table of*.
**vodka** (*vod'-kah*). A kind of Russian whiskey.
**Vogt's point** (*fōht*) [Paul Friedrich Emmanuel *Vogt*, German surgeon, 1847–1885]. The point selected by Vogt for trephining in cases of traumatic meningeal hemorrhage. It is found at the intersection of a horizontal line two fingerbreadths above the zygomatic arch, with a vertical line a thumb's breadth behind the ascending sphenofrontal process of the zygoma.
**Vohsen-Davidsohn's sign**. See *Davidsohn's sign*.
**voice** (*vois*) [*vox*, a voice]. The sounds, especially articulate sounds, produced by the vibration of the vocal bands and modified by the resonance organs. **v., change of**, in the transition period of youth the voice loses its treble quality and (sometimes irregularly) assumes the qualities of the adult voice.
**void** [ME., *voiden*, to void]. To evacuate.
**Voigt's boundary-lines** (*foit*) [Christian August *Voigt*, Austrian anatomist, 1809–1890]. The lines which divide the regions of distribution of two peripheral nerve-trunks.
**Voillemier's point**. A point on the linea alba

6 to 7 cm. below a line drawn between the two anterior superior spines of the ilium; suprapubic puncture of the bladder is made at this point in fat or edematous subjects.
**Voit's nucleus** (*foit*) [Carl von *Voit*, German physiologist, 1831-1908]. An accessory nucleus of the corpus dentatum in the cerebellum.
**vola** (*vo'-lah*) [L.]. The palm of the hand or the sole of the foot. **v. manus**, the palm of the hand. **v. pedis**, the sole of the foot.
**volar** (*vo'-lar*) [*vola*]. Pertaining to the palm or the sole.
**volatile** (*vol'-at-il*) [*volatilis*, from *volare*, to fly]. Passing into vapor at ordinary temperatures; evaporating. **v. alkali**, ammonia. **v. liniment**, ammonia liniment. **v. oils**. See *essential oils*.
**volatilization** (*vol-at-il-i-za'-shun*) [see *volatile*]. The act of volatilizing.
**volatilize** (*vol'-at-il-iz*) [see *volatile*]. To convert into vapor by means of heat; to pass into vapor.
**Volhard's solution** (*föl'-hart*) [J. *Volhard*, German chemist, 1834– ]. Decinormal solution of potassium sulphocyanate. **V.'s volumetric method**, a method for estimating halogens by means of ammonium sulphocyanate.
**volition** (*vo-lish'-un*) [*volitio*, will]. The will or determination to act.
**volitional** (*vo-lish'-un-al*) [*volitio*, will]. Pertaining to volition. **v. insanity**, insanity characterized by perversions of the will, or by abulia or hyperbulia.
**Volkmann's canals** (*fölk'-mahn*) [Alfred Wilhelm *Volkmann*, German physiologist, 1800-1877]. Small canals found in the circumferential lamellæ of bones and transmitting blood-vessels; they communicate with the Haversian canals.
**Volkmann's deformity** (*fölk'-mahn*) [Richard *Volkmann*, German surgeon, 1830-1889]. Congenital tibiotarsal dislocation. **V.'s spoon**, a sharp spoon for removing diseased tissue.
**volley** (*vol'-e*) [*volare*, to fly]. A series of artificially induced muscle-twitches.
**volsella** (*vol-sel'-ah*) [*vellere*, to pluck]. A forceps having one or more hooks at the end of each blade. Also called *vulsella*.
**volt** (*völt*) [Alessandro *Volta*, an Italian physicist, 1745-1827]. The unit of electromotive force, or the force sufficient to cause a current of one ampere to flow against a resistance of one ohm. **v.-ampere**, the amount of pressure developed by a current of one ampere having an electromotive force of one volt. Syn., *watt*.
**voltage** (*völt'-āj*). Electromotive strength measured in volts.
**voltagram** (*völt'-ah-gram*). A faradic battery so arranged as to produce an almost continuous current.
**voltaic** (*vol-ta'-ik*). Described by or named after Volta (see *volt*). **v. electricity**, galvanism. **v. irritability**, muscular irritability during galvanism.
**voltaism** (*völ'-ta-izm*). See *galvanism*.
**voltameter** (*völ-tam'-et-er*) [*volt*; μέτρον, a measure]. An instrument for ascertaining the electromotive force of a current in volts.
**voltammeter** (*völt'-am-me-ter*) [*volt*; *ampère*]. An instrument for estimating both volts and ampères.
**voltmeter** (*völt'-me-ter*) [see *voltameter*]. An instrument for measuring the voltage of an electric current.
**Voltolini's disease** (*vōl-to-le'-nē*) [Frederic Edward Rudolph *Voltolini*, German otologist and laryngologist, 1819-1889]. Primary labyrinthitis; an affection of childhood, characterized by meningitic symptoms, followed by deafness, deaf-mutism, and a staggering gait.
**Voltolini-Heryng's sign**. See *Heryng's sign*.
**volume** (*vol'-ūm*). In physics, the space which a substance fills. Cubic dimension. **v. index** (*of blood cells*), the average size of the red cells of an individual as compared with their normal size. **v., specific**, the molecular weight divided by the specific gravity.
**volumetric** (*vol ū met' rih*) [*volume;* μετρον, a measure]. Pertaining to measurement by volume. **v. analysis**. See *analysis, volumetric*.
**volumometer** (*vol-ū-mom'-e-ter*) [see *volumetric*]. An apparatus used for the purpose of measuring changes in volume.
**voluntary** (*vol'-un-ta-re*) [*voluntas*, will]. Under the control of the will; performed by an exercise of the will. **v. muscle**, striped muscle.
**voluntomotory** (*vol-un-to-mo'-to-re*) [*voluntary; motor*]. Pertaining to voluntary motion.

**volute** (*vo-lūt'*) [*voluta*, a spiral scroll]. Rolled up like a scroll; convoluted.
**volution** (*vol-ū'-shun*) [*voluta*, a spiral scroll]. A convolution; a gyrus.
**volvulus** (*vol'-vū-lus*) [*volvere*, to roll]. A twisting of the bowel upon itself so as to occlude the lumen, occurring most frequently in the sigmoid flexure.
**vomer** (*vo'-mer*) [L., a plowshare]. The thin plate of bone situated vertically between the nasal fossæ, and forming the posterior portion of the septum of the nose. **v., cartilaginous**, a cartilaginous plate that forms the anterior portion of the septum of the nose.
**vomerine** (*vo'-mer-in*) [*vomer*]. Pertaining to the vomer.
**vomerobasilar** (*vo-mer-o-bas'-il-ar*). Relating to the vomer and to the basal part of the cranium.
**vomica** (*vom'-ik-ah*) [*vomica*, an ulcer: *pl., vomicæ*]. 1. A cavity formed by the breaking down of tissue; especially a cavity in the lung. 2. [*vomere*, to vomit]. A collection of pus in the lungs or adjacent organs that may discharge through the bronchi and mouth. **v. laryngis**, perichondritis of the larynx.
**vomicose** (*vom'-ik-ōs*) [*vomica*]. Purulent; ulcerative.
**vomit** (*vom'-it*) [*vomere*, to vomit]. 1. To expel from the stomach by vomiting. 2. Vomited matter. **v., bilious**, vomit stained with bile. **v., black**, the characteristic vomit of yellow fever, a dark fluid consisting of blood and the contents of the stomach. **v., coffee-ground**, vomit consisting of broken-down blood and the contents of the stomach; it is frequently seen in carcinoma of the stomach.
**vomiting** (*vom'-it-ing*) [*vomit*]. The forcible ejection of the contents of the stomach through the mouth. **v., cyclic**, vomiting recurring at regular periods. **v., dry**, persistent nausea with attempts at vomiting, but with the ejection of nothing but gas. **v., pernicious**, a variety of vomiting occasionally seen in pregnancy and becoming at times so excessive as to threaten life. **v., stercoraceous**, the ejection of fecal matter in the vomit, usually due to intestinal obstruction.
**vomito negro** (*vo-me'-to na'-gro*) [Sp.]. Black vomit. Yellow fever.
**vomitory** (*vom'-it-or-e*) [*vomit*]. 1. Any agent that induces emesis. 2. A vessel to receive ejecta.
**vomiturition** (*vom-it-ū-rish'-un*) [*vomit*]. Ineffectual attempt at vomiting; retching.
**vomitus** (*vom'-it-us*) [*vomit*]. 1. Vomited matter. 2. The act of vomiting. **v. cruentus**, bloody vomit. **v. marinus**, seasickness. **v. matutinus**, morning sickness. **v. niger**, black vomit.
**von**. For names with this prefix see the name itself.
**vonulo** (*von'-ū-lo*). A bronchial disease seen in West Africa.
**voracious** (*vor-a'-shus*) [*vorare*, to devour]. Having an insatiable appetite or desire for food.
**vortex** (*vor'-teks*) [*vortex*, whirl: pl., *vortices*]. A structure having the appearance of being produced by a rotary motion about an axis. **v. of the heart**, a name applied to the spiral arrangement of the muscular fibres of the walls of the heart.
**vorticose** (*vor'-tik-ōs*) [*vortex*]. Whirling. **v. veins**. See *vena vorticosa*.
**vox** (*voks*) [L.]. The voice. **v. abscissa**, loss of voice. **v. capitis**, the upper register of the voice; falsetto voice. **v. cholerica**, a peculiar faint voice noted in the last stage of cholera. **v. rauca**, hoarse voice.
**voyeur** (*vwoy'-er*) [Fr. *voir*, to see]. One who indulges a desire to witness sexual intercourse.
**V. S.** Abbreviation for *volumetric solution*.
**vuerometer** (*vū-er-om'-et-er*) [Fr. *vue*, sight; μέτρον, a measure]. An apparatus for determining the distance of the eyes from each other.
**vulcanite** (*vul'-kan-it*) [*Vulcan*, the god of fire]. Vulcanized caoutchouc.
**vulcanize** (*vul'-kan-īz*) [see *vulcanite*]. To subject rubber to the process of vulcanization, a process wherein it is treated with sulphur at a high temperature, and thereby rendered either flexible or very hard (*vulcanite*).
**vulneral** (*vul'-ner-al*) [*vulnus*, a wound]. A proprietary salve for wounds and ulcers.
**vulnerary** (*vul'-ner-a-re*) [*vulnus*, a wound]. 1. Pertaining to wounds; healing wounds. 2. An agent useful in healing wounds.
**vulnus** (*vul'-nus*) [L.]. A wound.

**Vulpian's type** of progressive muscular atrophy. See *Aran-Duchenne's disease*.
**Vulpian-Prévost's law.** See *Prévost's symptom*.
**vulsella, vulsellum** (*vul-sel'-ah, vul-sel'-um*). See *volsella*.
**vultus** (*vul'-tus*) [L.]. The face, countenance, or looks.
**vulva** (*vul'-vah*) [*volvere*, to roll up]. The external organs of generation in the woman. **v. cerebri,** the anterior opening of the third ventricle of the brain. **v. connivens,** a form of vulva in which the labia majora are in close apposition. **v., garrulity of,** vaginal flatus. **v. hians,** the form of vulva in which the labia majora are gaping.
**vulval, vulvar** (*vul'-val, vul'-var*) [*vulva*]. Pertaining to the vulva.

**vulvismus** (*vul-vis'-mus*). See *vaginismus*.
**vulvitis** (*vul-vi'-tis*) [*vulva; ιτις,* inflammation]. Inflammation of the vulva.
**vulvo-** (*vul-vo-*) [*vulva*]. A prefix denoting pertaining to the vulva.
**vulvouterine** (*vul-vo-u'-ter-in*). Relating to the vulva and the uterus considered together.
**vulvovaginal** (*vul-vo-vaj'-in-al*) [*vulvo-; vagina*]. Pertaining to the vulva and the vagina. **v. gland,** a small gland situated on each side of the vulva near the vagina. Syn., *gland of Bartholin*.
**vulvovaginitis** (*vul-vo-vaj-in-i'-tis*) [*vulvo-; vagina; ιτις,* inflammation]. Inflammation of the vulva and of the vagina existing at the same time.
**vutrin** (*vū'-trin*). A concentrated powdered meat-extract.

# W

**W.** The chemical symbol of *tungsten (wolframium)*.
**wabain** (*wah'-bah-in*). A glucoside of waba, the root of *Carissa schimperi*; a cardiac stimulant and local anesthetic.
**wabran** (*wah'-bran*). Plantago.
**Wachendorff's membrane** (*vah'-ken-dorf*) [Eberhard Jacob von *Wachendorff*, Dutch anatomist, 18th century]. 1. The pupillary membrane which covers the pupil during fetal life. 2. Cell membrane.
**Wachsmuth's mixture** (*vahks'-moot*) [Hans *Wachsmuth*, German neurologist, 1872– ]. An anesthetic mixture of oil of turpentine 1 part, chloroform 5 parts.
**wadding** (*wod'-ing*). Common name for cotton wool or carded cotton in sheets.
**waddle** (*wod'-l*). To sway or rock from side to side in walking.
**Wade's balsam.** A compound tincture of benzoin.
**W.'s drops.** Same as *W.'s balsam*. **W.'s suppositories**, urethral suppositories containing iodoform bismuth subnitrate, chloral and morphine.
**wafer** (*wa'-fer*). A thin layer composed of moistened flour, and used to inclose powders that are taken internally.
**Wagner's corpuscles** (*vahg'-ner*) [Rudolf *Wagner*, German physiologist, 1805–1864]. See *Meissner's corpuscles*. **W.'s spot**, the germinal spot of the germinal vesicle. **W.'s tactile corpuscles.** Same as *W.'s corpuscles*.
**Wagner's migration theory** (*vahg'-ner*) [Moritz *Wagner*, German scientist, 1813–1887]. That new species of animals arise through the accommodation to surroundings of animals which have migrated or been transported.
**Wagstaffe's fracture** (*wag'-staff*) [William Warwick *Wagstaffe*, English surgeon, 1843–1910]. Separation of the internal malleolus.
**(von) Wahl's sign** (*vahl*) [Eduard von *Wahl*, German surgeon, 1833–1890]. 1. Distention of the bowel (local meteorism) above the point at which there exists an obstruction. 2. A scraping or blowing sound, synchronous with the cardiac impulse, heard over an arterial trunk immediately after the partial division, through injury, of the vessel.
**wahoo** (*wah-hoo'*). See *Euonymus*.
**waist.** The narrowest portion of the trunk above the hips.
**waistcoat, strait.** See *jacket, strait*.
**wakamba** (*wak-am'-bah*). A vegetable arrow poison, used in Zanzibar.
**Walcher's position** (*vahl'-ker*) [Gustav Adolf *Walcher*, German obstetrician, 1856– ]. A dorsal posture with the hips at the edge of the table and lower extremities hanging.
**Walcheren fever** (*vahl'-tsher-en*)[*Walcheren*, in the Netherlands]. A severe type of malarial fever.
**Waldenburg's apparatus** (*vahl'-den-boorg*) [Louis *Waldenburg*, German physician, 1837–1881]. An apparatus constructed on the principle of a gasometer, and used for compressing or rarefying air, which is inhaled, or into which the patient exhales.
**Waldeyer's fossa** (*vahl'-di-er*) [Heinrich Wilhelm Gottfried *Waldeyer*, German anatomist, 1836– ]. Mesentericoparietal fossa. See *Broesike's fossa*. **W.'s germinal epithelium**, the single layer of columnar epithelial cells covering the free surface of the ovary. **W.'s glands**, modified sudoriparous glands, located at the attached border of the tarsal plates of the eyelids. **W.'s plasma-cells.** See *cells, plasma-* (2). **W.'s sulcus**, the sulcus spiralis of the cochlea. **W.'s tonsillar ring**, the ring formed by the two facial tonsils, the pharyngeal tonsil, and smaller groups of adenoid follicles at the base of the tongue and behind the posterior pillars of the fauces. **W.'s vascular layer**, the internal or vascular layer of the ovary. **W.'s zonal layer**, Lissauer's tract.
**wale.** See *wheal*.
**Walker-Gordon milk.** A form of modified milk, prepared separately and specially for each case according to the prescription of a physician, and put up by a firm of this name.
**walking typhoid.** A mild grade of typhoid fever.
**wall-diseases.** Those due to the presence of saltpeter by the penetration into the body of the walls of houses of the bacilli of nitrification, making the houses cold and unwholesome, especially in damp localities.
**Waller's law.** See *Wallerian law*.
**Wallerian degeneration** (*wol-le'-re-an*) [Augustus Volney *Waller*, English physiologist, 1816–1870]. Degeneration of a nerve consecutive upon its section, the process consisting essentially in segmentation of the myelin and subsequent disappearance of the latter, together with the axis-cylinder. **W. law**, a nerve-fiber undergoes degenerative changes when it is separated from its trophic cells.
**wall-eye.** A colloquial name for leukoma of the cornea or for divergent strabismus.
**wall-teeth.** Molar teeth.
**walnut.** See *juglans*.
**Walter's ganglion.** See *Walther's ganglion*.
**Walther's arteriosonervous plexus** (*vahl'-ter*) [August Friedrich *Walther*, German anatomist, 1688–1746]. The cavernous plexus. **W.'s ducts**, the ducts of the accessory sublingual glands. **W.'s ganglion**, the ganglion impar or coccygeal ganglion. **W.'s oblique ligament**, the ligamentous band extending from the external malleolus inward to the posterior surface of the astragalus.
**wambles** (*wom'-bls*). Milk-sickness.
**wandering** (*wan'-der-ing*). 1. Moving about, as *wandering* cells. 2. Abnormally movable, as *wandering* spleen. **w. abscess**, one that points at a considerable distance from its real seat. **w. cell**, a leukocyte.
**wang** [ME., *wange*, cheek]. 1. The jaw, jaw-bone, or cheek-bone. 2. A cheek-tooth or grinder. **w.-tooth**, a cheek-tooth; a grinder or molar.
**waras.** See *warras*.
**Warburg's tincture.** An antiperiodic and diaphoretic mixture used in pernicious forms of malaria, consisting of: Aqueous extract of aloes, 28 gr.; rhubarb and angelica seed, each, 448 gr.; elecampane, saffron, and fennel, each, 224 gr.; gentian, zedoary root, cubeb, white agaric, camphor, and myrrh, each, 112 gr.; quinine sulphate, 1280 gr.; dilute alcohol, sufficient to make 8 pints. As originally made, it contained over 60 ingredients many of which are now unobtainable.
**ward.** A division or room of a hospital. **w.-carriage**, an apparatus on wheels for holding surgical material, instruments, etc.
**Ward's paste.** Confection of black pepper.
**Wardrop's disease** [James *Wardrop*, English surgeon, 1782–1869]. Onychia maligna. **W.'s operation**, ligation of an artery beyond an aneurism.
**warehouseman's itch.** Palmar eczema occurring among the workmen in warehouses.
**war fever.** A synonym of *typhus fever*.
**Waring's system** [George Edward *Waring*, American sanitary engineer, 1833–1898]. A system of sewage disposal by means of sub-surface irrigation. It should be called "Moule's System," because originated by the late Rev. Henry *Moule*, an English clergyman, of the last century.
**warm-blooded.** A term applied to animals that maintain a uniform temperature whatever the changes in the surrounding medium.
**warming plaster.** See *plaster, warming*.
**warras** (*war'-as*) [Ind.]. A variety of kamala said to be obtained from *Flemingia grahamiana*, a leguminous tree of India; it is a teniafuge and useful in skin diseases, also used as a dye. Dose of powder 3 ʒss–v; of tincture 3 ʒ.
**Warren's fat-columns** (*war'-en*) [John Collins *Warren*, Boston surgeon, 1778–1856]. Slender

columns of fatty tissue passing from the subcutaneous adipose tissue to the base of the hair-follicles. They are well developed over the dorsum of the body, particularly near the median line.
**Warren's styptic.** The *lotio adstringens* (N. F.). W.'s test, Trommer's test.
**wart** [AS., *wearte*]. A hyperplasia of the papillæ of the skin, forming a small projection. See *verruca*. w., **anatomical,** w., **postmortem.** See *tubercle, anatomical*. w., **Peruvian.** See *verruca peruana*. w., **venereal,** condyloma acuminatum.
**Warthin's sign.** Accentuation of the pulmonary sound in acute pericarditis.
**warty.** Resembling a wart; covered with warts.
w.-**smallpox,** *hornpox*, a name given to those cases of smallpox in which the eruption does not develop beyond the papular stage. w. **tubercle.** See *verruca necrogenica*. w. **ulcer.** See *Marjolin's ulcer*.
**wash.** See *lotio*. w., **black,** lotio hydrargyri nigra; mild mercuric chloride in water and lime-water. w., **eye,** collyrium. w., **yellow,** lotio hydrargyri flava, mercuric chloride in lime-water.
**washerwoman's itch.** Eczema of the hands.
**washing soda.** Sodium carbonate, Na₂CO₃.
**washleather-skin.** A condition of the skin in which certain metals, especially silver, mark it with a black line.
**Wasmann's glands.** The peptic glands.
**Wassermann's syphilis test** (*vas'-er-mahn*) [August Wassermann, German bacteriologist, 1866– ]. 1. *Complement.* One to 10 dilution of fresh guinea-pig serum in normal (.85 %) salt solution. 2. *Antigen.* Alcoholic extract of a syphilitic organ or suspension of an organ in weak carbolic acid solution (1 %)—amount determined by standardization. 3. *Amboceptor.* Inactivated serum of rabbit which has been highly immunized against sheep red-cell by five or six injections of increasing amounts of sheep red-cells. The amboceptor is standardized by putting in each of a series of test-tubes 1 Cc. of complement and 1 Cc. of 5 % emulsion of sheep red-cells. Different amounts of the inactivated rabbit serum are added to the tubes, beginning with 0.01 Cc. to 0.1 Cc. The tubes are then incubated one hour. That in which complete hemolysis occurs contains just enough amboceptor to dissolve 1 Cc. of 5 % emulsion of sheep red-cells. Double this quantity is the amboceptor to be used. Suspected serum to be examined is drawn from a superficial vein with a glass syringe under strict aseptic precautions, 5 to 10 Cc. of blood being desirable, but 1 to 2 Cc. suffices. Clear in a centrifuge, then inactivate by heat for thirty minutes at 56° C. Will keep in ice-box for weeks. *Test.*—Put 1 Cc. of complement, 2 drops of suspected serum, about 0.1 Cc. of antigen in test-tube and incubate one hour at 37° C. Then add the amount of amboceptor, determined by standardization, and 1 Cc. of 5 % emulsion of sheep's red-cells suspended in normal salt solution and incubate again for one hour. Then place in ice-box for six hours. Complete hemolysis is indicated by a clear, burgundy-red solution, showing no precipitate. No hemolysis, by a solid opaque sediment of the unaffected sheep cells at the bottom of the tube, while the supernatant fluid is clear and colorless. Result: Hemolysis, no syphilis; syphilis, no hemolysis. The control test is the same except that the antigen is omitted.
**waster** (*wäst'-er*). 1. A child suffering from marasmus. 2. An animal affected with tuberculosis.
**wasting palsy.** See *progressive muscular atrophy*.
**watchmaker's cramp.** 1. An occupation neurosis, characterized by painful cramps of the muscles of the hands. 2. Also spasm of the orbicularis palpebrarum muscle, due to holding the lens.
**water** (*waw'-ter*). 1. Hydrogen monoxide, H₂O. Boils at 212° F. (100° C.), and freezes at 32° F. (0° C.). See *ammonia-aqua, chlorine-water,* etc. 2. Euphemism for urine. w.-**bag,** a rubber bag containing hot or cold water for topical application. w.-**bed.** See *bed, water-*. w.-**borne,** produced by contaminated drinking-water. w. **on the brain,** hydrocephalus. w.-**brash.** See *pyrosis.* w.-**cancer,** noma. w.-**cress,** the plant *Nasturtium officinale.* w. of **crystallization,** the water contained in certain crystals, to which their crystalline structure is due. w.-**cure.** See *hydrotherapeutics.* w.-**dressing,** treatment of ulcers or wounds by the topical application of water. w.-**glass,** a solution of sodium or potassium silicate. w.-**hammer pulse.** See *Corrigan's pulse.* w., **hard,**

water containing soluble calcium salts and not readily forming a lather with soap. w. of **hydration.** Same as w. of *crystallization.* w., **Javelle,** a solution of potassium hypochlorite, KClO. w., **mineral,** a natural water containing mineral substances in solution. w. of **Pagliari,** a preparation employed in France as a hemostatic, consisting of crystallized alum, 15 gr.; gum benzoin, 75 gr.; distilled water, 3 oz. w.-**pox,** chicken pox. w.-**whistling,** a metallic rale heard in pneumothorax.
**water-gurgle test.** The swallowing of water causes a gurgling sound heard on auscultation, in cases of stricture of the esophagus.
**waters.** The liquor amnii. **bag of w.,** the amnion. w., **false,** a discharge of fluid before labor.
**watery eye.** Epiphora.
**watt** (*wot*) [James *Watt*, Scotch engineer, 1736–1819]. See *volt-ampère.*
**wattmeter.** An instrument for measuring electrical power or activity in watts.
**wave** (*wāv*) [AS., *wafian*, to waver in mind]. 1. A movement in a body which is propagated with a continuous motion, each particle of the body vibrating through a fixed path, usually a closed curve. 2. One of the curves in a series of curves representing a wave-like motion. w.-**length,** the distance between corresponding points, usually the crests, of two adjacent waves. w., **tidal,** in the sphygmogram, the wave succeeding the percussion wave, and due to the volume of blood poured out from the heart reaching the arteries.
**wavy respiration.** Cog-wheel respiration, a type of breathing in which inspiration or expiration is jerky and interrupted.
**wax** (*waks*). See *cera.*
**waxing** (*waks'-ing*) [ME., *waxen*, to increase]. Increasing in size. w.-**kernels,** enlarged inguinal and submaxillary lymph-glands in children.
**waxy cast** (*waks'-t*). A tube-cast composed of amyloid or similar material. w. **degeneration,** amyloid degeneration. w. **kidney,** w. **liver,** amyloid kidney or liver.
**weak.** Not strong. w. **ankle,** a condition in which there is an abnormal relaxation of the ligaments of the ankle-joint, with such weakness of the leg-muscles as may allow the foot to bend involuntarily, either inward or outward, in the act of standing or walking. It is common in feeble children. w.-**minded,** having a feeble intellect. w. **sight,** asthenopia.
**weaken.** To reduce the strength.
**weakness.** Loss of strength. w., **inward,** leukorrhea.
**wean** (*wēn*). To cease to give suck to offspring at a period when the latter is capable of taking substantial food from external sources.
**weaning-brash.** Severe infantile diarrhea due to weaning.
**weasand** (*we'-zand*). The trachea.
**weavers' bottom.** A chronic inflammation of the bursa over the tuberosity of the ischium, due to pressure.
**web.** A woven fabric; a membrane-like structure; tela. The thin, soft tissue between any two adjacent fingers or toes, fixed to the knuckles. w. of the **brain,** the bindweb, or neuroglia. w., **choroid,** the velum interpositum. w.-**eye.** See *pterygium*. w.-**eyed,** affected with pterygium. w.-**fingered,** having the fingers united by web-like tissue. w.-**foot,** a foot whose toes are webbed. w.-**footed,** having webfeet. w.-**footedness,** web-foot. w.-**toed,** web-footed.
**webbed fingers,** w. **toes.** Union of adjacent fingers or toes by a thin band of tissue. w. **penis.** See *penis palmatus.*
**weber** (*web'-er*) [Wilhelm *Weber*, German physicist, 1804–1891]. 1. Same as *coulomb.* 2. Same as *ampère.*
**Weber's glands** (*va'-ber*) [Ernst Heinrich *Weber*, German anatomist, 1795–1878]. Racemose glands situated in the posterior portion of the tongue and opening by several orifices on its border. W.'s **law,** the increase of stimulus necessary to produce the smallest perceptible change in a sensation is proportionate to the strength of the stimulus already acting. W.'s **orbicular zone,** that portion of the iliofemoral ligament which forms a loop around the neck of the femur. W.'s **organ,** W.'s vesicle, the sinus pocularis of the male urethra. W.'s **paradox,** a muscle when so loaded as to be unable to contract may elongate. W.'s **pouch,** the prostatic vesicle. W.'s **suture,** a

fine groove or suture on the inner surface of the nasal process of the superior maxilla. **W.'s symptom**, **W.'s syndrome**, paralysis of the motor oculi nerve on the side of the lesion and of the facial and hypoglossal nerves and extremities on the opposite side; it corresponds anatomically to a lesion in the pedunculopontine or upper pontine region. **W.'s test** [Friedrich Eugen *Weber*, German otologist, 1832–1891]. 1. When a vibrating tuning-fork is placed upon the vertex or the middle of the forehead, the sound is perceived equally by both ears. If it is heard only in one ear, a lesion exists in this. 2. For sensation: Determination of the smallest distance at which the two points of a pair of compasses, applied simultaneously and lightly to the skin, can be recognized as two separate objects. **W.'s test for indican in urine**, heat to boiling 30 Cc. of the urine with an equal volume of hydrochloric acid and 1 to 3 drops of dilute nitric acid; when cold, shake the solution with ether. The ether will assume a red or violet color with a blue foam on it.
**Webster's condenser**. In microscopy, an apparatus consisting of two lenses, used for intensifying the light thrown on the object. **W.'s (Lady) pill**, a pill of aloes and mastic.
**wedge** (*wej*). An instrument used by dentists to separate adjoining teeth. **w.-bone**, an ossicle sometimes found at a vertebral joint.
**Wedl's vesicular cells**. Commonly found in the crystalline lens in cases of cataract, especially the senile and diabetic varieties.
**weed**. 1. Milk fever. 2. Lymphangitis in legs of a horse.
**Weeks' bacillus**. Koch-Weeks bacillus of acute conjunctivitis.
**weeping**. 1. Lacrymation. 2. Exudation or leakage of a fluid. Exuding; applied to raw or excoriated surfaces bathed with a moist discharge. **w. eczema**, moist eczema. **w. sinew**. See *ganglion* (2).
**Wegner's disease of bone** (*veg'-ner*) [Fridericus Rudolphus Georgius *Wegner*, German pathologist, 1843–    ]. Epiphyseal osteochondritis affecting infants with hereditary syphilis. **W.'s line**, an angular line separating the epiphyses and diaphysis of the long bones, in certain diseased condition of the fetus, such as syphilis. **W.'s sign**, in fetal syphilis the dividing-line between the epiphysis and diaphysis of long bones, which under normal conditions is delicate and rectilinear, appears as a broad, irregular, yellowish line.
**Weichselbaum's coccus** (*vikh'-sel-bowm*) [Anton *Weichselbaum*, Austrian pathologist, 1845–    ]. The *Diplococcus intracellularis meningitidis*.
**Weidel's reaction** (*vi'-del*). 1. **for xanthin bodies**. Evaporate to dryness on the water-bath a little of the substance dissolved in fresh chlorine-water containing nitric acid, Treat the residue to ammonia vapors under a bell-jar, and a red or violet coloration will be produced in the presence of xanthin bodies. 2. **for uric acid**. See *murexide test*.
**Weigert's fibrin-stain** (*vi'-gert*) [Karl *Weigert*, German pathologist, 1843–1905]. A stain for fibrin, consisting in the application of a solution of gentian-violet, then one of iodine in potassium iodide, followed by one of anilin oil and xylol. Fibrin is stained blue. **W.'s law**, loss of elements or parts in organic structures is likely to be followed by overcompensation in the reparative process. **W.'s method**. 1. A method of staining the myelin of nerve-fibers with hematoxylin. 2. A method of staining the neuroglia according to a modified Weigert's fibrin-method, after the tissue has been fixed in formalin, subjected to a mordant of copper acetate, acetic acid, and chrome-alum, and a reducing agent composed of potassium permanganate and a solution of sodium sulphite containing a chromogen derived from naphthalin.
**weight** (*wāt*). The force with which bodies tend to approach the earth's center. **w., atomic**. See *atomic weight*. **w.s and measures**, the U. S. standard unit of weight is the troy pound; the standard unit of

## TABLE OF WEIGHTS AND MEASURES.

### TROY WEIGHT.

1 pound = 22.816 cubic inches of distilled water at 62° F.

| Grains. | Dwt. | Ounce. | Pound. |
|---|---|---|---|
| 24 | = 1 | | |
| 480 | = 20 | = 1 | |
| 5760 | = 240 | = 12 | = 1 |

### AVOIRDUPOIS WEIGHT.

1 pound = 1.2153 pounds troy.

| Grains. | Drams. | Ounces. | Pound. |
|---|---|---|---|
| gr. 27.34375 | = dr. 1 | | |
| 437.5 | = 16 | = oz. 1 | |
| 7000 | = 256 | = 16 | = ℔. 1 |

### APOTHECARIES' WEIGHT.

| Grains. | Scruples. | Drams. | Troy Ounces. | Pound. |
|---|---|---|---|---|
| gr. 20 | = ℈ 1 | | | |
| 60 | = 3 | = ℨ 1 | | |
| 480 | = 24 | = 8 | = ℥ 1 | |
| 5760 | = 288 | = 96 | = 12 | = ℔. 1 |

### APOTHECARIES' MEASURE.

| Minims. | Fluidrams. | Fluidounces. | Pints. | Gallon. |
|---|---|---|---|---|
| ♏ 60 | = fℨ 1 | | | |
| 480 | = 8 | = f℥ 1 | | |
| 7,680 | = 128 | = 16 | = O. 1 | |
| 61,440 | = 1024 | = 128 | = 8 | = C. 1 |

### IMPERIAL MEASURE.

| Minims. | Fluidrams. | Fluidounces. | Pints. | Gallon. |
|---|---|---|---|---|
| 60 | = | | | |
| 480 | = 8 | = | | |
| 9,600 | = 160 | = 20 | = 1 | |
| 76,800 | = 1280 | = 160 | = 8 | = 1 |

### LIQUID OR WINE MEASURE.

1 gill = 7.2187 cubic inches.

| Gills. | Pints. | Quarts. | Gallons. | Hogsheads. | Pipes. | Tun. |
|---|---|---|---|---|---|---|
| 4 | = 1 | | | | | |
| 8 | = 2 | = 1 | | | | |
| 32 | = 8 | = 4 | = 1 | | | |
| 2016 | = 504 | = 252 | = 63 | = 1 | | |
| 4032 | = 1008 | = 504 | = 126 | = 2 | = 1 | |
| 8064 | = 2016 | = 1008 | = 252 | = 4 | = 2 | = 1 |

### DRY MEASURE.

1 gallon = 268.8 cubic inches.

| Pints. | Quarts. | Gallons. | Pecks. | Bushels. | Quarter. |
|---|---|---|---|---|---|
| 2 | = 1 | | | | |
| 8 | = 4 | = 1 | | | |
| 16 | = 8 | = 2 | = 1 | | |
| 64 | = 32 | = 8 | = 4 | = 1 | |
| 512 | = 256 | = 64 | = 32 | = 8 | = 1 |

### SOLID MEASURE.

| Cubic Inches. | Cubic Feet. | Cubic Yard. |
|---|---|---|
| 1,728 | = 1 | |
| 46,656 | = 27 | = 1 |

### LINEAR MEASURE.

| Inches. | Feet. | Yards. | Fathoms. | Perches. | Furlongs. | Mile. |
|---|---|---|---|---|---|---|
| 12 | = 1 | | | | | |
| 36 | = 3 | = 1 | | | | |
| 72 | = 6 | = 2 | = 1 | | | |
| 198 | = 16.5 | = 5.5 | = 2.75 | = 1 | | |
| 7,920 | = 660 | = 220 | = 110 | = 40 | = 1 | |
| 63,360 | = 5280 | = 1760 | = 880 | = 320 | = 8 | = 1 |

### SQUARE MEASURE.

| Square Inches. | Square Feet. | Square Yards. | Perches. | Roods. | Acre. |
|---|---|---|---|---|---|
| 144 | = 1 | | | | |
| 1,296 | = 9 | = 1 | | | |
| 39,204 | = 272.25 | = 30.25 | = 1 | | |
| 1,568,160 | = 10,890 | = 1210 | = 40 | = 1 | |
| 6,272,640 | = 43,560 | = 4840 | = 160 | = 4 | = 1 |

### METRIC WEIGHTS.

1 gram = 1 cubic centimeter of distilled water at 62° F.

| | Gram. | Troy Gr. | Avoir. Oz. |
|---|---|---|---|
| Milligram | = | .001 | = .01543 | |
| Centigram | = | .01 | = .15432 | |
| Decigram | = | .1 | = 1.54323 | |
| Gram | = | 1. | = 15.43235 | = .03528 |
| Decagram | = | 10. | | = .3528 |
| Hectogram | = | 100. | | = 3.52758 |
| Kilogram | = | 1,000. | | = 35.2758 |
| Myriogram | = | 10,000. | | |
| Quintal | = | 100,000. | | |
| Tonneau | = | 1,000,000. | | |

# WEIGHT

## COMPARATIVE VALUES OF APOTHECARIES' AND METRIC LIQUID MEASURES.

| Minims. | Cubic Centimeters. | Minims. | Cubic Centimeters. | Minims. | Cubic Centimeters. | Fluidounces. | Cubic Centimeters. | Fluidounces. | Cubic Centimeters. |
|---|---|---|---|---|---|---|---|---|---|
| 1 | 0.06 | 25 | 1.54 | 1 | 30.00 | 21 | 621.00 |   |   |
| 2 | 0.12 | 30 | 1.90 | 2 | 59.20 | 22 | 650.00 |   |   |
| 3 | 0.18 | 35 | 2.16 | 3 | 89.00 | 23 | 680.00 |   |   |
| 4 | 0.24 | 40 | 2.50 | 4 | 118.40 | 24 | 710.00 |   |   |
| 5 | 0.30 | 45 | 2.80 | 5 | 148.00 | 25 | 740.00 |   |   |
| 6 | 0.36 | 50 | 3.08 | 6 | 178.00 | 26 | 769.00 |   |   |
| 7 | 0.42 | 55 | 3.40 | 7 | 207.00 | 27 | 798.50 |   |   |
| 8 | 0.50 |   |   | 8 | 236.00 | 28 | 828.00 |   |   |
| 9 | 0.55 | Fluid- |   | 9 | 266.00 | 29 | 858.00 |   |   |
| 10 | 0.60 | rams. |   | 10 | 295.70 | 30 | 887.25 |   |   |
| 11 | 0.68 | 1 | 3.75 | 11 | 325.25 | 31 | 917.00 |   |   |
| 12 | 0.74 | 1¼ | 4.65 | 12 | 355.00 | 32 | 946.00 |   |   |
| 13 | 0.80 | 1½ | 5.60 | 13 | 385.00 | 48 | 1419.00 |   |   |
| 14 | 0.85 | 1¾ | 6.51 | 14 | 414.00 | 56 | 1655.00 |   |   |
| 15 | 0.92 | 2 | 7.50 | 15 | 444.00 | 64 | 1892.00 |   |   |
| 16 | 1.00 | 3 | 11.25 | 16 | 473.11 | 72 | 2128.00 |   |   |
| 17 | 1.05 | 4 | 15.00 | 17 | 503.00 | 80 | 2365.00 |   |   |
| 18 | 1.12 | 5 | 18.50 | 18 | 532.00 | 96 | 2839.00 |   |   |
| 19 | 1.17 | 6 | 22.50 | 19 | 562.00 | 112 | 3312.00 |   |   |
| 20 | 1.25 | 7 | 26.00 | 20 | 591.50 | 128 | 3785.00 |   |   |

## COMPARATIVE VALUES OF METRIC LIQUID AND APOTHECARIES' MEASURES.

| Cubic Centimeters. | Fluidounces. | Cubic Centimeters. | Fluidounces. | Cubic Centimeters. | Fluidounces. | Cubic Centimeters. | Fluiddrams. | Cubic Centimeters. | Minims. |
|---|---|---|---|---|---|---|---|---|---|
| 1000 | 33.81 | 400 | 13.53 | 25 | 6.76 | 4 | 64.8 |   |   |
| 900 | 30.43 | 300 | 10.14 | 10 | 2.71 | 3 | 48.6 |   |   |
| 800 | 27.05 | 200 | 6.76 | 9 | 2.43 | 2 | 32.4 |   |   |
| 700 | 23.67 | 100 | 3.38 | 8 | 2.16 | 1 | 16.23 |   |   |
| 600 | 20.29 | 75 | 2.53 | 7 | 1.89 | 0.09 | 1.46 |   |   |
| 500 | 16.90 | 50 | 1.69 | 6 | 1.62 | 0.07 | 1.14 |   |   |
| 473 | 16.00 | 30 | 1.01 | 5 | 1.35 | 0.05 | 0.81 |   |   |

## COMPARATIVE VALUES OF AVOIRDUPOIS AND METRIC WEIGHTS.

| Avoir. Ounces. | Grams. | Avoir. Ounces. | Grams. | Avoir. Pounds. | Grams. |
|---|---|---|---|---|---|
| 1/16 | 1.772 | 8 | 226.80 | 2 | 907.18 |
| 1/8 | 3.544 | 9 | 255.15 | 2.2 | 1000.00 |
| ¼ | 7.088 | 10 | 283.50 | 3 | 1360.78 |
| ½ | 14.175 | 11 | 311.84 | 4 | 1814.37 |
| 1 | 28.350 | 12 | 340.20 | 5 | 2267.96 |
| 2 | 56.700 | 13 | 368.54 | 6 | 2727.55 |
| 3 | 85.050 | 14 | 396.90 | 7 | 3175.14 |
| 4 | 113.400 | 15 | 425.25 | 8 | 3628.74 |
| 5 | 141.750 | Avoir. |   | 9 | 4082.33 |
| 6 | 170.100 | Pounds. |   | 10 | 4535.92 |
| 7 | 198.450 | 1 | 453.60 |   |   |

For Comparative Thermometry, see *thermometers*.

## COMPARATIVE VALUES OF STANDARD AND METRIC MEASURES OF LENGTH.

| Inches. | Centimeters. | Inches. | Centimeters. | Inches. | Millimeters. | Inches. | Millimeters. |
|---|---|---|---|---|---|---|---|
| 12 | 30.48 | 6 | 15.24 | 1/25 | 1.00 |   | 15.85 |
| 11 | 27.94 | 5 | 12.70 | 1/12 | 2.11 |   | 16.92 |
| 10 | 25.40 | 4 | 10.16 | 1/8 | 3.17 |   | 19.05 |
| 9 | 22.86 | 3 | 7.62 | ¼ | 6.35 |   | 21.15 |
| 8 | 20.32 | 2 | 5.08 | ⅓ | 8.46 |   | 22.19 |
| 7 | 17.78 | 1 | 2.54 | ½ | 12.70 |   | 23.28 |

## METRIC DRY AND LIQUID MEASURE.

|   | Liter. | U. S. Cu. In. |   | U. S. |
|---|---|---|---|---|
| Milliliter = | .001 = | .061 = | { Liquid | .00845 gill. |
|   |   |   | Dry | .0018 pint. |
| Centiliter = | .01 = | .61 = | { Liquid | .0845 gill. |
|   |   |   | Dry | .018 pint. |
| Deciliter = | .1 = | 6.1 = | { Liquid | .845 gill. |
|   |   |   | Dry | .18 pint. |
| Liter = | 1. = | 61.02 = | { Liquid | 1.057 quarts. |
|   |   |   | Dry | .908 quart. |
| Decaliter = | 10. | = 610.16 = | { Liquid | 2.641 gallons. |
|   |   | U. S. Cu. Ft. | Dry | 9.08 quarts. |
| Hectoliter = | 100. | = 3.531 = | { Liquid | 26.414 gallons. |
|   |   |   | Dry | 2.837 bushels. |
| Kiloliter = | 1,000. | = 35.31 = | { Liquid | 264.141 gallons. |
|   |   |   | Dry | 28.374 bushels. |
| Myrialiter = | 10,000. | = 353.1 = | { Liquid | 2641.4 gallons. |
|   |   |   | Dry | 283.7 bushels. |

## METRIC LINEAR MEASURE.

|   | Meter. | U. S. Inches. | Feet. | Yards. | Miles. |
|---|---|---|---|---|---|
| Millimeter* = | .001 = | .03937 = | .00328 |   |   |
| Centimeter† = | .01 = | .3937 = | .03280 |   |   |
| Decimeter = | .1 = | 3.937 = | .32807 = | .10936 |   |
| Meter = | 1. = | 39.3685 = | 3.2807 = | 1.0936 = |   |
| Decameter = | 10. = |   | 32.807 = | 10.936 = |   |
| Hectometer = | 100. = |   | 328.07 = | 109.36 = | .0621347 |
| Kilometer = | 1,000. = |   | 3,280.7 = | 1,093.6 = | .621347 |
| Myriameter = | 10,000. = |   | 32,807. = | 10,936. = | 6.213466 |

* Nearly 1/25 of an inch.   † Full ⅜ of an inch.

## METRIC SQUARE MEASURE.

|   | Sq. Meter. | U. S. Sq. In. | Sq. Ft. | Sq. Yds. | Acres. |
|---|---|---|---|---|---|
| Sq. Centimeter = | .0001 = | .155 |   |   |   |
| Sq. Decimeter = | .01 = | 15.5 = | .10763 = | .01196 |   |
| Centiare = | 1. = | 1,549.88 = | 10.763 = | 1.196 = | .00025 |
| Are = | 100. = | 154,988. = | 1,076.3 = | 119.6 = | .0247 |
| Hectare = | 10,000. = |   | 107,630. = | 11,959. = | 2.47 |
| Sq. Kilometer = |   |   | .38607 Sq. Mile |   | = 247. |
| Sq. Myriameter = |   |   | 38.607 Sq. Miles |   | = 24,708. |

TABLE FOR CONVERTING METRIC WEIGHTS INTO TROY WEIGHTS.

| Grams. | Exact Equivalents in Grains. | Approximate Equivalents in Troy Weights. | | | | Grams. | Exact Equivalents in Grains. | Approximate Equivalents in Troy Weights. | | | |
|---|---|---|---|---|---|---|---|---|---|---|---|
| | | Ounces. | Drams. | Scruples. | Grains. | | | Ounces. | Drams. | Scruples. | Grains. |
| 0.01 | 0.1543 | ....... | ....... | ....... | 1/6 | 12.0 | 185.188 | ....... | 3 | ....... | 5 1/5 |
| 0.02 | 0.3086 | ....... | ....... | ....... | 1/3 | 13.0 | 200.621 | ....... | 3 | 1 | 3/5 |
| 0.03 | 0.4630 | ....... | ....... | ....... | 6/13 | 14.0 | 216.053 | ....... | 3 | 1 | 16 |
| 0.04 | 0.6173 | ....... | ....... | ....... | 7/11 | 15.0 | 231.485 | ....... | 3 | 2 | 11 1/2 |
| 0.05 | 0.7717 | ....... | ....... | ....... | 3/4 | 16.0 | 246.918 | ....... | 4 | ....... | 6 9/10 |
| 0.06 | 0.9260 | ....... | ....... | ....... | 9/10 | 17.0 | 262.350 | ....... | 4 | 1 | 2 1/3 |
| 0.07 | 1.0803 | ....... | ....... | ....... | 1 | 18.0 | 277.782 | ....... | 4 | 1 | 17 4/5 |
| 0.08 | 1.2347 | ....... | ....... | ....... | 1 1/4 | 19.0 | 293.215 | ....... | 4 | 2 | 13 1/5 |
| 0.09 | 1.3890 | ....... | ....... | ....... | 1 1/3 | 20.0 | 308.647 | ....... | 5 | ....... | 8 3/5 |
| 0.1 | 1.543 | ....... | ....... | ....... | 1 1/2 | 21.0 | 324.079 | ....... | 5 | 1 | 4 1/10 |
| 0.2 | 3.086 | ....... | ....... | ....... | 3 | 22.0 | 339.512 | ....... | 5 | 1 | 19 1/2 |
| 0.3 | 4.630 | ....... | ....... | ....... | 4 3/5 | 23.0 | 354.944 | ....... | 5 | 2 | 5 |
| 0.4 | 6.173 | ....... | ....... | ....... | 6 1/5 | 24.0 | 370.376 | ....... | 6 | ....... | 10 2/5 |
| 0.5 | 7.716 | ....... | ....... | ....... | 7 3/4 | 25.0 | 385.809 | ....... | 6 | 1 | 5 4/5 |
| 0.6 | 9.259 | ....... | ....... | ....... | 9 1/4 | 26.0 | 401.241 | ....... | 6 | 2 | 1 1/4 |
| 0.7 | 10.803 | ....... | ....... | ....... | 10 4/5 | 27.0 | 416.673 | ....... | 6 | 2 | 16 3/4 |
| 0.8 | 12.346 | ....... | ....... | ....... | 12 1/3 | 28.0 | 432.106 | ....... | 7 | ....... | 12 1/10 |
| 0.9 | 13.889 | ....... | ....... | ....... | 14 | 29.0 | 447.538 | ....... | 7 | 1 | 7 1/2 |
| 1.0 | 15.432 | ....... | ....... | ....... | 15 1/2 | 30.0 | 462.970 | ....... | 7 | 2 | 3 |
| 2.0 | 30.865 | ....... | ....... | 1 | 10 4/5 | 31.0 | 478.403 | ....... | 7 | 2 | 18 2/5 |
| 3.0 | 46.297 | ....... | ....... | 2 | 6 1/4 | 32.0 | 493.835 | 1 | ....... | ....... | 13 4/5 |
| 4.0 | 61.729 | ....... | 1 | ....... | 1 3/4 | 40.0 | 617.294 | 1 | 2 | ....... | 17 3/10 |
| 5.0 | 77.162 | ....... | 1 | ....... | 17 1/6 | 45.0 | 694.456 | 1 | 3 | 1 | 10 1/2 |
| 6.0 | 92.594 | ....... | 1 | 1 | 12 3/5 | 50.0 | 771.617 | 1 | 4 | 2 | 11 3/5 |
| 7.0 | 108.026 | ....... | 1 | 2 | 8 | 60.0 | 925.941 | 1 | 7 | 1 | 6 |
| 8.0 | 123.459 | ....... | 2 | ....... | 3 1/2 | 70.0 | 1080.264 | 2 | 2 | ....... | 1/4 |
| 9.0 | 138.891 | ....... | 2 | ....... | 18 9/10 | 80.0 | 1234.588 | 2 | 4 | 1 | 14 3/5 |
| 10.0 | 154.323 | ....... | 2 | 1 | 14 1/3 | 90.0 | 1388.911 | 2 | 7 | ....... | 9 |
| 11.0 | 169.756 | ....... | 2 | 2 | 9 3/4 | 100.0 | 1543.235 | 3 | 1 | 2 | 3 1/4 |

TABLE FOR CONVERTING TROY WEIGHTS INTO METRIC WEIGHTS.

| Grains. | Grams. | Grains. | Grams. | Grains. | Grams. | Grains. | Grams. | Grains. | Grams. | Grains. | Grams. | Grains. | Grams. | Grains. | Grams. |
|---|---|---|---|---|---|---|---|---|---|---|---|---|---|---|---|
| 1/50 | 0.00130 | 2 | 0.1296 | 18 | 1.166 | 34 | 2.203 | 50 | 3.234 | 66 | 4.276 | 82 | 5.313 | 98 | 6.350 |
| 1/32 | 0.00202 | 3 | 0.1944 | 19 | 1.231 | 35 | 2.268 | 51 | 3.304 | 67 | 4.341 | 83 | 5.378 | 99 | 6.414 |
| 1/20 | 0.00324 | 4 | 0.2592 | 20 | 1.296 | 36 | 2.332 | 52 | 3.369 | 68 | 4.406 | 84 | 5.442 | 100 | 6.479 |
| 1/18 | 0.00360 | 5 | 0.3240 | 21 | 1.361 | 37 | 2.397 | 53 | 3.434 | 69 | 4.471 | 85 | 5.507 | 120 | 7.776 |
| 1/16 | 0.00405 | 6 | 0.3888 | 22 | 1.426 | 38 | 2.462 | 54 | 3.499 | 70 | 4.535 | 86 | 5.572 | 150 | 9.719 |
| 1/15 | 0.00432 | 7 | 0.4536 | 23 | 1.490 | 39 | 2.527 | 55 | 3.564 | 71 | 4.600 | 87 | 5.637 | 180 | 11.664 |
| 1/12 | 0.00540 | 8 | 0.5184 | 24 | 1.555 | 40 | 2.592 | 56 | 3.628 | 72 | 4.665 | 88 | 5.702 | 200 | 12.958 |
| 1/10 | 0.00648 | 9 | 0.5832 | 25 | 1.620 | 41 | 2.656 | 57 | 3.693 | 73 | 4.730 | 89 | 5.766 | 480 | 31.103 |
| 1/8 | 0.00810 | 10 | 0.6480 | 26 | 1.685 | 42 | 2.721 | 58 | 3.758 | 74 | 4.795 | 90 | 5.831 | 500 | 32.396 |
| 1/6 | 0.01080 | 11 | 0.7130 | 27 | 1.749 | 43 | 2.786 | 59 | 3.823 | 75 | 4.859 | 91 | 5.896 | 600 | 38.875 |
| 1/5 | 0.01296 | 12 | 0.7776 | 28 | 1.814 | 44 | 2.851 | 60 | 3.888 | 76 | 4.924 | 92 | 5.961 | 700 | 45.354 |
| 1/4 | 0.01620 | 13 | 0.8424 | 29 | 1.869 | 45 | 2.916 | 61 | 3.952 | 77 | 4.989 | 93 | 6.026 | 800 | 51.833 |
| 1/3 | 0.02160 | 14 | 0.9072 | 30 | 1.944 | 46 | 2.980 | 62 | 4.017 | 78 | 5.054 | 94 | 6.090 | 900 | 58.313 |
| 1/2 | 0.03240 | 15 | 0.9720 | 31 | 2.009 | 47 | 3.045 | 63 | 4.082 | 79 | 5.118 | 95 | 6.155 | 960 | 62.207 |
| 3/4 | 0.04860 | 16 | 1.037 | 32 | 2.073 | 48 | 3.110 | 64 | 4.147 | 80 | 5.183 | 96 | 6.220 | 1000 | 64.792 |
| 1 | 0.0648 | 17 | 1.102 | 33 | 2.138 | 49 | 3.175 | 65 | 4.211 | 81 | 5.248 | 97 | 6.285 | | |

liquid measure is the Winchester wine gallon, containing 231 cubic inches. The imperial gallon, adopted by Great Britain, contains 277.274 cubic inches. The standard unit of the U. S. and British linear measure is the yard. The actual standard of length of the U. S. is a brass scale 82 inches long in the U. S. Treasury Department. The yard is between the twenty-seventh and the sixty-third inch of this scale. See *tables*, pages 947 to 949. **w., equivalent,** is the weight of an element which can combine with a unit weight of hydrogen or other univalent element. **w., molecular.** See *molecular weight*. **w., specific.** Same as *gravity, specific*.

**Weil's disease** (*vīl*) [Adolf *Weil*, German physician, 1848– ]. An infectious disease somewhat resembling typhoid fever, accompanied by muscular pain and grave disturbance of the digestive organs. **W.'s syndrome,** unilateral hyperesthesia of the muscles, nerve-trunks, and bones, sometimes seen in cases of pulmonary tuberculosis.

**Weiland's test.** For the determination of binocular fixation: A vertical bar is interposed between the eyes and the letters to be read.

**Weill's sign** (*vīl*) [Edmond *Weill*, French physician]. Absence of chest expansion in the subclavicular region, noticed on the affected side in very severe lobar pneumonia.

**Weir Mitchell's disease, W. M.'s treatment.** See under *Mitchell*.

**Weir's operation** (*wēr*) [Robert Fulton *Weir*, New York surgeon, 1838– ]. See *appendicostomy*.

**Weiss' reflex** (*vīs*) [Leopold *Weiss*, German oculist, 1849– ]. A curvilinear reflex on the nasal side of the optic disc, regarded as a prodromal sign of myopia.

**Weiss' sign** (*vīs*) [Nathan *Weiss*, German physician]. Contraction of the facial muscles upon light percussion; it is noticed in tetany, neurasthenia, hysteria, and exophthalmic goiter. Syn., *facialis phenomenon*.

**Weissmann's bundle** (*vīs'-man*). The aggregation of striped muscular fibers of a neuromuscular spindle.

**Weismann's theory of heredity** (*vīs'-man*) [August Friedrich Leopold *Weismann*, German biologist, 1834– ]. The theory of continuity of the germ-plasm, and the non-inheritance of acquired characters.

**Weitbrecht's cartilage** (*vīt'-brekt*) [Josias Weitbrecht, German anatomist, 1702–1747]. A fibro-cartilaginous lamella frequently found interposed between the articular surfaces of the acromioclavicular joint. **W.'s foramen,** one in the capsule of the shoulder-joint, through which the synovial membrane communicates with the bursa lining the under surface of the tendon of the subscapularis muscle. **W.'s ligament,** a rounded, fibrous bundle, extending from the outer portion of the coronoid process to the inner border of the radius, above the bicipital tuberosity. **W.'s retinacula,** flat bands lying on the neck of the femur and formed by the deeper fibers of the capsular ligament, which are reflected upward along the neck to be attached nearer to the head.

**Welch's bacillus** [William Henry Welch, American pathologist, 1850– ]. *Bacillus ærogenes capsulatus.*

**Wells' facies** [Sir Thomas Spencer Wells, English gynecologist, 1818–1897]. The facies of ovarian disease.

**welt.** See *wheal.*

**Weltmerism** (*welt'-mer-izm*) [Samuel A. Weltmer]. A method of treatment by suggestion.

**wen.** A sebaceous cyst.

**Wender's test for glucose** (*ven'-der*) [Neumann Wender, Austrian chemist]. Make a solution of 1 part methylene-blue in 3000 parts of distilled water. On rendering this solution alkaline with potassium hydroxide and heating with a glucose solution it becomes decolorized.

**Wenzell's test** (*vent'-sel*) [William Theodore Wenzell, American physician, 1829– ]. A test for strychnine. One part of potassium permanganate in 2,000 of sulphuric acid is added to the suspected fluid; a color reaction is given if strychnine is present.

**Wenz's method.** A method of removing the proteid from a fluid. Saturate the solutions with ammonium sulphate, and all the proteids except peptones will be precipitated, and may be filtered off.

**Weppen's test.** (1) *For morphine:* sugar, sulphuric acid, and bromine are added to the suspected solution; a red color indicates the presence of morphine. (2) *For veratrine:* sugar and sulphuric acid are added to the suspected solution; a blue or green or yellow color indicates the presence of veratrine.

**Werlhof's disease** (*verl'-hof*) [Paul Gottlieb Werlhof, a German physician, 1699–1767]. Purpura hæmorrhagica.

**Wernekinck's commissure** (*ver'-ne-kink*) [Friedrich Christian Wernekinck, German anatomist, 1798–1835]. The decussating fibers of the middle cerebellar peduncle.

**Wernicke's aphasia** (*ver'-ne-keh*) [Karl Wernicke, German neurologist, 1848–1905]. Cortical sensory phasia. **W.'s area,** same as W.'s triangle, *q. v.* **W.'s center,** the auditory word-center in the posterior third of the first temporosphenoid convolution. **W.'s convolution,** the first temporosphenoid convolution. **W.'s disease,** polioencephalitis acuta hæmorrhagica; acute superior encephalitis. **W.'s fibers.** See *Gratiolet's optic radiation.* **W.'s field.** See *W.'s triangle.* **W.'s fissure,** a nearly vertical fissure sometimes seen to divide the parietal and temporal lobes from the occipital lobe. **W.'s reaction, W.'s sign,** A reaction obtained in some cases of hemianopia in which a pencil of light thrown on the blind side of the retina gives rise to no movement in the iris, but thrown upon the normal side, produces contraction of the iris. It indicates that the lesion producing the hemianopia is situated at or anterior to the geniculate bodies. **W.'s triangle,** a triangular area formed by the decussation, at various angles, of the radiating fibers of Gratiolet with the fibers proceeding from the external geniculate body and pulvinar; it occupies the extreme posterior segment of the capsula.

**Wertheim's ointment** (*vert'-hīm*) [Gustav Wertheim, Austrian physician, 1822–1888]. An ointment used in treating chloasma. It consists of ammoniated chloride of mercury and bismuth, each two drams, and glycerin ointment one ounce.

**Wertheim's operation** (*vert'-hīm*) [Ernst Wertheim, Austrian gynecologist, 1864– ]. A plastic operation for procidentia uteri, and cystocele.

**Westbrook's operation.** Cardicentesis.

**Westphal's nucleus** (*vest'-fahl*) [Karl Friedrich Otto Westphal, German neurologist, 1833–1890]. The nucleus of origin of a part of the trochlear nerve-fibers; it is situated posteriorly to the trochlear nucleus proper. **W.'s paradoxic contraction,** tonic contraction of the anterior muscles of the leg (especially the tibialis anticus) on passive flexion of the foot; it is occasionally seen in multiple sclerosis, paralysis agitans, tabes, alcoholism, and hysteria. **W.'s sign,** absence of the patellar reflex; it occurs in lesions of the spinal cord at the level of the reflex center (*e. g.,* tabes dorsalis, paretic dementia), neuritis, certain cases of cerebellar disease, etc. **W.'s zone,** a zone in the posterior column of the lumbar spinal cord, which is bounded externally by the inner side of the posterior horn, internally by an imaginary anteroposterior line drawn through the point at which the posterior horn turns inward, and posteriorly by the periphery of the cord. It contains the afferent fibers concerned in the patellar reflex mechanism.

**Westphal-Erb's sign.** See *Westphal's sign.*

**Westphal-Piltz's reflex** [Alexander Karl Otto Westphal, German neurologist, 1863– ; Alexander Piltz, Austrian neurologist, 1871– ]. See *Gifford's reflex.*

**wet.** Not dry; moist. **w. brain,** an excessively serous condition of the brain. **w. cupping,** cupping combined with scarification, whereby some blood is drawn. **w.-dream,** seminal emission during sleep, generally accompanying an erotic dream. **w. nurse,** a woman who suckles the child of another. **w. pack,** a means of reducing temperature by wrapping a patient in a wet sheet and covering with dry blankets. **w. scald,** eczema in sheep. **w. tetter,** weeping eczema.

**Wetzel's test for CO in blood** [*vet'-sel*] (Georg Wetzel, German physician, 1871– ]. Add to the blood 4 volumes of water and treat with 3 volumes of a 1% tannic acid solution. In the presence of carbon monoxide the blood becomes carmine red; normal blood gradually becomes gray.

**Weyl's reaction for creatinin** (*vīl*) [Theodor Weyl, German chemist, 1851– ]. Add to the creatinin solution a few drops of a dilute solution of sodium nitroprusside, and then, drop by drop, a few drops of sodium hydroxide. A ruby-red coloration results, quickly changing to yellow again.

**wharl** (*hwarl*) [A variety of *whirl*]. The uvular or rattling utterance of the *r.* sound.

**Wharton's duct** (*hwar'-tun*) [Thomas Wharton, English anatomist, 1610–1673]. The duct of the submaxillary gland. **W.'s jelly,** the gelatinous embryonic connective tissue of the umbilical cord.

**wheal** (*hwēl*) [AS., *hwēle*]. A whitish or pinkish elevation, developing suddenly upon the skin, and lasting usually but a short time. Wheals are produced by urticaria, the bites of insects, or the sting of a nettle. Syn., *pomphus; urtica.* /w.-**worm,** the *acarus scabiei,* or itch insect.

**wheat** (*hwēt*). See *triticum.*

**wheatena** (*hwēt-ē'-nah*). An artificial food said to contain all the elements of the wheat-berry except the husk. The starch granules have already been ruptured by heat, and only a few moments' cooking is necessary.

**Wheatstone's bridge** (*hwēt'-stōn*) [Charles Wheatstone, English physicist, 1802–1875]. An instrument for measuring electrical resistance.

**Wheelhouse's operation** (*hwēl'-house*) [Claudius Galen Wheelhouse, English surgeon, ]. A perineal incision through the urethra for stricture; external urethrotomy.

**wheeze** (*hwēz*). To breathe hard; to breathe with difficulty and with an audible whistling sound.

**wheezing** (*hwēez'-ing*). The half-stertorous, sibilant sound occasionally observed in the breathing of persons affected with croup, asthma, or coryza.

**whelk** (*hwelk*). A protuberance upon the face, due to alcoholism; acne rosacea; a pimple.

**whetstone crystals.** Peculiar crystals of xanthin found in urine.

**whettle-bones.** The vertebræ.

**whey** (*hwā*). The liquid part of milk separating from the curd. **alum-w.,** whey separated by stirring milk with a lump of alum; a popular remedy for sore eyes. **wine-whey,** a whey prepared by adding Rhine wine one part to hot milk four parts, and straining. **w.-cure,** the administration of whey as a method of treating certain diseases.

**whiff** (*hwif*) [origin obscure]. A puff of air, w., oral, a peculiar sound heard during expiration from the open mouth, principally in cases of thoracic aneurysm.

**whip-worm** (*hwip'-wurm*) The *Trichocephalus dispar*.
**whirl** (*hwurl*) [ME., *whirlen*, to whirl]. To revolve rapidly. **w.-bone.** 1. The head of the femur. 2. The patella.
**whiskey, whisky** (*hwis'-ke*). See *spiritus frumenti*. **w.-nose.** See *acne rosacea*.
**whisper** (*hwis'-per*). A low, soft, sibilant sound produced by the passage of the breath through the glottis without throwing the vocal cords into vibration.
**whispered bronchophony.** See *bronchophony*.
**whispering pectoriloquy.** See *pectoriloquy, whispering*.
**whistle.** A sound produced by forcing the breath through the contracted lips. **w., Galton's.** See *Galton's whistle*.
**white** (*hwit*) [AS., *hwit*]. 1. Having a color produced by reflection of all the rays of the spectrum; opposed to black. 2. Any white substance, as *white of egg*. **w. arsenic,** arsenic trioxide. **w. cell, w. corpuscle,** the leukocyte. **w. commissure,** the anterior commissure of the spinal cord. **w. gangrene,** gangrene with anemia of the tissues. **w. lead,** basic lead carbonate. **w. leg.** See *Phlegmasia alba dolens*. **w. leprosy,** vitiligo. **w. line,** the linea alba. **w. matter,** the part of the brain and spinal cord consisting of medullated nerve-fibers, and having a white color. **w. mustard,** sinapis alba. See under *mustard*. **w., pearl-,** bismuth subnitrate. **w., permanent,** a commercial name for barium sulphate. **w. pneumonia.** See *pneumonia, white*. **w. precipitate,** hydrargyrum ammoniatum. See *mercury, ammoniated*. **w. softening,** softening of nerve-substance in which the affected area presents a whitish color, due to fatty degeneration following anemia. **w. substance of Schwann,** the myelin sheath of medullated nerve-fibers. **w. swelling.** See *swelling, white*. **w. vitriol,** zinc sulphate.
**White's disease.** Keratosis follicularis.
**White's operation** [J. William *White*, American surgeon, 1850– ]. Castration for cure of enlarged prostate.
**Whitehead's operation** [Walter *Whitehead*, English surgeon, 1840–1913]. 1. Excision of the tongue, through the mouth. 2. Excision of hemorrhoids, by removal of a circular strip of mucous membrane around the anus, including the tumors.
**whites** (*hwits*). See *leukorrhea*.
**whiting.** Prepared chalk or white clay; purified calcium carbonate.
**whitlow** (*hwit'-lo*). See *paronychia*. **w., melanotic,** a form of melanotic sarcoma simulating whitlow in appearance. **w., painless.** See *Morvan's disease*.
**whoop** (*hoop*). The inspiratory crow which precedes or occurs during a paroxysm in whooping-cough.
**whooping-cough** (*hoo'-ping-kof*). An infectious disease characterized by catarrhal inflammation of the air-passages and peculiar paroxysms of cough ending in a loud whooping inspiration. It is most frequent in children, and is probably due to a specific microorganism.
**whorl** (*hworl*). 1. A spiral turn, in general. 2. The spiral turn of the external fibers of the heart where they join the inner fibers.
**Whytt's disease** (*hwit*) [Robert *Whytt*, Scottish physician, 1714–1766]. Hydrocephalus internus: a collection of fluid in the cerebral ventricles.
**Wichmann's asthma** (*vik'-mahn*) [Johann Ernst *Wichmann*, German physician, 1740–1802]. Laryngismus stridulus. Kopp's asthma.
**Wickersheimer's fluid** (*vik'-ers-hi-mer*). A fluid employed for the preservation of anatomical specimens, consisting of arsenic trioxide, sodium chloride, potassium sulphate, carbonate, and nitrate, dissolved in a mixture of glycerol, methyl-alcohol, and water.
**wicking** (*wik'-ing*). Loosely twisted unspun cotton or wool; it is employed in packing cavities.
**Widal's reaction or test,** (*ve'-dal*) [Fernand *Widal*, French physician, 1862– ]. The addition of a few drops of a recent culture of *Bacillus typhi abdominalis* (Eberth) to the serum of a typhoid-fever patient causes an agglutination and loss of movement of the bacilli.
**Widmer's sign.** The temperature in the right axilla is higher than that in the left axilla; found in appendicitis.
**Wigger's ergotin.** An alcoholic extract prepared from ergot deprived of fixed oil by means of ether.
**Wilde's cords** [Sir William Robert Willis *Wilde*,

Irish surgeon, 1815–1876]. The transverse fibers of the callosum. **W.'s incision.** See *Wilde's operation*. **W.'s luminous triangle.** See *Politzer's luminous cone*. **W.'s operation.** For mastoid or cerebral abscess; the bone is exposed from the base to the apex of the mastoid process, one-half inch behind the auricle, and, if necessary, the bone is opened with a drill, gouge, or trephine.
**Wildermuth's ear** (*vil'-der-moot*) [Hermann A. *Wildermuth*, German neurologist, 1852– ]. A congenital deformity of the ear consisting in a prominence of the anthelix, the helix being turned downward.
**Wilkinson's disease.** Paralysis agitans.
**Wilks' disease** [Sir Samuel *Wilks*, English physician, 1824–1911]. Chronic parenchymatous nephritis. **W.'s kidney,** the large white kidney. **W.'s symptom-complex.** See *Erb's disease*.
**Willan's leprosy** [Robert *Willan*, English physician, 1757–1812]. Psoriasis. **W.'s lupus,** lupus vulgaris.
**Willard's disease.** See *lupus vulgaris*.
**Williams' sign** [Charles *Williams*, English physician, 1838–1889]. Diminished inspiratory expansion on the left side in adherent pericardium. **W.'s tracheal sound,** the peculiar resonance sometimes found in the second intercostal space in cases of very large pleural effusion. It is a dull tympanic resonance, becoming higher on opening the mouth, and arising from the vibrations of air in a large bronchus surrounded by compressed lung.
**Williamson's blood-test for diabetes.** Place in a narrow test-tube 40 Cmm. of water and 20 Cmm. of blood; add 1 Cc. of an aqueous solution of methylene-blue (1 : 6000) and 40 Cmm. of solution of potassium hydroxide. Place the tube in a water-pot, which is kept boiling. From the blood of a diabetic patient the blue color disappears in four minutes and becomes yellow. In blood that is not diabetic the blue color remains.
**Willis' accessory nerve** [Thomas *Willis*, English anatomist, 1621–1675]. The spinal accessory nerve. **W.'s arteries,** the anterior and posterior communicating arteries of the brain. **W., circle of,** the arterial anastomosis at the base of the brain, formed by the anterior communicating artery between the anterior cerebral arteries, the internal carotids and middle and posterior cerebral arteries, and the posterior communicating arteries. **W.'s cords.** 1. Fibrous trabeculæ stretching across the lower angle of the superior longitudinal sinus. 2. See *Wilde's cords*. **W.'s disease,** diabetes mellitus. **W.'s glands,** the corpora albicantia. **W.'s ophthalmic branch,** the ophthalmic division of the fifth cranial nerve. **W.'s paracusis,** increased hearing power in the presence of a loud noise. **W.'s valve.** See *valve of Vieussens*.
**Willock's respiratory jacket** (*wil'-ok*). A jacket used in pulmonary emphysema.
**willow** (*wil'-o*) [AS., *welig*]. *Salix*, a genus of trees of the order *Salicaceæ*, the bark and leaves of which contain salicin. *Salix alba* and *Salix nigra* have been used in spermatorrhea, neuralgia, and malaria.
**Wilson's disease** [William James Erasmus *Wilson*, English dermatologist, 1809–1884]. General exfoliative dermatitis. **W.'s lichen,** lichen planus. **W.'s muscle,** a nonconstant fasciculus of the compressor urethræ which is attached to the body of the pubis near the symphysis.
**Wilson-Brocq's disease.** See *Wilson's disease*.
**Winckel's disease.** (*ving'-kel*) [Franz Karl Ludwig Wilhelm von *Winckel*, German obstetrician, 1837– ]. An epidemic disease of children, characterized by cyanosis, jaundice, and hemoglobinuria.
**Winckler's bodies.** Spherical masses seen in syphilitic tissues.
**Winckler's test for free HCl in the gastric juice.** Mix a few drops of the filtered gastric juice in a porcelain capsule with a few drops of a 5% alcoholic solution of alphanaphthol to which 0.5 to 1% of glucose has been added. On heating gently a bluish-violet zone appears, which darkens rapidly.
**windage** (*win'-dāj*). The compression of air said to be produced by the passage of a bullet or other similar missile close to the body, and to give rise to an injury called *wind-contusion*.
**wind-contusion.** Windage.
**windlass, Spanish.** A form of tourniquet consisting of a handkerchief tied about a part and twisted by means of a stick.
**window** (*win'-do*): 1. An aperture in a wall for the admission of light and air. 2. A small aperture in a

bone. See *fenestra*. **w. resection,** submucous resection of part of nasal septum.
**windpipe** (*wind'-pīp*). See *trachea*.
**wind-pox.** Chickenpox.
**wine** (*wīn*) [L., *vinum*]. 1. The fermented juice of the grape. 2. A solution of a medicinal substance in wine. Wines consist chiefly of water and alcohol, the latter varying from 6 to 22% (from 10 to 14% in the official wines). In addition they contain volatile oil, enanthic ether, grape-sugar, traces of glycerol, coloring-matter, tannic, malic, phosphoric, carbonic, and acetic acids, potassium bitartrate, and calcium tartrate. Wine is used as a beverage in most civilized countries, and in medicine as a stimluant like alcohol. **w., red** (*vinum rubrum*, U. S. P.), the fermented juice of fresh colored grapes. The most important varieties are claret, Bordeaux, and port. **w., white** (*vinum album*, U. S. P.), the fermented juice of grapes freed from seeds, stems, and skins. The most important varieties are sherry (*vinum xericum*), Madeira, Catawba, etc.
**wineglass.** A measure holding nearly two fluid ounces.
**wine-press of Herophilus.** See *torcular*.
**Winiwarter's operation** (*vin'-e-var-ter*) [Alexander von *Winiwarter*, German surgeon, 1848– ]. Cholecystenterostomy, in two stages. In the first the gall-bladder is united to the upper portion of the jejunum and the parts are fixed to the parietal peritoneum; in the second, after about five days, the bowel is incised and a communication is established between the latter and the gall-bladder.
**wing.** See *ala*. **w. of Ingrassias,** the alæ of the sphenoid.
**wink.** To open and close the eyelids quickly.
**Winslow's foramen** (*wins'-lo*) [Jacob Benignus *Winslow*, Danish anatomist, 1669–1760]. An aperture between the liver and stomach, bounded in front by the portal vein, hepatic artery and duct, behind, by the inferior vena cava, below by the hepatic artery, and above, by the liver. It is forned by folds of the peritoneum, and establishes communication between the greater and lesser cavities of the peritoneum. **W.'s ligament,** the ligamentum posticum Winslowii, the posterior ligament of the knee-joint. **W.'s pancreas,** the lesser pancreas. **W.'s pouch,** the gastrohepatic omentum. **W.'s stars,** capillary whorls which form the beginning of the vorticose veins of the choroid.
**wintera,** winter's bark (*win'-ter-ah*). The bark of *Drimys winteri*; it is aromatic and tonic, used in scurvy.
**wintergreen.** See *gaultheria*.
**winter-itch.** See *pruritus hiemalis*.
**Winternitz's sound** (*vin'-ter-nitz*) [Wilhelm *Winternitz*, Austrian physician, 1835– ]. A double current catheter through which water may circulate, and so heat or cold may be applied to urethra or prostate.
**Wintrich's change of pitch** (*vin'-trik*) [Anton *Wintrich*, German physician, 1812–1882]. The tympanitic sound of pneumothorax and of cavities communicating freely with a bronchus becomes higher in pitch when the mouth is opened and lower when the mouth is closed.
**wiring.** Securing by means of wire the fragments of a broken bone.
**Wirsung, canal or duct of** (*vēr'-soong*) [Johann Georg *Wirsung*, Bavarian anatomist, –1643]. The excretory duct of the pancreas.
**wiry** (*wi'-re*). Resembling wire; tough and flexible. **w. pulse.** See *pulse*.
**wisdom-tooth.** The last molar tooth, which is the last of all the teeth to appear.
**wismol.** A proprietary dusting powder.
**Wistar's pyramids.** See *Bertin, bones of*.
**witch-hazel.** See *hamamelis*.
**witherite** (*with'-ur-ite*) [W. *Withering*, English physician, 1741–1799]. Native barium carbonate.
**withers** (*with'-ers*). The ridge above the shoulders of the horse, formed by the spinous processes of the first eight or ten thoracic vertebræ.
**Witz's test for HCl in the contents of the stomach.** An aqueous solution of methyl-violet (strength 0.025%) is first colored blue, then green, and finally decolorized by dilute inorganic acids.
**Wladimiroff's operation** (*vla-de'-me-rof*). Tarsectomy; Mikulicz's operation. The heel portion of the foot, consisting of the astragalus, os calcis, and the soft parts covering them, is removed; the articular surfaces of the tibia, fibula, cuboid, and scaphoid are sawn off and the foot is brought into a straight line with the leg.
**Woehler's rings.** See *Meyer's rings*.
**Woelde's triangle** (*vel'-der*). Politzer's luminous cone.
**Woelfler's operation** (*vel'-fler*) [Anton *Woelfler*, Austrian surgeon, 1850– ]. Gastroenterostomy.
**Woillez's disease** (*vwah-la'*) [Eugene Joseph *Woillez*, French physician, 1811–1882]. Acute pulmonary congestion.
**Wolff's law.** Every change in the static relations of a bone leads not only to a corresponding change of internal structure, but also to a change of external form and physiological function.
**Wolf-Eisner reaction or test** (*vulf-īz'-ner*) [Alfred *Wolff-Eisner*, German physician]. Calmette's ophthalmo-reaction.
**Wolffian body** [Kaspar Friedrich *Wolff*, German anatomist, 1733–1794]. The mesonephron, an organ of embryonic life situated on each side of the vertebral column and consisting of a series of convoluted tubes opening into a lateral duct, which is connected with the common cloaca of the alimentary and genito-urinary tracts. It disappears toward the end of the second month, leaving as a vestige the parovarium. **W. cyst,** a cyst of the broad ligament of the uterus, believed to be developed from vestiges of the Wolffian body. **W. duct,** the mesonephric duct; an embryonic duct of the mesonephron formed by longitudinal fission of part of the segmental duct. In the male it becomes the vas deferens; in the female it almost entirely disappears. **W. ridge,** a protuberance from which the W. body is developed. **W. tubules,** small tubes joining the Wolffian duct at right angles.
**wolf-flaps.** Whole skin-flaps without pedicles.
**wolfram.** Tungsten.
**wolframate.** Tungstate.
**wolfsbane.** Aconite.
**womb** (*woom*) [ME., *woombe*]. The uterus.
**wood.** The hard part of trees; the part within the bark. **w.-alcohol,** methyl-alcohol. **w., flour,** sawdust, used in surgical dressings. **w.-naphtha,** same as **w.-alcohol.** **w.-oil.** See *gurjun balsam*. **w.-sorrel,** *Oxalis acetosella*, a low, tender pubescent herb of North America, Europe, Asia, and northern Africa. It contains potassium binoxalate, which is sometimes obtained from it and sold as salt of sorrel. It has refrigerant and antiscorbutic qualities. **w.-spirit,** methyl-alcohol. **w.-sugar,** xylose. **-tar.** See *tar*. **w.-vinegar,** vinegar obtained by thewdry distillation of wood. **w.-wool,** prepared fibers of wood, used mainly as a surgical dressing.
**Woodbridge treatment** [John Eliot *Woodbridge*, American physician, 1845–1901]. Treatment of typhoid fever by intestinal antisepsis and elimination.
**wooden tongue.** See *actinomycosis*.
**wool-fat.** See *lanolin*.
**woolsorter's disease.** Anthrax.
**woorara** (*woo-rar'-ah*). See *curara*.
**word-blindness.** See under *aphasia*.
**word-deafness.** See *deafness, psychic*.
**working distance.** In a microscope the distance between the object and the objective.
**worm** (*wurm*). A member of the class *Vermes*, of the division *Invertebrata*, especially one parasitic in man or animals. **w., bladder-,** the *Tænia echinococcus*. See *tape-worm, dog-.* **w., guinea-,** **w., medina-,** the *Filaria medinensis*. See under *filaria*. **w., pin-, w., seat-, w., thread-.** See *oxyuris*. **w., whip-,** the *Trichocephalus dispar.*.
**Wormian bone** (*uw'-me-an*) [Olaus *Worm*, a Danish* physician, 1588–1654]. Any one of the small supernumerary bones found in the sutures of the skull.
**Wormley's test** [Theodore George *Wormley*, American chemist, 1826–1897]. A color reaction for alkaloids, made by treating the suspected solution with an alcoholic solution of picric acid, or with a dilute iodine-potassium-iodide solution.
**Worm-Mueller's test for sugar** [Jacob *Worm-Mueller*, Norwegian physician, 1834–1889]. A mixture of a 1.5 to 2.5% solution of cupric sulphate and an alkaline solution of Rochelle salt is added to the urine; on boiling, a yellowish precipitate of copper suboxide is formed.
**wormseed** (*wurm'-sēd*). 1. See *chenopodium*. 2. See *santonica*.
**wormwood.** See *absinthium*.
**Worremberg's apparatus.** An apparatus for polarization by reflection.

**worsted test.** See *Holmgren's test*.
**Woulfe's bottles** [Peter *Woulfe*, English chemist, 1727–1803]. An apparatus consisting of a series of two or three necked bottles connected by suitable tubes and used for washing gases or saturating liquids therewith.
**wound** (*woond*) [AS., *wund*]. A solution of continuity of an external or internal surface of the body. **w., contused,** one produced by a blunt body. **w., incised,** one caused by a cutting instrument. **w., lacerated,** one in which the tissues are torn. **w., open,** one having a free external opening. **w., penetrating,** one that pierces the walls of a cavity or enters into an organ. **w., poisoned,** one in which septic materials are introduced. **w., punctured,** one made by a pointed instrument. **w., subcutaneous,** one with a very small external opening in the skin.
**wreath.** In biology, applied to a stage in karyokinesis, as the *mother-wreath*, *daughter-wreath*.
**Wreden's test** (*wra'-den*) [Robert Robertovich *Wreden*, Russian otologist, 1837–1893]. A test of live-birth. It consists in the absence of gelatinous matter from the middle ear; this is regarded as a proof that the fetus was born alive and has breathed.
**Wright's method** [Sir Almroth Edward *Wright*, English bacteriologist, 1861– ]. The opsonic method. **W.'s vaccine,** antityphoid vaccine.
**Wright's solution** [James Homer *Wright*, American pathologist, 1869– ]. Sodium citrate, 0.5; sodium chloride, 3.0; distilled water, 100.
**wrightine** (*ri'-tēn*), C₃₄H₄₀N₂. An astringent and anthelmintic alkaloid from rnoessi bark. *Wrightia zeylanica.*
**Wrisberg's ansa memorabilis** (*ris'-berg*) [Heinrich August *Wrisberg*, German anatomist, 1739–1808]. A loop formed by the right semilunar ganglion and the anastomosis of the right pneumogastric and great splanchnic nerves. **W.'s cartilages,** the cuneiform cartilages, one on each side of the fold of membrane stretching from the arytenoid cartilage to the epiglottis. **W.'s ganglion.** 1. A ganglion frequently found in the superficial cardiac plexus at the point of union of the lower cervical cardiac branch of the left pneumogastric with the upper cardiac nerve of the sympathetic of the left side. 2. Intumescentia semilunaris. See *ganglion, Gasserian*. **W.'s lingula,** the filaments connecting the sensory and motor roots

of the trigeminus. **W.'s nerve.** 1. A small branch of the brachial plexus supplying the skin of the arm. See *cutaneous, lesser internal,* under *nerve*. 2. A small nerve arising from the medulla oblongata and coursing between the facial and auditory nerves. **W.'s pars intermedia.** See *W.'s nerve* (2).
**wrist** (*rist*) [AS.]. The part joining the forearm and the hand. See *carpus*. **w.-clonus.** See *reflex*. **wrist. w.-drop,** a paralysis of the extensor muscles of the wrist and fingers causing a dropping of the hand. **w.-joint,** the articulation between the forearm and the hand; the radio-carpal articulation.
**writer's cramp.** An occupation-neurosis occurring in those who write a great deal, and characterized by painful spasm of the fingers when an effort at writing is made.
**writing hand.** A peculiar position assumed by the hand in paralysis agitans.
**wry-neck** (*ri'-nek*). See *torticollis*.
**Wunderlich's law** or **curve** (*voon'-der-lik*) [Carl Reinhold *Wunderlich*, German physician, 1815–1867]. The ascending oscillations of the temperature-curve in typhoid fever.
**Wundt's tetanus.** A prolonged tetanic contraction induced in a frog's muscle by injury or the passage of a strong current.
**wurali** (*woo-rah'-le*). Same as *curara*.
**Wurster's test for hydrogen dioxide.** Paper saturated with a solution of tetramethylparaphenylendiamine turns blue-violet with hydrogen dioxide. **W.'s test for tyrosine.** 1. Treat a boiling aqueous solution of tyrosin drop by drop with a 1% acetic acid and a sodium nitrite solution. A red coloration results. 2. Add some dry quinone to a hot aqueous solution of tyrosine. A deep ruby-red coloration results, lasting for 24 hours, and then changing to brown.
**Wutzer's operation** (*vūt'-zer*). An operation for the radical cure of inguinal hernia.
**Wyeth's operation** [John Allan *Wyeth*, American surgeon, 1845– ]. A method of bloodless amputation of the hip or shoulder; hemorrhage is controlled by long pins and a strong elastic band or tube.
**Wylie's operation** [Walter Gill *Wylie*, American gynecologist, 1848– ]: Intra-abdominal shortening of the round ligaments of the uterus.
**Wyman's strap.** An arrangement of straps for holding a violently insane person in bed.

# X

**X.** Symbol for the decimal scale of potency or dilution, used by the homeopaths.
**xanol** (*zan'-ol*). Trade name of a preparation of sodium-caffeine salicylate.
**xanthæmatin.** See *xanthematin*.
**xanthaline** (*zan'-thal-ēn*) [ξανθός, yellow], $C_{17}H_{25}N_2$-$O_5$. An alkaloid from opium; a white, crystalline substance forming yellow salts.
**xanthamide** (*zan'-tham-id*), $C_2H_7NSO_4$. A crystalline substance derived from xanthic acid.
**xanthate** (*zan'-thāt*) [ξανθός, yellow]. A salt of xanthic acid.
**xanthein** (*zan'-the-in*) [ξανθός, yellow]. The yellow coloring matter of plants; it is soluble in water, thus differing from xanthin.
**xanthelasma** (*zan-thel-az'-mah*) [ξανθός, yellow; ἔλασμα, a metal plate]. See *xanthoma*.
**xanthelasmoidea** (*zan-thel-az-mo-id'-e-ah*) [*xanthelasma*; εἶδος, form]. Synonym of *urticaria pigmentosa*.
**xanthematin, xanthæmatin** (*zan-them'-at-in*) [ξανθός, yellow; *hematin*]. A bitter yellow substance obtained by dissolving hematin in dilutenitric acid.
**xanthene** (*zan'-thēn*) [ξανθός, yellow]. A yellow, mixture obtained from persulphocyanic acid.
**xanthic** (*zan'-thik*) [*xanthin*]. 1. Yellow. 2. Pertaining to xanthin. **x. acid**, an ester of thiosulphocarbonic acid; it is an unstable, colorless oil which decomposes at 25° C. into carbon disulphide and alcohol. **x. calculus**, urinary calculus composed mainly of xanthin. **x. oxide**, an ingredient of stony formations; xanthin.
**xanthin** (*zan'-thin*) [ξανθός, yellow], $C_5H_4N_4O_2$. A nonpoisonous leukomaine found in nearly all the tissues and liquids of the animal economy, and also in many plants; it occurs in minute quantities in urine, also in guano. It is formed in the decomposition of nuclein by dilute acids. According to Ross, it is an auxetic in cancer. It is a colorless powder, almost insoluble in cold water, but readily soluble in dilute acids and alkalies, and acts as a muscle-stimulant, especially to the heart. For tests see *Hoppe-Seyler* and *Weidel*. **x. bases**, alloxuric bases.
**xanthinoxidase** (*zan-thin-oks'-id-ās*) [*xanthin*; *oxidase*]. An oxidizing ferment which converts xanthin and hypoxanthin into uric acid.
**xanthinuria** (*zan-thin-ū'-re-ah*) [*xanthin*; *urine*]. The presence of xanthin in excess in the urine.
**Xanthium** (*zan'-the-um*). Clotbur; a genus of *Compositæ*.
**xanthiuria** (*zan-the-ū'-re-ah*). See *xanthinuria*.
**xantho-** (*zan-tho-*) [ξανθός, yellow]. A prefix meaning yellow.
**xanthochroia** (*zan-tho-kro'-e-ah*) [ξανθός, yellow; χροιά, skin]. Yellow discoloration of the skin.
**xanthochromia** (*zan-tho-kro'-me-ah*) [*xantho-*; χρῶμα, a color]. 1. A yellowish discoloration of the skin. 2. The yellow hemorrhagic discoloration of the cerebrospinal fluid, diagnostic of hemorrhage of the spinal cord.
**xanthochrous** (*zan-tho-kro'-us*) [ξανθός, yellow; χροιά, skin]. Yellow-skinned.
**xanthocreatinine** (*zan-tho-kre-at'-in-in*) [*xantho-*; *creatinine*], $C_5H_{10}N_4O$. A leukomaine found in muscle, crystallizing in yellow crystals and resembling creatinine. It produces depression, somnolence, fatigue, frequent defecation, and vomiting.
**xanthocyanopia, xanthocyanopsia** (*zan-tho-si-an-o'-pe-ah, zan-tho-si-an-op'-se-ah*) [*xantho-*; κύανος, blue; ὄψις, sight]. A defect of color-vision in which yellow and blue are perceived, while red is imperceptible.
**xanthocystin** (*zan-tho-sis'-tin*) [ξανθός, yellow; κύστις, bladder]. A nitrogenous substance found in the whitish tubercles in the mucosæ and organs of a corpse.
**xanthocyte** (*zan'-tho-sīt*) [*xantho-*; κύτος, a cell]. A cell secreting a yellow pigment.
**xanthoderma, xanthodermia** (*san-tho-der'-mah, zan-tho-der'-me-ah*) [*xantho-*; δέρμα, skin]. A yellow discoloration of the skin.
**xanthodont, xanthodontous** (*zan'-tho-doni, zan-tho-don'-tus*) [*xantho-*; ὀδούς, tooth]. Having yellow teeth.
**xanthogen** (*zan'-tho-jen*). Same as *xanthein*.
**xanthogenic acid** (*zan-tho-jen'-ik*). Same as *xanthic acid*.
**xanthoglobulin** (*zan-tho-glob'-ū-lin*) [ξανθός, yellow; *globulin*]. Same as *hypoxanthin*. A yellow substance found in the liver and pancreas.
**xanthokyanopy** (*zan-tho-ki-an'-o-pe*) [ξανθός, yellow; κύανος, blue; ὤψ, eye]. Red-green blindness, with undiminished spectrum. See *xanthocyanopsia*.
**xantholin** (*zan'-tho-lin*). Same as *santonica*.
**xanthoma** (*zan-tho'-mah*) [*xantho-*; ὄμα, tumor]. Xanthelasma; a newgrowth of the skin occurring as flat or slightly raised patches or nodules from a pin-head to a bean in size, and of a yellowish color. The flat lesions (*xanthoma planum*) usually occur about the eyelids; the elevated or tubercular variety (*xanthoma tuberculatum*; *xanthoma tuberosum*) on the neck, trunk, and extrémities. Histologically the lesions consist of connective tissue undergoing a partial fatty degeneration. **x. diabeticorum**, a rare disease of the skin associated with diabetes mellitus, the lesions of which are denser and firmer than those of true xanthoma, and are dull red, discrete, and solid, with a yellowish point at the apex. **x. glycosuricum**, that marked by grape-sugar or pentose in the urine. **x. multiplex**, a form occurring usually in women about middle life. **x. planum**, **x. palpebrarum**, the commoner form of xanthoma, usually occurring on the eyelids. **x. tuberculatum**, **x. tuberosum**, a form marked by tubercular lesions on the extensor surfaces of the extremities and on parts exposed to pressure.
**xanthomatosis** (*zan-tho-mat-o'-sis*) [ξανθός, yellow; ὄμα, tumor]. Xanthoma of so marked a type as to indicate a special diathesis.
**xanthomatous** (*zan-tho'-mat-us*) [ξανθός, yellow; ὄμα, tumor]. Of the nature of, or affected with xanthoma.
**xanthomelanous** (*san-tho-mel'-an-us*) [ξανθός, yellow; μέλας, black]. Having yellow or olive skin and black hair.
**xanthone** (*zan'-thōn*). Same as *bromeione*.
**xanthopathy** (*zan-thop'-ath-e*). See *xanthoderma*.
**xanthophane** (*zan'-tho-fān*) [*xantho-*; φαίνειν, to show]. A yellow pigment found in the retinal cones.
**xanthophose** (*zan'-tho-foz*) [*xantho-*; φῶς, light]. A yellow phose.
**xanthophyll** (*zan'-tho-fil*) [ξανθός, yellow; φύλλον, leaf]. The yellow pigment of plants, developed in the leaves.
**xanthopia** (*zan-tho'-pe-ah*). Same as *xanthopsia*.
**xanthopicrin** (*zan-tho-pik'-rin*) [ξανθός, yellow; πικρός, bitter]. A yellowish coloring matter, derived from the bark of *Xanthoxylum caribæum*. Same as *berberine*.
**xanthoplasty** (*zan'-tho-plas-te*) [ξανθός, yellow; πλάσσειν, to form]. A plastic operation for xanthoderma.
**xanthoproteic** (*zan-tho-pro-te'-ik*) [*xanthoprotein*]. Derived from or related to xanthoprotein. **x. acid**, $C_8H_{20}O_{24}N_4$, a non-crystallizable acid, produced by decomposing proteins with nitric acid. **x. reaction**, the deep-orange color obtained by the addition of ammonia to proteids that have been heated with strong nitric acid.
**xanthoprotein** (*zan-tho-pro'-te-in*) [*xantho-*; *protein*]. A yellowish substance formed from proteids by the action of nitric acid.
**xanthopsia** (*san-thop'-se-ah*) [*xantho-*; ὄψις, vision]. Yellow vision; the condition in which objects look yellow. It sometimes accompanies jaundice.
**xanthopsydracia** (*san-thop-se-dra'-she-ah*) [*xanthos*; ψύδραξ, a pimple]. The occurrence on the skin of yellow pimples or pustules.
**xanthopuccine** (*zan-tho-puk'-seen*) [*xanthos*; *puccoon*]. An alkaloid found in hydrastis.
**xanthorrhea, xanthorrhœa** (*san-thor-e'-ah*) [ξανθός,

yellow; ῥέων, to flow]. An acrid, purulent, yellow discharge from the vagina.
**Xanthorrhiza** (zan-tho-ri'-zah) [ξανθός, yellow; ῥίζα, a root]. A genus of ranunculaceous plants. **X. apiifolia**, yellow root, a bitter tonic.
**xanthosis** (zan-tho'-sis) [ξανθός, yellow]. The yellow pigmentation sometimes observed in carcinoma and degenerating tissues.
**xanthous** (zan'-thus) [ξανθός, yellow]. Having a yellow skin or complexion.
**xanthoxylene** (zan-thok'-sil-ēn) [ξανθός, yellow; ξύλον, wood]. C₁₀H₁₆. A colorless, volatile oil obtained from the fruit of *Xanthoxylum alatum*.
**xanthoxylin** (zan-thok'-sil-in) [ξανθός, yellow; ξύλον, wood]. 1. A precipitate from a tincture of *Xanthoxylum fraxineum*, stimulant, styptic, tonic, sialagogue. Dose 1-2 grains. 2. A stearopten from the volatile oil of *Xanthoxylum piperitum*.
**xanthoxylum** (zan-thoks'-il-um) [xantho-; ξύλον, wood]. Prickly ash, a genus of trees of the order *Rutaceæ*. The dried bark of **X.** *americanum* or of *Fagara clava-herculis* yields the *xanthoxylum* of the U. S. P.; it contains a crystalline principle, *xanthoxylin*, and is irritant, stimulant, and slightly diaphoretic. It is used in chronic rheumatism and as an emmenagogue. Dose 30 grains (2.0 Gm.). **x.**, **fluidextract of** (*fluidextractum xanthoxyli*, U. S. P.). Dose 30 minims (2.0 Cc.).
**xanthuria** (zan-thū'-re-ah). See *xanthinuria*.
**xanthylic** (zan-thil'-ik) [*xanthine*; ὕλη, matter]. Pertaining to xanthine.
**xaxa** (saks'-ah). Acetyl-salicylic acid.
**xaxaquin** (saks'-ak-win). Trade name of a preparation of acetyl salicylic acid. **x.-bases.** Alloxur bases.
**x-disease.** A condition of general malaise, with abnormal sensitiveness to cold, disturbances of digestion, respiration and cardiac action; its origin is unknown.
**Xe.** Chemical symbol of *xenon*.
**xenarthral** (zen-ar'-thral) [ξένος, strange; ἄρθρον, joint]. Strangely jointed.
**xenembole** (zen-em'-bo-le) [ξένος, foreign; ἐμβολή, a throwing in]. Same as *xenenthesis*.
**xenenthesis** (zen-en'-the-sis) [ξένος, foreign; ἐν, in; θέσις, a placing]. The introduction of a foreign body into the organism.
**xeno-** (zen-o-) [ξένος, strange]. A prefix meaning strange or foreign.
**xenogenesis** (zen-o-jen'-es-is). See *heterogenesis*.
**xenogenetic, xenogenic** (zen-o-jen-et'-ik, zen-o-jen'-ik) [ξένος, strange; γεννᾶν, to produce]. Pertaining to xenogenesis.
**xenogenous** (zen-oj'-en-us) [ξένος, foreign; γεννᾶν, to produce]. Caused by a foreign body.
**xenogeny** (zen-oj'-en-e) [ξένος, strange; γένος, kind]. Same as *xenogenesis*.
**xenomenia** (zen-o-me'-ne-ah) [xeno-; μηνιαῖα, menses]. Vicarious menstruation.
**xenon** (zen'-on) [ξένος, strange]. A gaseous element found in the atmosphere; atomic weight 130.2; symbol Xe.
**xenophthalmia** (zen-of-thal'-me-ah) [*xeno-*; ὀφθαλμός, eye]. Conjunctivitis due to injury.
**xenosite** (zen'-o-sīt) [*xeno-*; σῖτος, food]. A parasite in an intermediate condition, organ, or host.
**xeransis** (ze-ran'-sis) [ξήρανσις, a drying up]. The drying up or desiccation of a part or of a drug.
**xerantic** (ze-ran'-tik) [see *xeransis*]. Having desiccative properties; drying.
**xeraphion** (ze-raf'-e-on) [ξηρόν, dry]. 1. A medicine to check discharges. 2. A medicine to be taken dry.
**xerasia** (zer-a'-ze-ah) [ξηρός, dry]. A disease of the hair marked by cessation of growth and excessive dryness.
**xerium** (ze'-re-um). See *xeraphion*.
**xero-** (zer-o-) [ξηρός, dry]. A prefix meaning dry.
**xerocollyrium** (ze-ro-kol-ir'-e-um) [*xero-*; κολλύριον, collyrium]. A dry collyrium; an eye-salve.
**xeroderma, xerodermia** (zer-o-der'-mah, zer-o-der'-me-ah) [*xero-*; δέρμα, skin]. 1. An abnormal dryness of the skin. 2. A disease characterized by dryness and harshness of the skin, discoloration, and a fine scaly desquamation; by some it is considered a mild form of ichthyosis. **x. pigmentosum**, a rare disease of the skin usually beginning in childhood, and characterized by disseminated pigment-spots, telangiectasis, atrophy of muscles, and contraction of the skin, generally followed by the development of ulcers, and ending in death. Syn., *angioma pigmentosum atrophicum*; *atrophoderma pigmentosum*; *Kaposi's disease*; *melanosis lenticularis progressiva*.
**xeroform** (zer'-o-form). Tribromphenol-bismuth: an odorless, neutral powder containing 49.5% of bismuth oxide and 50% of tribromphenol. It is an internal antiseptic, and is recommended as a specific against Asiatic cholera. Dose 7½ gr. (0.5 Gm.).
**xeroma** (zer-o'-mah). See *xerophthalmia*.
**xeromycteria** (ze-ro-mik-te'-re-ah) [*xero-*; μυκτήρ, the nose]. Lack of moisture in the nasal passages.
**xeronosus** (zer-on'-o-sus) [*xero-*; νόσος, disease]. A condition of dryness of the skin.
**xerophagia** (ze-ro-faj'-e-ah) [*xero-*; φαγεῖν, to eat]. The use of dry or desiccated food.
**xerophagy** (ze-rof'-aj-e). See *xerophagia*.
**xerophthalmia** (zer-of-thal'-me-ah) [*xero-*; ὀφθαλμός, eye]. A dry and thickened condition of the conjunctiva, sometimes following chronic conjunctivitis or disease of the lacrimal apparatus.
**xerosis** (ze-ro'-sis) [ξηρός, dry]. A state of dryness, especially of the skin (see *xeroderma*) or of the conjunctiva (see *xerophthalmia*). **x. epithelialis, x. infantilis, x. triangularis**, xerophthalmia marked by a lusterless, grayish-white, foamy, greasy, very persistent deposit on the conjunctiva.
**xerostomia** (zer-o-sto'-me-ah) [*xero-*; στόμα, mouth]. Dry mouth, a peculiar condition characterized by suppression of the secretion of the salivary and buccal glands.
**xerotes** (zer-o'-tēz) [ξηρότης, dryness]. Dryness; a dry habit of the body.
**xerotic** (zer-ot'-ik). Characterized by xerosis; dry.
**xerotripsis** (zer-o-trip'-sis) [*xero-*; τρίβειν, to rub]. Dry friction.
**xinol** (si'-nol). See *zinol*.
**xiphi-, xipho-** (zif-i-, xif-o-) [ξίφος, sword]. Prefixes signifying relating to the xiphoid cartilage.
**xiphicostal.** See *xiphocostal*.
**xiphisternum** (zif-is-ter'-num) [*xiphi-*; *sternum*]. The xiphoid cartilage.
**xiphocardia.** Pain in the ensiform cartilage.
**Xiphoid** (zif'-oid) [ξίφος, sword; εἶδος, like]. Swordshaped; ensiform. **x. appendix, x. cartilage, x. process.** The third piece, or ensiform process, of the sternum; it becomes osseous in mature age.
**xiphopagus** (zif-op'-ag-us) [*xipho-*; πάγος, fixed]. A double monster united by the xiphoid cartilages.
**x-knee.** Knock-knee.
**x-leg.** Genu valgum.
**x-gram.** Skiagram, radiogram.
**x-graph.** Skiagraph, radiograph.
**x-ray photography.** Synonym of *skiagraphy*.
**x-rays.** See *rays*, *Roentgen-*.
**xylem** (zi'-lem) [ξύλον, wood]. The inner part of the vascular bundle in a plant stem.
**xyli** (zi'-lēn). See *xylol*.
**xylenin, xylenobacillin** (si'-len-in, zi-len-o-bas'-il-in). See *zylenin*.
**xylenol** (zi'-len-ol). A colorless, crystalline substance resembling phenol. It occurs in three isomeric forms (ortho-, meta-, and paraxylenol).
**xylidene** (zi'-lid-ēn) [ξύλον, wood]. C₈H₁₁NH₂. A methylated homologue of anilin; used for the preparation of pigments.
**xylo-** (zi-lo-) [ξύλον, wood]. A prefix meaning pertaining to or derived from wood.
**xylobalsamum** (zi-lo-bol'-sam-um) [*xylon*; *balsam*]. Balm of Gilead.
**xylochloral** (zi-lo-klo'-ral). A crystalline compound of xylose and chloral, prepared by heating with hydrochloric acid; used as a hypnotic.
**xylogene** (zi'-lo-jēn) [ξύλον, wood; γεννᾶν, to produce]. A woody substance found in vegetable cellwalls.
**xyloidin** (zi-loi'-din) [*xylo-*; εἶδος, like], C₆H₉NO₇. A white, inflammable substance obtained by the action of nitric acid upon starch or various forms of woody fiber.
**xylol** (zi'-lol) [*xylo-*; *oleum*, oil], C₈H₁₀. Dimethyl-

# XYLOMA 956 XYSTER

benzene, a volatile hydrocarbon used in microscopy as a clearing-agent and as a solvent for Canada balsam. It has been used in small-pox. Dose 3 to 10 minims.

**xyloma** (*zi-lo'-mah*) [ξύλον, wood; -ομα, tumor]. A woody tumor found on trees or plants.

**xylon** (*zi'-lon*) [ξύλον, wood]. 1. Wood. 2. The cotton plant. 3. A substance identical with woodcellulose.

**xylonite** (*zi'-lon-īt*) [ξύλον, wood]. Celluloid.

**xylose** (*zi'-lōs*) [ξύλον, wood], $C_5H_{10}O_5$. A glucose obtained by boiling wood gum (beechwood, jute, etc.) with dilute acids.

**xylostein** (*zi-los'-te-in*) [ξύλον, wood; ὀστέον, bone]. A bitter glucoside obtained from the berries of *Lonicera xylosteum*. It is purgative and emetic.

**xylostyptic ether** (*zi-lo-stip'-tik*). Styptic collodion.

**xylotherapy** (*zi-lo-ther'-a-pe*) [ξύλον, wood; θεραπεία, therapy]. Medical treatment by the application of certain woods.

**xyol** (*zi'-ol*). Trade name of a preparation of green soap and formaldehyde.

**xysma** (*zis'-mah*) [ξύσμα, scrapings]. The flocculent pseudomembrane sometimes seen in the stools in diarrhea.

**xyster** (*zis'-ter*) [ξυστήρ, a rasp]. A surgeon's raspatory or scraping instrument.

# Y

**Yaba bark.** The bark of *Andira excelsa.*
**yabine** (*yab'-ēn*). An amorphous bitter alkaloid obtained from the bark of *Andira excelsa.*
**Yakimoff's test** (*yah'-kem-off*). For *atoxyl:* a little atoxyl is warmed in a test-tube; the faintest trace of a yellow discoloration denotes impurity.
**yam.** The esculent root of several varieties of *Dioscorea;* also, incorrectly, a coarse variety of the sweet potato. wild y. See *dioscorea.*
**y.-angle.** The angle between the radius fixus and a line joining the lambda and the inion.
**yaourt** (*yowrt*) [Turk.]. An oriental fermented drink prepared from milk.
**yard.** 1. A measure used in the United States and in England, equal to three feet. 2. The penis.
**yard-sitting,** in *massage,* sitting with the arms stretched out laterally and horizontally.
**yarrow** (*ya'-ro*). See *Achillea.*
**yava-skin** (*yah'-vah-skin*). Same as *elephantiasis, q. v.*
**yawey.** Affected with yaws.
**yawn** [AS., *ganian,* to yawn]. To gape, to open the mouth widely.
**yawning** (*yaw'-ning*). An involuntary stretching of the muscles accompanied by a deep inspiration, occurring during the drowsy state preceding the onset of sleep.
**yaw-root.** Stillingia.
**yaws** (*yaws*). See *frambesia.*
**Yb.** The chemical symbol of *ytterbium.*
**Y.-cartilage.** The cartilage occupying the triradiate fissure in the immature socket of the hipjoint.
**yeast** (*yēst*) [AS., *gist*]. The name applied to various species of *Saccharomyces.* Yeast acts as a ferment, producing the alcoholic fermentation. y., **beer-, y., brewer's,** the *cerevisiæ fermentum* of the B. P., produced by *Saccharomyces cerevisiæ.* It is used as a stimulant and locally as a poultice and deodorant to gangrenous ulcers. **y.-poultice** (*cataplasma fermenti,* B. P.), a poultice containing yeast.
**yelk.** See *yolk.*
**yellow** (*yel'-o*) [AS., *geolo*]. Of a color like that of gold; producing such a color. **y. fever,** an acute infectious disease of tropical and subtropical regions of America, and due to a specific organism, probably a protozoon, disseminated by the *Stegomyia fasciata.* After a period of incubation varying from a few hours to several days the disease begins with a chill and pain in the head, back, and limbs. The temperature rises rapidly to from 103° to 105° F., vomiting occurs, the bowels are constipated, the urine scanty and albuminous. A remission follows, after which, in severe cases, the temperature rises to its original height, jaundice develops, and the vomited material becomes dark from the presence of blood (*black vomit*). Hemorrhages may occur from the intestinal mucous membrane. The disease is very fatal, death occurring in the typhoid state or from uremia. **y.-jack.** Same as *yellow fever.* **y. precipitate,** yellow oxide of mercury. **y.-root.** See *hydrastis.* **y.-softening.** Cerebral softening with yellow discoloration. **y. spot,** the macula lutea. **y. wash.** See *wash, yellow.*
**yerba** (*yer'-bah*). An herb. **y. sagrada,** *Lantana brasiliensis;* it is antipyretic. **y. santa.** See *eriodictyon.*
**yerbine** (*yer'-bēn*) [Sp., *yerba,* herb]. An alkaloid resembling caffeine, derived from *Ilex paraguayensis.*
**Yersin Roux serum** (*yer-san'*) [Alexandre *Yersin,*

French surgeon, 1863– ]. A prophylactic and curative serum used in the treatment of plague.
**-yl** [ὕλη, matter, stuff]. A termination used in chemistry to denote a radical.
**-ylene.** A termination used in chemistry to denote a bivalent hydrocarbon radical.
**Y ligament.** The iliofemoral ligament.
**yohimbé bark** (*yo-him'-ba*). The bark of *Corynanthe yohimbé,* a tree of the Cameroon region.
**yohimbine** (*yo-him'-bēn*). A mixture of alkaloids from the bark of *Corynanthe yohimbé,* used as an aphrodisiac. Dose of the hydrochloride $\frac{1}{30}-\frac{1}{8}$ gr. (0.0032–0.01 Gm.) 3 times daily; and as a local anesthetic in 1% solution.
**yoke-bone.** The malar bone.
**yolk** (*yōk*). 1. The nutritive part of an ovum. 2. The yellow portion of an egg as distinguished from the white. **y.-cells** or **y.-granules,** the elements composing the yolk. **y.-food.** See *deutoplasm.* **y., formative,** the active living portion of the protoplasm of an ovum, with the nucleus it incloses. **y.-sac,** the larger of the two globes formed by the blastodermic membrane in the early development of the embryo, and containing the food of the embryo. **y.-stalk,** the umbilical duct. **y. of wool,** suint.
**Young-Helmholtz theory of color-vision** [Thomas *Young,* English physicist, 1773–1829; Hermann Ludwig Ferdinand *Helmholtz,* German physicist, 1821–1894]. Color-vision depends upon the presence in the retina of three different sets of fibers, which respond to stimulation by a sensation of red, green, or violet respectively. All other colors are simply combinations of the three primary colors. The excitation of any one set is a matter of wave-length. The longest waves excite the red, the shortest the violet, and those of intermediate length the green fibers.
**Young's rule** [Thomas *Young,* English physicist, 1773–1829]. A rule of dosage in children. The dose is obtained by adding 12 to the age and dividing the result by the age, and making the quotient the denominator of a fraction the numerator of which is 1. The fraction represents the proportion of the adult dose to be given to the child.
**youth** (*ūth*). The period between childhood and maturity.
**Ys.** Abbreviation for the *yellow spot* of the retina.
**Yt.** Chemical symbol of *yttrium.*
**ytterbium** (*it-tur'-be-um*) [*Ytterby,* in Sweden]. A rare metal, having the symbol Yb, and atomic weight 172.
**yttria** (*it'-re-ah*). Yttrium oxide.
**yttrium** (*it'-re-um*). [*Ytterby,* in Sweden]. A rare metallic element. Symbol Yt, atomic weight 89.
**Yucca** (*yuk'-ah*) [Am. Ind.]. A genus of liliaceous plants. *Y. filamentosa,* Adam's needle, of the southern United States, is diuretic; its tincture is employed in uretinitis.
**Yvon's coefficient** (*e-vorn*(*g*)) [Paul *Yvon,* French physician]. The ratio existing between the amount of urea and the phosphates in the urine, represented by $\frac{1}{4}$. **Y's test.** 1. *For acetanilide in urine:* extract with chloroform and then heat the residue with mercurous nitrate; a green color denotes the presence of acetanilide. 2. *For alkaloids:* add a mixture of bismuth subnitrate, potassium iodide, hydrochloric acid, and water, to the suspected solution; a red color denotes the presence of an alkaloid.
**Yzquierdo's bacillus** (*ĕs-ke-ār'-do*) [Vincente *Yzquierdo,* histologist in Santiago, Chile]. A bacillus which is supposed to be the cause of *Verruga peruana,* or Carrion's disease.

# Z

**zacatilla** (*zak-ah-teel'-yah*). The best quality of cochineal.

**Zaglas' ligament** (*tsah-glah'*). The portion of the posterior sacroiliac ligament that extends from the posterior superior spinous process of the ilium downward to the side of the sacrum. Z.'s perpendicular external muscle, the vertical fibers of the tongue, which, decussating with the transverse fibers and the insertions of the geniohyoglossus, curve outward in each half of the tongue.

**Zahn's ribs.** The whitish, transverse markings often formed on the surface of a thrombus by the extremities of the columns of blood-platelets and leukocytes.

**zakavaska.** The name given in Russia to the grains used as a ferment to produce kephir or kumiss.

(von). **Zaleski's hepatin.** See *ferratin, Schmiedeberg's*. Z.'s test for CO in the blood, add to 2 Cc. of the blood to be tested an equal volume of water and three drops of a one-third saturated copper sulphate solution. In the presence of carbon monoxide a brick-red precipitate is obtained, while normal blood gives a greenish-brown precipitate.

**Zambesi ulcer** (*zam-be'-ze*) [*Zambesi*, river in East Africa]. An ulcer occurring on the foot or leg, found only in laborers near the Zambesi river, and supposed to be due to a spirillum.

**zanaloin** (*san-al'-o-in*) [*Zanzibar; aloin*]. The aloin derived from Zanzibar aloes; said to be the same as socaloin.

**Zander's system** (*tsan'-der*) [Jonas Gustaf *Zander*, Swedish physician, 1835– ]. Passive movement by means of special apparatus.

**Zang's space** [Christoph Bonifacius Zang, German surgeon, 1772–1835]. The space between the two tendons of origin of the sternomastoid in the supraclavicular fossa.

**zanol** (*san'-ol*). Trade name of a preparation containing sodium-caffeine salicylate; it is said to be a diuretic and to have vasomotor properties.

**zanzolin** (*san'-zo-lin*). A proprietary mixture of pyrethrum flowers, *Chrysanthemum coronopifolium*, and valerian root, *Valeriana officinalis*, used to combat mosquitoes.

**Zappert's chamber** or **cell** (*tsap'-pert*, Austrian physician, 1867– ). A chamber for counting blood corpuscles, like Thoma's cell.

**zarathan** (*tsar-ath'-an*). Scirrhous hardening of the breast.

**zea** (*ze'-ah*) [ζέα, a sort of grain]. A genus of grasses. The fresh styles and stigmas of Z. *mays*, maize, Indian corn, constitute the *zea* of the U. S. P. It has been used as a diuretic in cystitis, gonorrhea, and cardiac dropsy.

**zean** (*se'-an*) [*zea*]. A highly concentrated fluid extract of corn-silk; a diuretic and urinary antiseptic.

**zedoary** (*sed'-o-a-re*). The rhizome of several species of *Curcuma*. It resembles ginger, but is less agreeable and is seldom used in medicine.

**zein** (*tse'-in*) [*zea*]. A yellowish, soft, insipid protein obtained from maize.

**Zeisel's test for colchicine** (*tsi'-sel*). Dissolve the suspected substance in hydrochloric acid, then boil with ferric chloride, and shake with chloroform; a brown or dark red precipitate indicates the presence of colchicine.

**zeism** (*tse'-ism*) [*zea*]. Pellagra.

**zeismus** (*ze-is'-mus*) [*zea*]. Pellagra (believed by some to be due to a diet of maize).

**Zeiss' glands** (*tsis*). See *Zeissian glands*.

**Zeissian glands.** The sebaceous glands of the eyelashes. Z. sty, hordeolum externum; a sty produced by suppuration of one of the Zeissian glands.

**Zeller's test for melanin in urine.** Treat the urine with bromine water, and in the presence of melanin a yellow precipitate is formed which gradually changes to black.

**Zellner's test-paper.** Prepare by applying the coloring-matter employed as indicator, say fluorescein in solution, upon an underground, for which a neutral black coloring-matter is used. The fluorescein shows the minutest traces of alkali by a greenish color.

**zelotypia** (*ze-lo-tip'-e-ah*) [ζῆλος, zeal; τύπτειν, to strike]. Morbid or monomaniacal zeal in any pursuit.

**zematol** (*zem'-at-ol*). A proprietary ointment said to contain oil of betula, zinc oxide and ichthyol.

**zematone** (*zem'-at-ōn*). A proprietary remedy for asthma said to consist of extractives of *Datura stramonium, Hyoscyamus niger*, each, 8 parts; *Grindelia robusta*, 15 parts; *Solanum nigrum*, 4 parts; *Atropa belladonna*, 6 parts; white agaric, 5 parts; poppy capsules, 5 parts; and potassium nitrate, 22 parts.

**Zenker's crystals** (*zeng'-ker*) [Friedrich Albert Zenker, German pathologist, 1825–1808]. See *Charcot's crystals*. Z.'s degeneration, Z.'s disease of muscles, waxy or hyaline degeneration of muscles occurring in acute infectious diseases, especially in typhoid fever. Z.'s paralysis, paresis and disturbance of sensation in the lower extremities, the external popliteal nerve being most involved; it is caused by frequent and prolonged kneeling or squatting. Z.'s solution, a fixing agent, containing mercuric chloride 5, potassium bichromate 2.5, sodium sulphate 1, and water 100.

**Zenkerism** (*zeng'-ker-ism*) [see *Zenker's degeneration*]. The condition of Zenker's degeneration.

**zeoscope** (*ze'-o-skōp*) [ζέειν, to boil; σκοπεῖν, to view]. An apparatus for determining the alcoholic strength of a liquid by means of its boiling-point.

**zero** (*se'-ro*). 1. Any character denoting absence of quantity. 2. The point from which thermometers are graduated.

**zerumbet** (*ze-rum'-bet*) [E. Ind.]. An E. Indian drug or spice, by some asserted to be the same as cassimuniar; probably the rhizome of *Zingiber zerumbet*. It resembles ginger; little used.

**zestocausis** (*zes-to-kaw'-sis*) [ζεστός, boiling; καῦσις, a burning]. Cauterization with an instrument heated by steam; atmocausis.

**zestocautery** (*zes-to-kaw'-ter-e*). A double-channeled intrauterine catheter, the outer unfenestrated tube of which is heated by steam and acts as a cautery.

**zibet** (*sib'-et*). A variety of civet produced by *Viverra zibetha*, an animal of South and East Asia. It was formerly used as a substitute for musk.

**Ziehl-Neelsen method** (*tsēl'-nāl'-sen*) [Franz Ziehl, German bacteriologist; Friedrich Carl Adolf *Neelsen*, German pathologist, 1854–1894]. A method of staining tubercle bacilli with Ziehl's solution.

**Ziehl's solution** (*tsēl*) [Franz *Ziehl*, German bacteriologist]. A fluid employed to stain (lepra and) tubercle bacilli: It consists of a 5 % aqueous solution of phenol, with one-tenth its volume of a saturated alcoholic solution of fuchsin. Heat the specimen in this for three minutes, and the entire specimen will be stained red. Decolorize with 20 or 30 % of nitric acid, and the tubercle bacilli alone will retain the stain.

**Ziemssen's motor points** (*tsēm'-sen*) [Hugo von Ziemssen, German physician, 1829–1902]. Points of election in electrization of muscles; they correspond to the places of entrance of the motor nerves into the muscles.

**zimb** (*sim*) [Ar. *zimb*, a fly]. A gadfly of the genus *Pangonia*, found in East Africa; it bites man and beast and is believed to transmit disease.

**Zimmerlin's type of progressive muscular atrophy** (*tsim'-mer-lin*) [Franz *Zimmerlin*, Swiss physician]. The scapulohumeral type, distinguished from Erb's type by the absence of secondary lipomatosis.

**Zimmermann's corpuscles**, or **granules** (*tsim'-merman*) [Karl Wilhelm *Zimmermann*, German histologist, 1861– ]. See *Bizzozero's blood-platelets*.

**Zimmermann's decoction.** A decoction made of rhubarb 30 grains, potassium bitartrate 4 drams, barley 4 drams, and water 16 ounces; it is sweetened with syrup and used as a cathartic.

**zimphen** (*sim'-fen*). Sodium metaoxycyanocinnamate; used as a gastro-intestinal stimulant and tonic in 5 to 10 grain doses (0.33–0.66).

**zinc, zincum.** A bluish-white metal (*zincum*, U. S. P.) having a specific gravity of 7.12, an atomic weight of 65.37, and a valence of 2. Symbol, Zn. In nature it occurs in two principal forms, as a sulphide, called *blende*, and as a carbonate and silicate, termed *ealamine*. When melted and poured into water it becomes granular (*zincum granulatum*, B. P.). Zinc is used to prepare zinc sulphate and zinc chloride, and for generating hydrogen. The compounds of zinc are poisonous, and the slow ingestion of it produces a chronic intoxication resembling, but less severe than, that produced by lead. z. **acetate** (*zinci acetas*, U. S. P., B. P.), Zn(C₂H₃O₂)₂.- 2H₂O, is used locally in ophthalmia and gonorrhea in solutions of from 1–2 gr. to the oz. (0.065–0.13 Gm. to 30 Cc.) of water. z. **bromide** (*zinci bromidum*, U. S. P.), ZnBr₂, has been used in epilepsy. Dose 1–2 gr. (0.065–0.13 Gm.). z., **butter of.** See *zinc chloride*. z. **carbolate**, a white, antiseptic powder, slightly soluble in water or alcohol; used as a surgical dusting-powder. z. **carbonate, precipitated** (*zinci carbonas præcipitatas*, U. S. P.; *zinci carbonas*, B. P.) is used generally in the form of prepared calamine, as a dusting-powder on excoriated surfaces, or in the form of a cerate. z. **chloride** (*zinci chloridum*, U. S. P., B. P.), ZnCl₂, is used chiefly as an escharotic in carcinoma and spreading ulcers, as an injection in gonorrhea, and as an astringent in conjunctivitis. It is also employed as a disinfectant and for preserving anatomical preparations. z. **chloride, solution of** (*liquor zinci chloridi*, U. S. P., B. P.), used as disinfectant and preservative. *Burnett's disinfecting fluid* is a solution of zinc chloride. z., **flowers of, zinc oxide.** z. **iodide** (*zinci iodidum*, U. S. P.), ZnI₂, has been used in chorea, scrofula, and hysteria, and locally as an astringent, like the chloride. Dose ½–2 gr. (0.032–0.13 Gm.). z. **oxide** (*zinci oxidum*, U. S. P., B. P.), ZnO, is an amorphous white powder, and is used internally in chorea, epilepsy, whooping-cough, and gastrointestinal catarrh; locally, as a desiccant to excoriated surfaces, in the form of powder or ointment. Dose 2–8 gr. (0.13–0.52 Gm.). z. **oxide, ointment of** (*unguentum zinci oxidi*, U. S. P.; *unguentum zinci*, B. P.), an ointment composed of zinc oxide and benzoinated lard. z. **permanganate**, Zn(MnO₄)₂+6H₂O, used in aqueous solution (1 : 4000) as injection in gonorrhea and in 1 : 1000 or 2 : 1000 solution as eye-lotion. It is incompatible with all combustible or easily oxidizable substances. z. **phenolsulphonate** (*zinci phenolsulphonas*, U. S. P.), Zn(C₆H₄OHS)₂+8H₂O, used as an antiseptic and astringent. Dose 2 gr. (0.13 Gm.). z. **stearate** (*zinci stearas*, U. S. P.), a very fine white powder, tasteless, and having a slight odor, resembling that of fat. z. **sulphate** (*zinci sulphas*, U. S. P., B. P.), ZnSO₄+7H₂O, white vitriol, is tonic, astringent, and emetic. It is used in gastric catarrh, as an emetic, and locally in ophthalmia, gonorrhea, leukorrhea, and as a caustic in cases of ulcer, condyloma, etc. In overdoses it is a gastrointestinal irritant. Dose ¼–½ gr. (0.016–0.032 Gm.); as an emetic, 10–30 gr. (0.65–2.0 Gm.). z. **valerate** (*zinci valeras*, U. S. P.)· Zn(C₅H₉O₂)₂ . 2H₂O, is used in neuralgia, epilepsy, hysteria, and diabetes insipidus. Dose 1–2 gr. (0.065–0.13 Gm.). z.-**white**, zinc oxide.

**zinci** (*sin'-ki*). Genitive of zincum.

**zincoid** (*sin'-koid*) [*zincum; εἶδος*, form]. 1. Resembling zinc. 2. The positive plate in a battery.

**zincum** (*sin'-kum*). See *zinc*.

**zingiber** (*zin'-jib-er*) [ζιγγίβερις, ginger]. Ginger, a genus of plants of the *Zingiberaceæ*. The rhizome of *Z. officinale* is the *zingiber* of the U. S. P.; it contains a volatile oil, and is used as a stimulant and carminative in dyspepsia, flatulence, and intestinal atony; externally it is rubefacient. Dose 10–20 gr. (0.65–1.3 Gm.). **zingiberis, fluidextractum** (U. S. P.), fluidextract of ginger. Dose 10–20 min. (0.65–1.3 Cc.). **zingiberis, oleoresina** (U. S. P.), oleoresin of ginger. Dose ½ grain (0.030 Gm.). **zingiberis, syrupus** (U. S. P., B. P.), syrup of ginger. Dose 4 dr. (16 Cc.). **zingiberis, tinctura** (U. S. P., B. P.), tincture of ginger. Dose 30 minims (2 Cc.). Ginger also enters into the composition of *pulvis aromaticus* (U. S. P.); *pulvis cinnamomi compositus* (B. P.), *pulvis rhei compositus* (U. S. P., B. P.), and *fluidextractum aromaticum* (U. S. P.).

**zingiberin** (*zin-jib'-er-in*). The oleoresin of ginger.

**Zinn's artery** (*tsin*) [Johann Gottfried Zinn, German anatomist, 1727–1759]. The central artery of the retina. Z.'s **circle**, the plexus formed by small branches of the ciliary arteries within the fibrous layer of the sclera at the entrance of the optic nerve. Z.'s **ligament.** See *Z.'s ring*. Z.'s **membrane**, the anterior layer of the iris. Z.'s **ring**, Z.'s **tendon**, the circular fibrous sheath formed by the common tendon of the internal, external, and inferior rectus muscles. Z.'s **zonula**, Z.'s **zonule**, zonula ciliaris, the suspensory ligament of the crystalline lens. It is a thin, transparent membrane covering the ciliary processes and extending to the anterior capsule. A portion lies above the processes in folds, that covering the process being smooth.

**zinol** (*zi'-nol*). A mixture of zinc acetate and aluminol; used in solution in gonorrhea.

**Zionist** (*zi'-on-ist*). A follower of the faith-healer, Dowie.

**zirconia** (*zir-ko'-ne-ah*). Zirconium oxide, ZrO₃.

**zirconium** (*zir-ko'-ne-um*) [Pers. *zargun*, gold-colored]. A metallic element (symbol Zr; atomic weight 90.6), resembling titanium and silicon, and soluble in aqua regia and hydrofluoric acid. It is obtained from a mineral called zircon.

**Zittmann's decoction** (*tsit'-man*) [Johann Friedrich Zittmann, German physician, 1671–1757]. A drink used in old, obstinate cases of syphilis. It consists of sarsaparilla, 12½ oz.; water, 325 troy. oz.; alum and sugar, each, 6 dr.; anise and fennel, each, 4 dr.; senna, 3 oz.; licorice root, 1½ oz.

**Zn.** The chemical symbol for zinc.

**zoamylin** (*zo-am'-il-in*) [ζωή, life; *amylum*, starch]. Glycogen.

**zoanthropy** (*zo-an'-thro-pe*) [ζῶον, animal; ἄνθρωπος, a man]. A form of insanity in which the person imagines himself transformed into or inhabited by an animal.

**zoarium** (*zo-ar'-e-um*) [ζῴάριον, dim. of ζῶον, an animal: *pl.*, *zoaria*]. In biology, the composite structure formed by repeated gemination in the *Polyzoa*.

**zodiophilous** (*zo-de-off'-il-us*) [ζῴδιον, dim. of ζῶον, animal; φιλεῖν, to love]. In biology applied to plants which are frequented by animals and pollinated by their agency.

**Zoellner's lines** (*tsel'-ner*) [Johann Karl Friedrich Zoellner, German physicist, 1834–1882]. A device

ZOELLNER'S LINES.

to illustrate false estimates of direction or parallelism by intersecting lines crossing parallel lines at a certain angle.

**zoescope** (*zo-e-skōp*). See *stroboscope*.

**zoetic** (*zo-et'-ik*) [ζωή, life]. Vital, pertaining to life.

**zoetrope** (*ωʹ-e-ū ōp*) [ζωή, life; τρέπειν, to turn]. A stroboscope.

**zoiatria** (*zo-e-a'-tre-ah*) [ζῶον, animal; *ιατρός*, physician or surgeon]. The art and science of veterinary surgery.

**zoiatrics** (*zo-i-at'-riks*). See *zoiatria*.

**zoic** (*zo'-ik*) [ζωϊκός, of animals]. In biology, of or pertaining to living organisms; relating especially to animal life.

**zoism** (*zo'-izm*) [ζωή, life]. The doctrine or theory

that life is the manifestation of the operations of a peculiar vital principle; the doctrine of vital force.
**zomakyne** (*zo'-mak-ĭn*). A proprietary antipyretic substance.
**zomol** (*zo'-mol*) [ζωμός, meat-juice]. The plasma of raw beef. Evaporated to dryness it is used as a concentrated food. Dose 1.50 gr. (10 Gm.) daily.
**zomotherapy** (*zo-mo-ther'-ap-e*) [ζωμός, meat-juice; θεραπεία, therapy]. Treatment of tuberculosis by means of a raw meat diet; the meat, finely hashed or scraped, is given in daily doses of 6 oz. (200 Gm.) with soup, etc.
**zona** (*zo'-nah*) [L.]. 1. A belt or girdle. 2. See *herpes zoster*. z. **arcuata**, the inner zone of the basilar membrane, extending from the lower edge of the spiral groove of the cochlea to the external edge of the base of the outer rods of Corti. z. **cartilaginea**, the limbus of the spiral lamina. z. **ciliaris**, the ciliary processes collectively. z. **denticulata**, the inner zone of the basilar membrane, together with the limbus of the spiral lamina. z. **fasciculata**, the central portion of the cortex of the suprarenal capsule, composed of tube-like transverse bands. z. **glomerulosa**, a part of the cortical portion of the suprarenal capsule, having a net-like appearance on section, situated near the surface of the organ. z. **incerta**, the anterior portion of the reticular formation under the optic thalamus. z. **ophthalmica**, herpes zoster along the course of the ophthalmic division of the fifth nerve. z. **orbicularis**, a thickening of the capsular ligament around the acetabulum. z. **pectinata**, the outer portion of the basilar membrane, extending from the rods of Corti to the spiral ligament. z. **pellucida**, the thick, solid, elastic envelope of the ovum, corresponding to the cell-wall of a cell. Syn., *vitelline membrane*. z. **perforata**, the lower edge of the spiral groove of the cochlea. z. **tecta**, the inner portion of the basilar membrane, bearing the organ of Corti. z. **terminalis**. See *terma*.
**zonal** (*zo'-nal*) [*zona*, zone]. Pertaining to a zone, or to the disease called zona or to a girdle or to a band-like structure.
**zonary** (*zo'-nar-e*) [*zona*, zone]. Characterized by, or pertaining to a zone. z. **placenta**, a placenta which occupies a broad band around the chorion; found in carnivora.
**zonate** (*zo'-nāt*) [*zona*]. Marked with concentric bands.
**zone** (*zōn*). See *zona*. z., **cornuradicular**, the external part of Burdach's column, abutting on the middle third of the internal border of the posterior horn, and representing approximately the posterior root-zone. z., **entry**, the parts along the posterior horns of gray matter of the spinal cord where the posterior roots enter the cord. z., **hypnogenous**, an area or tract, pressure upon which induces sleep. z., **hysterogenous**, a region, as the ovarian or submammary region, where pressure in hysterical women calls forth an hysterical attack. z., **neogenic**, the subcapsular layer of the kidney, so-called because it is the one in which the most active processes are going on. z., **radiary**, a layer in the cortical gray matter of the brain characterized by radiating nerve-fibers. z., **sclerotic**, a condition occurring in iritis, marked by a ring of anastomoses of the cornea, which perforate the sclerotic and anastomose with those of the iris and choroid. z. **supra-radiary**, the layer of cortex immediately above the radiary zone.
**zonesthesia, zonæsthesia** (*zon-es-the'-ze-ah*). See *girdle-pain*.
**zonular** (*son'-ū-lar*) [*zonule*]. Pertaining to or in the shape of a zone or band. z. **cataract**, a cataract forming alternate layers.
**zonule** (*zon'-ūl*) [*zonula*, a little zone]. A small band. z. of Zinn. See *Zinn's zonula*. z. **ciliaris**. See *Zinn's zonula*.
**zonulitis** (*zon-ū-li'-tis*). Inflammation of Zinn's zonule.
**zoo-** (*zo-o-*) [ζῶον, an animal]. A prefix meaning animal or pertaining to an animal.
**zooamilin** (*zo-o-am'-il-in*) [*zoo-*; *amylum*, starch]. Glycogen, amyloid.
**zooamylon** (*zo-o-am'-il-on*) [*zoo-*; *amylum*, starch]. The ternary substance allied to starch and glycogen found in the cytoplasm of certain sporozoa.
**zoobiology** (*zo-o-bi-ol'-o-je*) [*zoo-*; *biology*]. Animal biology.
**zoobiotism** (*zo-o-bi'-ot-ism*) [ζῶον, animal; βίος, life]. Same as biotics.

**zooblast** (*zo'-o-blast*). [ζῶον, animal; βλαστός, germ]. An animal cell.
**zoochemia, zoochemistry** (*zo-o-ke'-me-ah*, *zo-o-kem'-is-tre*) [*zoo-*; χημεία, chemistry]. The chemistry of animal life and tissues.
**zoocyst** (*zo'-o-sist*) [*zoo-*; κύστις, cell]. A variety of encysted rhizopods resembling a sporocyst, except in the thickness and number of the protective layers.
**zoocytium** (*zo-o-sit'-e-um*) [*zoo-*; κύτος, cavity]. The gelatinous matrix secreted by certain infusoria.
**zoodermic** (*zo-o-der'-mik*). Pertaining to or taken from the skin of some animal other than man; applied to a form of skin-grafting.
**zoodynamics** (*zo-o-di-nam'-iks*) [*zoo-*; δύναμις, power]. Animal physiology.
**zoogamete** (*zo-o-gam'-ēt*) [ζῶον, animal; γαμετή, a wife]. In biology, a gamete or sexual spore endowed with the power of locomotion.
**zoogamy** (*zo-og'-am-e*) [ζῶον, animal; γάμος, marriage]. In biology, the sexual generation of animals; copulation, conjugation, mating.
**zoogenesis** (*zo-o-jen'-es-is*) [*zoo-*; γεννᾶν, to beget]. The generation of animal forms.
**zoogenous** (*zo-oj'-en-us*) [see *zoogenesis*]. Developed or derived from animals.
**zooglea, zoogloea** (*zo-og'-le-ah*) [*zoo-*; γλοιός, a sticky substance]. A stage in the life-history of certain bacteria in which they lie embedded in a gelatinous matrix.
**zoogonia** (*zo-o-gon'-e-ah*) [ζῶον, animal; γονή, generation]. Viviparous generation.
**zoograft** (*zo'-o-graft*) [*zoo-*; *graft*]. A graft of tissue derived from an animal.
**zoografting** (*zo'-o-graft-ing*) [see *zoograft*]. Grafting with tissue taken from the lower animals.
**zoography** (*zo-og'-ra-fe*) [ζῶον, animal; γράφειν, to write]. A descriptive treatise on the distribution of animals.
**zooid** (*zo'-oid*) [ζῶον, animal; εἶδος, form]. 1. Animal-like, resembling an animal. 2. A zoophyte. 3. An animal cell which can exist or move independently.
**zoolak** (*zoo'-lak*). A commercial name for matzoon.
**zoolite, zoolith** (*zo'-ol-īt*, *-ith*) [ζῶον, animal; λίθος, stone]. A fossil animal, or any part of it.
**zoology** (*zo-ol'-o-je*) [*zoo-*; λόγος, a treatise]. That branch of biology treating of the form, nature, and classification of animals.
**zoomagnetism** (*zo-o-mag'-net-ism*). Animal magnetism.
**zoometry** (*zo-om'-et-re*) [ζῶον, animal; μέτρον, measure]. The measurement of the proportionate lengths or sizes of the parts of animals.
**zoonite** (*zo'-on-īt*) [ζῶον, animal]. In biology, one of the segments or somites, metameres, or arthromeres of which an articulate animal is composed.
**zoonomia, zoonomy** (*zo-o-no'-me-ah*, *zo-on'-o-me*) [*zoo-*; νόμος, law]. The principles or laws of animal life; zoöbiology.
**zooparasite** (*zo-o-par'-as-īt*) [*zoo-*; *parasite*]. An animal parasite.
**zoopathology** (*zo-o-path-ol'-o-je*) [*zoo-*; *pathology*]. The science of the diseases of animals.
**zoopery** (*zo-op'-er-e*) [ζῶον, an animal; πειράειν, to experiment]. Experimentation upon animals.
**zoophagous** (*zo-of'-ag-us*) [*zoo-*; φαγεῖν, to eat]. Subsisting on animal food.
**zoopharmacology** (*zo-o-far-ma-kol'-o-je*) [ζῶον, animal; *pharmacology*]. Veterinary pharmacology.
**zoophilism** (*zo-of'-il-ism*) [ζῶον, animal; φιλεῖν, to love]. The love of animals; it is usually immoderate, and toward certain animals, illustrated in the fanaticism of antivivisection.
**zoophobia** (*zo-o-fo'-be-ah*) [*zoo-*; φόβος, fear]. Morbid dread of certain animals.
**zoophysiology** (*zo-o-fiz-e-ol'-o-je*) [ζῶον, animal; *physiology*]. Animal physiology.
**zoophyte** (*zo'-o-fīt*) [*zoo-*; φυτόν, a plant]. A member of the lower invertebrates.
**zooplasty** (*zo'-o-plas-te*) [ζῶον, animal; πλάσσειν, to form]. The surgical transfer of zoografts; the transplantation of tissue from any of the lower animals to man.
**zoopsia** (*zo-op'-se-ah*) [ζῶον, animal; ὄψις, a vision]. The seeing of animals, as an illusion or as an hallucination in a dream.
**zoopsychology** (*zo-o-si-kol'-o-je*) [ζῶον, animal; ψυχή, soul, mind; λόγος, science]. The science of the mental activities of lower animals.

**zooscopy** (zo-os'-ko-pe) [ζῷον, animal; σκοπεῖν, to see]. The hallucinatory appearance of animal forms.
**zoosperm** (zo'-o-sperm). See *spermatozoon*.
**zoospore** (zo'-o-spōr) [zoo-; σπόρος, seed]. A motile spore.
**zootechnics, zootechny** (zo-o-tek'-niks, zo'-o-tek-ne) [ζῷον, animal; τέχνη, art]. The science of breeding and domesticating animals.
**zootherapy** (zo-o-ther'-a-pe) [ζῷον, animal; θεραπεία, therapy]. Veterinary therapeutics.
**zootomist** (zo-ot'-o-mist) [zoo-; τομή, a cutting]. One who dissects animals; a comparative anatomist.
**zootomy** (zo-ot'-o-me) [see *zootomist*]. The dissection of animals.
**zootoxin** (zo-o-tok'-sin) [ζῷον, animal; τοξικόν, poison]. A toxin or poison of animal origin.
**zootrophic** (zo-o-trof'-ik) [ζῷον, animal; τρέφειν, to nourish]. Pertaining to animal alimentation.
**zootrophotoxism** (zo-o-trof-o-toks'-ism) [zoo-; τροφή, nourishment; τοξικόν, poison]. Poisoning with infected animal food. z., **gastric**, z., **intestinal**, that occurring through ingestion of spoiled flesh, milk, or cheese, and marked by cholera nostras, colic, diarrhea, fever, cramps, progressing to collapse and cyanosis. z., **tropeinic**, due to ingestion of poisonous sausage and salted fish, accompanied by symptoms similar to those of tropeinism.
**zoster** (zos'-ter) [ζωστήρ, a girdle]. An acute inflammatory painful disease, consisting of grouped vesicles corresponding in distribution to the course of the cutaneous nerves. See *herpes zoster*. z. **auricularis**, a form affecting the ear. z. **brachialis**, a form affecting the arm or forearm. z. **ophthalmicus**, an eruption in the course of the ophthalmic division of the fifth nerve.
**zosteriform** (zos-ter'-if-orm). Resembling zoster.
**Zouchlos' test for albumin.** A reagent consisting of 10 % of potassium sulphocyanide solution and 20 parts of acetic acid, added drop by drop to an albumin solution, produces a marked cloudiness.
**Zr.** Chemical symbol of zirconium.
**Zuckerkandl's convolution** (tsook'-er-kan-dl) [Emil Zuckerkandl, Austrian anatomist, 1849–1910]. The gyrus subcallosus; the peduncle of the callosum; it is located in the mesal aspect of the cerebrum and extends from the chiasm to the rostrum. Z.'s **dehiscences**, small gaps sometimes existing in the papyraceous lamina of the ethmoid bone, and bringing the lining membrane of the latter in contact with the dura. They are not pathological. Z.'s **vein**, a small branch of the ethmoid veins through which the veins of the lateral wall of the nose communicate with the cerebral veins.
**Zwanck's pessary** (tswank'). A pessary with two wings.
**Zwenger's test for cholesterin.** See *Liebermann-Buchard's test*.
**zygal** (zi'-gal) [ζυγόν, a yoke]. Yoked; applied to cerebral fissures consisting of two pairs of branches connected by a stem.
**zygapophysis** (zi-gap-of'-is-is) [zygon; apophysis]. The articular process of a vertebra.
**zygion** (zij'-e-on). A craniometric point at either end of the zygomatic diameter.
**zygolabialis** (zi-go-la-be-a'-lis) [zygon; labium, a lip]. The zygomaticus minor. See under *muscle*.
**zygoma** (zi-go'-mah) [ζύγωμα, the cheek-bone]. 1. The arch formed by the union of the zygomatic process of the temporal bone and the malar bone. 2. The malar bone.
**zygomatic** (zi-go-mat'-ik) [zygoma]. Pertaining to the zygoma. z. **arch**, the zygoma.
**zygomatico-** (zi-go-mat-ik-o-) [zygoma]. A prefix meaning relating to the zygoma.
**zygomaticoangularis** (zi-go-mat-ik-o-an-gū-lar'-is). Pertaining to the zygoma and the angle of the eye.
**zygomaticoauricular** (zi-go-mat-ik-o-aw-rik'-ū-lar) [zygomatico-; auricularis, of the ear]. Pertaining to the zygoma and the ear.
**zygomaticoauricularis** (zi-go-mat-ik-o-aw-rik-ū-la'-ris) [see *zygomaticoauricular*]. The attrahens aurem muscle. See under *muscle*.
**zygomaticofacial** (zi-go-mat-ik-o-fa'-shal) [zygomatico-; facies, face]. Pertaining to the zygoma and the face.
**zygomaticofrontal** (zi-go-mat-ik-o-fron'-tal). Pertaining to the zygoma and the frontal bone.
**zygomaticomaxillary** (zi-go-mat-ik-o-max'-il-la-re). Pertaining to the zygoma and the maxilla.

**zygomaticoorbital** (zi-go-mat-ik-o-or'-bit-al). Pertaining to the zygoma and the orbit.
**zygomaticosphenoid** (zi-go-mat-ik-o-sphen'-oid). Pertaining to the zygoma and the sphenoid bone.
**zygomaticotemporal** (zi-go-mat-ik-o-tem'-po-ral) [zygomatico-; temporal]. Pertaining to the zygoma and the temporal bone or fossa.
**zygomaticus** (zi-go-mat'-ik-us) [zygoma]. One of several small subcutaneous muscles arising from or in relation with the zygoma. See under *muscle*.
**zygomaxillary** (zi-go-maks'-il-a-re). See *zygomaticomaxillary*.
**zygomycetes** (zi-go-mi-se'-tēz) [ζυγόν, a yoke; μύκης, fungus]. A group of fungi characterized by sexual reproduction through the union of two similar gametes (zygospores).
**zygon** (zi'-gon) [ζυγόν, yoke]. In the cerebrum, the bar that connects the two pairs of branches of a zygal fissure.
**zygoneure** (zi'-go-nūr) [zygon; νεῦρον, nerve]. A nerve-cell joining other nerve-cells.
**zygosis** (zi-go'-sis) [ζυγόν, yoke]. The process of asexual reproduction by conjugation or fusion of two protoplasmic bodies or gametes.
**zygosperm** (zi'-go-spurm) [ζυγόν, yoke; σπέρμα, seed]. Same as *zygospore*.
**zygospore** (zi'-go-spōr) [ζυγόν, yoke; σπορά, seed]. The spore resulting from the zygosis or conjugation of two protoplasmic bodies or gametes.
**zygote** (zi'-gōt). Same as *zygospore*.
**zylenin, zylenobacillin** (zi'-len-in, zi-len-o-bas'-il-in). A toxin from tubercle bacilli.
**zylonite** (zi'-lo-nīt) [ξύλον, wood]. Celluloid.
**zymase** (zi'-mās). 1. See *microzyme*. 2. The unorganized ferment or enzyme to which the fermentative activity of the yeast-cell is due. z., **Buchner's**, that expressed from dried yeast; yeast-cell plasma.
**zyme** (zīm) [ζύμη, leaven]. An organized ferment. Cf. *enzyme*.
**zymic** (zi'-mik) [zyme]. Of or pertaining to organized ferments.
**zymin** (zi'-min) [zyme]. 1. A pancreatic preparation used in the treatment of diabetes mellitus. See *zyme*. 2. Sterile dried yeast; mixed with sugar and water it is used as an application in leukorrhea of gonorrheal origin.
**zyminized** (zi'-min-īzd) [zyme]. A term applied to milk in which a fermentive change has been induced, comparable to peptonization.
**zymo-** (zi-mo-) [ζύμη, leaven]. A prefix meaning pertaining to or produced by fermentation.
**zymocide** (zi'-mo-sīd) [ζύμη, leaven; cædere, to kill]. A proprietary disinfectant.
**zymogen** (zi'-mo-jen) [zymo-; γεννᾶν, to produce]. The substance existing in the glands secreting a digestive juice, and which, when set free, splits into a ferment (enzyme) and a proteid.
**zymogenic** (zi-mo-jen'-ik) [zymogen]. 1. Causing fermentation. 2. Pertaining to or producing a zymogen.
**zymohydrolysis** (zi-mo-hi-drol'-is-is) [zymo-; ὕδωρ, water; λύειν, to loosen]. Hydrolysis produced by the cleavage action of enzymes.
**zymoid** (zi'-moid) [zymo-; εἶδος, like]. Resembling an organized ferment.
**zymoidin** (zi-moid'-in). A proprietary wound antiseptic said to be a mixture of oxides of zinc, bismuth, and aluminum with iodine, boric acid, salicylic acid, phenol, gallic acid, etc.
**zymology** (zi-mol'-o-je) [zymo-; λόγος, science]. The science dealing with fermentation.
**zymolysis** (zi-mol'-is-is). See *zymosis* (1).
**zymolytic** (zi-mo-lit'-ik) [ζύμη, leaven; λύσις, loosening]. Due to, attended with, or relating to zymolysis; zymotic.
**zymoma** (zi-mo'-mah) [ζυμοῦν, to make to ferment]. Any ferment, fermented mixture, or culture.
**zymometer** (zi-mom'-et-er) [zymo-; μέτρον, a measure]. An instrument for measuring fermentation.
**zymophore** (zi'-mo-fōr) [ζύμη, leaven; φορεῖν, to bear]. The active part of an enzyme, that which bears the ferment.
**zymophoric, zymophorous** (zi-mo-for'-ik, zi-mof'-or-us) [zymo-; φορεῖν, to bear]. Exerting a fermentive action; bearing specific fermentive properties.
**zymophyte** (zi'-mo-fīt) [zymo-; φυτόν, a plant]. A microorganism producing fermentation.
**zymoplastic** (zi-mo-plas'-tik) [ζύμη, leaven; πλάσσειν, to form]. Ferment-producing.

**zymose** (*zi'-mōs*). See *enzyme*.
**zymosimeter** (*zi-mo-sim'-et-er*). See *zymometer*.
**zymosis** (*zi-mo'-sis*) [*zyme*]. 1. Fermentation, the result of the vital activity of certain microorganisms, organized ferments, or zymes. 2. The condition of one affected with a zymotic disease. 3. An infectious disease. z. **gastrica**, organacidia gastrica in the stomach of growing, sporulating, budding yeast.
**zymosthenic** (*zi-mos-then'-ik*) [*zymo-*; σθένος, strength]. An agent which increases the functional activity of an enzyme.
**zymotechnic** (*zi-mo-tek'-nik*) [ζύμη, leaven; τέχνη, art]. The art of inducing and conducting zymotic processes in connection with vivification, acetification, etc.
**zymotic** (*zi-mot'-ik*) [*zymosis*]. Pertaining to zymosis; produced by zymosis. z. **disease**, an infectious disease.
**zymotoxic** (*zi-mo-toks'-ik*) [*zymo-*; τοξικόν, poison]. In the side-chain theory, relating to the hemolytic action of the toxophore group.
**zymurgy** (*zi'-mur-je*) [ζύμη, leaven; ἔργον, work]. That department of technological chemistry which treats of the scientific principles of wine-making, brewing, and distilling, and the preparation of yeast and vinegar, in which processes fermentation plays the principal part.

*From the*

**BRITISH MEDICAL JOURNAL**

"Dr. George M. Gould is the Johnson of medical lexicography. His various dictionaries, adapted to the needs of student, practitioner and scholar have had a commercial success that of itself is sufficient to prove their practical usefulness."

"Dr. George M. Gould is not only the senior and most prolific of medical lexicographers, but the most successful from the publisher's point of view. The success has been due to a combination of accuracy and good judgment."

CPSIA information can be obtained
at www.ICGtesting.com
Printed in the USA
LVHW021456050219
606472LV00019B/1026/P